CHILTON®

ASIAN
SERVICE MANUAL
2012 EDITION
VOLUME III
INFINITI
NISSAN

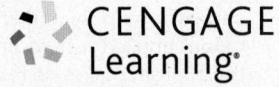

CENGAGE
Learning·

Australia • Brazil • Japan • Korea • Mexico • Singapore • Spain • United Kingdom • United States

CENGAGE Learning

CHILTON®
Asian Service Manual
2012 Edition
Volume III
Infiniti, Nissan

Vice President,
Technology & Trades Professional
Business Unit:
  Gregory L. Clayton

Publisher:
  David Koontz

Director of Marketing:
  Beth A. Lutz

Senior Production Director:
  Wendy Troeger

Production Manager:
  Sherondra Thedford

Senior Marketing Manager:
  Jennifer Barbic

Associate Marketing Manager:
  Rachael Torres

Chilton Content Specialist:
  Paula Baillie

Graphical Designer:
  Melinda Fantozzi

Art Director:
  Benjamin Gleeksman

Sr. Content Project Manager:
  William Tubbert

Senior Editor:
  Christine L. Sheeky
  Eugene F. Hannon Jr., A.S.E.
  Ryan Lee Price

Editors:
  David G. Olson, A.S.E.
  Kyla White
  Lance Williams
  Sherry Burdette
  Nicholas D'Andrea
  Maureen Lazarz

For product information and technology assistance, contact us at
Professional & Career Group customer Support, 1-800-648-7450.
For permission to use material from this text or product,
submit all requests online at
www.cengage.com/permissions.
Further permissions questions can be e-mailed to
permissionrequest@cengage.com

ISBN-13: 978-1-2854-7107-5
ISBN-10: 1-2854-7107-5
ISSN: 2161-8755

Chilton
5 Maxwell Drive
Clifton Park, NY 12065-2919
USA

Chilton products are represented in Canada by Nelson Education, Ltd.

NOTICE TO THE READER

Printed in the United States of America
1 2 3 4 5 6 7 XX 16 15 14 13

Contents

Model Index

USING THIS INFORMATION

Organization

To find where a particular model section or procedure is located, look in the Table of Contents. Main topics are listed with the page number on which they may be found. Following the main topics is an alphabetical listing of all of the procedures within the section and their page numbers.

Manufacturer and Model Coverage

This product covers 2011-2012 Asian models that are produced in sufficient quantities to warrant coverage, and which have technical content available from the vehicle manufacturers before our publication date. Although this information is as complete as possible at the time of publication, some manufacturers may make changes which cannot be included here. While striving for total accuracy, the publisher cannot assume responsibility for any errors, changes, or omissions that may occur in the compilation of this data.

Part Numbers and Special Tools

Part numbers and special tools are recommended by the publisher and vehicle manufacturer to perform specific jobs. Before substituting any part or tool for the one recommended, you must be completely satisfied that neither your personal safety, nor the performance of the vehicle will be endangered.

ACKNOWLEDGEMENT

The publisher would like to express appreciation to the following vehicle manufacturers for their assistance in producing this manual: Nissan North America, including Infiniti and Nissan Divisions. No further reproduction or distribution of the material in this manual is allowed without the expressed written permission of the vehicle manufacturers and the publisher.

PRECAUTIONS

Before servicing any vehicle, please be sure to read all of the following precautions, which deal with personal safety, prevention of component damage, and important points to take into consideration when servicing a motor vehicle:
- Always wear safety glasses or goggles when drilling, cutting, grinding or prying.
- Steel-toed work shoes should be worn when working with heavy parts. Pockets should not be used for carrying tools. A slip or fall can drive a screwdriver into your body.
- Work surfaces, including tools and the floor should be kept clean of grease, oil or other slippery material.
- When working around moving parts, don't wear loose clothing. Long hair should be tied back under a hat or cap, or in a hair net.
- Always use tools only for the purpose for which they were designed. Never pry with a screwdriver.
- Keep a fire extinguisher and first aid kit handy.
- Always properly support the vehicle with approved stands or lift.
- Always have adequate ventilation when working with chemicals or hazardous material.
- Carbon monoxide is colorless, odorless and dangerous. If it is necessary to operate the engine with vehicle in a closed area such as a garage, always use an exhaust collector to vent the exhaust gases outside the closed area.

- When draining coolant, keep in mind that small children and some pets are attracted by ethylene glycol antifreeze, and are quite likely to drink any left in an open container, or in puddles on the ground. This will prove fatal in sufficient quantity. Always drain the coolant into a sealable container.
- To avoid personal injury, do not remove the coolant pressure relief cap while the engine is operating or hot. The cooling system is under pressure; steam and hot liquid can come out forcefully when the cap is loosened slightly. Failure to follow these instructions may result in personal injury. The coolant must be recovered in a suitable, clean container for reuse. If the coolant is contaminated it must be recycled or disposed of correctly.
- When carrying out maintenance on the starting system be aware that heavy gauge leads are connected directly to the battery. Make sure the protective caps are in place when maintenance is completed. Failure to follow these instructions may result in personal injury.
- Do not remove any part of the engine emission control system. Operating the engine without the engine emission control system will reduce fuel economy and engine ventilation. This will weaken engine performance and shorten engine life. It is also a violation of Federal law.
- Due to environmental concerns, when the air conditioning system is drained, the refrigerant must be collected using refrigerant recovery/recycling equipment. Federal law requires that refrigerant be recovered into appropriate recovery equipment and the process be conducted by qualified technicians who have been certified by an approved organization, such as MACS, ASI, etc. Use of a recovery machine dedicated to the appropriate refrigerant is necessary to reduce the possibility of oil and refrigerant incompatibility concerns. Refer to the instructions provided by the equipment manufacturer when removing refrigerant from or charging the air conditioning system.
- Always disconnect the battery ground when working on or around the electrical system.
- Batteries contain sulfuric acid. Avoid contact with skin, eyes, or clothing. Also, shield your eyes when working near batteries to protect against possible splashing of the acid solution. In case of acid contact with skin or eyes, flush immediately with water for a minimum of 15 minutes and get prompt medical attention. If acid is swallowed, call a physician immediately. Failure to follow these instructions may result in personal injury.
- Batteries normally produce explosive gases. Therefore, do not allow flames, sparks or lighted substances to come near the battery. When charging or working near a battery, always shield your face and protect your eyes. Always provide ventilation. Failure to follow these instructions may result in personal injury.

• When lifting a battery, excessive pressure on the end walls could cause acid to spew through the vent caps, resulting in personal injury, damage to the vehicle or battery. Lift with a battery carrier or with your hands on opposite corners. Failure to follow these instructions may result in personal injury.

• Observe all applicable safety precautions when working around fuel. Whenever servicing the fuel system, always work in a well-ventilated area. Do not allow fuel spray or vapors to come in contact with a spark, open flame, or excessive heat (a hot drop light, for example). Keep a dry chemical fire extinguisher near the work area. Always keep fuel in a container specifically designed for fuel storage; also, always properly seal fuel containers to avoid the possibility of fire or explosion. Do not smoke or carry lighted tobacco or open flame of any type when working on or near any fuel-related components.

• Fuel injection systems often remain pressurized, even after the engine has been turned OFF. The fuel system pressure must be relieved before disconnecting any fuel lines. Failure to do so may result in fire and/or personal injury.

• The evaporative emissions system contains fuel vapor and condensed fuel vapor. Although not present in large quantities, it still presents the danger of explosion or fire. Disconnect the battery ground cable from the battery to minimize the possibility of an electrical spark occurring, possibly causing a fire or explosion if fuel vapor or liquid fuel is present in the area. Failure to follow these instructions can result in personal injury.

• The EPA warns that prolonged contact with used engine oil may cause a number of skin disorders, including cancer! You should make every effort to minimize your exposure to used engine oil. Protective gloves should

be worn when changing oil. Wash your hands and any other exposed skin areas as soon as possible after exposure to used engine oil. Soap and water, or waterless hand cleaner should be used.

• Some vehicles are equipped with an air bag system, often referred to as a Supplemental Restraint System (SRS) or Supplemental Inflatable Restraint (SIR) system. The system must be disabled before performing service on or around system components, steering column, instrument panel components, wiring and sensors. Failure to follow safety and disabling procedures could result in accidental air bag deployment, possible personal injury and unnecessary system repairs.

• Always wear safety goggles when working with, or around, the air bag system. When carrying a non-deployed air bag, be sure the bag and trim cover are pointed away from your body. When placing a non-deployed air bag on a work surface, always face the bag and trim cover upward, away from the surface. This will reduce the motion of the module if it is accidentally deployed.

• Electronic modules are sensitive to electrical charges. The ABS module can be damaged if exposed to these charges.

• Brake pads and shoes may contain asbestos, which has been determined to be a cancer-causing agent. Never clean brake surfaces with compressed air. Avoid inhaling brake dust. Clean all brake surfaces with a commercially available brake cleaning fluid.

• When replacing brake pads, shoes, discs or drums, replace them as complete axle sets.

• When servicing drum brakes, disassemble and assemble one side at a time, leaving the remaining side intact for reference.

• Brake fluid often contains polyglycol ethers and polyglycols. Avoid contact with the

eyes and wash your hands thoroughly after handling brake fluid. If you do get brake fluid in your eyes, flush your eyes with clean, running water for 15 minutes. If eye irritation persists, or if you have taken brake fluid internally, immediately seek medical assistance.

• Clean, high quality brake fluid from a sealed container is essential to the safe and proper operation of the brake system. You should always buy the correct type of brake fluid for your vehicle. If the brake fluid becomes contaminated, completely flush the system with new fluid. Never reuse any brake fluid. Any brake fluid that is removed from the system should be discarded. Also, do not allow any brake fluid to come in contact with a painted or plastic surface; it will damage the paint.

• Never operate the engine without the proper amount and type of engine oil; doing so will result in severe engine damage.

• Timing belt maintenance is extremely important! Many models utilize an interference- type, non freewheeling engine. If the timing belt breaks, the valves in the cylinder head may strike the pistons, causing potentially serious (also time-consuming and expensive) engine damage.

• Disconnecting the negative battery cable on some vehicles may interfere with the functions of the on-board computer system(s) and may require the computer to undergo a relearning process once the negative battery cable is reconnected.

• Steering and suspension fasteners are critical parts because they affect performance of vital components and systems and their failure can result in major service expense. They must be replaced with the same grade or part number or an equivalent part if replacement is necessary. Do not use a replacement part of lesser quality or substitute design. Torque values must be used as specified during reassembly.

INFINITI

EX35 • EX35 Journey

SPECIFICATIONS AND MAINTENANCE CHARTS

ENGINE AND VEHICLE IDENTIFICATION

		Engine					Model Year	
Code	Liters (cc)	Cu. In.	Cyl.	Fuel Sys.	Engine Type	Eng. Mfg.	Code ①	Year
VQ35HR	3.5 (3498)	213.45	6	EFI	DOHC	Nissan	B	2011
							C	2012

① 10th position of VIN

71075_EX35_C0001

GENERAL ENGINE SPECIFICATIONS

All measurements are given in inches.

Year	Model	Engine Disp. Liters (cc)	Engine ID	Fuel System Type	Net Horsepower @ rpm	Net Torque @ rpm (ft. lbs.)	Bore x Stroke (in.)	Compression Ratio	Oil Pressure @ rpm
2011	EX35/EX35 Journey	3.5 (3498)	VQ35HR	EFI	297@6800	253@4800	95.5x81.4	10.6:0	43@2000
2012	EX35/EX35 Journey	3.5 (3498)	VQ35HR	EFI	297@6800	253@4800	95.5x81.4	10.6:0	43@2000

71075_EX35_C0002

ENGINE TUNE-UP SPECIFICATIONS

Year	Engine Displacement Liters	Engine ID	Spark Plug Gap (in.)	Ignition Timing (deg.) ① AT	Fuel Pump (psi)	Idle Speed (rpm) AT	Valve Clearance ② Intake	Valve Clearance ② Exhaust
2011	3.5	VQ35HR	0.043	11-21 B	51	600-700	0.010-0.013	0.011-0.015
2012	3.5	VQ35HR	0.043	11-21 B	51	600-700	0.010-0.013	0.011-0.015

B: Before top dead center

① Under the following conditions:

A/C switch; OFF

Electric load; OFF (Lights, heater fan & rear window defogger)

Steering wheel; Kept in straight-ahead position

② Engine cold - approximately 68°F (20°C)

71075_EX35_C0003

CAPACITIES

Year	Model	Engine Displacement Liters	Engine ID	Engine Oil with Filter (qts.)	Transmission (pts.) Auto.	Transfer Case (pts.)	Fuel Tank (gal.)	Cooling System (qts.)
2011	EX35	3.5	VQ35HR	5.0	19.5	2.3	20.0	9.1
2012	EX35	3.5	VQ35HR	5.0	19.5	2.3	20.0	9.1

NOTE: All capacities are approximate. Add fluid gradually and ensure a proper fluid level is obtained.

71075_EX35_C0004

FLUID SPECIFICATIONS

Year	Model	Engine Disp. Liters	Engine Oil	Auto. Trans.	Drive Axle	Transfer Case	Power Steering Fluid	Brake Master Cylinder	Cooling System
2011	EX35/EX35 Journey	3.5	5W-30	①	②	③	④	⑤	⑥
2012	EX35/EX35 Journey	3.5	5W-30	①	②	③	④	⑤	⑥

DOT: Department Of Transpotation

① Genuine NISSAN Matic S ATF

② Genuine NISSAN Differential Oil Hypoid Super GL-5 80W-90 or API GL-5 Viscosity SAE 80W-90

③ Genuine NISSAN Matic J ATF or equivalent (if available)

④ Genuine NISSAN PSF or equivalent

⑤ Genuine NISSAN Super Heavy Duty Brake Fluid or equivalent DOT 3

⑥ Genuine NISSAN Long Life Antifreeze/Coolant (blue) or equivalent

71075_EX35_C0005

VALVE SPECIFICATIONS

Year	Engine Displacement Liters	Engine ID	Seat Angle (deg.)	Face Angle (deg.)	Spring Test Pressure (lbs. @ in.)	Spring Free-Length (in.)	Spring Installed Height (in.)	Stem-to-Guide Clearance (in.) Intake	Stem-to-Guide Clearance (in.) Exhaust	Stem Diameter (in.) Intake	Stem Diameter (in.) Exhaust
2011	3.5	VQ35HR	45.15-45.45	44.23-45.08	84-95 @1.071	1.726	1.457	0.0008-0.0021	0.0012-0.0022	0.2348-0.2354	0.2347-0.2350
2012	3.5	VQ35HR	45.15-45.45	44.23-45.08	84-95 @1.071	1.726	1.457	0.0008-0.0021	0.0012-0.0022	0.2348-0.2354	0.2347-0.2350

71075_EX35_C0006

CAMSHAFT SPECIFICATIONS
All measurements in inches unless noted

Year	Engine Displacement Liters	Engine Code	Journal Diameter	Brg. Oil Clearance	Shaft End-play	Runout	Journal Bore	Lobe Height Intake	Lobe Height Exhaust
2011	3.5	VQ35HR	①	②	0.0045-0.0074	0.0008 ③	NA	1.8057-1.8132	1.8061-1.8136
2012	3.5	VQ35HR	①	②	0.0045-0.0074	0.0008 ③	NA	1.8057-1.8132	1.8061-1.8136

NA: Not Available

① No. 1: 1.0211-1.0218 in.
   No. 2, 3, 4: 0.9230-0.9238 in.

② No. 1: 0.0018-0.0034 in.
   No. 2, 3, 4: 0.0014-0.0030 in.

③ 0.0020 maximum runout

71075_EX35_C0007

CRANKSHAFT AND CONNECTING ROD SPECIFICATIONS
All measurements are given in inches.

Year	Engine Displacement Liters	Engine ID	Crankshaft Main Brg. Journal Dia.	Crankshaft Main Brg. Oi Clearance	Crankshaft Shaft End-play	Crankshaft Thrust on No.	Connecting Rod Journal Diameter	Connecting Rod Oil Clearance	Connecting Rod Side Clearance
2011	3.5	VQ35HR	①	0.0014-0.0018	0.0039-0.0098	3	②	0.0016-0.0021	0.0079-0.0138
2012	3.5	VQ35HR	①	0.0014-0.0018	0.0039-0.0098	3	②	0.0016-0.0021	0.0079-0.0138

① Depends on the grade of the crankshaft. The mominal range is: 2.5581-2.5580 in.

② Grade A: 2.2441-2.2441
   Grade B: 2.2441-2.2442
   Grade C: 2.2442-2.2442

71075_EX35_C0008

PISTON AND RING SPECIFICATIONS
All measurements are given in inches.

Year	Engine Displacement Liters	Engine ID	Piston Clearance	Ring Gap Top Compression	Ring Gap Bottom Compression	Ring Gap Oil Control	Ring Side Clearance Top Compression	Ring Side Clearance Bottom Compression	Ring Side Clearance Oil Control
2011	3.5	VQ35HR	0.0004-0.0012	0.0091-0.0130	0.0130-0.0189	0.0067-0.0185	0.0016-0.0031	0.0012-0.0028	0.0022-0.0061
2012	3.5	VQ35HR	0.0004-0.0012	0.0091-0.0130	0.0130-0.0189	0.0067-0.0185	0.0016-0.0031	0.0012-0.0028	0.0022-0.0061

71075_EX35_C0009

TORQUE SPECIFICATIONS
All readings in ft. lbs.

Year	Engine Disp. Liters	Engine ID	Cylinder Head Bolts	Main Bearing Bolts	Rod Bearing Bolts	Crankshaft Damper Bolts	Flywheel Bolts	Manifold Intake	Exhaust	Spark Plugs	Oil Pan Drain Plug
2011	3.5	VQ35HR	①	②	③	④	N/A	⑤	⑥	14	25
2012	3.5	VQ35HR	①	②	③	④	N/A	⑤	⑥	14	25

NOTE: Dip main bearing bolts, crankshaft damper bolt, and flywheel bolts in clean engine oil prior to tightening.

N/A: Not available

① Step 1: Tighten all in sequence to 77 ft. lbs.

  Step 2: Completely loosen all in reverse sequence

  Step 3: Tighten in sequence to 30 ft. lbs.

  Step 4: Tighten 95 degress clockwise

  Step 5: Tighten another 95 degress clockwise

② Step 1: Tighten in sequence to 10 ft. lbs.

  Step 2: Tighten in sequence to 26 ft. lbs.

  Step 3: Tighten 90 degrees

③ Step 1: Tighten to 21 ft. lbs.

  Step 2: Completely loosen all in reverse sequence

  Step 3: Tighten to 18 ft. lbs.

  Step 4: Tighten another 90 degrees clockwise

④ Step 1: Tighten to 69 ft. lbs.

  Step 2: Tighten 90 degrees

⑤ Step 1: Tighten to 60 inch lbs.

  Step 2: Tighten to 19 ft. lbs.

⑥ Tighten to 51 inch lbs.

71075_EX35_C0010

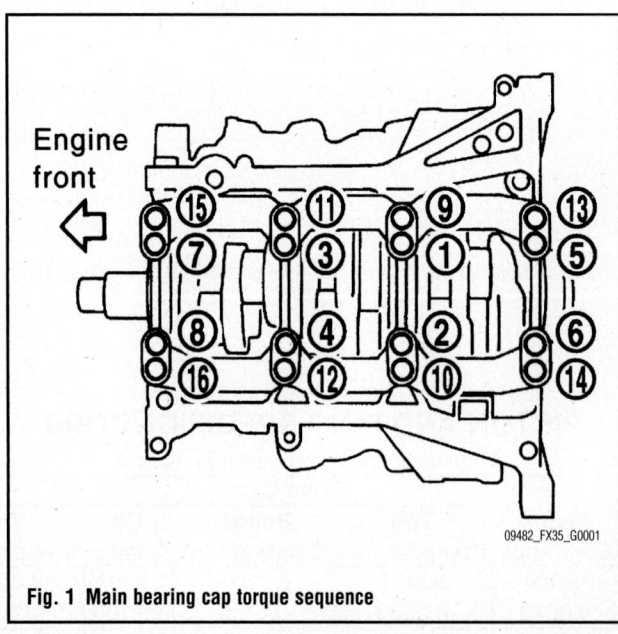

09482_FX35_G0001

Fig. 1 Main bearing cap torque sequence

WHEEL ALIGNMENT

Year	Model		Caster Range (+/-Deg.)	Caster Preferred Setting (Deg.)	Camber Range (+/-Deg.)	Camber Preferred Setting (Deg.)	Toe-in (in.)
2011	EX35	F	①	①	①	①	0.040+/-0.080
		R	NA	NA	①	①	0.114+/-0.228
2012	EX35	F	①	①	①	①	0.040+/-0.080
		R	NA	NA	①	①	0.114+/-0.228

NOTE: Measurements are given for unladen vehicle: fuel, engine coolant, and fluid levels are full. Spare tire, jack, hand tools, and mats are in designated positions.

NA: Not applicable

① 2WD: Camber Preferred Setting; Front - 0.08, Rear - -0.58

  Camber Range; Front - 0.55, Rear - 0.50

  Caster Preferred Setting; Front - 4.25,

  Caster Range; Front - 0.75

  4WD: Camber Preferred Setting; Front - -0.33, Rear - -0.58

  Camber Range; Front - 0.75, Rear - 0.50

  Caster Preferred Setting; Front - 4.17

  Caster Range; Front - 0.75

71075_EX35_C0011

TIRE, WHEEL AND BALL JOINT SPECIFICATIONS

Year	Model	OEM Tires Standard	OEM Tires Optional ①	Tire Pressures (psi) Front	Tire Pressures (psi) Rear	Wheel Size	Ball Joint Inspection	Lug Nut (ft. lbs.)
2011	EX35	225/55VR18	245/45VR19	33	33	8.0J x 18	②	80
2012	EX35	225/55VR18	245/45VR19	33	33	8.0J x 18	②	80

OEM: Original Equipment Manufacturer

PSI: Pounds Per Square Inch

NA: Information not available

① Available Only on the Journey

② Replace if any measurable axial end play is found.

71075_EX35_C0012

BRAKE SPECIFICATIONS

All measurements in inches unless noted

Year	Model		Brake Disc Original Thickness	Brake Disc Minimum Thickness	Brake Disc Max. Runout	Minimum Pad/Lining Thickness Front	Minimum Pad/Lining Thickness Rear	Brake Caliper Bracket Bolts (ft. lbs.)	Brake Caliper Mounting Bolts (ft. lbs.)
2011	EX35	F	1.772	1.024	0.0014	0.079	—	NA	98
		R	1.687	0.551	0.0022	—	0.079	NA	62
2012	EX35	F	1.772	1.024	0.0014	0.079	—	NA	98
		R	1.687	0.551	0.0022	—	0.079	NA	62

F: Front

R: Rear

NA: Information not available

71075_EX35_C0013

SCHEDULED MAINTENANCE INTERVALS
INFINITI—EX35 & EX35 Journey

TO BE SERVICED	TYPE OF SERVIC	VEHICLE MILEAGE INTERVAL (x1000)														
		7.5	15	22.5	30	37.5	45	52.5	60	67.5	75	82.5	90	97.5	105	120
Accessory drive belts ①	S/I								✓							✓
Air cleaner element (engine)	R				✓				✓				✓			✓
Air conditioner system	S/I	Inspect system operation annually														
Automatic transaxle fluid	S/I		✓		✓		✓		✓		✓		✓		✓	✓
Brake lines, hoses, cables, and connections	S/I		✓		✓		✓		✓		✓		✓		✓	✓
Brake pads, calipers, & rotors	S/I		✓		✓		✓		✓		✓		✓		✓	✓
Differential gear oil	S/I		✓		✓		✓		✓		✓		✓		✓	✓
Driveshafts and CV-boots	S/I		✓		✓		✓		✓		✓		✓		✓	✓
Engine coolant	R								✓				✓			✓
Engine oil and filter	R	✓	✓	✓	✓	✓	✓	✓	✓	✓	✓	✓	✓	✓	✓	✓
Exhaust pipe connections, muffler, and suspension bolts	S/I				✓				✓				✓			✓
EVAP vapor lines	S/I				✓				✓				✓			✓
Fuel lines and connections	S/I				✓				✓				✓			✓
In-cabin microfilter	S/I		✓		✓		✓		✓		✓		✓		✓	✓
Spark plugs (Platinum-tipped)	R	105,000 miles (under normal usage)														
Steering system	S/I				✓				✓				✓			✓
Suspension system	S/I				✓				✓				✓			✓
Transfer case fluid	S/I		✓		✓		✓		✓		✓		✓		✓	✓
Valve clearance	S/I	Whenever valve noise increases														

R: Replace   S/I: Service or Inspect

① Replace if worn or damaged or if the auto-tensioner has reached its limit (V8)

FREQUENT OPERATION MAINTENANCE (SEVERE SERVICE)

If a vehicle is operated under any of the following conditions it is considered severe service:

- Extremely dusty areas.

- 50% or more of the vehicle operation is in 90°F (32°C) or higher temperatures, or constant operation in temperatures below 32°F (0°C).

- Prolonged idling (vehicle operation in stop and go traffic).

- Frequent short running periods (engine does not warm to normal operating temperatures).

- Police, taxi, delivery usage, or trailer towing usage.

Automatic transaxle fluid (and filter), transfer case fluid, and differential gear oil: check every 15,000 miles, replace every 30,000 miles

Brake pads, calipers & rotors: service or inspect every 7,500 miles

Driveshafts and CV-boots inspect every 7,500 miles

Exhaust system inspect every 7,500 miles

Oil and oil filter: change every 3,750 miles

Steering system and suspension components inspect for looseness and damage every 7,500 miles

PRECAUTIONS

Before servicing any vehicle, please be sure to read all of the following precautions, which deal with personal safety, prevention of component damage, and important points to take into consideration when servicing a motor vehicle:

• Never open, service or drain the radiator or cooling system when the engine is hot; serious burns can occur from the steam and hot coolant.

• Observe all applicable safety precautions when working around fuel. Whenever servicing the fuel system, always work in a well-ventilated area. Do not allow fuel spray or vapors to come in contact with a spark, open flame, or excessive heat (a hot drop light, for example). Keep a dry chemical fire extinguisher near the work area. Always keep fuel in a container specifically designed for fuel storage; also, always properly seal fuel containers to avoid the possibility of fire or explosion. Refer to the additional fuel system precautions later in this section.

• Fuel injection systems often remain pressurized, even after the engine has been turned **OFF**. The fuel system pressure must be relieved before disconnecting any fuel lines. Failure to do so may result in fire and/or personal injury.

• Brake fluid often contains polyglycol ethers and polyglycols. Avoid contact with the eyes and wash your hands thoroughly after handling brake fluid. If you do get brake fluid in your eyes, flush your eyes with clean, running water for 15 minutes. If eye irritation persists, or if you have taken brake fluid internally, IMMEDIATELY seek medical assistance.

• The EPA warns that prolonged contact with used engine oil may cause a number of skin disorders, including cancer. You should make every effort to minimize your exposure to used engine oil. Protective gloves should be worn when changing oil. Wash your hands and any other exposed skin areas as soon as possible after exposure to used engine oil. Soap and water, or waterless hand cleaner should be used.

• All new vehicles are now equipped with an air bag system, often referred to as a Supplemental Restraint System (SRS) or Supplemental Inflatable Restraint (SIR) system. The system must be disabled before performing service on or around system components, steering column, instrument panel components, wiring and sensors. Failure to follow safety and disabling procedures could result in accidental air bag deployment, possible personal injury and unnecessary system repairs.

• Always wear safety goggles when working with, or around, the air bag system. When carrying a non-deployed air bag, be sure the bag and trim cover are pointed away from your body. When placing a non-deployed air bag on a work surface, always face the bag and trim cover upward, away from the surface. This will reduce the motion of the module if it is accidentally deployed. Refer to the additional air bag system precautions later in this section.

• Clean, high quality brake fluid from a sealed container is essential to the safe and proper operation of the brake system. You should always buy the correct type of brake fluid for your vehicle. If the brake fluid becomes contaminated, completely flush the system with new fluid. Never reuse any brake fluid. Any brake fluid that is removed from the system should be discarded. Also, do not allow any brake fluid to come in contact with a painted surface; it will damage the paint.

• Never operate the engine without the proper amount and type of engine oil; doing so WILL result in severe engine damage.

• Timing belt maintenance is extremely important. Many models utilize an interference-type, non-freewheeling engine. If the timing belt breaks, the valves in the cylinder head may strike the pistons, causing potentially serious (also time-consuming and expensive) engine damage. Refer to the maintenance interval charts for the recommended replacement interval for the timing belt, and to the timing belt section for belt replacement and inspection.

• Disconnecting the negative battery cable on some vehicles may interfere with the functions of the on-board computer system(s) and may require the computer to undergo a relearning process once the negative battery cable is reconnected.

• When servicing drum brakes, only disassemble and assemble one side at a time, leaving the remaining side intact for reference.

• Only an MVAC-trained, EPA-certified automotive technician should service the air conditioning system or its components.

BRAKES

ANTI-LOCK BRAKE SYSTEM

PRECAUTIONS

• Certain components within the ABS system are not intended to be serviced or repaired individually.

• Do not use rubber hoses or other parts not specifically specified for and ABS system. When using repair kits, replace all parts included in the kit. Partial or incorrect repair may lead to functional problems and require the replacement of components.

• Lubricate rubber parts with clean, fresh brake fluid to ease assembly. Do not use shop air to clean parts; damage to rubber components may result.

• Use only DOT 3 brake fluid from an unopened container.

• If any hydraulic component or line is removed or replaced, it may be necessary to bleed the entire system.

• A clean repair area is essential. Always clean the reservoir and cap thoroughly before removing the cap. The slightest amount of dirt in the fluid may plug an orifice and impair the system function. Perform repairs after components have been thoroughly cleaned; use only denatured alcohol to clean components. Do not allow ABS components to come into contact with any substance containing mineral oil; this includes used shop rags.

• The Anti-Lock control unit is a microprocessor similar to other computer units in the vehicle. Ensure that the ignition switch is **OFF** before removing or installing controller harnesses. Avoid static electricity discharge at or near the controller.

• If any arc welding is to be done on the vehicle, the control unit should be unplugged before welding operations begin.

WHEEL SPEED SENSORS

REMOVAL & INSTALLATION

Front

See Figure 2.

1. Be careful with the following when removing sensor.

    a. Do not twist sensor harness as much as possible, when removing it. Pull sensors out without pulling sensor harness.

    b. Be careful to avoid damaging sensor edges or rotor teeth. Remove wheel sensor first before removing front or rear wheel hub. This is to avoid damage to sensor wiring and loss of sensor function.

To install:

2. Be careful with the following when installing wheel sensor. Tighten installation bolts to the specified torques. Refer to the accompanying illustration.

a. When installing, make sure there is no foreign material such as iron chips on and in the mounting hole of the wheel sensor. Make sure no foreign material has been caught in the sensor rotor. Remove any foreign material and clean the mount.

b. When installing wheel sensor, be sure to press rubber grommets in until they lock at locations shown above in the figure. When installed, harness must not be twisted.

c. When you see the harness of the wheel sensor from the front side of the vehicle ensure that the white lines as shown in the accompanying illustration are not twisted.

Rear

See Figure 3.

1. Be careful with the following when removing sensor.

a. Do not twist sensor harness as much as possible, when removing it. Pull sensors out without pulling sensor harness.

b. Be careful to avoid damaging sensor edges or rotor teeth. Remove wheel sensor first before removing front or rear wheel hub. This is to avoid damage to sensor wiring and loss of sensor function.

To install:

2. Be careful with the following when installing wheel sensor. Tighten installation bolts to 120 inch lbs. (13 Nm).

a. When installing, make sure there is no foreign material such as iron chips on and in the mounting hole of the wheel sensor. Make sure no foreign material has been caught in the sensor rotor. Remove any foreign material and clean the mount.

b. When installing a rear LH wheel sensor, be sure to pass the wheel sensor harness under the breather hose.

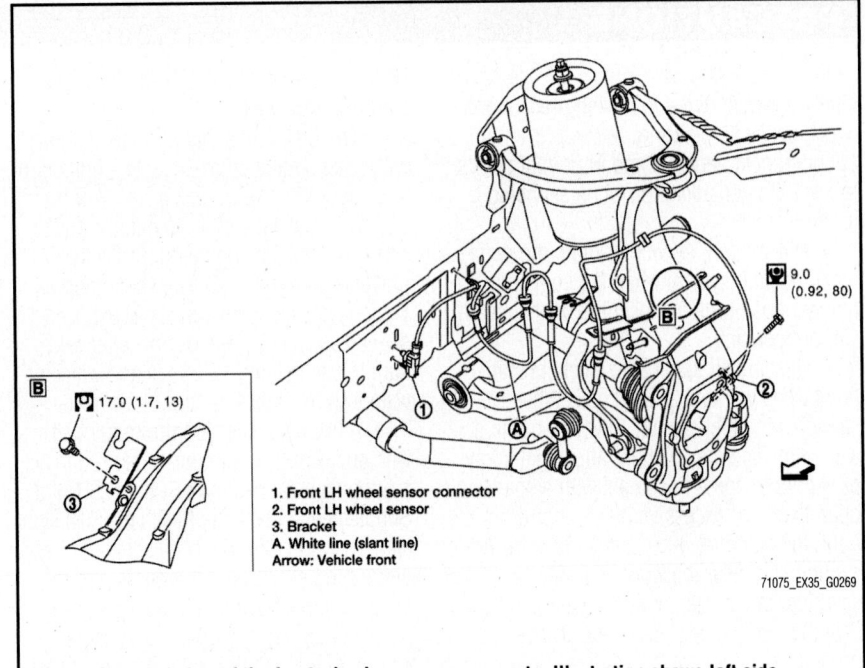

1. Front LH wheel sensor connector
2. Front LH wheel sensor
3. Bracket
A. White line (slant line)
Arrow: Vehicle front

71075_EX35_G0269

Fig. 2 Exploded view of the front wheel sensor components. Illustration shows left side. Right side is the mirror image

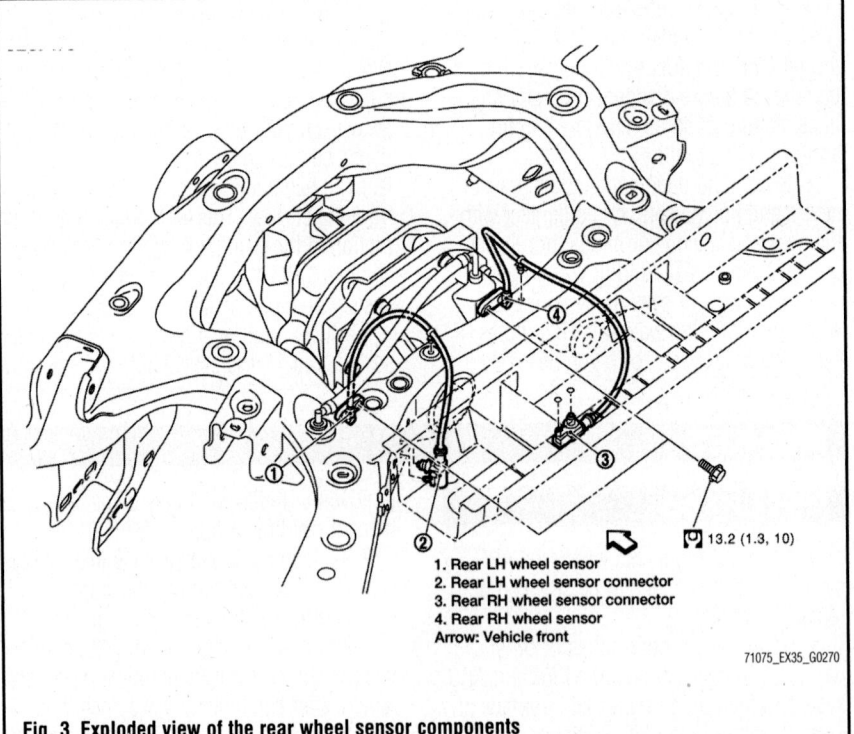

1. Rear LH wheel sensor
2. Rear LH wheel sensor connector
3. Rear RH wheel sensor connector
4. Rear RH wheel sensor
Arrow: Vehicle front

71075_EX35_G0270

Fig. 3 Exploded view of the rear wheel sensor components

BRAKES   **BLEEDING THE BRAKE SYSTEM**

BLEEDING PROCEDURE

MANUAL

When any part of the hydraulic system has been disconnected for repair or replacement, air may get into the lines and cause spongy pedal action (because air can be compressed and brake fluid cannot). To correct this condition, it is necessary to bleed the hydraulic system so to be sure all air is purged.

When bleeding the brake system, bleed one brake cylinder at a time, beginning at the cylinder with the longest hydraulic line (farthest from the master cylinder) first. ALWAYS keep the master cylinder reservoir filled with brake fluid during the bleeding operation. Never use brake fluid that has been drained from the hydraulic system, no matter how clean it is.

The primary and secondary hydraulic brake systems are separate and are bled independently. During the bleeding operation, do not allow the reservoir to run dry. Keep the master cylinder reservoir filled with brake fluid.

1. Clean all dirt from around the master cylinder fill cap, remove the cap and fill the master cylinder with brake fluid until the level is within ¼ in. (6mm) of the top edge of the reservoir.
2. Clean the bleeder screws at all 4 wheels. The bleeder screws are located on the top of the brake calipers.
3. Attach a length of rubber hose over the bleeder screw and place the other end of the hose in a glass jar, submerged in brake fluid.

4. Open the bleeder screw ½–¾ turn. Have an assistant slowly depress the brake pedal.

❊❊ CAUTION

Brake fluid contains polyglycol ethers and polyglycols. Avoid contact with the eyes and wash your hands thoroughly after handling brake fluid. If you do get brake fluid in your eyes, flush your eyes with clean, running water for 15 minutes. If eye irritation persists, or if you have taken brake fluid internally, IMMEDIATELY seek medical assistance.

5. Close the bleeder screw and tell your assistant to allow the brake pedal to return slowly. Continue this process to purge all air from the system.
6. When bubbles cease to appear at the end of the bleeder hose, close the bleeder screw and remove the hose. Tighten the bleeder screw to the proper torque.
7. Check the master cylinder fluid level and add fluid accordingly. Do this after bleeding each wheel.
8. Repeat the bleeding operation at the remaining 3 wheels, ending with the one closet to the master cylinder.
9. Fill the master cylinder reservoir to the proper level.

BLEEDING THE MASTER CYLINDER

Bench Bleeding

1. Before servicing the vehicle, refer to the Precautions Section.

2. If removed from the vehicle, clamp the master cylinder in a vise with soft-jaw caps.
3. Attach the special tools for bleeding the master cylinder in the following fashion:
   a. Thread the bleeder tube adapters into the primary and secondary outlet ports of the master cylinder and tighten the adapters.
   b. Thread a bleeder tube into each adapter and tighten the tube nuts.
   c. Flex each bleeder tube and place the open ends into the neck of the master cylinder reservoir. Position the open ends of the tubes into the reservoir so their outlets are below the surface of the brake fluid in the reservoir when filled.

➡**Make sure the ends of the bleeder tubes stay below the surface of the brake fluid in the reservoir at all times during the bleeding procedure.**

4. Fill the brake fluid reservoir with fresh brake fluid (DOT 3).
5. Using an appropriately sized wooden dowel as a pushrod, slowly press the pistons inward discharging brake fluid through the bleeder tubes, then release the pressure, allowing the pistons to return to the released position. Repeat this several times until all air bubbles are expelled from the master cylinder bore and bleeder tubes.
6. Remove the bleeder tubes and adapters from the master cylinder and plug the master cylinder outlet ports.
7. Install the fill cap on the reservoir.
8. Remove the master cylinder from the vise.
9. Install the master cylinder on the vehicle.

BRAKES   **FRONT DISC BRAKES**

BRAKE CALIPERS

INSPECTION

AFTER INSTALLATION
1. Check a drag of front disc brake. If any drag is found, follow the procedure described below.
2. Remove brake pads.
3. Press the pistons.

❊❊ WARNING

Never damage the piston boot.

➡**When replacing a pad with new one, check a brake fluid level in the reservoir tank because brake**

fluid returns to master cylinder reservoir tank when pressing piston in.

➡**Use a disc brake piston tool to easily press piston.**

4. Install brake pads.
5. Depress the brake pedal several times.
6. Check a drag of front disc brake again. If any drag is found, disassemble the cylinder body.
7. Burnish contact surface between disc rotors and brake pads according after refinishing or replacing disc rotor.

REMOVAL & INSTALLATION
See Figures 4 and 5.

➡**Clean any dust from the brake caliper and brake pads with a vacuum dust collector. Never blow with compressed air.**

➡**Never depress the brake pedal. Brake fluid may splash while removing the brake hose.**

1. Remove tires with power tool.
2. Fix the disc rotor using wheel nuts.
3. Drain brake fluid.

➡**Never spill or splash brake fluid on the disc rotor.**

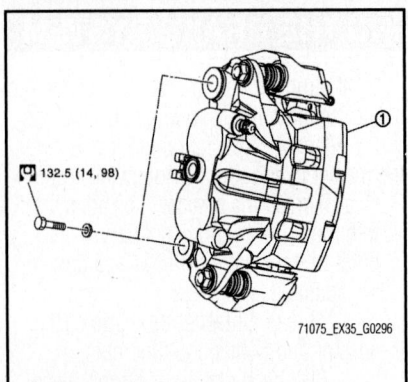

Fig. 4 Front disc brake caliper assembly (1)

4. Remove union bolt and copper washer, and disconnect brake hose from caliper assembly.

5. Remove torque member mounting bolts, and remove brake caliper assembly.

➡**Never drop brake pad and caliper assembly.**

6. Remove disc rotor.

➡**Put matching marks on the wheel hub and bearing assembly and the disc rotor before removing the disc rotor.**

➡**Never drop disc rotor.**

To install:

➡**Clean any dust from the brake caliper and brake pads with a vacuum dust collector. Never blow with compressed air.**

➡**Never depress the brake pedal. Brake fluid may splash while removing the brake hose.**

7. Install disc rotor.

➡**Align the matching marks that have been made during removal when reusing the disc rotor.**

8. Install the brake caliper assembly to the vehicle and tighten the torque member mounting bolts to the specified torque.

➡**Never spill or splash any grease and moisture on the brake caliper assembly mounting face, threads, mounting bolts and washers. Wipe out any grease and moisture.**

9. Install brake hose and copper washers to brake caliper assembly, and tighten union bolts to 13 ft. lbs. (18.2 Nm)

➡**Never reuse copper washer.**

10. Refill with new brake fluid and perform the air bleeding.

➡**Never reuse drained brake fluid.**

➡**Never spill or splash brake fluid on the disc rotor.**

11. Check a drag of front disc brake. If any drag is found, refer to Inspection in Front Disc Brakes in the Brakes Section.

BRAKE PADS

ADJUSTMENT

See Figure 6.

INSPECTION

Check brake pad wear thickness from an inspection hole on cylinder body. Check using a scale if necessary.

ADJUSTMENT

➡**Burnish contact surfaces between pads according to the following procedure after refinishing or replacing pads, or if a soft pedal occurs at very low mileage.**

➡**Be careful of vehicle speed because the brake does not operate firmly/securely until pads and disc rotor are securely fitted.**

➡**Only perform this procedure under safe road and traffic conditions. Use extreme caution.**

1. Drive vehicle on straight, flat road.

2. Depress brake pedal with the power to stop vehicle within 3 to 5 seconds until the vehicle stops.

3. Drive without depressing brake for a few minutes to cool the brake.

4. Repeat steps 1 to 3 until pad and disc rotor are securely fitted.

INSPECTION

INSPECTION AFTER REMOVAL

1. Replace the shims and the shim covers if rust is excessively attached.

INSPECTION AFTER INSTALLATION

2. Check a drag of front disc brake. If any drag is found, follow the procedure described below.

3. Remove brake pads.

4. Press the pistons.

❊❊ WARNING

Never damage the piston boot.

➡**When replacing a pad with new one, check a brake fluid level in the reservoir tank because brake fluid returns to master cylinder reservoir tank when pressing piston in.**

➡**Use a disc brake piston tool to easily press piston.**

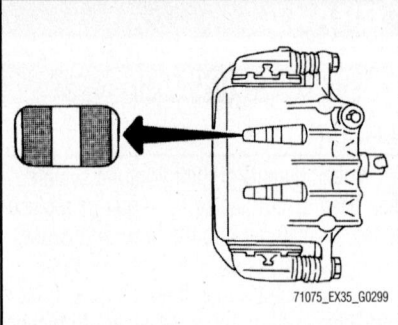

Fig. 6 Check brake pad wear thickness from an inspection hole on cylinder body. Check using a scale if necessary

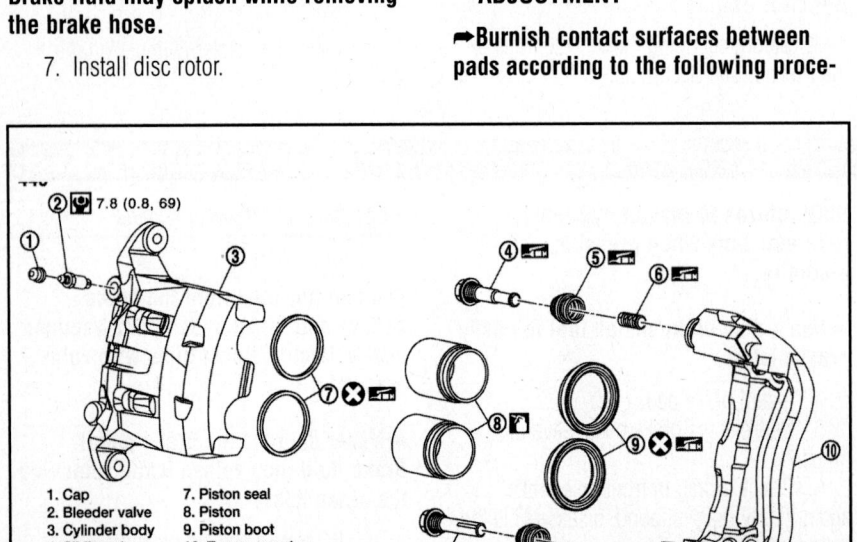

1. Cap
2. Bleeder valve
3. Cylinder body
4. Sliding pin
5. Sliding pin boot
6. Bushing
7. Piston seal
8. Piston
9. Piston boot
10. Torque member
Grease Gun: Apply rubber grease.
Oil/Fluid Gun: Apply brake fluid.

Fig. 5 Exploded view of the front disc brake caliper assembly components

5. Install brake pads.

6. Depress the brake pedal several times.

7. Check a drag of front disc brake again. If any drag is found, disassemble the cylinder body.

8. Burnish contact surfaces after refinishing or replacing brake pads, or if a soft pedal occurs at very low mileage.

REMOVAL & INSTALLATION

See Figures 7 and 8.

➡**Clean any dust from the brake caliper and brake pads with a vacuum dust collector. Never blow with compressed air.**

➡**Never depress the brake pedal while removing the brake pads because the piston may pop out.**

➡**Never spill or splash brake fluid on the disc rotor.**

1. Remove tires with power tool.
2. Remove lower sliding pin bolt.
3. Suspend the cylinder body with suitable wire so that the brake hose will not stretch. Then remove the brake pads, shims, shim covers and pad retainers from the torque member.

➡**Never deform the pad retainer when removing the pad retainer from the torque member.**

✳✳ WARNING

Never damage the piston boot.

➡**Never drop the brake pads, shims, and the shim covers.**

➡**Remember each position of the removed brake pads.**

To install:

➡**Clean any dust from the brake caliper and brake pads with a vacuum dust collector. Never blow with compressed air.**

➡**Never depress the brake pedal while removing the brake pads or the cylinder body because the piston may pop out.**

➡**Never spill or splash brake fluid on the disc rotor.**

4. Apply Copper based brake grease to the pad retainers before installing it to the torque member if the pad retainers has been removed.

➡**Securely assemble the pad retainers so that it will not be lifted up from the torque member.**

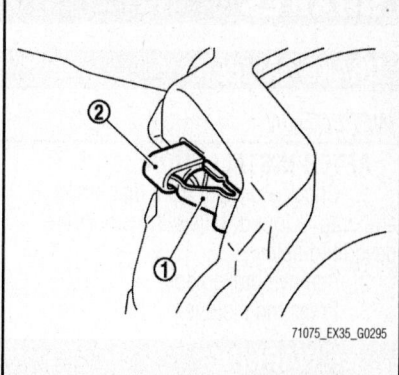

71075_EX35_G0295

Fig. 8 Both inner and outer pads have a pad return system on the pad retainer. Install pad return lever (1) securely to pad wear sensor (2).

➡**Never deform the pad retainers.**

5. Apply Copper based brake grease to the mating faces between the shims and the shim covers and install them to the brake pad.

➡**Always replace the shims together with the shim covers when replacing the brake pad.**

6. Install the brake pads to the torque member.

➡**Both inner and outer pads have a pad return system on the pad retainer. Install pad return lever securely to pad wear sensor.**

7. Install cylinder body to torque member.

✳✳ WARNING

Never damage the piston boot.

➡**When replacing brake pad with new one, check a brake fluid level in the reservoir tank because brake fluid returns to master cylinder reservoir tank when pressing piston in.**

➡**Use a disc brake piston tool to easily press piston.**

8. Install the lower sliding pin bolt and tighten it to the specified torque.

9. Depress the brake pedal several times to check that no drag feel is present for the front disc brake. Refer to Inspection in Front Disc Brakes in the Brakes Section.

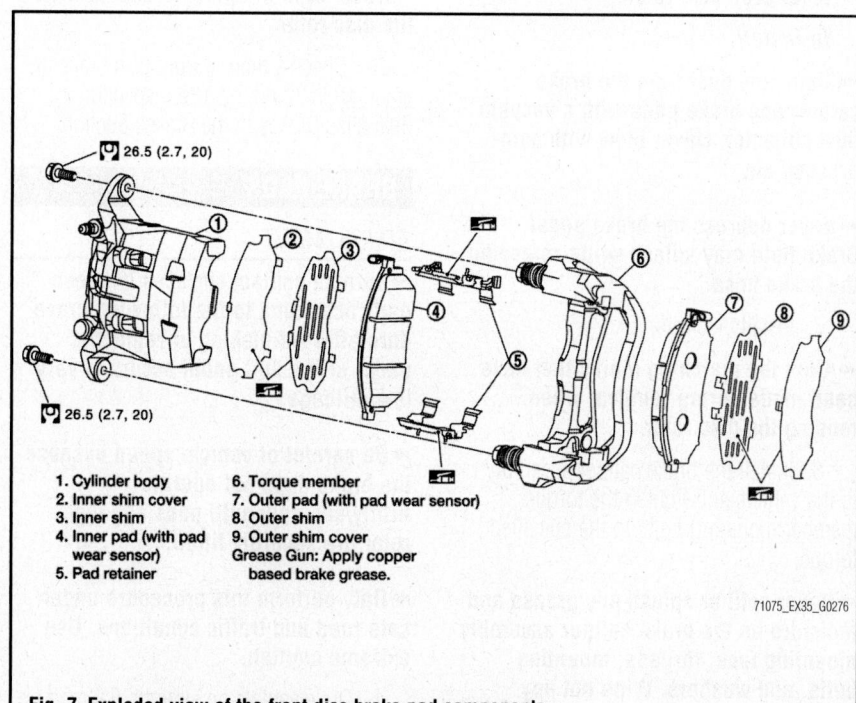

26.5 (2.7, 20)

26.5 (2.7, 20)

1. Cylinder body
2. Inner shim cover
3. Inner shim
4. Inner pad (with pad wear sensor)
5. Pad retainer
6. Torque member
7. Outer pad (with pad wear sensor)
8. Outer shim
9. Outer shim cover
    Grease Gun: Apply copper based brake grease.

71075_EX35_G0276

Fig. 7 Exploded view of the front disc brake pad components

BRAKES   **REAR DISC BRAKES**

BRAKE CALIPERS

INSPECTION

AFTER INSTALLATION

1. Check a drag of front disc brake. If any drag is found, follow the procedure described below.
2. Remove brake pads.
3. Press the pistons.

✳ WARNING

Never damage the piston boot.

➡**When replacing a pad with new one, check a brake fluid level in the reservoir tank because brake fluid returns to master cylinder reservoir tank when pressing piston in.**

➡**Use a disc brake piston tool to easily press piston.**

4. Install brake pads.
5. Depress the brake pedal several times.
6. Check a drag of front disc brake again. If any drag is found, disassemble the cylinder body.
7. Burnish contact surface between disc rotors and brake pads according after refinishing or replacing disc rotor.

REMOVAL & INSTALLATION

See Figure 9.

➡**Clean any dust from the brake caliper and brake pads with a vacuum dust collector. Never blow with compressed air.**

➡**Never depress the brake pedal. Brake fluid may splash while removing the brake hose.**

1. Remove tires with power tool.
2. Fix the disc rotor using wheel nuts.
3. Drain brake fluid.

➡**Never spill or splash brake fluid on the disc rotor.**

4. Remove union bolt and copper washers, and disconnect brake hose from caliper assembly.
5. Remove torque member mounting bolts, and remove brake caliper assembly.

➡**Never drop brake pad and caliper assembly.**

6. Remove disc rotor.

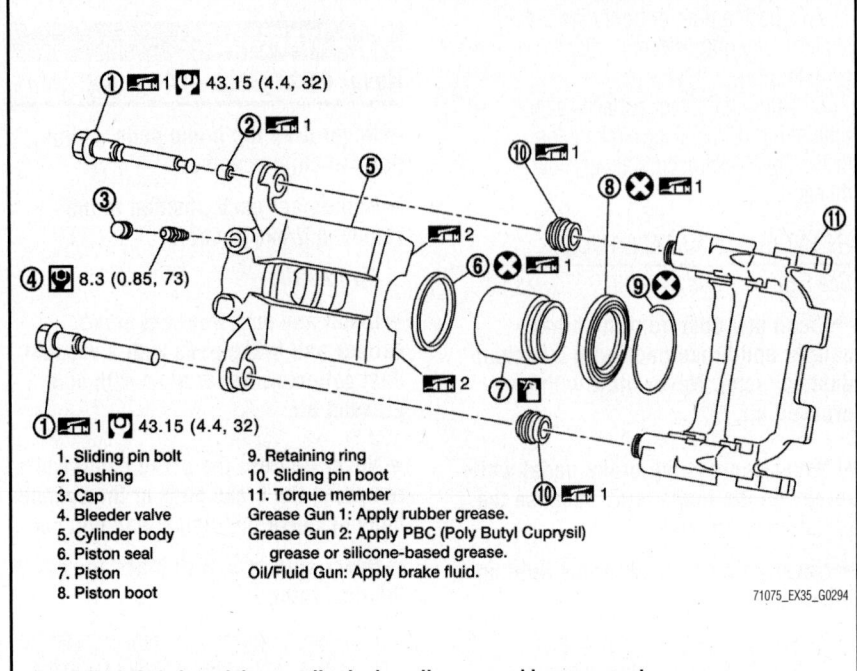

1. Sliding pin bolt
2. Bushing
3. Cap
4. Bleeder valve
5. Cylinder body
6. Piston seal
7. Piston
8. Piston boot
9. Retaining ring
10. Sliding pin boot
11. Torque member
Grease Gun 1: Apply rubber grease.
Grease Gun 2: Apply PBC (Poly Butyl Cuprysil) grease or silicone-based grease.
Oil/Fluid Gun: Apply brake fluid.

71075_EX35_G0294

Fig. 9 Exploded view of the rear disc brake caliper assembly components

➡**Put matching marks on the wheel hub and bearing assembly and the disc rotor before removing the disc rotor.**

➡**Never drop disc rotor.**

To install:

➡**Clean any dust from the brake caliper and brake pads with a vacuum dust collector. Never blow with compressed air.**

➡**Never depress the brake pedal. Brake fluid may splash while removing the brake hose.**

7. Install disc rotor.

➡**Align the matching marks that have been made during removal when reusing the disc rotor.**

8. Install the brake caliper assembly to the vehicle and tighten the torque member mounting bolts to the specified torque.

➡**Never spill or splash any grease and moisture on the brake caliper assembly mounting face, threads, mounting bolts, and washers. Wipe out any grease and moisture.**

9. Install brake hose and copper washers to brake caliper assembly, and tighten union bolts to the specified torque.

10. Refill with new brake fluid and perform the air bleeding.

➡**Never reuse drained brake fluid.**

➡**Never spill or splash brake fluid on the disc rotor.**

11. Check a drag of front disc brake. If any drag is found, refer to Inspection in Rear Disc Brakes in the Brakes Section.

BRAKE PADS

ADJUSTMENT

➡**Burnish contact surfaces between pads according to the following procedure after refinishing or replacing pads, or if a soft pedal occurs at very low mileage.**

➡**Be careful of vehicle speed because the brake does not operate firmly/securely until pads and disc rotor are securely fitted.**

➡**Only perform this procedure under safe road and traffic conditions. Use extreme caution.**

1. Drive vehicle on straight, flat road.
2. Depress brake pedal with the power to stop vehicle within 3 to 5 seconds until the vehicle stops.
3. Drive without depressing brake for a few minutes to cool the brake.

4. Repeat steps 1 to 3 until pad and disc rotor are securely fitted.

INSPECTION

AFTER REMOVAL

1. Replace the shims and the shim covers if rust is excessively attached.

AFTER INSTALLATION

2. Check a drag of front disc brake. If any drag is found, follow the procedure described below.

3. Remove brake pads.

4. Press the pistons.

✳✳ WARNING

Never damage the piston boot.

➡**When replacing a pad with new one, check a brake fluid level in the reservoir tank because brake fluid returns to master cylinder reservoir tank when pressing piston in.**

➡**Use a disc brake piston tool to easily press piston.**

5. Install brake pads.

6. Depress the brake pedal several times.

7. Check a drag of front disc brake again. If any drag is found, disassemble the cylinder body.

8. Burnish contact surfaces after refinishing or replacing brake pads, or if a soft pedal occurs at very low mileage.

REMOVAL & INSTALLATION

See Figure 10.

➡**Clean any dust from the brake caliper and brake pads with a vacuum dust collector. Never blow with compressed air.**

➡**Never depress the brake pedal while removing the brake pads or the cylinder body because the piston may pop out.**

➡**Never spill or splash brake fluid on the disc rotor.**

1. Remove tires with power tool.

2. Remove the upper sliding pin bolt.

3. Suspend the cylinder body with a wire so that the brake hose will not stretch. Remove the brake pads, shims, shim cover and pad retainers from the torque member.

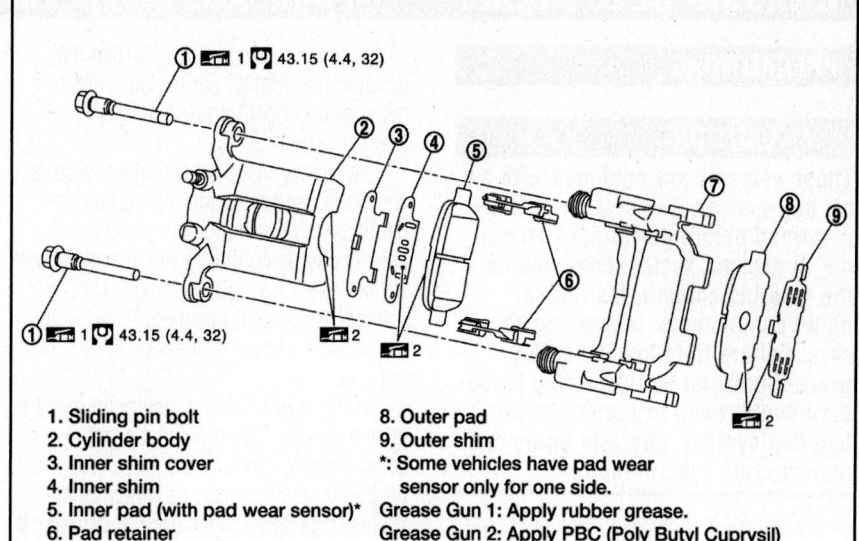

1. Sliding pin bolt
2. Cylinder body
3. Inner shim cover
4. Inner shim
5. Inner pad (with pad wear sensor)*
6. Pad retainer
7. Torque member
8. Outer pad
9. Outer shim
*: Some vehicles have pad wear sensor only for one side.
Grease Gun 1: Apply rubber grease.
Grease Gun 2: Apply PBC (Poly Butyl Cuprysil) grease or silicone-based grease.

71075_EX35_G0277

Fig. 10 Exploded view of the rear disc brake pad components

➡**Never deform the pad retainers if removing the pad retainers.**

✳✳ WARNING

Never damage the piston boot.

➡**Never drop the brake pad, shims, and the shim cover.**

➡**Remember each position of removed brake pads.**

To install:

➡**Clean any dust from the brake caliper and brake pads with a vacuum dust collector. Never blow with compressed air.**

➡**Never depress the brake pedal while removing the brake pads or the cylinder body because the piston may pop out.**

➡**Never spill or splash brake fluid on the disc rotor.**

4. Apply PBC (Poly Butyl Cuprysil) grease or silicone-based grease to the pad retainers before installing it to the torque member if the pad retainers has been removed.

➡**Securely assemble the pad retainers so that it will not be lifted up from the torque member.**

➡**Never deform the pad retainers.**

5. Apply PBC (Poly Butyl Cuprysil) grease or silicone-based grease to the mating faces between the shims and the shim cover and install them to the brake pad.

➡**Always replace the shims together with the shim cover when replacing the brake pad.**

6. Install cylinder body and brake pads to torque member.

✳✳ WARNING

Never damage the piston boot.

➡**When of replacing brake pad with new one, check a brake fluid level in the reservoir tank because brake fluid returns to master cylinder reservoir tank when pressing piston in.**

➡**Use a disc brake piston tool to easily press piston.**

7. Install the upper sliding pin bolt and tighten it to the specified torque.

8. Depress the brake pedal several times to check that no drag feel is present for the rear disc brake. Refer Inspection in Rear Disc Brakes in the Brakes Section.

CHASSIS ELECTRICAL **AIR BAGS (SUPPLEMENTAL RESTRAINT SYSTEM)**

PRECAUTIONS

✳ CAUTION

These vehicles are equipped with an air bag system. The system must be disarmed before performing service on, or around, system components, the steering column, instrument panel components, wiring and sensors. Failure to follow the safety precautions and the disarming procedure could result in accidental air bag deployment, possible injury and unnecessary system repairs.

Disconnect and isolate the battery negative cable before beginning any airbag system component diagnosis, testing, removal, or installation procedures. Allow system capacitor to discharge for two minutes before beginning any component service. This will disable the airbag system. Failure to disable the airbag system may result in accidental airbag deployment, personal injury, or death.

Do not place an intact undeployed airbag face down on a solid surface. The airbag will propel into the air if accidentally deployed and may result in personal injury or death.

When carrying or handling an undeployed airbag, the trim side (face) of the airbag should be pointing towards the body to minimize possibility of injury if accidental deployment occurs. Failure to do this may result in personal injury or death.

Replace airbag system components with OEM replacement parts. Substitute parts may appear interchangeable, but internal differences may result in inferior occupant protection. Failure to do so may result in occupant personal injury or death.

Wear safety glasses, rubber gloves, and long sleeved clothing when cleaning powder residue from vehicle after an airbag deployment. Powder residue emitted from a deployed airbag can cause skin irritation. Flush affected area with cool water if irritation is experienced. If nasal or throat irritation is experienced, exit the vehicle for fresh air until the irritation ceases. If irritation continues, see a physician.

Do not use a replacement airbag that is not in the original packaging. This may result in improper deployment, personal injury, or death.

The factory installed fasteners, screws and bolts used to fasten airbag components have a special coating and are specifically designed for the airbag system. Do not use substitute fasteners. Use only original equipment fasteners listed in the parts catalog when fastener replacement is required.

During, and following, any child restraint anchor service, due to impact event or vehicle repair, carefully inspect all mounting hardware, tether straps, and anchors for proper installation, operation, or damage. If a child restraint anchor is found damaged in any way, the anchor must be replaced. Failure to do this may result in personal injury or death.

Deployed and non-deployed airbags may or may not have live pyrotechnic material within the airbag inflator.

Do not dispose of driver/passenger/curtain airbags or seat belt tensioners unless you are sure of complete deployment. Refer to the Hazardous Substance Control System for proper disposal.

Dispose of deployed airbags and tensioners consistent with state, provincial, local, and federal regulations.

After any airbag component testing or service, do not connect the battery negative cable. Personal injury or death may result if the system test is not performed first.

If the vehicle is equipped with the Occupant Classification System (OCS), do not connect the battery negative cable before performing the OCS Verification Test using the scan tool and the appropriate diagnostic information. Personal injury or death may result if the system test is not performed properly.

Never replace both the Occupant Restraint Controller (ORC) and the Occupant Classification Module (OCM) at the same time. If both require replacement, replace one, then perform the Airbag System test before replacing the other.

Both the ORC and the OCM store Occupant Classification System (OCS) calibra-

tion data, which they transfer to one another when one of them is replaced. If both are replaced at the same time, an irreversible fault will be set in both modules and the OCS may malfunction and cause personal injury or death.

If equipped with OCS, the Seat Weight Sensor is a sensitive, calibrated unit and must be handled carefully. Do not drop or handle roughly. If dropped or damaged, replace with another sensor. Failure to do so may result in occupant injury or death.

If equipped with OCS, the front passenger seat must be handled carefully as well. When removing the seat, be careful when setting on floor not to drop. If dropped, the sensor may be inoperative, could result in occupant injury, or possibly death.

If equipped with OCS, when the passenger front seat is on the floor, no one should sit in the front passenger seat. This uneven force may damage the sensing ability of the seat weight sensors. If sat on and damaged, the sensor may be inoperative, could result in occupant injury, or possibly death.

DISARMING THE SYSTEM

1. Before servicing the vehicle, refer to the Precautions Section.
2. Turn the ignition switch to **OFF**.
3. Disconnect the negative battery cable and isolate it from accidental reconnection. Insulate the cable end with high-quality electrical tape or a similar non-conductive wrapping.
4. Wait at least 3 minutes for the system capacitor to discharge before performing any service. The air bag system is designed to retain enough voltage to deploy the air bag for a short period of time after the battery has been disconnected.

ARMING THE SYSTEM

1. Before servicing the vehicle, refer to the Precautions Section.
2. Reconnect the negative battery cable.
3. To confirm proper system operation, turn the ignition switch to the **ON** position. The SRS indicator light should light for at least 7 seconds and then go off.

DRIVELINE

AUTOMATIC TRANSMISSION

DRAIN & REFILL

See Figures 11 and 12.

➡**Use only Genuine NISSAN Matic S ATF. Never mix with other ATF.**

❋❋ WARNING

Using ATF other than Genuine NISSAN Matic S ATF will cause deterioration in driveability and A/T durability, and may damage the A/T, which is not covered by the warranty.

➡**When filling ATF, be careful not to scatter heat generating parts such as exhaust.**

  1. Step 1
    a. Install the O-ring (315268E000) to the charging pipe (310811EA5A).
  2. Step 2
    a. Use CONSULT to check that the ATF temperature is 104°F (40°C) or less.
    b. Lift up the vehicle.
    c. Remove the drain plug from the oil pan, and then drain the ATF.
    d. When the ATF starts to drip, tem-

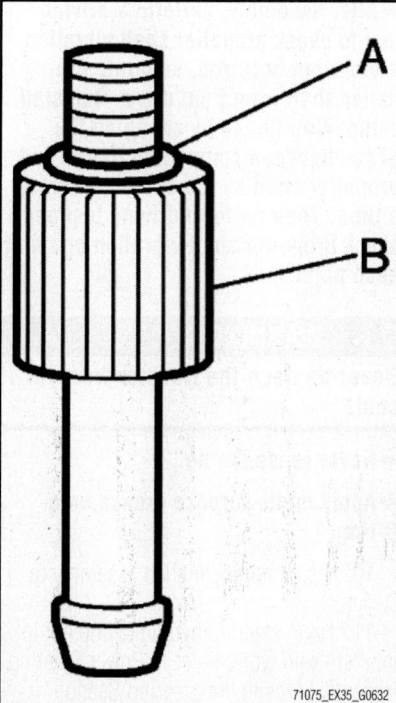

Fig. 11 Install the O-ring (315268E000) (A) to the charging pipe (310811EA5A) (B).

71075_EX35_G0632

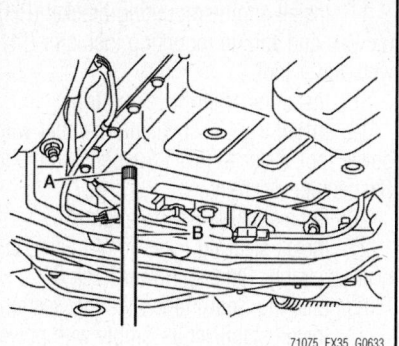

71075_EX35_G0633

Fig. 12 Install the charging pipe (A) to the overflow plug hole. Install the bucket pump hose (B) to the charging pipe.

porarily tighten the drain plug to the oil pan.

➡**Never replace drain plug and drain plug gasket with new ones yet.**

    e. Remove overflow plug from oil pan.
    f. Install the charging pipe to the overflow plug hole.

➡**Tighten the charging pipe by hand.**

    g. Install the bucket pump hose to the charging pipe.

➡**Insert the bucket pump hose all the way to the end of the charging pipe.**

    h. Fill approximately 3 qts. of the ATF.
    i. Remove the bucket pump hose to remove the charging pipe, and then temporarily tighten the overflow plug to the oil pan.

➡**Quickly perform the procedure to avoid ATF leakage from the oil pan.**

    j. Lift down the vehicle.
    k. Start the engine and wait for approximately 3 minutes.
    l. Stop the engine.
  3. Step 3
    a. Repeat "Step 2".
  4. Final Step
    a. Use CONSULT to check that the ATF temperature is 104°F (40°C) or less.
    b. Lift up the vehicle.
    c. Remove the drain plug from the oil pan, and then drain the ATF.
    d. When the ATF starts to drop, tighten the drain plug to the oil pan to 25 ft. lbs. (34 Nm).

➡**Never reuse drain plug and drain plug gasket.**

    e. Remove overflow plug from oil pan.
    f. Install the charging pipe to the overflow plug hole.

➡**Tighten the charging pipe by hand.**

    g. Install the bucket pump hose to the charging pipe.

➡**Insert the bucket pump hose all the way to the end of the charging pipe.**

    h. Fill approximately 3 qts. of the ATF.
    i. Remove the bucket pump hose to remove the charging pipe, and then temporarily tighten the overflow plug to the oil pan.

➡**Quickly perform the procedure to avoid ATF leakage from the oil pan.**

    j. Lift down the vehicle.
    k. Start the engine.
    l. Make the ATF temperature approximately 104°F (40°C).

➡**The ATF level is greatly affected by the temperature. Always check the ATF temperature on "ATF TEMP 1"of "Data Monitor" using CONSULT.**

    m. Park vehicle on level surface and set parking brake.
    n. Shift the selector lever through each gear position. Leave selector lever in ".
    o. Lift up the vehicle when the ATF temperature reaches 104°F (40°C), and then remove the overflow plug from the oil pan.
    p. When the ATF starts to drop, tighten the overflow plug to the oil pan to 108 inch lbs. (12.4 Nm).

➡**Never reuse overflow plug.**

FLUID INSPECTION

  1. Check transaxle surrounding area (oil seal and plug etc.) for fluid leakage.
  2. If anything is found, repair or replace damaged parts and adjust A/T fluid level.

FLUID RECOMMENDATIONS

  The manufacturer recommends Genuine NISSAN Matic S ATF.

FRONT FINAL DRIVE

REMOVAL & INSTALLATION

  1. Remove both front drive shaft.
  2. Remove front crossbar with power tool.
  3. Separate steering outer socket and

steering knuckle. Refer to Power Rack & Pinion Steering Gear in the Steering Section.

4. Remove side shaft.

5. Remove three way catalyst (right bank) with power tool.

6. Remove front propeller shaft. Refer to Front Drive Axle in the Driveline Section.

7. Separate power steering solenoid valve connector.

8. Separate power steering hydraulic line. Refer to Power Rack & Pinion Steering Gear in the Steering Section.

9. Remove stabilizer assembly with power tool. Refer to Front Suspension in the Suspension Section.

10. Separate steering lower joint and steering gear assembly. Refer to Power Rack & Pinion Steering Gear in the Steering Section.

11. Set a suitable jack to engine.

12. Remove front suspension member with power tool. Refer to Front Suspension in the Suspension Section.

13. Remove breather hose and tube.

14. Remove engine mounting bracket (RH) (Lower) and engine mounting insulator (RH) with power tool.

15. Remove final drive assembly mounting bolts with power tool and separate front final drive assembly from engine.

To install:

➡**When installing the side shaft, apply multi-purpose grease to contact surface of side shaft and side shaft oil seal.**

➡**Tighten mounting bolts in the order described below when installing front final drive assembly: side of gear carrier, upper side of gear carrier, part of carrier cover.**

➡**Align the mating faces of gear carrier and oil pan for installation.**

➡**When installing breather hose and tube, refer to the accompanying illustration.**

➡**Make sure there are no pinched or restricted areas on the breather hose caused by bending or winding when installing it.**

- Make sure the paint mark is facing up.
- Securely install the hose until it seats the rounded portion of the tube (front final drive side).
- Securely install the hose until it to paint mark of the tube (vehicle rear side).
- Face the bend of the breather hose to the engine.

16. Install front final drive assembly to engine, and then install final drive assembly mounting bolts with power tool.

17. Install engine mounting bracket (RH) (Lower) and engine mounting insulator (RH) with power tool.

18. Install breather hose and tube.

19. Install front suspension member with power tool. Refer to Front Suspension in the Suspension Section.

20. Set a suitable jack to engine.

21. Install steering lower joint to steering gear assembly. Refer to Power Rack & Pinion Steering Gear in the Steering Section.

22. Install stabilizer assembly with power tool. Refer to Front Suspension in the Suspension Section.

23. Install power steering hydraulic line. Refer to Power Rack & Pinion Steering Gear in the Steering Section.

24. Install power steering solenoid valve connector.

25. Install front propeller shaft. Refer to Front Drive Axle in the Driveline Section.

26. Install three way catalyst (right bank) with power tool.

27. Install side shaft.

28. Install steering outer socket and steering knuckle. Refer to Power Rack & Pinion Steering Gear in the Steering Section.

29. Install front crossbar with power tool.

30. Install both front drive shaft.

➡**When oil leaks while removing final drive assembly, check oil level after the installation.**

Oil Level Inspection

OIL LEAKAGE

1. Make sure that oil is not leaking from final drive assembly or around it.

OIL LEVEL

2. Remove filler plug and check oil level from filler plug mounting hole as shown in the accompanying illustration.

> ❋❋ **CAUTION**
>
> **Never start engine while checking oil level.**

3. Set a gasket on filler plug and install it on final drive assembly. Refer to the accompanying illustration.

➡**Never reuse gasket.**

FRONT PROPELLER SHAFT

REMOVAL & INSTALLATION

1. Shift the transmission to the neutral position, and then release the parking brake.

2. Remove engine undercover with a power tool.

3. Remove front cross bar.

4. Remove the three-way catalyst (right bank) with a power tool.

5. Put matching mark onto propeller shaft flange yoke and final drive companion flange.

> ❋❋ **WARNING**
>
> **For matching mark, use paint. Never damage propeller shaft flange and final drive companion flange.**

6. Remove the propeller shaft assembly fixing bolts.

7. Remove propeller shaft assembly from the front final drive and transfer.

> ❋❋ **WARNING**
>
> **Never damage the transfer front oil seal.**

8. Hang steering hydraulic line not to interfere with work. Refer to Power Steering Hoses & Lines in the Steering Section.

9. Remove propeller shaft assembly from O-ring.

To install:

➡**Align matching mark to install propeller shaft assembly to final drive companion flange.**

➡**After assembly, perform a driving test to check propeller shaft vibration. If vibration occurred, separate propeller shaft from final drive. Reinstall companion flange by changing the phase between companion flange and propeller shaft by the one bolt hole at a time. Then perform driving test and check propeller shaft vibration again at each point.**

> ❋❋ **WARNING**
>
> **Never damage the transfer front oil seal.**

➡**Never reuse O-ring.**

➡**Apply multi-purpose grease onto O-ring.**

10. Install propeller shaft assembly to O-ring.

11. Hang steering hydraulic line not to interfere with work. Refer to Power Steering Hoses & Lines in the Steering Section.

> ❋❋ **WARNING**
>
> **Never damage the transfer front oil seal.**

12. Install propeller shaft assembly to the front final drive and transfer.

13. Install the propeller shaft assembly fixing bolts.

✳✳ WARNING

For matching mark, use paint. Never damage propeller shaft flange and final drive companion flange.

14. Align matching mark on propeller shaft flange yoke and final drive companion flange.

15. Install the three-way catalyst (right bank) with a power tool.

16. Install front cross bar.

17. Install engine undercover with a power tool.

18. Shift the transmission to the park position, and then engage the parking brake.

FRONT HALFSHAFT

REMOVAL & INSTALLATION

Left Side—AWD Vehicles

See Figures 13 and 14.

1. Remove tires with power tool.

2. Remove wheel sensor and sensor harness.

➡**Never pull on wheel sensor harness.**

3. Remove brake hose bracket.

4. Remove caliper assembly. Hang caliper assembly in a place where it will not interfere with work. Refer to Front Disc Brakes in the Brakes Section.

➡**Never depress brake pedal while brake caliper is removed.**

5. Remove disc rotor.

6. Remove cotter pin, and then loosen wheel hub lock nut with a power tool. Refer to Front Drive Axle in the Driveline Section.

7. Patch wheel hub lock nut with a piece of wood. Hammer the wood to disengage wheel hub and bearing assembly from drive shaft.

✳✳ WARNING

Never place drive shaft joint at an extreme angle. Also be careful not to overextend slide joint.

➡**Never allow drive shaft to hang down without support for joint sub-assembly, shaft and the other parts.**

➡**Use suitable puller if wheel hub and bearing assembly and drive shaft cannot be separated even after performing the above procedure.**

8. Remove wheel hub lock nut.

9. Remove steering outer socket. Refer to Power Rack & Pinion Steering Gear in the Steering Section.

10. Remove drive shaft from wheel hub and bearing assembly.

11. Remove mounting bolts, and then remove drive shaft from vehicle.

To install:

12. Install drive shaft to vehicle, and then install mounting bolts.

13. Install drive shaft to wheel hub and bearing assembly.

14. Install steering outer socket. Refer to Power Rack & Pinion Steering Gear in the Steering Section.

15. Install wheel hub lock nut.

➡**Use suitable puller if wheel hub and bearing assembly and drive shaft cannot be separated even after performing the above procedure.**

➡**Never allow drive shaft to hang down without support for joint sub-assembly, shaft and the other parts.**

✳✳ WARNING

Never place drive shaft joint at an extreme angle. Also be careful not to overextend slide joint.

16. Patch wheel hub lock nut with a piece of wood. Hammer the wood to engage wheel hub and bearing assembly to drive shaft.

17. Tighten wheel hub lock nut with a power tool, and then install cotter pin. Refer to Front Drive Axle in the Driveline Section.

18. Install disc rotor.

➡**Never depress brake pedal while brake caliper is removed.**

19. Install caliper assembly. Refer to Front Disc Brakes in the Brakes Section.

Fig. 13 Patch wheel hub lock nut with a piece of wood. Hammer the wood to disengage wheel hub and bearing assembly from drive shaft

71075_EX35_G0634

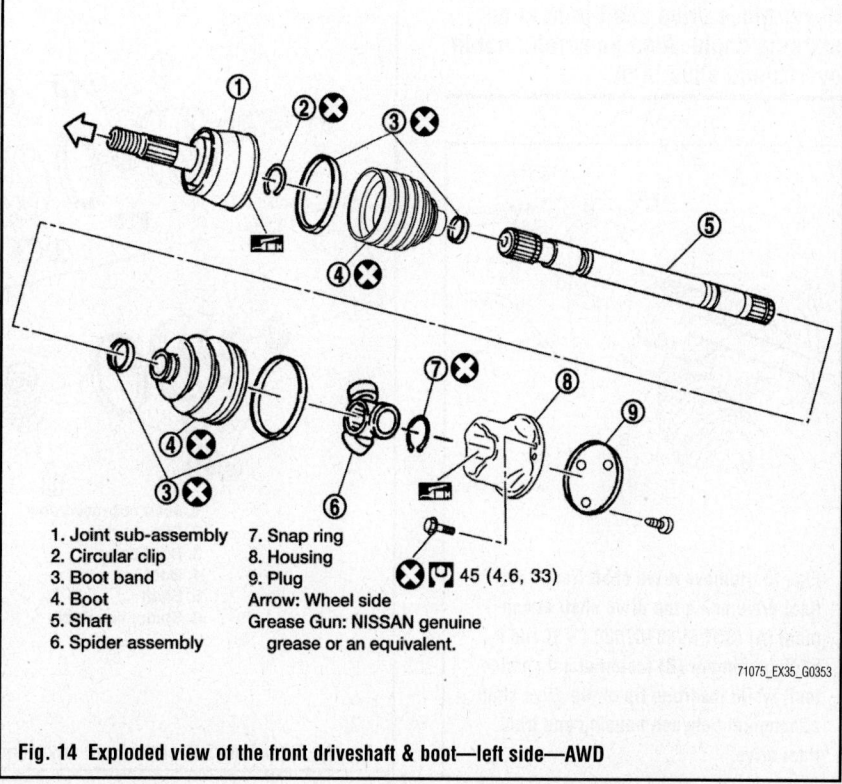

1. Joint sub-assembly
2. Circular clip
3. Boot band
4. Boot
5. Shaft
6. Spider assembly
7. Snap ring
8. Housing
9. Plug
Arrow: Wheel side
Grease Gun: NISSAN genuine grease or an equivalent.

45 (4.6, 33)

71075_EX35_G0353

Fig. 14 Exploded view of the front driveshaft & boot—left side—AWD

20. Install brake hose bracket.

➡ **Never pull on wheel sensor harness.**

21. Install wheel sensor and sensor harness.
22. Install tires with power tool.

Right Side—AWD Vehicles

See Figures 13, 15 through 17.

1. Remove tires with power tool.
2. Remove wheel sensor and sensor harness.

➡ **Never pull on wheel sensor harness.**

3. Remove brake hose bracket.
4. Remove caliper assembly. Hang caliper assembly in a place where it will not interfere with work. Refer to Front Disc Brakes in the Brakes Section.

➡ **Never depress brake pedal while brake caliper is removed.**

5. Remove disc rotor.
6. Remove cotter pin, and then loosen wheel hub lock nut with a power tool. Refer to Front Drive Axle in the Driveline Section.
7. Patch wheel hub lock nut with a piece of wood. Hammer the wood to disengage wheel hub and bearing assembly from drive shaft.

➡ **WARNING**

Never place drive shaft joint at an extreme angle. Also be careful not to overextend slide joint.

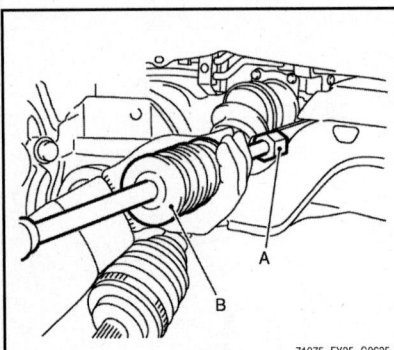

71075_EX35_G0635

Fig. 15 Remove drive shaft from front final drive using the drive shaft attachment (A) [SST:KV40107500 (–)] and a sliding hammer (B) (commercial service tool) while inserting tip of the drive shaft attachment between housing and front final drive

➡ **Never allow drive shaft to hang down without support for joint sub-assembly, shaft and the other parts.**

➡ **Use suitable puller if wheel hub and bearing assembly and drive shaft cannot be separated even after performing the above procedure.**

8. Remove wheel hub lock nut.
9. Remove steering outer socket. Refer to Power Rack & Pinion Steering Gear in the Steering Section.
10. Remove drive shaft from wheel hub and bearing assembly.
11. Remove drive shaft from front final drive using the drive shaft attachment [SST:KV40107500 (–)] and a sliding hammer (commercial service tool) while inserting tip of the drive shaft attachment between housing and front final drive.

❄❄ **WARNING**

Never place drive shaft joint at an extreme angle when removing drive shaft. Also be careful not to overextend slide joint.

To install:

➡ **Always replace final drive oil seal with new one when installing drive shaft. Refer to Front Drive Axle in the Driveline Section.**

❄❄ **WARNING**

Place the protector [SST:KV38107900 (–)] onto final drive to prevent damage to the oil seal while inserting drive shaft. Slide drive shaft sliding joint and tap with a hammer to install securely.

❄❄ **WARNING**

Never place drive shaft joint at an extreme angle when removing drive shaft. Also be careful not to overextend slide joint.

12. Install drive shaft to front final drive using the drive shaft attachment [SST:KV40107500 (–)] and a sliding hammer (commercial service tool) while inserting tip of the drive shaft attachment between housing and front final drive.
13. Install drive shaft to wheel hub and bearing assembly.
14. Install steering outer socket. Refer to Power Rack & Pinion Steering Gear in the Steering Section.
15. Install wheel hub lock nut.

➡ **Use suitable puller if wheel hub and bearing assembly and drive shaft cannot be separated even after performing the above procedure.**

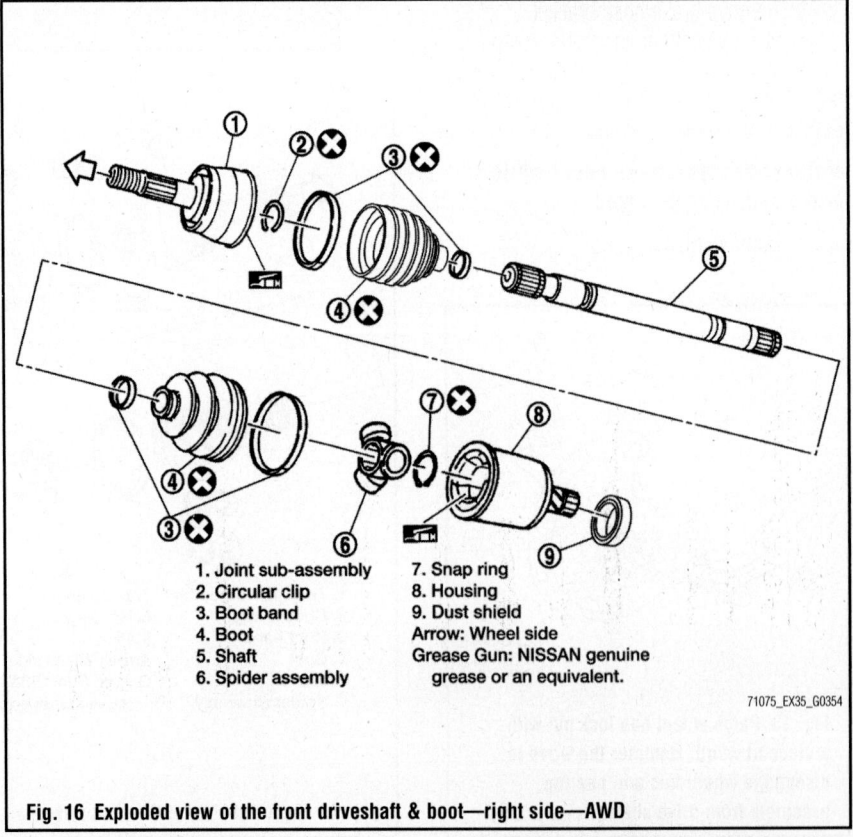

1. Joint sub-assembly
2. Circular clip
3. Boot band
4. Boot
5. Shaft
6. Spider assembly
7. Snap ring
8. Housing
9. Dust shield
Arrow: Wheel side
Grease Gun: NISSAN genuine grease or an equivalent.

71075_EX35_G0354

Fig. 16 Exploded view of the front driveshaft & boot—right side—AWD

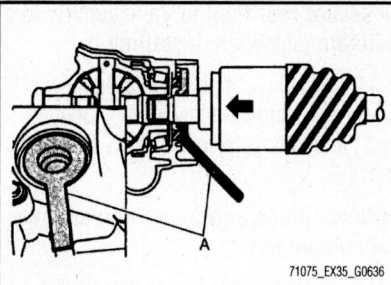

Fig. 17 Place the protector (A) [SST:KV38107900 (–)] onto final drive to prevent damage to the oil seal while inserting drive shaft. Slide drive shaft sliding joint and tap with a hammer to install securely

➡️Never allow drive shaft to hang down without support for joint sub-assembly, shaft and the other parts.

❄️ WARNING

Never place drive shaft joint at an extreme angle. Also be careful not to overextend slide joint.

16. Patch wheel hub lock nut with a piece of wood. Hammer the wood to engage wheel hub and bearing assembly to drive shaft.

17. Tighten wheel hub lock nut with a power tool, and then install cotter pin. Refer to Front Drive Axle in the Driveline Section.

18. Install disc rotor.

➡️**Never depress brake pedal while brake caliper is removed.**

19. Install caliper assembly. Refer to Front Disc Brakes in the Brakes Section.

20. Install brake hose bracket.

➡️**Never pull on wheel sensor harness.**

21. Install wheel sensor and sensor harness.

22. Install tires with power tool.

REAR HALFSHAFT

REMOVAL & INSTALLATION

See Figure 18.

1. Remove tire and wheel assembly.

2. Remove cotter pin and adjusting cap, then loosen wheel hub lock nut with power tool.

3. Put matching mark on drive shaft and wheel hub and bearing assembly.

4. Remove center muffler.

5. Patch wheel hub lock nut with a piece of wood. Hammer the wood to disengage wheel hub and bearing assembly from drive shaft.

❄️ WARNING

Never place drive shaft joint at an extreme angle. Also be careful not to overextend slide joint.

➡️Never allow drive shaft to hang down without support for counterpart such as joint sub-assembly, and other parts.

➡️Using a suitable puller if wheel hub and bearing assembly and drive shaft cannot be separated even after performing the above procedure.

6. Remove wheel hub lock nut.

7. Remove mounting bolts between side flange and drive shaft.

To install:

➡️Clean the matching surface of wheel hub lock nut and wheel hub and bearing assembly.

➡️Never apply lubricating oil to these matching surface.

➡️Clean the matching surface of drive shaft and wheel hub and bearing assembly. And then apply paste [service parts (440037S000)] to surface of joint sub-assembly of drive shaft.

➡️Apply paste to cover entire flat surface of joint sub-assembly of drive shaft.

➡️When installing drive shaft, change the drive shaft and wheel hub and bearing assembly matching marks put at the removal step by 180 degree.

➡️Use the following torque range for tightening the wheel hub lock nut. Tighten to 74-77 ft. lbs. (100-105 Nm).

➡️Since the drive shaft is assembled by press-fitting, use the tightening torque range for the wheel hub lock nut.

➡️Be sure to use torque wrench to tighten the wheel hub lock nut. Never use a power tool.

➡️Wheel hub lock nut tightening torque does not over torque for avoiding axle noise, and does not less than torque for avoiding looseness.

➡️Perform the final tightening of each of parts under unladen conditions, which were removed when removing wheel hub and bearing assembly and axle housing.

➡️When installing the spring washer, face the identification paint mark to the wheel hub and bearing assembly side.

➡️When installing the adjusting cap, check that there must be no play.

➡️Never reuse cotter pin, wheel hub lock nut, spring washer, and bushing.

8. Install mounting bolts between side flange and drive shaft.

9. Install wheel hub lock nut.

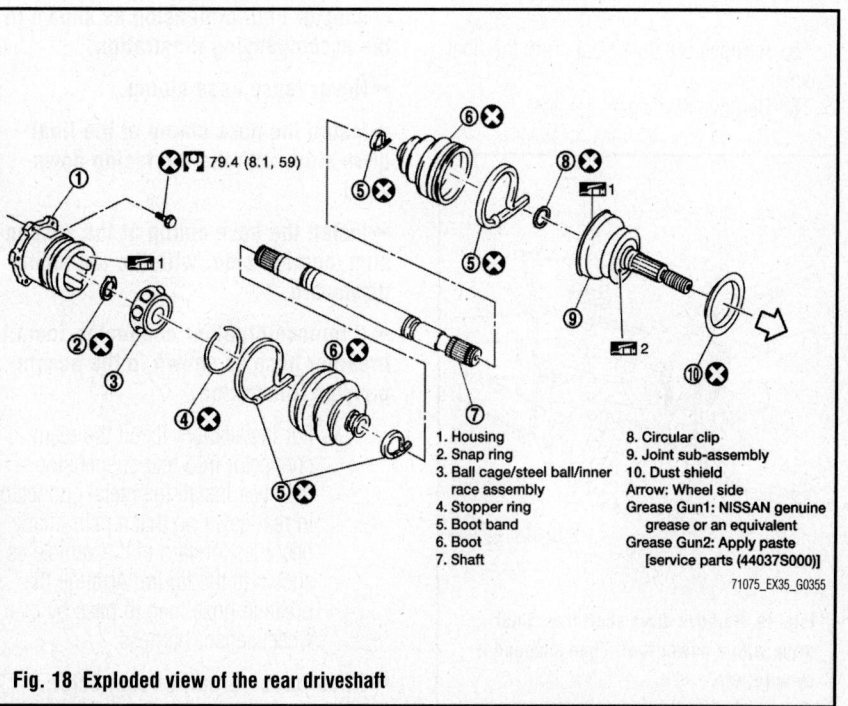

1. Housing
2. Snap ring
3. Ball cage/steel ball/inner race assembly
4. Stopper ring
5. Boot band
6. Boot
7. Shaft
8. Circular clip
9. Joint sub-assembly
10. Dust shield
Arrow: Wheel side
Grease Gun1: NISSAN genuine grease or an equivalent
Grease Gun2: Apply paste [service parts (44037S000)]

Fig. 18 Exploded view of the rear driveshaft

➡Using a suitable puller if wheel hub and bearing assembly and drive shaft cannot be separated even after performing the above procedure.

➡Never allow drive shaft to hang down without support for counterpart such as joint sub-assembly, and other parts.

❋❋ WARNING

Never place drive shaft joint at an extreme angle. Also be careful not to overextend slide joint.

10. Patch wheel hub lock nut with a piece of wood. Hammer the wood to engage wheel hub and bearing assembly to drive shaft.

11. Install center muffler.

12. Align matching mark on drive shaft and wheel hub and bearing assembly.

13. Install cotter pin and adjusting cap, then loosen wheel hub lock nut with power tool.

14. Install tire and wheel assembly.

REAR FINAL DRIVE

REMOVAL & INSTALLATION

See Figures 19 through 26.

1. Remove center muffler with a power tool.

2. Remove stabilizer bar with a power tool.

3. Remove rear propeller shaft from the final drive.

4. Remove drive shaft from final drive with a power tool. Then suspend it by wire, etc.

5. Remove breather hose from the final drive.

6. Remove rear wheel sensor.

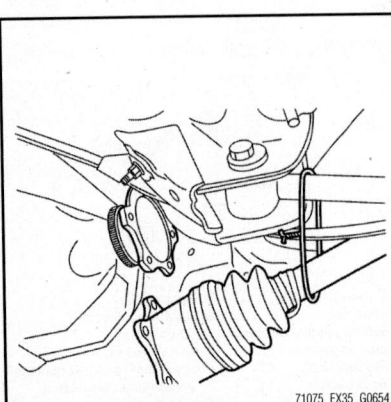

Fig. 19 Remove drive shaft from final drive with a power tool. Then suspend it by wire, etc.

71075_EX35_G0655

Fig. 20 Set a suitable jack to rear final drive assembly. 2WD final drive shown, AWD is similar

7. Set a suitable jack to rear final drive assembly.

➡Never place a jack on the rear cover (aluminum case).

8. Remove the mounting bolts and nuts connecting to the suspension member, and remove rear final drive assembly with a power tool.

➡Secure rear final drive assembly to a suitable jack while removing it.

To install:

➡Check that there are no pinched or restricted areas on the breather hose caused by bending or winding when installing it.

➡Install the breather hose to breather connector until dimension as shown in the accompanying illustration.

➡Never reuse hose clamp.

➡Install the hose clamp at the final drive side, with the tab facing downward.

➡Install the hose clamp at the suspension member side, with the tab facing downward.

➡If remove breather connector, install breather hose as shown in the accompanying illustration.

• For installation, insert the resin connector into rear suspension member. Install the metal connector in rear cover so that a paint mark becomes forward of the vehicle as shown in the figure. Arrange the breather hose then to pass by over wheel sensor harness.

➡Never reuse breather connector.

➡Secure rear final drive assembly to a suitable jack while installing it.

9. Install rear final drive assembly with power tool, and then install the mounting bolts and nuts connecting to the suspension member.

➡Never place a jack on the rear cover (aluminum case).

10. Set a suitable jack to rear final drive assembly.

11. Install rear wheel sensor.

12. Install breather hose to the final drive.

13. Install drive shaft to final drive with a power tool.

14. Install rear propeller shaft to the final drive.

15. Install stabilizer bar with a power tool.

16. Install center muffler with a power tool.

➡When oil leaks while removing final drive assembly, check oil level after the installation.

Oil Level Inspection

See Figures 27 and 28.

OIL LEAKAGE

1. Make sure that oil is not leaking from final drive assembly or around it.

OIL LEVEL

2. Remove filler plug and check oil level from filler plug mounting hole as shown in the accompanying illustration.

❋❋ CAUTION

Never start engine while checking oil level.

3. Set a gasket on filler plug and install it on final drive assembly. Refer to the accompanying illustration.

➡Never reuse gasket.

REAR PROPELLER SHAFT

REMOVAL & INSTALLATION

2WD Vehicles

See Figures 29 through 35.

1. Shift the transmission to the neutral position, and then release the parking brake.

2. Remove the floor reinforcement.

3. Remove the center muffler with power tool.

4. Remove the heat plate.

5. Put matching marks onto propeller shaft rubber coupling and final drive companion flange.

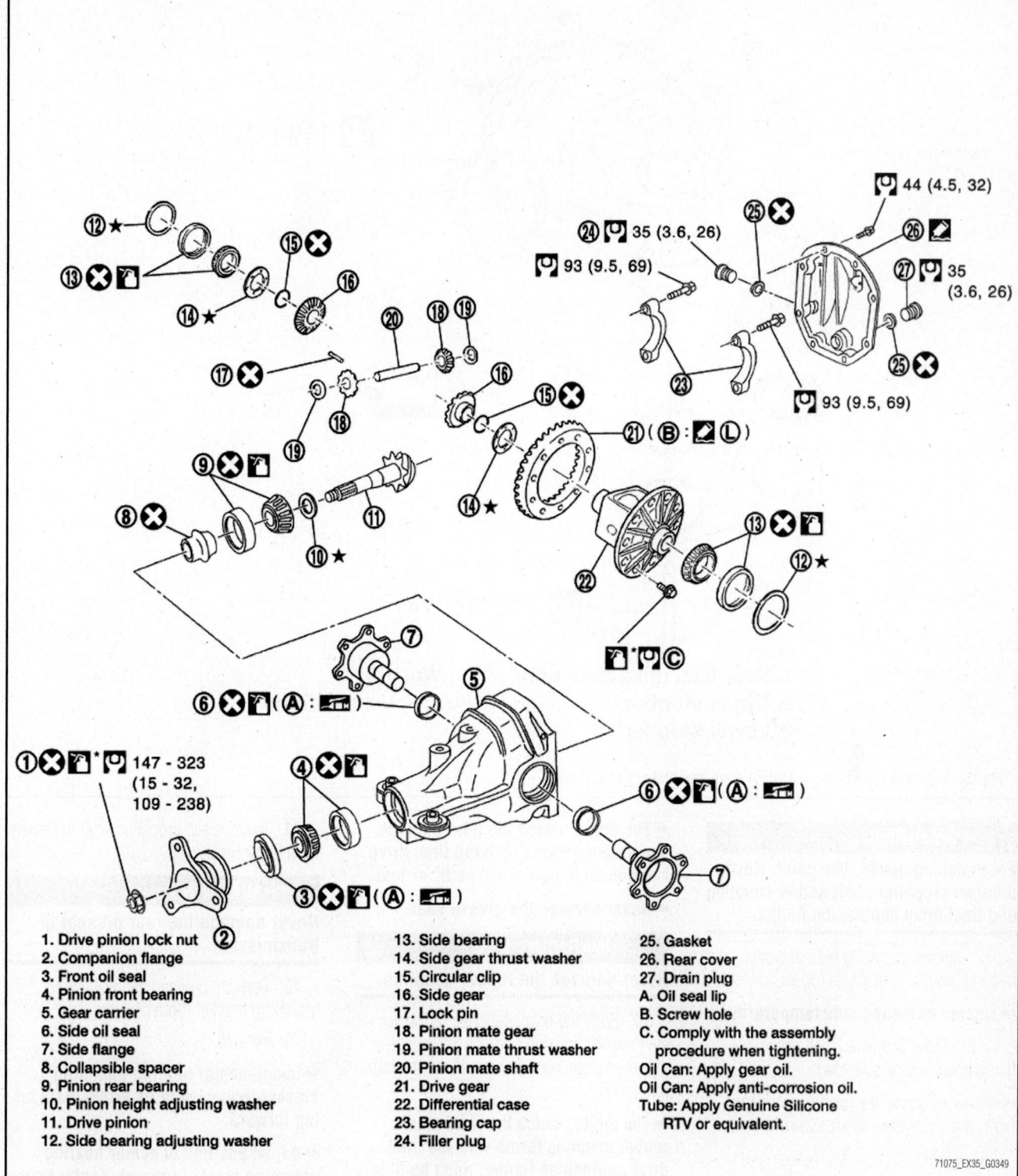

1. Drive pinion lock nut
2. Companion flange
3. Front oil seal
4. Pinion front bearing
5. Gear carrier
6. Side oil seal
7. Side flange
8. Collapsible spacer
9. Pinion rear bearing
10. Pinion height adjusting washer
11. Drive pinion
12. Side bearing adjusting washer
13. Side bearing
14. Side gear thrust washer
15. Circular clip
16. Side gear
17. Lock pin
18. Pinion mate gear
19. Pinion mate thrust washer
20. Pinion mate shaft
21. Drive gear
22. Differential case
23. Bearing cap
24. Filler plug
25. Gasket
26. Rear cover
27. Drain plug
A. Oil seal lip
B. Screw hole
C. Comply with the assembly procedure when tightening.
Oil Can: Apply gear oil.
Oil Can: Apply anti-corrosion oil.
Tube: Apply Genuine Silicone RTV or equivalent.

71075_EX35_G0349

Fig. 21 Exploded view of the rear final drive assembly (1 of 2)–2WD

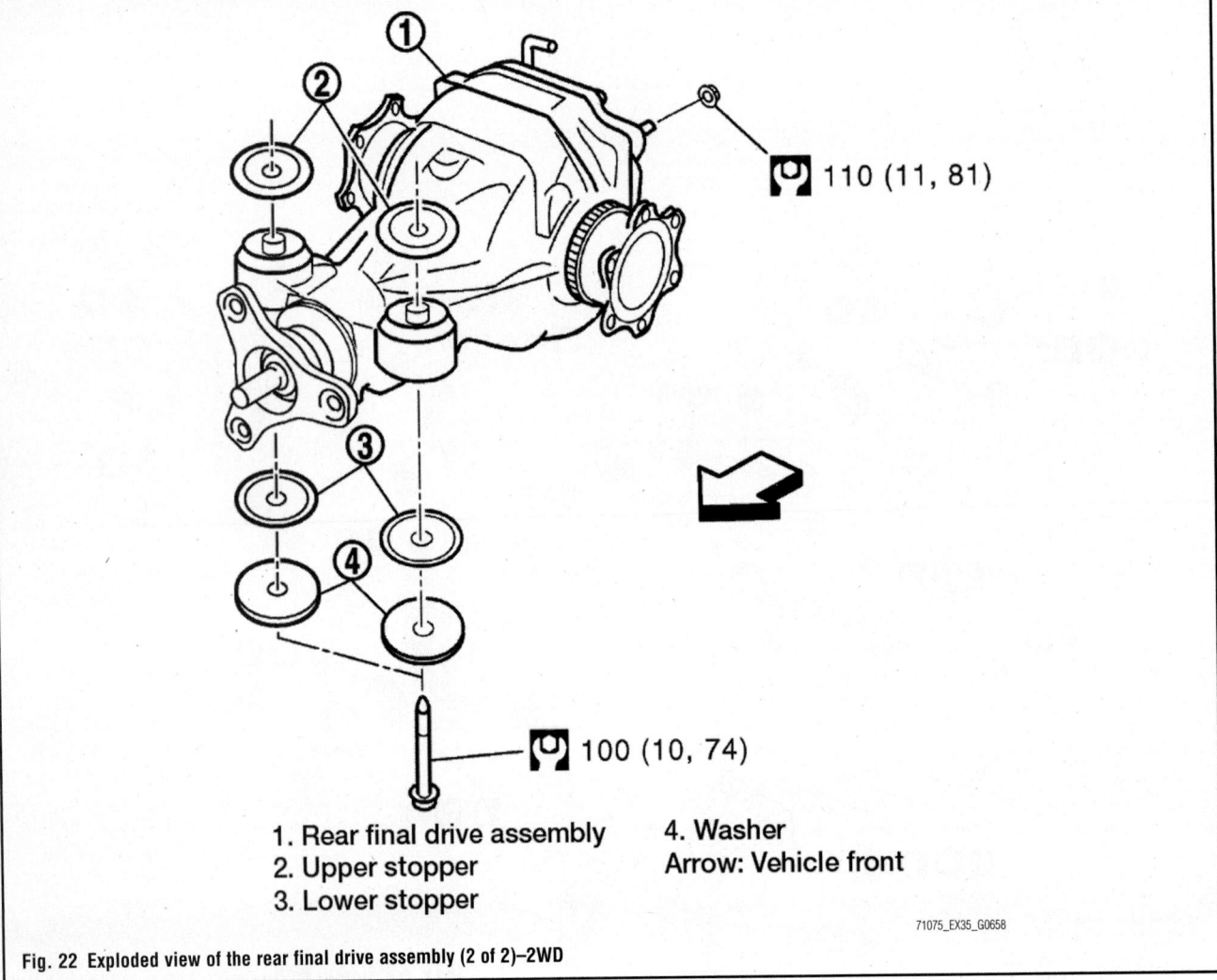

110 (11, 81)

100 (10, 74)

1. Rear final drive assembly
2. Upper stopper
3. Lower stopper

4. Washer
Arrow: Vehicle front

71075_EX35_G0658

Fig. 22 Exploded view of the rear final drive assembly (2 of 2)–2WD

✳✳ WARNING

For matching marks, use paint. Never damage propeller shaft rubber coupling and final drive companion flange.

6. Loosen mounting nuts of center bearing mounting brackets (upper/lower).

➡**Tighten mounting nuts temporarily.**

7. Remove propeller shaft assembly fixing bolts and nuts (arrow).

➡**Never remove the rubber coupling from the propeller shaft assembly.**

8. Slightly separate the rubber coupling from the final drive companion flange.

✳✳ WARNING

Never damage the final drive companion flange and rubber coupling.

9. Remove center bearing mounting bracket fixing nuts.

➡**The angle, which the third axis rubber coupling forms with the final drive companion flange, must be 5° or less.**

➡**Never damage the grease seal.**

✳✳ WARNING

Never damage the rubber coupling.

10. Slide the propeller shaft in the vehicle forward direction slightly. Separate the propeller shaft from the final drive companion flange.

➡**The angle, which the third axis rubber coupling forms with the final drive companion flange, must be 5°or less.**

✳✳ WARNING

Never damage the grease seal.

✳✳ WARNING

Never damage the rubber coupling.

11. Remove the propeller shaft assembly from the vehicle.

✳✳ WARNING

Never damage the rear oil seal of transmission.

12. Remove clip and center bearing mounting bracket (upper/lower).

To install:

➡**Install center bearing mounting bracket (upper) with its arrow mark facing forward.**

➡**Adjust position of center bearing mounting bracket (upper), center bearing mounting bracket (lower) sliding back and forth to prevent play in thrust direction of center bearing insulator. Install center bearing mounting bracket (upper/lower) to vehicle.**

➡**Align matching marks to install propeller shaft rubber coupling to final drive companion flange.**

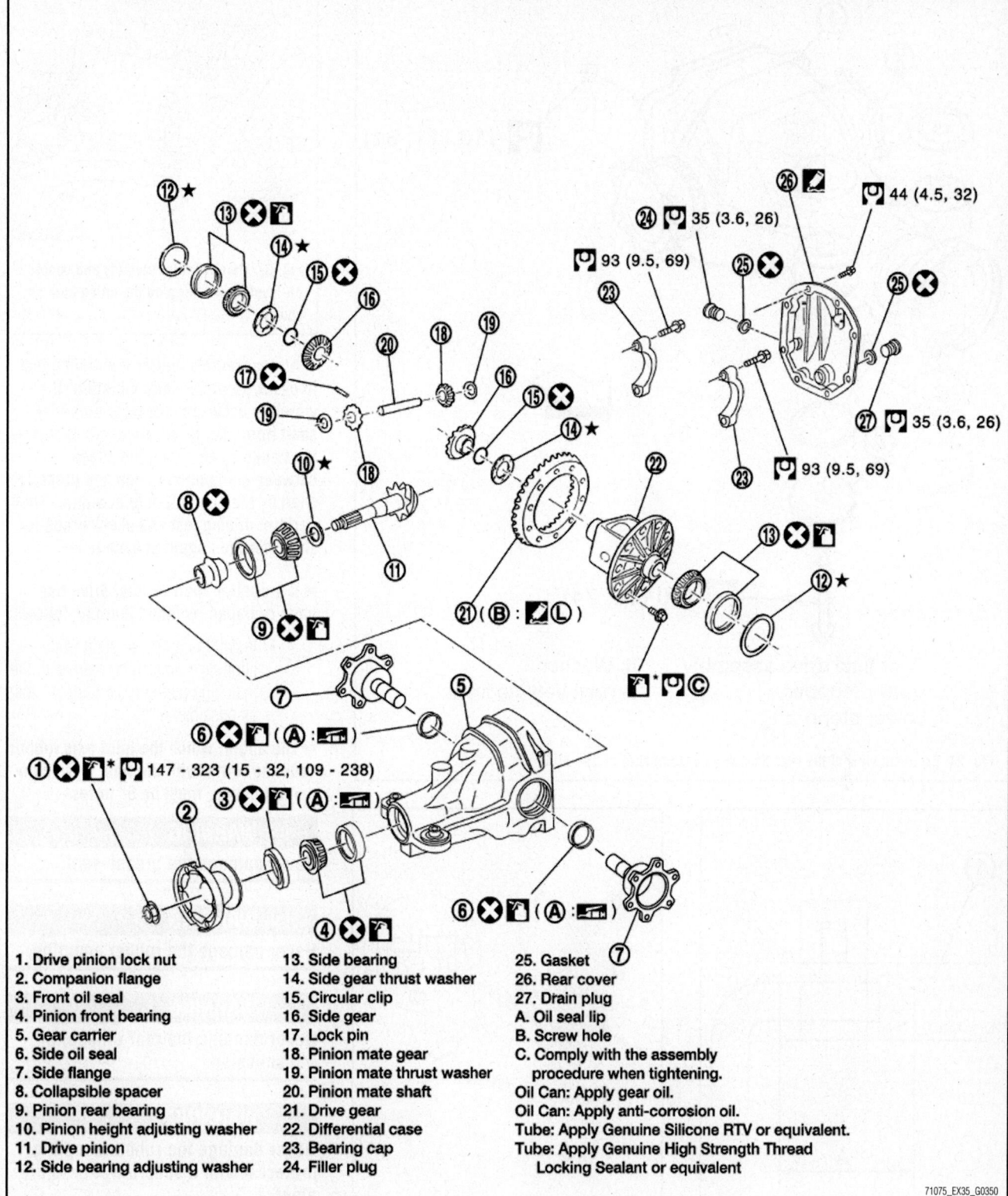

1. Drive pinion lock nut
2. Companion flange
3. Front oil seal
4. Pinion front bearing
5. Gear carrier
6. Side oil seal
7. Side flange
8. Collapsible spacer
9. Pinion rear bearing
10. Pinion height adjusting washer
11. Drive pinion
12. Side bearing adjusting washer

13. Side bearing
14. Side gear thrust washer
15. Circular clip
16. Side gear
17. Lock pin
18. Pinion mate gear
19. Pinion mate thrust washer
20. Pinion mate shaft
21. Drive gear
22. Differential case
23. Bearing cap
24. Filler plug

25. Gasket
26. Rear cover
27. Drain plug
A. Oil seal lip
B. Screw hole
C. Comply with the assembly
   procedure when tightening.
Oil Can: Apply gear oil.
Oil Can: Apply anti-corrosion oil.
Tube: Apply Genuine Silicone RTV or equivalent.
Tube: Apply Genuine High Strength Thread
   Locking Sealant or equivalent

71075_EX35_G0350

Fig. 23 Exploded view of the rear final drive assembly (1 of 2)—AWD

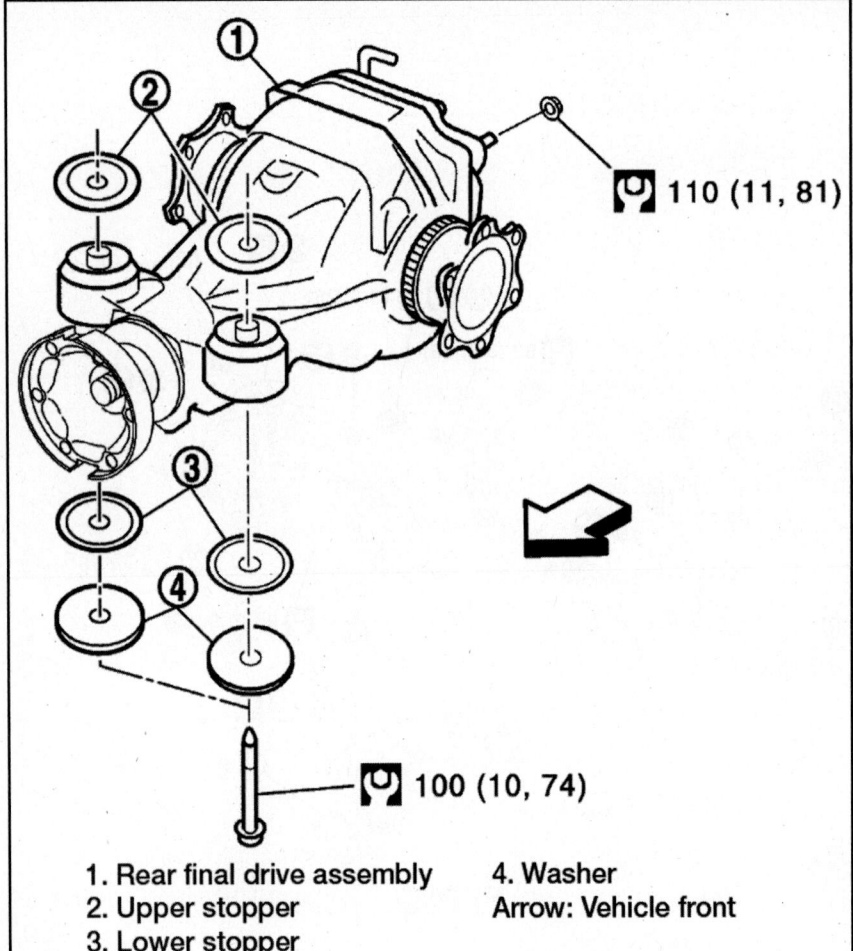

110 (11, 81)

100 (10, 74)

1. Rear final drive assembly
2. Upper stopper
3. Lower stopper
4. Washer
Arrow: Vehicle front

71075_EX35_G0659

Fig. 24 Exploded view of the rear final drive assembly (2 of 2)—AWD

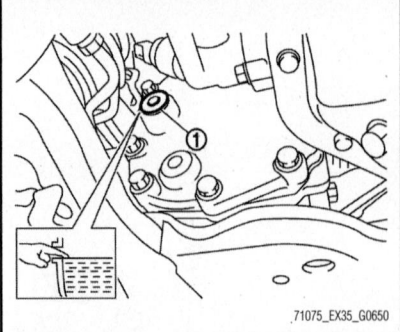

71075_EX35_G0650

Fig. 27 Remove filler plug (1) and check oil level from filler plug mounting hole as shown

➡️After assembly, perform a driving test to check propeller shaft vibration. If vibration occurred, separate propeller shaft from final drive. Reinstall companion flange by changing the phase between companion flange and propeller shaft by the one bolt hole at a time. Then perform driving test and check propeller shaft vibration again at each point.

➡️If propeller shaft or final drive has been replaced, connect them as follows:

- Install the propeller shaft while aligning its matching mark with the matching mark on the joint as close as possible.

➡️The angle, which the third axis rubber coupling forms with the final drive companion flange, must be 5° or less.

❄❄ WARNING
Never damage the grease seal.

❄❄ WARNING
Never damage the rubber coupling.

❄❄ WARNING
Never damage the rear oil seal of transmission.

❄❄ WARNING
Never damage the rubber coupling, protect it with a shop towel or equivalent.

13. Install clip and center bearing mounting bracket (upper/lower).

❄❄ WARNING
Never damage the rear oil seal of transmission.

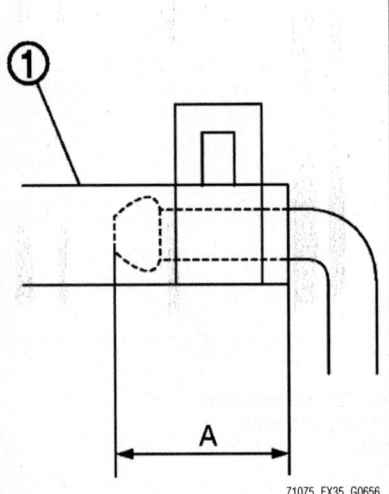

71075_EX35_G0656

Fig. 25 Install the breather hose (1) to breather connector until dimension (A) shown. A: Final drive side : 0.79 inch (20 mm), Suspension member side: 0.807 inch (20.5 mm)

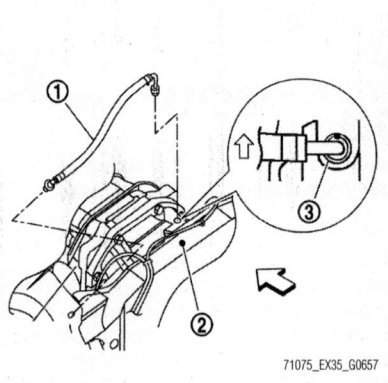

71075_EX35_G0657

Fig. 26 If remove breather connector, install breather hose (1) as shown. For installation, insert the resin connector into rear suspension member (2). Install the metal connector (3) in rear cover so that a paint mark becomes forward of the vehicle as shown.

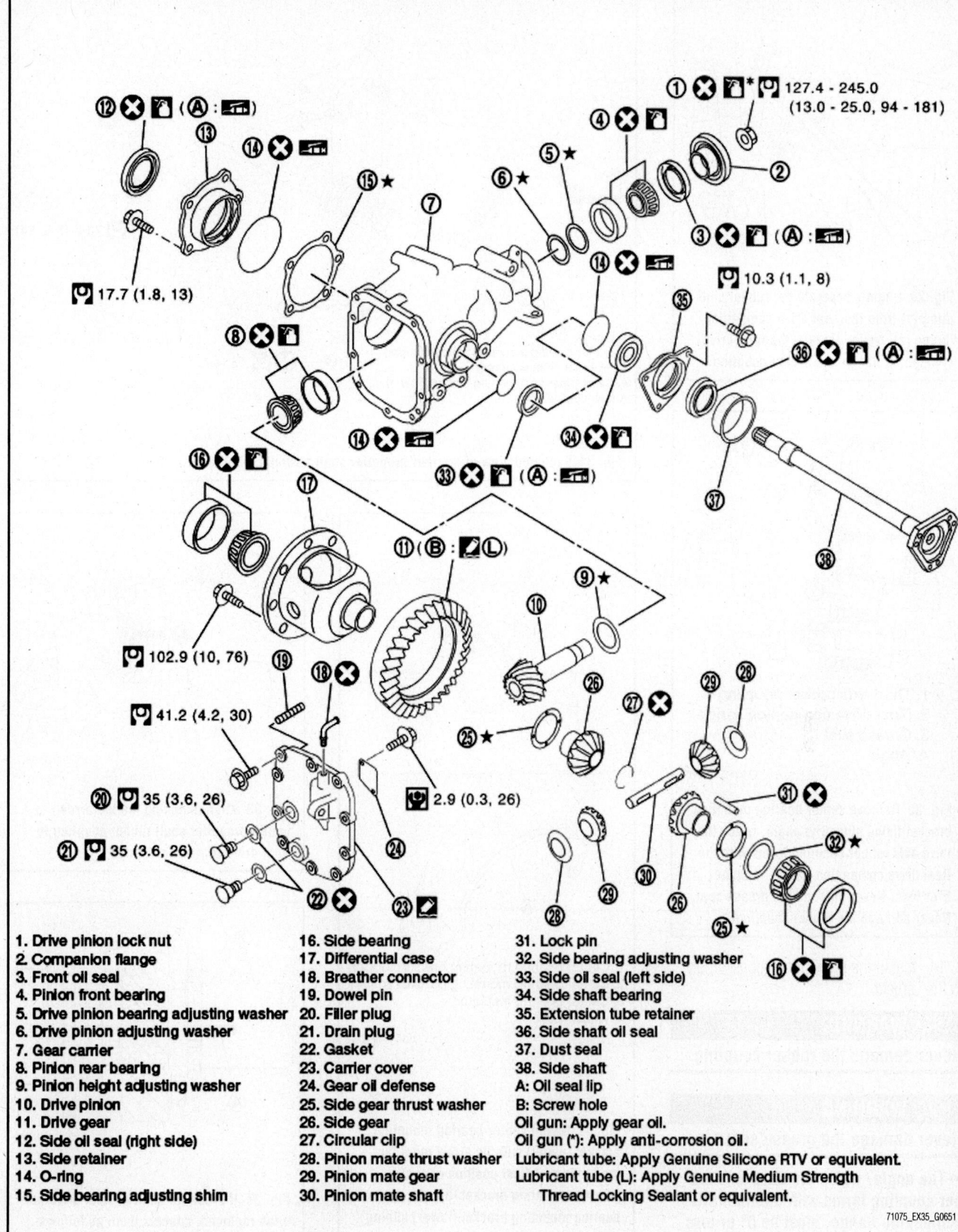

1. Drive pinion lock nut
2. Companion flange
3. Front oil seal
4. Pinion front bearing
5. Drive pinion bearing adjusting washer
6. Drive pinion adjusting washer
7. Gear carrier
8. Pinion rear bearing
9. Pinion height adjusting washer
10. Drive pinion
11. Drive gear
12. Side oil seal (right side)
13. Side retainer
14. O-ring
15. Side bearing adjusting shim

16. Side bearing
17. Differential case
18. Breather connector
19. Dowel pin
20. Filler plug
21. Drain plug
22. Gasket
23. Carrier cover
24. Gear oil defense
25. Side gear thrust washer
26. Side gear
27. Circular clip
28. Pinion mate thrust washer
29. Pinion mate gear
30. Pinion mate shaft

31. Lock pin
32. Side bearing adjusting washer
33. Side oil seal (left side)
34. Side shaft bearing
35. Extension tube retainer
36. Side shaft oil seal
37. Dust seal
38. Side shaft
A: Oil seal lip
B: Screw hole
Oil gun: Apply gear oil.
Oil gun (*): Apply anti-corrosion oil.
Lubricant tube: Apply Genuine Silicone RTV or equivalent.
Lubricant tube (L): Apply Genuine Medium Strength
   Thread Locking Sealant or equivalent.

71075_EX35_G0651

Fig. 28 Set a gasket on filler plug and install it on final drive assembly

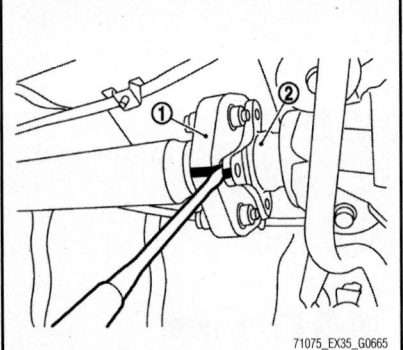

Fig. 29 Slightly separate the rubber coupling (1) from the final drive companion flange (2). Never damage the final drive companion flange and rubber coupling

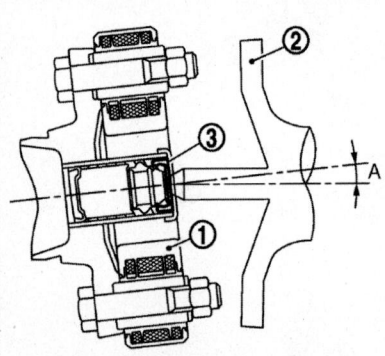

1. Third axis rubber coupling
2. Final drive companion flange
3. Grease seal
A. Angle

71075_EX35_G0666

Fig. 30 Remove center bearing mounting bracket fixing nuts. The angle, which the third axis rubber coupling forms with the final drive companion flange, must be 5°or less. Never damage the grease seal. Never damage the rubber coupling

14. Install the propeller shaft assembly to the vehicle.

※※ WARNING
Never damage the rubber coupling.

※※ WARNING
Never damage the grease seal.

➡The angle, which the third axis rubber coupling forms with the final drive companion flange, must be 5° or less.

15. Slide the propeller shaft in the vehicle backward direction slightly. Engage the propeller shaft to the final drive companion flange.

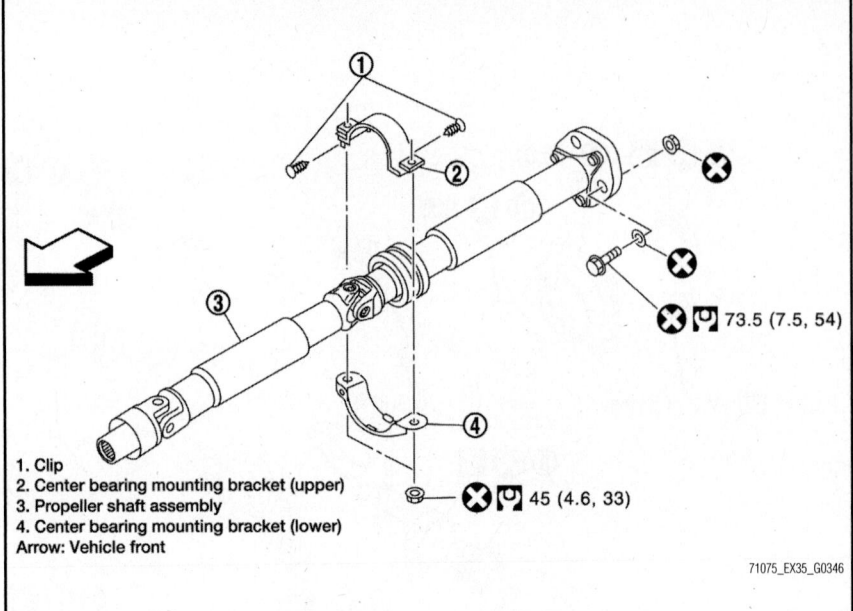

1. Clip
2. Center bearing mounting bracket (upper)
3. Propeller shaft assembly
4. Center bearing mounting bracket (lower)
Arrow: Vehicle front

73.5 (7.5, 54)

45 (4.6, 33)

71075_EX35_G0346

Fig. 31 Exploded view of the rear propeller shaft assembly–3S80A-R

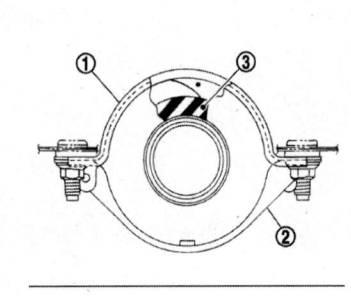

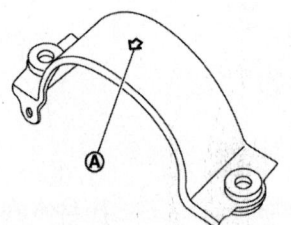

1. Center bearing mounting bracket (upper)
2. Center bearing mounting bracket (lower)
3. Center bearing insulator
A. Arrow mark

71075_EX35_G0667

Fig. 32 Install center bearing mounting bracket (upper) with its arrow mark facing forward. Adjust position of center bearing mounting bracket (upper), center bearing mounting bracket (lower) sliding back and forth to prevent play in thrust direction of center bearing insulator. Install center bearing mounting bracket (upper/lower) to vehicle

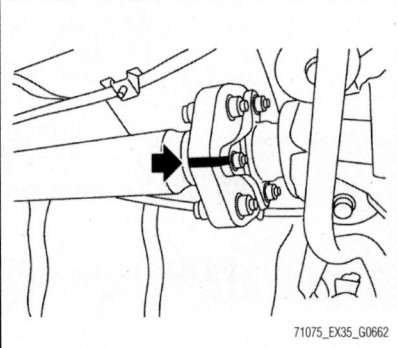

71075_EX35_G0662

Fig. 33 Align matching marks (arrow) to install propeller shaft rubber coupling to final drive companion flange

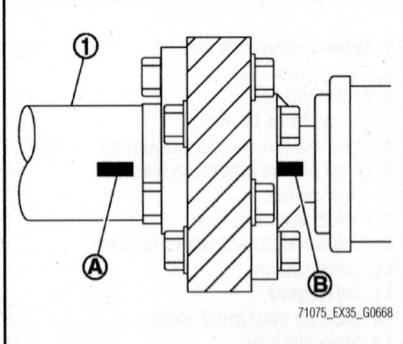

71075_EX35_G0668

Fig. 34 If propeller shaft or final drive has been replaced, connect them as follows: Install the propeller shaft (1) while aligning its matching mark (A) with the matching mark (B) on the joint as close as possible

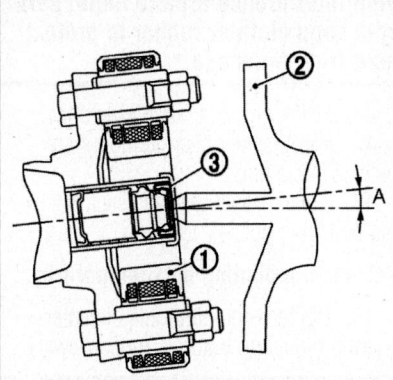

1. Third axis rubber coupling
2. Final drive companion flange
3. Grease seal
A. Angle

71075_EX35_G0666

Fig. 35 The angle, which the third axis rubber coupling forms with the final drive companion flange, must be 5°or less. Never damage the grease seal. Never damage the rubber coupling. Never damage the rear oil seal of transmission. Never damage the rubber coupling, protect it with a shop towel or equivalent.

✳✳ WARNING

Never damage the rubber coupling.

➡**Never damage the grease seal.**

➡**The angle, which the third axis rubber coupling forms with the final drive companion flange, must be 5° or less.**

16. Install center bearing mounting bracket fixing nuts.

✳✳ WARNING

Never damage the final drive companion flange and rubber coupling.

17. Install the rubber coupling to the final drive companion flange.

➡**Never remove the rubber coupling from the propeller shaft assembly.**

18. Install propeller shaft assembly fixing bolts and nuts (arrow).

➡**Tighten mounting nuts temporarily.**

19. Tighten mounting nuts of center bearing mounting brackets (upper/lower).

✳✳ WARNING

For matching marks, use paint. Never damage propeller shaft rubber coupling and final drive companion flange.

20. Align matching marks on propeller shaft rubber coupling and final drive companion flange.
21. Install the heat plate.
22. Install the center muffler with power tool.
23. Install the floor reinforcement.
24. Shift the transmission to the park position, and then engage the parking brake.

AWD Vehicles

See Figures 36 through 40.

1. Shift the transmission to the neutral position, and release the parking brake.
2. Remove the floor reinforcement.
3. Remove the center muffler with power tool.
4. Remove the heat plate.
5. Put matching marks on propeller shaft flange yoke and transfer companion flange.

✳✳ WARNING

For matching marks, use paint. Never damage propeller shaft flange yoke and transfer companion flange.

6. Put matching marks on propeller shaft rebro joint and final drive companion flange.

✳✳ WARNING

For matching marks, use paint. Never damage propeller shaft rebro joint and final drive companion flange.

7. Loosen mounting nuts of center bearing mounting brackets (upper/lower).

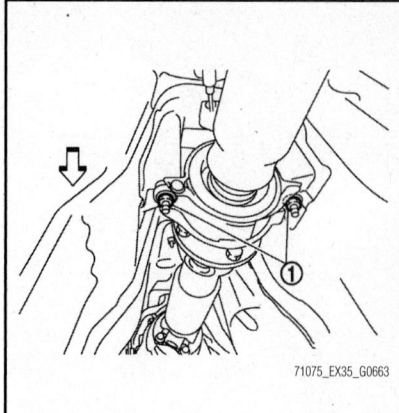

71075_EX35_G0663

Fig. 36 Loosen mounting nuts (1) of center bearing mounting brackets (upper/lower), Arrow=Vehicle front

➡**Tighten mounting nuts temporarily.**

8. Remove propeller shaft assembly fixing bolts and nuts.
9. Remove center bearing mounting bracket fixing nuts.
10. Remove propeller shaft assembly.

✳✳ WARNING

If constant velocity joint was bent during propeller shaft assembly removal, installation, or transportation, its boot may be damaged. Wrap boot interference area to metal part with shop cloth or rubber to protect boot from breakage.

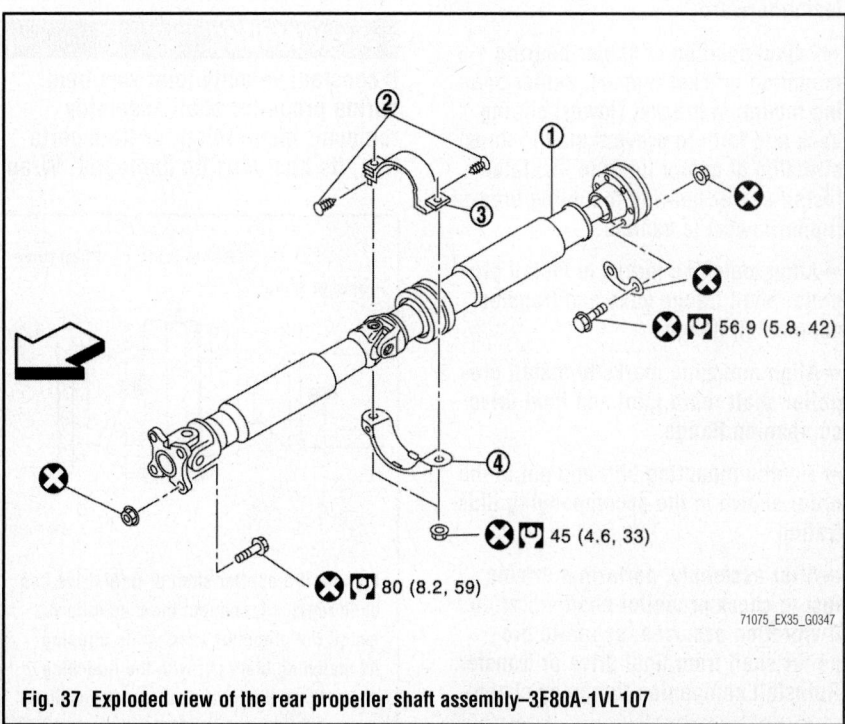

Fig. 37 Exploded view of the rear propeller shaft assembly–3F80A-1VL107

71075_EX35_G0347

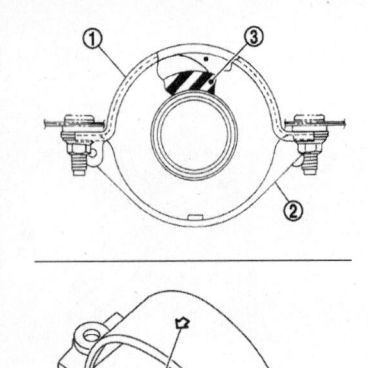

1. Center bearing mounting bracket (upper)
2. Center bearing mounting bracket (lower)
3. Center bearing insulator
A. Arrow mark

71075_EX35_G0667

Fig. 38 Install center bearing mounting bracket (upper) with its arrow mark facing forward. Adjust position of center bearing mounting bracket (upper), center bearing mounting bracket (lower) sliding back and forth to prevent play in thrust direction of center bearing insulator. Install center bearing mounting bracket (upper/lower) to vehicle.

To install:

➡Install center bearing mounting bracket (upper) with its arrow mark facing forward.

➡Adjust position of center bearing mounting bracket (upper), center bearing mounting bracket (lower) sliding back and forth to prevent play in thrust direction of center bearing insulator. Install center bearing mounting bracket (upper/lower) to vehicle.

➡Align matching marks to install propeller shaft flange yoke and transfer companion flange.

➡Align matching marks to install propeller shaft rebro joint and final drive companion flange.

➡Tighten mounting bolt and nut in the order shown in the accompanying illustration.

➡After assembly, perform a driving test to check propeller shaft vibration. If vibration occurred, separate propeller shaft from final drive or transfer. Reinstall companion flange by chang-

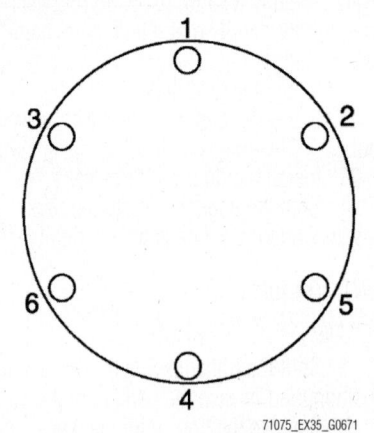

71075_EX35_G0671

Fig. 39 Tighten mounting bolt and nut in the order shown

ing the phase between companion flange and propeller shaft by the one bolt hole at a time. Then perform driving test and check propeller shaft vibration again at each point.

➡If propeller shaft or final drive has been replaced, connect them as follows:

- Install the propeller shaft while aligning its matching mark with the matching mark on the joint as close as possible.

❋❋ WARNING
Avoid damaging the rebro joint boot, protect it with a shop towel or equivalent.

❋❋ WARNING
If constant velocity joint was bent during propeller shaft assembly removal, installation, or transportation, its boot may be damaged. Wrap

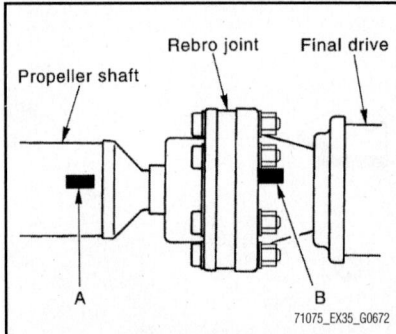

71075_EX35_G0672

Fig. 40 If propeller shaft or final drive has been replaced, connect them as follows: Install the propeller shaft while aligning its matching mark (A) with the matching mark (B) on the joint as close as possible

boot interference area to metal part with shop cloth or rubber to protect boot from breakage.

11. Install propeller shaft assembly.
12. Install center bearing mounting bracket fixing nuts.
13. Install propeller shaft assembly fixing bolts and nuts.

➡**Loosen mounting nuts temporarily.**

14. Tighten mounting nuts of center bearing mounting brackets (upper/lower).

❋❋ WARNING
For matching marks, use paint. Never damage propeller shaft rebro joint and final drive companion flange.

15. Align matching marks on propeller shaft rebro joint and final drive companion flange.

❋❋ WARNING
For matching marks, use paint. Never damage propeller shaft flange yoke and transfer companion flange.

16. Align matching marks on propeller shaft flange yoke and transfer companion flange.
17. Install the heat plate.
18. Install the center muffler with power tool.
19. Install the floor reinforcement.
20. Shift the transmission to the park position, and engage the parking brake.

TRANSFER CASE

DRAIN & REFILL

See Figure 41.

Draining
1. Run the vehicle to warm up the transfer unit sufficiently.
2. Stop the engine, and remove the drain plug to drain the transfer fluid.
3. Set a new gasket onto the drain plug, and install it on the transfer and tighten to 26 ft. lbs. (35 Nm).

➡**Never reuse gasket.**

Refilling
4. Remove filler plug and gasket. Then fill fluid up to mounting hole for the filler plug.

➡**Carefully fill the fluid. (Fill up for approximately 3 minutes.)**

5. Leave the vehicle for 3 minutes, and check the fluid level again.

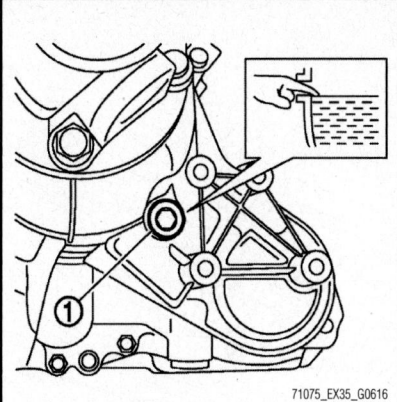

Fig. 41 Remove filler plug (1) and gasket. Then fill fluid up to mounting hole for the filler plug

6. Set a new gasket onto filler plug, and install it on transfer and tighten to 26 ft. lbs. (35 Nm).

➡**Never reuse gasket.**

FLUID LEVEL CHECK

See Figure 42.

1. Remove filler plug and gasket. Then check that fluid is filled up from mounting hole for the filler plug.

☀☀ CAUTION

Never start engine while checking fluid level.

2. Set a new gasket onto filler plug, and install it on transfer and tighten to 26 ft. lbs. (35 Nm).

➡**Never reuse gasket.**

FLUID RECOMMENDATIONS

The manufacturer recommends Genuine NISSAN Matic J ATF. Using transfer fluid other than Genuine NISSAN Matic J ATF will cause deterioration in driveability and transfer durability, and may damage the transfer, which is not

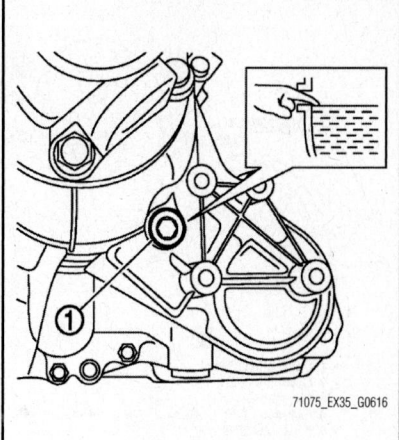

Fig. 42 Remove filler plug (1) and gasket. Then check that fluid is filled up from mounting hole for the filler plug

covered by the INFINITI new vehicle limited warranty.

ENGINE COOLING

ENGINE COOLANT

DRAIN & REFILL

Draining

Note the following:
• Never change engine coolant when the engine is hot to avoid being scalded.
• Wrap a thick cloth around radiator cap and carefully remove radiator cap. First, turn radiator cap a quarter of a turn to release built-up pressure. Then turn radiator cap all the way.

1. Open radiator drain plug at the bottom of radiator, and then remove radiator cap. When draining all of engine coolant in the system, open water drain plugs on cylinder block.

2. Remove reservoir tank if necessary, and drain engine coolant and clean reservoir tank before installing.

3. Check drained engine coolant for contaminants such as rust, corrosion or discoloration. If contaminated, flush the engine cooling system.

Refilling

1. Remove engine cover.
2. Install reservoir tank if removed, and radiator drain plug.

☀☀ WARNING

Be sure to clean drain plug and install with new O-ring. If water drain plugs on cylinder block are removed, close and tighten them.

3. Check that each hose clamp has been firmly tightened.
4. Remove air relief plug on radiator left side.
5. Fill radiator, and reservoir tank if removed, to specified level.
• Pour engine coolant through engine coolant filler neck slowly, wait a minute to allow air in system to escape.
• Use Genuine NISSAN Long Life Antifreeze/Coolant or equivalent mixed with water (distilled or demineralized).
6. When engine coolant overflows air relief hole on radiator, install air relief plug with new O-ring.
7. Repeat step 5.
8. Install radiator cap.
9. Warm up engine until opening thermostat. Standard for warming-up time is approximately 10 minutes at 3,000 rpm.
• Check thermostat opening condition by touching radiator hose (lower) to see a flow of warm water.

☀☀ WARNING

Watch water temperature gauge so as not to overheat engine.

10. Stop the engine and cool down to less than approximately 50°C (122°F).

• Cool down using fan to reduce the time.
• If necessary, refill radiator up to filler neck with engine coolant.
11. Refill reservoir tank to "MAX" level line with engine coolant.
12. Repeat steps 8 through 11 two or more times with radiator cap installed until engine coolant level no longer drops.
13. Check cooling system for leakage with engine running.
14. Warm up the engine, and check for sound of engine coolant flow while running engine from idle up to 3,000 rpm with heater temperature controller set at several position between "COOL" and "WARM".
• Sound may be noticeable at heater unit.
15. Repeat step 14 three times.
16. If sound is heard, bleed air from cooling system by repeating step 5, and steps from 8 to 15 until engine coolant level no longer drops.
17. Check that the reservoir tank cap is tightened.

ELECTRIC ENGINE FAN

REMOVAL & INSTALLATION

See Figure 43.

1. Remove engine under cover with power tool.
2. Drain engine coolant.
3. Remove reservoir tank.

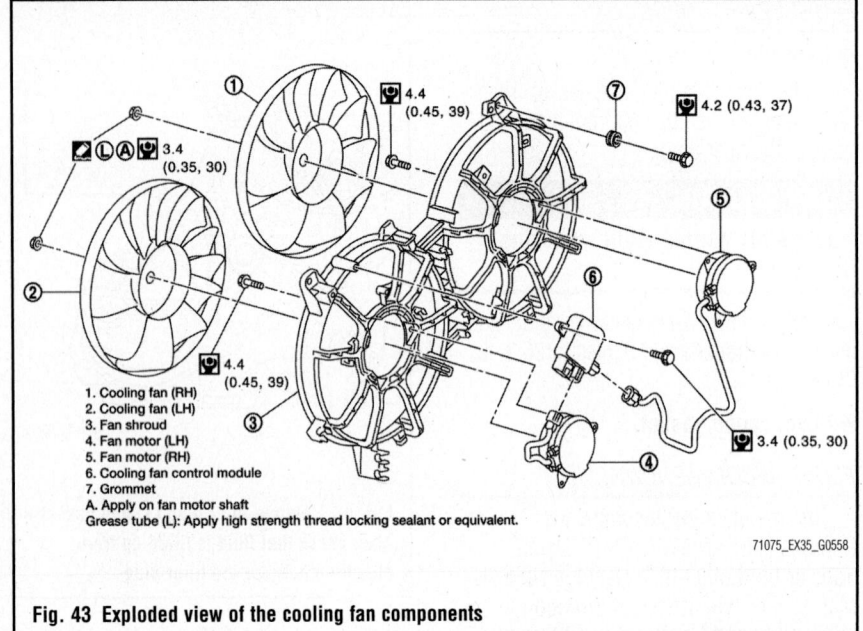

1. Cooling fan (RH)
2. Cooling fan (LH)
3. Fan shroud
4. Fan motor (LH)
5. Fan motor (RH)
6. Cooling fan control module
7. Grommet
A. Apply on fan motor shaft
Grease tube (L): Apply high strength thread locking sealant or equivalent.

71075_EX35_G0558

Fig. 43 Exploded view of the cooling fan components

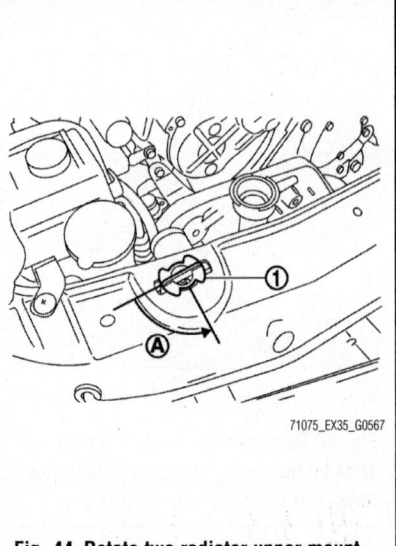

71075_EX35_G0567

Fig. 44 Rotate two radiator upper mount brackets (1) 90° (A) in direction (counter-clockwise) as shown, and remove them

4. Remove air cleaner case (LH and RH).

5. Remove mounting bolt from high pressure flexible hose bracket.

6. Remove radiator hose (upper).

7. Disconnect harness connector from cooling fan control module, and move harness to aside.

8. Remove cooling fan assembly.

✳✳ WARNING

Be careful not to damage or scratch on radiator core.

To install:

➡Note the following, and install in the reverse order of removal.

✳✳ WARNING

Only use genuine parts for cooling fan mounting bolt and observe the specified torque (to prevent radiator from being damaged).

✳✳ WARNING

Be careful not to damage or scratch on radiator core.

9. Install cooling fan assembly.

10. Connect harness connector to cooling fan control module, and move harness to aside.

11. Install radiator hose (upper).

12. Install mounting bolt to high pressure flexible hose bracket.

13. Install air cleaner case (LH and RH).

14. Install reservoir tank.

15. Fill engine coolant.

16. Install engine under cover with power tool.

RADIATOR

REMOVAL & INSTALLATION

See Figures 44 through 48.

Never remove radiator cap when engine is hot. Serious burns could occur from high-pressure engine coolant escaping from water outlet (front). Wrap a thick cloth around the cap. Slowly turn it a quarter of a turn to release built-up pressure. Carefully remove radiator cap by turning it all the way.

1. Remove the following parts:
   • Engine under cover with power tool.
   • Engine cover:
   • Air cleaner case (RH and LH):
   • Reservoir tank:
   • Hood lock cover, hood lock stay assembly and horn.
2. Remove condenser.
3. Drain engine coolant from radiator.

➡**Perform this step when the engine is cold.**

➡**Never spill engine coolant on drive belt.**

4. Disconnect A/T fluid cooler hoses from radiator.
   a. Install blind plug to avoid leakage of A/T fluid.
5. Remove radiator hoses (upper and lower) and reservoir tank hose.

➡**Be careful not to allow engine coolant to contact drive belt.**

➡**Never loosen radiator water inlet pipe mounting screw. If loosened, replace radiator.**

6. Remove cooling fan assembly.

✳✳ WARNING

Never damage or scratch radiator core when removing.

7. Rotate two radiator upper mount brackets 90° in direction (counterclockwise) as shown in the accompanying illustration, and remove them.

8. Remove radiator as follows:

✳✳ WARNING

Be careful not to damage radiator core.

   a. Lift up and pull the radiator forward, and then remove the mounting rubber (lower) from the radiator core support.

   b. Remove radiator from front of radiator core support.

To install:

➡Note the following, and install in the reverse order of removal.

➡**Replace water hose clamp if it is removed.**

➡**Use genuine mounting bolts for the cooling fan assembly and strictly observe the tightening torque. (Breakage prevention for radiator)**

➡**Insert the radiator hose all the way to the stopper or by 1.30 in. (33 mm) (hose without a stopper).**

For the orientation of the hose clamp pawl

Radiator hose	Hose end	Paint mark	Position of hose clamp*
Radiator hose (upper)	Radiator side	Upper	A
	Engine side	Upper	B
Radiator hose (lower)	Radiator side	Lower	C
	Engine side	Right side	D

*Refer to the illustrations for the specific position each hose clamp tab.

71075_EX35_G0571

Fig. 45 Hose clamp tab specific position (1 of 2)

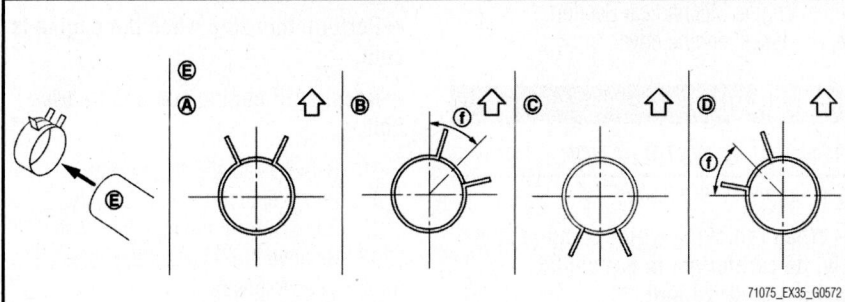

71075_EX35_G0572

Fig. 46 Hose clamp tab specific position (2 of 2). View E (E), 45°(F), Arrow=Vehicle upper

➡For the orientation of the hose clamp pawl, refer to the accompanying illustration.

➡The angle created by the hose clamp pawl and the specified line must be within ±30° as shown in the accompanying illustration.

➡To install hose clamps, check that the dimension from the end of the paint mark on the radiator hose to the hose clamp is within the reference value.

9. Install radiator as follows:
    a. Install radiator to front of radiator core support.

※※ **WARNING**

Be careful not to damage radiator core.

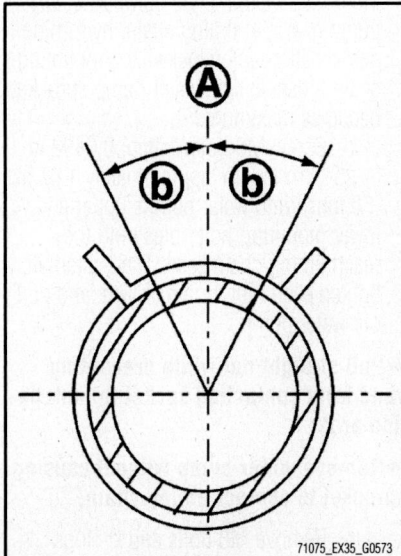

71075_EX35_G0573

Fig. 47 The angle (b) created by the hose clamp pawl and the specified line (A) must be within ±30° as shown

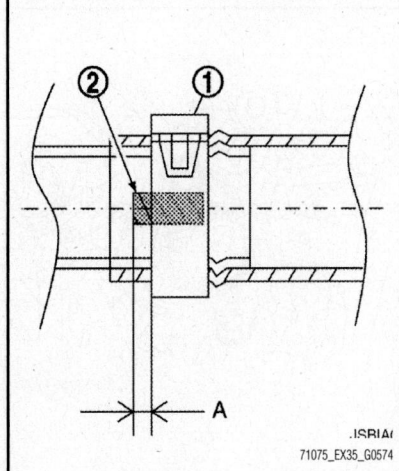

71075_EX35_G0574

Fig. 48 To install hose clamps (1), check that the dimension [(A) Dimension "A" (−1) − (+1) mm] from the end of the paint mark (2) on the radiator hose to the hose clamp is within the reference value

b. Lower down and push the radiator backward, and then install the mounting rubber (lower) to the radiator core support.

※※ **WARNING**

Never damage or scratch radiator core when removing.

10. Rotate two radiator upper mount brackets 90° in opposite direction (clockwise) as shown in the accompanying illustration, and install them.
11. Install cooling fan assembly.

➡Be careful not to allow engine coolant to contact drive belt.

➡Never loosen radiator water inlet pipe mounting screw. If loosened, replace radiator.

12. Install radiator hoses (upper and lower) and reservoir tank hose.
13. Connect A/T fluid cooler hoses to radiator.
    a. Remove blind plug that was installed to avoid leakage of A/T fluid.

➡Perform this step when the engine is cold.

➡Never spill engine coolant on drive belt.

14. Fill engine coolant to radiator.
15. Install condenser.
16. Install the following parts:
    • Hood lock cover, hood lock stay assembly and horn.
    • Reservoir tank:
    • Air cleaner case (RH and LH):
    • Engine cover:
    • Engine under cover with power tool.

➡Check for leakage of engine coolant using the radiator cap tester adapter (commercial service tool) and the radiator cap tester (commercial service tool).

➡Start and warm up the engine. Visually check that there is no leakage of engine coolant and A/T fluid.

THERMOSTAT & WATER INLET ASSEMBLY

REMOVAL & INSTALLATION

See Figure 49.

1. Remove engine cover.
2. Remove air duct and air cleaner case assembly (LH). Refer to Air Intake Systems in the Engine Mechanical Section.

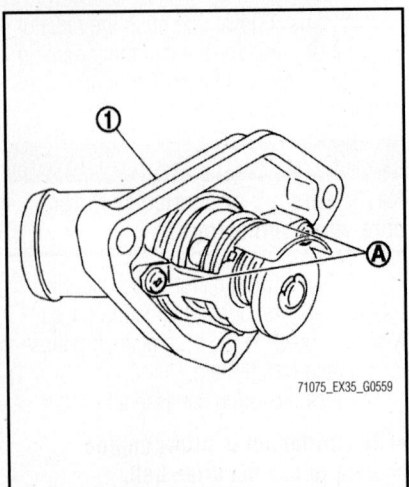

Fig. 49 Remove water inlet and thermostat assembly (1).

3. Remove reservoir tank.
4. Remove engine undercover with power tool.
5. Drain engine coolant from radiator drain plug at the bottom of radiator.

➡**Perform this step when the engine is cold.**

➡**Never spill engine coolant on drive belt.**

6. Disconnect radiator hose (lower).
7. Disconnect intake valve timing control solenoid valve harness connector (bank 2), and remove intake valve timing control solenoid valve.
8. Remove water inlet and thermostat assembly.

✳ WARNING

Never disassemble water inlet and thermostat assembly. Replace them as a unit, if necessary.

To install:

➡Note the following, and install in the reverse order of removal.

➡Be careful not to spill engine coolant over engine room. Use rag to absorb engine coolant.

✳ WARNING

Never disassemble water inlet and thermostat assembly. Replace them as a unit, if necessary.

9. Install water inlet and thermostat assembly (1).
10. Connect intake valve timing control solenoid valve harness connector (bank 2), and install intake valve timing control solenoid valve.

11. Connect radiator hose (lower).

➡**Never spill engine coolant on drive belt.**

➡**Perform this step when the engine is cold.**

12. Fill engine coolant to radiator drain plug at the bottom of radiator.
13. Install engine undercover with power tool.
14. Install reservoir tank.
15. Install air duct and air cleaner case assembly (LH). Refer to Air Intake Systems in the Engine Mechanical Section.
16. Install engine cover.

WATER PUMP

REMOVAL & INSTALLATION
See Figures 50 through 52.

➡**When removing water pump assembly, be careful not to get engine coolant on drive belt.**

➡**Water pump cannot be disassembled and should be replaced as a unit.**

➡**After installing water pump, connect hose and clamp securely, then check for leakage using the radiator cap tester (commercial service tool) and the radiator cap tester adapter (commercial service tool).**

1. Remove engine cover.
2. Release the fuel pressure.
3. Disconnect the battery cable from the negative terminal.

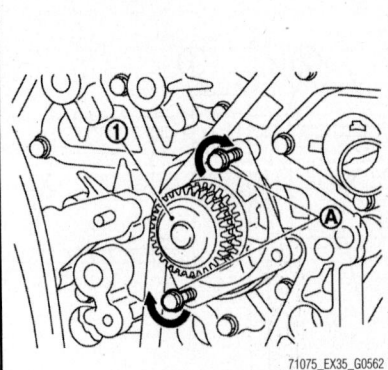

Fig. 50 Screw M8 bolts (A) [pitch: 0.0492 in (1.25 mm) length: approximately 1.97 in. (50 mm)] into water pumps upper and lower mounting bolt holes until they reach timing chain case. Then, alternately tighten each bolt for a half turn, and pull out water pump (1).

4. Remove air duct and air cleaner case assembly (RH and LH).
5. Remove reservoir tank.
6. Separate engine harness removing their brackets from front timing chain case.
7. Remove engine undercover with power tool.
8. Drain engine oil.

➡**Perform this step when the engine is cold.**

➡**Never spill engine oil on drive belt.**

9. Drain engine coolant from radiator.

➡**Perform this step when the engine is cold.**

➡**Never spill engine coolant on drive belt.**

10. Remove radiator hose (lower).
11. Remove cooling fan assembly.
12. Remove front timing chain case.
13. Remove timing chain tensioner (primary) as follows:
    a. Remove lower mounting bolt.
    b. Loosen upper mounting bolt slowly, and then turn chain tensioner (primary) on the upper mounting bolt so that plunger is fully expanded.

➡**Even if plunger is fully expanded, it is not dropped from the body of timing chain tensioner (primary).**

    c. Remove upper mounting bolt, and then remove timing chain tensioner (primary).
14. Remove water pump as follows:
    a. Remove three water pump mounting bolts. Secure a gap between water pump gear and timing chain, by turning crankshaft counterclockwise until timing chain looseness on water pump sprocket becomes maximum.
    b. Screw M8 bolts [pitch: 0.0492 in (1.25 mm) length: approximately 1.97 in. (50 mm)] into water pumps upper and lower mounting bolt holes until they reach timing chain case. Then, alternately tighten each bolt for a half turn, and pull out water pump.

➡**Pull straight out while preventing vane from contacting socket in installation area.**

➡**Remove water pump without causing sprocket to contact timing chain.**

    c. Remove M8 bolts and O-rings from water pump.

➡**Never disassemble water pump.**

15. Check for badly rusted or corroded water pump body assembly.

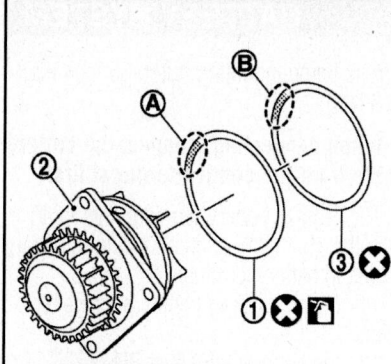

1. Engine oil to O-ring
2. Water pump
3. Engine coolant to O-ring

A. O-ring with yellow paint mark
B. O-ring with light blue paint mark

71075_EX35_G0563

Fig. 51 Install new O-rings to water pump. Apply engine oil to O-ring and engine coolant to O-ring as shown

16. Check for rough operation due to excessive end play.

17. If anything is found, replace water pump.

To install:

18. Install new O-rings to water pump.

a. Apply engine oil to O-ring (1) and engine coolant to O-ring (3) as shown in the accompanying illustration.

b. Locate O-ring with yellow paint mark (A) to front side.

c. Locate O-ring with light blue paint mark (B) to rear side.

19. Install water pump.

➡**Never allow cylinder block to nip O-rings when installing water pump.**

➡**Check timing chain and water pump sprocket are engaged.**

➡**Insert water pump by tightening mounting bolts alternately and evenly.**

20. Install timing chain tensioner (primary) as follows:

a. Turn crankshaft clockwise so that timing chain on the timing chain tensioner (primary) side is loose.

b. Pull plunger stopper tab up (or turn lever downward) so as to remove plunger stopper tab from the ratchet of plunger.

➡**Plunger stopper tab and lever are synchronized.**

c. Push plunger into the inside of tensioner body.

d. Hold plunger in the fully compressed position by engaging plunger stopper tab with the tip of ratchet.

e. To secure lever, insert stopper pin through hole of lever into tensioner body hole.

• The lever parts and the tab are synchronized. Therefore, the plunger will be secured under this condition.

➡**Illustration shows the example of 1.2 mm (0.047 in) diameter thin screwdriver being used as the stopper pin.**

f. Install timing chain tensioner (primary).

• Remove dust and foreign material completely from backside of timing chain tensioner (primary) and from installation area of rear timing chain case.

g. Remove stopper pin.

h. Check again that timing chain and water pump sprocket are engaged.

21. Install in the reverse order of removal for remaining parts.

a. After starting engine, let idle for three minutes, then rev engine up to 3,000 rpm under no load to purge air from the high-pressure chamber of chain

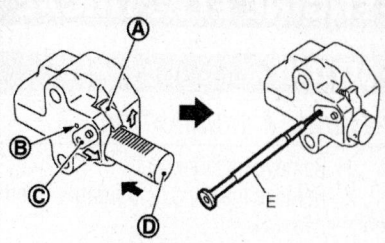

A. Plunger stopper tab
B. Tensioner body hole
C. Plunger stopper tab and lever
D. Ratchet of plunger
E. Stopper pin

71075_EX35_G0564

Fig. 52 Install timing chain tensioner (primary). Turn crankshaft clockwise so that timing chain on the timing chain tensioner (primary) side is loose. Pull plunger stopper tab up (or turn lever downward) so as to remove plunger stopper tab from the ratchet of plunger. Plunger stopper tab and lever are synchronized. Push plunger into the inside of tensioner body. Hold plunger in the fully compressed position by engaging plunger stopper tab with the tip of ratchet. To secure lever, insert stopper pin through hole of lever into tensioner body hole

tensioner. Engine may produce a rattling noise. This indicates that air still remains in the chamber and is not a matter of concern.

22. Check that the reservoir tank cap is tightened.

23. Check for leakage of engine coolant using the radiator cap tester adapter (commercial service tool) and the radiator cap tester (commercial service tool).

24. Start and warm up the engine. Visually check that there is no leakage of engine coolant.

ENGINE ELECTRICAL   BATTERY SYSTEM

BATTERY

REMOVAL & INSTALLATION

1. Remove battery cover.
2. Remove the clips, and remove hood ledge cover RH.
3. Remove cowl top cover RH.
4. Remove cover of battery positive terminal.
5. Loosen battery terminal nuts, and disconnect both battery cables from battery terminals.

➡**When disconnecting, disconnect the battery cable from the negative terminal first.**

6. Remove battery fix frame mounting nuts and battery fix frame.
7. Remove battery.

To install:

➡**When connecting, connect the battery cable to the positive terminal first.**

8. Install battery.
9. Install battery fix frame mounting nuts and battery fix frame. Tighten the battery fix frame mounting nut to 35 inch lbs. (3.9 Nm).

➡**When connecting, connect the battery cable from to negative terminal first.**

10. Tighten battery terminal nuts to 48 inch lbs. (5.4 Nm), and connect both battery cables to battery terminals.
11. Install cover of battery positive terminal.
12. Install cowl top cover RH.
13. Install the clips, and install hood ledge cover RH.
14. Install battery cover.

ENGINE ELECTRICAL   CHARGING SYSTEM

➡Disconnecting the negative battery cable on some vehicles may interfere with the functions of the on board computer system. The computer may undergo a relearning process once the negative battery cable is reconnected.

ALTERNATOR

REMOVAL & INSTALLATION
See Figure 53.

➡Disconnecting the negative battery cable on some vehicles may interfere

with the functions of the on board computer system. The computer may undergo a relearning process once the negative battery cable is reconnected.

1. Before servicing the vehicle, refer to the Precautions Section.

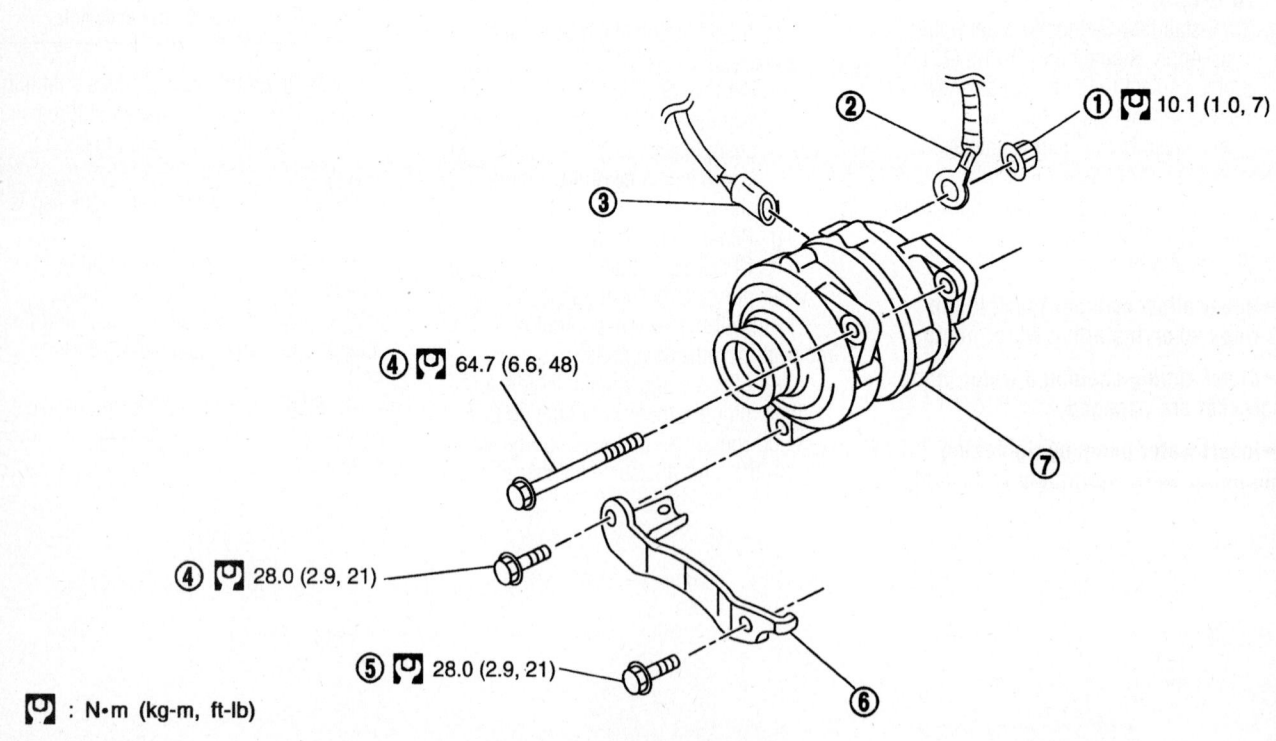

① 🔧 10.1 (1.0, 7)

④ 🔧 64.7 (6.6, 48)

④ 🔧 28.0 (2.9, 21)

⑤ 🔧 28.0 (2.9, 21)

🔧 : N•m (kg-m, ft-lb)

1. B terminal nut
4. Alternator mounting bolt
7. Alternator
2. Alternator B terminal harness
5. Alternator stay mounting bolt
3. Alternator connector
6. Alternator stay

67162-FX35-G100

Fig. 53 Exploded view of alternator mounting

2. Disconnect the negative battery cable.

3. Remove engine front undercover.

4. Remove the alternator and power steering oil pump belt.

5. Disconnect the alternator connector. Remove the **B** terminal nut.

6. Remove the harness clip and water hose bracket from the alternator.

7. For 2WD models:

   a. Remove the oil pressure switch harness clip from the alternator stay.

   b. Disconnect the oil pressure switch connector.

8. Remove the alternator stay mounting bolts and alternator stay.

9. Remove the alternator mounting bolt.

10. Remove the alternator assembly, by lowering it out of the bottom of the engine compartment.

To install:

11. Reposition the alternator in place on the engine and tighten the mounting bolts. Tighten the long alternator bolt 48 ft. lbs. (65 Nm) and the short alternator bracket bolts to 21 ft. lbs. (28 Nm).

12. Tighten the **B** terminal nut carefully to 84 inch lbs. (10 Nm).

13. For 2WD models:

   a. Connect the oil pressure switch connector.

   b. Install the oil pressure switch harness clip to the alternator stay.

14. Reconnect the alternator wiring.

15. Install the accessory drive belt.

16. Install the front engine undercover.

17. Reconnect the negative battery cable.

ENGINE ELECTRICAL   IGNITION SYSTEM

➡**Disconnecting the negative battery cable on some vehicles may interfere with the functions of the on board computer system. The computer may undergo a relearning process once the negative battery cable is reconnected.**

FIRING ORDERS

See Figure 54.

IGNITION COIL(S)

REMOVAL & INSTALLATION

See Figure 55.

1. Before servicing the vehicle, refer to the Precautions Section.

2. Remove the engine cover.

3. Remove the air duct (for ignition coil of left bank side).

4. Move aside the wiring harness, wiring harness bracket, and hoses located above the ignition coil.

5. Disconnect the wiring harness connector from the ignition coil.

6. Remove the ignition coil retaining bolt.

7. Remove the ignition coil.

❊❊ WARNING

Do not subject the ignition coils to excessive shock or vibration.

To install:

8. Install the ignition coil on the engine. Tighten the retaining bolt to 62 inch lbs. (7 Nm).

9. Reconnect the wiring harness to the coil.

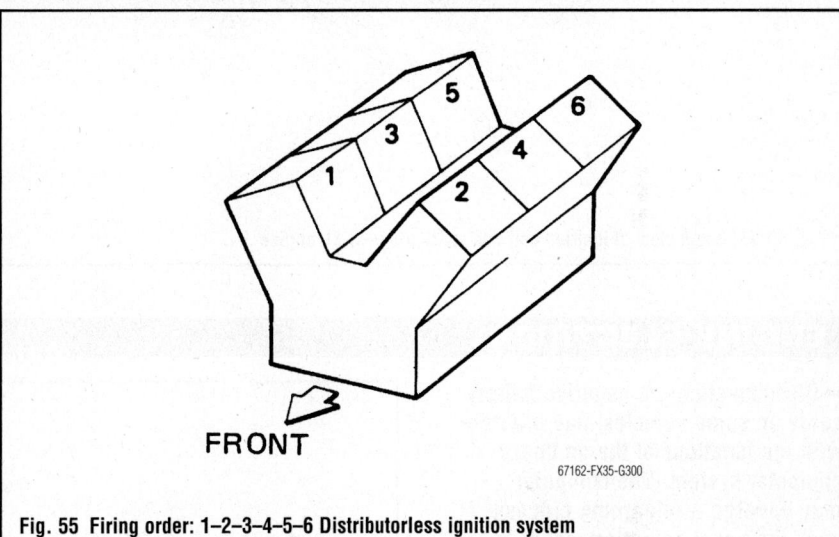

Fig. 55 Firing order: 1–2–3–4–5–6 Distributorless ignition system

67162-FX35-G300

10. Reposition the wiring harness, bracket and hoses.

11. Install the air duct and the engine cover.

IGNITION TIMING

INSPECTION & ADJUSTMENT

The ignition timing is controlled by the Electronic Control Module (ECM). No adjustment is necessary or possible.

SPARK PLUGS

REMOVAL & INSTALLATION

See Figure 55.

➡**Disconnecting the negative battery cable on some vehicles may interfere with the functions of the on board computer system. The computer may undergo a relearning process once the negative battery cable is reconnected.**

1. Disconnect the negative battery cable.

2. Remove the engine cover.

3. Remove the ignition coil retaining bolt.

4. Remove the ignition coil.

5. Remove the spark plug using a spark plug socket and wrench.

To install:

6. Be sure the spark plug gap is to specification: 0.043 in.

7. Carefully install the spark plug and torque to specification: 18 ft. lbs. (25 Nm).

8. Install the ignition coil, torque the retaining bolt to 62 inch lbs. (7 Nm).

9. Install the engine cover.

10. Connect the negative battery cable.

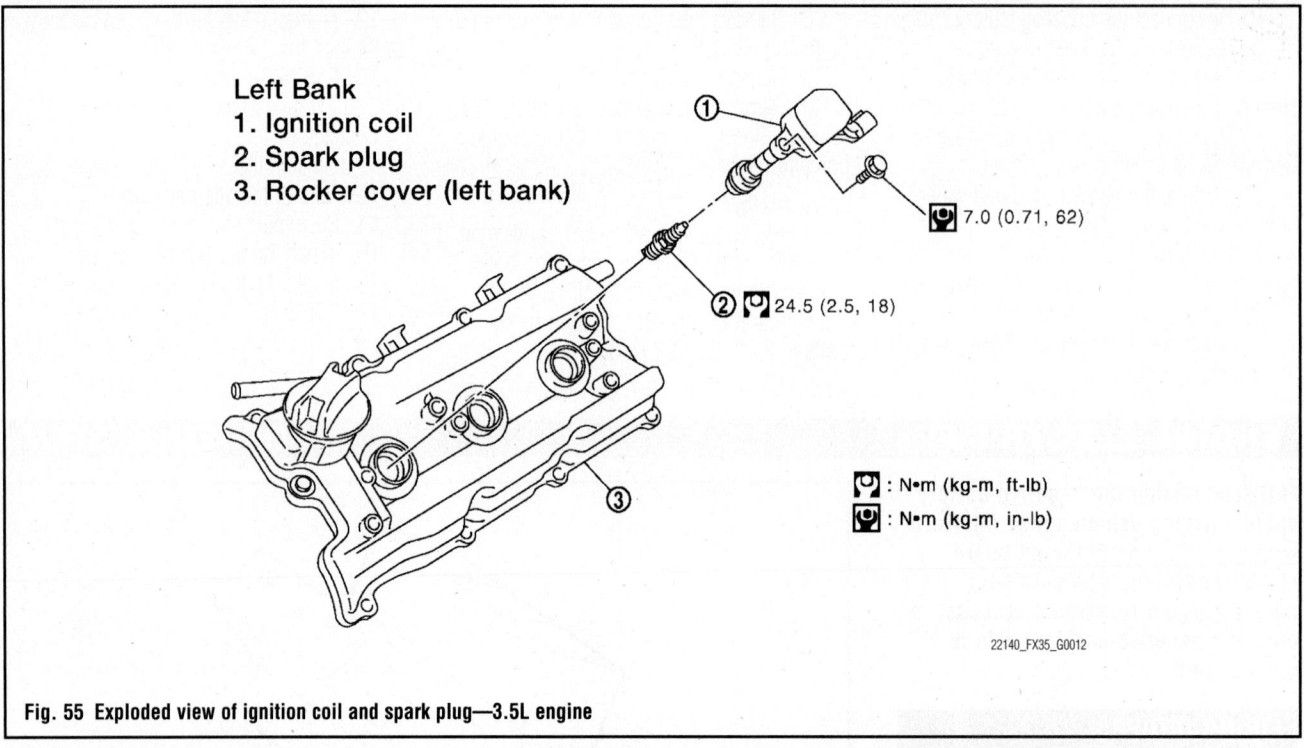

Left Bank
1. Ignition coil
2. Spark plug
3. Rocker cover (left bank)

7.0 (0.71, 62)

24.5 (2.5, 18)

: N•m (kg-m, ft-lb)
: N•m (kg-m, in-lb)

22140_FX35_G0012

Fig. 55 Exploded view of ignition coil and spark plug—3.5L engine

ENGINE ELECTRICAL

STARTING SYSTEM

➡Disconnecting the negative battery cable on some vehicles may interfere with the functions of the on board computer system. The computer may undergo a relearning process once the negative battery cable is reconnected.

STARTER

REMOVAL & INSTALLATION

See Figures 56 and 57.

1. Before servicing the vehicle, refer to the Precautions Section.
2. Disconnect the negative battery cable.
3. Remove the engine rear undercover.
4. Remove the starter electrical wires.
5. Remove the starter retaining bolts.
6. On 2WD vehicles, remove the harness clip bracket.
7. Remove the starter from its mounting.

To install:

8. Installation is the reverse of the removal procedure.

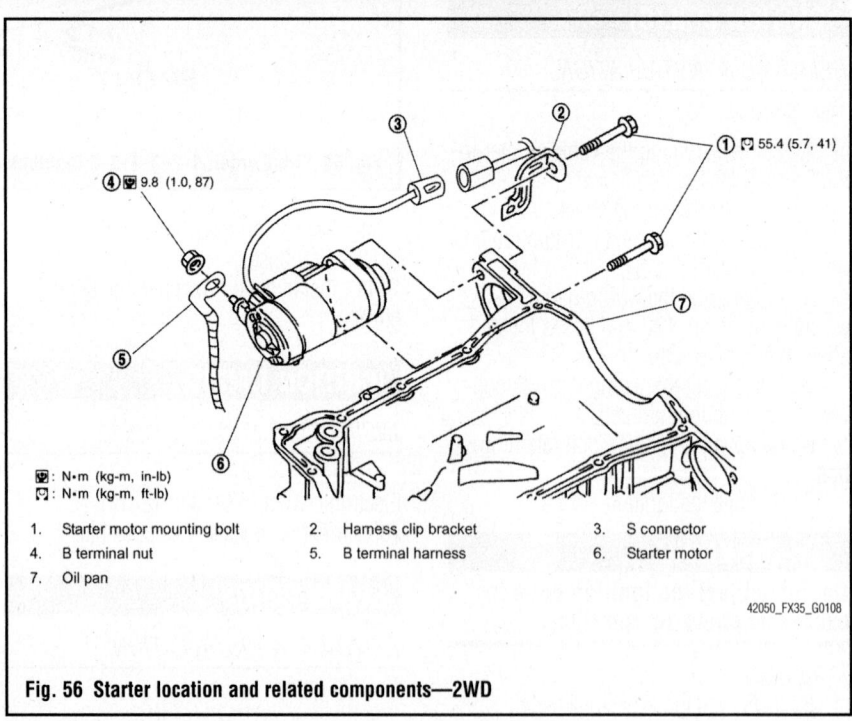

4. 9.8 (1.0, 87)

3.

2.

1. 55.4 (5.7, 41)

5.

6.

7.

: N•m (kg-m, in-lb)
: N•m (kg-m, ft-lb)

1. Starter motor mounting bolt
2. Harness clip bracket
3. S connector
4. B terminal nut
5. B terminal harness
6. Starter motor
7. Oil pan

42050_FX35_G0108

Fig. 56 Starter location and related components—2WD

9. Tighten the retaining bolts to 41 ft. lbs. (55 Nm).

10. Tighten the terminal nut to 87 inch lbs. (10 Nm).

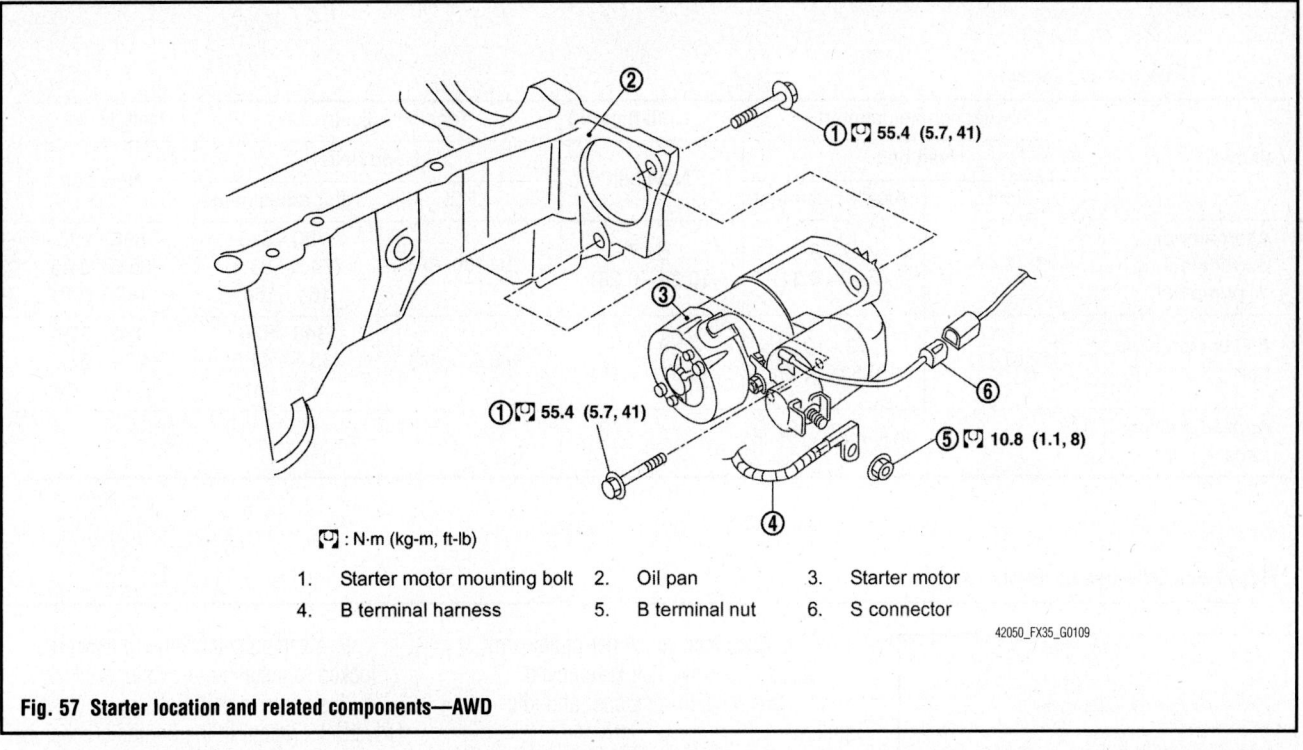

☑ : N·m (kg-m, ft-lb)

1. Starter motor mounting bolt
2. Oil pan
3. Starter motor
4. B terminal harness
5. B terminal nut
6. S connector

42050_FX35_G0109

Fig. 57 Starter location and related components—AWD

ENGINE MECHANICAL

ACCESSORY DRIVE BELT SYSTEM

ADJUSTMENT

Air Conditioning Compressor Belt

See Figures 58 and 59.

1. Before servicing the vehicle, refer to the Precautions Section.
2. Disconnect the negative battery cable.
3. Remove the engine undercover.
4. Loosen the idler pulley locknut and adjust the tension by turning the adjusting bolt.
5. Adjust the belt deflection/tension using the following table.
6. Tighten the locknut to 26 ft. lbs. (35 Nm).

Alternator And Power Steering Belt

See Figures 58 and 59.

1. Before servicing the vehicle, refer to the Precautions Section.
2. Disconnect the negative battery cable.
3. Remove the engine undercover.
4. Loosen the idler pulley locknut (A) and adjust the tension by turning the adjusting bolt (B).
5. Adjust the belt deflection/tension using the following table.
6. Tighten the locknut (A) to 26 ft. lbs. (35 Nm).

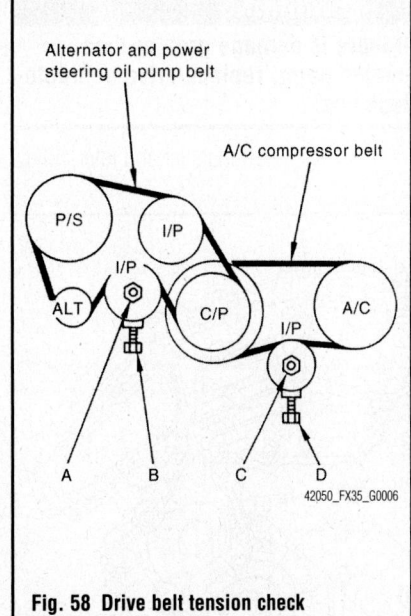

Alternator and power steering oil pump belt

A/C compressor belt

42050_FX35_G0006

Fig. 58 Drive belt tension check

BELT ROUTINGS

See Figure 60.

Refer to the accompanying illustration for belt routing.

INSPECTION

Inspect the drive belt for signs of glazing or cracking. A glazed belt will be perfectly smooth from slippage, while a good belt will have a slight texture of fabric visible. Cracks will usually start at the inner edge of the belt and run outward. All worn or damaged drive belts should be replaced immediately.

REMOVAL & INSTALLATION

Drive Belt

1. Before servicing the vehicle, refer to the Precautions Section.
2. Disconnect the negative battery cable.
3. Remove the engine undercover.
4. Remove the alternator and power steering belt.
5. Remove the air conditioning compressor belt.

To install:

6. Installation is the reverse of removal.
7. Be sure not to get grease or oil on the belts.
8. Make sure the drive belts are correctly engaged with the pulley groove.
9. Torque and adjust the drive belts to specification. Refer to Accessory Drive Belts, Adjustment.

Drive Belt Auto Tensioner And Idler Pulley

See Figure 61.

1. Remove drive belt. Refer to Accessory Drive Belt System in the Engine Mechanical Section.

Belt Deflection and Tension

Items	Deflection adjustment			Tension adjustment		
	Used belt		Unit: mm (in)	Used belt		Unit: N (kg, lb)
	Limit	After adjustment	New belt	Limit	After adjustment	New belt
Alternator and power steering oil pump belt	12 (0.47)	7 - 8 (0.28 - 0.31)	6 - 7 (0.24 - 0.28)	294 (30, 66)	730 - 818 (74.5 - 83.4, 164 - 184)	838 - 926 (85.5 - 94.5, 188 - 208)
A/C compressor belt	12 (0.47)	9 - 10 (0.35 - 0.39)	8 - 9 (0.31 - 0.35)	196 (20, 44)	348 - 436 (35.5 - 44.5, 78 - 98)	470 - 559 (47.9 - 57.0, 106 - 126)
Applied pushing force	98 N (10 kg, 22 lb)			—		

22140_FX35_G0014

Fig. 59 Belt deflection and tension table

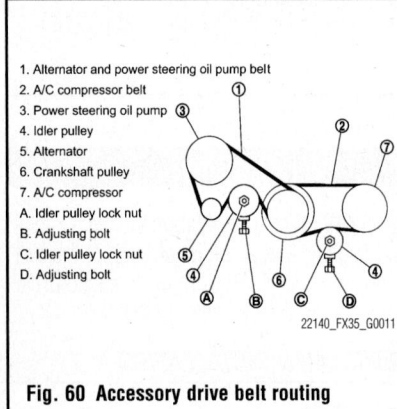

1. Alternator and power steering oil pump belt
2. A/C compressor belt
3. Power steering oil pump
4. Idler pulley
5. Alternator
6. Crankshaft pulley
7. A/C compressor
A. Idler pulley lock nut
B. Adjusting bolt
C. Idler pulley lock nut
D. Adjusting bolt

22140_FX35_G0011

Fig. 60 Accessory drive belt routing

a. Keep auto-tensioner pulley arm locked after drive belt is removed.
2. Remove auto-tensioner and idler pulley.

a. Keep auto-tensioner pulley arm locked to remove auto-tensioner.

To install:

✳✳ WARNING

If there is damage greater than peeled paint, replace drive belt auto-tensioner.

3. Install auto-tensioner and idler pulley.

a. Keep auto-tensioner pulley arm locked to install auto-tensioner
4. Install drive belt. Refer to Accessory Drive Belt System in the Engine Mechanical Section.

a. Keep auto-tensioner pulley arm locked after drive belt is install.

AIR CLEANER

REMOVAL & INSTALLATION

Air Cleaner And Air Duct Assembly
See Figure 62.

➥**Mass air flow sensor is removable under the car-mounted condition.**

1. Disconnect mass air flow sensor harness connector.
2. Disconnect PCV hose.
3. Remove air cleaner case with mass air flow sensor and air duct, disconnecting each joints.

a. Add marks if necessary for easier installation.
4. Remove mass air flow sensor from air cleaner case if necessary.

➥**Handle mass air flow sensor with the following cares.**

a. Never shock mass air flow sensor.
b. Never disassemble mass air flow sensor.
c. Never touch mass air flow sensor.

To install:
5. Install mass air flow sensor to air cleaner case if necessary.

➥**Handle mass air flow sensor with the following cares.**

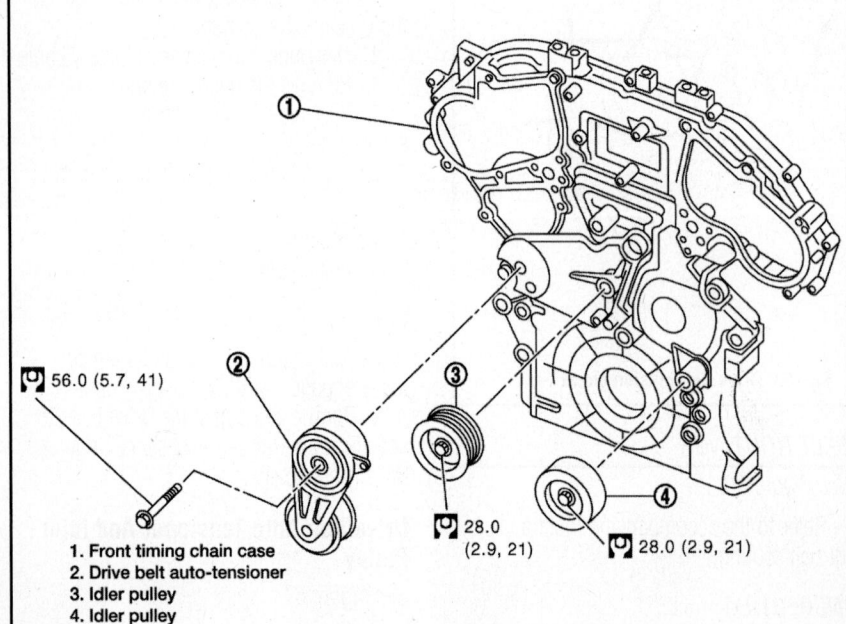

56.0 (5.7, 41)

28.0 (2.9, 21)

28.0 (2.9, 21)

1. Front timing chain case
2. Drive belt auto-tensioner
3. Idler pulley
4. Idler pulley

71075_EX35_G0366

Fig. 61 Exploded view of the drive belt auto tensioner and idler pulley components

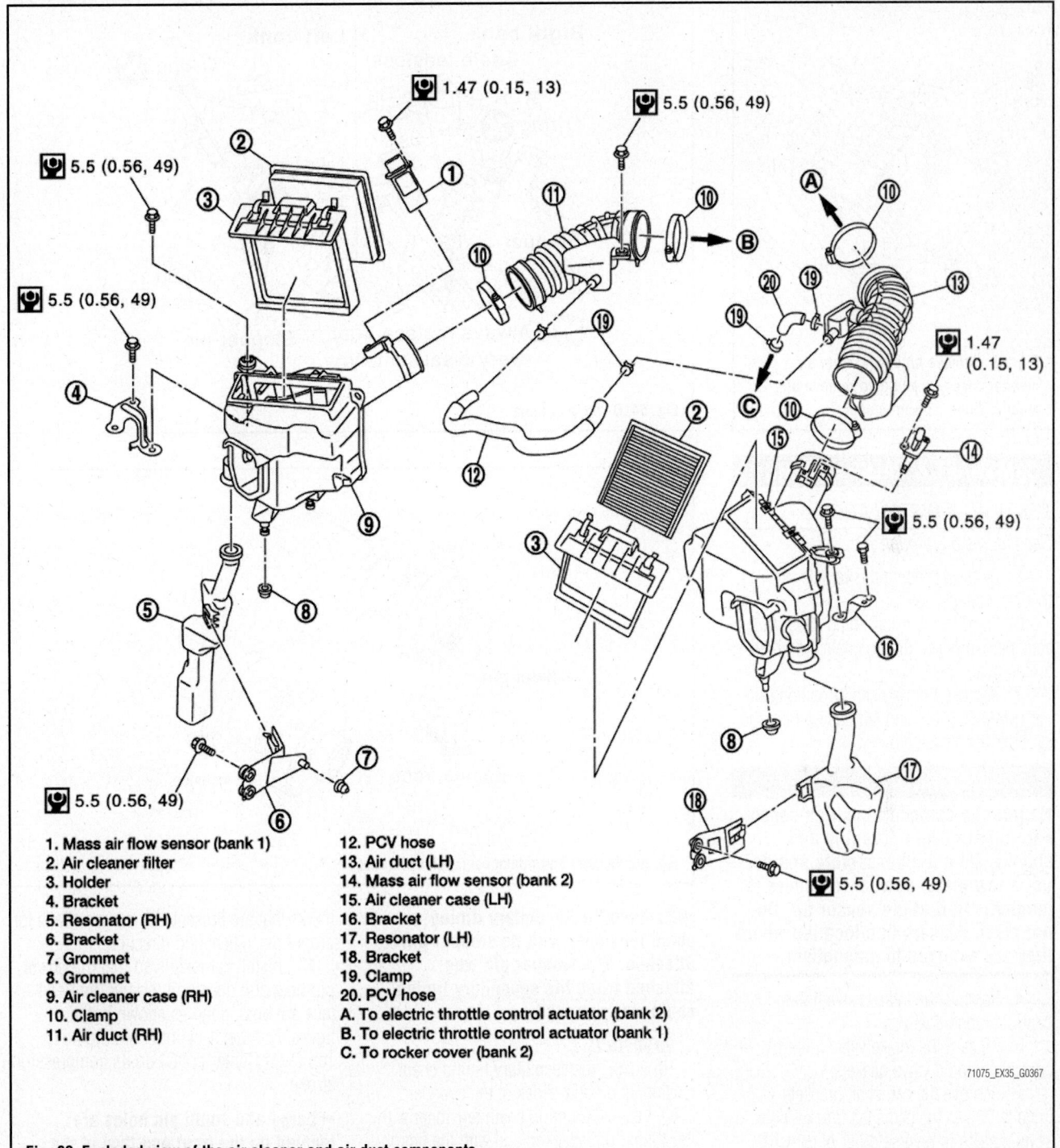

Fig. 62 Exploded view of the air cleaner and air duct components

1. Mass air flow sensor (bank 1)
2. Air cleaner filter
3. Holder
4. Bracket
5. Resonator (RH)
6. Bracket
7. Grommet
8. Grommet
9. Air cleaner case (RH)
10. Clamp
11. Air duct (RH)
12. PCV hose
13. Air duct (LH)
14. Mass air flow sensor (bank 2)
15. Air cleaner case (LH)
16. Bracket
17. Resonator (LH)
18. Bracket
19. Clamp
20. PCV hose
A. To electric throttle control actuator (bank 2)
B. To electric throttle control actuator (bank 1)
C. To rocker cover (bank 2)

71075_EX35_G0367

a. Never shock mass air flow sensor.

b. Never disassemble mass air flow sensor.

c. Never touch mass air flow sensor.

6. Install air cleaner case with mass air flow sensor and air duct, connecting each joints.

a. Add marks if necessary for easier installation.

7. Connect PCV hose.

8. Connect mass air flow sensor harness connector.

9. Inspect air duct and resonator assembly for crack or tear.

a. If anything found, replace air duct and resonator assembly.

Air Cleaner Filter

See Figure 63.

1. Unhook clips.
2. Remove holder from air cleaner case,

and then remove air cleaner filter from holder.

To install:

➡️**Install the air cleaner filter by aligning the seal with the notch of air cleaner case.**

3. Install air cleaner filter to holder, and then install holder to air cleaner case.

4. Hook clips.

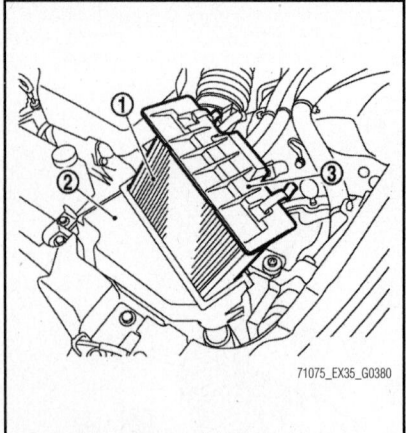

Fig. 63 Remove holder (3) from air cleaner case (2), and then remove air cleaner filter (1) from holder

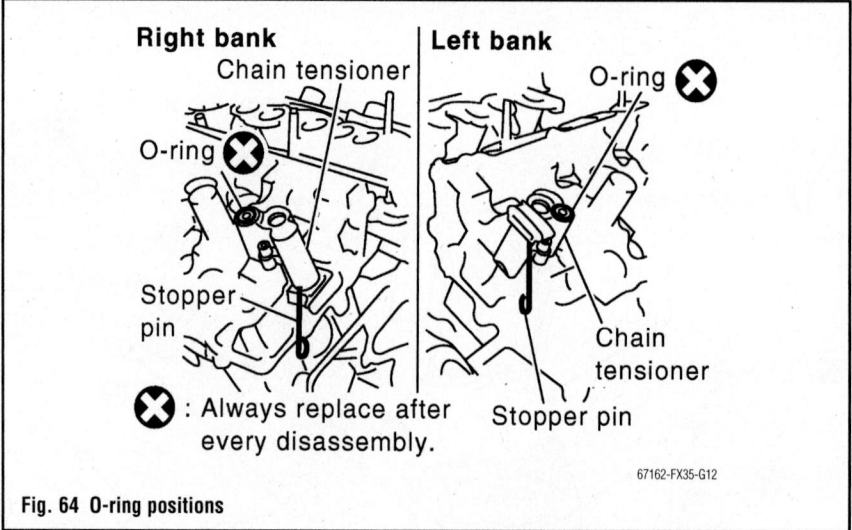

Fig. 64 O-ring positions

CAMSHAFT & VALVE LIFTERS

REMOVAL & INSTALLATION

See Figures 64 through 70.

1. Before servicing the vehicle, refer to the Precautions Section.

2. Remove the front timing chain case, camshaft sprocket, timing chain, and rear timing chain case.

3. Remove the Camshaft Position sensor (PHASE) (right and left banks) from the cylinder head back side.

✳✳ WARNING

Handle the camshaft position sensor carefully to avoid dropping and shocks. Do not disassemble and do not allow metal powder to adhere to magnetic part at the sensor tip. Do not place sensors in a location where they are exposed to magnetism.

4. Remove the intake valve timing control solenoid valves.

5. Discard the intake valve timing control solenoid valve gaskets.

6. Remove the camshaft brackets. Equally loosen the camshaft bracket bolts in several steps in reverse order of camshaft bearing cap mounting bolt tightening sequence shown in the figure.

➡**Mark the camshafts, camshaft brackets, and bolts so they are placed in the same position and direction at installation.**

7. Remove the camshaft(s).

8. Remove the valve lifters. Identify installation positions and store them without mixing them up.

9. Remove the secondary timing chain tensioner from the cylinder head.

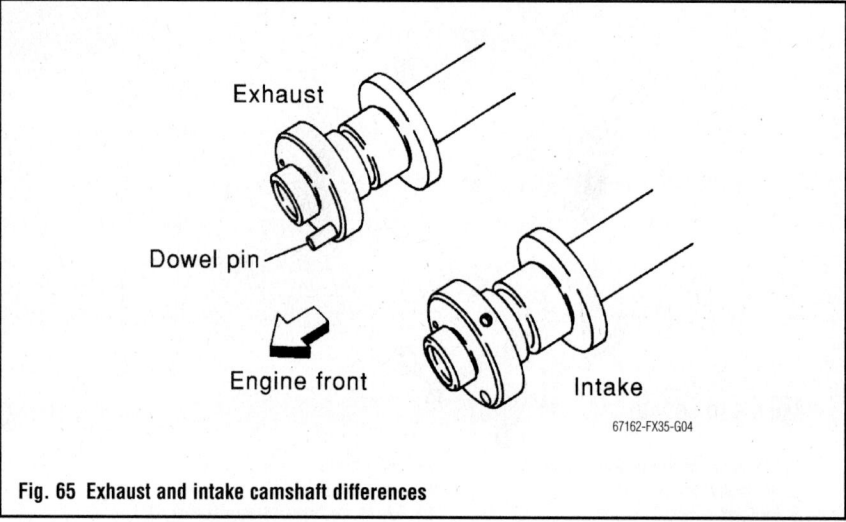

Fig. 65 Exhaust and intake camshaft differences

➡**Remove the secondary timing chain tensioner with its stopper pin attached. The stopper pin was attached when the secondary timing chain was removed.**

To install:

10. Install the secondary timing chain tensioners on both sides of the cylinder head. Install the timing chain tensioner with its stopper pin attached and with the timing chain tensioner sliding part facing downward on the right-side cylinder head while the sliding part is facing upward on the left-side cylinder head.

11. Install new O-rings as shown in the figure.

12. Install valve lifters in the original positions.

13. Install the camshafts. Install the camshaft with the dowel pin attached to its front end face on the exhaust side.

14. Follow the identification marks made during removal, or follow the identification

marks that are present on new camshafts for proper placement and direction.

15. Install camshafts so that the dowel pin hole and dowel pin on the front end face are positioned as shown in the figure. Number 1 cylinder should be in Top Dead Center (TDC) on its compression stroke.

➡**Large and small pin holes are located on the front end face of the intake camshaft at intervals of 180°. Face the small diameter side pin hole upward (in cylinder head upper face direction). Though the camshaft does not stop at the portion as shown in the figure, for the placement of the cam nose, it is generally accepted that the camshaft is placed in the same direction of the figure.**

16. Install the camshaft brackets. Remove foreign material completely from the camshaft bracket backside and from the cylinder head installation face. Install the

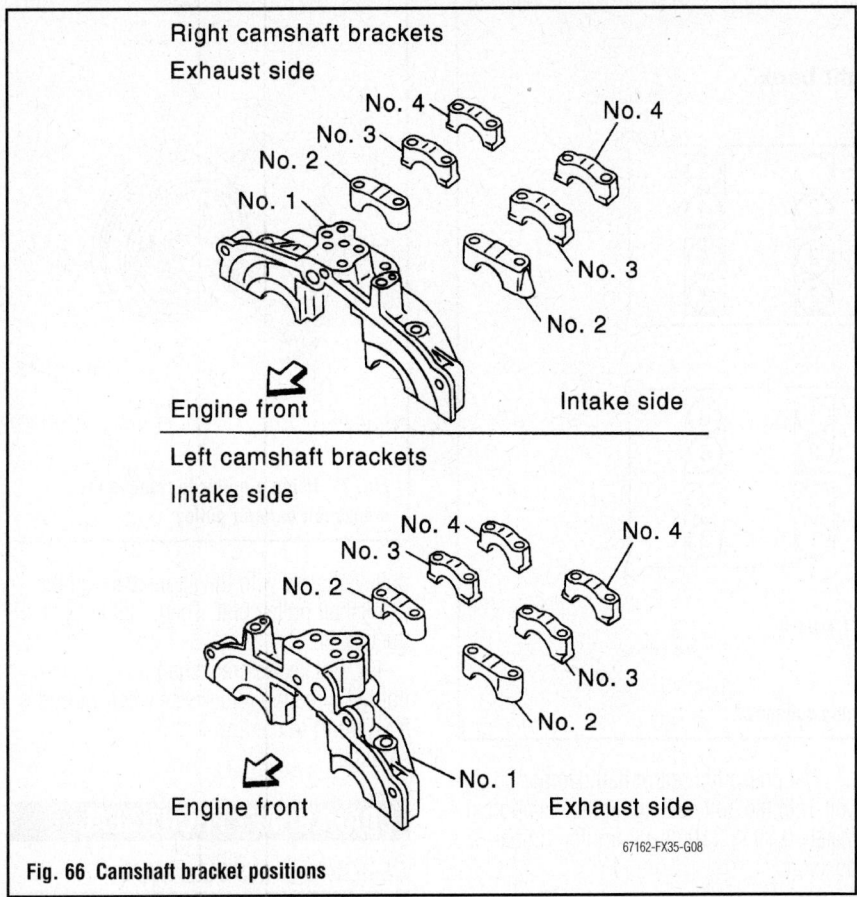

Right camshaft brackets
Exhaust side

Engine front

Intake side

Left camshaft brackets
Intake side

Engine front

Exhaust side

67162-FX35-G08

Fig. 66 Camshaft bracket positions

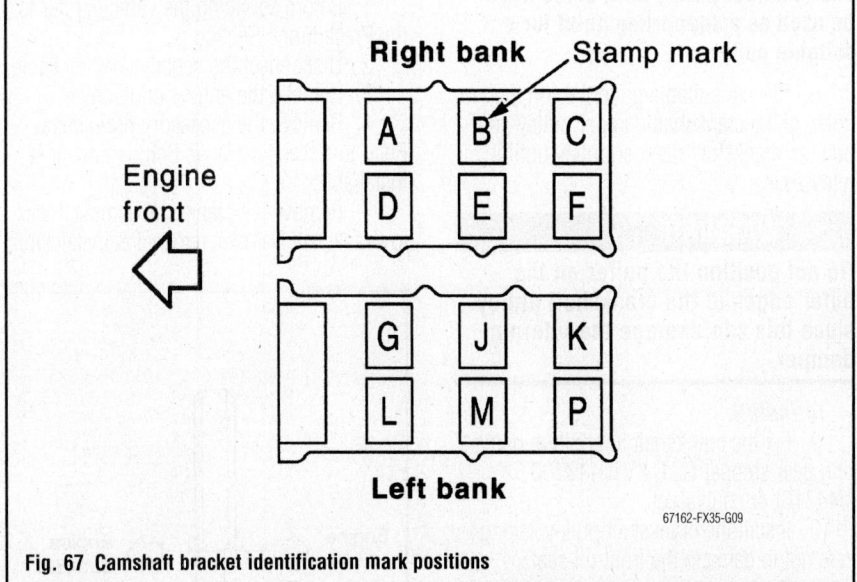

Right bank Stamp mark

Engine front

| A | B | C |
| D | E | F |

| G | J | K |
| L | M | P |

Left bank

67162-FX35-G09

Fig. 67 Camshaft bracket identification mark positions

camshaft bracket in the original position and direction as shown in figure.

17. Install camshaft brackets (numbers 2–4) aligning the stamp marks as shown in the figure.

➥**There are no identification marks indicating left and right for camshaft bracket (number 1).**

18. Apply liquid gasket to mating surface of camshaft bracket (number 1) as shown on both right and left banks. Use Genuine RTV Silicone Sealant or equivalent.

19. Tighten the camshaft bracket bolts in the following steps, in numerical order as shown.

a. Tighten numbers 7–10 to 12 inch lbs. (2 Nm).

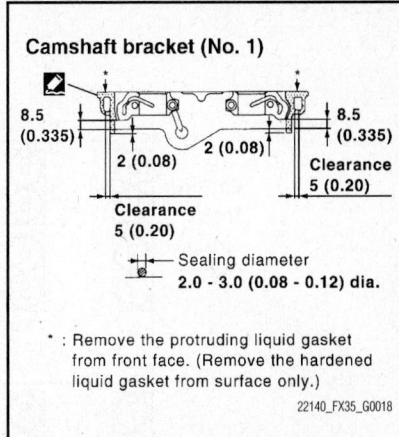

Camshaft bracket (No. 1)

8.5 (0.335)   8.5 (0.335)

2 (0.08)   2 (0.08)

Clearance 5 (0.20)

Clearance 5 (0.20)

Sealing diameter
2.0 - 3.0 (0.08 - 0.12) dia.

* : Remove the protruding liquid gasket from front face. (Remove the hardened liquid gasket from surface only.)

22140_FX35_G0018

Fig. 68 Apply liquid gasket to mating surface of camshaft bracket (number 1) as shown

b. Tighten numbers 1–6 to 12 inch lbs. (2 Nm).

c. Tighten numbers 1–10 to 48 inch lbs. (6 Nm).

d. Tighten numbers 1–10 in numerical order to 96 inch lbs. (10 Nm).

➥**After tightening the mounting bolts of camshaft brackets (number 1), be sure to wipe off excessive liquid gasket from the mating surface of the rocker cover and the mating surface of the rear timing chain case.**

20. Measure difference in levels between the front end faces of the number 1 camshaft bracket and the cylinder head. If the measurement is outside the specified range, re-install the camshaft and camshaft brackets.

➥**Camshaft bracket and cylinder head standard is: -0.0055–0.0055 in. (-0.14–0.14mm).**

a. Measure 2 positions (both intake and exhaust side) for a single bank.

b. If the measured value is out of the standard, re-install the camshaft bracket (number 1).

21. Inspect and adjust the valve clearance.

22. The remainder of installation is the reverse of removal.

CRANKSHAFT DAMPER PULLEY

REMOVAL & INSTALLATION

See Figure 71.

1. Before servicing the vehicle, refer to the Precautions Section.

2. Remove the front engine undercover.

3. Remove the accessory drive belt. Refer to Accessory Drive Belts, removal & installation.

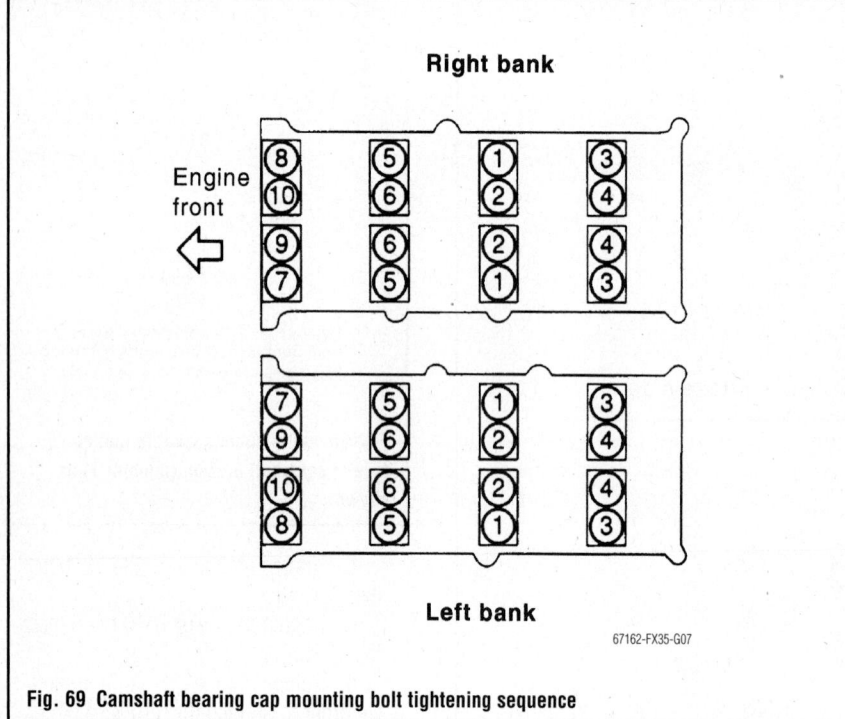

Right bank

Engine front

Left bank

67162-FX35-G07

Fig. 69 Camshaft bearing cap mounting bolt tightening sequence

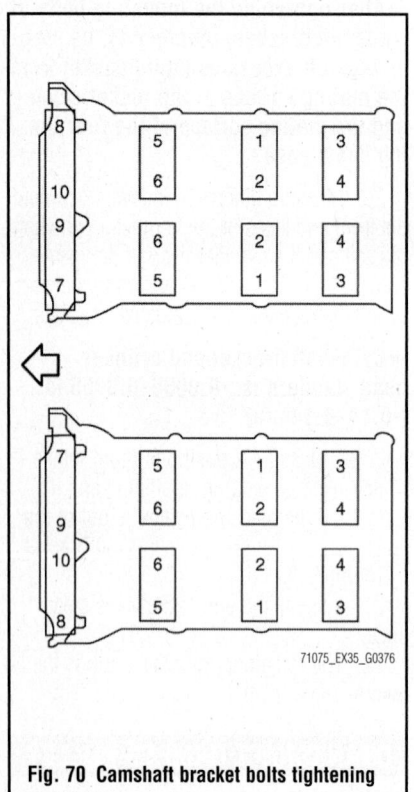

71075_EX35_G0376

Fig. 70 Camshaft bracket bolts tightening sequence

4. For 2WD models, remove the rear cover plate.

5. For AWD models, remove the starter motor.

6. Set the ring gear stopper SST: KV10117700 (J44716), or equivalent, to hold the crankshaft in position.

7. Loosen the crankshaft damper pulley bolt until the bolt seating surface is approximately 0.39 in. (10mm) from its original position.

➡ **Do not completely remove the crankshaft damper pulley bolt, since it will be used as a supporting point for a suitable puller.**

8. Place a suitable puller tab on the holes of the crankshaft damper pulley and pull the crankshaft damper pulley until it releases.

❋❋ WARNING

Do not position the puller on the outer edges of the crankshaft pulley since this can damage the internal damper.

To install:

9. Fix the crankshaft in position using ring gear stopper SST: KV10117700 (J44716) or equivalent.

10. Install the crankshaft pulley, taking care not to damage the front oil seal.

❋❋ WARNING

When press-fitting the crankshaft pulley with a plastic hammer, tap on its center portion, not on the circumference.

11. Tighten the crankshaft bolt to 33 ft. lbs. (44 Nm).

12. Put a paint mark on the crankshaft

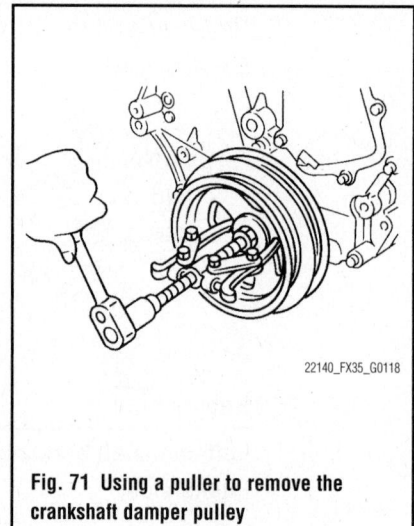

22140_FX35_G0118

Fig. 71 Using a puller to remove the crankshaft damper pulley

pulley aligned with the angle mark on the crankshaft pulley bolt. Then, further tighten the bolt 90°.

13. Rotate the crankshaft pulley in the normal direction (clockwise when viewed from front) to confirm that it turns smoothly.

CRANKSHAFT FRONT SEAL

REMOVAL & INSTALLATION

See Figure 72.

1. Before servicing the vehicle, refer to the Precautions Section.

2. Disconnect the negative battery cable.

3. Remove the engine undercover.

4. Remove the accessory drive belts. Refer to Accessory Drive Belts, removal & installation.

5. Remove the crankshaft damper. Refer to Crankshaft Damper, removal & installation.

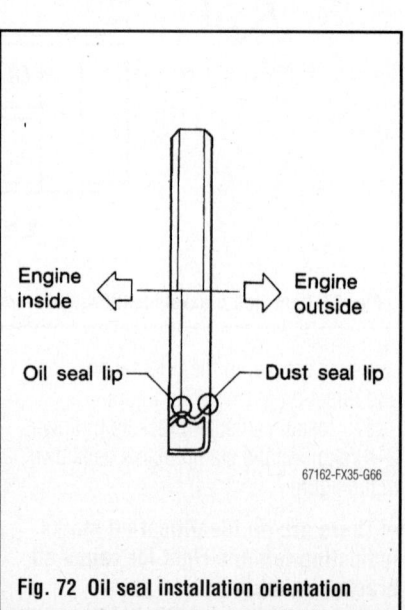

Engine inside

Engine outside

Oil seal lip

Dust seal lip

67162-FX35-G66

Fig. 72 Oil seal installation orientation

6. Remove the crankshaft front oil seal using a suitable tool.

✳✳ WARNING

Be careful not to damage the front timing chain case or the crankshaft.

To install:

7. Apply new engine oil to the oil seal lip and the dust seal lip of new crankshaft front oil seal.

8. Install the crankshaft front oil seal.

➡**The oil seal must be oriented as shown in the figure.**

9. Using a suitable drift, press-fit until the height of front oil seal is level with the mounting surface.

➡**A suitable drift should have an outer diameter of 2.36 in. (60mm), and an inner diameter of 1.97 in. (50mm). Make sure the garter spring is in position and the seal lips are not inverted.**

✳✳ WARNING

Be careful not to damage front timing chain case or the crankshaft. Press-fit straight and avoid causing burrs or tilting the oil seal during installation.

10. Installation continues in the reverse order of the removal procedure.

CYLINDER HEAD

REMOVAL & INSTALLATION

See Figures 73 through 76.

1. Before servicing the vehicle, refer to the Precautions Section.

2. Remove the camshafts. Refer to Camshaft and Valve Lifters, removal & installation.

3. Temporarily support the front suspension member to support the engine.

➡**Temporary support means that the engine is adequately stable although the weight supported by the hoist may be released. The front suspension member is removed and the cylinder head is hung by the hoist with an engine slinger installed.**

4. Release the hoist from hanging, then remove the engine slinger.

5. Remove the fuel tube and fuel injector assembly.

6. Remove the intake manifold.

7. Remove the exhaust manifold.

8. Remove the water inlet and thermostat housing.

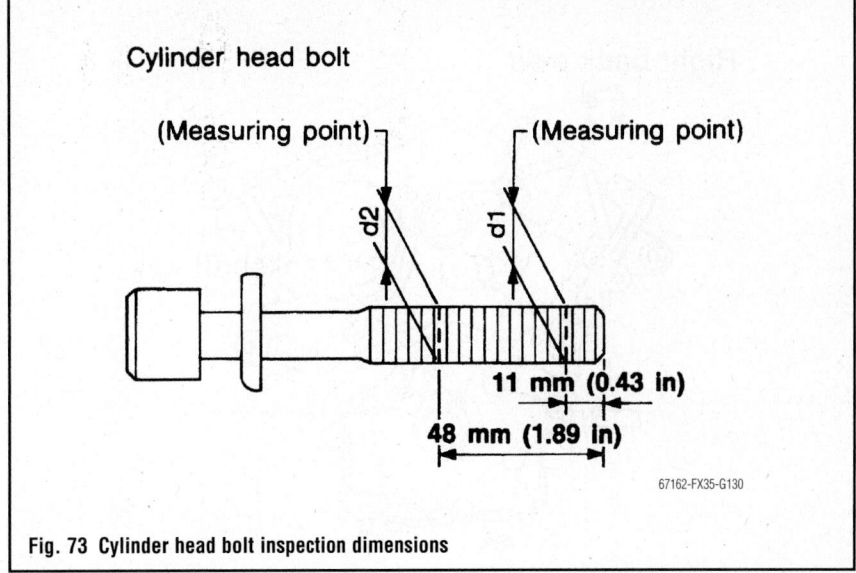

Fig. 73 Cylinder head bolt inspection dimensions

9. Remove the water outlet and water piping.

10. Loosen the cylinder head bolts in the reverse of the tightening order.

11. Remove the cylinder head.

12. Remove the cylinder head gaskets.

To install:

13. Inspect the cylinder head bolt diameters. The cylinder head bolts are tightened by the plastic zone tightening method. Whenever the size difference between d1 and d2 exceeds the limit, replace the bolt with a new one. The specification for d1 minus d2 for the cylinder head bolts is 0.0043 in. (0.11mm).

➡**If the reduction of the outer diameter appears in a position other than at d2, use it as the d2 point value.**

14. Check the cylinder head for distortion. Using a scraper, wipe off oil, scale, gasket, sealant, and carbon deposits from the surface of the cylinder head. At each of several locations on the bottom surface of the cylinder head, measure distortion in 6 directions. If cylinder head distortion exceeds the recommended limit of 0.004 in. (0.1mm), replace the cylinder head.

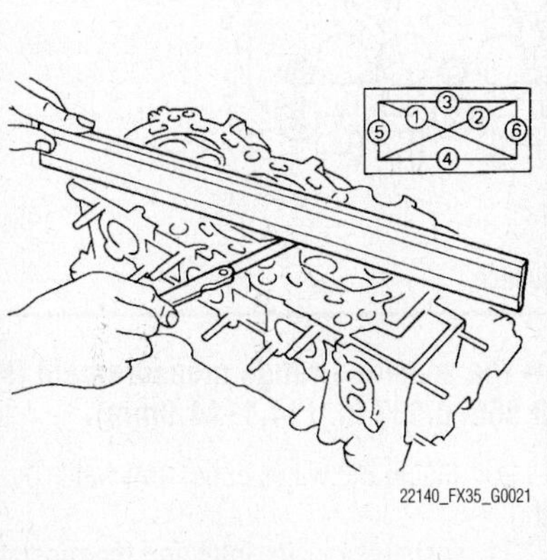

✳✳ WARNING

Do not allow gasket fragments to enter engine oil or engine coolant passages.

15. Install a new cylinder head gasket.

16. Turn the crankshaft until number 1 piston is set at Top Dead Center (TDC) on the compression stroke. The crankshaft key should line up with the right bank cylinder center line.

17. Install the cylinder head.

18. Install and tighten the cylinder head bolts in the proper order:

  a. Apply new engine oil to the threads and seat surfaces of the cylinder head bolts.

  b. Tighten all bolts to 72 ft. lbs. (98 Nm).

  c. Completely loosen all bolts in the reverse order of the tightening sequence.

  d. Retighten all bolts to 29 ft. lbs. (39 Nm).

  e. Turn all bolts 90° clockwise (angle tightening).

  f. Turn all bolts 90° clockwise again (angle tightening).

✳✳ WARNING

Check and confirm the tightening angle by using an angle wrench, or equivalent, and a cylinder head bolt wrench (commercial service tool). Avoid tightening the bolts with a visual inspection only.

Fig. 74 Checking cylinder head distortion

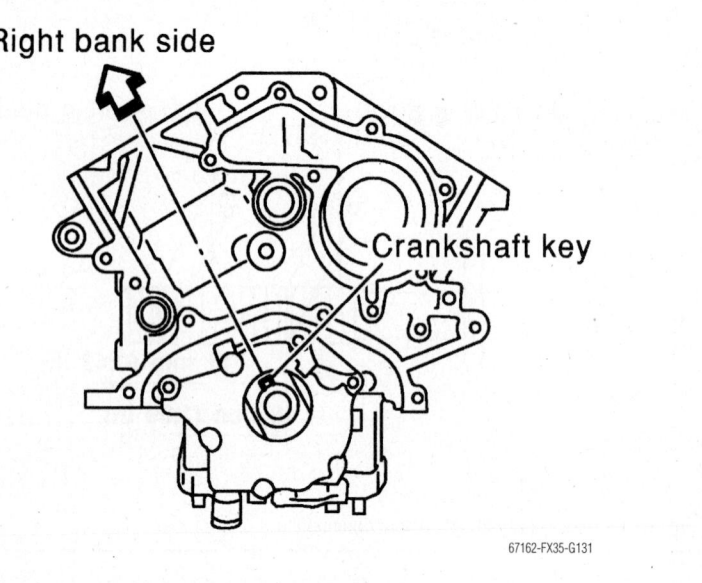

Fig. 75 Crankshaft positioning for cylinder head installation

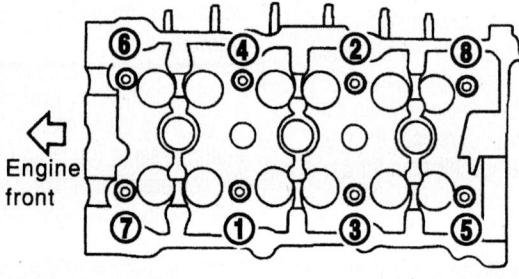

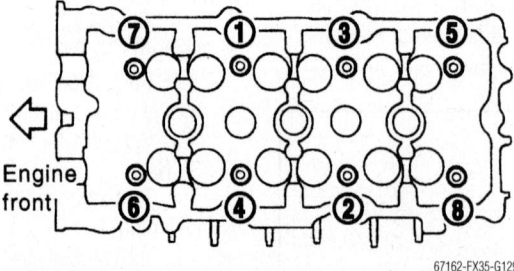

Fig. 76 Cylinder head mounting bolt tightening sequence

19. After installing the cylinder head, measure the distance between the front end faces of the cylinder block and the cylinder head (left and right banks). If the measurement is outside the specified range, re-install the cylinder head.

➡**The specified range measurement is 0.555–0.587 in. (14.1–14.9mm).**

20. Install the water outlet and water piping.

21. Install the water inlet and thermostat housing.

22. Install the exhaust manifold.

23. Install the intake manifold.

24. Install the fuel tube and fuel injector assembly.

25. Install the camshafts. Refer to Camshaft and Valve Lifters, removal & installation.

ENGINE COVER

REMOVAL & INSTALLATION

See Figure 77.

1. Loosen mounting bolts and nuts in the reverse order of the tightening sequence, and then remove engine cover.

✳✳ WARNING

Never damage or scratch engine cover when installing or removing.

To install:

2. Install engine cover, and then tighten mounting bolts and nuts in numerical order as shown the accompanying illustration.

✳✳ WARNING

Never damage or scratch engine cover when installing or removing.

ENGINE OIL & FILTER

OIL LEVEL CHECK

See Figure 78.

➡**Before starting engine, put vehicle horizontally and check the engine oil level. If engine is already started, stop it and allow 10 minutes before checking.**

1. Pull out oil level gauge and wipe it clean.

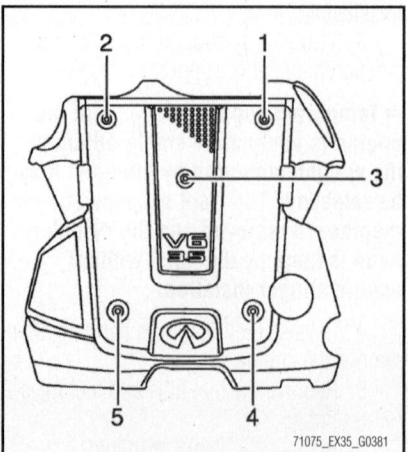

Fig. 77 Loosen mounting bolts and nuts in the reverse order of the tightening sequence, and then remove engine cover

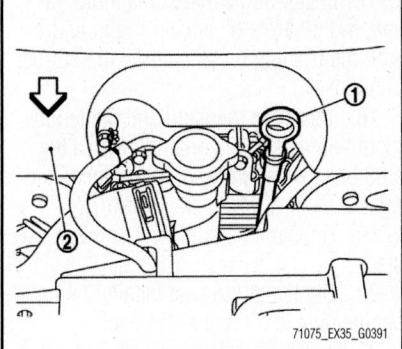

Fig. 78 When checking the engine oil level, insert oil level gauge (1) with its tip aligned with oil level gauge guide

2. Insert oil level gauge and check the engine oil level is within the range.
3. If it is out of range, adjust it.

➡When checking the engine oil level, insert oil level gauge with its tip aligned with oil level gauge guide.

OIL & FILTER CHANGE

Draining

See Figures 79 and 80.

➡Oil filter is provided with relief valve. Use genuine NISSAN oil filter or equivalent.

✳✳ CAUTION

Be careful not to get burn yourself, as engine oil may be hot.

✳✳ CAUTION

Prolonged and repeated contact with used engine oil may cause skin cancer. Try to avoid direct skin contact

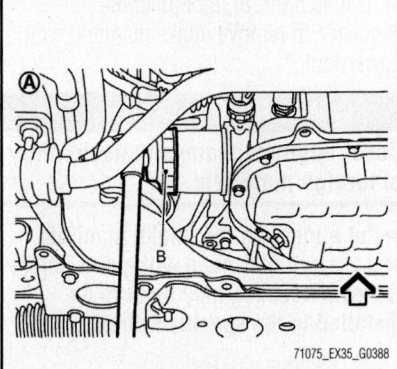

Fig. 79 Using oil filter wrench [SST: KV10115801 (J-38956)] (B), remove oil filter–2WD (A), Arrow=Engine front

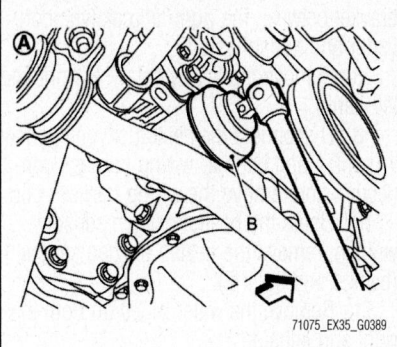

Fig. 80 Using oil filter wrench [SST: KV10115801 (J-38956)] (B), remove oil filter—AWD (A), Arrow=Engine front

with used engine oil. If skin contact is made, wash thoroughly with soap or hand cleaner as soon as possible.

➡When removing, prepare a shop cloth to absorb any engine oil leakage or spillage.

✳✳ WARNING

Never allow engine oil to adhere to drive belt.

✳✳ WARNING

Completely wipe off any engine oil that adheres to engine and vehicle.

1. Warm up the engine, and check for engine oil leakage from engine components.
2. Stop the engine and wait for 10 minutes.
3. Loosen oil filler cap.
4. Remove undercover with power tool.
5. Remove drain plug and then drain engine oil.
6. Using oil filter wrench [SST: KV10115801 (J-38956)], remove oil filter.

Refilling

See Figure 81.

1. Install drain plug with new washer.

✳✳ CAUTION

Be sure to clean drain plug and install with new washer.

2. Remove foreign materials adhering to oil filter installation surface.
3. Apply engine oil to the oil seal contact surface of new oil filter.
4. Screw oil filter manually until it touches the installation surface, then tighten it by 2/3 turn, or tighten to 13 ft. lbs. (17.7 Nm).

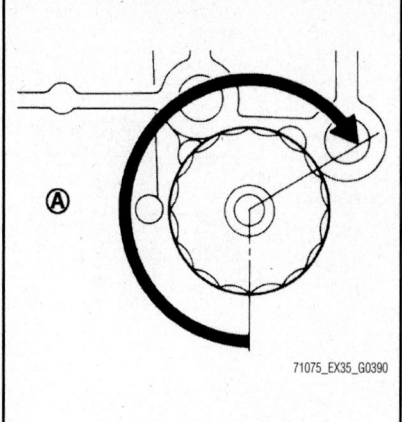

Fig. 81 Screw oil filter manually until it touches the installation surface, then tighten it by 2/3 turn (A), or tighten to 156 inch lbs. (17.7 Nm)

5. Refill with new engine oil.

➡When filling engine oil, never pull out oil level gauge.

➡The refill capacity depends on the engine oil temperature and drain time. Use these specifications for reference only.

➡Always use oil level gauge to determine the proper amount of engine oil in engine.

6. Warm up the engine and check area around drain plug and oil filter for engine oil leakage.
7. Stop the engine and wait for 10 minutes.
8. Check the engine oil level.

EXHAUST MANIFOLD

REMOVAL & INSTALLATION

See Figures 82 and 83.

1. Before servicing the vehicle, refer to the Precautions Section.

✳✳ CAUTION

Perform the work when the exhaust and cooling system have completely cooled down.

2. Remove the engine cover.
3. Remove the air cleaner case and air duct.
4. Remove the front and rear engine undercover and front cross bar.
5. Disconnect the heated oxygen sensor 2 wiring harness connectors (bank 1 and bank 2).
6. Using a heated oxygen sensor

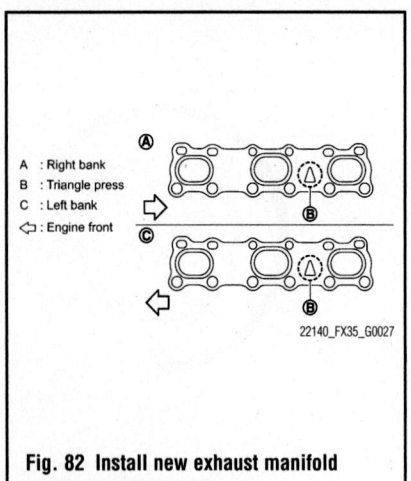

A : Right bank
B : Triangle press
C : Left bank
⇦ : Engine front

22140_FX35_G0027

Fig. 82 Install new exhaust manifold gaskets in the direction shown

wrench, remove the heated oxygen sensors (bank 1 and bank 2).

☀ WARNING

Be careful not to damage heated oxygen sensor. Discard any heated oxygen sensor which has been dropped from a height of more than 20 in. (51cm) onto a hard surface such as a concrete floor; replace with a new sensor.

7. Remove the exhaust mounting bracket between the right/left catalytic converter and transmission.

8. Remove the 3-way catalyst (right and left bank).

9. Disconnect the heated oxygen sensor 1 (bank 1 and bank 2) wiring harness connectors and remove the wiring harness clip.

10. Using the heated oxygen sensor wrench, remove the heated oxygen sensor 1 (bank 1 and bank 2).

11. Remove the water pipes on both the right and left side.

12. Remove the exhaust manifold cover (right and left bank).

13. Loosen the mounting nuts in the reverse order of the tightening sequence shown in the illustration.

14. Remove the exhaust manifold.

15. Remove the exhaust manifold gaskets.

➡**Cover all engine openings to avoid entry of foreign materials.**

To install:

16. Check the surface distortion of the exhaust manifold mating surface with a straightedge and feeler gauge. If it exceeds the limit, replace the exhaust manifold. Limit of surface distortion: 0.012 in. (0.3mm).

17. Install new exhaust manifold gaskets in the direction shown in the figure with the triangle press mark in the correct position.

18. Install the manifold and tighten the mounting nuts in the order shown. If the stud bolts were removed, install them and tighten them to 11 ft. lbs. (15 Nm).

19. Tighten nuts number 1 and 2 in two steps.

20. Tighten all exhaust manifold nuts-to-engine nuts to 22 ft. lbs. (31 Nm).

21. Install the exhaust manifold cover (right and left bank).

22. Install the water pipes on both the right and left side.

23. Install the heated oxygen sensor 1 (bank 1 and bank 2).

24. Install the 3-way catalyst (right and left bank).

25. Install the exhaust mounting bracket between the right/left catalytic converter and transmission.

26. Install the heated oxygen sensors 2 (bank 1 and bank 2).

27. Reconnect the heated oxygen sensor wiring harness connectors.

28. Install the front and rear engine undercover and front cross bar.

29. Install the air cleaner case and air duct.

30. Install the engine cover.

INTAKE MANIFOLD

REMOVAL & INSTALLATION
See Figures 84 through 86.

1. Release fuel pressure.

2. Remove intake manifold collector.

3. Remove fuel tube and fuel injector assembly.

4. Remove harness bracket.

5. Loosen mounting bolts and nuts in reverse order of the tightening sequence to remove intake manifold with power tool.

☀ WARNING

Cover engine openings to avoid entry of foreign materials.

➡**Put a mark on the intake manifold and the cylinder head with paint before removal because they need to be installed in the specified direction.**

6. Remove gaskets.

a. Check the surface distortion of the intake manifold mating surface with a straightedge and a feeler gauge.

b. If it exceeds the limit, replace intake manifold.

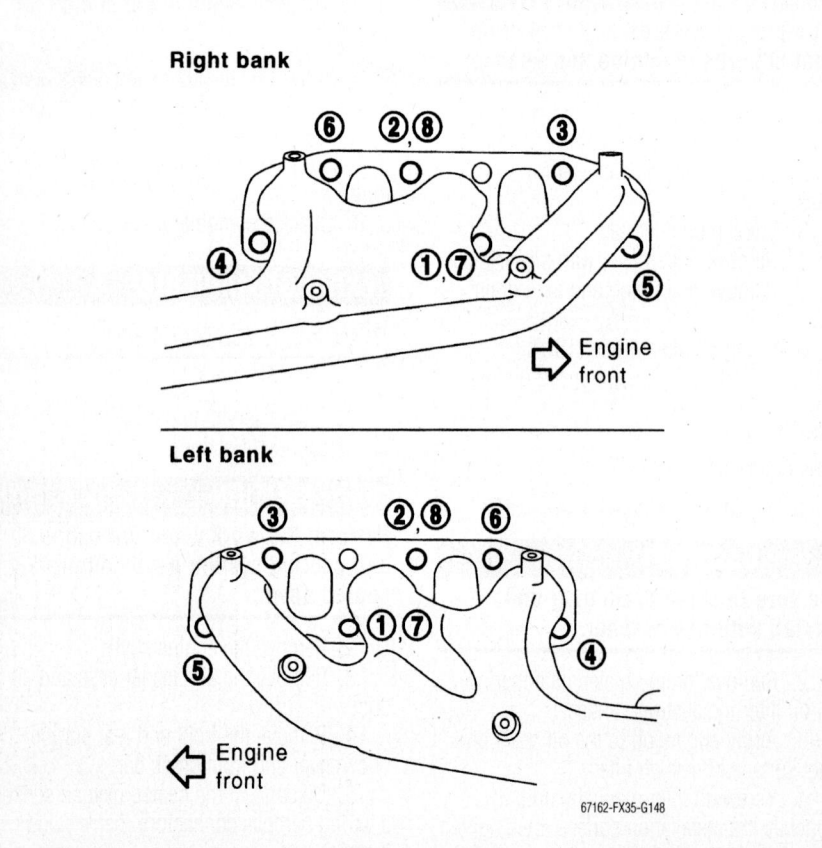

Right bank

Left bank

67162-FX35-G148

Fig. 83 Exhaust manifold mounting bolt tightening sequence

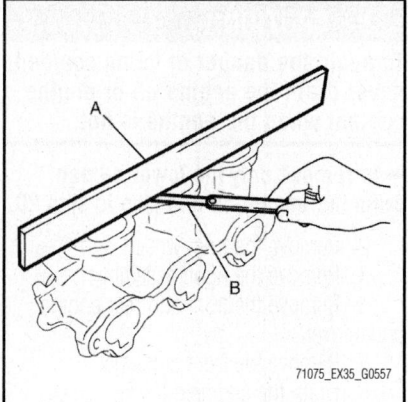

Fig. 84 Check the surface distortion of the intake manifold mating surface with a straightedge (A) and a feeler gauge (B)

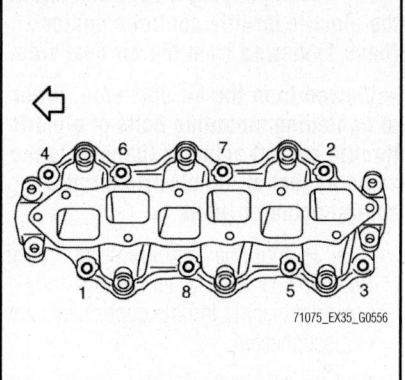

Fig. 86 Tighten mounting bolts and nuts in order as shown to install intake manifold with power tool (Arrow=Engine front)

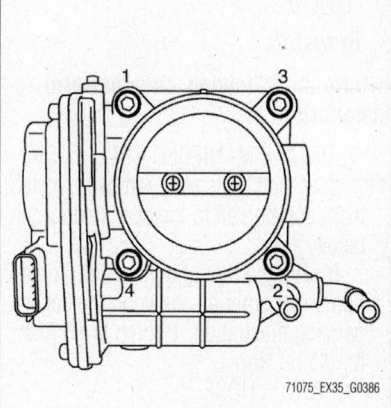

Fig. 87 Tighten mounting bolts and nuts in numerical order as shown

To install:

➡Note the following, and install in the reverse order of removal.

7. Intake Manifold

   a. If stud bolts were removed, install them and tighten bolts to 96 inch lbs. (10.8 Nm)

   b. Tighten all mounting bolts and nuts to the specified torque in two or more steps in numerical order as shown in the accompanying illustration.

   c. Tighten bolts to:
   • 1st step: 60 inch lbs. (7.4 Nm)
   • 2nd step and after: 19 ft. lbs. (25.5 Nm)

➡Install intake manifold with the marks (put on the intake manifold and the cylinder head before removal) aligned.

   8. Install gaskets.

   9. Tighten mounting bolts and nuts in order as shown in the accompanying illustration to install intake manifold with power tool.

   10. Install harness bracket.

   11. Install fuel tube and fuel injector assembly.

   12. Install intake manifold collector. Refer to Intake

   13. Release fuel pressure.

INTAKE MANIFOLD COLLECTOR

REMOVAL & INSTALLATION

See Figures 87 and 88.

Never drain engine coolant when the engine is hot to avoid the danger of being scalded.

   1. Remove engine cover with power tool.
   2. Remove air cleaner case and air duct (RH and LH).
   3. Remove electric throttle control actuator as follows:

   a. Drain engine coolant, or when water hoses are disconnected, attach plug to prevent engine coolant leakage.

Perform this step when engine is cold.

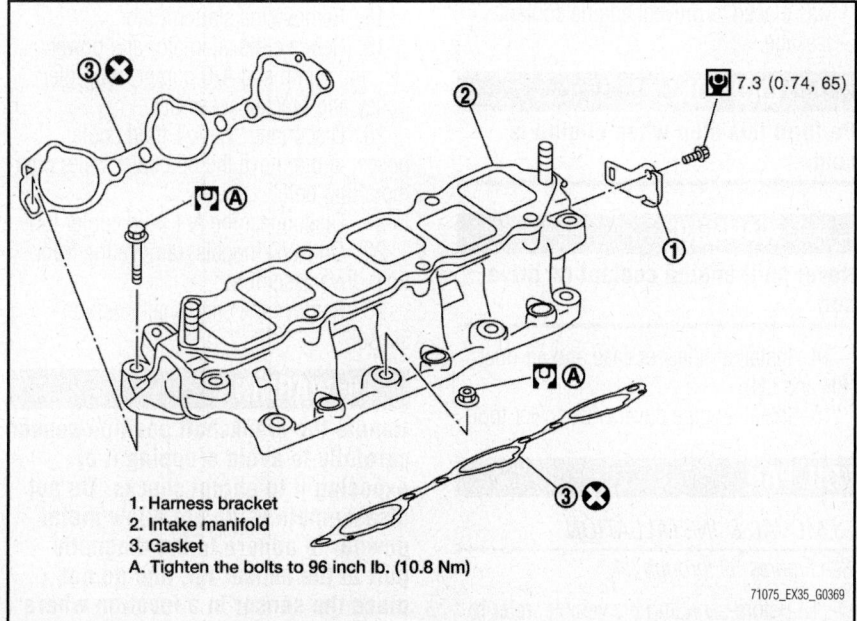

1. Harness bracket
2. Intake manifold
3. Gasket
A. Tighten the bolts to 96 inch lb. (10.8 Nm)

Fig. 85 Exploded view of the intake manifold components

Fig. 88 The illustration shows the electric throttle control actuator (bank 1) viewed from the air duct side. Viewed from the air duct side, order of tightening mounting bolts of electric throttle control actuator (bank 2) is the same as that of the electric throttle control actuator (bank 1).

✳✳ WARNING

Never spill engine coolant on drive belt.

b. Disconnect water hoses from electric throttle control actuator. When engine coolant is not drained from radiator, attach plug to water hoses to prevent engine coolant leakage.

c. Disconnect harness connector.

d. Loosen mounting bolts in reverse order of the tightening sequence.

➡ **When removing only intake manifold collector, move electric throttle control actuator without disconnecting the water hose.**

➡ **The accompanying illustration shows the electric throttle control actuator (bank 1) viewed from the air duct side.**

➡ **Viewed from the air duct side, order of loosening mounting bolts of electric throttle control actuator (bank 2) is the same as that of the electric throttle control actuator (bank 1).**

✳✳ WARNING

Handle carefully to avoid any shock to electric throttle control actuator.

4. Disconnect vacuum hose, PCV hose and EVAP hose from intake manifold collector.

5. Remove EVAP canister purge volume control solenoid valve and EVAP tube assembly from intake manifold collector.

6. Loosen mounting bolts and nuts with power tool in the reverse order of the tightening sequence to remove intake manifold collector.

To install:

➡ **Note the following, then perform installation.**

7. INTAKE MANIFOLD COLLECTOR

a. If stud bolts were removed, install them and tighten to the specified torque below.

b. Tighten mounting bolts and nuts in numerical order as shown in the accompanying illustration. Tighten to 96 inch lbs. (10.8 Nm).

8. WATER HOSE

a. Insert hose by 1.06 to 1.26 in. (27 to 32 mm) from connector end.

b. Clamp hose at location of 0.12 to 0.28 in. (3 to 7 mm) from hose end.

9. ELECTRIC THROTTLE CONTROL ACTUATOR (BANK 1 AND BANK 2)

a. Tighten in numerical order as shown in the accompanying illustration.

➡ **The accompanying illustration shows the electric throttle control actuator (bank 1) viewed from the air duct side.**

➡ **Viewed from the air duct side, order of tightening mounting bolts of electric throttle control actuator (bank 2) is the same as that of the electric throttle control actuator (bank 1).**

b. Perform the "Throttle Valve Closed Position Learning" when harness connector of electric throttle control actuator is disconnected.

➡ **Perform the "Idle Air Volume Learning" and "Throttle Valve Closed Position Learning" when electric throttle control actuator is replaced.**

10. Tighten mounting bolts and nuts with power tool in the order as shown in the accompanying illustration to install intake manifold collector.

11. Install EVAP canister purge volume control solenoid valve and EVAP tube assembly to intake manifold collector.

12. Connect vacuum hose, PCV hose and EVAP hose to intake manifold collector.

13. Install electric throttle control actuator as follows:

a. Tighten mounting bolts in order as shown in the accompanying illustration.

b. Connect harness connector.

c. Connect water hoses to electric throttle control actuator. When engine coolant is not drained from radiator, attach plug to water hoses to prevent engine coolant leakage.

d. Fill engine coolant, or when water hoses are connected, remove plug that was placed to prevent engine coolant leakage.

✳✳ CAUTION

Perform this step when engine is cold.

✳✳ WARNING

Never spill engine coolant on drive belt.

14. Install air cleaner case and air duct (RH and LH).

15. Install engine cover with power tool.

OIL PAN

REMOVAL & INSTALLATION

See Figures 89 through 94.

1. Before servicing the vehicle, refer to the Precautions Section.

✳✳ CAUTION

To avoid the danger of being scalded, never drain the engine oil or engine coolant when the engine is hot.

➡ **To remove only the lower oil pan, drain the engine oil and skip to step 26.**

2. Remove the front wheels and tires.

3. Remove the hood assembly.

4. Remove the front and rear engine undercover.

5. Remove the front cross bar.

6. Drain the engine oil.

7. Drain the engine coolant.

8. Remove the engine cover.

9. Remove the air hose from the air duct to the mass air flow and the electric throttle control actuator side.

10. Remove the alternator, power steering pump, and A/C compressor belt.

11. On AWD models, remove the front left and right halfshafts, and side shaft.

12. Remove the engine rear lower slinger, and install the engine rear slinger tool SST: 10006 31U00 (or equivalent) to hold the engine assembly in position. Tighten the engine rear slinger tool mounting bolts to 21 ft. lbs. (28 Nm).

13. Remove the front suspension member.

14. On AWD models, remove the engine mounting bracket, lower engine mounting bracket and insulator.

15. On AWD models, remove the front driveshaft.

16. On AWD models, remove the oil filter and oil filter bracket.

17. Remove the alternator stay.

18. Remove the starter motor.

19. Remove the alternator and power steering pump and A/C compressor idler pulley and bracket assembly.

20. Disconnect the A/T fluid cooler hoses, and remove the oil cooler water pipe mounting bolt.

21. Disconnect the A/T fluid cooler tube.

22. On AWD models, remove the front final drive assembly. .

23. Remove the crankshaft position sensor.

✳✳ WARNING

Handle the crankshaft position sensor carefully to avoid dropping it or exposing it to abrupt shocks. Do not disassemble it, do not allow metal powder to adhere to the magnetic part at the sensor tip, and do not place the sensor in a location where it may be exposed to magnetism.

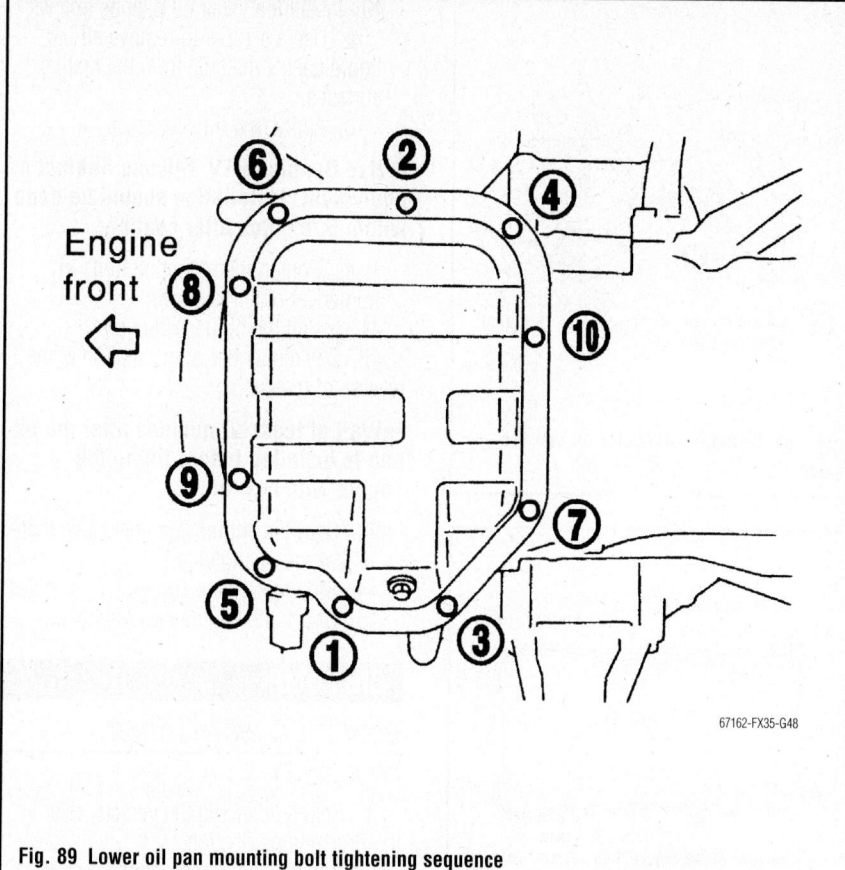

Fig. 89 Lower oil pan mounting bolt tightening sequence

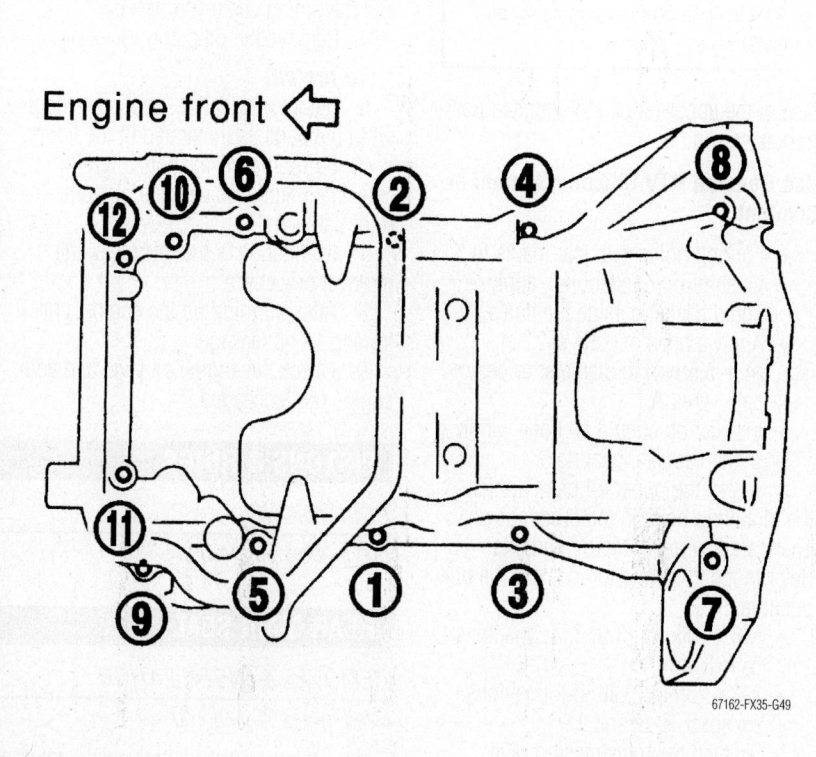

Fig. 90 Upper oil pan mounting bolt tightening sequence

24. Remove the oil filter, as necessary.
25. Remove the oil cooler, as necessary.
26. Remove the oil pan (lower), as follows:

    a. Loosen the mounting bolts in the reverse order of the tightening sequence.

    b. Insert a seal cutter SST: KV10111100 (J37228) or equivalent, between the upper oil pan and lower oil pan.

    c. Slide the seal cutter by tapping on the side of the tool with a hammer.

    d. Remove the lower oil pan.

✴✴ WARNING

Be careful not to damage the mating surface. Do not use a flat-bladed screwdriver as this could damage the mating surfaces.

27. Remove the oil strainer.
28. Remove the transmission joint bolts which pass through the upper oil pan.
29. On 2WD models, remove the rear cover plate.
30. Loosen the upper oil pan bolts in the reverse order of the tightening sequence.
31. Insert a seal cutter SST: KV10111100 (J37228) between the upper oil pan and cylinder block. Slide the seal cutter by tapping on the side of the tool with a hammer.
32. Remove the upper oil pan.

✴✴ WARNING

Be careful not to damage the mating surface. Do not use a flat-bladed screwdriver as this could damage the mating surfaces.

33. Remove the O-rings from the bottom of the cylinder block and oil pump.
34. Remove the oil pan gaskets.
35. For AWD models, remove the axle pipe from the upper oil pan using a suitable drift, if necessary.
36. Clean the oil strainer, if necessary.

To install:

37. On AWD models, install the axle pipe to the oil pan, if removed:

    a. Lubricate the O-ring groove of the axle pipe, O-ring, and O-ring joint of the oil pan with new engine oil.

    b. Install the axle pipe to the oil pan (upper) from the axle pipe flange side (left side) using a suitable drift with an outer diameter of 1.7 to 2.2 in. (43 to 57mm).

➡**Insert the axle pipe with care to prevent the O-ring from sliding.**

38. Install the upper oil pan, as follows:
    a. Use a scraper to remove the old liquid gasket from all mating surfaces.

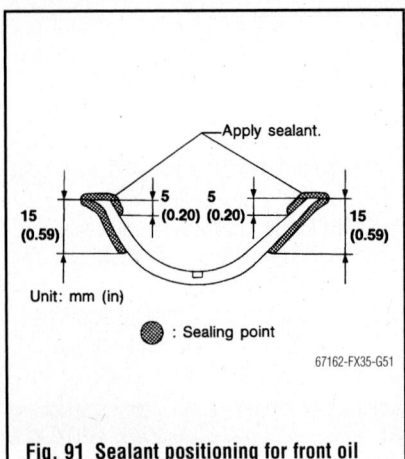

Fig. 91 Sealant positioning for front oil pan seal installation

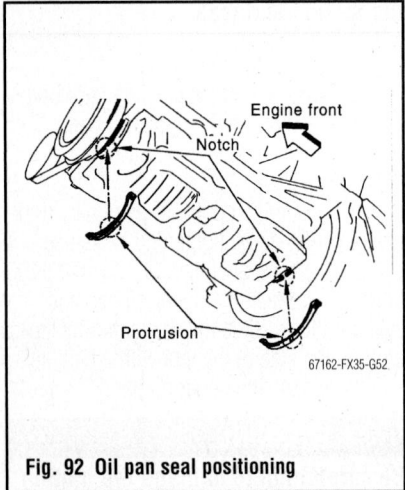

Fig. 92 Oil pan seal positioning

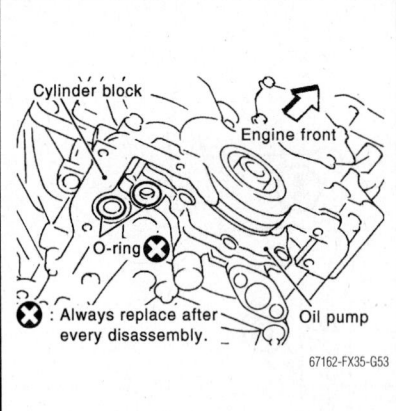

Fig. 93 O-ring locations for oil pan service

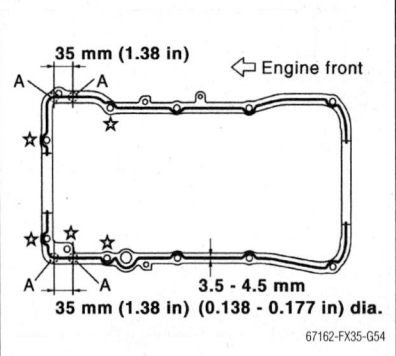

Fig. 94 Liquid gasket positioning for oil pan installation

Remove old liquid gasket from the mating surface of the cylinder block, and the bolt holes and threads.

✳✳ WARNING

Do not scratch or damage the mating surfaces when cleaning off the old liquid gasket material.

b. Apply liquid gasket to the oil pan gaskets as shown.

➡**Use Genuine RTV Silicone Sealant or equivalent.**

c. Install the new gasket. Align the protrusion of the oil pan gasket with the notches of the front timing chain case and rear oil seal retainer.

d. Install the oil pan gasket with the smaller arc to the front timing chain case side.

e. Install new O-rings on the cylinder block and oil pump.

f. Apply a continuous bead of liquid gasket to the cylinder block mating sur-

face of the upper oil pan to a limited portion as shown.

➡**Use Genuine RTV Silicone Sealant or equivalent.**

- For bolt holes with star marks in illustration (5 locations), apply liquid gasket outside the holes.
- Apply a bead of 0.18–0.22 in. (4.5–5.5mm) in diameter to designated area **A**.
- Installation should be done within 5 minutes after coating.

g. Install the upper oil pan. Tighten the mounting bolts in the order shown. There are 2 types of mounting bolts. Refer to the following for locating the bolt positions:

- M8 x 100mm (3.97 in.): positions 5, 7, 8, and 11
- M8 x 25mm (0.98 in.): positions except 5, 7, 8, and 11

h. Tighten the transmission joint bolts.

39. Install the oil strainer onto the oil pump.

40. Install the lower oil pan, as follows:

a. Use a scraper to remove all old liquid gasket material from the mating surfaces.

b. Apply new liquid gasket.

➡**Use Genuine RTV Silicone Sealant or equivalent. Installation should be done within 5 minutes after coating.**

c. Tighten the mounting bolts in numerical order as shown.

41. Install the oil pan drain plug.

42. The remainder of installation is the reverse of removal.

➡**Wait at least 30 minutes after the oil pan is installed before filling the engine with new oil.**

43. Start the engine and check that there is no leakage of engine oil.

44. Stop the engine and wait 10 minutes.

45. Check the engine oil level again.

OIL PUMP

REMOVAL & INSTALLATION

See Figure 95.

1. Before servicing the vehicle, refer to the Precautions Section.

2. Remove the oil pan (lower and upper) and the oil strainer.

3. Remove the front timing chain case and the timing chain (primary).

4. Remove the oil pump assembly.

To install:

5. Before installation, apply new engine oil to the parts as illustrated in the figure.

6. For pump installation, align the crankshaft flat faces with the oil pump inner rotor flat faces.

7. Installation is the reverse of the removal procedure.

8. After warming up the engine, check for engine oil leakage.

9. Check the engine oil level and add engine oil, as needed.

PISTONS & RINGS

POSITIONING

See Figures 96 through 98.

REAR MAIN SEAL

REMOVAL & INSTALLATION

See Figures 99 and 100.

1. Before servicing the vehicle, refer to the Precautions Section.

2. Remove the upper oil pan.

3. Remove the transmission assembly.

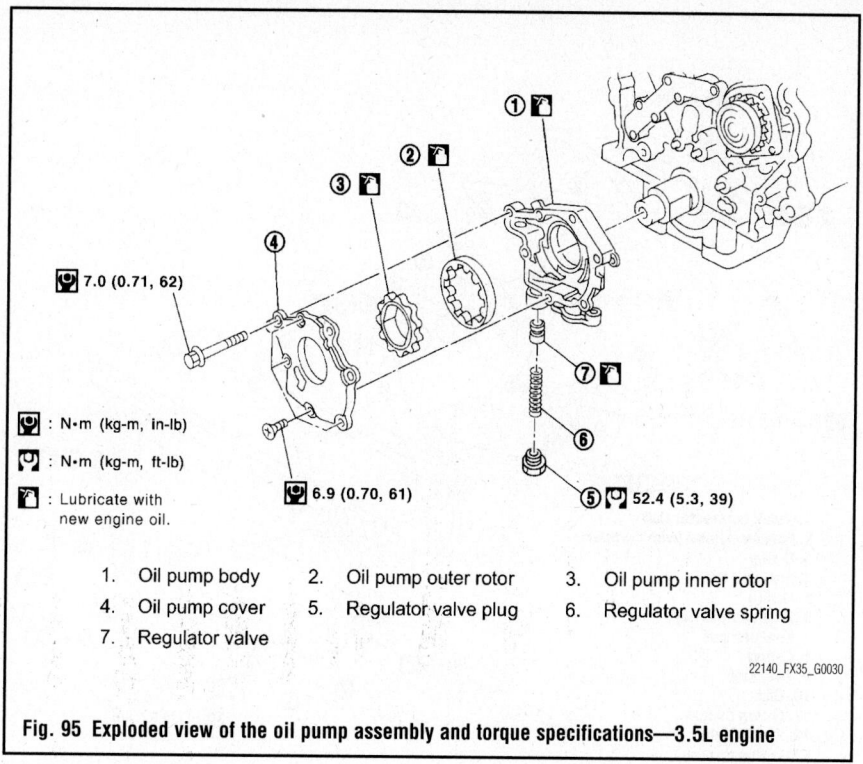

⊙ : N·m (kg-m, in-lb)

⊙ : N·m (kg-m, ft-lb)

🛢 : Lubricate with new engine oil.

⊙ 7.0 (0.71, 62)

⊙ 6.9 (0.70, 61)

⑤ ⊙ 52.4 (5.3, 39)

1. Oil pump body
2. Oil pump outer rotor
3. Oil pump inner rotor
4. Oil pump cover
5. Regulator valve plug
6. Regulator valve spring
7. Regulator valve

22140_FX35_G0030

Fig. 95 Exploded view of the oil pump assembly and torque specifications—3.5L engine

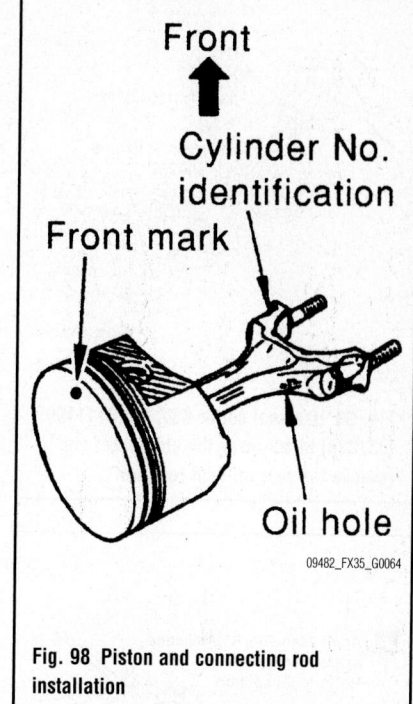

Front

Cylinder No. identification

Front mark

Oil hole

09482_FX35_G0064

Fig. 98 Piston and connecting rod installation

4. Remove the drive plate.

a. Install ring gear stopper SST: KV1011770 (J44716), or equivalent, and remove the mounting bolts in a diagonal order.

b. Carefully remove the drive plate.

※※ WARNING

Do not disassemble the drive plate. Never place the drive plate with the signal plate facing down. When handling the signal plate, take care not to damage or scratch it. Handle the signal plate in a manner that prevents it from becoming magnetized.

5. Use seal cutter SST: KV10111100 (J37228), or equivalent, to cut away the old liquid gasket material and remove the rear oil seal retainer.

※※ WARNING

Be careful not to damage the mounting surfaces.

➡The rear oil seal and retainer form a single part and are handled as one assembly.

To install:

6. Remove the old liquid gasket from the mating surface of the cylinder block and oil pan using a scraper.

7. Apply new engine oil to the oil and dust seal lips.

8. Apply liquid gasket to the rear oil seal retainer as illustrated.

➡Use Genuine RTV Silicone Sealant or equivalent. Installation should be done within 5 minutes after coating, otherwise the liquid gasket may not seal properly.

9. Install the rear oil seal retainer onto the cylinder block.

➡Make sure the garter spring is in position and the seal lips are not inverted.

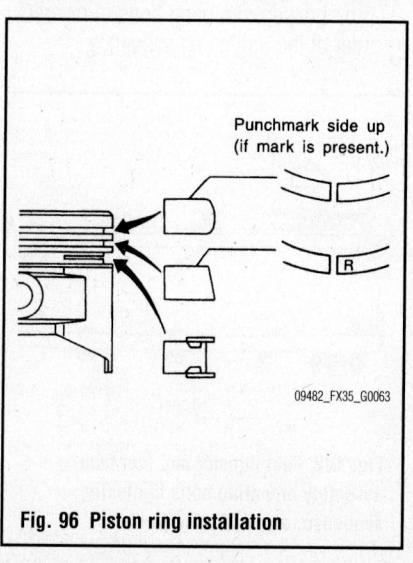

Punchmark side up (if mark is present.)

R

09482_FX35_G0063

Fig. 96 Piston ring installation

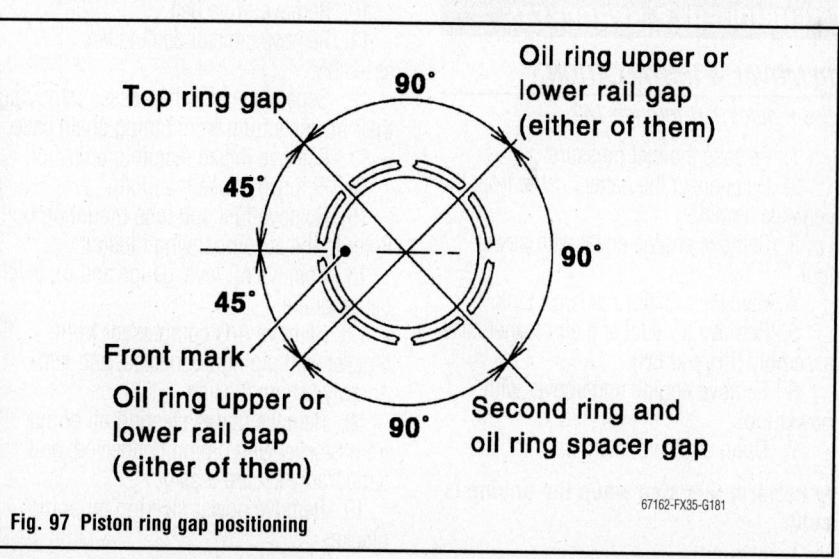

Top ring gap

90°

Oil ring upper or lower rail gap (either of them)

45°

90°

45°

90°

Front mark
Oil ring upper or lower rail gap (either of them)

90°

Second ring and oil ring spacer gap

67162-FX35-G181

Fig. 97 Piston ring gap positioning

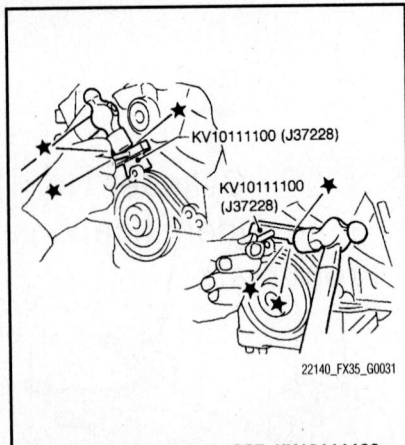

Fig. 99 Use seal cutter SST: KV10111100 (J37228) to cut away the old gasket and remove the rear oil seal retainer

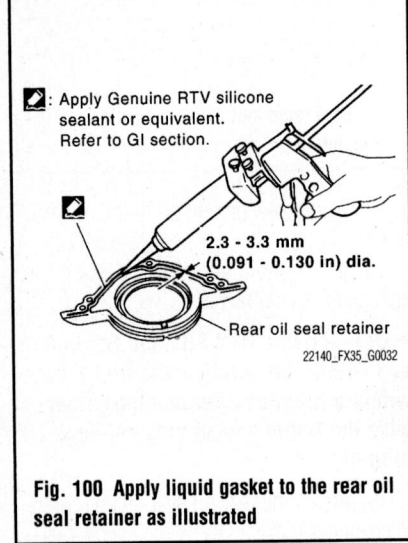

Fig. 100 Apply liquid gasket to the rear oil seal retainer as illustrated

10. The remainder of installation is the reverse of the removal procedure.

TIMING CHAIN COVER, CHAIN, TENSIONER, & SPROCKETS

REMOVAL & INSTALLATION

See Figures 101 through 142.

1. Release the fuel pressure.
2. Disconnect the battery cable from the negative terminal.
3. Remove engine cover with power tool.
4. Remove radiator reservoir tank.
5. Remove air duct and air cleaner case assembly (RH and LH).
6. Remove engine undercover with power tool.
7. Drain engine coolant from radiator.

➥**Perform this step when the engine is cold.**

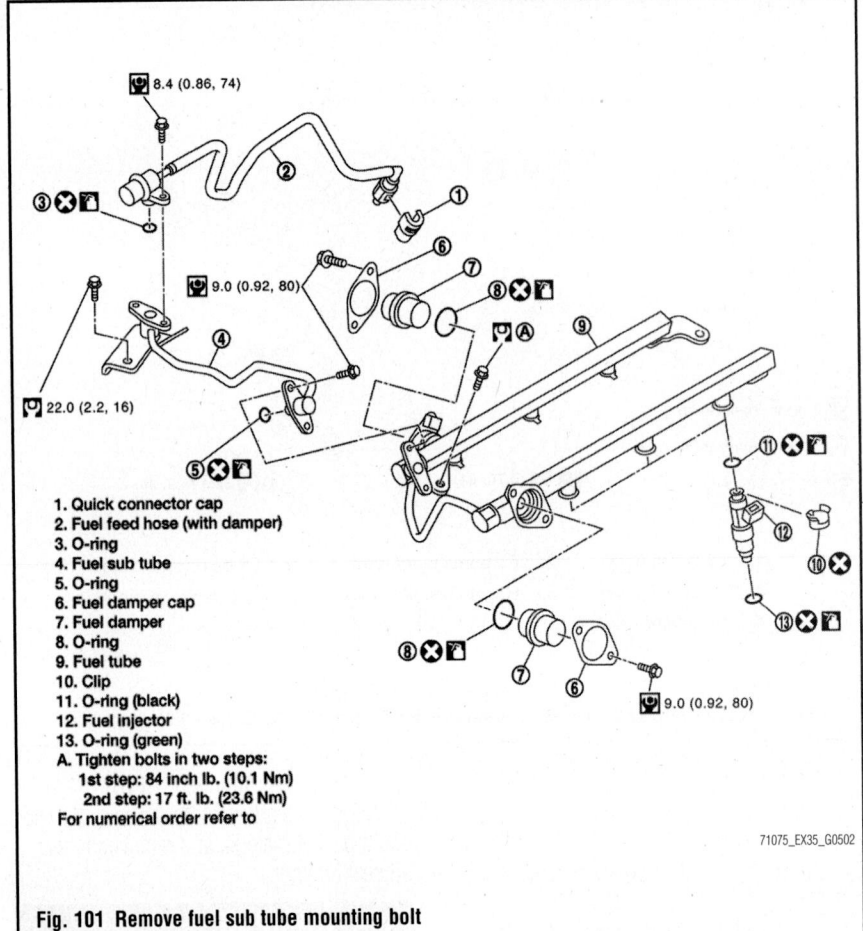

1. Quick connector cap
2. Fuel feed hose (with damper)
3. O-ring
4. Fuel sub tube
5. O-ring
6. Fuel damper cap
7. Fuel damper
8. O-ring
9. Fuel tube
10. Clip
11. O-ring (black)
12. Fuel injector
13. O-ring (green)
A. Tighten bolts in two steps:
    1st step: 84 inch lb. (10.1 Nm)
    2nd step: 17 ft. lb. (23.6 Nm)
For numerical order refer to

Fig. 101 Remove fuel sub tube mounting bolt

➥**Never spill engine coolant on drive belt.**

8. Remove radiator hose (upper and lower).
9. Drain engine oil.

➥**Perform this step when the engine is cold.**

➥**Never spill engine oil on drive belt.**

10. Remove drive belt.
11. Remove radiator cooling fan assembly.
12. Separate engine harnesses removing their brackets from front timing chain case.
13. Remove intake manifold collector.
14. Remove intake manifold.
15. Remove fuel sub tube mounting bolt. Refer to the accompanying illustration.
16. Remove oil level gauge and oil level gauge guide.
17. Remove A/C compressor from bracket with piping connected, and temporarily secure it aside.
18. Remove power steering oil pump from bracket with piping connected, and temporarily secure it aside.
19. Remove power steering oil pump bracket.

20. Remove idler pulley, auto tensioner and bracket.
21. Remove alternator and alternator bracket.
22. Remove water outlet (front) and water piping.
23. Remove valve timing control covers (bank 1 and bank 2) and gasket as follows:
    a. Disconnect valve timing control harness connector.
    b. Loosen mounting bolts in reverse order of the tightening sequence.

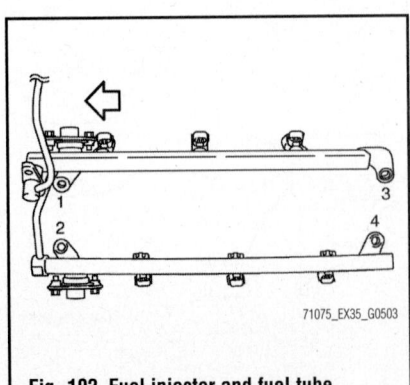

Fig. 102 Fuel injector and fuel tube assembly mounting bolts tightening sequence, arrow=Engine front

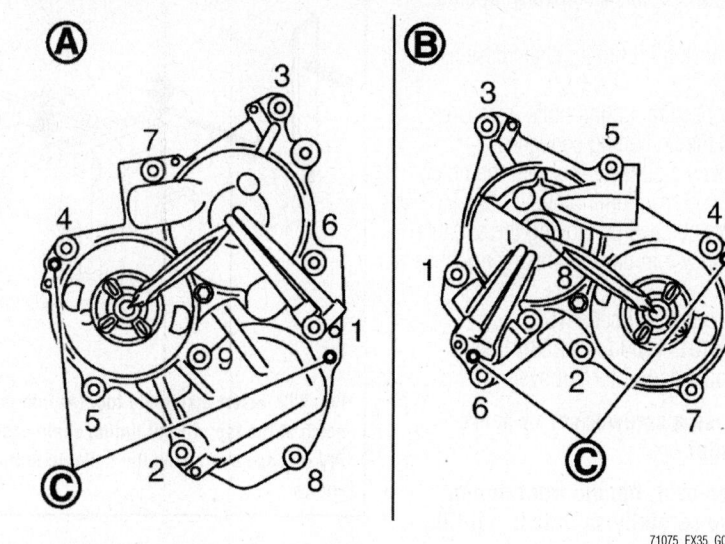

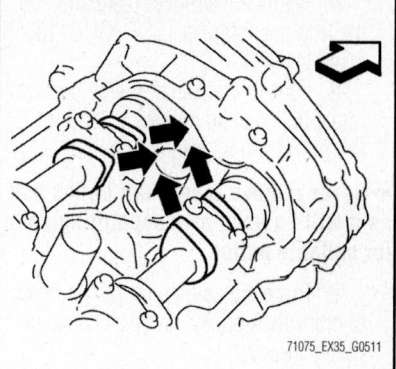

Fig. 106 Check that intake and exhaust cam noses on No. 1 cylinder (engine front side of bank 1) are located as shown. If not, turn crankshaft one revolution (360 degrees) and align as shown

Fig. 103 Remove valve timing control covers (bank 1 (A) and bank 2 (B)) and gasket, Dowel pin hole (C)

➡**Shaft is internally jointed with camshaft sprocket (INT) center hole. When removing, keep it horizontal until it is completely disconnected.**

c. Shaft is engaged with intake side camshaft sprocket center hole on inside. pull straight out so as not to tilt until the joint is disengaged.

• The mating surface of magnet retarder may be fitted with the exhaust side camshaft sprocket via the engine oil. Open valve timing control cover carefully.

• If the mating surface of magnet retarder is fitted with the camshaft sprocket, open the cover within the range that the load is not applied to the harness. And then, remove it so as to prevent magnet retarder from dropping.

❄❄ WARNING

Be careful not to damage magnet retarder.

➡**When carrying valve timing control cover, face the magnet retarder side up to prevent the cover from falling from magnet retarder.**

➡**Never remove magnet retarder from valve timing control cover. (Disassembly prohibited parts)**

24. Remove rocker covers (bank 1 and bank 2).
25. Obtain No. 1 cylinder at TDC of its compression stroke as follows:
a. Rotate crankshaft pulley clockwise to align timing mark (grooved line without color) with timing indicator.
b. Check that intake and exhaust cam noses on No. 1 cylinder (engine front side of bank 1) are located as shown in the accompanying illustration.
• If not, turn crankshaft one revolution (360 degrees) and align as shown in the accompanying illustration.
26. Remove crankshaft pulley as follows:
a. Remove front cross bar. Refer to Suspension Member in Front Suspension in the Suspension Section.
b. Remove power steering pipe mounting bolt.

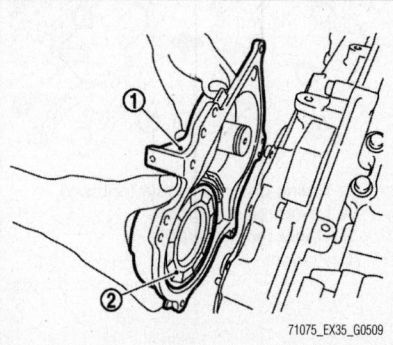

Fig. 104 The mating surface of magnet retarder (2) may be fitted with the exhaust side camshaft sprocket via the engine oil. Open valve timing control cover (1) carefully

Fig. 105 Rotate crankshaft pulley clockwise to align timing mark (grooved line without color) with timing indicator

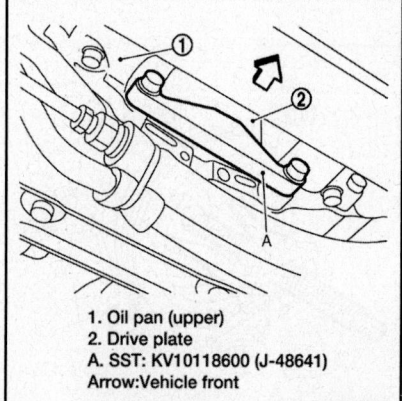

1. Oil pan (upper)
2. Drive plate
A. SST: KV10118600 (J-48641)
Arrow:Vehicle front

Fig. 107 Remove rear cover plate and set the ring gear stopper [SST: KV10118600 (J-48641)] (A).

c. Remove rear cover plate and set the ring gear stopper [SST: KV10118600 (J-48641)].

d. Loosen crankshaft pulley bolt and rotate bolt seating surface at 0.39 in. (10 mm) from its original position.

➡**Never remove crankshaft pulley bolt as it will be used as a supporting point for suitable puller.**

e. Place suitable puller tab on holes of crankshaft pulley, and pull crankshaft pulley through.

✳✳ WARNING

Never put suitable puller tab on crankshaft pulley periphery, as this will damage internal damper.

27. Remove oil pan (lower). Refer to Oil Pan in the Engine Mechanical Section.

28. Loosen two mounting bolts in front of oil pan (upper) with power tool in reverse order as shown in the accompanying illustration.

29. Remove front timing chain case as follows:

a. Loosen mounting bolts in reverse order of the tightening sequence.

b. Insert a suitable tool into the notch at the top of front timing chain case.

c. Pry off case by moving the suitable tool as shown in the accompanying illustration.

• Use the seal cutter [SST: KV10111100 (J-37228)] to cut liquid gasket for removal.

➡**Never use a screwdriver or something similar.**

➡**After removal, handle front timing chain case carefully so it does not tilt, cant, or warp under a load.**

30. Remove front oil seal from front timing chain case using a suitable tool.

a. Use a screwdriver for removal.

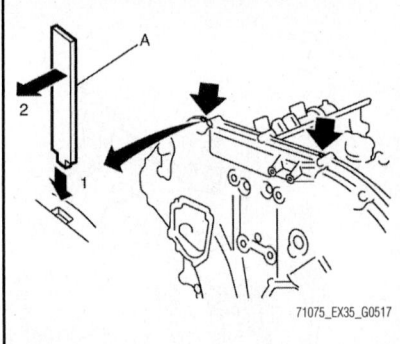

Fig. 112 Insert a suitable tool (A) into the notch at the top of front timing chain case. Pry off case by moving the suitable tool as shown

✳✳ WARNING

Be careful not to damage front timing chain case.

31. Remove O-rings from rear timing chain case.

32. Remove timing chain tensioner (primary) as follows:

a. Remove lower mounting bolt.

b. Loosen upper mounting bolt slowly, and then turn timing chain tensioner (primary) on the upper mounting bolt so that plunger is fully expanded.

➡**Even if plunger is fully expanded, it is not dropped from the body of timing chain tensioner (primary).**

Fig. 108 Loosen crankshaft (1) pulley bolt and rotate bolt seating surface at 0.39 in. (10 mm) from its original position.

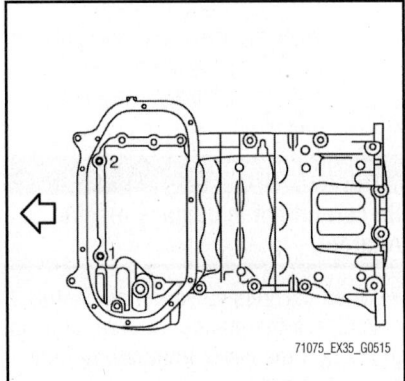

Fig. 110 Loosen two mounting bolts in front of oil pan (upper) with power tool in reverse order as shown

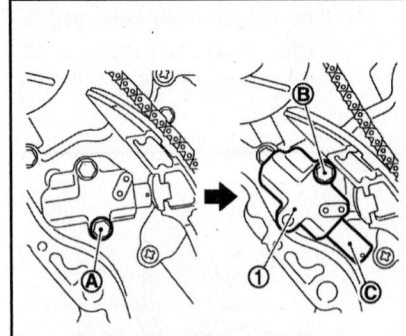

1. Timing chain tensioner (primary)
A. Lower mounting bolt
B. Upper mounting bolt
C. Plunger

Fig. 113 Remove timing chain tensioner (primary). Remove lower mounting bolt. Loosen upper mounting bolt slowly, and then turn timing chain tensioner (primary) on the upper mounting bolt so that plunger is fully expanded

Fig. 109 Place suitable puller tab on holes of crankshaft pulley, and pull crankshaft pulley through

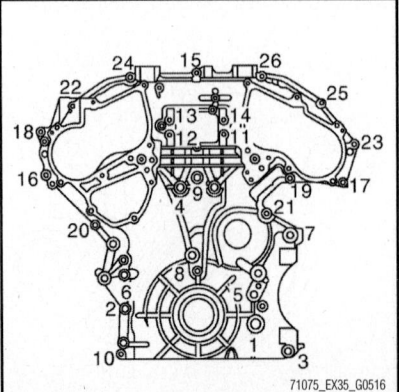

Fig. 111 Remove front timing chain case. Loosen mounting bolts in reverse order as shown

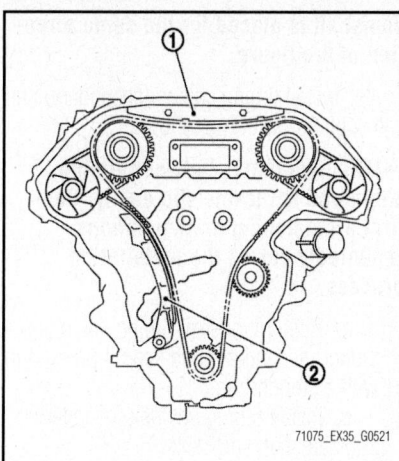

Fig. 114 Remove internal chain guide (1), and slack guide (2)

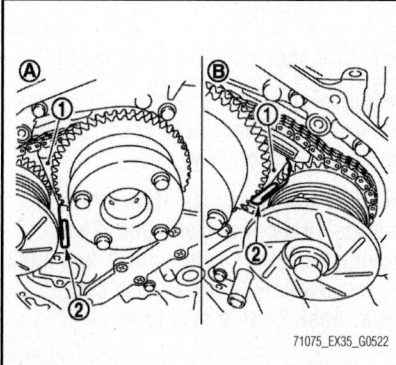

Fig. 115 Remove timing chain (secondary) and camshaft sprockets. Attach suitable stopper pin to the timing chain tensioners (secondary)

c. Remove upper mounting bolt, and then remove timing chain tensioner (primary).

33. Remove internal chain guide, and slack guide.

34. Remove timing chain (primary) and crankshaft sprocket.

➡**After removing timing chain tensioner (primary), never turn crankshaft and camshaft separately, or valves will strike the piston heads.**

35. Remove timing chain (secondary) and camshaft sprockets as follows:

a. Attach suitable stopper pin to the timing chain tensioners (secondary).

➡**Use approximately 0.5 mm (0.02 in) dia. hard metal pin as a stopper pin.**

➡**For removal of timing chain tensioners (secondary), refer to Camshaft in the Engine Mechanical Section**

Fig. 116 Secure the hexagonal portion of camshaft using a wrench to loosen mounting bolts. Never loosen the mounting bolts with securing anything other than the camshaft hexagonal portion or with tensioning the timing chain

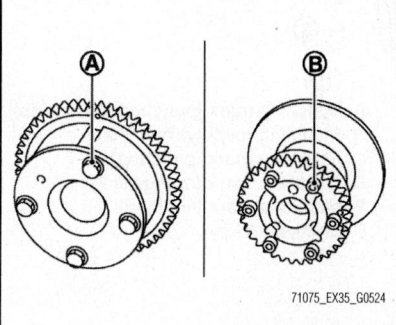

Fig. 117 Remove timing chain (secondary) together with camshaft sprockets. Never disassemble. [Never loosen bolts (A) and (B) as shown

[Removing camshaft bracket (No. 1) is required.].

b. Remove camshaft sprocket mounting bolts (INT and EXH).

• Secure the hexagonal portion of camshaft using a wrench to loosen mounting bolts.

➡**Never loosen the mounting bolts with securing anything other than the camshaft hexagonal portion or with tensioning the timing chain.**

c. Remove timing chain (secondary) together with camshaft sprockets.

➡**Never disassemble. [Never loosen bolts as shown in the accompanying illustration.]**

36. Remove timing chain tensioners (secondary) from cylinder head as follows, if necessary.

a. Remove camshaft brackets (No. 1). Refer to Camshaft in the Engine Mechanical Section.

b. Remove timing chain tensioners (secondary) with a stopper pin attached.

37. Use a scraper to remove all traces of old liquid gasket from front and rear timing chain cases and oil pan (upper), and liquid gasket mating surfaces.

✳✳ WARNING

Be careful not to allow gasket fragments to enter oil pan.

38. Remove old liquid gasket from bolt hole and thread.

Inspection After Removal

39. Timing Chain

40. Check for cracks and any excessive wear at link plates and roller links of timing chain. Replace timing chain if necessary.

To install:

➡**The below figure shows the relationship between the matching mark on each timing chain and that on the corresponding sprocket, with the components installed**

41. Install timing chain tensioners (secondary) to cylinder head as follows if removed.

42. Check that dowel pin and crankshaft key are located as shown in the figure. (No. 1 cylinder at compression TDC).

a. Camshaft dowel pin: At cylinder head upper face side in each bank.

b. Crankshaft key: At cylinder head side of bank 1.

➡**Though camshaft does not stop at the position as shown in the accompanying illustration, for the placement of cam noses, it is generally accepted**

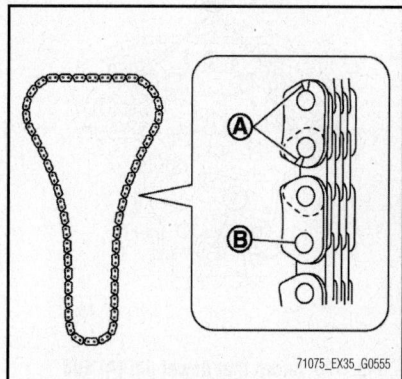

Fig. 118 Check for cracks (A) and any excessive wear (B) at link plates and roller links of timing chain. Replace timing chain if necessary

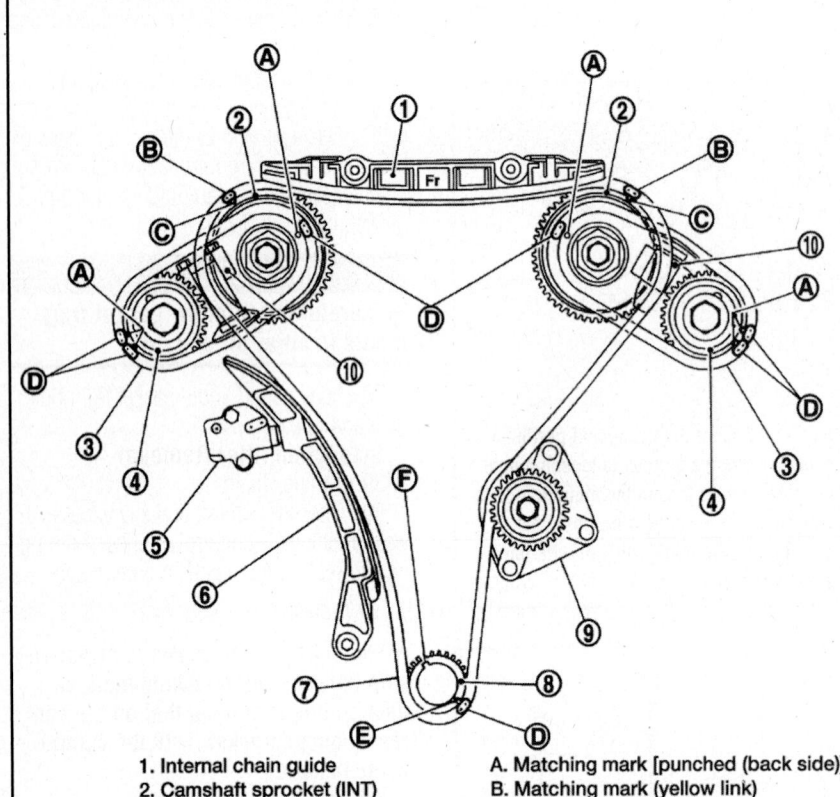

1. Internal chain guide
2. Camshaft sprocket (INT)
3. Timing chain (secondary)
4. Camshaft sprocket (EXH)
5. Timing chain tensioner (primary)
6. Slack guide
7. Timing chain (primary)
8. Crankshaft sprocket
9. Water pump
10. Timing chain tensioner (secondary)

A. Matching mark [punched (back side)]
B. Matching mark (yellow link)
C. Matching mark (punched)
D. Matching mark (orange link)
E. Matching mark (notched)
F. Crankshaft key

71075_EX35_G0527

Fig. 119 The illustration shows the relationship between the matching mark on each timing chain and that on the corresponding sprocket, with the components installed

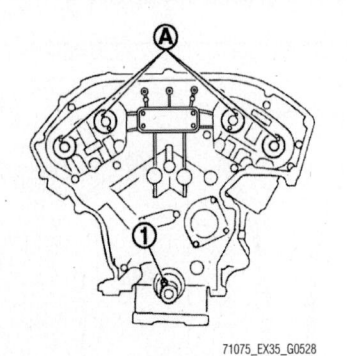

71075_EX35_G0528

Fig. 120 Check that dowel pin (A) and crankshaft key (1) are located as shown in the figure. (No. 1 cylinder at compression TDC). Camshaft dowel pin: At cylinder head upper face side in each bank. Crankshaft key: At cylinder head side of bank 1.

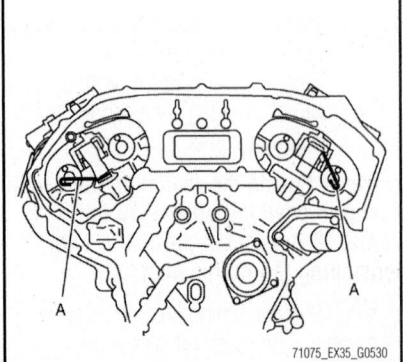

71075_EX35_G0530

Fig. 121 Push plunger of timing chain tensioner (secondary) and keep it pressed in with a stopper pin (A)

camshaft is placed for the same direction of the figure.

43. Install timing chains (secondary) and camshaft sprockets as follows:

➡**Matching marks between timing chain and sprockets slip easily. Confirm all matching mark positions repeatedly during the installation process.**

   a. Push plunger of timing chain tensioner (secondary) and keep it pressed in with a stopper pin.
   b. Install timing chains (secondary) and camshaft sprockets.
- Align the matching marks on timing chain (secondary) (orange link) with the ones on intake and exhaust camshaft sprockets (punched), and install them.

➡**Figure shows bank 1 (rear view).**

➡**Matching marks for camshaft sprockets are on the back side of camshaft sprockets (secondary).**

➡**There are two types of matching marks, circle and oval types. They should be used for the bank 1 and bank 2, respectively. Bank 1 : Use circle type, Bank 2 : Use oval type.**

- Align dowel pin camshafts with the groove or dowel hole on sprockets, and install them.
- On the intake side, align dowel pin on camshaft front end with pin groove on the back side of camshaft sprocket, and install them.
- On the exhaust side, align dowel pin on camshaft front end with pin hole on camshaft sprocket, and install them.
- In case that positions of each matching mark and each dowel pin are not fit on matching parts, make fine adjustment to the position holding the hexagonal portion on camshaft with wrench or equivalent.
- Mounting bolts for camshaft sprockets must be tightened in the next step. Tightening them by hand is enough to prevent the dislocation of dowel pins.
- Check the matching marks (punched) on each camshaft sprocket are positioned on the matching marks (orange link) on timing chain (secondary).

➡**Matching mark (punched) in the figure is for checking loose at this step.**

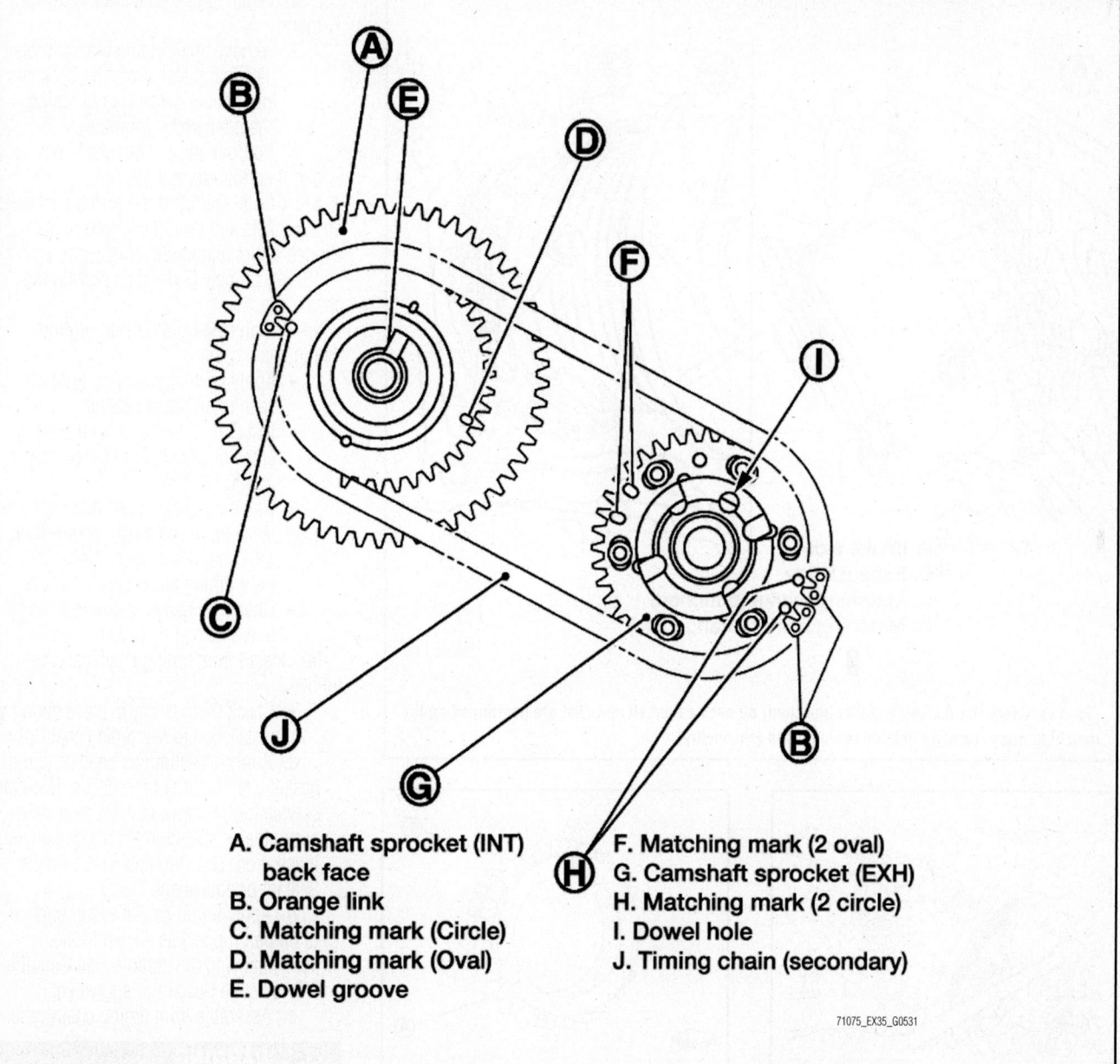

A. Camshaft sprocket (INT)
   back face
B. Orange link
C. Matching mark (Circle)
D. Matching mark (Oval)
E. Dowel groove

F. Matching mark (2 oval)
G. Camshaft sprocket (EXH)
H. Matching mark (2 circle)
I. Dowel hole
J. Timing chain (secondary)

71075_EX35_G0531

Fig. 122 Install timing chains (secondary) and camshaft sprockets. Align the matching marks on timing chain (secondary) (orange link) with the ones on intake and exhaust camshaft sprockets (punched), and install them.

c. After confirming the matching marks are aligned, tighten camshaft sprocket mounting bolts.
- Secure camshaft using a wrench at the hexagonal portion to tighten mounting bolts.

d. Pull stopper pins out from timing chain tensioners (secondary).

44. Install timing chain (primary) as follows:

a. Install crankshaft sprocket.
- Check the matching marks on crankshaft sprocket face the front of the engine.

b. Install timing chain (primary).
- Install timing chain (primary) so the matching mark (punched) on camshaft sprocket (INT) is aligned with the yellow link on timing chain, while the matching mark (notched) on crankshaft sprocket is aligned with the orange link one on timing chain, as shown in the accompanying illustration.
- When it is difficult to align matching marks of timing chain (primary) with each sprocket, gradually turn camshaft using wrench on the

hexagonal portion to align it with the matching marks.
- During alignment, be careful to prevent dislocation of matching mark alignments of timing chains (secondary).

45. Install internal chain guide, slack guide.

➡**Never overtighten slack guide mounting bolts. It is normal for a gap to exist under the bolt seats when mounting bolts are tightened to the specification.**

46. Install the timing chain tensioner (primary) with the following procedure:

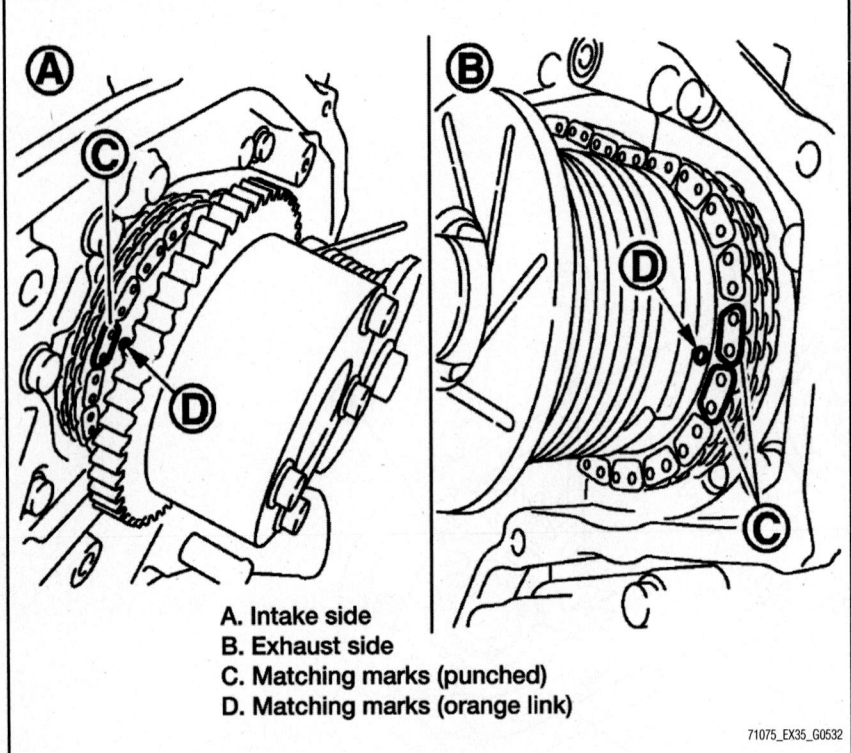

A. Intake side
B. Exhaust side
C. Matching marks (punched)
D. Matching marks (orange link)

71075_EX35_G0532

Fig. 123 Check the matching marks (punched) on each camshaft sprocket are positioned on the matching marks (orange link) on timing chain (secondary).

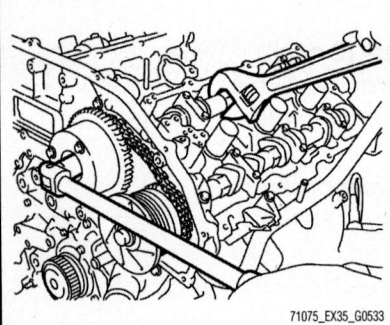

71075_EX35_G0533

Fig. 124 After confirming the matching marks are aligned, tighten camshaft sprocket mounting bolts. Secure camshaft using a wrench at the hexagonal portion to tighten mounting bolts.

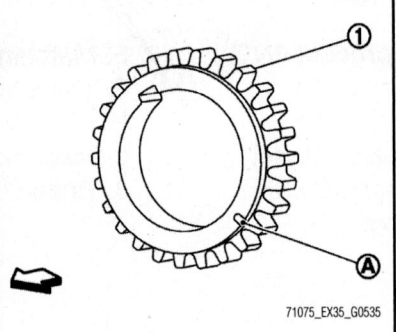

71075_EX35_G0535

Fig. 125 Install timing chain (primary) as follows: Install crankshaft sprocket (1). Check the matching marks (A) (front side) on crankshaft sprocket face the front of the engine (Arrow).

a. Pull plunger stopper tab up (or turn lever downward) so as to remove plunger stopper tab from the ratchet of plunger.

➡**Plunger stopper tab and lever are synchronized.**

b. Push plunger into the inside of tensioner body.

c. Hold plunger in the fully compressed position by engaging plunger stopper tab with the tip of ratchet.

d. To secure lever, insert stopper pin through hole of lever into tensioner body hole.

• The lever parts and the plunger stopper tab are synchronized. Therefore, the plunger will be secured under this condition.

➡**The illustration shows the example of 0.047 in (1.2 mm) diameter thin screwdriver being used as the stopper pin.**

e. Install timing chain tensioner (primary).

• Remove any dirt and foreign materials completely from the back and the mounting surfaces of timing chain tensioner (primary).

f. Pull out stopper pin after installing, and then release plunger.

47. Check again that the matching marks on sprockets and timing chain have not slipped out of alignment.

48. Install new O-rings on rear timing chain case.

49. Install new front oil seal on front timing chain case.

• Apply new engine oil to both oil seal lip and dust seal lip.

• Install it so that each seal lip is oriented as shown in the accompanying illustration.

• Using a suitable drift [outer diameter: 2.36 in (60 mm)], press-fit oil seal until it becomes flush with front timing chain case end face.

• Check the garter spring is in position and seal lip is not inverted.

50. Install front timing chain case as follows:

• Check O-rings stay in place during installation to rear timing chain case.

a. Apply a continuous bead of liquid gasket with the tube presser (commercial service tool) to front timing chain case back side as shown in the accompanying illustration. Use Genuine RTV Silicone Sealant or equivalent.

b. Apply liquid gasket to top surface of oil pan (upper) as shown in the accompanying illustration. Use Genuine RTV silicone Sealant or equivalent.

c. Assemble front timing chain case.

✳✳ WARNING

Be careful not to damage front oil seal by interference with front end of crankshaft.

➡**Attaching should be done within 5 minutes after liquid gasket application.**

d. Install front timing chain case as to fit its dowel pin hole together dowel pin on rear timing chain case.

e. Tighten mounting bolts to the specified torque in numerical order as shown in the accompanying illustration. M10 bolts : 1, 2, 3, 4, 5, 6, 7: 41 ft. lbs. (55.0 Nm), M6 bolts : Except the above: 108 inch lbs. (12.7 Nm)

• There are two types of mounting bolts. Refer to the following for locating bolts.

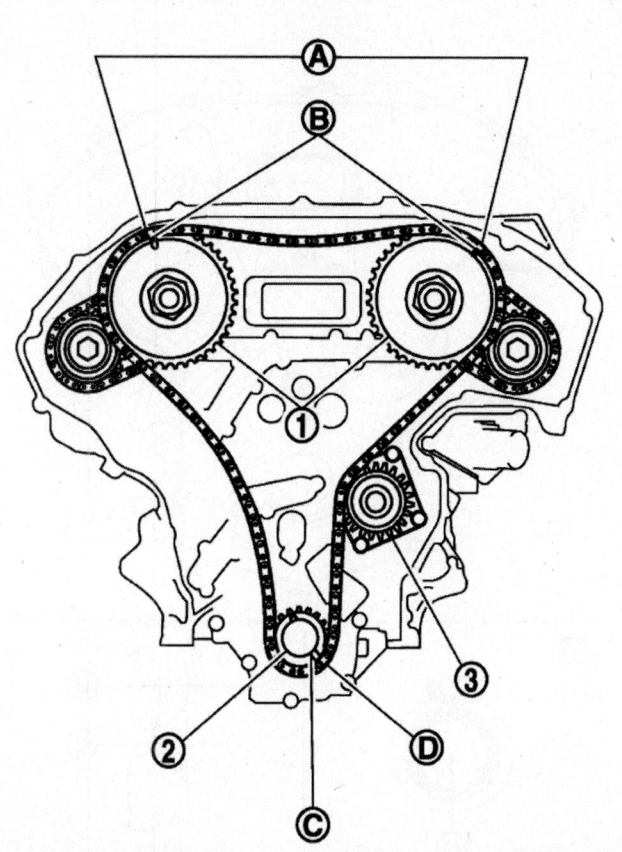

1. Camshaft sprocket (INT) (1)
2. Crankshaft sprocket
3. Water pump
A. Yellow link

B. Matching mark (punched)
C. Matching mark (notched)
D. Orange link

71075_EX35_G0536

Fig. 126 Install timing chain (primary) so the matching mark (punched) on camshaft sprocket (INT) is aligned with the yellow link on timing chain, while the matching mark (notched) on crankshaft sprocket is aligned with the orange link one on timing chain, as shown

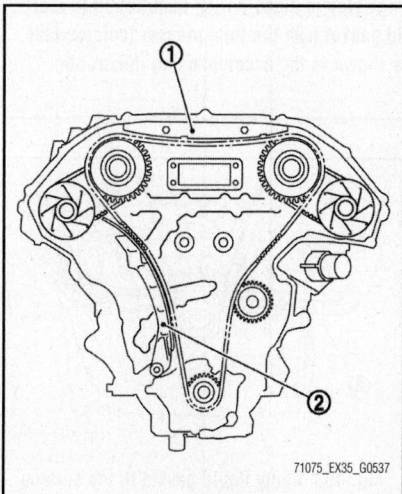

71075_EX35_G0537

Fig. 127 Install internal chain guide (1), slack guide (2)

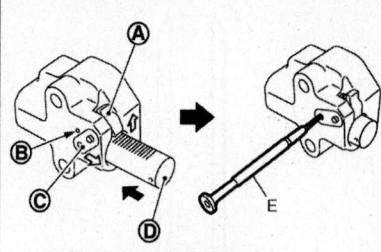

A. Plunger stopper tab
B. Tensioner body hole
C. Plunger stopper tab and lever
D. Ratchet of plunger
E. Stopper pin

71075_EX35_G0539

Fig. 128 Install the timing chain tensioner (primary).Pull plunger stopper tab up (or turn lever downward) so as to remove plunger stopper tab from the ratchet of plunger. Plunger stopper tab and lever are synchronized. Push plunger into the inside of tensioner body. Hold plunger in the fully compressed position by engaging plunger stopper tab with the tip of ratchet. To secure lever, insert stopper pin through hole of lever into tensioner body hole.

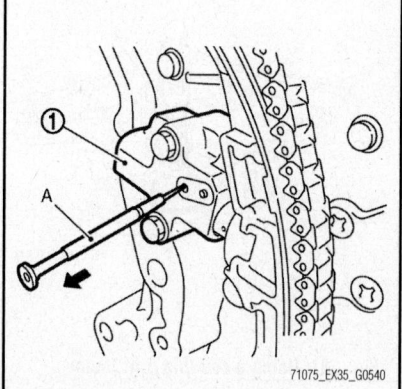

71075_EX35_G0540

Fig. 129 Install timing chain tensioner (primary) (1). Pull out stopper pin (A) after installing, and then release plunger

f. After all bolts are tightened, retighten them to the specified torque in numerical order shown in the accompanying illustration.

➡**Be sure to wipe off any excessive liquid gasket leaking on surface mating with oil pan (upper).**

g. Install two mounting bolts in front of oil pan (upper) in numerical order shown in the accompanying illustration.
51. Install valve timing control covers (bank 1 and bank 2) as follows:
a. Install new seal rings in shaft grooves.

➡**When replacing seal ring, replace all rings with new ones.**

b. To check the joint between dowel pins and dowel pin holes, check the looseness in the axle direction by pushing the circumferential looseness (between dowel pins and dowel pin holes) by twisting in the circumferential direction.

➡**Always perform this procedure when removing because the gap between dowel pins and dowel pin holes may not be caused on purpose.**

c. Install valve timing control cover with new gasket to front timing chain case.

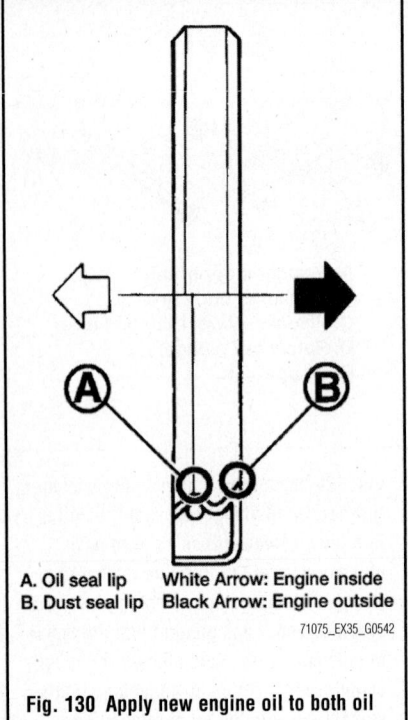

A. Oil seal lip White Arrow: Engine inside
B. Dust seal lip Black Arrow: Engine outside

71075_EX35_G0542

Fig. 130 Apply new engine oil to both oil seal lip and dust seal lip

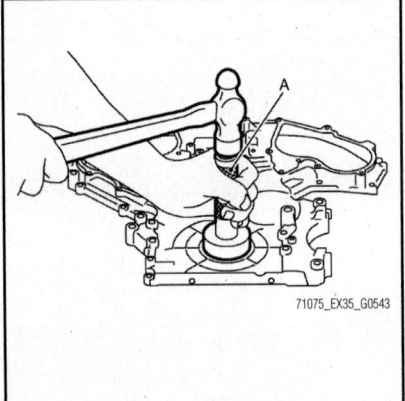

71075_EX35_G0543

Fig. 131 Using a suitable drift [outer diameter: 2.36 in (60 mm)] (A), press-fit oil seal until it becomes flush with front timing chain case end face.

➥Never face the magnet retarder side down to prevent magnet retarder from dropping.

➥Check the mating surface of magnet retarder and the drum of exhaust side camshaft sprocket for foreign materials.

➥Align the center of both shaft holes of the shaft and the intake side camshaft sprocket, and then insert them.

➥Be careful not to drop the seal ring from the shaft groove.

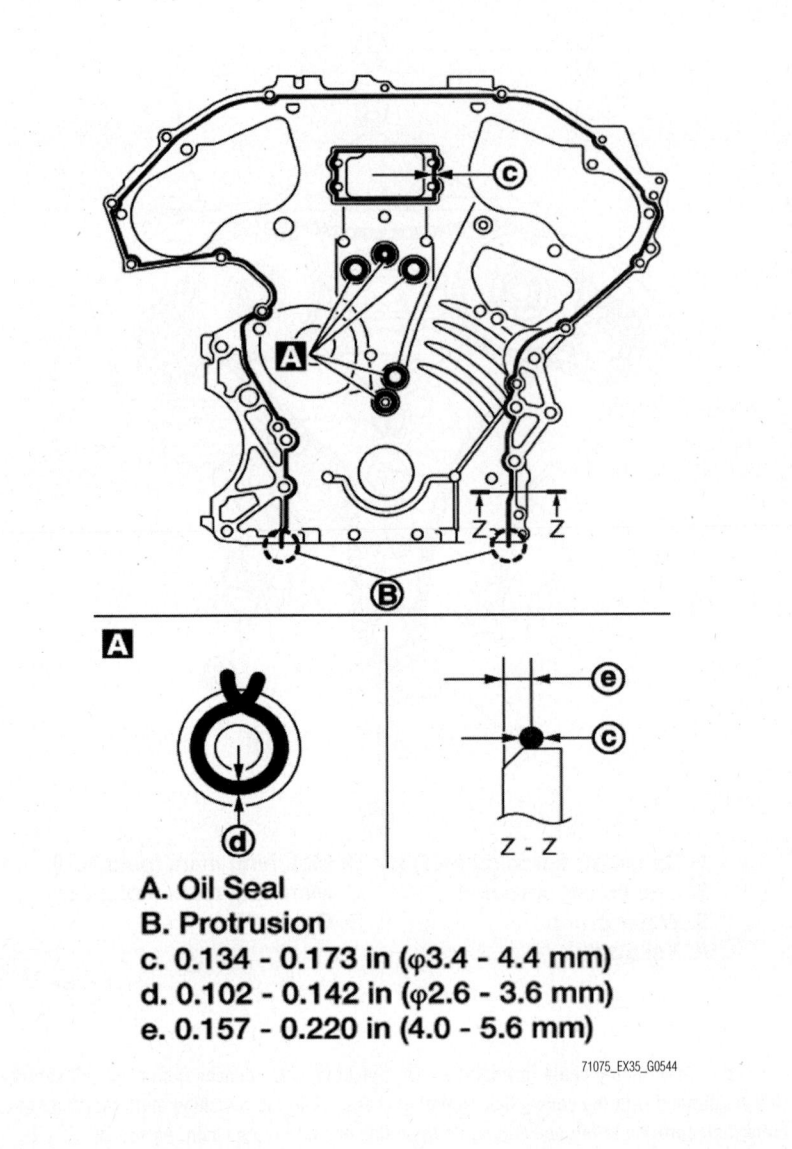

A. Oil Seal
B. Protrusion
c. 0.134 - 0.173 in (φ3.4 - 4.4 mm)
d. 0.102 - 0.142 in (φ2.6 - 3.6 mm)
e. 0.157 - 0.220 in (4.0 - 5.6 mm)

71075_EX35_G0544

Fig. 132 Install front timing chain case. Check O-rings stay in place during installation to rear timing chain case. Apply a continuous bead of liquid gasket with the tube presser (commercial service tool) to front timing chain case back side as shown in the accompanying illustration. Use Genuine RTV Silicone Sealant or equivalent

➥When setting the valve timing control cover in position by hand, if valve timing control cover is not contacting with the front timing chain case, the dowel pin of magnet retarder may not be aligned with the dowel pin holes of cover. In this case, return to step "b".

d. Being careful not to move seal ring from the installation groove, align dowel pins on front timing chain case with holes to install valve timing control covers.

e. Tighten mounting bolts in numerical order as shown in the accompanying illustration. Tighten bolts to 96 inch lbs. (11.3 Nm)

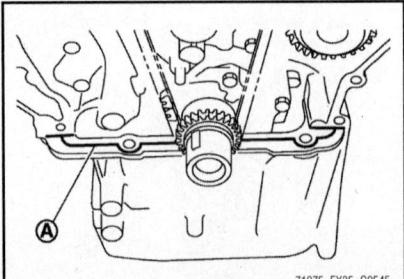

71075_EX35_G0545

Fig. 133 Apply liquid gasket to top surface of oil pan (upper) as shown. Use Genuine RTV silicone Sealant or equivalent. A : 0.157 - 0.197 in (φ4.0 - 5.0 mm)

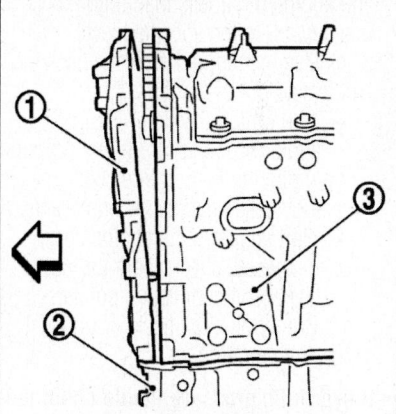

1. Front timing chain case
2. Oil pan (upper)
3. Cylinder block
Arrow: Engine front

71075_EX35_G0546

Fig. 134 Assemble front timing chain case

- After all bolts are tightened, tighten No. 1 bolt to the specified torque again.

52. Install oil pan (lower). Refer to Oil Pan in the Engine Mechanical Section.

53. Install rocker covers (bank 1 and bank 2). Refer to the accompanying illustration.

54. Install crankshaft pulley as follows:

a. Fix crankshaft using the ring gear stopper [SST: KV10118600 (J-48641)].

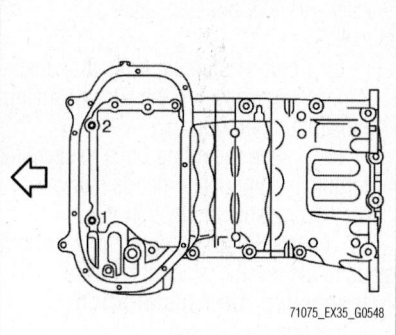

Fig. 136 Install two mounting bolts in front of oil pan (upper) in numerical order shown (Arrow=Engine front)

b. Install crankshaft pulley, taking care not to damage front oil seal.

- When press-fitting crankshaft pulley with plastic hammer, tap on its center portion (not circumference).

c. Tighten crankshaft pulley bolt to 33 ft. lbs. (44 Nm).

d. Place a matching mark on crankshaft pulley aligning with the matching mark of crankshaft pulley bolt. Tighten the bolt 90 degrees (one marks).

e. Rotate crankshaft pulley in normal direction (clockwise when viewed from front) to confirm it turns smoothly.

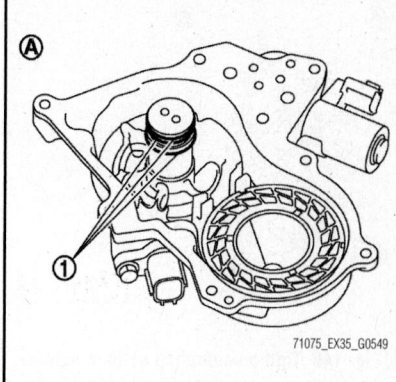

71075_EX35_G0549

Fig. 137 Install valve timing control covers (bank 1 and bank 2 (A)). Install new seal rings (1) in shaft grooves

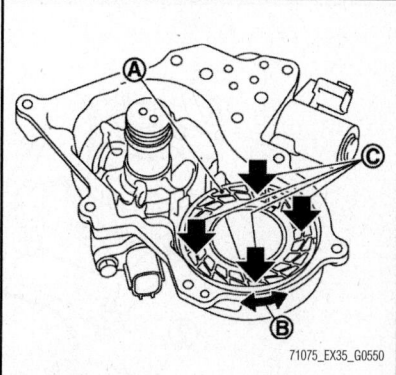

71075_EX35_G0550

Fig. 138 To check the joint between dowel pins and dowel pin holes, check the looseness in the axle direction by pushing the circumferential looseness (between dowel pins and dowel pin holes) by twisting in the circumferential direction. Mating surface of magnet retarder (A), Moves slightly (B), Not shaken (C)

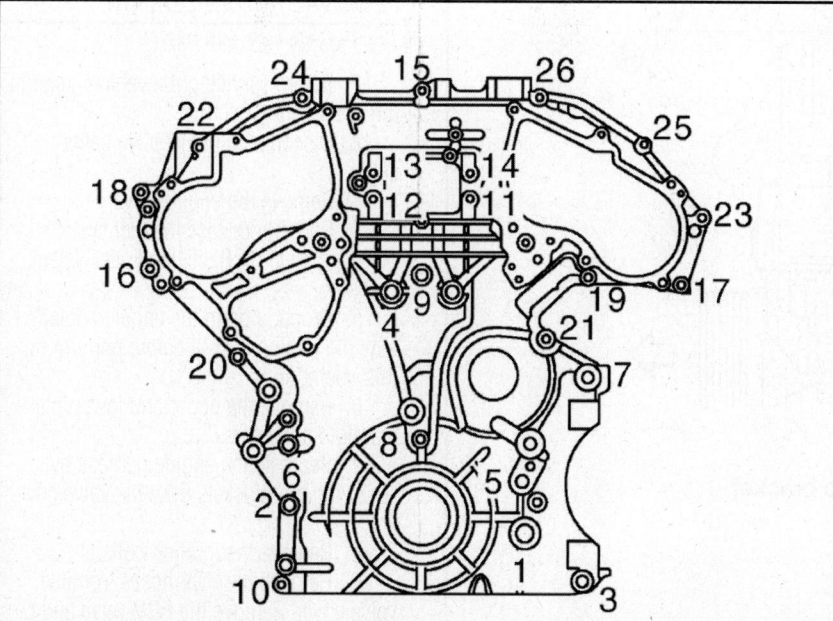

71075_EX35_G0547

Fig. 135 Tighten mounting bolts to the specified torque in numerical order as shown. There are two types of mounting bolts. Refer to the following for locating bolts. After all bolts are tightened, retighten them to the specified torque in numerical order shown

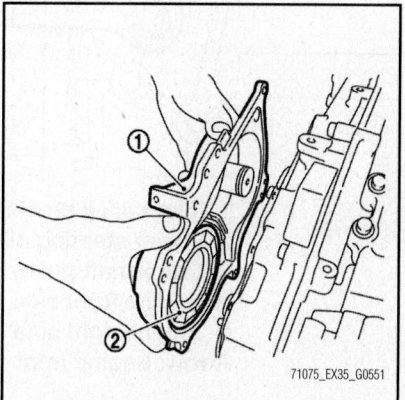

71075_EX35_G0551

Fig. 139 Install valve timing control cover (1) with new gasket to front timing chain case, magnet retarder (2)

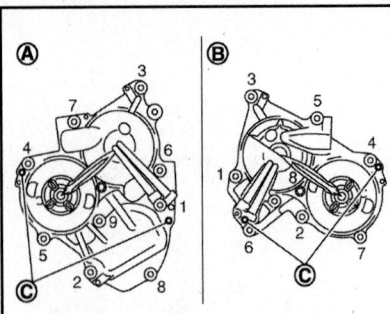

71075_EX35_G0552

Fig. 140 Tighten mounting bolts in numerical order as shown. After all bolts are tightened, tighten No. 1 bolt to the specified torque again. Bank 1 (A), Bank 2 (B), Dowel pin hole (C)

55. Install power steering oil pump bracket and idler pulley bracket as follows:

a. Tighten mounting bolts in numerical order as shown in the accompanying illustration (temporarily).

b. Tighten mounting bolts to specified torque in numerical order as shown in the accompanying illustration.

56. For the following operations, perform steps in the reverse order of removal.

Inspection After Installation

Inspection for Leakage

57. The following are procedures for checking fluids leakage, lubricates leakage.

a. Before starting engine, check oil/fluid levels including engine coolant and engine oil. If less than required quantity, fill to the specified level.

b. Use procedure below to check for fuel leakage.

- Turn ignition switch "ON"(with engine stopped). With fuel pressure applied to fuel piping, check for fuel leakage at connection points.
- Start engine. With engine speed increased, check again for fuel leakage at connection points.

c. Run engine to check for unusual noise and vibration.

➡ **If hydraulic pressure inside chain tensioner drops after removal/installation, slack in guide may generate a pounding noise during and just after the engine start. However, this does not indicate an unusualness. Noise will stop after hydraulic pressure rises.**

d. Warm up engine thoroughly to check there is no leakage of fuel, or any oil/fluids including engine oil and engine coolant.

e. Bleed air from lines and hoses of applicable lines, such as in cooling system.

f. After cooling down engine, again check oil/fluid levels including engine oil and engine coolant. Refill to the specified level, if necessary.

VALVE COVERS

REMOVAL & INSTALLATION

See Figures 143 and 144.

1. Before servicing the vehicle, refer to the Precautions Section.

2. Disconnect the negative battery cable.

3. Remove the engine cover.

4. Properly release the fuel system pressure. Refer to Relieving Fuel System Pressure.

5. Properly drain the engine coolant. Be sure the engine is cold before performing this operation.

6. Remove the upper and lower intake manifold collectors.

7. Separate the engine harness by removing the brackets from the valve covers.

8. Remove the ignition coils.

9. Remove the PCV hoses from the valve cover. Remove the PCV valve and O-ring from the valve cover on the right bank, if necessary.

10. Remove the oil filler cap and oil catcher from the valve cover on the left bank, if necessary.

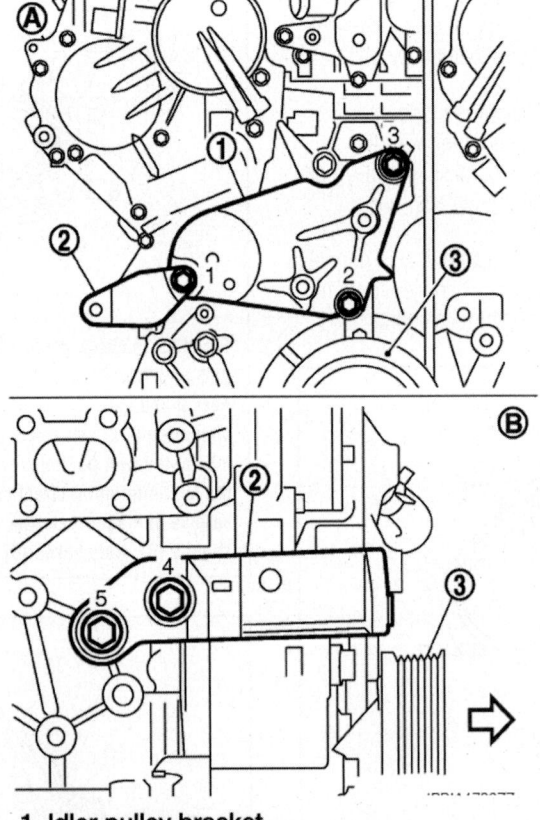

1. Idler pulley bracket
2. Power steering oil pump bracket
3. Crankshaft pulley
A. Engine front side
B. Engine right side
Arrow: Engine front

71075_EX35_G0554

Fig. 141 Install power steering oil pump bracket and idler pulley bracket as follows: Tighten mounting bolts in numerical order as shown in the accompanying illustration (temporarily).

Summary of the inspection items:

Items		Before starting engine	Engine running	After engine stopped
Engine coolant		Level	Leakage	Level
Engine oil		Level	Leakage	Level
Transmission / transaxle fluid	AT & CVT Models	Leakage	Level / Leakage	Leakage
	MT Models	Level / Leakage	Leakage	Level / Leakage
Other oils and fluids*		Level	Leakage	Level
Fuel		Leakage	Leakage	Leakage
Exhaust gases		—	Leakage	—

*: Power steering fluid, brake fluid, etc.

71075_EX35_G0529

Fig. 142 Summary of the inspection items

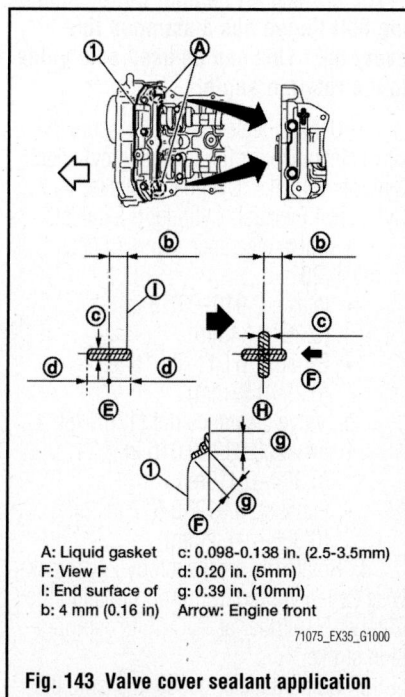

A: Liquid gasket
F: View F
I: End surface of
b: 4 mm (0.16 in)
c: 0.098-0.138 in. (2.5-3.5mm)
d: 0.20 in. (5mm)
g: 0.39 in. (10mm)
Arrow: Engine front

71075_EX35_G1000

Fig. 143 Valve cover sealant application (left side shown)

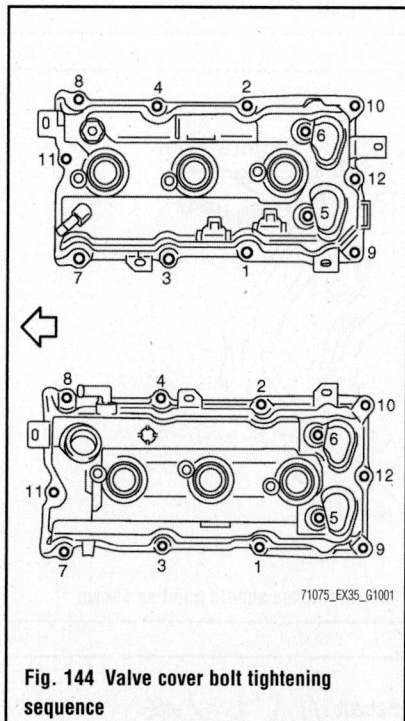

71075_EX35_G1001

Fig. 144 Valve cover bolt tightening sequence

11. Loosen the valve cover retaining bolts. Remove the bolts in the reverse order of the tightening sequence illustrated.

12. Remove the valve covers from the engine.

To install:

13. Use a scraper and remove all the old gasket material from the cylinder head and camshaft bracket (No. 1).

※※ WARNING

Be sure not to scratch the mating surfaces with the scraper when removing the gasket material.

14. Apply genuine RTV silicone sealant, or equivalent, to the joints of the valve cover, cylinder head, and camshaft bracket (No. 1) as shown in the illustration.

15. Refer to figure **a** in the illustration to apply liquid gasket to the joint part of the camshaft (No. 1) and cylinder head.

16. Refer to figure **b** in the illustration to apply liquid gasket to the figure **a** squarely.

17. Install a new valve cover gasket to the valve cover.

18. Install the valve cover. Torque the retaining bolts to 17 inch lbs. (2 Nm) and then to 74 inch lbs. (8 Nm), in the proper sequence as shown in the illustration.

➡**Be sure that the gasket has not dropped from the grove prior to installation.**

19. Continue the installation in the reverse order of the removal procedure.

20. When installing the PCV hose, insert it 0.98–1.18 in. (25–30mm) from the connector end.

21. When installing the PCV hose between the right and left valve covers, be sure the identification paint mark is facing upward (right cover side).

VALVE LASH (CLEARANCE) ADJUSTMENT

ADJUSTMENT

See Figures 145 through 149.

1. Before servicing the vehicle, refer to the Precautions Section.

Perform inspection after removal, installation, or replacement of camshaft or valve-related parts, or if there is unusual engine condition regarding valve clearance.

2. Remove the right and left rocker covers.

3. Set the number 1 cylinder at Top Dead Center (TDC) of its compression stroke, as follows:

a. Rotate the crankshaft pulley clockwise until the timing mark (grooved line without color) is aligned with the timing indicator.

b. Make sure the number 1 cylinder intake and exhaust cam noses are facing inward and upward from the cylinder head, as shown. If they are not positioned as shown, rotate the crankshaft pulley 360° clockwise until they are appropriately positioned.

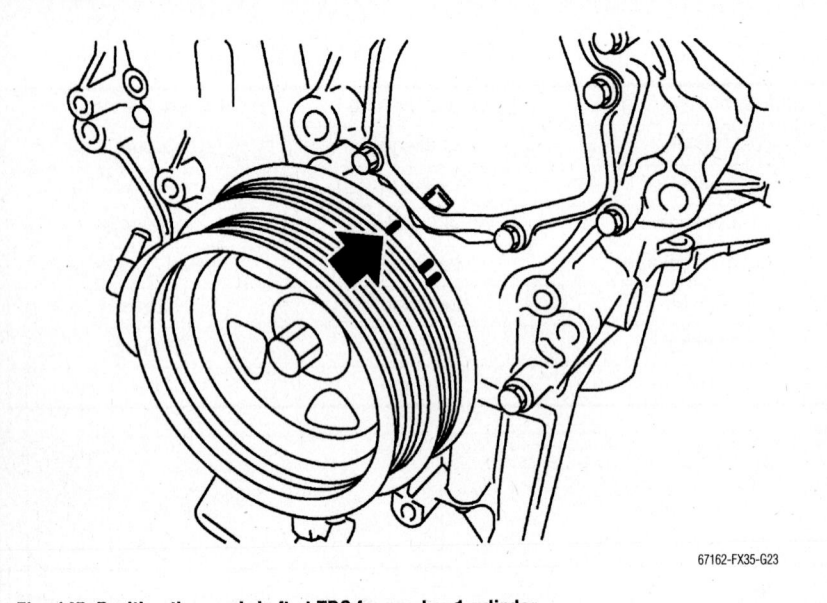

67162-FX35-G23

Fig. 145 Position the crankshaft at TDC for number 1 cylinder

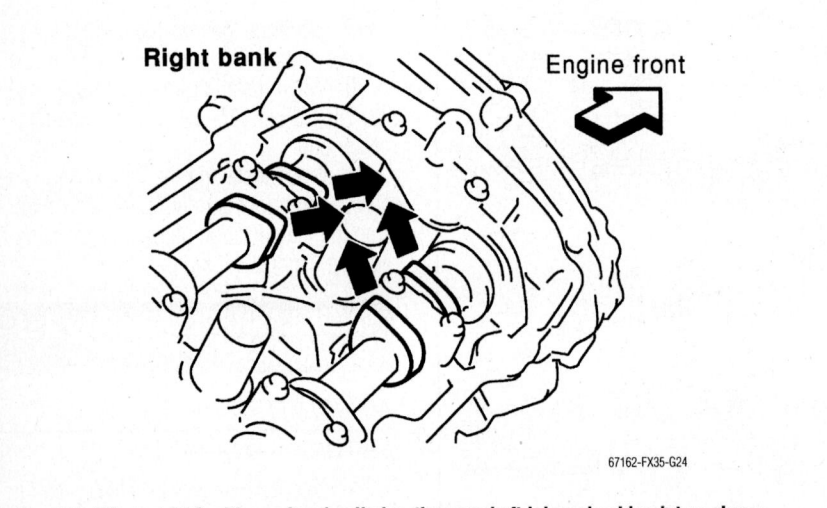

Right bank

Engine front

67162-FX35-G24

Fig. 146 When at TDC with number 1 cylinder, the camshaft lobes should point as shown

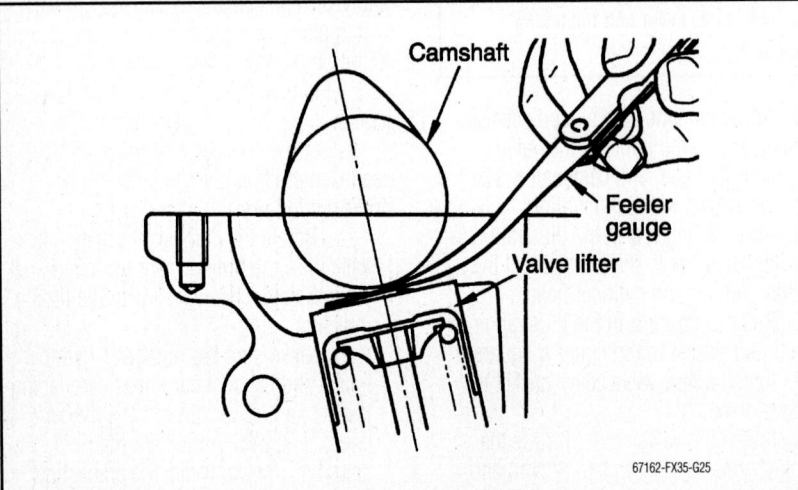

Camshaft

Feeler gauge

Valve lifter

67162-FX35-G25

Fig. 147 Measure the valve lash with the camshaft lobe positioned as shown

4. Using a feeler gauge, measure the valve lash clearance as illustrated.

5. Measure the valve clearance for the following cylinders: Cylinder 1 Intake, Cylinder 2 Exhaust, Cylinder 3 Exhaust, Cylinder 6 Intake:

   a. Valve clearance cold (68°F/20°C):
- Intake: 0.010–0.013 in. (0.26–0.34mm)
- Exhaust: 0.011–0.015 in. (0.29–0.37mm)

   b. Valve clearance hot (176°F/80°C):
- Intake: 0.012–0.016 in. (0.304–0.416mm)
- Exhaust: 0.012–0.017 in. (0.308–0.432mm)

6. Rotate the crankshaft by 240°clockwise (when viewed from front) to align the number 3 cylinder at TDC of its compression stroke.

➡**The crankshaft damper pulley mounting bolt flange has a stamped line every 60°. This can be used as a guide to the rotation angle.**

7. Using a feeler gauge, measure the valve clearance for the following cylinders: Cylinder 2 Intake, Cylinder 3 Intake, Cylinder 4 Exhaust, Cylinder 5 Exhaust:

   a. Valve clearance standard cold (68°F/20°C):
- Intake: 0.010–0.013 in. (0.26–0.34mm)
- Exhaust: 0.011–0.015 in. (0.29–0.37mm)

   b. Valve clearance hot (176°F/80°C):
- Intake: 0.012–0.016 in. (0.304–0.416mm)
- Exhaust: 0.012–0.017 in. (0.308–0.432mm)

8. Rotate the crankshaft by 240°clockwise (when viewed from front) to align the number 5 cylinder at TDC of its compression stroke.

9. Using a feeler gauge, measure the valve clearance for the following cylinders: Cylinder 1 Exhaust, Cylinder 4 Intake, Cylinder 5 Intake, Cylinder 6 Exhaust:

   a. Valve clearance standard cold (68°F/20°C):
- Intake: 0.010–0.013 in. (0.26–0.34mm)
- Exhaust: 0.011–0.015 in. (0.29–0.37mm)

   b. Valve clearance hot (176°F/80°C):
- Intake: 0.012–0.016 in. (0.304–0.416mm)
- Exhaust: 0.012–0.017 in. (0.308–0.432mm)

➡**If the inspection was carried out with a cold engine, make sure the values**

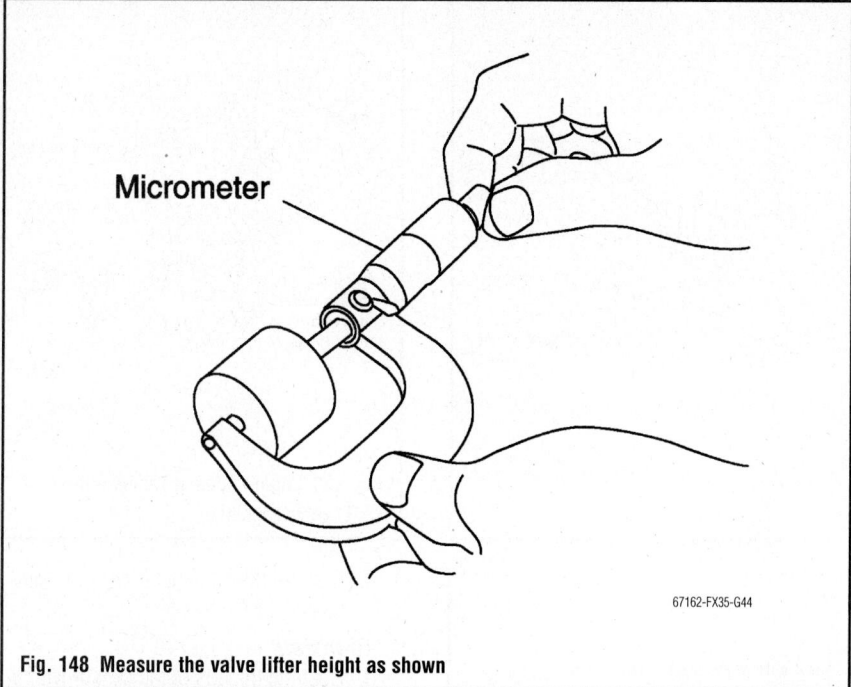

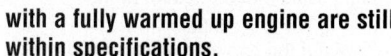

Fig. 148 Measure the valve lifter height as shown

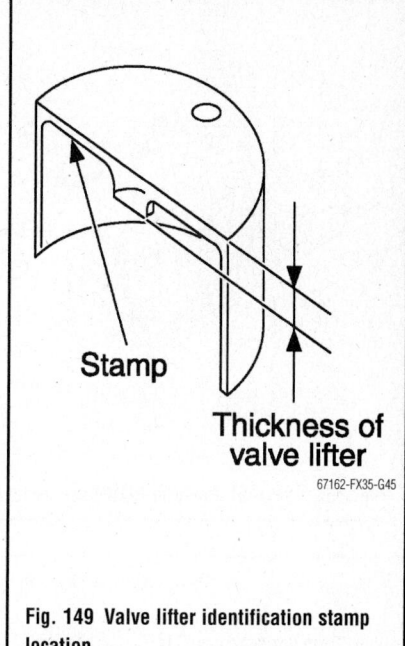

Fig. 149 Valve lifter identification stamp location

with a fully warmed up engine are still within specifications.

10. For all valve lifters that are found to be outside the specified range, perform the following steps.

  a. Perform adjustment depending on selected head thickness of valve lifter.

  b. The specified valve lifter thickness is the dimension at normal temperatures. Ignore dimensional differences caused by temperature. Use the specifications for hot engine condition to adjust.

11. Remove the camshaft.

12. Remove the valve lifters at the locations that are outside the standard.

13. Measure the center thickness of the removed valve lifters with a micrometer.

14. Use the following equation to calculate valve lifter thickness for the replacement lifters.

➡**Valve lifter thickness calculation: thickness of replacement valve lifter = t1 + (C1 - C2). t1 = Thickness of removed valve lifter, C1 = measured valve clearance, C2 = standard valve clearance.**

The thickness of a new valve lifter can be identified by stamp marks on the reverse side (inside the cylinder). Stamp mark 788U or 788R indicates 0.3102 in. (7.88mm) in thickness.

➡**Two types of stamp marks are used for parallel setting and for manufac-**

turer identification. Available thicknesses of valve lifters include 27 sizes covering a range of 0.3102–0.3307 in. (7.88–8.40mm) in steps of 0.0008 in. (0.02mm).

15. Install the selected valve lifter(s).

16. Install the camshaft.

17. Manually turn crankshaft pulley a few turns.

18. Make sure the valve clearances for the cold engine are within specifications by referring to the specified values.

19. After completing the repair, check valve clearances again with the specifications for a warmed engine. Make sure the values are within specifications.

ENGINE PERFORMANCE & EMISSION CONTROLS

CAMSHAFT POSITION SENSOR

LOCATION
See Figure 150.

Refer to the accompanying illustration.

REMOVAL & INSTALLATION

1. Before servicing the vehicle, refer to the Precautions Section.

2. Disconnect the negative battery cable.

3. Disconnect the connector from the Camshaft Position (CMP) sensor.

4. Remove the bolt that retains the CMP sensor.

5. Remove the CMP sensor.

To install:

6. Installation is the reverse of the removal procedure.

7. Tighten the bolt that retains the CMP sensor.

8. Connect the sensor connector.

CRANKSHAFT POSITION SENSOR

LOCATION
See Figure 151.

Refer to the accompanying illustration.

REMOVAL & INSTALLATION

1. Before servicing the vehicle, refer to the Precautions Section.

2. Disconnect the negative battery cable.

3. Disconnect the connector from the sensor.

4. Remove the bolt that retains the sensor in place.

5. Remove the sensor from its mounting.

To install:

6. Installation is the reverse of the removal procedure.

7. Tighten the sensor retaining bolt.

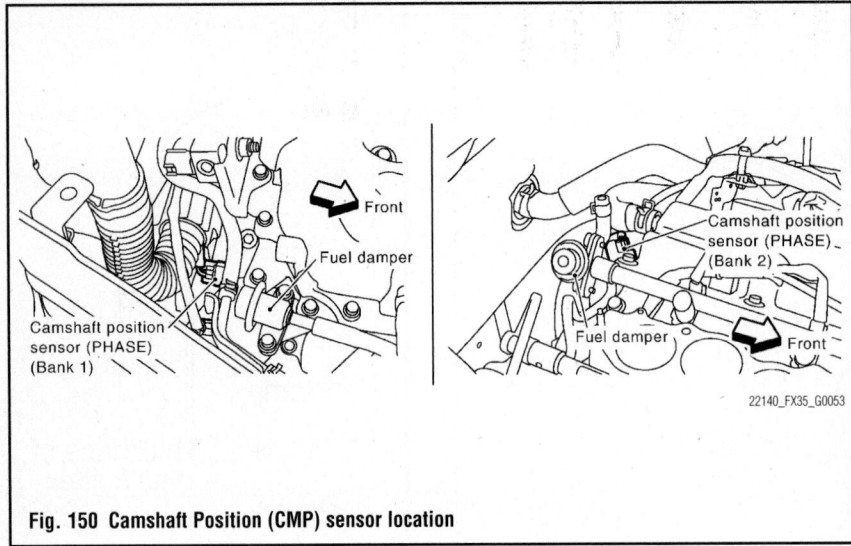

Fig. 150 Camshaft Position (CMP) sensor location

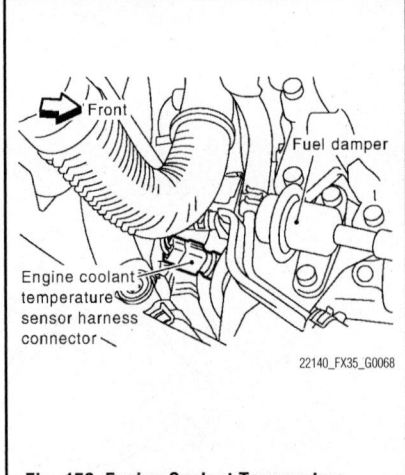

Fig. 153 Engine Coolant Temperature (ECT) sensor location

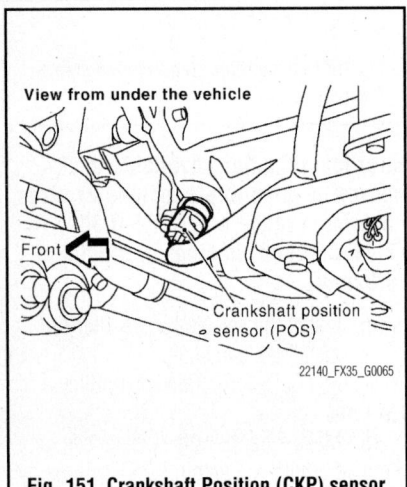

Fig. 151 Crankshaft Position (CKP) sensor location

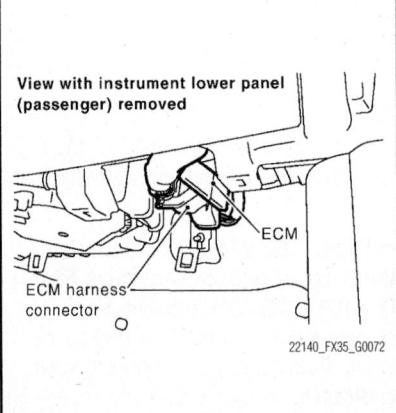

Fig. 152 Electronic Control Module (ECM) location

ELECTRONIC CONTROL MODULE

LOCATION

See Figure 152.

Refer to the accompanying illustration.

REMOVAL & INSTALLATION

1. Before servicing the vehicle, refer to the Precautions Section.
2. Turn the ignition switch **OFF**.
3. Disconnect the negative battery cable from the battery.
4. Disconnect the Electronic Control Module (ECM) connectors.
5. Remove the ECM mounting bolts and remove the ECM from the vehicle.

   To install:
6. Installation is the reverse of the removal.
7. Tighten the ECM mounting bolts.

※※ WARNING

When replacing the ECM, be careful to use the right part number, as damage to the injection system could occur.

ENGINE COOLANT TEMPERATURE SENSOR

LOCATION

See Figure 153.

Refer to the accompanying illustration.

REMOVAL & INSTALLATION

1. Before servicing the vehicle, refer to the Precautions Section.
2. Drain the coolant to a level below the bottom of the sensor.
3. Disconnect the ground cable from the battery and then remove the sensor connector.

4. Remove the coolant temperature sensor.

To install:
5. Coat the threads of the sensor with a suitable sealant and thread into the housing.
6. Tighten the sensor.
7. Refill the cooling system to the proper level.
8. Attach the electrical connector to the sensor securely.
9. Connect the negative battery cable.

HEATED OXYGEN SENSOR (HO2S)

LOCATION

See Figures 154 and 155.

Refer to the accompanying illustrations.

REMOVAL & INSTALLATION

※※ CAUTION

The temperature of the exhaust system is extremely high after the engine has been run. To prevent personal injury, allow the exhaust system to cool before removing the sensor from the exhaust system.

1. Before servicing the vehicle, refer to the Precautions Section.
2. Disconnect the negative battery cable.
3. Raise and safely support the vehicle, as needed.
4. Detach the electrical connector from the oxygen sensor.
5. Using an oxygen sensor socket, remove the Heated Oxygen Sensor (HO2S).

   To install:
6. If installing the old HO2S sensor, coat the threads with anti-seize compound. New

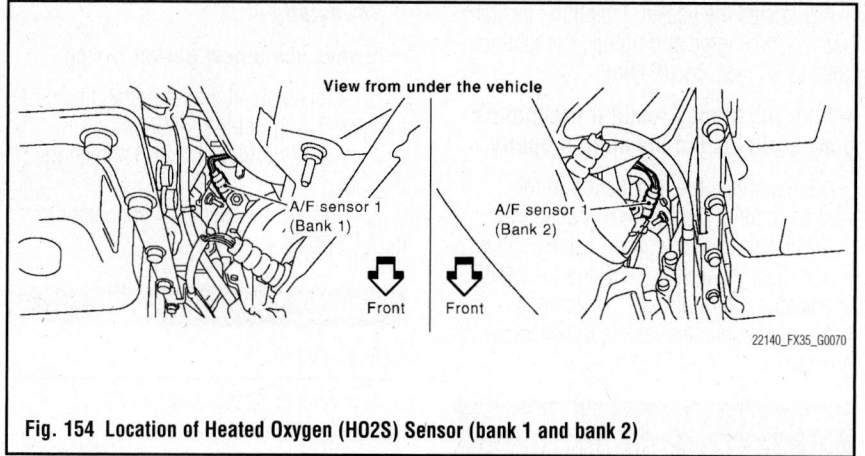

Fig. 154 Location of Heated Oxygen (HO2S) Sensor (bank 1 and bank 2)

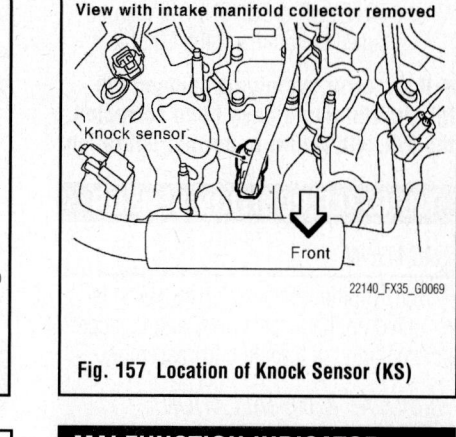

Fig. 157 Location of Knock Sensor (KS)

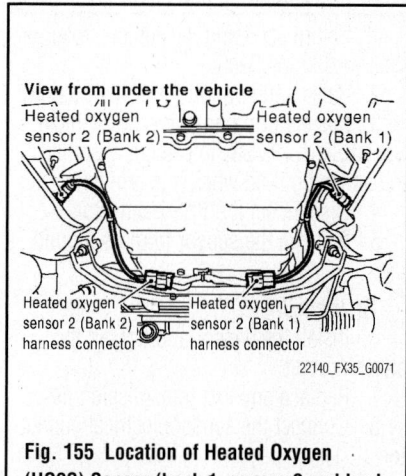

Fig. 155 Location of Heated Oxygen (HO2S) Sensor (bank 1, sensor 2 and bank 2, sensor 2

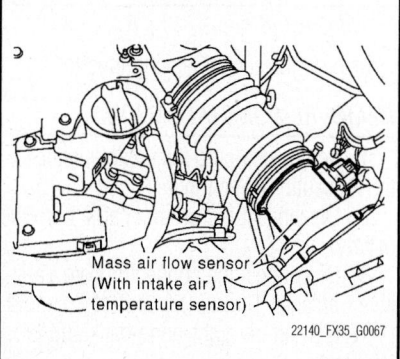

Fig. 156 Location of Mass Air Flow (MAF) and Intake Air Temperature (IAT) sensor—3.5L engine

sensors are already coated. Take care not to contaminate the oxygen sensor probe with the anti-seize compound.

7. Install the oxygen sensor. Using the correct tool, tighten the sensor.

8. Attach the wiring to the sensor.

9. Connect the negative battery cable.

INTAKE AIR TEMPERATURE SENSOR

LOCATION

See Figure 156.

Refer to the accompanying illustration.

REMOVAL & INSTALLATION

1. Before servicing the vehicle, refer to the Precautions Section.

2. Disconnect the negative battery cable.

3. Disconnect the connector from the sensor.

4. Remove the sensor retaining screws.

5. Remove the sensor from its mounting.

To install:

6. Installation is the reverse of the removal procedure.

7. Handle the sensor assembly carefully, protecting it from impact, extremes of temperature and/or exposure to shop chemicals.

KNOCK SENSOR (KS)

LOCATION

See Figure 157.

Refer to the accompanying illustration.

REMOVAL & INSTALLATION

1. Before servicing the vehicle, refer to the Precautions Section.

2. Disconnect the negative battery cable.

3. Disconnect the sensor connector.

4. Remove the sensor from its mounting.

To install:

5. Installation is the reverse of the removal procedure.

6. Tighten the sensor to 16 ft. lbs. (21 Nm).

MALFUNCTION INDICATOR LIGHT

RESET PROCEDURE

1. Proper operation of the Malfunction Indicator Light (MIL):
   - The MIL will illuminate with the ignition switch ON and the engine OFF
   - The MIL will turn OFF when the engine is started
   - The MIL will remain ON if the self-diagnostic system has detected a malfunction
   - The MIL may turn OFF if the malfunction is no longer present
   - If the MIL is illuminated and then the engine stalls, the MIL will remain illuminated as long as the ignition switch is ON
   - If the MIL is not illuminated and the engine stalls, the MIL will not illuminate until the ignition switch is cycled OFF, then ON

2. Resetting the MIL:
   - The control module turns OFF the MIL after 3 consecutive ignition cycles that the diagnostic system runs and does not fail
   - The control module turns OFF the MIL after a current Diagnostic Trouble Code (DTC) clears when the diagnostic cycle runs and passes
   - There may still be a history of DTC's stored in the system. These will clear after 40 consecutive warm-up cycles, if no failures are reported by any other related diagnostic system
   - Manual resetting of the MIL and any DTC stored in the system, requires the use of an OBD2 scan tool connected to the Data Link Connector (DLC) for communication with the vehicle. Follow the

instructions of the scan tool for both retrieval and resetting of DTC's. The CONSULT-III® can be used to command the MIL off.

➡**If the error symptoms causing the MIL to illuminate have been corrected, the MIL will return to normal operation.**

THROTTLE POSITION SENSOR

LOCATION

The Throttle Position Sensor (TPS) is mounted on the throttle body and is incorporated into the throttle body assembly.

REMOVAL & INSTALLATION

The throttle position sensor is an integral part of the throttle body.

1. Before servicing the vehicle, refer to the Precautions Section.
2. Properly relieve the fuel system pressure.
3. Drain the engine coolant.
4. Remove the air intake hose.
5. Remove the battery.
6. Disconnect the throttle position sensor connector.
7. Disconnect the water hose connection.
8. Remove the throttle body retaining bolts.
9. Remove the throttle body from the engine.
10. Discard the gasket.

To install:

✳ WARNING

Do not loosen the retaining screws for the resin cover of the throttle body assembly. If the screws are loosened, the sensor incorporated in the resin cover becomes misaligned and the throttle body may not work properly.

11. Align the recess on the intake manifold plenum with the projection of the throttle body gasket.

12. Install the gasket. Install the throttle body to the engine and tighten the retaining bolts to 80 inch lbs. (9 Nm).

➡**Poor idling may result if the throttle body gasket is not installed properly.**

13. Continue the installation in the reverse order of the removal procedure.
14. Connect the negative battery cable.
15. Turn the ignition **ON** and then **OFF**, and keep it off for at least 10 seconds.
16. Complete the vehicle initialization procedure.

VARIABLE CAMSHAFT TIMING OIL CONTROL SOLENOID

LOCATION

See Figure 158.

Refer to the accompanying illustration.

REMOVAL & INSTALLATION

1. Before servicing the vehicle, refer to the Precautions Section.
2. Disconnect the ground cable from the battery.
3. Disconnect the connector from Variable Camshaft Timing Oil Control Solenoid (VCTOCS) on the right-hand bank and/or left-hand bank.
4. Remove the bolt retaining the VCTOCS.
5. Remove the VCTOCS.

To install:

➡**Always use a new gasket/O-ring.**

6. Apply a small amount of engine oil to the new O-ring of the VCTOCS.
7. Install the VCTOCS and tighten the mounting bolt.
8. Connect the electrical connector to the VCTOCS.

VEHICLE SPEED SENSOR

LOCATION

The Vehicle Speed Sensor (VSS) is installed on the transmission.

REMOVAL & INSTALLATION

1. Before servicing the vehicle, refer to the Precautions Section.
2. Raise and support the vehicle safely.
3. Place a drip pan below the Vehicle Speed Sensor (VSS) to catch any spilled transmission fluid when it is removed.
4. Disconnect the VSS connector.
5. Remove the sensor from its mounting.

To install:

6. Install the sensor and tighten the attaching bolt.
7. Replace any lost transmission fluid.
8. Connect the sensor electrical connector.

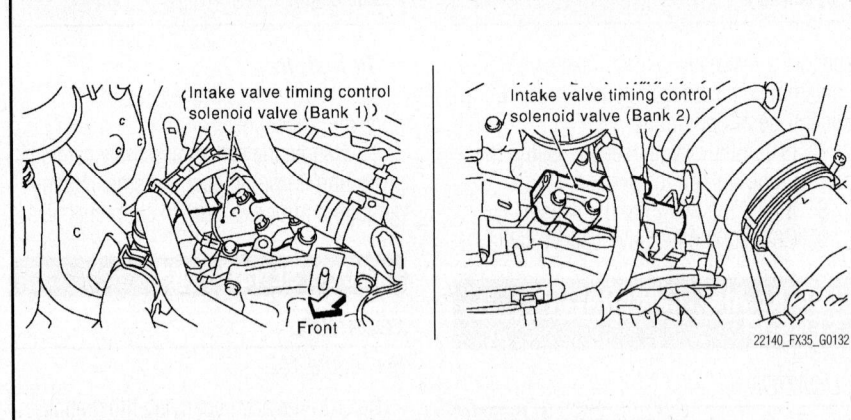

Fig. 158 Location of the variable camshaft timing oil control solenoid valves

FUEL SYSTEM SERVICE PRECAUTIONS

Safety is the most important factor when performing not only fuel system maintenance but any type of maintenance. Failure to conduct maintenance and repairs in a safe manner may result in serious personal injury or death. Maintenance and testing of the vehicle's fuel system components can be accomplished safely and effectively by adhering to the following rules and guidelines.

• To avoid the possibility of fire and personal injury, always disconnect the negative battery cable unless the repair or test procedure requires that battery voltage be applied.

• Always relieve the fuel system pressure prior to disconnecting any fuel system component (injector, fuel rail, pressure regulator, etc.), fitting or fuel line connection. Exercise extreme caution whenever relieving fuel system pressure to avoid exposing skin, face and eyes to fuel spray. Please be advised that fuel under pressure may penetrate the skin or any part of the body that it contacts.

• Always place a shop towel or cloth around the fitting or connection prior to loosening to absorb any excess fuel due to spillage. Ensure that all fuel spillage (should it occur) is quickly removed from engine surfaces. Ensure that all fuel soaked cloths or towels are deposited into a suitable waste container.

• Always keep a dry chemical (Class B) fire extinguisher near the work area.

• Do not allow fuel spray or fuel vapors to come into contact with a spark or open flame.

• Always use a back-up wrench when loosening and tightening fuel line connection fittings. This will prevent unnecessary stress and torsion to fuel line piping.

• Always replace worn fuel fitting O-rings with new Do not substitute fuel hose or equivalent where fuel pipe is installed.

Before servicing the vehicle, make sure to also refer to the precautions in the beginning of this section as well.

RELIEVING FUEL SYSTEM PRESSURE

With The Consult-III® Tool

1. Before servicing the vehicle, refer to the Precautions Section.
2. Turn the ignition switch **ON**.

3. Perform the "FUEL PRESSURE RELEASE" in "WORK SUPPORT" mode with the CONSULT-III®.
4. Start the engine.
5. After engine stalls, crank it over 2–3 times to release all fuel pressure.
6. Turn the ignition switch **OFF** .

Without The Consult-III® Tool

See Figure 159.

1. Before servicing the vehicle, refer to the Precautions Section.
2. Remove the fuel pump fuse located in IPDM E/R.
3. Start the engine.
4. After the engine stalls, crank it over 2–3 times to release all fuel pressure.
5. Turn the ignition switch **OFF** .
6. Reinstall the fuel pump fuse after servicing the fuel system.

FUEL FILTER

REMOVAL & INSTALLATION

The fuel delivery system integrates the fuel filter with the in-tank fuel pump. To service this filter, remove the fuel pump. Refer to Fuel Pump in the Fuel Systems Section.

FUEL LEVEL SENDING UNIT

REMOVAL & INSTALLATION

See Figures 160 through 162.

The fuel level sending unit detects a fuel level in the fuel tank and transmits a signal to the combination meter. The combination meter sends the fuel level sensor signal to the ECM through CAN communication line.

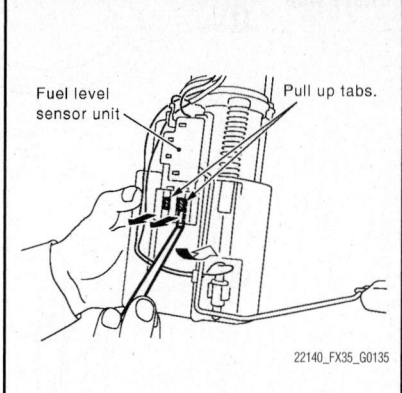

Fig. 160 **Using a suitable tool, pull up the tab points on the fuel level sensor unit to release the lock**

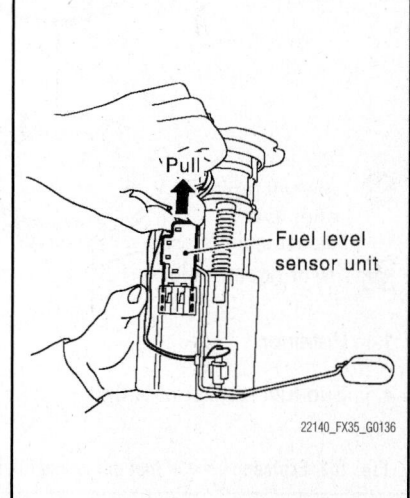

Fig. 161 **Slide the fuel level sensor unit out in the direction shown by the arrow**

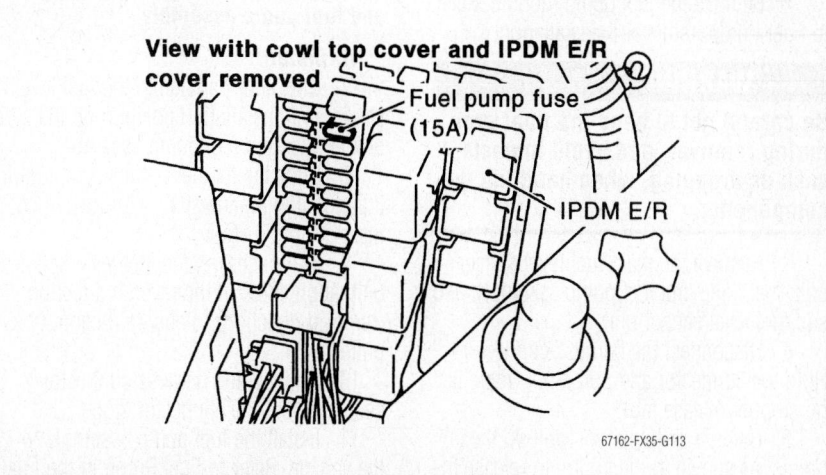

Fig. 159 **Fuel pump fuse location for fuel pressure release**

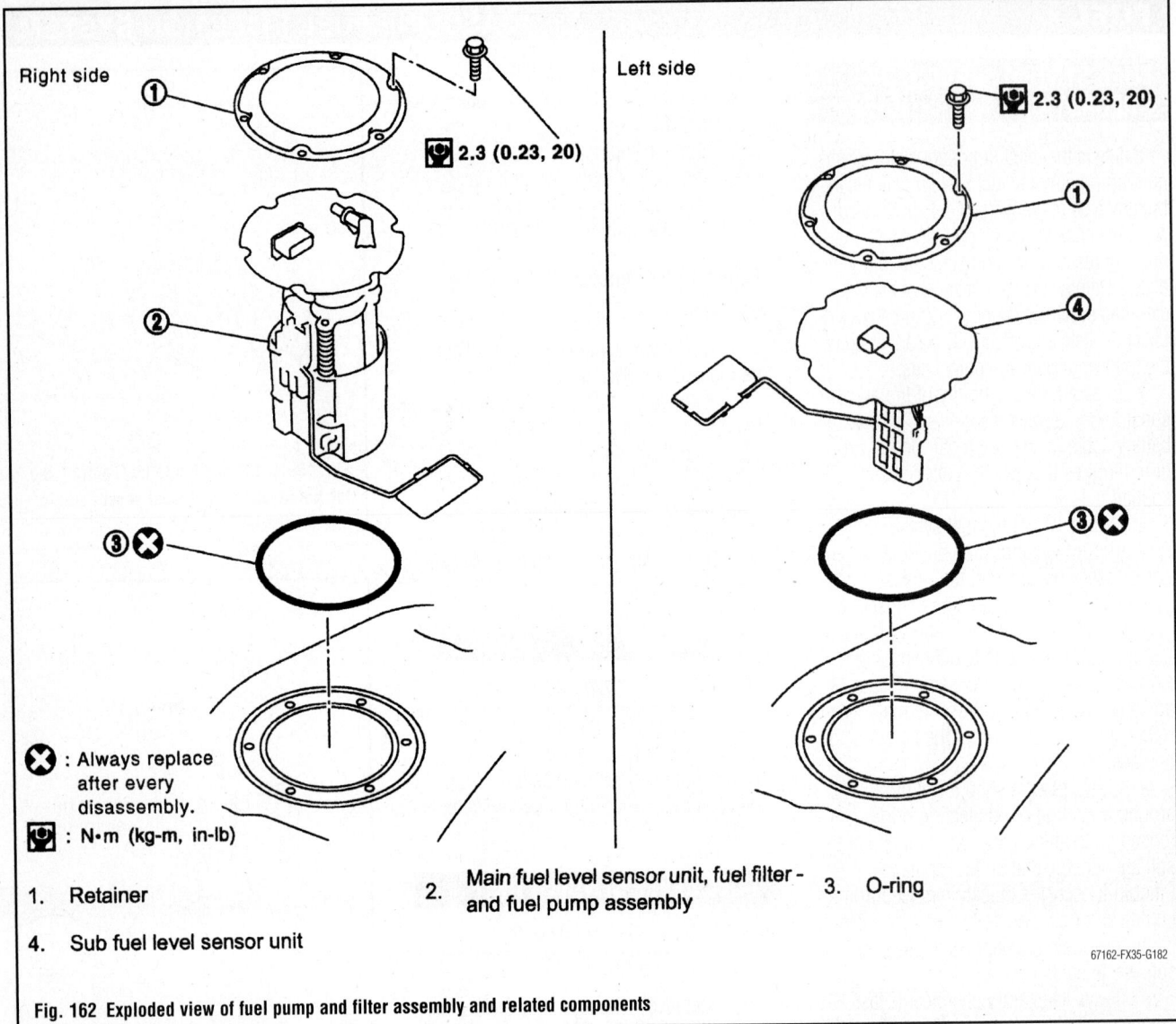

Right side

2.3 (0.23, 20)

Left side

2.3 (0.23, 20)

❌ : Always replace after every disassembly.

⚙ : N•m (kg-m, in-lb)

1. Retainer

2. Main fuel level sensor unit, fuel filter - and fuel pump assembly

3. O-ring

4. Sub fuel level sensor unit

67162-FX35-G182

Fig. 162 Exploded view of fuel pump and filter assembly and related components

This component is removed along with the fuel pump.

1. Before servicing the vehicle, refer to the Precautions Section.

2. Remove the fuel pump module. Refer to Fuel Pump, removal & installation.

✳✳ WARNING

Be careful not to bend the float arm during removal, and avoid impacts, such as dropping, when handling the components.

3. Remove the main fuel level sensor unit, fuel filter, and fuel pump assembly, and sub fuel level sensor unit.

4. Disconnect the harness connector. Hold the connector and pull it out (there is no stopper release tab).

5. Using a suitable tool, pull up the tab points, as shown in the figure, to release the lock. Be careful not to damage it.

6. After the fixing tabs are disengaged, slide the fuel level sensor unit out in the direction shown by the arrow.

➡ **Do not disassemble the fuel filter and fuel pump assembly.**

To install:

7. Check for damage to the fuel level sensor unit installation position on the side of fuel filter and fuel pump assembly.

8. Slide the fuel level sensor unit until it aligns into the installation groove, then insert it until it stops.

9. After inserting the fuel level sensor unit, apply force in the reverse direction (removal direction) to ensure it cannot be pulled out.

10. Connect the harness connector securely until the connector stops.

11. Install the fuel pump assembly to the vehicle. Refer to Fuel Pump in the Fuel Systems Section.

FUEL PRESSURE REGULATOR

REMOVAL & INSTALLATION

The fuel pressure regulator is built in to the fuel pump. Refer to Fuel Pump in the Fuel Systems Section.

FUEL PUMP

REMOVAL & INSTALLATION

See Figures 162 through 164.

1. Before servicing the vehicle, refer to the Precautions Section.

2. Check the fuel level on the fuel gauge. If the fuel gauge indicates full or almost full, drain the fuel from the fuel tank until the gauge indicates a level near ¾ of a tank.

✳✳ CAUTION

Fuel will be spilled when removing the main and sub fuel level sensor

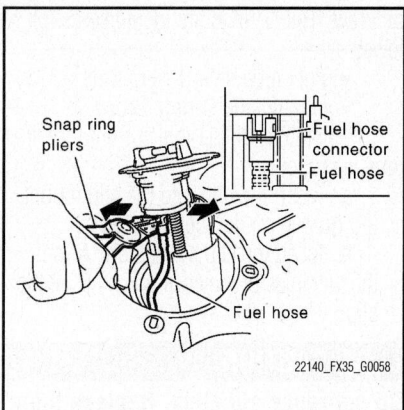

Fig. 163 Raise the fuel pump assembly and using snapring pliers, remove the fuel hose connector

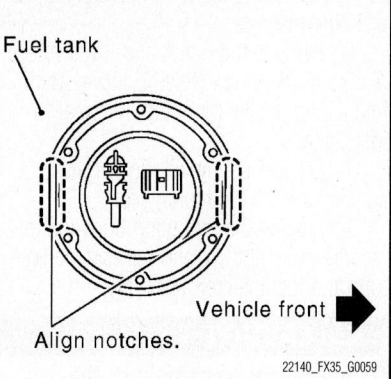

Fig. 164 Install the fuel pump retainer so that its notch becomes parallel with the notch on the fuel tank

units if the level of the fuel in the tank is higher than about ¾ of a tank.

3. In the case that the fuel pump does not operate, perform the following procedure:

a. Insert a hose of less than 1 in. (25mm) in diameter into the fuel filler tube through the fuel filler opening to draw fuel from the fuel filler tube.

b. Disconnect the fuel filler hose from the fuel filler tube.

c. Insert the fuel tube into the fuel tank through the fuel filler hose to draw the fuel from the fuel tank.

4. Release the fuel pressure from the fuel lines.

5. Open the fuel filler lid.

6. Open the filler cap and release the pressure inside the fuel tank.

7. Remove the rear seat cushion, as follows:

a. Pull the lock at the front bottom of the seat cushion forward (1 for each side).

b. Pull the seat cushion upward to release the retaining wire from the plastic hook.

c. Pull the seat cushion forward to remove.

8. Lift up the floor carpet, then remove the inspection hole cover for the main and sub fuel level sensor units by turning the retaining clips clockwise by 90°.

9. Disconnect the wiring harness connector and fuel feed tube.

10. Disconnect the fuel line quick connector, as follows:

a. Hold the sides of the connector, push in the tabs and pull out the tube.

➥**If the quick connector sticks to the tube of the main fuel level sensor unit,**

push and pull the quick connector several times until they start to move. Then, disconnect them by pulling.

When dealing with the fuel line quick connector, heed the following:

• The quick connector can be disconnected when the tabs are completely depressed. Do not twist it more than necessary

• Do not use any tools to disconnect the quick connector

• Keep the resin tube away from heat. Be especially careful when welding near the resin tube

• Prevent acidic liquid such as battery electrolyte from getting on the resin tube

• Do not bend or twist the resin tube during connection and disconnection

• Do not remove the remaining retainer on the hard tube (or the equivalent) except when the resin tube or retainer is replaced

• When the resin tube or hard tube (or the equivalent) is replaced, also replace the retainer with a new one

• To keep the connecting portion clean and to avoid damage and foreign materials, cover them completely with plastic bags or a similar material.

✳✳ **WARNING**

Make sure to not bend the float arm during removal, and avoid impacts, such as falling, when handling components.

11. Remove the main fuel level sensor unit, fuel filter, and fuel pump assembly, and sub fuel level sensor unit, as follows:

a. Remove the main fuel sensor unit retainer.

b. Raise the main fuel level sensor

unit, fuel filter and fuel pump assembly, and using snapring pliers, remove the fuel hose connector.

✳✳ **WARNING**

Be careful not to damage the fuel hose connector by expanding it excessively.

12. Removal of sub fuel level sensor unit:

a. Remove the sub fuel level sensor unit retainer.

b. Raise and release the sub fuel level sensor unit.

To install:

13. Installation is the reverse of removal.

14. When installing the fuel hose connectors insert them fully until a click sound of full stopper engagement is heard.

15. Install the fuel pump retainer so that its notch becomes parallel with the notch on the fuel tank.

16. Tighten the retainer mounting bolts evenly to 20 inch lbs. (2 Nm).

FUEL RAIL & INJECTORS

REMOVAL & INSTALLATION

See Figures 165 through 171.

1. Before servicing the vehicle, refer to the Precautions Section.

2. Remove the engine cover.

3. Relieve the fuel pressure.

4. Remove the fuel feed hose (with damper) from the fuel sub-tube.

➥**There is no fuel return route.**

✳✳ **CAUTION**

While the hoses are disconnected, plug them to prevent fuel from draining. Also, do not separate the damper and hose.

5. When separating the fuel feed hose (with damper) and the centralized underfloor piping connection, disconnect the quick connector, as follows:

a. Remove the quick connector cap from the quick connector connection on the right member side.

b. Disconnect the fuel feed hose (with damper) from the bracket hose clamp.

➥**Disconnect the quick connector by using quick connector release tool SST: J-45488, or equivalent.**

c. With the sleeve side of the quick connector release facing the quick con-

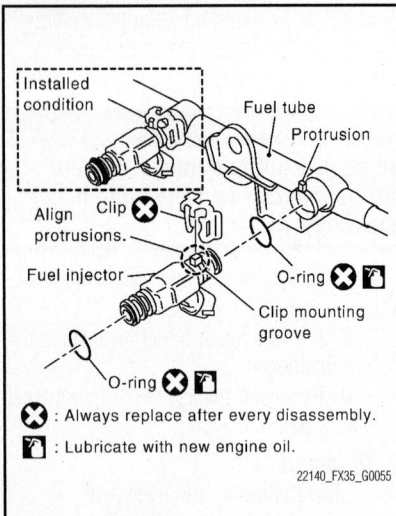

Fig. 165 Removing the fuel injector from the fuel rail (tube)

nector, install the quick connector release onto the centralized under-floor piping.

d. Insert the quick connector release into the quick connector until the sleeve contacts and goes no further. Hold the quick connector release at that position.

> **⁂ CAUTION**
>
> **Inserting the quick connector release hard will not disconnect the quick connector. Hold the quick connector release where it contacts and goes no further.**

e. Draw and pull out the quick connector straight from the centralized under-floor piping.

When disconnecting the fuel line, heed the following:

• Pull the quick connector holding **A** position as shown in the figure. Do not pull it with lateral force applied. The O-ring inside the quick connector may be damaged.

• Prepare a container and cloth beforehand as fuel will leak out

• Avoid fire and sparks

• Keep parts away from all heat sources. Especially, be careful when welding is performed

• Do not expose the parts to battery electrolyte or other acids

• Do not bend or twist the connection between the quick connector and the fuel feed hose (with damper) during installation/removal

• To keep the connecting portion clean and to avoid damage from foreign materials,

cover them completely with plastic bags or a similar material

6. Remove the upper and lower intake manifold collectors. Refer to Intake Manifold, Upper Intake Manifold, removal & installation.

7. Disconnect the wiring harness connector from the fuel injector.

8. Loosen the mounting bolts in the reverse order shown, and remove the fuel tube and fuel injector assembly.

> **⁂ CAUTION**
>
> **Do not tilt the assembly or the remaining fuel may leak.**

9. Remove the fuel injectors from the fuel rail (tube) with the following procedure:

a. Open and remove the clip.

b. Remove the fuel injector from the fuel tube by pulling straight.

During injector removal, heed the following items:

• Be careful with the remaining fuel that leaks from the fuel tube

• Be careful not to damage the injector nozzles during removal

• Do not bump or drop the fuel injectors

• Do not disassemble the fuel injectors

10. Remove the fuel sub-tube and fuel damper.

To install:

When handling all O-rings in this procedure, heed the following:

• Handle the O-ring with bare hands. Never wear gloves

• Lubricate the O-ring with new engine oil

• Do not clean the O-ring with solvent

• Make sure that the O-ring and its mating part are free of foreign material

• When installing the O-ring, be careful not to scratch it with a tool or fingernails

• Be careful not to twist or stretch the O-ring

• Insert the O-ring straight into the fuel tube

11. Install the fuel damper and fuel sub-tube.

12. Insert the fuel damper and fuel sub-tube straight into the fuel tube.

13. Tighten the mounting bolts.

14. After tightening the mounting bolts, make sure that there is no gap between the flange and fuel tube.

15. Install O-rings onto the fuel injector—the upper and lower O-rings are

different. The O-rings are identified as follows:

• Fuel tube side O-ring: Blue
• Nozzle side O-ring: Brown

16. Install each fuel injector onto the fuel tube, as follows:

a. Insert the clip into the clip mounting groove on the fuel injector.

b. Insert the clip so that lug "A" of the fuel injector matches notch "A" of the clip.

> **⁂ CAUTION**
>
> **Do not reuse old clips. Replace them with new ones. Be careful to keep the clip from interfering with the O-ring. If interference occurs, replace the O-ring with a new one.**

c. Insert the fuel injector into the fuel tube, matching it to the axial center, with the clip attached. Insert the fuel injector so that lug "B" of fuel tube matches notch " the clip. Make sure that the fuel tube flange is securely fixed in the groove on the clip.

d. Make sure that installation is complete by checking that the fuel injector does not rotate or come off.

17. Install the fuel tube and fuel injector assembly onto the intake manifold, and tighten the mounting bolts in 2 steps as shown.

a. 1st Step: 84 inch lbs. (10 Nm).
b. 2nd Step: 17 ft. lbs. (24 Nm).

> **⁂ WARNING**
>
> **Be careful not to let the tips of the injector nozzles come into contact with other parts.**

18. Connect the injector sub-wiring harness.

19. Install the upper and lower intake manifold collectors. Refer to Intake Manifold, Upper Intake Manifold, removal & installation.

20. Install the fuel sub-tube on the rear end of the lower intake manifold collector.

21. Connect the fuel feed hose and damper. After tightening the mounting bolts, make sure that there is no gap between the flange and the fuel sub-tube.

22. Connect the quick connector between the fuel feed hose and centralized under-floor piping connection with the following procedure:

a. Check the connection for damage and foreign materials.

b. Align the connector with the tube, then insert the connector straight into the tube until a click is heard.

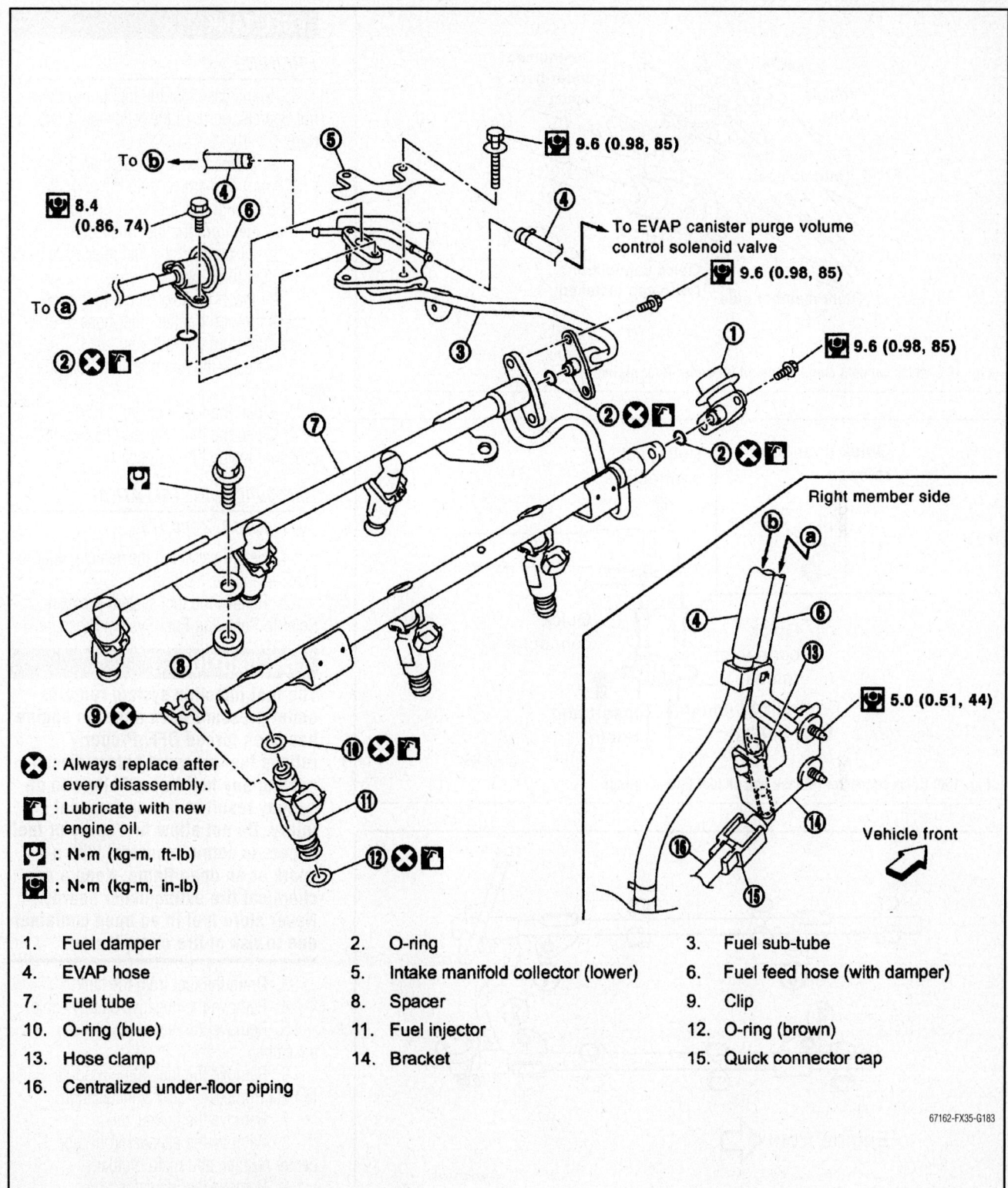

8.4 (0.86, 74)

To ⓑ

To ⓐ

9.6 (0.98, 85)

To EVAP canister purge volume control solenoid valve

9.6 (0.98, 85)

9.6 (0.98, 85)

Right member side

5.0 (0.51, 44)

Vehicle front

❌ : Always replace after every disassembly.

🛢 : Lubricate with new engine oil.

🔧 : N•m (kg-m, ft-lb)

🔧 : N•m (kg-m, in-lb)

1. Fuel damper	2. O-ring	3. Fuel sub-tube
4. EVAP hose	5. Intake manifold collector (lower)	6. Fuel feed hose (with damper)
7. Fuel tube	8. Spacer	9. Clip
10. O-ring (blue)	11. Fuel injector	12. O-ring (brown)
13. Hose clamp	14. Bracket	15. Quick connector cap
16. Centralized under-floor piping		

67162-FX35-G183

Fig. 166 Exploded view of the fuel injector and rail assembly

c. After connecting the quick connector, visually confirm that the 2 retainer tabs are connected to the connector, then pull the tube and connector to make sure they are securely connected.

d. Install quick connector cap to quick connector connection.

e. Install the quick connector cap with the arrow on the surface facing in the direction of the quick connector (fuel feed hose side).

➡**If the cap cannot be installed smoothly, the quick connector may not**

have been installed correctly. Check the connection again.

f. Secure the fuel feed hose to the clamp.

23. The remainder of installation is the reverse of removal.

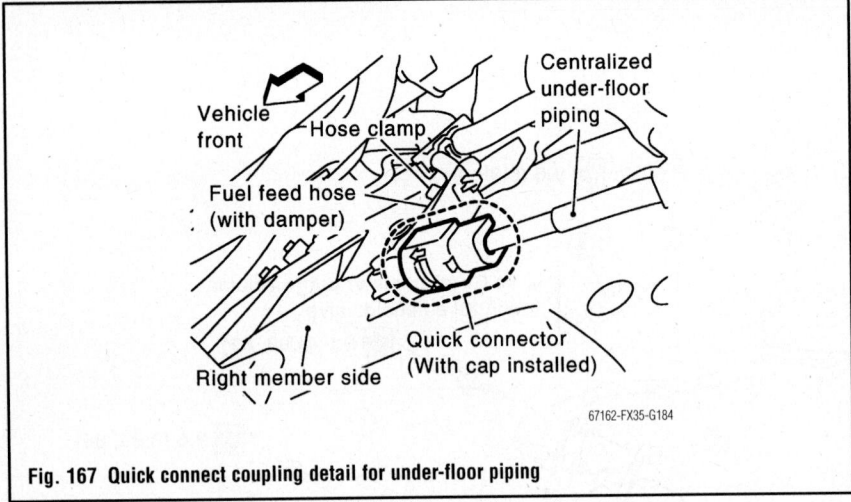

Fig. 167 Quick connect coupling detail for under-floor piping

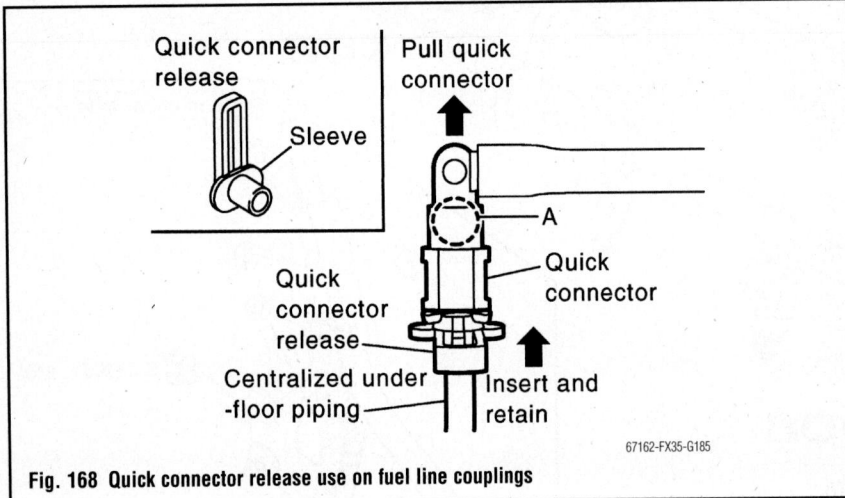

Fig. 168 Quick connector release use on fuel line couplings

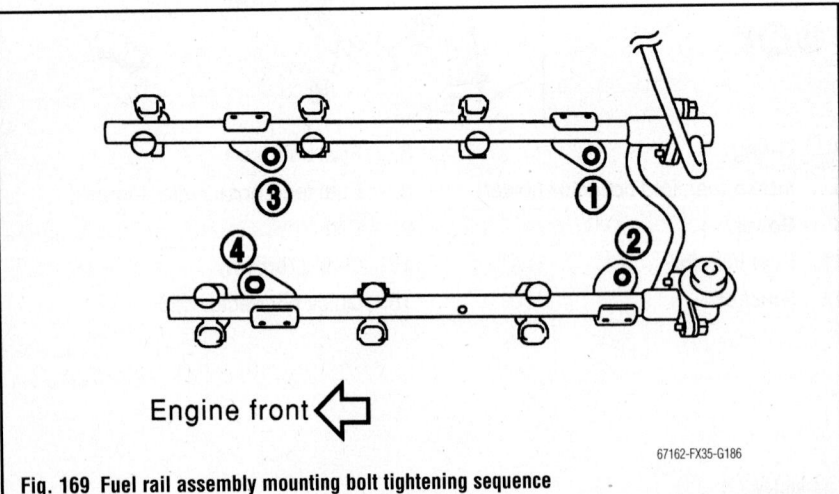

Fig. 169 Fuel rail assembly mounting bolt tightening sequence

24. Perform the following once installation is complete.

a. After installing the fuel tubes, make sure there is no fuel leakage at the connections.

b. Apply fuel pressure to the fuel lines by turning the ignition switch **ON** with the engine **OFF**. Then check for fuel leaks at all connections.

c. Start the engine, and while holding it at a high RPM, check for fuel leaks at all connections.

➡Use mirrors to check hard-to-see connections.

FUEL TANK

DRAINING

1. In the case that the fuel pump does not operate, perform the following procedure:

a. Insert a hose of less than 1 in. (25mm) in diameter into the fuel filler tube through the fuel filler opening to draw fuel from the fuel filler tube.

b. Disconnect the fuel filler hose from the fuel filler tube.

c. Insert the fuel tube into the fuel tank through the fuel filler hose to draw the fuel from the fuel tank.

2. Release the fuel pressure from the fuel lines.

3. Open the fuel filler lid.

4. Open the filler cap and release the pressure inside the fuel tank.

REMOVAL & INSTALLATION

See Figures 172 and 173.

1. Before servicing the vehicle, refer to the Precautions Section.

2. Relieve the fuel system pressure. Refer to Relieving Fuel System Pressure.

✴✴ CAUTION

The fuel injection system remains under pressure even after the engine has been turned OFF. Properly relieve fuel pressure before disconnecting any fuel lines. Failure to do so may result in fire or personal injury. Do not allow fuel spray or fuel vapors to come in contact with a spark or an open flame. Keep a dry chemical fire extinguisher nearby. Never store fuel in an open container due to risk of fire or explosion.

3. Drain the fuel from the tank.
4. Remove the negative battery cable.
5. Remove the rear seat cushion assembly.
6. Remove the fuel pump module. Refer to Fuel Pump, removal & installation.
7. Remove the tunnel stay.
8. Remove the exhaust front tube, center muffler, and main muffler.
9. Remove the insulator.
10. Remove the propeller shaft.
11. Remove the parking rear brake cables.
12. Compress the coil springs and remove them.
13. Remove the rear suspension assembly.
14. Remove the fuel tank protector.
15. Disconnect the fuel filler hose, vent

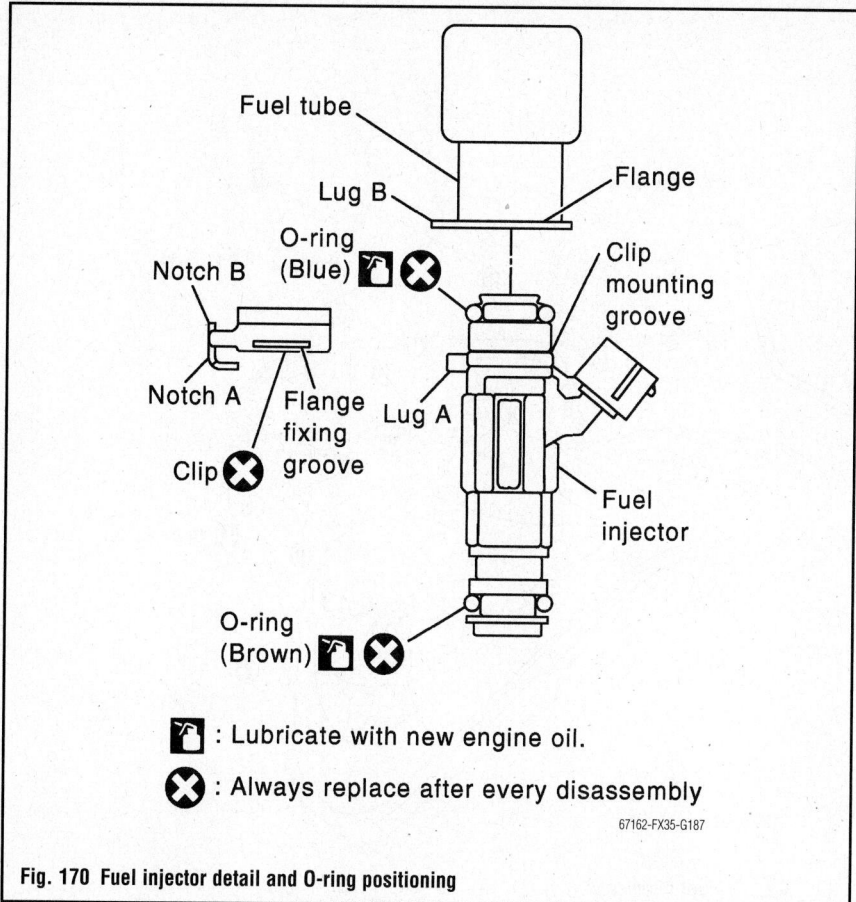

: Lubricate with new engine oil.

: Always replace after every disassembly

67162-FX35-G187

Fig. 170 Fuel injector detail and O-ring positioning

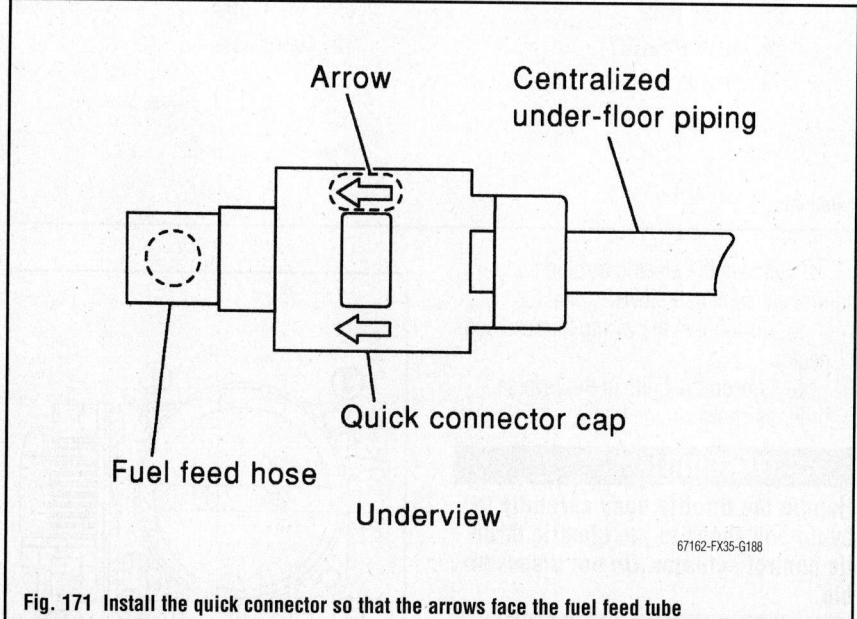

67162-FX35-G188

Fig. 171 Install the quick connector so that the arrows face the fuel feed tube

hose, and EVAP hoses at the fuel tank side.

16. Support the lower part of fuel tank with a transmission jack.

➡**Support the fuel tank in a position that fuel tank mounting bands do not engage.**

17. Remove the fuel tank mounting bands.

18. Help support the fuel tank and lower the transmission jack carefully to remove the fuel tank.

➡**Make sure that all the connection points have been disconnected. Con-**

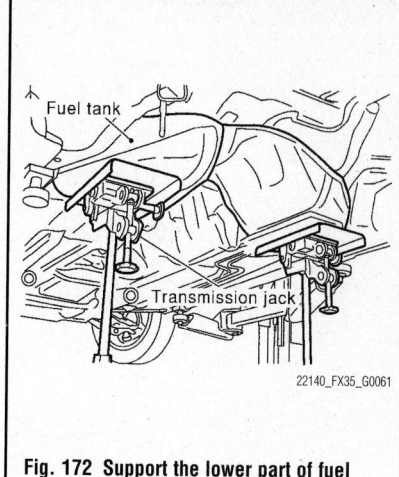

22140_FX35_G0061

Fig. 172 Support the lower part of fuel tank with a transmission jack

firm there is no interference with the vehicle while lowering the fuel tank.

19. Remove fuel filler tube protector and fuel filler tube, if necessary.

To install:

20. Installation is the reverse of the removal procedure.

21. Tighten the retaining bolts and fasteners to specification as illustrated.

22. Perform the following once installation is complete.

a. After installing the fuel tubes, make sure there is no fuel leakage at the connections.

b. Apply fuel pressure to the fuel lines by turning the ignition switch **ON** with the engine **OFF**. Then check for fuel leaks at all connections.

c. Start the engine, and while holding it at a high RPM, check for fuel leaks at all connections.

➡**Use mirrors to check hard-to-see connections.**

THROTTLE BODY

REMOVAL & INSTALLATION

See Figures 174 and 175.

1. Before servicing the vehicle, refer to the Precautions Section.

2. Disconnect the negative battery cable.

3. Remove the engine cover.

4. Disconnect and plug the water hoses from the intake manifold collector (upper).

➡**Do not spill engine coolant on the drive belts.**

5. Remove the air cleaner case and air duct, as follows:

a. Remove the air duct (inlet).

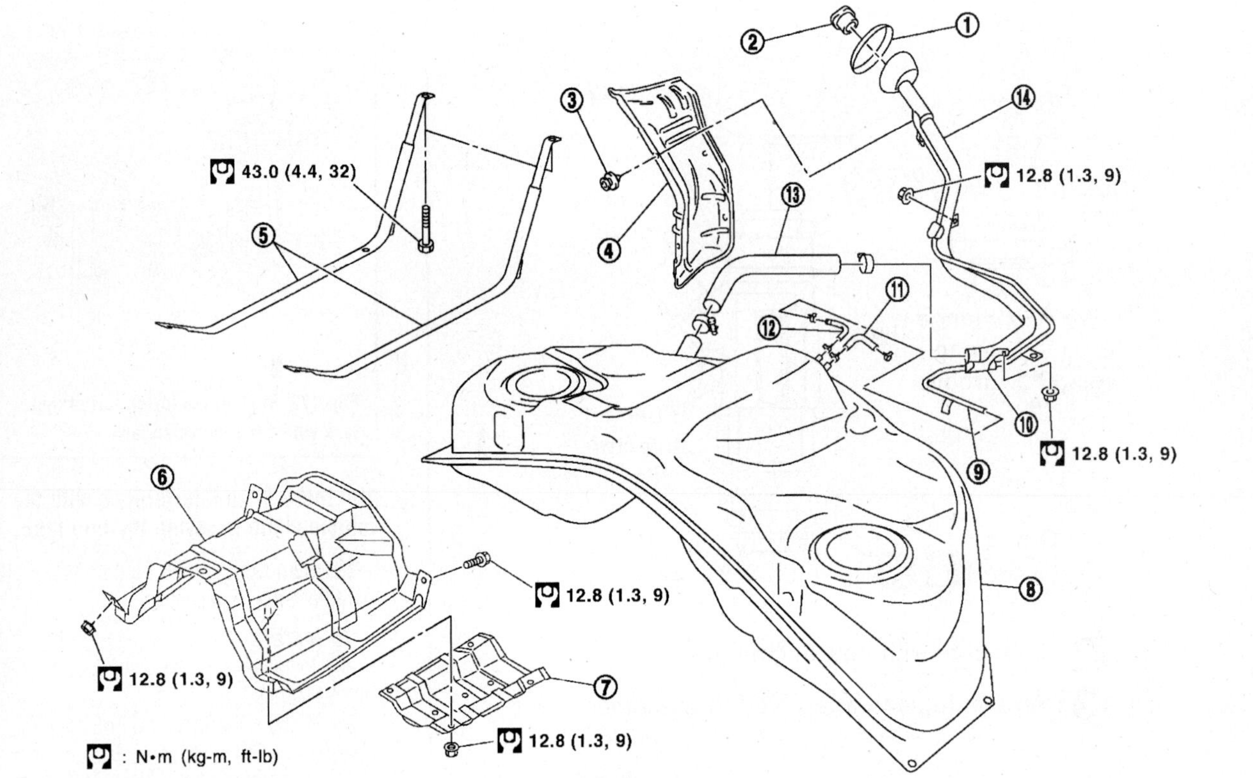

43.0 (4.4, 32)

12.8 (1.3, 9)

12.8 (1.3, 9)

12.8 (1.3, 9)

12.8 (1.3, 9)

12.8 (1.3, 9)

: N•m (kg-m, ft-lb)

1.	Grommet	2.	Fuel filler cap	3.	Clip
4.	Fuel filler tube protector	5.	Fuel tank mounting band	6.	Fuel tank protector
7.	Insulator	8.	Fuel tank	9.	Vent tube
10.	Vent hose	11.	EVAP hose	12.	Vent hose
13.	Fuel filler hose	14.	Fuel filler tube		

22140_FX35_G0060

Fig. 173 Exploded view of fuel tank and related components

b. Disconnect the mass air flow sensor wiring harness connector.

c. Remove the air cleaner case/mass air flow sensor assembly and the air duct/resonator assembly disconnecting them at the joints.

➡**Add match marks as necessary for easier installation.**

d. Remove the mass air flow sensor from air cleaner case.

❊❊ WARNING

Handle the mass air flow sensor with care. Do not expose it to harsh vibration or shock. Do not disassemble it or touch its sensor.

e. Remove the resonator in the fender, lifting the left fender protector.

6. Remove the electric throttle body control actuator, as follows:

a. Disconnect the wiring harness connector.

b. Loosen the bolts in the reverse order as shown in the figure.

❊❊ WARNING

Handle the throttle body carefully to avoid any shock to the electric throttle control actuator. Do not disassemble.

To install:

7. Install the electric throttle control actuator. Tighten the bolts to 75 inch lbs. (9 Nm).

8. Install the mass air flow sensor in the air cleaner case.

9. Install the air cleaner case and air duct.

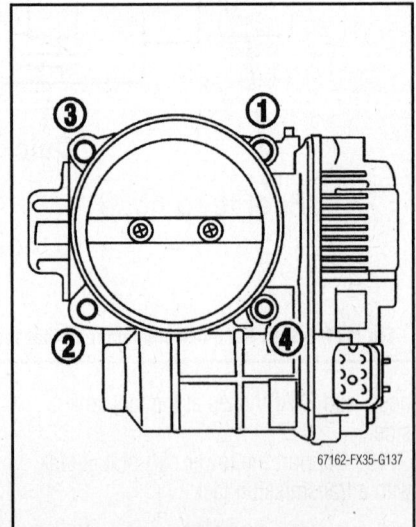

67162-FX35-G137

Fig. 174 Throttle body mounting bolt tightening sequence

10. Reconnect the hoses to the intake manifold collector (upper).
11. Install the engine cover.
12. Connect the negative battery cable.
13. Perform the following drivability adjustments:
14. Perform the "Throttle Valve Closed Position Learning" procedure (below) when the wiring harness connector of the electric throttle control actuator is disconnected, or

perform the "Idle Air Volume Learning" and "Throttle Valve Closed Position Learning" procedures (below) when the electric throttle control actuator is replaced.

Throttle Valve Closed Position Learning

1. Before servicing the vehicle, refer to the Precautions Section.
The Throttle Valve Closed Position

Learning procedure is an operation for the ECM to relearn the fully closed position of the throttle valve by monitoring the throttle position sensor output signal. It must be performed each time the wiring harness connector of the electric throttle control actuator or ECM is disconnected.

2. Make sure that accelerator pedal is fully released.
3. Turn ignition switch **ON**.

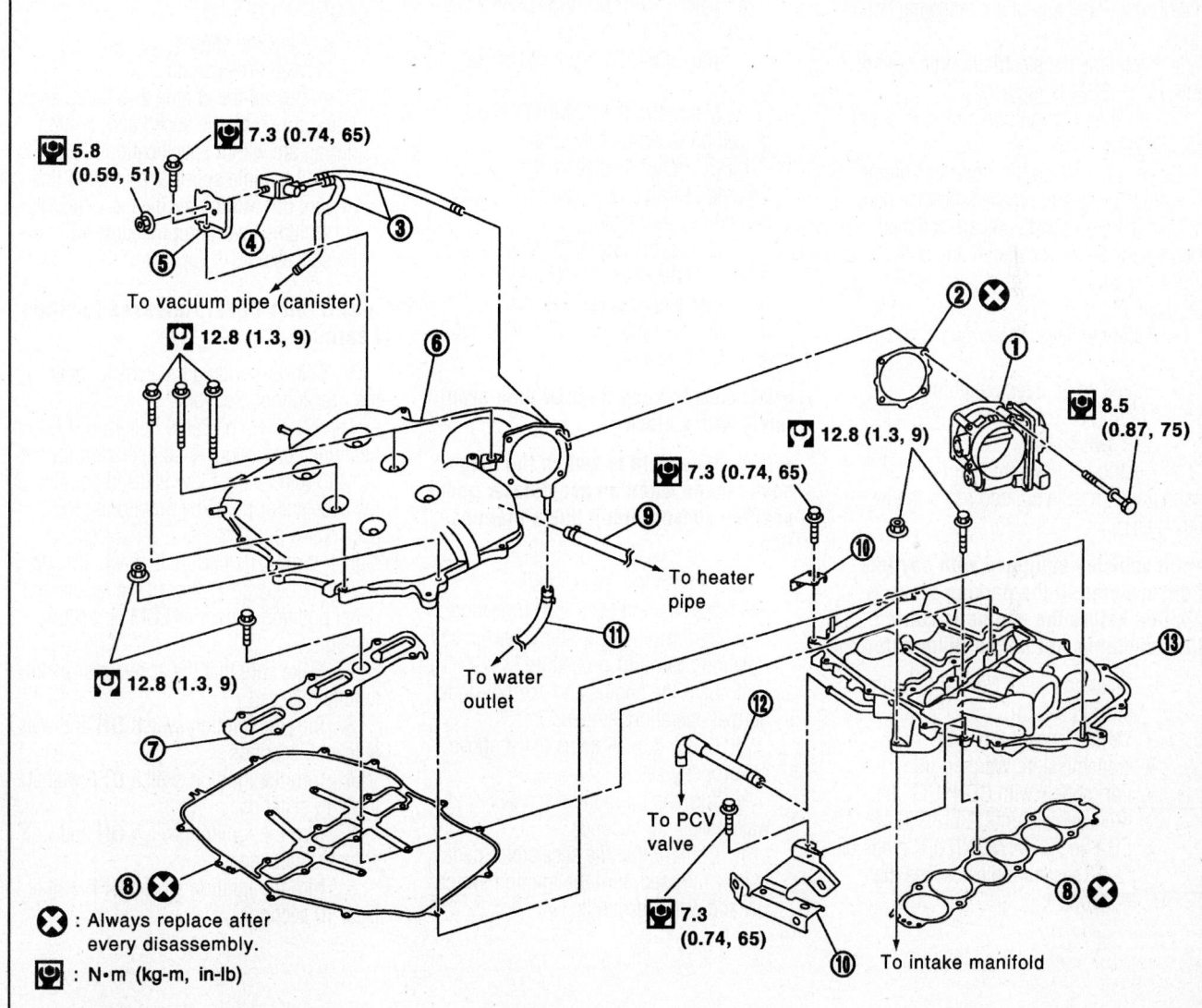

1. Electric throttle control actuator
2. Gasket
3. Vacuum hose
4. EVAP canister purge volume control solenoid valve
5. Bracket
6. Intake manifold collector (upper)
7. Intake manifold collector cover
8. Gasket
9. Water hose
10. Bracket
11. Water hose
12. PCV hose
13. Intake manifold collector (lower)

Fig. 175 Exploded view of the upper and lower halves of the upper intake manifold

67162-FX35-G136

4. Turn ignition switch **OFF** wait at least 10 seconds. Make sure that throttle valve moves during the above 10 seconds by confirming the operating sound.

Idle Air Volume Learning

1. Before servicing the vehicle, refer to the Precautions Section.

Idle Air Volume Learning is an operation to learn the idle air volume that keeps each engine within the specific range. It must be performed under any of the following conditions:

- Each time the electric throttle control actuator or ECM is replaced
- Idle speed or ignition timing is out of specification

Before performing the "Idle Air Volume Learning" procedure, make sure that all of the following conditions are satisfied. Learning will be cancelled if any of the following conditions are missed for even a moment.

- Battery voltage: More than 12.9 volts (at idle)
- Engine coolant temperature: 158–212°F (70–100°C)
- PNP switch: ON
- Electric load switch: OFF (air conditioner, headlamp, and rear window defogger)

➡**On vehicles equipped with daytime light systems, if the parking brake is applied before the engine is started, the headlamp will not be illuminated.**

- Steering wheel: Neutral (straight-ahead position)
- Vehicle speed: Stopped
- Transmission: Warmed-up
- For models with CONSULT-III®, drive vehicle until "FLUID TEMP SE 1"in "DATA MONITOR" mode of "A/T" system indicates less than 0.9 volts.

- For models without CONSULT-III®, drive vehicle for 10 minutes.

2. If using the CONSULT-III® tool, perform the following:

a. Perform the "Accelerator Pedal Released Position Learning" procedure.

b. Perform the "Throttle Valve Closed Position Learning" procedure.

c. Start the engine and warm it up to normal operating temperature.

d. Check that all items listed above are properly set.

e. Select "IDLE AIR VOL LEARN" in "WORK SUPPORT" mode.

f. Touch "START" and wait 20 seconds.

g. Make sure that "CMPLT" is displayed on CONSULT-III®screen. If "CMPLT" is not displayed, the Idle Air Volume Learning procedure will not be carried out successfully.

h. Rev up the engine 2–3 times and make sure that idle speed and ignition timing are within specifications.

3. If NOT using the CONSULT-III® tool, perform the following:

➡**It is best to keep track of time accurately with a clock.**

➡**It is impossible to switch the diagnostic mode when an accelerator pedal position sensor circuit has a malfunction.**

a. Perform the "Accelerator Pedal Released Position Learning" procedure.

b. Perform the "Throttle Valve Closed Position Learning" procedure.

c. Start the engine and warm it up to normal operating temperature.

d. Check that all items listed above are properly set.

e. Turn the ignition switch OFF and wait at least 10 seconds.

f. Confirm that the accelerator pedal is fully released, turn the ignition switch ON and wait 3 seconds.

➡**Repeat the following 2 steps quickly 5 times within 5 seconds.**

g. Fully depress the accelerator pedal.

h. Fully release the accelerator pedal.

i. Wait 7 seconds, fully depress the accelerator pedal and keep it for approx. 20 seconds until the MIL stops blinking and remains ON.

j. Fully release the accelerator pedal within 3 seconds after the MIL turned ON.

k. Start the engine and let it idle.

l. Wait 20 seconds.

m. Rev up the engine 2–3 times and make sure that idle speed and ignition timing are within specifications.

n. If the idle speed and ignition timing are not within specification, the Idle Air Volume Learning procedure will not be successful.

Accelerator Pedal Released Position Learning

1. Before servicing the vehicle, refer to the Precautions Section.

The "Accelerator Pedal Released Position Learning" procedure is an operation for the ECM to relearn the fully released position of the accelerator pedal by monitoring the accelerator pedal position sensor output signal. It must be performed each time the wiring harness connector of the accelerator pedal position sensor or ECM is disconnected.

2. Make sure that the accelerator pedal is fully released.

3. Turn the ignition switch **ON** and wait at least 2 seconds.

4. Turn the ignition switch **OFF** wait at least 10 seconds.

5. Turn the ignition switch **ON** and wait at least 2 seconds.

6. Turn the ignition switch **OFF** wait at least 10 seconds.

HEATING & AIR CONDITIONING SYSTEM

BLOWER MOTOR

REMOVAL & INSTALLATION

See Figures 176 and 177.

❈❈ CAUTION

Before servicing, or working around, the SRS system, turn the ignition switch OFF, disconnect both battery cables and wait at least 3 minutes. When servicing, or working around, the SRS system, do not work directly in front of the air bag module.

1. Before servicing the vehicle, refer to the Precautions Section.
2. Disconnect the negative battery cable.
3. Remove the lower instrument panel, passenger side.
4. Disconnect the blower motor electrical connector.
5. Remove the blower motor retaining screws.
6. Remove the blower motor from its mounting.

To install:

7. Installation is the reverse of the removal procedure.

HEATER CORE

REMOVAL & INSTALLATION

See Figures 177 through 182.

1. Before servicing the vehicle, refer to the Precautions Section.
2. Use approved refrigerant collecting equipment to discharge refrigerant.

❈❈ CAUTION

Make sure the engine is cold before draining the coolant.

3. Drain the coolant from the cooling system.
4. Remove the cowl top cover.
5. Remove the 2 high-pressure pipe mounting clips.
6. Remove the low-pressure flexible hose bracket mounting bolts.
7. Disconnect the evaporator-side one touch joint, as follows:

   a. Set disconnect SST: 9253089908 (high-pressure side) and SST: 9253089916 (low-pressure side) on the A/C piping.

   b. Slide the disconnect tool toward the front of the vehicle until it clicks.

   c. Slide the A/C pipe toward the front of the vehicle front and disconnect it.

➡**Seal the connection opening of the pipe with a cap or vinyl tape to avoid exposure to atmosphere.**

8. Remove the electronic control throttle assembly.

9. Disconnect the 2 heater hoses from the heater core.
10. Remove the instrument panel assembly, as follows:

   a. Remove the front kicking plate on both sides of the vehicle.

   b. Remove dash side finisher plastic nuts, then remove the dash side finisher.

   c. Pull to the inside of the vehicle, disengage the metal clips and remove the front pillar garnish.

   d. To remove the A/T Select Lever Knob, pull down the knob cover. Remove the lock-pin of the select lever knob. Then, lift up the select lever knob and remove it.

   e. Insert a remover into the side between the gaps of the instrument clock finisher and pull back to the side.

   f. Disconnect the clips and wiring harness connector, then remove the instrument clock finisher.

   g. Insert a remover into the side between the gaps of the A/T console finisher and remove it by lifting the A/T console finisher.

   h. Disconnect the wiring harness connector.

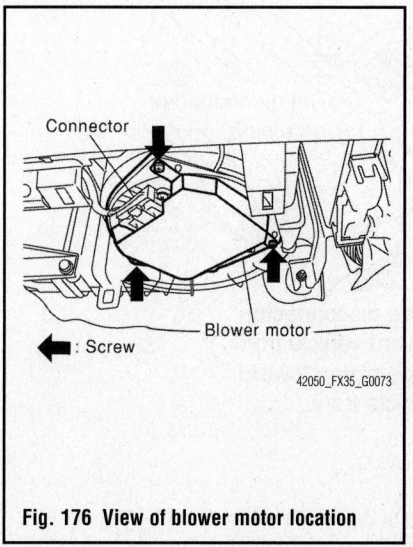

42050_FX35_G0073

Fig. 176 View of blower motor location

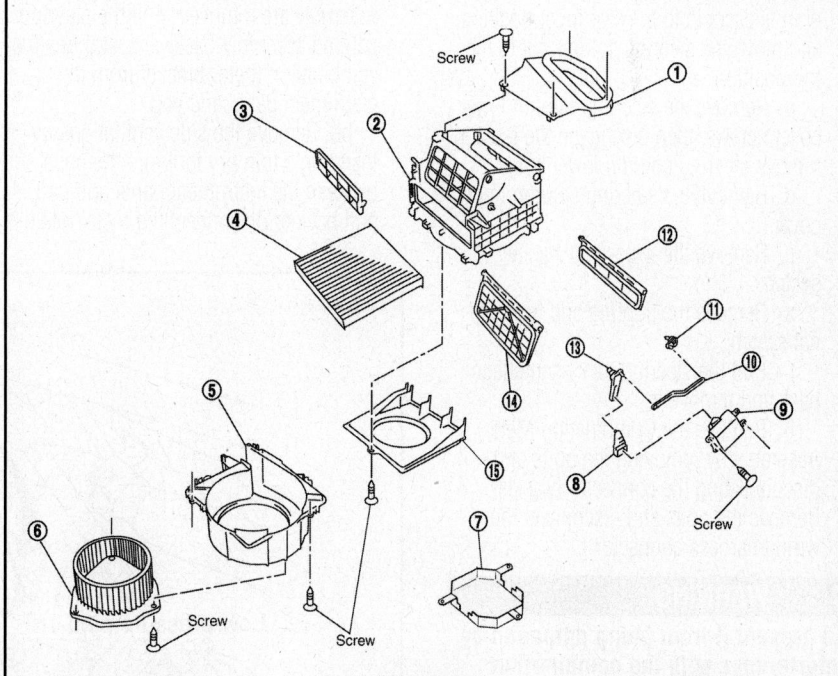

1.	Adapter	2.	Intake upper case	3.	Filter cover
4.	In-cabin microfilter	5.	Intake lower case	6.	Blower motor assembly
7.	Motor cover	8.	Intake door lever 2	9.	Intake door motor
10.	Intake door link	11.	Intake door lever 3	12.	Intake door 2
13.	Intake door lever 1	14.	Intake door 1	15.	Intake bell mouth

42050_FX35_G0072

Fig. 177 Blower unit and related components

i. Remove the console finisher screws.

j. Remove the console finishers.

k. To remove the center console, remove the mounting screws, then remove the console sub-wiring harness.

✳✳ WARNING

When removing console, be careful not to pull the wiring harness.

l. Remove the instrument lower cover by pulling down on the front instrument lower cover and disconnecting clips. Pull it horizontally, and remove it from the lower cover pawls.

m. Remove the instrument passenger lower panel screws, disconnect the wiring harness connector, and remove the lower panel.

n. Remove the instrument driver lower panel bolt and screws, detach the data link connector, pull to disengage the clip and pawl by removing panel in a horizontal direction. Then, disconnect the in-vehicle sensor and all electrical parts. Remove the grommet and remove the hood lock cable.

o. Remove the steering column front lower cover screw, disengage the tab, and then remove the steering column front lower cover. Move the steering column telescopic to the rear most position, and move the steering column tilt to the top position.

p. Remove the steering column lower cover screws, then disengage the tab and remove steering column lower cover.

q. Remove the steering column upper cover.

r. Remove the wiper and washer switch.

s. Remove the lighting and turn signal switch.

t. Pull the steering lock escutcheon back and remove it.

u. Remove the Combination Meter Assembly by removing the bolts and disconnecting the connector bracket. Remove the bolts and disconnect the wiring harness connector.

✳✳ WARNING

To prevent it from being damaged by interference with the combination meter assembly, protect the combination meter assembly with cloths.

v. Remove the instrument panel side panel screws, then pull the panels to the side, disconnect the clip and pawls, and remove the instrument side panels.

Perform for both right-hand and left-hand panels.

w. To remove the cluster lid, insert a pry tool into the gap between the instrument panel and pad, pull back towards you, and disconnect the metal clips. Then, disconnect the wiring harness connectors, and remove the cluster lid.

➡ **Cover surroundings with cloth to avoid making scratches or causing damage.**

x. Remove the display unit and audio unit by removing the screws, disconnecting the wiring harness connector, and removing the display unit and audio unit.

✳✳ CAUTION

The unit is heavy, so be careful not to pinch your fingers when working.

y. Insert a thin pry tool into the gaps between the front defroster grille and instrument panel and pad, lift the front defroster grille upward, and remove the front defroster grille. Perform this task for both right-hand and left-hand grilles.

z. Remove the combination meter bracket bolts and remove the bracket from the vehicle.

aa. Once the mounting bolts of the wiring harness clip and steering column assembly are removed, pull the steering column assembly backward, and free the combination meter bracket from the instrument panel and pad.

bb. Remove the side ventilations by inserting a thin pry tool into the gaps between the instrument panel and pad, pull back to disconnect the metal retain-

ing clips. Then, disconnect the door mirror switch wiring harness connectors, and remove the side ventilations.

cc. Remove the instrument panel and pad by removing the bolts and screws. Then, remove the front passenger air bag module, disconnect the wiring harness connectors, and remove the instrument panel and pad from the passenger door opening.

11. Remove the blower unit, as follows:

a. Remove the ECM with the bracket attached.

b. Disconnect the intake door motor connector and blower fan motor connector.

c. Remove the wiring harness clip from the blower unit.

d. Remove the mounting bolt and screws from the blower unit.

✳✳ CAUTION

Move the blower unit rightward, and remove the locating pin and joint. Then, remove the blower unit downward.

e. Remove the blower unit.

12. Remove the instrument stays (driver-side and passenger-side).

13. Remove the mounting bolts from the heater and cooling unit.

14. Disconnect the drain hose.

15. Remove the ventilator ducts, defroster nozzle and ducts.

16. Remove the steering member mounting bolts, nut and wiring harness clips.

17. Remove the steering member.

18. Remove the mounting screws and then remove the heater pipe cover.

19. Remove the heater pipe bracket.

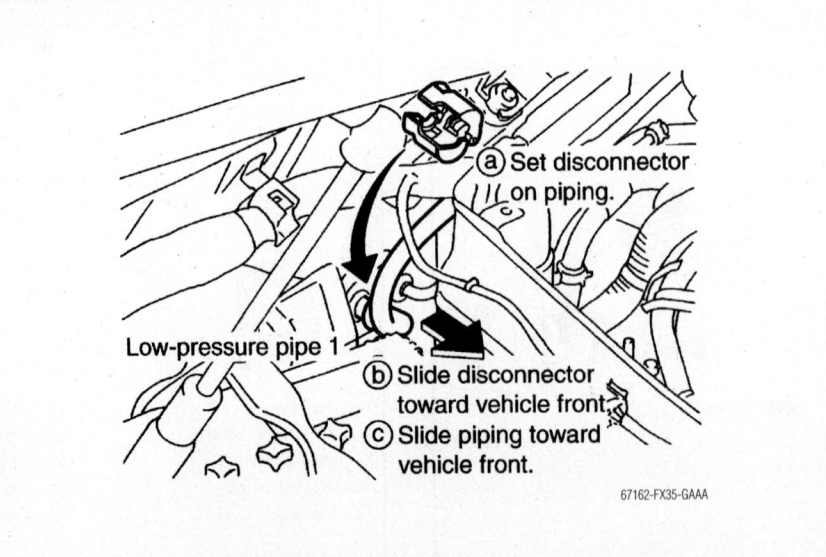

67162-FX35-GAAA

Fig. 178 Slide the disconnect tool toward the front of the vehicle until it clicks

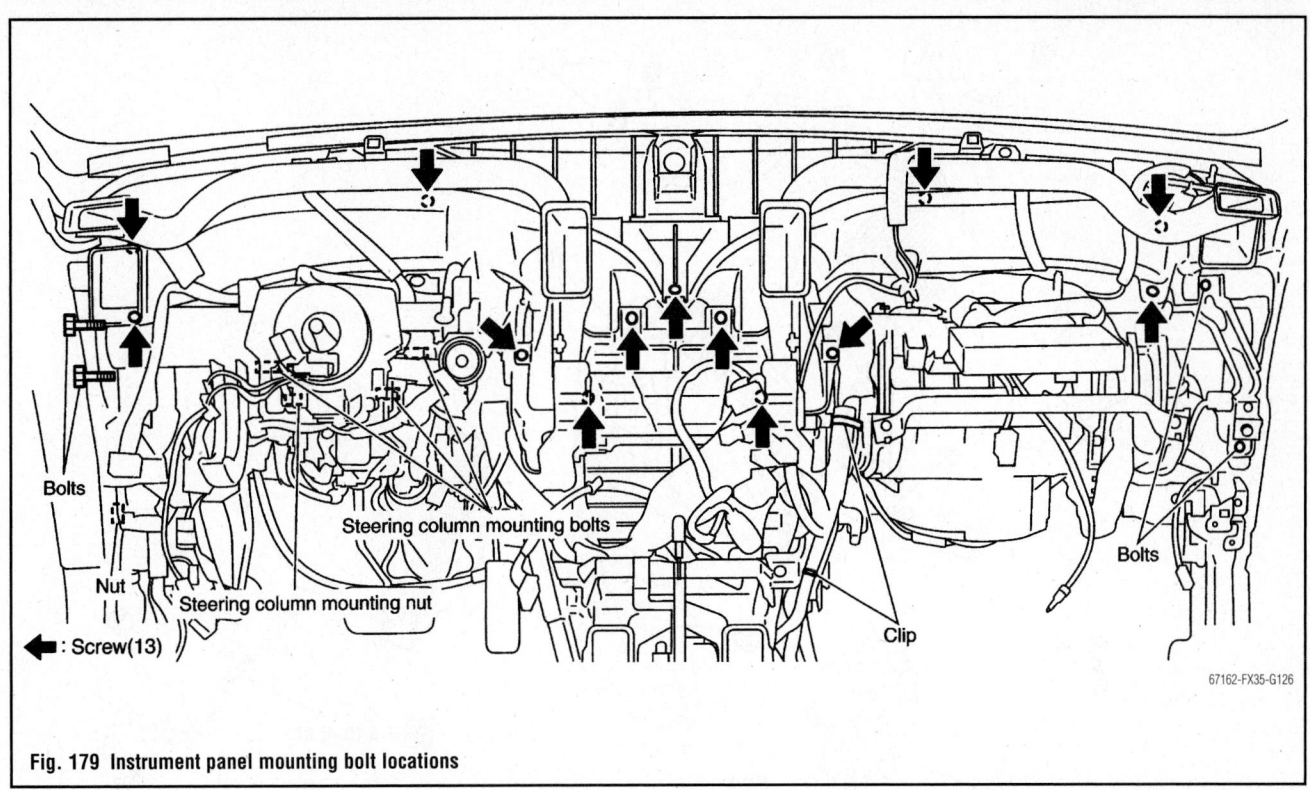

Fig. 179 Instrument panel mounting bolt locations

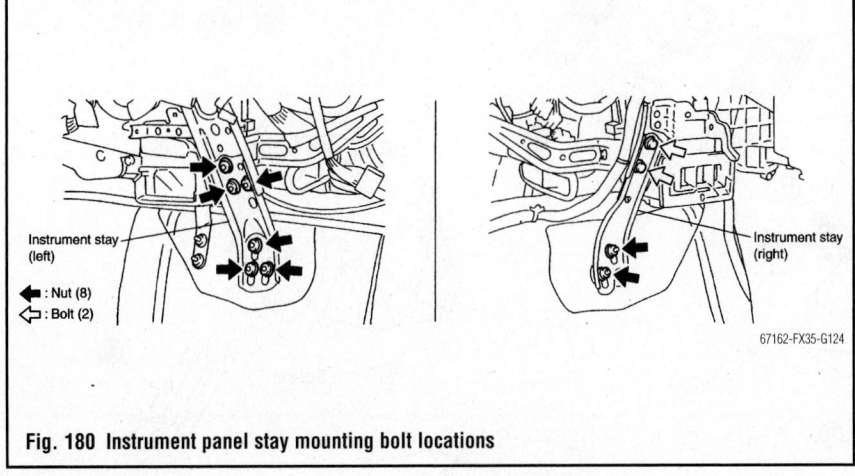

Fig. 180 Instrument panel stay mounting bolt locations

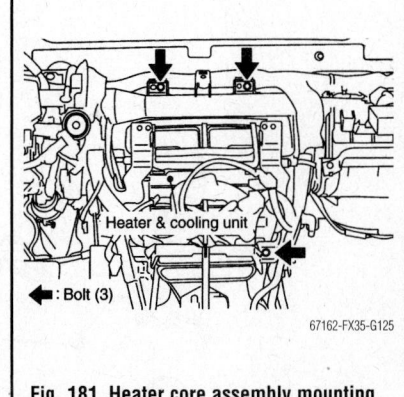

Fig. 181 Heater core assembly mounting bolt locations

20. Slide the heater core (shown in the figure) toward the driver's side.

21. Remove the heater core.

To install:

22. Install the heater core into the heating/cooling unit.

23. Install the heater and cooling unit. Tighten the mounting bolts to 60 inch lbs. (7 Nm).

24. Install the steering member.

25. Install the steering member mounting bolts, nut and wiring harness clips. Tighten the steering member mounting bolts to 108 inch lbs. (12 Nm).

26. Install the ventilator ducts, defroster nozzle and ducts.

27. Reconnect the drain hose.

28. Install the mounting bolts for the heater and cooling unit.

29. Install the instrument stays (driver-side and passenger-side).

✳✳ CAUTION

Make sure the locating pin and joint are securely inserted.

30. Install the blower unit, as follows:

 a. Install the blower unit.

 b. Install the mounting bolt and screws for the blower unit.

 c. Install the wiring harness clip for the blower unit.

 d. Reconnect the intake door motor connector and blower fan motor connector.

 e. Install the ECM with its bracket attached.

31. Install the instrument panel assembly, as follows:

 a. Install the front passenger air bag module and instrument panel and pad.

 b. Install the side ventilations.

 c. Install the combination meter bracket and reinstall the steering column assembly.

 d. Install the front defroster grilles.

 e. Install the display unit and audio unit.

 f. Install the cluster lid.

 g. Install the instrument side panels.

 h. Install the combination meter assembly.

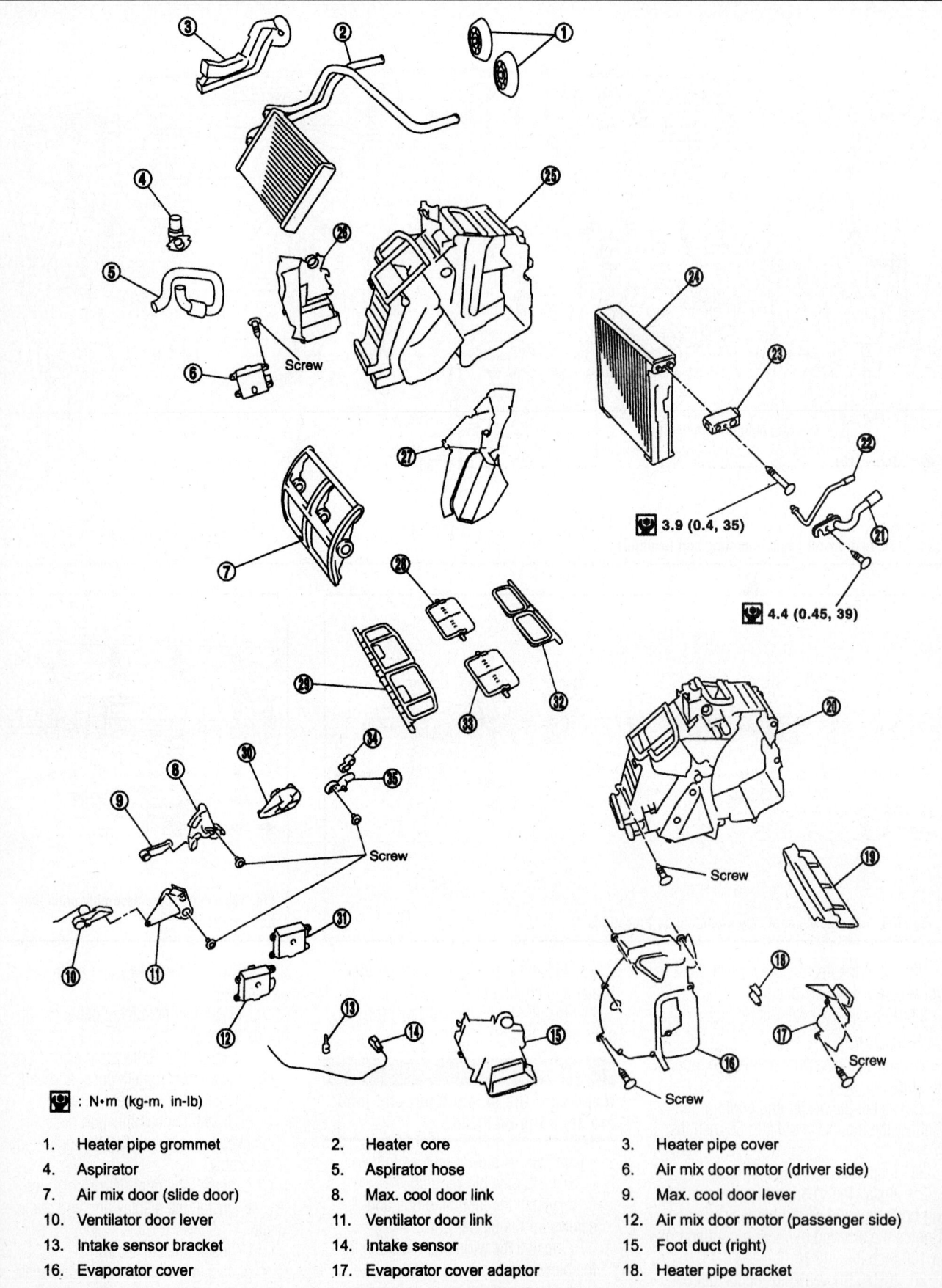

Fig. 182 Exploded view of the Heater & Cooling Unit, which contains the heater core

1.	Heater pipe grommet	2.	Heater core	3.	Heater pipe cover
4.	Aspirator	5.	Aspirator hose	6.	Air mix door motor (driver side)
7.	Air mix door (slide door)	8.	Max. cool door link	9.	Max. cool door lever
10.	Ventilator door lever	11.	Ventilator door link	12.	Air mix door motor (passenger side)
13.	Intake sensor bracket	14.	Intake sensor	15.	Foot duct (right)
16.	Evaporator cover	17.	Evaporator cover adaptor	18.	Heater pipe bracket

67162-FX35-G127

i. Install the steering lock escutcheon.

j. Install the lighting and turn signal switch.

k. Install the wiper and washer switch

l. Install the steering column upper cover.

m. Install the steering column lower cover.

n. Install the steering column front lower cover.

o. Install the instrument driver lower panel.

p. Reattach the data link connector.

q. Reconnect the in-vehicle sensor and all electrical parts.

r. Install the grommet and hood lock cable.

s. Install the instrument passenger lower panel.

t. Install the instrument lower cover.

u. Connect the center console sub-wiring harness.

v. Install the console finishers.

w. Install the console finisher screws.

x. Reconnect the wiring harness connector.

y. Install the A/T console finisher.

z. Install instrument clock finisher.

aa. Install the A/T select lever knob.

bb. Install the front pillar garnish.

cc. Install the dash side finisher and plastic nuts.

dd. Install the front kicking plate on both sides of the vehicle.

32. Reconnect the two heater hoses to the heater core.

33. Install the electronic control throttle assembly.

➡**Replace the O-rings for A/C piping with new ones, then apply compressor oil to them when installing them.**

34. Reconnect the evaporator-side one touch joint. The connection point for the female-side piping is thin, so when inserting the male-side piping, take care not to deform the female-side piping. Slowly insert it in the axial direction. Insert the one-touch joint connection point securely until it clicks. After the piping has been connected, pull on the male-side piping by hand to make sure the piping does not come off.

35. Install the low-pressure flexible hose bracket mounting bolts.

36. Install the 2 high-pressure pipe mounting clips.

37. Install the cowl top cover.

38. Fill the cooling system with the proper amount and type of fluid.

39. Recharge the vehicle A/C system and check for leaks.

STEERING

POWER RACK & PINION STEERING GEAR

REMOVAL & INSTALLATION

See Figures 183 through 186.

1. Before servicing the vehicle, refer to the Precautions Section.

❋❋ CAUTION

The spiral cable may snap due to steering operation if the steering column is separated from the steering gear assembly. Therefore fix the steering wheel with a string to avoid turning it too far.

2. Set the wheels in the straight-ahead position.

3. Raise and safely support the vehicle.

4. Remove the tires from vehicle.

5. Remove the undercover.

6. Confirm the slit of the lower joint fits with the projection on the rear cover cap marking the position on the steering gear assembly nearly fits with the projection on the rear cover cap.

7. Remove the cotter pin at steering outer socket, then loosen the mounting nut.

8. Use a ball joint remover to remove the steering outer socket from the steering knuckle. Be careful not to damage the ball joint boot.

➡**Temporarily tighten the mounting nut to prevent damage to the threads and to prevent the ball joint remover from coming off.**

9. Remove the high-pressure side and low-pressure side oil pipes from the steering gear assembly, then drain the fluid from the pipes.

10. Remove the mounting bolt of the steering hydraulic pipe bracket from the steering gear assembly.

11. Remove the lower side mounting bolt of the lower joint.

12. Remove the mounting bolts of the steering gear assembly, then remove the steering gear assembly from the vehicle.

To install:

13. Installation is the reverse of the removal procedure.

➡**Refer to component parts location and do not reuse non-reusable parts.**

14. After installation, check wheel alignment.

15. After adjusting wheel alignment, adjust the neutral position of the steering angle sensor.

16. When the steering wheel is set in the straight ahead direction, confirm the slit of the lower joint fits with the projection on the rear cover cap, and that the marking position on steering gear assembly nearly fits with the projection on rear cover cap.

17. Bleed all air from the power steering system, as follows:

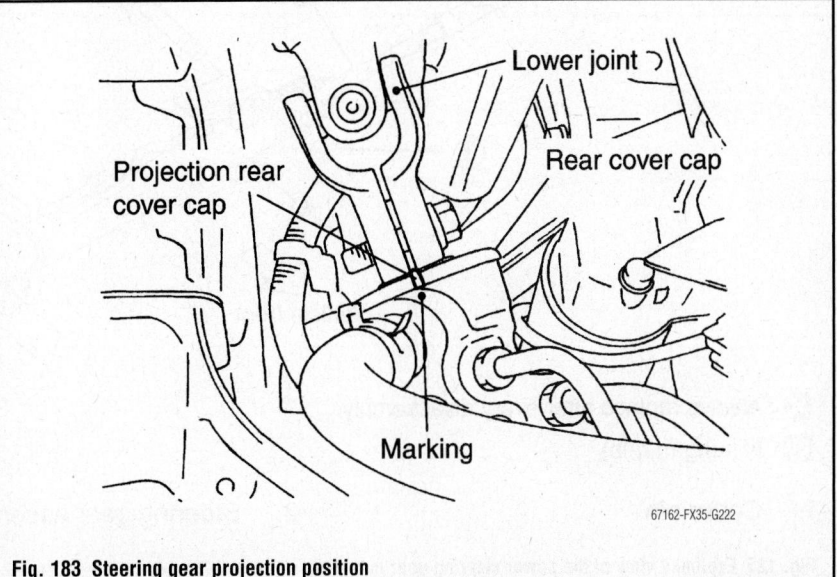

Projection rear cover cap

Lower joint

Rear cover cap

Marking

67162-FX35-G222

Fig. 183 Steering gear projection position

Fig. 184 Remove the mounting bolts of the steering gear assembly

a. Stop the engine, then turn the steering wheel fully to the right and left several times.

➡**Do not allow the steering fluid reservoir tank level to drop below the low-level line. Check the tank frequently and add fluid as needed.**

b. Run the engine at idle speed. Turn the steering wheel fully to the right and then fully to the left, and keep hold for about 3 seconds. Then check whether any fluid leaks have occurred.

c. Repeat sub-step b several times at about 3 second intervals.

✳✳ WARNING

Do not hold the steering wheel in the locked position for more than 10 seconds. There is the possibility that the oil pump may be damaged.

d. Check for air bubbles and cloudiness in the fluid.

e. If air bubbles and/or cloudiness don't fade, stop the engine, and stop air bleeding until the air bubbles and cloudiness fade.

f. Perform until all bubbles and cloudiness are gone.

g. Stop the engine and check the fluid level.

Incomplete air bleeding causes the following. When this happens, bleed the system again:

• Generation of air bubbles in the reservoir tank

• Generation of clicking noise in the oil pump

• Excessive buzzing in the oil pump

➡**When the vehicle is stationary or while the steering wheel is being turned slowly, some noise may be heard from the oil pump or gear. This noise is normal.**

18. Check that the steering wheel turns smoothly when it is turned several times fully to the end of the left and right.

POWER STEERING PUMP

BLEEDING

1. Before servicing the vehicle, refer to the Precautions Section.

2. Stop the engine.

3. Turn the steering wheel fully to the right and left several times.

➡**Do not allow the fluid level in the reservoir tank to go below the MIN level line. Check and add fluid as needed.**

4. Run the engine at idle speed. Turn the steering wheel fully to the right and then fully to the left. Hold for about 3 seconds. Check for fluid leakage.

5. Repeat the above step several times at 3 second intervals.

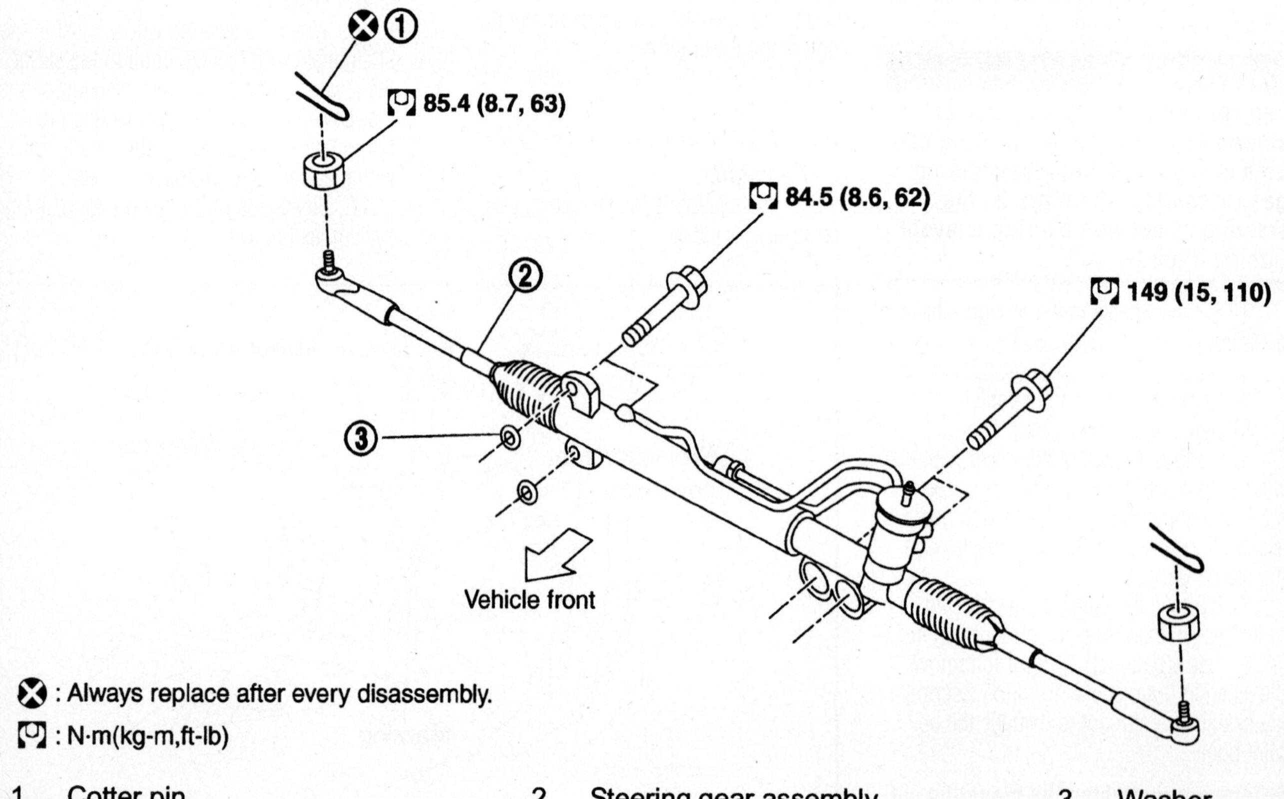

⊗ : Always replace after every disassembly.

🔧 : N·m(kg-m,ft-lb)

1. Cotter pin 2. Steering gear assembly 3. Washer

Fig. 185 Exploded view of the power steering gear mounting

✳✳ WARNING

Do not hold the steering wheel in the locked position for more than 10 seconds. Damage to the pump may occur.

6. Check for air bubbles or cloudy fluid. If found, repeat the bleeding procedure.

7. Stop the engine and check the fluid level. Correct as required.

REMOVAL & INSTALLATION

See Figure 186.

1. Before servicing the vehicle, refer to the Precautions Section.

2. Disconnect the negative battery cable.

3. Remove the undercover.

4. Remove the drive belt. Refer to Accessory Drive Belts, removal & installation.

5. Drain the power steering fluid from the reservoir tank into a suitable container. Properly discard the used fluid.

6. Remove the high pressure and the low pressure lines from the power steering fluid pump.

7. Remove the pump mounting bolts.

8. Remove the pump from the vehicle.

To install:

9. Installation is the reverse of the removal procedure.

10. Bleed the power steering system.

11. Adjust the belt tension

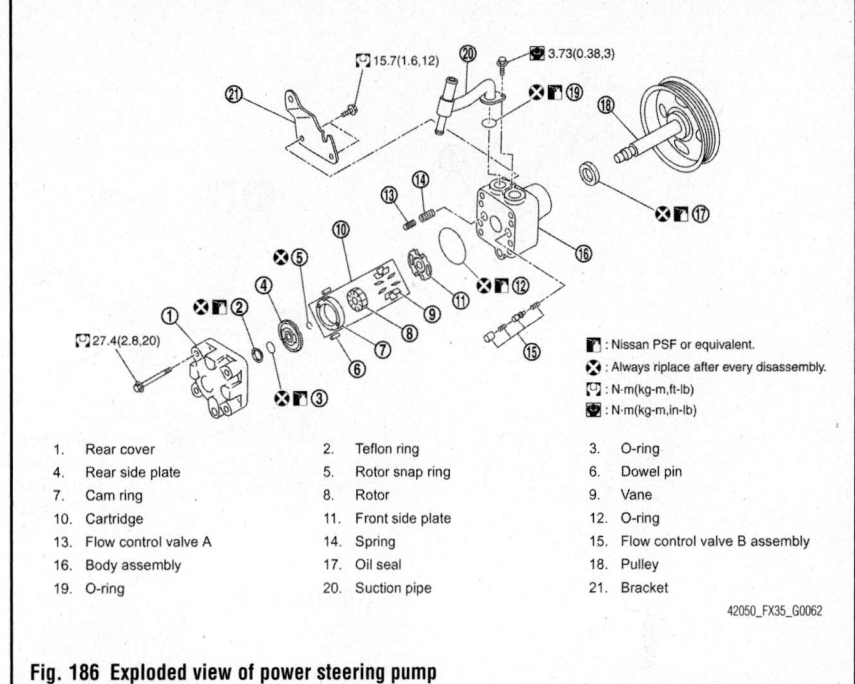

1. Rear cover
2. Teflon ring
3. O-ring
4. Rear side plate
5. Rotor snap ring
6. Dowel pin
7. Cam ring
8. Rotor
9. Vane
10. Cartridge
11. Front side plate
12. O-ring
13. Flow control valve A
14. Spring
15. Flow control valve B assembly
16. Body assembly
17. Oil seal
18. Pulley
19. O-ring
20. Suction pipe
21. Bracket

🅟 : Nissan PSF or equivalent.
✖ : Always replace after every disassembly.
🔧 : N·m(kg-m,ft-lb)
🔧 : N·m(kg-m,in-lb)

42050_FX35_G0062

Fig. 186 Exploded view of power steering pump

POWER STEERING FLUID

FLUID RECOMMENDATIONS

NISSAN recommends Genuine NISSAN PSF or equivalent (1.25 qt.)

FLUID LEVEL CHECK

1. Check fluid level with engine stopped.
2. Ensure that fluid level is between MIN and MAX.

3. Fluid levels at HOT and COLD are different. Do not confuse them.

➡**The fluid level should not exceed the MAX line. Excessive fluid causes fluid leakage from the cap.**

✳✳ WARNING

Never reuse drained power steering fluid.

SUSPENSION   FRONT SUSPENSION

COIL SPRINGS AND SHOCK ABSORBER

REMOVAL & INSTALLATION

See Figures 187 through 190.

1. Remove tires with power tool.
2. Remove wheel sensor and harness connector from shock absorber.

✳✳ WARNING

Never pull on wheel sensor harness.

3. Remove brake hose bracket.
4. Remove stabilizer connecting rod with power tool. Refer to the accompanying illustration.
5. Remove shock absorber from transverse link with power tool (AWD Only).
6. Separate upper link from steering knuckle. Refer to Power Rack & Pinion Steering Gear in the Steering Section.

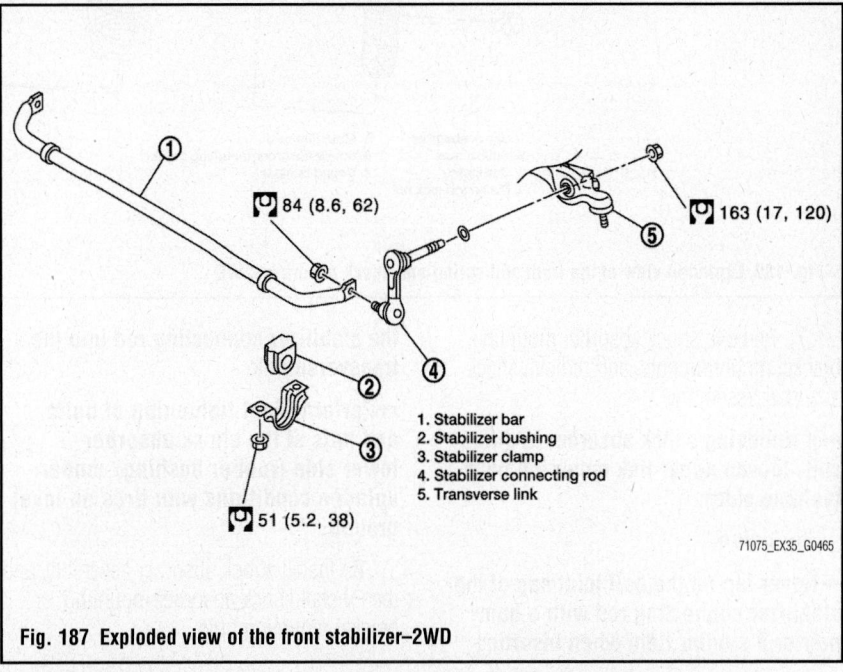

🔧 84 (8.6, 62)
🔧 163 (17, 120)
🔧 51 (5.2, 38)

1. Stabilizer bar
2. Stabilizer bushing
3. Stabilizer clamp
4. Stabilizer connecting rod
5. Transverse link

71075_EX35_G0465

Fig. 187 Exploded view of the front stabilizer–2WD

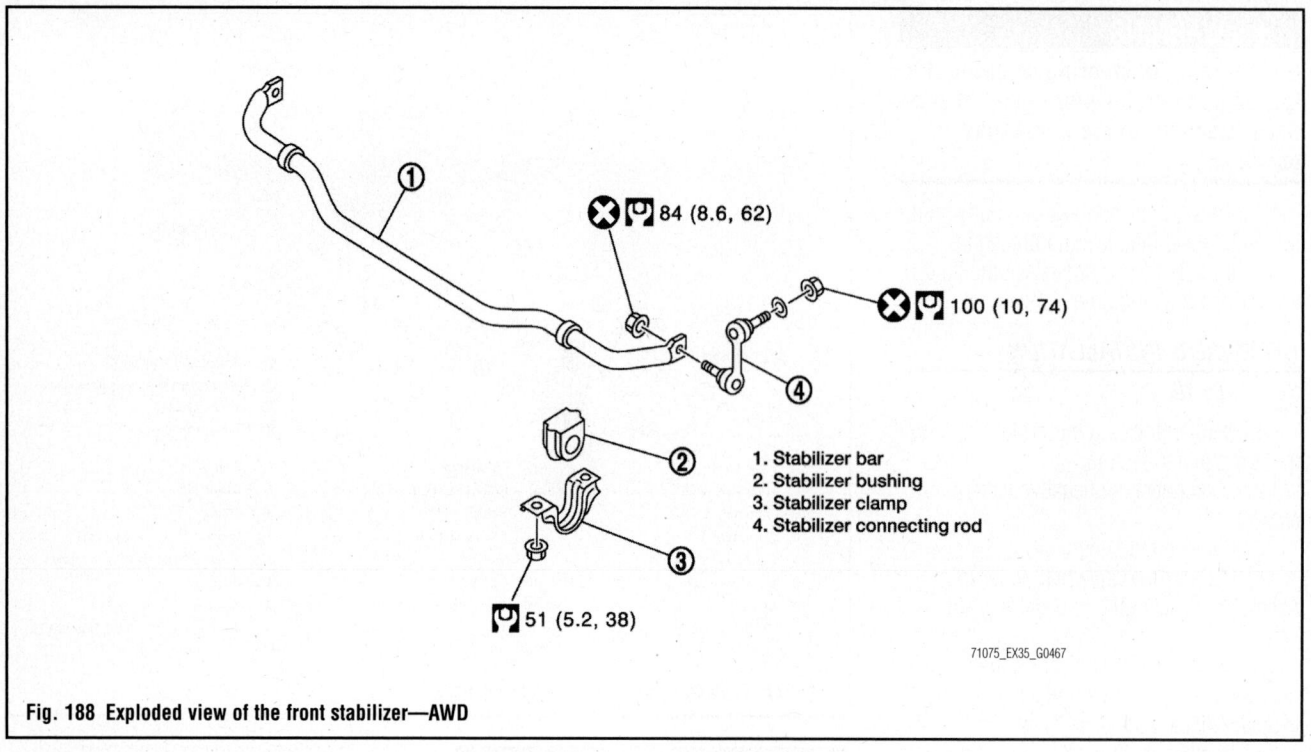

Fig. 188 Exploded view of the front stabilizer—AWD

84 (8.6, 62)

100 (10, 74)

1. Stabilizer bar
2. Stabilizer bushing
3. Stabilizer clamp
4. Stabilizer connecting rod

51 (5.2, 38)

71075_EX35_G0467

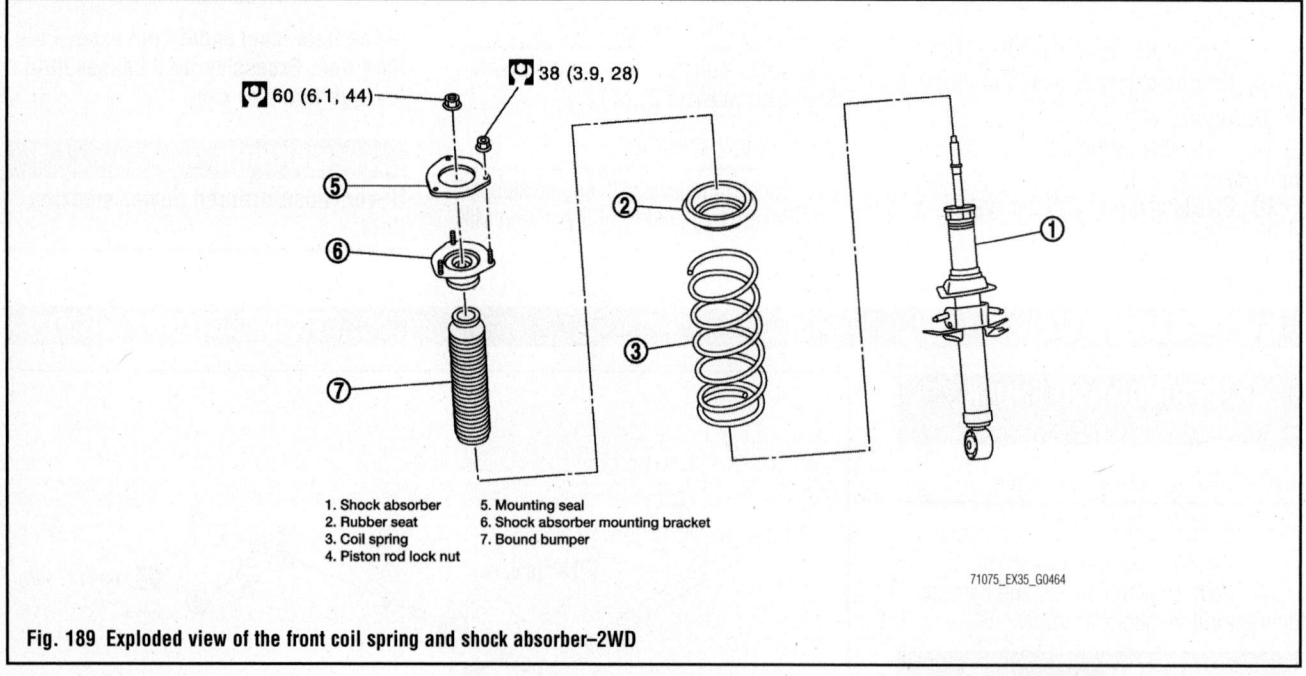

60 (6.1, 44)   38 (3.9, 28)

1. Shock absorber   5. Mounting seal
2. Rubber seat   6. Shock absorber mounting bracket
3. Coil spring   7. Bound bumper
4. Piston rod lock nut

71075_EX35_G0464

Fig. 189 Exploded view of the front coil spring and shock absorber–2WD

7. Remove shock absorber mounting bracket mounting nuts, and remove shock absorber assembly.

➡If removing shock absorber is difficult, loosen upper link mounting bolts (vehicle side).

To install:

➡Never tap on the ball joint cap of the stabilizer connecting rod with a hammer or a similar item when inserting

the stabilizer connecting rod into the transverse link.

➡Perform final tightening of bolts and nuts at the shock absorber lower side (rubber bushing), under unladen conditions with tires on level ground.

8. Install shock absorber assembly, and then install shock absorber mounting bracket mounting nuts.

9. Install upper link to steering knuckle. Refer to Power Rack & Pinion Steering Gear in the Steering Section.

10. Install shock absorber to transverse link with power tool (AWD Only).

11. Install stabilizer connecting rod with power tool. Refer to the accompanying illustration.

12. Install brake hose bracket.

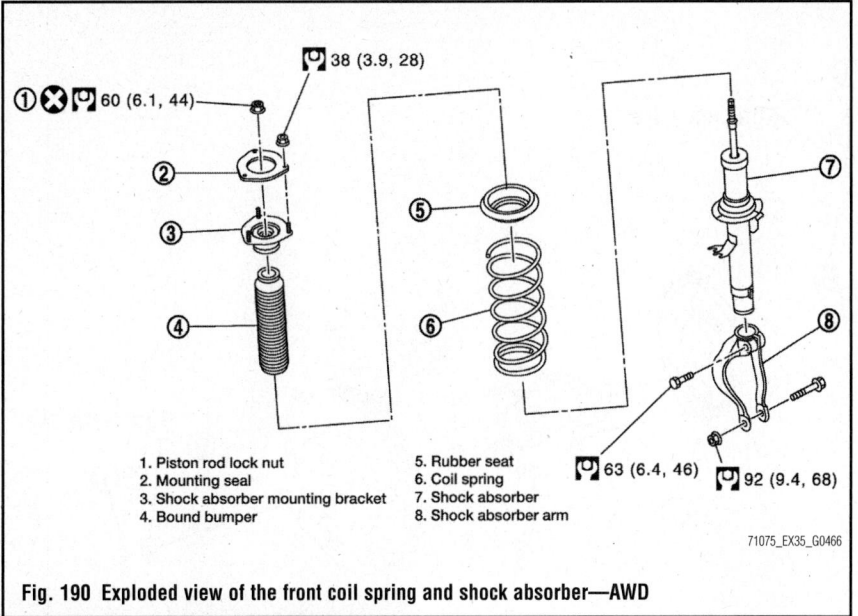

1. Piston rod lock nut
2. Mounting seal
3. Shock absorber mounting bracket
4. Bound bumper
5. Rubber seat
6. Coil spring
7. Shock absorber
8. Shock absorber arm

71075_EX35_G0466

Fig. 190 Exploded view of the front coil spring and shock absorber—AWD

✳✳ WARNING

Never pull on wheel sensor harness.

13. Install wheel sensor and harness connector to shock absorber.
14. Install tires with power tool.

CONTROL LINKS

REMOVAL & INSTALLATION

1. Before servicing the vehicle, refer to the Precautions Section.
2. Remove the tires from the vehicle.
3. Remove the undercover.
4. Remove the stabilizer control link lower nut and separate the stabilizer bar and stabilizer control link (the control link is also called the stabilizer connecting rod).
5. Remove the stabilizer control link upper nut.
6. Remove the stabilizer control link.

To install:

7. Check the stabilizer control link for cracks or damage, and replace if necessary.
8. Install in the reverse order of removal.
9. Tighten the stabilizer control link nuts to 75 ft. lbs. (102 Nm).
10. Check the wheel alignment and adjust as necessary.

KNUCKLE & SPINDLE

REMOVAL & INSTALLATION

See Figures 191 and 192.

1. Before servicing the vehicle, refer to the Precautions Section.
2. Raise and safely support the vehicle.

---

3. Remove the appropriate wheel.
4. Remove the brake caliper. Support it in a place where it will not interfere with work.

➡**Avoid depressing brake pedal while brake caliper is removed.**

5. Remove the disc rotor.

---

6. Remove the wheel sensor from the wheel hub and bearing assembly.

➡**Do not pull on wheel sensor wiring harness.**

7. Remove the cotter pin from the steering outer socket, then loosen the mounting nut.
8. Use a ball joint remover to separate the steering outer socket from the steering knuckle. Be careful not to damage the ball joint boot.

➡**Temporarily tighten the mounting nut to prevent damage to the threads and to prevent the ball joint remover from coming off.**

9. Remove the cotter pin at the transverse link, then loosen the mounting nut.
10. Use a ball joint remover to separate the transverse link from the steering knuckle. Be careful not to damage ball joint boot.

✳✳ CAUTION

Temporarily tighten the mounting nut to prevent damage to the threads and to prevent the ball joint remover from coming off.

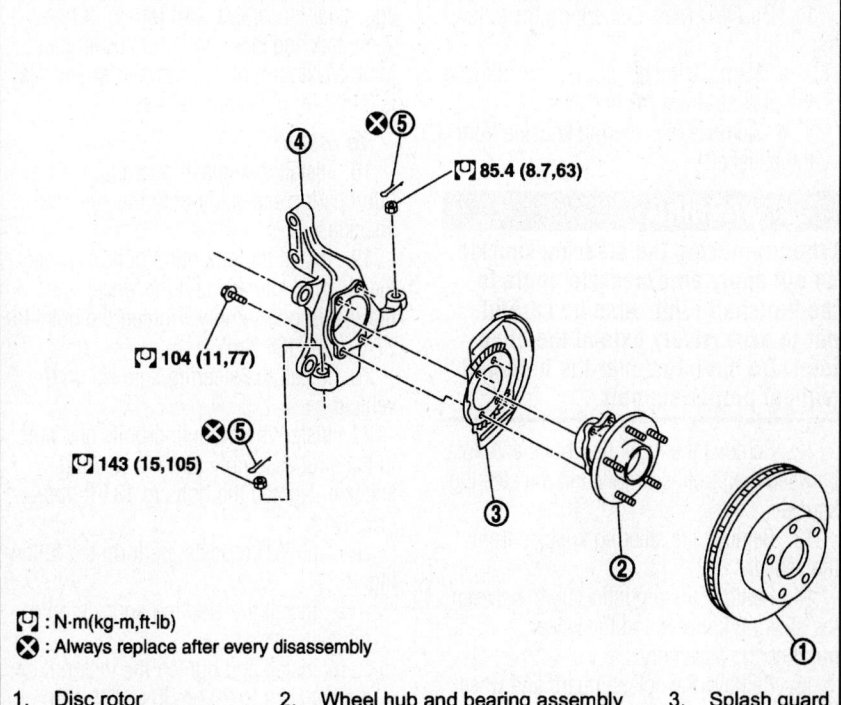

☐ : N·m(kg-m,ft-lb)
✖ : Always replace after every disassembly

1. Disc rotor
2. Wheel hub and bearing assembly
3. Splash guard
4. Steering knuckle
5. Cotter pin

67162-FX35-G228

Fig. 191 Exploded view of front wheel hub mounting—2WD models

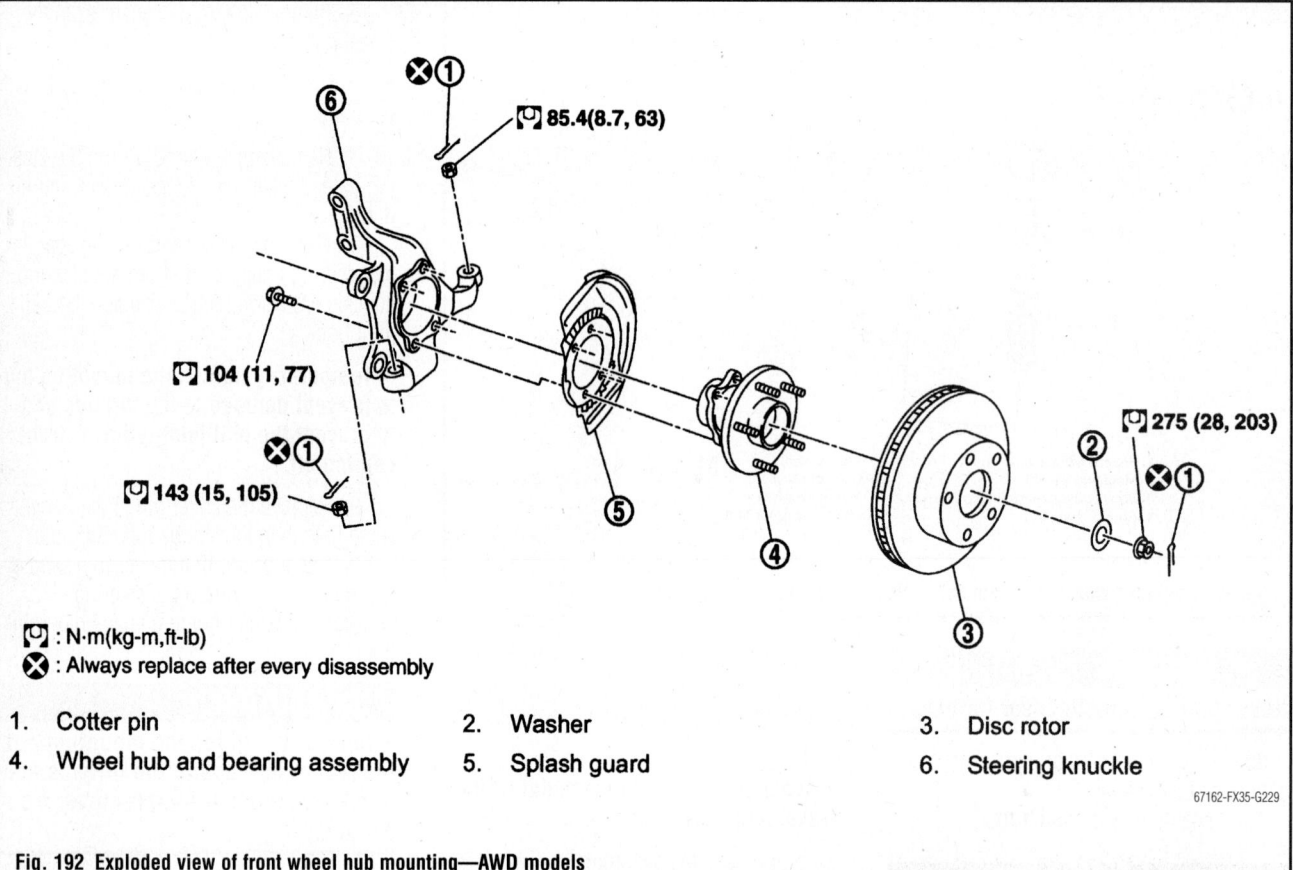

[🔧] : N·m(kg-m,ft-lb)
[✕] : Always replace after every disassembly

1. Cotter pin	2. Washer	3. Disc rotor	
4. Wheel hub and bearing assembly	5. Splash guard	6. Steering knuckle	

67162-FX35-G229

Fig. 192 Exploded view of front wheel hub mounting—AWD models

11. On AWD models, perform the following:

a. Remove the cotter pin, then remove the lock nut from the halfshaft.

b. Remove the steering knuckle from the halfshaft.

> **⚠ WARNING**
>
> **When removing the steering knuckle, do not apply an excessive angle to the halfshaft joint. Also be careful not to excessively extend the slide joint. Do not hang over the halfshaft without proper support.**

12. Remove the mounting bolts and nuts between the strut assembly and the steering knuckle.

13. Remove the steering knuckle from the vehicle.

14. Remove the mounting bolts between the steering knuckle and the wheel hub/bearing assembly.

15. Remove the splash guard and wheel hub/bearing assembly from the steering knuckle.

16. Check for deformities, cracks and damage on all parts and replace if necessary.

17. Inspect the ball joint for boot breakage, axial looseness, and torque of transverse link and steering outer socket ball joint. Maximum of allowable axial end play is 0.002 in. (0.05mm) or less.

To install:

18. Install the splash guard and wheel hub/bearing assembly onto the steering knuckle.

19. Install the mounting bolts between the steering knuckle and the wheel hub/bearing assembly. Tighten the bolts to 77 ft. lbs. (104 Nm).

20. Install the steering knuckle on the vehicle.

21. Install the mounting bolts and nuts to the strut assembly and the steering knuckle. Tighten the bolts to 134 ft. lbs. (182 Nm).

22. On AWD models, perform the following:

a. Install the steering knuckle onto the halfshaft.

b. Install and tighten the lock nut on the halfshaft to 203 ft. lbs. (275 Nm). Install a new cotter pin.

23. Reattach the transverse link to the steering knuckle. Tighten the ball joint nut to 105 ft. lbs. (143 Nm).

24. Install a new cotter pin.

25. Reconnect the steering outer socket to the steering knuckle. Tighten the ball joint nut to 63 ft. lbs. (86 Nm).

26. Install a new cotter pin.

27. Install the wheel sensor onto the wheel hub and bearing assembly.

28. Install the disc rotor.

29. Install the brake caliper.

30. Install the wheel.

31. Check wheel alignment.

32. After adjusting wheel alignment, adjust the neutral position of the steering angle sensor.

33. Check the installation condition of the wheel sensor wiring harness.

LOWER BALL JOINTS

REMOVAL & INSTALLATION

The front lower control arm ball joints are not separately replaceable from the control arms themselves. If the joints are found to be defective, the entire assembly must be replaced.

LOWER CONTROL ARMS

REMOVAL & INSTALLATION

1. Before servicing the vehicle, refer to the Precautions Section.

2. Raise and safely support the vehicle.

3. Remove the wheels from vehicle.

4. Remove the undercover.

5. Remove the front cross bar.

6. Remove the cotter pin at the lower control arm (also called the transverse link), then loosen the mounting nut.

7. Use a ball joint remover to remove the transverse link from the steering knuckle. Be careful not to damage the ball joint boot.

➡**Temporarily tighten the mounting nut to prevent damage to the threads and to prevent the ball joint remover from coming off.**

8. Remove the mounting bolts which are at the back of the transverse link (mounting part with body), then separate the transverse link.

9. Remove the mounting bolts which are at the front of the transverse link (mounting part with the front suspension member), then separate the transverse link.

10. Remove the transverse link from the vehicle.

11. Check transverse link and bushing for deformation, cracks, or damage. If any non-standard condition is found, replace it.

12. Check the boot of the ball joint for cracks, or other damage, and also for grease leakage. If any non-standard condition is found, replace it.

13. Manually move ball stud to confirm it moves smoothly with no binding.

➡**Before measurement, move ball joint at least ten times by hand to check for smooth movement.**

14. Hook a spring scale onto the ball stud tip. Confirm that the spring scale measurement value is within specifications when the ball stud begins to move. If it is outside the specified range, replace the transverse link assembly.

➡**Swing torque specification: Less than 5–43 inch lbs. (1–5 Nm), measure value of spring scale: less than 5–43 inch lbs. (1–5 Nm).**

15. Attach the mounting nut onto the ball stud. Check that the rotating torque is within specifications with a preload gauge . If it is outside the specified range, replace transverse link assembly.

➡**Rotating torque specification: Less than 5–43 inch lbs. (1–5 Nm).**

16. Move the tip of ball joint in axial direction to check for looseness. If it is outside the specified range, replace transverse link assembly.

➡**Axial end play specification: 0.004 in. (0.1mm).**

To install:

17. Install the transverse link on the vehicle.

18. Install and tighten the mounting bolts at the front of the transverse link (mounting part with the front suspension member) to 89 ft. lbs. (120 Nm), then install and tighten the mounting bolts which are at the back of the transverse link (mounting part with body) to 118 ft. lbs. (160 Nm).

19. Reattach the transverse link to the steering knuckle, and tighten the ball joint nut to 105 ft. lbs. (142 Nm). Be careful not to damage the ball joint boot.

20. Install a new cotter pin.

21. Install the front cross bar. Tighten the inner 2 bolts on each end to 33 ft. lbs. (45 Nm) and the outer 2 bolts on each end to 41 ft. lbs. (55 Nm).

22. Install the undercover.

23. Install the tire.

24. Check the wheel alignment.

25. After adjusting wheel alignment, adjust the neutral position of the steering angle sensor.

STABILIZER BAR (SWAY BAR) & LINKS

REMOVAL & INSTALLATION

Stabilizer Bar

See Figure 193.

1. Before servicing the vehicle, refer to the Precautions Section.

2. Raise and safely support the vehicle.

3. Remove the wheels.

4. Remove the fixing bolts and remove the stabilizer connecting rod mount bracket from the suspension arm.

5. Remove the lower side fixing nut on the stabilizer connecting rod and remove the stabilizer connecting rod from the stabilizer bar.

6. Remove the fixing nuts on the stabilizer clamps and remove the stabilizer from the vehicle.

7. Check the stabilizer bar, stabilizer bushings, stabilizer clamps, stabilizer connecting rod, and stabilizer connecting rod mounting bracket for any deformation, cracks, or damage. Replace if necessary.

To install:

8. Refer to the exploded view for tightening torques. Installation is the reverse of removal.

➡**Do not reuse non-reusable parts during assembly.**

9. The stabilizer bar uses pillow ball type connecting rod, position the ball joint with the case on pillow ball head parallel to the stabilizer bar.

10. When the bushing and clamp are installed to the stabilizer bar, position the bushing and clamp inside of the side slip prevention clamp

STRUTS (MACPHERSON STRUTS)

OVERHAUL

See Figures 194 through 196.

1. Before servicing the vehicle, refer to the Precautions Section.

✳✳ WARNING

Make sure the piston rod on the strut is not damaged when removing the components from the strut assembly.

2. Install the strut attachment SST: ST35652000 or equivalent to the strut and secure it in a vise. When installing the strut attachment to the strut, wrap a shop towel around the strut to protect it from damage.

3. Using a spring compressor, compress the coil spring between the spring upper seat and spring lower seat (on the strut) until the coil spring is free.

✳✳ CAUTION

Be sure the spring compressor is securely attached to the coil spring before compressing the spring.

4. After making sure the coil spring is free between the spring upper seat and spring lower seat of the strut, remove the piston rod locknut.

5. Remove the mounting insulator, mounting insulator bracket, mounting bearing, spring upper seat, spring upper rubber seat, and bound bumper. Then, remove the coil spring and spring lower rubber seat from the strut.

6. Gradually release the spring compressor, and remove the coil spring.

✳✳ CAUTION

Loosen the spring compressor while making sure the coil spring attachment position does not move.

7. Remove the strut attachment from the strut.

To assemble:

8. Check the strut for deformation, cracks, damage, and replace if necessary.

9. Check the piston rod for damage,

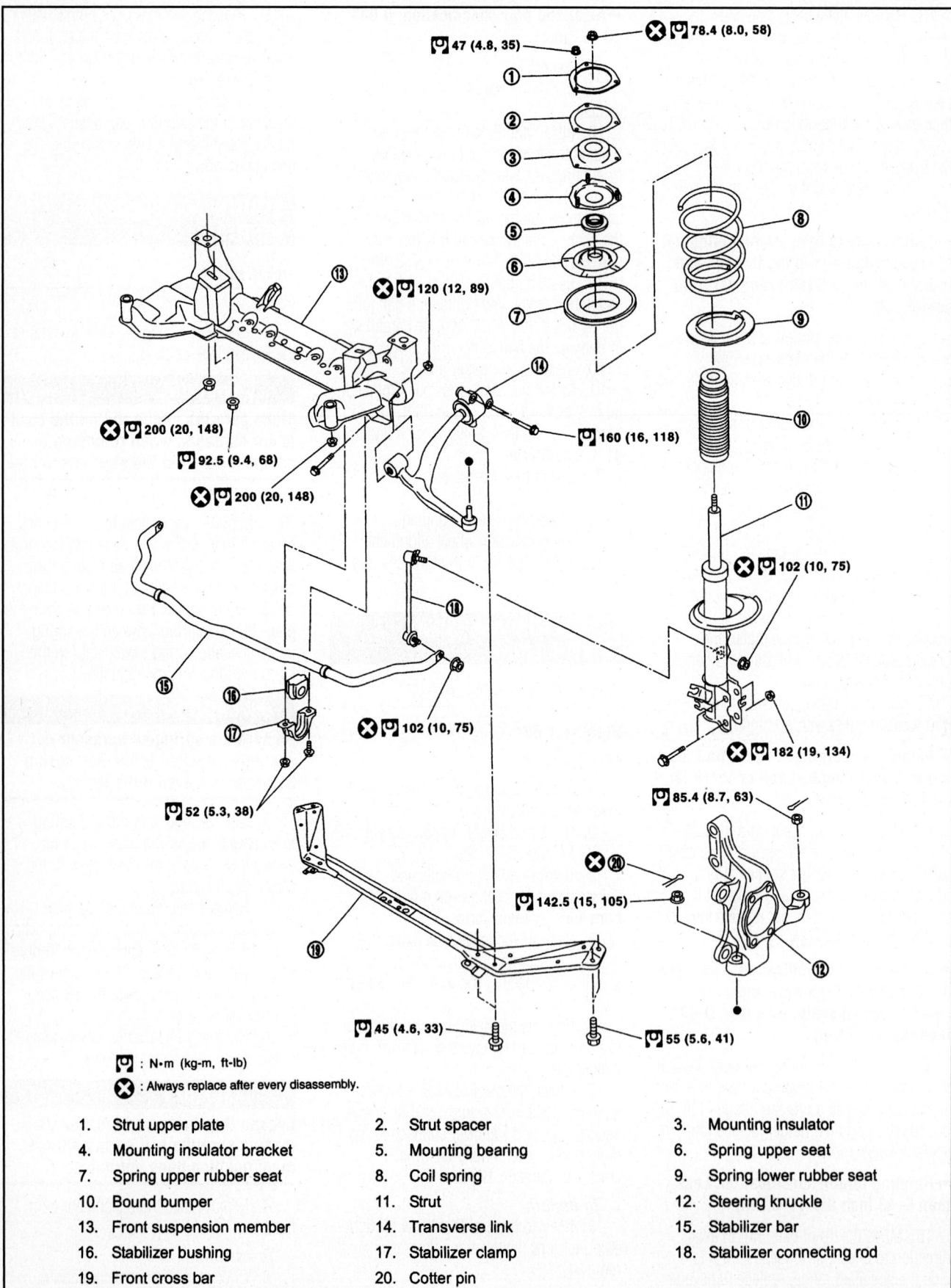

Ⓒ 47 (4.8, 35)
✖ Ⓒ 78.4 (8.0, 58)
✖ Ⓒ 120 (12, 89)
Ⓒ 160 (16, 118)
✖ Ⓒ 200 (20, 148)
Ⓒ 92.5 (9.4, 68)
✖ Ⓒ 200 (20, 148)
✖ Ⓒ 102 (10, 75)
✖ Ⓒ 102 (10, 75)
✖ Ⓒ 182 (19, 134)
Ⓒ 85.4 (8.7, 63)
✖ ⓴
Ⓒ 142.5 (15, 105)
Ⓒ 52 (5.3, 38)
Ⓒ 45 (4.6, 33)
Ⓒ 55 (5.6, 41)

Ⓒ : N•m (kg-m, ft-lb)

✖ : Always replace after every disassembly.

1. Strut upper plate	2. Strut spacer	3. Mounting insulator
4. Mounting insulator bracket	5. Mounting bearing	6. Spring upper seat
7. Spring upper rubber seat	8. Coil spring	9. Spring lower rubber seat
10. Bound bumper	11. Strut	12. Steering knuckle
13. Front suspension member	14. Transverse link	15. Stabilizer bar
16. Stabilizer bushing	17. Stabilizer clamp	18. Stabilizer connecting rod
19. Front cross bar	20. Cotter pin	

67162-FX35-G223

Fig. 193 Exploded view of the front suspension components

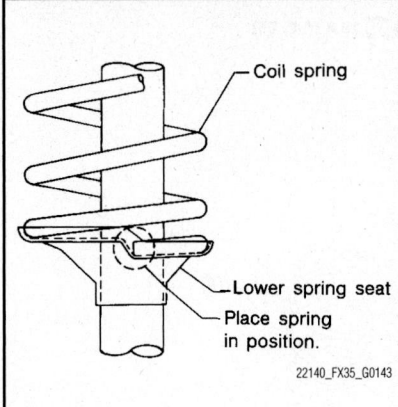

Fig. 194 Face the tube side of the coil spring downward and align the lower end to the spring rubber seat as shown

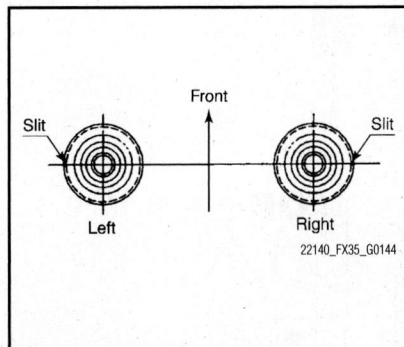

Fig. 195 The installation position of spring upper seat is as shown

uneven wear or distortion, and replace if necessary.

10. Check the welded and sealed areas for oil leakage, and replace if necessary.

11. Check the mounting insulator for cracks and rubber parts for wear. Replace them if necessary.

12. Check the coil spring for cracks, wear or damage, and replace if necessary.

➡**Make sure the piston rod on the strut is not damaged when attaching the components to the strut.**

13. Install the strut attachment to the strut and fix it in a vise. When installing the strut attachment to strut, wrap a shop cloth around strut to protect it from damage.

14. Compress the coil spring using a spring compressor and install it onto the strut.

15. Face the tube side of the coil spring downward. Align the lower end to the spring rubber seat as shown. Be sure the spring compressor is securely attached to the coil spring before compressing the coil spring.

16. Apply soapy water to the bound

bumper and insert it into the mounting insulator. Do not use machine oil.

17. Install the mounting insulator bracket, mounting bearing, bound bumper, spring upper seat, spring upper rubber seat, and spring lower rubber seat.

➡**The installation position of spring upper seat is as shown.**

18. Fix the mounting insulator and tighten the strut upper plate nuts to 35 ft. lbs. (47 Nm).

19. Install a new piston rod locknut and tighten to 58 ft. lbs. (78 Nm).

✳✳ WARNING

Be sure not to deform the mounting insulator bracket.

20. Gradually release the spring compressor and allow the tension to be placed on the strut assembly. Loosen the spring compressor while making sure the coil spring attachment position does not move.

21. Remove the strut spring compressor attachment from the strut.

22. Install the strut assembly to the vehicle.

REMOVAL & INSTALLATION
See Figure 197.

1. Before servicing the vehicle, refer to the Precautions Section.

2. Raise and safely support the vehicle.

3. Make an alignment marking on the camber adjusting bolt and strut for approximate installation alignment later.

4. Remove the wheels from the vehicle.

5. Remove the brake hose lock plate. Then, remove the brake hose from the strut assembly.

6. Remove the wheel sensor wiring harness from the strut assembly.

✳✳ WARNING

Do not pull on the wheel sensor wiring harness.

7. Remove the stabilizer connecting rod upper nut, separate the stabilizer connecting rod and strut assembly.

8. Remove the attaching bolts and nuts between the strut assembly and the steering knuckle.

9. Remove the mounting nuts on the mounting insulator bracket, then remove the strut upper plate, strut spacer and the strut from the vehicle.

To install:

➡**Attach strut upper plate as shown in the accompanying illustration.**

10. Install the strut upper plate, strut spacer, and the strut onto the vehicle.

11. Install and tighten the mounting nuts on the mounting insulator bracket to 35 ft. lbs. (47 Nm).

12. Install and tighten the attaching bolts and nuts between the strut assembly and the steering knuckle to 134 ft. lbs. (182 Nm).

➡**Use the matchmarks made earlier to approximate the front end alignment during installation of the strut assembly.**

13. Reattach the stabilizer connecting rod to the strut and tighten the upper nut to 75 ft. lbs. (102 Nm).

14. Install the wheel sensor wiring harness onto the strut assembly.

15. Install the brake hose and the hose lock plate.

16. Install the wheel and tire to the vehicle.

17. After installation, check wheel alignment.

18. After adjusting the wheel alignment, adjust the neutral position of steering angle sensor.

19. Double-check to ensure that the wheel sensor wiring harness is properly routed.

SUSPENSION MEMBER

REMOVAL & INSTALLATION
See Figures 198 and 199.

1. Remove tires with power tool.

2. Remove under cover with power tool.

3. Remove suspension member stays (2WD), front cross bar (AWD) with power tool.

4. Separate steering gear assembly and lower joint. Refer to Power Rack & Pinion Gear in the Steering Section.

5. Remove steering outer sockets from steering knuckles. Refer to Power Rack & Pinion Gear in the Steering Section.

6. Remove wheel sensors and sensor harness from steering knuckles. Refer to Anti-Lock Brakes System in the Brakes Section.

7. Remove shock absorbers from transverse links (AWD Only). Refer to Coil Springs And Shock Absorber in Front Suspension in the Suspension Section.

8. Remove stabilizer connecting rods and stabilizer bar. Refer to Front Suspension in the Suspension Section.

9. Install engine slinger, and then hoist engine.

10. Remove transverse link from front

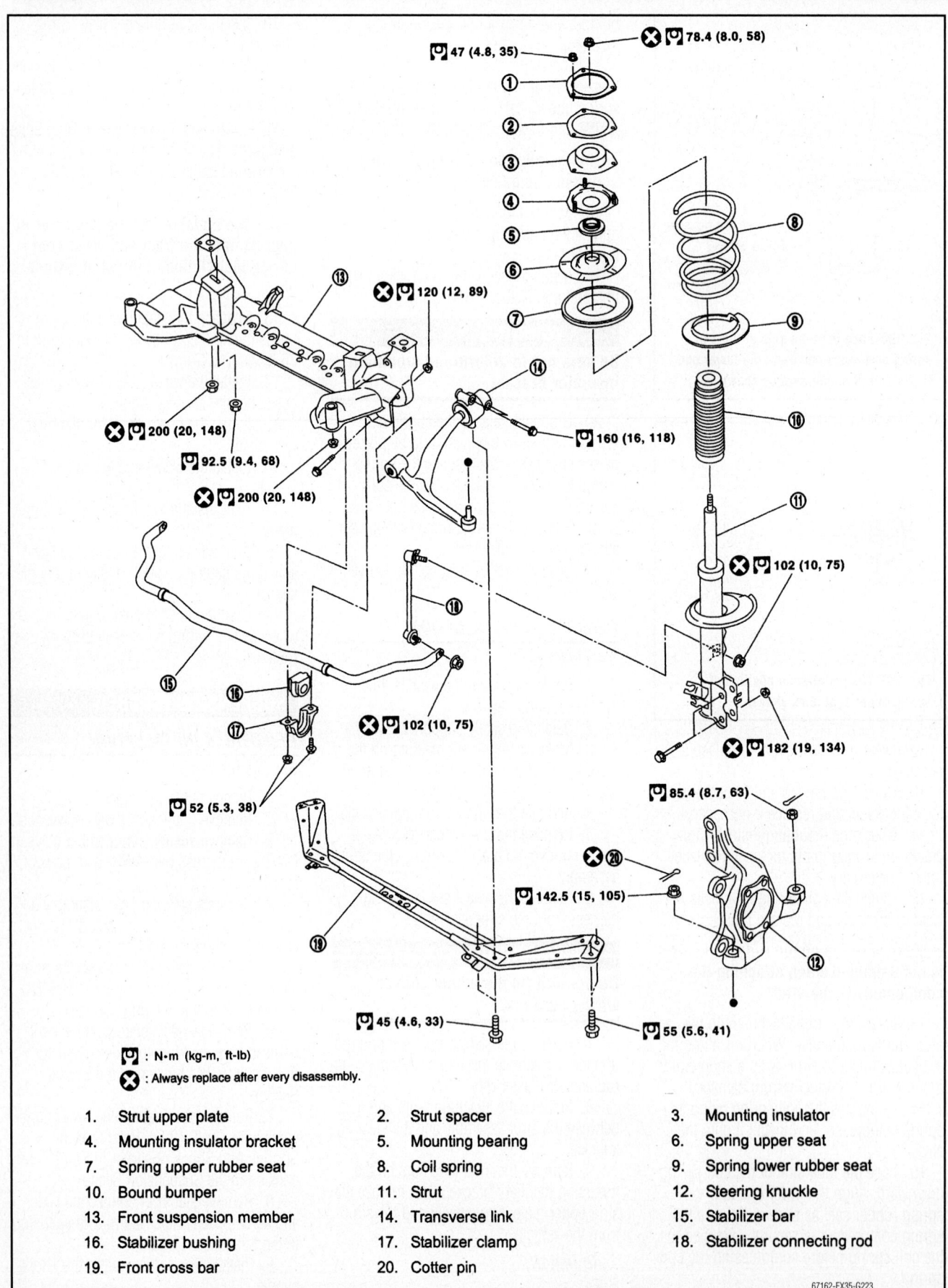

47 (4.8, 35)

78.4 (8.0, 58)

120 (12, 89)

200 (20, 148)

92.5 (9.4, 68)

200 (20, 148)

160 (16, 118)

102 (10, 75)

102 (10, 75)

182 (19, 134)

52 (5.3, 38)

85.4 (8.7, 63)

142.5 (15, 105)

45 (4.6, 33)

55 (5.6, 41)

: N·m (kg-m, ft-lb)

: Always replace after every disassembly.

1.	Strut upper plate	2.	Strut spacer	3.	Mounting insulator
4.	Mounting insulator bracket	5.	Mounting bearing	6.	Spring upper seat
7.	Spring upper rubber seat	8.	Coil spring	9.	Spring lower rubber seat
10.	Bound bumper	11.	Strut	12.	Steering knuckle
13.	Front suspension member	14.	Transverse link	15.	Stabilizer bar
16.	Stabilizer bushing	17.	Stabilizer clamp	18.	Stabilizer connecting rod
19.	Front cross bar	20.	Cotter pin		

67162-FX35-G223

Fig. 196 Exploded view of the front suspension components

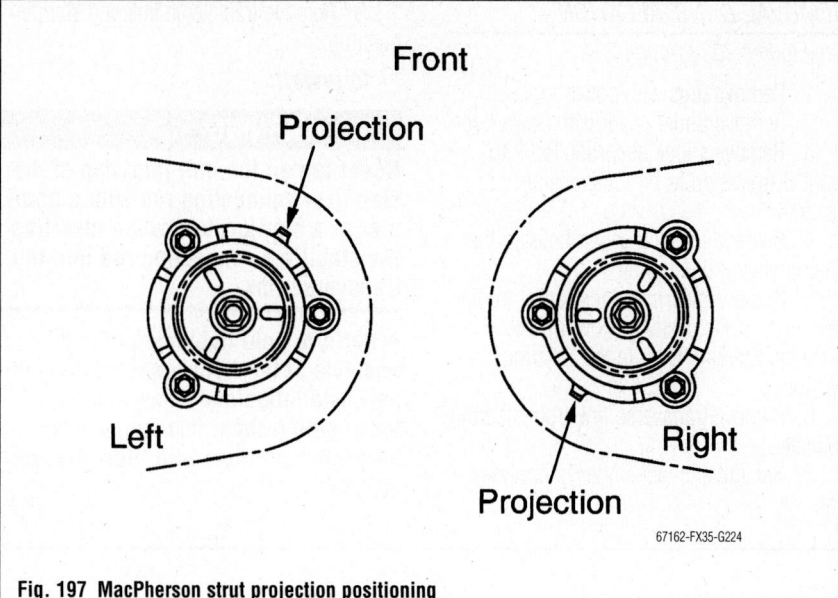

Fig. 197 MacPherson strut projection positioning

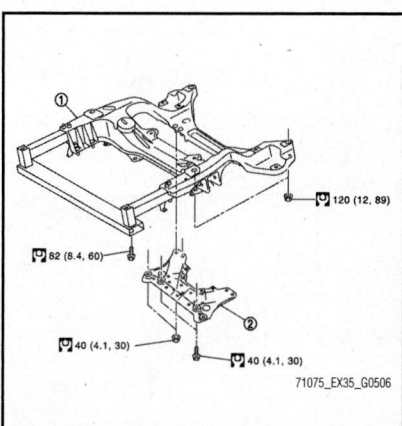

Fig. 198 Exploded view of the front suspension member (1), suspension member stay (2)–2WD

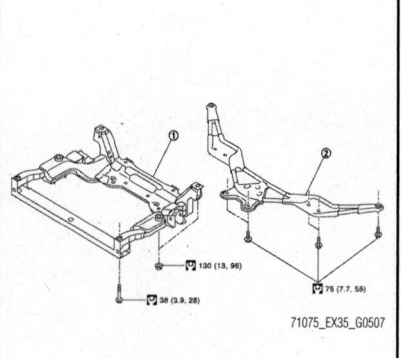

Fig. 199 Exploded view of the front suspension member (1), front cross bar (2)—AWD

suspension member. Refer to Transverse Link in Front Suspension in the Suspension Section.

11. Remove steering hydraulic piping bracket and steering gear from front suspension member. Refer to Power Steering Hoses & Lines in the Steering Section.

12. Set suitable jack front suspension member.

13. Remove mounting nuts between engine mounting insulator and from suspension member.

14. Remove mounting bolts and nuts of front suspension member with power tool.

15. Gradually lower jack to remove front suspension assembly from vehicle.

To install:

➡**Perform final tightening of bolts and nuts at the vehicle installation position**

(rubber bushing), under unladen condition with tires on level ground.

16. Gradually raise jack to install front suspension assembly to vehicle.

17. Install mounting bolts and nuts of front suspension member with power tool.

18. Install mounting nuts between engine mounting insulator and to suspension member.

19. Set suitable jack front suspension member.

20. Install steering gear to front suspension member, and then install Remove steering hydraulic piping bracket. Refer to Power Steering Hoses & Lines in the Steering Section.

21. Install transverse link to front suspension member. Refer to Transverse Link in Front Suspension in the Suspension Section.

22. Remove engine slinger, and then lower engine.

23. Install stabilizer connecting rods and stabilizer bar. Refer to Front Suspension in the Suspension Section.

24. Install shock absorbers to transverse links (AWD Only). Refer to Coil Springs And Shock Absorber in Front Suspension in the Suspension Section.

25. Install wheel sensors and sensor harness to steering knuckles. Refer to Anti-Lock Brakes System in the Brakes Section.

26. Install steering outer sockets to steering knuckles. Refer to Power Rack & Pinion Gear in the Steering Section.

27. Attach steering gear assembly and lower joint. Refer to Power Rack & Pinion Gear in the Steering Section.

28. Install suspension member stays (2WD), front cross bar (AWD) with power tool.

29. Install under cover with power tool.

30. Install tires with power tool.

TRANSVERSE LINK

INSPECTION

Inspection After Removal

See Figures 200 and 201.

Appearance

1. Check the following items, and replace the part if necessary:

a. Transverse link and bushing for deformation, cracks or damage.

b. Ball joint boot for cracks or other damage, and also for grease leakage.

Ball Joint Inspection

2. Manually move ball stud to confirm it moves smoothly with no binding.

Swing Torque Inspection

3. Move the ball stud at least ten times by hand to check for smooth movement.

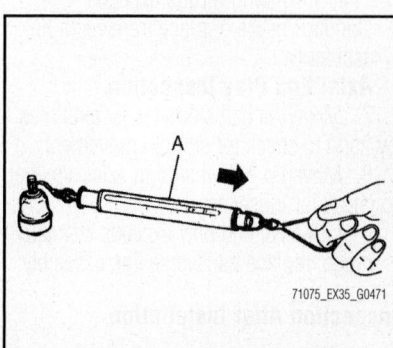

Fig. 200 Hook a spring balance (A) at cotter pin mounting hole. Confirm spring balance measurement value is within specifications when ball stud begins moving.

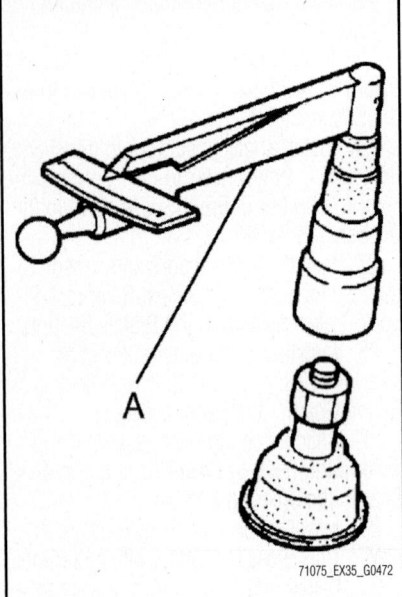

Fig. 201 Attach mounting nut to ball stud. Make sure that rotating torque is within specifications with a preload gauge (A) [SST: 3127S000 (J-25765-A)]

4. Hook a spring balance at cotter pin mounting hole. Confirm spring balance measurement value is within specifications when ball stud begins moving.

   a. If swing torque exceeds standard range, replace transverse link assembly.

Rotating Torque Inspection

5. Move the ball stud at least ten times by hand to check for smooth movement.

6. Attach mounting nut to ball stud. Make sure that rotating torque is within specifications with a preload gauge [SST: 3127S000 (J-25765-A)].

   a. If rotating torque exceeds standard range, replace transverse link assembly.

Axial End Play Inspection

7. Move the ball stud at least ten times by hand to check for smooth movement.

8. Move tip of ball stud in axial direction to check for looseness.

   a. If axial end play exceeds standard range, replace transverse link assembly.

Inspection After Installation

1. Check wheel sensor harness for proper connection. Refer to Front Disc Brakes in the Brakes Section.

2. Check wheel alignment. Refer to Specifications in the Suspension Section.

3. Adjust neutral position of steering angle sensor.

REMOVAL & INSTALLATION

See Figures 202 through 204.

1. Remove tires with power tool.
2. Remove under cover with power tool.
3. Remove shock absorber. Refer to Front Suspension in the Suspension Section.
4. Remove front crossbar. Refer to the accompanying illustration.
5. Remove steering outer socket from steering knuckle. Refer to Power Rack & Pinion Steering Gear in the Steering Section.
6. Remove transverse link from steering knuckle.
7. Set suitable jack under transverse link.
8. Remove transverse link and stopper bushing.

   To install:

✳✳ WARNING

Never tap on the ball joint cap of the stabilizer connecting rod with a hammer or a similar item when inserting the stabilizer connecting rod into the transverse link.

➡**Perform final tightening of bolts and nuts at the front suspension member installation and shock absorber lower side (rubber bushing), under unladen conditions with tires on level ground.**

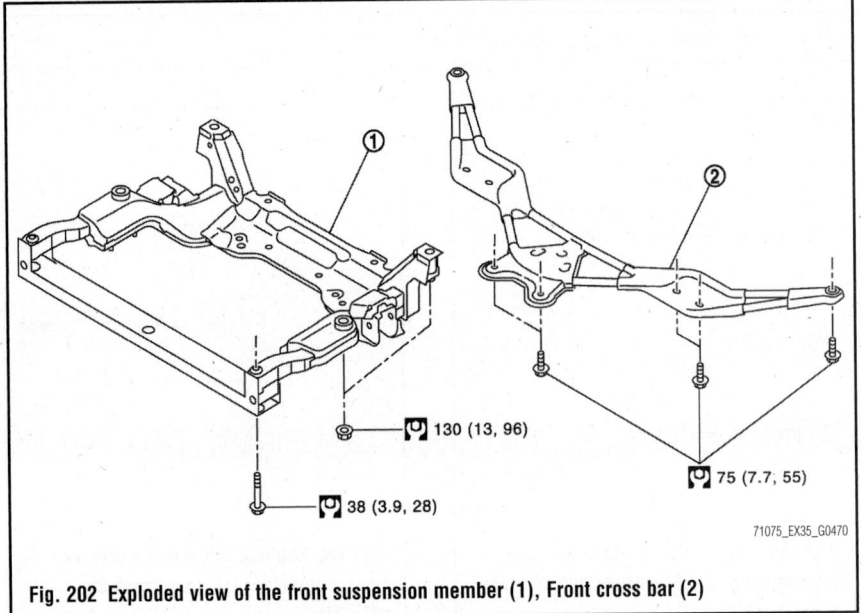

Fig. 202 Exploded view of the front suspension member (1), Front cross bar (2)

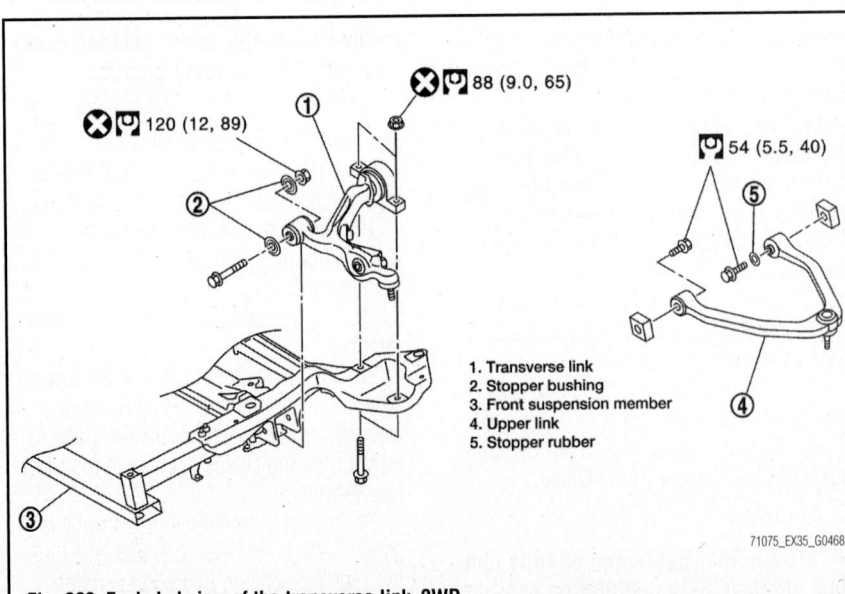

1. Transverse link
2. Stopper bushing
3. Front suspension member
4. Upper link
5. Stopper rubber

Fig. 203 Exploded view of the transverse link–2WD

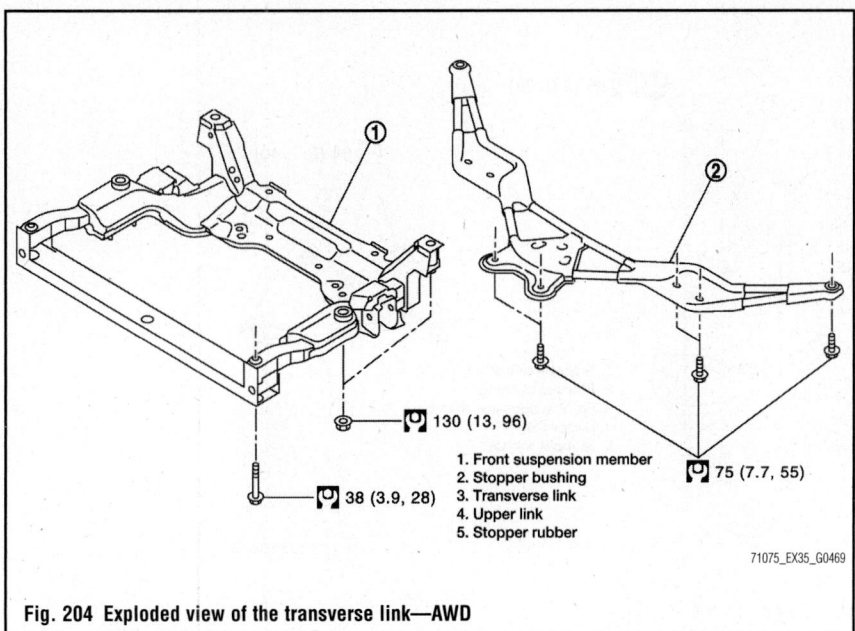

130 (13, 96)

38 (3.9, 28)

75 (7.7, 55)

1. Front suspension member
2. Stopper bushing
3. Transverse link
4. Upper link
5. Stopper rubber

71075_EX35_G0469

Fig. 204 Exploded view of the transverse link—AWD

9. Install transverse link and stopper bushing.
10. Remove jack from under transverse link.
11. Install transverse link to steering knuckle.
12. Install steering outer socket to steering knuckle. Refer to Power Rack & Pinion Steering Gear in the Steering Section.
13. Install front crossbar. Refer to the accompanying illustration.
14. Install shock absorber. Refer to Front Suspension in the Suspension Section.
15. Install under cover with power tool.
16. Install tires with power tool.

UPPER LINK

INSPECTION

Inspection After Removal

See Figure 205.

Appearance
1. Check the following items, and replace the part if necessary.
   a. Upper link and bushing for deformation, cracks or damage.
   b. Ball joint boot for cracks or other damage, and also for grease leakage.
Ball Joint Inspection
2. Manually move ball stud to confirm it moves smoothly with no binding.
Swing Torque Inspection
3. Move the ball stud at least ten times by hand to check for smooth movement.
4. Hook a spring balance at cutout on ball stud. Confirm spring balance measurement value is within specifications when ball stud begins moving.

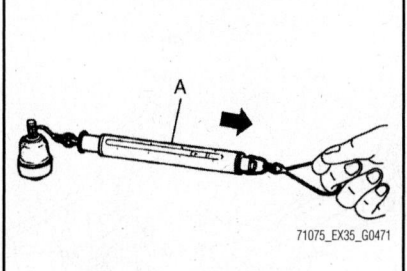

A

71075_EX35_G0471

Fig. 205 Hook a spring balance (A) at cutout on ball stud. Confirm spring balance measurement value is within specifications when ball stud begins moving

   a. f swing torque exceeds standard range, replace upper link assembly.
Axial End Play Inspection
5. Move the ball stud at least ten times by hand to check for smooth movement.
6. Move tip of ball stud in axial direction to check for looseness.
   a. If axial end play exceeds standard range, replace upper link assembly.

Inspection After Installation
1. Check wheel sensor harness for proper connection.
2. Check wheel alignment. Refer to Specifications in the Suspension Section.
3. Adjust neutral position of steering angle sensor.

REMOVAL & INSTALLATION

See Figures 206 and 207.

1. Remove tires with power tool.
2. Remove shock absorber. Refer to

Front Suspension in the Suspension Section.
3. Remove upper link from steering knuckle. Refer to Power Rack & Pinion Steering Gear in the Steering Section
4. Remove upper link and stopper rubber.

To install:

➡**Perform final tightening of bolts and nuts at the vehicle installation position (rubber bushing), under unladen conditions with tires on level ground.**

5. Install upper link and stopper rubber.
6. Install upper link from steering knuckle. Refer to Power Rack & Pinion Steering Gear in the Steering Section
7. Install shock absorber. Refer to Front Suspension in the Suspension Section.
8. Install tires with power tool.

WHEEL HUBS & BEARINGS

REMOVAL & INSTALLATION

See Figures 208 and 209.

1. Before servicing the vehicle, refer to the Precautions Section.
2. Raise and safely support the vehicle.
3. Remove the appropriate wheel.
4. Remove wheel sensor and sensor harness.

✳✳ WARNING

Never pull on wheel sensor harness.

5. Remove brake hose bracket. Refer
6. Remove the brake caliper. Support it in a place where it will not interfere with work.

➡**Avoid depressing the brake pedal while brake caliper is removed.**

7. Remove the disc rotor.
8. Remove the wheel sensor from the wheel hub and bearing assembly.

✳✳ WARNING

Do not pull on wheel sensor wiring harness.

9. Remove the cotter pin from the steering outer socket, then loosen the mounting nut.
10. Use a ball joint remover to separate the steering outer socket from the steering knuckle. Be careful not to damage the ball joint boot. Temporarily tighten the mounting nut to prevent damage to the threads and to prevent the ball joint remover from coming off.

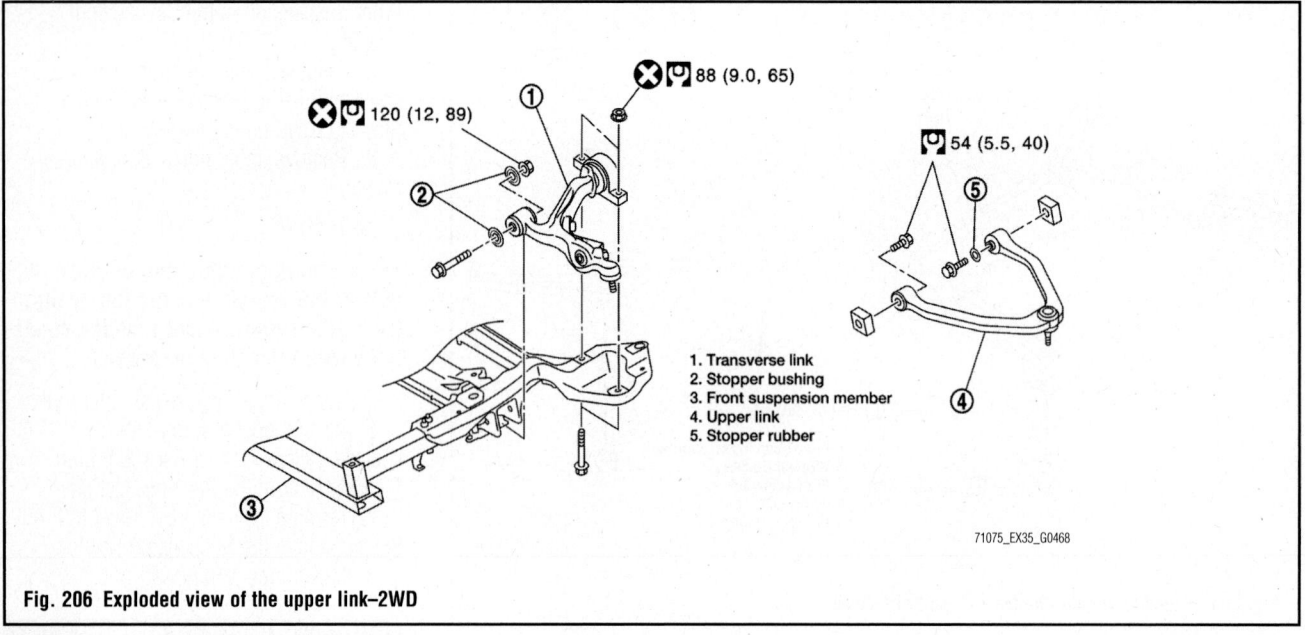

1. Transverse link
2. Stopper bushing
3. Front suspension member
4. Upper link
5. Stopper rubber

71075_EX35_G0468

Fig. 206 Exploded view of the upper link–2WD

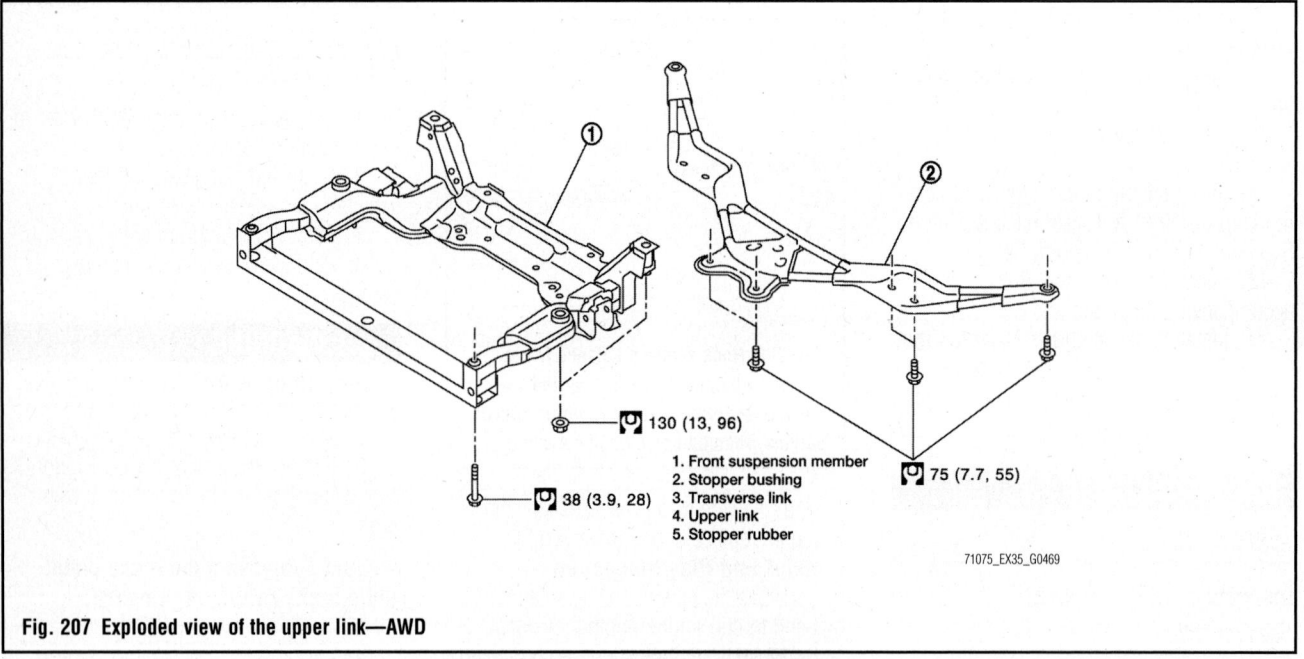

1. Front suspension member
2. Stopper bushing
3. Transverse link
4. Upper link
5. Stopper rubber

71075_EX35_G0469

Fig. 207 Exploded view of the upper link—AWD

11. Remove the cotter pin at the lower control arm (also known as the transverse link), then loosen the mounting nut.

12. Use a ball joint remover to separate the transverse link from the steering knuckle. Be careful not to damage ball joint boot. Temporarily tighten the mounting nut to prevent damage to the threads and to prevent the ball joint remover from coming off.

13. On AWD models, perform the following:

    a. Remove the cotter pin, then remove the lock nut from the halfshaft.

    b. Remove the steering knuckle from the halfshaft.

✳✳ WARNING

When removing steering knuckle, do not apply an excessive angle to the halfshaft joint. Also be careful not to excessively extend the slide joint. Do not hang the halfshaft without proper support.

14. Remove the mounting bolts and nuts between the strut assembly and the steering knuckle.

15. Remove the steering knuckle from the vehicle.

16. Remove the mounting bolts between the steering knuckle and the wheel hub/bearing assembly.

17. Remove the splash guard and wheel hub/bearing assembly from the steering knuckle.

18. Check for deformities, cracks, and damage on all parts and replace if necessary.

19. Inspect the ball joint for boot breakage, axial looseness, and torque of transverse link and steering outer socket ball joint. Maximum of allowable axial end play is 0.002 in. (0.05mm) or less.

To install:

20. Install the splash guard and wheel hub/bearing assembly onto the steering knuckle.

21. Install the mounting bolts between

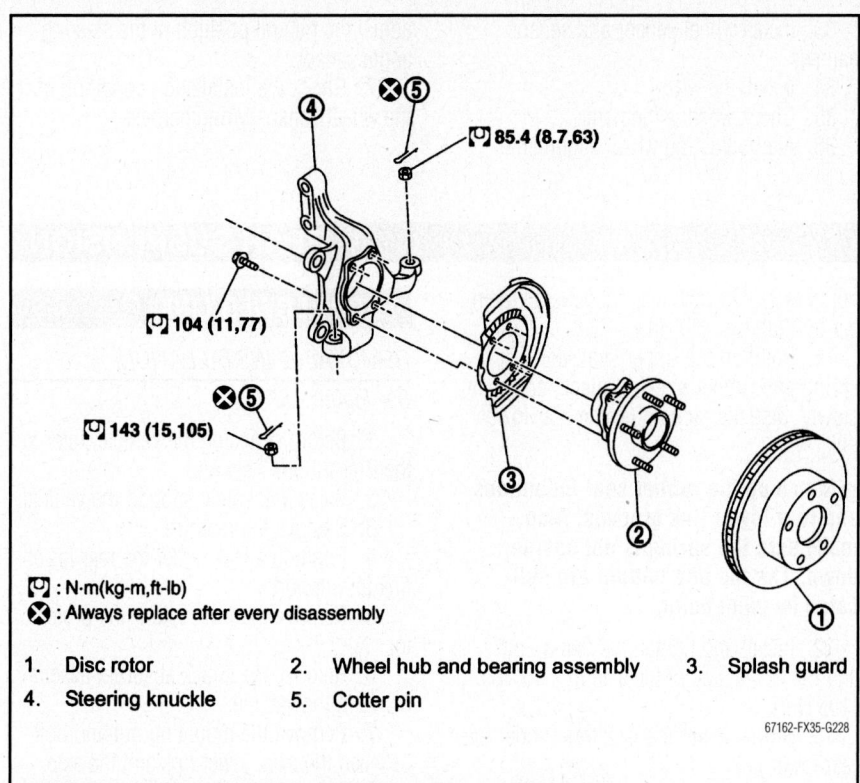

1 : N·m(kg-m,ft-lb)

⊗ : Always replace after every disassembly

1. Disc rotor
2. Wheel hub and bearing assembly
3. Splash guard
4. Steering knuckle
5. Cotter pin

67162-FX35-G228

Fig. 208 Exploded view of front wheel hub mounting—2WD models

the steering knuckle and the wheel hub/bearing assembly. Tighten the bolts to 77 ft. lbs. (104 Nm).

22. Install the steering knuckle on the vehicle.

23. Install the mounting bolts and nuts to the strut assembly and the steering knuckle. Tighten the bolts to 134 ft. lbs. (182 Nm).

24. On AWD models, perform the following:

　a. Install the steering knuckle onto the halfshaft.

　b. Install and tighten the lock nut on the halfshaft to 203 ft. lbs. (275 Nm). Install a new cotter pin.

25. Reattach the transverse link to the steering knuckle. Tighten the ball joint nut to 105 ft. lbs. (143 Nm).

26. Install a new cotter pin.

27. Reconnect the steering outer socket to the steering knuckle. Tighten the ball joint nut to 63 ft. lbs. (86 Nm).

28. Install a new cotter pin.

29. Install the wheel sensor onto the wheel hub and bearing assembly.

30. Install the disc rotor.

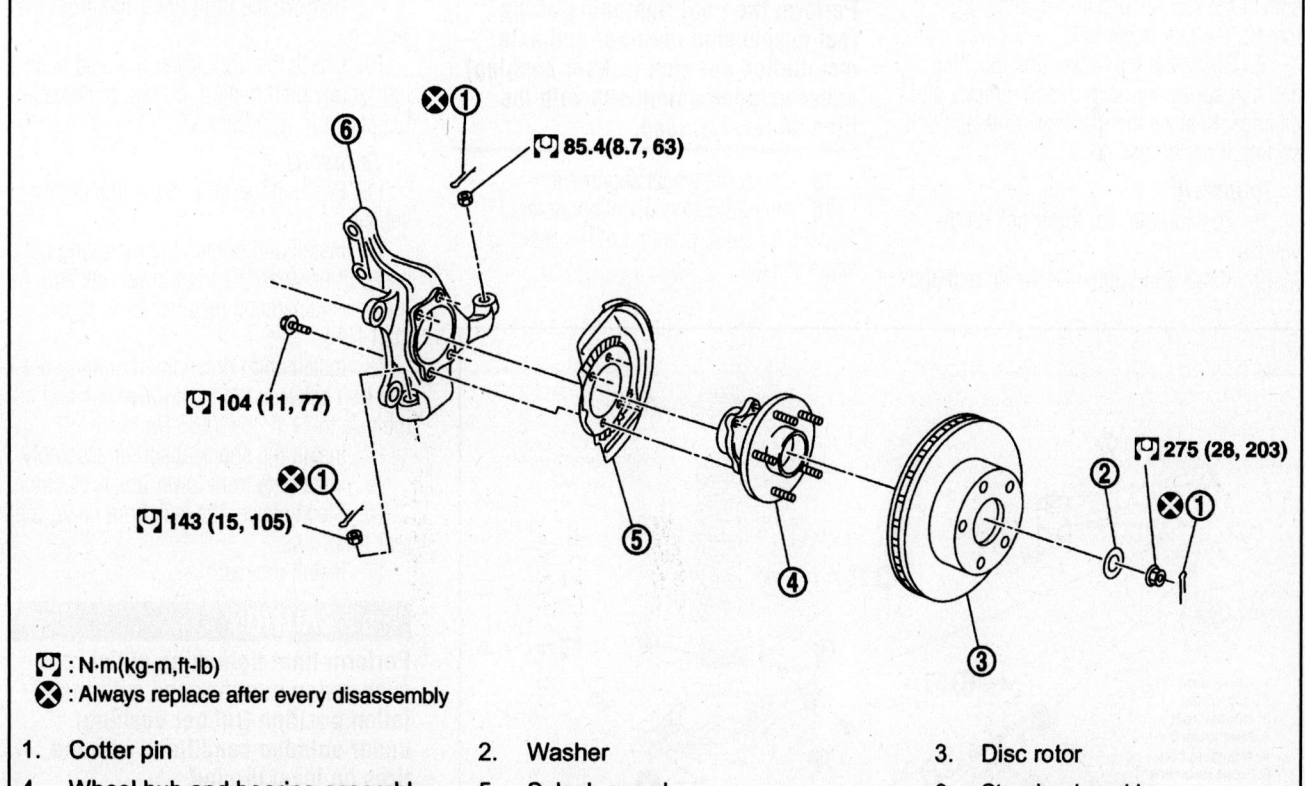

1 : N·m(kg-m,ft-lb)

⊗ : Always replace after every disassembly

1. Cotter pin
2. Washer
3. Disc rotor
4. Wheel hub and bearing assembly
5. Splash guard
6. Steering knuckle

67162-FX35-G229

Fig. 209 Exploded view of front wheel hub mounting—AWD models

31. Install the brake caliper.
32. Install brake hose bracket.

33. Install wheel sensor and sensor harness.
34. Install the wheel.
35. Check wheel alignment.
36. After adjusting wheel alignment, adjust the neutral position of the steering angle sensor.
37. Check the installation condition of the wheel sensor wiring harness.

SUSPENSION

COIL SPRINGS

REMOVAL & INSTALLATION

See Figure 210.

1. Before servicing the vehicle, refer to the Precautions Section.
2. Raise and safely support the vehicle.
3. Remove the rear tire.
4. Position a jack under the rear lower link for support.
5. Loosen the fixing bolt and nut of the rear lower link in the side of the suspension member, and then remove the fixing bolt and nut in the side of the axle.
6. Slowly lower the jack, then remove the upper seat, coil spring and rubber sheet from the rear lower link.
7. Remove the fixing bolt and nut in the side of the rear suspension member to remove the rear lower link.
8. Check the rear lower link, bushing, and coil spring for deformation, cracks, and damage. Replace the rear lower link and coil spring, if necessary.

To install:

9. Position the rear lower link on the vehicle.
10. Install and tighten the fixing bolt and nut in the side of the rear suspension member to 48 ft. lbs. (65 Nm).
11. Position the upper seat, coil spring and rubber sheet in place, and then slowly raise the jack under the rear lower link.

➡**Match up the rubber seat indentions and rear lower link grooves. Also, make sure the spring is not upside down. The top and bottom are indicated by paint color.**

12. Install and tighten the fixing bolt and nut in the side of the axle to 77 ft. lbs. (105 Nm).
13. Slowly lower the jack from under the rear lower link.
14. Install the rear tire.

❊❊ CAUTION

Perform the final tightening of the rear suspension member and axle installation position (rubber bushing) under unladen conditions with the tires on level ground.

15. Check the wheel alignment.
16. After adjusting wheel alignment, adjust the neutral position of the steering angle sensor.

REAR SUSPENSION

FRONT LOWER LINK

REMOVAL & INSTALLATION

See Figure 210.

1. Before servicing the vehicle, refer to the Precautions Section.
2. Raise and safely support the vehicle.
3. Remove the rear tire.
4. Position a jack under the rear lower link for support.
5. Remove the front lower link protector.
6. Remove the shock absorber assembly from the vehicle.
7. Remove the mounting nut and bolt between the front lower link and the axle.
8. Remove the mounting nut and bolt between the front lower link and the rear suspension member.
9. Remove the front lower link from the vehicle.
10. Check the front lower link and bushing for any deformation, cracks, or damage. Replace it if necessary.

To install:

11. Position the front lower link on the vehicle.
12. Install and tighten the mounting nut and bolt between the front lower link and the rear suspension member to 74 ft. lbs. (101 Nm).
13. Install and tighten the mounting nut and bolt between the front lower link and the axle to 77 ft. lbs. (105 Nm).
14. Install the shock absorber assembly.
15. Install the front lower link protector.
16. Slowly lower the jack from under the rear lower link.
17. Install the rear tire.

❊❊ CAUTION

Perform final tightening of the rear suspension member and axle installation position (rubber bushing) under unladen conditions with the tires on level ground.

18. Check the wheel alignment.
19. After adjusting wheel alignment, adjust the neutral position of the steering angle sensor.

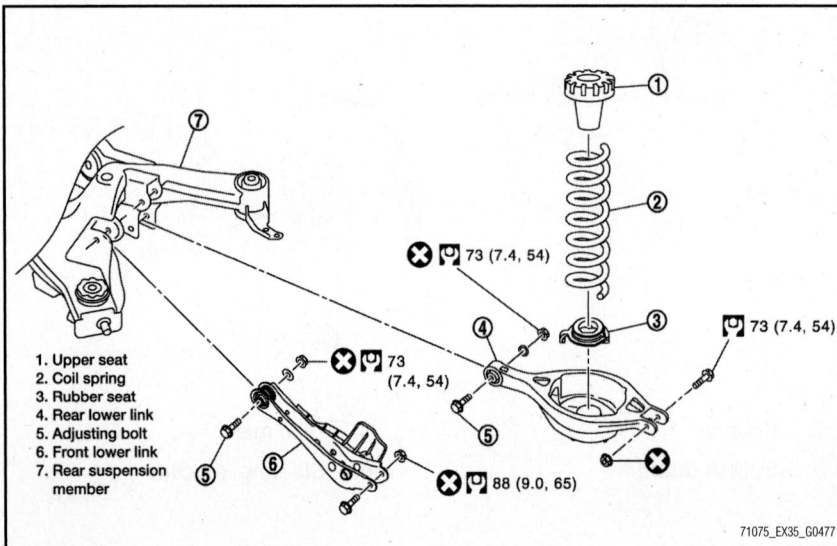

1. Upper seat
2. Coil spring
3. Rubber seat
4. Rear lower link
5. Adjusting bolt
6. Front lower link
7. Rear suspension member

❌ ⊔ 73 (7.4, 54)

⊔ 73 (7.4, 54)

❌ ⊔ 73 (7.4, 54)

❌ ⊔ 88 (9.0, 65)

71075_EX35_G0477

Fig. 210 Exploded view of the coil spring components

RADIUS ROD

REMOVAL & INSTALLATION

See Figure 211.

1. Remove tires with power tool.
2. Remove radius rod mounting bolt and nut (axle housing side).
3. Remove radius rod mounting bolt (rear suspension member side), and remove radius rod.

To install:

➡**Perform final tightening of rear suspension member and axle installation position (rubber bushing), under unladen conditions with tires on level ground.**

4. Install radius rod, and then install radius rod mounting bolt (rear suspension member side).
5. Install radius rod mounting bolt and nut (axle housing side).
6. Install tires with power tool.

REAR LATERAL LINKS

REMOVAL & INSTALLATION

See Figure 210.

Rear Lower Link

1. Before servicing the vehicle, refer to the Precautions Section.
2. Raise and safely support the vehicle.
3. Remove the rear tire.
4. Position a jack under the rear lower link for support.
5. Loosen the fixing bolt and nut of the rear lower link in the side of the suspension member and then remove the fixing bolt and nut in the side of the axle.
6. Slowly lower the jack, then remove the upper seat, coil spring, and rubber sheet from the rear lower link.

7. Remove the fixing bolt and nut in the side of the rear suspension member to remove the rear lower link.
8. Check the rear lower link, bushing, and coil spring for deformation, cracks, and damage. Replace the rear lower link and coil spring, if necessary.

To install:

9. Position the rear lower link on the vehicle.
10. Install and tighten the fixing bolt and nut in the side of the rear suspension member to 48 ft. lbs. (65 Nm).
11. Position the upper seat, coil spring and rubber sheet in place, and then slowly raise the jack under the rear lower link.

➡**Match up the rubber seat indentions and rear lower link grooves. Also, make sure the spring is not upside down. The top and bottom are indicated by paint color.**

12. Install and tighten the fixing bolt and nut in the side of the axle to 77 ft. lbs. (105 Nm).
13. Slowly lower the jack from under the rear lower link.
14. Install the rear tire.

➡**Perform final tightening of the rear suspension member and axle installation position (rubber bushing) under unladen conditions with the tires on level ground.**

15. Check the wheel alignment.

16. After adjusting wheel alignment, adjust the neutral position of the steering angle sensor.

REAR SUSPENSION MEMBER

REMOVAL & INSTALLATION

See Figure 212.

1. Remove tires with power tool.
2. Remove radius rod.
3. Remove caliper assemblies. Hang caliper assembly in a place where it will not interfere with work.

❈❈ WARNING

Avoid depressing brake pedal while brake caliper is removed.

4. Remove disc rotors.
5. Remove wheel sensors and sensor harness from rear suspension member and suspension arms.
6. Remove height sensor harness from rear suspension member (with xenon head lamp).
7. Remove center muffler.
8. Remove stabilizer bar.
9. Remove drive shafts.
10. Remove propeller shaft.
11. Remove final drive.
12. Remove parking brake cable mounting bolts and separate parking brake cable from vehicle and rear suspension member.
13. Remove shock absorber mounting bolts (lower side).

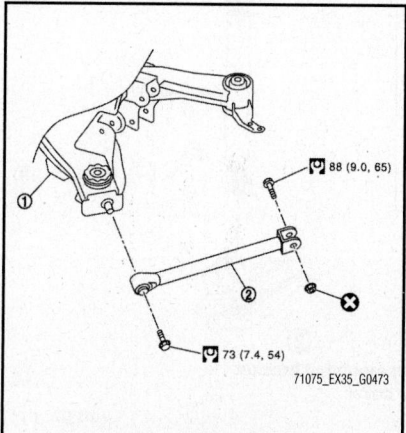

88 (9.0, 65)

73 (7.4, 54)

71075_EX35_G0473

Fig. 211 Exploded view of the radius rod (2), rear suspension member (1)

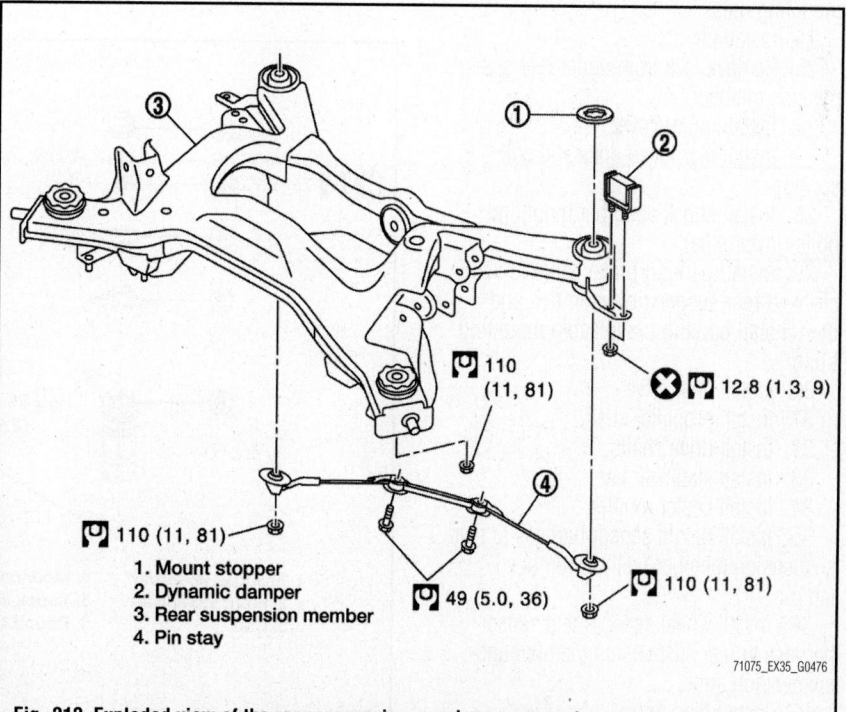

110 (11, 81)

❌ 12.8 (1.3, 9)

110 (11, 81)

49 (5.0, 36)

110 (11, 81)

1. Mount stopper
2. Dynamic damper
3. Rear suspension member
4. Pin stay

71075_EX35_G0476

Fig. 212 Exploded view of the rear suspension member components

14. Remove rear lower links and coil springs.

15. Remove under cover.

16. Set suitable jack under rear suspension member.

17. Remove pin stay.

18. Remove rear suspension member mounting nuts.

19. Slowly lower jack, then remove rear suspension member, suspension arms, front lower links, wheel hub and housings from vehicle as a unit.

20. Remove mounting bolts and nuts, then remove suspension arms, front lower links, wheel hub and housings from rear suspension member.

To install:

➡ **Perform the final tightening of each of parts under unladen conditions, which were removed when removing rear suspension assembly.**

➡ **Check wheel sensor harness for proper connection.**

➡ **Never reuse cotter pin.**

21. Install suspension arms, front lower links, wheel hub and housings to rear suspension member, and then install mounting bolts and nuts.

22. Slowly lower jack, then install rear suspension member, suspension arms, front lower links, wheel hub and housings to vehicle as a unit.

23. Install rear suspension member mounting nuts.

24. Install pin stay.

25. Remove jack from under rear suspension member.

26. Install under cover.

27. Install rear lower links and coil springs.

28. Install shock absorber mounting bolts (lower side).

29. Install parking brake cable to vehicle, and rear suspension member, and then install parking brake cable mounting bolts.

30. Install final drive.

31. Install propeller shaft.

32. Install drive shafts.

33. Install stabilizer bar.

34. Install center muffler.

35. Install height sensor harness to rear suspension member (with xenon head lamp).

36. Install wheel sensors and sensor harness to rear suspension member and suspension arms.

37. Install disc rotors.

38. Install caliper assemblies.

39. Install radius rod.

40. Install tires with power tool.

SHOCK ABSORBERS

REMOVAL & INSTALLATION

See Figure 213.

1. Before servicing the vehicle, refer to the Precautions Section.

2. Raise and safely support the vehicle.

3. Remove the rear tire.

4. Position a jack or equivalent support under the rear lower link.

5. Remove the fixing bolt in the lower side of the shock absorber assembly.

6. Remove the attaching nuts in the upper side of the shock absorber assembly and remove the shock absorber assembly from the vehicle.

7. Check the shock absorber assembly for deformation, cracks, or damage, and replace if necessary.

8. Check the piston rod for damage, uneven wear, or distortion, and replace if necessary.

9. Check the welded and sealed areas for oil leakage, and replace if necessary.

To install:

10. Position the shock absorber assembly on the vehicle.

11. Install and tighten the attaching nuts in the upper side of the shock absorber assembly to 22 ft. lbs. (30 Nm).

12. Install a new upper shock absorber mounting nut and tighten to 33 ft. lbs. (45 Nm).

13. Install the fixing bolt in the lower side of the shock absorber assembly and tighten until snug.

14. Remove the jack or equivalent support from under the rear lower link.

15. Install the tire.

16. With the weight of the vehicle resting on the suspension (empty vehicle), tighten the shock absorber lower fixing bolt to 66 ft. lbs. (89 Nm).

17. Check the wheel alignment.

18. After adjusting wheel alignment, adjust the neutral position of the steering angle sensor.

TESTING

1. Before servicing the vehicle, refer to the Precautions Section.

2. Check for oil leakage around seals and welds.

3. Move the piston rod up and down to check if it operates smoothly without any binding.

STABILIZER BAR (SWAY BAR)

REMOVAL & INSTALLATION

See Figure 214.

1. Remove center muffler.

2. Remove under cover.

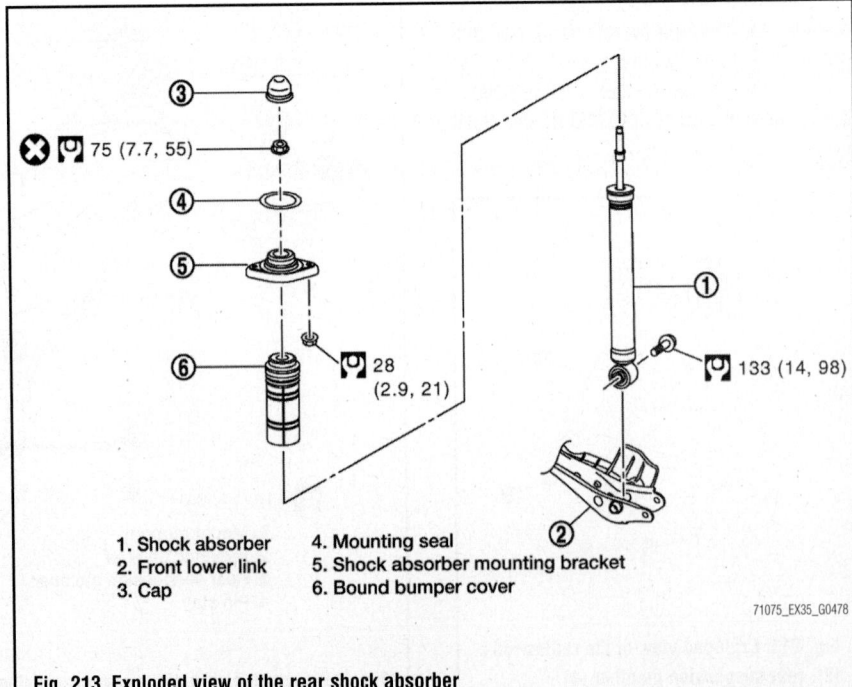

③ ❌ 🔧 75 (7.7, 55)

🔧 28 (2.9, 21)

① 🔧 133 (14, 98)

1. Shock absorber
2. Front lower link
3. Cap
4. Mounting seal
5. Shock absorber mounting bracket
6. Bound bumper cover

71075_EX35_G0478

Fig. 213 Exploded view of the rear shock absorber

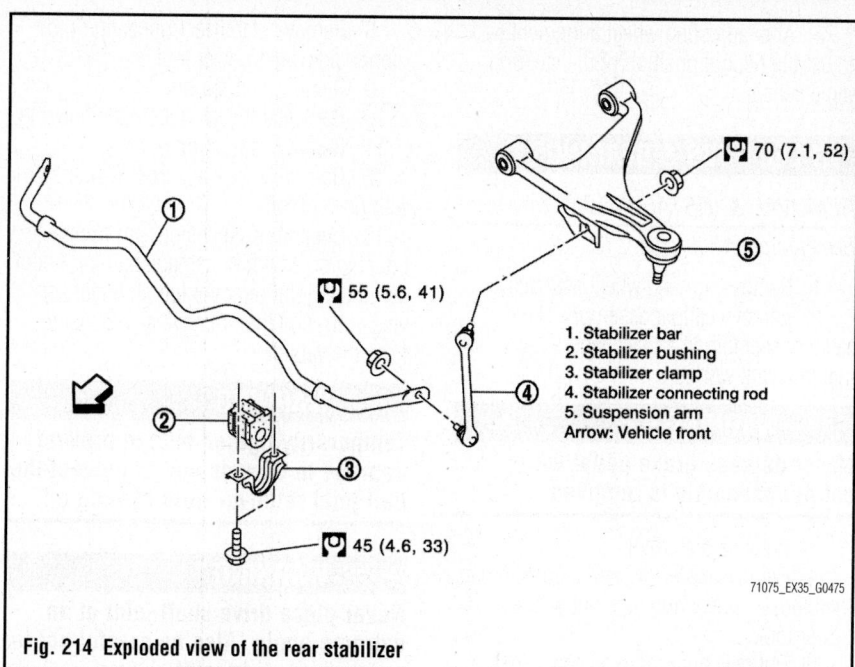

Fig. 214 Exploded view of the rear stabilizer

1. Stabilizer bar
2. Stabilizer bushing
3. Stabilizer clamp
4. Stabilizer connecting rod
5. Suspension arm
Arrow: Vehicle front

70 (7.1, 52)
55 (5.6, 41)
45 (4.6, 33)
71075_EX35_G0475

3. Remove stabilizer connecting rod mounting nuts (lower side), and remove stabilizer connecting rods from stabilizer bar.

4. Remove stabilizer connecting rod mounting nuts (upper side), and remove stabilizer connecting rods from stabilizer connecting rod mounting brackets.

5. Remove mounting nuts on stabilizer clamps and remove stabilizer bar.

To install:

➡ **Tighten the mounting nut to the specified torque while holding a hexagonal part of stabilizer connecting rod side.**

6. Install stabilizer bar, and then install mounting nuts on stabilizer clamps.

7. Install stabilizer connecting rods to stabilizer connecting rod mounting brackets, and then install stabilizer connecting rod mounting nuts (upper side).

8. Install stabilizer connecting rods to stabilizer bar, and then install stabilizer connecting rod mounting nuts (lower side).

9. Install under cover.

10. Install center muffler.

SUSPENSION ARM

REMOVAL & INSTALLATION

See Figure 215.

1. Remove tires with power tool.

2. Remove radius rod.

3. Remove caliper assembly. Hang torque member in a place where it will not interfere with work.

4. Set suitable jack under axle assembly to relieve the coil spring tension.

5. Remove stabilizer connecting rod.

6. Remove drive shaft.

7. Remove height sensor (with xenon head lamp).

8. Remove cotter pin of suspension arm ball joint, and loosen nut.

9. Remove suspension arm mounting bolts and nuts (rear suspension member side).

10. Use the ball joint remover to remove suspension arm from axle housing. Be careful not to damage ball joint boot.

✳✳ WARNING

Tighten temporarily mounting nut to prevent damage to threads and to

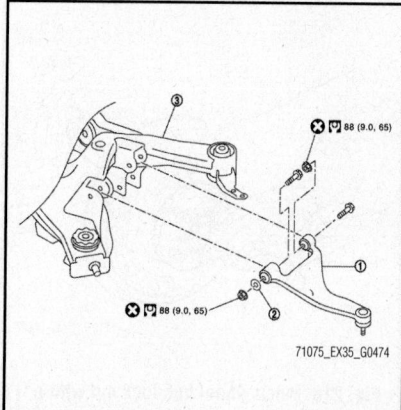

Fig. 215 Exploded view of the suspension arm (1), stopper rubber (2), rear suspension member (3)

88 (9.0, 65)
88 (9.0, 65)
71075_EX35_G0474

prevent ball joint remover from coming off.

11. Remove suspension arm.

12. Remove stabilizer connecting rod mounting bracket.

To install:

➡ **Perform final tightening of rear suspension member installation position (rubber bussing), under unladen conditions with tires on level ground.**

➡ **Never reuse cotter pin.**

13. Install stabilizer connecting rod mounting bracket.

14. Remove suspension arm.

✳✳ WARNING

Tighten temporarily mounting nut to prevent damage to threads and to prevent ball joint remover from coming off.

15. Use the ball joint remover to install suspension arm to axle housing. Be careful not to damage ball joint boot.

16. Install suspension arm mounting bolts and nuts (rear suspension member side).

17. Install cotter pin of suspension arm ball joint, and loosen nut.

18. Install height sensor (with xenon head lamp).

19. Install drive shaft.

20. Install stabilizer connecting rod.

21. Remove jack from under axle assembly to engage the coil spring tension.

22. Install caliper assembly. Hang torque member in a place where it will not interfere with work.

23. Install radius rod.

24. Install tires with power tool.

UPPER CONTROL ARMS

REMOVAL & INSTALLATION

1. Before servicing the vehicle, refer to the Precautions Section.

2. Raise and safely support the vehicle.

3. Remove the rear tire.

4. Remove the stabilizer connecting rod mounting bracket from the suspension arm.

5. Remove the halfshaft from the vehicle.

6. Remove the cotter pin of the suspension arm ball joint, and loosen the nut.

7. Use a ball joint remover or suitable tool to remove the suspension arm from the axle. Be careful not to damage the ball joint boot.

➡️**Temporarily tighten the mounting nut to prevent damage to the threads and to prevent the ball joint remover from coming off.**

8. Remove the fixing nuts and bolts between the suspension arm and the rear suspension member.

9. Remove the suspension arm from the vehicle.

10. Check the suspension arm and bushing for deformation, cracks, or damage. If any non-standard condition is found, replace it.

11. Check the boot of the ball joint for cracks or damage and also for grease leakage.

12. Manually move the ball stud to confirm it moves smoothly with no binding.

➡️**Before measuring, move ball joint at least 10 times by hand to check for smooth movement.**

13. Hook a spring scale at the cotter pin mounting hole. Confirm the spring scale measurement value is within 2–15 lbs. (10–66 N) when the ball joint stud begins moving. If it is outside the specified range, replace the suspension arm assembly.

14. Attach the mounting nut to the ball stud. Make sure the rotating torque is within 5–30 inch lbs. (1–3 Nm) with a preload gauge . If it is outside the specified range, replace the suspension arm assembly.

15. Move the tip of the ball joint in the axial direction to check for looseness. If it is outside the specified range of 0 in. (0mm), replace the suspension arm assembly.

To install:

16. Install the suspension arm onto the vehicle.

17. Install the fixing nuts and bolts between the suspension arm and the rear suspension member. Tighten them to 53 ft. lbs. (73 Nm).

18. Reattach the suspension arm ball joint to the axle. Tighten the ball joint nut to 96 ft. lbs. (130 Nm).

19. Install a new cotter pin.

20. Install the halfshaft.

21. Install the stabilizer connecting rod mounting bracket onto the suspension arm, and tighten the 2 mounting bolts to 41 ft. lbs. (55 Nm).

22. Install the rear tire.

➡️**Do not reuse non-reusable parts.**

23. Perform the final tightening of the rear suspension member installation position (rubber bushing) under unladen conditions with the tires on level ground.

24. Check wheel alignment.

25. After adjusting wheel alignment, adjust the neutral position of the steering angle sensor.

WHEEL HUBS & BEARINGS

REMOVAL & INSTALLATION

See Figures 216 through 218.

1. Remove tire and wheel assembly.

2. Remove caliper assembly. Hang caliper assembly in a place where it will not interfere with work.

❊❊ WARNING

Never depress brake pedal while caliper assembly is removed.

3. Remove disc rotor.

4. Remove cotter pin and adjusting cap, then loosen wheel hub lock nut with a power tool.

5. Put matching mark on drive shaft and wheel hub and bearing assembly.

6. Patch wheel hub lock nut with a piece of wood. Hammer the wood to disengage wheel hub and bearing assembly from drive shaft. Take out the wheel hub lock nut.

➡️**Never place drive shaft joint at an extreme angle. Also be careful not to overextend slide joint.**

➡️**Never allow drive shaft to hang down without support for counterpart such as joint sub-assembly, and other parts.**

➡️**Use a suitable puller, if wheel hub and bearing assembly and drive shaft cannot be separated even after performing the above procedure.**

7. Remove parking brake shoe and parking brake cable from back plate.

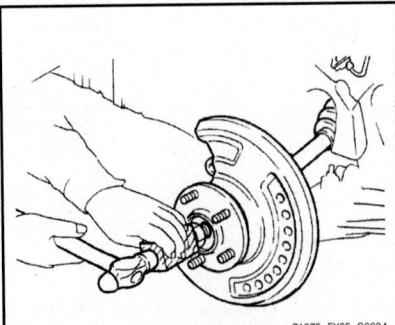

71075_EX35_G0634

Fig. 216 Patch wheel hub lock nut with a piece of wood. Hammer the wood to disengage wheel hub and bearing assembly from drive shaft. Take out the wheel hub lock nut.

8. Remove stabilizer connecting rod (upper side) with power tool.

9. Remove coil spring.

10. Set suitable jack under axle housing.

11. Remove radius rod.

12. Remove shock absorber (lower side) with power tool.

13. Separate suspension arm from axle housing so as not to damage ball joint boot using ball joint remover (commercial service tool), and then remove axle housing from the vehicle.

❊❊ WARNING

Temporarily tighten nuts to prevent damage to threads and to prevent the ball joint remover from coming off.

❊❊ WARNING

Never place drive shaft joint at an extreme angle. Also be careful not to overextend slide joint.

❊❊ WARNING

Never allow drive shaft to hang down without support for counterpart such as joint sub-assembly, and other parts.

14. Remove front lower link (axle housing side).

15. Remove rear lower link (axle housing side).

16. Remove the wheel hub and bearing assembly.

17. Remove anchor block mounting nuts, and then remove anchor block and back plate from axle housing.

To install:

➡️**Clean the matching surface of wheel hub lock nut and wheel hub and bearing assembly.**

❊❊ WARNING

Never apply lubricating oil to these matching surface.

➡️**Clean the matching surface of drive shaft and wheel hub and bearing assembly. And then apply paste (0.10 oz.) [service parts (440037S000)] to surface of joint sub-assembly of drive shaft.**

➡️**Apply paste to cover entire flat surface of joint sub-assembly of drive shaft.**

➡️**When installing drive shaft, change the drive shaft and wheel hub and**

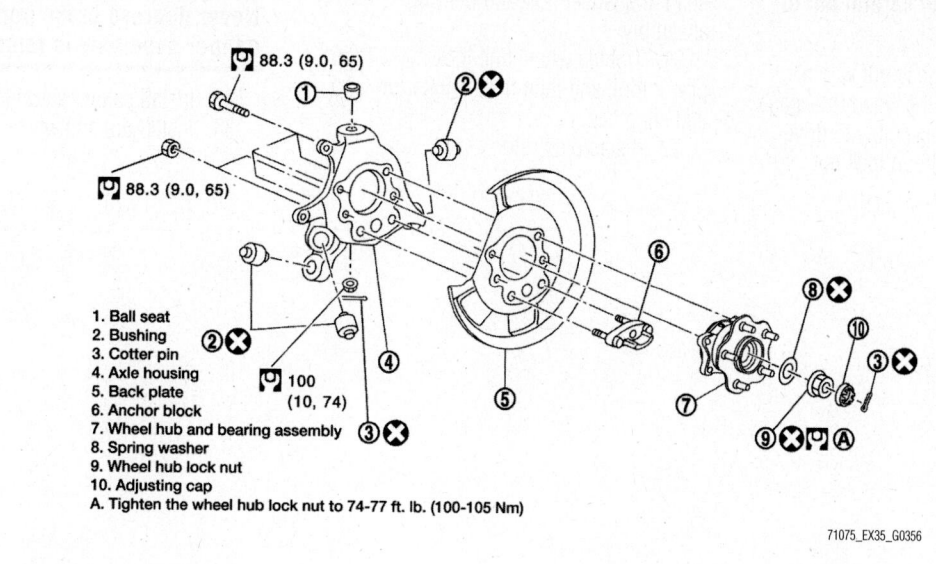

1. Ball seat
2. Bushing
3. Cotter pin
4. Axle housing
5. Back plate
6. Anchor block
7. Wheel hub and bearing assembly
8. Spring washer
9. Wheel hub lock nut
10. Adjusting cap
A. Tighten the wheel hub lock nut to 74-77 ft. lb. (100-105 Nm)

88.3 (9.0, 65)

88.3 (9.0, 65)

100 (10, 74)

71075_EX35_G0356

Fig. 217 Exploded view of the rear wheel hub and housing

bearing assembly matching marks put at the removal step by 180 degree.

➡Use the following torque range for tightening the wheel hub lock nut. Tighten to 74-77 ft. lbs. (100-105 Nm).

➡Since the drive shaft is assembled by press-fitting, use the tightening torque range for the wheel hub lock nut. Tighten to 74-77 ft. lbs. (100-105 Nm).

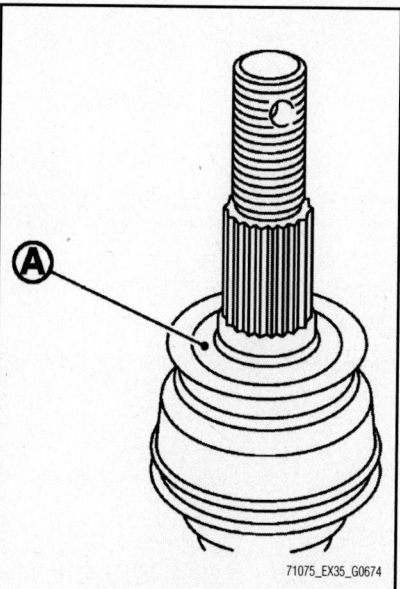

71075_EX35_G0674

Fig. 218 Clean the matching surface of drive shaft and wheel hub and bearing assembly. And then apply paste (0.10 oz.) [service parts (440037S000)] to surface (A) of joint sub-assembly of drive shaft.

➡Be sure to use torque wrench to tighten the wheel hub lock nut. Never use a power tool.

➡Wheel hub lock nut tightening torque does not over torque for avoiding axle noise, and does not less than torque for avoiding looseness.

➡Perform the final tightening of each of parts under unladen conditions, which were removed when removing wheel hub and bearing assembly and axle housing.

➡When installing the spring washer, face the identification paint mark to the wheel hub and bearing assembly side.

➡When installing the adjusting cap, check that there must be no play.

➡Never reuse cotter pin, wheel hub lock nut, spring washer, and bushing.

18. Install anchor block and back plate to axle housing, and then install anchor block mounting nuts.
19. Install the wheel hub and bearing assembly.
20. Install rear lower link (axle housing side).
21. Install front lower link (axle housing side).

✳✳ WARNING

Never allow drive shaft to hang down without support for counterpart such

as joint sub-assembly, and other parts.

✳✳ WARNING

Never place drive shaft joint at an extreme angle. Also be careful not to overextend slide joint.

✳✳ WARNING

Temporarily tighten nuts to prevent damage to threads and to prevent the ball joint remover from coming off.

22. Install axle housing to the vehicle, and then install suspension arm to axle housing so as not to damage ball joint boot using ball joint remover (commercial service tool).
23. Install shock absorber (lower side) with power tool.
24. Install radius rod.
25. Set suitable jack under axle housing.
26. Install coil spring.
27. Install stabilizer connecting rod (upper side) with power tool.
28. Install parking brake shoe and parking brake cable from back plate.

➡Use a suitable puller, if wheel hub and bearing assembly and drive shaft cannot be separated even after performing the above procedure.

➡Never allow drive shaft to hang down without support for counterpart such as joint sub-assembly, and other parts.

➡**Never place drive shaft joint at an extreme angle. Also be careful not to overextend slide joint.**

29. Patch wheel hub lock nut with a piece of wood. Hammer the wood to engage wheel hub and bearing assembly to drive shaft. Take out the wheel hub lock nut.

30. Align matching mark on drive shaft and wheel hub and bearing assembly.

31. Tighten wheel hub lock nut with a power tool, and then install cotter pin and adjusting cap.

32. Install disc rotor.

❄❄ WARNING

Never depress brake pedal while caliper assembly is removed.

33. Install caliper assembly.
34. Install tire and wheel assembly.

SPECIFICATIONS AND MAINTENANCE CHARTS

ENGINE AND VEHICLE IDENTIFICATION

	Engine							Model Year	
Code ①	Liters (cc)	Cu. In.	Cyl.	Fuel Sys.	Engine Type	Eng. Mfg.		Code ②	Year
VQ35HR	3.5 (3,498)	213.5	6	MFI	DOHC	Nissan		B	2011
VK50VE	5.0 (5,026)	306.7	8	MFI	DOHC	Nissan		C	2012

MFI: Multiport Fuel Injection

DOHC: Double Overhead Camshafts

① The VQ35HR engine code is stamped on the right rear of the engine block near the flywheel.

The VK50VE engine code is stamped on the front of the engine block near the water outlet.

② 10th position of the Vehicle Identification Number (VIN)

71075_FX35_C0001

GENERAL ENGINE SPECIFICATIONS

All measurements are given in inches.

Year	Model	Engine Displacement Liters (cc)	Engine ID/VIN	Fuel System Type	Net Horsepower @ rpm	Net Torque @ rpm (ft. lbs.)	Bore x Stroke (in.)	Com-pression Ratio	Oil Pressure @ rpm
2011	FX35	3.5 (3,498)	VQ35HR	MFI	303@6,800	262@4,800	3.76 x 3.21	10.6:1	43 psi@2,000
	FX50	5.0 (5,026)	VK50VE	MFI	390@6,500	369@4,400	3.76 x 3.45	10.9:1	43 psi@2,000
2012	FX35	3.5 (3,498)	VQ35HR	MFI	303@6,800	262@4,800	3.76 x 3.21	10.6:1	43 psi@2,000
	FX50	5.0 (5,026)	VK50VE	MFI	390@6,500	369@4,400	3.76 x 3.45	10.9:1	43 psi@2,000

MFI: Multiport Fuel Injection

71075_FX35_C0002

ENGINE TUNE-UP SPECIFICATIONS

Year	Engine Displacement Liters	Engine ID/VIN	Spark Plug Gap (in.)	Ignition Timing (deg. BTDC) MT	AT ①	Fuel Pump (psi) ②	Idle Speed (rpm) MT	AT ①	Valve Clearance (in.) Intake ③	Exhaust ③
2011	3.5	VQ35HR	0.043	N/A	10-20	51	N/A	625-725	0.010-0.013	0.011-0.015
	5.0	VK50VE	0.043	N/A	10-20	51	N/A	600-700	0.010-0.013	0.011-0.015
2012	3.5	VQ35HR	0.043	N/A	10-20	51	N/A	625-725	0.010-0.013	0.011-0.015
	5.0	VK50VE	0.043	N/A	10-20	51	N/A	600-700	0.010-0.013	0.011-0.015

NOTE: The Vehicle Emission Control Information label often reflects specification changes made during production.

The label figures must be used if they differ from those in this chart.

BTDC: Before Top Dead Center

N/A: Not Applicable

① Under no load condition (in P or N position)

② System pressure at idle

③ With engine cold

71075_FX35_C0003

CAPACITIES

Year	Model	Engine Displacement Liters	Engine ID/VIN	Engine Oil with Filter (qts.)	Transaxle (pts.) Auto. ①	Transaxle (pts.) Manual	Drive Axle (pts.) Front	Drive Axle (pts.) Rear	Transfer Case (pts.)	Fuel Tank (gal.)	Cooling System (qts.) ②
2011	FX35	3.5	VQ35HR	5.2	19.4	N/A	1.4	3.0	2.1	23.8	9.7
	FX50	5.0	VK50VE	7.1	24.0	N/A	1.4	3.7	2.1	23.8	11.6
2012	FX35	3.5	VQ35HR	5.2	19.4	N/A	1.4	3.0	2.1	23.8	9.7
	FX50	5.0	VK50VE	7.1	24.0	N/A	1.4	3.7	2.1	23.8	11.6

NOTE: All capacities are approximate. Add fluid gradually and ensure a proper fluid level is obtained.

N/A: Not Applicable

① Drain and refill

② With reservoir tank at "MAX" level

71075_FX35_C0004

FLUID SPECIFICATIONS

Year	Model	Engine Disp. Liters	Engine Oil	Manual Trans.	Auto. Trans.	Drive Axle Front	Drive Axle Rear	Transfer Case	Power Steering Fluid	Brake Master Cylinder	Cooling System
2011	FX35	3.5	5W-30	N/A	①	②	③	④	⑤	⑥	⑦
	FX50	5.0	5W-30	N/A	①	②	③	④	⑤	⑥	⑦
2012	FX35	3.5	5W-30	N/A	①	②	③	④	⑤	⑥	⑦
	FX50	5.0	5W-30	N/A	①	②	③	④	⑤	⑥	⑦

N/A: Not Applicable

DOT: Department Of Transportation

① Genuine NISSAN Matic S ATF

② Genuine NISSAN Differential Oil Hypoid Super GL-5 80W-90 or API GL-5, Viscosity SAE 80W-90

   For hot climates, viscosity SAE 90 is suitable for ambient temperatures above 32°F (0°C)

③ VQ35HR without towing package: Genuine NISSAN Differential Oil Hypoid Super GL-5 80W-90 or API GL-5 Viscosity SAE 80W-90

   VK50VE and VQ35HR with towing package: API GL-5 Synthetic gear oil, Viscosity SAE 75W-90

④ Genuine NISSAN Matic J ATF

⑤ Genuine NISSAN PSF or equivalent

⑥ Genuine NISSAN Super Heavy Duty Brake Fluid or equivalent DOT 3

⑦ Pre-diluted Genuine NISSAN Long Life Antifreeze/Coolant (blue) or equivalent

71075_FX35_C0005

VALVE SPECIFICATIONS

Year	Engine Displacement Liters	Engine ID/VIN	Seat Angle (deg.)	Spring Test Pressure (lbs. @ in.)	Spring Free-Length (in.)	Spring Installed Height (in.)	Stem-to-Guide Clearance (in.)		Stem Diameter (in.)	
							Intake	Exhaust	Intake	Exhaust
2011	3.5	VQ35HR	45.25-45.75	113-127@ 1.055	1.7264	1.4567	0.0008-0.0021	0.0012-0.0022	0.2348-0.2354	0.2347-0.2350
	5.0	VK50VE	45.25-45.75	①	②	③	0.0008-0.0021	0.0012-0.0025	0.2348-0.2354	0.2344-0.2350
2012	3.5	VQ35HR	45.25-45.75	113-127@ 1.055	1.7264	1.4567	0.0008-0.0021	0.0012-0.0022	0.2348-0.2354	0.2347-0.2350
	5.0	VK50VE	45.25-45.75	①	②	③	0.0008-0.0021	0.0012-0.0025	0.2348-0.2354	0.2344-0.2350

① Intake: 137-156 lbs. @ 1.1350 inches
Exhaust: 83-96 lbs. @ 1.0098 inches

② Intake: 1.9169 inches
Exhaust: 1.8642 inches

③ Intake: 1.6693 inches
Exhaust: 1.3957 inches

71075_FX35_C0006

CAMSHAFT SPECIFICATIONS
All measurements in inches unless noted

Year	Engine Displacement Liters	Engine Code	Journal Diameter	Brg. Oil Clearance	Shaft End-play	Runout	Journal Bore	Cam Height	
								Intake	Exhaust
2011	3.5	VQ35HR	①	②	0.0045-0.0074	0.0008	③	1.8057-1.8132	1.8061-1.8136
	5.0	VK50VE	④	⑤	0.0045-0.0074	0.0008	⑥	NS	1.7904-1.7978
2012	3.5	VQ35HR	①	②	0.0045-0.0074	0.0008	③	1.8057-1.8132	1.8061-1.8136
	5.0	VK50VE	④	⑤	0.0045-0.0074	0.0008	⑥	NS	1.7904-1.7978

NS: Not Specified

① No. 1: 1.0211-1.0218 inches
No. 2, 3, 4: 0.9230-0.9238 inch

② No. 1: 0.0018-0.0034 inches
No. 2, 3, 4: 0.0014-0.0030 inch

③ Camshaft bracket inner diameter
No. 1: 1.0236-1.0244 inches
No. 2, 3, 4: 0.9252-0.9260 inch

④ No. 1: 1.0211-1.0218 inches
No. 2, 3, 4, 5: 1.0217-1.0224 inches

⑤ No. 1: 0.0018-0.0034 inch
No. 2, 3, 4, 5: 0.0012-0.0028 inch

⑥ VVEL ladder assembly bracket inner diameter (EXH side): 1.0236-1.0244 inches

71075_FX35_C0007

CRANKSHAFT AND CONNECTING ROD SPECIFICATIONS

All measurements are given in inches.

| Year | Engine Disp. Liters | Engine ID/VIN | Crankshaft | | | | Connecting Rod | | |
			Main Brg. Journal Dia.	Main Brg. Oil Clearance	Shaft End-play	Thrust on No.	Journal Diameter	Oil Clearance	Side Clearance
2011	3.5	VQ35HR	2.5571-2.5581 ①	0.0014-0.0018	0.0039-0.0098	3	2.2441-2.2446 ①	0.0016-0.0021	0.0079-0.0138
	5.0	VK50VE	②	0.0014-0.0018	0.0039-0.0102	3	2.2441-2.2446 ①	0.0016-0.0021	0.0079-0.0138
2012	3.5	VQ35HR	2.5571-2.5581 ①	0.0014-0.0018	0.0039-0.0098	3	2.2441-2.2446 ①	0.0016-0.0021	0.0079-0.0138
	5.0	VK50VE	②	0.0014-0.0018	0.0039-0.0102	3	2.2441-2.2446 ①	0.0016-0.0021	0.0079-0.0138

① Variance depending on diameter Grade

② Journals No. 1 and 5: 2.5173-2.5183 inches (depending on Grade)

  Journals No. 2, 3, 4: 2.5174-2.5182 inches (depending on Grade)

71075_FX35_C0008

PISTON AND RING SPECIFICATIONS

All measurements are given in inches.

| Year | Engine Displacement Liters | Engine ID/VIN | Piston Clearance | Ring Gap | | | Ring Side Clearance | | |
				Top Compression	Bottom Compression	Oil Control	Top Compression	Bottom Compression	Oil Control
2011	3.5	VQ35HR	0.0004-0.0012	0.0091-0.0130	0.0130-0.0189	0.0067-0.0185	0.0016-0.0031	0.0012-0.0028	0.0022-0.0061
	5.0	VK50VE	0.0004-0.0012	0.0091-0.0130	0.0130-0.0189	0.0067-0.0185	0.0016-0.0031	0.0012-0.0028	0.0022-0.0061
2012	3.5	VQ35HR	0.0004-0.0012	0.0091-0.0130	0.0130-0.0189	0.0067-0.0185	0.0016-0.0031	0.0012-0.0028	0.0022-0.0061
	5.0	VK50VE	0.0004-0.0012	0.0091-0.0130	0.0130-0.0189	0.0067-0.0185	0.0016-0.0031	0.0012-0.0028	0.0022-0.0061

71075_FX35_C0009

TORQUE SPECIFICATIONS
All readings in ft. lbs.

Year	Engine Disp. Liters	Engine ID/VIN	Cylinder Head Bolts	Main Bearing Bolts	Rod Bearing Bolts	Crankshaft Damper Bolts	Flywheel Bolts	Manifold Intake	Manifold Exhaust	Spark Plugs	Oil Pan Drain Plug
2011	3.5	VQ35HR	①	②	③	④	65	⑤	22	15	25
	5.0	VK50VE	⑥	⑦	⑧	⑨	65	8	21	15	25
2012	3.5	VQ35HR	①	②	③	④	65	⑤	22	15	25
	5.0	VK50VE	⑥	⑦	⑧	⑨	65	8	21	15	25

① Apply engine oil to bolts, refer to procedure for tightening sequen

Step 1: Tighten to 77 ft. lbs.

Step 2: Loosen bolts completely in reverse sequence

Step 3: Tighten to 30 ft. lbs.

Step 4: Tighten 95 degrees

Step 5: Tighten another 95 degrees

② Apply engine oil to bolts, refer to procedure for tightening sequen

Step 1: Tighten bolts 17-26 to 18 ft. lbs.

Step 2: Repeat Step 1

Step 3: Tighten bolts 1-16 to 26 ft. lbs.

Step 4: Tighten bolts 1-16 another 90 degrees

③ Apply engine oil to bolts

Step 1: Tighten to 21 ft. lbs.

Step 2: Loosen bolts completely

Step 3: Tighten to 18 ft. lbs.

Step 4: Tighten 90 degrees

④ Step 1: Tighten to 33 ft. lbs.

Step 2: Tighten 90 degrees

⑤ Refer to procedure for tightening sequence

Step 1: Tighten to 66 inch lbs.

Step 2: Tighten to 19 ft. lbs.

⑥ Apply engine oil to bolts, refer to procedure for tightening sequence

Step 1: Tighten to 30 ft. lbs.

Step 2: Tighten 75 degrees

Step 3: Loosen bolts completely in reverse sequence

Step 4: Tighten to 30 ft. lbs.

Step 5: Tighten 65 degrees

Step 6: Tighten another 65 degrees

⑦ Apply engine oil to bolts, refer to procedure for tightening sequence

Step 1: Tighten M12 bolts 1-10 to 40 ft. lbs.

Step 2: Tighten M9 bolts 11-20 to 14 ft. lbs.

Step 3: Tighten M12 bolts 1-10 another 90 degrees

Step 4: Tighten M9 bolts 11-20 another 90 degrees

Step 5: Tighten M10 bolts 21-30 to 36 ft. lbs.

⑧ Apply engine oil to bolt

Step 1: Tighten to 21 ft. lbs.

Step 2: Loosen bolts completely

Step 3: Tighten to 18 ft. lbs.

Step 4: Tighten 90 degrees

⑨ Apply engine oil to bolt

Step 1: Tighten to 116 ft. lbs.

Step 2: Tighten 90 degrees

71075_FX35_C0010

WHEEL ALIGNMENT

Year	Model		Caster Range (+/-Deg.)	Caster Preferred Setting (Deg.)	Camber Range (+/-Deg.)	Camber Preferred Setting (Deg.)	Toe-in (in.)
2011	FX35	F	0.75	3.67	0.75	-0.33	0.07 +/- 0.04
		R	N/A	N/A	0.50	-1.16	0.12 +/- 0.12
	FX50	F	0.75	3.67	0.75	-0.33	0.07 +/- 0.04
		R	N/A	N/A	0.50	-1.16	0.12 +/- 0.12
2012	FX35	F	0.75	3.67	0.75	-0.33	0.07 +/- 0.04
		R	N/A	N/A	0.50	-1.16	0.12 +/- 0.12
	FX50	F	0.75	3.67	0.75	-0.33	0.07 +/- 0.04
		R	N/A	N/A	0.50	-1.16	0.12 +/- 0.12

NOTE: Measurements given for an unladen vehicle with fuel, coolant, and engine oil full; spare tire, jack, hand tools, and mats in designated positions.

N/A: Not Applicable

F: Front

R: Rear

71075_FX35_C0011

TIRE, WHEEL AND BALL JOINT SPECIFICATIONS

| Year | Model | OEM Tires | | Tire Pressures (psi) | | Wheel Size | | Ball Joint Inspection | Lug Nut (ft. lbs.) |
		Standard	Optional	Front	Rear	Standard	Optional		
2011	FX35	P265/60R18	P265/50R20	①	①	18 x 8.0	20 x 8.0	②	80
	FX50	P265/50R20	P265/45R21	①	①	20 x 8.0	21 x 9.5	②	80
2012	FX35	P265/60R18	P265/50R20	①	①	18 x 8.0	20 x 8.0	②	80
	FX50	P265/50R20	P265/45R21	①	①	20 x 8.0	21 x 9.5	②	80

OEM: Original Equipmer PSI: Pounds Per Square Inch

① Always refer to the owner's manual and/or vehicle label: conventional tires should be inflated to 33 psi

② Replace if any measurable axial end play is found.

71075_FX35_C0012

BRAKE SPECIFICATIONS
All measurements in inches unless noted

| Year | Model | | Brake Disc | | | Pad/Lining Thickness | | Brake Caliper | |
			Original Thickness	Minimum Thickness	Max. Runout	Standard	Limit	Torque Member Bolts (ft. lbs.)	Guide Pin Bolts (ft. lbs.)
2011	FX35	F	1.339	1.260	0.0014	0.433	0.079	91	34
		R	0.630	0.551	0.0022	0.335	0.079	62	32
	FX50	F	1.260	1.181	0.0014	0.433	0.079	91	NS
		R	0.787	0.709	0.0022	0.374	0.079	62	NS
2012	FX35	F	1.339	1.260	0.0014	0.433	0.079	91	34
		R	0.630	0.551	0.0022	0.335	0.079	62	32
	FX50	F	1.260	1.181	0.0014	0.433	0.079	91	NS
		R	0.787	0.709	0.0022	0.374	0.079	62	NS

NS: Not Specified

F: Front

R: Rear

71075_FX35_C0013

SCHEDULED MAINTENANCE INTERVALS
INFINITI—FX35 & FX50

TO BE SERVICED	TYPE OF SERVIC	VEHICLE MILEAGE INTERVAL (x1000)														
		7.5	15	22.5	30	37.5	45	52.5	60	67.5	75	82.5	90	97.5	105	120
Accessory drive belts ①	S/I								✓							✓
Air cleaner element (engine)	R				✓				✓				✓			✓
Air conditioner system	S/I	Inspect system operation annually														
Automatic transaxle fluid	S/I		✓		✓		✓		✓		✓		✓		✓	✓
Brake lines, hoses, cables, and connections	S/I		✓		✓		✓		✓		✓		✓		✓	✓
Brake pads, calipers, & rotors	S/I		✓		✓		✓		✓		✓		✓		✓	✓
Differential gear oil	S/I		✓		✓		✓		✓		✓		✓		✓	✓
Driveshafts and CV-boots	S/I		✓		✓		✓		✓		✓		✓		✓	✓
Engine coolant	R								✓				✓			✓
Engine oil and filter	R	✓	✓	✓	✓	✓	✓	✓	✓	✓	✓	✓	✓	✓	✓	✓
Exhaust pipe connections, muffler, and suspension bolts	S/I				✓				✓				✓			✓
EVAP vapor lines	S/I				✓				✓				✓			✓
Fuel lines and connections	S/I				✓				✓				✓			✓
In-cabin microfilter	S/I		✓		✓		✓		✓		✓		✓		✓	✓
Spark plugs (Platinum-tipped)	R	105,000 miles (under normal usage)														
Steering system	S/I				✓				✓				✓			✓
Suspension system	S/I				✓				✓				✓			✓
Transfer case fluid	S/I		✓		✓		✓		✓		✓		✓		✓	✓
Valve clearance	S/I	Whenever valve noise increases														

R: Replace   S/I: Service or Inspect

① Replace if worn or damaged or if the auto-tensioner has reached its limit (V8)

FREQUENT OPERATION MAINTENANCE (SEVERE SERVICE)

If a vehicle is operated under any of the following conditions it is considered severe service:

- Extremely dusty areas.

- 50% or more of the vehicle operation is in 90°F (32°C) or higher temperatures, or constant operation in temperatures below 32°F (0°C).

- Prolonged idling (vehicle operation in stop and go traffic).

- Frequent short running periods (engine does not warm to normal operating temperatures).

- Police, taxi, delivery usage, or trailer towing usage.

Automatic transaxle fluid (and filter), transfer case fluid, and differential gear oil: check every 15,000 miles, replace every 30,000 miles

Brake pads, calipers & rotors: service or inspect every 7,500 miles

Driveshafts and CV-boots inspect every 7,500 miles

Exhaust system inspect every 7,500 miles

Oil and oil filter: change every 3,750 miles

Steering system and suspension components inspect for looseness and damage every 7,500 miles

PRECAUTIONS

Before servicing any vehicle, please be sure to read all of the following precautions, which deal with personal safety, prevention of component damage, and important points to take into consideration when servicing a motor vehicle:

• Never open, service or drain the radiator or cooling system when the engine is hot; serious burns can occur from the steam and hot coolant.

• Observe all applicable safety precautions when working around fuel. Whenever servicing the fuel system, always work in a well-ventilated area. Do not allow fuel spray or vapors to come in contact with a spark, open flame, or excessive heat (a hot drop light, for example). Keep a dry chemical fire extinguisher near the work area. Always keep fuel in a container specifically designed for fuel storage; also, always properly seal fuel containers to avoid the possibility of fire or explosion. Refer to the additional fuel system precautions later in this section.

• Fuel injection systems often remain pressurized, even after the engine has been turned **OFF**. The fuel system pressure must be relieved before disconnecting any fuel lines. Failure to do so may result in fire and/or personal injury.

• Brake fluid often contains polyglycol ethers and polyglycols. Avoid contact with the eyes and wash your hands thoroughly after handling brake fluid. If you do get brake fluid in your eyes, flush your eyes with clean, running water for 15 minutes. If eye irritation persists, or if you have taken brake fluid internally, IMMEDIATELY seek medical assistance.

• The EPA warns that prolonged contact with used engine oil may cause a number of skin disorders, including cancer. You should make every effort to minimize your exposure to used engine oil. Protective gloves should be worn when changing oil. Wash your hands and any other exposed skin areas as soon as possible after exposure to used engine oil. Soap and water, or waterless hand cleaner should be used.

• All new vehicles are now equipped with an air bag system, often referred to as a Supplemental Restraint System (SRS) or Supplemental Inflatable Restraint (SIR) system. The system must be disabled before performing service on or around system components, steering column, instrument panel components, wiring and sensors. Failure to follow safety and disabling procedures could result in accidental air bag deployment, possible personal injury and unnecessary system repairs.

• Always wear safety goggles when working with, or around, the air bag system. When carrying a non-deployed air bag, be sure the bag and trim cover are pointed away from your body. When placing a non-deployed air bag on a work surface, always face the bag and trim cover upward, away from the surface. This will reduce the motion of the module if it is accidentally deployed. Refer to the additional air bag system precautions later in this section.

• Clean, high quality brake fluid from a sealed container is essential to the safe and proper operation of the brake system. You should always buy the correct type of brake fluid for your vehicle. If the brake fluid becomes contaminated, completely flush the system with new fluid. Never reuse any brake fluid. Any brake fluid that is removed from the system should be discarded. Also, do not allow any brake fluid to come in contact with a painted surface; it will damage the paint.

• Never operate the engine without the proper amount and type of engine oil; doing so WILL result in severe engine damage.

• Timing belt maintenance is extremely important. Many models utilize an interference-type, non-freewheeling engine. If the timing belt breaks, the valves in the cylinder head may strike the pistons, causing potentially serious (also time-consuming and expensive) engine damage. Refer to the maintenance interval charts for the recommended replacement interval for the timing belt, and to the timing belt section for belt replacement and inspection.

• Disconnecting the negative battery cable on some vehicles may interfere with the functions of the on-board computer system(s) and may require the computer to undergo a relearning process once the negative battery cable is reconnected.

• When servicing drum brakes, only disassemble and assemble one side at a time, leaving the remaining side intact for reference.

• Only an MVAC-trained, EPA-certified automotive technician should service the air conditioning system or its components.

BRAKES

GENERAL INFORMATION

PRECAUTIONS

• Certain components within the ABS system are not intended to be serviced or repaired individually.

• Do not use rubber hoses or other parts not specifically specified for and ABS system. When using repair kits, replace all parts included in the kit. Partial or incorrect repair may lead to functional problems and require the replacement of components.

• Lubricate rubber parts with clean, fresh brake fluid to ease assembly. Do not use shop air to clean parts; damage to rubber components may result.

• Use only DOT 3 brake fluid from an unopened container.

• If any hydraulic component or line is removed or replaced, it may be necessary to bleed the entire system.

• A clean repair area is essential. Always clean the reservoir and cap thoroughly before removing the cap. The slightest amount of dirt in the fluid may plug an orifice and impair the system function. Perform repairs after components have been thoroughly cleaned; use only denatured alcohol to clean components. Do not allow ABS components to come into contact with any substance containing mineral oil; this includes used shop rags.

• The Anti-Lock control unit is a microprocessor similar to other computer units in the vehicle. Ensure that the ignition switch is **OFF** before removing or installing controller harnesses. Avoid static electricity discharge at or near the controller.

ANTI-LOCK BRAKE SYSTEM (ABS)

• If any arc welding is to be done on the vehicle, the control unit should be unplugged before welding operations begin.

WHEEL SPEED SENSORS

REMOVAL & INSTALLATION

See Figures 1 and 2.

At this time, the manufacturer does not provide a specific removal and installation procedure for this component, refer to the illustration as required.

Before servicing the vehicle, refer to the Precautions Section.

Note the following:

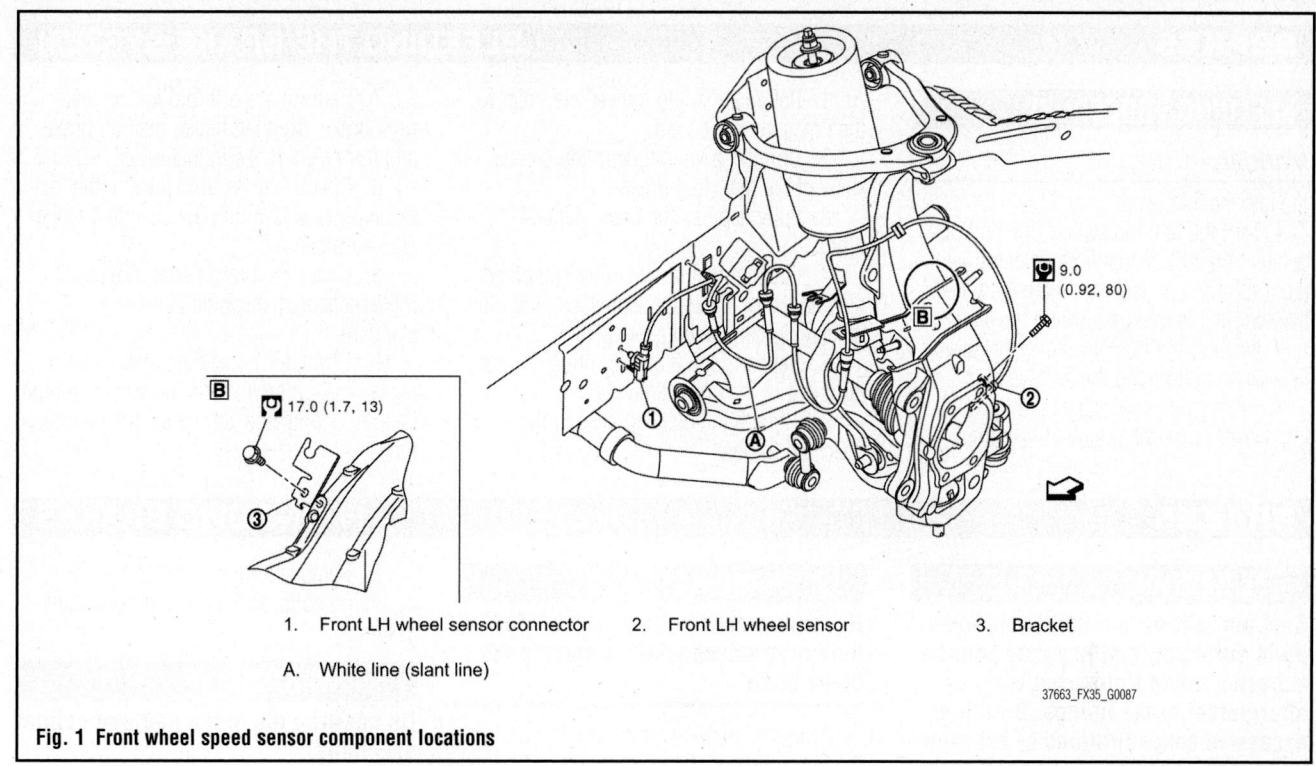

1. Front LH wheel sensor connector   2. Front LH wheel sensor   3. Bracket

A. White line (slant line)

37663_FX35_G0087

Fig. 1 Front wheel speed sensor component locations

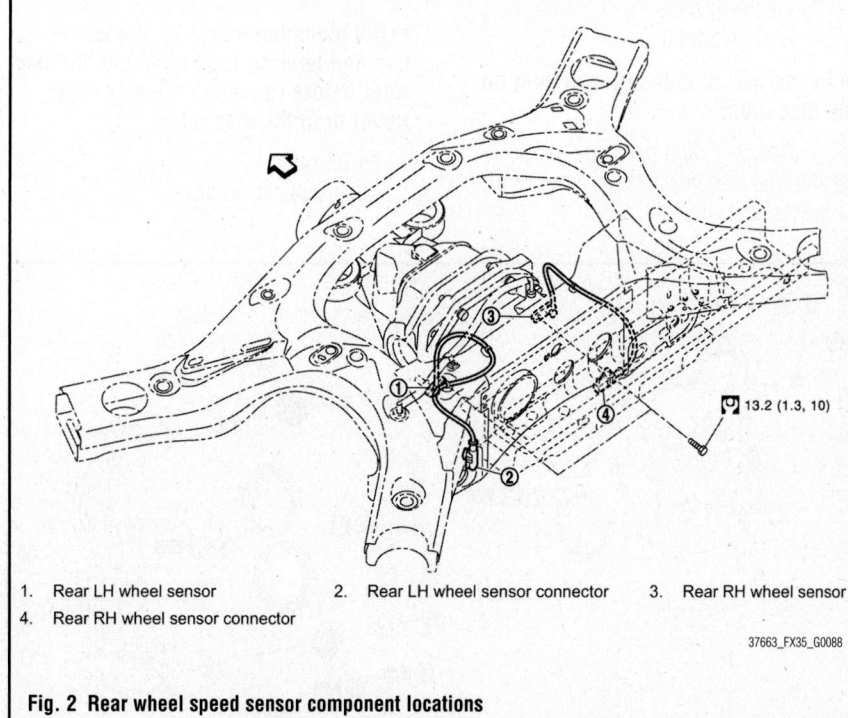

1. Rear LH wheel sensor   2. Rear LH wheel sensor connector   3. Rear RH wheel sensor
4. Rear RH wheel sensor connector

37663_FX35_G0088

Fig. 2 Rear wheel speed sensor component locations

• Refrain from twisting the sensor harness as much as possible, when removing it. Pull sensors out without pulling the sensor harness.

• Be careful to avoid damaging the sensor edges or rotor teeth. Remove the wheel sensor first before removing the front or rear wheel hub. This is to avoid damage to the sensor wiring and loss of sensor function.

• When installing, make sure there is no foreign material such as iron chips on and in the mounting hole of the wheel sensor. Make sure no foreign material has been caught in the sensor rotor. Remove any foreign material and clean the mount.

• When installing the wheel sensor, be sure to press the rubber grommets in until they lock at the locations shown. When installed, the harness must not be twisted.

• When you see the harness of the wheel sensor from the front side of the vehicle, ensure that the white lines are not twisted.

BRAKES

BLEEDING THE BRAKE SYSTEM

BLEEDING PROCEDURE

MANUAL

Note the following:
• Turn the ignition switch OFF and disconnect the ABS actuator and electric unit (control unit) connector or the battery negative terminal before performing the work
• Monitor the fluid level in the reservoir tank while performing the air bleeding
• Always use new brake fluid for refilling. Never reuse the drained brake fluid

1. Before servicing the vehicle, refer to the Precautions Section.
2. Connect a vinyl tube to the bleeder valve of the rear right brake.
3. Fully depress the brake pedal 4–5 times.
4. Loosen the bleeder valve and bleed air with the brake pedal depressed, and then quickly tighten the bleeder valve.
5. Repeat steps 3 and 4 until all of the air is out of the brake line.
6. Tighten the bleeder valve to the specified torque.

7. Perform steps 2 to 6 for the rear right brake, front left brake, rear left brake, and front right brake in that order.
8. Check that the fluid level in the reservoir tank is within the specified range after air bleeding.
9. Check the brake pedal. Adjust it if the measurement value is not the standard.
10. Check for brake fluid leakage from the master cylinder mounting face, reservoir tank mounting face, and brake tube connections.

BRAKES

FRONT DISC BRAKES

✳✳ CAUTION

Dust and dirt accumulating on brake parts during normal use may contain asbestos fibers from production or aftermarket brake linings. Breathing excessive concentrations of asbestos fibers can cause serious bodily harm. Exercise care when servicing brake parts. Do not sand or grind brake lining unless equipment used is designed to contain the dust residue. Do not clean brake parts with compressed air or by dry brushing. Cleaning should be done by dampening the brake components with a fine mist of water, then wiping the brake components clean with a dampened cloth. Dispose of cloth and all residue containing asbestos fibers in an impermeable container with the appropriate label. Follow practices prescribed by the Occupational Safety and Health Administration (OSHA) and the Environmental Protection Agency (EPA) for the handling, processing, and disposing of dust or debris that may contain asbestos fibers.

BRAKE CALIPERS

REMOVAL & INSTALLATION

2 Piston Type

See Figure 3.

✳✳ CAUTION

Clean any dust from the brake caliper and brake pads with a vacuum dust collector. Never blow with compressed air.

✳✳ WARNING

Never depress the brake pedal. Brake fluid may splash while removing the brake hose.

1. Before servicing the vehicle, refer to the Precautions Section.
2. Remove tires and wheels.
3. Fix the disc rotor using wheel nuts.
4. Drain brake fluid.

➡**Never spill or splash brake fluid on the disc rotor.**

5. Remove union bolt and copper washer, and disconnect brake hose from caliper assembly.

6. Remove torque member mounting bolts, and remove brake caliper assembly.

✳✳ WARNING

Do not drop the brake pad and caliper assembly.

7. Remove disc rotor.

➡**Put matching marks on the wheel hub and bearing assembly and the disc rotor before removing the disc rotor. Never drop the disc rotor.**

To install:
8. Install disc rotor.

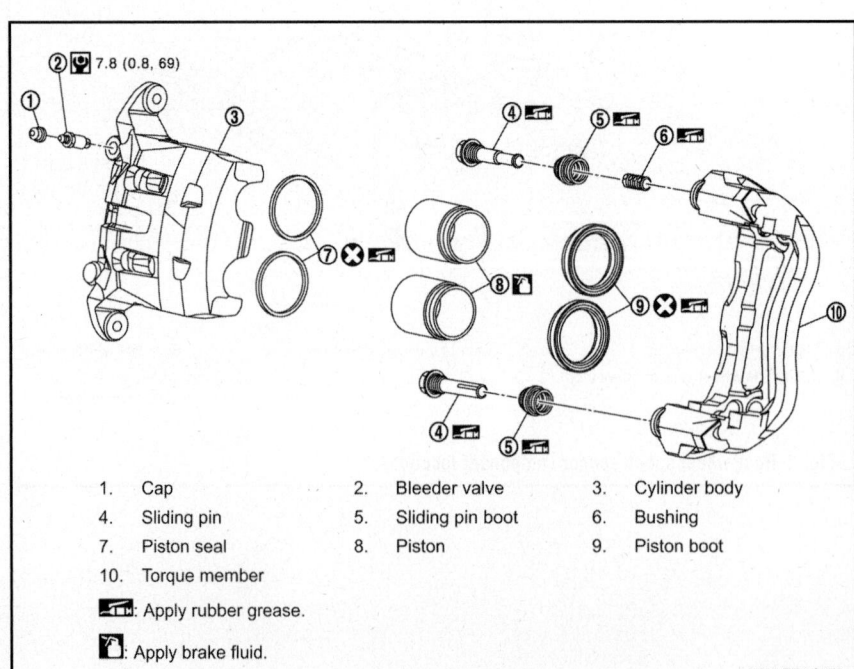

1. Cap	2. Bleeder valve	3. Cylinder body
4. Sliding pin	5. Sliding pin boot	6. Bushing
7. Piston seal	8. Piston	9. Piston boot
10. Torque member		

🔧 : Apply rubber grease.

🛢 : Apply brake fluid.

37663_FX35_G0093

Fig. 3 Exploded view of brake caliper assembly—2 piston type

➡**Align the matching marks that have been made during removal when reusing the disc rotor.**

9. Install the brake caliper assembly to the vehicle and tighten the torque member mounting bolts to the specified torque.

➡**Never spill or splash any grease and moisture on the brake caliper assembly mounting face, threads, mounting bolts and washers. Wipe out any grease and moisture.**

10. Install brake hose and copper washers to brake caliper assembly, and tighten union bolts to the specified torque.

➡**Never reuse copper washer.**

11. Refill with new brake fluid and perform the air bleeding.

✳✳ WARNING

Never reuse drained brake fluid. Never spill or splash brake fluid on the disc rotor.

12. Check the drag of the front disc brake.

4 Piston Type

See Figure 4.

✳✳ CAUTION

Clean any dust from the brake caliper and brake pads with a vacuum dust collector. Never blow with compressed air.

✳✳ WARNING

Never depress the brake pedal. Brake fluid may splash while removing the brake hose.

1. Before servicing the vehicle, refer to the Precautions Section.
2. Remove tires and wheels.
3. Fix the disc rotor using wheel nuts.
4. Drain brake fluid.

➡**Never spill or splash brake fluid on the disc rotor.**

5. Loosen the flare nut with a flare nut wrench and separate the brake tube from caliper.
   Note the following:
   • Cover flare nut wrench with a cloth as not to damage the caliper
   • Never scratch the flare nut and the brake tube
   • Never bend sharply, twist or strongly pull out the brake tube

• Cover open end of brake tube when disconnecting to prevent entrance of dirt
6. Remove caliper mounting bolts, and remove caliper.

✳✳ WARNING

Never drop brake pad and caliper.

7. Remove disc rotor.

➡**Put matching marks on the wheel hub and bearing assembly and the disc rotor before removing the disc rotor. Never drop disc rotor.**

To install:

8. Install disc rotor.

➡**Align the matching marks that have been made during removal when reusing the disc rotor.**

9. Install the brake caliper assembly to the vehicle and tighten the torque member mounting bolts to the specified torque.

➡**Never spill or splash any grease and moisture on the brake caliper assembly mounting face, threads, mounting bolts and washers. Wipe out any grease and moisture.**

10. Tighten the flare nut to the specified torque with a flare nut crowfoot and a torque wrench.

✳✳ WARNING

Cover crowfoot with a cloth as not to damage the caliper. Never scratch the flare nut and the brake tube.

11. Refill with new brake fluid and perform the air bleeding.

✳✳ WARNING

Never reuse drained brake fluid. Never spill or splash brake fluid on the disc rotor and caliper.

12. Check the drag of the front disc brake.

BRAKE PADS

REMOVAL & INSTALLATION

2 Piston Type

See Figures 5 and 6.

✳✳ CAUTION

Clean any dust from the brake caliper and brake pads with a vacuum dust collector. Never blow with compressed air.

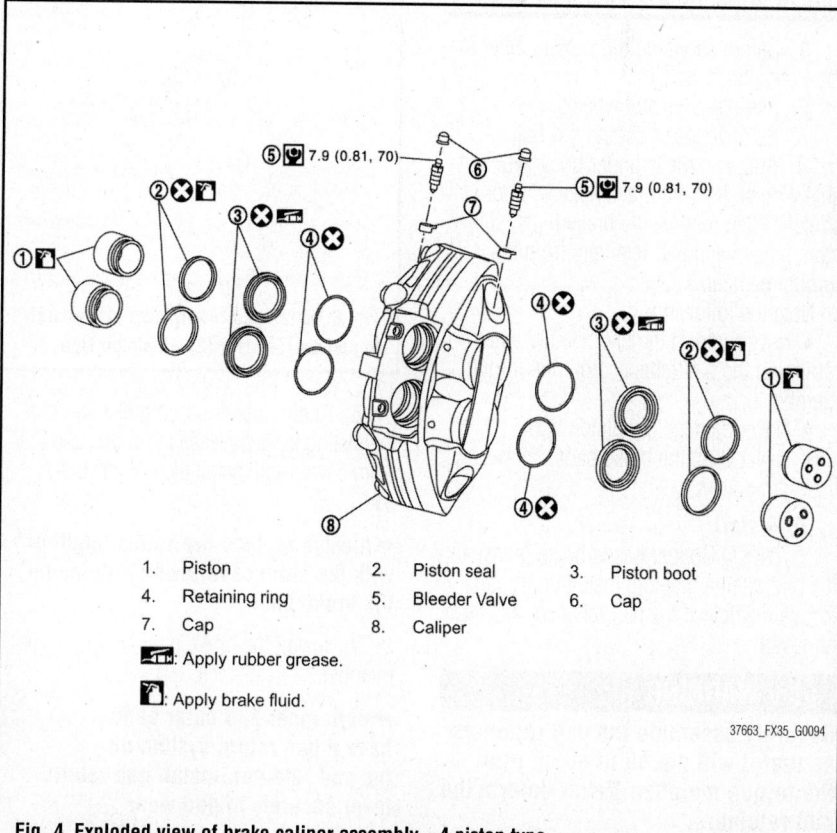

1.	Piston	2.	Piston seal	3.	Piston boot
4.	Retaining ring	5.	Bleeder Valve	6.	Cap
7.	Cap	8.	Caliper		

🔧: Apply rubber grease.

🛢: Apply brake fluid.

37663_FX35_G0094

Fig. 4 Exploded view of brake caliper assembly—4 piston type

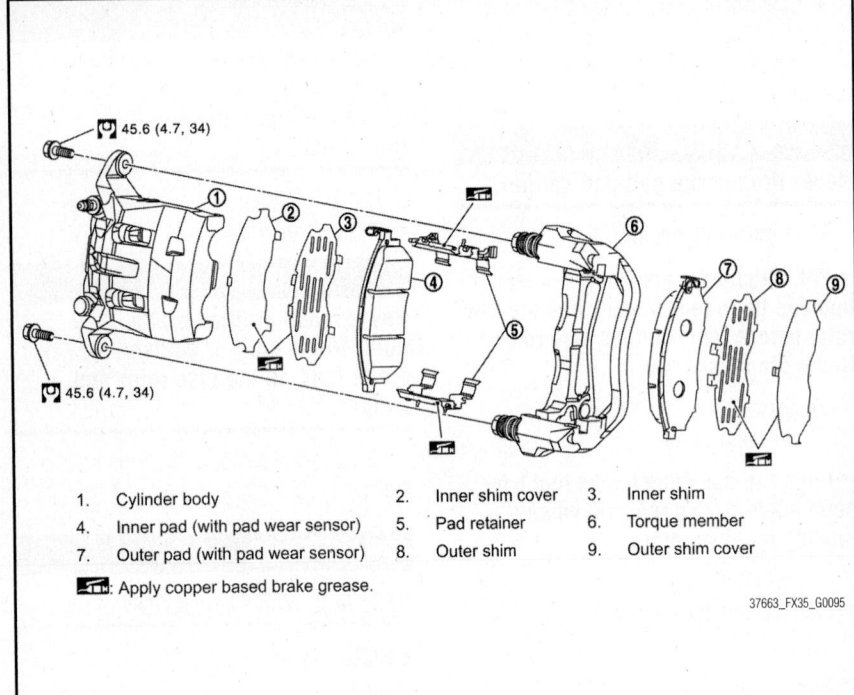

45.6 (4.7, 34)

45.6 (4.7, 34)

1. Cylinder body
2. Inner shim cover
3. Inner shim
4. Inner pad (with pad wear sensor)
5. Pad retainer
6. Torque member
7. Outer pad (with pad wear sensor)
8. Outer shim
9. Outer shim cover

: Apply copper based brake grease.

37663_FX35_G0095

Fig. 5 Exploded view of front brake pad and caliper assembly—2 piston type

> **⁂ WARNING**
>
> **Never depress the brake pedal while removing the brake pads because the piston may pop out. Never spill or splash brake fluid on the disc rotor.**

1. Before servicing the vehicle, refer to the Precautions Section.
2. Remove tires and wheels.
3. Remove lower sliding pin bolt.
4. Suspend the cylinder body with suitable wire so that the brake hose will not stretch. Then remove the brake pads, shims, shim covers and pad retainers from the torque member.

Note the following:
• Never deform the pad retainer when removing the pad retainer from the torque member
• Never damage the piston boot
• Never drop the brake pads, shims, and the shim covers

To install:

5. Apply Copper based brake grease to the pad retainers before installing it to the torque member if the pad retainers has been removed.

> **⁂ WARNING**
>
> **Securely assemble the pad retainers so that it will not be lifted up from the torque member. Never deform the pad retainers.**

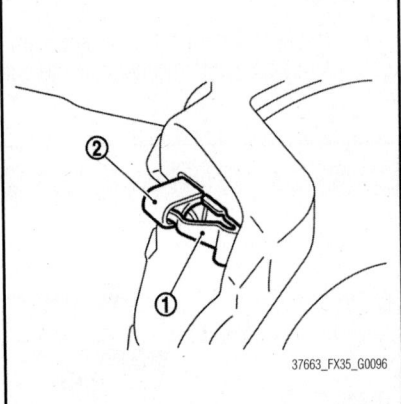

37663_FX35_G0096

Fig. 6 Install pad return lever (1) securely to pad wear sensor (2)—2 piston type

6. Apply Copper based brake grease to the mating faces between the shims and the shim covers and install them to the brake pad.

➡**Always replace the shims together with the shim covers when replacing the brake pad.**

7. Install the brake pads to the torque member.

➡**Both inner and outer pads have a pad return system on the pad retainer. Install pad return lever securely to pad wear sensor.**

8. Install cylinder body to torque member.

Note the following:
• Never damage the piston boot
• When replacing brake pad with new one, check a brake fluid level in the reservoir tank because brake fluid returns to master cylinder reservoir tank when pressing piston in

➡**Use a disc brake piston tool to easily press piston.**

9. Install the lower sliding pin bolt and tighten it to the specified torque.
10. Depress the brake pedal several times to check that no drag feel is present for the front disc brake.

4 Piston Type

See Figures 7 through 9.

> **⁂ CAUTION**
>
> **Clean any dust from the brake caliper and brake pads with a vacuum dust collector. Never blow with compressed air.**

> **⁂ WARNING**
>
> **Never depress the brake pedal while removing the brake pads because the piston may pop out. Never spill or splash brake fluid on the disc rotor and caliper.**

1. Before servicing the vehicle, refer to the Precautions Section.
2. Remove tires and wheels.
3. Remove clips from pad pins.
4. Remove pad pins while holding down cross spring, then remove cross spring from caliper.
5. Using pliers, remove brake pads and shims from caliper.

> **⁂ WARNING**
>
> **Do not damage the piston boot. Never drop the brake pads and shims.**

To install:

6. Apply copper based brake grease to the mating faces between the brake pads and shims, and install shims to the brake pad.

➡**Always replace the shims together when replacing the brake pad.**

7. Apply copper based brake grease to the mating faces between the brake pads and caliper.

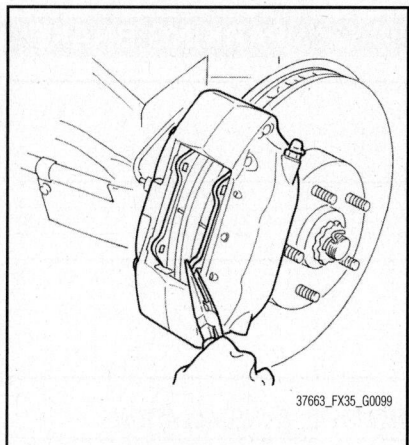

Fig. 7 Using pliers, remove brake pads and shims from caliper—4 piston type

8. Install brake pads to caliper. Note the following:
• Never damage the piston boot
• In the case of replacing a pad with new

one, check a brake fluid level in the reservoir tank because brake fluid returns to master cylinder reservoir tank when pressing piston in

➡Use a disc brake piston tool to easily press piston.

9. Install upper pad pin from the inner side, then install firmly to the outer side through the hole in the top of brake pad.

10. Place the top of cross spring over the upper pad pin, press in the cross spring, install lower pad pin from the inner side to the outer side, and secure cross spring.

11. Install clips to the pad pins.

☼☼ CAUTION

If clip is not fully attached, pad pin or brake pad could fall out while vehicle is in motion.

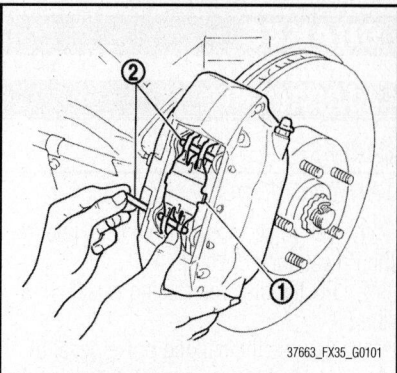

Fig. 9 Place the top of cross spring (1) over the upper pad pin (2), press in the cross spring, install lower pad pin from the inner side to the outer side, and secure cross spring—4 piston type

12. Depress the brake pedal several times to check that no drag feel is present for the front disc brake.

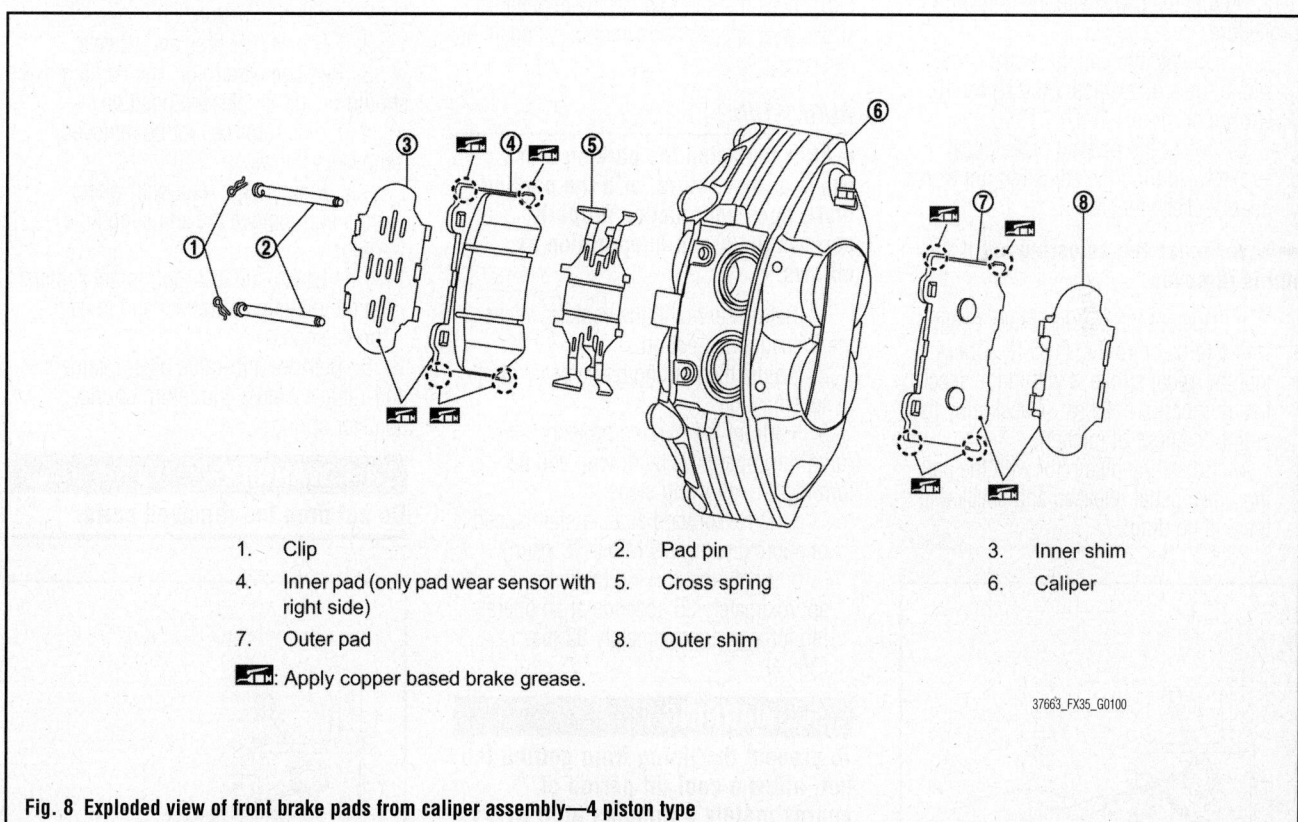

1. Clip
2. Pad pin
3. Inner shim
4. Inner pad (only pad wear sensor with right side)
5. Cross spring
6. Caliper
7. Outer pad
8. Outer shim

🔧: Apply copper based brake grease.

Fig. 8 Exploded view of front brake pads from caliper assembly—4 piston type

BRAKES

PARKING BRAKE

ADJUSTMENTS

CABLES

See Figures 10 and 11.

1. Before servicing the vehicle, refer to the Precautions Section.

2. Fix the disc rotor using wheel nuts.

3. Release the parking brake pedal by turning the adjusting nut with a deep socket wrench and loosening the cable.

4. Remove the adjusting hole plug from the disc rotor. Turn the adjuster in the direction as shown in the figure using a suitable tool until the disc rotor is locked.

5. Turn back the adjuster 5 or 6 notches from the locked position.

6. Rotate the disc rotor to check that there is no drag. Install the adjusting hole plug.

7. Adjust the cable with the following procedure.

   a. Operate the parking brake pedal with a force of 110 lbs. (490 N) for 10 strokes or more.

   b. Adjust the parking brake pedal stroke by turning the adjusting nut with a deep socket wrench.

➡ **Never reuse the adjusting nut if the nut is removed.**

   c. Operate the parking brake pedal with a force of 44 lbs. (196 N). Check that the pedal stroke is within the specified number of notches. (Check it by listening to clicks of ratchet).

   d. Rotate the disc rotor with the parking brake pedal released and check that there is no drag.

Item	Standard
Number of notches [under force of 196 N (20 kg, 44 lb)]	2 – 3 notches
Number of notches when brake warning lamp turns ON	1 notch

71075_FX35_G0036

Fig. 11 Parking brake control pedal stroke table

PARKING BRAKE SHOES

ADJUSTMENTS

See Figure 11.

1. Before servicing the vehicle, refer to the Precautions Section.

2. Adjust the parking brake pedal stroke to specification.

3. After the break-in operation (burnishing), check that the parking brake pedal stroke is at specification and adjust again as necessary.

BURNISHING

➡**After replacing the parking brake shoes or disc rotors, or if the parking brake does not function properly, perform the break-in operation as follows.**

1. Before servicing the vehicle, refer to the Precautions Section.

2. Adjust the parking brake pedal stroke to specification.

3. Perform the parking brake break-in (drag run) operation by driving and performing the following steps:

   a. Drive forward at a constant speed of approximately 19 MPH (30 km/h).

   b. Apply the parking brake for approximately 35 seconds at an operating force at approximately 82 lbs. (365 N).

✳✳ WARNING

To prevent the lining from getting too hot, allow a cool off period of approximately 5 minutes after every break-in operation. Do not perform excessive break-in operations, because this may cause uneven or early the wear of lining.

4. After the break-in operation, check that the parking brake pedal stroke is at specification and adjust again as necessary.

REMOVAL & INSTALLATION

See Figures 12 through 14.

✳✳ CAUTION

Clean any dust from the parking brake shoes and back plates with a vacuum dust collector. Never blow with compressed air.

1. Before servicing the vehicle, refer to the Precautions Section.

2. Remove rear tires and wheels.

3. Remove disc rotor. The parking brake should be in the released position.

4. If disc rotor cannot be removed, remove as follows:

   a. Fix the disc rotor with wheel nuts and remove the adjusting hole plug.

   b. Using suitable tool, rotate adjuster in the direction to retract and loosen brake shoe.

5. Remove anti-rattle pins, retainers, anti-rattle springs, and return spring, adjuster spring.

✳✳ WARNING

Do not drop the removed parts.

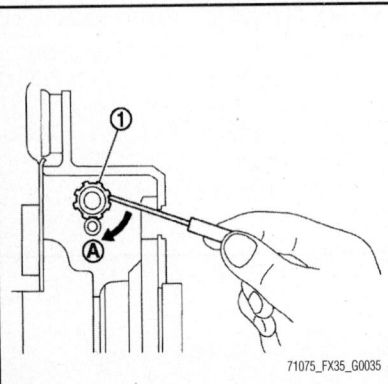

71075_FX35_G0035

Fig. 10 Turn the adjuster (1) in the direction (A) shown using a suitable tool until the disc rotor is locked

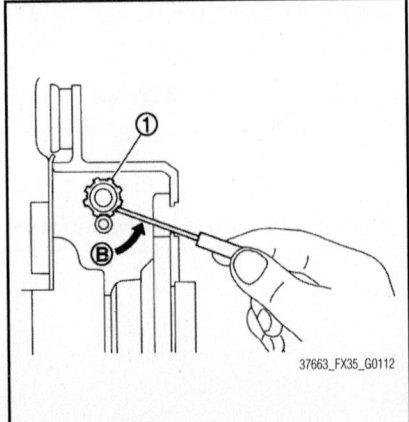

37663_FX35_G0112

Fig. 12 Using suitable tool, rotate adjuster (1) in the direction (B) to retract and loosen brake shoe

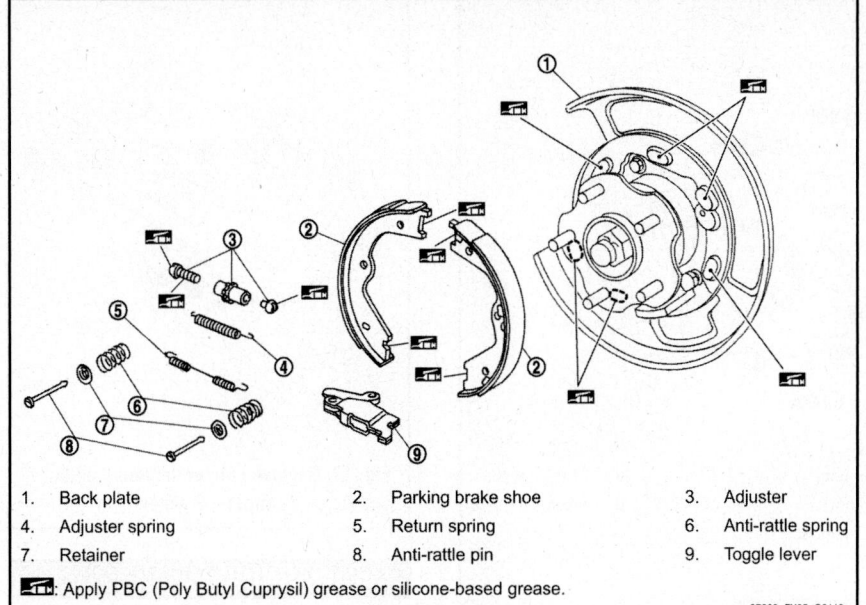

1. Back plate
2. Parking brake shoe
3. Adjuster
4. Adjuster spring
5. Return spring
6. Anti-rattle spring
7. Retainer
8. Anti-rattle pin
9. Toggle lever

⊞: Apply PBC (Poly Butyl Cuprysil) grease or silicone-based grease.

37663_FX35_G0113

Fig. 13 Exploded view of parking brake shoe assembly

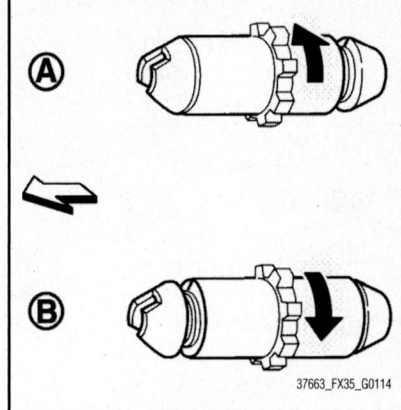

37663_FX35_G0114

Fig. 14 Assemble adjusters so that threaded part is expanded when rotating it in the direction shown by arrow

6. Remove parking brake shoes, adjuster assembly, and toggle lever.
Note the following:
• The parking brake shoes for the front wheels are made of different materials from those for the rear wheels. Never misidentify them when removing or replacing

To install:
7. Installation is the reverse of the removal procedure.
8. Apply Poly Butyl Cuprysil (PBC) grease or silicone-based grease to the back plate and brake shoe.
9. Assemble adjusters so that threaded part is expanded when rotating it in the direction shown by arrow.
10. Shorten adjuster by rotating it.
11. When disassembling apply PBC grease or silicone-based grease to threads.
12. Check brake shoe sliding surface and drum inner surface for grease. Wipe it off if it adheres on the surfaces.

BRAKES

❋❋ CAUTION

Dust and dirt accumulating on brake parts during normal use may contain asbestos fibers from production or aftermarket brake linings. Breathing excessive concentrations of asbestos fibers can cause serious bodily harm. Exercise care when servicing brake parts. Do not sand or grind brake lining unless equipment used is designed to contain the dust residue. Do not clean brake parts with compressed air or by dry brushing. Cleaning should be done by dampening the brake components with a fine mist of water, then wiping the brake components clean with a dampened cloth. Dispose of cloth and all residue containing asbestos fibers in an impermeable container with the appropriate label. Follow practices prescribed by the Occupational Safety and Health Administration (OSHA) and the Environmental Protection Agency (EPA) for the handling, processing, and disposing of dust or debris that may contain asbestos fibers.

BRAKE CALIPERS

REMOVAL & INSTALLATION

1 Piston Type
See Figures 15 and 16.

❋❋ CAUTION

Clean any dust from the brake caliper and brake pads with a vacuum dust collector. Never blow with compressed air.

❋❋ WARNING

Never depress the brake pedal. Brake fluid may splash while removing the brake hose.

1. Before servicing the vehicle, refer to the Precautions Section.
2. Remove tires with power tool.
3. Fix the disc rotor using wheel nuts.
4. Drain brake fluid.

REAR DISC BRAKES

➡**Never spill or splash brake fluid on the disc rotor.**

5. Remove union bolt and copper washers, and disconnect brake hose from caliper assembly.
6. Remove torque member mounting bolts, and remove brake caliper assembly.

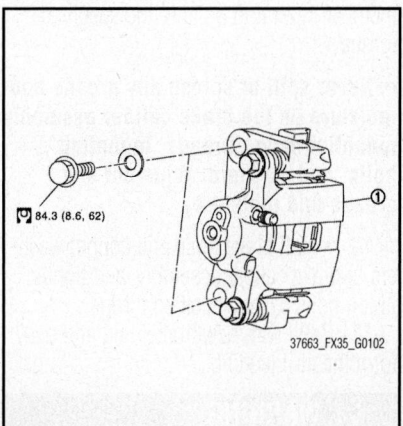

84.3 (8.6, 62)

37663_FX35_G0102

Fig. 15 Remove torque member mounting bolts, and remove brake caliper assembly (1)—1 piston type

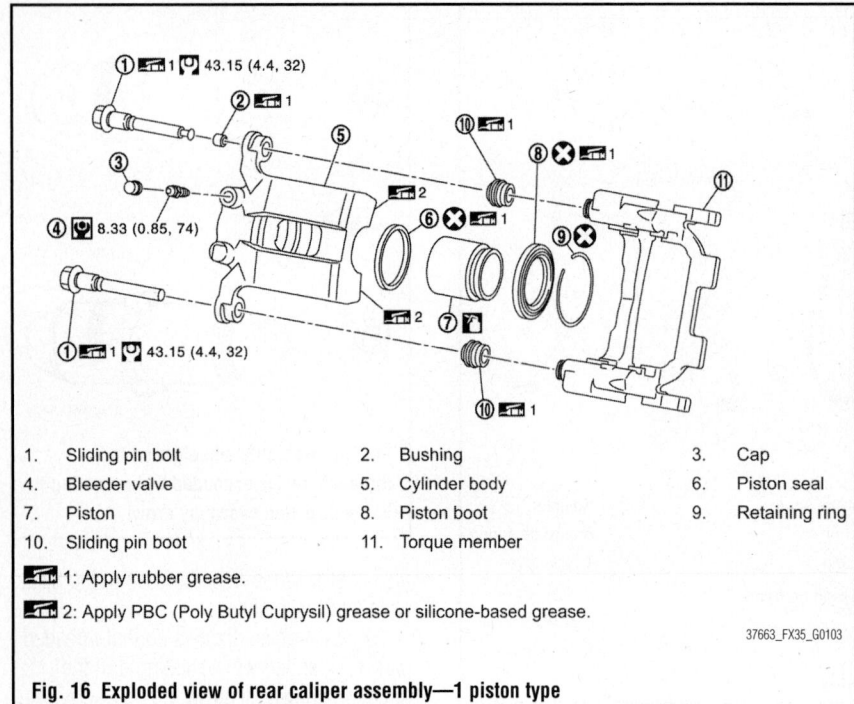

1. Sliding pin bolt
2. Bushing
3. Cap
4. Bleeder valve
5. Cylinder body
6. Piston seal
7. Piston
8. Piston boot
9. Retaining ring
10. Sliding pin boot
11. Torque member

1: Apply rubber grease.

2: Apply PBC (Poly Butyl Cuprysil) grease or silicone-based grease.

37663_FX35_G0103

Fig. 16 Exploded view of rear caliper assembly—1 piston type

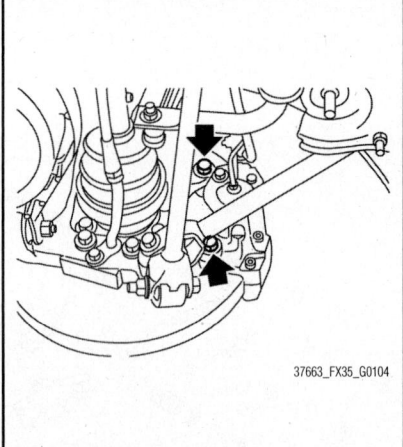

37663_FX35_G0104

Fig. 17 Remove caliper mounting bolts, and remove caliper—2 piston type

❊❊ WARNING

Do not drop brake pad and caliper assembly.

7. Remove disc rotor.

➡Put matching marks on the wheel hub and bearing assembly and the disc rotor before removing the disc rotor. Never drop the disc rotor.

To install:

8. Install disc rotor.

➡Align the matching marks that have been made during removal when reusing the disc rotor.

9. Install the brake caliper assembly to the vehicle and tighten the torque member mounting bolts to the specified torque.

➡Never spill or splash any grease and moisture on the brake caliper assembly mounting face, threads, mounting bolts, and washers. Wipe out any grease and moisture.

10. Install brake hose and copper washers to brake caliper assembly, and tighten union bolts to the specified torque.

11. Refill with new brake fluid and perform the air bleeding.

❊❊ WARNING

Never reuse drained brake fluid. Never spill or splash brake fluid on the disc rotor.

12. Check the drag of the rear disc brake.

2 Piston Type

See Figures 17 and 18.

❊❊ CAUTION

Clean any dust from the brake caliper and brake pads with a vacuum dust collector. Never blow with compressed air.

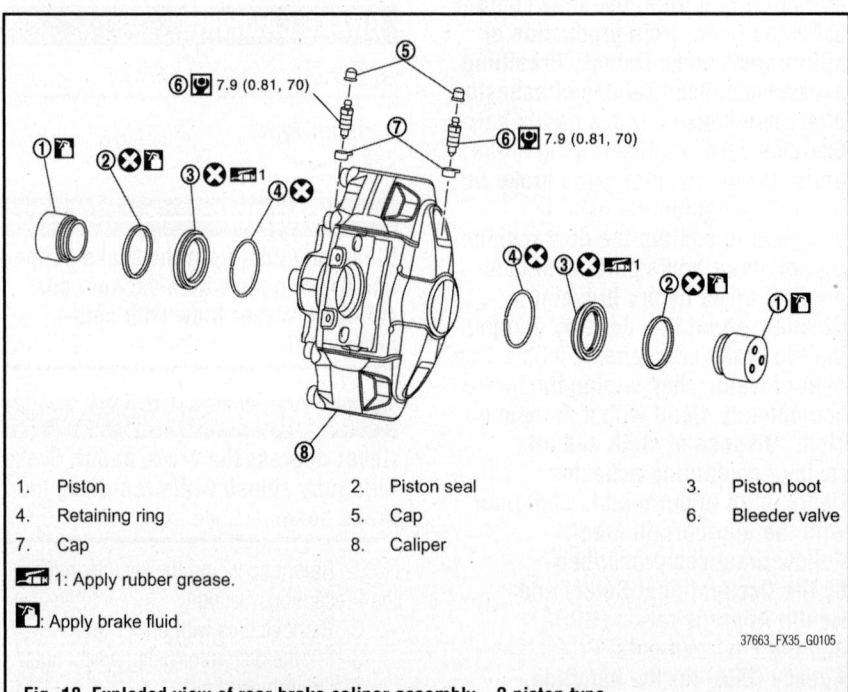

1. Piston
2. Piston seal
3. Piston boot
4. Retaining ring
5. Cap
6. Bleeder valve
7. Cap
8. Caliper

1: Apply rubber grease.

: Apply brake fluid.

37663_FX35_G0105

Fig. 18 Exploded view of rear brake caliper assembly—2 piston type

❊❊ WARNING

Never depress the brake pedal. Brake fluid may splash while removing the brake hose and brake tube.

1. Before servicing the vehicle, refer to the Precautions Section.
2. Remove tires and wheels.
3. Fix the disc rotor using wheel nuts.
4. Drain brake fluid.

➡Never spill or splash brake fluid on the disc rotor.

5. Loosen the flare nut with a flare nut wrench and separate the brake tube from caliper.

Note the following:

- Cover flare nut wrench with a cloth as not to damage the caliper
- Never scratch the flare nut and the brake tube
- Never bend sharply, twist or strongly pull out the brake tube
- Cover open end of brake tube when disconnecting to prevent entrance of dirt

6. Remove brake hose mounting bolt.

7. Remove caliper mounting bolts, and remove caliper.

❋❋ WARNING

Do not drop brake pad and caliper.

8. Remove disc rotor.

➡ **Put matching marks on the wheel hub and bearing assembly and the disc rotor before removing the disc rotor. Never drop disc rotor.**

To install:

9. Install disc rotor.

➡ **Align the matching marks that have been made during removal when reusing the disc rotor.**

10. Install the brake caliper to the vehicle and tighten the caliper mounting bolts to the specified torque.

➡ **Never spill or splash any grease and moisture on the caliper mounting face, threads, mounting bolts and washers. Wipe out any grease and moisture.**

11. Install the brake hose mounting bolt to the specified torque.

12. Tighten the flare nut to the specified torque with a flare nut crowfoot and a torque wrench.

❋❋ WARNING

Cover crowfoot with a cloth as not to damage the caliper. Never scratch the flare nut and the brake tube.

13. Refill with new brake fluid and perform the air bleeding.

❋❋ WARNING

Never reuse drained brake fluid. Never spill or splash brake fluid on the disc rotor.

14. Check the drag of the rear disc brake.

BRAKE PADS

REMOVAL & INSTALLATION

1 Piston Type

See Figure 19.

❋❋ CAUTION

Clean any dust from the brake caliper and brake pads with a vacuum dust collector. Never blow with compressed air.

❋❋ WARNING

Never depress the brake pedal while removing the brake pads or the cylinder body because the piston may pop out. Never spill or splash brake fluid on the disc rotor.

1. Before servicing the vehicle, refer to the Precautions Section.

2. Remove tires with power tool.

3. Remove the upper sliding pin bolt.

4. Suspend the cylinder body with a wire so that the brake hose will not stretch. Remove the brake pads, shims, shim cover and pad retainers from the torque member.

Note the following:
- Never deform the pad retainers if removing the pad retainers
- Never damage the piston boot
- Never drop the brake pad, shims, and the shim cover

To install:

5. Apply Poly Butyl Cuprysil (PBC) grease or silicone-based grease to the pad retainers before installing it to the torque

member if the pad retainers has been removed.

❋❋ WARNING

Securely assemble the pad retainers so that it will not be lifted up from the torque member. Never deform the pad retainers.

6. Apply PBC grease or silicone-based grease to the mating faces between the shims and the shim cover.

➡ **Always replace the shims together with the shim cover when replacing the brake pad.**

7. Apply PBC grease or silicone-based grease to the mating faces between the brake pads and pad retainers and install them to the brake pads.

8. Install cylinder body and brake pads to torque member.

Note the following:
- Never damage the piston boot
- When replacing brake pad with new one, check a brake fluid level in the reservoir tank because brake fluid returns to master cylinder reservoir tank when pressing piston in

➡ **Use a disc brake piston tool to easily press piston.**

9. Install the upper sliding pin bolt and tighten it to the specified torque.

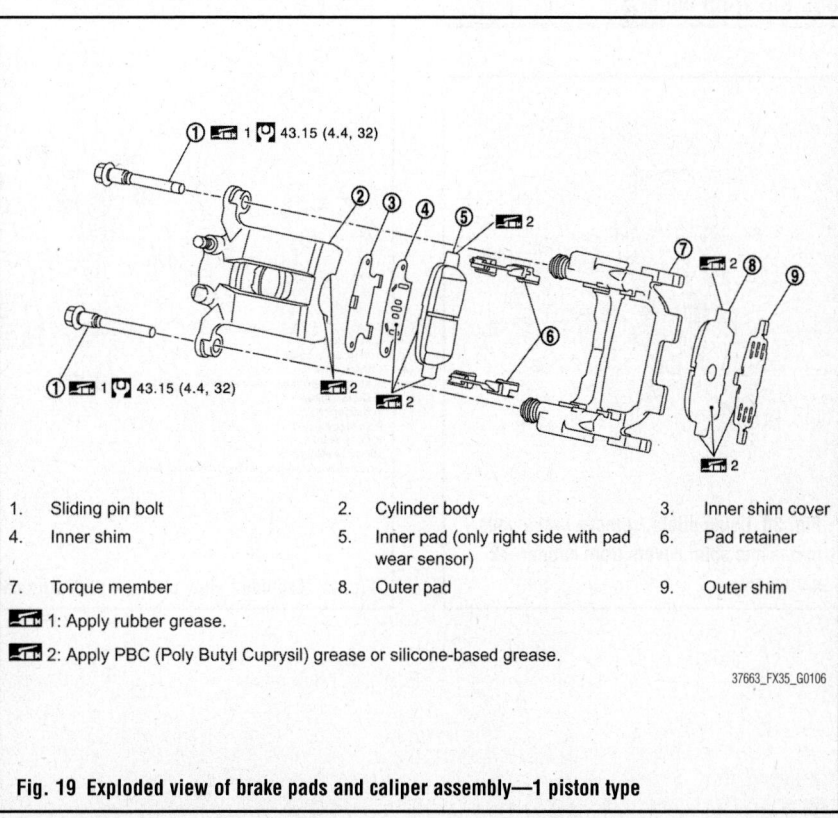

1. Sliding pin bolt
2. Cylinder body
3. Inner shim cover
4. Inner shim
5. Inner pad (only right side with pad wear sensor)
6. Pad retainer
7. Torque member
8. Outer pad
9. Outer shim

1: Apply rubber grease.

2: Apply PBC (Poly Butyl Cuprysil) grease or silicone-based grease.

37663_FX35_G0106

Fig. 19 Exploded view of brake pads and caliper assembly—1 piston type

10. Depress the brake pedal several times to check that no drag feel is present for the rear disc brake.

2 Piston Type

See Figures 20 through 22.

※※ CAUTION

Clean any dust from the brake caliper and brake pads with a vacuum dust collector. Never blow with compressed air.

※※ WARNING

Never depress the brake pedal while removing the brake pads or the cylinder body because the piston may pop out. Never spill or splash brake fluid on the disc rotor and caliper.

1. Before servicing the vehicle, refer to the Precautions Section.
2. Remove tires with power tool.
3. Remove clips from pad pins.
4. Remove pad pins while holding down cross spring, then remove cross spring from caliper.
5. Using pliers, remove brake pads, shims and shim covers from caliper.

※※ WARNING

Do not damage the piston boot. Never drop the brake pad, shims, and the shim cover.

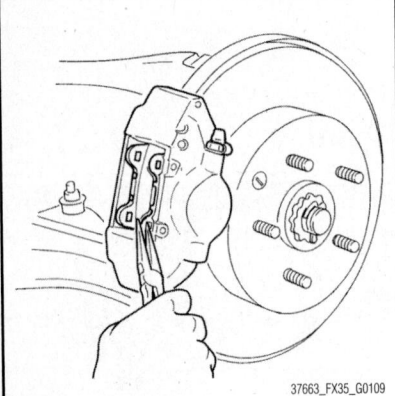

Fig. 20 Using pliers, remove brake pads, shims and shim covers from caliper—2 piston type

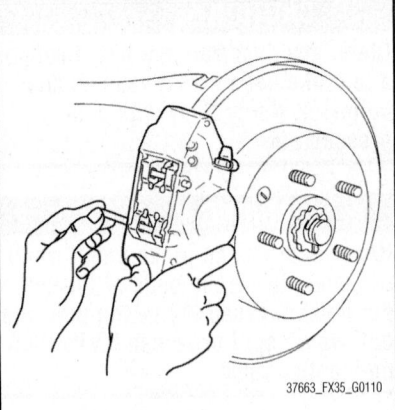

Fig. 21 Installing pad pins and cross spring—2 piston type

To install:

6. Apply copper based brake grease to the mating faces between the brake pads, shims and shim cover, and install shims and shim cover to the brake pad.

➡ Always replace the shims together when replacing the brake pad.

7. Apply copper based brake grease to the mating faces between the brake pads and caliper.
8. Apply copper based brake grease to the mating faces between the brake pads and pad pins.

9. Apply copper based brake grease to the mating faces between the brake pads and cross spring.
10. Install brake pads to caliper. Note the following:
- Never damage the piston boot
- In the case of replacing a pad with new one, check a brake fluid level in the reservoir tank because brake fluid returns to master cylinder reservoir tank when pressing piston in

➡ Use a disc brake piston tool to easily press piston.

11. Install upper pad pin from the inner side, then install firmly to the outer side through the hole in the top of brake pad.
12. Place the top of cross spring over the upper pad pin, press in the cross spring, install lower pad pin from the inner side to the outer side, and secure cross spring.
13. Install clips to the pad pins.

※※ CAUTION

If clip is not fully attached, pad pin or brake pad could fall out while vehicle is in motion.

14. Depress the brake pedal several times to check that no drag feel is present for the rear disc brake.

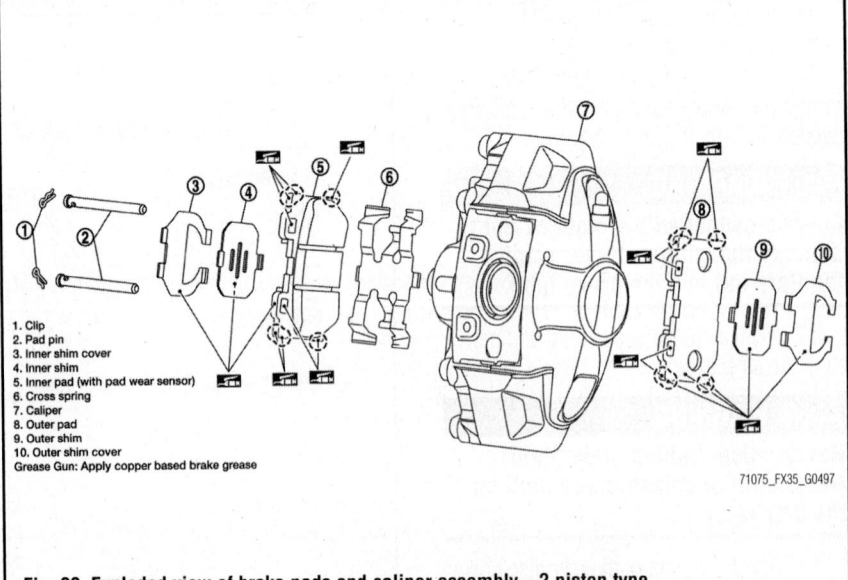

1. Clip
2. Pad pin
3. Inner shim cover
4. Inner shim
5. Inner pad (with pad wear sensor)
6. Cross spring
7. Caliper
8. Outer pad
9. Outer shim
10. Outer shim cover
Grease Gun: Apply copper based brake grease

Fig. 22 Exploded view of brake pads and caliper assembly—2 piston type

CHASSIS ELECTRICAL | **AIR BAGS (SUPPLEMENTAL RESTRAINT SYSTEM)**

All new vehicles are now equipped with an air bag system, often referred to as a Supplemental Restraint System (SRS) or Supplemental Inflatable Restraint (SIR) system. The system must be disabled before performing service on or around system components, steering column, instrument panel components, wiring, and sensors. Failure to follow safety and disabling procedures could result in accidental air bag deployment, possible personal injury, and unnecessary system repairs.

❄❄ WARNING

Always wear safety goggles when working with, or around, the air bag system. When carrying a non-deployed air bag, be sure the bag and trim cover are pointed away from your body. When placing a non-deployed air bag on a work surface, always face the bag and trim cover upward, away from the surface. This will reduce the motion of the module if it is accidentally deployed. Refer to air bag system precautions.

PRECAUTIONS

Disconnect and isolate the battery negative cable before beginning any airbag system component diagnosis, testing, removal, or installation procedures. Allow the system capacitor to discharge for 3 minutes before beginning any component service. This will disable the airbag system. Failure to disable the airbag system may result in accidental airbag deployment, personal injury, or death.

Replace airbag system components with OEM replacement parts. Substitute parts may appear interchangeable, but internal differences may result in inferior occupant protection. Failure to do so may result in occupant personal injury or death.

Wear safety glasses, rubber gloves, and long sleeved clothing when cleaning powder residue from vehicle after an airbag deployment. Powder residue emitted from a deployed airbag can cause skin irritation. Flush affected area with cool water if irritation is experienced. If nasal or throat irritation is experienced, exit the vehicle for fresh air until the irritation ceases. If irritation continues, see a physician.

Do not use a replacement airbag that is not in the original packaging. This may result in improper deployment, personal injury, or death.

The factory installed fasteners, screws and bolts used to fasten airbag components have a special coating and are specifically designed for the airbag system. Do not use substitute fasteners. Use only original equipment fasteners listed in the parts catalog when fastener replacement is required.

Deployed and non-deployed airbags may or may not have live pyrotechnic material within the airbag inflator. Handle with care.

Do not dispose of driver/passenger/curtain airbags or seat belt tensioners unless you are sure of complete deployment.

Dispose of deployed airbags and tensioners consistent with state, provincial, local, and federal regulations.

After any airbag component service, do not immediately connect the battery negative cable. Personal injury or death may result if the system test is not performed first.

To avoid rendering the SRS inoperative, which could increase the risk of personal injury or death in the event of a collision which would result in air bag inflation, all maintenance must be performed by an authorized NISSAN/INFINITI dealer.

Improper maintenance, including incorrect removal and installation of the SRS, can lead to personal injury caused by unintentional activation of the system.

Do not use electrical test equipment on any circuit related to the SRS unless instructed to in this Service Manual. SRS wiring harnesses can be identified by yellow and/or orange harnesses or harness connectors.

When working near the Airbag Diagnosis Sensor Unit or other Airbag System sensors with the Ignition ON or engine running, DO NOT use air or electric power tools or strike near the sensor(s) with a hammer. Heavy vibration could activate the sensor(s) and deploy the air bag(s), possibly causing serious injury.

When using air or electric power tools or hammers, always switch the Ignition OFF, disconnect the battery, and wait at least 3 minutes before performing any service.

Do not use electrical test equipment to check SRS circuits unless instructed to in this Service Manual.

Before servicing the SRS, turn the ignition switch OFF, disconnect both battery cables and wait at least 3 minutes. For approximately 3 minutes after the cables are removed, it is still possible for the air bag

and seat belt pretensioner to deploy. Therefore, do not work on any SRS connectors or wires until at least 3 minutes have passed.

The diagnosis sensor unit must always be installed with arrow marks pointing towards the front of the vehicle for proper operation. Also check the diagnosis sensor unit for cracks, deformities, or rust before installation and replace as required.

The spiral cable must be aligned with the neutral position since its rotations are limited. Do not turn the steering wheel and column after removal of the steering gear.

Handle the air bag module carefully. Always place driver and front passenger air bag modules with the pad side facing upward and the seat mounted front side air bag module standing with the stud bolt side facing down.

Conduct a self-diagnosis to check the entire SRS for proper function after replacing any components.

After the air bag inflates, the front instrument panel assembly should be replaced if damaged.

Always replace the instrument panel pad following a front passenger air bag deployment.

DISARMING THE SYSTEM

❄❄ CAUTION

All SRS electrical wiring harnesses and connectors are covered with YELLOW outer insulation. Do not use electrical test equipment on any circuit related to the SRS (air bag) sensors. When installing SRS components, always install with the arrow marks facing the front of the vehicle.

1. Before servicing the vehicle, refer to the Precautions Section.
2. Turn the ignition switch to the **OFF** position.
3. Disconnect both battery cables starting with the negative cable first.
4. Wait at least 3 minutes after the cables are disconnected. Be sure to insulate the battery terminal ends.

ARMING THE SYSTEM

1. Before servicing the vehicle, refer to the Precautions Section.

2. Make sure that the removed components are installed and/or the disconnected connectors are connected properly.

3. Turn the ignition switch to the **OFF** position.

4. Connect both battery cables starting with the positive cable first.

5. The SRS or air bag system is equipped with a self-diagnostic operation after turning the ignition key to the ON or START position.

a. The AIR BAG warning lamp will illuminate for 7 seconds.

b. After 7 seconds, the AIR BAG lamp will extinguish if no malfunction is detected.

c. If the AIR BAG lamp does not extinguish after 7 seconds, check the SRS self-diagnostic system for a malfunction.

DRIVETRAIN

AUTOMATIC TRANSMISSION

DRAIN & REFILL

See Figures 23 through 25.

Note the following:

• Use only Genuine NISSAN Matic S ATF. Never mix with other ATF

• Using ATF other than Genuine NISSAN Matic S ATF will cause deterioration in driveability and A/T durability, and may damage the A/T, which is not covered by the INFINITI new vehicle limited warranty

• When filling ATF, be careful not to spill fluid on heat generating parts such as the exhaust

• Before servicing the vehicle, refer to the Precautions Section

1. Step 1: Install the O-ring (315268E000) to the charging pipe (310811EA5A).

2. Step 2:

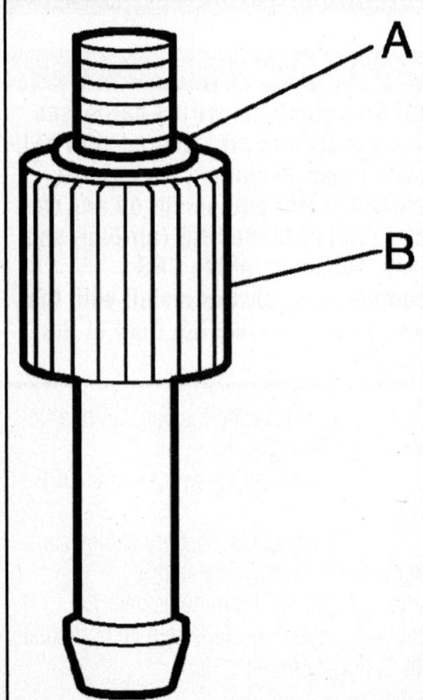

a. Use CONSULT to check that the ATF temperature is 104°F (40°C) or less.

b. Raise and safely support the vehicle.

c. Remove the drain plug from the oil pan, and then drain the ATF.

d. When the ATF starts to drip, temporarily tighten the drain plug to the oil pan.

➡**Do not replace the drain plug and drain plug gasket with new ones at this point.**

e. Remove overflow plug from oil pan.

f. Install the charging pipe to the overflow plug hole.

❋❋ WARNING

Tighten the charging pipe by hand.

g. Install the bucket pump hose to the charging pipe.

➡**Insert the bucket pump hose all the way to the end of the charging pipe.**

h. Fill approximately 3.2 quarts (3 liters) of the ATF.

i. Remove the bucket pump hose to remove the charging pipe, and then temporarily tighten the overflow plug to the oil pan.

➡**Quickly perform the procedure to avoid ATF leakage from the oil pan.**

j. Lower the vehicle.

k. Start the engine and wait for approximately 3 minutes.

l. Stop the engine.

3. Step 3: Repeat "Step 2".

4. Final Step:

a. Use CONSULT to check that the ATF temperature is 104°F (40°C) or less.

b. Raise and safely support the vehicle.

c. Remove the drain plug from the oil pan, and then drain the ATF.

d. When the ATF starts to drip, tighten the drain plug to the oil pan to the specified torque.

➡**Do not reuse the drain plug and drain plug gasket.**

e. Remove overflow plug from oil pan.

f. Install the charging pipe to the overflow plug hole.

❋❋ WARNING

Tighten the charging pipe by hand.

g. Install the bucket pump hose to the charging pipe.

➡**Insert the bucket pump hose all the way to the end of the charging pipe.**

h. Fill approximately 3.2 quarts (3 liters) of the ATF.

i. Remove the bucket pump hose to remove the charging pipe, and then temporarily tighten the overflow plug to the oil pan.

➡**Quickly perform the procedure to avoid ATF leakage from the oil pan.**

j. Lower the vehicle.

k. Start the engine.

l. Make the ATF temperature approximately 104°F (40°C).

➡**The ATF level is greatly affected by the temperature. Always check the ATF temperature on "ATF TEMP 1" of "Data Monitor" using CONSULT.**

71075_FX35_G0095

Fig. 23 O-ring (315268E000) (A) on the charging pipe (310811EA5A) (B) shown

71075_FX35_G0096

Fig. 24 Install the charging pipe (A) to the overflow plug hole. Install the bucket pump hose (B) to the charging pipe

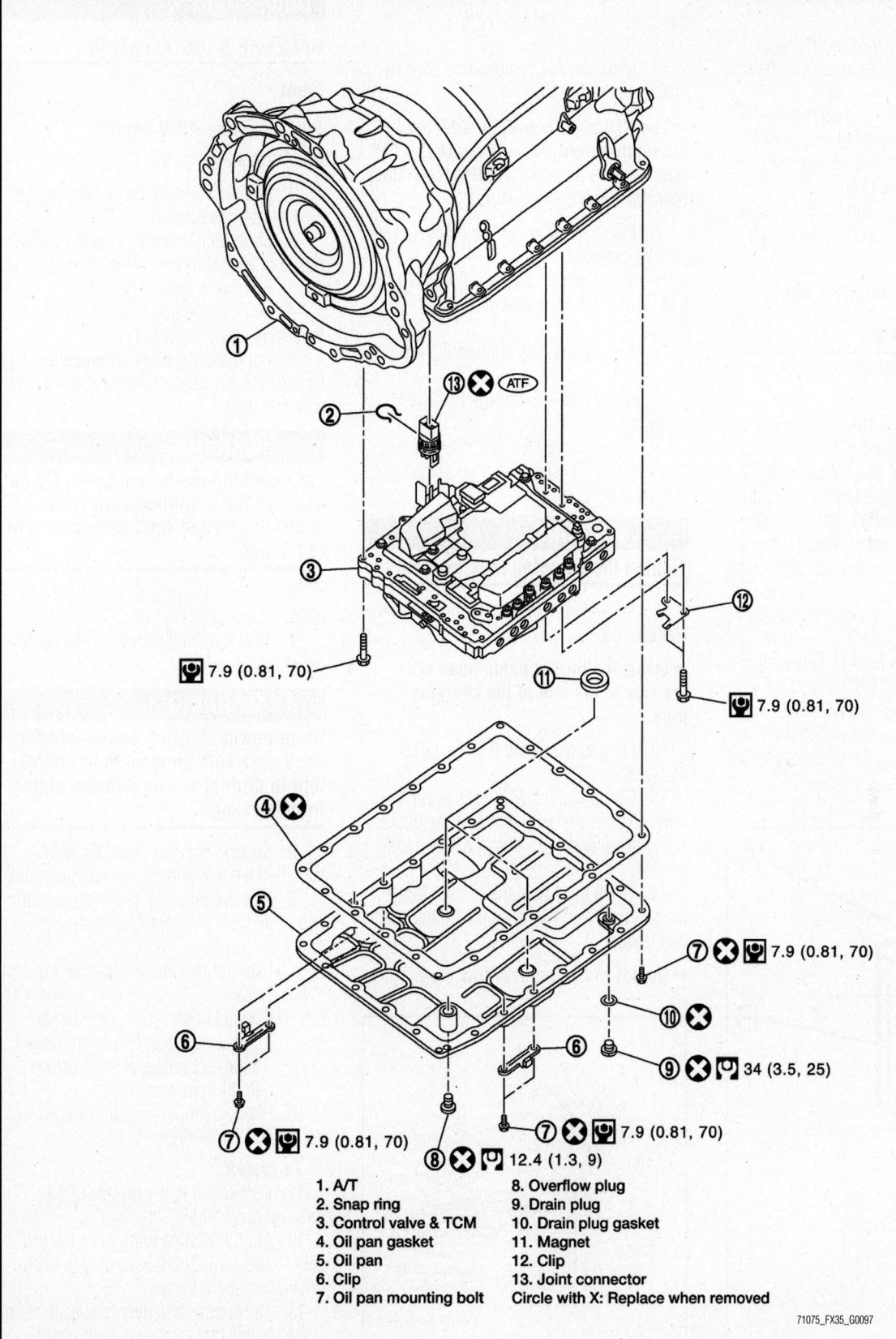

7.9 (0.81, 70)

7.9 (0.81, 70)

7.9 (0.81, 70)

34 (3.5, 25)

7.9 (0.81, 70)

12.4 (1.3, 9)

1. A/T
2. Snap ring
3. Control valve & TCM
4. Oil pan gasket
5. Oil pan
6. Clip
7. Oil pan mounting bolt

8. Overflow plug
9. Drain plug
10. Drain plug gasket
11. Magnet
12. Clip
13. Joint connector
Circle with X: Replace when removed

71075_FX35_G0097

Fig. 25 Transmission control valve and TCM component locations

m. Park vehicle on level surface and set parking brake.

n. Shift the selector lever through each gear position. Leave selector lever in "P" position.

o. Raise and safely support the vehicle when the ATF temperature reaches 104°F (40°C), and then remove the overflow plug from the oil pan.

p. When the ATF starts to drip, tighten the overflow plug to the oil pan to the specified torque.

➡**Do not reuse the overflow plug.**

FLUID LEVEL CHECK

See Figures 26 through 28.

Note the following:
• Use only Genuine NISSAN Matic S ATF. Never mix with other ATF
• Using ATF other than Genuine NISSAN Matic S ATF will cause deterioration in driveability and A/T durability, and may damage the A/T, which is not covered by the INFINITI new vehicle limited warranty
• When filling ATF, be careful not to spill fluid on heat generating parts such as the exhaust system
• Always maintain the ATF temperature within between 95°F (35°C) and 113°F (45°C) while checking with CONSULT when the ATF level adjustment is performed

1. Before servicing the vehicle, refer to the Precautions Section.

2. Install the O-ring (315268E000) to the charging pipe (310811EA5A).

3. Start the engine.

4. Make the ATF temperature approximately 104°F (40°C).

➡**The ATF level is greatly affected by the temperature. Always check the ATF temperature on "ATF TEMP 1" of "Data Monitor" using CONSULT.**

5. Park vehicle on level surface and set parking brake.

6. Shift the selector lever through each gear position. Leave selector lever in "P" position.

7. Raise and safely support the vehicle.

8. Check the ATF leakage from transmission.

9. Remove overflow plug from oil pan.

10. Install the charging pipe to the overflow plug hole.

✳✳ WARNING

Tighten the charging pipe by hand.

11. Install the bucket pump hose to the charging pipe.

➡**Insert the bucket pump hose all the way to the end of the charging pipe.**

12. Fill approximately 0.5 quart (0.5 liters) of the ATF.

13. Check that the ATF leaks when removing the charging pipe and the bucket pump hose. If the ATF does not leak, refill the ATF.

14. When the ATF starts to drip, tighten the overflow plug to the oil pan to the specified torque.

➡**Do not reuse the overflow plug.**

DRIVESHAFT

REMOVAL & INSTALLATION

Front

3.5L Engine—AWD Model

See Figures 29 and 30.

1. Before servicing the vehicle, refer to the Precautions Section.

2. Shift the transmission to the neutral position, and then release the parking brake.

3. Remove engine undercover.

4. Remove exhaust front tube and three-way catalyst (bank 1).

5. Put matching mark on propeller shaft (driveshaft) flange yoke and final drive companion flange.

✳✳ WARNING

For matching mark, use paint. Do not damage the propeller shaft (driveshaft) flange and final drive companion flange.

6. Remove the propeller shaft (driveshaft) assembly fixing bolts.

7. Move steering hydraulic line not to interfere with work.

✳✳ WARNING

Wrap power steering piping interference area with shop cloth or equivalent to protect power steering piping from breakage.

8. Support transfer assembly with a jack, remove rear engine mounting member.

9. Remove propeller shaft (driveshaft) assembly from the front final drive and transfer.

• Do not damage the transfer front oil seal
• Wrap transmission interference area with shop cloth or equivalent to protect propeller shaft (driveshaft) from breakage

10. Remove propeller shaft (driveshaft) assembly from O-ring.

To install:

11. Installation is the reverse of the removal procedure.

12. Align matching mark to install propeller shaft (driveshaft) assembly to final drive companion flange.

13. Tighten the fasteners to specification.

14. Inspect the propeller shaft (driveshaft) assembly after installation.

a. Perform a driving test to check propeller shaft (driveshaft) vibration.

![Fig. 26 O-ring diagram]

71075_FX35_G0095

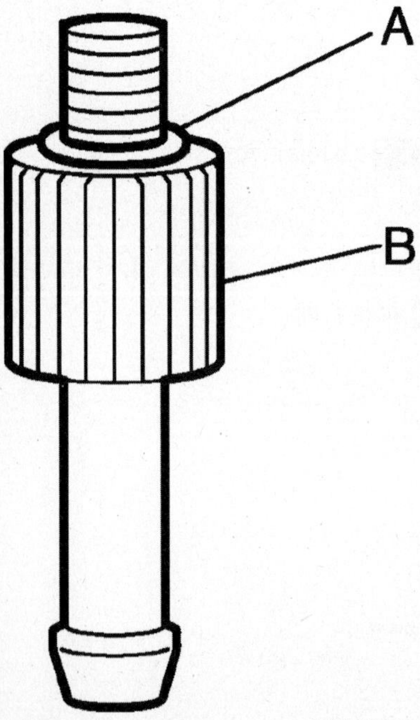

Fig. 26 O-ring (315268E000) (A) on the charging pipe (310811EA5A) (B) shown

71075_FX35_G0096

Fig. 27 Install the charging pipe (A) to the overflow plug hole. Install the bucket pump hose (B) to the charging pipe

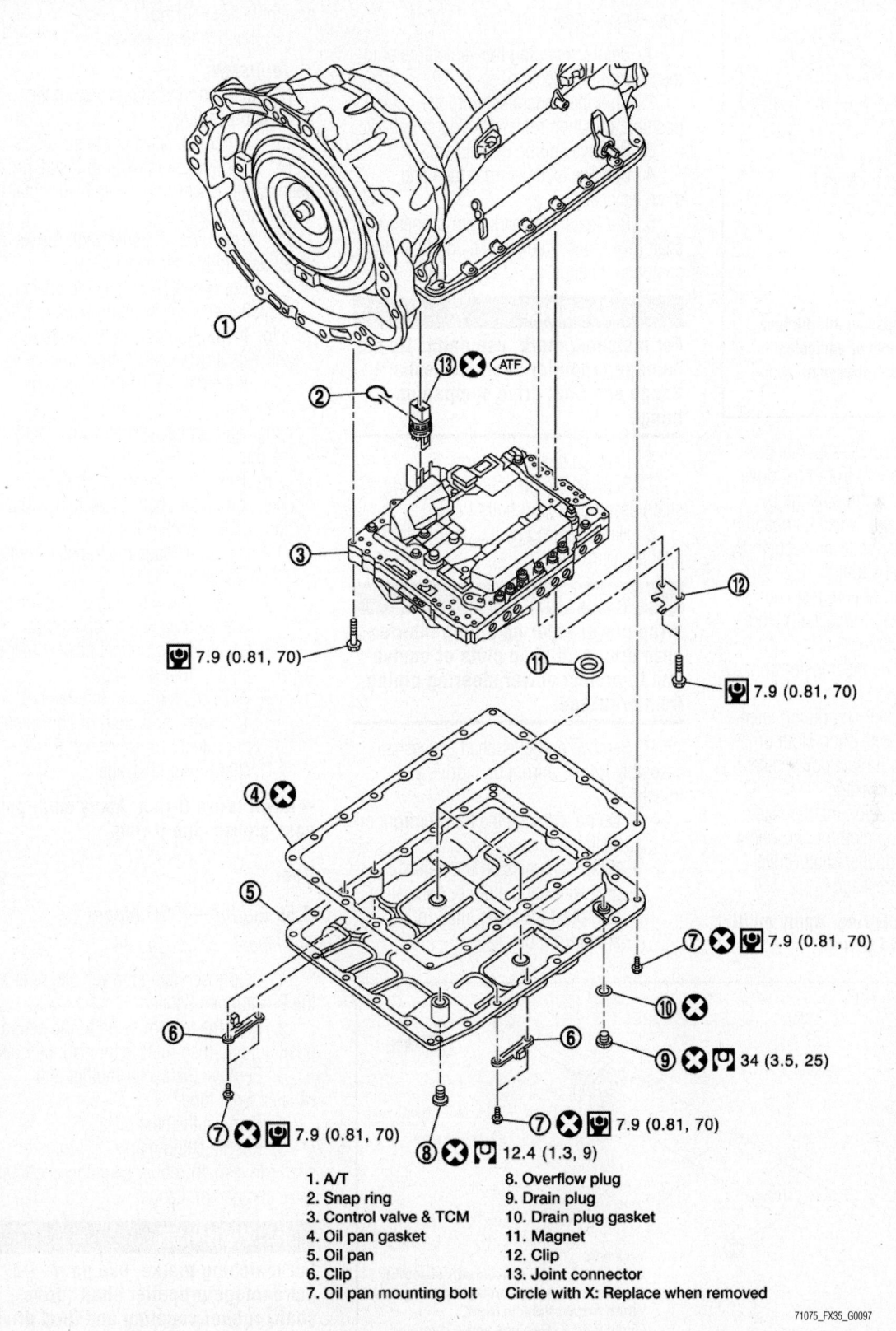

7.9 (0.81, 70)

7.9 (0.81, 70)

7.9 (0.81, 70)

34 (3.5, 25)

7.9 (0.81, 70)

7.9 (0.81, 70)

12.4 (1.3, 9)

1. A/T	8. Overflow plug
2. Snap ring	9. Drain plug
3. Control valve & TCM	10. Drain plug gasket
4. Oil pan gasket	11. Magnet
5. Oil pan	12. Clip
6. Clip	13. Joint connector
7. Oil pan mounting bolt	Circle with X: Replace when removed

71075_FX35_G0097

Fig. 28 Transmission control valve and TCM component locations

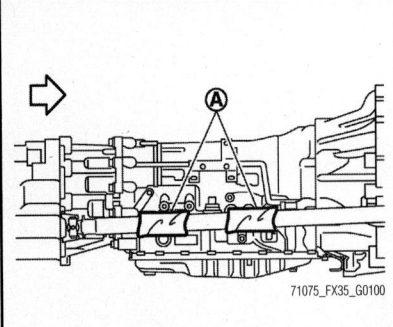

Fig. 29 Wrap transmission interference area (A) with shop cloth or equivalent to protect propeller shaft (driveshaft) from breakage

b. If vibration occurs, separate propeller shaft (driveshaft) from final drive.

c. Reinstall companion flange by changing the phase between companion flange and propeller shaft (driveshaft) by the one bolt hole at a time.

d. Then perform driving test and check propeller shaft (driveshaft) vibration again at each point.

- Do not damage the transfer front oil seal
- Wrap power steering piping interference area with shop cloth or equivalent to protect power steering piping from breakage
- Wrap transmission interference area with shop cloth or equivalent to protect propeller shaft (driveshaft) from breakage

➡ **Do not reuse the O-ring. Apply multipurpose grease onto O-ring.**

5.0L Engine

See Figures 29 and 31.

1. Before servicing the vehicle, refer to the Precautions Section.

2. Shift the transmission to the neutral position, and then release the parking brake.

3. Remove engine undercover.

4. Remove exhaust front tube and three-way catalyst.

5. Put matching mark onto propeller shaft (driveshaft) flange yoke and final drive companion flange.

❋❋ WARNING

For matching mark, use paint. Do not damage propeller shaft (driveshaft) flange and final drive companion flange.

6. Remove heat insulator.

7. Remove the propeller shaft (driveshaft) assembly fixing bolts.

8. Hang steering hydraulic line not to interfere with work.

❋❋ WARNING

Wrap power steering piping interference area with shop cloth or equivalent to protect power steering piping from breakage.

9. Remove propeller shaft (driveshaft) assembly from the front final drive and transfer.

- Do not damage the transfer front oil seal
- Wrap transmission interference area with shop cloth or equivalent to protect propeller shaft (driveshaft) from breakage

10. Remove propeller shaft (driveshaft) assembly from O-ring.

11. Remove heat bracket.

To install:

12. Installation is the reverse of the removal procedure.

13. Tighten the fasteners to specification.

14. Align matching mark to install propeller shaft (driveshaft) assembly to final drive companion flange.

15. Inspect the propeller shaft (driveshaft) assembly after installation.

a. Perform a driving test to check propeller shaft (driveshaft) vibration.

b. If vibration occurs, separate propeller shaft (driveshaft) from final drive.

c. Reinstall companion flange by changing the phase between companion flange and propeller shaft (driveshaft) by the one bolt hole at a time.

d. Then perform driving test and check propeller shaft (driveshaft) vibration again at each point.

- Do not damage the transfer front oil seal
- Wrap power steering piping interference area with shop cloth or equivalent to protect power steering piping from breakage
- Wrap transmission interference area with shop cloth or equivalent to protect propeller shaft (driveshaft) from breakage

➡ **Never reuse O-ring. Apply multi-purpose grease onto O-ring.**

Rear

3.5L Engine—2WD Model

See Figures 32 through 34.

1. Before servicing the vehicle, refer to the Precautions Section.

2. Shift the transmission to the neutral position, and then release the parking brake.

3. Remove the center muffler and exhaust front tube.

4. Remove the heat plate.

5. Put matching marks on propeller shaft (driveshaft) rubber coupling and final drive companion flange.

❋❋ WARNING

For matching marks, use paint. Do not damage propeller shaft (driveshaft) rubber coupling and final drive companion flange.

6. Loosen mounting nuts of center bearing mounting brackets (upper/lower).

➡ **Tighten mounting nuts temporarily.**

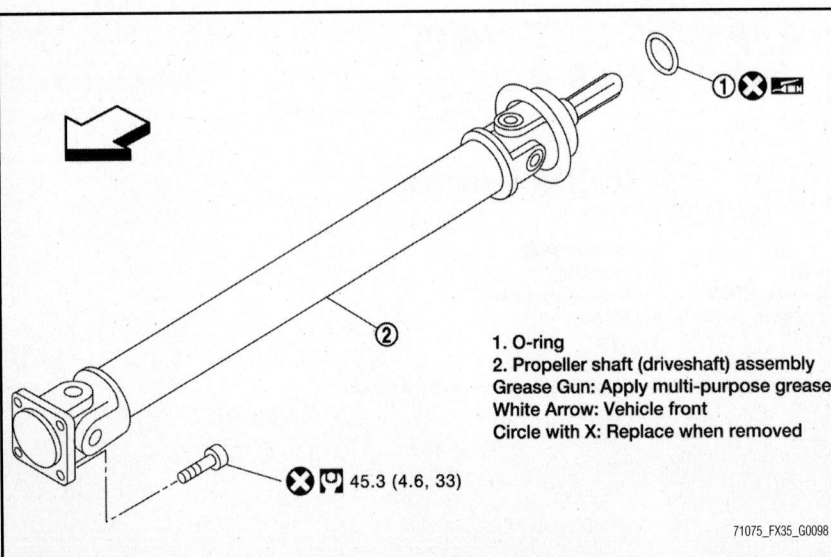

1. O-ring
2. Propeller shaft (driveshaft) assembly
Grease Gun: Apply multi-purpose grease
White Arrow: Vehicle front
Circle with X: Replace when removed

❌ ⌷ 45.3 (4.6, 33)

Fig. 30 Front propeller shaft (driveshaft) assembly component locations—3.5L engine

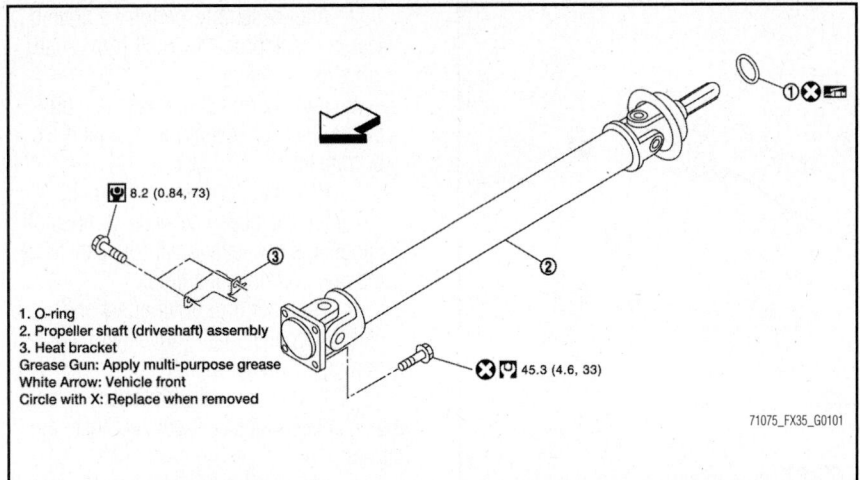

Fig. 31 Front propeller shaft (driveshaft) assembly component locations—5.0L engine

1. O-ring
2. Propeller shaft (driveshaft) assembly
3. Heat bracket
Grease Gun: Apply multi-purpose grease
White Arrow: Vehicle front
Circle with X: Replace when removed

71075_FX35_G0101

7. Remove propeller shaft (driveshaft) assembly fixing bolts and nuts.

※ WARNING

Do not remove the rubber coupling from the propeller shaft (driveshaft) assembly.

8. Slightly separate the rubber coupling from the final drive companion flange.

※ WARNING

Do not damage the final drive companion flange and rubber coupling.

9. Remove center bearing mounting bracket fixing nuts.
   - Never bend the rubber coupling above an angle of 0–4°
   - Do not damage the grease seal
10. Slide the propeller shaft (driveshaft) in the vehicle forward direction slightly. Separate the propeller shaft (driveshaft)

from the final drive companion flange.

※ WARNING

Do not damage the grease seal. Never damage the rubber coupling.

11. Remove the propeller shaft (driveshaft) assembly from the vehicle.

※ WARNING

Do not damage the rear oil seal of transmission.

12. Remove clip and center bearing mounting bracket (upper/lower).

To install:

13. Installation is the reverse of the removal procedure.

14. Install center bearing mounting bracket (upper) with its arrow mark facing forward.

15. Adjust position of center bearing mounting bracket (upper), center bearing mounting bracket (lower) sliding back and forth to prevent play in thrust direction of center bearing insulator. Install center bearing mounting bracket (upper/lower) to vehicle.

16. Align matching marks to install propeller shaft (driveshaft) rubber coupling to final drive companion flange.

17. Tighten the fasteners to specification.

18. Perform inspection after installation.

   a. After assembly, perform a driving test to check propeller shaft (driveshaft) vibration.

   b. If vibration occurs, separate propeller shaft (driveshaft) from final drive.

   c. Reinstall companion flange by changing the phase between companion flange and propeller shaft (driveshaft) by the one bolt hole at a time.

   d. Then perform driving test and check propeller shaft (driveshaft) vibration again at each point.

19. If propeller shaft (driveshaft) or final drive has been replaced, connect them as follows:

   a. Install the propeller shaft (driveshaft) while aligning its matching mark with the matching mark on the joint as close as possible.

※ WARNING

Do not damage the grease seal or rubber coupling. Do not damage the rear oil seal of transmission. Do not damage the rubber coupling, protect it with a shop towel or equivalent.

3.5L Engine—AWD Model

See Figures 35 and 36.

1. Before servicing the vehicle, refer to the Precautions Section.

2. Shift the transmission to the neutral position, and release the parking brake.

3. Remove the center muffler and exhaust front tube.

4. Remove the heat plate.

5. Put matching marks on propeller shaft (driveshaft) flange yoke and transfer companion flange.

※ WARNING

For matching marks, use paint. Do not damage propeller shaft (driveshaft) flange yoke and transfer companion flange.

6. Put matching marks on propeller shaft (driveshaft) joint and final drive companion flange.

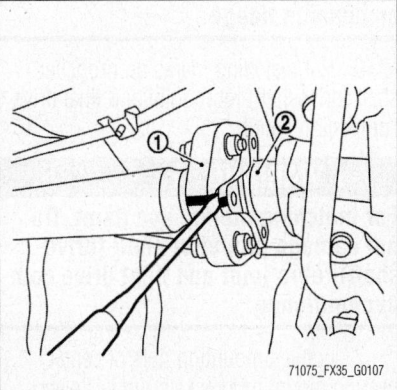

71075_FX35_G0107

Fig. 32 Slightly separate the rubber coupling (1) from the final drive companion flange (2)

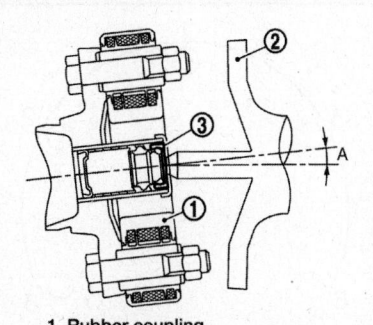

1. Rubber coupling
2. Final drive companion flange
3. Grease seal.
A. Never bend rubber coupling above 0 – 4°

71075_FX35_G0108

Fig. 33 Remove center bearing mounting bracket fixing nuts

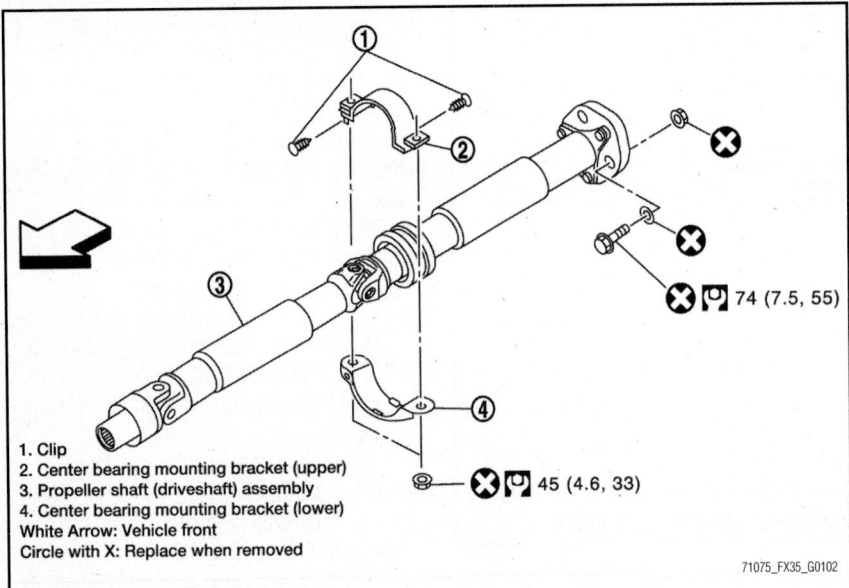

1. Clip
2. Center bearing mounting bracket (upper)
3. Propeller shaft (driveshaft) assembly
4. Center bearing mounting bracket (lower)
White Arrow: Vehicle front
Circle with X: Replace when removed

74 (7.5, 55)

45 (4.6, 33)

71075_FX35_G0102

Fig. 34 Rear propeller shaft (driveshaft) assembly component locations—3.5L engine (2WD model)

✳✳ WARNING

For matching marks, use paint. Do not damage propeller shaft (driveshaft) joint and final drive companion flange.

7. Loosen mounting nuts of center bearing mounting brackets (upper/lower).

➡**Tighten mounting nuts temporarily.**

8. Remove propeller shaft (driveshaft) assembly fixing bolts and nuts.

9. Remove center bearing mounting bracket fixing nuts.

10. Remove propeller shaft (driveshaft) assembly.

✳✳ WARNING

Do not damage the rear oil seal of transmission.

✳✳ WARNING

If constant velocity joint was bent during propeller shaft (driveshaft) assembly removal, installation, or transportation, its boot may be damaged. Wrap boot interference area to metal part with shop cloth or equivalent to protect boot from breakage.

11. Remove clip and center bearing mounting bracket (upper/lower).

To install:

12. Installation is the reverse of the removal procedure.

13. Tighten the fasteners to specification.

14. Install center bearing mounting bracket (upper) with its arrow mark facing forward.

15. Adjust position of center bearing mounting bracket (upper), center bearing mounting bracket (lower) sliding back and forth to prevent play in thrust direction of center bearing insulator. Install center bearing mounting bracket (upper/lower) to vehicle.

16. Align matching marks to install propeller shaft (driveshaft) flange yoke and transfer companion flange.

17. Align matching marks to install propeller shaft (driveshaft) joint and final drive companion flange.

18. Tighten mounting bolt and nut in the order shown in the figure.

19. Perform inspection after installation.

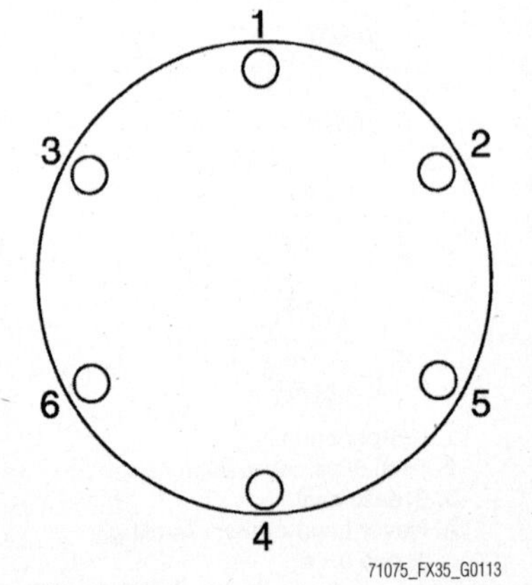

71075_FX35_G0113

Fig. 35 Tighten mounting bolt and nut in the order shown

a. After assembly, perform a driving test to check propeller shaft (driveshaft) vibration.

b. If vibration occurs, separate propeller shaft (driveshaft) from final drive or transfer.

c. Reinstall companion flange by changing the phase between companion flange and propeller shaft (driveshaft) by the one bolt hole at a time.

d. Then perform driving test and check propeller shaft (driveshaft) vibration again at each point.

20. If propeller shaft (driveshaft) or final drive has been replaced, connect them as follows:

a. Install the propeller shaft (driveshaft) while aligning its matching mark with the matching mark on the joint as close as possible.

✳✳ WARNING

Avoid damaging the rebro joint boot, protect it with a shop cloth or equivalent.

5.0L Engine

See Figures 35 and 37.

1. Before servicing the vehicle, refer to the Precautions Section.

2. Shift the transmission to the neutral position, and release the parking brake.

3. Remove exhaust front tube and center muffler.

4. Remove the heat plate.

5. Put matching marks on propeller shaft (driveshaft) rubber coupling and transfer companion flange.

✳✳ WARNING

For matching marks, use paint. Do not damage propeller shaft (driveshaft) rubber coupling and transfer companion flange.

6. Put matching marks on propeller shaft (driveshaft) rebro joint and final drive companion flange.

✳✳ WARNING

For matching marks, use paint. Do not damage propeller shaft (driveshaft) rebro joint and final drive companion flange.

7. Loosen mounting nuts of center bearing mounting brackets (upper/lower).

➡**Tighten mounting nuts temporarily.**

8. Remove propeller shaft (driveshaft) assembly fixing bolts and nuts.

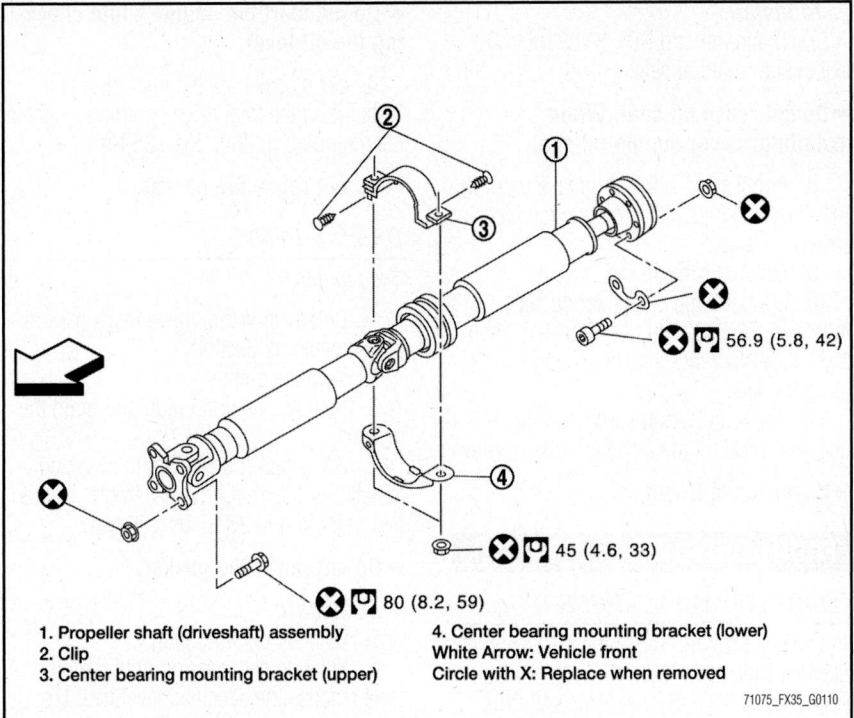

1. Propeller shaft (driveshaft) assembly
2. Clip
3. Center bearing mounting bracket (upper)
4. Center bearing mounting bracket (lower)
White Arrow: Vehicle front
Circle with X: Replace when removed

71075_FX35_G0110

Fig. 36 Rear propeller shaft (driveshaft) assembly component locations—3.5L engine (AWD model)

✳✳ WARNING

Never remove the rubber coupling from the propeller shaft (driveshaft) assembly.

9. Slightly separate the rubber coupling from transfer companion flange.

✳✳ WARNING

Do not damage transfer companion flange and rubber coupling.

10. Remove center bearing mounting bracket fixing nuts.
- Never bend the rubber coupling above an angle of 0–4°
- Do not damage the grease seal

11. Remove propeller shaft (driveshaft) assembly.

✳✳ WARNING

If the constant velocity joint was bent during propeller shaft (driveshaft) assembly removal, installation, or transportation, its boot may be damaged. Wrap boot interference area to metal part with shop cloth or equivalent to protect boot from breakage.

12. Remove clip and center bearing mounting bracket (upper/lower).

To install:

13. Installation is the reverse of the removal procedure.
14. Tighten the fasteners to specification.
15. Install center bearing mounting bracket (upper) with its arrow mark facing forward.

16. Adjust position of center bearing mounting bracket (upper), center bearing mounting bracket (lower) sliding back and forth to prevent play in thrust direction of center bearing insulator. Install center bearing mounting bracket (upper/lower) to vehicle.

17. Align matching marks to install propeller shaft (driveshaft) flange yoke and transfer companion flange.

18. Align matching marks to install propeller shaft (driveshaft) joint and final drive companion flange.

19. Tighten mounting bolt and nut in the order shown in the figure.

20. If propeller shaft (driveshaft) or final drive has been replaced, connect them as follows:

   a. Install the propeller shaft (driveshaft) while aligning its matching mark with the matching mark on the joint as close as possible.

21. Perform inspection after installation.

   a. After assembly, perform a driving test to check propeller shaft (driveshaft) vibration.

   b. If vibration occurs, separate propeller shaft (driveshaft) from final drive or transfer.

   c. Reinstall companion flange by changing the phase between companion flange and propeller shaft (driveshaft) by the one bolt hole at a time.

   d. Then perform driving test and check propeller shaft (driveshaft) vibration again at each point.

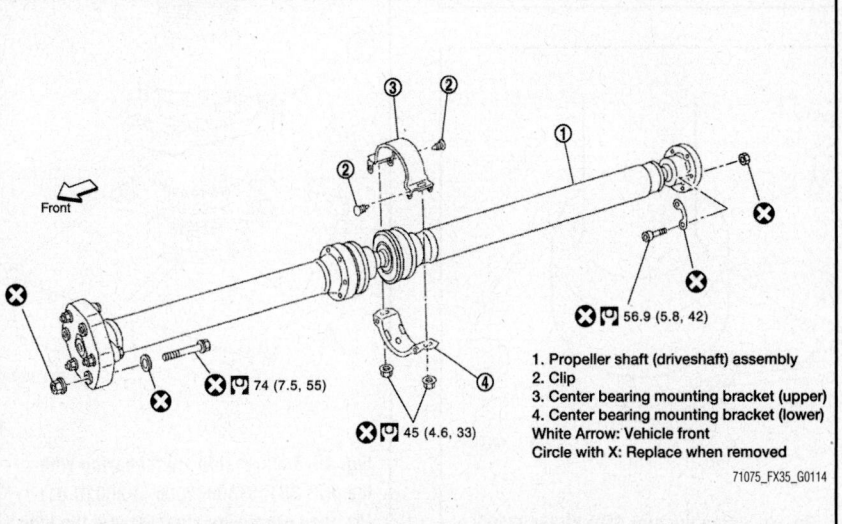

1. Propeller shaft (driveshaft) assembly
2. Clip
3. Center bearing mounting bracket (upper)
4. Center bearing mounting bracket (lower)
White Arrow: Vehicle front
Circle with X: Replace when removed

71075_FX35_G0114

Fig. 37 Rear propeller shaft (driveshaft) assembly component locations—5.0L engine

FRONT AXLE SHAFT, BEARING & SEAL

REMOVAL & INSTALLATION

See Figures 38 through 42.

1. Before servicing the vehicle, refer to the Precautions Section.
2. Hold extension tube retainer with puller, then press outside shaft using a press.
3. Remove side shaft oil seal from extension tube retainer with a suitable tool.

✳✳ WARNING

Do not damage extension tube retainer.

4. Remove side shaft bearing from extension tube retainer.
5. Remove O-ring from extension tube retainer.
6. Remove dust seal from side shaft.

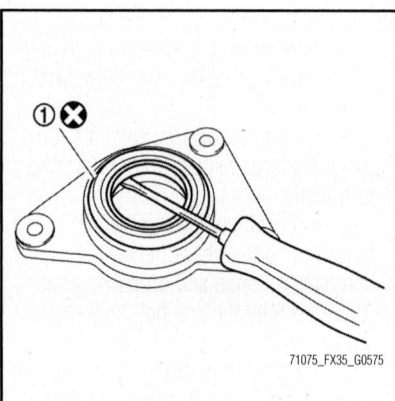

Fig. 38 Remove side shaft oil seal (1) from extension tube retainer with a suitable tool

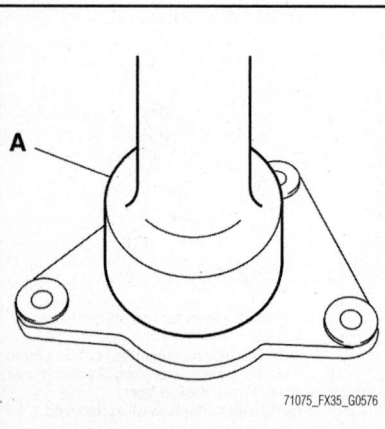

Fig. 39 Using the drift SST: KV38100200 (—) (A) to install side shaft oil seal

To install:

7. Using the drift SST: KV38100200, install side shaft oil seal.

➡**Do not reuse oil seal. When installing, never incline oil seal.**

8. Apply multi-purpose grease onto oil seal lips, and gear oil onto the circumference of oil seal.
9. Install dust seal.
10. Support side shaft bearing with the drift SST: ST30032000 (J-26010-01), then press side shaft into the side shaft bearing using a press.
11. Apply multi-purpose grease to O-ring, and install it to extension tube retainer.

➡**Never reuse O-ring.**

FRONT DRIVE AXLE

FLUID RECOMMENDATIONS

The manufacturer recommends Genuine NISSAN Differential Oil Hypoid Super GL-5 80W-90 or API GL-5, Viscosity SAE 80W-90. For hot climates, viscosity SAE 90 is suitable for ambient temperatures above 32°F (0°C).

LEVEL CHECK

See Figure 43.

1. Before servicing the vehicle, refer to the Precautions Section.
2. Make sure that oil is not leaking from final drive assembly or around it.
3. Remove filler plug and check oil level from filler plug mounting hole as shown in the figure.

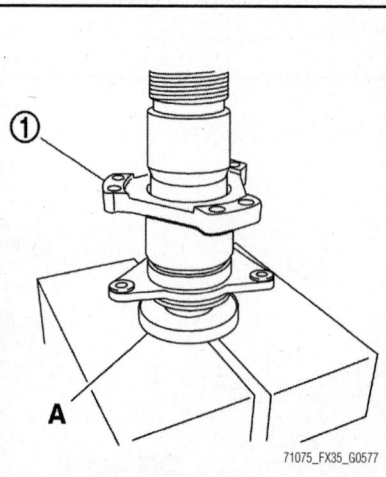

Fig. 40 Support side shaft bearing with the drift SST: ST30032000 (J-26010-01) (A), then press side shaft (1) into the side shaft bearing using a press

➡**Do not start the engine while checking the oil level.**

4. Set a gasket on the filler plug and install it on the final drive assembly. Tighten the filler plug to 26 ft. lbs. (35 Nm).

➡**Do not reuse the gasket.**

DRAIN & REFILL

See Figures 43 and 44.

1. Before servicing the vehicle, refer to the Precautions Section.
2. Stop the engine.
3. Remove the drain plug and drain the gear oil.
4. Set a gasket on the drain plug and install it to the final drive assembly. Tighten the drain plug to 26 ft. lbs. (35 Nm).

➡**Do not reuse the gasket.**

To install:

5. Remove the filler plug.
6. Fill with new gear oil until the oil level reaches the specified level near the filler plug mounting hole.
7. After refilling oil, check the oil level. Set a gasket to filler plug, then install it to final drive assembly. Tighten the filler plug to 26 ft. lbs. (35 Nm).

➡**Do not reuse the gasket.**

FRONT HALFSHAFT

REMOVAL & INSTALLATION

Left Side

See Figures 45 and 46.

1. Before servicing the vehicle, refer to the Precautions Section.
2. Remove tires and wheels.
3. Remove wheel sensor and sensor harness.

✳✳ WARNING

Never pull on wheel sensor harness.

4. Remove brake hose bracket.
5. Remove caliper assembly mounting bolts. Hang caliper assembly in a place where it will not interfere with work.

➡**Do not depress brake pedal while brake caliper is removed.**

6. Remove disc rotor.
7. Remove cotter pin, and then loosen wheel hub lock nut.
8. Tap the wheel hub lock nut with a piece of wood. Hammer the wood to disengage wheel hub and bearing assembly from halfshaft.

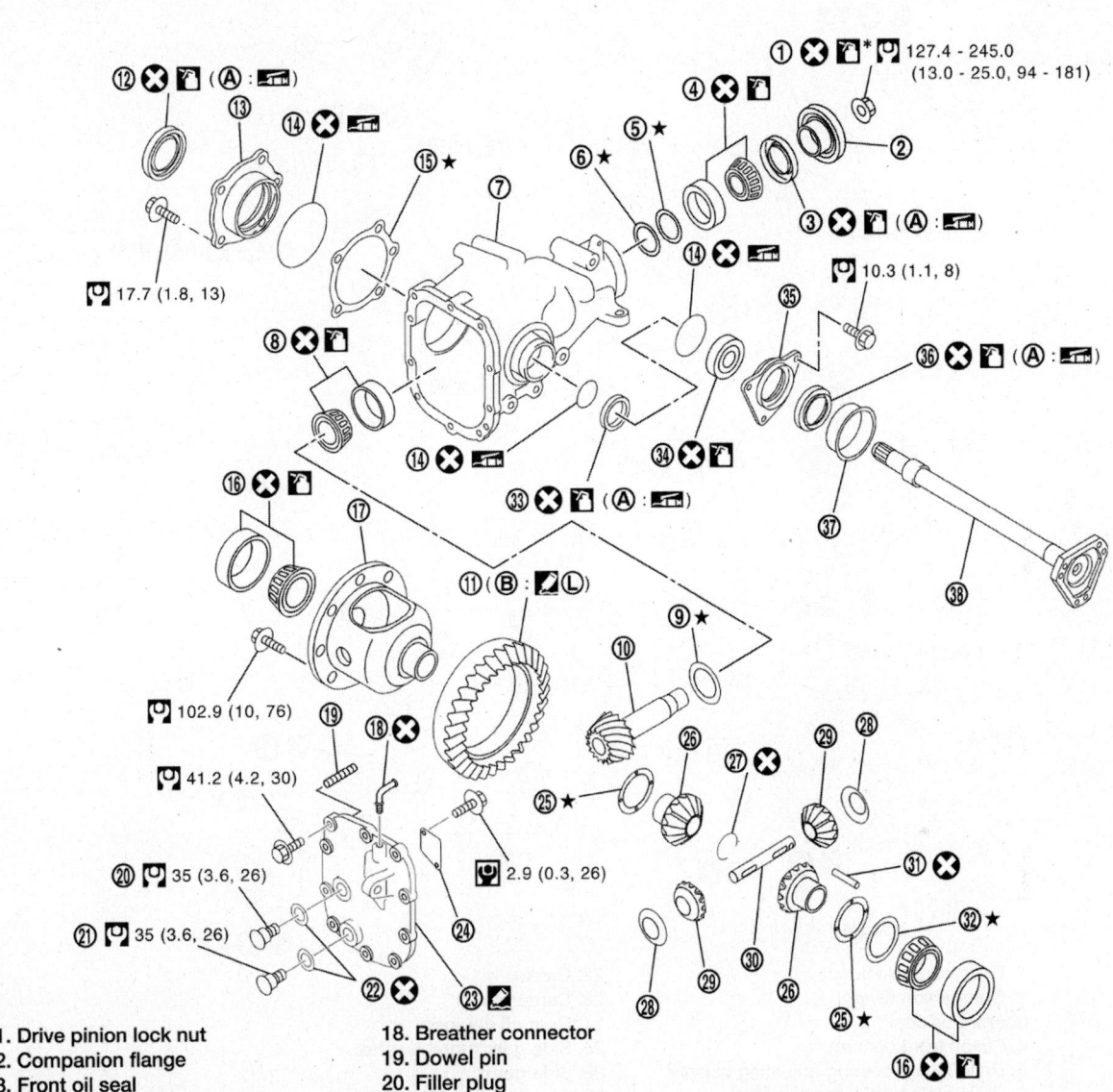

1. Drive pinion lock nut
2. Companion flange
3. Front oil seal
4. Pinion front bearing
5. Drive pinion bearing adjusting washer
6. Drive pinion adjusting washer
7. Gear carrier
8. Pinion rear bearing
9. Pinion height adjusting washer
10. Drive pinion
11. Drive gear
12. Side oil seal (right side)
13. Side retainer
14. O-ring
15. Side bearing adjusting shim
16. Side bearing
17. Differential case

18. Breather connector
19. Dowel pin
20. Filler plug
21. Drain plug
22. Gasket
23. Carrier cover
24. Gear oil defense
25. Side gear thrust washer
26. Side gear
27. Circular clip
28. Pinion mate thrust washer
29. Pinion mate gear
30. Pinion mate shaft
31. Lock pin
32. Side bearing adjusting washer
33. Side oil seal (left side)
34. Side shaft bearing

35. Extension tube retainer
36. Side shaft oil seal
37. Dust seal
38. Side shaft
A: Oil seal lip
B: Screw hole
Oil Can: Apply gear oil
Oil Can with Star: Apply anti-corrosion oil
Grease Gun: Apply multi-purpose grease
Caulk Tube: Apply Genuine Silicone RTV or equivalent
Caulk Tube with L: Apply Genuine Medium Strength
    Thread Locking Sealant or equivalent

71075_FX35_G0572

Fig. 41 Exploded view of front final drive component locations—3.5L engine

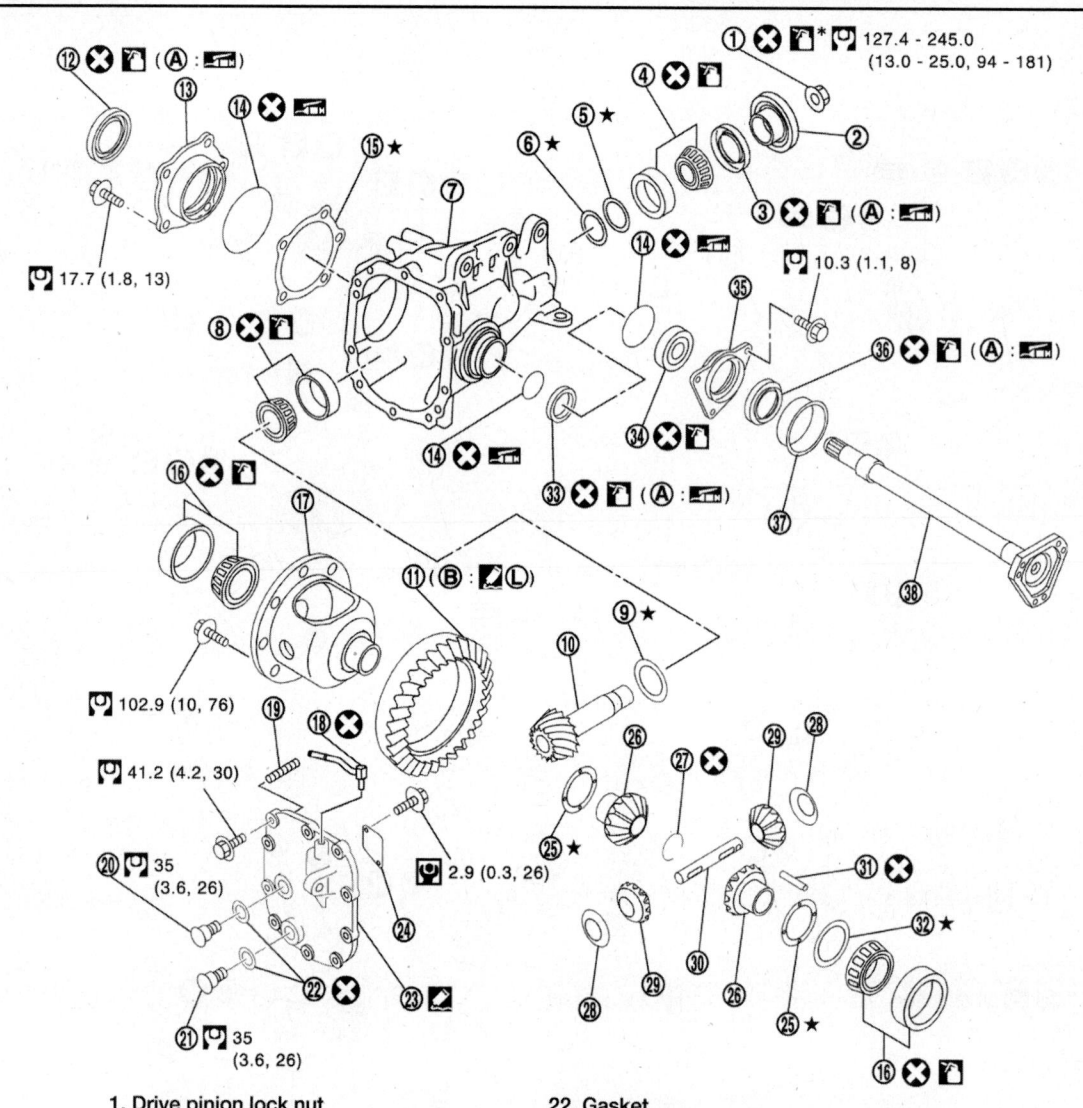

1. Drive pinion lock nut
2. Companion flange
3. Front oil seal
4. Pinion front bearing
5. Drive pinion bearing adjusting washer
6. Drive pinion adjusting washer
7. Gear carrier
8. Pinion rear bearing
9. Pinion height adjusting washer
10. Drive pinion
11. Drive gear
12. Side oil seal (right side)
13. Side retainer
14. O-ring
15. Side bearing adjusting shim
16. Side bearing
17. Differential case
18. Breather connector
19. Dowel pin
20. Filler plug
21. Drain plug

22. Gasket
23. Carrier cover
24. Gear oil defense
25. Side gear thrust washer
26. Side gear
27. Circular clip
28. Pinion mate thrust washer
29. Pinion mate gear
30. Pinion mate shaft
31. Lock pin
32. Side bearing adjusting washer
33. Side oil seal (left side)
34. Side shaft bearing
35. Extension tube retainer
36. Side shaft oil seal
37. Dust seal
38. Side shaft
A: Oil seal lip
B: Screw hole
Oil Can: Apply gear oil
Oil Can with Star: Apply anti-corrosion oil
Grease Gun: Apply multi-purpose grease
Caulk Tube: Apply Genuine Silicone RTV or equivalent
Caulk Tube with L: Apply Genuine Medium Strength
Thread Locking Sealant or equivalent

71075_FX35_G0573

Fig. 42 Exploded view of front final drive component locations—5.0L engine

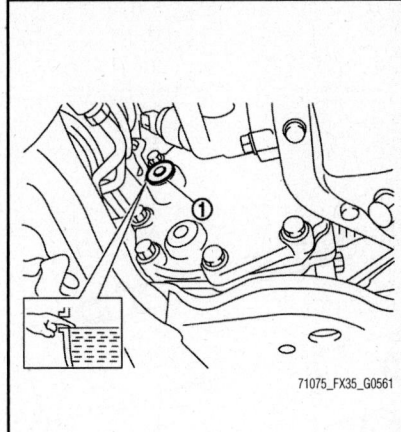

Fig. 43 View of filler plug (1) and proper oil level from filler plug mounting hole

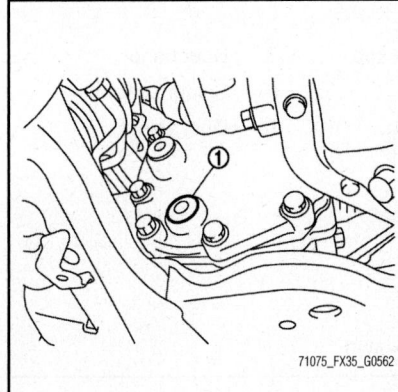

Fig. 44 Remove the drain plug (1) and drain the gear oil

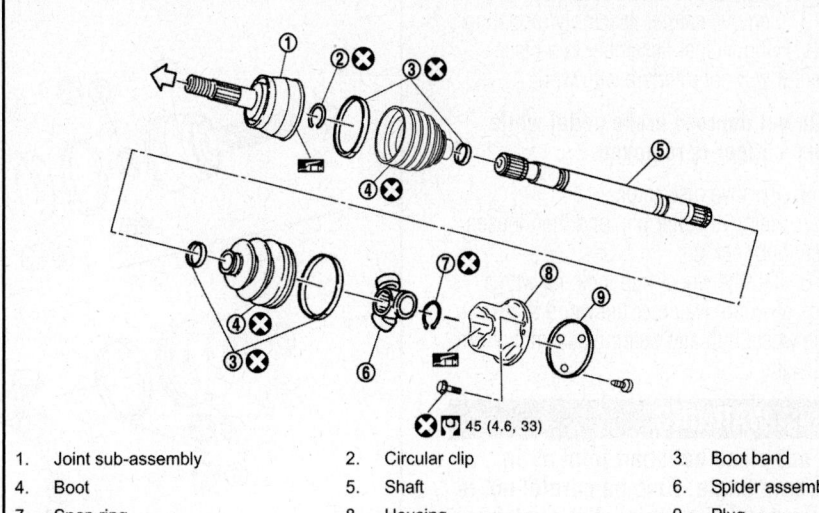

1.	Joint sub-assembly	2.	Circular clip	3.	Boot band
4.	Boot	5.	Shaft	6.	Spider assembly
7.	Snap ring	8.	Housing	9.	Plug

⟨⇦⟩: Wheel side

▓▓▓: NISSAN genuine grease or an equivalent.

37663_FX35_G0166

Fig. 45 Exploded view of front left side drive axle—3.5L engine

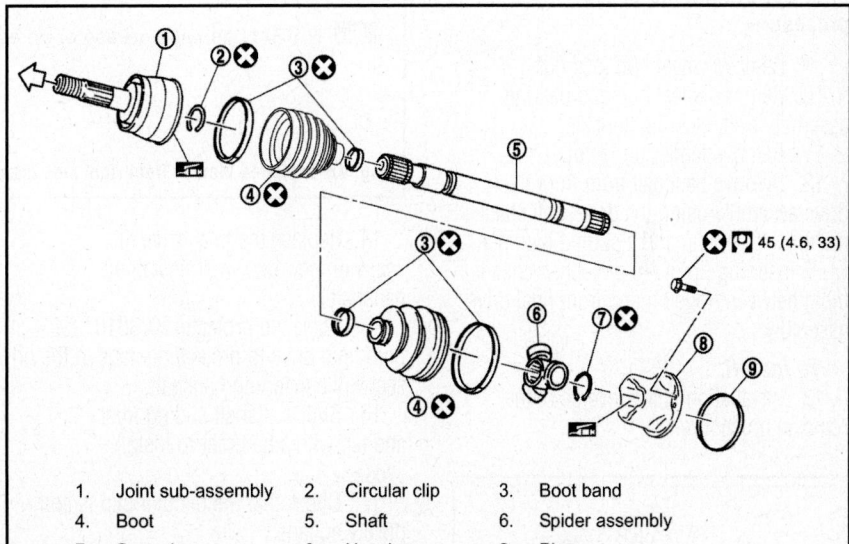

1.	Joint sub-assembly	2.	Circular clip	3.	Boot band
4.	Boot	5.	Shaft	6.	Spider assembly
7.	Snap ring	8.	Housing	9.	Plug

⟨⇦⟩: Wheel side

37663_FX35_G0167

Fig. 46 Exploded view of front left side drive axle—5.0L engine

---

❋❋❋ **WARNING**

Do not place halfshaft joint at an extreme angle. Also be careful not to overextend slide joint. Never allow halfshaft to hang down without support for joint sub-assembly, shaft and the other parts.

➡**Use suitable puller if wheel hub and halfshaft cannot be separated even after performing the above procedure.**

9. Remove wheel hub lock nut.
10. Remove steering outer socket (tie rod end).
11. Separate upper link from steering knuckle.
12. Remove halfshaft from wheel hub and bearing assembly.
13. Remove shock absorber from vehicle.
14. Remove under cover.
15. Remove mounting bolts, and then remove halfshaft from the front final drive assembly.

To install:
16. Installation is the reverse of the removal procedure.
17. Install halfshaft. Tighten the wheel hub lock nut to 92 ft. lbs. (125 Nm).

➡**Be sure to use torque wrench to tighten the wheel hub lock nut. Never use a power tool. Do not reuse cotter pin.**

Right Side

See Figures 47 through 49.

1. Before servicing the vehicle, refer to the Precautions Section.
2. Remove tires and wheels.
3. Remove wheel sensor and sensor harness.

❋❋❋ **WARNING**

Do not pull on wheel sensor harness.

4. Remove brake hose bracket.

5. Remove caliper assembly mounting bolts. Hang caliper assembly in a place where it will not interfere with work.

➡**Do not depress brake pedal while brake caliper is removed.**

6. Remove disc rotor.

7. Remove cotter pin, and then loosen wheel hub lock nut.

8. Tap the wheel hub lock nut with a piece of wood. Hammer the wood to disengage wheel hub and bearing assembly from halfshaft.

✳✳ WARNING

Do not place halfshaft joint at an extreme angle. Also be careful not to overextend slide joint. Never allow halfshaft to hang down without support for joint sub-assembly, shaft and the other parts.

➡**Use suitable puller if wheel hub and halfshaft cannot be separated even after performing the above procedure.**

9. Remove wheel hub lock nut.

10. Remove wheel hub and bearing assembly from steering knuckle.

11. Remove fender protector.

12. Remove halfshaft from front final drive assembly using the driveshaft attachment KV40107500 and a sliding hammer while inserting tip of the driveshaft attachment between housing and front final drive assembly.

To install:

13. Installation is the reverse of the removal procedure.

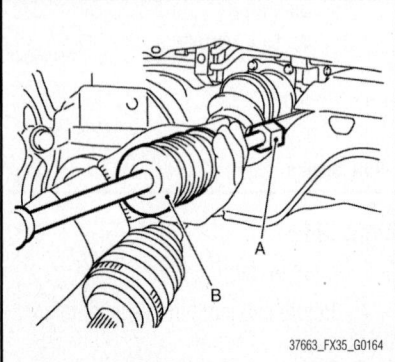

Fig. 47 Remove halfshaft from front drive assembly using the driveshaft attachment KV40107500 (A) and a sliding hammer (B) while inserting tip of the driveshaft attachment between housing and front final drive assembly

37663_FX35_G0164

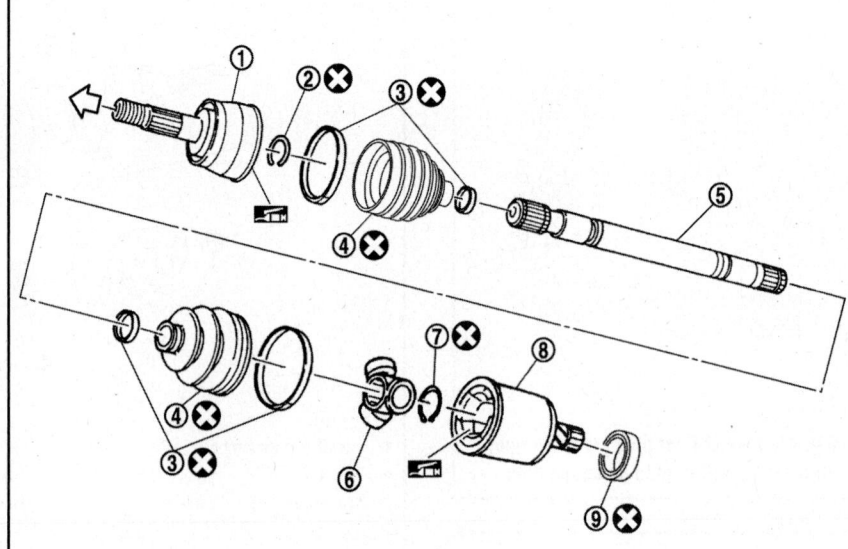

1.	Joint sub-assembly	2.	Circular clip	3.	Boot band
4.	Boot	5.	Shaft	6.	Spider assembly
7.	Snap ring	8.	Housing	9.	Dust shield

⬅: Wheel side

▥: NISSAN genuine grease or an equivalent.

37663_FX35_G0168

Fig. 48 Exploded view of front right side drive axle—AWD

14. Replace the final drive oil seal with new one when installing halfshaft.

15. Place the protector KV38107900 onto final drive to prevent damage to the oil seal while inserting halfshaft.

16. Slide halfshaft sliding joint and tap with a hammer to install securely.

17. Check that the circular clip is completely engaged.

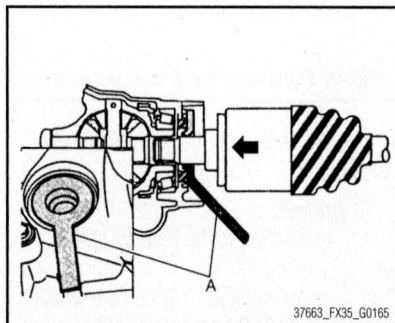

37663_FX35_G0165

Fig. 49 Place the protector KV38107900 (A) onto final drive to prevent damage to the oil seal while inserting halfshaft

REAR DRIVE AXLE

FLUID RECOMMENDATIONS

For 3.5L engine without towing package—The manufacturer recommends Genuine NISSAN Differential Oil Hypoid Super GL-5 80W-90 or API GL-5, Viscosity SAE 80W-90.

For 3.5L engine and 5.0L engine with towing package—The manufacturer recommends API GL-5 Synthetic gear oil, Viscosity SAE 75W-90.

LEVEL CHECK

3.5L Engine—Rear Differential (R200)

See Figure 50.

1. Before servicing the vehicle, refer to the Precautions Section.

2. Make sure that oil is not leaking from final drive assembly or around it.

3. Remove filler plug and check oil level from filler plug mounting hole as shown in the figure.

➡**Never start engine while checking oil level.**

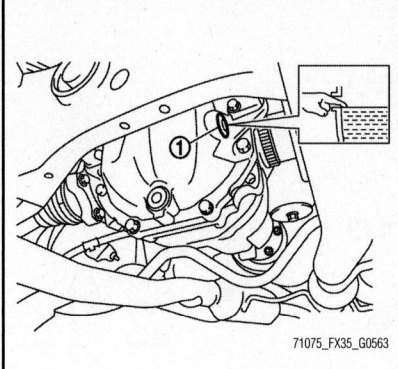

Fig. 50 View of filler plug (1) and proper oil level from filler plug mounting hole—rear differential (R200)

4. Set a gasket on filler plug and install it on final drive assembly. Tighten the filler plug to 26 ft. lbs. (35 Nm).

➡ **Do not reuse gasket.**

5.0L Engine—Rear Differential (R230)

See Figure 51.

1. Before servicing the vehicle, refer to the Precautions Section.
2. Make sure that differential gear oil is not leaking from the rear final drive assembly or around it.
3. Check the differential gear oil level from the filler plug hole as shown.

➡ **Do not start engine while checking differential gear oil level.**

4. Install the filler plug with a new gasket on it to the rear final drive assembly. Tighten to 26 ft. lbs. (35 Nm).

➡ **Do not reuse gasket.**

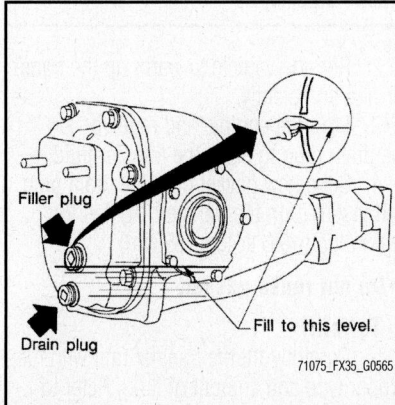

Fig. 51 View of drain plug, filler plug, and proper oil level from filler plug mounting hole—rear differential (R230)

DRAIN & REFILL

3.5L Engine—Rear Differential (R200)

See Figures 50 and 52.

1. Before servicing the vehicle, refer to the Precautions Section.
2. Stop the engine.
3. Remove the drain plug and drain the gear oil.
4. Set a gasket on drain plug and install it to final drive assembly and tighten to 26 ft. lbs. (35 Nm).

➡ **Do not reuse gasket.**

To install:

5. Remove the filler plug.
6. Fill with new gear oil until the oil level reaches the specified level near the filler plug mounting hole.
7. After refilling oil, check oil level.
8. Set a gasket to filler plug, then install it to final drive assembly. Tighten the filler plug to 26 ft. lbs. (35 Nm).

➡ **Do not reuse gasket.**

5.0L Engine—Rear Differential (R230)

See Figure 51.

1. Before servicing the vehicle, refer to the Precautions Section.
2. Stop the engine.
3. Remove the drain plug and gasket from the rear final drive assembly to drain the differential gear oil.
4. Install the drain plug with a new gasket to the rear final drive assembly. Tighten to 26 ft. lbs. (35 Nm).

➡ **Do not reuse gasket.**

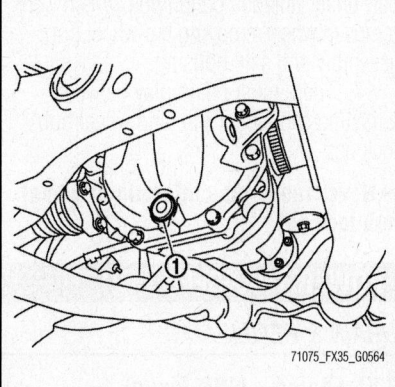

Fig. 52 Remove the drain plug (1) and drain the gear oil—rear differential (R200)

To install:

5. Remove the filler plug and gasket from the rear final drive assembly.
6. Fill the rear final drive assembly with new differential gear oil until the level reaches the specified level near the filler plug hole.
7. Install the filler plug with a new gasket on it to the rear final drive assembly. Tighten to 26 ft. lbs. (35 Nm).

➡ **Do not reuse gasket.**

REAR HALFSHAFT

REMOVAL & INSTALLATION

See Figures 53 through 55.

1. Before servicing the vehicle, refer to the Precautions Section.
2. Remove tires and wheels.
3. Remove cotter pin and adjusting cap, then loosen wheel hub lock nut.
4. Put matching mark on halfshaft and wheel hub and bearing assembly.

✳✳ **WARNING**

Use paint or similar substance for matching marks. Do not scratch the surface.

5. Remove center muffler.
6. Tap the wheel hub lock nut with a piece of wood. Hammer the wood to disengage wheel hub and bearing assembly from halfshaft.

✳✳ **WARNING**

Do not place the halfshaft joint at an extreme angle. Also be careful not to overextend slide joint. Never allow halfshaft to hang down without support for counterpart such as joint sub-assembly, and other parts.

➡ **Use a suitable puller if wheel hub and bearing assembly and halfshaft cannot be separated even after performing the above procedure.**

7. Remove wheel hub lock nut.
8. Remove mounting bolts between side flange and halfshaft.
9. Remove halfshaft.

To install:

10. Installation is the reverse of the removal procedure.
11. Clean the matching surface of wheel hub lock nut and wheel hub and bearing assembly.

➡ **Do not apply lubricating oil to these matching surface.**

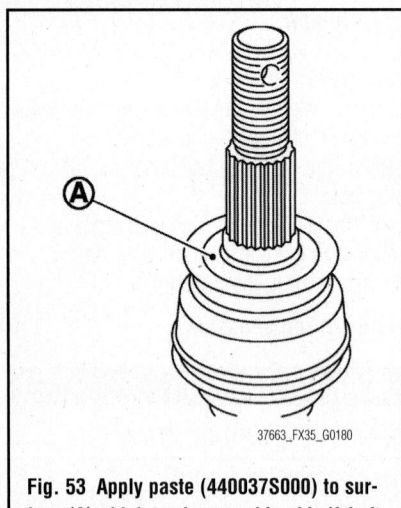

Fig. 53 Apply paste (440037S000) to surface (A) of joint sub-assembly of halfshaft

12. Clean the matching surface of halfshaft and wheel hub and bearing assembly and then apply paste (440037S000) to the surface of the joint sub-assembly of the halfshaft.

➡**Apply paste to cover entire flat surface of joint sub-assembly of halfshaft.**

• Amount of paste: 0.04–0.10 ounce (1.0–3.0 grams)

13. Tighten the wheel hub lock nut to 74–77 ft. lbs. (100–105 Nm).

✳✳ WARNING

Since the halfshaft is assembled by press-fitting, use the tightening torque range for the wheel hub lock nut. Be sure to use torque wrench to tighten the wheel hub lock nut. Never use a power tool. Do not reuse the hub lock nut.

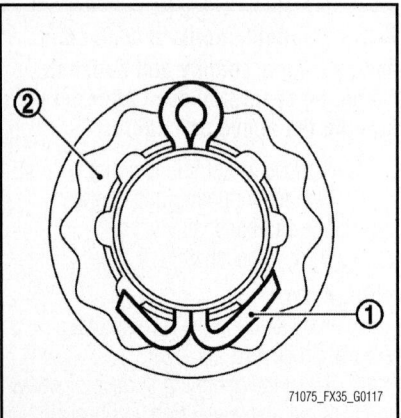

Fig. 54 When installing a cotter pin (1) and adjusting cap (2), securely bend the basal portion to prevent rattles

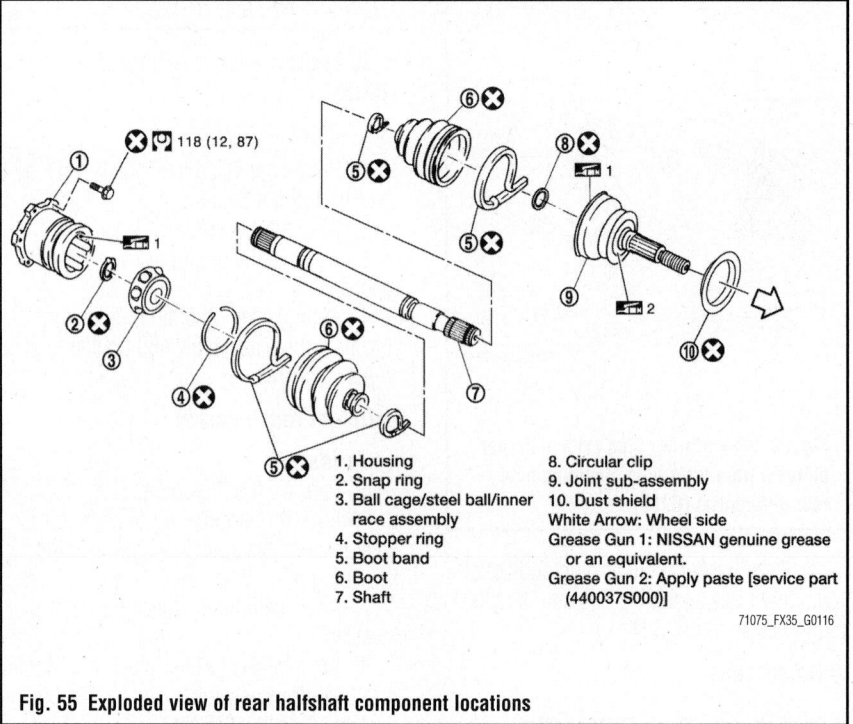

1. Housing
2. Snap ring
3. Ball cage/steel ball/inner race assembly
4. Stopper ring
5. Boot band
6. Boot
7. Shaft
8. Circular clip
9. Joint sub-assembly
10. Dust shield
White Arrow: Wheel side
Grease Gun 1: NISSAN genuine grease or an equivalent
Grease Gun 2: Apply paste [service part (440037S000)]

Fig. 55 Exploded view of rear halfshaft component locations

➡**Do not over-torque the wheel hub lock nut to avoid axle noise or under-torque the wheel hub lock nut to avoid looseness.**

➡**When installing the spring washer, face the identification paint mark to the wheel hub and bearing assembly side. Never reuse spring washer.**

14. Align the matching marks that were made during removal if reusing the disc rotor.

15. When installing a cotter pin and adjusting cap, securely bend the basal portion to prevent rattles.

➡**Never reuse the cotter pin.**

16. Perform the final tightening of each part under unladen conditions, which were removed when removing the wheel hub assembly and axle housing.

17. There must be no play between adjusting cap, cotter pin, and wheel hub lock nut.

➡**Never reuse the cotter pin or wheel hub lock nut.**

POWER TRANSFER UNIT

DRAIN & REFILL

3.5L Engine—AWD Model
See Figures 56 and 57.

1. Before servicing the vehicle, refer to the Precautions Section.

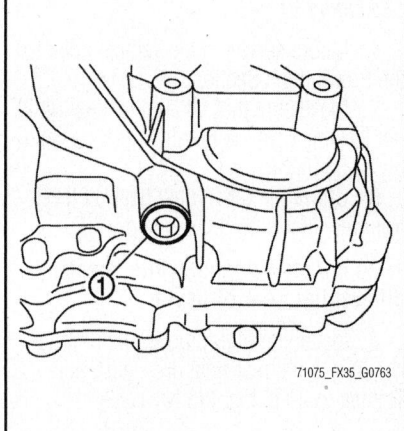

Fig. 56 Transfer unit drain plug (1) location—3.5L engine

2. Run the vehicle to warm up the transfer unit sufficiently.

3. Stop the engine, and remove the drain plug to drain the transfer fluid.

4. Set a new gasket onto the drain plug, and install it on the transfer. Tighten the drain plug to 26 ft. lbs. (35 Nm).

➡**Do not reuse gasket.**

To refill:

5. Carefully fill the transfer unit with the proper type and amount of fluid. Refer to Fluid Recommendations.

6. Leave the vehicle for 3 minutes, and check the fluid level again.

7. Set a new gasket onto filler plug, and

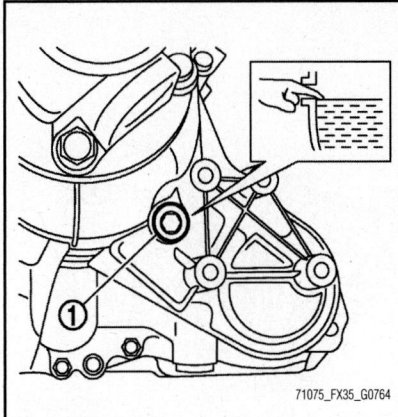

Fig. 57 Transfer unit filler plug (1) location and proper fluid level illustrated—3.5L engine

install it on transfer. Tighten the filler plug to 26 ft. lbs. (35 Nm).

5.0L Engine

See Figures 58 and 59.

1. Before servicing the vehicle, refer to the Precautions Section.
2. Run the vehicle to warm up the transfer unit sufficiently.
3. Stop the engine, and remove the drain plug to drain the transfer fluid.
4. Set a new gasket onto the drain plug, and install it on the transfer. Tighten the drain plug to 26 ft. lbs. (35 Nm).

➡**Do not reuse gasket.**

To refill:

5. Carefully fill the transfer unit with the proper type and amount of fluid. Refer to Fluid Recommendations.
6. Leave the vehicle for 3 minutes, and check the fluid level again.

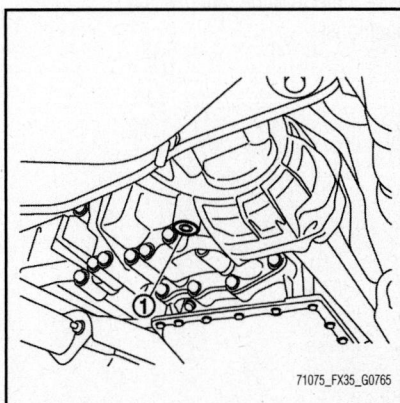

Fig. 58 Transfer unit drain plug (1) location—5.0L engine

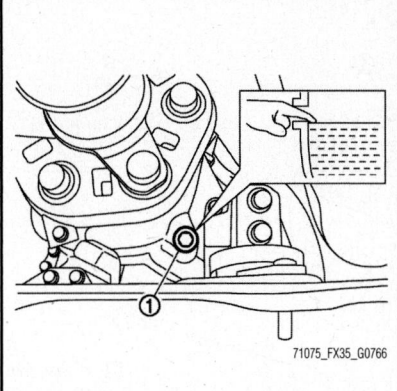

Fig. 59 Transfer unit filler plug (1) location and proper fluid level illustrated—5.0L engine

7. Set a new gasket onto filler plug, and install it on transfer. Tighten the filler plug to 26 ft. lbs. (35 Nm).

➡**Do not reuse gasket.**

FLUID LEVEL CHECK

See Figures 57 and 59.

1. Before servicing the vehicle, refer to the Precautions Section.
2. Check transfer surrounding area (oil seal, drain plug, and filler plug etc.) for fluid leakage.
3. Remove filler plug and gasket.
4. Check that fluid is filled up to mounting hole for the filler plug.

➡**Do not start engine while checking fluid level.**

5. Set a new gasket onto filler plug, and install it on transfer unit. Tighten the filler plug to 26 ft. lbs. (35 Nm).

➡**Do not reuse gasket.**

FLUID RECOMMENDATIONS

The manufacturer recommends Genuine NISSAN Matic J ATF for transfer unit fluid.

✳✳ WARNING

Using transfer fluid other than Genuine NISSAN Matic J ATF may cause deterioration in driveability and transfer durability, and may damage the transfer, which is not covered by the INFINITI new vehicle limited warranty.

➡**Approximate amount of transfer fluid required for drain/refill: 2 ⅛ pints (1.0L).**

REMOVAL & INSTALLATION

3.5L Engine—AWD Model

See Figures 60 through 62.

1. Before servicing the vehicle, refer to the Precautions Section.
2. Remove rear propeller shaft (driveshaft).
3. Remove front propeller shaft (driveshaft).
4. Disconnect AWD solenoid harness connector and separate harness from transfer assembly.
5. Remove transfer air breather hose.
6. Remove control rod.
7. Support transfer assembly and transmission assembly with a jack.
8. Remove rear engine mounting member and engine mounting insulator.
9. Lower jack to the position where the top transfer mounting bolts can be removed.
10. Remove transfer mounting bolts and separate transfer from transmission.

✳✳ WARNING

Secure transfer assembly and transmission assembly to a jack.

To install:

11. Installation is the reverse of the removal procedure.
12. When installing the transfer to the transmission, install the mounting bolts following the standard illustrated, tighten bolts to the specified torque.
13. Torque all transfer assembly attaching bolts to 27 ft. lbs. (37 Nm).
14. When installing transfer air breather hose, make sure there are no pinched or restricted areas on the transfer air breather hose caused by bending or winding.

   a. Set transfer air breather hose of transmission side with the paint mark facing upward, and insert air breather hose to air breather tube until hose end reaches the tube bend portion.

   b. Be sure to insert air breather hose of transfer side to air breather tube until hose end reaches the tube bend portion.

   c. Be sure to attach air breather hose in parts of transmission and transfer.

15. Check for fluid leakage, fluid level, and the A/T positions.

5.0L Engine

See Figures 60 and 61.

1. Before servicing the vehicle, refer to the Precautions Section.

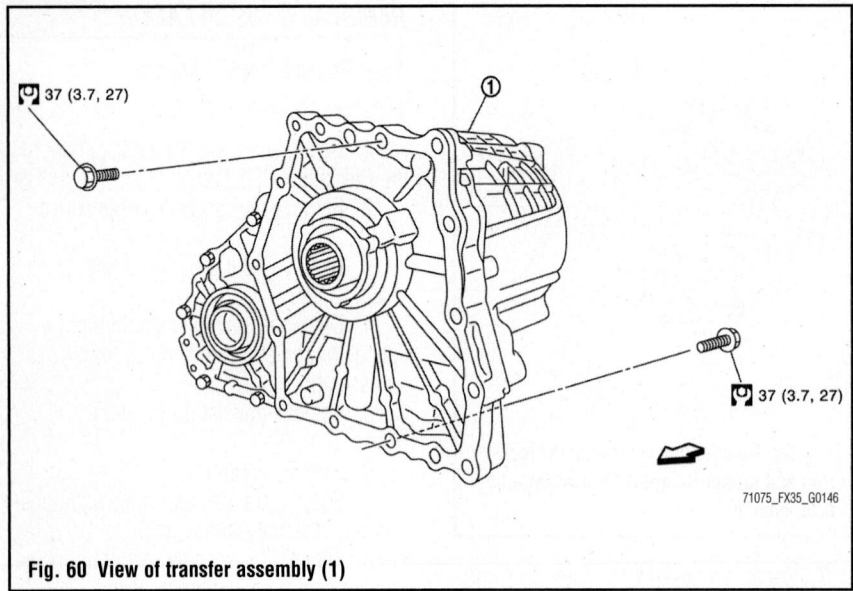

Fig. 60 View of transfer assembly (1)

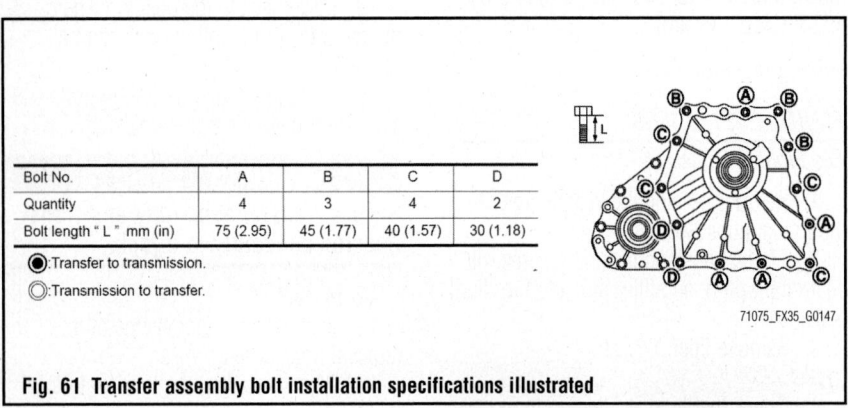

Bolt No.	A	B	C	D
Quantity	4	3	4	2
Bolt length " L " mm (in)	75 (2.95)	45 (1.77)	40 (1.57)	30 (1.18)

●:Transfer to transmission.

○:Transmission to transfer.

Fig. 61 Transfer assembly bolt installation specifications illustrated

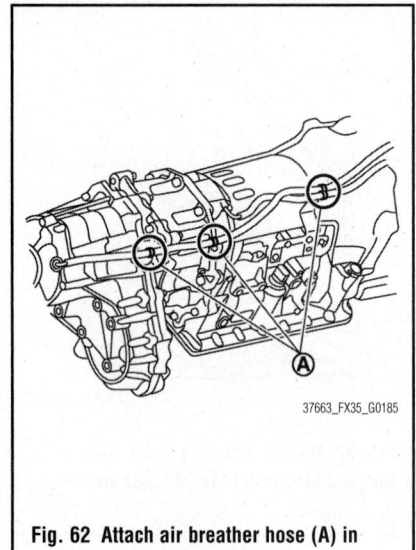

Fig. 62 Attach air breather hose (A) in parts of transmission and transfer

2. Remove the transmission assembly from the vehicle.

3. Remove the transfer air breather hose.

4. Remove the rear engine mounting member and engine mounting insulator.

5. Support the transfer assembly with a jack.

6. Remove the transfer mounting bolts and separate the transfer from the transmission.

✷✷ WARNING

Secure the transfer assembly and transmission assembly to a jack.

To install:

7. Installation is the reverse of the removal procedure.

8. When installing the transfer to the transmission, install the mounting bolts following the standard illustrated, and tighten the bolts to 27 ft. lbs. (37 Nm).

9. When installing the transfer air breather hose, make sure there are no pinched or restricted areas on the transfer air breather hose caused by bending or winding.

a. Set the transfer air breather hose with the paint mark facing upward.

b. Be sure to insert the air breather hose of the transmission side to the air breather tube until the hose end reaches the tube bend portion.

c. Be sure to insert the air breather hose of the transfer side to the air breather tube until the hose end reaches the tube bend portion.

d. Be sure to fix the air breather hose in the parts of the transmission and transfer assembly.

10. After the installation, check fluid level, fluid leakage, and the A/T positions.

ENGINE COOLING

ENGINE COOLANT

BLEEDING

Refer to Drain & Refill Procedure.

DRAIN & REFILL

3.5L Engine

Draining

See Figure 63.

❋❋ CAUTION

To avoid being scalded, never change the coolant when the engine is hot. Wrap a thick cloth around the cap and carefully remove the cap. First, turn the cap a quarter of a turn to release built-up pressure. Then turn the cap all the way.

1. Before servicing the vehicle, refer to the Precautions Section.

2. Connect drain hose.
   a. Use general-purpose hose.
   b. Drain hose dimensions:
   - Internal diameter of hose: 0.59–0.63 inch (15–16mm)
   - Length of hose: 5.7 inches (145mm)

3. Open radiator drain plug at the bottom of radiator, and then remove radiator cap.

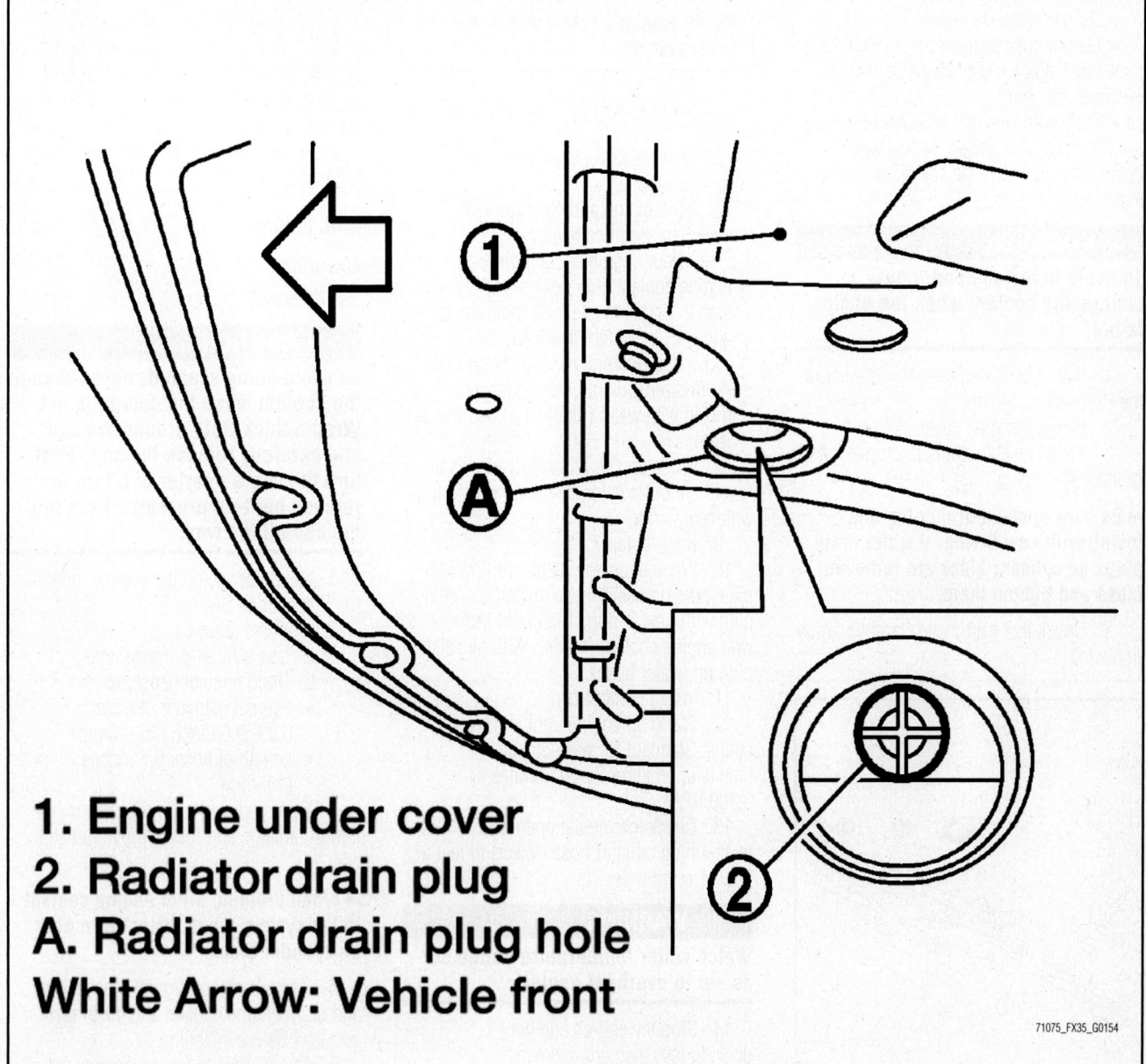

1. Engine under cover
2. Radiator drain plug
A. Radiator drain plug hole
White Arrow: Vehicle front

71075_FX35_G0154

Fig. 63 Open radiator drain plug at the bottom of radiator

➡**When draining all of engine coolant in the system, open water drain plugs on cylinder block.**

4. Remove reservoir tank as necessary, and drain engine coolant and clean reservoir tank before installing.

5. Check drained engine coolant for contaminants such as rust, corrosion or discoloration. If contaminated, flush the engine cooling system.

6. Disconnect the drain hose.

Refilling

See Figures 64 and 65.

Note the following:
- Do not reuse O-rings
- Do not put additive such as waterleak preventive, since it may cause cooling waterway clogging
- Refill with Genuine NISSAN Long Life Antifreeze/Coolant (blue), or equivalent quality, mixed with water (distilled or demineralized).

❋❋ CAUTION

To avoid being scalded, never change the coolant when the engine is hot.

1. Before servicing the vehicle, refer to the Precautions Section.

2. Remove engine cover.

3. Install reservoir tank if removed, and radiator drain plug.

➡**Be sure to clean drain plug and install with new O-ring. If water drain plugs on cylinder block are removed, close and tighten them.**

4. Check that each hose clamp is firmly tightened.

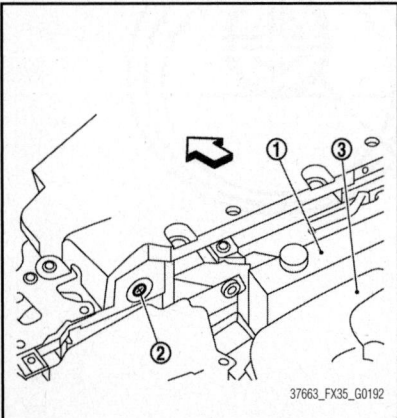

Fig. 64 View of air relief plug (2) on radiator (1) left side and engine cover (3)

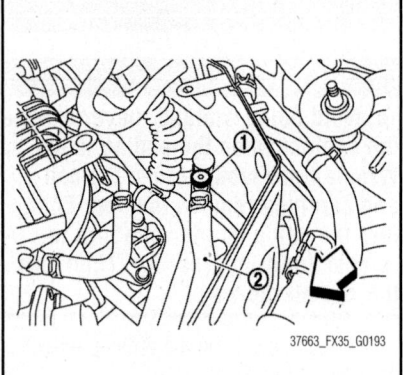

Fig. 65 Remove air relief plug (1) on heater hose (2)

5. Remove air relief plug on radiator left side.

6. Remove air relief plug on heater hose, if equipped.

7. Fill radiator, and reservoir tank if removed, to specified level.

   a. Pour engine coolant through engine coolant filler neck slowly at less than 2 ⅛ quarts (2 liters) a minute to allow air in system to escape.

   b. Use Genuine NISSAN Long Life Antifreeze/Coolant or an equivalent mixed with water (distilled or demineralized).

8. When engine coolant overflows air relief hole on radiator, install air relief plug with new O-ring.

9. Repeat step 7.

10. When engine coolant overflows air relief hole on heater hose, install air relief plug with new O-ring. Then refill radiator with engine coolant (models with air relief plug on heater hose).

11. Install radiator cap.

12. Warm up engine until thermostat opens. Standard for warming-up time is approximately 10 minutes at 3,000 RPM.

13. Check thermostat opening condition by touching radiator hose (lower) to feel a flow of warm water.

❋❋ WARNING

Watch water temperature gauge so as not to overheat engine.

14. Stop the engine and cool down to less than approximately 122°F (50°C).

   a. Cool down using fan to reduce the time.

   b. If necessary, refill radiator up to filler neck with engine coolant.

15. Refill reservoir tank to "MAX" level line with engine coolant.

16. Repeat steps 11–13 two or more times with radiator cap installed until engine coolant level no longer drops.

17. Check cooling system for leakage with engine running.

18. Warm up the engine, and check for sound of engine coolant flow while running engine from idle up to 3,000 RPM with heater temperature controller set at several position between "COOL" and "WARM".

➡**Sound may be heard from the heater unit.**

19. Repeat step 17 three times.

20. If sound is heard, bleed air from cooling system by repeating step 7, and steps from 11–18 until engine coolant level no longer drops.

21. Check that the reservoir tank cap is tightened.

5.0L Engine

Draining

See Figure 66.

❋❋ CAUTION

To avoid being scalded, never change the coolant when the engine is hot. Wrap a thick cloth around the cap and carefully remove the cap. First, turn the cap a quarter of a turn to release built-up pressure. Then turn the cap all the way.

1. Before servicing the vehicle, refer to the Precautions Section.

2. Connect drain hose.

   a. Use general-purpose hose.

   b. Drain hose dimensions:
- Internal diameter of hose: 0.59–0.63 inch (15–16mm)
- Length of hose: 5.7 inches (145mm)

3. Open radiator drain plug at the bottom of radiator, and then remove radiator cap.

➡**When draining all of engine coolant in the system, open water drain plug on cylinder block.**

4. Remove reservoir tank if necessary, and drain engine coolant and clean reservoir tank before installing.

5. Check drained engine coolant for contaminants such as rust, corrosion or discoloration. If contaminated, flush the engine cooling system.

6. Disconnect the drain hose.

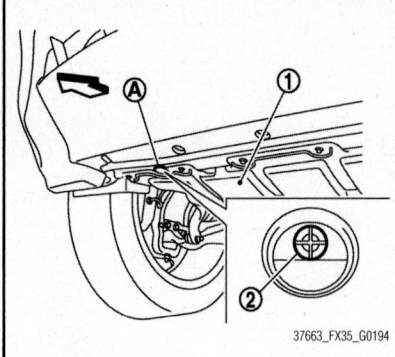

37663_FX35_G0194

Fig. 66 Open radiator drain plug (2) at the bottom of radiator and then remove radiator cap; access is through the drain plug hole (A) in the engine under cover (1)

Refilling

See Figures 67 and 68.

Note the following:
- Do not reuse O-rings
- Do not put additive such as waterleak preventive, since it may cause cooling waterway clogging
- Refill with Genuine NISSAN Long Life Antifreeze/Coolant (blue), or equivalent quality, mixed with water (distilled or demineralized).

※※ CAUTION

To avoid being scalded, never change the coolant when the engine is hot.

1. Before servicing the vehicle, refer to the Precautions Section.
2. Remove engine cover and engine room cover (LH).
3. Install reservoir tank if removed, and radiator drain plug.

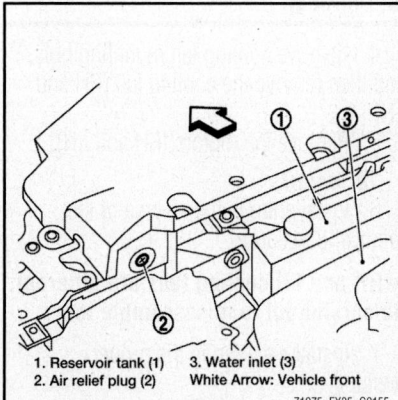

1. Reservoir tank (1)   3. Water inlet (3)
2. Air relief plug (2)   White Arrow: Vehicle front
71075_FX35_G0155

Fig. 67 Remove the air relief plug on the radiator left side

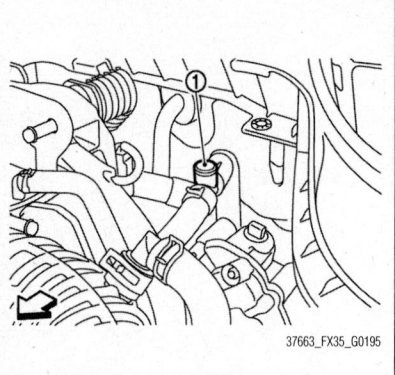

37663_FX35_G0195

Fig. 68 Remove the air relief plug (1) on the heater hose

➡Be sure to clean drain plug and install with new O-ring. If water drain plug on cylinder block is removed, close and tighten it to 11 inch lbs. (1 Nm).

4. Check that each hose clamp is firmly tightened.
5. Remove the air relief plug on the radiator left side.
6. Remove the air relief plug on the heater hose.
7. Fill water inlet, and reservoir tank if removed, to specified level.
  a. Pour engine coolant through engine coolant filler neck slowly at less than 2 ⅛ quarts (2 liters) a minute to allow air in system to escape.
  b. Use Genuine NISSAN Long Life Antifreeze/Coolant, or an equivalent, mixed with water (distilled or demineralized).
8. When engine coolant overflows air relief hole on radiator, install air relief plug with new O-ring.
9. Repeat step 7.
10. When engine coolant overflows air relief hole on heater hose, install air relief plug with new O-ring. Then refill radiator with engine coolant.
11. Install radiator cap.
12. Warm up engine until opening thermostat. Standard for warming-up time is approximately 10 minutes at 3,000 RPM.

➡Check thermostat opening condition by touching radiator hose (lower) to feel a flow of warm water.

※※ WARNING

Watch water temperature gauge so as not to overheat engine.

13. Stop the engine and cool down to less than approximately 122°F (50°C).

  a. Cool down using fan to reduce the time.
  b. If necessary, refill radiator up to filler neck with engine coolant.
14. Refill reservoir tank to "MAX" level line with engine coolant.
15. Repeat steps 11–14 two or more times with radiator cap installed until engine coolant level no longer drops.
16. Check cooling system for leakage with engine running.
17. Warm up the engine, and check for sound of engine coolant flow while running engine from idle up to 3,000 RPM with heater temperature controller set at several position between "COOL" and "WARM".

➡Sound may be heard from the heater unit.

18. Repeat step 17 three times.
19. If sound is heard, bleed air from cooling system by repeating step 7, and steps from 11–18 until engine coolant level no longer drops.
20. Check that the reservoir tank cap is tightened.

LEVEL CHECK

See Figure 69.

※※ CAUTION

If the radiator cap is removed when the engine and radiator are still hot, extremely hot fluid and steam may spew out under pressure. This can cause severe burns.

1. Wait until the engine coolant temperature has lowered before removing the radiator cap.
2. When the engine is cool, check the coolant level in the reservoir.

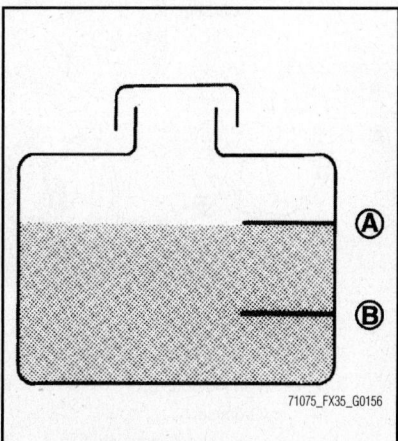

71075_FX35_G0156

Fig. 69 Check if the reservoir tank coolant level is within the MIN (B) to MAX (A) range when the engine is cool

a. Normal coolant level should be between the MAX mark and the MIN mark on the reservoir.

b. If the coolant level is below the MIN mark, remove the reservoir cap and add the proper coolant to the reservoir to bring the coolant level up to the MAX mark.

➡It is not necessary to the remove radiator cap to check the coolant level, unless the coolant reservoir is empty or very low.

ELECTRIC ENGINE FAN

REMOVAL & INSTALLATION

See Figures 70 and 71.

1. Before servicing the vehicle, refer to the Precautions Section.
2. Drain engine coolant.
3. Remove reservoir tank.
4. Remove air cleaner case (bank 1 and bank 2).
5. Remove mounting bolt from high pressure flexible hose bracket.
6. Remove the upper radiator hose.
7. Disconnect harness connector from cooling fan control module, and move harness to aside.
8. Remove cooling fan assembly.

※※ WARNING

Be careful not to damage or scratch the radiator core.

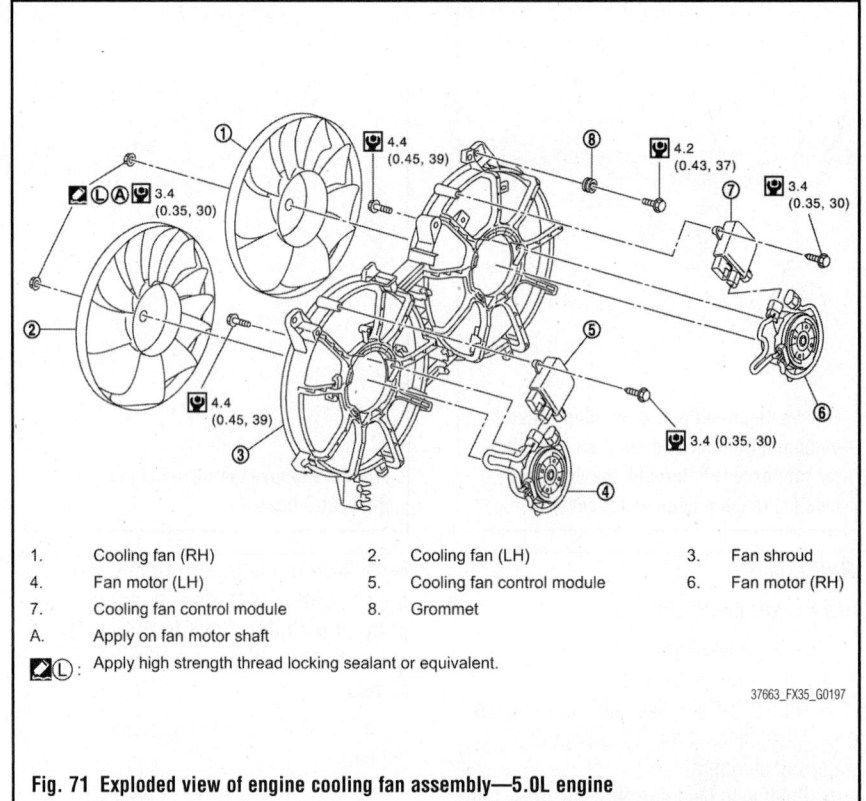

1.	Cooling fan (RH)	2.	Cooling fan (LH)	3.	Fan shroud
4.	Fan motor (LH)	5.	Cooling fan control module	6.	Fan motor (RH)
7.	Cooling fan control module	8.	Grommet		
A.	Apply on fan motor shaft				
🖊Ⓛ:	Apply high strength thread locking sealant or equivalent.				

37663_FX35_G0197

Fig. 71 Exploded view of engine cooling fan assembly—5.0L engine

To install:

9. Installation is the reverse of the removal procedure.
10. Only use genuine parts for cooling fan mounting bolt and observe the specified torque (to prevent radiator from being damaged).

FAN SHROUD

REMOVAL & INSTALLATION

See Figures 70 and 71.

1. Before servicing the vehicle, refer to the Precautions Section.
2. Disconnect harness connector from cooling fan control module.
3. Remove cooling fan control module from cooling fan assembly.

※※ WARNING

Handle carefully to avoid dropping and impact.

4. Remove cooling fan mounting nuts, and then remove the cooling fan (RH and LH).
5. Remove fan motors (RH and LH).

To install:

6. Installation is the reverse of the removal procedure.

➡**RH and LH cooling fans are different. Be careful not to misassemble them.**

7. Install each fan in the proper position.
   • Right side: 9 blades
   • Left side: 11 blades
8. Secure the harness tightly to the fan shroud to prevent the fan rotation area from being slack.

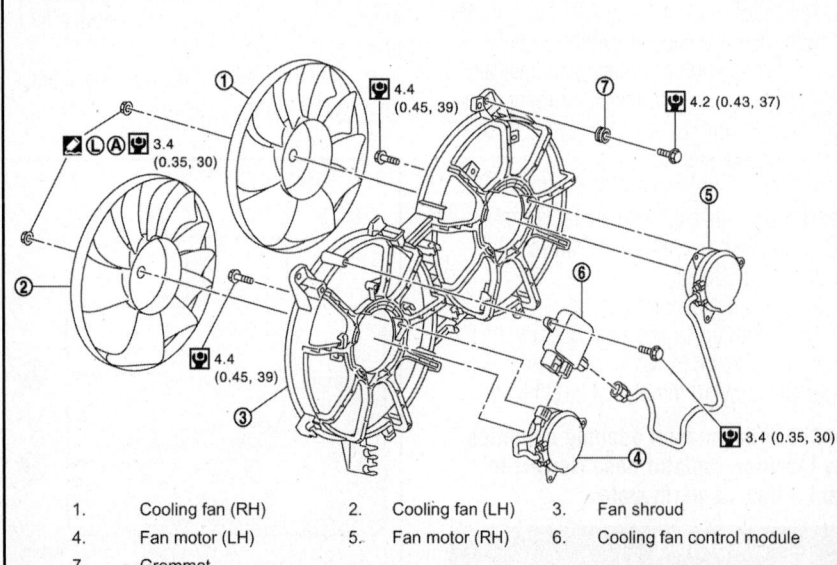

1.	Cooling fan (RH)	2.	Cooling fan (LH)	3.	Fan shroud
4.	Fan motor (LH)	5.	Fan motor (RH)	6.	Cooling fan control module
7.	Grommet				
A.	Apply on fan motor shaft				

37663_FX35_G0196

Fig. 70 Exploded view of engine cooling fan assembly—3.5L engine

9. After installation, check that fan motors operate normally.

➡**Cooling fans are controlled by the cooling fan control module.**

RADIATOR

REMOVAL & INSTALLATION

See Figures 72 through 75.

❄❄ **CAUTION**

Never remove radiator cap when engine is hot. Serious burns could occur from high-pressure engine coolant escaping from water outlet (front). Wrap a thick cloth around the cap. Slowly turn it a quarter of a turn to release built-up pressure. Carefully remove radiator cap by turning it all the way.

1. Before servicing the vehicle, refer to the Precautions Section.
2. Remove the following parts:
   • Engine under cover
   • Engine cover
   • Air cleaner case
   • Air duct (inlet)
   • Hood lock stay assembly and horn
3. Remove condenser.
4. Drain engine coolant from radiator.

❄❄ **CAUTION**

Perform this step when the engine is cold.

❄❄ **WARNING**

Do not spill engine coolant on drive belt.

5. Disconnect A/T fluid cooler hoses from radiator.

➡**Install blind plug to avoid leakage of A/T fluid.**

6. Remove radiator hoses (upper and lower) and reservoir tank hose.
Note the following:
   • Be careful not to allow engine coolant to contact drive belt
   • Never loosen radiator water inlet pipe mounting screw. If loosened, replace radiator
7. Rotate two radiator upper mount brackets 90° in direction shown, and remove them.
8. Remove radiator as per the following:

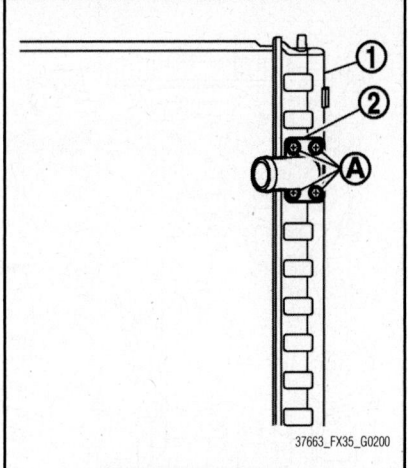

37663_FX35_G0200

Fig. 72 Never loosen radiator water inlet pipe mounting screw (A). If loosened, replace radiator (1); radiator inlet pipe (2)

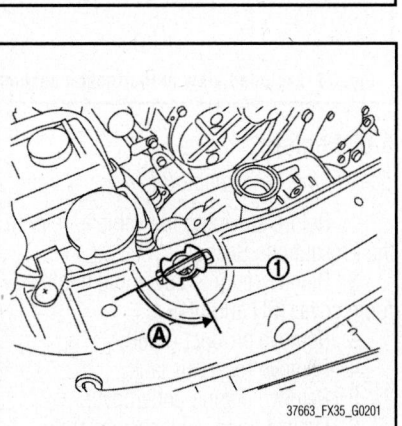

37663_FX35_G0201

Fig. 73 Rotate two radiator upper mount brackets (1) 90 degrees in direction (A) as shown

❄❄ **WARNING**

Be careful not to damage radiator core.

a. Lift up and pull the radiator forward, and then remove the mounting rubber (lower) from the radiator core support.
b. Remove radiator from front of radiator core support.

To install:

9. Installation is the reverse of the removal procedure.
10. Do not reuse O-rings.
11. Replace water hose clamp if it is removed.
12. Use genuine mounting bolts for the cooling fan assembly and strictly observe the tightening torque (breakage prevention for radiator).

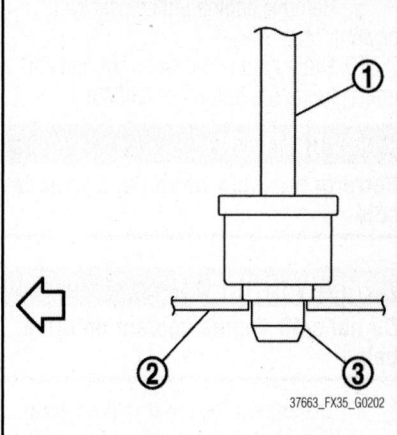

37663_FX35_G0202

Fig. 74 Lift up and pull the radiator (1) forward, and then remove the mounting rubber (lower) (3) from the radiator core support (2)

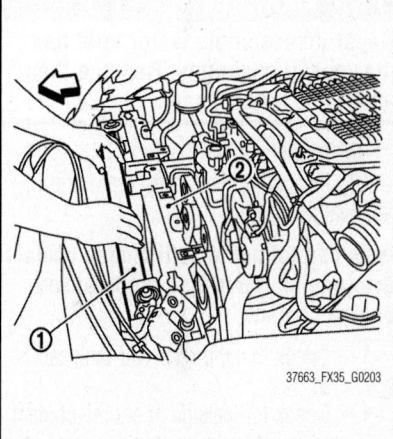

37663_FX35_G0203

Fig. 75 Remove radiator (1) from front of radiator core support (2)

13. Install hoses and clamps in original positions.
14. Check for leakage of engine coolant using the radiator cap tester adapter and the radiator cap tester.
15. Start and warm up the engine. Visually check that there is no leakage of engine coolant or A/T fluid.

THERMOSTAT

REMOVAL & INSTALLATION

3.5L Engine

See Figures 76 and 77.

1. Before servicing the vehicle, refer to the Precautions Section.
2. Remove engine cover.
3. Remove air duct and air cleaner case assembly (bank 2).
4. Remove reservoir tank.

5. Remove engine undercover with power tool.

6. Drain engine coolant from radiator drain plug at the bottom of radiator.

❊❊ CAUTION

Perform this step when the engine is cold.

❊❊ WARNING

Do not spill engine coolant on drive belt.

7. Disconnect the lower radiator hose.

8. Disconnect intake valve timing control solenoid valve harness connector (bank 2), and remove intake valve timing control solenoid valve.

9. Remove water inlet and thermostat assembly.

❊❊ WARNING

Never disassemble water inlet and thermostat assembly. Replace them as a unit, if necessary.

To install:

10. Installation is the reverse of the removal procedure.

➡ **Be careful not to spill engine coolant over engine room. Use rag to absorb engine coolant.**

11. Check that the reservoir tank cap is tightened.

12. Check for leakage of engine coolant using the radiator cap tester adapter and the radiator cap tester.

13. Start and warm up the engine. Visually check that there is no leakage of engine coolant.

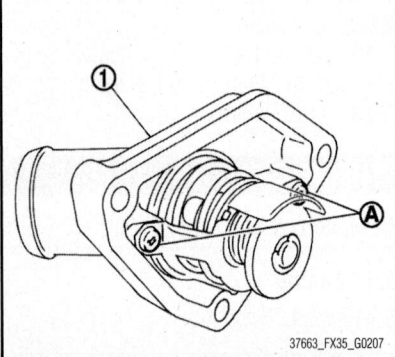

Fig. 76 Remove water inlet and thermostat assembly (1); never loosen these screws (A)

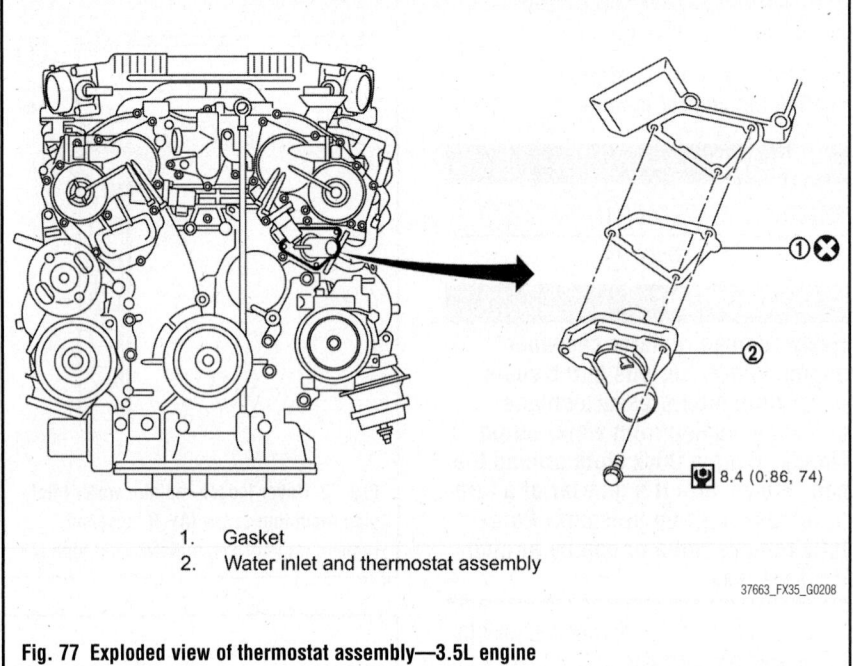

1. Gasket
2. Water inlet and thermostat assembly

37663_FX35_G0208

Fig. 77 Exploded view of thermostat assembly—3.5L engine

5.0L Engine

See Figures 78 through 80.

1. Before servicing the vehicle, refer to the Precautions Section.

2. Remove engine cover and engine room cover (RH and LH).

3. Remove air duct (inlet).

4. Remove reservoir tank.

5. Remove engine undercover.

6. Drain engine coolant from drain plugs on radiator and cylinder block.

❊❊ CAUTION

Perform this step when engine is cold.

❊❊ WARNING

Do not spill engine coolant on drive belts.

7. Disconnect the upper and lower radiator hoses.

8. Remove intake manifold.

9. Remove water suction pipe and water suction hose.

10. Remove water inlet and thermostat.

11. Remove water connector, heater pipes and heater hoses.

12. Remove thermostat housing.

To install:

13. Installation is the reverse of the removal procedure.

➡ **Be careful not to spill engine coolant over engine room. Use rag to absorb engine coolant.**

14. Install thermostat with the whole circumference of each flange part fit securely inside rubber ring.

15. Install thermostat with jiggle valve facing upwards. The position deviation may be within the range of 20°.

16. Apply a neutral detergent to O-rings, then quickly insert the insertion parts of the water connector and heater pipe into the installation holes.

17. Check that the reservoir tank cap is tightened.

18. Check for leakage of engine coolant using the radiator cap tester adapter and the radiator cap tester.

19. Start and warm up the engine. Visually check that there is no leakage of engine coolant.

WATER PUMP

REMOVAL & INSTALLATION

3.5L Engine

See Figures 81 through 85.

Note the following:

• When removing water pump assembly, be careful not to get engine coolant on drive belt

• Water pump cannot be disassembled and should be replaced as a unit

• After installing water pump, connect hose and clamp securely, then check for leakage using the radiator cap tester and the radiator cap tester adapter

1. Before servicing the vehicle, refer to the Precautions Section.

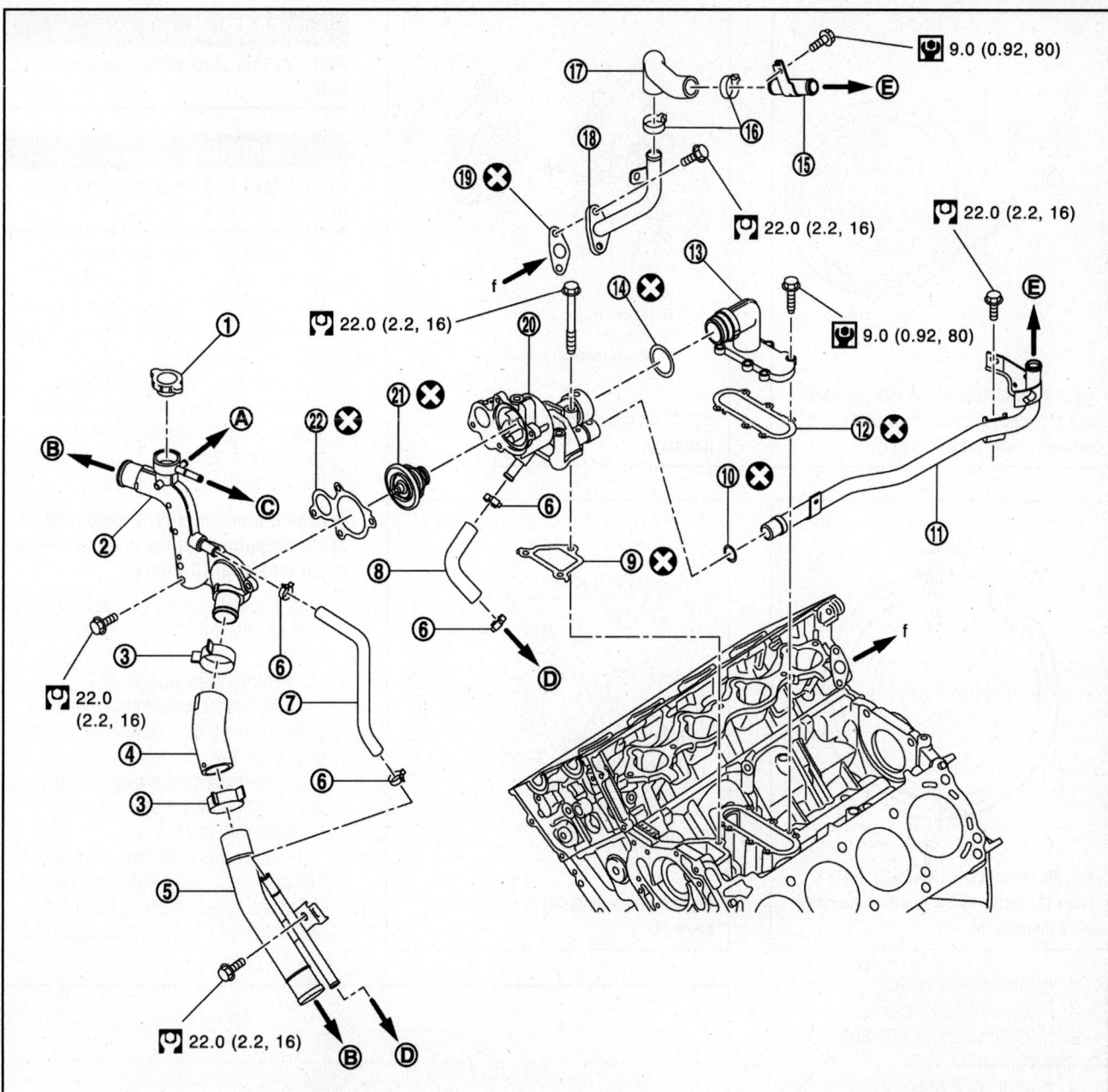

1.	Radiator cap	2.	Water inlet	3.	Clamp
4.	Water suction hose	5.	Water suction pipe	6.	Clamp
7.	Water hose	8.	Water hose	9.	Gasket
10.	O-ring	11.	Heater pipe	12.	Gasket
13.	Water connector	14.	O-ring	15.	Water pipe
16.	Clamp	17.	Water hose	18.	Water pipe
19.	Gasket	20.	Thermostat housing	21.	Thermostat
22.	Gasket				
A.	To electric throttle control actuator	B.	To radiator	C.	To reservoir tank
D.	To oil cooler	E.	To heater		

37663_FX35_G0199

Fig. 78 Exploded view showing thermostat assembly—5.0L engine

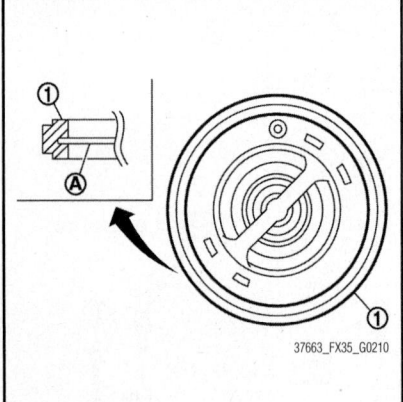

Fig. 79 Install thermostat with the whole circumference of each flange part (A) fit securely inside rubber ring (1)

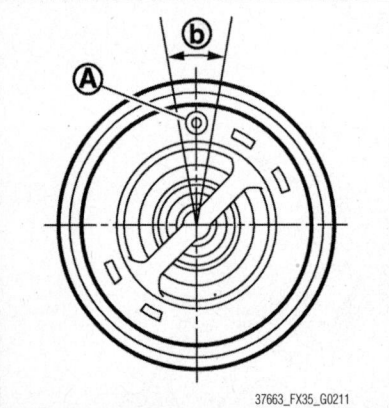

Fig. 80 Install thermostat with jiggle valve (A) facing upwards within the range of 20 degrees (b)

2. Remove engine cover.

3. Release the fuel pressure.

4. Disconnect the battery cable from the negative terminal.

5. Remove air duct and air cleaner case assembly (bank 1 and bank 2).

6. Remove reservoir tank.

7. Separate engine harness removing their brackets from front timing chain case.

8. Remove engine undercover.

9. Drain engine oil.

✲✲ CAUTION

Perform this step when the engine is cold.

✲✲ WARNING

Do not spill engine oil on drive belt.

10. Drain engine coolant from radiator.

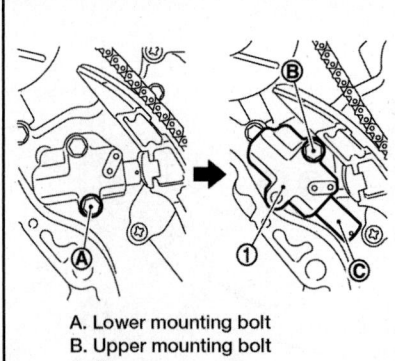

A. Lower mounting bolt
B. Upper mounting bolt
C. Plunger
1. Chain tensioner (primary)

Fig. 81 Remove timing chain tensioner (primary)

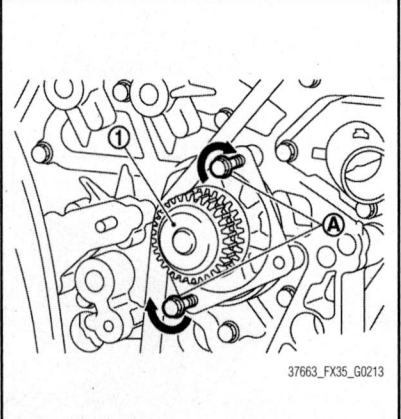

Fig. 82 Screw bolts (A) into the water pump (1)

✲✲ CAUTION

Perform this step when the engine is cold.

✲✲ WARNING

Do not spill engine coolant on drive belt.

11. Remove the lower radiator hose.

12. Remove cooling fan assembly.

13. Remove front timing chain case.

14. Remove timing chain tensioner (primary):

    a. Remove lower mounting bolt.

    b. Loosen upper mounting bolt slowly, and then turn chain tensioner (primary) on the upper mounting bolt so that plunger is fully expanded.

➡**Even if plunger is fully expanded, it is not dropped from the body of timing chain tensioner (primary).**

    c. Remove upper mounting bolt, and then remove timing chain tensioner (primary).

15. Remove water pump:

    a. Remove 3 water pump mounting bolts. Secure a gap between water pump gear and timing chain, by turning crankshaft counterclockwise until timing chain looseness on water pump sprocket becomes maximum.

    b. Screw M8 bolts, approximately 2 inches (50mm) long, into water pump upper and lower mounting bolt holes until they reach timing chain case. Then,

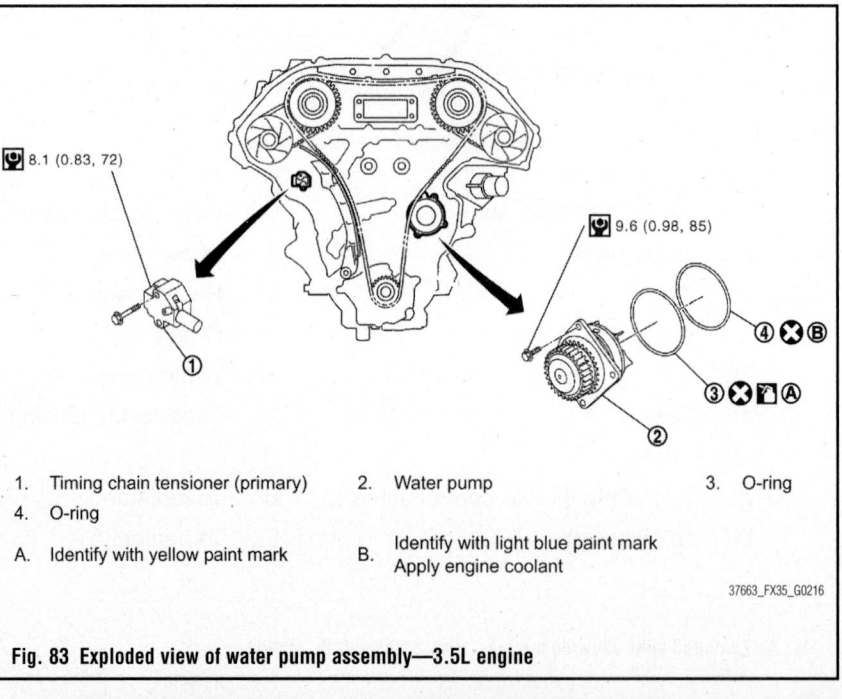

8.1 (0.83, 72)

9.6 (0.98, 85)

1. Timing chain tensioner (primary)
2. Water pump
3. O-ring
4. O-ring

A. Identify with yellow paint mark
B. Identify with light blue paint mark Apply engine coolant

Fig. 83 Exploded view of water pump assembly—3.5L engine

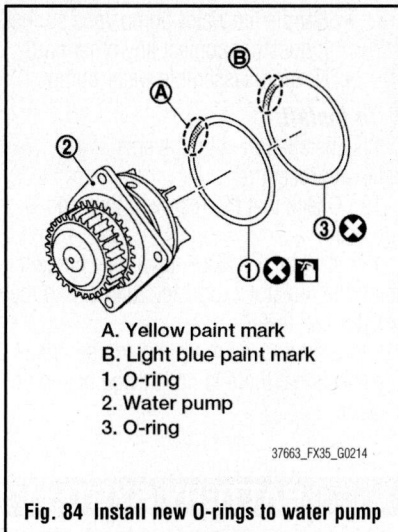

A. Yellow paint mark
B. Light blue paint mark
1. O-ring
2. Water pump
3. O-ring

37663_FX35_G0214

Fig. 84 Install new O-rings to water pump

alternately tighten each bolt for a half turn, and pull out water pump (1). Note the following:
• Pull straight out while preventing vane from contacting socket in installation area
• Remove water pump without causing sprocket to contact timing chain
   a. Remove M8 bolts and O-rings from water pump.

✳✳ WARNING

Do not disassemble the water pump.

To install:

16. Install new O-rings to water pump. Do not reuse O-rings.
   a. Apply engine oil to O-ring and engine coolant to O-ring.
   b. Locate O-ring with yellow paint mark to front side.
   c. Locate O-ring with light blue paint mark to rear side.
17. Install water pump.

✳✳ WARNING

Do not allow cylinder block to nip O-rings when installing water pump.

   a. Check timing chain and water pump sprocket are engaged.
   b. Insert water pump by tightening mounting bolts alternately and evenly.
18. Install timing chain tensioner (primary):
   a. Turn crankshaft clockwise so that timing chain on the timing chain tensioner (primary) side is loose.
   b. Pull plunger stopper tab up (or turn lever downward) so as to remove plunger stopper tab from the ratchet of plunger.

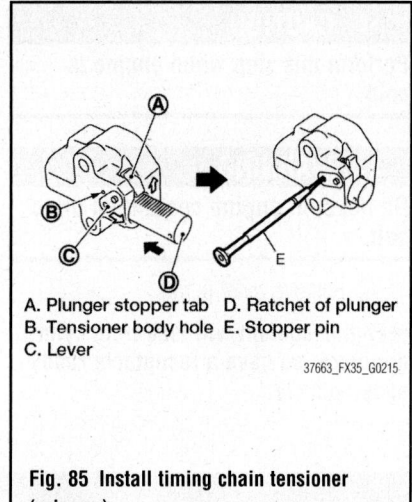

A. Plunger stopper tab   D. Ratchet of plunger
B. Tensioner body hole   E. Stopper pin
C. Lever

37663_FX35_G0215

Fig. 85 Install timing chain tensioner (primary)

➡ **Plunger stopper tab and lever are synchronized.**

   c. Push plunger into the inside of tensioner body.
   d. Hold plunger in the fully compressed position by engaging plunger stopper tab with the tip of ratchet.
   e. To secure lever, insert stopper pin through hole of lever into tensioner body hole.

➡ **The lever parts and the tab are synchronized. Therefore, the plunger will be secured under this condition.**

   f. Install timing chain tensioner (primary).

➡ **Remove dust and foreign material completely from backside of timing**

chain tensioner (primary) and from installation area of rear timing chain case.

   g. Remove stopper pin.
   h. Check again that timing chain and water pump sprocket are engaged.
19. Install in the reverse order of removal for remaining parts.
20. After starting engine, let idle for 3 minutes, then rev engine up to 3,000 RPM under no load to purge air from the high-pressure chamber of chain tensioner. Engine may produce a rattling noise. This indicates that air still remains in the chamber and is not a matter of concern.
21. Check that the reservoir tank cap is tightened.
22. Check for leakage of engine coolant using the radiator cap tester adapter and the radiator cap tester.
23. Start and warm up the engine. Visually check that there is no leakage of engine coolant.

5.0L Engine

See Figure 86.

Note the following:
• When removing water pump assembly, be careful not to get engine coolant on drive belts
• Water pump cannot be disassembled and should be replaced as a unit
• After installing water pump, connect hose and clamp securely, then check for leakage using the radiator cap tester and the radiator cap tester adapter

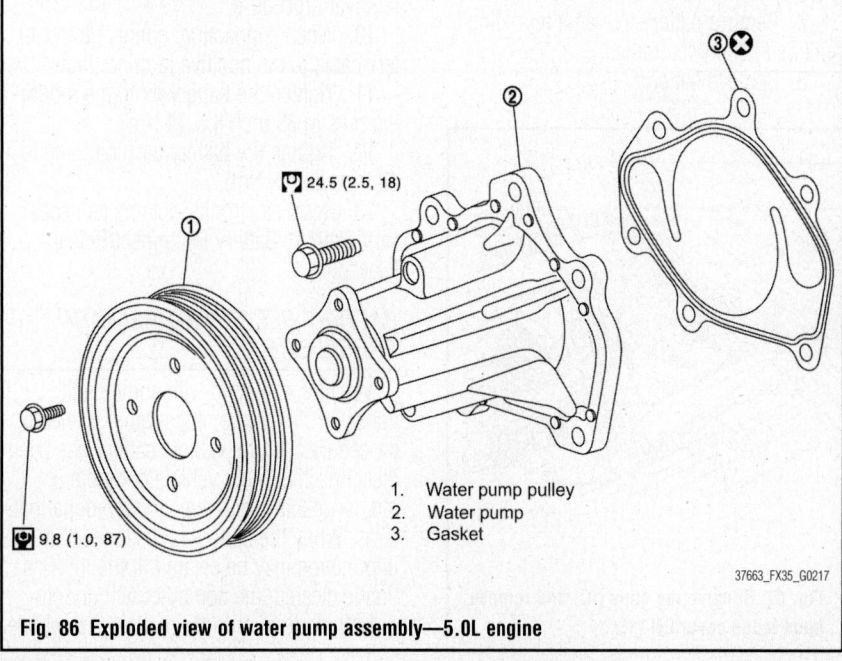

24.5 (2.5, 18)

9.8 (1.0, 87)

1. Water pump pulley
2. Water pump
3. Gasket

37663_FX35_G0217

Fig. 86 Exploded view of water pump assembly—5.0L engine

1. Before servicing the vehicle, refer to the Precautions Section.

2. Remove the engine undercover.

3. Remove the engine cover, and engine room covers (RH and LH).

4. Remove the air duct (inlet).

5. Remove the engine coolant reservoir tank.

6. Loosen water pump pulley mounting bolts.

7. Remove alternator, water pump and A/C compressor belt.

8. Remove water pump pulley.

9. Drain engine coolant from drain plugs on radiator and cylinder block.

✳✳ CAUTION

Perform this step when engine is cold.

✳✳ WARNING

Do not spill engine coolant on drive belt.

10. Remove water pump.

➡**Engine coolant will leak from cylinder block, so have a receptacle ready under vehicle.**

- Handle the water pump vane so that it does not contact any other parts
- Never disassemble water pump

To install:

11. Installation is the reverse of the removal procedure.

12. Check that the reservoir tank cap is tightened.

13. Check for leakage of engine coolant using the radiator cap tester adapter and the radiator cap tester.

14. Start and warm up the engine. Visually check that there is no leakage of engine coolant.

ENGINE ELECTRICAL

BATTERY

REMOVAL & INSTALLATION

See Figures 87 and 88.

1. Before servicing the vehicle, refer to the Precautions Section.

2. Remove battery cover.

3. Remove the clips and remove hood ledge cover RH.

4. Remove cowl top cover RH.

5. Remove cover of battery positive terminal.

6. Loosen battery terminal nuts and disconnect both battery cables from battery terminals.

✳✳ CAUTION

When disconnecting, disconnect the battery cable from the negative terminal first.

7. Remove battery fix frame mounting nuts and battery fix frame.

8. Remove battery.

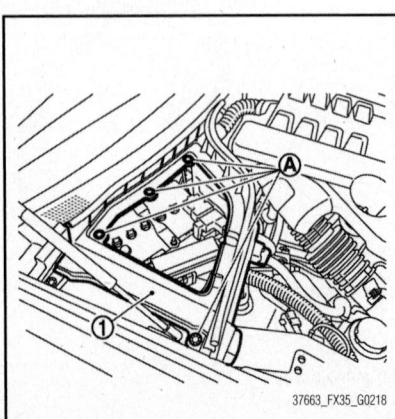

Fig. 87 Remove the clips (A), and remove hood ledge cover RH (1)

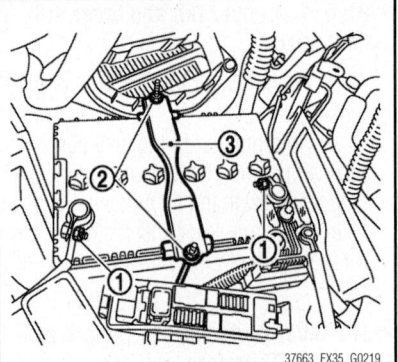

37663_FX35_G0219

Fig. 88 Loosen battery terminal nuts (1); remove battery fix frame mounting nuts (2) and battery fix frame (3)

To install:

9. Installation is the reverse of the removal procedure.

10. When connecting, connect the battery cable to the positive terminal first.

11. Tighten the battery fix frame mounting nuts to 35 inch lbs. (4 Nm).

12. Tighten the battery terminal nuts to 48 inch lbs. (5 Nm).

13. Reset electronic systems as necessary. Refer to Battery Reconnect/Relearn Procedure.

BATTERY RECONNECT/RELEARN PROCEDURE

Vehicles equipped with engine and transaxle computers may require a relearn procedure after the vehicle battery has been disconnected. Most vehicle computers memorize and store vehicle operational patterns. When the battery is disconnected, the information may be cleared. If the information is cleared, the computer will go into default mode in order to operate the vehicle.

BATTERY SYSTEM

The vehicle computer will relearn operational patterns each time the vehicle is restarted. The relearning process may take up to 40 or more key cycles.

When a specific engine component is replaced, a relearn procedure may be required. If the relearn procedure is not performed, the vehicle may exhibit the following:

- Harsh or poor shift quality
- Poor fuel mileage
- Hesitation or stumble
- Unstable idle or stalling
- Lean or rich running conditions

If an accessory component was replaced, a relearn procedure may also be required. The following systems and components may not work properly without a relearn procedure:

- Anti-theft system
- Steering system
- Power window system
- Power sunroof system

Automatic Drive Positioner System

See Figure 89.

Before servicing the vehicle, refer to the Precautions Section.

➡**Notice that disconnecting the battery when detected DTCs are present will erase the DTC memory.**

System Initialization

Always perform the initialization when the battery terminal is disconnected or the driver seat control unit is replaced. The entry/exit assist function will not operate normally if initialization is not performed.

1. Step 1: CHOOSE METHOD
   a. There are 2 initialization methods.
   b. Which method do you use?

Function	Condition	Procedure
Memory (Seat, steering, mirror)	Erased	Perform storing
Entry/exit assist	OFF	Perform initialization
		Set slide amount[*1]
Intelligent Key interlock	Erased	Perform storing
Seat synchronization	OFF	—

[*1]: Default value is 40mm.

71075_FX35_G0184

Fig. 89 Each function of the automatic drive positioner system is reset to the following condition when a battery terminal is disconnected

- With door switch, GO TO Step 2.
- With vehicle speed, GO TO Step 4.
2. Step 2: A-1
   a. Turn ignition switch from ACC to OFF position.
   b. GO TO Step 3.
3. Step 3: A-2
   a. Driver door switch
   - Is ON (open)—OFF (close)—ON (open).
   - END
4. Step 4: B-1
   a. Drive the vehicle at more than 16 MPH (25 km/h).
   b. END
5. Go to System Setting.

System Setting

See Figure 90.

The settings of the automatic driving positioner system can be changed, using CONSULT, the display unit in the center of the instrument panel, and the set switch. Always check the settings before and after disconnecting the battery terminal or replacing the driver seat control unit.
   1. Step 1: CHOOSE METHOD
      a. There are 3 ways of setting the system.

   b. Which method do you choose?
   - With set switch, GO TO Step 2.
   - With CONSULT, GO TO Step 4.
2. Step 2: WITH SET SWITCH (STEP 1)
   a. Turn ignition switch OFF.
   b. Push setting button and hold for more than 10 seconds, then confirm blinking of the memory switch indicator.
   - Entry/exit assist (seat/steering column) are ON: Memory switch indicator blinks 2 times
   - Entry/exit assist (seat/steering column) are OFF: Memory switch indicator blinks once
   - GO TO Step 3.
3. Step 3: WITH SET SWITCH (STEP 2)
   a. Turn ignition switch ACC
   b. Push setting button and hold for more than 10 seconds, then confirm blinking of the memory switch indicator.
   - Seat synchronization is ON: Memory switch indicator blinks 2 times
   - Seat synchronization is OFF: Memory switch indicator blinks once
   - END
4. Step 4: WITH CONSULT (STEP 1)
   - Select "Work support"
   - GO TO Step 5.

5. Step 5: WITH CONSULT (STEP 2)
   a. Select "EXIT SEAT SLIDE SETTING", "EXIT TILT SETTING" or "SEAT SLIDE VOLUME SET" then touch display to change between ON and OFF.
   - EXIT SEAT SLIDE SETTING: Entry/exit assist (seat)
   - EXIT TILT SETTING: Entry/exit assist (steering column)
   b. Then touch "OK".
   c. END
6. Go to Memory Storage.

Memory Storage

Always perform the memory storage when the battery terminal is disconnected or the driver seat control unit is replaced. The memory function and Intelligent Key interlock function will not operate normally if memory storage is not performed.

Two positions for the driver seat, steering column and outside mirror can be stored for memory operation by the following procedure.
   1. Step 1
      a. Shift A/T selector lever to P position.
      b. GO TO Step 2.
   2. Step 2
      a. Turn ignition switch ON.
      b. GO TO Step 3.
   3. Step 3
      a. Adjust driver seat, steering column and outside mirror position manually.
      b. GO TO Step 4.
   4. Step 4
      a. Push set switch.

- Memory indicator for which driver seat position is already retained in memory is illuminated for 5 seconds
- Memory indicator for which driver seat position is not retained in memory is illuminated for 0.5 second
   b. Push the memory switch (1 or 2) for at least 1 second within 5 seconds after pushing the set switch.
   - To enter driver seat positions into blank memory, memory indicator will be turned on for 5 seconds
   - To modify driver seat positions, memory indicator will be turned OFF for 0.5 second, then turned ON for 5 seconds
   - If memory is stored in the same memory switch, the previous memory will be deleted
   c. Do you need to link Intelligent Key?
   - If YES, GO TO Step 6.

				×: Applicable		
Item	Content	CONSULT	Set switch	Factory setting
Amount of seat sliding for entry/exit assist	The amount of seat sliding for entry/exit assist can be selected from 3 items. [40mm/80mm/150mm]	x	—	40mm
Entry/exit assist (seat)	Entry/exit assist (seat) can be selected: ON (operated) – OFF (not operated)	x	x	ON
Entry/exit assist (steering column)	Entry/exit assist (steering column) can be selected: ON (operated) – OFF (not operated)	x	x	ON
Seat synchronization	Seat synchronization can be selected: ON (operated) – OFF (not operated)	—	x	OFF
Reset custom settings	All settings can be set to default (factory setting).	—	—	—

71075_FX35_G0185

Fig. 90 Setting the automatic driving positioner system table

- If NO, GO TO Step 5.
5. Step 5
   a. Confirm the operation of each part with memory operation.
   b. END
6. Step 6
   a. Push the Intelligent Key unlock button within 5 seconds after pushing memory switch (while the memory indicator is turned ON).
   b. GO TO Step 7.
7. Step 7
   a. Confirm the operation of each part with memory operation and Intelligent Key interlock operation.
   b. END

Power Window Control System

When the battery negative terminal is disconnected, the initialization is necessary.

If any of the following operations are performed, the initialization is necessary as well as when the negative terminal of battery is disconnected.

- Power supply to the power window main switch or power window motor is cut off by the removal of battery terminal or if the battery fuse is blown
- Disconnection and connection of power window main switch harness connector
- Removal and installation of motor from regulator assembly
- Operation of regulator assembly as an independent unit
- Removal and installation of glass
- Removal and installation of door glass run

The following specified operations cannot be performed under the non-initialized condition.

- Auto-up operation
- Anti-pinch function

Initialization Procedure

1. Before servicing the vehicle, refer to the Precautions Section.
2. Disconnect the battery negative terminal or power window main switch connector. Reconnect it after a minute or more.
3. Turn ignition switch ON.
4. Operate power window switch to fully open the window. (This operation is unnecessary if the window is already fully open).
5. Continue pulling the power window switch UP (AUTO-UP operation). Even after glass stops at the fully closed position, keep pulling the switch for 3 seconds or more.
6. Initializing procedure is completed.
7. Inspect anti-pinch function. Refer to Check Anti-Pinch Function.

Check Anti-Pinch Function

1. Before servicing the vehicle, refer to the Precautions Section.
2. Fully open the door window.
3. Place a piece of wood near fully closed position.
4. Close door glass completely with AUTO-UP.
   a. Check that glass lowers for approximately 5.9 inches (150mm) or for 2 seconds without the pinching piece of wood and stops.
   b. Check that glass does not rise when operating the power window main switch while lowering.
   Note the following:
- Perform initial setting when auto-up operation or anti-pinch function does not operate normally
- Check that AUTO-UP operates before inspection when system initialization is performed
- Never check with hands or other body parts because they may be pinched
- It may switch to fail-safe mode if open/close operation is performed continuously without fully closing. Perform initial setting in that situation
- Finish initial setting. Otherwise, next operation cannot be done
5. Check the Auto-up operation.
6. Check the Anti-pinch function.

Predictive Course Line Center Position Adjustment

Adjust the center position of the predictive course line of the rear view monitor if it is shifted.

1. Before servicing the vehicle, refer to the Precautions Section.
2. Drive the vehicle straight ahead 328 ft. (100m) or more at a speed of 19 MPH (30 km/h) or more.
3. END.

Automatic Back Door System

When the battery is disconnected from the negative terminal, it is necessary to perform initial setting to operate automatic back door control system normally.

1. Before servicing the vehicle, refer to the Precautions Section.
2. Fully close the back door manually. (When back door is already fully closed, this operation is not necessary).
3. Perform automatic back door open/close operation of back door.
4. Check for noise or malfunctioning during operation.
5. Check that hazard lamp blinks and warning buzzer operates.

❋❋ CAUTION

Do not touch the back door, or allow foreign materials to be pinched in the back door, when performing the automatic back door open/close operation of the back door, until it is in the fully closed or fully open position.

6. END.

ENGINE ELECTRICAL | **CHARGING SYSTEM**

ALTERNATOR

REMOVAL & INSTALLATION

3.5L Engine—2WD Model

See Figure 91.

1. Before servicing the vehicle, refer to the Precautions Section.
2. Disconnect the battery cable from the negative terminal.
3. Remove engine front undercover.
4. Remove drive belt.
5. Remove the splash guard (RH).
6. Disconnect alternator connector.
7. Remove "B" terminal nut.
8. Remove the harness bracket bolts.
9. Remove oil pressure switch harness clip from alternator stay.
10. Disconnect oil pressure switch connector and oil temperature sensor connector.
11. Remove alternator mounting bolt and alternator stay mounting bolt, then remove alternator stay.
12. Remove alternator mounting bolt.
13. Move power steering oil pump hose upward.
14. Remove alternator assembly downward to remove from the vehicle.

To install:

15. Installation is the reverse of the removal procedure.

> ☀ **WARNING**
>
> **Be sure to tighten "B" terminal nut carefully.**

16. Install alternator, and check tension of drive belt.

➡ **The power generation voltage variable control system that controls the** power generation voltage of the alternator has been adopted. Therefore, an inspection should be performed after replacing the alternator to make sure that the system operates normally.

17. Inspect alternator operation.

3.5L Engine—AWD Model

See Figures 92 and 93.

1. Before servicing the vehicle, refer to the Precautions Section.
2. Disconnect the battery cable from the negative terminal.
3. Remove air cleaner case.
4. Disconnect power steering oil pressure sensor connector.
5. Remove the clip from the harness bracket and "B" terminal harness from the clip.
6. Remove engine undercover.
7. Remove drive belt.

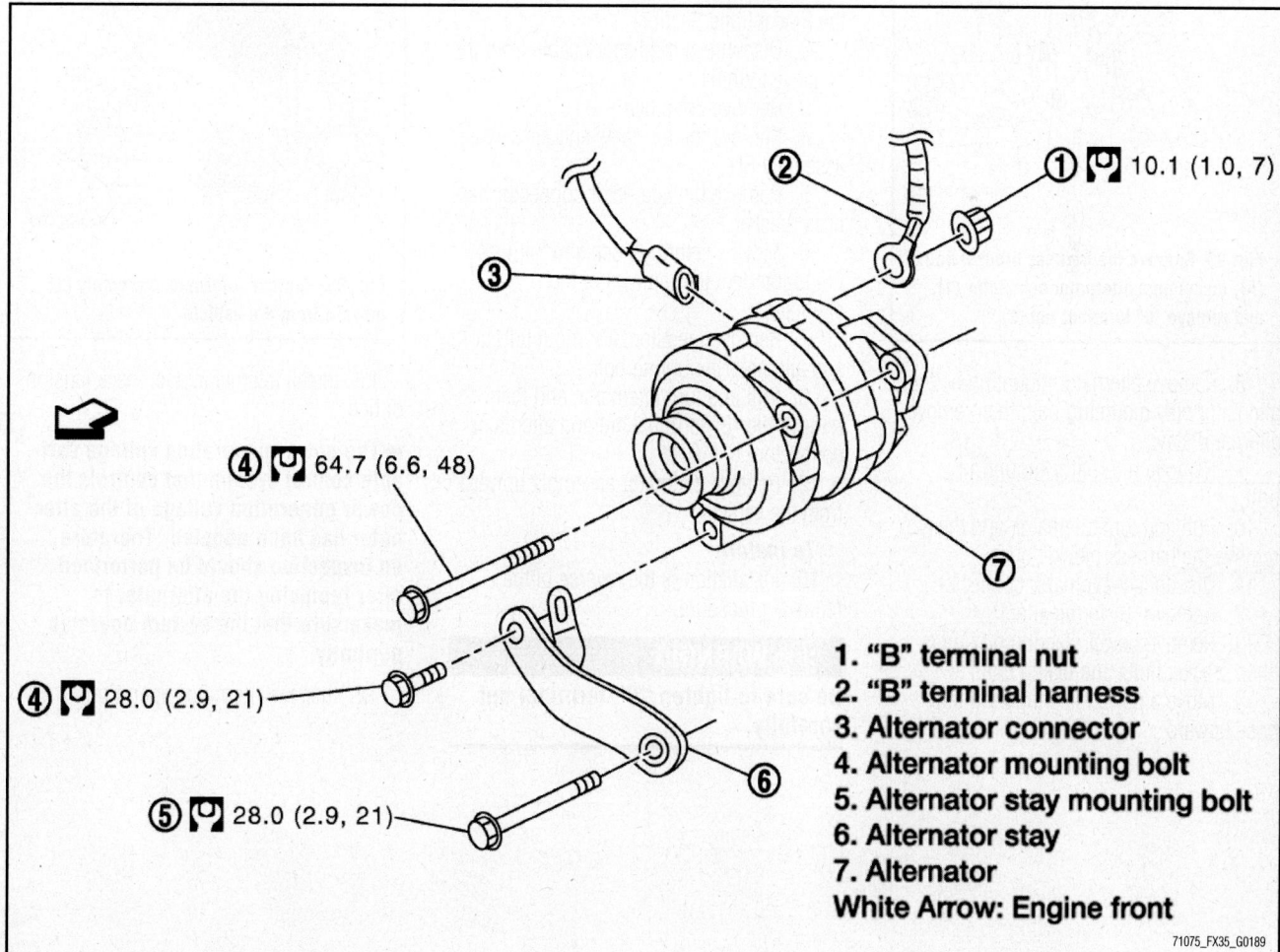

1. "B" terminal nut
2. "B" terminal harness
3. Alternator connector
4. Alternator mounting bolt
5. Alternator stay mounting bolt
6. Alternator stay
7. Alternator
White Arrow: Engine front

① 🔧 10.1 (1.0, 7)
④ 🔧 64.7 (6.6, 48)
④ 🔧 28.0 (2.9, 21)
⑤ 🔧 28.0 (2.9, 21)

Fig. 91 Exploded view of alternator—3.5L engine

71075_FX35_G0189

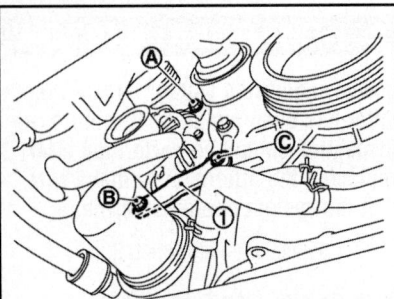

A. Alternator mounting bolt
B. Alternator mounting bolt
C. Alternator stay mounting bolt
1. Alternator stay

37663_FX35_G0224

Fig. 92 Remove alternator mounting bolt and alternator stay mounting bolt, then remove alternator stay

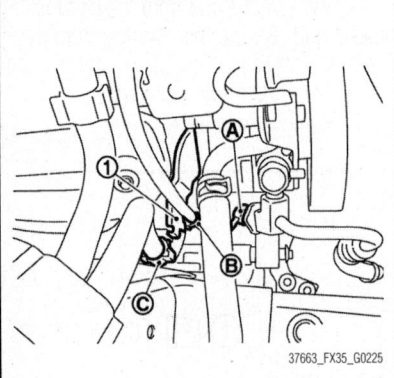

37663_FX35_G0225

Fig. 93 Remove the harness bracket bolts (A), disconnect alternator connector (1), and remove "B" terminal nut (2)

8. Remove alternator mounting bolt and alternator stay mounting bolt, then remove alternator stay.

9. Remove alternator mounting bolt.

10. Pull and turn alternator, and then remove the harness bracket bolts.

11. Disconnect alternator connector.

12. Remove "B" terminal nut.

13. Remove power steering oil pump hose bracket bolts and clamp bolts.

14. Move a power steering oil pump hose upward.

15. Remove alternator assembly downward from the vehicle.

To install:

16. Installation is the reverse of the removal procedure.

❄❄ WARNING

Be sure to tighten "B" terminal nut carefully.

17. Install alternator, and check tension of belt.

➡**The power generation voltage variable control system that controls the power generation voltage of the alternator has been adopted. Therefore, an inspection should be performed after replacing the alternator to make sure that the system operates normally.**

18. Inspect alternator operation.

5.0L Engine

See Figures 94 and 95.

1. Before servicing the vehicle, refer to the Precautions Section.

2. Disconnect the battery cable from the negative terminal.

3. Remove drive belt.

4. Remove the air ducts and air cleaner assembly RH.

5. Remove the alternator connector harness bracket.

6. Move a steering hose and harness not to interfere the removal of the alternator.

7. Remove the alternator mounting bolt and alternator mounting bolt.

8. Pull and turn alternator, and then remove the "B" terminal nut and alternator connector.

9. Remove alternator assembly upward from the vehicle.

To install:

10. Installation is the reverse of the removal procedure.

❄❄ WARNING

Be sure to tighten "B" terminal nut carefully.

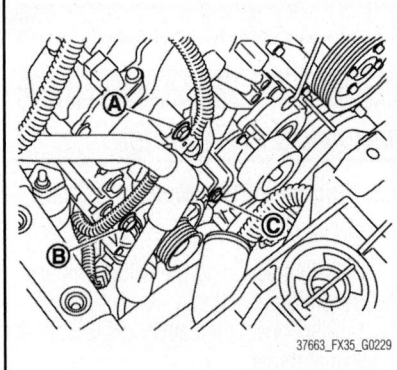

37663_FX35_G0229

Fig. 94 Remove the alternator connector harness bracket (A), the alternator mounting bolt (B) and alternator mounting bolt (C)

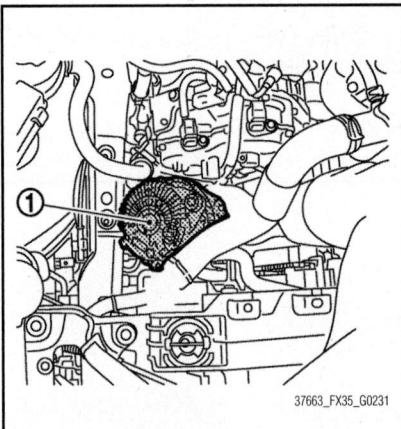

37663_FX35_G0231

Fig. 95 Remove alternator assembly (1) upward from the vehicle

11. Install alternator, and check tension of belt.

➡**The power generation voltage variable control system that controls the power generation voltage of the alternator has been adopted. Therefore, an inspection should be performed after replacing the alternator to make sure that the system operates normally.**

12. Inspect alternator operation.

ENGINE ELECTRICAL   **IGNITION SYSTEM**

FIRING ORDERS

See Figures 96 and 97.

The firing order for the 3.5L engine is 1–2–3–4–5–6.

The firing order for the 5.0L engine is 1–8–7–3–6–5–4–2.

IGNITION COIL(S)

REMOVAL & INSTALLATION

3.5L Engine

See Figure 98.

1. Before servicing the vehicle, refer to the Precautions Section.
2. Remove the engine cover.
3. Remove the air cleaner case and air duct.
4. Remove the intake manifold collector.

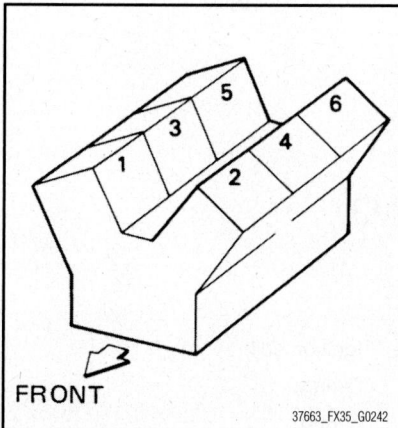

FRONT

37663_FX35_G0242

Fig. 96 3.5L engine cylinder number locations

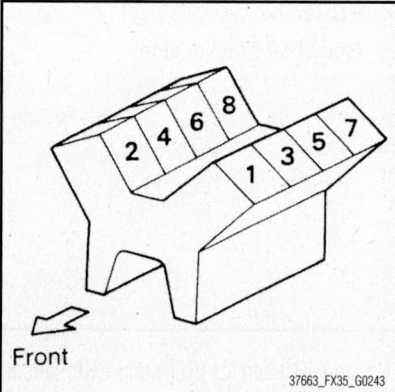

Front

37663_FX35_G0243

Fig. 97 5.0L engine cylinder number locations

5. Disconnect PCV hose from valve cover.
6. Remove camshaft position sensor (PHASE) and exhaust valve timing control position sensor (bank 1 and bank 2), as needed.
- Handle carefully to avoid dropping and shocks
- Never disassemble
- Never allow metal powder to adhere to magnetic part at sensor tip
- Never place sensors in a location where they are exposed to magnetism

7. Remove PCV valve and O-ring from valve cover, if necessary.
8. Remove oil filler cap from valve cover, if necessary.
9. Remove ignition coil.

✳✳ WARNING

Do not impact ignition coil.

To install:

10. Installation is the reverse of the removal procedure.
11. Tighten the fasteners to specification.

5.0L Engine

See Figures 99 and 100.

1. Before servicing the vehicle, refer to the Precautions Section.
2. Remove the engine cover and engine room cover (RH and LH).
3. Remove the air cleaner case and air duct.
4. Remove the fuel feed hose.
5. Disconnect the PCV hose from the rocker cover.
6. Remove the ignition coil.

✳✳ WARNING

Do not impact ignition coil.

➥**Installation position of ignition coil depends on cylinder position.**

To install:

7. Installation is the reverse of the removal procedure.
8. Install ignition coil.

➥**Be sure to install ignition coils marked with an identification mark on cylinder No. 7 and 8.**

9. Tighten the fasteners to specification.

IGNITION TIMING

INSPECTION & ADJUSTMENT

➥**The ignition timing is not adjustable. If not within specifications, further diagnostic inspection is required. The following procedure is for viewing the ignition timing setting.**

Visually check the air cleaner, intake hoses, ducts, Exhaust Gas Recirculation (EGR) valve operation and electrical connections prior to the adjustment of the ignition timing. Correct or repair any problem as required. Be sure to inspect the throttle valve and Throttle Position (TP) sensor for proper operation.

1. Before servicing the vehicle, refer to the Precautions Section.
2. Locate the timing marks on the crankshaft pulley and the front of the engine.
3. Clean the timing marks.

➥**The ignition timing specification is 10–20° Before Top Dead Center (BTDC).**

4. Using chalk or white paint, color the mark on the crankshaft pulley and the mark on the scale that will indicate the correct timing when aligned with the notch on the crankshaft pulley.
5. Attach a tachometer to the engine.
6. Attach a timing light to the engine to number 1 cylinder ignition wire.
7. Turn all electrical equipment and accessories **OFF**.
8. Check to be sure all of the wires clear the fan, then, start the engine and allow it to reach normal operating temperatures.
9. Block the front wheels and set the parking brake. Shift the transmission into **NEUTRAL**. Do not stand in front of the vehicle when making adjustments.
10. Perform the following procedures:
   a. Race the engine at 2,000 RPM for about 2 minutes under a no-load condition; be sure all of the accessories are turned **OFF**.
   b. Perform on board engine diagnostics and repair any fault code.
   c. Race the engine at 2,000 RPM for about 2 minutes under a no-load condition.
   d. Turn the engine **OFF** and disconnect the TP sensor.
   e. Start and race the engine 2 to 3 times under no-load, then run the engine at idle speed.

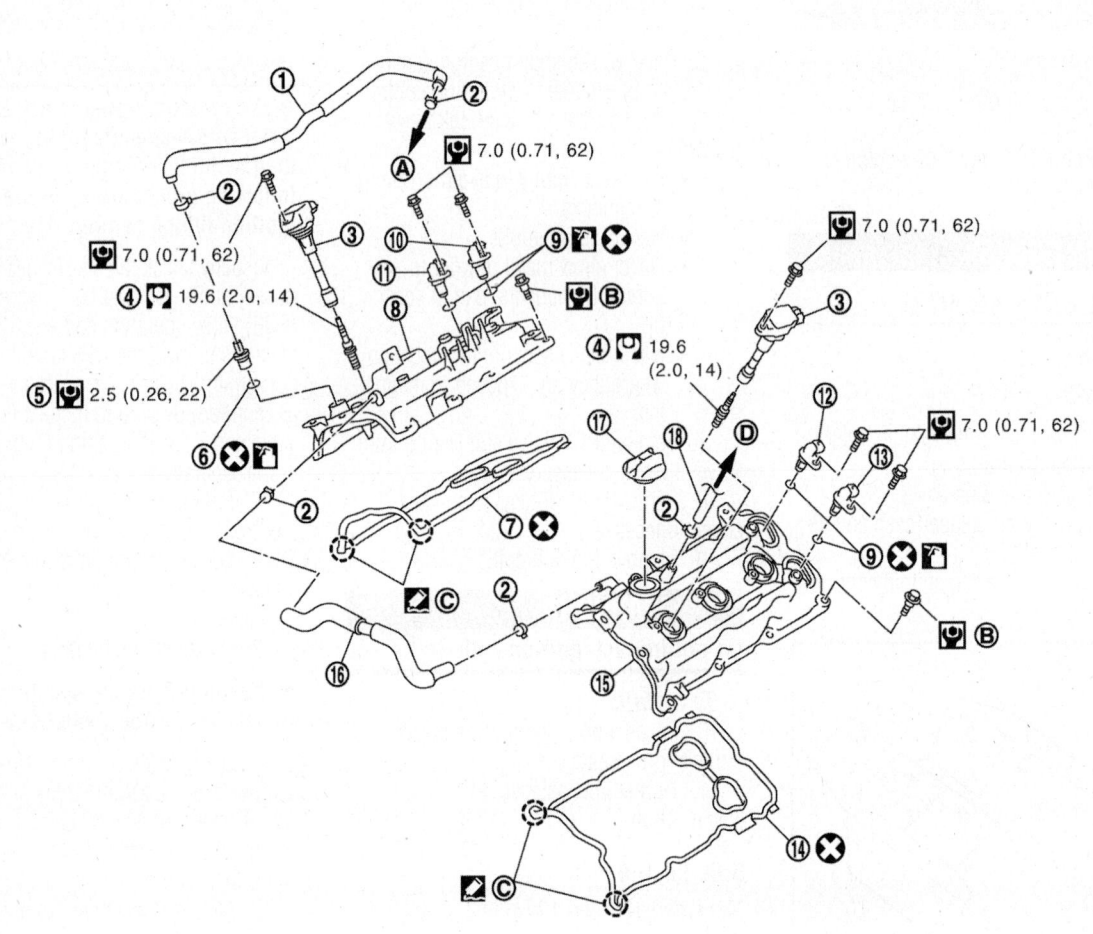

1. PCV hose
2. Clamp
3. Ignition coil
4. Spark plug
5. PCV valve
6. O-ring
7. Rocker cover gasket (bank 1)
8. Rocker cover (bank 1)
9. O-ring
10. Camshaft position sensor (PHASE) (bank 1)
11. Exhaust valve timing control position sensor (bank 1)
12. Camshaft position sensor (PHASE) (bank 2)
13. Exhaust valve timing control position sensor (bank 2)
14. Rocker cover gasket (bank 2)
15. Rocker cover (bank 2)
16. PCV hose
17. Oil filler cap
18. PCV hose
A. To intake manifold collector
C. Camshaft bracket side
D. To air duct

37663_FX35_G0236

Fig. 98 Exploded view of ignition coil, spark plug and valve cover—3.5L engine

11. Aim the timing light at the timing marks. If the marks on the pulley and the engine are aligned when the light flashes, the timing is correct. Turn the engine **OFF** and remove the tachometer and the timing light. If the marks are not in alignment, proceed with the following steps:
   a. Turn the engine **OFF**.
   b. Check the Camshaft Position (CMP) sensor (PHASE), Crankshaft Position (CKP) sensor (REF) and CKP sensor (POS). Replace if necessary.
   c. If the ignition timing is still not correct, substitute a known good Electronic Control Module (ECM).

A
37663_FX35_G0241

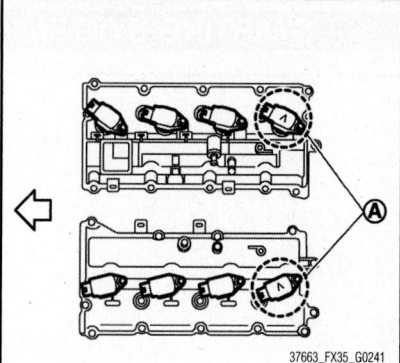

Fig. 99 Install ignition coil marked with an identification mark (A) on cylinder No. 7 and 8

➡**The ECM may be the cause of the problem, but this is rarely the case.**

12. Turn the engine **OFF** and remove the tachometer and the timing light.

SPARK PLUGS

REMOVAL & INSTALLATION

3.5L Engine

See Figure 98.

1. Before servicing the vehicle, refer to the Precautions Section.
2. Remove engine cover.
3. Remove air duct.
4. Remove electric throttle control actuator.
5. Remove ignition coil. Refer to Ignition Coil(s), removal & installation.
6. Remove spark plug with a spark plug wrench (commercial service tool).
7. Installation is the reverse of the removal procedure.

5.0L Engine

See Figure 100.

1. Before servicing the vehicle, refer to the Precautions Section.
2. Remove engine cover.
3. Remove ignition coil. Refer to Ignition Coil(s), removal & installation.
4. Remove spark plug with a spark plug wrench (commercial service tool).
5. Installation is the reverse of the removal procedure.

7.0 (0.71, 62)
2.5 (0.26, 22)
19.6 (2.0, 14)
19.6
(2.0, 14)
7.0 (0.71, 62)
1. Clamp
2. PCV hose
3. Oil filler cap
4. Oil catcher
5. Ignition coil (No. 1 - 6)
6. Spark plug
7. Rocker cover (bank 2)
8. Rocker cover gasket (bank 2)
9. Rocker cover gasket (bank 1)
10. Rocker cover (bank 1)
11. Ignition coil (No. 7, 8)
12. PCV hose
13. PCV hose
14. Clamp
15. PCV valve
16. O-ring
17. PCV hose
A. To air duct (bank 2)
B. Comply with the installation procedure when tightening
C. To air duct (bank 1)
D. To intake manifold
71075_FX35_G0191

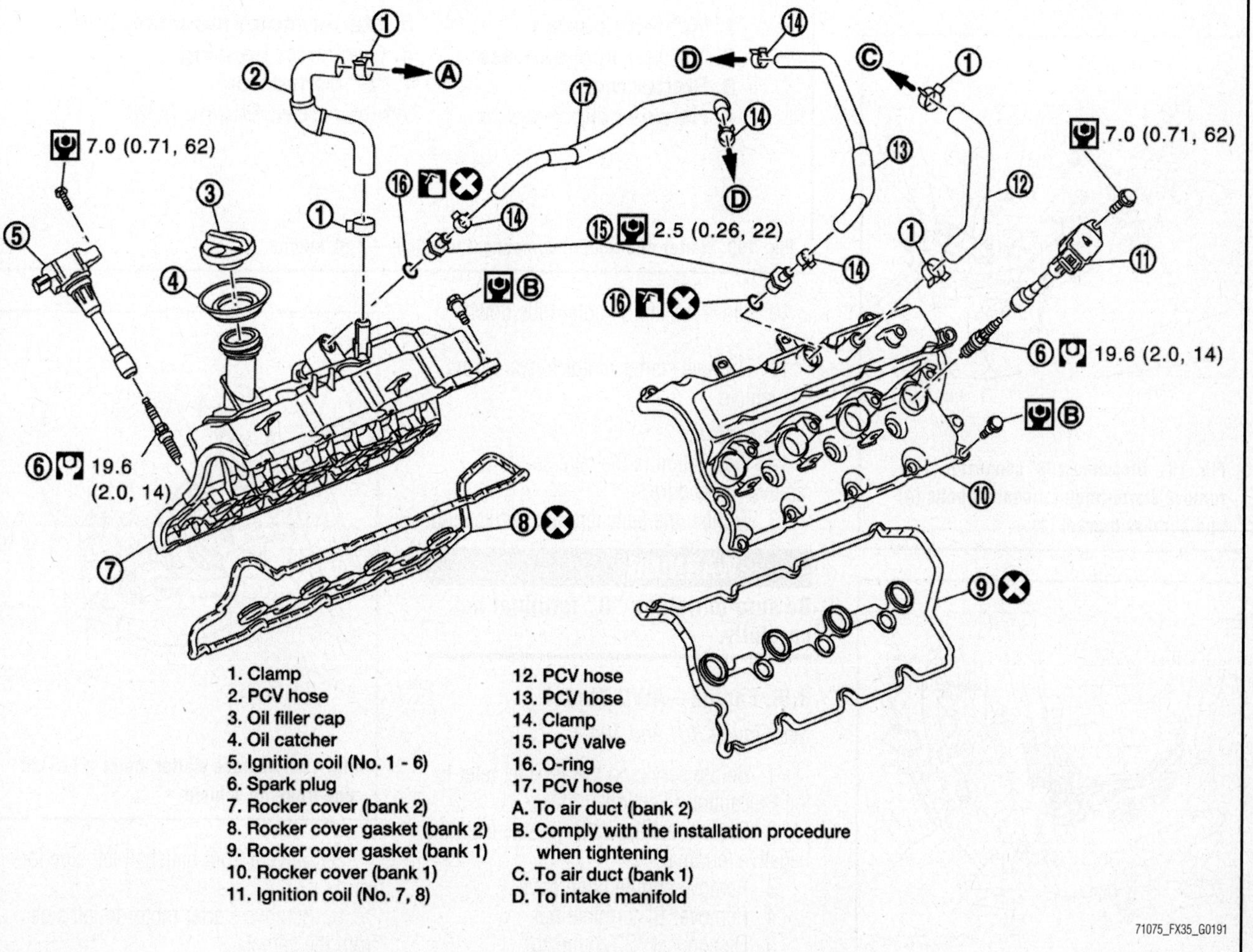

Fig. 100 Exploded view of ignition coil, spark plug, and rocker cover component locations—5.0L engine

STARTER

REMOVAL & INSTALLATION

3.5L Engine—2WD Model

See Figures 101 through 103.

1. Before servicing the vehicle, refer to the Precautions Section.
2. Disconnect the battery cable from the negative terminal.
3. Remove engine undercover.
4. Remove "B" terminal nut.
5. Disconnect "S" connector.
6. Remove starter motor mounting bolts and harness bracket.
7. Remove compressor bracket bolts.
8. Remove compressor bracket.
9. Remove A/T fluid cooler tube clip bolts and bracket.

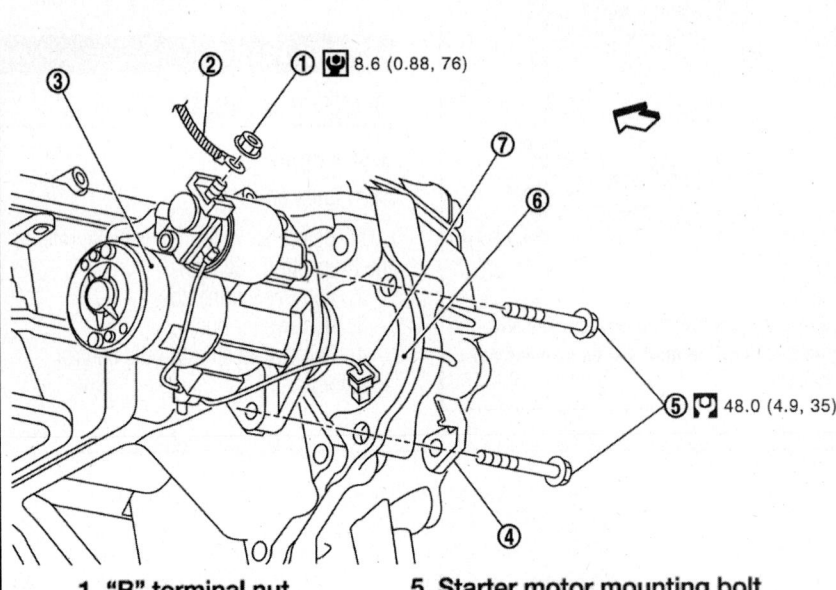

1. "B" terminal nut
2. "B" terminal harness
3. Starter motor
4. Harness clip bracket
5. Starter motor mounting bolt
6. Converter housing
7. "S" connector

White Arrow: Engine front

71075_FX35_G0195

Fig. 103 Starter and related component locations—3.5L engine

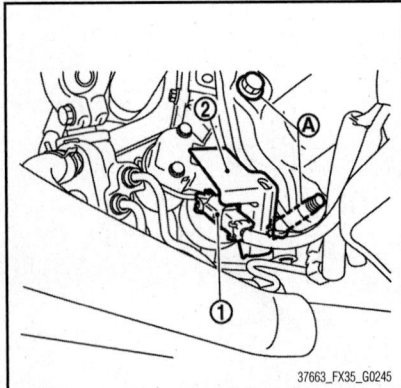

37663_FX35_G0245

Fig. 101 Disconnect "S" connector (1); remove starter motor mounting bolts (A) and harness bracket (2)

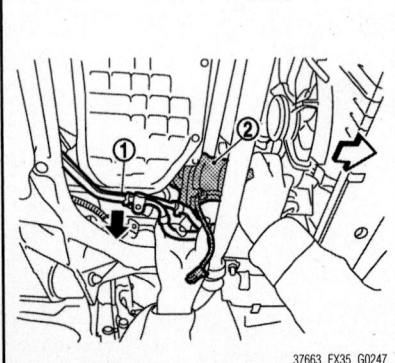

37663_FX35_G0247

Fig. 102 Move A/T fluid cooler tube (1) downward; remove starter motor (2) forward

10. Move A/T fluid cooler tube downward.
11. Remove starter motor forward from the vehicle.

To install:

12. Installation is the reverse of the removal procedure.
13. Tighten the fasteners to specification.

✸✸ WARNING

Be sure to tighten "B" terminal nut carefully.

3.5L Engine—AWD Model

See Figures 101 and 104.

1. Before servicing the vehicle, refer to the Precautions Section.
2. Disconnect the battery cable from the negative terminal.
3. Remove engine undercover.
4. Remove "B" terminal nut.
5. Disconnect "S" connector.
6. Remove starter motor mounting bolts and harness bracket.
7. Remove front halfshaft left side housing bolts.

37663_FX35_G0248

Fig. 104 Remove starter motor (1) to left side from the vehicle

8. Move a front halfshaft left side forward.
9. Remove starter motor to left side from the vehicle.

To install:

10. Installation is the reverse of the removal procedure.
11. Tighten the fasteners to specification.

✳✳ WARNING

Be sure to tighten "B" terminal nut carefully.

5.0L Engine

See Figure 105.

1. Before servicing the vehicle, refer to the Precautions Section.

2. Disconnect the battery cable from the negative terminal.

3. Remove engine cover.

4. Remove intake manifold. Refer to Intake Manifold, removal & installation.

5. Remove "B" terminal nut.

6. Disconnect "S" connector.

7. Remove starter motor mounting bolts.

8. Remove starter motor upward from the vehicle.

To install:

9. Installation is the reverse of the removal procedure.

10. Tighten the fasteners to specification.

✳✳ WARNING

Be sure to tighten "B" terminal nut carefully.

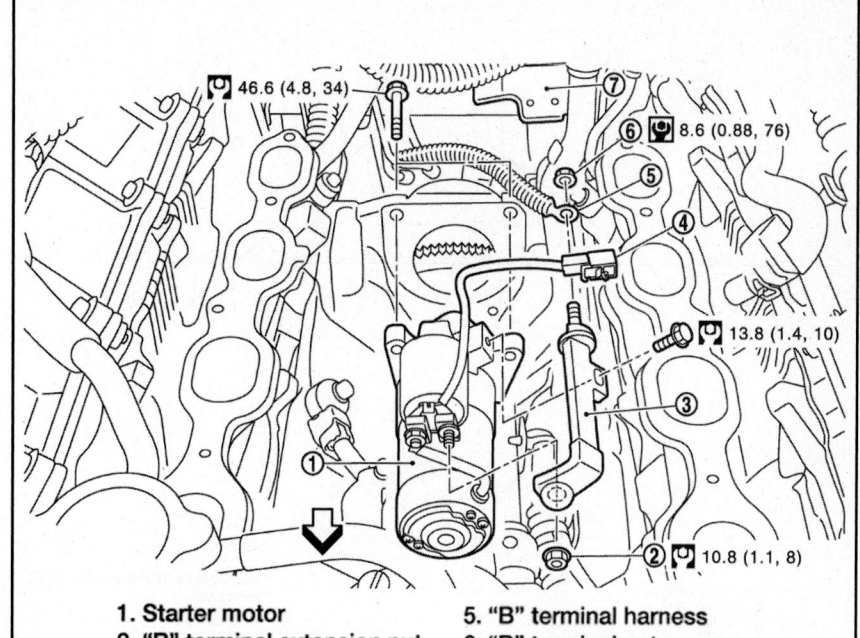

1. Starter motor
2. "B" terminal extension nut
3. "B" terminal extension
4. "S" connector
5. "B" terminal harness
6. "B" terminal nut
7. "S" connector bracket
White Arrow: Engine front

71075_FX35_G0196

Fig. 105 Starter and related component locations—5.0L engine

ENGINE MECHANICAL

ACCESSORY DRIVE BELT SYSTEM

ADJUSTMENT

Belt tension is not manually adjustable; it is automatically adjusted by the drive belt auto-tensioner. If the drive belt is out of adjustment, replace the drive belt as needed.

BELT ROUTINGS

See Figures 106 and 107.

INSPECTION

See Figures 106 through 107.

✳✳ CAUTION

Inspect and check the drive belts with the engine OFF.

1. Before servicing the vehicle, refer to the Precautions Section.

2. Check that the indicator of the drive belt auto-tensioner is within the possible use range.

➡**Check the drive belt auto-tensioner indication when the engine is cold.**

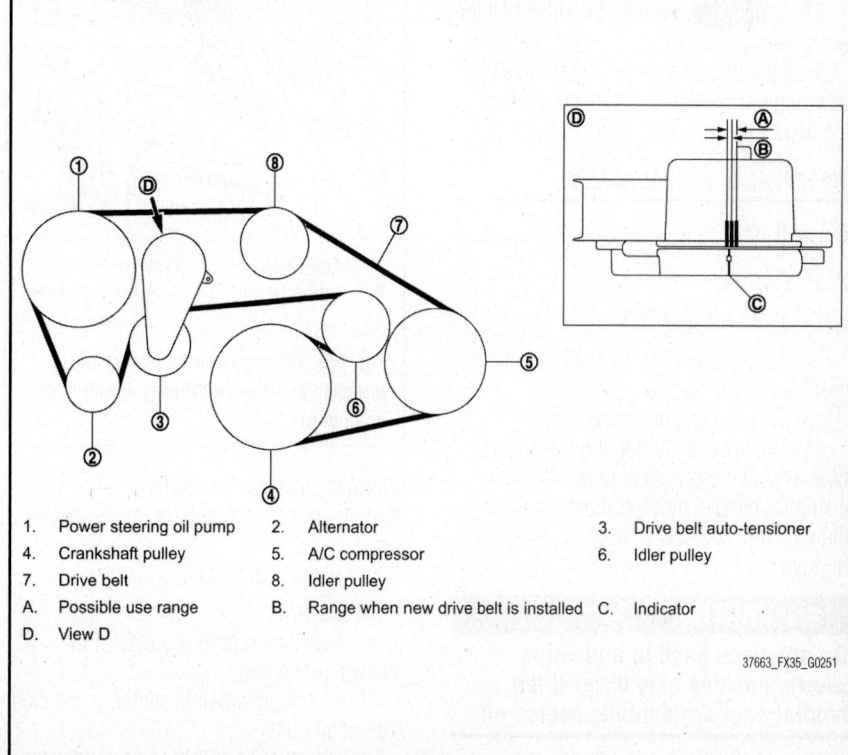

1.	Power steering oil pump	2.	Alternator	3.	Drive belt auto-tensioner
4.	Crankshaft pulley	5.	A/C compressor	6.	Idler pulley
7.	Drive belt	8.	Idler pulley		
A.	Possible use range	B.	Range when new drive belt is installed	C.	Indicator
D.	View D				

37663_FX35_G0251

Fig. 106 Accessory belt routing—3.5L engine

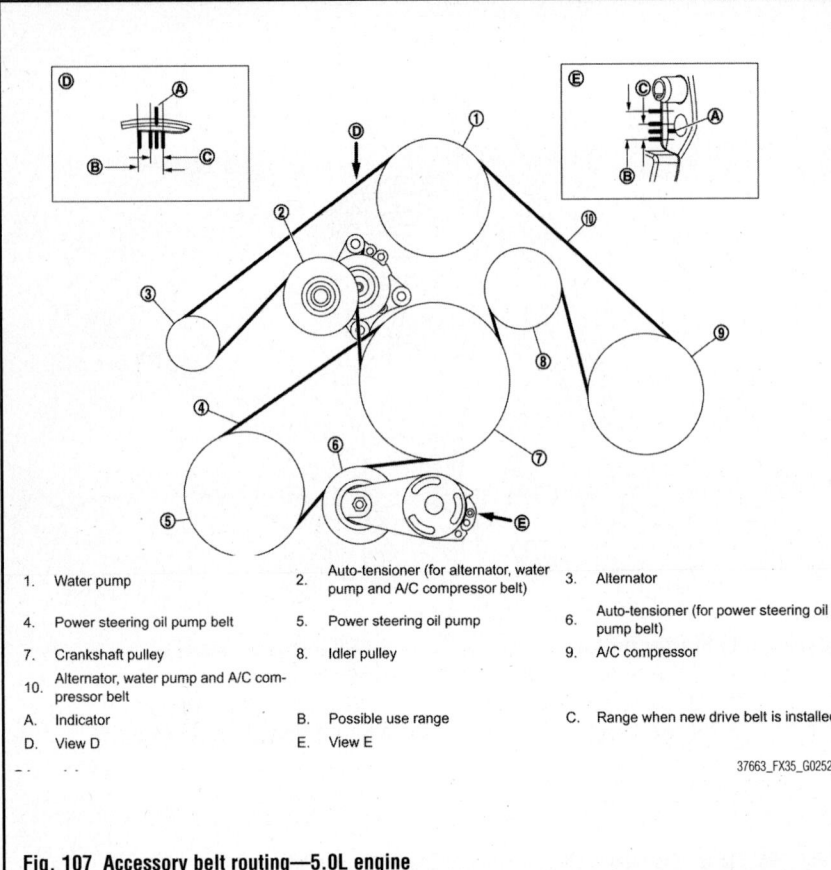

1.	Water pump	2.	Auto-tensioner (for alternator, water pump and A/C compressor belt)	3.	Alternator
4.	Power steering oil pump belt	5.	Power steering oil pump	6.	Auto-tensioner (for power steering oil pump belt)
7.	Crankshaft pulley	8.	Idler pulley	9.	A/C compressor
10.	Alternator, water pump and A/C compressor belt				
A.	Indicator	B.	Possible use range	C.	Range when new drive belt is installed
D.	View D	E.	View E		

37663_FX35_G0252

Fig. 107 Accessory belt routing—5.0L engine

3. When a new drive belt is installed, the indicator should be within the new drive belt range.

4. Visually check entire the drive belt for wear, damage, or cracks.

5. If the indicator is out of the possible use range or the belt is damaged, replace the drive belt.

REMOVAL & INSTALLATION

Drive Belt

3.5L Engine

See Figures 106 and 108.

1. Before servicing the vehicle, refer to the Precautions Section.

2. Remove engine undercover.

3. While securely holding the square hole in pulley center of auto tensioner with a spinner handle, move spinner handle in the direction of arrow (loosening direction of drive belt).

※ CAUTION

Do not place hand in a location where pinching may occur if the holding tool accidentally comes off.

4. Under the above condition, insert a metallic bar of approximately 0.24 inches

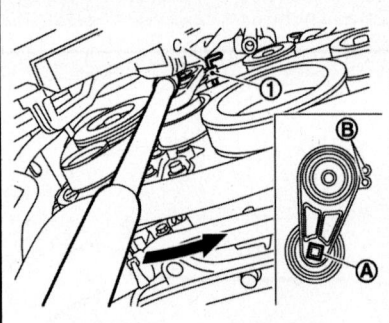

A. Square hole
B. Holding boss
C. Wrench
1. Auto-tensioner

37663_FX35_G0253

Fig. 108 Move spinner handle in the direction of arrow (loosening direction of drive belt)

(6mm) in diameter through the holding boss to lock auto-tensioner pulley arm.

5. Remove drive belt.

To install:

6. Installation is the reverse of the removal procedure.

7. Check drive belt is securely installed around all pulleys.

8. Check drive belt is correctly engaged with the pulley groove.

9. Check that engine oil or engine coolant are not adhered to the drive belt or pulley groove.

10. Turn crankshaft pulley clockwise several times to equalize tension between each pulley, and then confirm tension of drive belt at indicator (notch on fixed side) is within the possible use range

5.0L Engine—Alternator, Water Pump, and A/C Compressor Belt

See Figures 107 and 109.

1. Before servicing the vehicle, refer to the Precautions Section.

2. Remove air duct (inlet).

3. Remove reservoir tank.

4. With box wrench, and while securely holding the hexagonal part in pulley center of auto tensioner, move wrench handle in the direction of loosening belt.

※ CAUTION

Do not place hand in a location where pinching may occur if the holding tool accidentally comes off.

※ WARNING

Never loosen the hexagonal part in center of auto tensioner pulley. If turned clockwise, the complete auto tensioner must be replaced as a unit, including the pulley.

5. Under the above condition, insert a metallic bar of approximately 0.24 inches (6mm) in diameter through the holding boss to lock auto tensioner pulley arm.

➡**Leave the auto tensioner pulley arm locked until the belt is installed again.**

6. Remove the alternator, water pump, and A/C compressor drive belt.

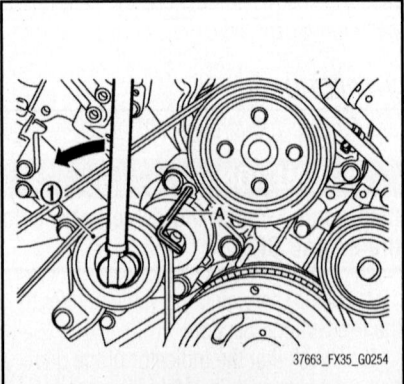

37663_FX35_G0254

Fig. 109 Move wrench handle in the direction of arrow (loosening direction of belt) and place a holding pin (A) in position

To install:

7. Installation is the reverse of the removal procedure.

8. Check drive belts are securely installed around all pulleys.

9. Check drive belts are correctly engaged with the pulley groove.

10. Check that engine oil or engine coolant are not adhered to the drive belt or pulley groove.

11. Turn crankshaft pulley clockwise several times to equalize tension between each pulley, and then confirm tension of drive belts at indicator (notch on fixed side) is within the possible use range.

5.0L Engine—Power Steering Oil Pump Belt

See Figures 107 and 110.

1. Before servicing the vehicle, refer to the Precautions Section.

2. Remove engine undercover.

3. Remove alternator, water pump, and A/C compressor drive belt.

4. With a box wrench, and while securely holding the hexagonal part in pulley center of auto tensioner, move wrench handle in the direction of loosening the belt.

✳✳ CAUTION

Do not place hand in a location where pinching may occur if the holding tool accidentally comes off.

✳✳ WARNING

Never loosen the hexagonal part in center of auto tensioner pulley. If turned clockwise, the complete auto tensioner must be replaced as a unit, including the pulley.

---

5. Under the above condition, insert a metallic bar of approximately 0.24 inches (6mm) in diameter through the holding boss to lock auto tensioner pulley arm.

➡ **Leave auto tensioner pulley arm locked until belt is installed again.**

6. Remove power steering oil pump drive belt.

To install:

7. Installation is the reverse of the removal procedure.

8. Check drive belts are securely installed around all pulleys.

9. Check drive belts are correctly engaged with the pulley groove.

10. Check that engine oil or engine coolant are not adhered to the drive belt or pulley groove.

11. Turn crankshaft pulley clockwise several times to equalize tension between each pulley, and then confirm tension of drive belts at indicator (notch on fixed side) is within the possible use range.

Drive Belt Tensioner And Idler Pulley

3.5L Engine

See Figures 106 and 111.

1. Before servicing the vehicle, refer to the Precautions Section.

2. Remove drive belt.

➡ **Keep auto-tensioner pulley arm locked after drive belt is removed.**

3. Remove auto-tensioner and idler pulley. Keep auto-tensioner pulley

arm locked to install or remove auto-tensioner.

To install:

4. Installation is the reverse of the removal procedure.

5. Tighten the fasteners to specification.

➡ **If there is damage greater than peeled paint, replace drive belt auto-tensioner.**

5.0L Engine

See Figures 107 and 112.

➡ **The complete drive belt auto-tensioner must be replaced as a unit, including the pulley.**

1. Before servicing the vehicle, refer to the Precautions Section.

2. Remove the drive belts.

➡ **Keep the auto-tensioner pulley arm locked after the drive belt is removed.**

3. Remove the drive belt auto-tensioners. Keep the auto-tensioner pulley arm locked to install or remove the auto-tensioner.

✳✳ WARNING

Never loosen the hexagonal part in the center of the drive belt auto tensioner pulley. If turned clockwise, the complete drive belt auto tensioner must be replaced as a unit, including the pulley.

---

4. Remove the idler pulley.

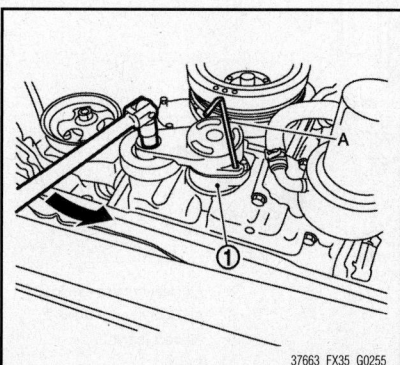

37663_FX35_G0255

Fig. 111 Move wrench handle in the direction of arrow (loosening direction of belt) on the belt tensioner (1) and place a holding pin (A) in position

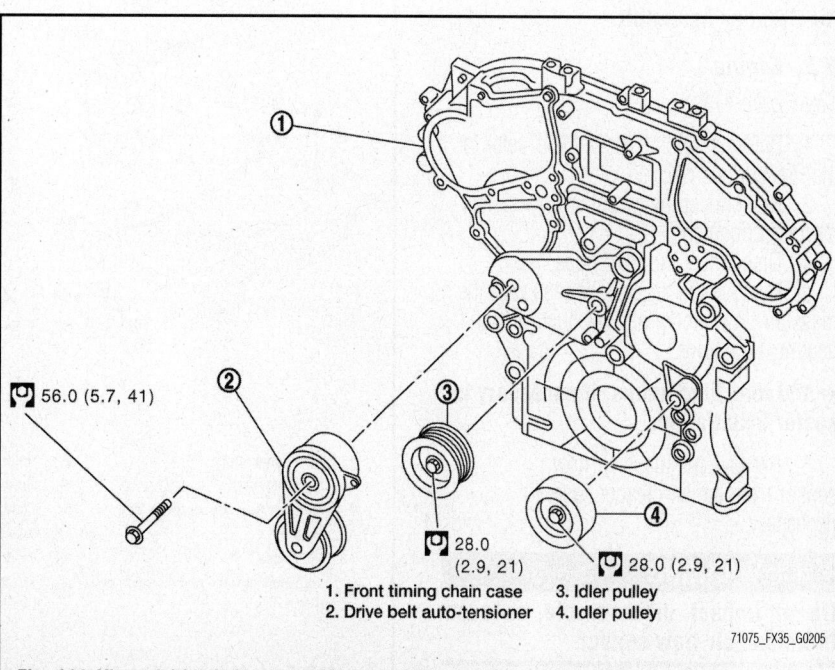

56.0 (5.7, 41)

28.0 (2.9, 21)   28.0 (2.9, 21)

1. Front timing chain case
2. Drive belt auto-tensioner
3. Idler pulley
4. Idler pulley

71075_FX35_G0205

Fig. 114 View of drive belt auto tensioner and idler pulley components—3.5L engine

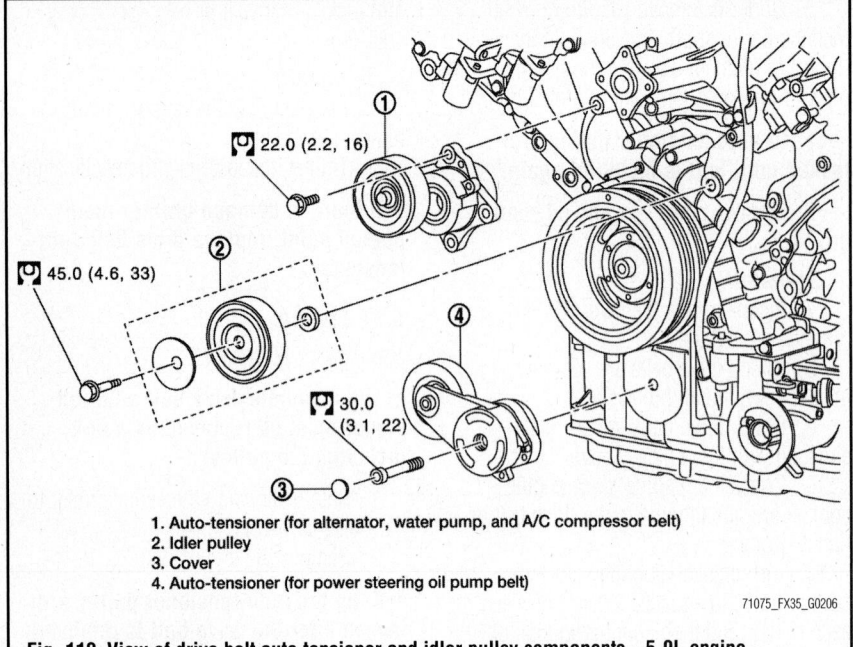

1. Auto-tensioner (for alternator, water pump, and A/C compressor belt)
2. Idler pulley
3. Cover
4. Auto-tensioner (for power steering oil pump belt)

71075_FX35_G0206

Fig. 112 View of drive belt auto tensioner and idler pulley components—5.0L engine

To install:

5. Installation is the reverse of the removal procedure.

6. Tighten the fasteners to specification.

> ⁕⁕ **WARNING**
>
> **Never swap the pulley between the new and the old drive belt auto tensioner.**

AIR INTAKE SYSTEM

REMOVAL & INSTALLATION

Air Cleaner Assembly

3.5L Engine

See Figure 113.

1. Before servicing the vehicle, refer to the Precautions Section.

2. Disconnect the mass air flow sensor harness connector.

3. Disconnect the PCV hose.

4. Remove the air cleaner case with the mass air flow sensor and air duct, disconnecting each joint.

➠**Add matching marks, if necessary for easier installation.**

5. Remove the mass air flow sensor from the air cleaner case, if necessary.

> ⁕⁕ **WARNING**
>
> **Do not impact, disassemble, or touch the mass air flow sensor.**

To install:

6. Installation is the reverse of the removal procedure.

7. Tighten the fasteners to specification.

8. Align marks. Attach each joint. Screw clamps firmly.

• Tighten the clamps to 40 inch lbs. (5 Nm)

9. Inspect the air duct and resonator assembly for cracks or tears. If damage is found, replace the air duct and resonator assembly.

5.0L Engine

See Figure 114.

1. Before servicing the vehicle, refer to the Precautions Section.

2. Remove engine cover and engine room cover (RH and LH).

3. Remove air duct (inlet).

4. Disconnect mass air flow sensor harness connector.

5. Disconnect PCV hose.

6. Remove air cleaner case & mass air flow sensor assembly and air duct by disconnecting their joints.

➠**Add matching marks, if necessary for easier installation.**

7. Remove mass air flow sensor from air cleaner case, if necessary.

> ⁕⁕ **WARNING**
>
> **Do not impact, disassemble, or touch the mass air flow sensor.**

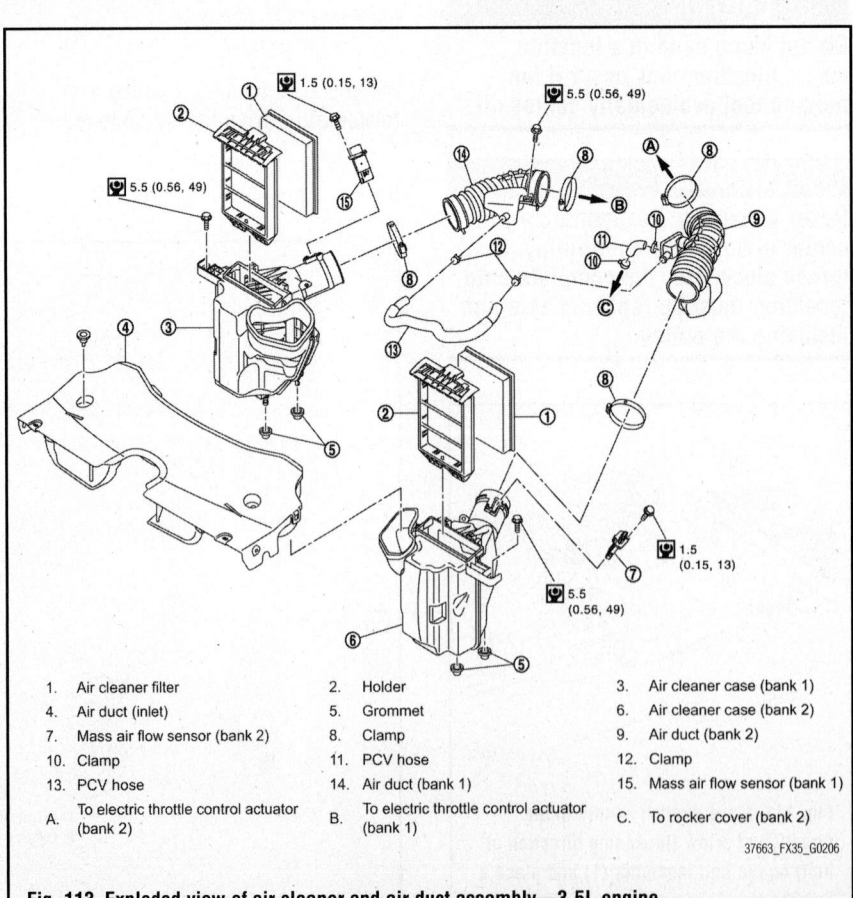

1. Air cleaner filter
2. Holder
3. Air cleaner case (bank 1)
4. Air duct (inlet)
5. Grommet
6. Air cleaner case (bank 2)
7. Mass air flow sensor (bank 2)
8. Clamp
9. Air duct (bank 2)
10. Clamp
11. PCV hose
12. Clamp
13. PCV hose
14. Air duct (bank 1)
15. Mass air flow sensor (bank 1)
A. To electric throttle control actuator (bank 2)
B. To electric throttle control actuator (bank 1)
C. To rocker cover (bank 2)

37663_FX35_G0206

Fig. 113 Exploded view of air cleaner and air duct assembly—3.5L engine

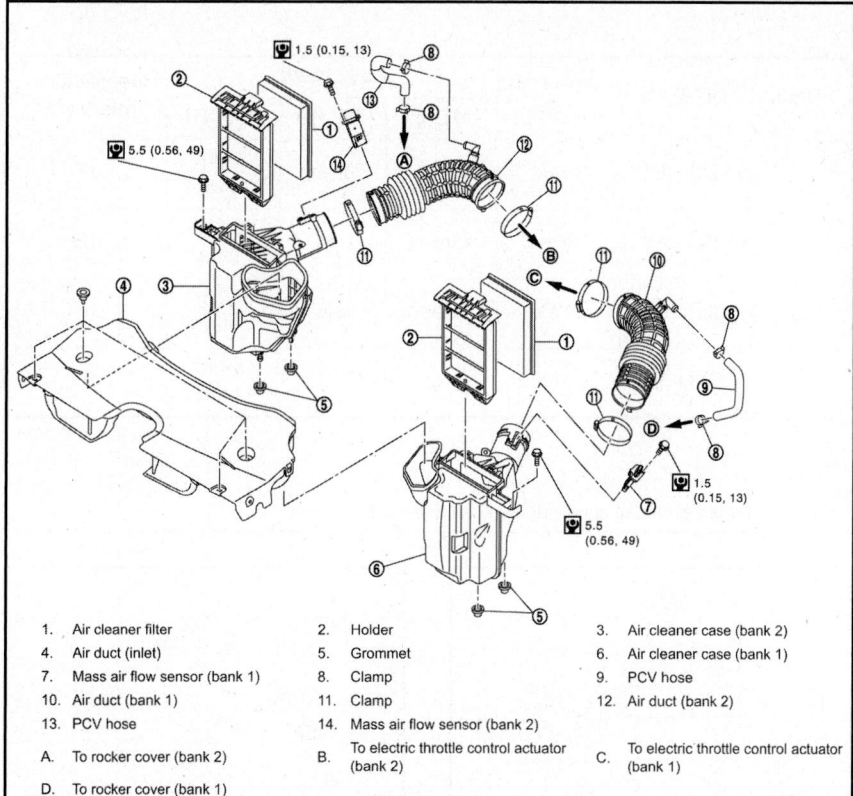

1. Air cleaner filter	2. Holder	3. Air cleaner case (bank 2)
4. Air duct (inlet)	5. Grommet	6. Air cleaner case (bank 1)
7. Mass air flow sensor (bank 1)	8. Clamp	9. PCV hose
10. Air duct (bank 1)	11. Clamp	12. Air duct (bank 2)
13. PCV hose	14. Mass air flow sensor (bank 2)	
A. To rocker cover (bank 2)	B. To electric throttle control actuator (bank 2)	C. To electric throttle control actuator (bank 1)
D. To rocker cover (bank 1)		

37663_FX35_G0209

Fig. 114 Exploded view of air cleaner and air duct assembly—5.0L engine

To install:

8. Installation is the reverse of the removal procedure.

9. Tighten the fasteners to specification.

10. Align marks. Attach each joint. Screw clamps firmly.

- Tighten the clamps to 40 inch lbs. (5 Nm)

11. Inspect the air duct and resonator assembly for cracks or tears. If damage is found, replace the air duct and resonator assembly.

Air Filter Element

3.5L Engine

See Figure 115.

1. Before servicing the vehicle, refer to the Precautions Section.

2. Unhook clips from holder of air cleaner case.

3. Remove holder from air cleaner case, and then remove air cleaner filter from holder.

To install:

4. Installation is the reverse of the removal procedure.

5. Install the air cleaner filter by aligning the seal with the notch of the air cleaner case.

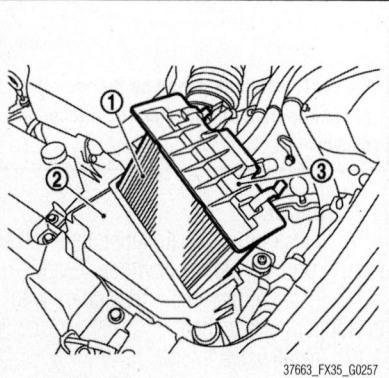

37663_FX35_G0257

Fig. 115 Remove holder (3) from air cleaner case (2), and then remove air cleaner filter (1) from holder

5.0L Engine

See Figure 116.

1. Before servicing the vehicle, refer to the Precautions Section.

2. Unhook clips from holder of air cleaner case.

3. Remove the air cleaner filter from the air cleaner case.

To install:

4. Installation is the reverse of the removal procedure.

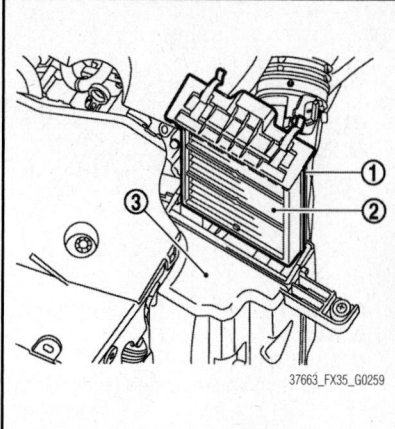

37663_FX35_G0259

Fig. 116 View of air cleaner filter (2), air cleaner case (3), and holder (1)—5.0L engine

5. Install the air cleaner filter by aligning the seal with the notch of the air cleaner case.

CAMSHAFT & LIFTERS

REMOVAL & INSTALLATION

3.5L Engine

See Figures 117 through 126.

1. Before servicing the vehicle, refer to the Precautions Section.

2. Remove front timing chain case, camshaft sprocket and timing chain.

3. Remove fuel sub tube.

4. Remove camshaft sensor bracket. Loosen camshaft sensor bracket

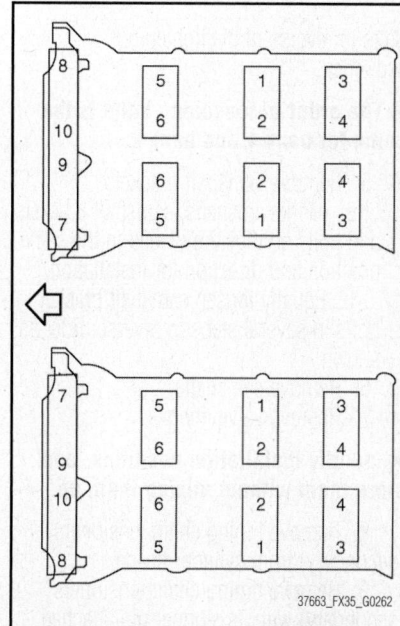

37663_FX35_G0262

Fig. 117 Camshaft bracket bolt tightening sequence. Loosen in the reverse order

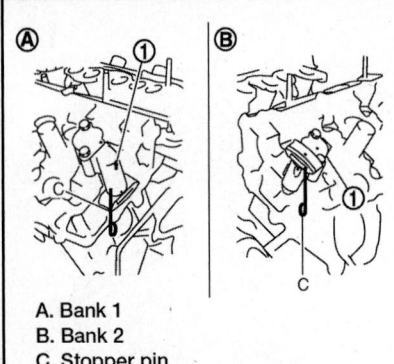

A. Bank 1
B. Bank 2
C. Stopper pin
1. Timing chain tensioners (secondary)

37663_FX35_G0263

Fig. 118 Remove timing chain tensioners (secondary) from cylinder heads

Bank	INT/EXH	Dowel pin (1)	Paint marks			Identification mark (C)
			M1 (E)	M2 (F)	M3 (D)	
1	EXH (B)	Yes	No	Green	Light blue	1F
	INT (A)	Yes	Green	No	Light blue	1E
2	INT (A)	Yes	Green	No	Light blue	1G
	EXH (B)	Yes	No	Green	Light blue	1H

37663_FX35_G0265

Fig. 120 Table describing camshaft identification marks

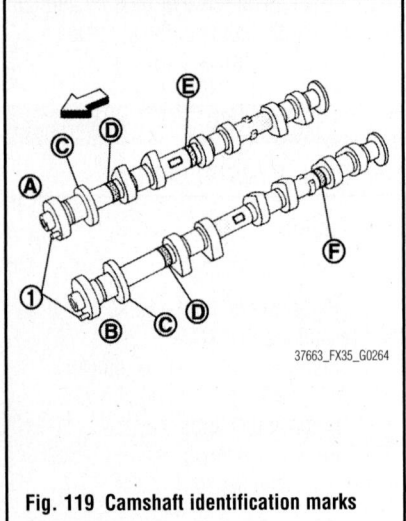

37663_FX35_G0264

Fig. 119 Camshaft identification marks

37663_FX35_G0266

Fig. 121 Install camshaft so that dowel pin (A) on front end face are positioned as shown

A. No. 1
B. No. 2
C. No. 3
D. No. 4
E. Camshaft brackets (bank 1)
F. Exhaust side
G. Intake side
H. Camshaft brackets (bank 2)
I. Intake side
J. Exhaust side

37663_FX35_G0267

Fig. 122 Install camshaft brackets

bolts in reverse of the tightening sequence.

➡**The order of loosening bolts is the same for bank 1 and bank 2.**

5. Remove camshaft brackets.
    a. Mark camshafts, camshaft brackets and bolts so they are placed in the same position and direction for installation.
    b. Equally loosen camshaft bracket bolts in several steps in reverse order as shown.
6. Remove camshafts.
7. Remove valve lifters.

➡**Identify installation positions, and store them without mixing them up.**

8. Remove timing chain tensioners (secondary) from cylinder heads.
9. Remove timing chain tensioners (secondary) with its stopper pin attached.

➡**Stopper pin should be attached when timing chain (secondary) is removed.**

To install:
10. Install timing chain tensioners (secondary) on both sides of cylinder head.
11. Install timing chain tensioners with its stopper pin attached.
12. Install valve lifters.

➡**Install in the original position.**

13. Install camshafts.
    a. Follow identification marks made during removal, or follow the identification marks that are present on new camshafts for proper placement and direction.
    b. Install camshaft so that dowel pin on front end face are positioned as shown (No. 1 cylinder TDC on its compression stroke).

➡**Though camshaft does not stop at the portion as shown, for the placement of cam nose, it is generally accepted camshaft is placed for the same direction of the figure.**

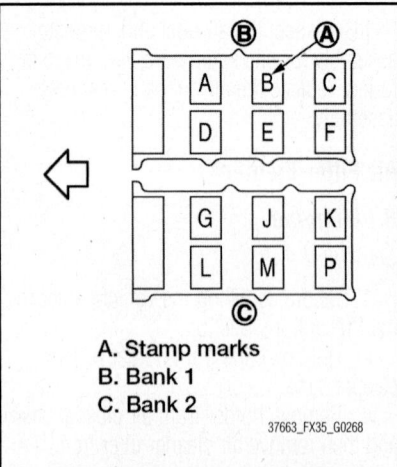

A. Stamp marks
B. Bank 1
C. Bank 2

37663_FX35_G0268

Fig. 123 Install camshaft brackets (No. 2 to 4) aligning the stamp marks as shown

14. Install camshaft brackets.
    a. Remove foreign material completely from camshaft bracket backside and from cylinder head installation face.

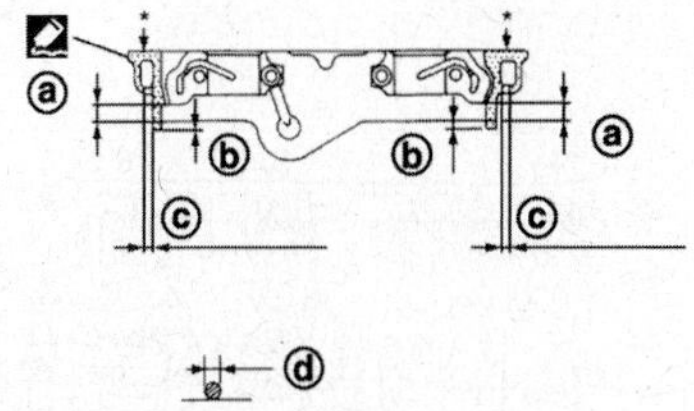

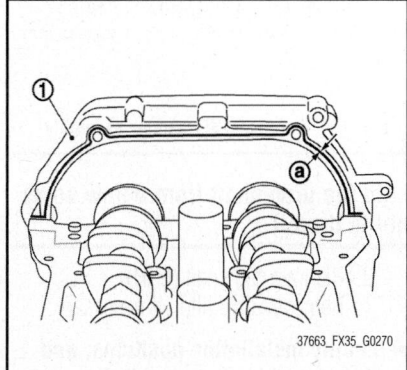

a. 0.335 inches (8.5 mm)
b. 0.08 inches (2 mm)
c. Clearance 0.20 inches (5 mm)
d. 0.098 inches (2.5 mm)
* Apply liquid gasket to rear timing chain side

37663_FX35_G0269

Fig. 124 Apply liquid gasket to mating surface of camshaft bracket (No. 1) as shown

37663_FX35_G0270

Fig. 125 Apply liquid gasket to camshaft bracket (No. 1) contact surface on the rear timing chain case backside as shown

b. Install camshaft bracket in original position and direction as shown.
c. Install camshaft brackets (No. 2 to 4) aligning the stamp marks as shown.

➡**There are no identification marks indicating bank 1 and bank 2 for camshaft bracket (No. 1).**

d. Apply liquid gasket to mating surface of camshaft bracket (No. 1) as shown on both bank 1 and bank 2.
e. Apply liquid gasket to camshaft bracket (No. 1) contact surface on the rear timing chain case backside as shown on both bank 1 and bank 2.

➡**For camshaft bracket (No. 1) near installation position, install it without disturbing the liquid gasket applied to the surfaces.**

15. Tighten camshaft bracket bolts in the following steps, in numerical order as shown.

37663_FX35_G0261

Fig. 126 Camshaft sensor bracket bolt tightening sequence

a. Tighten No. 7–10 to 18 inch lbs. (2 Nm) in numerical order as shown.
b. Tighten No. 1–6 to 18 inch lbs. (2 Nm) in numerical order as shown.
c. Tighten No. 1–10 to 53 inch lbs. (6 Nm) in numerical order as shown.
d. Tighten No. 1–10 to 89 inch lbs. (10 Nm) in numerical order as shown.
16. Install camshaft sensor bracket.
17. Tighten camshaft sensor bracket bolts in numerical order.

➡**The order of tightening bolts is the same for bank 1 and bank 2.**

18. Inspect and adjust the valve clearance.

19. Continue installation in the reverse of the removal procedure.

5.0L Engine
See Figures 127 through 145.

✳✳ WARNING

A high degree of precision is required for a valve on the intake side. Never remove the valve related parts unless necessary.

Note the following:
• As for replacement of parts on the intake side as shown in the exploded view, replace VVEL ladder assembly and cylinder head assembly
• VVEL ladder assembly cannot be replaced as a single part, because it is machined together with cylinder head assembly

➡**The figure shows an example of bank 1.**

✳✳ WARNING

Never loosen adjusting bolts and mounting bolts (black color) of VVEL ladder assembly. If loosened, the stroke of cam lift becomes out of adjustment. In such case, replacement of VVEL ladder assembly and cylinder head assembly is required.

1. Before servicing the vehicle, refer to the Precautions Section.

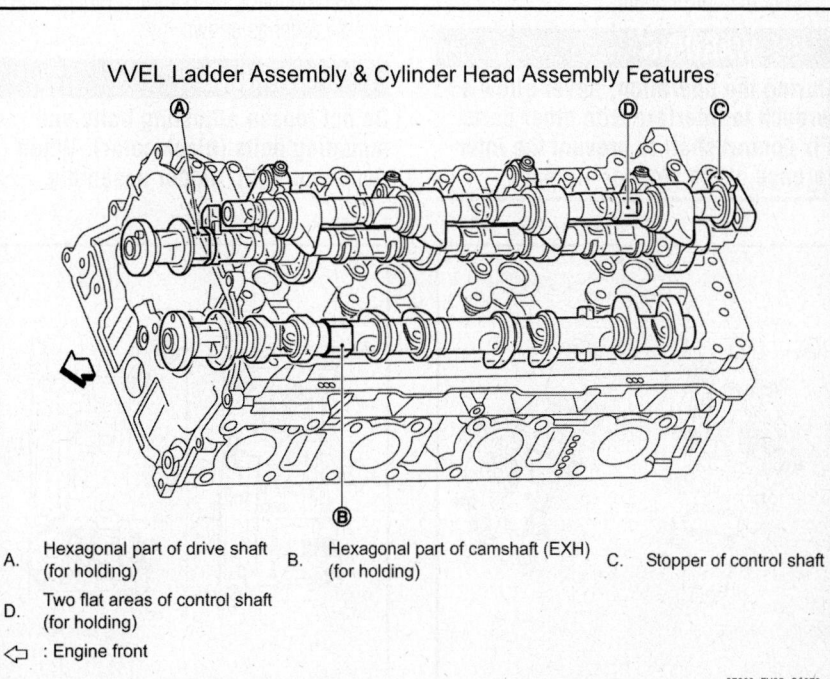

VVEL Ladder Assembly & Cylinder Head Assembly Features

A. Hexagonal part of drive shaft (for holding)
B. Hexagonal part of camshaft (EXH) (for holding)
C. Stopper of control shaft
D. Two flat areas of control shaft (for holding)
⇦ : Engine front

37663_FX35_G0272

Fig. 127 VVEL Ladder Assembly & Cylinder Head Assembly Features

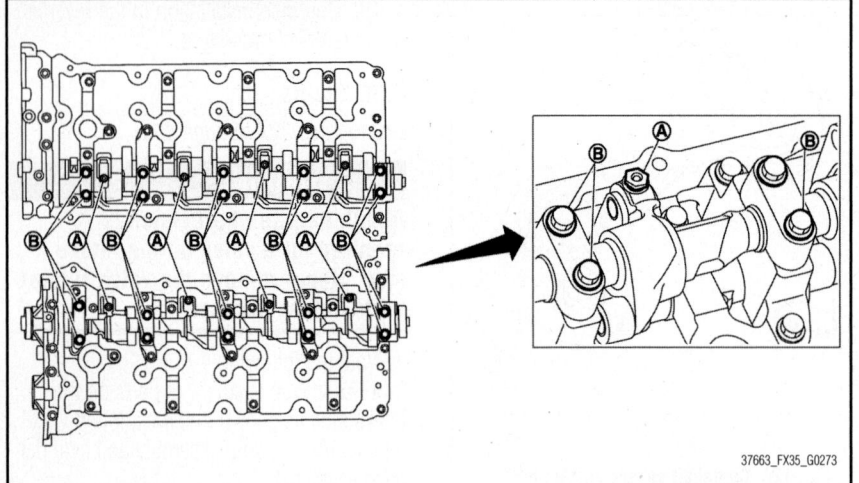

Fig. 128 Never loosen adjusting bolts (A) and mounting bolts (black color) (B) of VVEL ladder assembly

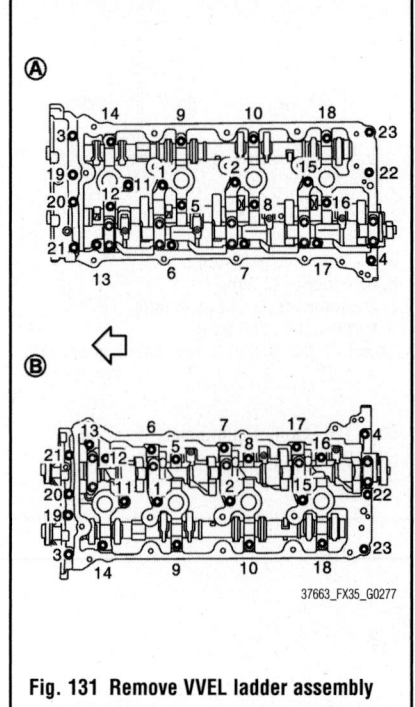

Fig. 131 Remove VVEL ladder assembly

2. Remove valve covers (bank 1 and bank 2).

3. Remove VVEL actuator sub assembly as per the following:

✳✳ WARNING

VVEL actuator sub assembly and VVEL control shaft position sensor are not reusable. Never remove them unless required.

a. Remove VVEL control shaft position sensor.

b. Hold the 2 flat areas of control shaft with a wrench to remove mounting bolts of control shaft.

✳✳ WARNING

During the operation, never allow a wrench to interfere with other parts. Fix control shaft to prevent the interference of the stopper surface.

c. Remove VVEL actuator sub assembly. Loosen mounting bolts in the reverse order as shown.

➡**When removing, prepare for oil spills. When installing, be careful with VVEL actuator sub assembly (bank 1) mounting bolt No. 4 because its length is different.**

d. Remove actuator bracket (rear). Loosen mounting bolts in the reverse order as shown.

4. Remove front cover, camshaft sprockets, and timing chains.

5. Remove VVEL ladder assembly. Loosen mounting bolts (gold color) in the reverse order as shown.

✳✳ WARNING

Do not loosen adjusting bolts and mounting bolts (black color). When removing VVEL ladder assembly,

hold the driveshaft from below so as not to drop it.

6. Remove camshaft (EXH).

7. Remove valve lifter, if necessary.

➡**Identify installation positions, and store them without mixing them up.**

To install:

8. Install valve lifters in their original positions.

9. Install camshaft (EXH).

➡**Distinction between camshaft (EXH) is performed with the identification mark.**

10. Install VVEL ladder assembly as per the following:

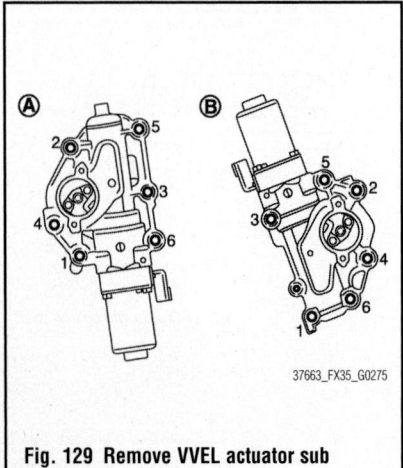

Fig. 129 Remove VVEL actuator sub assembly

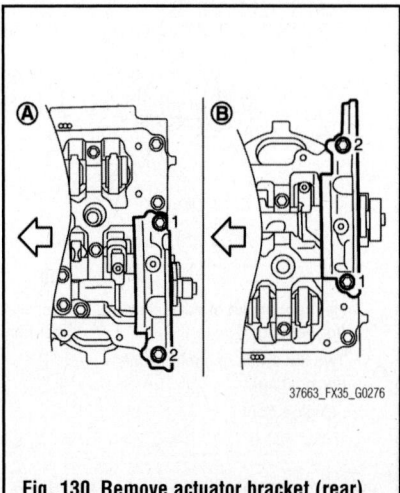

Fig. 130 Remove actuator bracket (rear)

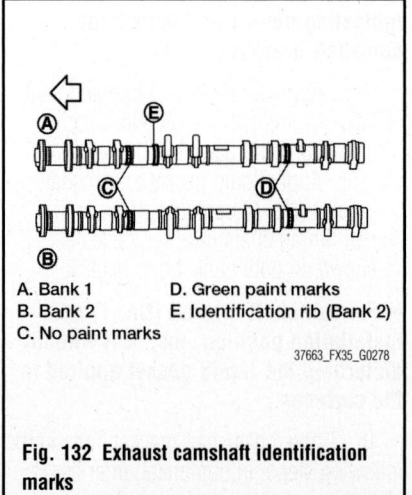

A. Bank 1
B. Bank 2
C. No paint marks
D. Green paint marks
E. Identification rib (Bank 2)

Fig. 132 Exhaust camshaft identification marks

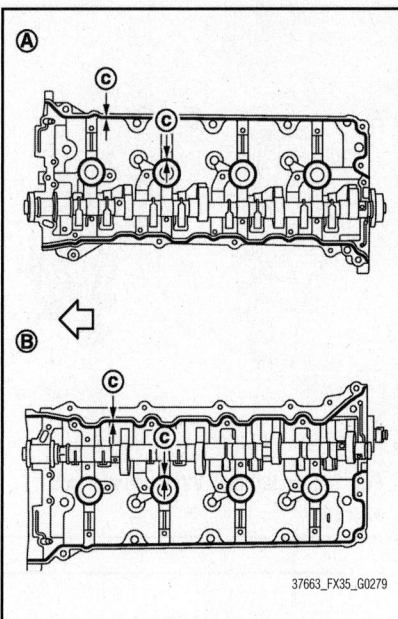

Fig. 133 Apply a continuous bead of liquid gasket (C) to the cylinder heads Bank 1 (A) and Bank 2 (B) as shown

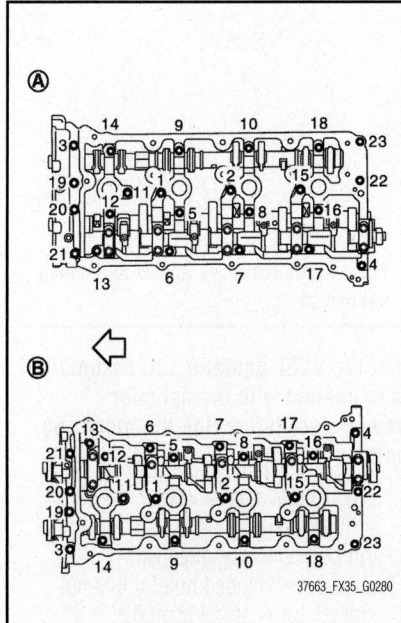

Fig. 134 Tighten mounting bolts in the following steps, in numerical order as shown

a. Apply a continuous bead of liquid gasket with tube presser to the cylinder head as shown.

b. Tighten mounting bolts in the following steps, in numerical order as shown.

➡**Do not reuse washers.**

- Step 1: Tighten bolts to 18 inch lbs. (2 Nm)

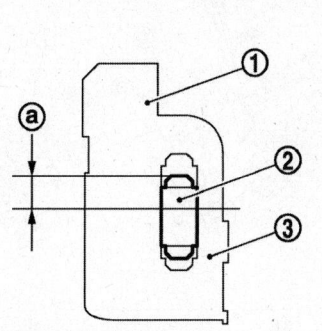

1. Actuator bracket
2. Dowel pins
3. VVEL ladder assembly
a. Bead of liquid gasket
   0.157-0.236 inches (4.0-6.0 mm)

Fig. 135 Refer to the figure to replace new dowel pins, if removed

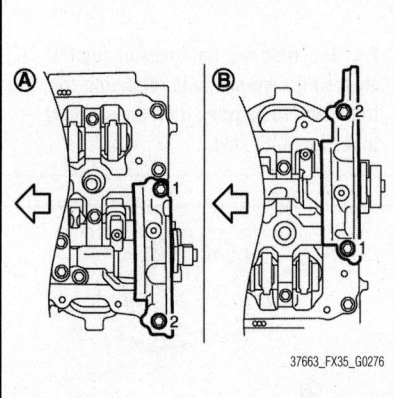

Fig. 136 Tighten mounting bolts in the following steps, in numerical order as shown

- Step 2: Tighten bolts to 53 inch lbs. (6 Nm)
- Step 3: Tighten bolts 89 inch lbs. (10 Nm)

11. Install camshaft sprockets and timing chains.

12. Install actuator bracket (rear) as per the following:

a. Refer to the figure to replace new dowel pins, if removed.

b. Apply a continuous bead of liquid gasket with tube presser to the actuator bracket (rear) as shown.

➡**Do not apply gasket to the oil passage.**

c. Tighten mounting bolts in the following steps, in numerical order as shown.

➡**Do not reuse washers.**

- Step 1: Tighten bolts to 1 ft. lbs. (2 Nm)
- Step 2: Tighten bolts to 4 ft. lbs. (6 Nm)
- Step 3: Tighten bolts to 23 ft. lbs. (31 Nm)

13. Install new VVEL actuator sub assembly as per the following:

➡**Regarding replacement, because VVEL actuator sub assembly and VVEL control shaft position sensor are controlled on a one-on-one basis, replace them as a set.**

➡**VVEL actuator arm is factory-fixed at 10° from the small lift with a holding jig. The holding jig is supplied in the new VVEL actuator sub assembly.**

✳✳ WARNING

Do not disassemble VVEL actuator sub assembly. Never loosen actuator motor mounting bolts shown. Never impact VVEL actuator sub assembly.

a. Move control shaft to the position of small lift stopper.

b. The position where a part of the stopper of control shaft contacts VVEL ladder bracket.

✳✳ WARNING

Be careful not to damage the stopper surface.

c. If control shaft cannot be moved, set crankshaft in position referring to the information below (to displace cam nose).

- Bank 1: Turn 360° from No. 1 cylinder at TDC

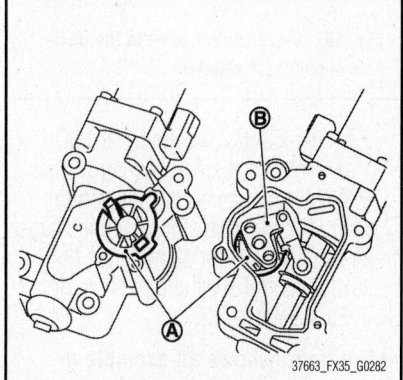

Fig. 137 VVEL actuator arm (B) is factory-fixed at 10° from the small lift with a holding jig (A)

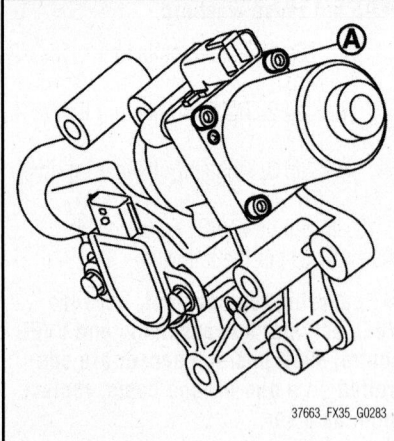

Fig. 138 Never loosen actuator motor mounting bolts (A) shown

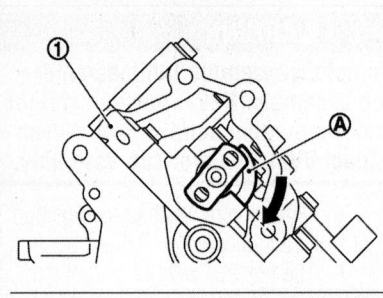

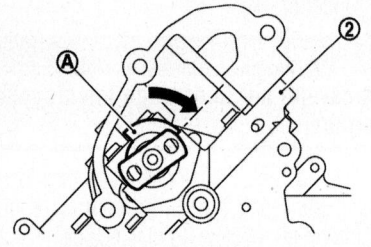

1. VVEL ladder assembly (bank 2)
2. VVEL ladder assembly (bank 1)
A. Stopper of control shaft
Arrow: Small lift side

Fig. 139 Move control shaft to the position of small lift stopper

• Bank 2: No. 1 cylinder at TDC
  d. Hold 2 flat areas of control shaft with a wrench, and rotate the control shaft (10° from the stopper) to the large lift side. (This is for aligning the bolt hole of control shaft and the hole of VVEL actuator arm).

➡The figure shows an example of bank 2.

  e. Apply a continuous bead of liquid gasket with tube presser to the VVEL actuator sub assembly as shown.

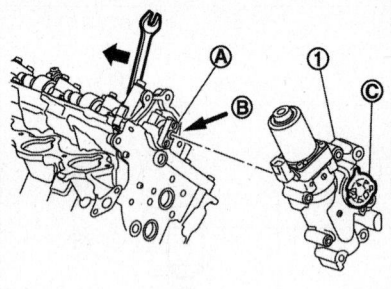

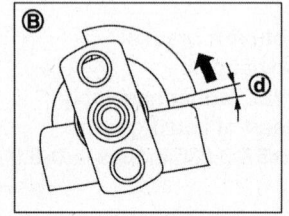

1. VVEL actuator sub assembly (bank 2)
A. Control shaft
B. View B
C. Holding jig
d. 10 degrees
Arrow: Large lift side

Fig. 140 Hold two flat areas of control shaft with a wrench, and rotate the control shaft (10 degrees from the stopper) to the large lift side

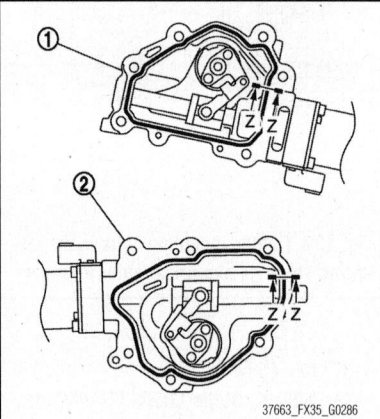

Fig. 141 Apply a continuous bead of liquid gasket to the VVEL actuator sub assembly as shown

➡Do not apply gasket to the oil passage.

  f. Install new VVEL actuator sub assembly. Tighten mounting bolts in the following step, in numerical order as shown.

➡When installing, be careful with VVEL actuator sub assembly (bank 1) mounting bolt No. 4 because its length is different. Be sure to check

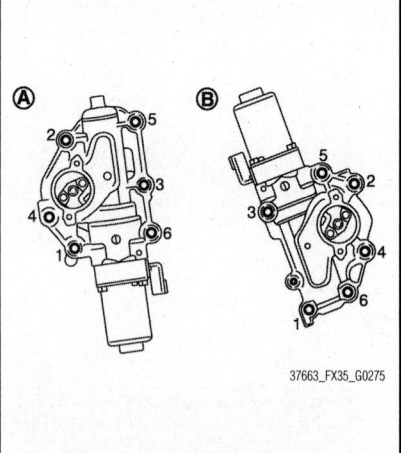

Fig. 142 Install new VVEL actuator sub assembly

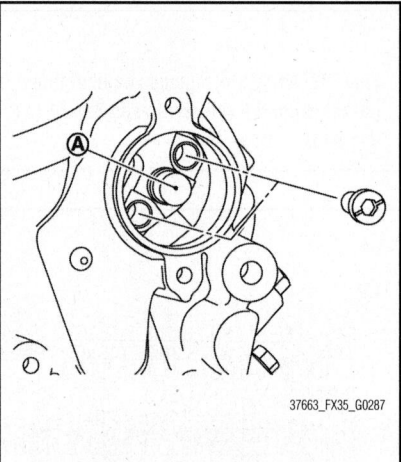

Fig. 143 Do not cause impact to the magnet part (A)

that the VVEL actuator sub assembly is in contact with the cylinder head before tightening the mounting bolts.

  g. Remove holding jig.
  h. Check that VVEL actuator arm bolt hole is aligned with control shaft tapped hole. If it is not aligned, turn control shaft for alignment.

❈❈ WARNING

Do not cause impact to the magnet part.

  i. Hold 2 flat areas of control shaft with a wrench to tighten mounting bolts of control shaft.

❈❈ WARNING

During the operation, never allow a wrench to interfere with other parts. Hold control shaft to prevent

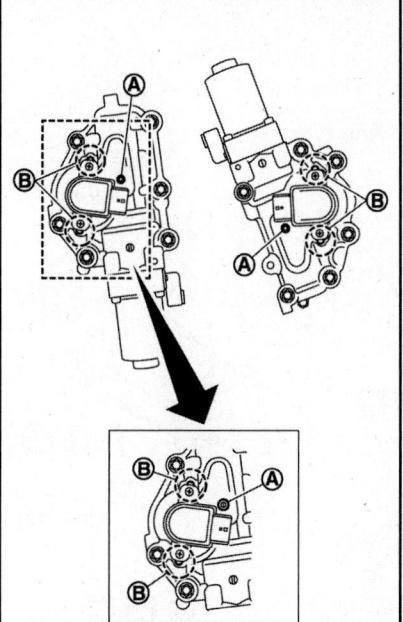

Fig. 144 Align matching marks (B) of VVEL control shaft position sensor and upper housing. Face connector toward matching mark (A)

the interference of the stopper surface.

14. Install new VVEL control shaft position sensor as per the following:

➡**Regarding replacement, because VVEL actuator sub assembly and VVEL control shaft position sensor are controlled on a one-on-one basis, replace them as a set.**

a. Apply engine oil to O-ring or contact surface of O-ring.
b. Align matching marks of VVEL control shaft position sensor and upper housing. Face connector toward matching mark.
c. Temporarily tighten bolt.
d. Adjust VVEL control shaft position sensor after setting the engine assembly in the vehicle.

➡**Be sure to adjust VVEL control shaft position sensor.**

e. After adjusting VVEL control shaft position sensor, tighten bolts to the specified torque.
15. Install actuator cover.
16. Inspect the valve clearance.
17. Installation continues in the reverse of the removal procedure.

CRANKSHAFT DAMPER (BALANCER)

REMOVAL & INSTALLATION

3.5L Engine
See Figures 146 through 149.

1. Before servicing the vehicle, refer to the Precautions Section.
2. Remove the accessory drive belt.
3. Remove front cross bar.
4. Remove power steering pipe mounting bolt.
5. Remove rear cover plate and set the ring gear stopper KV10118600 (J-48641) as shown.
6. Loosen crankshaft pulley bolt and rotate bolt seating surface at 0.39 inches (10mm) from its original position.

➡**Do not remove the crankshaft pulley bolt as it will be used as a supporting point for the suitable puller.**

7. Place suitable puller tab on holes of crankshaft pulley, and pull crankshaft pulley loose.

❋❋ **WARNING**

Never put suitable puller tab on crankshaft pulley periphery, as this will damage internal damper.

To install:
8. Installation is the reverse of the removal procedure.
9. Fix crankshaft using the ring gear stopper KV10118600 (J-48641).
10. Install crankshaft pulley, taking care not to damage front oil seal.

❋❋ **WARNING**

When press-fitting crankshaft pulley with plastic hammer, tap on its center portion (not circumference).

11. Tighten crankshaft pulley bolt.
a. Step 1: Tighten to 33 ft. lbs. (44 Nm).
b. Step 2: Tighten the bolt 90°.
12. Rotate crankshaft pulley in normal direction (clockwise when viewed from front) to confirm it turns smoothly.

5.0L Engine
See Figures 146 and 150.

1. Before servicing the vehicle, refer to the Precautions Section.
2. Remove the engine undercover.
3. Remove the accessory drive belt(s).
4. Remove the cooling fan assembly.
5. Remove the front cross bar.

6. Remove rear plate cover.
7. Set the ring gear stopper KV10119200 (J-49277) as shown.
8. Loosen crankshaft pulley bolt, and then pull crankshaft pulley with both hands to remove it.

➡**Do not remove the crankshaft pulley bolt. Keep the loosened crankshaft pulley bolt in place to protect the removed crankshaft pulley from dropping.**

To install:
9. Installation is the reverse of the removal procedure.
10. Fix the crankshaft with the ring gear stopper KV10119200 (J-49277).
11. Install crankshaft pulley, taking care not to damage front oil seal.
12. Apply engine oil onto threaded parts of crankshaft pulley bolt and seating area.
13. Lightly tapping its center with a plastic hammer, insert crankshaft pulley.

❋❋ **WARNING**

Never tap crankshaft pulley on the side surface where belt is installed (outer circumference).

14. Tighten crankshaft pulley bolt.
a. Step 1: Tighten the pulley bolt to 116 ft. lbs. (157 Nm).
b. Step 2: Tighten an additional 90°.
15. Rotate crankshaft pulley in normal direction (clockwise when viewed from engine front) to confirm it turns smoothly.

CRANKSHAFT FRONT SEAL

REMOVAL & INSTALLATION

3.5L Engine
See Figure 151.

1. Before servicing the vehicle, refer to the Precautions Section.
2. Remove the engine undercover.
3. Remove the drive belt. Refer to Accessory Drive Belt System, Drive Belt, removal & installation.
4. Remove the crankshaft pulley. Refer to Crankshaft Damper (Balancer), removal & installation.
5. Remove front oil seal using a suitable tool.

❋❋ **WARNING**

Be careful not to damage front timing chain case and crankshaft.

To install:
6. Apply new engine oil to both oil seal lip and dust seal lip of new front oil seal.
7. Install front oil seal.

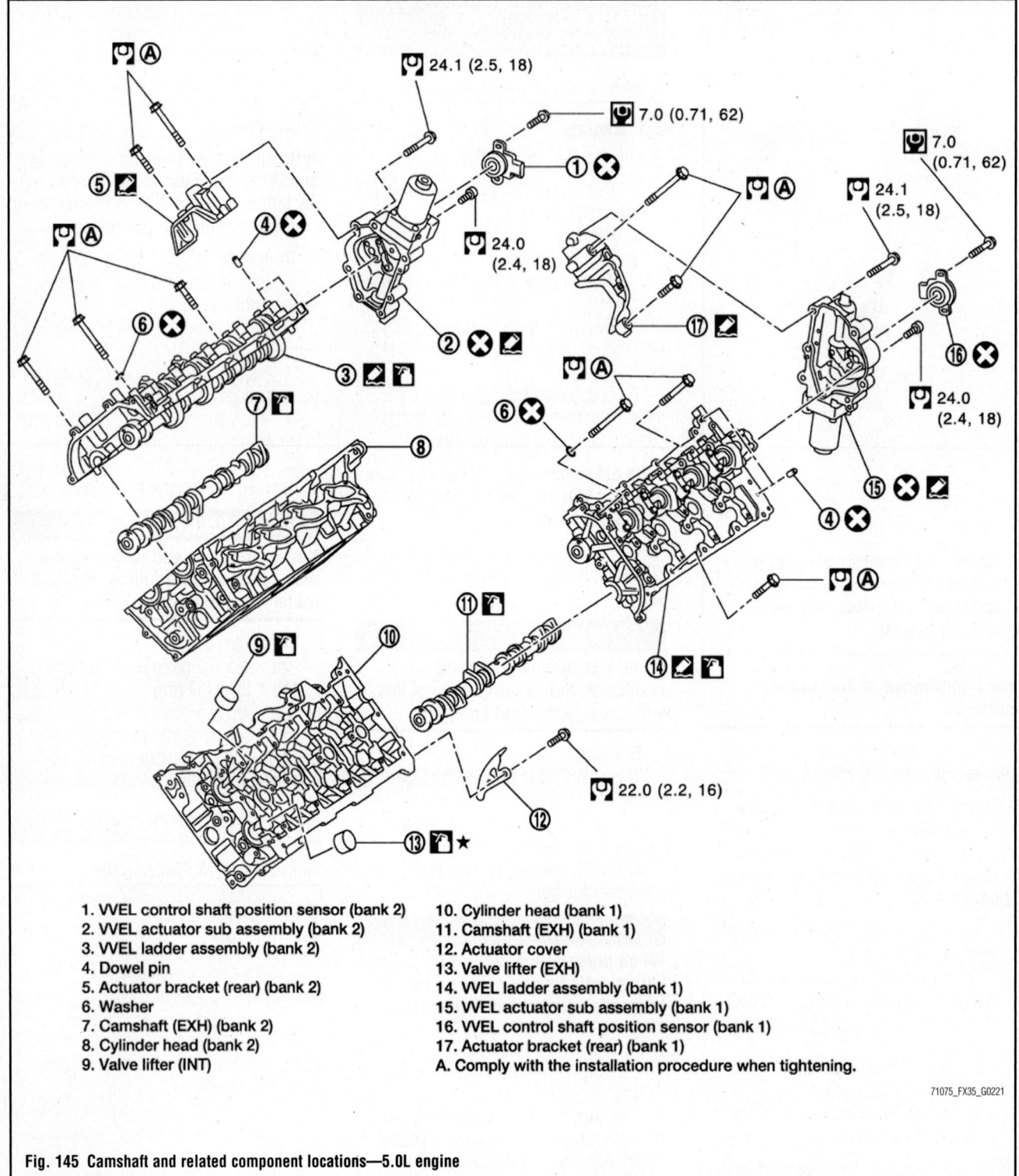

1. VVEL control shaft position sensor (bank 2)
2. VVEL actuator sub assembly (bank 2)
3. VVEL ladder assembly (bank 2)
4. Dowel pin
5. Actuator bracket (rear) (bank 2)
6. Washer
7. Camshaft (EXH) (bank 2)
8. Cylinder head (bank 2)
9. Valve lifter (INT)
10. Cylinder head (bank 1)
11. Camshaft (EXH) (bank 1)
12. Actuator cover
13. Valve lifter (EXH)
14. VVEL ladder assembly (bank 1)
15. VVEL actuator sub assembly (bank 1)
16. VVEL control shaft position sensor (bank 1)
17. Actuator bracket (rear) (bank 1)
A. Comply with the installation procedure when tightening.

71075_FX35_G0221

Fig. 145 Camshaft and related component locations—5.0L engine

a. Install front oil seal so that each seal lip is oriented as shown.

b. Using a suitable drift, press-fit until the height of front oil seal is level with the mounting surface.

- Suitable drift: outer diameter 2.36 inches (60mm), inner diameter 2 inches (50mm)

c. Check the garter spring is in position and seal lips are not inverted.

❉❉ WARNING

Be careful not to damage front timing chain case and crankshaft. Press-fit straight and avoid causing burrs or tilting the oil seal.

8. Installation continues in the reverse of the removal procedure.

5.0L Engine

See Figures 146 and 151.

1. Before servicing the vehicle, refer to the Precautions Section.

2. Remove the engine undercover.

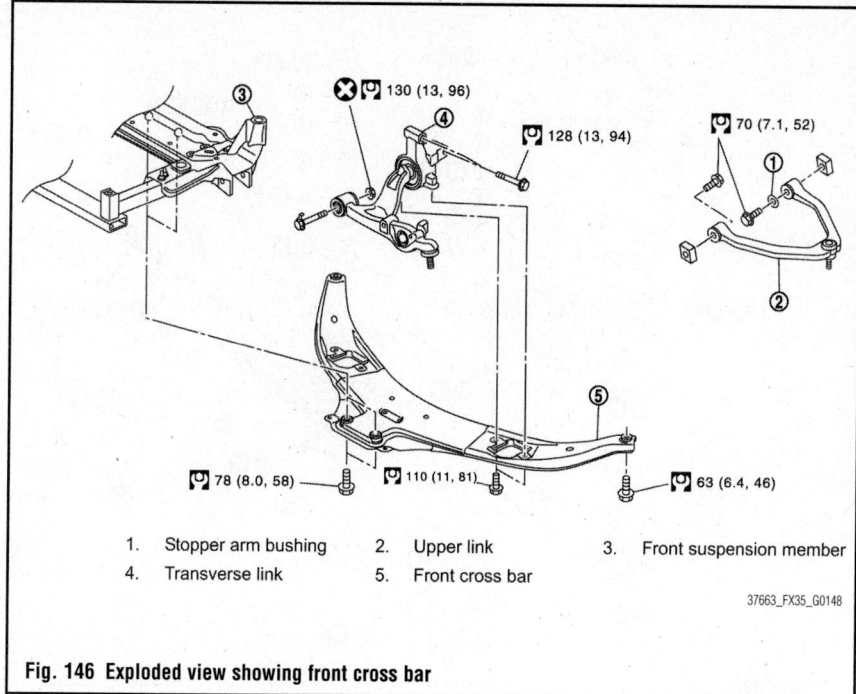

1. Stopper arm bushing
2. Upper link
3. Front suspension member
4. Transverse link
5. Front cross bar

37663_FX35_G0148

Fig. 146 Exploded view showing front cross bar

37663_FX35_G0292

Fig. 147 Remove rear cover plate and set the ring gear stopper (A) as shown

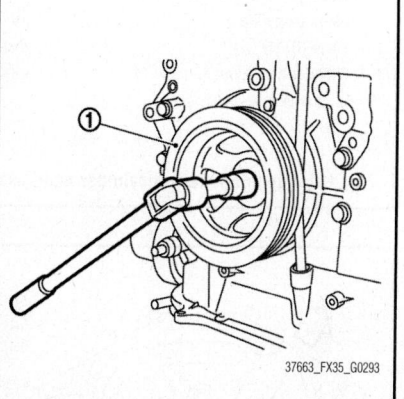

37663_FX35_G0293

Fig. 148 Loosen crankshaft pulley bolt and rotate bolt seating surface at 0.39 inches (10mm) from its original position

37663_FX35_G0294

Fig. 149 Place suitable puller tab on holes of crankshaft pulley, and pull crankshaft pulley loose

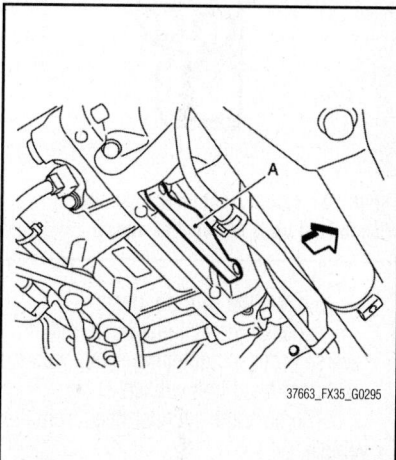

37663_FX35_G0295

Fig. 150 Set the ring gear stopper (A) as shown

CYLINDER HEAD

REMOVAL & INSTALLATION

3.5L Engine

See Figures 152 through 156.

3. Remove the drive belts. Refer to Accessory Drive Belt System, Drive Belt, removal & installation.

4. Remove the cooling fan assembly.

5. Remove the front cross bar.

6. Remove the crankshaft pulley. Refer to Crankshaft Damper (Balancer), removal & installation.

7. Remove front oil seal using a suitable tool.

❊❊ WARNING

Be careful not to damage front cover and crankshaft.

To install:

8. Install front oil seal on front cover.

a. Apply new engine oil to both oil seal lip and dust seal lip.

b. Install it so that each seal lip is oriented as shown.

❊❊ WARNING

Be careful not to scratch or make burrs on circumference of oil seal.

c. Using a suitable drift with outer diameter: 2.20 inches (56mm), press-fit oil seal until it becomes flush with front cover end face.

d. Check the garter spring is in position and seal lips are not inverted.

9. Installation continues in the reverse of the removal procedure.

1. Before servicing the vehicle, refer to the Precautions Section.

2. Remove the intake manifold collector.

3. Remove the valve covers, ignition coils, and spark plugs.

4. Remove the fuel tube and fuel injector assembly.

5. Remove the intake manifold.

6. Remove the exhaust manifold.

7. Remove the water inlet and thermostat assembly.

8. Remove the water outlets (front and rear), water pipe and heater pipe.

9. Remove the timing chain.

10. Remove the rear timing chain case.

11. Remove the camshafts.

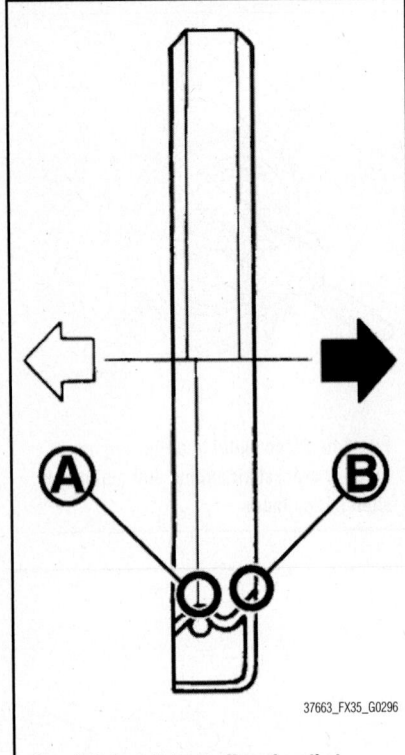

Fig. 151 Install front oil seal so that each seal lip is oriented as shown

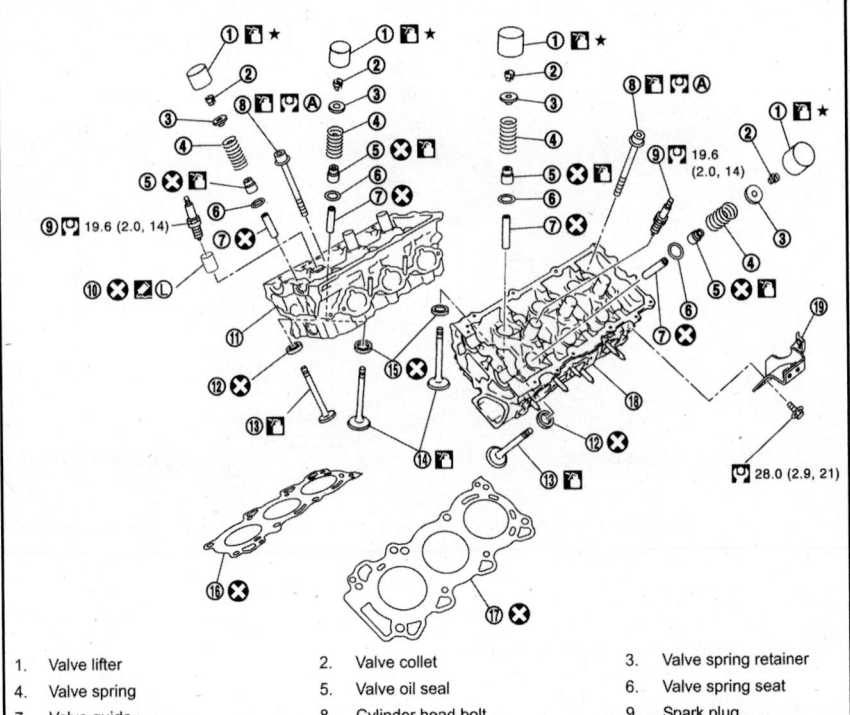

1. Valve lifter
2. Valve collet
3. Valve spring retainer
4. Valve spring
5. Valve oil seal
6. Valve spring seat
7. Valve guide
8. Cylinder head bolt
9. Spark plug
10. Spark plug tube
11. Cylinder head (bank 1)
12. Valve seat (EXH)
13. Valve (EXH)
14. Valve (INT)
15. Valve seat (INT)
16. Cylinder head gasket (bank 1)
17. Cylinder head gasket (bank 2)
18. Cylinder head (bank 2)
19. Engine rear lower slinger

Fig. 152 Exploded view of cylinder head assembly—3.5L engine

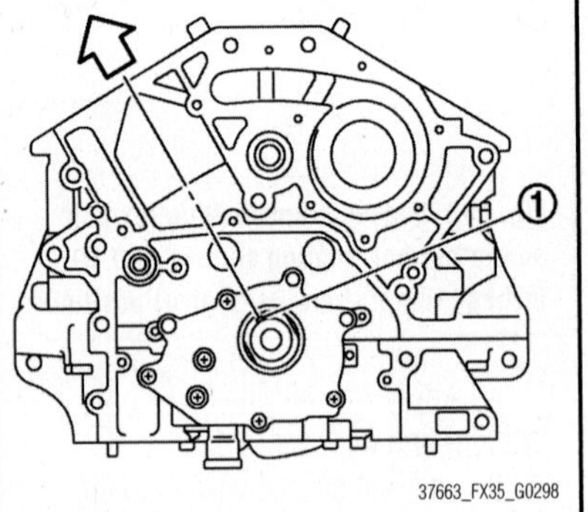

12. Remove the cylinder head.

a. Loosen the cylinder head bolts in reverse order of tightening sequence with a cylinder head bolt wrench.

b. Remove the cylinder head from the vehicle.

c. Remove cylinder head gaskets.

To install:

13. Install new cylinder head gaskets.

14. Turn crankshaft until No. 1 piston is set at Top Dead Center (TDC). The crankshaft key should line up with the bank 1 cylinder center line as shown.

15. Install cylinder head following the steps below and tighten the cylinder head bolts in numerical order as shown with a cylinder head bolt wrench.

➡**If the cylinder head bolts are re-used, check their outer diameters before installation. Before installing the cylinder head, inspect the cylinder head distortion.**

a. Apply new engine oil to threads and seat surfaces of cylinder head bolts.

b. Tighten all cylinder head bolts to 77 ft. lbs. (105 Nm) in numerical order.

c. Completely loosen all cylinder head bolts in reverse of tightening order.

d. Tighten all cylinder head bolts to 30 ft. lbs. (40 Nm) in numerical order.

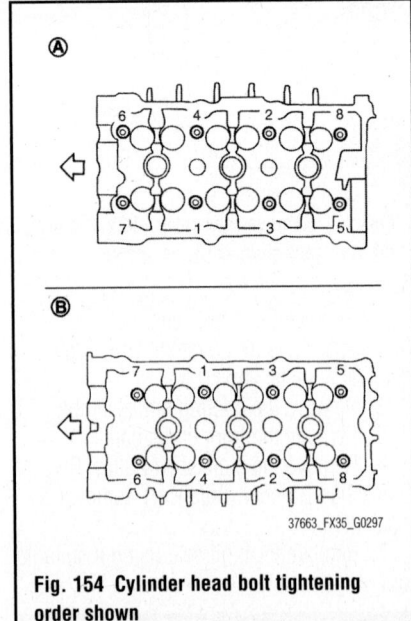

Fig. 153 Turn crankshaft until No. 1 piston is set at TDC

e. Tighten all cylinder head bolts (clockwise) 95° in numerical order.

❊❊ WARNING

Check the tightening angle by using the angle wrench KV10112100 (BT8653-A). Never make judgment by visual inspection. Check tightening

Fig. 154 Cylinder head bolt tightening order shown

angle indicated on the angle wrench indicator plate.

f. Tighten all cylinder head bolts again (clockwise) 95° in numerical order.

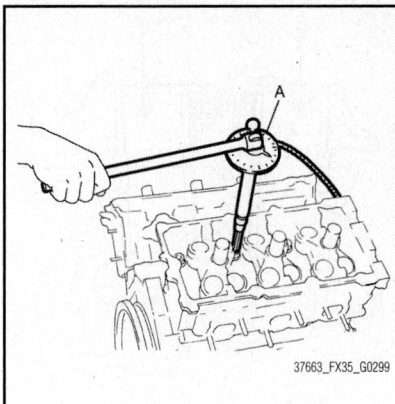

Fig. 155 Check the tightening angle by using the angle wrench KV10112100 (BT8653-A) (A)

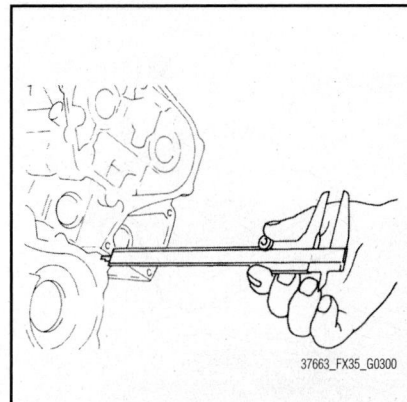

Fig. 156 Measure distance between front end faces of cylinder block and cylinder head (bank 1 and bank 2)

16. After installing cylinder head, measure distance between front end faces of cylinder block and cylinder head (bank 1 and bank 2). If measured value is out of the standard of 0.555–0.587 inch (14.1–14.9mm), reinstall the cylinder head.

17. Install valve lifters in their original positions.

18. Install spark plug with spark plug wrench.

19. Installation continue in the reverse of the removal procedure.

20. Before starting engine, check oil/fluid levels including engine coolant and engine oil. If less than required quantity, fill to the specified level.

21. Use procedure below to check for fuel leakage:

a. Turn ignition switch "ON" (with engine stopped). With fuel pressure applied to fuel piping, check for fuel leakage at connection points.

b. Start engine. With engine speed increased, check again for fuel leakage at connection points.

22. Run engine to check for unusual noise and vibration.

23. Warm up engine thoroughly to check there is no leakage of fuel, exhaust gases, or any oil/fluids including engine oil and engine coolant.

24. Bleed air from lines and hoses of applicable lines, such as in cooling system.

25. After cooling down engine, again check oil/fluid levels including engine oil and engine coolant. Refill to the specified level, if necessary.

5.0L Engine

See Figures 157 through 159.

1. Before servicing the vehicle, refer to the Precautions Section.

2. Remove the valve covers, ignition coils, and spark plugs.

3. Remove the intake manifold.

4. Remove the exhaust manifolds.

5. Remove the water inlet and thermostat housing.

6. Remove the water pipe and heater pipe.

7. Remove the timing chain.

8. Remove the camshaft (EXH) and VVEL ladder assembly.

9. Remove the cylinder heads.

a. Use TORX® socket and power tool.

b. Loosen the mounting bolts in reverse order of tightening sequence.

10. Remove cylinder head gaskets.

To install:

11. Install new cylinder head gaskets.

12. Install cylinder head.

➡If the cylinder head bolts are re-used, check their outer diameters before installation. Before installing the cylinder head, inspect the cylinder head distortion.

13. Tighten the cylinder head bolts in numerical order as shown using TORX® socket.

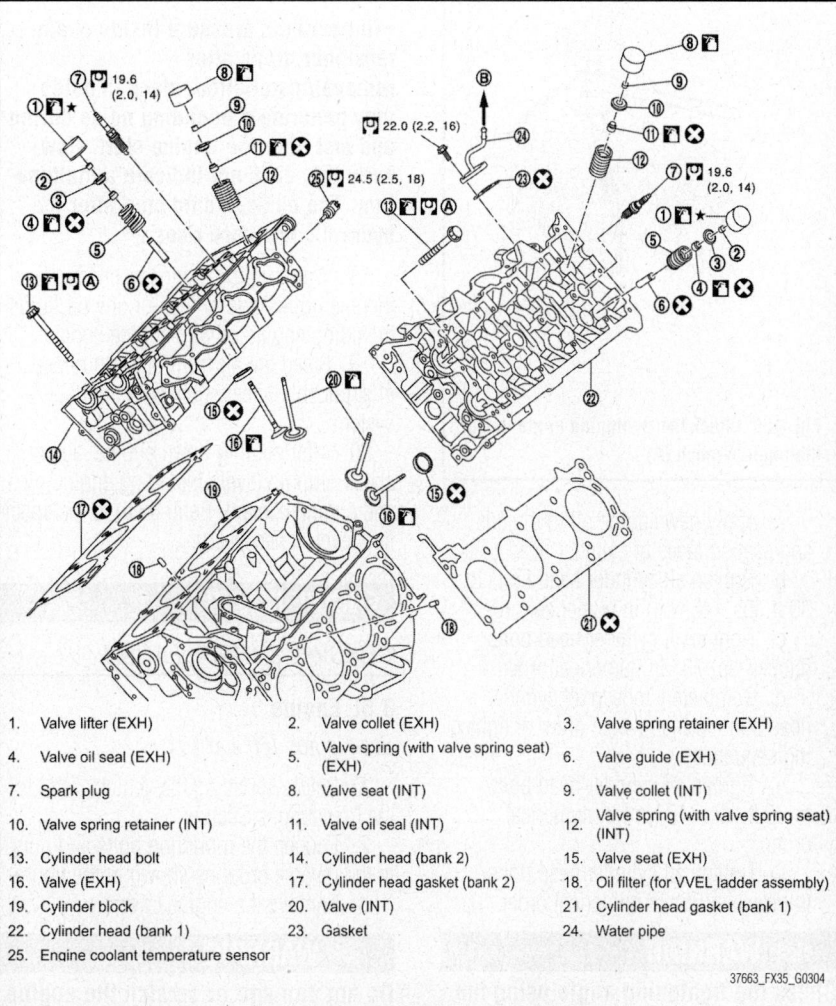

1.	Valve lifter (EXH)	2.	Valve collet (EXH)	3.	Valve spring retainer (EXH)
4.	Valve oil seal (EXH)	5.	Valve spring (with valve spring seat) (EXH)	6.	Valve guide (EXH)
7.	Spark plug	8.	Valve seat (INT)	9.	Valve collet (INT)
10.	Valve spring retainer (INT)	11.	Valve oil seal (INT)	12.	Valve spring (with valve spring seat) (INT)
13.	Cylinder head bolt	14.	Cylinder head (bank 2)	15.	Valve seat (EXH)
16.	Valve (EXH)	17.	Cylinder head gasket (bank 2)	18.	Oil filter (for VVEL ladder assembly)
19.	Cylinder block	20.	Valve (INT)	21.	Cylinder head gasket (bank 1)
22.	Cylinder head (bank 1)	23.	Gasket	24.	Water pipe
25.	Engine coolant temperature sensor				

Fig. 157 Exploded view of cylinder head assembly—5.0L engine

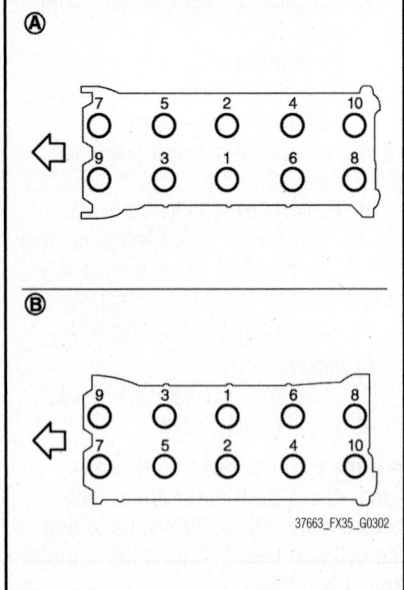

Fig. 158 Tighten cylinder head bolts in numerical order as shown

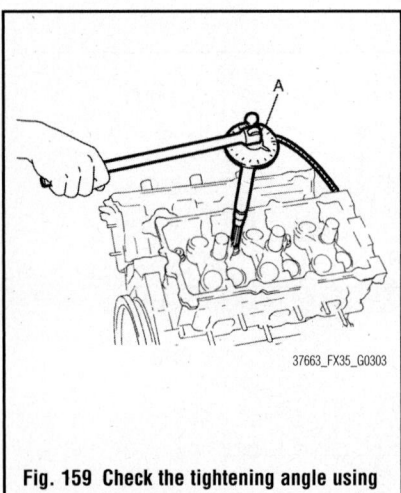

Fig. 159 Check the tightening angle using the angle wrench (A)

a. Apply new engine oil to threads and seat surfaces of cylinder head bolts.

b. Tighten all cylinder head bolts to 30 ft. lbs. (40 Nm) in numerical order.

c. Tighten all cylinder head bolts (clockwise) 75° in numerical order.

d. Completely loosen all cylinder head bolts in the reverse order of tightening sequence.

e. Tighten all cylinder head bolts to 30 ft. lbs. (40 Nm) in numerical order.

f. Tighten all cylinder head bolts (clockwise) 65° in numerical order.

> ❊❊ **WARNING**
>
> **Check the tightening angle using the angle wrench KV10112100 (BT8653-**

A). Never make judgment by visual inspection. Check tightening angle indicated on the angle wrench indicator plate.

g. Tighten all cylinder head bolts again (clockwise) 65° in numerical order.

14. Install the valve lifters in their original positions.

15. Installation continues in the reverse of the removal procedure.

16. Before starting engine, check oil/fluid levels including engine coolant and engine oil. If any are less than the required quantity, fill them to the specified level.

17. Follow the procedure below to check for fuel leakage.

a. Turn ignition switch to the "ON" position (with engine stopped). With fuel pressure applied to fuel piping, check for fuel leakage at connection points.

b. Start engine. With engine speed increased, check again for fuel leakage at connection points.

c. Run engine to check for unusual noise and vibration.

➡ **If hydraulic pressure inside chain tensioner drops after removal/installation, slack in guide may generate a pounding noise during and just after the engine start. However, this does not indicate a malfunction. The noise should stop after the hydraulic pressure rises.**

18. Warm up the engine to check that there is no leakage of fuel, or any oil/fluids including engine oil and engine coolant.

19. Bleed the air from lines and hoses of applicable lines, such as in cooling system.

20. After cooling down engine, again check oil/fluid levels including engine oil and engine coolant. Refill them to the specified level, if necessary.

ENGINE COVER

REMOVAL & INSTALLATION

3.5L Engine

See Figures 160 and 161.

1. Before servicing the vehicle, refer to the Precautions Section.

2. Loosen the mounting bolts and nuts in the reverse order as shown in the figure.

3. Remove the engine cover.

> ❊❊ **WARNING**
>
> **Do not damage or scratch the engine cover when installing or removing.**

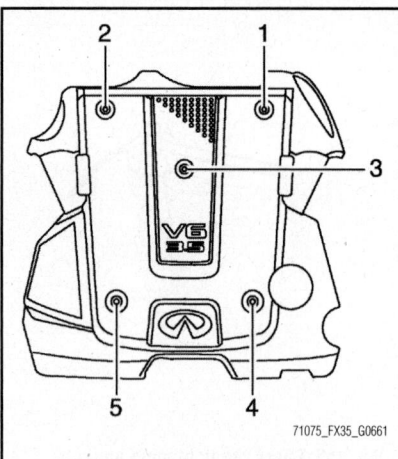

Fig. 160 Engine cover mounting bolt and nut tightening order—3.5L engine

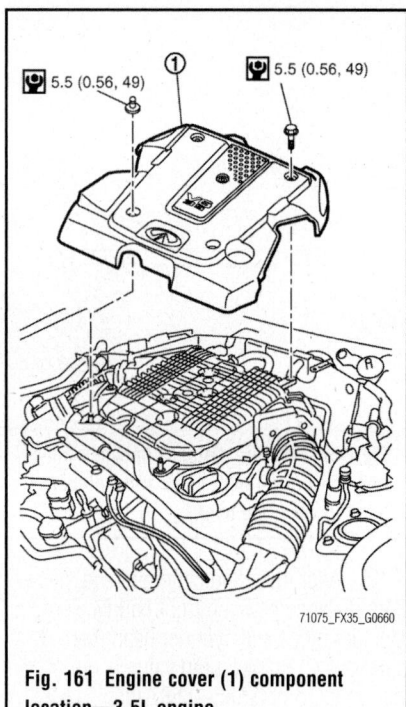

Fig. 161 Engine cover (1) component location—3.5L engine

To install:

4. Install the engine cover.

5. Tighten the mounting bolts and nuts in numerical order as shown in the figure.

5.0L Engine

See Figures 162 and 163.

> ❊❊ **WARNING**
>
> **Do not damage or scratch the engine cover when installing or removing.**

1. Before servicing the vehicle, refer to the Precautions Section.

2. Remove clip, and remove engine room cover (RH and LH).

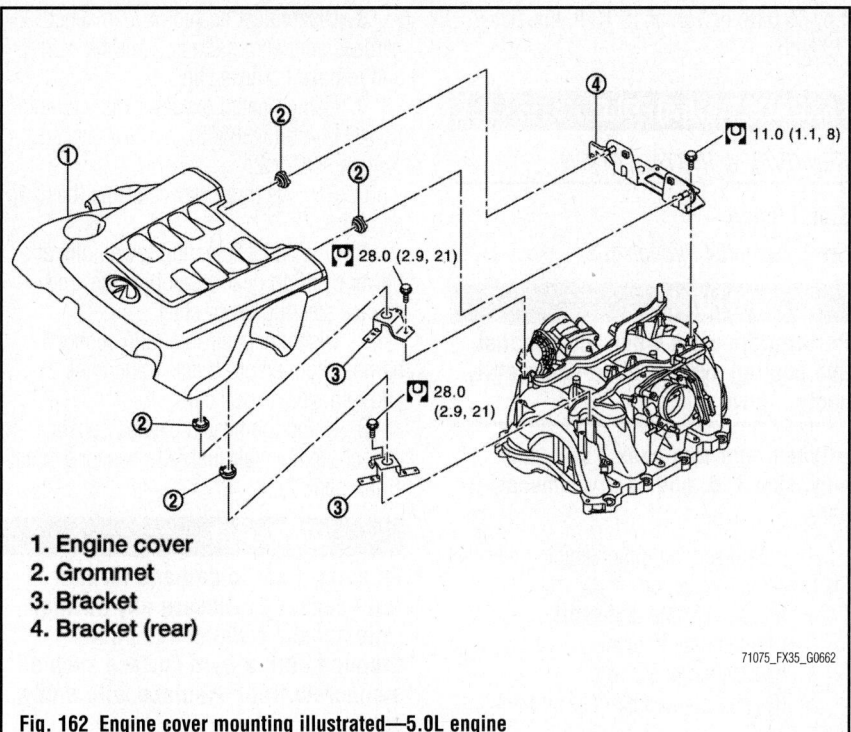

1. Engine cover
2. Grommet
3. Bracket
4. Bracket (rear)

71075_FX35_G0662

Fig. 162 Engine cover mounting illustrated—5.0L engine

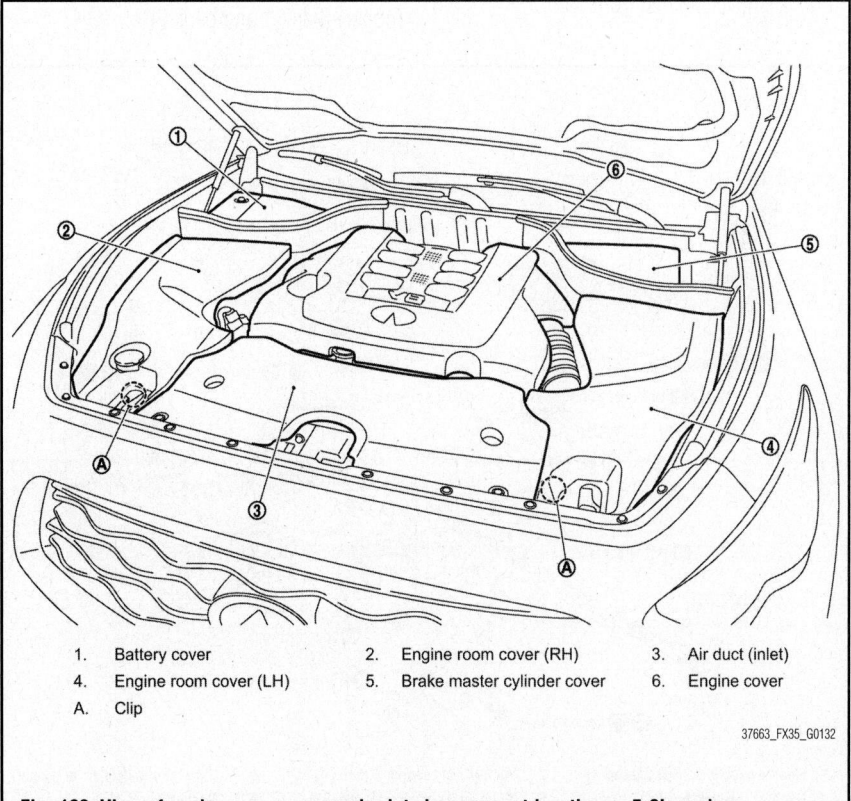

1.	Battery cover	2.	Engine room cover (RH)	3.	Air duct (inlet)
4.	Engine room cover (LH)	5.	Brake master cylinder cover	6.	Engine cover
A.	Clip				

37663_FX35_G0132

Fig. 163 View of engine room cover and related component locations—5.0L engine

3. Remove engine cover as per the following:

    a. Front side: Lift and remove from fitting.

    b. Rear side: Pull forward and remove from fitting.

4. Remove battery cover and brake master cylinder cover, if necessary.

5. Remove air duct (inlet).

6. Installation is the reverse of the removal procedure.

ENGINE OIL & FILTER

OIL LEVEL CHECK

1. Before servicing the vehicle, refer to the Precautions Section.

2. Before starting the engine, check the oil level. If the engine is already started, stop it and allow 10 minutes before checking.

3. Check that the oil level is within the range as indicated on the dipstick.

4. If it is out of range, add oil as necessary until the dipstick indicates the correct level.

5. Check the engine oil for a white milky appearance or excessive contamination.

- If the engine oil becomes milky, it is highly probable that it is contaminated with engine coolant. Repair or replace damaged parts.

OIL & FILTER CHANGE

✲✲ CAUTION

Prolonged and repeated contact with used engine oil may cause skin cancer. Try to avoid direct skin contact with used oil. If skin contact is made, wash thoroughly with soap or hand cleaner as soon as possible. Wear protective clothing, including impervious gloves where practicable. Do not use gasoline, kerosene, diesel fuel, gas oil, thinners, or solvents for cleaning skin. Where there is a risk of eye contact, eye protection should be worn, for example, chemical goggles or face shields; in addition an eye wash facility should be provided.

✲✲ CAUTION

Hot oil can scald.

➡**Use only engine oil with the American Petroleum Institute (API) Certified For Gasoline Engines "Starburst" symbol. It is highly recommended to use SAE 5W-30 oil.**

Draining

1. Before servicing the vehicle, refer to the Precautions Section.

2. Warm up the engine, and check for engine oil leakage from engine components.

3. Stop the engine and wait for 10–15 minutes.

4. Loosen oil filler cap.

5. Remove undercover.

6. Remove drain plug and then drain engine oil.

Refilling

1. Before servicing the vehicle, refer to the Precautions Section.

2. Install drain plug with new washer. Tighten the drain plug to 25 ft. lbs. (34 Nm).

➡ **Be sure to clean drain plug and install with new washer.**

3. Refill with new engine oil.

   a. When filling engine oil, do not pull out oil level gauge.

   b. The refill capacity depends on the engine oil temperature and drain time. Use these specifications for reference only.

   c. Always use oil level gauge to determine the proper amount of engine oil in engine.

   d. For 3.5L engine:
   - With oil filter change: 5.2 quarts (4.9L) oil
   - Without oil filter change: 4.9 quarts (4.6L) oil

   e. For 5.0L engine:
   - With oil filter change: 7.1 quarts (6.7L) oil
   - Without oil filter change: 6.1 quarts (5.8L) oil

4. Warm up the engine and check area around drain plug and oil filter for engine oil leakage.

5. Stop the engine and wait for 10–15 minutes.

6. Check the engine oil level.

Filter Replacement

Note the following:

- Oil filter is provided with relief valve. Use genuine NISSAN oil filter or an equivalent

- Be careful not to get burned when engine and engine oil may be hot

- When removing, prepare a shop cloth to absorb any engine oil leakage or spillage

- Never allow engine oil to adhere to drive belt

- Completely wipe off any engine oil that adheres to engine and vehicle

1. Before servicing the vehicle, refer to the Precautions Section.

2. Remove engine undercover.

3. Using oil filter wrench KV10115801 (J38956), remove oil filter.

To install:

4. Remove foreign matter adhering to oil filter installation surface.

5. Apply engine oil to the oil seal contact surface of new oil filter.

6. Screw oil filter manually until it touches the installation surface, then tighten

it by ⅔ turn, or tighten to 13 ft. lbs. (18 Nm).

EXHAUST MANIFOLD

REMOVAL & INSTALLATION

3.5L Engine

See Figures 164 through 166.

✳ CAUTION

Perform the work when the exhaust and cooling system have completely cooled down.

➡ **When removing bank 1 side parts only, step 3, 6, and 11 are unnecessary.**

1. Before servicing the vehicle, refer to the Precautions Section.

2. Remove engine undercover.

3. Drain engine coolant.

4. Remove engine cover.

5. Remove air cleaner case and air duct.

6. Remove heater pipe and water hose.

7. Remove exhaust front tube.

8. Disconnect heated oxygen sensor 2 harness connectors (bank 1 and bank 2) and remove harness clip.

9. Using heated oxygen sensor wrench KV10114400 (J-38365), removal heated oxygen sensor 2.

10. Remove three way catalysts (bank 1 and bank 2).

11. Disconnect steering lower joint at power steering gear assembly side, and release steering lower shaft.

12. Disconnect air fuel ratio sensor 1 harness connectors (bank 1 and bank 2) and remove harness clip.

13. Using the heated oxygen sensor wrench, remove air fuel ratio sensor 1 (bank 1 and bank 2).

✳ WARNING

Be careful not to damage air fuel ratio sensor 1. Discard any air fuel ratio sensor 1 which has been dropped onto a hard surface such as a concrete floor. Replace with a new sensor.

14. Remove exhaust manifold cover (upper) (bank 1 and bank 2).

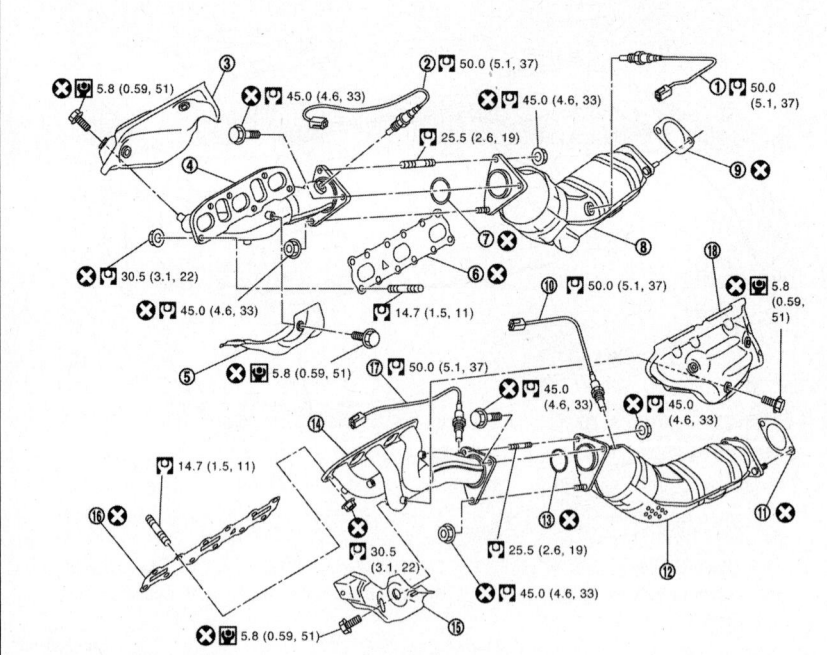

1.	Heated oxygen sensor 2 (bank 1)	2.	Air fuel ratio sensor 1 (bank 1)	3.	Exhaust manifold cover (upper)
4.	Exhaust manifold (bank 1)	5.	Exhaust manifold cover (lower)	6.	Gasket
7.	Ring gasket	8.	Three way catalyst (bank 1)	9.	Gasket
10.	Heated oxygen sensor 2 (bank 2)	11.	Gasket	12.	Three way catalyst (bank 2)
13.	Ring gasket	14.	Exhaust manifold (bank 2)	15.	Exhaust manifold cover (lower)
16.	Gasket	17.	Air fuel ratio sensor 1 (bank 2)	18.	Exhaust manifold cover (upper)

37663_FX35_G0290

Fig. 164 Exploded view of exhaust manifold assembly—3.5L engine

15. Loosen mounting nuts in the reverse order of the tightening sequence.

➡ **Disregard the numerical order No. 7 and 8 in removal.**

16. Remove gaskets.

➡ **Cover engine openings to avoid entry of foreign materials.**

To install:

17. Installation is the reverse of the removal procedure.

18. Install exhaust manifold gasket in direction shown. (Follow same procedure for both banks).

19. If stud bolts were removed, install them and tighten to 11 ft. lbs. (15 Nm).

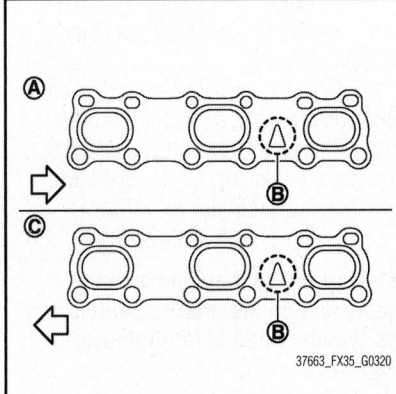

Fig. 165 Install exhaust manifold gasket in direction shown

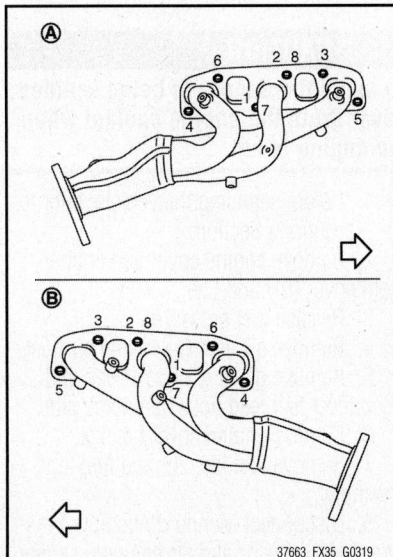

Fig. 166 Install exhaust manifold and tighten mounting nuts in numerical order as shown

20. Install exhaust manifold and tighten mounting nuts in numerical order as shown.

➡ **Tighten nuts No. 1 and 2 in two steps. The numerical order No. 7 and 8 shows second step.**

21. Before installing a new sensor, clean exhaust system threads using heated oxygen sensor thread cleaner tool and apply anti-seize lubricant.

✲✲ WARNING

Do not over-torque sensors. Doing so may cause damage to the sensors, resulting in the "MIL" coming on.

5.0L Engine

See Figures 167 through 169.

1. Before servicing the vehicle, refer to the Precautions Section.

2. Remove heated oxygen sensor 2.

➡ **Heated oxygen sensor 2 is not reusable. Never remove heated oxygen sensor 2 unless this is required.**

3. Using the heated oxygen sensor wrench KV10114400 (J-38365), remove heated oxygen sensor 2.

➡ **The heated oxygen sensor 2 is removable under vehicle mounted condition. The figure shows an example of bank 1.**

4. Remove three way catalyst (bank 1 and bank 2).

5. Remove air fuel ratio sensor 1 as per the following:

➡ **Air fuel ratio sensor 1 is not reusable. Never remove air fuel ratio sensor 1 unless this is required.**

- Using the heated oxygen sensor wrench KV10114400 (J-38365), remove air fuel ratio sensor 1

➡ **The air fuel ration sensor 1 is removable under vehicle-mounted condition.**

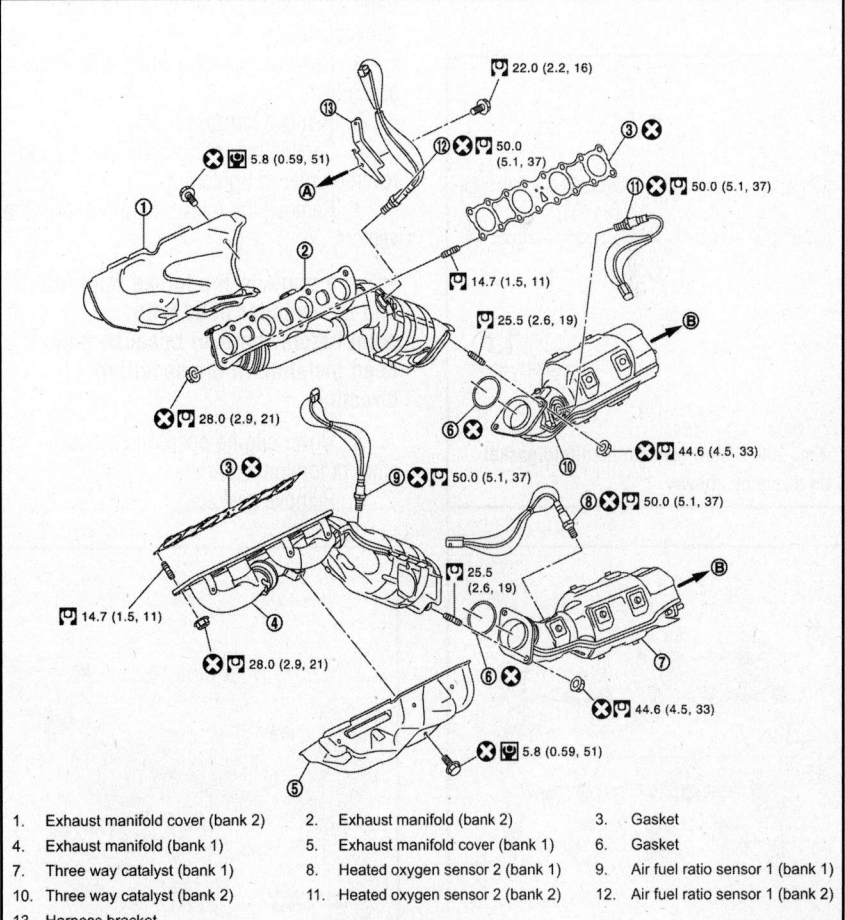

1. Exhaust manifold cover (bank 2)
2. Exhaust manifold (bank 2)
3. Gasket
4. Exhaust manifold (bank 1)
5. Exhaust manifold cover (bank 1)
6. Gasket
7. Three way catalyst (bank 1)
8. Heated oxygen sensor 2 (bank 1)
9. Air fuel ratio sensor 1 (bank 1)
10. Three way catalyst (bank 2)
11. Heated oxygen sensor 2 (bank 2)
12. Air fuel ratio sensor 1 (bank 2)
13. Harness bracket
A. To cylinder head (bank 2)
B. To exhaust front tube

37663_FX35_G0291

Fig. 167 Exploded view of exhaust manifold assembly—5.0L engine

6. Remove exhaust manifold. Loosen nuts in the reverse order of the tightening sequence.

➡Disregard No. 9 to No. 12 when loosening.

7. Remove exhaust manifold gaskets.

➡Cover engine openings to avoid entry of foreign materials.

To install:

8. Installation is the reverse of the removal procedure.

9. Install exhaust manifold gasket in direction shown.

10. Tighten mounting nuts in numerical order as shown.

➡Tighten mounting nuts No. 1 to 8 in two steps. The numerical order No. 9 to No. 12 shown are second steps.

11. Before installing a new sensors, clean exhaust system threads using oxygen sensor thread cleaner (J-43897-18 or J-43897-12), and apply anti-seize lubricant.

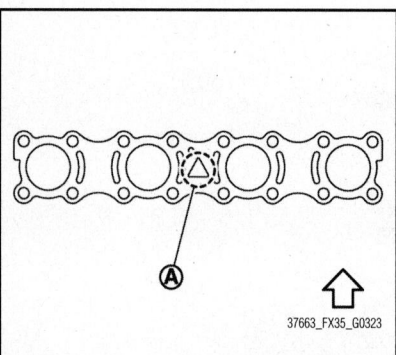

Fig. 168 Install exhaust manifold gasket in direction shown

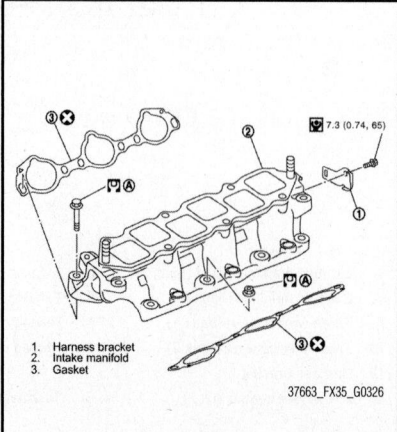

Fig. 169 Exhaust manifold bolt tightening sequence

➡Sensors are not reusable. Replace them with a new one after removal. When replacing them, handle with care not to impact them. When installing the new sensors, set the heated oxygen sensor wrench KV10114400(J-38365) in the hexagonal part to tighten them.

✳✳ WARNING

Never over-torque sensors. Doing so may cause damage to the sensors, resulting in "MIL" coming on.

INTAKE MANIFOLD

REMOVAL & INSTALLATION

3.5L Engine

See Figures 170 and 171.

1. Before servicing the vehicle, refer to the Precautions Section.
2. Release fuel pressure.
3. Remove intake manifold collector. Refer to Intake Manifold Collector, removal & installation.
4. Remove fuel tube and fuel injector assembly.
5. Remove harness bracket.
6. Loosen mounting bolts and nuts in reverse order of tightening.
7. Remove the intake manifold from the vehicle.

➡Put a mark on the intake manifold and the cylinder head with paint before removal because they need installed in the specified direction.

8. Cover engine openings to avoid entry of foreign materials.
9. Remove gaskets.

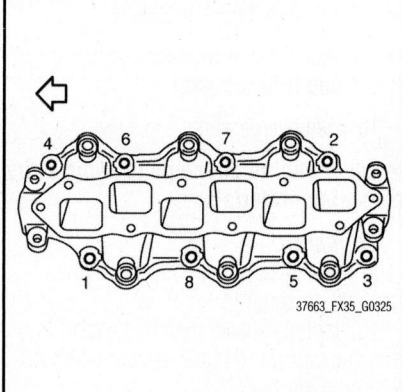

Fig. 171 Tighten all mounting bolts and nuts in two steps in numerical order shown

To install:

10. Installation is the reverse of the removal procedure.

11. If stud bolts were removed, install them and tighten to 97 inch lbs. (11 Nm).

12. Tighten all mounting bolts and nuts in 2 steps in numerical order as shown.
   a. Step 1: 66 inch lbs. (7 Nm).
   b. Step 2: 19 ft. lbs. (26 Nm).

➡Install intake manifold with the marks (put on the intake manifold and the cylinder head before removal) aligned.

13. Install intake manifold collector. Refer to Intake Manifold Collector, removal & installation.

5.0L Engine

See Figures 172 and 173.

✳✳ CAUTION

To avoid the danger of being scalded, never drain the engine coolant when the engine is hot.

1. Before servicing the vehicle, refer to the Precautions Section.
2. Remove engine cover and engine room cover (RH and LH).
3. Release fuel pressure.
4. Remove air duct (inlet) and air duct.
5. Remove quick connector cap and disconnect fuel feed hose on engine side.
6. Remove engine cover bracket.
7. Remove fuel injector and fuel tube assembly.
8. Disconnect Manifold Absolute Pressure (MAP) sensor and air fuel ratio sensor 1 (bank 1) harness connector.
9. Remove vacuum tank, EVAP service port hose and EVAP canister purge control solenoid valve.

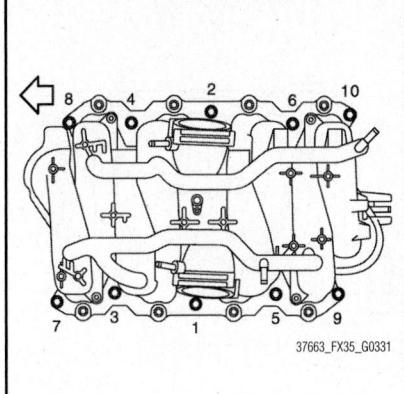

Fig. 172 Intake Manifold bolt tightening sequence

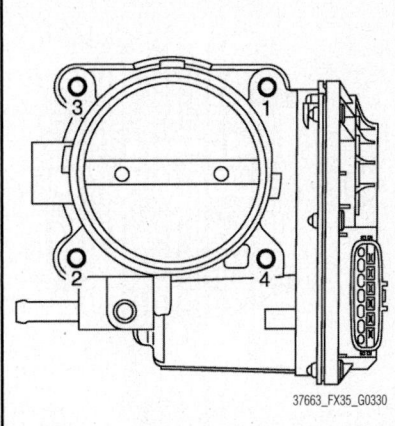

Fig. 173 Electric throttle control actuator tightening sequence

10. Disconnect PCV hoses and vacuum hose from intake manifold. Add matching marks as necessary for easier installation.

11. Drain engine coolant from radiator.

✳✳ CAUTION

Perform this step when the engine is cold. Never spill engine coolant on drive belts.

➡**When removing only intake manifold, move electric throttle control actuator without disconnecting the water hoses.**

12. Remove electric throttle control actuator. Loosen mounting bolts in reverse order of tightening.

➡**The figure shows the electric throttle control actuator (bank 1) viewed from the air duct side. Viewed from the air duct side, the order of loosening mounting bolts of electric throttle control actuator (bank 1) is the same as that of the electric throttle control actuator (bank 2).**

✳✳ WARNING

Handle carefully to avoid any impact to electric throttle control actuator. Never disassemble.

13. Remove intake manifold. Loosen mounting bolts in reverse order of tightening sequence.

14. Remove intake manifold gaskets.

➡**Cover engine openings to avoid entry of foreign materials.**

15. Remove Manifold Absolute Pressure (MAP) sensor, if necessary.

✳✳ WARNING

Handle carefully to avoid any impact to Manifold Absolute Pressure (MAP) sensor.

16. Remove acoustic absorbent material.

To install:

17. Installation is the reverse of the removal procedure.

18. Tighten the intake manifold fasteners in the numerical order as shown.

19. Tighten the electric throttle control actuator in the numerical order as shown.

➡**The figure shows the electric throttle control actuator (bank 1) viewed from the air duct side. Viewed from the air duct side, the order of tightening mounting bolts of electric throttle control actuator (bank 1) is the same as that of the electric throttle control actuator (bank 2).**

20. Perform the "Throttle Valve Closed Position Learning" when harness connector of electric throttle control actuator is disconnected.

21. Perform the "Idle Air Volume Learning" and "Throttle Valve Closed Position Learning" when electric throttle control actuator is replaced.

Throttle Valve Closed Position Learning

Throttle Valve Closed Position Learning is a function of ECM to learn the fully closed position of the throttle valve by monitoring the throttle position sensor output signal. It must be performed each time the harness connector of the electric throttle control actuator or ECM is disconnected.

1. Before servicing the vehicle, refer to the Precautions Section.

2. Check that accelerator pedal is fully released.

3. Turn ignition switch ON.

4. Turn ignition switch OFF and wait at least 10 seconds.

➡**Check that throttle valve moves during the above 10 seconds by confirming the operating sound.**

Idle Air Volume Learning

Idle Air Volume Learning is a function of ECM to learn the idle air volume that keeps engine idle speed within the specific range. It must be performed under the following conditions:

• Each time the electric throttle control actuator or ECM is replaced

• Idle speed or ignition timing is out of the specification

1. Before servicing the vehicle, refer to the Precautions Section.

2. Check that all of the following conditions are satisfied.

➡**Learning will be cancelled if any of the following conditions are missed for even a moment.**

    a. Battery voltage: More than 12.9 V (at idle).

    b. Engine coolant temperature: 158–221°F (70–105°C).

    c. Selector lever position: P or N.

    d. Electric load switch: OFF. (Air conditioner, headlamp, rear window defogger).

➡**On vehicles equipped with daytime light systems, if the parking brake is applied before the engine is started the headlamp will not illuminate.**

    e. Steering wheel: Neutral (Straight-ahead position).

    f. Vehicle speed: Stopped.

    g. Transmission: Warmed-up.

3. With CONSULT-III:

• Drive vehicle until "ATF TEMP 2" in "DATA MONITOR" mode of "A/T" system indicates less than 0.9 V.

4. Without CONSULT-III:

• Drive vehicle for 10 minutes.

With Consult-III

1. Before servicing the vehicle, refer to the Precautions Section.

2. Perform Accelerator Pedal Released Position Learning.

3. Perform Throttle Valve Closed Position Learning.

4. Start engine and warm it up to normal operating temperature.

5. Select "IDLE AIR VOL LEARN" in "WORK SUPPORT" mode.

6. Touch "START" and wait 20 seconds.

Without Consult-III

Note the following:

• It is better to count the time accurately with a clock

• It is impossible to switch the diagnostic mode when an accelerator pedal position sensor circuit has a malfunction

1. Before servicing the vehicle, refer to the Precautions Section.

2. Perform Accelerator Pedal Released Position Learning.

3. Perform Throttle Valve Closed Position Learning.

4. Start engine and warm it up to normal operating temperature.

5. Turn ignition switch OFF and wait at least 10 seconds.

6. Confirm that accelerator pedal is fully released, turn ignition switch ON and wait 3 seconds.

7. Repeat the following procedure quickly 5 times within 5 seconds.

   a. Fully depress the accelerator pedal.

   b. Fully release the accelerator pedal.

8. Wait 7 seconds, fully depress the accelerator pedal for approx. 20 seconds until the MIL stops blinking and turns ON.

9. Fully release the accelerator pedal within 3 seconds after the MIL turns ON.

10. Start engine and let it idle.

11. Wait 20 seconds.

12. Rev up engine 2–3 times and check that idle speed and ignition timing are within the specifications.

13. Check the following:

   a. Check that throttle valve is fully closed.

   b. Check PCV valve operation.

   c. Check that downstream of throttle valve is free from air leakage.

INTAKE MANIFOLD COLLECTOR

REMOVAL & INSTALLATION

3.5L Engine

See Figures 174 and 175.

✳✳ CAUTION

Never drain engine coolant when the engine is hot to avoid the danger of being scalded.

1. Before servicing the vehicle, refer to the Precautions Section.

2. Remove engine cover.

3. Remove air cleaner case and air duct.

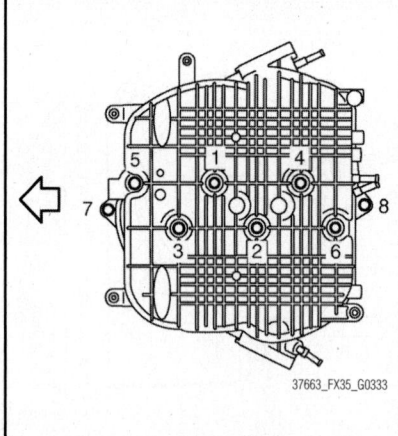

Fig. 174 Intake manifold collector mounting bolts and nuts tightening sequence

37663_FX35_G0333

4. Remove electric throttle control actuator as per the following:

   a. Drain engine coolant, or when water hoses are disconnected, attach plug to prevent engine coolant leakage.

✳✳ CAUTION

Perform this step when engine is cold. Never spill engine coolant on drive belt.

   b. Disconnect water hoses from electric throttle control actuator. When engine coolant is not drained from radiator, attach plug to water hoses to prevent engine coolant leakage.

   c. Disconnect harness connector.

   d. Loosen mounting bolts in the reverse order of tightening.

➡**When removing only intake manifold collector, move electric throttle control actuator without disconnecting the water hose. The figure shows the electric throttle control actuator (bank 1) viewed from the air duct side. Viewed from the air duct side, order of loosening mounting bolts of electric throttle control actuator (bank 2) is the same as that of the electric throttle control actuator (bank 1).**

✳✳ WARNING

Handle carefully to avoid any impact to electric throttle control actuator.

5. Disconnect vacuum hose, PCV hose and EVAP hose from intake manifold collector.

6. Remove EVAP canister purge volume control solenoid valve and EVAP tube

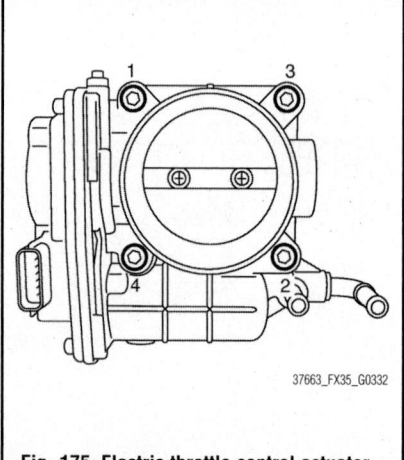

Fig. 175 Electric throttle control actuator tightening sequence

37663_FX35_G0332

assembly from intake manifold collector.

7. Loosen mounting bolts and nuts in the reverse order of tightening sequence.

8. Remove the intake manifold collector.

To install:

9. Installation is the reverse of the removal procedure.

10. If stud bolts were removed, install them and tighten to 97 inch lbs. (11 Nm).

11. Tighten mounting bolts and nuts in numerical order as shown.

12. To install the water hose, perform the following:

   a. Insert hose by 1.06–1.26 inches (27–32mm) from connector end.

   b. Clamp hose at location of 0.12–0.28 inch (3–7mm) from hose end.

13. To install the electric throttle control actuator (bank 1 & 2). Tighten in numerical order as shown.

➡**The figure shows the electric throttle control actuator (bank 1) viewed from the air duct side. Viewed from the air duct side, order of tightening mounting bolts of electric throttle control actuator (bank 2) is the same as that of the electric throttle control actuator (bank 1).**

14. Perform the "Throttle Valve Closed Position Learning" when harness connector of electric throttle control actuator is disconnected. Refer to Intake Manifold, Throttle Valve Closed Position Learning.

15. Perform the "Idle Air Volume Learning" and "Throttle Valve Closed Position Learning" when electric throttle control actuator is replaced. Refer to Intake Manifold, Idle Air Volume Learning.

OIL PAN

REMOVAL & INSTALLATION

3.5L Engine

Lower Oil Pan and Oil Strainer

See Figures 176 through 179.

✳✳ CAUTION

To avoid the danger of being scalded, never drain engine oil when engine is hot.

1. Before servicing the vehicle, refer to the Precautions Section.
2. Remove engine undercover.
3. Drain engine oil.
4. Remove lower oil pan as per the following:
   a. Loosen mounting bolts in reverse order of tightening sequence.
   b. Insert the seal cutter KV10111100 (J-37228) between upper oil pan and lower oil pan.

✳✳ WARNING

Be careful not to damage the mating surfaces. Never insert a screwdriver, this will damage the mating surfaces.

   c. Slide the seal cutter by tapping on the side of tool with a hammer.
5. Remove lower oil pan.
6. Remove oil strainer.

To install:

7. Install oil strainer.
8. Install lower oil pan as per the following:
   a. Use scraper to remove old liquid gasket from mating surfaces. Remove old liquid gasket from the bolt holes and thread.

✳✳ WARNING

Do not scratch or damage the mating surfaces when cleaning off old liquid gasket.

   b. Apply a continuous bead of liquid gasket to the lower oil pan as shown.

➡**Attaching must be done within 5 minutes after coating.**

   c. Install lower oil pan. Tighten mounting bolts in numerical order as shown.
9. Install oil pan drain plug.
10. Installation continues in the reverse of the removal procedure.

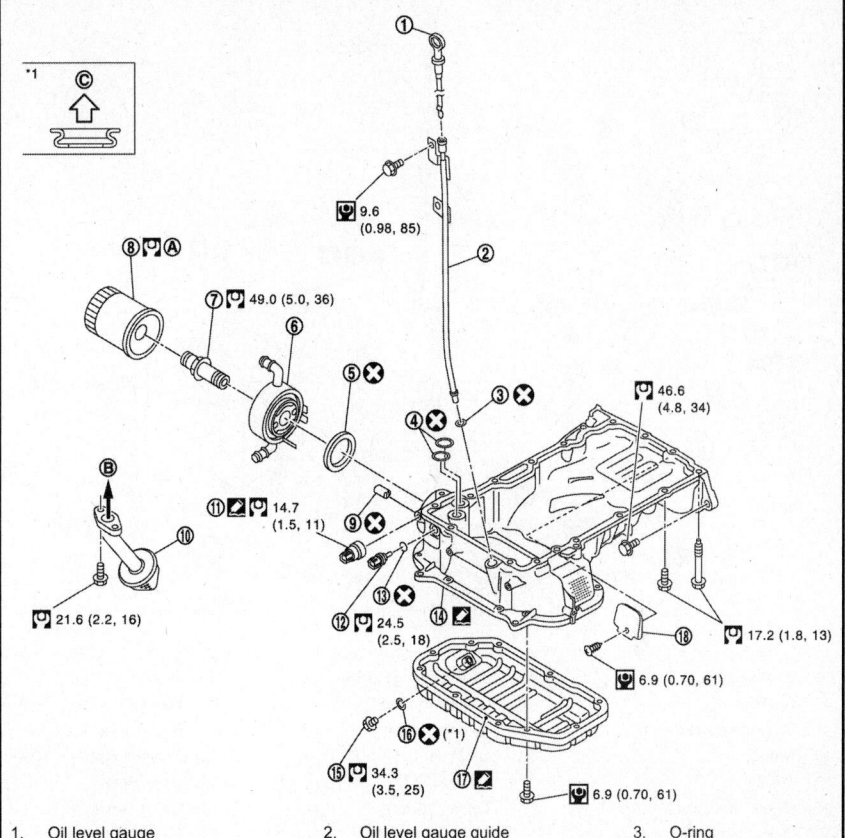

1.	Oil level gauge	2.	Oil level gauge guide	3.	O-ring
4.	O-ring	5.	O-ring	6.	Oil cooler
7.	Connector bolt	8.	Oil filter	9.	Relief valve
10.	Oil strainer	11.	Oil pressure switch	12.	Oil temperature sensor
13.	Washer	14.	Oil pan (upper)	15.	Drain plug
16.	Drain plug washer	17.	Oil pan (lower)	18.	Rear plate cover

37663_FX35_G0338

Fig. 176 Exploded view of oil pan assembly—3.5L engine (2WD model)

➡**Wait at least 30 minutes after oil pan is installed before adding engine oil.**

11. Check the engine oil level and adjust engine oil.
12. Start engine, and check there is no leakage of engine oil.
13. Stop engine and wait for 10 minutes.
14. Check the engine oil level again.

Upper Oil Pan (2WD)

See Figures 176, 180 and 181.

✳✳ CAUTION

Never drain engine oil when the engine is hot to avoid the danger of being scalded.

1. Before servicing the vehicle, refer to the Precautions Section.
2. Remove oil level gauge, oil pressure switch and oil temperature sensor.
3. Remove lower oil pan.

4. Remove oil strainer.
5. Loosen the mounting bolts in the reverse order of tightening sequence.
6. Insert the seal cutter KV10111100 (J-37228) between upper oil pan and lower cylinder block. Slide seal cutter by tapping on the side of tool with a hammer.
7. Remove upper oil pan.

✳✳ WARNING

Be careful not to damage the mating surfaces. Never insert a screwdriver, this will damage the mating surfaces.

8. Remove O-rings from bottom of lower cylinder block and oil pump.

To install:

9. Use a scraper to remove old liquid gasket from mating surfaces.

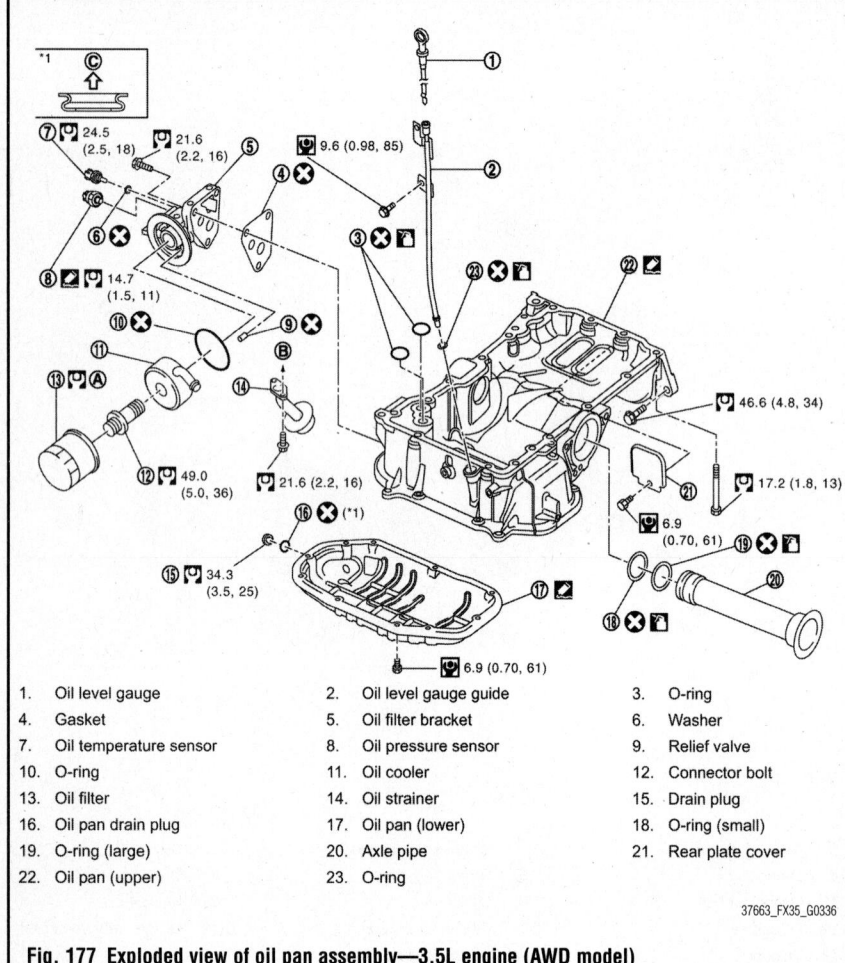

1. Oil level gauge
2. Oil level gauge guide
3. O-ring
4. Gasket
5. Oil filter bracket
6. Washer
7. Oil temperature sensor
8. Oil pressure sensor
9. Relief valve
10. O-ring
11. Oil cooler
12. Connector bolt
13. Oil filter
14. Oil strainer
15. Drain plug
16. Oil pan drain plug
17. Oil pan (lower)
18. O-ring (small)
19. O-ring (large)
20. Axle pipe
21. Rear plate cover
22. Oil pan (upper)
23. O-ring

37663_FX35_G0336

Fig. 177 Exploded view of oil pan assembly—3.5L engine (AWD model)

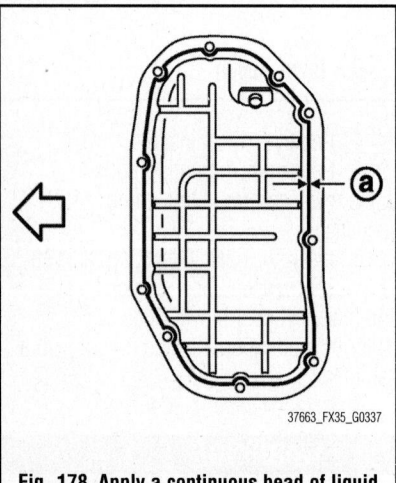

37663_FX35_G0337

Fig. 178 Apply a continuous bead of liquid gasket to the lower oil pan as shown

✳✳ WARNING

Do not scratch or damage the mating surfaces when cleaning off old liquid gasket.

10. Remove old liquid gasket from mating surface of lower cylinder block.

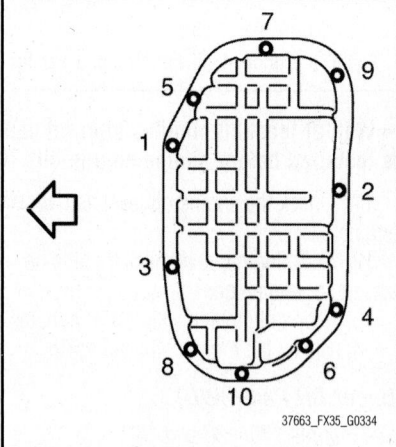

37663_FX35_G0334

Fig. 179 Lower oil pan mounting bolt tightening sequence

11. Remove old liquid gasket from the bolt holes and threads.

12. Install new O-rings on the bottom of lower cylinder block and oil pump.

13. Apply a continuous bead of liquid gasket to the cylinder block mating surface of upper oil pan as shown.

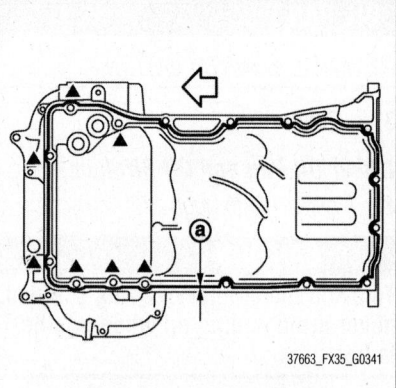

37663_FX35_G0341

Fig. 180 Apply a continuous bead of liquid gasket (a) to the cylinder block mating surface of upper oil pan as shown

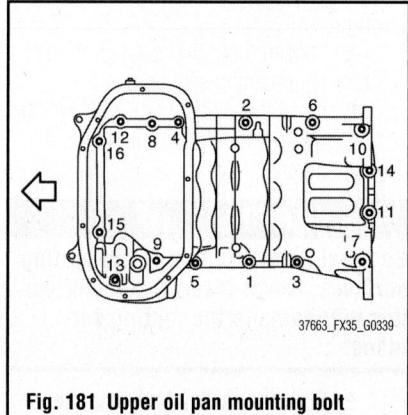

37663_FX35_G0339

Fig. 181 Upper oil pan mounting bolt tightening sequence

14. For bolt holes with triangle marks (7 locations), apply liquid gasket outside the holes.

➥**Attaching must be done within 5 minutes after coating.**

15. Install upper oil pan. Avoid misalignment of O-rings.

16. Tighten mounting bolts in numerical order as shown.

➥**There are 2 types of mounting bolts. The 3.5 inch (90mm) bolts are used in holes 7, 10 and 13. All other holes use 1 inch (25mm) bolts.**

17. Install oil strainer to oil pump.

18. Install lower oil pan.

19. Install oil pan drain plug.

20. Installation continues in the reverse of the removal procedure.

➥**Wait at least 30 minutes after oil pan is installed to add engine oil.**

21. Check the engine oil level and adjust engine oil.

22. Start engine, and check there is no leakage of engine oil.

23. Stop engine and wait for 10 minutes.

24. Check the engine oil level again.

Upper Oil Pan (AWD)

See Figures 177, 182 and 183.

❊❊ CAUTION

Never drain engine oil when the engine is hot to avoid the danger of being scalded.

1. Before servicing the vehicle, refer to the Precautions Section.

2. Remove oil level gauge, oil pressure switch and oil temperature sensor.

3. Remove oil filter bracket.

4. Remove lower oil pan.

5. Remove oil strainer.

6. Loosen mounting bolts in the reverse order of the tightening sequence.

7. Insert the seal cutter KV10111100 (J-37228) between upper oil pan and lower cylinder block. Slide seal cutter by tapping on the side of tool with a hammer.

8. Remove upper oil pan.

❊❊ WARNING

Be careful not to damage the mating surfaces. Never insert a screwdriver, this will damage the mating surfaces.

9. Remove O-rings from bottom of lower cylinder block and oil pump.

10. Remove axle pipe, if necessary.

To install:

11. Install axle pipe to upper oil pan, if removed.

a. Install axle pipe to upper oil pan from axle pipe flange side (left side).

➡**Insert it with care to prevent O-ring from sliding.**

12. Use a scraper to remove old liquid gasket from mating surfaces.

❊❊ WARNING

Do not scratch or damage the mating surfaces when cleaning off old liquid gasket.

13. Remove old liquid gasket from mating surface of lower cylinder block.

14. Remove old liquid gasket from the bolt holes and threads.

15. Install new O-rings on the bottom of lower cylinder block and oil pump.

16. Apply a continuous bead of liquid gasket to the cylinder block mating surface of upper oil pan as shown.

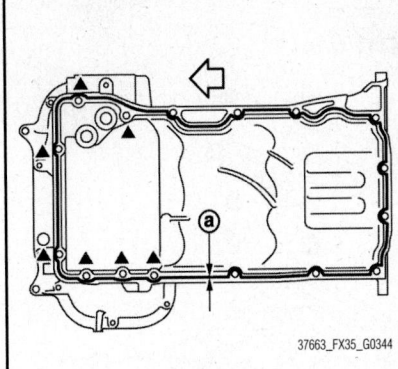

Fig. 182 Apply a continuous bead of liquid gasket to the cylinder block mating surface of upper oil pan as shown

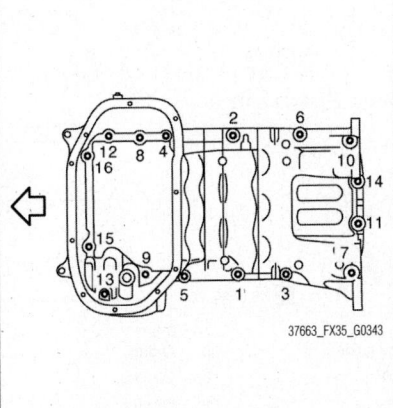

Fig. 183 Tighten mounting bolts in numerical order as shown

➡**For bolt holes with triangle marks (7 locations), apply liquid gasket outside the holes. Attaching must be done within 5 minutes after coating.**

17. Install upper oil pan. Install avoiding misalignment of O-rings.

18. Tighten mounting bolts in numerical order as shown.

19. There are 3 types of mounting bolts:

a. 1 inch (25mm) bolts: Bolt positions 3, 6, 8, 9, 11, 12, 14, 15, and 16.

b. 2 inches (50mm) bolt: Bolt position 2.

c. 3.5 inches (90mm) bolts: Bolt positions 1, 4, 5, 7, 10, and 13.

20. Install oil strainer to oil pump.

21. Install lower oil pan.

22. Install oil pan drain plug.

23. Installation continues in the reverse of the removal procedure.

➡**Wait at least 30 minutes after oil pan is installed to add engine oil.**

24. Check the engine oil level and adjust engine oil.

25. Start engine, and check there is no leakage of engine oil.

26. Stop engine and wait for 10 minutes.

27. Check the engine oil level again.

5.0L Engine

Lower Oil Pan

See Figures 184 through 186.

❊❊ CAUTION

To avoid the danger of being scalded, never drain engine oil when engine is hot.

1. Before servicing the vehicle, refer to the Precautions Section.

2. Drain engine oil.

3. Remove lower oil pan as per the following:

a. Loosen mounting bolts in reverse order of the tightening sequence.

b. Insert the seal cutter KV10111100 (J-37228) between upper oil pan and lower oil pan.

❊❊ WARNING

Be careful not to damage the mating surfaces. Never insert a screwdriver. This damages the mating surfaces.

c. Slide the seal cutter by tapping on the side of tool with a hammer.

4. Remove lower oil pan.

5. Remove oil strainer.

To install:

6. Install oil strainer.

7. Use scraper to remove old liquid gasket from mating surfaces.

8. Remove old liquid gasket from the bolt holes and thread.

❊❊ WARNING

Never scratch or damage the mating surfaces when cleaning off old liquid gasket.

9. Apply a continuous bead of liquid gasket to the lower oil pan as shown.

➡**Attaching must be done within 5 minutes after coating.**

10. Install lower oil pan.

11. Tighten mounting bolts in numerical order as shown.

12. Install oil pan drain plug.

13. Installation continues in the reverse of the removal procedure.

➡**Wait at least 30 minutes after oil pan is installed before adding engine oil.**

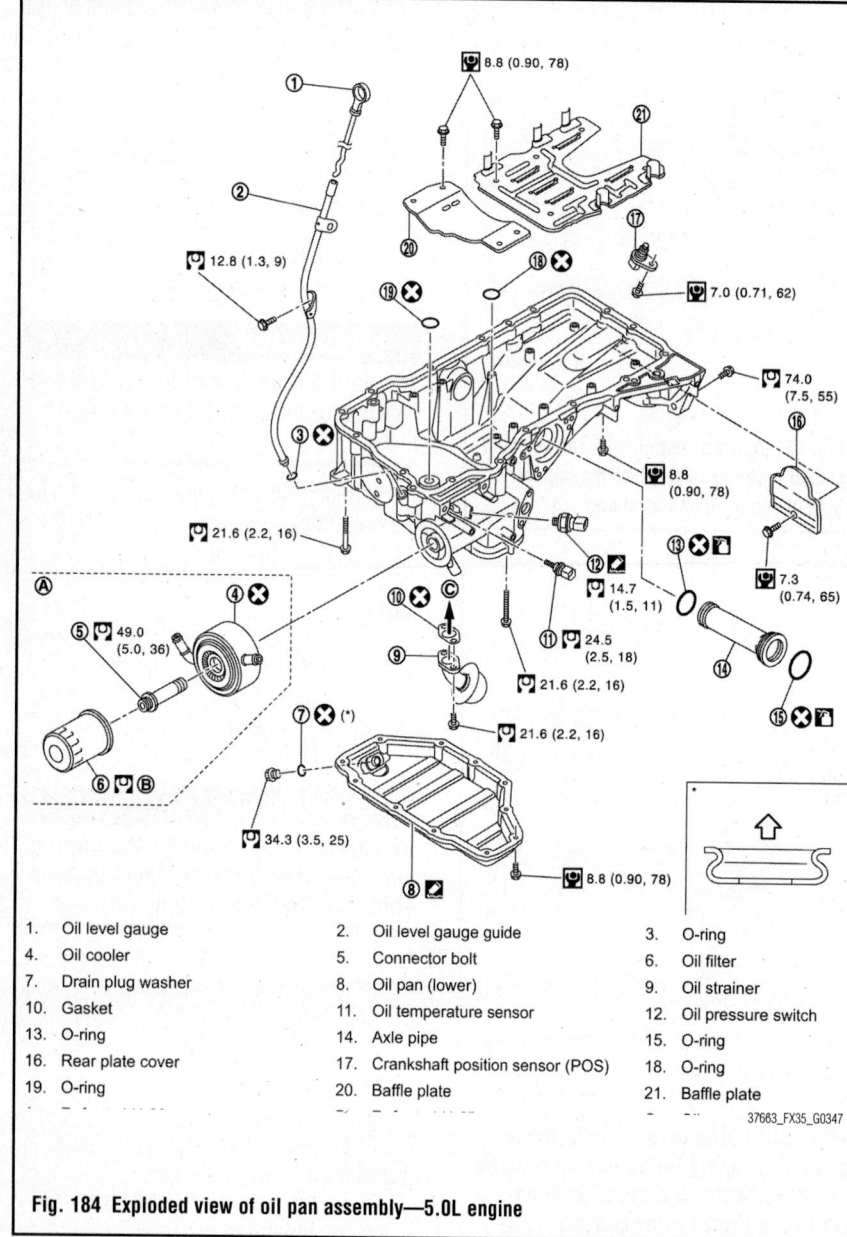

Fig. 184 Exploded view of oil pan assembly—5.0L engine

1. Oil level gauge
2. Oil level gauge guide
3. O-ring
4. Oil cooler
5. Connector bolt
6. Oil filter
7. Drain plug washer
8. Oil pan (lower)
9. Oil strainer
10. Gasket
11. Oil temperature sensor
12. Oil pressure switch
13. O-ring
14. Axle pipe
15. O-ring
16. Rear plate cover
17. Crankshaft position sensor (POS)
18. O-ring
19. O-ring
20. Baffle plate
21. Baffle plate

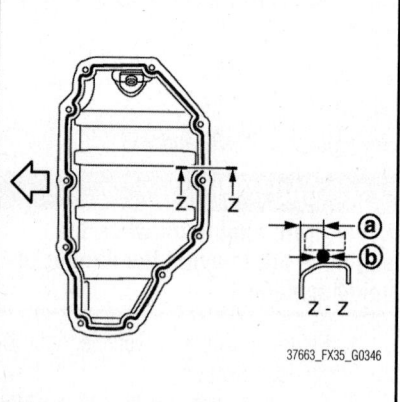

Fig. 185 Apply a continuous bead of liquid gasket to the lower oil pan as shown

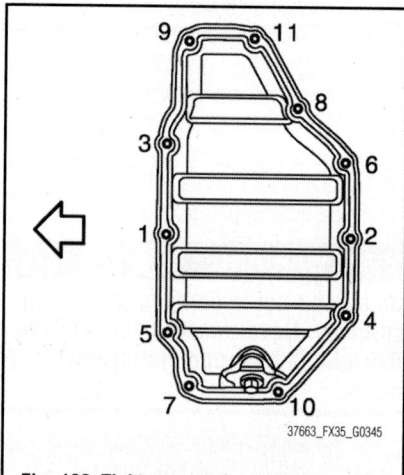

Fig. 186 Tighten mounting bolts in numerical order as shown

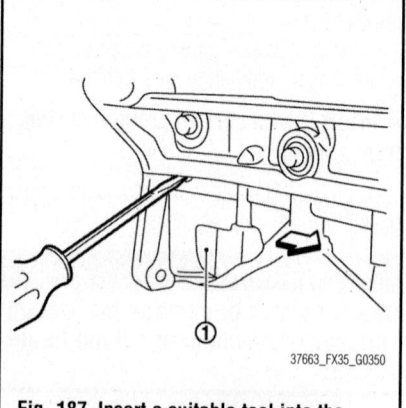

Fig. 187 Insert a suitable tool into the notch at upper oil pan (1) as shown

14. Check the engine oil level and adjust engine oil.

15. Start engine, and check there is no leakage of engine oil.

16. Stop engine and wait for 15 minutes.

17. Check the engine oil level again.

Upper Oil Pan

See Figures 187 through 189.

❈❈ CAUTION

To avoid the danger of being scalded, never drain engine oil when engine is hot.

1. Before servicing the vehicle, refer to the Precautions Section.

2. Remove oil filter.

3. Remove oil cooler.

4. Remove A/C compressor and A/C compressor bracket.

5. Remove oil level gauge and oil level gauge guide.

6. Remove oil pressure switch and oil temperature sensor if necessary.

7. Remove rear plate cover.

8. Remove lower oil pan.

9. Remove oil strainer.

10. Loosen the upper oil pan mounting bolts in the reverse order of the tightening sequence.

➡ **Disregard No. 12, 17 when loosening.**

11. Insert a suitable tool into the notch at upper oil pan as shown.

12. Pry off case by using a suitable tool.

❈❈ WARNING

Be careful not to damage the mating surfaces.

13. Remove O-ring from bottom of cylinder block and oil pump.

14. Remove baffle plate, if necessary.

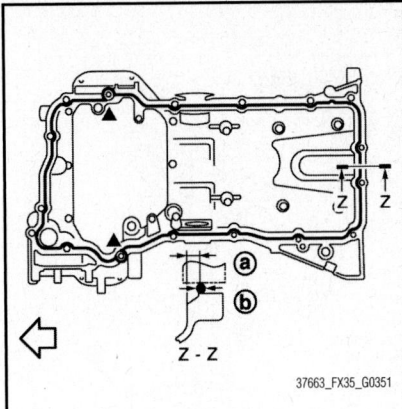

Fig. 188 Apply a continuous bead of liquid gasket to the cylinder block mating surfaces of upper oil pan as shown

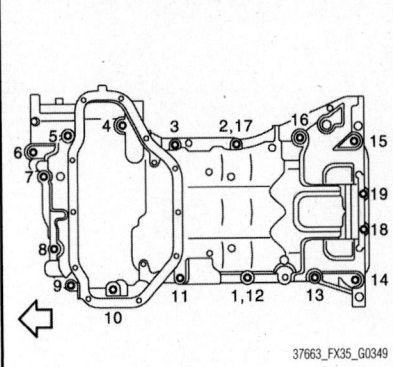

Fig. 189 Upper oil pan mounting bolt tightening sequence

15. Remove axle pipe from upper oil pan, if necessary.

16. Remove axle pipe from upper oil pan using a suitable drift.

To install:

17. Install axle pipe to upper oil pan, if removed.

  a. Lubricate O-ring groove of axle pipe, O-ring, and O-ring joint of oil pan with new engine oil.

  b. Install axle pipe to upper oil pan from halfshaft (LH) side.

➡**Insert it with care to prevent O-ring from sliding.**

18. Use a scraper to remove old liquid gasket from mating surfaces.

19. Also remove the old liquid gasket from mating surface of cylinder block.

20. Remove old liquid gasket from the bolt holes and threads.

✹✹ WARNING

Never scratch or damage the mating surfaces when cleaning off old liquid gasket.

21. Install new O-rings on the bottom of cylinder block and oil pump.

22. Apply a continuous bead of liquid gasket to the cylinder block mating surfaces of upper oil pan as shown.

➡**Attaching must be done within 5 minutes after coating.**

23. Tighten mounting bolts in numerical order as shown. Install avoiding misalignment of O-rings.

➡**Tighten mounting bolts No. 1 and 2 in two steps. The numerical order No. 12 and 17 shown as second steps.**

24. There are 3 types of mounting bolts.
  a. 1.18 inches (30mm) bolts: Bolt positions 18, 19.
  b. 3.94 inches (100mm) bolts: Bolt positions 3, 4, 5, 7, 10, 11, 14, and 15.
  c. 1.77 inches (45mm) bolts: Bolt positions all others except above.

25. Tighten transmission joint bolts.
  a. Install rear plate cover.

26. Install oil strainer.

27. Install lower oil pan.

28. Install in the reverse order of removal.

➡**Wait at least 30 minutes after oil pan is installed before adding engine oil.**

29. Check the engine oil level and adjust engine oil.

30. Start engine, and check there is no leakage of engine oil.

31. Stop engine and wait for 15 minutes.

32. Check the engine oil level again.

OIL PUMP

REMOVAL & INSTALLATION

3.5L Engine

See Figure 190.

1. Before servicing the vehicle, refer to the Precautions Section.

2. Remove lower oil pan and oil strainer.

3. Remove upper oil pan.

4. Remove front timing chain case and primary timing chain.

5. Remove oil pump assembly.

6. Installation is the reverse of the removal procedure.

5.0L Engine

See Figures 191 through 193.

1. Before servicing the vehicle, refer to the Precautions Section.

2. Remove lower oil pan and oil strainer.

3. Remove upper oil pan.

4. Remove front timing chain cover.

5. Remove oil pump drive chain.
  a. Push oil pump drive chain tensioner.
  b. Insert a stopper pin into the body hole.
  c. Hold the 2 flat parts of oil pump shaft, and then loosen the oil pump sprocket (oil pump side) nut.

6. Remove oil pump.

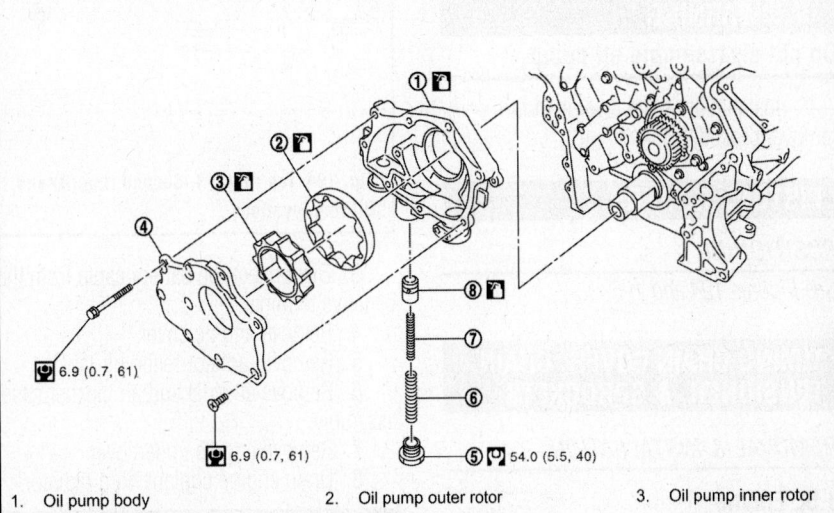

1. Oil pump body	2. Oil pump outer rotor	3. Oil pump inner rotor
4. Oil pump cover	5. Regulator valve plug	6. Regulator valve spring
7. Regulator valve spring	8. Regulator valve	

Fig. 190 Exploded view of oil pump assembly—3.5L engine

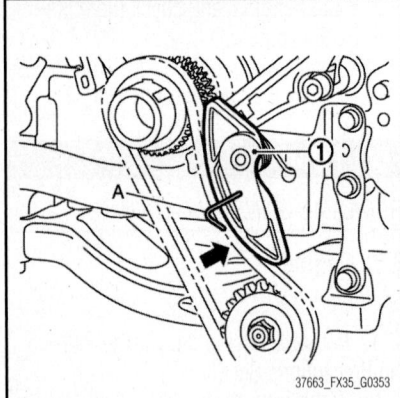

Fig. 191 Push oil pump drive chain tensioner (1); Insert a stopper pin (A) into the body hole

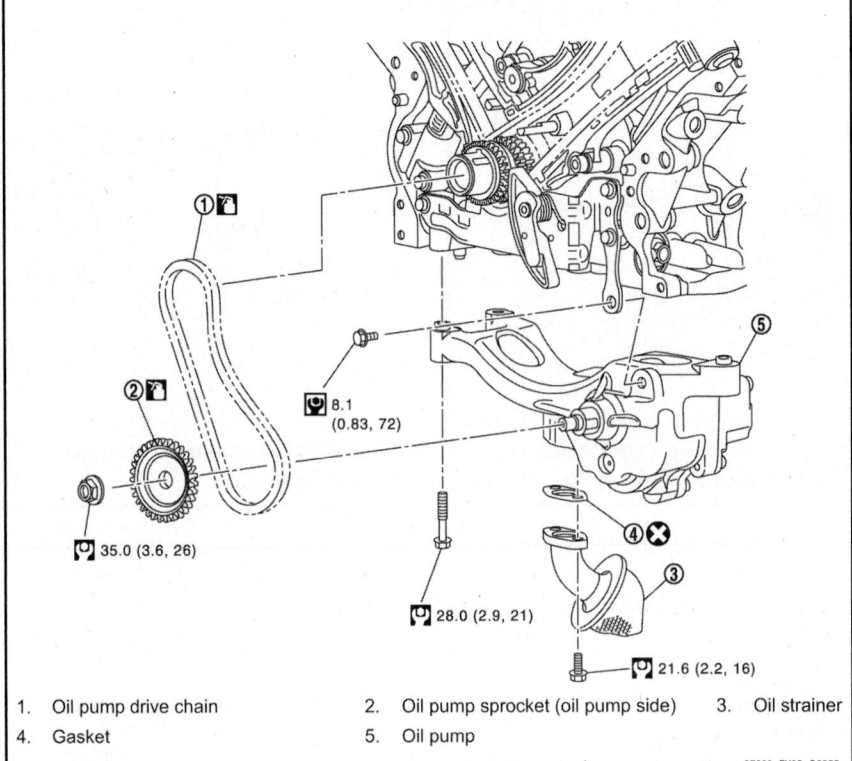

1. Oil pump drive chain
2. Oil pump sprocket (oil pump side)
3. Oil strainer
4. Gasket
5. Oil pump

Fig. 193 Exploded view of oil pump assembly—5.0L engine

Fig. 192 Hold the two flat parts of oil pump shaft, and then loosen the oil pump sprocket (1) (oil pump side) nut

✳✳ WARNING

Do not disassemble oil pump.

7. Installation is the reverse of the removal procedure.

PISTONS & RINGS

POSITIONING

See Figures 194 and 195.

TIMING CHAIN COVER, CHAIN, TENSIONER, & SPROCKETS

REMOVAL & INSTALLATION

3.5L Engine

See Figures 196 through 223.

1. Before servicing the vehicle, refer to the Precautions Section.
2. Release the fuel pressure.

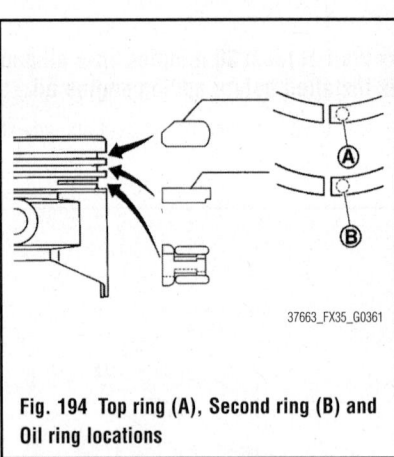

Fig. 194 Top ring (A), Second ring (B) and Oil ring locations

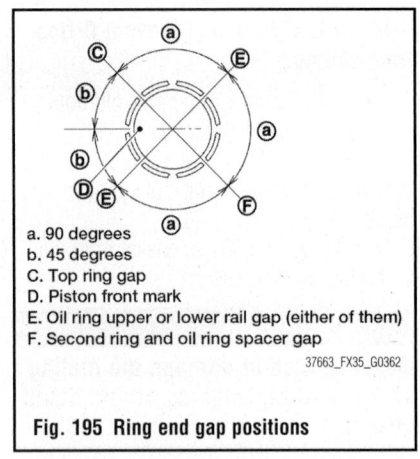

a. 90 degrees
b. 45 degrees
C. Top ring gap
D. Piston front mark
E. Oil ring upper or lower rail gap (either of them)
F. Second ring and oil ring spacer gap

Fig. 195 Ring end gap positions

3. Disconnect the battery cable from the negative terminal.
4. Remove engine cover.
5. Remove radiator reservoir tank.
6. Remove air duct and air cleaner case assembly.
7. Remove engine undercover.
8. Drain engine coolant from radiator.

✳✳ CAUTION

Perform this step when the engine is cold. Never spill engine coolant on drive belt.

9. Remove upper and lower radiator hoses.
10. Drain engine oil.

✳✳ CAUTION

Perform this step when the engine is cold. Never spill engine oil on drive belt.

11. Remove drive belt.
12. Remove radiator cooling fan assembly.
13. Separate engine harnesses removing their brackets from front timing chain case.

14. Remove intake manifold collector.

15. Remove intake manifold.

16. Remove oil level gauge and oil level gauge guide.

17. Remove A/C compressor from bracket with piping connected, and temporarily secure it aside.

18. Remove power steering oil pump from bracket with piping connected, and temporarily secure it aside.

19. Remove power steering oil pump bracket.

20. Remove idler pulley, auto tensioner and bracket.

21. Remove alternator and alternator bracket.

22. Remove water outlet (front) and water piping.

23. Remove valve timing control covers (bank 1 and bank 2) and gasket as per the following:

    a. Disconnect valve timing control harness connector

    b. Loosen mounting bolts in reverse order as shown.

➡**Shaft is internally jointed with camshaft sprocket (INTAKE) center hole. When removing, keep it horizontal until it is completely disconnected.**

    c. Shaft is engaged with intake side camshaft sprocket center hole on inside. Pull straight out so as not to tilt until the joint is disengaged.

    d. The mating surface of magnet retarder may be fitted with the exhaust side camshaft sprocket via the engine oil. Open valve timing control cover carefully.

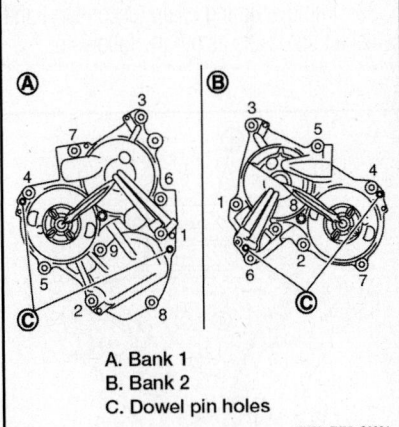

A. Bank 1
B. Bank 2
C. Dowel pin holes

37663_FX35_G0364

Fig. 196 Loosen mounting bolts in reverse order as shown

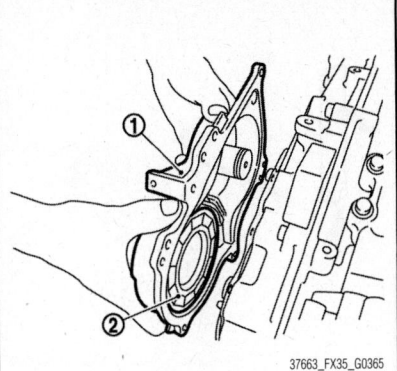

37663_FX35_G0365

Fig. 197 View of mating surface of magnet retarder (2) and valve timing control cover (1)

    e. If the mating surface of magnet retarder is fitted with the camshaft sprocket, open the cover within the range that the load is not applied to the harness. And then, remove it so as to prevent magnet retarder from dropping.

❊❊ WARNING

Be careful not to damage magnet retarder. When carrying valve timing control cover, face the magnet retarder side up to prevent the cover from falling from magnet retarder. Never remove magnet retarder from valve timing control cover.

24. Remove valve covers (bank 1 and bank 2).

37663_FX35_G0366

Fig. 198 Rotate crankshaft pulley clockwise to align timing mark (grooved line without color) with timing indicator

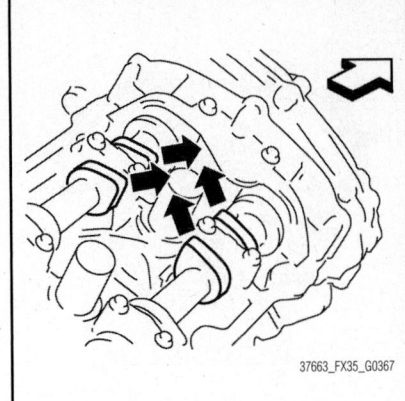

37663_FX35_G0367

Fig. 199 Check that intake and exhaust cam noses on No. 1 cylinder (engine front side of bank 1) are located as shown

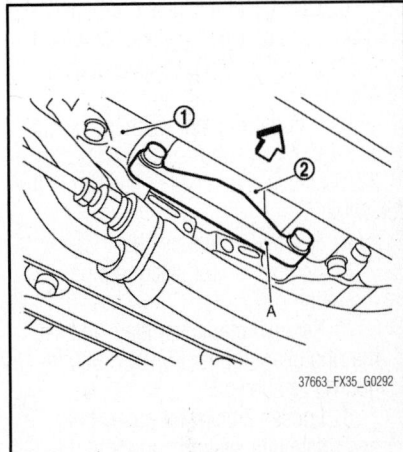

37663_FX35_G0292

Fig. 200 Remove rear cover plate and set the ring gear stopper (A) as shown

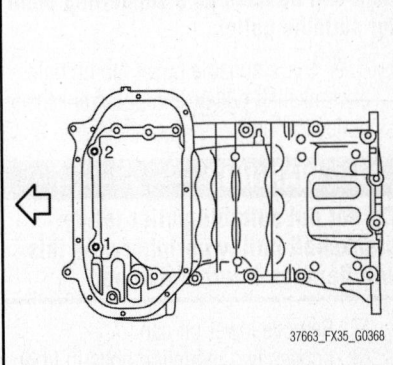

37663_FX35_G0368

Fig. 201 Loosen two mounting bolts in front of upper oil pan in reverse order as shown

25. Obtain No. 1 cylinder at TDC of its compression stroke as per the following:

    a. Rotate crankshaft pulley clockwise to align timing mark (grooved line without color) with timing indicator.

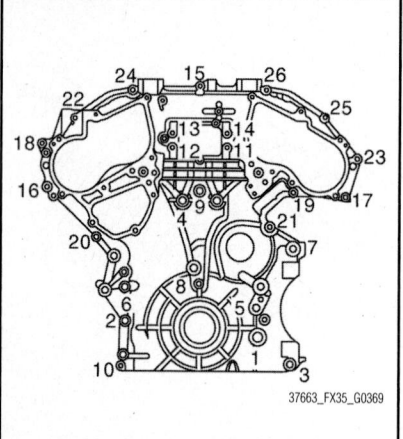

Fig. 202 Loosen mounting bolts in reverse order as shown

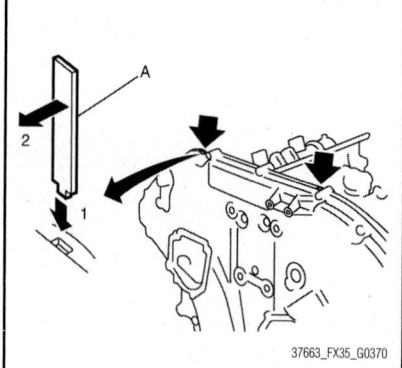

Fig. 203 Insert a suitable tool (A) into the notch at the top of front timing chain case as shown

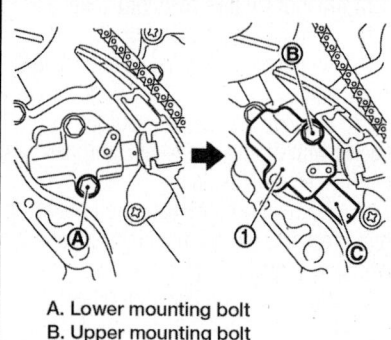

A. Lower mounting bolt
B. Upper mounting bolt
C. Plunger
1. Timing chain tensioner (primary)

Fig. 205 Remove timing chain tensioner (primary)

b. Check that intake and exhaust cam noses on No. 1 cylinder (engine front side of bank 1) are located as shown.

c. If not, turn crankshaft one revolution (360°) and align as shown.

26. Remove crankshaft pulley as per the following:

a. Remove front cross bar.

b. Remove power steering pipe mounting bolt.

c. Remove rear cover plate and set the ring gear stopper KV10118600 (J-48641) as shown.

d. Loosen crankshaft pulley bolt and rotate bolt seating surface at 0.39 inch (10mm) from its original position.

➡**Do not remove crankshaft pulley bolt as it will be used as a supporting point for suitable puller.**

e. Place suitable puller tab on holes of crankshaft pulley, and pull crankshaft pulley through.

☀☀ WARNING

Do not put suitable puller tab on crankshaft pulley periphery, as this will damage internal damper.

27. Remove lower oil pan.

28. Loosen two mounting bolts in front of upper oil pan in reverse order as shown.

29. Remove front timing chain case as per the following:

a. Loosen mounting bolts in reverse order as shown.

b. Insert a suitable tool into the notch at the top of front timing chain case as shown.

c. Pry off case by moving the suitable tool as shown.

d. Use the seal cutter KV10111100 (J-37228) to cut liquid gasket for removal.

☀☀ WARNING

Do not use a screwdriver or similar tool. After removal, handle front timing chain case carefully so it does not tilt, cant, or warp under a load.

30. Remove front oil seal from front timing chain case using a suitable tool.

☀☀ WARNING

Be careful not to damage front timing chain case.

31. Remove O-rings from rear timing chain case.

32. Remove timing chain tensioner (primary) as per the following:

a. Remove lower mounting bolt.

b. Loosen upper mounting bolt slowly, and then turn timing chain ten-

sioner (primary) on the upper mounting bolt so that plunger is fully expanded.

➡**Even if plunger is fully expanded, it is not dropped from the body of timing chain tensioner (primary).**

c. Remove upper mounting bolt, and then remove timing chain tensioner (primary).

33. Remove internal chain guide, and slack guide.

34. Remove timing chain (primary) and crankshaft sprocket.

☀☀ WARNING

After removing timing chain tensioner (primary), never turn crankshaft and camshaft separately, or valves will strike the piston heads.

35. Remove timing chain (secondary) and camshaft sprockets as per the following:

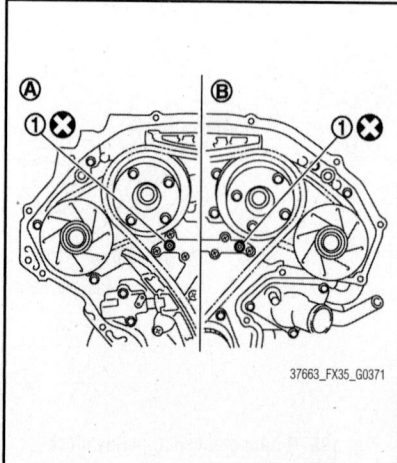

Fig. 204 Remove O-rings (1) from rear timing chain case

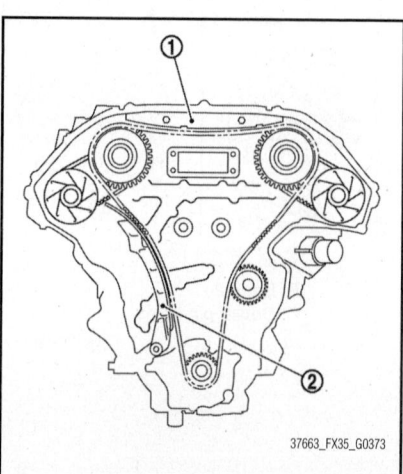

Fig. 206 Remove internal chain guide (1), and slack guide (2)

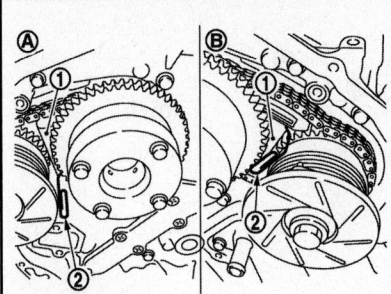

A. Bank 1
B. Bank 2
1. Timing chain tensioners (secondary)
2. Stopper pin

37663_FX35_G0374

Fig. 207 Remove timing chain (secondary) and camshaft sprockets

a. Attach suitable stopper pin to the timing chain tensioners (secondary).

➡**Use a hard metal pin as a stopper pin. Removing of camshaft bracket (No. 1) is required.**

b. Remove camshaft sprocket mounting bolts (INT and EXH).

c. Secure the hexagonal portion of camshaft using a wrench to loosen mounting bolts.

d. Remove timing chain (secondary) together with camshaft sprockets.

➡**Do not loosen the mounting bolts with securing anything other than the camshaft hexagonal portion or with tensioning the timing chain. Never disassemble. Never loosen bolts as shown.**

36. Remove timing chain tensioners (secondary) from cylinder head as per the following, if necessary.

a. Remove camshaft brackets (No. 1).

b. Remove timing chain tensioners (secondary) with a stopper pin attached.

37. Use a scraper to remove all traces of old liquid gasket from front and rear timing chain cases and upper oil pan, and liquid gasket mating surfaces.

➡**Be careful not to allow gasket fragments to enter oil pan.**

37663_FX35_G0375

Fig. 209 Secure the hexagonal portion of camshaft using a wrench to loosen mounting bolts

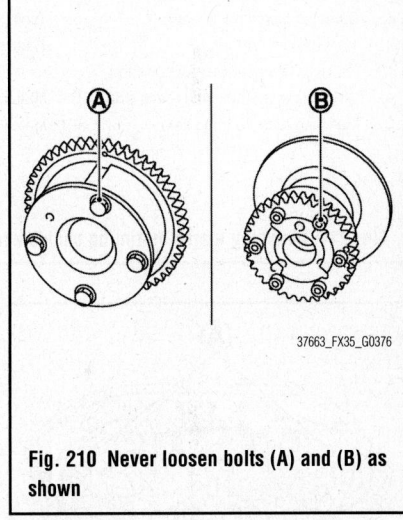

37663_FX35_G0376

Fig. 210 Never loosen bolts (A) and (B) as shown

38. Remove old liquid gasket from bolt hole and thread.

To install:

➡**For installation, refer to figure that shows the relationship between the matching mark on each timing chain and that on the corresponding sprocket, with the components installed.**

39. Install timing chain tensioners (secondary) to cylinder head if removed.

40. Check that dowel pin and crankshaft key are located as shown (No. 1 cylinder at compression TDC).

➡**Though camshaft does not stop at the position as shown, for the placement of cam noses, it is generally accepted camshaft is placed the same direction as the figure.**

41. Install timing chains (secondary) and camshaft sprockets as per the following:

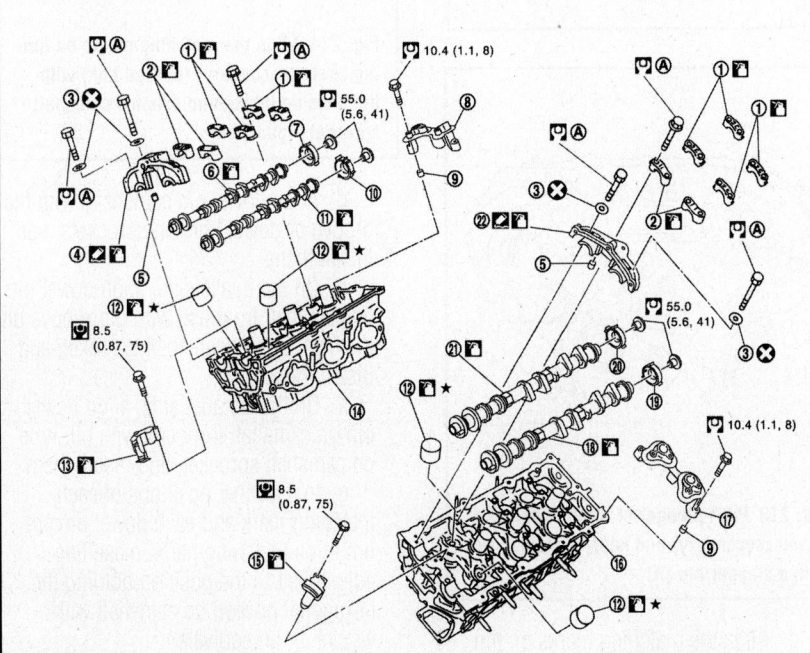

1. Camshaft bracket (No. 3, 4)	2. Camshaft bracket (No. 2)	3. Seal washer
4. Camshaft bracket (No. 1) (bank 1)	5. Dowel pin	6. Camshaft (EXH) (bank 1)
7. Camshaft signal plate (EXH)	8. Camshaft sensor bracket (bank 1)	9. Dowel pin
10. Camshaft signal plate (INT)	11. Camshaft (INT) (bank 1)	12. Valve lifter
13. Timing chain tensioner (secondary) (bank 1)	14. Cylinder head (bank 1)	15. Timing chain tensioner (secondary) (bank 2)
16. Cylinder head (bank 2)	17. Camshaft sensor bracket (bank 2)	18. Camshaft (EXH) (bank 2)
19. Camshaft signal plate (EXH)	20. Camshaft signal plate (INT)	21. Camshaft (INT) (bank 2)
22. Camshaft bracket (No. 1) (bank 2)		

37663_FX35_G0271

Fig. 208 Camshaft and related component locations—3.5L engine

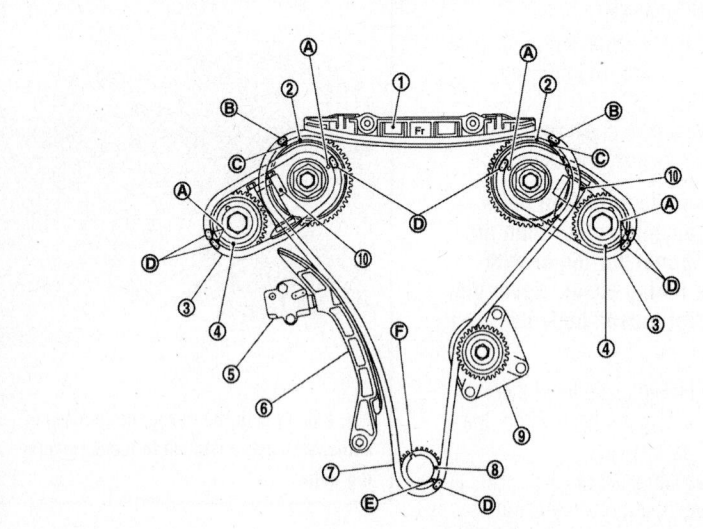

1. Internal chain guide
2. Camshaft sprocket (INT)
3. Timing chain (secondary)
4. Camshaft sprocket (EXH)
5. Timing chain tensioner (primary)
6. Slack guide
7. Timing chain (primary)
8. Crankshaft sprocket
9. Water pump
10. Timing chain tensioner (secondary)

A. Matching mark [punched (back side)]
B. Matching mark (yellow link)
C. Matching mark (punched)
D. Matching mark (orange link)
E. Matching mark (notched)
F. Crankshaft key

37663_FX35_G0377

Fig. 211 Matching marks on timing chains and sprockets

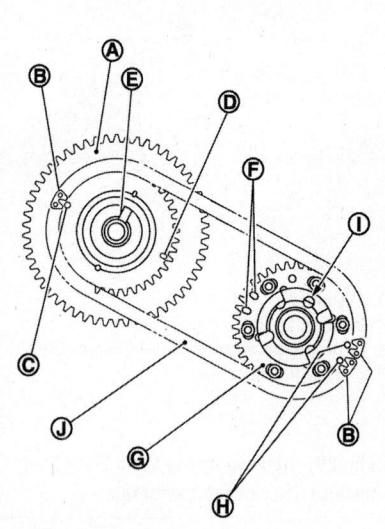

A. Camshaft sprocket (INT) back face
B. Orange link
C. Matching mark (Circle)
D. Matching mark (Oval)
E. Dowel groove
F. Matching mark (2 oval)
G. Camshaft sprocket (EXH) back face
H. Matching mark (2 circle)
I. Dowel hole
J. Timing chain (secondary)

37663_FX35_G0380

Fig. 214 Align the matching marks on timing chain (secondary) (orange link) with the ones on intake and exhaust camshaft sprockets (punched)

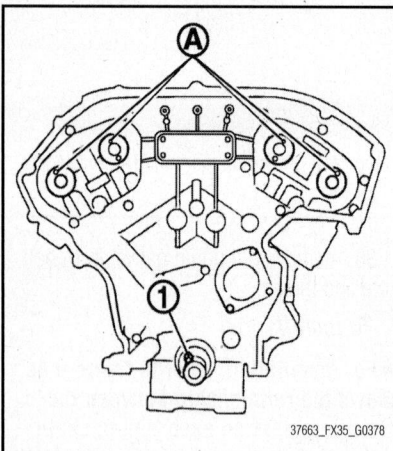

37663_FX35_G0378

Fig. 212 Check that dowel pin (A) and crankshaft key (1) are located as shown

✵✵ WARNING

Matching marks between timing chain and sprockets slip easily. Confirm all matching mark positions repeatedly during the installation process.

a. Push plunger of timing chain tensioner (secondary) and keep it pressed in with a stopper pin.

b. Install timing chains (secondary) and camshaft sprockets.

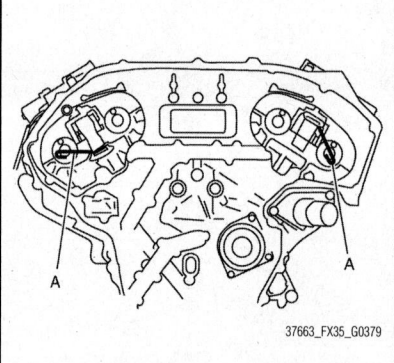

37663_FX35_G0379

Fig. 213 Push plunger of timing chain tensioner (secondary) and keep it pressed in with a stopper pin (A)

c. Align the matching marks on timing chain (secondary) (orange link) with the ones on intake and exhaust camshaft sprockets (punched), and install them.

➡**Figure shows bank 1 (rear view). Matching marks for camshaft sprockets are on the back side of camshaft sprockets (secondary). There are 2 types of matching marks, circle and oval types. They should be used for the bank 1 and bank 2, respectively.**

d. Align dowel pin camshafts with the groove or dowel hole on sprockets, and install them.

e. On the intake side, align dowel pin on camshaft front end with pin groove on the back side of camshaft sprocket, and install them.

f. On the exhaust side, align dowel pin on camshaft front end with pin hole on camshaft sprocket, and install them.

g. In case that positions of each matching mark and each dowel pin are not fit on matching parts, make fine adjustment to the position holding the hexagonal portion on camshaft with wrench or an equivalent.

h. Mounting bolts for camshaft sprockets must be tightened in the next step. Tightening them by hand is enough to prevent the dislocation of dowel pins.

i. Check the matching marks (punched) on each camshaft sprocket are positioned on the matching marks (orange link) on timing chain (secondary).

➡**Matching mark (punched) is for checking loose at this step.**

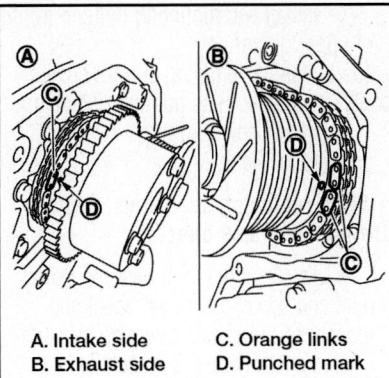

A. Intake side C. Orange links
B. Exhaust side D. Punched mark

37663_FX35_G0381

Fig. 215 Check the matching marks (punched) on each camshaft sprocket are positioned on the matching marks (orange link) on timing chain (secondary)

j. After confirming the matching marks are aligned, tighten camshaft sprocket mounting bolts.

k. Secure camshaft using a wrench at the hexagonal portion to tighten mounting bolts.

l. Pull stopper pins out from timing chain tensioners (secondary).

42. Install timing chain (primary) as per the following:

a. Install crankshaft sprocket. Check the matching marks on crankshaft sprocket face the front of the engine.

b. Install timing chain (primary).

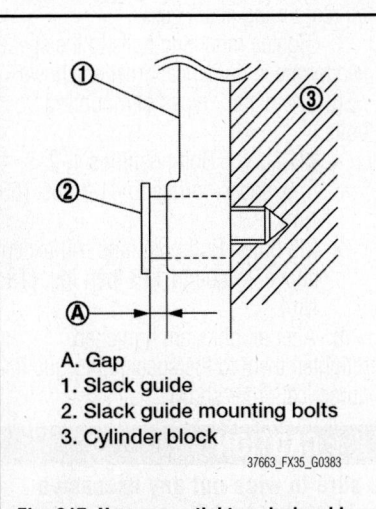

A. Yellow link
B. Punched mark
C. Notched mark
D. Orange link
1. Camshaft sprocket (INTAKE)
2. Crankshaft sprocket

37663_FX35_G0382

Fig. 216 Install timing chain (primary)

c. Install timing chain (primary) so the matching mark (punched) on camshaft sprocket (INT) is aligned with the yellow link on timing chain, while the matching mark (notched) on crankshaft sprocket is aligned with the orange link one on timing chain, as shown.

d. When it is difficult to align matching marks of timing chain (primary) with each sprocket, gradually turn camshaft using wrench on the hexagonal portion to align it with the matching marks.

e. During alignment, be careful to prevent dislocation of matching mark alignments of timing chains (secondary).

43. Install internal chain guide and slack guide.

✸✸ WARNING

Never over-tighten slack guide mounting bolts. It is normal for a gap to exist under the bolt seats when mounting bolts are tightened to the specification.

44. Install the timing chain tensioner (primary) with the following procedure:

a. Pull plunger stopper tab up (or turn lever downward) so as to remove plunger stopper tab from the ratchet of plunger.

➡**Plunger stopper tab and lever are synchronized.**

b. Push plunger into the inside of tensioner body.

c. Hold plunger in the fully compressed position by engaging plunger stopper tab with the tip of ratchet.

d. To secure lever, insert stopper pin through hole of lever into tensioner body hole.

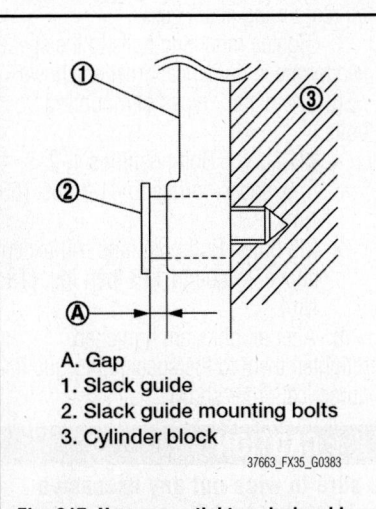

A. Gap
1. Slack guide
2. Slack guide mounting bolts
3. Cylinder block

37663_FX35_G0383

Fig. 217 Never over-tighten slack guide mounting bolts

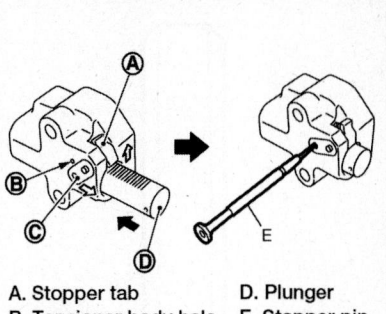

A. Stopper tab D. Plunger
B. Tensioner body hole E. Stopper pin
C. Lever

37663_FX35_G0384

Fig. 218 Install the timing chain tensioner (primary)

e. The lever parts and the plunger stopper tab are synchronized. Therefore, the plunger will be secured under this condition.

➡**Figure shows the example of thin screwdriver being used as the stopper pin.**

f. Install timing chain tensioner (primary).

➡**Remove any dirt and foreign materials completely from the back and the mounting surfaces of timing chain tensioner (primary).**

g. Pull out stopper pin after installing, and then release plunger.

45. Check again that the matching marks on sprockets and timing chain have not slipped out of alignment.

46. Install new O-rings on rear timing chain case.

47. Install new front oil seal on front timing chain case.

a. Apply new engine oil to both oil seal lip and dust seal lip.

b. Install it so that each seal lip is oriented as shown.

c. Using a suitable drift, press-fit oil seal until it becomes flush with front timing chain case end face.

d. Check the garter spring is in position and seal lip is not inverted.

48. Install front timing chain case as per the following:

a. Check O-rings stay in place during installation to rear timing chain case.

b. Apply a continuous bead of liquid gasket to the back side of the front timing chain case as shown.

c. Apply liquid gasket to top surface of upper oil pan as shown.

d. Install front timing chain case.

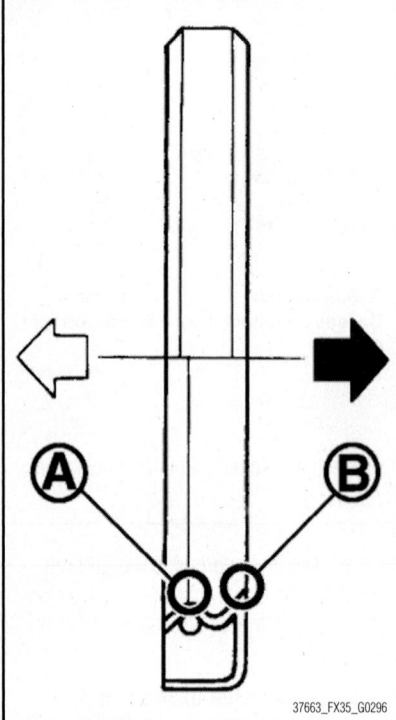

Fig. 219 Install it so that each seal lip is oriented as shown

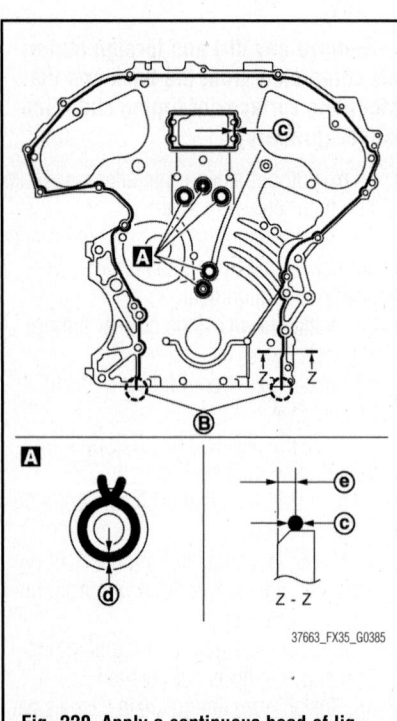

Fig. 220 Apply a continuous bead of liquid gasket to the back side of the front timing chain case

✳✳ WARNING

Be careful not to damage front oil seal by interference with front end of crankshaft. Attaching must be done

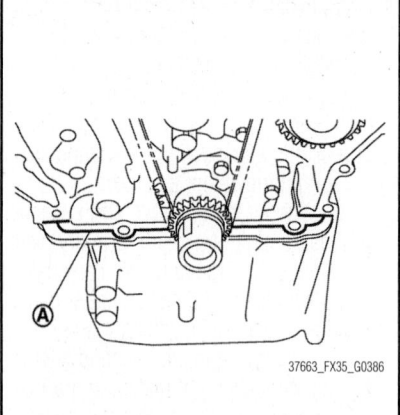

Fig. 221 Apply liquid gasket to top surface of upper oil pan as shown

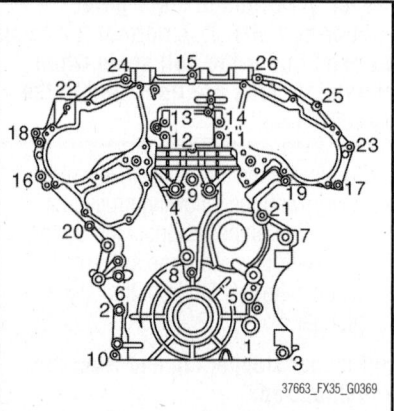

Fig. 222 Tighten mounting bolts in numerical order as shown

within 5 minutes after liquid gasket application.

  e. Install front timing chain case as to fit its dowel pin hole together dowel pin on rear timing chain case.

  f. Tighten mounting bolts to the specified torque in numerical order as shown.

  g. There are 2 types of mounting bolts.

- M10 bolts: Bolt positions 1, 2, 3, 4, 5, 6, and 7, torque to 41 ft. lbs. (55 Nm)
- M6 bolts: Bolt positions: All except above, torque to 115 inch lbs. (13 Nm)

  h. After all bolts are tightened, retighten them to the specified torque in numerical order shown.

✳✳ CAUTION

Be sure to wipe out any excessive liquid gasket leaking on surface mating with upper oil pan.

  i. Install two mounting bolts in front of upper oil pan.

49. Install valve timing control covers (bank 1 and bank 2) as per the following:

  a. Install new seal rings in shaft grooves for Bank 2.

➡**When replacing seal rings, replace all rings with new ones.**

  b. To check the joint between dowel pins and dowel pin holes, check the looseness in the axle direction by pushing the circumferential looseness (between dowel pins and dowel pin holes) by twisting in the circumferential direction.

➡**Always perform this procedure when removing because the gap between dowel pins and dowel pin holes may occur.**

  c. Install valve timing control cover with new gasket to front timing chain case.

✳✳ WARNING

Never face the magnet retarder side down to prevent magnet retarder from dropping.

  d. Check the mating surface of magnet retarder and the drum of exhaust side camshaft sprocket for foreign materials.

  e. Align the center of both shaft holes of the shaft and the intake side camshaft sprocket, and then insert them.

  f. Be careful not to drop the seal ring from the shaft groove.

  g. When setting the valve timing control cover in position by hand, if valve timing control cover is not contacting with the front timing chain case, the dowel pin of magnet retarder may not be aligned with the dowel pin holes of cover.

  h. Being careful not to move seal ring from the installation groove, align dowel pins on front timing chain case with holes to install valve timing control covers.

  i. Tighten mounting bolts in numerical order as shown.

  j. After all bolts are tightened, tighten No. 1 bolt to the specified torque again.

50. Install lower oil pan.

51. Install valve covers (bank 1 and bank 2).

52. Install crankshaft pulley as per the following:

  a. Secure crankshaft using the ring gear stopper KV10118600 (J-48641).

  b. Install crankshaft pulley, taking care not to damage front oil seal.

c. When press-fitting crankshaft pulley with plastic hammer, tap on its center portion (not circumference).

d. Tighten crankshaft pulley bolt.
- Step 1: Tighten to 33 ft. lbs. (44 Nm)
- Step 2: Tighten the bolt 90°

53. Rotate crankshaft pulley in normal direction (clockwise when viewed from front) to confirm it turns smoothly.

54. Install drive belt auto-tensioner bracket and power steering oil pump bracket as per the following:

a. Install drive belt auto-tensioner bracket, and tighten mounting bolts No. 2, 3 (temporarily).

b. Tighten mounting bolts No. 2, 3 to specified torque.

c. Install power steering oil pump bracket, and tighten mounting bolts No. 1, 4, 5 (temporarily).

d. Tighten mounting bolts No. 1 to specified torque.

e. Tighten mounting bolts No. 4, 5 to specified torque.

55. For the following operations, perform steps in the reverse order of removal.

56. Before starting engine, check oil/fluid levels including engine coolant and engine oil. If less than required quantity, fill to the specified level.

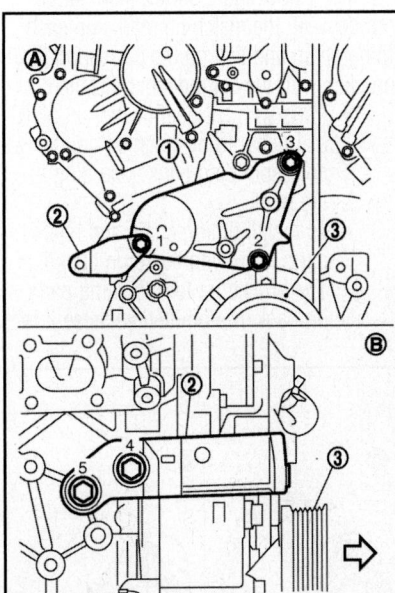

A. Engine front side
B. engine right side
1. Drive belt auto-tensioner
2. Power steering oil pump bracket
3. Crankshaft pulley

37663_FX35_G0388

Fig. 223 Install drive belt auto-tensioner bracket and power steering oil pump bracket

57. Use procedure below to check for fuel leakage.

a. Turn ignition switch "ON" (with engine stopped). With fuel pressure applied to fuel piping, check for fuel leakage at connection points.

b. Start engine. With engine speed increased, check again for fuel leakage at connection points.

58. Run engine to check for unusual noise and vibration.

➡ **If hydraulic pressure inside chain tensioner drops after removal/installation, slack in guide may generate a pounding noise during and just after the engine start. However, this does not indicate a problem. The noise should stop after the hydraulic pressure rises.**

59. Warm up engine thoroughly to check there is no leakage of fuel, or any oil/fluids including engine oil and engine coolant.

60. Bleed air from lines and hoses of applicable lines, such as in cooling system.

61. After cooling down engine, again check oil/fluid levels including engine oil and engine coolant. Refill to the specified level, if necessary.

5.0L Engine

See Figures 224 through 245.

1. Before servicing the vehicle, refer to the Precautions Section.

2. Remove auto tensioners and idler pulley.

3. Remove oil level gauge and oil level gauge guide.

4. Remove alternator bracket and alternator stay.

5. Remove camshaft position sensors.

※ WARNING

Handle carefully to avoid dropping and shocks. Never disassemble. Do not allow metal powder to adhere to magnetic part at sensor tip. Never place sensors in a location where they are exposed to magnetism.

6. Remove valve timing control cover as per the following:

a. Disconnect valve timing control solenoid valve harness connector.

b. Loosen mounting bolts in the reverse order of the tightening sequence.

※ WARNING

Exercise care not to damage mating surfaces. Shaft is internally jointed with camshaft sprocket center hole.

When removing, keep it horizontal until it is completely disconnected.

7. Remove valve timing control solenoid valve (INT and EXH), if necessary.

➡ **Valve timing control solenoid valve is not reusable. Never remove it unless required.**

8. Remove O-rings from front cover.

9. Remove valve covers.

10. Obtain No. 1 cylinder at TDC of its compression stroke.

11. Remove crankshaft pulley.

12. Remove water pump pulley.

13. Remove lower oil pan and oil strainer.

14. Remove upper oil pan.

15. Remove front cover as per the following:

a. Loosen mounting bolts in reverse order of the tightening sequence.

b. Insert a suitable tool into the notch at front cover. Pry off case by moving a suitable tool.

※ WARNING

Exercise care not to damage mating surfaces. After removal, handle front cover carefully so it does not tilt, cant, or warp under a load.

16. Remove front oil seal from front cover using suitable tool.

※ WARNING

Be careful not to damage front cover.

17. Remove O-rings from cylinder heads and cylinder block.

18. Remove oil filter (for valve timing control solenoid valve), if necessary.

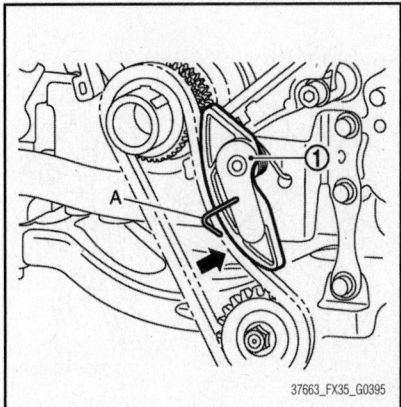

37663_FX35_G0395

Fig. 224 Push oil pump drive chain tensioner (1); insert a stopper pin (A) into the body hole

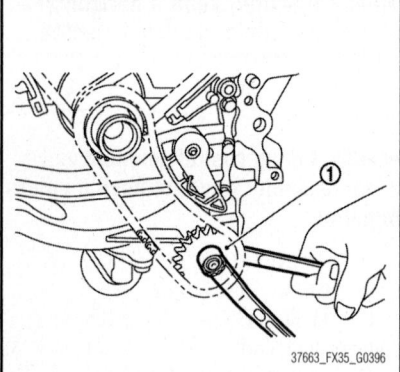

Fig. 225 Hold the 2 flat parts of oil pump shaft, and then loosen the oil pump sprocket (1) (oil pump side) nut

19. Remove timing chain tensioner cover from front cover, if necessary. Use seal cutter KV10111100 (J-37228) to cut liquid gasket for removal.

20. Remove oil pump drive chain as per the following:

    a. Push oil pump drive chain tensioner.

    b. Insert a stopper pin into the body hole.

    c. Hold the 2 flat parts of oil pump shaft, and then loosen the oil pump sprocket (oil pump side) nut.

➡Secure the oil pump unit shaft with the 2 flat parts.

21. Remove oil pump drive chain tensioner.

22. Remove timing chain tensioner (bank 1) as per the following:

➡To remove timing chain and related parts, start with those on bank 1. The procedure for removing parts on bank

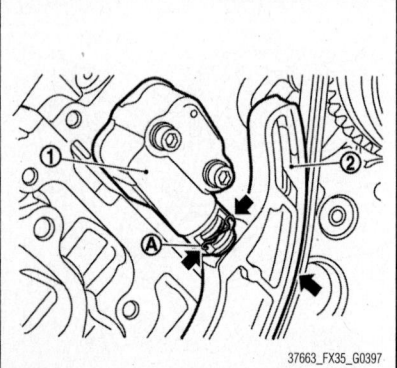

Fig. 226 Timing chain tensioner (bank 1) (1): Push both sides of spring (A) against spring tension, and then press in plunger with a slack guide (2)

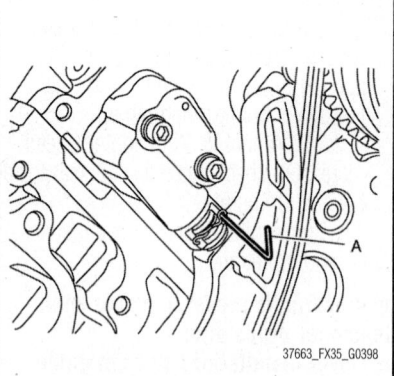

Fig. 227 Insert a stopper pin (A) into the body hole

2 is omitted because it is the same as that for bank 1.

    a. Push both sides of spring against spring tension, and then press in plunger with a slack guide.

    b. Insert a stopper pin into the body hole, and then fix it with the plunger pushed in.

23. Remove tension guide and slack guide.

24. Remove timing chain and crankshaft sprocket.

✳✳ WARNING

After removing timing chain, never turn crankshaft and camshaft separately, or valves will strike the piston head.

25. Remove camshaft sprocket (INT) and (EXH) as per the following:

    a. Exhaust side: Secure the hexagonal portion of camshaft (EXH)

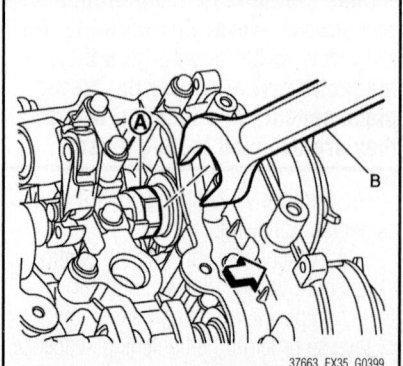

Fig. 228 Secure the hexagonal portion (located in between journal No. 1 and journal No. 2) of driveshaft (A) using a wrench (B) to loosen mounting bolt

using a wrench to loosen mounting bolt.

    b. Intake side: Secure the hexagonal portion (located in between journal No. 1 and journal No. 2) of driveshaft using a wrench to loosen mounting bolt.

➡The figure shows an example of bank 2.

✳✳ WARNING

Never loosen the mounting bolt by securing anything other than the camshaft (driveshaft) hexagonal portion or with tensioning the timing chain. When holding the hexagonal part of camshaft (driveshaft) with a wrench, be careful not to allow the wrench to cause interference with other parts. Do not disassemble camshaft sprocket. Never loosen the bolts on the sprockets.

26. Use scraper to remove all traces of old liquid gasket from front cover and opposite mating surfaces. Remove old liquid gasket from bolt hole and thread.

To install:

➡Do not reuse O-rings.

Note the following:

• The below figure shows the relationship between the matching marks on each timing chain and that on the corresponding sprocket, with the components installed.

    • Parts with an identification mark (R or L) should be installed on the corresponding bank according to the mark.

    • To install timing chain and related parts, start with those on bank 2. The procedure for installing parts on bank 1 is omitted because it is

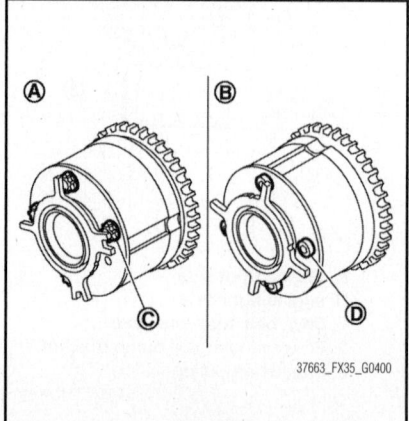

Fig. 229 Never disassemble camshaft sprocket. Never loosen bolts (C) or (D) as shown

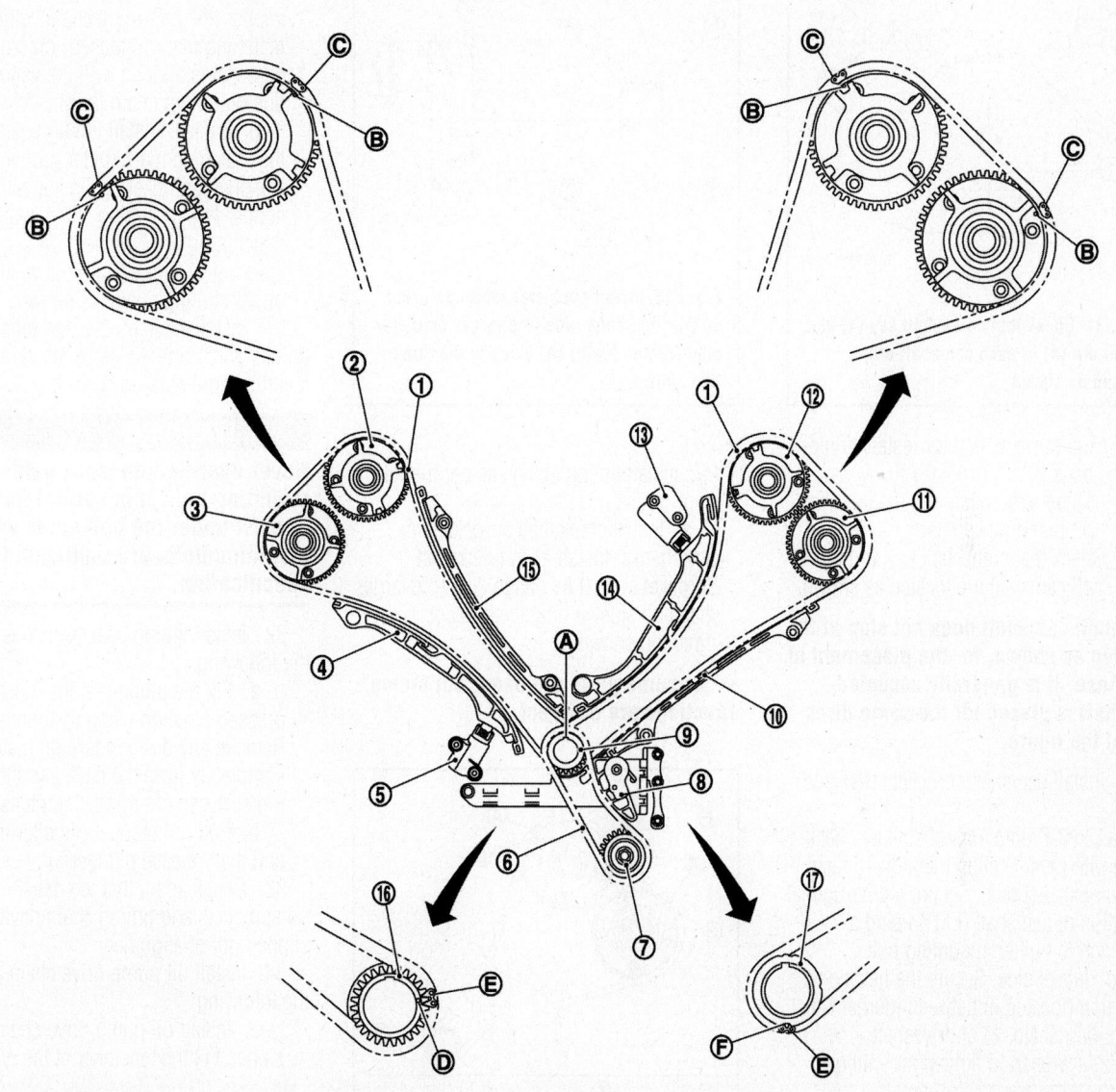

1. Timing chain
2. Camshaft sprocket (INT) (bank 2)
3. Camshaft sprocket (EXH) (bank 2)
4. Slack guide (bank 2)
5. Timing chain tensioner (bank 2)
6. Oil pump drive chain
7. Oil pump sprocket (oil pump side)
8. Oil pump drive chain tensioner
9. Oil pump sprocket (crankshaft side)
10. Tension guide (bank 1)
11. Camshaft sprocket (EXH) (bank 1)
12. Camshaft sprocket (INT) (bank 1)
13. Timing chain tensioner (bank 1)
14. Slack guide (bank 1)
15. Tension guide (bank 2)
16. Crankshaft sprocket (bank 2 side)
17. Crankshaft sprocket (bank 1 side)
A. Crankshaft key
B. Matching mark (outer groove)
C. Matching mark (copper link)
D. Matching mark (punched)
E. Matching mark (yellow link)
F. Matching mark (notched)

37663_FX35_G0401

Fig. 230 This figure shows the relationship between the matching marks on each timing chain and that on the corresponding sprocket, with the components installed

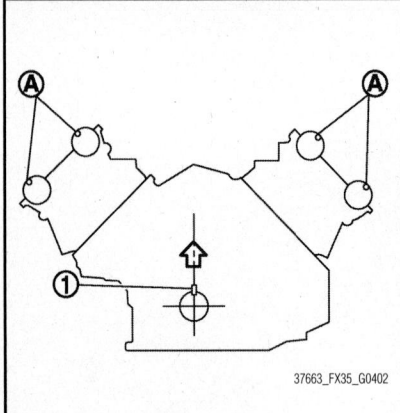

Fig. 231 Check that crankshaft key (1) and dowel pin (A) of each camshaft are located as shown

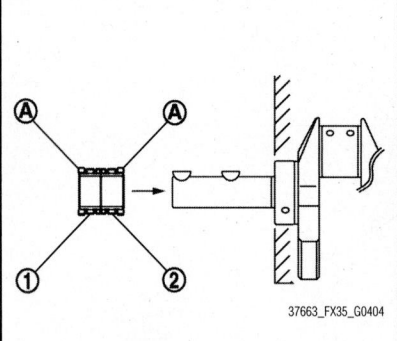

Fig. 233 Install each crankshaft sprocket so that its flange side (the larger diameter side without teeth) (A) faces in the direction shown

the same as that for installation on bank 2.

- There is no matching mark in the oil pump related parts.

27. Check that crankshaft key and dowel pin of each camshaft are located as shown.

➡**Though camshaft does not stop at the position as shown, for the placement of cam nose, it is generally accepted camshaft is placed for the same direction of the figure.**

28. Install camshaft sprockets (INT and EXH).

a. Install onto correct side by checking with identification mark on surface.

b. Exhaust side: Secure the hexagonal portion of camshaft (EXH) using a wrench to tighten mounting bolt.

c. Intake side: Secure the hexagonal portion (located in between journal No. 1 and journal No. 2) of driveshaft using a wrench to tighten mounting

bolt.

29. Install timing chains as per the following:

a. Install crankshaft sprockets for both banks. Install each crankshaft sprocket so that its flange side (the larger diameter side without teeth) faces in the direction shown.

➡**The same parts are used, but facing directions are different.**

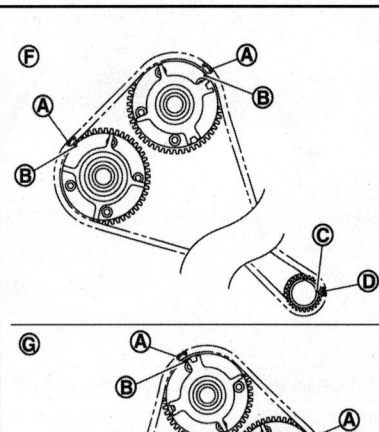

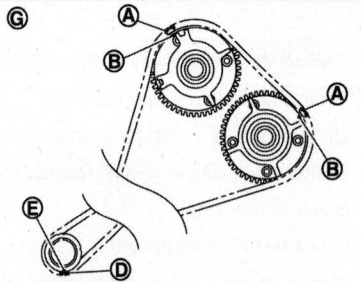

A. Copper link
B. Matching mark (outer groove)
C. Matching mark (punched)
D. Yellow link
E. Matching mark (notched)
F. Bank 2
G. Bank 1

Fig. 234 Install timing chains

30. Install timing chains.

a. Bank 2: Install timing chain so that the matching mark (outer groove) on camshaft sprocket is aligned with the copper link on timing chain, while the matching mark (punched) on crankshaft sprocket is aligned with the yellow link one on timing chain, as shown.

b. Bank 1: Install timing chain so that the matching mark (outer groove) on camshaft sprocket is aligned with the copper link on timing chain, while the matching mark (notched) on crankshaft sprocket is aligned with the yellow link one on timing chain, as shown.

31. Install slack guides and tension guides onto correct side by checking with identification mark on surface.

✳✳ WARNING

Never over-tighten slack guide mounting bolt. It is normal for a gap to exist under the bolt seats when mounting bolts are tightened to the specification.

32. Install timing chain tensioner as per the following:

a. Fix the plunger at the most compressed position using a stopper pin (A). Remove any dirt and foreign materials completely from the back and the mounting surfaces of timing chain tensioner.

b. Pull out stopper pin after installing, and then release plunger.

33. Check again that the matching marks on sprockets and timing chain have not slipped out of alignment.

34. Install oil pump drive chain as per the following:

a. Install oil pump drive chain tensioner. Fix the tensioner at the most

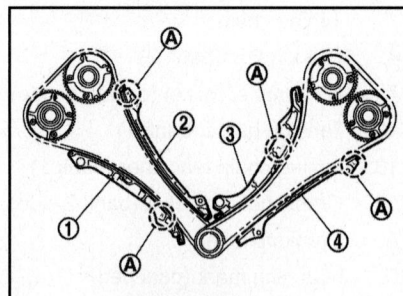

A. Identification mark
1. Slack guide (bank 2)
2. Tension guide (bank 2)
3. Slack guide (bank 1)
4. Tension guide (bank 1)

Fig. 235 Install slack guides and tension guides onto correct side

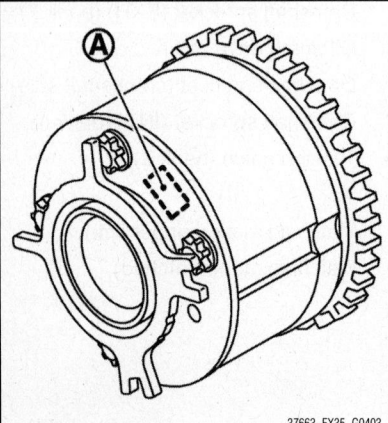

Fig. 232 Install onto correct side by checking with identification mark (A) on surface

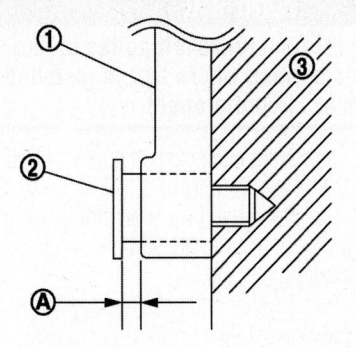

A. Gap
1. Slack guide
2. Slack guide mounting bolt
3. Cylinder block

37663_FX35_G0407

Fig. 236 Never over-tighten slack guide mounting bolt

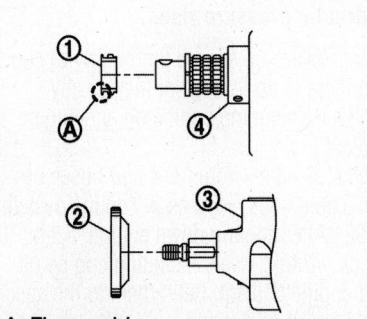

A. Flange side
1. Oil pump sprocket (crankshaft side)
2. Oil pump sprocket (oil pump side)
3. Oil pump
4. Crankshaft

37663_FX35_G0408

Fig. 237 Install each oil pump sprocket so that its flange side (the larger diameter side without teeth) faces in the direction shown

compressed position using a stopper pin and then install it.

b. Install the oil pump sprocket (crankshaft side), oil pump sprocket (oil pump side) and oil pump drive chain at the same time. Install each oil pump sprocket so that its flange side (the larger diameter side without teeth) faces in the direction shown.

➡**There are no matching marks on the oil pump related parts.**

c. Hold the 2 flat parts of oil pump shaft, and then tighten the oil pump sprocket (oil pump side) nut. Secure the oil pump shaft with the 2 flat parts.

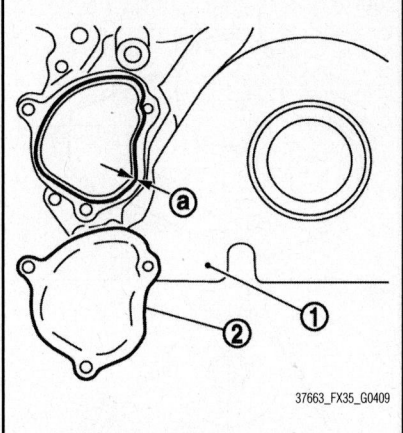

37663_FX35_G0409

Fig. 238 Install timing chain tensioner cover (2) to front cover (1)

d. Pull out the stopper pin after installing the oil pump drive chain. Check that the tension is applied to the oil pump drive chain after installing.

35. Install front oil seal on front cover.

36. Install timing chain tensioner cover to front cover. Apply a continuous bead of liquid gasket to front cover as shown.

37. Install oil filter (for valve timing control solenoid valve) in the direction shown, if removed. Check that the oil filter does not protrude from the upper surface of front cover after installation.

38. Install front cover as per the following:

a. Install new O-ring onto cylinder heads and cylinder block.

b. Apply a continuous bead of liquid gasket to front cover as shown.

c. Check again that the matching marks on timing chain and that on each sprocket are aligned. Then, install front cover.

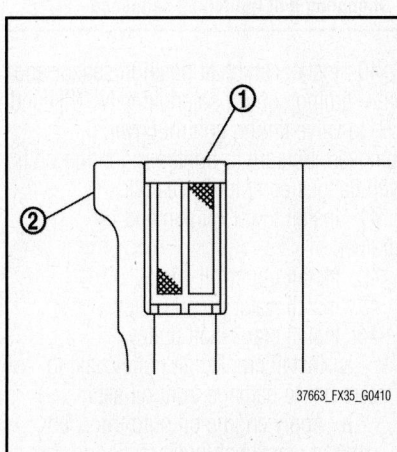

37663_FX35_G0410

Fig. 239 Install oil filter (1) in the direction shown; front cover (2)

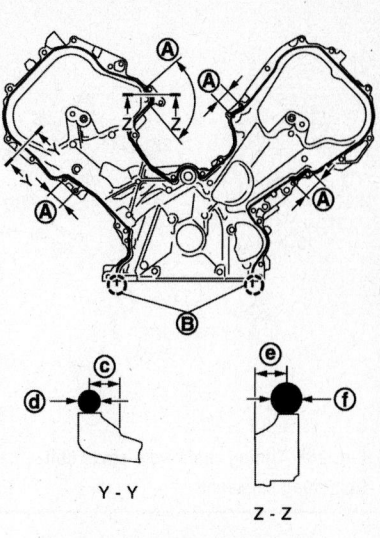

A. Junction between cylinder block and cylinder head
B. Protrusion
c. 0.169-0.209 inches (4.3-5.3 mm)
d. 0.134-0.173 inches (3.4-4.4 mm)
e. 0.157-0.220 inches (4.0-5.6 mm)
f. 0.189-0.228 inches (4.8-5.8 mm)

37663_FX35_G0411

Fig. 240 Apply a continuous bead of liquid gasket to front cover as shown

✳✳ **WARNING**

Be careful not to damage front oil seal by interference with front end of crankshaft.

d. Tighten mounting bolts in numerical order as shown.

➡**There are 3 types of mounting bolts.**

- Bolt A: 0.79 inch (20mm)
- Bolt B: 1.77 inches (45mm)
- Bolt C: 3.15 inches (80mm)

e. After all mounting bolts are tightened, retighten them in numerical order as shown.

➡**Be sure to wipe out any excessive liquid gasket leaking onto surface mating with oil pan.**

39. Install valve timing control cover as per the following:

a. Install new O-rings on front cover.

b. Install new seal rings in shaft grooves.

➡**When replacing seal ring, replace all rings with new ones.**

c. Apply a continuous bead of liquid gasket to valve timing control covers as shown.

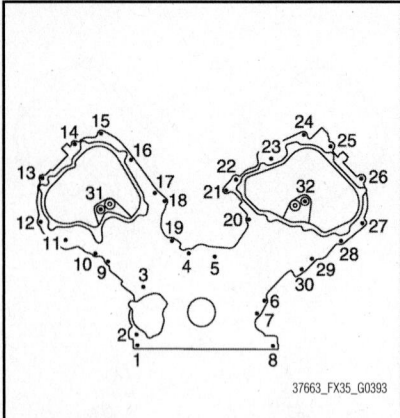

Fig. 241 Timing chain front cover bolt tightening sequence

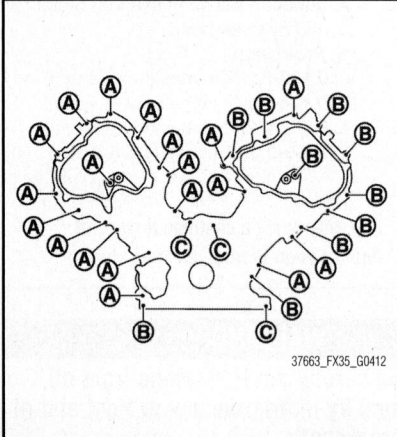

Fig. 242 Proper bolt locations (A, B, C)

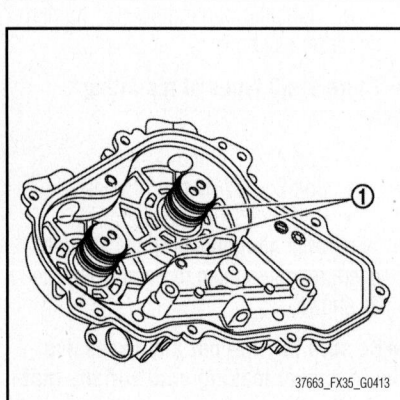

Fig. 243 Install new seal rings (1) in shaft grooves

d. Being careful not to move seal ring from the installation groove, align dowel pins on front cover with dowel pin holes to install valve timing control covers.

e. Tighten mounting bolts in numerical order as shown.

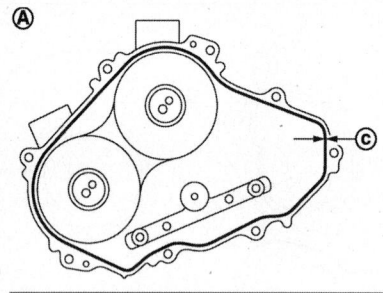

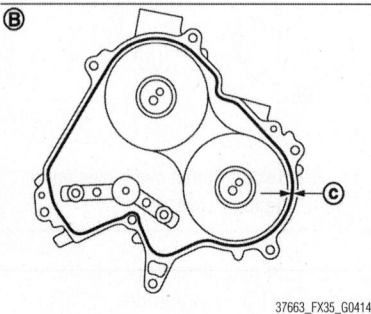

Fig. 244 Apply a continuous bead of liquid gasket to valve timing control covers as shown

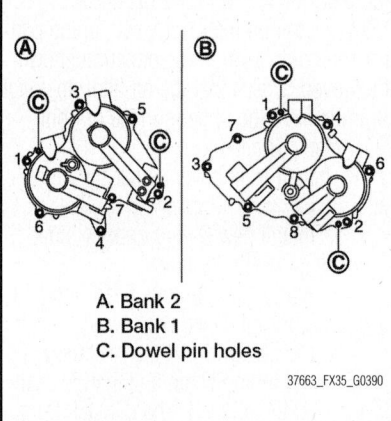

A. Bank 2
B. Bank 1
C. Dowel pin holes

Fig. 245 Valve timing control cover mounting bolt tightening sequence

40. Install camshaft position sensor and valve timing control solenoid valve (RH and LH) to valve timing control cover, if removed. Be sure to tighten mounting bolts with flanges completely seated.

41. Install lower oil pan and oil strainer.

42. Install upper oil pan.

43. Install water pump pulley.

44. Install crankshaft pulley.

a. Install crankshaft pulley, taking care not to damage front oil seal.

b. Apply engine oil onto threaded parts of crankshaft pulley bolt and seating area. Lightly tapping its center with plastic hammer, insert crankshaft pulley.

✲✲ WARNING

Never tap crankshaft pulley on the side surface where belt is installed (outer circumference).

c. Tighten crankshaft pulley bolt to 116 ft. lbs. (157 Nm).

d. Put a paint mark on crankshaft pulley aligning with angle mark on crankshaft pulley bolt.

e. Tighten crankshaft pulley bolt (clockwise) 90°.

45. Rotate crankshaft pulley in normal direction (clockwise when viewed from engine front) to confirm it turns smoothly.

46. Installation continues in the reverse of the removal procedure.

47. Before starting engine, check oil/fluid levels including engine coolant and engine oil. If any are less than the required quantity, fill them to the specified level.

48. Follow the procedure below to check for fuel leakage.

a. Turn ignition switch to the "ON" position (with engine stopped). With fuel pressure applied to fuel piping, check for fuel leakage at connection points.

b. Start engine. With engine speed increased, check again for fuel leakage at connection points.

49. Run engine to check for unusual noise and vibration.

➡ **If hydraulic pressure inside chain tensioner drops after removal/installation, slack in guide may generate a pounding noise during and just after the engine start. However, this does not indicate a malfunction. The noise should stop after the hydraulic pressure rises.**

50. Warm up engine thoroughly to check that there is no leakage of fuel, or any oil/fluids including engine oil and engine coolant.

51. Bleed air from lines and hoses of applicable lines, such as in cooling system.

52. After cooling down engine, again check oil/fluid levels including engine oil and engine coolant. Refill them to the specified level, if necessary.

VALVE (ROCKER) COVER

REMOVAL & INSTALLATION

3.5L Engine

See Figures 246 and 247.

1. Before servicing the vehicle, refer to the Precautions Section.

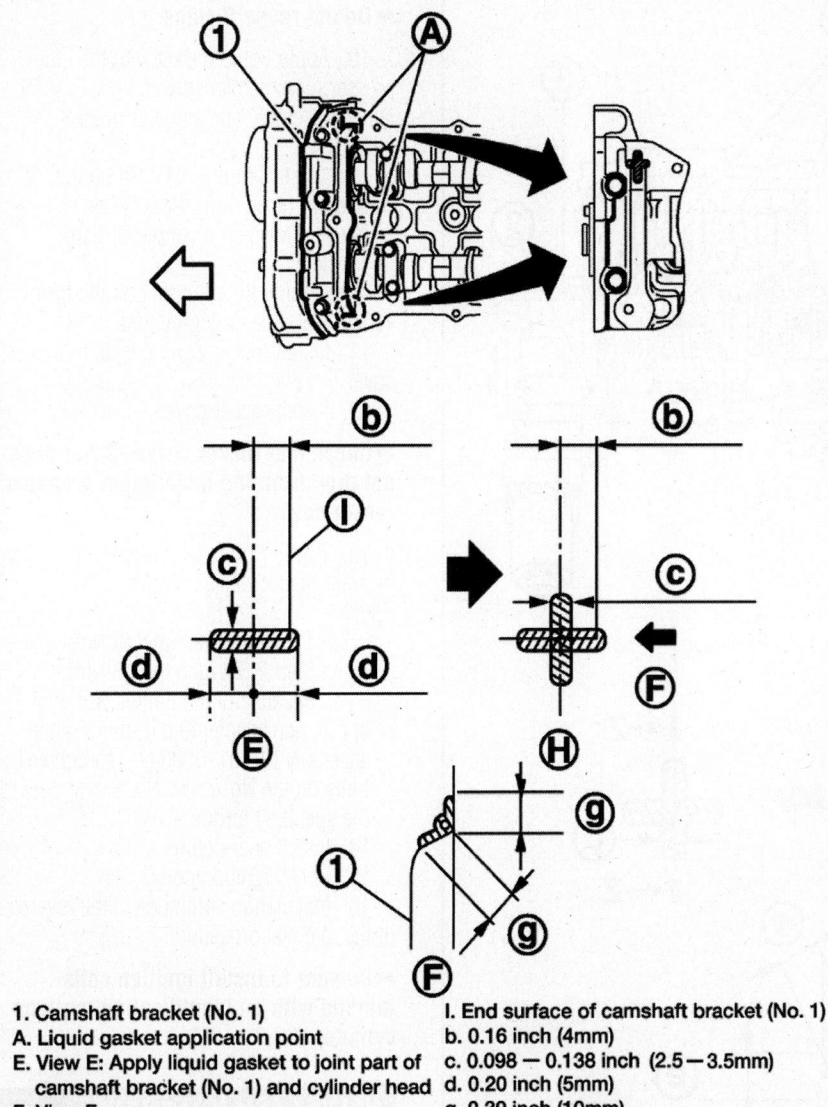

1. Camshaft bracket (No. 1)
A. Liquid gasket application point
E. View E: Apply liquid gasket to joint part of camshaft bracket (No. 1) and cylinder head
F. View F
H. View H: Apply liquid gasket in 90°

I. End surface of camshaft bracket (No. 1)
b. 0.16 inch (4mm)
c. 0.098 — 0.138 inch (2.5 — 3.5mm)
d. 0.20 inch (5mm)
g. 0.39 inch (10mm)
White Arrow: Engine front

71075_FX35_G0238

Fig. 246 Apply liquid gasket to the position shown

2. Remove ignition coils. Refer to Ignition Coil(s), removal & installation.

3. Remove PCV valve and O-ring from rocker cover, if necessary.

4. Remove oil filler cap from rocker cover, if necessary.

5. Remove harness clips on the rocker cover.

6. Loosen mounting bolts in reverse order of tightening sequence.

7. Remove rocker cover gasket from rocker cover.

8. Use scraper to remove all traces of liquid gasket from cylinder head and camshaft bracket (No. 1).

✳✳ WARNING

Never scratch or damage the mating surface when cleaning off old liquid gasket.

To install:

➡**Do not reuse O-rings.**

9. Apply liquid gasket to the position shown in the figure.

➡**Use Genuine RTV silicone sealant or an equivalent.**

a. Refer to figure to apply liquid gasket to joint part of camshaft bracket (No. 1) and cylinder head.

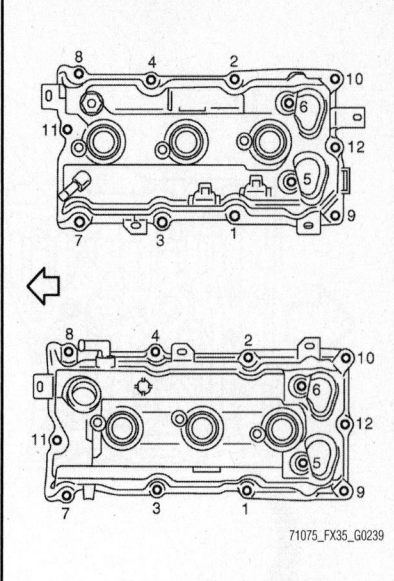

Fig. 247 Valve (rocker) cover bolt tightening sequence and direction of engine front (White Arrow)—3.5L engine

b. Refer to figure to apply liquid gasket in 90° to figure.

10. Install rocker cover gasket to rocker cover.

11. Install rocker cover.

➡**Be sure rocker cover gasket has not dropped from the installation groove of the rocker cover.**

12. Tighten bolts in 2 steps separately in numerical order as shown in the figure.
   a. Step 1: 18 inch lbs. (2 Nm).
   b. Step 2: 73 inch lbs. (8 Nm).

13. Installation continues in the reverse of the removal procedure.

5.0L Engine

See Figures 248 through 250.

1. Before servicing the vehicle, refer to the Precautions Section.

2. Remove ignition coils. Refer to Ignition Coil(s), removal & installation.

3. Disconnect the PCV hose from the rocker cover.

➡**Installation position of ignition coil depends on cylinder position.**

4. Remove spark plugs.

5. Remove rocker cover. Loosen bolts in reverse order of the tightening sequence.

6. Remove rocker cover gasket from rocker cover.

7. Use scraper to remove all traces of liquid gasket from cylinder head and VVEL ladder assembly.

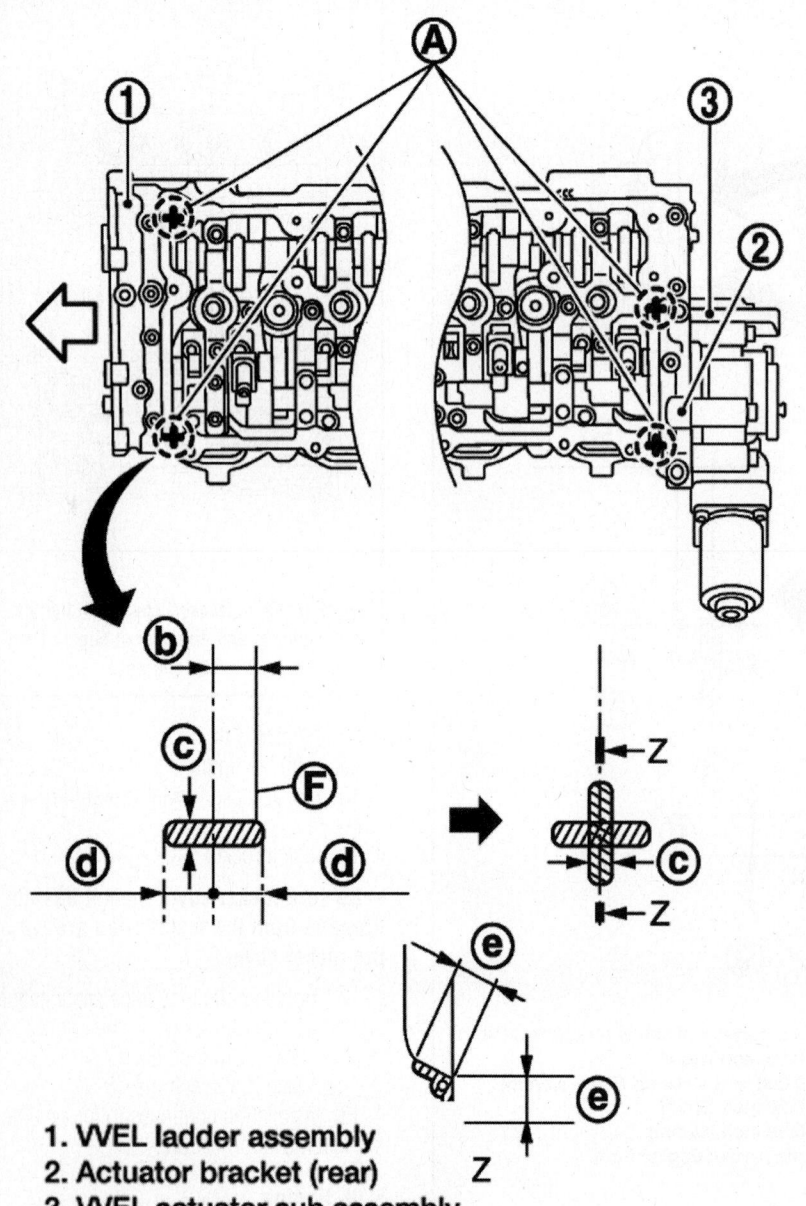

1. VVEL ladder assembly
2. Actuator bracket (rear)
3. VVEL actuator sub assembly
A. Liquid gasket application point
F. End surface of VVEL ladder assembly
b. 0.16 inch (4mm)
c. 0.098 – 0.138 inch
   (2.5 – 3.5mm)
d. 0.20 inch (5mm)
e. 0.39 inch (10mm)
White Arrow: Engine front

71075_FX35_G0240

Fig. 248 Apply liquid gasket to VVEL ladder assembly and actuator bracket (rear)

※※ WARNING

Do not scratch or damage the mating surface when cleaning off old liquid gasket.

---

8. Remove PCV valve from rocker cover, if necessary.

9. Remove oil filler cap and oil catcher from rocker cover, if necessary.

To install:

➡**Do not reuse O-rings.**

10. Apply liquid gasket with the tube presser (commercial service tool) to VVEL ladder assembly and actuator bracket (rear).

   a. Use Genuine RTV Silicone Sealant or an equivalent. (The figure shows an example of bank 1 side).

   b. Apply liquid gasket on the front and rear side of engine first.

11. Install rocker cover gasket to rocker cover.

12. Install rocker cover.

➡**Check that rocker cover gasket does not drop from the installation groove of rocker cover.**

13. Tighten bolts in 2 steps separately in numerical order as shown in the figure.

   a. Step 1: 18 inch lbs. (2 Nm).

   b. Step 2: 73 inch lbs. (8 Nm).

   c. Because of the limited working space, use adapter and torque wrench assembly KV10119300 (-) to tighten bolts on the No. 7 and No. 8 cylinders to the specified torque.

14. Install spark plugs.

15. Install ignition coils.

16. Installation continues in the reverse of the removal procedure.

➡**Be sure to install ignition coils marked with an identification mark on cylinder No. 7 and 8.**

VALVE LASH (CLEARANCE) ADJUSTMENT

ADJUSTMENT

3.5L Engine

See Figures 251 and 252.

Perform adjustment depending on selected head thickness of valve lifter.

1. Before servicing the vehicle, refer to the Precautions Section.

2. Measure the valve clearance. Refer to Inspection procedure.

3. Remove camshaft.

4. Remove valve lifters at the locations that are out of the standard.

   a. Standard intake valve clearance:

   • Cold: 0.010–0.013 inch (0.26–0.34mm)

   • Hot: 0.012–0.016 inch (0.304–0.416mm)

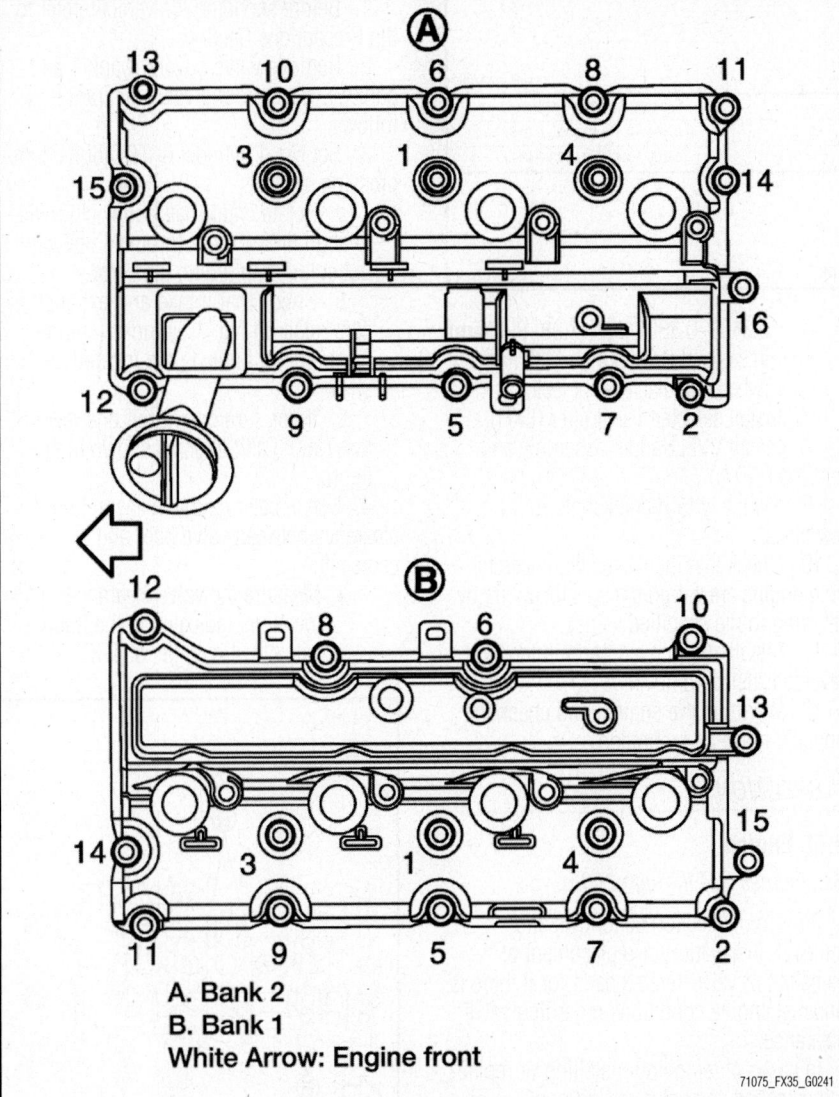

A. Bank 2
B. Bank 1
White Arrow: Engine front

71075_FX35_G0241

Fig. 249 Valve (rocker) cover bolt tightening sequence—5.0L engine

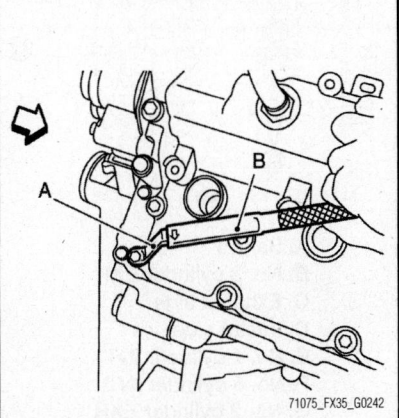

71075_FX35_G0242

Fig. 250 Using adapter (A) and torque wrench (B) assembly KV10119300 (-) to tighten bolts on the No. 7 and No. 8 cylinders. Engine front shown (White Arrow)

b. Standard exhaust valve clearance:
• Cold: 0.011–0.015 inch (0.29–0.37mm)
• Hot: 0.012–0.017 inch (0.308–0.432mm)

5. Measure the center thickness of the removed valve lifters with a micrometer.

6. Use the equation below to calculate valve lifter thickness for replacement.

a. Valve lifter thickness calculation:
$t = t1 + (C1 - C2)$.

b. t = Valve lifter thickness to be replaced.

c. $t1$ = Removed valve lifter thickness.

d. $C1$ = Measured valve clearance.

e. $C2$ = Standard valve clearance.

• Thickness of new valve lifter can be identified by stamp marks on the reverse side (inside the cylinder)

• Stamp mark 788 indicates 0.3102 inch (7.88mm) in thickness

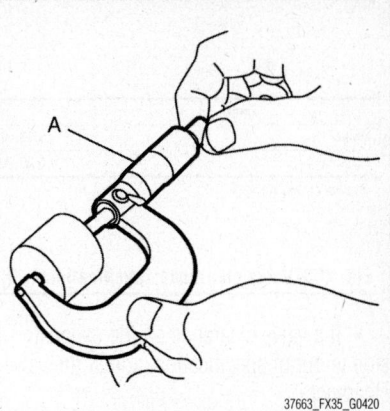

37663_FX35_G0420

Fig. 251 Measure the center thickness of the removed valve lifters with a micrometer (A)

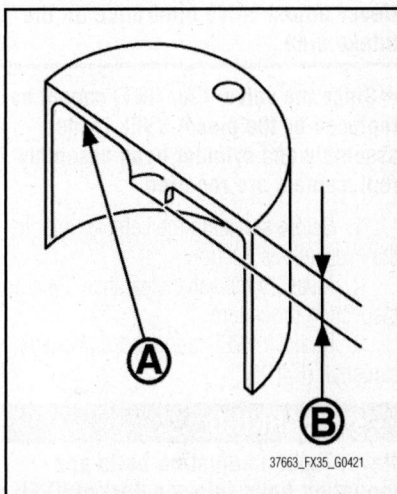

37663_FX35_G0421

Fig. 252 Valve lifter stamp mark (A) and measured thickness (B)

• Available thickness of valve lifter: 27 sizes with range 0.3102–0.3307 inch (7.88–8.40mm) in steps of 0.0008 inch (0.02mm) (when manufactured at factory)

7. Install selected valve lifter.

8. Install camshaft.

9. Manually turn crankshaft pulley a few turns.

10. Check that the valve clearances for cold engine are within the specifications by referring to the specified values.

11. Install all removal parts in the reverse order of removal.

12. Warm up the engine, and check for unusual noise and vibration.

5.0L Engine

See Figures 251 through 253.

Note the following:
• Perform adjustment depending on selected head thickness of valve lifter (EXH)

Items	Cold	Hot* (reference data)
		Unit: mm (in)
Intake	0.26 - 0.34 (0.010 - 0.013)	0.304 - 0.416 (0.012 - 0.016)
Exhaust	0.29 - 0.37 (0.011 - 0.015)	0.308 - 0.432 (0.012 - 0.017)

*: Approximately 80°C (176°F)

71075_FX35_G0243

Fig. 253 Valve clearance specifications—5.0L engine

- If a valve clearance on the exhaust side is out of specification, adjust the valve clearance
- If a valve clearance on the intake side is out of specification, replace VVEL ladder assembly and cylinder head assembly.

✳✳ WARNING

Never adjust valve clearance on the intake side.

➡Since the valve lifter (INT) cannot be replaced by the piece, VVEL ladder assembly and cylinder head assembly replacement are required.

1. Before servicing the vehicle, refer to the Precautions Section.
2. Measure the valve clearance. Refer to Inspection procedure.
3. Remove VVEL ladder assembly and camshaft (EXH).

✳✳ WARNING

Never loosen adjusting bolts and mounting bolts (black color) of VVEL ladder assembly.

4. Remove valve lifter (EXH) at the locations that are out of the standard.
5. Measure the center thickness of the removed valve lifters (EXH) with a micrometer (A).
6. Use the equation below to calculate valve lifter (EXH) thickness for replacement.
   a. Valve lifter (EXH) thickness calculation: $t = t1 + (C1 - C2)$.
   b. t = Valve lifter (EXH) thickness to be replaced.
   c. $t1$ = Removed valve lifter (EXH) thickness.
   d. $C1$ = Measured valve clearance.
   e. $C2$ = Standard valve clearance.
- Thickness of new valve lifter (EXH) can be identified by stamp marks on the reverse side (inside the cylinder)
- Stamp mark 788 indicates 0.3102 inch (7.88mm) in thickness.
- Available thickness of valve lifter (EXH): 27 sizes with range

0.3102–0.3307 inch (7.88–8.40mm) in steps of 0.0008 inch (0.02mm) (when manufactured at factory)

7. Install selected valve lifter (EXH).
8. Install VVEL ladder assembly and camshaft (EXH).
9. Manually turn crankshaft pulley a few turns.
10. Check that the valve clearances for cold engine are within the specifications by referring to the specified values.
11. Install all removed parts in the reverse order of removal.
12. Warm up the engine, and check for unusual noise and vibration.

INSPECTION

3.5L Engine

See Figures 254 through 258.

Perform inspection as follows after removal, installation or replacement of camshaft or valve-related parts, or if there is unusual engine conditions regarding valve clearance.

In cases of removing/installing or replacing camshaft and valve related parts, or of unusual engine conditions due to changes in valve clearance (found malfunctions during starting, idling or causing noise), perform inspection as follows.

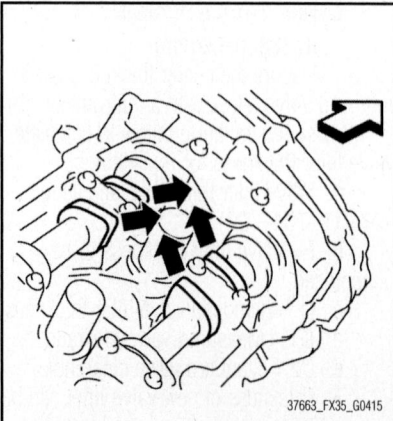

37663_FX35_G0415

Fig. 254 Check that intake and exhaust cam nose on No. 1 cylinder (engine front side of bank 1) are located as shown

1. Before servicing the vehicle, refer to the Precautions Section.
2. Remove valve covers (bank 1 and bank 2). Measure the valve clearance as follows.
3. Set No. 1 cylinder at TDC of its compression stroke.
   a. Rotate crankshaft pulley clockwise to align timing mark (grooved line without color) with timing indicator.
   b. Check that intake and exhaust cam nose on No. 1 cylinder (engine front side of bank 1) are located as shown.
   c. If not, turn crankshaft one revolution (360°) and align as shown in the figure.
4. Use a feeler gauge, measure the clearance between valve lifter and camshaft.

- Measure the valve clearances at locations indicated in the figure with No. 1 cylinder at TDC.

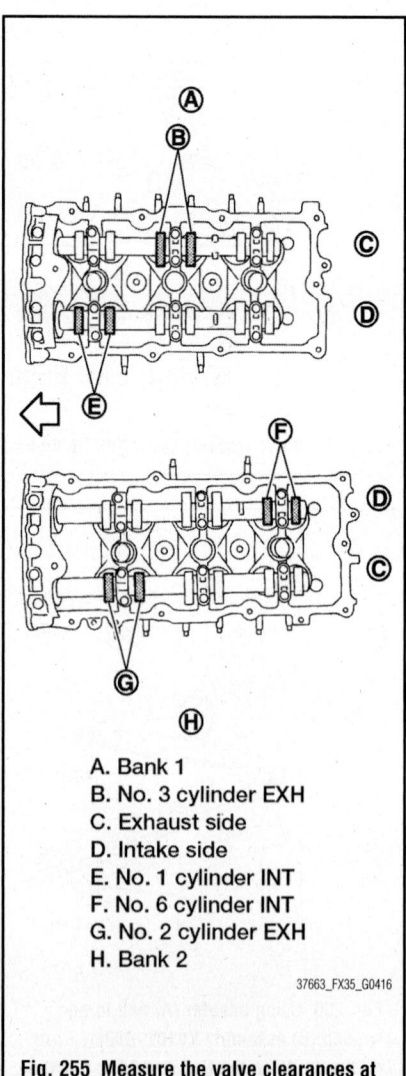

A. Bank 1
B. No. 3 cylinder EXH
C. Exhaust side
D. Intake side
E. No. 1 cylinder INT
F. No. 6 cylinder INT
G. No. 2 cylinder EXH
H. Bank 2

37663_FX35_G0416

Fig. 255 Measure the valve clearances at locations indicated in the figure

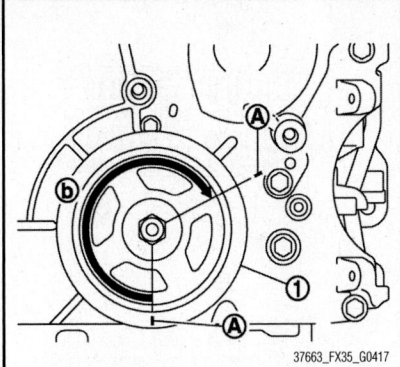

Fig. 256 Mark (A) a position 240 degrees (b) from a corner of the hexagonal part of crankshaft pulley (1) mounting bolt as shown

5. Rotate crankshaft 240° clockwise (when viewed from engine front) to align No. 3 cylinder at TDC its compression stroke.

   a. Mark a position 240° from a corner of the hexagonal part of crankshaft pulley mounting bolt as shown. Use the hexagonal part as a guide.

   b. Measure the valve clearances at locations indicated in the figure with No. 3 cylinder at TDC.

6. Rotate crankshaft 240° clockwise (when viewed from engine front) to align No. 5 cylinder at TDC of compression stroke.

   a. Mark a position 240° from a corner of the hexagonal part of crankshaft pulley mounting bolt. Use the hexagonal part as a guide.

   b. Measure the valve clearances at locations indicated in the figure with No. 5 cylinder at TDC.

7. Perform adjustment if the measured value is out of the standard.

   a. Standard intake valve clearance:
   • Cold: 0.010–0.013 inches (0.26–0.34mm)
   • Hot: 0.012–0.016 inch (0.304–0.416mm)

   b. Standard exhaust valve clearance:
   • Cold: 0.011–0.015 inch (0.29–0.37mm)
   • Hot: 0.012–0.017 inch (0.308–0.432mm)

5.0L Engine

Intake Side

Inspect at the removal and installation of VVEL ladder assembly or valve-related parts, or at the occurrence of malfunction (poor starting, idle malfunction, unusual

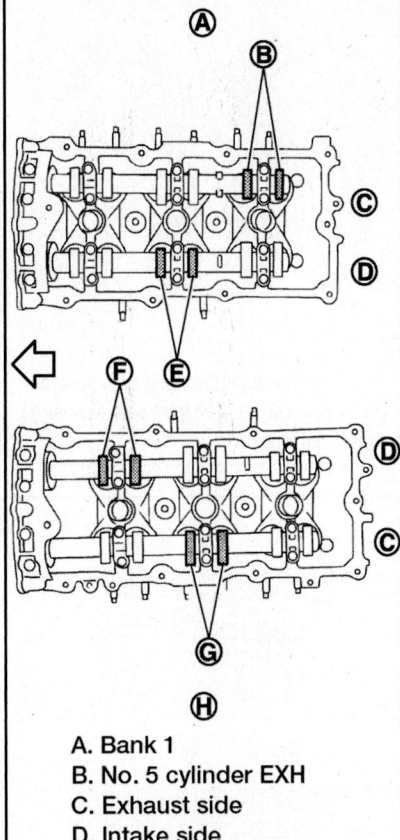

A. Bank 1
B. No. 5 cylinder EXH
C. Exhaust side
D. Intake side
E. No. 3 cylinder INT
F. No. 2 cylinder INT
G. No. 4 cylinder EXH
H. Bank 2

Fig. 257 Measure the valve clearances at locations indicated in the figure

noise) due to aged deterioration in valve clearance.

➡**Valve clearance check on the intake side is not required after replacing the VVEL ladder assembly and cylinder head assembly with a new one. (Install new VVEL ladder assembly and cylinder head assembly in factory-shipped condition because it is factory-adjusted and inspected.)**

➡**VVEL ladder assembly cannot be replaced as a single part, because it is machined together with cylinder head assembly.**

Exhaust Side

See Figures 259 through 265.

Inspect at the removal, installation, and replacement of camshaft (EXH) or valve-related parts, or at the occurrence of malfunction (poor starting, idle malfunction,

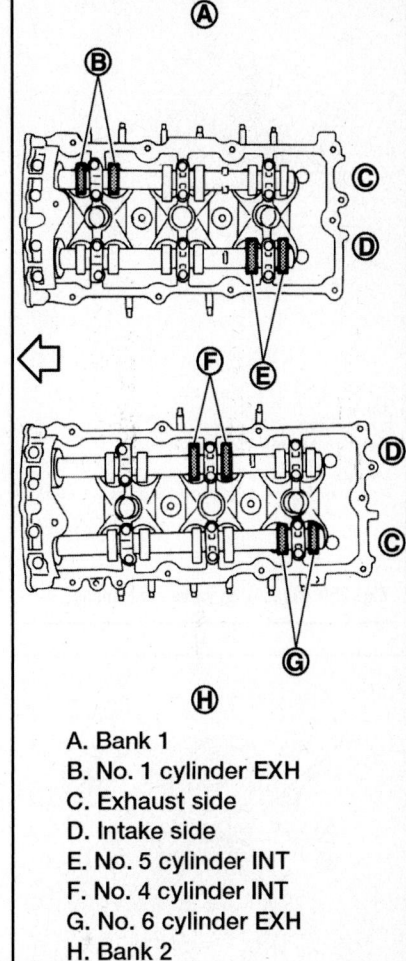

A. Bank 1
B. No. 1 cylinder EXH
C. Exhaust side
D. Intake side
E. No. 5 cylinder INT
F. No. 4 cylinder INT
G. No. 6 cylinder EXH
H. Bank 2

Fig. 258 Measure the valve clearances at locations indicated in the figure

unusual noise) due to aged deterioration in valve clearance.

1. Before servicing the vehicle, refer to the Precautions Section.

2. Remove rocker covers (bank 1 and bank 2).

3. Measure the valve clearance as per the following:

   a. Use the feeler gauge of curved-tip. This allows the feeler gauge to access the clearance between camshaft (drive-shaft) nose and valve lifter with ease.

➡**Be sure to note the following points when measuring valve clearance on the intake side.**

   b. Before measuring, check that the position of driveshaft nose is within the angle shown.

   c. Refer to the figure for the insertion direction of the feeler gauge since the direction depends on the bank.

4. Set No. 1 cylinder at TDC of its compression stroke.

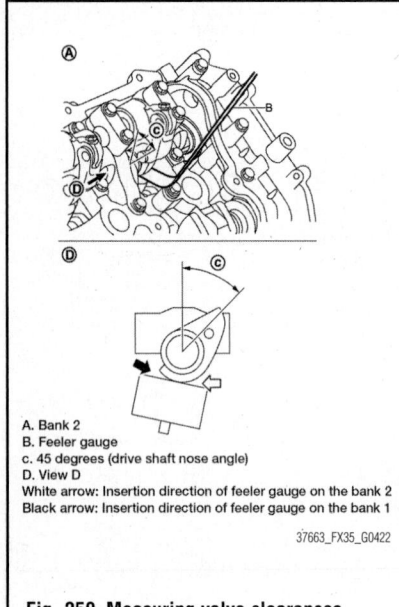

A. Bank 2
B. Feeler gauge
c. 45 degrees (drive shaft nose angle)
D. View D
White arrow: Insertion direction of feeler gauge on the bank 2
Black arrow: Insertion direction of feeler gauge on the bank 1

37663_FX35_G0422

Fig. 259 Measuring valve clearances

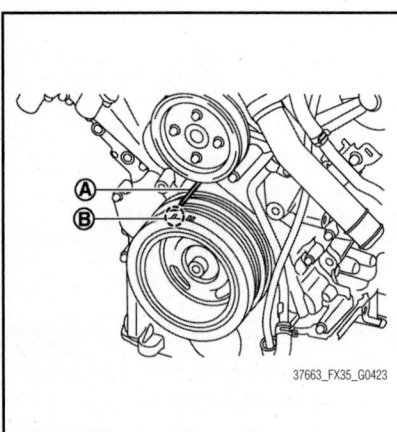

37663_FX35_G0423

Fig. 260 Rotate crankshaft pulley clockwise to align timing mark (grooved line without color) (B) with timing indicator (A)

a. Rotate crankshaft pulley clockwise to align timing mark (grooved line without color) with timing indicator.

b. Check that exhaust cam nose on No. 1 cylinder (engine front side of bank 1) is located as shown.

c. If not, turn crankshaft one revolution (360°) and align as shown.

d. Measure the valve clearances at locations indicated in the figure with No. 1 cylinder at compression TDC.

➡**To measure valve clearance of No. 1 cylinder INT valve (front side), insert feeler gauge from the front side of the control shaft bracket or camshaft (EXH) side.**

5. Rotate crankshaft 270° clockwise (when viewed from engine front) to align No. 3 cylinder at TDC its compression stroke.

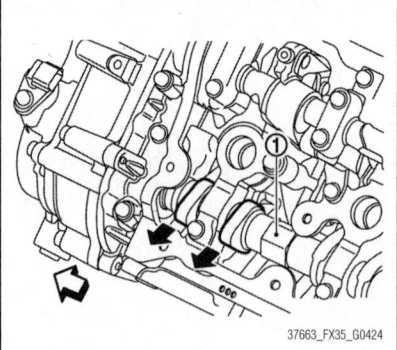

37663_FX35_G0424

Fig. 261 Check that exhaust cam nose on No. 1 cylinder (engine front side of bank 1) is located as shown

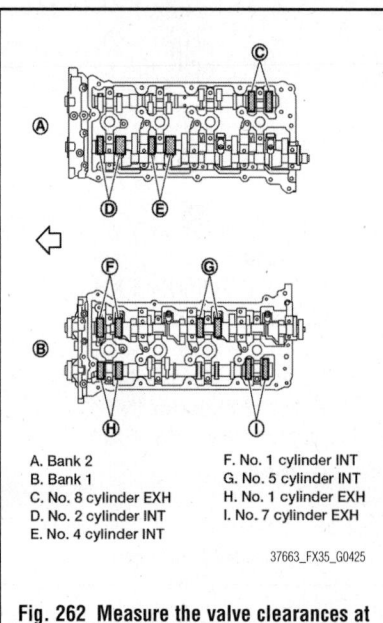

A. Bank 2
B. Bank 1
C. No. 8 cylinder EXH
D. No. 2 cylinder INT
E. No. 4 cylinder INT
F. No. 1 cylinder INT
G. No. 5 cylinder INT
H. No. 1 cylinder EXH
I. No. 7 cylinder EXH

37663_FX35_G0425

Fig. 262 Measure the valve clearances at locations indicated in the figure with No. 1 cylinder at compression TDC

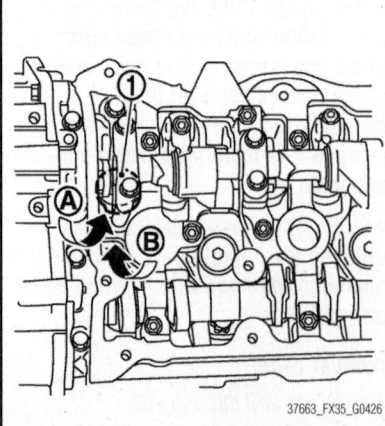

37663_FX35_G0426

Fig. 263 Insert feeler gauge from the front side (A) of the control shaft bracket or camshaft (EXH) side (B)

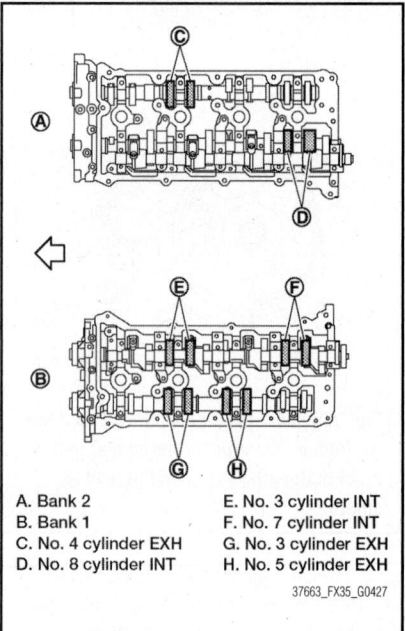

A. Bank 2
B. Bank 1
C. No. 4 cylinder EXH
D. No. 8 cylinder INT
E. No. 3 cylinder INT
F. No. 7 cylinder INT
G. No. 3 cylinder EXH
H. No. 5 cylinder EXH

37663_FX35_G0427

Fig. 264 Measure the valve clearances at locations indicated in the figure with No. 3 cylinder at compression TDC

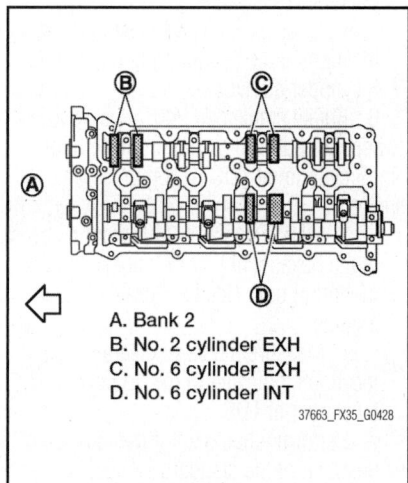

A. Bank 2
B. No. 2 cylinder EXH
C. No. 6 cylinder EXH
D. No. 6 cylinder INT

37663_FX35_G0428

Fig. 265 Measure the valve clearances at locations indicated in the figure with No. 6 cylinder at compression TDC

➡**Crankshaft pulley mounting bolt flange has an angle mark every 90°. They can be used as a guide to rotation angle.**

a. Measure the valve clearances at locations indicated in the figure with No. 3 cylinder at compression TDC.

6. Rotate crankshaft 90° clockwise (when viewed from engine front) to align No. 6 cylinder at TDC of compression stroke.

➡**Crankshaft pulley mounting bolt flange has an angle mark every 90°.**

They can be used as a guide to rotation angle.

a. Measure the valve clearances at locations indicated in the figure with No. 6 cylinder at compression TDC.

7. Perform adjustment or replacement if the measured value is out of the standard.

a. If a valve clearance on the exhaust side is out of specification, adjust the valve clearance.

b. If a valve clearance on the intake side is out of specification, replace VVEL ladder assembly and cylinder head assembly.

> **✳✳ WARNING**
>
> **Never adjust valve clearance on the intake side.**

➡ **Since the valve lifter (INT) cannot be replaced by the piece, VVEL ladder assembly and cylinder head assembly replacement are required.**

ENGINE PERFORMANCE & EMISSION CONTROLS

CAMSHAFT POSITION (CMP) SENSOR

LOCATION

See Figures 266 and 267.

REMOVAL & INSTALLATION

See Figures 266 and 267.

1. Before servicing the vehicle, refer to the Precautions Section.
2. Turn ignition switch OFF.
3. Loosen the mounting bolt of the sensor.
4. Disconnect camshaft position sensor harness connector.
5. Remove the sensor.

To install:

6. Installation is the reverse of removal procedure.
7. Tighten the fasteners to specification.

CRANKSHAFT POSITION (CKP) SENSOR

LOCATION

See Figures 268 and 269.

REMOVAL & INSTALLATION

See Figures 268 and 269.

1. Before servicing the vehicle, refer to the Precautions Section.
2. Turn ignition switch OFF.
3. Loosen the fixing bolt of the sensor.
4. Disconnect crankshaft position sensor harness connector.
5. Remove the sensor.

To install:

6. Installation is the reverse of removal procedure.
7. Tighten the fasteners to specification.

ELECTRONIC CONTROL MODULE (ECM)

LOCATION

See Figure 270.

REMOVAL & INSTALLATION

See Figure 270.

1. Before servicing the vehicle, refer to the Precautions Section.
2. Turn the ignition switch **OFF**.
3. Disconnect the negative battery cable from the battery.
4. Disconnect the Electronic Control Module (ECM) connectors.
5. Remove the ECM mounting bolts and remove the ECM from the vehicle.

To install:

6. Installation is the reverse of the removal procedure.
7. Tighten the ECM mounting olts.

> **✳✳ WARNING**
>
> **When replacing the ECM, be careful to use the right part number, as damage to the injection system could occur.**

RESET

When replacing the ECM, the following procedure must be performed in order.
• Initialization/Registration of NVIS (NATS) System and Ignition Key IDs
• VIN Registration
• Accelerator Pedal Released Position Learning
• Throttle Valve Closed Position Learning
• Idle Air Volume Learning
• Exhaust Valve Timing Control Learning

Initialization/Registration of NVIS (NATS) System and Ignition Key IDs

Performing the following procedure can automatically perform re-communication of the ECM and BCM, but only when the ECM has been replaced with a new one. (New one means an ECM which has never been energized on-board).

When registering new Key IDs or replacing the ECM that is not brand new, refer to the CONSULT Operation Manual.

If multiple keys are attached to the key holder, separate them before work.

Distinguish keys with unregistered key ID from those with registered ID.

1. Before servicing the vehicle, refer to the Precautions Section.
2. Insert the registered Intelligent Key.

• To perform this step, use the key that has been used before performing ECM replacement

3. Turn the ignition switch to ON.
4. Maintain the ignition switch in ON position for at least 5 seconds.
5. Turn the ignition switch to OFF.
6. Start the engine.
   a. Can the engine be started?
   • If yes, the procedure is completed.
   • If no, initialize the control unit. Refer to the CONSULT Operation Manual.

VIN Registration

VIN Registration is an operation to register the VIN in the ECM. It must be performed each time the ECM is replaced.

➡ **An accurate VIN which is registered in the ECM may be required for inspection and maintenance procedures.**

1. Before servicing the vehicle, refer to the Precautions Section.
2. Check the VIN of the vehicle and note it.
3. With CONSULT tool,
   a. Turn the ignition switch ON with the engine stopped.
   b. Select "VIN REGISTRATION" in "WORK SUPPORT" mode.
   c. Follow the instruction of the CONSULT display.

Accelerator Pedal Released Position Learning

Accelerator Pedal Released Position Learning is an operation to learn the fully released position of the accelerator pedal by

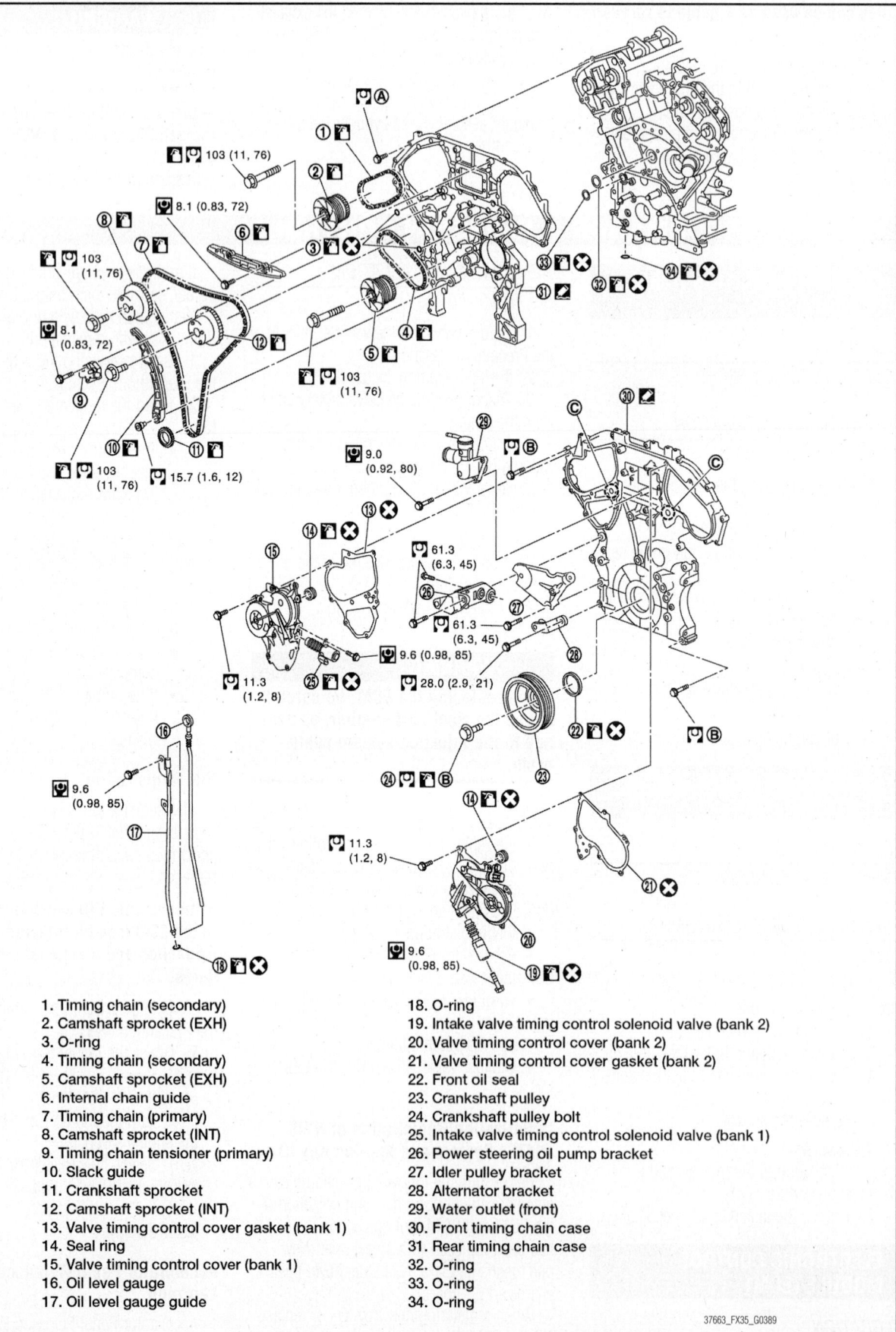

Fig. 266 Exploded view of timing chain assembly—3.5L engine

1. Timing chain (secondary)
2. Camshaft sprocket (EXH)
3. O-ring
4. Timing chain (secondary)
5. Camshaft sprocket (EXH)
6. Internal chain guide
7. Timing chain (primary)
8. Camshaft sprocket (INT)
9. Timing chain tensioner (primary)
10. Slack guide
11. Crankshaft sprocket
12. Camshaft sprocket (INT)
13. Valve timing control cover gasket (bank 1)
14. Seal ring
15. Valve timing control cover (bank 1)
16. Oil level gauge
17. Oil level gauge guide
18. O-ring
19. Intake valve timing control solenoid valve (bank 2)
20. Valve timing control cover (bank 2)
21. Valve timing control cover gasket (bank 2)
22. Front oil seal
23. Crankshaft pulley
24. Crankshaft pulley bolt
25. Intake valve timing control solenoid valve (bank 1)
26. Power steering oil pump bracket
27. Idler pulley bracket
28. Alternator bracket
29. Water outlet (front)
30. Front timing chain case
31. Rear timing chain case
32. O-ring
33. O-ring
34. O-ring

37663_FX35_G0389

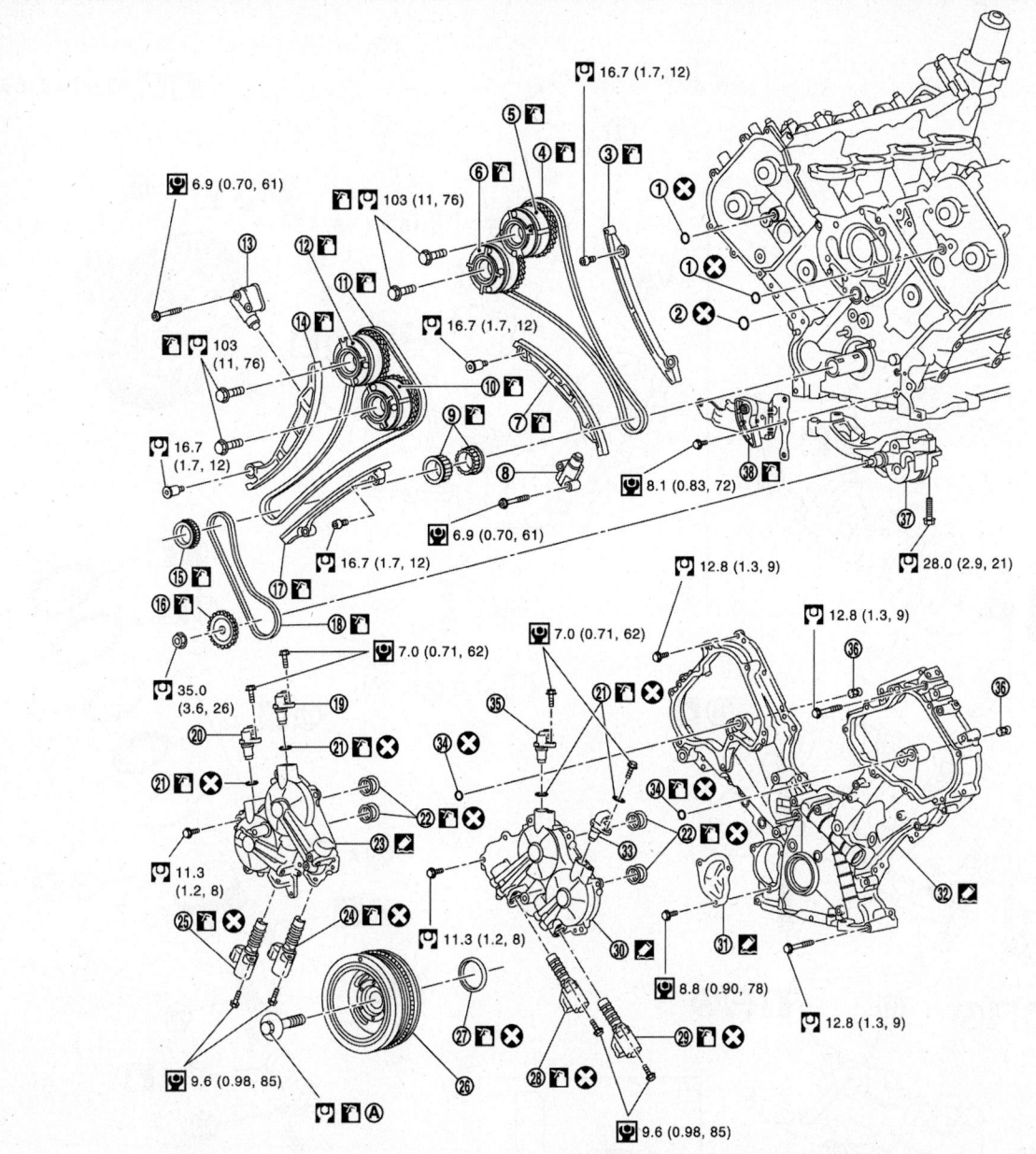

1. O-ring
2. O-ring
3. Tension guide (bank 2)
4. Timing chain (bank 2)
5. Camshaft sprocket (INT) (bank 2)
6. Camshaft sprocket (EXH) (bank 2)
7. Slack guide (bank 2)
8. Timing chain tensioner (bank 2)
9. Crankshaft sprocket
10. Camshaft sprocket (EXH) (bank 1)
11. Timing chain (bank 1)
12. Camshaft sprocket (INT) (bank 1)
13. Timing chain tensioner (bank 1)
14. Slack guide (bank 1)
15. Oil pump sprocket (crankshaft side)
16. Oil pump sprocket (oil pump side)
17. Tension guide (bank 1)
18. Oil pump drive chain
19. Camshaft position sensor (INT) (bank 2)
20. Camshaft position sensor (EXH) (bank 2)
21. O-ring
22. Seal ring
23. Valve timing control cover (bank 2)
24. Intake valve timing control solenoid valve (bank 2)
25. Exhaust valve timing control solenoid valve (bank 2)
26. Crankshaft pulley
27. Front oil seal
28. Intake valve timing control solenoid valve (bank 1)
29. Exhaust valve timing control solenoid valve (bank 1)
30. Valve timing control cover (bank 1)
31. Timing chain tensioner cover
32. Front cover
33. Camshaft position sensor (EXH) (bank 1)
34. O-ring
35. Camshaft position sensor (INT) (bank 1)
36. Oil filter (for valve timing control solenoid valve)
37. Oil pump
38. Oil pump drive chain tensioner

37663_FX35_G0391

Fig. 267 Exploded view of timing chain assembly—5.0L engine

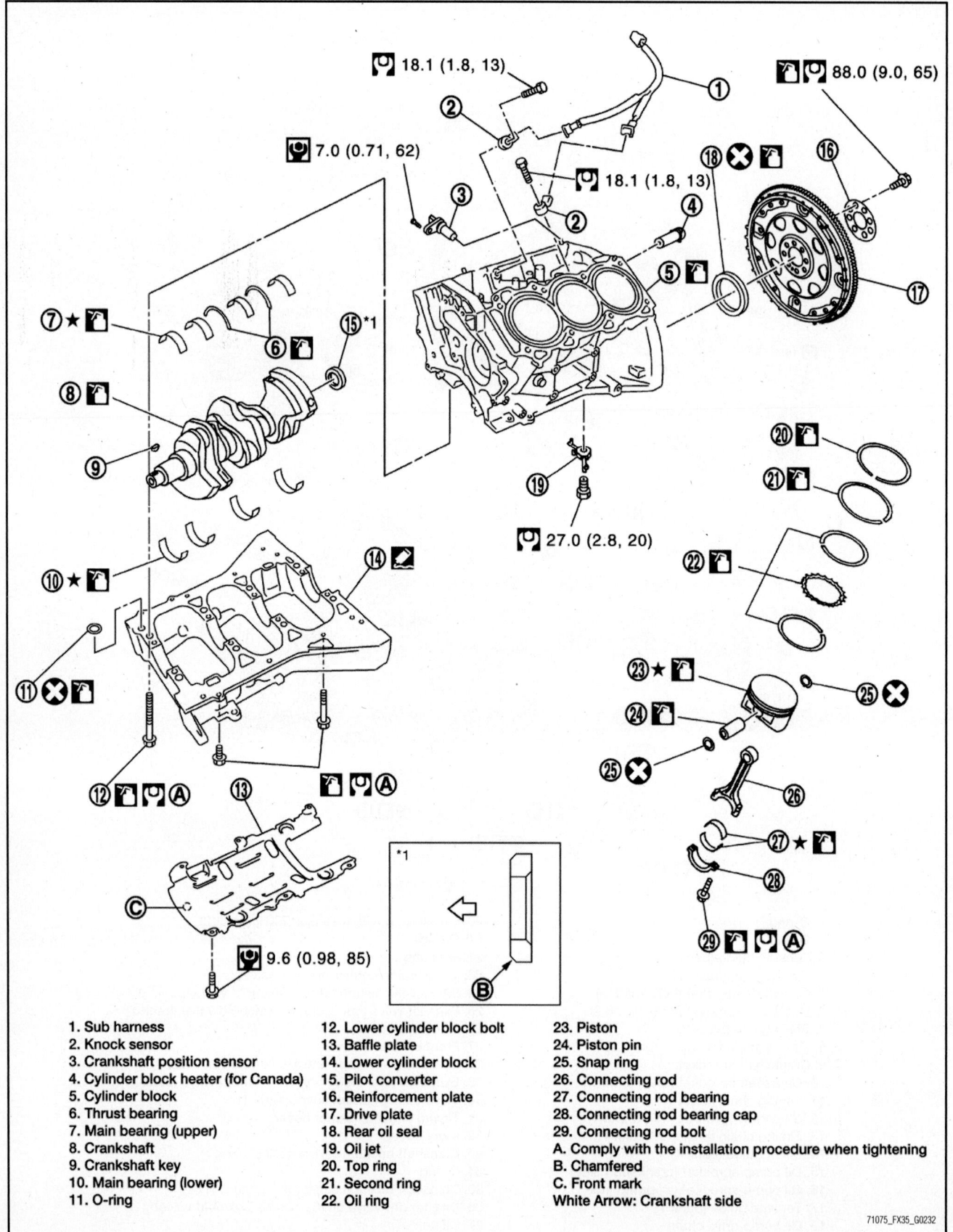

18.1 (1.8, 13)

88.0 (9.0, 65)

7.0 (0.71, 62)

18.1 (1.8, 13)

27.0 (2.8, 20)

9.6 (0.98, 85)

*1

1. Sub harness
2. Knock sensor
3. Crankshaft position sensor
4. Cylinder block heater (for Canada)
5. Cylinder block
6. Thrust bearing
7. Main bearing (upper)
8. Crankshaft
9. Crankshaft key
10. Main bearing (lower)
11. O-ring
12. Lower cylinder block bolt
13. Baffle plate
14. Lower cylinder block
15. Pilot converter
16. Reinforcement plate
17. Drive plate
18. Rear oil seal
19. Oil jet
20. Top ring
21. Second ring
22. Oil ring
23. Piston
24. Piston pin
25. Snap ring
26. Connecting rod
27. Connecting rod bearing
28. Connecting rod bearing cap
29. Connecting rod bolt
A. Comply with the installation procedure when tightening
B. Chamfered
C. Front mark
White Arrow: Crankshaft side

71075_FX35_G0232

Fig. 268 Cylinder block and related component locations—3.5L engine

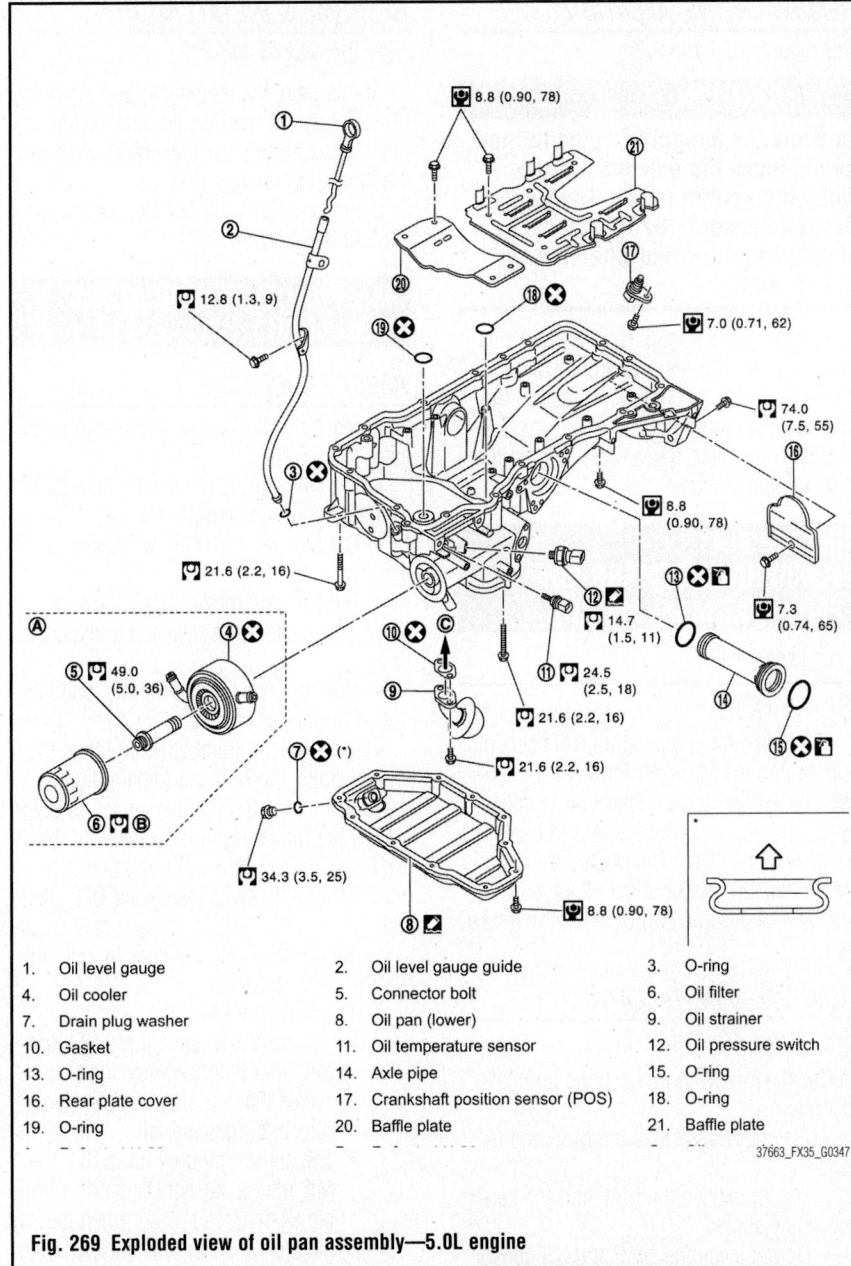

1. Oil level gauge
2. Oil level gauge guide
3. O-ring
4. Oil cooler
5. Connector bolt
6. Oil filter
7. Drain plug washer
8. Oil pan (lower)
9. Oil strainer
10. Gasket
11. Oil temperature sensor
12. Oil pressure switch
13. O-ring
14. Axle pipe
15. O-ring
16. Rear plate cover
17. Crankshaft position sensor (POS)
18. O-ring
19. O-ring
20. Baffle plate
21. Baffle plate

37663_FX35_G0347

Fig. 269 Exploded view of oil pan assembly—5.0L engine

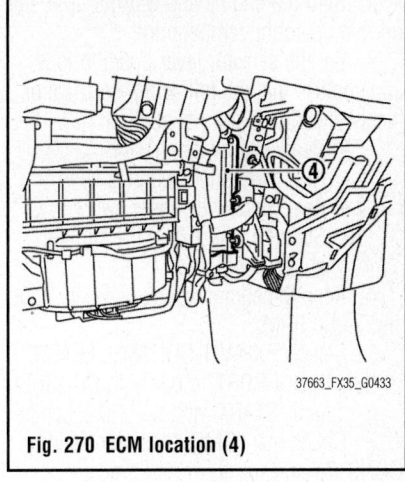

37663_FX35_G0433

Fig. 270 ECM location (4)

4. Follow the instructions on the CON-SULT display.

5. Turn the ignition switch OFF and wait at least 10 seconds.

6. Check that the throttle valve moves during the above 10 seconds by confirming the operating sound.

Without Consult Tool

1. Before servicing the vehicle, refer to the Precautions Section.

2. Start the engine.
- Engine coolant temperature should be 77°F (25°C) or less before the engine starts

3. Warm up the engine.
- Raise the engine coolant temperature until it reaches 149°F (65°C) or more

4. Turn the ignition switch OFF and wait at least 10 seconds.

5. Check that the throttle valve moves during the above 10 seconds by confirming the operating sound.

Idle Air Volume Learning

Refer to Intake Manifold, Idle Air Volume Learning procedure.

Exhaust Valve Timing Control Learning

Exhaust Valve Timing Control Learning is a function of the ECM to learn the characteristic of the exhaust valve timing control magnet retarder by comparing the target angle of the exhaust camshaft with the actual retarded angle of the exhaust camshaft. It must be performed each time the exhaust valve timing control magnet retarder is disconnected or replaced, or the ECM is replaced.

With Consult Tool

1. Before servicing the vehicle, refer to the Precautions Section.

---

monitoring the accelerator pedal position sensor output signal. It must be performed each time the harness connector of the accelerator pedal position sensor or ECM is disconnected.

1. Before servicing the vehicle, refer to the Precautions Section.

2. Check that the accelerator pedal is fully released.

3. Turn the ignition switch ON and wait at least 2 seconds.

4. Turn the ignition switch OFF and wait at least 10 seconds.

5. Turn the ignition switch ON and wait at least 2 seconds.

6. Turn the ignition switch OFF and wait at least 10 seconds.

Throttle Valve Closed Position Learning

Throttle Valve Closed Position Learning is an operation to learn the fully closed position of the throttle valve by monitoring the throttle position sensor output signal. It must be performed each time the harness connector of the electric throttle control actuator or the ECM is disconnected or the electric throttle control actuator is cleaned.

With Consult Tool

1. Before servicing the vehicle, refer to the Precautions Section.

2. Turn the ignition switch ON.

3. Select "CLSD THL POS LEARN" in "WORK SUPPORT" mode.

2. Start the engine and warm it up to the normal operating temperature.

3. Set the selector lever position to N and confirm that the following electrical or mechanical loads are not applied.
- Headlamp switch is OFF
- Air conditioner switch is OFF
- Rear window defogger switch is OFF
- Steering wheel is in the straight-ahead position

4. Keep the engine speed between 1,800 and 2,000 RPM.

5. Select "EXH V/T CONTROL LEARN" in "WORK SUPPORT" mode with CONSULT.

6. Touch "START" and wait 20 seconds.

7. Check that "CMPLT" is displayed on CONSULT screen.

Without Consult Tool

1. Before servicing the vehicle, refer to the Precautions Section.

2. Start the engine and warm it up to the normal operating temperature.

3. Set the selector lever position to N and confirm that the following electrical or mechanical loads are not applied.
- Headlamp switch is OFF
- Air conditioner switch is OFF
- Rear window defogger switch is OFF
- Steering wheel is in the straight-ahead position

4. Keep the engine speed between 1,800 and 2,000 RPM for 20 seconds.

ENGINE COOLANT TEMPERATURE (ECT) SENSOR

LOCATION

See Figures 271 through 273.

REMOVAL & INSTALLATION

See Figures 271 and 273.

1. Before servicing the vehicle, refer to the Precautions Section.

2. Turn ignition switch OFF.

3. Disconnect engine coolant temperature sensor harness connector.

4. Remove engine coolant temperature sensor.

To install:

5. Installation is the reverse of the removal procedure.

6. Tighten the fasteners to specification.

HEATED OXYGEN SENSOR (HO2S)

LOCATION

See Figures 274 and 275.

REMOVAL & INSTALLATION

See Figures 274 and 275.

✳✳ CAUTION

To avoid the danger of being burned, do not touch the exhaust system when the system is hot. The Heated Oxygen Sensor (HO2S) removal should be performed when the system is cool.

At this time, the manufacturer does not provide a specific removal and installation procedure for this component, refer to the illustration as required.

Before servicing the vehicle, refer to the Precautions Section.

INTAKE AIR TEMPERATURE/MASS AIR FLOW (IAT/MAF) SENSOR

LOCATION

See Figures 276 and 277.

The Intake Air Temperature (IAT) sensor is built into the Mass Air Flow (MAF) sensor. The sensor detects intake air temperature in the stream of intake air and transmits a signal to the ECM. The Mass Air Flow (MAF) sensor measures the intake flow rate by measuring a part of the entire intake flow.

REMOVAL & INSTALLATION

See Figures 276 and 277.

Use the following precautions for this procedure.
- Do not disassemble the MAF and IAT sensor
- Do not expose the MAF and IAT sensor to any shock
- Do not clean the MAF and IAT sensor
- If the MAF and IAT sensor has been dropped, it should be replaced
- Do not blow compressed air through the MAF and IAT sensor
- Do not place a finger or any other object into the MAF and IAT sensor. Malfunction may occur

At this time, the manufacturer does not provide specific removal and installation procedures for this component, refer to the illustration as required.

Before servicing the vehicle, refer to the Precautions Section.

KNOCK SENSOR (KS)

LOCATION

See Figures 278 and 279.

REMOVAL & INSTALLATION

See Figures 278 and 279.

At this time, the manufacturer does not provide specific removal and installation procedures for this component(s), refer to the illustration as required.

Before servicing the vehicle, refer to the Precautions Section.

MALFUNCTION INDICATOR LIGHT (MIL)

RESET PROCEDURE

Proper operation of the Malfunction Indicator Light (MIL):
- The MIL will illumine with the ignition switch ON and the engine OFF
- The MIL will turn OFF when the engine is started
- The MIL will remain ON if the self-diagnostic system has detected a malfunction
- The MIL may turn OFF if the malfunction is no longer present
- If the MIL is illuminated and then the engine stalls, the MIL will remain illuminated as long as the ignition switch is ON
- If the MIL is not illuminated and the engine stalls, the MIL will not illuminate until the ignition switch is cycled OFF, then ON

1. Before servicing the vehicle, refer to the Precautions Section.

2. Resetting the MIL:
- The control module turns OFF the MIL after 3 consecutive ignition cycles that the diagnostic system runs and does not fail
- The control module turns OFF the MIL after a current Diagnostic Trouble Code (DTC) clears when the diagnostic cycle runs and passes
- There may still be a history of DTCs stored in the system. These will clear after 40 consecutive warm-up cycles, if no failures are reported by any other related diagnostic system
- Manual resetting of the MIL and any DTC stored in the system, requires the use of an OBD2 scan tool connected to the Data Link Connector (DLC) for communication with the vehicle. Follow the instructions of the scan tool for both retrieval and resetting of DTCs. The scan tool can be used to command the MIL off.

➡ **If the error symptoms causing the MIL to illuminate have been corrected,**

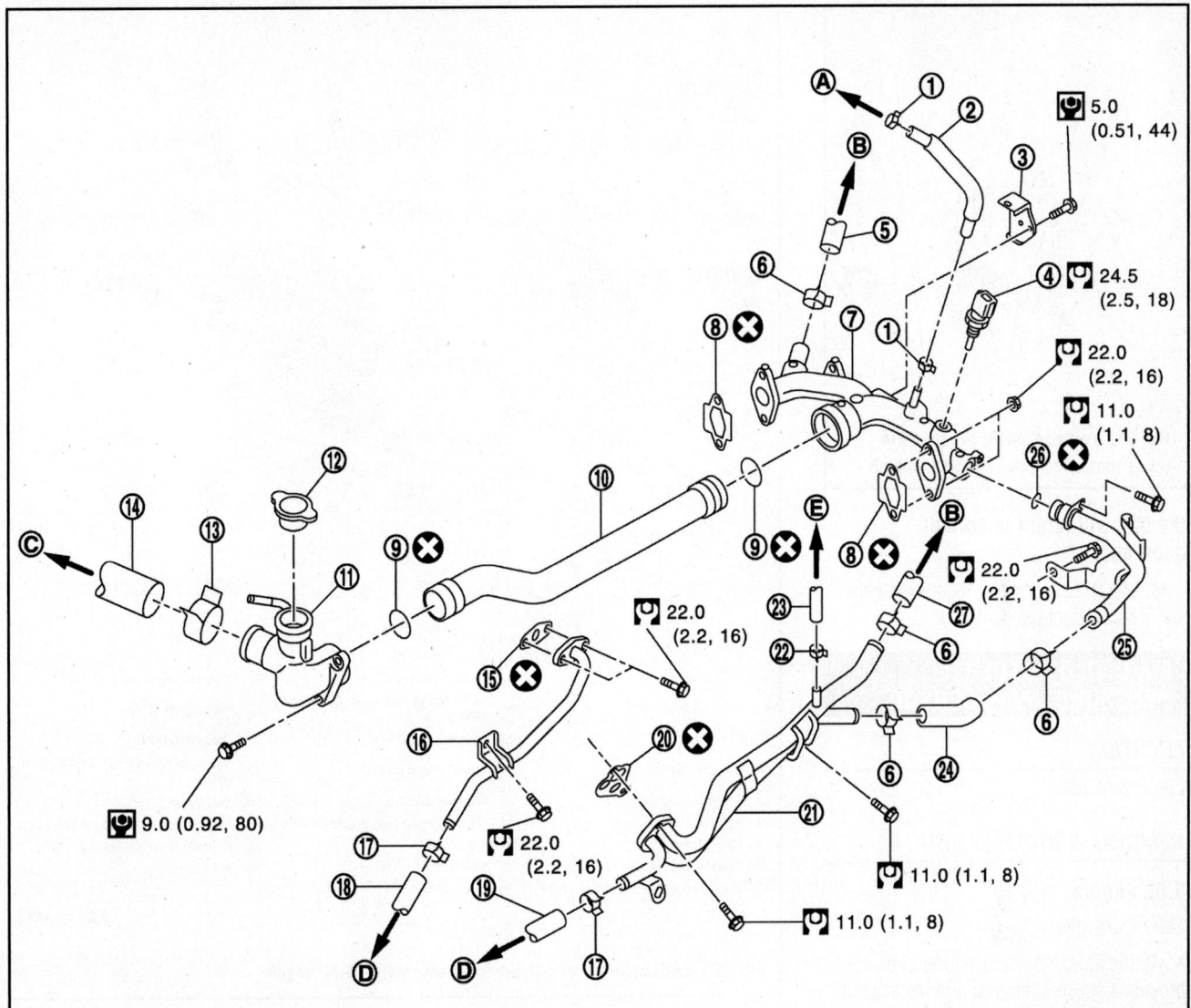

1. Clamp
2. Water hose
3. Harness bracket
4. Engine coolant temperature sensor
5. Heater hose
6. Clamp
7. Water outlet (rear)
8. Gasket
9. O-ring
10. Water outlet pipe
11. Water outlet (front)
12. Radiator cap
13. Clamp
14. Radiator hose (upper)
15. Gasket
16. Water pipe
17. Clamp
18. Water hose
19. Water hose
20. Gasket
21. Heater pipe
22. Clamp
23. Water hose
24. Water hose
25. Water bypass pipe
26. O-ring
27. Heater hose
A. To electric throttle control actuator (bank 1)
B. To heater core
C. To radiator
D. To oil cooler
E. To electric throttle control actuator (bank 2)

Fig. 271 Exploded view of water hoses and piping—3.5L engine

37663_FX35_G0198

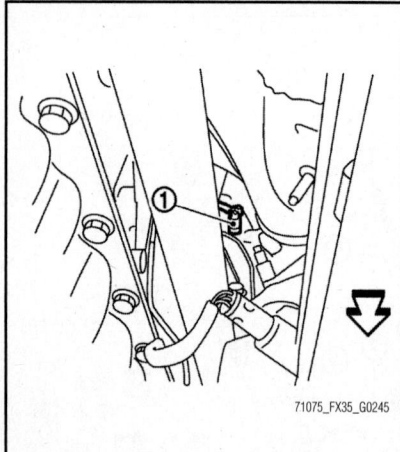

Fig. 272 Engine Coolant Temperature (ECT) sensor location (1)—5.0L engine

the MIL will return to normal operation.

3. If a DTC is present, record the code and troubleshoot the fault.

MANIFOLD ABSOLUTE PRESSURE (MAP) SENSOR

LOCATION

See Figure 280.

REMOVAL & INSTALLATION

5.0L Engine

See Figure 280.

At this time, the manufacturer does not provide a specific removal and installation procedure for this component, refer to the illustration as required.

Before servicing the vehicle, refer to the Precautions Section.

THROTTLE CONTROL ACTUATOR

LOCATION

The electric throttle control actuators are located on the intake manifold collector on the 3.5L engine and on the intake manifold of the 5.0L engine.

REMOVAL & INSTALLATION

3.5L Engine

See Figure 281.

⁂ **CAUTION**

Never drain engine coolant when the engine is hot to avoid the danger of being scalded.

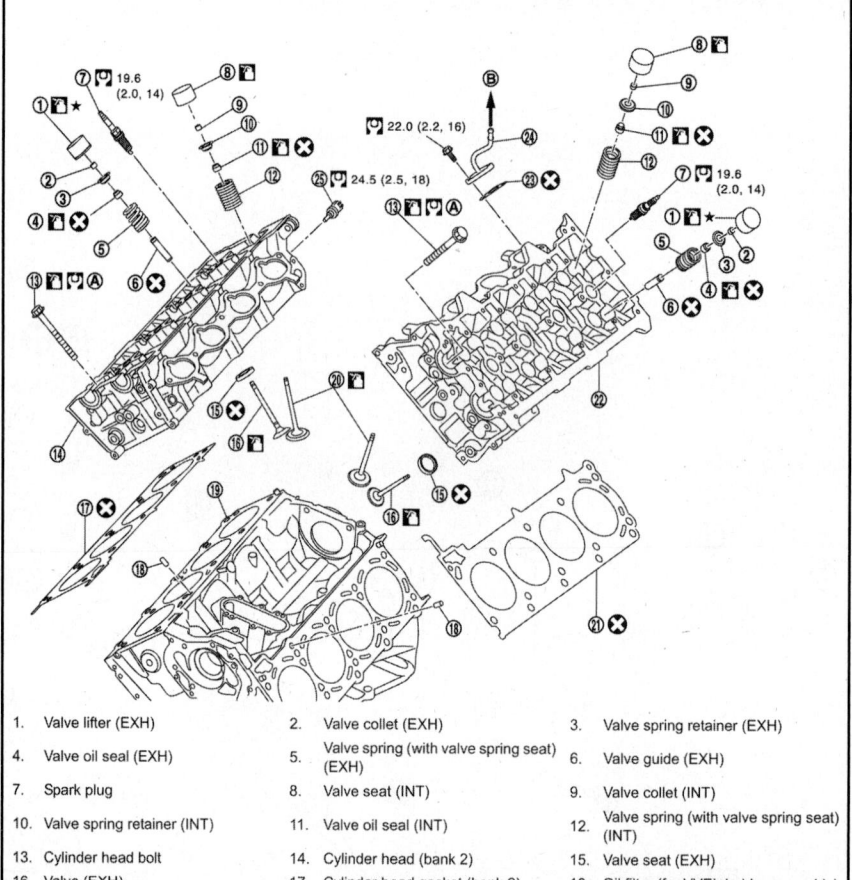

1.	Valve lifter (EXH)	2.	Valve collet (EXH)	3.	Valve spring retainer (EXH)
4.	Valve oil seal (EXH)	5.	Valve spring (with valve spring seat) (EXH)	6.	Valve guide (EXH)
7.	Spark plug	8.	Valve seat (INT)	9.	Valve collet (INT)
10.	Valve spring retainer (INT)	11.	Valve oil seal (INT)	12.	Valve spring (with valve spring seat) (INT)
13.	Cylinder head bolt	14.	Cylinder head (bank 2)	15.	Valve seat (EXH)
16.	Valve (EXH)	17.	Cylinder head gasket (bank 2)	18.	Oil filter (for VVEL ladder assembly)
19.	Cylinder block	20.	Valve (INT)	21.	Cylinder head gasket (bank 1)
22.	Cylinder head (bank 1)	23.	Gasket	24.	Water pipe
25.	Engine coolant temperature sensor				

Fig. 273 Exploded view of cylinder head assembly—5.0L engine

1. A/F sensor 1 (bank 2)
2. Heated oxygen sensor 2 (bank 2)
3. Heated oxygen sensor 2 (bank 2) harness connector
4. Heated oxygen sensor 2 (bank 1) harness connector
5. Heated oxygen sensor 2 (bank 1)
6. A/F sensor 1 (bank 1)
White Arrow: Vehicle front

Fig. 274 Heated Oxygen Sensor (HO2S) and A/F sensor component locations—3.5L engine

1. Before servicing the vehicle, refer to the Precautions Section.
2. Remove engine cover.
3. Remove air cleaner case and air duct.

4. Remove electric throttle control actuator as per the following:

a. Drain engine coolant, or when water hoses are disconnected, attach plug to prevent engine coolant leakage.

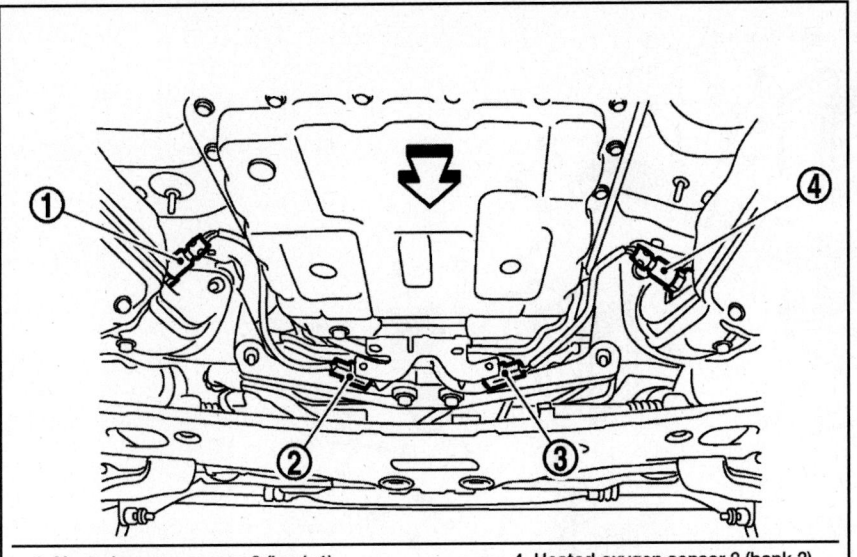

1. Heated oxygen sensor 2 (bank 1)
2. Heated oxygen sensor 2 (bank 1) harness connector
3. Heated oxygen sensor 2 (bank 2) harness connector
4. Heated oxygen sensor 2 (bank 2)
White Arrow: Vehicle front

71075_FX35_G0252

Fig. 275 Heated Oxygen Sensor (HO2S) component locations—5.0L engine

a. Insert hose by 1.06–1.26 inches (27–32mm) from connector end.
b. Clamp hose at location of 0.12–0.28 inch (3–7mm) from hose end.
7. To install the electric throttle control actuator (bank 1 & 2). Tighten in numerical order as shown.

➡The figure shows the electric throttle control actuator (bank 1) viewed from the air duct side. Viewed from the air duct side, order of tightening mounting bolts of electric throttle control actuator (bank 2) is the same as that of the electric throttle control actuator (bank 1).

8. Perform the "Throttle Valve Closed Position Learning" when harness connector of electric throttle control actuator is disconnected. Refer to Intake Manifold, Throttle Valve Closed Position Learning.

9. Perform the "Idle Air Volume Learning" and "Throttle Valve Closed Position Learning" when electric throttle control actuator is replaced. Refer to Intake Manifold, Idle Air Volume Learning.

✳✳ CAUTION

Perform this step when engine is cold. Never spill engine coolant on drive belt.

b. Disconnect water hoses from electric throttle control actuator. When engine coolant is not drained from radiator, attach plug to water hoses to prevent engine coolant leakage.
c. Disconnect harness connector.
d. Loosen mounting bolts in the reverse order of tightening.

➡When removing only intake manifold collector, move electric throttle control actuator without disconnecting the water hose. The figure shows the electric throttle control actuator (bank 1) viewed from the air duct side. Viewed from the air duct side, order of loosening mounting bolts of electric throttle control actuator (bank 2) is the same as that of the electric throttle control actuator (bank 1).

✳✳ WARNING

Handle carefully to avoid any impact to electric throttle control actuator.

To install:

5. Installation is the reverse of the removal procedure.
6. To install the water hose, perform the following:

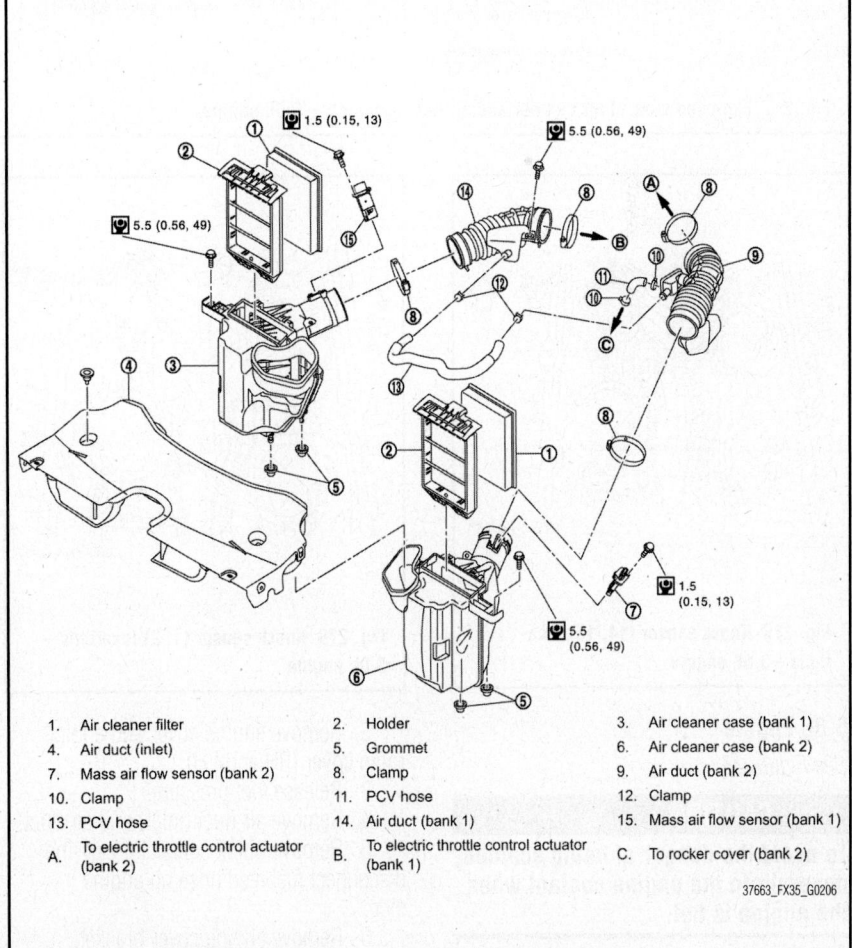

1. Air cleaner filter
2. Holder
3. Air cleaner case (bank 1)
4. Air duct (inlet)
5. Grommet
6. Air cleaner case (bank 2)
7. Mass air flow sensor (bank 2)
8. Clamp
9. Air duct (bank 2)
10. Clamp
11. PCV hose
12. Clamp
13. PCV hose
14. Air duct (bank 1)
15. Mass air flow sensor (bank 1)
A. To electric throttle control actuator (bank 2)
B. To electric throttle control actuator (bank 1)
C. To rocker cover (bank 2)

37663_FX35_G0206

Fig. 276 Exploded view of air cleaner and air duct assembly—3.5L engine

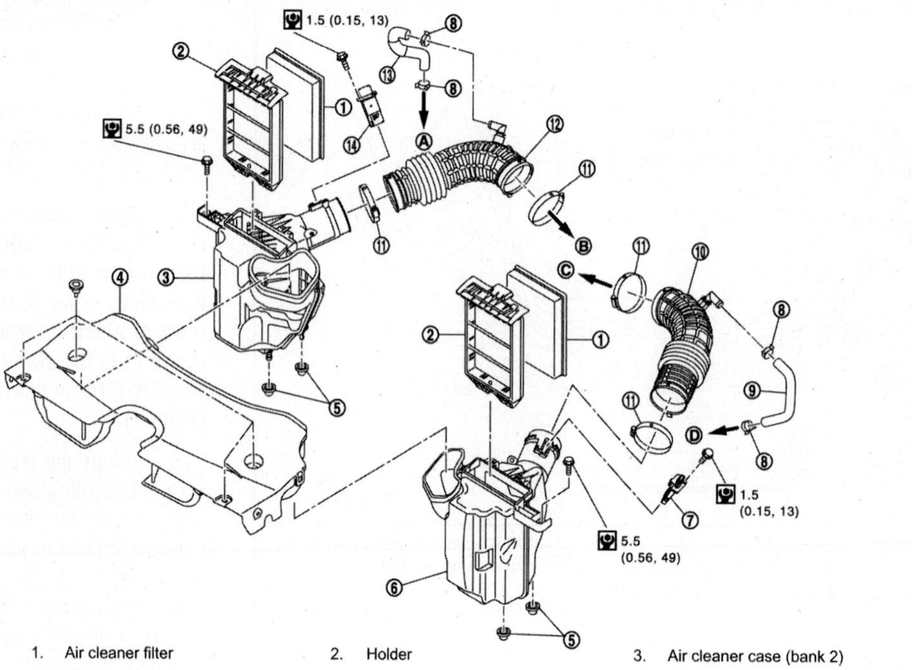

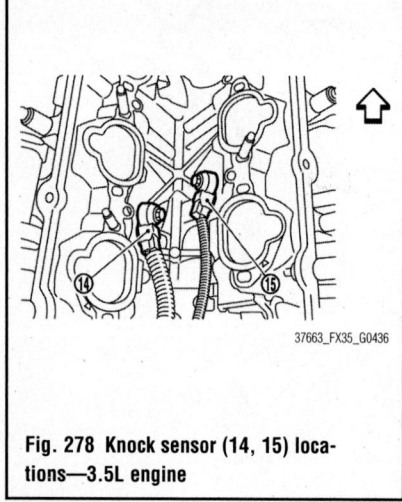

Fig. 277 Exploded view of air cleaner and air duct assembly—5.0L engine

1.	Air cleaner filter	2.	Holder	3.	Air cleaner case (bank 2)
4.	Air duct (inlet)	5.	Grommet	6.	Air cleaner case (bank 1)
7.	Mass air flow sensor (bank 1)	8.	Clamp	9.	PCV hose
10.	Air duct (bank 1)	11.	Clamp	12.	Air duct (bank 2)
13.	PCV hose	14.	Mass air flow sensor (bank 2)		
A.	To rocker cover (bank 2)	B.	To electric throttle control actuator (bank 2)	C.	To electric throttle control actuator (bank 1)
D.	To rocker cover (bank 1)				

37663_FX35_G0209

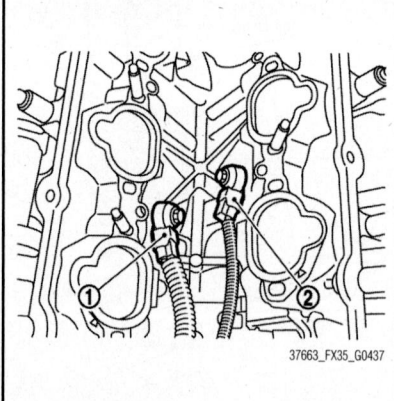

37663_FX35_G0436

Fig. 278 Knock sensor (14, 15) locations—3.5L engine

37663_FX35_G0437

Fig. 279 Knock sensor (1, 2) locations—5.0L engine

5.0L Engine

See Figure 282.

✳✳ CAUTION

To avoid the danger of being scalded, never drain the engine coolant when the engine is hot.

1. Before servicing the vehicle, refer to the Precautions Section.

2. Remove engine cover and engine room cover (RH and LH).
3. Release fuel pressure.
4. Remove air duct (inlet) and air duct.
5. Remove quick connector cap and disconnect fuel feed hose on engine side.
6. Remove engine cover bracket.
7. Remove fuel injector and fuel tube assembly.

8. Disconnect Manifold Absolute Pressure (MAP) sensor and air fuel ratio sensor 1 (bank 1) harness connector.
9. Remove vacuum tank, EVAP service port hose and EVAP canister purge control solenoid valve.
10. Disconnect PCV hoses and vacuum hose from intake manifold. Add matching marks as necessary for easier installation.
11. Drain engine coolant from radiator.

✳✳ CAUTION

Perform this step when the engine is cold. Never spill engine coolant on drive belts.

12. Remove electric throttle control actuator. Loosen mounting bolts in reverse order of the tightening sequence.

✳✳ WARNING

Handle carefully to avoid any impact to electric throttle control actuator. Never disassemble.

To install:

13. Installation is the reverse of the removal procedure.

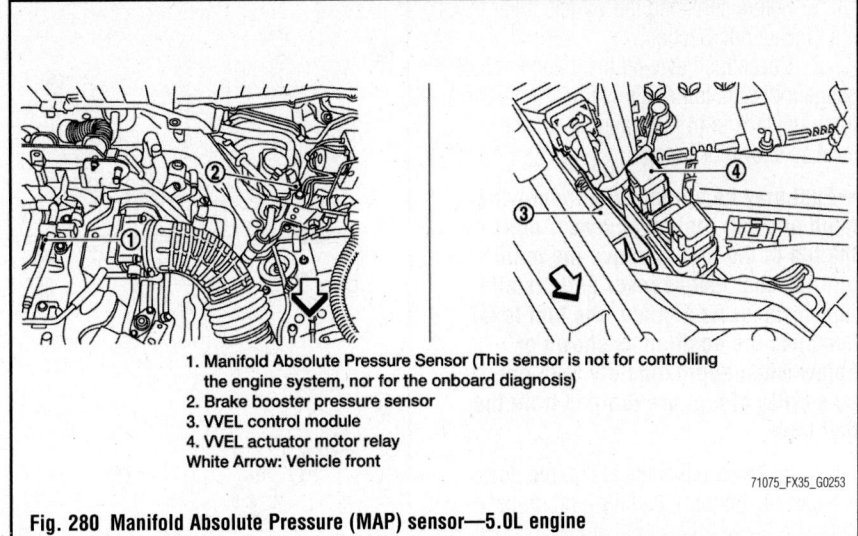

1. Manifold Absolute Pressure Sensor (This sensor is not for controlling the engine system, nor for the onboard diagnosis)
2. Brake booster pressure sensor
3. VVEL control module
4. VVEL actuator motor relay
White Arrow: Vehicle front

71075_FX35_G0253

Fig. 280 Manifold Absolute Pressure (MAP) sensor—5.0L engine

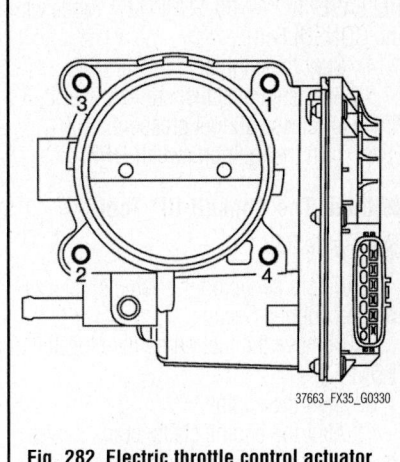

37663_FX35_G0330

Fig. 282 Electric throttle control actuator tightening sequence

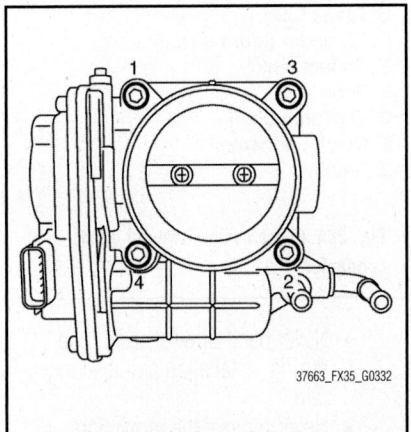

37663_FX35_G0332

Fig. 281 Electric throttle control actuator tightening sequence

➡The figure shows the electric throttle control actuator (bank 1) viewed from the air duct side. Viewed from the air duct side, the order of loosening mounting bolts of electric throttle control actuator (bank 1) is the same as that of the electric throttle control actuator (bank 2).

14. Perform the "Throttle Valve Closed Position Learning" when harness connector of electric throttle control actuator is disconnected. Refer to Intake Manifold, Throttle Valve Closed Position Learning.

15. Perform the "Idle Air Volume Learning" and "Throttle Valve Closed Position Learning" when electric throttle

control actuator is replaced. Refer to Intake Manifold, Idle Air Volume Learning.

THROTTLE POSITION SENSOR (TPS)

LOCATION

The electric throttle body includes a throttle control motor and Throttle Position Sensors (TPS). The TPS responds to the throttle valve movement. The TPS has two sensors.

REMOVAL & INSTALLATION

Refer to Throttle Control Actuator, removal & installation, when servicing this component.

FUEL   GASOLINE FUEL INJECTION SYSTEM

FUEL SYSTEM SERVICE PRECAUTIONS

Safety is the most important factor when performing not only fuel system maintenance but any type of maintenance. Failure to conduct maintenance and repairs in a safe manner may result in serious personal injury or death. Maintenance and testing of the vehicle's fuel system components can be accomplished safely and effectively by adhering to the following rules and guidelines.

• To avoid the possibility of fire and personal injury, always disconnect the negative battery cable unless the repair or test procedure requires that battery voltage be applied.

• Always relieve the fuel system pressure prior to disconnecting any fuel system component (injector, fuel rail, pressure reg-

ulator, etc.), fitting or fuel line connection. Exercise extreme caution whenever relieving fuel system pressure to avoid exposing skin, face and eyes to fuel spray. Please be advised that fuel under pressure may penetrate the skin or any part of the body that it contacts.

• Always place a shop towel or cloth around the fitting or connection prior to loosening to absorb any excess fuel due to spillage. Ensure that all fuel spillage (should it occur) is quickly removed from engine surfaces. Ensure that all fuel soaked cloths or towels are deposited into a suitable waste container.

• Always keep a dry chemical (Class B) fire extinguisher near the work area.

• Do not allow fuel spray or fuel vapors to come into contact with a spark or open flame.

• Always use a back-up wrench when loosening and tightening fuel line connection fittings. This will prevent unnecessary stress and torsion to fuel line piping.

• Always replace worn fuel fitting O-rings with new Do not substitute fuel hose or equivalent where fuel pipe is installed.

Before servicing the vehicle, make sure to also refer to the precautions in the beginning of this section as well.

RELIEVING FUEL SYSTEM PRESSURE

With The Consult-III® Tool

1. Before servicing the vehicle, refer to the Precautions Section.
2. Turn the ignition switch **ON**.

3. Perform the "FUEL PRESSURE RELEASE" in "WORK SUPPORT" mode with the CONSULT-III®.

4. Start the engine.

5. After engine stalls, crank it over 2–3 times to release all fuel pressure.

6. Turn the ignition switch **OFF**.

Without The Consult-III® Tool

See Figure 283.

1. Before servicing the vehicle, refer to the Precautions Section.

2. Remove the fuel pump fuse located in IPDM E/R.

3. Start the engine.

4. After the engine stalls, crank it over 2–3 times to release all fuel pressure.

5. Turn the ignition switch **OFF**.

6. Reinstall the fuel pump fuse after servicing the fuel system.

FUEL FILTER

REMOVAL & INSTALLATION

The fuel filter is installed in the fuel pump assembly inside the fuel tank. Replace the fuel filter (or fuel pump assembly) with a new one periodically. Refer to Fuel Pump Assembly, removal & installation.

FUEL LEVEL SENDING UNIT

REMOVAL & INSTALLATION

The fuel level sensor unit is an integral part of the fuel pump assembly. Refer to the Fuel Pump Assembly, removal & installation when servicing this component.

FUEL PUMP ASSEMBLY

REMOVAL & INSTALLATION

See Figures 284 through 288.

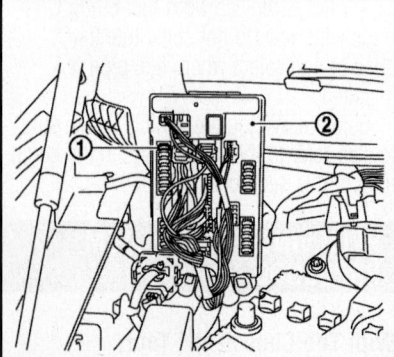

Fig. 283 Remove fuel pump fuse (1) located in IPDM E/R (2)

1. Before servicing the vehicle, refer to the Precautions Section.

2. Check fuel level on fuel gauge. If fuel gauge indicates full or almost full, drain fuel from fuel tank until fuel gauge indicates level as shown or below.

→**Fuel may be spilled when removing main and sub fuel level sensor units for the top of the fuel is above the main and sub fuel level sensor unit installation surface. As a guide, the fuel level becomes the position as shown or below when approximately 5.25 gallons (20L) of fuel are drained from the fuel tank.**

3. In the case that the fuel pump does not operate, perform the following procedure.

a. Insert a hose of less than 1 inch (25mm) in diameter into the fuel filler tube through the fuel filler opening to draw the fuel from the fuel filler tube.

b. Disconnect the fuel filler hose from the fuel filler tube.

c. Insert the fuel tube into the fuel tank through the fuel filler hose to draw the fuel from the fuel tank.

4. Release the fuel pressure from the fuel lines.

5. Open fuel filler lid.

6. Open filler cap and release the pressure inside fuel tank.

7. Remove rear seat cushion.

8. Peel off floor carpet, then remove inspection hole cover units by turning clips clockwise by 90°.

9. Disconnect harness connector and fuel feed tube.

a. Hold the sides of connector, push in tabs and pull out fuel feed tube.

b. If quick connector sticks to tube of main fuel level sensor unit, push and pull quick connector several times until they start to move. Then disconnect them by pulling.

✲✲ WARNING

Quick connector can be disconnected when the tabs are completely depressed. Never twist it more than necessary.

- Never use any tools to disconnected quick connector.
- Keep resin tube away from heat. Be especially careful when welding near the resin tube.
- Prevent acid liquid such as battery electrolyte, etc. from getting on resin tube.

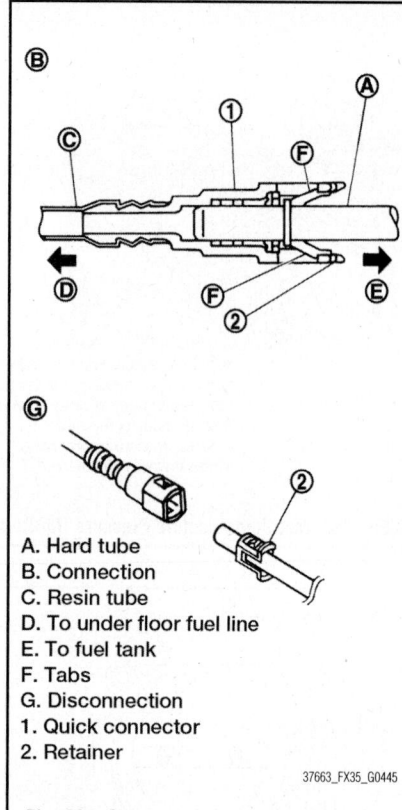

A. Hard tube
B. Connection
C. Resin tube
D. To under floor fuel line
E. To fuel tank
F. Tabs
G. Disconnection
1. Quick connector
2. Retainer

37663_FX35_G0445

Fig. 284 Cross section view of quick connector

- Never bend or twist resin tube during installation and disconnection.
- Never remove the remaining retainer on hard tube (or the equivalent) except when resin tube or retainer is replaced.
- When resin tube or hard tube (or the equivalent) is replaced, also replace retainer with new one.
- To keep the connecting portion clean and to avoid damage and foreign materials, cover them completely with plastic bags or something similar.

10. Removal of main fuel level sensor unit, fuel filter and fuel pump assembly:

a. Remove retainer.

b. Raise main fuel level sensor unit, fuel filter and fuel pump assembly, and disconnect quick connector. Push in tabs and pull out fuel tube.

✲✲ WARNING

Do not bend float arm during removal. Avoid impacts such as falling when handling components.

11. Removal of sub fuel level sensor unit:

Fig. 285 Push in tabs (1) and pull out fuel tube (2)

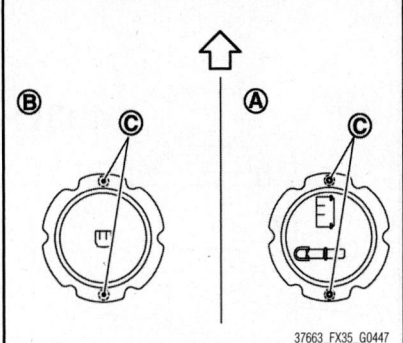

Fig. 287 Face main and sub fuel level sensor units as shown, and install them with the knock pin (C) on back aligned with pin hole on fuel tank; right side (A), left side (B)

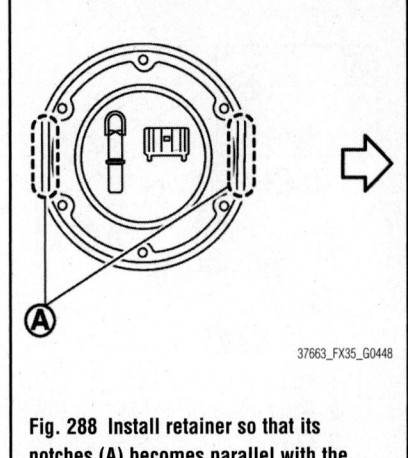

Fig. 288 Install retainer so that its notches (A) becomes parallel with the notch on fuel tank

a. Remove retainer.

b. Raise and release sub fuel level sensor unit to remove.

To install:

12. Installation is the reverse of the removal procedure.

13. Install fuel tube.

14. Face main and sub fuel level sensor units as shown, and install them with the knock pin on back aligned with pin hole on fuel tank.

15. Install retainer so that its notch becomes parallel with the notch on fuel tank.

16. Tighten retainer mounting bolts evenly.

17. Check the quick connector connection for damage or any foreign materials.

18. Align the connector with the tube, then insert the connector straight into the tube until a click sound is heard.

19. After connecting, check that the connection is secure by following method.

a. Pull the tube and the connector to check they are securely connected.

b. Visually confirm that the two retainer tabs are connected to the connector.

20. Turn ignition switch "ON" (with engine stopped), then check connections for leakage by applying fuel pressure to fuel piping.

21. Start engine and let it idle and check that there is no fuel leakage at the fuel system connections.

FUEL RAIL & INJECTORS

REMOVAL & INSTALLATION

3.5L Engine

See Figures 289 through 293.

Note the following:
- Be sure to work in a well-ventilated area and furnish workshop with a CO2 fire extinguisher
- Never smoke while servicing fuel system. Keep open flames and sparks away from the work area
- Never drain engine coolant when the engine is hot to avoid the danger of being scalded
- Do not remove or disassemble parts unless instructed to do so in the procedure

1. Before servicing the vehicle, refer to the Precautions Section.

2. Release fuel pressure.

3. Disconnect battery cable from the negative terminal.

4. Remove engine cover.

5. Remove air cleaner case and air duct.

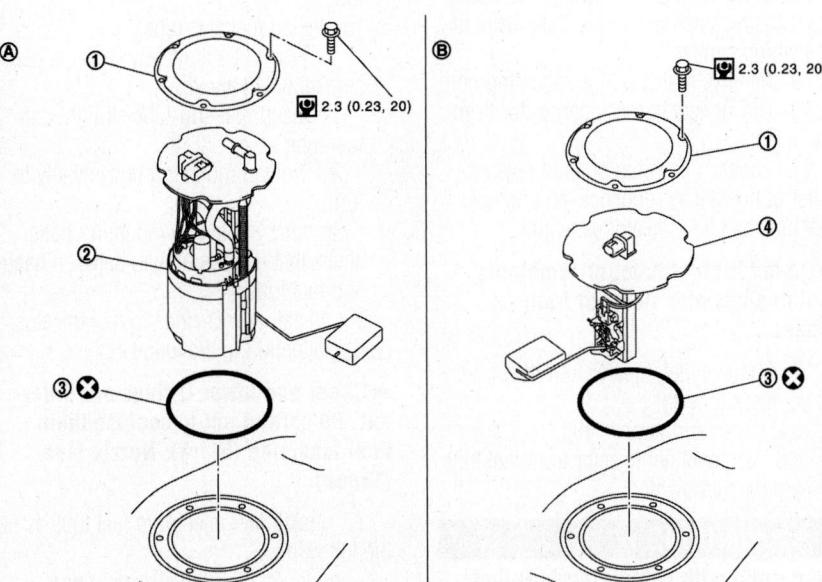

1. Retainer
2. Main fuel level sensor unit, fuel filter and fuel pump assembly
3. O-ring
4. Sub fuel level sensor unit
A. Right side
B. Left side

Fig. 286 Exploded view of fuel level sensor unit, fuel filter and fuel pump assembly

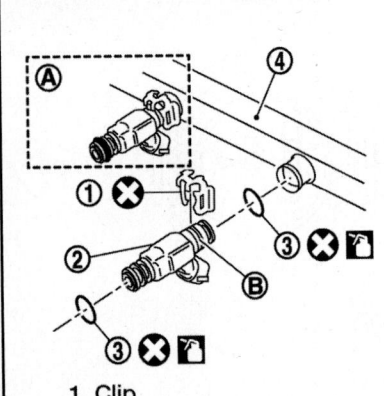

1. Clip
2. Fuel injector
3. O-ring
4. Fuel tube
A. Installed condition
B. Clip mounting groove

37663_FX35_G0452

Fig. 289 Remove fuel injector from fuel tube

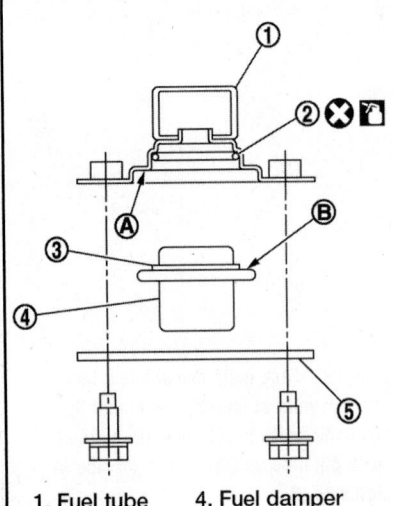

1. Fuel tube
2. O-ring
3. Spacer
4. Fuel damper
5. Fuel damper cap

37663_FX35_G0453

Fig. 290 Exploded view of fuel damper installation

6. Remove intake manifold collector. Refer to Intake Manifold Collector, removal & installation.

7. Remove fuel feed hose (with damper) from fuel sub-tube and remove harness bracket.

➡**There is no fuel return route. While hoses are disconnected, plug them to prevent fuel from draining. Never separate damper and hose.**

8. When separating fuel feed hose (with damper) and centralized under-floor piping connection, disconnect quick connector as per the following:

a. Remove quick connector cap from quick connector connection on right member side.

b. Disconnect fuel feed hose (with damper) from bracket hose clamp.

c. Push in retainer tabs.

d. Draw and pull out quick connector straight from centralized under-floor piping.

Note the following:

• Pull quick connector holding position (at 90°)

• Never pull with lateral force applied. O-ring inside quick connector may be damaged

• Prepare container and cloth beforehand as fuel will leak out

• Avoid fire and sparks

• Keep parts away from heat source. Especially, be careful when welding is performed

• Never expose parts to battery electrolyte or other acids

• Never bend or twist connection between quick connector and fuel feed hose (with damper) during installation/removal

• To keep the connecting portion clean and to avoid damage and foreign materials, cover them completely with plastic bags or something similar

9. Remove fuel sub tube mounting bolt.

10. Disconnect harness connector from fuel injector.

11. Loosen mounting bolts in reverse order of tightening sequence, and remove fuel tube and fuel injector assembly.

➡**Do not tilt fuel tube, or remaining fuel in pipes may flow out from pipes.**

12. Remove fuel injector from fuel tube as per the following:

a. Open and remove clip.

b. Remove fuel injector from fuel tube by pulling straight.

✳✳ WARNING

Be careful with remaining fuel that may go out from fuel tube. Be careful not to damage injector nozzles during removal. Never bump or drop fuel injector. Do not disassemble fuel injector.

13. Remove fuel sub-tube and fuel damper, if necessary.

To install:

➡**Do not reuse the O-rings.**

14. Install fuel damper as per the following:

a. Install new O-ring to fuel tube as shown.

➡**Handle O-ring with bare hands. Never wear gloves. Lubricate O-ring with new engine oil. Never clean O-ring with solvent. Check that O-ring and its mating part are free of foreign material. When installing O-ring, be careful not to scratch it with tool or fingernails. Also, be careful not to twist or stretch O-ring. If O-ring was stretched while it was being attached, never insert it quickly into fuel tube. Insert new O-ring straight into fuel tube. Never twist it.**

b. Install spacer to fuel damper.

c. Insert fuel damper straight into fuel tube.

➡**Insert straight, checking sure that the axis is lined up. Insert fuel damper at 29 lbs. (130 N) or less to prevent damage to the parts. Insert fuel damper until B is touching A of fuel tube.**

d. Tighten bolts evenly in turn. After tightening bolts, check that there is no gap between fuel damper cap and fuel tube.

15. Install fuel sub-tube.

a. When handling new O-rings, use care as noted above.

b. Insert fuel sub-tube straight into fuel tube.

c. Tighten mounting bolts evenly in turn.

d. After tightening mounting bolts, check that there is no gap between flange and fuel tube.

16. Install new O-rings to fuel injector, paying attention to the following.

➡**Upper and lower O-rings are different. Be careful not to confuse them. Fuel tube side (Black); Nozzle side (Green).**

17. Install fuel injector to fuel tube as per the following:

a. Insert clip into clip mounting groove on fuel injector.

✳✳ CAUTION

Never reuse clip. Replace it with a new one. Be careful to keep clip from interfering with O-ring. If interference occurs, replace O-ring.

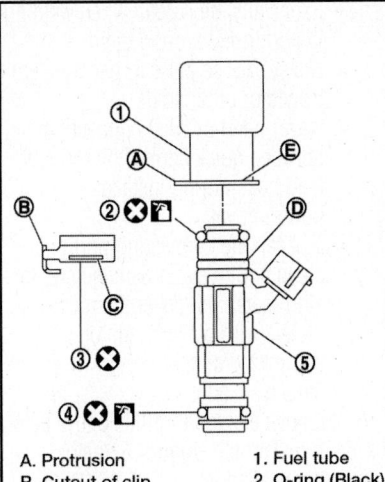

A. Protrusion
B. Cutout of clip
C. Flange fixing groove
D. Clip mounting groove
E. Fuel tube flange

1. Fuel tube
2. O-ring (Black)
3. Clip
4. O-ring (Green)
5. Fuel injector

37663_FX35_G0454

Fig. 291 Install fuel injector to fuel tube

b. Insert fuel injector into fuel tube with clip attached.
- Insert it while matching it to the axial center.
- Insert fuel injector so that protrusion of fuel tube matches cutout of clip.
- Check that fuel tube flange is securely fixed in flange fixing groove on clip.

c. Check that installation is complete by checking that fuel injector does not rotate or come off. Check that protrusions of fuel injectors are aligned with cutouts of clips after installation.

18. Install fuel tube and fuel injector assembly to intake manifold. Tighten

mounting bolts in 2 steps in numerical order as shown.
  a. Step 1: 89 inch lbs. (10 Nm).
  b. Step 2: 17 ft. lbs. (24 Nm).

❋❋ WARNING

Be careful not to let tip of injector nozzle come in contact with other parts.

19. Connect injector sub-harness.
20. Install fuel sub tube mounting bolt.
21. Connect fuel feed hose (with damper).
  a. Handling procedure of O-ring is the same as that of fuel damper and fuel sub-tube.
  b. Insert fuel damper straight into fuel sub-tube.
  c. Tighten mounting bolts evenly in turn.
  d. After tightening mounting bolts,

check that there is no gap between flange and fuel sub-tube.
22. Connect quick connector between fuel feed hose (with damper) and centralized under-floor piping connection as per the following:
  a. Check no foreign substances are deposited in and around centralized under-floor piping and quick connector, and no damage on them.
  b. Thinly apply new engine oil around centralized under-floor piping from tip end to spool end.
  c. Align center to insert quick connector straightly into centralized under-floor piping. Visually confirm that the 2 retainer tabs are connected to the quick connector.

➡**Carefully align center to avoid inclined insertion to prevent damage to O-ring inside quick connector. Insert**

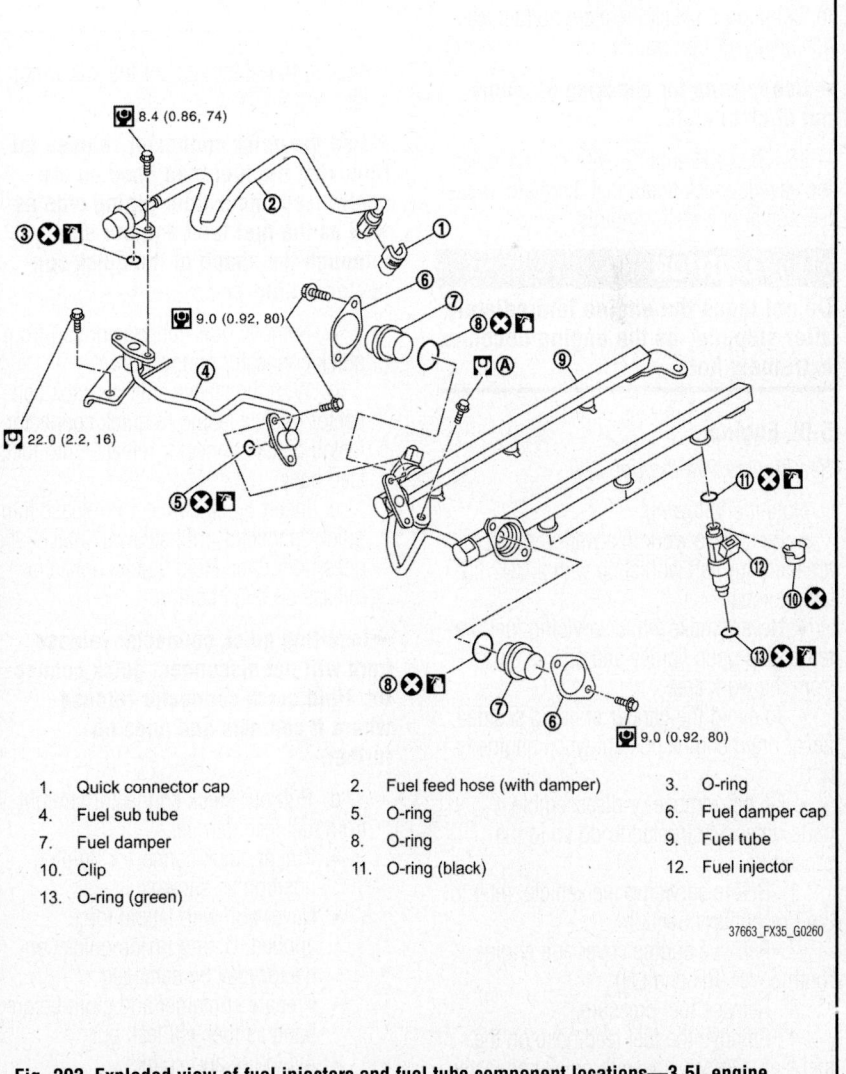

1. Quick connector cap
2. Fuel feed hose (with damper)
3. O-ring
4. Fuel sub tube
5. O-ring
6. Fuel damper cap
7. Fuel damper
8. O-ring
9. Fuel tube
10. Clip
11. O-ring (black)
12. Fuel injector
13. O-ring (green)

37663_FX35_G0260

Fig. 293 Exploded view of fuel injectors and fuel tube component locations—3.5L engine

Fig. 292 Fuel tube and fuel injector assembly bolt tightening sequence

37663_FX35_G0451

until you hear a "click" sound and actually feel the engagement. To avoid misidentification of engagement with a similar sound, be sure to perform the next step.

    d. Pull quick connector by hand holding position. Check it is completely engaged (connected) so that it does not come out from centralized under-floor piping.

    e. Install quick connector cap to quick connector connection. Install quick connector cap with arrow on surface facing the direction of quick connector (fuel feed hose side).

➡**If quick connector cap cannot be installed smoothly, quick connector may have not been installed correctly. Check the connection again.**

23. Installation continues in the reverse of the removal procedure.

24. Turn ignition switch "ON" (with the engine stopped). With fuel pressure applied to fuel piping, check there are no fuel leakage at connection points.

➡**Use mirrors for checking at points out of clear sight.**

25. Start the engine. With engine speed increased, check again that there are no fuel leakage at connection points.

✳✳ CAUTION

Do not touch the engine immediately after stopped, as the engine becomes extremely hot.

5.0L Engine

See Figures 294 through 299.

Note the following:
- Be sure to work in a well-ventilated area and furnish workshop with a CO_2 fire extinguisher
- Never smoke while servicing fuel system. Keep open flames and sparks away from the work area
- To avoid the danger of being scalded, never drain engine coolant when engine is hot
- Do not remove or disassemble parts unless instructed to do so in the procedure

1. Before servicing the vehicle, refer to the Precautions Section.

2. Remove engine cover and engine room cover (RH and LH).

3. Release fuel pressure.

4. Remove the fuel feed hose on the fuel feed damper side with quick connector

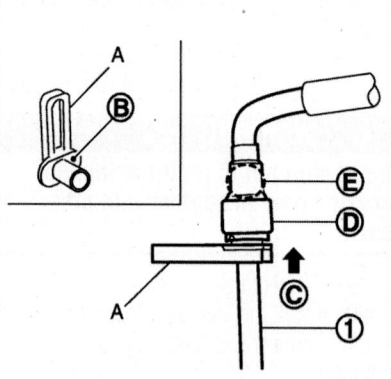

A. Quick connector release
B. Sleeve side
C. Insert and retain
D. Quick connector
E. Holding position
1. Fuel feed damper

37663_FX35_G0457

Fig. 294 Using a quick connector release tool

release J-45488 tool as per the followings steps.

➡**Use the quick connector release for removing the fuel feed hose on the centralized under-floor piping side as well as the fuel feed damper side although the shape of the quick connector is different.**

    a. Remove quick connector cap from quick connector connection.

    b. With the sleeve side of quick connector release facing to quick connector, install quick connector release onto fuel feed hose.

    c. Insert quick connector release into quick connector until sleeve contacts and goes no further. Hold quick connector release on that position.

➡**Inserting quick connector release hard will not disconnect quick connector. Hold quick connector release where it contacts and goes no further.**

    d. Pull out quick connector straight from fuel feed damper.
- Pull at quick connector holding position as shown
- Never pull with lateral force applied. O-ring inside quick connector may be damaged
- Prepare container and cloth beforehand as fuel will leak out
- Avoid fire and sparks
- Keep parts away from heat source

Especially, be careful when welding is performed around them
- Never expose parts to battery electrolyte or other acids
- Never bend or twist connection between quick connector and fuel feed hose during installation/removal
- To keep the connecting portion clean and to avoid damage and foreign materials, cover them completely with plastic bags or something similar

5. Remove air duct.

6. Remove electric throttle control actuator. Refer to Throttle Control Actuator, removal & installation.

7. Remove fuel hose (center).

➡**The procedure for removing the quick connector is the same as for removing the fuel feed damper.**

➡**Disconnect quick connector by using quick connector release J-45488, not by picking out retainer tabs.**

8. Remove fuel tube and fuel injector assembly, loosening the bolts in reverse of tightening sequence.

✳✳ CAUTION

Do not tilt it, or remaining fuel in pipes may flow out from pipes.

9. Remove fuel injector from fuel tube as per the following:
    a. Open and remove clip.
    b. Remove fuel injector from fuel tube by pulling straight.

✳✳ WARNING

Be careful with remaining fuel that may go out from fuel tube. Be careful not to damage injector nozzles during removal. Never bump or drop fuel injector. Do not disassemble fuel injector.

10. Disconnect sub harness connector from fuel injectors.

11. Remove fuel damper and fuel feed damper, if necessary.

    To install:

➡**Do not reuse O-rings.**

12. Install fuel damper as per the following:

    a. Install new O-ring to fuel tube (bank 1). When handling new O-ring, pay attention to the following caution items:
- Handle O-ring with bare hands. Never wear gloves

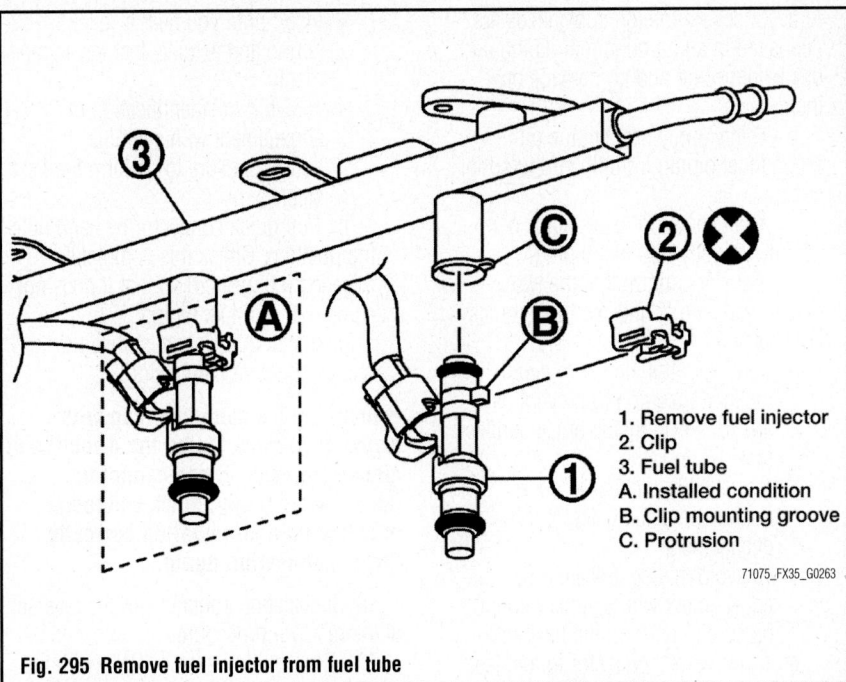

1. Remove fuel injector
2. Clip
3. Fuel tube
A. Installed condition
B. Clip mounting groove
C. Protrusion

71075_FX35_G0263

Fig. 295 Remove fuel injector from fuel tube

- Lubricate O-ring with new engine oil
- Never clean O-ring with solvent
- Check that O-ring and its mating part are free of foreign material
- When installing O-ring, be careful not to scratch it with tool or fingernails. Also be careful not to twist or stretch O-ring. If O-ring was stretched while it was being

attached, never insert it quickly into fuel tube
- Insert new O-ring straight into fuel tube. Never decenter or twist it

   b. Install spacer to fuel damper. Insert fuel damper straight into fuel tube (bank 1).
- Insert straight, check that the axis is lined up
- Insert fuel damper at 29 lbs. (130 N) or less to prevent damage to the parts
- Insert fuel damper until the rim reaches the cap flange

   c. Tighten mounting bolts evenly in turn. After tightening mounting bolts, check that there is no gap between flange and fuel tube (bank 1).

13. Install fuel feed damper.

   a. Handling procedure of O-ring is the same as that of fuel damper.

   b. Insert fuel feed damper straight into fuel tube (bank 2).

➡**Insert fuel feed damper at 33 lbs. (147 N) or less to prevent damage to the parts.**

   c. Tighten mounting bolts evenly in turn. After tightening mounting bolts, check that there is no gap between flange and fuel tube (bank 2).

14. Install new O-rings to fuel injector paying attention to the following caution.
- Upper and lower O-ring are different. Be careful not to confuse them
- Fuel tube side O-ring (Black); Nozzle side O-ring (Green)

- Handle O-ring with bare hands. Never wear gloves
- Lubricate O-ring with new engine oil
- Never clean O-ring with solvent
- Check that O-ring and its mating part are free of foreign material
- When installing O-ring, be careful not to scratch it with tool or fingernails. Also be careful not to twist or stretch O-ring. If O-ring was stretched while it was being attached, never insert it quickly into fuel tube
- Insert O-ring straight into fuel injector. Never decenter or twist it

15. Install fuel injector to fuel tube as per the following:

   a. Insert clip into clip mounting groove on fuel injector. Insert clip so that protrusion of fuel injector matches cutout of clip.

➡**Never reuse clip. Replace it with a new one. Be careful to keep clip from interfering with O-ring. If interference occurs, replace O-ring.**

   b. Insert fuel injector into fuel tube with clip attached.
- Insert it while matching it to the axial center
- Insert fuel injector so that protrusion of fuel tube matches cutout of clip

Fig. 296

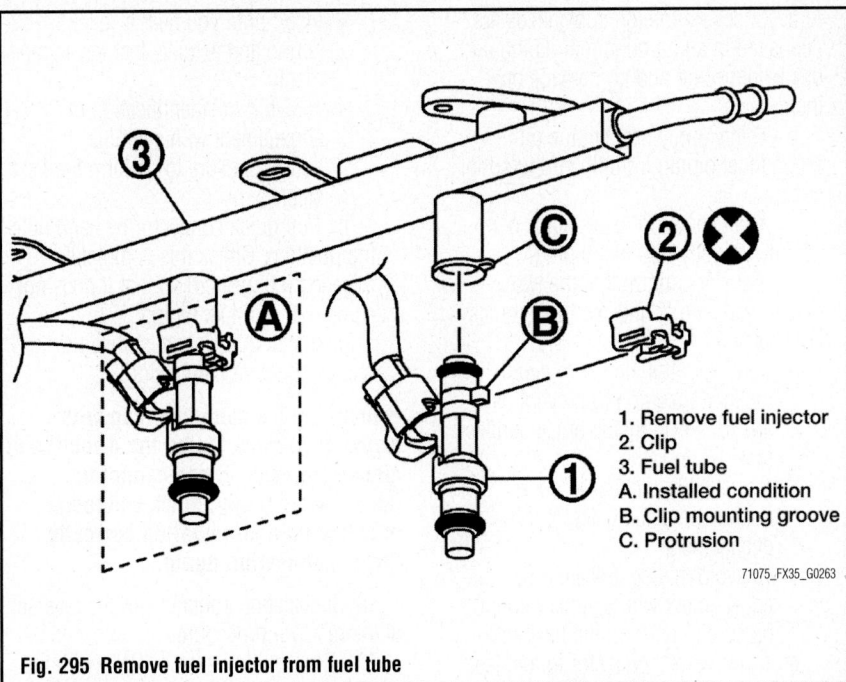

A. Spacer
B. Cap flange
C. Rim

1. Fuel damper cap
2. Fuel tube (Bank 1)
3. O-ring
4. Fuel damper

37663_FX35_G0460

Fig. 296 Installing fuel damper

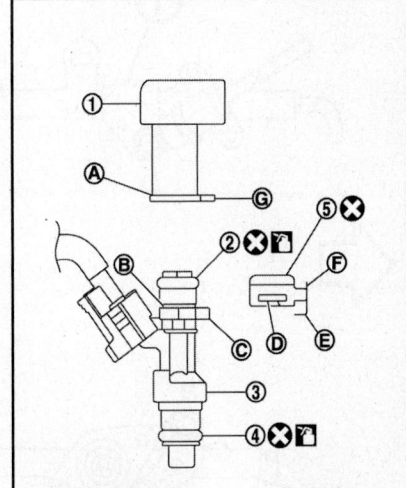

A. Fuel tube flange
B. Clip mounting groove
C. Protrusion
D. Flange fixing groove
E. Cutout of clip
F. Cutout of clip
G. Protrusion

1. Fuel tube
2. O-ring (Black)
3. Fuel injector
4. O-ring (Green)
5. Clip

37663_FX35_G0461

Fig. 297 Install fuel injector to fuel tube

- Check that fuel tube flange is securely fixed in flange fixing groove on clip

➡ **Insert fuel injector at 33 lbs. (147 N) or less to prevent damage to the parts.**

c. Check that installation is complete by checking that fuel injector does not rotate or come off.

➡ **Check that protrusions of fuel injectors and fuel tube are aligned with cutouts of clips after installation.**

16. Install fuel tube and fuel injector assembly. Tighten in 2 steps.
  a. Step 1: 89 inch lbs. (10 Nm)
  b. Step 2: 17 ft. lbs. (24 Nm)

✳✳ WARNING

Be careful not to let tip of injector nozzle come in contact with other parts. Insert fuel injector at 33 lbs. (147 N) or less to prevent damage to the parts.

17. Install quick connecters as per the following:

➡ **Unless otherwise indicated, the installation to the engine side and centralized under-floor piping side is exactly alike.**

a. Check no foreign substances are deposited in and around fuel piping and quick connector, and no damage on them.

b. Thinly apply new engine oil around fuel piping from tip end to spool end.

c. Align center to insert quick connector straightly into fuel piping.

d. Visually confirm that the two retainer tabs are connected to the quick connector.

- Carefully align center to avoid inclined insertion to prevent damage to O-ring inside quick connector
- Insert until you hear a "click" sound and actually feel the engagement
- To avoid misidentification of engagement with a similar sound, be sure to perform the next step

e. Insert quick connector to fuel feed damper piping until top spool is completely inside quick connector and 2nd level spool exposes just below quick connector.

- Carefully align center to avoid inclined insertion to prevent damage to O-ring inside quick connector

- Insert until you hear a "click" sound and actually feel the engagement
- To avoid misidentification of engagement with a similar sound, be sure to perform the next step

f. Pull quick connector by hand holding position. Check it is completely engaged (connected) so that it does not come out from fuel piping.

g. Install quick connector cap to quick connector connection.

➡ **Install quick connector cap with arrow on surface facing the direction of quick connector. If cap cannot be installed smoothly, quick connector may have not be installed correctly. Check connection again.**

18. Installation continues in the reverse of the removal procedure.

19. Turn ignition switch "ON" (with the engine stopped). With fuel pressure applied to fuel piping, check that there is no fuel leakage at connection points.

➡ **Use mirrors for checking at points out of clear sight.**

20. Start the engine. With engine speed increased, check again that there is no fuel leakage at connection points.

✳✳ CAUTION

Never touch the engine immediately after it is stopped because the engine is extremely hot.

FUEL TANK

DRAINING

1. Before servicing the vehicle, refer to the Precautions Section.

2. Insert fuel tubing of less than 1 inch (25mm) diameter into the fuel filler tube through the fuel filler opening to drain fuel from the fuel filler tube.

3. Disconnect the fuel filler hose from the fuel filler tube.

4. Insert fuel tubing into the fuel tank through the fuel filler hose to drain fuel from the fuel tank.

REMOVAL & INSTALLATION

See Figures 300 through 302.

1. Before servicing the vehicle, refer to the Precautions Section.

2. Check fuel level on fuel gauge. If fuel gauge indicates full or almost full, drain fuel from fuel tank until fuel gauge indicates level as shown or below.

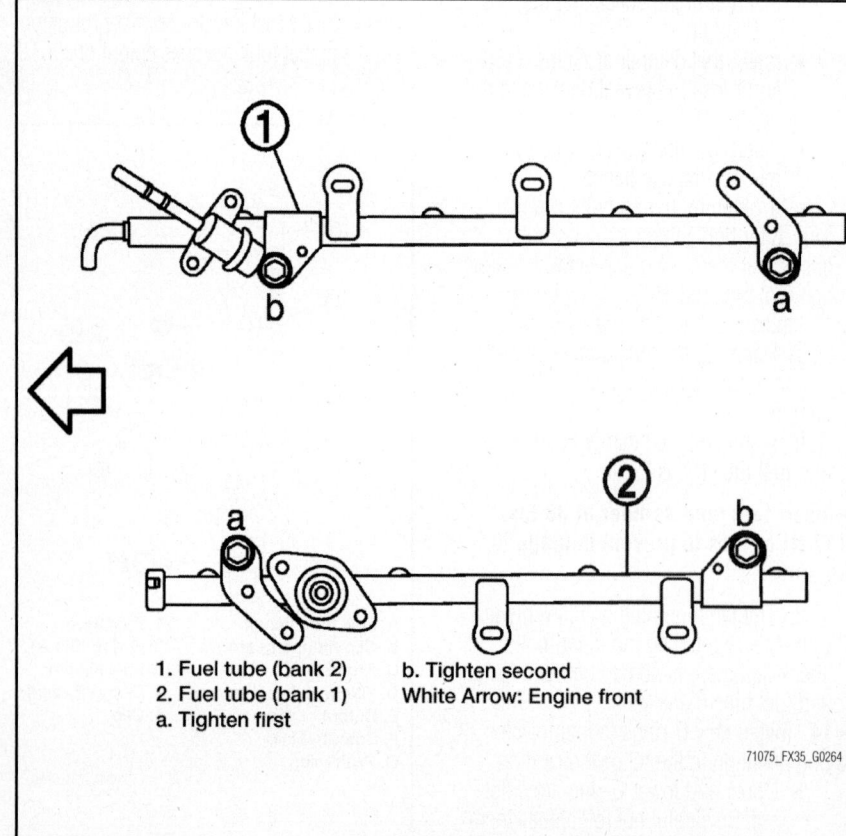

1. Fuel tube (bank 2)
2. Fuel tube (bank 1)
a. Tighten first
b. Tighten second
White Arrow: Engine front

71075_FX35_G0264

Fig. 298 Fuel tube and fuel injector assembly bolt tightening sequence

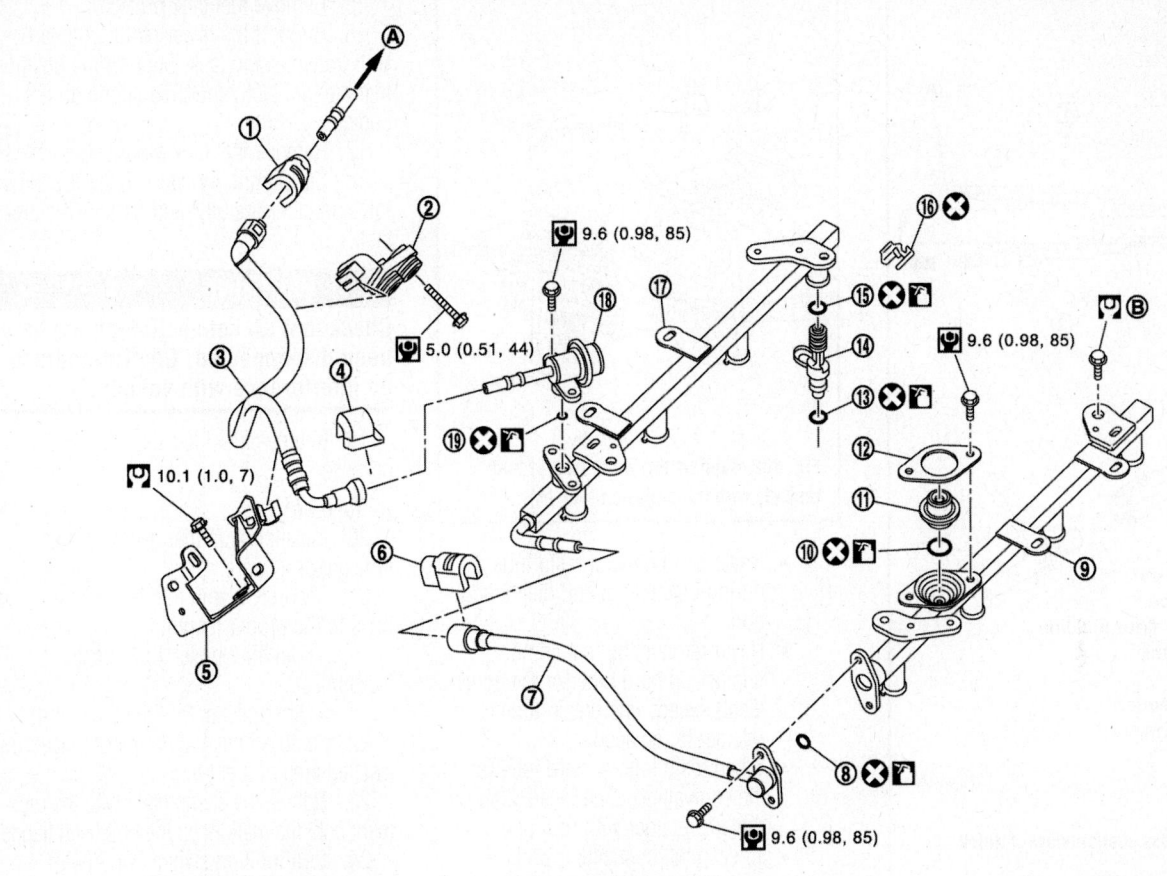

1. Quick connector cap
2. Fuel hose bracket
3. Fuel feed hose
4. Quick connector cap
5. Fuel hose bracket
6. Quick connector cap
7. Fuel hose (center)
8. O-ring
9. Fuel tube (bank 1)
10. O-ring
11. Fuel damper
12. Fuel damper cap
13. O-ring (green)
14. Fuel injector
15. O-ring (black)
16. Clip
17. Fuel tube (bank 2)
18. Fuel feed damper
19. O-ring

37663_FX35_G0237

Fig. 299 Exploded view of the fuel injectors and related component locations—5.0L engine

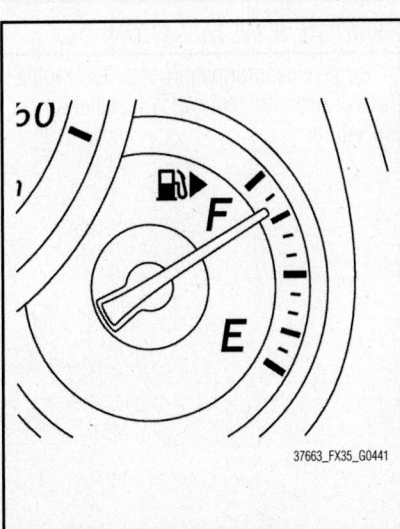

37663_FX35_G0441

Fig. 300 Fuel gauge level shown

➡**Fuel may be spilled when removing main and sub fuel level sensor units for the top of the fuel is above the main and sub fuel level sensor unit installation surface. As a guide, the fuel level becomes the position as shown or below when approximately 5.25 gallons (20L) of fuel are drained from the fuel tank.**

3. In the case that the fuel pump does not operate, perform the following procedure.

a. Insert a hose of less than 1 inch (25mm) in diameter into the fuel filler tube through the fuel filler opening to draw the fuel from the fuel filler tube.

b. Disconnect the fuel filler hose from the fuel filler tube.

c. Insert the fuel tube into the fuel tank through the fuel filler hose to draw the fuel from the fuel tank.

4. Release the fuel pressure from the fuel lines.

5. Open fuel filler lid.

6. Open filler cap and release the pressure inside fuel tank.

7. Remove rear seat cushion.

8. Peel off floor carpet, then remove inspection hole cover units by turning clips clockwise by 90°.

9. Disconnect harness connector and fuel feed tube.

a. Hold the sides of connector, push in tabs and pull out fuel feed tube.

b. If quick connector sticks to tube of main fuel level sensor unit, push and pull quick connector several times until they

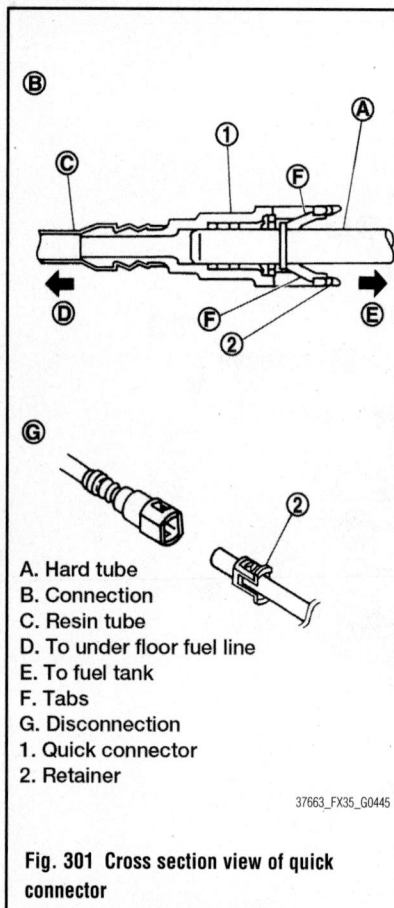

A. Hard tube
B. Connection
C. Resin tube
D. To under floor fuel line
E. To fuel tank
F. Tabs
G. Disconnection
1. Quick connector
2. Retainer

37663_FX35_G0445

Fig. 301 Cross section view of quick connector

start to move. Then disconnect them by pulling.

- Quick connector can be disconnected when the tabs are completely depressed. Never twist it more than necessary
- Never use any tools to disconnected quick connector
- Keep resin tube away from heat. Be especially careful when welding near the resin tube
- Prevent acid liquid such as battery electrolyte, etc. from getting on resin tube

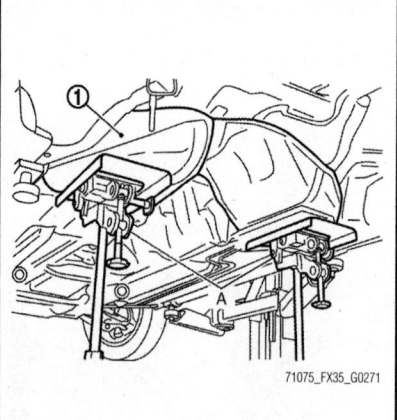

71075_FX35_G0271

Fig. 302 Support the lower part of fuel tank (1) with transmission jack (A)

- Never bend or twist resin tube during installation and disconnection
- Never remove the remaining retainer on hard tube (or the equivalent) except when resin tube or retainer is replaced
- When resin tube or hard tube (or the equivalent) is replaced, also replace retainer with new one
- To keep the connecting portion clean and to avoid damage and foreign materials, cover them completely with plastic bags or something similar

10. Remove exhaust front tube, center muffler and main muffler.
11. Remove propeller shaft (driveshaft).
12. Remove rear parking brake cables.
13. Remove rear suspension member assembly.

➡**For this service, the halfshaft, final drive, and rear suspension member are not required to be separated from one another during removal.**

14. Disconnect fuel filler hose, EVAP tube and vent hose.

15. Remove fuel tank protector.
16. Support the lower part of fuel tank with transmission jack. Support the position that the fuel tank mounting bands do not engage.
17. Remove fuel tank mounting bands.
18. Supporting by hand, lower the transmission jack carefully, and remove the fuel tank.

✳✳ WARNING

Check that all connection points have been disconnected. Confirm there is no interference with vehicle.

19. Remove fuel filler tube, if necessary.

To install:
20. Installation is the reverse of the removal procedure.
21. Securely clamp fuel hoses and insert hose to the proper length.
    a. Fuel filler hose: 1.38 inches (35mm).
    b. Other hoses: 0.98 inch (25mm).
22. Be sure hose clamp is not placed on swelled area of fuel tube.
23. Tighten the clamp hand with the top mark until the mark is on the bolt head flange.
24. Turn ignition switch "ON" (with engine stopped), and check connections for leakage by applying fuel pressure to fuel piping.
25. Start engine and rev it up and check there is no fuel leakage at the fuel system tube and hose connections.
26. After removing/installing rear suspension assembly, check to adjust wheel alignment.

THROTTLE BODY

REMOVAL & INSTALLATION

For service information, refer to Throttle Control Actuator, removal & installation procedure.

HEATING & AIR CONDITIONING SYSTEM

BLOWER MOTOR

REMOVAL & INSTALLATION

See Figure 303.

1. Before servicing the vehicle, refer to the Precautions Section.
2. Remove instrument lower cover RH.
3. Disconnect blower motor connector.
4. Remove mounting screws, and then remove blower motor.
5. Installation is the reverse of the removal procedure.

HEATER CORE

REMOVAL & INSTALLATION

See Figure 304.

1. Before servicing the vehicle, refer to the Precautions Section.
2. Remove heater and cooling unit assembly. Refer to Heater & Cooling Unit, removal & installation.
3. Remove mounting screws, and then remove heater pipe cover.
4. Remove mounting screws, and then remove foot duct (left).
5. Slide heater core to leftward.

To install:

6. Installation is the reverse of the removal procedure.
7. Fill the cooling system with the proper type and amount of fluid. Check for any leakage.

HEATER & COOLING UNIT

REMOVAL & INSTALLATION

See Figures 305 through 310.

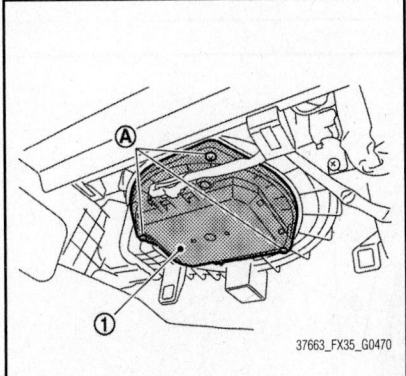

Fig. 303 Blower motor (1) and mounting screw (A) component locations

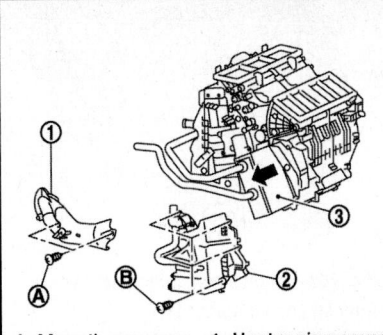

A. Mounting screws
B. Mounting screws
1. Heater pipe cover
2. Foot duct (left)
3. Heater core

37663_FX35_G0488

Fig. 304 Exploded view showing heater core removal

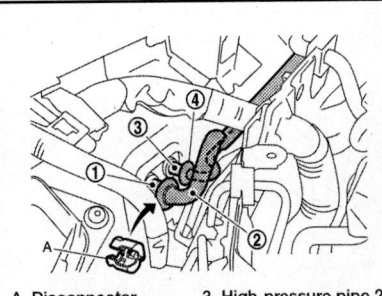

A. Disconnector
1. Low-pressure pipe 1
2. Low-pressure pipe 2
3. High-pressure pipe 2
4. High-pressure pipe 1

37663_FX35_G0489

Fig. 305 Disconnect one-touch joint between low-pressure pipe 1 and low-pressure pipe 2

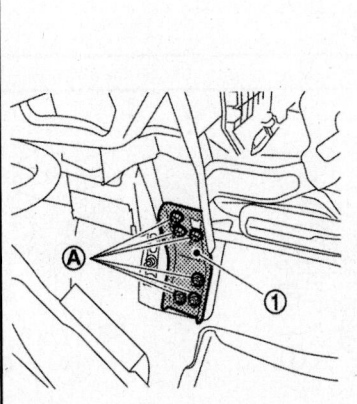
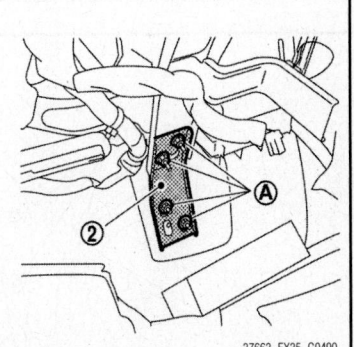

37663_FX35_G0490

Fig. 306 Remove mounting nuts (A), and then remove instrument stay (left) (1) and instrument stay (right) (2)

1. Before servicing the vehicle, refer to the Precautions Section.
2. Set the temperature at full cold.
3. Disconnect the battery cable from the negative terminal.
4. Use refrigerant collecting equipment (for HFC-134a) to discharge the refrigerant.
5. Drain engine coolant from cooling system.
6. Remove cowl top cover.
7. Remove engine cover.
8. Disconnect one-touch joint between low-pressure pipe 1 and low-pressure pipe 2 with Disconnector SST: 9253089916.

➡**Cap or wrap the joint of the A/C piping with suitable material such as vinyl tape to avoid the entry of air.**

9. Disconnect one-touch joint between high-pressure pipe 1 and high-pressure pipe 2 with Disconnector SST: 9253089908.

➡**Cap or wrap the joint of the A/C piping with suitable material such as vinyl tape to avoid the entry of air.**

10. Remove clamps and then disconnect heater hoses.
11. Remove instrument panel assembly.
12. Remove defroster nozzle and adaptor duct.
13. Remove blower unit assembly.
14. Remove mounting nuts, and then remove instrument stay (left) and instrument stay (right).

15. Disconnect drain hose from heater & cooling unit assembly.

16. Remove mounting bolts from steering member.

17. Remove mounting bolts from steering member.

18. Remove ground bolts from steering member.

19. Remove steering column mounting nuts and bolts.

20. Remove harness connector and clips of vehicle harness from steering member.

21. Remove mounting bolts from steering member.

22. Remove steering member mounting bolt.

23. Remove steering member, and then remove heater & cooling unit assembly.

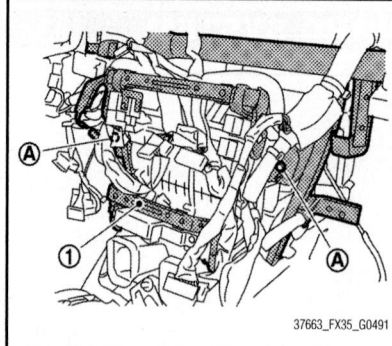

37663_FX35_G0491

Fig. 307 Remove mounting bolts (A) from steering member (1)

To install:

24. Installation is the reverse of the removal procedure.

25. Replace O-rings with new ones. Then apply compressor oil to them when installing.

❄❄ WARNING

Female-side piping connection is thin and easy to deform. Slowly insert the male-side piping straight in axial direction.

26. Insert piping securely until a clicks is heard.

27. After piping connection is completed, pull male-side piping by hand to make sure that connection does not come loose.

28. Recharge the refrigerant. Check for leakage when recharging refrigerant.

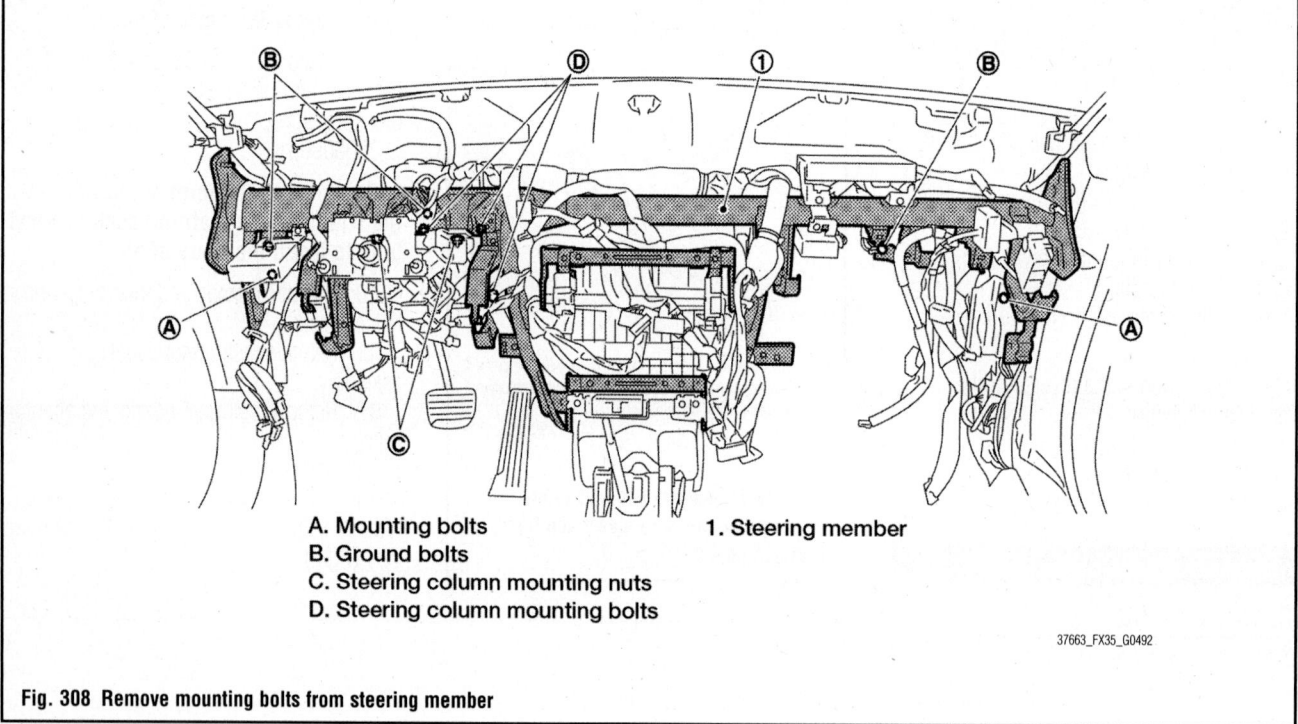

A. Mounting bolts
B. Ground bolts
C. Steering column mounting nuts
D. Steering column mounting bolts

1. Steering member

37663_FX35_G0492

Fig. 308 Remove mounting bolts from steering member

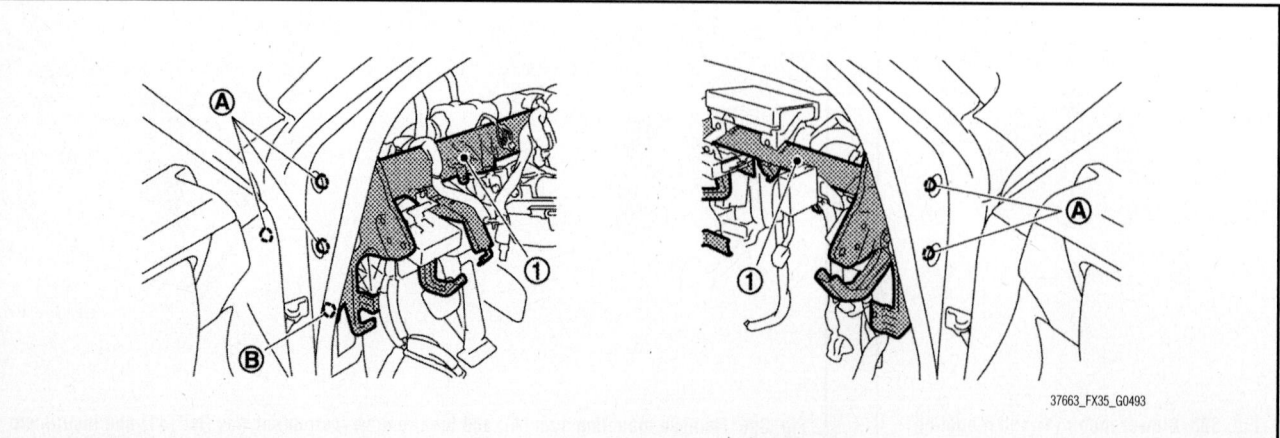

37663_FX35_G0493

Fig. 309 Remove mounting bolts (A) and steering member mounting bolt (B) from steering member (1)

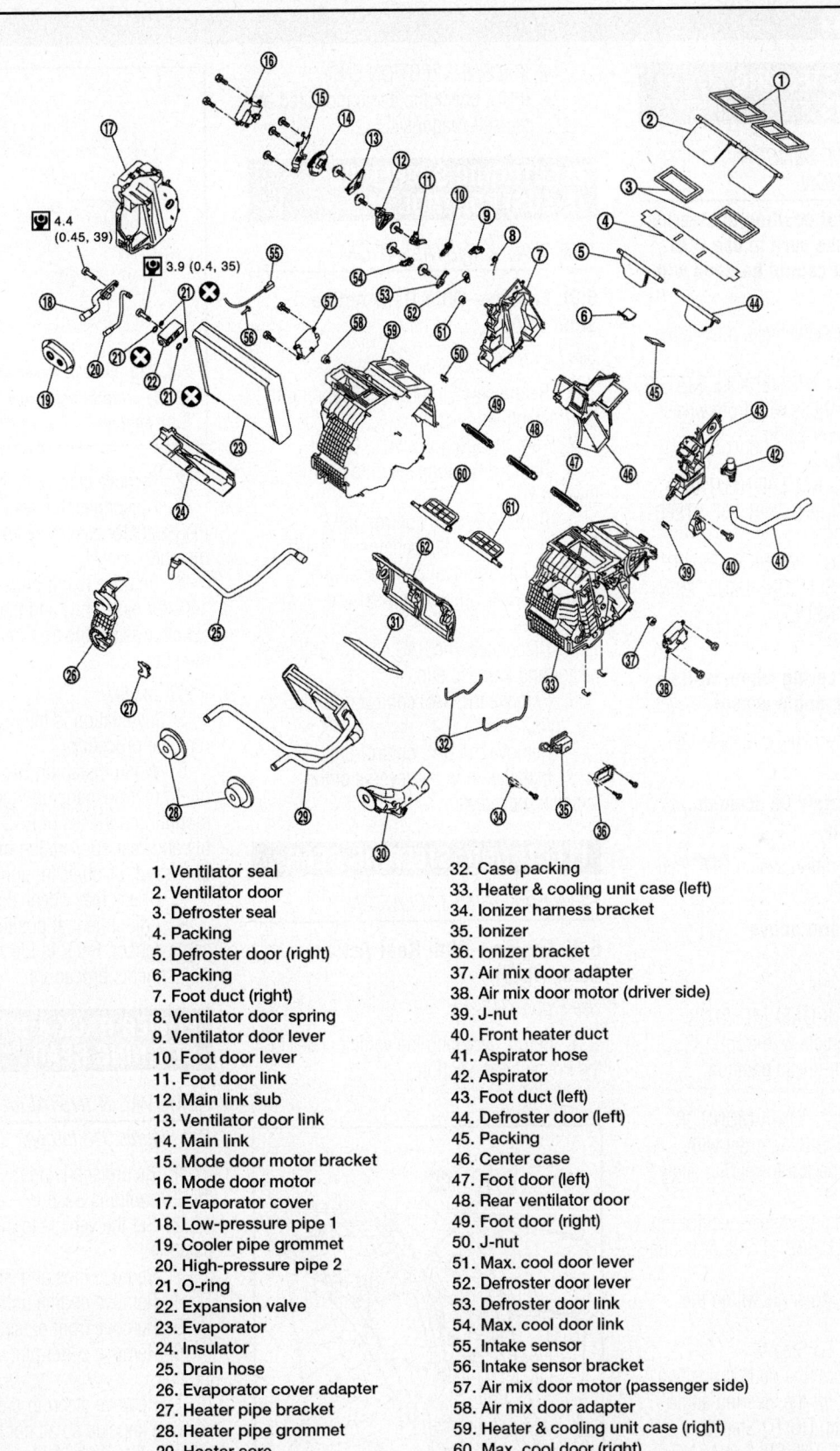

1. Ventilator seal
2. Ventilator door
3. Defroster seal
4. Packing
5. Defroster door (right)
6. Packing
7. Foot duct (right)
8. Ventilator door spring
9. Ventilator door lever
10. Foot door lever
11. Foot door link
12. Main link sub
13. Ventilator door link
14. Main link
15. Mode door motor bracket
16. Mode door motor
17. Evaporator cover
18. Low-pressure pipe 1
19. Cooler pipe grommet
20. High-pressure pipe 2
21. O-ring
22. Expansion valve
23. Evaporator
24. Insulator
25. Drain hose
26. Evaporator cover adapter
27. Heater pipe bracket
28. Heater pipe grommet
29. Heater core
30. Heater pipe cover
31. Packing

32. Case packing
33. Heater & cooling unit case (left)
34. Ionizer harness bracket
35. Ionizer
36. Ionizer bracket
37. Air mix door adapter
38. Air mix door motor (driver side)
39. J-nut
40. Front heater duct
41. Aspirator hose
42. Aspirator
43. Foot duct (left)
44. Defroster door (left)
45. Packing
46. Center case
47. Foot door (left)
48. Rear ventilator door
49. Foot door (right)
50. J-nut
51. Max. cool door lever
52. Defroster door lever
53. Defroster door link
54. Max. cool door link
55. Intake sensor
56. Intake sensor bracket
57. Air mix door motor (passenger side)
58. Air mix door adapter
59. Heater & cooling unit case (right)
60. Max. cool door (right)
61. Max. cool door (left)
62. Air mix door (Slide door)

37663_FX35_G0494

Fig. 310 Exploded view of heating and cooling unit

STEERING

ELECTRONIC STEERING ADJUSTMENTS

STEERING ANGLE SENSOR NEUTRAL POSITION

➡**To adjust neutral position of steering angle sensor, make sure to use CONSULT. Adjustment cannot be done without CONSULT.**

Before servicing the vehicle, refer to the Precautions Section.

1. Step 1: ALIGN THE VEHICLE STATUS
   a. Stop the vehicle with front wheels in straight-ahead position.
   b. GO TO Step 2.
2. Step 2: PERFORM THE NEUTRAL POSITION ADJUSTMENT FOR THE STEERING ANGLE SENSOR
   a. Select "ABS", "WORK SUPPORT" and "ST ANGLE SENSOR ADJUSTMENT" in order with CONSULT.
   b. Select "START".

➡**Do not touch steering wheel while adjusting steering angle sensor.**

   c. After approximately 10 seconds, select "END".

➡**After approximately 60 seconds, it ends automatically.**

   d. Turn the ignition switch OFF, then turn it ON again.

➡**Be sure to perform above operation.**

   e. GO TO Step 3.
3. Step 3: CHECK DATA MONITOR
   a. Run the vehicle with front wheels in straight-ahead position, then stop.
   b. Select "ABS", "DATA MONITOR" and "STR ANGLE SIG" in order with CONSULT, and check the steering angle sensor signal.
   • Steering angle sensor specification (STR ANGLE SIG): 0 plus or minus 2.5°
   c. Is the steering angle within the specified range?
   • If YES, GO TO Step 4.
   • If NO, perform the neutral position adjustment for the steering angle sensor again, GO TO Step 1.
4. Step 4: ERASE THE SELF-DIAGNOSIS MEMORY
   a. Erase the self-diagnosis memories for "ABS", "ENGINE" and "ICC/ADAS" with CONSULT.
   b. Are the memories erased?

   • If YES, INSPECTION END.
   • If NO, check the items indicated by the self-diagnosis.

ELECTRONIC SUSPENSION (E-SUS) CONTROL UNIT

REMOVAL & INSTALLATION

5.0L Engine—With Rear Active Steer (RAS)

See Figure 311.

1. Before servicing the vehicle, refer to the Precautions Section.
2. Turn the ignition switch OFF.
3. Remove the luggage side finisher lower (LH).
4. Remove E-SUS control unit. Refer to Electronic Suspension (E-SUS) Control Unit, removal & installation under the Rear Suspension service procedures.
5. Disconnect the RAS control unit connector and harness clip.
6. Remove the RAS control unit mounting bolts.
7. Remove the RAS control unit.
8. Installation is the reverse of the removal procedure.

ELECTRONIC STEERING GEAR

REMOVAL & INSTALLATION

5.0L Engine—With Rear Active Steer (RAS)

See Figure 312.

1. Before servicing the vehicle, refer to the Precautions Section.

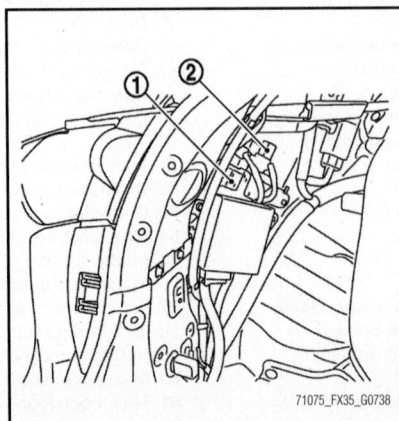

Fig. 311 View of RAS control unit (1) and RAS motor relay (2)—5.0L engine

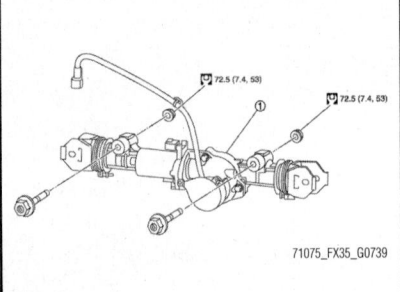

71075_FX35_G0739

Fig. 312 View of the Rear Active Steer (RAS) actuator assembly (1) component— 5.0L engine

2. Remove coil spring and rear lower link.
3. Disconnect harness connector from RAS actuator assembly and rear suspension member.
4. Remove fixing bolts and nuts of RAS actuator assembly, and then remove RAS actuator assembly from rear suspension member.

To install:

5. Installation is the reverse of the removal procedure.
6. When installing RAS actuator assembly to rear suspension member, check the mounting surfaces of RAS actuator assembly and rear suspension member for oil, dirt, sand, or other foreign materials.
7. Check rear wheel alignment.
8. Adjust neutral position of steering angle sensor. Refer to Electronic Steering, Adjustments procedure.

POWER RACK & PINION STEERING GEAR

REMOVAL & INSTALLATION

See Figures 313 through 318.

1. Before servicing the vehicle, refer to the Precautions Section.
2. Set the vehicle to the straight-ahead position.
3. Remove tires and wheels.
4. Remove engine under cover.
5. Remove front cross bar.
6. Remove cotter pin, and then loosen the nut.
7. Remove steering outer socket from steering knuckle so as not to damage ball joint boot using suitable ball joint remover.

❈❈ WARNING

Temporarily tighten the nut to prevent damage to threads and to prevent the

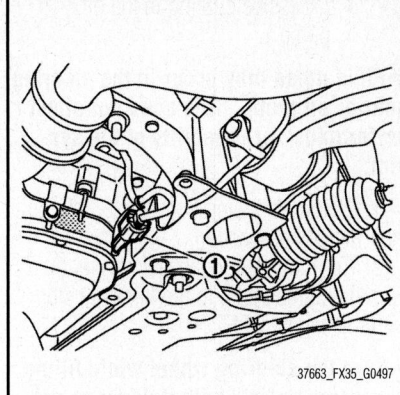

Fig. 313 Remove power steering solenoid valve harness connector (1) and harness clip

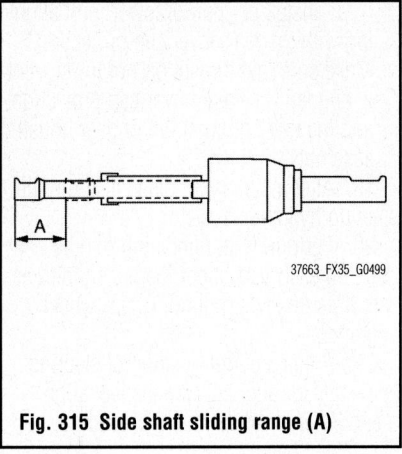

37663_FX35_G0499

Fig. 315 Side shaft sliding range (A)

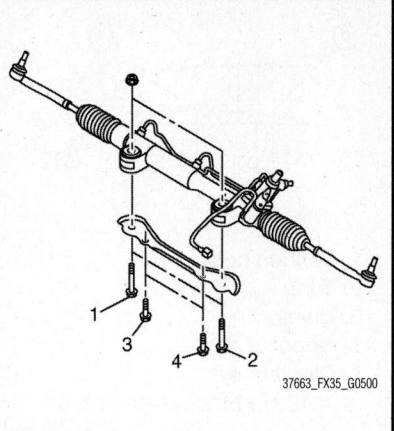

37663_FX35_G0500

Fig. 316 Tighten the mounting bolts in the order shown

ball joint remover from suddenly coming off.

8. Remove high pressure piping and low pressure piping of hydraulic piping, and then drain power steering fluid.
9. Remove power steering solenoid valve harness connector and harness clip.
10. Remove steering hydraulic piping bracket (5.0L engine).
11. Remove rack stay.
12. Remove lower joint mounting bolt (steering gear side).

13. Separate the lower shaft from the steering gear assembly by sliding the side shaft.

✳✳ WARNING
The spiral cable may be cut if the steering wheel turns while separating the steering column assembly and steering gear assembly. Be sure to secure the steering wheel to avoid turning.

✳✳ WARNING
When removing the lower joint, do not insert a tool, such as a screw-

driver, into the yoke groove to pull out the lower joint. In case of the violation of the above, replace the lower joint with a new one.

14. Remove the steering gear assembly.

To install:
15. Installation is the reverse of the removal procedure.
16. Tighten the mounting bolts in the order shown when installing the steering gear assembly.

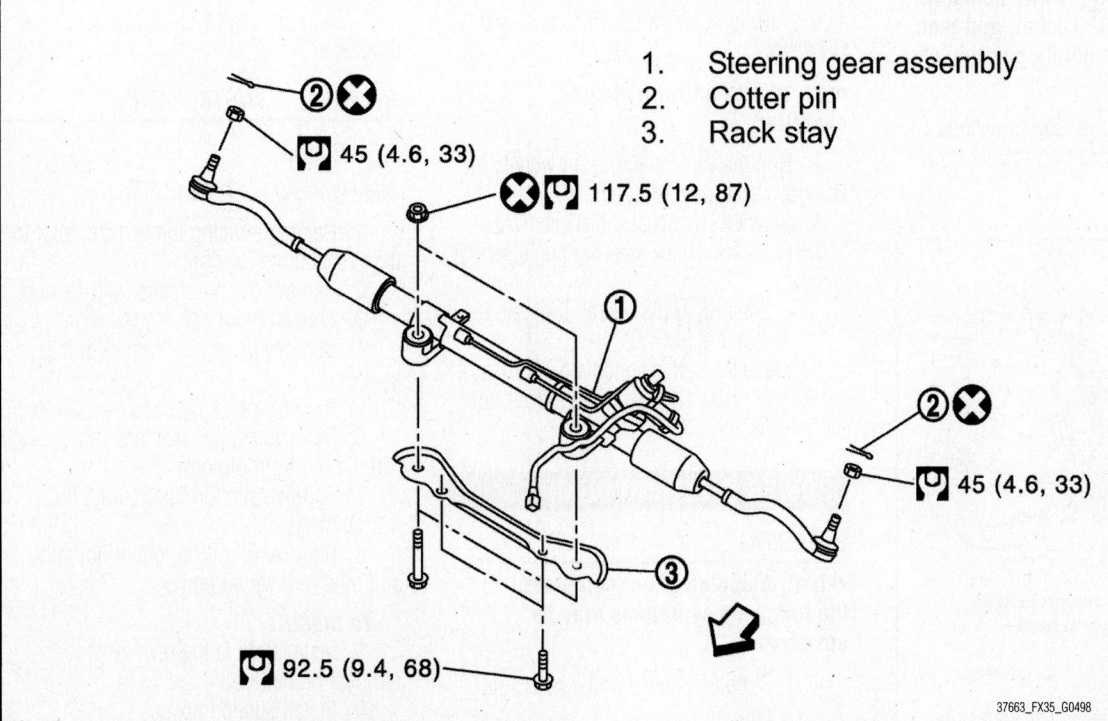

1. Steering gear assembly
2. Cotter pin
3. Rack stay

45 (4.6, 33)

117.5 (12, 87)

45 (4.6, 33)

92.5 (9.4, 68)

37663_FX35_G0498

Fig. 314 Steering gear and linkage component locations

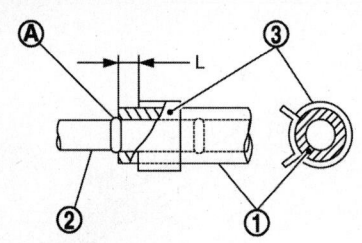

1. Suction hose
2. Tube
3. Clamp
A. Spool of tube
L. Clearance:
   0.12-0.31 inches (3-8 mm)

37663_FX35_G0501

Fig. 317 Install the suction hoses according to the figure

➡**Do not reuse the steering gear assembly mounting nut.**

17. Install the suction hoses.

➡**Never apply fluid to the hose and tube. Insert hose securely until it contacts spool of tube. Leave clearance when installing clamp.**

18. When installing lower joint to steering gear assembly, follow the procedure listed below.

   a. Set rack of steering gear in the neutral position.

➡**To get the neutral position of rack, turn gear-sub assembly and measure the distance of inner socket, and then measure the intermediate position of the distance.**

   b. Align rear cover cap projection with the marking position of gear housing assembly.

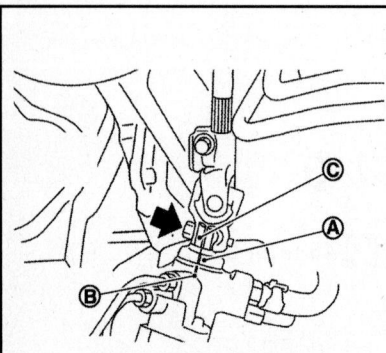

A. Rear cover cap projection
B. Gear housing assembly
C. Lower joint

37663_FX35_G0502

Fig. 318 Align rear cover cap projection with the marking position of gear housing assembly

c. Install slit part of lower joint aligning with the rear cover cap projection. Make sure that the slit part of lower joint is aligned with rear cover cap projection and the marking position of gear housing assembly.

19. After installation, bleed air from the steering hydraulic system.

20. Perform final tightening of nuts and bolts on each part under loaded conditions with tires on level ground. Check wheel alignment.

21. Adjust neutral position of steering angle sensor after checking wheel alignment. Refer to Adjustments procedure.

ADJUSTMENTS

Steering Angle Sensor

1. Before servicing the vehicle, refer to the Precautions Section.

2. Stop the vehicle with front wheels in straight-ahead position.

3. On the CONSULT-III screen, touch "WORK SUPPORT" and "ST ANGLE SENSOR ADJUSTMENT" in order.

4. Touch "START".

➡**Do not touch steering wheel while adjusting steering angle sensor.**

5. After approximately 10 seconds, touch "END".

➡**After approximately 60 seconds, it ends automatically.**

6. Turn ignition switch OFF, then turn it ON again.

➡**Be sure to perform above operation.**

7. Run the vehicle with front wheels in straight-ahead position, then stop.

8. Select "STR ANGLE SIG" in "DATA MONITOR" and check steering angle sensor signal.

   • Steering angle signal: 0 plus or minus 2.5°

9. Erase the self-diagnosis memories of the ABS actuator and electric unit (control unit), ECM and ICC.

POWER STEERING PUMP

BLEEDING

➡**If air bleeding is not complete, the following symptoms may be observed:**

   • Bubbles created in the reservoir tank
   • Clicking noise heard from the oil pump

   • Excessive buzzing in the oil pump

➡**Fluid noise may occur in the steering gear or oil pump. This does not affect performance or durability of the system.**

1. Before servicing the vehicle, refer to the Precautions Section.

2. Turn the steering wheel several times from the full left stop to the full right stop with the engine OFF.

➡**Turn the steering wheel while filling the reservoir tank with fluid so as not to lower the fluid level below the MIN line.**

3. Start the engine and hold the steering wheel at each lock position for 3 seconds at idle to check for fluid leakage.

4. Repeat several times at approximately 3 second intervals.

✳✳ WARNING

Do not hold the steering wheel in a locked position for more than 10 seconds. There is the possibility that the oil pump may be damaged.

5. Check the fluid for bubbles and white contamination.

6. Stop the engine if bubbles and white contamination do not drain out. Perform steps above after waiting until the bubbles and white contamination drain out.

7. Stop the engine, and then check the fluid level.

REMOVAL & INSTALLATION

3.5L Engine

See Figures 317, 319 and 320.

1. Before servicing the vehicle, refer to the Precautions Section.

2. Remove the air cleaner and air duct.

3. Loosen drive belt.

4. Remove drive belt from oil pump pulley.

5. Remove pressure sensor connector.

6. Remove copper washers and eye bolt (drain fluid from piping).

7. Remove suction hose (drain fluid from piping).

8. Remove oil pump mounting bolts, and then remove oil pump.

To install:

9. Installation is the reverse of the removal procedure.

10. Install suction hoses.

➡**Do not apply fluid to the hose and tube. Insert hose securely until it con-**

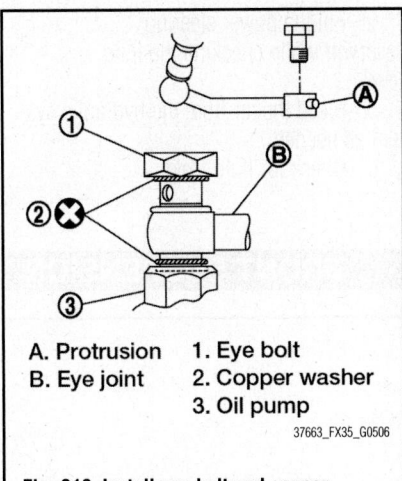

A. Protrusion
B. Eye joint
1. Eye bolt
2. Copper washer
3. Oil pump

37663_FX35_G0506

Fig. 319 Install eye bolt and copper washer to oil pump

tacts spool of tube. Leave clearance when installing clamp.

11. Install eye bolt and copper washer to oil pump.

➡**Do not reuse copper washer.**

12. Apply power steering fluid to and around copper washers.

13. Install eye bolt with eye joint (assembled to high pressure hose) protrusion facing with pump side cutout, tighten by hand, then tighten to 43 ft. lbs. (59 Nm).

14. Securely insert harness connector to pressure sensor.

15. Adjust belt tension.

16. Check fluid level and fluid leakage.

17. Air bleed the hydraulic system after the installation. Refer to Power Steering Pump, Bleeding.

5.0L Engine

See Figures 317, 319 and 321.

1. Before servicing the vehicle, refer to the Precautions Section.

2. Drain power steering fluid from reservoir tank.

3. Remove the engine under cover from vehicle.

4. Remove drive belt from oil pump pulley.

5. Remove pressure sensor connector.

6. Remove joint mounting nut.

7. Remove suction hose (drain fluid from piping).

8. Remove oil pump mounting bolts, and then remove oil pump.

To install:

9. Installation is the reverse of the removal procedure.

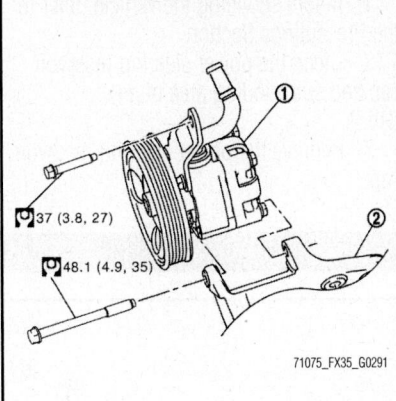

71075_FX35_G0291

Fig. 320 View of power steering oil pump (1) and bracket (2)—3.5L engine

10. Install suction hoses according to the figure shown.

➡**Do not apply fluid to the hose and tube. Insert hose securely until it contacts spool of tube. Leave clearance when installing clamp.**

11. Install eye bolt and copper washer to oil pump.

➡**Do not reuse copper washer.**

12. Apply power steering fluid to and around copper washers.

13. Install eye bolt with eye joint (assembled to high pressure hose) protrusion facing with pump side cutout, tighten by hand, then tighten to 43 ft. lbs. (59 Nm).

14. Securely insert harness connector to pressure sensor.

15. Install drive belt.

16. Check fluid level and fluid leakage.

17. Air bleed the hydraulic system after the installation. Refer to Power Steering Pump, Bleeding.

POWER STEERING FLUID

FLUID RECOMMENDATIONS

Recommended fluid is Genuine Nissan PSF or equivalent power steering fluid.

FLUID LEVEL CHECK

See Figure 322.

1. Before servicing the vehicle, refer to the Precautions Section.

2. Check the fluid level with the engine stopped.

3. Make sure that the fluid level is between the MIN and MAX markings.

 a. Fluid levels at HOT and COLD are different. Do not confuse them.
 • HOT Fluid temperature: 122–176°F (50–80°C)
 • COLD Fluid temperature: 32–86°F (0–30°C)
 b. The fluid level should not exceed the MAX line. Excessive fluid may cause fluid leakage from the cap.

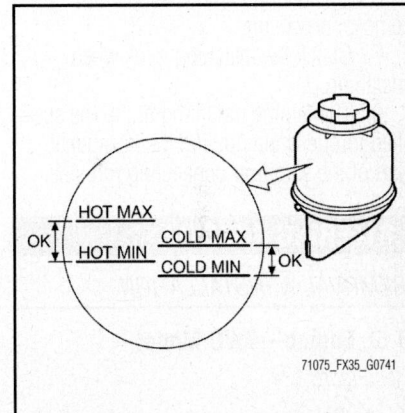

71075_FX35_G0741

Fig. 322 HOT and COLD fluid temperature levels indicated for the power steering fluid reservoir

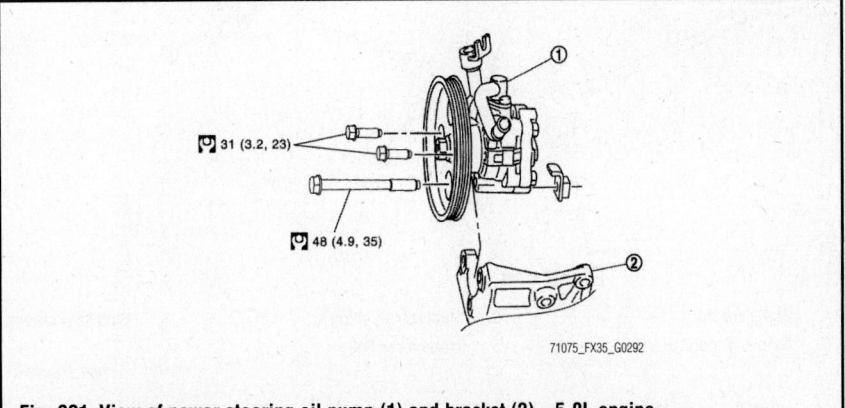

71075_FX35_G0292

Fig. 321 View of power steering oil pump (1) and bracket (2)—5.0L engine

FLUID FILL PROCEDURE

See Figure 322.

1. Before servicing the vehicle, refer to the Precautions Section.
2. Clean the power steering reservoir cap and surrounding area of any debris.
3. Remove the power steering reservoir cap.

4. Fill the power steering reservoir while checking the fluid level.
5. Bleed the air from the hydraulic system as needed.
6. Check for fluid leaks.

SUSPENSION

STABILIZER BAR

REMOVAL & INSTALLATION

See Figure 323.

1. Before servicing the vehicle, refer to the Precautions Section.
2. Remove under cover.
3. Remove stabilizer connecting rod.

➡**Apply a matching mark to identify the installation position.**

4. Remove stabilizer clamp and stabilizer bushing.
5. Remove stabilizer bar.

To install:

6. Installation is the reverse of the removal procedure.
7. Check the matching mark when installing.
8. Tighten the mounting nut to the specified torque while holding the hexagonal part of the stabilizer connecting rod side.

STEERING KNUCKLE

REMOVAL & INSTALLATION

3.5L Engine—2WD Model

See Figure 324.

1. Before servicing the vehicle, refer to the Precautions Section.

2. Remove tires and wheels.
3. Remove wheel sensor and sensor harness.

FRONT SUSPENSION

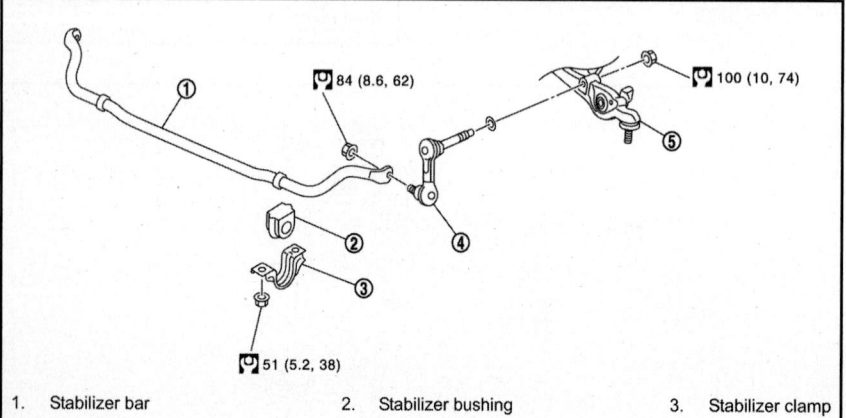

Fig. 324 Wheel hub and steering knuckle component locations—2WD vehicles

1. Steering knuckle	2. Ball seat	3. Cotter pin
4. Splash guard	5. Wheel hub and bearing assembly	

55 (5.6, 41)
88.3 (9.0, 65)
156 (16, 115)

37663_FX35_G0516

4. Remove brake hose bracket.
5. Remove caliper assembly. Hang caliper assembly in a place where it will not interfere with work.

➡**Do not depress brake pedal while brake caliper is removed.**

6. Remove disc rotor.
7. Remove wheel hub and bearing assembly, and then remove splash guard.
8. Remove steering outer socket.
9. Remove cotter pin of transverse link and steering knuckle, and then loosen nut.
10. Separate steering knuckle from upper link.
11. Separate steering knuckle from transverse link so as not to damage ball joint boot using the ball joint remover, and remove steering knuckle.

84 (8.6, 62)   100 (10, 74)
51 (5.2, 38)

1. Stabilizer bar	2. Stabilizer bushing	3. Stabilizer clamp
4. Stabilizer connecting rod	5. Transverse link	

37663_FX35_G0518

Fig. 323 Front stabilizer bar and related component locations

To install:

12. Installation is the reverse of the removal procedure.

13. Perform the final tightening of each of the parts under unladen conditions.

➡**Do not reuse the cotter pin.**

14. Check wheel sensor harness for proper connection.

15. Check the wheel alignment.

16. Adjust neutral position of steering angle sensor.

AWD Vehicles

See Figure 325.

1. Before servicing the vehicle, refer to the Precautions Section.

2. Remove tires and wheels.

3. Remove wheel sensor and sensor harness.

✳ WARNING

Do not pull on wheel sensor harness.

4. Remove brake hose bracket.

5. Remove caliper assembly. Hang caliper assembly in a place where it will not interfere with work.

➡**Do not depress brake pedal while brake caliper is removed.**

6. Remove disc rotor.

7. Remove cotter pin, and then loosen wheel hub lock nut.

8. Patch wheel hub lock nut with a piece of wood. Hammer the wood to disengage wheel hub and bearing assembly from halfshaft.

✳ WARNING

Never place halfshaft joint at an extreme angle. Also be careful not to overextend slide joint. Do not allow halfshaft to hang down without support for or joint sub-assembly, shaft and the other parts.

➡**Use suitable puller, if wheel hub and bearing assembly and halfshaft cannot be separated even after performing the above procedure.**

9. Remove wheel hub lock nut.

10. Remove wheel hub and bearing assembly, and then remove splash guard.

11. Remove steering outer socket (tie rod end).

12. Remove cotter pin of transverse link and steering knuckle, and then loosen nut.

13. Separate steering knuckle from upper link.

14. Separate steering knuckle link from transverse so as not to damage ball joint boot using the ball joint remover, and remove steering knuckle.

✳ WARNING

Temporarily tighten the nut to prevent damage to threads and to prevent the ball joint remover from suddenly coming off.

To install:

15. Installation is the reverse of the removal procedure.

16. Perform the final tightening of each of the parts under unladen conditions.

17. Install halfshaft using tightening torque of wheel hub lock nut.

✳ WARNING

Be sure to use torque wrench to tighten the wheel hub lock nut. Do not use a power tool.

➡**Do not reuse the cotter pin.**

18. Check wheel sensor harness for proper connection.

19. Check the wheel alignment.

20. Adjust neutral position of steering angle sensor.

STRUTS

REMOVAL & INSTALLATION

3.5L Engine—2WD Model

See Figure 326.

1. Before servicing the vehicle, refer to the Precautions Section.

2. Remove tires and wheels.

3. Remove wheel sensor and harness connector from shock absorber (strut).

✳ WARNING

Do not pull on wheel sensor harness.

4. Remove brake hose bracket.

5. Remove shock absorber from transverse link.

6. Separate upper link from steering knuckle.

7. Remove shock absorber mounting bracket mounting nuts, and remove shock absorber (strut) assembly.

➡**If removing shock absorber is difficult, loosen upper link mounting bolts (vehicle side).**

To install:

8. Note the following, and install in the reverse order of removal.

9. Never tap on the ball joint cap of the stabilizer connecting rod with a hammer or a similar item when inserting the stabilizer connecting rod into the transverse link.

10. Perform final tightening of bolts and nuts at the shock absorber lower side (rubber bushing), under unladen conditions with tires on level ground.

AWD Vehicles

See Figures 327 through 329.

1. Before servicing the vehicle, refer to the Precautions Section.

2. Remove engine cover.

3. Remove front fender protector.

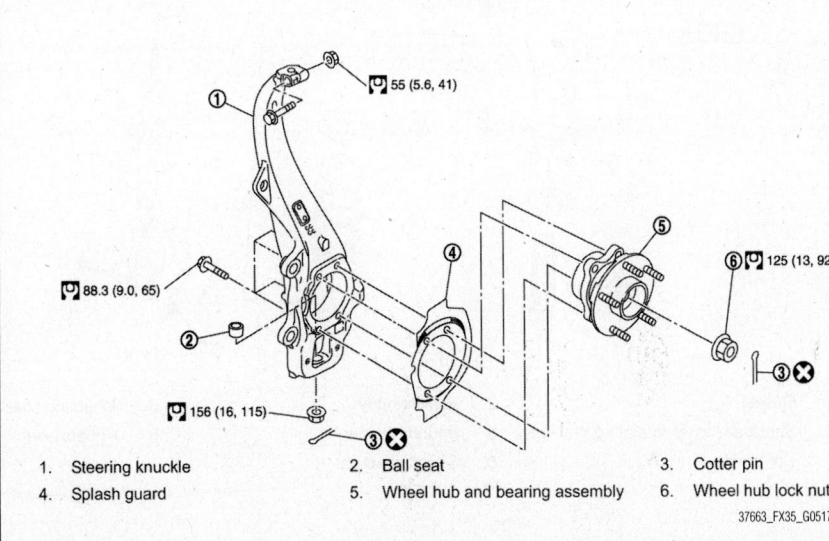

55 (5.6, 41)

88.3 (9.0, 65)

156 (16, 115)

125 (13, 92)

| 1. | Steering knuckle | 2. | Ball seat | 3. | Cotter pin |
| 4. | Splash guard | 5. | Wheel hub and bearing assembly | 6. | Wheel hub lock nut |

37663_FX35_G0517

Fig. 325 Wheel hub and steering knuckle component locations—AWD vehicles

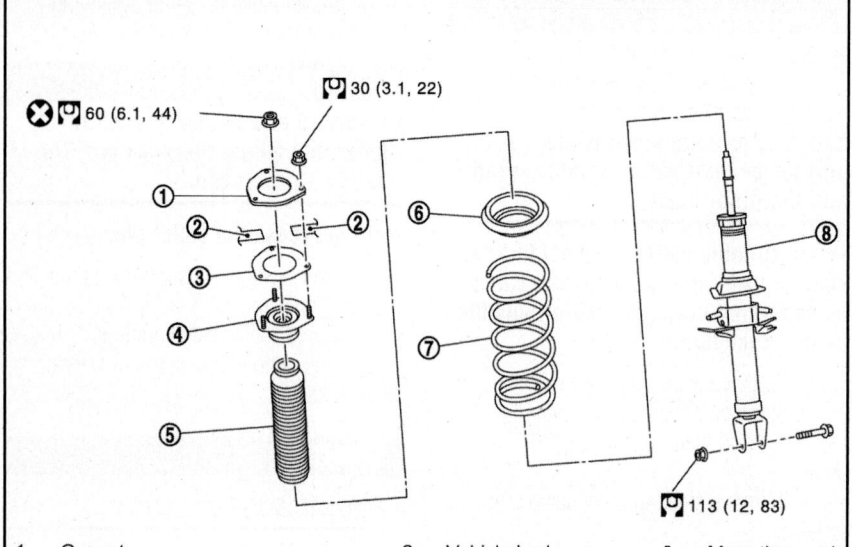

1. Gusset
2. Vehicle body
3. Mounting seal
4. Shock absorber mounting bracket
5. Bound bumper
6. Rubber seat
7. Coil spring
8. Shock absorber

37663_FX35_G0519

Fig. 326 Exploded view of strut assembly—2WD vehicle

4. Remove tires and wheels.

5. Remove wheel sensor and harness connector from vehicle.

☀☀ WARNING

Do not pull on wheel sensor harness.

6. With Continuous Damping Control (CDC).

a. Remove shock absorber actuator harness connector.

b. Remove front wheel vertical G sensor.

7. Remove brake hose bracket.

8. Remove stabilizer connecting rod.

9. Remove wheel hub lock nut.

10. Remove shock absorber (strut) from transverse link.

11. Separate upper link from steering knuckle.

12. Separate halfshaft from wheel hub and bearing assembly.

13. Remove shock absorber (strut) assembly.

➡If removing shock absorber (strut) is difficult, loosen upper link mounting bolts (vehicle side).

To install:

14. Note the following, and install in the reverse order of removal.

15. Never tap on the ball joint cap of the stabilizer connecting rod with a hammer or a similar item when inserting the stabilizer connecting rod into the transverse link.

16. Perform final tightening of bolts and nuts at the shock absorber lower side (rubber bushing), under unladen conditions with tires on level ground.

TRANSVERSE LINK

REMOVAL & INSTALLATION

See Figure 330.

1. Before servicing the vehicle, refer to the Precautions Section.

2. Remove the tire and wheel assembly.

3. Remove the shock absorber.

4. Temporarily install upper link and steering knuckle.

5. Remove front cross bar.

6. Remove transverse link from steering knuckle.

7. Set suitable jack under transverse link.

8. Remove transverse link and stopper bushings.

➡If removing the transverse link mounting bolt (front side) is difficult,

1. Front wheel vertical G sensor
2. Bracket
3. Front strut

37663_FX35_G0521

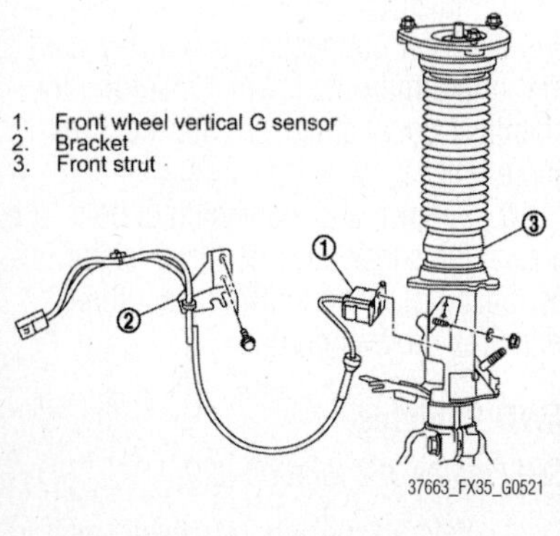

Fig. 327 Remove front wheel vertical G sensor

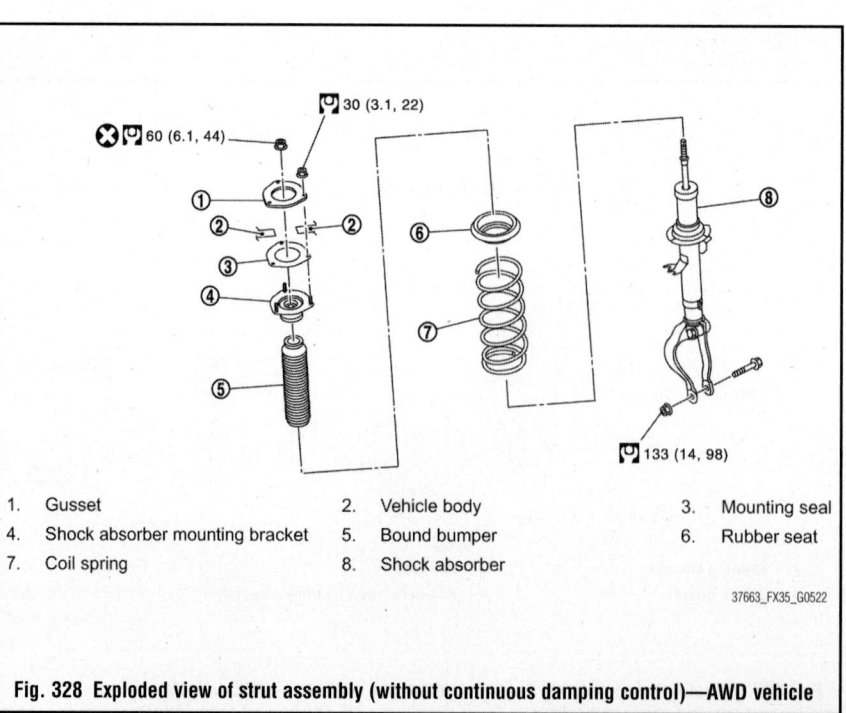

1. Gusset
2. Vehicle body
3. Mounting seal
4. Shock absorber mounting bracket
5. Bound bumper
6. Rubber seat
7. Coil spring
8. Shock absorber

37663_FX35_G0522

Fig. 328 Exploded view of strut assembly (without continuous damping control)—AWD vehicle

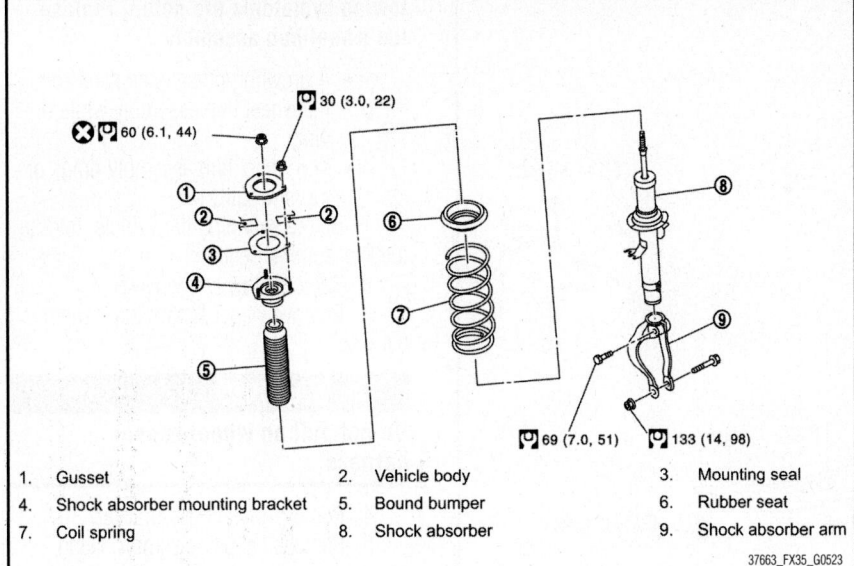

1. Gusset
2. Vehicle body
3. Mounting seal
4. Shock absorber mounting bracket
5. Bound bumper
6. Rubber seat
7. Coil spring
8. Shock absorber
9. Shock absorber arm

37663_FX35_G0523

Fig. 329 Exploded view of strut assembly (with continuous damping control)—AWD vehicle

rotate the steering wheel and remove the steering outer socket.

To install:

9. Installation is the reverse of the removal procedure.

> ❋❋ **WARNING**
>
> **Do not tap on the ball joint cap of the stabilizer connecting rod with a hammer or a similar item when inserting the stabilizer connecting rod into the transverse link.**

10. Perform final tightening of bolts and nuts at the front suspension member installation and shock absorber lower side (rubber bushing), under unladen conditions with tires on level ground.

11. Check shock absorber actuator harness connector for proper connection (with Continuous Damping Control).
12. Check wheel sensor harness for proper connection.
13. Check wheel alignment.

UPPER LINK

REMOVAL & INSTALLATION

See Figure 331.

1. Before servicing the vehicle, refer to the Precautions Section.
2. Remove tires and wheels.
3. Remove shock absorber (strut).
4. Remove upper link and stopper arm bushing.

To install:

5. Installation is the reverse of the removal procedure.
6. Perform final tightening of bolts and nuts at the vehicle installation position (rubber bushing), under unladen conditions with tires on level ground.
7. Check wheel sensor harness for proper connection.
8. Check wheel alignment.

WHEEL HUBS & BEARINGS

REMOVAL & INSTALLATION

3.5L Engine—2WD Model
See Figure 332.

➡**The wheel hub assembly does not require maintenance. If any of the following symptoms are noted, replace the wheel hub assembly.**

- A growling noise is emitted from the wheel hub assembly while driving
- The wheel hub assembly drags or turns roughly

1. Before servicing the vehicle, refer to the Precautions Section.
2. Remove tires and wheels.
3. Remove wheel sensor and sensor harness.

> ❋❋ **WARNING**
>
> **Do not pull on wheel sensor harness.**

4. Remove brake hose bracket.
5. Remove caliper assembly. Hang caliper assembly in a place where it will not interfere with work.

➡**Do not depress brake pedal while brake caliper is removed.**

6. Remove disc rotor.
7. Remove wheel hub and bearing assembly, and then remove splash guard.
8. Remove steering outer socket.
9. Remove cotter pin of transverse link and steering knuckle, and then loosen nut.
10. Separate steering knuckle from upper link.
11. Separate steering knuckle from transverse link so as not to damage ball joint boot using the ball joint remover, and remove steering knuckle.

> ❋❋ **WARNING**
>
> **Temporarily tighten the nut to prevent damage to threads and to prevent the ball joint remover from suddenly coming off.**

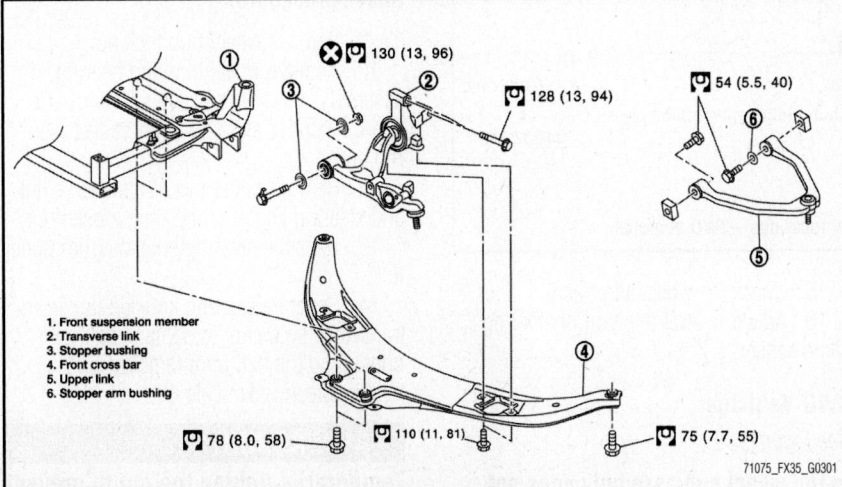

1. Front suspension member
2. Transverse link
3. Stopper bushing
4. Front cross bar
5. Upper link
6. Stopper arm bushing

71075_FX35_G0301

Fig. 330 Transverse link and related component locations

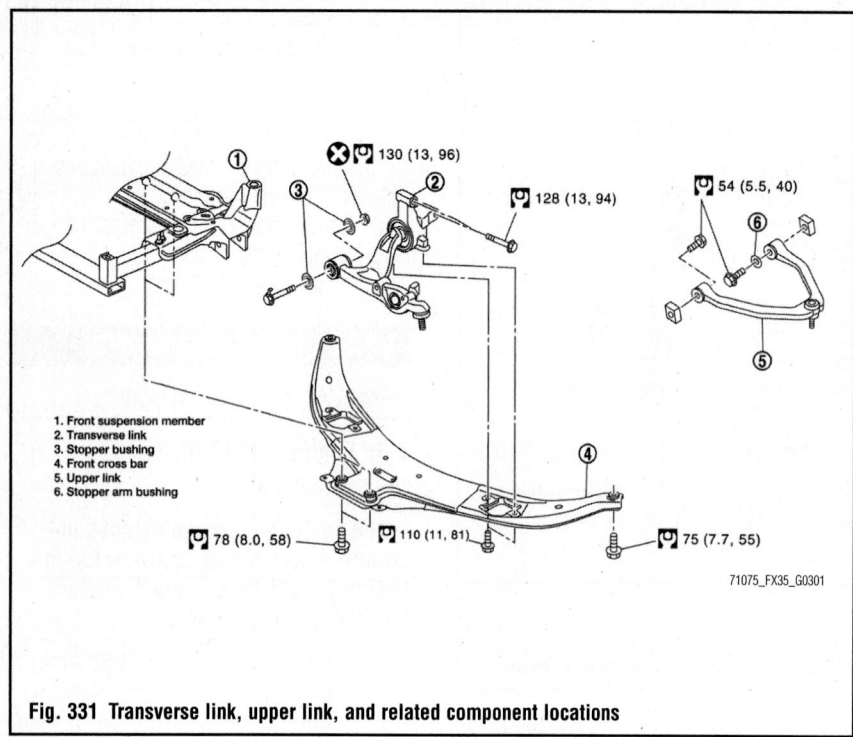

Fig. 331 Transverse link, upper link, and related component locations

1. Front suspension member
2. Transverse link
3. Stopper bushing
4. Front cross bar
5. Upper link
6. Stopper arm bushing

130 (13, 96)

128 (13, 94)

54 (5.5, 40)

78 (8.0, 58)

110 (11, 81)

75 (7.7, 55)

71075_FX35_G0301

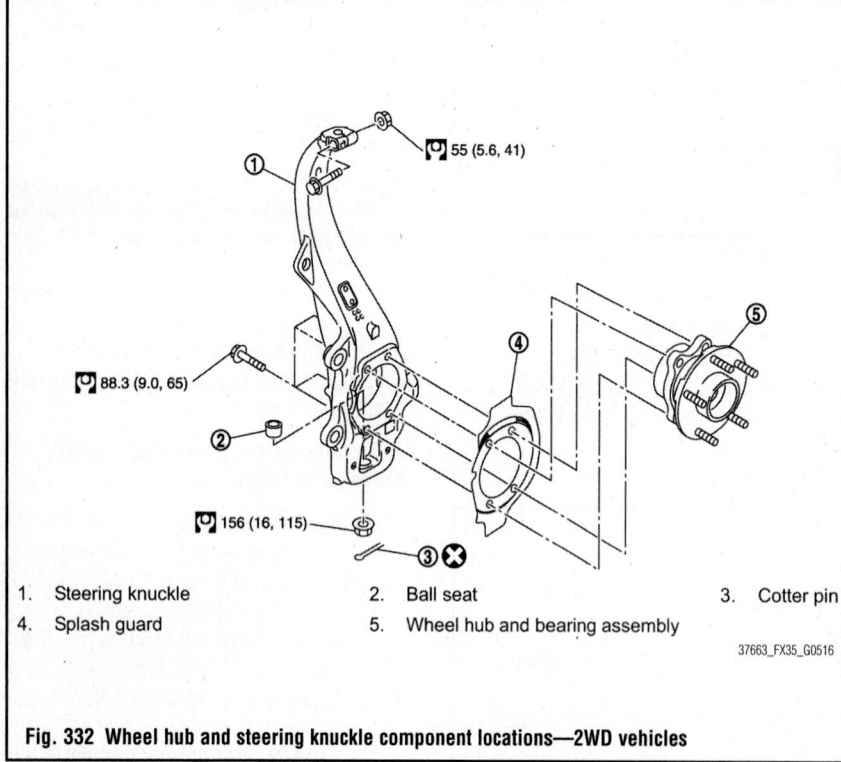

Fig. 332 Wheel hub and steering knuckle component locations—2WD vehicles

55 (5.6, 41)

88.3 (9.0, 65)

156 (16, 115)

1. Steering knuckle
2. Ball seat
3. Cotter pin
4. Splash guard
5. Wheel hub and bearing assembly

37663_FX35_G0516

To install:

12. Installation is the reverse of the removal procedure.

13. Perform the final tightening of each of the parts under unladen conditions.

➡️ **Do not reuse the cotter pin.**

14. Check wheel sensor harness for proper connection.

15. Check the wheel alignment.

16. Adjust neutral position of steering angle sensor.

AWD Vehicles

See Figure 333.

➡️ **The wheel hub assembly does not require maintenance. If any of the fol-**

lowing symptoms are noted, replace the wheel hub assembly.

- A growling noise is emitted from the wheel hub assembly while driving
- The wheel hub assembly drags or turns roughly

1. Before servicing the vehicle, refer to the Precautions Section.

2. Remove tires and wheels.

3. Remove wheel sensor and sensor harness.

✳✳ WARNING

Do not pull on wheel sensor harness.

4. Remove brake hose bracket.

5. Remove caliper assembly. Hang caliper assembly in a place where it will not interfere with work.

➡️ **Do not depress brake pedal while brake caliper is removed.**

6. Remove disc rotor.

7. Remove cotter pin, and then loosen wheel hub lock nut.

8. Patch wheel hub lock nut with a piece of wood. Hammer the wood to disengage wheel hub and bearing assembly from halfshaft.

✳✳ WARNING

Never place halfshaft joint at an extreme angle. Also be careful not to overextend slide joint. Do not allow halfshaft to hang down without support for or joint sub-assembly, shaft and the other parts.

➡️ **Use suitable puller, if wheel hub and bearing assembly and halfshaft cannot be separated even after performing the above procedure.**

9. Remove wheel hub lock nut.

10. Remove wheel hub and bearing assembly, and then remove splash guard.

11. Remove steering outer socket (tie rod end).

12. Remove cotter pin of transverse link and steering knuckle, and then loosen nut.

13. Separate steering knuckle from upper link.

14. Separate steering knuckle link from transverse so as not to damage ball joint boot using the ball joint remover, and remove steering knuckle.

✳✳ WARNING

Temporarily tighten the nut to prevent damage to threads and to prevent the

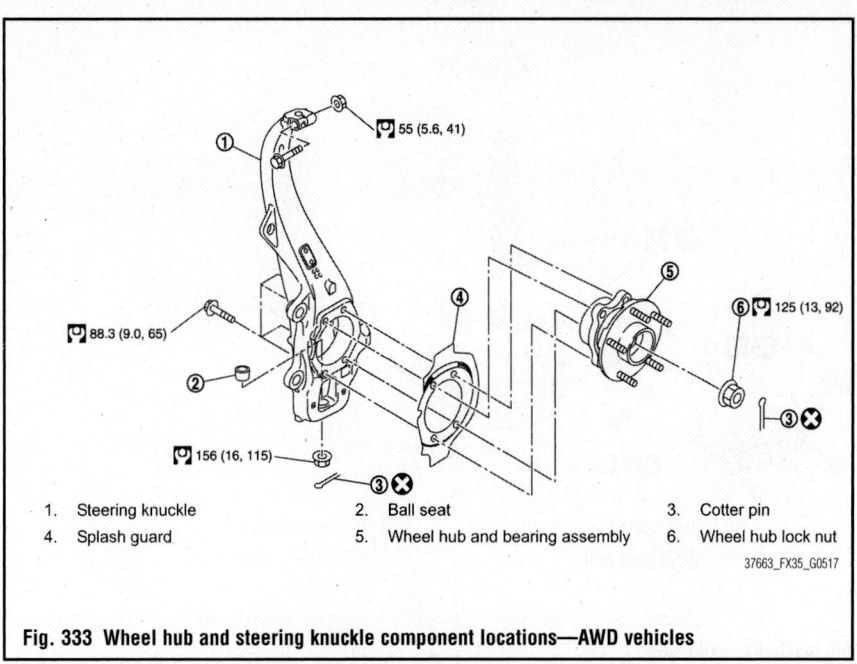

1. Steering knuckle
2. Ball seat
3. Cotter pin
4. Splash guard
5. Wheel hub and bearing assembly
6. Wheel hub lock nut

37663_FX35_G0517

Fig. 333 Wheel hub and steering knuckle component locations—AWD vehicles

ball joint remover from suddenly coming off.

To install:

15. Installation is the reverse of the removal procedure.

16. Perform the final tightening of each of the parts under unladen conditions.

17. Install halfshaft using tightening torque of wheel hub lock nut.

☀ WARNING

Be sure to use torque wrench to tighten the wheel hub lock nut. Do not use a power tool.

➡ **Do not reuse the cotter pin.**

18. Check wheel sensor harness for proper connection.

19. Check the wheel alignment.

20. Adjust neutral position of steering angle sensor.

SUSPENSION

REAR SUSPENSION

COIL SPRINGS

REMOVAL & INSTALLATION

See Figures 334 through 336.

1. Before servicing the vehicle, refer to the Precautions Section.

2. Remove tires and wheels.

3. Set suitable jack under rear lower link to relieve the coil spring tension.

4. Loosen rear lower link mounting nuts [rear suspension member side (without Rear Active Steer (RAS) or RAS actuator assembly (with RAS)], and remove rear lower link mounting bolts and nuts (axle housing side).

5. Slowly lower jack, then remove upper seat, coil spring and rubber sheet from rear lower link.

6. Remove rear lower link mounting nuts and adjusting bolts [rear suspension member side (without RAS) or RAS actuator assembly (with RAS)] and remove rear lower link.

To install:

7. Note the following, and install in the reverse order of removal.

8. Match up rubber seat indentions and rear lower link grooves and attach.

9. Install coil spring by aligning the lower end of the large diameter side to the step between the rubber seat and the rear lower link.

➡ **Make sure spring is not upside down.**

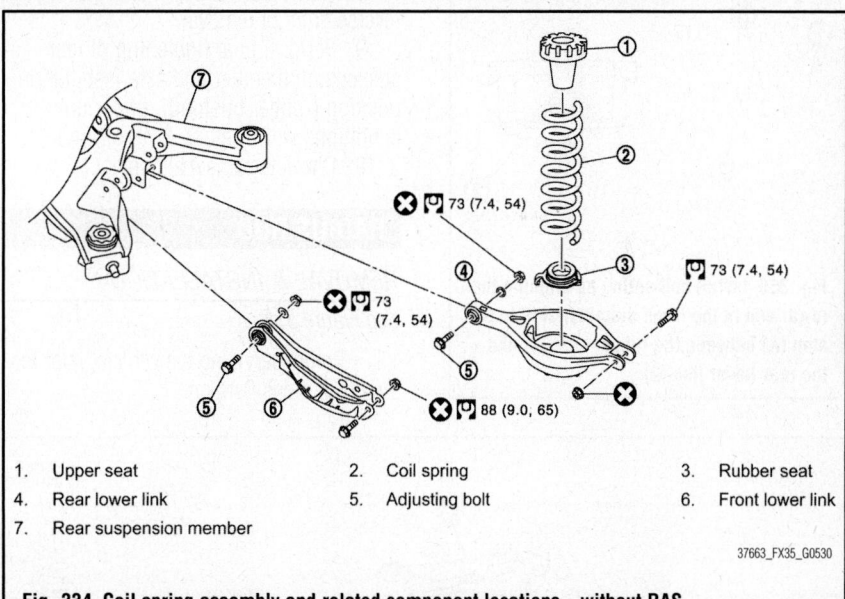

1. Upper seat
2. Coil spring
3. Rubber seat
4. Rear lower link
5. Adjusting bolt
6. Front lower link
7. Rear suspension member

37663_FX35_G0530

Fig. 334 Coil spring assembly and related component locations—without RAS

10. Perform the final tightening of rear suspension member and axle housing rubber bushing position under unladen condition with tires on level ground.

11. Check the wheel alignment.

12. Adjust neutral position of steering angle sensor.

FRONT LOWER LINK

REMOVAL & INSTALLATION

See Figures 335 and 337.

1. Before servicing the vehicle, refer to the Precautions Section.

2. Remove tire and wheels.

3. Set suitable jack under axle assembly to relieve the coil spring tension.

4. Remove shock absorber mounting bolts from front lower link.

5. Remove front lower link mounting bolts and nuts from axle housing.

6. Remove stabilizer clamp and stabilizer bushing.

7. Remove front lower link mounting bolts and nuts from rear suspension member, and remove front lower link.

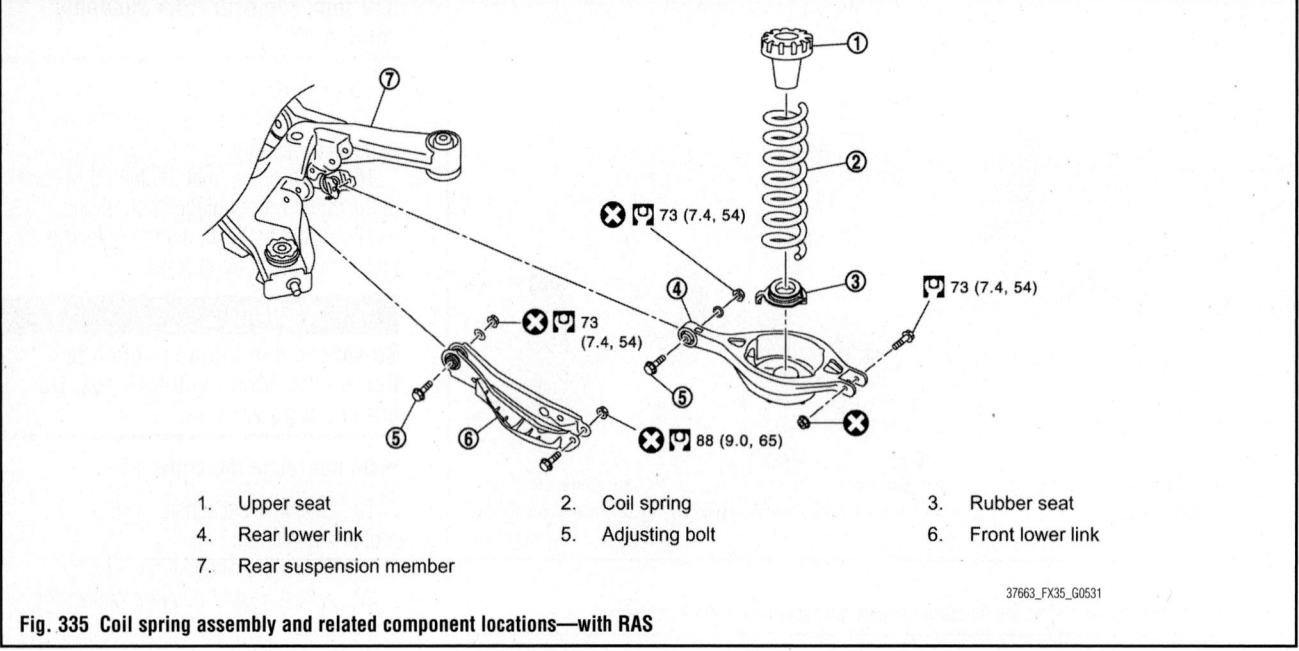

1. Upper seat
2. Coil spring
3. Rubber seat
4. Rear lower link
5. Adjusting bolt
6. Front lower link
7. Rear suspension member

37663_FX35_G0531

Fig. 335 Coil spring assembly and related component locations—with RAS

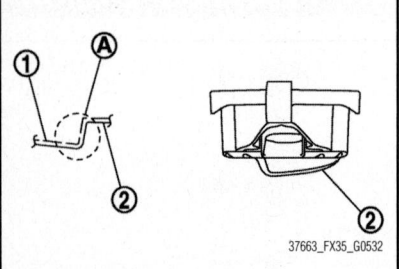

37663_FX35_G0532

Fig. 336 Install coil spring by aligning the lower end of the large diameter side to the step (A) between the rubber seat (1) and the rear lower link (2)

To install:

8. Note the following, and install in the reverse order of removal.

9. Perform final tightening of rear suspension member and axle installation position (rubber bushing), under unladen conditions with tires on level ground.

10. Check the wheel alignment.

RADIUS ROD

REMOVAL & INSTALLATION

See Figure 338.

1. Before servicing the vehicle, refer to the Precautions Section.

2. Remove tire and wheels.

3. Remove radius rod mounting bolt and nut (axle housing side).

4. Remove radius rod mounting bolt (rear suspension member side), and remove radius rod.

To install:

5. Note the following, and install in the reverse order of removal.

6. Perform final tightening of rear suspension member and axle installation position (rubber bushing), under unladen conditions with tires on level ground.

7. Check the wheel alignment.

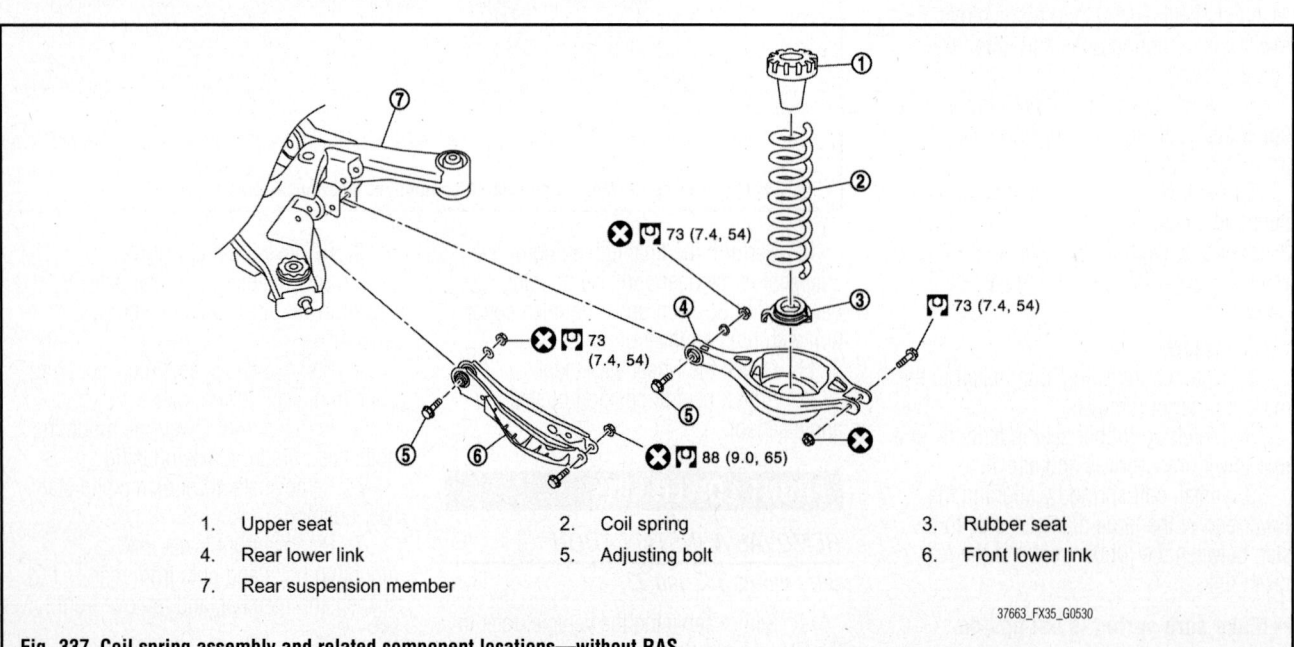

1. Upper seat
2. Coil spring
3. Rubber seat
4. Rear lower link
5. Adjusting bolt
6. Front lower link
7. Rear suspension member

37663_FX35_G0530

Fig. 337 Coil spring assembly and related component locations—without RAS

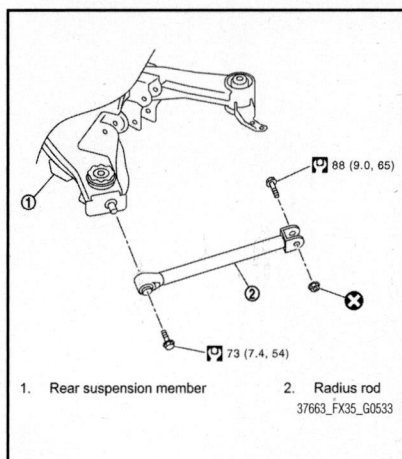

Fig. 338 Exploded view of radius rod assembly

1. Rear suspension member   2. Radius rod
37663_FX35_G0533

REAR SUSPENSION MEMBER

REMOVAL & INSTALLATION

See Figures 339 and 340.

1. Before servicing the vehicle, refer to the Precautions Section.
2. Remove tires and wheels.
3. Remove center muffler.
4. Remove radius rod.
5. Remove caliper assembly. Hang caliper assembly in a place where it will not interfere with work.

➡**Avoid depressing brake pedal while brake caliper is removed.**

6. Remove disc rotor.
7. Remove wheel sensor harness from rear suspension member.
8. Remove height sensor harness from rear suspension member (with xenon head lamp).

9. Remove shock absorber actuator harness connector (with Continuous Damping Control (CDC)).
10. Remove stabilizer bar.
11. Remove halfshaft.
12. Remove propeller shaft (driveshaft).
13. Remove final drive.
14. Remove parking brake cable mounting bolt and separate parking brake cable from vehicle and rear suspension member.
15. Remove shock absorber mounting bolts (lower side).
16. Remove rear lower link and coil spring.
17. Remove RAS actuator assembly (with RAS).
18. Set suitable jack under rear suspension member.
19. Remove pin stay.
20. Remove dynamic dampers (5.0L engine).
21. Remove tunnel stay.
22. Remove rear suspension member stay.
23. Slowly lower jack, then remove rear suspension member, suspension arm, front lower link, wheel hub and housing from vehicle as a unit.
24. Remove mounting bolts and nuts, then remove suspension arm, front lower link, wheel hub and housing from rear suspension member.

To install:

25. Note the following, and install in the reverse order of the removal.
26. Perform the final tightening of each of parts under unladen conditions, which were removed when removing rear suspension assembly.

27. Check wheel sensor harness for proper connection.

➡**Do not reuse cotter pin.**

28. Check shock absorber actuator harness connector for proper connection (with CDC).
29. Adjust parking brake operation (stroke).
30. Check wheel alignment.

SHOCK ABSORBERS

REMOVAL & INSTALLATION

See Figures 341 and 342.

1. Before servicing the vehicle, refer to the Precautions Section.
2. Remove tires and wheels.
3. Remove shock absorber actuator harness connector (with Continuous Damping Control (CDC)).
4. Set suitable jack under axle assembly to relieve the coil spring tension.
5. Remove shock absorber (lower side).
6. Gradually lower the jack to remove it from rear lower link.
7. Remove shock absorber assembly mounting nuts (upper side), and then remove shock absorber assembly.

To install:

8. Note the following, and install in the reverse order of removal.
9. Perform final tightening of bolts and nuts at the shock absorber lower side (rubber bushing), under unladen conditions with tires on level ground.
10. Check shock absorber actuator harness connector for proper connection (with CDC).
11. Check wheel alignment.

STABILIZER BAR

REMOVAL & INSTALLATION

See Figure 343.

1. Before servicing the vehicle, refer to the Precautions Section.
2. Remove center muffler.
3. Remove under cover.
4. Remove stabilizer connecting rod mounting nuts (lower side), and remove stabilizer connecting rod from stabilizer bar.
5. Remove stabilizer connecting rod mounting nuts (upper side), and remove stabilizer connecting rod from stabilizer connecting rod mounting bracket.
6. Remove mounting bolts or nuts on stabilizer clamp, and remove stabilizer bar.

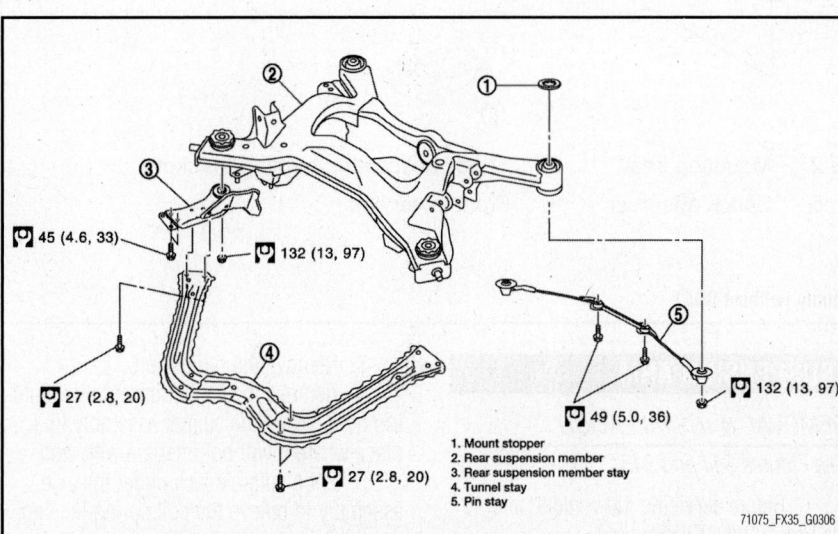

1. Mount stopper
2. Rear suspension member
3. Rear suspension member stay
4. Tunnel stay
5. Pin stay
71075_FX35_G0306

Fig. 339 Rear suspension member and related component locations—3.5L engine

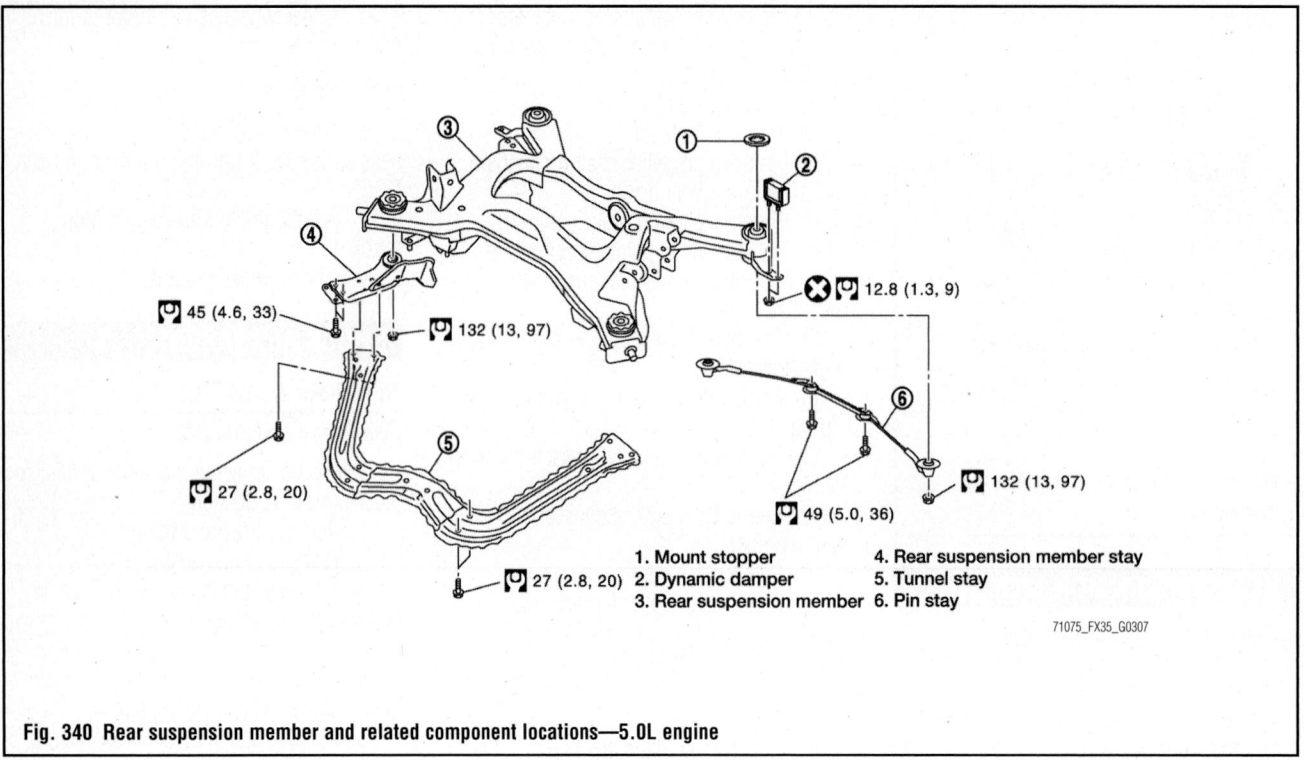

Fig. 340 Rear suspension member and related component locations—5.0L engine

1. Mount stopper
2. Dynamic damper
3. Rear suspension member
4. Rear suspension member stay
5. Tunnel stay
6. Pin stay

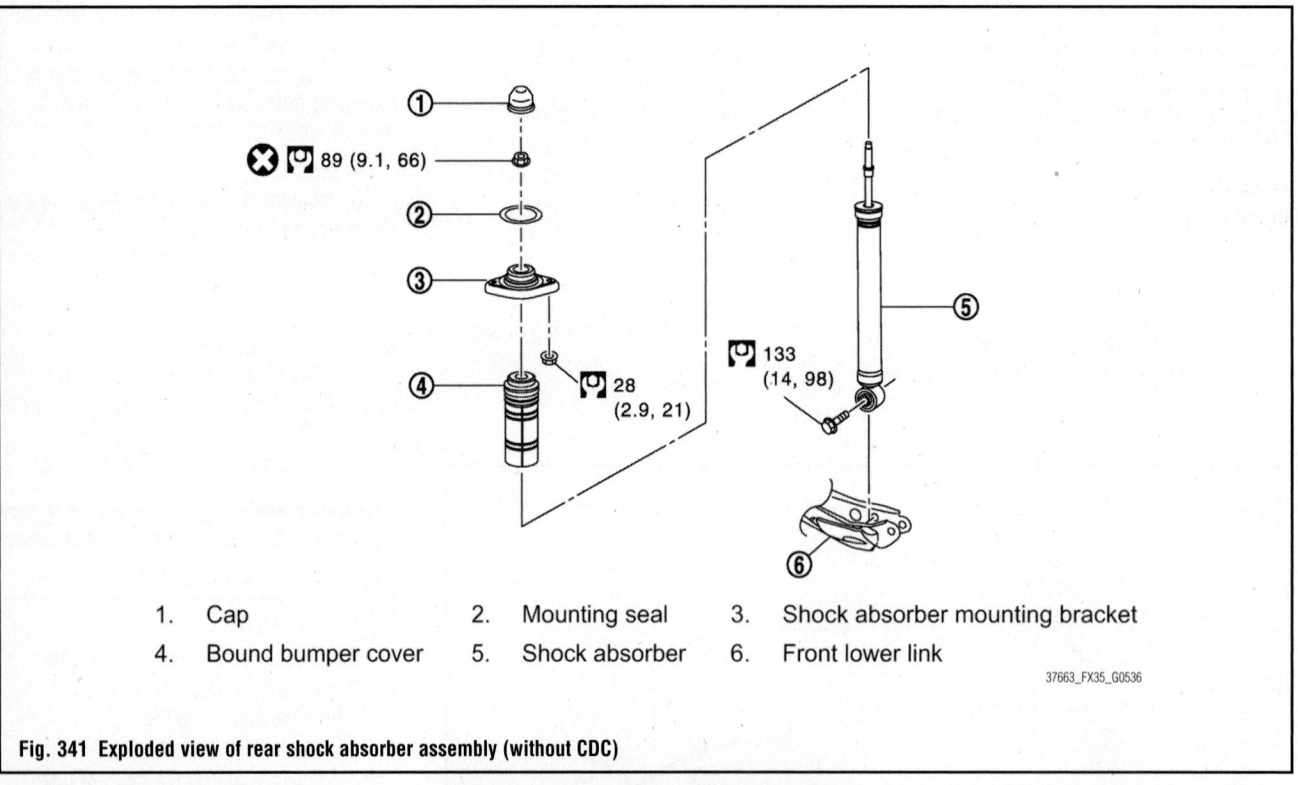

1. Cap
2. Mounting seal
3. Shock absorber mounting bracket
4. Bound bumper cover
5. Shock absorber
6. Front lower link

Fig. 341 Exploded view of rear shock absorber assembly (without CDC)

7. Remove stabilizer connecting rod mounting bracket.

To install:

8. Note the following, and install in the reverse order of removal.

9. Tighten the mounting nut to the specified torque while holding a hexagonal part of stabilizer connecting rod side.

SUSPENSION ARM

REMOVAL & INSTALLATION

See Figures 344 and 345.

1. Before servicing the vehicle, refer to the Precautions Section.

2. Remove the tire and wheels.

3. Remove the radius rod.

4. Remove the caliper assembly mounting bolts. Hang the caliper assembly in a place where it will not interfere with work.

5. Set a suitable jack under the axle assembly to relieve the coil spring tension.

6. Remove the stabilizer connecting rod.

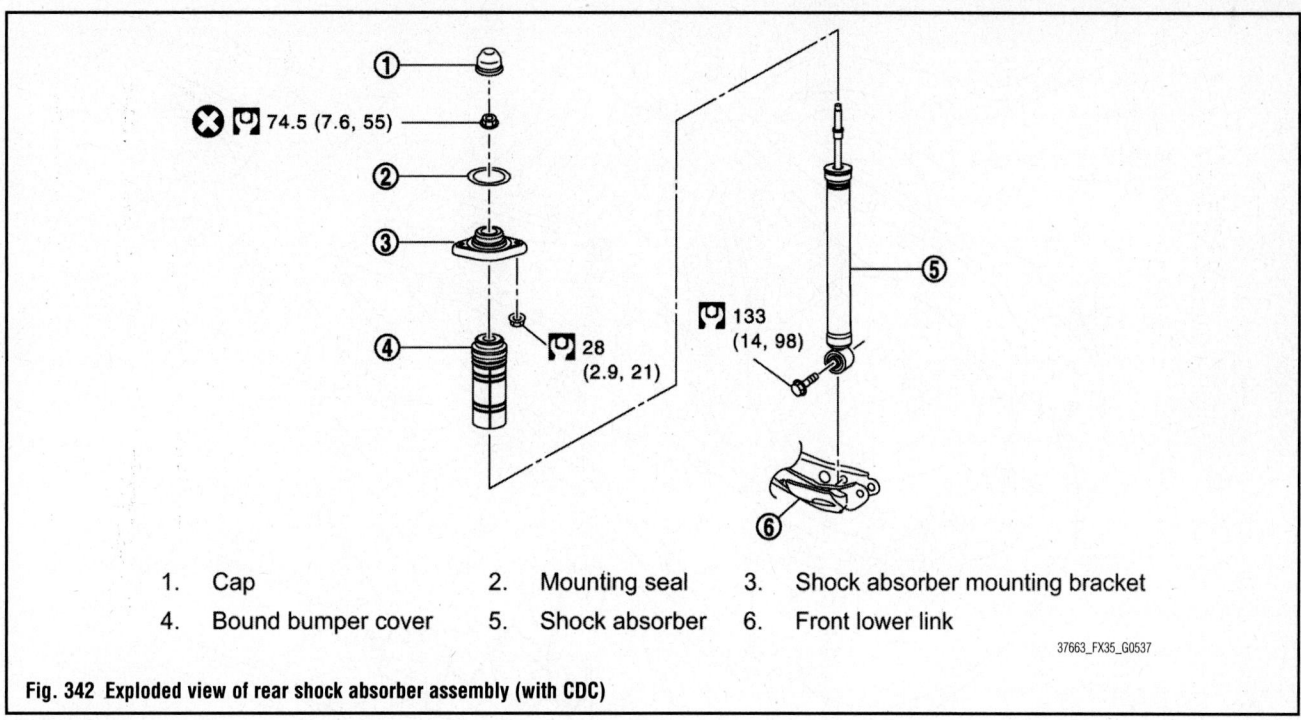

1. Cap
2. Mounting seal
3. Shock absorber mounting bracket
4. Bound bumper cover
5. Shock absorber
6. Front lower link

37663_FX35_G0537

Fig. 342 Exploded view of rear shock absorber assembly (with CDC)

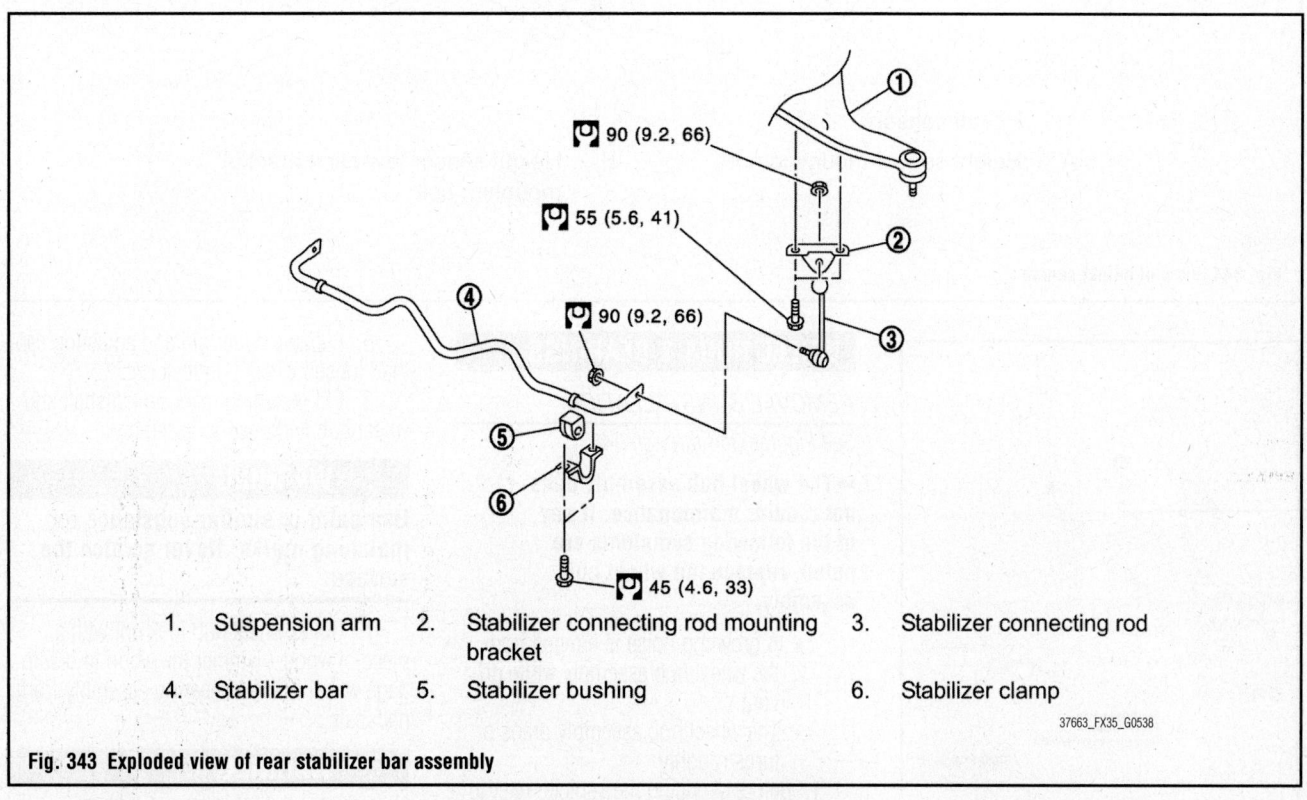

1. Suspension arm
2. Stabilizer connecting rod mounting bracket
3. Stabilizer connecting rod
4. Stabilizer bar
5. Stabilizer bushing
6. Stabilizer clamp

37663_FX35_G0538

Fig. 343 Exploded view of rear stabilizer bar assembly

7. Remove the halfshaft.

8. Remove the height sensor (with xenon head lamp).

9. Remove the cotter pin of the suspension arm ball joint, and loosen the nut.

10. Remove the suspension arm mounting bolts and nuts (rear suspension member side).

11. Use the ball joint remover to remove the suspension arm from the axle housing. Be careful not to damage the ball joint boot.

✳✳ WARNING

Temporarily tighten the mounting nut to prevent damage to the threads and to prevent the ball joint remover from coming off.

12. Remove the suspension arm.

13. Remove the stabilizer connecting rod mounting bracket.

To install:

14. Note the following and, install in the reverse order of removal.

15. Perform the final tightening of the rear suspension member installation posi-

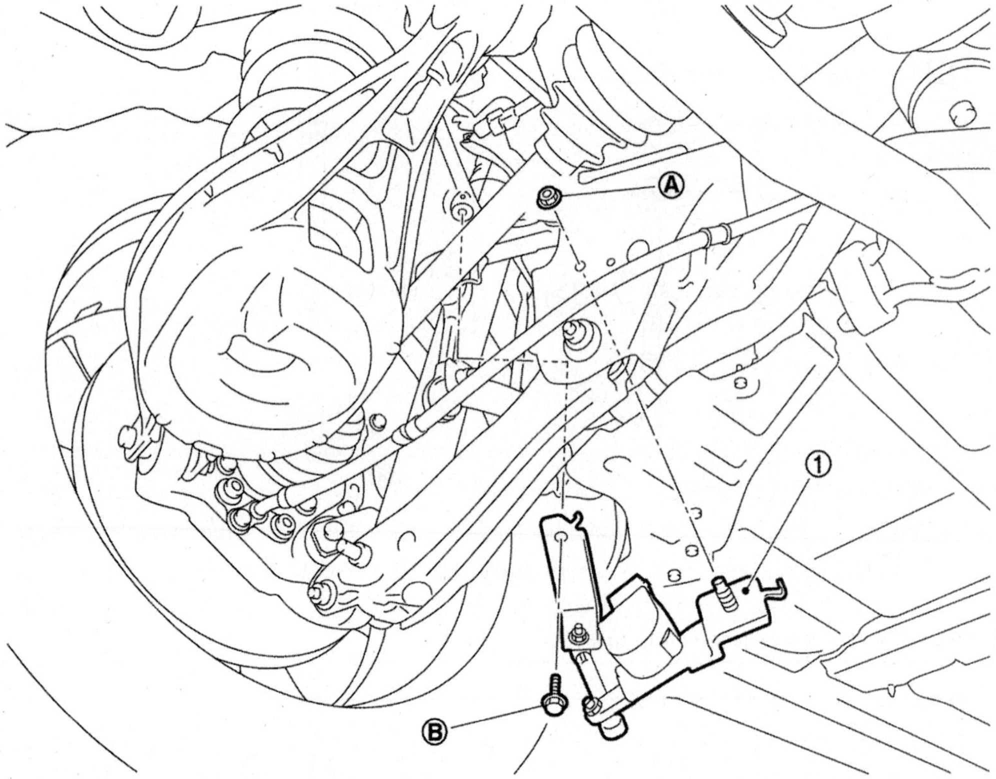

1. Height sensor

A Height sensor mounting nut

B. Height sensor lever link bracket mounting bolt

37663_FX35_G0534

Fig. 344 View of height sensor

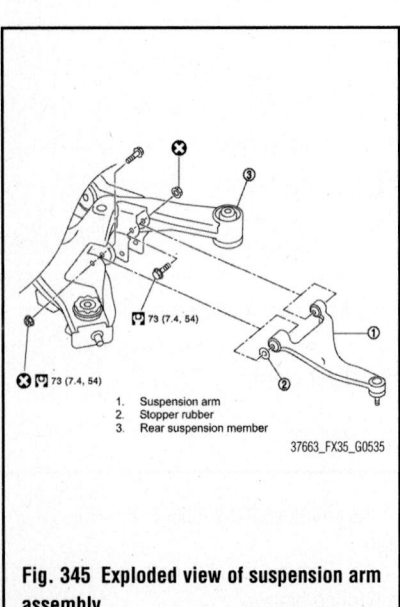

1. Suspension arm
2. Stopper rubber
3. Rear suspension member

37663_FX35_G0535

Fig. 345 Exploded view of suspension arm assembly

tion (rubber bussing), under unladen conditions with the tires on level ground.

16. Check the wheel alignment.

WHEEL HUBS & BEARINGS

REMOVAL & INSTALLATION

See Figures 346 through 348.

➡The wheel hub assembly does not require maintenance. If any of the following symptoms are noted, replace the wheel hub assembly.

- A growling noise is emitted from the wheel hub assembly while driving
- The wheel hub assembly drags or turns roughly

1. Before servicing the vehicle, refer to the Precautions Section.

2. Remove tire and wheels.

3. Remove caliper assembly. Hang caliper assembly in a place where it will not interfere with work.

➡Do not depress brake pedal while caliper assembly is removed.

4. Remove disc rotor.

5. Remove cotter pin and adjusting cap, then loosen wheel hub lock nut.

6. Put matching mark on halfshaft and wheel hub and bearing assembly.

✳✳ WARNING

Use paint or similar substance for matching marks. Never scratch the surface.

7. Cover wheel hub lock nut with a piece of wood. Hammer the wood to disengage wheel hub and bearing assembly from halfshaft.

✳✳ WARNING

Never place halfshaft joint at an extreme angle. Also be careful not to overextend slide joint. Do not allow halfshaft to hang down without support for counterpart such as joint sub-assembly, and other parts.

➡Use a suitable puller, if wheel hub and bearing assembly and halfshaft

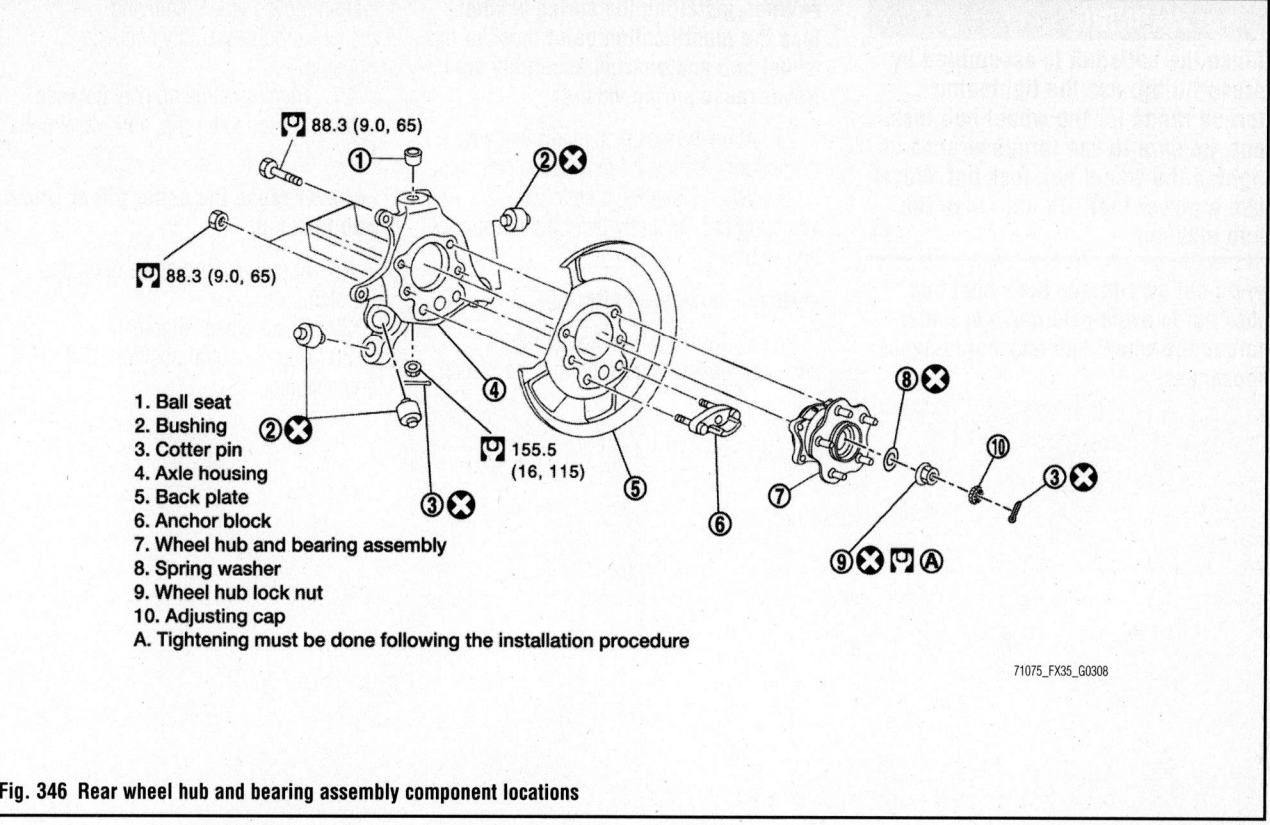

1. Ball seat
2. Bushing
3. Cotter pin
4. Axle housing
5. Back plate
6. Anchor block
7. Wheel hub and bearing assembly
8. Spring washer
9. Wheel hub lock nut
10. Adjusting cap
A. Tightening must be done following the installation procedure

88.3 (9.0, 65)

88.3 (9.0, 65)

155.5 (16, 115)

71075_FX35_G0308

Fig. 346 Rear wheel hub and bearing assembly component locations

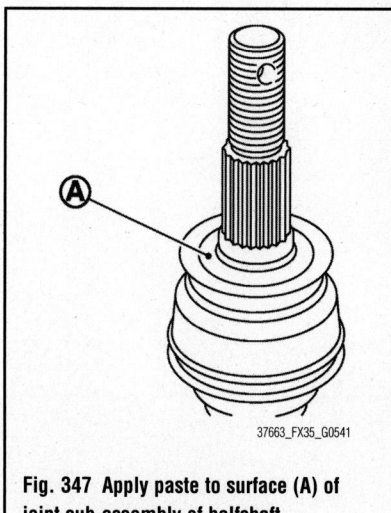

37663_FX35_G0541

Fig. 347 Apply paste to surface (A) of joint sub-assembly of halfshaft

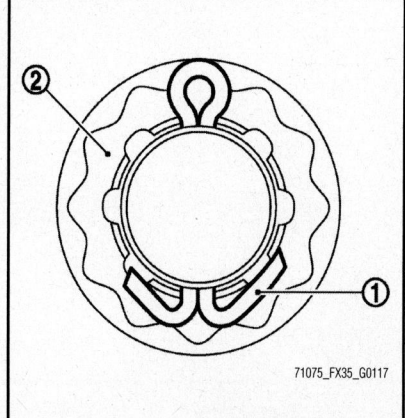

71075_FX35_G0117

Fig. 348 When installing a cotter pin (1) and adjusting cap (2), securely bend the basal portion to prevent rattles

cannot be separated even after performing the above procedure.

8. Remove wheel hub lock nut.
9. Remove parking brake shoe and parking brake cable from back plate.
10. Remove stabilizer connecting rod (upper side).
11. Remove coil spring.
12. Set suitable jack under axle housing.
13. Remove radius rod.
14. Remove shock absorber (lower side).
15. Separate suspension arm from axle housing so as not to damage ball joint boot using ball joint remover, and then remove axle housing from the vehicle.

⁜ WARNING

Temporarily tighten nuts to prevent damage to threads and to prevent the ball joint remover from coming off. Never place halfshaft joint at an extreme angle. Also be careful not to overextend slide joint. Never allow halfshaft to hang down without support for counterpart such as joint sub-assembly, and other parts.

16. Remove front lower link (axle housing side).
17. Remove rear lower link (axle housing side).
18. Remove wheel hub and bearing assembly.
19. Remove anchor block mounting nuts, and then remove anchor block and back plate from axle housing.

To install:
20. Note the following, and install in the reverse order of removal.
21. Clean the matching surface of wheel hub lock nut and wheel hub and bearing assembly.

➡Do not apply lubricating oil to these matching surface.

22. Clean the matching surface of halfshaft and wheel hub and bearing assembly. And then apply paste (440037S000) to surface of joint sub-assembly of halfshaft.

➡Apply paste to cover entire flat surface of joint sub-assembly of halfshaft.

• Amount of paste: 0.04–0.10 ounce (1.0–3.0 grams)
23. Tighten the wheel hub lock nut to 74–77 ft. lbs. (100–105 Nm).

❊❊ WARNING

Since the halfshaft is assembled by press-fitting, use the tightening torque range for the wheel hub lock nut. Be sure to use torque wrench to tighten the wheel hub lock nut. Never use a power tool. Do not reuse the hub lock nut.

➡Do not over-torque the wheel hub lock nut to avoid axle noise or under-torque the wheel hub lock nut to avoid looseness.

➡When installing the spring washer, face the identification paint mark to the wheel hub and bearing assembly side. Never reuse spring washer.

24. Align the matching marks that were made during removal if reusing the disc rotor.

25. When installing a cotter pin and adjusting cap, securely bend the basal portion to prevent rattles.

➡Never reuse the cotter pin.

26. Perform the final tightening of each part under unladen conditions, which were removed when removing the wheel hub assembly and axle housing.

27. There must be no play between adjusting cap, cotter pin, and wheel hub lock nut.

➡Never reuse the cotter pin or wheel hub lock nut.

28. Adjust parking brake operation (stroke).

29. Check wheel alignment.

30. Adjust neutral position of steering angle sensor.

INFINITI

G25 • G37 • G37X

3

SPECIFICATIONS AND MAINTENANCE CHARTS

ENGINE AND VEHICLE IDENTIFICATION

Engine							Model Year	
Code ①	Liters (cc)	Cu. In.	Cyl.	Fuel Sys.	Engine Type	Eng. Mfg.	Code ②	Year
VQ37VHR	3.7 (3696)	225	6	MFI	DOHC	Nissan	B	2011
VQ25HR	2.5 (2496)	152	6	MFI	DOHC	Nissan	C	2012

MFI: Multi-port Fuel Injection

DOHC: Double Overhead Camshaft

① The engine VIN "C" is the 4th position of the Vehicle Identification Number (VIN)

② 10th position of the Vehicle Identification Number (VIN)

71075_IG37_C0001

GENERAL ENGINE SPECIFICATIONS

Year	Model	Engine Displacement Liters	Engine ID	Net Horsepower @ rpm	Net Torque @ rpm (ft. lbs.)	Bore x Stroke (in.)	Compression Ratio	Oil Pressure @ rpm
2011	G37	3.7	VQ37VHR	①	②	3.760X3386	11.0:1	43@2000
	G25	2.5	VQ25HR	218 @ 6400	187 @ 4800	3.35X2.89	10.3	43@2000
2012	G37	3.7	VQ37VHR	①	②	3.760X3386	11.0:1	43@2000
	G25	2.5	VQ25HR	218 @ 6400	187 @ 4800	3.35X2.89	10.3	43@2000

① Sedan: 328@7000. Coupe: 330@7000. Convertible: 325@7000.

② Sedan: 269@5200. Coupe: 270@5200. Convertible: 267@5200.

71075_IG37_C0002

ENGINE TUNE-UP SPECIFICATIONS

Year	Model	Engine ID	Spark Plug Gap (in.)	Ignition Timing (deg.) MT	Ignition Timing (deg.) AT	Fuel Pump (psi) ①	Idle Speed (rpm) MT	Idle Speed (rpm) AT ②	Valve Clearance (in.) Intake ③	Valve Clearance (in.) Exhaust ③
2011	G37	VQ37VHR	0.043	④	④	51	600-700	600-700	0.010-0.013	0.011-0.015
	G25	VQ25HR	0.043	⑤	⑤	51	NA	600-700	0.010-0.013	0.011-0.015
2012	G37	VQ37VHR	0.043	④	④	51	600-700	600-700	0.010-0.013	0.011-0.015
	G25	VQ25HR	0.043	⑤	⑤	51	NA	600-700	0.010-0.013	0.011-0.015

NOTE: The Vehicle Emission Control Information label often reflects specification changes made during production.

The label figures must be used if they differ from those in this chart.

NA: Not Available

B: Before top dead center

① System pressure at idle with vacuum hose connected; should increase to 43 psi when disconnected

② Automatic transmission in park or neutral

③ Engine cold

④ 5-15 degrees BTDC

⑤ 12-16 degrees BTDC

71075_IG37_C0003

CAPACITIES

Year	Model	Engine ID	Engine Displacement Liters	Engine Oil with Filter (qts.)	Transmission (pts.)		Drive Axle Rear (pts.)	Fuel Tank (gal.)	Cooling System (qts.)
					Man	Auto.			
2011	G37	VQ37VHR	3.7	5.0	6.0	19.5	①	20.0	②
	G25	VQ25HR	2.5	5.0	6.0	19.5	①	20.0	③
2012	G37	VQ37VHR	3.7	5.0	6.0	19.5	①	20.0	②
	G25	VQ25HR	2.5	5.0	6.0	19.5	①	20.0	③

NOTE: All capacities are approximate. Add fluid gradually and check to be sure a proper fluid level is obtained.

① Front: 1 38. Rear: 3.00

② Automatic transmission: 9.00. Manual transmission: 9.18

③ 2WD: 8.75. AWD: 9.25

71075_IG37_C0004

FLUID SPECIFICATIONS

Year	Model	Engine ID	Engine Oil	Auto. Trans.	Manual Trans.	Drive Axle		Transfer Case	Power Steering Fluid	Brake Master Cylinder	Cooling System
						Front	Rear				
2011	G37	VQ37VHR	5W-30	①	②	③	④ ⑤ ⑥	⑦	⑧	DOT 3	⑨
	G25	VQ25HR	5W-30	①	②	③	④ ⑤ ⑥	⑦	⑧	DOT 3	⑨
2012	G37	VQ37VHR	5W-30	①	②	③	④ ⑤ ⑥	⑦	⑧	DOT 3	⑨
	G25	VQ25HR	5W-30	①	②	③	④ ⑤ ⑥	⑦	⑧	DOT 3	⑨

DOT: Department Of Transportation

① Nissan Matic S ATF. Using non approved fluid will cause deterioration in driveability and transmission damage.

② Nissan (MTF) HQ MULTI 75W-85 or API GL-4, Viscosity 75W-85

③ Nissan differential oil hypoid super GL-5 80W-90 or API GL-5 viscosity SAE 80W-90 gear oil.

④ Sedan and Coupe: Nissan differential oil hypoid super GL-5 80W-90 or API GL-5 viscosity SAE 80W-90 gear oil, except if equipped with 7A/T 2WD.

⑤ Sedan and Coupe: Nissan differential synthetic 75W-90 or API GL-5 synthetic gear oil, viscosity SAE 75W-90 gear oil, if equipped with 7A/T 2WD.

⑥ Convertible: Nissan differential oil hypoid super GL-5 80W-90 or API GL-5 viscosity SAE 80W-90 gear oil, if equipped with M/T. Nissan differential synthetic 75W-90 or API GL-5 synthetic gear oil, viscosity SAE 75W-90 gear oil, if equipped with A/T.

⑦ Nissan Matic J ATF. Using non approved fluid will cause deterioration in driveability and transfer case damage.

⑧ Nissan Power Steering Fluid

⑨ Nissan Long Life antifreeze or equivalent

71075_IG37_C0013

VALVE SPECIFICATIONS

Year	Engine ID	Engine Displacement Liters	Seat Angle (deg.)	Face Angle (deg.)	Spring Test Pressure (lbs. @ in.)	Spring Installed Height (in.)	Stem-to-Guide Clearance (in.)		Stem Diameter (in.)	
							Intake	Exhaust	Intake	Exhaust
2011	VQ37VHR	3.7	45.15-45.45	45	①	②	0.0008-0.0021	0.0012-0.0022	0.2348-0.2354	0.2347-0.2350
	VQ25HR	2.5	③	45	37-42@ 1.055	1.4567	0.0008-0.0021	0.0012-0.0022	0.2348-0.2354	0.2347-0.2350
2012	VQ37VHR	3.7	45.15-45.45	45	①	②	0.0008-0.0021	0.0012-0.0022	0.2348-0.2354	0.2347-0.2350
	VQ25HR	2.5	③	45	37-42@ 1.055	1.4567	0.0008-0.0021	0.0012-0.0022	0.2348-0.2354	0.2347-0.2350

① Intake (upon installation): 43-48 lbs. @ 1.6102 in.

Intake (valve open): 187-211 lbs. @ 1.1051 in.

Exhaust (upon installation): 37-42 lbs. @ 1.4567 in.

Exhaust (valve open): 113-127 lbs. @ 1.0551 in.

② Intake: 1.7976. Exhaust: 1.7264.

③ Angle "a1": 60 degrees

Angle "a2": 88.45-90.15 degrees

Angle "a3": 120 degrees

71075_IG37_C0005

CAMSHAFT AND BEARING SPECIFICATIONS CHART
All measurements are given in inches.

Year	Engine Displ. Liters	Engine ID	Journal Dia.	Brg. Oil Clearance	Shaft End-play	Runout	Journal-to-Bore Clearance	Lobe Lift	
								Intake	Exhaust
2011	3.7	VQ37VHR	① ③	② ③	0.0045-0.0074	0.0008	NA	NA	④
	2.5	VQ25HR	⑤	⑥	0.0045-0.0074	0.0008	NA	NA	④
2012	3.7	VQ37VHR	① ③	② ③	0.0045-0.0074	0.0008	NA	1.7585-1.7659	1.7628-1.7703
	2.5	VQ25HR	⑤	⑥	0.0045-0.0074	0.0008	NA	NA	④

NA: Not Available

① Front No. 1: 1.0211- 1.0218

No. 2, 3, 4: 0.9230- 0.9238

② Front No. 1: 0.0018- 0.0034

No. 2, 3, 4: 0.0014- 0.0030

③ Specification is for exhaust

④ Bank 1: 1.7722- 1.7797

Bank 2: 1.8400- 1.8474

⑤ No. 1: 1.0236-1.0244

No. 2, 3, 4: 0.9252-0.9260

⑥ No. 1: 0.0018-0.0034

No. 2, 3, 4: 0.0014-0.0030

71075_IG37_C0012

CRANKSHAFT AND CONNECTING ROD SPECIFICATIONS

All measurements are given in inches.

Year	Engine Displacement Liters	Engine ID	Crankshaft				Connecting Rod		
			Main Brg. Journal Dia.	Main Brg. Oil Clearance	Shaft End-play	Thrust on No.	Journal Diameter	Oil Clearance	Side Clearance
2011	3.7	VQ37VHR	①	0.0014- 0.0018	0.0039- 0.0098	3	②	0.0016- 0.0021	0.0079- 0.0138
	2.5	VQ25HR	③	0.0014- 0.0018	0.0039- 0.0098	3	②	0.0016- 0.0021	0.0079- 0.0138
2012	3.7	VQ37VHR	①	0.0014- 0.0018	0.0039- 0.0098	3	②	0.0016- 0.0021	0.0079- 0.0138
	2.5	VQ25HR	③	0.0014- 0.0018	0.0039- 0.0098	3	②	0.0016- 0.0021	0.0079- 0.0138

① There are 24 different grades, ranging from 2.5581- 2.5571

② There are 13 different grades, ranging from 2.2441- 2.2446. Specification is for rod big end diameter (without bearing).

③ There are 24 different grades, ranging from 2.5194-2.5203

④ There are 13 different grades, ranging from 2.0866-2.0871

71075_IG37_C0008

PISTON AND RING SPECIFICATIONS

All measurements are given in inches.

Year	Engine Displ. Liters	Engine ID	Piston Clearance	Ring Gap			Ring Side Clearance		
				Top Compression	Bottom Compression	Oil Control	Top Compression	Bottom Compression	Oil Control
2011	3.7	VQ37VHR	0.0004- 0.0012	0.0091- 0.0130	0.0091- 0.0130	0.0067- 0.0185	0.0016- 0.0031	0.0012- 0.0028	0.0022- 0.0061
	2.5	VQ25HR	0.0004- 0.0012	0.0079- 0.0118	0.0122- 0.0181	0.0079- 0.0197	0.0018- 0.0031	0.0012- 0.0028	0.0018- 0.0049
2012	3.7	VQ37VHR	0.0004- 0.0012	0.0091- 0.0130	0.0091- 0.0130	0.0067- 0.0185	0.0016- 0.0031	0.0012- 0.0028	0.0022- 0.0061
	2.5	VQ25HR	0.0004- 0.0012	0.0079- 0.0118	0.0122- 0.0181	0.0079- 0.0197	0.0018- 0.0031	0.0012- 0.0028	0.0018- 0.0049

71075_IG37_C0007

TORQUE SPECIFICATIONS

All readings in ft. lbs.

Year	Engine Displacement Liters	Engine ID	Cylinder Head Bolts	Main Bearing Bolts	Rod Bearing Bolts	Crankshaft Damper Bolts	Flywheel Bolts	Manifold		Spark Plugs	Oil Drain Plug
								Intake	Exhaust		
2011	3.7	VQ37VHR	①	②	③	④	65	⑤	⑥	18	25
	2.5	VQ25HR	⑦	26	⑧	NA	65	⑨	⑥	18	25
2012	3.7	VQ37VHR	①	②	③	④	65	⑤	⑥	18	25
	2.5	VQ25HR	⑦	26	⑧	NA	65	⑨	⑥	18	25

① Step 1: 77 ft. lbs.
Step 2: Loosen bolts completely
Step 3: 30 ft. lbs.
Step 4: Tighten an additional 95 degrees
Step 5: Tighten an additional 95 degrees

② Step 1: 18 ft. lbs.
Step 2: 26 ft.lbs.
Step 3: Tighten an additional 90 degrees

③ Step 1: 21 ft. lbs.
Step 2: Loosen bolts completely
Step 3: 18 ft.lbs.
Step 4: Tighten an additional 90 degrees

④ Step 1: 33 ft. lbs.
Step 2: Tighten an additional 90 degrees

⑤ Step 1: 5 ft. lbs.
Step 2: 19 ft. lbs.

⑥ Step 1: 11 ft. lbs. in 2 steps

⑦ Step 1: 72 ft. lbs.
Step 2: Loosen bolts completely
Step 3: 28 ft. lbs.
Step 4: Tighten an additional 90 degrees
Step 5: Tighten an additional 90 degrees

⑧ 14 + 90 degrees

⑨ Step 1: 5 ft. lbs.
Step 2: 21 ft. lbs.

71075_IG37_C0006

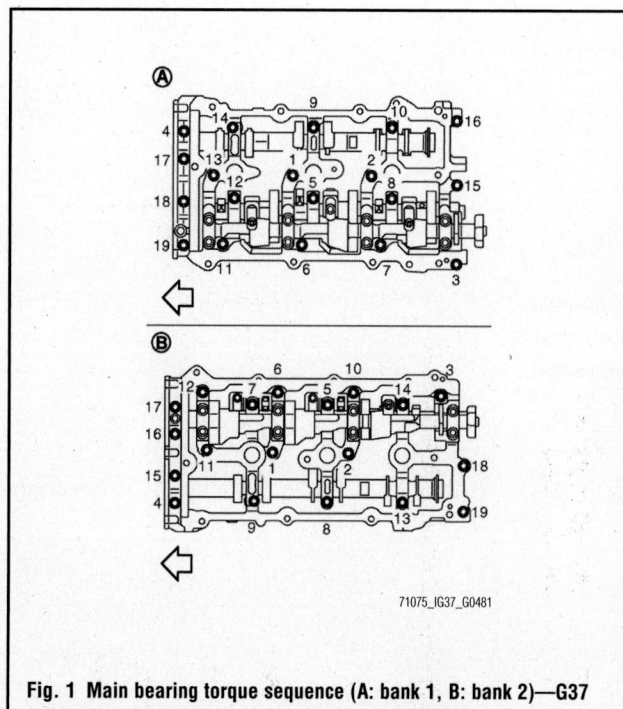

Fig. 1 Main bearing torque sequence (A: bank 1, B: bank 2)—G37

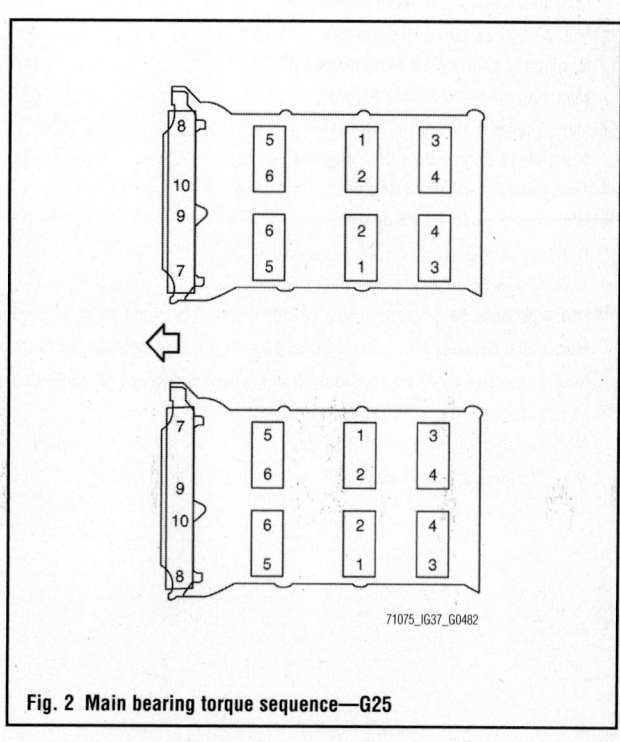

Fig. 2 Main bearing torque sequence—G25

WHEEL ALIGNMENT

Year	Model		Caster Range (+/-Deg.)	Caster Preferred Setting (Deg.)	Camber Range (+/-Deg.)	Camber Preferred Setting (Deg.)	Toe-in (in.)
2011	G37/G25 sedan	F	①	①	②	②	③
		R	—	—	④	④	⑤
	G37 coupe	F	⑥	⑥	⑦	⑦	③
		R	—	—	⑧	⑧	⑤
	G37 convertible	F	⑨	⑨	⑩	⑩	③
		R	—	—	⑧	⑧	⑤
2012	G37/G25 sedan	F	①	①	②	②	③
		R	—	—	④	④	⑤
	G37 coupe	F	⑥	⑥	⑦	⑦	③
		R	—	—	⑧	⑧	⑤
	G37 convertible	F	⑨	⑨	⑩	⑩	③
		R	—	—	⑧	⑧	⑤

① Min: 3 degrees 50' (3.83 degrees) except sport models. 3 degrees 55' (3.92 degrees) sport models.

   Nominal: 4 degrees 35' (4.58 degrees) except sport models. 4 degrees 40' (4.67 degrees) sport models.

   Max: 5 degrees 20' (5.33 degrees) except sport models. 5 degrees 25' (5.42 degrees) sport models.

② Min: -1 degree 05' (-1.08 degrees)

   Nominal: -0 degree 20' (-0.33 degrees)

   Max: 0 degree 25' (0.42 degrees)

③ Min: 0mm (0 inch) distance

   Nominal: In 1mm (0.04 inch) distance

   Max: In 2mm (0.08 inch) distance

④ Min: -1 degrees 20' (-1.33 degrees) 2WD except sport models and AWD. -1 degree 25' (-1.42 degrees) 2WD sport models and AWD.

   Nominal: -0 degrees 50' (-0.83 degree) 2WD except sport models and AWD. -0 degrees 55' (-0.92 degrees) 2WD sport models and AWD.

   Max: -0 degrees 20' (-0.33 degrees) 2WD except sport models and AWD. -0 degrees 25' (-0.42 degrees) 2WD sport models and AWD.

⑤ Min: 0mm (0 inch) distance

   Nominal: In 2.8mm (0.110 inch) distance

   Max: In 5.6mm (0.220 inch) distance

⑥ Min: 3 degrees 30' (3.50 degrees)

   Nominal: 4 degrees 15' (4.25 degrees)

   Max: 5 degrees 00' (5.00 degrees)

⑦ Min: -1 degree 10' (-1.16 degrees)

   Nominal: -0 degree 25' (-0.42 degrees)

   Max: 0 degree 20' (0.33 degrees)

⑧ Min: -1 degree 45' (-1.75 degrees)

   Nominal: -1 degree 15' (-1.25 degrees)

   Max: -0 degree 45' (-0.75 degrees)

⑨ Min: 4 degrees 05' (4.08 degrees) 18 inch wheel. 4 degrees 10' (4.17 degrees) 19 inch wheel.

   Nominal: 4 degrees 50' (4.83 degrees) 18 inch wheel. 4 degrees 55' (4.92 degrees) 19 inch wheel

   Max: 5 degrees 35' (5.58 degrees) 18 inch wheel. 5 degrees 40' (5.66 degrees) 19 inch wheel.

⑩ Min: -1 degree 10' (-1.16 degrees)

   Nominal: -0 degree 25' (-0.41 degrees)

   Max: 0 degree 20' (0.33 degrees)

71075_IG37_C0009

TIRE, WHEEL AND BALL JOINT SPECIFICATIONS

Year	Model	OEM Tires		Tire Pressures (psi)		Wheel Size	Lug Nut Torque (ft. lbs.)
		Standard	Optional	Front	Rear		
2011	G37/G25 sedan	①	①	33	33	NA	NA
	G37 coupe	②	②	③	③	NA	NA
	G37 convertible	④	④	⑤	⑤	NA	NA
2012	G37/G25 sedan	①	①	33	33	NA	NA
	G37 coupe	②	②	③	③	NA	NA
	G37 convertible	④	④	⑤	⑤	NA	NA

Note: If specification differes from vehicle placard, use specification given on vehicle placard.

NA: Not Available

OEM: Original Equipment Manufacturer

PSI: Pounds Per Square Inch

① P225/55R17, P225/50R18, 225/50R18, 245/45R18

② P225/50R18, 225/45R19, 245/40R19

③ P225/50R18: 33. 225/45R19 and 245/40R19: 35.

④ P225/50R18, P245/45R18, 225/45R19, 245/40R19. On 225/45R19 and 245/40R19 XL indicates extra load (reinforced) tire.

⑤ P225/50R18 and P245/45R18: 38. 225/45R19 and 245/40R19: 39.

71075_IG37_C0010

BRAKE SPECIFICATIONS
All measurements in inches unless noted

Year	Model		Brake Disc			Minimum Lining Thickness		Brake Caliper	
			Original Thickness	Minimum Thickness	Maximum Run-out	Front	Rear	Bracket Bolts (ft. lbs.)	Mounting Bolts (ft. lbs.)
2011	G37/G25 sedan	F	①	②	0.0014	0.079	0.079	NA	NA
		R	③	②	0.0022	0.079	0.079	NA	NA
	G37 coupe	F	④	②	0.0014	0.079	0.079	NA	NA
		R	⑤	②	0.0022	0.079	0.079	NA	NA
	G37 convertible	F	⑥	②	0.0014	0.079	0.079	NA	NA
		R	⑦	②	0.0022	0.079	0.079	NA	NA
2012	G37/G25 sedan	F	①	②	0.0014	0.079	0.079	NA	NA
		R	③	②	0.0022	0.079	0.079	NA	NA
	G37 coupe	F	④	②	0.0014	0.079	0.079	NA	NA
		R	⑤	②	0.0022	0.079	0.079	NA	NA
	G37 convertible	F	⑥	②	0.0014	0.079	0.079	NA	NA
		R	⑦	②	0.0022	0.079	0.079	NA	NA

NA: Not Available

① Rotor outer diameter x thickness: 2 piston caliper: 12.60x1.102. 4 piston caliper: 13.98x1.260.

② Thickness variation measured at 8 positions: 0.0006

③ Rotor outer diameter x thickness: 1 piston caliper: 12.13x0.630. 2 piston caliper: 13.78x0.787.

④ Rotor outer diameter x thickness: 1 piston caliper: 12.99x1.260. 2 piston caliper: 12.60x1.102. 4 piston caliper: 13.98x1.260.

⑤ Rotor outer diameter x thickness: 1 piston caliper: 12.99x0.630. 1 piston caliper: 12.13x0.630 (if used with 2 piston front caliper). 2 piston caliper: 13.78x0.787.

⑥ Rotor outer diameter x thickness: 1 piston caliper: 12.99x0.630. 4 piston caliper: 13.98x1.260.

⑦ Rotor outer diameter x thickness: 1 piston caliper: 12.99x0.630. 2 piston caliper: 13.78x0.787.

71075_IG37_C0011

SCHEDULED MAINTENANCE INTERVALS
INFINITI G37/G25

TO BE SERVICED	TYPE OF	VEHICLE MILEAGE INTERVAL (x1000)												
		3.75	7.5	15	22.5	30	37.5	45	52.5	60	67.5	75	82.5	90
Engine oil & filter	R	✓	✓	✓	✓	✓	✓	✓	✓	✓	✓	✓	✓	✓
Brake lines & cables	I			✓		✓		✓		✓		✓		✓
Brake pads& rotors	L/I			✓		✓		✓		✓		✓		✓
Driveshaft boots & propeller shaft (4WD)	I					✓				✓				
Automatic transmission, final drive oil & transfer case	I			✓		✓		✓		✓				
LSD gear oil	I			✓		✓		✓		✓		✓		✓
Front wheel bearing grease (4WD)	R					✓				✓				
Air cleaner filter	R					✓				✓				✓
Engine coolant	R									✓				✓
Exhaust system	I						✓	✓	✓	✓				
Spark plugs	R	Replace every 105,000 miles												
Drive belt(s)	I			✓		✓		✓		✓		✓	.	✓
Cabin air filter	R							✓						✓
Exhaust system	I		✓			✓				✓				✓
Fuel lines	I		✓			✓				✓				
Steering gear (box) & linkage, axle & suspension parts	I					✓				✓				✓
Transfer case	I					✓				✓				✓
Tire rotation			✓	✓	✓	✓	✓	✓	✓	✓	✓	✓	✓	✓
Vapor lines	S/I					✓				✓				✓

R: Replace   S/I: Service or Inspect

FREQUENT OPERATION MAINTENANCE (SEVERE SERVICE)

If a vehicle is operated under any of the following conditions it is considered severe service:

- Extremely dusty areas.
- 50% or more of the vehicle operation is in 32°C (90°F) or higher temperatures, or constant operation in temperatures below 0°C (32°F).
- Prolonged idling (vehicle operation in stop and go traffic).
- Frequent short running periods (engine does not warm to normal operating temperatures).
- Police, taxi, delivery usage or trailer towing usage.

Oil & oil filter: change every 3750 miles.

Brake pads & discs: service or inspect every 7500 miles.

Driveshaft boots: service or inspect every 7500 miles.

Exhaust system: service or inspect every 7500 miles.

Steering gear & linkage, axle & suspension parts: service or inspect every 7500 miles.

Steering linkage ball joints & front suspension ball joints: service or inspect every 7500 miles.

Air cleaner filter: service or inspect every 15,000 miles.

Final drive oil: Change every 30000 miles if towing a trailer.

Transfer case fluid: Change every 30000 miles if towing a trailer.

PRECAUTIONS

Before servicing any vehicle, please be sure to read all of the following precautions, which deal with personal safety, prevention of component damage, and important points to take into consideration when servicing a motor vehicle:

• Never open, service or drain the radiator or cooling system when the engine is hot; serious burns can occur from the steam and hot coolant.

• Observe all applicable safety precautions when working around fuel. Whenever servicing the fuel system, always work in a well-ventilated area. Do not allow fuel spray or vapors to come in contact with a spark, open flame, or excessive heat (a hot drop light, for example). Keep a dry chemical fire extinguisher near the work area. Always keep fuel in a container specifically designed for fuel storage; also, always properly seal fuel containers to avoid the possibility of fire or explosion. Refer to the additional fuel system precautions later in this section.

• Fuel injection systems often remain pressurized, even after the engine has been turned **OFF**. The fuel system pressure must be relieved before disconnecting any fuel lines. Failure to do so may result in fire and/or personal injury.

• Brake fluid often contains polyglycol ethers and polyglycols. Avoid contact with the eyes and wash your hands thoroughly after handling brake fluid. If you do get brake fluid in your eyes, flush your eyes with clean, running water for 15 minutes. If eye irritation persists, or if you have taken

brake fluid internally, IMMEDIATELY seek medical assistance.

• The EPA warns that prolonged contact with used engine oil may cause a number of skin disorders, including cancer. You should make every effort to minimize your exposure to used engine oil. Protective gloves should be worn when changing oil. Wash your hands and any other exposed skin areas as soon as possible after exposure to used engine oil. Soap and water, or waterless hand cleaner should be used.

• All new vehicles are now equipped with an air bag system, often referred to as a Supplemental Restraint System (SRS) or Supplemental Inflatable Restraint (SIR) system. The system must be disabled before performing service on or around system components, steering column, instrument panel components, wiring and sensors. Failure to follow safety and disabling procedures could result in accidental air bag deployment, possible personal injury and unnecessary system repairs.

• Always wear safety goggles when working with, or around, the air bag system. When carrying a non-deployed air bag, be sure the bag and trim cover are pointed away from your body. When placing a non-deployed air bag on a work surface, always face the bag and trim cover upward, away from the surface. This will reduce the motion of the module if it is accidentally deployed. Refer to the additional air bag system precautions later in this section.

• Clean, high quality brake fluid from a sealed container is essential to the safe and

proper operation of the brake system. You should always buy the correct type of brake fluid for your vehicle. If the brake fluid becomes contaminated, completely flush the system with new fluid. Never reuse any brake fluid. Any brake fluid that is removed from the system should be discarded. Also, do not allow any brake fluid to come in contact with a painted surface; it will damage the paint.

• Never operate the engine without the proper amount and type of engine oil; doing so WILL result in severe engine damage.

• Timing belt maintenance is extremely important. Many models utilize an interference-type, non-freewheeling engine. If the timing belt breaks, the valves in the cylinder head may strike the pistons, causing potentially serious (also time-consuming and expensive) engine damage. Refer to the maintenance interval charts for the recommended replacement interval for the timing belt, and to the timing belt section for belt replacement and inspection.

• Disconnecting the negative battery cable on some vehicles may interfere with the functions of the on-board computer system(s) and may require the computer to undergo a relearning process once the negative battery cable is reconnected.

• When servicing drum brakes, only disassemble and assemble one side at a time, leaving the remaining side intact for reference.

• Only an MVAC-trained, EPA-certified automotive technician should service the air conditioning system or its components.

BRAKES   ANTI-LOCK BRAKE SYSTEM (ABS)

GENERAL INFORMATION

PRECAUTIONS

• Certain components within the ABS system are not intended to be serviced or repaired individually.

• Do not use rubber hoses or other parts not specifically specified for and ABS system. When using repair kits, replace all parts included in the kit. Partial or incorrect repair may lead to functional problems and require the replacement of components.

• Lubricate rubber parts with clean, fresh brake fluid to ease assembly. Do not use shop air to clean parts; damage to rubber components may result.

• Use only DOT 3 brake fluid from an unopened container.

• If any hydraulic component or line is removed or replaced, it may be necessary to bleed the entire system.

• A clean repair area is essential. Always clean the reservoir and cap thoroughly before removing the cap. The slightest amount of dirt in the fluid may plug an orifice and impair the system function. Perform repairs after components have been thoroughly cleaned; use only denatured alcohol to clean components. Do not allow ABS components to come into contact with any substance containing mineral oil; this includes used shop rags.

• The Anti-Lock control unit is a microprocessor similar to other computer units in the vehicle. Ensure that the ignition switch is **OFF** before removing or installing con-

troller harnesses. Avoid static electricity discharge at or near the controller.

• If any arc welding is to be done on the vehicle, the control unit should be unplugged before welding operations begin.

WHEEL SPEED SENSORS

REMOVAL & INSTALLATION

See Figures 3 through 5.

At this time the manufacturer does not provide removal and installation procedures for this component. The following procedure is a guideline and may differ from the vehicle you are servicing.

➡**Whenever the negative battery cable is disconnected the following components will require resetting. The Auto-**

matic temperature control system, Automatic drive positioner, Power window control, Sunroof system, Sunshade system, Rear view monitor, Idle Air Volume Learning, Steering Angle Sensor Neutral Position, Audio presets and Navigation. You will need the CONSULT-III diagnostic tool, or equivalent. Follow the directions on the screen of the tool, as needed.

1. Before servicing the vehicle, refer to the Precautions Section.

➡️If working near and/or around the SRS system and components, be sure to disable the SRS system. After disabling the system wait three minutes or more before servicing the vehicle.

2. Disconnect the negative battery cable.

3. Raise and support the vehicle safely.

4. Remove the tire and wheel assembly, as required.

5. Never twist or bend sensor harness when removing it.

6. Pull the wheel sensor out without pulling the harness.

7. Be careful not to damage the sensor edges or rotor teeth.

8. Remove the sensor first, before removing the wheel hub and bearing assembly.

To install:

➡️Be sure to use new fasteners, as required.

9. Installation is the reverse of the removal procedure.

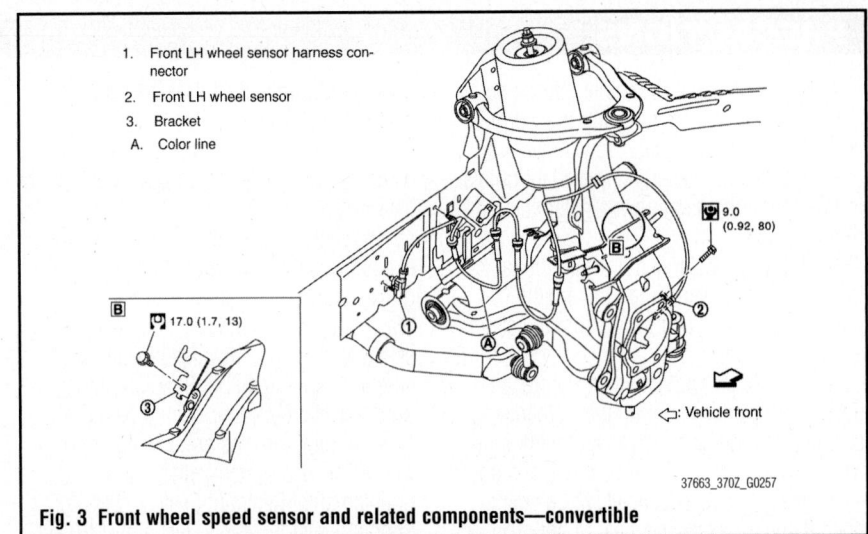

1. Front LH wheel sensor harness connector
2. Front LH wheel sensor
3. Bracket
A. Color line

9.0 (0.92, 80)

17.0 (1.7, 13)

⟻ : Vehicle front

37663_370Z_G0257

Fig. 3 Front wheel speed sensor and related components—convertible

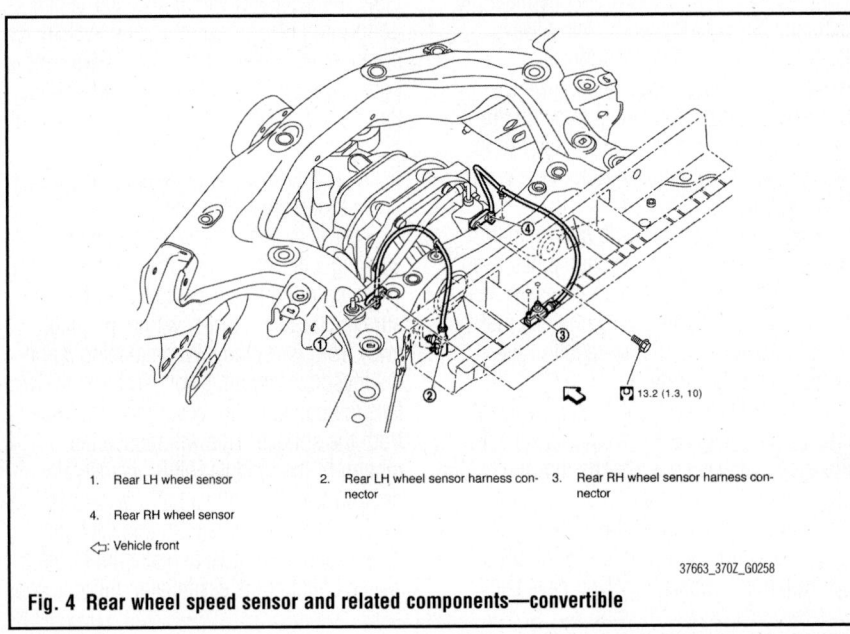

1. Rear LH wheel sensor
2. Rear LH wheel sensor harness connector
3. Rear RH wheel sensor harness connector
4. Rear RH wheel sensor
⟻ : Vehicle front

13.2 (1.3, 10)

37663_370Z_G0258

Fig. 4 Rear wheel speed sensor and related components—convertible

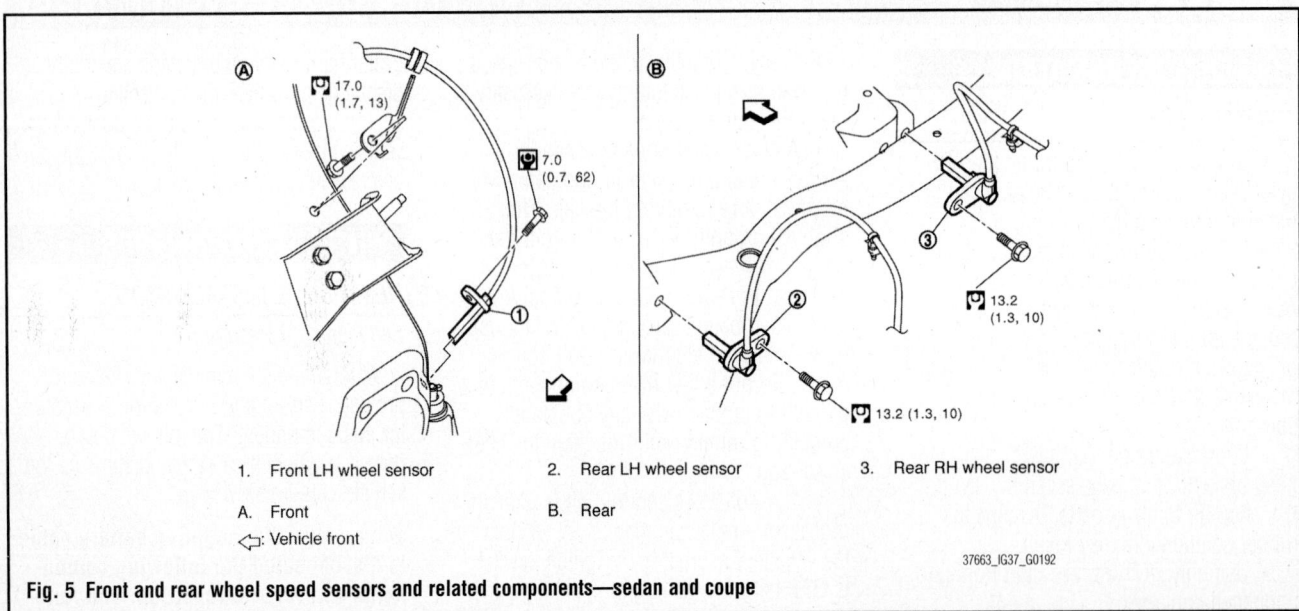

Ⓐ 17.0 (1.7, 13)

7.0 (0.7, 62)

Ⓑ 13.2 (1.3, 10)

13.2 (1.3, 10)

1. Front LH wheel sensor
2. Rear LH wheel sensor
3. Rear RH wheel sensor
A. Front
B. Rear
⟻ : Vehicle front

37663_IG37_G0192

Fig. 5 Front and rear wheel speed sensors and related components—sedan and coupe

BLEEDING PROCEDURE

MANUAL

⁂ CAUTION

Turn the ignition switch OFF and disconnect the ABS actuator and electric unit (control unit) connector or the battery negative terminal before performing the work.

⁂ WARNING

Monitor the fluid level in the reservoir tank while performing the air bleeding

⁂ WARNING

Always use new brake fluid for refilling. Never reuse the drained brake fluid.

1. Connect a vinyl tube to the bleeder valve of the rear right brake.
2. Fully depress the brake pedal 4 to 5 times.
3. Loosen the bleeder valve and bleed air with the brake pedal depressed, and then quickly tighten the bleeder valve.
4. Repeat steps 2 and 3 until all of the air is out of the brake line.
5. Tighten the bleeder valve to the specified torque.
6. Check that the fluid level in the reservoir tank is within the specified range after air bleeding.
7. Check each item of brake pedal. Adjust it if the measurement value is not the standard.

BLEEDING THE ABS SYSTEM

➡Whenever the negative battery cable is disconnected the following components will require resetting. The Automatic temperature control system, Automatic drive positioner, Power window control, Sunroof system, Sunshade system, Rear view monitor, Idle Air Volume Learning, Steering Angle Sensor Neutral Position, Audio presets and Navigation. You will need the CONSULT-III diagnostic tool, or equivalent. Follow the directions on the screen of the tool, as needed.

1. Before servicing the vehicle, refer to the Precautions Section.

➡If working near and/or around the SRS system and components, be sure to disable the SRS system. After disabling the system wait three minutes or more before servicing the vehicle.

2. Disconnect the negative battery cable.

➡Turn the ignition switch off and disconnect the ABS actuator and electric control unit connector, or the negative battery cable before performing the work.

➡Monitor the fluid level in the reservoir while performing the work. Always use new brake fluid. Be sure to use the proper grade and type fluid.

➡As required, cover the crowfoot and flare nut wrench with a shop towel to prevent damage to the front four piston type caliper and rear two piston type caliper.

3. Connect a vinyl tube to the bleeder valve of the right rear brake.
4. Fully depress the brake pedal four or five times.
5. Loosen the bleeder valve and bleed the air with the brake pedal depressed, quickly tighten the bleeder valve.
6. Repeat the above step until all air is expelled out of the brake line.
7. Tighten the bleeder valve.
8. Perform the above procedure to the brakes in the following sequence.
9. Right rear brake, left front brake, left rear brake and right front brake.
10. Check and refill the master cylinder, as required.
11. Be sure to perform the reconnect/relearn procedures.

FLUID FILL PROCEDURE

➡Whenever the negative battery cable is disconnected the following components will require resetting. The Automatic temperature control system, Automatic drive positioner, Power window control, Sunroof system, Sunshade system, Rear view monitor, Idle Air Volume Learning, Steering Angle Sensor Neutral Position, Audio presets and Navigation. You will need the CONSULT-III diagnostic tool, or equivalent. Follow the directions on the screen of the tool, as needed.

1. Before servicing the vehicle, refer to the Precautions Section.

➡If working near and/or around the SRS system and components, be sure to disable the SRS system. After disabling the system wait three minutes or more before servicing the vehicle.

➡Turn the ignition switch off and disconnect the ABS actuator and electric control unit connector, or the negative battery cable before performing the work.

➡Cover the crowfoot and flare nut wrench with a shop towel to prevent damage to the front four piston type caliper and rear two piston type caliper.

2. Check that there is no foreign material in the reservoir tank or around it. Never reuse used fluid.
3. Loosen the bleeder valve.
4. Slowly depress the brake pedal to the full stroke. Release the pedal.
5. Repeat at intervals of two or three seconds until all brake fluid is discharged.
6. Close the bleeder valve with the brake pedal depressed.
7. Repeat the above on each wheel.
8. Bleed the brake system.
9. Be sure to perform the reconnect/relearn procedures.

BRAKE CALIPERS

REMOVAL & INSTALLATION

1-Piston Type

> ✳✳ **WARNING**
>
> **Clean any dust from the brake caliper and brake pads with a vacuum dust collector. Never blow with compressed air.**

> ✳✳ **CAUTION**
>
> **Never depress the brake pedal. Brake fluid may splash while removing the brake hose.**

1. Remove tires with power tool.
2. Fix the disc rotor using wheel nuts.
3. Drain brake fluid.

> ✳✳ **WARNING**
>
> **Never spill or splash brake fluid on the disc rotor.**

4. Remove union bolt and copper washer, and disconnect brake hose from caliper assembly.
5. Remove torque member mounting bolts, and remove brake caliper assembly.

> ✳✳ **WARNING**
>
> **Never drop brake pads and caliper assembly.**

6. Remove disc rotor.

> ✳✳ **WARNING**
>
> **Put matching marks on the wheel hub and bearing assembly and the disc rotor before removing the disc rotor.**

> ✳✳ **WARNING**
>
> **Never drop disc rotor.**

To install:

> ✳✳ **WARNING**
>
> **Clean any dust from the brake caliper and brake pads with a vacuum dust collector. Never blow with compressed air.**

> ✳✳ **WARNING**
>
> **Never depress the brake pedal. Brake fluid may splash while removing the brake hose.**

7. Install disc rotor.

> ✳✳ **WARNING**
>
> **Align the matching marks that have been made during removal when reusing the disc rotor.**

8. Install the brake caliper assembly to the vehicle and tighten the torque member mounting bolts to the specified torque.

> ✳✳ **WARNING**
>
> **Never spill or splash any grease and moisture on the brake caliper assembly mounting face, threads, mounting bolts and washers. Wipe out any grease and moisture.**

9. Install brake hose and copper washers to brake caliper assembly, and tighten union bolts to the specified torque.
10. Refill with new brake fluid and perform the air bleeding.

> ✳✳ **WARNING**
>
> **Never reuse drained brake fluid.**

> ✳✳ **WARNING**
>
> **Never spill or splash brake fluid on the disc rotor.**

11. Check the front disc brakes for drag.

4-Piston Type

> ✳✳ **WARNING**
>
> **Clean any dust from the brake caliper and brake pads with a vacuum dust collector. Never blow with compressed air.**

> ✳✳ **WARNING**
>
> **Never depress the brake pedal. Brake fluid may splash while removing the brake hose and brake tube.**

1. Remove tires with power tool.
2. Fix the disc rotor using wheel nuts.
3. Drain brake fluid.

> ✳✳ **WARNING**
>
> **Never spill or splash brake fluid on the disc rotor and caliper.**

4. Loosen the flare nut with a flare nut wrench and separate the brake tube from caliper.

> ✳✳ **WARNING**
>
> **Cover flare nut wrench with a cloth as not to damage the caliper.**

> ✳✳ **WARNING**
>
> **Never scratch the flare nut and the brake tube.**

> ✳✳ **WARNING**
>
> **Never bend sharply, twist or strongly pull out the brake tube.**

> ✳✳ **WARNING**
>
> **Cover open end of brake tube when disconnecting to prevent entrance of dirt.**

5. Remove caliper mounting bolts, and remove caliper.

> ✳✳ **WARNING**
>
> **Never drop brake pad and caliper.**

6. Remove disc rotor.

> ✳✳ **WARNING**
>
> **Put matching marks on the wheel hub and bearing assembly and the disc rotor before removing the disc rotor.**

> ✳✳ **WARNING**
>
> **Never drop disc rotor.**

To install:

> ✳✳ **WARNING**
>
> **Clean any dust from the brake caliper and brake pads with a vacuum dust collector. Never blow with compressed air.**

> ✳✳ **WARNING**
>
> **Never depress the brake pedal. Brake fluid may splash while removing the brake hose and brake tube.**

7. Install disc rotor.

> ✳✳ **WARNING**
>
> **Align the matching marks that have been made during removal when reusing the disc rotor.**

8. Install the brake caliper to the vehicle and tighten the caliper mounting bolts to the specified torque.

9. Tighten the flare nut to the specified torque with a flare nut torque wrench.

10. Refill with new brake fluid and perform the air bleeding.

11. Check the front disc brakes for drag.

BRAKE PADS

REMOVAL & INSTALLATION

1-Piston Type

See Figures 6 and 7.

1. Remove tires with power tool.
2. Remove the protector and location pin.

➡**Make sure to note the installed positions of the brake pads.**

3. Suspend the cylinder body with suitable wire so that the brake hose will not stretch. Then remove the pad return springs and brake pads from the torque member.

To install:

4. Apply bentonite noise damping brake grease to the pad retainers before it to installing the torque member if the pad retainers has been removed.
5. Securely assemble the pad retainers so that it will not be lifted up from the torque member. Be careful not to deform the pad retainers.
6. Install the brake pads to the torque member.

➡**The brake pads are directional. Never mistake the direction.**

7. Install the pad return spring to the brake pad.

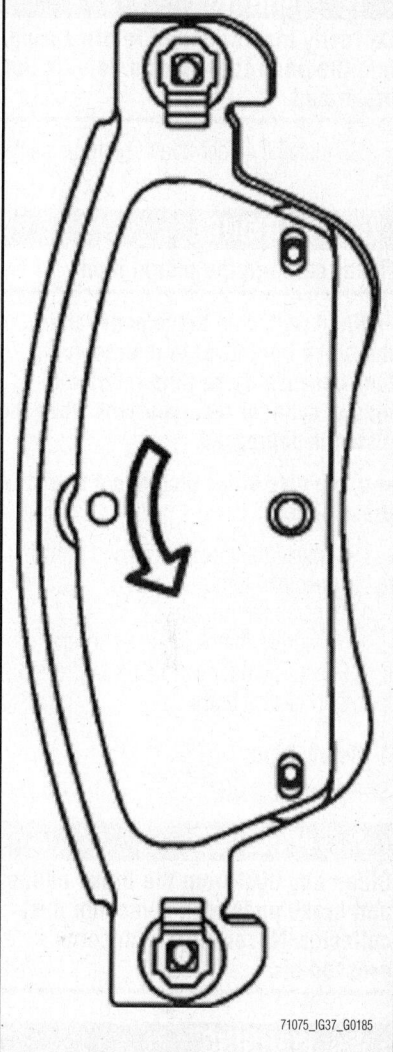

71075_IG37_G0185

Fig. 6 Direction of disc rotor rotation (Forward direction: arrow)

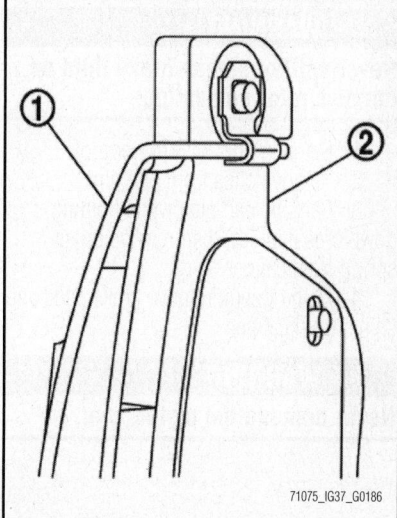

71075_IG37_G0186

Fig. 7 Installing the pad return spring (1) to the brake pad (2)

✖✖ WARNING

Correctly insert the pad return spring into the pad return spring hole on the brake pad.

8. Install cylinder body to torque member.

✖✖ WARNING

Never damage the piston boot.

➡When replacing brake pads, check the brake fluid level in the reservoir tank because brake fluid returns to master cylinder reservoir tank when the piston is depressed.

➡Use a disc brake piston tool to easily press piston.

9. Install the location pin and tighten it to the specified torque.
10. Install the protector.
11. Depress the brake pedal several times to check that no drag feel is present for the front disc brake.

4-Piston Type

See Figures 8 and 9.

✖✖ WARNING

Clean any dust from the brake caliper and brake pads with a vacuum dust collector. Never blow with compressed air.

✖✖ WARNING

Never depress the brake pedal while removing the brake pads because the piston may pop out.

✖✖ WARNING

Never spill or splash brake fluid on the disc rotor and caliper.

1. Remove tires with power tool.
2. Remove clips from pad pins.
3. Remove pad pins while holding down cross spring, then remove cross spring from caliper.
4. Using pliers, remove brake pads and shims from caliper.

✖✖ WARNING

Never damage the piston boot.

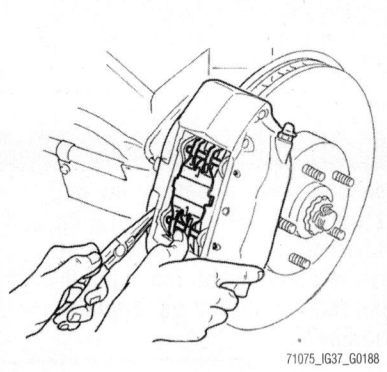

Fig. 8 Removing pad pins while holding down cross spring, then removing cross spring from caliper

✖✖ WARNING

Never drop the brake pads, shims.

✖✖ WARNING

Remember each position of the removed brake pads.

To install:

✖✖ WARNING

Clean any dust from the brake caliper and brake pads with a vacuum dust collector. Never blow with compressed air.

✖✖ WARNING

Never depress the brake pedal while removing the brake pads because the piston may pop out.

✖✖ WARNING

Never spill or splash brake fluid on the disc rotor and caliper.

5. Apply copper based brake grease to the mating faces between the brake pads and shims, and install shims to the brake pad.

➡**Always replace the shims together when replacing the brake pad.**

6. Apply copper based brake grease to the mating faces between the brake pads and caliper.
7. Install brake pads to caliper.

✖✖ WARNING

Be careful not to damage the piston boot.

➡When replacing brake pads, check the brake fluid level in the reservoir tank because brake fluid returns to master cylinder reservoir tank when the piston is depressed.

➡Use a disc brake piston tool to easily depress the piston.

8. Install upper pad pin from the inner side, then install firmly to the outer side through the hole in the top of brake pad.
9. Place the top of cross spring (1) over the upper pad pin (2), press in the cross spring, install lower pad pin from the inner side to the outer side, and secure cross spring.
10. Install clips to the pad pins.

✖✖ CAUTION

If clip is not fully attached, pad pin or brake pad could fall out while vehicle is in motion.

11. Depress the brake pedal several times to check that no drag feel is present for the front disc brake.

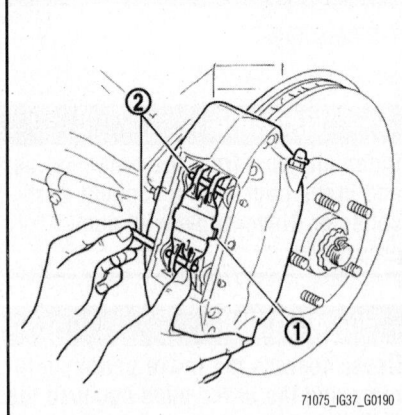

Fig. 9 Placing top of cross spring (1) over the upper pad pin (2), pressing in the cross spring and installing the lower pad pin from the inner side to the outer side, and securing the cross spring

BRAKES   **REAR DISC BRAKES**

BRAKE CALIPERS

REMOVAL & INSTALLATION

1 Piston Type

See Figure 10.

> ❋❋ **WARNING**
>
> **Clean any dust from the brake caliper and brake pads with a vacuum dust collector. Never blow with compressed air.**

> ❋❋ **WARNING**
>
> **Never depress the brake pedal. Brake fluid may splash while removing the brake hose.**

1. Remove tires with power tool.
2. Fix the disc rotor using wheel nuts.
3. Drain brake fluid.

> ❋❋ **WARNING**
>
> **Never spill or splash brake fluid on the disc rotor.**

4. Remove union bolt and copper washers, and disconnect brake hose from caliper assembly.
5. Remove torque member mounting bolts, and remove brake caliper assembly.

> ❋❋ **WARNING**
>
> **Never drop brake pads and caliper assembly.**

6. Remove disc rotor.

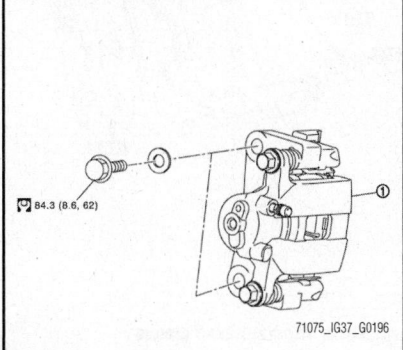

84.3 (8.6, 62)

Fig. 10 Removing and installing the rear brake caliper—1 piston type

> ❋❋ **WARNING**
>
> **Never drop disc rotor.**

To install:

> ❋❋ **WARNING**
>
> **Clean any dust from the brake caliper and brake pads with a vacuum dust collector. Never blow with compressed air.**

> ❋❋ **WARNING**
>
> **Never depress the brake pedal. Brake fluid may splash while removing the brake hose.**

7. Install disc rotor.

> ❋❋ **WARNING**
>
> **Align the matching marks that have been made during removal when reusing the disc rotor.**

8. Install the brake caliper assembly to the vehicle and tighten the torque member mounting bolts to the specified torque.

> ❋❋ **WARNING**
>
> **Never spill or splash any grease and moisture on the brake caliper assembly mounting face, threads, mounting bolts, and washers. Wipe out any grease and moisture.**

9. Install brake hose and copper washers to brake caliper assembly, and tighten union bolts to the specified torque.
10. Refill with new brake fluid and perform the air bleeding.

> ❋❋ **WARNING**
>
> **Never reuse drained brake fluid.**

> ❋❋ **WARNING**
>
> **Never spill or splash brake fluid on the disc rotor.**

11. Check a drag of rear disc brake.

2 Piston Type

See Figure 11.

> ❋❋ **WARNING**
>
> **Clean any dust from the brake caliper and brake pads with a vacuum dust**

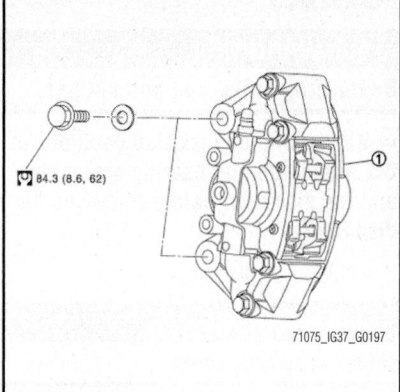

84.3 (8.6, 62)

71075_IG37_G0197

Fig. 11 Removing and installing the rear brake caliper—2 piston type

collector. Never blow with compressed air.

> ❋❋ **WARNING**
>
> **Never depress the brake pedal. Brake fluid may splash while removing the brake hose and brake tube.**

1. Remove tires with power tool.
2. Fix the disc rotor using wheel nuts.
3. Drain brake fluid.

> ❋❋ **WARNING**
>
> **Never spill or splash brake fluid on the disc rotor.**

4. Loosen the flare nut with a flare nut wrench and separate the brake tube from caliper.

> ❋❋ **WARNING**
>
> **Cover flare nut wrench with a cloth as not to damage the caliper.**

> ❋❋ **WARNING**
>
> **Never scratch the flare nut and the brake tube.**

> ❋❋ **WARNING**
>
> **Never bend sharply, twist or strongly pull out the brake tube.**

> ❋❋ **WARNING**
>
> **Cover open end of brake tube when disconnecting to prevent entrance of dirt.**

5. Remove brake hose mounting bolt.
6. Remove caliper mounting bolts, and remove caliper.

⁂ WARNING

Never drop brake pad and caliper.

➡Matchmark the installed positions of the wheel hub and bearing assembly and the disc rotor before removing the disc rotor.

7. Remove disc rotor.

⁂ WARNING

Never drop disc rotor.

To install:

⁂ WARNING

Clean any dust from the brake caliper and brake pads with a vacuum dust collector. Never blow with compressed air.

⁂ WARNING

Never depress the brake pedal. Brake fluid may splash while removing the brake hose.

8. Install disc rotor.
9. Align the matching marks that have been made during removal when reusing the disc rotor.
10. Install the brake caliper to the vehicle and tighten the caliper mounting bolts to the specified torque.

⁂ WARNING

Never spill or splash any grease and moisture on the caliper mounting face, threads, mounting bolts and washers. Wipe out any grease and moisture.

11. Install the brake hose mounting bolt to the specified torque.
12. Tighten the flare nut to the specified torque with a flare nut torque wrench.

⁂ WARNING

Cover crowfoot with a cloth as not to damage the caliper.

⁂ WARNING

Never scratch the flare nut and the brake tube.

13. Refill with new brake fluid and perform the air bleeding.

⁂ WARNING

Never reuse drained brake fluid.

⁂ WARNING

Never spill or splash brake fluid on the disc rotor.

14. Check a drag of rear disc brake.

BRAKE PADS

REMOVAL & INSTALLATION

1 Piston Type
See Figure 12.

⁂ WARNING

Clean any dust from the brake caliper and brake pads with a vacuum dust collector. Never blow with compressed air.

⁂ WARNING

Never depress the brake pedal while removing the brake pads or the cylinder body because the piston may pop out.

⁂ WARNING

Never spill or splash brake fluid on the disc rotor.

1. Remove tires with power tool.
2. Remove the upper sliding pin bolt.
3. Suspend the cylinder body with suitable wire so that the brake hose will not stretch. Remove the brake pads, shims, shim cover and pad retainers from the torque member.

⁂ WARNING

Never deform the pad retainers if removing the pad retainers.

⁂ WARNING

Never damage the piston boot.

⁂ WARNING

Never drop the brake pad, shims, and the shim cover.

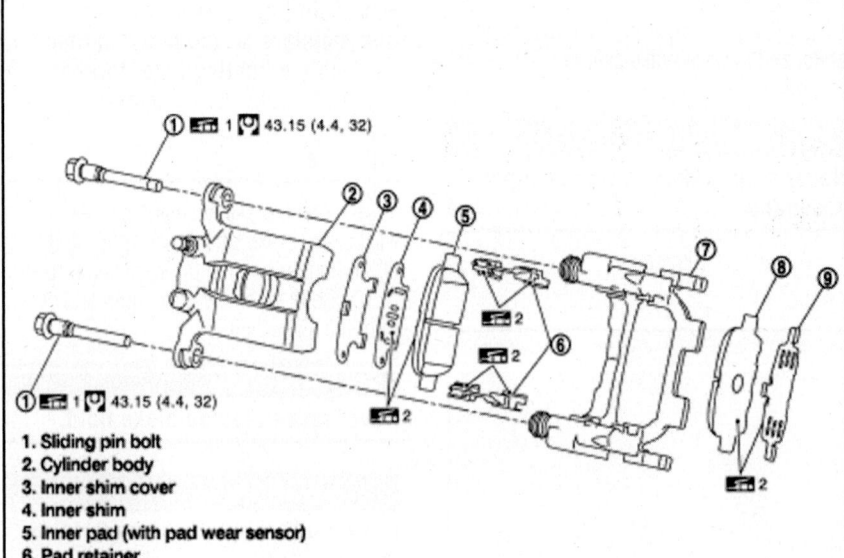

1. Sliding pin bolt
2. Cylinder body
3. Inner shim cover
4. Inner shim
5. Inner pad (with pad wear sensor)
6. Pad retainer
7. Torque member
8. Outer pad
9. Outer shim
Grease applicator: 1: Apply rubber grease.
Grease applicator: 2: Apply PBC (Poly Butyl Cuprysil) grease or silicone-based grease.

71075_IG37_G0198

Fig. 12 Rear brake pad exploded view—1 piston type

⁂ WARNING

Remember each position of the removed brake pads.

To install:

⁂ WARNING

Clean any dust from the brake caliper and brake pads with a vacuum dust collector. Never blow with compressed air.

⁂ WARNING

Never depress the brake pedal while removing the brake pads or the cylinder body because the piston may pop out.

⁂ WARNING

Never spill or splash brake fluid on the disc rotor.

4. Apply PBC (Poly Butyl Cuprysil) grease or silicone-based grease to the pad retainers before installing it to the torque member if the pad retainers has been removed.

⁂ WARNING

Securely assemble the pad retainers so that it will not be lifted up from the torque member.

⁂ WARNING

Never deform the pad retainers.

5. Apply PBC (Poly Butyl Cuprysil) grease or silicone-based grease to the mating faces between the brake pads and shims, and install shims and shim cover to brake pads.

➡Always replace the shims together with the shim cover when replacing the brake pad.

6. Install cylinder body and brake pads to torque member.

⁂ WARNING

Never damage the piston boot.

➡When replacing brake pads, check the brake fluid level in the reservoir tank because brake fluid returns to master cylinder reservoir tank when the piston is depressed.

➡Use a disc brake piston tool to easily compress the piston.

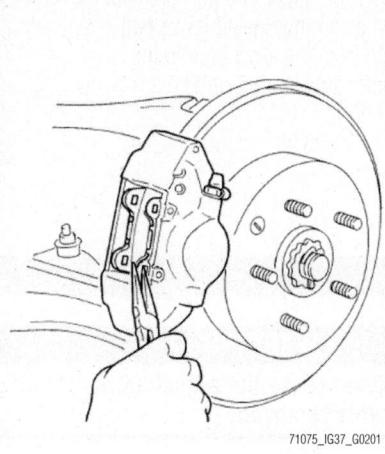

71075_IG37_G0201

Fig. 13 Removing brake pads, shims and shim covers from caliper

7. Install the upper sliding pin bolt and tighten it to the specified torque.

8. Depress the brake pedal several times to check that no drag feel is present for the rear disc brake.\

2 Piston Type

See Figure 13.

⁂ WARNING

Clean any dust from the brake caliper and brake pads with a vacuum dust collector. Never blow with compressed air.

⁂ WARNING

Never depress the brake pedal while removing the brake pads or the cylinder body because the piston may pop out.

⁂ WARNING

Never spill or splash brake fluid on the disc rotor and caliper.

1. Remove tires with power tool.

2. Remove clips from pad pins.

3. Remove pad pins while holding down cross spring, then remove cross spring from caliper.

4. Using pliers, remove brake pads, shims and shim covers from caliper.

⁂ WARNING

Never damage the piston boot.

⁂ WARNING

Never drop the brake pad, shims, and the shim cover.

⁂ WARNING

Remember each position of the removed brake pads.

To install:

⁂ WARNING

Clean any dust from the brake caliper and brake pads with a vacuum dust collector. Never blow with compressed air.

⁂ WARNING

Never depress the brake pedal while removing the brake pads because the piston may pop out.

⁂ WARNING

Never spill or splash brake fluid on the disc rotor and caliper.

5. Apply copper based brake grease to the mating faces between the brake pads, shims and shim cover, and install shims and shim cover to the brake pad.

⁂ WARNING

Always replace the shims together when replacing the brake pad.

6. Apply copper based brake grease to the mating faces between the brake pads and caliper.

7. Apply copper based brake grease to the mating faces between the brake pads and pad pins.

8. Apply copper based brake grease to the mating faces between the brake pads and cross spring.

9. Install brake pads to caliper.

⁂ WARNING

Never damage the piston boot.

➡When replacing brake pads, check the brake fluid level in the reservoir tank because brake fluid returns to master cylinder reservoir tank when the piston is depressed.

➡ **Use a disc brake piston tool to easily compress the piston.**

10. Install upper pad pin from the inner side, then install firmly to the outer side through the hole in the top of brake pad.

11. Place the top of cross spring over the upper pad pin, press in the cross spring, install lower pad pin from the inner side to the outer side, and secure cross spring.

12. Install clips to the pad pins.

✳✳ CAUTION

If clip is not fully attached, pad pin or brake pad could fall out while vehicle is in motion.

13. Depress the brake pedal several times to check that no drag feel is present for the rear disc brake.

BRAKES

ADJUSTMENTS

CONTROL ASSEMBLY

Pedal Type

See Figure 14.

1. Fix the disc rotor using wheel nuts.

2. Release the parking brake pedal by turning the adjusting nut with a deep socket wrench and loosening the cable.

3. Remove the adjusting hole plug from the disc rotor. Turn the adjuster (1) in the direction (A) as shown in the figure using a suitable tool until the disc rotor is locked.

4. Turn back the adjuster 5 or 6 notches from the locked position.

5. Rotate the disc rotor to check that there is no drag. Install the adjusting hole plug.

6. Adjust the cable with the following procedure.

   a. Operate the parking brake pedal with a force of 110 lbs. (490 N) for 10 strokes or more.

   b. Adjust the parking brake pedal stroke by turning the adjusting nut with a deep socket wrench.

✳✳ WARNING

Never reuse the adjusting nut if the nut is removed.

   c. Operate the parking brake pedal with a force of 44 lb. (196 N). Check that the pedal stroke is within the specified number of notches. (Check it by listening to clicks of ratchet.)

   d. Rotate the disc rotor with the parking brake pedal released and check that there is no drag.

Lever Type

See Figure 14.

1. Fix the disc rotor using wheel nuts.

2. Release the parking brake lever by turning the adjusting nut with a deep socket wrench and loosening the cable.

3. Remove the adjusting hole plug from the disc rotor. Turn the adjuster in the direction shown using a suitable tool until the disc rotor is locked.

4. Turn back the adjuster 5 or 6 notches from the locked position.

5. Rotate the disc rotor to check that there is no drag. Install the adjusting hole plug.

6. Adjust the cable with the following procedure.

   a. Operate the parking brake lever with a force of 66 lbs. (294 N) for 10 strokes or more.

   b. Adjust the parking brake lever stroke by turning the adjusting nut with a deep socket wrench.

✳✳ WARNING

Never reuse the adjusting nut if the nut is removed.

   c. Operate the parking brake lever with a force of 44 lbs. (196 N). Check that the lever stroke is within the specified number of notches.

PARKING BRAKE

(Check it by listening to clicks of ratchet.)

   d. Rotate the disc rotor with the parking brake lever released and check that there is no drag.

PARKING BRAKE SHOES

ADJUSTMENTS

1. Adjust parking brake pedal (pedal type) or parking brake lever (lever type) stroke.

2. Perform parking brake break-in (drag on) operation by driving vehicle under the following conditions:

   - Drive forward
   - Vehicle speed: Approx. 19 MPH (30 km/h) set (constant and forward)
   - Parking brake operating force: 71.6 lbs. (318.5 N) set contact
   - Time: Approx. 35 sec.

✳✳ WARNING

To prevent lining from getting too hot, allow a cool off period of approximately 5 minutes after every break-in operation.

3. After the break-in procedure, check parking brake pedal (pedal type) or parking brake lever (lever type) stroke of parking brake.

✳✳ WARNING

If it is out of the specification, adjust again.

REMOVAL & INSTALLATION

See Figures 15 and 16.

✳✳ WARNING

Clean any dust from the parking brake shoes and back plates with a vacuum dust collector. Never blow with compressed air.

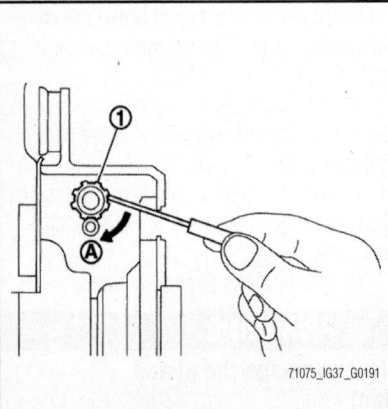

71075_IG37_G0191

Fig. 14 Turning adjuster (1) in direction (A) until the disc rotor is locked

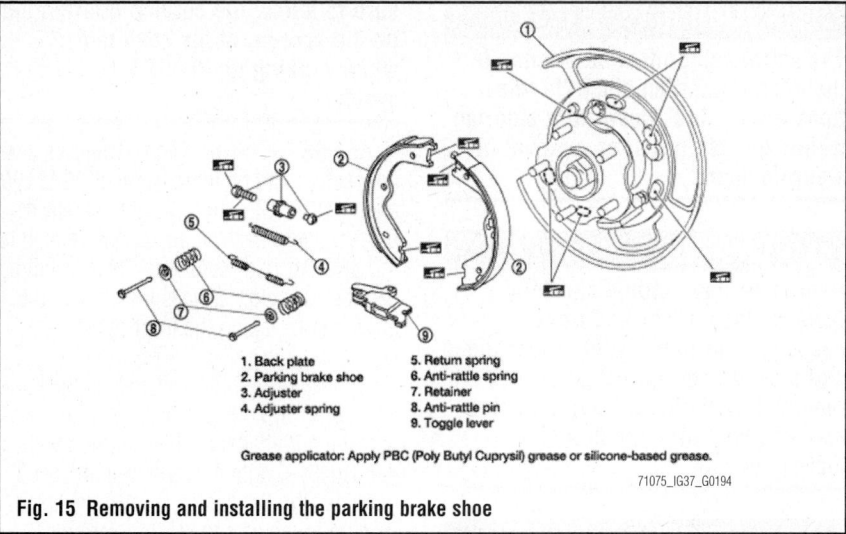

1. Back plate
2. Parking brake shoe
3. Adjuster
4. Adjuster spring
5. Return spring
6. Anti-rattle spring
7. Retainer
8. Anti-rattle pin
9. Toggle lever

Grease applicator: Apply PBC (Poly Butyl Cuprysil) grease or silicone-based grease.

71075_IG37_G0194

Fig. 15 Removing and installing the parking brake shoe

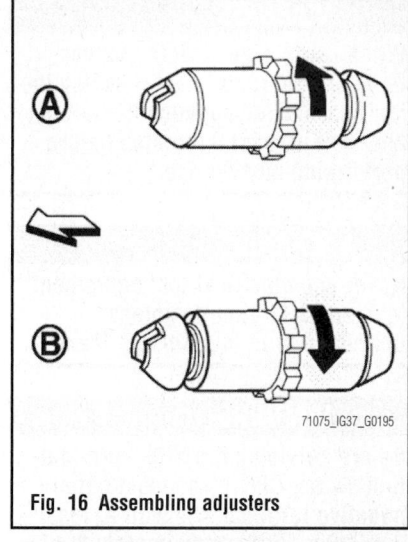

71075_IG37_G0195

Fig. 16 Assembling adjusters

1. Remove rear tires.
2. Remove disc rotor.

✳✳ WARNING

Parking brake completely in the released position.

3. If disc rotor cannot be removed, remove as follows:
   a. Fix the disc rotor with wheel nuts and remove the adjusting hole plug.
   b. Using suitable tool, rotate adjuster (1) in the direction (B) to retract and loosen brake shoe.
4. Remove anti-rattle pins, retainers, anti-rattle springs, and return spring, adjuster spring.

✳✳ WARNING

Never drop the removed parts.

5. Remove parking brake shoes, adjuster assembly, and toggle lever.

✳✳ WARNING

The parking brake shoes for the front wheels are made of different materials from those for the rear wheels. Never misidentify them when removing.

✳✳ WARNING

Never drop the removed parts.

   To install:
6. To install, reverse the removal procedure.
7. Apply PBC (Poly Butyl Cuprysil) grease or silicone-based grease to the back plate and brake shoe.

✳✳ WARNING

The parking brake shoes for the front wheels are made of different materials from those for the rear wheels. Never misidentify them when removing and replacing.

8. Assemble adjusters so that threaded part is expanded when rotating it in the direction shown by arrow.
9. Shorten adjuster by rotating it.
10. When disassembling apply PBC (Poly Butyl Cuprysil) grease or silicone-based grease to threads.
11. Check brake shoe sliding surface and drum inner surface for grease. Wipe it off if it adhere on the surfaces.

CHASSIS ELECTRICAL   AIR BAGS (SUPPLEMENTAL RESTRAINT SYSTEM)

PRECAUTIONS

The Supplemental Restraint System such as "AIR BAG" and "SEAT BELT PRE-TENSIONER", used along with a front seat belt, helps to reduce the risk or severity of injury to the driver and front passenger for certain types of collision. This system includes seat belt switch inputs and dual stage front air bag modules. The SRS system uses the seat belt switches to determine the front air bag deployment, and may only deploy one front air bag, depending on the severity of a collision and whether the front occupants are belted or unbelted. Information necessary to service the system safely is included in the "SRS AIR BAG" and "SEAT BELT".

✳✳ WARNING

Always observe the following items for preventing accidental activation.

✳✳ CAUTION

Improper maintenance, including incorrect removal and installation of the SRS, can lead to personal injury caused by unintentional activation of the system. For removal of Spiral Cable and Air Bag Module, see "SRS AIR BAG".

✳✳ WARNING

Never use electrical test equipment on any circuit related to the SRS unless instructed. SRS wiring

harnesses can be identified by yellow and/or orange harnesses or harness connectors.

PRECAUTIONS WHEN USING POWER TOOLS (AIR OR ELECTRIC) AND HAMMERS

✳✳ WARNING

When working near the Air Bag Diagnosis Sensor Unit or other Air Bag System sensors with the ignition ON or engine running, never use air or electric power too ls or strike near the sensor(s) with a hammer. Heavy vibration could activate the sensor(s) and deploy the air bag(s), possibly causing serious injury.

✳✳ WARNING

When using air or electric power tools or hammers, always switch the ignition OFF, disconnect the battery, and wait at least 3 minutes before performing any service.

✳✳ WARNING

Never use electrical test equipment to check SRS circuits unless instructed to in this Service Manual.

✳✳ WARNING

Before servicing the SRS, turn ignition switch OFF, disconnect battery negative terminal and wait at least 3 minutes. For approximately 3 minutes after the cables are removed, it is still possible for the air bag and seat belt pretensioner to deploy. Therefore, never work on any SRS connectors or wires until at least 3 minutes have elapsed.

✳✳ WARNING

Diagnosis sensor unit must always be installed with their arrow marks "⇐"pointing towards the front of the vehicle for proper operation. Also check diagnosis sensor unit for cracks, deformities or rust before installation and replace as required.

✳✳ WARNING

The spiral cable must be aligned in the neutral position since its rotations are limited. Never turn steering wheel and column after removal of steering gear.

✳✳ CAUTION

Handle air bag module carefully. Always place driver and front passenger air bag modules with the pad side facing upward and seat mounted front side air bag module standing with the stud bolt side facing down.

✳✳ WARNING

Conduct self-diagnosis to check entire SRS for proper function after replacing any components.

✳✳ WARNING

Always replace instrument panel pad following front passenger air bag deployment.

DISARMING THE SYSTEM

✳✳ WARNING

Servicing the SRS system will require the use of the CONSULT-III scan tool, or equivalent. Be sure to follow the service information on the screen, of the scan tool, when working on the SRS system.

All SRS electrical wiring harnesses and connectors can be identified with YELLOW and or ORANGE color. Do not use electrical test equipment on any circuit related to the SRS (air bag) sensors. When installing SRS components, always install with the arrow marks facing the front of the vehicle.

To disarm the SRS system turn the ignition switch to **OFF** position. Then, disconnect the both battery cables starting with the negative cable first and wait at least 3 minutes after the cables are disconnected. Be sure to insulate the battery terminal ends.

ARMING THE SYSTEM

To arm the SRS system turn the ignition switch to **OFF** position. Connect the both battery cables starting with the positive cable first.

The SRS or air bag system is equipped with a self-diagnostic operation. After turning the ignition key to the ON or START position, the AIR BAG warning lamp will illuminate for 7 seconds. After 7 seconds, the AIR BAG lamp will extinguish if no malfunction is detected. If the AIR BAG lamp does not extinguish after 7 seconds, check the SRS self-diagnostic system for a malfunction.

DRIVETRAIN

AUTOMATIC TRANSMISSION

DRAIN & REFILL

Draining

1. Start the engine and let it run to warm up transmission.
2. Stop the engine.
3. Remove drain plug and gasket from transmission case and then drain gear oil.
4. Set a gasket on drain plug and install it to transmission case.

➡**Never reuse gasket.**

5. Tighten drain plug to 25 ft. lbs. (34.5 Nm).

Refilling

1. Remove filler plug and gasket from transmission case.

2. Fill with new gear oil to transmission as shown in the figure.

✳✳ WARNING

Never reuse drained gear oil.

3. After refilling gear oil, check the oil level.
4. Set a gasket on filler plug and then install it to transmission case.

➡**Never reuse gasket.**

5. Tighten filler plug to 25 ft. lbs. (34.5 Nm).

FLUID INSPECTION

Make sure that gear oil is not leaking from transmission or around it.

FLUID LEVEL CHECK

See Figure 17.

1. Remove filler plug (1) and gasket from transmission case.
2. Check the oil level from filler plug mounting hole.

✳✳ WARNING

Never start engine while checking oil level.

3. Set a gasket on filler plug and then install to transmission case.

✳✳ WARNING

Never reuse gasket.

4. Tighten filler plug to 25 ft. lbs. (34.5 Nm).

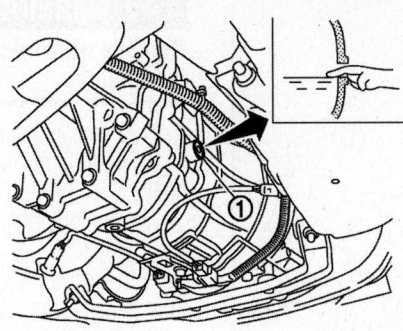

Fig. 17 Removing filler plug (1) and checking oil level front filler plug mounting hole

FLUID RECOMMENDATIONS

Genuine NISSAN Matic S ATF is recommended.

✳✳ WARNING

Using automatic transmission fluid other than Genuine NISSAN Matic S ATF will cause deterioration in drive-ability and automatic transmission durability, and may damage the automatic transmission, which is not covered by the INFINITI new vehicle limited warranty.

MANUAL TRANSMISSION

DRAIN & REFILL

Draining

1. Start the engine and let it run to warm up transmission.
2. Stop the engine.
3. Remove drain plug and gasket from transmission case and then drain gear oil.
4. Set a gasket on drain plug and install it to transmission case.
5. Never reuse gasket.
6. Tighten drain plug to 25 ft. lbs. (34.5 Nm).

Refilling

1. Remove filler plug and gasket from transmission case.
2. Fill with new gear oil to transmission.

✳✳ WARNING

Never reuse drained gear oil.

3. After refilling gear oil, check the oil level.
4. Set a gasket on filler plug and then install it to transmission case.

✳✳ WARNING

Never reuse gasket.

5. Tighten filler plug to 25 ft. lbs. (34.5 Nm).

FLUID LEVEL CHECK

See Figure 18.

1. Remove filler plug (1) and gasket from transmission case.
2. Check the oil level from filler plug mounting hole as shown.

✳✳ WARNING

Never start engine while checking oil level.

✳✳ WARNING

Set a gasket on filler plug and then install it to transmission case.

➡**Never reuse gasket.**

3. Tighten filler plug to 25 ft. lbs. (34.5 Nm).

FLUID RECOMMENDATIONS

Genuine NISSAN Manual Transmission Fluid (MTC) HQ Multi 75W-85 or API GL-4 Viscosity SAE 75W-85 is recommended.

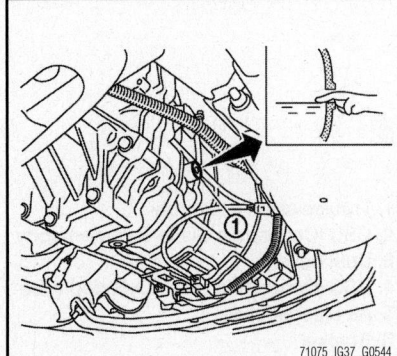

Fig. 18 Removing filler plug (1) and gasket from transmission case

CLUTCH

FLUID LEVEL CHECK

1. Check that the fluid level in the reservoir tank is within the specified range (MAX – MIN lines).
2. Visually check for any fluid leakage around the reservoir tank.
3. Check the clutch system for any leakage if the fluid level is extremely low (lower than MIN).

FLUID RECOMMENDATIONS

Genuine NISSAN Super Heavy Duty Brake Fluid or equivalent DOT 3 (US FMVSS No. 116).

➡**Available through an INFINITI dealer.**

BLEEDING PROCEDURE

➡**Monitor clutch fluid level in reservoir tank to make sure it does not empty.**

✳✳ WARNING

Keep painted surface on the body or other parts free of clutch fluid. If it spills, wipe up immediately and wash the affected area with water.

➡**Do not use a vacuum assist or any other type of power bleeder on this system. Use of vacuum assist or power bleeder will not purge all the air from the system.**

1. Fill master cylinder reservoir tank with new clutch fluid.

✳✳ WARNING

Never reuse drained clutch fluid.

2. Connect a transparent vinyl hose to air bleeder valve.
3. Depress clutch pedal slowly and fully several times at an interval of 2 to 3 seconds and hold it.
4. With clutch pedal depressed, loosen air bleeder valve to release air.
5. Tighten air bleeder valve.
6. Release clutch pedal and wait for 5 seconds.
7. Repeat steps 3 to 6 until no bubbles can be observed in clutch fluid.
8. Tighten air bleeder valve to the specified torque.
9. Check that the fluid level in the reservoir tank is within the specified range after air bleeding.

FLUID FILL PROCEDURE

> ⁕⁕ **WARNING**
>
> **Keep painted surface on the body or other parts free of clutch fluid. If it spills, wipe up immediately and wash the affected area with water.**

1. Check that there is no foreign material in reservoir tank and then fill with new clutch fluid.

> ⁕⁕ **WARNING**
>
> **Never reuse drained clutch fluid.**

2. Loosen air bleeder valve, slowly depress clutch pedal to the full stroke and then release clutch pedal.

3. Repeat this operation at intervals of 2 or 3 seconds until new clutch fluid is discharged.

4. Tighten air bleeder valve with the clutch pedal depressed.

5. Perform the air bleeding.

CONCENTRIC SLAVE (RELEASE) CYLINDER

REMOVAL & INSTALLATION

See Figures 19 through 24.

> ⁕⁕ **WARNING**
>
> **Never reuse CSC (Concentric Slave Cylinder) body and CSC tube. Because CSC slides back to the original position every time when removing transmission assembly. At this timing, dust on the sliding parts may damage a seal of CSC and may cause clutch fluid leakage.**

Fig. 19 Removing mounting bolt (arrow)

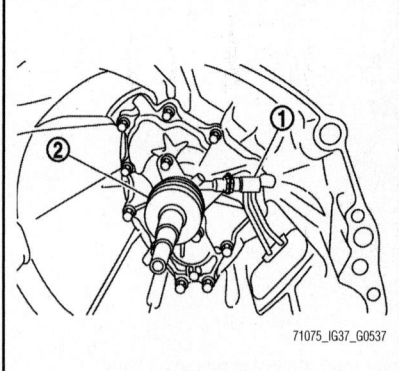

Fig. 20 Removing CSC tube (1) from the CSC body (2)

> ⁕⁕ **WARNING**
>
> **Never disassemble CSC body.**

> ⁕⁕ **WARNING**
>
> **Keep painted surface on the body or other parts free of clutch fluid. If it spills, wipe up immediately and wash the affected area with water.**

> ⁕⁕ **WARNING**
>
> **Remove transmission assembly from the engine.**

1. Remove mounting bolt (arrow).

2. Pull up the lock pin of the CSC body.

3. Pull out the CSC tube (1) from the CSC body (2).

4. Remove CSC tube and dust cover from transmission case.

5. Remove air bleeder valve and bracket from CSC tube.

6. Remove CSC body from transmission case.

To install:

7. Install CSC body to transmission case and then tighten mounting bolts (arrow) to the specified torque.

> ⁕⁕ **WARNING**
>
> **Never reuse CSC body.**

> ⁕⁕ **WARNING**
>
> **Never insert and operate CSC body because piston and stopper of CSC body components may fall off.**

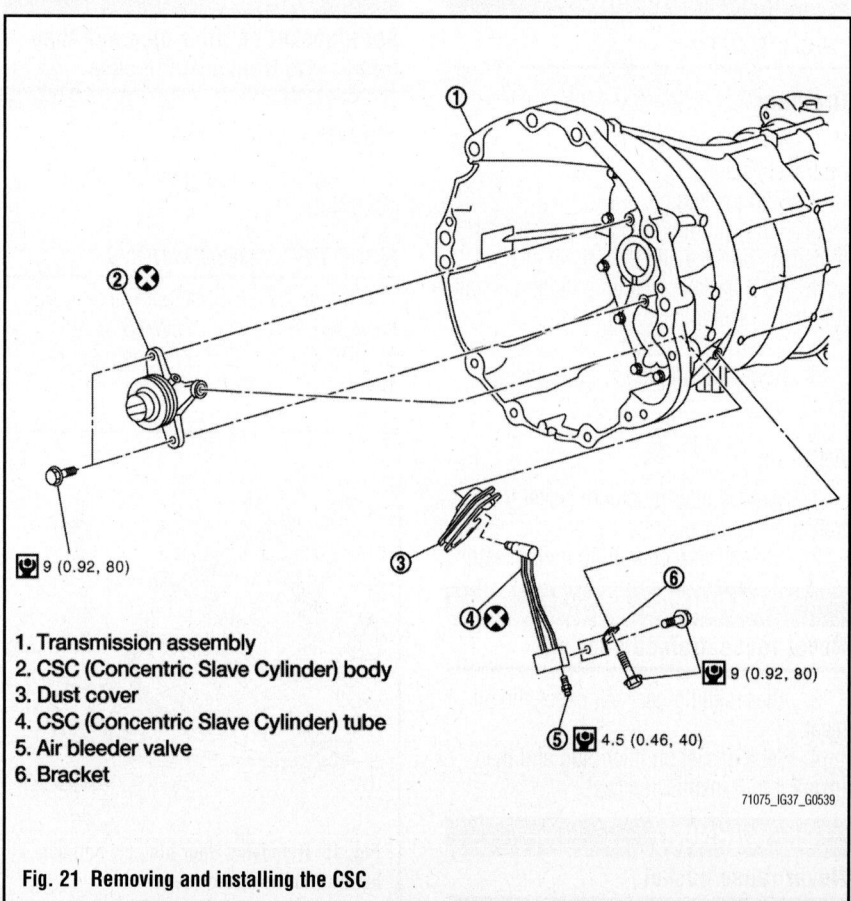

1. Transmission assembly
2. CSC (Concentric Slave Cylinder) body
3. Dust cover
4. CSC (Concentric Slave Cylinder) tube
5. Air bleeder valve
6. Bracket

Fig. 21 Removing and installing the CSC

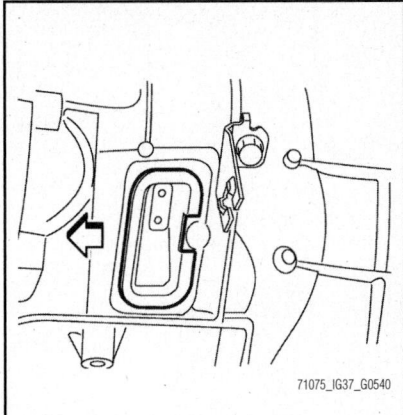

Fig. 22 Installing dust cover to transmission case (arrow: vehicle front)

8. Install dust cover to transmission case.

✳✳ WARNING
Be careful with the orientation of dust cover.

9. Insert CSC tube to dust cover.

✳✳ WARNING
Never reuse CSC tube.

✳✳ WARNING
Never damage O-ring of CSC tube.

10. Press down the lock pin of the CSC body.
11. Insert the CSC tube into the connector of the CSC body until it clicks.
12. Install bracket and mounting bolts and then tighten mounting bolts to the specified torque in the numerical order as shown.

Fig. 23 Identifying the mounting bolt tightening sequence

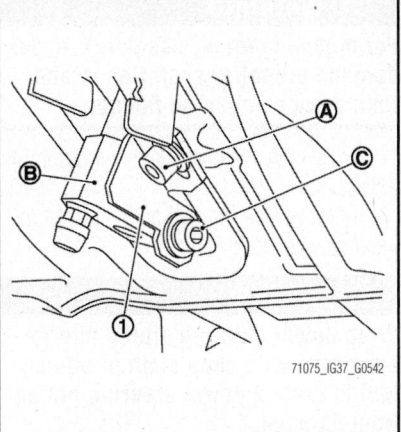

Fig. 24 Checking CSC tube (B), bracket (1) mounting bolt (A and C)

✳✳ WARNING
Check that CSC tube's (B) and bracket (1) are fit tightly before tightening the mounting bolt (A).

✳✳ WARNING
Tighten the mounting bolt within the range of bracket's mounting hole.

✳✳ WARNING
After replacing the CSC tube, the mounting bolt (C) is still temporary tightening. Never forget tightening the mounting bolt.

13. Install air bleeder valve to CSC tube and then tighten air bleeder valve to the specified torque.
14. Install transmission assembly to the engine.

DRIVESHAFT

REMOVAL & INSTALLATION

Front

2S56A—Coupe
See Figures 25 and 26.

1. Shift the transmission to the neutral position, and then release the parking brake.
2. Remove engine undercover with a power tool.
3. Remove exhaust front tube and three-way catalyst (bank 1).
4. Put matching mark on propeller shaft flange yoke and final drive companion flange.

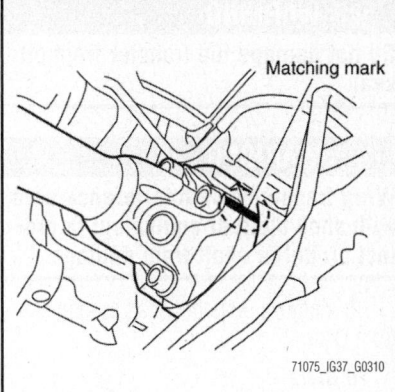

Fig. 25 Putting matching mark on propeller shaft flange yoke and final drive companion flange

✳✳ WARNING
For matching mark, use paint. Never damage propeller shaft flange and final drive companion flange.

5. Remove the propeller shaft assembly fixing bolts.
6. Move steering hydraulic line not to interfere with work.

✳✳ WARNING
Wrap power steering piping interference area with shop cloth or equivalent to protect power steering piping from damage.

7. Support transfer assembly with a jack, remove rear engine mounting member
8. Remove propeller shaft assembly from the front final drive and transfer.

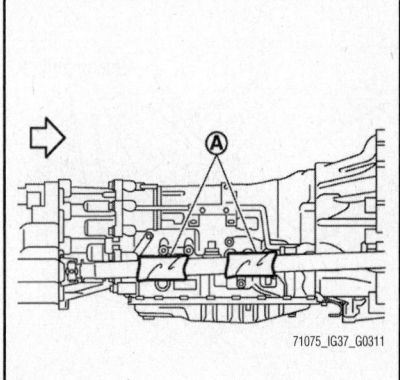

Fig. 26 Identifying transmission interference area (A)

※※ WARNING

Do not damage the transfer front oil seal.

※※ WARNING

Wrap transmission interference area with shop cloth or equivalent to protect propeller shaft from damage.

9. Remove propeller shaft assembly from O-ring.

To install:

Note the following, and install in the reverse order of removal.

10. Align matching mark to install propeller shaft assembly to final drive companion flange.

11. Perform inspection after installation.

12. Never damage the transfer front oil seal.

13. Wrap power steering piping interference area with shop cloth or equivalent to protect power steering piping from damage.

14. Wrap transmission interference area (A) with shop cloth or equivalent to protect propeller shaft from damage.

15. Never reuse O-ring.

16. Apply multi-purpose grease onto O-ring.

2S56A—Sedan

See Figures 27 through 29.

1. Shift the transmission to the neutral position, and then release the parking brake.

2. Remove engine undercover with a power tool.

3. Remove exhaust front tube and three-way catalyst (bank 1).

4. Put matching mark on propeller shaft flange yoke and final drive companion flange.

※※ WARNING

For matching mark, use paint. Never damage propeller shaft flange and final drive companion flange.

5. Remove the propeller shaft assembly fixing bolts.

6. Move steering hydraulic line not to interfere with work.

※※ WARNING

Wrap power steering piping interference area with shop cloth or equivalent to protect power steering piping from damage.

7. Support transfer assembly with a jack, remove rear engine mounting member.

8. Remove propeller shaft assembly from the front final drive and transfer.

※※ WARNING

Do not damage the transfer front oil seal.

※※ WARNING

Wrap transmission interference area (A) with shop cloth or equivalent to protect propeller shaft from damage.

9. Remove propeller shaft assembly from O-ring.

To install:

Note the following, and install in the reverse order of removal.

10. Align matching mark to install propeller shaft assembly to final drive companion flange.

11. After assembly, perform a driving test to check propeller shaft vibration. If vibration occurred, separate propeller shaft

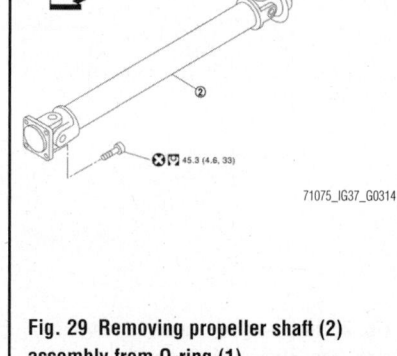

Fig. 29 Removing propeller shaft (2) assembly from O-ring (1)

from final drive. Reinstall companion flange after rotating it by 90, 180, 270 degrees. Then perform driving test and check propeller shaft vibration again at each point.

※※ WARNING

Do not damage the transfer front oil seal.

※※ WARNING

Wrap power steering piping interference area with shop cloth or equivalent to protect power steering piping from damage.

※※ WARNING

Wrap transmission interference area (A) with shop cloth or equivalent to protect propeller shaft from damage.

※※ WARNING

Never reuse O-ring.

※※ WARNING

Apply multi-purpose grease onto O-ring.

Rear

3S80A & 3S80A-R

See Figures 30 through 36.

1. Move the A/T selector lever to N position and release the parking brake.

2. Remove the floor reinforcement.

3. Remove the center muffler with power tool.

4. Remove the heat plate.

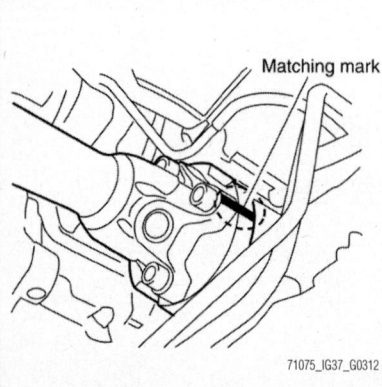

Fig. 27 Identifying matching mark on propeller shaft flange yoke and final drive companion flange

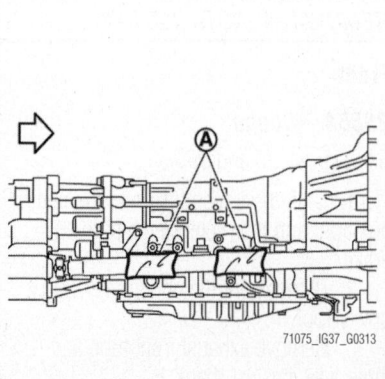

Fig. 28 Identifying transmission interference area (A)

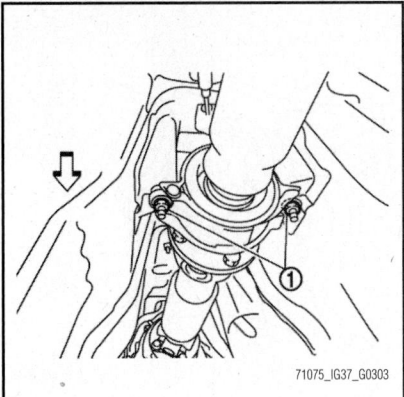

Fig. 30 Loosening mounting nuts (1) of center bearing mounting brackets

5. Put matching marks on propeller shaft rubber coupling with final drive companion flange.

※ **WARNING**

For matching marks, use paint. Never damage propeller shaft rubber coupling and final drive companion flange.

6. Loosen mounting nuts of center bearing mounting brackets.

※ **WARNING**

Tighten mounting nuts temporarily.

7. Remove propeller shaft assembly fixing bolts and nuts.

※ **WARNING**

Never remove the rubber coupling from the propeller shaft assembly.

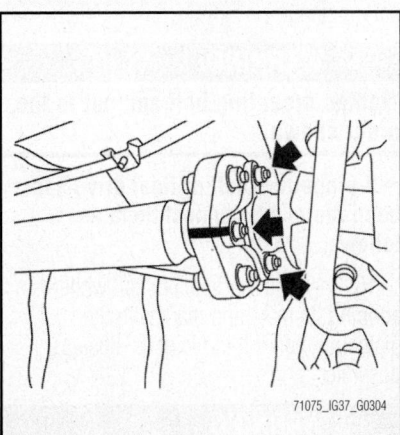

Fig. 31 Removing propeller shaft assembly fixing bolts and nuts

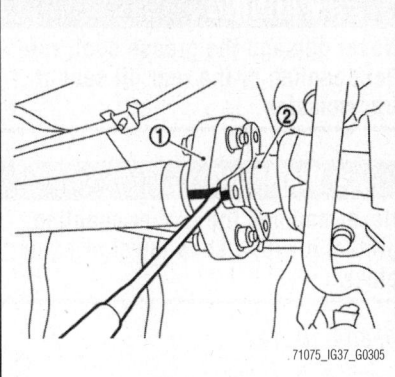

Fig. 32 Separating rubber coupling (1) from the final drive companion flange (2)

8. Slightly separate the rubber coupling from the final drive companion flange.

※ **WARNING**

Never damage the final drive companion flange and rubber coupling.

9. Remove center bearing mounting bracket fixing nuts.

※ **WARNING**

The angle, which the third axis rubber coupling forms with the final drive companion flange, must be 5° or less.

※ **WARNING**

Never damage the grease seal or rubber coupling.

10. Slide the propeller shaft in the vehicle forward direction slightly. Separate the propeller shaft from the final drive companion flange.

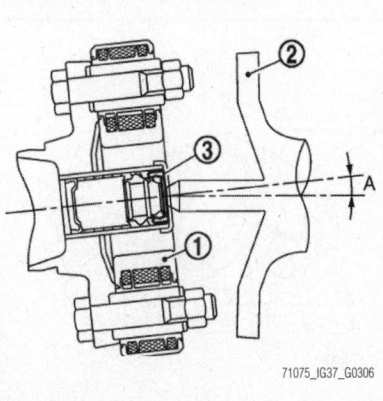

Fig. 33 Identifying the angle (A), which the third axis rubber coupling (1) forms with the final drive companion flange (2) and the grease seal (3)

peller shaft from the final drive companion flange.

※ **WARNING**

The angle, which the third axis rubber coupling forms with the final drive companion flange, must be 5° or less.

※ **WARNING**

Do not damage the grease seal or rubber coupling.

11. Remove the propeller shaft assembly from the vehicle.

※ **WARNING**

Do not damage the rear oil seal of transmission.

To install:

Note the following, and install in the reverse order of removal.

- Install center bearing mounting bracket (upper) with its arrow mark facing forward.
- Adjust position of center bearing mounting bracket (upper) and center bearing mounting bracket (lower) sliding back and forth to prevent play in thrust direction of center bearing insulator. Install bracket to vehicle.
- Align matching marks to install propeller shaft rubber coupling to final drive companion flange.
- Perform inspection after installation.
- If propeller shaft or final drive has been replaced, connect them as follows:

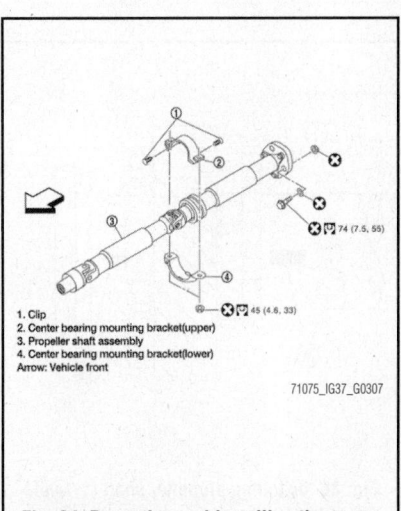

1. Clip
2. Center bearing mounting bracket(upper)
3. Propeller shaft assembly
4. Center bearing mounting bracket(lower)
Arrow: Vehicle front

Fig. 34 Removing and installing the rear propeller shaft—3S80A and 3S80A-R

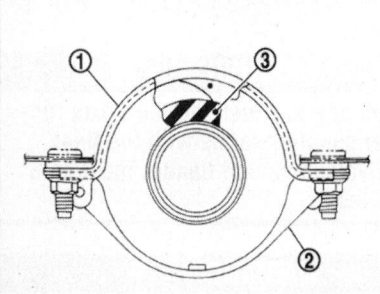

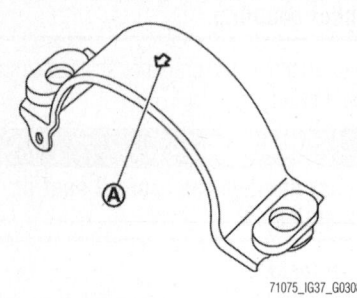

71075_IG37_G0308

Fig. 35 Installing center bearing mounting bracket (upper (1) with arrow mark (A) facing forward and positioning center bearing mounting bracket (upper) (1) and (lower) (2) to prevent play in thrust direction of center bearing insulator (3)

12. Install the propeller shaft (1) while aligning its matching mark (A) with the matching mark (B) on the joint as close as possible.

⁂ WARNING

The angle, which the third axis rubber coupling forms with the final drive companion flange, must be 5° or less.

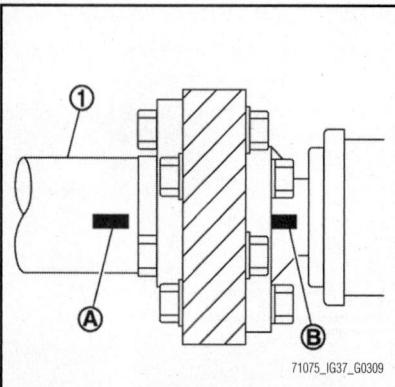

71075_IG37_G0309

Fig. 36 Installing propeller shaft (1) while aligning its matching mark (A) with the matching mark (B) on the joint

⁂ WARNING

Never damage the grease seal, rubber coupling or the rear oil seal of transmission.

⁂ WARNING

Never damage the rubber coupling, protect it with a shop towel or equivalent.

3F80A-1VL107

See Figures 37 through 41.

1. Shift the transmission to the neutral position, and release the parking brake.
2. Remove the center muffler and exhaust front tube with the power tool.
3. Remove the heat plate.
4. Using paint, put matching marks on propeller shaft flange yoke and transfer companion flange.
5. Using paint, put matching marks on propeller shaft rebro joint and final drive companion flange.
6. Loosen mounting nuts of center bearing mounting brackets (upper/lower).

⁂ WARNING

Tighten mounting nuts temporarily.

7. Remove propeller shaft assembly fixing bolts and nuts.
8. Remove center bearing mounting bracket fixing nuts.
9. Remove propeller shaft assembly being careful not to damage the transmission rear oil seal.

➡If constant velocity joint was bent during propeller shaft assembly removal, installation, or transportation, its boot may be

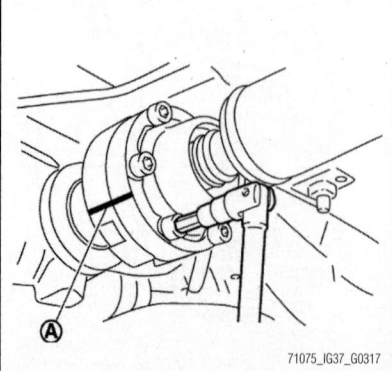

71075_IG37_G0317

Fig. 37 Putting marks (A) on propeller shaft rebro joint and final drive companion flange

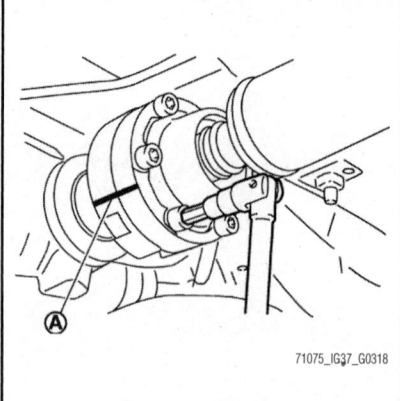

71075_IG37_G0318

Fig. 38 Loosening the mounting nuts (1) of center bearing mounting brackets (upper/lower)

damaged. Wrap boot interference area to metal part with shop cloth or equivalent to protect boot from breakage.

10. Remove clip and center bearing mounting bracket (upper/lower).

To install:
Note the following, and install in the reverse order of removal.

11. Install center bearing mounting bracket (upper) with its arrow mark facing forward.
12. Adjust position of center bearing mounting bracket (upper), center bearing mounting bracket (lower) sliding back and forth to prevent play in the thrust direction of center bearing insulator. Install center bearing mounting bracket (upper/lower) to vehicle.
13. Align matching marks to install propeller shaft flange yoke and transfer companion flange.
14. Align matching marks to install propeller shaft rebro joint and final drive companion flange.

⁂ WARNING

Tighten mounting bolt and nut in the order shown.

➡If propeller shaft or final drive has been replaced, connect them as follows:

15. Install the propeller shaft while aligning its matching mark with the matching mark on the joint as close as possible.

⁂ WARNING

Avoid damaging the rebro joint boot, protect it with a shop cloth or equivalent.

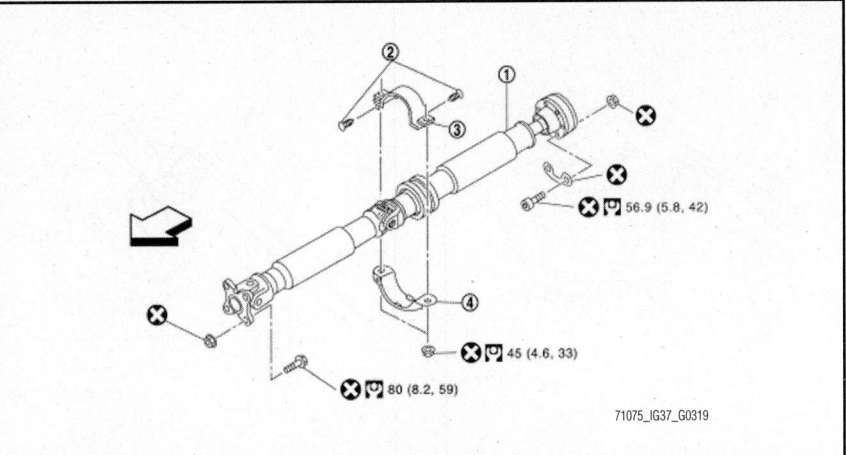

Fig. 39 Locating propeller shaft (1), clip (2), center bearing mounting bracket (upper), and center bearing mounting bracket (lower) (arrow: vehicle front)

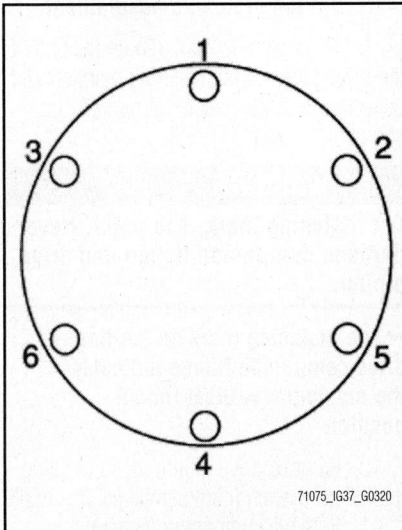

Fig. 40 Bolt tightening sequence

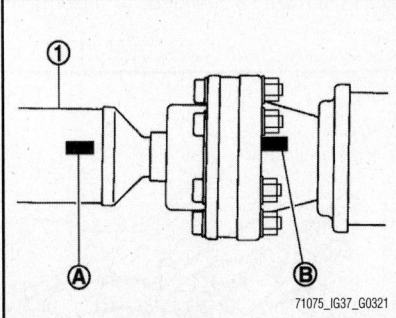

Fig. 41 Installing propeller shaft (1) while aligning its matching mark (A) with the matching mark (B) on the joint as close as possible

REAR HALFSHAFT

REMOVAL & INSTALLATION

See Figure 42.

Follow the directions on the screen of the tool, as needed.

1. Before servicing the vehicle, refer to the Precautions Section.

➡If working near and/or around the SRS system and components, be sure to disable the SRS system. After disabling the system wait three minutes or more before servicing the vehicle.

2. Disconnect the negative battery cable.

3. Raise and support the vehicle safely.

4. Remove the tire and wheel assemblies.

5. Remove and discard the cotter pin.

6. Loosen the wheel hub locknut.

7. Matchmark the halfshaft and wheel hub and bearing.

8. On convertible remove the diag. brace.

9. Remove the main muffler and center muffler, on coupe and sedan.

10. Remove the center muffler, convertible.

11. Patch wheel hub locknut with a piece of wood. Hammer the wood to disengage the wheel hub and bearing assembly from the halfshaft.

➡Whenever the negative battery cable is disconnected the following components will require resetting. The Automatic temperature control system, Automatic drive positioner, Power window control, Sunroof system, Sunshade system, Rear view monitor, Idle Air Volume Learning, Steering Angle Sensor Neutral Position, Audio presets and Navigation. You will need the CONSULT-III diagnostic tool, or equivalent.

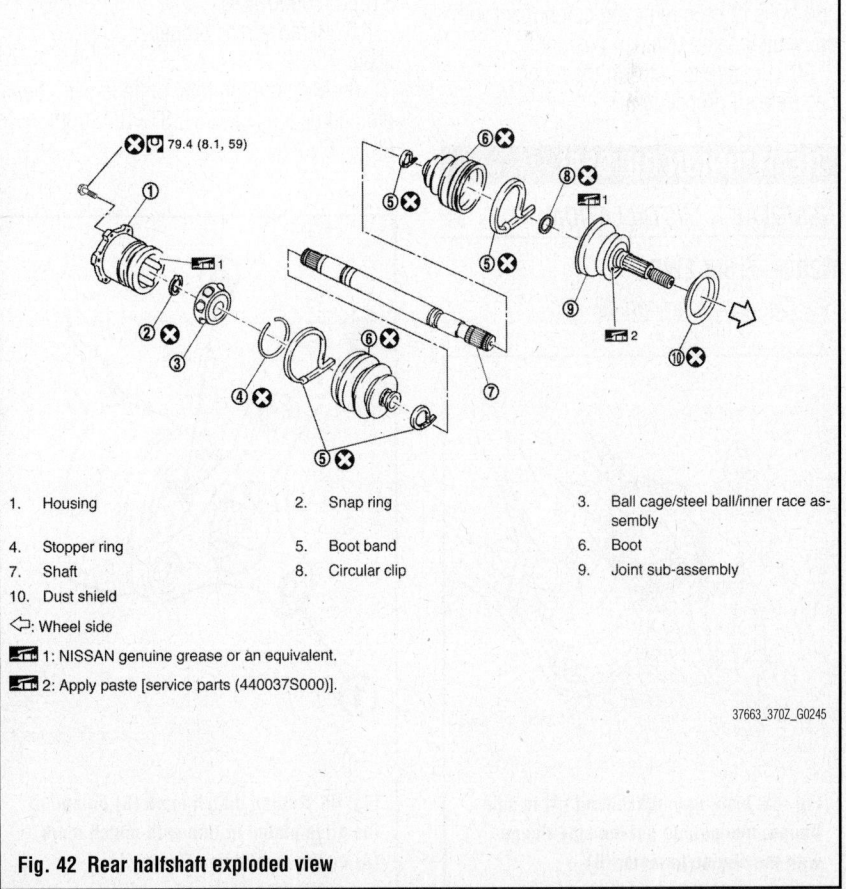

1. Housing	2. Snap ring	3. Ball cage/steel ball/inner race assembly
4. Stopper ring	5. Boot band	6. Boot
7. Shaft	8. Circular clip	9. Joint sub-assembly
10. Dust shield		

◁: Wheel side

▦ 1: NISSAN genuine grease or an equivalent.

▦ 2: Apply paste [service parts (440037S000)].

Fig. 42 Rear halfshaft exploded view

➡Never position the halfshaft at an extreme angle. Do not overextend the slide joint. Properly support the halfshaft, do not allow it to hang unsupported.

12. Use a suitable puller if the wheel hub and bearing assembly and halfshaft cannot be separated even after performing the above procedure.

13. Remove the wheel locknut.

14. Remove the mounting bolts between the side flange and the halfshaft.

To install:

➡Be sure to use new fasteners, as required.

15. Installation is the reverse of the removal procedure.

16. Clean the matching surface of the halfshaft and wheel hub and bearing assembly.

17. Apply paste part number 440037S000 or equivalent to the surface of the sub joint assembly of the halfshaft.

➡Apply the paste, about 0.04–0.10 ounce) to cover the entire flat surface of the sub joint assembly of the halfshaft.

18. The wheel hub locknut tightening specification is 136 ft. lbs. (185 Nm).

19. Perform a final tightening of nuts and bolts of each removed component with the vehicle in an unladen position.

20. Be sure to perform the reconnect/relearn procedures.

REAR PINION OIL SEAL

REMOVAL & INSTALLATION

R200—Front 2WD

See Figures 43 through 49.

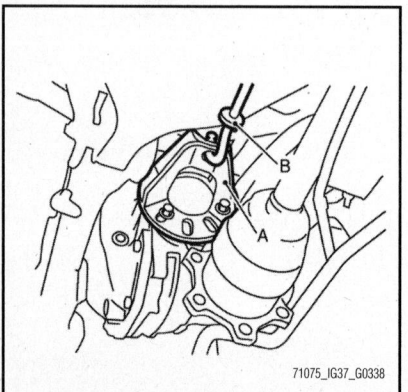

Fig. 43 Installing attachment (A) to side flange, and pulling out the side flange with the sliding hammer (B)

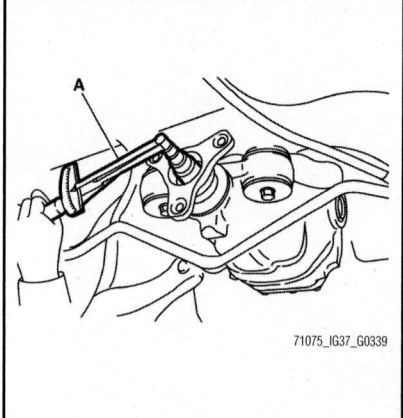

Fig. 44 Measuring the total preload with the preload gauge (A)

1. Drain gear oil.

2. Make a judgment if a collapsible spacer replacement is required.

3. Remove center muffler with a power tool.

4. Remove rear wheel sensor.

5. Remove drive shaft from final drive. Then suspend it by wire, etc.

6. Install attachment (A: attachment [SST: KV40104100]) to side flange, and then pull out the side flange with the sliding hammer (B: sliding hammer [SST: ST362310000]).

7. Remove rear propeller shaft.

8. Measure the total preload with the preload gauge (A) [SST: ST3127S000 (J-25765-A)].

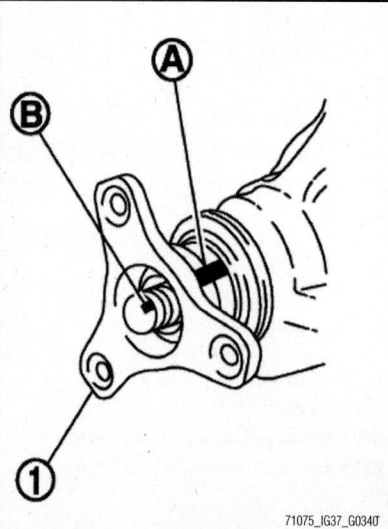

Fig. 45 Putting match mark (B) on end of the drive pinion in line with match mark (A) on the companion flange (1)

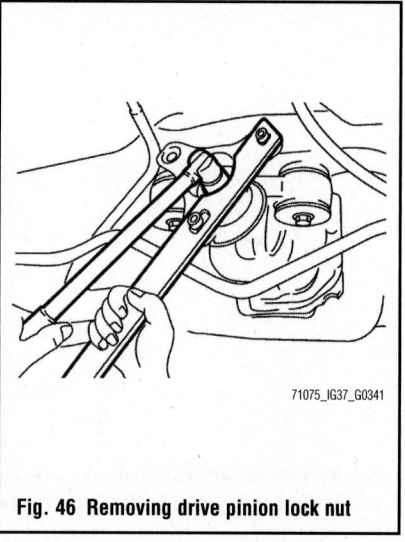

Fig. 46 Removing drive pinion lock nut

➡Record the preload measurement.

9. Put matching mark (B) on the end of the drive pinion. The matching mark should be in line with the matching mark (A) on companion flange (1).

❋❋ WARNING

For matching mark, use paint. Never damage companion flange and drive pinion.

➡The matching mark on the final drive companion flange indicates the maximum vertical runout position.

10. Remove drive pinion lock nut using the flange wrench (commercial service tool).

11. Remove companion flange using pullers (commercial service tool).

12. Remove front oil seal using the puller (A) [SST: KV381054S0 (J-34286)].

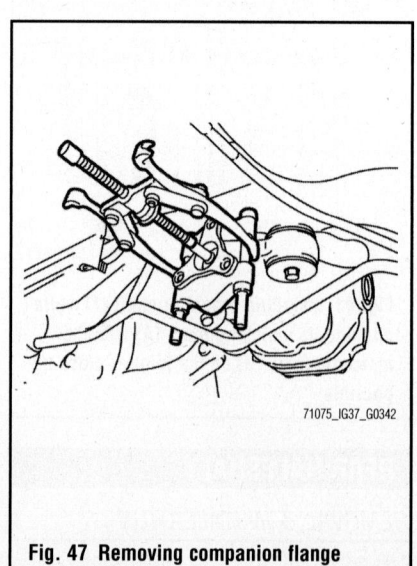

Fig. 47 Removing companion flange

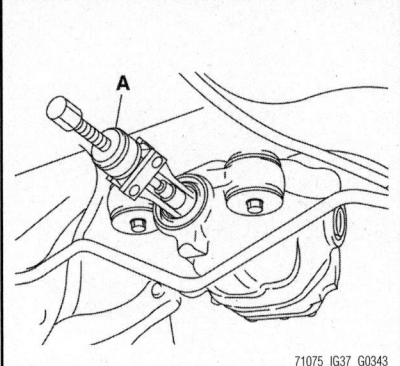

Fig. 48 Removing front oil seal using the puller (A)

71075_IG37_G0343

To install:

13. Apply multi-purpose grease to front oil seal lips.

14. Install front oil seal using the drift (A) [SST: ST30720000 (J- 25405)].

➡️**Do not reuse oil seal.**

❄ WARNING

Never incline oil seal when installing.

15. Align the matching mark of drive pinion with the matching mark of companion flange, and then install the companion flange.

16. Apply anti-corrosion oil to the thread and seat of new drive pinion lock nut, and temporarily tighten drive pinion lock nut to drive pinion, using flange wrench (commercial service tool).

❄ WARNING

Never reuse drive pinion lock nut.

17. Tighten drive pinion lock nut within the limits of specified torque so as to keep the pinion bearing preload within a standard values, using preload gauge [SST: ST3127S000 (J-25765-A)].

❄ WARNING

Adjust to the lower limit of the drive pinion lock nut tightening torque first.

❄ WARNING

If the preload torque exceeds the specified value, replace collapsible spacer and tighten it again to adjust. Never loosen drive pinion lock nut to adjust the preload torque.

18. Set a dial indicator vertically to the tip of the drive pinion.

19. Rotate drive pinion to check for runout.

   a. If the runout value is still outside of the limit after the phase has been changed, possible causes are an assembly malfunction of drive pinion and pinion bearing and malfunction of pinion bearing. Check for these items and repair if necessary.

20. Make a stamping for identification of front oil seal replacement frequency.

❄ WARNING

Make a stamping after replacing front oil seal.

21. Install rear propeller shaft.

22. Install side flange with the following procedure.

   a. Attach the protector [SST: KV38107900 (J-39352)] to side oil seal.

   b. After the side flange is inserted and the serrated part of side gear has engaged the serrated part of flange, remove the protector.

   c. Put a suitable drift on the center of side flange, then drive it until sound changes.

➡️**When installation is completed, driving sound of the side flange turns into a sound that seems to affect the whole final drive.**

   d. Confirm that the dimension of the side flanges (1) installation measurement (A) in the figure comes into the following.

23. Install drive shaft.

24. Install rear wheel sensor.

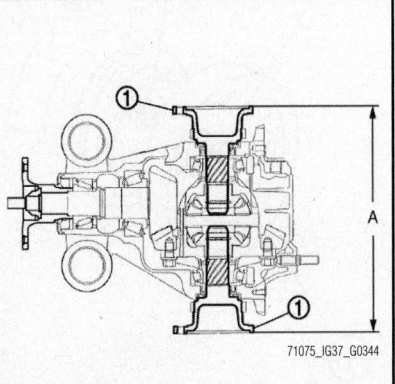

Fig. 49 Confirming the dimension of the side flanges (1) installation measurement (A)

71075_IG37_G0344

25. Install center muffler.

26. Refill gear oil to the final drive and check oil level.

27. Check the final drive for oil leakage.

R200—Front AWD

See Figures 50 through 55.

❄ WARNING

Verify identification stamp of replacement frequency put in the lower part of gear carrier to determine replacement for collapsible spacer when replacing front oil seal. Refer to "Identification stamp of replacement frequency of front oil seal". If collapsible spacer replacement is necessary, remove final drive assembly and disassemble it to replace front oil seal and collapsible spacer.

➡️**The reuse of collapsible spacer is prohibited in principle. However, it is reusable on a one-time basis only in cases when replacing front oil seal.**

Identification stamp of replacement frequency of front oil seal:

• The diagonally shaded area in the figure shows stamping point for replacement frequency of front oil seal.

• The following table shows if collapsible spacer replacement is needed before replacing front oil seal.

When collapsible spacer replacement is required, disassemble final drive assembly to replace collapsible spacer and front oil seal.

Stamp: collapsible spacer replacement:

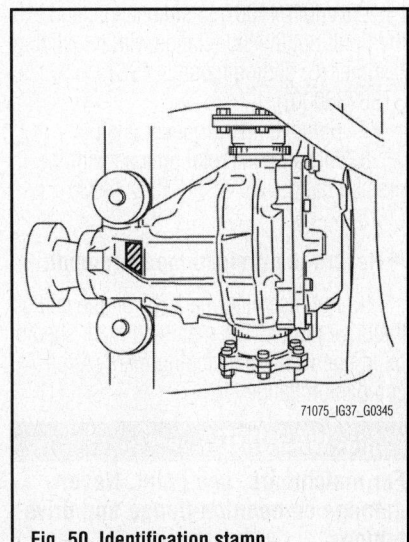

71075_IG37_G0345

Fig. 50 Identification stamp

Stamp before stamping	Stamping on the far right	Stamping
No stamp	0	0
"0" (Front oil seal was replaced once.)	1	01
"01" (Collapsible spacer and front oil seal were replaced last time.)	0	010
"0" is on the far right. (Only front oil seal was replaced last time.)	1	...01
"1" is on the far right. (Collapsible spacer and front oil seal were replaced last time.)	0	...010

71075_IG37_G0346

Fig. 51 Stamping frequency

- No stamp: Not required
- "0" or "0" on the far right stamp: Required
- "01" or "1" on the far right side of stamp: Not required

❋❋ WARNING

Make a stamping after replacing front oil seal.

❋❋ WARNING

After replacing front oil seal, make a stamping on the stamping point in accordance with the table below in order to identify replacement frequency.

1. Drain gear oil.
2. Make a judgment if a collapsible spacer replacement is required.
3. Remove center muffler with a power tool.
4. Remove rear wheel sensor.
5. Remove drive shaft from final drive. Then suspend it by wire, etc.
6. Install attachment (A: attachment [SST: KV40104100]) to side flange, and then pull out the side flange with the sliding hammer (B: sliding hammer [SST: ST36230000]).
7. Remove rear propeller shaft.
8. Measure the total preload with the preload gauge (A) [SST: ST3127S000 (J-25765-A)].

➡Record the preload measurement.

9. Put matching mark (B) on the end of the drive pinion. The matching mark should be in line with the matching mark (A) on companion flange (1).

❋❋ WARNING

For matchmark, use paint. Never damage companion flange and drive pinion.

➡The matching mark on the final drive companion flange indicates the maximum vertical runout position.

10. Remove drive pinion lock nut using the flange wrench (commercial service tool).
11. Remove companion flange using puller (commercial service tool).
12. Remove front oil seal using the puller (A) [SST: KV381054S0 (J-34286)].

To install:

13. Apply multi-purpose grease to front oil seal lips.
14. Install front oil seal using the drift (A) [SST: ST30720000 (J-25405)].

❋❋ WARNING

Never reuse oil seal.

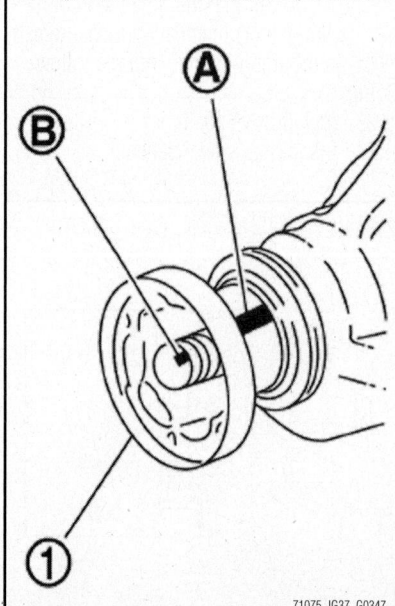

71075_IG37_G0347

Fig. 52 Identifying match mark (B) on the end of the drive pinion and (A) on the companion flange (1)

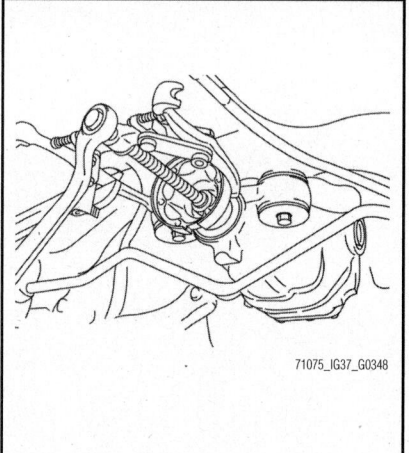

71075_IG37_G0348

Fig. 53 Removing companion flange using puller

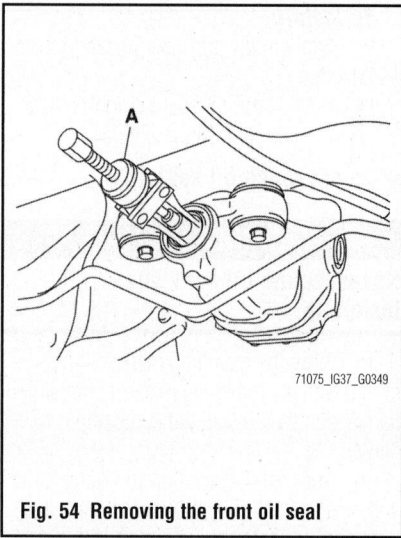

71075_IG37_G0349

Fig. 54 Removing the front oil seal

❋❋ WARNING

Never incline oil seal when installing.

15. Align the matching mark of drive pinion with the matching mark of companion flange, and then install the companion flange.
16. Apply anti-corrosion oil to the thread and seat of new drive pinion lock nut, and temporarily tighten drive pinion lock nut to drive pinion, using flange wrench (commercial service tool).

❋❋ WARNING

Never reuse drive pinion lock nut.

17. Tighten drive pinion lock nut within the limits of specified torque so as to keep the pinion bearing preload within a standard values, using preload gauge [SST: ST3127S000 (J-25765-A)].

Standard Total Preload Torque:

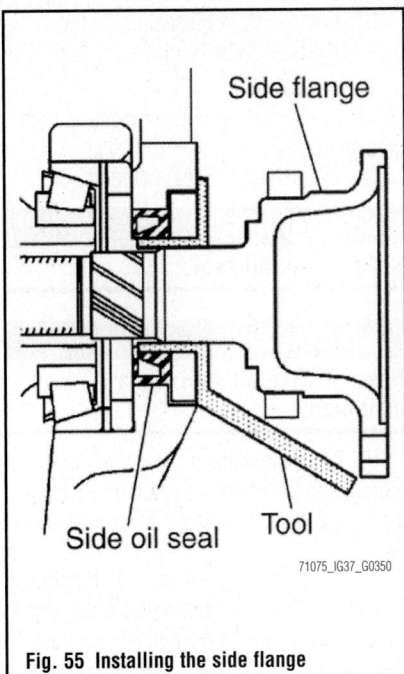

Fig. 55 Installing the side flange

- A value that add 0.9-3.5 inch lbs. (0.1-0.4 Nm) to the measured value before removing.

※ WARNING

Adjust to the lower limit of the drive pinion lock nut tightening torque first.

※ WARNING

If the preload torque exceeds the specified value, replace collapsible spacer and tighten it again to adjust. Never loosen drive pinion lock nut to adjust the preload torque.

18. Fit a test indicator to the inner side of companion flange (socket diameter).
19. Rotate companion flange to check for runout.
   Limit: Companion flange runout:
   - Companion flange face runout: 0.0031 in. (0.08mm)
   - Inner side of the companion flange runout: 0.0031 in. (0.08mm)
20. If the runout value is outside the runout limit, follow the procedure below to adjust.
   a. Check for runout while changing the phase between companion flange and drive pinion by 90° step, and search for the position where the runout is the minimum.
   b. If the runout value is still outside of the limit after the phase has been changed, possible cause will be an assembly malfunction of drive pinion and

pinion bearing and malfunction of pinion bearing. Check for these items and repair if necessary.
   c. If the runout value is still outside of the limit after the check and repair, replace companion flange.
21. Make a stamping for identification of front oil seal replacement frequency. Refer to "Identification stamp of replacement frequency of front oil seal".

※ WARNING

Make a stamping after replacing front oil seal.

22. Install rear propeller shaft.
23. Install side flange with the following procedure.
   a. Attach the protector [SST: KV38107900 (J-39352)] to side oil seal.
   b. After the side flange is inserted and the serrated part of side gear has engaged the serrated part of flange, remove the protector.
   c. Put a suitable drift on the center of side flange, then drive it until sound changes.

➡ **When installation is completed, driving sound of the side flange turns into a sound that seems to affect the whole final drive.**

   d. Confirm that the dimension of the side flanges installation measurement (A) in the figure comes into the following.
   - Standard A: 12.83-12.91 inch (326-328 mm)
24. Install drive shaft.
25. Install rear wheel sensor.
26. Install center muffler.
27. Refill gear oil to the final drive and check oil level.
28. Check the final drive for oil leakage.

R200V—Front 2WD (M/T)

See Figures 56 through 59.

※ WARNING

Verify identification stamp of replacement frequency put in the lower part of gear carrier to determine replacement for collapsible spacer when replacing front oil seal. Refer to "Identification stamp of replacement frequency of front oil seal". If collapsible spacer replacement is necessary, remove final drive assembly and disassemble it to replace front oil seal and collapsible spacer.

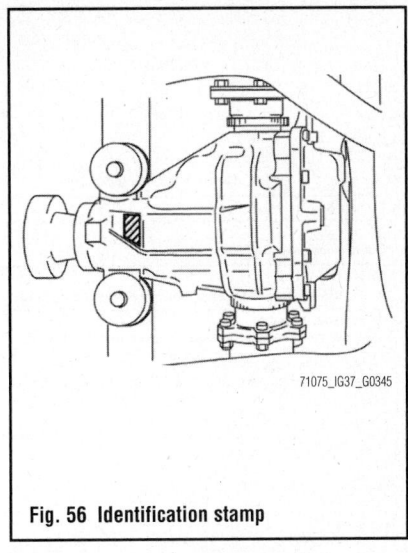

Fig. 56 Identification stamp

➡ **The reuse of collapsible spacer is prohibited in principle. However, it is reusable on a one-time basis only in cases when replacing front oil seal.**

   Identification stamp of replacement frequency of front oil seal:
   - The diagonally shaded area in the figure shows stamping point for replacement frequency of front oil seal.
   - The following table shows if collapsible spacer replacement is needed before replacing front oil seal.
   When collapsible spacer replacement is required, disassemble final drive assembly to replace collapsible spacer and front oil seal.
   Stamp: collapsible spacer replacement:
   - No stamp: Not required
   - "0"or "0"on the far right stamp: Required
   - "01"or "1"on the far right side of stamp: Not required

※ WARNING

Make a stamping after replacing front oil seal.

※ WARNING

After replacing front oil seal, make a stamping on the stamping point in accordance with the table below in order to identify replacement frequency.

1. Drain gear oil.
2. Make a judgment if a collapsible spacer replacement is required.
3. Remove center muffler with a power tool.

Stamp before stamping	Stamping on the far right	Stamping
No stamp	0	0
"0" (Front oil seal was replaced once.)	1	01
"01" (Collapsible spacer and front oil seal were replaced last time.)	0	010
"0" is on the far right. (Only front oil seal was replaced last time.)	1	...01
"1" is on the far right. (Collapsible spacer and front oil seal were replaced last time.)	0	...010

71075_IG37_G0346

Fig. 57 Stamping frequency

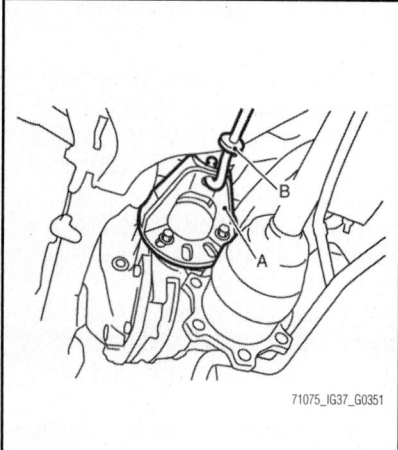

71075_IG37_G0351

Fig. 58 Installing attachment (A) to side flange, and pulling out side flange with the sliding hammer (B)

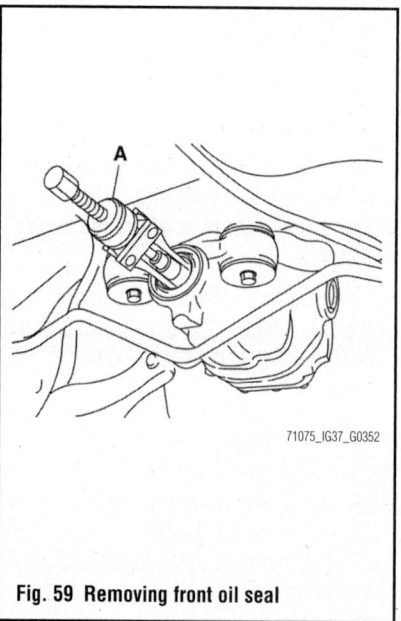

71075_IG37_G0352

Fig. 59 Removing front oil seal

4. Remove rear wheel sensor.
5. Remove drive shaft from final drive. Then suspend it by wire, etc.
6. Install attachment (A: attachment [SST: KV40104100])) to side flange, and then pull out the side flange with the sliding hammer (B: sliding hammer [SST: ST36230000]).

➡**Circular clip installation position: Final drive side**

7. Remove propeller shaft.
8. Measure the total preload with the preload gauge (A) [SST: ST3127S000 (J-25765-A)].

➡**Record the preload measurement.**

9. Put matching mark on the end of the drive pinion. The matching mark should be in line with the matching mark on companion flange.

※※ WARNING

For matching mark, use paint. Never damage companion flange and drive pinion.

➡**The matching mark on the final drive companion flange indicates the maximum vertical runout position.**

10. Remove drive pinion lock nut using the flange wrench (commercial service tool).
11. Remove companion flange using a puller (commercial service tool).

12. Remove front oil seal using the puller (A) [SST: KV381054S0 (J- 34286)].

To install:

13. Apply multi-purpose grease to front oil seal lips.
14. Install front oil seal using the drift.

※※ WARNING

Never reuse oil seal.

※※ WARNING

Never incline oil seal when installing.

15. Align the matching mark of drive pinion with the matching mark of companion flange, and then install the companion flange.
16. Apply anti-corrosion oil to the thread and seat of new drive pinion lock nut, and temporarily tighten drive pinion lock nut to drive pinion, using flange wrench (commercial service tool).

※※ WARNING

Never reuse drive pinion lock nut.

17. Tighten drive pinion lock nut within the limits of specified torque so as to keep the pinion bearing preload within a standard values, using [SST: ST3127S000 (J-25765-A)].

※※ WARNING

Adjust to the lower limit of the drive pinion lock nut tightening torque first.

※※ WARNING

If the preload torque exceeds the specified value, replace collapsible spacer and tighten it again to adjust. Never loosen drive pinion lock nut to adjust the preload torque.

Total preload torque:
• A value that add 0.01-0.04 kg m (0.1-0.4 Nm)

18. Fit a dial indicator onto the companion flange face (inner side of the propeller shaft mounting bolt holes).
19. Rotate the companion flange to check for runout.
20. Fit a test indicator to the inner side of the companion flange (socket diameter).
21. Rotate the companion flange to check for runout.
22. If the runout value is outside the repair limit, follow the procedure below to adjust.

a. Check for runout while changing the phase between companion flange and drive pinion gear by 90° step, and search for the position where the runout is the minimum.

b. If the runout value is still outside of the limit after the phase has been changed, possible causes are be an assembly malfunction of drive pinion and pinion bearing and malfunction of pinion bearing. Check for these items and repair if necessary.

c. If the runout value is still outside of the limit after the check and repair, replace companion flange.

23. Make a stamping for identification of front oil seal replacement frequency.

❄❄ WARNING

Make a stamping after replacing front oil seal.

24. Install propeller shaft.
25. Install side flange with the following procedure.

a. Attach the protector [SST: KV38107900 (J-39352)] to side oil seal.

b. After the side flange is inserted and the serrated part of side gear has engaged the serrated part of flange, remove the protector.

c. Put a suitable drift on the center of side flange, then drive it until sound changes.

➡ **When installation is completed, driving sound of the side flange turns into a sound that seems to affect the whole final drive.**

d. Confirm that the dimension of the side flange installation (Measurement A) in the figure comes into the following. Measurement A: 12.83-12.91 inch (326-328 mm)
26. Install drive shaft.
27. Install rear wheel sensor.
28. Install center muffler.
29. Refill gear oil to the final drive and check oil level.
30. Check the final drive for oil leakage.

R200V—Front 2WD (A/T)

See Figures 57, 60 and 61.

❄❄ WARNING

Verify identification stamp of replacement frequency put in the lower part of gear carrier to determine replace-

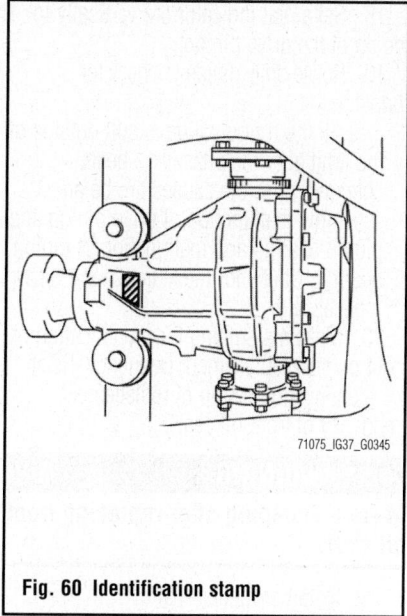

Fig. 60 Identification stamp

ment for collapsible spacer when replacing front oil seal. Refer to "Identification stamp of replacement frequency of front oil seal". If collapsible spacer replacement is necessary, remove final drive assembly and disassemble it to replace front oil seal and collapsible spacer.

➡ **The reuse of collapsible spacer is prohibited in principle. However, it is reusable on a one-time basis only in cases when replacing front oil seal.**

Identification stamp of replacement frequency of front oil seal:

• The diagonally shaded area in the figure shows stamping point for replacement frequency of front oil seal.

• The following table shows if collapsible spacer replacement is needed before replacing front oil seal.

When collapsible spacer replacement is required, disassemble final drive assembly to replace collapsible spacer and front oil seal.

Stamp: collapsible spacer replacement:

• No stamp: Not required
• "0" or "0" on the far right stamp: Required
• "01" or "1" on the far right side of stamp: Not required

❄❄ WARNING

Make a stamping after replacing front oil seal.

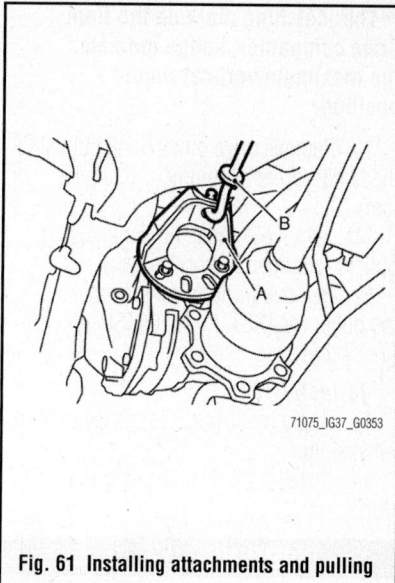

Fig. 61 Installing attachments and pulling out side flange

❄❄ WARNING

After replacing front oil seal, make a stamping on the stamping point in accordance with the table below in order to identify replacement frequency.

1. Drain gear oil.
2. Make a judgment if a collapsible spacer replacement is required.
3. Remove center muffler with a power tool.
4. Remove rear wheel sensor.
5. Remove drive shaft from final drive with a power tool. Then suspend it by wire, etc.
6. Install attachment (A: attachment [SST: KV40104100]) to side flange, and then pull out the side flange with the sliding hammer (B: sliding hammer [SST: ST36230000]).

➡ **Circular clip installation position: Final drive side**

7. Remove propeller shaft.
8. Measure the total preload with the preload gauge (A) [SST: ST3127S000 (J-25765-A)].

➡ **Record the preload measurement.**

9. Put matching mark on the end of the drive pinion. The matching mark should be in line with the matching mark on companion flange.

❄❄ WARNING

For matching mark, use paint. Never damage companion flange and drive pinion.

➡ **The matching mark on the final drive companion flange indicates the maximum vertical runout position.**

10. Remove drive pinion lock nut using the flange wrench (commercial service tool).

11. Remove companion flange using pullers (commercial service tool).

12. Remove front oil seal using the puller (A) [SST: KV381054S0 (J- 34286)].

To install:

13. Apply multi-purpose grease to front oil seal lips.

14. Install front oil seal using the drift.

✳✳ WARNING
Never reuse oil seal.

✳✳ WARNING
Never incline oil seal when installing.

15. Align the matching mark of drive pinion with the matching mark of companion flange, and then install the companion flange.

16. Apply anti-corrosion oil to the thread and seat of new drive pinion lock nut, and temporarily tighten drive pinion lock nut to drive pinion, using flange wrench (commercial service tool).

✳✳ WARNING
Never reuse drive pinion lock nut.

17. Tighten drive pinion lock nut within the limits of specified torque so as to keep the pinion bearing preload within a standard values, using preload gauge [SST: ST3127S000 (J-25765-A)].

Total preload torque:
• A value that add 0.01-0.04 kg m (0.1-0.4 Nm) to the measured value when removing.

✳✳ WARNING
Adjust to the lower limit of the drive pinion lock nut tightening torque first.

✳✳ WARNING
If the preload torque exceeds the specified value, replace collapsible spacer and tighten it again to adjust. Never loosen drive pinion lock nut to adjust the preload torque.

18. Set a dial indicator (A) vertically to the tip of the drive pinion.

19. Rotate drive pinion to check for runout.

a. If the runout value is still outside of the limit after the phase has been changed, possible causes are be an assembly malfunction of drive pinion and pinion bearing and malfunction of pinion bearing. Check for these items and repair if necessary.

20. Make a stamping for identification of front oil seal replacement frequency. Refer to "Identification stamp of replacement frequency of front oil seal".

✳✳ WARNING
Make a stamping after replacing front oil seal.

21. Install propeller shaft.

22. Install side flange with the following procedure.

a. Attach the protector [SST: KV38107900 (J-39352)] to side oil seal.

b. After the side flange is inserted and the serrated part of side gear has engaged the serrated part of flange, remove the protector.

c. Put a suitable drift on the center of side flange, then drive it until sound changes.

➡ **When installation is completed, driving sound of the side flange turns into a sound that seems to affect the whole final drive.**

d. Confirm that the dimension of the side flange installation (Measurement A) in the figure comes into the following.
A: 12.83-12.91 inch (326-328 mm)

23. Install drive shaft.

24. Install rear wheel sensor.

25. Install center muffler.

26. Refill gear oil to the final drive and check oil level.

27. Check the final drive for oil leakage.

R200V—AWD
See Figures 62 through 67.

✳✳ WARNING
Verify identification stamp of replacement frequency put in the lower part of gear carrier to determine replacement for collapsible spacer when replacing front oil seal. Refer to "Identification stamp of replacement frequency of front oil seal". If collapsible spacer replacement is necessary, remove final drive assembly

and disassemble it to replace front oil seal and collapsible spacer.

➡ **The reuse of collapsible spacer is prohibited in principle. However, it is reusable on a one-time basis only in cases when replacing front oil seal.**

Identification stamp of replacement frequency of front oil seal:
• The diagonally shaded area in the figure shows stamping point for replacement frequency of front oil seal.
• The following table shows if collapsible spacer replacement is needed before replacing front oil seal.

When collapsible spacer replacement is required, disassemble final drive assembly to replace collapsible spacer and front oil seal.

Stamp: collapsible spacer replacement:
• No stamp: Not required
• "0" or "0" on the far right stamp: Required
• "01" or "1" on the far right side of stamp: Not required

✳✳ WARNING
Make a stamping after replacing front oil seal.

✳✳ WARNING
After replacing front oil seal, make a stamping on the stamping point in accordance with the table below in order to identify replacement frequency.

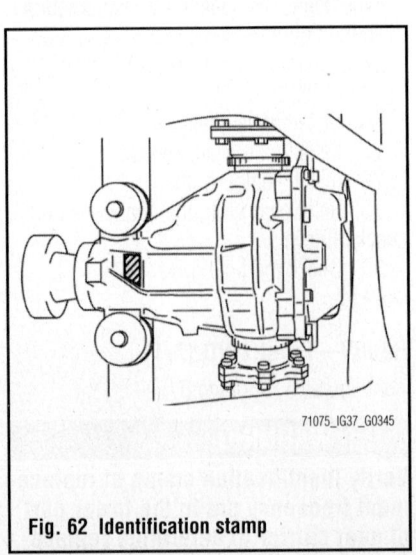

71075_IG37_G0345

Fig. 62 Identification stamp

Stamp before stamping	Stamping on the far right	Stamping
No stamp	0	0
"0" (Front oil seal was replaced once.)	1	01
"01" (Collapsible spacer and front oil seal were replaced last time.)	0	010
"0" is on the far right. (Only front oil seal was replaced last time.)	1	...01
"1" is on the far right. (Collapsible spacer and front oil seal were replaced last time.)	0	...010

71075_IG37_G0346

Fig. 63 Stamping frequency

1. Drain gear oil.

2. Make a judgment if a collapsible spacer replacement is required.

3. Remove center muffler with a power tool.

4. Remove rear wheel sensor.

5. Remove drive shaft from final drive with a power tool. Then suspend it by wire, etc.

6. Install attachment (A: attachment [SST: KV40104100]) to side flange, and then pull out of the side flange with the sliding hammer (B: sliding hammer [SST: ST36230000]).

➡**Circular clip installation position. Final drive side.**

7. Remove propeller shaft.

8. Measure the total preload with the preload gauge (A: SST ST3127S000).

9. Put matching mark (B) on the end of the drive pinion. The matching mark should be in line with the matching mark (A) on companion flange (1).

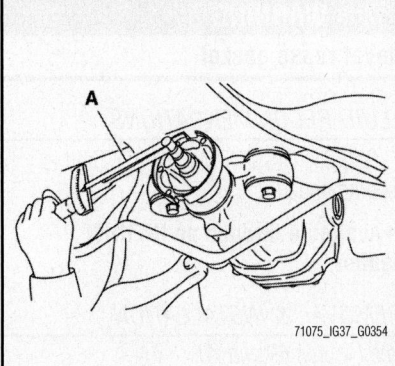

71075_IG37_G0354

Fig. 64 Measuring preload with preload gauge (A)

✳✳ **WARNING**

For matching mark, use paint. Never damage companion flange and drive pinion.

➡**The matching mark on the final drive companion flange indicates the maximum vertical runout position.**

10. Remove drive pinion lock nut using the flange wrench (commercial service tool).

11. Remove companion flange using a puller (commercial service tool).

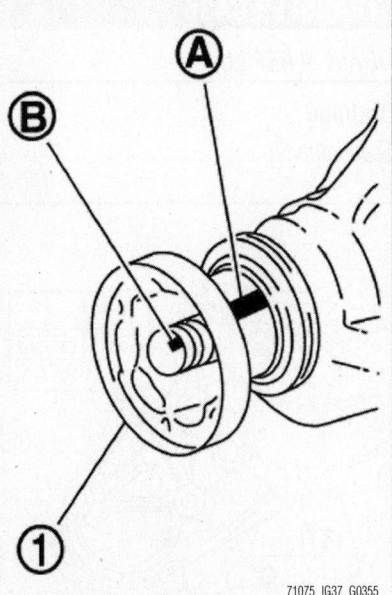

71075_IG37_G0355

Fig. 65 Putting matching mark (B) on the end of the drive pinion and (A) on the companion flange (1)

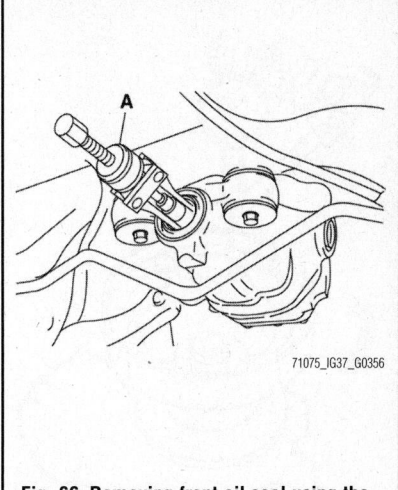

71075_IG37_G0356

Fig. 66 Removing front oil seal using the puller (A)

12. Remove front oil seal using the puller (A: SST KV381054S0).

To install:

13. Apply multi-purpose grease to front oil seal lips.

14. Install front oil seal using the drift (A) [SST: ST30720000 (J-25405)].

✳✳ **WARNING**

Never reuse oil seal.

✳✳ **WARNING**

Never incline oil seal when installing.

15. Align the matching mark of drive pinion with the matching mark of companion flange, and then install the companion flange.

16. Apply anti-corrosion oil to the thread and seat of new drive pinion lock nut, and temporarily tighten drive pinion lock nut to drive pinion, using flange wrench (commercial service tool).

✳✳ **WARNING**

Never reuse drive pinion lock nut.

17. Tighten drive pinion lock nut within the limits of specified torque so as to keep the pinion bearing preload within a standard values, using preload gauge [SST: ST3127S000 (J-25765-A)].

✳✳ **WARNING**

Adjust to the lower limit of the drive pinion lock nut tightening torque first.

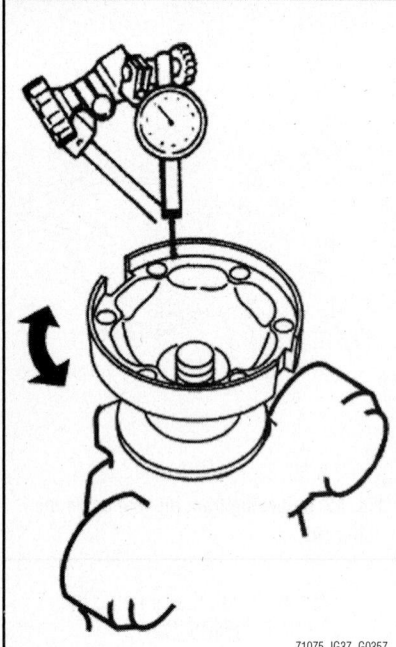

71075_IG37_G0357

Fig. 67 Fitting test indicator to the inner side of companion flange (socket diameter)

❋❋ **WARNING**

If the preload torque exceeds the specified value, replace collapsible spacer and tighten it again to adjust. Never loosen drive pinion lock nut to adjust the preload torque.

18. Fit a test indicator to the inner side of companion flange (socket diameter).

19. Rotate the companion flange to check for runout.

20. If the runout value is outside the repair limit, follow the procedure below to adjust.

a. Check for runout while changing the phase between companion flange and drive pinion gear by 90° step, and search for the position where the runout is the minimum.

b. If the runout value is still outside of the limit after the phase has been changed, possible causes are be an assembly malfunction of drive pinion and pinion bearing and malfunction of pinion bearing. Check for these items and repair if necessary.

c. If the runout value is still outside of the limit after the check and repair, replace companion flange.

21. Make a stamping for identification of front oil seal replacement frequency.

❋❋ **WARNING**

Make a stamping after replacing front oil seal.

22. Install propeller shaft.

23. Install side flange with the following procedure.

a. Attach the protector [SST: KV38107900 (J-39352)] to side oil seal.

b. After the side flange is inserted and the serrated part of side gear has engaged the serrated part of flange, remove the protector.

c. Put a suitable drift on the center of side flange, then drive it until sound changes.

➡**When installation is completed, driving sound of the side flange turns into a sound that seems to affect the whole final drive.**

d. Confirm that the dimension of the side flange installation (Measurement A) in the figure comes into the following.

• Measurement A: 12.83-12.91 inch (326-328 mm)

24. Install drive shaft.

25. Install rear wheel sensor.

26. Install center muffler.

27. Refill gear oil to the final drive and check oil level.

28. Check the final drive for oil leakage.

TRANSFER CASE

DRAIN & REFILL

Draining

See Figure 68.

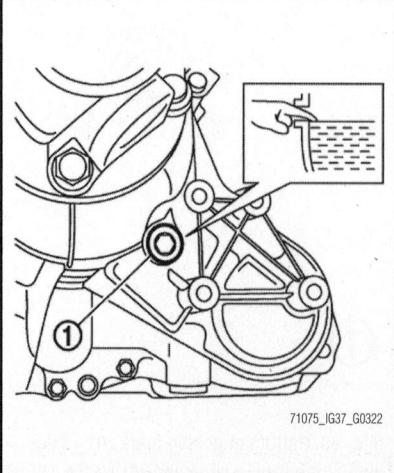

71075_IG37_G0322

Fig. 68 Locating the filler plug (1)

1. Run the vehicle to warm up the transfer unit sufficiently.

2. Stop the engine, and remove the drain plug to drain the transfer fluid.

3. Set a new gasket onto the drain plug, and install it on the transfer and tighten to 26 ft. lbs. (35 Nm).

❋❋ **WARNING**

Never reuse gasket.

Refilling

1. Remove the filler plug and gasket. Then fill fluid up to mounting hole for the filler plug.

➡**Fluid capacity (approx.) 2 ⅛ pt.**

❋❋ **WARNING**

Carefully fill the fluid. (Fill up for approximately 3 minutes.)

2. Leave the vehicle for 3 minutes, and check the fluid level again.

3. Set a new gasket onto filler plug, and install it on transfer and tighten to 26 ft. lbs. (35 Nm).

❋❋ **WARNING**

Never reuse gasket.

FLUID LEVEL CHECK

See Figure 68.

1. Remove filler plug and gasket. Then check that fluid is filled up from mounting hole for the filler plug.

❋❋ **WARNING**

Never start engine while checking fluid level.

2. Set a new gasket onto filler plug, and install it on transfer and tighten to 26 ft. lbs.

❋❋ **WARNING**

Never reuse gasket.

FLUID RECOMMENDATIONS

Genuine NISSAN Matic J ATF fluid is recommended.

➡**Available through an INFINITI dealer.**

REMOVAL & INSTALLATION

See Figures 69 and 70.

➡**Whenever the negative battery cable is disconnected the following components will require resetting. The Automatic temperature control system,**

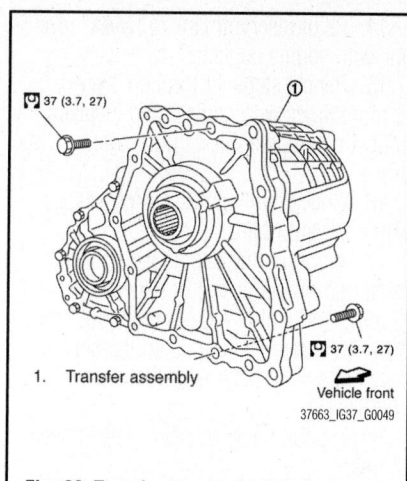

Fig. 69 Transfer case and related components

1. Transfer assembly

Vehicle front

37663_IG37_G0049

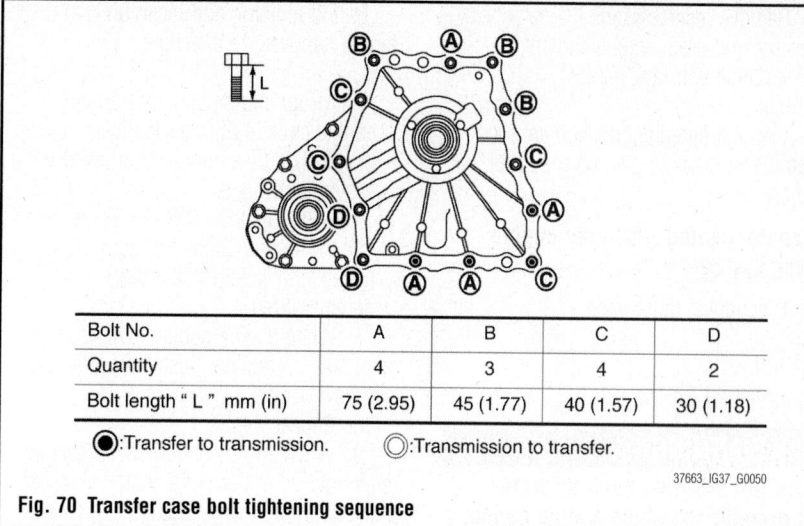

Bolt No.	A	B	C	D
Quantity	4	3	4	2
Bolt length " L " mm (in)	75 (2.95)	45 (1.77)	40 (1.57)	30 (1.18)

●:Transfer to transmission. ○:Transmission to transfer.

37663_IG37_G0050

Fig. 70 Transfer case bolt tightening sequence

Automatic drive positioner, Power window control, Sunroof system, Sunshade system, Rear view monitor, Idle Air Volume Learning, Steering Angle Sensor Neutral Position, Audio presets and Navigation. You will need the CONSULT-III diagnostic tool, or equivalent. Follow the directions on the screen of the tool, as needed.

1. Before servicing the vehicle, refer to the Precautions Section.

→If working near and/or around the SRS system and components, be sure to disable the SRS system. After disabling the system wait three minutes or more before servicing the vehicle.

2. Disconnect the negative battery cable.

3. Raise and safely support the vehicle.

4. Drain the transfer case. Be sure to properly dispose of use fluid.

5. Remove the rear driveshaft.

6. Remove the front driveshaft.

7. Disconnect the AWD solenoid harness connector. Separate the harness from the transfer case assembly.

8. Remove the air breather hose.

9. Remove the control rod.

10. Support the transfer case using a suitable jack.

11. Remove the rear mounting member and engine mounting insulator.

12. Lower the jack to a position where the top transfer case mounting bolts can be removed.

13. Remove the transfer case mounting bolts and separate the transfer case from the transmission.

→Be sure to secure the transfer case and transmission assembly to the jack.

To install:

→Be sure to use new fasteners, as required.

14. Installation is the reverse of the removal procedure.

15. Be sure to fill the transfer case with the proper grade and type fluid.

16. Be sure to perform the reconnect/relearn procedures.

ENGINE COOLING

ENGINE COOLANT

DRAIN & REFILL

Draining

See Figures 71 and 72.

※ WARNING

To avoid being scalded, never change engine coolant when the engine is hot.

※ CAUTION

Wrap a thick cloth around radiator cap and carefully remove radiator cap. First, turn radiator cap a quarter of a turn to release built-up pressure. Then turn radiator cap all the way.

1. Connect drain hose.

→Use a general-purpose hose with a width (A) of 0.59-0.63 inch (15-16 mm) and a length (B) of 5.17 inch (145 mm).

2. Open radiator drain plug (2) at the bottom of radiator, and then remove radiator cap.

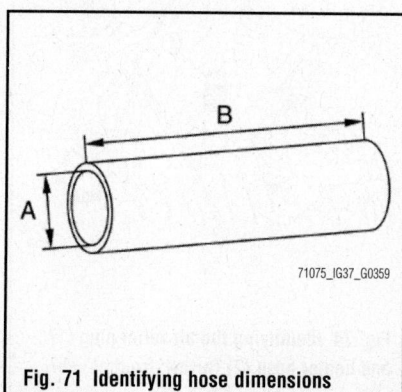

71075_IG37_G0359

Fig. 71 Identifying hose dimensions

→When draining all of engine coolant in the system, open water drain plugs on cylinder block.

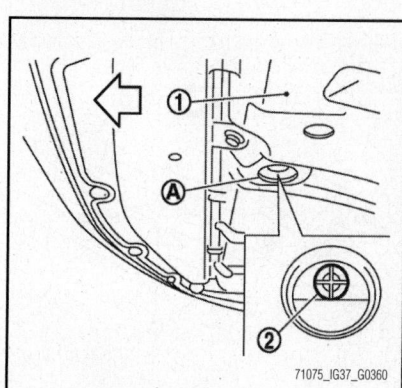

71075_IG37_G0360

Fig. 72 Identifying radiator drain plug (2), engine under cover (1) and radiator drain plug hole (A)(arrow: vehicle front)

3. Remove reservoir tank if necessary, and drain engine coolant and clean reservoir tank before installing.

4. Check drained engine coolant for contaminants such as rust, corrosion or discoloration.

➡**If contaminated, flush the engine cooling system.**

5. Disconnect drain hose.

Refilling

See Figures 73 and 74.

❋❋ WARNING

Do not put additive such as water leak preventive, since it may cause cooling waterway clogging.

➡**Use pre-diluted Genuine NISSAN Long Life Antifreeze/Coolant (blue) or equivalent.**

1. Remove air cleaner case (LH).

2. Install reservoir tank if removed, and radiator drain plug.

❋❋ WARNING

Be sure to clean drain plug and install with new O-ring.

➡**If water drain plugs on cylinder block are removed, close and tighten them.**

3. Check that each hose clamp has been firmly tightened.

4. Remove air relief plug on radiator left side.

5. Remove air relief plug on heater hose.

6. Fill radiator, and reservoir tank if removed, to specified level.

a. Pour engine coolant through engine coolant filler neck slowly of less than 2-1/8 qts a minute to allow air in system to escape.

b. Use Genuine NISSAN Long Life Antifreeze/Coolant or equivalent mixed with water (distilled or demineralized).

7. When engine coolant overflows air relief hole on radiator, install air relief plug with new O-ring.

8. Repeat step 6.

9. When engine coolant overflows air relief hole on heater hose, install air relief plug with new O-ring. Tighten to 11 inch lbs. (1.2 Nm). Then refill radiator with engine coolant.

10. Install air cleaner case (LH).

11. Install radiator cap.

12. Warm up engine until opening thermostat. Standard for warming-up time is approximately 10 minutes at 3,000 rpm.

a. Check thermostat opening condition by touching radiator hose (lower) to see a flow of warm water.

❋❋ WARNING

Watch water temperature gauge so as not to overheat engine.

13. Stop the engine and cool down to less than approximately 122°F (50°C).

a. Cool down using fan to reduce the time.

b. If necessary, refill radiator up to filler neck with engine coolant.

14. Refill reservoir tank to "MAX" level line with engine coolant.

15. Repeat steps 11 through 14 two or more times with radiator cap installed until engine coolant level no longer drops.

16. Check cooling system for leakage with engine running.

17. Warm up the engine, and check for sound of engine coolant flow while running engine from idle up to 3,000 rpm with heater temperature controller set at several position between "COOL" and "WARM".

a. Sound may be noticeable at heater unit.

18. Repeat step 17 three times.

19. If sound is heard, bleed air from cooling system by repeating step 6, and steps from 11 to 18 until engine coolant level no longer drops.

20. Check that the reservoir tank cap is tightened.

LEVEL CHECK

See Figure 75.

1. Check if the reservoir tank engine coolant level is within the "MIN" to "MAX" when the engine is cool.

2. Adjust the engine coolant level if necessary.

3. Check that the reservoir tank cap is tightened.

ELECTRIC ENGINE FAN

REMOVAL & INSTALLATION

See Figure 76.

1. Remove reservoir tank.

2. Remove air cleaner case (LH).

3. Disconnect harness connector from

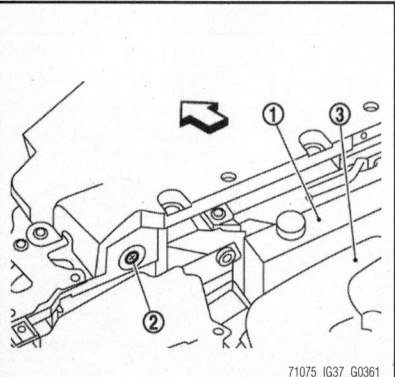

Fig. 73 Identifying the reservoir tank (1), air relief plug (2) and engine cover (3) (arrow: vehicle front)

71075_IG37_G0361

Fig. 74 Identifying the air relief plug (1) and heater hose (2) (arrow: front of vehicle)

71075_IG37_G0362

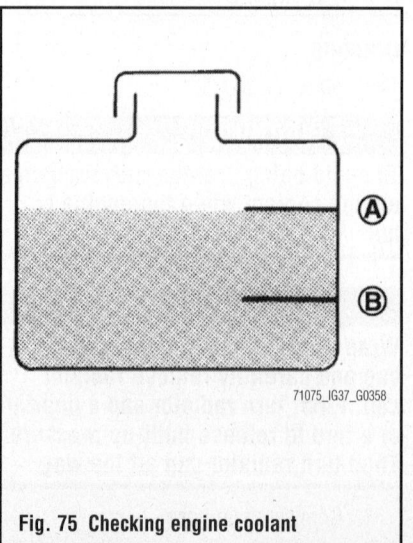

Fig. 75 Checking engine coolant

71075_IG37_G0358

cooling fan control module, and move harness to aside.

4. Remove engine under cover with power tool.

5. Remove cooling fan assembly from under the vehicle.

✳✳ WARNING

Be careful not to damage or scratch on radiator core.

To install:

6. Note the following, and install in the reverse order of removal.

a. Only use genuine parts for cooling fan mounting bolt and observe the specified torque (to prevent core support from being damaged).

RADIATOR

REMOVAL & INSTALLATION

See Figures 77 through 83.

✳✳ CAUTION

Never remove radiator cap when engine is hot. Serious burns could occur from high-pressure engine coolant escaping from water inlet (front). Wrap a thick cloth around the cap. Slowly turn it a quarter of a turn to release built-up pressure. Carefully remove radiator cap by turning it all the way.

1. Remove the following parts:
   a. Engine under cover with power tool.

b. Engine cover.
c. Air cleaner case (RH and LH).
d. Reservoir tank.
e. Radiator core support ornament, radiator core support center.
f. Horn.
g. Hood lock.

2. Remove condenser pipe assembly.

3. Drain engine coolant from radiator.

✳✳ CAUTION

Perform this step when the engine is cold.

✳✳ WARNING

Never spill engine coolant on drive belt.

4. Disconnect A/T fluid cooler hoses from radiator. (A/T models)
   a. Install blind plug to avoid leakage of A/T fluid.

5. Remove radiator hoses (upper and lower) and reservoir tank hose.

✳✳ WARNING

Be careful not to allow engine coolant to contact drive belt.

6. Remove cooling fan assembly.

✳✳ WARNING

Never damage or scratch radiator & condenser assembly core when removing.

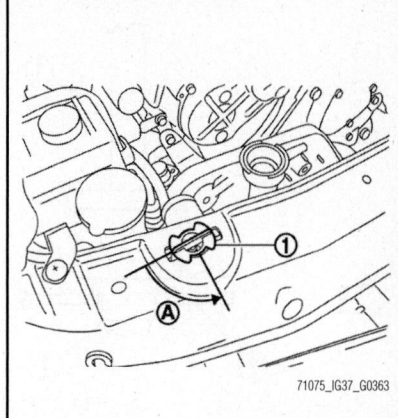

71075_IG37_G0363

Fig. 77 Rotating two radiator upper mount brackets (1) and turning (A) 90° counter-clockwise

7. Rotate two radiator upper mount brackets 90 degrees in direction shown, and remove them.

✳✳ WARNING

Be careful not to damage radiator & condenser assembly core.

8. Remove radiator & condenser assembly as follows:
   a. Lift up and pull the radiator & condenser assembly forward, and then remove the mounting rubber (lower) from the radiator core support.
   b. Remove radiator & condenser assembly from front of radiator core support.

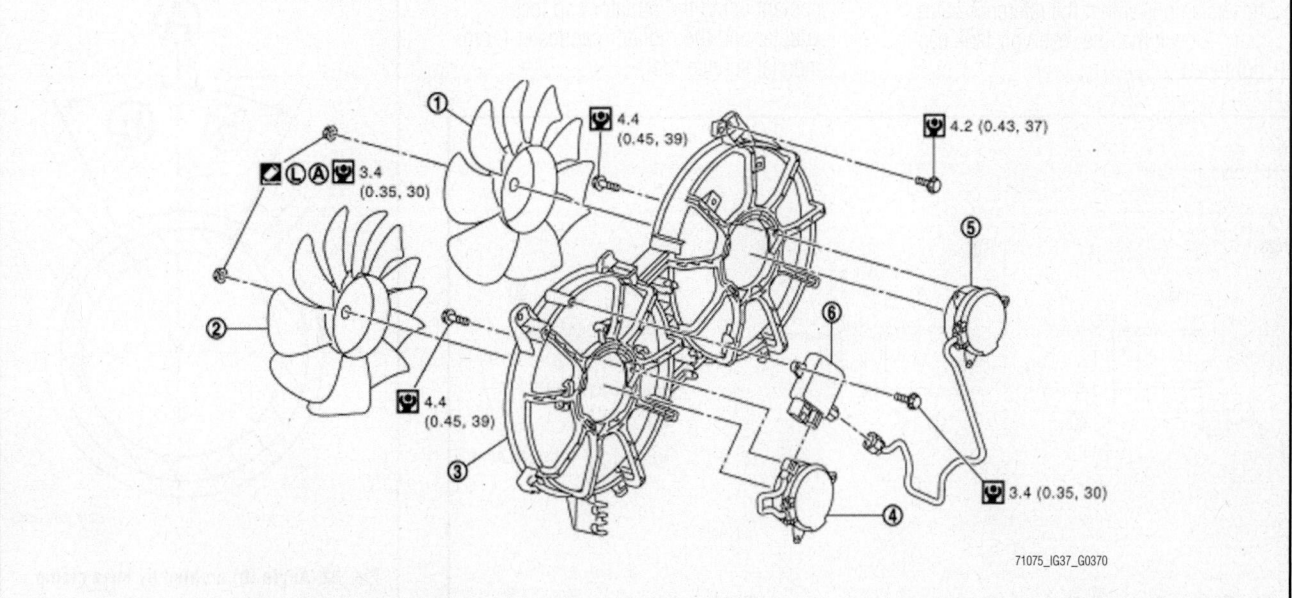

71075_IG37_G0370

Fig. 76 Exploded view of cooling fan and related components

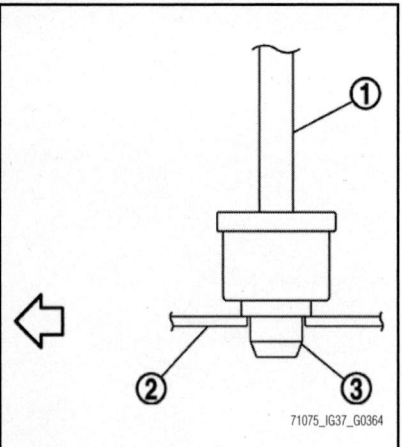

Fig. 78 Lifting radiator and condenser assembly (1) forward, and removing the mounting rubber (lower) (3) from the radiator core support (2)

Radiator hose	Hose end	Paint mark	Position of hose clamp*
Radiator hose (upper)	Radiator side	Upper	A
	Engine side	Upper	B
Radiator hose (lower)	Radiator side	Lower	C
	Engine side	Right side	D

*Refer to the illustrations for the specific position each hose clamp tab.

Fig. 80 Hose positioning specifications (f: 45°) (1 of 2)

To install:

9. Note the following, and install the reverse order of removal:

   a. Replace water hose clamp if it is removed.

   b. Use genuine mounting bolts for the cooling fan assembly and strictly observe the tightening torque. (Breakage prevention for radiator).

   c. Insert the radiator hose (1) all the way to the stopper (2) by 1.3 inch (33 mm) (hose without stopper).

   d. The angle (b) created by the hose clamp pawl and the specified line (A) must be within ±30°.

   e. To install hose clamps (1), check that the dimension (A) from the end of the paint mark (2) on the radiator hose to the hose clamp is within the reference value.

   f. Check that the reservoir tank cap is tightened.

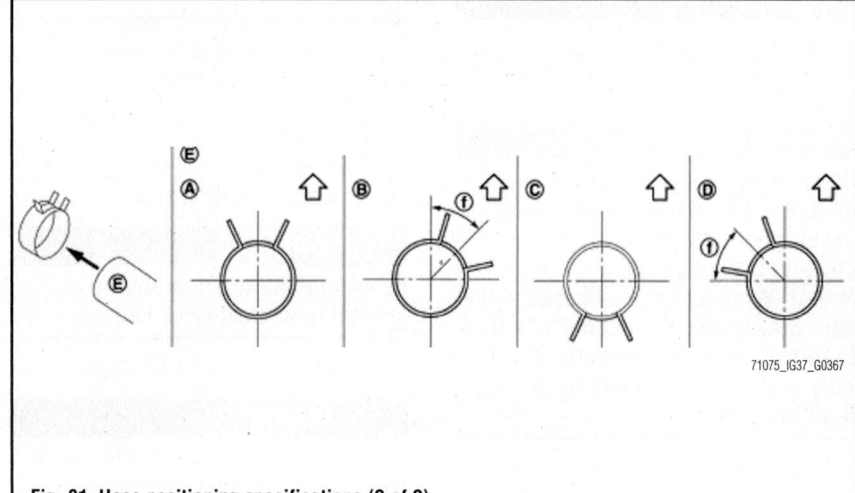

Fig. 81 Hose positioning specifications (2 of 2)

   g. Check for leakage of engine coolant using the radiator cap tester adapter and the radiator cap tester (commercial service tool).

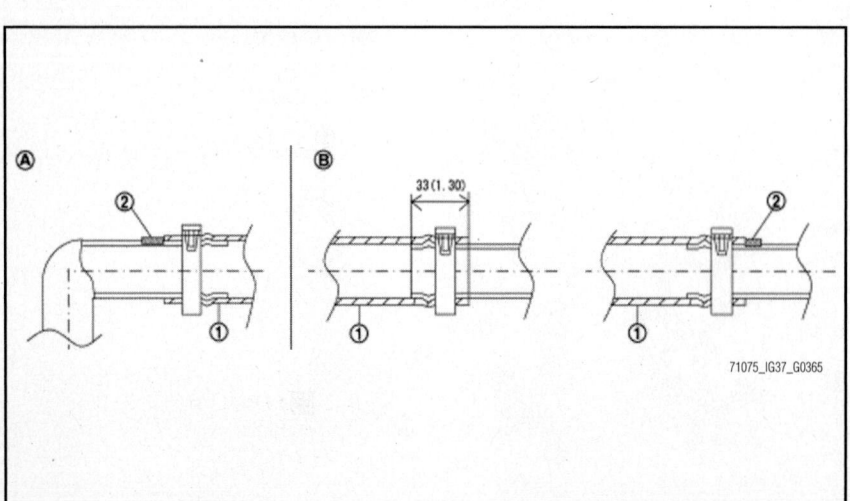

Fig. 79 Inserting radiator hose (1) all the way to the stopper (2) (A: radiator side, B: engine side)

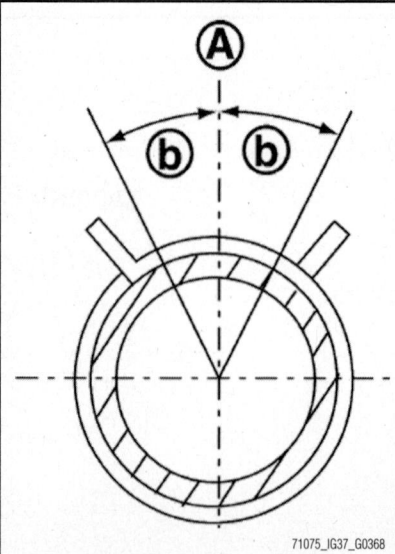

Fig. 82 Angle (b) created by hose clamp pawl and specified line (A) must be within ±30°

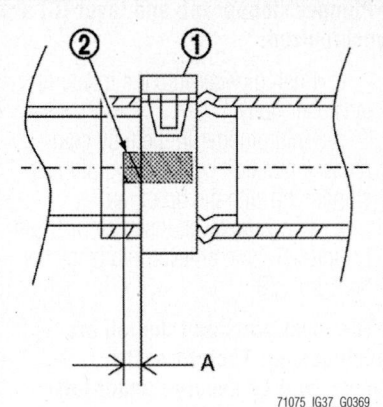

Fig. 83 Installing hose clamps (1) and checking dimension (A) from the end of the paint mark (2)

h. Start and warm up the engine. Visually check that there is no leakage of engine coolant and A/T fluid (A/T models).

THERMOSTAT

REMOVAL & INSTALLATION

See Figure 84.

1. Remove engine cover.
2. Remove air duct and air cleaner case assembly (LH).
3. Remove reservoir tank.
4. Remove engine undercover with power tool.
5. Drain engine coolant from radiator drain plug at the bottom of radiator.

➥Perform this step when the engine is cold.

✳✳ WARNING
Never spill engine coolant on drive belt.

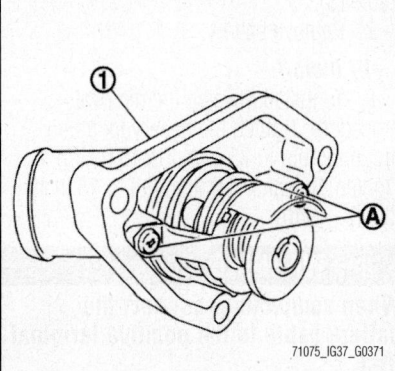

Fig. 84 Removing water inlet and thermostat assembly (1) (A: DO NOT loosen these screws)

6. Disconnect radiator hose (lower).
7. Disconnect intake valve timing control valve harness connector (LH), and remove intake valve timing control solenoid.
8. Remove water inlet and thermostat assembly (1).

✳✳ WARNING
Never disassemble water inlet and thermostat assembly. Replace them as a unit, if necessary.

To install:
Note the following, and install in the reverse order of removal.

➥Be careful not to spill engine coolant over engine room. Use rag to absorb engine coolant.

WATER PUMP

REMOVAL & INSTALLATION
See Figures 85 through 87.

✳✳ WARNING
When removing water pump assembly, be careful not to get engine coolant on drive belt.

➥Water pump cannot be disassembled and should be replaced as a unit.

✳✳ WARNING
After installing water pump, connect hose and clamp securely, then check for leakage using the radiator cap tester and the radiator cap tester adapter (commercial service tool).

1. Remove engine cover.
2. Release the fuel pressure.

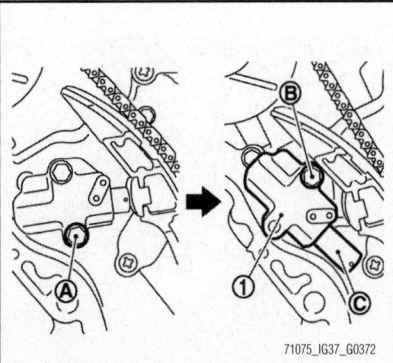

Fig. 85 Removing timing chain primary tensioner (1) by removing lower mounting bolt (A), loosening upper mounting bolt (B) and expanding plunger (C)

3. Disconnect the battery cable from the negative terminal.
4. Remove air duct and air cleaner case assembly (RH and LH).
5. Remove reservoir tank.
6. Separate engine harness removing their brackets from front timing chain case.
7. Remove engine undercover with power tool.
8. Drain engine oil.

✳✳ WARNING
Perform this step when the engine is cold.

✳✳ WARNING
Never spill engine oil on drive belt.

9. Drain engine coolant from radiator.
10. Perform this step when the engine is cold.
11. Never spill engine coolant on drive belt.
12. Remove cooling fan assembly.
13. Remove radiator hose (upper and lower).
14. Remove front timing chain case.
15. Remove timing chain tensioner (primary) as follows:
   a. Remove lower mounting bolt.
   b. Loosen upper mounting bolt slowly, and then turn chain tensioner (primary) on the upper mounting bolt so that plunger is fully expanded.

➥Even if plunger is fully expanded, it is not dropped from the body of timing chain tensioner (primary).

   c. Remove upper mounting bolt, and then remove timing chain tensioner (primary).
   d. Remove water pump as follows:
   e. Remove three water pump mounting bolts. Secure a gap between water pump gear and timing chain, by turning crankshaft counterclockwise until timing chain looseness on water pump sprocket becomes maximum.
   f. Screw M8 bolts (A) [pitch: 0.049 inch (1.25 mm) length: approx. 1.97 inch (50 mm)] into water pumps upper and lower mounting bolt holes until they reach timing chain case. Then, alternately tighten each bolt for a half turn, and pull out water pump.

✳✳ WARNING
Pull straight out while preventing vane from contacting socket in installation area.

Fig. 86 Identifying M8 bolts (A) and water pump (1)

⁂ WARNING

Remove water pump without causing sprocket to contact timing chain.

g. Remove M8 bolts and O-rings from water pump.

⁂ WARNING

Never disassemble water pump.

To install:

16. Install new O-rings to water pump.

   a. Apply engine oil to O-ring (1) and engine coolant to O-ring (3).
   b. Locate O-ring with yellow paint mark (A) to front side.
   c. Locate O-ring with light blue paint mark (B) to rear side.

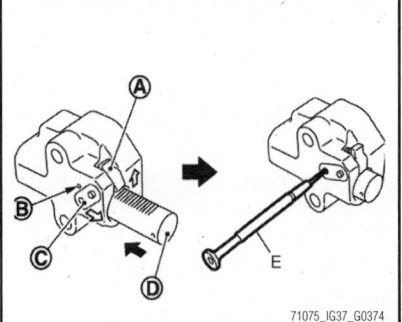

Fig. 87 Identifying the plunger stopper tab (A), tensioner body hole (B), plunger stopper tab and lever (C), plunger (D) and stopper pin (E)

17. Install water pump.

⁂ WARNING

Never allow cylinder block to nip O-rings when installing water pump.

   a. Check timing chain and water pump sprocket are engaged.
   b. Insert water pump by tightening mounting bolts alternately and evenly.
18. Install timing chain tensioner (primary) as follows:
   a. Turn crankshaft clockwise so that timing chain on the timing chain tensioner (primary) side is loose.
   b. Pull plunger stopper tab (A) up (or turn lever downward) so as to remove plunger stopper tab from the ratchet of plunger (D).

➡ **Plunger stopper tab and lever (C) are synchronized.**

   c. Push plunger into the inside of tensioner body.
   d. Hold plunger in the fully compressed position by engaging plunger stopper tab with the tip of ratchet.
   e. To secure lever, insert stopper pin (E) through hole of lever into tensioner body hole (B).

➡ **The lever parts and the tab are synchronized. Therefore, the plunger will be secured under this condition.**

➡ **Figure shows the example of 1.2 mm (0.047 in) diameter thin screwdriver being used as the stopper pin.**

   f. Install timing chain tensioner (primary).
   g. Remove dust and foreign material completely from backside of timing chain tensioner (primary) and from installation area of rear timing chain case.
   h. Remove stopper pin.
   i. Check again that timing chain and water pump sprocket are engaged.
19. Install in the reverse order of removal for remaining parts.
   a. After starting engine, let idle for three minutes, then rev engine up to 3,000 rpm under no load to purge air from the high-pressure chamber of chain tensioner. Engine may produce a rattling noise. This indicates that air still remains in the chamber and is not a matter of concern.

ENGINE ELECTRICAL

BATTERY

REMOVAL & INSTALLATION

See Figure 88.

1. Remove battery cover.
2. Remove the clips, and remove hood ledge cover (RH).
3. Remove cowl top cover (RH).
4. Remove cover of battery positive terminal.
5. Loosen battery terminal nuts (1), and disconnect both battery cables from battery terminals.

⁂ WARNING

When disconnecting, disconnect the battery cable from the negative terminal first.

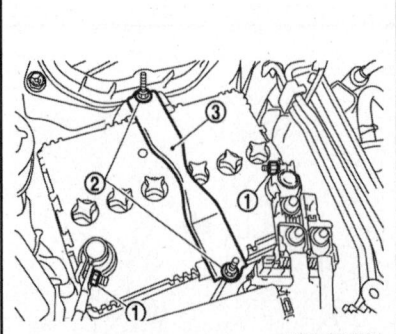

Fig. 88 Identifying battery terminal nuts (1), fix frame mounting nuts (2) and battery fix frame (3)

BATTERY SYSTEM

6. Remove battery fix frame mounting nuts (2) and battery fix frame (3).
7. Remove battery.

To install:

8. To install, reverse the removal procedure. Tighten the battery fix frame mounting nut to 35 inch lbs. (3.9 Nm). Tighten the batter terminal nut to 48 inch lbs. (5.4 Nm).

⁂ WARNING

When connecting, connect the battery cable to the positive terminal first.

➡ Reset electronic systems as necessary.

ENGINE ELECTRICAL **CHARGING SYSTEM**

ALTERNATOR

REMOVAL & INSTALLATION

2WD Vehicles

See Figures 89 and 90.

1. Disconnect the battery cable from the negative terminal.

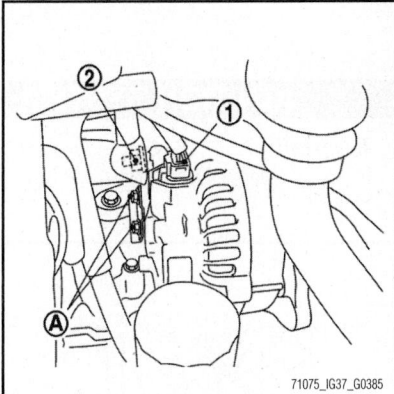

Fig. 89 Disconnecting alternator connector (1), removing "B" terminal nut (2) and harness bracket bolts (A)

2. Remove engine front undercover, using power tools.

3. Remove radiator cooling fan assembly.

4. Remove drive belt.

5. Disconnect alternator connector (1).

6. Remove "B" terminal nut (2).

7. Remove the harness bracket bolts (A).

8. Remove oil pressure switch harness clip (C) from alternator stay (1).

9. Disconnect oil pressure switch connector (D) and oil temperature sensor connector (E).

10. Remove alternator mounting bolt (B) and alternator stay mounting bolt (F) using power tools, then remove alternator stay.

11. Remove alternator mounting bolt (A), using power tools.

12. Remove alternator assembly downward from the vehicle.

To install:

✳✳ WARNING

Be sure to tighten "B" terminal nut carefully.

13. Install alternator, and check tension of belt.

14. For this model, the power generation voltage variable control system that controls the power generation voltage of the alternator has been adopted. Therefore, the power generation voltage variable control system operation inspection should be performed after replacing the alternator, and then make sure that the system operates normally.

AWD Vehicles

See Figures 91 through 93.

1. Disconnect the battery cable from the negative terminal.

2. Remove air cleaner case (RH).

3. Remove the clip (B) from the harness bracket (1) and "B" terminal harness from the clip (C).

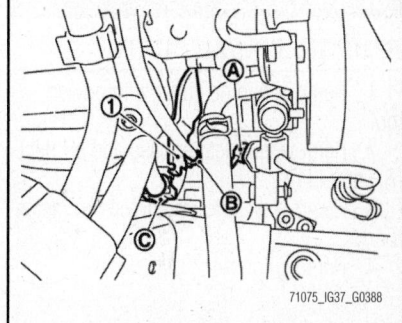

Fig. 91 Removing clip (B) from harness bracket (1) and "B" terminal harness from the clip (C) and disconnecting pressure sensor connector (A)

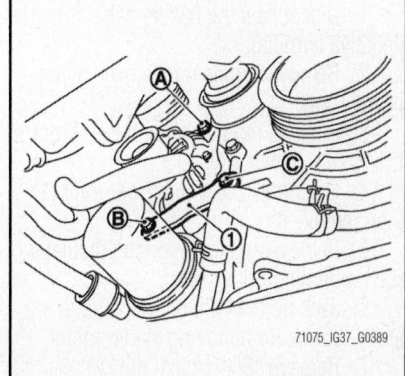

Fig. 92 Removing alternator mounting bolt (B) and alternator stay mounting bolt (C), then alternator stay (1) and alternator mounting bolt (A)

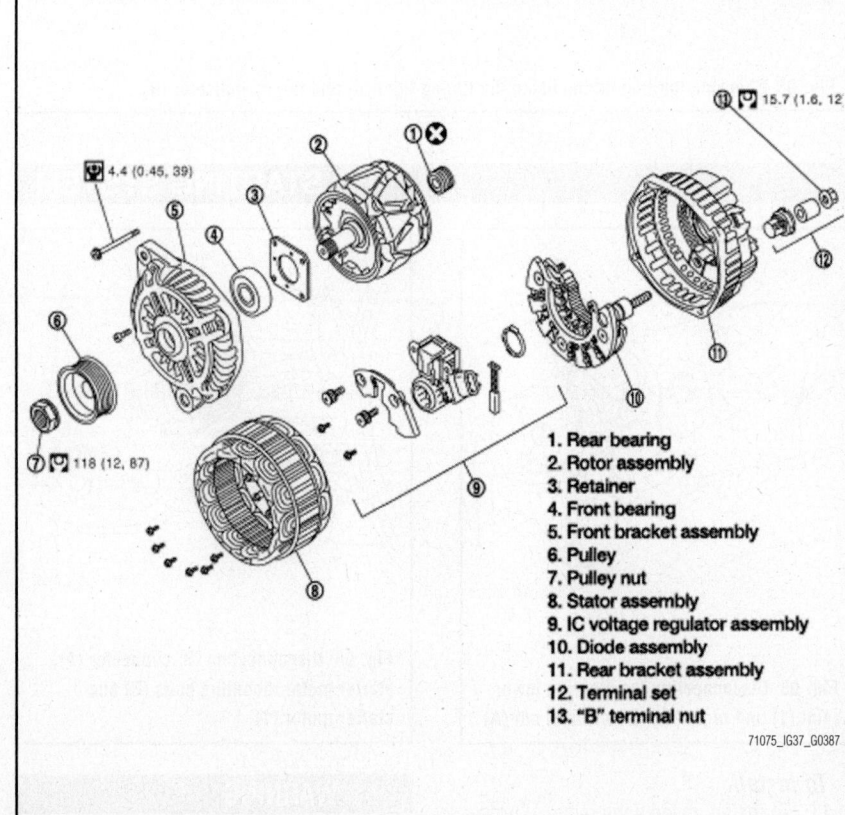

1. Rear bearing
2. Rotor assembly
3. Retainer
4. Front bearing
5. Front bracket assembly
6. Pulley
7. Pulley nut
8. Stator assembly
9. IC voltage regulator assembly
10. Diode assembly
11. Rear bracket assembly
12. Terminal set
13. "B" terminal nut

Fig. 90 Exploded view of alternator and related components—2WD

4. Disconnect pressure sensor connector (A).

5. Remove engine undercover, using power tools.

6. Remove radiator cooling fan assembly.

7. Remove drive belt.

8. Remove alternator mounting bolt (B) and alternator stay mounting bolt (C) using power tools, then remove alternator stay (1).

9. Remove alternator mounting bolt (A), using power tools.

10. Pull and turn alternator, and then remove the harness bracket bolts (A).

11. Disconnect alternator connector (1).

12. Remove "B" terminal nut (2).

13. Remove alternator assembly downward from the vehicle.

To install:

※※ **WARNING**

Be sure to tighten "B" terminal nut carefully.

14. Install alternator, and check tension of belt.

15. For this model, the power generation voltage variable control system that controls the power generation voltage of the alternator has been adopted. Therefore, the power generation voltage variable control system operation inspection should be performed after replacing the alternator, and then make sure that the system operates normally.

ENGINE ELECTRICAL | IGNITION SYSTEM

IGNITION TIMING

INSPECTION & ADJUSTMENT

See Figure 94.

1. Attach timing light to loop wire.
2. Check ignition timing.

SPARK PLUGS

REMOVAL & INSTALLATION

1. Remove engine cover with power tool.
2. Remove air cleaner case and air duct (RH and LH).
3. Remove electric throttle control actuator.
4. Remove ignition coil.

5. Remove spark plug with a spark plug wrench (commercial service tool 0.55 inch [14 mm]).

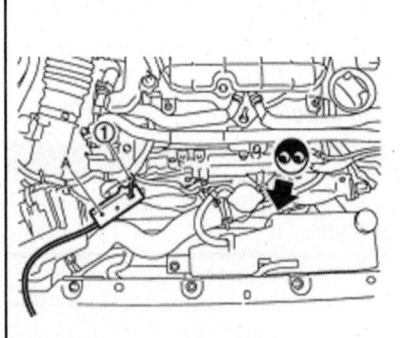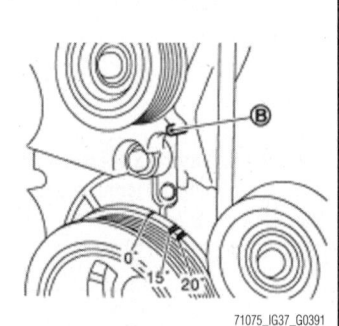

71075_IG37_G0391

Fig. 94 Checking ignition timing using the timing light (A) and timing indicator (B)

To install:
To install reverse the removal procedure.

ENGINE ELECTRICAL | STARTING SYSTEM

STARTER

REMOVAL & INSTALLATION

See Figures 95 and 96.

1. Disconnect the battery cable from the negative terminal.

2. Remove engine undercover, using power tools.

3. Remove road wheel and tire (Front LH), using power tools.

4. Disconnect steering lower joint (1), then remove it.

5. Remove engine mounting insulator (LH) mounting nut (Lower).

6. Jack up the engine front side to create clearance for removing starter motor.

7. Remove "B" terminal nut (A).

8. Disconnect "S" connector (A).

9. Remove starter motor mounting bolts (B), using power tools.

10. Remove starter motor (1) from the side of the vehicle.

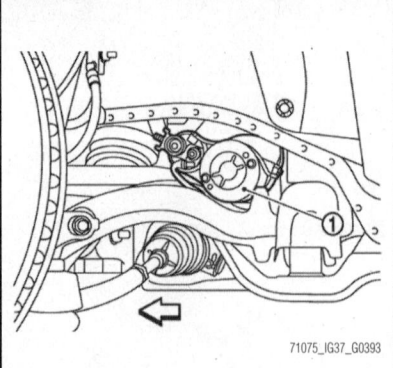

71075_IG37_G0392

Fig. 95 Disconnecting the steering lower joint (1) and removing "B" terminal nut (A)

To install:
11. To install, reverse the removal procedure.

71075_IG37_G0393

Fig. 96 Disconnecting "S" connector (A), starter motor mounting bolts (B) and starter motor (1)

※※ **WARNING**

Be sure to tighten "B" terminal nut carefully.

ENGINE MECHANICAL

ACCESSORY DRIVE BELT SYSTEM

BELT ROUTING

See Figure 97.

Refer to the accompanying illustration.

INSPECTION

❊❊ WARNING

Be sure to perform the this step when engine is stopped.

1. Check that the indicator (notch on fixed side) of drive belt auto-tensioner is within the possible use range (A).

➡**Check the drive belt auto-tensioner indication when the engine is cold.**

➡**When new drive belt is installed, the indicator (notch on fixed side) should be within the range.**

2. Visually check the entire drive belt for wear, damage or crack.

3. If the indicator (notch on fixed side) is out of the possible use range or belt is damaged, replace drive belt.

REMOVAL & INSTALLATION

Drive Belt

See Figure 98.

1. Remove engine undercover with power tool.

2. While securely holding the square hole (A) in pulley center of auto tensioner (1) with a spinner handle, move spinner handle in the direction of arrow (loosening direction of drive belt).

❊❊ CAUTION

Never place hand in a location where pinching may occur if the holding tool accidentally comes off.

3. Under the above condition, insert a metallic bar of approximately 0.24 inch

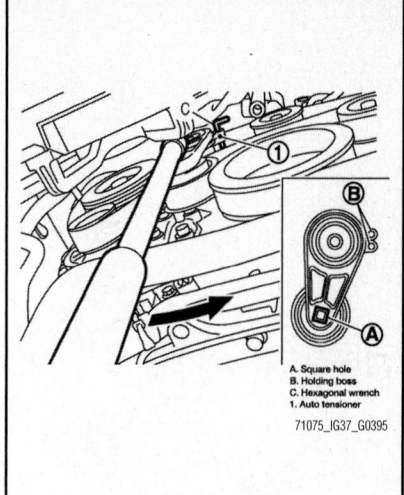

A. Square hole
B. Holding boss
C. Hexagonal wrench
1. Auto tensioner

71075_IG37_G0395

Fig. 98 Removing the drive belt

(6 mm) in diameter [hexagonal wrench (C) shown as example in the figure] through the holding boss (B) to lock auto-tensioner pulley arm.

4. Remove drive belt.

To install:

5. Note the following, and install in the reverse order of removal:

   a. Check drive belt is securely installed around all pulleys.

   b. Check drive belt is correctly engaged with the pulley groove.

   c. Check for engine oil and engine coolant are not adhered drive belt and pulley groove.

Drive Belt Tensioner & Idler Pulley

See Figure 99.

1. Remove drive belt:

   a. Keep auto tensioner pulley arm locked after drive belt is removed.

2. Remove auto tensioner and idler pulley:

   a. Keep auto tensioner pulley arm locked to install or remove auto tensioner.

To install:

3. To install, reverse the removal procedure.

❊❊ WARNING

If there is damage greater than peeled paint, replace drive belt auto tensioner.

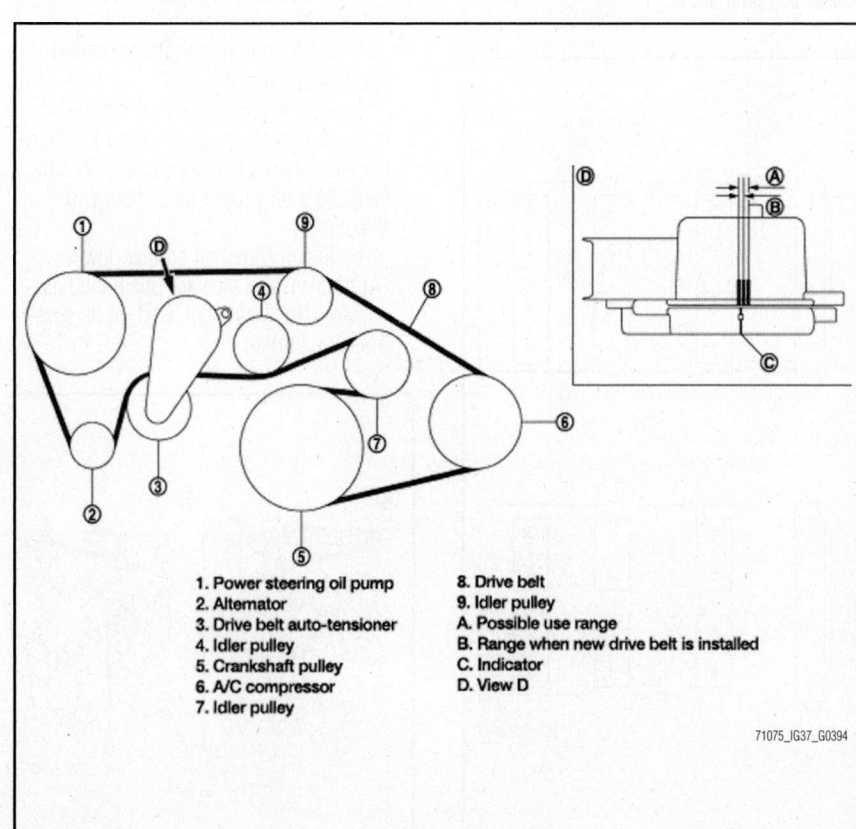

1. Power steering oil pump
2. Alternator
3. Drive belt auto-tensioner
4. Idler pulley
5. Crankshaft pulley
6. A/C compressor
7. Idler pulley
8. Drive belt
9. Idler pulley
A. Possible use range
B. Range when new drive belt is installed
C. Indicator
D. View D

71075_IG37_G0394

Fig. 97 Drive belt routing

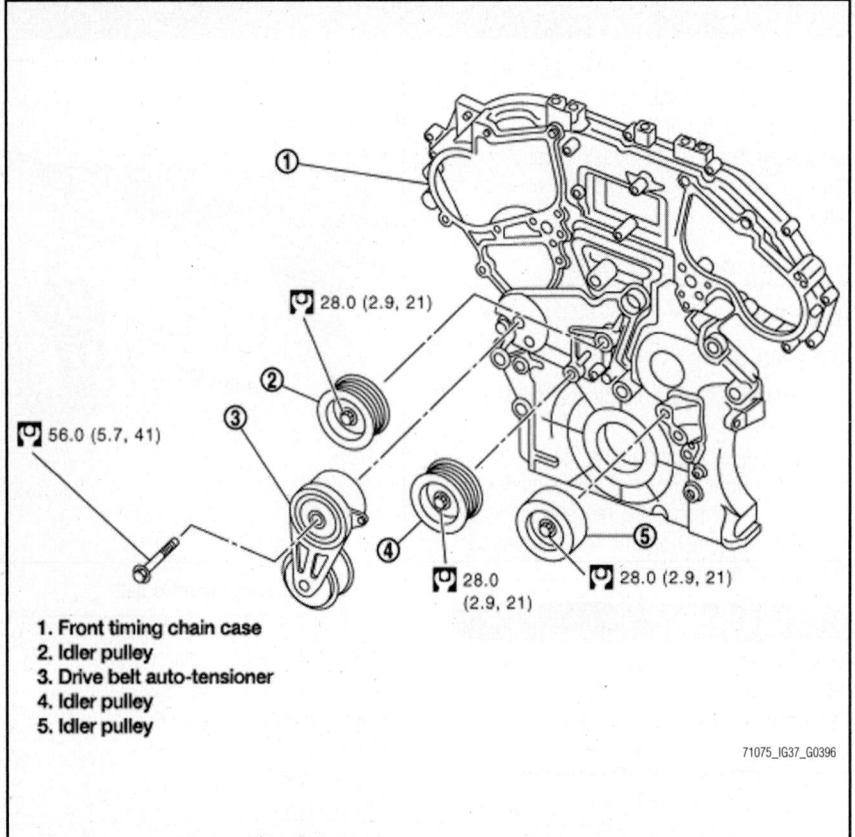

28.0 (2.9, 21)

56.0 (5.7, 41)

28.0 (2.9, 21)

28.0 (2.9, 21)

1. Front timing chain case
2. Idler pulley
3. Drive belt auto-tensioner
4. Idler pulley
5. Idler pulley

71075_IG37_G0396

Fig. 99 Removing and installing drive belt auto tensioner and idler pulley

CAMSHAFT & BEARINGS

REMOVAL & INSTALLATION

G25

See Figures 100 through 110.

1. Remove front timing chain case, camshaft sprocket and timing chain.
2. Remove fuel sub tube.

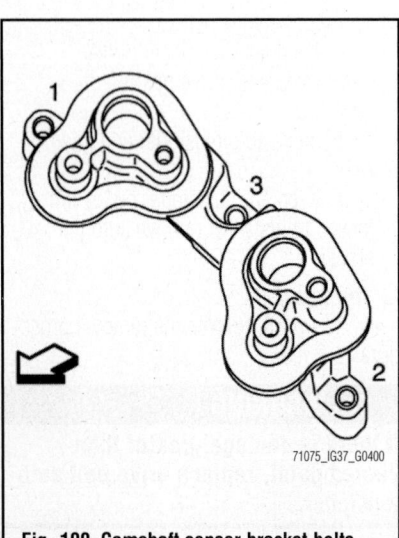

71075_IG37_G0400

Fig. 100 Camshaft sensor bracket bolts

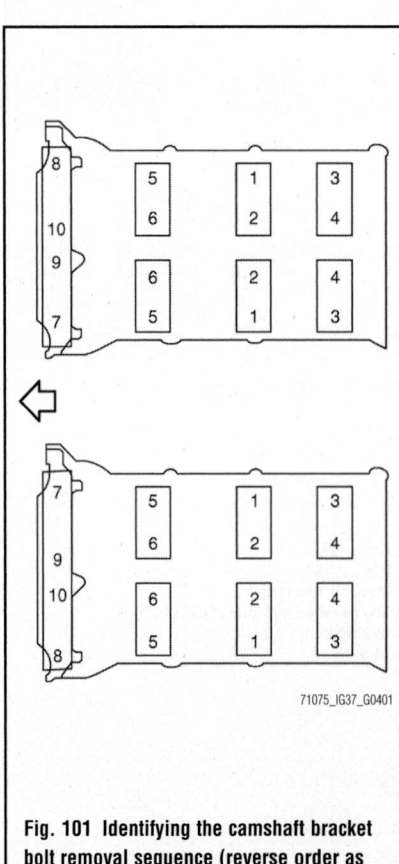

71075_IG37_G0401

Fig. 101 Identifying the camshaft bracket bolt removal sequence (reverse order as shown)

3. Loosen camshaft sensor bracket bolts in reverse order as shown.

➡ **The order of loosening bolts is the same for bank 1 and bank 2.**

4. Remove camshaft brackets.
   a. Mark camshafts, camshaft brackets and bolts so they are placed in the same position and direction for installation.
   b. Equally loosen camshaft bracket bolts in several steps in reverse order as shown.
5. Remove camshaft.
6. Remove valve lifter.
   a. Identify installation positions, and store them without mixing them up.
7. Remove timing chain tensioners (secondary) (1) from cylinder head.
   a. Remove timing chain tensioners (secondary) with its stopper pin (C) attached.

➡ **Stopper pin should be attached when timing chain (secondary) is removed.**

To install:

8. Install timing chain tensioners (secondary) on both sides of cylinder head.
   a. Install timing chain tensioners (1) with its stopper pin (C) attached.
9. Install valve lifter.
   a. Install it in the original position.
10. Install camshafts.
   a. Follow your identification marks made during removal, or follow the identification marks that are present on new camshafts for proper placement and direction.
   b. Install camshaft so that dowel pin (A) on front end face are positioned as shown. (No. 1 cylinder TDC on its compression stroke).

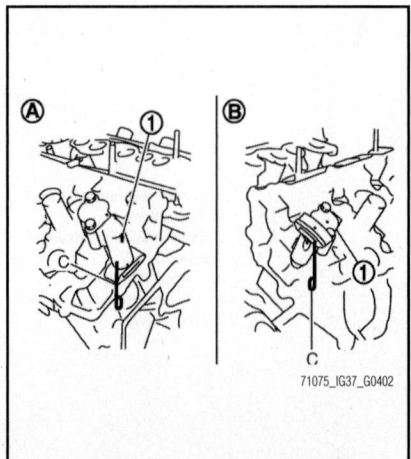

71075_IG37_G0402

Fig. 102 Removing timing chain tensioners (secondary)(1) with stopper pin (C) attached (A: bank 1, B: bank 2)

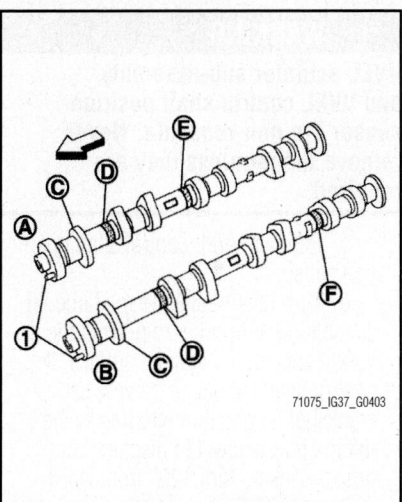

Fig. 103 Installing the camshaft (1 of 2)

71075_IG37_G0403

Bank	INT/EXH	Dowel pin (1)	Paint marks M1 (E)	Paint marks M2 (F)	Paint marks M3 (D)	Identification mark (C)
1	EXH (B)	Yes	No	Yellow	Light blue	1K
1	INT (A)	Yes	Yellow	No	Light blue	1J
2	INT (A)	Yes	Yellow	No	Light blue	1L
2	EXH (B)	Yes	No	Yellow	Light blue	1M

71075_IG37_G0404

Fig. 104 Installing the camshaft (2 of 2)

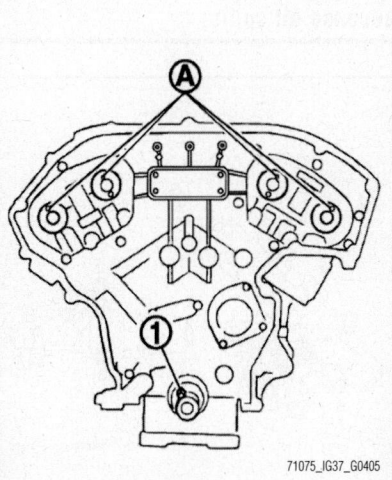

71075_IG37_G0405

Fig. 105 Installing camshaft so dowel pin (A) on front end face are positioned as shown (1: crankshaft key)

➡Though camshaft does not stop at the portion as shown, for the placement of cam nose, it is generally accepted camshaft is placed for the same direction of the figure.

11. Install camshaft brackets.

   a. Remove foreign material completely from camshaft bracket backside and from cylinder head installation face.

   b. Install camshaft bracket in original position and direction as shown.

   c. Install camshaft brackets (No. 2 to 4) aligning the stamp marks (A) as shown.

➡There are no identification marks indicating bank 1 and bank 2 for camshaft bracket (No. 1).

   d. Apply liquid gasket to mating surface of camshaft bracket (No. 1) as shown on both bank 1 and bank 2.

➡Use Genuine RTV Silicone Sealant or equivalent.

   e. Apply liquid gasket to camshaft bracket (No. 1) contact surface on the rear timing chain case backside as shown on both bank 1 and bank 2.

➡Use Genuine RTV Silicone Sealant or equivalent.

❈❈ WARNING

For camshaft bracket (No. 1) near installation position, and install it without disturbing the liquid gasket applied to the surfaces.

12. Tighten camshaft bracket bolts in the following steps, in numerical order as shown.

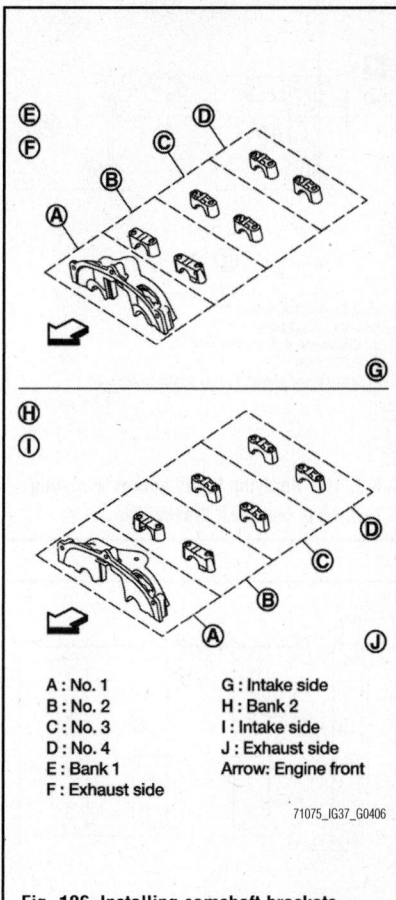

A : No. 1
B : No. 2
C : No. 3
D : No. 4
E : Bank 1
F : Exhaust side
G : Intake side
H : Bank 2
I : Intake side
J : Exhaust side
Arrow: Engine front

71075_IG37_G0406

Fig. 106 Installing camshaft brackets

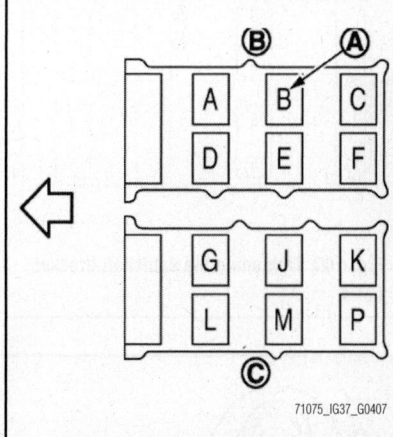

71075_IG37_G0407

Fig. 107 Installing camshaft brackets (No. 2 to 4) aligning the stamp marks (A) as shown (B: Bank 1, C: Bank 2, arrow: engine front)

   a. Tighten No. 7 to 10 to 1 ft. lb. (1.96 Nm) in numerical order as shown.

   b. Tighten No. 1 to 6 to 1 ft. lb. (1.96 Nm) in numerical order as shown.

   c. Tighten No. 1 to 10 to 4 ft. lbs. (5.88 Nm) in numerical order as shown.

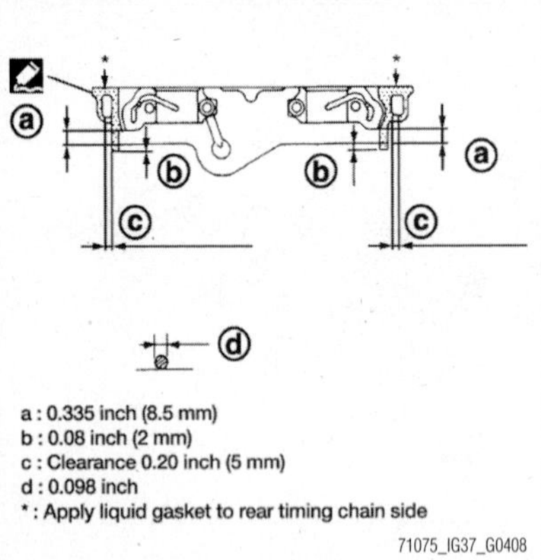
a
b
c
d
a : 0.335 inch (8.5 mm)
b : 0.08 inch (2 mm)
c : Clearance 0.20 inch (5 mm)
d : 0.098 inch
* : Apply liquid gasket to rear timing chain side
71075_IG37_G0408

Fig. 108 Applying liquid gasket to mating surface of camshaft bracket (No. 1)

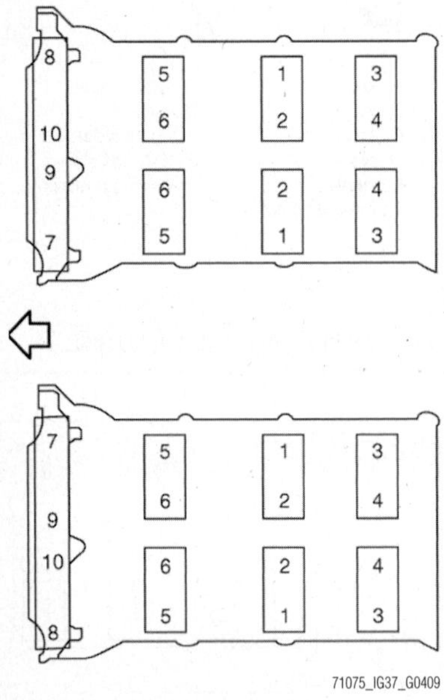
71075_IG37_G0409

Fig. 109 Tightening the camshaft bracket bolts

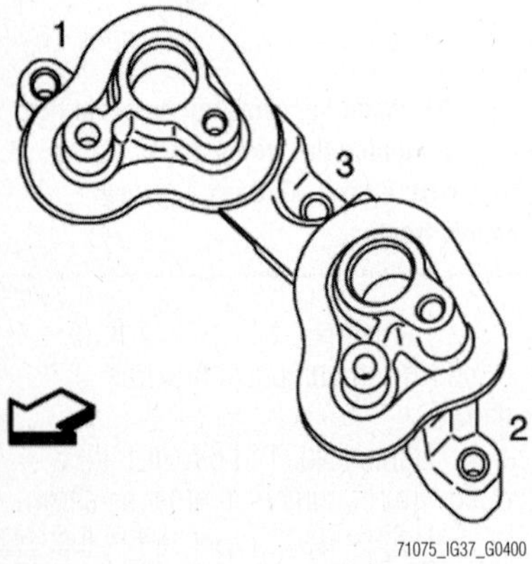
1
3
2
71075_IG37_G0400

Fig. 110 Camshaft sensor bracket bolts

d. Tighten No. 1 to 10 to 8 ft. lbs. (10.4 Nm) in numerical order as shown.

13. Tighten camshaft sensor bracket bolts in numerical order as shown.

➡The order of tightening bolts is the same for bank 1 and bank 2.

14. Inspect and adjust the valve clearance.

15. Install in the reverse order of removal after this step.

G37

See Figures 111 through 125.

⁂ WARNING

Never loosen adjusting bolts (A) and mounting bolts (black color) (B) of VVEL ladder assembly. If loosened, the stroke of cam lift becomes out of adjustment. In such case, replacement of VVEL ladder assembly and cylinder head assembly is required.

➡VVEL ladder assembly cannot be replaced as a single part, because it is machined together with cylinder head assembly.

1. Remove rocker covers (bank 1 and bank 2).

2. Remove VVEL actuator sub-assembly as follows.

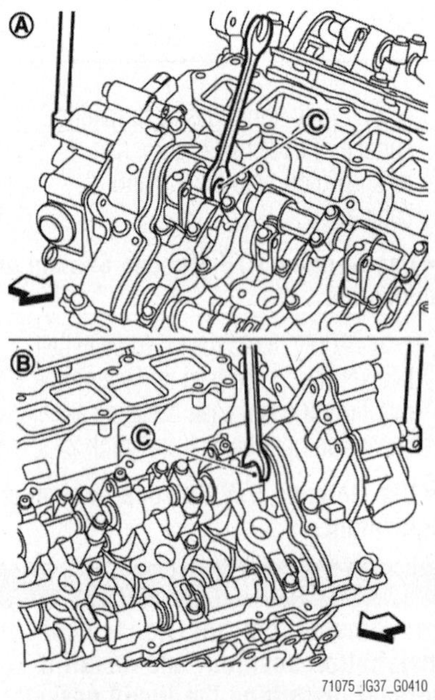
A
C
B
C
71075_IG37_G0410

Fig. 111 Fixing two flat areas (C) of control shaft with a wrench to remove mounting bolts of control shaft (A: Bank 1, B: Bank 2, arrow: engine front)

⁂ WARNING

VVEL actuator sub-assembly and VVEL control shaft position sensor are non-reusable. Never remove them unless they are required.

a. Remove VVEL control shaft position sensor.

b. Turn control shaft to the large lift side and fix it in order to prevent the interference of the stopper surface. If control shaft cannot be moved, set crankshaft in position referring to the information below. (To displace cam nose). Bank 1: Turn 120° from No. 1 cylinder at TDC. Bank 2: No. 1 cylinder at TDC.

c. Fix two flat areas (C) of control shaft with a wrench to remove mounting bolts of control shaft.

⁂ WARNING

During the operation, never allow a wrench to interfere with other parts.

⁂ WARNING

Fix control shaft to prevent the interference of the stopper surface.

d. Remove VVEL actuator sub-assembly. Loosen mounting bolts in the reverse order as shown.

⁂ WARNING

When removing, prepare wastes because oil spills.

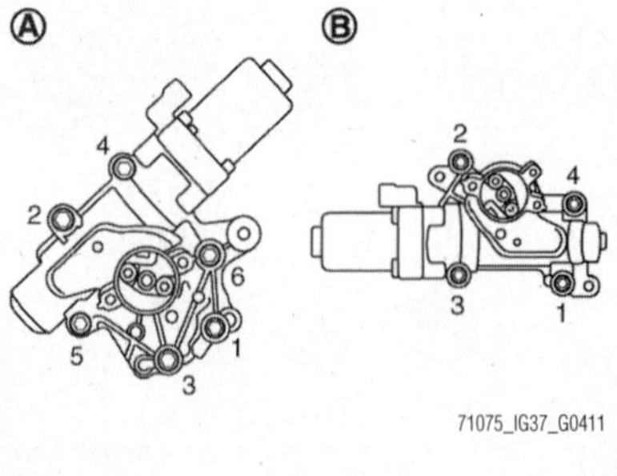
A
B
71075_IG37_G0411

Fig. 112 Removing Bank 1 (A) and Bank 2 (B) VVEL actuator sub-assembly mounting bolts

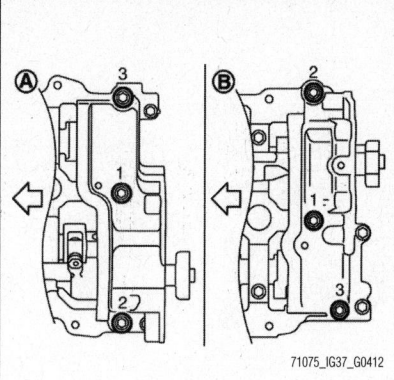

Fig. 113 Removing Bank 1 (A) and Bank 2 (B) actuator rear bracket (arrow: engine front)

※ WARNING

When installing, be careful with VVEL actuator sub-assembly (bank 2) mounting bolt No. 1 because its length is different.

   e. Remove actuator bracket (rear). Loosen mounting bolts in the reverse order as shown in the figure.
  3. Remove front timing chain case, camshaft sprockets, and timing chain.
  4. Remove rear timing chain case.
  5. Remove VVEL ladder assembly.

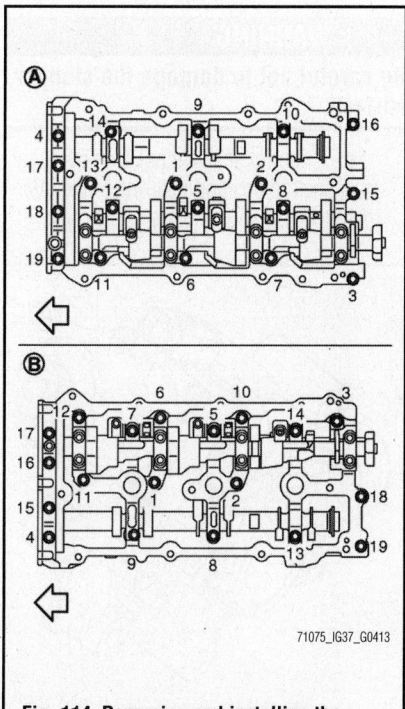

Fig. 114 Removing and installing the VVEL ladder assembly (A: Bank 1, B: Bank 2, arrow: engine front)

   a. Loosen mounting bolts (gold color) in the reverse order as shown.

※ WARNING

Never loosen adjusting bolts and mounting bolts (black color).

※ WARNING

When removing VVEL ladder assembly, hold the drive shaft from below so as not to drop it.

  6. Remove camshaft (EXH).
  7. Remove valve lifter.
   a. Identify installation positions, and store them without mixing them up.
  8. Remove timing chain tensioners (secondary) (1) from cylinder head.
   a. Remove timing chain tensioners (secondary) with its stopper pin (C) attached.

➡**Stopper pin should be attached when timing chain (secondary) is removed.**

  9. Remove oil filter from cylinder head, if necessary.

To install:

  10. Install timing chain tensioners (secondary) (1) on both sides of cylinder head.
   a. Install timing chain tensioner with its stopper pin attached.
   b. Install timing chain tensioner with sliding part facing downward on cylinder head (bank 1), and with sliding part facing upward on cylinder head (bank 2).
  11. Install oil filter, if removed.
   a. Do not project from the cylinder head surface.
  12. Install valve lifter.
   a. Install it in the original position.
  13. Install camshaft (EXH).

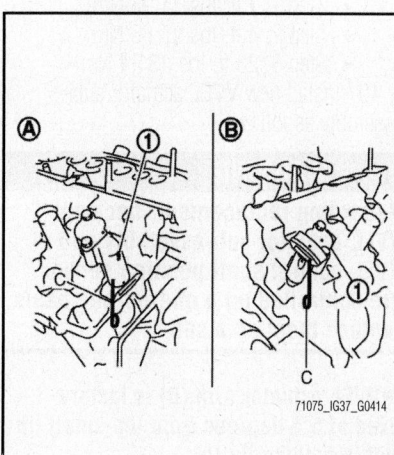

Fig. 115 Removing timing chain tensioners (secondary: 1, A: Bank 1, B: Bank 2)

   a. Distinction between camshaft (EXH) (bank 1 and bank 2) is performed with the identification mark.
  14. Install VVEL ladder assembly as follows:
   a. Apply a continuous bead of liquid gasket with tube presser (commercial service tool) to the cylinder head as shown. Use Genuine RTV Silicone Sealant or equivalent.
   b. Tighten mounting bolts in the following step, in numerical order as shown.
   c. Tighten bolts in numerical order as shown in the figure in three steps.
   • Step 1: 1 ft. lb. (1.96 Nm)
   • Step 2: 4 ft. lbs. (5.88 Nm)
   • Step 3: 8 ft. lbs. (10.4 Nm)
  15. Measure difference in levels between front end faces of VVEL ladder assembly and cylinder head.
   a. Measure two positions (both intake and exhaust side) for a single bank.
   b. If the measured value is out of the standard, re-install VVEL ladder assembly.
  16. Install rear timing chain case.
  17. Install camshaft sprockets and timing chains.
  18. Install actuator bracket (rear) as follows:

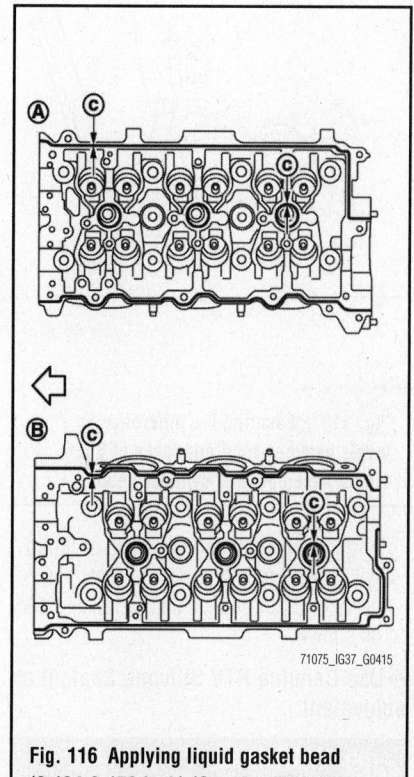

Fig. 116 Applying liquid gasket bead (0.134-0.173 inch) (Genuine RTV silicone sealant or equivalent) to Bank 1 (A) and Bank 2 (B)

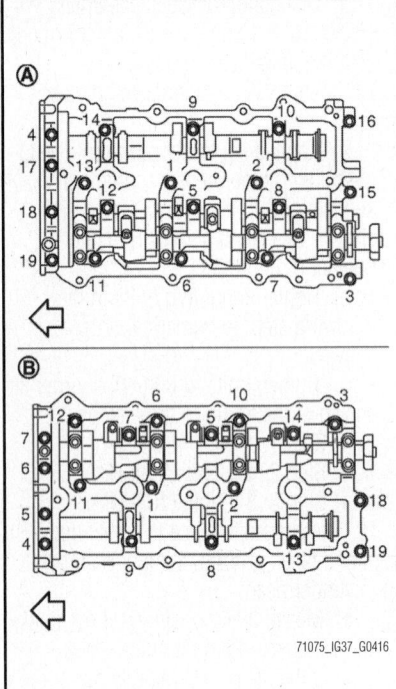

Fig. 117 Tightening mounting bolts in the order shown (A: Bank 1, B: Bank 2, arrow: engine front)

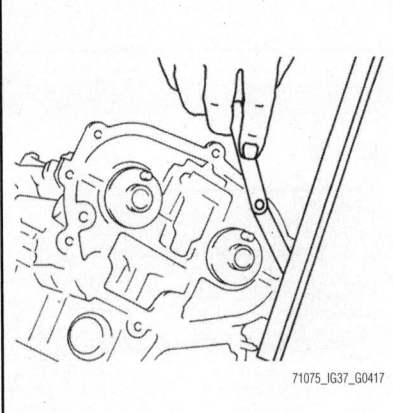

Fig. 118 Measuring the difference in levels between front end faces of VVEL ladder assembly and cylinder head

a. Apply a continuous bead of liquid gasket with tube presser (commercial service tool) to the actuator bracket (rear) as shown.

➥Use Genuine RTV Silicone Sealant or equivalent.

⁂ WARNING

Never apply gasket to the oil passage.

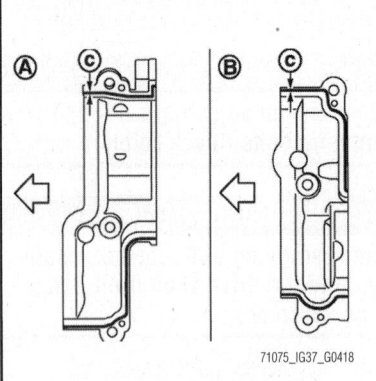

Fig. 119 Applying bead of liquid gasket (0.134-0.173 inch) to Bank 1 (A) and Bank 2 (B) (arrow: engine front)

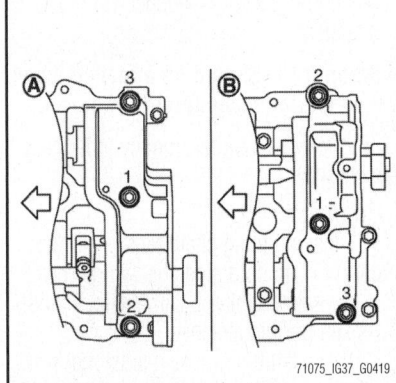

Fig. 120 Mounting bolt tightening sequence

b. Tighten mounting bolts in the following steps, in numerical order as shown.
c. Tighten bolts in three steps in numerical order as shown in the figure.
• Step 1: 1 ft. lbs. (1.96 Nm)
• Step 2: 4 ft. lbs. (5.88 Nm)
• Step 3: 23 ft. lbs. (31.4 Nm)
19. Install new VVEL actuator sub-assembly as follows:

⁂ WARNING

Regarding replacement, because VVEL actuator sub-assembly and VVEL control shaft position sensor are controlled on a one-on-one basis, replace them as a set.

➥VVEL actuator arm (B) is factory-fixed at 5.5 degrees from the small lift with a holding jig (A).

➥The holding jig is supplied in the new VVEL actuator sub-assembly.

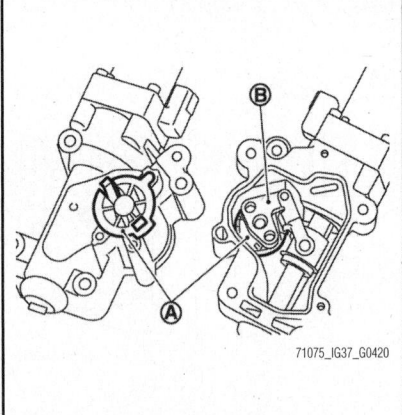

Fig. 121 VVEL actuator arm (B) is factory-fixed at 5.5° from the small lift with a holding jig (A)

⁂ WARNING

Never disassemble VVEL actuator sub-assembly. [Never loosen actuator motor mounting bolts (A) shown]

⁂ WARNING

Never shock VVEL actuator sub-assembly.

a. Move control shaft to the position of small lift stopper.
b. The position where a part of the stopper of control shaft contacts VVEL ladder bracket.

⁂ WARNING

Be careful not to damage the stopper surface.

c. If control shaft cannot be moved, set crankshaft in position referring to the information below. (To displace cam nose).

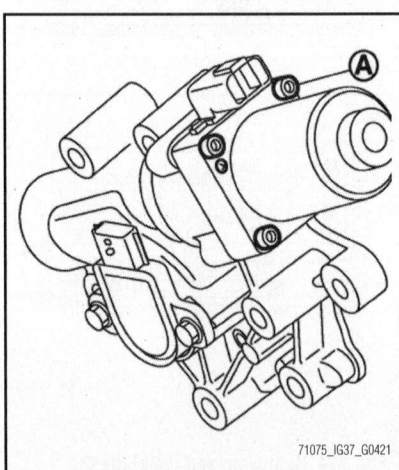

Fig. 122 Never loosen actuator motor mounting bolts (A)

Fig. 123 Holding 2 flat areas of control shaft with a wrench and rotating the control shaft (A) (5.5°(b) from the stopper) to the large lift side (1: VVEL actuator sub-assembly Bank 1)

- Bank 1: Turn 120° from No. 1 cylinder TDC
- Bank 2: No. 1 cylinder at TDC

d. Hold two flat areas of control shaft with a wrench, and rotate the control shaft (5.5 degrees from the stopper) to the large lift side. (This is for aligning the bolt hole of control shaft and the hole of VVEL actuator arm.)

e. Apply a continuous bead of liquid gasket with tube presser (commercial service tool) to the VVEL actuator sub-assembly as shown. Use Genuine RTV Silicone Sealant or equivalent.

❊❊ WARNING

Never apply gasket to the oil passage.

f. Install new VVEL actuator sub-assembly. Tighten mounting bolts in the following step, in numerical order as shown.

❊❊ WARNING

When installing, be careful with VVEL actuator sub-assembly (bank 2) mounting bolt No. 1 because its length is different.

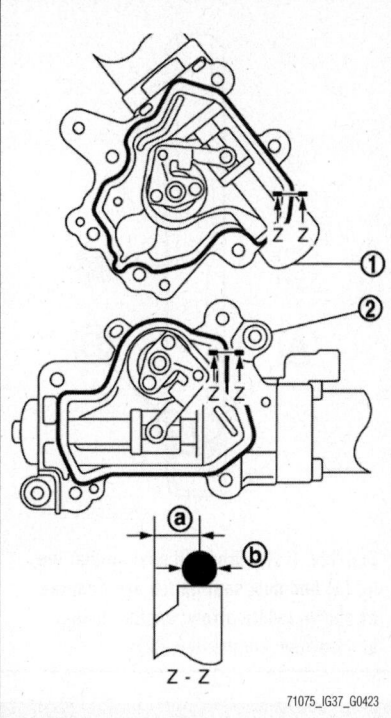

Fig. 124 Applying liquid gasket (a: 0.157-0.220 inch, b: 0.134-0.173 inch) to VVEL actuator sub-assembly (1: bank 2, 2: bank 1)

❊❊ WARNING

Be sure to check that the VVEL actuator sub-assembly is in contact with the cylinder head before tightening the mounting bolts.

g. Remove holding jig.
h. Check that VVEL actuator arm bolt hole is aligned with control shaft tapped hole. If it is not aligned, turn control shaft for alignment.
i. Fix two flat areas of control shaft with a wrench to install mounting bolts of control shaft.

❊❊ WARNING

During the operation, never allow a wrench to interfere with other parts.

❊❊ WARNING

Fix control shaft to prevent the interference of the stopper surface.

20. Install new VVEL control shaft position sensor as follows:

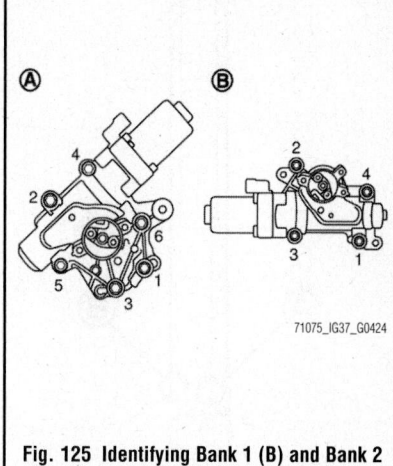

Fig. 125 Identifying Bank 1 (B) and Bank 2 (A) mounting bolts tightening sequence

❊❊ WARNING

Regarding replacement, because VVEL actuator sub-assembly and VVEL control shaft position sensor are controlled on a one-on-one basis, replace them as a set.

a. Apply engine oil to O-ring or contact surface of O-ring.
b. Align matching marks (B) of VVEL control shaft position sensor and upper housing.
c. Face connector toward matching mark (A).
d. Temporarily tighten bolt.
e. Adjust VVEL control shaft position sensor after setting the engine assembly in the vehicle.

❊❊ WARNING

Be sure to adjust VVEL control shaft position sensor.

f. After adjusting VVEL control shaft position sensor, tighten bolts to the specified torque.
21. Inspect the valve clearance.
22. Install in the reverse order of removal after this step.

CRANKSHAFT FRONT SEAL

REMOVAL & INSTALLATION

G25

See Figure 126.

1. Remove the following parts:
- Engine undercover with power tool.
- Drive belt
- Crankshaft pulley
2. Remove front oil seal using a suitable tool.

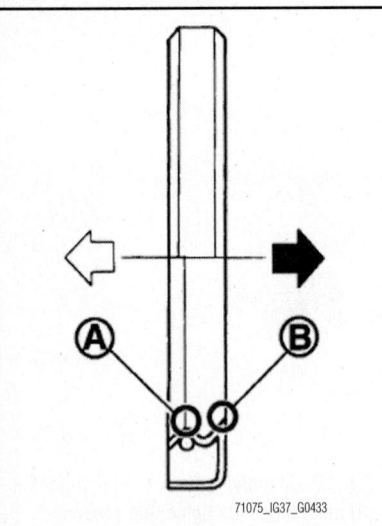

Fig. 126 Installing front oil seal so that the front oil seal lip (A) and dust seal lip (B) are oriented as shown (white arrow: engine inside, black arrow: outside)

❄❄ WARNING

Be careful not to damage front timing chain case and crankshaft.

To install:

3. Apply new engine oil to both oil seal lip and dust seal lip of new front oil seal.

4. Install front oil seal.

   a. Install front oil seal so that each seal lip is oriented as shown.

   b. Using a suitable drift, press-fit until the height of front oil seal is level with the mounting surface.

   c. Suitable drift: outer diameter 2.36 inch (60 mm), inner diameter 1.97 inch (50 mm).

   d. Check the garter spring is in position and seal lips not inverted

❄❄ WARNING

Be careful not to damage front timing chain case and crankshaft.

   e. Press-fit straight and avoid causing burrs or tilting oil seal.

5. Install in the reverse order of removal after this step.

G37

See Figure 127.

1. Remove the following parts:
   • Engine undercover with power tool
   • Drive belt
   • Crankshaft pulley

2. Remove front oil seal using a suitable tool.

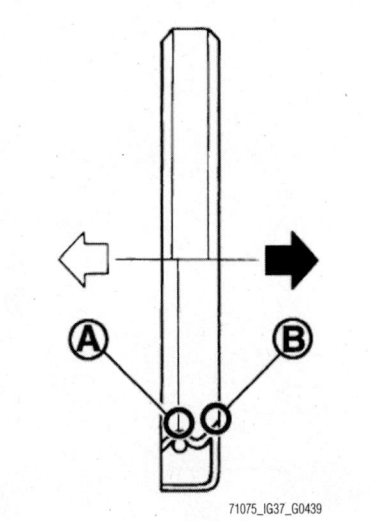

Fig. 127 Install front oil seal so that the lip (A) and dust seal lip (B) are oriented as shown (white arrow: engine inside, black arrow: engine outside)

❄❄ WARNING

Be careful to damage front timing chain case and crankshaft.

To install:

3. Apply new engine oil to both oil seal lip and dust seal lip of new front oil seal.

4. Install front oil seal.

   a. Install front oil seal so that each seal lip is oriented as shown.

   b. Using a suitable drift, press-fit until the height of front oil seal is level with the mounting surface.

   c. Suitable drift: outer diameter 2.36 inch (60 mm), inner diameter 1.97 inch (50 mm).

5. Check that the garter spring is in position and seal lips are not inverted.

   a. Be careful not to damage front timing chain case and crankshaft.

   b. Press-fit straight and avoid causing burrs or tilting oil seal.

   c. Install in the reverse order of removal after this step.

CRANKSHAFT PULLEY

REMOVAL & INSTALLATION

G25

See Figures 128 through 131.

1. Remove rear cover plate and set the ring gear stopper as shown.

   a. Loosen crankshaft pulley bolt

and rotate bolt seating surface at 0.39 inch (10 mm) from its original position.

❄❄ WARNING

Never remove crankshaft pulley bolt as it will be used as a supporting point for suitable puller.

   b. Place suitable puller tab on holes of crankshaft pulley, and pull crankshaft pulley through.

❄❄ WARNING

Never put suitable puller tab on crankshaft pulley periphery, as this will damage internal damper.

To install:

2. Fix crankshaft using the ring gear stopper [SST: KV10118600].

3. Install crankshaft pulley, taking care not to damage front oil seal.

   a. When press-fitting crankshaft pulley with plastic hammer, tap on its center portion (not circumference).

4. Tighten crankshaft pulley bolt to 33 ft. lbs. (44.1 Nm).

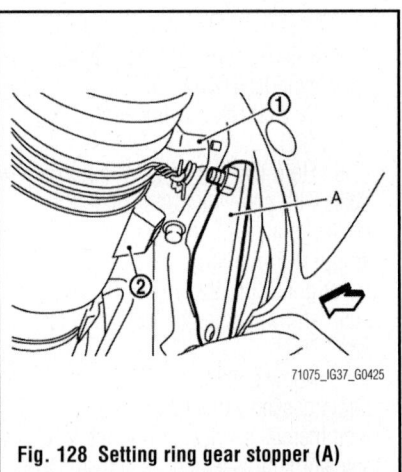

Fig. 128 Setting ring gear stopper (A)

Fig. 129 Loosening the crankshaft pulley (1) bolt

Fig. 130 Placing suitable puller tab on holes of crankshaft pulley, and pull crankshaft pulley through

Fig. 132 Setting ring gear stopper (A)

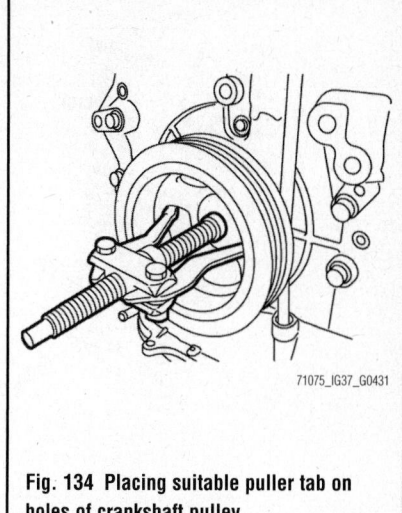

Fig. 134 Placing suitable puller tab on holes of crankshaft pulley

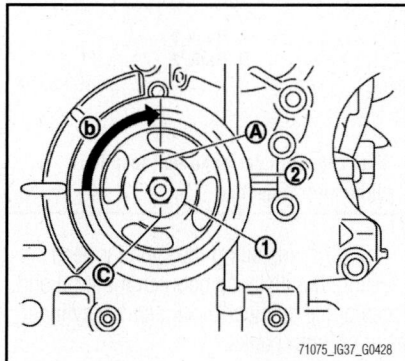

Fig. 131 Placing match mark (A) on crankshaft pulley (2) aligning with match mark (C) of crankshaft pulley bolt (1) and tighten bolt 90° (B)

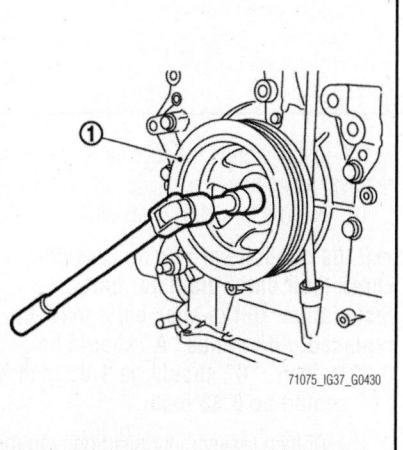

Fig. 133 Loosening crankshaft pulley (1) bolt

5. Place a matching mark (A) on crankshaft pulley (2) aligning with the matching mark (C) of crankshaft pulley bolt (1). Tighten the bolt 90 degrees (one marks) (b).

6. Rotate crankshaft pulley in normal direction (clockwise when viewed from front) to confirm it turns smoothly.

G37

See Figures 131 through 134.

1. Remove rear cover plate and set the ring gear stopper.

2. Loosen crankshaft pulley bolt and rotate bolt seating surface at 0.39 inch (10 mm) from its original position.

> ✳✳ **WARNING**
>
> **Never remove crankshaft pulley bolt because it is used as a supporting point for suitable puller.**

3. Placing suitable puller tab on holes of crankshaft pulley, and pull crankshaft pulley through.

> ✳✳ **WARNING**
>
> **Never put suitable puller tab on crankshaft pulley periphery, because this damages internal damper.**

To install:

4. Fix crankshaft using the ring gear stopper.

5. Install crankshaft pulley, taking care not to damage front oil seal.

 a. When press-fitting crankshaft pulley with plastic hammer, tap on its center portion (not circumference).

6. Tighten crankshaft pulley bolt to 33 ft. lbs. (44.1 Nm).

7. Tighten the bolt 90 degrees (one mark) (b).

 a. Place a matching mark (A) on crankshaft pulley (2) aligning with the matching (C) of crankshaft pulley bolt (1).

8. Rotate crankshaft pulley in normal direction (clockwise when viewed from front) to confirm it turns smoothly.

CYLINDER HEAD

REMOVAL & INSTALLATION

See Figures 135 through 138.

At this time the manufacturer provides service information for this component with the engine removed from the vehicle and positioned in a suitable holding fixture.

➡ **Whenever the negative battery cable is disconnected the following components will require resetting. The Automatic temperature control system, Automatic drive positioner, Power window control, Sunroof system, Sunshade system, Rear view monitor, Idle Air Volume Learning, Steering Angle Sensor Neutral Position, Audio presets and Navigation. You will need the CONSULT-III diagnostic tool, or equivalent. Follow the directions on the screen of the tool, as needed.**

1. Before servicing the vehicle, refer to the Precautions Section.

➡ **If working near and/or around the SRS system and components, be sure to disable the SRS system. After disabling the system wait three minutes or more before servicing the vehicle.**

2. Disconnect the negative battery cable.

3. Remove the engine and position it in a suitable holding fixture.

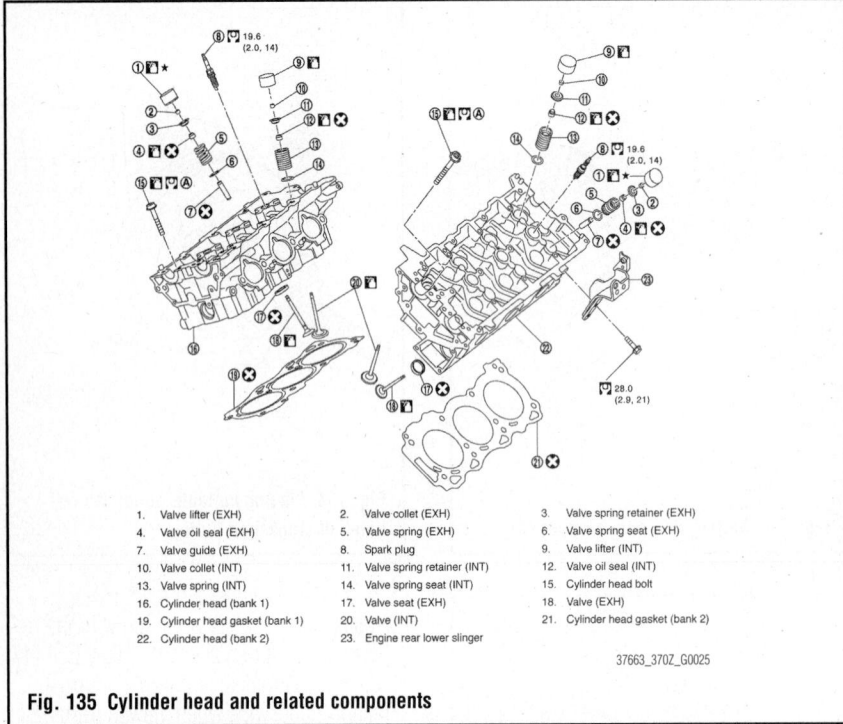

1. Valve lifter (EXH)
2. Valve collet (EXH)
3. Valve spring retainer (EXH)
4. Valve oil seal (EXH)
5. Valve spring (EXH)
6. Valve spring seat (EXH)
7. Valve guide (EXH)
8. Spark plug
9. Valve lifter (INT)
10. Valve collet (INT)
11. Valve spring retainer (INT)
12. Valve oil seal (INT)
13. Valve spring (INT)
14. Valve spring seat (INT)
15. Cylinder head bolt
16. Cylinder head (bank 1)
17. Valve seat (EXH)
18. Valve (EXH)
19. Cylinder head gasket (bank 1)
20. Valve (INT)
21. Cylinder head gasket (bank 2)
22. Cylinder head (bank 2)
23. Engine rear lower slinger

37663_370Z_G0025

Fig. 135 Cylinder head and related components

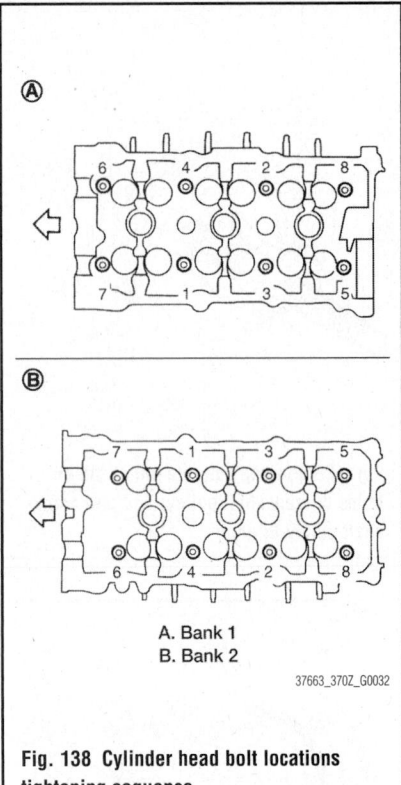

A. Bank 1
B. Bank 2

37663_370Z_G0032

Fig. 138 Cylinder head bolt locations tightening sequence

4. Remove the intake manifold collector.

5. Remove the fuel tube and fuel injector assembly.

6. Remove the intake manifold.

7. Remove the valve covers.

8. Remove the exhaust manifold.

9. Remove the water inlet and thermostat assembly. Remove the water pipe and heater pipe assemblies.

10. Remove the timing chain.

11. Remove the Camshaft.

12. Remove the cylinder head retaining bolts in the reverse order of the tightening sequence.

13. Remove the cylinder head gaskets. Discard the gaskets.

To install:

➡Be sure to use new fasteners, as required.

14. Installation is the reverse of the removal procedure.

15. Be sure to use new gaskets.

➡If the old bolts are being reused check their outer diameter before installation. Out of spec bolts must be replaced. "B" minus "A" should be 0.0071 inch. "C" should be 1.89 inch. "D" should be 0.43 inch.

16. Tighten the cylinder head bolts in the proper sequence and to specification. Coat the bolt threads with clean engine oil before installation.

17. Specification is 77 ft. lbs. (105 Nm) first pass. Completely loosen all bolts, in the reverse order of the tightening sequence. Tighten all bolts to 30 ft. lbs. (40.0 Nm) in the proper sequence. Finally turn all bolts 95 degrees clockwise (angle tightening) in the proper sequence.

18. After installing the cylinder head measure the distance between the front end faces of the cylinder block and the cylinder head on both banks.

19. Specification should be 0.555–0.587 inch (14.1–14.9 mm).

20. Continue the installation in the reverse order of the removal procedure.

21. Be sure to perform the reconnect/relearn procedures.

ENGINE COVER

REMOVAL & INSTALLATION

See Figures 139 and 140.

1. Loosen mounting bolts and nuts in the reverse order as shown in the figure, and then remove engine cover.

✳✳ WARNING

Never damage or scratch engine cover when installing or removing.

To install:

Install engine cover, and then tighten mounting bolts and nuts in numerical order as shown.

✳✳ WARNING

Never damage or scratch engine cover when installing or removing.

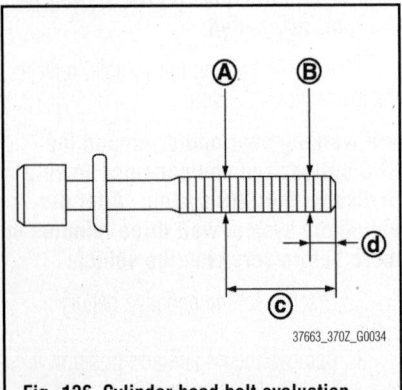

37663_370Z_G0034

Fig. 136 Cylinder head bolt evaluation

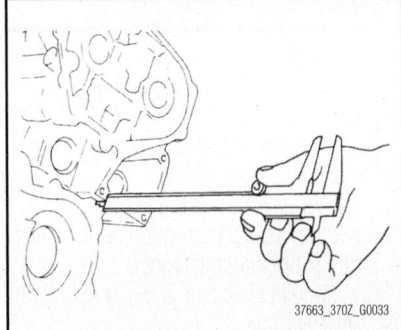

37663_370Z_G0033

Fig. 137 Cylinder head to cylinder block measurement check

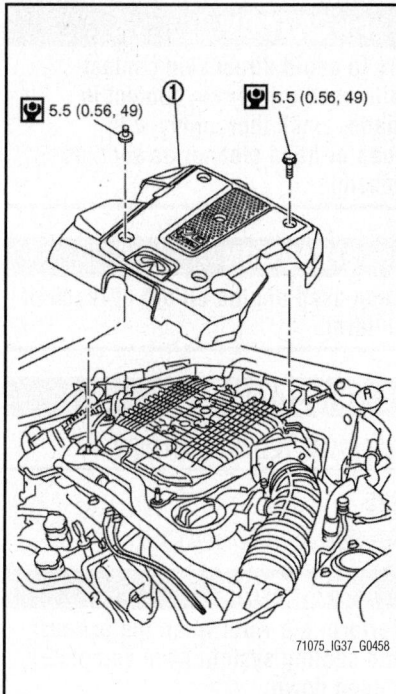

Fig. 139 Removing and installing the engine cover

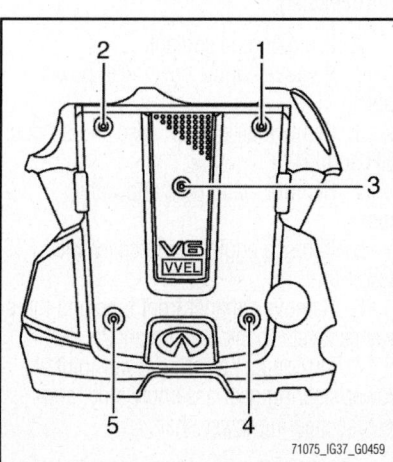

Fig. 140 Engine cover bolt removal and installation sequence

ENGINE OIL & FILTER

OIL LEVEL CHECK

See Figures 141 and 142.

1. Park the vehicle on a level surface and apply the parking brake.
2. Run the engine until it reaches operating temperature.
3. Turn off the engine. Wait more than 15 minutes for the oil to drain back into the oil pan.
4. Remove the dipstick and wipe it clean. Reinsert it all the way.
5. Remove the dipstick again and check

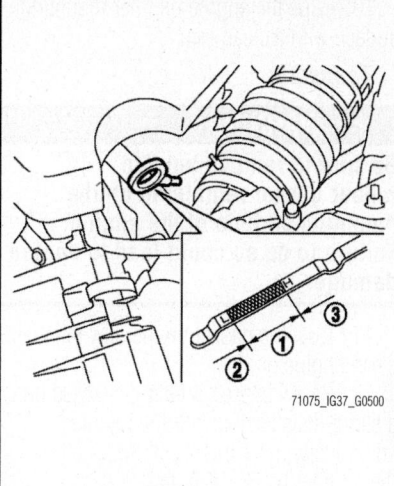

Fig. 141 Checking oil level—G25

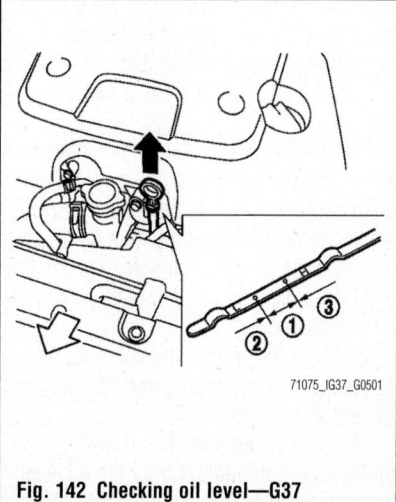

Fig. 142 Checking oil level—G37

the oil level. It should be within the range *1. If the oil level is below *2, remove the oil filler cap and pour recommended oil through the opening. Do not overfill *3.

6. Recheck oil level with the dipstick. It is normal to add some oil between oil maintenance intervals or during the break in period, depending on the severity of operating conditions.

> ✽✽ **WARNING**
>
> **Oil level should be checked regularly. Operating the engine with an insufficient amount of oil can damage the engine, and such damage is not covered by warranty.**

OIL & FILTER CHANGE

See Figures 143 through 146.

> ✽✽ **CAUTION**
>
> **Be careful not to get burned when engine and engine oil may be hot.**

> ✽✽ **WARNING**
>
> **When removing, prepare a shop cloth to absorb any engine oil leakage or spillage.**

> ✽✽ **WARNING**
>
> **Never allow engine oil to adhere to drive belt.**

> ✽✽ **WARNING**
>
> **Completely wipe off any engine oil that adheres to engine and vehicle.**

1. Park the vehicle on a level surface and apply the parking brake.
2. Run the engine until it reaches operating temperature.
3. Turn the engine off and wait more than 15 minutes.
4. Raise and support the vehicle using a suitable floor jack and safety jack stands.
   a. Place the safety jack stands under the vehicle jack-up points.
   b. A suitable adapter should be attached to the jack stand saddle.
5. Remove the plastic engine undercover.
   a. Remove the small plastic clip at the center point of the undercover.
   b. Then remove the other bolts that hold the undercover in place.
6. Place a large drain pan under the drain plug.
7. Remove the oil filler cap.
8. Remove the drain plug *1 with a wrench and completely drain the oil.

> ✽✽ **CAUTION**
>
> **Be careful not to burn yourself, as the engine oil is hot.**

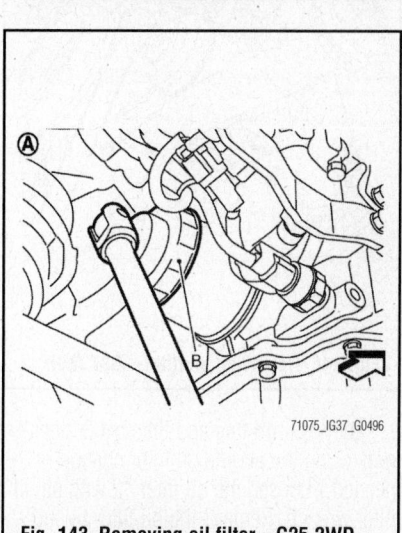

Fig. 143 Removing oil filter—G25 2WD

Fig. 144 Removing oil filter—G25 AWD

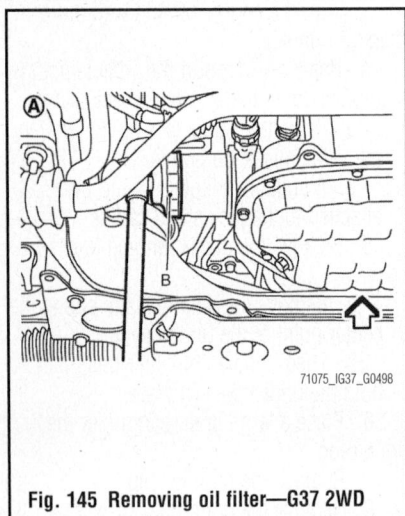

Fig. 145 Removing oil filter—G37 2WD

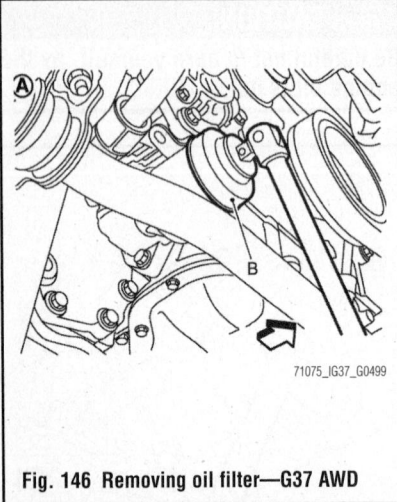

Fig. 146 Removing oil filter—G37 AWD

9. (Perform this and the next 3 steps only when the engine oil filter change is needed.) Loosen the oil filter *2 with an oil filter wrench. Remove the oil filter by turning it by hand.

10. Wipe the engine oil filter mounting surface with a clean rag.

To install:

✳✳ WARNING

Be sure to remove any old rubber gasket remaining on the mounting surface of the engine. Failure to do so could lead to engine damage.

11. Coat the gasket on the new filter with clean engine oil.

12. Screw in the oil filter clockwise until a slight resistance is felt, then tighten additionally more than 2/3 turn. Tighten the oil filter to 11–15 ft. lbs. (14.7–20.5 Nm).

13. Clean and re-install the drain plug with a new washer. Securely tighten the drain plug with a wrench. Tighten the drain plug to 22–29 ft. lbs. (29–39 Nm). Do not use excessive force.

14. Refill engine with recommended oil and install the oil filler cap securely. The drain and refill capacity depends on the oil temperature and drain time. Use these specifications for reference only. Always use the dipstick to determine the proper amount of oil in the engine.

15. Start the engine and check for leakage around the drain plug and the oil filter. Correct as required.

16. Turn the engine off and wait more than 15 minutes. Check the oil level with the dipstick. Add engine oil if necessary.

17. Install the engine undercover into position as the following steps.

   a. Pull the center of the small plastic clip out.

   b. Hold the engine undercover into position.

   c. Insert the clip through the undercover into the hole in the frame, then push the center of the clip in to lock the clip in place.

   d. Install the other bolts that hold the undercover in place. Be careful not to strip the bolts or over-tighten them.

18. Lower the vehicle carefully to the ground.

19. Dispose of waste oil and filter properly.

✳✳ CAUTION

Prolonged and repeated contact with used engine oil may cause skin cancer.

✳✳ CAUTION

Try to avoid direct skin contact with used oil. If skin contact is made, wash thoroughly with soap or hand cleaner as soon as possible.

✳✳ CAUTION

Keep used engine oil out of reach of children.

EXHAUST MANIFOLD

REMOVAL & INSTALLATION

G25

See Figures 147 through 150.

✳✳ CAUTION

Perform the work when the exhaust and cooling system have completely cooled down.

➡When removing bank 1 side parts only, steps 1 and 4 are unnecessary.

1. Drain engine coolant.
2. Remove engine cover with power tool.
3. Remove air cleaner case and air duct (RH and LH).
4. Remove water pipe and water hose.
5. Remove engine undercover with power tool.
6. Remove exhaust front tube and three way catalysts (bank 1 and bank 2).
7. Disconnect steering lower joint at power steering gear assembly side, and release steering lower shaft.

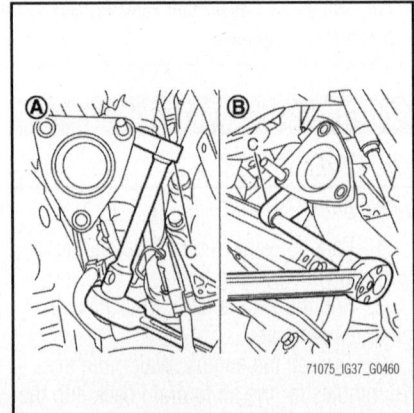

Fig. 147 Using heated oxygen sensor wrench (C) to remove air fuel ration sensor 1 (A: Bank 2, B: Bank 1)

8. Disconnect air fuel ratio sensor 1 (bank 1 and bank 2) harness connectors and remove harness clip.

9. Using the heated oxygen sensor wrench [SST: KV10114400] (C), remove air fuel ratio sensor 1 (bank 1 and bank 2).

※※ WARNING

Be careful not to damage air fuel ratio sensor 1.

※※ WARNING

Discard any air fuel ratio sensor 1 that has been dropped onto a hard surface such as a concrete floor. Replace with a new sensor.

10. Remove exhaust manifold cover (upper) (bank 1 and bank 2).

11. Loosen mounting nuts in the reverse order as shown in the figure to remove exhaust manifold.

➡**Disregard the numerical order No. 7 and 8 in removal.**

12. Remove gaskets.

※※ WARNING

Cover engine openings to avoid entry of foreign materials.

To install:

Note the following, and install in the reverse order or removal.

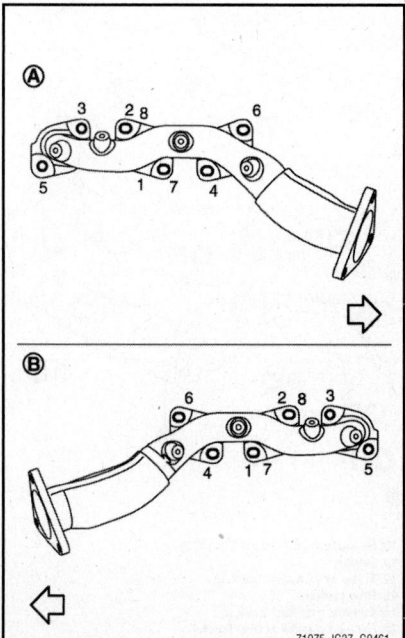

Fig. 148 Mounting nut removal and installation sequence (A: Bank 1, B: Bank: 2, arrow: Engine front)

71075_IG37_G0461

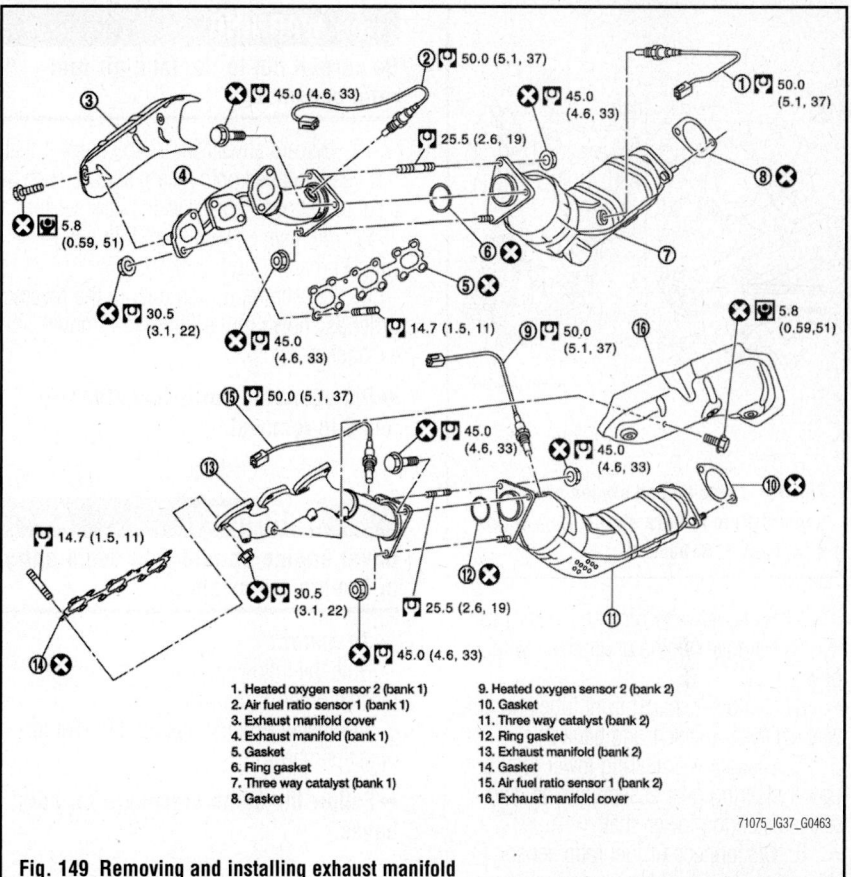

1. Heated oxygen sensor 2 (bank 1)
2. Air fuel ratio sensor 1 (bank 1)
3. Exhaust manifold cover
4. Exhaust manifold (bank 1)
5. Gasket
6. Ring gasket
7. Three way catalyst (bank 1)
8. Gasket
9. Heated oxygen sensor 2 (bank 2)
10. Gasket
11. Three way catalyst (bank 2)
12. Ring gasket
13. Exhaust manifold (bank 2)
14. Gasket
15. Air fuel ratio sensor 1 (bank 2)
16. Exhaust manifold cover

71075_IG37_G0463

Fig. 149 Removing and installing exhaust manifold

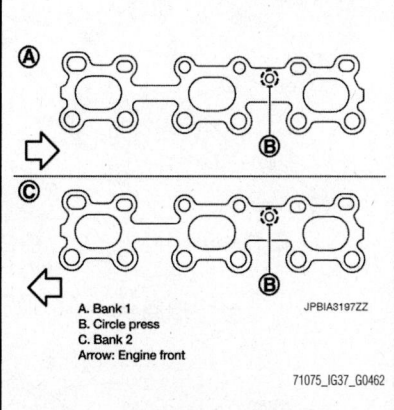

A. Bank 1
B. Circle press
C. Bank 2
Arrow: Engine front

JPBIA3197ZZ

71075_IG37_G0462

Fig. 150 Installing exhaust manifold gasket

13. Install the exhaust manifold gasket in direction shown. (Follow the same procedure for both banks.)

14. If stud bolts were removed, install then and tighten to specification.

15. Install exhaust manifold and tighten mounting bolts in numerical order as shown.

➡**Tighten nuts No. 1 and 2 in two steps. The numerical order No. 7 and 8 shows the second step.**

※※ WARNING

Before installing a new air fuel ratio sensor 1, clean exhaust system threads using heated oxygen sensor thread cleaner tool (commercial service tool) and apply anti-seize lubricant.

※※ WARNING

Never apply excessive torque to air fuel ratio sensor 1. Doing so may cause damage to air fuel ratio sensor 1, resulting in the "MIL" illuminating.

G37

See Figures 151 through 154.

※※ WARNING

Perform the work when the exhaust and cooling system have completely cooled down.

➡**When removing bank 1 side parts only, steps 1, 4 and 7 are unnecessary.**

1. Drain engine coolant.
2. Remove engine cover with power tool.
3. Remove air cleaner case and air duct (RH and LH).

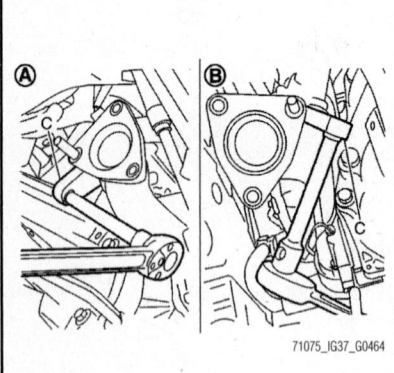

Fig. 151 Using heated oxygen sensor wrench (C) to remove air fuel ratio sensor 1 (A: bank 1, B: bank 2)

4. Remove water pipe and water hose.

5. Remove engine undercover with power tool.

6. Remove exhaust front tube and three way catalysts (bank 1 and bank 2).

7. Disconnect steering lower joint at power steering gear assembly side, and release steering lower shaft.

8. Disconnect air fuel ratio sensor 1 (bank 1 and bank 2) harness connectors and remove harness clip.

9. Using the heated oxygen sensor wrench [SST: KV10114400 (J-38365)] (C), remove air fuel ratio sensor 1 (bank 1 and bank 2).

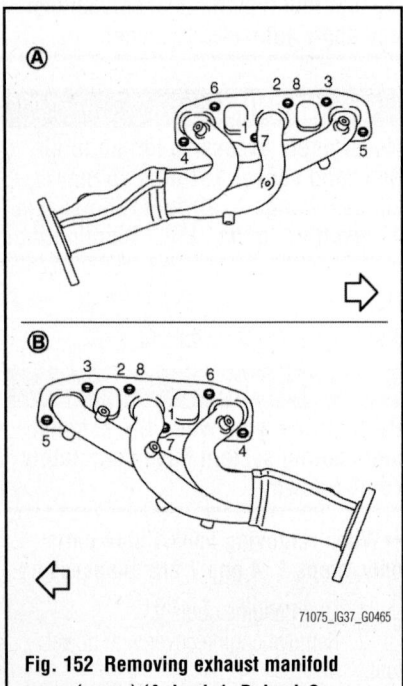

Fig. 152 Removing exhaust manifold cover (upper) (A: bank 1, B: bank 2, arrow: engine front)

> ❋❋ **WARNING**
>
> **Be careful not to damage air fuel ratio sensor 1.**

10. Discard any air fuel ratio sensor 1 that has been dropped onto a hard surface such as a concrete floor. Replace with a new sensor.

11. Remove exhaust manifold cover (upper) (bank 1 and bank 2).

12. Loosen mounting nuts in the reverse order as shown in the figure to remove exhaust manifold.

➡ **Disregard the numerical order No. 7 and 8 in removal.**

13. Remove gaskets.

> ❋❋ **WARNING**
>
> **Cover engine openings to avoid entry of foreign materials.**

To install:

Note the following, and install in the reverse order of removal.

14. Install exhaust manifold gasket in direction shown.

➡ **Follow the same procedure for both banks.**

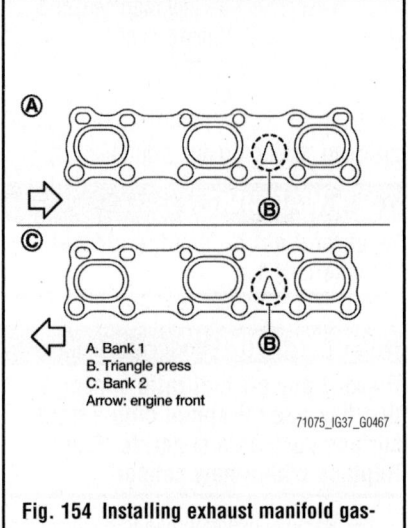

A. Bank 1
B. Triangle press
C. Bank 2
Arrow: engine front

Fig. 154 Installing exhaust manifold gasket

15. If stud bolts were removed, install them and tighten to the specification.

16. Install exhaust manifold and tighten mounting bolts in numerical order.

➡ **Tighten nuts No. 1 and 2 in two steps. The numerical order No. 7 and 8 shows the second step.**

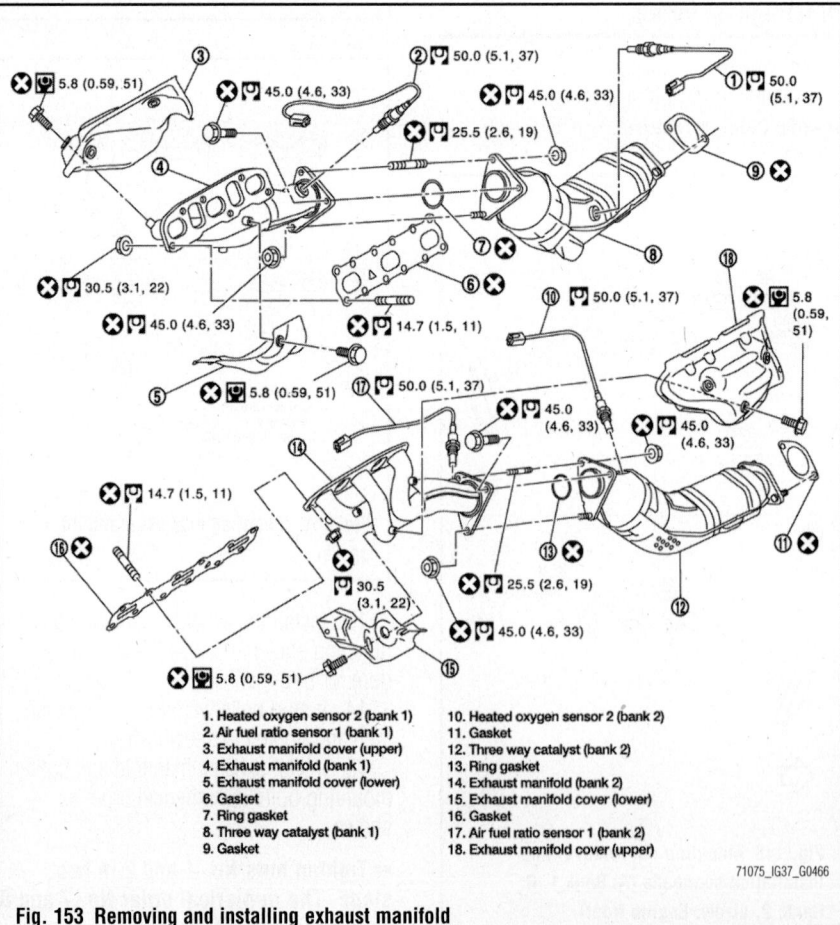

1. Heated oxygen sensor 2 (bank 1)
2. Air fuel ratio sensor 1 (bank 1)
3. Exhaust manifold cover (upper)
4. Exhaust manifold (bank 1)
5. Exhaust manifold cover (lower)
6. Gasket
7. Ring gasket
8. Three way catalyst (bank 1)
9. Gasket
10. Heated oxygen sensor 2 (bank 2)
11. Gasket
12. Three way catalyst (bank 2)
13. Ring gasket
14. Exhaust manifold (bank 2)
15. Exhaust manifold cover (lower)
16. Gasket
17. Air fuel ratio sensor 1 (bank 2)
18. Exhaust manifold cover (upper)

Fig. 153 Removing and installing exhaust manifold

> ※※ **WARNING**
>
> Before installing a new air fuel ratio sensor 1, clean exhaust system threads using heated oxygen sensor thread cleaner tool (commercial service tool) and apply anti-seize lubricant.

> ※※ **WARNING**
>
> Never apply excessive torque to air fuel ratio sensor 1. Doing so may cause damage to air fuel ratio sensor 1, resulting in the MIL illuminating.

INTAKE MANIFOLD

REMOVAL & INSTALLATION

G25

See Figures 155 through 159.

1. Release fuel pressure.

> ※※ **WARNING**
>
> Never drain engine coolant when the engine is hot to avoid the danger of being scalded.

2. Remove engine cover with power tool.

3. Remove air cleaner case and air duct (RH and LH).

4. Remove electric throttle control actuator as follows:

   a. Drain engine coolant. When water hoses are disconnected, attach plug to prevent engine coolant leakage.

> ※※ **WARNING**
>
> Perform this step when engine is cold.

> ※※ **WARNING**
>
> Never spill engine coolant on drive belt.

   b. Disconnect water hoses from electric throttle control actuator. When engine coolant is not drained from radiator, attach plug to water hoses to prevent engine coolant leakage.

   c. Disconnect harness connector.

   d. Loosen mounting bolts in reverse order as shown.

➡When removing only intake manifold collector, move electric throttle control actuator without disconnecting the water hose.

➡The figure shows the electric throttle control actuator (bank 1) viewed from the air duct side.

➡Viewed from the air duct side, the order of loosening mounting bolts of electric throttle control actuator (bank 2) is the same as that of the electric throttle control actuator (bank 1).

> ※※ **WARNING**
>
> Handle carefully to avoid any shock to electric throttle control actuator.

5. Disconnect vacuum hose, PCV hose and EVAP hose from intake manifold collector.

6. Remove EVAP canister purge volume control solenoid valve and EVAP tube assembly from intake manifold collector.

7. Loosen mounting bolts and nuts with power tool in the reverse order as shown to remove intake manifold collector.

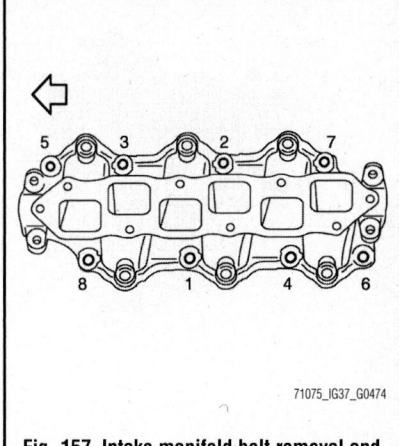

Fig. 157 Intake manifold bolt removal and installation sequence

8. Remove fuel tube and fuel injector assembly.

9. Loosen mounting bolts in reverse order as shown to remove intake manifold with power tool.

> ※※ **WARNING**
>
> Cover engine openings to avoid entry of foreign materials.

> ※※ **WARNING**
>
> Put a mark on the intake manifold and the cylinder head with paint before removal because they need to be installed in the specified direction.

10. Remove gaskets.

To install:

Note the following, and install in the reverse order of removal.

11. If stud bolts were removed, install them and tighten to 8 ft. lbs. (10.8 Nm).

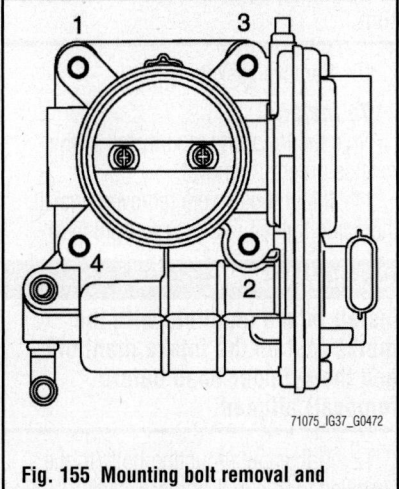

Fig. 155 Mounting bolt removal and installation sequence

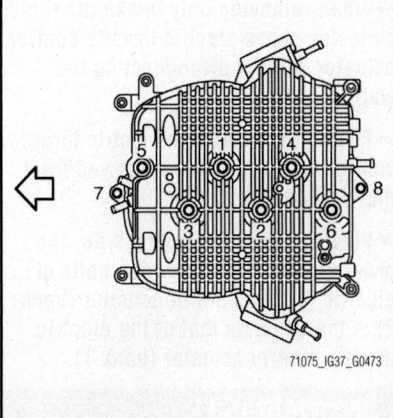

Fig. 156 Removing and installing the intake manifold collector

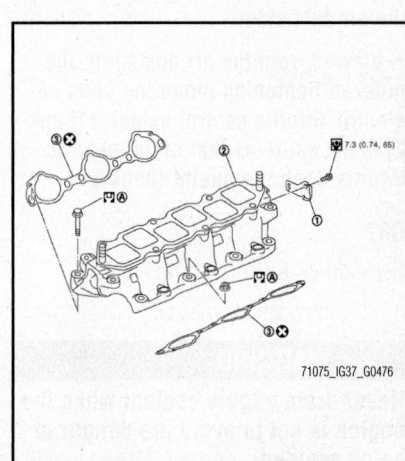

Fig. 158 Removing and installing the intake manifold (2), harness bracket (1) and gasket (3)

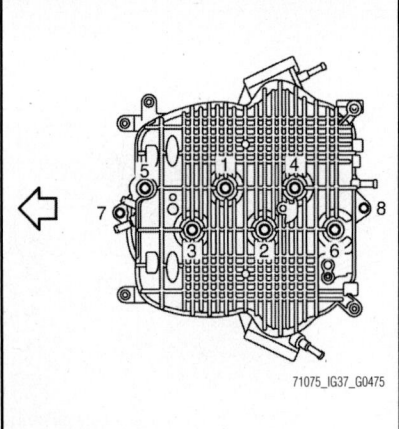

71075_IG37_G0475

Fig. 159 Identifying bolt tightening sequence

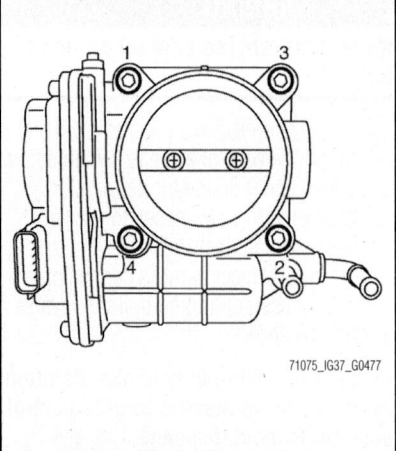

71075_IG37_G0477

Fig. 160 Mounting bolt removal and installation sequence

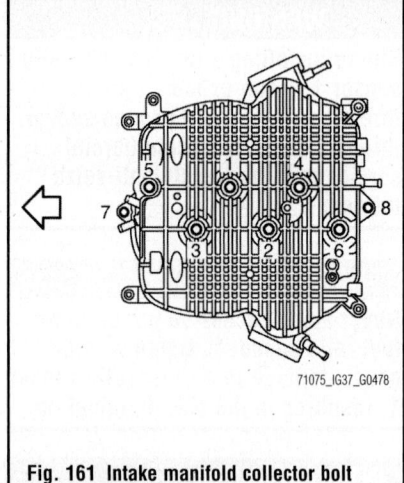

71075_IG37_G0478

Fig. 161 Intake manifold collector bolt removal and installation sequence

※※ WARNING

Install intake manifold with the marks (put on the intake manifold and the cylinder head before removal) aligned.

12. Tighten all mounting bolts to the specified torque in 2 or more steps in numerical order as shown.
- Step 1: 5 ft. lbs. (7.4 Nm)
- Step 2: 21 ft. lbs. (29 Nm)

13. Install the intake manifold collector.

a. If stud bolts were removed, install them and tighten to 8 ft. lbs. (10.8 Nm).

b. Tighten mounting bolts and nuts in numerical order as shown.

14. Insert water hose by 1.06-1.26 inch (27-32 mm) from connector end.

15. Clamp the hose at location of 0.12-0.28 (3-7 mm) from connector end.

16. Tighten in numerical order.

➡The figure shows the electric throttle control actuator (bank 1) viewed from the air duct side.

➡Viewed from the air duct side, the order of tightening mounting bolts of electric throttle control actuator (bank 2) is the same as that of the electric throttle control actuator (bank 1).

G37

See Figures 160 through 163.

1. Release fuel pressure.

※※ WARNING

Never drain engine coolant when the engine is hot to avoid the danger of being scalded.

2. Remove engine cover with power tool.

3. Remove air cleaner case and air duct (RH and LH).

4. Remove electric throttle control actuator as follows:

a. Drain engine coolant. When water hoses are disconnected, attach plug to prevent engine coolant leakage.

※※ WARNING

Perform this step when engine is cold.

※※ WARNING

Never spill engine coolant on drive belt.

b. Disconnect water hoses from electric throttle control actuator. When engine coolant is not drained from radiator, attach plug to water hoses to prevent engine coolant leakage.

c. Disconnect harness connector.

d. Loosen mounting bolts in REVERSE order as shown.

➡When removing only intake manifold collector, move electric throttle control actuator without disconnecting the water hose.

➡The figure shows the electric throttle control actuator (bank 1) viewed from the air duct side.

➡Viewed from the air duct side, the order of loosening mounting bolts of electric throttle control actuator (bank 2) is the same as that of the electric throttle control actuator (bank 1).

※※ WARNING

Handle carefully to avoid any shock to electric throttle control actuator.

5. Disconnect vacuum hose, PCV hose and EVAP hose from intake manifold collector.

6. Remove EVAP canister purge volume control solenoid valve and EVAP tube assembly from intake manifold collector.

7. Loosen mounting bolts and nuts with power tool in the reverse order as shown to remove intake manifold collector

8. Remove fuel tube and fuel injector assembly.

9. Loosen mounting bolts in reverse order as shown in the figure to remove intake manifold with power tool.

※※ WARNING

Cover engine openings to avoid entry of foreign materials.

※※ WARNING

Put a mark on the intake manifold and the cylinder head with paint before removal because they need to be installed in the specified direction.

10. Remove gaskets.

To install:

Note the following, and install in the reverse order of removal.

11. If stud bolts were removed, install them and tighten to 8 ft. lbs. (10.8 Nm).

※※ WARNING

Install intake manifold with the marks (put on the intake manifold and the cylinder head before removal) aligned.

12. Tighten all mounting bolts to the specified torque in 3 or more steps in numerical order.

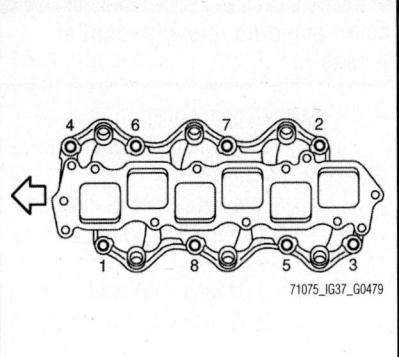

Fig. 162 Identifying intake manifold bolt removal and installation sequence

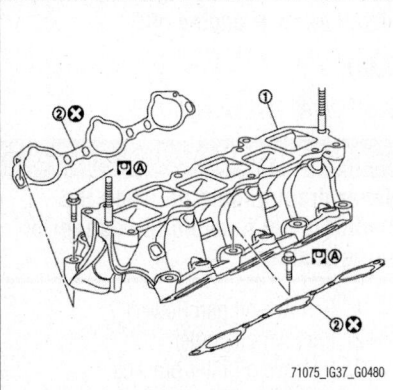

Fig. 163 Removing and installing the intake manifold (1) and gasket (2)

- Step 1: 5 ft. lbs. (7.4 Nm)
- Step 2: 19 ft. lbs. (25.5 Nm)

13. If the stud bolts on the intake manifold collector were removed, install and tighten them to 8 ft. lbs. (10.8 Nm).

14. Tighten the mounting bolts and nuts in numerical order.

15. Insert the water hose by 1.06-1.26 inch (27-32 mm) from connector end.

16. Clamp hose at location of 0.12-0.28 inch (3-7 mm) from hose end.

17. Tighten the electric throttle control actuator (Bank 1 and 2) in numerical order as shown.

➡ **The figure shows the electric throttle control actuator (bank 1) viewed from the air duct side.**

➡ **Viewed from the air duct side, the order of tightening mounting bolts of electric throttle control actuator (bank 2) is the same as that of the electric throttle control actuator (bank 1).**

OIL PAN

REMOVAL & INSTALLATION

G25

Lower

See Figure 164.

✳✳ CAUTION

Never drain engine oil when the engine is hot to avoid the danger of being scalded.

1. Remove engine undercover with power tool.
2. Drain engine oil.
3. Remove oil pan (lower) as follows:
    a. Loosen mounting bolts in reverse order as shown to remove.
    b. Insert the seal cutter between oil pan (upper) and oil pan (lower).

✳✳ WARNING

Be careful not to damage the mating surfaces.

✳✳ WARNING

Never insert a screwdriver. This damages the mating surfaces.

    c. Slide the seal cutter by tapping on the side of tool with a hammer. Remove oil pan (lower).

To install:

4. use scraper to remove oil liquid gasket from mating surfaces.
    a. Remove oil liquid gasket from the bolt holes and thread.

✳✳ WARNING

Never scratch or damage the mating surfaces when cleaning off oil liquid gasket.

5. Apply a continuous bead of liquid gasket with the tube presser to the oil pan (lower) about 0.157-0.197 inch wide.

➡ **Use Genuine RTV Silicone Sealant or equivalent.**

✳✳ WARNING

Attaching should be done within 5 minutes after coating.

6. Install oil pan (lower).
7. Tighten mounting bolts in numerical order as shown.
8. Install the oil pan drain plug.

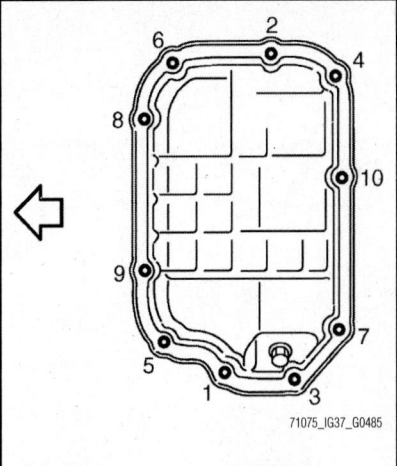

Fig. 164 Lower oil pan mounting bolt removal and installation sequence

9. Install in the reverse order of removal after this step.

➡ **Wait at least 30 minutes after oil pan is installed before pouring engine oil.**

Upper and Oil Strainer

See Figures 165 through 169.

✳✳ CAUTION

Never drain oil when the engine is hot to avoid the danger of being scalded.

1. Remove oil pan (lower).
2. Remove oil strainer.
3. Loosen mounting bolts in the reverse order as shown.
    a. Insert the seal cutter between oil pan (upper) and lower cylinder block. Slide seal cutter by tapping on the side of tool with a hammer. Remove oil pan (upper).

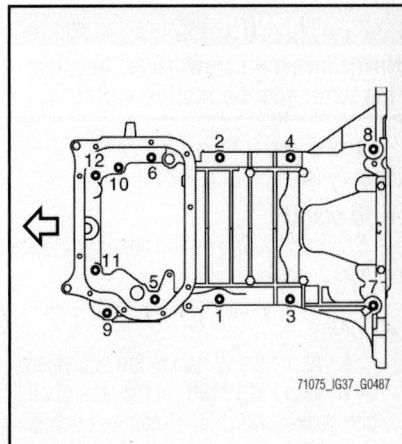

Fig. 165 Mounting bolt removal and installation sequence

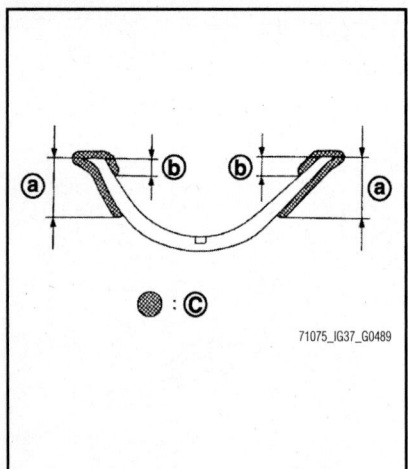

Fig. 166 Liquid gasket application position (C) (a: 0.59 inch, b: 0.20 inch)

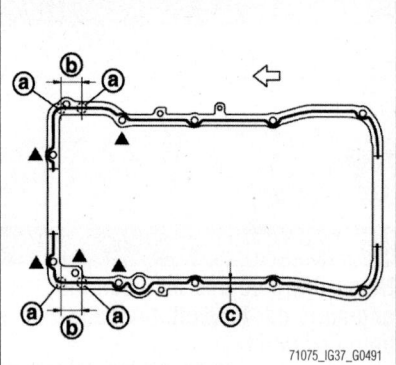

Fig. 168 Applying liquid gasket to the cylinder block mating surface of oil pan upper

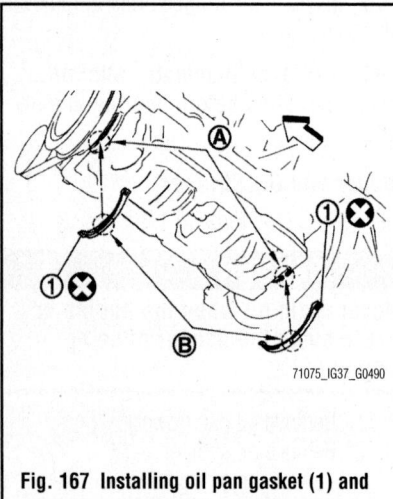

Fig. 167 Installing oil pan gasket (1) and aligning protrusion (B) with the notches

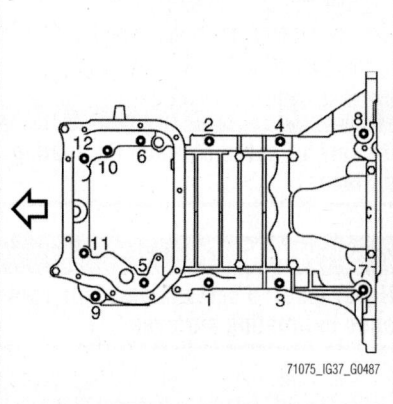

Fig. 169 Mounting bolt removal and installation sequence

✳✳ WARNING

Be careful not to damage the mating surfaces.

✳✳ WARNING

Never insert a screwdriver, because this damages the mating surfaces.

4. Remove O-rings from bottom of lower cylinder block and oil pump.

To install:

5. Install O-ring on the bottom of lower cylinder block and oil pump.

6. Install oil pan gaskets (both front and rear).

　a. Apply liquid gasket (an equivalent of Three Bond 1218B) to the area of oil pan gasket shown in a seamless single layer.

　b. To install oil pan gasket (1), align the protrusion (B) with the notches (A) of the front timing chain case and the rear oil seal retainer.

　c. Install the oil pan gasket with smaller arc to the front timing chain case side.

7. Apply a continuous bead of liquid gasket with the tube presser (commercial service tool) to the cylinder block mating surface of oil pan (upper) to a limited portion as shown.

➡**Use Genuine RTV Silicone Sealant or equivalent.**

✳✳ WARNING

For bolt holes with triangle marks (5 locations), apply liquid gasket outside the holes.

✳✳ WARNING

Attaching should be done within 5 minutes after coating.

✳✳ WARNING

Install avoiding misalignment of O-rings.

8. Install oil pan (upper).
　a. Tighten mounting bolts in numerical order as shown.
　b. There are three types of mounting bolts. Refer to the following for locating bolts.
　　• M8 x 3.94 inch (100 mm): 5, 7, 11
　　• M8 x 0.98 inch (25 mm): Except 5, 7, 11
9. Install oil strainer to oil pump.
10. Install oil pan (lower).
11. Install oil pan drain plug.
12. Install in the reverse order of removal after this step.

➡**At least 30 minutes after oil pan is installed, pour engine oil.**

AWD

See Figures 170 through 172.

✳✳ CAUTION

Never drain engine oil when the engine is hot to avoid the danger of being scalded.

1. Remove oil pan (lower).
2. Remove oil cooler.
3. Remove oil filter bracket.
4. Remove oil strainer.
5. Loosen mounting bolts in the reverse order as shown with power tool to remove.

　a. Insert the seal cutter [SST: KV10111100] between oil pan (upper) and lower cylinder block. Slide seal cutter by tapping on the side of tool with a hammer. Remove oil pan (upper).

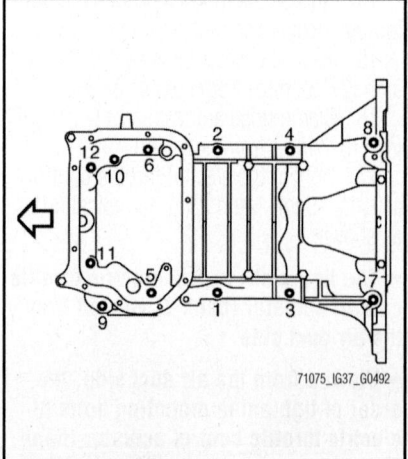

Fig. 170 Mounting bolt removal and installation sequence

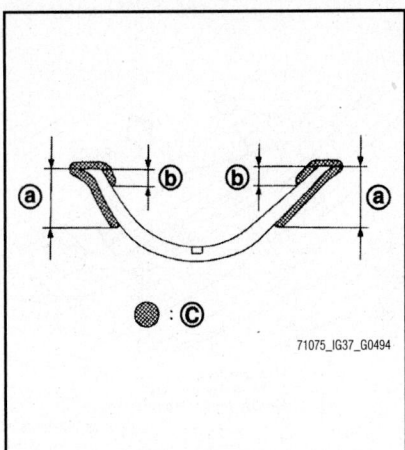

Fig. 171 Liquid gasket application position (C) (a: 0.59 inch, b: 0.20 inch)

> ❊❊ **WARNING**
>
> **Be careful not to damage the mating surfaces.**

> ❊❊ **WARNING**
>
> **Never insert a screwdriver, because this damages the mating surfaces.**

6. Remove O-rings from bottom of lower cylinder block and oil pump.

7. Remove axle pipe, if necessary.

   a. Remove axle pipe from oil pan (upper) using a suitable drift (A) [outer diameter: 37 mm (1.46 in)].

To install:

8. Install axle pipe (3) to oil pan (upper), if removed.

   a. Lubricate O-ring groove of axle pipe, O-rings and O-ring joint of oil pan with new engine oil.

   b. Install axle pipe to oil pan (upper) from axle pipe flange side (left side) using a suitable drift (outer diameter: 1.69–2.24 inch.

> ❊❊ **WARNING**
>
> **Insert it with care to prevent O-ring from sliding.**

9. Install O-ring on the bottom of lower cylinder block and oil pump.

10. Install oil pan gaskets (both front and rear).

   a. To install, oil pan gasket, align the protrusion with the notches of the front timing chain case and the rear oil seal retainer.

   b. Install the oil pan gasket with smaller arc to the front timing chain case side.

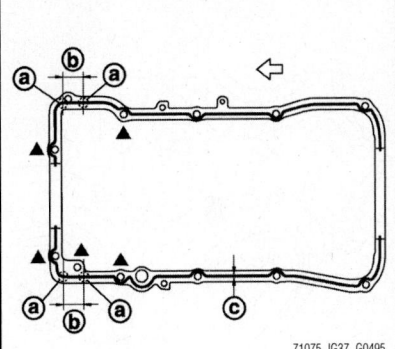

Fig. 172 Applying liquid gasket to the cylinder block mating surface of oil pan (upper)

11. Apply a continuous bead of liquid gasket with the tube presser (commercial service tool) to the cylinder block mating surface of oil pan (upper) to a limited portion as shown.

➡**Use Genuine RTV Silicone Sealant or equivalent.**

> ❊❊ **WARNING**
>
> **For bolt holes with marks (5 locations), apply liquid gasket outside the holes.**

> ❊❊ **WARNING**
>
> **Attaching should be done within 5 minutes after coating.**

12. Install oil pan (upper).

> ❊❊ **WARNING**
>
> **Install avoiding misalignment of O-rings.**

   a. Tighten mounting bolts in numerical order as shown.

   b. There are three types of mounting bolts.

   • M8 x 3.94 inch (100 mm): 5, 7, 11
   • M8 x 0.98 inch (25 mm): Except 5, 7, 11

13. Install oil strainer to oil pump.

14. Install oil pan (lower).

15. Install oil pan drain plug.

16. Install in the reverse order of removal after this step.

➡**At least 30 minutes after oil pan is installed, pour engine oil.**

PISTONS & RINGS

POSITIONING

See Figure 173.

TIMING CHAIN COVER, CHAIN, TENSIONER, & SPROCKETS

REMOVAL & INSTALLATION

Front Cover

See Figures 174 through 182.

➡**Whenever the negative battery cable is disconnected the following components will require resetting. The Automatic temperature control system, Automatic drive positioner, Power window control, Sunroof system, Sunshade system, Rear view monitor, Idle Air Volume Learning, Steering Angle Sensor Neutral Position, Audio presets and Navigation. You will need the CONSULT-III diagnostic tool, or equivalent. Follow the directions on the screen of the tool, as needed.**

1. Before servicing the vehicle, refer to the Precautions Section.

➡**If working near and/or around the SRS system and components, be sure to disable the SRS system. After disabling the system wait three minutes or more before servicing the vehicle.**

2. Relieve the fuel system pressure.
3. Disconnect the negative battery cable.
4. Remove the engine undercover.
5. Drain the engine coolant. Be sure to properly dispose of used coolant.
6. Drain the engine oil. Be sure to properly dispose of used oil.

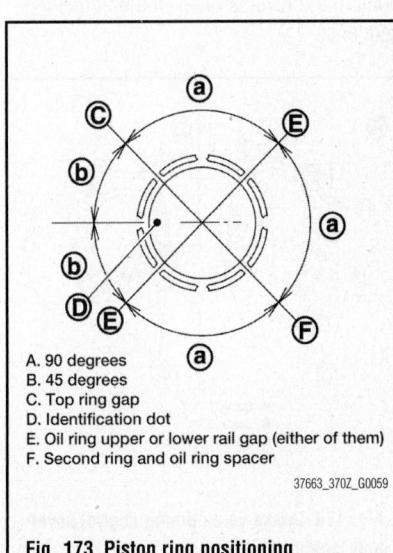

A. 90 degrees
B. 45 degrees
C. Top ring gap
D. Identification dot
E. Oil ring upper or lower rail gap (either of them)
F. Second ring and oil ring spacer

Fig. 173 Piston ring positioning

7. Remove the engine cover.

8. Remove the reservoir tank.

9. Remove the air cleaner case assembly.

10. Remove the upper and lower radiator hoses.

11. Remove the radiator cooling fan assembly.

12. Remove the drive belt.

13. Separate the engine harnesses by removing their brackets from the timing chain cover.

14. Remove the intake manifold collector.

15. Remove the fuel sub mounting bolt.

16. Remove the oil level gauge and guide.

17. Remove the air conditioning compressor from the bracket. Secure it to the side. Do not discharge the refrigerant or disconnect the refrigerant hoses.

18. Remove the power steering fluid pump from the bracket with the hoses connected, secure it to the side. Remove the power steering oil pump bracket.

19. Remove the idler pulley, drive belt auto tensioner and bracket.

20. Remove the alternator and alternator bracket.

21. Remove the front water outlet.

22. Remove the Camshaft Position (CMP) sensor.

➡ **Do not drop the sensor. Never disassemble it. Never allow metal powder to adhere to the magnetic portion of the sensor. Never store the sensor where it is exposed to magnetism.**

23. To remove the intake valve timing control covers and gasket, disconnect the intake valve timing control solenoid valve harness connector. Loosen the mounting bolts in the reverse order of the tightening sequence.

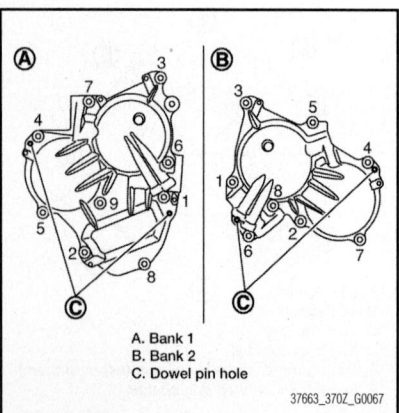

A. Bank 1
B. Bank 2
C. Dowel pin hole

37663_370Z_G0067

Fig. 174 Intake valve timing control cover bolt locations

37663_370Z_G0068

Fig. 175 TDC alignment (1 of 2)

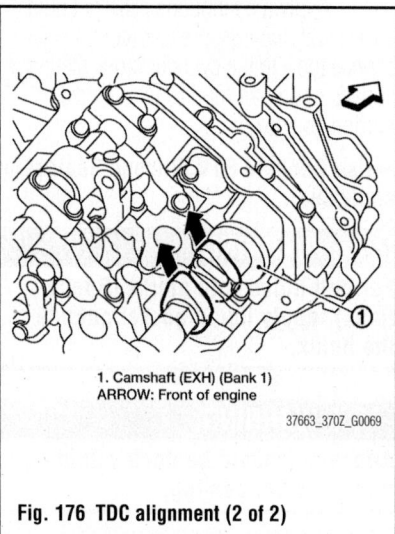

1. Camshaft (EXH) (Bank 1)
ARROW: Front of engine

37663_370Z_G0069

Fig. 176 TDC alignment (2 of 2)

➡ **The shaft is internally jointed with the camshaft sprocket (INT) center hole. When removing, keep it horizontal until it is completely disconnected.**

24. The shaft is engaged with the camshaft sprocket (INT) center hole on the inside. Pull straight out so that it does not tilt until the joint is disengaged.

25. Remove the intake valve timing control solenoid valve, if necessary.

➡ **This valve is not reusable. Never remove it unless required.**

26. Remove the valve covers.

27. To position the engine to TDC on the compression stroke, rotate the crankshaft pulley clockwise to align the timing mark (grooved line without color) with the timing indicator. Check that the exhaust cam noses on No. 1 cylinder (engine front side of bank 1) is located as shown in the illustration. If not turn the crankshaft on complete revolution and align.

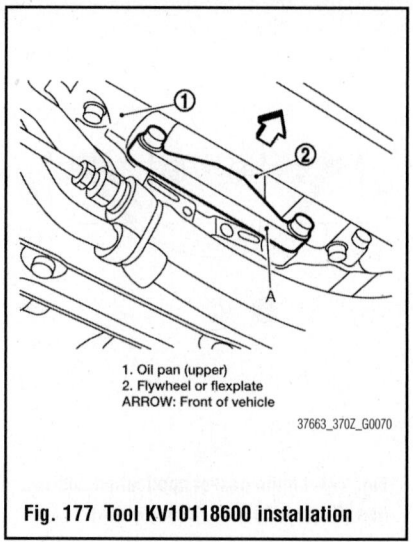

1. Oil pan (upper)
2. Flywheel or flexplate
ARROW: Front of vehicle

37663_370Z_G0070

Fig. 177 Tool KV10118600 installation

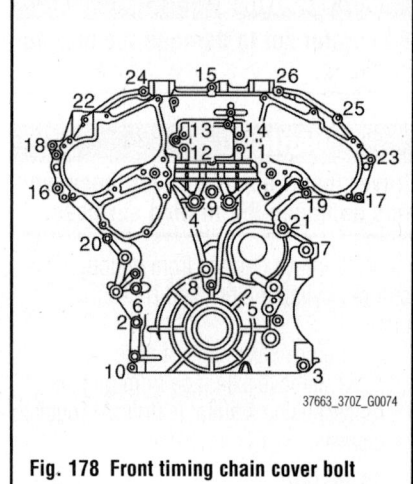

37663_370Z_G0074

Fig. 178 Front timing chain cover bolt locations

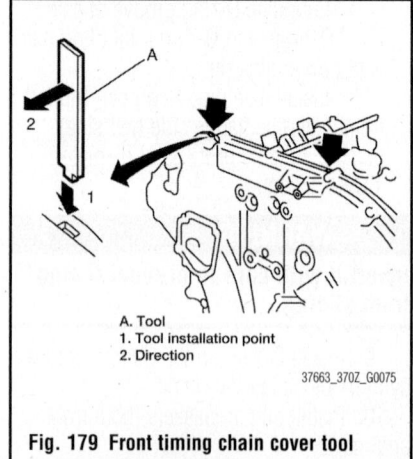

A. Tool
1. Tool installation point
2. Direction

37663_370Z_G0075

Fig. 179 Front timing chain cover tool installation and removal

28. Remove the crankshaft pulley.

29. Remove the lower oil pan.

30. Loosen the mounting bolts in the front of the upper oil pan in the reverse order of the installation sequence.

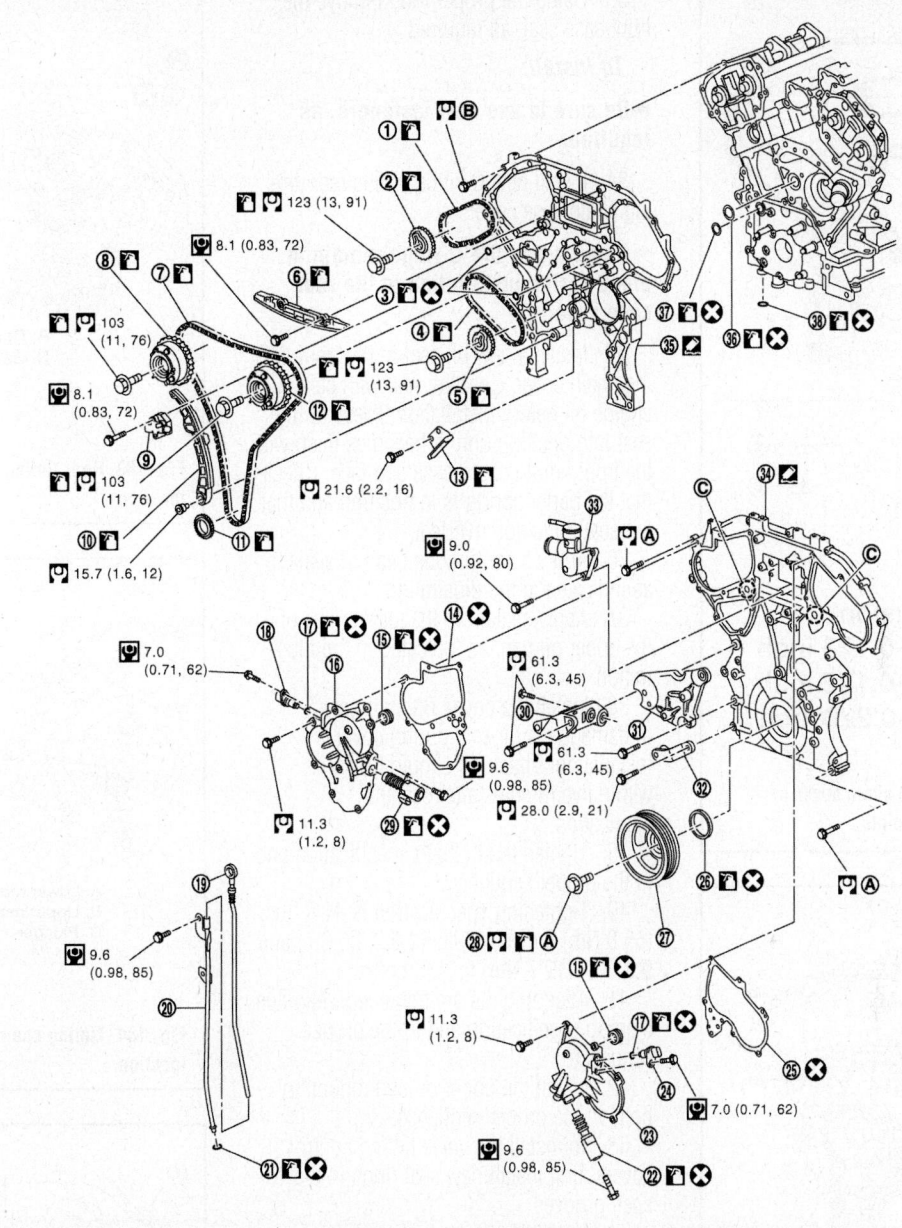

Fig. 180 Timing chain case and related components

1.	Timing chain (secondary) (bank 1)	2.	Camshaft sprocket (EXH) (bank 1)	3.	O-ring
4.	Timing chain (secondary) (bank 2)	5.	Camshaft sprocket (EXH) (bank 2)	6.	Internal chain guide
7.	Timing chain (primary)	8.	Camshaft sprocket (INT) (bank 1)	9.	Timing chain tensioner (primary)
10.	Slack guide	11.	Crankshaft sprocket	12.	Camshaft sprocket (INT) (bank 2)
13.	Tension guide	14.	Intake valve timing control cover gasket (bank 1)	15.	Seal ring
16.	Intake valve timing control cover (bank 1)	17.	O-ring	18.	Camshaft position sensor (PHASE) (bank 1)
19.	Oil level gauge	20.	Oil level gauge guide	21.	O-ring
22.	Intake valve timing control solenoid valve (bank 2)	23.	Intake valve timing control cover (bank 2)	24.	Camshaft position sensor (PHASE) (bank 2)
25.	Intake valve timing control cover gasket (bank 2)	26.	Front oil seal	27.	Crankshaft pulley
28.	Crankshaft pulley bolt	29.	Intake valve timing control solenoid valve (bank 1)	30.	Power steering oil pump bracket
31.	Idler pulley bracket	32.	Alternator bracket	33.	Water outlet (front)
34.	Front timing chain case	35.	Rear timing chain case	36.	O-ring
37.	O-ring	38.	O-ring		
				C.	Oil filter

37663_370Z_G0010

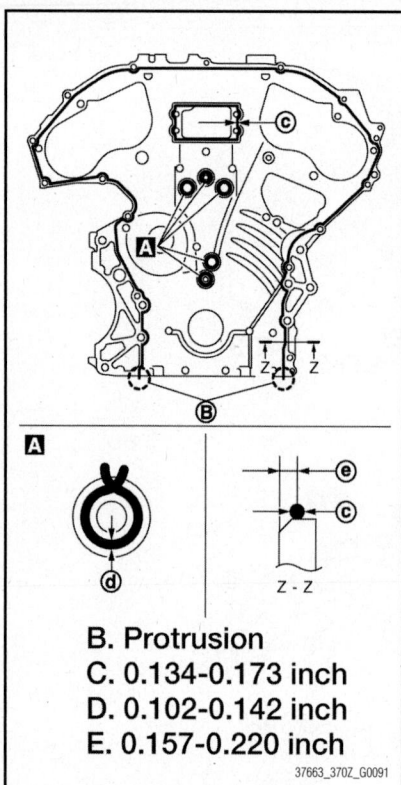

B. Protrusion
C. 0.134-0.173 inch
D. 0.102-0.142 inch
E. 0.157-0.220 inch

37663_370Z_G0091

Fig. 181 Front timing chain cover sealant application points

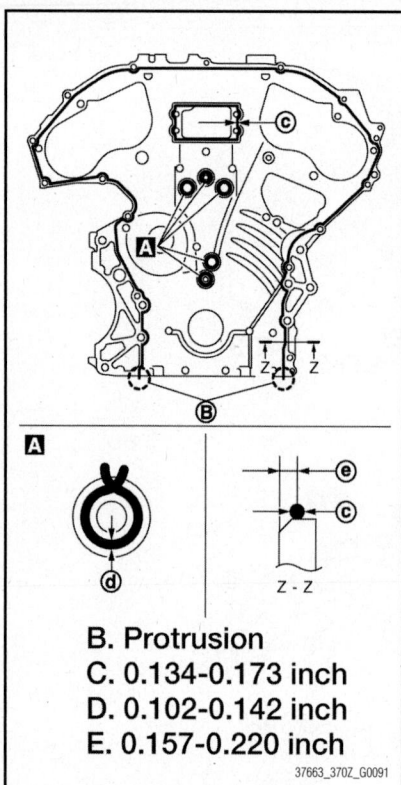

37663_370Z_G0074

Fig. 182 Front timing chain cover bolt locations and tightening sequence

31. To remove the case cover, loosen the mounting bolts in the reverse order of the installation sequence.

32. Insert a suitable tool (KV10111100) into the notch at the top of the front timing chain cover, as shown in the illustration. Pry off the case by moving the tool, as shown in the illustration.

➡️**Never use a screwdriver or similar item. After removal handle the cover carefully so it does not tilt, cant or warp under load.**

33. Using the proper tool, remove the front case seal, as required.

To install:

➡️**Be sure to use new fasteners, as required.**

34. Install new O-rings on the rear timing chain case cover.

➡️**Be sure that the O-rings remain in place during installation to the rear case cover.**

35. Install a new oil seal in the front timing chain cover. Coat the seal with clean engine oil before installation. Press fit the seal into position until it becomes flush with the front timing chain case end face. Check that the garter spring is in position and that the seal lip is not inverted.

36. Apply a continuous bead of sealant as indicated in the illustration.

37. Apply sealant to the top surface of the upper oil pan, as indicated in the illustration.

38. Install the cover. Be sure not to damage the oil seal during cover installation. Attaching should be done within five minutes after sealant application.

39. Tighten the bolts to specification and in the proper sequence.

40. Tightening specification is 41 ft. lbs. (55.0 Nm) for M10 bolts (1,2,3,4,5,6,7) and 9 ft. lbs. (12.7 Nm) for M6 bolts.

41. After all bolts are tightened, retighten them to specification and in the proper sequence.

42. Install the upper oil pan mounting bolts in the proper sequence.

43. To install the valve timing control covers, first install new seal rings in the shaft grooves.

44. Install the covers, using new gaskets.

➡️**Align the center of both shaft holes of the camshaft sprocket (INT) and the shaft and then insert them. Be careful not to drop the seal ring from the shaft groove.**

45. Tighten the mounting bolts in the sequence shown in the illustration.

46. Continue the installation in the reverse order of the removal procedure.

47. Be sure to perform the reconnect/relearn procedures.

Chain & Sprocket

See Figures 183 through 199.

➡️**Whenever the negative battery cable is disconnected the following compo-**

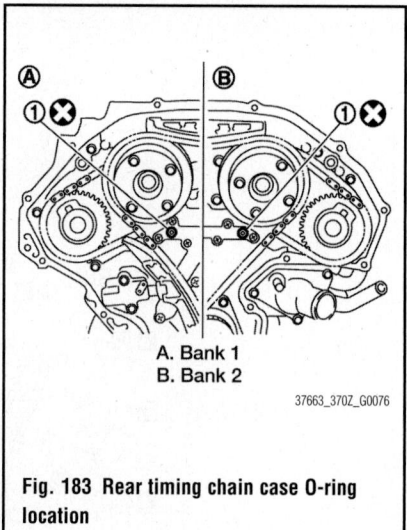

A. Bank 1
B. Bank 2

37663_370Z_G0076

Fig. 183 Rear timing chain case O-ring location

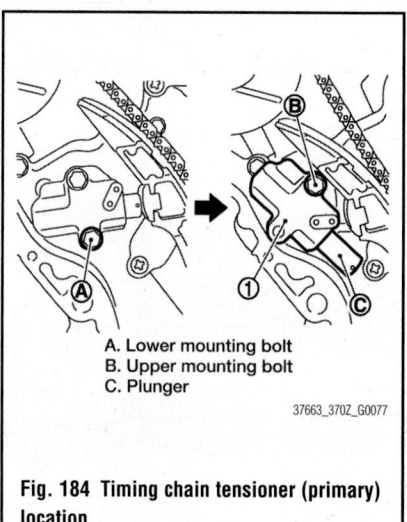

A. Lower mounting bolt
B. Upper mounting bolt
C. Plunger

37663_370Z_G0077

Fig. 184 Timing chain tensioner (primary) location

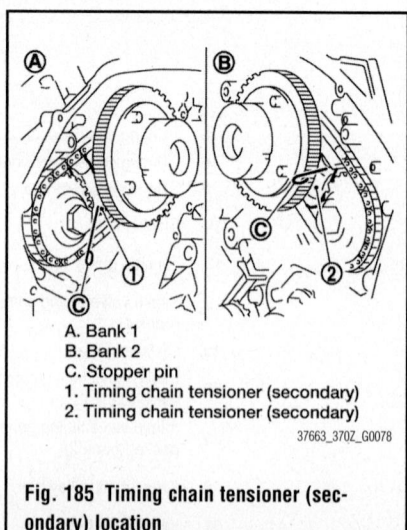

A. Bank 1
B. Bank 2
C. Stopper pin
1. Timing chain tensioner (secondary)
2. Timing chain tensioner (secondary)

37663_370Z_G0078

Fig. 185 Timing chain tensioner (secondary) location

nents will require resetting. The Automatic temperature control system, Automatic drive positioner, Power window control, Sunroof system, Sun-

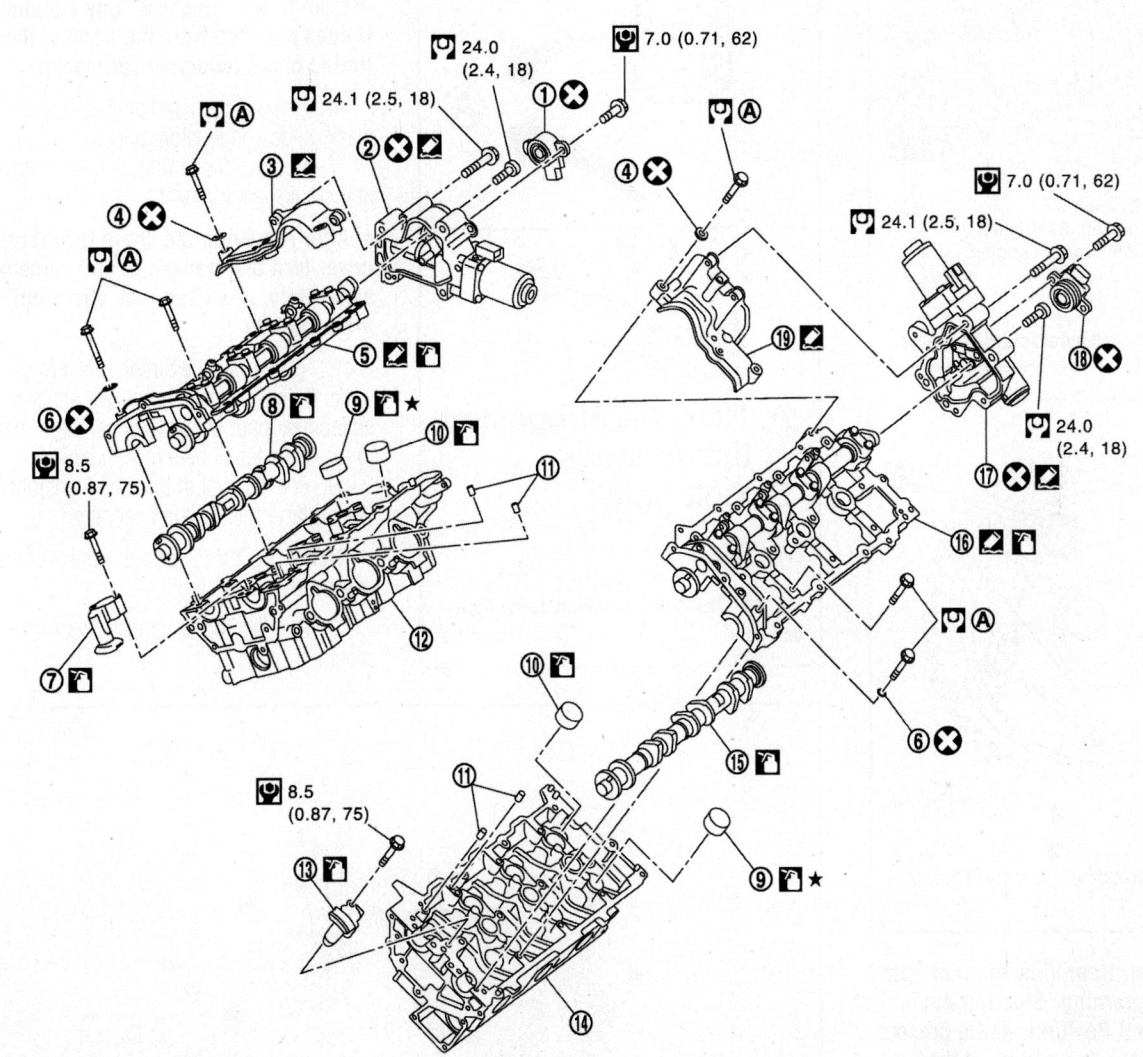

1. VVEL control shaft position sensor (bank 1)
2. VVEL actuator sub assembly (bank 1)
3. Actuator bracket (rear) (bank 1)
4. Washer
5. VVEL ladder assembly (bank 1)
6. Washer
7. Timing chain tensioner (secondary) (bank 1)
8. Camshaft (EXH) (bank 1)
9. Valve lifter (EXH)
10. Valve lifter (INT)
11. Oil filter
12. Cylinder head (bank 1)
13. Timing chain tensioner (secondary) (bank 2)
14. Cylinder head (bank 2)
15. Camshaft (EXH) (bank 2)
16. VVEL ladder assembly (bank 2)
17. VVEL actuator sub assembly (bank 2)
18. VVEL control shaft position sensor (bank 2)
19. Actuator bracket (rear) (bank 2)

37663_370Z_G0004

Fig. 186 Camshaft and related components (VVEL ladder assembly)

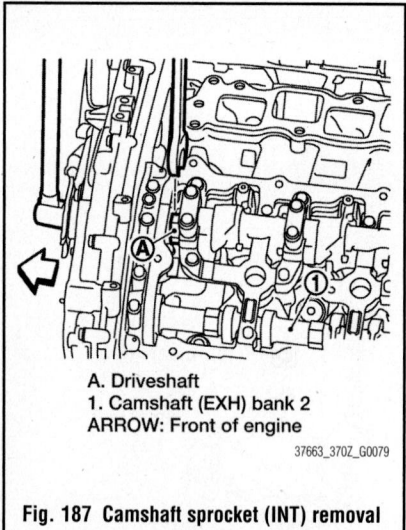

A. Driveshaft
1. Camshaft (EXH) bank 2
ARROW: Front of engine

37663_370Z_G0079

Fig. 187 Camshaft sprocket (INT) removal

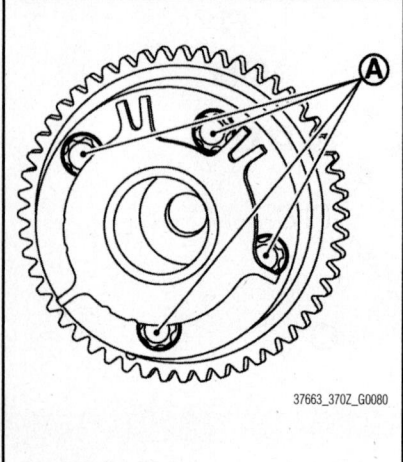

37663_370Z_G0080

Fig. 188 Camshaft sprocket (INT) bolt "A" location

shade system, Rear view monitor, Idle Air Volume Learning, Steering Angle Sensor Neutral Position, Audio presets and Navigation. You will need the CONSULT-III diagnostic tool, or equivalent. Follow the directions on the screen of the tool, as needed.

1. Before servicing the vehicle, refer to the Precautions Section.

➡ **If working near and/or around the SRS system and components, be sure to disable the SRS system. After disabling the system wait three minutes or more before servicing the vehicle.**

2. Disconnect the negative battery cable.
3. Remove the timing chain cover.
4. Remove the O-ring from the rear timing chain case.
5. To remove the timing chain tensioner (primary), remove the lower mounting bolt. Loosen the upper mounting bolt slowly and then turn the timing chain

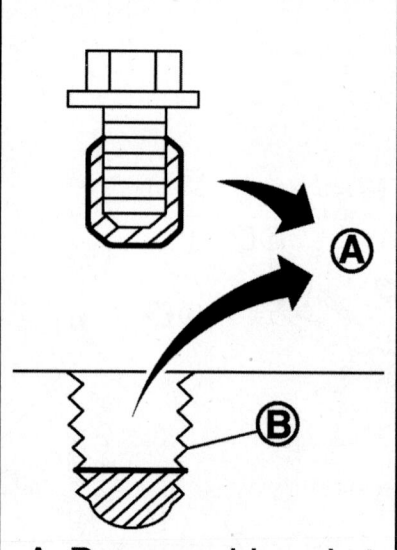

A. Remove old gasket that is stuck
B. Bolt hole

37663_370Z_G0081

Fig. 189 Removing gasket material from bolt hole

tensioner (primary) on the upper mounting bolt so that the plunger is fully expanded. Remove the upper mounting bolt and then remove the timing chain tensioner (primary).

➡ **Even if the plunger is fully expanded, it does not drop from the body of the timing chain tensioner (primary).**

6. Remove the internal chain guide, slack guide and tension guide.
7. Remove the timing chain (primary) and crankshaft sprocket.

➡ **After removing the chain (primary), never turn the crankshaft and camshaft separately, or valves may strike the piston heads.**

8. To remove the timing chain (secondary) and camshaft sprockets attach a suitable stopper pin (0.020 inch hard metal) to the chain tensioners (secondary).
9. For removal of the chain tensioners (secondary) refer to the illustration.

➡ **Removing VVEL ladder assembly is required.**

10. Remove the camshaft sprocket (EXH) mounting bolt.

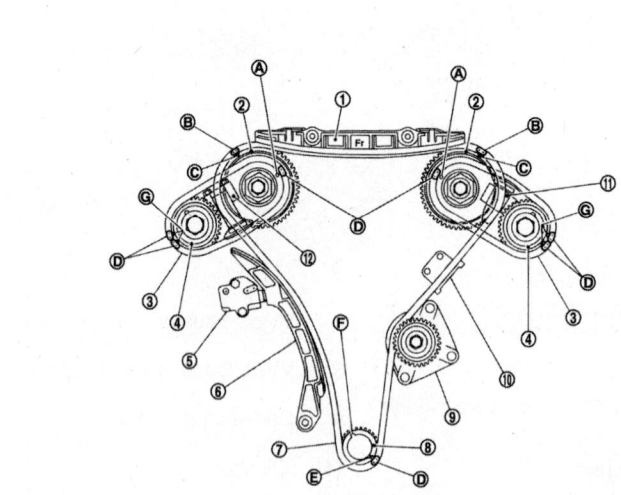

1.	Internal chain guide	2.	Camshaft sprocket (INT)	3.	Timing chain (secondary)
4.	Camshaft sprocket (EXH)	5.	Timing chain tensioner (primary)	6.	Slack guide
7.	Timing chain (primary)	8.	Crankshaft sprocket	9.	Water pump
10.	Tension guide	11.	Timing chain tensioner (secondary) (bank 2)	12.	Timing chain tensioner (secondary) (bank 1)
A.	Matching mark [punched (back side)]	B.	Matching mark (yellow link)	C.	Matching mark (punched)
D.	Matching mark (orange link)	E.	Matching mark (notched)	F.	Crankshaft key

37663_370Z_G0082

Fig. 190 Timing chains and sprockets (relationship) alignment

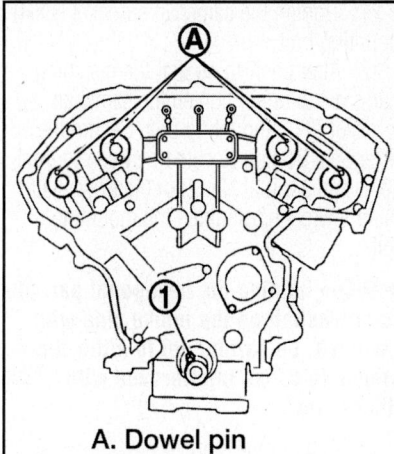

A. Dowel pin
1. Crankshaft key

37663_370Z_G0083

Fig. 191 Camshafts and crankshaft positioning

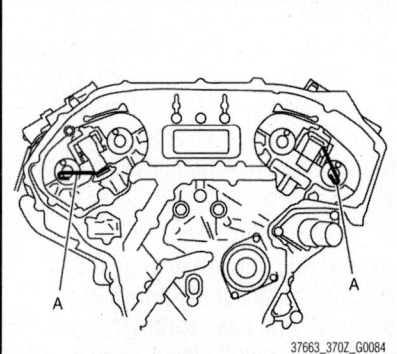

37663_370Z_G0084

Fig. 192 Stopper pin installation and timing chain tensioner (secondary) plunger location

➥Secure the hexagonal portion of the camshaft (EXH) using a wrench to loosen mounting bolt. Never loosen the mounting bolt by securing anything other than camshaft (EXH) hexagonal portion or with tensioning the timing chain.

11. Remove the camshaft sprocket (INT) mounting bolt.

➥Secure the hexagonal portion (located between journal No. 1 and journal No. 2) of the driveshaft using a wrench to loosen the mounting bolt. Never loosen the mounting bolt by securing anything other than the driveshaft hexagonal portion or with tensioning the timing chain. When holding the hexagonal part of the driveshaft on the intake side with a wrench, be careful not to allow the wrench to cause interference with other parts.

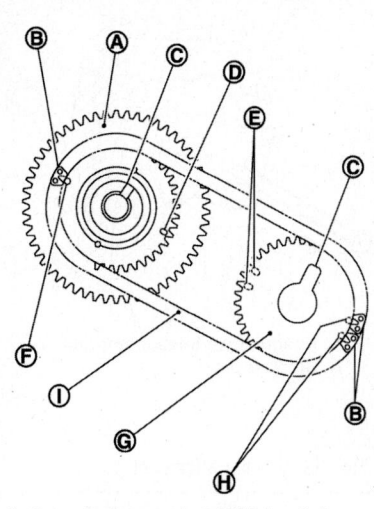

A. Camshaft sprocket (INT) back face
B. Orange link
C. Dowel groove
D. Matching mark (oval)
E. Matching mark (2 oval: on front face)
F. Matching mark (circle)
G. Camshaft sprocket (EXH) back face
H. Matching mark (2 oval: on front face)
I. Timing chain (secondary)

37663_370Z_G0085

Fig. 193 Timing chain tensioner (secondary) and camshaft sprockets—bank 1 rear view

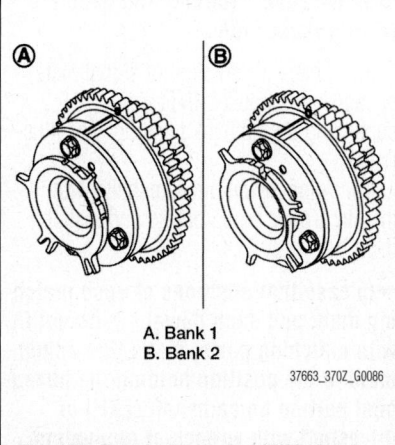

A. Bank 1
B. Bank 2

37663_370Z_G0086

Fig. 194 Camshaft sprocket (INT) signal plate orientation

➥Never disassemble the camshaft sprocket (INT). Never loosen bolts (A) as shown in the illustration.

12. Remove the timing chain (secondary) along with the camshaft sprockets.

13. Remove all traces of gasket material from the front and rear timing chain covers. Be sure to remove all material from the bolt holes and threads. See illustration

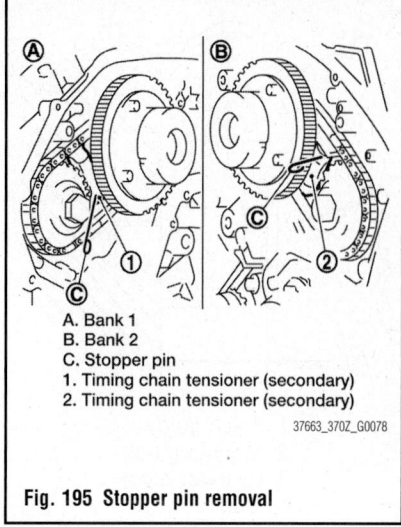

A. Bank 1
B. Bank 2
C. Stopper pin
1. Timing chain tensioner (secondary)
2. Timing chain tensioner (secondary)

37663_370Z_G0078

Fig. 195 Stopper pin removal

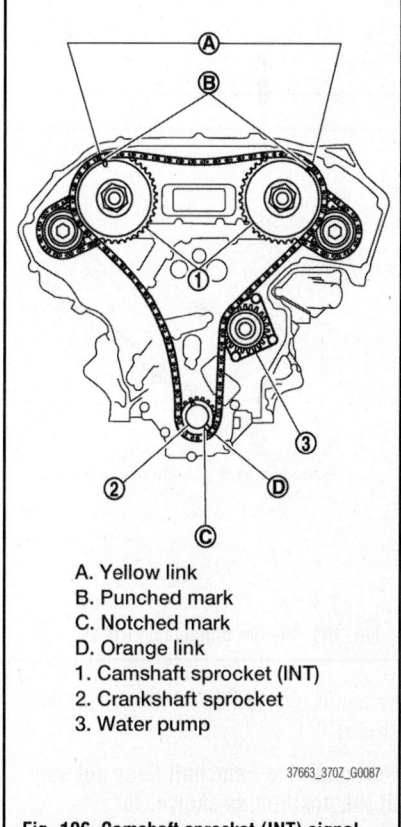

A. Yellow link
B. Punched mark
C. Notched mark
D. Orange link
1. Camshaft sprocket (INT)
2. Crankshaft sprocket
3. Water pump

37663_370Z_G0087

Fig. 196 Camshaft sprocket (INT) signal plate orientation

To install:

➥Be sure to use new fasteners, as required.

➥See illustration that shows the relationship between the matching mark on each timing chain and that on the corresponding sprocket with components installed.

14. Check that the dowel pin and crankshaft key are located as shown in the illus-

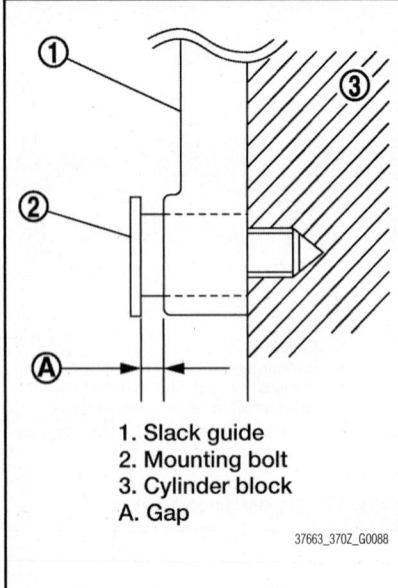

1. Slack guide
2. Mounting bolt
3. Cylinder block
A. Gap

37663_370Z_G0088

Fig. 197 Slack guide mounting bolt installation

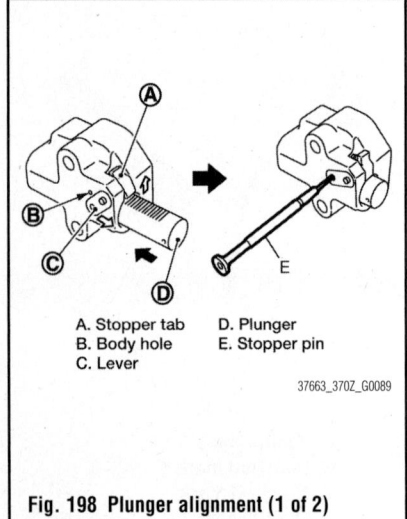

A. Stopper tab   D. Plunger
B. Body hole   E. Stopper pin
C. Lever

37663_370Z_G0089

Fig. 198 Plunger alignment (1 of 2)

tration (engine at TDC on the compression stroke).

➡ Though the camshaft does not stop at the position as shown, for placement of cam noses, it is generally accepted that the camshaft is placed in the same direction as that of the illustration.

➡ Matching marks between the chain and sprockets slip easily. Confirm all matching mark positions repeatedly during the installation process.

15. To install the timing chains (secondary) and camshaft sprockets, push the plunger of the timing chain tensioner (secondary) and keep it pressed in with a stopper pin.

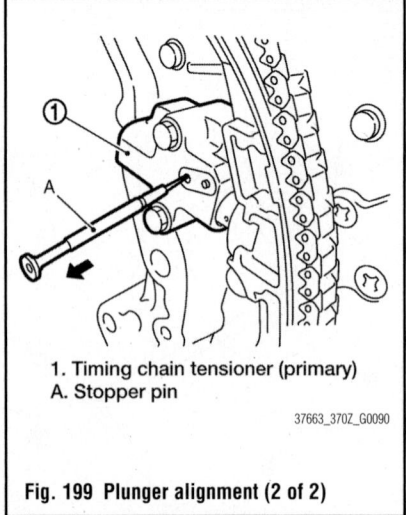

1. Timing chain tensioner (primary)
A. Stopper pin

37663_370Z_G0090

Fig. 199 Plunger alignment (2 of 2)

16. Install the timing chains (secondary) and camshaft sprockets. See illustration.

➡ Illustration shows bank 1 (rear view)

17. Align the matching marks on the chain (secondary) (orange link) with the ones on the intake and exhaust camshaft sprockets (punched), and install them.

➡ Matching marks for camshaft sprockets (INT) are on the back side of the camshaft sprockets (secondary). There are two types of matching marks, the circle and the oval. They should be used for bank 1 (circle) and bank 2 (oval) respectively.

18. Shape (orientation of signal plate) of camshaft sprocket (INT) varies depending on the bank position. See illustration.

19. Align dowel pin camshafts with the pin groove on sprockets and install them.

➡ In case that positions of each matching mark and each dowel pin do not fit with matching parts, make fine adjustment to the position holding the hexagonal portion on camshaft (EXH) or driveshaft with wrench or equivalent tool.

20. Mounting bolts for camshaft sprockets must be tightened in the next step. Tightening by hand is sufficient to prevent the dislocation of the dowel pins.

➡ It may be difficult to visibly check the dislocation of the matching marks during and after installation. To make the matching easier make a matching mark on the top of the sprocket teeth and its extended line in advance using paint.

21. Tighten the camshaft sprocket (EXH) mounting bolt.

22. After confirming that the matching marks are aligned, tighten the camshaft sprocket (INT) mounting bolt. Secure the hexagonal portion (located between journal No. 1 and Journal No. 2) of the driveshaft using a wrench to tighten the mounting bolt.

➡ When holding the hexagonal part of the driveshaft on the intake side with a wrench, be careful not to allow the wrench to cause interference with other parts.

23. Pull out the stopper pins from the timing chain tensioners (secondary).

24. To install the timing chain (primary), install the crankshaft sprocket.

➡ Be sure that the matching marks on the crankshaft sprocket face the front of the engine.

25. Install the timing chain (primary) so that the matching mark (punched) on the camshaft sprocket (INT) is aligned with the yellow link on the timing chain while the matching mark (notched) on the crankshaft sprocket is aligned with the orange link on the timing chain. See illustration.

➡ When it is difficult to align the matching marks of the timing chain (primary) with each sprocket, gradually turn the driveshaft sing a wrench on the hexagonal portion to align it with the matching marks.

26. Install the internal chain guide, slack guide and tension guide.

➡ Never overtighten the slack guide mounting bolt. It is normal for a gap to exist under the bolt seats when mounting bolts are tightened to specification.

27. To install the timing chain tensioner (primary) pull the plunger stopper tab up (or turn lever downward) so as to remove the plunger stopper tab from the ratchet of the plunger. Note that the plunger stopper tab and lever are synchronized.

28. Push the plunger into the inside of the tensioner body. Hold the plunger in the fully compressed position by engaging the plunger stopper tab with the tip of the ratchet.

29. To secure the lever, insert the stopper pin through the hole of the lever into the tensioner body hole. The lever parts and the plunger stopper tab are synchronized, therefore the plunger is secured under this condition.

➡The illustration shows a suitable tool of 0.047 inch diameter being used as a stopper pin.

30. Pull out the stopper pin after installing and release the plunger.

31. Check again that the matching marks on the sprockets and the timing chain have not slipped out of alignment.

32. Install the timing chain cover.

33. Be sure to perform the reconnect/relearn procedures.

VALVE COVERS

REMOVAL & INSTALLATION

See Figures 200 through 202.

➡Whenever the negative battery cable is disconnected the following components will require resetting. The Automatic temperature control system, Automatic drive positioner, Power window control, Sunroof system, Sunshade system, Rear view monitor, Idle Air Volume Learning, Steering Angle Sensor Neutral Position, Audio presets and Navigation. You will need the CON-

SULT-III diagnostic tool, or equivalent. Follow the directions on the screen of the tool, as needed.

1. Before servicing the vehicle, refer to the Precautions Section.

➡If working near and/or around the SRS system and components, be sure to disable the SRS system. After disabling the system wait three minutes or more before servicing the vehicle.

2. Disconnect the negative battery cable.

3. Remove the engine cover.

4. Remove the air cleaner case and air duct.

5. Remove the intake manifold collector.

6. Disconnect the PCV hose from the cover.

7. Remove the PCV valve and O-ring, if necessary.

8. Remove the oil filler cap from the cover, if necessary.

9. Remove the ignition coil. Never shock the coil.

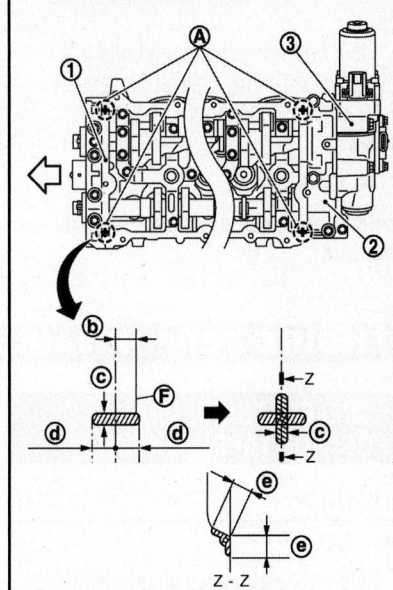

2. Actuator bracket (rear)
3. VVEL actuator sub assembly
A. Sealant application point
F. End surface of VVEL sub assembly
B. 0.16 inch
C. 0.098-0.138 inch
D. 0.20 inch
E 0.39 inch
ARROW: Front of engine

37663_370Z_G0065

Fig. 201 Valve cover sealant application points

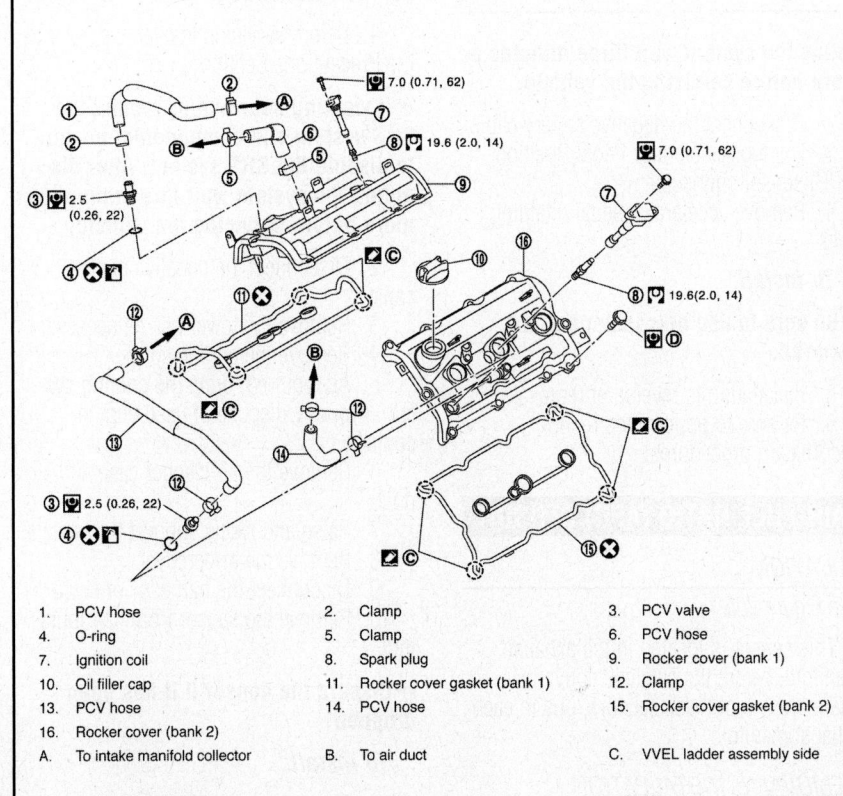

1.	PCV hose	2.	Clamp	3.	PCV valve
4.	O-ring	5.	Clamp	6.	PCV hose
7.	Ignition coil	8.	Spark plug	9.	Rocker cover (bank 1)
10.	Oil filler cap	11.	Rocker cover gasket (bank 1)	12.	Clamp
13.	PCV hose	14.	PCV hose	15.	Rocker cover gasket (bank 2)
16.	Rocker cover (bank 2)				
A.	To intake manifold collector	B.	To air duct	C.	VVEL ladder assembly side

37663_370Z_G0007

Fig. 200 Valve covers and related components

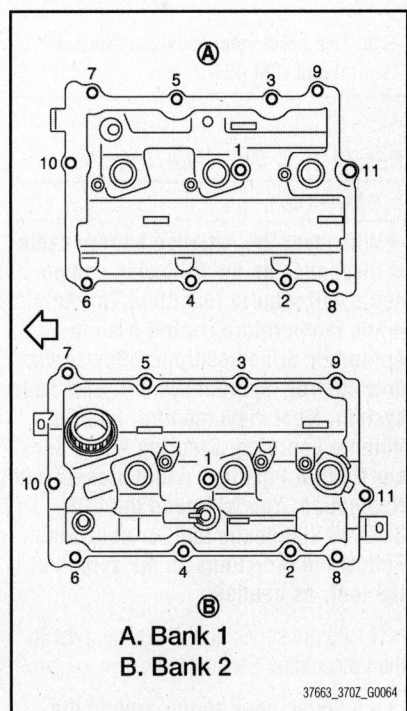

A. Bank 1
B. Bank 2

37663_370Z_G0064

Fig. 202 Valve cover bolt locations and tightening sequence

10. Remove the harness clips on the cover.

11. Loosen the mounting bolts in the reverse order of the installation sequence.

12. Remove the cover. Discard the gasket.

To install:

➡**Be sure to use new fasteners, as required.**

13. Apply liquid gasket, as shown in the illustration.

➡**Refer to the illustration to apply sealant to the joint part of the VVEL ladder assembly and cylinder head. Apply sealant in ninety degrees to illustration.**

14. Install the valve cover gasket to the valve cover.

15. Install the rocker cover.

16. Tighten the bolts in two steps and in the proper sequence.

17. Tightening specification is 18 inch lbs. (2.0 Nm), first pass and then 73 inch lbs. (8.3 Nm), second pass.

18. Continue the installation in the reverse order of the removal procedure.

19. Be sure to perform the reconnect/relearn procedures.

ENGINE PERFORMANCE & EMISSION CONTROLS

ACCELERATOR PEDAL POSITION (APP) SENSOR

LOCATION

See Figure 203.

Refer to the accompanying illustration.

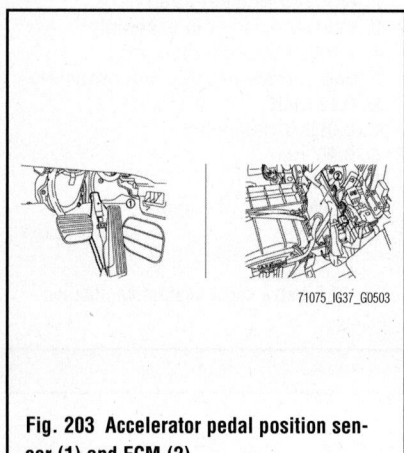

71075_IG37_G0503

Fig. 203 Accelerator pedal position sensor (1) and ECM (2)

REMOVAL & INSTALLATION

See Figure 204.

➡**Whenever the negative battery cable is disconnected the following components will require resetting. The Automatic temperature control system, Automatic drive positioner, Power window control, Sunroof system, Sunshade system, Rear view monitor, Idle Air Volume Learning, Steering Angle Sensor Neutral Position, Audio presets and Navigation. You will need the CONSULT-III diagnostic tool, or equivalent. Follow the directions on the screen of the tool, as needed.**

1. Before servicing the vehicle, refer to the Precautions Section.

➡**If working near and/or around the SRS system and components, be sure to disable the SRS system. After dis-**

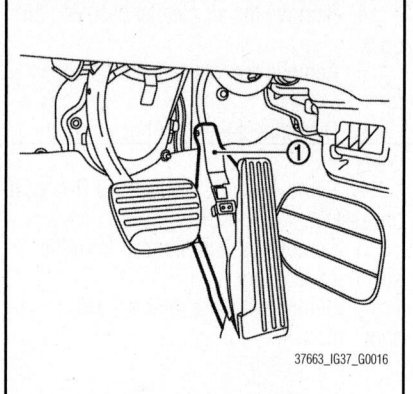

37663_IG37_G0016

Fig. 204 Accelerator position sensor and related components

abling the system wait three minutes or more before servicing the vehicle.

2. Disconnect the negative battery cable.
3. Unplug Accelerator Pedal Position (APP) sensor connector.
4. Remove accelerator pedal retaining bolts.

To install:

➡**Be sure to use new fasteners, as required.**

5. Installation is reverse of removal.
6. Be sure to perform the reconnect/relearn procedures.

AIR-FUEL RATIO (AFR) SENSOR

LOCATION

See Figure 205.

This sensor is located in the exhaust manifold, before the catalytic converter. There are two of these sensors, one in each exhaust manifold.

REMOVAL & INSTALLATION

See Figures 205 and 206.

At this time the manufacturer does not provide removal and installation procedures

for this component. The following procedure is a guideline and may differ from the vehicle you are servicing.

➡**Whenever the negative battery cable is disconnected the following components will require resetting. The Automatic temperature control system, Automatic drive positioner, Power window control, Sunroof system, Sunshade system, Rear view monitor, Idle Air Volume Learning, Steering Angle Sensor Neutral Position, Audio presets and Navigation. You will need the CONSULT-III diagnostic tool, or equivalent. Follow the directions on the screen of the tool, as needed.**

1. Before servicing the vehicle, refer to the Precautions Section.

➡**If working near and/or around the SRS system and components, be sure to disable the SRS system. After disabling the system wait three minutes or more before servicing the vehicle.**

2. Disconnect the negative battery cable.
3. Remove the tower bar, as necessary.
4. Remove the engine cover.
5. As required, drain the cooling system. Properly dispose of used engine coolant.
6. Remove the air cleaner case and air duct.
7. Raise and safely support the vehicle.
8. Remove the undercover.
9. Disconnect the harness connector.
10. Remove the sensor from its mounting.

➡**Discard the sensor if it has been dropped.**

To install:

➡**Be sure to use new fasteners, as required.**

11. Installation is the reverse of the removal procedure.

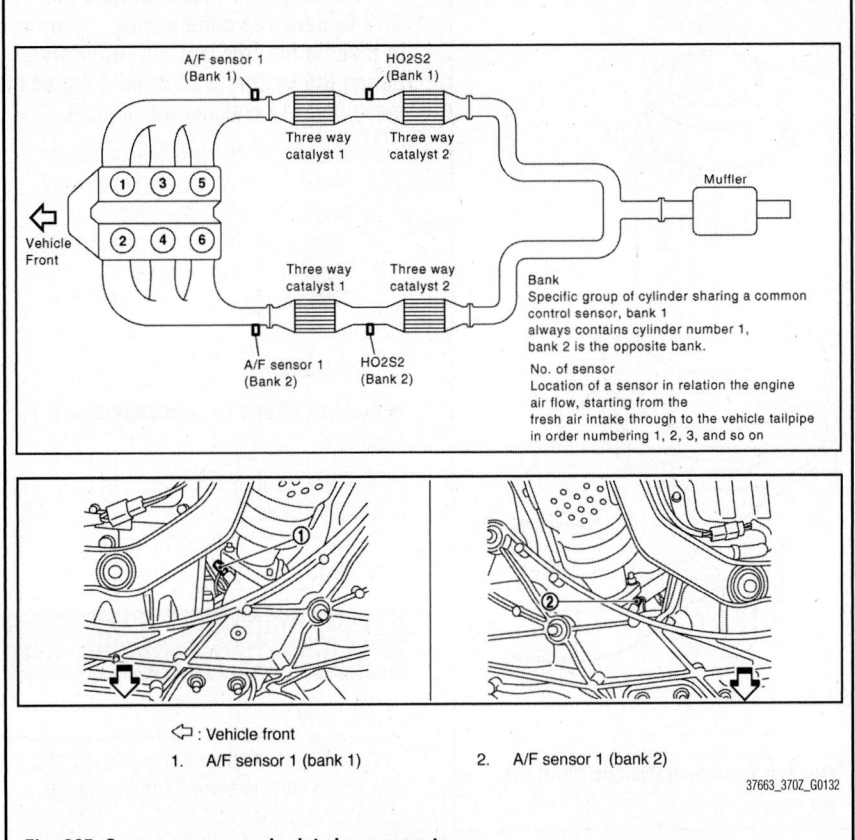

: Vehicle front
1. A/F sensor 1 (bank 1)   2. A/F sensor 1 (bank 2)

37663_370Z_G0132

Fig. 205 Oxygen sensors and related components

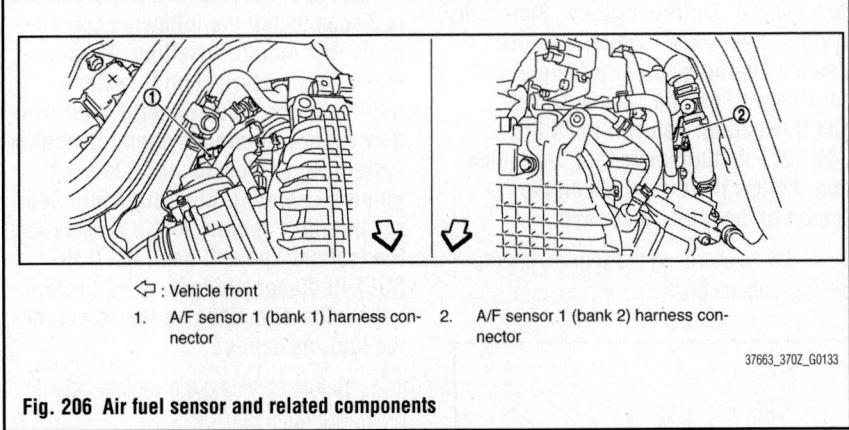

: Vehicle front
1. A/F sensor 1 (bank 1) harness connector   2. A/F sensor 1 (bank 2) harness connector

37663_370Z_G0133

Fig. 206 Air fuel sensor and related components

12. Before installing a new sensor coat the threads with an approved anti-seize lubricant.

13. Be sure to perform the reconnect/ relearn procedures.

CAMSHAFT POSITION (CMP) SENSOR

LOCATION

See Figure 207.

Refer to the accompanying illustration.

REMOVAL & INSTALLATION

See Figures 207 through 208.

At this time the manufacturer does not provide removal and installation procedures for this component. The following procedure is a guideline and may differ from the vehicle you are servicing.

➡**Whenever the negative battery cable is disconnected the following components will require resetting. The Automatic temperature control system, Automatic drive positioner, Power window control, Sunroof system, Sunshade system, Rear view monitor, Idle Air Volume Learning, Steering Angle Sensor Neutral Position, Audio presets and Navigation. You will need the CONSULT-III diagnostic tool, or equivalent. Follow the directions on the screen of the tool, as needed.**

1. Before servicing the vehicle, refer to the Precautions Section.

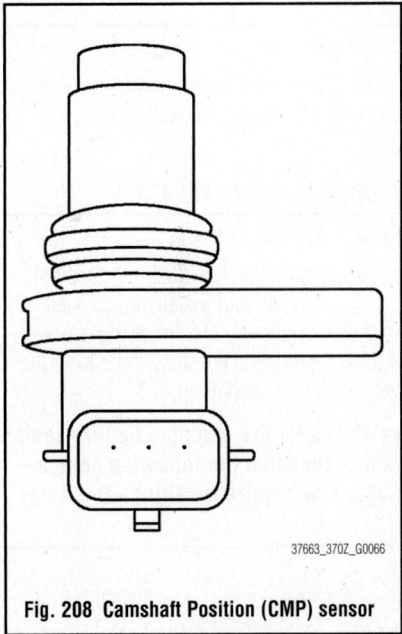

37663_370Z_G0066

Fig. 208 Camshaft Position (CMP) sensor

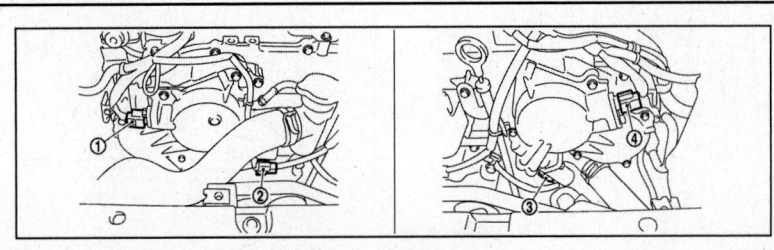

1. Camshaft position sensor (PHASE) (bank 1)
2. Intake valve timing control solenoid valve (bank 1) harness connector
3. Intake valve timing control solenoid valve (bank 2) harness connector
4. Camshaft position sensor (PHASE) (bank 2)

37663_370Z_G0135

Fig. 207 Camshaft Position (CMP) sensor location

➡️If working near and/or around the SRS system and components, be sure to disable the SRS system. After disabling the system wait three minutes or more before servicing the vehicle.

2. Disconnect the negative battery cable.
3. Remove the necessary components in order to gain access to the sensor.
4. Disconnect the electrical connector.
5. Remove the retaining bolt.
6. Remove the component from the vehicle.

To install:

➡️Be sure to use new fasteners, as required.

7. Installation is the reverse of the removal procedure.
8. Be sure to perform the reconnect/relearn procedures.

CRANKSHAFT POSITION (CKP) SENSOR

LOCATION
See Figure 209.

This sensor is located on the cylinder block facing the gear teeth on the signal plate.

REMOVAL & INSTALLATION
See Figure 210.

At this time the manufacturer does not provide removal and installation procedures for this component. The following procedure is a guideline and may differ from the vehicle you are servicing.

➡️Whenever the negative battery cable is disconnected the following components will require resetting. The Auto-

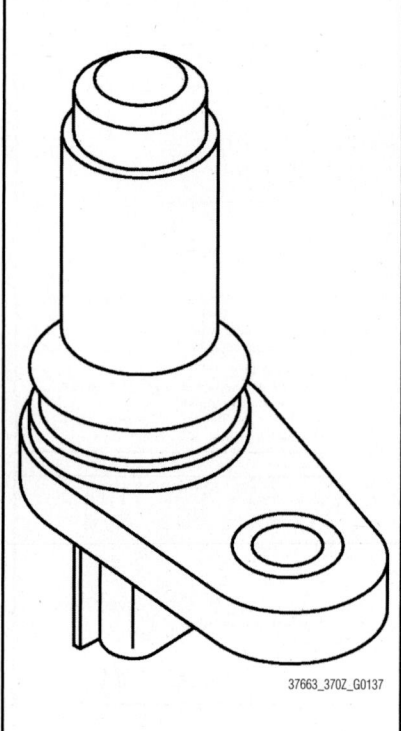

Fig. 210 Crankshaft Position (CKP) sensor

matic temperature control system, Automatic drive positioner, Power window control, Sunroof system, Sunshade system, Rear view monitor, Idle Air Volume Learning, Steering Angle Sensor Neutral Position, Audio presets and Navigation. You will need the CONSULT-III diagnostic tool, or equivalent. Follow the directions on the screen of the tool, as needed.

1. Before servicing the vehicle, refer to the Precautions Section.

➡️If working near and/or around the SRS system and components, be sure to disable the SRS system. After disabling the system wait three minutes or more before servicing the vehicle.

2. Disconnect the negative battery cable.
3. Remove the necessary components in order to gain access to the sensor.
4. Disconnect the electrical connector.
5. Remove the retaining bolt.
6. Remove the component from the vehicle.

To install:

➡️Be sure to use new fasteners, as required.

7. Installation is the reverse of the removal procedure.
8. Be sure to perform the reconnect/relearn procedures.

ELECTRONIC CONTROL MODULE (ECM)

LOCATION

The ECM is located on the passenger's side of the vehicle behind the instrument assist lower panel.

REMOVAL & INSTALLATION

➡️Whenever the negative battery cable is disconnected the following components will require resetting. The Automatic temperature control system, Automatic drive positioner, Power window control, Sunroof system, Sunshade system, Rear view monitor, Idle Air Volume Learning, Steering Angle Sensor Neutral Position, Audio presets and Navigation. You will need the CONSULT-III diagnostic tool, or equivalent. Follow the directions on the screen of the tool, as needed.

1. Before servicing the vehicle, refer to the Precautions Section.

➡️If working near and/or around the SRS system and components, be sure to disable the SRS system. After disabling the system wait three minutes or more before servicing the vehicle.

2. Disconnect the negative battery cable.
3. Remove the passenger's side instrument lower panel.
4. Disconnect electrical connectors.
5. Remove the component.

To install:

➡️Be sure to use new fasteners, as required.

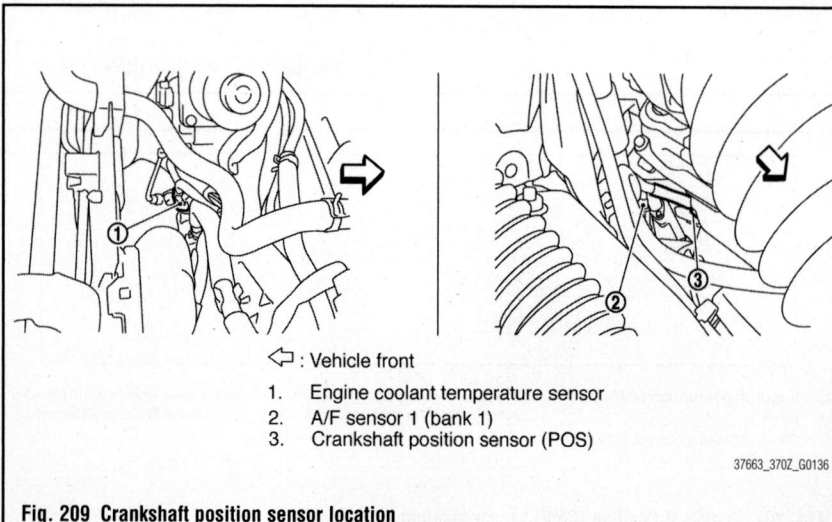

⇦ : Vehicle front
1. Engine coolant temperature sensor
2. A/F sensor 1 (bank 1)
3. Crankshaft position sensor (POS)

Fig. 209 Crankshaft position sensor location

6. Installation is the reverse of the removal procedure.

7. Be sure to perform the reconnect/relearn procedures.

ENGINE COOLANT TEMPERATURE (ECT) SENSOR

LOCATION

See Figure 211.

Refer to the accompanying illustration.

REMOVAL & INSTALLATION

See Figure 212.

At this time the manufacturer does not provide removal and installation procedures for this component. The following procedure is a guideline and may differ from the vehicle you are servicing.

➡Whenever the negative battery cable is disconnected the following components will require resetting. The Automatic temperature control system, Automatic drive positioner, Power window control, Sunroof system, Sunshade system, Rear view monitor, Idle Air Volume Learning, Steering Angle Sensor Neutral Position, Audio presets and Navigation. You will need the CONSULT-III diagnostic tool, or equivalent. Follow the directions on the screen of the tool, as needed.

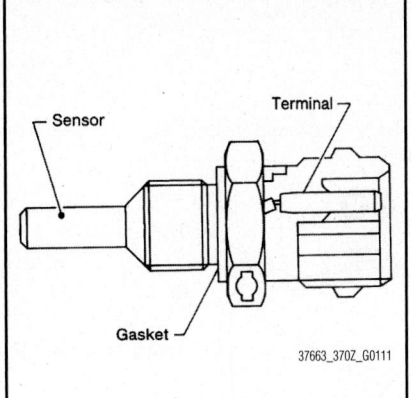

Fig. 212 Engine Coolant Temperature (ECT) sensor

1. Before servicing the vehicle, refer to the Precautions Section.

➡If working near and/or around the SRS system and components, be sure to disable the SRS system. After disabling the system wait three minutes or more before servicing the vehicle.

2. Disconnect the negative battery cable.

3. Drain the cooling system. Properly dispose of used engine coolant.

4. Remove the necessary components to gain access to the sensor.

5. Disconnect the electrical sensor.

6. Remove the sensor from its mounting.

To install:

➡Be sure to use new fasteners, as required.

7. Installation is the reverse of the removal procedure.

8. Be sure to perform the reconnect/relearn procedures.

EVAPORATIVE (EVAP) CANISTER

LOCATION

The EVAP canister is located under the vehicle near the fuel tank.

REMOVAL & INSTALLATION

See Figure 213.

➡Whenever the negative battery cable is disconnected the following components will require resetting. The Automatic temperature control system, Automatic drive positioner, Power window control, Sunroof system, Sunshade system, Rear view monitor, Idle Air Volume Learning, Steering Angle Sensor Neutral Position, Audio presets and Navigation. You will need the CONSULT-III diagnostic tool, or equivalent. Follow the directions on the screen of the tool, as needed.

1. Before servicing the vehicle, refer to the Precautions Section.

➡If working near and/or around the SRS system and components, be sure to disable the SRS system. After disabling the system wait three minutes or more before servicing the vehicle.

2. Disconnect the negative battery cable.

3. Raise and support the vehicle safely.

4. Remove the canister retaining bolt.

5. Remove the canister from its mounting.

➡The canister vent control valve and system pressure sensor can be removed without removing the canister.

To install:

➡Be sure to use new fasteners, as required.

6. Installation is the reverse of the removal procedure.

7. Be sure to perform the reconnect/relearn procedures.

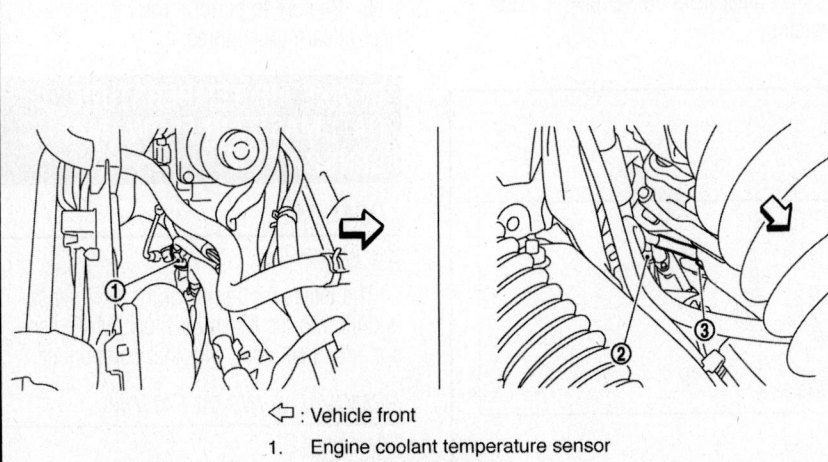

⇦ : Vehicle front

1. Engine coolant temperature sensor
2. A/F sensor 1 (bank 1)
3. Crankshaft position sensor (POS)

Fig. 211 Engine Coolant Temperature (ECT) sensor location

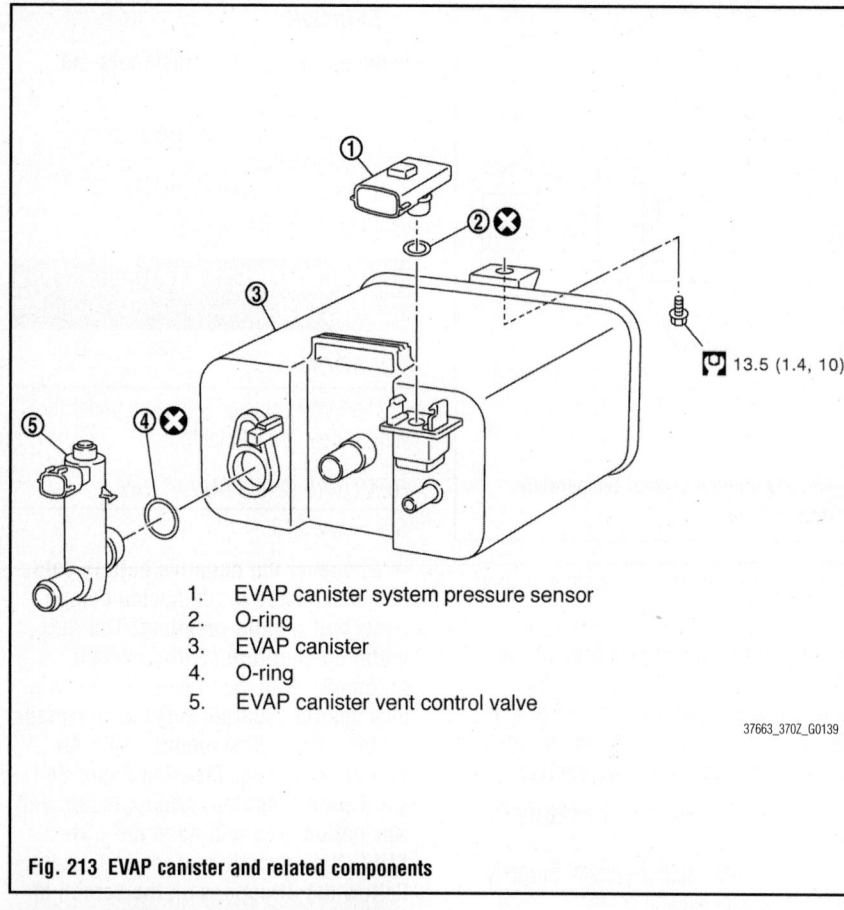

1. EVAP canister system pressure sensor
2. O-ring
3. EVAP canister
4. O-ring
5. EVAP canister vent control valve

13.5 (1.4, 10)

37663_370Z_G0139

Fig. 213 EVAP canister and related components

HEATED OXYGEN SENSOR (HO2S)

LOCATION

See Figures 214 and 215.

This sensor is located in the exhaust stream, after the catalytic converter. Two sensors are used, one for each cylinder bank.

REMOVAL & INSTALLATION

See Figure 214.

At this time the manufacturer does not provide removal and installation procedures for this component. The following procedure is a guideline and may differ from the vehicle you are servicing.

➡ **Whenever the negative battery cable is disconnected the following components will require resetting. The Automatic temperature control system, Automatic drive positioner, Power window control, Sunroof system, Sunshade system, Rear view monitor, Idle Air Volume Learning, Steering Angle Sensor Neutral Position, Audio presets and Navigation. You will need the CONSULT-III diagnostic tool, or equivalent. Follow the directions on the screen of the tool, as needed.**

1. Before servicing the vehicle, refer to the Precautions Section.

➡ **If working near and/or around the SRS system and components, be sure to disable the SRS system. After disabling the system wait three minutes or more before servicing the vehicle.**

2. Disconnect the negative battery cable.
3. Raise and safely support the vehicle.
4. Remove the undercover.
5. Disconnect the harness connector.
6. Remove the sensor from its mounting.

➡ **Discard the sensor if it has been dropped.**

To install:

➡ **Be sure to use new fasteners, as required.**

7. Installation is the reverse of the removal procedure.
8. Before installing a new sensor coat the threads with an approved anti-seize lubricant.
9. Be sure to perform the reconnect/relearn procedures.

INTAKE AIR TEMPERATURE (IAT)/MASS AIR FLOW (MAF) SENSOR

LOCATION

See Figure 216.

The Intake Air Temperature (IAT) sensor is built into the Mass Air Flow (MAF) sensor and is serviced with that component.

REMOVAL & INSTALLATION

See Figure 216.

At this time the manufacturer does not provide removal and installation procedures for this component. The following procedure is a guideline and may differ from the vehicle you are servicing.

➡ **Whenever the negative battery cable is disconnected the following compo-**

⇦ : Vehicle front

1. Heated oxygen sensor (bank 2)
2. Heated oxygen sensor 2 (bank 2) harness connector
3. Heated oxygen sensor 2 (bank 1)
4. Heated oxygen sensor 2 (bank 1) harness connector

37663_370Z_G0134

Fig. 214 Heated oxygen sensor and related components

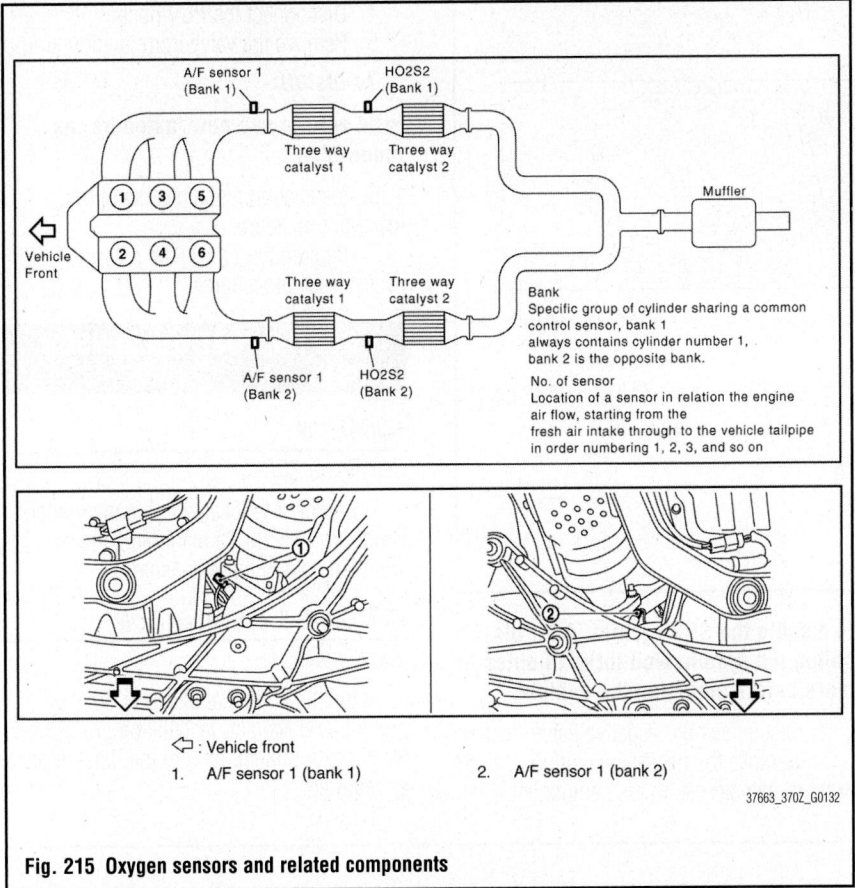

⇦ : Vehicle front

1. A/F sensor 1 (bank 1)   2. A/F sensor 1 (bank 2)

37663_370Z_G0132

Fig. 215 Oxygen sensors and related components

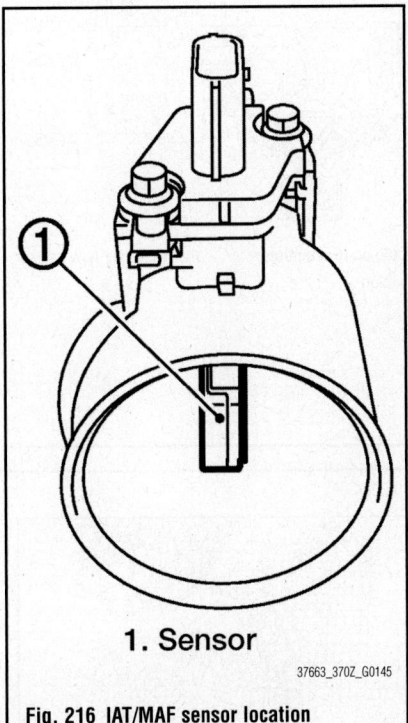

1. Sensor

37663_370Z_G0145

Fig. 216 IAT/MAF sensor location

nents will require resetting. The Automatic temperature control system, Automatic drive positioner, Power window control, Sunroof system, Sunshade

system, Rear view monitor, Idle Air Volume Learning, Steering Angle Sensor Neutral Position, Audio presets and Navigation. You will need the CONSULT-III diagnostic tool, or equivalent. Follow the directions on the screen of the tool, as needed.

1. Before servicing the vehicle, refer to the Precautions Section.

➡ **If working near and/or around the SRS system and components, be sure to disable the SRS system. After disabling the system wait three minutes or more before servicing the vehicle.**

2. Disconnect the negative battery cable.

3. Disconnect the harness connector from the IAT/MAF sensor.

4. Disconnect the tube clamp at the electric throttle control actuator and at the fresh air intake tube.

5. Remove air cleaner to electric throttle control actuator tube, air cleaner case (upper) with the IAT/MAF sensor attached.

6. Remove IAT/MAF sensor from air cleaner case (upper), as necessary.

7. Remove resonator in the fender, lifting left fender protector, as necessary.

To install:

➡ **Be sure to use new fasteners, as required.**

8. Installation is the reverse of the removal procedure.

9. Be sure to perform the reconnect/relearn procedures.

KNOCK SENSOR (KS)

LOCATION

See Figure 219.

The knock sensors are located under the intake manifold. There are two of them.

REMOVAL & INSTALLATION

See Figure 219.

At this time the manufacturer does not provide removal and installation procedures for this component. The following procedure is a guideline and may differ from the vehicle you are servicing.

➡ **Whenever the negative battery cable is disconnected the following components will require resetting. The Automatic temperature control system, Automatic drive positioner, Power window control, Sunroof system, Sunshade system, Rear view monitor, Idle Air Volume Learning, Steering Angle Sensor Neutral Position, Audio presets and Navigation. You will need the CONSULT-III diagnostic tool, or equivalent. Follow the directions on the screen of the tool, as needed.**

1. Before servicing the vehicle, refer to the Precautions Section.

➡ **If working near and/or around the SRS system and components, be sure to disable the SRS system. After disabling the system wait three minutes or more before servicing the vehicle.**

2. Disconnect the negative battery cable.

3. Remove the intake manifold collector.

4. Remove the intake manifold.

5. Disconnect the sensor electrical connector.

6. Remove the sensor from its mounting.

To install:

➡ **Be sure to use new fasteners, as required.**

7. Installation is the reverse of the removal procedure.

8. Be sure to perform the reconnect/relearn procedures.

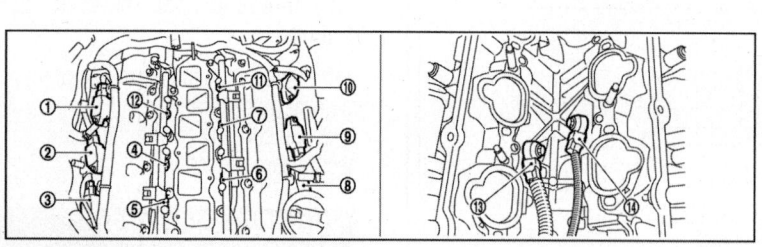

1. Ignition coil No.5 (with power transistor)
2. Ignition coil No.3 (with power transistor)
3. Ignition coil No.1 (with power transistor)
4. Fuel injector No.3
5. Fuel injector No.1
6. Fuel injector No.2
7. Fuel injector No.4
8. Ignition coil No.2 (with power transistor)
9. Ignition coil No.4 (with power transistor)
10. Ignition coil No.6 (with power transistor)
11. Fuel injector No.6
12. Fuel injector No.5
13. Knock sensor (bank 1)
14. Knock sensor (bank 2)

37663_370Z_G0144

Fig. 219 Knock sensor location

MALFUNCTION INDICATOR LIGHT

RESET PROCEDURE

Clearing diagnostic trouble codes resets the MIL.

POSITIVE CRANKCASE VENTILATION (PCV) VALVE

LOCATION

See Figure 220.

Refer to the accompanying illustration.

REMOVAL & INSTALLATION

At this time the manufacturer does not provide removal and installation procedures for this component. The following procedure is a guideline and may differ from the vehicle you are servicing.

➡Whenever the negative battery cable is disconnected the following components will require resetting. The Automatic temperature control system, Automatic drive positioner, Power window control, Sunroof system, Sunshade system, Rear view monitor, Idle Air Volume Learning, Steering Angle Sensor Neutral Position, Audio presets and Navigation. You will need the CONSULT-III diagnostic tool, or equivalent. Follow the directions on the screen of the tool, as needed.

1. Before servicing the vehicle, refer to the Precautions Section.

➡If working near and/or around the SRS system and components, be sure

to disable the SRS system. After disabling the system wait three minutes or more before servicing the vehicle.

2. Disconnect the negative battery cable.
3. Remove the necessary components in order to gain access to the component.

4. Disconnect the PCV hose.
5. Remove the valve from its mounting.

To install:

➡Be sure to use new fasteners, as required.

6. Installation is the reverse of the removal procedure.
7. Be sure to perform the reconnect/relearn procedures.

THROTTLE POSITION SENSOR (TPS)

LOCATION

See Figure 221.

This sensor is located in the main hose connecting the intake manifold collector . There are two of these sensors.

REMOVAL & INSTALLATION

See Figure 221.

At this time the manufacturer does not provide removal and installation procedures for this component, refer to the illustration as required.

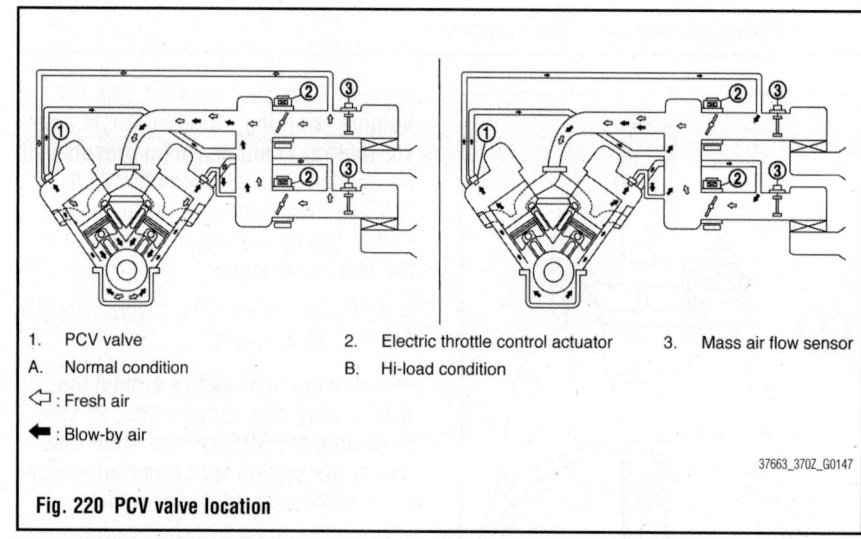

1. PCV valve
2. Electric throttle control actuator
3. Mass air flow sensor
A. Normal condition
B. Hi-load condition
◁ : Fresh air
◀ : Blow-by air

37663_370Z_G0147

Fig. 220 PCV valve location

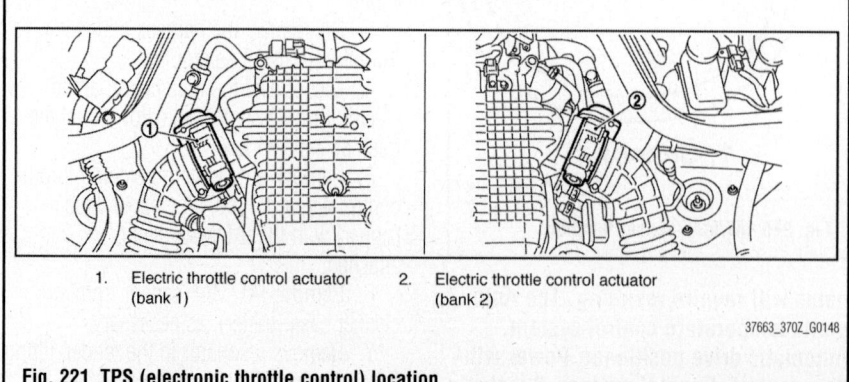

1. Electric throttle control actuator (bank 1)
2. Electric throttle control actuator (bank 2)

37663_370Z_G0148

Fig. 221 TPS (electronic throttle control) location

FUEL | GASOLINE FUEL INJECTION SYSTEM

FUEL SYSTEM SERVICE PRECAUTIONS

Safety is the most important factor when performing not only fuel system maintenance but any type of maintenance. Failure to conduct maintenance and repairs in a safe manner may result in serious personal injury or death. Maintenance and testing of the vehicle's fuel system components can be accomplished safely and effectively by adhering to the following rules and guidelines.

• To avoid the possibility of fire and personal injury, always disconnect the negative battery cable unless the repair or test procedure requires that battery voltage be applied.

• Always relieve the fuel system pressure prior to disconnecting any fuel system component (injector, fuel rail, pressure regulator, etc.), fitting or fuel line connection. Exercise extreme caution whenever relieving fuel system pressure to avoid exposing skin, face and eyes to fuel spray. Please be advised that fuel under pressure may penetrate the skin or any part of the body that it contacts.

• Always place a shop towel or cloth around the fitting or connection prior to loosening to absorb any excess fuel due to spillage. Ensure that all fuel spillage (should it occur) is quickly removed from engine surfaces. Ensure that all fuel soaked cloths or towels are deposited into a suitable waste container.

• Always keep a dry chemical (Class B) fire extinguisher near the work area.

• Do not allow fuel spray or fuel vapors to come into contact with a spark or open flame.

• Always use a back-up wrench when loosening and tightening fuel line connection fittings. This will prevent unnecessary stress and torsion to fuel line piping.

• Always replace worn fuel fitting O-rings with new Do not substitute fuel hose or equivalent where fuel pipe is installed.

Before servicing the vehicle, make sure to also refer to the precautions in the beginning of this section as well.

RELIEVING FUEL SYSTEM PRESSURE

See Figure 223.

➡Whenever the negative battery cable is disconnected the following components will require resetting. The Auto-

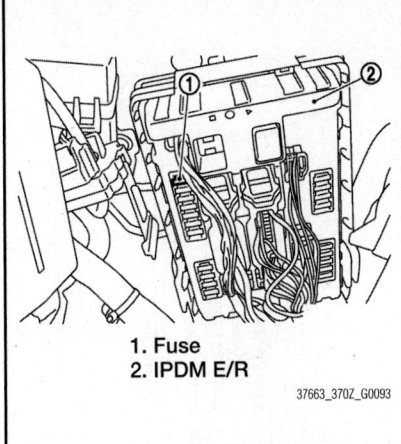

1. Fuse
2. IPDM E/R

37663_370Z_G0093

Fig. 223 Fuel pump fuse location

matic temperature control system, Automatic drive positioner, Power window control, Sunroof system, Sunshade system, Rear view monitor, Idle Air Volume Learning, Steering Angle Sensor Neutral Position, Audio presets and Navigation. You will need the CONSULT-III diagnostic tool, or equivalent. Follow the directions on the screen of the tool, as needed.

1. Before servicing the vehicle, refer to the Precautions Section.

➡If working near and/or around the SRS system and components, be sure to disable the SRS system. After disabling the system wait three minutes or more before servicing the vehicle.

2. Remove the fuel pump fuse, located in the IPDM E/R.
3. Start the engine.
4. After the engine stalls, crank it two or three times to release all fuel pressure.
5. Turn the ignition switch off.
6. Reinstall the fuel pump fuse after servicing the fuel system.
7. Be sure to perform the reconnect/relearn procedures, as required.

FUEL FILTER

REMOVAL & INSTALLATION

The fuel filter is attached to the fuel pump assembly. The fuel pump must be removed before the filter can be serviced.

FUEL LEVEL SENDING UNIT

REMOVAL & INSTALLATION

See Figure 224.

➡Whenever the negative battery cable is disconnected the following components will require resetting. The Automatic temperature control system, Automatic drive positioner, Power window control, Sunroof system, Sunshade system, Rear view monitor, Idle Air Volume Learning, Steering Angle Sensor Neutral Position, Audio presets and Navigation. You will need the CONSULT-III diagnostic tool, or equivalent. Follow the directions on the screen of the tool, as needed.

1. Before servicing the vehicle, refer to the Precautions Section.

➡If working near and/or around the SRS system and components, be sure to disable the SRS system. After disabling the system wait three minutes or more before servicing the vehicle.

2. Properly relieve the fuel system pressure.
3. Disconnect the negative battery cable.
4. Remove the fuel pump module.
5. Discard the gasket.
6. Service the fuel level sending unit, as required.

➡This component cannot be disassembled and should be replaced as a unit.

To install:

➡Be sure to use new fasteners, as required.

7. Installation is the reverse of the removal procedure.
8. Be sure to perform the reconnect/relearn procedures.

FUEL PUMP MODULE

REMOVAL & INSTALLATION

See Figures 225 through 231.

➡Whenever the negative battery cable is disconnected the following components will require resetting. The Automatic temperature control system, Automatic drive positioner, Power window control, Sunroof system, Sunshade system, Rear view monitor, Idle Air Volume Learning, Steering Angle Sensor Neutral Position, Audio presets and Navigation. You will need the

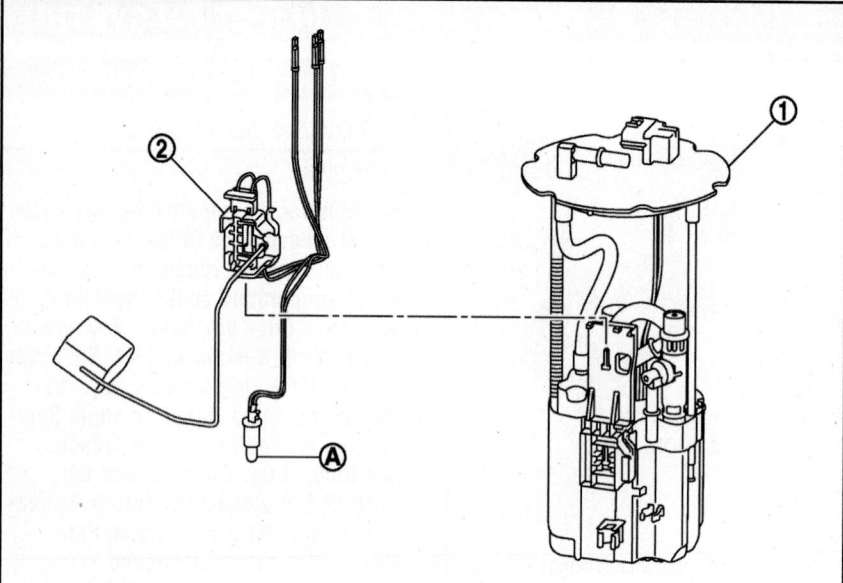

1. Fuel filter and fuel pump assembly
2. Main fuel level sensor unit
A. Fuel temp sensor

37663_370Z_G0095

Fig. 224 Fuel level sending unit and related components

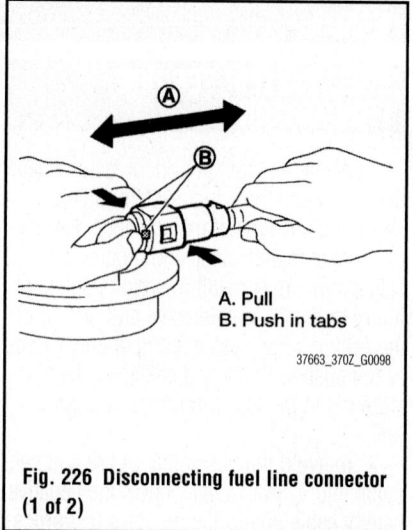

A. Pull
B. Push in tabs

37663_370Z_G0098

Fig. 226 Disconnecting fuel line connector (1 of 2)

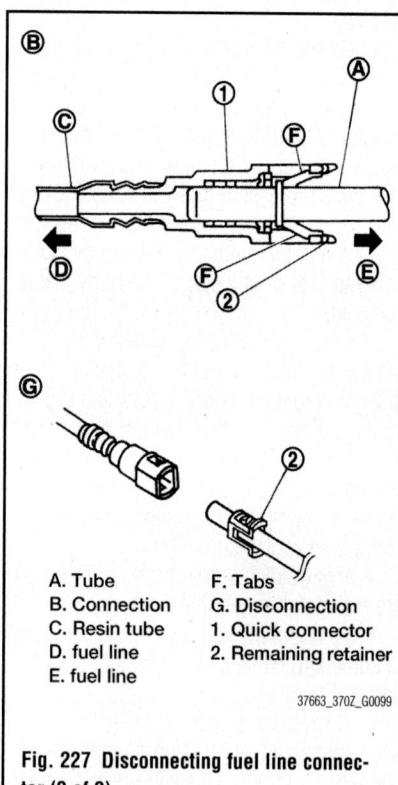

A. Tube
B. Connection
C. Resin tube
D. fuel line
E. fuel line
F. Tabs
G. Disconnection
1. Quick connector
2. Remaining retainer

37663_370Z_G0099

Fig. 227 Disconnecting fuel line connector (2 of 2)

CONSULT-III diagnostic tool, or equivalent. Follow the directions on the screen of the tool, as needed.

1. Before servicing the vehicle, refer to the Precautions Section.

➡️If working near and/or around the SRS system and components, be sure to disable the SRS system. After disabling the system wait three minutes or more before servicing the vehicle.

2. Disconnect the negative battery cable.

3. Drain the fuel tank to an acceptable level. If the fuel level indicates more than the level shown in the illustration, full or almost full, drain the fuel from the tank until the fuel gauge indicates a level as shown in the illustration.

➡️Because fuel will be spilled when removing the main and sub fuel level sensor units for the top of the fuel is above the main and sub fuel level sensor units installed surface. As a guide, fuel level becomes the position as shown in the illustration when approximately 3 3/8 gallons of fuel are removed from the tank.

4. Properly relieve the fuel system pressure.

5. Remove the fuel tank cap.

6. Remove the rear seat cushion.

7. Peel off the floor carpet. Remove the inspection hole cover.

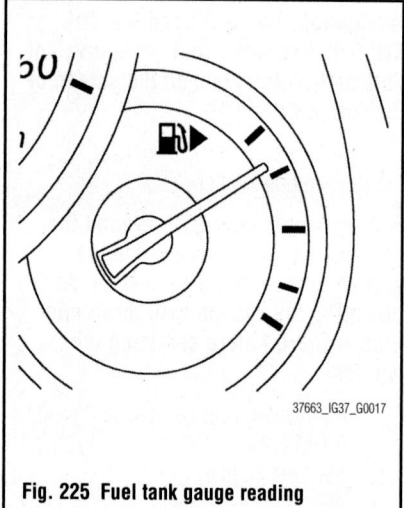

37663_IG37_G0017

Fig. 225 Fuel tank gauge reading

➡️Right side for main fuel level sensor, fuel filter and fuel pump assembly. Left side for sub fuel level sensor unit.

8. Disconnect and fuel feed tube. Disconnect the harness connector.

9. Disconnect the quick connector. Hold the sides of the connector, push in the tabs and pull out the fuel feed tube.

➡️The quick connector can be disconnected when the tabs are completely depressed. Never twist it more than necessary. Never use tools to disconnect the quick connector. Cover the fuel line openings to prevent dirt from entering the fuel system.

10. To remove the main fuel level sensor unit, fuel filter and fuel pump assembly, remove the retainer. Raise the unit and disconnect the quick connector.

11. To remove the sub fuel level sensor, remove the retainer. Raise the component and remove it.

To install:

➡️Be sure to use new fasteners, as required.

12. Installation is the reverse of the removal procedure.

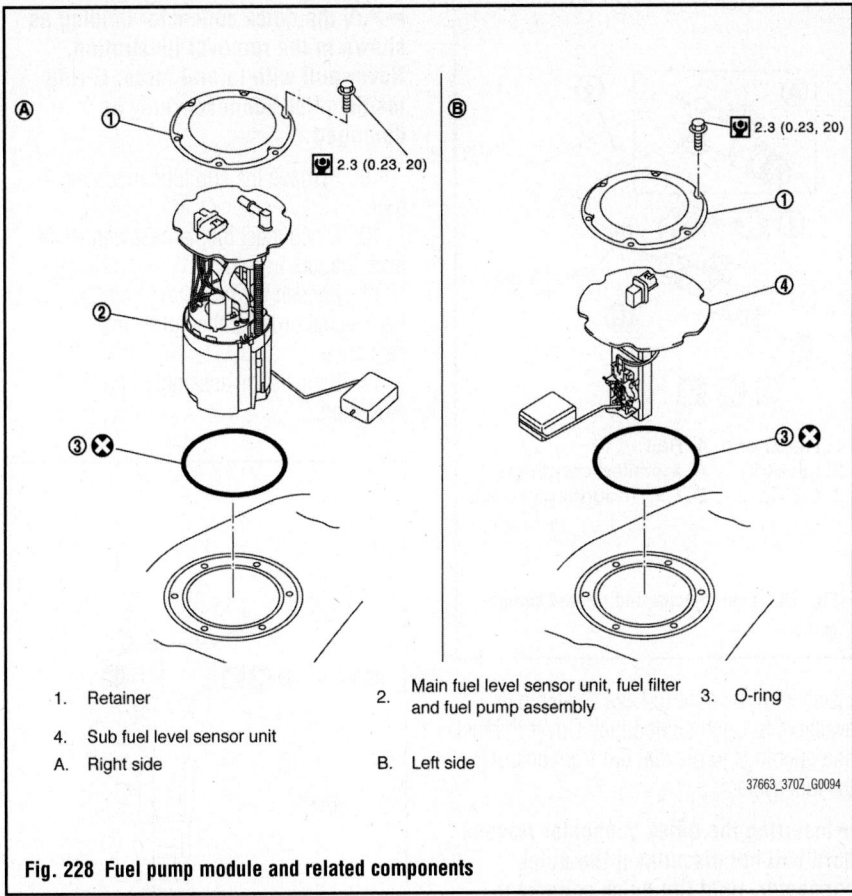

1. Retainer
2. Main fuel level sensor unit, fuel filter and fuel pump assembly
3. O-ring
4. Sub fuel level sensor unit
A. Right side
B. Left side

37663_370Z_G0094

Fig. 228 Fuel pump module and related components

13. When installing, face the units as shown in the illustration and install them with the knock pun on back aligned with the pin hole on the fuel tank.

14. Install the retainer so that its notch becomes parallel with the notch on the fuel tank. Tighten the retainer bolts evenly.

15. Install the fuel pump fuse, if removed.

16. Turn the ignition switch ON (with the engine stopped), check all connections for fuel leakage, correct as required.

17. Start the engine and let it idle, check for fuel leaks. Correct as required.

18. Be sure to perform the reconnect/relearn procedures.

FUEL RAIL & INJECTORS

REMOVAL & INSTALLATION

See Figures 231 through 236.

➡Whenever the negative battery cable is disconnected the following components will require resetting. The Automatic temperature control system, Automatic drive positioner, Power window control, Sunroof system, Sunshade system, Rear view monitor, Idle Air Volume Learning, Steering Angle Sensor Neutral Position, Audio presets and Navigation. You will need the CONSULT-III diagnostic tool, or equivalent. Follow the directions on the screen of the tool, as needed.

1. Before servicing the vehicle, refer to the Precautions Section.

➡If working near and/or around the SRS system and components, be sure to disable the SRS system. After disabling the system wait three minutes or more before servicing the vehicle.

2. Properly relieve the fuel system pressure.

3. Disconnect the negative battery cable.

4. Remove the engine cover.

5. Remove the air cleaner assembly.

6. Remove the intake manifold collector.

7. Remove the fuel feed hose (with damper) from the fuel sub tube. Remove the harness bracket.

➡There is no fuel return route. Plug the lines to prevent fuel leakage. Never separate the damper and the hose.

➡When separating the fuel feed hose (with damper) and centralized under floor piping connection, disconnect the quick connector, as shown in the illustrations. Disconnect the quick connector by using a quick connector tool, J-45488 or equivalent.

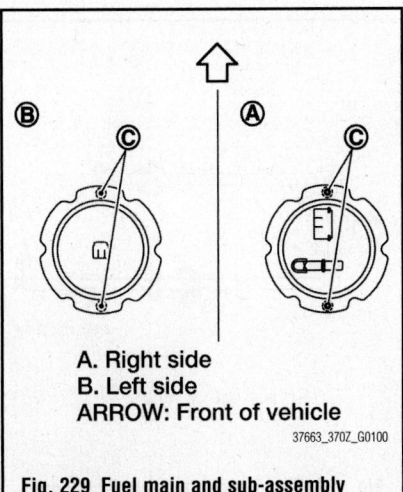

A. Right side
B. Left side
ARROW: Front of vehicle

37663_370Z_G0100

Fig. 229 Fuel main and sub-assembly installation alignment

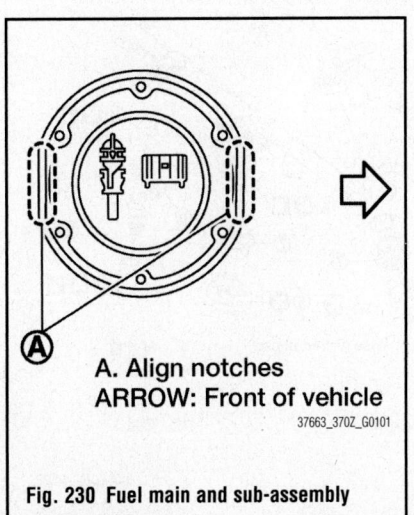

A. Align notches
ARROW: Front of vehicle

37663_370Z_G0101

Fig. 230 Fuel main and sub-assembly retainer alignment

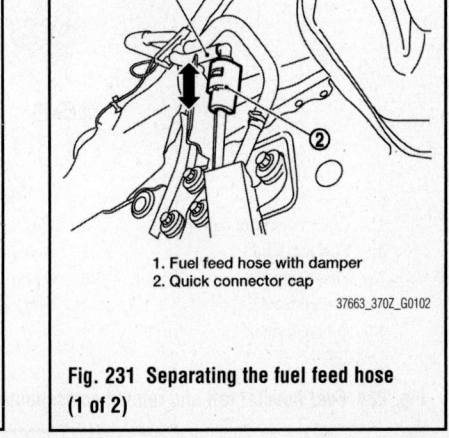

1. Fuel feed hose with damper
2. Quick connector cap

37663_370Z_G0102

Fig. 231 Separating the fuel feed hose (1 of 2)

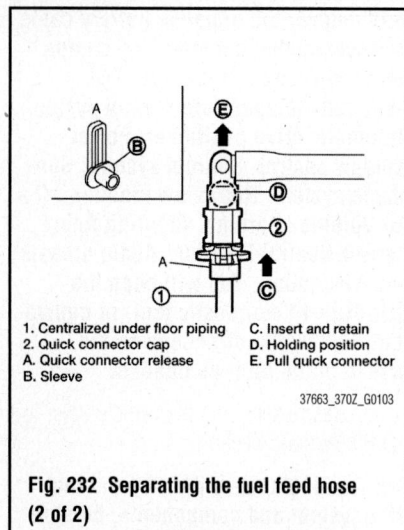

1. Centralized under floor piping   C. Insert and retain
2. Quick connector cap   D. Holding position
A. Quick connector release   E. Pull quick connector
B. Sleeve

37663_370Z_G0103

Fig. 232 Separating the fuel feed hose (2 of 2)

8. To disconnect the quick connector from the centralized under floor piping, with the sleeve side of the quick connector release facing the quick connector, install the quick connector release onto the centralized floor piping. Insert the quick connector release into the quick connector, until the sleeve contacts and stops. Hold the quick connector and release on that position. Draw and pull out the quick connector straight from the centralized under floor piping. Never bend or twist the connection between the quick connector and fuel feed hose (with damper) during removal and/or

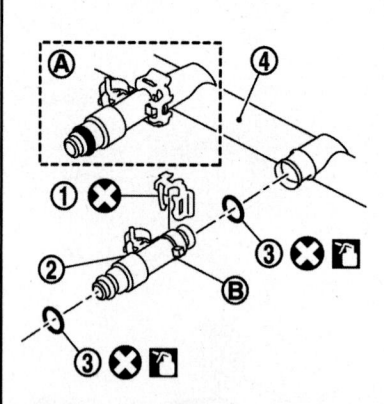

1. Retainer   4. Rail
2. Injector   A. Installed condition
3. O-ring   B. Clip mounting groove

37663_370Z_G0105

Fig. 233 Fuel injector and related components

installation. Be sure to have a catch pan available to catch spilled fuel. Cover the fuel line openings to prevent dirt from entering the fuel system.

➡**Inserting the quick connector release hard will not disconnect the quick connector. Hold the quick connector release where it contacts and goes no further.**

➡**Pull the quick connector holding as shown in the removal illustration. Never pull with lateral force. O-ring inside quick connector may be damaged.**

9. Remove the sub tube mounting bolt.

10. Disconnect the harness connector from the fuel injector.

11. Loosen the mounting bolts in the reverse order of the tightening sequence.

12. Remove the assembly from its mounting.

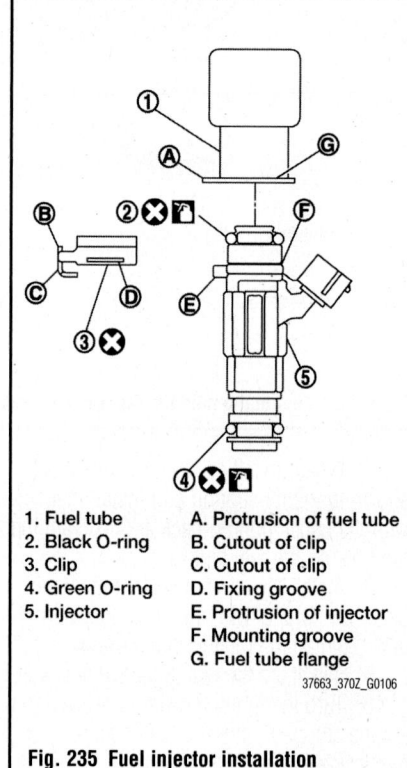

1. Fuel tube   A. Protrusion of fuel tube
2. Black O-ring   B. Cutout of clip
3. Clip   C. Cutout of clip
4. Green O-ring   D. Fixing groove
5. Injector   E. Protrusion of injector
     F. Mounting groove
     G. Fuel tube flange

37663_370Z_G0106

Fig. 235 Fuel injector installation

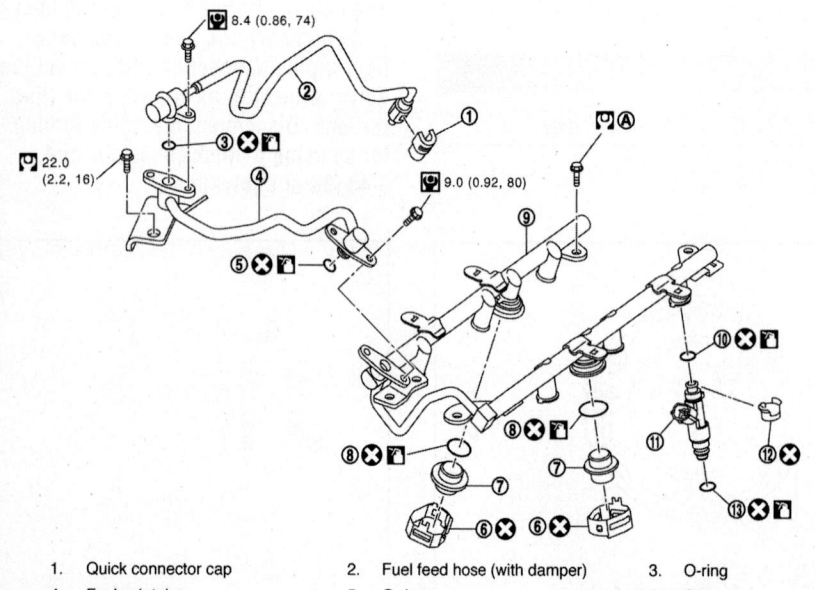

1. Quick connector cap   2. Fuel feed hose (with damper)   3. O-ring
4. Fuel sub tube   5. O-ring   6. Clip
7. Fuel damper   8. O-ring   9. Fuel tube
10. O-ring (black)   11. Fuel injector   12. Clip
13. O-ring (green)

37663_370Z_G0027

Fig. 234 Fuel injector rail and related components

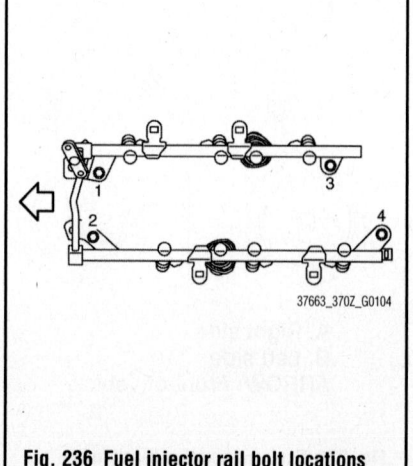

37663_370Z_G0104

Fig. 236 Fuel injector rail bolt locations and tightening sequence

13. Remove the fuel injector from the rail, as required. Discard the O-rings.

14. Do not remove the fuel sub tube and fuel damper.

To install:

➡ **Be sure to use new fasteners, as required.**

15. Install new O-rings on the injector, if removed. Lubricate the O-ring with clean engine oil prior to installation.

➡ **Fuel tube side O-ring is black, nozzle side is green.**

16. To install the injector, insert a new retaining clip into the mounting groove of the injector. Never reuse the old retaining clip.

17. Do not remove the fuel sub tube and fuel damper.

Insert the injector into the fuel tube, with the clip attached.

➡ **Insert it while matching it to the axial center. Insert the injector so that the protrusion of the fuel tube matches the cutout of the clip. Check that the tube flange is securely fixed in the flange fixing groove on the clip. Check that the injector does not rotate. See illustration.**

18. Install the fuel injector rail and injectors. Tighten to specification and in the proper sequence.

19. Tightening specification is 7 ft. lbs. (10.1 Nm), first pass. 17 ft. lbs. (23.6 Nm), second pass.

20. Continue the installation in the reverse order of the removal procedure.

21. Be sure to perform the reconnect/relearn procedures.

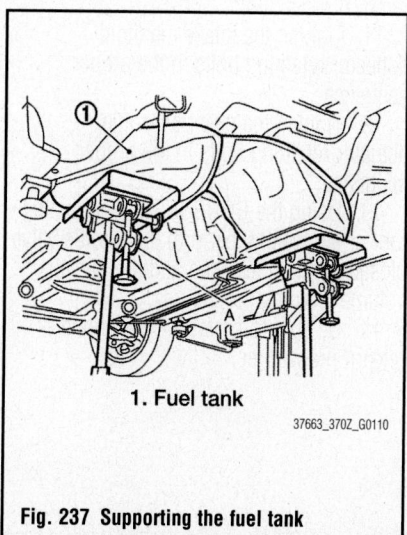

1. Fuel tank

37663_370Z_G0110

Fig. 237 Supporting the fuel tank

FUEL TANK

REMOVAL & INSTALLATION

See Figures 225 through 227, 237 and 238.

➡ **Whenever the negative battery cable is disconnected the following components will require resetting. The Automatic temperature control system, Automatic drive positioner, Power window control, Sunroof system, Sunshade system, Rear view monitor, Idle Air Volume Learning, Steering Angle Sensor Neutral Position, Audio presets and Navigation. You will need the CONSULT-III diagnostic tool, or equivalent. Follow the directions on the screen of the tool, as needed.**

1. Before servicing the vehicle, refer to the Precautions Section.

➡ **If working near and/or around the SRS system and components, be sure to disable the SRS system. After disabling the system wait three minutes or more before servicing the vehicle.**

2. Disconnect the negative battery cable.

3. Drain the fuel tank to an acceptable level. If the fuel level indicates more than the level shown in the illustration, full or almost full, drain the fuel from the tank until the fuel gauge indicates a level as shown in the illustration.

➡ **Because fuel will be spilled when removing the main and sub fuel level sensor units for the top of the fuel is above the main and sub fuel level sensor units installed surface. As a guide, fuel level becomes the position as shown in the illustration when approximately 3 3/8 gallons of fuel are removed from the tank.**

4. Properly relieve the fuel system pressure.

5. Remove the fuel tank cap.

6. Remove the rear seat cushion.

7. Peel off the floor carpet. Remove the inspection hole cover.

➡ **Right side for main fuel level sensor, fuel filter and fuel pump assembly. Left side for sub fuel level sensor unit.**

8. Disconnect and fuel feed tube. Disconnect the harness connector.

9. Disconnect the quick connector. Hold the sides of the connector, push in the tabs and pull out the fuel feed tube.

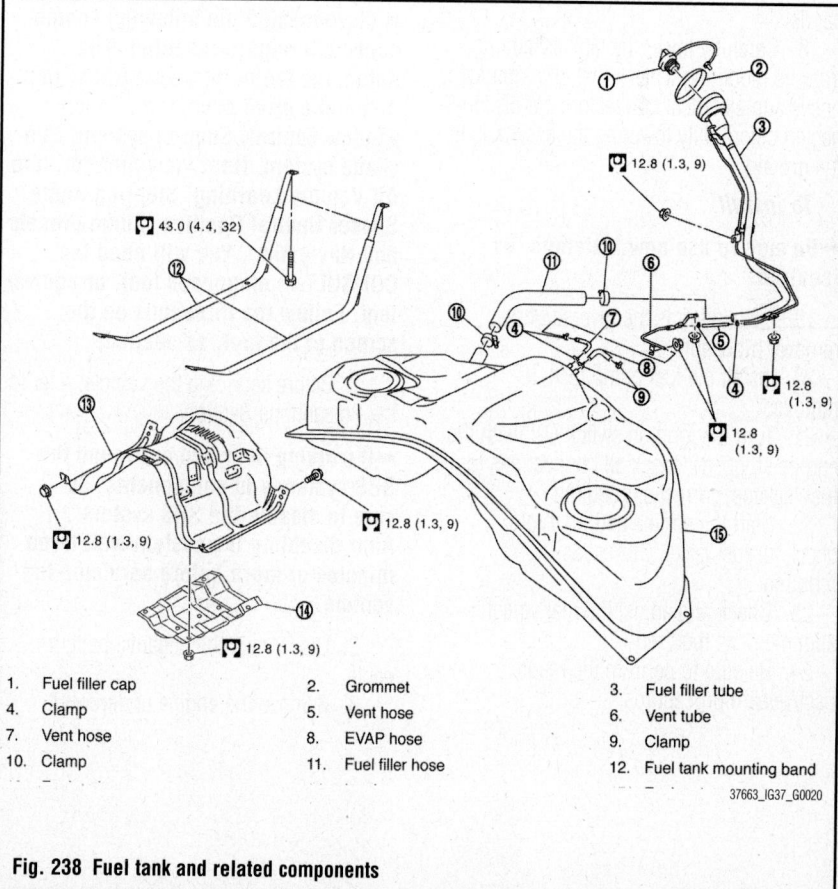

1.	Fuel filler cap	2.	Grommet	3.	Fuel filler tube
4.	Clamp	5.	Vent hose	6.	Vent tube
7.	Vent hose	8.	EVAP hose	9.	Clamp
10.	Clamp	11.	Fuel filler hose	12.	Fuel tank mounting band

37663_IG37_G0020

Fig. 238 Fuel tank and related components

➡The quick connector can be disconnected when the tabs are completely depressed. Never twist it more than necessary. Never use tools to disconnect the quick connector. Cover the fuel line openings to prevent dirt from entering the fuel system.

10. Remove the exhaust front tube, center muffler and main muffler.

11. Remove the driveshaft.

12. Remove the parking brake cables.

13. Remove the rear suspension member assembly.

➡For this service, halfshaft, final drive and rear suspension member are not required to be separated from one another during removal.

14. Disconnect the fuel filler hose, vent hose and EVAP hose at the tube side.

15. Remove the fuel tank protector, if equipped.

16. Properly support the lower part of the fuel tank, with a suitable jack.

➡Support the position that the fuel tank retaining straps do not engage.

17. Remove the fuel tank mounting bands.

18. Carefully lower the tank assembly from its mounting. Check that all required hoses and electrical connectors are disconnected before fully lowering the assembly to the ground.

To install:

➡Be sure to use new fasteners, as required.

19. Installation is the reverse of the removal procedure.

20. Install the fuel pump fuse, if removed.

21. Turn the ignition switch ON (with the engine stopped), check all connections for fuel leakage, correct as required.

22. Start the engine and let it idle, check for fuel leaks. Correct as required.

23. Check and adjust the rear wheel alignment, as required.

24. Be sure to perform the reconnect/relearn procedures.

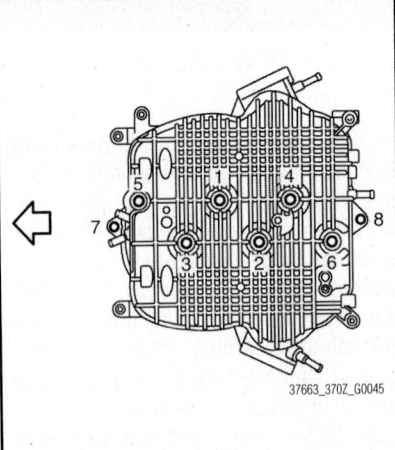

Fig. 239 Intake manifold collector bolt tightening sequence

THROTTLE BODY

REMOVAL & INSTALLATION

See Figures 239 and 240.

At this time the manufacturer does not provide removal and installation procedures for this component. The following procedure is a guideline and may differ from the vehicle you are servicing.

➡Whenever the negative battery cable is disconnected the following components will require resetting. The Automatic temperature control system, Automatic drive positioner, Power window control, Sunroof system, Sunshade system, Rear view monitor, Idle Air Volume Learning, Steering Angle Sensor Neutral Position, Audio presets and Navigation. You will need the CONSULT-III diagnostic tool, or equivalent. Follow the directions on the screen of the tool, as needed.

1. Before servicing the vehicle, refer to the Precautions Section.

➡If working near and/or around the SRS system and components, be sure to disable the SRS system. After disabling the system wait three minutes or more before servicing the vehicle.

2. Disconnect the negative battery cable.

3. Remove the engine undercover.

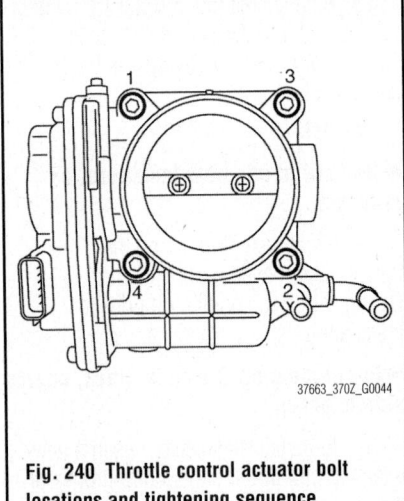

Fig. 240 Throttle control actuator bolt locations and tightening sequence

4. Drain the engine coolant. Be sure to properly dispose of used coolant.

5. Remove the engine cover.

6. Remove the air cleaner assembly.

7. To remove the electric throttle control actuator, Disconnect the water hoses from the component. Disconnect the harness connector.

8. Loosen the mounting bolts in the reverse order of the tightening sequence.

9. Remove the component from its mounting.

➡When removing only the intake manifold collector, move the electric throttle control actuator without disconnecting the water hose. Handel the component carefully to avoid shock damage.

To install:

➡Be sure to use new fasteners, as required.

10. Installation is the reverse of the removal procedure.

11. Tighten the intake manifold collector retaining bolts in the proper sequence.

12. Tighten the electric throttle actuator retaining bolts in the proper sequence.

13. Using the CONSULT-III diagnostic tool or equivalent, perform the throttle valve closed position learning and the idle air volume learning procedures.

14. Be sure to perform the reconnect/relearn procedures.

HEATING & AIR CONDITIONING SYSTEM

BLOWER MOTOR

REMOVAL & INSTALLATION
See Figure 241.

➡Whenever the negative battery cable is disconnected the following components will require resetting. The Automatic temperature control system, Automatic drive positioner, Power window control, Sunroof system, Sunshade system, Rear view monitor, Idle Air Volume Learning, Steering Angle Sensor Neutral Position, Audio presets and Navigation. You will need the CONSULT-III diagnostic tool, or equivalent. Follow the directions on the screen of the tool, as needed.

1. Before servicing the vehicle, refer to the Precautions Section.

➡If working near and/or around the SRS system and components, be sure to disable the SRS system. After disabling the system wait three minutes or more before servicing the vehicle.

2. Disconnect the negative battery cable.
3. As necessary remove the instrument lower cover.
4. Disconnect the blower motor electrical connector.
5. Remove the blower motor retaining screws.
6. Remove the component from its mounting.

To install:

➡Be sure to use new fasteners, as required.

7. Installation is the reverse of the removal procedure.
8. Be sure to perform the reconnect/relearn procedures.

HEATER CORE

REMOVAL & INSTALLATION
See Figure 242.

➡Whenever the negative battery cable is disconnected the following components will require resetting. The Automatic temperature control system, Automatic drive positioner, Power window control, Sunroof system, Sunshade system, Rear view monitor, Idle Air Volume Learning, Steering Angle Sensor Neutral Position, Audio presets and Navigation. You will need the CONSULT-III diagnostic tool, or equivalent. Follow the directions on the screen of the tool, as needed.

1. Before servicing the vehicle, refer to the Precautions Section.

➡If working near and/or around the SRS system and components, be sure to disable the SRS system. After disabling the system wait three minutes or more before servicing the vehicle.

2. Disconnect the negative battery cable.
3. Remove the heating and cooling unit assembly.
4. Remove the mounting screws and remove the heater pipe cover.
5. Remove the mounting screws and remove the left foot duct.
6. Slide the heater core leftward and remove it.

To install:

➡Be sure to use new fasteners, as required.

7. Installation is the reverse of the removal procedure.
8. Be sure to use new O-rings, coated with clean refrigerant oil prior to installation.
9. Properly charge the air conditioning system.
10. Properly refill the cooling system.
11. Start the engine and check the system for proper operation and refrigerant leakage.
12. Using the CONSULT-III diagnostic tool, or equivalent, perform the 4WAS front actuator adjustment if equipped with 4WAS.
13. Be sure to perform the reconnect/relearn procedures.

HEATER AND COOLING UNIT

REMOVAL & INSTALLATION
See Figures 243 through 247.

➡Whenever the negative battery cable is disconnected the following components will require resetting. The Automatic temperature control system, Automatic drive positioner, Power window control, Sunroof system, Sunshade system, Rear view monitor, Idle Air Volume Learning, Steering Angle Sensor Neutral Position, Audio presets and Navigation. You will need the CONSULT-III diagnostic tool, or equivalent. Follow the directions on the screen of the tool, as needed.

1. Before servicing the vehicle, refer to the Precautions Section.

➡If working near and/or around the SRS system and components, be sure to disable the SRS system. After disabling the system wait three minutes or more before servicing the vehicle.

2. Disconnect the negative battery cable.
3. Properly discharge the air conditioning system
4. Drain the cooling system. Be sure to properly dispose of used coolant.
5. Remove the cowl top cover.
6. Using tool SST: J-45815 disconnect and plug the one touch connectors at the housing.

1. Blower motor
A. Mounting screws

37663_370Z_G0195

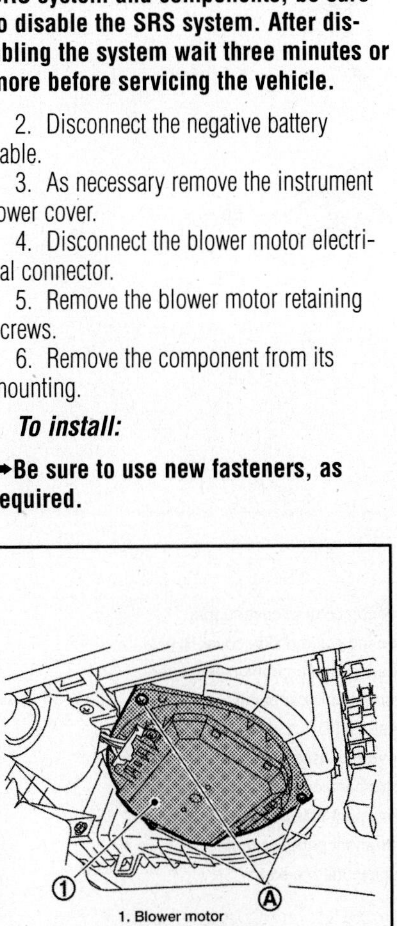

Fig. 241 Blower motor location

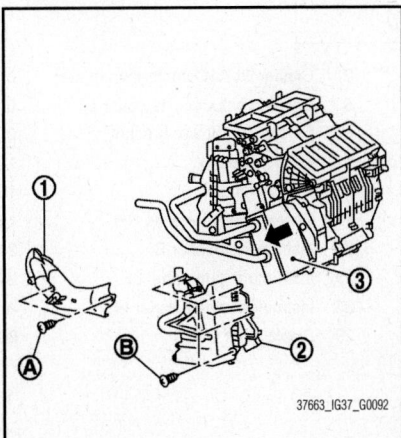

37663_IG37_G0092

Fig. 242 HVAC heater core and related components

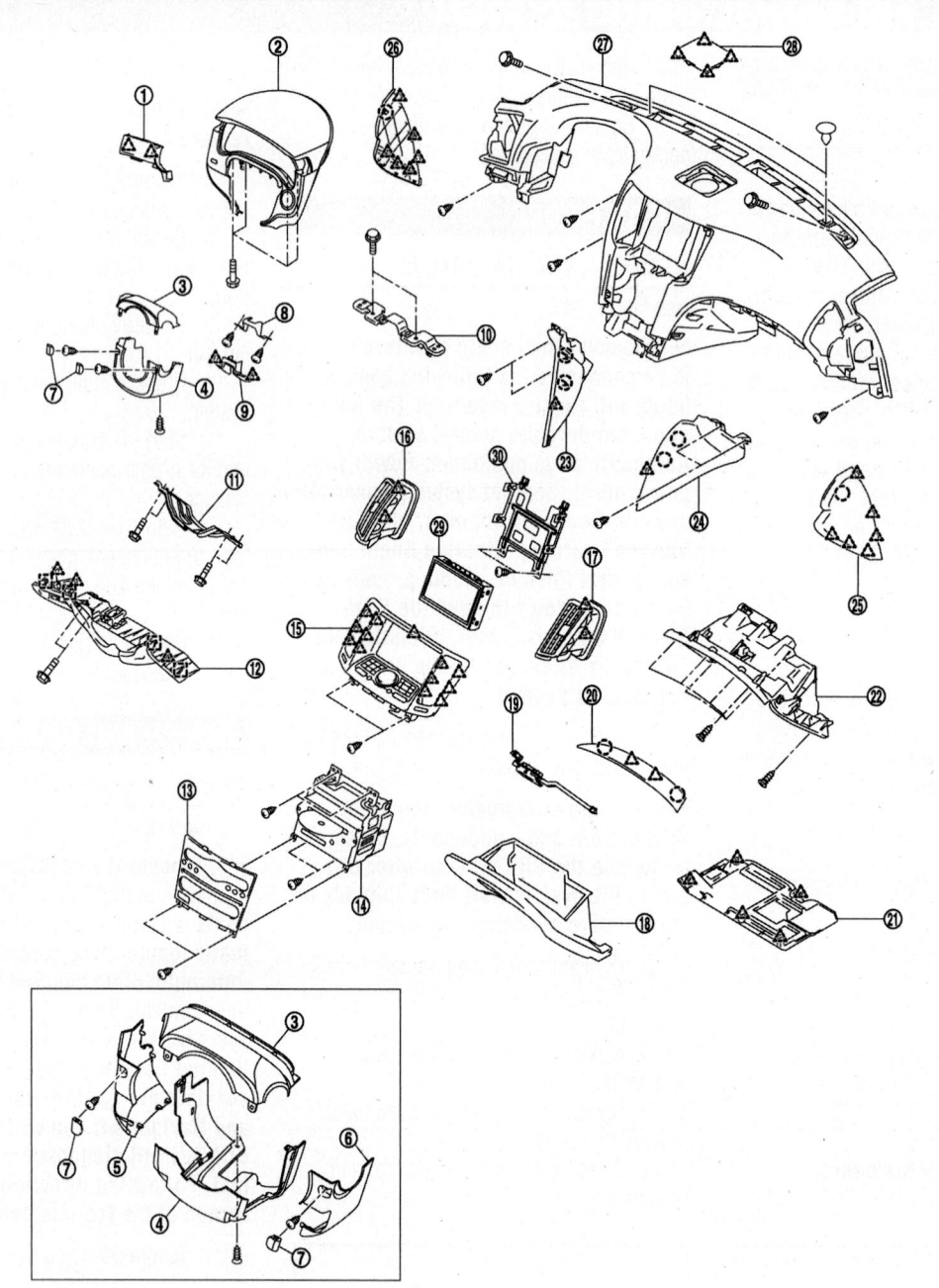

1.	Instrument finisher A	2.	Cluster lid A (Combination meter)	3.	Steering column cover upper
4.	Steering column cover lower	5.	Steering column side cover LH	6.	Steering column side cover RH
7.	Steering column mask	8.	Cluster lid A lower bracket	9.	Steering column front lower cover
10.	Meter bracket	11.	Knee protector	12.	Instrument lower panel LH
13.	Cluster lid C	14.	AV control unit	15.	Cluster lid D
16.	Center ventilator grille LH	17.	Center ventilator grille RH	18.	Glove box assembly
19.	Glove box lock assembly	20.	Instrument finisher B	21.	Instrument lower cover
22.	Instrument lower panel RH	23.	Instrument side panel LH	24.	Instrument side panel RH
25.	Instrument side finisher RH	26.	Instrument side finisher LH	27.	Instrument panel assembly
28.	Center speaker grille	29.	Display unit	30.	Display unit bracket

◯ : Clip

△ : Pawl

⬚ : Metal clip

37663_IG37_G0093

Fig. 243 Instrument panel and related components

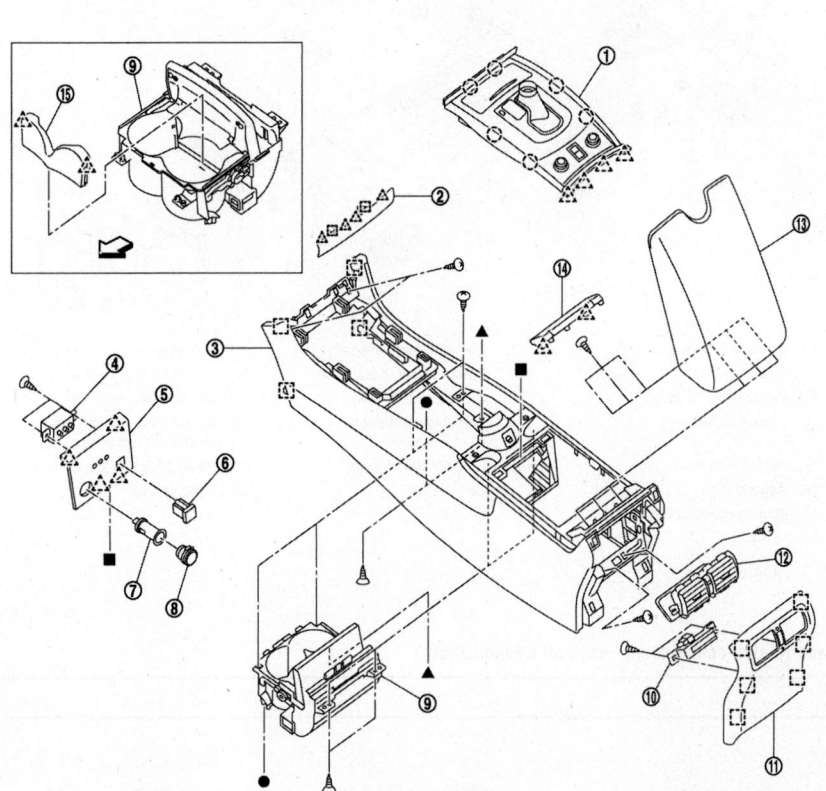

1. Console finisher assembly
2. Instrument finisher E
3. Center console assembly
4. Auxiliary input jacks (if equipped)
5. Console center finisher
6. USB connector (if equipped)
7. Power socket Inner case
8. Socket knob
9. Cup holder assembly
10. Inside key antenna
11. Console rear finisher
12. Rear ventilator grille (if equipped)
13. Console lid assembly
14. Console mask
15. Cup separator

⬡ : Clip

△ : Pawl

⬜ : Metal clip

◁ : Vehicle front

37663_IG37_G0095

Fig. 244 Center console and related components—automatic transmission

7. Remove the clamps and disconnect the heater hoses. Plug the hoses.

8. Remove the instrument panel assembly.

9. Remove the blower unit.

10. Remove the clips of the vehicle harness from the steering member.

11. Remove the instrument left and right stay.

12. Remove the drain hose.

13. Remove the mounting bolts from the heater and cooling unit assembly.

14. Remove the front defroster nozzle, side defroster nozzle and ventilator duct.

15. Remove the steering column mounting bolts and nuts. Remove the steering member mounting bolts. Remove the steering member.

16. Remove the heater and cooling unit assembly.

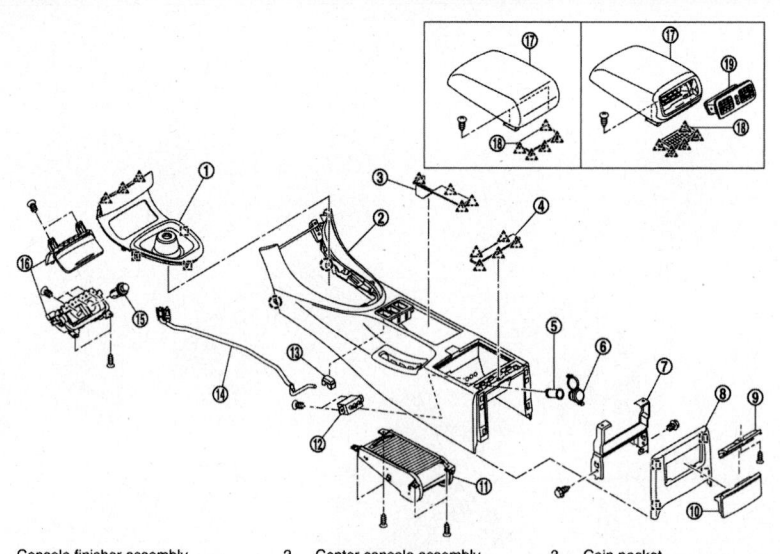

1. Console finisher assembly	2. Center console assembly	3. Coin pocket
4. Console mask	5. Power socket (if equipped)	6. Socket knob (if equipped)
7. Console rear bracket	8. Console rear finisher	9. Console ashtray bracket
10. Console ashtray	11. Cup holder assembly	12. Auxiliary input jacks (if equipped) or USB connector (if equipped)
13. Switch hole mask	14. Console sub harness	15. Power socket or Cigarette lighter
16. Ashtray (front) or Coin pocket	17. Console lid	18. Console mask
19. Rear ventilator grille		

◯ : Clip

△ : Pawl

37663_IG37_G0096

Fig. 245 Center console and related components—manual transmission

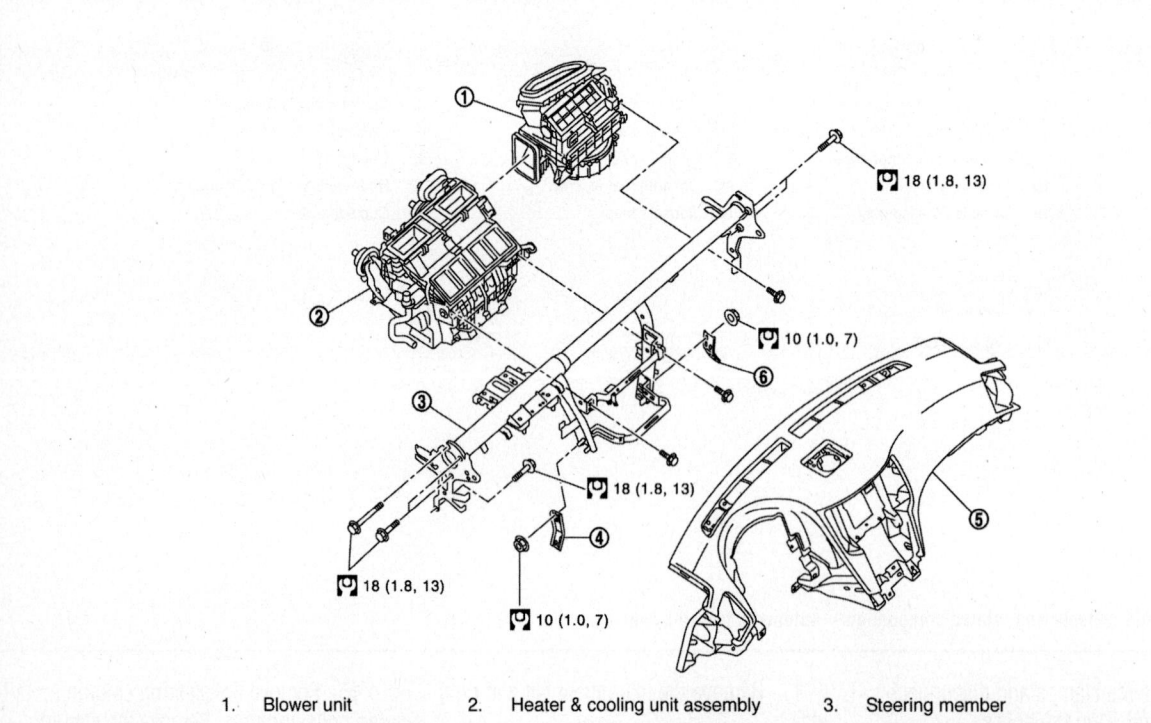

1. Blower unit	2. Heater & cooling unit assembly	3. Steering member
4. Instrument stay (left)	5. Instrument panel assembly	6. Instrument stay (right)

37663_IG37_G0078

Fig. 246 HVAC blower unit and related components

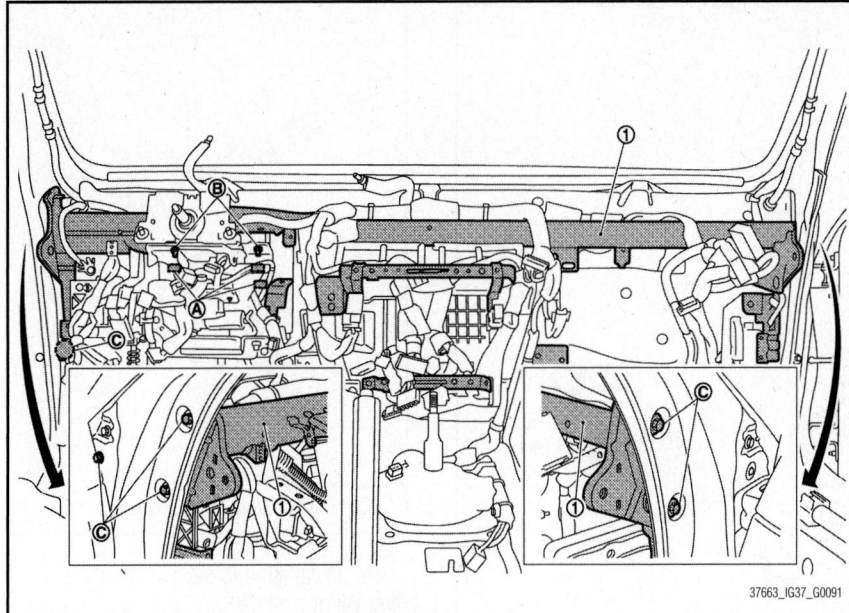

Fig. 247 HVAC heating and cooling unit removal points

To install:

➡ **Be sure to use new fasteners, as required.**

17. Installation is the reverse of the removal procedure.

18. Be sure to use new O-rings, coated with clean refrigerant oil prior to installation.

19. Properly charge the air conditioning system.

20. Properly refill the cooling system.

21. Start the engine and check the system for proper operation and refrigerant leakage.

22. Using the CONSULT-III diagnostic tool, or equivalent, perform the 4WAS front actuator adjustment if equipped with 4WAS.

23. Be sure to perform the reconnect/relearn procedures.

STEERING

ELECTRONIC STEERING CONTROL UNIT

REMOVAL & INSTALLATION
See Figure 248.

1. Remove glove box assembly.
2. Remove power steering control unit screws.
3. Remove power steering control unit (1).
4. Disconnect power steering control unit connector.

To install:

5. To install, reverse removal procedure.

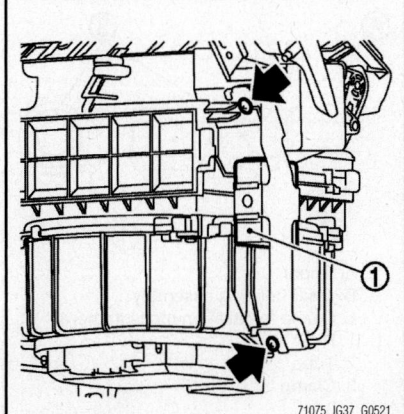

Fig. 248 Removing and installing power steering control unit (1)

POWER RACK & PINION STEERING GEAR

REMOVAL & INSTALLATION
See Figures 249 through 254.

➡ **Whenever the negative battery cable is disconnected the following components will require resetting. The Automatic temperature control system, Automatic drive positioner, Power window control, Sunroof system, Sunshade system, Rear view monitor, Idle Air Volume Learning, Steering Angle Sensor Neutral Position, Audio presets and Navigation. You will need the CONSULT-III diagnostic tool, or equivalent. Follow the directions on the screen of the tool, as needed.**

1. Before servicing the vehicle, refer to the Precautions Section.

➡ **If working near and/or around the SRS system and components, be sure to disable the SRS system. After disabling the system wait three minutes or more before servicing the vehicle.**

2. Disconnect the negative battery cable.

3. Position the front tires in the straight ahead position.

4. Perform the 4WAS front actuator neutral position adjustment, using the CONSULT-III diagnostic tool, or equivalent, if equipped with 4WAS.

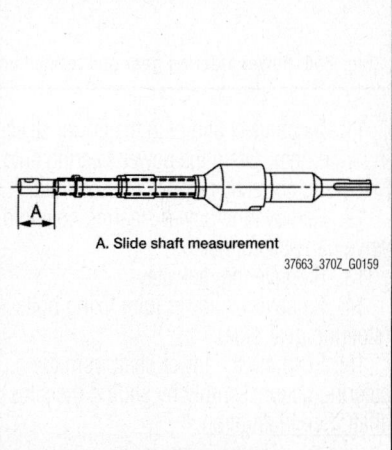

A. Slide shaft measurement

Fig. 249 Power steering gear slide shaft measurement

5. Raise and support the vehicle safely.

6. Remove the tire and wheel assemblies.

7. Remove the front suspension member stay, except AWD.

8. Remove the front cross bar, AWD.

9. Remove the cotter pin and loosen the locknut.

10. Remove the outer steering socket from the steering knuckle, so as not to damage the ball joint boot, using a suitable removal tool.

➡ **Temporarily tighten the nut to prevent damage to the threads and prevent the ball joint remover from sudden drop.**

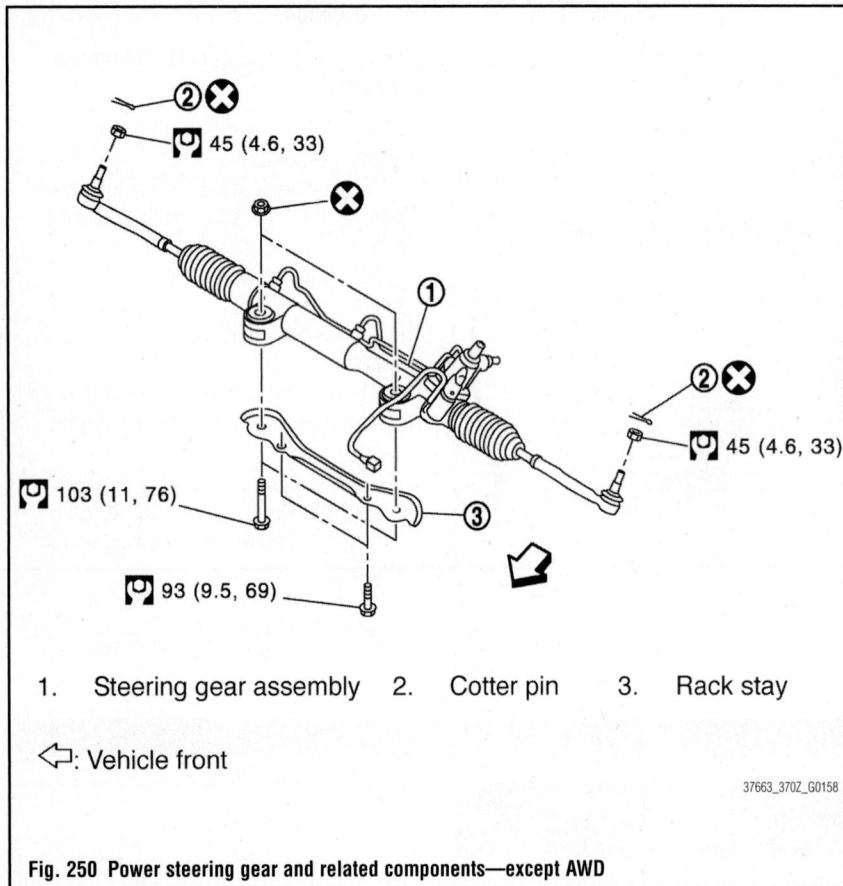

1.   Steering gear assembly   2.   Cotter pin   3.   Rack stay

⟵: Vehicle front

37663_370Z_G0158

Fig. 250 Power steering gear and related components—except AWD

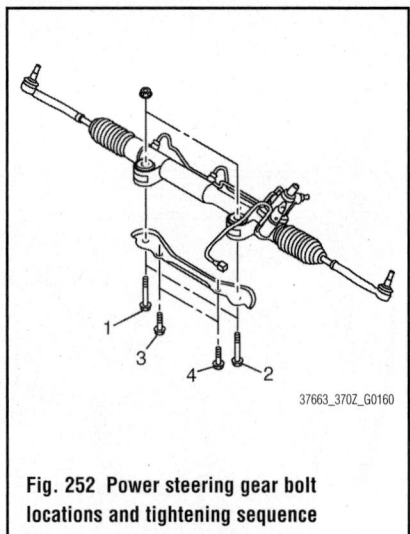

37663_370Z_G0160

Fig. 252 Power steering gear bolt locations and tightening sequence

11.   Disconnect and plug the power steering fluid lines. Drain the power steering fluid. Be sure to correctly dispose of used fluid.

12.   Remove the power steering solenoid valve harness connector.

13.   Remove the rack stay.

14.   Remove the lower joint fixing bolts (steering gear side).

15.   Separate the lower shaft from the steering gear assembly by sliding the slide shaft. See illustration.

➡**The spiral cable may be cut if the steering wheel turns while separating the steering column assembly and steering gear assembly. Always lock the steering wheel using string to avoid turning.**

16.   On AWD, position a suitable jack to the transmission assembly. Remove the mounting nuts and bolts on the lower side of the shock absorber arm and remove the shock from the transverse link. Position a suitable jack to the front suspension member. Remove the mounting bolts and nuts of the steering gear assembly. Remove the mounting nuts of the engine mounting insulator. Remove the mounting nuts of the front suspension member. Position a suitable jack and slowly lower it to the position

where the steering gear assembly can be removed. Support the gear assembly so it will not drop.

17.   Remove the steering gear retaining bolts, except AWD.

18.   Remove the gear assembly from its mounting.

To install:

➡**Be sure to use new fasteners, as required.**

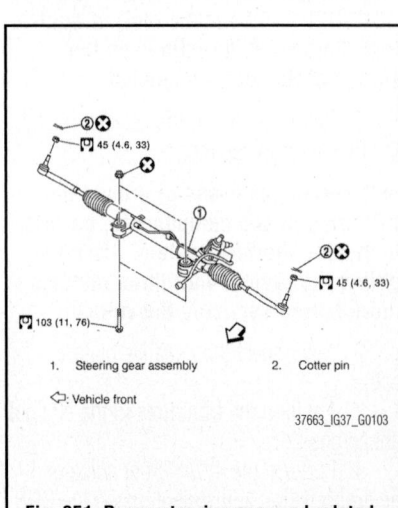

1.   Steering gear assembly   2.   Cotter pin

⟵: Vehicle front

37663_IG37_G0103

Fig. 251 Power steering gear and related components—AWD

19.   Installation is the reverse of the removal procedure.

➡**The spiral cable may be cut if the steering wheel turns while separating the steering column assembly and steering gear assembly. Always lock the steering wheel using string to avoid turning.**

20.   Tighten the mounting bolts in the order shown in the illustration.

21.   Tighten step one, temporary and in step two final.

22.   When installing the suction hoses, refer to the illustration.

➡**Never apply fluid to the hose and the tube. Insert the hose securely until it contacts the spool of the tube. Install the clamp at the hose (0.12–0.31 inch from the edge of the hose.**

23.   To install the lower joint to the steering gear, see the illustration. Position the

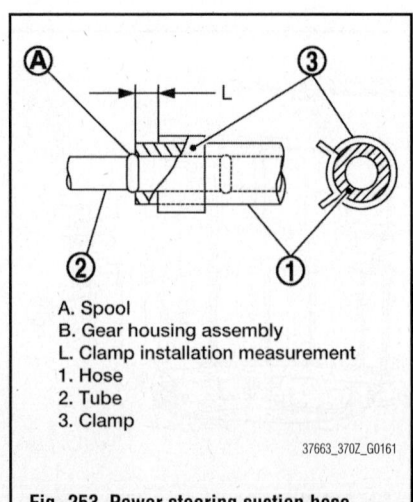

A. Spool
B. Gear housing assembly
L. Clamp installation measurement
1. Hose
2. Tube
3. Clamp

37663_370Z_G0161

Fig. 253 Power steering suction hose installation

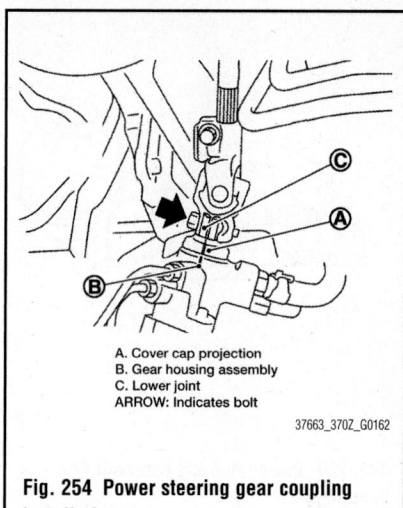

A. Cover cap projection
B. Gear housing assembly
C. Lower joint
ARROW: Indicates bolt

37663_370Z_G0162

Fig. 254 Power steering gear coupling installation

rack of the steering gear in the neutral position.

➡To get to the neutral position turn the sub gear assembly and measure the distance of the inner socket, then measure the intermediate position and distance. Align the rear cover cap projection with the marking position of the gear housing assembly. Install the slip part of the lower joint aligning with the rear cover cap projection. Make sure that the slit part of the lower joint is aligned with the rear cover cap protection and the marking position of the gear housing assembly. Make sure that there is no clearance between the lower joint, gear housing assembly and mounting bolt.

24. Fill the system. Bleed the system.
25. Perform a final tightening of nuts and bolts of each removed component with the vehicle in an unladen position.
26. Check and adjust the wheel alignment, as required.
27. Adjust the neutral position of the steering angle sensor using the CONSULT-III diagnostic tool, or equivalent, after checking the wheel alignment.
28. Perform the 4WAS front actuator neutral position adjustment, using the CONSULT-III diagnostic tool, or equivalent, if equipped with 4WAS.
29. Be sure to perform the reconnect/relearn procedures.

POWER STEERING PUMP

REMOVAL & INSTALLATION

See Figures 255 and 256.

➡Whenever the negative battery cable is disconnected the following compo-

nents will require resetting. The Automatic temperature control system, Automatic drive positioner, Power window control, Sunroof system, Sunshade system, Rear view monitor, Idle Air Volume Learning, Steering Angle Sensor Neutral Position, Audio presets and Navigation. You will need the CONSULT-III diagnostic tool, or equivalent. Follow the directions on the screen of the tool, as needed.

1. Before servicing the vehicle, refer to the Precautions Section.

➡If working near and/or around the SRS system and components, be sure to disable the SRS system. After disabling the system wait three minutes or more before servicing the vehicle.

2. Disconnect the negative battery cable.
3. Drain the power steering fluid from the reservoir. Be sure to properly dispose of used fluid.
4. Remove the air cleaner assembly.
5. Loosen the drive belt.
6. Remove the belt from the steering pump.
7. Remove the copper washers and eye bolt. Drain fluid from their piping's.
8. Remove the suction hose. Drain fluid from their piping's.
9. Remove the pump retaining bolts.
10. Remove the pump from the vehicle.

To install:

➡Be sure to use new fasteners, as required.

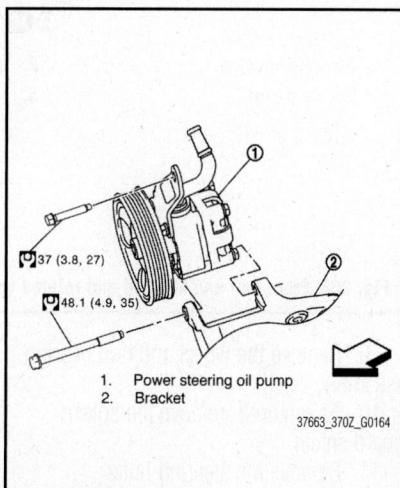

1. Power steering oil pump
2. Bracket

37663_370Z_G0164

Fig. 255 Power steering pump and related components

11. Installation is the reverse of the removal procedure.
12. When installing the suction hoses see note below.

➡Never apply fluid to the hose and the tube. Insert the hose securely until it contacts the spool of the tube. Install the clamp at the hose (0.12–0.31 inch from the edge of the hose.

13. When installing the eye bolt and copper washer to the pump, see illustration.

➡Never reuse the cooper washer. Apply clean power steering fluid around the washers, then install the eye bolt. Install the eye bolt with the eye joint (assembled to the high pressure hose) protrusion facing with pump side cutout, and then tighten it to specification.

14. Adjust the belt tension.
15. Fill the system with the proper grade and type fluid. Bleed the system.
16. Check for fluid leakage, correct as required.

BLEEDING

1. Before servicing the vehicle, refer to the Precautions Section.
2. Stop engine, and then turn steering wheel fully to right and left several times.
3. Do not allow steering fluid reservoir tank to go below the low-level line. Check tank frequenter and add fluid as needed.
4. Run engine at idle speed. Turn steering wheel fully to the right and then fully to the left, and keep for about 3 seconds. Then check whether a fluid leakage has occurred.

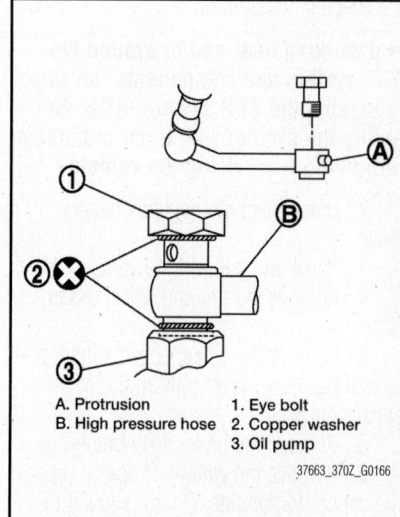

A. Protrusion
B. High pressure hose
1. Eye bolt
2. Copper washer
3. Oil pump

37663_370Z_G0166

Fig. 256 Power steering pump eye bolt and cooper washer installation

5. Repeat the 2nd procedure several times at about three seconds interval Check generation of air bubbles and cloud in fluid.

6. If air bubbles and the cloud don't fade, stop engine, hold air bleeding until air bubbles and the cloud fade. Perform the 2nd and the 3rd procedures again.

7. Stop engine, check fluid level.

POWER STEERING FLUID

FLUID FILL PROCEDURE

See Figure 257.

1. Stop the engine before performing the fluid level check.

2. Ensure that the fluid level is between the MAX range and the MIN level.

➡ **Because fluid level differs within the HOT range and the COLD range, check carefully.**

3. Hot specification temperature is 122–176 degrees F. Cold specification temperature is 32—86 degrees F.

4. Do not overfill the MAX level.

5. Do not reuse old power steering fluid.

6. Be sure to use the proper grade and type replacement fluid.

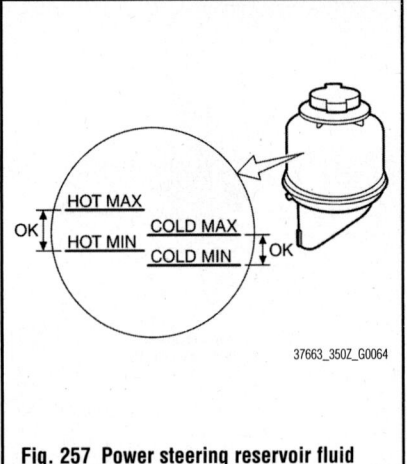

Fig. 257 Power steering reservoir fluid markings

SUSPENSION   FRONT SUSPENSION

KNUCKLE & SPINDLE

REMOVAL & INSTALLATION

See Figures 258, and 259.

➡ **Whenever the negative battery cable is disconnected the following components will require resetting. The Automatic temperature control system, Automatic drive positioner, Power window control, Sunroof system, Sunshade system, Rear view monitor, Idle Air Volume Learning, Steering Angle Sensor Neutral Position, Audio presets and Navigation. You will need the CONSULT-III diagnostic tool, or equivalent. Follow the directions on the screen of the tool, as needed.**

1. Before servicing the vehicle, refer to the Precautions Section.

➡ **If working near and/or around the SRS system and components, be sure to disable the SRS system. After disabling the system wait three minutes or more before servicing the vehicle.**

2. Disconnect the negative battery cable.

3. Raise and support the vehicle safely.

4. Remove the tire and wheel assemblies.

5. Remove the wheel speed sensor and sensor harness. Never pull on the wheel sensor harness.

6. Remove the brake hose bracket.

7. Remove the caliper. Properly support the caliper to the side. Do not allow it to hang by the brake hose. Never depress the brake pedal with the caliper removed.

8. Remove the brake rotor.

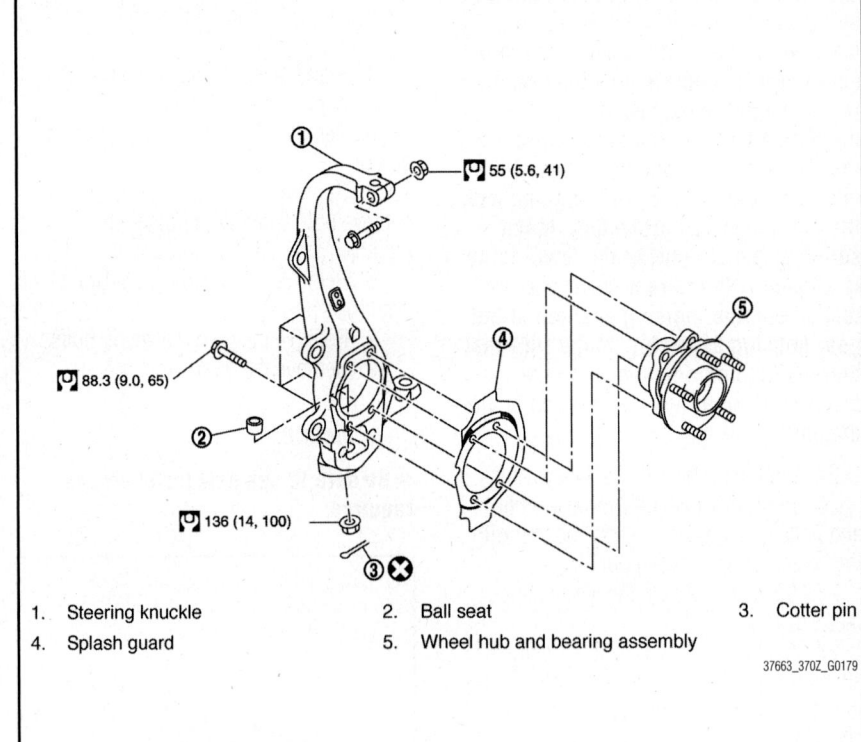

1. Steering knuckle
4. Splash guard
2. Ball seat
5. Wheel hub and bearing assembly
3. Cotter pin

37663_370Z_G0179

Fig. 258 Front hub and knuckle and related components—except AWD

9. Remove the wheel and hub bearing assembly.

10. As required, remove the splash guard shield.

11. Remove the steering outer socket.

12. Remove the cotter pin of the transverse link and steering knuckle, and then loosen nut.

13. Separate the upper link from the steering knuckle.

14. Separate the transverse link from the steering knuckle, using a ball joint removal tool. Remove the steering knuckle.

➡ **Temporarily tighten the nut to prevent damage to the threads and to prevent the removal tool from suddenly coming off.**

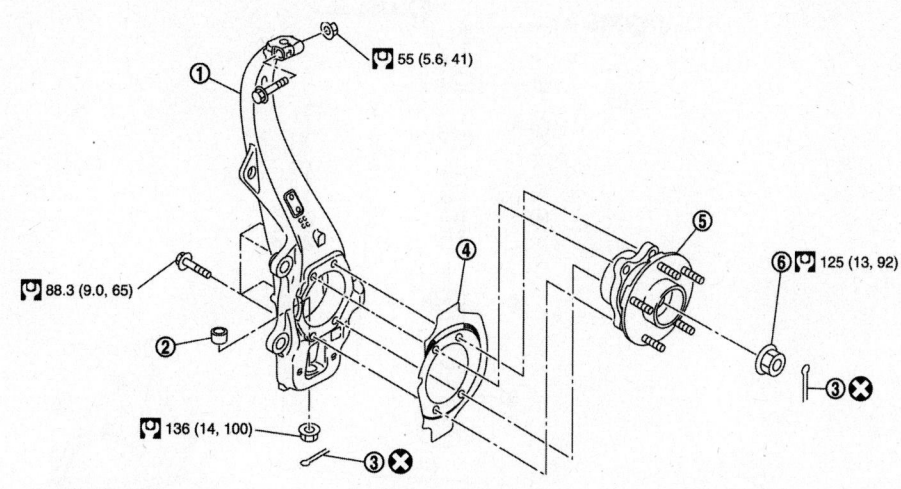

1. Steering knuckle
4. Splash guard
2. Ball seat
5. Wheel hub and bearing assembly
3. Cotter pin
6. Wheel hub lock nut

37663_IG37_G0118

Fig. 259 Front hub and knuckle and related components—AWD

To install:

➡ **Be sure to use new fasteners, as required.**

15. Installation is the reverse of the removal procedure.
16. Be sure to use new cotter pins.
17. Perform a final tightening of nuts and bolts of each removed component with the vehicle in an unladen position.
18. Check wheel speed sensor for proper operation.
19. Adjust the steering angle sensor neutral position, using the CONSULT-III diagnostic tool, or equivalent.
20. Check and adjust the front alignment, as required.
21. Be sure to perform the reconnect/relearn procedures.

STABILIZER BAR (SWAY BAR) & LINKS

REMOVAL & INSTALLATION

Stabilizer Bar

See Figure 260.

1. Remove tires with power tool.
2. Remove under cover with power tool.
3. Remove stabilizer connecting rod.

⁜ WARNING

Apply a matching mark to identify the installation position.

4. Remove the stabilizer clamp and stabilizer bushing.
5. Remove the stabilizer bar.

To install:
6. To install, reverse the removal procedure.

➡ **Check the mounting mark when installing.**

➡ **Tighten the mounting nut to the specified torque while holding a hexagonal part of stabilizer connecting rod side.**

STRUTS

REMOVAL & INSTALLATION

See Figures 261 through 263.

➡ **Whenever the negative battery cable is disconnected the following components will require resetting. The Automatic temperature control system, Automatic drive positioner, Power window control, Sunroof system, Sunshade system, Rear view monitor, Idle Air Volume Learning, Steering Angle**

Sensor Neutral Position, Audio presets and Navigation. You will need the CONSULT-III diagnostic tool, or equivalent. Follow the directions on the screen of the tool, as needed.

1. Before servicing the vehicle, refer to the Precautions Section.

➡ **If working near and/or around the SRS system and components, be sure to disable the SRS system. After disabling the system wait three minutes or more before servicing the vehicle.**

2. Disconnect the negative battery cable.
3. Raise and safely support the vehicle.
4. Remove the tire and wheel assemblies, as necessary.
5. Remove the wheel sensor and harness connector from the shock absorber, except AWD.
6. Remove the brake hose bracket, except AWD.
7. Remove the stabilizer connecting rod.
8. Remove the halfshaft, AWD.
9. Separate the upper link from the steering knuckle. Remove the shock absorber assembly mounting nuts.

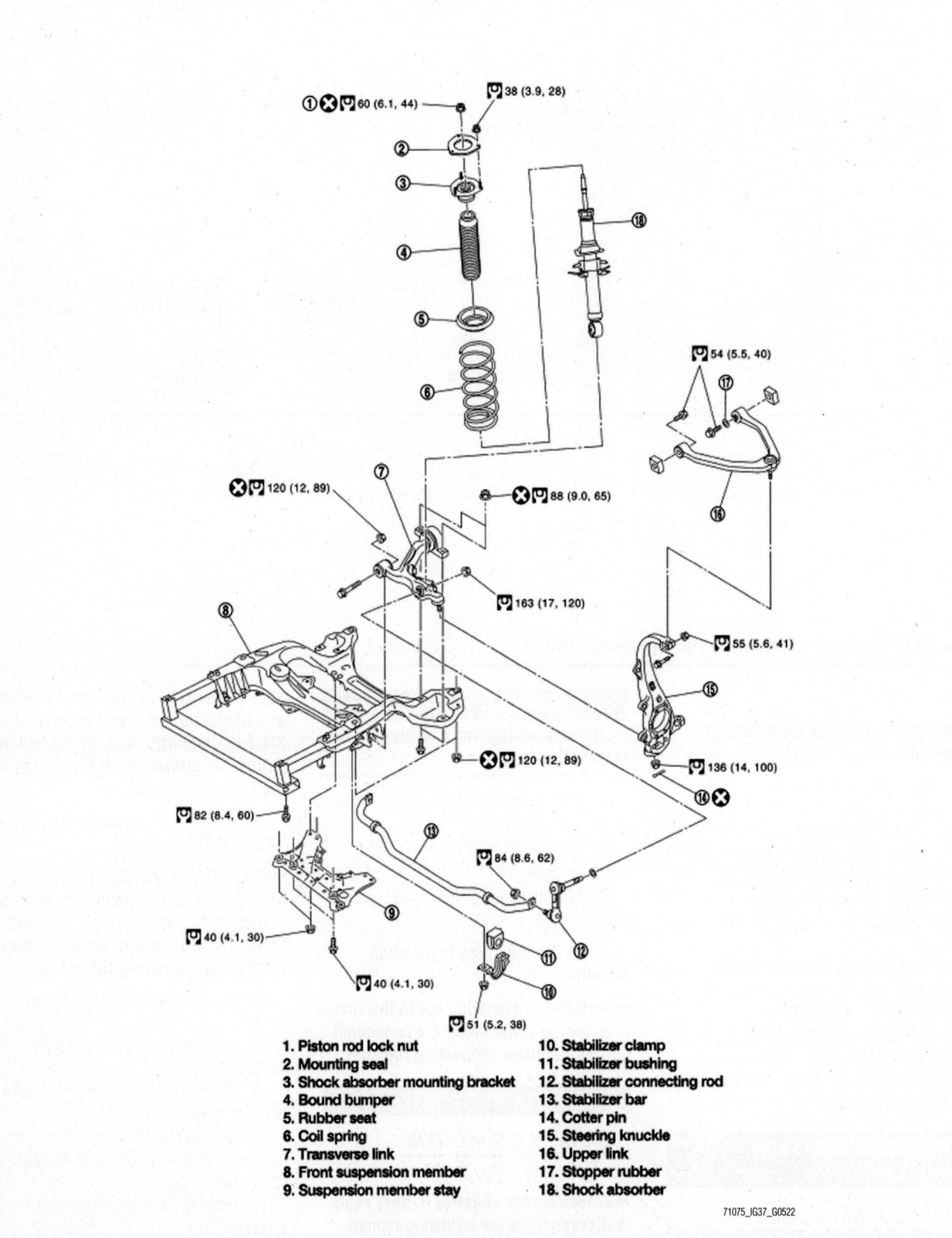

1. Piston rod lock nut
2. Mounting seal
3. Shock absorber mounting bracket
4. Bound bumper
5. Rubber seat
6. Coil spring
7. Transverse link
8. Front suspension member
9. Suspension member stay
10. Stabilizer clamp
11. Stabilizer bushing
12. Stabilizer connecting rod
13. Stabilizer bar
14. Cotter pin
15. Steering knuckle
16. Upper link
17. Stopper rubber
18. Shock absorber

71075_IG37_G0522

Fig. 260 Removing and installing the stabilizer bar

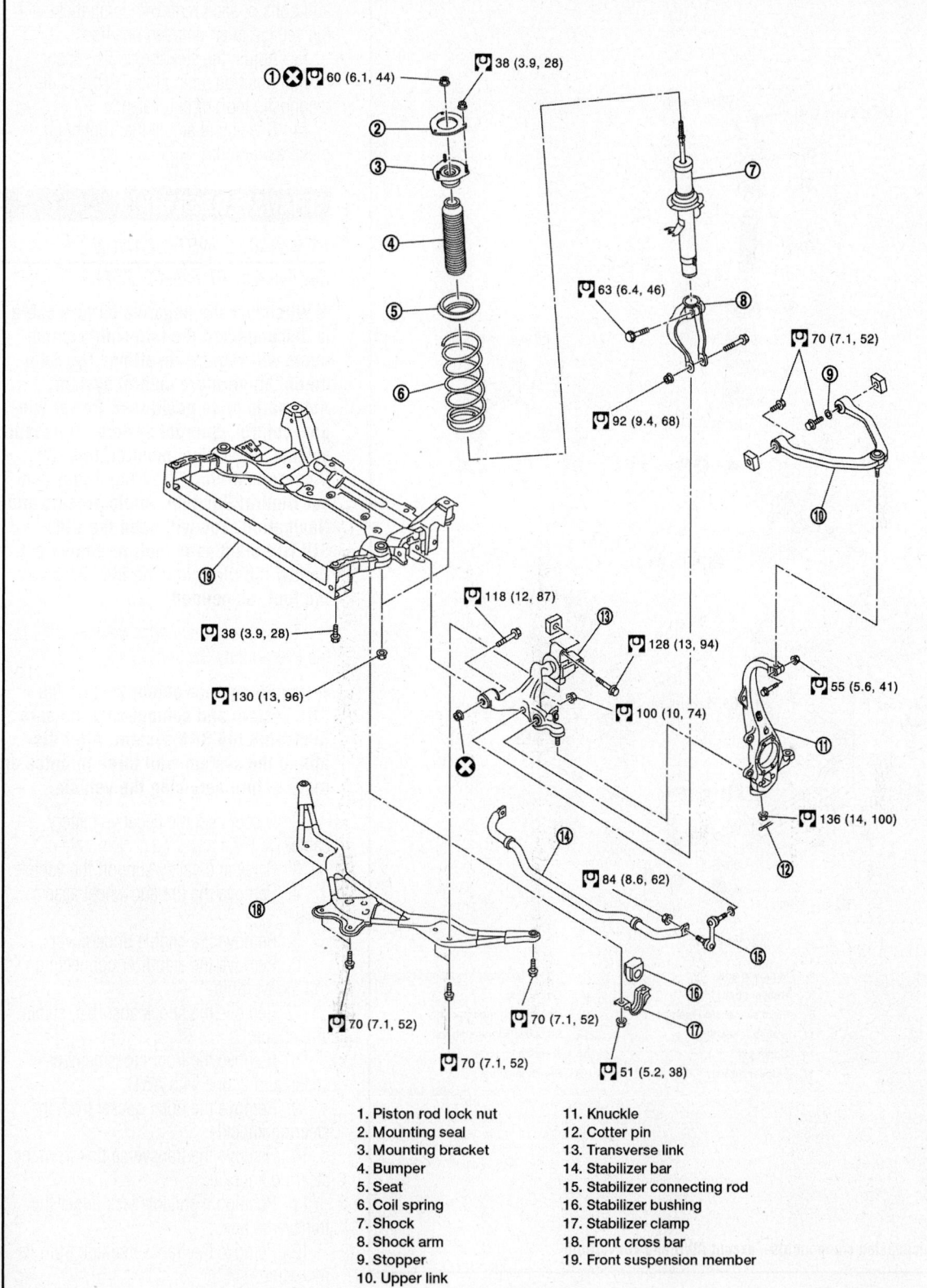

1. Piston rod lock nut
2. Mounting seal
3. Mounting bracket
4. Bumper
5. Seat
6. Coil spring
7. Shock
8. Shock arm
9. Stopper
10. Upper link
11. Knuckle
12. Cotter pin
13. Transverse link
14. Stabilizer bar
15. Stabilizer connecting rod
16. Stabilizer bushing
17. Stabilizer clamp
18. Front cross bar
19. Front suspension member

37663_IG37_G0009

Fig. 261 Front suspension components—AWD

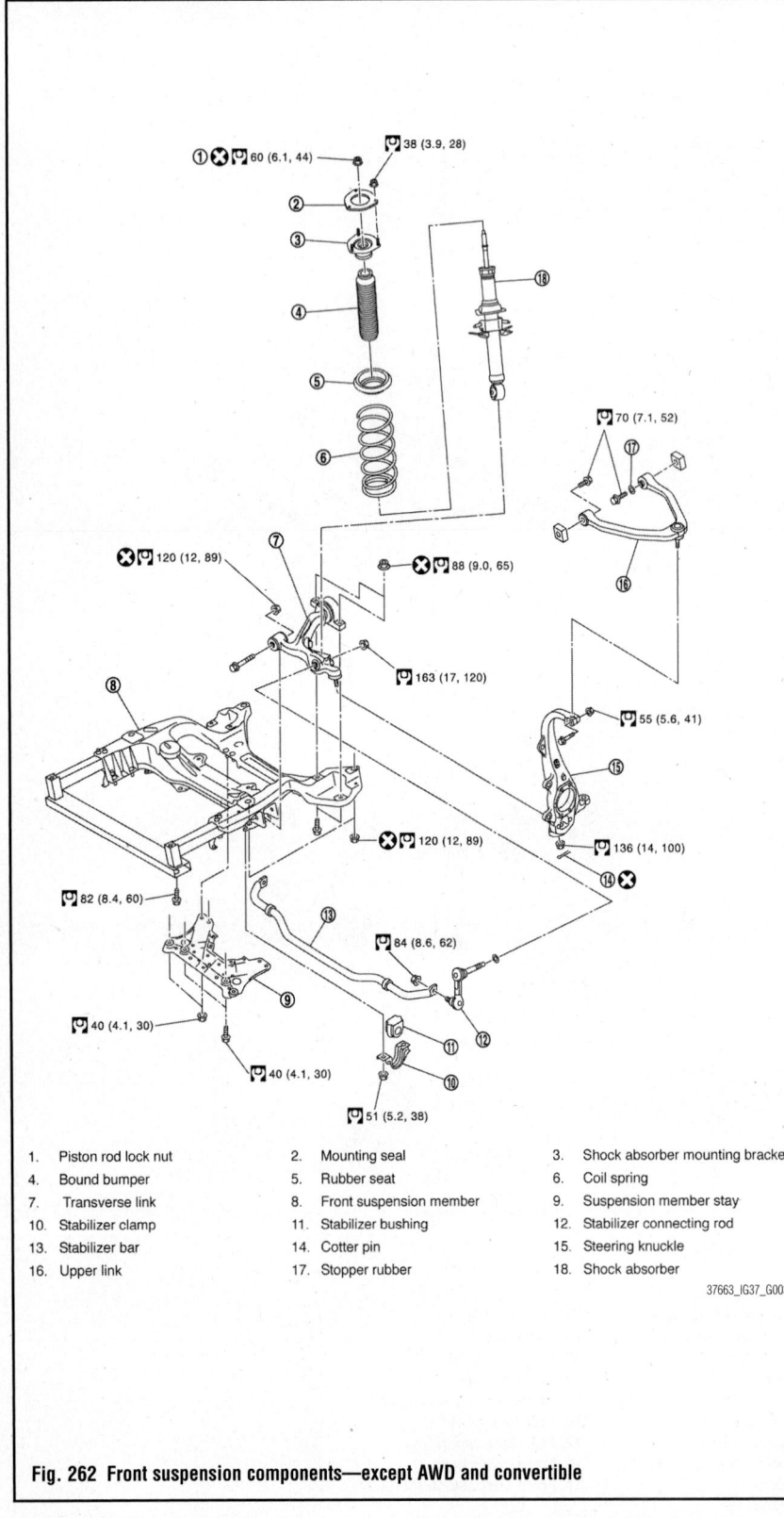

Fig. 262 Front suspension components—except AWD and convertible

1.	Piston rod lock nut	2.	Mounting seal	3.	Shock absorber mounting bracket
4.	Bound bumper	5.	Rubber seat	6.	Coil spring
7.	Transverse link	8.	Front suspension member	9.	Suspension member stay
10.	Stabilizer clamp	11.	Stabilizer bushing	12.	Stabilizer connecting rod
13.	Stabilizer bar	14.	Cotter pin	15.	Steering knuckle
16.	Upper link	17.	Stopper rubber	18.	Shock absorber

37663_IG37_G0035

10. Remove the component from the vehicle.

To install:

→Be sure to use new fasteners, as required.

11. Installation is the reverse of the removal procedure.

→**Never tap on the ball joint cap of the stabilizer connecting rod with a hammer when inserting the stabilizer connecting rod into the transverse link.**

12. Perform a final tightening of nuts and bolts of each removed component with the vehicle in an unladen position.

13. Adjust the steering angle sensor neutral position, using the CONSULT-III diagnostic tool, or equivalent.

14. Check and adjust the front alignment, as required.

TRANSVERSE LINK

REMOVAL & INSTALLATION

See Figures 261 and 262, 264.

→**Whenever the negative battery cable is disconnected the following components will require resetting. The Automatic temperature control system, Automatic drive positioner, Power window control, Sunroof system, Sunshade system, Rear view monitor, Idle Air Volume Learning, Steering Angle Sensor Neutral Position, Audio presets and Navigation. You will need the CONSULT-III diagnostic tool, or equivalent. Follow the directions on the screen of the tool, as needed.**

1. Before servicing the vehicle, refer to the Precautions Section.

→**If working near and/or around the SRS system and components, be sure to disable the SRS system. After disabling the system wait three minutes or more before servicing the vehicle.**

2. Disconnect the negative battery cable.

3. Raise and safely support the vehicle.

4. Remove the tire and wheel assemblies.

5. Remove the engine undercover.

6. Remove the stabilizer connecting rod, convertible.

7. Remove the shock absorber, sedan and coupe.

8. Remove the front crossmember, sedan and coupe with AWD.

9. Remove the outer socket from the steering knuckle.

10. Remove the transverse link from the steering knuckle.

11. Position a suitable jack under the transverse link.

12. Remove the transverse link from its mounting.

To install:

→Be sure to use new fasteners, as required.

13. Installation is the reverse of the removal procedure.

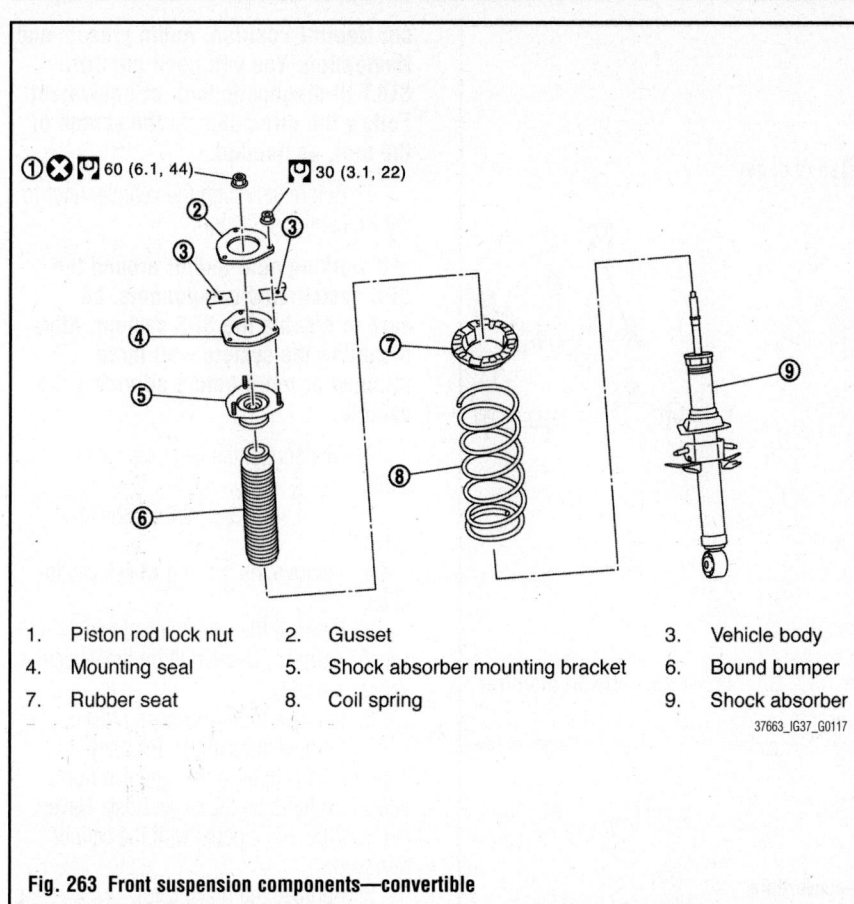

1. Piston rod lock nut   2. Gusset   3. Vehicle body
4. Mounting seal   5. Shock absorber mounting bracket   6. Bound bumper
7. Rubber seat   8. Coil spring   9. Shock absorber

37663_IG37_G0117

Fig. 263 Front suspension components—convertible

Volume Learning, Steering Sensor Neutral Position, Audio presets and Navigation. You will need the CONSULT-III diagnostic tool, or equivalent. Follow the directions on the screen of the tool, as needed.

1. Before servicing the vehicle, refer to the Precautions Section.

➡️If working near and/or around the SRS system and components, be sure to disable the SRS system. After disabling the system wait three minutes or more before servicing the vehicle.

2. Disconnect the negative battery cable.

3. Raise and safely support the vehicle.

4. Remove the tire and wheel assemblies.

5. Remove the shock absorber.

6. Remove the upper link from the steering knuckle.

7. Remove the upper link and stopper rubber.

To install:

➡️Be sure to use new fasteners, as required.

➡️Never tap on the ball joint cap of the stabilizer connecting rod with a hammer when inserting the stabilizer connecting rod into the transverse link.

14. Perform a final tightening of nuts and bolts of each removed component with the vehicle in an unladen position.

15. Check wheel speed sensor for proper operation.

16. Adjust the steering angle sensor neutral position, using the CONSULT-III diagnostic tool, or equivalent.

17. Check and adjust the front alignment, as required.

18. Be sure to perform the reconnect/relearn procedures.

UPPER LINK

REMOVAL & INSTALLATION

See Figures 261 and 262, 265.

➡️Whenever the negative battery cable is disconnected the following components will require resetting. The Automatic temperature control system, Automatic drive positioner, Power window control, Sunroof system, Sunshade system, Rear view monitor, Idle Air

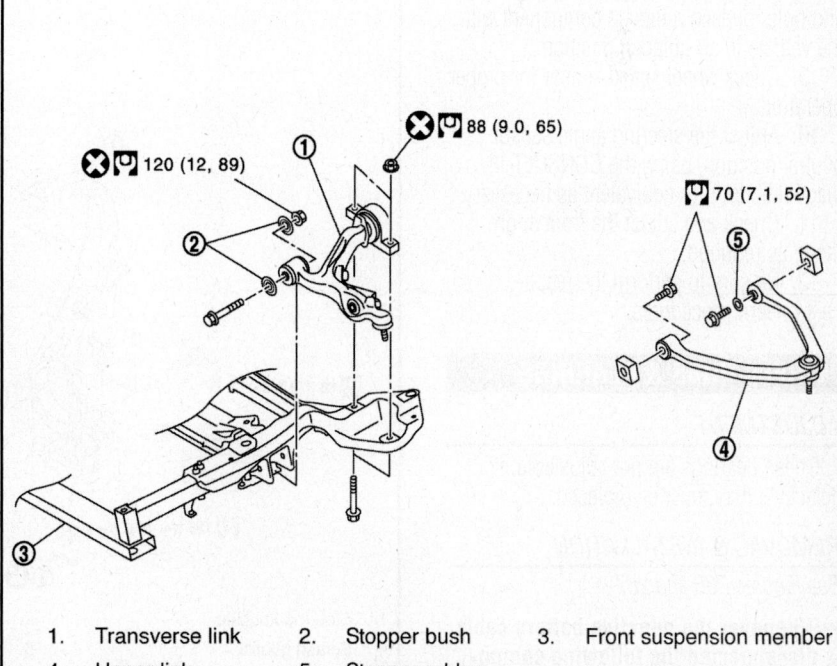

1. Transverse link   2. Stopper bush   3. Front suspension member
4. Upper link   5. Stopper rubber

37663_370Z_G0182

Fig. 264 Front transverse link and related components—convertible

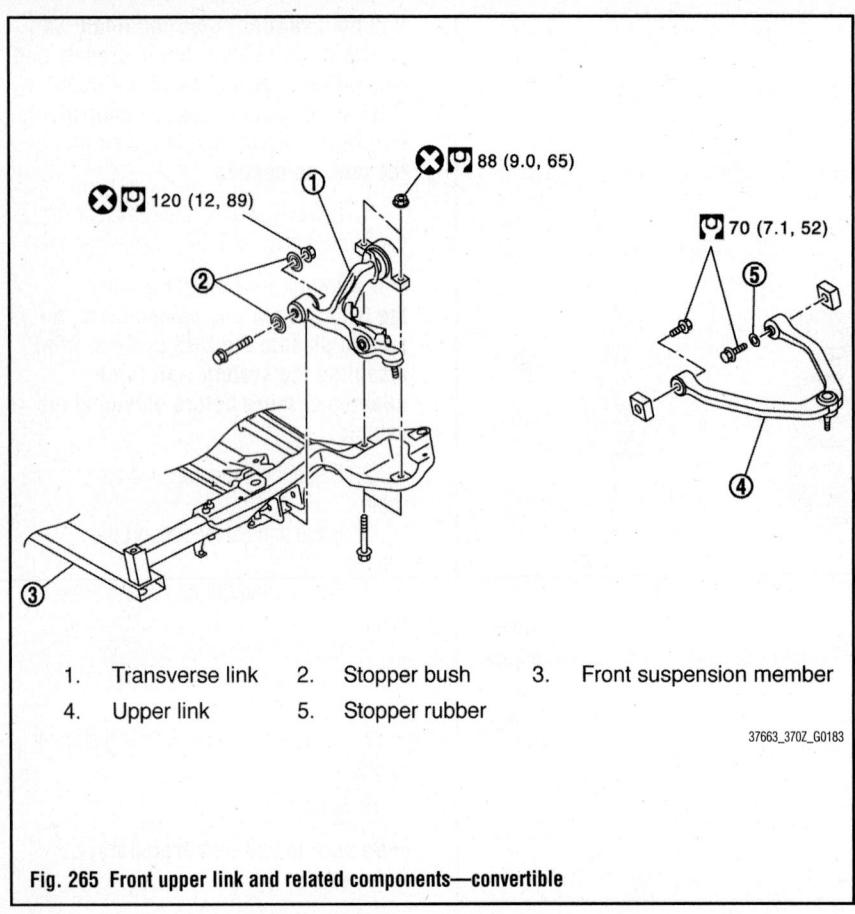

1.	Transverse link	2.	Stopper bush	3.	Front suspension member
4.	Upper link	5.	Stopper rubber		

37663_370Z_G0183

Fig. 265 Front upper link and related components—convertible

8. Perform a final tightening of nuts and bolts of each removed component with the vehicle in an unladen position.

9. Check wheel speed sensor for proper operation.

10. Adjust the steering angle sensor neutral position, using the CONSULT-III diagnostic tool, or equivalent as necessary.

11. Check and adjust the front alignment, as required.

12. Be sure to perform the reconnect/relearn procedures.

WHEEL HUBS & BEARINGS

ADJUSTMENT

These bearings are not adjustable. If defective, they must be replaced.

REMOVAL & INSTALLATION

See Figures 266 and 267.

➡Whenever the negative battery cable is disconnected the following components will require resetting. The Automatic temperature control system, Automatic drive positioner, Power window control, Sunroof system, Sunshade system, Rear view monitor, Idle Air Volume Learning, Steering Angle Sen-

sor Neutral Position, Audio presets and Navigation. You will need the CONSULT-III diagnostic tool, or equivalent. Follow the directions on the screen of the tool, as needed.

1. Before servicing the vehicle, refer to the Precautions Section.

➡If working near and/or around the SRS system and components, be sure to disable the SRS system. After disabling the system wait three minutes or more before servicing the vehicle.

2. Disconnect the negative battery cable.

3. Raise and support the vehicle safely.

4. Remove the tire and wheel assemblies.

5. Remove the wheel speed sensor and sensor harness. Never pull on the wheel sensor harness.

6. Remove the brake hose bracket.

7. Remove the caliper. Properly support the caliper to the side. Do not allow it to hang by the brake hose. Never depress the brake pedal with the caliper removed.

8. Remove the brake rotor.

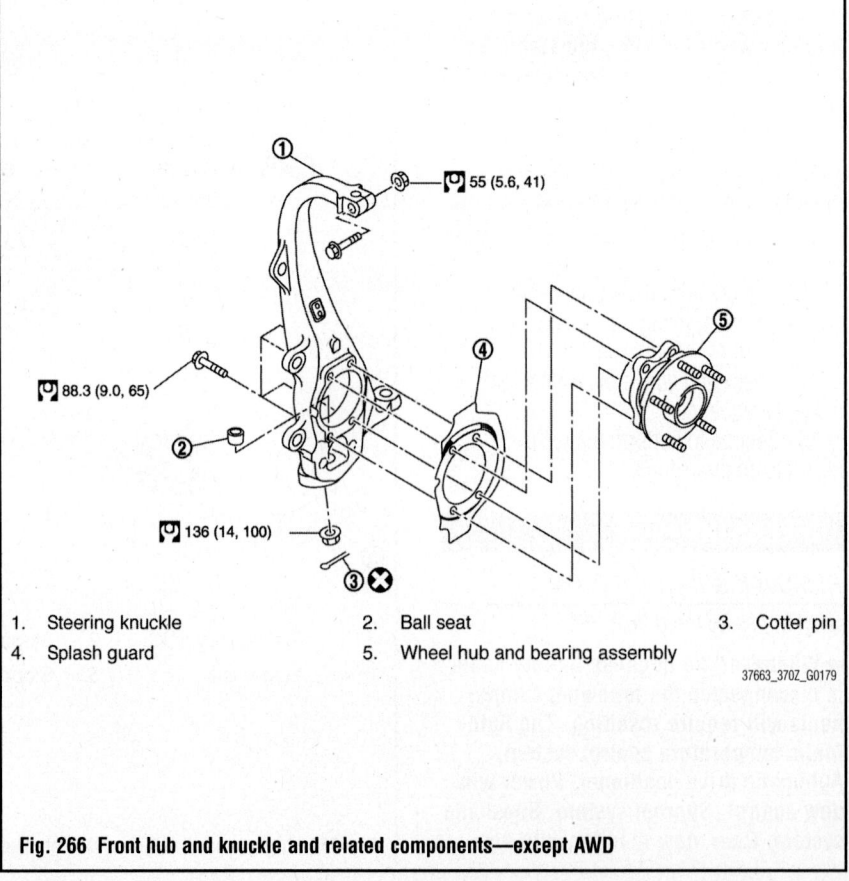

1.	Steering knuckle	2.	Ball seat	3.	Cotter pin
4.	Splash guard	5.	Wheel hub and bearing assembly		

37663_370Z_G0179

Fig. 266 Front hub and knuckle and related components—except AWD

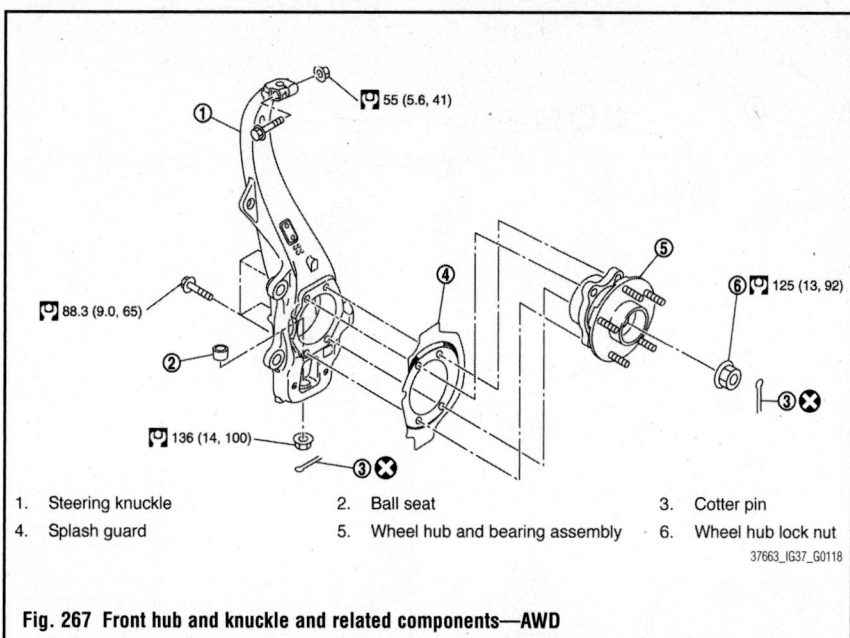

1. Steering knuckle
2. Ball seat
3. Cotter pin
4. Splash guard
5. Wheel hub and bearing assembly
6. Wheel hub lock nut

37663_IG37_G0118

Fig. 267 Front hub and knuckle and related components—AWD

9. Remove the wheel and hub bearing assembly.

10. As required, remove the splash guard shield.

To install:

➡️**Be sure to use new fasteners, as required.**

11. Installation is the reverse of the removal procedure.

12. Perform a final tightening of nuts and bolts of each removed component with the vehicle in an unladen position.

13. Check wheel speed sensor for proper operation.

14. Adjust the steering angle sensor neutral position, using the CONSULT-III diagnostic tool, or equivalent.

15. Check and adjust the front alignment, as required.

SUSPENSION

FRONT LINK

REMOVAL & INSTALLATION

Lower

See Figures 268 through 270.

➡️**Whenever the negative battery cable is disconnected the following components will require resetting. The Automatic temperature control system, Automatic drive positioner, Power window control, Sunroof system, Sunshade system, Rear view monitor, Idle Air Volume Learning, Steering Angle Sensor Neutral Position, Audio presets and Navigation. You will need the CONSULT-III diagnostic tool, or equivalent. Follow the directions on the screen of the tool, as needed.**

1. Before servicing the vehicle, refer to the Precautions Section.

➡️**If working near and/or around the SRS system and components, be sure to disable the SRS system. After disabling the system wait three minutes or more before servicing the vehicle.**

2. Disconnect the negative battery cable.
3. Raise and support the vehicle safely.
4. Remove the tire and wheel assemblies.
5. Position a suitable jack under the rear axle assembly to relieve tension on the coil spring.
6. Remove the lower link retaining bolts.

7. Remove the component from its mounting.

To install:

➡️**Be sure to use new fasteners, as required.**

8. Installation is the reverse of the removal procedure.

9. Perform a final tightening of nuts and bolts of each removed component with the vehicle in an unladen position.

10. Check wheel speed sensor for proper operation.

11. Adjust the steering angle sensor neutral position, using the CONSULT-III diagnostic tool, or equivalent.

12. Check and adjust the front alignment, as required.

13. Be sure to perform the reconnect/relearn procedures.

RADIUS ARM

REMOVAL & INSTALLATION

See Figures 268 and 269, 271.

➡️**Whenever the negative battery cable is disconnected the following components will require resetting. The Automatic temperature control system, Automatic drive positioner, Power window control, Sunroof system, Sunshade system, Rear view monitor, Idle Air Volume Learning, Steering Angle Sensor Neutral Position, Audio presets and Navigation. You will need the CONSULT-III diagnostic tool, or equivalent.**

REAR SUSPENSION

Follow the directions on the screen of the tool, as needed.

1. Before servicing the vehicle, refer to the Precautions Section.

➡️**If working near and/or around the SRS system and components, be sure to disable the SRS system. After disabling the system wait three minutes or more before servicing the vehicle.**

2. Disconnect the negative battery cable.
3. Raise and support the vehicle safely.
4. Remove the tire and wheel assemblies.
5. Remove the radius rod retaining nuts.
6. Remove the component from the vehicle.

To install:

➡️**Be sure to use new fasteners, as required.**

7. Installation is the reverse of the removal procedure.

8. Perform a final tightening of nuts and bolts of each removed component with the vehicle in an unladen position.

9. Check wheel speed sensor for proper operation.

10. Adjust the steering angle sensor neutral position, using the CONSULT-III diagnostic tool, or equivalent.

11. Check and adjust the front alignment, as required.

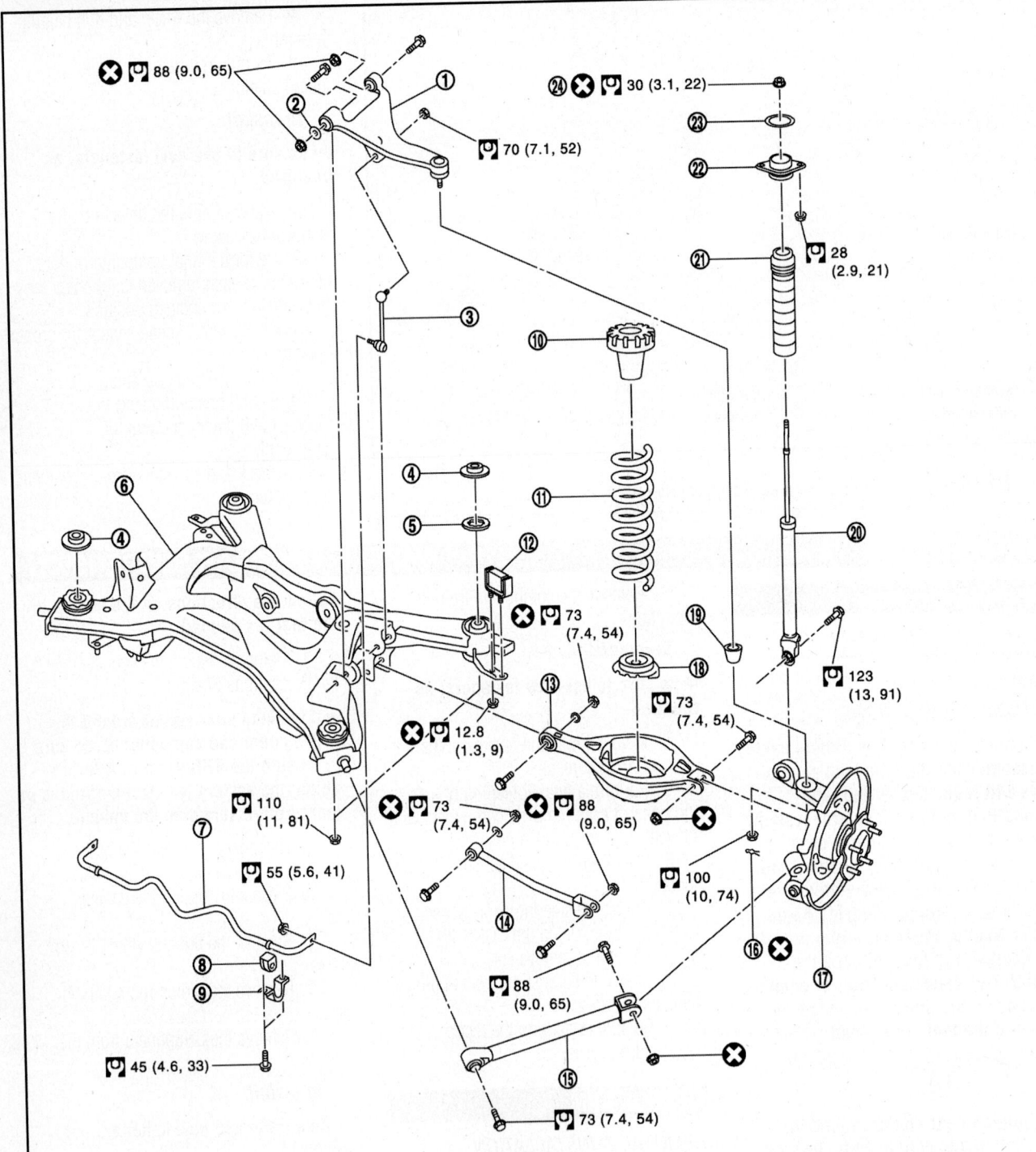

88 (9.0, 65)

70 (7.1, 52)

30 (3.1, 22)

28 (2.9, 21)

73 (7.4, 54)

12.8 (1.3, 9)

73 (7.4, 54)

123 (13, 91)

110 (11, 81)

55 (5.6, 41)

73 (7.4, 54)

88 (9.0, 65)

100 (10, 74)

88 (9.0, 65)

45 (4.6, 33)

73 (7.4, 54)

1. Suspension arm	2. Stopper rubber	3. Stabilizer connecting rod
4. Upper stopper (AWD model)	5. Mount stopper	6. Suspension member
7. Stabilizer bar	8. Stabilizer bushing	9. Stabilizer clamp
10. Upper seat	11. Coil spring	12. Damper assembly (AWD model)
13. Rear lower link	14. Front lower link	15. Radius rod
16. Cotter pin	17. Axle assembly	18. Rubber seat
19. Ball seat	20. Shock absorber	21. Bound bumper cover
22. Shock absorber mounting bracket	23. Mounting seal	24. Piston rod lock nut

37663_IG37_G0119

Fig. 268 Rear suspension components—sedan

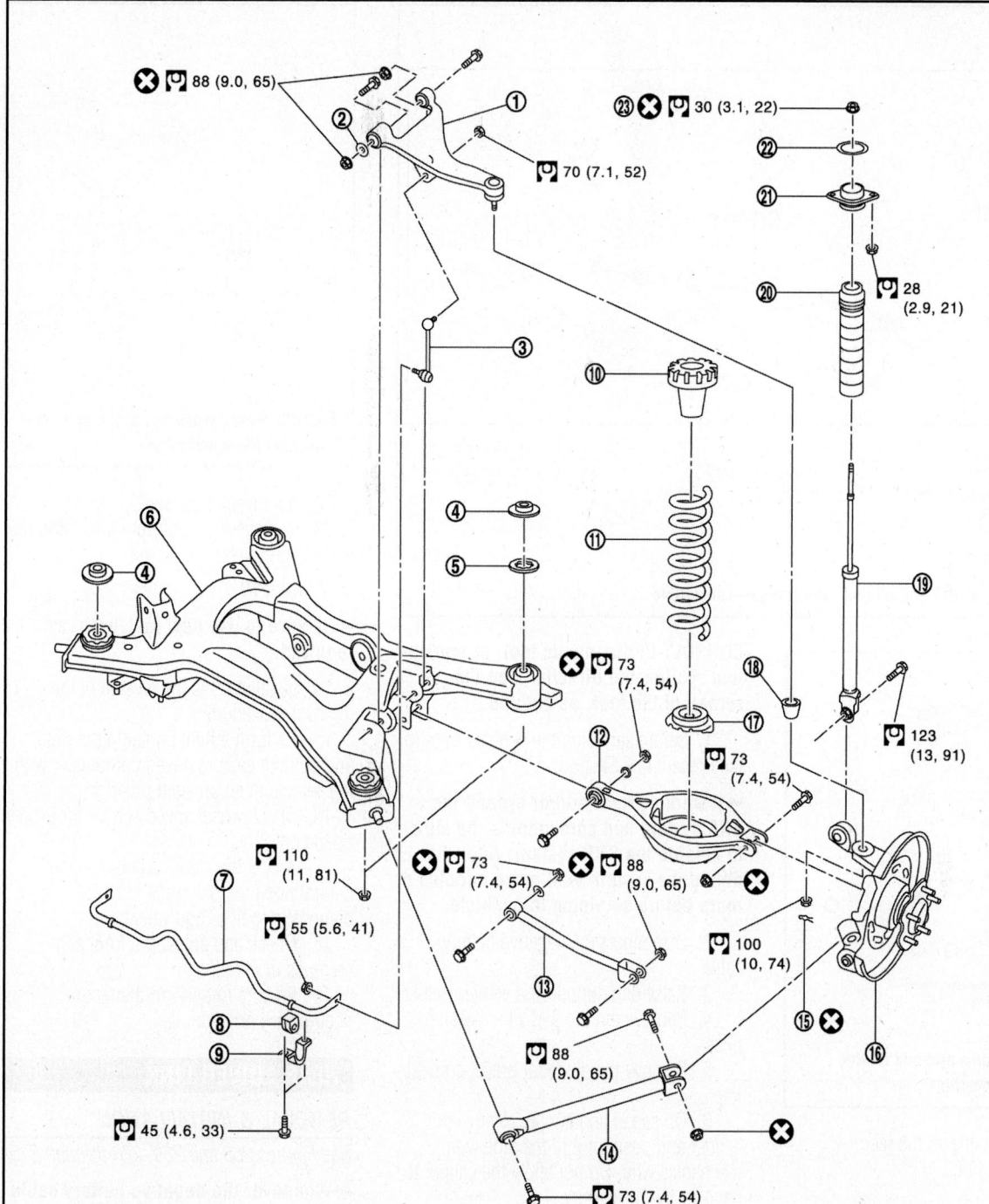

88 (9.0, 65)

70 (7.1, 52)

30 (3.1, 22)

28 (2.9, 21)

73 (7.4, 54)

73 (7.4, 54)

123 (13, 91)

110 (11, 81)

55 (5.6, 41)

73 (7.4, 54)

88 (9.0, 65)

100 (10, 74)

88 (9.0, 65)

45 (4.6, 33)

73 (7.4, 54)

1. Suspension arm
2. Stopper rubber
3. Stabilizer connecting rod
4. Upper stopper (AWD model)
5. Mount stopper
6. Suspension member
7. Stabilizer bar
8. Stabilizer bushing
9. Stabilizer clamp
10. Upper seat
11. Coil spring
12. Rear lower link
13. Front lower link
14. Radius rod
15. Cotter pin
16. Axle assembly
17. Rubber seat
18. Ball seat
19. Shock absorber
20. Bound bumper cover
21. Shock absorber mounting bracket
22. Mounting seal
23. Piston rod lock nut

37663_IG37_G0121

Fig. 269 Rear suspension components—coupe

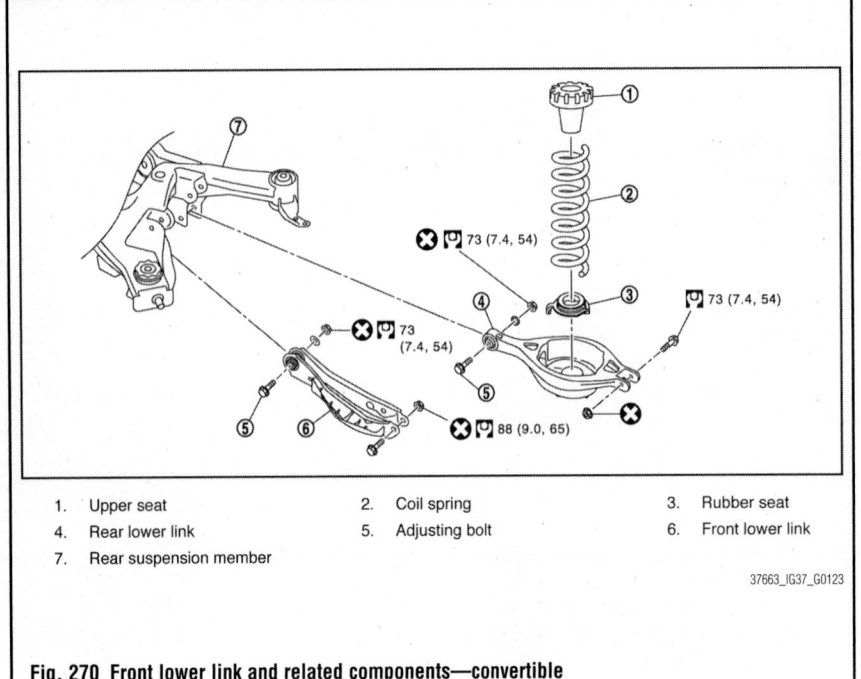

1. Upper seat
2. Coil spring
3. Rubber seat
4. Rear lower link
5. Adjusting bolt
6. Front lower link
7. Rear suspension member

37663_IG37_G0123

Fig. 270 Front lower link and related components—convertible

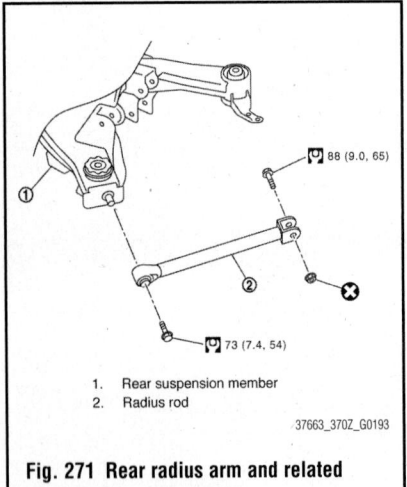

1. Rear suspension member
2. Radius rod

37663_370Z_G0193

Fig. 271 Rear radius arm and related components—convertible

12. Be sure to perform the reconnect/relearn procedures.

REAR ARM

REMOVAL & INSTALLATION

See Figures 268 and 269, 272.

➡Whenever the negative battery cable is disconnected the following components will require resetting. The Automatic temperature control system, Automatic drive positioner, Power window control, Sunroof system, Sunshade system, Rear view monitor, Idle Air Volume Learning, Steering Angle Sensor Neutral Position, Audio presets and Navigation. You will need the

CONSULT-III diagnostic tool, or equivalent. Follow the directions on the screen of the tool, as needed.

1. Before servicing the vehicle, refer to the Precautions Section.

➡If working near and/or around the SRS system and components, be sure to disable the SRS system. After disabling the system wait three minutes or more before servicing the vehicle.

2. Disconnect the negative battery cable.
3. Raise and support the vehicle safely.
4. Remove the tire and wheel assemblies.
5. Remove the diagonal brace, convertible.
6. On sedan and coupe, remove the caliper and position it to the side with mechanics wire. Do not allow the caliper to hang by the brake hose.
7. Remove the stabilizer connecting rod.
8. Remove the halfshaft.
9. Remove the cotter pin of the suspension arm ball joint, and loosen the nut.
10. Remove the suspension arm (rear suspension member side).
11. Use a ball joint removal tool to remove the suspension arm from the axle housing.

➡Be careful not to damage the ball joint boot. Temporarily tighten the nut to prevent damage to the threads and to prevent the tool from coming off.

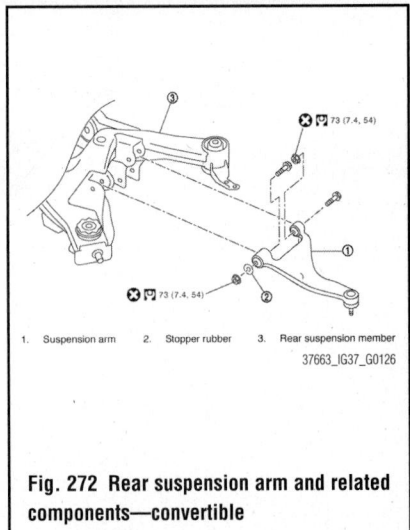

1. Suspension arm
2. Stopper rubber
3. Rear suspension member

37663_IG37_G0126

Fig. 272 Rear suspension arm and related components—convertible

12. Remove the suspension arm.
13. Remove the stabilizer connecting rod mounting bracket.

To install:

➡Be sure to use new fasteners, as required.

14. Installation is the reverse of the removal procedure.
15. Perform a final tightening of nuts and bolts of each removed component with the vehicle in an unladen position.
16. Check wheel speed sensor for proper operation.
17. Adjust the steering angle sensor neutral position, using the CONSULT-III diagnostic tool, or equivalent.
18. Check and adjust the front alignment, as required.
19. Be sure to perform the reconnect/relearn procedures.

REAR LOWER LINK

REMOVAL & INSTALLATION

See Figures 268 and 269, 273 through 275.

➡Whenever the negative battery cable is disconnected the following components will require resetting. The Automatic temperature control system, Automatic drive positioner, Power window control, Sunroof system, Sunshade system, Rear view monitor, Idle Air Volume Learning, Steering Angle Sensor Neutral Position, Audio presets and Navigation. You will need the CONSULT-III diagnostic tool, or equivalent. Follow the directions on the screen of the tool, as needed.

1. Before servicing the vehicle, refer to the Precautions Section.

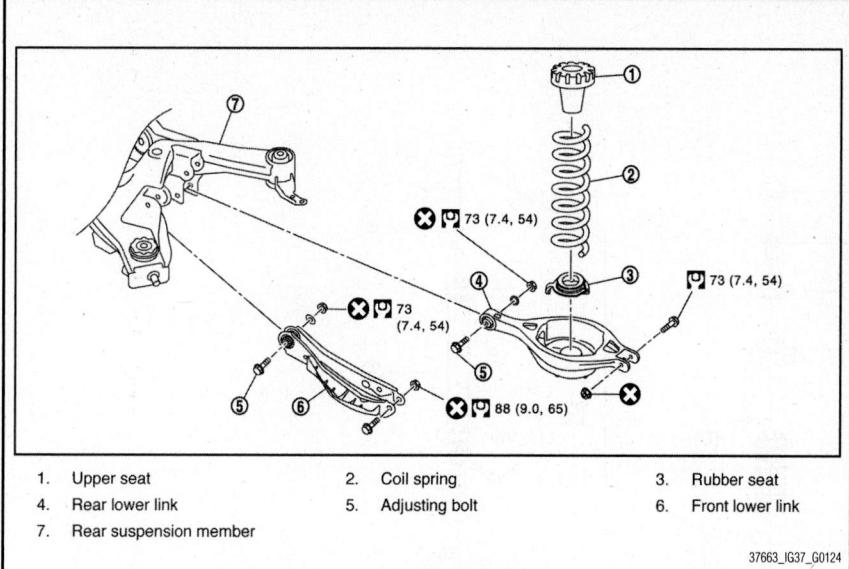

1. Upper seat
2. Coil spring
3. Rubber seat
4. Rear lower link
5. Adjusting bolt
6. Front lower link
7. Rear suspension member

37663_IG37_G0124

Fig. 273 Rear lower link and related components—convertible

➡**If working near and/or around the SRS system and components, be sure to disable the SRS system. After disabling the system wait three minutes or more before servicing the vehicle.**

2. Disconnect the negative battery cable.

3. Raise and support the vehicle safely.

4. Remove the tire and wheel assemblies.

5. Position a suitable jack under the rear lower link to relieve the coil spring tension.

6. Loosen the rear lower link mounting nuts (rear suspension member side).

7. Remove the rear lower link (axle housing side).

8. Slowly lower the jack and remove the upper seat, coil spring and rubber sheet from the rear lower link.

9. Remove the rear lower link.

To install:

➡**Be sure to use new fasteners, as required.**

10. Installation is the reverse of the removal procedure.

11. Be sure that the upper seat is attached as indicated in the illustration.

➡**Make sure that the projecting parts of the floor panel is securely fitted with the upper seat tab.**

12. Match up the rubber seat indentations and the rear lower link grooves and attach.

13. Install the coil spring by aligning the lower end of the large diameter side to the step between the rubber seat and the rear lower link.

➡**Make sure that the spring is not upside down. The top and bottom are indicated by paint color.**

14. Perform a final tightening of nuts and bolts of each removed component with the vehicle in an unladen position.

15. Check wheel speed sensor for proper operation.

16. Adjust the steering angle sensor neutral position, using the CONSULT-III diagnostic tool, or equivalent.

17. Check and adjust the front alignment, as required.

SHOCK ABSORBERS

REMOVAL & INSTALLATION

See Figures 268 and 269, 276.

➡**Whenever the negative battery cable is disconnected the following components will require resetting. The Automatic temperature control system, Automatic drive positioner, Power window control, Sunroof system, Sunshade system, Rear view monitor, Idle Air Volume Learning, Steering Angle Sensor Neutral Position, Audio presets and Navigation. You will need the CONSULT-III diagnostic tool, or equivalent. Follow the directions on the screen of the tool, as needed.**

1. Before servicing the vehicle, refer to the Precautions Section.

➡**If working near and/or around the SRS system and components, be sure to disable the SRS system. After disabling the system wait three minutes or more before servicing the vehicle.**

2. Disconnect the negative battery cable.

3. Raise and support the vehicle safely.

4. Remove the tire and wheel assemblies.

5. Position a suitable jack under the rear axle assembly to relieve the coil spring tension.

6. Gradually lower the jack and separate the shock absorber (lower side) from the axle housing.

7. Remove the shock absorber mounting nuts (upper side), and then remove the shock absorber.

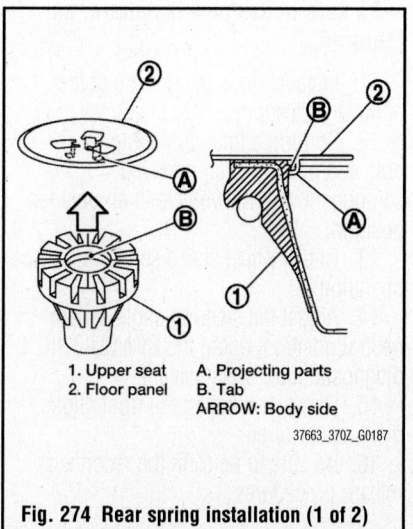

1. Upper seat
2. Floor panel
A. Projecting parts
B. Tab
ARROW: Body side

37663_370Z_G0187

Fig. 274 Rear spring installation (1 of 2)

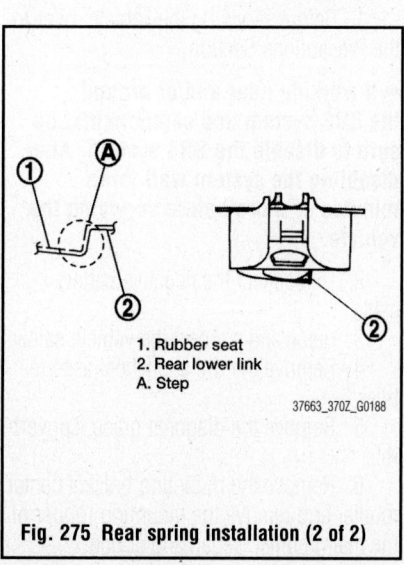

1. Rubber seat
2. Rear lower link
A. Step

37663_370Z_G0188

Fig. 275 Rear spring installation (2 of 2)

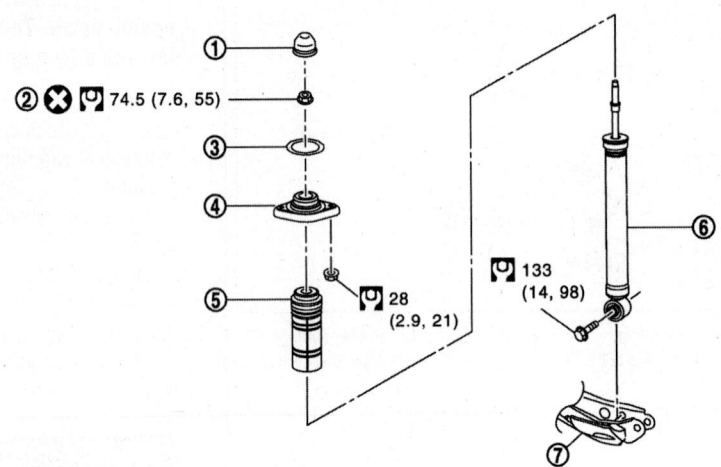

1. Cap
2. Piston rod lock nut
3. Mounting seal
4. Shock absorber mounting bracket
5. Bound bumper cover
6. Shock absorber
7. Front lower link

37663_IG37_G0125

Fig. 276 Rear shock absorber and related components—convertible

To install:

→Be sure to use new fasteners, as required.

8. Installation is the reverse of the removal procedure.

9. Perform a final tightening of nuts and bolts of each removed component with the vehicle in an unladen position.

10. Check wheel speed sensor for proper operation.

11. Adjust the steering angle sensor neutral position, using the CONSULT-III diagnostic tool, or equivalent.

12. Check and adjust the front alignment, as required.

13. Be sure to perform the reconnect/relearn procedures.

STABILIZER BAR

REMOVAL & INSTALLATION

See Figures 268 and 269, 277.

→Whenever the negative battery cable is disconnected the following components will require resetting. The Automatic temperature control system,

Automatic drive positioner, Power window control, Sunroof system, Sunshade system, Rear view monitor, Idle Air Volume Learning, Steering Angle Sensor Neutral Position, Audio presets and Navigation. You will need the CONSULT-III diagnostic tool, or equivalent. Follow the directions on the screen of the tool, as needed.

1. Before servicing the vehicle, refer to the Precautions Section.

→If working near and/or around the SRS system and components, be sure to disable the SRS system. After disabling the system wait three minutes or more before servicing the vehicle.

2. Disconnect the negative battery cable.

3. Raise and support the vehicle safely.

4. Remove the tire and wheel assemblies.

5. Remove the diagonal brace, convertible.

6. Remove the mounting bracket center muffler and remove the mounting rubber of the main muffler, sedan and coupe.

7. Remove the stabilizer connecting rods.

8. Remove the stabilizer clamps. Remove the stabilizer bushings.

9. Remove the stabilizer bar.

10. Remove the stabilizer connecting rod mounting brackets from the suspension arm.

To install:

→Be sure to use new fasteners, as required.

11. Installation is the reverse of the removal procedure.

12. Perform a final tightening of nuts and bolts of each removed component with the vehicle in an unladen position.

13. Check wheel speed sensor for proper operation.

14. Adjust the steering angle sensor neutral position, using the CONSULT-III diagnostic tool, or equivalent.

15. Check and adjust the front alignment, as required.

16. Be sure to perform the reconnect/relearn procedures.

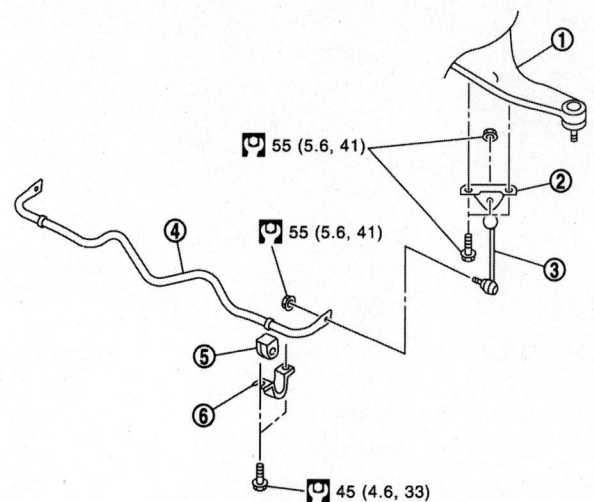

1. Suspension arm	2. Stabilizer connecting rod mounting bracket	3. Stabilizer connecting rod
4. Stabilizer bar	5. Stabilizer bushing	6. Stabilizer clamp

37663_370Z_G0191

Fig. 277 Rear stabilizer bar and related components—convertible

WHEEL HUBS & BEARINGS

ADJUSTMENT

These bearings are not adjustable. If defective, they must be replaced.

REMOVAL & INSTALLATION

Except Convertible

See Figure 278.

➡Whenever the negative battery cable is disconnected the following components will require resetting. The Automatic temperature control system, Automatic drive positioner, Power window control, Sunroof system, Sunshade system, Rear view monitor, Idle Air Volume Learning, Steering Angle Sensor Neutral Position, Audio presets and Navigation. You will need the CONSULT-III diagnostic tool, or equivalent. Follow the directions on the screen of the tool, as needed.

1. Before servicing the vehicle, refer to the Precautions Section.

➡If working near and/or around the SRS system and components, be sure to disable the SRS system. After disabling the system wait three minutes or more before servicing the vehicle.

2. Disconnect the negative battery cable.
3. Raise and support the vehicle safely.
4. Remove the tire and wheel assembly.
5. Remove the caliper. Properly support the caliper to the side. Do not allow it to hang by the brake hose. Never depress the brake pedal with the caliper removed.
6. Remove the brake rotor.
7. Remove the cotter pin and adjusting cap. Loosen the wheel hub locknut.
8. Matchmark the halfshaft and the wheel hub and bearing assembly.
9. Remove the locknut.
10. Remove the cotter pin, then loosen the suspension arm mounting nut of the axle housing.
11. Remove the wheel hub and bearing assembly.

To install:

➡Be sure to use new fasteners, as required.

12. Installation is the reverse of the removal procedure.
13. Use the matchmarks to align removed components.
14. Clean the matching surface of the halfshaft and wheel hub and bearing assembly.
15. Apply paste part number 440037S000 or equivalent to the surface of the sub joint assembly of the halfshaft.

➡Apply the paste, about 0.04–0.10 ounce) to cover the entire flat surface of the sub joint assembly of the halfshaft.

16. The wheel hub locknut tightening specification is 136 ft. lbs. (185 Nm).
17. Perform a final tightening of nuts and bolts of each removed component with the vehicle in an unladen position.
18. Check wheel speed sensor for proper operation.
19. Adjust the steering angle sensor

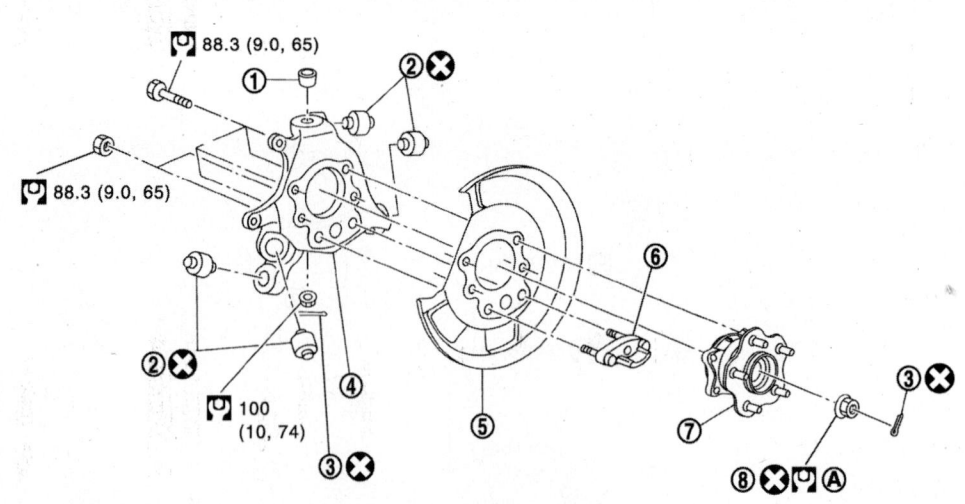

1. Ball seat
4. Axle housing
7. Wheel hub and bearing assembly

2. Bushing
5. Back plate
8. Wheel hub lock nut

3. Cotter pin
6. Anchor block

37663_370Z_G0192

Fig. 278 Rear wheel bearing and hub assembly and related components

neutral position, using the CONSULT-III diagnostic tool, or equivalent.

20. Check and adjust the front alignment, as required.

Convertible

See Figure 278.

➡Whenever the negative battery cable is disconnected the following components will require resetting. The Automatic temperature control system, Automatic drive positioner, Power window control, Sunroof system, Sunshade system, Rear view monitor, Idle Air Volume Learning, Steering Angle Sensor Neutral Position, Audio presets and Navigation. You will need the CONSULT-III diagnostic tool, or equivalent. Follow the directions on the screen of the tool, as needed.

1. Before servicing the vehicle, refer to the Precautions Section.

➡If working near and/or around the SRS system and components, be sure to disable the SRS system. After disabling the system wait three minutes or more before servicing the vehicle.

2. Disconnect the negative battery cable.

3. Raise and support the vehicle safely.

4. Remove the tire and wheel assembly.

5. Remove the caliper. Properly

support the caliper to the side. Do not allow it to hang by the brake hose. Never depress the brake pedal with the caliper removed.

6. Remove the brake rotor.

7. Remove the cotter pin and adjusting cap. Loosen the wheel hub locknut.

8. Matchmark the halfshaft and the wheel hub and bearing assembly.

9. Remove the locknut and spring washer.

10. Remove the parking brake shoe and parking brake cable.

11. Remove the stabilizer connecting rod (upper side).

12. Remove the coil spring.

13. Properly position a suitable jack under the axle housing.

14. Remove the radius rod.

15. Remove the shock absorber (lower side).

16. Remove the front lower link (axle housing side). Remove the rear lower link (axle housing side)

17. Separate the axle housing from the suspension arm, using the proper tool and remove the axle housing.

➡Be careful not to damage the ball joint boot. Temporarily tighten the nut to prevent damage to the threads and to prevent the tool from coming off.

➡Never place the halfshaft at an extreme angle. Be careful not to overextend the slide joint. Never allow

the halfshaft to hang down with proper support.

18. Remove the wheel hub and bearing assembly.

To install:

➡Be sure to use new fasteners, as required.

19. Installation is the reverse of the removal procedure.

20. Use the matchmarks to align removed components.

21. Clean the matching surface of the halfshaft and wheel hub and bearing assembly.

22. Apply paste part number 440037S000 or equivalent to the surface of the sub joint assembly of the halfshaft.

➡Apply the paste, about 0.04–0.10 ounce) to cover the entire flat surface of the sub joint assembly of the halfshaft.

23. The wheel hub locknut tightening specification is 136 ft. lbs. (185 Nm).

24. Perform a final tightening of nuts and bolts of each removed component with the vehicle in an unladen position.

25. Check wheel speed sensor for proper operation.

26. Adjust the steering angle sensor neutral position, using the CONSULT-III diagnostic tool, or equivalent.

27. Check and adjust the front alignment, as required.

INFINITI

M37 • M56

SPECIFICATIONS AND MAINTENANCE CHARTS

ENGINE AND VEHICLE IDENTIFICATION

	Engine						Model Year	
Code	Liters (cc)	Cu. In.	Cyl.	Fuel Sys.	Engine Type	Eng. Mfg.	Code ①	Year
VQ37VHR	3.7 (3700)	225	6	MFI	DOHC	Nissan	B	2011
VK56VD	5.6 (5600)	341	8	MFI	DOHC	Nissan	C	2012

MFI: Multi-Port Fuel Injection

DOHC: Double Overhead Camshaft

① 10th digit of the Vehicle Identification Number (VIN)

71075_IM37_C0001

GENERAL ENGINE SPECIFICATIONS

Year	Model	Engine Displacement Liters	Engine ID	Net Horsepower @ rpm	Net Torque @ rpm (ft. lbs.)	Bore x Stroke (in.)	Compression Ratio	Oil Pressure @ rpm
2011	M37/37X	3.7	VQ37VHR	330@6800	262@4800	3.76x3.386	11.5:1	43@2000
	M56/56X	5.6	VK56VD	420@6000	417@4400	3.858x3.622	11.5:1	43@2000
2012	M37/37X	3.7	VQ37VHR	330@6800	262@4800	3.76x3.386	11.5:1	43@2000
	M56/56X	5.6	VK56VD	420@6000	417@4400	3.858x3.622	11.5:1	43@2000

X denotes AWD

71075_IM37_C0002

ENGINE TUNE-UP SPECIFICATIONS

Year	Engine Displacement Liters	Engine ID	Spark Plug Gap (in.)	Ignition Timing (deg.) MT	Ignition Timing (deg.) AT	Fuel Pump (psi)	Idle Speed (rpm) MT	Idle Speed (rpm) AT	Valve Clearance Intake	Valve Clearance Exhaust
2011	3.7	VQ37VHR	0.043	NA	15 ①	34	NA	600-700	②	③
	5.6	VK56VD	0.043	NA	12 ①	34	MA	600-700	②	③
2012	3.7	VQ37VHR	0.043	NA	15 ①	34	MA	600-700	②	③
	5.6	VK56VD	0.043	NA	12 ①	34	NA	600-700	②	③

NA: Not Available

NOTE: The Vehicle Emission Control Information label often reflects specification changes made during production.

The label figures must be used if they differ from those in this chart.

① +/-5° Before top dead center

② Intake 0.010 - 0.013 Cold

③ Exhaust 0.011 - 0.015 Cold

71075_IM37_C0003

CAPACITIES

Year	Model	Engine Displacement Liters	Engine ID	Engine Oil with Filter (qts.)	Transmission (pts.) Manual	Transmission (pts.) Auto.	Transfer Case (pts.)	Drive Axle (pts.)	Fuel Tank (gal.)	Cooling System (qts.)
2011	M37	3.7	VQ37VHR	5.125	—	①	2.6	②	20	9.0
	M56	5.6	VK56VD	5.75	—	①	2.6	②	20	11.5
2012	M37	3.7	VQ37VHR	5.125	—	①	2.6	②	20	9.0
	M56	5.6	VK56VD	5.75	—	①	2.6	②	20	11.5

NOTE: All capacities are approximate. Add fluid gradually and check to be sure a proper fluid level is obtained.

① 5A/T: 21-5/8 pts.
  6A/T: 19-1/2 pts.

② Front Axle 1.38 (AWD) -Rear Axle 3.00

71075_IM37_C0005

FLUID SPECIFICATIONS

Year	Model	Engine Displ. Liters	Engine Oil	Auto. Trans.	Drive Axle Front	Drive Axle Rear	Transfer Case	Power Steering Fluid	Brake Master Cylinder	Cooling System
2011	M37	3.7	①	②	GL-5 80W-90	GL-5 80W-90	③	NISSAN PSF	DOT 3	④
	M56	5.6	①	②	GL-5 80W-90	GL-5 80W-90	③	NISSAN PSF	DOT 3	④
2012	M37	3.7	①	②	GL-5 80W-90	GL-5 80W-90	③	NISSAN PSF	DOT 3	④
	M56	5.6	①	②	GL-5 80W-90	GL-5 80W-90	③	NISSAN PSF	DOT 3	④

DOT: Department Of Transpotation

① API Certification Mark 5W-30

② Genuine NISSAN Matic S ATF

③ Genuine NISSAN Matic J ATF

④ NISSAN Long Life Antifreeze/ Coolant or equivalent

71075_IM37_C0004

VALVE SPECIFICATIONS

Year	Engine Displacement Liters	Engine ID	Seat Angle (deg.)	Face Angle (deg.)	Spring Test Pressure (lbs. @ in.)	Spring Free Height (in.)	Stem-to-Guide Clearance (in.) Intake	Stem-to-Guide Clearance (in.) Exhaust	Stem Diameter (in.) Intake	Stem Diameter (in.) Exhaust
2011	3.7	VQ37VHR	①	NA	NA	1.7979	0.0008-0.0021	0.0012-0.0022	0.2348-0.2354	0.2347-0.2350
	5.6	VK56VD	①	NA	NA	②	0.0008-0.0018	0.0012-0.0022	0.2348-0.2354	0.2344-0.2350
2012	3.7	VQ37VHR	①	NA	NA	1.7976	0.0008-0.0021	0.0012-0.0022	0.2348-0.2354	0.2347-0.2350
	5.6	VK56VD	①	NA	NA	②	0.0008-0.0018	0.0012-0.0022	0.2348-0.2354	0.2344-0.2350

NA: Not Available

① 45 degrees, 15 minutes to 45 degrees, 45 minutes

② Intake: 1.9169 inches
  Exhaust: 1.8642

71075_IM37_C0006

CAMSHAFT AND BEARING SPECIFICATIONS

All measurements are given in inches.

Year	Engine Displacement Liters	Engine VIN	Journal Diameter	Brg. Oil Clearance	Shaft End-play	Runout	Journal Bore	Lobe Lift	
								Intake	Exhaust
2011	3.7	VQ37VHR	①	②	0.0045-0.0074	0.0045-0.0074	NA	1.8057-1.8132	1.8061-1.8136
	5.6	VK56VD	③	④	0.0045-0.0074	0.0059	1.0236-1.0244	1.7663-1.7738	1.7293-1.7368
2012	3.7	VQ37VHR	①	②	0.0045-0.0074	0.0045-0.0074	NA	1.8057-1.8132	1.8061-1.8136
	5.6	VK56VD	③	④	0.0045-0.0074	0.0059	1.0236-1.0244	1.7663-1.7738	1.7293-1.7368

NA: Information not available

Note: (Oil clearance) = (Camshaft bracket inner diameter) – (Camshaft journal diameter).

① No. 1: 1.0211 - 1.0218 inches

   No 2,3 and 4 - 0.9230-0.9238 inches

② No. 1: 0.0018 - 0.0034 inches

   No 2,3,4 and 5 - 0.0014-0.0030 inches

③ No. 1: 1.0212 - 1.0218 inches

   No 2,3 and 4: 1.218-1.0224 inches

④ No. 1: 0.0018 - 0.0033 inches

   No 2,3,4 and 5 - 0.0012-0.0027 inches

71075_IM37_C0007

CRANKSHAFT AND CONNECTING ROD SPECIFICATIONS

All measurements represent standard values and are given in inches.

Year	Engine Displacement Liters	Engine ID	Crankshaft				Connecting Rod		
			Main Brg. Journal Dia.	Main Brg. Oil Clearance	Shaft End-play	Thrust on No.	Journal Diameter	Oil Clearance	Side Clearance
2011	3.7	VQ37VHR	2.5571-2.5581	0.0014-0.0018	0.0039-0.0098	3	NA	0.0016-0.0021	NA
	5.6	VK56VD	2.5173-2.5183	①	0.0039-0.0098	3	NA	0.0008-0.0018	NA
2012	3.7	VQ37VHR	2.5571-2.5581	0.0014-0.0018	0.0039-0.0098	3	NA	0.0016-0.0021	NA
	5.6	VK56VD	2.5173-2.5183	①	0.0039-0.0098	3	NA	0.0008-0.0018	NA

① Nos. 1 and 5: 0.00004-0.0004 in.

   Nos. 2, 3 and 4: 0.0003-0.0007 in.

71075_IM37_C0009

PISTON AND RING SPECIFICATIONS
All measurements are given in inches.

Year	Engine Disp. Liters	Engine ID	Piston Clearance	Ring Gap			Ring Side Clearance		
				Top Compression	Bottom Compression	Oil Control	Top Compression	Bottom Compression	Oil Control
2011	3.7	VQ37VHR	0.0004-0.0012	0.0091-0.0130	0.0130-0.0189	0.0067-0.0189	0.0016-0.0031	0.0012-0.0028	0.0022-0.0061
	5.6	VK56VD	0.0004-0.0012	0.0087-0.0126	0.0087-0.0126	0.0079-0.0197	0.0018-0.0031	0.0012-0.0028	0.0026-0.0053
2012	3.7	VQ37VHR	0.0004-0.0012	0.0091-0.0130	0.0130-0.0189	0.0067-0.0189	0.0016-0.0031	0.0012-0.0028	0.0022-0.0061
	5.6	VK56VD	0.0004-0.0012	0.0087-0.0126	0.0087-0.0126	0.0079-0.0197	0.0018-0.0031	0.0012-0.0028	0.0026-0.0053

71075_IM37_C0008

TORQUE SPECIFICATIONS
All readings in ft. lbs.

Year	Engine Displacement Liters	Engine ID	Cylinder Head Bolts	Main Bearing Bolts	Rod Bearing Bolts	Crankshaft Damper Bolts	Flywheel Bolts	Manifold		Spark Plugs	Oil Pan Drain Plug
								Intake	Exhaust		
2009	3.7	VQ37VHR	①	②	③	33	65	④	22	14	25
	5.6	VK56VD	⑤	⑥	⑦	⑧	65	⑨ ⑩	21	18	25
2010	3.7	VQ37VHR	①	②	③	33	65	④	22	14	25
	5.6	VK56VD	⑤	⑥	⑦	⑧	65	⑨ ⑩	21	18	25

① Step 1: 77 ft. lbs.

Step 2: Loosen bolts completely

Step 3: 30 ft. lbs.

Step 4: Tighten an additional 95 degrees

Step 5: Again, tighten an additional 95 degrees.

Step 6: After installing cylinder head, measure
   distance between front end faces of cylinder
   block and cylinder head (left and right banks)
   Standard : 0.555 - 0.587 in (14.1 - 14.9 mm)

② Step 1: Bolts 17-26: 18 ft. lbs.

Step 2: Repeat Step 1

Step 3: Bolts 1-16: 26 ft. lbs.

Step 4: Bolts 1-16: Additional 90 degrees

③ Step 1: Tighten to 21 ft. lbs.

Step 2: Completely loosen rod bolts

Step 3: Tighten to 18 ft. lbs.

Step 4: Additional 90 degrees

④ Step 1: Tighten stud bolts to 8 ft. lbs.

Step 2: Tighten to 5 ft. lbs.

Step 3: Tighten, again, to 19 ft. lbs.

⑤ Step 1: 30 ft.lbs.

Step 2: Additional 75 degrees

Step 3: Loosen bolts completely

Step 3: 30 ft. lbs.

Step 4: Tighten an additional 90 degrees.

⑥ Step 1: M12 bolts 1-10: 40 ft. lbs.

Step 2: M9 bolts 11-20: 14 ft. lbs.

Step 3: M12 bolts: Tighten an additional 90 degrees

Step 4: M9 bolts: Tighten an additional 90 degrees

Step 5: M10 side bolts 21-30: 36 ft. lbs.

⑦ Step 1: Tighten to 21 ft. lbs.

Step 2: Completely loosen rod bolts

Step 3: Tighten to 18 ft. lbs.

Step 4: Additional 90 degrees

⑧ Step 1: 69 ft. lbs.

Step 2: Tighten an additional 90 degrees

⑨ Step 1: Tighten Upper Manifold bolts to 8 ft. lbs.

⑩ Step 1: Tighten stud bolts to 8 ft. lbs.

Step 2: Tighten to 11 ft. lbs.

Step 3: Tighten, again, to 21 ft. lbs.

71075_IM37_C0010

WHEEL ALIGNMENT

| Year | Model | | Caster | | Camber | | Toe-in |
			Range (+/-Deg.)	Preferred Setting (Deg.)	Range (+/-Deg.)	Preferred Setting (Deg.)	(Deg.)
2011	M37	F	0.75	①	0.75	-0.25	0.05 +/- 0.05
		R	NA	NA	②	③	0.12 +0.11/- 0.12
	M37X	F	0.75	3.83	0.75	-0.25	0.04 +/- 0.04
	AWD	R	NA	NA	0.50	-0.17	0.12 +0.11/- 0.12
	M56	F	0.75	①	0.75	-0.25	0.05 +/- 0.05
		R	NA	NA	②	③	0.12 +0.11/- 0.12
	M56X	F	0.75	3.83	0.75	-0.25	0.04 +/- 0.04
	AWD	R	NA	NA	0.50	-0.17	0.12 +0.11/- 0.12
2012	M37	F	0.75	①	0.75	-0.25	0.05 +/- 0.05
		R	NA	NA	②	③	0.12 +0.11/- 0.12
	M37X	F	0.75	3.83	0.75	-0.25	0.04 +/- 0.04
	AWD	R	NA	NA	0.50	-0.17	0.12 +0.11/- 0.12
	M56	F	0.75	①	0.75	-0.25	0.05 +/- 0.05
		R	NA	NA	②	③	0.12 +0.11/- 0.12
	M56X	F	0.75	3.83	0.75	-0.25	0.04 +/- 0.04
	AWD	R	NA	NA	0.50	-0.17	0.12 +0.11/- 0.12

Note: Measure wheel alignment under unladen conditions.

"Unladen conditions" means that fuel, engine coolant, and lubricant are full. Spare tire, jack, hand tools and mats are in designated position.

① 18" Wheels 4.50°

19" Wheels 4.58°

② 18" Wheels: 40°

19" Wheels: 50°

③ Rear 18" Wheels -67°

Rear 19" Wheels -83°

71075_IM37_C0011

TIRE, WHEEL AND BALL JOINT SPECIFICATIONS

Year	Model	OEM Tires Standard	OEM Tires Optional	Tire Pressures (psi) Front	Tire Pressures (psi) Rear	Wheel Size	Ball Joint Inspection	Lug Nut (ft. lbs.)
2011	M37	P245/50R18 99V	P245/40R20 95W	33	33	Std: 8.0-JJ	①	80
	M37 AWD	P245/50R18 99V	NA	33	33	Opt: 8,5-JJ		
	M56	P245/50R18 99V	P245/40R20 95W	33	33	Std: 8.0-JJ	①	80
	M56 AWD	P245/50R18 99V	NA	33	33	Opt: 8.5-JJ		
2012	M37	P245/50R18 99V	P245/40R20 95W	33	33	Std: 8.0-JJ	①	80
	M37 AWD	P245/50R18 99V	NA	33	33	Opt: 8,5-JJ		
	M45	P245/50R18 99V	P245/40R20 95W	33	33	Std: 8.0-JJ	①	80
	M56 AWD	P245/50R18 99V	NA	33	33	Opt: 8.5-JJ		

OEM: Original Equipment Manufacturer

PSI: Pounds Per Square Inch

STD: Standard

OPT: Optional

NA: Not Available

① Replace is any measureable movement is found.

71075_IM37_C0012

BRAKE SPECIFICATIONS
All measurements in inches unless noted

Year	Model	Front Brake Disc Original Thickness	Front Brake Disc Minimum Thickness	Front Brake Disc Maximum Run-out	Rear Brake Disc Original Thickness	Rear Brake Disc Minimum Thickness	Rear Brake Disc Maximum Run-out	Minimum Lining Thickness Front	Minimum Lining Thickness Rear	Brake Caliper Bracket Bolts (ft. lbs.)	Brake Caliper Mounting Bolts (ft. lbs.)
2011	M37	NA	①	0.0014	NA	②	0.0022	0.079	0.079	③	④
	M56	NA	①	0.0014	NA	②	0.0022	0.079	0.079	③	④
2012	M37	NA	①	0.0014	NA	②	0.0022	0.079	0.079	③	④
	M56	NA	①	0.0014	NA	②	0.0022	0.079	0.079	③	④

NA: Not Avialble

① 2 Piston type: 1.024 inches

   4 Piston type: 1.181 inches

② 1 Piston type: 0.551 inches

   2 Piston type: 0.709 inches

③ Front: 20

   Rear: 32

④ Front: 98

   Rear: 62

71075_IM37_C0013

SCHEDULED MAINTENANCE INTERVALS
Infiniti— M37 & M56

TO BE SERVICED	TYPE OF SERVICE	VEHICLE MILEAGE INTERVAL (x1000)												
		7.5	15	22.5	30	37.5	45	52.5	60	67.5	75	82.5	90	97.5
Engine oil & filter	R	✓	✓	✓	✓	✓	✓	✓	✓	✓	✓	✓	✓	✓
Automatic transaxle fluid ①	S/I		✓		✓		✓		✓		✓		✓	
Brake lines & cables	S/I		✓		✓		✓		✓		✓		✓	
Brake pads & discs	S/I		✓		✓		✓		✓		✓		✓	
Differential gear oil	S/I		✓		✓		✓		✓		✓		✓	
Driveshaft boots	S/I		✓		✓		✓		✓		✓		✓	
Active suspension fluid ②	S/I		✓		✓		✓		✓		✓		✓	
In-cabin microfilter	R		✓		✓		✓		✓					
Air cleaner filter ③	R								✓					
Exhaust system	S/I		✓		✓		✓		✓		✓		✓	
Fuel lines	S/I								✓					
Steering gear & linkage, axle & suspension parts	S/I		✓		✓		✓		✓		✓		✓	
Vapor lines	S/I				✓				✓				✓	
Engine coolant ④	R								✓				✓	
Spark plugs(Platinum-tipped type) ⑤	R	Replace every 105,000 miles												
Drive belts ⑥	S/I								✓					

R: Replace   S/I: Service or Inspect

① If towing a trailer, using a camper or car-top carrier, or driving on rough or muddy roads, CHANGE oil every 30,000 miles or 24 months.

② Replace at 60,000 miles (if not previously replaced).

③ If operating in dusty conditions, more frequent maintenance may be required.

④ After 60,000 miles or 48 months, replace coolant every 30,000 miles or 24 months.

⑤ Platinum-tipped spark plugs should be changed every 105,000 miles.

⑥ After 60,000 miles or 48 months, inspect every 15,000 miles or 12 months. Replace belts if found damaged.

FREQUENT OPERATION MAINTENANCE (SEVERE SERVICE)

If a vehicle is operated under any of the following conditions it is considered severe service:

- Extremely dusty areas.

- 50% or more of the vehicle operation is in 90°F or higher temperatures, or constant operation in temperatures below 32°F.

- Prolonged idling (vehicle operation in stop and go traffic).

- Frequent short running periods (engine does not warm to normal operating temperatures).

- Police, taxi, delivery or trailer towing usage.

Oil & oil filter: change every 3750 miles.

Brake pads & discs: service or inspect every 7500 miles.

Driveshaft boots: service or inspect every 7500 miles

Exhaust system: service or inspect every 7500 miles.

Steering gear, linkage, axle & suspension ball joints: service or inspect every 7500 miles.

Steering linkage, ball joints & front suspension ball joints: service or inspect every 7500 miles.

71075_IM37_C0014

PRECAUTIONS

Before servicing any vehicle, please be sure to read all of the following precautions, which deal with personal safety, prevention of component damage, and important points to take into consideration when servicing a motor vehicle:

• Never open, service or drain the radiator or cooling system when the engine is hot; serious burns can occur from the steam and hot coolant.

• Observe all applicable safety precautions when working around fuel. Whenever servicing the fuel system, always work in a well-ventilated area. Do not allow fuel spray or vapors to come in contact with a spark, open flame, or excessive heat (a hot drop light, for example). Keep a dry chemical fire extinguisher near the work area. Always keep fuel in a container specifically designed for fuel storage; also, always properly seal fuel containers to avoid the possibility of fire or explosion. Refer to the additional fuel system precautions later in this section.

• Fuel injection systems often remain pressurized, even after the engine has been turned **OFF**. The fuel system pressure must be relieved before disconnecting any fuel lines. Failure to do so may result in fire and/or personal injury.

• Brake fluid often contains polyglycol ethers and polyglycols. Avoid contact with the eyes and wash your hands thoroughly after handling brake fluid. If you do get brake fluid in your eyes, flush your eyes with clean, running water for 15 minutes. If eye irritation persists, or if you have taken brake fluid internally, IMMEDIATELY seek medical assistance.

• The EPA warns that prolonged contact with used engine oil may cause a number of skin disorders, including cancer. You should make every effort to minimize your exposure to used engine oil. Protective gloves should be worn when changing oil. Wash your hands and any other exposed skin areas as soon as possible after exposure to used engine oil. Soap and water, or waterless hand cleaner should be used.

• All new vehicles are now equipped with an air bag system, often referred to as a Supplemental Restraint System (SRS) or Supplemental Inflatable Restraint (SIR) system. The system must be disabled before performing service on or around system components, steering column, instrument panel components, wiring and sensors. Failure to follow safety and disabling procedures could result in accidental air bag deployment, possible personal injury and unnecessary system repairs.

• Always wear safety goggles when working with, or around, the air bag system. When carrying a non-deployed air bag, be sure the bag and trim cover are pointed away from your body. When placing a non-deployed air bag on a work surface, always face the bag and trim cover upward, away from the surface. This will reduce the motion of the module if it is accidentally deployed. Refer to the additional air bag system precautions later in this section.

• Clean, high quality brake fluid from a sealed container is essential to the safe and proper operation of the brake system. You should always buy the correct type of brake fluid for your vehicle. If the brake fluid becomes contaminated, completely flush the system with new fluid. Never reuse any brake fluid. Any brake fluid that is removed from the system should be discarded. Also, do not allow any brake fluid to come in contact with a painted surface; it will damage the paint.

• Never operate the engine without the proper amount and type of engine oil; doing so WILL result in severe engine damage.

• Timing belt maintenance is extremely important. Many models utilize an interference-type, non-freewheeling engine. If the timing belt breaks, the valves in the cylinder head may strike the pistons, causing potentially serious (also time-consuming and expensive) engine damage. Refer to the maintenance interval charts for the recommended replacement interval for the timing belt, and to the timing belt section for belt replacement and inspection.

• Disconnecting the negative battery cable on some vehicles may interfere with the functions of the on-board computer system(s) and may require the computer to undergo a relearning process once the negative battery cable is reconnected.

• When servicing drum brakes, only disassemble and assemble one side at a time, leaving the remaining side intact for reference.

• Only an MVAC-trained, EPA-certified automotive technician should service the air conditioning system or its components.

BRAKES

GENERAL INFORMATION

PRECAUTIONS

• Certain components within the ABS system are not intended to be serviced or repaired individually.

• Do not use rubber hoses or other parts not specifically specified for and ABS system. When using repair kits, replace all parts included in the kit. Partial or incorrect repair may lead to functional problems and require the replacement of components.

• Lubricate rubber parts with clean, fresh brake fluid to ease assembly. Do not use shop air to clean parts; damage to rubber components may result.

• Use only DOT 3 brake fluid from an unopened container.

• If any hydraulic component or line is removed or replaced, it may be necessary to bleed the entire system.

• A clean repair area is essential. Always clean the reservoir and cap thoroughly before removing the cap. The slightest amount of dirt in the fluid may plug an orifice and impair the system function. Perform repairs after components have been thoroughly cleaned; use only denatured alcohol to clean components. Do not allow ABS components to come into contact with any substance containing mineral oil; this includes used shop rags.

• The Anti-Lock control unit is a microprocessor similar to other computer units in the vehicle. Ensure that the ignition switch

ANTI-LOCK BRAKE SYSTEM (ABS)

is **OFF** before removing or installing controller harnesses. Avoid static electricity discharge at or near the controller.

• If any arc welding is to be done on the vehicle, the control unit should be unplugged before welding operations begin.

SPEED SENSORS

REMOVAL & INSTALLATION

Front

See Figure 1.

1. Remove tires with power tool.
2. Remove the fender protector (front).
3. Remove front wheel sensor from steering knuckle.

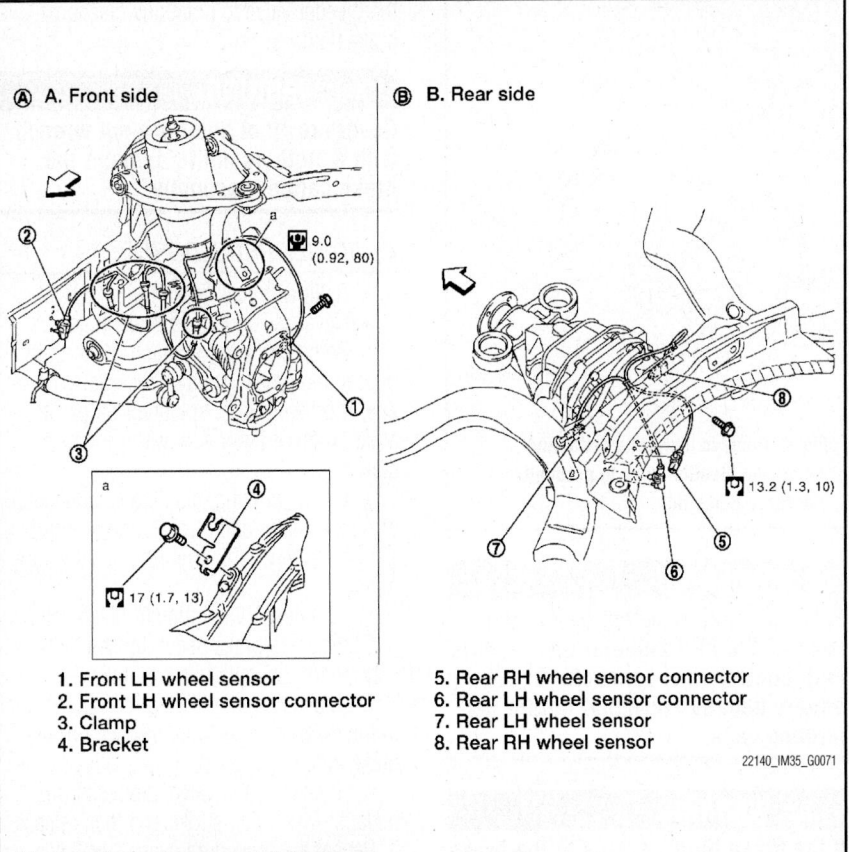

A. Front side

B. Rear side

9.0
(0.92, 80)

13.2 (1.3, 10)

a

17 (1.7, 13)

1. Front LH wheel sensor
2. Front LH wheel sensor connector
3. Clamp
4. Bracket

5. Rear RH wheel sensor connector
6. Rear LH wheel sensor connector
7. Rear LH wheel sensor
8. Rear RH wheel sensor

22140_IM35_G0071

Fig. 1 ABS wheel sensors exploded view

> **⁂ WARNING**
> **Never rotate and never pull front wheel sensor as much as possible, when pulling out.**

4. Remove front wheel sensor harness from the vehicle.
5. Never twist or pull front wheel sensor harness, when removing.

6. Note the following, and install in the reverse order of the removal.
   a. Check that there is no foreign material like iron powder or damage on inner surface of front wheel sensor mounting hole of steering knuckle and sensor rotor. Install after cleaning when there are foreign material like iron powder, or replace when there is a malfunction.

b. Never twist front wheel sensor harness when installing front wheel sensor. Check that grommet is fully inserted to bracket. Check that front wheel sensor harness is not twisted after installation.

> **⁂ WARNING**
> **Check that front wheel sensor identification line faces toward the vehicle front.**

Rear

See Figure 1.

1. Remove rear wheel sensor from rear final drive.

> **⁂ WARNING**
> **Never rotate or pull rear wheel sensor as much as possible, when pulling out.**

2. Remove rear wheel sensor harness from the vehicle.

> **⁂ WARNING**
> **Never twist and never pull rear wheel sensor harness, when removing.**

3. Note the following, and install in the reverse order of removal.
   a. Check that there is no foreign material like iron powder or damage on inner surface of rear wheel sensor mounting hole of rear final drive and sensor rotor. Install after cleaning when there are foreign material like iron powder, or replace when there is a malfunction.

BRAKES

BLEEDING PROCEDURE

BLEEDING PROCEDURE

> **⁂ WARNING**
> **Use of any other than the approved DOT 3 brake fluid will cause permanent damage to brake components and will render the brakes inoperative. Failure to follow these instructions may result in personal injury.**

• Brake fluid contains polyglycol ethers and polyglycols. Avoid contact with eyes. Wash hands thoroughly after handling. If brake fluid contacts eyes, flush eyes with running water for 15 minutes. Get medical attention if irritation persists. If taken internally, drink water and induce vomiting. Get medical attention immediately. Failure to follow these instructions may result in personal injury.

> **⁂ WARNING**
> **Do not allow the brake master cylinder reservoir to run dry during the bleeding operation. Keep the master cylinder reservoir filled with the specified brake fluid. Never reuse the brake fluid that has been drained from the hydraulic system.**

BLEEDING THE BRAKE SYSTEM

> **⁂ WARNING**
> **Brake fluid is harmful to painted and plastic surfaces. If brake fluid is spilled onto a painted or plastic surface, immediately wash it with water.**

> **⁂ WARNING**
> **When any part of the hydraulic system has been disconnected or a new component is installed, air may enter the system, causing spongy brake pedal action. This requires the bleeding of the hydraulic system after it has been correctly connected.**

1. Connect a vinyl tube to rear right brake caliper bleed valve.

2. Fully depress brake pedal 4 or 5 times.

3. With brake pedal depressed, loosen bleed valve to bleed air in brake line, and then tighten it immediately.

4. Repeat steps 2 and 3 until all of the air is out of the brake line.

5. Tighten the bleed valve to the specified torque.

6. From step 1 to 5, with master cylinder reservoir tank filled at least half way, bleed air from brake hydraulic line bleed valves in the following order:
- Rear right brake
- Front left brake
- Rear left brake
- Front right brake

7. Check that the fluid level in the reservoir tank is within the specified range after air bleeding.

8. Check each item of brake pedal. Adjust it if the measurement value is not the standard.

DRAINING

See Figure 2.

> **※ WARNING**
>
> **Never spill or splash brake fluid on painted surfaces. Brake fluid may seriously damage paint. Wipe it off immediately and wash with water if it gets on a painted surface.**

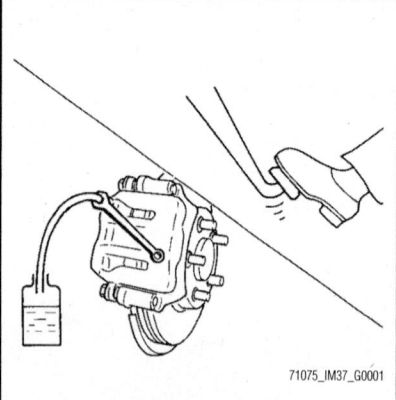

71075_IM37_G0001

Fig. 2 Depress the brake pedal and loosen the bleeder valve to gradually discharge brake fluid

> **※ WARNING**
>
> **Turn the ignition switch OFF and disconnect the ABS actuator and electric unit (control unit) connector or the battery negative terminal before performing work.**

> **※ WARNING**
>
> **If the brake fluid adheres to the brake caliper assembly and disc rotor, quickly wipe it off.**

1. Connect a vinyl tube to the bleed valve.

2. Depress the brake pedal and loosen

the bleeder valve to gradually discharge brake fluid.

> **※ WARNING**
>
> **Cover crowfoot and flare nut wrench with a cloth as not to damage the brake caliper assembly.**

FLUID FILL PROCEDURE

- Refill with new brake fluid "DOT 3".
- Never reuse drained brake fluid.
- Be careful not to splash brake fluid on painted areas; it may cause paint damage. If brake fluid is splashed on painted areas, wash it away with water immediately.
- Before working, disconnect connectors of ABS actuator and electric unit (control unit) or battery cable from the negative terminal.

1. Connect a vinyl tube to bleed valve.

2. Depress brake pedal, loosen bleed valve, and gradually remove brake fluid.

3. Make sure there is no foreign material in the reservoir tank, and refill with new brake fluid.

4. Loosen bleed valve, depress brake pedal slowly to full stroke and then release it. Repeat the procedure every 2 or 3 seconds until the new brake fluid comes out, then close the bleed valve while depressing the pedal. Repeat the same work for each wheel.

5. Bleed air.

BRAKES

> **※ WARNING**
>
> **Dust and dirt accumulating on brake parts during normal use may contain asbestos fibers from production or aftermarket brake linings. Breathing excessive concentrations of asbestos fibers can cause serious bodily harm. Exercise care when servicing brake parts. Do not sand or grind brake lining unless equipment used is designed to contain the dust residue. Do not clean brake parts with compressed air or by dry brushing. Cleaning should be done by dampening the brake components with a fine mist of water, then wiping the brake components clean with a dampened cloth. Dispose of cloth and all residue containing asbestos fibers in an impermeable container with the appropriate label. Follow practices prescribed by the Occupational Safety**

and Health Administration (OSHA) and the Environmental Protection Agency (EPA) for the handling, processing, and disposing of dust or debris that may contain asbestos fibers.

BRAKE CALIPER

REMOVAL & INSTALLATION

> **※ WARNING**
>
> **Clean any dust from the brake caliper and brake pads with a vacuum dust collector. Never blow with compressed air.**

> **※ WARNING**
>
> **Never spill or splash brake fluid on painted surfaces. Brake fluid may seriously damage paint. Wipe it off**

FRONT DISC BRAKES

immediately and wash with water if it gets on a painted surface.

> **※ WARNING**
>
> **Never depress the brake pedal while removing the brake pads because the piston may pop out.**

> **※ WARNING**
>
> **If the brake fluid or grease adheres to the brake caliper assembly and disc rotor, quickly wipe it off.**

2 Piston Type

See Figure 3.

1. Remove tires with power tool.

2. Fix the disc rotor using wheel nuts.

3. Drain brake fluid.

4. Remove union bolt and copper

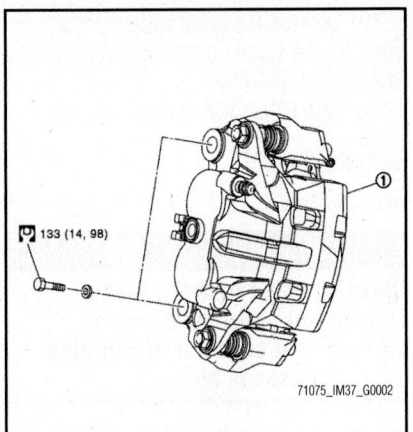

133 (14, 98)

71075_IM37_G0002

Fig. 3 Removing the brake caliper assembly (1)—2 Piston Type

washer, and separate brake hose from brake caliper assembly.

5. Remove torque member mounting bolts, and remove brake caliper assembly.

※※ WARNING

Never drop brake pad and brake caliper assembly.

6. Remove disc rotor. Refer to this section.

To install:

※※ WARNING

Clean any dust from the brake caliper and brake pads with a vacuum dust collector. Never blow with compressed air.

※※ WARNING

Never spill or splash brake fluid on painted surfaces. Brake fluid may seriously damage paint. Wipe it off immediately and wash with water if it gets on a painted surface.

- Never depress the brake pedal while removing the brake pads because the piston may pop out.
- If the brake fluid or grease adheres to the brake caliper assembly and disc rotor, quickly wipe it off.

7. Install disc rotor.

8. Install the brake caliper assembly to the steering knuckle and tighten the torque member mounting bolts to the specified torque.

※※ WARNING

Never spill or splash any grease and moisture on the brake caliper assem-

bly mounting face, threads, mounting bolts and washers. Wipe out any grease and moisture.

9. Install brake hose and copper washers to brake caliper assembly.

10. Refill with new brake fluid and perform the air bleeding.

11. Check a drag of front disc brake.

12. Install tires with power tool.

4 Piston Type

See Figure 4.

1. Remove tires with power tool.

2. Fix the disc rotor using wheel nuts.

3. Drain brake fluid.

4. Loosen the flare nut with a flare nut wrench and separate the brake tube from caliper.

5. Remove brake caliper assembly mounting bolts, and remove brake caliper assembly.

※※ WARNING

Never drop brake pad and caliper assembly.

6. Remove disc rotor.

To install:

※※ WARNING

Clean any dust from the brake caliper and brake pads with a vacuum dust collector. Never blow with compressed air.

※※ WARNING

Never spill or splash brake fluid on painted surfaces. Brake fluid may seriously damage paint. Wipe

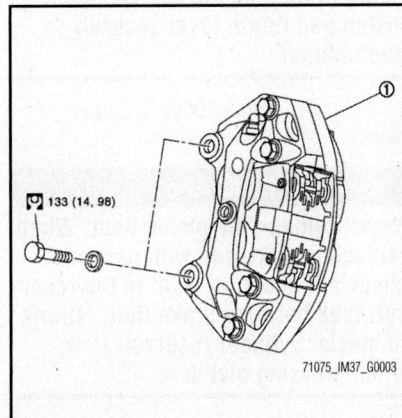

133 (14, 98)

71075_IM37_G0003

Fig. 4 Removing the brake caliper assembly (1)—2 Piston Type

it off immediately and wash with water if it gets on a painted surface.

- Never depress the brake pedal while removing the brake pads because the piston may pop out.
- If the brake fluid or grease adheres to the brake caliper assembly and disc rotor, quickly wipe it off.
- Cover crowfoot and flare nut wrench with a cloth as not to damage the brake caliper assembly.

7. Install disc rotor.

8. Install the brake caliper assembly to the steering knuckle and tighten the brake caliper assembly mounting bolts to the specified torque.

※※ WARNING

Never spill or splash any grease and moisture on the brake caliper assembly mounting face, threads, mounting bolts and washers. Wipe out any grease and moisture.

9. Install brake tube to brake caliper assembly.

10. Refill with new brake fluid and perform the air bleeding.

11. Check a drag of front disc brake.

12. Install tires with power tool.

DISC BRAKE PADS

REMOVAL & INSTALLATION

2 Piston Type
See Figure 5.

※※ WARNING

Clean any dust from the brake caliper and brake pads with a vacuum dust collector. Never blow with compressed air.

※※ WARNING

Never depress the brake pedal while removing the brake pads because the piston may pop out. If the brake fluid or grease adheres to the brake caliper assembly and disc rotor, quickly wipe it off.

1. Remove tires with power tool.

2. Remove lower sliding pin bolt.

3. Suspend the cylinder body with suitable wire so that the brake hose will not stretch. Then remove the brake pads from the torque member.

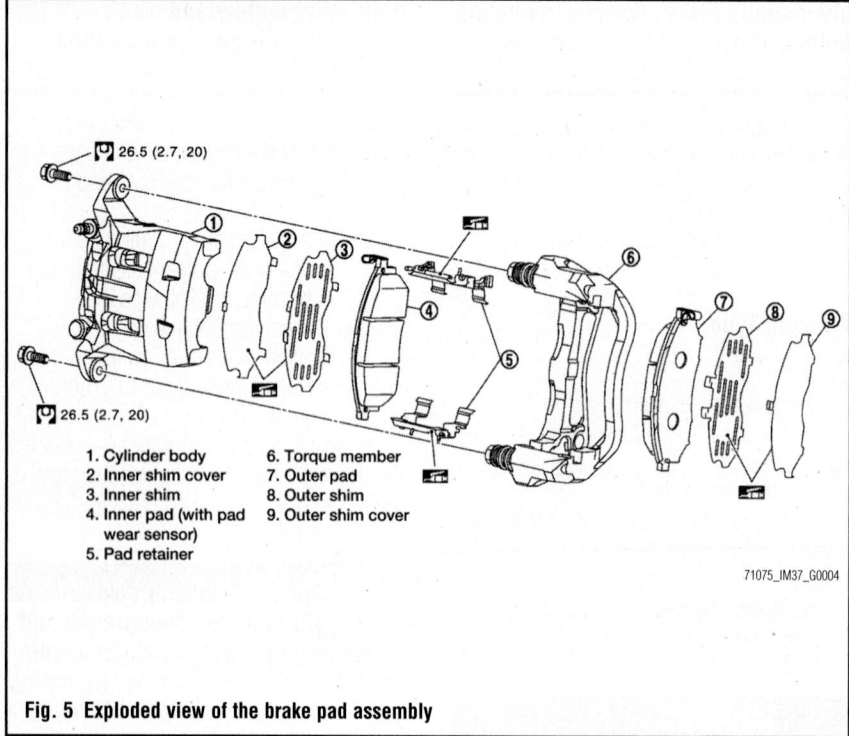

26.5 (2.7, 20)

26.5 (2.7, 20)

1. Cylinder body
2. Inner shim cover
3. Inner shim
4. Inner pad (with pad wear sensor)
5. Pad retainer
6. Torque member
7. Outer pad
8. Outer shim
9. Outer shim cover

71075_IM37_G0004

Fig. 5 Exploded view of the brake pad assembly

- Never deform the pad retainer when removing the pad retainer from the torque member.
- Never damage the piston boot.
- Never drop the brake pads, shims and shim covers.
- Remember each position of the removed brake pads.

4. Perform inspection after removal.

To install:

✳✳ WARNING
Clean any dust from the brake caliper and brake pads with a vacuum dust collector. Never blow with compressed air.

✳✳ WARNING
Never depress the brake pedal while removing the brake pads because the piston may pop out. If the brake fluid or grease adheres to the brake caliper assembly and disc rotor, quickly wipe it off.

5. Apply copper based brake grease to the pad retainers before installing it to the torque member if the pad retainers has been removed.

✳✳ WARNING
Securely assemble the pad retainers so that it will not be lifted up from

the torque member. **Never deform the pad retainers.**

6. Apply copper based brake grease to the matching faces between the shim and shim cover, and install shim and shim cover to the brake pad.

✳✳ WARNING
Always replace the shims and shim covers when replacing the brake pad.

7. Install the brake pads to the torque member.

✳✳ WARNING
Both inner and outer pads have a pad return system on the pad retainer. Install pad return lever securely to pad retainer.

8. Install cylinder body to torque member.

✳✳ WARNING
Never damage the piston boot. When replacing brake pad with new one, check a brake fluid level in the reservoir tank because brake fluid returns to master cylinder reservoir tank when pressing piston in.

➡ Use a disc brake piston tool to easily press piston.

9. Install the lower sliding pin bolt and tighten it to the specified torque.

10. Depress the brake pedal several times to check that no drag feel is present for the front disc brake.

11. Install tires with power tool.

4 Piston Type
See Figure 6.

✳✳ WARNING
Clean any dust from the brake caliper and brake pads with a vacuum dust collector. Never blow with compressed air.

✳✳ WARNING
Never depress the brake pedal while removing the brake pads because the piston may pop out. If the brake fluid or grease adheres to the brake caliper assembly and disc rotor, quickly wipe it off.

1. Remove tires with power tool.
2. Remove clips from pad pins with suitable pliers.
3. Remove pad pins with suitable pliers, while holding down cross spring, then remove cross spring from caliper.
4. Remove brake pads and shims from caliper with suitable pliers.
   - Never damage the piston boot
   - Never drop the brake pads and shims
   - Remember each position of the removed brake pads
5. Perform inspection after removal.

To install:

✳✳ WARNING
Clean any dust from the brake caliper and brake pads with a vacuum dust collector. Never blow with compressed air.

✳✳ WARNING
Never depress the brake pedal while removing the brake pads because the piston may pop out. If the brake fluid or grease adheres to the brake caliper assembly and disc rotor, quickly wipe it off.

6. Apply copper based brake grease to the matching faces (A) between the brake pad and shim, and install shim to the brake pad.

✳✳ WARNING
Always replace the shims when replacing the brake pad.

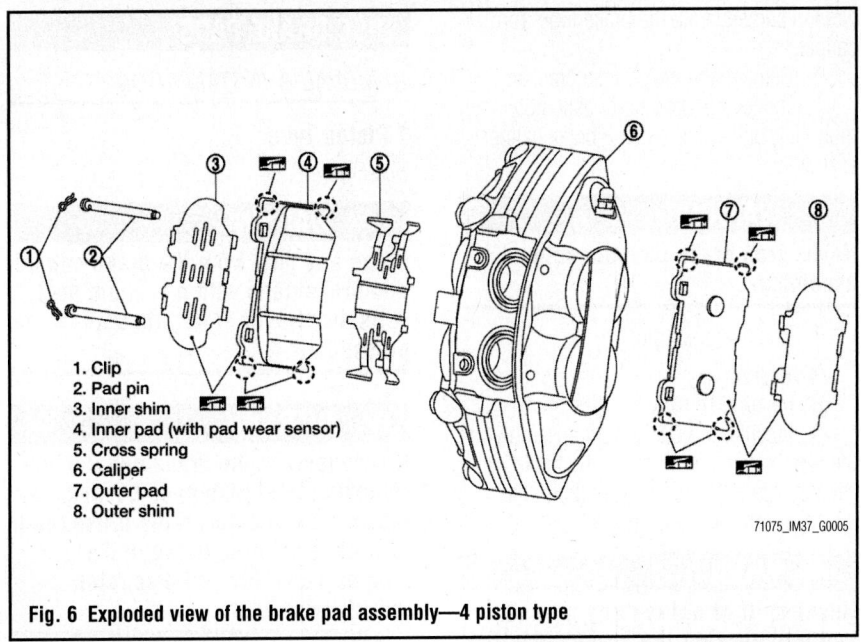

1. Clip
2. Pad pin
3. Inner shim
4. Inner pad (with pad wear sensor)
5. Cross spring
6. Caliper
7. Outer pad
8. Outer shim

71075_IM37_G0005

Fig. 6 Exploded view of the brake pad assembly—4 piston type

7. Apply copper based brake grease to the matching faces (B) between the brake pad and caliper.
8. Install the brake pads to the caliper.

> ✳✳ **WARNING**
>
> Never damage the piston boot. When replacing brake pad with new one,

check a brake fluid level in the reservoir tank because brake fluid returns to master cylinder reservoir tank when pressing piston in.

➡ Use a disc brake piston tool to easily press piston.

9. Install upper pad pin from the inner side, then install firmly to the outer side through the hole in the top of brake pad.
10. Place the top of cross spring (1) over the upper pad pin (2), press in the cross spring, install lower pad pin from the inner side to the outer side, and secure cross spring.
11. Install clips to the pad pins.

> ✳✳ **WARNING**
>
> If clip is not fully attached, pad pin or brake pad could fall out while vehicle is in motion.

12. Depress the brake pedal several times to check that no drag feel is present for the front disc brake.
13. Install tires with power tool.

BRAKES

> ✳✳ **WARNING**
>
> Dust and dirt accumulating on brake parts during normal use may contain asbestos fibers from production or aftermarket brake linings. Breathing excessive concentrations of asbestos fibers can cause serious bodily harm. Exercise care when servicing brake parts. Do not sand or grind brake lining unless equipment used is designed to contain the dust residue. Do not clean brake parts with compressed air or by dry brushing. Cleaning should be done by dampening the brake components with a fine mist of water, then wiping the brake components clean with a dampened cloth. Dispose of cloth and all residue containing asbestos fibers in an impermeable container with the appropriate label. Follow practices prescribed by the Occupational Safety and Health Administration (OSHA) and the Environmental Protection Agency (EPA) for the handling, processing, and disposing of dust or debris that may contain asbestos fibers.

BRAKE CALIPER

REMOVAL & INSTALLATION

1 Piston Type
See Figure 7.

> ✳✳ **WARNING**
>
> Clean any dust from the brake caliper and brake pads with a vacuum dust collector. Never blow with compressed air.

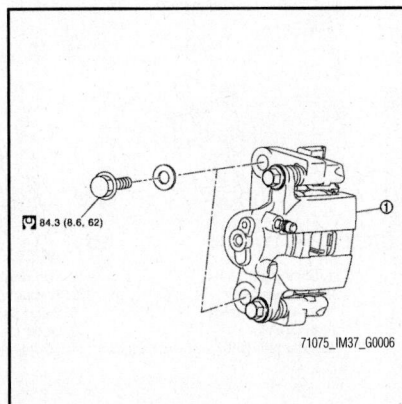

🔧 84.3 (8.6, 62)

71075_IM37_G0006

Fig. 7 Removing the rear caliper (1)—1 piston type

REAR DISC BRAKES

> ✳✳ **WARNING**
>
> Never spill or splash brake fluid on painted surfaces. Brake fluid may seriously damage paint. Wipe it off immediately and wash with water if it gets on a painted surface. Never depress the brake pedal while removing the brake pads because the piston may pop out. If the brake fluid or grease adheres to the brake caliper assembly and disc rotor, quickly wipe it off.

1. Remove tires with power tool.
2. Fix the disc rotor using wheel nuts.
3. Drain brake fluid.
4. Remove union bolt and copper washer, and separate brake hose from caliper assembly.
5. Remove torque member mounting bolts, and remove brake caliper assembly.

> ✳✳ **WARNING**
>
> Never drop brake pad and caliper assembly.

6. Remove disc rotor.

To install:
7. Install disc rotor.
8. Install the brake caliper assembly to the axle housing and tighten the torque member mounting bolts to the specified torque.

✳✳ WARNING

Never spill or splash any grease and moisture on the brake caliper assembly mounting face, threads, mounting bolts and washers. Wipe out any grease and moisture.

9. Install brake hose and copper washers to brake caliper assembly.

10. Refill with new brake fluid and perform the air bleeding.

11. Check a drag of rear disc brake.

12. Install tires with power tool.

2 Piston Type

See Figure 8.

✳✳ WARNING

Clean any dust from the brake caliper and brake pads with a vacuum dust collector. Never blow with compressed air.

✳✳ WARNING

Never spill or splash brake fluid on painted surfaces. Brake fluid may seriously damage paint. Wipe it off immediately and wash with water if it gets on a painted surface. Never depress the brake pedal while removing the brake pads because the piston may pop out. If the brake fluid or grease adheres to the brake caliper assembly and disc rotor, quickly wipe it off.

1. Remove tires with power tool.

2. Fix the disc rotor using wheel nuts.

3. Drain brake fluid.

4. Loosen the flare nut with a flare nut

wrench and separate the brake tube from caliper.

5. Remove the brake hose bracket.

6. Remove brake caliper assembly mounting bolts, and remove brake caliper assembly.

✳✳ WARNING

Never drop brake pad and caliper assembly.

7. Remove disc rotor.

To install:

8. Install disc rotor.

9. Install the brake caliper assembly to the axle housing and tighten the brake caliper assembly mounting bolts to the specified torque.

✳✳ WARNING

Never spill or splash any grease and moisture on the brake caliper assembly mounting face, threads, mounting bolts and washers. Wipe out any grease and moisture.

10. Install brake hose bracket (caliper side).

11. Install brake tube to brake caliper assembly.

12. Refill with new brake fluid and perform the air bleeding.

13. Check a drag of rear disc brake.

14. Install tires with power tool.

DISC BRAKE PADS

REMOVAL & INSTALLATION

1 Piston Type

See Figure 9.

✳✳ WARNING

Clean any dust from the brake caliper and brake pads with a vacuum dust collector. Never blow with compressed air.

✳✳ WARNING

Never depress the brake pedal while removing the brake pads because the piston may pop out. If the brake fluid or grease adheres to the brake caliper assembly and disc rotor, quickly wipe it off.

1. Remove tires with power tool.

2. Remove upper sliding pin bolt.

3. Suspend the cylinder body with suitable wire so that the brake hose will not stretch. Then remove the brake pads from the torque member.

✳✳ WARNING

Never deform the pad retainer when removing the pad retainer from the torque member.

- Never damage the piston boot.
- Never drop the brake pads, shims and shim covers.

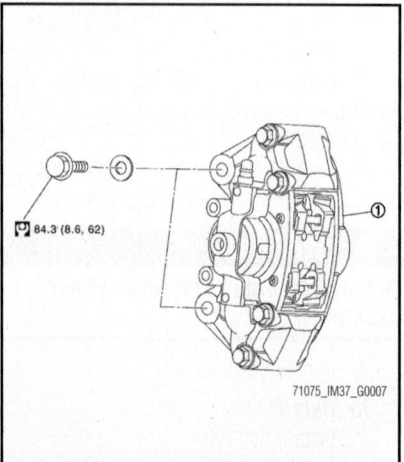

84.3 (8.6, 62)

71075_IM37_G0007

Fig. 8 Removing the rear caliper (1)—2 piston type

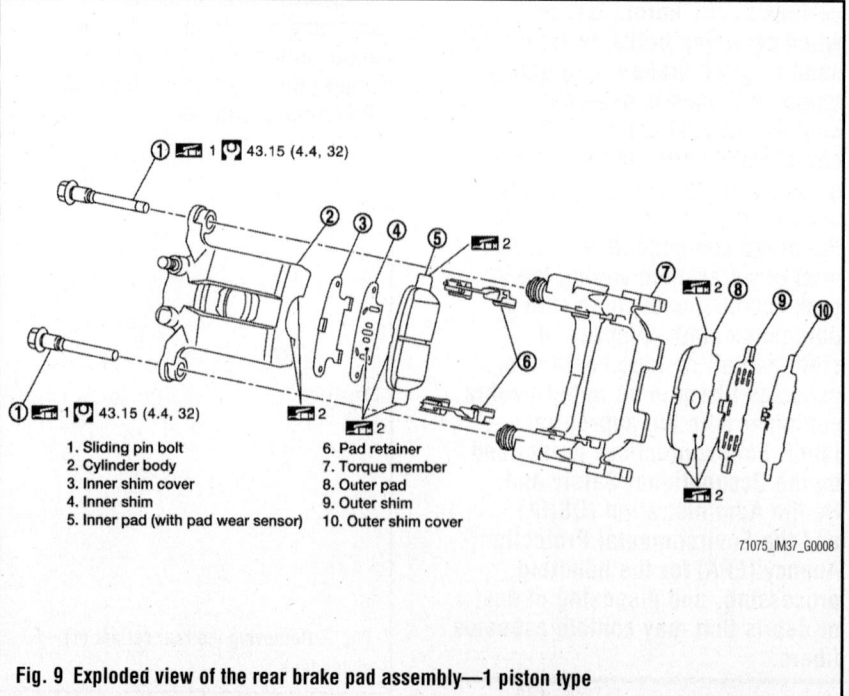

① ⟦ 1 ⟧ 43.15 (4.4, 32)

① ⟦ 1 ⟧ 43.15 (4.4, 32)

1. Sliding pin bolt
2. Cylinder body
3. Inner shim cover
4. Inner shim
5. Inner pad (with pad wear sensor)
6. Pad retainer
7. Torque member
8. Outer pad
9. Outer shim
10. Outer shim cover

71075_IM37_G0008

Fig. 9 Exploded view of the rear brake pad assembly—1 piston type

- Remember each position of the removed brake pads.
4. Perform inspection after removal.

To install:

> ✳✳ **WARNING**
>
> **Clean any dust from the brake caliper and brake pads with a vacuum dust collector. Never blow with compressed air.**

> ✳✳ **WARNING**
>
> **Never depress the brake pedal while removing the brake pads because the piston may pop out. If the brake fluid or grease adheres to the brake caliper assembly and disc rotor, quickly wipe it off.**

5. Install the torque member if the pad retainers has been removed.

> ✳✳ **WARNING**
>
> **Securely assemble the pad retainers so that it will not be lifted up from the torque member. Never deform the pad retainers.**

6. Apply PBC (Poly Butyl Cuprysil) grease or silicone-based grease to the matching faces (A) between the brake pad and shim, and install shim and shim cover to the brake pad.

> ✳✳ **WARNING**
>
> **Always replace the shims and shim covers when replacing the brake pad.**

7. Apply PBC (Poly Butyl Cuprysil) grease or silicone-based grease to the matching faces between the brake pad and pad retainer, and install brake pad to the torque member.
8. Apply PBC (Poly Butyl Cuprysil) grease or silicone-based grease to the pawls part of cylinder body, and install cylinder body to the torque member.

> ✳✳ **WARNING**
>
> **Never damage the piston boot. When replacing brake pad with new one, check a brake fluid level in the reservoir tank because brake fluid returns to master cylinder reservoir tank when pressing piston in.**

➡ Use a disc brake piston tool to easily press piston.

9. Apply rubber grease to the sliding pin bolt, install the upper sliding pin bolt and tighten it to the specified torque.
10. Depress the brake pedal several times to check that no drag feel is present for the front disc brake.
11. Install tires with power tool.

2 Piston Type

See Figures 10 and 11.

> ✳✳ **WARNING**
>
> **Clean any dust from the brake caliper and brake pads with a vacuum dust collector. Never blow with compressed air.**

> ✳✳ **WARNING**
>
> **Never depress the brake pedal while removing the brake pads because the piston may pop out. If the brake fluid or grease adheres to the brake caliper assembly and disc rotor, quickly wipe it off.**

1. Remove tires with power tool.
2. Remove clips from pad pins with suitable pliers.
3. Remove pad pins with suitable pliers, while holding down cross spring, then remove cross spring from caliper.
4. Remove brake pads, shims and shim covers from caliper with suitable pliers.

> ✳✳ **WARNING**
>
> **Never deform the pad retainer when removing the pad retainer from the torque member.**

- Never damage the piston boot.
- Never drop the brake pads, shims and shim covers.
- Remember each position of the removed brake pads.
5. Perform the inspection after removal.

> ✳✳ **WARNING**
>
> **Clean any dust from the brake caliper and brake pads with a vacuum dust collector. Never blow with compressed air.**

> ✳✳ **WARNING**
>
> **Never depress the brake pedal while removing the brake pads because the piston may pop out. If the brake fluid or grease adheres to the brake caliper assembly and disc rotor, quickly wipe it off.**

6. Apply copper based brake grease to the matching faces between the shim and shim cover, and install shim and shim cover to the brake pad.

> ✳✳ **WARNING**
>
> **Always replace the shims and shim covers when replacing the brake pad.**

7. Apply copper based brake grease to the following matching faces:
   a. Between the brake pad and caliper.
   b. Between the brake pad and pad pin.

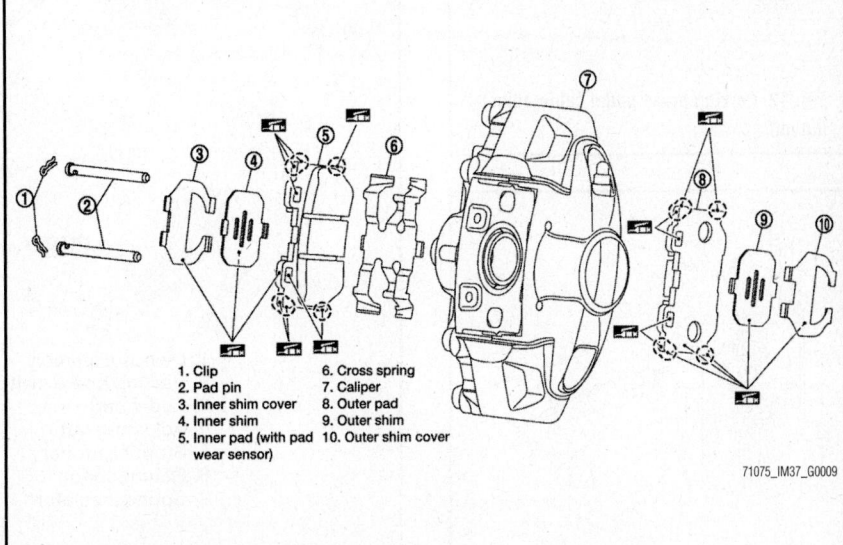

1. Clip
2. Pad pin
3. Inner shim cover
4. Inner shim
5. Inner pad (with pad wear sensor)
6. Cross spring
7. Caliper
8. Outer pad
9. Outer shim
10. Outer shim cover

71075_IM37_G0009

Fig. 10 Exploded view of the rear brake pad assembly—2 piston type

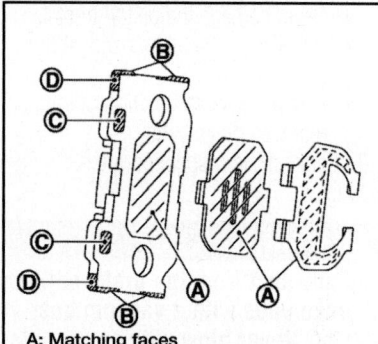

A: Matching faces
B: Between the brake pad and caliper
C: Between the brake pad and pad pin
D: Between the brake pad and cross spring

71075_IM37_G0010

Fig. 11 Locating grease areas for application

c. Between the brake pad and cross spring.

8. Install the brake pads to the caliper.

※※ WARNING

Never damage the piston boot. When replacing brake pad with new one, check a brake fluid level in the reservoir tank because brake fluid returns to master cylinder reservoir tank when pressing piston in.

➡**Use a disc brake piston tool to easily press piston.**

9. Install upper pad pin from the inner side, then install firmly to the outer side through the hole in the top of brake pad.

10. Place the top of cross spring over the upper pad pin, press in the cross spring, install lower pad pin from the inner side to the outer side, and secure cross spring.

a. Install clips to the pad pins.

※※ WARNING

If clip is not fully attached, pad pin or brake pad could fall out while vehicle is in motion.

11. Depress the brake pedal several times to check that no drag feel is present for the front disc brake.

12. Install tires with power tool.

BRAKES   PARKING BRAKE

PARKING BRAKE CABLES

ADJUSTMENT

See Figures 12 through 14.

1. To perform adjustment operations, remove rear tires from vehicle.

2. Insert a deep socket wrench onto adjusting nut.

3. Rotate adjusting nut to fully loosen cable, and then release parking brake pedal.

4. Secure disc rotor to hub using wheel nut so as not to tilt disc rotor.

5. Remove adjuster hole plug installed on the disc rotor.

6. Turn the adjuster in direction "A"

using a flat-bladed screwdriver as shown, until disc rotor is locked.

7. Turn the adjuster in the opposite direction by 5 or 6 notches after locking.

8. Rotate disc rotor to make sure that there is no drag. Install the adjuster hole plug.

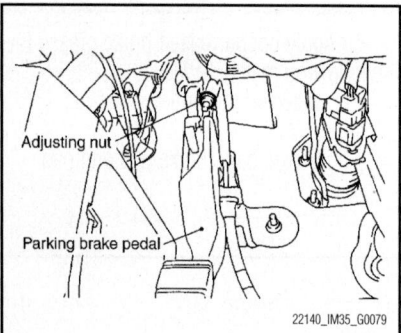

22140_IM35_G0079

Fig. 12 Parking brake pedal cable adjusting nut

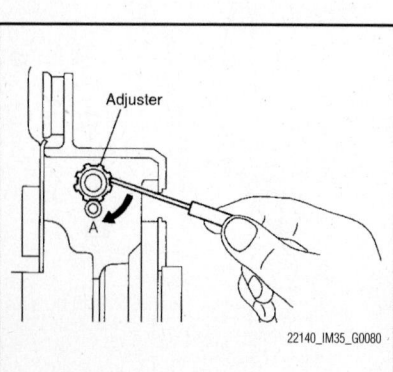

22140_IM35_G0080

Fig. 13 Parking brake pedal cable adjustment

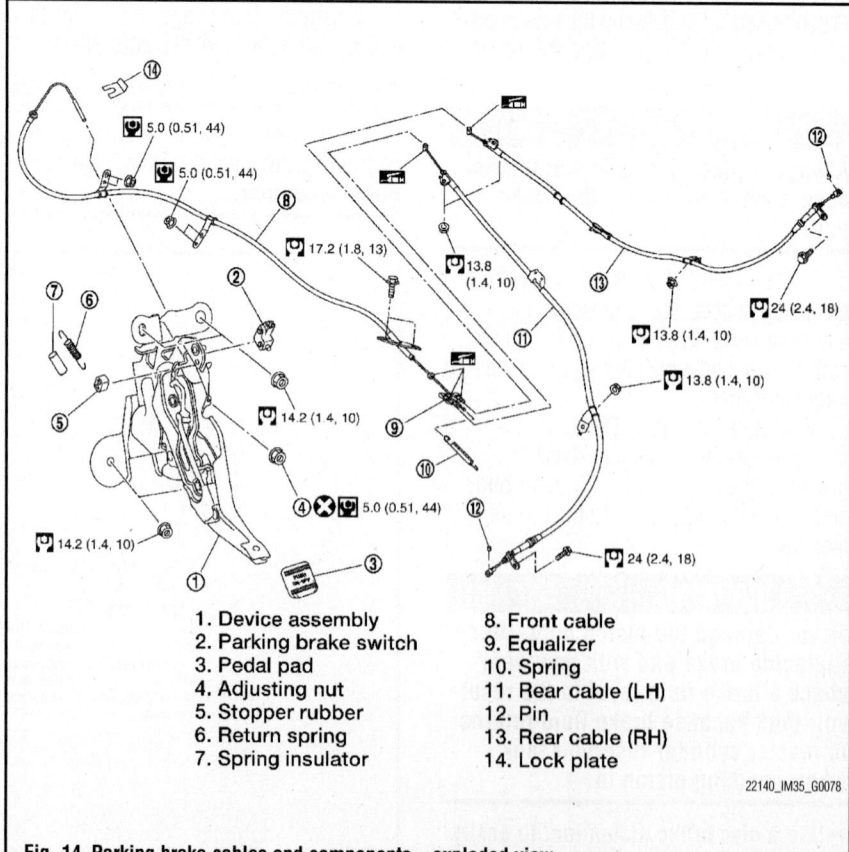

1. Device assembly
2. Parking brake switch
3. Pedal pad
4. Adjusting nut
5. Stopper rubber
6. Return spring
7. Spring insulator
8. Front cable
9. Equalizer
10. Spring
11. Rear cable (LH)
12. Pin
13. Rear cable (RH)
14. Lock plate

22140_IM35_G0078

Fig. 14 Parking brake cables and components—exploded view

9. Adjust parking brake cable with the following procedure.

- Operate parking brake pedal 10 or more times with the force of 110 ft. lbs. (150 Nm).
- Rotate adjusting nut to adjust parking brake pedal stroke using a deep socket wrench.

➡️**Do not reuse adjusting nut after removing it.**

- Operate parking brake pedal with a force of 44 lbs. (60 N), make sure the pedal stroke is within the specified number of notches. (Check it by listening and counting ratchet clicks.)

➡️**Pedal stroke 3–4 notches**

➡️**Make sure that there is no drag on rear brake with parking brake pedal completely released.**

PARKING BRAKE SHOES

REMOVAL & INSTALLATION

See Figures 15 and 16.

➡️**Clean brakes with a vacuum dust collector to minimize the hazard of air borne particles or other materials.**

✳✳ WARNING

Clean dust on disc rotor and back plate using a vacuum dust collector. Do not blow with compressed air.

➡️**Put matching marks on both disc rotor and wheel hub when removing disc rotor.**

1. Remove rear tires from vehicle.
2. Remove disc rotor with parking brake pedal completely in the released position See "Removal and Installation of Brake Caliper Assembly" in this section.
3. Remove the parking brake shoes (refer to the illustration).

To install:
4. Installation is the reverse of the removal procedure.

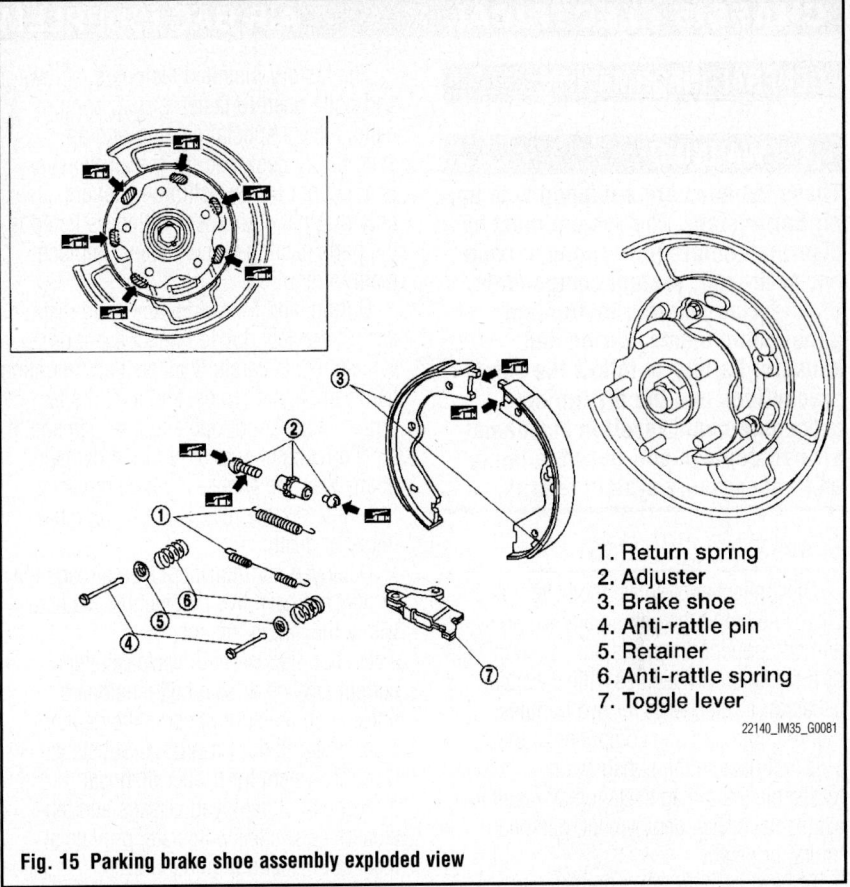

1. Return spring
2. Adjuster
3. Brake shoe
4. Anti-rattle pin
5. Retainer
6. Anti-rattle spring
7. Toggle lever

22140_IM35_G0081

Fig. 15 Parking brake shoe assembly exploded view

➡️**The orientation of the adjuster is different for the left and right sides. Refer to the illustration for proper installation.**

5. Assemble adjusters so that threaded part is expanded when rotating it in the direction shown by arrow.
6. Shorten adjuster by rotating it.
7. Check shoe sliding surface and drum inner surface for grease.
8. Wipe it off if it adheres on the surfaces.
9. Adjust the parking brake.

ADJUSTMENT

See Parking brake cable adjustment in this section.

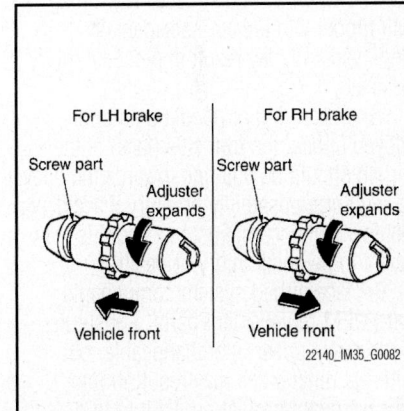

For LH brake   For RH brake
Screw part   Screw part
Adjuster   Adjuster
expands   expands
Vehicle front   Vehicle front

22140_IM35_G0082

Fig. 16 Parking brake shoe adjuster orientation

GENERAL INFORMATION

❋❋ WARNING

These vehicles are equipped with an air bag system. The system must be disarmed before performing service on, or around, system components, the steering column, instrument panel components, wiring and sensors. Failure to follow the safety precautions and the disarming procedure could result in accidental air bag deployment, possible injury and unnecessary system repairs.

SERVICE PRECAUTIONS

Disconnect and isolate the battery negative cable before beginning any airbag system component diagnosis, testing, removal, or installation procedures. Allow system capacitor to discharge for two minutes before beginning any component service. This will disable the airbag system. Failure to disable the airbag system may result in accidental airbag deployment, personal injury, or death.

Do not place an intact undeployed airbag face down on a solid surface. The airbag will propel into the air if accidentally deployed and may result in personal injury or death.

When carrying or handling an undeployed airbag, the trim side (face) of the airbag should be pointing towards the body to minimize possibility of injury if accidental deployment occurs. Failure to do this may result in personal injury or death.

Replace airbag system components with OEM replacement parts. Substitute parts may appear interchangeable, but internal differences may result in inferior occupant protection. Failure to do so may result in occupant personal injury or death.

Wear safety glasses, rubber gloves, and long sleeved clothing when cleaning powder residue from vehicle after an airbag deployment. Powder residue emitted from a deployed airbag can cause skin irritation. Flush affected area with cool water if irritation is experienced. If nasal or throat irritation is experienced, exit the vehicle for fresh air until the irritation ceases. If irritation continues, see a physician.

Do not use a replacement airbag that is not in the original packaging. This may result in improper deployment, personal injury, or death.

The factory installed fasteners, screws and bolts used to fasten airbag components have a special coating and are specifically designed for the airbag system. Do not use substitute fasteners. Use only original equipment fasteners listed in the parts catalog when fastener replacement is required.

During, and following, any child restraint anchor service, due to impact event or vehicle repair, carefully inspect all mounting hardware, tether straps, and anchors for proper installation, operation, or damage. If a child restraint anchor is found damaged in any way, the anchor must be replaced. Failure to do this may result in personal injury or death.

Deployed and non-deployed airbags may or may not have live pyrotechnic material within the airbag inflator.

Do not dispose of driver/passenger/ curtain airbags or seat belt tensioners unless you are sure of complete deployment. Refer to the Hazardous Substance Control System for proper disposal.

Dispose of deployed airbags and tensioners consistent with state, provincial, local, and federal regulations.

After any airbag component testing or service, do not connect the battery negative cable. Personal injury or death may result if the system test is not performed first.

If the vehicle is equipped with the Occupant Classification System (OCS), do not connect the battery negative cable before performing the OCS Verification Test using the scan tool and the appropriate diagnostic information. Personal injury or death may result if the system test is not performed properly.

Never replace both the Occupant Restraint Controller (ORC) and the Occupant Classification Module (OCM) at the same time. If both require replacement, replace one, then perform the Airbag System test before replacing the other.

Both the ORC and the OCM store Occupant Classification System (OCS) calibration data, which they transfer to one another when one of them is replaced. If both are replaced at the same time, an irreversible fault will be set in both modules and the OCS may malfunction and cause personal injury or death.

If equipped with OCS, the Seat Weight Sensor is a sensitive, calibrated unit and must be handled carefully. Do not drop or handle roughly. If dropped or damaged, replace with another sensor. Failure to do so may result in occupant injury or death.

If equipped with OCS, the front passenger seat must be handled carefully as well. When removing the seat, be careful when setting on floor not to drop. If dropped, the sensor may be inoperative, could result in occupant injury, or possibly death.

If equipped with OCS, when the passenger front seat is on the floor, no one should sit in the front passenger seat. This uneven force may damage the sensing ability of the seat weight sensors. If sat on and damaged, the sensor may be inoperative, could result in occupant injury, or possibly death.

DISARMING THE SYSTEM

All Air Bag electrical wiring harnesses and connectors are covered with **YELLOW** outer insulation. Do not use electrical test equipment on any circuit related to the Air Bag sensors. When installing Air Bag components, always install with the arrow marks facing the front of the vehicle.

1. Before servicing the vehicle, refer to the precautions in the beginning of this section.
2. Turn the ignition switch to the **OFF** position.
3. Disconnect both battery cables starting with the negative cable first and wait at least 3 minutes after the cables are disconnected. Be sure to insulate the battery terminal ends.

ARMING THE SYSTEM

1. Before servicing the vehicle, refer to the precautions in the beginning of this section.
2. Turn the ignition switch to the **OFF** position.
3. Connect both battery cables starting with the positive cable first.
4. The Air Bag or Air Bag system is equipped with a self-diagnostic operation. After turning the ignition key to the **ON** or **START** position, the **AIR BAG** warning lamp will illuminate for 7 seconds. After 7 seconds, the **AIR BAG** lamp will extinguish if no malfunction is detected. If the **AIR BAG** lamp does not extinguish after 7 seconds, check the Air Bag self-diagnostic system for a malfunction.

CLOCKSPRING CENTERING

See Figure 17.

❋❋ WARNING

The spiral cable may snap by steering operation if the cable is installed in an improper position.

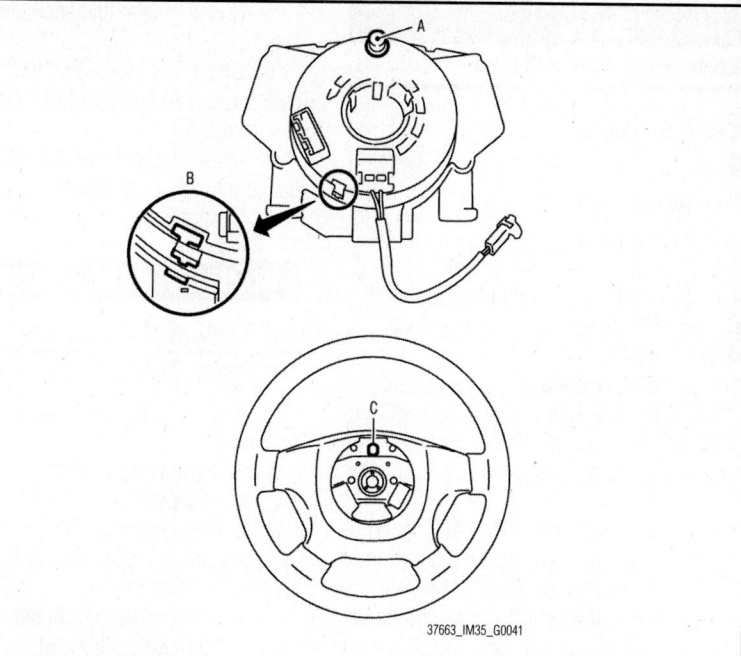

Fig. 17 Adjust the spiral cable locating pin (shown as A) to the steering wheel locating pin hole (shown as C)

The neutral position is set as follows.

1. Turn quietly the spiral cable clockwise to the end position.

2. Then turn it counterclockwise (about 2 and half turns) and stop turning at the point on which the stopper insertion holes are in the same position.

3. The service part is installed in the neutral position by the stopper and can be set without adjusting after the stopper is removed.

> ※※ **WARNING**
>
> **Never turn the spiral cable rashly and also beyond the limit number of turns. (These will cause cable snap.)**

4. Adjust the spiral cable locating pin to the steering wheel locating pin hole.

5. Secure the air bag harness with the harness attaching hook.

DRIVE TRAIN

AUTOMATIC TRANSMISSION FLUID

DRAIN AND REFILL

With 3.7L Engine

See Figure 18.

➡If Genuine NISSAN Matic S ATF is not available, Genuine NISSAN Matic J ATF may also be used.

> ※※ **WARNING**
>
> **Using ATF other than Genuine NISSAN Matic S ATF will cause deterioration in driveability and A/T durability, and may damage the A/T, which is not covered by the INFINITI new vehicle limited warranty.**

➡When filling ATF, be careful not to scatter heat generating parts such as exhaust.

1. Step 1
   a. Install the O-ring (315268E000) to the charging pipe (310811EA5A).
2. Step 2
   a. Use CONSULT-III to check that the ATF temperature is 104° F (40° C) or less.
   b. Raise the vehicle.
   c. Remove the drain plug from the oil pan, and then drain the ATF.

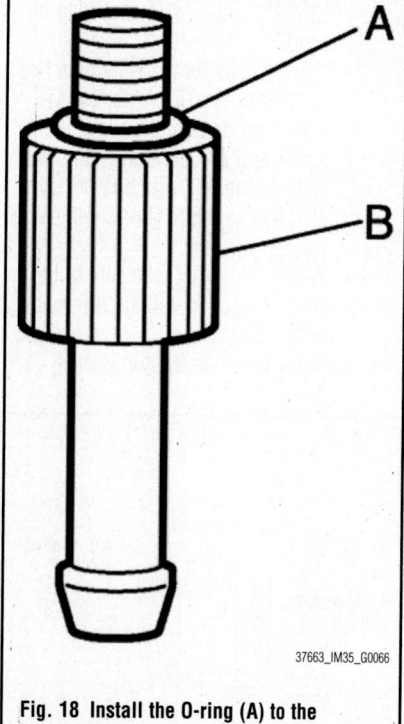

Fig. 18 Install the O-ring (A) to the charging pipe (B)

d. When the ATF starts to drip, temporarily tighten the drain plug to the oil pan.

➡Never replace drain plug and drain plug gasket with new ones yet.

e. Remove overflow plug from oil pan.
   f. Install the charging pipe to the overflow plug hole.

> ※※ **WARNING**
>
> **Tighten the charging pipe by hand.**

g. Install the bucket pump hose to the charging pipe.

> ※※ **WARNING**
>
> **Insert the bucket pump hose all the way to the end of the charging pipe.**

h. Fill approximately 3⅛ qts. (3 liters) of the ATF.
   i. Remove the bucket pump hose to remove the charging pipe, and then temporarily tighten the overflow plug to the oil pan.

> ※※ **WARNING**
>
> **Quickly perform the procedure to avoid ATF leakage from the oil pan.**

j. Lower the vehicle.
   k. Start the engine and wait for approximately 3 minutes.
   l. Stop the engine.
3. Step 3
   a. Repeat "Step 2".
4. Final Step

a. Use CONSULT-III to check that the ATF temperature is 104° F (40° C) or less.

b. Raise the vehicle.

c. Remove the drain plug from the oil pan, and then drain the ATF.

5. When the ATF starts to drip, tighten the drain plug to the oil pan to the specified torque.

✳✳ WARNING

Never reuse drain plug and drain plug gasket.

a. Remove overflow plug from oil pan.

b. Install the charging pipe to the overflow plug hole.

✳✳ WARNING

Tighten the charging pipe by hand.

c. Install the bucket pump hose to the charging pipe.

✳✳ WARNING

Insert the bucket pump hose all the way to the end of the charging pipe.

d. Fill approximately 3⅛ qts. (3 liters) of the ATF.

e. Remove the bucket pump hose to remove the charging pipe, and then temporarily tighten the overflow plug to the oil pan.

✳✳ WARNING

Quickly perform the procedure to avoid ATF leakage from the oil pan.

f. Lower the vehicle.

g. Start the engine.

6. Make the ATF temperature approximately 104° F (40° C).

➥**The ATF level is greatly affected by the temperature. Always check the ATF temperature on "ATF TEMP 1" of "Data Monitor" using CONSULT-III.**

a. Park vehicle on level surface and set parking brake.

b. Shift the selector lever through each gear position. Leave selector lever in "P" position.

c. Raise the vehicle when the ATF temperature reaches 104° F (40° C), and then remove the overflow plug from the oil pan.

d. When the ATF starts to drip, tighten the overflow plug to the oil pan to the specified torque.

✳✳ WARNING

Never reuse overflow plug.

With 5.6L Engine

See Figure 19.

1. Warm up ATF.

2. Stop engine.

3. Loosen the level gauge bolt.

4. Drain ATF from drain plug and refill with new ATF. Always refill same volume with drained ATF.

a. To replace the ATF, pour in new ATF at the A/T fluid charging pipe with the engine idling and at the same time drain the old ATF from the radiator cooler hose return side.

b. When the color of the ATF coming out is about the same as the color of the new ATF, the replacement is complete. The amount of new ATF to use should be 30 to 50% increase of the stipulated amount.

✳✳ WARNING

- If Genuine NISSAN Matic S ATF is not available, Genuine NISSAN Matic J ATF may also be used.
- Using ATF other than Genuine NISSAN Matic S ATF or Matic J ATF will cause deterioration in driveability and A/T durability, and may damage the A/T, which is not covered by the INFINITI new vehicle limited warranty.
- When filling ATF, take care not to scatter heat generating parts such as exhaust.
- Never reuse drain plug gasket.

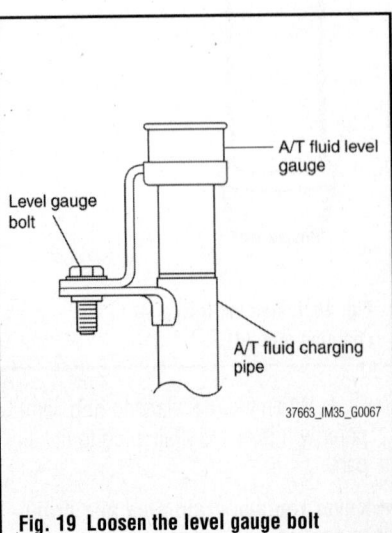

Fig. 19 Loosen the level gauge bolt

5. Run engine at idle speed for 5 minutes.

6. Check A/T fluid level and condition. If ATF is still dirty, repeat step 2. through 5.

7. Install the removed A/T fluid level gauge into A/T fluid charging pipe.

8. Tighten the level gauge bolt.

TRANSFER CASE ASSEMBLY

REMOVAL & INSTALLATION

See Figures 20 through 22.

1. For 3.7L Engines, perform the following:

a. Remove rear driveshaft.

b. Remove front driveshaft

c. Disconnect AWD solenoid harness connector and separate harness from transfer assembly.

d. Remove transfer breather hose.

e. Remove control rod.

f. Transfer assembly and transmission assembly with a jack.

✳✳ WARNING

Secure transfer assembly and transmission assembly to a jack.

g. Remove rear engine mounting member and engine mounting insulator with power tool.

h. Lower jack to the position where the top transfer mounting bolts can be removed.

i. Remove transfer mounting bolts with power tool and separate transfer from transmission.

2. For 5.6L engines, perform the following:

a. Remove transmission assembly from the vehicle.

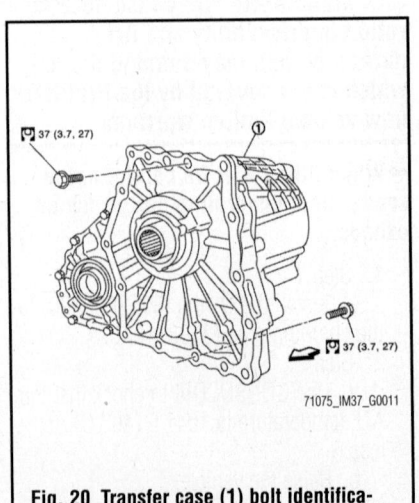

Fig. 20 Transfer case (1) bolt identification and torque value

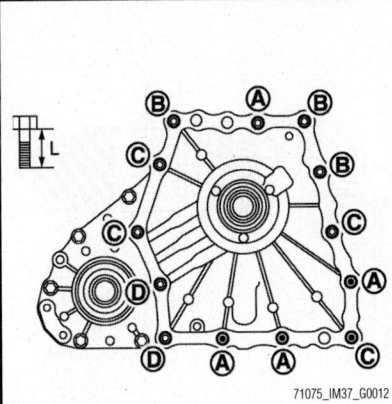

Fig. 21 Bolt installation pattern for the transfer case to transmission

b. Remove transfer air breather hose.

c. Remove rear engine mounting member and engine mounting insulator with power tool.

d. Support transfer assembly with a jack.

e. Remove transfer mounting bolts with power tool and separate transfer from transmission.

✳✳ WARNING

Secure the transfer assembly and transmission assembly to a jack.

To install:

3. Note the following, and install in the reverse order of removal.

a. When installing the transfer to the transmission, install the mounting bolts and tighten bolts to the specified torque.

4. When installing transfer breather hose, make sure there are no pinched or restricted areas on the transfer breather hose caused by bending or winding.

a. Set transfer breather hose of transmission side with the paint mark facing upward, and insert breather hose to

breather tube until hose end reaches the tube bend R portion.

b. Be sure to insert air breather hose of transfer side to air breather tube until hose end reaches the tube bend R portion.

Be sure to fix air breather hose in parts of transmission and transfer.

5. After the installation, check the fluid level, fluid leakage and the A/T positions.

FRONT DRIVESHAFT

REMOVAL & INSTALLATION

See Figures 23 and 24.

1. Remove engine undercover.
2. If necessary, remove heat bracket.
3. Remove the three way catalyst (right bank).
4. Put matching marks onto propeller shaft flange yoke and final drive companion flange.

✳✳ WARNING

For matching marks, use paint. Never damage propeller shaft flange and companion flange.

5. Remove the propeller shaft mounting bolts.

6. Remove propeller shaft from the front final drive and transfer.

To install:

7. Note the following, install in the reverse order of removal.

8. Align matching marks to install propeller shaft to final drive companion flange, and then tighten to specified torque.

✳✳ WARNING

Never reuse the bolts.

✳✳ WARNING

Never damage the transfer front oil seal.

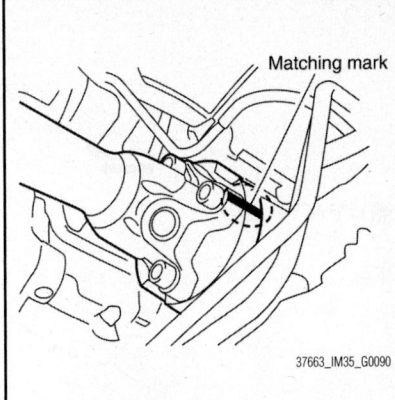

Fig. 23 Put matching marks onto propeller shaft flange yoke and final drive companion flange

- Wrap the power steering piping interference area with shop cloth or equivalent to protect power steering piping from breakage.
- Wrap the transmission interference area with shop cloth or equivalent to protect propeller shaft from breakage.
- Never reuse O-ring.
- Apply multi-purpose grease onto O-ring.

9. After assembly, perform a driving test to check propeller shaft vibration. If vibration occurred, separate propeller shaft from final drive or transfer. Reinstall companion flange after rotating it by 90, 180, 270 degrees. Then perform driving test and check propeller shaft vibration again at each point.

FRONT HALFSHAFT

REMOVAL & INSTALLATION

See Figures 25 through 27.

1. Remove the tires and wheels from vehicle.
2. Remove wheel sensor from steering knuckle.

✳✳ WARNING

Do not pull on wheel sensor harness.

3. Remove brake hose bracket.
4. Remove torque member mounting bolts. Hang torque member in a place where it will not interfere with work.

✳✳ WARNING

Do not depress brake pedal while brake caliper is removed.

5. Remove disc rotor.

Bolt No.	A	B	C	D
Quantity	4	3	4	2
Bolt length " L " mm (in)	75 (2.95)	45 (1.77)	40 (1.57)	30 (1.18)

●: Transfer to transmission.

◯: Transmission to transfer.

71075_IM37_G0013

Fig. 22 Bolt installation chart for transfer case assemblies

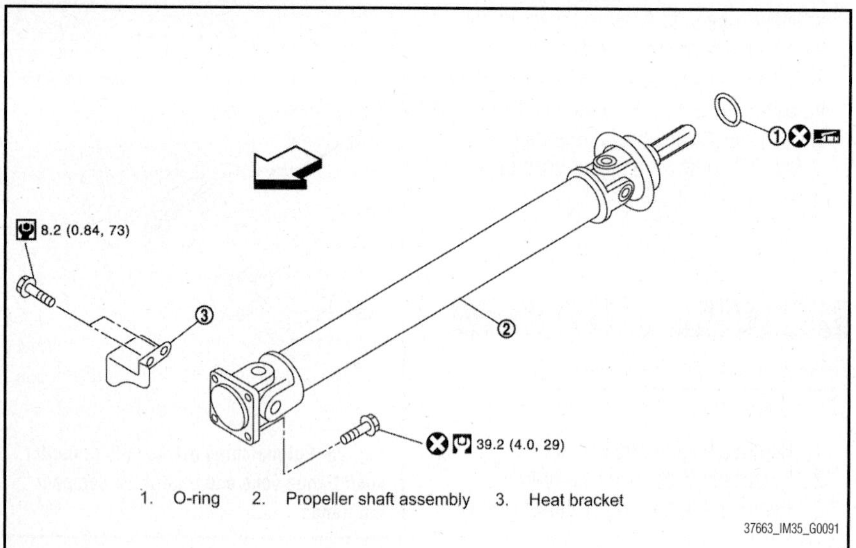
8.2 (0.84, 73)
39.2 (4.0, 29)
1. O-ring 2. Propeller shaft assembly 3. Heat bracket
37663_IM35_G0091

Fig. 24 Exploded view of front propeller shaft

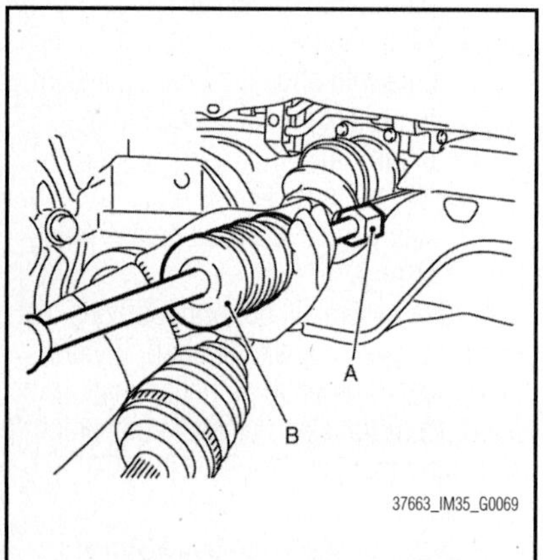
A
B
37663_IM35_G0069

Fig. 25 Remove drive shaft from front final drive using the drive shaft attachment (A) and a sliding hammer (B)

6. Remove cotter pin, then loosen hub lock nut.

7. Separate wheel hub and bearing assembly from drive shaft by lightly tapping the end with a hammer and a wood block, and then remove hub lock nut.

WARNING

Do not place drive shaft joint at an extreme angle. Also be careful not to overextend slide joint. Do not allow drive shaft to hang down without support for housing (or joint sub-assembly), shaft and the other parts.

➡Use a puller if wheel hub and drive shaft cannot be separated even after performing the above procedure.

8. Remove cotter pin, and then loosen the nut.

9. Remove steering outer socket from steering knuckle so as not to damage ball joint boot using the ball joint remover.

WARNING

Temporarily tighten the nut to prevent damage to threads and to prevent the ball joint remover from suddenly coming off.

10. Remove drive shaft from wheel hub and bearing assembly.

11. Remove mounting nuts and bolts, and then remove shock absorber arm.

12. Remove drive shaft from front final drive. (Right side)

a. Remove drive shaft from front final drive using the drive shaft attachment KV40107500 and a sliding hammer while inserting tip of the drive shaft attachment between housing and front final drive.

WARNING

Never place drive shaft joint at an extreme angle when removing drive shaft. Also be careful not to overextend slide joint.

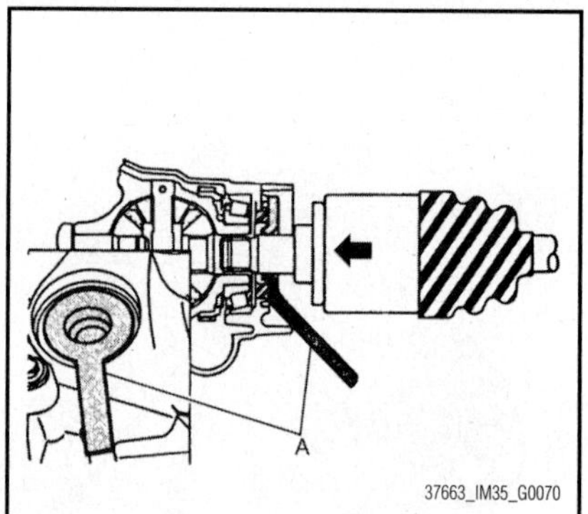
A
37663_IM35_G0070

Fig. 27 Place the protector (A) onto front final drive to prevent damage to the oil seal while inserting drive shaft

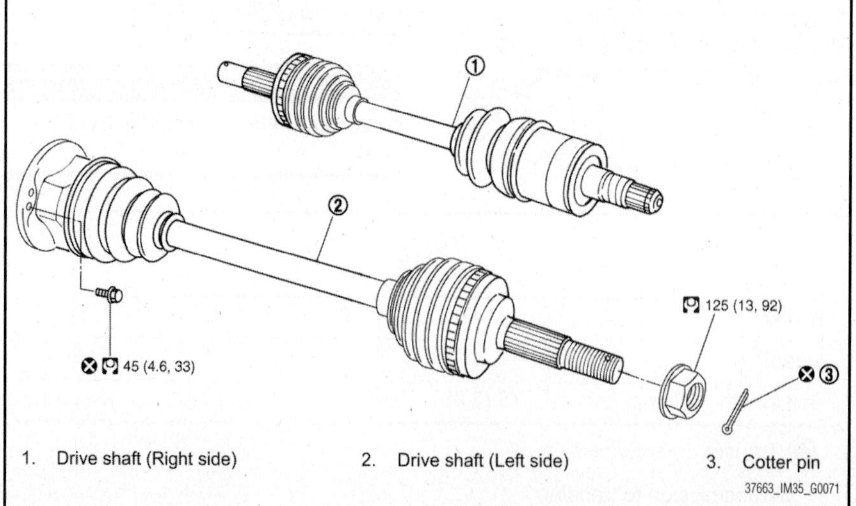
45 (4.6, 33)
125 (13, 92)
1. Drive shaft (Right side) 2. Drive shaft (Left side) 3. Cotter pin
37663_IM35_G0071

Fig. 26 Left and right front drive shafts

13. Remove mounting nuts and bolts, and then remove drive shaft from vehicle. (Left side)

To install:

> ❋❋ **WARNING**
>
> **Always replace transaxle side oil seal with new one when installing drive shaft.**

14. Installation is the reverse order of removal.

15. Place the protector KV38107900 onto front final drive to prevent damage to the oil seal while inserting drive shaft.

16. Slide drive shaft sliding joint and tap with a hammer to install securely. (Right side)

REAR AXLE FLUID

DRAIN & REFILL

Draining

See Figure 28.

1. Stop engine.
2. Remove drain plug and drain gear oil.
3. Set a gasket on drain plug and install it to final drive assembly and tighten to the specified torque.

> ❋❋ **WARNING**
>
> **Never reuse gasket.**

Refilling

See Figure 29.

1. Remove filler plug. Fill with new gear oil until oil level reaches the level of the filler plug mounting hole.

2. After refilling oil, check oil level. Set a gasket to filler plug, then install it to final drive assembly.

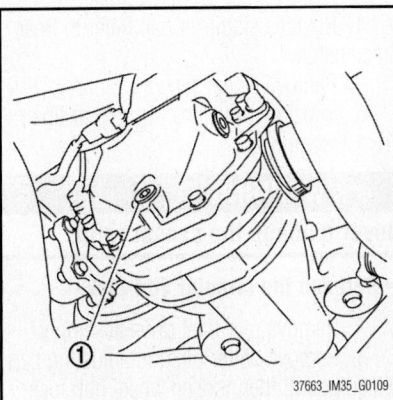

Fig. 28 Remove drain plug (1) and drain gear oil

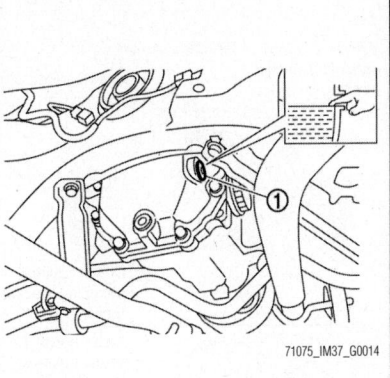

71075_IM37_G0014

Fig. 29 Remove filler plug (1) and fill with new gear oil

> ❋❋ **WARNING**
>
> **Never reuse gasket.**

REAR DRIVESHAFT

REMOVAL & INSTALLATION

See Figures 30 through 34.

1. Move the A/T select lever to N position and release the parking brake.
2. Remove the floor reinforcement.
3. Remove the center muffler.
4. Matching marks:
   a. For 3.7L engine RWD models, put matching marks on propeller shaft rebro joint with final drive companion flange.

> ❋❋ **WARNING**
>
> **For matching marks, use paint. Never damage propeller shaft rebro joint and companion flange.**

   b. For 5.6L engine RWD models, put matching marks on propeller shaft rubber coupling with transmission companion flange and on rebro joint with final drive companion flange.
   c. For AWD models, put matching marks on propeller shaft flange yoke with transfer companion flange and on rebro joint with final drive companion flange.

> ❋❋ **WARNING**
>
> **For matching marks, use paint. Never damage propeller shaft flange yoke, rebro joint and companion flanges.**

5. Loosen, but do not remove mounting nuts of center bearing mounting brackets.
6. Remove propeller shaft mounting bolts and nuts.
7. Remove center bearing mounting bracket mounting nuts.
8. Remove propeller shaft.

> ❋❋ **WARNING**
>
> **If constant velocity joint was bent during propeller shaft assembly removal, installation, or transportation, its boot may be damaged. Wrap boot interference area to metal part with shop cloth or rubber to protect boot from breakage.**

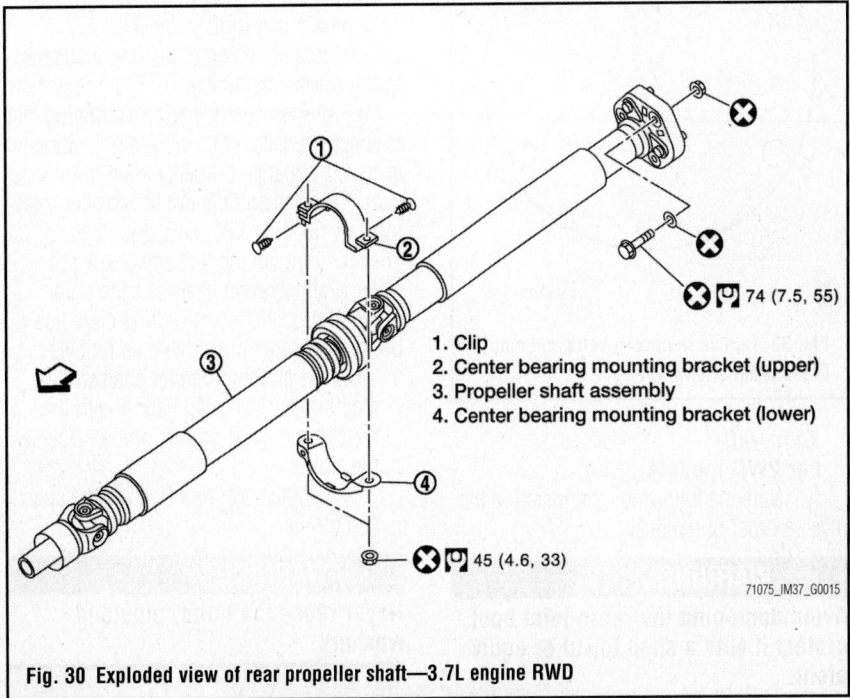

1. Clip
2. Center bearing mounting bracket (upper)
3. Propeller shaft assembly
4. Center bearing mounting bracket (lower)

74 (7.5, 55)

45 (4.6, 33)

71075_IM37_G0015

Fig. 30 Exploded view of rear propeller shaft—3.7L engine RWD

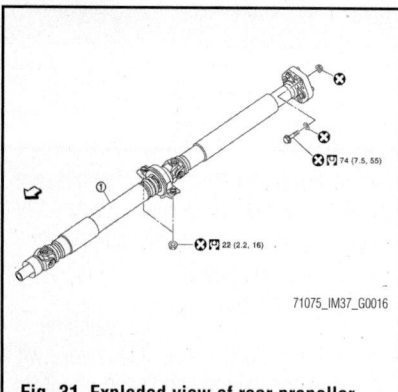

Fig. 31 Exploded view of rear propeller (1) shaft—5.6L engine RWD

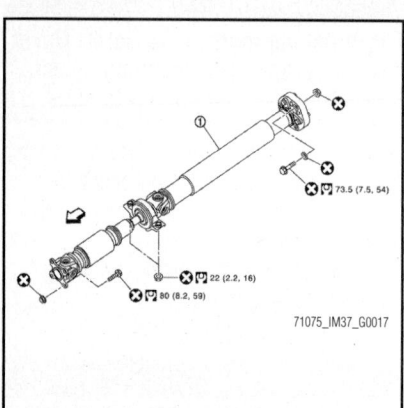

Fig. 32 Exploded view of rear propeller shaft (1)—3.7L and 5.6L engine AWD

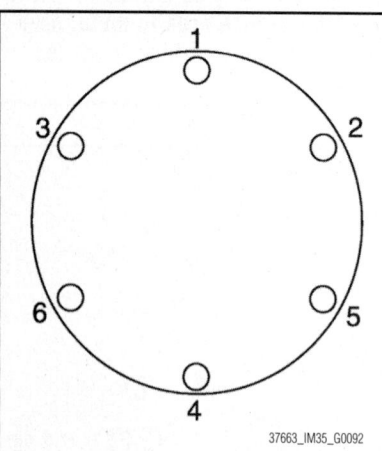

Fig. 33 Tighten mounting bolts and nuts in the order shown

To install:
For 2WD models

9. Note the following, and install in the reverse order of removal.

※ WARNING

Avoid damaging the rebro joint boot, protect it with a shop towel or equivalent.

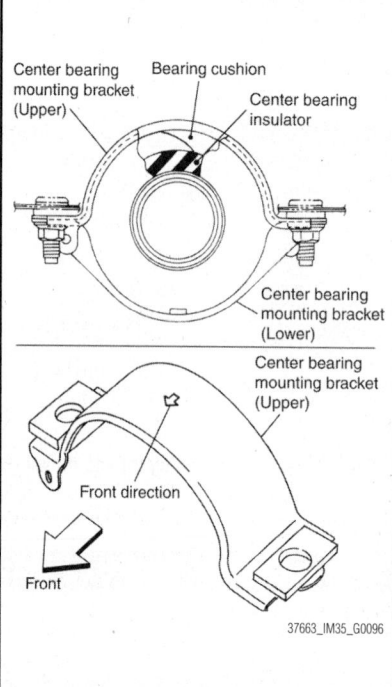

Fig. 34 Install center bearing mounting bracket (Upper) with its arrow mark facing forward

10. Tighten mounting bolts and nuts in the order shown.

11. Align matching marks to install propeller shaft to final drive and transfer (AWD models only) companion flanges, and then tighten to specified torque.

12. Install center bearing mounting bracket (Upper) with its arrow mark facing forward.

13. Adjust position of mounting bracket sliding back and forth to prevent play in thrust direction of center bearing insulator. Install bracket to vehicle.

14. After assembly, perform a driving test to check propeller shaft vibration. If vibration occurred, separate propeller shaft from final drive. Reinstall companion flange after rotating it by 60, 120, 180, 240, 300 degrees. Then perform driving test and check propeller shaft vibration again at each point.

15. If propeller shaft or final drive has been replaced, connect them as follows:

a. Install the propeller shaft while aligning its matching mark A with the matching mark B on the joint as close as possible.

b. Tighten the joint bolts to the specified torque.

※ WARNING

Never reuse the bolts, nuts and washers.

For AWD Models

16. Note the following, and install in the reverse order of removal.

a. Align the matching marks to install propeller shaft flange yoke and transfer companion flange.

b. Align the matching marks to install propeller shaft rubber coupling to final drive companion flange.

※ WARNING

The angle is third axis rubber coupling (1) forms with the final drive companion flange (2). Never bend rubber coupling above the angle. A : 0 - 4°

- Never damage the grease seal.
- Never damage the rubber coupling.
- Never damage the rear oil seal of transmission.
- Never damage the rubber coupling; protect it with a shop towel or equivalent.
- Perform inspection after installation.
- If propeller shaft or final drive has been replaced, connect them as follows:

17. Install the propeller shaft while aligning its matching mark (A) with the matching mark (B) on the joint as close as possible.

※ WARNING

Never damage rubber coupling.

REAR HALFSHAFT

REMOVAL & INSTALLATION

See Figures 35 and 36.

1. Remove tires and wheels.
2. Remove center muffler.
3. Remove rear propeller shaft.
4. Remove stabilizer bar. Refer to Rear Suspension.
5. Remove wheel sensor.
6. Separate drive shaft from rear final drive assembly.

※ WARNING

Never damage the sensor rotor.

→**Release the circular clip lock.**

7. Remove rear final drive assembly.
8. Remove cotter pin and adjusting cap (if equipped), then loosen wheel hub lock nut.
9. Remove stabilizer connecting rod mounting bracket mounting bolt and free stabilizer connecting rod.

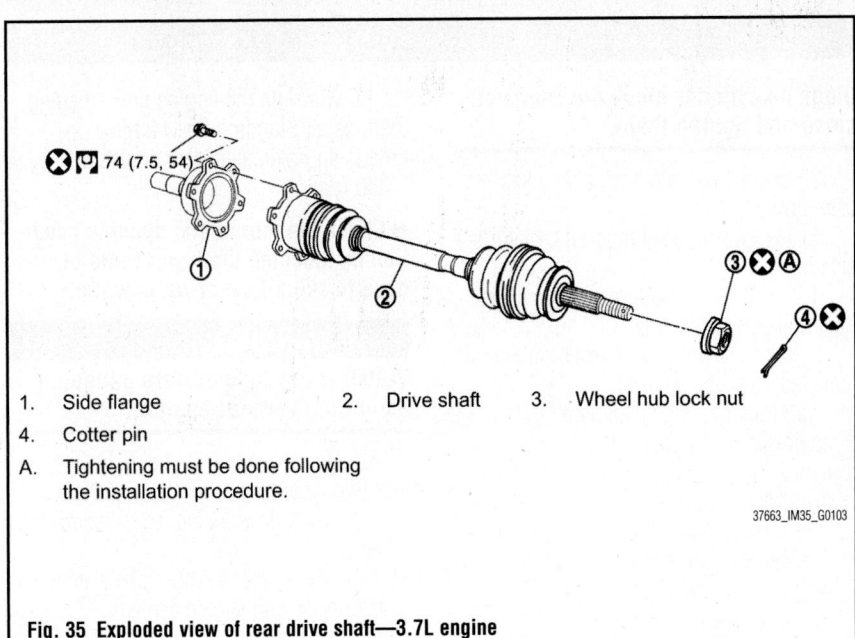

1. Side flange
2. Drive shaft
3. Wheel hub lock nut
4. Cotter pin
A. Tightening must be done following the installation procedure.

❌ 🔧 74 (7.5, 54)

③❌Ⓐ
④❌

37663_IM35_G0103

Fig. 35 Exploded view of rear drive shaft—3.7L engine

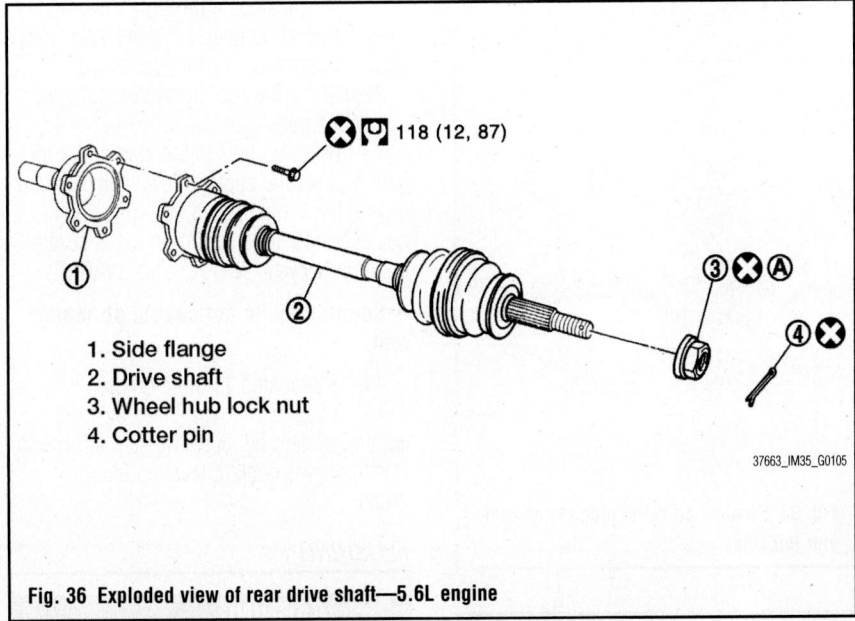

❌ 🔧 118 (12, 87)

③❌Ⓐ
④❌

1. Side flange
2. Drive shaft
3. Wheel hub lock nut
4. Cotter pin

37663_IM35_G0105

Fig. 36 Exploded view of rear drive shaft—5.6L engine

10. Separate the wheel hub and bearing assembly from drive shaft by lightly tapping the end with a hammer and wood block, and then remove wheel hub lock nut and spring washer (if equipped).

✴✴ WARNING

Do not place drive shaft joint at an extreme angle. Also be careful not to overextend slide joint. Do not allow drive shaft to hang down without support for counterpart such as joint sub-assembly, and other parts.

➡ Using a puller if the wheel hub and bearing assembly and drive shaft can-not be separated even after performing the above procedure.

11. Remove the mounting bolts between side flange and drive shaft.

To install:
12. Note the following, and install in the reverse order of removal.
13. Clean the matching surface of wheel hub lock nut and wheel hub and bearing assembly.

✴✴ WARNING

Never apply lubricating oil to these matching surface.

14. Clean the matching surface of drive shaft and wheel hub and bearing assembly. And then apply paste 440037S000 to surface (A) of joint sub-assembly of drive shaft.

✴✴ WARNING

Apply paste to cover entire flat surface of joint sub-assembly of drive shaft.

15. Use the following torque range for tightening the wheel hub lock nut.
   a. Tighten to 74–77 ft. lbs. (100–105 Nm).

✴✴ WARNING

Since the drive shaft is assembled by press-fitting, use the tightening torque range for the wheel hub lock nut. Be sure to use torque wrench to tighten the wheel hub lock nut. Never use a power tool.

➡ Wheel hub lock nut tightening torque does not over torque for avoiding axle noise, and does not less than torque for avoiding looseness.

16. Perform the final tightening of each of parts under unladen conditions, which were removed when removing wheel hub and bearing assembly and axle housing.
17. When installing the spring washer, face the identification paint mark to the wheel hub and bearing assembly side. (With adjusting cap and spring washer for wheel hub lock nut)
18. When installing the adjusting cap, check that there must be no play. (With adjusting cap and spring washer for wheel hub lock nut)
19. Never reuse cotter pin, wheel hub lock nut, spring washer (if equipped).
20. Final Drive Side:
   a. Replace rear final drive side oil seal.
   b. Place the protector (A) [SST: KV38106700 (J-34296)] (VQ37VHR) or [SST: KV38105500 (J-33904)] (VK56VD) onto final drive to prevent damage to the oil seal while inserting drive shaft. Slide drive shaft sliding joint and tap with a hammer to install securely.

✴✴ WARNING

Check that circular clip is completely engaged.

ENGINE COOLING

ENGINE COOLANT

DRAIN & REFILL PROCEDURE

Draining Engine Coolant

See Figure 37.

✳✳ WARNING

To avoid being scalded, never change engine coolant when the engine is hot. Wrap a thick cloth around radiator cap and carefully remove radiator cap. First, turn radiator cap a quarter of a turn to release built-up pressure. Then turn radiator cap all the way.

1. Remove air duct (inlet).
2. Open radiator drain plug at the bottom of radiator, and then remove radiator cap.

➡ **When draining all of engine coolant in the system, open water drain plugs on cylinder block.**

3. Remove the reservoir tank if necessary, and drain engine coolant and clean reservoir tank before installing.
4. Check drained engine coolant for contaminants such as rust, corrosion or discoloration. If contaminated, flush the engine cooling system.

Refilling Engine Coolant

See Figures 38 and 39.

1. Install the reservoir tank if removed, and radiator drain plug.

✳✳ WARNING

Be sure to clean drain plug and install with new O-ring. If water drain

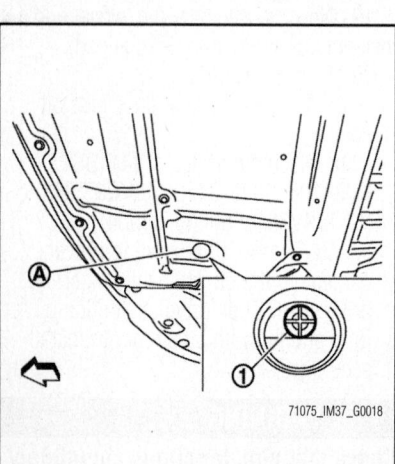

Fig. 37 Locate radiator drain plug hole (A) and open radiator drain plug (1)

---

plugs on cylinder block are removed, close and tighten them.

2. Check that each hose clamp has been firmly tightened.
3. Remove air relief plug on radiator left side.
4. For 5.6L Engines, remove air relief plug on heater hose side.
5. Fill the radiator, and reservoir tank if removed, to specified level.
   a. Pour the engine coolant through engine coolant filler neck slowly of less than 2 ⅛ qts. (2 L) a minute to allow air in system to escape.
   b. Use Genuine NISSAN Long Life Antifreeze/Coolant or equivalent mixed with water (distilled or demineralized).
6. When the engine coolant overflows the air relief hole on heater hose, install air relief plug with new O-ring.
7. Install the radiator cap.

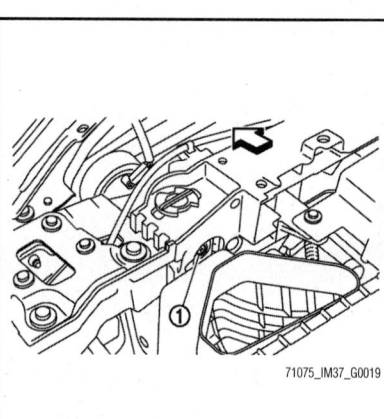

Fig. 38 Remove air relief plug (1) on radiator left side

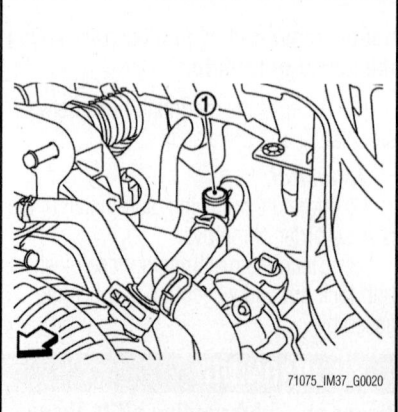

Fig. 39 Remove air relief plug (1) on heater hose side—5.6L Engines

---

8. Warm up the engine until opening thermostat. Standard for warming-up time is approximately 10 minutes at 3,000 rpm.

➡ **Check the thermostat opening condition by touching the lower radiator hose to see a flow of warm water.**

✳✳ WARNING

Watch water temperature gauge so as not to overheat engine.

9. Stop the engine and cool down to less than approximately 122° F (50° C).
   a. Cool down using fan to reduce the time.
   b. If necessary, refill radiator up to filler neck with engine coolant.
10. Refill the reservoir tank to "MAX" level line with engine coolant.
11. Repeat steps 4 through 7 two or more times with radiator cap installed until engine coolant level no longer drops.
12. Check the cooling system for leakage with engine running.
13. Warm up the engine, and check for sound of engine coolant flow while running engine from idle up to 3,000 rpm with heater temperature controller set at several position between "COOL" and "WARM".

➡ **Sound may be noticeable at heater unit.**

14. Repeat step 11 three times.
15. If sound is heard, bleed air from cooling system by repeating step 4 through 7 until engine coolant level no longer drops.

FLUSHING

✳✳ WARNING

To avoid being scalded, never change engine coolant when the engine is hot. Wrap a thick cloth around radiator cap and carefully remove radiator cap. First, turn radiator cap a quarter of a turn to release built-up pressure. Then turn radiator cap all the way.

1. Install reservoir tank if removed, and radiator drain plug.

✳✳ WARNING

Be sure to clean drain plug and install with new O-ring. If water drain plugs on cylinder block are removed, close and tighten them.

2. Remove air relief plug on the radiator.

3. Fill radiator with water until water spills from the air relief hole, then close air relief plug. Fill radiator and reservoir tank with water and reinstall radiator cap.

4. Run the engine and warm it up to normal operating temperature.

5. Rev the engine two or three times under no-load.

6. Stop the engine and wait until it cools down.

7. Drain water from the system.

8. Repeat steps 1 through 7 until clear water begins to drain from radiator.

ENGINE FAN

REMOVAL & INSTALLATION

See Figures 40 and 41.

1. For 3.7L Engines, remove reservoir tank and drain hose.

2. Remove air cleaner case (LH) and air duct (inlet).

3. Disconnect harness connector from cooling fan control module, and move harness to aside.

4. Remove harness clips.

5. Remove A/T oil cooler tube from fan shroud.

6. Remove the cooling fan assembly from under the vehicle.

❊❊ WARNING

Be careful not to damage or scratch on radiator core.

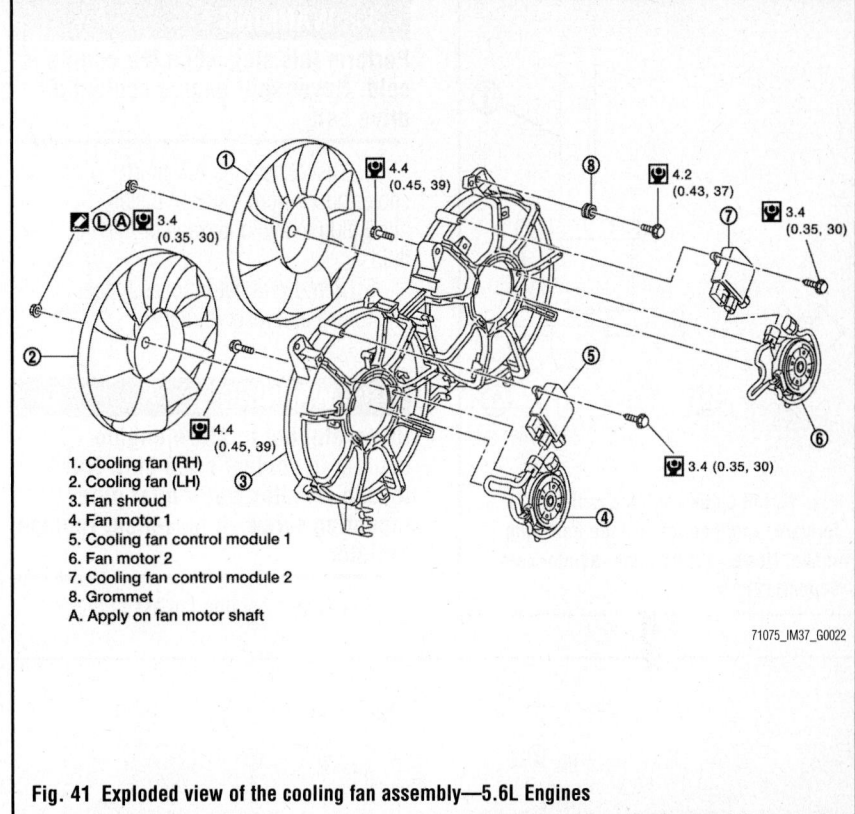

1. Cooling fan (RH)
2. Cooling fan (LH)
3. Fan shroud
4. Fan motor 1
5. Cooling fan control module 1
6. Fan motor 2
7. Cooling fan control module 2
8. Grommet
A. Apply on fan motor shaft

71075_IM37_G0022

Fig. 41 Exploded view of the cooling fan assembly—5.6L Engines

7. Note the following, and install in the reverse order of removal.

8. Only use genuine parts for cooling fan mounting bolt and observe the specified torque (to prevent core support from being damaged).

RADIATOR

REMOVAL & INSTALLATION

Radiator

See Figures 42 and 43.

❊❊ WARNING

Never remove radiator cap when engine is hot. Serious burns could occur from high-pressure engine coolant escaping from radiator. Wrap a thick cloth around the cap. Slowly turn it a quarter of a turn to release built-up pressure. Carefully remove radiator cap by turning it all the way.

1. Remove the following parts:
   a. Engine undercover with power tool.
   b. Engine cover.
   c. Air cleaner case (RH and LH) and air duct (inlet).
   d. For 5.6L Engines: the hood lock say assembly and horn.
   e. Reservoir tank.
   f. Radiator core support ornament, radiator core support center.
   g. For 3.7L Engines: the horn.

2. Remove the condenser pipe assembly.

3. Drain engine coolant from radiator.

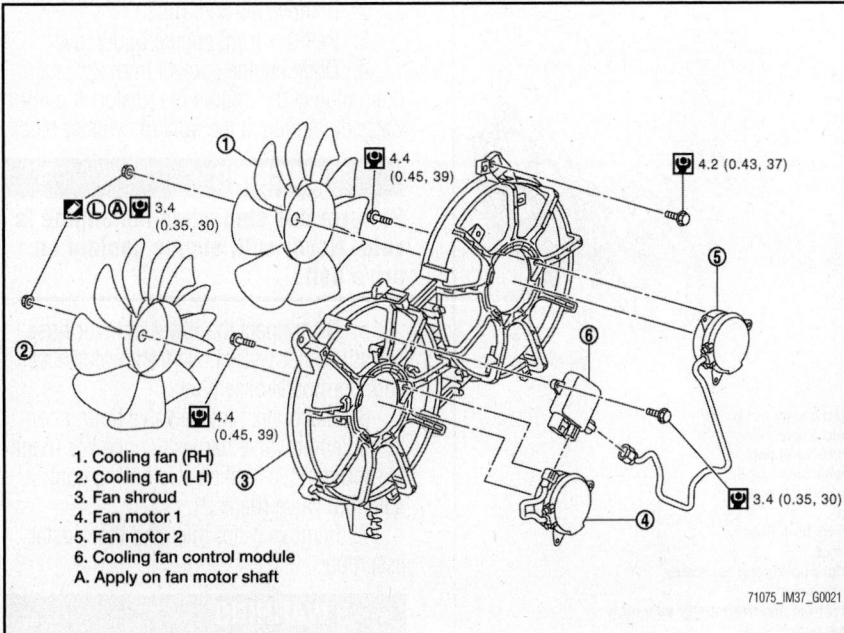

1. Cooling fan (RH)
2. Cooling fan (LH)
3. Fan shroud
4. Fan motor 1
5. Fan motor 2
6. Cooling fan control module
A. Apply on fan motor shaft

71075_IM37_G0021

Fig. 40 Exploded view of the cooling fan assembly—3.7L Engines

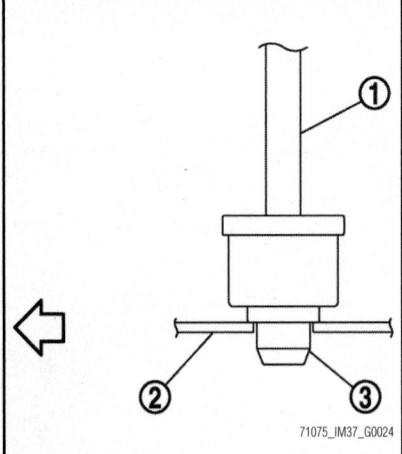

Fig. 42 Lift up and pull the radiator (1) forward, and then remove the mounting rubber (lower) (3) from the radiator core support (2)

71075_IM37_G0024

✳✳ WARNING

Perform this step when the engine is cold. Never spill engine coolant on drive belt.

4. Disconnect the A/T fluid cooler hoses from radiator. Install blind plug to avoid leakage of A/T fluid.

5. Remove radiator hoses (upper and lower) and reservoir tank hose.

✳✳ WARNING

Be careful not to allow engine coolant to contact drive belt. Never loosen radiator water inlet pipe mounting screw. If loosened, replace radiator.

6. Remove cooling fan assembly.

✳✳ WARNING

Never damage or scratch radiator core when removing.

7. Rotate two radiator upper mount brackets 90 degrees counter clockwise, and remove them.

8. Remove condenser.

9. Remove radiator as follows:

✳✳ WARNING

Be careful not to damage radiator core.

10. Lift up and pull the radiator (1) forward, and then remove the mounting rubber (lower) (3) from the radiator core support (2).

To install:

11. Install in the reverse order of removal.

12. Check for leakage of engine coolant using the radiator cap tester adapter and the radiator cap tester.

13. Start and warm up the engine. Visually check that there is no leakage of engine coolant and A/T fluid.

THERMOSTAT

REMOVAL & INSTALLATION

3.7L Engine

See Figures 44 and 45.

1. Remove engine room cover (RH and LH).

2. Remove air duct (inlet).

3. Remove front engine undercover.

4. Drain engine coolant from radiator drain plug at the bottom of radiator, and from water drain plug at the front of cylinder block.

✳✳ WARNING

Perform this step when the engine is cold. Never spill engine coolant on drive belt.

5. Disconnect the lower radiator hose and oil cooler water hose from water inlet and thermostat assembly.

6. Disconnect intake valve timing control solenoid valve harness connector (bank 2), and remove intake valve timing control solenoid valve (bank 2).

7. Remove water inlet and thermostat assembly.

✳✳ WARNING

Never disassemble water inlet and thermostat assembly. Replace them as a unit, if necessary.

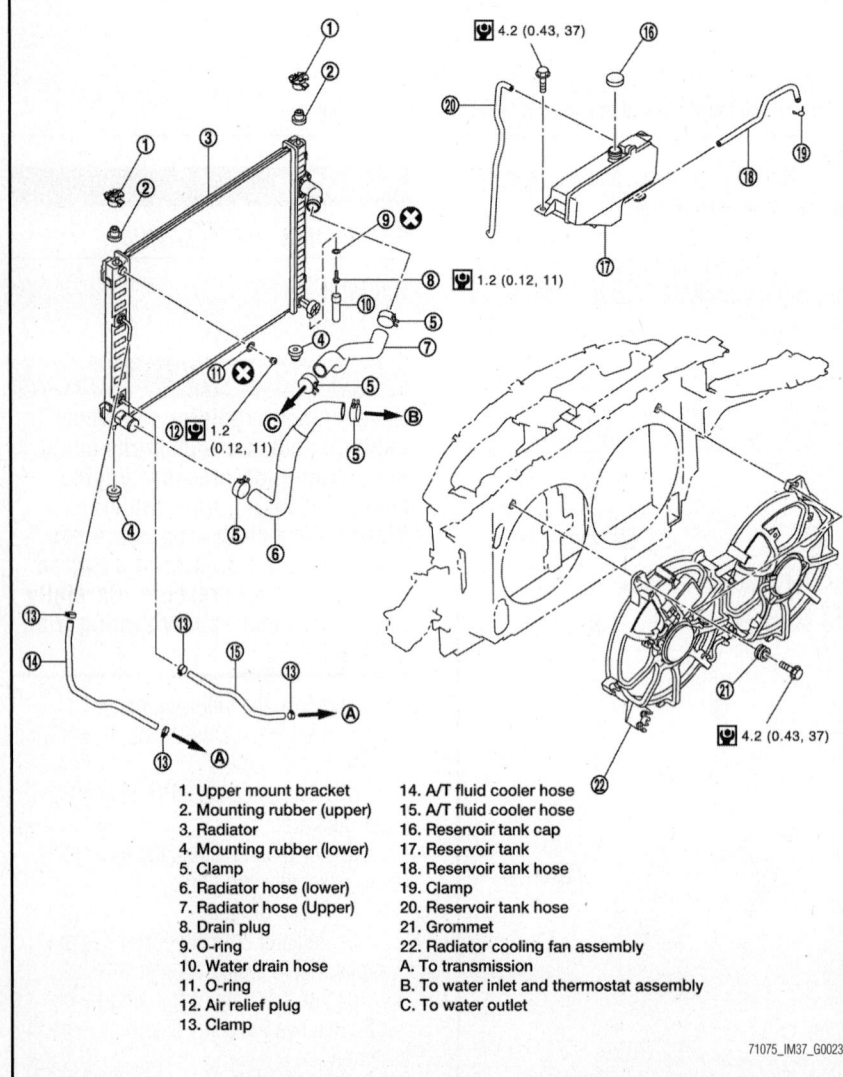

1. Upper mount bracket
2. Mounting rubber (upper)
3. Radiator
4. Mounting rubber (lower)
5. Clamp
6. Radiator hose (lower)
7. Radiator hose (Upper)
8. Drain plug
9. O-ring
10. Water drain hose
11. O-ring
12. Air relief plug
13. Clamp
14. A/T fluid cooler hose
15. A/T fluid cooler hose
16. Reservoir tank cap
17. Reservoir tank
18. Reservoir tank hose
19. Clamp
20. Reservoir tank hose
21. Grommet
22. Radiator cooling fan assembly
A. To transmission
B. To water inlet and thermostat assembly
C. To water outlet

71075_IM37_G0023

Fig. 43 Exploded view of radiator assembly—3.7L and 5.6L engines

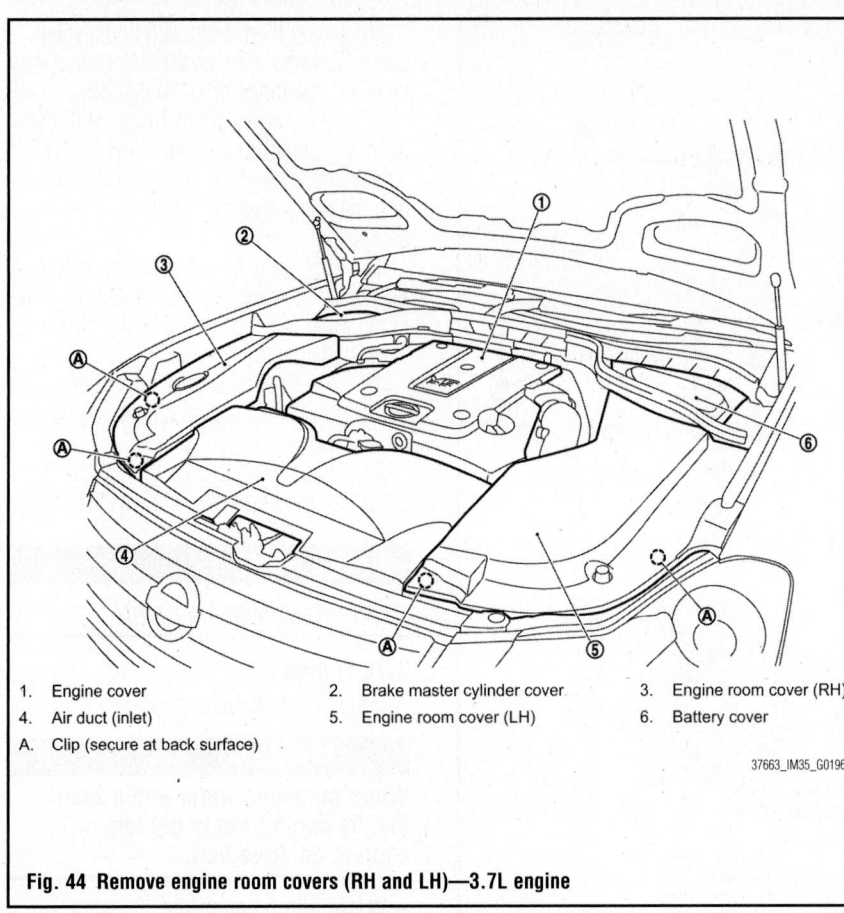

1. Engine cover
4. Air duct (inlet)
A. Clip (secure at back surface)

2. Brake master cylinder cover
5. Engine room cover (LH)

3. Engine room cover (RH)
6. Battery cover

37663_IM35_G0196

Fig. 44 Remove engine room covers (RH and LH)—3.7L engine

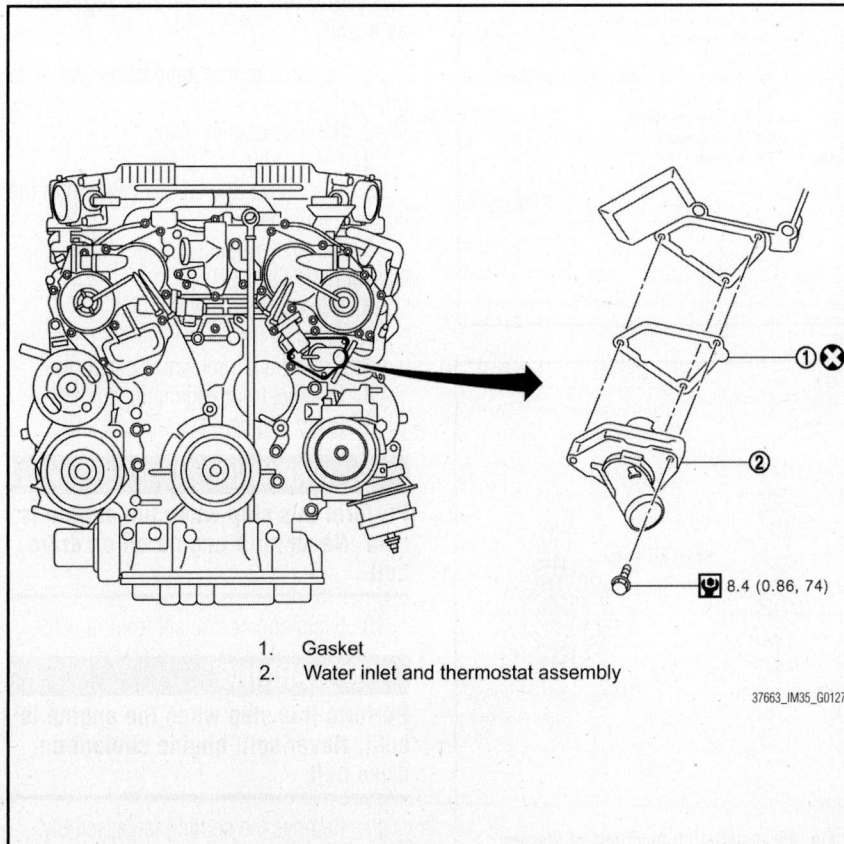

8.4 (0.86, 74)

1. Gasket
2. Water inlet and thermostat assembly

37663_IM35_G0127

Fig. 45 Exploded view of thermostat assembly

To install:

8. Note the following, and install in the reverse order of removal.

⁂ WARNING

Be careful not to spill engine coolant over engine room. Use rag to absorb engine coolant.

9. Check for leakage of engine coolant using the radiator cap tester adapter and the radiator cap tester.
10. Start and warm up the engine. Visually check that there is no leakage of engine coolant.

5.6L Engine

See Figures 46 through 48.

1. Remove front engine cover.
2. Remove air duct (inlet).
3. Remove reservoir tank.
4. Remove engine undercover.
5. Drain engine coolant from drain plugs on radiator and both side of cylinder block.

⁂ WARNING

Perform this step when engine is cold. Never spill engine coolant on drive belts.

6. Disconnect the upper radiator hose from thermostat housing.
7. Remove water suction pipe and water suction hose.
8. Remove the intake manifolds.
   a. Remove the following parts: Move injector harness to the position without the hindrance for work.
   b. Harness connector
   c. Harness clip
9. Remove fuel tube insulator.
10. Remove fuel feed tube (pump side) and fuel feed tube (bank side).
11. Remove water inlet and thermostat.
12. Remove water connector, heater pipes and heater hoses.
13. Remove thermostat housing.

To install:

14. Note the following, and install in the reverse order of removal.

⁂ WARNING

Be careful not to spill engine coolant over engine room. Use rag to absorb engine coolant.

15. Install thermostat and water control valve with the whole circumference of each flange part fit securely inside rubber ring.

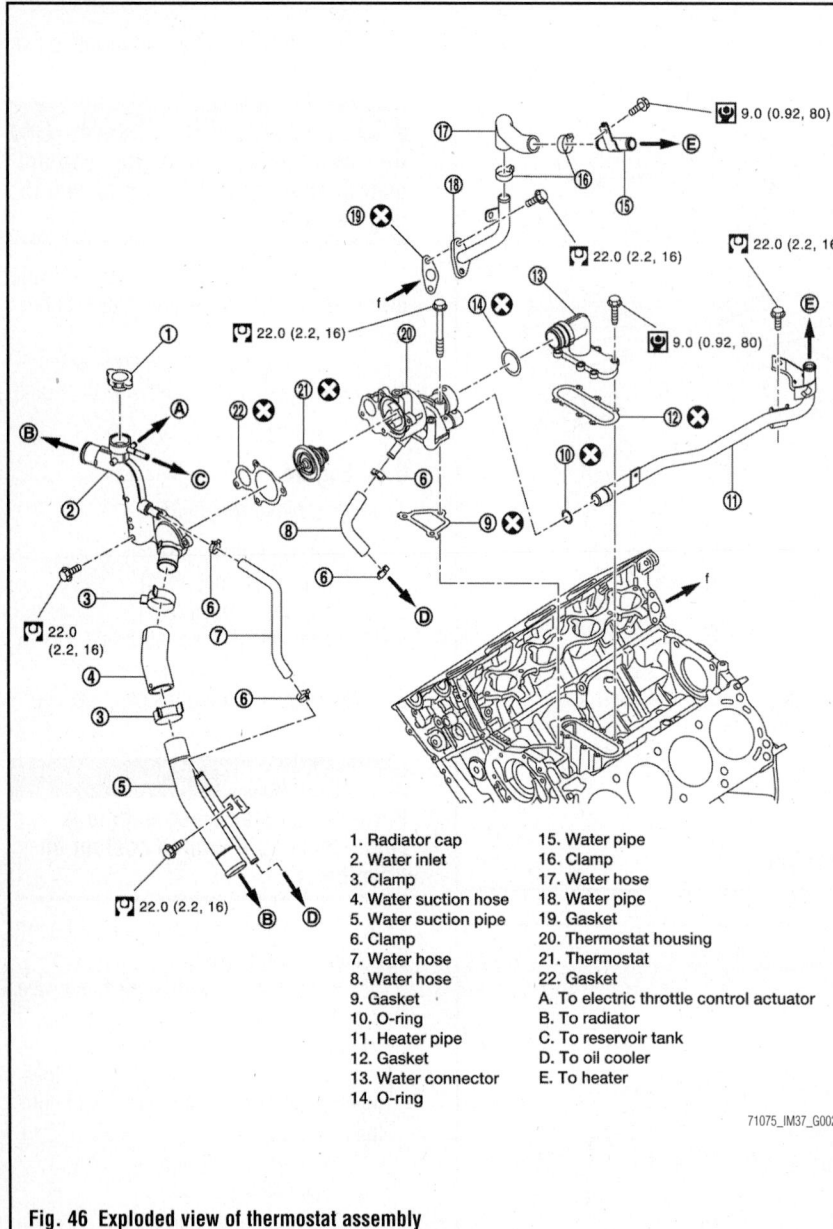

1. Radiator cap
2. Water inlet
3. Clamp
4. Water suction hose
5. Water suction pipe
6. Clamp
7. Water hose
8. Water hose
9. Gasket
10. O-ring
11. Heater pipe
12. Gasket
13. Water connector
14. O-ring
15. Water pipe
16. Clamp
17. Water hose
18. Water pipe
19. Gasket
20. Thermostat housing
21. Thermostat
22. Gasket
A. To electric throttle control actuator
B. To radiator
C. To reservoir tank
D. To oil cooler
E. To heater

71075_IM37_G0025

Fig. 46 Exploded view of thermostat assembly

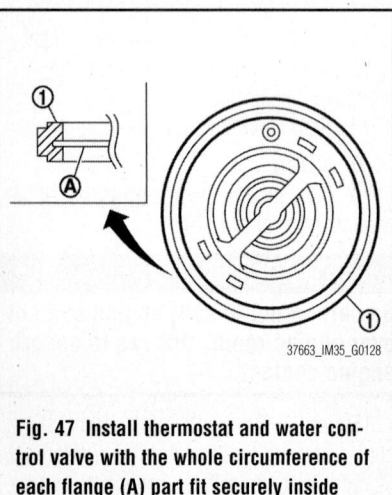

37663_IM35_G0128

Fig. 47 Install thermostat and water control valve with the whole circumference of each flange (A) part fit securely inside rubber ring (1)

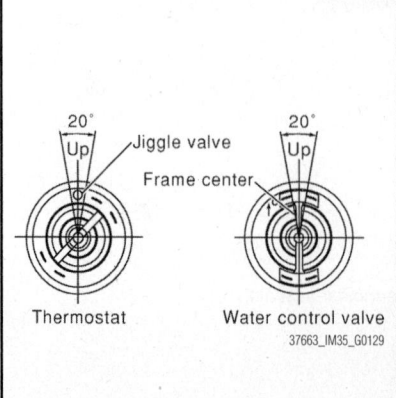

37663_IM35_G0129

Fig. 48 Installation positions of thermostat and water control valve

16. Install thermostat with jiggle valve facing upwards. (The position deviation may be within the range of +/-10 degrees)

17. Install water control valve with the up-mark facing up and the frame center part facing upwards. (The position deviation may be within the range of +/-10 degrees)

18. First apply a neutral detergent to O-rings, then quickly insert the insertion parts of the water outlet pipe and heater pipe into the installation holes.

19. Check for leakage of engine coolant using radiator cap tester adapter and radiator cap tester.

20. Start and warm up engine. Visually check if there is no leakage of engine coolant.

WATER PUMP

REMOVAL & INSTALLATION

3.7L Engine

See Figures 49 through 52.

❄❄ WARNING

When removing water pump assembly, be careful not to get engine coolant on drive belt.

➡**Water pump cannot be disassembled and should be replaced as a unit.**

1. Remove engine room cover (RH and LH).
2. Remove engine cover.
3. Release the fuel pressure.
4. Disconnect the battery cable from the negative terminal.
5. Remove air duct and air cleaner case assembly (RH and LH).
6. Remove the reservoir tank.
7. Separate the engine harness removing their brackets from front timing chain case.
8. Remove front engine undercover.
9. Drain engine oil.

❄❄ WARNING

Perform this step when the engine is cold. Never spill engine oil on drive belt.

10. Drain engine coolant from radiator.

❄❄ WARNING

Perform this step when the engine is cold. Never spill engine coolant on drive belt.

11. Remove the cooling fan assembly.
12. Remove the upper and lower radiator hoses.

13. Remove front timing chain case.

14. Remove timing chain tensioner (primary) as follows:

   a. Remove lower mounting bolt.

   b. Loosen upper mounting bolt slowly, and then turn chain tensioner (primary) on the upper mounting bolt so that plunger is fully expanded.

➡**Even if plunger is fully expanded, it is not dropped from the body of timing chain tensioner (primary).**

   c. Remove upper mounting bolt, and then remove timing chain tensioner (primary).

15. Remove water pump as follows:

   a. Remove three water pump mounting bolts. Secure a gap between water pump gear and timing chain, by turning crankshaft counterclockwise until timing chain looseness on water pump sprocket becomes maximum.

   b. Screw bolts into water pumps upper and lower mounting bolt holes until they reach timing chain case. Then, alternately tighten each bolt for a half turn, and pull out water pump.

❋❋ **WARNING**

Pull straight out while preventing vane from contacting socket in installation area. Remove water pump without causing sprocket to contact timing chain.

   c. Remove M8 bolts and O-rings from water pump.

❋❋ **WARNING**

Never disassemble water pump.

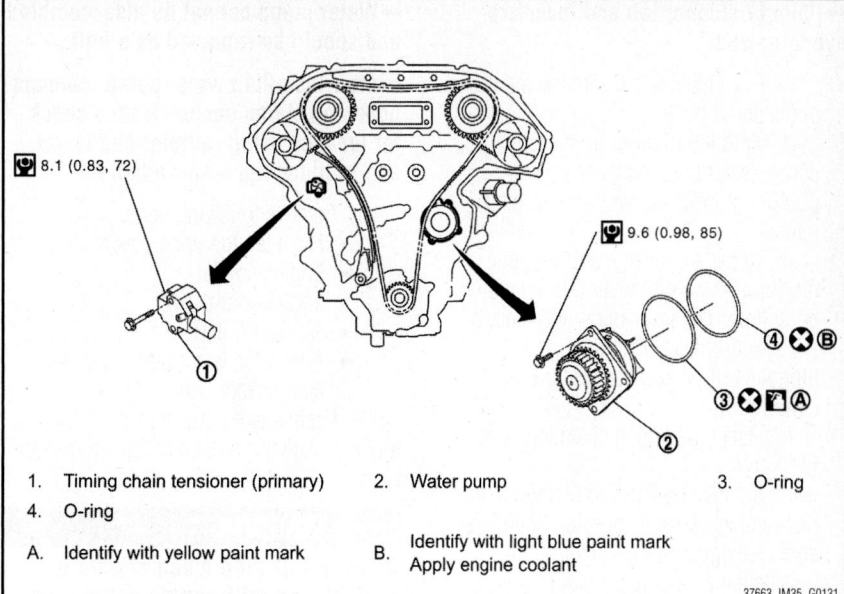

1. Timing chain tensioner (primary)   2. Water pump   3. O-ring
4. O-ring
A. Identify with yellow paint mark   B. Identify with light blue paint mark
                                           Apply engine coolant

37663_IM35_G0131

Fig. 50 Exploded view of water pump assembly

To install:

16. Install new O-rings to water pump.

   a. Apply engine oil to O-ring and engine coolant to O-ring as shown.

   b. Locate O-ring with yellow paint mark to front side.

   c. Locate O-ring with light blue paint mark to rear side.

17. Install water pump.

❋❋ **WARNING**

Never allow cylinder block to nip O-rings when installing water pump.

   a. Check timing chain and water pump sprocket are engaged.

   b. Insert water pump by tightening mounting bolts alternately and evenly.

18. Install timing chain tensioner (primary) as follows:

   a. Turn crankshaft clockwise so that timing chain on the timing chain tensioner (primary) side is loose.

   b. Pull the plunger stopper tab up (or turn lever downward) so as to remove plunger stopper tab from the ratchet of plunger.

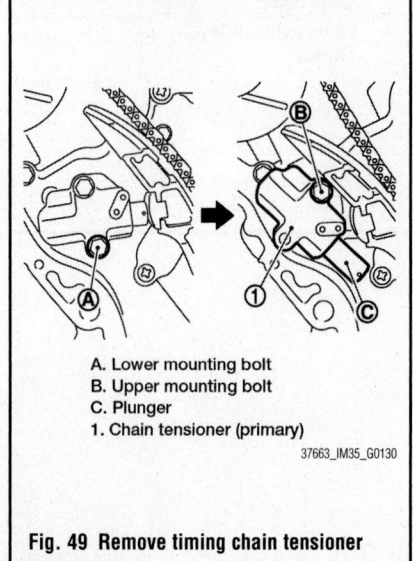

A. Lower mounting bolt
B. Upper mounting bolt
C. Plunger
1. Chain tensioner (primary)

37663_IM35_G0130

Fig. 49 Remove timing chain tensioner (primary)

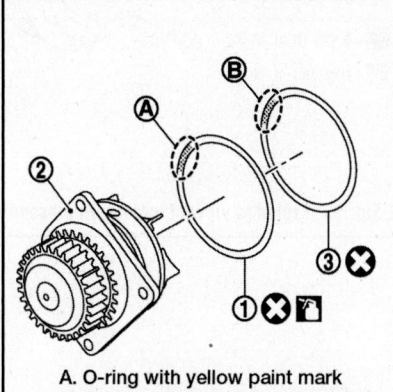

A. O-ring with yellow paint mark
B. O-ring with light blue paint mark
1. Apply engine oil to this O-ring
2. Water pump
3. Apply engine coolant to this O-ring

37663_IM35_G0132

Fig. 51 Install new O-rings to water pump

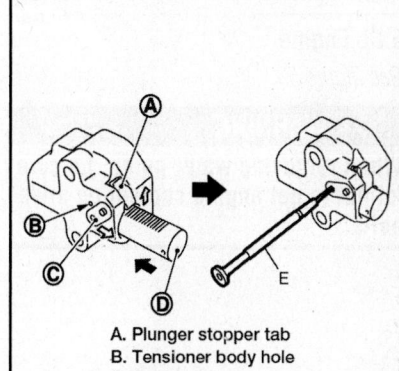

A. Plunger stopper tab
B. Tensioner body hole
C. Lever
D. Plunger
E. Stopper pin

37663_IM35_G0133

Fig. 52 Install timing chain tensioner (primary)

➡**Plunger stopper tab and lever are synchronized.**

    c. Push the plunger into the inside of tensioner body.

    d. Hold the plunger in the fully compressed position by engaging plunger stopper tab with the tip of ratchet.

    e. To secure lever, insert stopper pin through hole of lever into tensioner body hole. The lever parts and the tab are synchronized. Therefore, the plunger will be secured under this condition.

    f. Install the timing chain tensioner (primary).

    g. Remove the dust and foreign material completely from backside of timing chain tensioner (primary) and from installation area of rear timing chain case.

    h. Remove stopper pin.

    i. Check again that timing chain and water pump sprocket are engaged.

19. Install in the reverse order of removal for remaining parts.

➡**After starting engine, let idle for three minutes, then rev engine up to 3,000 rpm under no load to purge air from the high-pressure chamber of chain tensioner. Engine may produce a rattling noise. This indicates that air still remains in the chamber and is not a matter of concern.**

20. Check for leakage of engine coolant using the radiator cap tester adapter and the radiator cap tester.

21. Start and warm up the engine. Visually check that there is no leakage of engine coolant.

5.6L Engine

See Figure 53.

❉❉ WARNING

When removing water pump, be careful not to get engine coolant on drive belts.

➡**Water pump cannot be disassembled and should be replaced as a unit.**

➡**After installing water pump, connect hose and clamp securely, then check for leakage using radiator cap tester and radiator cap tester adapter.**

1. Remove following parts:
- Front engine undercover
- Engine cover
- Engine room cover
- Air duct (inlet)
- Alternator, water pump and A/C compressor belt

2. Drain engine coolant from drain plugs on radiator and both side of cylinder block.

❉❉ WARNING

Perform this step when engine is cold. Never spill engine coolant on drive belts.

3. Remove water pump pulley.
4. Remove water pump.

➡**Engine coolant will leakage from cylinder block, so have a receptacle ready under vehicle.**

❉❉ WARNING

Handle the water pump vane so that it never contact any other parts. Never disassemble water pump.

5. Visually check that there is no significant dirt or rusting on water pump body and vane.

6. Check there is no looseness in vane shaft, and that it turns smoothly when rotated by hand.

7. If anything is found, replace water pump.

To install:

8. Install in the reverse order of removal.

9. Check for leakage of engine coolant using radiator cap tester adapter and radiator cap tester.

10. Start and warm up engine. Visually check if there is no leakage of engine coolant.

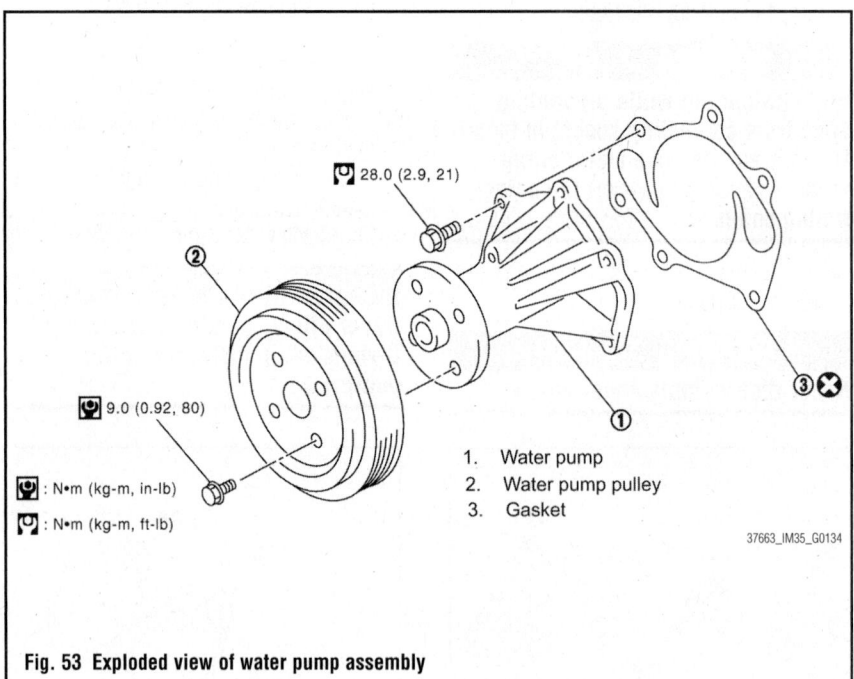

28.0 (2.9, 21)

9.0 (0.92, 80)

: N•m (kg-m, in-lb)

: N•m (kg-m, ft-lb)

1. Water pump
2. Water pump pulley
3. Gasket

37663_IM35_G0134

Fig. 53 Exploded view of water pump assembly

BATTERY

REMOVAL & INSTALLATION

See Figure 54.

1. Remove the battery cover.
2. Remove the cowl top cover (RH).
3. Remove the clips, and remove hood ledge cover RH.
4. Loosen battery terminal nuts, and disconnect both battery cables from battery terminals.

> ### ✳✳ WARNING
> **When disconnecting, disconnect the battery cable from the negative terminal first.**

5. Remove the battery holder frame mounting nuts and battery holder frame.
6. Remove the battery.

To install:

7. Installation is the reverse order of removal.

➡**Locate the battery at the outside of the vehicle in the battery tray when installing the battery. Check that the positive terminal cap opens and closes.**

> ### ✳✳ WARNING
> **When connecting, connect the battery cable to the positive terminal first.**

➡**Reset electronic systems as necessary.**

BATTERY RECONNECT/ RELEARN PROCEDURE

Automatic Drive Positioner Initialization

After reconnecting battery cable, perform initialization procedure A or B. If initialization has not been performed, EXITING OPERATION will not operate.

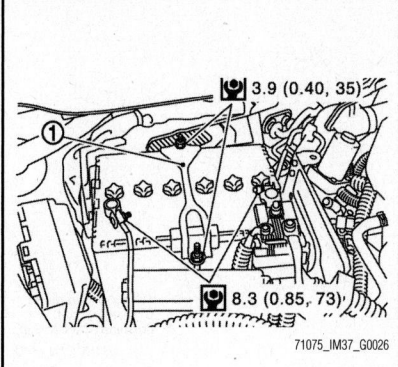

Fig. 54 Remove the battery hold down clamp (1)

Procedure A

1. Turn ignition switch from ACC to OFF position.
2. Driver door switch is ON (open), then OFF (close), then ON (open).
3. END

Procedure B

1. Drive the vehicle at more than 25 km/h (16 MPH).
2. END

Power Window Initialization

Perform the initialization when the following operations are performed or when the auto up operation is not performed:

• When the power supply to the power window main switch, power window sub-switch or each power window motor is cut off by the removal of battery terminal or the battery fuse is blown.

  • Disconnection and connection of power window main switch or each power window sub-switch harness connector.
  • Removal and installation of regulator assembly.

• Removal and installation of motor from regulator assembly.
• Operation of regulator assembly as an independent unit.
• Removal and installation of glass.
• Removal and installation of door glass run.

1. Disconnect the minus terminal of battery or disconnect power window switch's harness connector temporarily, then reconnect after at least 1 minute.
2. Turn ignition switch ON.
3. Open the window to its full width by operating the power window switch. (Exclude this procedure if the window is already fully opened)
4. Fully draw the power window switch in up direction (auto close position) and hold, keep holding the switch even when window is completely closed and then release after 3 second has passed.
5. Inspection of the anti-pinch system function.

➡**Initialization may be cancelled with continuous opening and closing operation. In this case, initialize the system.**

Power Sunroof Initialization

If the battery is disconnected, the sunroof motor connector is disconnected sunroof does not close or open automatically, use the following procedure to return sunroof operation to normal.

1. Close the sunroof if it is not in the closed position. It may be necessary to repeatedly push the switch to close the sunroof.
2. Press the SLIDE OPEN/TILT DOWN switch for approximately 1 second or more.
3. Initialization procedure is completed. Confirm proper operation of the sunroof (slide open, slide close, tilt up, tilt down.)

ENGINE ELECTRICAL   **CHARGING SYSTEM**

ALTERNATOR

REMOVAL & INSTALLATION

3.7L Engine, RWD Models

See Figures 55 through 57.

1. Before servicing the vehicle, refer to the service precautions.
2. Disconnect the battery cable from the negative terminal.
3. Remove engine front undercover.
4. Remove the alternator and power steering oil pump belt.
5. Disconnect the alternator connector.
6. Remove "B" terminal nut.
7. Remove the harness bracket bolts.
8. Remove oil pressure switch harness clip from alternator stay.
9. Disconnect oil pressure switch connector.
10. Remove alternator mounting bolt and alternator stay mounting bolt, then remove alternator stay.
11. Remove alternator mounting bolt.
12. Remove alternator assembly downward from the vehicle

To install:
13. Installation is the reverse order of removal.
14. Tighten the alternator mounting bolt (C) to 48 ft. lbs. (64.7 Nm).
15. Tighten the alternator stay mounting bolts (A and B) to 21 ft. lbs. (28 Nm).

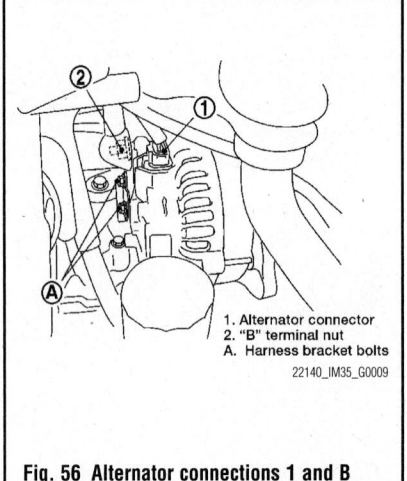

1. Alternator connector
2. "B" terminal nut
A. Harness bracket bolts

22140_IM35_G0009

Fig. 56 Alternator connections 1 and B and harness bracket bolt locations

16. Connect the B terminal; and alternator connector.

✻✻ WARNING

Check the tension of the belt.

➡**For this model, the power generation voltage variable control system that controls the power generation voltage of the alternator has been adopted. Therefore, the power generation voltage variable control system operation inspection should be performed after replacing the alternator, and then make sure that the system operates normally.**

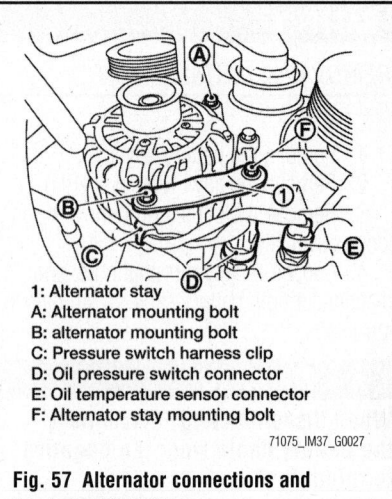

1: Alternator stay
A: Alternator mounting bolt
B: alternator mounting bolt
C: Pressure switch harness clip
D: Oil pressure switch connector
E: Oil temperature sensor connector
F: Alternator stay mounting bolt

71075_IM37_G0027

Fig. 57 Alternator connections and bracket bolt locations

3.7L Engine, AWD Models

See Figures 58 through 60.

1. Disconnect the battery cable from the negative terminal.
2. Remove air duct (inlet).
3. Remove air cleaner case RH.
4. Remove the terminal B harness from harness clamp.
5. Remove harness clip from harness bracket.
6. Disconnect the alternator connector.
7. Remove engine under cover.
8. Remove drive belt.
9. Remove alternator mounting bolt (B) and alternator stay mounting bolt (C), and then remove alternator stay (1).
10. Remove alternator mounting bolt (A).
11. Remove alternator from engine and laterally rotate to a position so that terminal B nut (A) is visible.

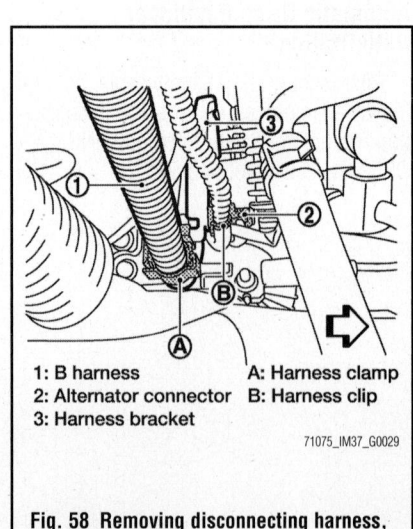

1: B harness   A: Harness clamp
2: Alternator connector B: Harness clip
3: Harness bracket

71075_IM37_G0029

Fig. 58 Removing disconnecting harness, clips and connectors

① 10.1 (1.0, 7)

④ 64.7 (6.6, 48)

④ 28.0 (2.9, 21)

⑤ 28.0 (2.9, 21)

1. Terminal B nut
2. Terminal B harness
3. Alternator connector
4. Alternator mounting bolt
5. Alternator stay mounting bolt
6. Alternator stay
7. Alternator

71075_IM37_G0028

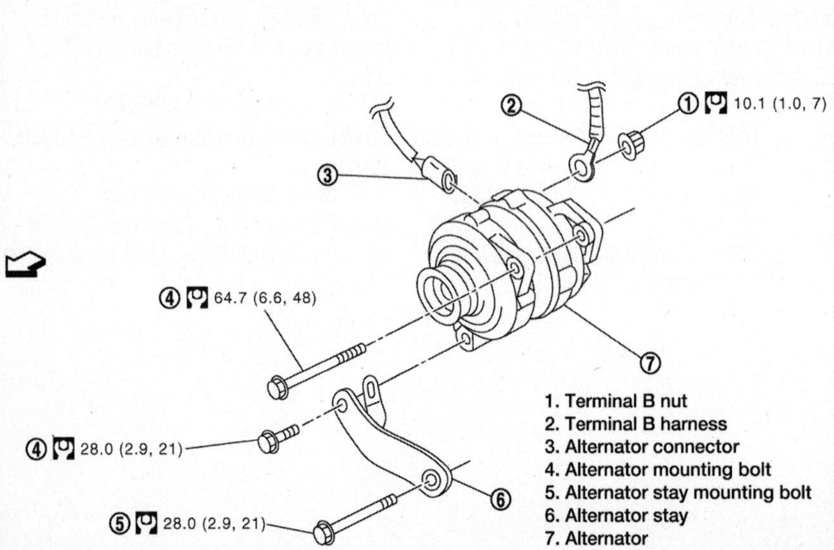

Fig. 55 Alternator and components exploded view—3.7L engine

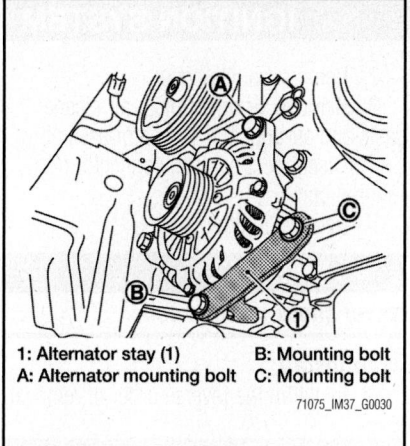

1: Alternator stay (1)   B: Mounting bolt
A: Alternator mounting bolt   C: Mounting bolt

71075_IM37_G0030

Fig. 59 Removing the alternator and components

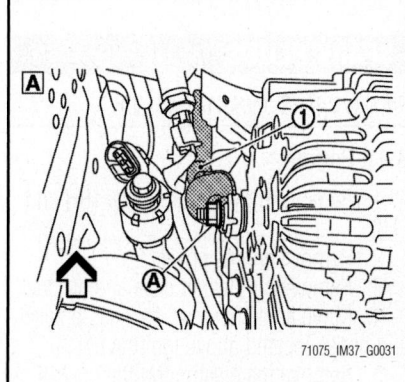

71075_IM37_G0031

Fig. 60 Removing the terminal B nut (A) and harness (1)

✳✳ WARNING

Be careful not to damage engine oil filter.

12. Remove terminal B nut, and then remove terminal B harness (1).

To install:

13. Note the following item, and then install in the reverse order of removal.

14. Be sure to tighten terminal B nut carefully.

15. Install alternator, and check tension of belt.

➡For this model, the power generation voltage variable control system that controls the power generation voltage of the alternator has been adopted. Therefore, the power generation voltage variable control system operation inspection should be per-

formed after replacing the alternator, and then make sure that the system operates normally.

5.6L Engine

See Figures 61 and 62.

1. Disconnect the battery cable from the negative terminal.

2. Remove air duct (inlet) and air cleaner case (bank 2).

3. Remove drive belt.

4. Remove the mounting bolts. Move power steering suction hose and power steering high pressure piping and secure work space.

5. Remove harness bracket mounting bolt.

6. Disconnect VDC harness connector.

7. Move harness together with harness brackets and, and secure work space.

8. Remove engine under cover.

9. Disconnect the alternator connector.

10. Remove terminal B nut, and then remove terminal B harness.

11. Remove alternator mounting bolt lower.

12. Remove the alternator mounting bolt upper.

13. Remove alternator assembly upward from the vehicle.

To install:

14. Note the following item, and then install in the reverse order of removal.

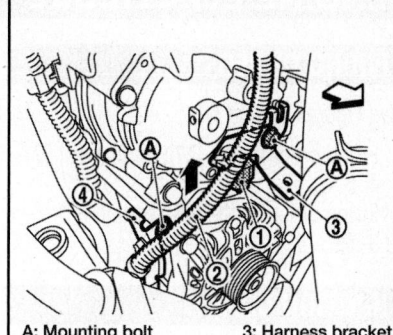

A: Mounting bolt   3: Harness bracket
1: VDC harness connector 4: Harness bracket
2: Harness

71075_IM37_G0032

Fig. 61 Removing the alternator—5.6L Engines

15. Be sure to tighten terminal B nut carefully.

16. Install alternator, and check tension of belt.

➡For this model, the power generation voltage variable control system that controls the power generation voltage of the alternator has been adopted. Therefore, the power generation voltage variable control system operation inspection should be performed after replacing the alternator, and then make sure that the system operates normally.

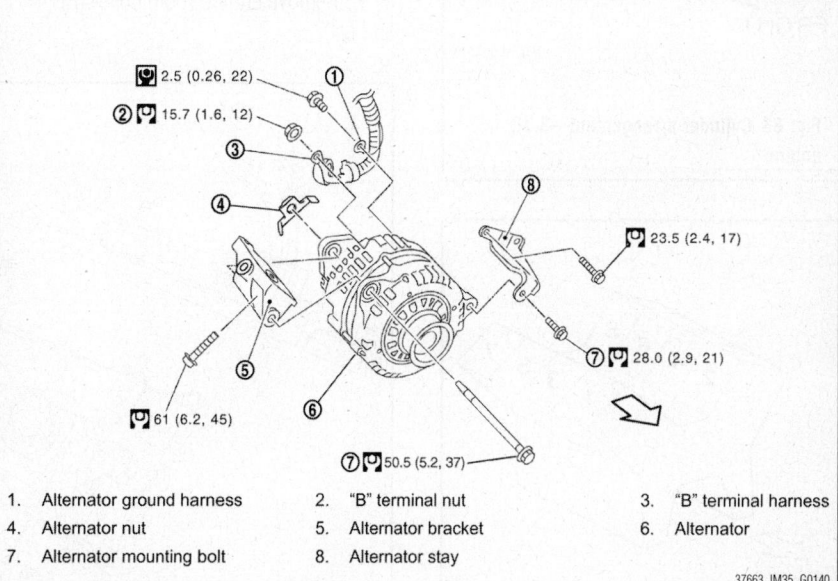

1. Alternator ground harness
2. "B" terminal nut
3. "B" terminal harness
4. Alternator nut
5. Alternator bracket
6. Alternator
7. Alternator mounting bolt
8. Alternator stay

37663_IM35_G0140

Fig. 62 Exploded view of alternator assembly

ENGINE ELECTRICAL

IGNITION SYSTEM

FIRING ORDER

See Figures 63 and 64.

Firing order for the 3.7L engine is 1–2–3–4–5–6.

Firing order for the 5.6L engine is 1–8–7–3–6–5–4–2.

IGNITION COIL

REMOVAL & INSTALLATION

3.7L Engine

See Figure 65.

1. Remove engine room cover (RH and LH).
2. Remove engine cover.
3. Remove air cleaner case and air duct.
4. Move aside harness, harness bracket, and hoses located above ignition coil.

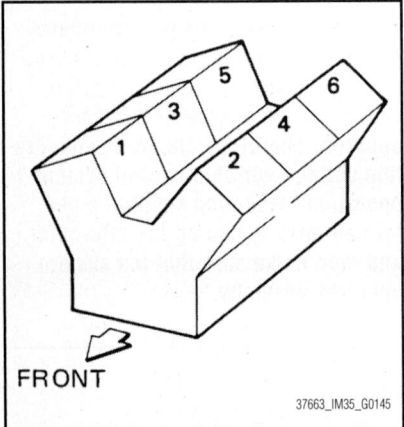

Fig. 63 Cylinder arrangement—3.7L engine

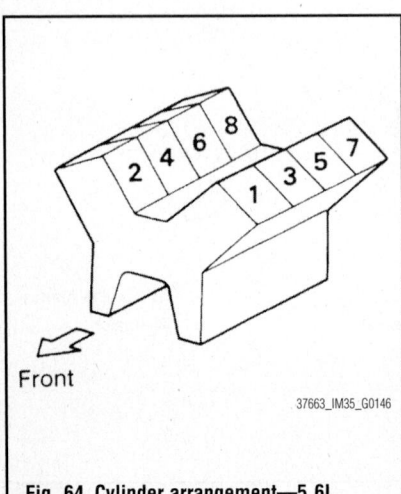

Fig. 64 Cylinder arrangement—5.6L engine

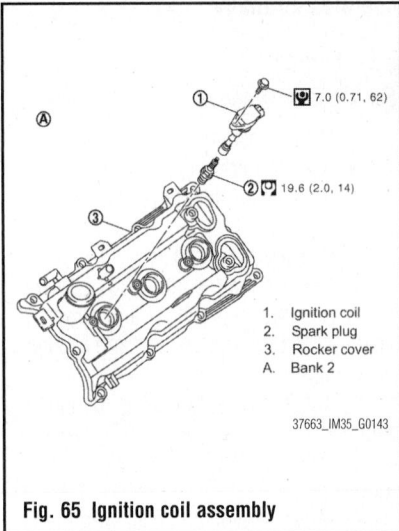

1. Ignition coil
2. Spark plug
3. Rocker cover
A. Bank 2

37663_IM35_G0143

Fig. 65 Ignition coil assembly

5. Remove the electric throttle control actuator.
6. Disconnect harness connector from ignition coil.
7. Remove the ignition coil.

✳✳ WARNING

Never shock ignition coil.

To install:

8. Install in the reverse order of removal.

5.6L Engine

See Figure 66.

1. Remove engine room cover (RH and LH).

2. Remove engine cover.
3. Remove air duct (inlet), air cleaner case, and air duct and resonator assembly.
4. Disconnect harness connector from ignition coil.
5. Remove ignition coil.

✳✳ WARNING

Never shock ignition coil.

To install:

6. Install in the reverse order of removal.

IGNITION TIMING

ADJUSTMENT

The ignition timing is controlled by the ECM. No adjustment is possible or necessary.

SPARK PLUGS

REMOVAL & INSTALLATION

3.7L Engine

1. Remove engine room cover (RH and LH).
2. Remove engine cover.
3. Remove air cleaner case and air duct.
4. Move aside harness, harness bracket, and hoses located above ignition coil.
5. Remove the electric throttle control actuator.
6. Disconnect harness connector from ignition coil.
7. Remove the ignition coil.

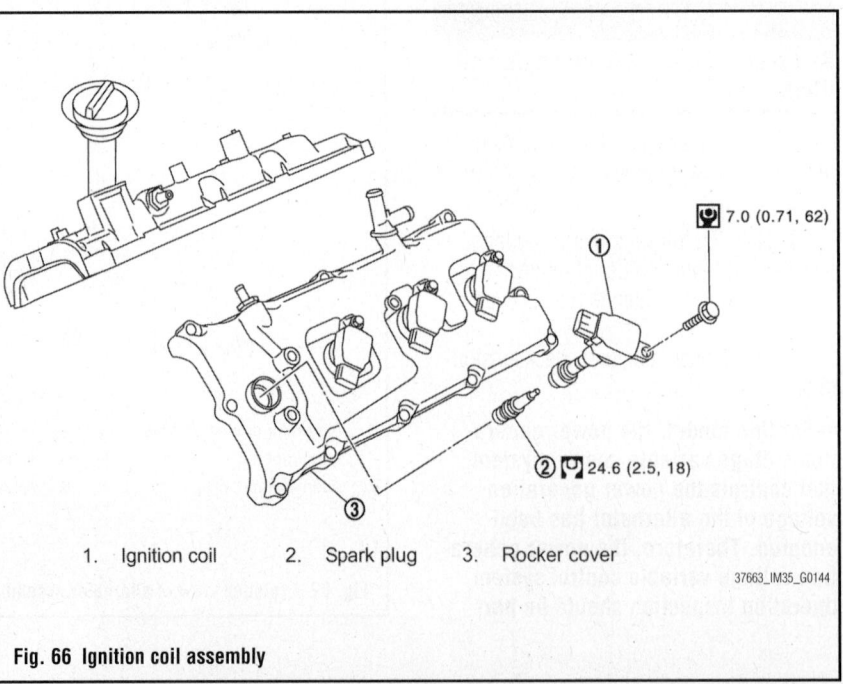

1. Ignition coil   2. Spark plug   3. Rocker cover

37663_IM35_G0144

Fig. 66 Ignition coil assembly

Never shock ignition coil.

8. Remove spark plug with a spark plug wrench, 0.55 inch (commercial service tool).

To install:

9. Install in the reverse order of removal.

ENGINE ELECTRICAL

STARTER

REMOVAL & INSTALLATION

3.7L Engine

2WD Models

See Figure 67.

1. Disconnect the battery cable from the negative terminal.
2. Remove front undercover.
3. Remove road wheel and tire (Front LH).
4. Disconnect steering lower joint, then remove it.
5. Remove the terminal B nut.
6. Disconnect the connector S.
7. Remove the starter motor mounting bolts, using power tools.
8. Remove the starter motor from the side of the vehicle.

To install:

9. Note the following item, and install in the reverse order of removal.

5.6L Engine

1. Remove engine room cover (RH and LH).
2. Remove engine cover.
3. Remove air duct (inlet), air cleaner case, and air duct and resonator assembly.
4. Disconnect harness connector from ignition coil.
5. Remove ignition coil.

10. Be sure to tighten terminal B nut carefully.

4WD Models

See Figure 67.

1. Disconnect the battery cable from the negative terminal.
2. Remove front undercover.
3. Remove road wheel and tire (Front LH).
4. Disconnect steering lower joint, then remove it.
5. Remove the terminal B nut.
6. Disconnect the connector S.
7. Remove the starter motor mounting bolts, using power tools.
8. Remove exhaust mounting bracket.
9. Remove the starter motor downward from the vehicle.

To install:

10. Note the following item, and install in the reverse order of removal.

Never shock ignition coil.

6. Remove spark plug with a spark plug wrench, 0.55 inch (commercial service tool).

To install:

7. Install in the reverse order of removal.

STARTING SYSTEM

11. Be sure to tighten terminal B nut carefully.

5.6L Engine

See Figure 68.

1. Disconnect the battery cable from the negative terminal.
2. Remove the engine cover.
3. Remove the intake manifold.
4. Remove the terminal B nut.
5. Disconnect the connector S.
6. Remove the starter motor mounting bolts, using power tools.
7. Remove the starter motor upward from the vehicle.

To install:

8. Note the following item, and install in the reverse order of removal.
9. Be sure to tighten terminal B nut carefully.

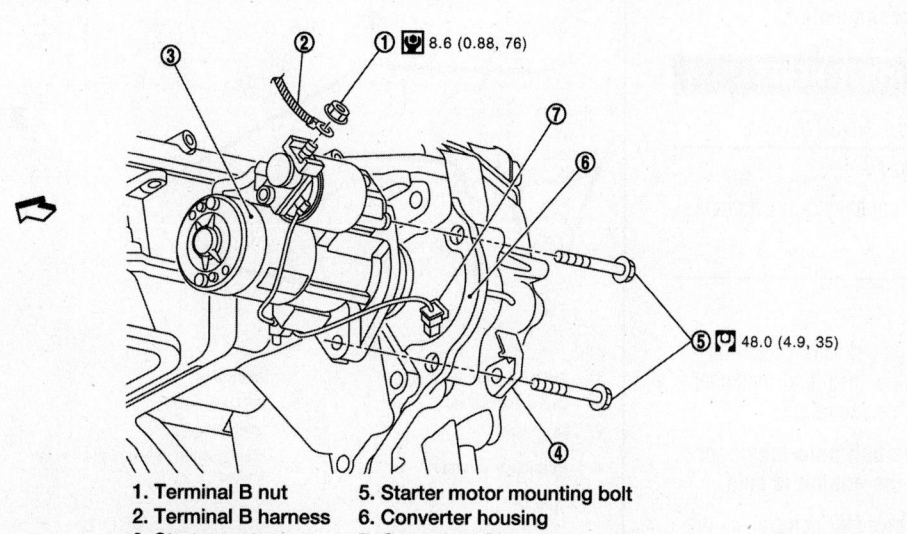

1. Terminal B nut
2. Terminal B harness
3. Starter motor
4. Harness clip bracket
5. Starter motor mounting bolt
6. Converter housing
7. Connector S

71075_IM37_G0033

Fig. 67 View of the starter assembly—3.7L engine

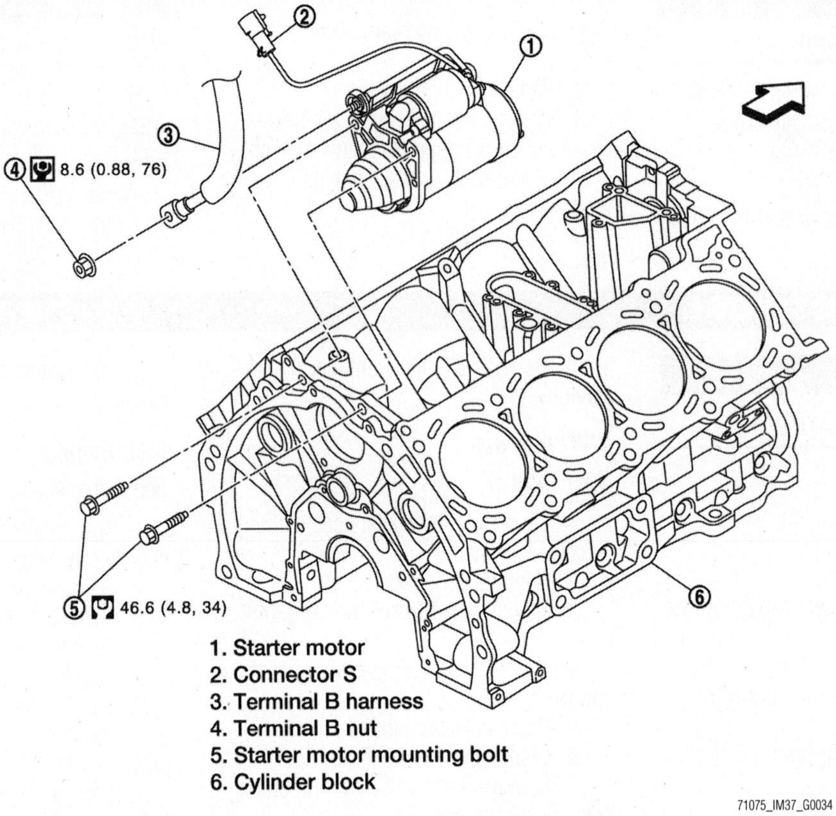

8.6 (0.88, 76)

46.6 (4.8, 34)

1. Starter motor
2. Connector S
3. Terminal B harness
4. Terminal B nut
5. Starter motor mounting bolt
6. Cylinder block

71075_IM37_G0034

Fig. 68 View of the starter assembly—5.6L Engine

ENGINE MECHANICAL

➡**Disconnecting the negative battery cable may interfere with the functions of the on board computer systems and may require the computer to undergo a relearning process, once the negative battery cable is reconnected.**

ACCESSORY DRIVE BELTS

ACCESSORY BELT ROUTING

See Figures 69 and 70.

Refer to the accompanying illustrations.

INSPECTION

3.7L Engine

1. Check that the indicator (notch on mounted side) of drive belt auto-tensioner is within the possible use range.

➡**Check the drive belt auto-tensioner indication when the engine is cold.**

2. When new drive belt is installed, the indicator (notch on mounted side) should be within the range.

3. Visually check entire drive belt for wear, damage or cracks.

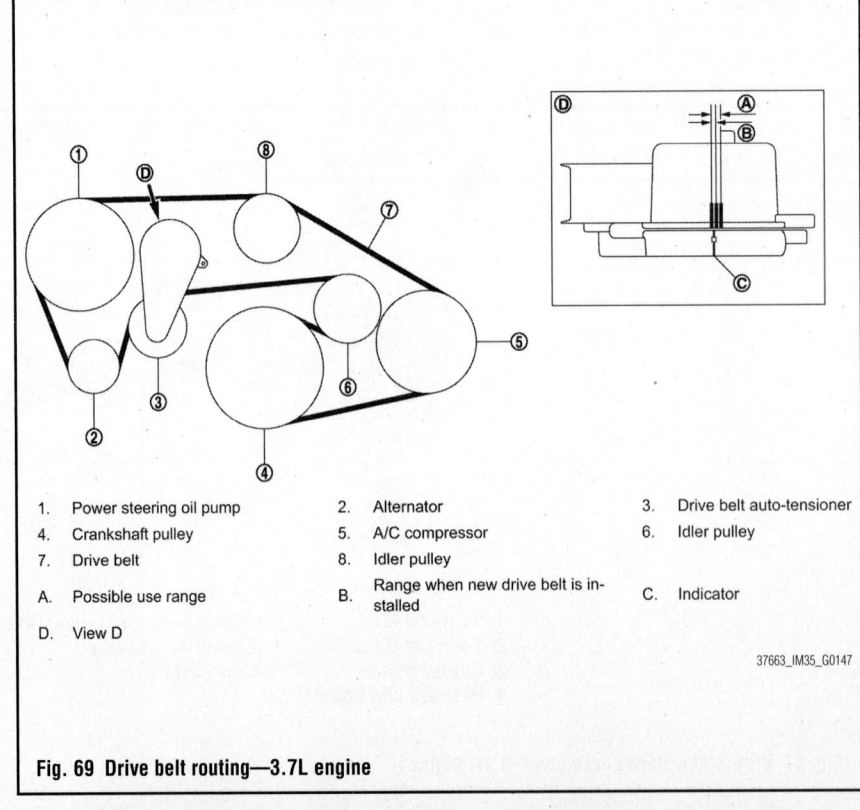

1. Power steering oil pump	2. Alternator	3. Drive belt auto-tensioner
4. Crankshaft pulley	5. A/C compressor	6. Idler pulley
7. Drive belt	8. Idler pulley	
A. Possible use range	B. Range when new drive belt is installed	C. Indicator
D. View D		

37663_IM35_G0147

Fig. 69 Drive belt routing—3.7L engine

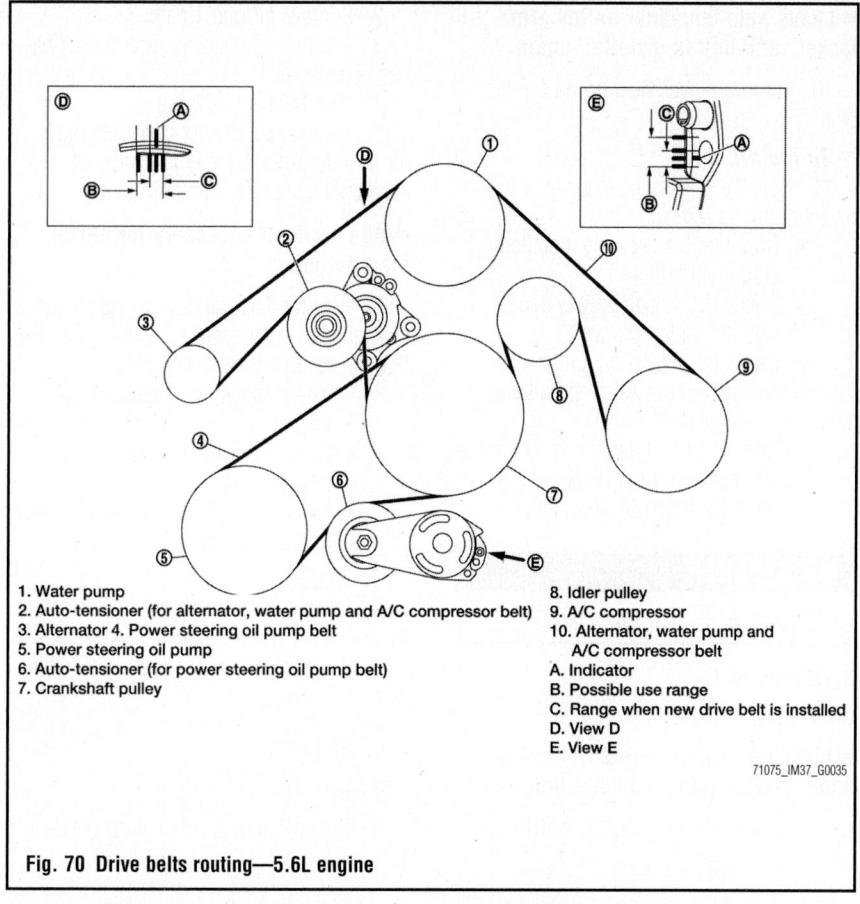

1. Water pump
2. Auto-tensioner (for alternator, water pump and A/C compressor belt)
3. Alternator **4.** Power steering oil pump belt
5. Power steering oil pump
6. Auto-tensioner (for power steering oil pump belt)
7. Crankshaft pulley

8. Idler pulley
9. A/C compressor
10. Alternator, water pump and A/C compressor belt
A. Indicator
B. Possible use range
C. Range when new drive belt is installed
D. View D
E. View E

71075_IM37_G0035

Fig. 70 Drive belts routing—5.6L engine

4. If the indicator (notch on mounted side) is out of the possible use range or belt is damaged, replace drive belt.

5.6L Engine

1. Remove air duct (inlet) when inspecting drive belt for alternator, water pump and A/C compressor.
2. Remove front engine undercover when inspecting power steering oil pump belt.
3. Check that indicator (single line notch) of each auto tensioner is within the allowable working range (between three line notches).

➡**Check auto tensioner indication when engine is cold.**

4. The indicator notch is located on the moving side of auto tensioner for alternator, water pump and A/C compressor belt, while it is found on the mounted side for power steering oil pump belt.
5. Visually check entire belt for wear, damage or cracks.
6. If the indicator is out of allowable working range or belt is damaged, replace belt.

ADJUSTMENT

Belt tension is not necessary, as it is automatically adjusted by drive belt auto-tensioner.

REMOVAL & INSTALLATION

3.7L Engine

See Figure 71.

1. Remove front engine undercover.
2. While securely holding the square hole in pulley center of auto tensioner with a spinner handle, move spinner handle in the direction of arrow (loosening direction of drive belt).

❈❈ WARNING

Avoid placing hand in a location where pinching may occur if the holding tool accidentally comes off.

3. Under the above condition, insert a metallic bar through the holding boss to lock auto-tensioner pulley arm.
4. Remove drive belt.

To install:

➡**Note the following, and install in the reverse order of removal.**

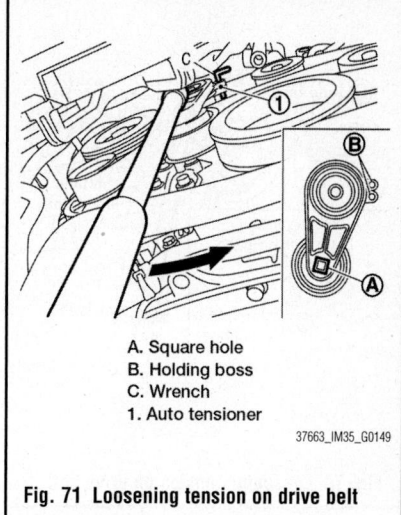

A. Square hole
B. Holding boss
C. Wrench
1. Auto tensioner

37663_IM35_G0149

Fig. 71 Loosening tension on drive belt

- Check drive belt is securely installed around all pulleys.
- Check drive belt is correctly engaged with the pulley groove.
- Check for engine oil and engine coolant are not adhered drive belt and pulley groove.

5. Turn crankshaft pulley clockwise several times to equalize tension between each pulley, and then confirm tension of drive belt at indicator (notch on mounted side) is within the possible use range.

5.6L Engine

See Figures 72 and 73.

ALTERNATOR, WATER PUMP AND A/C COMPRESSOR BELT

1. Remove air duct (inlet).
2. With box wrench, and while securely holding the hexagonal part in pulley center of auto tensioner, move wrench handle in the direction of arrow (loosening direction of tensioner).

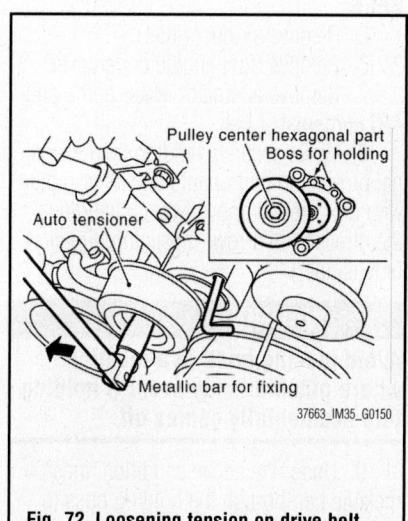

Pulley center hexagonal part
Boss for holding
Auto tensioner
Metallic bar for fixing

37663_IM35_G0150

Fig. 72 Loosening tension on drive belt

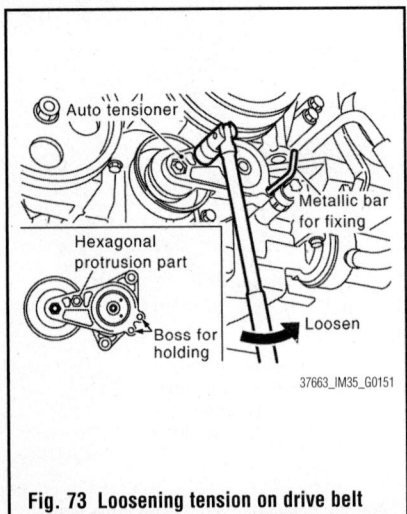

Fig. 73 Loosening tension on drive belt

※※ WARNING

Avoid placing hand in a location where pinching may occur if the holding tool accidentally comes off.

※※ WARNING

Never loosen the hexagonal part in center of drive belt auto tensioner pulley (Never turn it clockwise). If turned clockwise, the complete drive belt auto tensioner must be replaced as a unit, including the pulley.

3. Under the above condition, insert a metallic bar through the holding boss to lock auto tensioner pulley arm.

➡**Leave auto tensioner pulley arm locked until belt is installed again.**

4. Remove alternator, water pump and A/C compressor belt.

POWER STEERING OIL PUMP BELT

5. Remove air duct (inlet).
6. Remove front engine undercover.
7. Remove alternator, water pump and A/C compressor belt.
8. While securely holding the hexagonal protrusion part of auto tensioner pulley with box wrench, move wrench handle in the direction of arrow (loosening direction of tensioner).

※※ WARNING

Avoid placing hand in a location where pinching may occur if holding tool accidentally comes off.

9. Under the above condition, insert a metallic bar through the holding boss to lock auto tensioner pulley arm.

➡**Leave auto tensioner pulley arm locked until belt is installed again.**

10. Remove power steering oil pump belt.

To install:

11. Note the following, and install in the reverse order of removal:
- Check belt is securely installed around all pulleys.
- Check belt is correctly engaged with the pulley groove.
- Check for engine oil and engine coolant are not adhered belt and pulley groove.
- Check that belt tension is within the allowable working range, using indicator notch on auto tensioner.

AIR CLEANER

REMOVAL & INSTALLATION

3.7L Engine
See Figure 74.

➡**Mass air flow sensor is removable under the car-mounted condition.**

1. Remove engine room cover (RH and LH).

2. Remove air duct (inlet).
3. Disconnect mass air flow sensor harness connector.
4. Disconnect PCV hose.
5. Remove air cleaner case with mass air flow sensor assembly and air duct assembly disconnecting their joints.

➡**Add marks if necessary for easier installation.**

6. Remove mass air flow sensor from air cleaner case if necessary. Handle the mass air flow sensor with care, noting the following:
- Never shock the mass air flow sensor.
- Never disassemble the mass air flow sensor.
- Never touch the sensor of the mass air flow sensor.

To install:

7. Note the following, and install in the reverse order of removal.
8. Align marks. Attach each joint. Screw clamps firmly.

5.6L Engine
See Figure 75.

1. Remove engine room cover (RH and LH).

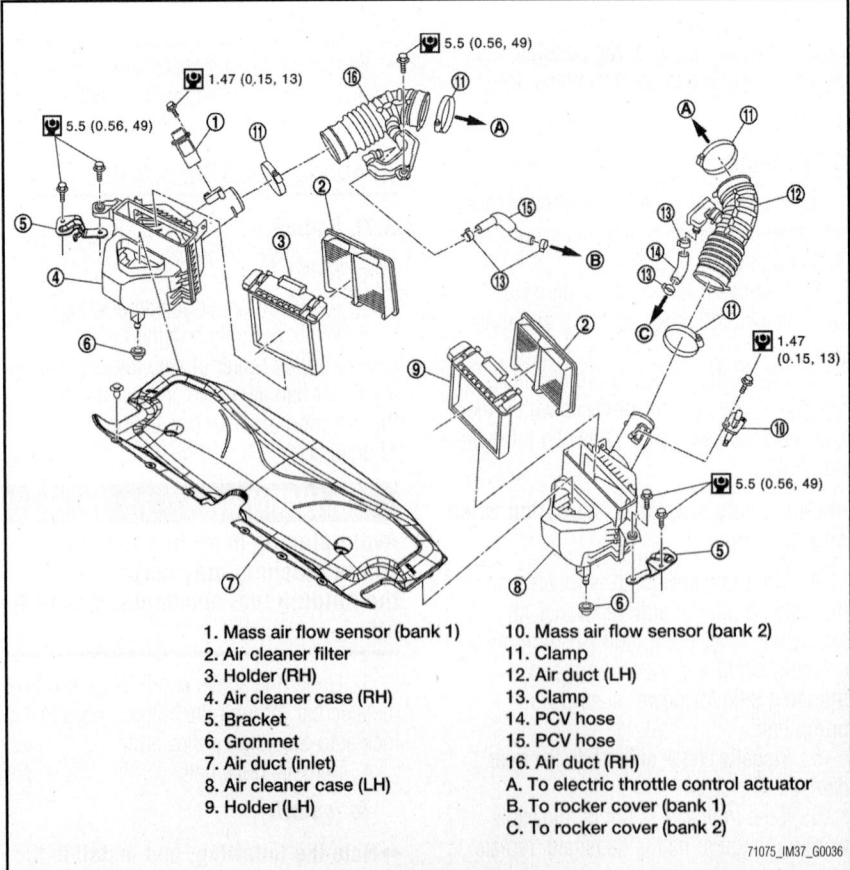

1. Mass air flow sensor (bank 1)
2. Air cleaner filter
3. Holder (RH)
4. Air cleaner case (RH)
5. Bracket
6. Grommet
7. Air duct (inlet)
8. Air cleaner case (LH)
9. Holder (LH)
10. Mass air flow sensor (bank 2)
11. Clamp
12. Air duct (LH)
13. Clamp
14. PCV hose
15. PCV hose
16. Air duct (RH)
A. To electric throttle control actuator
B. To rocker cover (bank 1)
C. To rocker cover (bank 2)

Fig. 74 Exploded view of air cleaner and air duct assembly

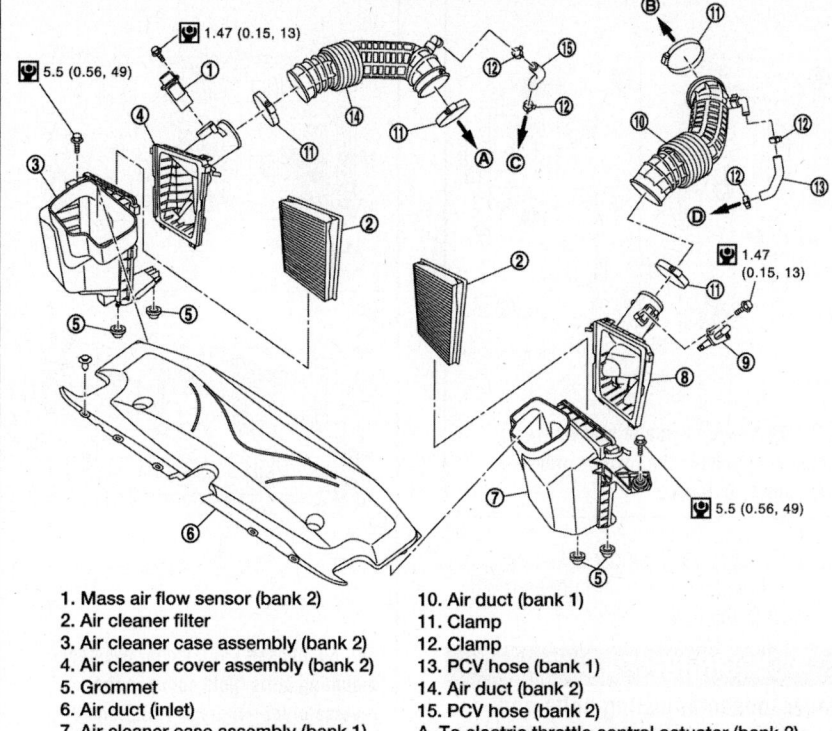

1. Mass air flow sensor (bank 2)
2. Air cleaner filter
3. Air cleaner case assembly (bank 2)
4. Air cleaner cover assembly (bank 2)
5. Grommet
6. Air duct (inlet)
7. Air cleaner case assembly (bank 1)
8. Air cleaner cover assembly (bank 1)
9. Mass air flow sensor (bank1)
10. Air duct (bank 1)
11. Clamp
12. Clamp
13. PCV hose (bank 1)
14. Air duct (bank 2)
15. PCV hose (bank 2)
A. To electric throttle control actuator (bank 2)
B. To electric throttle control actuator (bank 1)
C. To rocker cover (bank 2)
D. To rocker cover (bank 1)

71075_IM37_G0037

Fig. 75 Exploded view of air cleaner and air duct assembly

2. Disconnect harness connector from mass air flow sensor.

3. Disconnect vacuum hose and PCV hose.

4. Remove air duct (inlet), air cleaner case and mass air flow sensor assembly, and air duct and resonator assembly disconnecting their joints.

➡**Add marks if necessary for easier installation.**

5. Remove mass air flow sensor from air cleaner case if necessary. Handle the mass air flow sensor with care, noting the following:

- Never shock mass air flow sensor.
- Never disassemble mass air flow sensor.
- Never touch the sensor of the mass air flow sensor.

6. Inspect air duct and resonator assembly for crack or tear. If anything found, replace air duct and resonator assembly.

To install:

7. Note the following, and install in the reverse order of removal.

8. Align marks. Attach each joint. Screw clamps firmly.

FILTER/ELEMENT REPLACEMENT

3.7L Engine

See Figure 76.

1. Unhook clips.
2. Remove air cleaner filter from air cleaner case.

To install:

3. Note the following, and install in the reverse order of removal.

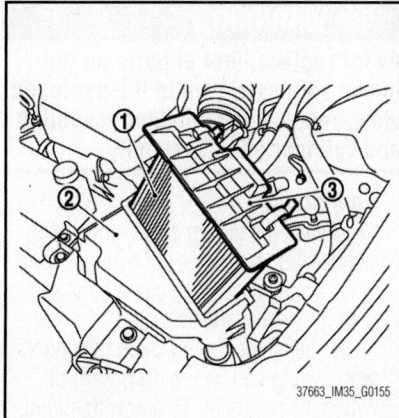

37663_IM35_G0155

Fig. 76 Remove air cleaner filter (1) from air cleaner case (2); holder (3)

4. Install the air cleaner filter by aligning the seal with the notch of air cleaner case.

5.6L Engine

See Figure 77.

1. Remove engine room cover (LH).
2. Unhook clips.
3. Remove holder and air cleaner filter assembly from air cleaner case.

To install:

4. Note the following, and install in the reverse order of removal.

5. Install the air cleaner filter by aligning the seal with the notch of air cleaner case.

CAMSHAFT AND VALVE LIFTERS

REMOVAL & INSTALLATION

3.7L Engine

See Figures 78 through 86.

✳✳ WARNING

Never loosen adjusting bolts and mounting bolts (black color) of VVEL ladder assembly. If loosened, the stroke of cam lift becomes out of adjustment. In such case, replacement of VVEL ladder assembly and cylinder head assembly is required.

➡**VVEL ladder assembly cannot be replaced as a single part, because it is machined together with cylinder head assembly.**

1. Remove rocker covers (bank 1 and bank 2).

2. Remove VVEL actuator sub assembly as follows:

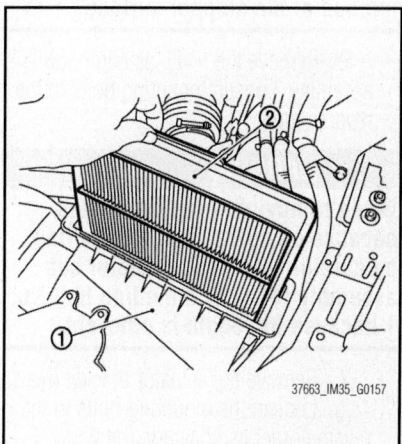

37663_IM35_G0157

Fig. 77 Remove holder and air cleaner filter assembly (2) from air cleaner case (1)

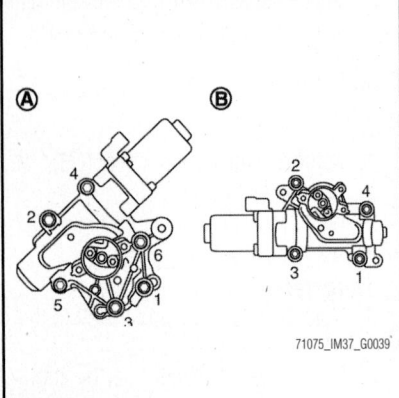

Fig. 78 Loosen VVEL actuator sub assembly mounting bolts in the reverse order—A: bank 2, B: Bank 1

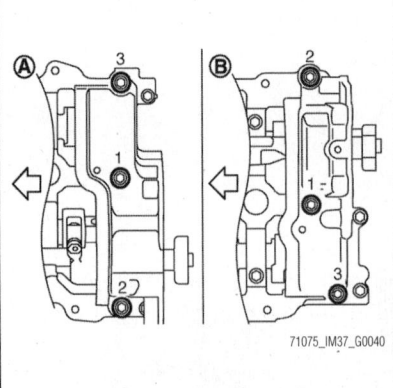

Fig. 79 Loosen actuator bracket (rear) mounting bolts in the reverse order—A: bank 1, B: Bank 2

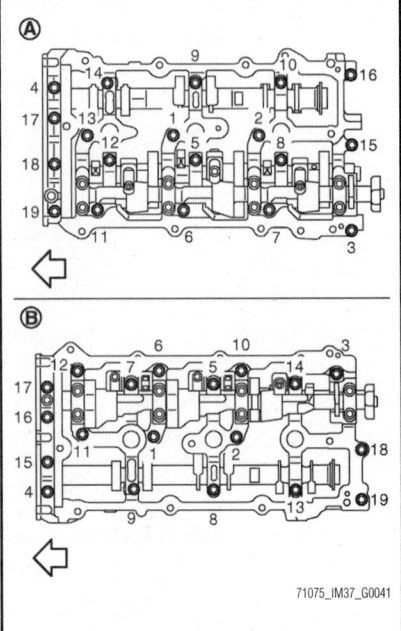

Fig. 80 Loosen VVEL ladder assembly mounting bolts (gold color) in the reverse order—A: bank 1, B: Bank 2

✳✳ WARNING

VVEL actuator sub assembly and VVEL control shaft position sensor are not reusable. Never remove them unless they are required.

   a. Remove VVEL control shaft position sensor.

   b. Turn control shaft to the large lift side and fix it in order to prevent the interference of the stopper surface.

   c. If control shaft cannot be moved, set crankshaft in position referring to the information below. (To displace cam nose)

   d. Fix two flat areas of control shaft with a wrench to remove mounting bolts of control shaft.

✳✳ WARNING

During the operation, never allow a wrench to interfere with other parts. Fix control shaft to prevent the interference of the stopper surface.

   e. Remove the VVEL actuator sub assembly. Loosen mounting bolts in the reverse order.

✳✳ WARNING

When removing, prepare wastes because oil spills. When installing, be careful with VVEL actuator sub assembly (bank 2) mounting bolt No. 1 because its length is different.

   f. Remove the actuator bracket (rear).

   g. Loosen the mounting bolts in the reverse order as shown in the figure.

  3. Remove front timing chain case, camshaft sprockets, and timing chain.

  4. Remove rear timing chain case.

  5. Remove VVEL ladder assembly.

   a. Loosen the mounting bolts (gold color) in the reverse order.

✳✳ WARNING

Never loosen adjusting bolts and mounting bolts (black color). When removing VVEL ladder assembly, hold the drive shaft from below so as not to drop it.

  6. Remove the camshaft (EXH).

  7. Remove valve lifter.

  8. Identify installation positions, and store them without mixing them up.

  9. Remove the timing chain tensioners (secondary) from cylinder head.

  10. Remove timing chain tensioners (secondary) with its stopper pin (C) attached.

➡**Stopper pin should be attached when timing chain (secondary) is removed.**

  11. Remove oil filter from cylinder head, if necessary.

✳✳ WARNING

As for replacement of parts on the intake side as shown in the exploded view, replace VVEL ladder assembly and cylinder head assembly.

To install:

  12. Install the timing chain tensioners (secondary) on both sides of cylinder head.

   a. Install the timing chain tensioner with its stopper pin attached.

   b. Install the timing chain tensioner with sliding part facing downward on cylinder head (bank 1), and with sliding part facing upward on cylinder head (bank 2).

  13. Install oil filter, if removed. Do not project from the cylinder head surface.

  14. Install valve lifter. Install it in the original position.

  15. Install the camshaft (EXH).

   a. Distinction between camshaft (EXH) (bank 1 and bank 2) is performed with the identification mark.

  16. Install VVEL ladder assembly as follows:

   a. Apply a continuous bead of liquid gasket with tube presser (commercial service tool) to the cylinder head.

   b. Use Genuine RTV Silicone Sealant or equivalent.

   c. Tighten mounting bolts in numerical order in the following steps:

- First pass: 1 ft. lbs. (1.96 Nm)
- Second pass: 4 ft. lbs. (5.88 Nm)
- Third pass: 8 ft. lbs. (10 Nm)

  17. Measure difference in levels between front end faces of VVEL ladder assembly and cylinder head. (_0.0055 to 0.0055 inches)

   a. Measure two positions (both intake and exhaust side) for a single bank.

   b. If the measured value is out of the standard, re-install VVEL ladder assembly.

  18. Install rear timing chain case.

  19. Install the camshaft sprockets and timing chains.

  20. Install the actuator bracket (rear) as follows:

   a. Apply a continuous bead of liquid gasket with tube presser (commercial service tool) to the actuator bracket (rear).

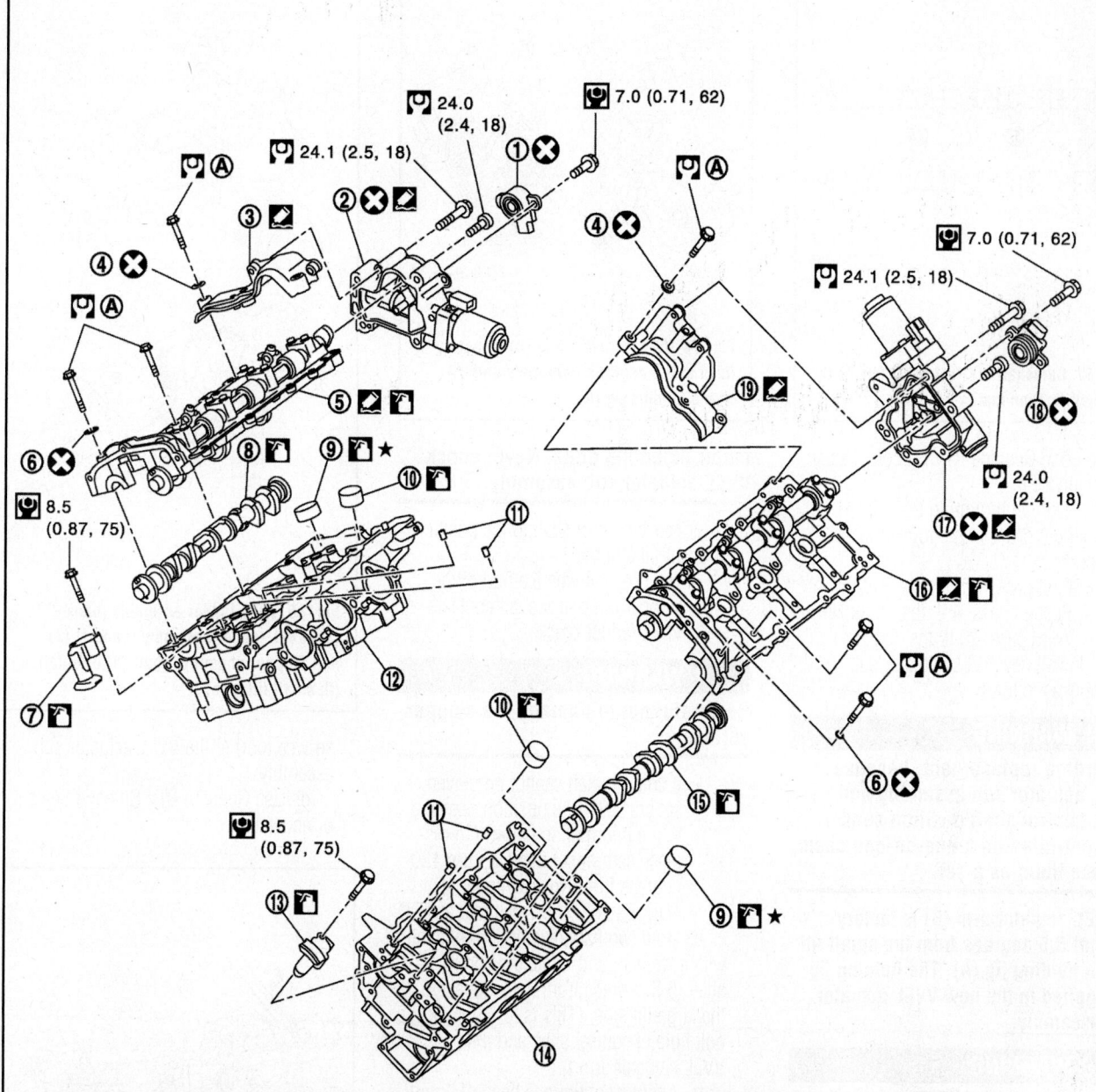

1. VVEL control shaft position sensor (bank 1)
2. VVEL actuator sub assembly (bank 1)
3. Actuator bracket (rear) (bank 1)
4. Washer
5. VVEL ladder assembly (bank 1)
6. Washer
7. Timing chain tensioner (secondary) (bank 1)
8. Camshaft (EXH) (bank 1)
9. Valve lifter (EXH)
10. Valve lifter (INT)
11. Oil filter
12. Cylinder head (bank 1)
13. Timing chain tensioner (secondary) (bank 2)
14. Cylinder head (bank 2)
15. Camshaft (EXH) (bank 2)
16. VVEL ladder assembly (bank 2)
17. VVEL actuator sub assembly (bank 2)
18. VVEL control shaft position sensor (bank 2)
19. Actuator bracket (rear) (bank 2)
A. Comply with the assembly procedure when tightening.

71075_IM37_G0038

Fig. 81 Exploded view of the camshaft assembly—3.7L Engine

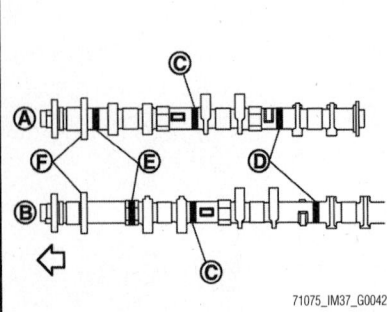

Fig. 82 Camshaft (EXH) (bank 1 and bank 2) identification marks—1 of 2

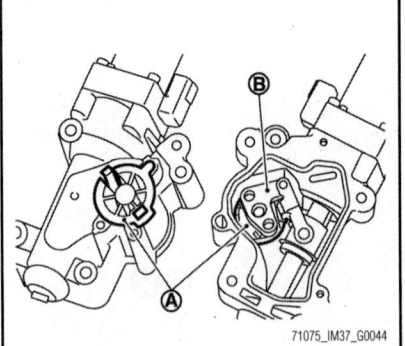

Fig. 84 VVEL actuator arm (B) is factory-fixed at 5.5 degrees from the small lift with a holding jig (A)

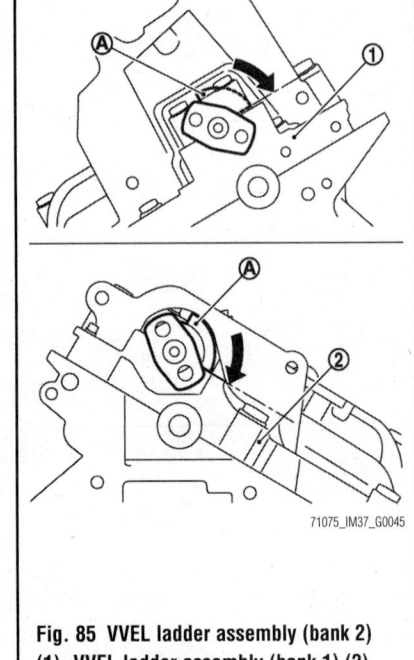

Fig. 85 VVEL ladder assembly (bank 2) (1), VVEL ladder assembly (bank 1) (2) and Stopper of control shaft (A) orientation (fl: small lift side)

b. Use Genuine RTV Silicone Sealant or equivalent.

c. Tighten mounting bolts in numerical order in the following steps:

- First pass: 1 ft. lbs. (1.96 Nm)
- Second pass: 4 ft. lbs. (5.88 Nm)
- Third pass: 23 ft. lbs. (31 Nm)

21. Install new VVEL actuator sub assembly as follows:

☀☀ WARNING

Regarding replacement, because VVEL actuator sub assembly and VVEL control shaft position sensor are controlled on a one-on-one basis, replace them as a set.

→VVEL actuator arm (B) is factory-fixed at 5.5 degrees from the small lift with a holding jig (A). The holding jig is supplied in the new VVEL actuator sub assembly.

☀☀ WARNING

Never disassemble VVEL actuator sub assembly. Never loosen actuator

motor mounting bolts. Never shock VVEL actuator sub assembly.

a. Move control shaft to the position of small lift stopper.

- The position where a part of the stopper of control shaft contacts VVEL ladder bracket.

☀☀ WARNING

Be careful not to damage the stopper surface.

- If control shaft cannot be moved, set crankshaft in position referring to the information below. (To displace cam nose) Bank 1: Turn 120 degrees from No. 1 cylinder at TDC, Bank 2: No. 1 cylinder at TDC

b. Hold two flat areas of control shaft with a wrench, and rotate the control shaft (5.5 degrees from the stopper) to the large lift side. (This is for aligning the bolt hole of control shaft and the hole of VVEL actuator arm.)

c. Apply a continuous bead of liquid gasket with tube presser (commercial

service tool) to the VVEL actuator sub assembly.

d. Use Genuine RTV Silicone Sealant or equivalent.

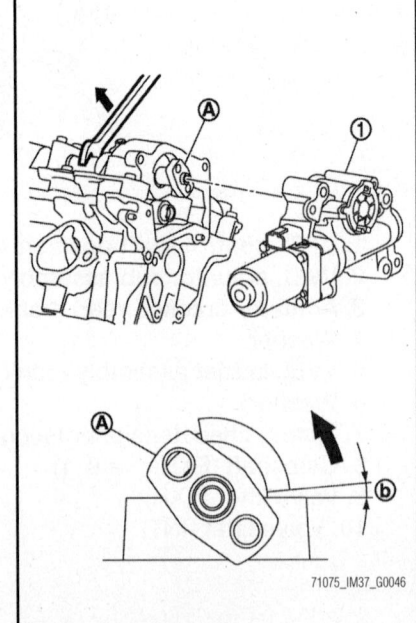

Fig. 86 VVEL actuator sub assembly (bank 1) (1), Control shaft (A), 5.5 degrees (fl: Large lift side) alignment

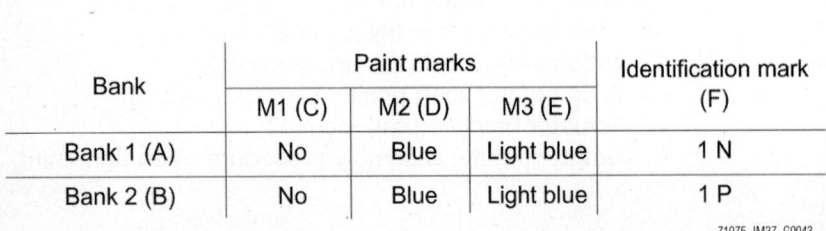

Bank	Paint marks			Identification mark (F)
	M1 (C)	M2 (D)	M3 (E)	
Bank 1 (A)	No	Blue	Light blue	1 N
Bank 2 (B)	No	Blue	Light blue	1 P

Fig. 83 Camshaft (EXH) (bank 1 and bank 2) identification marks—2 of 2

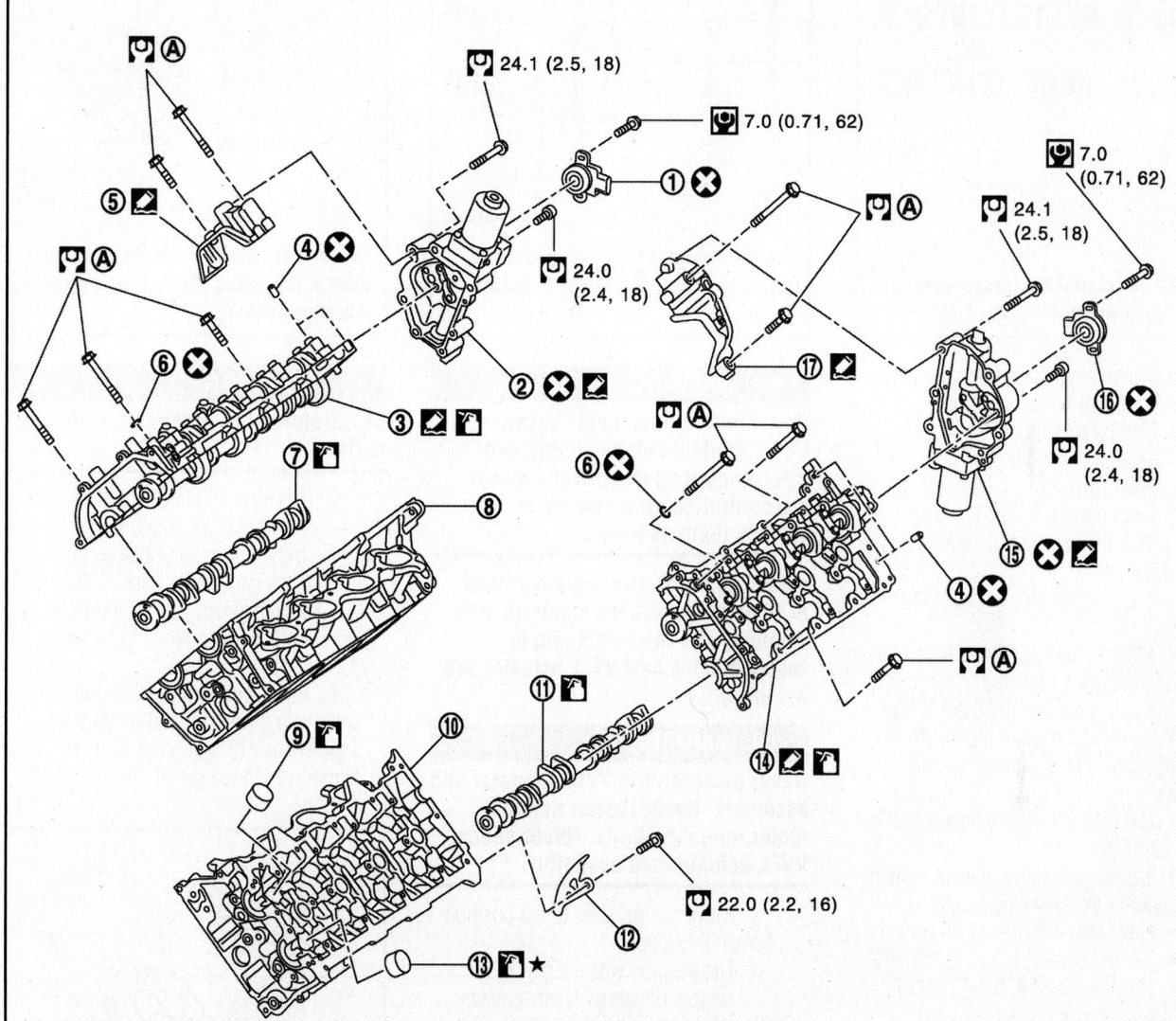

1. VVEL control shaft position sensor (bank 2)
2. VVEL actuator sub assembly (bank 2)
3. VVEL ladder assembly (bank 2)
4. Dowel pin
5. Actuator bracket (rear) (bank 2)
6. Washer
7. Camshaft (EXH) (bank 2)
8. Cylinder head (bank 2)
9. Valve lifter (INT)

10. Cylinder head (bank 1)
11. Camshaft (EXH) (bank 1)
12. Actuator cover
13. Valve lifter (EXH)
14. VVEL ladder assembly (bank 1)
15. VVEL actuator sub assembly (bank 1)
16. VVEL control shaft position sensor (bank 1)
17. Actuator bracket (rear) (bank 1)
A. Comply with the installation procedure when tightening.

71075_IM37_G0052

Fig. 92 Exploded view of the camshaft assembly—5.6L Engine

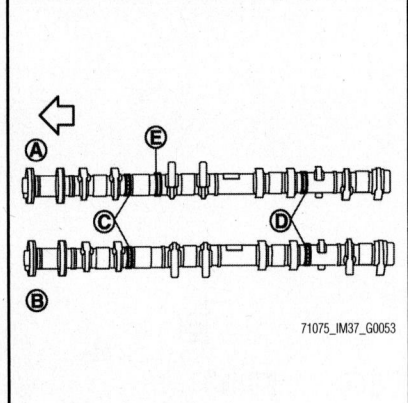

Fig. 93 Camshaft (EXH) (bank 1 and bank 2) identification marks—1 of 2

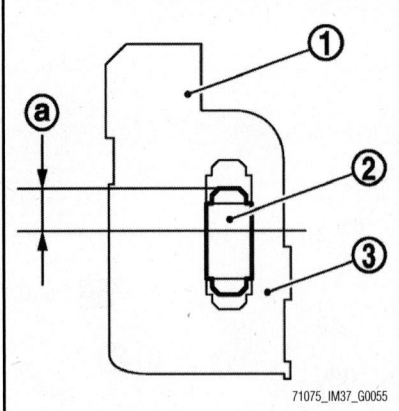

Fig. 95 Actuator bracket (1), dowel pins (2), VVEL ladder (3) and 0.157 - 0.236 inches (a)

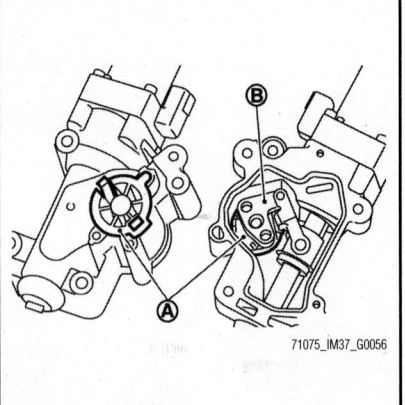

Fig. 96 VVEL actuator arm (B) is factory-fixed at 10 degrees from the small lift with a holding jig (A)

c. Tighten mounting bolts in numerical order in the following steps:
- First pass: 1 ft. lbs. (1.96 Nm)
- Second pass: 4 ft. lbs. (5.88 Nm)
- Third pass: 8 ft. lbs. (10 Nm)

15. Install the camshaft sprockets and timing chains.

16. Install the actuator bracket (rear) as follows:

a. Refer to the figure to replace new dowel pins (2), if removed.

b. Apply a continuous bead of liquid gasket with tube presser (commercial service tool) to the actuator bracket (rear).

c. Use Genuine RTV Silicone Sealant or equivalent.

d. Tighten mounting bolts in numerical order in the following steps:
- First pass: 1 ft. lbs. (1.96 Nm)
- Second pass: 4 ft. lbs. (5.88 Nm)
- Third pass: 23 ft. lbs. (31 Nm)

17. Install new VVEL actuator sub assembly as follows:

❋❋ WARNING

Regarding replacement, because VVEL actuator sub assembly and VVEL control shaft position sensor are controlled on a one-on-one basis, replace them as a set.

➡**VVEL actuator arm is factory-fixed at 10 degrees from the small lift with a holding jig. The holding jig is supplied in the new VVEL actuator sub assembly.**

❋❋ WARNING

Never disassemble VVEL actuator sub assembly. Never loosen actuator motor mounting bolts. Never shock VVEL actuator sub assembly.

a. Move control shaft to the position of small lift stopper.
- The position where a part of the stopper of control shaft contacts VVEL ladder bracket.

❋❋ WARNING

Be careful not to damage the stopper surface.

- If control shaft cannot be moved, set crankshaft in position referring to the information below. (To displace cam nose) Bank 1: Turn 360 degrees from No. 1 cylinder at TDC, Bank 2: No. 1 cylinder at TDC

b. Hold two flat areas of control shaft with a wrench, and rotate the control shaft (5.5 degrees from the stopper) to the large lift side. (This is

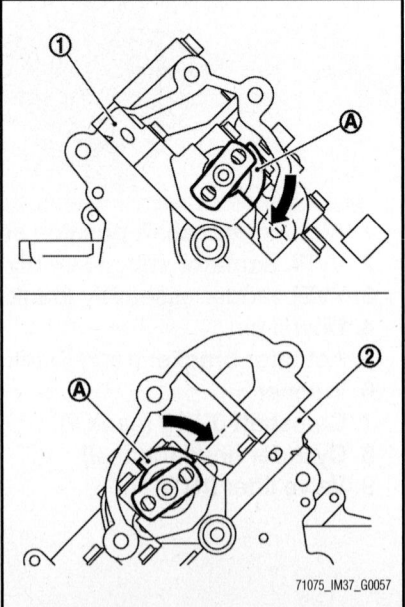

Fig. 97 VVEL ladder assembly (bank 2) (1), VVEL ladder assembly (bank 1) (2) and Stopper of control shaft (A) orientation (fl: small lift side)

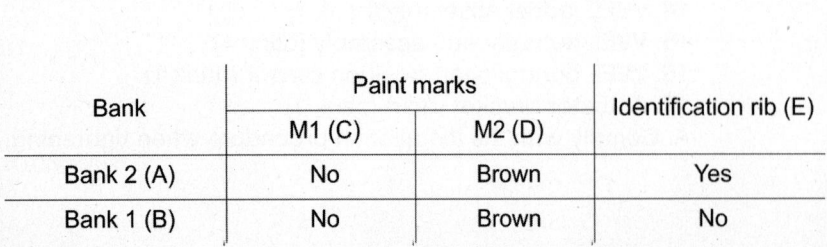

Bank	Paint marks		Identification rib (E)
	M1 (C)	M2 (D)	
Bank 2 (A)	No	Brown	Yes
Bank 1 (B)	No	Brown	No

Fig. 94 Camshaft (EXH) (bank 1 and bank 2) identification marks—2 of 2

for aligning the bolt hole of control shaft and the hole of VVEL actuator arm.)

➡**The figure shows an example of bank 2.**

c. Apply a continuous bead of liquid gasket with tube presser (commercial service tool) to the VVEL actuator sub assembly.

d. Use Genuine RTV Silicone Sealant or equivalent.

⁜ **WARNING**

Never apply gasket to the oil passage.

e. Install the new VVEL actuator sub assembly. Tighten mounting bolts, in numerical order.

⁜ **WARNING**

When installing, be careful with VVEL actuator sub assembly (bank 1) mounting bolt No. 4 because its length is different. Be sure to check that the VVEL actuator sub assembly is in contact with the cylinder head before tightening the mounting bolts.

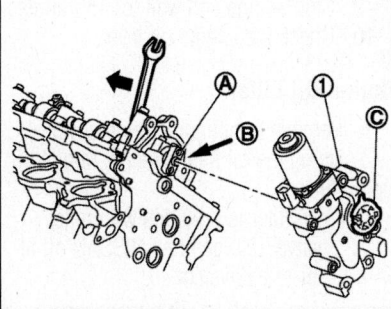

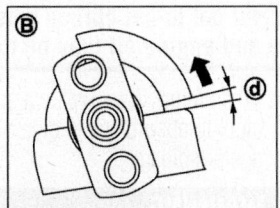

1: VVEL actuator sub assembly (bank 2)
A: Control shaft
B: View B
C: Holding jig
D: 10 degrees
⊠: Large lift side

71075_IM37_G0058

Fig. 98 Hold two flat areas of control shaft with a wrench, and rotate the control shaft

f. Remove the holding jig.

g. Check that VVEL actuator arm bolt hole is aligned with control shaft tapped hole. If it is not aligned, turn control shaft for alignment.

⁜ **WARNING**

Never give an impact to the magnet part.

h. Fix two flat areas of control shaft with a wrench to install mounting bolts of control shaft.

⁜ **WARNING**

During the operation, never allow a wrench to interfere with other parts. Fix control shaft to prevent the interference of the stopper surface.

18. Install new VVEL control shaft position sensor as follows:

⁜ **WARNING**

Regarding replacement, because VVEL actuator sub assembly and VVEL control shaft position sensor are controlled on a one-on-one basis, replace them as a set.

a. Apply engine oil to O-ring or contact surface of O-ring.

b. Align matching marks of VVEL control shaft position sensor and upper housing.

c. Face connector toward matching mark.

d. Temporarily tighten bolt.

e. Adjust VVEL control shaft position sensor after setting the engine assembly in the vehicle.

⁜ **WARNING**

Be sure to adjust VVEL control shaft position sensor.

f. After adjusting VVEL control shaft position sensor, tighten bolts to the specified torque.

19. Inspect the valve clearance.

20. Install in the reverse order of removal after this step.

CATALYTIC CONVERTER

REMOVAL & INSTALLATION

Refer to Exhaust Manifold.

CRANKSHAFT FRONT SEAL

REMOVAL & INSTALLATION

3.7L Engine
See Figure 99.

1. Remove the following parts:
   - Front engine undercover
   - Drive belt
   - Radiator cooling fan assembly
   - Crankshaft pulley
2. Remove front oil seal.

⁜ **WARNING**

Be careful not to damage front timing chain case and crankshaft.

To install:
3. Apply engine oil to both oil seal lip and dust seal lip of new front oil seal.
4. Install front oil seal.
   a. Install front oil seal so that each seal lip is oriented as shown.
   b. Using suitable drift, press-fit until the height of front oil seal is level with the mounting surface.

⁜ **WARNING**

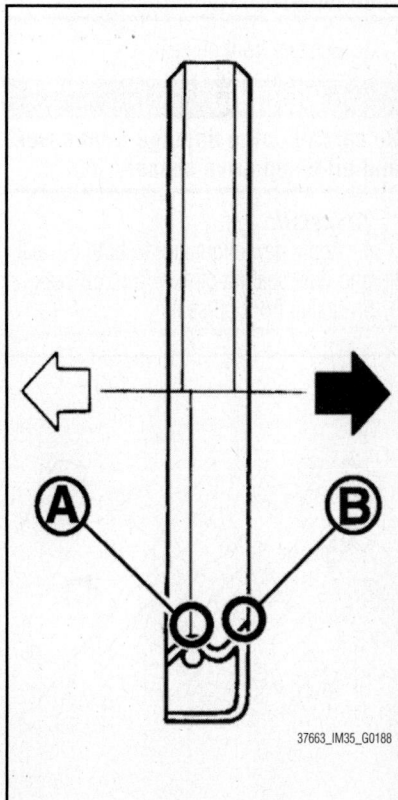

37663_IM35_G0188

Fig. 99 Install front oil seal so that each seal lip is oriented as shown

- Be careful not to damage front timing chain case and crankshaft.
- Press-fit straight and avoid causing burrs or tilting oil seal.

5. Install in the reverse order of removal after this step.

5.6L Engine

See Figures 100 and 101.

1. Remove the following parts:
   - Front engine undercover
   - Radiator
   - Drive belts
   - Rear plate cover
2. Remove crankshaft pulley as follows:
   a. Set ring gear stopper.
   b. Loosen crankshaft pulley bolt, and then pull crankshaft pulley with both hands to remove it.

> ※※ **WARNING**
>
> **Never remove crankshaft pulley bolt. Keep loosened crankshaft pulley bolt in place to protect removed crankshaft pulley from dropping.**

> ※※ **WARNING**
>
> **Never remove balance weight (inner hexagon bolt) at the front of crankshaft pulley.**

3. Remove front oil seal.

> ※※ **WARNING**
>
> **Be careful not to damage front cover and oil pump drive spacer.**

To install:

4. Apply new engine oil to both oil seal lip and dust seal lip of new front oil seal.
5. Install front oil seal.

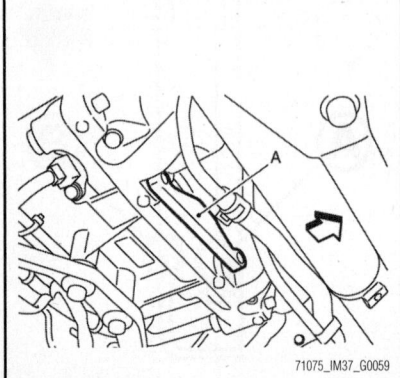

Fig. 100 Set ring gear stopper [SST: KV10119200 (J-49277)] (A)

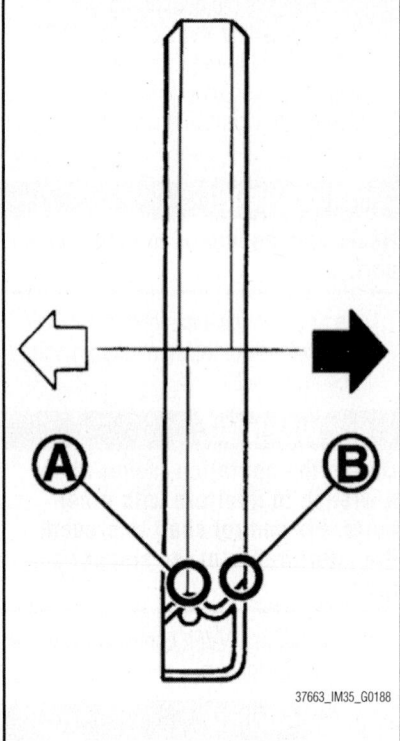

Fig. 101 Install front oil seal so that each seal lip is oriented as shown

a. Install front oil seal so that each seal lip is oriented as shown.
b. Using suitable drift, press fit until the height of front oil seal is level with the mounting surface.
c. Check the garter spring is in position and seal lips not inverted.

> ※※ **WARNING**
>
> **Be careful not to damage front cover and oil pump drive spacer.**

> ※※ **WARNING**
>
> **Press fit straight and avoid causing burrs or tilting oil seal.**

6. Install in the reverse order of removal.

ENGINE OIL & FILTER

REPLACEMENT

Engine Oil

> ※※ **WARNING**
>
> **Be careful not to burn yourself, as engine oil may be hot.**

> ※※ **WARNING**
>
> **Prolonged and repeated contact with used engine oil may cause skin cancer. Try to avoid direct skin contact with used engine oil. If skin contact is made, wash thoroughly with soap or hand cleaner as soon as possible.**

1. Warm up engine, put vehicle horizontally and check for engine oil leakage from engine components.
2. Stop engine and wait for 15 minutes.
3. Loosen oil filler cap.
4. Remove mounting bolts, and then pull down the rear of front engine undercover and secure it using clip.
5. Remove drain plug and then drain engine oil.
6. Install drain plug with new drain plug washer.

> ※※ **WARNING**
>
> **Be sure to clean drain plug and install with new drain plug washer.**

7. Refill with new engine oil:
   - The refill capacity depends on the engine oil temperature and drain time.
   - Always use oil level gauge to determine the proper amount of engine oil in engine.
8. Warm up engine and check area around drain plug and oil filter for oil leakage.
9. Stop engine and wait for 15 minutes.
10. Check the engine oil level.

Engine Oil Filter

1. Remove front engine undercover.
2. Using the oil filter wrench, remove the oil filter:
   - Oil filter is provided with relief valve. Use genuine NISSAN oil filter or equivalent.

> ※※ **WARNING**
>
> **Be careful not to get burned when engine and engine oil may be hot.**

- When removing, prepare a shop cloth to absorb any engine oil leakage or spillage.

> ※※ **WARNING**
>
> **Never allow engine oil to adhere to drive belts.**

- Completely wipe off any engine oil that adhere to engine and vehicle.

To install:

3. Remove foreign materials adhering to oil filter installation surface.
4. Apply new engine oil to the oil seal circumference of the new oil filter.

5. Screw oil filter manually until it touches the installation surface, then tighten it by ⅔ turn. Or tighten to specification.

6. Check the engine oil level.

7. Start engine, and check there is no leakage of engine oil.

8. Stop engine and wait for 15 minutes.

9. Check the engine oil level and adjust engine oil.

EXHAUST MANIFOLD

REMOVAL & INSTALLATION

3.7L Engine

See Figures 102 through 106.

✳✳ WARNING

Perform the work when the exhaust and cooling system have completely cooled down.

➡**When removing bank 1 side parts only, step 3, 10 and 11 are unnecessary.**

1. Remove engine room cover (RH and LH).

2. Remove engine cover.

3. Drain engine coolant.

✳✳ WARNING

Perform this step when engine is cold. Never spill engine coolant on drive belt.

4. Remove air cleaner case and air duct.

5. Remove front and rear undercover.

6. Disconnect heated oxygen sensor harness connectors.

✳✳ WARNING

Be careful not to damage heated oxygen sensor 2. Discard any heated

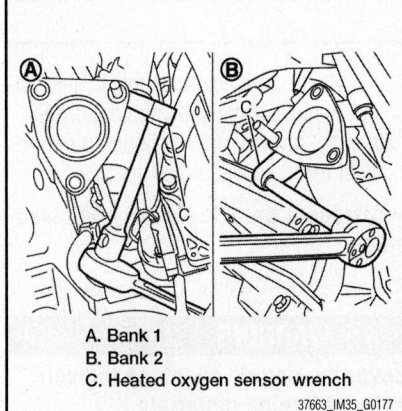

A. Bank 1
B. Bank 2
C. Heated oxygen sensor wrench

37663_IM35_G0177

Fig. 102 Disconnect harness connector and remove air fuel ratio sensor 1 on both banks

oxygen sensor 2 which has been dropped onto a hard surface such as a concrete floor. Replace with a new sensor.

7. Remove exhaust mounting bracket between three way catalysts (bank 1 and bank 2) and transmission.

8. Remove exhaust front tube and three way catalysts (bank 1 and bank 2).

9. Disconnect harness connector and remove air fuel ratio sensor 1 on both banks using heated oxygen sensor wrench KV10114400 (J38365).

10. Put marks to identify installation positions of each air fuel ratio sensor 1.

✳✳ WARNING

Be careful not to damage air fuel ratio sensor 1. Discard any air fuel ratio sensor 1 which has been dropped onto a hard surface such as a concrete floor. Replace with a new sensor.

11. Disconnect steering lower joint at power steering gear assembly side, and release steering lower shaft.

12. Remove water bypass pipe and heater pipe.

13. Remove exhaust manifold cover.

14. Loosen mounting nuts in reverse order as shown to remove exhaust manifold.

➡**Disregard the numerical order No. 7 and 8 in removal.**

15. Remove gaskets.

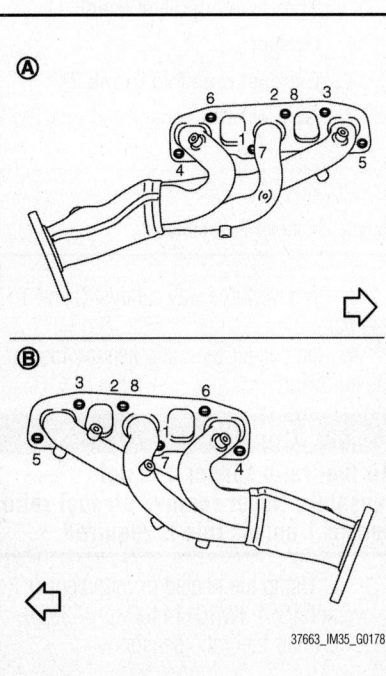

37663_IM35_G0178

Fig. 103 Loosen mounting nuts in reverse order as shown

✳✳ WARNING

Cover engine openings to avoid entry of foreign materials.

To install:

16. Note the following, and install in the reverse order of removal.

17. Install exhaust manifold gasket in direction shown.

18. If the exhaust manifold stud bolts were removed, install them and tighten to 11 ft. lbs. (15 Nm).

19. Install exhaust manifold retainers in numerical order as shown.

➡**Tighten nuts No. 1 and 2 in two steps. The numerical order No. 7 and 8 shown second step.**

✳✳ WARNING

Before installing a new air fuel ratio sensor 1 and heated oxygen sensor 2, clean exhaust system threads using oxygen sensor thread cleaner (J-43897-18 or J43897-12) and apply anti-seize lubricant.

✳✳ WARNING

Never over torque air fuel ratio sensor 1 and heated oxygen sensor 2. Doing so may cause damage to air fuel ratio sensor 1 and heated oxygen sensor 2, resulting in the "MIL" coming on.

5.6L Engine

See Figures 107 and 108.

✳✳ WARNING

Perform the work when the exhaust and cooling system have completely cooled down.

➡**For 2WD models, the exhaust manifold on the bank 2 side can be removed without removing the engine.**

1. Remove heated oxygen sensor 2.

✳✳ WARNING

Heated oxygen sensor 2 is not reusable. Never remove heated oxygen sensor 2 unless this is required.

2. Using the heated oxygen sensor wrench [SST: KV10114400 (J-38365)], remove heated oxygen sensor 2.

➡**The heated oxygen sensor 2 is removable under vehicle mounted condition**

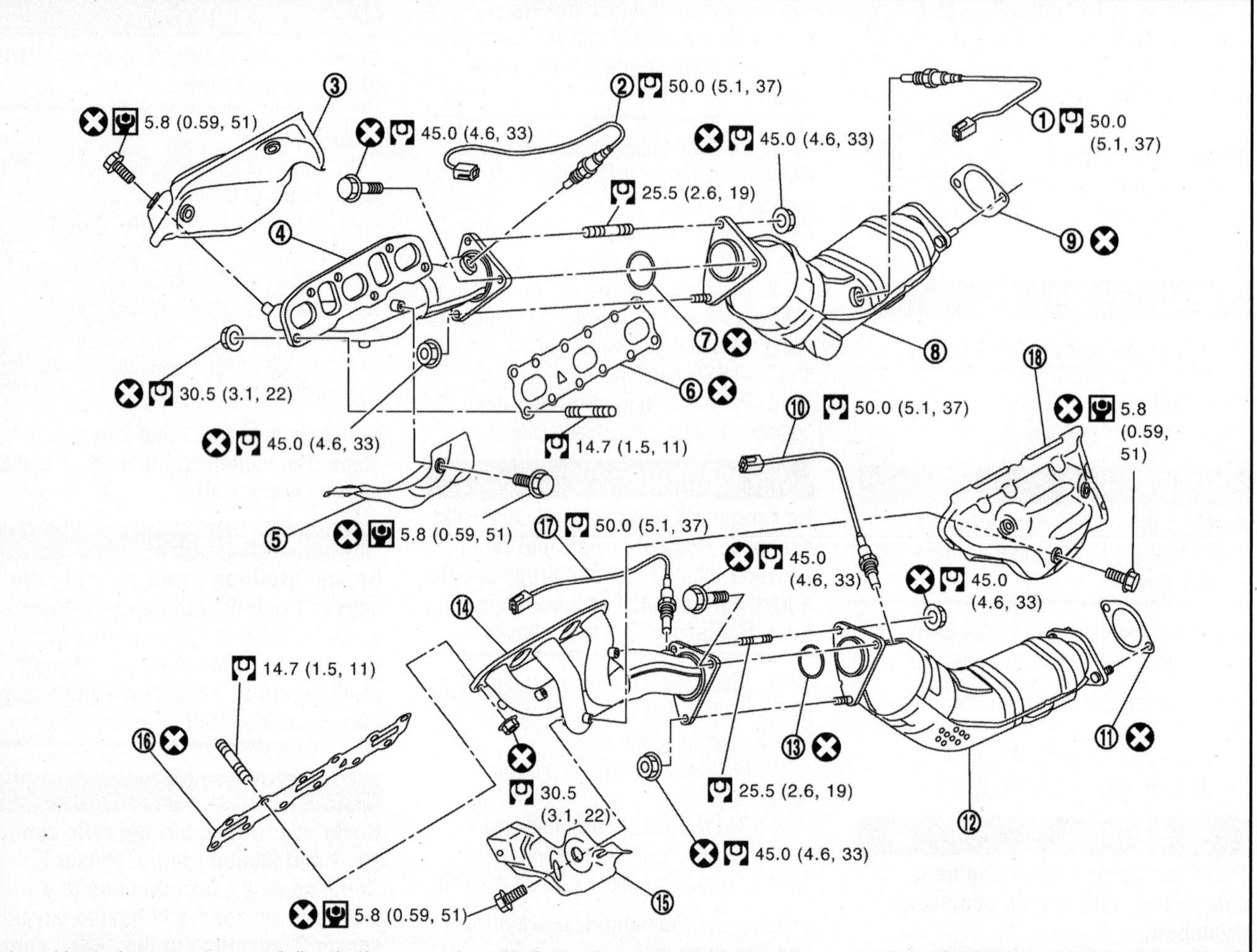

⊗ 🔧 5.8 (0.59, 51)

② 🔧 50.0 (5.1, 37)

⊗ 🔧 45.0 (4.6, 33)

⊗ 🔧 45.0 (4.6, 33)

① 🔧 50.0 (5.1, 37)

🔧 25.5 (2.6, 19)

③

④

⑨ ⊗

⊗ 🔧 30.5 (3.1, 22)

⑦ ⊗

⑥ ⊗

⑧

⑱

⊗ 🔧 5.8 (0.59, 51)

⊗ 🔧 45.0 (4.6, 33)

🔧 14.7 (1.5, 11)

⑩ 🔧 50.0 (5.1, 37)

⑤ ⊗ 🔧 5.8 (0.59, 51)

⑰ 🔧 50.0 (5.1, 37)

⊗ 🔧 45.0 (4.6, 33)

⊗ 🔧 45.0 (4.6, 33)

🔧 14.7 (1.5, 11)

⑭

⊗ 🔧 30.5 (3.1, 22)

🔧 25.5 (2.6, 19)

⑬ ⊗

⑪ ⊗

⑯ ⊗

⑫

⊗ 🔧 45.0 (4.6, 33)

⑮

⊗ 🔧 5.8 (0.59, 51)

1. Heated oxygen sensor 2 (bank 1)	2. Air fuel ratio sensor 1 (bank 1)	3. Exhaust manifold cover (upper)
4. Exhaust manifold (bank 1)	5. Exhaust manifold cover (lower)	6. Gasket
7. Ring gasket	8. Three way catalyst (bank 1)	9. Gasket
10. Heated oxygen sensor 2 (bank 2)	11. Gasket	12. Three way catalyst (bank 2)
13. Ring gasket	14. Exhaust manifold (bank 2)	15. Exhaust manifold cover (lower)
16. Gasket	17. Air fuel ratio sensor 1 (bank 2)	18. Exhaust manifold cover (upper)

37663_IM35_G0180

Fig. 104 Exploded view of exhaust manifold and catalytic converter assembly

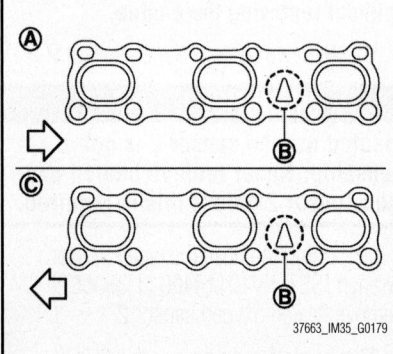

Ⓐ

Ⓒ

37663_IM35_G0179

Fig. 105 Install exhaust manifold gasket in direction shown

3. Remove three way catalyst (bank 1 and bank 2).

4. Remove air fuel ratio sensor 1as per the following:

❋❋ WARNING

Air fuel ratio sensor 1 is not reusable. Never remove air fuel ratio sensor 1 unless this is required.

   a. Using the heated oxygen sensor wrench [SST: KV10114400 (J-38365)], remove air fuel ratio sensor 1.

➡**The air fuel ration sensor 1 is removable under vehicle-mounted condition.**

5. Remove exhaust manifold cover.

6. Remove exhaust manifold. Loosen nuts in the reverse order of figure to remove exhaust manifold with power tool.

➡**Disregard No. 9 to No. 12 when loosening.**

7. Remove exhaust manifold gaskets.

❋❋ WARNING

Cover the engine openings to avoid entry of foreign materials.

To install:

8. Note the following, and install in the reverse order of removal.

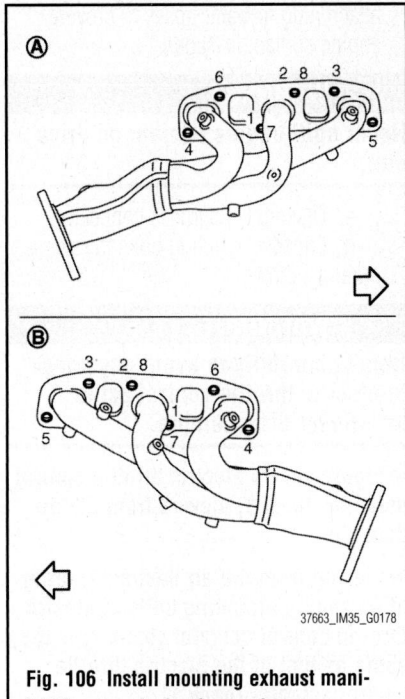

**Fig. 106 Install mounting exhaust mani-
fold in numerical order as shown**

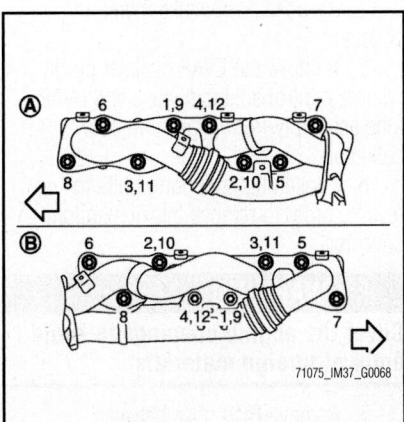

**Fig. 107 Loosen nuts in the reverse order
to remove exhaust manifold and three way
catalyst (left bank)**

9. Install exhaust manifold gasket with
its directional protrusion set upward.

10. Install exhaust manifold and tighten
mounting nuts in numerical.

➡**Tighten the mounting nuts No. 1 to 4
in two steps. The numerical order No.
9 to 12 shown second steps.**

❊❊ WARNING

**Before installing a new sensors,
clean exhaust system threads using
oxygen sensor thread cleaner (commercial service tool: J-43897-18 or
J-43897-12), and apply anti-seize
lubricant (commercial service tool).**

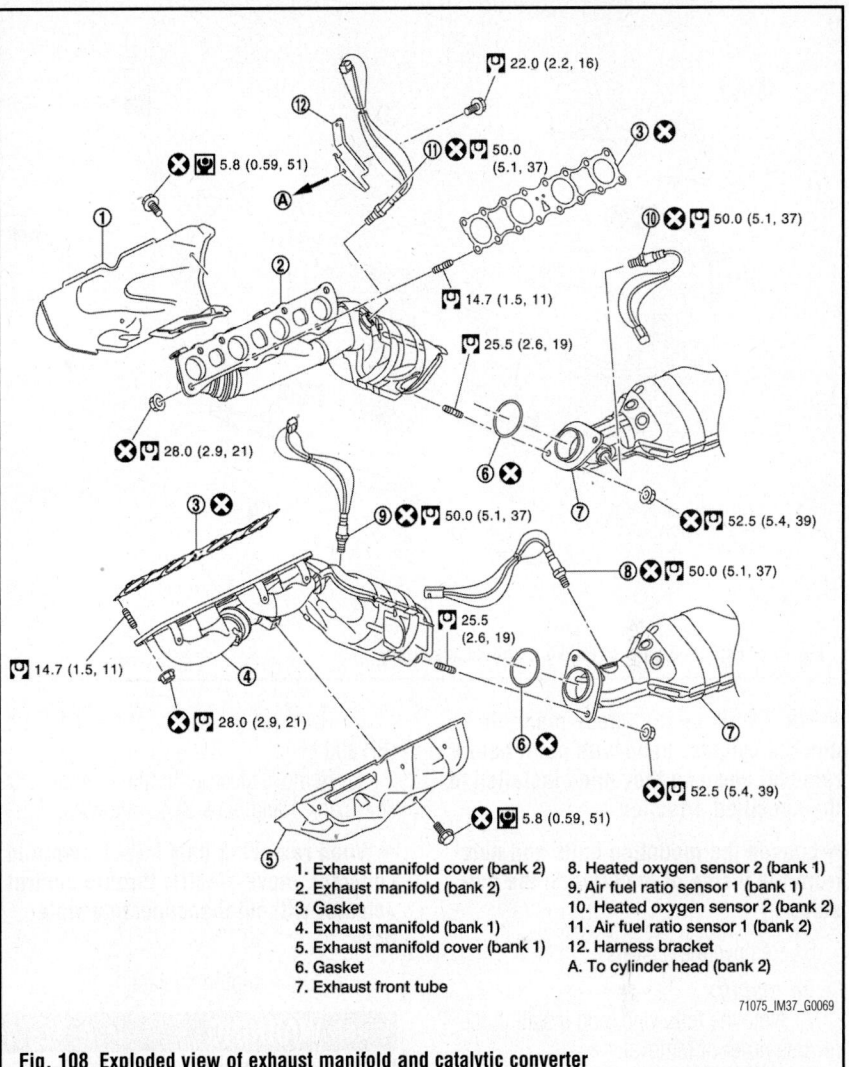

1. Exhaust manifold cover (bank 2)
2. Exhaust manifold (bank 2)
3. Gasket
4. Exhaust manifold (bank 1)
5. Exhaust manifold cover (bank 1)
6. Gasket
7. Exhaust front tube
8. Heated oxygen sensor 2 (bank 1)
9. Air fuel ratio sensor 1 (bank 1)
10. Heated oxygen sensor 2 (bank 2)
11. Air fuel ratio sensor 1 (bank 2)
12. Harness bracket
A. To cylinder head (bank 2)

Fig. 108 Exploded view of exhaust manifold and catalytic converter

• Sensors are not reusable. Replace
them with a new one after removal.
When replacing them, handle with
care not to impact on them.

• When installing the new sensors,
set the heated oxygen sensor
wrench [SST: KV10114400
(J-38365)] in the hexagonal part
to tighten the them.

• Never over torque sensors. Doing
so may cause damage to the sensors, resulting in "MIL" coming on.

• Prevent rust preventives from
adhering to the sensor body.

INTAKE MANIFOLD

REMOVAL & INSTALLATION

3.7L Engine

See Figures 109 and 110.

1. Release fuel pressure.
2. Remove intake manifold collector.

3. Remove fuel tube and fuel injector
assembly.

4. Remove harness bracket.

5. Loosen mounting nuts and bolts in
reverse order as shown to remove intake
manifold.

➡**Cover the engine openings to avoid
entry of foreign materials.**

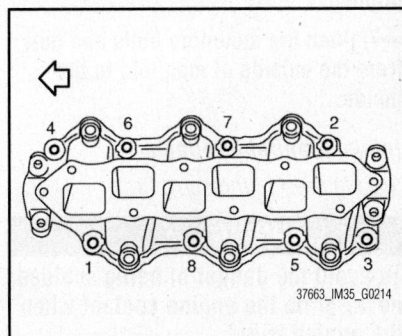

**Fig. 109 Loosen mounting nuts and bolts
in reverse order as shown**

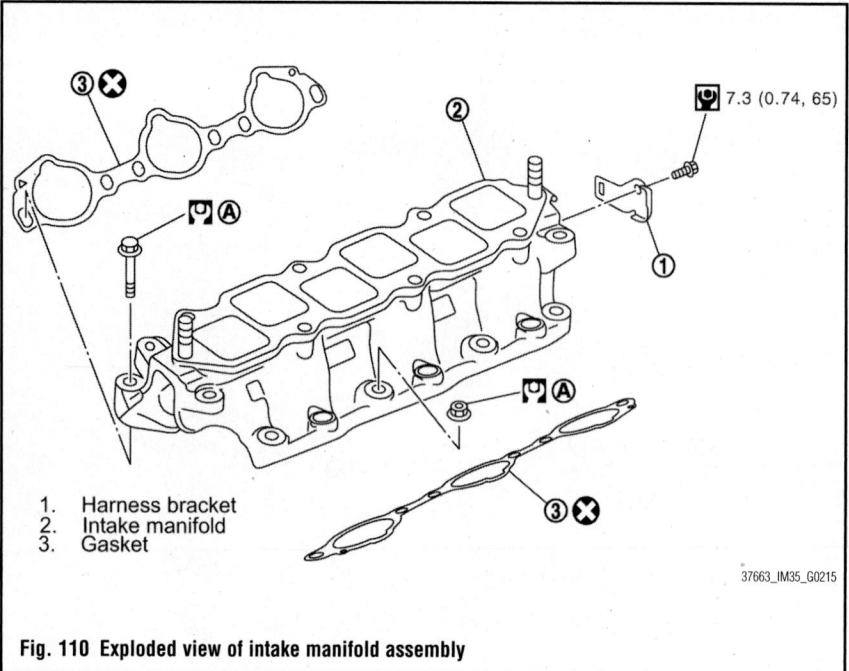

1. Harness bracket
2. Intake manifold
3. Gasket

37663_IM35_G0215

Fig. 110 Exploded view of intake manifold assembly

➡ **Put a mark on the intake manifold and the cylinder head with paint before removal because they need installed in the specified direction.**

➡ **Loosen the mounting bolts and nuts from the inside of manifold to the outside.**

6. Remove the gaskets.

To install:

7. Note the following, and install in the reverse order of removal.

8. If stud bolts were removed, install them and tighten to 8 ft. lbs. (11 Nm).

9. Tighten all the mounting nuts and bolts to the specified torque in two or more steps in numerical order.
   a. Step 1: 5 ft. lbs. (7 Nm)
   b. Step 2: 19 ft. lbs. (26 Nm)

➡ **Install intake manifold with the marks (put on the intake manifold and the cylinder head before removal) aligned.**

➡ **Tighten the mounting bolts and nuts from the outside of manifold to the inside.**

Intake Manifold Collector

See Figures 111 through 113.

⚖ **WARNING**

To avoid the danger of being scalded, never drain the engine coolant when the engine is hot.

1. Remove engine room cover (RH and LH).

2. Remove air cleaner case and air duct (RH and LH).

3. Remove electric throttle control actuator (bank 1 and bank 2) as follows:

➡ **When removing only intake manifold collector, move electric throttle control actuator without disconnecting water hose.**

   a. Drain engine coolant.

⚖ **WARNING**

Perform this step when engine is cold.

   b. Disconnect water hoses from electric throttle control actuator. When engine coolant is not drained from radiator,

attach plug to water hoses to prevent engine coolant leakage.

⚖ **WARNING**

Never spill engine coolant on drive belt.

   c. Disconnect harness connector.
   d. Loosen mounting bolts in reverse order as shown.

⚖ **WARNING**

Handle carefully to avoid any shock to electric throttle control actuator.>Never disassemble.

➡ **Figure shows electric throttle control actuator (bank 1) viewed from the air duct side.**

➡ **Viewed from the air duct side, order of loosening mounting bolts of electric throttle control actuator (bank 2) is the same as that of the electric throttle control actuator (bank 1).**

4. Disconnect vacuum hose, PCV hose and EVAP hose from intake manifold collector.

5. Remove the EVAP canister purge volume control solenoid valve and EVAP tube assembly from intake manifold collector.

6. Loosen the mounting bolts in reverse order to remove intake manifold collector.

⚖ **WARNING**

Cover the engine openings to avoid entry of foreign materials.

7. Remove PCV hose between intake manifold collector and rocker cover (bank 1).

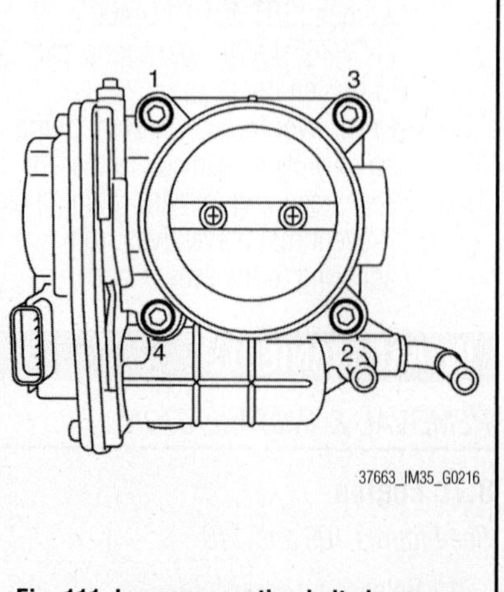

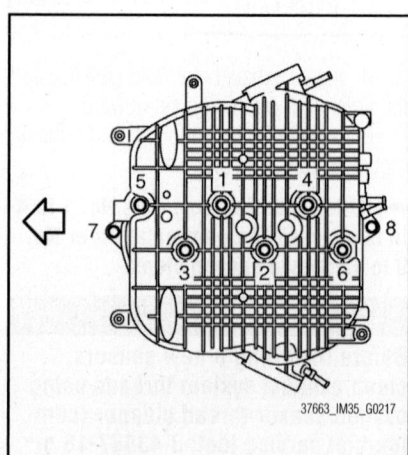

37663_IM35_G0216

Fig. 111 Loosen mounting bolts in reverse order as shown

37663_IM35_G0217

Fig. 112 Loosen mounting bolts in reverse order as shown

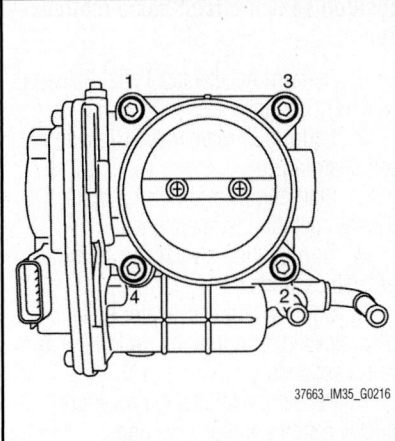

Fig. 113 Tighten mounting bolts in numerical order as shown

To install:

8. Note the following, and install in the reverse order of removal.

9. If intake manifold collector stud bolts were removed, install them and tighten to 8 ft. lbs. (11 Nm). Tighten mounting bolts in numerical order.

➥**Tighten mounting bolts to secure gasket and intake manifold collector.**

10. Insert water hose by 1.06 to 1.26 inches (27 to 32 mm) from connector end.

11. Clamp hose at location of 0.12 to 0.28 inches (3 to 7 mm) from hose end.

12. Install the electric throttle control actuator in the reverse order of removal.

13. Tighten the mounting bolts in numerical order as shown.

※※ WARNING

Handle carefully to avoid any shock to electric throttle control actuator. Never disassemble.

➥**The figure shows the electric throttle control actuator (bank 1) viewed from the air duct side. Viewed from the air duct side, order of tightening mounting bolts of electric throttle control actuator (bank 2) is the same as that of the electric throttle control actuator (bank 1).**

14. Perform the "Throttle Valve Closed Position Learning" when harness connector of electric throttle control actuator is disconnected.

15. Perform the "Idle Air Volume Learning" and "Throttle Valve Closed Position Learning" when electric throttle control actuator is replaced.

5.6L Engine

See Figures 114 and 115.

※※ WARNING

To avoid the danger of being scalded, never drain the engine coolant when the engine is hot.

1. Remove engine cover.
2. Remove air duct (inlet) and air duct.
3. Disconnect harness connector and harness bracket (bank 2 rear side).
4. Remove engine cover bracket.
5. Disconnect the air flow sensor harness connector.
6. Remove the air flow sensor harness connector clip from intake manifold.
7. Remove vacuum tank, EVAP service port hose and EVAP canister purge control solenoid valve.
8. Disconnect PCV hoses and vacuum hose from intake manifold. Add matching marks as necessary for easier installation.
9. Drain engine coolant from radiator.

※※ WARNING

Perform this step when the engine is cold. Never spill engine coolant on drive belts.

➥**When removing only intake manifold, move electric throttle control actuator without disconnecting the water hoses.**

10. Remove the electric throttle control actuator. Refer to Engine Performance.

※※ WARNING

Handle carefully to avoid any impact to electric throttle control actuator. Never disassemble.

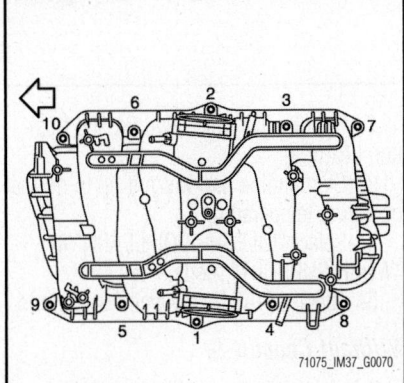

Fig. 114 Loosen the mounting bolts in reverse order

11. Remove engine harness clip on rocker cover.
12. Remove water pipe mounting bolts.
13. Remove water pipe mounting bolts (intake manifold back side).
14. Remove high pressure fuel pump insulator.
15. Remove harness bracket (intake manifold back side).
16. Remove the intake manifold.
17. Loosen the mounting bolts in reverse order as shown in the figure.
18. Remove the intake manifold gaskets.

※※ WARNING

Cover the engine openings to avoid entry of foreign materials.

19. Remove manifold absolute pressure (MAP) sensor, if necessary.

※※ WARNING

Handle carefully to avoid any impact to manifold absolute pressure (MAP) sensor.

20. Remove the acoustic absorbent.

To install:

21. Note the following item, and install in the reverse order of removal.

IDLE AIR VOLUME LEARNING

Idle Air Volume Learning is a function of ECM to learn the idle air volume that keeps engine idle speed within the specific range. It must be performed under the following conditions:

• Each time electric throttle control actuator or ECM is replaced.
• Idle speed or ignition timing is out of specification.

Preconditioning

Check that all of the following conditions are satisfied. Learning will be cancelled if any of the following conditions are missed for even a moment.

• Battery voltage: More than 12.9 V (At idle)
• Engine coolant temperature: 158–221° F (70–105° C)
• Selector lever: P or N
• Electric load switch: OFF (Air conditioner, headlamp, rear window defogger)

➥**On vehicles equipped with daytime light systems, if the parking brake is applied before the engine is started the headlamp will not illuminate.**

• Steering wheel: Neutral (Straight-ahead position)

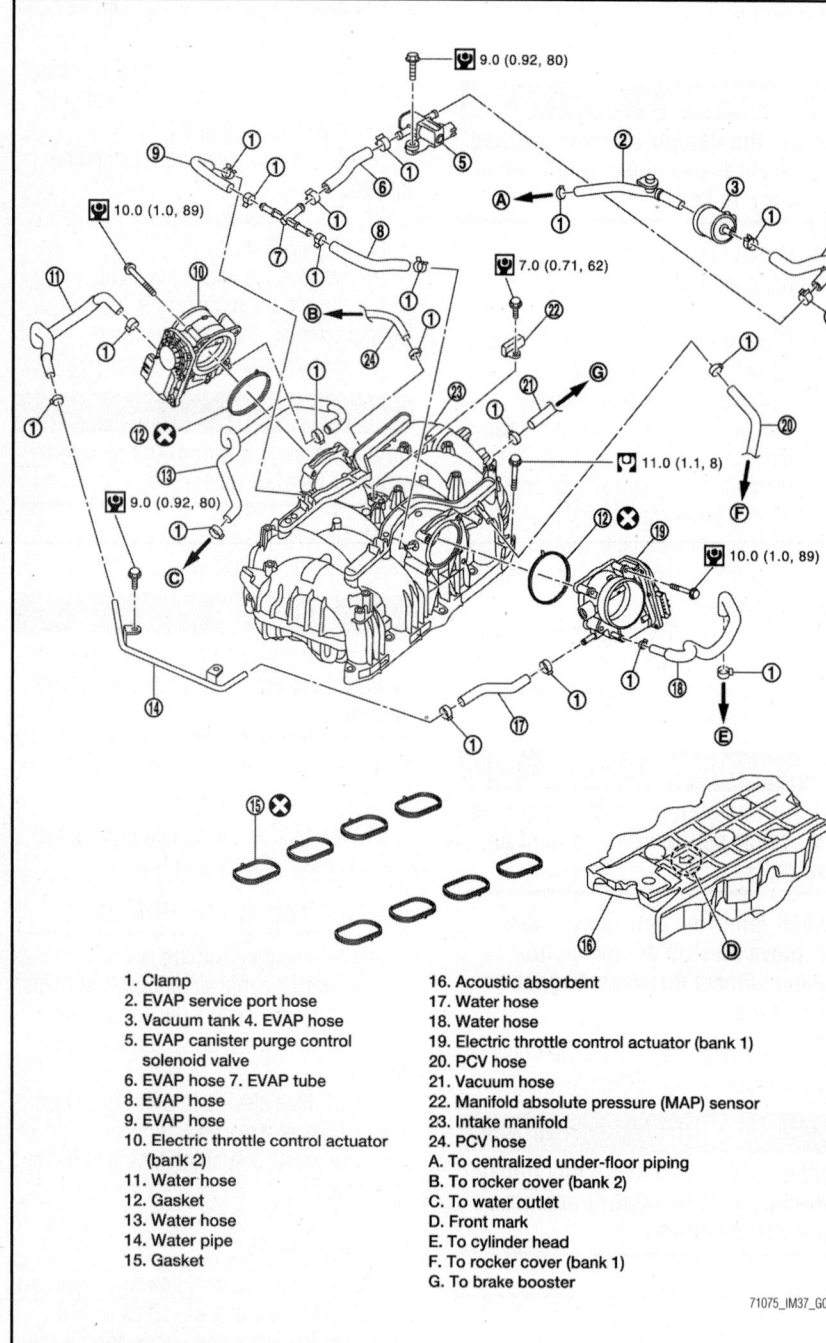

Fig. 115 Exploded view of the intake manifold

1. Clamp
2. EVAP service port hose
3. Vacuum tank 4. EVAP hose
5. EVAP canister purge control solenoid valve
6. EVAP hose 7. EVAP tube
8. EVAP hose
9. EVAP hose
10. Electric throttle control actuator (bank 2)
11. Water hose
12. Gasket
13. Water hose
14. Water pipe
15. Gasket
16. Acoustic absorbent
17. Water hose
18. Water hose
19. Electric throttle control actuator (bank 1)
20. PCV hose
21. Vacuum hose
22. Manifold absolute pressure (MAP) sensor
23. Intake manifold
24. PCV hose
A. To centralized under-floor piping
B. To rocker cover (bank 2)
C. To water outlet
D. Front mark
E. To cylinder head
F. To rocker cover (bank 1)
G. To brake booster

71075_IM37_G0071

- Vehicle speed: Stopped
- Transmission: Warmed-up
- With CONSULT-III: Drive vehicle until "ATF TEMP SE 1" in "DATA MONITOR" mode of "A/T" system indicates less than 0.9 V.
- Without CONSULT-III: Drive vehicle for 10 minutes.

Perform Idle Air Volume Learning

With Consult-III

1. Perform Accelerator Pedal Released Position Learning.

position sensor circuit has a malfunction.

1. Perform Accelerator Pedal Released Position Learning.
2. Perform Throttle Valve Closed Position Learning.
3. Start engine and warm it up to normal operating temperature.
4. Turn ignition switch OFF and wait at least 10 seconds.
5. Confirm that accelerator pedal is fully released, turn ignition switch ON and wait 3 seconds.
6. Repeat the following procedure quickly 5 times within 5 seconds.
   a. Fully depress the accelerator pedal.
   b. Fully release the accelerator pedal.
7. Wait 7 seconds, fully depress the accelerator pedal for approx. 20 seconds until the MIL stops blinking and turns ON.
8. Fully release the accelerator pedal within 3 seconds after the MIL turns ON.
9. Start engine and let it idle.
10. Wait 20 seconds.
11. Rev up the engine 2 or 3 times and check that idle speed and ignition timing are within the specifications.

THROTTLE VALVE CLOSED POSITION LEARNING

Throttle Valve Closed Position Learning is a function of ECM to learn the fully closed position of the throttle valve by monitoring the throttle position sensor output signal. It must be performed each time harness connector of electric throttle control actuator or ECM is disconnected.

1. Check that accelerator pedal is fully released.
2. Turn ignition switch ON.
3. Turn ignition switch OFF and wait at least 10 seconds.
4. Check that throttle valve moves during the above 10 seconds by confirming the operating sound.

OIL PAN

REMOVAL & INSTALLATION

3.7L Engine

RWD Models

See Figures 116 through 123.

✳✳ WARNING

To avoid the danger of being scalded, never drain engine oil when engine is hot.

➡To remove the upper oil pan, remove engine assembly first. When removing

2. Perform Throttle Valve Closed Position Learning.
3. Start engine and warm it up to normal operating temperature.
4. Select "IDLE AIR VOL LEARN" in "WORK SUPPORT" mode.
5. Touch "START" and wait 20 seconds.

Without Consult-III

➡It is better to count the time accurately with a clock.

➡It is impossible to switch the diagnostic mode when an accelerator pedal

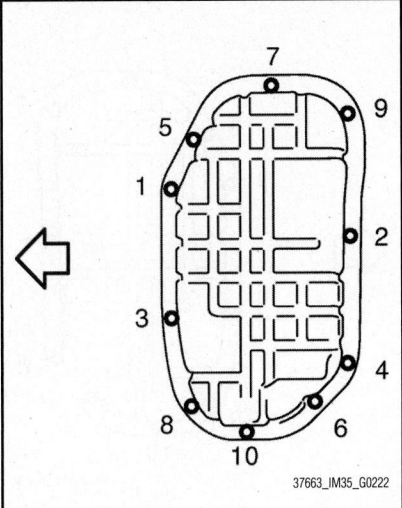

Fig. 116 Loosen mounting bolts in reverse order as shown

the lower oil pan only, remove engine assembly is not necessary. **Perform steps 1, 2 and 10.**

1. Remove front and rear undercover.
2. Drain engine oil.

❊❊ WARNING

Perform this step when engine is cold. Never spill engine oil on drive belt.

3. Remove engine assembly from the vehicle, and separate front suspension member and transmission from engine.
4. Lift the engine with hoist, and mount it onto widely use engine stand.
5. Remove the alternator.
6. Remove the starter motor.
7. Remove the idler pulley and bracket assembly.

8. Remove oil filter, if necessary.
9. Remove oil temperature sensor, if necessary.
10. Remove the lower oil pan as follows:
   a. Loosen the mounting bolts in reverse order as shown to remove the lower oil pan.
   b. Insert seal cutter KV10111100 (J37228) between the upper oil pan and the lower oil pan.

❊❊ WARNING

Be careful not to damage the mating surfaces. Never insert screwdriver, this will damage the mating surface.

11. Remove oil strainer.
12. Remove rear cover plate.
13. Loosen mounting bolts in reverse order as shown to remove the upper oil pan.
14. Insert seal cutter KV10111100 (J37228) between the upper oil pan and cylinder block. Slide seal cutter by tapping on the side of tool with hammer.

❊❊ WARNING

Be careful not to damage mating surfaces. Never insert screwdriver, this will damage the mating surface.

15. Remove the upper oil pan.
16. Remove O-rings from bottom of lower cylinder block and oil pump.

To install:

17. Install the upper oil pan as follows:
   a. Use scraper to remove old liquid gasket from mating surfaces.
   b. Also remove the old liquid gasket from mating surface of lower cylinder block.
   c. Remove old liquid gasket from the bolt holes and threads.

❊❊ WARNING

Never scratch or damage the mating surfaces when cleaning off old liquid gasket.

   d. Install new O-rings on the bottom of lower cylinder block and oil pump.
   e. Apply a continuous bead of liquid gasket to the lower cylinder block mating surface of the upper oil pan as shown.

❊❊ WARNING

For bolt holes B (7 locations), apply liquid gasket outside the holes. Attaching should be done within 5 minutes after coating.

   f. Install the upper oil pan.

❊❊ WARNING

Install avoiding misalignment of both oil pan gasket and O-rings.

   g. Tighten mounting bolts in numerical order as shown.

➡**Bolts measuring 3.62 inches (92 mm) are used at bolt positions 7, 10, and 13. The bolts measuring 1.0 inches (25 mm) are used in all other bolt positions.**

   h. Tighten transmission joint bolts.
18. Install oil strainer to oil pump.

➡**Apply locking sealant to the thread of mounting bolts.**

19. Install the lower oil pan as follows:
   a. Use scraper (A) to remove old liquid gasket from mating surfaces.
   b. Also remove old liquid gasket from mating surface of the upper oil pan.
   c. Remove old liquid gasket from the bolt holes and thread.

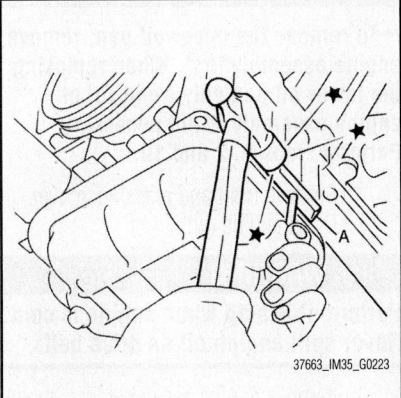

Fig. 117 Insert seal cutter (A) between the upper oil pan and the lower oil pan

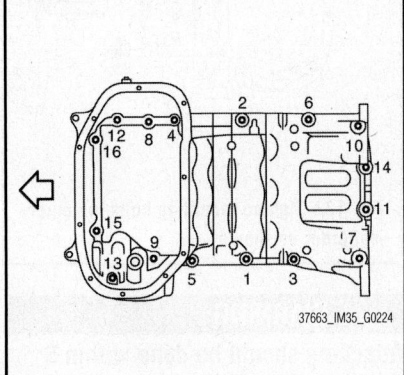

Fig. 118 Loosen mounting bolts in reverse order as shown

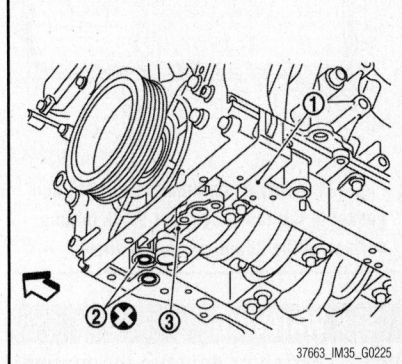

Fig. 119 Remove O-rings (2) from bottom of lower cylinder block (1) and oil pump (3)

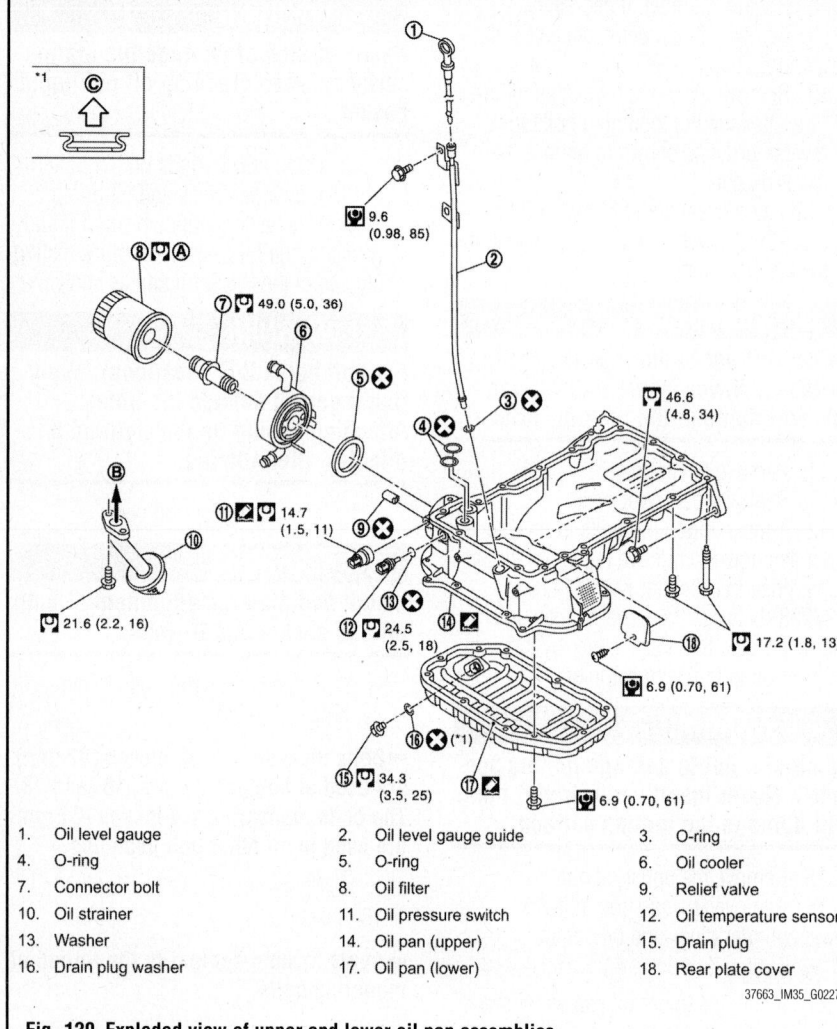

1. Oil level gauge
2. Oil level gauge guide
3. O-ring
4. O-ring
5. O-ring
6. Oil cooler
7. Connector bolt
8. Oil filter
9. Relief valve
10. Oil strainer
11. Oil pressure switch
12. Oil temperature sensor
13. Washer
14. Oil pan (upper)
15. Drain plug
16. Drain plug washer
17. Oil pan (lower)
18. Rear plate cover

37663_IM35_G0227

Fig. 120 Exploded view of upper and lower oil pan assemblies

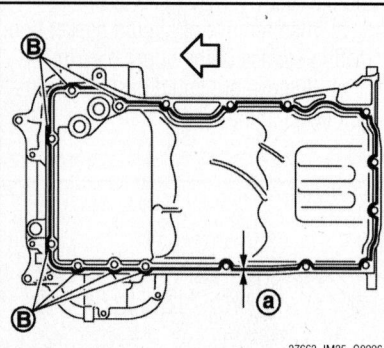

37663_IM35_G0226

Fig. 121 Apply a continuous bead of liquid gasket to the lower cylinder block mating surface of the upper oil pan as shown

❊❊ WARNING

Never scratch or damage the mating surfaces when cleaning off old liquid gasket.

d. Apply a continuous bead of liquid gasket to the lower oil pan.

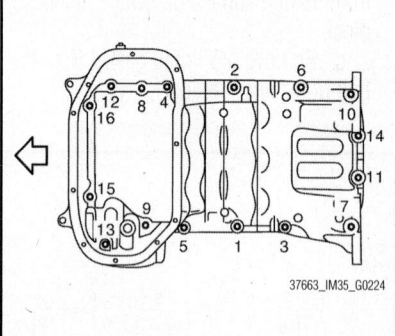

37663_IM35_G0224

Fig. 122 Tighten mounting bolts in numerical order as shown

❊❊ WARNING

Attaching should be done within 5 minutes after coating.

e. Install the lower oil pan.
f. Tighten mounting bolts in numerical order as shown.

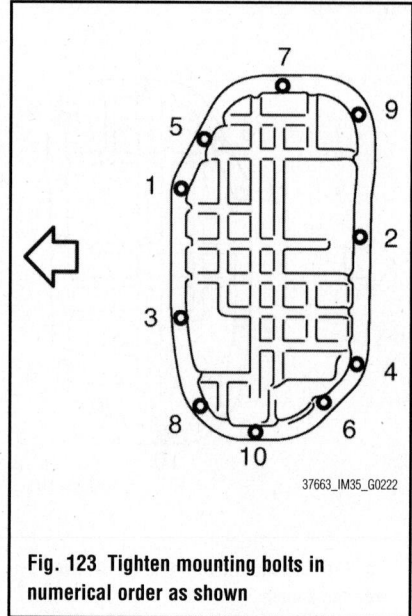

37663_IM35_G0222

Fig. 123 Tighten mounting bolts in numerical order as shown

20. Install oil pan drain plug.
21. Install in the reverse order of removal after this step.

➤At least 30 minutes after oil pan is installed, pour engine oil.

22. Check engine oil level and adjust engine oil.
23. Start engine, and check there is no leakage of engine oil.
24. Stop engine and wait for 10 minutes.
25. Check engine oil level again.

3.7L Engine

AWD Models

See Figures 124 through 132.

❊❊ WARNING

To avoid the danger of being scalded, never drain engine oil when engine is hot.

➤To remove the upper oil pan, remove engine assembly first. When removing the lower oil pan only, removal of engine assembly is not necessary. Perform steps 1, 2 and 10.

1. Remove front and rear undercover.
2. Drain engine oil.

❊❊ WARNING

Perform this step when engine is cold. Never spill engine oil on drive belt.

3. Remove engine assembly from the vehicle, and separate front suspension member and transmission from engine.
4. Lift the engine with hoist, and mount it onto widely use engine stand.

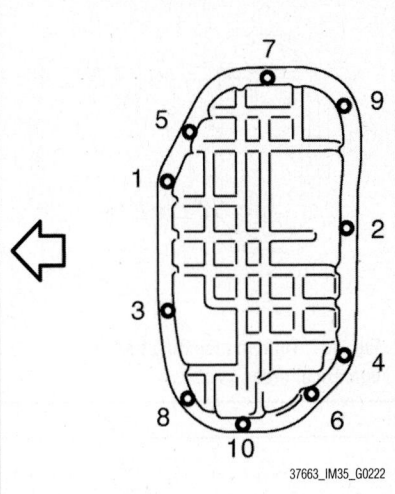

Fig. 124 Loosen mounting bolts in reverse order as shown

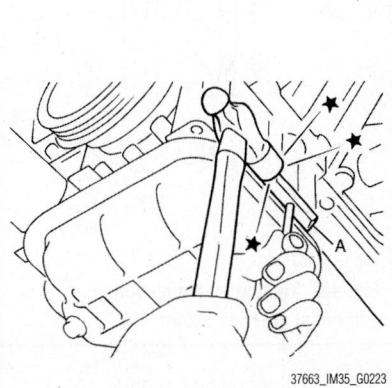

Fig. 125 Insert seal cutter (A) between the upper oil pan and the lower oil pan

5. Remove alternator.

6. Remove starter motor.

7. Remove idler pulley and bracket assembly.

8. Remove oil filter, if necessary.

9. Remove oil temperature sensor, if necessary.

10. Remove the lower oil pan as follows:

a. Loosen mounting bolts in reverse order as shown to remove the lower oil pan.

b. Insert seal cutter KV10111100 (J37228) between the upper oil pan and the lower oil pan.

✻✻ WARNING

Be careful not to damage the mating surfaces. Never insert screwdriver, this will damage the mating surface.

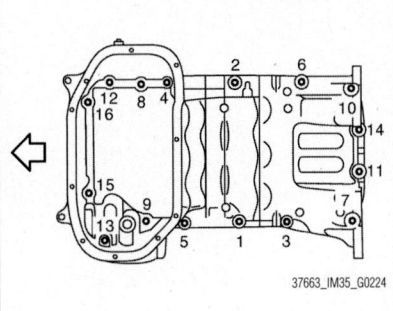

Fig. 126 Loosen mounting bolts in reverse order as shown

11. Remove oil strainer.

12. Remove rear cover plate.

13. Loosen mounting bolts in reverse order as shown to remove the upper oil pan.

14. Insert seal cutter KV10111100 (J37228) between the upper oil pan and cylinder block. Slide seal cutter by tapping on the side of tool with hammer.

15. Remove the upper oil pan.

✻✻ WARNING

Be careful not to damage mating surfaces. Never insert screwdriver, this will damage the mating surface.

16. Remove O-rings from bottom of lower cylinder block and oil pump.

17. Remove axle pipe, if necessary. Remove axle pipe from the upper oil pan using a suitable drift.

To install:

18. Install axle pipe to the upper oil pan, if removed.

a. Lubricate O-ring groove of axle pipe, O-rings, and O-ring joint of oil pan with new engine oil.

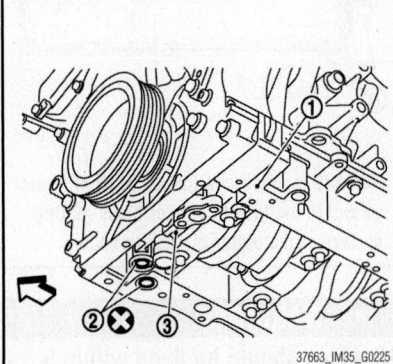

Fig. 127 Remove O-rings (2) from bottom of lower cylinder block (1) and oil pump (3)

b. Install axle pipe to the upper oil pan from axle pipe flange side (left side) using a suitable drift.

✻✻ WARNING

Insert it with care to prevent O-ring from sliding.

19. Install the upper oil pan as follows:

a. Use scraper to remove old liquid gasket from mating surfaces. Also remove the old liquid gasket from mating surface of lower cylinder block. Remove old liquid gasket from the bolt holes and threads.

✻✻ WARNING

Never scratch or damage the mating surfaces when cleaning off old liquid gasket.

b. Install new O-rings on the bottom of lower cylinder block and oil pump.

c. Apply a continuous bead of liquid gasket to the lower cylinder block mating surface of the upper oil pan as shown.

✻✻ WARNING

For bolt holes B (7 locations), apply liquid gasket outside the holes. Attaching should be done within 5 minutes after coating.

d. Install the upper oil pan.

✻✻ WARNING

Install avoiding misalignment of both oil pan gasket and O-rings.

e. Tighten mounting bolts in numerical order as shown.

f. There are three types of mounting bolts. Refer to the following for locating bolts:

- 1 inches (25 mm) bolts used in positions 3, 6, 8, 9, 11, 12, 14, 15, 16
- 2 inches (50 mm) bolt used in position 2
- 3.5 inches (90 mm) bolts used in positions 1, 4, 5, 7, 10, 13

g. Tighten transmission joint bolts.

20. Install oil strainer to oil pump. Apply locking sealant to the thread of mounting bolts.

21. Install the lower oil pan as follows:

a. Use scraper to remove old liquid gasket from mating surfaces. Also remove old liquid gasket from mating surface of the upper oil pan. Remove old liquid gasket from the bolt holes and thread.

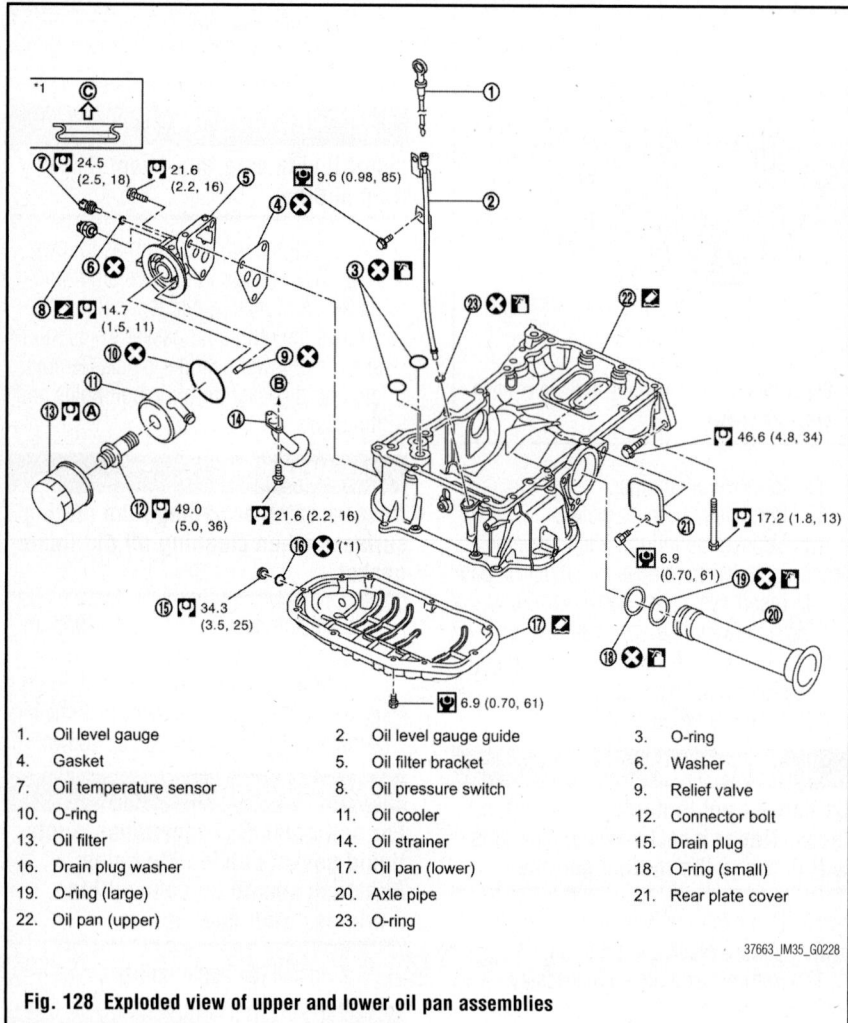

Fig. 128 Exploded view of upper and lower oil pan assemblies

1. Oil level gauge
2. Oil level gauge guide
3. O-ring
4. Gasket
5. Oil filter bracket
6. Washer
7. Oil temperature sensor
8. Oil pressure switch
9. Relief valve
10. O-ring
11. Oil cooler
12. Connector bolt
13. Oil filter
14. Oil strainer
15. Drain plug
16. Drain plug washer
17. Oil pan (lower)
18. O-ring (small)
19. O-ring (large)
20. Axle pipe
21. Rear plate cover
22. Oil pan (upper)
23. O-ring

37663_IM35_G0228

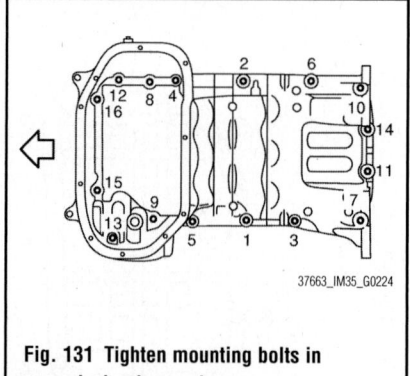

Fig. 131 Tighten mounting bolts in numerical order as shown

37663_IM35_G0224

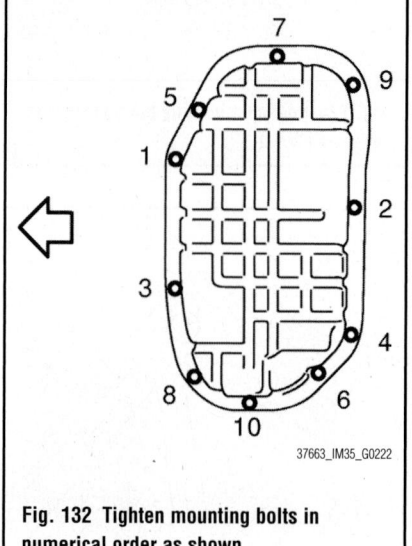

Fig. 132 Tighten mounting bolts in numerical order as shown

37663_IM35_G0222

22. Install oil pan drain plug.
23. Install in the reverse order of removal after this step.

➡At least 30 minutes after oil pan is installed, pour engine oil.

24. Check engine oil level and adjust engine oil.
25. Start engine, and check there is no leakage of engine oil.
26. Stop engine and wait for 10 minutes.
27. Check engine oil level again.

5.6L Engine

See Figures 133 through 135.

✳✳ WARNING

To avoid the danger of being scalded, never drain engine oil when engine is hot.

1. Drain engine oil.
2. Remove oil filter.
3. Remove oil cooler.
4. Remove the A/C compressor.

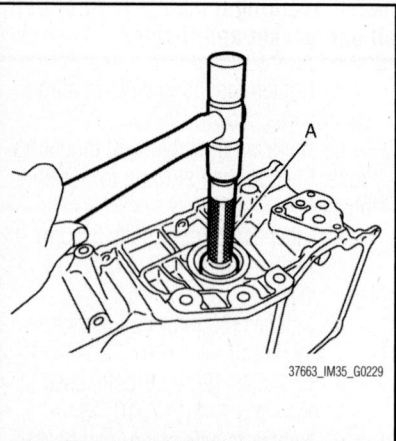

37663_IM35_G0229

Fig. 129 Remove axle pipe from the upper oil pan using a suitable drift (A)

✳✳ WARNING

Never scratch or damage the mating surfaces when cleaning off old liquid gasket.

b. Apply a continuous bead of liquid gasket to the lower oil pan.

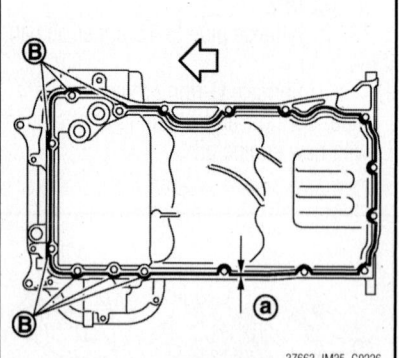

37663_IM35_G0226

Fig. 130 Apply a continuous bead of liquid gasket to the lower cylinder block mating surface of the upper oil pan as shown

✳✳ WARNING

Attaching should be done within 5 minutes after coating.

c. Install the lower oil pan.
d. Tighten mounting bolts in numerical order as shown.

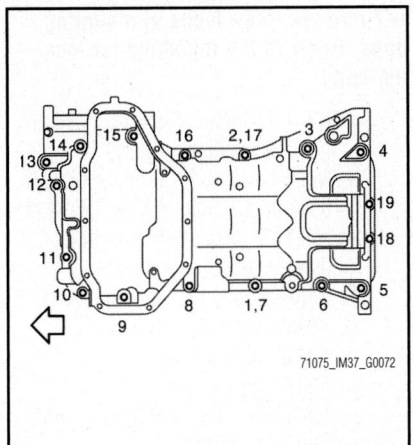

Fig. 133 Upper oil pan bolt removal and installation sequence

5. Remove engine oil level gauge and engine oil level gauge guide.

6. Remove engine oil pressure sensor and engine oil temperature sensor if necessary.

7. Remove rear plate cover.

8. Remove power steering oil pump bracket.

9. Remove power steering belt tensioner pulley.

10. Remove front final drive assembly (AWD models).

11. Remove oil pan (lower). Refer to exploded view illustrations.

12. Remove oil strainer. Refer to exploded view illustrations.

13. Remove oil pan (upper) as per the following:

a. Remove the transmission mounting bolts.

b. Loosen the mounting bolts in the reverse order with power tool to remove.

➡**The oil pan (upper) removal order for AWD models is the same as the one for RWD models.**

➡**Disregard No. 7, 17 when loosening.**

c. Insert a suitable tool into the notch at oil pan (upper) (1). Pry off case by using a suitable tool.

⁑⁑ **WARNING**

Be careful not to damage the mating surfaces.

14. Remove O-ring from bottom of cylinder block and oil pump.

15. Remove baffle plate, if necessary.

16. Remove axle pipe from oil pan (upper), if necessary (AWD models).

➡**Pull the axle pipe from oil pan (upper) using a suitable drift.**

To install:

17. Install oil strainer.

18. Install axle pipe to oil pan, if removed. (AWD models)

a. Lubricate O-ring groove of axle pip, O-ring, and O-ring joint of oil pan with new engine oil.

b. Right/left O-ring diameters differ from each other. O-ring with identification paint mark is installed on front drive shaft (left) installing side.

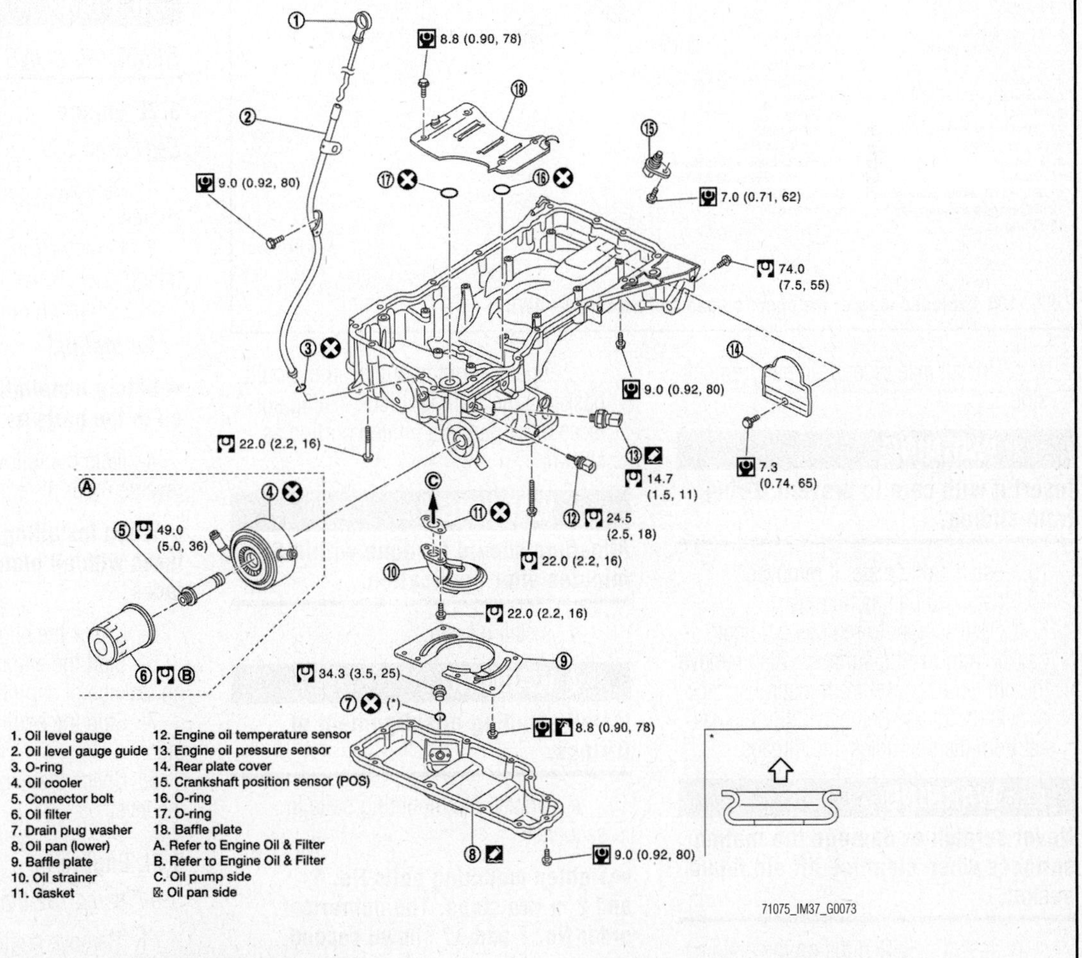

1. Oil level gauge
2. Oil level gauge guide
3. O-ring
4. Oil cooler
5. Connector bolt
6. Oil filter
7. Drain plug washer
8. Oil pan (lower)
9. Baffle plate
10. Oil strainer
11. Gasket
12. Engine oil temperature sensor
13. Engine oil pressure sensor
14. Rear plate cover
15. Crankshaft position sensor (POS)
16. O-ring
17. O-ring
18. Baffle plate
A. Refer to Engine Oil & Filter
B. Refer to Engine Oil & Filter
C. Oil pump side
Oil pan side

Fig. 134 Exploded view of the upper oil pan and assembly—2WD

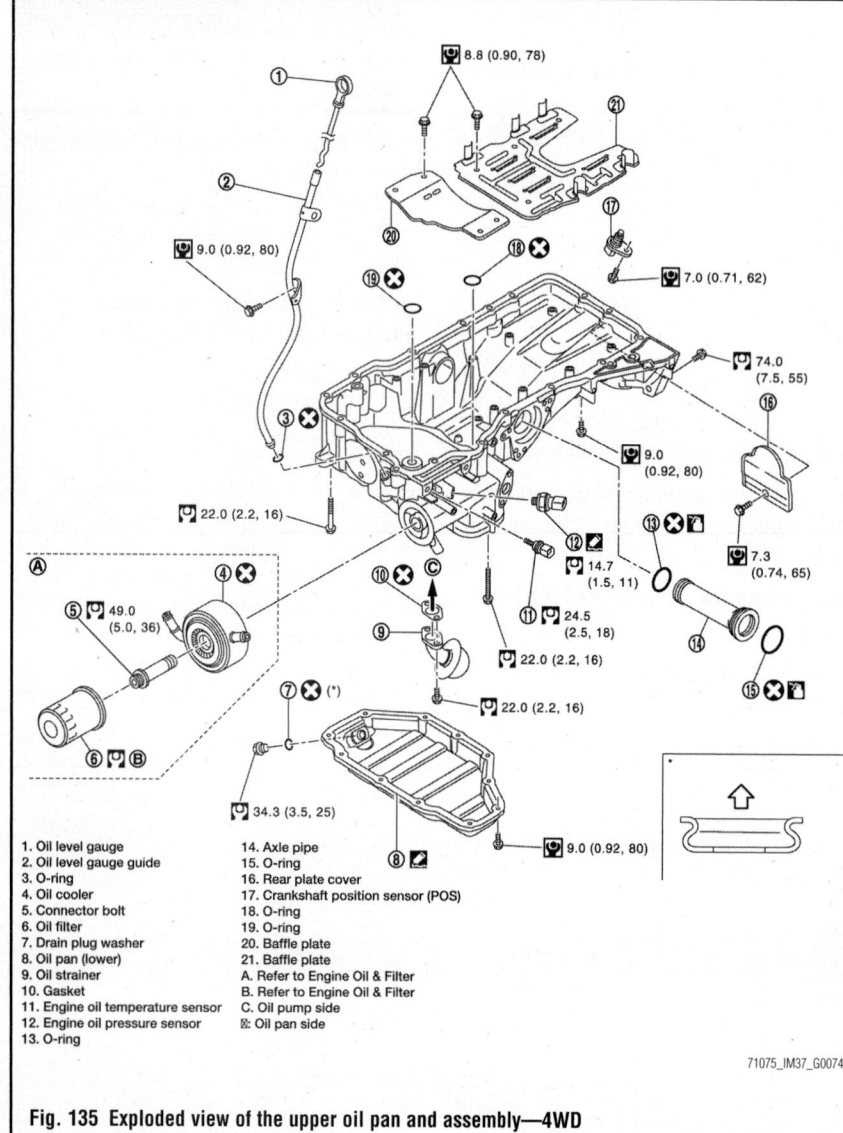

Fig. 135 Exploded view of the upper oil pan and assembly—4WD

1. Oil level gauge
2. Oil level gauge guide
3. O-ring
4. Oil cooler
5. Connector bolt
6. Oil filter
7. Drain plug washer
8. Oil pan (lower)
9. Oil strainer
10. Gasket
11. Engine oil temperature sensor
12. Engine oil pressure sensor
13. O-ring
14. Axle pipe
15. O-ring
16. Rear plate cover
17. Crankshaft position sensor (POS)
18. O-ring
19. O-ring
20. Baffle plate
21. Baffle plate
A. Refer to Engine Oil & Filter
B. Refer to Engine Oil & Filter
C. Oil pump side
▣: Oil pan side

71075_IM37_G0074

c. Install axle pipe to oil pan from left side.

✱✱ WARNING

Insert it with care to prevent O-ring from sliding.

19. Install baffle plate, if removed.
20. Install oil pan as follows:
   a. Use scraper to remove old liquid gasket from mating surfaces. Also remove the old liquid gasket from mating surface of cylinder block. Remove old liquid gasket from the bolt holes and threads.

✱✱ WARNING

Never scratch or damage the mating surfaces when cleaning off old liquid gasket.

   b. Install new O-rings on the bottom of cylinder block and oil pump.

c. Apply a continuous bead of liquid gasket to the cylinder block mating surfaces of oil pan to a limited portion as shown.

✱✱ WARNING

Attaching should be done within 5 minutes after application.

   d. Install oil pan.

✱✱ WARNING

Install avoiding misalignment of O-rings.

   e. Tighten the mounting bolts in sequence.

➡Tighten mounting bolts No. 1 and 2 in two steps. The numerical order No. 7 and 17 shown second steps.

➡There are three types of mounting bolts. Refer to the following for locating bolts:

   • M6 1.18 inches (30 mm) bolts are used in positions 18 and 19
   • M8 4.0 inches (100 mm) bolts are used in positions 4,5, 9, 12, 14, 15
   • M8 1.77 inches (45 mm) bolts are used in all other positions
   f. Tighten transmission joint bolts.
   g. Install rear plate cover.
   h. Install oil strainer.
   i. Install oil pan (lower).
   j. Install the oil pan drain plug with new drain plug washer.
21. Install in the reverse order of removal after this step.

➡At least 30 minutes after oil pan is installed, pour engine oil.

22. Check engine oil level and adjust engine oil.
23. Start engine, and check there is no leakage of engine oil.
24. Stop engine and wait for 15 minutes.
25. Check engine oil level again.

OIL PUMP

REMOVAL & INSTALLATION

3.7L Engine

See Figure 136.

1. Remove the lower oil pan and oil strainer.
2. Remove front timing chain case and timing chain (primary).
3. Remove oil pump assembly.

   To install:

➡Before installation, apply new engine oil to the parts as shown.

4. Note the following, and install in the reverse order of removal.

➡When installing, align crankshaft flat faces with oil pump inner rotor flat faces.

5. Check the engine oil level.
6. Start the engine, and check there is no leakage of engine oil.
7. Stop the engine and wait for 10 minutes.
8. Check the engine oil level and adjust the level.

5.6L Engine

See Figures 137 through 139.

1. Remove engine assembly from vehicle.

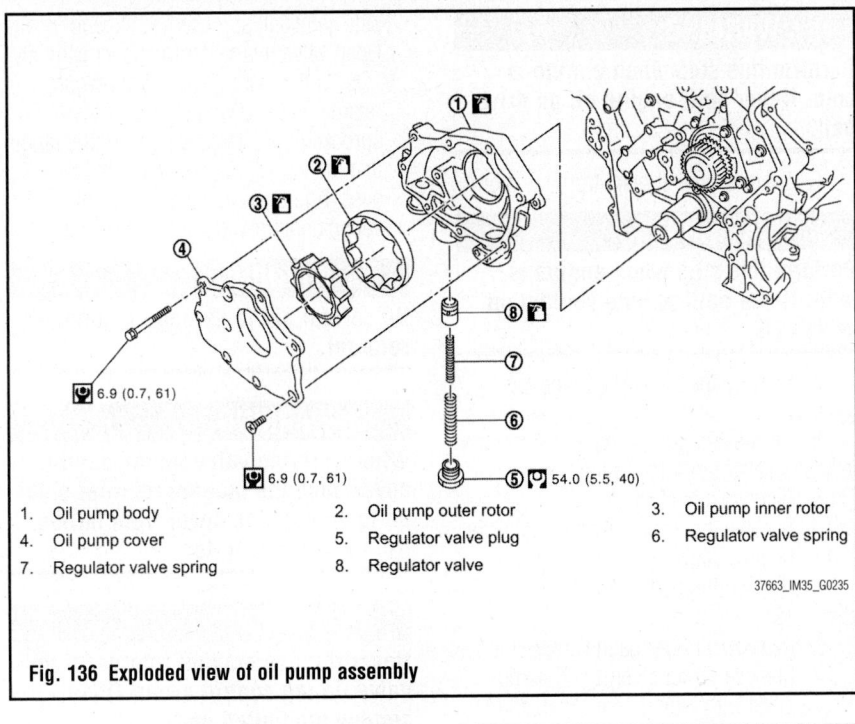

6.9 (0.7, 61)

6.9 (0.7, 61)

⑤ 54.0 (5.5, 40)

1.	Oil pump body	2.	Oil pump outer rotor	3.	Oil pump inner rotor
4.	Oil pump cover	5.	Regulator valve plug	6.	Regulator valve spring
7.	Regulator valve spring	8.	Regulator valve		

37663_IM35_G0235

Fig. 136 Exploded view of oil pump assembly

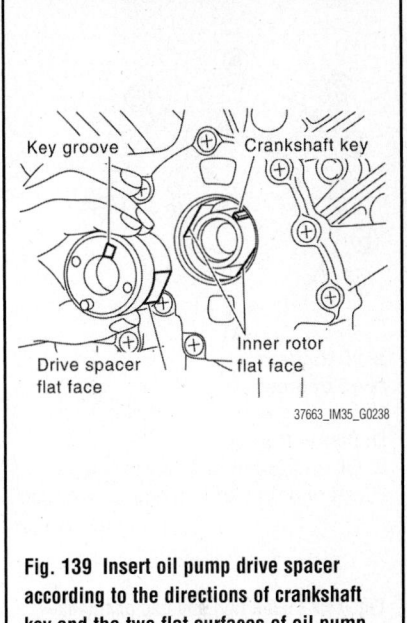

Key groove   Crankshaft key

Drive spacer
flat face   Inner rotor
flat face

37663_IM35_G0238

Fig. 139 Insert oil pump drive spacer according to the directions of crankshaft key and the two flat surfaces of oil pump inner rotor

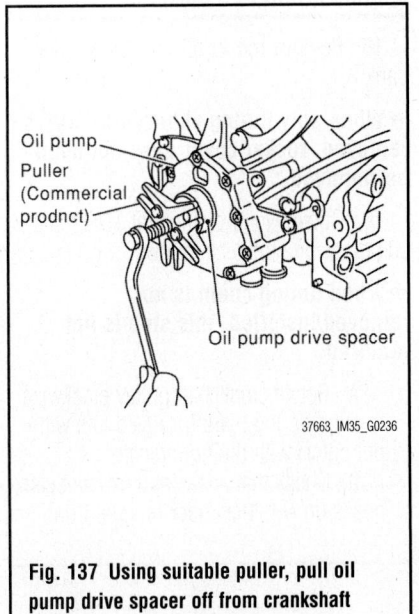

Oil pump

Puller
(Commercial
prodnct)

Oil pump drive spacer

37663_IM35_G0236

Fig. 137 Using suitable puller, pull oil pump drive spacer off from crankshaft

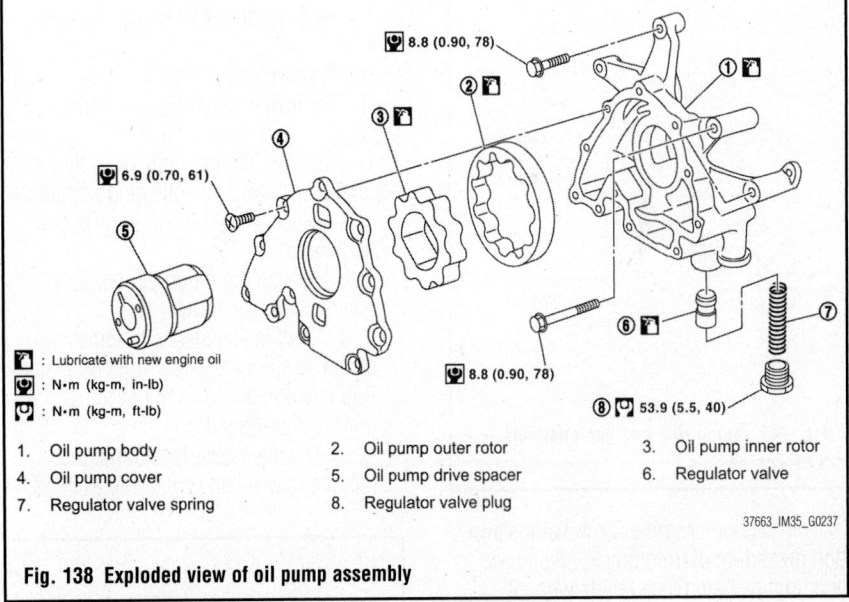

8.8 (0.90, 78)

6.9 (0.70, 61)

8.8 (0.90, 78)

⑧ 53.9 (5.5, 40)

: Lubricate with new engine oil

: N•m (kg-m, in-lb)

: N•m (kg-m, ft-lb)

1.	Oil pump body	2.	Oil pump outer rotor	3.	Oil pump inner rotor
4.	Oil pump cover	5.	Oil pump drive spacer	6.	Regulator valve
7.	Regulator valve spring	8.	Regulator valve plug		

37663_IM35_G0237

Fig. 138 Exploded view of oil pump assembly

2. Remove front cover.

3. Using suitable puller, pull oil pump drive spacer off from crankshaft.

4. Remove oil pump.

To install:

5. Install the oil pump.

6. Install oil pump drive spacer as follows:

a. Insert oil pump drive spacer according to the directions of crankshaft key and the two flat surfaces of oil pump inner rotor.

➡**If the positional relationship does not allow the insertion, rotate oil pump inner rotor with a finger to allow spacer.**

b. After confirming that the position of each part is in correct condition to allow for spacer, force fit spacer by lightly tapping with plastic hammer until it contacts and does not go further.

7. Install in the reverse order of removal after this step.

8. Check the engine oil level.

9. Start engine, and check there is no leakage of engine oil.

10. Stop engine and wait for 15 minutes.

11. Check the engine oil level and adjust engine oil.

PISTON AND RING

POSITIONING

See Figures 140 and 141.

TIMING CHAIN FRONT COVER

REMOVAL & INSTALLATION

3.7L Engine

See Figures 142 through 156.

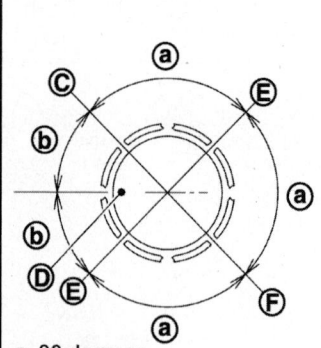

a. 90 degrees
b. 45 degrees
C. Top ring gap
D. Piston front mark
E. Oil ring upper or lower rail gap
F. Second ring and oil ring spacer gap

37663_IM35_G0250

Fig. 140 Piston ring end gap positions—3.7L engine

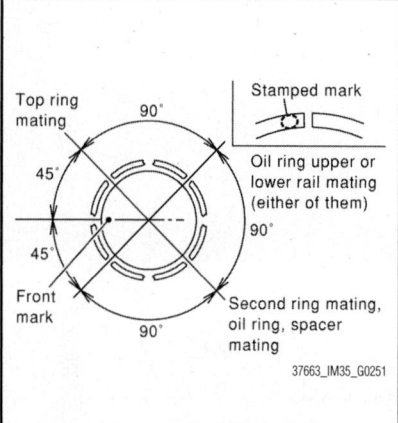

37663_IM35_G0251

Fig. 141 Piston ring end gap positions—5.6L engine

This section describes removal/installation procedure of front timing chain case and timing chain related parts without removing the upper oil pan on vehicle.

When the upper oil pan needs to be removed or installed, or when rear timing chain case is removed or installed, remove the upper and lower oil pans first. Then remove front timing chain case, timing chain related parts, and rear timing chain case in this order, and install in the reverse order of removal.

1. Remove engine cover.
2. Remove the front and rear under covers.
3. Release the fuel pressure.
4. Disconnect the battery cable from the negative terminal.
5. Drain engine oil.

※ WARNING

Perform this step when engine is cold. Never spill engine oil on drive belt.

6. Drain engine coolant from radiator.

※ WARNING

Perform this step when engine is cold. Never spill engine coolant on drive belt.

7. Remove the radiator cooling fan assembly.
8. Separate engine harnesses removing their brackets from front timing chain case.
9. Remove drive belt.
10. Remove the intake manifold collector.
11. Remove harness bracket and fuel sub tube mounting bolt on front timing chain case.
12. Remove oil level gauge and guide.
13. Remove power steering oil pump from bracket with piping connected, and temporarily secure it to aside.
14. Remove power steering oil pump bracket.
15. Remove the alternator.
16. Remove water outlet and water piping.
17. Remove left and right valve timing control covers with the following procedure.
    a. Disconnect valve timing control harness connector.
    b. Loosen the mounting bolts in reverse order as shown.
    c. Shaft is engaged with intake side camshaft sprocket center hole on inside. Pull straight out so as not to tilt until the joint is disengaged.
    d. The mating surface of magnet retarder may be fitted with the exhaust

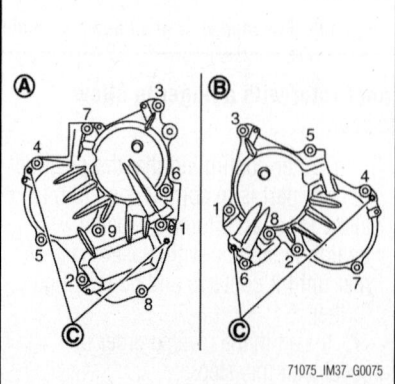

71075_IM37_G0075

Fig. 142 Loosen mounting bolts in reverse order as shown

---

side camshaft sprocket via the engine oil. Open valve timing control cover carefully.
    e. If the mating surface of magnet retarder is fitted with the camshaft sprocket, open the cover within the range that the load is not applied to the harness. And then, remove it so as to prevent magnet retarder from dropping.

※ WARNING

Be careful not to damage magnet retarder.

※ WARNING

When carrying valve timing control cover, face the magnet retarder side up to prevent the cover from falling from magnet retarder.

※ WARNING

Never remove magnet retarder from valve timing control cover. (Disassembly prohibited parts)

18. Remove rocker covers (bank 1 and bank 2).

➡**When only timing chain (primary) is removed, rocker cover does not need to be removed.**

19. Obtain No. 1 cylinder at TDC of its compression stroke as follows:

➡**When timing chain is not removed/installed, this step is not required.**

    a. Rotate crankshaft pulley clockwise to align timing mark (grooved line without color) with timing indicator.
    b. Check that intake and exhaust cam noses on No. 1 cylinder (engine front

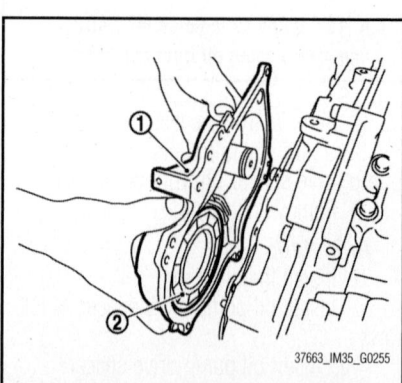

37663_IM35_G0255

Fig. 143 The mating surface of magnet retarder (2) may be fitted with the exhaust side camshaft sprocket via the engine oil; open valve timing control cover (1) carefully

Fig. 144 Rotate crankshaft pulley clockwise to align timing mark (grooved line without color) with timing indicator

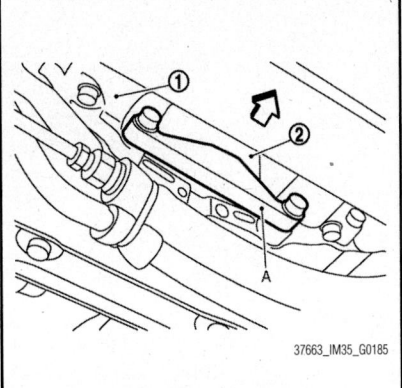

Fig. 146 Remove rear cover plate and set ring gear stopper (A) as shown; upper oil pan (1); drive plate (2)

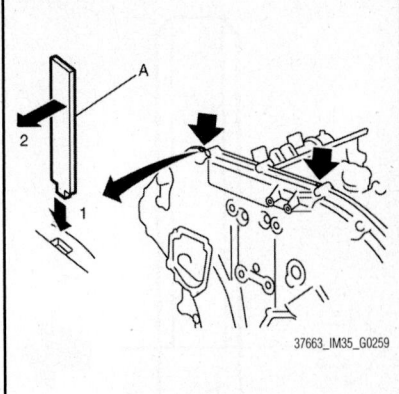

Fig. 148 Insert suitable tool (A) into the notch at the top of the front timing chain case as shown

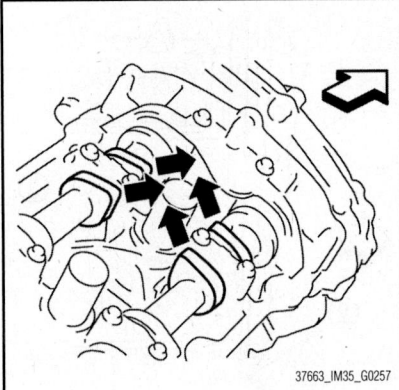

Fig. 145 Check that intake and exhaust cam noses on No. 1 cylinder (engine front side of bank 1) are located as shown

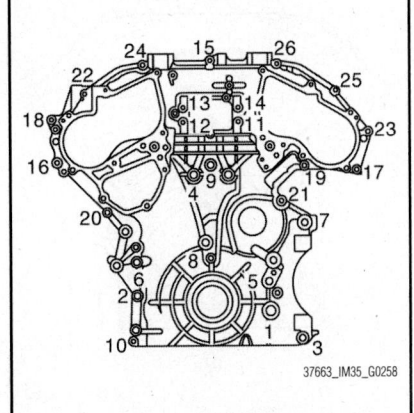

Fig. 147 Loosen mounting bolts in reverse order as shown

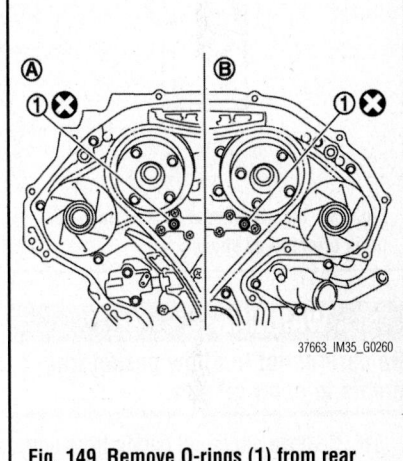

Fig. 149 Remove O-rings (1) from rear timing chain case

side of bank 1) are located as shown. If not, turn crankshaft one revolution (360 degrees) and align as shown.

➡When only timing chain (primary) is removed, rocker cover does not need to be removed. To check that No. 1 cylinder is at its compression TDC, remove front timing chain case first. Then check mating marks on camshaft sprockets.

20. Remove crankshaft pulley as follows:
   a. Remove rear cover plate and set ring gear stopper KV10118600 (J-48641) (A) as shown.
   b. Loosen crankshaft pulley bolt and locate bolt seating surface as 0.39 inches (10 mm) from its original position.

✳✳ WARNING

Never remove crankshaft pulley bolt as it will be used as a supporting point for suitable puller.

   c. Place suitable puller tab on holes of crankshaft pulley, and pull crankshaft pulley through.

✳✳ WARNING

Never put suitable puller tab on crankshaft pulley periphery, as this will damage internal damper.

21. Remove the lower oil pan.
22. Loosen two mounting bolts in front of the upper oil pan.
23. Remove front timing chain case as follows:
   a. Loosen the mounting bolts in reverse order as shown.
   b. Insert the suitable tool into the notch at the top of the front timing chain case as shown.
   c. Pry off the case by moving tool as shown.

➡Use the seal cutter KV10111100 (J37228) to cut liquid gasket for removal.

• Never use screwdriver or something similar.
• After removal, handle front timing chain case carefully so it never tilt, cant, or warp under a load.

24. Remove O-rings from rear timing chain case.
25. Remove front oil seal from front timing chain case.

✳✳ WARNING

Be careful not to damage front timing chain case.

26. Remove the timing chain and related parts.
27. Use scraper to remove all traces of old liquid gasket from front and rear timing chain cases and the upper oil pan, and liquid gasket mating surfaces.

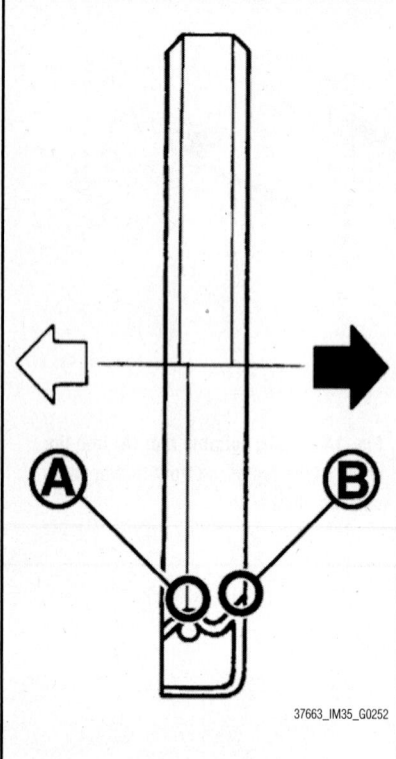

Fig. 150 Install oil seal so that each seal lip is oriented as shown

※※ WARNING

Be careful not to allow gasket fragments to enter oil pan.

28. Remove old liquid gasket from bolt holes and threads.

To install:
29. Install the timing chain and related parts.
30. Install new O-rings on rear timing chain case.
31. Install new front oil seal on the front timing chain case.
  a. Apply new engine oil to both oil seal lip (A) and dust seal lip (B).
  b. Install it so that each seal lip is oriented as shown.
32. Hammer dowel pins (right and left) into front timing chain case up to a point close to taper in order to shorten protrusion length.
  a. Using suitable drift, press-fit oil seal until it becomes flush with front timing chain case end face.
  b. Check the garter spring is in position and seal lip is not inverted.
33. Install front timing chain case as follows:
  a. Apply a continuous bead of liquid gasket to front timing chain case back side as shown.

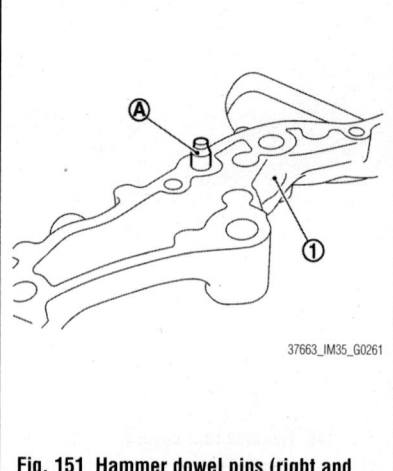

Fig. 151 Hammer dowel pins (right and left) (A) into front timing chain case (1)

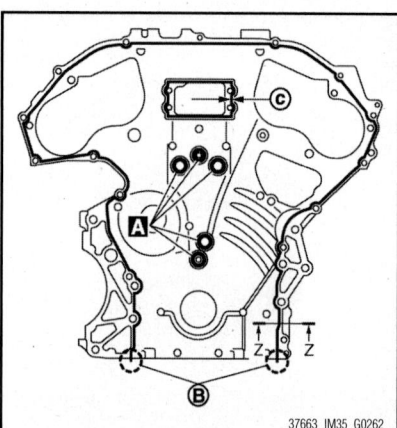

Fig. 152 Apply a continuous bead of liquid gasket to front timing chain case back side as shown

  b. Apply the liquid gasket to top surface of the upper oil pan as shown.
34. Assemble front timing chain case as follows:
  a. Fit lower end of front timing chain case tightly onto top face of the upper oil pan. From the fitting point, make entire front timing chain case contact rear timing chain case completely.

※※ WARNING

Be careful not to damage front oil seal by interference with front end of crankshaft.

※※ WARNING

Attaching should be done within 5 minutes after liquid gasket application.

  b. Install front timing chain case as to fit its dowel pin hole together dowel pin on rear timing chain case.

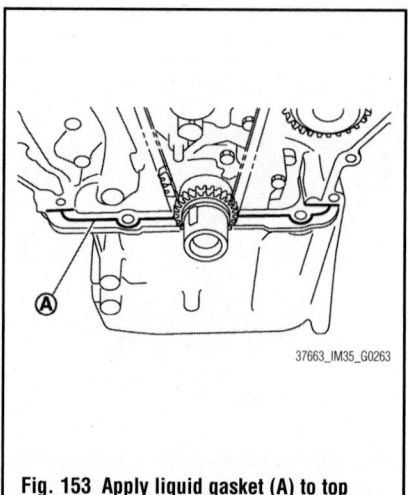

Fig. 153 Apply liquid gasket (A) to top surface of the upper oil pan as shown

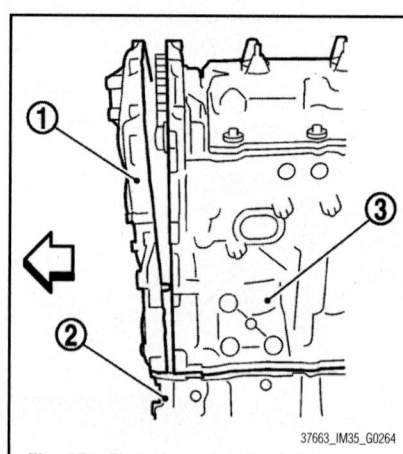

Fig. 154 Fit lower end of front timing chain case (1) tightly onto top face of the upper oil pan (2); cylinder block (3)

  c. Tighten the mounting bolts to the specified torque in numerical order as shown.

➡**There are two types of mounting bolts. Refer to the following for locating bolts.**

  d. M10 bolts are used in bolt positions 1 through 7. Tighten to 41 ft. lbs. (55 Nm).
  e. M6 bolts are used in all other bolts positions. Tighten to 9 ft. lbs. (13 Nm).
  f. After all bolts tightened, retighten them to the specified torque in numerical order.
35. Install two mounting bolts in front of the upper oil pan.
36. Install the lower oil pan.
37. Install right and left valve timing control covers as follows.
  a. Install new seal rings in shaft grooves.

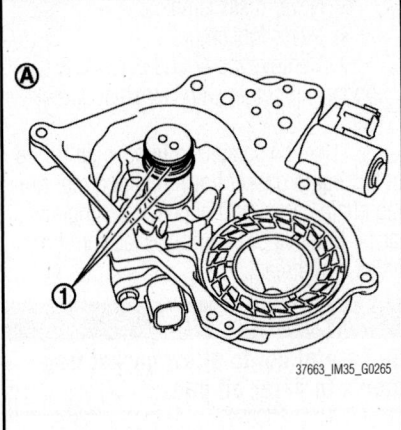

Fig. 155 Install new seal rings (1) in shaft grooves

> ※ WARNING
>
> **When replacing seal rings, replace all rings with new ones.**

b. Install the intake valve timing control cover with new gasket to front timing chain case.

> ※ WARNING
>
> **Align the center of both shaft holes of the shaft and the intake side camshaft sprocket, and then insert them.**

> ※ WARNING
>
> **Be careful not to drop the seal ring from the shaft groove.**

c. Being careful not to move seal ring from the installation groove, align dowel pins on front timing chain case with dowel pin holes to install intake valve timing control covers.

d. Tighten the mounting bolts in numerical order.

e. After all bolt are tightened, tighten No. 1 bolt to the specified torque again.

38. Install the oil pan lower.

39. Install the valve covers.

40. Install crankshaft pulley as follows:

a. Secure crankshaft using ring gear stopper KV10118600 (J-48641).

b. Install the crankshaft pulley, taking care not to damage front oil seal.

➡ **When press-fitting crankshaft pulley with plastic hammer, tap on its center portion (not circumference).**

c. Tighten the crankshaft pulley bolt to 33 ft. lbs. (44 Nm).

d. Tighten the bolt 90 degrees (one

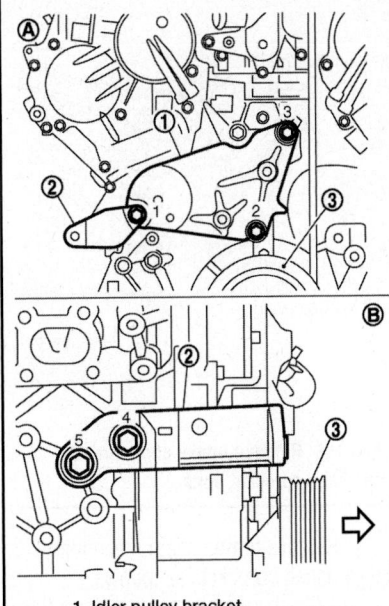

1. Idler pulley bracket
2. Power steering oil pump bracket
3. Crankshaft pulley
A. Engine front side
B. Engine right side

Fig. 156 Install power steering oil pump bracket and idler pulley bracket

mark). Place a matching mark on crankshaft pulley aligning with the matching of crankshaft pulley bolt.

e. Rotate crankshaft pulley in normal direction (clockwise when viewed from front) to confirm it turns smoothly.

41. Install power steering oil pump bracket and idler pulley bracket as follows:

a. Tighten mounting bolts in numerical order as shown. (temporarily)

b. Tighten mounting bolts to specified torque in numerical order.

42. For the following operations, perform steps in the reverse order of removal.

43. Before starting engine, check oil/fluid levels including engine coolant and engine oil. If less than required quantity, fill to the specified level.

44. Run engine to check for unusual noise and vibration.

➡ **If hydraulic pressure inside timing chain tensioner drops after removal/installation, slack in the guide may generate a pounding noise during and just after engine start. However, this is normal. Noise will stop after hydraulic pressure rises.**

45. Warm up engine thoroughly to check there is no leakage of exhaust gases, or any

oil/fluids including engine oil and engine coolant.

46. Bleed air from lines and hoses of applicable lines, such as in cooling system.

47. After cooling down engine, again check oil/fluid levels including engine oil and engine coolant. Refill to the specified level, if necessary.

TIMING CHAIN & SPROCKETS

REMOVAL & INSTALLATION

3.7L Engine

See Figures 157 through 169.

1. Remove the Timing Chain Front Cover.

2. Remove O-ring from rear timing chain case.

3. Remove timing chain tensioner (primary) as follows:

a. Remove lower mounting bolt.

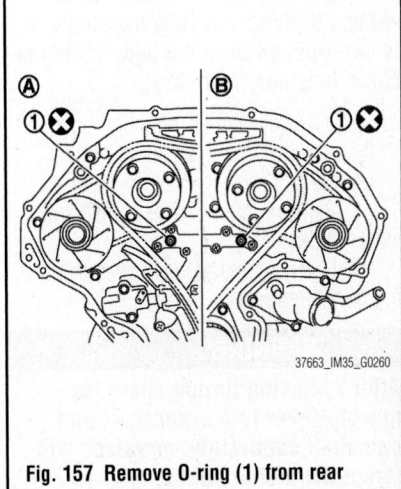

Fig. 157 Remove O-ring (1) from rear timing chain case

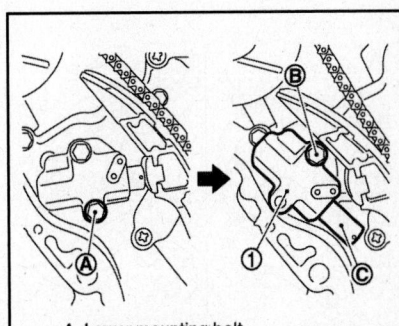

A. Lower mounting bolt
B. Upper mounting bolt
C. Plunger
1. Timing chain tensioner (primary)

Fig. 158 Remove timing chain tensioner (primary)

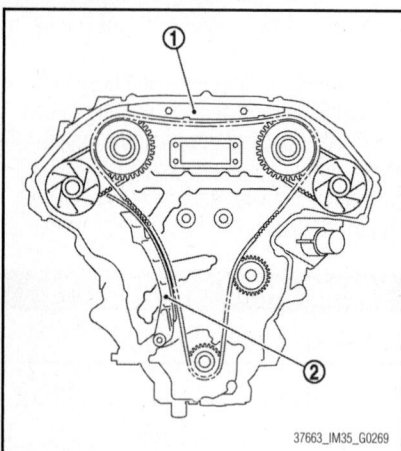

Fig. 159 Remove internal chain guide (1) and slack guide (2)

b. Loosen upper mounting bolt slowly, and then turn timing chain tensioner (primary) on the mounting bolt so that plunger is fully expanded.

➡**Even if plunger is fully expanded, it is not dropped from the body of timing chain tensioner (primary).**

c. Remove upper mounting bolt, and then remove timing chain tensioner (primary).

4. Remove internal chain guide and slack guide.

5. Remove timing chain (primary) and crankshaft sprocket.

❊❊ WARNING

After removing timing chain (primary), never turn crankshaft and camshaft separately, or valves will strike the piston heads.

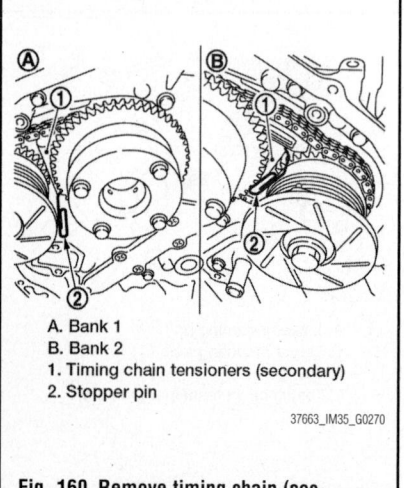

A. Bank 1
B. Bank 2
1. Timing chain tensioners (secondary)
2. Stopper pin

37663_IM35_G0270

Fig. 160 Remove timing chain (secondary) and camshaft sprockets

37663_IM35_G0271

Fig. 161 Remove intake and exhaust camshaft sprocket bolts

6. Remove timing chain (secondary) and camshaft sprockets as follows:

a. Attach suitable stopper pin to the right and left timing chain tensioners (secondary).

➡**Use a hard metal pin as a stopper pin.**

b. Remove intake and exhaust camshaft sprocket bolts. Secure the hexagonal portion of camshaft using wrench to loosen mounting bolts.

❊❊ WARNING

Never loosen mounting bolts with securing anything other than the camshaft hexagonal portion or with tensioning the timing chain.

c. Remove timing chain (secondary) together with camshaft sprockets:
• Handle carefully to avoid any shock to camshaft sprocket.

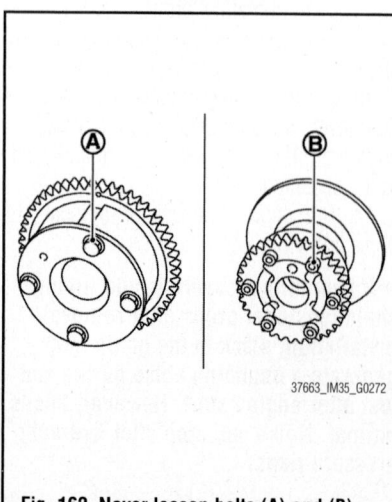

37663_IM35_G0272

Fig. 162 Never loosen bolts (A) and (B) as shown

• Never disassemble.
• Never loosen bolts.

d. Remove the timing chain (secondary) together with camshaft sprockets.

7. Use the scraper to remove all traces of old liquid gasket from front and rear timing chain cases, and opposite mating surfaces. Remove old liquid gasket from bolt holes and threads.

❊❊ WARNING

Be careful not to allow gasket fragments to enter oil pan.

8. Check for cracks and any excessive wear at link plates and roller links of timing chain. Replace timing chain if necessary.

To install:

9. Note the following: The below figure shows the relationship between the mating mark on each timing chain and that on the corresponding sprocket, with the components installed.

10. Check that dowel pin and crankshaft key are located as shown. (No. 1 cylinder at compression TDC)

➡**Though camshaft does not stop at the position as shown, for the placement of cam nose, it is generally accepted camshaft is placed for the same direction as the figure.**

11. Install timing chains (secondary) and camshaft sprockets as follows:

❊❊ WARNING

Matching marks between timing chain and sprockets slip easily. Confirm all matching mark positions repeatedly during the installation process.

a. Push plunger of timing chain tensioner (secondary) and keep it pressed in with stopper pin.

b. Install timing chains (secondary) and camshaft sprockets.

➡**Figure shows bank 1 (rear view).**

c. Align the mating marks on timing chain (secondary) (orange link) with the ones on intake and exhaust camshaft sprockets (punched), and install them.

➡**Mating marks for intake camshaft sprocket are on the back side of camshaft sprocket (secondary). There are two types of mating marks, circle and oval types. They should be used for the bank 1 (Circle Type) and bank 2 (Oval Type), respectively.**

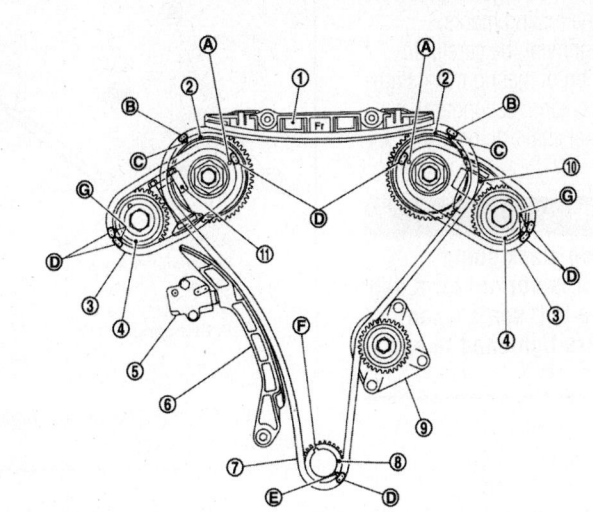

1.	Internal chain guide	2.	Camshaft sprocket (INT)	3.	Timing chain (secondary)
4.	Camshaft sprocket (EXH)	5.	Timing chain tensioner (primary)	6.	Slack guide
7.	Timing chain (primary)	8.	Crankshaft sprocket	9.	Water pump
10.	Timing chain tensioner (secondary) (bank 2)	11.	Timing chain tensioner (secondary) (bank 1)		
A.	Mating mark [punched (back side)]	B.	Mating mark (colored link)	C.	Mating mark (punched)
D.	Mating mark (colored link)	E.	Mating mark (notched)	F.	Crankshaft key

37663_IM35_G0277

Fig. 163 Timing chain marks orientation

d. Shape (orientation of signal plate) of camshaft sprocket (INT) varies depending on the bank position.

e. Align dowel pin hole on the small diameter side of the camshaft front end with dowel pin on the back side of camshaft sprockets, and install them.

f. In case that positions of each mating mark and each dowel pin are not fit on mating parts, make fine adjustment to the position holding the hexagonal por-

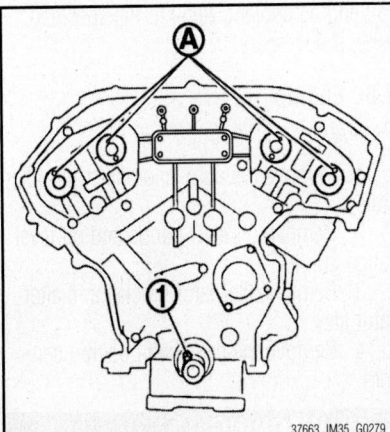

37663_IM35_G0279

Fig. 164 Check that dowel pin (A) and crankshaft key (1) are located as shown

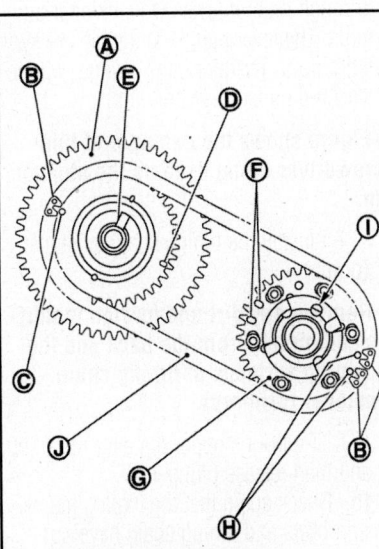

A. Camshaft sprocket (INT) back face
B. Colored link
C. Mating mark (circle)
D. Mating mark (oval)
E. Dowel groove
F. Mating mark (2 oval)
G. Camshaft sprocket (EXH) back face
H. Mating mark (2 circle)
I. Dowel pin hole
J. Timing chain (secondary)

37663_IM35_G0280

Fig. 165 Install timing chains (secondary) and camshaft sprockets

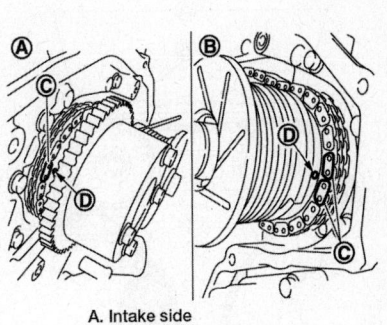

A. Intake side
B. Exhaust side
C. Mating marks (punched)
D. Mating marks (orange link)

37663_IM35_G0281

Fig. 166 Check the mating marks (punched) on each camshaft sprocket are positioned on the mating marks (orange link) on timing chain (secondary)

tion on camshaft with wrench or equivalent.

g. Mounting bolts for camshaft sprockets must be tightened in the step "d". Tightening them by hand is enough to prevent the dislocation of dowel pins.

h. It may be difficult to visually check the dislocation of matching marks during and after installation. To make the matching easier, make a matching mark on the top of sprocket teeth and its extended line in advance with paint.

i. Tighten the camshaft sprocket mounting bolts. Secure camshaft using wrench at the hexagonal portion to tighten mounting bolts.

j. After confirming the matching marks are aligned, tighten camshaft sprocket (INT) mounting bolt. Secure the hexagonal portion (located in between journal No.1 and journal No. 2) of drive shaft (A) using a wrench to tighten mounting bolt.

✷✷ WARNING

When holding the hexagonal part of drive shaft on the intake side with a wrench, be careful not to allow the wrench to cause interference with other parts.

k. Pull stopper pins out from timing chain tensioners (secondary).

12. Install timing chain (primary) as follows:

a. Install the crankshaft sprocket. Check the mating marks on crankshaft sprocket face the front of engine.

b. Install the timing chain (primary).

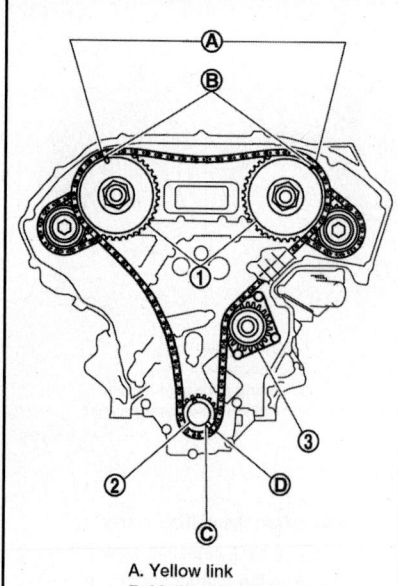

A. Yellow link
B. Mating mark (punched)
C. Mating mark (notched)
D. Orange link
1. Camshaft sprocket INT
2. Crankshaft sprocket
3. Water pump

37663_IM35_G0282

Fig. 167 Install timing chain (primary)

c. Install timing chain (primary) so the mating mark (punched) on camshaft sprocket (INT) is aligned with the yellow link on timing chain, while the mating mark (notched) on crankshaft sprocket is aligned with the orange link on timing chain, as shown.

d. When it is difficult to align mating marks of timing chain (primary) with

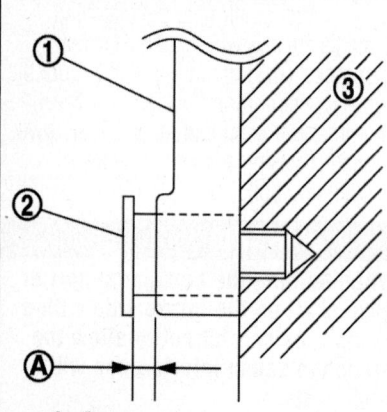

A. Gap
1. Slack guide
2. Slack guide mounting bolts
3. Cylinder block

37663_IM35_G0283

Fig. 168 It is normal for a gap to exist under the bolt seats when mounting bolts are tightened

each sprocket, gradually turn camshaft using wrench on the hexagonal portion to align it with the mating marks.

e. During alignment, be careful to prevent dislocation of mating mark alignments of timing chains (secondary).

13. Install internal chain guide and slack guide.

✳✳ WARNING

Never over tighten slack guide mounting bolts. It is normal for a gap to exist under the bolt seats when mounting bolts are tightened to specification.

14. Install the timing chain tensioner (primary) with the following procedure:

a. Pull plunger stopper tab up (or turn lever downward) so as to remove plunger stopper tab from the ratchet of plunger.

➡**Plunger stopper tab and lever are synchronized.**

b. Push plunger into the inside of tensioner body.

c. Hold plunger in the fully compressed position by engaging plunger stopper tab with the tip of ratchet.

d. To secure lever, insert stopper pin through hole of lever into tensioner body hole. The lever parts and the tab are synchronized. Therefore, the plunger will be secured under this condition.

➡**Figure shows the example of thin screwdriver being used as the stopper pin.**

e. Install the timing chain tensioner (primary).

➡**Remove any dirt and foreign materials completely from the back and the mounting surfaces of timing chain tensioner (primary).**

f. Pull out stopper pin after installing, and then release plunger.

15. Check again that the mating marks on sprockets and timing chain have not slipped out of alignment.

16. Install new O-ring on rear timing chain case.

17. For the remaining operations, perform steps in the reverse order of removal.

18. Before starting the engine, check oil/fluid levels including engine coolant and engine oil. If less than required quantity, fill to the specified level.

19. Use procedure below to check for fuel leakage.

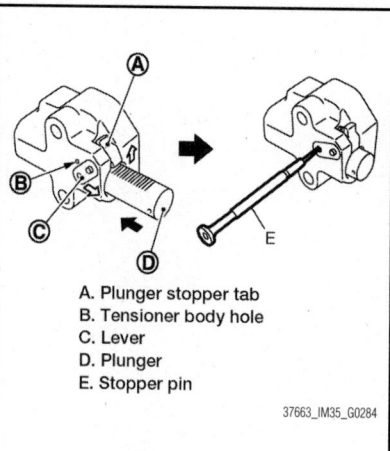

A. Plunger stopper tab
B. Tensioner body hole
C. Lever
D. Plunger
E. Stopper pin

37663_IM35_G0284

Fig. 169 Installing the timing chain tensioner (primary)

a. Turn ignition switch "ON" (with engine stopped). With fuel pressure applied to fuel piping, check for fuel leakage at connection points.

b. Start the engine. With engine speed increased, check again for fuel leakage at connection points.

20. Run engine to check for unusual noise and vibration.

➡**If hydraulic pressure inside timing chain tensioner drops after removal/installation, slack in the guide may generate a pounding noise during and just after engine start. However, this is normal. Noise will stop after hydraulic pressure rises.**

21. Warm up the engine thoroughly to check there is no leakage of fuel, exhaust gases, or any oil/fluids including engine oil and engine coolant.

22. Bleed air from lines and hoses of applicable lines, such as in cooling system.

23. After cooling down engine, again check oil/fluid levels including engine oil and engine coolant. Refill to the specified level, if necessary.

5.6L Engine

See Figures 170 through 180.

1. Remove the auto tensioners and idler pulley.

2. Remove oil level gauge and oil level gauge guide.

3. Remove alternator bracket and alternator stay.

4. Remove the camshaft position sensors.

✳✳ WARNING

Handle carefully to avoid dropping and shocks. Never disassemble. Never allow metal powder to adhere

to magnetic part at sensor tip. **Never place sensors in a location where they are exposed to magnetism.**

5. Remove the high pressure fuel pump and lifter.

✳✳ WARNING

After removing lifter, replace lifter with a new one.

6. Remove water suction hose and water suction pipe.
7. Remove valve timing control cover as per the following:
   a. Disconnect valve timing control solenoid valve harness connector.
   b. Loosen the mounting bolts in the reverse order.

✳✳ WARNING

Exercise care to not to damage mating the surfaces. Shaft is internally jointed with camshaft sprocket center hole. When removing, keep it horizontal until it is completely disconnected.

8. Remove the valve timing control solenoid valve (INT and EXH), if necessary.

✳✳ WARNING

Valve timing control solenoid valve is not reusable. Never remove it unless required.

9. Remove the fuel pump connector protector.
10. Remove O-rings from front cover.
11. Remove rocker cover.
12. Obtain the No. 1 cylinder at the TDC of its compression stroke.
13. Remove oil cooler tube and hose.
14. Remove air compressor bracket.
15. Remove the crankshaft pulley.
16. Remove water pump pulley.

17. Remove oil pan (lower) and oil strainer.
18. Remove oil pan (upper).
19. Remove front cover as per the following:
   a. Loosen the mounting bolts, and then remove camshaft bracket.
   b. Loosen the mounting bolts in reverse order as shown in the figure.
   c. Insert a suitable tool into the notch at front cover. Pry off case by moving a suitable tool.

✳✳ WARNING

Exercise care to not to damage the mating surfaces. After removal, handle front cover carefully so it does not tilt, cant, or warp under a load.

20. Remove front oil seal from front cover using suitable tool. Use screwdriver for removal.

✳✳ WARNING

Be careful not to damage front cover.

21. Remove O-rings from cylinder heads and cylinder block.
22. Remove the oil filter (for valve timing control solenoid valve), if necessary.
23. Remove the timing chain tensioner cover from front cover, if necessary. Use seal cutter [SST: KV10111100 (J-37228)] to cut liquid gasket for removal.
24. Remove oil pump drive chain as per the following:
   a. Push the oil pump drive chain tensioner.
   b. Insert a stopper pin into the body hole.
   c. Hold the two flat parts of oil pump shaft, and then loosen the oil pump sprocket (oil pump side) nut.

✳✳ WARNING

Secure the oil pump unit shaft with the two flat parts.

25. Remove the oil pump drive chain tensioner.
26. Remove timing chain tensioner (bank 1) as per the following:

➡**To remove timing chain and related parts, start with those on bank 1. The procedure for removing parts on bank 2 is omitted because it is the same as that for bank 1.**

   a. Push both sides of spring against spring tension, and then press in plunger with a slack guide.
   b. Insert a stopper pin into the body hole, and then fix it with the plunger pushed in.
27. Remove high pressure fuel pump camshaft.
28. Remove tension guide and slack guide.
29. Remove the timing chain and crankshaft sprocket.

✳✳ WARNING

After removing the timing chain, never turn crankshaft and camshaft separately, or valves will strike the piston head.

30. Secure the hexagonal portion (located in between journal No.1 and journal No. 2) of drive shaft (A) using a wrench (B) to loosen mounting bolt.

✳✳ WARNING

Never loosen the mounting bolt by securing anything other than the camshaft (drive shaft) hexagonal portion or with tensioning the timing chain. When holding the hexagonal part of camshaft (drive shaft) with a wrench, be careful not to allow the wrench to cause interference with other parts.

✳✳ WARNING

Never disassemble camshaft sprocket. Never loosen bolts.

31. Use scraper to remove all traces of old liquid gasket from front cover and opposite mating surfaces. Remove old liquid gasket from bolt hole and thread.

To install:

➡**The figure shows the relationship between the mating mark on each**

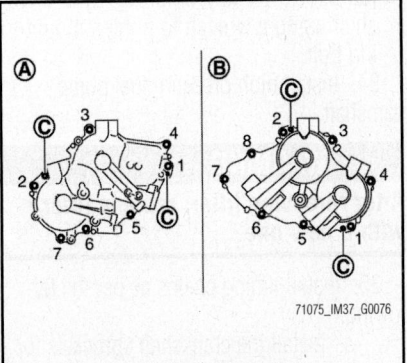

Fig. 170 Valve timing control cover removal and installation sequence

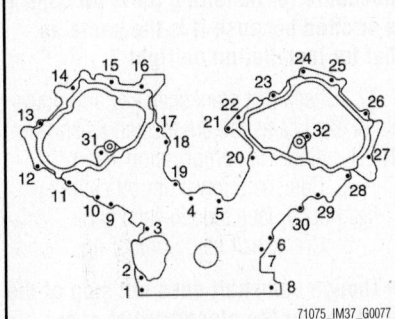

Fig. 171 Front cover removal and installation sequence

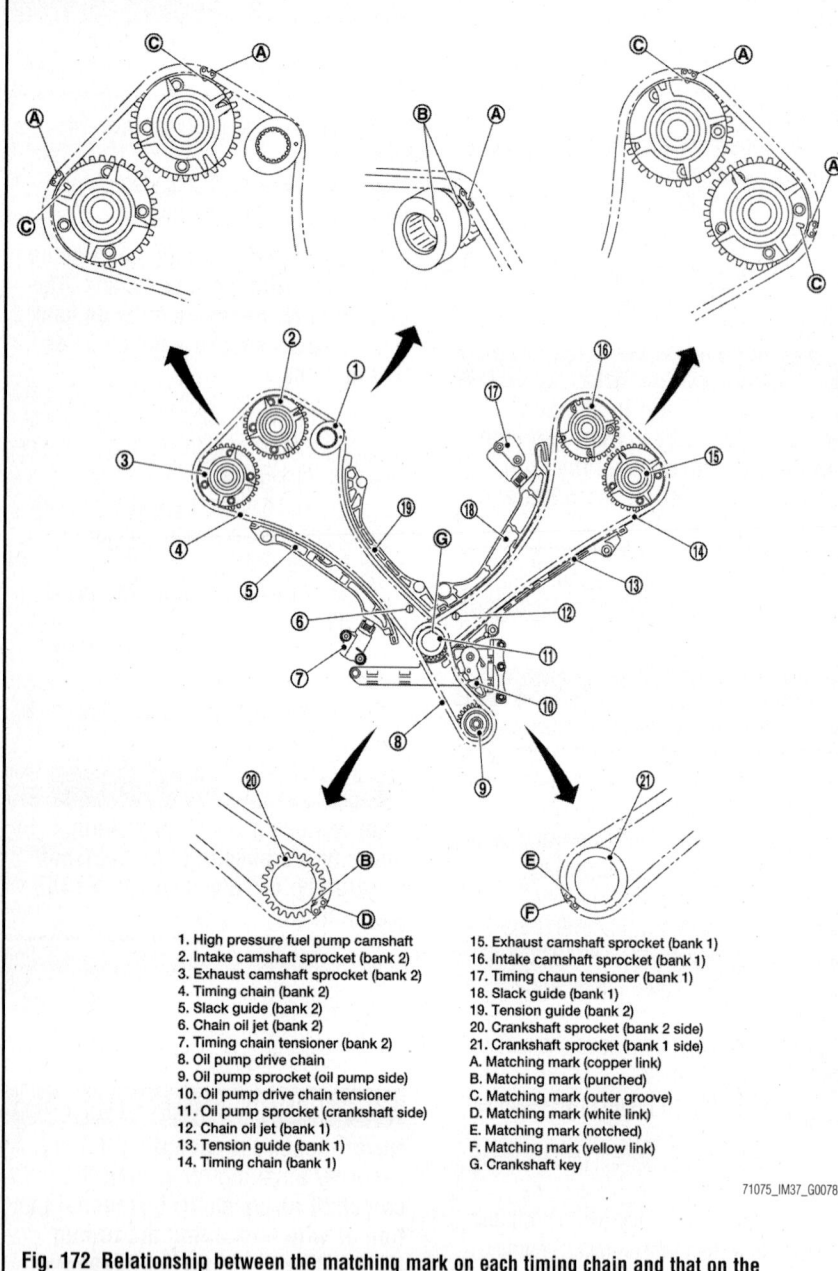

1. High pressure fuel pump camshaft
2. Intake camshaft sprocket (bank 2)
3. Exhaust camshaft sprocket (bank 2)
4. Timing chain (bank 2)
5. Slack guide (bank 2)
6. Chain oil jet (bank 2)
7. Timing chain tensioner (bank 2)
8. Oil pump drive chain
9. Oil pump sprocket (oil pump side)
10. Oil pump drive chain tensioner
11. Oil pump sprocket (crankshaft side)
12. Chain oil jet (bank 1)
13. Tension guide (bank 1)
14. Timing chain (bank 1)
15. Exhaust camshaft sprocket (bank 1)
16. Intake camshaft sprocket (bank 1)
17. Timing chaun tensioner (bank 1)
18. Slack guide (bank 1)
19. Tension guide (bank 2)
20. Crankshaft sprocket (bank 2 side)
21. Crankshaft sprocket (bank 1 side)
A. Matching mark (copper link)
B. Matching mark (punched)
C. Matching mark (outer groove)
D. Matching mark (white link)
E. Matching mark (notched)
F. Matching mark (yellow link)
G. Crankshaft key

71075_IM37_G0078

Fig. 172 Relationship between the matching mark on each timing chain and that on the corresponding sprocket, with the components installed

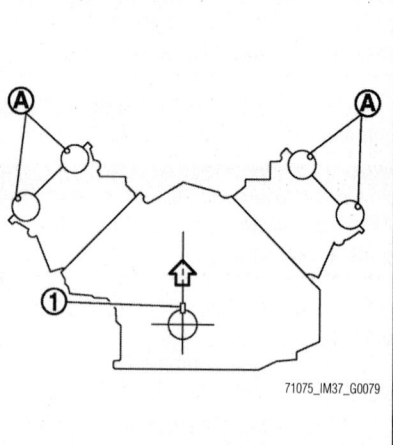

71075_IM37_G0079

Fig. 173 Check that crankshaft key (1) and dowel pin (A) of each camshaft are located as shown

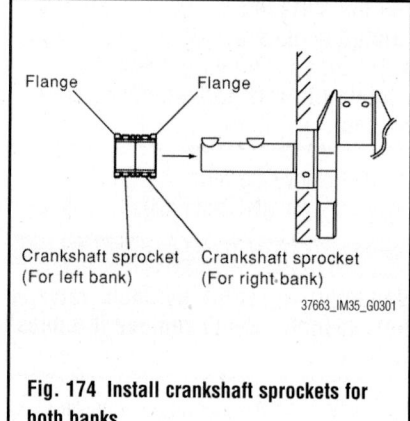

37663_IM35_G0301

Fig. 174 Install crankshaft sprockets for both banks

timing chain and that on the corresponding sprocket, with the components installed.

➡**Parts with an identification mark (R or L) should be installed on the corresponding bank according to the mark.**

The following parts have an identification mark:

- Camshaft sprocket (INT)
- Camshaft sprocket (EXH)
- Chain tension guide
- Chain slack guide

➡**To install timing chain and related parts, start with those on bank 2. The** procedure for installing parts on bank 1 is omitted because it is the same as that for installation on right 2.

32. Check that crankshaft key and dowel pin of each camshaft are located as shown. (No. 1 cylinder at compression TDC)

a. Camshaft dowel pin: At cylinder head upper face side in each bank.

b. Crankshaft key: Straight up.

➡**Though camshaft does not stop at the position, for the placement of cam nose, it is generally accepted camshaft is placed for the same direction.**

33. Install the camshaft sprockets.

a. Install onto the correct side by checking with the identification mark on the surface.

b. Exhaust side: Secure the hexagonal portion of camshaft (EXH) using a wrench to tighten mounting bolt.

c. Intake side: Secure the hexagonal portion (located in between journal No.1 and journal No. 2) of drive shaft using a wrench to tighten mounting bolt.

34. Install high pressure fuel pump camshaft.

✴✴ WARNING
After removing lifter, replace lifter with a new one.

35. Install timing chains as per the following:

a. Install the crankshaft sprockets for both banks.

b. Install each crankshaft sprocket so that its flange side (the larger diameter

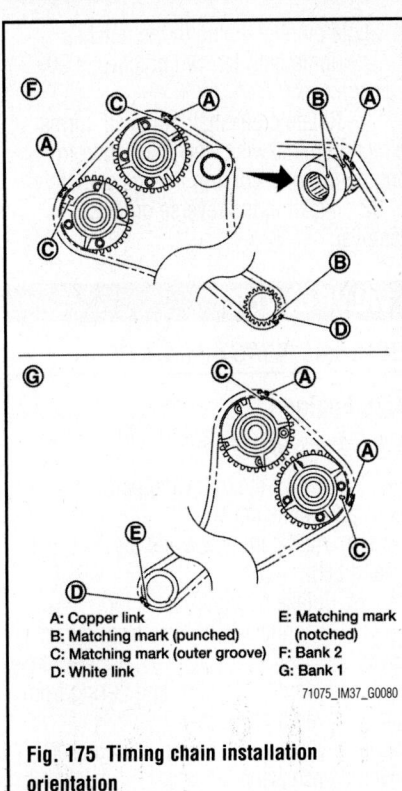

A: Copper link
B: Matching mark (punched)
C: Matching mark (outer groove)
D: White link
E: Matching mark (notched)
F: Bank 2
G: Bank 1

71075_IM37_G0080

Fig. 175 Timing chain installation orientation

side without teeth) faces in the right direction.

➡**The same parts are used but facing directions are different.**

36. Install the timing chains.

a. Bank 2: Install timing chain so that the matching mark (punched) on high pressure fuel pump camshaft and the matching mark (outer groove) on camshaft sprocket is aligned with the copper link on timing chain, while the matching mark (punched) on crankshaft sprocket is aligned with the white link one on timing chain.

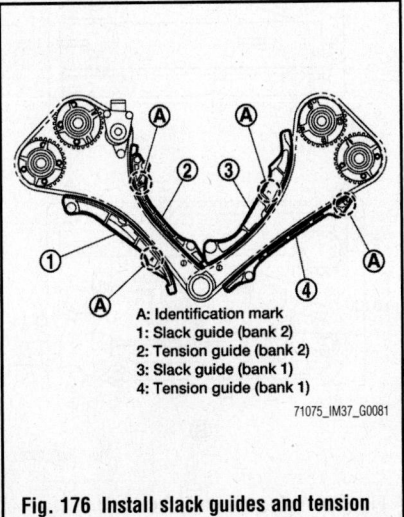

A: Identification mark
1: Slack guide (bank 2)
2: Tension guide (bank 2)
3: Slack guide (bank 1)
4: Tension guide (bank 1)

71075_IM37_G0081

Fig. 176 Install slack guides and tension guides

b. Bank 1: Install timing chain so that the matching mark (outer groove) on camshaft sprocket is aligned with the copper link on timing chain, while the matching mark (notched) on crankshaft sprocket is aligned with the yellow link one on timing chain.

37. Install slack guides and tension guides onto correct side by checking with identification mark (A) on surface.

✳✳ WARNING

Never overtighten slack guide mounting bolt. It is normal for a gap to exist under the bolt seats when mounting bolt are tightened to the specification.

38. Install timing chain tensioner as per the following:

a. Fix the plunger at the most compressed position using a stopper pin. Remove any dirt and foreign materials completely from the back and the mounting surfaces of timing chain tensioner.

b. Pull out stopper pin after installing, and then release plunger.

39. Check again that the matching marks on sprockets and timing chain have not slipped out of alignment.

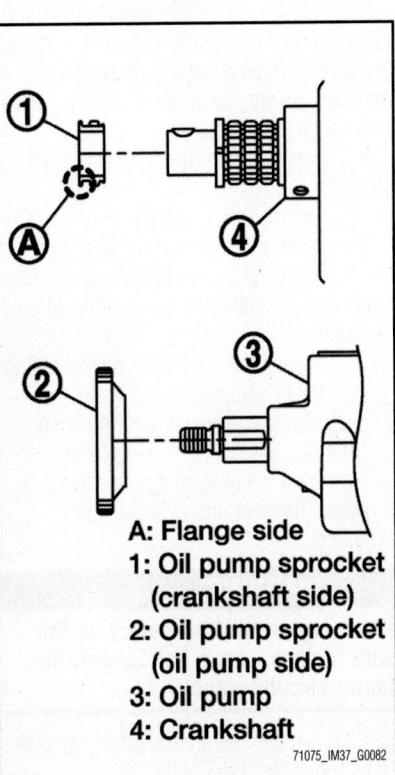

A: Flange side
1: Oil pump sprocket (crankshaft side)
2: Oil pump sprocket (oil pump side)
3: Oil pump
4: Crankshaft

71075_IM37_G0082

Fig. 177 Install the oil pump sprocket (crankshaft side), oil pump sprocket (oil pump side)

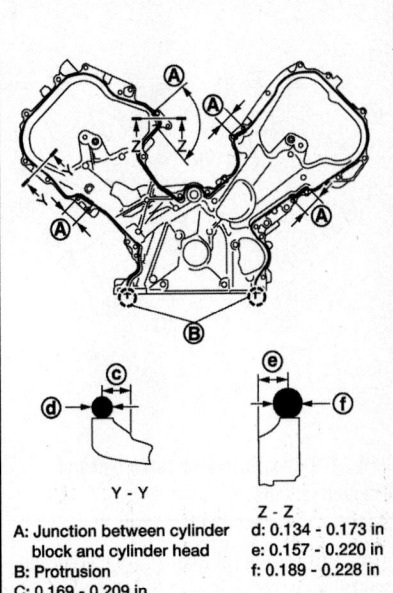

A: Junction between cylinder block and cylinder head
B: Protrusion
C: 0.169 - 0.209 in

d: 0.134 - 0.173 in
e: 0.157 - 0.220 in
f: 0.189 - 0.228 in

71075_IM37_G0083

Fig. 178 Apply a continuous bead of liquid gasket with tube presser (commercial service tool) to front cover

40. Install the oil pump drive chain as per the following:

a. Install the oil pump drive chain tensioner. Fix the tensioner at the most compressed position using a stopper pin and then install it.

b. Install the oil pump sprocket (crankshaft side), oil pump sprocket (oil pump side) and oil pump drive chain at the same time. Install each oil pump sprocket so that its flange side (the larger diameter side without teeth) (A) faces in the direction shown in the figure.

➡**There is no matching mark in the oil pump related parts.**

c. Hold the two flat parts of oil pump shaft, and then tighten the oil pump sprocket (oil pump side) nut.

✳✳ WARNING

Secure the oil pump shaft with the two flat parts.

d. Securely pull out the stopper pin after installing the oil pump drive chain. Check that the tension is applied to the oil pump drive chain after installing.

41. Install front oil seal on front cover.

42. Install the timing chain tensioner cover to front cover. Apply a continuous bead of liquid gasket with tube presser (commercial service tool) to front cover. Use Genuine RTV Silicone Sealant or an equivalent.

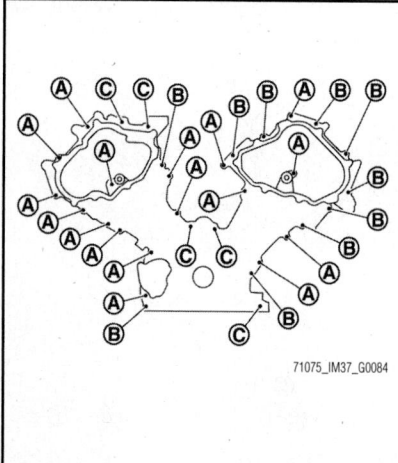

Fig. 179 Identifying the three types of mounting bolts

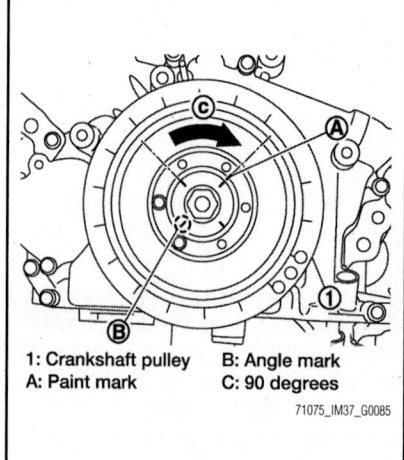

1: Crankshaft pulley   B: Angle mark
A: Paint mark   C: 90 degrees

71075_IM37_G0085

Fig. 180 Crankshaft alignment

43. Install the oil filter (for valve timing control solenoid valve) in the direction shown in the figure, if removed.

a. Check that the oil filter does not protrude from the upper surface of front cover after installation.

44. Install front cover as per the following:

a. Install new O-ring (1), (2) onto cylinder heads and cylinder block.

b. Apply a continuous bead of liquid gasket with tube presser (commercial service tool) to front cover. Use Genuine RTV Silicone Sealant or equivalent.

c. Check again that the matching marks on timing chain and that on each sprocket are aligned. Then, install front cover.

✳✳ WARNING
Be careful not to damage front oil seal by interference with front end of crankshaft.

d. Tighten the mounting bolts in numerical order. There are three types of mounting bolts.

45. Install valve timing control cover as per the following:

➡Both of fuel pump connector protector cannot be installed after installing valve timing control cover. Therefore, install fuel pump connector protector in advance, if it is being removed.

a. Install new O-rings on front cover.
b. Install new seal rings in shaft grooves.

✳✳ WARNING
When replacing seal ring, replace all rings with new ones.

c. Apply a continuous bead of liquid gasket with tube presser (commercial service tool) to valve timing control covers. Use Genuine RTV Silicone Sealant or equivalent.

d. Being careful not to move seal ring from the installation groove, align dowel pins on front cover with dowel pin holes to install valve timing control covers.

e. Tighten mounting bolts in numerical order.

46. Install camshaft position sensor and valve timing control solenoid valve (RH and LH) to valve timing control cover, if removed. Be sure to tighten mounting bolts with flanges completely seated.

47. Install oil pan (lower) and oil strainer.

48. Install oil pan (upper).

49. Install water pump pulley.

50. Install the crankshaft pulley. Fix the crankshaft as instructed in the removal procedure.

a. Install the crankshaft pulley, taking care not to damage front oil seal.

b. Apply engine oil onto threaded parts of crankshaft pulley bolt and seating area. Lightly tapping its center with plastic hammer, insert crankshaft pulley.

✳✳ WARNING
Never tap crankshaft pulley on the side surface where belt is installed (outer circumference).

c. Tighten the crankshaft pulley bolt to 116 ft. lbs. (157 Nm).

d. Put a paint mark on crankshaft pulley aligning with angle mark on crankshaft pulley bolt.

e. Tighten the crankshaft pulley bolt (clockwise) to 90°. Check the tightening

angle by referencing to the notches. The angle between two notches is 90 degrees.

51. Rotate crankshaft pulley in normal direction (clockwise when viewed from engine front) to confirm it turns smoothly.

52. Install in the reverse order of removal.

VALVE COVERS

REMOVAL & INSTALLATION

3.7L Engine

See Figures 181 and 182.

1. Remove the following parts:
a. Engine cover.
b. Air cleaner case and air duct (RH and LH).
c. Intake manifold collector.

2. Disconnect PCV hose from rocker cover.

3. Remove PCV valve and O-ring from rocker cover, if necessary.

4. Remove oil filler cap from rocker cover, if necessary.

5. Remove the ignition coil.

✳✳ WARNING
Never shock the ignition coil.

6. Remove harness clips on the rocker cover.

7. Loosen the mounting bolts with power tool in reverse order.

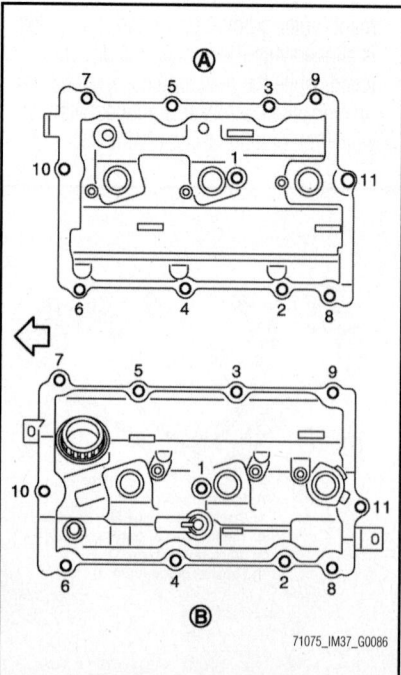

71075_IM37_G0086

Fig. 181 Valve cover removal and installation sequence

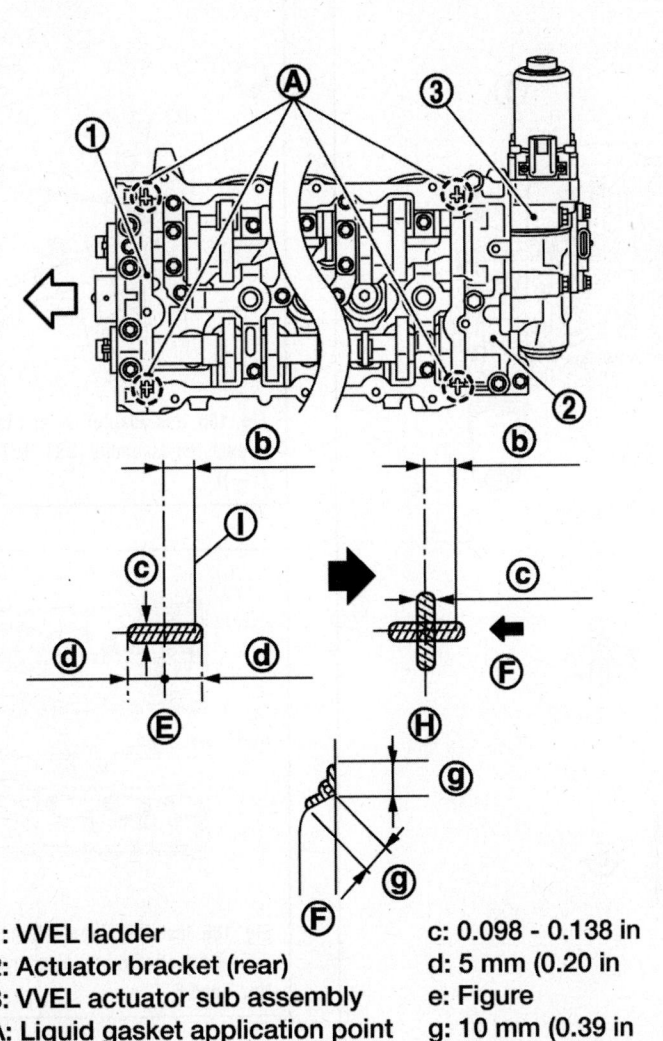

1: VVEL ladder
2: Actuator bracket (rear)
3: VVEL actuator sub assembly
A: Liquid gasket application point
F: View F
I: End surface of VVEL ladder assembly
b: 0.16 in

c: 0.098 - 0.138 in
d: 5 mm (0.20 in
e: Figure
g: 10 mm (0.39 in
h: Figure
⊠: Engine front

71075_IM37_G0087

Fig. 182 Apply liquid gasket to the positions shown

8. Remove rocker cover gasket from rocker cover.

9. Use scraper to remove all traces of liquid gasket from cylinder head and VVEL ladder assembly.

✳✳ WARNING

Never scratch or damage the mating surface when cleaning off old liquid gasket.

To install:

10. Apply liquid gasket to the position shown in the figure with the following procedure:

a. Refer to the figure to apply the liquid gasket to joint part of VVEL ladder assembly and cylinder head.

b. Refer to the figure to apply the liquid gasket. Use Genuine RTV Silicone Sealant or equivalent.

11. Install rocker cover gasket to rocker cover.

12. Install rocker cover. Check that rocker cover gasket does not drop from the installation groove of rocker cover.

13. Tighten bolts in two steps separately in numerical order.

a. 1st step: 18 inch lbs. (2 Nm)
b. 2nd step: 73 inch lbs. (8 Nm)

14. Install in the reverse order of removal after this step.

5.6L Engine

See Figures 183 through 186.

1. Engine cover.

2. Air duct (inlet), air cleaner case assembly and air duct. Remove harness clip and harness bracket, and move engine harness.

3. Remove the ignition coil.

✳✳ WARNING

Never impact it.

➡**Installation position of ignition coil depends on cylinder position.**

4. Disconnect PCV hose from rocker cover.

5. Remove spark plugs.

6. Remove the rocker cover, refer to following.

a. Bank 1: Discharge refrigerant from A/C circuit.

b. Remove low pressure flexible hose compressor side.

c. Bank 2: Remove EVAP hose.

d. Release fuel pressure.

e. Disconnect the battery cable from the negative terminal.

f. Remove water hose.

g. Disconnect low fuel pressure sensor harness connector and quick connector.

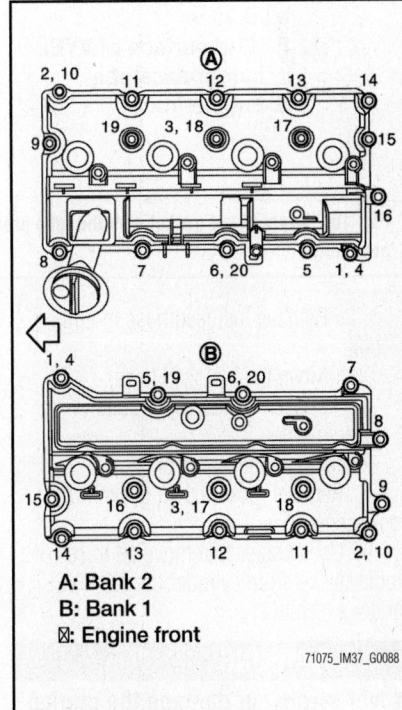

A: Bank 2
B: Bank 1
⊠: Engine front

71075_IM37_G0088

Fig. 183 Valve cover removal and installation bolt sequence

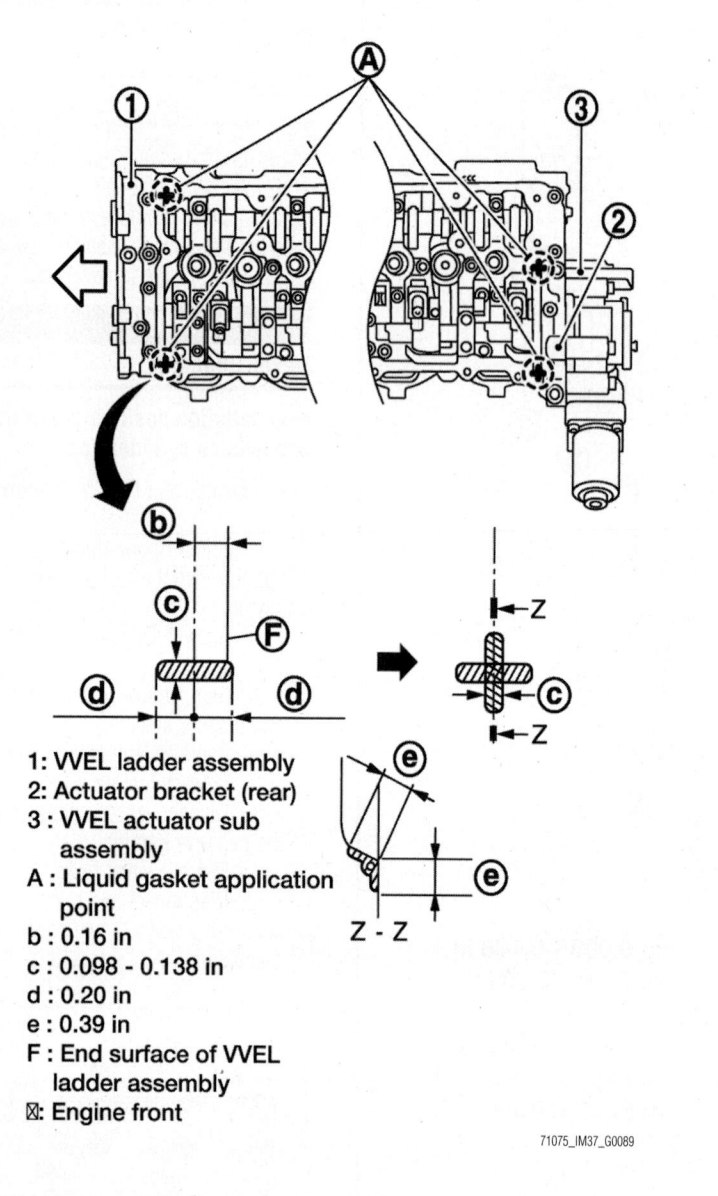

1: VVEL ladder assembly
2: Actuator bracket (rear)
3 : VVEL actuator sub
    assembly
A : Liquid gasket application
    point
b : 0.16 in
c : 0.098 - 0.138 in
d : 0.20 in
e : 0.39 in
F : End surface of VVEL
    ladder assembly
⊠: Engine front

71075_IM37_G0089

Fig. 184 Apply liquid gasket with the tube presser (commercial service tool) to VVEL ladder assembly

h. Remove fuel feed hose mounting bolt.

i. Move the fuel feed hose.

j. Loosen rocker cover bolts in reverse order.

7. Remove rocker cover.

8. Remove rocker cover gasket from rocker cover.

9. Use scraper to remove all traces of liquid gasket from cylinder head & VVEL ladder assembly.

> ❋❋ **WARNING**

Never scratch or damage the mating surface when cleaning off old liquid gasket.

10. Remove PCV valve from rocker cover, if necessary.

11. Remove oil filler cap and oil catcher from rocker cover, if necessary.

To install:

12. Apply the liquid gasket with the tube presser (commercial service tool) to the VVEL ladder assembly and to the actuator bracket (rear). Use Genuine RTV Silicone Sealant or an equivalent.

➡The figure shows an example of bank 1 side. Apply liquid gasket on the front and rear side of engine first. 0.20 inches + 0.20 inches side as shown in the figure.

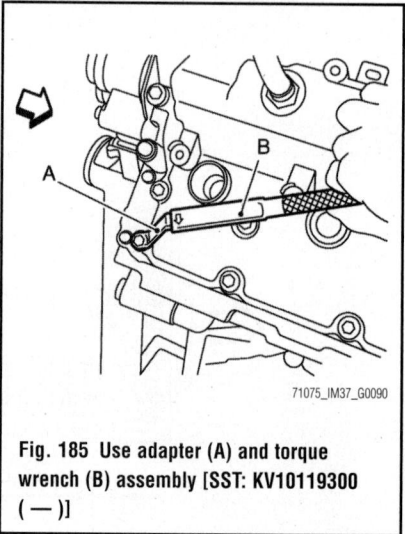

Fig. 185 Use adapter (A) and torque wrench (B) assembly [SST: KV10119300 (—)]

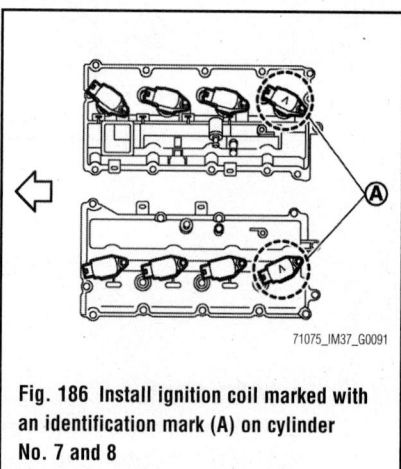

71075_IM37_G0091

Fig. 186 Install ignition coil marked with an identification mark (A) on cylinder No. 7 and 8

Install rocker cover gasket to rocker cover.

13. Install rocker cover. Check that rocker cover gasket does not drop from the installation groove of rocker cover.

14. Tighten bolts in two steps separately in numerical order. Because of the limited working space, use adapter and torque wrench assembly [SST: KV10119300 (—)] to tighten bolts (on the No.7 and No. 8 cylinders) to the specified torque.

15. Install spark plug.

16. Install ignition coil. Install ignition coil marked with an identification mark on cylinder No. 7 and 8.

17. Install in the reverse order of removal.

> **VALVE LASH**

ADJUSTMENT

See Figures 187 and 188.

➡Perform adjustment depending on selected head thickness of valve lifter.

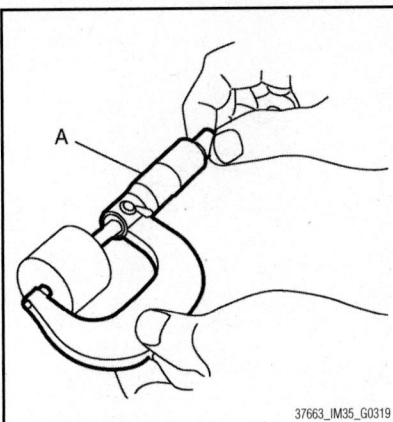

37663_IM35_G0319

Fig. 187 Measure the center thickness of removed valve lifters with a micrometer (A)

1. Measure the valve clearance.
2. Remove camshafts.
3. Remove valve lifters at the locations that are out of the standard.
4. Measure the center thickness of removed valve lifters with a micrometer.
5. Use the equation below to calculate valve lifter thickness for replacement:

- Valve lifter thickness calculation:
  $t = t1 + (C1 - C2)$
- t = Valve lifter thickness to be replaced
- $t1$ = Removed valve lifter thickness
- $C1$ = Measured valve clearance
- $C2$ = Standard valve clearance: Intake: 0.012 inches (0.30 mm); Exhaust: 0.013 inches (0.33 mm)

➡ **Thickness of new valve lifter can be identified by stamp marks on the reverse side (inside the cylinder). Stamp mark 788 indicates 0.3102 inches (7.88 mm) in thickness. Available thickness of valve lifter: 27 sizes with range 7.88 to 8.40 mm (0.3102 to 0.3307 inches) in steps of 0.02 mm (0.0008 inches) (when manufactured at factory).**

6. Install selected valve lifter.
7. Install camshaft.
8. Manually turn crankshaft pulley a few turns.
9. Check that the valve clearances for cold engine are within the specifications by referring to the specified values.

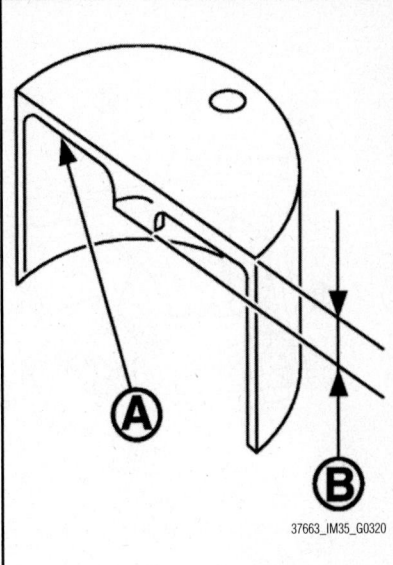

37663_IM35_G0320

Fig. 188 Thickness of new valve lifter (B) can be identified by stamp marks (A) on the reverse side (inside the cylinder)

10. Install all removal parts in the reverse order of removal.
11. Warm up the engine, and check for unusual noise and vibration.

ENGINE PERFORMANCE & EMISSION CONTROLS

COMPONENT LOCATIONS

See Figures 189 through 192.

CAMSHAFT POSITION (CMP) SENSOR

LOCATION

Refer to component locations.

REMOVAL & INSTALLATION

See Figure 193.

✳ WARNING

Handle carefully to avoid dropping and shocks. Never disassemble. Never allow metal powder to adhere to magnetic part at sensor tip. Never place sensors in a location where they are exposed to magnetism.

1. Turn ignition switch OFF.
2. Loosen the mounting bolt of the sensor.
3. Disconnect the camshaft position sensor harness connector.
4. Remove the sensor.
5. Installation is the reverse of removal.

CRANKSHAFT POSITION (CKP) SENSOR

LOCATION

The Crankshaft Position (CKP) sensor is located on the A/T near the rear of the engine. Refer to Component Locations illustrations.

ELECTRONIC CONTROL MODULE (ECM)

LOCATION

Refer to Component Location illustrations.

REMOVAL & INSTALLATION

See Figures 194 and 195.

1. Remove the instrument lower cover.
2. Remove the dash side finisher (LH).
3. Remove the ECM cover bolts and remove ECM cover.
4. Disconnect ECM harness connectors.
5. Remove ECM bracket bolt (A)
6. Slide the ECM bracket (1) upward and then remove ECM bracket with ECM.
7. Remove ECM bracket bolts (B) and separate ECM (2) and ECM bracket.

8. Install in the reverse order of removal.

✳ WARNING

Must be perform additional service when replacing ECM.

ENGINE COOLANT TEMPERATURE (ECT) SENSOR

REMOVAL & INSTALLATION

1. Turn ignition switch OFF.
2. Disconnect engine coolant temperature sensor harness connector.
3. Remove engine coolant temperature sensor.
4. Installation is the reverse of removal.

HEATED OXYGEN (HO2S) SENSOR

REMOVAL & INSTALLATION

See Figures 196 and 197.

1. Remove heated oxygen sensor 2 as follows:
   a. Using heated oxygen sensor wrench KV10114400 (J38365), removal heated oxygen sensor 2.

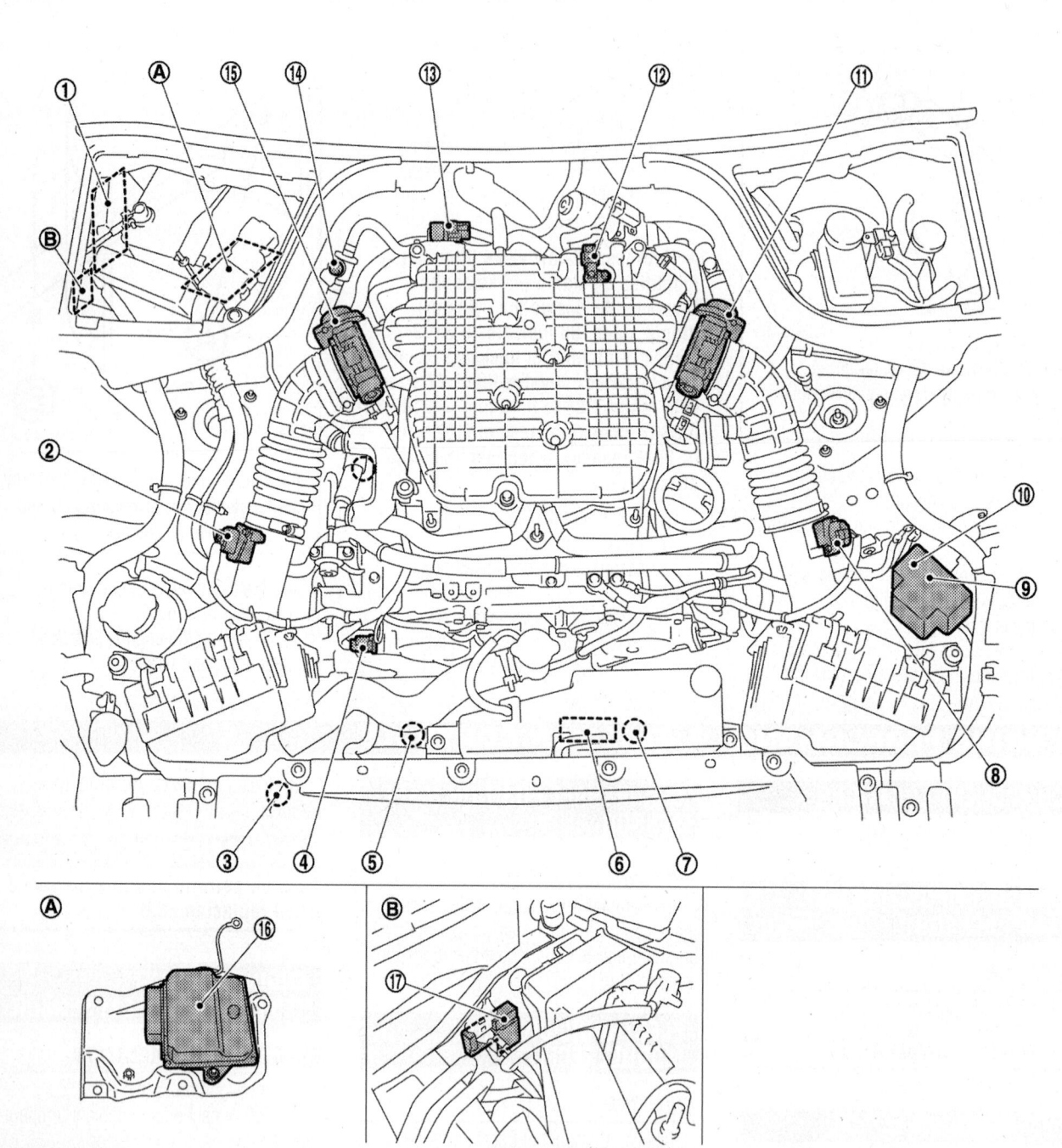

1. **IPDM E/R**
2. **Mass air flow sensor (with intake air temperature sensor) (bank 1)**
3. **Refrigerant pressure sensor**
4. **Camshaft position sensor (PHASE) (bank 1)**
5. **Cooling fan motor-2**
6. **Cooling fan control module**
7. **Cooling fan motor-1**
8. **Mass air flow sensor (bank 2)**
9. **VVEL actuator motor relay**
10. **Cooling fan relay**
11. **Electric throttle control actuator (bank 2)**
12. **Manifold absolute pressure (MAP) sensor**
13. **EVAP canister purge volume control solenoid valve**
14. **EVAP service port**
15. **Electric throttle control actuator (bank 1)**
16. **VVEL control module**
17. **Battery current sensor (with battery temperature sensor)**
A. **Under the battery tray (View with upside-down)**
B. **Body side in battery case**

71075_IM37_G0092

Fig. 189 Engine control component locations—3.7L Engine—1 of 2

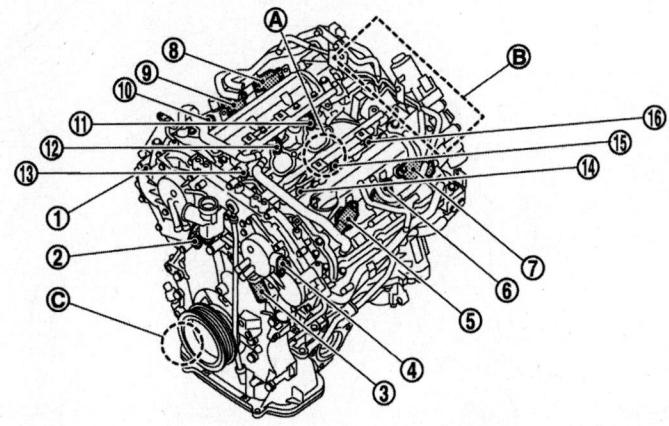

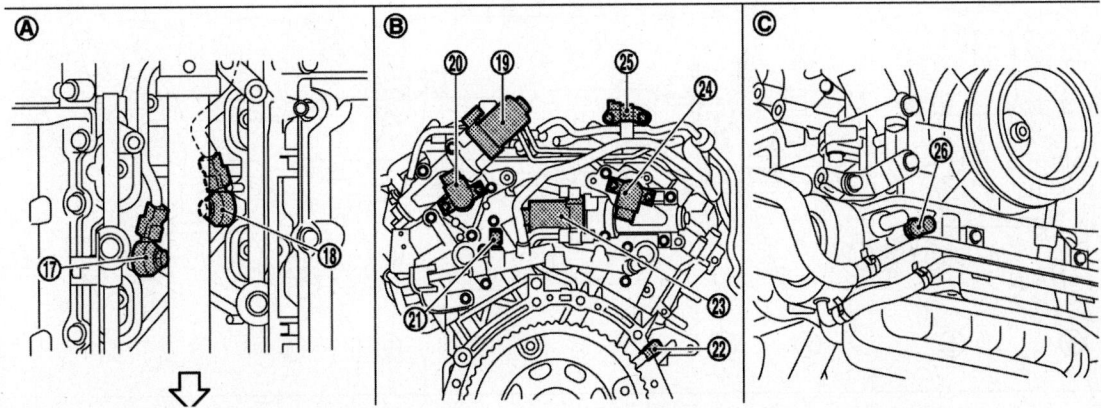

1. Camshaft position sensor (PHASE) (bank 1)
2. Intake valve timing control solenoid valve (bank 1)
3. Intake valve timing control solenoid valve (bank 2)
4. Camshaft position sensor (PHASE) (bank 2)
5. Ignition coil (with power transistor) and spark plug (No.2 cylinder)
6. Ignition coil (with power transistor) and spark plug (No.4 cylinder)
7. Ignition coil (with power transistor) and spark plug (No.6 cylinder)
8. Ignition coil (with power transistor) and spark plug (No.5 cylinder)
9. Ignition coil (with power transistor) and spark plug (No.3 cylinder)
10. Ignition coil (with power transistor and spark plug (No.1 cylinder)
11. Fuel injector (No.5 cylinder)
12. Fuel injector (No.3 cylinder)
13. Fuel injector (No.1 cylinder)
14. Fuel injector (No.2 cylinder)
15. Fuel injector (No.4 cylinder)
16. Fuel injector (No.6 cylinder)
17. Knock sensor (bank 1)
18. Knock sensor (bank 2)
19. VVEL actuator motor (bank 2)
20. VVEL control shaft position sensor (bank 2)
21. Engine coolant temperature sensor
22. Crankshaft position sensor (POS)
23. VVEL actuator motor (bank 1)
24. VVEL control shaft position sensor (bank 1)
25. EVAP canister purge volume control solenoid valve
26. Engine oil temperature sensor
A. Top view of the engine (View with intake manifold removed)
B. Rear view of the engine
C. Front view of the engine
Arrow indicates front of engine

71075_IM37_G0093

Fig. 190 Engine control component locations—3.7L Engine—2 of 2

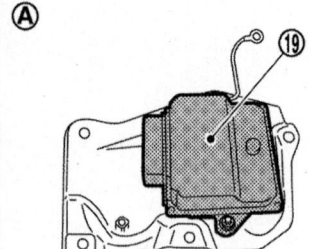

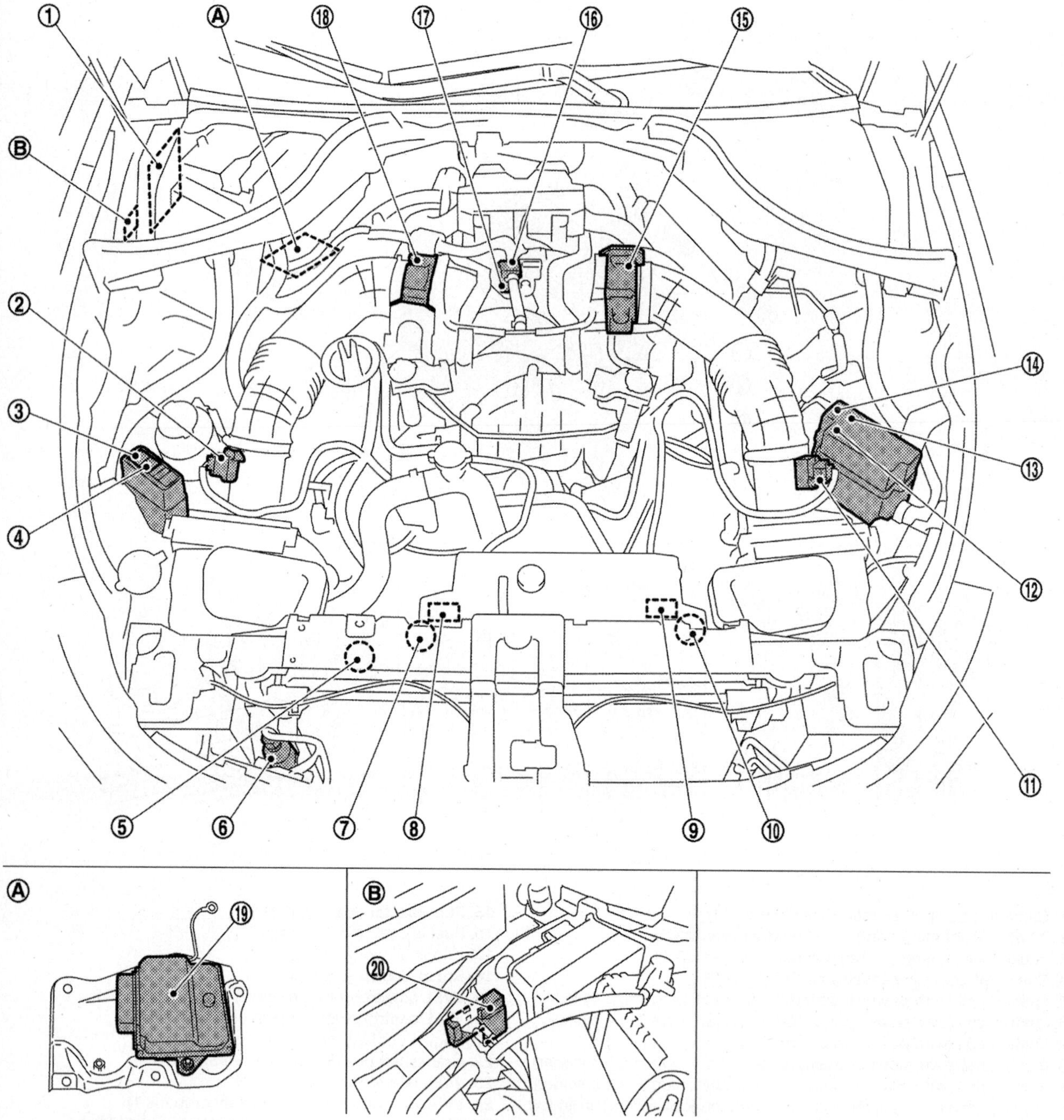

1. IPDM E/R
2. Mass air flow sensor
3. Cooling fan relay 2
4. Injector relay 2
5. Power steering pressure sensor
6. Refrigerant pressure sensor
7. Cooling fan motor 2
8. Cooling fan control module 2
9. Cooling fan control module 1
10. Cooling fan motor 1
11. Mass air flow sensor (with intake air temperature sensor) (bank 1)

12. Injector relay 1
13. VVEL actuator motor relay
14. Cooling fan relay 1
15. Electric throttle control actuator (bank 1)
16. EVAP canister purge volume control solenoid valve
17. Manifold absolute pressure (MAP) sensor
18. Electric throttle control actuator (bank 2) 19. VVEL control module
20. Battery current sensor (with battery temperature sensor)
A. Under the battery tray (View with upside-down)
B. Body side in battery case

71075_IM37_G0094

Fig. 191 Engine control component locations—5.6L Engine—1 of 2

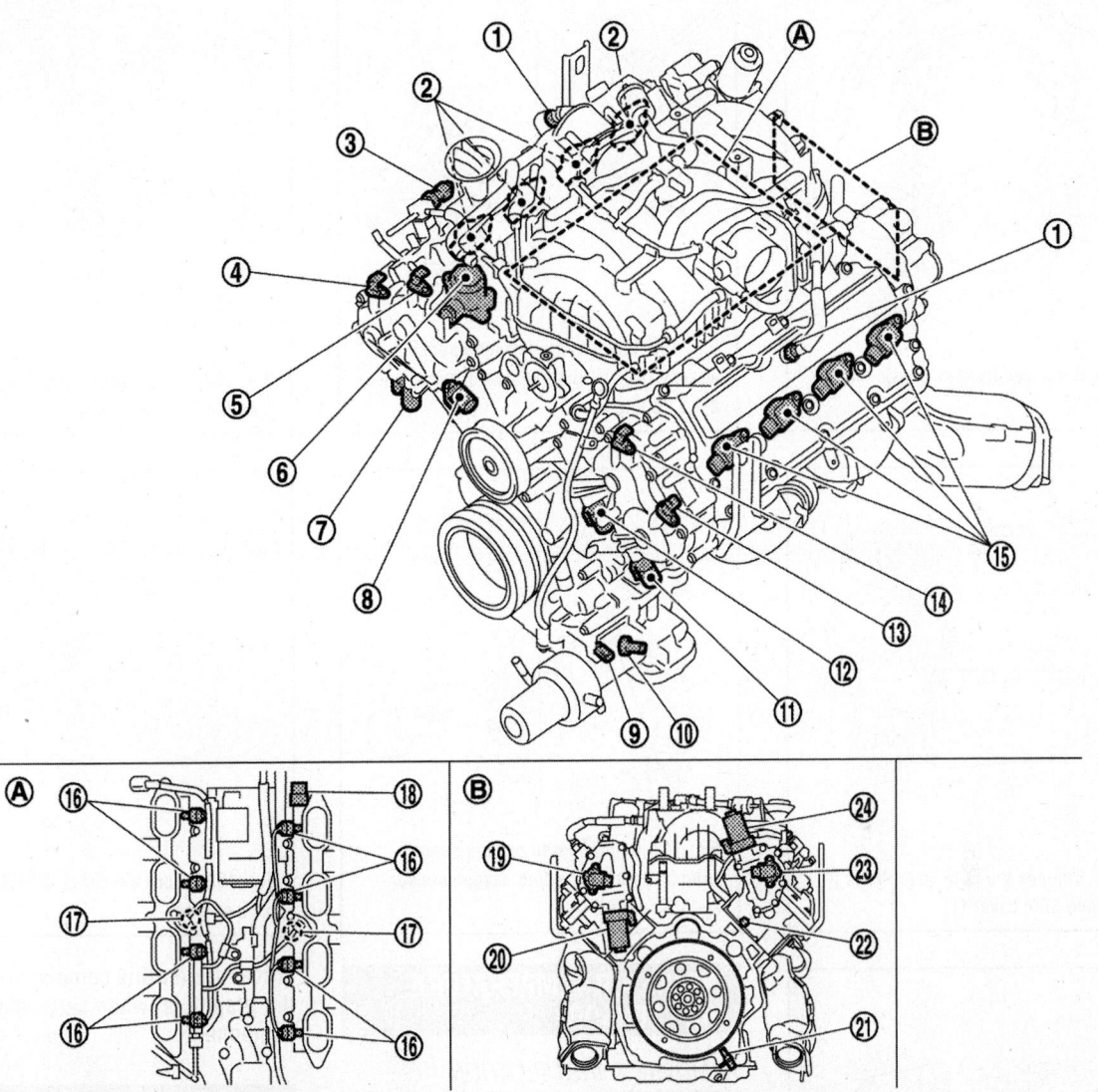

1. Positive crankcase ventilation (PCV) valve
2. Ignition coil (with power transistor) and spark plug (bank 2)
3. Low fuel pressure sensor 4. Exhaust valve timing control position sensor (bank 2)
5. Camshaft position sensor (bank 2)
6. High pressure fuel pump
7. Exhaust valve timing control solenoid valve (bank 2)
8. Intake valve timing control solenoid valve (bank 2)
9. Engine oil temperature sensor
10. Engine oil pressure sensor
11. Exhaust valve timing control solenoid valve (bank 1)
12. Intake valve timing control solenoid valve (bank 1)
13. Exhaust valve timing control position sensor (bank 1)
14. Camshaft position sensor (bank 1)

15. Ignition coil (with power transistor) and spark plug (bank 1)
16. Fuel injector
17. Knock sensor
18. Fuel rail pressure sensor
19. VVEL control shaft position sensor (bank 1)
20. VVEL actuator motor (bank 1)
21. Crankshaft position sensor
22. Engine coolant temperature sensor
23. VVEL control shaft position sensor (bank 2)
24. VVEL actuator motor (bank 2)
A. Top view of the engine (View with intake manifold is removed)
B. Rear view of the engine

71075_IM37_G0095

Fig. 192 Engine control component locations—5.6L Engine—2 of 2

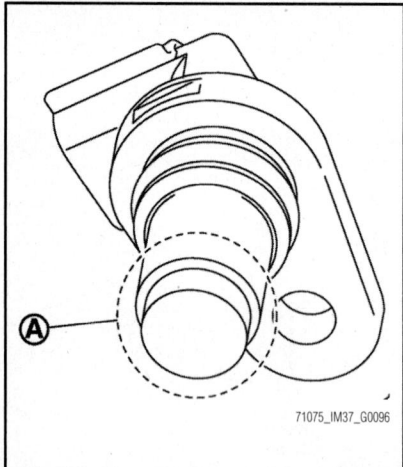

Fig. 193 Keep free from magnetic materials (A)

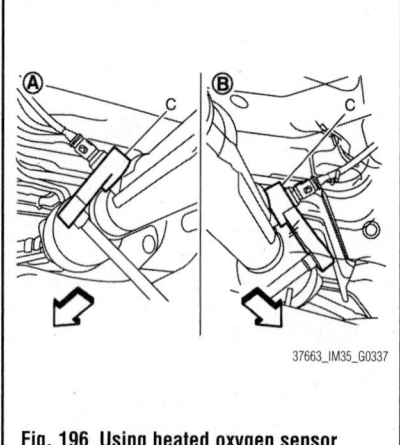

Fig. 196 Using heated oxygen sensor wrench (C), removal heated oxygen sensor 2—3.7L engine

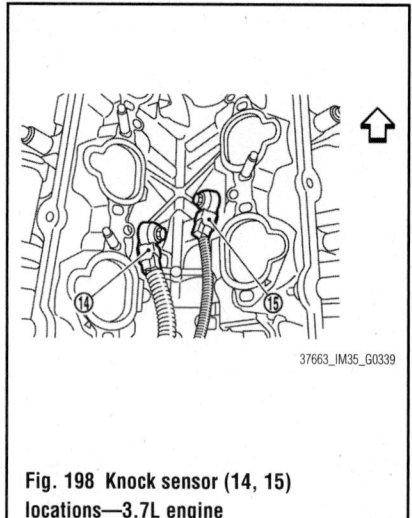

Fig. 198 Knock sensor (14, 15) locations—3.7L engine

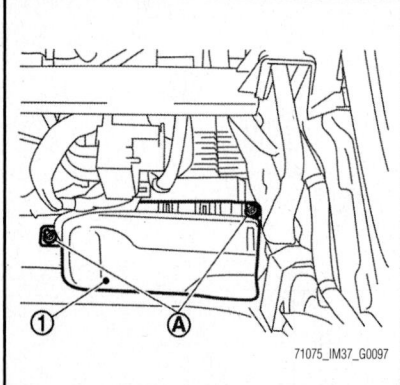

Fig. 194 Remove the ECM cover bolts (A) and remove ECM cover (1)

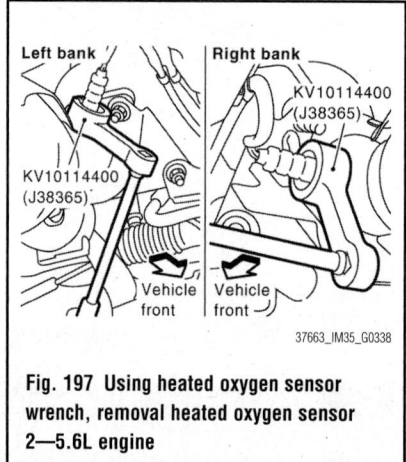

Fig. 197 Using heated oxygen sensor wrench, removal heated oxygen sensor 2—5.6L engine

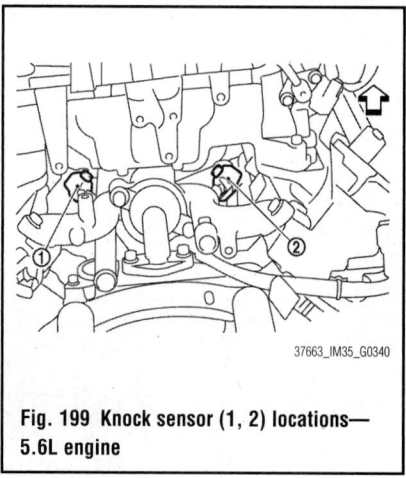

Fig. 199 Knock sensor (1, 2) locations—5.6L engine

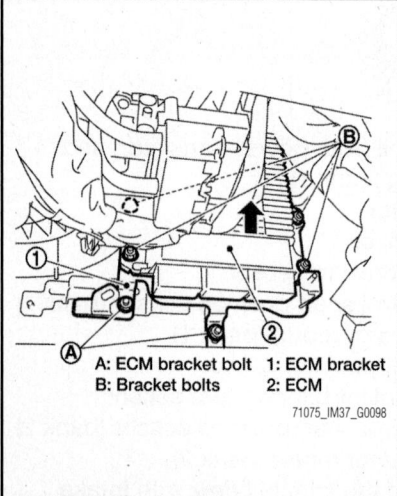

A: ECM bracket bolt 1: ECM bracket
B: Bracket bolts 2: ECM

Fig. 195 Removing the ECM

✷✷ WARNING

Be careful not to damage heated oxygen sensor 2.

2. Installation is the reverse of removal.

INTAKE AIR TEMPERATURE (IAT) SENSOR

REMOVAL & INSTALLATION

The Intake Air Temperature (IAT) sensor is an integral part of the Mass Air Flow (MAF) sensor/Intake Air Temperature (IAT) sensor assembly which is mounted on the air intake duct. Refer to the Mass Air Flow (MAF) section for information regarding servicing this component.

KNOCK SENSOR (KS)

LOCATION
See Figures 198 and 199.

The knock sensors are located in the valley of the cylinder block beneath the intake manifold.

REMOVAL & INSTALLATION

➡The manufacturer does not provide a specific Removal and Installation

procedure for this component. Refer to the graphic(s) when servicing this component.

MASS AIR FLOW (MAF) SENSOR

LOCATION

Refer to Component Location illustrations.

REMOVAL & INSTALLATION

Refer to Air Cleaner in Engine Mechanical.

THROTTLE POSITION SENSOR (TPS)

LOCATION

The Throttle Position Sensor is an integral part of the electric throttle control actuator.

REMOVAL & INSTALLATION

Refer to Intake Manifold in Engine Mechanical.

FUEL SYSTEM SERVICE PRECAUTIONS

Safety is the most important factor when performing not only fuel system maintenance but any type of maintenance. Failure to conduct maintenance and repairs in a safe manner may result in serious personal injury or death. Maintenance and testing of the vehicle's fuel system components can be accomplished safely and effectively by adhering to the following rules and guidelines.

• To avoid the possibility of fire and personal injury, always disconnect the negative battery cable unless the repair or test procedure requires that battery voltage be applied.

• Always relieve the fuel system pressure prior to disconnecting any fuel system component (injector, fuel rail, pressure regulator, etc.), fitting or fuel line connection. Exercise extreme caution whenever relieving fuel system pressure to avoid exposing skin, face and eyes to fuel spray. Please be advised that fuel under pressure may penetrate the skin or any part of the body that it contacts.

• Always place a shop towel or cloth around the fitting or connection prior to loosening to absorb any excess fuel due to spillage. Ensure that all fuel spillage (should it occur) is quickly removed from engine surfaces. Ensure that all fuel soaked cloths or towels are deposited into a suitable waste container.

• Always keep a dry chemical (Class B) fire extinguisher near the work area.

• Do not allow fuel spray or fuel vapors to come into contact with a spark or open flame.

• Always use a back-up wrench when loosening and tightening fuel line connection fittings. This will prevent unnecessary stress and torsion to fuel line piping.

• Always replace worn fuel fitting O-rings with new Do not substitute fuel hose or equivalent where fuel pipe is installed.

Before servicing the vehicle, make sure to also refer to the precautions in the beginning of this section as well.

RELIEVING FUEL SYSTEM PRESSURE

With Consult-III

1. Turn ignition switch ON.
2. Perform "FUEL PRESSURE RELEASE" in "WORK SUPPORT" mode with CONSULT-III.

3. Start engine.
4. After engine stalls, crank it 2 or 3 times to release all fuel pressure.
5. Turn ignition switch OFF.

Without Consult-III

See Figure 200.

1. Remove fuel pump fuse located in IPDM E/R.
2. Start engine.
3. After engine stalls, crank it 2 or 3 times to release all fuel pressure.
4. Turn ignition switch OFF.
5. Reinstall fuel pump fuse after servicing fuel system.

FUEL LEVEL SENSOR UNIT, FUEL FILTER AND FUEL PUMP ASSEMBLY

REMOVAL & INSTALLATION

See Figures 201 through 206.

1. Check fuel level on fuel gauge. If fuel gauge indicates more than the level as shown (full or almost full), drain fuel from fuel tank until fuel gauge indicates level as shown or below.

➡**Because fuel will be spilled when removing main and sub fuel level sensor units for the top of the fuel is above the main and sub fuel level sensor units installation surface. As a guide, fuel level becomes the position as shown or below when approximately 5¼ US gal. (20 L) of fuel are drained from fuel tank.**

2. In a case that fuel pump does not operate, perform the following procedure.

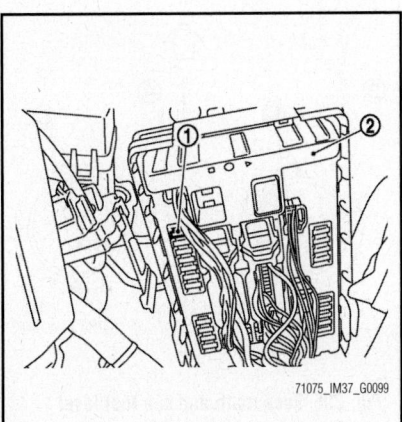

Fig. 200 Remove fuel pump fuse (1) located in IPDM E/R (2)

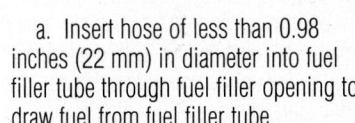

37663_IM35_G0356

Fig. 201 Check fuel level on fuel gauge

a. Insert hose of less than 0.98 inches (22 mm) in diameter into fuel filler tube through fuel filler opening to draw fuel from fuel filler tube.

b. Disconnect fuel filler hose from fuel filler tube.

c. Insert fuel tube into fuel tank through fuel filler hose to draw fuel from fuel tank.

3. Release the fuel pressure from the fuel lines.

4. Open fuel filler lid.

5. Open filler cap and release the pressure inside fuel tank.

6. Remove rear seat cushion.

7. Peel off floor carpet, then remove inspection hole cover units by turning clips clockwise by 90 degrees.

8. Disconnect harness connector and fuel feed tube.

9. Disconnect quick connector as follows:

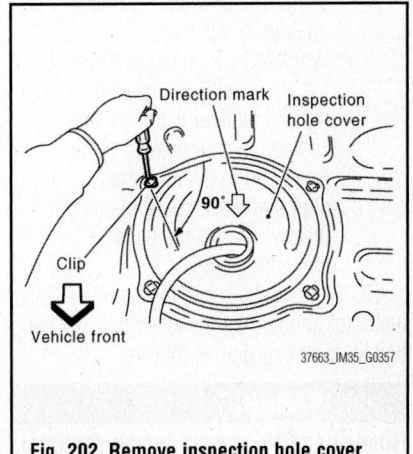

37663_IM35_G0357

Fig. 202 Remove inspection hole cover units by turning clips clockwise by 90 degrees

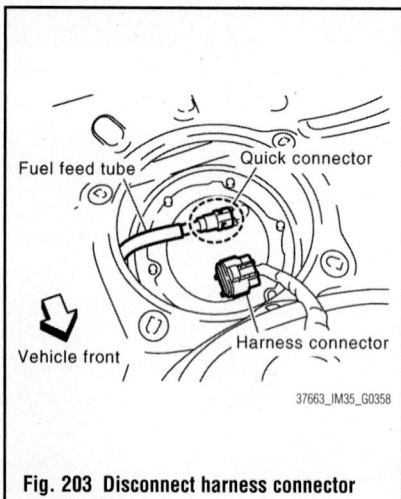

Fig. 203 Disconnect harness connector and fuel feed tube

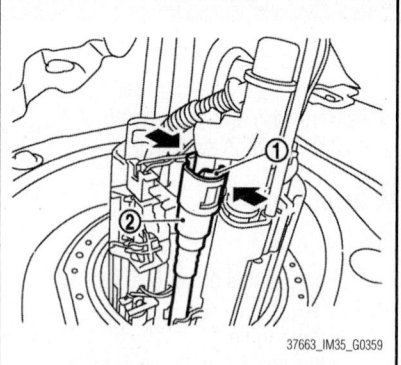

Fig. 204 Raise main fuel level sensor unit, fuel filter and fuel pump assembly, and disconnect quick connector by pushing in tabs (1) and pulling out fuel tube (2)

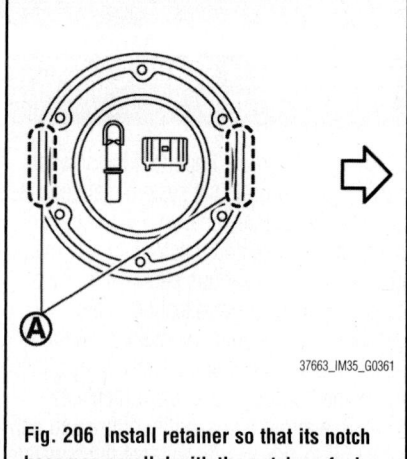

Fig. 206 Install retainer so that its notch becomes parallel with the notch on fuel tank

  a. Hold the sides of connector, push in tabs and pull out fuel feed tube.

  b. If quick connector sticks to tube of main fuel level sensor unit, push and pull quick connector several times until they start to move. Then disconnect them by pulling.

  • Quick connector can be disconnected when the tabs are completely depressed. Never twist it more than necessary.

  • Never use any tools to disconnected quick connector.

  • Keep resin tube away from heat. Be especially careful when welding near the resin tube.

  • Prevent acid liquid such as battery electrolyte, etc. from getting on resin tube.

  • Never bend or twist resin tube during installation and disconnection.

  • Never remove the remaining retainer on hard tube (or the equivalent) except when resin tube or retainer is replaced.

  • When resin tube or hard tube (or the equivalent) is replaced, also replace retainer with new one.

  • To keep the connecting portion clean and to avoid damage and foreign materials, cover them completely with plastic bags or something similar.

10. Remove main fuel level sensor unit, fuel filter and fuel pump assembly, and sub fuel level sensor unit as follows:

  a. Removal of main fuel level sensor unit, fuel filter and fuel pump assembly, and sub fuel level sensor unit:

  • Remove retainer.

  • Raise main fuel level sensor unit, fuel filter and fuel pump assembly, and disconnect quick connector by pushing in tabs and pulling out fuel tube.

  b. Removal of sub fuel level sensor unit:

  • Remove retainer.

  • Raise and release sub fuel level sensor unit to remove.

To install:

11. Note the following, and install in the reverse order of removal.

  a. Main and Sub Fuel Level Sensor Unit:

  • Face main and sub fuel level sensor units as shown, and install

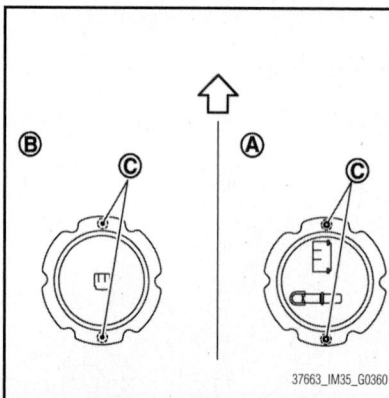

Fig. 205 Face main and sub fuel level sensor units as shown, and install them with the knock pin (C) on back aligned with pin hole on fuel tank; Right side (A), Left side (B)

them with the knock pin on back aligned with pin hole on fuel tank.

  • Install retainer so that its notch becomes parallel with the notch on fuel tank.

  • Tighten retainer mounting bolts evenly.

  b. Connect quick connector as follows:

  • Check the connection for damage or any foreign materials.

  • Align the connector with the tube, then insert the connector straight into the tube until a click sound is heard.

  • After connecting, check that the connection is secure by following method.

  • Pull the tube and the connector to check they are securely connected.

  • Visually confirm that the two retainer tabs are connected to the connector.

12. Turn ignition switch "ON" (with engine stopped), then check connections for leakage by applying fuel pressure to fuel piping.

13. Start engine and let it idle and check there are no fuel leakage at the fuel system connections.

FUEL INJECTORS

REMOVAL & INSTALLATION

3.7L Engine

See Figures 207 through 212.

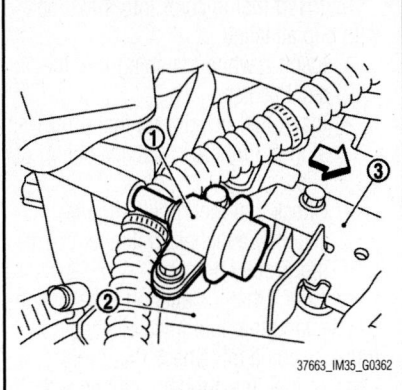

Fig. 207 Remove fuel feed hose (with damper) (1) from fuel sub-tube (2) and harness bracket (3)

✳✳ WARNING

Never smoke while servicing fuel system. Keep open flames and sparks away from the work area.

✳✳ WARNING

To avoid the danger of being scalded, never drain engine coolant when engine is hot.

1. Remove engine room cover (RH and LH).
2. Remove engine cover.
3. Release fuel pressure.
4. Drain engine coolant, or when water hoses are disconnected, attach plug to prevent engine coolant leakage.

✳✳ WARNING

Perform this step when engine is cold.

5. Remove intake manifold collector.
6. Remove fuel feed hose (with damper) from fuel sub-tube and harness bracket.

➡**There is no fuel return route.**

✳✳ WARNING

While hoses are disconnected, plug them to prevent fuel from draining. Never separate fuel damper and fuel feed hose.

7. When separating fuel feed hose (with damper) and centralized under-floor piping connection, disconnect quick connector as follows:
   a. Remove quick connector cap from quick connector connection on right member side.

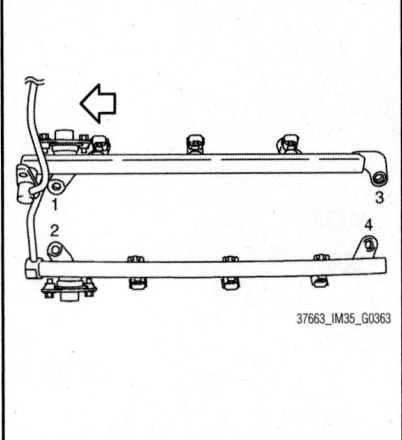

Fig. 208 Loosen mounting bolts in reverse order as shown

   b. Disconnect fuel feed hose (with damper) from bracket hose clamp.
   c. Push in retainer tabs.
   d. Draw and pull out quick connector straight from centralized under-floor piping.
   • Never pull with lateral force applied. O-ring inside quick connector may be damaged.
   • Prepare container and cloth beforehand because fuel will leakage out.
   • Avoid fire and sparks.
   • Keep parts away from heat source. Especially, be careful when welding is performed around them.
   • Never expose parts to battery electrolyte or other acids.
   • Never bend or twist connection between quick connector and fuel feed hose (with damper) during installation/removal.
   • To keep clean the connecting portion and to avoid damage and foreign materials, cover them completely with plastic bags, etc. or something similar.

8. Disconnect harness connector from fuel injector.
9. Loosen mounting bolts in reverse order as shown, and remove fuel tube and fuel injector assembly.

✳✳ WARNING

Never tilt it, or remaining fuel in pipes may flow out from pipes.

10. Remove fuel injector from fuel tube as follows:
   a. Open and remove clip.
   b. Remove fuel injector from fuel tube by pulling straight.
   • Be careful with remaining fuel that may go out from fuel tube.

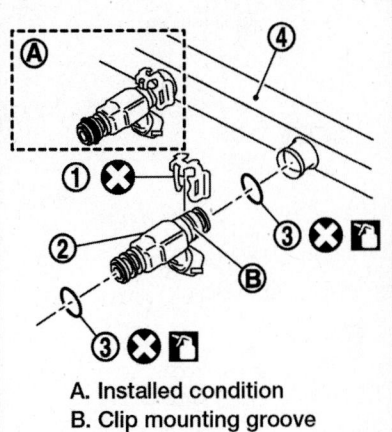

A. Installed condition
B. Clip mounting groove
1. Clip
2. Fuel injector
3. O-ring
4. Fuel tube

Fig. 209 Remove fuel injector from fuel tube

   • Be careful not to damage injector nozzles during removal.
   • Never bump or drop fuel injector.
   • Never disassemble fuel injector.

11. Remove fuel sub-tube and fuel damper, if necessary.

To install:

12. Install fuel damper as follows:
   a. Install new O-ring to fuel tube as shown.
   • Handle O-ring with bare hands. Never wear gloves.
   • Lubricate O-ring with new engine oil.
   • Never clean O-ring with solvent.
   • Check that O-ring and its mating part are free of foreign material.
   • When installing O-ring, be careful not to scratch it with tool or fingernails. Also be careful not to twist or stretch O-ring. If O-ring was stretched while it was being attached, never insert it quickly into fuel tube.
   • Insert O-ring straight into fuel tube. Never decenter or twist it.
   b. Install spacer to fuel damper.
   c. Insert fuel damper straight into fuel tube.

✳✳ WARNING

Insert straight, checking that the axis is lined up. Never pressure-fit with excessive force. Insert fuel damper unit is touching of fuel tube.

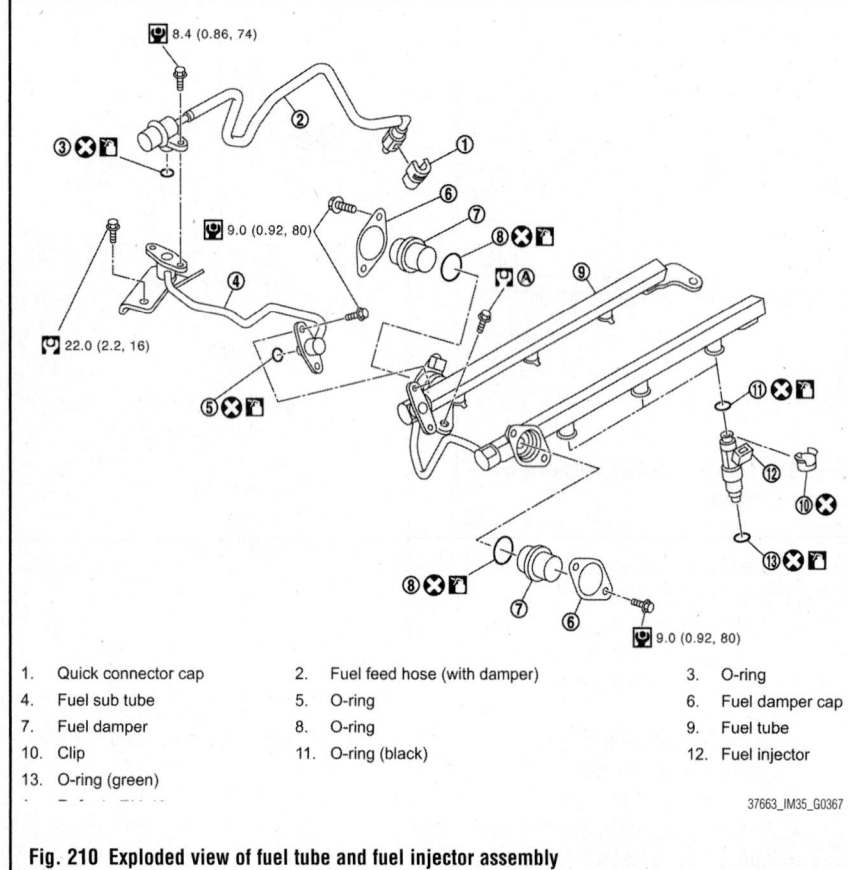

Fig. 210 Exploded view of fuel tube and fuel injector assembly

1. Quick connector cap	2. Fuel feed hose (with damper)	3. O-ring
4. Fuel sub tube	5. O-ring	6. Fuel damper cap
7. Fuel damper	8. O-ring	9. Fuel tube
10. Clip	11. O-ring (black)	12. Fuel injector
13. O-ring (green)		

37663_IM35_G0367

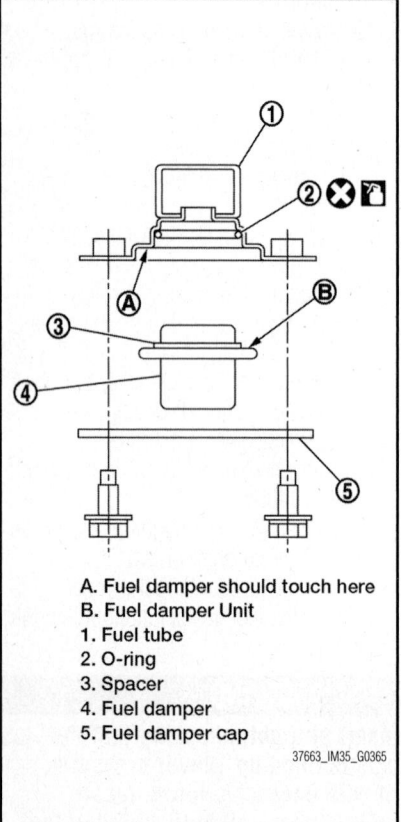

A. Fuel damper should touch here
B. Fuel damper Unit
1. Fuel tube
2. O-ring
3. Spacer
4. Fuel damper
5. Fuel damper cap

37663_IM35_G0365

Fig. 211 Install fuel damper

d. Tighten bolts evenly in turn.

➠**After tightening bolts, check that there is no gap between fuel damper cap and fuel tube.**

13. Install fuel sub-tube.
   a. Insert fuel sub-tube straight into fuel tube.
   b. Tighten mounting bolts evenly in turn.
   c. After tightening mounting bolts, check that there is no gap between flange and fuel sub-tube.
14. Install O-rings to fuel injector, paying attention to the following:
   • Upper and lower O-ring are different. Be careful not to confuse them.
   • Fuel tube side is black
   • Nozzle side is green
15. Install fuel injector to fuel tube as follows:
   a. Insert clip into clip mounting groove on fuel injector.

❄❄ WARNING

Never reuse clip. Replace it with a new one. Be careful to keep clip from interfering with O-ring. If interference occurs, replace O-ring.

b. Insert fuel injector into fuel tube with clip attached.
   • Insert it while matching it to the axial center.
   • Insert fuel injector so that protrusion of fuel tube matches cutout of clip.
   • Check that fuel tube flange is securely attached in flange mounting groove on clip.
   c. Check that installation is complete by checking that fuel injector does not rotate or come off. Check that protrusions of fuel injectors are aligned with cutouts of clips after installation.
16. Install fuel tube and fuel injector assembly to intake manifold.

❄❄ WARNING

Be careful not to let tip of injector nozzle come in contact with other parts.

17. Tighten mounting bolts in two steps in numerical order as shown.
   a. Step 1: 7 ft. lbs. (10 Nm)
   b. Step 2: 17 ft. lbs. (24 Nm)
18. Connect fuel injector harness connector.
19. Connect fuel feed hose (with damper).
   a. Handling procedure of O-ring is the same as that of fuel damper and fuel sub-tube.
   b. Insert fuel damper straight into fuel sub-tube.
   c. Tighten mounting bolts evenly in turn.
   d. After tightening mounting bolts, check that there is no gap between flange and fuel sub-tube.
20. Connect quick connector between fuel feed hose (with damper) and centralized under-floor piping connection as follows:
   a. Check no foreign substances are deposited in and around centralized under-floor piping and quick connector, and no damage on them.
   b. Thinly apply new engine oil around centralized under-floor piping from tip end to spool end.
   c. Align center to insert quick connector straightly into centralized under-floor piping. Insert quick connector to centralized under-floor piping until top spool is completely inside quick connector, and 2nd level spool exposes right below quick connector.
   • Hold align center to avoid inclined insertion to prevent to O-ring inside quick connector.
   • Insert until you hear a "click" sound and actually feel the engagement.

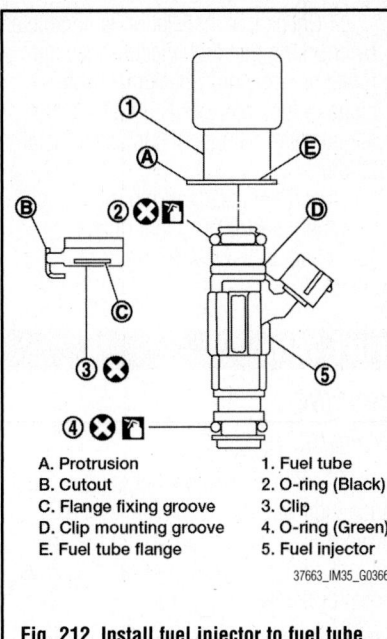

A. Protrusion
B. Cutout
C. Flange fixing groove
D. Clip mounting groove
E. Fuel tube flange

1. Fuel tube
2. O-ring (Black)
3. Clip
4. O-ring (Green)
5. Fuel injector

37663_IM35_G0366

Fig. 212 Install fuel injector to fuel tube

- To avoid misidentification of engagement with a similar sound, be sure to perform the next step.

d. Pull quick connector by hand holding position. Check it is completely engaged (connected) so that it does not come out from centralized under-floor piping.

e. Install quick connector cap to quick connector connection. Install quick connector cap with arrow on surface facing in direction of quick connector (fuel feed hose side).

❋ WARNING
If quick connector cap cannot be installed smoothly, quick connector may have not been installed correctly. Check the connection again.

21. Install in the reverse order of removal after this step.

22. Turn ignition switch "ON" (with engine stopped). With fuel pressure applied to fuel piping, check for fuel leakage at connection points.

➡**Use mirrors for checking at points out of clear sight.**

23. Start engine. With engine speed increased, check again for fuel leakage at connection points.

❋ WARNING
Never touch engine immediately after stopped, as engine becomes extremely hot.

5.6L Engine
See Figures 213 through 218.

❋ WARNING
Be sure to work in a well-ventilated area and furnish workshop with a CO2 fire extinguisher.

❋ WARNING
Never smoke while servicing the fuel system. Keep open flames and sparks away from the work area.

❋ WARNING
To avoid the danger of being scalded, never drain engine coolant when engine is hot.

1. Release fuel pressure.
2. Remove the intake manifold.
3. Remove the acoustic absorbent.
4. Remove fuel feed tube (pump side) and fuel feed tube (bank side).

❋ WARNING
Never reuse fuel feed tube.

5. Remove fuel rail (bank 1) and fuel rail (bank 2).
6. Disconnect harness connector from fuel injectors.
7. Remove fuel injector from cylinder head as per the following:

❋ WARNING
Be careful with remaining fuel that may go out from fuel tube. Be careful not to damage injector nozzles during removal. Never bump or drop fuel injector. Never disassemble fuel injector.

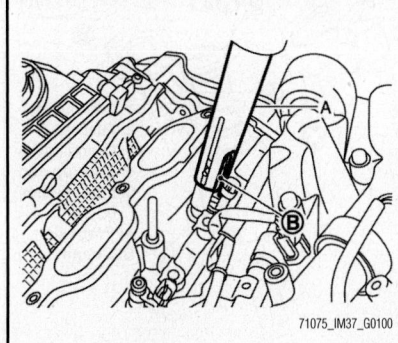

71075_IM37_G0100

Fig. 213 Install an injector remover [SST:KV10119600 (—)] (A) to the injector connector side so that cutout (B) of injector remover faces the injector connector side

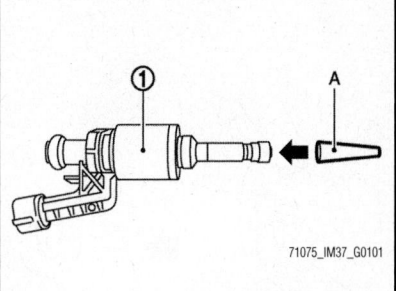

71075_IM37_G0101

Fig. 214 Install an injector seal drift set [SST: KV101197S0 (—)] (A) to fuel injector (1)

a. Remove the injector holder.

b. Install an injector remover to the injector connector side so that cutout of injector remover faces the injector connector side.

c. Hook the pawl portion of injector remover to groove portion of injector.

d. Press down body portion of injector remover until it contacts cylinder head.

e. Tighten injector remover clockwise and remove injector from cylinder head.

f. Cut Teflon seal (1) while pinching it. Be careful not to damage injector.

g. Remove the insulator from mounting hole of fuel injector of cylinder head.

To install:
8. Install seal ring to fuel injector as per the following:

❋ WARNING
Handle seal ring with bare hands. Never wear gloves. Never apply engine oil to seal ring. Never clean seal ring with solvent.

a. Install an injector seal drift set to fuel injector.

b. Set seal ring to injector seal drift set .

c. Straightly insert seal ring, which is set in precedent step, to fuel injector and install.

❋ WARNING
Be careful that seal ring does not exceed the groove portion of fuel injector.

d. Insert injector seal drift set to injector and rotate clockwise and counterclockwise by 90° while pressing seal ring to fit it.

➡**Compress seal ring, because this operation is for rectifying stretch of**

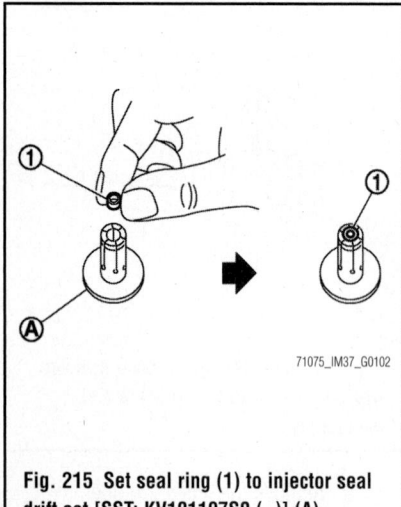

Fig. 215 Set seal ring (1) to injector seal drift set [SST: KV101197S0 (-)] (A)

seal ring caused by installation and for preventing sticking when inserting injector into cylinder head.

9. Install O-ring and back up ring to fuel injector. When handing new O-ring and back up ring, paying attention to the following caution items:

✷✷ WARNING

Handle O-ring with bare hands. Never wear gloves. Lubricate O-ring with new engine oil. Never clean O-ring with solvent. Check that O-ring and its mating part are free of foreign material. When installing O-ring, be careful not to scratch it with tool or fingernails. Also be careful not to twist or stretch O-ring. If O-ring was stretched while it was being attached, never insert it quickly into fuel rail. Insert new O-ring straight into fuel rail. Never de-center or twist it. Always install the backup ring in the right direction as instructed.

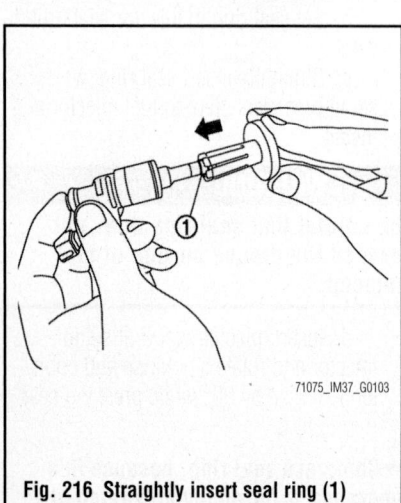

Fig. 216 Straightly insert seal ring (1)

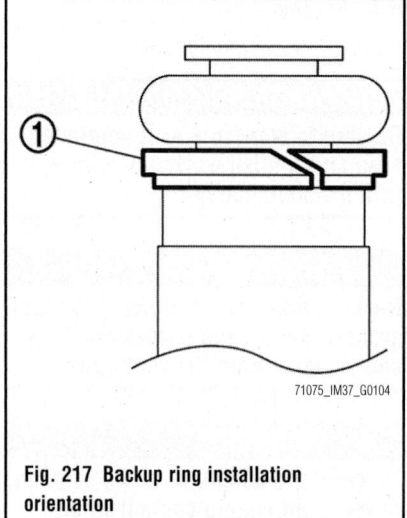

Fig. 217 Backup ring installation orientation

10. Install fuel injector (1) to fuel rail (2) as per the following:
　　a. Install fuel injector holder (5) to fuel injector.

✷✷ WARNING

Never reuse holder. Replace it with a new one. Be careful to keep fuel injector holder from interfering with O-ring. If interference occurs, replace O-ring.

　　b. Insert fuel injector into fuel rail with fuel injector holder attached. Insert it while matching it to the axial center. Insert so that protrusion (A) of fuel injector is aligned to cutout (B).

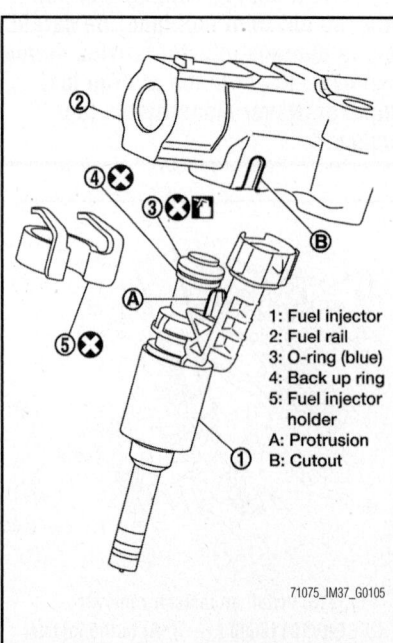

1: Fuel injector
2: Fuel rail
3: O-ring (blue)
4: Back up ring
5: Fuel injector holder
A: Protrusion
B: Cutout

Fig. 218 Installing the fuel injector to the fuel rail

　　c. Check that installation is complete by checking that fuel injector does not rotate or come off. Check that protrusions of fuel injectors and fuel rail are aligned with cutouts of clips after installation.

11. Insert the insulator into mounting hole of fuel injector of cylinder head.

12. Install in the reverse order of removal.

FUEL TANK

DRAINING

See Figure 219.

1. Check fuel level on fuel gauge. If fuel gauge indicates more than the level as shown (full or almost full), drain fuel from fuel tank until fuel gauge indicates level as shown or below.

➡**Because fuel will be spilled when removing main and sub fuel level sensor units for the top of the fuel is above the main and sub fuel level sensor units installation surface. As a guide, fuel level becomes the position as shown or below when approximately 5¼ US gal. (20 L) of fuel are drained from fuel tank.**

2. In a case that fuel pump does not operate, perform the following procedure.
　　a. Insert hose of less than 0.98 inches (22 mm) in diameter into fuel filler tube through fuel filler opening to draw fuel from fuel filler tube.
　　b. Disconnect fuel filler hose from fuel filler tube.
　　c. Insert fuel tube into fuel tank through fuel filler hose to draw fuel from fuel tank.

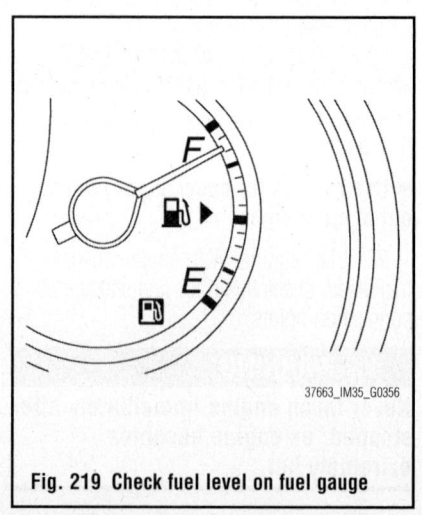

Fig. 219 Check fuel level on fuel gauge

REMOVAL & INSTALLATION

See Figures 220 through 223.

1. Release the fuel pressure from the fuel lines.

2. Open the fuel filler lid.

3. Open filler cap and release the pressure inside fuel tank.

4. Remove rear seat cushion.

5. Peel off floor carpet, then remove inspection hole cover units by turning clips clockwise by 90 degrees.

6. Disconnect harness connector and fuel feed tube.

7. Disconnect quick connector as follows:

a. Hold the sides of connector, push in tabs and pull out fuel feed tube.

b. If quick connector sticks to tube of main fuel level sensor unit, push and pull quick connector several times until they start to move. Then disconnect them by pulling.

- Quick connector can be disconnected when the tabs are completely depressed. Never twist it more than necessary.
- Never use any tools to disconnected quick connector.
- Keep resin tube away from heat. Be especially careful when welding near the resin tube.
- Prevent acid liquid such as battery electrolyte, etc. from getting on resin tube.
- Never bend or twist resin tube during installation and disconnection.
- Never remove the remaining retainer on hard tube (or the equivalent) except when resin tube or retainer is replaced.
- When resin tube or hard tube (or the equivalent) is replaced, also replace retainer with new one.
- To keep the connecting portion clean and to avoid damage and foreign materials, cover them completely with plastic bags or something similar.

8. Remove exhaust front tube, center muffler, and main muffler.

9. Remove propeller shaft.

10. Remove parking rear brake cables.

11. Remove rear suspension assembly.

➡**For this service, drive shaft, final drive, and rear suspension member are required not to be separate one another during removal.**

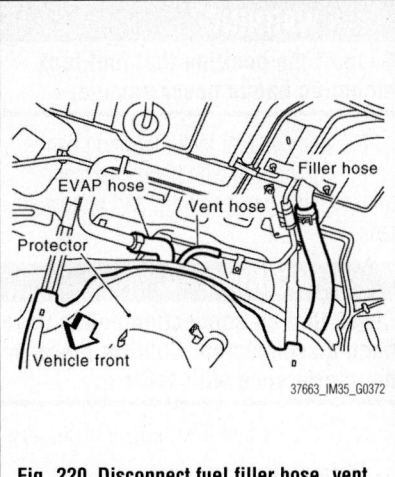

Fig. 220 Disconnect fuel filler hose, vent hose, and EVAP hoses at fuel tank side

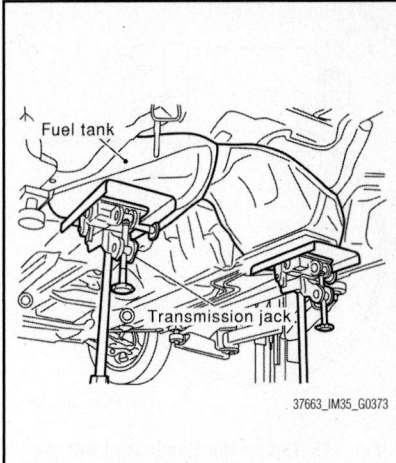

Fig. 221 Support the lower part of fuel tank with transmission jack

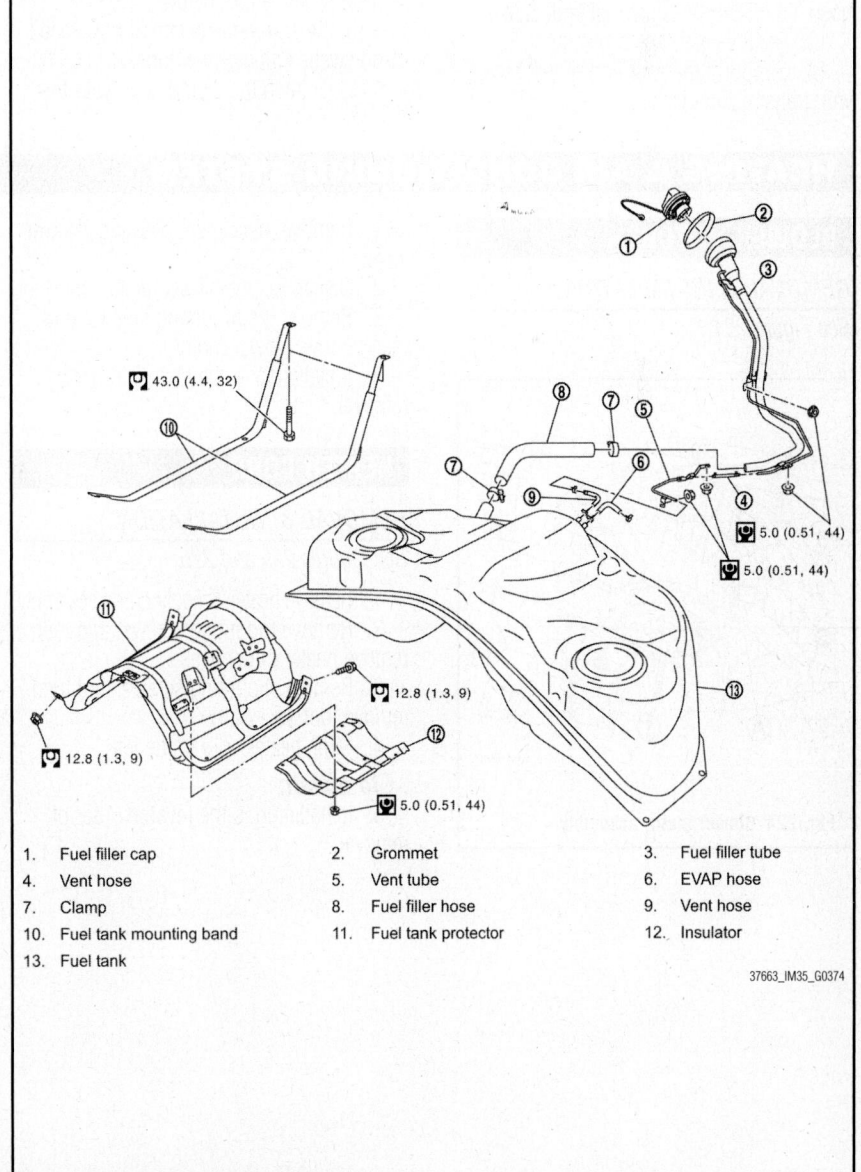

1. Fuel filler cap	2. Grommet	3. Fuel filler tube
4. Vent hose	5. Vent tube	6. EVAP hose
7. Clamp	8. Fuel filler hose	9. Vent hose
10. Fuel tank mounting band	11. Fuel tank protector	12. Insulator
13. Fuel tank		

Fig. 222 Exploded view of fuel tank assembly

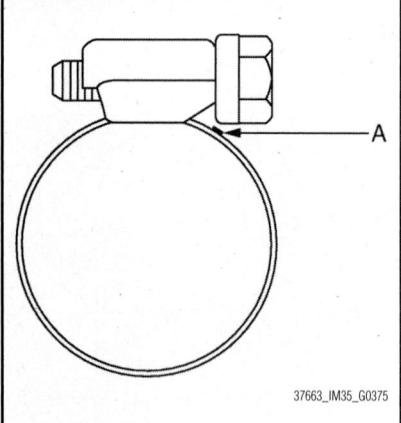

Fig. 223 Tighten the clamp band with the top mark (A) until the mark is on the bolt head flange

12. Disconnect fuel filler hose, vent hose, and EVAP hoses at fuel tank side.

13. Remove fuel tank protector.

14. Support the lower part of fuel tank with transmission jack.

✸✸ WARNING
Support the position that fuel tank mounting bands never engage.

15. Remove fuel tank mounting bands.

16. Supporting with hands, descend transmission jack carefully, and remove fuel tank.

✸✸ WARNING
Check that all connection points have been disconnected. Confirm there is no interference with vehicle.

17. Remove fuel filler tube, if necessary.

To install:

18. Note the following, and install in the reverse order of removal.

   a. Surely clamp fuel hoses and insert hose to the length below.

   b. Be sure hose clamp is not placed on swelled area of fuel tube.

   c. Tighten the clamp band with the

top mark until the mark is on the bolt head flange.

19. Turn ignition switch "ON" (with engine stopped), and check connections for leakage by applying fuel pressure to fuel piping.

20. Start engine and rev it up and check there are no fuel leakage at the fuel system tube and hose connections.

21. After removing/installing rear suspension assembly, check to adjust wheel alignment and then, adjust neutral position of steering angle sensor.

IDLE SPEED

ADJUSTMENT

Idle speed is controlled by the ECM. No adjustment is necessary or possible.

THROTTLE BODY

REMOVAL & INSTALLATION

Refer to Intake Manifold.

HEATING & AIR CONDITIONING SYSTEM

BLOWER MOTOR

REMOVAL & INSTALLATION
See Figure 224.

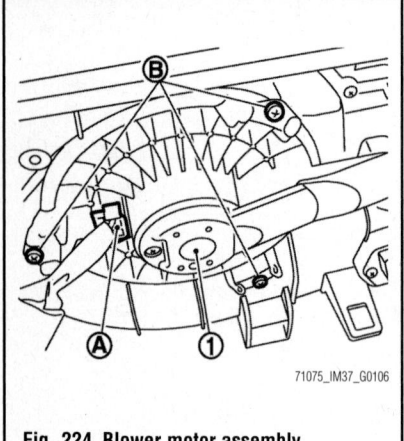

Fig. 224 Blower motor assembly

1. Remove instrument passenger lower cover.

2. Disconnect the blower motor connector.

3. Remove the mounting screws, and then remove blower motor.

4. Installation is the reverse order of removal.

HEATER CORE

REMOVAL & INSTALLATION
See Figures 225 and 226.

1. Remove heater & cooling unit assembly.

2. Remove mounting screws, and then remove heater pipe cover.

3. Remove mounting screws, and then remove foot duct (left).

4. Slide heater core to the left.

To install:

5. Installation is the reverse order of removal.

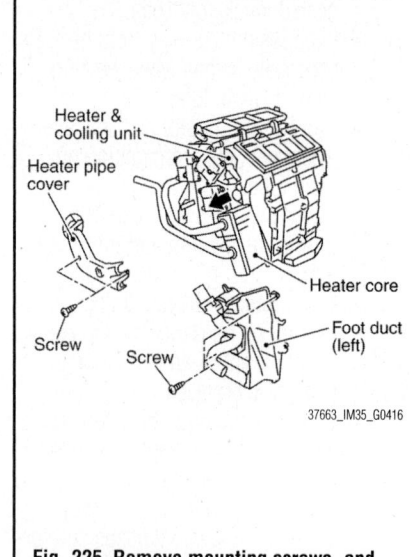

Fig. 225 Remove mounting screws, and then remove foot duct (left)

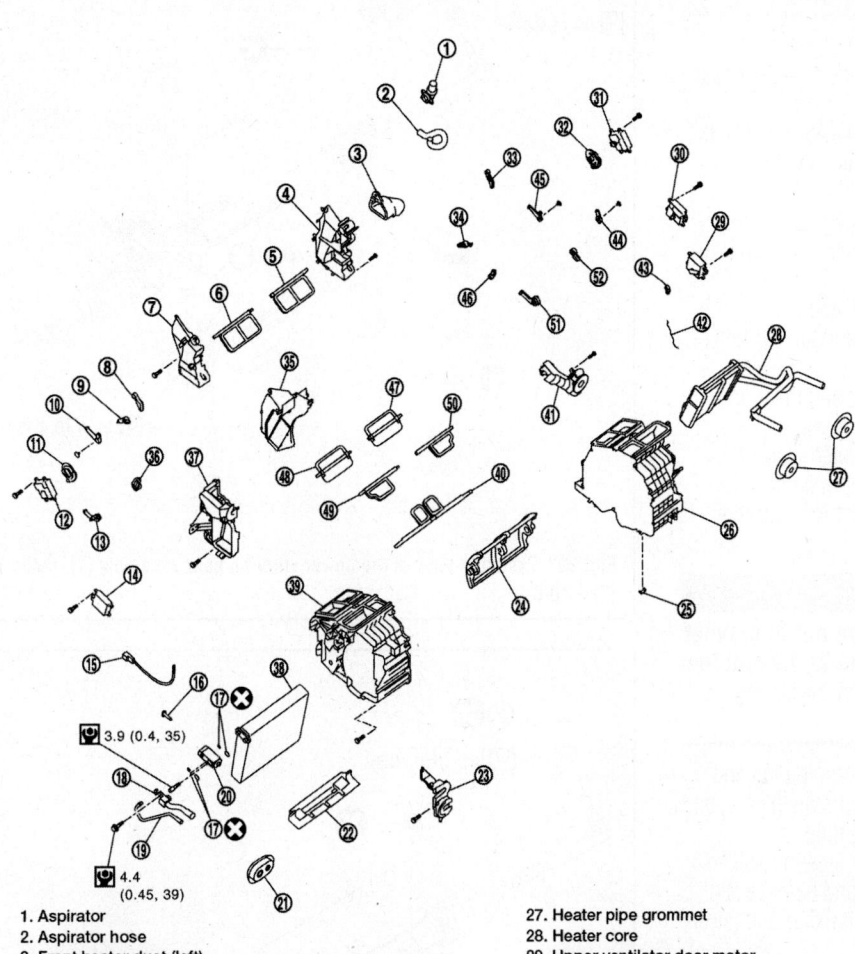

3.9 (0.4, 35)

4.4 (0.45, 39)

1. Aspirator
2. Aspirator hose
3. Front heater duct (left)
4. Foot duct (left)
5. Ventilator door (left)
6. Ventilator door (right)
7. Foot duct (right)
8. Main link sub (right)
9. Ventilator door lever (right)
10. Ventilator door link (right)
11. Main link (right) 12. Mode door motor (passenger side)
13. Max. cool door link (right)
14. Air mix door motor (passenger side)
15. Intake sensor
16. Intake sensor bracket
17. O-ring
18. Low-pressure pipe 1
19. High-pressure pipe 2
20. Expansion valve
21. Cooler pipe grommet
22. Insulator
23. Evaporator cover adapter
24. Air mix door (Slide door)
25. Clip
26. Heater & cooling unit case (left)

27. Heater pipe grommet
28. Heater core
29. Upper ventilator door motor
30. Air mix door motor (driver side)
31. Mode door motor (driver side)
32. Main link (left)
33. Main link sub (left)
34. Ventilator door lever (left)
35. Center case
36. Max. cool door lever (right)
37. Evaporator cover
38. Evaporator
39. Heater & cooling unit case (right)
40. Upper ventilator door
41. Heater pipe cover
42. Upper ventilator door rod
43. Upper ventilator door lever
44. Defroster door link
45. Ventilator door link (left)
46. Max. cool door lever (left)
47. Max. cool door (left)
48. Max. cool door (right)
49. Defroster door (right)
50. Defroster door (left)
51. Max. cool door link (left)
52. Defroster door lever

37663_IM35_G0408

Fig. 226 Exploded view of heater and cooling unit assembly

STEERING

POWER STEERING GEAR

REMOVAL & INSTALLATION

See Figures 227 and 228.

1. Set the vehicle to the straight-ahead position.
2. For vehicles with 4WAS, perform the adjustment before removal, as outlined in this section.
3. Remove tires from vehicle.
4. Remove the suspension member stay.
5. Remove cotter pin, and then loosen
6. Remove lower side mounting bolt of lower joint.
7. Remove cotter pin, and then loosen the nut.
8. Remove steering outer socket from steering knuckle so as not to damage ball joint boot using a ball joint remover (commercial service tool).

✳✳ WARNING

Temporarily tighten the nut to prevent damage to threads and to prevent the ball joint remover from suddenly coming off.

9. Remove high pressure piping and low pressure piping of hydraulic piping, and then drain power steering fluid.
10. Remove the power steering solenoid valve harness connector and harness clip.
11. Remove lower joint fixing bolt (steering gear side).
12. Separate the lower joint from the steering gear assembly by sliding the side shaft.

✳✳ WARNING

Place a matching mark on both lower joint and steering gear assembly before removing lower joint. When removing lower joint, never insert a tool, such as a screwdriver, into the yoke groove to pull out the lower joint. In case of the violation of the above, replace lower joint with a new one. Spiral cable may be cut if steering wheel turns while separating lower joint and steering gear assembly. Be sure to secure steering wheel using string to avoid turning.

13. Remove rack stay.
14. Remove the steering gear assembly mounting bolts, and nuts.
15. Remove the steering gear assembly.

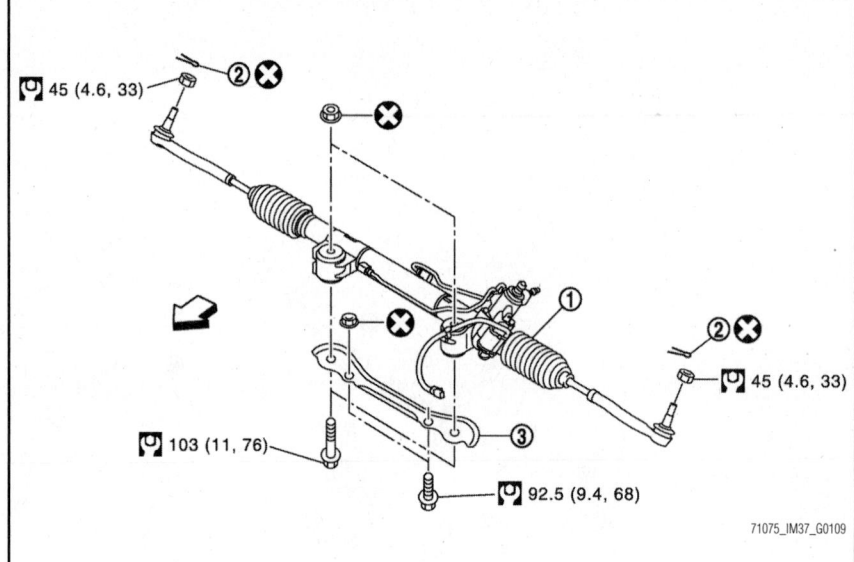

Fig. 227 Exploded view of the power steering gear assembly (1), cotter pin (2) and rack stay (3)—2WD

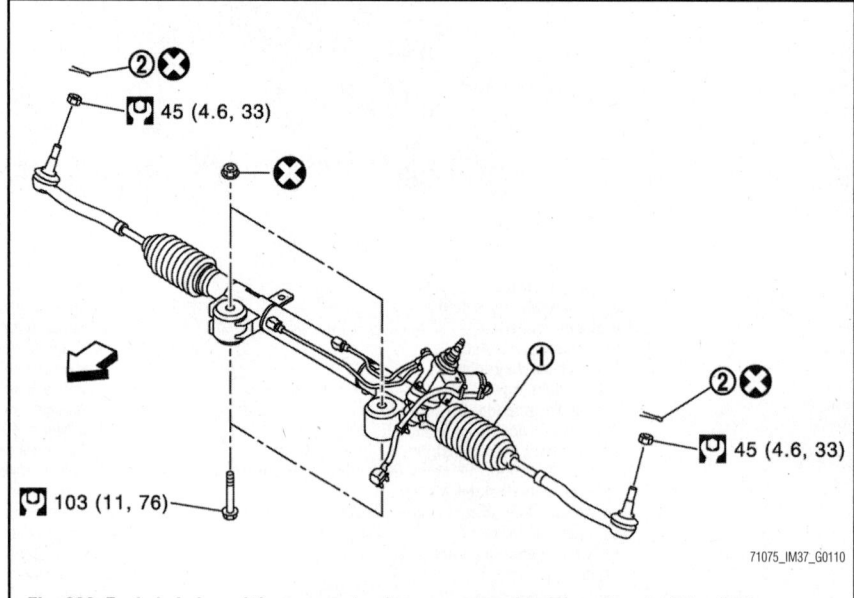

Fig. 228 Exploded view of the power steering gear assembly (1), cotter pin (2)—4WD

To install:

16. Installation is the reverse order of removal.
17. Tighten the outer mounting bolts then the inner when installing the steering gear assembly.

✳✳ WARNING

Never reuse the steering gear assembly mounting nut.

18. When installing the suction hoses:

a. Never apply fluid to the hose and tube.
b. Insert hose securely until it contacts spool of tube.
c. Leave the clearance when installing the clamp.

19. When installing lower joint to steering gear assembly, follow the procedure listed below. Set rack of steering gear in the neutral position.

➡**To get the neutral position of rack, turn gear-sub assembly and measure** the distance of inner socket, and then measure the intermediate position of the distance.

20. Align the rear cover cap projection with the marking position of gear housing assembly.

21. Install slit part of lower joint aligning with the projection of rear cover cap. Make sure that the slit part of lower joint is aligned with both the projection of rear cover cap and the marking position of gear housing assembly.

22. After installation, bleed air from the steering hydraulic system.

23. Perform final tightening of nuts and bolts on each part under unladen conditions with tires on level ground when removing steering gear assembly. Check wheel alignment.

24. Adjust the neutral position of steering angle sensor after checking wheel alignment.

25. Make sure that steering wheel operates smoothly by turning several times from full left stop to full right stop.

4WAS ADJUSTMENT

1. WAS Front Actuator Adjustment

➡**A CONSULT diagnostic tool is required for this procedure.**

1. Turn the ignition switch ON.

> ※※ **WARNING**
>
> **Never start the engine.**

2. Steer 30° leftward slowly. Steer 30° rightward and return the steering wheel to the straight-ahead position.

3. Perform the steering angle sensor neutral position adjustment.

4. Turn the ignition switch OFF.

5. Perform Active Test (Slow Mode).

2. Perform Active Test (Slow Mode)

➡**A CONSULT diagnostic tool is required for this procedure.**

1. Start the engine.

> ※※ **WARNING**
>
> **Never drive the vehicle.**

2. Select "SLOW MODE" item on "ACTIVE TEST" for "4WAS(FRONT)".

3. Perform "MODE START" of "ACTIVE TEST".

4. Steer the steering wheel leftward slowly until the turning stops.

5. Steer the steering wheel rightward slowly until the turning stops.

6. Is "OK" indicated on both right and left on "SLOW MODE"?

   a. If YES>> GO TO 3. Perform Self-Diagnosis.

3. Perform Self-Diagnosis (4WAS Front Control Unit)

➡**A CONSULT diagnostic tool is required for this procedure.**

➡**Detect DTC "C1671" when replacing 4WAS front control unit or performing 4WAS front actuator adjustment. DTC "C1671" becomes past record if 4WAS front actuator adjustment is completed normally.**

1. Is any error system detected?

   a. If YES>> Check the error system.

   b. If NO>> GO TO 4.

4. Erase Error History

➡**A CONSULT diagnostic tool is required for this procedure.**

1. Erase the memory of self-diagnostic result for "4WAS (FRONT)" and "4WAS(MAIN)/RAS/HICAS".>> END

POWER STEERING PUMP

REMOVAL & INSTALLATION

See Figures 229 through 232.

1. Drain power steering fluid from reservoir tank.

2. Remove the undercover from vehicle.

3. Loosen drive belt.

4. Remove copper washers and eye bolt (drain fluid from their pipings).

5. Remove suction hose (drain fluid from their pipings).

6. Remove oil pump mounting bolts, and then remove oil pump.

To install:

7. Note the following, and install in the reverse order of removal.

8. When installing suction hoses :

   a. Never apply fluid to the hose and tube.

   b. Insert hose securely until it contacts spool of tube.

   c. Leave clearance when installing the clamp.

9. When installing eye bolt and copper washer to oil pump:

   a. Never reuse copper washer.

   b. Apply power steering fluid to around copper washers, then install eye bolt.

   c. Install eye bolt with eye joint (assembled to high pressure hose) (B)

protrusion (A) facing with pump side cutout, and then tighten it to the specified torque after tightening by hand.

   d. Securely insert harness connector to pressure sensor.

10. Check the fluid level, fluid leakage and air bleeding hydraulic system after the installation.

BLEEDING

If air bleeding is not complete, the following symptoms can be observed.

- Bubbles are created in reservoir tank.

- Clicking noise can be heard from oil pump.

- Excessive buzzing in the oil pump.

➡**Fluid noise may occur in the steering gear or oil pump. This does not affect performance or durability of the system.**

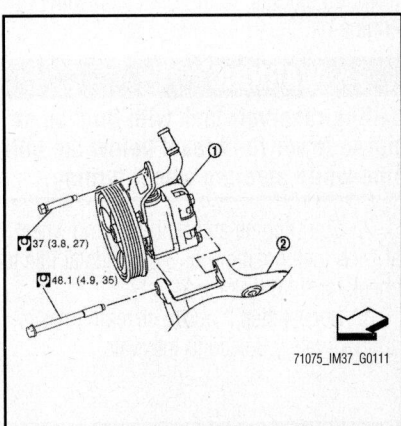

71075_IM37_G0111

Fig. 229 Removing and installing the power steering pump (1) and bracket (2)— 3.7L without 4WAS

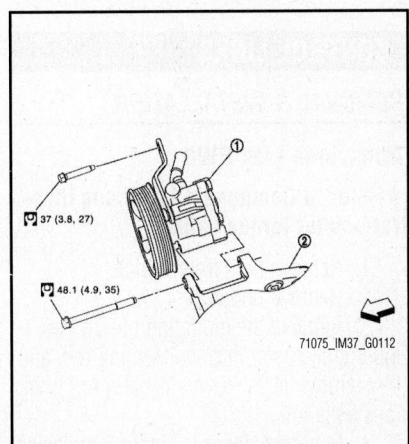

71075_IM37_G0112

Fig. 230 Removing and installing the power steering pump (1) and bracket (2)— 3.7L with 4WAS

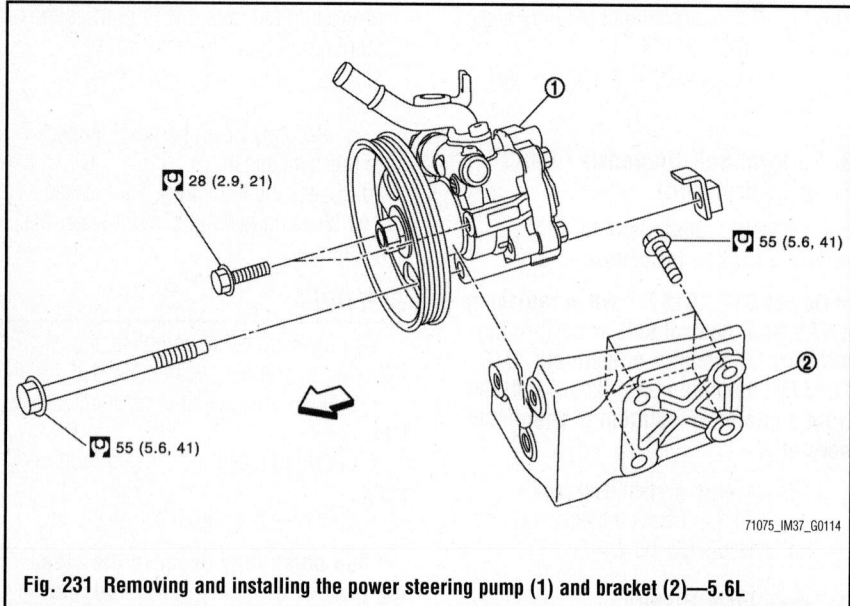

Fig. 231 Removing and installing the power steering pump (1) and bracket (2)—5.6L

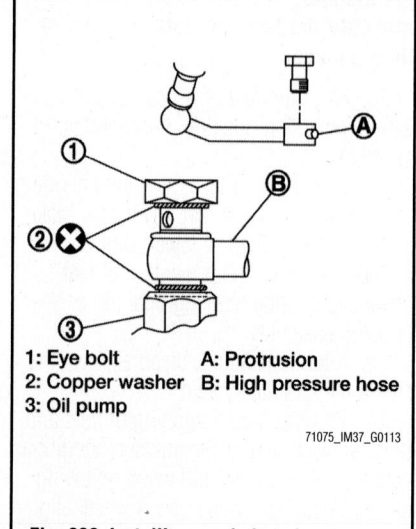

1: Eye bolt   A: Protrusion
2: Copper washer   B: High pressure hose
3: Oil pump

Fig. 232 Installing eye bolt and copper washer to oil pump

1. Turn steering wheel several times from full left stop to full right stop with engine off.

✳✳ WARNING

Filling reservoir tank with fluid so as not to lower fluid level below the MIN line while steering wheel turning.

2. Start engine and hold steering wheel at each lock position for 3 seconds at idle to check for fluid leakage.

3. Repeat step 2 above several times at approximately 3 second intervals.

✳✳ WARNING

Do not hold the steering wheel in a locked position for more than 10 seconds. (There is the possibility that oil pump may be damaged.)

4. Check fluid for bubbles and while contamination.

5. Stop engine if bubbles and white contamination do not drain out. Perform step 2 and 3 above after waiting until bubbles and white contamination drain out.

6. Stop the engine, and then check fluid level.

FLUID FILL PROCEDURE

1. Check fluid level with engine stopped.

2. Ensure that fluid level is between MIN and MAX.

3. Fluid levels at HOT and COLD are different. Do not confuse them.

✳✳ WARNING

The fluid level should not exceed the MAX line. Excessive fluid causes fluid leakage from the cap. Never reuse drained power steering fluid. Always use the specified fluid.

SUSPENSION

COMPONENT LOCATIONS

See Figures 233 and 234.

CONTROL LINKS

REMOVAL & INSTALLATION

Transverse Link RWD

➡**Refer to Component Locations illustrations for torque values.**

1. Remove tires from vehicle.
2. Remove undercover.
3. Remove the mounting nut on the upper side of stabilizer connecting rod, and then remove stabilizer connecting rod from transverse link.
4. Separate steering gear assembly and lower joint.
5. Remove rack stay.
6. Remove the mounting nut and bolt

on the lower side of shock absorber, and then remove shock absorber from transverse link.

7. Remove transverse link from steering knuckle.

8. Set jack under front suspension member.

9. Remove the mounting bolts of member bracket, and then remove member bracket from front suspension member.

10. Remove the mounting nut and bolts of member stay, and then remove member stay from front suspension member and vehicle.

11. Remove the mounting nut of front suspension member.

12. Gradually lower the suspension member to the position where transverse link mounting bolts is remove.

✳✳ WARNING

Be careful not to lower it too far. (Do not overload the links)

FRONT SUSPENSION

13. Remove mounting nut and bolts and stopper-arm bush, and then remove transverse link from vehicle.

To install:

14. Installation is the reverse order of removal.

15. Perform final tightening of bolts and nuts at the front suspension member installation position and the shock absorber lower side (rubber bushing) under unladen conditions with tires on level ground. Check wheel alignment.

16. Adjust neutral position of steering angle sensor after checking wheel alignment.

Transverse Link AWD

➡**Refer to Component Locations illustrations for torque values.**

1. Remove tires from vehicle.
2. Remove undercover.

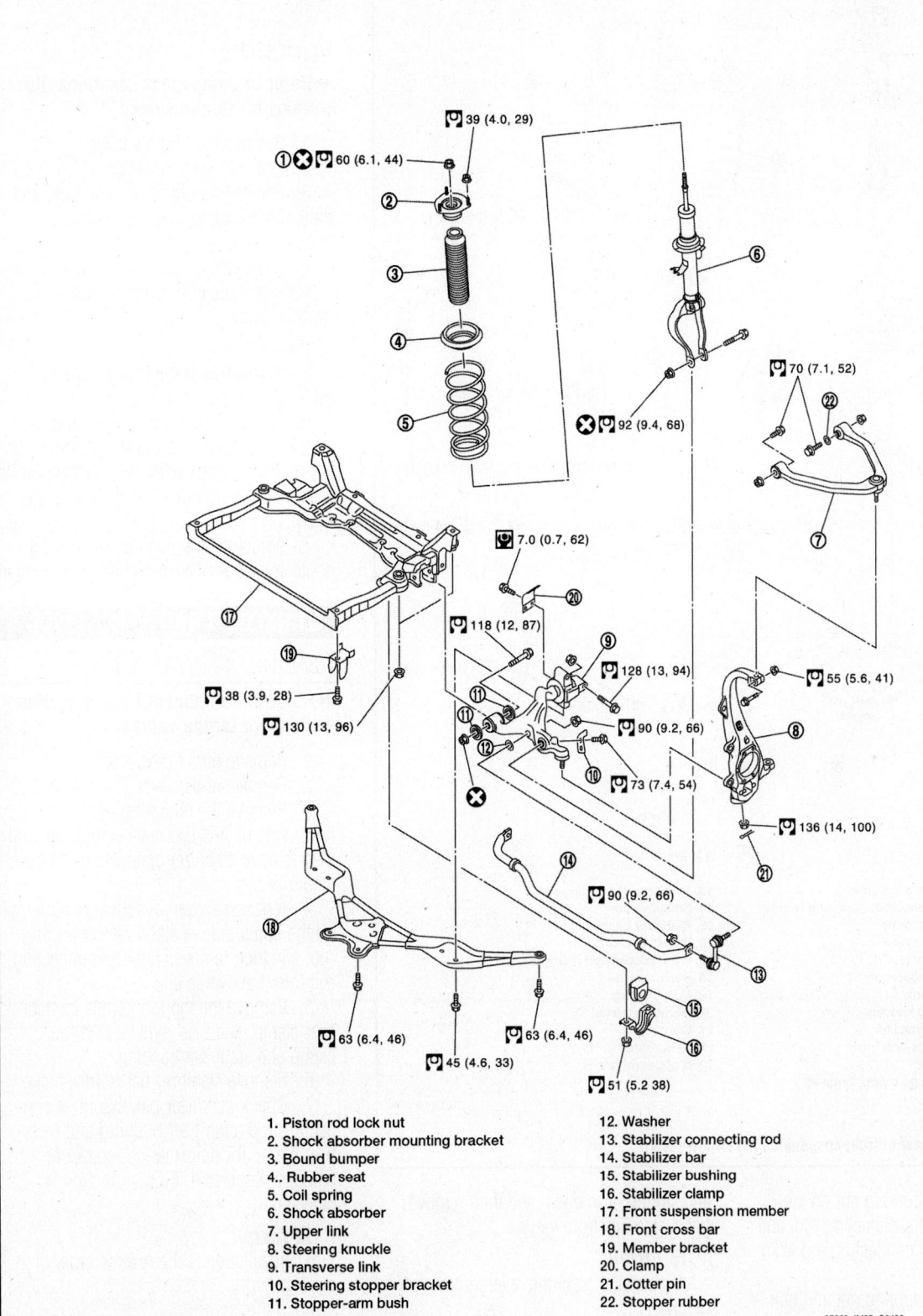

39 (4.0, 29)

60 (6.1, 44)

70 (7.1, 52)

92 (9.4, 68)

7.0 (0.7, 62)

118 (12, 87)

128 (13, 94)

55 (5.6, 41)

38 (3.9, 28)

90 (9.2, 66)

130 (13, 96)

73 (7.4, 54)

136 (14, 100)

90 (9.2, 66)

63 (6.4, 46)

63 (6.4, 46)

45 (4.6, 33)

51 (5.2 38)

1. Piston rod lock nut
2. Shock absorber mounting bracket
3. Bound bumper
4.. Rubber seat
5. Coil spring
6. Shock absorber
7. Upper link
8. Steering knuckle
9. Transverse link
10. Steering stopper bracket
11. Stopper-arm bush

12. Washer
13. Stabilizer connecting rod
14. Stabilizer bar
15. Stabilizer bushing
16. Stabilizer clamp
17. Front suspension member
18. Front cross bar
19. Member bracket
20. Clamp
21. Cotter pin
22. Stopper rubber

37663_IM35_G0436

Fig. 233 Exploded view of front suspension assembly—AWD

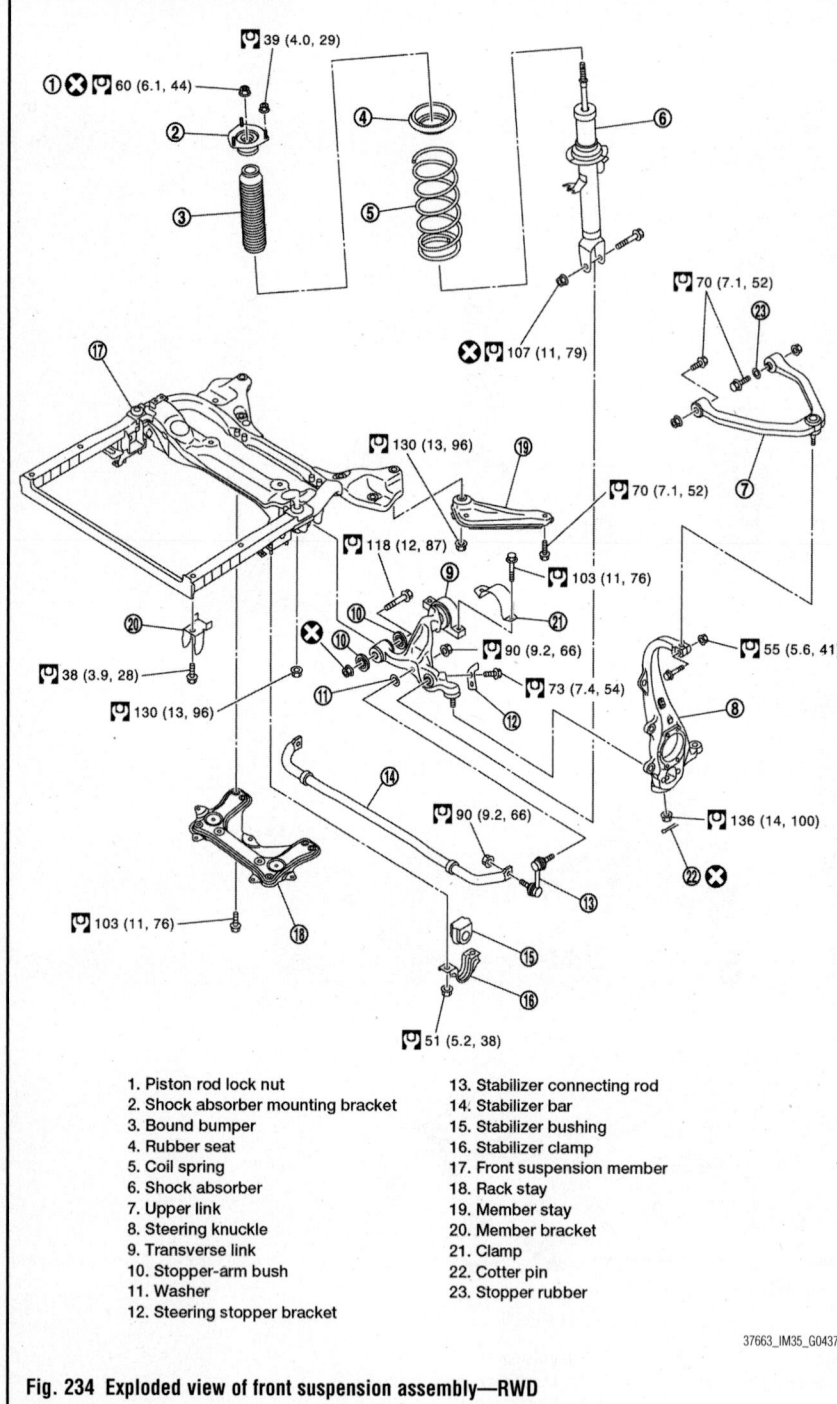

1. Piston rod lock nut
2. Shock absorber mounting bracket
3. Bound bumper
4. Rubber seat
5. Coil spring
6. Shock absorber
7. Upper link
8. Steering knuckle
9. Transverse link
10. Stopper-arm bush
11. Washer
12. Steering stopper bracket
13. Stabilizer connecting rod
14. Stabilizer bar
15. Stabilizer bushing
16. Stabilizer clamp
17. Front suspension member
18. Rack stay
19. Member stay
20. Member bracket
21. Clamp
22. Cotter pin
23. Stopper rubber

37663_IM35_G0437

Fig. 234 Exploded view of front suspension assembly—RWD

3. Remove the mounting nut on the upper side of stabilizer connecting rod, and then remove stabilizer connecting rod from transverse link.

4. Remove the mounting nut and bolt on the lower side of shock absorber arm, and then remove shock absorber arm from transverse link.

5. Remove front cross bar.

6. Remove transverse link from steering knuckle.

7. Remove mounting nuts and bolts

and stopper-arm bush, and then remove transverse link from vehicle.

To install:

8. Installation is the reverse order of removal.

9. Perform final tightening of bolts and nuts at the front suspension member installation position and the shock absorber lower side (rubber bushing) under unladen conditions with tires on level ground. Check wheel alignment.

10. Adjust neutral position of steering

angle sensor after checking wheel alignment.

Upper Link

→Refer to Component Locations illustrations for torque values.

1. Remove tires from vehicle.

2. Remove shock absorber.

3. Remove mounting nut and bolt, and then remove upper link from steering knuckle.

4. Remove mounting nuts and bolts, and then remove upper link and stopper rubber from vehicle.

To install:

5. Installation is the reverse order of removal.

6. Perform final tightening of bolts and nuts at the vehicle installation position (rubber bushing) under unladen conditions with tires on level ground. Check wheel alignment.

7. Adjust neutral position of steering angle sensor after checking wheel alignment

STABILIZER BAR

REMOVAL & INSTALLATION

→Refer to Component Locations illustrations for torque values.

1. Remove tires from vehicle.

2. Remove undercover.

3. Remove the mounting nut on the lower side of stabilizer connecting rod, and then remove stabilizer connecting rod from stabilizer bar.

4. If necessary remove the mounting nut on the upper side of stabilizer connecting rod, and then remove stabilizer connecting rod from transverse link.

5. Remove the mounting nuts of stabilizer clamp, and then remove stabilizer clamp and stabilizer bushing.

6. Remove stabilizer bar from vehicle.

7. Check stabilizer bar, stabilizer connecting rod, stabilizer bushing and stabilizer clamp for deformation, cracks or damage. Replace it if a malfunction is detected.

To install:

8. Installation is the reverse order of removal.

STEERING KNUCKLE

REMOVAL & INSTALLATION

See Figures 235 and 236.

→Refer to Component Locations illustrations for torque values.

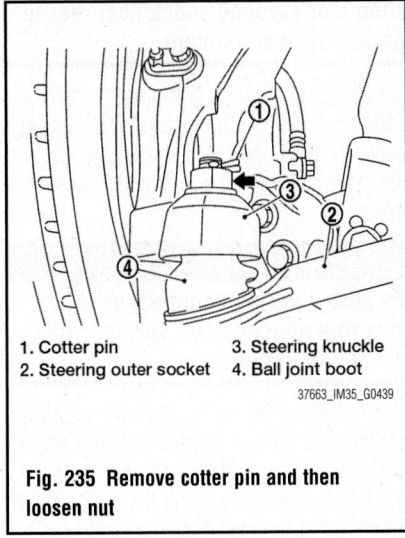

1. Cotter pin
2. Steering outer socket
3. Steering knuckle
4. Ball joint boot

37663_IM35_G0439

Fig. 235 Remove cotter pin and then loosen nut

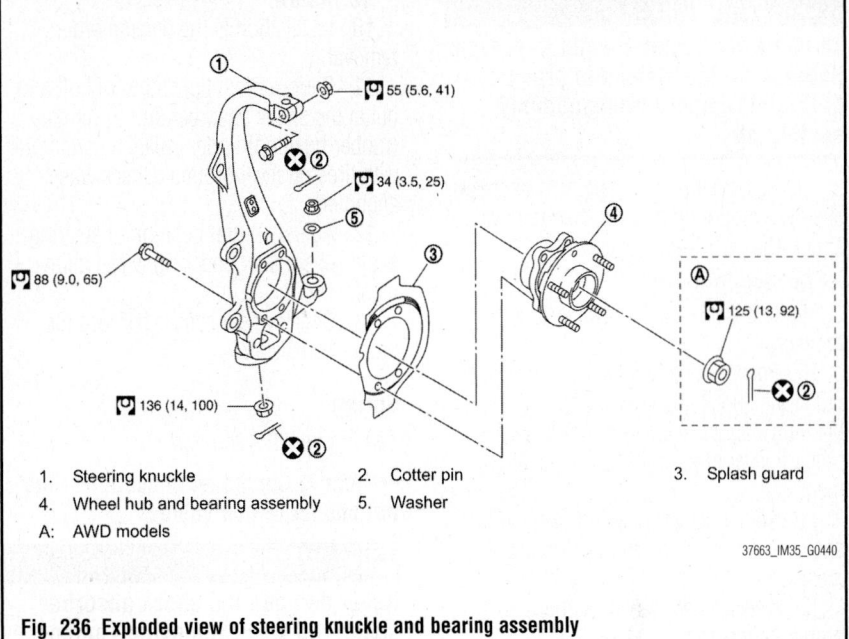

1. Steering knuckle
2. Cotter pin
3. Splash guard
4. Wheel hub and bearing assembly
5. Washer
A: AWD models

37663_IM35_G0440

Fig. 236 Exploded view of steering knuckle and bearing assembly

1. Remove tires from vehicle.
2. Remove wheel sensor from steering knuckle.

✳✳ WARNING

Do not pull on wheel sensor harness.

3. Remove brake hose bracket.
4. Remove torque member mounting bolts. Hang torque member in a place where it will not interfere with work.

✳✳ WARNING

Do not depress brake pedal while brake caliper is removed.

5. Put matching mark on disc rotor and wheel hub and bearing assembly, then remove disc rotor.
6. Remove cotter pin, then loosen hub lock nut. (AWD)
7. Separate wheel hub and bearing assembly from drive shaft by lightly tapping the end with a hammer and a wood block, and then remove hub lock nut. (AWD)

✳✳ WARNING

Do not place drive shaft joint at an extreme angle. Also be careful not to overextend slide joint. Do not allow drive shaft to hang down without support for housing (or joint sub-assembly), shaft and the other parts.

➡**Use a puller if wheel hub and bearing assembly and drive shaft cannot be separated even after performing the above procedure.**

8. Remove cotter pin and then loosen the nut.

9. Remove steering outer socket from steering knuckle so as not to damage ball joint boot using the ball joint remover.

✳✳ WARNING

Temporarily tighten the nut to prevent damage to threads and to prevent the ball joint remover from suddenly coming off.

10. Remove cotter pin of transverse link and steering knuckle, and then loosen nut.
11. Remove transverse link from steering knuckle so as not to damage ball joint boot using the ball joint remover.

✳✳ WARNING

Temporarily tighten the nut to prevent damage to threads and to prevent ball joint remover from suddenly coming off.

12. Remove mounting nut and bolt, and then remove steering knuckle from upper link.
13. Remove wheel hub and bearing assembly mounting bolts, and then remove splash guard and wheel hub and bearing assembly from steering knuckle.

To install:

14. Installation is the reverse order of the removal.
15. Perform the final tightening of each of parts under unladen conditions, which were removed when removing wheel hub and bearing assembly and steering knuckle. Check the wheel alignment.
16. Adjust neutral position of steering

angle sensor after checking the wheel alignment.
17. Check wheel sensor harness for proper connection.
18. Assemble disc rotor and wheel hub and bearing assembly by aligning each matching mark when installing disc rotor.

STRUT & SPRING ASSEMBLY

REMOVAL & INSTALLATION

Strut

RWD Models

➡**Refer to Component Locations illustrations for torque values.**

1. Remove tires from vehicle.
2. Remove harness of wheel sensor from shock absorber.

✳✳ WARNING

Do not pull on wheel sensor harness.

3. Remove brake hose bracket.
4. Remove the mounting nut on the upper side of stabilizer connecting rod, and then remove stabilizer connecting rod from transverse link.
5. Remove mounting nut and bolt on the lower side of shock absorber, and then remove shock absorber from transverse link.
6. Remove cotter pin of transverse link and steering knuckle, and then loosen nut.
7. Remove transverse link from steering knuckle so as not to damage ball joint boot using the ball joint remover.

⁂ **WARNING**

Temporarily tighten the nut to prevent damage to threads and to prevent ball joint remover from suddenly coming off.

8. Remove the mounting nuts of shock absorber mounting bracket, then remove shock absorber from vehicle.

To install:

9. Installation is the reverse order of removal.

10. Perform final tightening of bolt and nut at the shock absorber lower side (rubber bushing), under unladen conditions with tires on level ground. Check wheel alignment.

11. Adjust neutral position of steering angle sensor after checking wheel alignment.

12. Check wheel sensor harness for proper connection.

AWD Models

➡**Refer to Component Locations illustrations for torque values.**

1. Remove tires from vehicle.
2. Remove harness of wheel sensor from shock absorber.

⁂ **WARNING**

Do not pull on wheel sensor harness.

3. Remove brake hose bracket.
4. Remove the mounting nut on the upper side of stabilizer connecting rod, and then remove stabilizer connecting rod from transverse link.
5. Remove mounting nut and bolt on the lower side of shock absorber arm, and then remove shock absorber arm from transverse link.
6. Remove cotter pin of transverse link and steering knuckle, and then loosen nut.
7. Remove transverse link from steering knuckle so as not to damage ball joint boot using the ball joint remover.

⁂ **WARNING**

Temporarily tighten the nut to prevent damage to threads and to prevent ball joint remover from suddenly coming off.

8. Remove the mounting bolt on the upper side of shock absorber arm, and then remove shock absorber arm from shock absorber.
9. Remove the mounting nuts of shock absorber mounting bracket, then remove shock absorber from vehicle.

To install:

10. Installation is the reverse order of removal.

11. Perform final tightening of bolt and nut at the shock absorber arm lower side (rubber bushing) under unladen conditions with tires on level ground. Check wheel alignment.

12. Adjust neutral position of steering angle sensor after checking wheel alignment.

13. Check wheel sensor harness for proper connection.

Spring

See Figures 237 and 238.

➡**Refer to Component Locations illustrations for torque values.**

⁂ **WARNING**

Never damage the shock absorber piston rod when removing components from shock absorber.

1. Install shock absorber attachment to shock absorber and secure it in a vise.

⁂ **WARNING**

When installing the shock absorber attachment to shock absorber, wrap a shop cloth around shock absorber to protect it from damage.

2. Using a spring compressor (commercial service tool), compress coil spring between rubber seat and shock absorber until coil spring with a spring compressor is free.

⁂ **WARNING**

Be sure a spring compressor is securely attached coil spring. Compress coil spring.

3. Make sure coil spring with a spring compressor between rubber seat and shock absorber is free. And then remove piston rod lock nut while securing the piston rod tip so that piston rod does not turn.

⁂ **WARNING**

Start compressing the coil spring after checking that the spring compressor is completely attached.

4. Remove mounting seal, shock absorber mounting bracket, rubber seat, bound bumper from shock absorber.
5. After remove coil spring with a spring compressor, and then gradually release a spring compressor.

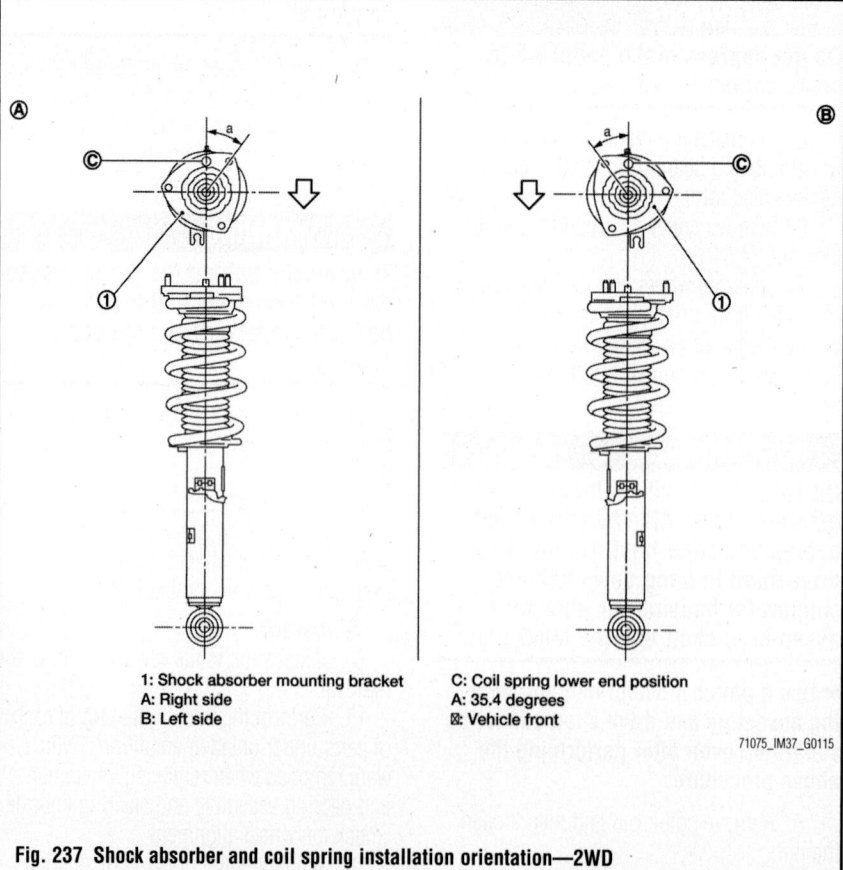

1: Shock absorber mounting bracket
A: Right side
B: Left side

C: Coil spring lower end position
A: 35.4 degrees
⬇: Vehicle front

71075_IM37_G0115

Fig. 237 Shock absorber and coil spring installation orientation—2WD

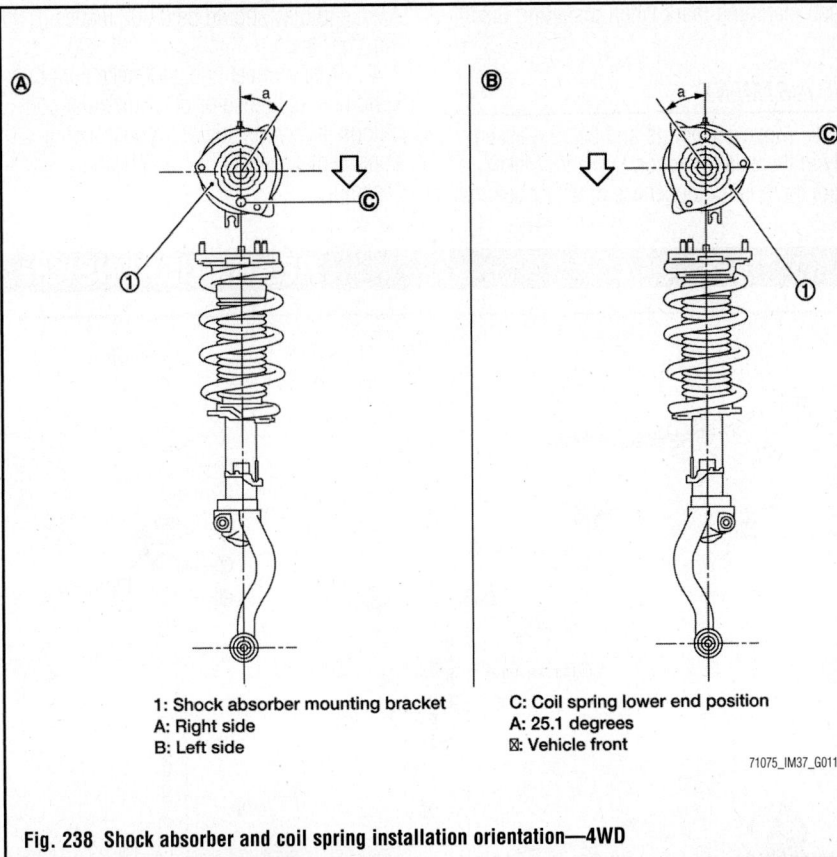

1: Shock absorber mounting bracket
A: Right side
B: Left side

C: Coil spring lower end position
A: 25.1 degrees
⬗: Vehicle front

71075_IM37_G0116

Fig. 238 Shock absorber and coil spring installation orientation—4WD

✳✳ WARNING

Loosen while making sure coil spring attachment position does not move.

6. Remove the shock absorber attachment from shock absorber.
7. Perform inspection after disassembly. Refer to Power Steering.

To install:

✳✳ WARNING

Never damage shock absorber piston rod when installing components from shock absorber.

8. Install shock absorber attachment to shock absorber and secure it in a vise.

✳✳ WARNING

When installing the shock absorber attachment to shock absorber, wrap a shop cloth around shock absorber to protect it from damage.

9. Compress coil spring using a spring compressor (commercial service tool), and install it onto shock absorber.

✳✳ WARNING

Install with the large-diameter side facing up and the small-diameter

side facing down. Be sure a spring compressor is securely attached to coil spring. Compress coil spring.

10. Install the shock absorber mounting bracket and rubber seat.
11. Apply soapy water to bound bumper.

✳✳ WARNING

Never use machine oil.

12. Insert bound bumper into shock absorber mounting bracket, and then install it to shock absorber together with rubber seat.
   a. Check that the lower end of the coil spring is positioned at the spring lower seat of the shock absorber.
13. Secure piston rod tip so that piston rod does not turn, then tighten piston rod lock nut with specified torque.

✳✳ WARNING

Never reuse piston rod lock nut.

14. Gradually release a spring compressor, and remove coil spring.

✳✳ WARNING

Loosen while making sure coil spring attachment position does not move.

15. Remove the shock absorber attachment from shock absorber.
16. Install the mounting seal to shock absorber mounting bracket.

WHEEL BEARINGS

REMOVAL & INSTALLATION

➡**Refer to Steering Knuckle for illustration.**

1. Remove tires from vehicle.
2. Remove wheel sensor from steering knuckle.

✳✳ WARNING

Do not pull on wheel sensor harness.

3. Remove brake hose bracket.
4. Remove torque member mounting bolts. Hang torque member in a place where it will not interfere with work.

✳✳ WARNING

Do not depress brake pedal while brake caliper is removed.

5. Put matching mark on disc rotor and wheel hub and bearing assembly, then remove disc rotor.
6. Remove cotter pin, then loosen hub lock nut. (AWD)
7. Separate wheel hub and bearing assembly from drive shaft by lightly tapping the end with a hammer and a wood block, and then remove hub lock nut. (AWD)

✳✳ WARNING

Do not place drive shaft joint at an extreme angle. Also be careful not to overextend slide joint. Do not allow drive shaft to hang down without support for housing (or joint sub-assembly), shaft and the other parts.

➡**Use a puller if wheel hub and bearing assembly and drive shaft cannot be separated even after performing the above procedure.**

8. Remove wheel hub and bearing assembly mounting bolts, and then remove splash guard and wheel hub and bearing assembly from steering knuckle.

To install:

9. Installation is the reverse order of the removal.
10. Perform the final tightening of each of parts under unladen conditions, which were removed when removing wheel hub and bearing assembly and steering knuckle. Check the wheel alignment.

11. Adjust neutral position of steering angle sensor after checking the wheel alignment.

12. Check wheel sensor harness for proper connection.

13. Assemble disc rotor and wheel hub and bearing assembly by aligning each matching mark when installing disc rotor.

ADJUSTMENT

1. Move wheel hub and bearing assembly in the axial direction by hand. Make sure there is no looseness of wheel bearing. Axial end play should be 0.002 inches (0.05 mm) or less.

2. Rotate wheel hub and make sure that is no unusual noise or other irregular conditions. If there is any of irregular conditions, replace wheel hub and bearing assembly

SUSPENSION

REAR SUSPENSION

COMPONENT LOCATIONS

See Figure 239.

CONTROL ARMS/LINKS

REMOVAL & INSTALLATION

Suspension Arm

See Figure 240.

➡**Refer to illustration for torque values.**

1. Remove the tires and wheels.

2. Set a jack under rear lower link to relieve the coil spring tension.

3. Remove the connecting rod mounting bracket from suspension arm.

4. Remove the mounting nuts and bolts between suspension arm and rear suspension member.

5. Remove cotter pin of suspension arm ball joint, and loosen nut.

6. Use a ball joint remover to remove suspension arm from axle. Be careful not to damage ball joint boot.

❄❄ WARNING

Temporarily tighten mounting nut to prevent damage to threads and to prevent ball joint remover from coming off.

7. Remove suspension arm and stopper rubber from vehicle.

To install:

8. Installation is the reverse order of removal.

❄❄ WARNING

Do not reuse non-reusable parts.

9. Perform the final tightening of rear suspension member installation position (rubber bushing) under unladen condition with tires on level ground.

10. Adjust the neutral position of steering angle sensor after checking the wheel alignment.

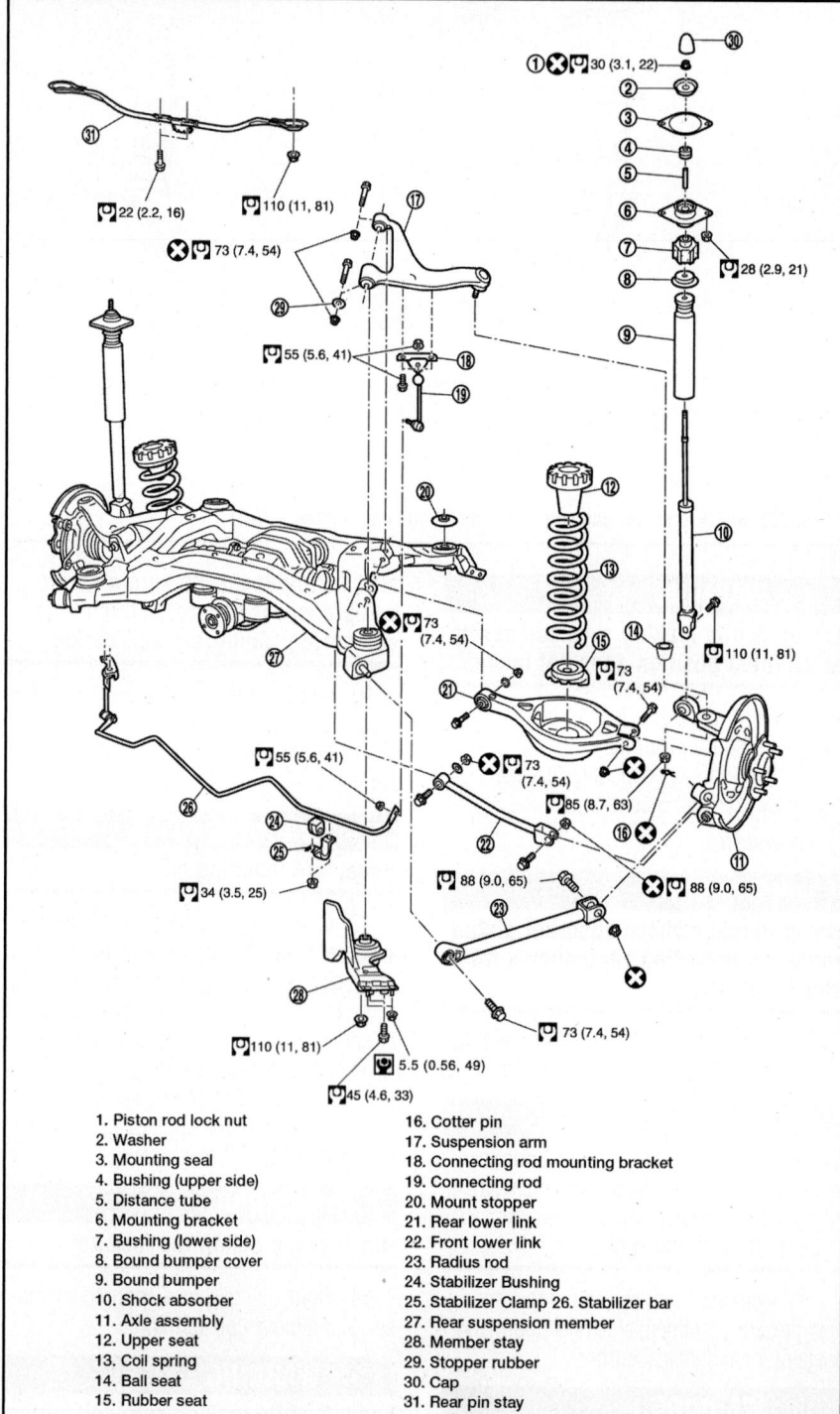

1. Piston rod lock nut
2. Washer
3. Mounting seal
4. Bushing (upper side)
5. Distance tube
6. Mounting bracket
7. Bushing (lower side)
8. Bound bumper cover
9. Bound bumper
10. Shock absorber
11. Axle assembly
12. Upper seat
13. Coil spring
14. Ball seat
15. Rubber seat
16. Cotter pin
17. Suspension arm
18. Connecting rod mounting bracket
19. Connecting rod
20. Mount stopper
21. Rear lower link
22. Front lower link
23. Radius rod
24. Stabilizer Bushing
25. Stabilizer Clamp 26. Stabilizer bar
27. Rear suspension member
28. Member stay
29. Stopper rubber
30. Cap
31. Rear pin stay

37663_IM35_G0438

Fig. 239 Exploded view of rear suspension assembly

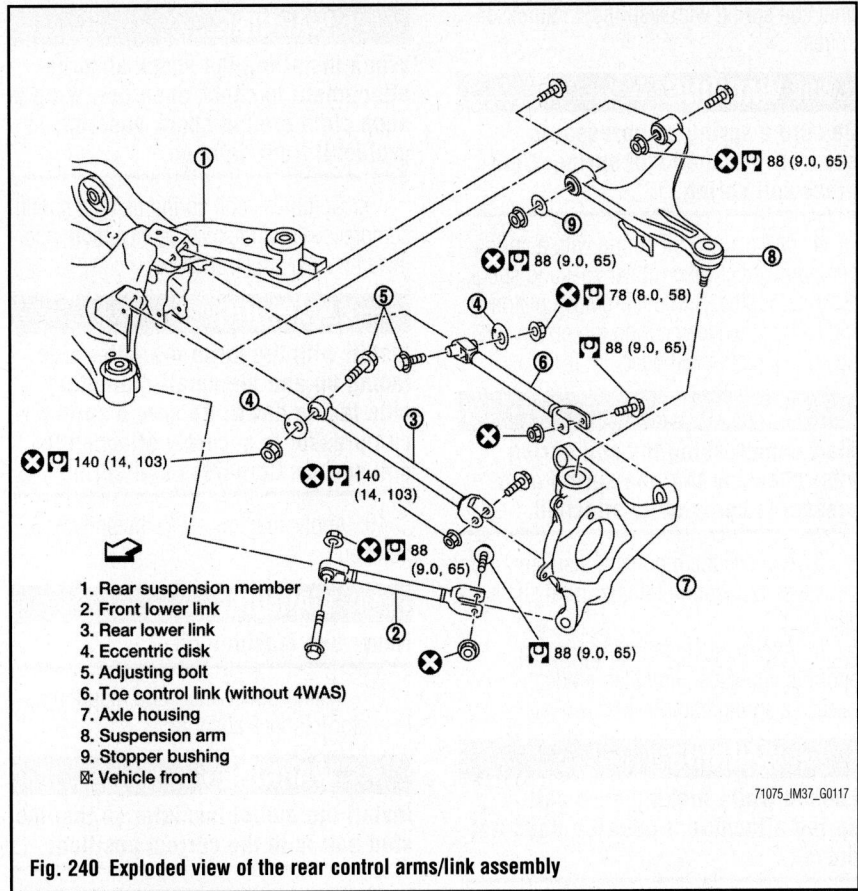

1. Rear suspension member
2. Front lower link
3. Rear lower link
4. Eccentric disk
5. Adjusting bolt
6. Toe control link (without 4WAS)
7. Axle housing
8. Suspension arm
9. Stopper bushing
⊠: Vehicle front

71075_IM37_G0117

Fig. 240 Exploded view of the rear control arms/link assembly

Front Lower Link

See Figure 240.

➡**Refer to illustration for torque values.**

1. Remove the tires and wheels.
2. Set a jack under rear lower link to relieve the coil spring tension.
3. Remove mounting nut and bolt between front lower link and rear suspension member.
4. Remove mounting nut and bolt between front lower link and axle.
5. Remove front lower link from vehicle.
6. Check front lower link and bushing for any deformation, cracks, or damage. Replace if there is any damage or excessive wear.

To install:
7. Installation is the reverse order of removal.

✳ WARNING

Do not reuse non-reusable parts.

8. Perform the final tightening of rear suspension member and axle installation position (rubber bushing) under unladen condition with tires on level ground. Check wheel alignment.

9. Adjust the neutral position of steering angle sensor after checking the wheel alignment.

Rear Lower Link

See Figure 240.

➡**Refer to illustration for torque values.**

1. Remove the tires and wheels.
2. Set a jack under rear lower link to relieve the coil spring tension.
3. Loosen mounting bolt and nut of rear lower link inside of suspension member, and then remove mounting bolt and nut inside of axle.
4. Slowly lower jack, then remove upper seat, coil spring and rubber sheet from rear lower link.
5. Remove the mounting bolt and nut inside of suspension member to remove.

To install:
6. Make sure that upper seat is attached as shown.
7. Make sure that the projecting parts on upper seat inside is securely fitted on the bracket tabs.
8. Match up rubber seat indentions and rear lower link grooves and attach.

✳ WARNING

Make sure spring is not upside down. The top and bottom are indicated by paint color.

9. Perform the final tightening of rear suspension member and axle installation position (rubber bushing) under unladen condition with tires on level ground.
10. Tighten rear suspension member to 53 ft. lbs. (72 Nm).
11. Check wheel alignment.

➡**Adjust the neutral position of steering angle sensor after checking the wheel alignment.**

Toe Control Link

See Figure 240.

➡**Refer to illustration for torque values.**

1. Remove tires with power tool.
2. Set suitable jack under axle housing.
3. Check that jack supporting status is stable.
4. Separate the shock absorber from axle housing.
5. Remove the eccentric disk, adjusting bolt, mounting bolt, and nut. Remove toe control link.
6. Perform inspection after removal.

To install:
7. Note the following, and install in the reverse order of removal.
   a. Perform final tightening of rear suspension member and axle installation position (rubber bushing), under unladen conditions with tires on level ground.
   b. Perform inspection after installation.

STABILIZER BAR

REMOVAL & INSTALLATION

➡**Refer to Component Locations illustrations for torque values.**

1. Remove the member stay.
2. Remove lower side mounting nut on stabilizer connecting rod and remove stabilizer connecting rod from stabilizer bar.
3. Remove mounting nut on stabilizer clamp and remove stabilizer from vehicle.
4. Check stabilizer bar, stabilizer bushings, stabilizer clamp, stabilizer connecting rod and stabilizer connecting rod mounting bracket for any deformation, crack or damage. Replace if there is any damage or excessive wear.

To install:
5. Installation is the reverse order of removal.

> ❋❋ **WARNING**
>
> Do not reuse non-reusable parts.

STRUT & SPRING ASSEMBLY

REMOVAL & INSTALLATION

Strut

➡ **Refer to Component Locations illustrations for torque values.**

1. Remove the tires and wheels from vehicle.
2. Set a jack under rear lower link to relieve the coil spring tension.
3. Remove shock absorber lower end bolt.
4. Remove shock absorber from axle housing.
5. Remove the rear parcel shelf finisher.
6. Remove the seat belt retractor.
7. Remove mounting insulator nuts, and then remove shock absorber assembly.
8. Check shock absorber assembly for deformation, cracks, damage, and replace if there is any damage or excessive wear.
9. Check welded and sealed areas for oil leakage, and replace if there are any defects noted.

To install:
10. Installation is the reverse order of removal.

> ❋❋ **WARNING**
>
> Do not reuse non-reusable parts.

11. Perform final tightening of shock absorber assembly lower side (rubber bushing) under unladen condition with tires on level ground. Check wheel alignment.
12. Adjust the neutral position of steering angle sensor after checking the wheel alignment.

Spring

See Figure 241.

1. Remove the gasket and cap from mounting insulator.
2. Install shock absorber attachment to shock absorber and secure it in a vise.

> ❋❋ **WARNING**
>
> When installing the shock absorber attachment to shock absorber, wrap a shop cloth around shock absorber to protect it from damage.

3. Using a spring compressor (commercial service tool), compress coil spring between rubber seat and shock absorber until coil spring with a spring compressor is free.

> ❋❋ **WARNING**
>
> Be sure a spring compressor is securely attached coil spring. Compress coil spring.

4. Make sure coil spring with a spring compressor between rubber seat and shock absorber is free. And then remove piston rod lock nut while securing the piston rod tip so that piston rod does not turn.

> ❋❋ **WARNING**
>
> Start compressing the coil spring after checking that the spring compressor is completely attached.

5. Remove the mounting insulator, rubber sheet, and bound bumper from shock absorber.
6. After remove coil spring with a spring compressor, and then gradually release a spring compressor.

> ❋❋ **WARNING**
>
> Loosen while making sure coil spring attachment position does not move.

7. Remove the shock absorber attachment from shock absorber.

To install:

> ❋❋ **WARNING**
>
> Never damage shock absorber piston rod when installing components from shock absorber.

8. Install shock absorber attachment to shock absorber and secure it in a vise.

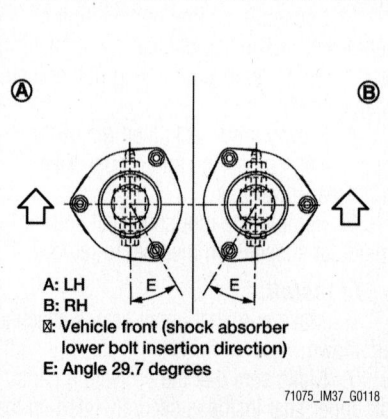

Ⓐ   Ⓑ

A: LH
B: RH
🡱: Vehicle front (shock absorber lower bolt insertion direction)
E: Angle 29.7 degrees

71075_IM37_G0118

Fig. 241 Mount insulator installation orientation

> ❋❋ **WARNING**
>
> When installing the shock absorber attachment to shock absorber, wrap a shop cloth around shock absorber to protect it from damage.

9. Compress coil spring using a spring compressor (commercial service tool), and install it onto shock absorber.

> ❋❋ **WARNING**
>
> Install with the large-diameter side facing up and the small-diameter side facing down. Be sure a spring compressor is securely attached to coil spring. Compress coil spring.

10. Apply soapy water to the bound bumper.

> ❋❋ **WARNING**
>
> Never use machine oil.

11. Install rubber sheet and mounting insulator to shock absorber.

> ❋❋ **WARNING**
>
> Install the mount insulator so that the stud bolt is in the correct position.

12. Secure piston rod tip so that piston rod does not turn, then tighten piston rod lock nut with specified torque.
13. Gradually release a spring compressor, and remove coil spring.

> ❋❋ **WARNING**
>
> Loosen while making sure coil spring attachment position does not move.

14. Remove the shock absorber attachment from shock absorber.
15. Install the gasket and cap to the mounting insulator.

WHEEL BEARINGS

REMOVAL & INSTALLATION

See Figures 242 through 244.

1. Remove tires and wheels from vehicle.
2. Remove rear brake caliper. Hang it in a place where it will not interfere with work.

> ❋❋ **WARNING**
>
> Do not depress brake pedal while brake caliper is removed.

3. Put matching mark on disc rotor and the wheel hub and bearing assembly then removing disc rotor.

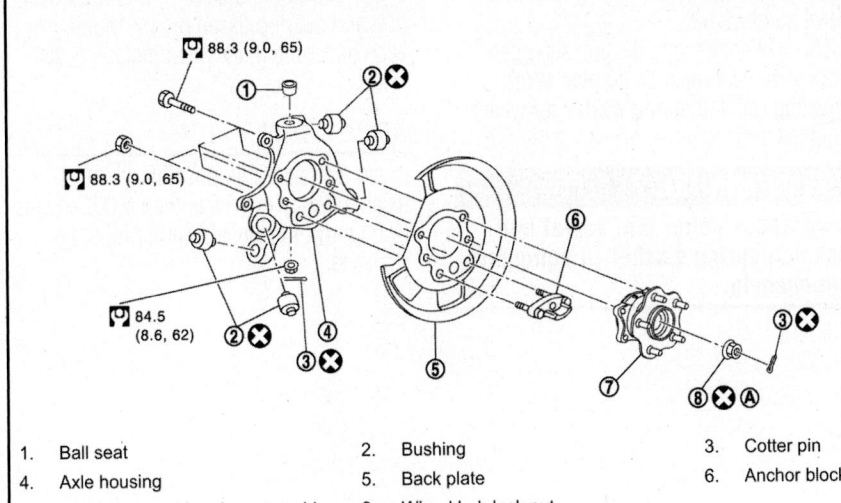

1. Ball seat
4. Axle housing
7. Wheel hub and bearing assembly

2. Bushing
5. Back plate
8. Wheel hub lock nut

3. Cotter pin
6. Anchor block

37663_IM35_G0444

Fig. 242 Exploded view of wheel hub and bearing assembly (Without adjusting cap and spring washer)

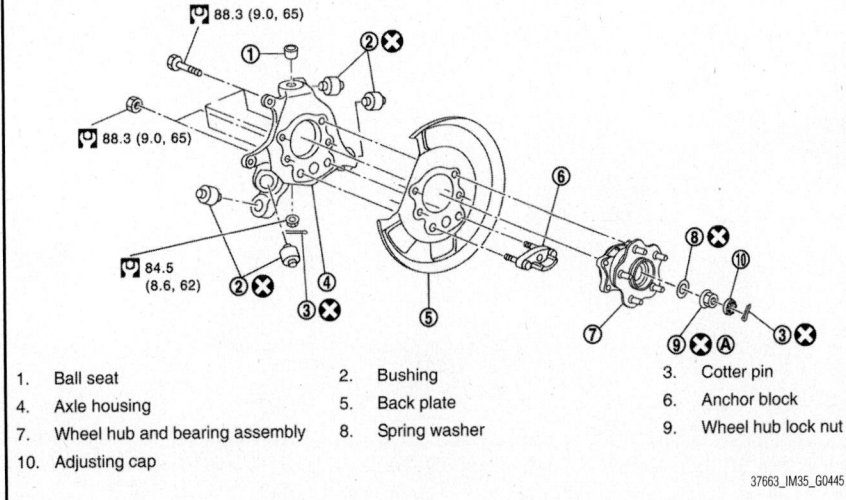

1. Ball seat
4. Axle housing
7. Wheel hub and bearing assembly
10. Adjusting cap

2. Bushing
5. Back plate
8. Spring washer

3. Cotter pin
6. Anchor block
9. Wheel hub lock nut

37663_IM35_G0445

Fig. 243 Exploded view of wheel hub and bearing assembly (With adjusting cap and spring washer)

4. Remove cotter pin and adjusting cap (if equipped), then loosen wheel hub lock nut.

5. Separate the wheel hub and bearing assembly from drive shaft by lightly tapping the end with a hammer and wood block, and then remove wheel hub lock nut and spring washer (if equipped).

✴✴ WARNING

Do not place drive shaft joint at an extreme angle. Also be careful not to overextend slide joint. Do not allow

drive shaft to hang down without support for housing (or joint sub-assembly), shaft and other parts.

➡**Use a puller, if the wheel hub and bearing assembly and drive shaft cannot be separated even after performing the above procedure.**

6. Remove the wheel hub and bearing assembly mounting bolts.

7. Remove the wheel hub and bearing assembly.

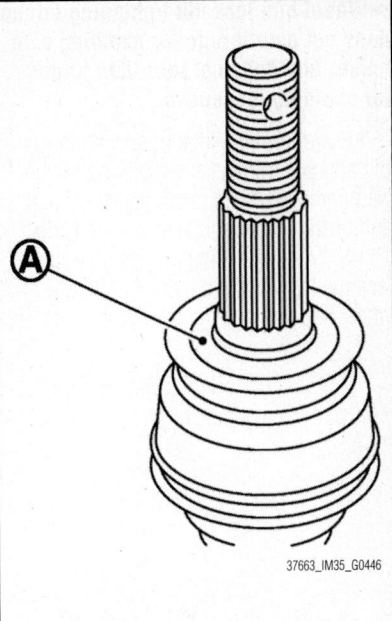

37663_IM35_G0446

Fig. 244 Apply paste to surface (A) of joint sub-assembly of drive shaft

To install:

8. Note the following, and install in the reverse order of removal.

9. Clean the matching surface of wheel hub lock nut and wheel hub and bearing assembly.

✴✴ WARNING

Never apply lubricating oil to these matching surface.

10. Clean the matching surface of drive shaft and wheel hub and bearing assembly.

11. Apply paste (440037S000) to surface of joint sub-assembly of drive shaft.

✴✴ WARNING

Apply paste to cover entire flat surface of joint sub-assembly of drive shaft.

12. Use the following torque range for tightening the wheel hub lock nut:
- Without adjusting cap and spring washer: 133 to 136 ft. lbs. (180 to 185 Nm)
- With adjusting cap and spring washer: 74 to 77 ft. lbs. (100 to 105 Nm)

✴✴ WARNING

Since the drive shaft is assembled by press-fitting, use the tightening torque range for the wheel hub lock nut. Be sure to use torque wrench to tighten the wheel hub lock nut. Never use a power tool.

➥**Wheel hub lock nut tightening torque does not over torque for avoiding axle noise, and does not less than torque for avoiding looseness.**

13. Perform the final tightening of each of parts under unladen conditions, which were removed when removing wheel hub and bearing assembly and axle housing.

14. When installing the spring washer, face the identification paint mark to the wheel hub and bearing assembly side.

(With adjusting cap and spring washer for wheel hub lock nut)

15. When installing the adjusting cap, check that there must be no play. (With adjusting cap and spring washer for wheel hub lock nut)

✳✳ WARNING

Never reuse cotter pin, wheel hub lock nut, spring washer (if equipped), and bushing.

16. Assemble disc rotor and the wheel hub and bearing assembly by aligning each matching mark when installing disc rotor.

ADJUSTMENT

No adjustment is possible. If the axial end play is greater than 0.002 inches (0.05 mm), the wheel bearing must be replaced.

INFINITI

QX56

5

SPECIFICATIONS AND MAINTENANCE CHARTS

ENGINE AND VEHICLE IDENTIFICATION

			Engine					Model Year	
Code ①	Liters (cc)	Cu. In.	Cyl.	Fuel Sys.	Engine	Eng. Mfg.		Code ②	Year
VK56VD	5.6 (5552)	338.8	8	MFI	DOHC	Nissan		B	2011
								C	2012

MFI: Multi-port Fuel Injection

DOHC: Double Overhead Camshafts

① Engine VIN: A

② 10th digit of the Vehicle Identification Number (VIN)

71075_QX56_C0001

GENERAL ENGINE SPECIFICATIONS

Year	Model	Engine Displacement Liters	Engine ID	Net Horsepower @ rpm	Net Torque @ rpm (ft. lbs.)	Bore x Stroke (in.)	Compression Ratio	Oil Pressure @ rpm
2011	QX56	5.6	VK56VD	400@5800	413@4000	3.86X3.62	10.8	43@2000
2012	QX56	5.6	VK56VD	400@5800	413@4000	3.86X3.62	10.8	43@2000

71075_QX56_C0002

ENGINE TUNE-UP SPECIFICATIONS

Year	Engine Displacement Liters	Engine ID	Spark Plug Gap (in.)	Ignition Timing	Fuel Pump (psi) ①	Idle Speed	Valve Clearance (in.) In.	Valve Clearance (in.) Ex.
2011	5.6	VK56VD	0.043	②	65	600-700	0.010-0.013	0.011-0.015
2012	5.6	VK56VD	0.043	②	65	600-700	0.010-0.013	0.011-0.015

NOTE: The Vehicle Emission Control Information label often reflects specification changes made during production. The label figures must be used if they differ from those in this chart.

① Approximate at idle

② 12 degrees +/- 2 degrees BTDC

71075_QX56_C0003

CAPACITIES

Year	Model	Engine Displacement Liters	Engine ID	Engine Oil with Filter (qts.)	Transmission (pts.)	Transfer Case (pts.)	Drive Axle		Fuel Tank (gal.)	Cooling System (qts.)
							Front (pts.)	Rear (pts.)		
2011	QX56	5.6	VK56VD	6.75	22.50	3.25	1.750	3.750	26.0	15.25
2012	QX56	5.6	VK56VD	6.75	22.50	3.25	1.750	3.750	26.0	15.25

NOTE: All capacities are approximate. Add fluid gradually and check to be sure a proper fluid level is obtained.

71075_QX56_C0004

FLUID SPECIFICATIONS

Year	Model	Engine Displ. Liters	Engine Oil	Auto. Trans.	Drive Axle		Transfer Case	Power Steering Fluid	Brake Master Cylinder	Cooling System
					Front	Rear				
2011	QX56	5.6	SAE 5W-30	①	②	②	③	④	⑤	⑥
2012	QX56	5.6	SAE 5W-30	①	②	②	③	④	⑤	⑥

① Genuine Nissan Matic S ATF fluid. Using non approved fluid may damage the transmission.

② Front: Nissan differential oil hypoid super GL-5 80W-90 or API GL-5 viscosity SAE 80W-90 gear oil. For hot climate viscosity SAE 90 may be used
   above 32 degrees F.ear: Nissan differential oil API GL-5 synthetic 75W-90 gear oil.

   Rear: Nissan differential oil API GL-5 synthetic 75W-90 gear oil.

③ Nissan ATX 90A. Using non approved fluid will damage the transfer case.

④ Nissan Power Steering Fluid. DEXRON VI type ATF may be used.

⑤ Nissan Super Heavy Duty DOT 3

⑥ Nissan Long Life antifreeze (BLUE) or equivalent

71075_QX56_C0013

VALVE SPECIFICATIONS

Year	Engine Displacement Liters	Engine ID	Seat Angle (deg.)	Face Angle (deg.)	Spring Test Pressure (lbs. @ in.)	Spring Installed Height (in.)	Stem-to-Guide Clearance (in.)		Stem Diameter (in.)	
							Intake	Exhaust	Intake	Exhaust
2011	5.6	VK56VD	45.15-45.45	45	①	②	0.0008-0.0021	0.0012-0.0025	0.2348-0.2354	0.2344-0.2350
2012	5.6	VK56VD	45.15-45.45	45	①	②	0.0008-0.0021	0.0012-0.0025	0.2348-0.2354	0.2344-0.2350

① Intake: 37-42@1.6140

   Exhaust: 37-42@1.3563

② Intake: 1.8614

   Exhaust: 1.8921

71075_QX56_C0006

CAMSHAFT SPECIFICATIONS

All measurements are given in inches.

Year	Engine Displ. Liters	Engine ID/VIN	Journal Dia.	Brg. Oil Clearance	Shaft End-play	Runout	Journal Bore	Lobe Height Intake	Lobe Height Exhaust
2011	5.6	VK56VD	1.0211-1.0218	0.0012-0.0028	0.0045-0.0074	0.0008	1.0236-1.0244	NA	1.7904-1.7978
2012	5.6	VK56VD	1.0211-1.0218	0.0012-0.0028	0.0045-0.0074	0.0008	1.0236-1.0244	NA	1.7746-1.7821

NA: not available

Note: Specifications are for exhaust camshafts only.

71075_QX56_C0007

CRANKSHAFT AND CONNECTING ROD SPECIFICATIONS

All measurements are given in inches.

Year	Engine Displ. Liters	Engine ID	Crankshaft Main Brg. Journal Dia.	Crankshaft Main Brg. Oil Clearance	Crankshaft Shaft End-play	Thrust on No.	Connecting Rod Journal Diameter	Connecting Rod Oil Clearance	Connecting Rod Side Clearance
2011	5.6	VK56VD	①	②	0.0039-0.0102	3	③	0.0008-0.0015	0.0079-0.0157
2012	5.6	VK56VD	①	②	0.0039-0.0102	3	③	0.0008-0.0015	0.0079-0.0157

① There are 24 different grades, ranging from 2.5182- 2.5174 for No. 1 and 5.

There are 24 different grades, ranging from 2.5182- 2.5173 for No. 2, 3 and 4.

② No. 1 and 5: 0.00004- 0.00043

No. 2, 3 and 4: 0.0003- 0.0007

③ There are 13 different grades, ranging from 2.2441- 2.2446. Specification is for rod bearing housing.

71075_QX56_C0005

PISTON AND RING SPECIFICATIONS

All measurements are given in inches.

Year	Engine Displacement Liters	Engine ID	Piston Clearance	Ring Gap Top Comp.	Ring Gap Bottom Comp.	Ring Gap Oil Control	Ring Side Clearance Top Comp.	Ring Side Clearance Bottom Comp.	Ring Side Clearance Oil Control
2011	5.6	VK56VD	0.0004-0.0012	0.0091-0.0110	0.0197-0.0256	0.0079-0.0236	0.0016-0.0031	0.0012-0.0028	0.0006-0.0073
2012	5.6	VK56VD	0.0004-0.0012	0.0091-0.0110	0.0197-0.0256	0.0079-0.0236	0.0016-0.0031	0.0012-0.0028	0.0006-0.0073

71075_QX56_C0008

TORQUE SPECIFICATIONS
All readings in ft. lbs.

Year	Engine Displacement Liters	Engine ID	Cylinder Head Bolts	Main Bearing Bolts	Rod Bearing Bolts	Crankshaft Damper Bolts	Flywheel Bolts	Manifold Intake	Manifold Exhaust	Spark Plugs	Oil Pan Drain Plug
2011	5.6	VK56VD	①	②	③	④	65	NA	25	18	25
2012	5.6	VK56VD	①	②	③	④	65	NA	25	18	25

NA: Not Available

① Step 1: 30 ft. lbs

   Step 2: +75 degrees clockwise

   Step 3: loosen in reverse order of tightening sequence

   Step 4: 30 ft. lbs.

   Step 5: +90 degrees clockwise

   Step 6: +90 degrees clockwise

② Step 1: cap bolts in order 1-10: 29 ft. lbs.

   Step 2: cap sub bolts in order 11-20: 22 ft. lbs.

   Step 3: cap bolts in order 1-10: +40 degrees

   Step 4: cap sub bolts in order 11-20: +30 degrees

   Step 5: side bolts in order 21-30: 36 ft. lbs.

③ Step 1: 21.7 ft. lbs.

   Step 2: loosen in reverse order of tightening sequence

   Step 3: 14.5 ft. lbs.

   Step 4: +90 degrees clockwise

④ Step 1: 151 ft. lbs.

   Step 2: +90 degrees

71075_QX56_C0009

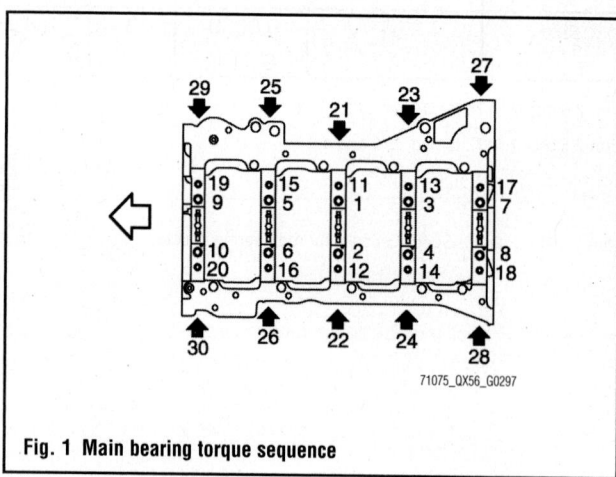

71075_QX56_G0297

Fig. 1 Main bearing torque sequence

WHEEL ALIGNMENT

| Year | Model | Caster | | Camber | | Toe-in (in.) |
		Range (+/-Deg.)	Preferred Setting (Deg.)	Range (+/-Deg.)	Preferred Setting (Deg.)	
2011	QX56	①	①	②	②	③
2012	QX56	①	①	②	②	③

NA - not available

① Left side: Minimum: 2 degrees 20' (2.34 degrees). Nominal: 3 degrees 05' (3.08 degrees). Maximum: 3 degrees 50' (3.83 degrees).

Right side: Minimum: 2 degrees 40' (2.67 degrees). Nominal: 3 degrees 25' (3.42 degrees). Maximum: 4 degree 10' (4.16 degrees).

Left and right difference: 0 degrees 45' (0.75 degrees) or less.

② Left side: Minimum: -0 degrees 45' (-0.75 degrees). Nominal: 0 degrees 00' (0.00 degrees). Maximum: 0 degrees 45' (0.75 degrees).

Right side: Minimum: -0 degrees 55' (-0.91 degrees). Nominal: -0 degrees 10' (-0.17 degrees). Maximum: 0 degree 35' (0.58 degrees).

Left and right difference: 0 degrees 33' (0.55 degrees) or less.

③ Toe-in distance: Minimum: 0.055 inch. Nominal: 0.094 inch. Maximum: 0.134 inch.

Toe angle: Minimum: 0 degrees 03' (0.05 degrees). Nominal: 0 degrees 05' (0.08 degrees). Maximum: 0 degree 07' (0.11 degrees).

71075_QX56_C0010

TIRE, WHEEL AND BALL JOINT SPECIFICATIONS

| Year | Model | OEM Tires | | Tire Pressures (psi) | | Wheel Size | Ball Joint Inspection | Lug Nut Torque (ft. lbs.) |
		Standard	Optional	Front	Rear			
2011	QX56	①	None	35	35	②	NA	98
2012	QX56	①	None	35	35	②	NA	98

OEM: Original Equipment Manufacturer

PSI: Pounds Per Square Inch

NA: not available

① P275/60R20 114H. P275/50R22 111H.

② 20x8.0 inch. 22x8.0 inch.

71075_QX56_C0011

BRAKE SPECIFICATIONS

All measurements in inches unless noted

| Year | Model | | Brake Disc | | | Minimum Pad Thickness | Brake Caliper | |
			Original Thickness	Minimum Thickness	Maximum Runout		Bracket Bolts (ft. lbs.)	Mounting Bolts (ft. lbs.)
2011	QX56	Front	1.181	NA	①	0.059	②	②
		Rear	0.787	NA	①	0.079	③	③
2012	QX56	Front	1.181	NA	①	0.059	②	②
		Rear	0.787	NA	①	0.079	③	③

NA: not available

① Maximum uneven wear measured at 8 positions: 0.0006. Runout limit, attached to vehicle: Front: 0.0021, Rear: 0.0020.

② Main bolt: 122 ft. lbs.

③ Cylinder body retaining bolt: 20 ft. lbs. Main bolt: 116 ft. lbs.

71075_QX56_C0012

SCHEDULED MAINTENANCE INTERVALS
INFINITI QX56

TO BE SERVICED	SERVICE	VEHICLE MILEAGE INTERVAL (x1000)												
		3.75	7.5	15	22.5	30	37.5	45	52.5	60	67.5	75	82.5	90
Engine oil & filter	R	✓	✓	✓	✓	✓	✓	✓	✓	✓	✓	✓	✓	✓
Brake lines & cables	I			✓		✓		✓		✓		✓		✓
Brake pads& rotors	L/I			✓		✓		✓		✓		✓		✓
Driveshaft boots & propeller shaft (4x4)	I					✓				✓				
Automatic transmission, final drive oil & transfer case	I			✓		✓		✓		✓				
LSD gear oil	I			✓		✓		✓		✓		✓		✓
Front wheel bearing grease (4x4)	R					✓				✓				
Air cleaner filter	R					✓				✓				✓
Engine coolant	R									✓				✓
Exhaust system	I						✓	✓	✓	✓				
Spark plugs	R	Replace every 105,000 miles												
Drive belt(s)	I			✓		✓		✓		✓		✓		✓
Cabin air filter	R							✓						✓
Exhaust system	I		✓			✓				✓				✓
Fuel lines	I		✓			✓				✓				✓
Steering gear (box) & linkage, axle & suspension parts	I					✓				✓				✓
Transfer case	I					✓				✓				✓
Tire rotation			✓	✓	✓	✓	✓	✓	✓	✓	✓	✓	✓	✓
Vapor lines	S/I					✓				✓				✓

R: Replace   S/I: Service or Inspect   L: Lubricate

FREQUENT OPERATION MAINTENANCE (SEVERE SERVICE)

If a vehicle is operated under any of the following conditions it is considered severe service:

- Extremely dusty areas.

- 50% or more of the vehicle operation is in 32°C (90°F) or higher temperatures, constant operation in temp. below 0°C (32°F).

- Prolonged idling (vehicle operation in stop and go traffic).

- Frequent short running periods (engine does not warm to normal operating temperatures).

- Police, taxi, delivery usage or trailer towing usage.

Oil & oil filter: replace every 3750 miles.

Brake pads, discs, drums & linings: service or inspect every 7500 miles.

Driveshaft boots & propeller shaft: service or inspect every 7500 miles.

Exhaust system: service or inspect every 7500 miles.

Final drive oil: Change every 30000 miles if towing a trailer.

Transfer case fluid: Change every 30000 miles if towing a trailer.

Steering gear (box) & linkage, (steering damper-4x4), axle & suspension parts: service or inspect every 7500 miles.

Steering linkage ball joints & front suspension ball joints: service or inspect every 7500 miles.

71075_QX56_C0014

PRECAUTIONS

Before servicing any vehicle, please be sure to read all of the following precautions, which deal with personal safety, prevention of component damage, and important points to take into consideration when servicing a motor vehicle:

• Never open, service or drain the radiator or cooling system when the engine is hot; serious burns can occur from the steam and hot coolant.

• Observe all applicable safety precautions when working around fuel. Whenever servicing the fuel system, always work in a well-ventilated area. Do not allow fuel spray or vapors to come in contact with a spark, open flame, or excessive heat (a hot drop light, for example). Keep a dry chemical fire extinguisher near the work area. Always keep fuel in a container specifically designed for fuel storage; also, always properly seal fuel containers to avoid the possibility of fire or explosion. Refer to the additional fuel system precautions later in this section.

• Fuel injection systems often remain pressurized, even after the engine has been turned **OFF**. The fuel system pressure must be relieved before disconnecting any fuel lines. Failure to do so may result in fire and/or personal injury.

• Brake fluid often contains polyglycol ethers and polyglycols. Avoid contact with the eyes and wash your hands thoroughly after handling brake fluid. If you do get brake fluid in your eyes, flush your eyes with clean, running water for 15 minutes. If eye irritation persists, or if you have taken brake fluid internally, IMMEDIATELY seek medical assistance.

• The EPA warns that prolonged contact with used engine oil may cause a number of skin disorders, including cancer. You should make every effort to minimize your exposure to used engine oil. Protective gloves should be worn when changing oil. Wash your hands and any other exposed skin areas as soon as possible after exposure to used engine oil. Soap and water, or waterless hand cleaner should be used.

• All new vehicles are now equipped with an air bag system, often referred to as a Supplemental Restraint System (SRS) or Supplemental Inflatable Restraint (SIR) system. The system must be disabled before performing service on or around system components, steering column, instrument panel components, wiring and sensors. Failure to follow safety and disabling procedures could result in accidental air bag deployment, possible personal injury and unnecessary system repairs.

• Always wear safety goggles when working with, or around, the air bag system. When carrying a non-deployed air bag, be sure the bag and trim cover are pointed away from your body. When placing a non-deployed air bag on a work surface, always face the bag and trim cover upward, away from the surface. This will reduce the motion of the module if it is accidentally deployed. Refer to the additional air bag system precautions later in this section.

• Clean, high quality brake fluid from a sealed container is essential to the safe and proper operation of the brake system. You should always buy the correct type of brake fluid for your vehicle. If the brake fluid becomes contaminated, completely flush the system with new fluid. Never reuse any brake fluid. Any brake fluid that is removed from the system should be discarded. Also, do not allow any brake fluid to come in contact with a painted surface; it will damage the paint.

• Never operate the engine without the proper amount and type of engine oil; doing so WILL result in severe engine damage.

• Timing belt maintenance is extremely important. Many models utilize an interference-type, non-freewheeling engine. If the timing belt breaks, the valves in the cylinder head may strike the pistons, causing potentially serious (also time-consuming and expensive) engine damage. Refer to the maintenance interval charts for the recommended replacement interval for the timing belt, and to the timing belt section for belt replacement and inspection.

• Disconnecting the negative battery cable on some vehicles may interfere with the functions of the on-board computer system(s) and may require the computer to undergo a relearning process once the negative battery cable is reconnected.

• When servicing drum brakes, only disassemble and assemble one side at a time, leaving the remaining side intact for reference.

• Only an MVAC-trained, EPA-certified automotive technician should service the air conditioning system or its components.

BRAKES

ANTI-LOCK BRAKE SYSTEM (ABS)

GENERAL INFORMATION

PRECAUTIONS

• Certain components within the ABS system are not intended to be serviced or repaired individually.

• Do not use rubber hoses or other parts not specifically specified for and ABS system. When using repair kits, replace all parts included in the kit. Partial or incorrect repair may lead to functional problems and require the replacement of components.

• Lubricate rubber parts with clean, fresh brake fluid to ease assembly. Do not use shop air to clean parts; damage to rubber components may result.

• Use only DOT 3 brake fluid from an unopened container.

• If any hydraulic component or line is removed or replaced, it may be necessary to bleed the entire system.

• A clean repair area is essential. Always clean the reservoir and cap thoroughly before removing the cap. The slightest amount of dirt in the fluid may plug an orifice and impair the system function. Perform repairs after components have been thoroughly cleaned; use only denatured alcohol to clean components. Do not allow ABS components to come into contact with any substance containing mineral oil; this includes used shop rags.

• The Anti-Lock control unit is a microprocessor similar to other computer units in the vehicle. Ensure that the ignition switch is **OFF** before removing or installing controller harnesses. Avoid static electricity discharge at or near the controller.

• If any arc welding is to be done on the vehicle, the control unit should be unplugged before welding operations begin.

SPEED SENSORS

REMOVAL & INSTALLATION

Front

See Figure 2.

➡**Replace the wheel hub/bearing assembly as an assembly when replacing the wheel sensor because the sensor cannot be disassembled.**

1. Before servicing the vehicle, refer to the Precautions Section.

➡**If working near and/or around the SRS system and components, be sure** to disable the SRS system. After disabling the system wait three minutes or more before servicing the vehicle.

➡Whenever the negative battery cable is disconnected the following components will require resetting. The air conditioning system, automatic drive positioner system, power window control system, around view monitor, automatic back door system, idle air volume learning, steering angle sensor neutral position, sunroof system, audio visual and navigation systems. Use the CONSULT-III diagnostic tool, or equivalent to perform the required resets. Follow the directions on the screen of the tool, as needed.

➡Before disconnecting the negative battery cable lower both the driver's side and passenger's side front windows. This will prevent any interference between the window edge and the vehicle when the door is opener/closed. The automatic window function will not work with the battery cable disconnected.

   2. Disconnect the negative battery cable.
   3. Raise and support the vehicle safely.
   4. Remove the tire and wheel assembly.
   5. Remove the hub and bearing assembly.

To install:

➡Be sure to use new fasteners, as required.

   6. Installation is the reverse of the removal procedure.
   7. Be sure to perform the reconnect/relearn procedures.

Rear

See Figure 3.

➡Replace the wheel hub/bearing assembly as an assembly when replacing the wheel sensor because the sensor cannot be disassembled.

   1. Before servicing the vehicle, refer to the Precautions Section.

➡If working near and/or around the SRS system and components, be sure to disable the SRS system. After disabling the system wait three minutes or more before servicing the vehicle.

➡Whenever the negative battery cable is disconnected the following components will require resetting. The air conditioning system, automatic drive positioner system, power window control system, around view monitor,

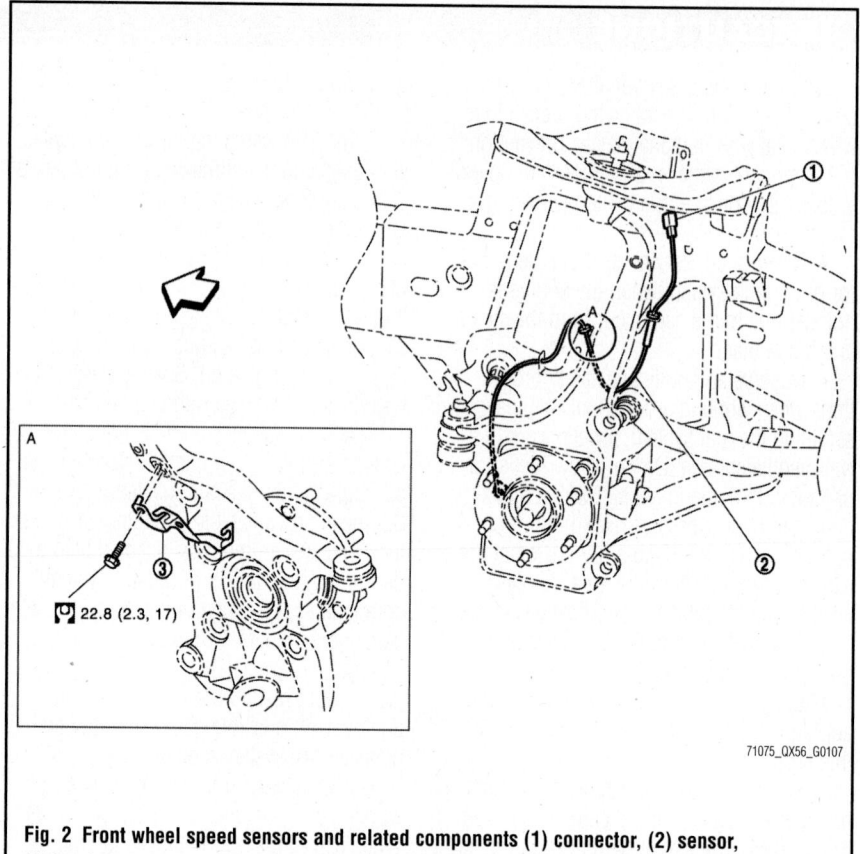

Fig. 2 Front wheel speed sensors and related components (1) connector, (2) sensor, (3) bracket

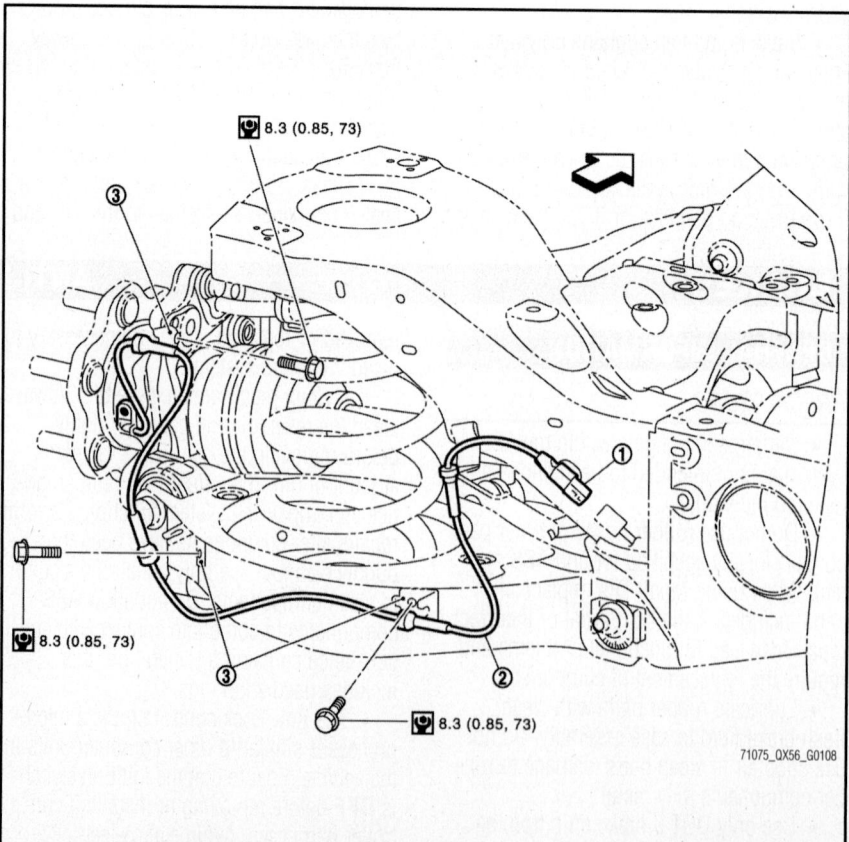

Fig. 3 Rear wheel speed sensors and related components (1) connector, (2) sensor, (3) bracket

automatic back door system, idle air volume learning, steering angle sensor neutral position, sunroof system, audio visual and navigation systems. Use the CONSULT-III diagnostic tool, or equivalent to perform the required resets. Follow the directions on the screen of the tool, as needed.

➡Before disconnecting the negative battery cable lower both the driver's

side and passenger's side front windows. This will prevent any interference between the window edge and the vehicle when the door is opener/closed. The automatic window function will not work with the battery cable disconnected.

    2. Disconnect the negative battery cable.
    3. Raise and support the vehicle safely.
    4. Remove the tire and wheel assembly.

    5. Remove the hub and bearing assembly.

  To install:

➡Be sure to use new fasteners, as required.

    6. Installation is the reverse of the removal procedure.
    7. Be sure to perform the reconnect/relearn procedures.

BRAKES

BLEEDING THE BRAKE SYSTEM

BLEEDING PROCEDURE

MOTOR ACCUMULATOR

➡Be sure that the master cylinder is full of clean fresh brake fluid before starting the bleeding process. Use only the recommended brake fluid when bleeding the system. Do not allow brake fluid to spill on painted surfaces as damage will occur.

➡The Nissan CONSULT III diagnostic tool, or equivalent is required to perform diagnostic this procedure.

➡Bleed the air in the following order: motor/accumulator, front right brake, front left brake, rear left brake, rear right brake.

➡The VDC warning lamp, ABS warning lamp and brake warning lamp turn ON and DTC code C118E may be detected in self-diagnosis result for ABS with the CONSULT-III diagnostic tool when the brake pedal is excessively operated, during this procedure. This is not a system malfunction as this occurs due to the temporary decrease in accumulator fluid pressure. The system returns to normal when the fluid pressure reaches a specified pressure with the ignition switch ON and the warning lamps will turn OFF. After these steps, clear the DTC code using the CONSULT-III diagnostic tool.

    1. Before servicing the vehicle, refer to the Precautions Section.

➡If working near and/or around the SRS system and components, be sure to disable the SRS system. After disabling the system wait three minutes or more before servicing the vehicle.

➡Whenever the negative battery cable is disconnected the following components will require resetting. The air conditioning system, automatic drive positioner system, power window

control system, around view monitor, automatic back door system, idle air volume learning, steering angle sensor neutral position, sunroof system, audio visual and navigation systems. Use the CONSULT-III diagnostic tool, or equivalent to perform the required resets. Follow the directions on the screen of the tool, as needed.

➡Before disconnecting the negative battery cable lower both the driver's side and passenger's side front windows. This will prevent any interference between the window edge and the vehicle when the door is opener/closed. The automatic window function will not work with the battery cable disconnected.

    2. Turn the ignition switch OFF.
    3. Depress the brake pedal twenty times or more.
    4. Check that there is no foreign material in the reservoir tank, and refill with clean fluid. Never reuse drained or used brake fluid.
    5. Turn the ignition switch ON. The motor will activate and stop.
    6. Turn the ignition switch OFF.
    7. Depress the brake pedal twenty or more times.
    8. Repeat steps 4 thru 6 an additional five times.
    9. Turn the ignition switch ON to check that the time between motor activation and motor stop is less than eighteen seconds. If longer, repeat steps 4 thru 8.

FRONT BRAKE

➡Be sure that the master cylinder is full of clean fresh brake fluid before starting the bleeding process. Use only the recommended brake fluid when bleeding the system. Do not allow brake fluid to spill on painted surfaces as damage will occur.

➡The Nissan CONSULT III diagnostic

tool, or equivalent is required to perform diagnostic this procedure.

➡Bleed the air in the following order: motor/accumulator, front right brake, front left brake, rear left brake, rear right brake.

➡The VDC warning lamp, ABS warning lamp and brake warning lamp turn ON and DTC code C118E may be detected in self-diagnosis result for ABS with the CONSULT-III diagnostic tool when the brake pedal is excessively operated, during this procedure. This is not a system malfunction as this occurs due to the temporary decrease in accumulator fluid pressure. The system returns to normal when the fluid pressure reaches a specified pressure with the ignition switch ON and the warning lamps will turn OFF. After these steps, clear the DTC code using the CONSULT-III diagnostic tool.

    1. Before servicing the vehicle, refer to the Precautions Section.

➡If working near and/or around the SRS system and components, be sure to disable the SRS system. After disabling the system wait three minutes or more before servicing the vehicle.

➡Whenever the negative battery cable is disconnected the following components will require resetting. The air conditioning system, automatic drive positioner system, power window control system, around view monitor, automatic back door system, idle air volume learning, steering angle sensor neutral position, sunroof system, audio visual and navigation systems. Use the CONSULT-III diagnostic tool, or equivalent to perform the required resets. Follow the directions on the screen of the tool, as needed.

➡Before disconnecting the negative battery cable lower both the driver's side and passenger's side front windows. This will prevent any interference between the window edge and the vehicle when the door is opener/closed. The automatic window function will not work with the battery cable disconnected.

2. Turn the ignition switch OFF.

3. Depress the brake pedal twenty times or more.

4. Check that there is no foreign material in the reservoir tank, and refill with clean fluid. Never reuse drained or used brake fluid.

5. Turn the ignition switch ON.

6. Connect a vinyl tube to the bleed valve.

7. Depress the brake pedal and loosen the bleeder valve.

8. Repeat above steps until all of the air is out of the brake line. Tighten the bleeder valve with the pedal depressed.

9. Check that no drag is present.

10. Check and adjust brake pedal, as required.

REAR BRAKE

➡Be sure that the master cylinder is full of clean fresh brake fluid before starting the bleeding process. Use only the recommended brake fluid when bleeding the system. Do not allow brake fluid to spill on painted surfaces as damage will occur.

➡The Nissan CONSULT III diagnostic tool, or equivalent is required to perform diagnostic this procedure.

➡Bleed the air in the following order: motor/accumulator, front right brake, front left brake, rear left brake, rear right brake.

➡The VDC warning lamp, ABS warning lamp and brake warning lamp turn ON and DTC code C118E may be detected in self-diagnosis result for ABS with the CONSULT-III diagnostic tool when the brake pedal is excessively operated, during this procedure. This is not a system malfunction as this occurs due to the temporary decrease in accumulator fluid pressure. The system returns to normal when the fluid pressure reaches a specified pressure with the ignition switch ON and the warning lamps will turn OFF. After these steps, clear the DTC code using the CONSULT-III diagnostic tool.

1. Before servicing the vehicle, refer to the Precautions Section.

➡If working near and/or around the SRS system and components, be sure to disable the SRS system. After disabling the system wait three minutes or more before servicing the vehicle.

➡Whenever the negative battery cable is disconnected the following components will require resetting. The air conditioning system, automatic drive positioner system, power window control system, around view monitor, automatic back door system, idle air volume learning, steering angle sensor neutral position, sunroof system, audio visual and navigation systems. Use the CONSULT-III diagnostic tool, or equivalent to perform the required resets. Follow the directions on the screen of the tool, as needed.

➡Before disconnecting the negative battery cable lower both the driver's side and passenger's side front windows. This will prevent any interference between the window edge and the vehicle when the door is opener/closed. The automatic window function will not work with the battery cable disconnected.

2. Turn the ignition switch OFF.

3. Depress the brake pedal twenty times or more.

4. Check that there is no foreign material in the reservoir tank, and refill with clean fluid. Never reuse drained or used brake fluid.

5. Turn the ignition switch ON.

6. Connect a vinyl tube to the bleed valve.

7. Depress the brake pedal and loosen the bleeder valve.

8. Depress and hold the brake pedal depression to discharge a 100cc of fluid before tightening the bleeder valve. Since fluid is conveyed by the motor, the pedal is not necessarily depressed.

9. Release the brake pedal.

10. Repeat above steps until all of the air is out of the brake line. Tighten the bleeder valve with the pedal depressed.

11. Check that no drag is present.

12. Check and adjust brake pedal, as required.

BRAKE FLUID

FLUID RECOMMENDATIONS

When adding/changing fluid to the brake system be sure to use DOT 3 (US FMVSS No. 116, or equivalent).

LEVEL CHECK

See Figure 4.

1. Before servicing the vehicle, refer to the Precautions Section.

➡If working near and/or around the SRS system and components, be sure to disable the SRS system. After disabling the system wait three minutes or more before servicing the vehicle.

➡Whenever the negative battery cable is disconnected the following components will require resetting. The air conditioning system, automatic drive positioner system, power window control system, around view monitor, automatic back door system, idle air volume learning, steering angle sensor neutral position, sunroof system, audio visual and navigation systems. Use the CONSULT-III diagnostic tool, or equivalent to perform the required resets. Follow the directions on the screen of the tool, as needed.

➡Before disconnecting the negative battery cable lower both the driver's side and passenger's side front windows. This will prevent any interference between the window edge and the vehicle when the door is opener/closed. The automatic window function will not work with the battery cable disconnected.

✳ CAUTION

Never check or adjust the fluid level with the ignition switch ON.

2. Check that there is no foreign material in the reservoir tank or around it. Never reuse used fluid.

3. Check that the fluid is within specification.

4. Correct fluid level, as required.

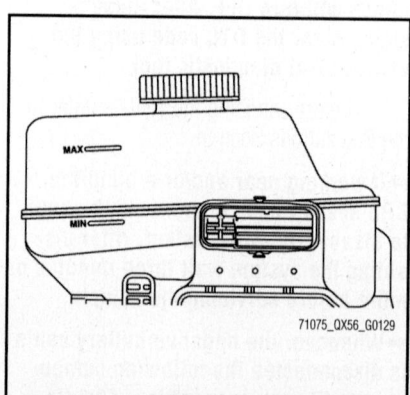

71075_QX56_G0129

Fig. 4 Fill the brake fluid reservoir tank between the MAX and MIN lines

BRAKE CALIPERS

REMOVAL & INSTALLATION

See Figure 5.

1. Before servicing the vehicle, refer to the Precautions Section.

➡ If working near and/or around the SRS system and components, be sure to disable the SRS system. After disabling the system wait three minutes or more before servicing the vehicle.

➡ Whenever the negative battery cable is disconnected the following components will require resetting. The air conditioning system, automatic drive positioner system, power window control system, around view monitor, automatic back door system, idle air volume learning, steering angle sensor neutral position, sunroof system, audio visual and navigation systems. Use the CONSULT-III diagnostic tool, or equivalent to perform the required resets. Follow the directions on the screen of the tool, as needed.

➡ Before disconnecting the negative battery cable lower both the driver's side and passenger's side front windows. This will prevent any interference between the window edge and the vehicle when the door is opener/closed. The automatic window function will not work with the battery cable disconnected.

2. Disconnect the negative battery cable.

3. Drain brake fluid as necessary.
4. Raise and safely support the vehicle.
5. Remove or disconnect the following:
   - Wheel and tire assembly
   - Union bolt
   - Disconnect and plug brake hose. Discard the washer.
   - Caliper mounting bolts
   - Brake caliper

To install:

➡ Be sure to use new fasteners, as required.

6. Installation is the reverse of the removal procedure.
7. Bleed the system, as required.
8. Be sure to perform the reconnect/relearn procedures.

BRAKE PADS

REMOVAL & INSTALLATION

See Figures 6 through 10.

1. Before servicing the vehicle, refer to the Precautions Section.

➡ If working near and/or around the SRS system and components, be sure to disable the SRS system. After disabling the system wait three minutes or more before servicing the vehicle.

➡ Whenever the negative battery cable is disconnected the following components will require resetting. The air conditioning system, automatic drive positioner system, power window control system, around view monitor,

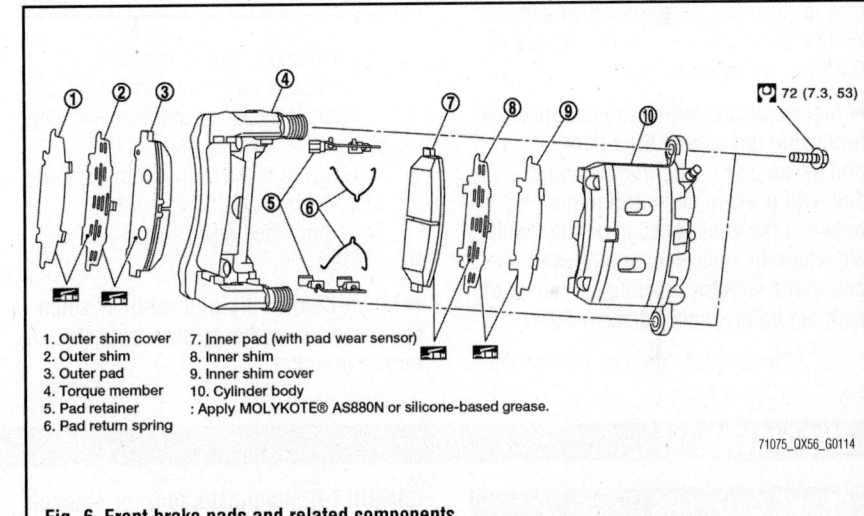

1. Outer shim cover
2. Outer shim
3. Outer pad
4. Torque member
5. Pad retainer
6. Pad return spring
7. Inner pad (with pad wear sensor)
8. Inner shim
9. Inner shim cover
10. Cylinder body
: Apply MOLYKOTE® AS880N or silicone-based grease.

71075_QX56_G0114

Fig. 6 Front brake pads and related components

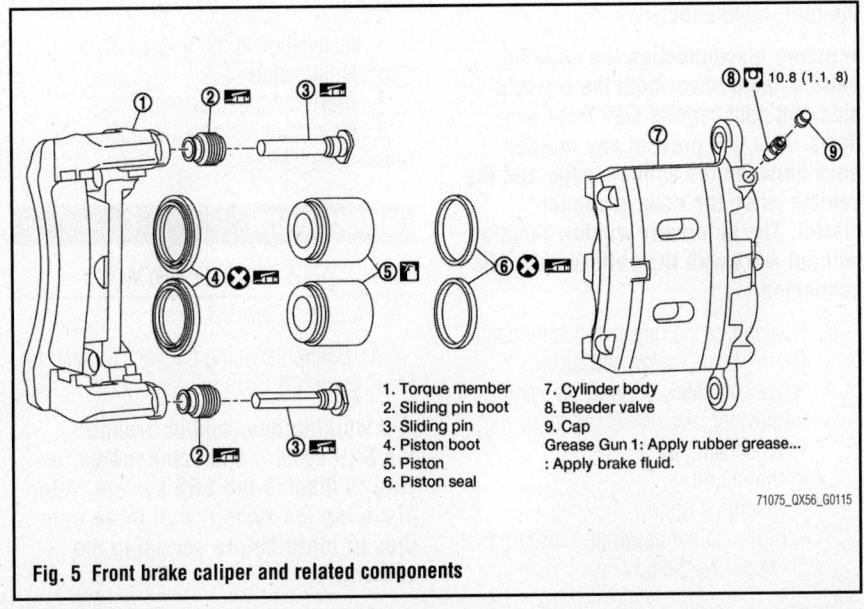

1. Torque member
2. Sliding pin boot
3. Sliding pin
4. Piston boot
5. Piston
6. Piston seal
7. Cylinder body
8. Bleeder valve
9. Cap
Grease Gun 1: Apply rubber grease...
: Apply brake fluid.

72 (7.3, 53)

10.8 (1.1, 8)

71075_QX56_G0115

Fig. 5 Front brake caliper and related components

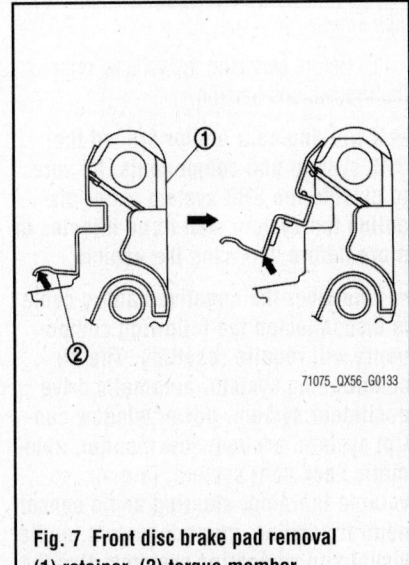

71075_QX56_G0133

Fig. 7 Front disc brake pad removal (1) retainer, (2) torque member

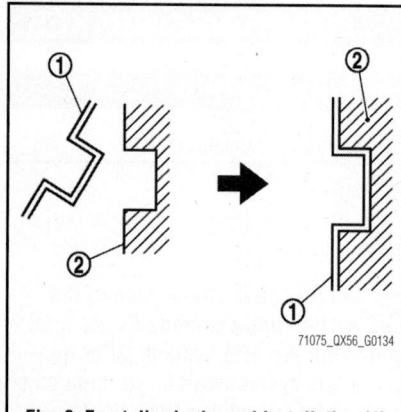

Fig. 8 Front disc brake pad installation (1) retainer, (2) torque member—1 of 3

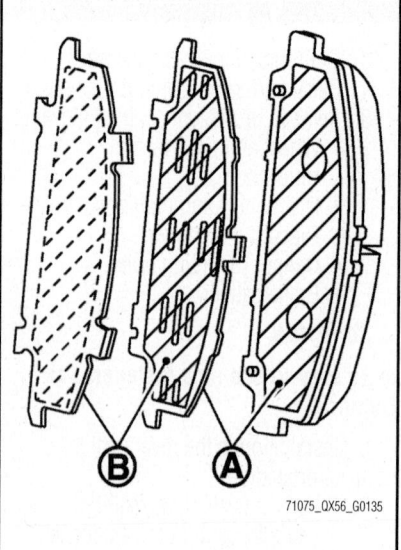

Fig. 9 Front disc brake pad installation (A and B) lube application points—2 of 3

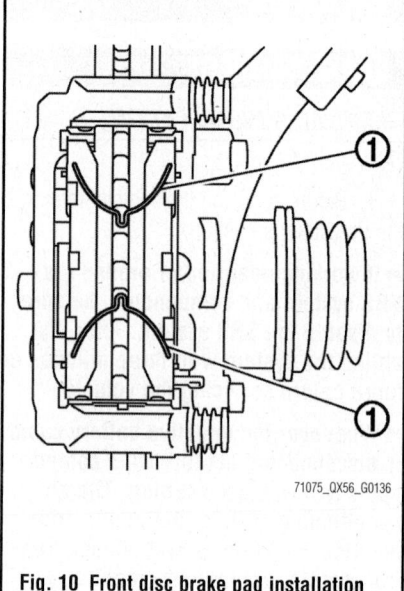

Fig. 10 Front disc brake pad installation (1) springs—3 of 3

automatic back door system, idle air volume learning, steering angle sensor neutral position, sunroof system, audio visual and navigation systems. Use the CONSULT-III diagnostic tool, or equivalent to perform the required resets. Follow the directions on the screen of the tool, as needed.

➡Before disconnecting the negative battery cable lower both the driver's side and passenger's side front windows. This will prevent any interference between the window edge and the vehicle when the door is opener/closed. The automatic window function will not work with the battery cable disconnected.

2. Disconnect the negative battery cable.

3. Drain brake fluid as necessary.
4. Raise and safely support the vehicle.
5. Remove the wheel and tire assembly.
6. Remove lower sliding pin bolt.
7. Suspend brake caliper with a remove and remove brake pad return springs.
8. Remove the brake pads, shims and shim covers.

➡Never deform the pad retainer when removing the pad retainers from the torque member.

To install:

➡Be sure to use new fasteners, as required.

9. Installation is the reverse of the removal procedure.
10. Lubricate required points with Molykote or a good silicone based grease.
11. Be sure to perform the reconnect/relearn procedures.

BRAKES

BRAKE CALIPERS

REMOVAL & INSTALLATION
See Figure 11.

1. Before servicing the vehicle, refer to the Precautions Section.

➡If working near and/or around the SRS system and components, be sure to disable the SRS system. After disabling the system wait three minutes or more before servicing the vehicle.

➡Whenever the negative battery cable is disconnected the following components will require resetting. The air conditioning system, automatic drive positioner system, power window control system, around view monitor, automatic back door system, idle air volume learning, steering angle sensor neutral position, sunroof system, audio visual and navigation systems. Use the

CONSULT-III diagnostic tool, or equivalent to perform the required resets. Follow the directions on the screen of the tool, as needed.

➡Before disconnecting the negative battery cable lower both the driver's side and passenger's side front windows. This will prevent any interference between the window edge and the vehicle when the door is opener/closed. The automatic window function will not work with the battery cable disconnected.

2. Disconnect the negative battery cable.
3. Drain brake fluid as necessary.
4. Raise and safely support the vehicle.
5. Remove or disconnect the following:
   • Wheel and tire assembly
   • Union bolt
   • Mounting bolts
   • Brake caliper assembly. Discard the copper washers.

REAR DISC BRAKES

To install:

➡Be sure to use new fasteners, as required.

6. Installation is the reverse of the removal procedure.
7. Bleed the brake system.
8. Be sure to perform the reconnect/relearn procedures.

BRAKE PADS

REMOVAL & INSTALLATION
See Figures 12 and 13.

1. Before servicing the vehicle, refer to the Precautions Section.

➡If working near and/or around the SRS system and components, be sure to disable the SRS system. After disabling the system wait three minutes or more before servicing the vehicle.

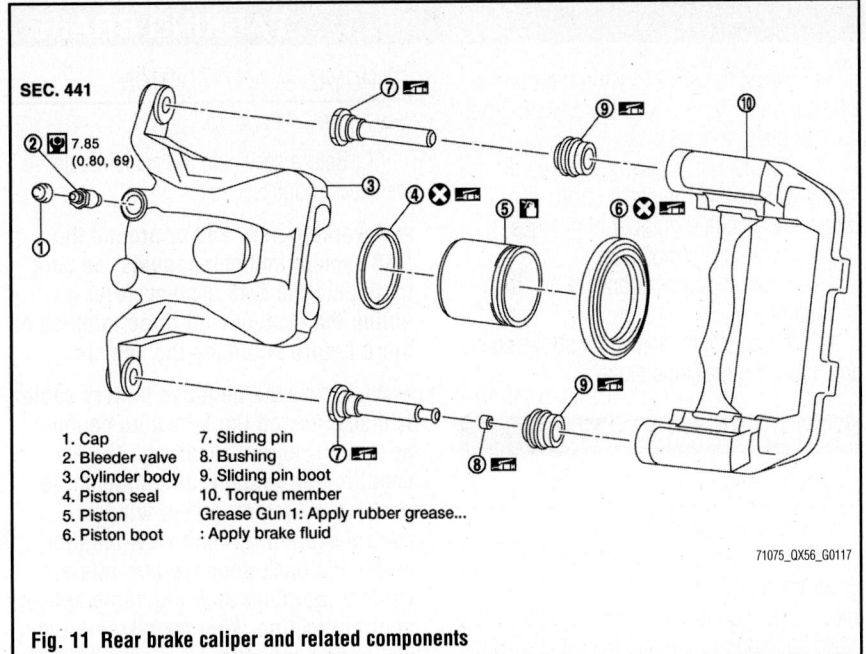

SEC. 441

1. Cap
2. Bleeder valve
3. Cylinder body
4. Piston seal
5. Piston
6. Piston boot
7. Sliding pin
8. Bushing
9. Sliding pin boot
10. Torque member
Grease Gun 1: Apply rubber grease...
: Apply brake fluid

71075_QX56_G0117

Fig. 11 Rear brake caliper and related components

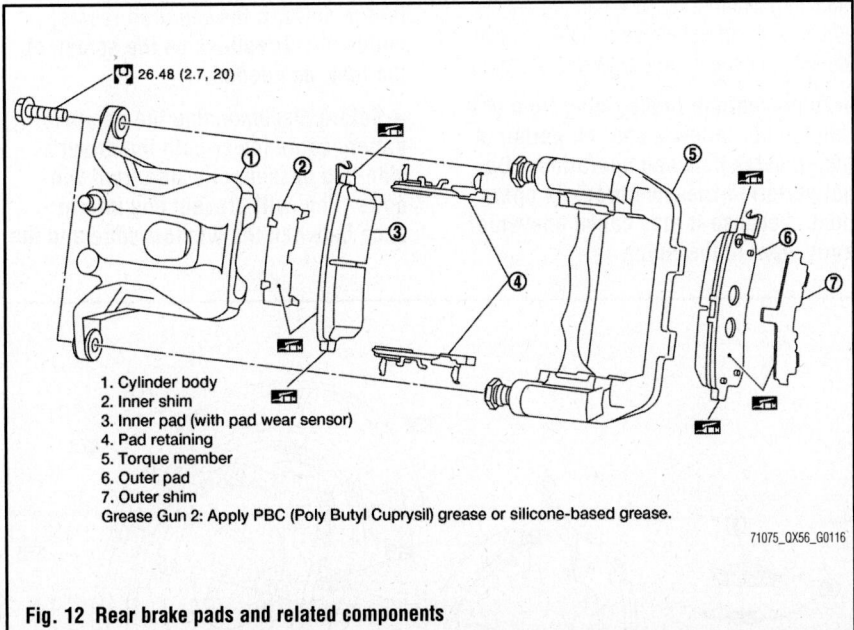

1. Cylinder body
2. Inner shim
3. Inner pad (with pad wear sensor)
4. Pad retaining
5. Torque member
6. Outer pad
7. Outer shim
Grease Gun 2: Apply PBC (Poly Butyl Cuprysil) grease or silicone-based grease.

71075_QX56_G0116

Fig. 12 Rear brake pads and related components

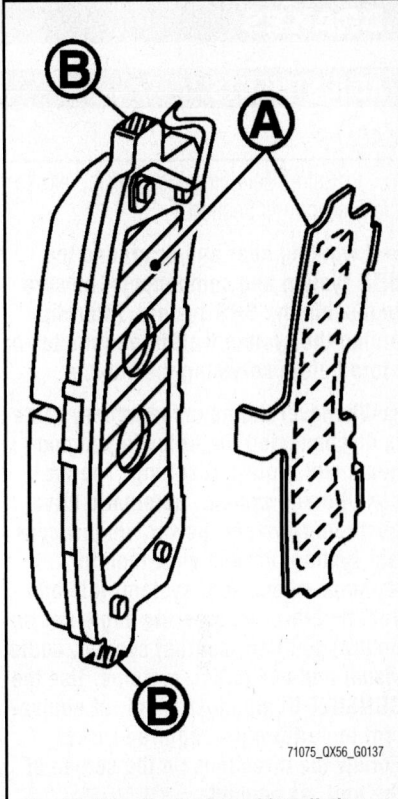

71075_QX56_G0137

Fig. 13 Rear disc brake pad installation (A and B) lube application points

→Whenever the negative battery cable is disconnected the following components will require resetting. The air conditioning system, automatic drive positioner system, power window control system, around view monitor, automatic back door system, idle air volume learning, steering angle sensor neutral position, sunroof system, audio visual and navigation systems. Use the CONSULT-III diagnostic tool, or equivalent to perform the required resets. Follow the directions on the screen of the tool, as needed.

→Before disconnecting the negative battery cable lower both the driver's side and passenger's side front windows. This will prevent any interference between the window edge and the vehicle when the door is opener/closed. The automatic window function will not work with the battery cable disconnected.

2. Disconnect the negative battery cable.

3. Drain brake fluid as necessary.

4. Raise and safely support the vehicle.

5. Remove the wheel and tire assembly.

6. Remove the upper pin sliding bolt.

7. Support the cylinder body, with mechanics wire.

8. Remove the pads, shims, cover and retainer.

To install:

→Be sure to use new fasteners, as required.

9. Installation is the reverse of the removal procedure.

10. Lubricate required points with poly butyl silicone or a good silicone based grease.

11. Be sure to perform the reconnect/relearn procedures.

BRAKES   **PARKING BRAKE**

ADJUSTMENTS

CABLES

1. Before servicing the vehicle, refer to the Precautions Section.

→**If working near and/or around the SRS system and components, be sure to disable the SRS system. After disabling the system wait three minutes or more before servicing the vehicle.**

→**Whenever the negative battery cable is disconnected the following components will require resetting. The air conditioning system, automatic drive positioner system, power window control system, around view monitor, automatic back door system, idle air volume learning, steering angle sensor neutral position, sunroof system, audio visual and navigation systems. Use the CONSULT-III diagnostic tool, or equivalent to perform the required resets. Follow the directions on the screen of the tool, as needed.**

→**Before disconnecting the negative battery cable lower both the driver's side and passenger's side front windows. This will prevent any interference between the window edge and the vehicle when the door is opener/closed. The automatic window function will not work with the battery cable disconnected.**

2. Disconnect the negative battery cable.
3. Raise and support the vehicle safely.
4. Remove the tire and wheel assembly.
5. Hold the rotor in place using at least two wheel lug nuts.
6. Release the parking brake pedal by turning the adjusting nut and loosening the cable.
7. Remove the plug from the rotor. Turn the adjuster in using a suitable tool until the rotor is locked.
8. Back out the adjuster five or six notches.
9. Rotate the rotor to make sure that there is no drag.
10. Adjust the parking brake pedal stroke again.
11. Check the rear disc brake.
12. To adjust the cable, temporarily adjust the cable so that the parking brake pedal operating force immediately before full stroke reaches 110 lbs. or more.
13. Maintain the force (110 lbs.) for thirty minutes or more.

14. Adjust the stroke by turning the adjusting nut. Never reuse the adjusting nut if it has been removed.
15. Operate the pedal with a force of 44 lbs. Check that the pedal stroke is within the specified number of notches. Specification is 6–7 notches.
16. Rotate the rotor to ensure that there is no drag.
17. If drag exists, adjust the stroke again and check the rear disc brake.

PARKING BRAKE SHOES

BURNISHING

Perform the parking brake burnishing operation by driving the vehicle forward under the following conditions: vehicle speed 51.6 mph forward direction, parking brake operating force 57.3 lbs set and apply time of 33 seconds. After parking brake burnishing operation, recheck parking brake adjustment, correct as required. Allow five minutes of cooling time between operations.

→**To prevent the brake lining from getting too hot, allow a cool off period of five minutes between operations. Do not perform excessive break-in operations, because it may cause uneven or early wear of the lining.**

REMOVAL & INSTALLATION

See Figures 14 and 15.

1. Before servicing the vehicle, refer to the Precautions Section.

→**If working near and/or around the SRS system and components, be sure to disable the SRS system. After disabling the system wait three minutes or more before servicing the vehicle.**

→**Whenever the negative battery cable is disconnected the following components will require resetting. The air conditioning system, automatic drive positioner system, power window control system, around view monitor, automatic back door system, idle air volume learning, steering angle sensor neutral position, sunroof system, audio visual and navigation systems. Use the CONSULT-III diagnostic tool, or equivalent to perform the required resets. Follow the directions on the screen of the tool, as needed.**

→**Before disconnecting the negative battery cable lower both the driver's side and passenger's side front windows. This will prevent any interference between the window edge and the**

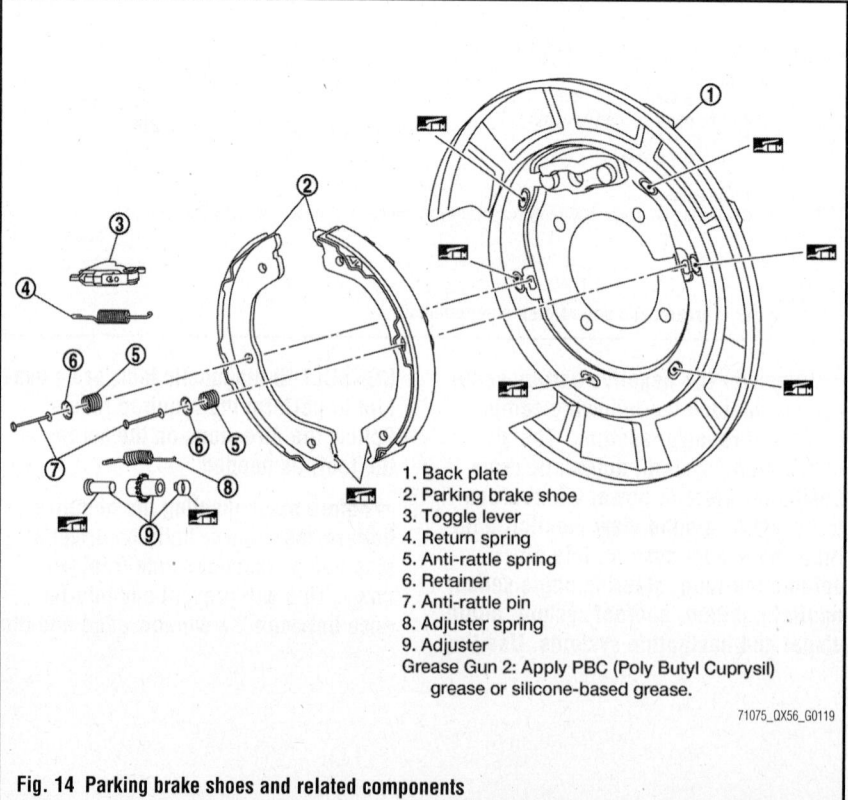

1. Back plate
2. Parking brake shoe
3. Toggle lever
4. Return spring
5. Anti-rattle spring
6. Retainer
7. Anti-rattle pin
8. Adjuster spring
9. Adjuster

Grease Gun 2: Apply PBC (Poly Butyl Cuprysil) grease or silicone-based grease.

71075_QX56_G0119

Fig. 14 Parking brake shoes and related components

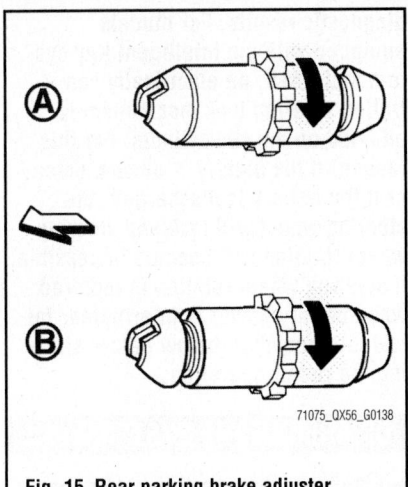

**Fig. 15 Rear parking brake adjuster
identification (A) right side, (B) left side**

vehicle when the door is opener/
closed. The automatic window function
will not work with the battery cable
disconnected.

2. Disconnect the negative battery
cable.
3. Raise and support the vehicle
safely.
4. Remove the tire and wheel
assembly.
5. Be sure that the parking brake lever
is in the released position.
6. Remove the rear disc rotor.
7. Remove the anti-rattle springs.
8. Remove the return spring. Remove
the adjuster spring.
9. Disconnect the parking brake cable
from the toggle lever.

10. Remove the shoes, adjuster assembly and toggle lever.

To install:

➡**Be sure to use new fasteners, as
required.**

11. Installation is the reverse of the
removal procedure.
12. Adjust the parking brake.
13. Perform the parking brake burnishing operation.
14. Assemble the adjusters so that the
threaded part is expanded when rotating it
in the direction of the arrow. See illustration.
15. Be sure to perform the reconnect/
relearn procedures.

CHASSIS ELECTRICAL

AIR BAGS (SUPPLEMENTAL RESTRAINT SYSTEM)

PRECAUTIONS

Disconnect and isolate the battery negative cable before beginning any airbag system component diagnosis, testing, removal, or installation procedures. Allow system capacitor to discharge for two minutes before beginning any component service. This will disable the airbag system. Failure to disable the airbag system may result in accidental airbag deployment, personal injury, or death.

Do not place an intact undeployed airbag face down on a solid surface. The airbag will propel into the air if accidentally deployed and may result in personal injury or death.

When carrying or handling an undeployed airbag, the trim side (face) of the airbag should be pointing towards the body to minimize possibility of injury if accidental deployment occurs. Failure to do this may result in personal injury or death.

Replace airbag system components with OEM replacement parts. Substitute parts may appear interchangeable, but internal differences may result in inferior occupant protection. Failure to do so may result in occupant personal injury or death.

Wear safety glasses, rubber gloves, and long sleeved clothing when cleaning powder residue from vehicle after an airbag deployment. Powder residue emitted from a deployed airbag can cause skin irritation. Flush affected area with cool water if irritation is experienced. If nasal or throat irritation is experienced, exit the vehicle for fresh air until the irritation ceases. If irritation continues, see a physician.

Do not use a replacement airbag that is

not in the original packaging. This may result in improper deployment, personal injury, or death.

The factory installed fasteners, screws and bolts used to fasten airbag components have a special coating and are specifically designed for the airbag system. Do not use substitute fasteners. Use only original equipment fasteners listed in the parts catalog when fastener replacement is required.

During, and following, any child restraint anchor service, due to impact event or vehicle repair, carefully inspect all mounting hardware, tether straps, and anchors for proper installation, operation, or damage. If a child restraint anchor is found damaged in any way, the anchor must be replaced. Failure to do this may result in personal injury or death.

Deployed and non-deployed airbags may or may not have live pyrotechnic material within the airbag inflator.

Do not dispose of driver/passenger/curtain airbags or seat belt tensioners unless you are sure of complete deployment. Refer to the Hazardous Substance Control System for proper disposal.

Dispose of deployed airbags and tensioners consistent with state, provincial, local, and federal regulations.

After any airbag component testing or service, do not connect the battery negative cable. Personal injury or death may result if the system test is not performed first.

If the vehicle is equipped with the Occupant Classification System (OCS), do not connect the battery negative cable before performing the OCS Verification Test using the scan tool and the appropriate diagnostic information. Personal injury or death may

result if the system test is not performed properly.

Never replace both the Occupant Restraint Controller (ORC) and the Occupant Classification Module (OCM) at the same time. If both require replacement, replace one, then perform the Airbag System test before replacing the other.

Both the ORC and the OCM store Occupant Classification System (OCS) calibration data, which they transfer to one another when one of them is replaced. If both are replaced at the same time, an irreversible fault will be set in both modules and the OCS may malfunction and cause personal injury or death.

If equipped with OCS, the Seat Weight Sensor is a sensitive, calibrated unit and must be handled carefully. Do not drop or handle roughly. If dropped or damaged, replace with another sensor. Failure to do so may result in occupant injury or death.

If equipped with OCS, the front passenger seat must be handled carefully as well. When removing the seat, be careful when setting on floor not to drop. If dropped, the sensor may be inoperative, could result in occupant injury, or possibly death.

If equipped with OCS, when the passenger front seat is on the floor, no one should sit in the front passenger seat. This uneven force may damage the sensing ability of the seat weight sensors. If sat on and damaged, the sensor may be inoperative, could result in occupant injury, or possibly death.

The Supplemental Restraint System (SRS) such as AIR BAG and SEAT BELT PRE-TENSIONER, used along with a front seat belt, helps to reduce the risk or severity of injury to the driver and front passenger for certain types of collision. This system includes seat belt switch inputs and dual stage front air bag modules. The SRS system uses the seat belt switches to determine the front air bag deployment, and may only deploy one front air bag, depending on the severity of a collision and whether the front occupants are belted or unbelted.

❋❋ CAUTION

Improper maintenance, including incorrect removal and installation of the SRS, can lead to personal injury caused by unintentional activation of the system. Do not use electrical test equipment on any circuit related to the SRS unless instructed. SRS wiring harnesses can be identified by yellow and/or orange harnesses or harness connectors.

❋❋ CAUTION

When working near the Airbag Diagnosis Sensor Unit or other Airbag System sensors with the Ignition ON or engine running, DO NOT use air or electric power tools or strike near the sensor(s) with a hammer. Heavy vibration could activate the sensor(s) and deploy the air bag(s), possibly causing serious injury. When using air or electric power tools or hammers, always switch the Ignition OFF, disconnect the battery, and wait at least 3 minutes before performing any service.

❋❋ CAUTION

Before removing the seat belt pre-tensioner assembly, turn the ignition switch OFF, disconnect both battery terminals and wait at least three minutes. For approximately three minutes after the battery terminals have been removed, it is still possible for the air bag and seat belt pre-tensioner to deploy. Therefore, do not attempt work on any SRS connectors or wires until at least three minutes have passed. After replacing or reinstalling seat belt pre-tensioner assembly, or reconnecting seat belt pre-tensioner assembly connector, make sure entire SRS operates properly.

➡Do not disassemble buckle or seat belt assembly. Replace anchor bolts if they are deformed or worn out. Never

oil tongue and buckle. If any component of seat belt assembly is questionable, do not repair. Replace the whole seat belt assembly. If webbing is cut, frayed, or damaged, replace seat belt assembly. When replacing seat belt assembly, use a genuine Nissan seat belt assembly.

➡After a collision inspect all seat belt assemblies including retractors and attaching hardware after any collision. Nissan recommends that all seat belt assemblies in use during a collision be replaced unless the collision was minor and the belts show no damage and continue to operate properly. Failure to do so could result in serious personal injury in an accident. Seat belt assemblies not in use during a collision should also be replaced if either damage or improper operation is noted. Seat belt pre-tensioner should be replaced even if the seat belts are not in use during a frontal collision in which the air bags are deployed. Replace any seat belt assembly (including anchor bolts) if the seat belt was in use at the time of a collision (except for minor collisions and the belts, retractors and buckles show no damage and continue to operate properly). The seat belt was damaged in an accident. (i.e., torn webbing, bent retractor or guide, etc.). The seat belt attaching point was damaged in an accident. Inspect the seat belt attaching area for damage or distortion and repair as necessary before installing a new seat belt assembly. Anchor bolts are deformed or worn out. The seat belt pre-tensioner should be replaced even if the seat belts are not in use during the collision in which the air bags are deployed.

➡Replace occupant classification system control unit and passenger front seat cushion as an assembly.

➡If the steering wheel was rotated after battery disconnect, perform the following: This Procedure is applied only to models with Intelligent Key system and NATS (NISSAN ANTI-THEFT SYSTEM). Remove and install all control units after disconnecting both battery cables with the ignition knob in the "LOCK" position. Always use CONSULT to perform self-diagnosis as a part of each function inspection after finishing work. If DTC is detected, perform trouble diagnosis according to self-

diagnostic results. For models equipped with the Intelligent Key system and NATS, an electrically controlled steering lock mechanism is adopted on the key cylinder. For this reason, if the battery is disconnected or if the battery is discharged, the steering wheel will lock and steering wheel rotation will become impossible. If steering wheel rotation is required when battery power is interrupted, follow the procedure below before starting the repair operation.

DISARMING THE SYSTEM

➡Whenever the negative battery cable is disconnected the following components will require resetting. The air conditioning system, automatic drive positioner system, power window control system, around view monitor, automatic back door system, idle air volume learning, steering angle sensor neutral position, sunroof system, audio visual and navigation systems. Use the CONSULT-III diagnostic tool, or equivalent to perform the required resets. Follow the directions on the screen of the tool, as needed.

➡Before disconnecting the negative battery cable lower both the driver's side and passenger's side front windows. This will prevent any interference between the window edge and the vehicle when the door is opener/ closed. The automatic window function will not work with the battery cable disconnected.

1. Before servicing the vehicle, refer to the Precautions Section.
2. Disconnect the negative battery cable.
3. Disconnect the positive battery cable.
4. Wait at least 3 minutes before working on the vehicle. The air bag system is designed to retain enough power to deploy the air bag for a short time after the battery has been disconnected.

ARMING THE SYSTEM

➡Once repair work has been completed, return the ignition switch to the LOCK position, before connecting the battery cables. At this time the steering lock mechanism will engage. Install the CONSULT-III diagnostic tool, or equivalent, and follow the directions on the screen of the tool and perform the self-diagnosis check.

DRIVE TRAIN

AUTOMATIC TRANSMISSION

DRAIN & REFILL

The automatic transmission uses Nissan approved Matic S ATF transmission fluid. If information differs from the owner's manual, use the data in the owner's manual.

At this time the manufacturer does not provide removal and installation procedures for this component. The following procedure is a guideline and may differ from the vehicle you are servicing.

1. Before servicing the vehicle, refer to the Precautions Section.

➡ **If working near and/or around the SRS system and components, be sure to disable the SRS system. After disabling the system wait three minutes or more before servicing the vehicle.**

➡ **Whenever the negative battery cable is disconnected the following components will require resetting. The air conditioning system, automatic drive positioner system, power window control system, around view monitor, automatic back door system, idle air volume learning, steering angle sensor neutral position, sunroof system, audio visual and navigation systems. Use the CONSULT-III diagnostic tool, or equivalent to perform the required resets. Follow the directions on the screen of the tool, as needed.**

➡ **Before disconnecting the negative battery cable lower both the driver's side and passenger's side front windows. This will prevent any interference between the window edge and the vehicle when the door is opener/closed. The automatic window function will not work with the battery cable disconnected.**

Use the CONSULT III diagnostic tool, or equivalent, to be sure that the transmission fluid temperature is below 104 F, before changing.

2. Disconnect the negative battery cable.
3. Stop the engine.
4. Raise and support the vehicle safely.
5. Loosen the drain plug.

➡ **Always replace the drain plug and drain plug gasket with a new one.**

6. Drain the fluid from the drain plug and refill with new fluid. Always refill same volume with drained fluid.

➡ To flush the old fluid from the oil coolers, pour new fluid into the charging pipe with the engine idling and at the same time drain the old fluid from the auxiliary cooler hose return line. When the color of the fluid coming out is about the same color as new fluid, the procedure is complete. The amount of new fluid used for flushing should be about 30–50 percent increase of the specified capacity.

7. Install the fluid level gauge.
8. Drive the vehicle until operating temperature is reached.
9. Check and correct the fluid level.

To install:

➡ **Be sure to use new fasteners, as required.**

10. Installation is the reverse of the removal procedure.
11. Be sure to fill the transmission with the proper grade and type transmission fluid.
12. Be sure to perform the reconnect/relearn procedures.

FLUID LEVEL CHECK

The automatic transmission uses Nissan approved Matic S ATF transmission fluid. If information differs from the owner's manual, use the data in the owner's manual.

1. Before servicing the vehicle, refer to the Precautions Section.

➡ **If working near and/or around the SRS system and components, be sure to disable the SRS system. After disabling the system wait three minutes or more before servicing the vehicle.**

➡ **Whenever the negative battery cable is disconnected the following components will require resetting. The air conditioning system, automatic drive positioner system, power window control system, around view monitor, automatic back door system, idle air volume learning, steering angle sensor neutral position, sunroof system, audio visual and navigation systems. Use the CONSULT-III diagnostic tool, or equivalent to perform the required resets. Follow the directions on the screen of the tool, as needed.**

➡ **Before disconnecting the negative battery cable lower both the driver's side and passenger's side front windows. This will prevent any interfer-**

ence between the window edge and the vehicle when the door is opener/closed. The automatic window function will not work with the battery cable disconnected.

2. Run the engine until operating temperature is reached.
3. Check for external fluid leakage.
4. Loosen the level gauge bolt.
5. Before driving, fluid level can be checked at fluid temperatures of 86–122 degrees F using the COLD range on the dipstick.
6. To COLD check the transmission, park the vehicle on a level surface and set the parking brake.
7. Start the engine and move the selector lever through each gear range. Leave the selector lever in the PARK (P) range.
8. Check the fluid level with the engine idling. Always use lint free paper, not a cloth to check the fluid level.
9. Re-insert the gauge. Remove the gauge and note the reading. As required add fluid. Do not overfill.
10. Drive the vehicle for about five minutes in urban areas.
11. Using the CONSULT-III tool, or equivalent, make the fluid temperature about 176 degrees F.
12. Recheck the fluid level, correct as required.

DRIVESHAFT

REMOVAL & INSTALLATION

Front

See Figure 16.

1. Before servicing the vehicle, refer to the Precautions Section.

➡ **If working near and/or around the SRS system and components, be sure to disable the SRS system. After disabling the system wait three minutes or more before servicing the vehicle.**

➡ **Whenever the negative battery cable is disconnected the following components will require resetting. The air conditioning system, automatic drive positioner system, power window control system, around view monitor, automatic back door system, idle air volume learning, steering angle sensor neutral position, sunroof system, audio visual and navigation systems. Use the CONSULT-III diagnostic tool, or** equivalent to perform the required resets. Follow the directions on the screen of the tool, as needed.

➡Before disconnecting the negative battery cable lower both the driver's side and passenger's side front windows. This will prevent any interference between the window edge and the vehicle when the door is opener/closed. The automatic window function will not work with the battery cable disconnected.

2. Disconnect the negative battery cable.

3. Position the selector lever in the N position.

4. Raise and support the vehicle safely. Remove the engine under cover, if equipped and as required.

5. Remove the air suspension protectors.

6. Remove the front suspension rear crossmember.

7. Matchmark the driveshaft and companion flange.

8. Remove the driveshaft. Discard the nuts.

To install:

➡Be sure to use new fasteners, as required.

9. Installation is the reverse of the removal procedure.

10. Check for vehicle vibration, correct as required.

11. Be sure to perform the reconnect/relearn procedures.

Rear

See Figures 17 and 18.

1. Before servicing the vehicle, refer to the Precautions Section.

➡If working near and/or around the SRS system and components, be sure to disable the SRS system. After disabling the system wait three minutes or more before servicing the vehicle.

➡Whenever the negative battery cable is disconnected the following components will require resetting. The air conditioning system, automatic drive positioner system, power window control system, around view monitor, automatic back door system, idle air volume learning, steering angle sensor neutral position, sunroof system, audio visual and navigation systems. Use the CONSULT-III diagnostic tool, or equivalent to perform the required resets.

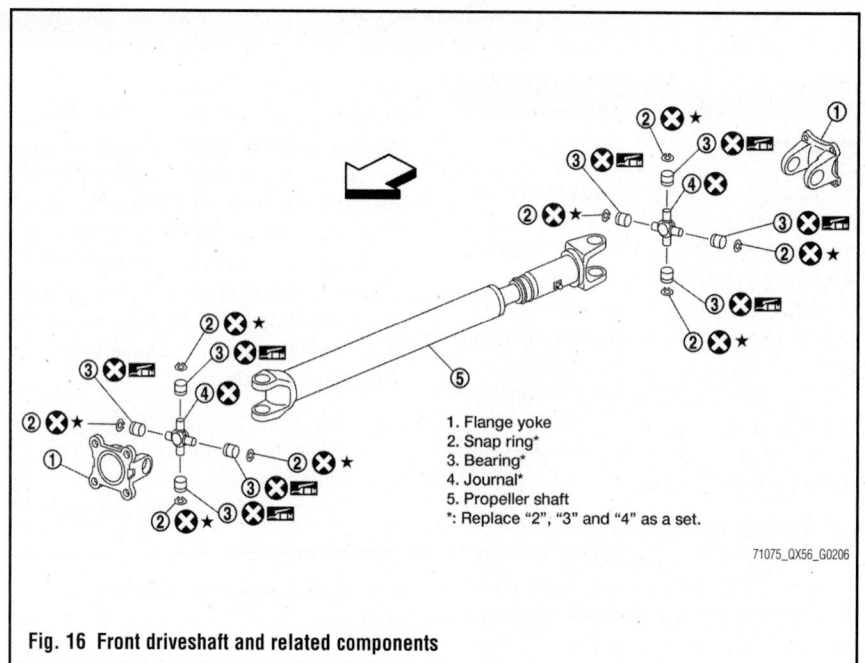

1. Flange yoke
2. Snap ring*
3. Bearing*
4. Journal*
5. Propeller shaft
*: Replace "2", "3" and "4" as a set.

71075_QX56_G0206

Fig. 16 Front driveshaft and related components

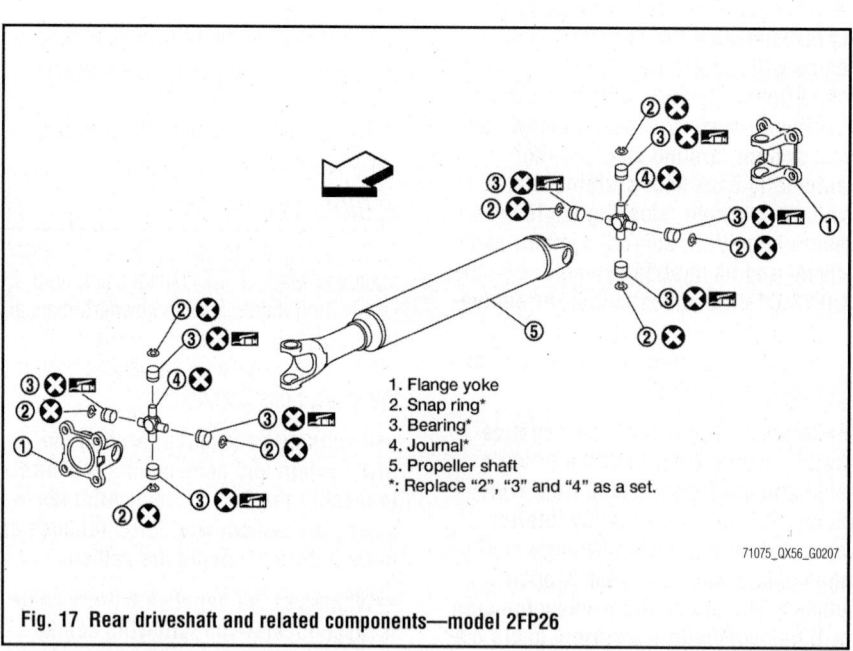

1. Flange yoke
2. Snap ring*
3. Bearing*
4. Journal*
5. Propeller shaft
*: Replace "2", "3" and "4" as a set.

71075_QX56_G0207

Fig. 17 Rear driveshaft and related components—model 2FP26

Follow the directions on the screen of the tool, as needed.

➡Before disconnecting the negative battery cable lower both the driver's side and passenger's side front windows. This will prevent any interference between the window edge and the vehicle when the door is opener/closed. The automatic window function will not work with the battery cable disconnected.

2. Disconnect the negative battery cable.

3. Position the selector lever in the N position.

4. Release the parking brake.

5. Raise and support the vehicle safely. Remove the engine under cover, if equipped and as required.

6. Matchmark the driveshaft and companion flange.

7. Remove the driveshaft. Discard the nuts.

To install:

➡Be sure to use new fasteners, as required.

8. Installation is the reverse of the removal procedure.

9. Check for vehicle vibration, correct as required.

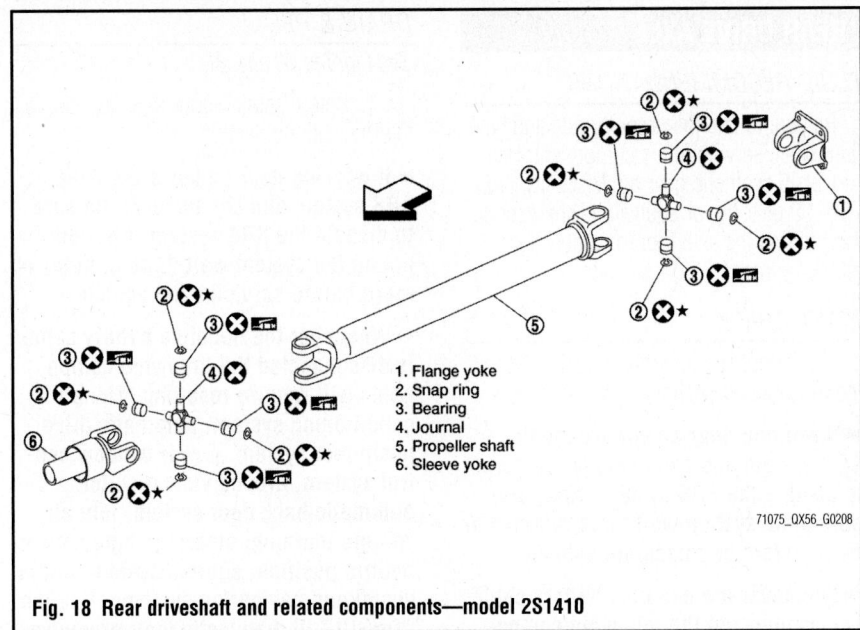

1. Flange yoke
2. Snap ring
3. Bearing
4. Journal
5. Propeller shaft
6. Sleeve yoke

71075_QX56_G0208

Fig. 18 Rear driveshaft and related components—model 2S1410

10. Be sure to perform the reconnect/relearn procedures.

DRIVESHAFT ANGLE MEASUREMENT

1. Before servicing the vehicle, refer to the Precautions Section.

➡If working near and/or around the SRS system and components, be sure to disable the SRS system. After disabling the system wait three minutes or more before servicing the vehicle.

➡Whenever the negative battery cable is disconnected the following components will require resetting. The air conditioning system, automatic drive positioner system, power window control system, around view monitor, automatic back door system, idle air volume learning, steering angle sensor neutral position, sunroof system, audio visual and navigation systems. Use the CONSULT-III diagnostic tool, or equivalent to perform the required resets. Follow the directions on the screen of the tool, as needed.

➡Before disconnecting the negative battery cable lower both the driver's side and passenger's side front windows. This will prevent any interference between the window edge and the vehicle when the door is opener/closed. The automatic window function will not work with the battery cable disconnected.

2. Disconnect the negative battery cable.
3. Raise and support the vehicle safely.

4. Matchmark the driveshaft and companion flange.
5. Remove the driveshaft. Discard the nuts.
6. Inspect the driveshaft runout, replace the driveshaft as necessary. Runout specification should be 0.004 inch (1.00mm) for 2FP26, and 0.0402 inch (1.02mm) for 2S1410.
7. While holding the flange yoke on one side, check the axial play of the joint. Journal axial play should be 0.0000 inch (0.00mm) for 2FP26 and 0.0008 inch (0.02mm) for 2S1410, repair or replace the journal parts.
8. Check the driveshaft for dents or cracks. If damage is detected, replace the driveshaft assembly.

To install:

➡Be sure to use new fasteners, as required.

9. Installation is the reverse of the removal procedure.
10. Check for vehicle vibration, correct as required.

❈❈ WARNING

Do not reuse the bolts and nuts. Always install new ones.

11. Be sure to perform the reconnect/relearn procedures.

FRONT DRIVE AXLE

FLUID RECOMMENDATIONS

Be sure to use the proper grade and type fluid when servicing this component. Use

API GL-5 viscosity SAE 80W-90 gear oil, or equivalent when servicing/refilling. Approximate capacity is 1 5/8 pints (0.75L).

LEVEL CHECK

See Figure 19.

1. Before servicing the vehicle, refer to the Precautions Section.

➡If working near and/or around the SRS system and components, be sure to disable the SRS system. After disabling the system wait three minutes or more before servicing the vehicle.

➡Whenever the negative battery cable is disconnected the following components will require resetting. The air conditioning system, automatic drive positioner system, power window control system, around view monitor, automatic back door system, idle air volume learning, steering angle sensor neutral position, sunroof system, audio visual and navigation systems. Use the CONSULT-III diagnostic tool, or equivalent to perform the required resets. Follow the directions on the screen of the tool, as needed.

➡Before disconnecting the negative battery cable lower both the driver's side and passenger's side front windows. This will prevent any interference between the window edge and the vehicle when the door is opener/closed. The automatic window function will not work with the battery cable disconnected.

2. Disconnect the negative battery cable.
3. Raise and support the vehicle safely, as required.
4. Position a catch pan under the assembly.

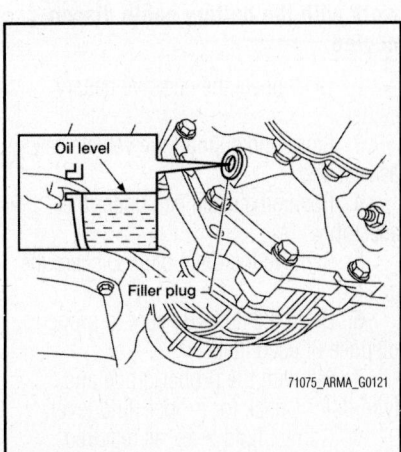

71075_ARMA_G0121

Fig. 19 Front drive axle plug location—checking

5. Remove the filler plug. Discard the gasket.

6. Check for proper fluid level, see illustration.

7. Correct fluid level, as required.

8. Install the filler plug, using a new gasket.

9. Tighten filler plug to 26 ft. lbs. (35 Nm).

DRAIN & REFILL

1. Before servicing the vehicle, refer to the Precautions Section.

➡ If working near and/or around the SRS system and components, be sure to disable the SRS system. After disabling the system wait three minutes or more before servicing the vehicle.

➡ Whenever the negative battery cable is disconnected the following components will require resetting. The air conditioning system, automatic drive positioner system, power window control system, around view monitor, automatic back door system, idle air volume learning, steering angle sensor neutral position, sunroof system, audio visual and navigation systems. Use the CONSULT-III diagnostic tool, or equivalent to perform the required resets. Follow the directions on the screen of the tool, as needed.

➡ Before disconnecting the negative battery cable lower both the driver's side and passenger's side front windows. This will prevent any interference between the window edge and the vehicle when the door is opener/closed. The automatic window function will not work with the battery cable disconnected.

2. Disconnect the negative battery cable.

3. Raise and support the vehicle safely, as required.

4. Position a catch pan under the assembly.

5. Remove the drain plug. Discard the gasket.

6. Drain the fluid. Be sure to properly dispose of used fluid.

7. Fill with the proper grade and type fluid. Check for proper fluid level.

8. Correct fluid level, as required.

9. Install the, using a new gasket.

10. Tighten checking plug to 26 ft. lbs. (35 Nm).

REAR DRIVE AXLE

FLUID RECOMMENDATIONS

Be sure to use the proper grade and type fluid when servicing this component. Use API GL-5 synthetic gear oil, Viscosity SAE 75W-90 gear oil, or equivalent when servicing/refilling the axle. Approximate capacity is 3 3.4 pints (1.75L).

LEVEL CHECK

1. Before servicing the vehicle, refer to the Precautions Section.

➡ If working near and/or around the SRS system and components, be sure to disable the SRS system. After disabling the system wait three minutes or more before servicing the vehicle.

➡ Whenever the negative battery cable is disconnected the following components will require resetting. The air conditioning system, automatic drive positioner system, power window control system, around view monitor, automatic back door system, idle air volume learning, steering angle sensor neutral position, sunroof system, audio visual and navigation systems. Use the CONSULT-III diagnostic tool, or equivalent to perform the required resets. Follow the directions on the screen of the tool, as needed.

➡ Before disconnecting the negative battery cable lower both the driver's side and passenger's side front windows. This will prevent any interference between the window edge and the vehicle when the door is opener/closed. The automatic window function will not work with the battery cable disconnected.

2. Disconnect the negative battery cable.

3. Raise and safely support the vehicle.

4. Position a catch pan under the assembly.

5. Remove the drain plug.

6. Discard the gasket.

7. Check the oil level, correct as required.

➡ Be sure to use the proper grade and type gear oil.

8. Check the oil level after refilling.

9. Using a new gasket install the plug. Tighten to specification: 26 ft. lbs. (35 Nm).

10. Be sure to perform the reconnect/relearn procedures.

DRAIN & REFILL

See Figures 20 and 21.

1. Before servicing the vehicle, refer to the Precautions Section.

➡ If working near and/or around the SRS system and components, be sure to disable the SRS system. After disabling the system wait three minutes or more before servicing the vehicle.

➡ Whenever the negative battery cable is disconnected the following components will require resetting. The air conditioning system, automatic drive positioner system, power window control system, around view monitor, automatic back door system, idle air volume learning, steering angle sensor neutral position, sunroof system, audio visual and navigation systems. Use the CONSULT-III diagnostic tool, or equivalent to perform the required resets. Follow the directions on the screen of the tool, as needed.

➡ Before disconnecting the negative battery cable lower both the driver's side and passenger's side front windows. This will prevent any interference between the window edge and the vehicle when the door is opener/closed. The automatic window function will not work with the battery cable disconnected.

2. Disconnect the negative battery cable.

3. Raise and safely support the vehicle.

4. Remove the drain plug and drain the fluid into a suitable container.

5. Discard the gasket.

6. Fill the unit with new gear oil until the oil level reaches the specified level near the filler plug mounting hole.

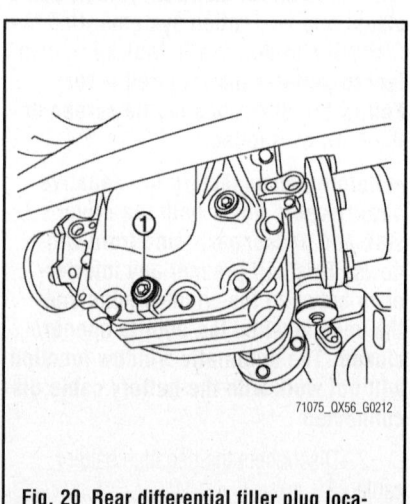

71075_QX56_G0212

Fig. 20 Rear differential filler plug location (1) plug—draining

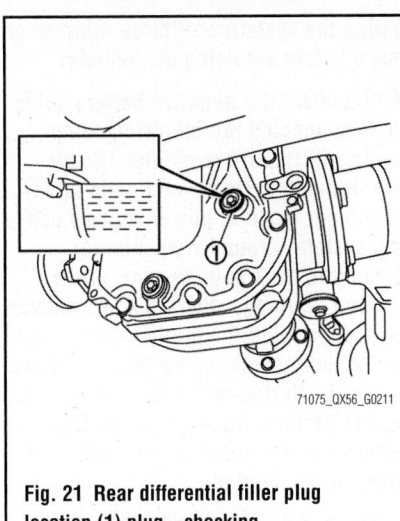

71075_QX56_G0211

Fig. 21 Rear differential filler plug location (1) plug—checking

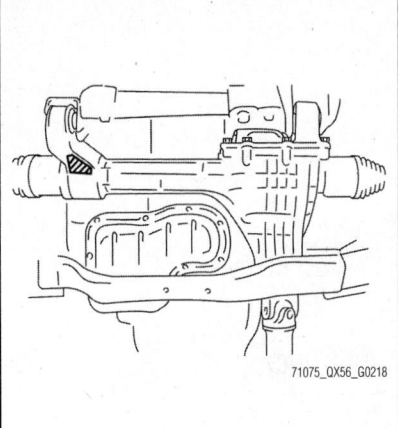

71075_QX56_G0218

Fig. 22 Front drive axle identification data location point

To install:

→Be sure to use new fasteners, as required.

7. Installation is the reverse of the removal procedure.

8. Using a new gasket install the drain plug. Tighten to specification: 26 ft. lbs. (35 Nm).

9. Check the oil level after refilling.

→Be sure to use the proper grade and type gear oil.

10. Be sure to perform the reconnect/relearn procedures.

FRONT PINION OIL SEAL

REMOVAL & INSTALLATION

See Figure 22.

1. Before servicing the vehicle, refer to the Precautions Section.

→If working near and/or around the SRS system and components, be sure to disable the SRS system. After disabling the system wait three minutes or more before servicing the vehicle.

→Whenever the negative battery cable is disconnected the following components will require resetting. The air conditioning system, automatic drive positioner system, power window control system, around view monitor, automatic back door system, idle air volume learning, steering angle sensor neutral position, sunroof system, audio visual and navigation systems. Use the CONSULT-III diagnostic tool, or equivalent to perform the required resets. Follow the directions on the screen of the tool, as needed.

→Before disconnecting the negative battery cable lower both the driver's side and passenger's side front windows. This will prevent any interference between the window edge and the vehicle when the door is opener/closed. The automatic window function will not work with the battery cable disconnected.

→Verify identification stamp of replacement frequency put in the lower part of the gear carrier to determine replacement for collapsible spacer when replacing the front oil seal. If collapsible spacer replacement is necessary, remove the final drive assembly and disassemble it to replace the seal and spacer. The reuse of the spacer is prohibited in principle, however it is reusable on a one time basis only in cases when replacing the front oil seal. See illustration (diagonally shaded area). No stamp: spacer replacement not required. 0 on far right of the stamp spacer: replacement required.)1 or 1 on the far right of the stamp: no spacer replacement required.

2. Disconnect the negative battery cable.

3. Raise and safely support the vehicle.

4. Drain the gear oil. Be sure to properly dispose of used oil.

5. Remove the halfshafts.

6. Remove the driveshaft.

7. Measure and record total preload.

8. Matchmark the end of the drive pinion and the companion flange.

9. Remove the retaining nut. Discard the nut. Always use a new one.

10. Remove the companion flange using the proper puller.

11. Remove the seal using a seal removal tool.

To install:

→Be sure to use new fasteners, as required.

12. Installation is the reverse of the removal procedure.

13. Be sure to perform the reconnect/relearn procedures.

FRONT HALFSHAFT

REMOVAL & INSTALLATION

See Figure 23.

1. Before servicing the vehicle, refer to the Precautions Section.

→If working near and/or around the SRS system and components, be sure to disable the SRS system. After disabling the system wait three minutes or more before servicing the vehicle.

→Whenever the negative battery cable is disconnected the following components will require resetting. The air conditioning system, automatic drive positioner system, power window control system, around view monitor, automatic back door system, idle air volume learning, steering angle sensor neutral position, sunroof system, audio visual and navigation systems. Use the CONSULT-III diagnostic tool, or equivalent to perform the required resets. Follow the directions on the screen of the tool, as needed.

→Before disconnecting the negative battery cable lower both the driver's side and passenger's side front windows. This will prevent any interference between the window edge and the vehicle when the door is opener/closed. The automatic window function will not work with the battery cable disconnected.

2. Disconnect the negative battery cable.

3. Raise and safely support the vehicle.

4. Remove the tire and wheel assembly.

5. Remove the caliper mounting bolts and position it to the side. Do not allow it to hang by the brake hose.

6. Remove the rotor.

7. Remove the wheel speed sensor harness. Remove the height sensor from the upper link, right side.

8. Remove the steering outer socket.

9. Position a suitable jack under the lower link.

10. Remove the shock lower mounting bolt.

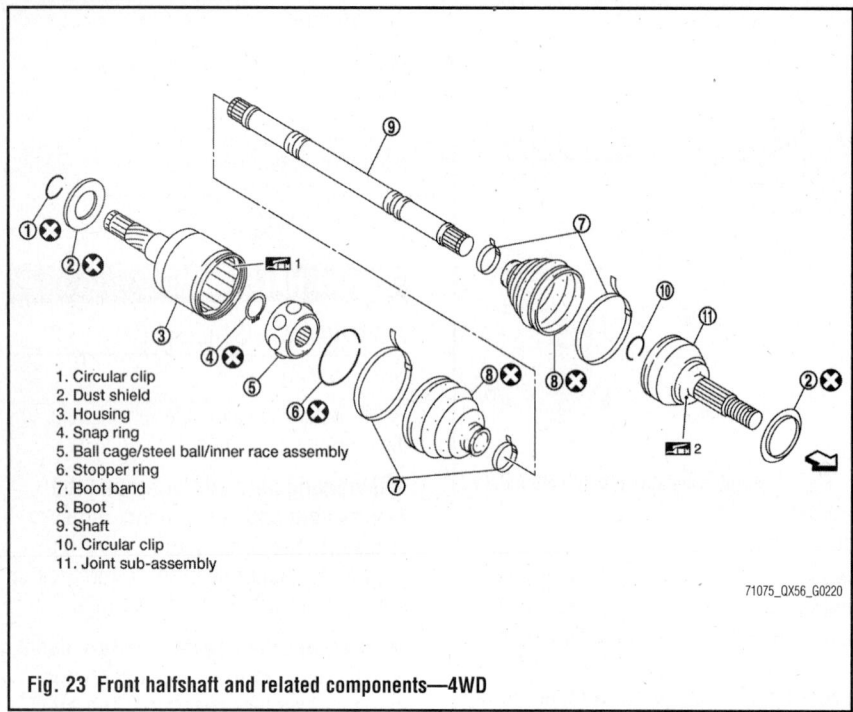

Fig. 23 Front halfshaft and related components—4WD

1. Circular clip
2. Dust shield
3. Housing
4. Snap ring
5. Ball cage/steel ball/inner race assembly
6. Stopper ring
7. Boot band
8. Boot
9. Shaft
10. Circular clip
11. Joint sub-assembly

71075_QX56_G0220

11. Separate the upper link from the steering knuckle, using the proper ball joint removal tool.

✳✳ CAUTION

Temporarily tighten the nut to prevent damage to the threads and to prevent the tool from coming off.

12. Remove the cotter pin and loosen the wheel hub locknut.

13. Tap the wheel hub locknut using a piece of wood. Hammer the wood to disengage the wheel hub and bearing assembly from the halfshaft.

➡**Never place the halfshaft joint at an extreme angle. Also be careful not to overextend the slide joint. Never allow the halfshaft to hang without being supported.**

14. Remove the wheel hub locknut.
15. Remove the fender protector.
16. Remove the halfshaft from the final drive unit using tool SST:KV40107500 and a sliding hammer.

To install:

➡**Be sure to use new fasteners, as required.**

17. Installation is the reverse of the removal procedure.
18. Always replace the final drive oil seal.
19. Be sure that the circlip is fully engaged.
20. Tighten the wheel hub locknut to 87–90 ft. lbs. (118-122 Nm).

21. Be sure to perform the reconnect/relearn procedures.

REAR HALFSHAFTS

REMOVAL & INSTALLATION

See Figure 24.

1. Before servicing the vehicle, refer to the Precautions Section.

➡**If working near and/or around the SRS system and components, be sure to disable the SRS system. After dis-**

abling the system wait three minutes or more before servicing the vehicle.

➡**Whenever the negative battery cable is disconnected the following components will require resetting. The air conditioning system, automatic drive positioner system, power window control system, around view monitor, automatic back door system, idle air volume learning, steering angle sensor neutral position, sunroof system, audio visual and navigation systems. Use the CONSULT-III diagnostic tool, or equivalent to perform the required resets. Follow the directions on the screen of the tool, as needed.**

➡**Before disconnecting the negative battery cable lower both the driver's side and passenger's side front windows. This will prevent any interference between the window edge and the vehicle when the door is opener/closed. The automatic window function will not work with the battery cable disconnected.**

2. Disconnect the negative battery cable.
3. Raise and support the vehicle safely.
4. Remove the tire and wheel assembly.
5. Remove the parking brake cable from the suspension member.
6. Remove the wheel height sensor harness.
7. Remove the height sensor from the lower link, both sides.
8. Position a suitable jack under the lower link, axle housing side.
9. Remove the coil spring.

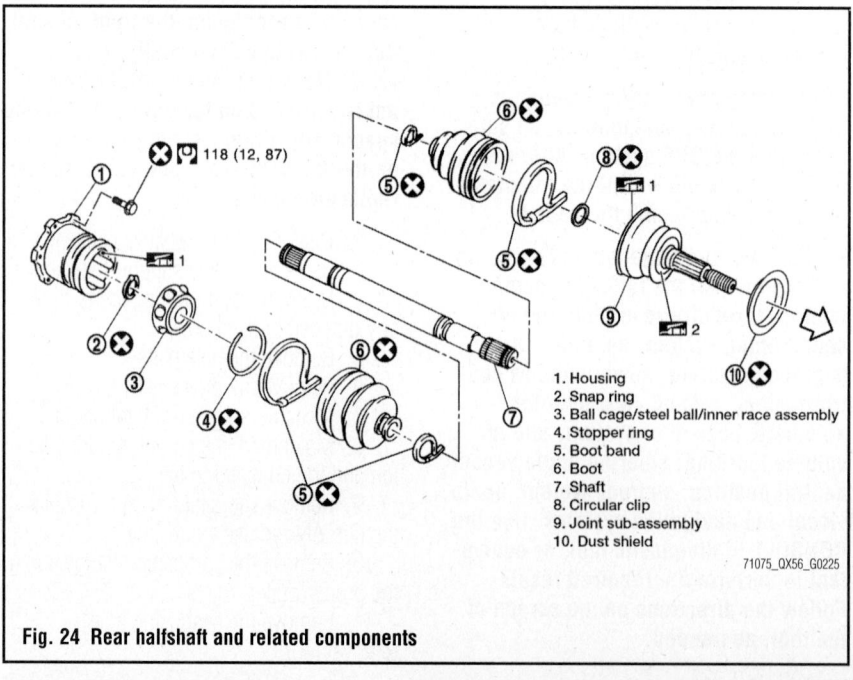

1. Housing
2. Snap ring
3. Ball cage/steel ball/inner race assembly
4. Stopper ring
5. Boot band
6. Boot
7. Shaft
8. Circular clip
9. Joint sub-assembly
10. Dust shield

71075_QX56_G0225

Fig. 24 Rear halfshaft and related components

10. Remove the cotter pin and adjusting cap. Loosen the wheel hub locknut.

11. Matchmark the halfshaft and wheel hub and bearing assembly.

12. Tap the wheel hub locknut using a piece of wood. Hammer the wood to disengage the wheel hub and bearing assembly from the halfshaft.

➡**Never place the halfshaft joint at an extreme angle. Also be careful not to overextend the slide joint. Never allow the halfshaft to hang without being supported.**

13. Remove the wheel hub locknut.

14. Remove the mounting bolts between the side flange and the halfshaft.

15. Remove the halfshaft from the vehicle.

To install:

➡**Be sure to use new fasteners, as required.**

16. Installation is the reverse of the removal procedure.

17. Clean the mating surfaces of the wheel hub locknut and wheel bearing and hub assembly.

➡**Never apply lubricating oil to these surfaces.**

18. Clean the mating surfaces of the halfshaft and wheel hub and bearing assembly. Apply paste, about 0.10 ounce, (part number 440037S000) to the surface of the sub-assembly of the halfshaft.

➡**Apply paste to cover the entire flat surface of the joint sub-assembly.**

19. Tighten the wheel hub locknut to 161–164 ft. lbs. (217–221 Nm).

20. Be sure to perform the reconnect/relearn procedures.

REAR PINION OIL SEAL

REMOVAL & INSTALLATION
See Figures 25 and 26.

1. Before servicing the vehicle, refer to the Precautions Section.

➡**If working near and/or around the SRS system and components, be sure to disable the SRS system. After disabling the system wait three minutes or more before servicing the vehicle.**

➡**Whenever the negative battery cable is disconnected the following components will require resetting. The air conditioning system, automatic drive positioner system, power window con-**

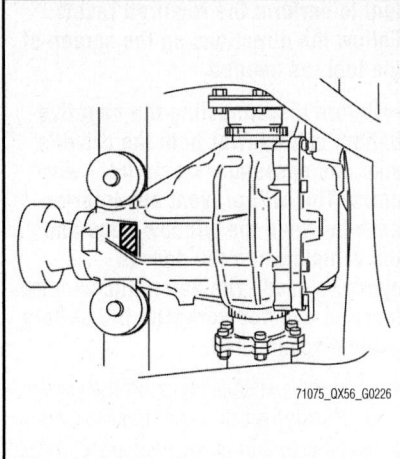

Fig. 25 Rear drive axle identification data location point

trol system, around view monitor, automatic back door system, idle air volume learning, steering angle sensor neutral position, sunroof system, audio visual and navigation systems. Use the CONSULT-III diagnostic tool, or equivalent to perform the required resets. Follow the directions on the screen of the tool, as needed.

➡**Before disconnecting the negative battery cable lower both the driver's side and passenger's side front windows. This will prevent any interference between the window edge and the vehicle when the door is opener/closed. The automatic window function**

will not work with the battery cable disconnected.

➡**Verify identification stamp of replacement frequency put in the lower part of the gear carrier to determine replacement for collapsible spacer when replacing the front oil seal. If collapsible spacer replacement is necessary, remove the final drive assembly and disassemble it to replace the seal and spacer. The reuse of the spacer is prohibited in principle, however it is reusable on a one time basis only in cases when replacing the front oil seal. See illustration (diagonally shaded area). No stamp: spacer replacement not required. 0 on far right of the stamp spacer: replacement required.)1 or 1 on the far right of the stamp: no spacer replacement required.**

2. Disconnect the negative battery cable.

3. Raise and safely support the vehicle.

4. Drain the gear oil. Be sure to properly dispose of used oil.

5. Remove the halfshafts.

6. Remove the driveshaft.

7. Measure and record total preload.

8. Matchmark the end of the drive pinion and the companion flange.

9. Remove the retaining nut. Discard the nut. Always use a new one.

10. Remove the companion flange using the proper puller.

11. Remove the seal using a seal removal tool.

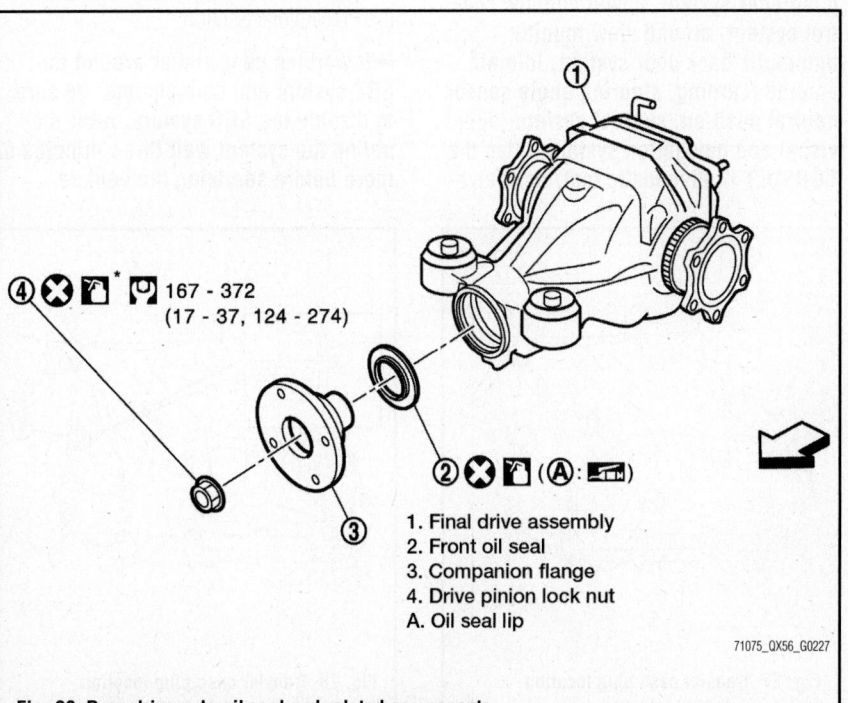

4 ⊗ 📦* 🔧 167 - 372
(17 - 37, 124 - 274)

2 ⊗ 📦 (Ⓐ: ▭)

1. Final drive assembly
2. Front oil seal
3. Companion flange
4. Drive pinion lock nut
A. Oil seal lip

Fig. 26 Rear drive axle oil seal and related components

To install:

➡Be sure to use new fasteners, as required.

12. Installation is the reverse of the removal procedure.

➡Check and adjust bearing preload. If preload torque exceeds the specified value, replace the collapsible spacer and tighten it again to adjust. Never loosen the drive pinion locknut to adjust bearing preload.

13. Be sure to perform the reconnect/relearn procedures.

TRANSFER CASE

DRAIN & REFILL
See Figure 27.

The transfer case uses Nissan transfer case fluid ATX90A. If information differs from the owner's manual, use the data in the owner's manual.

1. Before servicing the vehicle, refer to the Precautions Section.

➡If working near and/or around the SRS system and components, be sure to disable the SRS system. After disabling the system wait three minutes or more before servicing the vehicle.

➡Whenever the negative battery cable is disconnected the following components will require resetting. The air conditioning system, automatic drive positioner system, power window control system, around view monitor, automatic back door system, idle air volume learning, steering angle sensor neutral position, sunroof system, audio visual and navigation systems. Use the CONSULT-III diagnostic tool, or equiva-

lent to perform the required resets. Follow the directions on the screen of the tool, as needed.

➡Before disconnecting the negative battery cable lower both the driver's side and passenger's side front windows. This will prevent any interference between the window edge and the vehicle when the door is opener/closed. The automatic window function will not work with the battery cable disconnected.

2. Raise and safely support the vehicle.
3. Position a catch pan under the component. Be sure to properly dispose of used fluid.
4. Remove the transfer case drain plug. Drain the fluid. Discard the gasket.
5. Reinstall the drain plug using a new gasket.
6. Refill the unit with the proper grade and type fluid.
7. Check oil level and correct as required. Set a new gasket to the filler plug, then install it to the transfer case. Tighten the filler plug to 35 ft. lbs. (48 Nm).

➡Do not reuse the filler plug gasket.

FLUID LEVEL CHECK
See Figure 28.

The transfer case uses Nissan transfer case fluid ATX90A. If information differs from the owner's manual, use the data in the owner's manual.

1. Before servicing the vehicle, refer to the Precautions Section.

➡If working near and/or around the SRS system and components, be sure to disable the SRS system. After disabling the system wait three minutes or more before servicing the vehicle.

➡Whenever the negative battery cable is disconnected the following components will require resetting. The air conditioning system, automatic drive positioner system, power window control system, around view monitor, automatic back door system, idle air volume learning, steering angle sensor neutral position, sunroof system, audio visual and navigation systems. Use the CONSULT-III diagnostic tool, or equivalent to perform the required resets. Follow the directions on the screen of the tool, as needed.

➡Before disconnecting the negative battery cable lower both the driver's side and passenger's side front windows. This will prevent any interference between the window edge and the vehicle when the door is opener/closed. The automatic window function will not work with the battery cable disconnected.

2. Disconnect the negative battery cable.
3. Raise and safely support the vehicle.
4. Remove the transfer case checking plug. Discard the gasket.
5. Check oil level and correct as required. Set a new gasket to the filler plug, then install it to the transfer case. Tighten the filler plug to 35 ft. lbs. (48 Nm).

➡Do not reuse the filler plug gasket.

REMOVAL & INSTALLATION
See Figures 29 through 32.

1. Before servicing the vehicle, refer to the Precautions Section.

➡If working near and/or around the SRS system and components, be sure to disable the SRS system. After disabling the system wait three minutes or more before servicing the vehicle.

➡Whenever the negative battery cable is disconnected the following components will require resetting. The air conditioning system, automatic drive positioner system, power window control system, around view monitor, automatic back door system, idle air volume learning, steering angle sensor neutral position, sunroof system, audio visual and navigation systems. Use the CONSULT-III diagnostic tool, or equivalent to perform the required resets. Follow the directions on the screen of the tool, as needed.

➡Before disconnecting the negative battery cable lower both the driver's side and passenger's side front win-

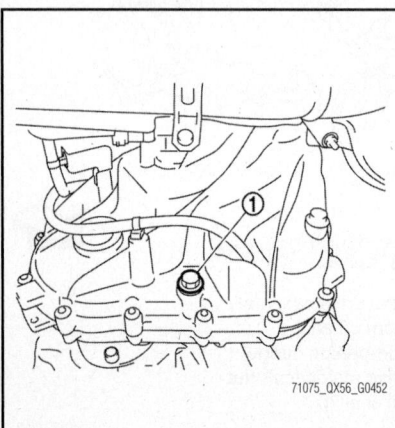

71075_QX56_G0452

**Fig. 27 Transfer case plug location
(1) plug—draining**

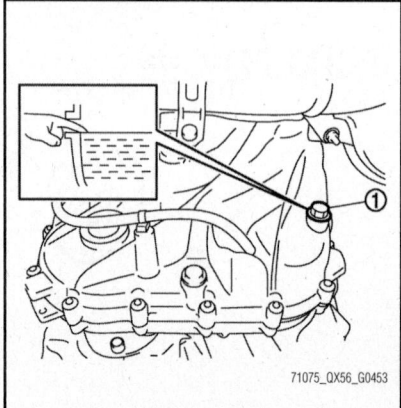

71075_QX56_G0453

**Fig. 28 Transfer case plug location
(1) plug—checking**

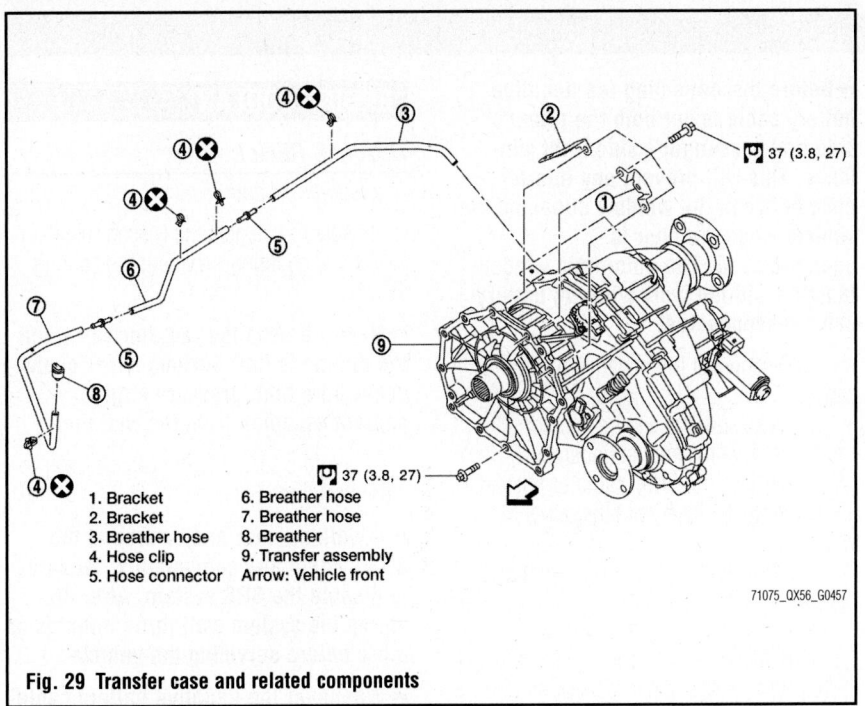

1. Bracket
2. Bracket
3. Breather hose
4. Hose clip
5. Hose connector
6. Breather hose
7. Breather hose
8. Breather
9. Transfer assembly
Arrow: Vehicle front

71075_QX56_G0457

Fig. 29 Transfer case and related components

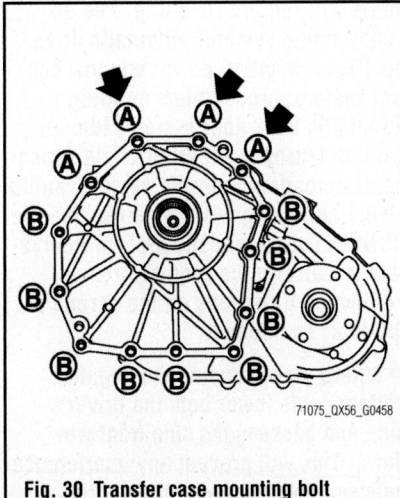

71075_QX56_G0458

Fig. 30 Transfer case mounting bolt locations (A) transfer to transmission, (B) transmission to transfer

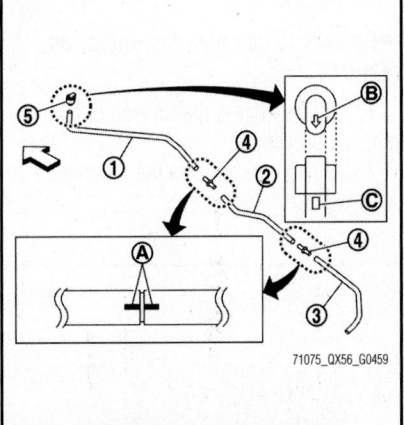

71075_QX56_G0459

Fig. 31 Transfer case breather hose installation (A) paint marks, (B) arrow, (C) mark, (1), (2) and (3) hoses, (4) connectors— 1 of 2

dows. This will prevent any interference between the window edge and the vehicle when the door is opener/closed. The automatic window function will not work with the battery cable disconnected.

2. Disconnect the negative battery cable.
3. Raise and support the vehicle safely.
4. Drain the transfer case. Be sure to properly dispose of used fluid.
5. Remove the rear driveshaft.
6. Remove the front driveshaft.
7. Disconnect the electrical connectors.
8. Remove the breather hose.
9. Remove the front right and left exhaust tubes.
10. Remove the main muffler.
11. Properly support the transfer case assembly using a service jack.
12. Remove the rear engine mounting member and engine mounting insulator.
13. Lower the jack to a position where the top transfer mounting bolts can be removed.
14. Remove the bolts and separate the transfer case from the transmission.
15. With the component securely secured to the service jack, remove it from the vehicle.

To install:

➡Be sure to use new fasteners, as required.

16. Installation is the reverse of the removal procedure.
17. Be sure to install the retaining bolts in the proper location. See illustration.

➡Be sure to install the breather hose correctly. See illustrations. Dimension D: 1.035 inch. Dimension E: 4.791 inch. Dimension F: 0.984 inch. Dimension G: 5.630 inch.

18. Be sure to perform the reconnect/relearn procedures.

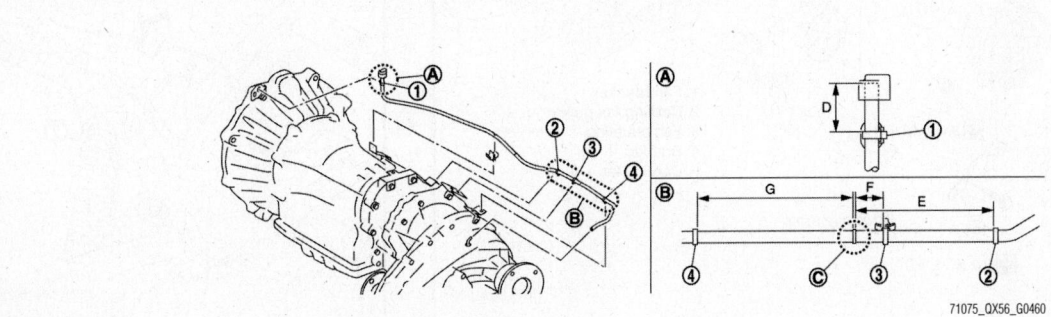

71075_QX56_G0460

Fig. 32 Transfer case breather hose installation (1) clip, (2), (3) and (4) clips, (C) joint—2 of 2

ENGINE COOLING

BELT-DRIVEN ENGINE FAN

REMOVAL & INSTALLATION

See Figure 33.

➡Never remove the radiator cap when the engine is hot. Serious burns could occur from high-pressure engine coolant escaping from the radiator.

1. Before servicing the vehicle, refer to the Precautions Section.

➡If working near and/or around the SRS system and components, be sure to disable the SRS system. After disabling the system wait three minutes or more before servicing the vehicle.

➡Whenever the negative battery cable is disconnected the following components will require resetting. The air conditioning system, automatic drive positioner system, power window control system, around view monitor, automatic back door system, idle air volume learning, steering angle sensor neutral position, sunroof system, audio visual and navigation systems. Use the CONSULT-III diagnostic tool, or equivalent to perform the required resets. Follow the directions on the screen of the tool, as needed.

➡Before disconnecting the negative battery cable lower both the driver's side and passenger's side front windows. This will prevent any interference between the window edge and the vehicle when the door is opener/closed. The automatic window function will not work with the battery cable disconnected.

2. Disconnect the negative battery cable.
3. Make sure the engine is cold.
4. Remove the engine undercover.
5. Remove the lower radiator shroud.
6. Move the reservoir tank out of the way.
7. Remove the radiator core support cover.
8. Remove the engine cover.
9. Remove the air cleaner assembly.
10. Remove the upper radiator shroud.
11. Remove the cooling fan.

To install:

➡Be sure to use new fasteners, as required.

12. Installation is the reverse of the removal procedure.
13. Be sure to perform the reconnect/relearn procedures.

ENGINE COOLANT

DRAIN & REFILL

See Figures 34 and 35.

Be sure to use genuine Nissan (blue) coolant when filling/servicing the cooling system.

➡Never remove the radiator cap when the engine is hot. Serious burns could occur from high-pressure engine coolant escaping from the radiator.

1. Before servicing the vehicle, refer to the Precautions Section.

➡If working near and/or around the SRS system and components, be sure to disable the SRS system. After disabling the system wait three minutes or more before servicing the vehicle.

➡Whenever the negative battery cable is disconnected the following components will require resetting. The air conditioning system, automatic drive positioner system, power window control system, around view monitor, automatic back door system, idle air volume learning, steering angle sensor neutral position, sunroof system, audio visual and navigation systems. Use the CONSULT-III diagnostic tool, or equivalent to perform the required resets. Follow the directions on the screen of the tool, as needed.

➡Before disconnecting the negative battery cable lower both the driver's side and passenger's side front windows. This will prevent any interference between the window edge and the vehicle when the door is opener/closed.

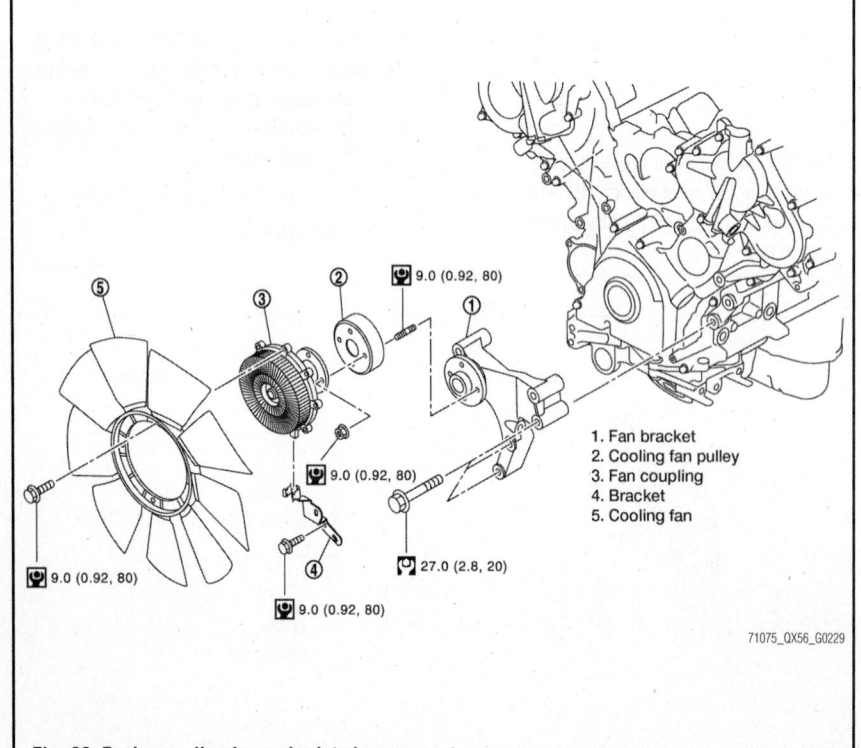

1. Fan bracket
2. Cooling fan pulley
3. Fan coupling
4. Bracket
5. Cooling fan

9.0 (0.92, 80)
27.0 (2.8, 20)

71075_QX56_G0229

Fig. 33 Engine cooling fan and related components

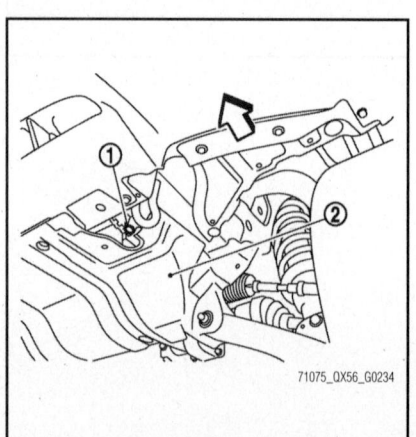

71075_QX56_G0234

Fig. 34 Radiator drain plug location (1) plug, (2) undercover

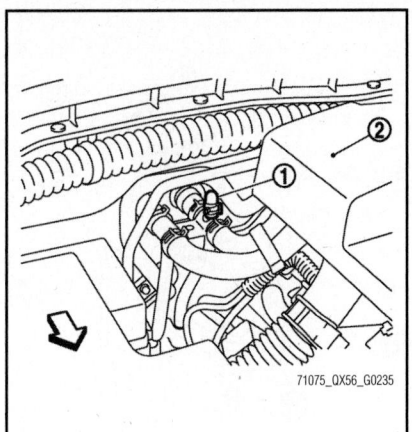

**Fig. 35 Radiator air relief plug location
(1) plug, (2) engine cover**

The automatic window function will not
work with the battery cable disconnected.

2. Disconnect the negative battery
cable.

3. Open the radiator drain plug.
Remove the radiator cap and reservoir cap.
Be sure to position a drain pan under the
radiator. Properly dispose of used coolant.

4. Remove the reservoir tank, if necessary. Drain the coolant and clean the tank.
If required, flush the cooling system.

To install:

➡**Be sure to use new fasteners, as
required.**

5. Installation is the reverse of the
removal procedure.

6. Install the drain plug, using a new
O-ring. Tighten to 17 ft. lbs. (23 Nm).

7. Remove the air relief plug, on the
heater hose.

8. Fill the radiator and reservoir. Pour
fluid slowly to allow air to escape.

9. Be sure to fill the cooling system
with the proper grade and type coolant.

10. When coolant overflows from the air
relief hole, install the plug using a clamp.
Refill the radiator.

11. Warm the engine until opening thermostat, less than 3000 RPM's. Watch temperature gauge as not to overheat engine.

12. Stop the engine and cool down to
less than 122 degrees F.

13. Refill the reservoir tank to the MAX
line, if necessary.

14. Repeat above two more times with
the reservoir tank cap installed until engine
the coolant level no longer drops.

15. Check the cooling system for leakage with the engine running. Correct as
required.

16. Warm the engine, and check for the

sound of engine coolant flow while running
the engine from idle up to 3000 rpm's with
the heater temperature controller set at several positions between COOL and WARM.
Repeat this step three times.

17. If sound is heard, bleed the air from
the cooling system by repeating the procedure.

18. Be sure to perform the reconnect/
relearn procedures.

LEVEL CHECK
See Figure 36.

Be sure to use genuine Nissan (blue)
coolant when filling/servicing the cooling
system.

Before servicing the vehicle, refer to the
Precautions Section.

➡**If working near and/or around the
SRS system and components, be sure
to disable the SRS system. After disabling the system wait three minutes or
more before servicing the vehicle.**

➡**Whenever the negative battery cable
is disconnected the following components will require resetting. The air
conditioning system, automatic drive
positioner system, power window control system, around view monitor,
automatic back door system, idle air
volume learning, steering angle sensor
neutral position, sunroof system, audio
visual and navigation systems. Use the
CONSULT-III diagnostic tool, or equivalent to perform the required resets.
Follow the directions on the screen of
the tool, as needed.**

➡**Before disconnecting the negative
battery cable lower both the driver's
side and passenger's side front win-**

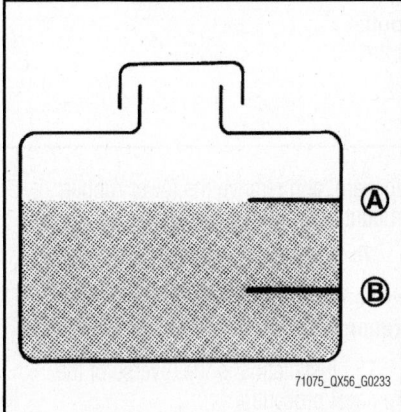

**Fig. 36 Coolant level MIN/MAX identifica-
tion**

dows. This will prevent any interference between the window edge and
the vehicle when the door is opener/
closed. The automatic window function
will not work with the battery cable
disconnected.

Check that the coolant reservoir tank is
level and within the MIN and MAX marks.
Adjust coolant level, as necessary.

RADIATOR

REMOVAL & INSTALLATION
See Figure 37.

✶✶ CAUTION

**Never remove the radiator cap when
the engine is hot. Serious burns could
occur from high-pressure engine
coolant escaping from the radiator.**

1. Before servicing the vehicle, refer to
the Precautions Section.

➡**If working near and/or around the
SRS system and components, be sure
to disable the SRS system. After disabling the system wait three minutes or
more before servicing the vehicle.**

➡**Whenever the negative battery cable
is disconnected the following components will require resetting. The air
conditioning system, automatic drive
positioner system, power window control system, around view monitor,
automatic back door system, idle air
volume learning, steering angle sensor
neutral position, sunroof system, audio
visual and navigation systems. Use the
CONSULT-III diagnostic tool, or equivalent to perform the required resets.
Follow the directions on the screen of
the tool, as needed.**

➡**Before disconnecting the negative
battery cable lower both the driver's
side and passenger's side front windows. This will prevent any interference between the window edge and
the vehicle when the door is opener/
closed. The automatic window function
will not work with the battery cable disconnected.**

2. Disconnect the negative battery
cable.

3. Remove the undercover.

4. Remove the engine cover.

5. Remove the air cleaner assembly.

6. Position a catch pan under the radiator. Drain the coolant. Be sure to properly
dispose of used coolant.

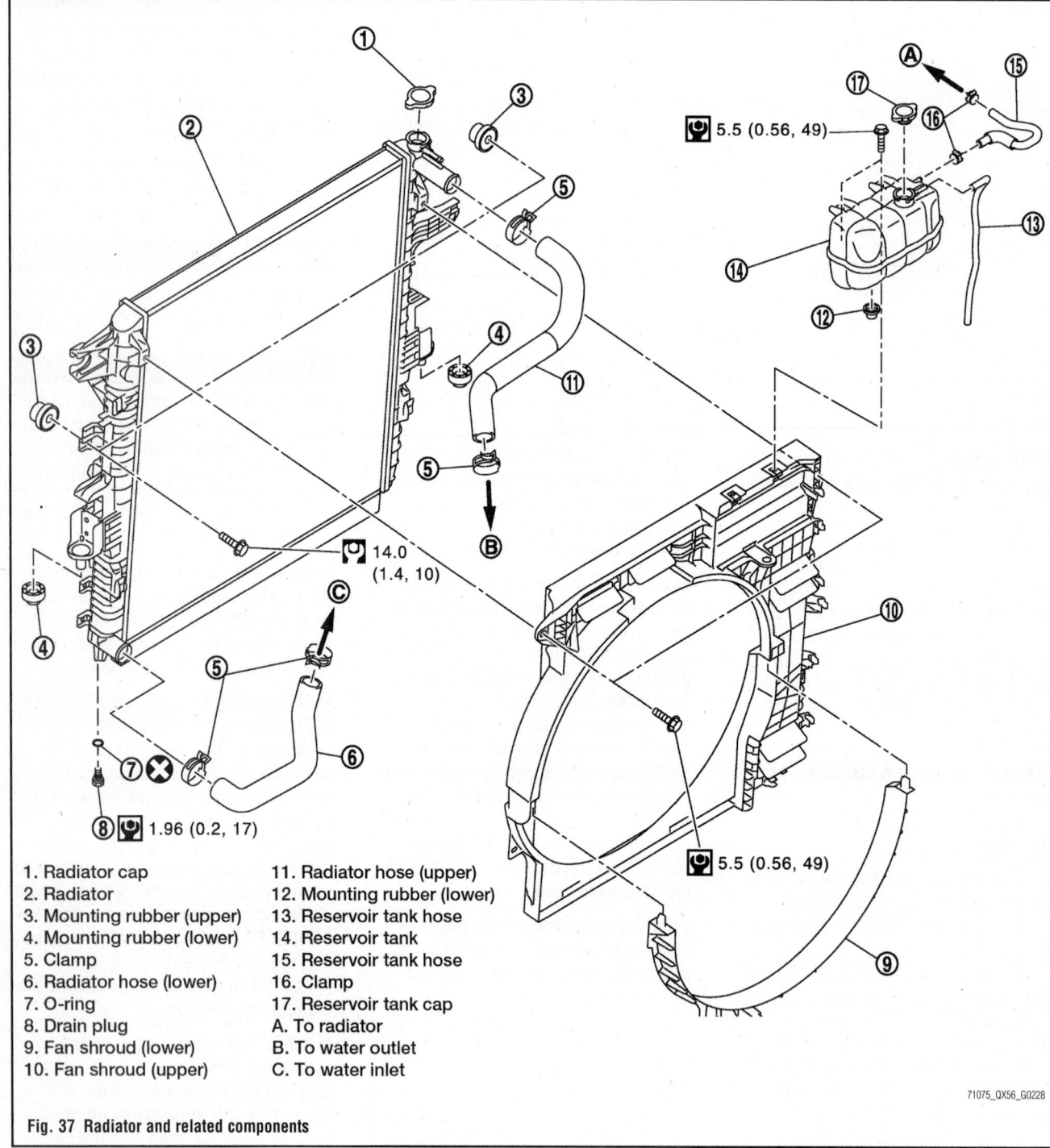

14.0 (1.4, 10)

5.5 (0.56, 49)

5.5 (0.56, 49)

1.96 (0.2, 17)

1. Radiator cap
2. Radiator
3. Mounting rubber (upper)
4. Mounting rubber (lower)
5. Clamp
6. Radiator hose (lower)
7. O-ring
8. Drain plug
9. Fan shroud (lower)
10. Fan shroud (upper)
11. Radiator hose (upper)
12. Mounting rubber (lower)
13. Reservoir tank hose
14. Reservoir tank
15. Reservoir tank hose
16. Clamp
17. Reservoir tank cap
A. To radiator
B. To water outlet
C. To water inlet

71075_QX56_G0228

Fig. 37 Radiator and related components

7. Remove the lower radiator hose. Remove the lower fan shroud.

8. Move the reservoir tank out of the way.

9. Remove the radiator core support cover.

10. Remove the upper fan shroud.

11. Remove the radiator grille.

12. Reposition the condenser out of the way.

13. Remove the upper radiator hose.

14. Carefully lift and pull the radiator forward, and remove the lower rubber mounting from the core support.

To install:

➡**Be sure to use new fasteners, as required.**

15. Installation is the reverse of the removal procedure.

16. Be sure to properly install radiator hoses.

17. Be sure to refill the cooling using the proper grade and type engine coolant.

18. Start the engine and check for leaks.

19. Start the engine and allow it to reach operation temperature. Recheck the coolant level, fill as required.

20. Be sure to perform the reconnect/relearn procedures.

THERMOSTAT

REMOVAL & INSTALLATION

See Figures 38 through 40.

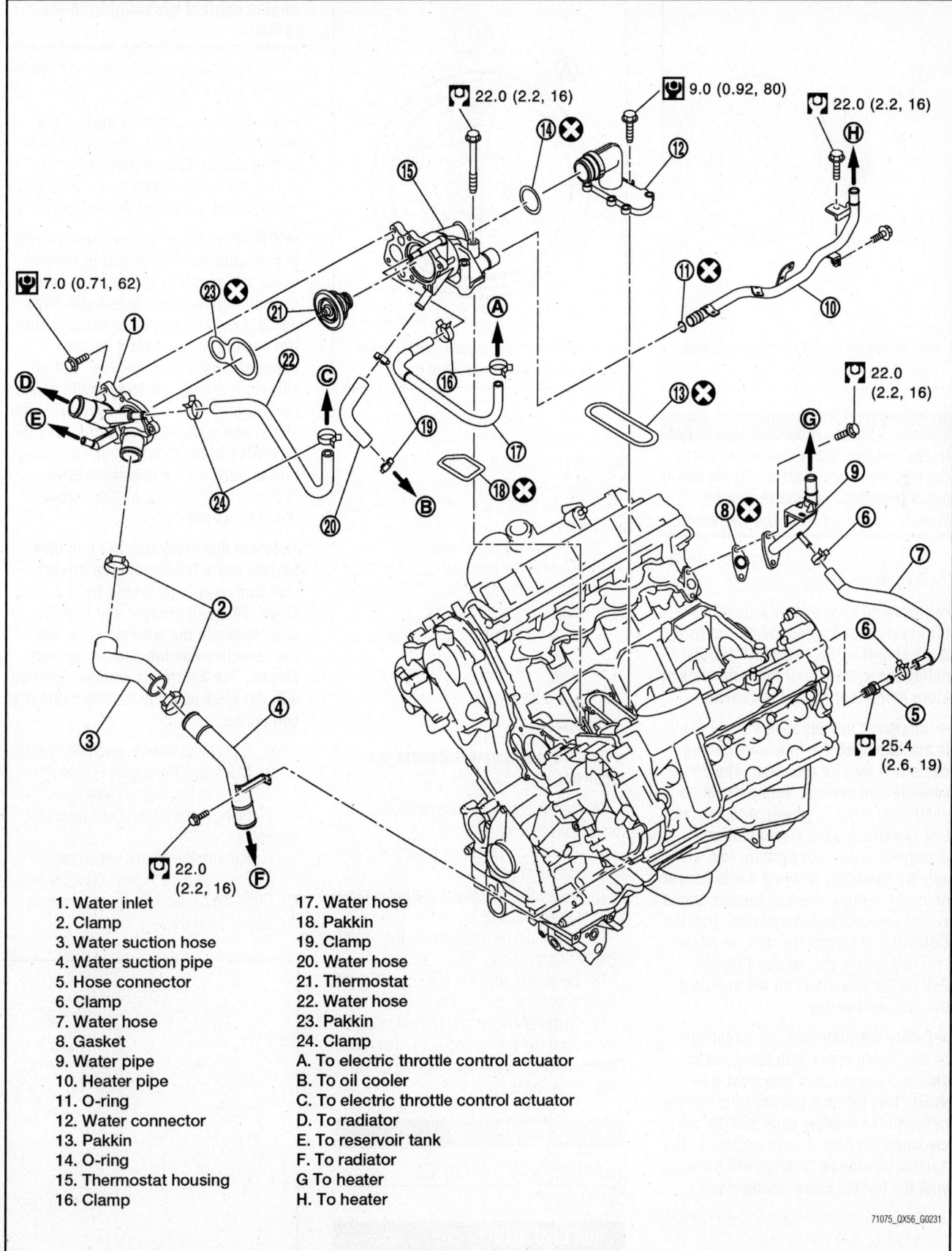

22.0 (2.2, 16)

9.0 (0.92, 80)

22.0 (2.2, 16)

7.0 (0.71, 62)

22.0 (2.2, 16)

22.0 (2.2, 16)

25.4 (2.6, 19)

22.0 (2.2, 16)

1. Water inlet
2. Clamp
3. Water suction hose
4. Water suction pipe
5. Hose connector
6. Clamp
7. Water hose
8. Gasket
9. Water pipe
10. Heater pipe
11. O-ring
12. Water connector
13. Pakkin
14. O-ring
15. Thermostat housing
16. Clamp
17. Water hose
18. Pakkin
19. Clamp
20. Water hose
21. Thermostat
22. Water hose
23. Pakkin
24. Clamp
A. To electric throttle control actuator
B. To oil cooler
C. To electric throttle control actuator
D. To radiator
E. To reservoir tank
F. To radiator
G To heater
H. To heater

71075_QX56_G0231

Fig. 38 Thermostat and related components

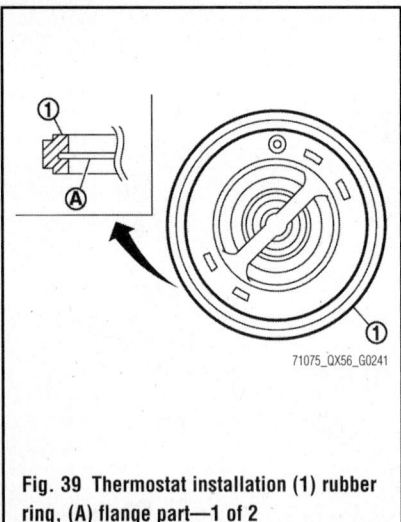

Fig. 39 Thermostat installation (1) rubber ring, (A) flange part—1 of 2

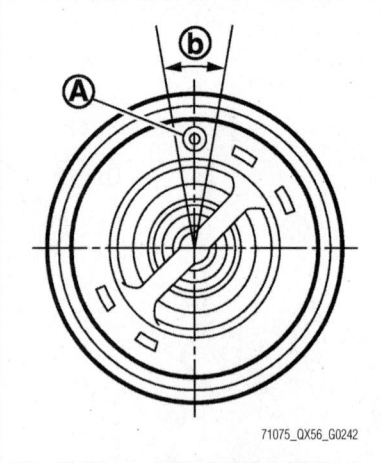

Fig. 40 Thermostat installation (A) jiggle valve, (B) degree range—2 of 2

❊❊ CAUTION

Never remove the radiator cap when the engine is hot. Serious burns could occur from high-pressure engine coolant escaping from the radiator.

1. Before servicing the vehicle, refer to the Precautions Section.

➡ If working near and/or around the SRS system and components, be sure to disable the SRS system. After disabling the system wait three minutes or more before servicing the vehicle.

➡ Whenever the negative battery cable is disconnected the following components will require resetting. The air conditioning system, automatic drive positioner system, power window control system, around view monitor, automatic back door system, idle air volume learning, steering angle sensor neutral position, sunroof system, audio visual and navigation systems. Use the CONSULT-III diagnostic tool, or equivalent to perform the required resets. Follow the directions on the screen of the tool, as needed.

➡ Before disconnecting the negative battery cable lower both the driver's side and passenger's side front windows. This will prevent any interference between the window edge and the vehicle when the door is opener/closed. The automatic window function will not work with the battery cable disconnected.

2. Disconnect the negative battery cable.
3. Make sure the engine is cold.
4. Remove the engine room cover.
5. Remove the air duct and resonator assembly.

6. Remove the engine undercover.
7. Position a catch pan under the radiator. Drain the coolant. Be sure to properly dispose of used coolant.
8. Disconnect the upper and lower radiator hoses.
9. Remove the intake manifold.
10. Remove the fuel feed tube.
11. Remove the water hose, water connector, heater pipes and hoses.
12. Remove the thermostat housing retaining bolts.
13. Remove the housing and the thermostat. Discard the gasket.

To install:

➡ Be sure to use new fasteners, as required.

14. Installation is the reverse of the removal procedure.
15. Be sure to use a new gasket. Do not reuse used O-rings.
16. Install the thermostat see illustration for proper positioning.
17. Install the thermostat with the jiggle valve facing upward.
18. Be sure to refill the cooling using the proper grade and type engine coolant.
19. Start the engine and check for leaks.
20. Start the engine and allow it to reach operation temperature. Recheck the coolant level, fill as required.

WATER PUMP

REMOVAL & INSTALLATION
See Figure 41.

❊❊ CAUTION

Never remove the radiator cap when the engine is hot. Serious burns could occur from high-pressure

engine coolant escaping from the radiator.

1. Before servicing the vehicle, refer to the Precautions Section.

➡ If working near and/or around the SRS system and components, be sure to disable the SRS system. After disabling the system wait three minutes or more before servicing the vehicle.

➡ Whenever the negative battery cable is disconnected the following components will require resetting. The air conditioning system, automatic drive positioner system, power window control system, around view monitor, automatic back door system, idle air volume learning, steering angle sensor neutral position, sunroof system, audio visual and navigation systems. Use the CONSULT-III diagnostic tool, or equivalent to perform the required resets. Follow the directions on the screen of the tool, as needed.

➡ Before disconnecting the negative battery cable lower both the driver's side and passenger's side front windows. This will prevent any interference between the window edge and the vehicle when the door is opener/ closed. The automatic window function will not work with the battery cable disconnected.

2. Disconnect the negative battery cable.
3. Make sure the engine is cold.
4. Remove the engine room cover.
5. Remove the air duct and resonator assembly.
6. Remove the engine undercover.
7. Position a catch pan under the radiator. Drain the coolant. Be sure to properly dispose of used coolant.

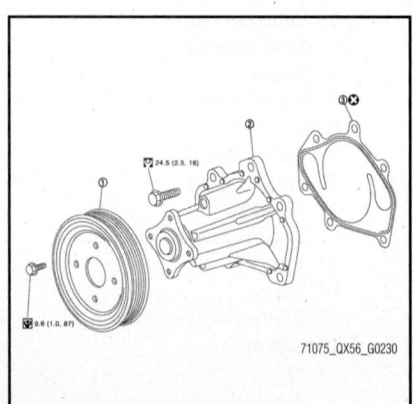

Fig. 41 Water pump and related components (1) pulley, (2) pump, (3) gasket

8. Remove the lower fan shroud.
9. Reposition the reservoir tank out of the way.
10. Remove the radiator core support cover.
11. Remove the upper fan shroud.
12. Remove the drive belt.
13. Remove the cooling fan assembly.
14. Remove the upper radiator hose.

15. Remove the water pump pulley.
16. Remove the water pump retaining bolts.
17. remove the water pump. Discard the gasket.

To install:

➡ **Be sure to use new fasteners, as required.**

18. Installation is the reverse of the removal procedure.
19. Refill the cooling system.
20. Start the engine and check for leaks.
21. Be sure to perform the reconnect/relearn procedures.

ENGINE ELECTRICAL

BATTERY SYSTEM

BATTERY

REMOVAL & INSTALLATION

See Figure 42.

1. Before servicing the vehicle, refer to the Precautions Section.

➡ **If working near and/or around the SRS system and components, be sure to disable the SRS system. After disabling the system wait three minutes or more before servicing the vehicle.**

➡ **Whenever the negative battery cable is disconnected the following components will require resetting. The air conditioning system, automatic drive positioner system, power window control system, around view monitor, automatic back door system, idle air volume learning, steering angle sensor neutral position, sunroof system, audio visual and navigation systems. Use the CONSULT-III diagnostic tool, or equivalent to perform the required resets. Follow the directions on the screen of the tool, as needed.**

➡ **Before disconnecting the negative battery cable lower both the driver's side and passenger's side front**

windows. This will prevent any interference between the window edge and the vehicle when the door is opener/closed. The automatic window function will not work with the battery cable disconnected.

2. Disconnect the negative battery cable.

➡ **Always disconnect the negative battery cable first.**

3. Remove the cover of the positive cable.
4. Disconnect the positive cable.
5. Remove the battery fix frame nuts and fix frame.
6. Remove the battery from its mounting.

To install:

➡ **Be sure to use new fasteners, as required.**

7. Installation is the reverse of the removal procedure.

➡ **Always connect the positive battery cable first.**

8. Be sure to perform the reconnect/relearn procedures.

BATTERY RECONNECT/ RELEARN PROCEDURE

➡ **Whenever the negative battery cable is disconnected the following components will require resetting. The air conditioning system, automatic drive positioner system, power window control system, around view monitor, automatic back door system, idle air volume learning, steering angle sensor neutral position, sunroof system, audio visual and navigation systems. Use the CONSULT-III diagnostic tool, or equivalent to perform the required resets. Follow the directions on the screen of the tool, as needed. The following procedures can be performed without the use of the diagnostic tool, however the use of a diagnostic tool is recommended.**

Throttle Valve Closed Position Learning

1. Start the engine.

➡ **Be sure the engine coolant temperature is 77 degrees F or less, before starting.**

2. Warm up the engine.

➡ **Raise the engine coolant temperature until it reaches 149 degrees F, or higher.**

3. Turn the ignition switch to the OFF position and wait ten seconds.
4. Check that the throttle valve moves during the above ten seconds by confirming the operating sound.

Idle Air Volume Learning

➡ **Before performing this procedure be sure that the following conditions are met. Drive the vehicle for ten minutes. Battery voltage is more than 12.9 volts at idle. Engine coolant temperature is between 158–221 degrees F. Selector lever is in either P or N. All accessories are OFF. The steering wheel is in the straight head position. The vehicle is stopped. The transmission is warmed up.**

➡ **It is better to count the time accurately using a clock. It is impossible to switch the diagnostic mode when an accelerator pedal position sensor circuit has a malfunction.**

1. Perform the accelerator pedal released position learning procedure.
2. Perform the throttle valve closed position learning procedure.
3. Start the engine and warm until operating temperature is reached.
4. Turn the ignition switch to the OFF position and wait ten seconds.
5. Confirm that the accelerator pedal is fully released, turn the ignition switch to the ON position and wait three seconds.
6. Repeat the following step quickly five times within five seconds.

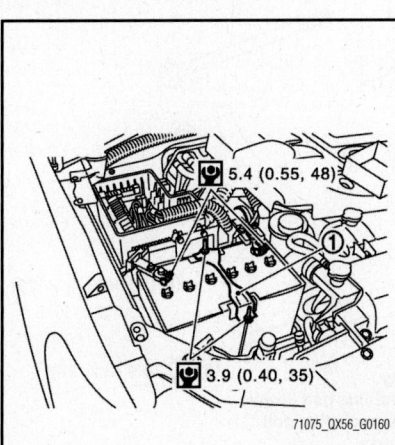

5.4 (0.55, 48)

3.9 (0.40, 35)

71075_QX56_G0160

Fig. 42 Battery and related components

7. Fully depress the accelerator pedal and then fully release the accelerator pedal.

8. Wait seven seconds, fully depress the accelerator pedal for about twenty seconds, wait until the MIL light stops blinking and turns ON.

9. Fully release the accelerator pedal within three seconds after the MIL light turns ON.

10. Start the engine and let it idle. Wait twenty seconds.

Accelerator Pedal Released Position Learning

1. Check that the accelerator pedal is fully released.

2. Turn the ignition switch to the ON position and wait at least two seconds.

3. Turn the ignition switch to the OFF position and wait at least ten seconds.

4. Turn the ignition switch to the ON position and wait at least two seconds.

5. Turn the ignition switch to the OFF position and wait at least ten seconds.

ENGINE ELECTRICAL

CHARGING SYSTEM

ALTERNATOR

REMOVAL & INSTALLATION

See Figures 43 and 44.

1. Before servicing the vehicle, refer to the Precautions Section.

➡ **If working near and/or around the SRS system and components, be sure to disable the SRS system. After disabling the system wait three minutes or more before servicing the vehicle.**

➡ **Whenever the negative battery cable is disconnected the following components will require resetting. The air conditioning system, automatic drive positioner system, power window control system, around view monitor, automatic back door system, idle air volume learning, steering angle sensor neutral position, sunroof system, audio visual and navigation systems. Use the CONSULT-III diagnostic tool, or equivalent to perform the required resets. Follow the directions on the screen of the tool, as needed.**

➡ **Before disconnecting the negative battery cable lower both the driver's side and passenger's side front windows. This will prevent any interference between the window edge and the vehicle when the door is opener/closed. The automatic window function will not work with the battery cable disconnected.**

2. Disconnect the negative battery cable.
3. Remove the engine under cover.
4. Remove the drive belt.
5. Disconnect the oil pressure switch connector and alternator connector.
6. Remove the harness bracket bolt.
7. Disconnect the alternator electrical connectors.
8. Remove the ground harness mounting bolt.
9. Remove the lower and upper alternator mounting bolts.
10. Remove the transmission fluid cooler tube from the fan shroud.

11. Move the transmission fluid cooler tube out of the way.

12. Remove the alternator assembly from underneath of the vehicle.

To install:

➡ **Be sure to use new fasteners, as required.**

13. Installation is the reverse of the removal procedure.

14. Tighten the alternator mounting bracket bolts in proper sequence. See illustration.

15. Using the CONSULT III diagnostic tool, perform the power generation voltage variable control system operation test.

16. Be sure to perform the reconnect/relearn procedures.

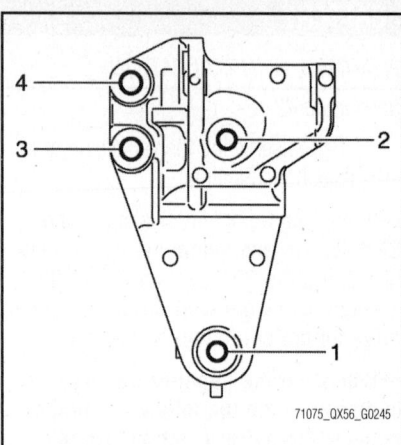

71075_QX56_G0245

Fig. 44 Alternator mounting bracket tightening sequence

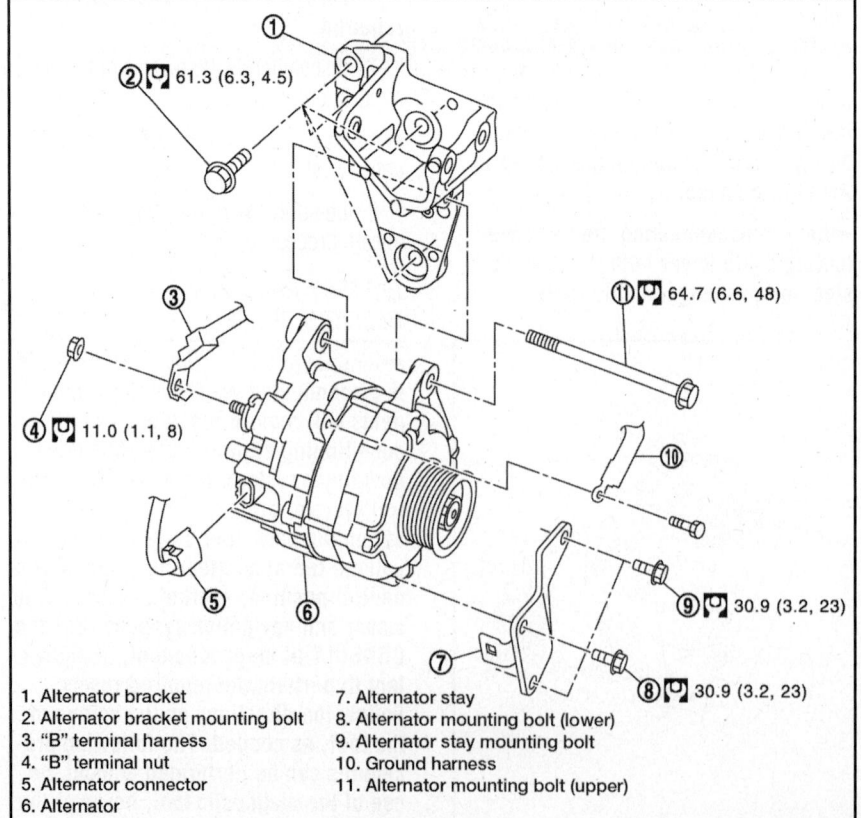

② ⟦ 61.3 (6.3, 4.5)

⑪ ⟦ 64.7 (6.6, 48)

④ ⟦ 11.0 (1.1, 8)

⑨ ⟦ 30.9 (3.2, 23)

⑧ ⟦ 30.9 (3.2, 23)

1. Alternator bracket
2. Alternator bracket mounting bolt
3. "B" terminal harness
4. "B" terminal nut
5. Alternator connector
6. Alternator
7. Alternator stay
8. Alternator mounting bolt (lower)
9. Alternator stay mounting bolt
10. Ground harness
11. Alternator mounting bolt (upper)

71075_QX56_G0244

Fig. 43 Alternator and related components

ENGINE ELECTRICAL

DISTRIBUTORLESS IGNITION SYSTEM

FIRING ORDERS

1–8–7–3–6–5–4–2

IGNITION COIL

REMOVAL & INSTALLATION

See Figures 45 and 46.

1. Before servicing the vehicle, refer to the Precautions Section.

➡If working near and/or around the SRS system and components, be sure to disable the SRS system. After disabling the system wait three minutes or more before servicing the vehicle.

➡Whenever the negative battery cable is disconnected the following components will require resetting. The air conditioning system, automatic drive positioner system, power window control system, around view monitor, automatic back door system, idle air volume learning, steering angle sensor neutral position, sunroof system, audio visual and navigation systems. Use the CONSULT-III diagnostic tool, or equivalent to perform the required resets. Follow the directions on the screen of the tool, as needed.

➡Before disconnecting the negative battery cable lower both the driver's side and passenger's side front windows. This will prevent any interference between the window edge and the vehicle when the door is opener/closed. The automatic window function

will not work with the battery cable disconnected.

2. Disconnect the negative battery cable.
3. Remove the engine room cover.
4. Remove the air cleaner assembly for access, as required.
5. Remove the ignition coil.

To install:

➡Be sure to use new fasteners, as required.

6. Installation is the reverse of the removal procedure.
7. Installed the coils marked (A) on cylinder number 5, 6, 7, and 8. See illustration.
8. Be sure to perform the reconnect/relearn procedures.

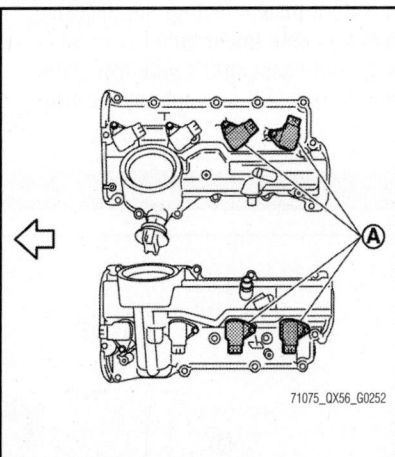

71075_QX56_G0252

Fig. 46 Ignition coil installation identification

IGNITION TIMING

INSPECTION & ADJUSTMENT

See Figure 47.

The ignition timing is controlled by the Powertrain Control Module (PCM). No adjustment is necessary or possible.

1. Before servicing the vehicle, refer to the Precautions Section.

➡If working near and/or around the SRS system and components, be sure to disable the SRS system. After disabling the system wait three minutes or more before servicing the vehicle.

➡Whenever the negative battery cable is disconnected the following components will require resetting. The air conditioning system, automatic drive positioner system, power window control system, around view monitor, automatic back door system, idle air volume learning, steering angle sensor neutral position, sunroof system, audio visual and navigation systems. Use the CONSULT-III diagnostic tool, or equivalent to perform the required resets. Follow the directions on the screen of the tool, as needed.

➡Before disconnecting the negative battery cable lower both the driver's side and passenger's side front windows. This will prevent any interference between the window edge and the vehicle when the door is opener/closed. The automatic window function will not work with the battery cable disconnected.

2. Remove the number one ignition coil.
3. Connect the number one ignition coil and spark plug with a suitable high tension wire.
4. Attach the timing light clamp to the wire.
5. Check the ignition timing.

SPARK PLUGS

REMOVAL & INSTALLATION

1. Before servicing the vehicle, refer to the Precautions Section.

➡If working near and/or around the SRS system and components, be sure to disable the SRS system. After disabling the system wait three minutes or more before servicing the vehicle.

➡Whenever the negative battery cable is disconnected the following

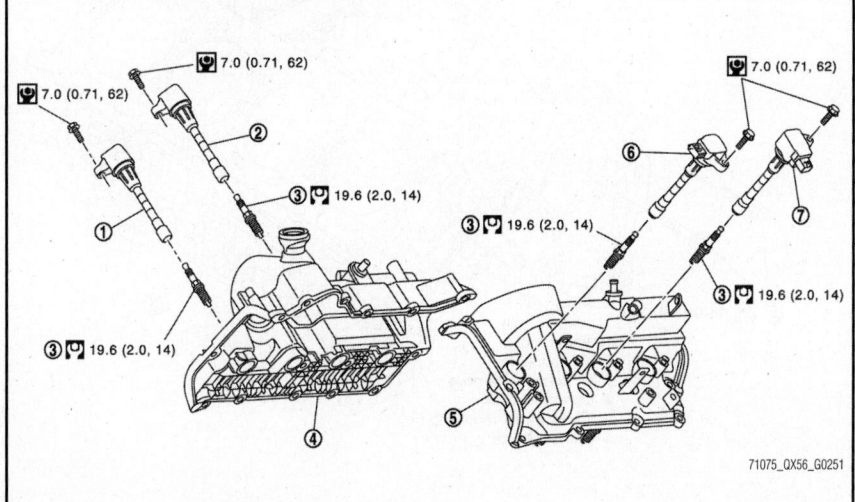

71075_QX56_G0251

Fig. 45 Ignition coil and related components

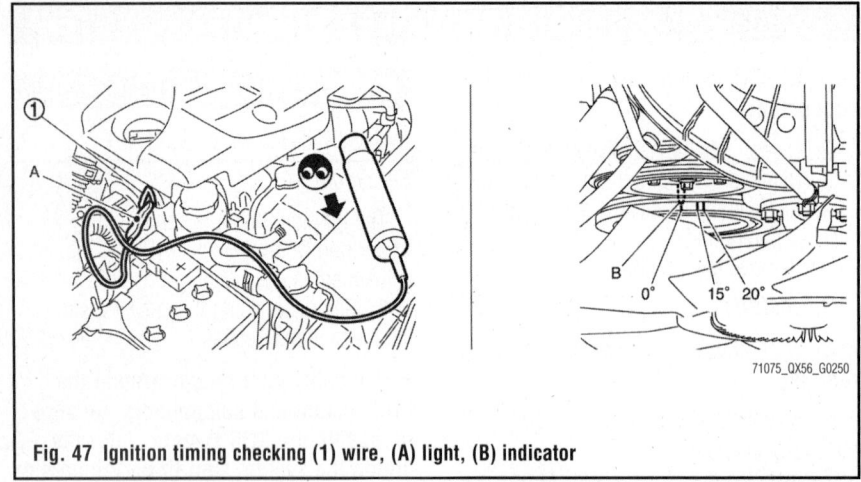

Fig. 47 Ignition timing checking (1) wire, (A) light, (B) indicator

components will require resetting. The air conditioning system, automatic drive positioner system, power window control system, around view monitor, automatic back door system, idle air volume learning, steering angle sensor neutral position, sunroof system, audio visual and navigation systems. Use the CONSULT-III diag-

nostic tool, or equivalent to perform the required resets. Follow the directions on the screen of the tool, as needed.

➡Before disconnecting the negative battery cable lower both the driver's side and passenger's side front windows. This will prevent any interfer-

ence between the window edge and the vehicle when the door is opener/closed. The automatic window function will not work with the battery cable disconnected.

2. Disconnect the negative battery cable.
3. Remove the engine room cover.
4. Remove the air cleaner assembly for access, as required.
5. Remove the ignition coil.
6. Remove the spark plug.

To install:

➡Be sure to use new fasteners, as required.

7. Be sure the spark plug gap is to specification (0.043 in.).
8. Carefully install the spark plug and torque to specification, 18 ft. lbs. (25 Nm).
9. Install the ignition coil, torque the retaining bolt to 80 inch lbs. (9 Nm).
10. Connect the harness coil.
11. Connect the negative battery cable.
12. Be sure to perform the reconnect/relearn procedures.

ENGINE ELECTRICAL

STARTER

REMOVAL & INSTALLATION
See Figure 48.

1. Before servicing the vehicle, refer to the Precautions Section.

➡If working near and/or around the SRS system and components, be sure to disable the SRS system. After disabling the system wait three minutes or more before servicing the vehicle.

➡Whenever the negative battery cable is disconnected the following components will require resetting. The air conditioning system, automatic drive positioner system, power window control system, around view monitor, automatic back door system, idle air volume learning, steering angle sensor neutral position, sunroof system, audio visual and navigation systems. Use the CONSULT-III diagnostic tool, or equivalent to perform the required resets. Follow the directions on the screen of the tool, as needed.

➡Before disconnecting the negative battery cable lower both the driver's side and passenger's side front windows. This will prevent any interference between the window edge and the

STARTING SYSTEM

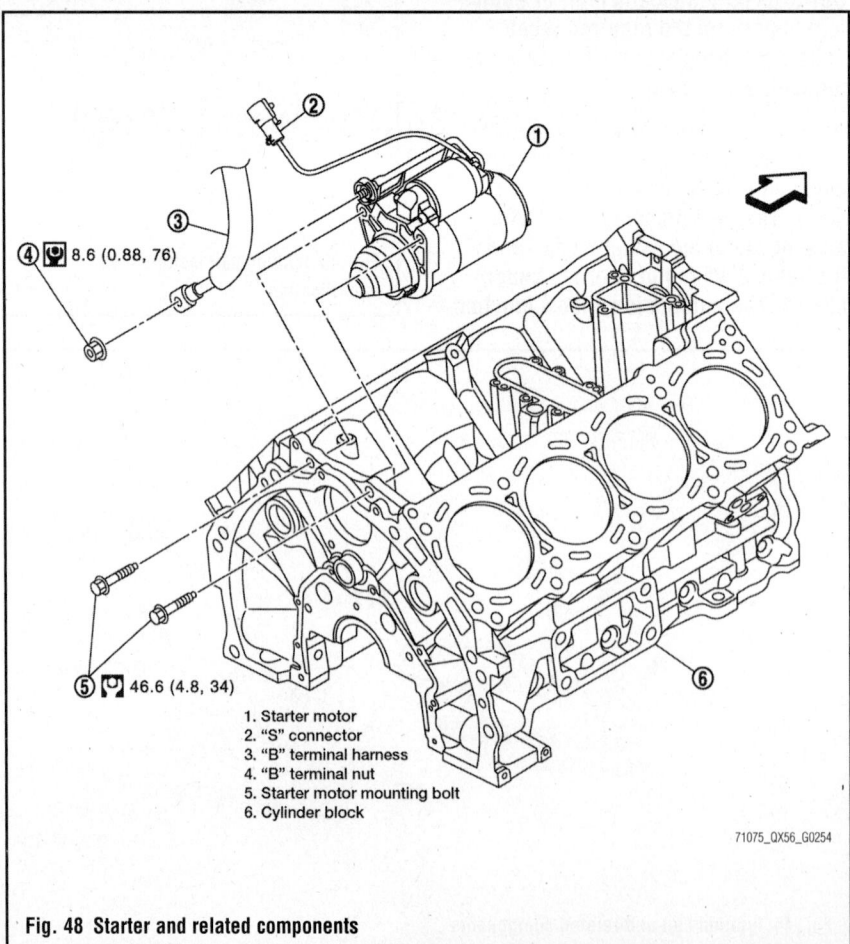

1. Starter motor
2. "S" connector
3. "B" terminal harness
4. "B" terminal nut
5. Starter motor mounting bolt
6. Cylinder block

Fig. 48 Starter and related components

vehicle when the door is opener/closed. The automatic window function will not work with the battery cable disconnected.

2. Disconnect the negative battery cable.
3. Remove the intake manifold.

4. Remove the starter harness connectors.
5. Remove the starter retaining bolts.
6. Remove the starter from its mounting.

To install:

➡ **Be sure to use new fasteners, as required.**

7. Installation is the reverse of the removal procedure.
8. Be sure to perform the reconnect/relearn procedures.

ENGINE MECHANICAL

ACCESSORY DRIVE BELT SYSTEM

ADJUSTMENT

Drive belt tension is not necessary, as it is automatically adjusted by the auto tensioner.

BELT ROUTINGS

See Figure 49.

INSPECTION

Inspect the drive belt for signs of glazing or cracking. A glazed belt will be perfectly smooth from slippage, while a good belt will have a slight texture of fabric visible. Cracks will usually start at the inner edge of the belt and run outward. All worn or damaged drive belts should be replaced immediately.

REMOVAL & INSTALLATION

See Figure 50.

1. Before servicing the vehicle, refer to the Precautions Section.

➡ **If working near and/or around the SRS system and components, be sure to disable the SRS system. After disabling the system wait three minutes or more before servicing the vehicle.**

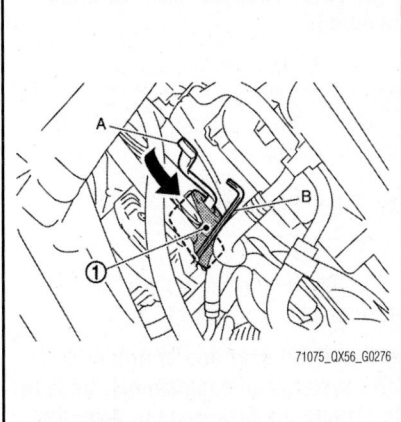

71075_QX56_G0276

Fig. 50 Drivebelt removal (1) pulley, (A) wrench, (B) bar

➡ **Whenever the negative battery cable is disconnected the following components will require resetting. The air conditioning system, automatic drive positioner system, power window control system, around view monitor, automatic back door system, idle air volume learning, steering angle sensor neutral position, sunroof system, audio visual and navigation systems. Use the CONSULT-III diagnostic tool, or equivalent to perform the required resets.**

Follow the directions on the screen of the tool, as needed.

➡ **Before disconnecting the negative battery cable lower both the driver's side and passenger's side front windows. This will prevent any interference between the window edge and the vehicle when the door is opener/ closed. The automatic window function will not work with the battery cable disconnected.**

2. Disconnect the negative battery cable.
3. Position the cooling system reservoir tank out of the way.
4. Remove the engine room cover and air cleaner assembly, as required.
5. Using a wrench, move the auto tensioner pulley. Position a bar thru the holding boss to lock the tensioner pulley arm. See illustration.

➡ **Never loosen the hexagonal part in the center of the tensioner pulley. Never turn clockwise, if turned clockwise the complete tensioner must be replaced as a unit, including the pulley.**

6. Remove the drive belt.

To install:

➡ **Be sure to use new fasteners, as required.**

7. Installation is the reverse of the removal procedure.
8. Be sure that the belt is securely installed around all pulleys.
9. Make sure that the belt tension is within the allowable working range, using the indicator notch on the auto tensioner.
10. Be sure to perform the reconnect/ relearn procedures.

AIR CLEANER

REMOVAL & INSTALLATION

Air Cleaner Assembly

See Figure 51.

1. Before servicing the vehicle, refer to the Precautions Section.

1. Drive belt
2. Power steering oil pump pulley
3. Alternator pulley
4. Crankshaft pulley
5. A/C compressor
6. Idler pulley
7. Cooling fan pulley
8. Water pump pulley
9. Drive belt auto-tensioner
A. Possible use range
B. Range when new drive belt is installed
C. Indicator
D. View D

71075_QX56_G0003

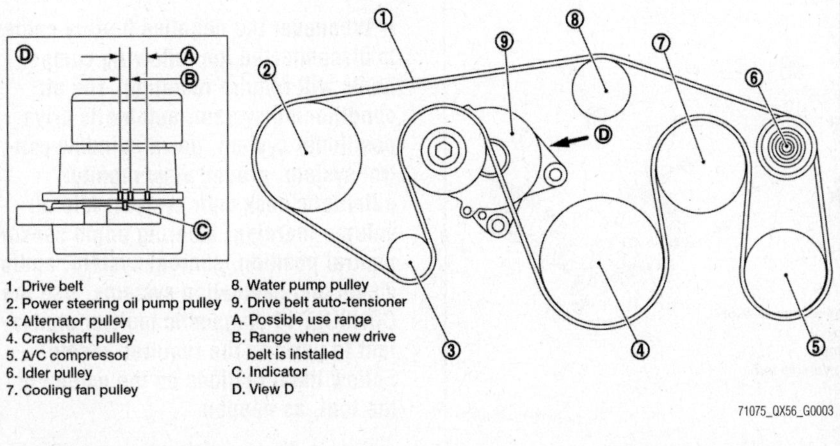

Fig. 49 Drivebelt routing

➡️If working near and/or around the SRS system and components, be sure to disable the SRS system. After disabling the system wait three minutes or more before servicing the vehicle.

➡️Whenever the negative battery cable is disconnected the following components will require resetting. The air conditioning system, automatic drive positioner system, power window control system, around view monitor, automatic back door system, idle air volume learning, steering angle sensor neutral position, sunroof system, audio visual and navigation systems. Use the CONSULT-III diagnostic tool, or equivalent to perform the required resets. Follow the directions on the screen of the tool, as needed.

➡️Before disconnecting the negative battery cable lower both the driver's side and passenger's side front windows. This will prevent any interference between the window edge and the vehicle when the door is opener/closed. The automatic window function will not work with the battery cable disconnected.

2. Disconnect the negative battery cable.
3. Remove the engine room cover.
4. Disconnect the mass air flow sensor harness connector. Remove the sensor from the case, if necessary.
5. Disconnect the PCV hose from the air duct.

6. Remove the air duct.
7. Remove the air filter.
8. Remove the lower case cover.
9. Remove the adapter.

➡️If removing the resonator, Raise and support the vehicle safely. Remove the left front tire and wheel assembly. Remove the fender protector.

To install:

➡️Be sure to use new fasteners, as required.

10. Installation is the reverse of the removal procedure.
11. Be sure to perform the reconnect/relearn procedures.

Air Filter Element
See Figure 51.

1. Before servicing the vehicle, refer to the Precautions Section.

➡️If working near and/or around the SRS system and components, be sure to disable the SRS system. After disabling the system wait three minutes or more before servicing the vehicle.

➡️Whenever the negative battery cable is disconnected the following components will require resetting. The air conditioning system, automatic drive positioner system, power window control system, around view monitor, automatic back door system, idle air volume learning, steering angle sensor

neutral position, sunroof system, audio visual and navigation systems. Use the CONSULT-III diagnostic tool, or equivalent to perform the required resets. Follow the directions on the screen of the tool, as needed.

➡️Before disconnecting the negative battery cable lower both the driver's side and passenger's side front windows. This will prevent any interference between the window edge and the vehicle when the door is opener/closed. The automatic window function will not work with the battery cable disconnected.

2. Disconnect the negative battery cable.
3. Unhook the clips and lift off the upper cover.
4. Remove the air cleaner filter.

To install:

➡️Be sure to use new fasteners, as required.

5. Installation is the reverse of the removal procedure.
6. Be sure to perform the reconnect/relearn procedures.

CAMSHAFT & BEARINGS

REMOVAL & INSTALLATION
See Figures 52 through 58.

➡️The following procedure is for the exhaust camshaft only. At this time the manufacturer does not provide information for the intake camshaft.

1. Before servicing the vehicle, refer to the Precautions Section.

➡️If working near and/or around the SRS system and components, be sure to disable the SRS system. After disabling the system wait three minutes or more before servicing the vehicle.

➡️Whenever the negative battery cable is disconnected the following components will require resetting. The air conditioning system, automatic drive positioner system, power window control system, around view monitor, automatic back door system, idle air volume learning, steering angle sensor neutral position, sunroof system, audio visual and navigation systems. Use the CONSULT-III diagnostic tool, or equivalent to perform the required resets. Follow the directions on the screen of the tool, as needed.

➡️Before disconnecting the negative battery cable lower both the driver's

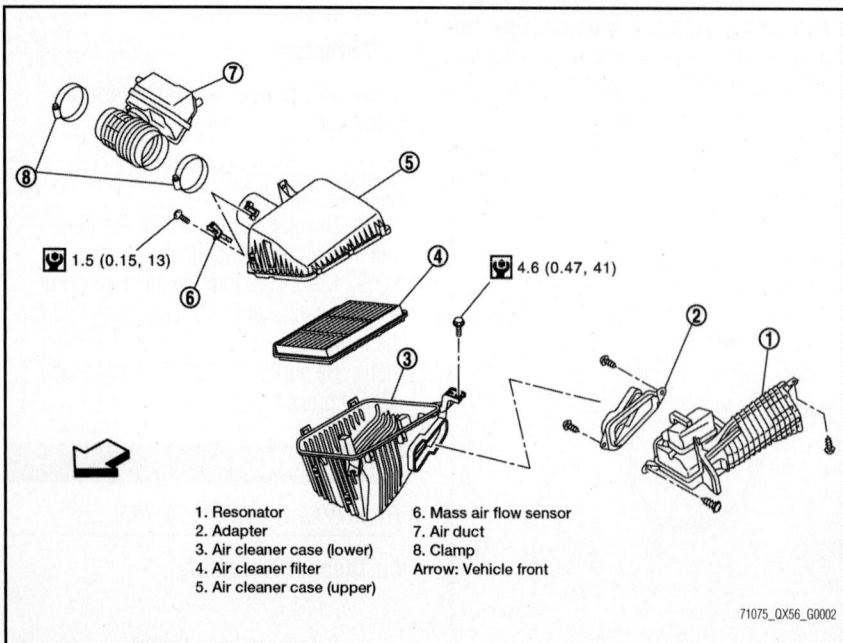

1.5 (0.15, 13)   4.6 (0.47, 41)

1. Resonator
2. Adapter
3. Air cleaner case (lower)
4. Air cleaner filter
5. Air cleaner case (upper)
6. Mass air flow sensor
7. Air duct
8. Clamp
Arrow: Vehicle front

71075_QX56_G0002

Fig. 51 Air cleaner assembly and related components

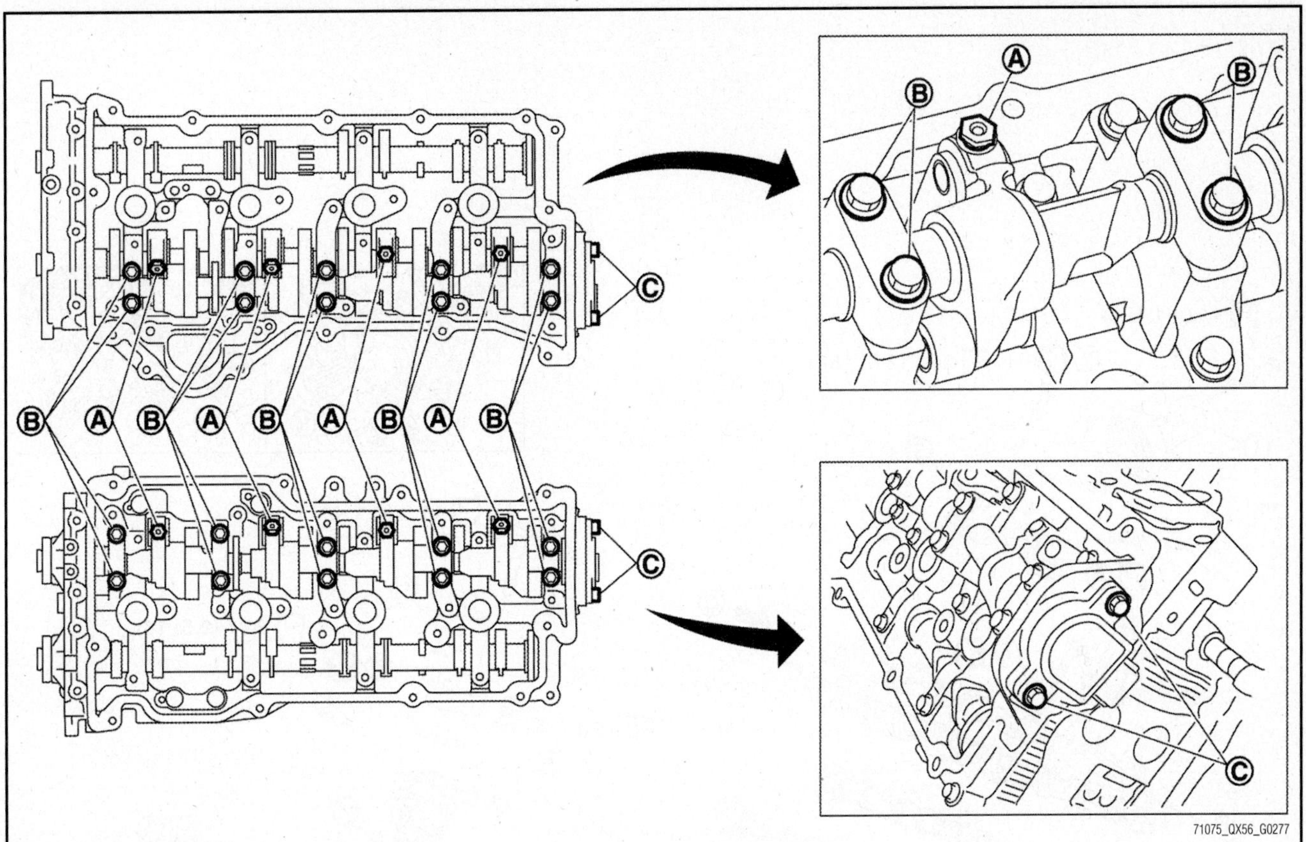

Fig. 52 VVEL ladder bolt location and identification

71075_QX56_G0277

side and passenger's side front windows. This will prevent any interference between the window edge and the vehicle when the door is opener/closed. The automatic window function will not work with the battery cable disconnected.

2. Disconnect the negative battery cable.

3. Remove the engine cover.

4. Remove the air cleaner assembly.

➡ Never loosen the adjusting bolts (A), mounting bolts (B) black color of the VVEL ladder assembly and the mounting bolts (C) of the VVEL control shaft position sensor. If loosened the stroke of the cam lift becomes out of adjustment. In such case, replacement of the VVEL ladder assembly and cylinder head assembly is required.

➡ Never loosen the mounting bolts (C) of the VVEL control shaft position sensor. VVEL control shaft position sensor mounting bolts are required to be loosened for adjustment only when using a new VVEL ladder assembly. See illustration.

5. Remove the VVEL actuator motor assembly.

➡ The VVEL ladder assembly cannot be replaced as a single part, because it is machined together with the cylinder head.

6. Remove the rocker cover.

7. Remove the VVEL actuator housing assembly.

8. Remove the front cover, camshaft sprockets and timing chains.

9. Remove the VVEL ladder assembly.

10. Loosen the mounting bolts (gold color) in the reverse order as shown in the illustration (A = bank 2, B = bank 1)

➡ Never loosen the adjusting and mounting bolts (black color. When removing the VVEL ladder assembly, hold the driveshaft from below as not to drop it.

11. Remove the exhaust camshaft.

To install:

➡ Be sure to use new fasteners, as required.

12. Installation is the reverse of the removal procedure.

➡ Do not reuse washers.

13. Install the exhaust camshaft.

➡ Camshaft identification is provided by identification mark. See illustration.

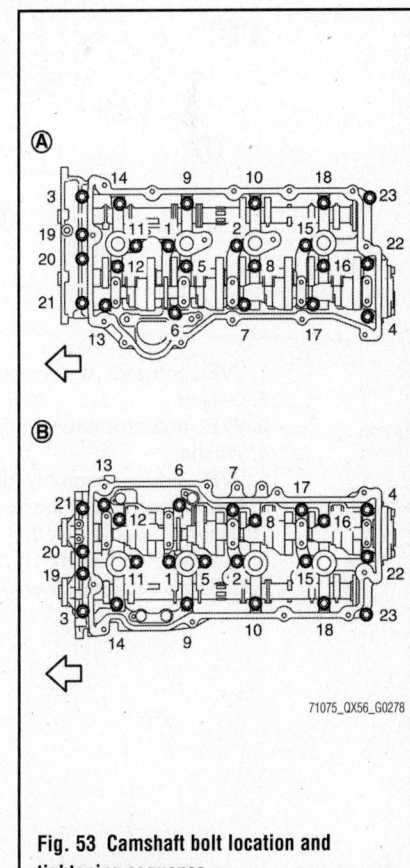

71075_QX56_G0278

Fig. 53 Camshaft bolt location and tightening sequence

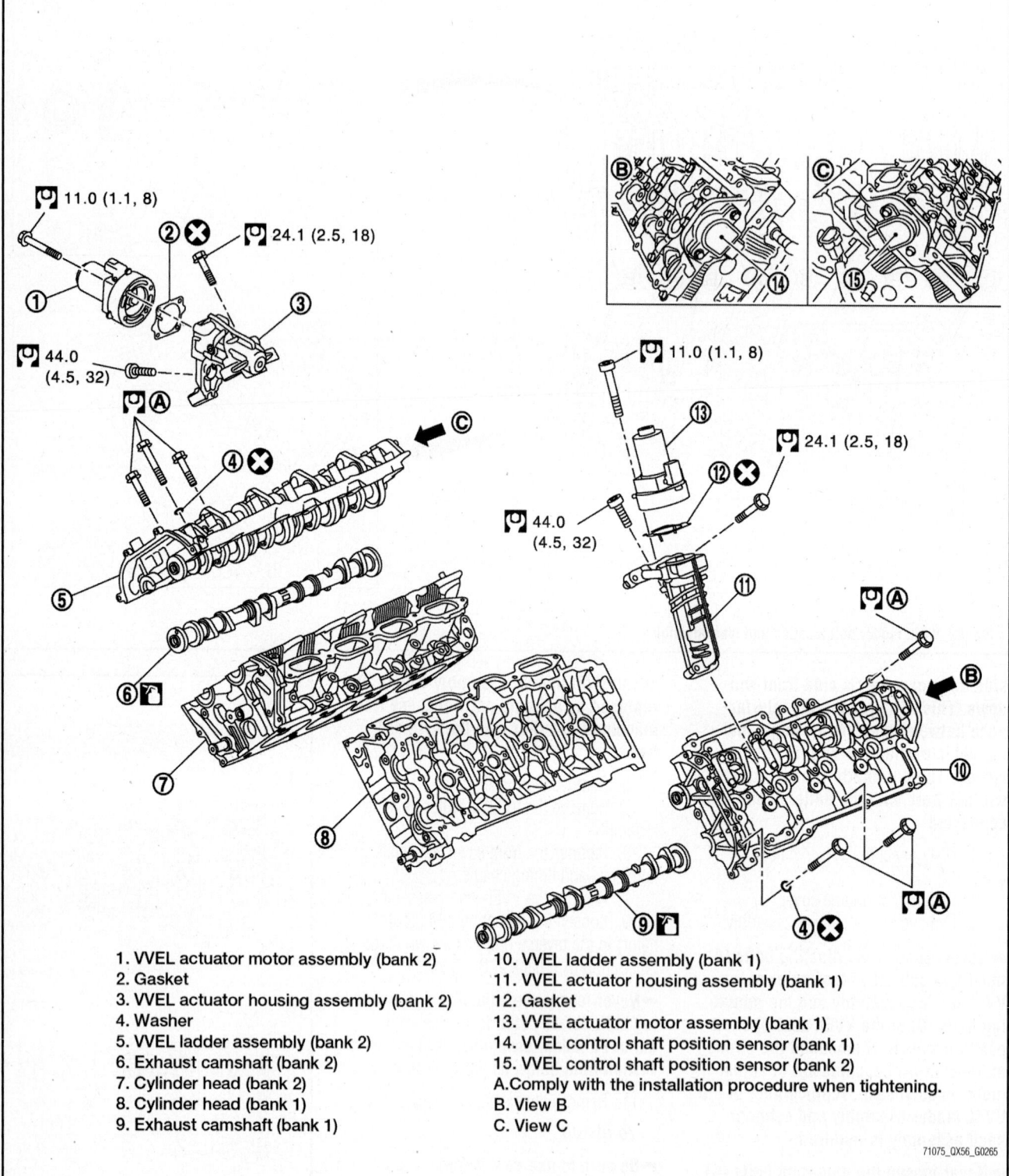

1. VVEL actuator motor assembly (bank 2)
2. Gasket
3. VVEL actuator housing assembly (bank 2)
4. Washer
5. VVEL ladder assembly (bank 2)
6. Exhaust camshaft (bank 2)
7. Cylinder head (bank 2)
8. Cylinder head (bank 1)
9. Exhaust camshaft (bank 1)
10. VVEL ladder assembly (bank 1)
11. VVEL actuator housing assembly (bank 1)
12. Gasket
13. VVEL actuator motor assembly (bank 1)
14. VVEL control shaft position sensor (bank 1)
15. VVEL control shaft position sensor (bank 2)
A. Comply with the installation procedure when tightening.
B. View B
C. View C

71075_QX56_G0265

Fig. 54 Camshaft and related components

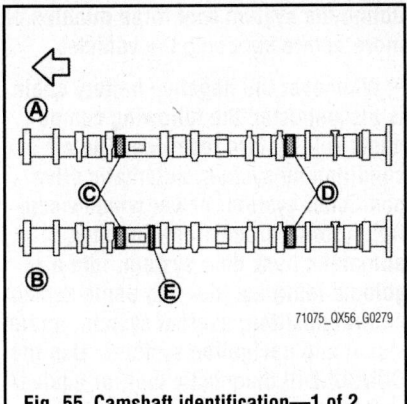

Fig. 55 Camshaft identification—1 of 2

Bank	Paint marks		Identification rib (E)
	M1 (C)	M2 (D)	
Bank 1 (A)	No	Purple	Yes
Bank 2 (B)	No	Purple	No

71075_QX56_G0280

Fig. 56 Camshaft identification—2 of 2

14. Apply a continuous bead (0.134–0.173 inch) of liquid sealer (RTV silicone sealant) to the VVEL ladder assembly. See illustration.

15. Tighten the mounting bolts in the proper sequence and to the following specification: first pass 1 ft. lb., second pass 4 ft. lbs., third pass 8 ft. lbs. Do not reuse washers. See illustration.

16. Install the timing chains.

17. Install the VVEL actuator housing assembly.

18. Inspect the valve clearance.

➡**When a new VVEL ladder assembly is used, Adjust the VVEL control shaft positioner using the CONSULT III diagnostic tool, or equivalent.**

19. Continue the installation in the reverse order of the removal procedure.

20. Be sure to perform the reconnect/relearn procedures.

CRANKSHAFT FRONT SEAL

REMOVAL & INSTALLATION

See Figures 59 and 60.

1. Before servicing the vehicle, refer to the Precautions Section.

➡**If working near and/or around the SRS system and components, be sure to disable the SRS system. After disabling the system wait three minutes or more before servicing the vehicle.**

➡**Whenever the negative battery cable is disconnected the following components will require resetting. The air conditioning system, automatic drive**

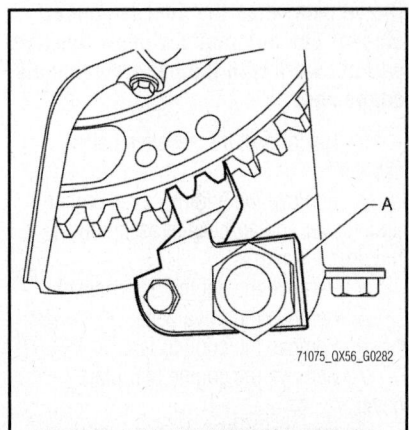

71075_QX56_G0282

Fig. 59 Tool SST:KV10120100 installation

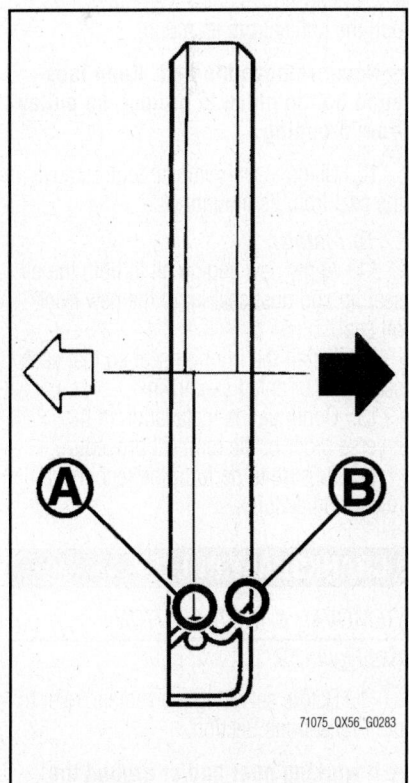

71075_QX56_G0283

Fig. 60 Crankshaft front oil seal installation (A) oil seal lip, (B) dust seal lip

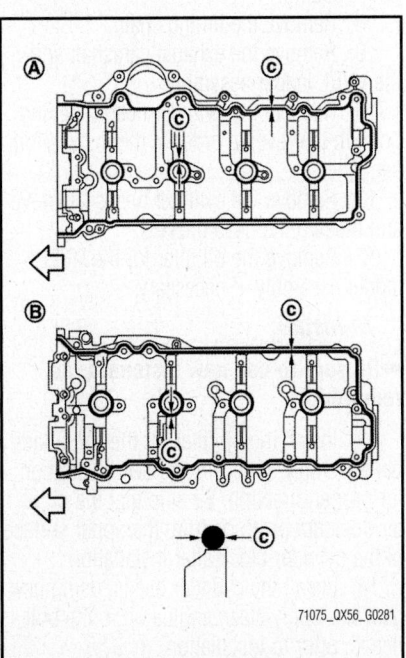

71075_QX56_G0281

Fig. 57 VVEL sealant application (A) bank 1, (B) bank 2, (C) 0.134-0.173 inch of sealant

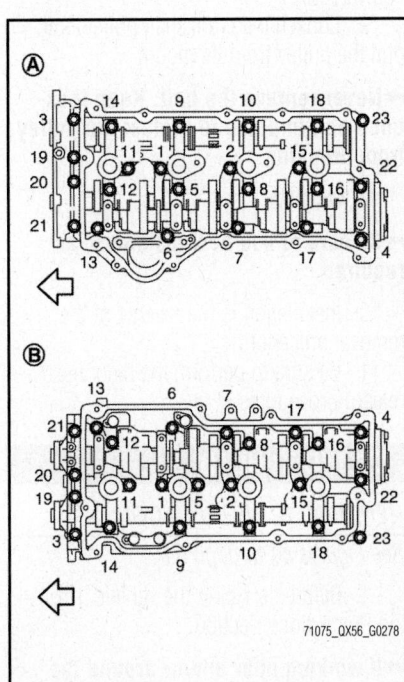

71075_QX56_G0278

Fig. 58 Camshaft bolt location and tightening sequence

positioner system, power window control system, around view monitor, automatic back door system, idle air volume learning, steering angle sensor neutral position, sunroof system, audio visual and navigation systems. Use the CONSULT-III diagnostic tool, or equivalent to perform the required resets. Follow the directions on the screen of the tool, as needed.

➡ Before disconnecting the negative battery cable lower both the driver's side and passenger's side front windows. This will prevent any interference between the window edge and the vehicle when the door is opener/closed. The automatic window function will not work with the battery cable disconnected.

2. Disconnect the negative battery cable.
3. Remove the engine cover, engine undercover and air cleaner assembly, as required for access.
4. Remove the engine under cover.
5. Remove the drive belt.
6. Remove the cooling fan.
7. Remove the engine rear plate cover.
8. Install tool SST: KV10120100, or equivalent.
9. Loosen the crankshaft pulley bolt, pull the pulley from its mount.

➡ Never remove the bolt. Keep loosened bolt in place to protect the pulley from dropping.

10. Using a seal removal tool, remove the seal from its mounting.

To install:

11. Apply new engine oil to both the oil seal lip and dust seal lip of the new front oil seal.
12. Install the front oil seal so that each seal lip is oriented as shown.
13. Continue the installation in the reverse order of the removal procedure.
14. Be sure to perform the reconnect/relearn procedures.

CRANKSHAFT PULLEY

REMOVAL & INSTALLATION

See Figure 59.

1. Before servicing the vehicle, refer to the Precautions Section.

➡ If working near and/or around the SRS system and components, be sure to disable the SRS system. After

disabling the system wait three minutes or more before servicing the vehicle.

➡ Whenever the negative battery cable is disconnected the following components will require resetting. The air conditioning system, automatic drive positioner system, power window control system, around view monitor, automatic back door system, idle air volume learning, steering angle sensor neutral position, sunroof system, audio visual and navigation systems. Use the CONSULT-III diagnostic tool, or equivalent to perform the required resets. Follow the directions on the screen of the tool, as needed.

➡ Before disconnecting the negative battery cable lower both the driver's side and passenger's side front windows. This will prevent any interference between the window edge and the vehicle when the door is opener/closed. The automatic window function will not work with the battery cable disconnected.

2. Disconnect the negative battery cable.
3. Remove the engine cover, engine undercover and air cleaner assembly, as required for access.
4. Remove the engine under cover.
5. Remove the drive belt.
6. Remove the cooling fan.
7. Remove the engine rear plate cover.
8. Install tool SST:KV10120100, or equivalent.
9. Loosen the crankshaft pulley bolt, pull the pulley from its mount.

➡ Never remove the bolt. Keep loosened bolt in place to protect the pulley from dropping.

To install:

➡ Be sure to use new fasteners, as required.

10. Installation is the reverse of the removal procedure.
11. Be sure to perform the reconnect/relearn procedures.

CYLINDER HEAD

REMOVAL & INSTALLATION

See Figures 61 through 64.

1. Before servicing the vehicle, refer to the Precautions Section.

➡ If working near and/or around the SRS system and components, be sure to disable the SRS system. After dis-

abling the system wait three minutes or more before servicing the vehicle.

➡ Whenever the negative battery cable is disconnected the following components will require resetting. The air conditioning system, automatic drive positioner system, power window control system, around view monitor, automatic back door system, idle air volume learning, steering angle sensor neutral position, sunroof system, audio visual and navigation systems. Use the CONSULT-III diagnostic tool, or equivalent to perform the required resets. Follow the directions on the screen of the tool, as needed.

➡ Before disconnecting the negative battery cable lower both the driver's side and passenger's side front windows. This will prevent any interference between the window edge and the vehicle when the door is opener/closed. The automatic window function will not work with the battery cable disconnected.

2. Disconnect the negative battery cable.
3. Remove the engine room cover and air cleaner assembly.
4. Remove the rocker cover. Remove the spark plugs.
5. Remove the intake manifold.
6. Remove the exhaust manifold.
7. Remove the thermostat and water inlet housing.
8. Remove the timing chain.
9. Remove the exhaust camshaft and the VVEL ladder assembly.
10. Remove the cylinder head retaining bolts in the reverse order of the installation sequence.
11. Remove the cylinder heads. Remove and discard the head gaskets.
12. Remove the oil filter for the VVEL ladder assembly, if necessary.

To install:

➡ Be sure to use new fasteners, as required.

13. Install the oil filter for the VVEL ladder assembly, if necessary. See illustration for proper direction. Be sure that the oil filter does not protrude from the upper surface of the cylinder block after installation.
14. Install the cylinder heads, using new gaskets. Apply clean engine oil to the bolt heads, prior to installation.

➡ If the head bolts are reused be sure to check their outer diameter before installation. Bolts are tightened by a

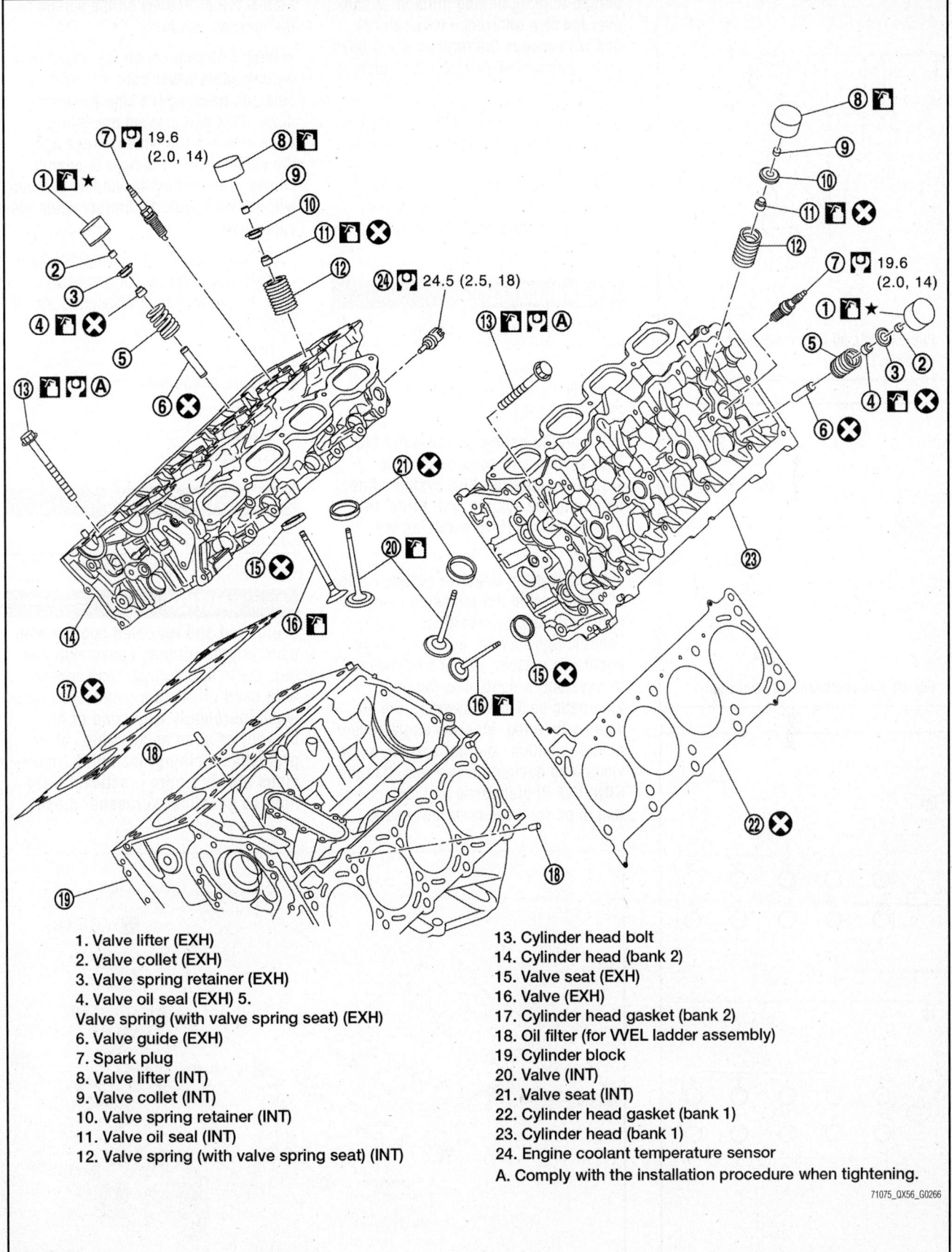

1. Valve lifter (EXH)
2. Valve collet (EXH)
3. Valve spring retainer (EXH)
4. Valve oil seal (EXH) 5.
Valve spring (with valve spring seat) (EXH)
6. Valve guide (EXH)
7. Spark plug
8. Valve lifter (INT)
9. Valve collet (INT)
10. Valve spring retainer (INT)
11. Valve oil seal (INT)
12. Valve spring (with valve spring seat) (INT)

13. Cylinder head bolt
14. Cylinder head (bank 2)
15. Valve seat (EXH)
16. Valve (EXH)
17. Cylinder head gasket (bank 2)
18. Oil filter (for VVEL ladder assembly)
19. Cylinder block
20. Valve (INT)
21. Valve seat (INT)
22. Cylinder head gasket (bank 1)
23. Cylinder head (bank 1)
24. Engine coolant temperature sensor

A. Comply with the installation procedure when tightening.

71075_QX56_G0266

Fig. 61 Cylinder head and related components

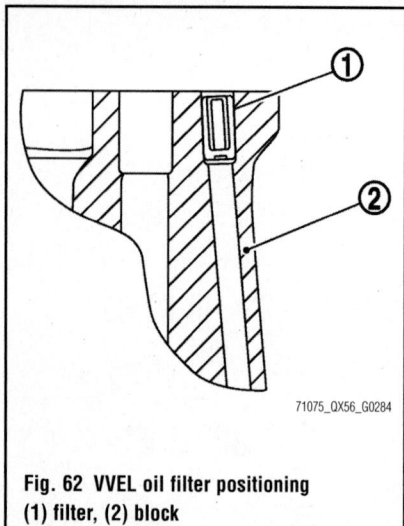

**Fig. 62 VVEL oil filter positioning
(1) filter, (2) block**

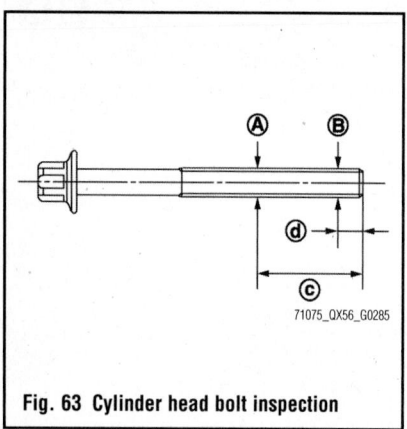

Fig. 63 Cylinder head bolt inspection

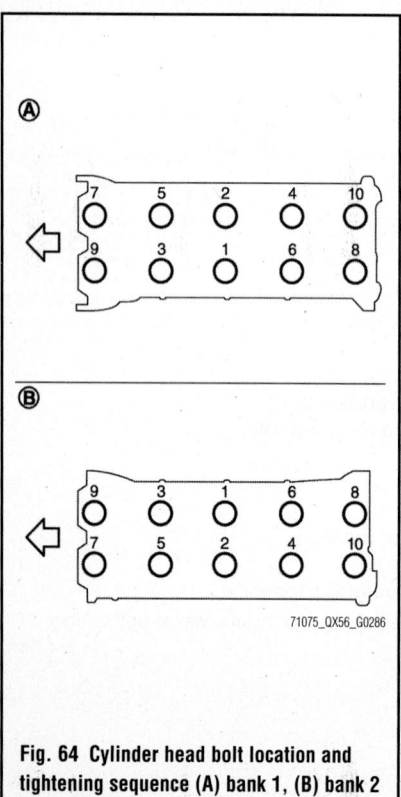

**Fig. 64 Cylinder head bolt location and
tightening sequence (A) bank 1, (B) bank 2**

plastic zone tightening method. Whenever the size difference between (B) and (A) exceeds the limit (B-A = 0.0071 inch) replace the bolt. C = 2.17 inch, D = 0.47 inch.

15. Tighten the head bolts in the proper sequence and to the proper torque specification.

16. Continue the installation in the reverse order of the removal procedure.

17. Be sure to perform the reconnect/relearn procedures.

ENGINE COVER

REMOVAL & INSTALLATION
See Figure 65.

1. Before servicing the vehicle, refer to the Precautions Section.

➡ If working near and/or around the SRS system and components, be sure to disable the SRS system. After disabling the system wait three minutes or more before servicing the vehicle.

➡ Whenever the negative battery cable is disconnected the following components will require resetting. The air conditioning system, automatic drive positioner system, power window control system, around view monitor, automatic back door system, idle air volume learning, steering angle sensor neutral position, sunroof system, audio visual and navigation systems. Use the CONSULT-III diagnostic tool, or equivalent to perform the required resets.

Follow the directions on the screen of the tool, as needed.

➡ Before disconnecting the negative battery cable lower both the driver's side and passenger's side front windows. This will prevent any interference between the window edge and the vehicle when the door is opener/closed. The automatic window function will not work with the battery cable disconnected.

2. Disconnect the negative battery cable.
3. Remove the cover retaining bolts.
4. Pull forward to disengage the snap fit mounts.

To install:

➡ Be sure to use new fasteners, as required.

5. Installation is the reverse of the removal procedure.

ENGINE OIL & FILTER

OIL LEVEL CHECK
See Figure 66.

✳✳ CAUTION

Prolonged and repeated contact with used engine oil may cause skin cancer. Try to avoid direct skin contact with used oil. If skin contact is made, wash thoroughly with soap or hand cleaner as soon as possible. Wear protective clothing, including impervious gloves where practicable. Do not use gasoline, kerosene, diesel

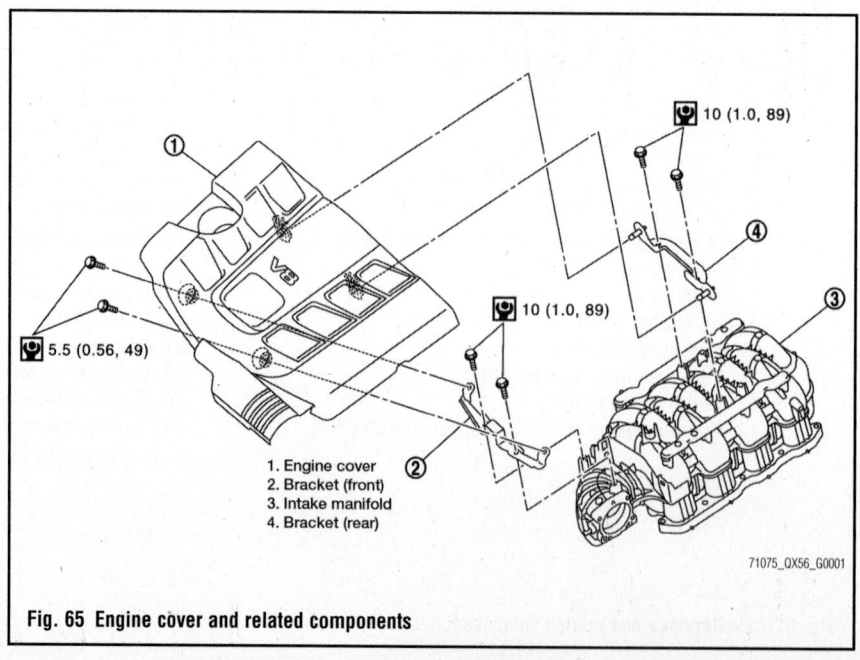

1. Engine cover
2. Bracket (front)
3. Intake manifold
4. Bracket (rear)

Fig. 65 Engine cover and related components

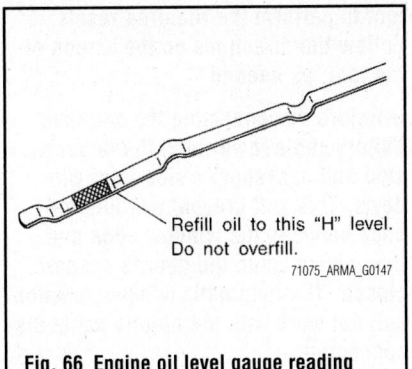

Refill oil to this "H" level.
Do not overfill.

71075_ARMA_G0147

Fig. 66 Engine oil level gauge reading

fuel, gas oil, thinners, or solvents for cleaning skin. Where there is a risk of eye contact, eye protection should be worn, for example, chemical goggles or face shields; in addition an eye wash facility should be provided.

➡Be sure to check the engine oil with the vehicle engine OFF and parked on a level surface. If the engine was running prior to checking the oil, turn it off and wait 10 minutes.

1. Before servicing the vehicle, refer to the Precautions Section.
2. Remove the oil level gauge and wipe it clean with a shop towel. Be sure to properly dispose of the used shop towel.
3. Insert the gauge into its mounting.
4. Remove the oil level gauge and check the reading. See illustration.
5. Correct, oil level as required. Be sure to use the proper grade and type engine oil.

OIL & FILTER CHANGE

1. Before servicing the vehicle, refer to the Precautions Section.

➡If working near and/or around the SRS system and components, be sure to disable the SRS system. After disabling the system wait three minutes or more before servicing the vehicle.

➡Whenever the negative battery cable is disconnected the following components will require resetting. The air conditioning system, automatic drive positioner system, power window control system, around view monitor, automatic back door system, idle air volume learning, steering angle sensor neutral position, sunroof system, audio visual and navigation systems. Use the CONSULT-III diagnostic tool, or equivalent to perform the required resets. Follow the directions on the screen of the tool, as needed.

➡Before disconnecting the negative battery cable lower both the driver's side and passenger's side front windows. This will prevent any interference between the window edge and the vehicle when the door is opener/closed. The automatic window function will not work with the battery cable disconnected.

➡Be sure that the engine is cold and the engine oil is cold.

2. Disconnect the negative battery cable.
3. Raise and safely support the vehicle.
4. Remove the undercover, if necessary.
5. Remove the oil drain plug. Discard the washer.
6. Drain the engine oil into a suitable container. Properly dispose of used engine oil.
7. Using oil filter wrench remove the oil filter. Discard the filter.

To install:

➡Be sure to use new fasteners, as required

8. Install the drain plug. Use a new washer and tighten to 25 ft. lbs. (34 Nm).
9. Coat the oil filter seal with clean engine oil prior to installation.

10. Do not over tighten the filter to its mounting. Tightening specification should be 13 ft. lbs. (17.7 Nm).
11. Fill the engine with the proper grade and type engine oil.
12. Start the engine and check for leaks, correct as required.

EXHAUST MANIFOLD

REMOVAL & INSTALLATION

See Figures 67 through 69.

1. Before servicing the vehicle, refer to the Precautions Section.

➡If working near and/or around the SRS system and components, be sure to disable the SRS system. After disabling the system wait three minutes or more before servicing the vehicle.

➡Whenever the negative battery cable is disconnected the following components will require resetting. The air conditioning system, automatic drive positioner system, power window control system, around view monitor, automatic back door system, idle air volume learning, steering angle sensor neutral position, sunroof system, audio visual and navigation systems. Use the

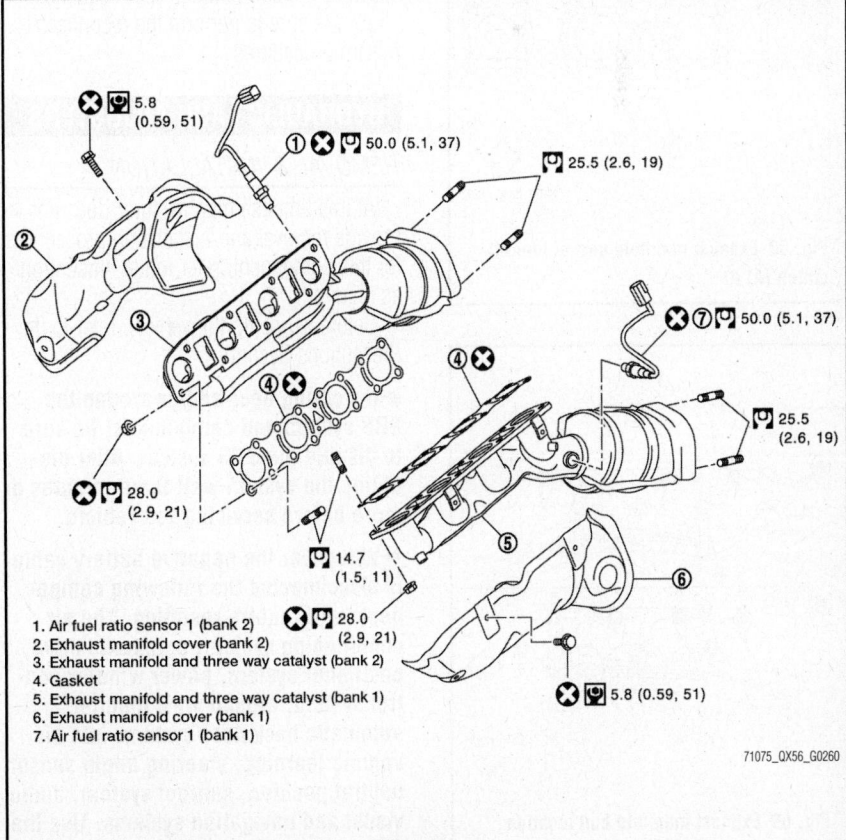

1. Air fuel ratio sensor 1 (bank 2)
2. Exhaust manifold cover (bank 2)
3. Exhaust manifold and three way catalyst (bank 2)
4. Gasket
5. Exhaust manifold and three way catalyst (bank 1)
6. Exhaust manifold cover (bank 1)
7. Air fuel ratio sensor 1 (bank 1)

71075_QX56_G0260

Fig. 67 Exhaust manifold and related components

CONSULT-III diagnostic tool, or equivalent to perform the required resets. Follow the directions on the screen of the tool, as needed.

➡️Before disconnecting the negative battery cable lower both the driver's side and passenger's side front windows. This will prevent any interference between the window edge and the vehicle when the door is opener/closed. The automatic window function will not work with the battery cable disconnected.

2. Properly discharge the air conditioning system.

3. Disconnect the negative battery cable.

4. Remove the engine under cover.

5. Remove the engine cover. Remove the air cleaner assembly.

6. Drain the radiator. Be sure to properly dispose of used coolant.

7. Remove the drive belt.

8. Remove the coolant reservoir tank.

9. Remove the power steering oil pump.

10. Remove the radiator.

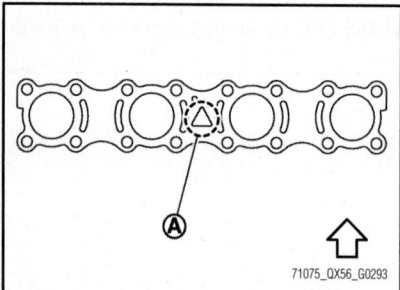

Fig. 68 Exhaust manifold gasket identification (A) ID

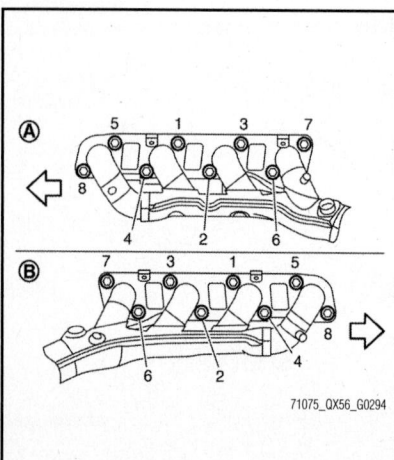

Fig. 69 Exhaust manifold bolt location and tightening sequence

11. Raise and safely support the vehicle. Remove the front tire and wheel assemblies.

12. Remove the air conditioning compressor. Plug the refrigerant lines.

13. Remove the alternator and bracket.

14. Remove the exhaust front tubes, both sides.

15. Remove the front driveshaft.

16. Disconnect the lower steering joint.

17. Remove the air/fuel sensors.

➡️Never remove sensors unless required, they are not reusable.

18. Remove the exhaust manifold cover.

19. Remove the oil level gauge guide.

20. Loosen the manifold retaining nuts in the reverse order of the tightening sequence. Remove the nuts.

21. Remove the exhaust manifold from the vehicle. Discard the gaskets.

To install:

➡️Be sure to use new fasteners, as required.

22. Installation is the reverse of the removal procedure.

23. Install new gaskets with the top of the triangular UP mark on it facing up and its coated face (black side) toward the exhaust manifold side.

24. Tighten the retaining nuts/bolts to specification and in the proper sequence.

25. Be sure to perform the reconnect/relearn procedures.

HYDRAULIC LASH ADJUSTERS

REMOVAL & INSTALLATION

At this time the manufacturer does not provide removal and installation procedures for this component, refer to the illustration as required.

Before servicing the vehicle, refer to the Precautions Section.

➡️If working near and/or around the SRS system and components, be sure to disable the SRS system. After disabling the system wait three minutes or more before servicing the vehicle.

➡️Whenever the negative battery cable is disconnected the following components will require resetting. The air conditioning system, automatic drive positioner system, power window control system, around view monitor, automatic back door system, idle air volume learning, steering angle sensor neutral position, sunroof system, audio visual and navigation systems. Use the CONSULT-III diagnostic tool, or equiva-

lent to perform the required resets. Follow the directions on the screen of the tool, as needed.

➡️Before disconnecting the negative battery cable lower both the driver's side and passenger's side front windows. This will prevent any interference between the window edge and the vehicle when the door is opener/closed. The automatic window function will not work with the battery cable disconnected.

INTAKE MANIFOLD

REMOVAL & INSTALLATION

See Figures 70 through 72.

1. Before servicing the vehicle, refer to the Precautions Section.

➡️If working near and/or around the SRS system and components, be sure to disable the SRS system. After disabling the system wait three minutes or more before servicing the vehicle.

➡️Whenever the negative battery cable is disconnected the following components will require resetting. The air conditioning system, automatic drive positioner system, power window control system, around view monitor, automatic back door system, idle air volume learning, steering angle sensor neutral position, sunroof system, audio visual and navigation systems. Use the CONSULT-III diagnostic tool, or equivalent to perform the required resets. Follow the directions on the screen of the tool, as needed.

➡️Before disconnecting the negative battery cable lower both the driver's side and passenger's side front windows. This will prevent any interference between the window edge and the vehicle when the door is opener/closed. The automatic window function will not work with the battery cable disconnected.

2. Disconnect the negative battery cable.

3. Drain the cooling system. Be sure to properly dispose of used coolant.

4. Remove the engine cover.

5. Remove the air cleaner assembly.

6. Disconnect the MAP sensor harness connector.

7. Remove the EVAP canister purge control solenoid valve.

8. Disconnect the PCV valve hoses from the manifold.

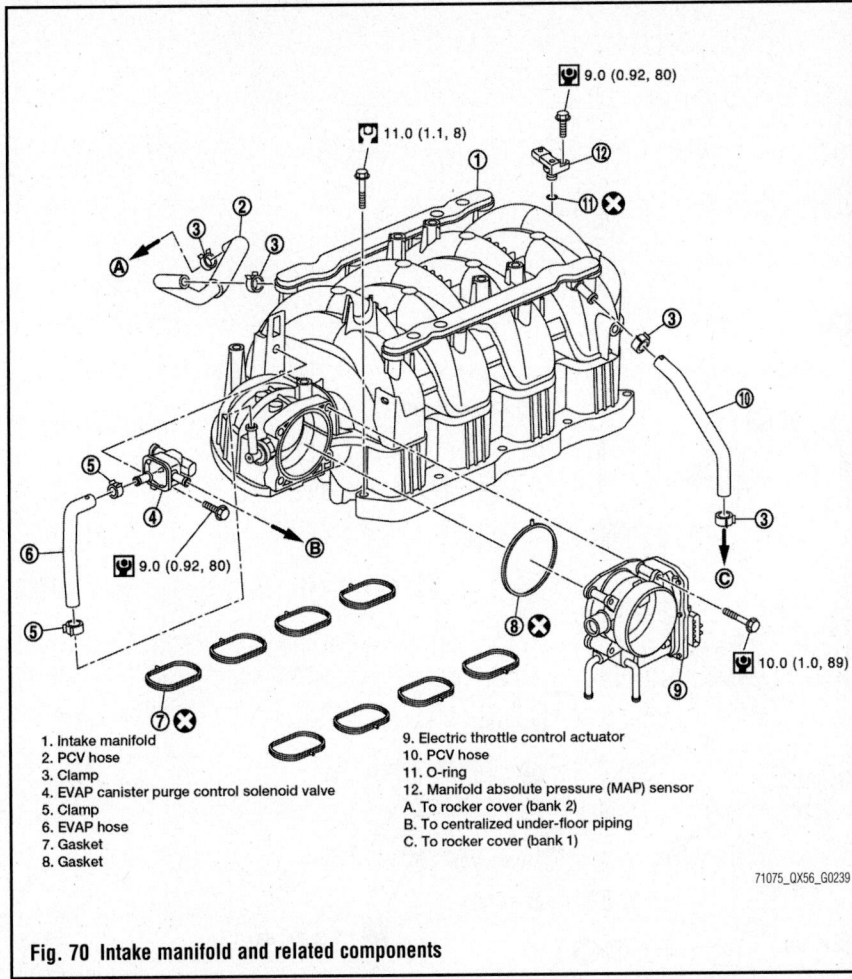

1. Intake manifold
2. PCV hose
3. Clamp
4. EVAP canister purge control solenoid valve
5. Clamp
6. EVAP hose
7. Gasket
8. Gasket
9. Electric throttle control actuator
10. PCV hose
11. O-ring
12. Manifold absolute pressure (MAP) sensor
A. To rocker cover (bank 2)
B. To centralized under-floor piping
C. To rocker cover (bank 1)

71075_QX56_G0239

Fig. 70 Intake manifold and related components

➡ **When removing only the manifold, move the electric throttle control actuator without disconnecting the water hoses.**

9. Remove the throttle control actuator retaining bolts. Loosen the retaining bolts in the reverse order of the tightening sequence. Remove the component.

➡ **Never disassemble the throttle control actuator. Handle this component carefully to avoid any impact.**

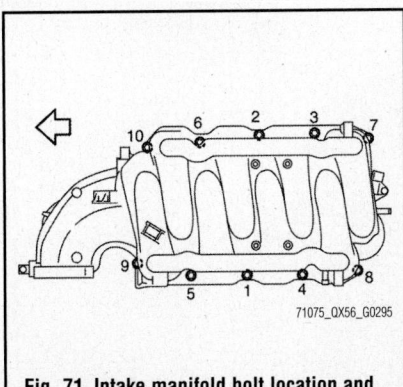

71075_QX56_G0295

Fig. 71 Intake manifold bolt location and tightening sequence

10. Remove the intake manifold retaining bolts in the reverse order of the tightening sequence.

11. Remove the intake manifold from the engine. Discard the gaskets. Discard the O-rings.

12. As required, remove the MAP sensor.

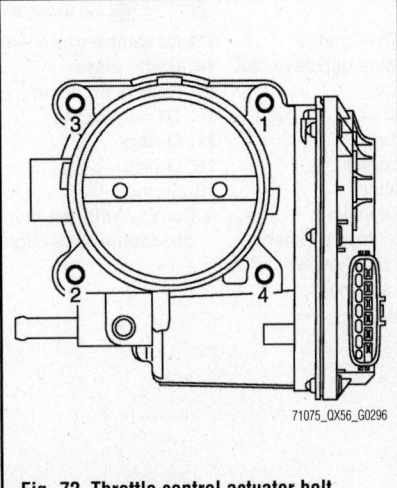

71075_QX56_G0296

Fig. 72 Throttle control actuator bolt location and tightening sequence

To install:

➡ **Be sure to use new fasteners, as required.**

13. Installation is the reverse of the removal procedure.

14. Be sure to use new gaskets and O-rings. Tighten the intake manifold bolts to the proper torque specification and in the proper order.

15. Tighten the throttle control actuator bolts in the proper sequence.

16. Be sure to perform the reconnect/relearn procedures.

OIL PAN

REMOVAL & INSTALLATION

Lower

See Figures 73 through 75.

1. Before servicing the vehicle, refer to the Precautions Section.

➡ **If working near and/or around the SRS system and components, be sure to disable the SRS system. After disabling the system wait three minutes or more before servicing the vehicle.**

➡ **Whenever the negative battery cable is disconnected the following components will require resetting. The air conditioning system, automatic drive positioner system, power window control system, around view monitor, automatic back door system, idle air volume learning, steering angle sensor neutral position, sunroof system, audio visual and navigation systems. Use the CONSULT-III diagnostic tool, or equivalent to perform the required resets. Follow the directions on the screen of the tool, as needed.**

➡ **Before disconnecting the negative battery cable lower both the driver's side and passenger's side front windows. This will prevent any interference between the window edge and the vehicle when the door is opener/closed. The automatic window function will not work with the battery cable disconnected.**

2. Disconnect the negative battery cable.

3. Raise and safely support the vehicle. Drain the engine oil. Be sure to properly dispose of used oil.

4. Remove the protector A and protector B.

5. Remove the front suspension rear crossmember.

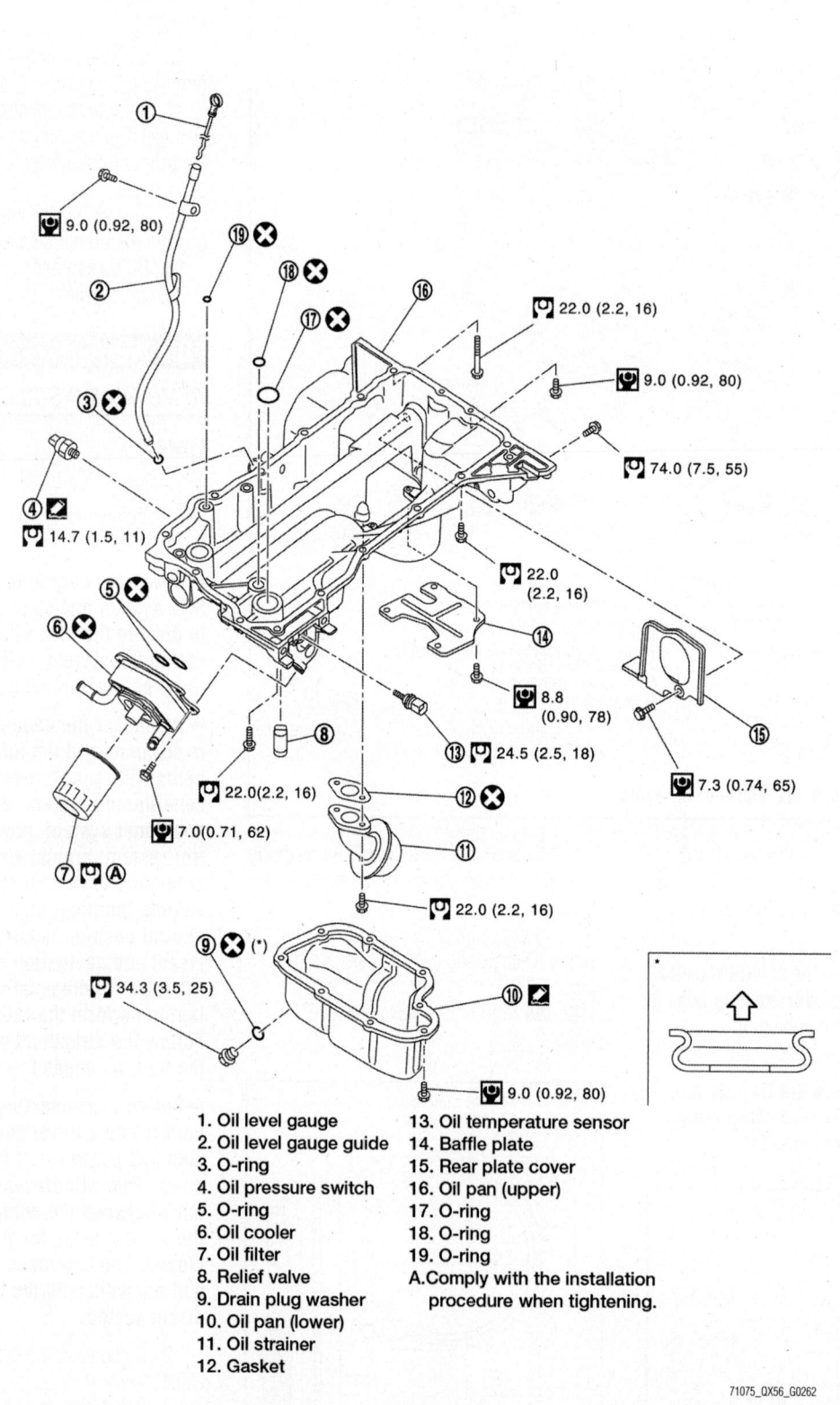

1. Oil level gauge
2. Oil level gauge guide
3. O-ring
4. Oil pressure switch
5. O-ring
6. Oil cooler
7. Oil filter
8. Relief valve
9. Drain plug washer
10. Oil pan (lower)
11. Oil strainer
12. Gasket
13. Oil temperature sensor
14. Baffle plate
15. Rear plate cover
16. Oil pan (upper)
17. O-ring
18. O-ring
19. O-ring
A. Comply with the installation procedure when tightening.

71075_QX56_G0262

Fig. 73 Lower oil pan/strainer and related components

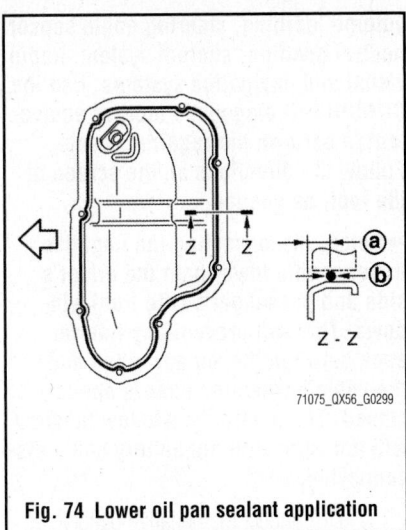

Fig. 74 Lower oil pan sealant application

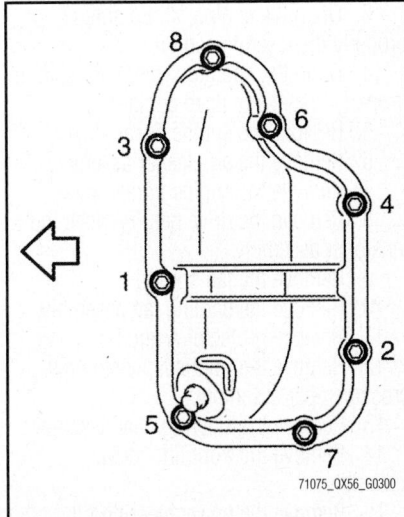

Fig. 75 Lower oil pan bolt location and tightening sequence

6. Remove the oil pan retaining bolts in the reverse order of the tightening sequence.

7. Remove the oil pan from its mounting using tool SST: KV10111100, or equivalent. Slide the seal cutter tool by tapping on the side of the tool with a hammer.

➡**Be careful not to damage the mating surfaces. Never use a screwdriver.**

To install:

➡**Be sure to use new fasteners, as required.**

8. Installation is the reverse of the removal procedure.

9. Be sure to use a new gasket and O-ring.

10. Apply a continuous bead of liquid gasket to the oil pan. A= 0.295–0.374 inch, B= 0.157–0.197 inch. Installation must be done within five minutes after coating. See illustration.

11. Tighten the retaining bolts to specification and in the proper sequence.

➡**Wait at least thirty minutes after pan installation before filling the engine with engine oil.**

12. Be sure to fill the engine with the proper grade and type engine oil.

13. Start the engine and check for leaks, correct as required.

14. Be sure to perform the reconnect/relearn procedures.

Upper

See Figures 76 through 78.

1. Before servicing the vehicle, refer to the Precautions Section.

➡**If working near and/or around the SRS system and components, be sure to disable the SRS system. After disabling the system wait three minutes or more before servicing the vehicle.**

➡**Whenever the negative battery cable is disconnected the following components will require resetting. The air conditioning system, automatic drive positioner system, power window control system, around view monitor, automatic back door system, idle air volume learning, steering angle sensor neutral position, sunroof system, audio visual and navigation systems. Use the CONSULT-III diagnostic tool, or equivalent to perform the required resets. Follow the directions on the screen of the tool, as needed.**

➡**Before disconnecting the negative battery cable lower both the driver's side and passenger's side front windows. This will prevent any interference between the window edge and the vehicle when the door is opener/closed. The automatic window function will not work with the battery cable disconnected.**

2. Disconnect the negative battery cable.

3. Raise and safely support the vehicle. Drain the engine oil. Be sure to properly dispose of used oil.

4. Remove the oil filter.

5. Remove the oil cooler.

6. Reposition the air conditioning compressor out of the way.

7. Remove the oil level gauge and guide. Remove the oil pressure and oil temperature switch, if necessary.

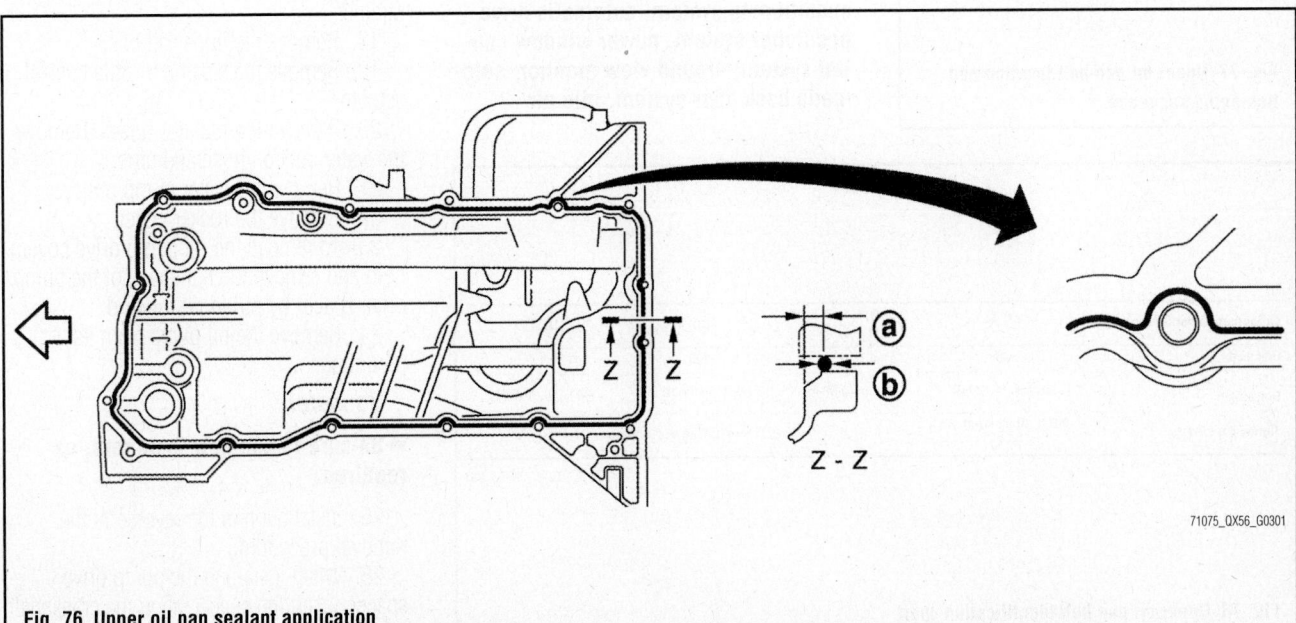

Fig. 76 Upper oil pan sealant application

8. Remove the rear plate cover.

9. Remove the protector A and protector B.

10. Remove the front suspension rear crossmember.

11. Remove the steering gear.

12. Remove the front final drive assembly.

13. Remove the lower oil pan.

14. Remove the oil strainer.

15. Remove the upper oil pan retaining bolts to the transmission assembly.

16. Remove the retaining bolts in the reverse order of the tightening sequence.

➡ **Disregard bolts 9 and 16 when loosening.**

17. Using a suitable flat bladed tool remove the upper oil pan from its mounting on the engine.

To install:

➡ **Be sure to use new fasteners, as required.**

18. Installation is the reverse of the removal procedure.

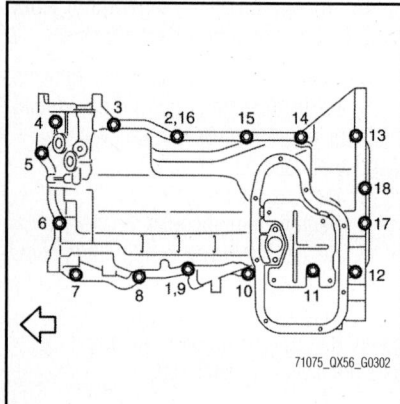

Fig. 77 Upper oil pan bolt location and tightening sequence

19. Be sure to use a new gasket and O-ring.

20. Apply a continuous bead of liquid gasket to the oil pan. A= 0.295–0.374 inch, B= 0.157–0.197 inch. Installation must be done within five minutes after coating. See illustration.

21. Tighten the retaining bolts to specification and in the proper sequence.

➡ **Tighten bolts 1 and 2 in two steps. The numerical order number 9 and 16 shown second steps, see illustration. There are four types of bolts.**

➡ **Wait at least thirty minutes after pan installation before filling the engine with engine oil.**

22. Be sure to fill the engine with the proper grade and type engine oil.

23. Start the engine and check for leaks, correct as required.

24. Be sure to perform the reconnect/relearn procedures.

OIL PUMP

REMOVAL & INSTALLATION

See Figures 79 and 80.

1. Before servicing the vehicle, refer to the Precautions Section.

➡ **If working near and/or around the SRS system and components, be sure to disable the SRS system. After disabling the system wait three minutes or more before servicing the vehicle.**

➡ **Whenever the negative battery cable is disconnected the following components will require resetting. The air conditioning system, automatic drive positioner system, power window control system, around view monitor, automatic back door system, idle air volume learning, steering angle sensor neutral position, sunroof system, audio visual and navigation systems. Use the CONSULT-III diagnostic tool, or equivalent to perform the required resets. Follow the directions on the screen of the tool, as needed.**

➡ **Before disconnecting the negative battery cable lower both the driver's side and passenger's side front windows. This will prevent any interference between the window edge and the vehicle when the door is opener/closed. The automatic window function will not work with the battery cable disconnected.**

2. Disconnect the negative battery cable.

3. Drain the engine oil. Be sure to properly dispose of used oil.

4. Drain the cooling system. Be sure to properly dispose of used coolant.

5. Remove the engine cover.

6. Remove the air cleaner assembly.

7. Remove the engine under cover.

8. Remove the drive belt. Remove the tensioner assembly.

9. Remove the fan shroud.

10. Remove the cooling fan assembly.

11. Remove protector A and B

12. Remove the front suspension rear crossmember assembly.

13. Remove the steering gear assembly.

14. Remove the front final drive assembly.

15. Remove the power steering oil pump.

16. Remove the alternator.

17. Remove the lower oil pan and strainer.

18. Remove the upper oil pan.

19. Remove the electric throttle control actuator.

20. Remove the radiator hoses. Remove the water suction hose and pipe.

21. Remove the water pump pulley.

22. Remove the rocker cover.

23. To remove the oil pump drive cover, hold and remove the flat space of the pump drive spacer by pulling it forward.

24. Remove the oil pump from its mounting.

To install:

➡ **Be sure to use new fasteners, as required.**

25. Installation is the reverse of the removal procedure.

26. When installing the pump drive spacer, insert the spacer, align the crankshaft

Order number for tightening	17, 18	2 (16), 3, 5, 6, 7, 8, 10, 11, 14, 15	1(9), 4	12, 13
Bolt size	M6	M8		
Bolt length	45 mm (1.77 in)	25 mm (0.98 in)	30.0 mm (1.18 in)	120 mm (4.72 in)
Tightening torque	9.0 N· (0.92 kg-m, 80 in-lb)	22.0 N·m (2.2 kg-m, 16 ft-lb)		

71075_QX56_G0303

Fig. 78 Upper oil pan bolt identification chart

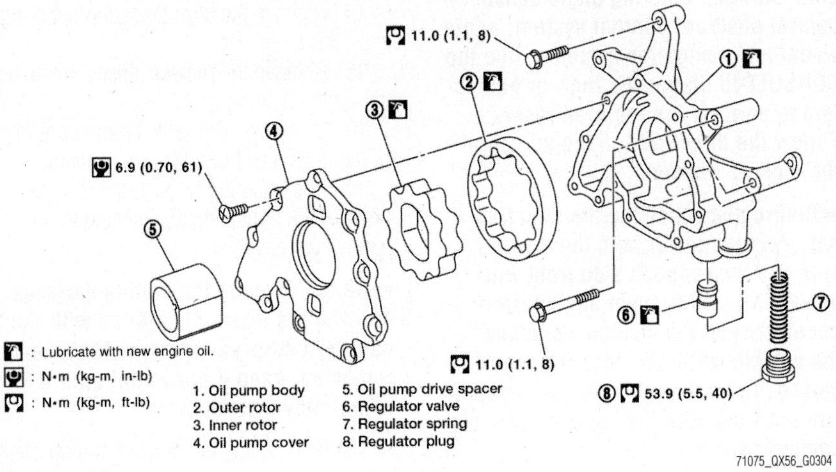

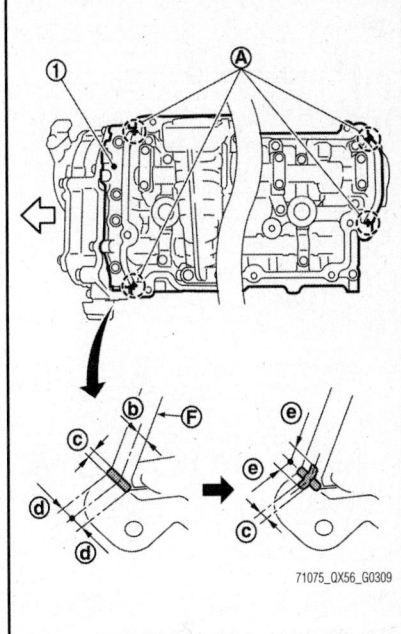

Fig. 79 Engine oil pump and related components

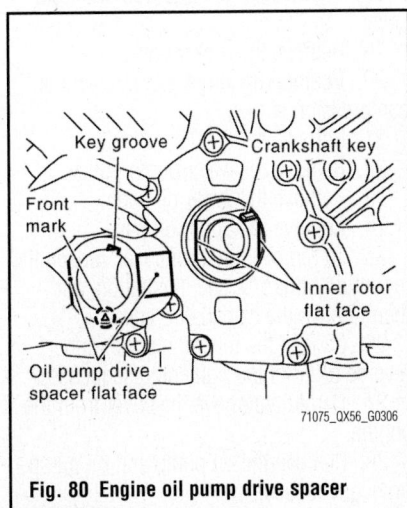

: Lubricate with new engine oil.

: N•m (kg-m, in-lb)

: N•m (kg-m, ft-lb)

1. Oil pump body
2. Outer rotor
3. Inner rotor
4. Oil pump cover
5. Oil pump drive spacer
6. Regulator valve
7. Regulator spring
8. Regulator plug

Fig. 82 Rocker cover sealant application (bank 1 shown)

Fig. 80 Engine oil pump drive spacer installation

key and the flat face of the inner rotor. If they are not aligned, rotate the pump inner rotor by hand. Be sure that each part is aligned and tap lightly until it reaches the end.

27. Be sure to perform the reconnect/relearn procedures.

PISTONS & RINGS

POSITIONING
See Figure 81.

ROCKER COVER

REMOVAL & INSTALLATION
See Figures 82 and 83.

1. Before servicing the vehicle, refer to the Precautions Section.

→If working near and/or around the SRS system and components, be sure to disable the SRS system. After disabling the system wait three minutes or more before servicing the vehicle.

→Whenever the negative battery cable is disconnected the following components will require resetting. The air conditioning system, automatic drive positioner system, power window control system, around view monitor, automatic back door system, idle air volume learning, steering angle sensor neutral position, sunroof system, audio visual and navigation systems. Use the CONSULT-III diagnostic tool, or equivalent to perform the required resets. Follow the directions on the screen of the tool, as needed.

→Before disconnecting the negative battery cable lower both the driver's

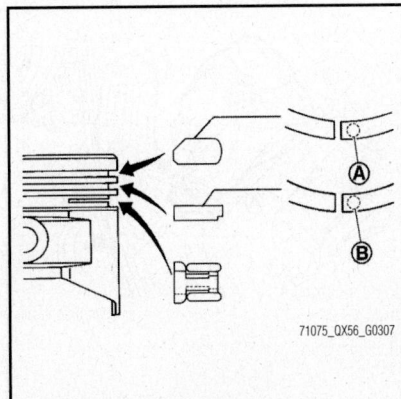

Fig. 81 Engine piston ring positioning (A) top ring, (B) second ring

side and passenger's side front windows. This will prevent any interference between the window edge and the vehicle when the door is opener/closed. The automatic window function will not work with the battery cable disconnected.

2. Disconnect the negative battery cable.

3. Remove the engine cover and bracket.

4. Remove the air cleaner assembly.

5. Reposition the oil level gauge guide.

6. Reposition the power steering pump reservoir bracket.

7. Reposition the EFAP canister purge control solenoid valve.

8. Reposition the fuel feeder hose.

9. Remove the VVEL actuator motor assembly.

10. Remove the ignition coil.

11. Remove the rocker cover retaining bolts in the reverse order of the tightening sequence.

12. Remove the rocker cover from the engine.

To install:

→Be sure to use new fasteners, as required.

13. Be sure to use a new gasket and O-ring.

14. Apply a continuous bead of liquid gasket to the component. Apply sealant on the front and rear side of the engine first.

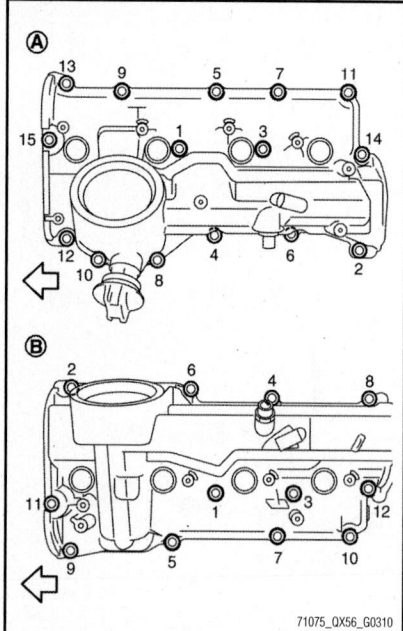

Fig. 83 Rocker cover bolt location and tightening sequence (A) bank 1, (B) bank 2

A= application point, B= 0.16 inch, C= 0.098–0.138 inch, D= 0.20 inch and E= 0.39 inch. Installation must be done within five minutes after coating. See illustration.

15. Tighten the retaining bolts in two steps and in the proper sequence. Step one 18 inch lbs. Step two 73 inch lbs.

➡**Because of limited working space to tighten bolts 7 and 8 you will need tool SST: KV10119300 or equivalent.**

16. Continue the installation in the reverse order of the removal procedure.

17. Be sure to perform the reconnect/relearn procedures.

TIMING CHAIN COVER, CHAIN, TENSIONER, & SPROCKETS

REMOVAL & INSTALLATION

See Figures 84 through 92.

1. Before servicing the vehicle, refer to the Precautions Section.

➡**If working near and/or around the SRS system and components, be sure to disable the SRS system. After disabling the system wait three minutes or more before servicing the vehicle.**

➡**Whenever the negative battery cable is disconnected the following components will require resetting. The air conditioning system, automatic drive positioner system, power window con-** trol system, around view monitor, automatic back door system, idle air volume learning, steering angle sensor neutral position, sunroof system, audio visual and navigation systems. Use the CONSULT-III diagnostic tool, or equivalent to perform the required resets. Follow the directions on the screen of the tool, as needed.

➡**Before disconnecting the negative battery cable lower both the driver's side and passenger's side front windows. This will prevent any interference between the window edge and the vehicle when the door is opener/closed. The automatic window function will not work with the battery cable disconnected.**

2. Properly release the fuel system pressure.

3. Disconnect the negative battery cable.

4. Remove the engine cover. Remove the air cleaner assembly.

5. Remove the engine undercover.

6. Drain the engine coolant. Properly dispose of used coolant.

7. Remove the fan shroud. Remove the fan bracket.

8. Remove the drive belt.

9. Remove the oil level gauge and gauge guide.

10. Reposition the power steering pump out of the way.

11. Remove the alternator.

12. Reposition the power steering reservoir out of the way. Remove the reservoir bracket.

13. Remove the Camshaft Position (CMP) sensors.

➡**Be careful not to drop the sensor. Do not leave it in an area where it can be exposed to magnetism. Never disas-** semble the sensor. Do not drop or shock the sensor.

14. Remove the high pressure fuel pump and lifter.

15. Remove the radiator hoses. Remove the water suction pipe.

16. To remove the valve timing control cover, disconnect the harness connector. Loosen the retaining bolts in the reverse order of the tightening sequence and remove the cover.

➡**Do not damage the mating surfaces. The shaft is internally jointed with the camshaft sprocket center hole. When removing, keep it horizontal until it is completely disconnected.**

17. Remove the intake valve timing control solenoid valve, both sides, if necessary.

➡**This valve is not reusable. Never remove it unless required.**

18. Remove the O-rings from the front cover.

19. Remove the rocker cover.

20. Position the engine at TDC on the compression stroke.

21. Remove the crankshaft pulley.

22. Remove the water pump pulley.

23. Remove the lower oil pan.

24. Remove the upper oil pan.

25. To remove the front cover, loosen the mounting bolts (A), see illustration, and then remove the camshaft bracket.

26. Loosen the retaining bolts in the reverse order of the tightening sequence.

27. Carefully remove the cover from the engine.

28. Remove the oil pump and oil pump drive gear spacer.

29. Remove the front oil seal from the front cover, using a flat bladed tool.

Fig. 84 Engine timing chain tensioner removal (A) spring, (2) guide—1 of 2

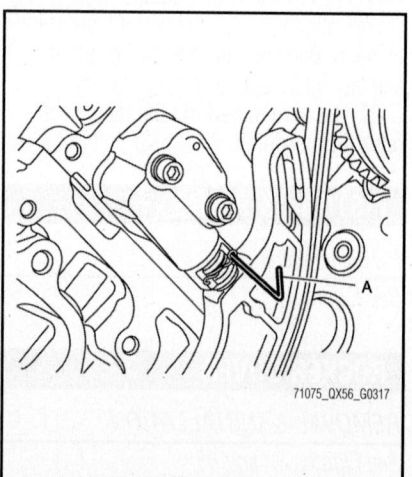

Fig. 85 Engine timing chain tensioner removal (A) stopper pin—2 of 2

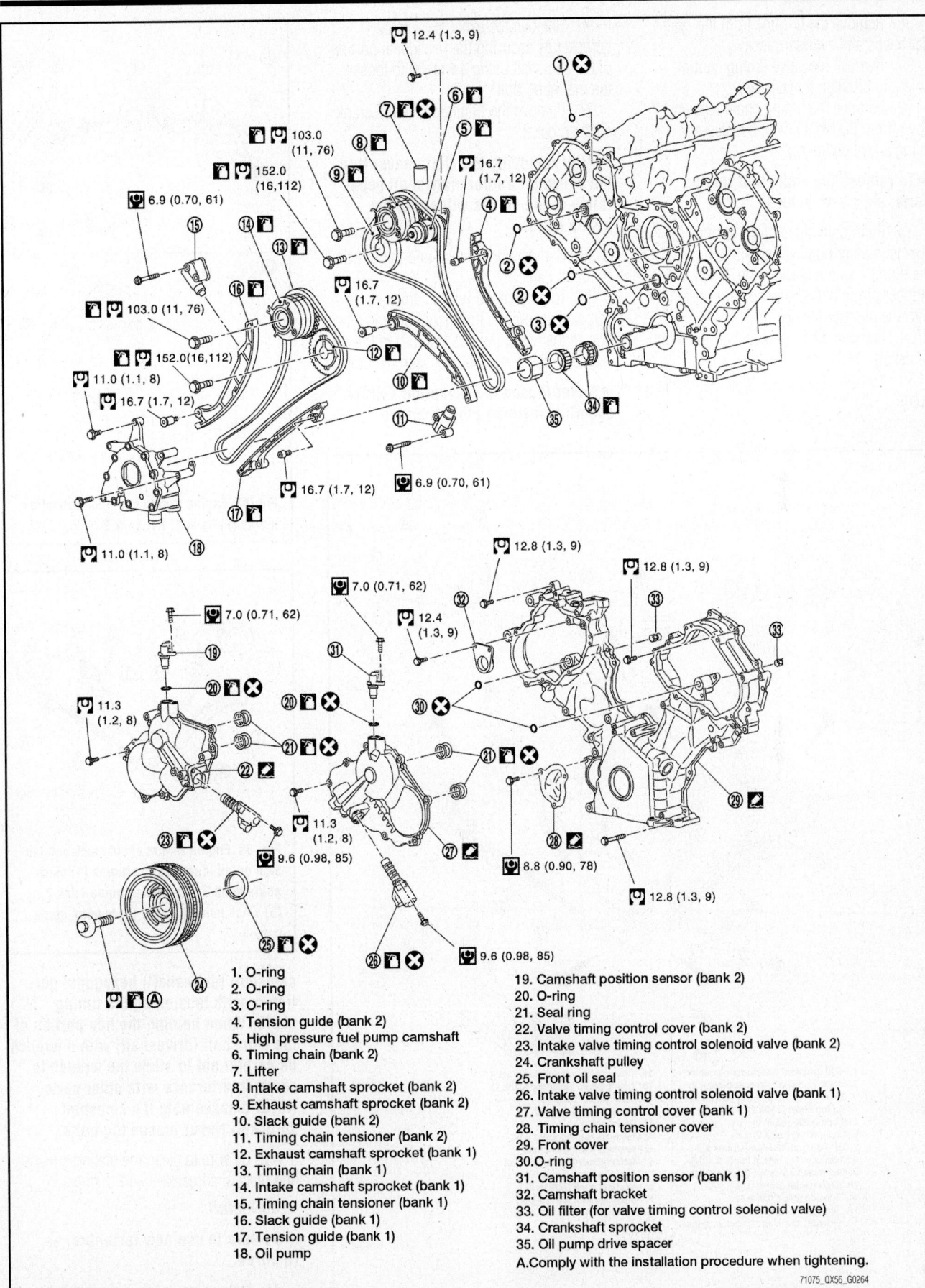

1. O-ring
2. O-ring
3. O-ring
4. Tension guide (bank 2)
5. High pressure fuel pump camshaft
6. Timing chain (bank 2)
7. Lifter
8. Intake camshaft sprocket (bank 2)
9. Exhaust camshaft sprocket (bank 2)
10. Slack guide (bank 2)
11. Timing chain tensioner (bank 2)
12. Exhaust camshaft sprocket (bank 1)
13. Timing chain (bank 1)
14. Intake camshaft sprocket (bank 1)
15. Timing chain tensioner (bank 1)
16. Slack guide (bank 1)
17. Tension guide (bank 1)
18. Oil pump

19. Camshaft position sensor (bank 2)
20. O-ring
21. Seal ring
22. Valve timing control cover (bank 2)
23. Intake valve timing control solenoid valve (bank 2)
24. Crankshaft pulley
25. Front oil seal
26. Intake valve timing control solenoid valve (bank 1)
27. Valve timing control cover (bank 1)
28. Timing chain tensioner cover
29. Front cover
30. O-ring
31. Camshaft position sensor (bank 1)
32. Camshaft bracket
33. Oil filter (for valve timing control solenoid valve)
34. Crankshaft sprocket
35. Oil pump drive spacer
A. Comply with the installation procedure when tightening.

71075_QX56_G0264

Fig. 86 Engine timing chain and related components

30. Remove the O-rings from the cylinder heads and cylinder block.

31. Remove the valve timing control solenoid oil filter, if necessary.

32. Remove the timing chain tensioner cover from the front cover, if necessary, using a seal cutter tool.

➡**To remove the chain and related parts, start with bank 1.**

33. Push both sides of the spring against the spring tension and then press the plunger with a slack guide. Insert a stopper pin into the body hole and retain it with the plunger pushed in.

34. Remove the high pressure fuel pump camshaft.

35. Remove the tension guide and slack guide.

36. Remove the exhaust camshaft sprocket by securing the hexagonal portion of the camshaft using a wrench to loosen the mounting bolt.

37. Remove the timing chain and crankshaft sprocket.

➡**After removing the chain never turn the camshaft and/or crankshaft separately or the valves will strike the piston head.**

38. Use the above steps and remove the chain on bank 2.

39. To remove the intake camshaft sprocket secure the hexagonal portion (located between journal 1 and 2) of the driveshaft using a wrench to loosen the bolt.

➡**Never loosen the mounting bolt by securing anything other than the**

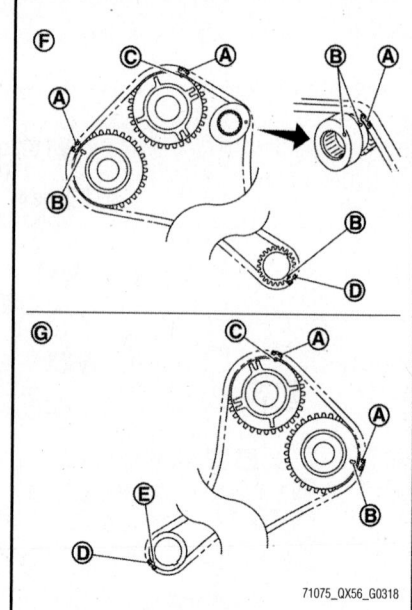

Fig. 88 Engine timing chain identification marks (F) bank 1, (G) bank 2

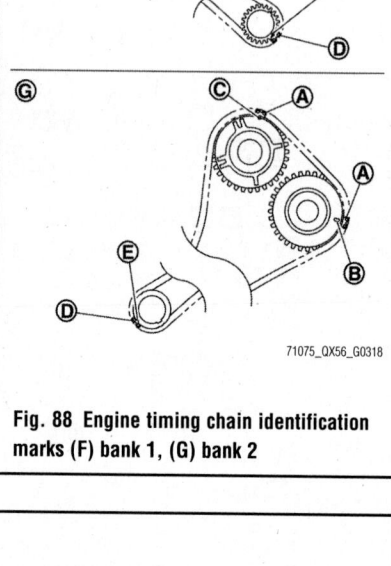

Fig. 89 Engine timing chain slack and tension guide identification marks (1) slack guide bank 2, (2) tension guide bank 2, (3) slack guide bank 1, (4) tension guide bank 1

camshaft (driveshaft) hexagonal portion or with tensioning the timing chain. When holding the hex portion of the camshaft (driveshaft) with a wrench be careful not to allow the wrench to cause interference with other parts. Never disassemble the camshaft sprocket. Never loosen the bolts.

40. Be sure to clean the mating surfaces and discard all gaskets and O-rings.

To install:

➡**Be sure to use new fasteners, as required.**

41. Installation is the reverse of the removal procedure.

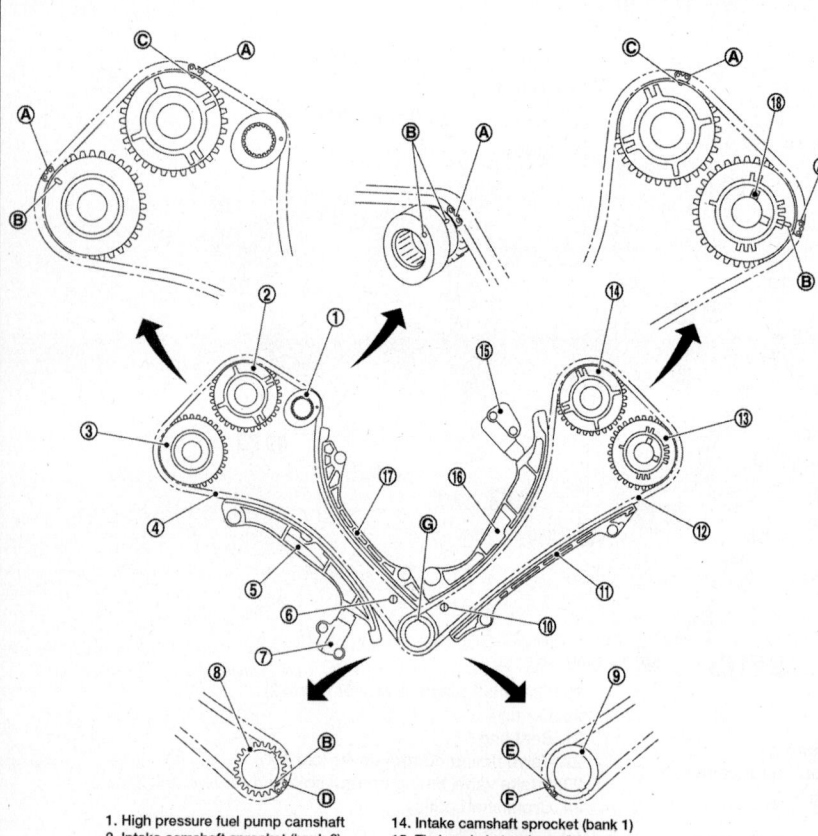

1. High pressure fuel pump camshaft
2. Intake camshaft sprocket (bank 2)
3. Exhaust camshaft sprocket (bank 2)
4. Timing chain (bank 2)
5. Slack guide (bank 2)
6. Chain oil jet (bank 2)
7. Timing chain tensioner (bank 2)
8. Crankshaft sprocket (bank 2 side)
9. Crankshaft sprocket (bank 1 side)
10. Chain oil jet (bank 1)
11. Tension guide (bank 1)
12. Timing chain (bank 1)
13. Exhaust camshaft sprocket (bank 1)
14. Intake camshaft sprocket (bank 1)
15. Timing chain tensioner (bank 1)
16. Slack guide (bank 1)
17. Tension guide (bank 2)
18. Dowel pin
A: Matching mark (copper link)
B: Matching mark (punched)
C: Matching mark (outer groove)
D: Matching mark (white link)
E: Matching mark (notched)
F: Matching mark (yellow link)
G: Crankshaft key

71075_QX56_G0311

Fig. 87 Engine timing chain alignment

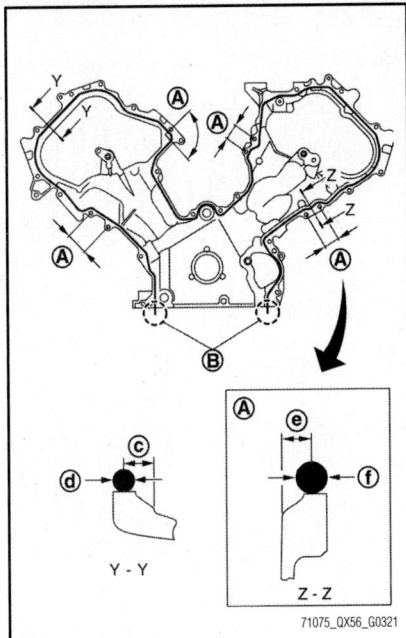

Fig. 90 Engine timing chain cover sealant application (B) protrusion

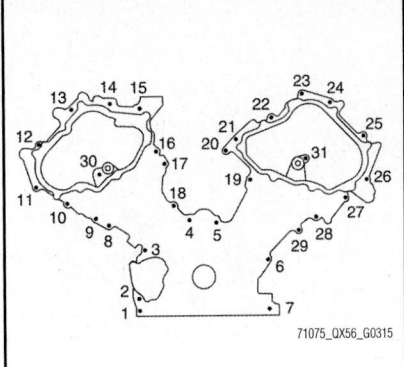

Fig. 91 Engine timing cover bolt location and tightening sequence

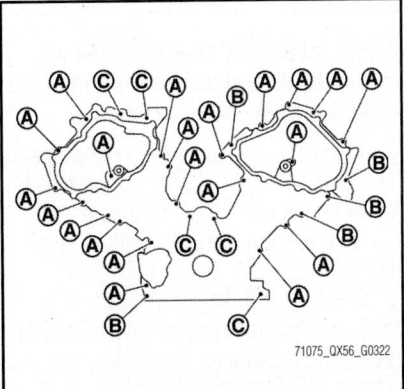

Fig. 92 Engine timing cover bolt size identification

➡The engine timing chain alignment illustration shows the relationship between the matching mark on each chain and on the corresponding sprocket, with the components installed. Parts with R and L should be installed on the corresponding bank according to the mark.

➡To install the timing chain and components, start with bank 2.

42. Be sure that the crankshaft key and dowel pin of each camshaft are in alignment (crankshaft key is straight up and camshaft dowel pin is face up at the cylinder head in each bank).

43. Install the camshaft sprockets. Be sure they are installed on the correct side by checking the identification mark on the surface.

44. Install the crankshaft sprocket so that its flange side (the large diameter without teeth) faces the proper direction.

45. Install the chain so that the matching mark B and the match mark outer groove C on the cam sprocket is aligned with the cooper link A on the chain, while the matching mark B (bank 1) E bank 2 on the crankshaft sprocket is aligned with the yellow link D on the chain. See illustration.

46. Install the slack guides and tension guides onto the correct side with the identification mark A on the surface. See illustration.

➡Never over tighten the slack guide mounting bolt. It is normal for a gap to exist under the bolt seats when the

mounting bolt is tightened to specification.

47. Continue the installation in the reverse order of the removal procedure.

48. When installing the timing chain tensioner cover to the timing cover, apply a continuous bead (0.134–0.173 inch) of liquid gasket to the front cover.

49. Apply a continuous bead of liquid gasket to the cover. c= 0.169–0.209 inch, d= 0.134–0.173 inch, e= 0.157–0.220 inch, f= 0.189–0.228 inch. See illustration.

50. Install the timing cover retaining bolts. There are three types of mounting bolts. Bolt A is 0.79 inch, bolt B is 1.97 inch and bolt C is 3.15 inch.

51. Be sure to perform the reconnect/relearn procedures.

VALVE LASH (CLEARANCE) ADJUSTMENT

ADJUSTMENT

See Figures 93 through 101.

1. Before servicing the vehicle, refer to the Precautions Section.

➡If working near and/or around the SRS system and components, be sure to disable the SRS system. After disabling the system wait three minutes or more before servicing the vehicle.

➡Whenever the negative battery cable is disconnected the following components will require resetting. The air conditioning system, automatic drive positioner system, power window control system, around view monitor, automatic back door system, idle air volume learning, steering angle sensor neutral position, sunroof system, audio visual and navigation systems. Use the CONSULT-III diagnostic tool, or equivalent to perform the required resets. Follow the directions on the screen of the tool, as needed.

➡Before disconnecting the negative battery cable lower both the driver's side and passenger's side front windows. This will prevent any interference between the window edge and the vehicle when the door is opener/closed. The automatic window function will not work with the battery cable disconnected.

➡Check valve clearance if the following applies. On the intake side after the removal and installation of the VVEL ladder assembly or valve related parts, or at the occurrence of malfunction (poor starting, poor idle or unusual noise) due to age deterioration in valve clearance. On the exhaust side after the removal and installation of the exhaust camshaft or valve related parts, or at the occurrence of malfunction (poor starting, poor idle or unusual noise) due to age deterioration in valve clearance.

➡Valve clearance check is not required on the intake side after replacing the VVEL ladder assembly and cylinder head with a new one. The VVEL ladder assembly cannot be replaced without replacing the cylinder head because it is machined together with the head.

2. Disconnect the negative battery cable.
3. Remove the VVEL actuator motor assembly.
4. Remove the rocker covers.
5. Remove the VVEL actuator housing assembly.
6. Measure the intake valve clearance using a feeler gauge.

➡Be sure to check the following points when measuring the intake side.

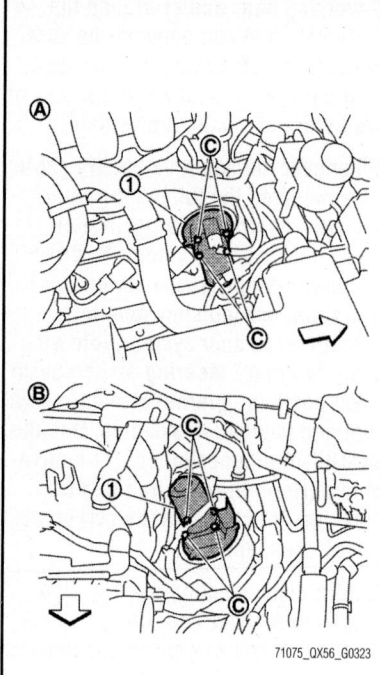

Fig. 93 VVEL actuator and related components (A) bank 1, (B) bank 2, (C) bolts, (1) actuator

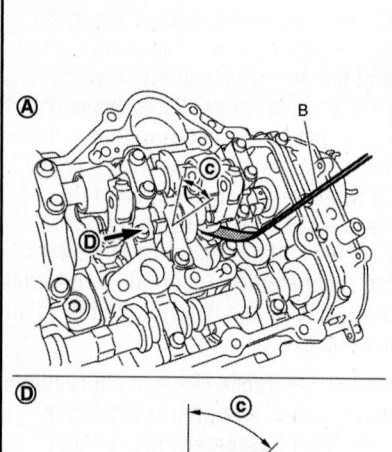

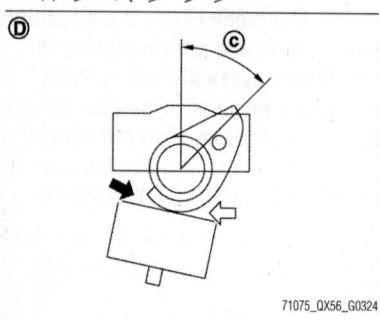

Fig. 94 Measuring valve clearance on the intake side (A) bank 2, (B) gauge, (c) 45 degrees, (D) view D

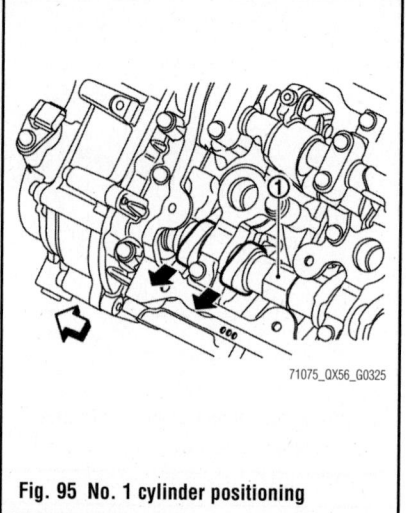

Fig. 95 No. 1 cylinder positioning

Before measuring check that the position of the driveshaft nose is within the angle shown in the illustration.

7. Position the engine at TDC on the compression stroke. Rotate the crankshaft pulley clockwise to align the timing mark (grooved line without color) with the timing indicator.

8. Check that the exhaust cam nose on No.1 cylinder (front engine side of bank 1) is properly positioned. If not rotate the crankshaft 360 degrees. See illustration.

9. Measure the valve clearance at the locations marked X. See illustration.

10. Measure the clearance of No.1 INT valve (front side).

11. Rotate the engine 270 degrees clockwise (when viewed from the front) to align No.3 cylinder at TDC. The crankshaft mounting bolt flange has an angle mark every 90 degrees. They can be used as a guide for rotation angle.

12. Measure the valve clearance at the locations marked X. See illustration.

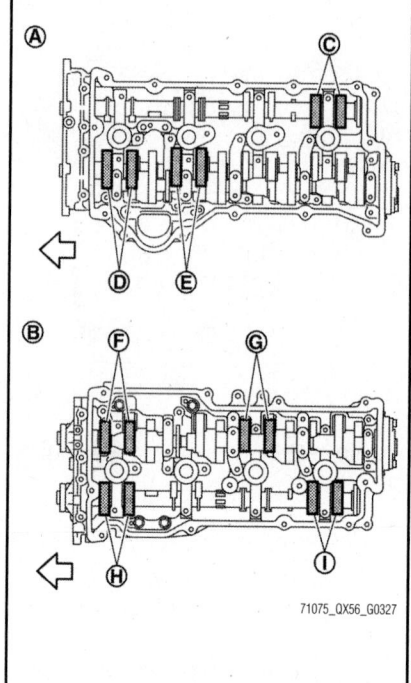

Fig. 97 Valve clearance measurement No.1—2 of 2

13. Rotate the engine 90 degrees clockwise (when viewed from the front) to align No.6 cylinder at TDC. The crankshaft mounting bolt flange has an angle mark every 90 degrees. They can be used as a guide for rotation angle.

14. Measure the valve clearance at the locations marked X. See illustration.

15. Adjust or replace components if measured value is not within specification. Never adjust valve clearance on the intake side.

16. Installation is the reverse of the removal procedure.

17. Be sure to perform the reconnect/relearn procedures.

Measuring position [bank 2 (A)]		No. 2 CYL.	No. 4 CYL.	No. 6 CYL.	No. 8 CYL.
No. 1 cylinder at compression TDC	EXH				× (C)
	INT	× (D)	× (E)		
Measuring position [bank 1 (B)]		No. 1 CYL.	No. 3 CYL.	No. 5 CYL.	No. 7 CYL.
No. 1 cylinder at compression TDC	INT	× (F)		× (G)	
	EXH	× (H)			× (I)

Fig. 96 Valve clearance measurement No.1—1 of 2

Measuring position [bank 2 (A)]		No. 2 CYL.	No. 4 CYL.	No. 6 CYL.	No. 8 CYL.
No. 3 cylinder at compression TDC	EXH		× (C)		
	INT				× (D)
Measuring position [bank 1 (B)]		No. 1 CYL.	No. 3 CYL.	No. 5 CYL.	No. 7 CYL.
No. 3 cylinder at compression TDC	INT		× (E)		× (F)
	EXH		× (G)	× (H)	

71075_QX56_G0328

Fig. 98 Valve clearance measurement No.3—1 of 2

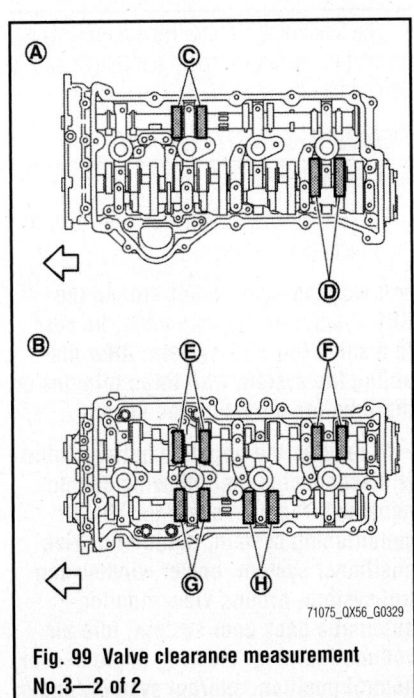

71075_QX56_G0329

Fig. 99 Valve clearance measurement No.3—2 of 2

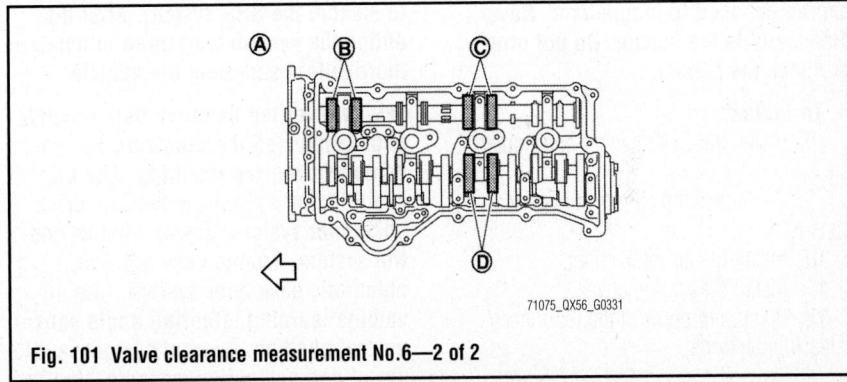

Measuring position [bank 2 (A)]		No. 2 CYL.	No. 4 CYL.	No. 6 CYL.	No. 8 CYL.
No. 6 cylinder at compression TDC	EXH	× (B)		× (C)	
	INT			× (D)	

71075_QX56_G0330

Fig. 100 Valve clearance measurement No.6—1 of 2

71075_QX56_G0331

Fig. 101 Valve clearance measurement No.6—2 of 2

ENGINE PERFORMANCE & EMISSION CONTROLS

CAMSHAFT POSITION (CMP) SENSOR

LOCATION

The Camshaft Position (CMP) sensors are located on the timing cover, facing the engine. There are two of them, one on each bank.

REMOVAL & INSTALLATION

See Figure 102.

At this time the manufacturer does not provide removal and installation procedures for this component. The following procedure is a guideline and may differ from the vehicle you are servicing.

1. Before servicing the vehicle, refer to the Precautions Section.

➡ If working near and/or around the SRS system and components, be sure to disable the SRS system. After disabling the system wait three minutes or more before servicing the vehicle.

➡ Whenever the negative battery cable is disconnected the following components will require resetting. The air conditioning system, automatic drive positioner system, power window control system, around view monitor, automatic back door system, idle air volume learning, steering angle sensor neutral position, sunroof system, audio visual and navigation systems. Use the CONSULT-III diagnostic tool, or equivalent to perform the required resets. Follow the directions on the screen of the tool, as needed.

➡ Before disconnecting the negative battery cable lower both the driver's side and passenger's side front windows. This will prevent any interference between the window edge and the vehicle when the door is opener/closed. The automatic window function will not work with the battery cable disconnected.

2. Disconnect the negative battery cable.

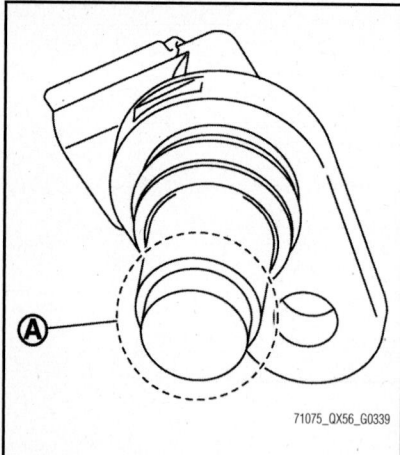

71075_QX56_G0339

Fig. 102 Camshaft Position (CMP) sensor (A) sensor

3. Remove the engine cover.
4. Remove the air cleaner assembly.
5. Disconnect the Camshaft Position (CMP) sensor connector.
6. Remove the bolt.
7. Remove the Camshaft Position (CMP) sensors.

➡Be careful not to drop the sensor. Do not leave it in an area where it can be exposed to magnetism. Never disassemble the sensor. Do not drop or shock the sensor.

To install:

8. Install the CMP sensor and tighten the bolt.
9. Reconnect the camshaft electrical sensor.
10. Install the air intake duct.
11. Install the engine cover.
12. Be sure to perform the reconnect/relearn procedures.

CRANKSHAFT POSITION (CKP) SENSOR

LOCATION

The Crankshaft Position (CKP) sensor is located on the transmission assembly facing the gear teeth (cogs) of the signal plate.

REMOVAL & INSTALLATION

See Figure 103.

At this time the manufacturer does not provide removal and installation procedures for this component. The following procedure is a guideline and may differ from the vehicle you are servicing.

1. Before servicing the vehicle, refer to the Precautions Section.

➡If working near and/or around the

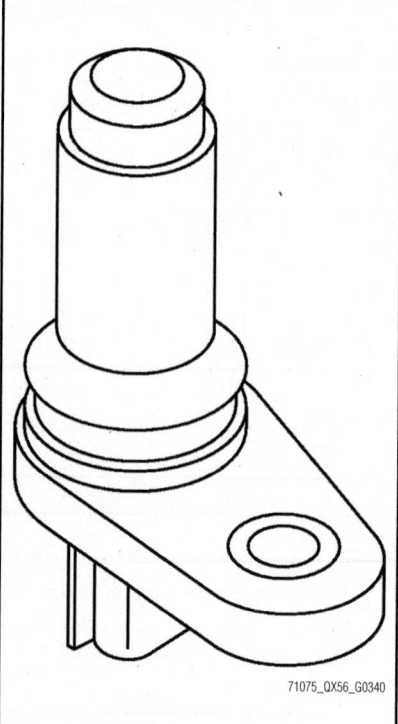

71075_QX56_G0340

Fig. 103 Crankshaft position sensor

SRS system and components, be sure to disable the SRS system. After disabling the system wait three minutes or more before servicing the vehicle.

➡Whenever the negative battery cable is disconnected the following components will require resetting. The air conditioning system, automatic drive positioner system, power window control system, around view monitor, automatic back door system, idle air volume learning, steering angle sensor neutral position, sunroof system, audio visual and navigation systems. Use the CONSULT-III diagnostic tool, or equivalent to perform the required resets. Follow the directions on the screen of the tool, as needed.

➡Before disconnecting the negative battery cable lower both the driver's side and passenger's side front windows. This will prevent any interference between the window edge and the vehicle when the door is opener/closed. The automatic window function will not work with the battery cable disconnected.

2. Disconnect the negative battery cable.
3. Raise and support the vehicle safely.
4. Disconnect the Crankshaft Position (CKP) sensor connector.
5. Remove the mounting bolt and CKP sensor.

➡Be careful not to drop the sensor. Do not leave it in an area where it can be exposed to magnetism. Never disassemble the sensor. Do not drop or shock the sensor.

To install:

6. Install the CKP sensor and tighten the mounting bolt.
7. Reconnect the CKP sensor connector.
8. Lower the vehicle.
9. Be sure to perform the reconnect/relearn procedures.

ELECTRONIC CONTROL MODULE (ECM)

LOCATION

The Electronic Control Module (ECM) is located in the engine room passenger side behind battery.

REMOVAL & INSTALLATION

See Figure 104.

1. Before servicing the vehicle, refer to the Precautions Section.

➡If working near and/or around the SRS system and components, be sure to disable the SRS system. After disabling the system wait three minutes or more before servicing the vehicle.

➡Whenever the negative battery cable is disconnected the following components will require resetting. The air conditioning system, automatic drive positioner system, power window control system, around view monitor, automatic back door system, idle air volume learning, steering angle sensor neutral position, sunroof system, audio visual and navigation systems. Use the

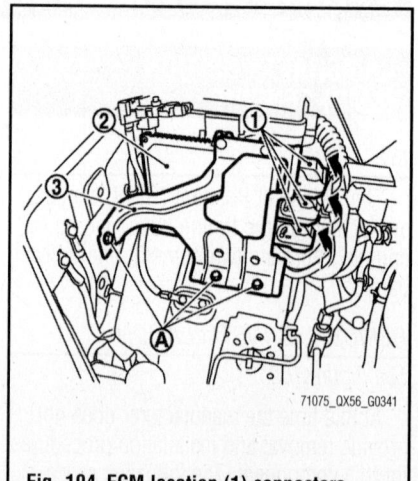

71075_QX56_G0341

Fig. 104 ECM location (1) connectors, (2) ECM, (3) bracket, (A) bolts

CONSULT-III diagnostic tool, or equivalent to perform the required resets. Follow the directions on the screen of the tool, as needed.

➡️Before disconnecting the negative battery cable lower both the driver's side and passenger's side front windows. This will prevent any interference between the window edge and the vehicle when the door is opener/closed. The automatic window function will not work with the battery cable disconnected.

2. Disconnect the negative battery cable.

3. Disconnect the positive battery cable and remove the battery. Remove the battery tray.

4. Carefully remove the Electronic Control Module (ECM) harness connectors.

5. Remove the ECM mounting bolts and the ECM.

To install:

6. Install the ECM and mounting bolts and tighten to 62 inch lbs. (7 Nm).

7. Carefully install the ECM harness connectors.

8. Install the battery.

9. Reconnect the battery cables.

10. Be sure to perform the reconnect/relearn procedures.

RESET

➡️The Nissan CONSULT III diagnostic tool, or equivalent is required to perform the resets used on this vehicle.

When replacing the ECM, the following procedures must be performed in the order listed.

- Initialize the immobilizer system
- Initialize the IVIS system
- Initialize the NATS system
- VIN Registration
- Accelerator Pedal Released Position Learning
- Throttle Valve Closed Position Learning
- Idle Air Volume Learning

ENGINE COOLANT TEMPERATURE (ECT) SENSOR

LOCATION

The Engine Coolant Temperature (ECT) sensor is mounted on the passenger's side cylinder head, at the rear of the engine.

REMOVAL & INSTALLATION

See Figure 105.

At this time the manufacturer does not provide removal and installation procedures

for this component. The following procedure is a guideline and may differ from the vehicle you are servicing.

⁕⁕ CAUTION

Never open, service or drain the radiator or cooling system when hot; serious burns can occur from the steam and hot coolant. Also, when draining engine coolant, keep in mind that cats and dogs are attracted to ethylene glycol antifreeze and could drink any that is left in an uncovered container or in puddles on the ground. This will prove fatal in sufficient quantities. Always drain coolant into a sealable container. Coolant should be reused unless it is contaminated or is several years old.

1. Before servicing the vehicle, refer to the Precautions Section.

➡️If working near and/or around the SRS system and components, be sure to disable the SRS system. After disabling the system wait three minutes or more before servicing the vehicle.

➡️Whenever the negative battery cable is disconnected the following components will require resetting. The air conditioning system, automatic drive positioner system, power window control system, around view monitor, automatic back door system, idle air volume learning, steering angle sensor neutral position, sunroof system, audio visual and navigation systems. Use the CONSULT-III diagnostic tool, or equivalent to perform the required resets. Follow the directions on the screen of the tool, as needed.

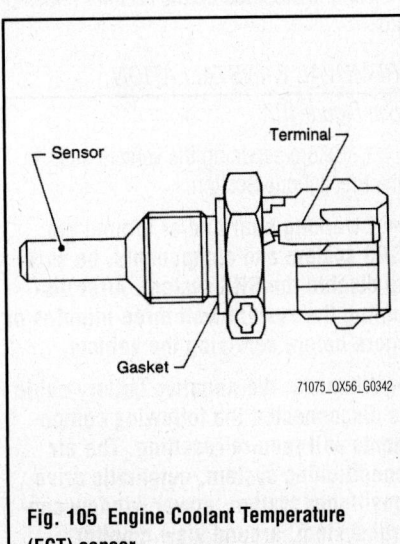

Fig. 105 Engine Coolant Temperature (ECT) sensor

Sensor

Terminal

Gasket

71075_QX56_G0342

➡️Before disconnecting the negative battery cable lower both the driver's side and passenger's side front windows. This will prevent any interference between the window edge and the vehicle when the door is opener/closed. The automatic window function will not work with the battery cable disconnected.

2. Disconnect the negative battery cable.

3. Remove the engine cover.

4. Remove the intake air duct.

5. Partially drain the cooling system.

6. Disconnect the harness connector.

7. Remove the Engine Coolant Temperature (ECT) sensor.

To install:

➡️Be sure to use new fasteners, as required.

8. Install the ECT sensor and carefully tighten.

9. Reconnect the harness connector.

10. Install the intake air duct.

11. Install the engine cover.

12. Refill the engine coolant.

13. Be sure to perform the reconnect/relearn procedures.

HEATED OXYGEN SENSOR (HO2S)

LOCATION

The Heated Oxygen (HO2S) sensors are located after the exhaust manifold converter assembly, in the lower part of the exhaust system.

REMOVAL & INSTALLATION

See Figure 106.

1. Before servicing the vehicle, refer to the Precautions Section.

➡️If working near and/or around the SRS system and components, be sure to disable the SRS system. After disabling the system wait three minutes or more before servicing the vehicle.

➡️Whenever the negative battery cable is disconnected the following components will require resetting. The air conditioning system, automatic drive positioner system, power window control system, around view monitor, automatic back door system, idle air volume learning, steering angle sensor neutral position, sunroof system, audio visual and navigation systems. Use the CONSULT-III diagnostic tool, or equiva-

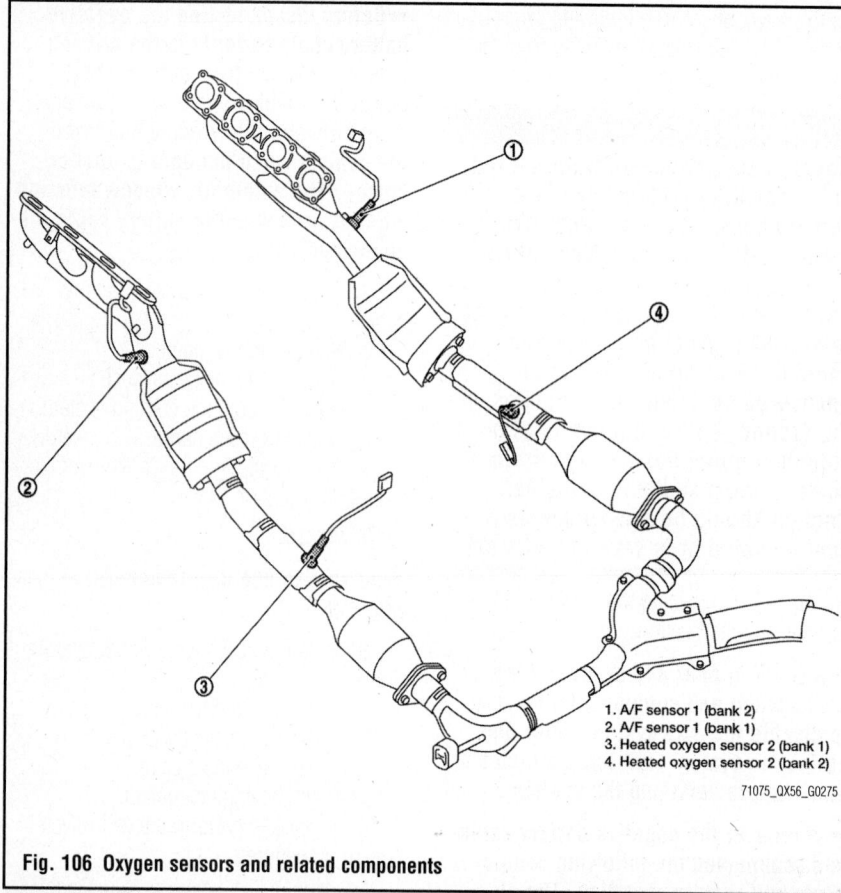

Fig. 106 Oxygen sensors and related components

1. A/F sensor 1 (bank 2)
2. A/F sensor 1 (bank 1)
3. Heated oxygen sensor 2 (bank 1)
4. Heated oxygen sensor 2 (bank 2)

71075_QX56_G0275

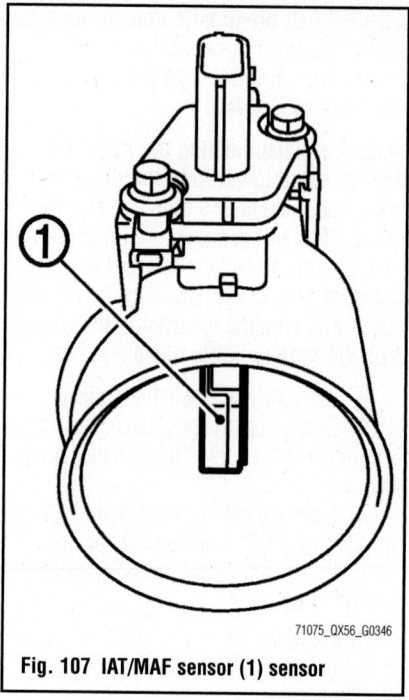

Fig. 107 IAT/MAF sensor (1) sensor

71075_QX56_G0346

lent to perform the required resets. Follow the directions on the screen of the tool, as needed.

➡Before disconnecting the negative battery cable lower both the driver's side and passenger's side front windows. This will prevent any interference between the window edge and the vehicle when the door is opener/closed. The automatic window function will not work with the battery cable disconnected.

2. Disconnect the negative battery cable.
3. Raise and safely support the vehicle.
4. Remove the engine undercover, as needed.
5. Unplug the Heated Oxygen (HO2S) sensor harness.
6. Using an O2 wrench remove the HO2S sensor.

➡Lower the exhaust in needed.

To install:

➡Be sure to use new fasteners, as required.

7. Install the HO2S sensor .
8. Install the harness connector.
9. Keep the harness connector and wiring away from exhaust system.

10. Be sure to perform the reconnect/relearn procedures.

INTAKE AIR TEMPERATURE (IAT)/MASS AIR FLOW (MAF) SENSOR

LOCATION

The Intake Air Temperature (IAT) sensor is integral to the Mass Air Flow (MAF) sensor, and is mounted on the air filter housing lid.

REMOVAL & INSTALLATION
See Figure 107.

1. Before servicing the vehicle, refer to the Precautions Section.

➡If working near and/or around the SRS system and components, be sure to disable the SRS system. After disabling the system wait three minutes or more before servicing the vehicle.

➡Whenever the negative battery cable is disconnected the following components will require resetting. The air conditioning system, automatic drive positioner system, power window control system, around view monitor, automatic back door system, idle air

volume learning, steering angle sensor neutral position, sunroof system, audio visual and navigation systems. Use the CONSULT-III diagnostic tool, or equivalent to perform the required resets. Follow the directions on the screen of the tool, as needed.

➡Before disconnecting the negative battery cable lower both the driver's side and passenger's side front windows. This will prevent any interference between the window edge and the vehicle when the door is opener/closed. The automatic window function will not work with the battery cable disconnected.

2. Disconnect the negative battery cable.
3. Remove the engine room cover.
4. Disconnect the sensor connector.
5. Remove the air cleaner case along with the sensor and air duct assembly.
6. Remove the sensor from the case.

To install:

➡Be sure to use new fasteners, as required.

7. Installation is the reverse of the removal procedure.
8. Be sure to perform the reconnect/relearn procedures.

KNOCK SENSOR (KS)

LOCATION

The Knock (KS) sensors are mounted under the intake manifold on the cylinder block.

REMOVAL & INSTALLATION

See Figure 108.

At this time the manufacturer does not provide removal and installation procedures for this component. The following procedure is a guideline and may differ from the vehicle you are servicing.

1. Before servicing the vehicle, refer to the Precautions Section.

➡ **If working near and/or around the SRS system and components, be sure to disable the SRS system. After disabling the system wait three minutes or more before servicing the vehicle.**

➡ **Whenever the negative battery cable is disconnected the following components will require resetting. The air conditioning system, automatic drive positioner system, power window control system, around view monitor, automatic back door system, idle air volume learning, steering angle sensor neutral position, sunroof system, audio visual and navigation systems. Use the CONSULT-III diagnostic tool, or equivalent to perform the required resets. Follow the directions on the screen of the tool, as needed.**

➡ **Before disconnecting the negative battery cable lower both the driver's side and passenger's side front windows. This will prevent any interference between the window edge and the vehicle when the door is opener/closed. The automatic window function will not work with the battery cable disconnected.**

2. Disconnect the negative battery cable.

3. Remove the engine cover.

4. Remove the air cleaner assembly.
5. Remove the intake manifold.
6. Disconnect the sensor connector.
7. Remove the sensor retaining bolt.
8. Remove the sensor from the vehicle.

To install:

➡ **Be sure to use new fasteners, as required.**

9. Installation is the reverse of the removal procedure.

10. Be sure to perform the reconnect/relearn procedures.

MALFUNCTION INDICATOR LIGHT

RESET PROCEDURE

Clearing diagnostic trouble codes resets MIL.

Proper operation of the Malfunction Indicator Light (MIL):

• The MIL will illumine with the ignition switch ON and the engine OFF
• The MIL will turn OFF when the engine is started
• The MIL will remain ON if the self-diagnostic system has detected a malfunction
• The MIL may turn OFF if the malfunction is no longer present
• If the MIL is illuminated and then the engine stalls, the MIL will remain illuminated as long as the ignition switch is ON
• If the MIL is not illuminated and the engine stalls, the MIL will not illuminate until the ignition switch is cycled OFF, then ON

1. Before servicing the vehicle, refer to the Precautions Section.

2. Resetting the MIL:
• The control module turns OFF the MIL after 3 consecutive ignition cycles that the diagnostic system runs and does not fail
• The control module turns OFF the MIL after a current Diagnostic Trouble Code (DTC) clears when the diagnostic cycle runs and passes
• There may still be a history of DTC's stored in the system. These will clear after 40 consecutive warm-up cycles, if no failures are reported by any other related diagnostic system
• Manual resetting of the MIL and any DTC stored in the system, requires the use of an OBD2 scan tool connected to the Data Link Connector (DLC) for communication with the vehicle. Follow the

instructions of the scan tool for both retrieval and resetting of DTC's. The scan tool can be used to command the MIL off.

➡ **If the error symptoms causing the MIL to illuminate have been corrected, the MIL will return to normal operation.**

3. If a DTC is present, record the code and troubleshoot the fault.

MANIFOLD ABSOLUTE PRESSURE (MAP) SENSOR

LOCATION

This sensor is located on the intake manifold collector.

REMOVAL & INSTALLATION

See Figure 109.

At this time the manufacturer does not provide removal and installation procedures for this component. The following procedure is a guideline and may differ from the vehicle you are servicing.

1. Before servicing the vehicle, refer to the Precautions Section.

➡ **If working near and/or around the SRS system and components, be sure to disable the SRS system. After disabling the system wait three minutes or more before servicing the vehicle.**

➡ **Whenever the negative battery cable is disconnected the following components will require resetting. The air conditioning system, automatic drive positioner system, power window control system, around view monitor, automatic back door system, idle air volume learning, steering angle sensor neutral position, sunroof system, audio**

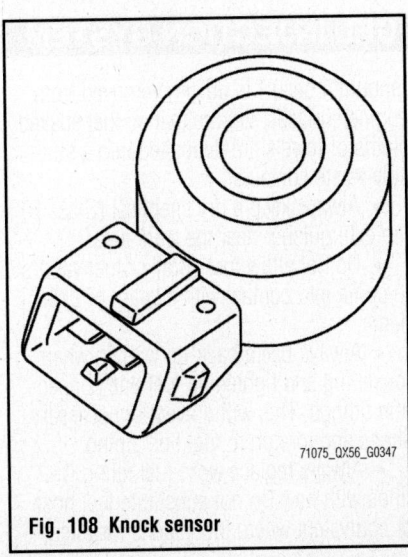

71075_QX56_G0347

Fig. 108 Knock sensor

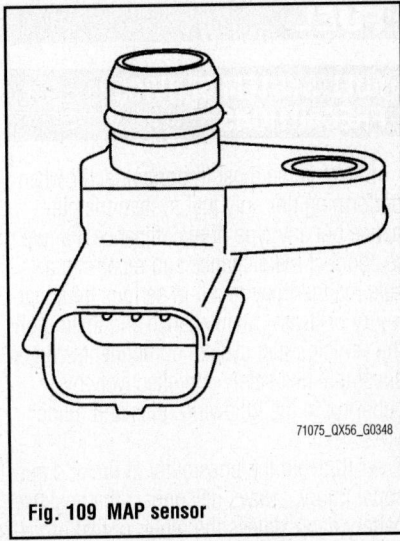

71075_QX56_G0348

Fig. 109 MAP sensor

visual and navigation systems. Use the CONSULT-III diagnostic tool, or equivalent to perform the required resets. Follow the directions on the screen of the tool, as needed.

→Before disconnecting the negative battery cable lower both the driver's side and passenger's side front windows. This will prevent any interference between the window edge and the vehicle when the door is opener/closed. The automatic window function will not work with the battery cable disconnected.

2. Disconnect the negative battery cable.
3. Remove the engine cover.
4. Remove the air cleaner assembly.
5. Disconnect the sensor connector.
6. Remove the sensor from its mounting.

To install:

→Be sure to use new fasteners, as required.

7. Installation is the reverse of the removal procedure.
8. Be sure to perform the reconnect/relearn procedures.

THROTTLE POSITION SENSOR

LOCATION

The Throttle Position (TPS) sensor is integral to the electric Throttle Control actuator. The Throttle Control actuator is mounted at the front of the intake manifold.

REMOVAL & INSTALLATION
See Figure 110.

At this time the manufacturer does not provide removal and installation procedures

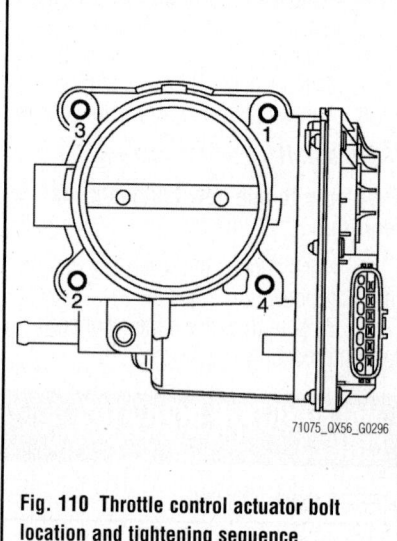

71075_QX56_G0296

Fig. 110 Throttle control actuator bolt location and tightening sequence

for this component. The following procedure is a guideline and may differ from the vehicle you are servicing.

1. Before servicing the vehicle, refer to the Precautions Section.

→If working near and/or around the SRS system and components, be sure to disable the SRS system. After disabling the system wait three minutes or more before servicing the vehicle.

→Whenever the negative battery cable is disconnected the following components will require resetting. The air conditioning system, automatic drive positioner system, power window control system, around view monitor, automatic back door system, idle air volume learning, steering angle sensor neutral position, sunroof system, audio visual and navigation

systems. Use the CONSULT-III diagnostic tool, or equivalent to perform the required resets. Follow the directions on the screen of the tool, as needed.

→Before disconnecting the negative battery cable lower both the driver's side and passenger's side front windows. This will prevent any interference between the window edge and the vehicle when the door is opener/closed. The automatic window function will not work with the battery cable disconnected.

2. Disconnect the negative battery cable.
3. Drain the cooling system, as required. Be sure to properly dispose of used engine coolant.
4. Remove the engine room cover. Remove the air intake duct.
5. Disconnect harness connector.
6. Disconnect water hoses.
7. Loosen the throttle body assembly mounting bolts in reverse order of the tightening sequence.

To install:

→Be sure to use new fasteners, as required.

8. Install the throttle body assembly with a new gasket.
9. Tighten the mounting bolts in sequence.
10. Reconnect the water hose.
11. Reconnect the harness connector.
12. Reconnect the air intake duct.
13. Fill the cooling system with the proper grade and type engine coolant.
14. Be sure to perform the reconnect/relearn procedures.

FUEL
GASOLINE FUEL INJECTION SYSTEM

FUEL SYSTEM SERVICE PRECAUTIONS

Safety is the most important factor when performing not only fuel system maintenance but any type of maintenance. Failure to conduct maintenance and repairs in a safe manner may result in serious personal injury or death. Maintenance and testing of the vehicle's fuel system components can be accomplished safely and effectively by adhering to the following rules and guidelines.

• To avoid the possibility of fire and personal injury, always disconnect the negative battery cable unless the repair or test proce-

dure requires that battery voltage be applied.

• Always relieve the fuel system pressure prior to disconnecting any fuel system component (injector, fuel rail, pressure regulator, etc.), fitting or fuel line connection. Exercise extreme caution whenever relieving fuel system pressure to avoid exposing skin, face and eyes to fuel spray. Please be advised that fuel under pressure may penetrate the skin or any part of the body that it contacts.

• Always place a shop towel or cloth around the fitting or connection prior to loosening to absorb any excess fuel due to spillage. Ensure that all fuel spillage

(should it occur) is quickly removed from engine surfaces. Ensure that all fuel soaked cloths or towels are deposited into a suitable waste container.

• Always keep a dry chemical (Class B) fire extinguisher near the work area.

• Do not allow fuel spray or fuel vapors to come into contact with a spark or open flame.

• Always use a back-up wrench when loosening and tightening fuel line connection fittings. This will prevent unnecessary stress and torsion to fuel line piping.

• Always replace worn fuel fitting O-rings with new Do not substitute fuel hose or equivalent where fuel pipe is installed.

Before servicing the vehicle, make sure to also refer to the precautions in the beginning of this section as well.

RELIEVING FUEL SYSTEM PRESSURE

WITH CONSULT-III®

1. Turn ignition switch **ON**.
2. Perform "FUEL PRESSURE RELEASE" in "WORK SUPPORT" mode with CONSULT-III®.
3. Start engine.
4. After engine stalls, turn over the engine two or three times to release all fuel pressure.
5. Turn ignition switch **OFF**.

WITHOUT CONSULT-III®

See Figure 111.

1. Before servicing the vehicle, refer to the Precautions Section.

➡If working near and/or around the SRS system and components, be sure to disable the SRS system. After disabling the system wait three minutes or more before servicing the vehicle.

➡Whenever the negative battery cable is disconnected the following components will require resetting. The air conditioning system, automatic drive positioner system, power window control system, around view monitor, automatic back door system, idle air volume learning, steering angle sensor neutral position, sunroof system, audio visual and navigation systems. Use the CONSULT-III diagnostic tool, or equivalent to perform the required resets. Follow the directions on the screen of the tool, as needed.

➡Before disconnecting the negative battery cable lower both the driver's side and passenger's side front windows. This will prevent any interference between the window edge and the vehicle when the door is opener/closed. The automatic window function will not work with the battery cable disconnected.

2. Remove fuel pump fuse located in IPDM E/R.
3. Start engine.
4. After engine stalls, turn over engine two or three times to release all fuel pressure.
5. Turn ignition switch **OFF**.
6. Disconnect the negative battery cable.
7. Reinstall fuel pump fuse after servicing fuel system.

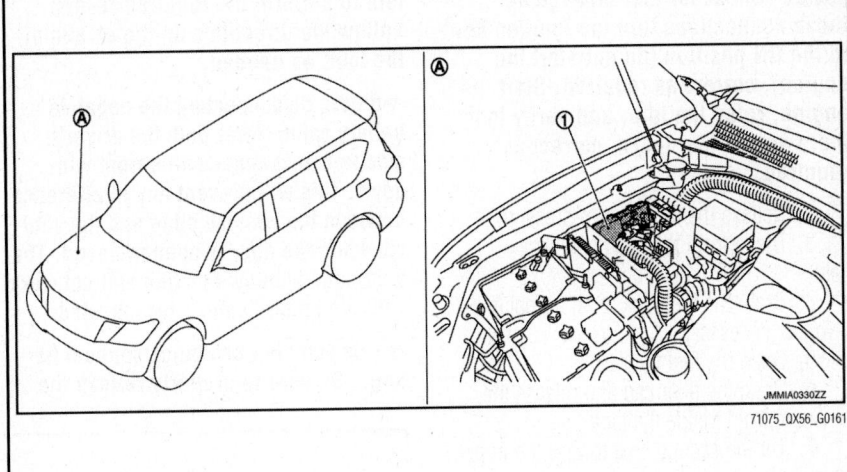

Fig. 111 IPDM/ER and related components (1) IPDM/ER, (A) underhood

8. Be sure to perform the reconnect/relearn procedures.

FUEL PUMP MODULE

REMOVAL & INSTALLATION

See Figure 112.

1. Before servicing the vehicle, refer to the Precautions Section.

➡If working near and/or around the SRS system and components, be sure to disable the SRS system. After disabling the system wait three minutes or more before servicing the vehicle.

➡Whenever the negative battery cable is disconnected the following components will require resetting. The air conditioning system, automatic drive positioner system, power window control system, around view monitor, automatic back door system, idle air volume learning, steering angle sensor neutral position, sunroof system, audio visual and navigation systems. Use the CONSULT-III diagnostic tool, or equivalent to perform the required resets. Follow the directions on the screen of the tool, as needed.

➡Before disconnecting the negative battery cable lower both the driver's side and passenger's side front windows. This will prevent any interference between the window edge and the vehicle when the door is opener/closed. The automatic window function will not work with the battery cable disconnected.

➡Be sure to check the fuel gauge indicator. Make sure that it reads not more than half a tank. If not, properly drain fuel until the gauge reads half a tank or less.

➡This vehicle uses quick connect fittings. Be sure to properly relieve the fuel system pressure before disconnecting any of these fittings. Always replace O-rings and clamps with new ones. Do not bend, twist or kink hoses when they are being removed or installed. Be sure that the clamp screw does not contact adjacent parts. When tightening the high pressure rubber hose clamp make sure the clamp end is 0.12 inch from the hose end. After connecting these fittings

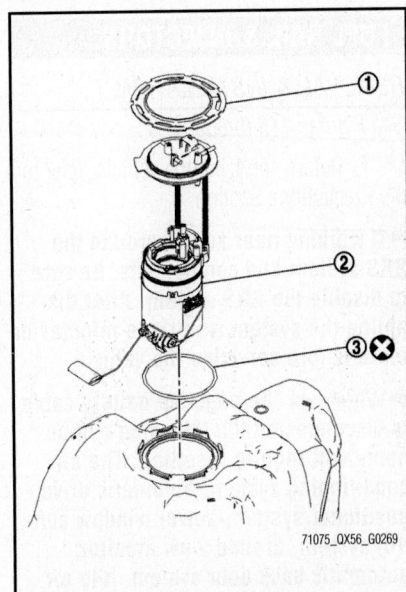

Fig. 112 Fuel pump module and related components (1) lock ring, (2) module, (3) O-ring

make sure that the connectors are secure. Check for fuel leakage at these connections turn the ignition key to the ON position (do not start the engine), correct as required. Start the engine, raise the idle, and verify that there are no fuel leaks, correct as required.

2. Relieve the fuel system pressure.
3. Disconnect the negative battery cable.
4. Drain the fuel tank to an acceptable level, as necessary.
5. Open the fuel filler lid.
6. Open the filler cap and release the pressure from inside the tank.
7. Tilt the second seat toward the front of the vehicle.
8. Peel off the floor carpet, then remove the inspection hole cover by turning the clips clockwise 90 degrees.
9. Disconnect the harness connector and all the fuel line tubes.
10. Disconnect the quick connectors.
11. Remove the lockring with tool SST: KV10119800, or equivalent, by turning the lock ring counterclockwise.
12. Remove the fuel pump module from the tank. Discard all O-rings and gaskets.

To install:

➡Be sure to use new fasteners, as required.

13. Installation is the reverse of the removal procedure.
14. Be sure to perform the reconnect/relearn procedures.

FUEL RAIL & INJECTORS

REMOVAL & INSTALLATION
See Figures 113 through 117.

1. Before servicing the vehicle, refer to the Precautions Section.

➡If working near and/or around the SRS system and components, be sure to disable the SRS system. After disabling the system wait three minutes or more before servicing the vehicle.

➡Whenever the negative battery cable is disconnected the following components will require resetting. The air conditioning system, automatic drive positioner system, power window control system, around view monitor, automatic back door system, idle air volume learning, steering angle sensor neutral position, sunroof system, audio visual and navigation systems. Use the

CONSULT-III diagnostic tool, or equivalent to perform the required resets. Follow the directions on the screen of the tool, as needed.

➡Before disconnecting the negative battery cable lower both the driver's side and passenger's side front windows. This will prevent any interference between the window edge and the vehicle when the door is opener/closed. The automatic window function will not work with the battery cable disconnected.

➡This vehicle uses quick connect fittings. Be sure to properly relieve the fuel system pressure before disconnecting any of these fittings. Always replace O-rings and clamps with new ones. Do not bend, twist or kink hoses when they are being removed or installed. Be sure that the clamp screw does not contact adjacent parts. When tightening the high pressure rubber hose clamp make sure the clamp end is 0.12 inch from the hose end. After connecting these fittings make sure that the connectors are secure. Check for fuel leakage at these connections turn the ignition key to the ON position (do not start the engine), correct as

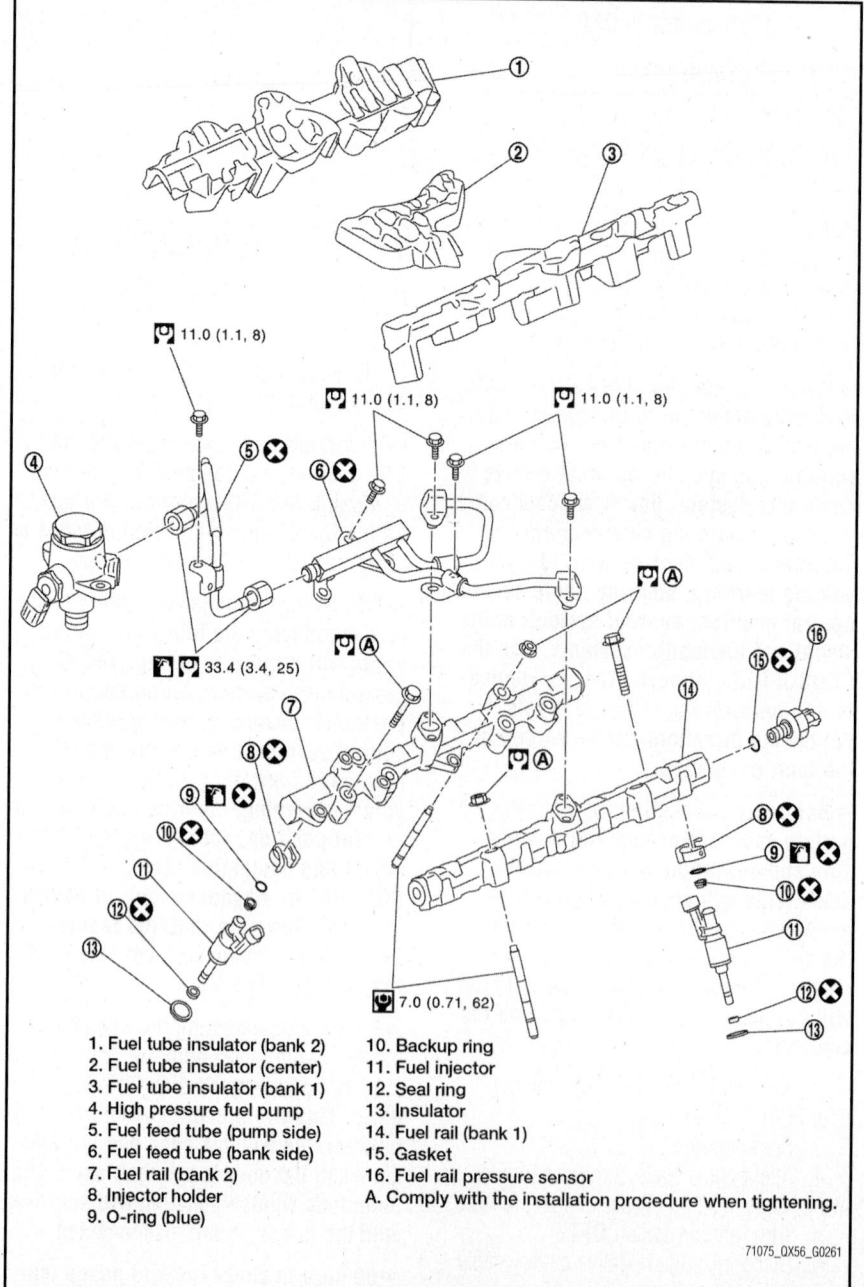

1. Fuel tube insulator (bank 2)
2. Fuel tube insulator (center)
3. Fuel tube insulator (bank 1)
4. High pressure fuel pump
5. Fuel feed tube (pump side)
6. Fuel feed tube (bank side)
7. Fuel rail (bank 2)
8. Injector holder
9. O-ring (blue)
10. Backup ring
11. Fuel injector
12. Seal ring
13. Insulator
14. Fuel rail (bank 1)
15. Gasket
16. Fuel rail pressure sensor
A. Comply with the installation procedure when tightening.

71075_QX56_G0261

Fig. 113 Fuel injectors/fuel rail and related components

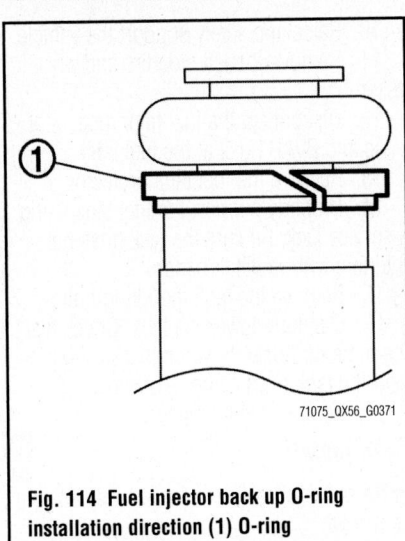

Fig. 114 Fuel injector back up O-ring installation direction (1) O-ring

required. **Start the engine, raise the idle, and verify that there are no fuel leaks, correct as required.**

2. Properly relieve the fuel system pressure.

3. Disconnect the negative battery cable.

4. Remove the engine cover.

5. Remove the air cleaner assembly.

6. Remove the intake manifold.

7. Remove the fuel feed tube, pump side and fuel feed tube, bank side.

➡**Never reuse the fuel feed tube, discard it.**

8. Disconnect the fuel injector harness connectors.

9. Remove the injector from the cylinder head. Discard the O-ring.

➡**Handle the injector carefully. Never disassemble it. Do not drop it. Discard the O-ring.**

10. To remove the injector holder install tool SST: KV10119600, or equivalent to the lower injector connector side so that the cutout of the tool faces the injector connector side. Hook the pawl portion of the tool to the groove portion of the injector. Press down on the body of the tool until it contacts the cylinder head. Tighten the tool clockwise and remove the injector. Cut the Teflon seal while pinching it. Be careful not to damage the injector. Remove the insulator from the mounting hole of the injector of the cylinder head.

To install:

➡**Be sure to use new fasteners, as required.**

11. Installation is the reverse of the removal procedure.

12. Install the seal guide (tool) to the injector. Position the seal ring to the tool. Straightly insert the seal ring to the injector and install. Be careful that the seal ring does not exceed the groove portion of the fuel injector. Insert drift, SST: KV10119710 or equivalent, to the injector and rotate clockwise and counterclockwise 90 degrees while pressing the seal ring to fit it. Compress the seal ring, because this operation is for rectifying stretch of seal ring caused by installation and preventing sticking when inserting the injector into the cylinder head.

13. Install the O-ring and back up O-ring on the injector. Coat the O-ring with clean engine oil.

➡**Always install the O-ring straight on the rail. Never de-center or twist it. Always install the backup O-ring in the right direction. See illustration.**

14. Install the fuel injector to the fuel rail.

➡**Never reuse the injector holder, replace it. Do not allow the injector to interfere with the O-ring, if it does replace the O-ring.**

15. Insert the injector into the rail. Be sure the injector does not rotate or come off. Check that the protrusions of the injectors and rail are aligned with the cutouts of the clips after installation.

16. Install the fuel rail and injector assembly to the cylinder head.

17. Tighten the retaining bolts in the proper sequence and to specification. Specification is 89 inch lbs., first pass and 15 ft. lbs., second pass.

18. Install the fuel feed tube, bank side to the fuel rail. Never reuse the fuel feed tube. Replace the O-ring and back up O-

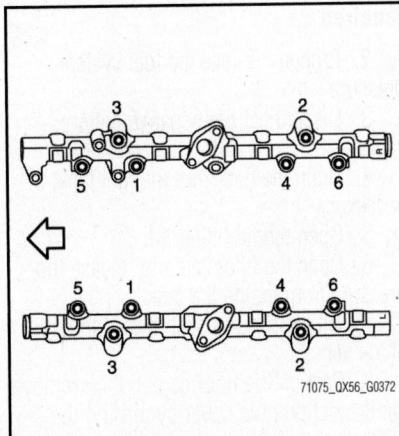

Fig. 115 Fuel injector/rail bolt location and tightening sequence

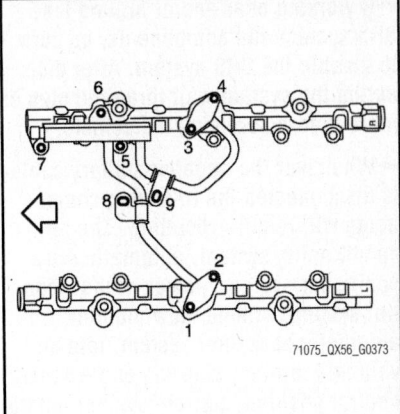

Fig. 116 Fuel feed tube (bank side) bolt location and tightening sequence

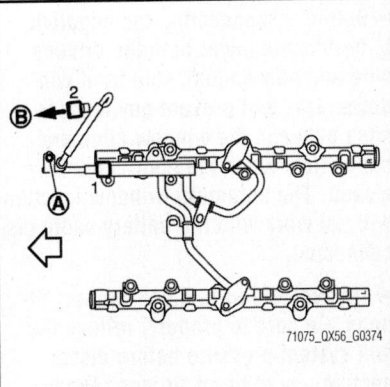

Fig. 117 Fuel feed tube (pump side) bolt location and tightening sequence (A) bolt, (B) to high pressure fuel pump

ring. Tighten the mounting bolts in the proper sequence.

19. Install the fuel feed tube, pump side to the fuel feed tube. Never reuse the fuel feed tube. Apply clean engine oil to the flare screw parts of the high pressure pump side and fuel feed tube, bank side. Manually tighten the two flare nuts without a tool until they are seated. Tighten bolt (A). Tighten the bolts in the proper sequence.

20. Continue the installation in the reverse order of the removal procedure.

21. Be sure to perform the reconnect/relearn procedures.

FUEL TANK

DRAINING

At this time the manufacturer does not provide service information for draining the fuel tank.

1. Before servicing the vehicle, refer to the Precautions Section.

➡If working near and/or around the SRS system and components, be sure to disable the SRS system. After disabling the system wait three minutes or more before servicing the vehicle.

➡Whenever the negative battery cable is disconnected the following components will require resetting. The air conditioning system, automatic drive positioner system, power window control system, around view monitor, automatic back door system, idle air volume learning, steering angle sensor neutral position, sunroof system, audio visual and navigation systems. Use the CONSULT-III diagnostic tool, or equivalent to perform the required resets. Follow the directions on the screen of the tool, as needed.

➡Before disconnecting the negative battery cable lower both the driver's side and passenger's side front windows. This will prevent any interference between the window edge and the vehicle when the door is opener/closed. The automatic window function will not work with the battery cable disconnected.

➡This vehicle uses quick connect fittings. Be sure to properly relieve the fuel system pressure before disconnecting any of these fittings. Always replace O-rings and clamps with new ones. Do not bend, twist or kink hoses when they are being removed or installed. Be sure that the clamp screw does not contact adjacent parts. When tightening the high pressure rubber hose clamp make sure the clamp end is 0.12 inch from the hose end. After connecting these fittings make sure that the connectors are secure. Check for fuel leakage at these connections turn the ignition key to the ON position (do not start the engine), correct as required. Start the engine, raise the idle, and verify that there are no fuel leaks, correct as required.

REMOVAL & INSTALLATION
See Figure 118.

1. Before servicing the vehicle, refer to the Precautions Section.

➡If working near and/or around the SRS system and components, be sure to disable the SRS system. After disabling the system wait three minutes or more before servicing the vehicle.

➡Whenever the negative battery cable is disconnected the following components will require resetting. The air conditioning system, automatic drive positioner system, power window control system, around view monitor, automatic back door system, idle air volume learning, steering angle sensor neutral position, sunroof system, audio visual and navigation systems. Use the CONSULT-III diagnostic tool, or equivalent to perform the required resets. Follow the directions on the screen of the tool, as needed.

➡Before disconnecting the negative battery cable lower both the driver's side and passenger's side front windows. This will prevent any interference between the window edge and the vehicle when the door is opener/closed. The automatic window function will not work with the battery cable disconnected.

➡This vehicle uses quick connect fittings. Be sure to properly relieve the fuel system pressure before disconnecting any of these fittings. Always replace O-rings and clamps with new ones. Do not bend, twist or kink hoses when they are being removed or installed. Be sure that the clamp screw does not contact adjacent parts. When tightening the high pressure rubber hose clamp make sure the clamp end is 0.12 inch from the hose end. After connecting these fittings make sure that the connectors are secure. Check for fuel leakage at these connections turn the ignition key to the ON position (do not start the engine), correct as required. Start the engine, raise the idle, and verify that there are no fuel leaks, correct as required.

2. Properly relieve the fuel system pressure.
3. Disconnect the negative battery cable.
4. Drain the fuel from the fuel tank, if necessary.
5. Open the fuel filler lid.
6. Open the filler cap and release the pressure from inside the tank.
7. Tilt the second seat toward the front of the vehicle.
8. Peel off the floor carpet, then remove the inspection hole cover by turning the clips clockwise 90 degrees.
9. Disconnect the harness connector and all the fuel line tubes.

10. Raise and safely support the vehicle.
11. Remove the left side tire and wheel assembly.
12. Disconnect the fuel filler tube, vent hose and EVAP hose at the fuel tank.
13. Remove the fuel tank protector.
14. Properly support the fuel tank using a service jack. Be sure the jack does not interfere with the tank bands.
15. Remove the tank mounting bolts.
16. Carefully lower the tank. Check that there are no hoses or wires that would prevent the tank from being removed.
17. Remove the fuel tank.

To install:

➡Be sure to use new fasteners, as required.

18. Installation is the reverse of the removal procedure.
19. Be sure to perform the reconnect/relearn procedures.

THROTTLE BODY

REMOVAL & INSTALLATION
See Figure 119.

At this time the manufacturer does not provide removal and installation procedures for this component. The following procedure is a guideline and may differ from the vehicle you are servicing.

1. Before servicing the vehicle, refer to the Precautions Section.

➡If working near and/or around the SRS system and components, be sure to disable the SRS system. After disabling the system wait three minutes or more before servicing the vehicle.

➡Whenever the negative battery cable is disconnected the following components will require resetting. The air conditioning system, automatic drive positioner system, power window control system, around view monitor, automatic back door system, idle air volume learning, steering angle sensor neutral position, sunroof system, audio visual and navigation systems. Use the CONSULT-III diagnostic tool, or equivalent to perform the required resets. Follow the directions on the screen of the tool, as needed.

➡Before disconnecting the negative battery cable lower both the driver's side and passenger's side front windows. This will prevent any interference between the window edge and the

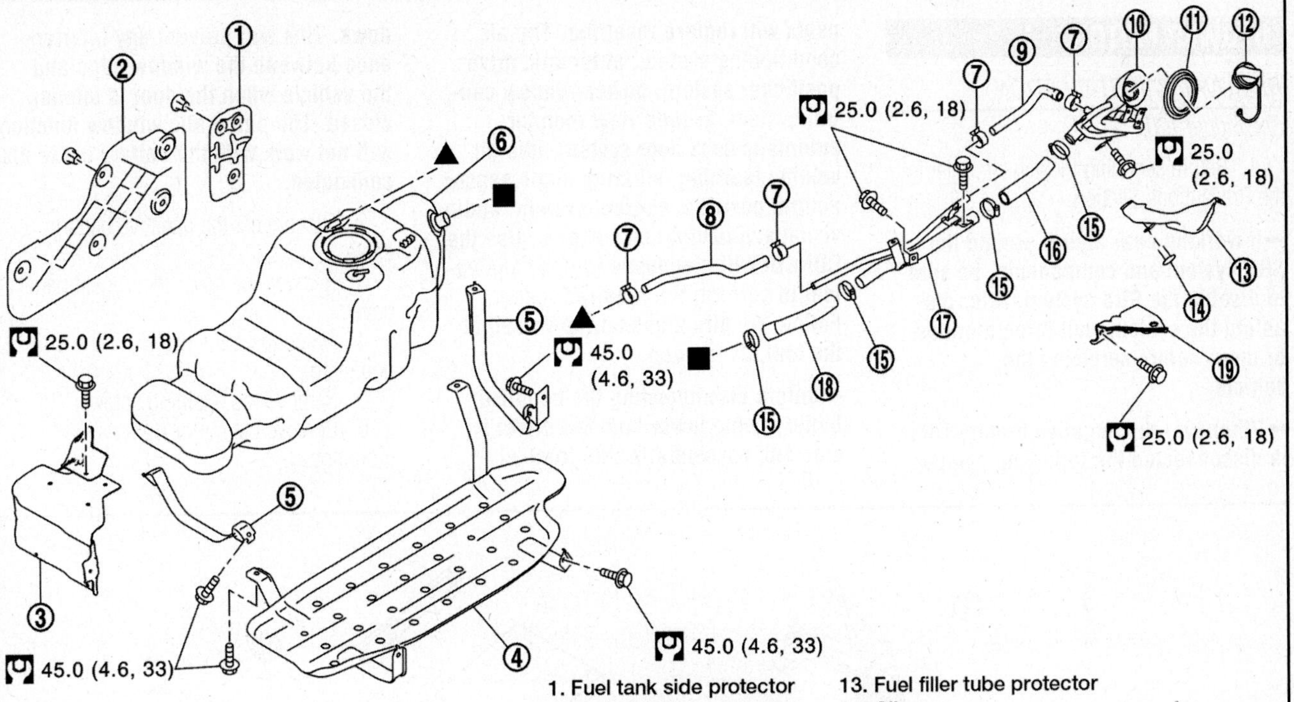

25.0 (2.6, 18)

25.0 (2.6, 18)

25.0 (2.6, 18)

25.0 (2.6, 18)

45.0 (4.6, 33)

45.0 (4.6, 33)

45.0 (4.6, 33)

1. Fuel tank side protector
2. Fuel tank side protector
3. Fuel tank front protector
4. Fuel tank protector
5. Fuel tank mounting band
6. Fuel tank
7. Clamp
8. Vent hose
9. Vent hose
10. Fuel filler tube
11. Grommet
12. Fuel filler cap
13. Fuel filler tube protector
14. Clip
15. Clamp
16. Fuel filler hose
17. Fuel filler tube
18. Fuel filler hose
19. Fuel filler tube protector
, : Indicates that the part is connected at
    points with same symbol in actual vehicle.

71075_QX56_G0270

Fig. 118 Fuel tank and related components

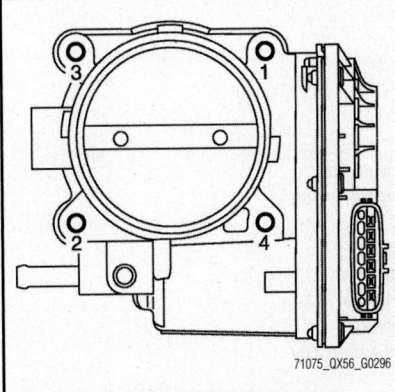

71075_QX56_G0296

Fig. 119 Throttle control actuator bolt location and tightening sequence

vehicle when the door is opener/ closed. The automatic window function will not work with the battery cable disconnected.

2. Disconnect the negative battery cable.

3. Drain the cooling system, as required. Be sure to properly dispose of used engine coolant.

4. Remove the engine room cover. Remove the air intake duct.

5. Disconnect harness connector.

6. Disconnect water hoses.

7. Loosen the throttle body assembly mounting bolts in reverse order of the tightening sequence.

To install:

➡Be sure to use new fasteners, as required.

8. Install the throttle body assembly with a new gasket.

9. Tighten the mounting bolts in sequence.

10. Reconnect the water hose.

11. Reconnect the harness connector.

12. Reconnect the air intake duct.

13. Fill the cooling system with the proper grade and type engine coolant.

14. Be sure to perform the reconnect/ relearn procedures.

HEATING & AIR CONDITIONING SYSTEM

BLOWER MOTOR

REMOVAL & INSTALLATION

See Figure 120.

1. Before servicing the vehicle, refer to the Precautions Section.

➡ If working near and/or around the SRS system and components, be sure to disable the SRS system. After disabling the system wait three minutes or more before servicing the vehicle.

➡ Whenever the negative battery cable is disconnected the following compo-

nents will require resetting. The air conditioning system, automatic drive positioner system, power window control system, around view monitor, automatic back door system, idle air volume learning, steering angle sensor neutral position, sunroof system, audio visual and navigation systems. Use the CONSULT-III diagnostic tool, or equivalent to perform the required resets. Follow the directions on the screen of the tool, as needed.

➡ Before disconnecting the negative battery cable lower both the driver's side and passenger's side front win-

dows. This will prevent any interference between the window edge and the vehicle when the door is opener/closed. The automatic window function will not work with the battery cable disconnected.

2. Disconnect the negative battery cable.
3. Remove the lower instrument panel cover.
4. Disconnect the motor connector.
5. Remove the retaining screws.
6. Remove the motor from its mounting.

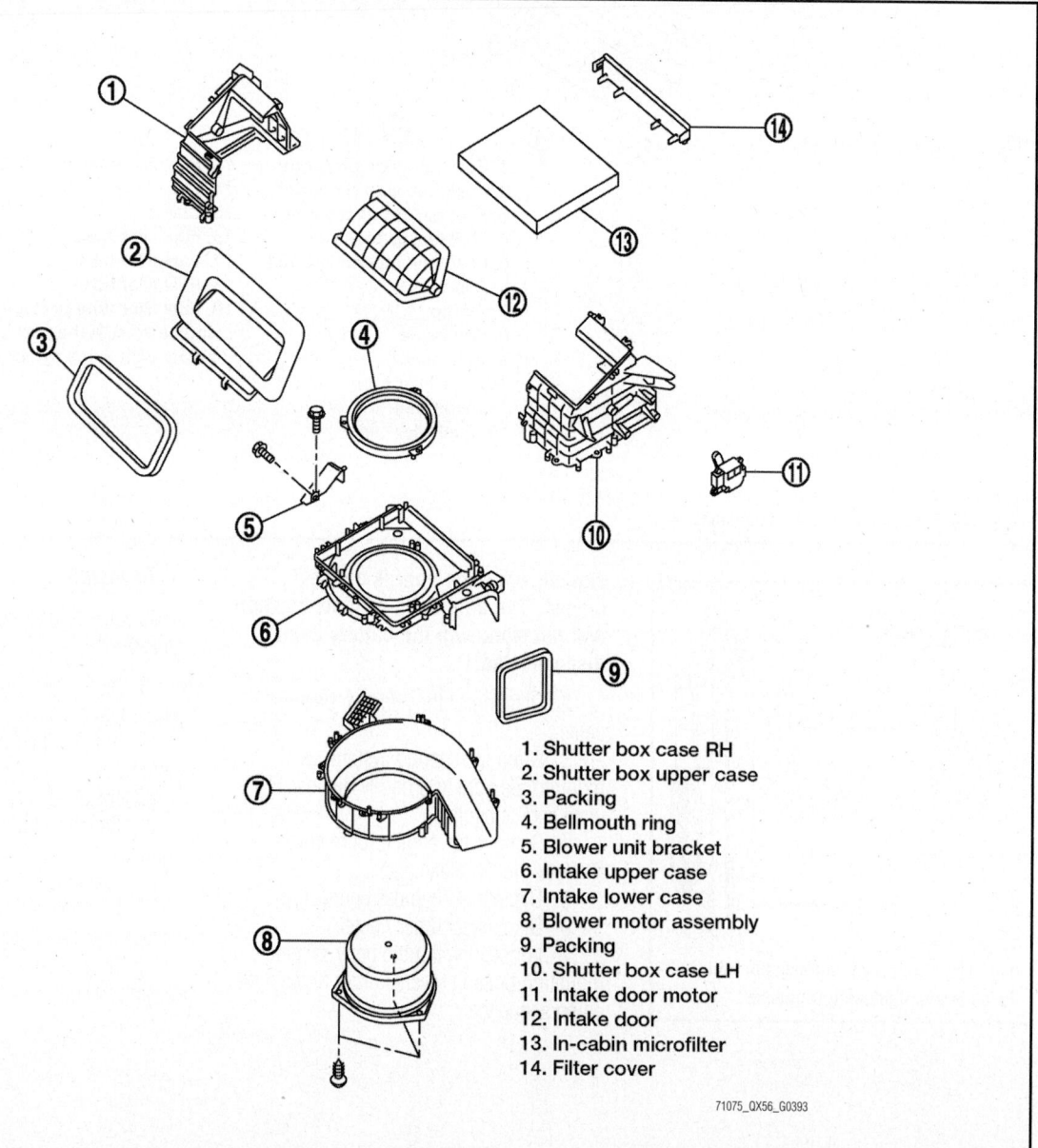

1. Shutter box case RH
2. Shutter box upper case
3. Packing
4. Bellmouth ring
5. Blower unit bracket
6. Intake upper case
7. Intake lower case
8. Blower motor assembly
9. Packing
10. Shutter box case LH
11. Intake door motor
12. Intake door
13. In-cabin microfilter
14. Filter cover

71075_QX56_G0393

Fig. 120 Front air conditioning/heating blower motor and related components

To install:

➡ **Be sure to use new fasteners, as required.**

7. Installation is the reverse of the removal procedure.

8. Properly recharge the system.

9. Be sure to perform the reconnect/relearn procedures.

HEATER/COOLING UNIT

REMOVAL & INSTALLATION

See Figures 121 through 123.

1. Before servicing the vehicle, refer to the Precautions Section.

➡ **If working near and/or around the SRS system and components, be sure to disable the SRS system. After disabling the system wait three minutes or more before servicing the vehicle.**

➡ **Whenever the negative battery cable is disconnected the following components will require resetting. The air conditioning system, automatic drive positioner system, power window control system, around view monitor,** automatic back door system, idle air volume learning, steering angle sensor neutral position, sunroof system, audio visual and navigation systems. Use the CONSULT-III diagnostic tool, or equivalent to perform the required resets. Follow the directions on the screen of the tool, as needed.

➡ **Before disconnecting the negative battery cable lower both the driver's side and passenger's side front windows. This will prevent any interference between the window edge and the vehicle when the door is opener/closed. The automatic window function will not work with the battery cable disconnected.**

➡ **Perform lubricant return operation before refrigeration system disassembly. However, if a large amount of lubricant or refrigerant leakage is detected never perform lubricant return operation procedure.**

2. Start the engine.

3. Set the engine speed to 1200 rpm, idling.

4. Turn the air conditioning ON.

5. Set the fan speed to MAX.

6. Set the intake door position to RECIRC.

7. Set the temperature to FULL cold.

8. Perform the lubricant return operation for about ten minutes.

9. Stop the engine.

10. The procedure is complete.

11. Properly discharge the air conditioning system.

12. Disconnect the negative battery cable.

13. Remove the engine cover.

14. Remove the air cleaner assembly.

15. Drain the cooling system. Be sure to properly dispose of used coolant.

16. Disconnect the heater hoses at the heater core.

17. Disconnect and plug the refrigerant lines from the expansion valve.

18. Remove the instrument panel assembly.

19. Remove the steering column mounting bolt and nuts.

20. Reposition the steering column assembly so it does not interfere with the procedure.

21. Disconnect the harness clips, harness connectors, ground bolts and brackets from the steering member. Move the vehicle harness out of the way.

22. Remove the mounting nuts and remove the right and left instrument stay.

23. Remove the heater and cooling unit assembly bolts.

24. Remove the mounting bolts and then remove the steering member from the vehicle.

25. Disconnect the drain hose from the heater and cooling unit.

26. Remove the component from the vehicle.

27. Remove the mounting bolts and disconnect the heater and cooling unit from the blower motor housing assembly.

To install:

➡ **Be sure to use new fasteners, as required.**

28. Installation is the reverse of the removal procedure.

29. Be sure to use new O-rings coated with clean refrigerant oil, as required.

30. Fill the cooling system with the proper grade and the coolant.

31. Properly recharge the A/C system.

32. Start the engine and check for leaks, correct as required.

33. Be sure to perform the reconnect/relearn procedures.

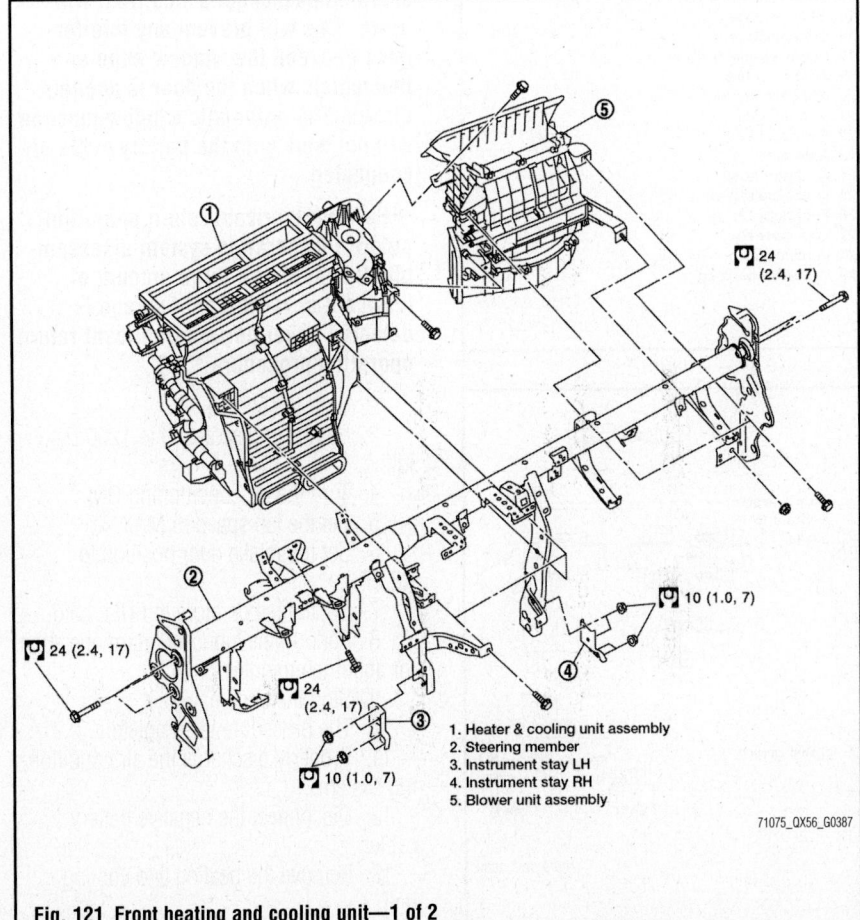

1. Heater & cooling unit assembly
2. Steering member
3. Instrument stay LH
4. Instrument stay RH
5. Blower unit assembly

71075_QX56_G0387

Fig. 121 Front heating and cooling unit—1 of 2

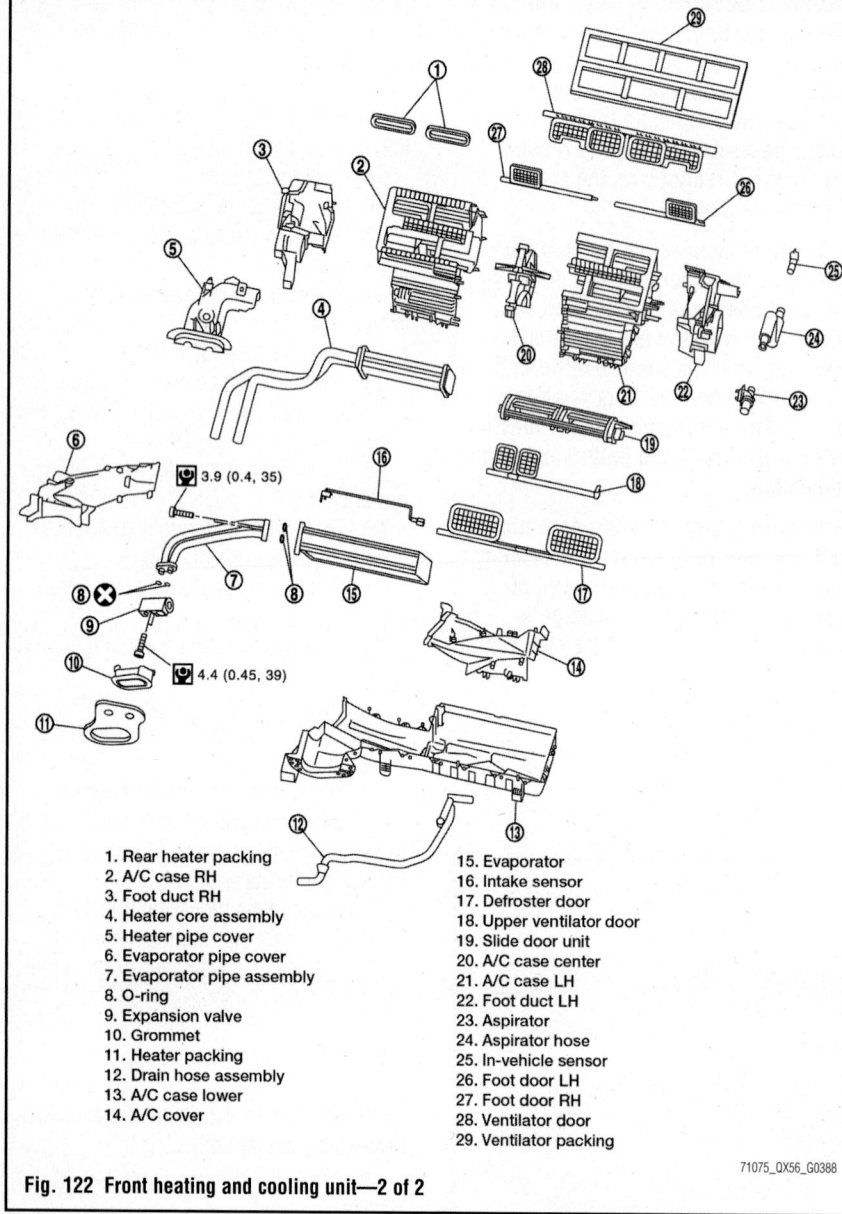

3.9 (0.4, 35)

4.4 (0.45, 39)

1. Rear heater packing
2. A/C case RH
3. Foot duct RH
4. Heater core assembly
5. Heater pipe cover
6. Evaporator pipe cover
7. Evaporator pipe assembly
8. O-ring
9. Expansion valve
10. Grommet
11. Heater packing
12. Drain hose assembly
13. A/C case lower
14. A/C cover
15. Evaporator
16. Intake sensor
17. Defroster door
18. Upper ventilator door
19. Slide door unit
20. A/C case center
21. A/C case LH
22. Foot duct LH
23. Aspirator
24. Aspirator hose
25. In-vehicle sensor
26. Foot door LH
27. Foot door RH
28. Ventilator door
29. Ventilator packing

71075_QX56_G0388

Fig. 122 Front heating and cooling unit—2 of 2

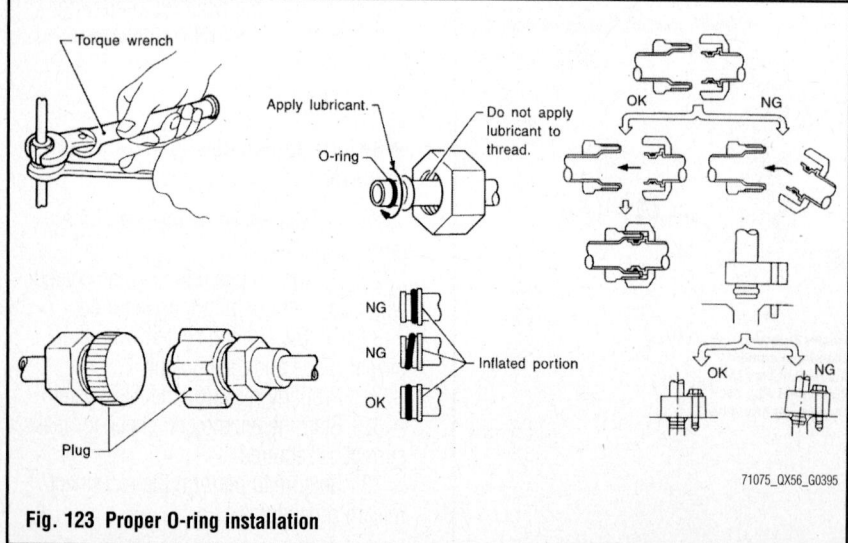

71075_QX56_G0395

Fig. 123 Proper O-ring installation

HEATER CORE

REMOVAL & INSTALLATION

See Figure 124.

1. Before servicing the vehicle, refer to the Precautions Section.

➡ If working near and/or around the SRS system and components, be sure to disable the SRS system. After disabling the system wait three minutes or more before servicing the vehicle.

➡ Whenever the negative battery cable is disconnected the following components will require resetting. The air conditioning system, automatic drive positioner system, power window control system, around view monitor, automatic back door system, idle air volume learning, steering angle sensor neutral position, sunroof system, audio visual and navigation systems. Use the CONSULT-III diagnostic tool, or equivalent to perform the required resets. Follow the directions on the screen of the tool, as needed.

➡ Before disconnecting the negative battery cable lower both the driver's side and passenger's side front windows. This will prevent any interference between the window edge and the vehicle when the door is opener/closed. The automatic window function will not work with the battery cable disconnected.

➡ Perform lubricant return operation before refrigeration system disassembly. However, if a large amount of lubricant or refrigerant leakage is detected never perform lubricant return operation procedure.

2. Start the engine.
3. Set the engine speed to 1200rpm, idling.
4. Turn the air conditioning ON.
5. Set the fan speed to MAX.
6. Set the intake door position to RECIRC.
7. Set the temperature to FULL cold.
8. Perform the lubricant return operation for about ten minutes.
9. Stop the engine.
10. The procedure is complete.
11. Properly discharge the air conditioning system.
12. Disconnect the negative battery cable.
13. Remove the heating and cooling unit.

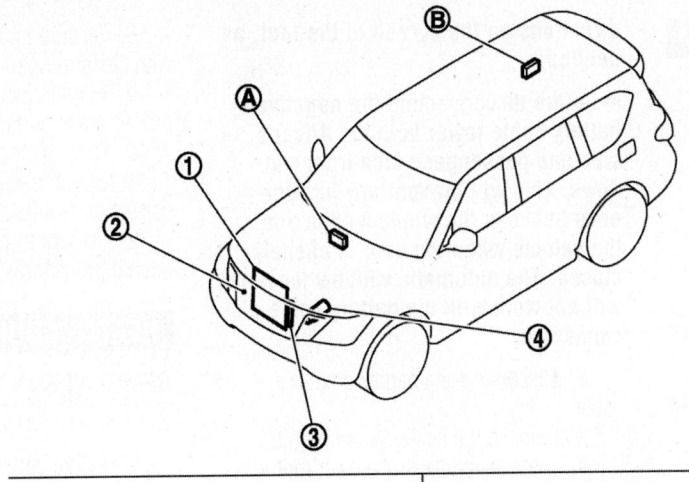

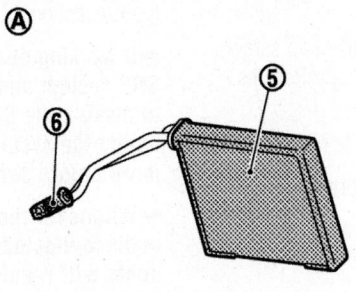
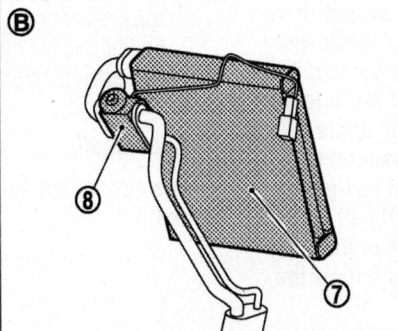

1. Condenser
2. Refrigerant pressure sensor
3. Liquid tank
4. Compressor
5. Front evaporator
6. Front expansion valve

7. Rear evaporator
8. Rear expansion valve
A. Built-in heater & cooling unit assembly
B. Built-in rear cooler unit assembly

71075_QX56_G0381

Fig. 124 Automatic air conditioning heater core and related components

14. Remove the mounting screws and remove the heater core pipe cover.

15. Remove the mode door motor.

16. Remove the main link, foot door link and foot door lever.

17. Remove the mounting screws and remove foot duct assist.

18. Pull the heater core from the heater and cooling unit.

To install:

➡**Be sure to use new fasteners, as required.**

19. Installation is the reverse of the removal procedure.

20. Be sure to use new O-rings coated with clean refrigerant oil, as required.

21. Fill the cooling system with the proper grade and the coolant.

22. Properly recharge the A/C system.

23. Start the engine and check for leaks, correct as required.

24. Be sure to perform the reconnect/relearn procedures.

AUXILIARY HEATING & AIR CONDITIONING

BLOWER MOTOR

REMOVAL & INSTALLATION

See Figures 125 and 126.

1. Before servicing the vehicle, refer to the Precautions Section.

➡ If working near and/or around the SRS system and components, be sure to disable the SRS system. After disabling the system wait three minutes or more before servicing the vehicle.

➡ Whenever the negative battery cable is disconnected the following components will require resetting. The air conditioning system, automatic drive positioner system, power window control system, around view monitor, automatic back door system, idle air volume learning, steering angle sensor neutral position, sunroof system, audio visual and navigation systems. Use the CONSULT-III diagnostic tool, or equivalent to perform the required resets. Follow the directions on the screen of the tool, as needed.

➡ Before disconnecting the negative battery cable lower both the driver's side and passenger's side front windows. This will prevent any interference between the window edge and the vehicle when the door is opener/closed. The automatic window function will not work with the battery cable disconnected.

2. Disconnect the negative battery cable.
3. Remove the heater/cooling unit.
4. Disconnect the rear blower motor electrical connector.
5. Remove the blower retaining screws.
6. Remove the blower motor from its mounting.

To install:

➡ Be sure to use new fasteners, as required.

7. Installation is the reverse of the removal procedure.

8. Be sure to use new O-rings coated with clean refrigerant oil, as required.
9. Fill the cooling system with the proper grade and the coolant.
10. Properly recharge the A/C system.
11. Start the engine and check for leaks, correct as required.
12. Be sure to perform the reconnect/relearn procedures.

HEATER/COOLING UNIT

REMOVAL & INSTALLATION

See Figures 125 through 127

1. Before servicing the vehicle, refer to the Precautions Section.

➡ If working near and/or around the SRS system and components, be sure to disable the SRS system. After disabling the system wait three minutes or more before servicing the vehicle.

➡ Whenever the negative battery cable is disconnected the following components will require resetting. The air

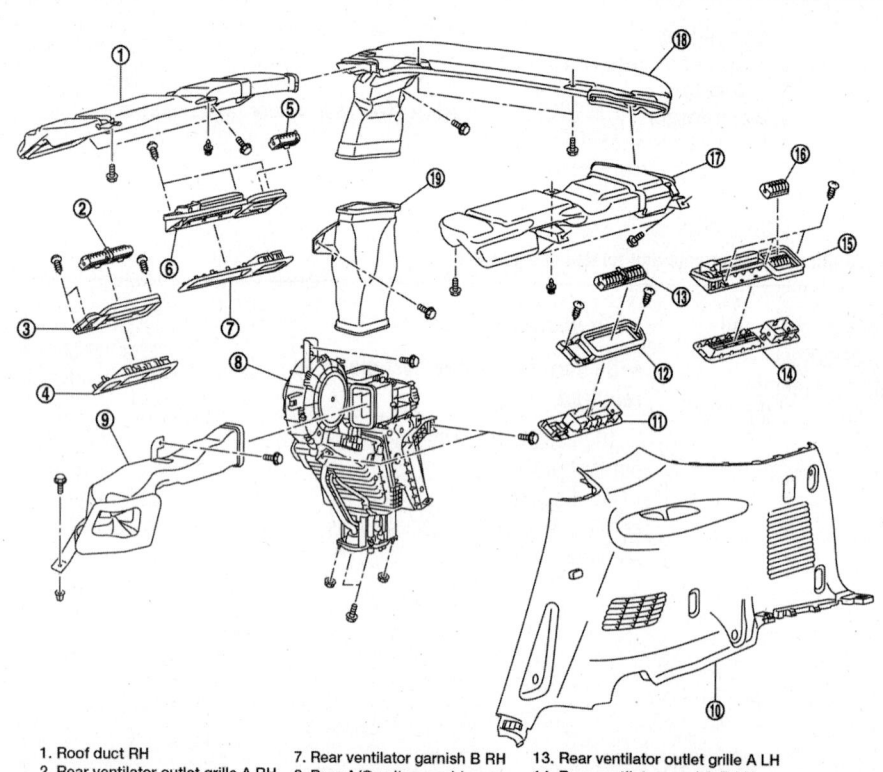

1. Roof duct RH
2. Rear ventilator outlet grille A RH
3. Rear ventilator base RH
4. Rear ventilator garnish A RH
5. Rear ventilator outlet grille B RH
6. Rear ventilator grille RH
7. Rear ventilator garnish B RH
8. Rear A/C unit assembly
9. Rear ventilator duct lower
10. Luggage side finisher RH
11. Rear ventilator garnish A LH
12. Rear ventilator base LH
13. Rear ventilator outlet grille A LH
14. Rear ventilator garnish B LH
15. Rear ventilator grille LH
16. Rear ventilator outlet grille B LH
17. Roof duct LH Roof duct center
18.
19. Rear ventilator duct upper

71075_QX56_G0389

Fig. 125 Rear heating and cooling unit—1 of 2

3. Disconnect the negative battery cable.

4. Drain the cooling system. Be sure to properly dispose of used coolant.

5. Remove the rear ventilator duct, upper and lower.

6. Remove the mounting bolt and disconnect the cooler unit pipe assembly from the rear cooler pipe assembly. Discard the O-rings. Cap the lines.

7. Remove the clamps and disconnect the heater pipe from the cooler pipe assembly.

8. Remove the vehicle tire jack assembly.

9. Remove the mounting bolts and nuts and remove the heater and cooling unit assembly.

To install:

➡ **Be sure to use new fasteners, as required.**

10. Installation is the reverse of the removal procedure.

11. Be sure to use new O-rings coated with clean refrigerant oil, as required.

12. Fill the cooling system with the proper grade and the coolant.

13. Properly recharge the A/C system.

14. Start the engine and check for leaks, correct as required.

15. Be sure to perform the reconnect/relearn procedures.

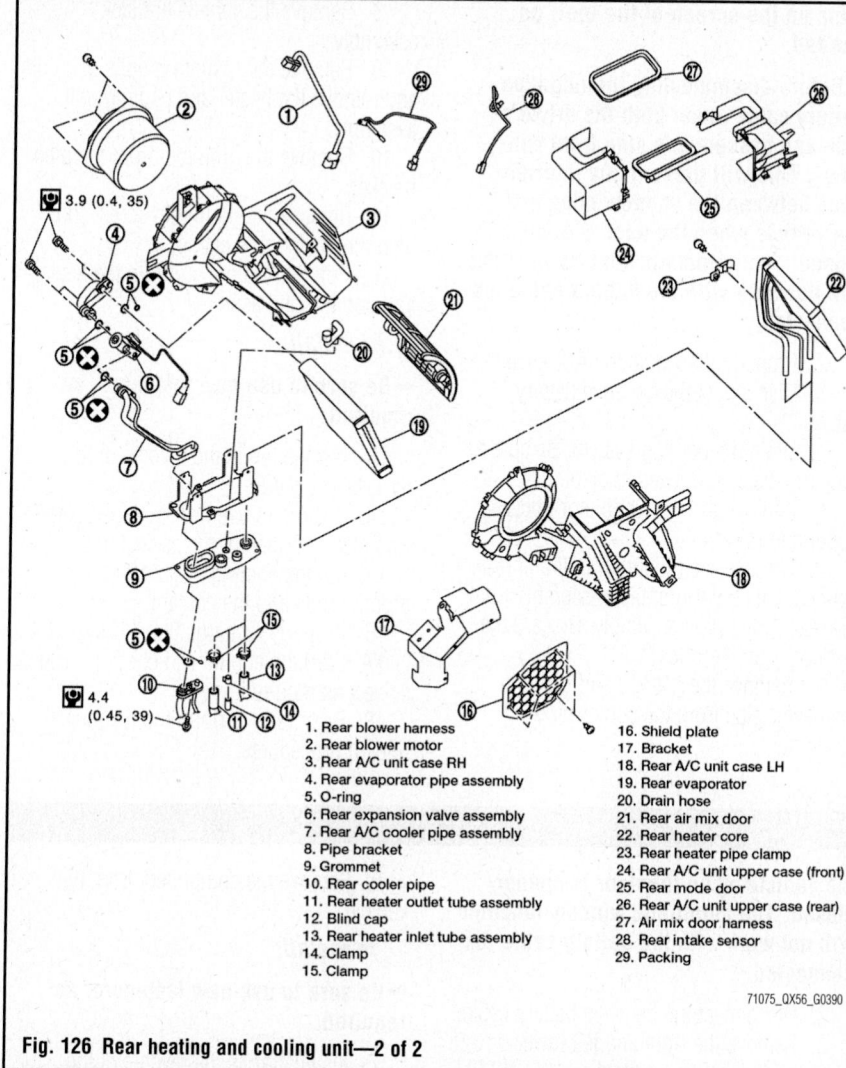

1. Rear blower harness
2. Rear blower motor
3. Rear A/C unit case RH
4. Rear evaporator pipe assembly
5. O-ring
6. Rear expansion valve assembly
7. Rear A/C cooler pipe assembly
8. Pipe bracket
9. Grommet
10. Rear cooler pipe
11. Rear heater outlet tube assembly
12. Blind cap
13. Rear heater inlet tube assembly
14. Clamp
15. Clamp
16. Shield plate
17. Bracket
18. Rear A/C unit case LH
19. Rear evaporator
20. Drain hose
21. Rear air mix door
22. Rear heater core
23. Rear heater pipe clamp
24. Rear A/C unit upper case (front)
25. Rear mode door
26. Rear A/C unit upper case (rear)
27. Air mix door harness
28. Rear intake sensor
29. Packing

3.9 (0.4, 35)
4.4 (0.45, 39)

71075_QX56_G0390

Fig. 126 Rear heating and cooling unit—2 of 2

conditioning system, automatic drive positioner system, power window control system, around view monitor, automatic back door system, idle air volume learning, steering angle sensor neutral position, sunroof system, audio visual and navigation systems. Use the CONSULT-III diagnostic tool, or equivalent to perform the required resets. Follow the directions on the screen of the tool, as needed.

➡ Before disconnecting the negative battery cable lower both the driver's side and passenger's side front windows. This will prevent any interference between the window edge and the vehicle when the door is opener/closed. The automatic window function will not work with the battery cable disconnected.

2. Properly discharge the A/C system.

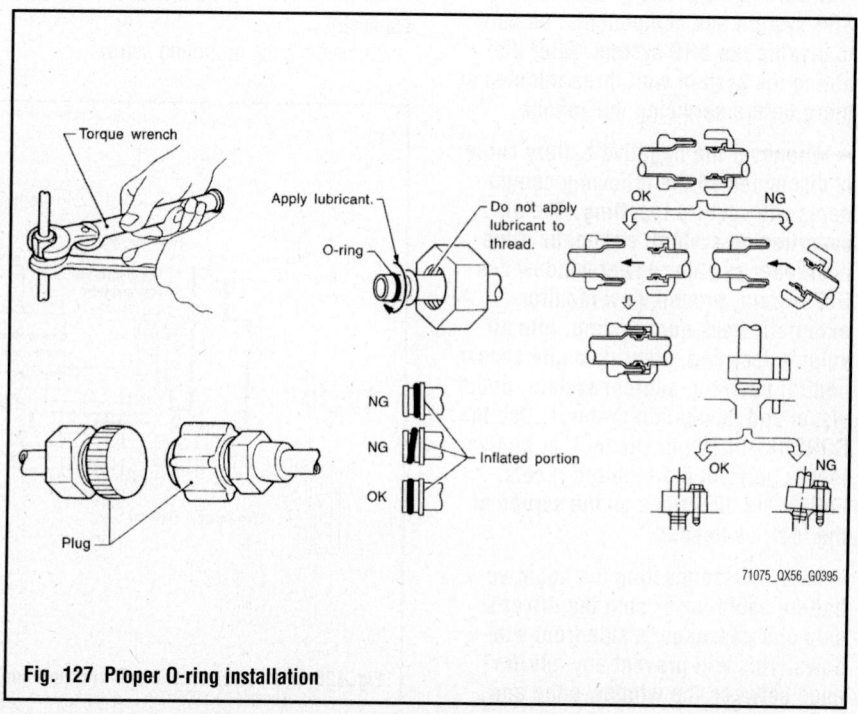

71075_QX56_G0395

Fig. 127 Proper O-ring installation

HEATER CORE

REMOVAL & INSTALLATION

See Figures 125 and 126.

1. Before servicing the vehicle, refer to the Precautions Section.

➡ If working near and/or around the SRS system and components, be sure to disable the SRS system. After disabling the system wait three minutes or more before servicing the vehicle.

➡ Whenever the negative battery cable is disconnected the following components will require resetting. The air conditioning system, automatic drive positioner system, power window control system, around view monitor, automatic back door system, idle air volume learning, steering angle sensor neutral position, sunroof system, audio visual and navigation systems. Use the CONSULT-III diagnostic tool, or equivalent to perform the required resets. Follow the directions on the screen of the tool, as needed.

➡ Before disconnecting the negative battery cable lower both the driver's side and passenger's side front windows. This will prevent any interference between the window edge and the vehicle when the door is opener/closed. The automatic window function will not work with the battery cable disconnected.

2. Properly discharge the A/C system.
3. Disconnect the negative battery cable.
4. Drain the cooling system. Be sure to properly dispose of used coolant.
5. Remove the rear ventilator duct, upper and lower.
6. Remove the mounting bolt and disconnect the cooler unit pipe assembly from the rear cooler pipe assembly. Discard the O-rings. Cap the lines.
7. Remove the clamps and disconnect the heater pipe from the cooler pipe assembly.

8. Remove the vehicle tire jack assembly.
9. Remove the mounting bolts and nuts and remove the heater and cooling unit assembly.
10. Remove the grommet from the pipe bracket.
11. Remove the mounting screw and then the heater pipe clamp.
12. Pull the heater core from the heater and cooling unit assembly.

To install:

➡ Be sure to use new fasteners, as required.

13. Installation is the reverse of the removal procedure.
14. Be sure to use new O-rings coated with clean refrigerant oil, as required.
15. Fill the cooling system with the proper grade and the coolant.
16. Properly recharge the A/C system.
17. Start the engine and check for leaks, correct as required.
18. Be sure to perform the reconnect/relearn procedures.

STEERING

EPS CONTROL UNIT

REMOVAL & INSTALLATION

See Figure 128.

1. Before servicing the vehicle, refer to the Precautions Section.

➡ If working near and/or around the SRS system and components, be sure to disable the SRS system. After disabling the system wait three minutes or more before servicing the vehicle.

➡ Whenever the negative battery cable is disconnected the following components will require resetting. The air conditioning system, automatic drive positioner system, power window control system, around view monitor, automatic back door system, idle air volume learning, steering angle sensor neutral position, sunroof system, audio visual and navigation systems. Use the CONSULT-III diagnostic tool, or equivalent to perform the required resets. Follow the directions on the screen of the tool, as needed.

➡ Before disconnecting the negative battery cable lower both the driver's side and passenger's side front windows. This will prevent any interference between the window edge and the vehicle when the door is opener/closed. The automatic window function will not work with the battery cable disconnected.

2. Disconnect the negative battery cable.
3. Remove the right and left lower instrument panel trim panels.
4. Disconnect the control unit electrical connector.
5. Remove the mounting screws.

6. Remove the component from the vehicle.

To install:

➡ Be sure to use new fasteners, as required.

7. Installation is the reverse of the removal procedure.
8. Be sure to perform the reconnect/relearn procedures.

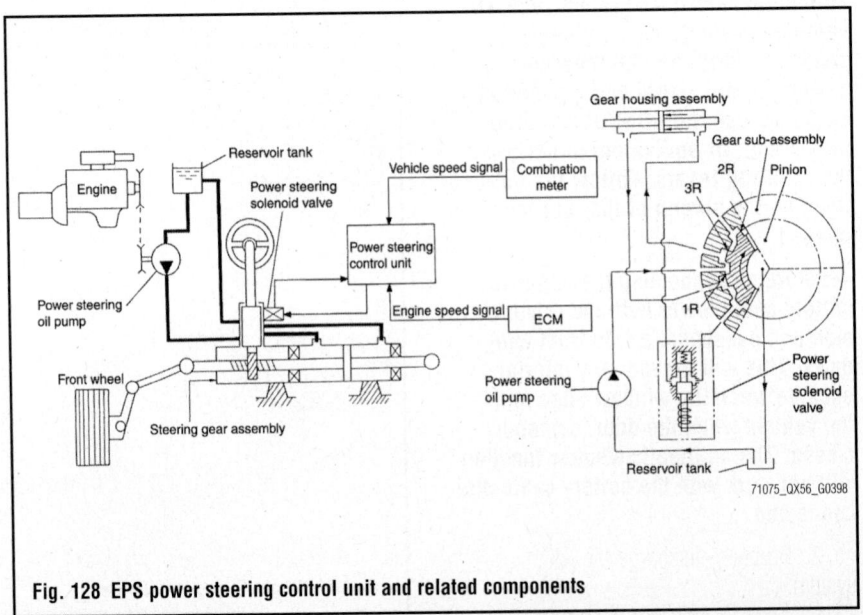

Fig. 128 EPS power steering control unit and related components

71075_QX56_G0398

POWER RACK & PINION STEERING GEAR

REMOVAL & INSTALLATION

See Figures 129 through 131.

1. Before servicing the vehicle, refer to the Precautions Section.

➡ **If working near and/or around the SRS system and components, be sure to disable the SRS system. After disabling the system wait three minutes or more before servicing the vehicle.**

➡ **Whenever the negative battery cable is disconnected the following components will require resetting. The air conditioning system, automatic drive positioner system, power window control system, around view monitor, automatic back door system, idle air volume learning, steering angle sensor neutral position, sunroof system, audio visual and navigation systems. Use the CONSULT-III diagnostic tool, or equivalent to perform the required resets. Follow the directions on the screen of the tool, as needed.**

➡ **Before disconnecting the negative battery cable lower both the driver's side and passenger's side front windows. This will prevent any interference between the window edge and the vehicle when the door is opener/closed. The automatic window function will not work with the battery cable disconnected.**

2. Ensure the wheels are in the straight-ahead position.

3. Disconnect the negative battery cable.

4. Raise and support the vehicle safely.

5. Remove the front tire and wheel assemblies.

6. Remove the front final drive unit, if equipped.

7. Remove the cotter pin and loosen the locknut.

8. Remove the steering outer socket from the knuckle, so as not to damage the ball joint using a ball joint removal tool.

➡ **Temporarily tighten the nut to prevent damage to the threads and to prevent the tool from suddenly coming off.**

9. Remove the high pressure piping and low pressure piping and drain the power steering fluid. Be sure to properly dispose of used fluid.

10. Remove the solenoid valve harness connector.

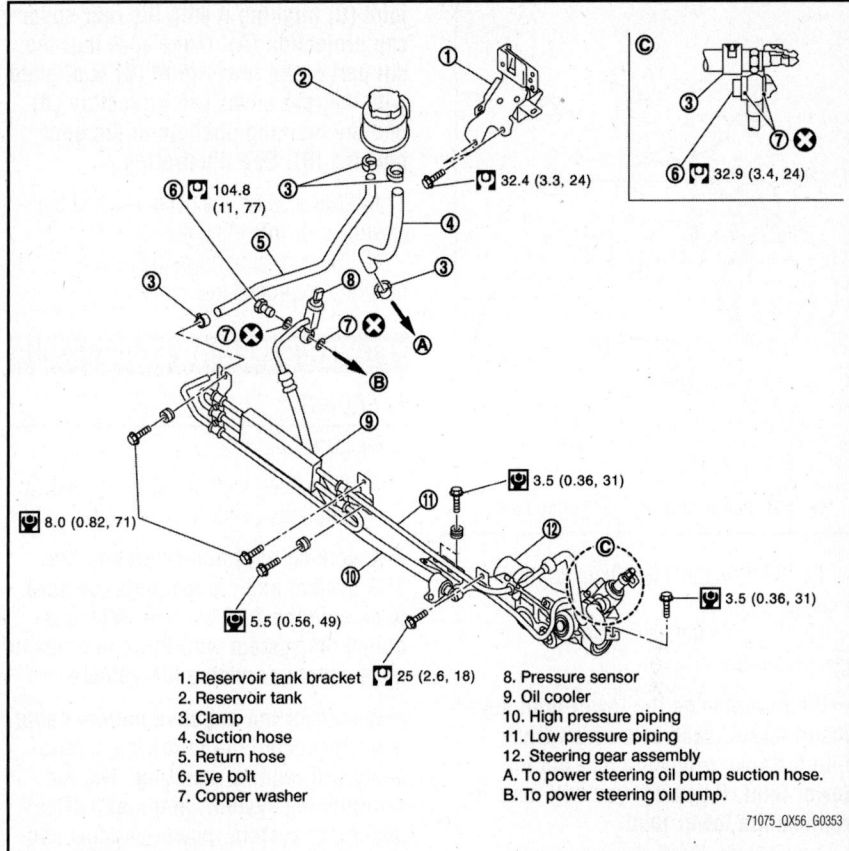

1. Reservoir tank bracket
2. Reservoir tank
3. Clamp
4. Suction hose
5. Return hose
6. Eye bolt
7. Copper washer
8. Pressure sensor
9. Oil cooler
10. High pressure piping
11. Low pressure piping
12. Steering gear assembly
A. To power steering oil pump suction hose.
B. To power steering oil pump.

71075_QX56_G0353

Fig. 129 Power steering gear and related components—engine speed sensitive system

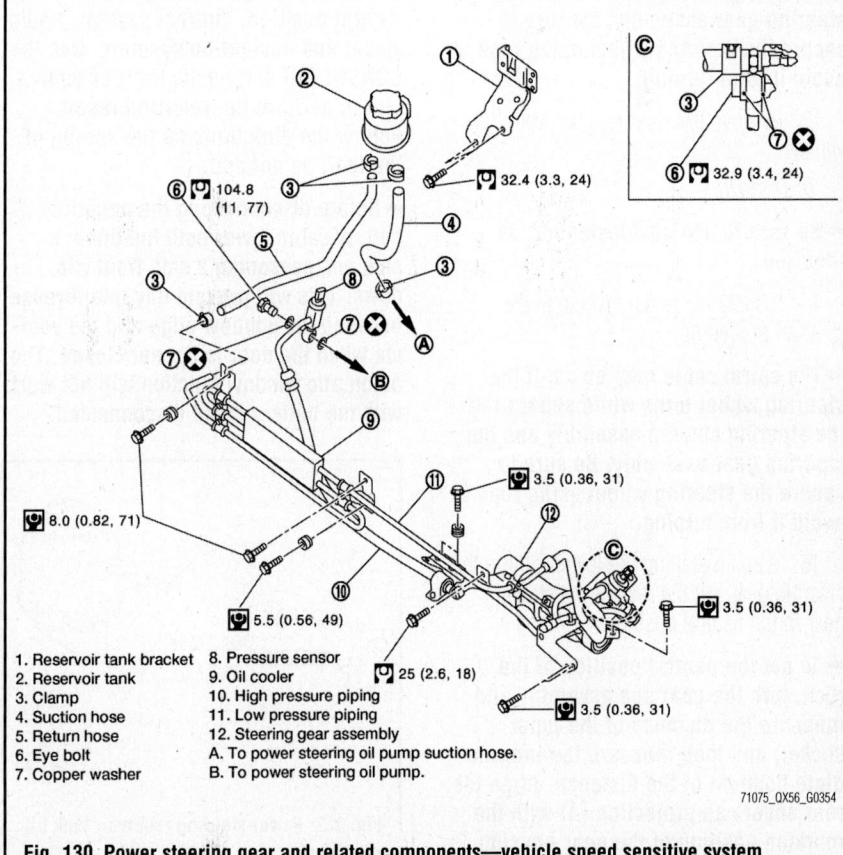

1. Reservoir tank bracket
2. Reservoir tank
3. Clamp
4. Suction hose
5. Return hose
6. Eye bolt
7. Copper washer
8. Pressure sensor
9. Oil cooler
10. High pressure piping
11. Low pressure piping
12. Steering gear assembly
A. To power steering oil pump suction hose.
B. To power steering oil pump.

71075_QX56_G0354

Fig. 130 Power steering gear and related components—vehicle speed sensitive system

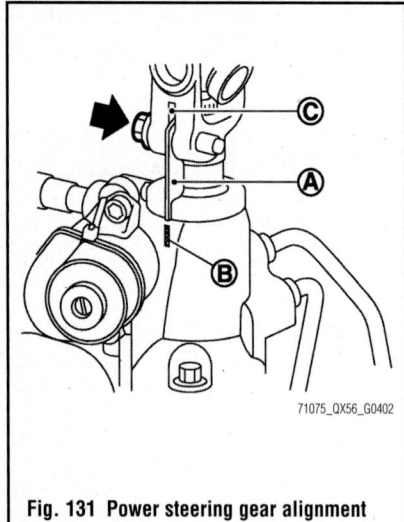

Fig. 131 Power steering gear alignment

11. Remove the lower fixing bolt (steering gear side).

12. Separate the lower joint from the gear assembly.

➡ **When removing the lower joint never insert a tool, such as a screwdriver, into the yoke groove to pull out the lower joint. If you do, you will have to replace the lower joint.**

➡ **The spiral cable may be cut if the steering wheel turns while separating the steering column assembly and the steering gear assembly. Be sure to secure the steering wheel using rope to avoid it from turning.**

13. Remove the steering gear from the vehicle.

To install:

➡ **Be sure to use new fasteners, as required.**

14. Installation is the reverse of the removal procedure.

➡ **The spiral cable may be cut if the steering wheel turns while separating the steering column assembly and the steering gear assembly. Be sure to secure the steering wheel using rope to avoid it from turning.**

15. When installing the lower joint to the steering gear, set the rack of the steering gear in the neutral position.

➡ **To get the neutral position of the rack, turn the gear sub assembly and measure the distance of the inner socket, and then measure the intermediate position of the distance. Align the rear cover cap projection (A) with the marking position of the gear housing (B). Install the slit part of the lower**

joint (C) aligning it with the rear cover cap projection (A). Make sure that the slit part of the lower joint (C) is aligned with the rear cover cap projection (A) and the marking position of the gear housing (B). See illustration.

16. Never reuse the cotter pin and the steering gear mounting nut.

17. Be sure to perform the reconnect/relearn procedures.

POWER STEERING PUMP

BLEEDING

See Figure 132.

1. Before servicing the vehicle, refer to the Precautions Section.

➡ **If working near and/or around the SRS system and components, be sure to disable the SRS system. After disabling the system wait three minutes or more before servicing the vehicle.**

➡ **Whenever the negative battery cable is disconnected the following components will require resetting. The air conditioning system, automatic drive positioner system, power window control system, around view monitor, automatic back door system, idle air volume learning, steering angle sensor neutral position, sunroof system, audio visual and navigation systems. Use the CONSULT-III diagnostic tool, or equivalent to perform the required resets. Follow the directions on the screen of the tool, as needed.**

➡ **Before disconnecting the negative battery cable lower both the driver's side and passenger's side front windows. This will prevent any interference between the window edge and the vehicle when the door is opener/closed. The automatic window function will not work with the battery cable disconnected.**

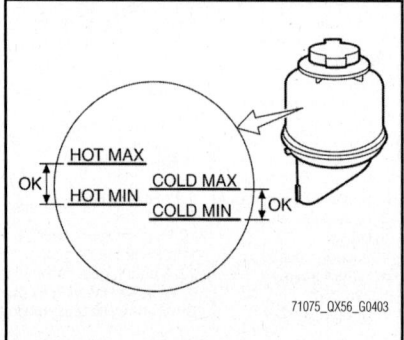

Fig. 132 Power steering reservoir tank fill markings

2. Stop the engine.

3. Turn the steering wheel fully to the right and left several times.

➡ **Do not allow the fluid level in the reservoir tank to go below the MIN level line. Check and add fluid as needed.**

4. Run the engine at idle speed. Turn the steering wheel fully to the right and then fully to the left. Hold for about three seconds. Check for fluid leakage.

5. Repeat the above step several times at three second intervals.

➡ **Do not hold the steering wheel in the locked position for more than ten seconds.**

6. Check for air bubbles or cloudy fluid. If found, repeat the bleeding procedure.

7. Stop the engine and check the fluid level. Correct as required.

REMOVAL & INSTALLATION

See Figures 133 through 135.

1. Before servicing the vehicle, refer to the Precautions Section.

➡ **If working near and/or around the SRS system and components, be sure to disable the SRS system. After disabling the system wait three minutes or more before servicing the vehicle.**

➡ **Whenever the negative battery cable is disconnected the following components will require resetting. The air conditioning system, automatic drive positioner system, power window control system, around view monitor, automatic back door system, idle air volume learning, steering angle sensor neutral position, sunroof system, audio visual and navigation systems. Use the CONSULT-III diagnostic tool, or equivalent to perform the required resets. Follow the directions on the screen of the tool, as needed.**

➡ **Before disconnecting the negative battery cable lower both the driver's side and passenger's side front windows. This will prevent any interference between the window edge and the vehicle when the door is opener/closed. The automatic window function will not work with the battery cable disconnected.**

2. Disconnect the negative battery cable.

3. Remove the engine cover.

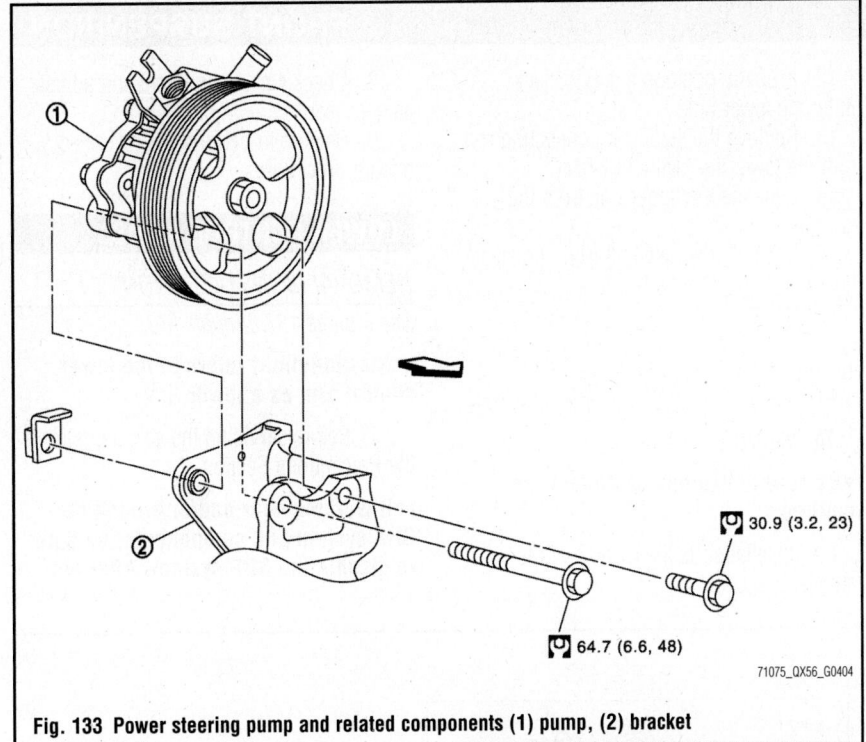

Fig. 133 Power steering pump and related components (1) pump, (2) bracket

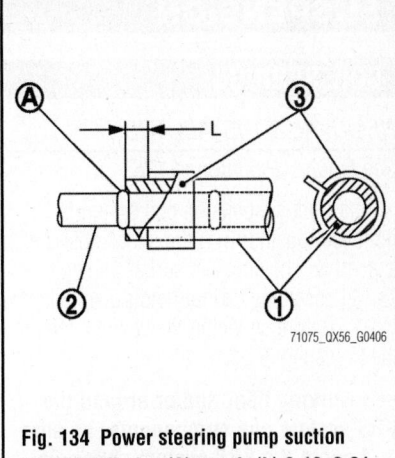

Fig. 134 Power steering pump suction hose installation (A) spool, (L) 0.12–0.31 inch, (1) hose, (2) tube

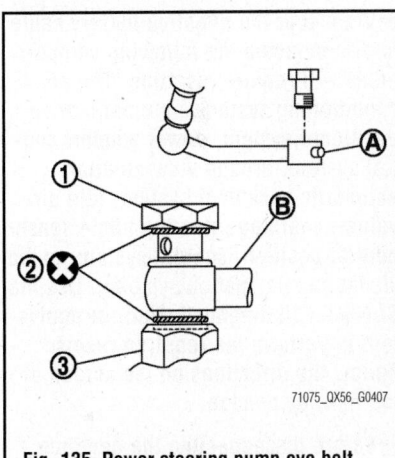

Fig. 135 Power steering pump eye bolt installation (A) protrusion, (B) hose, (1) eye bolt, (2) washer

4. Remove the air cleaner assembly.

5. Drain the power steering fluid. Be sure to properly dispose of used fluid.

6. Remove the coolant reservoir tank.

7. Remove the drive belt.

8. Remove the battery and battery tray.

9. Remove the pump hoses. Discard the washers.

10. Remove the pump retaining bolts.

11. Remove the pump from the engine.

To install:

➡ **Be sure to use new fasteners, as required.**

12. Installation is the reverse of the removal procedure.

13. Be sure to use new copper washers coated with clean power steering fluid.

14. Bleed the power steering system.

➡ **The drive belt tension is automatic and requires no adjustment.**

15. Be sure to perform the reconnect/relearn procedures.

FLUID FILL PROCEDURE

Use genuine Nissan power steering fluid (PSF), or equivalent when servicing the power steering system. The system capacity is 1 1/8 pints.

✳ CAUTION

Used fluid is considerably more dangerous than new fluid. Avoid skin contact with used fluid.

1. Before servicing the vehicle, refer to the Precautions Section.

➡ **If working near and/or around the SRS system and components, be sure to disable the SRS system. After disabling the system wait three minutes or more before servicing the vehicle.**

➡ **Whenever the negative battery cable is disconnected the following components will require resetting. The air conditioning system, automatic drive positioner system, power window control system, around view monitor, automatic back door system, idle air volume learning, steering angle sensor neutral position, sunroof system, audio visual and navigation systems. Use the CONSULT-III diagnostic tool, or equivalent to perform the required resets. Follow the directions on the screen of the tool, as needed.**

➡ **Before disconnecting the negative battery cable lower both the driver's side and passenger's side front windows. This will prevent any interference between the window edge and the vehicle when the door is opener/closed. The automatic window function will not work with the battery cable disconnected.**

2. Inspect the power steering fluid level in the power steering reservoir. Do allow the fluid to drop below the MIN marking.

3. Remove the power steering reservoir cap.

4. Add fluid, as necessary, referring to the scale on the reservoir tank.

- HOT range for fluid temperatures: 122–176°F (50–80°C)
- COLD range for fluid temperatures: 32–86°F (0–30°C)

➡ **Do not overfill the fluid. Do not reuse any used power steering fluid.**

SUSPENSION

COIL SPRINGS

REMOVAL & INSTALLATION

See Figures 136 through 138.

The front suspension coil spring is removed together with the shock absorber assembly. The shock absorber and coil spring assembly cannot be disassembled.

1. Before servicing the vehicle, refer to the Precautions Section.

➡ **If working near and/or around the SRS system and components, be sure to disable the SRS system. After disabling the system wait three minutes or more before servicing the vehicle.**

➡ **Whenever the negative battery cable is disconnected the following components will require resetting. The air conditioning system, automatic drive positioner system, power window control system, around view monitor, automatic back door system, idle air volume learning, steering angle sensor neutral position, sunroof system, audio visual and navigation systems. Use the CONSULT-III diagnostic tool, or equivalent to perform the required resets. Follow the directions on the screen of the tool, as needed.**

➡ **Before disconnecting the negative battery cable lower both the driver's side and passenger's side front windows. This will prevent any interference between the window edge and the vehicle when the door is opener/closed. The automatic window function will not work with the battery cable disconnected.**

2. Disconnect the negative battery cable.
3. Raise and support the vehicle safely.
4. Remove the tire and wheel assembly.
5. Release the pressure, if equipped with HBMC.
6. Remove the brake hose bracket from the knuckle.
7. Remove the brake caliper assembly mounting bolts. Remove the caliper. Do not disconnect the brake line. Do not allow the caliper to hang by the brake line. Reposition the caliper out of the work area.
8. Remove the brake rotor.
9. Remove the wheel speed sensor and harness connector.
10. Remove the steering outer socket from the knuckle.
11. Remove the height sensor from the upper link, right side.

12. Properly position a service jack under the lower link.
13. Remove the stabilizer connecting rod from the lower link, without HBMC.
14. Separate the upper link from the knuckle.
15. Separate the halfshaft from the steering knuckle, 4WD vehicles.
16. Remove the front tube assembly A and front tube assembly B from the shock absorber, if equipped with HBMC.
17. Remove the shock absorber assembly.

To install:

➡ **Be sure to use new fasteners, as required.**

18. Installation is the reverse of the removal procedure.

19. Check wheel alignment and adjust as necessary.
20. Be sure to perform the reconnect/relearn procedures.

LOWER CONTROL ARMS

REMOVAL & INSTALLATION

See Figures 137 through 139.

➡ **Nissan/Infiniti refers to the lower control arm as a lower link.**

1. Before servicing the vehicle, refer to the Precautions Section.

➡ **If working near and/or around the SRS system and components, be sure to disable the SRS system. After dis-**

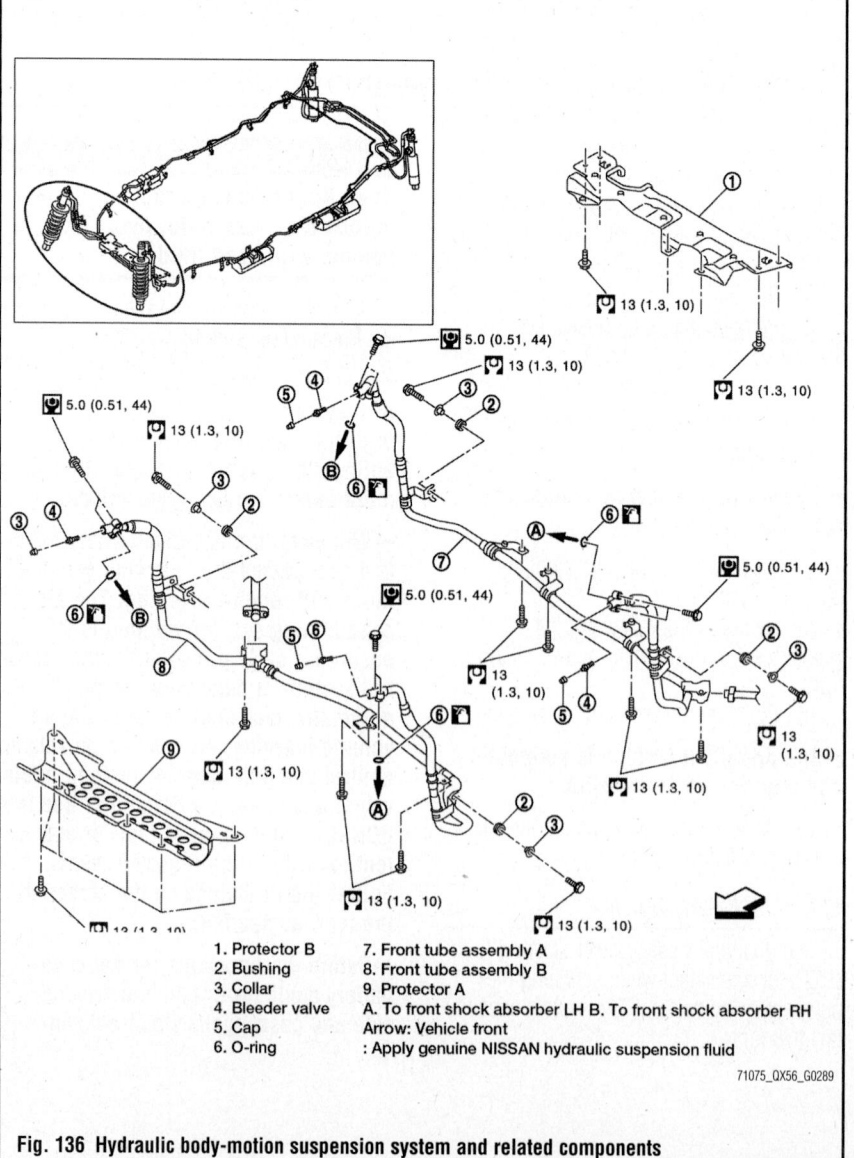

1. Protector B
2. Bushing
3. Collar
4. Bleeder valve
5. Cap
6. O-ring
7. Front tube assembly A
8. Front tube assembly B
9. Protector A
A. To front shock absorber LH B. To front shock absorber RH
Arrow: Vehicle front
: Apply genuine NISSAN hydraulic suspension fluid

71075_QX56_G0289

Fig. 136 Hydraulic body-motion suspension system and related components

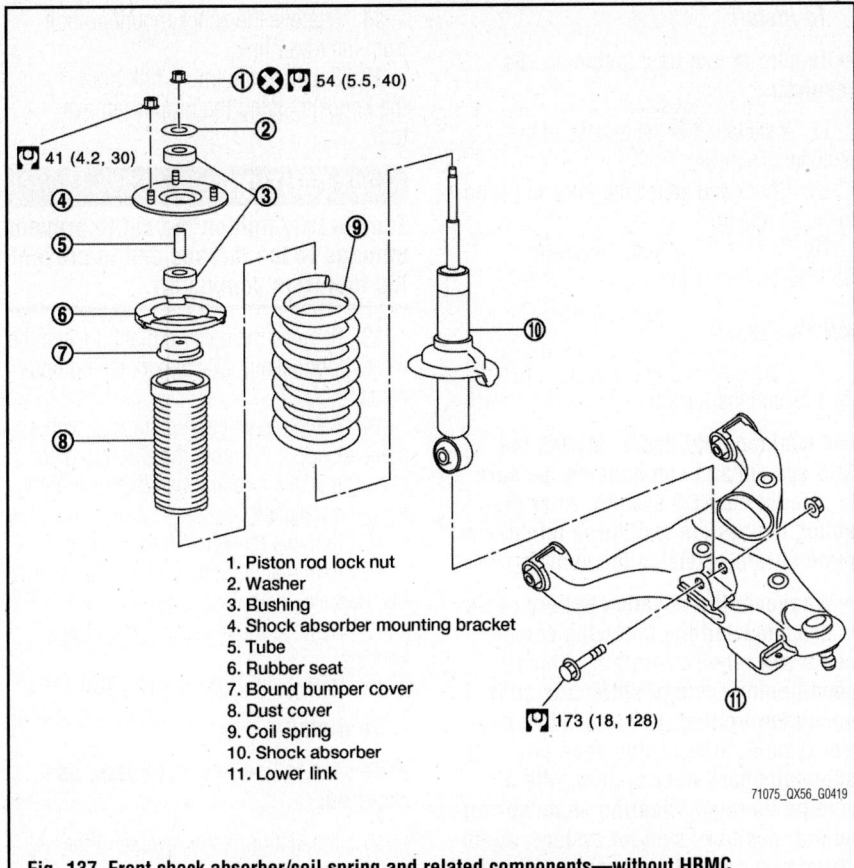

1. Piston rod lock nut
2. Washer
3. Bushing
4. Shock absorber mounting bracket
5. Tube
6. Rubber seat
7. Bound bumper cover
8. Dust cover
9. Coil spring
10. Shock absorber
11. Lower link

Fig. 137 Front shock absorber/coil spring and related components—without HBMC

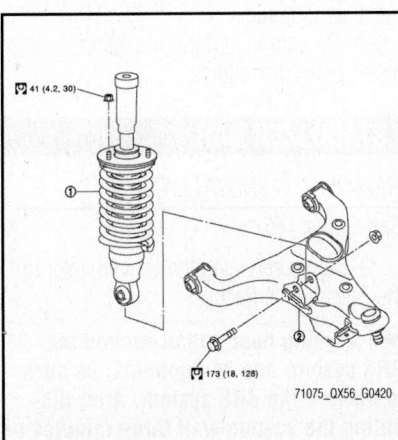

Fig. 138 Front shock absorber/coil spring and related components (1) spring, (2) link—with HBMC

abling the system wait three minutes or more before servicing the vehicle.

➡Whenever the negative battery cable is disconnected the following components will require resetting. The air conditioning system, automatic drive positioner system, power window control system, around view monitor, automatic back door system, idle air volume learning, steering angle sensor neutral position, sunroof system, audio

visual and navigation systems. Use the CONSULT-III diagnostic tool, or equivalent to perform the required resets. Follow the directions on the screen of the tool, as needed.

➡Before disconnecting the negative battery cable lower both the driver's side and passenger's side front windows. This will prevent any interference between the window edge and the vehicle when the door is opener/closed. The automatic window function will not work with the battery cable disconnected.

2. Disconnect the negative battery cable.
3. Raise and support the vehicle safely.
4. Remove the tire and wheel assembly.
5. Remove the brake hose bracket from the knuckle.
6. Remove the brake caliper assembly mounting bolts. Remove the caliper. Do not disconnect the brake line. Do not allow the caliper to hang by the brake line. Reposition the caliper out of the work area.
7. Remove the brake rotor.
8. Remove the wheel speed sensor and harness connector.
9. Remove the steering outer socket from the knuckle.

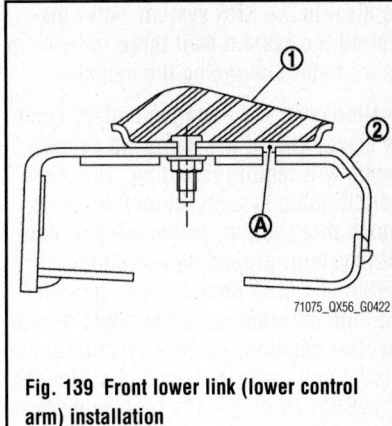

Fig. 139 Front lower link (lower control arm) installation

10. Remove the height sensor from the upper link, right side.
11. Properly position a service jack under the lower link.
12. Remove the stabilizer connecting rod from the lower link, without HBMC.
13. Remove the shock mounting bolt from the lower link.
14. Separate the upper link from the knuckle.
15. Separate the halfshaft from the steering knuckle, 4WD vehicles.
16. Remove the lower link from the steering knuckle.
17. Remove the front tube assembly A and front tube assembly B from the shock absorber, if equipped with HBMC. Remove the bracket.
18. Remove the adjusting bolts, nuts, eccentric discs, stopper rubber and lower link from the vehicle.

To install:

➡Be sure to use new fasteners, as required.

19. Installation is the reverse of the removal procedure.

➡Protrusion (A) of the bumper rubber (1) should be installed securely to hole of lower link (2). See illustration.

20. Check wheel alignment and adjust as necessary.
21. Be sure to perform the reconnect/relearn procedures.

STEERING KNUCKLE

REMOVAL & INSTALLATION

2WD Vehicles

1. Before servicing the vehicle, refer to the Precautions Section.

➡If working near and/or around the SRS system and components, be sure

to disable the SRS system. After disabling the system wait three minutes or more before servicing the vehicle.

➡Whenever the negative battery cable is disconnected the following components will require resetting. The air conditioning system, automatic drive positioner system, power window control system, around view monitor, automatic back door system, idle air volume learning, steering angle sensor neutral position, sunroof system, audio visual and navigation systems. Use the CONSULT-III diagnostic tool, or equivalent to perform the required resets. Follow the directions on the screen of the tool, as needed.

➡Before disconnecting the negative battery cable lower both the driver's side and passenger's side front windows. This will prevent any interference between the window edge and the vehicle when the door is opener/closed. The automatic window function will not work with the battery cable disconnected.

2. Disconnect the negative battery cable.

3. Raise and support the vehicle safely.

4. Remove the tire and wheel assembly.

5. Remove the brake hose bracket.

6. Remove the brake caliper assembly mounting bolts. Remove the caliper. Do not disconnect the brake line. Do not allow the caliper to hang by the brake line. Reposition the caliper out of the work area.

7. Remove the brake rotor.

8. Remove the wheel speed sensor and harness connector.

9. Remove the steering outer socket from the knuckle.

10. Properly position a service jack under the lower link.

11. Remove the shock mounting bolt from the lower link.

12. Separate the upper link from the knuckle, using the proper removal tool.

✳✳ CAUTION

Temporarily tighten the nut to prevent damage to the threads and to prevent the tool from coming off.

13. Remove the upper link mounting bolts and nuts. Remove the upper link from the vehicle.

14. Remove the wheel hub and bearing assembly, than remove the splash guard.

15. Remove the lower link from the knuckle.

16. Remove the steering knuckle.

To install:

➡**Be sure to use new fasteners, as required.**

17. Installation is the reverse of the removal procedure.

18. Check and adjust the front end alignment, as required.

19. Be sure to perform the reconnect/relearn procedures.

4WD Vehicles

1. Before servicing the vehicle, refer to the Precautions Section.

➡**If working near and/or around the SRS system and components, be sure to disable the SRS system. After disabling the system wait three minutes or more before servicing the vehicle.**

➡**Whenever the negative battery cable is disconnected the following components will require resetting. The air conditioning system, automatic drive positioner system, power window control system, around view monitor, automatic back door system, idle air volume learning, steering angle sensor neutral position, sunroof system, audio visual and navigation systems. Use the CONSULT-III diagnostic tool, or equivalent to perform the required resets. Follow the directions on the screen of the tool, as needed.**

➡**Before disconnecting the negative battery cable lower both the driver's side and passenger's side front windows. This will prevent any interference between the window edge and the vehicle when the door is opener/closed. The automatic window function will not work with the battery cable disconnected.**

2. Disconnect the negative battery cable.

3. Raise and support the vehicle safely.

4. Remove the tire and wheel assembly.

5. Remove the brake hose bracket.

6. Remove the brake caliper assembly mounting bolts. Remove the caliper. Do not disconnect the brake line. Do not allow the caliper to hang by the brake line. Reposition the caliper out of the work area.

7. Remove the brake rotor.

8. Remove the wheel speed sensor and harness connector.

9. Remove the steering outer socket from the knuckle.

10. Properly position a service jack under the lower link.

11. Remove the shock mounting bolt from the lower link.

12. Separate the upper link from the knuckle, using the proper removal tool.

✳✳ CAUTION

Temporarily tighten the nut to prevent damage to the threads and to prevent the tool from coming off.

13. Remove the cotter pin and loosen the wheel hub locknut. Matchmark the components.

14. Patch the wheel hub locknut with a piece of wood. Hammer the wood to disengage the wheel hub and bearing assembly from the halfshaft.

15. Remove the wheel hub locknut.

16. Remove the hub and bearing assembly. Remove the splash guard.

17. Remove the lower link from the knuckle.

18. Remove the steering knuckle.

To install:

➡**Be sure to use new fasteners, as required.**

19. Installation is the reverse of the removal procedure.

20. Check and adjust the front end alignment, as required.

21. Be sure to perform the reconnect/relearn procedures.

STABILIZER BAR

REMOVAL & INSTALLATION

See Figure 140.

1. Before servicing the vehicle, refer to the Precautions Section.

➡**If working near and/or around the SRS system and components, be sure to disable the SRS system. After disabling the system wait three minutes or more before servicing the vehicle.**

➡**Whenever the negative battery cable is disconnected the following components will require resetting. The air conditioning system, automatic drive positioner system, power window control system, around view monitor, automatic back door system, idle air volume learning, steering angle sensor neutral position, sunroof system, audio visual and navigation systems. Use the CONSULT-III diagnostic tool, or equivalent to perform the required resets. Follow the directions on the screen of the tool, as needed.**

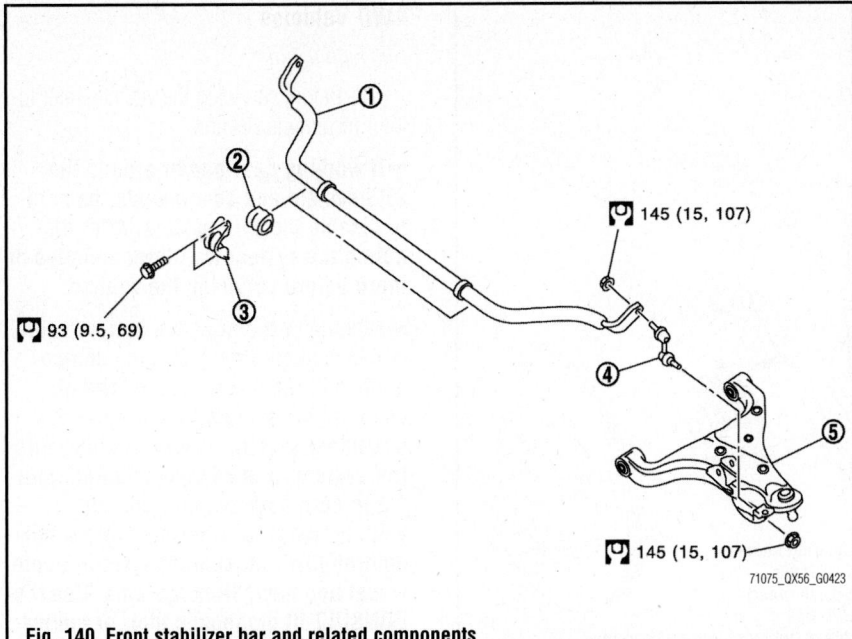

Fig. 140 Front stabilizer bar and related components

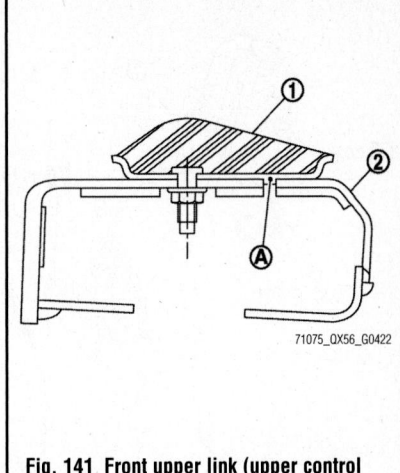

Fig. 141 Front upper link (upper control arm) installation

➡**Before disconnecting the negative battery cable lower both the driver's side and passenger's side front windows. This will prevent any interference between the window edge and the vehicle when the door is opener/closed. The automatic window function will not work with the battery cable disconnected.**

2. Disconnect the negative battery cable.

3. Raise and support the vehicle safely.

4. Matchmark and remove the stabilizer connecting rod.

5. Remove the stabilizer clamps and stabilizer bushing.

6. Remove the stabilizer bar.

To install:

➡**Be sure to use new fasteners, as required.**

7. Installation is the reverse of the removal procedure.

8. Be sure to perform the reconnect/relearn procedures.

UPPER BALL JOINTS

REMOVAL & INSTALLATION

At this time the manufacturer does not provide removal and installation procedures for this component. The upper ball joint is part of the upper control arm (upper link) assembly.

➡**Nissan/Infiniti refers to the upper control arm as a upper link.**

UPPER CONTROL ARMS

REMOVAL & INSTALLATION
See Figure 141.

➡**Nissan/Infiniti refers to the upper control arm as a upper link.**

1. Before servicing the vehicle, refer to the Precautions Section.

➡**If working near and/or around the SRS system and components, be sure to disable the SRS system. After disabling the system wait three minutes or more before servicing the vehicle.**

➡**Whenever the negative battery cable is disconnected the following components will require resetting. The air conditioning system, automatic drive positioner system, power window control system, around view monitor, automatic back door system, idle air volume learning, steering angle sensor neutral position, sunroof system, audio visual and navigation systems. Use the CONSULT-III diagnostic tool, or equivalent to perform the required resets. Follow the directions on the screen of the tool, as needed.**

➡**Before disconnecting the negative battery cable lower both the driver's side and passenger's side front windows. This will prevent any interference between the window edge and the vehicle when the door is opener/closed. The automatic window function will not work with the battery cable disconnected.**

2. Disconnect the negative battery cable.

3. Raise and support the vehicle safely.

4. Remove the tire and wheel assembly.

5. Remove the brake hose bracket from the knuckle.

6. Remove the brake caliper assembly mounting bolts. Remove the caliper. Do not disconnect the brake line. Do not allow the caliper to hang by the brake line. Reposition the caliper out of the work area.

7. Remove the brake rotor.

8. Remove the wheel speed sensor and harness connector.

9. Remove the height sensor from the upper link, right side.

10. Properly position a service jack under the lower link.

11. Separate the upper link from the knuckle, 4WD vehicles.

12. Remove the upper link mounting bolts and nuts. Remove the upper link from the vehicle.

To install:

➡**Be sure to use new fasteners, as required.**

13. Installation is the reverse of the removal procedure.

14. Check and adjust alignment, as required.

15. Be sure to perform the reconnect/relearn procedures.

WHEEL HUB & BEARING (SEALED UNIT)

REMOVAL & INSTALLATION

2WD Vehicles
See Figure 142.

1. Before servicing the vehicle, refer to the Precautions Section.

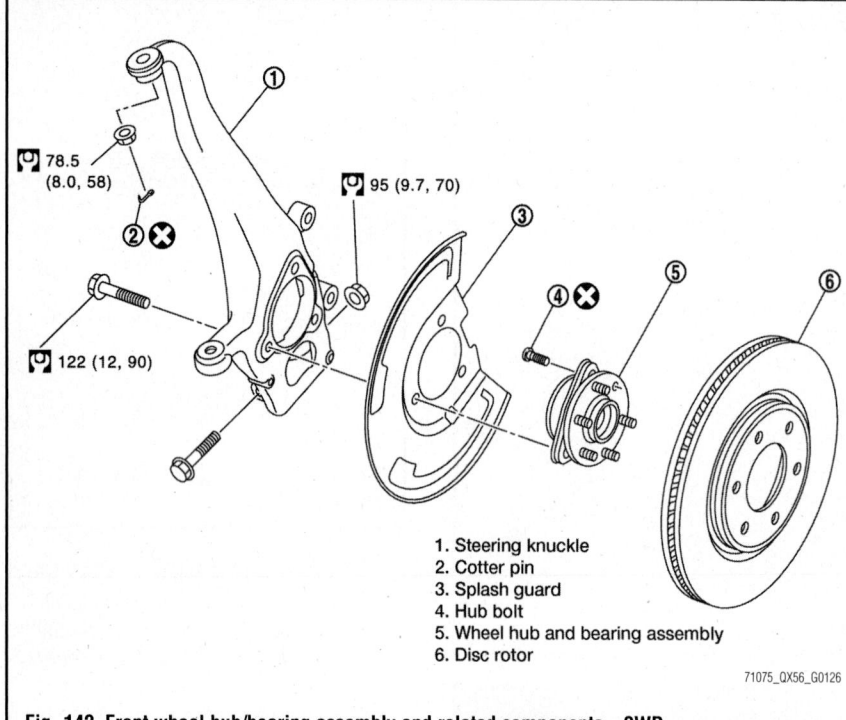

1. Steering knuckle
2. Cotter pin
3. Splash guard
4. Hub bolt
5. Wheel hub and bearing assembly
6. Disc rotor

71075_QX56_G0126

Fig. 142 Front wheel hub/bearing assembly and related components—2WD

➡If working near and/or around the SRS system and components, be sure to disable the SRS system. After disabling the system wait three minutes or more before servicing the vehicle.

➡Whenever the negative battery cable is disconnected the following components will require resetting. The air conditioning system, automatic drive positioner system, power window control system, around view monitor, automatic back door system, idle air volume learning, steering angle sensor neutral position, sunroof system, audio visual and navigation systems. Use the CONSULT-III diagnostic tool, or equivalent to perform the required resets. Follow the directions on the screen of the tool, as needed.

➡Before disconnecting the negative battery cable lower both the driver's side and passenger's side front windows. This will prevent any interference between the window edge and the vehicle when the door is opener/closed. The automatic window function will not work with the battery cable disconnected.

2. Disconnect the negative battery cable.
3. Raise and support the vehicle safely.
4. Remove the tire and wheel assembly.
5. Remove the brake hose bracket.
6. Remove the brake caliper assembly mounting bolts. Remove the caliper. Do not disconnect the brake line. Do not allow the caliper to hang by the brake line. Reposition the caliper out of the work area.
7. Remove the brake rotor.
8. Remove the wheel speed sensor and harness connector.
9. Remove the steering outer socket from the knuckle.
10. Properly position a service jack under the lower link.
11. Remove the shock mounting bolt from the lower link.
12. Separate the upper link from the knuckle, using the proper removal tool.

✸✸ CAUTION

Temporarily tighten the nut to prevent damage to the threads and to prevent the tool from coming off.

13. Remove the upper link mounting bolts and nuts. Remove the upper link from the vehicle.
14. Remove the wheel hub and bearing assembly, than remove the splash guard.

To install:

➡Be sure to use new fasteners, as required.

15. Installation is the reverse of the removal procedure.
16. Check and adjust the front end alignment, as required.
17. Be sure to perform the reconnect/relearn procedures.

4WD Vehicles

See Figure 143.

1. Before servicing the vehicle, refer to the Precautions Section.

➡If working near and/or around the SRS system and components, be sure to disable the SRS system. After disabling the system wait three minutes or more before servicing the vehicle.

➡Whenever the negative battery cable is disconnected the following components will require resetting. The air conditioning system, automatic drive positioner system, power window control system, around view monitor, automatic back door system, idle air volume learning, steering angle sensor neutral position, sunroof system, audio visual and navigation systems. Use the CONSULT-III diagnostic tool, or equivalent to perform the required resets. Follow the directions on the screen of the tool, as needed.

➡Before disconnecting the negative battery cable lower both the driver's side and passenger's side front windows. This will prevent any interference between the window edge and the vehicle when the door is opener/closed. The automatic window function will not work with the battery cable disconnected.

2. Disconnect the negative battery cable.
3. Raise and support the vehicle safely.
4. Remove the tire and wheel assembly.
5. Remove the brake hose bracket.
6. Remove the brake caliper assembly mounting bolts. Remove the caliper. Do not disconnect the brake line. Do not allow the caliper to hang by the brake line. Reposition the caliper out of the work area.
7. Remove the brake rotor.
8. Remove the wheel speed sensor and harness connector.
9. Remove the steering outer socket from the knuckle.
10. Properly position a service jack under the lower link.
11. Remove the shock mounting bolt from the lower link.
12. Separate the upper link from the knuckle, using the proper removal tool.

✸✸ CAUTION

Temporarily tighten the nut to prevent damage to the threads and to prevent the tool from coming off.

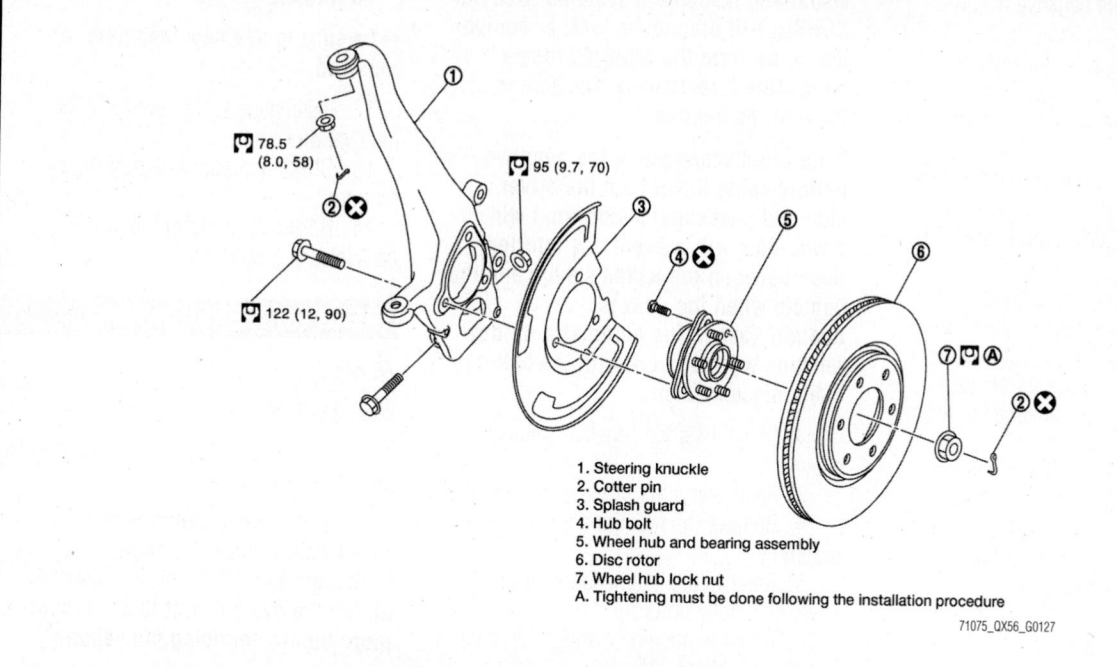

1. Steering knuckle
2. Cotter pin
3. Splash guard
4. Hub bolt
5. Wheel hub and bearing assembly
6. Disc rotor
7. Wheel hub lock nut
A. Tightening must be done following the installation procedure

71075_QX56_G0127

Fig. 143 Front wheel hub/bearing assembly and related components—4WD

13. Remove the cotter pin and loosen the wheel hub locknut. Matchmark the components.

14. Patch the wheel hub locknut with a piece of wood. Hammer the wood to disengage the wheel hub and bearing assembly from the halfshaft.

15. Remove the wheel hub locknut.

16. Remove the hub and bearing assembly. Remove the splash guard.

To install:

➡ **Be sure to use new fasteners, as required.**

17. Installation is the reverse of the removal procedure.

18. Check and adjust the front end alignment, as required.

19. Be sure to perform the reconnect/relearn procedures.

SUSPENSION

COIL SPRINGS

REMOVAL & INSTALLATION

See Figures 144 and 145.

1. Before servicing the vehicle, refer to the Precautions Section.

➡ **If working near and/or around the SRS system and components, be sure to disable the SRS system. After disabling the system wait three minutes or more before servicing the vehicle.**

➡ **Whenever the negative battery cable is disconnected the following components will require resetting. The air conditioning system, automatic drive positioner system, power window control system, around view monitor, automatic back door system, idle air volume learning, steering angle sensor neutral position, sunroof system, audio visual and navigation systems. Use the CONSULT-III diagnostic tool, or equivalent to perform**

the required resets. Follow the directions on the screen of the tool, as needed.

➡ **Before disconnecting the negative battery cable lower both the driver's side and passenger's side front win-**

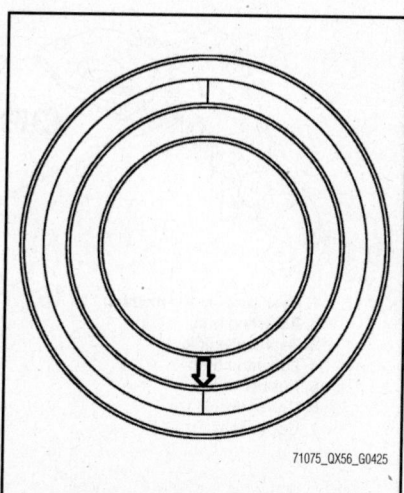

71075_QX56_G0425

Fig. 144 Rear spring installation—1 of 2

REAR SUSPENSION

dows. This will prevent any interference between the window edge and the vehicle when the door is opener/closed. The automatic window function will not work with the battery cable disconnected.

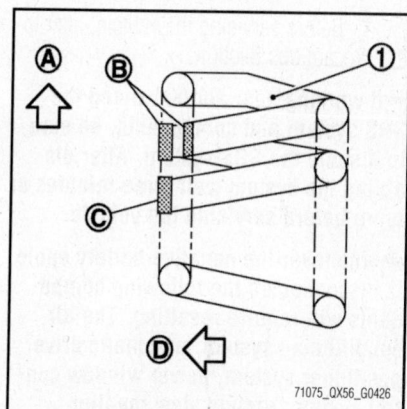

71075_QX56_G0426

Fig. 145 Rear spring installation
(A) upper side, (B) paint marks, (C) paint mark, (D) inside, (1) spring—2 of 2

2. Disconnect the negative battery cable.

3. Raise and support the vehicle safely.

4. Remove the rear tire and wheel assembly.

5. Remove the height sensor from the lower link, both sides.

6. Properly position a shop jack under the rear lower link.

7. Using a spring compressor tool, compress the spring between the rubber seal and the upper seat until the spring and the compressor tool are free. Be sure the tool is securely attached to the spring.

8. Remove the lower link mounting bolt, on the axle side.

9. Slowly lower the jack, then remove the upper seat, coil spring and rubber seat from the lower link.

To install:

➡ Be sure to use new fasteners, as required.

10. Installation is the reverse of the removal procedure.

➡ When installing the rubber seats for the coil spring, refer to the illustrations.

11. Check and adjust alignment, as required.

12. Be sure to perform the reconnect/relearn procedures.

LOWER CONTROL ARMS

REMOVAL & INSTALLATION

See Figure 146.

➡ Nissan/Infiniti refers to the lower control arm a lower link.

1. Before servicing the vehicle, refer to the Precautions Section.

➡ If working near and/or around the SRS system and components, be sure to disable the SRS system. After disabling the system wait three minutes or more before servicing the vehicle.

➡ Whenever the negative battery cable is disconnected the following components will require resetting. The air conditioning system, automatic drive positioner system, power window control system, around view monitor, automatic back door system, idle air volume learning, steering angle sensor neutral position, sunroof system, audio

visual and navigation systems. Use the CONSULT-III diagnostic tool, or equivalent to perform the required resets. Follow the directions on the screen of the tool, as needed.

➡ Before disconnecting the negative battery cable lower both the driver's side and passenger's side front windows. This will prevent any interference between the window edge and the vehicle when the door is opener/closed. The automatic window function will not work with the battery cable disconnected.

2. Disconnect the negative battery cable.

3. Raise and support the vehicle safely.

4. Remove the rear tire and wheel assembly.

5. Remove the height sensor from the lower link, both sides.

6. Properly position a shop jack under the rear lower link.

7. Using a spring compressor tool, compress the spring between the rubber seal and the upper seat until the spring and the compressor tool are free. Be sure the tool is securely attached to the spring.

8. Remove the lower link mounting bolt, on the axle side.

9. Slowly lower the jack, then remove the upper seat, coil spring and rubber seat from the lower link.

10. Remove the rear lower link mounting bolt, eccentric disc and adjusting bolt.

11. Remove the rear lower link.

To install:

➡ Be sure to use new fasteners, as required.

12. Installation is the reverse of the removal procedure.

13. Check and adjust alignment, as required.

14. Be sure to perform the reconnect/relearn procedures.

REAR SUSPENSION ARM

REMOVAL & INSTALLATION

See Figure 147.

1. Before servicing the vehicle, refer to the Precautions Section.

➡ If working near and/or around the SRS system and components, be sure to disable the SRS system. After disabling the system wait three minutes or more before servicing the vehicle.

➡ Whenever the negative battery cable is disconnected the following components will require resetting. The air conditioning system, automatic drive positioner system, power window control system, around view monitor, automatic back door system, idle air volume learning, steering angle sensor neutral position, sunroof system, audio visual and navigation systems. Use the CONSULT-III diagnostic tool, or equivalent to perform the required resets. Follow the directions on the screen of the tool, as needed.

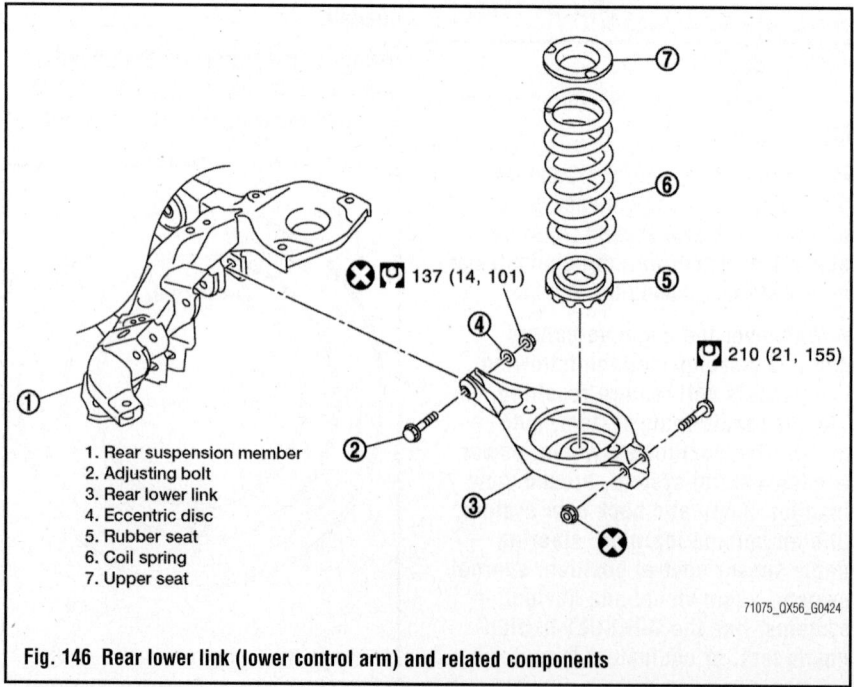

1. Rear suspension member
2. Adjusting bolt
3. Rear lower link
4. Eccentric disc
5. Rubber seat
6. Coil spring
7. Upper seat

137 (14, 101)

210 (21, 155)

71075_QX56_G0424

Fig. 146 Rear lower link (lower control arm) and related components

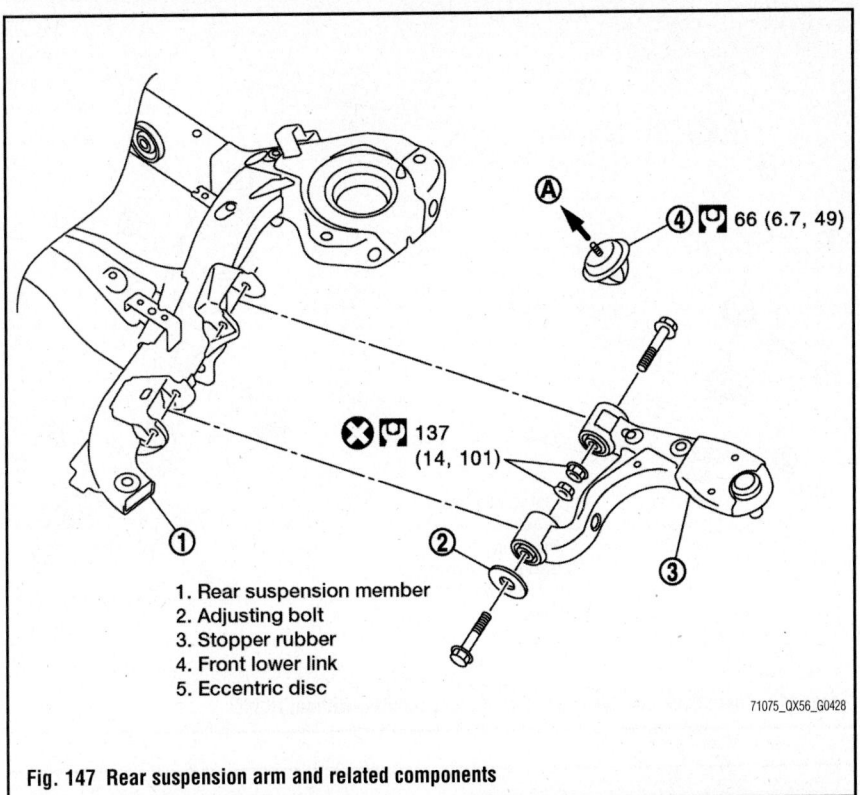

1. Rear suspension member
2. Adjusting bolt
3. Stopper rubber
4. Front lower link
5. Eccentric disc

71075_QX56_G0428

Fig. 147 Rear suspension arm and related components

➡️Before disconnecting the negative battery cable lower both the driver's side and passenger's side front windows. This will prevent any interference between the window edge and the vehicle when the door is opener/ closed. The automatic window function will not work with the battery cable disconnected.

2. Disconnect the negative battery cable.
3. Raise and support the vehicle safely.
4. Remove the tire and wheel assemblies.
5. Remove the rear suspension member.
6. Remove the bumper rubber from the frame.
7. Remove the suspension arm from the axle housing.
8. Remove the suspension arm retaining bolts, nuts and stopper rubber from the rear suspension member.
9. Remove the suspension arm.

To install:

➡️**Be sure to use new fasteners, as required.**

10. Installation is the reverse of the removal procedure.
11. Check and adjust the alignment, as required.
12. Be sure to perform the reconnect/relearn procedures.

REAR FRONT LOWER LINK

REMOVAL & INSTALLATION

See Figure 148.

1. Before servicing the vehicle, refer to the Precautions Section.

➡️**If working near and/or around the SRS system and components, be sure to disable the SRS system. After dis-**

abling the system wait three minutes or more before servicing the vehicle.

➡️Whenever the negative battery cable is disconnected the following components will require resetting. The air conditioning system, automatic drive positioner system, power window control system, around view monitor, automatic back door system, idle air volume learning, steering angle sensor neutral position, sunroof system, audio visual and navigation systems. Use the CONSULT-III diagnostic tool, or equivalent to perform the required resets. Follow the directions on the screen of the tool, as needed.

➡️Before disconnecting the negative battery cable lower both the driver's side and passenger's side front windows. This will prevent any interference between the window edge and the vehicle when the door is opener/ closed. The automatic window function will not work with the battery cable disconnected.

2. Disconnect the negative battery cable.
3. Raise and support the vehicle safely.
4. Remove the tire and wheel assemblies.
5. Position a suitable jack under the rear lower link.
6. Remove the shock bolts from the front lower link.
7. Remove the stabilizer connecting rod, vehicles without HBMC.

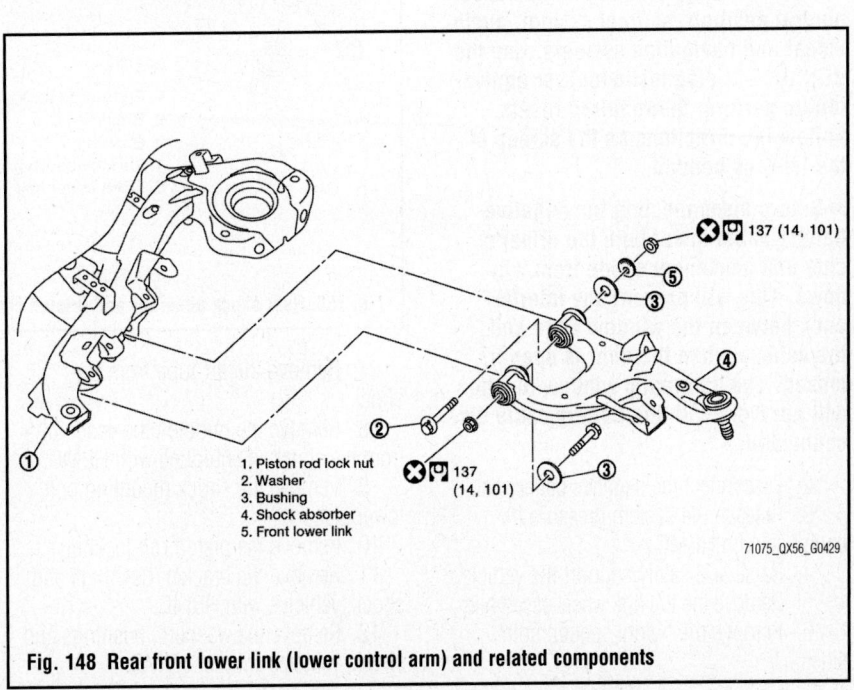

1. Piston rod lock nut
2. Washer
3. Bushing
4. Shock absorber
5. Front lower link

71075_QX56_G0429

Fig. 148 Rear front lower link (lower control arm) and related components

8. Remove the front lower link mounting bolts and nuts from the axle housing.

9. Remove the front lower link mounting bolt, nuts, stopper rubbers, eccentric disc and adjusting bolt from the rear suspension member.

10. Remove the front lower link.

To install:

➡ **Be sure to use new fasteners, as required.**

11. Installation is the reverse of the removal procedure.

12. Check and adjust the alignment, as required.

13. Be sure to perform the reconnect/relearn procedures.

SHOCK ABSORBERS

REMOVAL & INSTALLATION

See Figures 149 through 152.

1. Before servicing the vehicle, refer to the Precautions Section.

➡ **If working near and/or around the SRS system and components, be sure to disable the SRS system. After disabling the system wait three minutes or more before servicing the vehicle.**

➡ **Whenever the negative battery cable is disconnected the following components will require resetting. The air conditioning system, automatic drive positioner system, power window control system, around view monitor, automatic back door system, idle air volume learning, steering angle sensor neutral position, sunroof system, audio visual and navigation systems. Use the CONSULT-III diagnostic tool, or equivalent to perform the required resets. Follow the directions on the screen of the tool, as needed.**

➡ **Before disconnecting the negative battery cable lower both the driver's side and passenger's side front windows. This will prevent any interference between the window edge and the vehicle when the door is opener/closed. The automatic window function will not work with the battery cable disconnected.**

2. Disconnect the negative battery cable.

3. Release the system pressure if equipped with HBMC.

4. Raise and safely support the vehicle.

5. Remove the tire and wheel assemblies.

6. Remove the height sensor, both sides.

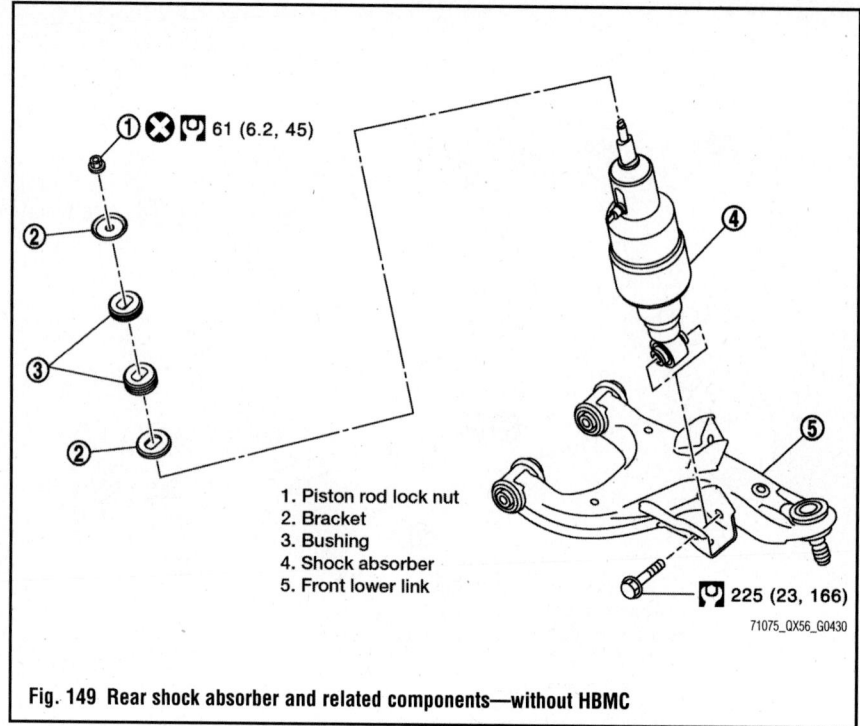

1. Piston rod lock nut
2. Bracket
3. Bushing
4. Shock absorber
5. Front lower link

61 (6.2, 45)

225 (23, 166)

71075_QX56_G0430

Fig. 149 Rear shock absorber and related components—without HBMC

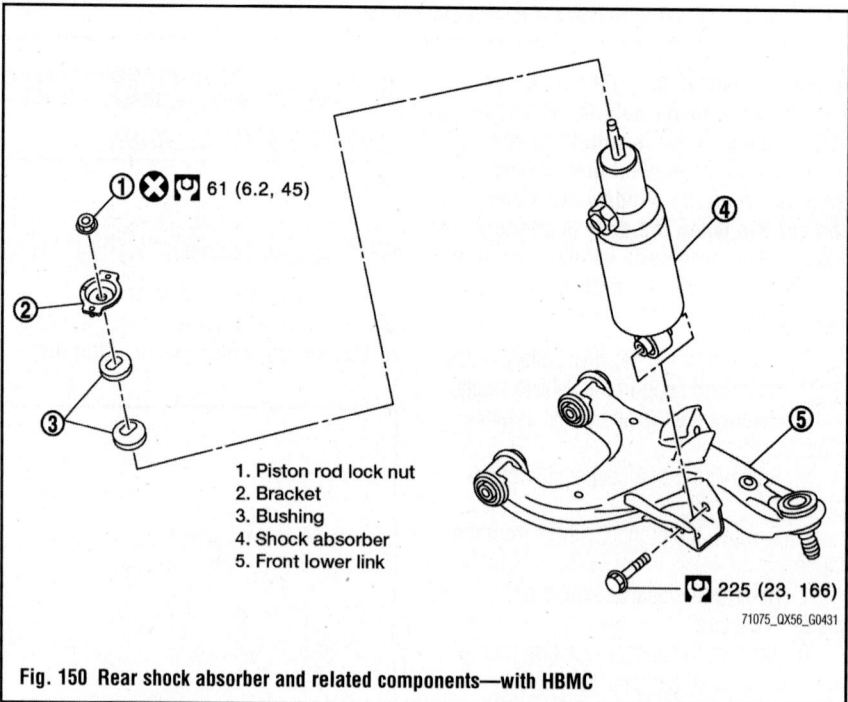

1. Piston rod lock nut
2. Bracket
3. Bushing
4. Shock absorber
5. Front lower link

61 (6.2, 45)

225 (23, 166)

71075_QX56_G0431

Fig. 150 Rear shock absorber and related components—with HBMC

7. Remove the air tube from the shock.

8. Remove the middle tube assembly from the shock, if equipped with HBMC.

9. Remove the shock mounting bolt, lower side.

10. Remove the piston rod locknut.

11. Remove the bracket, bushings and shock, vehicles with HBMC.

12. Remove the washers, bushings and shock, vehicles without HBMC.

To install:

➡ **Be sure to use new fasteners, as required.**

13. Installation is the reverse of the removal procedure.

➡ **On vehicles equipped with HBMC, when installing the bracket refer to the illustration. You will need holder tool SST: KV10109300.**

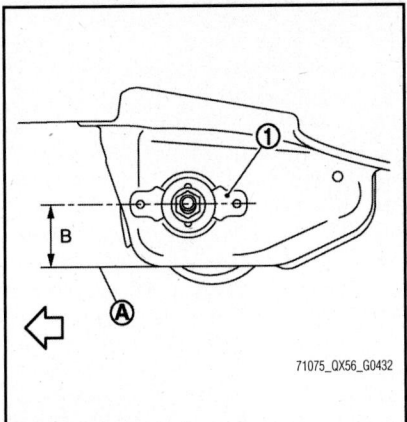

Fig. 151 Rear suspension bracket installation with HBMC (A) frame edge, (B) almost parallel, (1) bracket—1 of 2

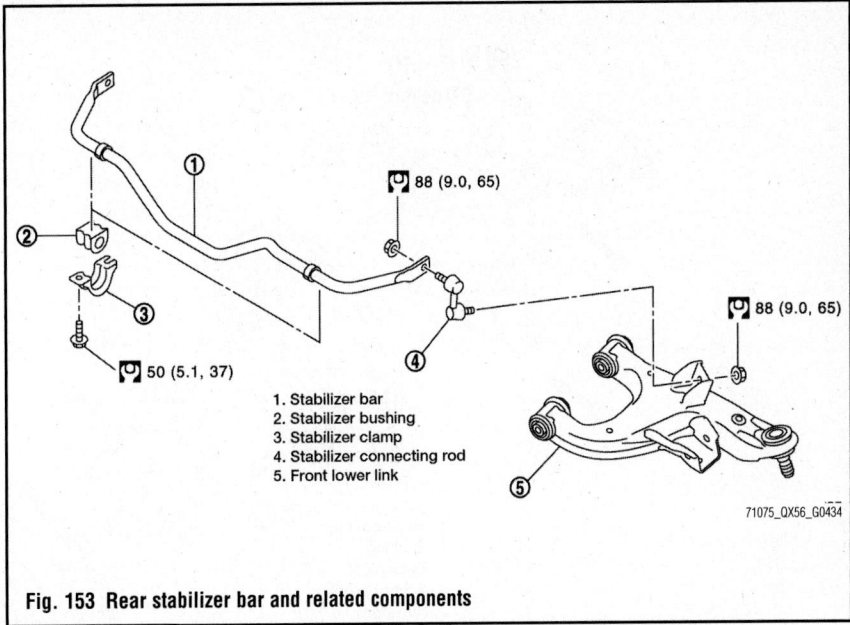

1. Stabilizer bar
2. Stabilizer bushing
3. Stabilizer clamp
4. Stabilizer connecting rod
5. Front lower link

Fig. 153 Rear stabilizer bar and related components

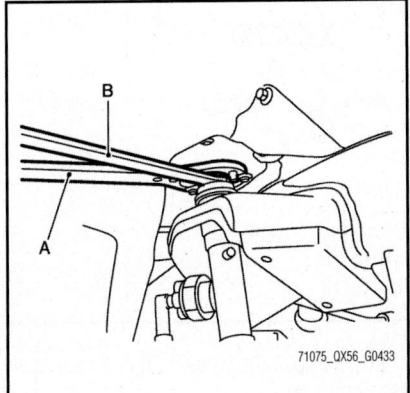

Fig. 152 Rear suspension bracket installation with HBMC (A) holder, (B) tool—2 of 2

14. Be sure to perform the reconnect/relearn procedures.

STABILIZER BAR (SWAY BAR) & LINKS

REMOVAL & INSTALLATION

See Figure 153.

1. Before servicing the vehicle, refer to the Precautions Section.

➡ **If working near and/or around the SRS system and components, be sure to disable the SRS system. After disabling the system wait three minutes or more before servicing the vehicle.**

➡ **Whenever the negative battery cable is disconnected the following components will require resetting. The air conditioning system, automatic drive positioner system, power window control system, around view monitor, automatic back door system, idle air volume learning, steering angle sensor neutral position, sunroof system, audio visual and navigation systems. Use the CONSULT-III diagnostic tool, or equivalent to perform the required resets. Follow the directions on the screen of the tool, as needed.**

➡ **Before disconnecting the negative battery cable lower both the driver's side and passenger's side front windows. This will prevent any interference between the window edge and the vehicle when the door is opener/closed. The automatic window function will not work with the battery cable disconnected.**

2. Disconnect the negative battery cable.
3. Raise and safely support the vehicle.
4. Disconnect the stabilizer connecting rods.
5. Remove the bar clamps. Remove the bar bushings.
6. Remove the stabilizer bar.

To install:

➡ **Be sure to use new fasteners, as required.**

7. Installation is the reverse of the removal procedure.
8. Be sure to perform the reconnect/relearn procedures.

WHEEL HUB & BEARING (SEALED UNIT)

REMOVAL & INSTALLATION

See Figure 154.

1. Before servicing the vehicle, refer to the Precautions Section.

➡ **If working near and/or around the SRS system and components, be sure to disable the SRS system. After disabling the system wait three minutes or more before servicing the vehicle.**

➡ **Whenever the negative battery cable is disconnected the following components will require resetting. The air conditioning system, automatic drive positioner system, power window control system, around view monitor, automatic back door system, idle air volume learning, steering angle sensor neutral position, sunroof system, audio visual and navigation systems. Use the CONSULT-III diagnostic tool, or equivalent to perform the required resets. Follow the directions on the screen of the tool, as needed.**

➡ **Before disconnecting the negative battery cable lower both the driver's side and passenger's side front windows. This will prevent any interference between the window edge and the vehicle when the door is opener/closed. The automatic window function will not work with the battery cable disconnected.**

2. Disconnect the negative battery cable.
3. Raise and support the vehicle safely.
4. Remove the tire and wheel assembly.
5. Remove the brake hose bracket.
6. Remove the brake caliper assembly mounting bolts. Remove the caliper. Do not disconnect the brake line. Do not allow the caliper to hang by the brake line. Reposition the caliper out of the work area.

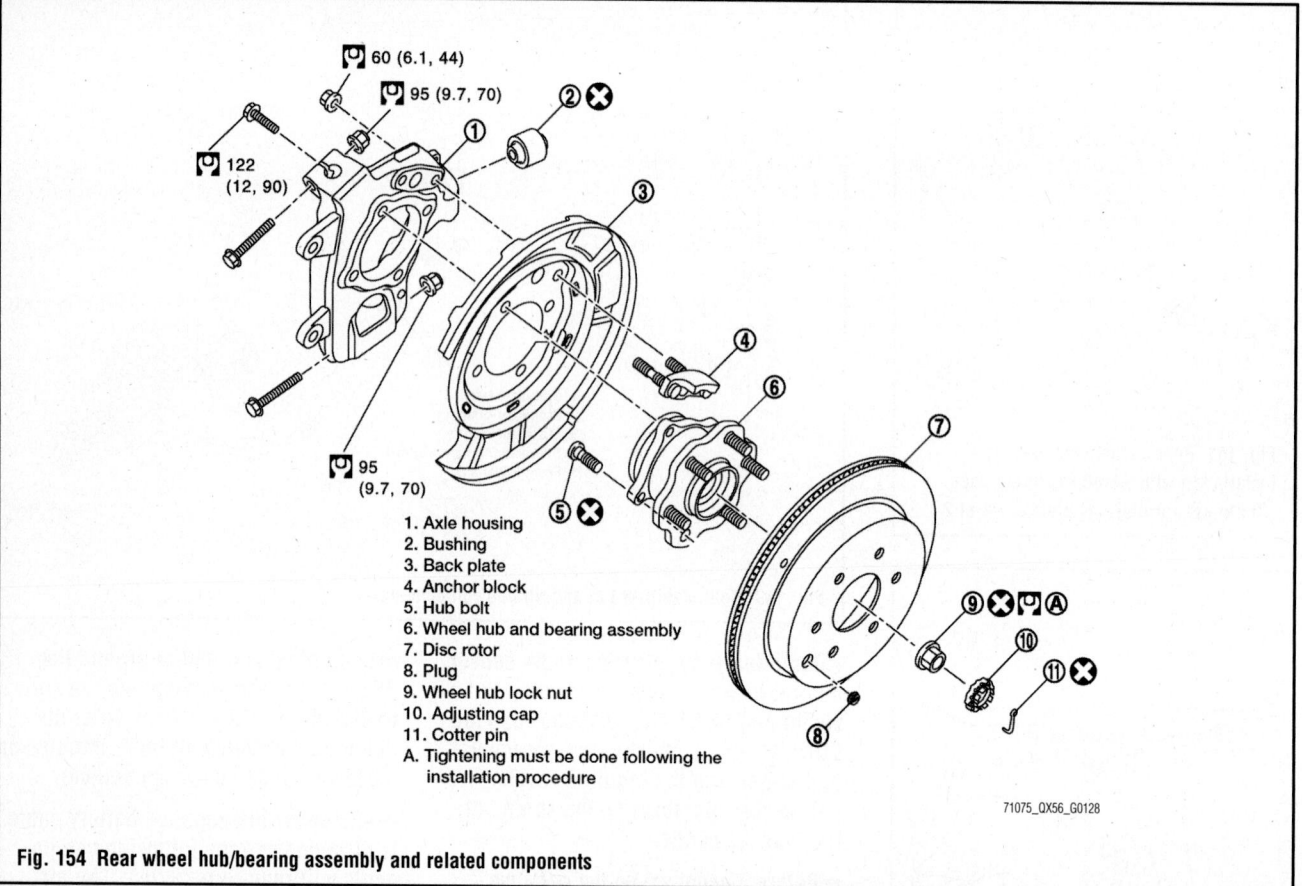

Fig. 154 Rear wheel hub/bearing assembly and related components

1. Axle housing
2. Bushing
3. Back plate
4. Anchor block
5. Hub bolt
6. Wheel hub and bearing assembly
7. Disc rotor
8. Plug
9. Wheel hub lock nut
10. Adjusting cap
11. Cotter pin
A. Tightening must be done following the installation procedure

71075_QX56_G0128

7. Remove the brake rotor.

8. Remove the wheel speed sensor and harness connector.

9. Remove the parking brake shoe and parking brake cable from the backing plate.

10. Remove the height sensor, both sides.

11. Properly position a suitable jack under the rear lower link.

12. Remove the rear lower link from the axle housing.

13. Remove the coil spring.

14. Remove the cotter pin and loosen the wheel hub locknut. Matchmark the components.

15. Patch the wheel hub locknut with a piece of wood. Hammer the wood to disengage the wheel hub and bearing assembly from the halfshaft.

16. Remove the wheel hub locknut.

17. Remove the suspension arm from the axle housing.

18. Remove the lower link from the axle housing.

19. Remove the axle housing.

20. Remove the wheel hub and bearing assembly.

To install:

➡**Be sure to use new fasteners, as required.**

21. Installation is the reverse of the removal procedure.

22. Check and adjust alignment, as required.

23. Be sure to perform the reconnect/relearn procedures.

INFINITI

Diagnostic Trouble Codes

DIAGNOSTIC TROUBLE CODES

OBD II VEHICLE APPLICATIONS

INFINITI

EX35
2011–2012
- 3.5L EFI . . . . . Engine Code: VQ35HR

FX35
2011–2012
- 3.5L EFI . . . . . Engine Code: VQ35HR

FX50
2011–2012
- 5.0L EFI . . . . . Engine Code: VK50VE

G25
2011–2012
- 2.5L EFI . . . . . Engine Code: VQ25HR

G37
2011–2012
- 3.7L EFI . . . . Engine Code: VQ37VHR

M37
2011–2012
- 3.7L EFI . . . . Engine Code: VQ37VHR

M56
2011–2012
- 5.6L EFI . . . . . Engine Code: VK56VD

QX56
2011–2012
- 5.6L EFI . . . . . Engine Code: VK56VD

OBD II Trouble Code List (P0XXX Codes)

DTC	Trouble Code Title and Conditions
DTC: P0011 **1T ECM, MIL:** Yes **Year:** 2011 **Model:** M56, QX56 **Engine:** 5.6L **Transmission:** All	**Intake Valve Timing Control Performance (Bank 1):** **NOTE: If DTC P0011 is displayed with DTC P0075, P0081, P1140 or P1145, first perform the trouble diagnosis for any other DTC before proceeding with P0011.** Condition A: The alignment of the intake valve timing control has been misregistered. Condition B: There is a gap between angle of target and phase-control angle degree.
DTC: P0011 **1T ECM, MIL:** Yes **Year:** 2011 **Model:** EX35, FX35, FX50, G37, M37 **Engine:** 3.5L, 3.7L, 5.0L **Transmission:** All	**Intake Valve Timing Control Performance (Includes Hybrid Models):** HYBRID MODELS CAUTION: Hybrid systems use very high-voltage battery systems. Before starting any service work involving the battery system, turn the ignition switch OFF and then remove the service plug from pocket in the trunk. After removing the service plug, wait 10 minutes before touching any of the high-voltage connectors and terminals. **NOTE: When the malfunction is detected, the ECM enters fail-safe mode.** **NOTE: If DTC P0011 is displayed with DTC P0075, P0081, P0524, P1111, or P1136, perform the appropriate trouble diagnosis first.** * There is a gap between angle of target and phase-control angle degree.
DTC: P0014 **1T ECM, MIL:** Yes **Year:** 2011 **Model:** EX35, FX35, FX50 **Engine:** 3.5L, 5.0L **Transmission:** All	**Exhaust Valve Timing (EVT) Control Performance (Bank 1):** **NOTE: If DTC P0014 or P0024 is displayed with DTC P0078, P0084, P1078 or P1084 first perform trouble diagnosis for respective DTC before proceeding with P0014.** There is a gap between angle of target and phase-control angle degree.
DTC: P0021 **1T ECM, MIL:** Yes **Year:** 2011 **Model:** EX35, FX35, FX50, G37, M37, M56, QX56 **Engine:** 3.5L, 3.7L, 5.0L, 5.6L **Transmission:** All	**Intake Valve Timing Control Performance (Bank 2):** * If DTC P0011 or P0021 is displayed with DTC P0075, P0081, P1111 or P1136, first perform trouble diagnosis for DTC P0075 or P0081. **NOTE: When the malfunction is detected, the ECM enters fail-safe mode.** * There is a gap between angle of target and phase-control angle degree. * When the malfunction is detected, the ECM enters fail-safe mode.
DTC: P0024 **1T ECM, MIL:** Yes **Year:** 2011 **Model:** EX35, FX35, FX50 **Engine:** 3.5L, 5.0L **Transmission:** All	**Exhaust Valve Timing (EVT) Control Performance (Bank 2):** **NOTE: If DTC P0014 or P0024 is displayed with DTC P0078, P0084, P1078 or P1084, first perform trouble diagnosis for the respective DTC before proceeding with P0024.** There is a gap between angle of target and phase-control angle degree.
DTC: P0031 **1T ECM, MIL:** Yes **Year:** 2011 **Model:** EX35, FX35, FX50, G37, M37, M56, QX56 **Engine:** 3.5L, 3.7L, 5.0L, 5.6L **Transmission:** All	**Air Fuel Ratio (A/F) Sensor 1 Heater Control Circuit Low (Includes Hybrid Models):** HYBRID MODELS CAUTION: Hybrid systems use very high-voltage battery systems. Before starting any service work involving the battery system, turn the ignition switch OFF and then remove the service plug from pocket in the trunk. After removing the service plug, wait 10 minutes before touching any of the high-voltage connectors and terminals. **NOTE: On and engines, this applies to Bank 1.** * The ECM performs ON/OFF duty control of the A/F sensor 1 heater corresponding to the engine operating condition to keep the temperature of A/F sensor 1 element at the specified range. * The current amperage in the A/F sensor 1 heater circuit is out of the normal range. (An excessively low voltage signal is sent to ECM through the A/F sensor heater.)
DTC: P0032 **1T ECM, MIL:** Yes **Year:** 2011 **Model:** EX35, FX35, FX50, G37, M37, M56, QX56 **Engine:** 3.5L, 3.7L, 5.0L, 5.6L **Transmission:** All	**Air Fuel Ratio (A/F) Sensor 1 Heater (Bank 1) Control Circuit High (Includes Hybrid Models):** HYBRID MODELS CAUTION: Hybrid systems use very high-voltage battery systems. Before starting any service work involving the battery system, turn the ignition switch OFF and then remove the service plug from pocket in the trunk. After removing the service plug, wait 10 minutes before touching any of the high-voltage connectors and terminals. **Note: On and engines, this applies to Bank 1.** * The ECM performs ON/OFF duty control of the A/F sensor 1 heater corresponding to the engine operating condition to keep the temperature of A/F sensor 1 element at the specified range. * The current amperage in the A/F sensor 1 heater circuit is out of the normal range. (An excessively high voltage signal is sent to ECM through the A/F sensor heater.)
DTC: P0037 **1T ECM, MIL:** Yes **Year:** 2011 **Model:** EX35, FX35, FX50, G37, M37, M56, QX56 **Engine:** 3.5L, 3.7L, 5.0L, 5.6L **Transmission:** All	**Heated Oxygen Sensor 2 Heater Control Circuit Low (Includes Hybrid Models):** HYBRID MODEL CAUTION: Hybrid systems use very high-voltage battery systems. Before starting any service work involving the battery system, turn the ignition switch OFF and then remove the service plug from pocket in the trunk. After removing the service plug, wait 10 minutes before touching any of the high-voltage connectors and terminals. **Note: On and engines, this applies to Bank 1.** The current amperage in the heated oxygen sensor 2 heater circuit is out of the normal range. (An excessively low voltage signal is sent to ECM via the heated oxygen sensor 2 heater.)

DTC	Trouble Code Title and Conditions
DTC: P0038 **1T ECM, MIL: Yes** **Year:** 2011 **Model:** EX35, FX35, FX50, G37, M37, M56, QX56 **Engine:** 3.5L, 3.7L, 5.0L, 5.6L **Transmission:** All	**Heated Oxygen Sensor 2 Heater Control Circuit High (Includes Hybrid Models):** HYBRID MODELS CAUTION: Hybrid systems use very high-voltage battery systems. Before starting any service work involving the battery system, turn the ignition switch OFF and then remove the service plug from pocket in the trunk. After removing the service plug, wait 10 minutes before touching any of the high-voltage connectors and terminals. **Note: On and engines, this applies to Bank 1.** The current amperage in the heated oxygen sensor 2 heater circuit is out of the normal range. (An excessively high voltage signal is sent to ECM via the heated oxygen sensor 2 heater.)
DTC: P0051 **1T ECM, MIL: Yes** **Year:** 2011 **Model:** EX35, FX35, FX50, G37, M37, M56, QX56 **Engine:** 3.5L, 3.7L, 5.0L, 5.6L **Transmission:** All	**Air Fuel Ratio (A/F) Sensor 1 (Bank 2) Heater Control Circuit Low:** **Note: On and engines, this applies to Bank 2.** The ECM performs ON/OFF duty control of the A/F sensor 1 heater corresponding to the engine operating condition to keep the temperature of A/F sensor 1 element at the specified range. The current amperage in the A/F sensor 1 heater circuit is out of the normal range. (An excessively low voltage signal is sent to ECM through the A/F sensor heater.)
DTC: P0052 **1T ECM, MIL: Yes** **Year:** 2011 **Model:** EX35, FX35, FX50, G37, M37, M56, QX56 **Engine:** 3.5L, 3.7L, 5.0L, 5.6L **Transmission:** All	**Air Fuel Ratio (A/F) Sensor 1 (Bank 2) Heater Control Circuit High:** **Note: On and engines, this applies to Bank 2.** The ECM performs ON/OFF duty control of the A/F sensor 1 heater corresponding to the engine operating condition to keep the temperature of A/F sensor 1 element at the specified range. The current amperage in the A/F sensor 1 heater circuit is out of the normal range. (An excessively high voltage signal is sent to ECM through the A/F sensor heater.)
DTC: P0057 **1T ECM, MIL: Yes** **Year:** 2011 **Model:** EX35, FX35, FX50, G37, M37, M56, QX56 **Engine:** 3.5L, 3.7L, 5.0L, 5.6L **Transmission:** All	**Heated Oxygen Sensor 2 (Bank 2) Heater Control Circuit Low:** **Note: On and engines, this applies to Bank 2.** The current amperage in the heated oxygen sensor 2 heater circuit is out of the normal range. (An excessively low voltage signal is sent to ECM via the heated oxygen sensor 2 heater.)
DTC: P0058 **1T ECM, MIL: Yes** **Year:** 2011 **Model:** EX35, FX35, FX50, G37, M37, M56, QX56 **Engine:** 3.5L, 3.7L, 5.0L, 5.6L **Transmission:** All	**Heated Oxygen Sensor 2 (Bank 2) Heater Control Circuit High:** **Note: On and engines, this applies to Bank 2.** The current amperage in the heated oxygen sensor 2 heater circuit is out of the normal range. (An excessively high voltage signal is sent to ECM via the heated oxygen sensor 2 heater.)
DTC: P0075 **1T ECM, MIL: Yes** **Year:** 2011 **Model:** EX35, FX35, FX50, G37, M37, M56, QX56 **Engine:** 3.5L, 3.7L, 5.0L, 5.6L **Transmission:** All	**Intake Valve Timing Control Solenoid Valve Circuit (Includes Hybrid Models):** HYBRID MODELS CAUTION: Hybrid systems use very high-voltage battery systems. Before starting any service work involving the battery system, turn the ignition switch OFF and then remove the service plug from pocket in the trunk. After removing the service plug, wait 10 minutes before touching any of the high-voltage connectors and terminals. **Note: On and, this IVT is on bank 1.** * An improper voltage is sent to the ECM via the intake valve timing control solenoid valve.
DTC: P0078 **1T ECM, MIL: Yes** **Year:** 2011 **Model:** EX35, FX35, FX50 **Engine:** 3.5L, 5.0L **Transmission:** All	**Exhaust Valve Timing (EVT) Control Magnet Retarder (Bank 1) Circuit:** An improper voltage is sent to the ECM via the exhaust valve timing control magnet retarder.
DTC: P0081 **1T ECM, MIL: Yes** **Year:** 2011 **Model:** EX35, FX35, FX50, G37, M37, M56, QX56 **Engine:** 3.5L, 3.7L, 5.0L, 5.6L **Transmission:** All	**Intake Valve Timing Control Solenoid Valve Circuit (Bank 2):** **Note: Applies to and engines only.** * An improper voltage is sent to the ECM via the intake valve timing control solenoid valve.

DTC	Trouble Code Title and Conditions
DTC: P0084 **1T ECM, MIL: Yes** **Year:** 2011 **Model:** EX35, FX35, FX50 **Engine:** 3.5L, 5.0L **Transmission:** All	**Exhaust Valve Timing (EVT) Timing Control (Magnet Retarder) Solenoid (Bank 2) Circuit:** An improper voltage is sent to the ECM via the exhaust valve timing control (magnet retarder).
DTC: P0101 **1T ECM, MIL: Yes** **Year:** 2011 **Model:** EX35, FX35, FX50, G37, M37, M56, QX56 **Engine:** 3.5L, 3.7L, 5.0L, 5.6L **Transmission:** All	**Mass Air Flow Sensor Circuit Range/Performance (Includes Hybrid Models):** HYBRID MODELS CAUTION: Hybrid systems use very high-voltage battery systems. Before starting any service work involving the battery system, turn the ignition switch OFF and then remove the service plug from pocket in the trunk. After removing the service plug, wait 10 minutes before touching any of the high-voltage connectors and terminals. * Condition A: A high voltage from the sensor is sent to ECM under light load driving condition. * Condition B: A low voltage from the sensor is sent to ECM under heavy load driving condition.
DTC: P0102 **1T ECM, MIL: Yes** **Year:** 2011 **Model:** EX35, FX35, FX50, G37, M37, M56, QX56 **Engine:** 3.5L, 3.7L, 5.0L, 5.6L **Transmission:** All	**Mass Air Flow Circuit Low Input (Includes Hybrid Models):** HYBRID MODELS CAUTION: Hybrid systems use very high-voltage battery systems. Before starting any service work involving the battery system, turn the ignition switch OFF and then remove the service plug from pocket in the trunk. After removing the service plug, wait 10 minutes before touching any of the high-voltage connectors and terminals. **NOTE: On and engines, this DTC is for Bank 1.** * An excessively low voltage from the sensor is sent to ECM.
DTC: P0103 **1T ECM, MIL: Yes** **Year:** 2011 **Model:** EX35, FX35, FX50, G37, M37, M56, QX56 **Engine:** 3.5L, 3.7L, 5.0L, 5.6L **Transmission:** All	**Mass Air Flow Sensor Circuit High Input (Includes Hybrid Models):** HYBRID MODELS CAUTION: Hybrid systems use very high-voltage battery systems. Before starting any service work involving the battery system, turn the ignition switch OFF and then remove the service plug from pocket in the trunk. After removing the service plug, wait 10 minutes before touching any of the high-voltage connectors and terminals. **NOTE: On and engines, this DTC is for Bank 1.** * An excessively high voltage from the sensor is sent to ECM.
DTC: P010A **1T ECM, MIL: Yes** **Year:** 2011 **Model:** G37, M37 **Engine:** 3.7L **Transmission:** All	**Manifold Absolute Pressure Sensor Circuit:** **NOTE: If DTC P010A is displayed with DTC P0643, first perform the trouble diagnosis for DTC P0643.** An excessively low voltage from the sensor is sent to ECM. An excessively high voltage from the sensor is sent to ECM.
DTC: P010B **1T ECM, MIL: Yes** **Year:** 2011 **Model:** EX35, FX35, FX50, G37, M37 **Engine:** 3.5L, 3.7L, 5.0L **Transmission:** All	**Mass Air Flow Sensor (Bank 2) Circuit/Range Performance:** Condition A: A high voltage from the sensor is sent to ECM under light load driving condition. Condition B: A low voltage from the sensor is sent to ECM under heavy load driving condition.
DTC: P010C **1T ECM, MIL: Yes** **Year:** 2011 **Model:** EX35, FX35, FX50, G37, M37 **Engine:** 3.5L, 3.7L, 5.0L **Transmission:** All	**Mass Air Flow Sensor (Bank 2) Circuit Low Input:** An excessively low voltage from the sensor is sent to ECM.
DTC: P010D **1T ECM, MIL: Yes** **Year:** 2011 **Model:** EX35, FX35, FX50, G37, M37 **Engine:** 3.5L, 3.7L, 5.0L **Transmission:** All	**Mass Air Flow Sensor (Bank 2) Circuit High Input:** An excessively high voltage from the sensor is sent to ECM.

DTC	Trouble Code Title and Conditions
DTC: P0112 **1T ECM, MIL: Yes** **Year:** 2011 **Model:** EX35, FX35, FX50, G37, M37, M56, QX56 **Engine:** 3.5L, 3.7L, 5.0L, 5.6L **Transmission:** All	**Intake Air Temperature Sensor Circuit Low Input (Includes Hybrid Models):** HYBRID MODELS CAUTION: Hybrid systems use very high-voltage battery systems. Before starting any service work involving the battery system, turn the ignition switch OFF and then remove the service plug from pocket in the trunk. After removing the service plug, wait 10 minutes before touching any of the high-voltage connectors and terminals. **NOTE: On and engines, this DTC is for Bank 1.** * An excessively low voltage from the sensor is sent to ECM.
DTC: P0113 **1T ECM, MIL: Yes** **Year:** 2011 **Model:** EX35, FX35, FX50, G37, M37, M56, QX56 **Engine:** 3.5L, 3.7L, 5.0L, 5.6L **Transmission:** All	**Intake Air Temperature Sensor Circuit High Input (Includes Hybrid Models):** HYBRID MODELS CAUTION: Hybrid systems use very high-voltage battery systems. Before starting any service work involving the battery system, turn the ignition switch OFF and then remove the service plug from pocket in the trunk. After removing the service plug, wait 10 minutes before touching any of the high-voltage connectors and terminals. **NOTE: On and engines, this DTC is for Bank 1.** * An excessively high voltage from the sensor is sent to ECM.
DTC: P0116 **1T ECM, MIL: Yes** **Year:** 2011 **Model:** EX35, FX35, FX50, G37, M37, M56, QX56 **Engine:** 3.5L, 3.7L, 5.0L, 5.6L **Transmission:** All	**Engine Coolant Temperature Sensor Circuit Range/Performance (Includes Hybrid Models):** HYBRID MODELS CAUTION: Hybrid systems use very high-voltage battery systems. Before starting any service work involving the battery system, turn the ignition switch OFF and then remove the service plug from pocket in the trunk. After removing the service plug, wait 10 minutes before touching any of the high-voltage connectors and terminals. **NOTE: If DTC P0116 is displayed with P0117 or P0118, first perform the trouble diagnosis for DTC P0117, P0118.** * Engine coolant temperature signal from engine coolant temperature sensor does not fluctuate, even when some time has passed after starting the engine with pre-warming up condition.
DTC: P0117 **1T ECM, MIL: Yes** **Year:** 2011 **Model:** EX35, FX35, FX50, G37, M37, M56, QX56 **Engine:** 3.5L, 3.7L, 5.0L, 5.6L **Transmission:** All	**Engine Coolant Temperature Circuit Low Input (Includes Hybrid Models):** HYBRID MODELS CAUTION: Hybrid systems use very high-voltage battery systems. Before starting any service work involving the battery system, turn the ignition switch OFF and then remove the service plug from pocket in the trunk. After removing the service plug, wait 10 minutes before touching any of the high-voltage connectors and terminals. * An excessively low voltage from the sensor is sent to ECM.
DTC: P0118 **1T ECM, MIL: Yes** **Year:** 2011 **Model:** EX35, FX35, FX50, G37, M37, M56, QX56 **Engine:** 3.5L, 3.7L, 5.0L, 5.6L **Transmission:** All	**Engine Coolant Temperature Sensor Circuit High Input (Includes Hybrid Models):** HYBRID MODELS CAUTION: Hybrid systems use very high-voltage battery systems. Before starting any service work involving the battery system, turn the ignition switch OFF and then remove the service plug from pocket in the trunk. After removing the service plug, wait 10 minutes before touching any of the high-voltage connectors and terminals. * An excessively high voltage from the sensor is sent to ECM.
DTC: P0122 **1T ECM, MIL: Yes** **Year:** 2011 **Model:** EX35, FX35, FX50, G37, M37, M56, QX56 **Engine:** 3.5L, 3.7L, 5.0L, 5.6L **Transmission:** All	**Throttle Position (TP) Sensor 2 Circuit Low Input (Includes Hybrid Models):** HYBRID MODELS CAUTION: Hybrid systems use very high-voltage battery systems. Before starting any service work involving the battery system, turn the ignition switch OFF and then remove the service plug from pocket in the trunk. After removing the service plug, wait 10 minutes before touching any of the high-voltage connectors and terminals. **NOTE: If this DTC is displayed with DTC P0510 or P0643, first perform the trouble diagnosis for DTC P0510 or P0643.** **NOTE: On and engines, this DTC is for Bank 1.** * An excessively low voltage from the TP sensor 2 is sent to ECM.
DTC: P0123 **1T ECM, MIL: Yes** **Year:** 2011 **Model:** EX35, FX35, FX50, G37, M37, M56, QX56 **Engine:** 3.5L, 3.7L, 5.0L, 5.6L **Transmission:** All	**Throttle Position Sensor 2 Circuit High Input (Includes Hybrid Models):** HYBRID MODELS CAUTION: Hybrid systems use very high-voltage battery systems. Before starting any service work involving the battery system, turn the ignition switch OFF and then remove the service plug from pocket in the trunk. After removing the service plug, wait 10 minutes before touching any of the high-voltage connectors and terminals. **NOTE: If DTC P0122 or P0123 is displayed with DTC P0510 or P0643, first perform the trouble diagnosis for DTC P0510 or P0643.** **NOTE: On 3.7L engine, this DTC is for Bank 1.** * An excessively high voltage from the TP sensor 2 is sent to ECM. * When the malfunction is detected, ECM enters fail-safe mode and the MIL lights up.
DTC: P0125 **1T ECM, MIL: Yes** **Year:** 2011 **Model:** EX35, FX35, FX50, G37, M37, M56, QX56 **Engine:** 3.5L, 3.7L, 5.0L, 5.6L **Transmission:** All	**Insufficient Engine Coolant Temperature for Closed Loop Fuel Control (Includes Hybrid Models):** HYBRID MODELS CAUTION: Hybrid systems use very high-voltage battery systems. Before starting any service work involving the battery system, turn the ignition switch OFF and then remove the service plug from pocket in the trunk. After removing the service plug, wait 10 minutes before touching any of the high-voltage connectors and terminals. * If DTC P0125 is displayed with P0116, P0117 or P0118, first perform the trouble diagnosis for the appropriate DTC, then proceed with P0125. * Voltage sent to ECM from the sensor is not practical, even when some time has passed after starting the engine. * Engine coolant temperature is insufficient for closed loop fuel control.

DTC	Trouble Code Title and Conditions
DTC: P0127 **1T ECM, MIL: Yes** **Year:** 2011 **Model:** EX35, FX35, FX50, G37, M37, M56, QX56 **Engine:** 3.5L, 3.7L, 5.0L, 5.6L **Transmission:** All	**Intake Air Temperature Too High (Includes Hybrid Models):** HYBRID MODELS CAUTION: Hybrid systems use very high-voltage battery systems. Before starting any service work involving the battery system, turn the ignition switch OFF and then remove the service plug from pocket in the trunk. After removing the service plug, wait 10 minutes before touching any of the high-voltage connectors and terminals. * Rationally incorrect voltage from the sensor is sent to ECM, compared with the voltage signal from engine coolant temperature sensor.
DTC: P0128 **1T ECM, MIL: Yes** **Year:** 2011 **Model:** EX35, FX35, FX50, G37, M37, M56, QX56 **Engine:** 3.5L, 3.7L, 5.0L, 5.6L **Transmission:** All	**Thermostat Function (includes Hybrid):** **NOTE: If DTC P0128 is displayed with DTC P0300, P0301, P0302, P0303, P0304, P0305, P0306, P0307 or P0308 first perform the trouble diagnosis for this DTC before continuing with P0128 diagnosis.** * The engine coolant temperature does not reach to specified temperature even though the engine has run long enough.
DTC: P0130 **1T ECM, MIL: Yes** **Year:** 2011 **Model:** EX35, FX35, FX50, G37, M37, M56, QX56 **Engine:** 3.5L, 3.7L, 5.0L, 5.6L **Transmission:** All	**Air Fuel Ratio (A/F) Sensor 1 Circuit (Includes Hybrid Models):** HYBRID MODELS CAUTION: Hybrid systems use very high-voltage battery systems. Before starting any service work involving the battery system, turn the ignition switch OFF and then remove the service plug from pocket in the trunk. After removing the service plug, wait 10 minutes before touching any of the high-voltage connectors and terminals. **NOTE: On and engines, this applies to Bank 1.** Condition A: The A/F signal computed by ECM from the A/F sensor 1 signal is constantly in the range other than approx. 2.2V. Condition B: The A/F signal computed by ECM from the A/F sensor 1 signal is constantly approx. 2.2V.
DTC: P0131 **1T ECM, MIL: Yes** **Year:** 2011 **Model:** EX35, FX35, FX50, G37, M37, M56, QX56 **Engine:** 3.5L, 3.7L, 5.0L, 5.6L **Transmission:** All	**Air Fuel Ratio (A/F) Sensor 1 Circuit Low Voltage (Includes Hybrid Models):** HYBRID MODELS CAUTION: Hybrid systems use very high-voltage battery systems. Before starting any service work involving the battery system, turn the ignition switch OFF and then remove the service plug from pocket in the trunk. After removing the service plug, wait 10 minutes before touching any of the high-voltage connectors and terminals. **NOTE: On and engines, this applies to Bank 1.** * To judge the malfunction, the diagnosis checks that the A/F signal computed by ECM from the A/F sensor 1 signal is not inordinately low. * The A/F signal computed by ECM from the A/F sensor 1 signal is constantly approx. 0V.
DTC: P0132 **1T ECM, MIL: Yes** **Year:** 2011 **Model:** EX35, FX35, FX50, G37, M37, M56, QX56 **Engine:** 3.5L, 3.7L, 5.0L, 5.6L **Transmission:** All	**Air Fuel Ratio (A/F) Sensor 1 Circuit High Voltage (Includes Hybrid Models):** HYBRID MODELS CAUTION: Hybrid systems use very high-voltage battery systems. Before starting any service work involving the battery system, turn the ignition switch OFF and then remove the service plug from pocket in the trunk. After removing the service plug, wait 10 minutes before touching any of the high-voltage connectors and terminals. **NOTE: On and engines, this applies to Bank 1.** * To judge the malfunction, the diagnosis checks that the A/F signal computed by ECM from the A/F sensor 1 signal is not inordinately low. * The A/F signal computed by ECM from the A/F sensor 1 signal is constantly approx. 5V. * An excessively high voltage signal is sent from the sensor.
DTC: P0133 **1T ECM, MIL: Yes** **Year:** 2011 **Model:** EX35, FX35, FX50, G37, M37, M56, QX56 **Engine:** 3.5L, 3.7L, 5.0L, 5.6L **Transmission:** All	**Air Fuel Ratio (A/F) Sensor 1 Circuit Slow Response (Includes Hybrid Models):** HYBRID MODELS CAUTION: Hybrid systems use very high-voltage battery systems. Before starting any service work involving the battery system, turn the ignition switch OFF and then remove the service plug from pocket in the trunk. After removing the service plug, wait 10 minutes before touching any of the high-voltage connectors and terminals. **NOTE: On and engines, this applies to Bank 1.** * To judge the malfunction of A/F sensor 1, this diagnosis measures response time of the A/F signal computed by ECM from the A/F sensor 1 signal. The time is compensated by engine operating (speed and load), fuel feedback control constant, and the A/F sensor 1 temperature index. Judgment is based on whether the compensated time (the A/F signal cycling time index) is inordinately long or not. * The response of the A/F signal computed by ECM from A/F sensor 1 signal takes more than the specified time.
DTC: P0137 **1T ECM, MIL: Yes** **Year:** 2011 **Model:** EX35, FX35, FX50, G37, M37, M56, QX56 **Engine:** 3.5L, 3.7L, 5.0L, 5.6L **Transmission:** All	**Heated Oxygen Sensor 2 Circuit Low Voltage (Includes Hybrid Models):** HYBRID MODELS CAUTION: Hybrid systems use very high-voltage battery systems. Before starting any service work involving the battery system, turn the ignition switch OFF and then remove the service plug from pocket in the trunk. After removing the service plug, wait 10 minutes before touching any of the high-voltage connectors and terminals. **NOTE: On and engines, this applies to Bank 1.** * The minimum voltage from the sensor is not reached to the specified voltage

DTC	Trouble Code Title and Conditions
DTC: P0138 **1T ECM, MIL: Yes** **Year:** 2011 **Model:** EX35, FX35, FX50, G37, M37, M56, QX56 **Engine:** 3.5L, 3.7L, 5.0L, 5.6L **Transmission:** All	**Heated Oxygen Sensor 2 Circuit High Voltage (Includes Hybrid Models):** HYBRID MODELS CAUTION: Hybrid systems use very high-voltage battery systems. Before starting any service work involving the battery system, turn the ignition switch OFF and then remove the service plug from pocket in the trunk. After removing the service plug, wait 10 minutes before touching any of the high-voltage connectors and terminals. **NOTE: On and engines, this applies to Bank 1.** * Condition A: An excessively high voltage from the sensor is sent to ECM, or, * Condition B: The minimum voltage from the sensor is not reached to the specified voltage.
DTC: P0139 **1T ECM, MIL: Yes** **Year:** 2011 **Model:** EX35, FX35, FX50, G37, M37, M56, QX56 **Engine:** 3.5L, 3.7L, 5.0L, 5.6L **Transmission:** All	**Heated Oxygen Sensor 2 Circuit Slow Response (Includes Hybrid Models):** HYBRID MODELS CAUTION: Hybrid systems use very high-voltage battery systems. Before starting any service work involving the battery system, turn the ignition switch OFF and then remove the service plug from pocket in the trunk. After removing the service plug, wait 10 minutes before touching any of the high-voltage connectors and terminals. **NOTE: On 4-cyl, sensor 2; on, sensor 2 bank 1** * It takes more time for the sensor to respond between rich and lean than the specified time.
DTC: P0150 **1T ECM, MIL: Yes** **Year:** 2011 **Model:** EX35, FX35, FX50, G37, M37, M56, QX56 **Engine:** 3.5L, 3.7L, 5.0L, 5.6L **Transmission:** All	**Air Fuel Ratio (A/F) Sensor 1 Bank 2 Circuit:** * The A/F signal computed by ECM from the A/F sensor 1 bank 2 signal is constantly in the range other than approx. 1.5V or 2.2V. * The A/F signal computed by ECM from the A/F sensor 1 bank 2 signal is constantly approx. 1.5V or 2.2V.
DTC: P0151 **1T ECM, MIL: Yes** **Year:** 2011 **Model:** EX35, FX35, FX50, G37, M37, M56, QX56 **Engine:** 3.5L, 3.7L, 5.0L, 5.6L **Transmission:** All	**Air Fuel Ratio (A/F) Sensor 1 Bank 2 Circuit Low Voltage:** * To judge the malfunction, the diagnosis checks that the A/F signal computed by ECM from the A/F sensor 1 signal is inordinately low. * The A/F signal computed by ECM from the A/F sensor 1 signal is constantly approx. 0V. * Maximum voltage is not reached.
DTC: P0152 **1T ECM, MIL: Yes** **Year:** 2011 **Model:** EX35, FX35, FX50, G37, M37, M56, QX56 **Engine:** 3.5L, 3.7L, 5.0L, 5.6L **Transmission:** All	**Air Fuel Ratio (A/F) Sensor 1 (Bank 2) Circuit High Voltage:** **NOTE: On and engines, this applies to Bank 2.** * Engine: After warming up. Maintaining engine speed at 2,000 rpm. * The A/F signal computed by ECM from the A/F sensor 1 signal is constantly approx. 5V. * An excessively high voltage from the sensor is sent to ECM.
DTC: P0153 **1T ECM, MIL: Yes** **Year:** 2011 **Model:** EX35, FX35, FX50, G37, M37, M56, QX56 **Engine:** 3.5L, 3.7L, 5.0L, 5.6L **Transmission:** All	**Air Fuel Ratio (A/F) Sensor 1 Bank 2 Circuit Slow Response:** The response of the A/F signal computed by ECM from A/F sensor 1 signal takes more than the specified time.
DTC: P0157 **1T ECM, MIL: Yes** **Year:** 2011 **Model:** EX35, FX35, FX50, G37, M37, M56, QX56 **Engine:** 3.5L, 3.7L, 5.0L, 5.6L **Transmission:** All	**Heated Oxygen Sensor 2 Bank 2 Circuit Low Voltage:** The maximum voltage from the sensor does not reach the specified voltage.
DTC: P0158 **1T ECM, MIL: Yes** **Year:** 2011 **Model:** EX35, FX35, FX50, G37, M37, M56, QX56 **Engine:** 3.5L, 3.7L, 5.0L, 5.6L **Transmission:** All	**Heated Oxygen Sensor 2 Bank 2 Circuit High Voltage:** * Condition A: An excessively high voltage from the sensor is sent to ECM, or, * Condition B: The minimum voltage from the sensor is not reached to the specified voltage.

DTC	Trouble Code Title and Conditions
DTC: P0159 **1T ECM, MIL: Yes** **Year:** 2011 **Model:** EX35, FX35, FX50, G37, M37, M56, QX56 **Engine:** 3.5L, 3.7L, 5.0L, 5.6L **Transmission:** All	**Heated Oxygen Sensor 2 Bank 2 Circuit Slow Response:** It takes more time for the sensor to respond between rich and lean than the specified time.
DTC: P0171 **1T ECM, MIL: Yes** **Year:** 2011 **Model:** EX35, FX35, FX50, G37, M37, M56, QX56 **Engine:** 3.5L, 3.7L, 5.0L, 5.6L **Transmission:** All	**Fuel Injection System Too Lean (Includes Hybrid Models):** HYBRID MODELS CAUTION: Hybrid systems use very high-voltage battery systems. Before starting any service work involving the battery system, turn the ignition switch OFF and then remove the service plug from pocket in the trunk. After removing the service plug, wait 10 minutes before touching any of the high-voltage connectors and terminals. **NOTE: On and engines, this applies to Bank 1.** * Fuel injection system does not operate properly. * The amount of mixture ratio compensation is too large. (The mixture ratio is too lean.)
DTC: P0172 **1T ECM, MIL: Yes** **Year:** 2011 **Model:** EX35, FX35, FX50, G37, M37, M56, QX56 **Engine:** 3.5L, 3.7L, 5.0L, 5.6L **Transmission:** All	**Fuel Injection System Too Rich (Includes Hybrid Models):** HYBRID MODELS CAUTION: Hybrid systems use very high-voltage battery systems. Before starting any service work involving the battery system, turn the ignition switch OFF and then remove the service plug from pocket in the trunk. After removing the service plug, wait 10 minutes before touching any of the high-voltage connectors and terminals. **NOTE: On and engines, this applies to Bank 1.** * Fuel injection system does not operate properly. * The amount of mixture ratio compensation is too large. (The mixture ratio is too rich.)
DTC: P0174 **1T ECM, MIL: Yes** **Year:** 2011 **Model:** EX35, FX35, FX50, G37, M37, M56, QX56 **Engine:** 3.5L, 3.7L, 5.0L, 5.6L **Transmission:** All	**Fuel Injection System Too Lean (Bank 2):** * Fuel injection system does not operate properly. * The amount of mixture ratio compensation is too large. (The mixture ratio is too lean.)
DTC: P0175 **1T ECM, MIL: Yes** **Year:** 2011 **Model:** EX35, FX35, FX50, G37, M37, M56, QX56 **Engine:** 3.5L, 3.7L, 5.0L, 5.6L **Transmission:** All	**Fuel Injection System Too Rich (Bank 2):** * Fuel injection system does not operate properly. * The amount of mixture ratio compensation is too large. (The mixture ratio is too rich.)
DTC: P0181 **1T ECM, MIL: Yes** **Year:** 2011 **Model:** EX35, FX35, FX50, G37, M37, M56, QX56 **Engine:** 3.5L, 3.7L, 5.0L, 5.6L **Transmission:** All	**Fuel Tank Temperature Sensor Circuit Range/Performance (Includes Hybrid Models):** HYBRID MODELS CAUTION: Hybrid systems use very high-voltage battery systems. Before starting any service work involving the battery system, turn the ignition switch OFF and then remove the service plug from pocket in the trunk. After removing the service plug, wait 10 minutes before touching any of the high-voltage connectors and terminals. * Rationally incorrect voltage from the sensor is sent to ECM, compared with the voltage signals from engine coolant temperature sensor and intake air temperature sensor.
DTC: P0182 **1T ECM, MIL: Yes** **Year:** 2011 **Model:** EX35, FX35, FX50, G37, M37, M56, QX56 **Engine:** 3.5L, 3.7L, 5.0L, 5.6L **Transmission:** All	**Fuel Tank Temperature Sensor Circuit Low Input (Includes Hybrid Models):** HYBRID MODELS CAUTION: Hybrid systems use very high-voltage battery systems. Before starting any service work involving the battery system, turn the ignition switch OFF and then remove the service plug from pocket in the trunk. After removing the service plug, wait 10 minutes before touching any of the high-voltage connectors and terminals. * An excessively low voltage from the sensor is sent to ECM.
DTC: P0183 **1T ECM, MIL: Yes** **Year:** 2011 **Model:** EX35, FX35, FX50, G37, M37, M56, QX56 **Engine:** 3.5L, 3.7L, 5.0L, 5.6L **Transmission:** All	**Fuel Tank Temperature Sensor Circuit High Input (Includes Hybrid Models):** HYBRID MODELS CAUTION: Hybrid systems use very high-voltage battery systems. Before starting any service work involving the battery system, turn the ignition switch OFF and then remove the service plug from pocket in the trunk. After removing the service plug, wait 10 minutes before touching any of the high-voltage connectors and terminals. An excessively high voltage from the sensor is sent to ECM.

DTC	Trouble Code Title and Conditions
DTC: P0196 **1T ECM, MIL: Yes** **Year:** 2011 **Model:** EX35, FX35, FX50, G37, M37 **Engine:** 3.5L, 3.7L, 5.0L **Transmission:** All	**Engine Oil Temperature (EOT) Sensor Range/Performance:** **NOTE: If DTC P0196 is displayed with P0197 or P0198, first perform the trouble diagnosis for DTC P0197, P0198.** * Rationally incorrect voltage from the sensor is sent to ECM, compared with the voltage signals from engine coolant temperature sensor and intake air temperature sensor.
DTC: P0197 **1T ECM, MIL: Yes** **Year:** 2011 **Model:** EX35, FX35, FX50, G37, M37 **Engine:** 3.5L, 3.7L, 5.0L **Transmission:** All	**Engine Oil Temperature (EOT) Sensor Circuit Low Input:** An excessively low voltage from the sensor is sent to ECM.
DTC: P0198 **1T ECM, MIL: Yes** **Year:** 2011 **Model:** EX35, FX35, FX50, G37, M37 **Engine:** 3.5L, 3.7L, 5.0L **Transmission:** All	**Engine Oil Temperature (EOT) Sensor Circuit High Input:** An excessively high voltage from the sensor is sent to ECM.
DTC: P0222 **1T ECM, MIL: Yes** **Year:** 2011 **Model:** EX35, FX35, FX50, G37, M37, M56, QX56 **Engine:** 3.5L, 3.7L, 5.0L, 5.6L **Transmission:** All	**Throttle Position (TP) Sensor 1 Circuit Low Input (Includes Hybrid Models):** HYBRID MODELS CAUTION: Hybrid systems use very high-voltage battery systems. Before starting any service work involving the battery system, turn the ignition switch OFF and then remove the service plug from pocket in the trunk. After removing the service plug, wait 10 minutes before touching any of the high-voltage connectors and terminals. **NOTE: If DTC P0222 or P0223 is displayed with DTC P0643, first perform the trouble diagnosis for DTC P0643.** **NOTE: On and engines, this DTC is for Bank 1.** * An excessively low voltage from the TP sensor 1 is sent to ECM.
DTC: P0223 **1T ECM, MIL: Yes** **Year:** 2011 **Model:** EX35, FX35, FX50, G37, M37, M56, QX56 **Engine:** 3.5L, 3.7L, 5.0L, 5.6L **Transmission:** All	**Throttle Position (TP) Sensor 1 Circuit High Input (Includes Hybrid Models):** HYBRID MODELS CAUTION: Hybrid systems use very high-voltage battery systems. Before starting any service work involving the battery system, turn the ignition switch OFF and then remove the service plug from pocket in the trunk. After removing the service plug, wait 10 minutes before touching any of the high-voltage connectors and terminals. **NOTE: If DTC P0222 or P0223 is displayed with DTC P0643, first perform the trouble diagnosis for DTC P0643.** **NOTE: On and engines, this DTC is for Bank 1.** * An excessively high voltage from the TP sensor 1 is sent to ECM.
DTC: P0225 **T ECM, MIL: Yes** **Year:** 2011 **Model:** FX50 **Engine:** 5.0L **Transmission:** All	**Closed Throttle Position Learning Performance (Bank 2):** **NOTE: DTC P0225 is displayed with another DTC for electric throttle control actuator. Perform the trouble diagnosis for the corresponding DTC.** Closed throttle position learning value is excessively low. Closed throttle position learning is not performed successfully, repeatedly.
DTC: P0227 **1T ECM, MIL: Yes** **Year:** 2011 **Model:** EX35, FX35, FX50, G37, M37 **Engine:** 3.5L, 3.7L, 5.0L **Transmission:** All	**Throttle Position Sensor 2 (Bank 2) Circuit Low Input:** **NOTE: If DTC P0122, P0123, P0227 or P0228 is displayed with DTC P0643, first perform the trouble diagnosis for DTC P0643.** An excessively low voltage from the TP sensor 2 is sent to ECM.
DTC: P0228 **1T ECM, MIL: Yes** **Year:** 2011 **Model:** EX35, FX35, FX50, G37, M37 **Engine:** 3.5L, 3.7L, 5.0L **Transmission:** All	**Throttle Position Sensor 2 (Bank 2) Circuit High Input:** **NOTE: If DTC P0122, P0123, P0227 or P0228 is displayed with DTC P0643, first perform the trouble diagnosis for DTC P0643.** An excessively high voltage from the TP sensor 2 is sent to ECM.

DTC	Trouble Code Title and Conditions
DTC: P0300 **1T ECM, MIL: Yes** **Year:** 2011 **Model:** EX35, FX35, FX50, G37, M37, M56, QX56 **Engine:** 3.5L, 3.7L, 5.0L, 5.6L **Transmission:** All	**Multiple Cylinder Misfire Detected (Includes Hybrid Models):** HYBRID MODELS CAUTION: Hybrid systems use very high-voltage battery systems. Before starting any service work involving the battery system, turn the ignition switch OFF and then remove the service plug from pocket in the trunk. After removing the service plug, wait 10 minutes before touching any of the high-voltage connectors and terminals. * Multiple cylinder misfire. * One Trip Detection Logic (Three Way Catalyst Damage) **Note: On the 1st trip, when a misfire condition occurs that can damage the three way catalyst (TWC) due to overheating, the MIL will blink.** * When a misfire condition occurs, the ECM monitors the CKP sensor (POS) signal every 200 engine revolutions for a change. * When the misfire condition decreases to a level that will not damage the TWC, the MIL will turn off. * If another misfire condition occurs that can damage the TWC on a second trip, the MIL will blink. * Two Trip Detection Logic (Exhaust quality deterioration) **Note: For misfire conditions that will not damage the TWC (but will affect vehicle emissions), the MIL will only light when the misfire is detected on a second trip. During this condition, the ECM monitors the CKP sensor signal every 1,000 engine revolutions.** * A misfire malfunction can be detected in any one cylinder or in multiple cylinders.
DTC: P0301 **1T ECM, MIL: Yes** **Year:** 2011 **Model:** EX35, FX35, FX50, G37, M37, M56, QX56 **Engine:** 3.5L, 3.7L, 5.0L, 5.6L **Transmission:** All	**No.1 Cylinder Misfire Detected (Includes Hybrid Models):** HYBRID MODELS CAUTION: Hybrid systems use very high-voltage battery systems. Before starting any service work involving the battery system, turn the ignition switch OFF and then remove the service plug from pocket in the trunk. After removing the service plug, wait 10 minutes before touching any of the high-voltage connectors and terminals. No. 1 cylinder misfires. 1. One Trip Detection Logic (Three Way Catalyst Damage) On the 1st trip, when a misfire condition occurs that can damage the three way catalyst (TWC) due to overheating, the MIL will blink. When a misfire condition occurs, the ECM monitors the CKP sensor (POS) signal every 200 engine revolutions for a change. When the misfire condition decreases to a level that will not damage the TWC, the MIL will turn off. 2. Two Trip Detection Logic (Exhaust quality deterioration) For misfire conditions that will not damage the TWC (but will affect vehicle emissions), the MIL will only light when the misfire is detected on a second trip. During this condition, the ECM monitors the CKP sensor signal every 1,000 engine revolutions. A misfire malfunction can be detected on any one cylinder or on multiple cylinders. If another misfire condition occurs that can damage the TWC on a second trip, the MIL will blink.
DTC: P0302 **1T ECM, MIL: Yes** **Year:** 2011 **Model:** EX35, FX35, FX50, G37, M37, M56, QX56 **Engine:** 3.5L, 3.7L, 5.0L, 5.6L **Transmission:** All	**No. 2 Cylinder Misfire Detected (Includes Hybrid Models):** HYBRID MODELS CAUTION: Hybrid systems use very high-voltage battery systems. Before starting any service work involving the battery system, turn the ignition switch OFF and then remove the service plug from pocket in the trunk. After removing the service plug, wait 10 minutes before touching any of the high-voltage connectors and terminals. * No. 2 cylinder misfires. 1. One Trip Detection Logic (Three Way Catalyst Damage) - On the 1st trip, when a misfire condition occurs that can damage the three way catalyst (TWC) due to overheating, the MIL will blink. - When a misfire condition occurs, the ECM monitors the CKP sensor (POS) signal every 200 engine revolutions for a change. - When the misfire condition decreases to a level that will not damage the TWC, the MIL will turn off. 2. Two Trip Detection Logic (Exhaust quality deterioration) - For misfire conditions that will not damage the TWC (but will affect vehicle emissions), the MIL will only light when the misfire is detected on a second trip. - During this condition, the ECM monitors the CKP sensor signal every 1,000 engine revolutions. - A misfire malfunction can be detected on any one cylinder or on multiple cylinders. - If another misfire condition occurs that can damage the TWC on a second trip, the MIL will blink.

DTC	Trouble Code Title and Conditions
DTC: P0303 **1T ECM, MIL: Yes** **Year:** 2011 **Model:** EX35, FX35, FX50, G37, M37, M56, QX56 **Engine:** 3.5L, 3.7L, 5.0L, 5.6L **Transmission:** All	**No. 3 Cylinder Misfire Detected (Includes Hybrid Models):** HYBRID MODELS CAUTION: Hybrid systems use very high-voltage battery systems. Before starting any service work involving the battery system, turn the ignition switch OFF and then remove the service plug from pocket in the trunk. After removing the service plug, wait 10 minutes before touching any of the high-voltage connectors and terminals. * No. 3 cylinder misfires. 1. One Trip Detection Logic (Three Way Catalyst Damage) - On the 1st trip, when a misfire condition occurs that can damage the three way catalyst (TWC) due to overheating, the MIL will blink. - When a misfire condition occurs, the ECM monitors the CKP sensor (POS) signal every 200 engine revolutions for a change. - When the misfire condition decreases to a level that will not damage the TWC, the MIL will turn off. 2. Two Trip Detection Logic (Exhaust quality deterioration) - For misfire conditions that will not damage the TWC (but will affect vehicle emissions), the MIL will only light when the misfire is detected on a second trip. - During this condition, the ECM monitors the CKP sensor signal every 1,000 engine revolutions. - A misfire malfunction can be detected on any one cylinder or on multiple cylinders. - If another misfire condition occurs that can damage the TWC on a second trip, the MIL will blink.
DTC: P0304 **1T ECM, MIL: Yes** **Year:** 2011 **Model:** EX35, FX35, FX50, G37, M37, M56, QX56 **Engine:** 3.5L, 3.7L, 5.0L, 5.6L **Transmission:** All	**No. 4 Cylinder Misfire Detected (Includes Hybrid Models):** HYBRID MODELS CAUTION: Hybrid systems use very high-voltage battery systems. Before starting any service work involving the battery system, turn the ignition switch OFF and then remove the service plug from pocket in the trunk. After removing the service plug, wait 10 minutes before touching any of the high-voltage connectors and terminals. * No. 4 cylinder misfires. * The misfire detection logic consists of the following two conditions. 1. One Trip Detection Logic (Three Way Catalyst Damage) - On the 1st trip, when a misfire condition occurs that can damage the three way catalyst (TWC) due to overheating, the MIL will blink. - When a misfire condition occurs, the ECM monitors the CKP sensor (POS) signal every 200 engine revolutions for a change. - When the misfire condition decreases to a level that will not damage the TWC, the MIL will turn off. - If another misfire condition occurs that can damage the TWC on a second trip, the MIL will blink. - When the misfire condition decreases to a level that will not damage the TWC, the MIL will remain on. - If another misfire condition occurs that can damage the TWC, the MIL will begin to blink again. 2. Two Trip Detection Logic (Exhaust quality deterioration) - For misfire conditions that will not damage the TWC (but will affect vehicle emissions), the MIL will only light when the misfire is detected on a second trip. - During this condition, the ECM monitors the CKP sensor signal every 1,000 engine revolutions. - A misfire malfunction can be detected on any one cylinder or on multiple cylinders.
DTC: P0305 **1T ECM, MIL: Yes** **Year:** 2011 **Model:** EX35, FX35, FX50, G37, M37, M56, QX56 **Engine:** 3.5L, 3.7L, 5.0L, 5.6L **Transmission:** All	**No. 5 Cylinder Misfire Detected:** * No. 5 cylinder misfires. * The misfire detection logic consists of the following two conditions. 1. One Trip Detection Logic (Three Way Catalyst Damage): - On the first trip, when a misfire condition occurs that can damage the three way catalyst (TWC) due to overheating, the MIL will blink. - When a misfire condition occurs, the ECM monitors the CKP sensor signal every 200 engine revolutions for a change. - When the misfire condition decreases to a level that will not damage the TWC, the MIL will turn off. - If another misfire condition occurs that can damage the TWC on a second trip, the MIL will blink. - When the misfire condition decreases to a level that will not damage the TWC, the MIL will remain on. - If another misfire condition occurs that can damage the TWC, the MIL will begin to blink again. 2. Two Trip Detection Logic (Exhaust quality deterioration): - For misfire conditions that will not damage the TWC (but will affect vehicle emissions), the MIL will only light when the misfire is detected on a second trip. - During this condition, the ECM monitors the CKP sensor signal every 1,000 engine revolutions. - A misfire malfunction can be detected in any one cylinder or in multiple cylinders.

DTC	Trouble Code Title and Conditions
DTC: P0306 **1T ECM, MIL: Yes** **Year:** 2011 **Model:** EX35, FX35, FX50, G37, M37, M56, QX56 **Engine:** 3.5L, 3.7L, 5.0L, 5.6L **Transmission:** All	**No. 6 Cylinder Misfire Detected:** * No. 6 cylinder misfires. * The misfire detection logic consists of the following two conditions. 1. One Trip Detection Logic (Three Way Catalyst Damage): - On the first trip, when a misfire condition occurs that can damage the three way catalyst (TWC) due to overheating, the MIL will blink. - When a misfire condition occurs, the ECM monitors the CKP sensor signal every 200 engine revolutions for a change. - When the misfire condition decreases to a level that will not damage the TWC, the MIL will turn off. - If another misfire condition occurs that can damage the TWC on a second trip, the MIL will blink. - When the misfire condition decreases to a level that will not damage the TWC, the MIL will remain on. - If another misfire condition occurs that can damage the TWC, the MIL will begin to blink again. 2. Two Trip Detection Logic (Exhaust quality deterioration): - For misfire conditions that will not damage the TWC (but will affect vehicle emissions), the MIL will only light when the misfire is detected on a second trip. - During this condition, the ECM monitors the CKP sensor signal every 1,000 engine revolutions. - A misfire malfunction can be detected in any one cylinder or in multiple cylinders.No. 5 cylinder misfires.
DTC: P0307 **1T ECM, MIL: Yes** **Year:** 2011 **Model:** FX50, M56, QX56 **Engine:** 5.0L, 5.6L **Transmission:** All	**No. 7 Cylinder Misfire Detected:** The misfire detection logic consists of the following two conditions: One Trip Detection Logic (Three Way Catalyst Damage) - On the 1st trip that a misfire condition occurs that can damage the three way catalyst (TWC) due to overheating, the MIL will blink. - When a misfire condition occurs, the ECM monitors the CKP sensor signal every 200 engine revolutions for a change. - When the misfire condition decreases to a level that will not damage the TWC, the MIL will turn off. - If another misfire condition occurs that can damage the TWC on a second trip, the MIL will blink. - When the misfire condition decreases to a level that will not damage the TWC, the MIL will remain on. - If another misfire condition occurs that can damage the TWC, the MIL will begin to blink again. Two Trip Detection Logic (Exhaust quality deterioration) - For misfire conditions that will not damage the TWC (but will affect vehicle emissions), the MIL will only light when the misfire is detected on a second trip. - During this condition, the ECM monitors the CKP sensor signal every 1,000 engine revolutions. - A misfire malfunction can be detected on any one cylinder or on multiple cylinders. No. 7 cylinder misfires.
DTC: P0308 **1T ECM, MIL: Yes** **Year:** 2011 **Model:** FX50, M56, QX56 **Engine:** 5.0L, 5.6L **Transmission:** All	**No. 8 Cylinder Misfire Detected:** The misfire detection logic consists of the following two conditions: One Trip Detection Logic (Three Way Catalyst Damage) - On the 1st trip that a misfire condition occurs that can damage the three way catalyst (TWC) due to overheating, the MIL will blink. - When a misfire condition occurs, the ECM monitors the CKP sensor signal every 200 engine revolutions for a change. - When the misfire condition decreases to a level that will not damage the TWC, the MIL will turn off. - If another misfire condition occurs that can damage the TWC on a second trip, the MIL will blink. - When the misfire condition decreases to a level that will not damage the TWC, the MIL will remain on. - If another misfire condition occurs that can damage the TWC, the MIL will begin to blink again. Two Trip Detection Logic (Exhaust quality deterioration) - For misfire conditions that will not damage the TWC (but will affect vehicle emissions), the MIL will only light when the misfire is detected on a second trip. - During this condition, the ECM monitors the CKP sensor signal every 1,000 engine revolutions. - A misfire malfunction can be detected on any one cylinder or on multiple cylinders. No. 8 cylinder misfires.
DTC: P0327 **1T ECM, MIL: Yes** **Year:** 2011 **Model:** EX35, FX35, FX50, G37, M37, M56, QX56 **Engine:** 3.5L, 3.7L, 5.0L, 5.6L **Transmission:** All	**Knock Sensor Circuit Low Input (Includes Hybrid Models):** HYBRID MODELS CAUTION: Hybrid systems use very high-voltage battery systems. Before starting any service work involving the battery system, turn the ignition switch OFF and then remove the service plug from pocket in the trunk. After removing the service plug, wait 10 minutes before touching any of the high-voltage connectors and terminals. **NOTE: On and engines, this applies to Bank 1.** An excessively low voltage from the sensor is sent to ECM.

DTC	Trouble Code Title and Conditions
DTC: P0328 **1T ECM, MIL: Yes** **Year:** 2011 **Model:** EX35, FX35, FX50, G37, M37, M56, QX56 **Engine:** 3.5L, 3.7L, 5.0L, 5.6L **Transmission:** All	**Knock Sensor Circuit High Input (Includes Hybrid Models):** HYBRID MODELS CAUTION: Hybrid systems use very high-voltage battery systems. Before starting any service work involving the battery system, turn the ignition switch OFF and then remove the service plug from pocket in the trunk. After removing the service plug, wait 10 minutes before touching any of the high-voltage connectors and terminals. **NOTE: On and engines, this applies to Bank 1.** An excessively high voltage from the sensor is sent to ECM.
DTC: P0332 **1T ECM, MIL: Yes** **Year:** 2011 **Model:** EX35, FX35, FX50, G37, M37, M56, QX56 **Engine:** 3.5L, 3.7L, 5.0L, 5.6L **Transmission:** All	**Knock Sensor (KS) Bank 2 Sensor Circuit Low Input:** An excessively low voltage from the sensor is sent to ECM.
DTC: P0333 **1T ECM, MIL: Yes** **Year:** 2011 **Model:** EX35, FX35, FX50, G37, M37, M56, QX56 **Engine:** 3.5L, 3.7L, 5.0L, 5.6L **Transmission:** All	**Knock Sensor (Bank 2) Circuit High Input:** An excessively high voltage from the sensor is sent to ECM.
DTC: P0335 **1T ECM, MIL: Yes** **Year:** 2011 **Model:** EX35, FX35, FX50, G37, M37, M56, QX56 **Engine:** 3.5L, 3.7L, 5.0L, 5.6L **Transmission:** All	**Crankshaft Position Sensor (POS) Circuit (Includes Hybrid Models):** HYBRID MODELS CAUTION: Hybrid systems use very high-voltage battery systems. Before starting any service work involving the battery system, turn the ignition switch OFF and then remove the service plug from pocket in the trunk. After removing the service plug, wait 10 minutes before touching any of the high-voltage connectors and terminals. * The crankshaft position sensor (POS) signal is not detected by the ECM during the first few seconds of engine cranking. * The proper pulse signal from the crankshaft position sensor (POS) is not sent to ECM while the engine is running. * The crankshaft position sensor (POS) signal is not in the normal pattern during engine running.
DTC: P0340 **1T ECM, MIL: Yes** **Year:** 2011 **Model:** EX35, FX35, FX50, G37, M37, M56, QX56 **Engine:** 3.5L, 3.7L, 5.0L, 5.6L **Transmission:** All	**Camshaft Position Sensor Circuit (Includes Hybrid Models):** HYBRID MODELS CAUTION: Hybrid systems use very high-voltage battery systems. Before starting any service work involving the battery system, turn the ignition switch OFF and then remove the service plug from pocket in the trunk. After removing the service plug, wait 10 minutes before touching any of the high-voltage connectors and terminals. **NOTE: On and engines, this applies to Bank 1.** **NOTE: If DTC P0340 is displayed with DTC P0643, first perform the trouble diagnosis for DTC P0643.** * The cylinder No. signal or proper position signal is not sent to ECM for the first few seconds during engine cranking. * The cylinder No. signal or proper position signal is not set to ECM during engine running. * The cylinder No. signal or proper position signal is not in the normal pattern during engine running.
DTC: P0345 **1T ECM, MIL: Yes** **Year:** 2011 **Model:** EX35, FX35, FX50, G37, M37 **Engine:** 3.5L, 3.7L, 5.0L **Transmission:** All	**Camshaft Position (CMP) Sensor Bank 2 Circuit:** **NOTE: If DTC P0340 or P0345 is displayed with DTC P0643, first perform the trouble diagnosis for DTC P0643.** * The cylinder No. signal is not sent to ECM for the first few seconds during engine cranking. * The cylinder No. signal is not sent to ECM during engine running. * The cylinder No. signal is not in the normal pattern during engine running.
DTC: P0420 **1T ECM, MIL: Yes** **Year:** 2011 **Model:** EX35, FX35, FX50, G37, M37, M56, QX56 **Engine:** 3.5L, 3.7L, 5.0L, 5.6L **Transmission:** All	**Catalyst System Efficiency Below Threshhold (Includes Hybrid Models):** HYBRID MODELS CAUTION: Hybrid systems use very high-voltage battery systems. Before starting any service work involving the battery system, turn the ignition switch OFF and then remove the service plug from pocket in the trunk. After removing the service plug, wait 10 minutes before touching any of the high-voltage connectors and terminals. **NOTE: On models with dual exhaust, this DTC refers to Bank 1.** * Three way catalyst (manifold) does not operate properly. * Three way catalyst (manifold) does not have enough oxygen storage capacity.
DTC: P0430 **1T ECM, MIL: Yes** **Year:** 2011 **Model:** EX35, FX35, FX50, G37, M37, M56, QX56 **Engine:** 3.5L, 3.7L, 5.0L, 5.6L **Transmission:** All	**Catalyst System Efficiency Below Threshhold (Bank 2):** * Three way catalyst (manifold) does not operate properly. * Three way catalyst (manifold) does not have enough oxygen storage capacity.

DTC	Trouble Code Title and Conditions
DTC: P0441 **1T ECM, MIL: Yes** **Year:** 2011 **Model:** EX35, FX35, FX50, G37, M37, M56, QX56 **Engine:** 3.5L, 3.7L, 5.0L, 5.6L **Transmission:** All	**EVAP Control System Incorrect Purge Flow (Includes Hybrid Models):** HYBRID MODELS CAUTION: Hybrid systems use very high-voltage battery systems. Before starting any service work involving the battery system, turn the ignition switch OFF and then remove the service plug from pocket in the trunk. After removing the service plug, wait 10 minutes before touching any of the high-voltage connectors and terminals. **NOTE: If DTC P0441 is displayed with other DTC such as P2122, P2123 P2127, P2128, P2138, first perform trouble diagnosis for other DTC.** * Under normal conditions (non-closed throttle), sensor output voltage indicates if pressure drop and purge flow are adequate. If not, a malfunction is determined. * EVAP control system does not operate properly – EVAP control system has a leak between intake manifold and EVAP control system pressure sensor.
DTC: P0442 **1T ECM, MIL: Yes** **Year:** 2011 **Model:** EX35, FX35, FX50, G37, M37, M56, QX56 **Engine:** 3.5L, 3.7L, 5.0L, 5.6L **Transmission:** All	**EVAP Control System Small Leak Detected (Negative Pressure) (Includes Hybrid Models):** **NOTE: If DTC P0442 is displayed with DTC P0456, first perform the trouble diagnosis for DTC P0456.** * EVAP control system has a leak, EVAP control system does not operate properly.
DTC: P0443 **2T CCM, MIL: Yes** **Year:** 2011 **Model:** M56, QX56 **Engine:** 5.6L **Transmission:** All	**EVAP Canister Purge Solenoid Circuit Malfunction:** Engine started, Engine running at cruise speed under light engine load, system voltage from 11-16v, and the PCM detected an unexpected voltage condition on the Purge solenoid circuit, or it detected an invalid EVAP signal present when the Purge solenoid was commanded "on" and "off".
DTC: P0443 **1T ECM, MIL: Yes** **Year:** 2011 **Model:** EX35, FX35, G37, M37, M56, QX56 **Engine:** 3.5L, 3.7L, 5.6L **Transmission:** All	**EVAP Canister Purge Volume Control Solenoid Valve (Includes Hybrid Models):** HYBRID MODELS CAUTION: Hybrid systems use very high-voltage battery systems. Before starting any service work involving the battery system, turn the ignition switch OFF and then remove the service plug from pocket in the trunk. After removing the service plug, wait 10 minutes before touching any of the high-voltage connectors and terminals. Condition A: The canister purge flow is detected during the vehicle is stopped while the engine is running, even when EVAP canister purge volume control solenoid valve is completely closed. Condition B: The canister purge flow is detected during the specified driving conditions, even when EVAP canister purge volume control solenoid valve is completely closed.
DTC: P0443 **1T ECM, MIL: Yes** **Year:** 2011 **Model:** FX35, FX50 **Engine:** 3.5L, 5.0L **Transmission:** All	**EVAP Canister Purge Volume Control Solenoid Valve:** The canister purge flow is detected during the specified driving conditions, even when EVAP canister purge volume control solenoid valve is completely closed.
DTC: P0444 **1T ECM, MIL: Yes** **Year:** 2011 **Model:** EX35, FX35, FX50, G37, M37, M56, QX56 **Engine:** 3.5L, 3.7L, 5.0L, 5.6L **Transmission:** All	**EVAP Canister Purge Volume Control Solenoid Valve Circuit Open (Includes Hybrid Models):** HYBRID MODELS CAUTION: Hybrid systems use very high-voltage battery systems. Before starting any service work involving the battery system, turn the ignition switch OFF and then remove the service plug from pocket in the trunk. After removing the service plug, wait 10 minutes before touching any of the high-voltage connectors and terminals. * An excessively low voltage signal is sent to ECM through the valve
DTC: P0445 **1T ECM, MIL: Yes** **Year:** 2011 **Model:** EX35, FX35, FX50, G37, M37, M56, QX56 **Engine:** 3.5L, 3.7L, 5.0L, 5.6L **Transmission:** All	**EVAP Canister Purge Volume Control Solenoid Valve Circuit Shorted (Includes Hybrid Models):** HYBRID MODELS CAUTION: Hybrid systems use very high-voltage battery systems. Before starting any service work involving the battery system, turn the ignition switch OFF and then remove the service plug from pocket in the trunk. After removing the service plug, wait 10 minutes before touching any of the high-voltage connectors and terminals. * An excessively high voltage signal is sent to ECM through the valve
DTC: P0447 **1T ECM, MIL: Yes** **Year:** 2011 **Model:** EX35, FX35, FX50, G37, M37, M56, QX56 **Engine:** 3.5L, 3.7L, 5.0L, 5.6L **Transmission:** All	**EVAP Canister Vent Control Valve Circuit Open (Includes Hybrid Models):** HYBRID MODELS CAUTION: Hybrid systems use very high-voltage battery systems. Before starting any service work involving the battery system, turn the ignition switch OFF and then remove the service plug from pocket in the trunk. After removing the service plug, wait 10 minutes before touching any of the high-voltage connectors and terminals. * An improper voltage signal is sent to ECM through EVAP canister vent control valve. * EVAP canister vent control valve remains open under specified driving conditions.

DTC	Trouble Code Title and Conditions
DTC: P0448 **1T ECM, MIL: Yes** **Year:** 2011 **Model:** EX35, FX35, FX50, G37, M37, M56, QX56 **Engine:** 3.5L, 3.7L, 5.0L, 5.6L **Transmission:** All	**EVAP Canister Vent Control Valve Closed (Includes Hybrid Models):** HYBRID MODELS CAUTION: Hybrid systems use very high-voltage battery systems. Before starting any service work involving the battery system, turn the ignition switch OFF and then remove the service plug from pocket in the trunk. After removing the service plug, wait 10 minutes before touching any of the high-voltage connectors and terminals. * EVAP canister vent control valve remains closed under specified driving conditions.
DTC: P0451 **1T ECM, MIL: Yes** **Year:** 2011 **Model:** EX35, FX35, FX50, G37, M37, M56, QX56 **Engine:** 3.5L, 3.7L, 5.0L, 5.6L **Transmission:** All	**EVAP Control System Pressure Sensor Performance (Includes Hybrid Models):** HYBRID MODELS CAUTION: Hybrid systems use very high-voltage battery systems. Before starting any service work involving the battery system, turn the ignition switch OFF and then remove the service plug from pocket in the trunk. After removing the service plug, wait 10 minutes before touching any of the high-voltage connectors and terminals. * ECM detects a sloshing signal from the EVAP control system pressure sensor
DTC: P0452 **1T ECM, MIL: Yes** **Year:** 2011 **Model:** EX35, FX35, FX50, G37, M37, M56, QX56 **Engine:** 3.5L, 3.7L, 5.0L, 5.6L **Transmission:** All	**EVAP Control System Pressure Sensor Low Input (Includes Hybrid Models):** HYBRID MODELS CAUTION: Hybrid systems use very high-voltage battery systems. Before starting any service work involving the battery system, turn the ignition switch OFF and then remove the service plug from pocket in the trunk. After removing the service plug, wait 10 minutes before touching any of the high-voltage connectors and terminals. * An excessively low voltage from the sensor is sent to ECM.
DTC: P0453 **1T ECM, MIL: Yes** **Year:** 2011 **Model:** EX35, FX35, FX50, G37, M37, M56, QX56 **Engine:** 3.5L, 3.7L, 5.0L, 5.6L **Transmission:** All	**EVAP Control System Pressure Sensor High Input (Includes Hybrid Models):** HYBRID MODELS CAUTION: Hybrid systems use very high-voltage battery systems. Before starting any service work involving the battery system, turn the ignition switch OFF and then remove the service plug from pocket in the trunk. After removing the service plug, wait 10 minutes before touching any of the high-voltage connectors and terminals. * An excessively high voltage from the sensor is sent to ECM.
DTC: P0455 **1T ECM, MIL: Yes** **Year:** 2011 **Model:** EX35, FX35, FX50, G37, M37, M56, QX56 **Engine:** 3.5L, 3.7L, 5.0L, 5.6L **Transmission:** All	**EVAP Control System Gross Leak Detected (Includes Hybrid Models):** * EVAP control system has a very large leak, such as fuel filler cap fell off. * EVAP control system does not operate properly. CAUTION: Never remove fuel filler cap during the DTC Confirmation Procedure.
DTC: P0456 **1T ECM, MIL: Yes** **Year:** 2011 **Model:** EX35, FX35, FX50, G37, M37, M56, QX56 **Engine:** 3.5L, 3.7L, 5.0L, 5.6L **Transmission:** All	**Evaporative Emission Control System Very Small Leak (Negative Pressure Check) (Includes Hybrid Models):** HYBRID MODELS CAUTION: Hybrid systems use very high-voltage battery systems. Before starting any service work involving the battery system, turn the ignition switch OFF and then remove the service plug from pocket in the trunk. After removing the service plug, wait 10 minutes before touching any of the high-voltage connectors and terminals. **NOTE: If ECM judges a leak which corresponds to a very small leak, the very small leak P0456 will be detected.** **NOTE: If ECM judges a leak equivalent to a small leak, EVAP small leak P0442 will be detected.** **NOTE: If ECM judges there are no leaks, the diagnosis will be OK.** * If DTC P0456 is displayed with DTC P0442, first perform the trouble diagnosis for DTC P0456. * This diagnosis detects very small leakage in the EVAP line between fuel tank and EVAP canister purge volume control solenoid valve, using the intake manifold vacuum in the same way as conventional EVAP small leakage diagnosis. **NOTE: If ECM judges a leakage which corresponds to a very small leakage, the very small leakage P0456 will be detected.** * If ECM judges a leakage equivalent to a small leakage, EVAP small leakage P0442 will be detected. * If ECM judges that there are no leakage, the diagnosis will be OK. * EVAP system has a very small leak. * EVAP system does not operate properly.
DTC: P0460 **1T ECM, MIL: Yes** **Year:** 2011 **Model:** EX35, FX35, FX50, G37, M37, M56, QX56 **Engine:** 3.5L, 3.7L, 5.0L, 5.6L **Transmission:** All	**Fuel Level Sensor Circuit Noise (Includes Hybrid Models):** HYBRID MODELS CAUTION: Hybrid systems use very high-voltage battery systems. Before starting any service work involving the battery system, turn the ignition switch OFF and then remove the service plug from pocket in the trunk. After removing the service plug, wait 10 minutes before touching any of the high-voltage connectors and terminals. **NOTE: If DTC P0461 is displayed with DTC UXXXX, first perform the trouble diagnosis for DTC UXXXX.** **NOTE: If DTC P0460 is displayed with DTC P0607, first perform the trouble diagnosis for DTC P0607.** * When the vehicle is parked, naturally the fuel level in the fuel tank is stable. It means that output signal of the fuel level sensor does not change. If ECM senses sloshing signal from the sensor, fuel level sensor malfunction is detected. * Even though the vehicle is parked, a signal being varied is sent from the fuel level sensor to ECM.

DTC	Trouble Code Title and Conditions
DTC: P0461 **1T ECM, MIL: Yes** **Year:** 2011 **Model:** EX35, FX35, FX50, G37, M37, M56, QX56 **Engine:** 3.5L, 3.7L, 5.0L, 5.6L **Transmission:** All	**Fuel Level Sensor Circuit Range/Performance (Includes Hybrid Models):** HYBRID MODELS CAUTION: Hybrid systems use very high-voltage battery systems. Before starting any service work involving the battery system, turn the ignition switch OFF and then remove the service plug from pocket in the trunk. After removing the service plug, wait 10 minutes before touching any of the high-voltage connectors and terminals. **NOTE: If DTC P0461 is displayed with DTC U1000 or U1001, first perform the trouble diagnosis for appropriate "U" code.** **NOTE: If DTC P0461 is displayed with DTC P0607, first perform the trouble diagnosis for DTC P0607.** * This diagnosis detects the fuel gauge malfunction of the gauge not moving even after a long distance has been driven. Driving long distances naturally affect fuel gauge level. * The output signal of the fuel level sensor does not change within the specified range even though the vehicle has been driven a long distance.
DTC: P0462 **1T ECM, MIL: Yes** **Year:** 2011 **Model:** EX35, FX35, FX50, G37, M37, M56, QX56 **Engine:** 3.5L, 3.7L, 5.0L, 5.6L **Transmission:** All	**Fuel Level Sensor Circuit Low Input (Includes Hybrid Models):** HYBRID MODELS CAUTION: Hybrid systems use very high-voltage battery systems. Before starting any service work involving the battery system, turn the ignition switch OFF and then remove the service plug from pocket in the trunk. After removing the service plug, wait 10 minutes before touching any of the high-voltage connectors and terminals. **NOTE: If DTC P0462 or P0463 is displayed with DTC UXXXX, first perform the trouble diagnosis for DTC UXXXX.** **NOTE: If DTC P0462 or P0463 is displayed with DTC P0607, first perform the trouble diagnosis for DTC P0607.** * An excessively low voltage from the sensor is sent to ECM.
DTC: P0463 **1T ECM, MIL: Yes** **Year:** 2011 **Model:** EX35, FX35, FX50, G37, M37, M56, QX56 **Engine:** 3.5L, 3.7L, 5.0L, 5.6L **Transmission:** All	**Fuel Level Sensor Circuit High Input (Includes Hybrid Models):** HYBRID MODELS CAUTION: Hybrid systems use very high-voltage battery systems. Before starting any service work involving the battery system, turn the ignition switch OFF and then remove the service plug from pocket in the trunk. After removing the service plug, wait 10 minutes before touching any of the high-voltage connectors and terminals. **NOTE: If DTC P0462 or P0463 is displayed with DTC UXXXX, first perform the trouble diagnosis for DTC UXXXX.** **NOTE: If DTC P0462 or P0463 is displayed with DTC P0607, first perform the trouble diagnosis for DTC P0607.** * An excessively high voltage from the sensor is sent to ECM.
DTC: P0500 **2T ECM, MIL: Yes** **Year:** 2011 **Model:** M56, QX56 **Engine:** 5.6L **Transmission:** All	**Vehicle Speed Sensor Circuit:** An almost 0 MPH signal from the vehicle speed sensor is sent to the ECM even when the vehicle is being driven.
DTC: P0506 **1T ECM, MIL: Yes** **Year:** 2011 **Model:** EX35, FX35, FX50, G37, M37, M56, QX56 **Engine:** 3.5L, 3.7L, 5.0L, 5.6L **Transmission:** All	**Idle Speed Control System RPM Lower Than Expected (Includes Hybrid Models):** HYBRID MODELS CAUTION: Hybrid systems use very high-voltage battery systems. Before starting any service work involving the battery system, turn the ignition switch OFF and then remove the service plug from pocket in the trunk. After removing the service plug, wait 10 minutes before touching any of the high-voltage connectors and terminals. **NOTE: If DTC P0506 is displayed with other DTC, first perform the trouble diagnosis for the other DTC.** * The idle speed is less than the target idle speed by 100 rpm or more.
DTC: P0507 **1T ECM, MIL: Yes** **Year:** 2011 **Model:** EX35, FX35, FX50, G37, M37, M56, QX56 **Engine:** 3.5L, 3.7L, 5.0L, 5.6L **Transmission:** All	**Idle Speed Control System RPM Higher Than Expected (Includes Hybrid Models):** HYBRID MODELS CAUTION: Hybrid systems use very high-voltage battery systems. Before starting any service work involving the battery system, turn the ignition switch OFF and then remove the service plug from pocket in the trunk. After removing the service plug, wait 10 minutes before touching any of the high-voltage connectors and terminals. **NOTE: If DTC P0507 is displayed with other DTC, first perform the trouble diagnosis for the other DTC.** * The idle speed is more than the target idle speed by 200 rpm or more.
DTC: P0524 **1T ECM, MIL: Yes** **Year:** 2011 **Model:** FX50, G37, M37 **Engine:** 3.7L, 5.0L **Transmission:** All	**Engine Oil Pressure Too Low:** Engine oil pressure is low because there is a gap between angle of target and phase-control angle.
DTC: P0550 **1T ECM** **Year:** 2011 **Model:** EX35, FX35, FX50, G37, M37, M56, QX56 **Engine:** 3.5L, 3.7L, 5.0L, 5.6L **Transmission:** All	**Power Steering Pressure Sensor Circuit:** The MIL will not illuminate for this diagnosis. **NOTE: If DTC P0550 is displayed with DTC P0643, first perform the trouble diagnosis for DTC P0643.** * An excessively low or high voltage from the sensor is sent to ECM.

DTC	Trouble Code Title and Conditions
DTC: P0555 **T ECM, MIL: Yes** **Year:** 2011 **Model:** G37, M37 **Engine:** 3.7L **Transmission:** All	**Brake Booster Pressure Sensor Circuit:** An excessively low voltage from the sensor is sent to ECM. An excessively high voltage from the sensor is sent to ECM.
DTC: P0603 **1T ECM, MIL: Yes** **Year:** 2011 **Model:** EX35, FX35, FX50, G37, M37, M56, QX56 **Engine:** 3.5L, 3.7L, 5.0L, 5.6L **Transmission:** All	**ECM Power Supply Circuit (Includes Hybrid Models):** HYBRID MODELS CAUTION: Hybrid systems use very high-voltage battery systems. Before starting any service work involving the battery system, turn the ignition switch OFF and then remove the service plug from pocket in the trunk. After removing the service plug, wait 10 minutes before touching any of the high-voltage connectors and terminals. * ECM back-up RAM system does not function properly.
DTC: P0605 **1T ECM, MIL: Yes** **Year:** 2011 **Model:** EX35, FX35, FX50, G37, M37, M56, QX56 **Engine:** 3.5L, 3.7L, 5.0L, 5.6L **Transmission:** All	**Engine Control Module (ECM) (Includes Hybrid Models):** HYBRID MODELS CAUTION: Hybrid systems use very high-voltage battery systems. Before starting any service work involving the battery system, turn the ignition switch OFF and then remove the service plug from pocket in the trunk. After removing the service plug, wait 10 minutes before touching any of the high-voltage connectors and terminals. A. ECM calculation function is malfunctioning. B. ECM EEP-ROM system is malfunctioning. C. ECM self shut-off function is malfunctioning.
DTC: P0607 **1T ECM, MIL: Yes** **Year:** 2011 **Model:** EX35, FX50, G37, M37, M56, QX56 **Engine:** 3.5L, 3.7L, 5.0L, 5.6L **Transmission:** All	**CAN Communication Bus (Includes Hybrid Models):** HYBRID MODELS CAUTION: Hybrid systems use very high-voltage battery systems. Before starting any service work involving the battery system, turn the ignition switch OFF and then remove the service plug from pocket in the trunk. After removing the service plug, wait 10 minutes before touching any of the high-voltage connectors and terminals. When detecting error during the initial diagnosis of CAN controller of ECM.
DTC: P0615 **T TCM, TCIL: Yes** **Year:** 2011 **Model:** EX35, FX35, FX50, G37, M37, M56, QX56 **Engine:** 3.5L, 3.7L, 5.0L, 5.6L **Transmission:** All	**Starter Relay Circuit:** * This is not an OBD-II self-diagnostic item * This DTC will set if the starter monitor value is OFF when the ignition switch is ON at the" P" and "N" positions.
DTC: P0643 **1T ECM, MIL: Yes** **Year:** 2011 **Model:** EX35, FX35, FX50, G37, M37, M56, QX56 **Engine:** 3.5L, 3.7L, 5.0L, 5.6L **Transmission:** All	**Sensor Power Supply Circuit Short (Includes Hybrid Models):** HYBRID MODELS CAUTION: Hybrid systems use very high-voltage battery systems. Before starting any service work involving the battery system, turn the ignition switch OFF and then remove the service plug from pocket in the trunk. After removing the service plug, wait 10 minutes before touching any of the high-voltage connectors and terminals. * ECM detects a voltage of power source for sensor is excessively low or high. **NOTE: When the malfunction is detected, ECM enters fail-safe mode and the MIL illuminates.** * ECM stops the electric throttle control actuator control, throttle valve is maintained at a fixed opening (approx. 5 degrees) by the return spring.
DTC: P0700 **T TCM, MIL: Yes,TCIL: Yes** **Year:** 2011 **Model:** EX35, M56, QX56 **Engine:** 3.5L, 5.6L **Transmission:** All	**TCM:** This is an OBD-II self-diagnostic item. Diagnostic trouble code P0700 is detected when the TCM is malfunctioning.
DTC: P0705 **T TCM, TCIL: Yes** **Year:** 2011 **Model:** EX35, FX35, FX50, G37, M37, M56, QX56 **Engine:** 3.5L, 3.7L, 5.0L, 5.6L **Transmission:** All	**Transmission Range Switch A:** * Transmission range switch 1 – 4 signals input with impossible pattern. * "P" position is detected from "N" position without any other position being detected in between.

DTC	Trouble Code Title and Conditions
DTC: P0710 **T TCM, TCIL: Yes** **Year:** 2011 **Model:** FX35, FX50, G37, M37 **Engine:** 3.5L, 3.7L, 5.0L **Transmission:** All	**Transmission Fluid Temperature Sensor A Circuit:** * Set DTC when the A/T fluid temperature sensor is −40 °C (−40 °F) or less for 5 seconds while driving the vehicle at the vehicle speed 10 km/h (7 MPH) or more. * Set DTC when the A/T fluid temperature sensor is 180 °C (356 °F) or more for 5 seconds.
DTC: P0717 **T TCM, TCIL: Yes** **Year:** 2011 **Model:** FX35, FX50, G37, M37 **Engine:** 3.5L, 3.7L, 5.0L **Transmission:** All	**Input/Turbine Speed Sensor A Circuit No Signal:** * The revolution of input speed sensor 1 and/or 2 is 270 rpm or less. * This is an OBD-II self-diagnostic item. * Diagnostic trouble code "P0717" is detected under the following conditions: - When TCM does not receive the proper voltage signal from the sensor. - When TCM detects an irregularity only at position of 4GR for input speed sensor 2.
DTC: P0717 **1T TCM, TCIL: Yes** **Year:** 2011 **Model:** EX35, M56, QX56 **Engine:** 3.5L, 5.6L **Transmission:** All	**Input Speed Sensor A (Turbine Revolution Sensor):** The input speed sensor detects input shaft rpm (revolutions per minute). It is located on the input side of the automatic transmission. Monitors revolution of sensor 1 and sensor 2 for non-standard conditions. This is an OBD-II self-diagnostic item. Diagnostic trouble code P0717 is detected under the following conditions: - When TCM does not receive the proper voltage signal from the sensor. - When TCM detects an irregularity only at position of 4th gear for input speed sensor 2.
DTC: P0720 **T TCM, TCIL: Yes** **Year:** 2011 **Model:** EX35 **Engine:** 3.5L **Transmission:** All	**Output Speed Sensor Circuit:** * Signal from vehicle speed sensor CVT [output speed sensor (secondary speed sensor)] is not input due to open or short circuit. * An unexpected signal is input during running. * After ignition switch is turned ON, unexpected signal input from vehicle speed signal before the vehicle starts moving.
DTC: P0720 **1T TCM, TCIL: Yes** **Year:** 2011 **Model:** M56, QX56 **Engine:** 5.6L **Transmission:** All	**Vehicle Speed Sensor A/T (Revolution Sensor/Output Speed Sensor):** * This is an OBD-II self-diagnostic item. * Diagnostic trouble code P0720 is detected under the following conditions: - When TCM does not receive the proper voltage signal from the sensor. - After ignition switch is turned "ON", irregular signal input from vehicle speed sensor MTR before the vehicle starts moving.
DTC: P0720 **T TCM, TCIL: Yes** **Year:** 2011 **Model:** FX35, FX50, G37, M37 **Engine:** 3.5L, 3.7L, 5.0L **Transmission:** All	**Output Speed Sensor Circuit:** * The output speed sensor recognizes that the vehicle speed is 5 km/h (3 MPH) or less even if the vehicle speed signal recognizes that the vehicle speed is 20 km/h (12 MPH) or more. (Only when starts after the ignition switch is turned ON.) * The vehicle speed recognized by the output speed sensor decelerates 36 km/h (23 MPH) or more during 60 msec when the output speed sensor recognizes that the vehicle speed is 36 km/h (23 MPH) or more and the vehicle speed signal recognizes that the vehicle speed is 24 km/h (15 MPH) or less. * The vehicle speed of output speed sensor decelerates 36 km/h (23 MPH) or more even if the vehicle speed of vehicle speed signal accelerates or decelerates 24 km/h (15 MPH) or less during 60 msec when the output speed sensor recognizes that the vehicle speed is 36 km/h (23 MPH) or more.
DTC: P0725 **T TCM, TCIL: Yes** **Year:** 2011 **Model:** EX35, FX35, FX50, G37, M37 **Engine:** 3.5L, 3.7L, 5.0L **Transmission:** All	**Engine Speed Signal:** TCM does not receive proper voltage signal from ECM.
DTC: P0725 **1T TCM, TCIL: Yes** **Year:** 2011 **Model:** FX35, M56, QX56 **Engine:** 3.5L, 5.6L **Transmission:** All	**Engine Speed Signal:** * The engine speed signal is sent from the ECM to the TCM. * Diagnostic trouble code P0725 is detected when TCM does not receive the engine speed signal or ignition signal (input by CAN communication) from ECM. * The engine speed signal is sent with the engine running and should closely match the tachometer reading.
DTC: P0729 **T TCM, TCIL: Yes** **Year:** 2011 **Model:** FX35, FX50, G37, M37 **Engine:** 3.5L, 3.7L, 5.0L **Transmission:** All	**Gear 6 Incorrect Ratio:** The gear ratio is: 0.914 or more 0.813 or less

DTC	Trouble Code Title and Conditions
DTC: P0730 **T TCM, TCIL: Yes** **Year:** 2011 **Model:** FX35, FX50, G37, M37 **Engine:** 3.5L, 3.7L, 5.0L **Transmission:** All	**Incorrect Gear Ratio:** The revolution of under drive sun gear is 8,000 rpm or more. **NOTE: Not detected when in "P" or "N" position and during a shift to "P" or "N" position.**
DTC: P0731 **T TCM, TCIL: Yes** **Year:** 2011 **Model:** EX35, FX35, FX50, G37, M37, M56, QX56 **Engine:** 3.5L, 3.7L, 5.0L, 5.6L **Transmission:** All	**A/T 1st Gear Function:** * This is an OBD-II self-diagnostic item. * Diagnostic trouble code P0731 is detected when TCM detects any inconsistency in the actual gear ratio.
DTC: P0732 **T TCM, TCIL: Yes** **Year:** 2011 **Model:** EX35, FX35, FX50, G37, M37, M56, QX56 **Engine:** 3.5L, 3.7L, 5.0L, 5.6L **Transmission:** All	**A/T 2nd Gear Function:** * This malfunction is detected when the A/T does not shift into 2GR position as instructed by TCM. This is not only caused by electrical malfunction (circuits open or shorted) but mechanical malfunction such as control valve sticking, improper solenoid valve operation. * This is an OBD-II self-diagnostic item. * Diagnostic trouble code P0732 is detected when TCM detects any inconsistency in the actual gear ratio.
DTC: P0733 **T TCM, TCIL: Yes** **Year:** 2011 **Model:** EX35, FX35, FX50, G37, M37, M56, QX56 **Engine:** 3.5L, 3.7L, 5.0L, 5.6L **Transmission:** All	**A/T 3rd Gear Function:** * This malfunction is detected when the A/T does not shift into 3GR position as instructed by TCM. This is not only caused by electrical malfunction (circuits open or shorted) but mechanical malfunction such as control valve sticking, improper solenoid valve operation. * This is an OBD-II self-diagnostic item. * Diagnostic trouble code P0733 is detected when TCM detects any inconsistency in the actual gear ratio.
DTC: P0734 **T TCM, TCIL: Yes** **Year:** 2011 **Model:** EX35, FX35, FX50, G37, M37, M56, QX56 **Engine:** 3.5L, 3.7L, 5.0L, 5.6L **Transmission:** All	**A/T 4th Gear Function:** * This malfunction is detected when the A/T does not shift into 4GR position as instructed by TCM. This is not only caused by electrical malfunction (circuits open or shorted) but mechanical malfunction such as control valve sticking, improper solenoid valve operation. * This is an OBD-II self-diagnostic item. * P0734 is detected when TCM detects any inconsistency in the actual gear ratio.
DTC: P0735 **T TCM, TCIL: Yes** **Year:** 2011 **Model:** EX35, FX35, FX50, G37, M37, M56, QX56 **Engine:** 3.5L, 3.7L, 5.0L, 5.6L **Transmission:** All	**A/T 5th Gear Function:** * This malfunction is detected when the A/T does not shift into 5GR position as instructed by TCM. This is not only caused by electrical malfunction (circuits open or shorted) but mechanical malfunction such as control valve sticking, improper solenoid valve operation. * This is an OBD-II self-diagnostic item. * Diagnostic trouble code P0735 is detected when TCM detects any inconsistency in the actual gear ratio.
DTC: P0740 **T TCM, TCIL: Yes** **Year:** 2011 **Model:** FX35, FX50, G37, M37 **Engine:** 3.5L, 3.7L, 5.0L **Transmission:** All	**Torque Converter Clutch Circuit - Open:** The torque converter clutch solenoid valve monitor value is 0.4 A or less when the torque converter clutch solenoid valve command value is more than 0.75 A.
DTC: P0740 **1T TCM, TCIL: Yes** **Year:** 2011 **Model:** EX35, FX35, M56, QX56 **Engine:** 3.5L, 5.6L **Transmission:** All	**Torque Converter Clutch Solenoid Valve:** * Diagnostic trouble code P0740 is detected under the following conditions: - TCM detects an improper voltage drop when it tries to operate the solenoid valve. - When TCM detects as irregular by comparing target value with monitor value.

DTC	Trouble Code Title and Conditions
DTC: P0744 **T TCM, TCIL: Yes** **Year:** 2011 **Model:** EX35, FX35, FX50, G37, M37, M56, QX56 **Engine:** 3.5L, 3.7L, 5.0L, 5.6L **Transmission:** All	**A/T TCC S/V Function (Lock-Up):** * This malfunction is detected when the A/T does not lock-up or does not shift to 5th gear. This is not only caused by electrical malfunction (circuits open or shorted) but also by mechanical malfunction such as control valve sticking, improper solenoid valve operation, etc. * This is an OBD-II self-diagnostic item. * Diagnostic trouble code P0744 is detected under the following conditions: - When A/T cannot perform lock-up even if electrical circuit is good. - When TCM detects as irregular by comparing difference value with slip rotation.
DTC: P0745 **T TCM, TCIL: Yes** **Year:** 2011 **Model:** FX35, G37, M37 **Engine:** 3.5L, 3.7L **Transmission:** All	**Pressure Control Solenoid A:** The line pressure solenoid valve monitor value is 0.4 A or less when the line pressure solenoid valve command value is more than 0.75 A.
DTC: P0745 **T TCM, TCIL: Yes** **Year:** 2011 **Model:** M56, QX56 **Engine:** 5.6L **Transmission:** All	**Pressure Control Solenoid A:** * The line pressure solenoid valve regulates the oil pump discharge pressure to suit the driving condition in response to a signal sent from the TCM. * This is an OBD-II self-diagnostic item. * Diagnostic trouble code P0745 is detected under the following conditions: - When TCM detects an improper voltage drop when it tries to operate the solenoid valve. - When TCM detects as irregular by comparing target value with monitor value.
DTC: P0750 **T TCM, TCIL: Yes** **Year:** 2011 **Model:** FX35, FX50, G37, M37 **Engine:** 3.5L, 3.7L, 5.0L **Transmission:** All	**Shift Solenoid A:** * The anti-interlock solenoid valve monitor value is ON when the anti-interlock solenoid valve command value is OFF. * The anti-interlock solenoid valve monitor value is OFF when the anti-interlock solenoid valve command value is ON.
DTC: P0775 **T TCM, TCIL: Yes** **Year:** 2011 **Model:** FX35, FX50, G37, M37 **Engine:** 3.5L, 3.7L, 5.0L **Transmission:** All	**Pressure Control Solenoid B:** The input clutch solenoid valve monitor value is 0.4 A or less when the input clutch solenoid valve command value is more than 0.75 A.
DTC: P0780 **T TCM, TCIL: Yes** **Year:** 2011 **Model:** FX35, FX50, G37, M37 **Engine:** 3.5L, 3.7L, 5.0L **Transmission:** All	**Shift Error:** * When shifting from 3GR to 4GR with the selector lever in "D" position, the gear ratio does not shift to 1.412 (gear ratio of 4GR). * When shifting from 5GR to 6GR or 6GR to 7GR, the engine speed exceeds the prescribed speed. * The shift change time from 4GR to 3GR is 0.2 second or less.
DTC: P0795 **T TCM, TCIL: Yes** **Year:** 2011 **Model:** FX35, FX50, G37, M37 **Engine:** 3.5L, 3.7L, 5.0L **Transmission:** All	**Pressure Control Solenoid C:** The front brake solenoid valve monitor value is 0.4 A or less when the front brake solenoid valve command value is more than 0.75 A.
DTC: P0850 **1T ECM, MIL: Yes** **Year:** 2011 **Model:** EX35, FX35, FX50, G37, M37, M56, QX56 **Engine:** 3.5L, 3.7L, 5.0L, 5.6L **Transmission:** All	**Park/Neutral Position Switch:** * CVT & M/T: When the shift lever position is P or N (CVT), Neutral (M/T), park/neutral position (PNP) switch is ON. ECM detects the position because the continuity of the line (the ON signal) exists. * A/T: The signal of the park/neutral position (PNP) switch does not change in the process of engine starting and driving.

OBD II Trouble Code List (P1XXX Codes)

DTC	Trouble Code Title and Conditions
DTC: P100A **1T ECM, MIL: Yes** **Year:** 2011 **Model:** FX50, G37, M37 **Engine:** 3.7L, 5.0L **Transmission:** All	**Variable Valve Event & Lift (VVEL) Response Malfunction (Bank 1):** **NOTE: If DTC P100A or P100B is displayed with DTC P1090 or P1093, first perform the trouble diagnosis for DTC P1090 or P1093.** Actual event response to target is poor.
DTC: P100B **1T ECM, MIL: Yes** **Year:** 2011 **Model:** FX50, G37, M37 **Engine:** 3.7L, 5.0L **Transmission:** All	**Variable Valve Event & Lift (VVEL) Response Malfunction (Bank 2):** **NOTE: If DTC P100A or P100B is displayed with DTC P1090 or P1093, first perform the trouble diagnosis for DTC P1090 or P1093.** Actual event response to target is poor.
DTC: P1078 **1T ECM, MIL: Yes** **Year:** 2011 **Model:** EX35, FX35, FX50 **Engine:** 3.5L, 5.0L **Transmission:** All	**Exhaust Valve Timing Control Position Sensor (Bank 1) Circuit:** **NOTE: If this DTC is displayed with DTC P0643, first perform the trouble diagnosis for DTC P0643.** * An excessively high or low voltage from the sensor is sent to ECM.
DTC: P1084 **1T ECM, MIL: Yes** **Year:** 2011 **Model:** EX35, FX35 **Engine:** 3.5L **Transmission:** All	**Exhaust Valve Timing Control Position Sensor (Bank 2) Circuit:** **NOTE: If this DTC is displayed with DTC P0643, first perform the trouble diagnosis for DTC P0643.** * An excessively high or low voltage from the sensor is sent to ECM.
DTC: P1087 **T ECM, MIL: Yes** **Year:** 2011 **Model:** FX50, G37, M37 **Engine:** 3.7L, 5.0L **Transmission:** All	**Variable Valve Event & Lift (VVEL) Small Event Angle Malfunction (Bank 1):** **NOTE: If DTC P1087 or P1088 is displayed with DTC P1090 or P1093, perform the diagnosis for P1090 or P1093 first.** The event angle of VVEL control shaft is always small.
DTC: P1088 **T ECM, MIL: Yes** **Year:** 2011 **Model:** FX50, G37, M37 **Engine:** 3.7L, 5.0L **Transmission:** All	**Variable Valve Event & Lift (VVEL) Small Event Angle Malfunction (Bank 2):** **NOTE: If DTC P1087 or P1088 is displayed with DTC P1090 or P1093, perform the diagnosis for P1090 or P1093 first.** The event angle of VVEL control shaft is always small.
DTC: P1089 **T , MIL: Yes** **Year:** 2011 **Model:** FX50, G37, M37 **Engine:** 3.7L, 5.0L **Transmission:** All	**Variable Valve Event & Lift (VVEL) Control Shaft Position Sensor (Bank 1) Circuit:** **NOTE: If DTC P1089 or P1092 is displayed with DTC P1608, first perform the trouble diagnosis for DTC P1608.** An excessively low voltage from the sensor is sent to VVEL control module. An excessively high voltage from the sensor is sent to VVEL control module. Rationally incorrect voltage is sent to VVEL control module compared with the signals from VVEL control shaft position sensor 1 and VVEL control shaft position sensor 2.
DTC: P1090 **T , MIL: Yes** **Year:** 2011 **Model:** FX50, G37, M37 **Engine:** 3.7L, 5.0L **Transmission:** All	**Variable Valve Event & Lift (VVEL) System Performance (Bank 1) :** **NOTE: If DTC P1090 or P1093 is displayed with DTC P1091, first perform the trouble diagnosis for DTC P1091.** Event angle difference between the actual and the target is detected. Abnormal current is sent to VVEL actuator motor.
DTC: P1091 **T ECM, MIL: Yes** **Year:** 2011 **Model:** FX50, G37, M37 **Engine:** 3.7L, 5.0L **Transmission:** All	**Variable Valve Event & Lift (VVEL) Actuator Motor Relay Circuit:** VVEL control module detects the VVEL actuator motor relay is stuck OFF. VVEL control module detects the VVEL actuator motor relay is stuck ON.

DTC	Trouble Code Title and Conditions
DTC: P1092 **T ECM, MIL: Yes** **Year:** 2011 **Model:** FX50, G37, M37 **Engine:** 3.7L, 5.0L **Transmission:** All	**Variable Valve Event & Lift (VVEL) Control Shaft Position Sensor (Bank 2) Circuit:** NOTE: If DTC P1089 or P1092 is displayed with DTC P1608, first perform the trouble diagnosis for DTC P1608. An excessively low voltage from the sensor is sent to VVEL control module. An excessively high voltage from the sensor is sent to VVEL control module. Rationally incorrect voltage is sent to VVEL control module compared with the signals from VVEL control shaft position sensor 1 and VVEL control shaft position sensor 2.
DTC: P1093 **T ECM, MIL: Yes** **Year:** 2011 **Model:** FX50, G37, M37 **Engine:** 3.7L, 5.0L **Transmission:** All	**Variable Valve Event & Lift (VVEL) System Performance (Bank 2) :** NOTE: If DTC P1090 or P1093 is displayed with DTC P1091, first perform the trouble diagnosis for DTC P1091 Event angle difference between the actual and the target is detected. Abnormal current is sent to VVEL actuator motor.
DTC: P1140 **1T ECM, MIL: Yes** **Year:** 2011 **Model:** M56, QX56 **Engine:** 5.6L **Transmission:** All	**Intake Valve Timing Control Position Sensor Circuit (Bank 1):** An excessively high or low voltage from the sensor is sent to ECM.
DTC: P1145 **1T ECM, MIL: Yes** **Year:** 2011 **Model:** M56, QX56 **Engine:** 5.6L **Transmission:** All	**Intake Valve Timing Control Position Sensor Circuit (Bank 2):** An excessively high or low voltage from the sensor is sent to ECM.
DTC: P1148 **1T ECM, MIL: Yes** **Year:** 2011 **Model:** EX35, FX35, FX50, G37, M37, M56, QX56 **Engine:** 3.5L, 3.7L, 5.0L, 5.6L **Transmission:** All	**Closed Loop Control Function (Includes Hybrid Models):** HYBRID MODELS CAUTION: Hybrid systems use very high-voltage battery systems. Before starting any service work involving the battery system, turn the ignition switch OFF and then remove the service plug from pocket in the trunk. After removing the service plug, wait 10 minutes before touching any of the high-voltage connectors and NOTE: On and engines, this applies to Bank 1, except 3.7L which is Bank 2. NOTE: DTC P1148 or P1168 is displayed with another DTC for A/F sensor 1. Perform the trouble diagnosis for the corresponding DTC. * The closed loop control function for bank 1 does not operate even when vehicle is being driven in the specified condition.
DTC: P1168 **1T ECM, MIL: Yes** **Year:** 2011 **Model:** EX35, FX35, FX50, G37, M37, M56, QX56 **Engine:** 3.5L, 3.7L, 5.0L, 5.6L **Transmission:** All	**Closed Loop Control Function (Bank 2):** NOTE: If DTC P1148 or P1168 is displayed with another DTC for air fuel ratio (A/F) sensor 2. Perform the trouble diagnosis for the corresponding DTC. * The closed loop control function for bank 2 does not operate even when vehicle is being driven in the specified condition.
DTC: P1211 **1T ECM** **Year:** 2011 **Model:** EX35, FX35, FX50, G37, M37, M56, QX56 **Engine:** 3.5L, 3.7L, 5.0L, 5.6L **Transmission:** All	**TCS Control Unit:** * Freeze frame data is not stored in the ECM for this self-diagnosis. * The MIL will not illuminate for this self-diagnosis. * ECM receives malfunction information from "ABS actuator and electric unit (control unit)".
DTC: P1212 **1T ECM, MIL: Yes** **Year:** 2011 **Model:** EX35, FX35, FX50, G37, M37, M56, QX56 **Engine:** 3.5L, 3.7L, 5.0L, 5.6L **Transmission:** All	**TCS Communication Line:** NOTE: If DTC P1212 is displayed with DTC UXXXX, first perform the trouble diagnosis for DTC UXXXX. NOTE: If DTC P1212 is displayed with DTC P0607, first perform the trouble diagnosis for DTC P0607. NOTE: Be sure to erase the malfunction information such as DTC not only for "ABS actuator and electric unit (control unit)" but also for ECM after TCS related repair. * Freeze frame data is not stored in the ECM for this self-diagnosis. * The MIL will not illuminate for this self-diagnosis. * ECM cannot receive the information from "ABS actuator and electric unit (control unit)".

DTC	Trouble Code Title and Conditions
DTC: P1217 **1T ECM, MIL: Yes** **Year:** 2011 **Model:** EX35, FX35, FX50, G37, M37, M56, QX56 **Engine:** 3.5L, 3.7L, 5.0L, 5.6L **Transmission:** All	**Engine Over Temperature (Overheat) (Includes Hybrid Models):** HYBRID MODELS CAUTION: Hybrid systems use very high-voltage battery systems. Before starting any service work involving the battery system, turn the ignition switch OFF and then remove the service plug from pocket in the trunk. After removing the service plug, wait 10 minutes before touching any of the high-voltage connectors and terminals. **NOTE: If DTC P1217 is displayed with DTC UXXXX, first perform the trouble diagnosis for DTC UXXXX.** **NOTE: If DTC P1217 is displayed with DTC P0607, first perform the trouble diagnosis for DTC P0607.** * The ECM controls cooling fan relays through CAN communication line. * Cooling fan does not operate properly (overheat). * Cooling fan system does not operate properly (overheat). * Engine coolant was not added to the system using the proper filling method. * Engine coolant is not within the specified range.
DTC: P1220 **1T ECM, MIL: Yes** **Year:** 2011 **Model:** FX50 **Engine:** 5.0L **Transmission:** All	**Fuel Pump Control Module (FPCM):** * An improper voltage signal from the FPCM, which is supplied to a point between the fuel pump and the dropping resistor, is detected by ECM. * During engine cranking, the signal voltage of the FPCM to the ECM is too low.
DTC: P1225 **1T ECM, MIL: Yes** **Year:** 2011 **Model:** EX35, FX35, FX50, G37, M37, M56, QX56 **Engine:** 3.5L, 3.7L, 5.0L, 5.6L **Transmission:** All	**Closed Throttle Position Learning Performance (Includes Hybrid Models):** HYBRID MODELS CAUTION: Hybrid systems use very high-voltage battery systems. Before starting any service work involving the battery system, turn the ignition switch OFF and then remove the service plug from pocket in the trunk. After removing the service plug, wait 10 minutes before touching any of the high-voltage connectors and terminals. **NOTE: For and, this DTC is for Bank 1.** * Closed throttle position learning value is excessively low.
DTC: P1226 **1T ECM, MIL: Yes** **Year:** 2011 **Model:** EX35, FX35, FX50, G37, M37, M56, QX56 **Engine:** 3.5L, 3.7L, 5.0L, 5.6L **Transmission:** All	**Closed Throttle Position Learning Performance (Includes Hybrid Models):** HYBRID MODELS CAUTION: Hybrid systems use very high-voltage battery systems. Before starting any service work involving the battery system, turn the ignition switch OFF and then remove the service plug from pocket in the trunk. After removing the service plug, wait 10 minutes before touching any of the high-voltage connectors and terminals. **NOTE: On abd, this DTC is for Bank 1.** * Closed throttle position learning is not performed successfully, repeatedly.
DTC: P1233 **T ECM, MIL: Yes** **Year:** 2011 **Model:** EX35, FX35, FX50, G37, M37 **Engine:** 3.5L, 3.7L, 5.0L **Transmission:** All	**Electric Throttle Control Performance (Bank 2):** **NOTE: If DTC P1233 or P2101 is displayed with DTC P1238, P1290, P2100 or 2119, first perform the trouble diagnosis for DTC P1238, P2119 or P1290, P2100.** Electric throttle control function does not operate properly
DTC: P1234 **1T ECM, MIL: Yes** **Year:** 2011 **Model:** EX35, FX35, FX50, G37, M37 **Engine:** 3.5L, 3.7L, 5.0L **Transmission:** All	**Closed Throttle Position Learning Performance (Bank 2):** Closed throttle position learning value is excessively low.
DTC: P1235 **1T ECM, MIL: Yes** **Year:** 2011 **Model:** EX35, FX35, FX50, G37, M37 **Engine:** 3.5L, 3.7L, 5.0L **Transmission:** All	**Closed Throttle Position Learning Performance (Bank 2):** Closed throttle position learning is not performed successfully, repeatedly.
DTC: P1236 **T ECM, MIL: Yes** **Year:** 2011 **Model:** EX35, FX35, FX50, G37, M37 **Engine:** 3.5L, 3.7L, 5.0L **Transmission:** All	**Throttle Control Motor (Bank 2) Circuit Short:** ECM detects short in both circuits between ECM and throttle control motor.

DTC	Trouble Code Title and Conditions
DTC: P1238 **T ECM, MIL: Yes** **Year:** 2011 **Model:** EX35, FX35, FX50, G37, M37 **Engine:** 3.5L, 3.7L, 5.0L **Transmission:** All	**Electrical Throttle Control Actuator (Bank 2):** Condition A: Electric throttle control actuator does not function properly due to the return spring malfunction. Condition B: Throttle valve opening angle in fail-safe mode is not in specified range. Condition CL ECM detect the throttle valve is stuck open.
DTC: P1239 **T ECM, MIL: Yes** **Year:** 2011 **Model:** EX35, FX35, FX50, G37, M37 **Engine:** 3.5L, 3.7L, 5.0L **Transmission:** All	**Throttle Position Sensor (Bank 2) Circuit Range/Performance:** Rationally incorrect voltage is sent to ECM compared with the signals from TP sensor 1 and TP sensor 2.
DTC: P1290 **T ECM, MIL: Yes** **Year:** 2011 **Model:** EX35, FX35, FX50, G37, M37 **Engine:** 3.5L, 3.7L, 5.0L **Transmission:** All	**Throttle Control Motor Relay Circuit Open (Bank 2):** ECM detects a voltage of power source for throttle control motor is excessively low.
DTC: P1421 **1T ECM, MIL: Yes** **Year:** 2011 **Model:** EX35, FX35, FX50, G37, M37, M56, QX56 **Engine:** 3.5L, 3.7L, 5.0L, 5.6L **Transmission:** All	**Cold Start Emission Reduction Strategy Monitoriing (Includes Hybrid Models):** HYBRID MODELS CAUTION: Hybrid systems use very high-voltage battery systems. Before starting any service work involving the battery system, turn the ignition switch OFF and then remove the service plug from pocket in the trunk. After removing the service plug, wait 10 minutes before touching any of the high-voltage connectors and terminals. **NOTE: If DTC P1421 is displayed with other DTC, first perform the trouble diagnosis for other DTC.** * ECM does not control ignition timing and engine idle speed properly when engine is started with pre-warming up condition.
DTC: P1446 **2T EVAP, MIL: Yes** **Year:** 2011 **Model:** M56, QX56 **Engine:** 5.6L **Transmission:** All	**EVAP Canister Vent Valve Malfunction (Closed):** EVAP canister vent control valve remains closed under specified driving conditions.
DTC: P1550 **1T ECM, MIL: Yes** **Year:** 2011 **Model:** EX35, FX35, FX50, G37, M37, M56, QX56 **Engine:** 3.5L, 3.7L, 5.0L, 5.6L **Transmission:** All	**Battery Current Sensor Circuit Range/Performance:** * The MIL will not illuminate for this diagnosis. **NOTE: If DTC P1550 is displayed with DTC P0643, first perform the trouble diagnosis for DTC P0643.** * The output voltage of the battery current sensor remains within the specified range while engine is running.
DTC: P1551 **1T ECM, MIL: Yes** **Year:** 2011 **Model:** EX35, FX35, FX50, G37, M37, M56, QX56 **Engine:** 3.5L, 3.7L, 5.0L, 5.6L **Transmission:** All	**Battery Current Sensor Circuit Low Input:** * The MIL will not illuminate for this diagnosis. **NOTE: If DTC P1551 or P1552 is displayed with DTC P0643, first perform the trouble diagnosis for DTC P0643.** * An excessively low voltage from the sensor is sent to ECM.
DTC: P1552 **1T ECM, MIL: Yes** **Year:** 2011 **Model:** EX35, FX35, FX50, G37, M37, M56, QX56 **Engine:** 3.5L, 3.7L, 5.0L, 5.6L **Transmission:** All	**Battery Current Sensor Circuit High Input:** * The MIL will not illuminate for this diagnosis. **NOTE: If DTC P1551 or P1552 is displayed with DTC P0643, first perform the trouble diagnosis for DTC P0643.** * An excessively high voltage from the sensor is sent to ECM.

DTC	Trouble Code Title and Conditions
DTC: P1553 **1T ECM, MIL: Yes** **Year:** 2011 **Model:** EX35, FX35, FX50, G37, M37, M56, QX56 **Engine:** 3.5L, 3.7L, 5.0L, 5.6L **Transmission:** All	**Battery Current Sensor Performance:** * The MIL will not illuminate for this diagnosis. **NOTE: If DTC P1553 is displayed with DTC P0643, first perform the trouble diagnosis for DTC P0643.** * The signal voltage transmitted from the sensor to ECM is higher than the amount of the maximum power generation.
DTC: P1554 **1T ECM, MIL: Yes** **Year:** 2011 **Model:** EX35, FX35, FX50, G37, M37, M56, QX56 **Engine:** 3.5L, 3.7L, 5.0L, 5.6L **Transmission:** All	**Battery Current Sensor Performance:** * The MIL will not illuminate for this diagnosis. **NOTE: If DTC P1554 is displayed with DTC P0643, first perform the trouble diagnosis for DTC P0643.** * The output voltage of the battery current sensor is lower than the specified value while the battery voltage is high enough.
DTC: P1564 **1T ECM** **Year:** 2011 **Model:** EX35, FX35, FX50, G37, M37, M56, QX56 **Engine:** 3.5L, 3.7L, 5.0L, 5.6L **Transmission:** All	**ASCD Steering Switch Malfunction (Includes Hybrid Models):** HYBRID MODELS CAUTION: Hybrid systems use very high-voltage battery systems. Before starting any service work involving the battery system, turn the ignition switch OFF and then remove the service plug from pocket in the trunk. After removing the service plug, wait 10 minutes before touching any of the high-voltage connectors and terminals. * This self-diagnosis has the one trip detection logic. * The MIL will not illuminate for this self-diagnosis. **NOTE: If DTC P1564 is displayed with DTC P0605, first perform the trouble diagnosis for DTC P0605.** * An excessively high voltage signal from the ASCD steering switch is sent to ECM. * ECM detects that input signal from the ASCD steering switch is out of the specified range. * ECM detects that the ASCD steering switch is stuck ON.
DTC: P1564 **T ECM, MIL: Yes** **Year:** 2011 **Model:** EX35, FX35, FX50, G37, M37, M56, QX56 **Engine:** 3.5L, 3.7L, 5.0L, 5.6L **Transmission:** All	**ICC Steering Switch:** **NOTE: If DTC P1564 is displayed with DTC P0605, first perform the trouble diagnosis for DTC P0605.** * An excessively high voltage signal from the ICC steering switch is sent to ECM. * ECM detects that input signal from the ICC steering switch is out of the specified range. * ECM detects that the ICC steering switch is stuck ON.
DTC: P1568 **T ECM, MIL: Yes** **Year:** 2011 **Model:** EX35, FX35, FX50, G37, M37, M56, QX56 **Engine:** 3.5L, 3.7L, 5.0L, 5.6L **Transmission:** All	**ICC Function:** **NOTE: If DTC P1568 is displayed with DTC UXXXX, first perform the trouble diagnosis for DTC UXXXX.** - If this DTC is displayed with DTC P0605 or P0607, first perform the trouble diagnosis for DTC P0605 or P0607. * ECM detects a difference between signals from ICC sensor integrated unit is out of specified range.
DTC: P1568 **1T CCM** **Year:** 2011 **Model:** FX35 **Engine:** 3.5L **Transmission:** All	**ASCD Command Valve Circuit Malfunction:** If DTC P1568 is displayed with DTC U1000 and/or U1001, diagnose the cause of the DTC U1000 and/or U1001 first. If DTC P1568 is displayed with DTC P0605, diagnose the cause of DTC P0605 first. Engine started, and PCM detected that the signals from the command valve unit (ICC unit) were out of range.
DTC: P1572 **1T ECM, MIL: Yes** **Year:** 2011 **Model:** EX35, FX35, FX50, G37, M37 **Engine:** 3.5L, 3.7L, 5.0L **Transmission:** All	**ACSD Brake Switch Malfunction (Includes Hybrid Models):** HYBRID MODELS CAUTION: Hybrid systems use very high-voltage battery systems. Before starting any service work involving the battery system, turn the ignition switch OFF and then remove the service plug from pocket in the trunk. After removing the service plug, wait 10 minutes before touching any of the high-voltage connectors and terminals. * This self-diagnosis has the one trip detection logic. * The MIL will not illuminate for this self-diagnosis. **NOTE: If DTC P1572 is displayed with DTC P0605, first perform the trouble diagnosis for DTC P0605.** * This self-diagnosis has the one trip detection logic. **NOTE: When malfunction A is detected, the DTC is not stored in ECM memory. And in that case, 1st trip DTC and 1st trip freeze frame data are displayed. 1st trip DTC is erased when ignition switch is turned OFF. And even when Malfunction A is detected in two consecutive trips, DTC is not stored in ECM memory.** * Malfunction A: When the vehicle speed is above 19 MPH, ON signals from the stop lamp switch and the ASCD brake switch are sent to ECM at the same time. * Malfunction B: ASCD brake switch signal is not sent to ECM for extremely long time while the vehicle is being driven.

DTC	Trouble Code Title and Conditions
DTC: P1572 **1T ECM, MIL: Yes** **Year:** 2011 **Model:** EX35, FX35, FX50, G37, M37, M56, QX56 **Engine:** 3.5L, 3.7L, 5.0L, 5.6L **Transmission:** All	**ICC Brake Switch:** **NOTE: If DTC P1572 is displayed with DTC P0605, first perform the trouble diagnosis for DTC P0605.** * This self-diagnosis has the one trip detection logic. When malfunction A is detected, DTC is not stored in ECM memory. And in that case, 1st trip DTC and 1st trip freeze frame data are displayed. 1st trip DTC is erased when ignition switch OFF. And even when malfunction A is detected in two consecutive trips, DTC is not stored in ECM memory. * Condition A: ON signals from the stop lamp switch and the ICC brake switch are sent to ECM at the same time. * Condition B: ICC brake switch signal is not sent to ECM for extremely long time while the vehicle is driving.
DTC: P1574 **1T ECM, MIL: Yes** **Year:** 2011 **Model:** EX35, FX35, FX50, G37, M37, M56, QX56 **Engine:** 3.5L, 3.7L, 5.0L, 5.6L **Transmission:** All	**ASCD Vehicle Speed Sensor Malfunction (Includes Hybrid Models):** HYBRID MODELS CAUTION: Hybrid systems use very high-voltage battery systems. Before starting any service work involving the battery system, turn the ignition switch OFF and then remove the service plug from pocket in the trunk. After removing the service plug, wait 10 minutes before touching any of the high-voltage connectors and terminals. * The MIL will not illuminate for this self-diagnosis. **NOTE: If DTC P1574 is displayed with DTC UXXXX, first perform the trouble diagnosis for DTC UXXXX.** **NOTE: If DTC P1574 is displayed with DTC P0500, P0605 and/or P0607, first perform the trouble diagnosis for these DTCs before continuing with DTC P1574.** * ECM detects a difference between two vehicle speed signals is out of the specified range.
DTC: P1574 **T ECM, MIL: Yes** **Year:** 2011 **Model:** EX35, FX35, FX50, G37, M37, M56, QX56 **Engine:** 3.5L, 3.7L, 5.0L, 5.6L **Transmission:** All	**ICC Vehicle Speed Sensor:** **NOTE: If DTC P1574 is displayed with DTC UXXXX, first perform the trouble diagnosis for DTC UXXXX.** **NOTE: If this DTC is displayed with DTC P0500, P0605 or P0607, first perform the trouble diagnosis for these other DTC(s) first.** * ECM detects a difference between two vehicle speed signals is out of the specified range.
DTC: P1606 **T ECM, MIL: Yes** **Year:** 2011 **Model:** FX50, G37, M37 **Engine:** 3.7L, 5.0L **Transmission:** All	**Variable Valve Event & Lift (VVEL) Control Module:** VVEL control module calculation function is malfunctioning. VVEL EEPROM system is malfunctioning.
DTC: P1607 **T ECM, MIL: Yes** **Year:** 2011 **Model:** FX50, G37, M37 **Engine:** 3.7L, 5.0L **Transmission:** All	**Variable Valve Event & Lift (VVEL) Control Module Circuit:** The internal circuit of the VVEL control module is malfunctioning.
DTC: P1608 **T ECM, MIL: Yes** **Year:** 2011 **Model:** FX50, G37, M37 **Engine:** 3.7L, 5.0L **Transmission:** All	**Variable Valve Event & Lift (VVEL) Sensor Power Supply Circuit:** VVEL control module detects a voltage of power source for sensor is excessively low or high.
DTC: P1610 **T BCM, MIL: Yes** **Year:** 2011 **Model:** EX35, FX35, FX50, G37, M37, M56, QX56 **Engine:** 3.5L, 3.7L, 5.0L, 5.6L **Transmission:** All	**Lock Mode:** * When the starting operation is carried out five or more times consecutively under the following conditions:
DTC: P1611 **T BCM** **Year:** 2011 **Model:** EX35, FX35, FX50, M56, QX56 **Engine:** 3.5L, 5.0L, 5.6L **Transmission:** All	**ID Discord – IMMU-ECM:** The ID verification results between BCM and ECM are NG. Registration is necessary. **NOTE: P1611 has the same meaning as B2192.**

DTC	Trouble Code Title and Conditions
DTC: P1612 **T BCM** **Year:** 2011 **Model:** EX35, FX35, FX50, G37, M37, M56, QX56 **Engine:** 3.5L, 3.7L, 5.0L, 5.6L **Transmission:** All	**Chain of ECM-IMMU:** NOTE: P1612 has the same meaning as B2193. Inactive communication between ECM and BCM **NOTE: If DTC P1612 is displayed with DTC U1000 (for BCM), first perform the trouble diagnosis for DTC U1000**
DTC: P1614 **T BCM** **Year:** 2011 **Model:** EX35, FX35, FX50, G37, M37, M56, QX56 **Engine:** 3.5L, 3.7L, 5.0L, 5.6L **Transmission:** All	**Chain of IMMU-ECM:** Inactive communication between key slot or NATS antenna amp. and BCM. Mechanical key is malfunctioning.
DTC: P1615 **T BCM** **Year:** 2011 **Model:** EX35, FX35, FX50, G37, M37, M56, QX56 **Engine:** 3.5L, 3.7L, 5.0L, 5.6L **Transmission:** All	**Difference of Key:** The ID verification results between BCM and Intelligent Key or Mechanical Key are NG. Registration is necessary.
DTC: P1705 **T TCM, TCIL: Yes** **Year:** 2011 **Model:** FX35, FX50, G37, M37, M56, QX56 **Engine:** 3.5L, 3.7L, 5.0L, 5.6L **Transmission:** All	**Accelerator Pedal Position (APP) Sensor Signal Circuit:** TCM detects improper accelerator pedal position signals received from ECM via CAN communication.
DTC: P1705 **1T TCM, TCIL: Yes** **Year:** 2011 **Model:** EX35 **Engine:** 3.5L **Transmission:** All	**Throttle Position/Accelerator Pedal Position Sensor Circuit:** Electric throttle control actuator consists of throttle control motor, accelerator pedal position sensor, throttle position sensor etc. The actuator sends a signal to the ECM, and ECM sends the signal to TCM with CAN communication. * This is not an OBD-II self-diagnostic item. * Diagnostic trouble code P1705 is detected when TCM does not receive the proper accelerator pedal position signals (input by CAN communication) from ECM.
DTC: P1710 **T TCM, TCIL: Yes** **Year:** 2011 **Model:** EX35, M56, QX56 **Engine:** 3.5L, 5.6L **Transmission:** All	**A/T Fluid Temperature Sensor Circuit:** * This is an OBD-II self-diagnostic item. * Diagnostic trouble code P1710 will be detected when TCM receives an excessively low or high voltage from the sensor. * A/T fluid temperature does not rise to the specified temperature while driving.
DTC: P1715 **1T TCM** **Year:** 2011 **Model:** EX35, FX50 **Engine:** 3.5L, 5.0L **Transmission:** All	**Input Speed Sensor (Primary Speed Sensor/TCM Output):** * The MIL will not illuminate for this self-diagnosis. **NOTE: If DTC P1715 is displayed with DTC UXXXX, first perform the trouble diagnosis for DTC UXXXX.** **NOTE: If DTC P1715 is displayed with DTC P0335, P0340, P0605 and/or P0607, first perform the trouble diagnosis for the appropriate DTC before proceeding with P1715 diagnosis.** * Sensor signal is different from the theoretical value calculated by ECM from secondary speed sensor signal and engine rpm signal.
DTC: P1721 **T TCM, TCIL: Yes** **Year:** 2011 **Model:** FX35, FX50, G37, M37 **Engine:** 3.5L, 3.7L, 5.0L **Transmission:** All	**Vehicle Speed Signal Circuit:** * The vehicle speed signal recognizes that the vehicle speed is 5 km/h (3 MPH) or less even if the output speed sensor recognizes that the vehicle speed is 20 km/h (12 MPH) or more. (Only when starts after the ignition switch is turned ON.) * The vehicle speed recognized by the vehicle speed signal decelerates 36 km/h (23 MPH) or more during 60 msec when the vehicle speed signal recognizes that the vehicle speed is 36 km/h (23 MPH) or more and the output speed sensor recognizes that the vehicle speed is 24 km/h (15 MPH) or less. * The vehicle speed of vehicle speed signal decelerates 36 km/h (23 MPH) or more even if the vehicle speed of output speed sensor accelerates or decelerates 24 km/h (15 MPH) or less during 60 msec when the vehicle speed sensor recognizes that the vehicle speed is 36 km/h (23 MPH) or more.

DTC	Trouble Code Title and Conditions
DTC: P1721 **T TCM, TCIL:** Yes **Year:** 2011 **Model:** M56, QX56 **Engine:** 5.6L **Transmission:** All	**Vehicle Speed Signal:** * This is not an OBD-II self-diagnostic item. * Diagnostic trouble code P1721 is detected when TCM does not receive the proper vehicle speed sensor MTR signal (input by CAN communication) from combination meter (unified meter and A/C amp).
DTC: P1721 **T TCM, TCIL:** Yes **Year:** 2011 **Model:** EX35 **Engine:** 3.5L **Transmission:** All	**Vehicle Speed Signal:** * Signal (CAN communication) from vehicle speed signal not input due to cut line or the like. * Unexpected signal input during running.
DTC: P1730 **T TCM, TCIL:** Yes **Year:** 2011 **Model:** FX35, FX50, G37, M37 **Engine:** 3.5L, 3.7L, 5.0L **Transmission:** All	**Interlock:** The output sensor detects the deceleration of 12 km/h (7 MPH) or more for 1 second.
DTC: P1730 **T TCM, TCIL:** Yes **Year:** 2011 **Model:** EX35 **Engine:** 3.5L **Transmission:** All	**A/T Interlock:** * This is an OBD-II self-diagnostic item. * Diagnostic trouble code P1730 is detected when TCM does not receive the proper voltage signal from the sensor and switch. * TCM monitors and compares gear position and conditions of each ATF pressure switch when gear is steady.
DTC: P1730 **2T PCM, MIL:** Yes, **TCIL:** Yes **Year:** 2011 **Model:** M56, QX56 **Engine:** 5.6L **Transmission:** All	**Problem in Shift Control System:** With the engine running and in Drive position allow transmission to shift to 5th gear. Shift solenoid A or D stuck OFF, Shift solenoid B stuck ON, Shift Valves A, B or D stuck.
DTC: P1731 **T TCM, TCIL:** Yes **Year:** 2011 **Model:** EX35, M56, QX56 **Engine:** 3.5L, 5.6L **Transmission:** All	**A/T 1st Engine Braking:** * This is not an OBD-II self-diagnostic item. * Diagnostic trouble code P1731 is detected under the following conditions. - When TCM does not receive the proper voltage signal from the sensor. - When TCM monitors each ATF pressure switch and solenoid monitor value, and detects as irregular when engine brake of 1st gear acts other than at "1" position.
DTC: P1734 **T TCM, TCIL:** Yes **Year:** 2011 **Model:** FX35, FX50, G37, M37 **Engine:** 3.5L, 3.7L, 5.0L **Transmission:** All	**7th Gear Incorrect Ratio:** DTC is set when any inconsistency is recognized in gear ratio.
DTC: P1752 **T TCM, TCIL:** Yes **Year:** 2011 **Model:** EX35, M56, QX56 **Engine:** 3.5L, 5.6L **Transmission:** All	**Input Clutch Solenoid Valve:** * This is an OBD-II self-diagnostic item. * Diagnostic trouble code P1752 is detected under the following conditions: - When TCM detects an improper voltage drop when it tries to operate the solenoid valve. - When TCM detects as irregular by comparing target value with monitor value.
DTC: P1757 **T TCM, TCIL:** Yes **Year:** 2011 **Model:** EX35, M56, QX56 **Engine:** 3.5L, 5.6L **Transmission:** All	**Front Brake Solenoid Valve:** * This is an OBD-II self-diagnostic item. * Diagnostic trouble code P1757 is detected under the following conditions: - When TCM detects an improper voltage drop when it tries to operate the solenoid valve. - When TCM detects as irregular by comparing target value with monitor value.

DTC	Trouble Code Title and Conditions
DTC: P1762 **T TCM, TCIL: Yes** **Year:** 2011 **Model:** EX35, M56, QX56 **Engine:** 3.5L, 5.6L **Transmission:** All	**Direct Clutch Solenoid Valve:** * This is an OBD-II self-diagnostic item. * Diagnostic trouble code P1762 will be detected under the following conditions: - When TCM detects an improper voltage drop when it tries to operate the solenoid valve. - When TCM detects as irregular by comparing target value with monitor value.
DTC: P1767 **T TCM, TCIL: Yes** **Year:** 2011 **Model:** EX35, M56, QX56 **Engine:** 3.5L, 5.6L **Transmission:** All	**High & Low Reverse Clutch Solenoid Valve:** This is an OBD-II self-diagnostic item. Diagnostic trouble code P1767 will be detected under the following conditions: - When TCM detects an improper voltage drop when it tries to operate the solenoid valve. - When TCM detects as irregular by comparing target value with monitor value.
DTC: P1772 **T TCM, TCIL: Yes** **Year:** 2011 **Model:** EX35, M56, QX56 **Engine:** 3.5L, 5.6L **Transmission:** All	**Low Coast Brake Solenoid Valve:** * This is an OBD-II self-diagnostic item. * Diagnostic trouble code P1772 will be set when the TCM detects an improper voltage drop when it tries to operate the solenoid valve.
DTC: P1774 **T TCM, TCIL: Yes** **Year:** 2011 **Model:** EX35, M56, QX56 **Engine:** 3.5L, 5.6L **Transmission:** All	**Low Coast Brake Solenoid Valve Function:** * This is an OBD-II self-diagnostic item. * Diagnostic trouble code P1774 will be detected under the following conditions: - TCM detects an improper voltage drop when it tries to operate the solenoid valve. - When TCM detects that actual gear ratio is irregular, and relation between gear position and condition of ATF pressure switch 2 is irregular during depressing accelerator pedal (other than during shift change). - When TCM detects that relation between gear position and condition of ATF pressure switch 2 is irregular during releasing accelerator pedal. (Other than during shift change)
DTC: P1805 **1T ECM** **Year:** 2011 **Model:** EX35, FX35, FX50, G37, M37, M56, QX56 **Engine:** 3.5L, 3.7L, 5.0L, 5.6L **Transmission:** All	**Brake Switch Signal Malfunction (Includes Hybrid Models):** HYBRID MODELS CAUTION: Hybrid systems use very high-voltage battery systems. Before starting any service work involving the battery system, turn the ignition switch OFF and then remove the service plug from pocket in the trunk. After removing the service plug, wait 10 minutes before touching any of the high-voltage connectors and terminals. * The MIL may not illuminate for this self-diagnosis. * A brake switch signal is not sent to ECM for extremely long time while the vehicle is being driven.
DTC: P1815 **T TCM, TCIL: Yes** **Year:** 2011 **Model:** EX35, FX35, FX50, G37, M37 **Engine:** 3.5L, 3.7L, 5.0L **Transmission:** All	**Manual Mode Switch Circuit:** * TCM monitors manual mode, non manual mode, up or down switch signal, and detects as irregular when impossible input pattern occurs 2 seconds or more. * Shift up/down signal of paddle shifter continuously remains ON for 60 seconds.

OBD II Trouble Code List (P2XXX Codes)

DTC	Trouble Code Title and Conditions
DTC: P2100 **1T ECM, MIL: Yes** **Year:** 2011 **Model:** EX35, FX35, FX50, G37, M37, M56, QX56 **Engine:** 3.5L, 3.7L, 5.0L, 5.6L **Transmission:** All	**Throttle Control Motor Relay Circuit is Open (Includes Hybrid Models):** HYBRID MODELS CAUTION: Hybrid systems use very high-voltage battery systems. Before starting any service work involving the battery system, turn the ignition switch OFF and then remove the service plug from pocket in the trunk. After removing the service plug, wait 10 minutes before touching any of the high-voltage connectors and terminals. **NOTE: On and, this DTC is for Bank 1.** * These self-diagnoses have the one trip detection logic. * ECM detects that the voltage of power source for throttle control motor is excessively low.
DTC: P2101 **1T ECM, MIL: Yes** **Year:** 2011 **Model:** EX35, FX35, FX50, G37, M37, M56, QX56 **Engine:** 3.5L, 3.7L, 5.0L, 5.6L **Transmission:** All	**Electric Throttle Control Performance (Includes Hybrid Models):** HYBRID MODELS CAUTION: Hybrid systems use very high-voltage battery systems. Before starting any service work involving the battery system, turn the ignition switch OFF and then remove the service plug from pocket in the trunk. After removing the service plug, wait 10 minutes before touching any of the high-voltage connectors and terminals. **NOTE: On and, this DTC refers to Bank 1.** **NOTE: If DTC P1233 or P2101 is displayed with DTC P1238, P1290, P2100 or 2119, first perform the trouble diagnosis for DTC P1238, P2119 or P1290, P2100.** * Electric throttle control function does not operate properly.

DTC	Trouble Code Title and Conditions
DTC: P2103 **1T ECM, MIL: Yes** **Year:** 2011 **Model:** EX35, FX35, FX50, G37, M37, M56, QX56 **Engine:** 3.5L, 3.7L, 5.0L, 5.6L **Transmission:** All	**Throttle Control Motor Relay Circuit is Short (Includes Hybrid Models):** HYBRID MODELS CAUTION: Hybrid systems use very high-voltage battery systems. Before starting any service work involving the battery system, turn the ignition switch OFF and then remove the service plug from pocket in the trunk. After removing the service plug, wait 10 minutes before touching any of the high-voltage connectors and terminals. * ECM detects the throttle control motor relay is stuck ON.
DTC: P2118 **1T ECM, MIL: Yes** **Year:** 2011 **Model:** EX35, FX35, FX50, G37, M37, M56, QX56 **Engine:** 3.5L, 3.7L, 5.0L, 5.6L **Transmission:** All	**Throttle Control Motor Circuit Short (Includes Hybrid Models):** HYBRID MODELS CAUTION: Hybrid systems use very high-voltage battery systems. Before starting any service work involving the battery system, turn the ignition switch OFF and then remove the service plug from pocket in the trunk. After removing the service plug, wait 10 minutes before touching any of the high-voltage connectors and terminals. **NOTE: On and, this DTC is for Bank 1.** * ECM detects short in both circuits between ECM and throttle control motor.
DTC: P2119 **1T ECM, MIL: Yes** **Year:** 2011 **Model:** EX35, FX35, FX50, G37, M37, M56, QX56 **Engine:** 3.5L, 3.7L, 5.0L, 5.6L **Transmission:** All	**Electric Throttle Control Actuator (Includes Hybrid Models):** HYBRID MODELS CAUTION: Hybrid systems use very high-voltage battery systems. Before starting any service work involving the battery system, turn the ignition switch OFF and then remove the service plug from pocket in the trunk. After removing the service plug, wait 10 minutes before touching any of the high-voltage connectors and terminals. **NOTE: When the malfunction is detected, ECM enters fail-safe mode and the MIL illuminates.** **NOTE: On and, this DTC is for Bank 1.** * Malfunction A: Electric throttle control actuator does not function properly due to the return spring malfunction. ECM controls the electric throttle actuator by regulating the throttle opening around the idle position. The engine speed will not rise more than 2,000 rpm. * Malfunction B: Throttle valve opening angle in fail-safe mode is not in specified range. ECM controls the electric throttle control actuator by regulating the throttle opening to 20 degrees or less. * Malfunction C: ECM detects the throttle valve is stuck open. While the vehicle is driving, it slows down gradually by fuel cut. After the vehicle stops, the engine stalls. * The engine can restart in N or P position (CVT), neutral (M/T), and engine speed will not exceed 1,000 rpm or more.
DTC: P2122 **1T ECM, MIL: Yes** **Year:** 2011 **Model:** EX35, FX35, FX50, G37, M37, M56, QX56 **Engine:** 3.5L, 3.7L, 5.0L, 5.6L **Transmission:** All	**Accelerator Pedal Position Sensor 1 Circuit Low Input:** **NOTE: If DTC P2122 or P2123 is displayed with DTC P0643, first perform the trouble diagnosis for DTC P0643.** * An excessively low voltage from the APP sensor 1 is sent to ECM.
DTC: P2123 **1T ECM, MIL: Yes** **Year:** 2011 **Model:** EX35, FX35, FX50, G37, M37, M56, QX56 **Engine:** 3.5L, 3.7L, 5.0L, 5.6L **Transmission:** All	**Accelerator Pedal Position Sensor 1 Circuit High Input:** **NOTE: If DTC P2122 or P2123 is displayed with DTC P0643, first perform the trouble diagnosis for DTC P0643.** * An excessively high voltage from the APP sensor 1 is sent to ECM. * When the malfunction is detected, ECM enters fail-safe mode and the MIL illuminates.
DTC: P2127 **1T ECM, MIL: Yes** **Year:** 2011 **Model:** EX35, FX35, FX50, G37, M37, M56, QX56 **Engine:** 3.5L, 3.7L, 5.0L, 5.6L **Transmission:** All	**Accelerator Pedal Position Sensor 2 Circuit Low Input:** * An excessively low voltage from the APP sensor 2 is sent to ECM. * When the malfunction is detected, ECM enters fail-safe mode and the MIL illuminates.
DTC: P2128 **1T ECM, MIL: Yes** **Year:** 2011 **Model:** EX35, FX35, FX50, G37, M37, M56, QX56 **Engine:** 3.5L, 3.7L, 5.0L, 5.6L **Transmission:** All	**Accelerator Pedal Position Sensor 2 Circuit High Input:** * An excessively high voltage from the APP sensor 2 is sent to ECM. * When the malfunction is detected, ECM enters fail-safe mode and the MIL illuminates.

DTC	Trouble Code Title and Conditions
DTC: P2132 **1T ECM, MIL: Yes** **Year:** 2011 **Model:** EX35, FX35, FX50, G37, M37 **Engine:** 3.5L, 3.7L, 5.0L **Transmission:** All	**Throttle Position Sensor 1 (Bank 2) Circuit Low Input:** An excessively low voltage from the TP sensor 1 is sent to ECM.
DTC: P2133 **1T ECM, MIL: Yes** **Year:** 2011 **Model:** EX35, FX35, FX50, G37, M37 **Engine:** 3.5L, 3.7L, 5.0L **Transmission:** All	**Throttle Position Sensor 1 (Bank 2) Circuit High Input:** An excessively high voltage from the TP sensor 1 is sent to ECM.
DTC: P2135 **1T ECM, MIL: Yes** **Year:** 2011 **Model:** EX35, FX35, FX50, G37, M37, M56, QX56 **Engine:** 3.5L, 3.7L, 5.0L, 5.6L **Transmission:** All	**Throttle Position Sensor Circuit Range/Performance (Includes Hybrid Models):** HYBRID MODELS CAUTION: Hybrid systems use very high-voltage battery systems. Before starting any service work involving the battery system, turn the ignition switch OFF and then remove the service plug from pocket in the trunk. After removing the service plug, wait 10 minutes before touching any of the high-voltage connectors and terminals. **NOTE: If DTC P2135 is displayed with DTC P0643, first perform the trouble diagnosis for DTC P0643.** **NOTE: On and, this DTC refers to Bank 1.** * Rationally incorrect voltage is sent to ECM compared with the signals from TP sensor 1 and TP sensor 2. * When the malfunction is detected, the ECM enters fail-safe mode and the MIL illuminates.
DTC: P2138 **1T ECM, MIL: Yes** **Year:** 2011 **Model:** EX35, FX35, FX50, G37, M37, M56, QX56 **Engine:** 3.5L, 3.7L, 5.0L, 5.6L **Transmission:** All	**Accelerator Pedal Position Sensor Circuit Range/Performance:** **NOTE: If DTC P2138 is displayed with DTC P0643, first perform the trouble diagnosis for DTC P0643.** * Rationally incorrect voltage is sent to ECM compared with the signals from APP sensor 1 and APP sensor 2. * When the malfunction is detected, ECM enters fail-safe mode and the MIL illuminates.
DTC: P2713 **T TCM, TCIL: Yes** **Year:** 2011 **Model:** FX35, FX50, G37, M37 **Engine:** 3.5L, 3.7L, 5.0L **Transmission:** All	**Pressure Control Solenoid D:** The high and low reverse clutch solenoid valve monitor value is 0.4 A or less when the high and low reverse clutch solenoid valve command value is more than 0.75 A.
DTC: P2722 **T TCM, TCIL: Yes** **Year:** 2011 **Model:** FX35, FX50, G37, M37 **Engine:** 3.5L, 3.7L, 5.0L **Transmission:** All	**Pressure Control Solenoid E:** The low brake solenoid valve monitor value is 0.4 A or less when the low brake solenoid valve command value is more than 0.75 A.
DTC: P2731 **T TCM, TCIL: Yes** **Year:** 2011 **Model:** FX35, FX50, G37, M37 **Engine:** 3.5L, 3.7L, 5.0L **Transmission:** All	**Pressure Control Solenoid F:** The 2346 brake solenoid valve monitor value is 0.4 A or less when the 2346 brake solenoid valve command value is more than 0.75 A.
DTC: P2807 **T TCM, TCIL: Yes** **Year:** 2011 **Model:** FX35, FX50, G37, M37 **Engine:** 3.5L, 3.7L, 5.0L **Transmission:** All	**Pressure Control Solenoid G:** The direct clutch solenoid valve monitor value is 0.4 A or less when the direct clutch solenoid valve command value is more than 0.75 A.

DTC	Trouble Code Title and Conditions
DTC: P2A00 **1T ECM, MIL: Yes** **Year:** 2011 **Model:** EX35, FX35, FX50, G37, M37, M56, QX56 **Engine:** 3.5L, 3.7L, 5.0L, 5.6L **Transmission:** All	**Air Fuel (A/F) Sensor 1 Circuit Range/Performance (Includes Hybrid Models):** HYBRID MODELS CAUTION: Hybrid systems use very high-voltage battery systems. Before starting any service work involving the battery system, turn the ignition switch OFF and then remove the service plug from pocket in the trunk. After removing the service plug, wait 10 minutes before touching any of the high-voltage connectors and terminals. **NOTE: On and engines, this applies to Bank 1.** * To judge the malfunction, the A/F signal computed by ECM from the A/F sensor 1 signal is monitored not to be shifted to LEAN side or RICH side. * The output voltage computed by ECM from the A/F sensor 1 signal is shifted to the lean side for a specified period. * The A/F signal computed by ECM from the A/F sensor 1 signal is shifted to the rich side for a specified period.
DTC: P2A03 **1T ECM, MIL: Yes** **Year:** 2011 **Model:** EX35, FX35, FX50, G37, M37, M56, QX56 **Engine:** 3.5L, 3.7L, 5.0L, 5.6L **Transmission:** All	**Air Fuel (A/F) Ratio Sensor 1 Circuit Range/Performance:** **NOTE: On and engines, this applies to Bank 2.** * The output voltage computed by ECM from the A/F sensor 1 signal is shifted to the lean side for a specified period. * The A/F signal computed by ECM from the A/F sensor 1 signal is shifted to the rich side for a specified period. * To judge the malfunction, the A/F signal computed by ECM from the A/F sensor 1 signal is monitored so it will not shift to LEAN side or RICH side.

NISSAN

370Z

SPECIFICATIONS AND MAINTENANCE CHARTS

ENGINE AND VEHICLE IDENTIFICATION

		Engine						Model Year	
Code ①	Liters (cc)	Cu. In.	Cyl.	Fuel Sys.	Engine Type	Eng. Mfg.		Code ②	Year
VQ37VHR	3696	225	6	MFI	DOHC	Nissan		B	2011
								C	2012

MFI: Multi-port Fuel Injection

DOHC: Double Overhead Camshaft

① The Engine Code is stamped on the engine block near the starter.

② 10th position of the Vehicle Identification Number (VIN)

71075_370Z_C0001

GENERAL ENGINE SPECIFICATIONS

Year	Model	Engine Displacement Liters	Engine Series (ID)	Net Horsepower @ rpm	Net Torque @ rpm (ft. lbs.)	Bore x Stroke (in.)	Com- pression Ratio	Oil Pressure @ rpm
2011	370Z	3.7	VQ37VHR	332@7000	270@5200	3.760X3386	11.0:1	43@2000
2012	370Z	3.7	VQ37VHR	332@7000	270@5200	3.760X3386	11.0:1	43@2000

71075_370Z_C0002

ENGINE TUNE-UP SPECIFICATIONS

Year	Model	Engine Displacement Liters (ID)	Spark Plug Gap (in.)	Ignition Timing (deg.) MT	AT	Fuel Pump (psi) ①	Idle Speed (rpm) MT	AT ②	Valve Clearance (in.) Intake ③	Exhaust ③
2011	370Z	3.7 (VQ37VHR)	0.043	④	④	51	600-700	600-700	0.010-0.013	0.011-0.015
2012	370Z	3.7 (VQ37VHR)	0.043	④	④	51	600-700	600-700	0.010-0.013	0.011-0.015

NOTE: The Vehicle Emission Control Information label often reflects specification changes made during production.

The label figures must be used if they differ from those in this chart.

NA: Not Available

B: Before top dead center

① System pressure at idle with vacuum hose connected; should increase to 43 psi when disconnected

② Automatic transmission in park or neutral

③ Engine cold

④ 5-15 degrees BTDC

71075_370Z_C0003

CAPACITIES

Year	Model	Engine ID	Engine Displacement Liters	Engine Oil with Filter (qts.)	Transmission (pts.)		Drive Axle Rear (pts.)	Fuel Tank (gal.)	Cooling System (qts.)
					Man	Auto.			
2011	370Z	VQ37VHR	3.7	5.0	6.0	19.5	3.0	19.0	9.0
2012	370Z	VQ37VHR	3.7	5.0	6.0	19.5	3.0	19.0	9.0

NOTE: All capacities are approximate. Add fluid gradually and check to be sure a proper fluid level is obtained.

71075_370Z_C0004

FLUID SPECIFICATIONS

Year	Model	Engine Displacement Liters	Engine ID	Engine Oil	Auto. Trans.	Manual Trans.	Power Steering Fluid	Brake Master Cylinder
2011	370Z	3.7	VQ37VHR	5W-30	①	②	③	DOT 3
2012	370Z	3.7	VQ37VHR	5W-30	①	②	③	DOT 3

DOT: Department Of Transportation

① Nissan Matic S ATF

② Nissan (MTF) HQ MULTI 75W-85 or API GL-4, Viscosity 75W-85

③ Nissan PSF or equivalent. DEXRON VI type ATF may be used

71075_370Z_C0013

VALVE SPECIFICATIONS

Year	Engine Displacement Liters	Engine ID	Seat Angle (deg.)	Face Angle (deg.)	Spring Test Pressure (lbs. @ in.)	Spring Installed Height (in.)	Stem-to-Guide Clearance (in.)		Stem Diameter (in.)	
							Intake	Exhaust	Intake	Exhaust
2011	3.7	VQ37VHR	45.15-45.45	45	①	②	0.0008-0.0021	0.0012-0.0022	0.2348-0.2354	0.2347-0.2350
2012	3.7	VQ37VHR	45.15-45.45	45	①	②	0.0008-0.0021	0.0012-0.0022	0.2348-0.2354	0.2347-0.2350

① Intake: 43-48 lb at 1.6102 inch. Exhaust: 37-42 lb at 1.4567 inch.

② Intake: 1.7976. Exhaust: 1.7264.

71075_370Z_C0005

CAMSHAFT AND BEARING SPECIFICATIONS CHART

All measurements are given in inches.

Year	Engine Displ. Liters	Engine ID	Journal Dia.	Brg. Oil Clearance	Shaft End-play	Runout	Lobe Lift Intake	Lobe Lift Exhaust
2011	3.7	VQ37VHR	① ③	② ③	0.0045-0.0074	0.0008	NA	③ ④
2012	3.7	VQ37VHR	① ③	② ③	0.0045-0.0074	0.0008	NA	③ ④

NA: Not Available

① Front No. 1: 1.0211- 1.0218

  No. 2, 3, 4: 0.9230- 0.9238

② Front No. 1: 0.0018- 0.0034

  No. 2, 3, 4: 0.0014- 0.0030

③ Specification is for exhaust camshaft.

④ 1.7722- 1.7797 bank one. 1.8400- 1.8474 bank two.

71075_370Z_C0014

CRANKSHAFT AND CONNECTING ROD SPECIFICATIONS

All measurements are given in inches.

Year	Engine Displacement Liters	Engine ID	Crankshaft Main Brg. Journal Dia.	Crankshaft Main Brg. Oil Clearance	Crankshaft Shaft End-play	Crankshaft Thrust on No.	Connecting Rod Journal Diameter	Connecting Rod Oil Clearance	Connecting Rod Side Clearance
2011	3.7	VQ37VHR	2.5581-2.5580*	0.0014-0.0018	0.0039-0.0098	3	2.1241-2.1241*	0.0016-0.0021	0.0079-0.0138
2012	3.7	VQ37VHR	2.5581-2.5580*	0.0014-0.0018	0.0039-0.0098	3	2.1241-2.1241*	0.0016-0.0021	0.0079-0.0138

* Based upon grade. Grade "A" specification. Specification for other grades will differ.

71075_370Z_C0008

PISTON AND RING SPECIFICATIONS

All measurements are given in inches.

Year	Engine Displ. Liters	Engine ID	Piston Clearance	Ring Gap Top Compression	Ring Gap Bottom Compression	Ring Gap Oil Control	Ring Side Clearance Top Compression	Ring Side Clearance Bottom Compression	Ring Side Clearance Oil Control
2011	3.7	VQ37VHR	0.0004-0.0012	0.0091-0.0130	0.0091-0.0130	0.0067-0.0185	0.0016-0.0031	0.0012-0.0028	0.0022-0.0061
2012	3.7	VQ37VHR	0.0004-0.0012	0.0091-0.0130	0.0091-0.0130	0.0067-0.0185	0.0016-0.0031	0.0012-0.0028	0.0022-0.0061

71075_370Z_C0007

TORQUE SPECIFICATIONS

All readings in ft. lbs.

Year	Engine Displacement Liters	Engine ID/VIN	Cylinder Head Bolts	Main Bearing Bolts	Rod Bearing Bolts	Crankshaft Damper Bolts	Flywheel Bolts	Manifold Intake	Manifold Exhaust	Spark Plugs	Oil Drain Plug
2011	3.7	VQ37VHR	①	②	③	④	65	⑤	⑥	18	25
2012	3.7	VQ37VHR	①	②	③	④	65	⑤	⑥	18	25

① Step 1: 77 ft. lbs.

   Step 2: Loosen bolts completely

   Step 3: 30 ft. lbs.

   Step 4: Tighten an additional 95 degrees

   Step 5: Tighten an additional 95 degrees

② Step 1: 18 ft. lbs.

   Step 2: 26 ft.lbs.

   Step 3: Tighten an additional 90 degrees

③ Step 1: 21 ft. lbs.

   Step 2: Loosen bolts completely

   Step 3: 18 ft.lbs.

   Step 4: Tighten an additional 90 degrees

④ Step 1: 33 ft. lbs.

   Step 2: Tighten an additional 90 degrees

⑤ Step 1: 5 ft. lbs.

   Step 2: 19 ft. lbs.

⑥ Step 1: 11 ft. lbs. in 2 steps

71075_370Z_C0006

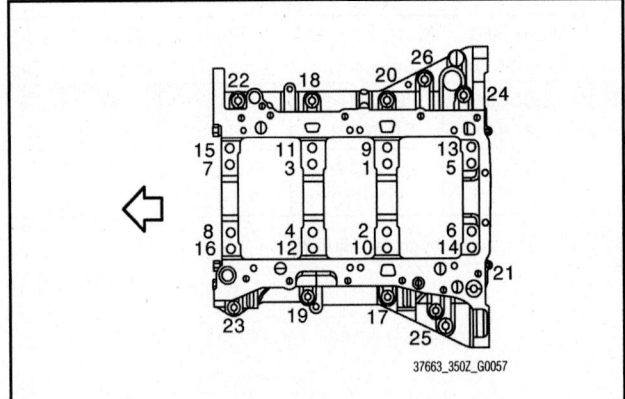

37663_350Z_G0057

Fig. 1 Main bearing (lower cylinder block bolts) tightening sequence

WHEEL ALIGNMENT

Year	Model		Caster Range (+/-Deg.)	Caster Preferred Setting (Deg.)	Camber Range (+/-Deg.)	Camber Preferred Setting (Deg.)	Toe-in (in.)
2011	370Z	F	①	①	②	②	③
		R	—	—	④	④	⑤ ⑥
2012	370Z	F	①	①	②	②	③
		R	—	—	④	④	⑤ ⑥

① Min 4 degrees 25' (4.42 degrees)
Max 5 degrees 55' (5.91 degrees)
② Min -1 degree 25' (-1.41 degrees)
Max 0 degree 40' (0.08 degrees)
③ Min 0.04 inch
Nominal 0.08 inch
Max 0.11 inch

④ Min -2 degrees 10' (-2.16 degrees)
Max -1 degrees 10' (-1.17 degrees)
⑤ Min 0.079 inch 18 inch wheel
Nominal 0.150 inch 18 inch wheel
Max 0.221 inch 18 inch wheel
⑥ Min 0.079 inch 19 inch wheel
Nominal 0.146 inch 19 inch wheel
Max 0.213 inch 19 inch wheel

71075_370Z_C0009

TIRE, WHEEL AND BALL JOINT SPECIFICATIONS

Year	Model		OEM Tires Standard	OEM Tires Optional	Tire Pressures (psi) Front	Tire Pressures (psi) Rear	Wheel Size	Lug Nut Torque (ft. lbs.)
2011	370Z	Front	225/50W5R18	245/40WR19	①	②	NA	80
		Rear	245/45WR18	275/35WR19	①	②	NA	80
2012	370Z	Front	225/50W5R18	245/40WR19	①	②	NA	80
		Rear	245/45WR18	275/35WR19	①	②	NA	80

Note: If specification differs from vehicle placard, use specification given on vehicle placard.

NA: Not Available

OEM: Original Equipment Manufacturer

PSI: Pounds Per Square Inch

① See vehicle placard for specification

71075_370Z_C0010

BRAKE SPECIFICATIONS
All measurements in inches unless noted

Year	Model		Brake Disc Original Thickness	Brake Disc Minimum Thickness	Brake Disc Maximum Run-out	Minimum Lining Thickness Front	Minimum Lining Thickness Rear	Brake Caliper Bracket Bolts (ft. lbs.)	Brake Caliper Mounting Bolts (ft. lbs.)
2011	370Z ①	F	1.1024	NA	0.0014	0.079	0.079	NA	NA
		R	0.5510	NA	0.0022	0.079	0.079	NA	NA
	370Z ②	F	1.1810	NA	0.0014	0.079	0.079	NA	NA
		R	0.7090	NA	0.0022	0.079	0.079	NA	NA
2012	370Z ①	F	1.1024	NA	0.0014	0.079	0.079	NA	NA
		R	0.5510	NA	0.0022	0.079	0.079	NA	NA
	370Z ②	F	1.1810	NA	0.0014	0.079	0.079	NA	NA
		R	0.7090	NA	0.0022	0.079	0.079	NA	NA

NA: Not Available

① One piston type rear. Two piston type front.
② Two piston type rear. Four piston type front.

71075_370Z_C0011

SCHEDULED MAINTENANCE INTERVALS
NISSAN—370Z

TO BE SERVICED	TYPE OF SERVICE	VEHICLE MILEAGE INTERVAL (x1000)												
		7.5	15	22.5	30	37.5	45	52.5	60	67.5	75	82.5	90	97.5
Engine oil & filter	R	✓	✓	✓	✓	✓	✓	✓	✓	✓	✓	✓	✓	✓
Brake lines & cables	S/I		✓		✓		✓		✓		✓		✓	
Brake pads & discs	S/I		✓		✓		✓		✓		✓		✓	
Cabin Filter	R		✓		✓		✓		✓		✓		✓	
Driveshaft boots	S/I		✓		✓		✓		✓		✓		✓	
Exhaust system	S/I				✓				✓				✓	
Transmission fluid	S/I		✓		✓		✓		✓		✓		✓	
Air cleaner filter	R				✓				✓				✓	
Spark plugs (except platinum)	R				✓				✓				✓	
Spark plugs (platinum tip) exc. Nismo 370Z	R	replace every 105,000 miles												
Spark plugs (platinum tip) Nismo 370Z	R								✓					
Steering gear & linkage, axle & suspension parts	S/I				✓				✓				✓	
Engine coolant	R								✓				✓	
Drive belts	S/I								✓		✓		✓	
Fuel lines	S/I								✓					
Vapor lines	S/I								✓					

R: Replace   S/I: Service or Inspect

FREQUENT OPERATION MAINTENANCE (SEVERE SERVICE)

If a vehicle is operated under any of the following conditions it is considered severe service:

- Extremely dusty areas.

- 50% or more of the vehicle operation is in 32°C (90°F) or higher temperatures, or constant operation in temperatures below 0°C (32°F).

- Prolonged idling (vehicle operation in stop and go traffic).

- Frequent short running periods (engine does not warm to normal operating temperatures).

- Police, taxi, delivery usage or trailer towing usage.

Oil & oil filter: change every 3750 miles.

Brake pads & discs: service or inspect every 7500 miles.

Driveshaft boots: service or inspect every 7500 miles.

Exhaust system: service or inspect every 7500 miles.

Steering gear & linkage, axle & suspension parts: service or inspect every 7500 miles.

Steering linkage ball joints & front suspension ball joints: service or inspect every 7500 miles.

Air cleaner filter: service or inspect every 15,000 miles.

71075_370Z_C0012

PRECAUTIONS

Before servicing any vehicle, please be sure to read all of the following precautions, which deal with personal safety, prevention of component damage, and important points to take into consideration when servicing a motor vehicle:

• Never open, service or drain the radiator or cooling system when the engine is hot; serious burns can occur from the steam and hot coolant.

• Observe all applicable safety precautions when working around fuel. Whenever servicing the fuel system, always work in a well-ventilated area. Do not allow fuel spray or vapors to come in contact with a spark, open flame, or excessive heat (a hot drop light, for example). Keep a dry chemical fire extinguisher near the work area. Always keep fuel in a container specifically designed for fuel storage; also, always properly seal fuel containers to avoid the possibility of fire or explosion. Refer to the additional fuel system precautions later in this section.

• Fuel injection systems often remain pressurized, even after the engine has been turned **OFF**. The fuel system pressure must be relieved before disconnecting any fuel lines. Failure to do so may result in fire and/or personal injury.

• Brake fluid often contains polyglycol ethers and polyglycols. Avoid contact with the eyes and wash your hands thoroughly after handling brake fluid. If you do get brake fluid in your eyes, flush your eyes with clean, running water for 15 minutes. If eye irritation persists, or if you have taken brake fluid internally, IMMEDIATELY seek medical assistance.

• The EPA warns that prolonged contact with used engine oil may cause a number of skin disorders, including cancer. You should make every effort to minimize your exposure to used engine oil. Protective gloves should be worn when changing oil. Wash your hands and any other exposed skin areas as soon as possible after exposure to used engine oil. Soap and water, or waterless hand cleaner should be used.

• All new vehicles are now equipped with an air bag system, often referred to as a Supplemental Restraint System (SRS) or Supplemental Inflatable Restraint (SIR) system. The system must be disabled before performing service on or around system components, steering column, instrument panel components, wiring and sensors. Failure to follow safety and disabling procedures could result in accidental air bag deployment, possible personal injury and unnecessary system repairs.

• Always wear safety goggles when working with, or around, the air bag system. When carrying a non-deployed air bag, be sure the bag and trim cover are pointed away from your body. When placing a non-deployed air bag on a work surface, always face the bag and trim cover upward, away from the surface. This will reduce the motion of the module if it is accidentally deployed. Refer to the additional air bag system precautions later in this section.

• Clean, high quality brake fluid from a sealed container is essential to the safe and proper operation of the brake system. You should always buy the correct type of brake fluid for your vehicle. If the brake fluid becomes contaminated, completely flush the system with new fluid. Never reuse any brake fluid. Any brake fluid that is removed from the system should be discarded. Also, do not allow any brake fluid to come in contact with a painted surface; it will damage the paint.

• Never operate the engine without the proper amount and type of engine oil; doing so WILL result in severe engine damage.

• Timing belt maintenance is extremely important. Many models utilize an interference-type, non-freewheeling engine. If the timing belt breaks, the valves in the cylinder head may strike the pistons, causing potentially serious (also time-consuming and expensive) engine damage. Refer to the maintenance interval charts for the recommended replacement interval for the timing belt, and to the timing belt section for belt replacement and inspection.

• Disconnecting the negative battery cable on some vehicles may interfere with the functions of the on-board computer system(s) and may require the computer to undergo a relearning process once the negative battery cable is reconnected.

• When servicing drum brakes, only disassemble and assemble one side at a time, leaving the remaining side intact for reference.

• Only an MVAC-trained, EPA-certified automotive technician should service the air conditioning system or its components.

BRAKES   ANTI-LOCK BRAKE SYSTEM

BLEEDING THE ABS SYSTEM

1. Before servicing the vehicle, refer to the Precautions Section.

➡ **If working near and/or around the SRS system and components, be sure to disable the SRS system. After disabling the system wait three minutes or more before servicing the vehicle.**

➡ **Before disconnecting the battery, lower both the driver's and passenger's windows. This will prevent any interference between the window edge and the vehicle when the door is opened or closed. During normal operation the window slightly raises and lowers automatically to prevent any window to vehicle interference. The automatic window function will not work with the battery disconnected.**

➡ **Turn the ignition switch off and disconnect the ABS actuator and electric control unit connector, or the negative battery cable before performing the work.**

➡ **Monitor the fluid level in the reservoir while performing the work. Always use new brake fluid. Be sure to use the proper grade and type fluid.**

➡ **Cover the crowfoot and flare nut wrench with a shop towel to prevent damage to the front four piston type** caliper and rear two piston type caliper.

2. Connect a vinyl tube to the bleeder valve of the right rear brake.

3. Fully depress the brake pedal four or five times.

4. Loosen the bleeder valve and bleed the air with the brake pedal depressed, quickly tighten the bleeder valve.

5. Repeat the above step until all air is expelled out of the brake line.

6. Tighten the bleeder valve.

7. Perform the above procedure to the brakes in the following sequence.

8. Right rear brake, left front brake, left rear brake and right front brake.

9. Check and refill the master cylinder, as required.

WHEEL SPEED SENSORS

REMOVAL & INSTALLATION

See Figures 2 and 3.

At this time the manufacturer does not provide removal and installation procedures for this component. The following procedure is a guideline and may differ from the vehicle you are servicing.

1. Before servicing the vehicle, refer to the Precautions Section.

➡**If working near and/or around the SRS system and components, be sure to disable the SRS system. After disabling the system wait three minutes or more before servicing the vehicle.**

➡**Before disconnecting the battery, lower both the driver's and passenger's windows. This will prevent any interference between the window edge and the vehicle when the door is opened or closed. During normal operation the window slightly raises and lowers automatically to prevent any window to vehicle interference. The automatic window function will not work with the battery disconnected.**

2. Raise and support the vehicle safely.

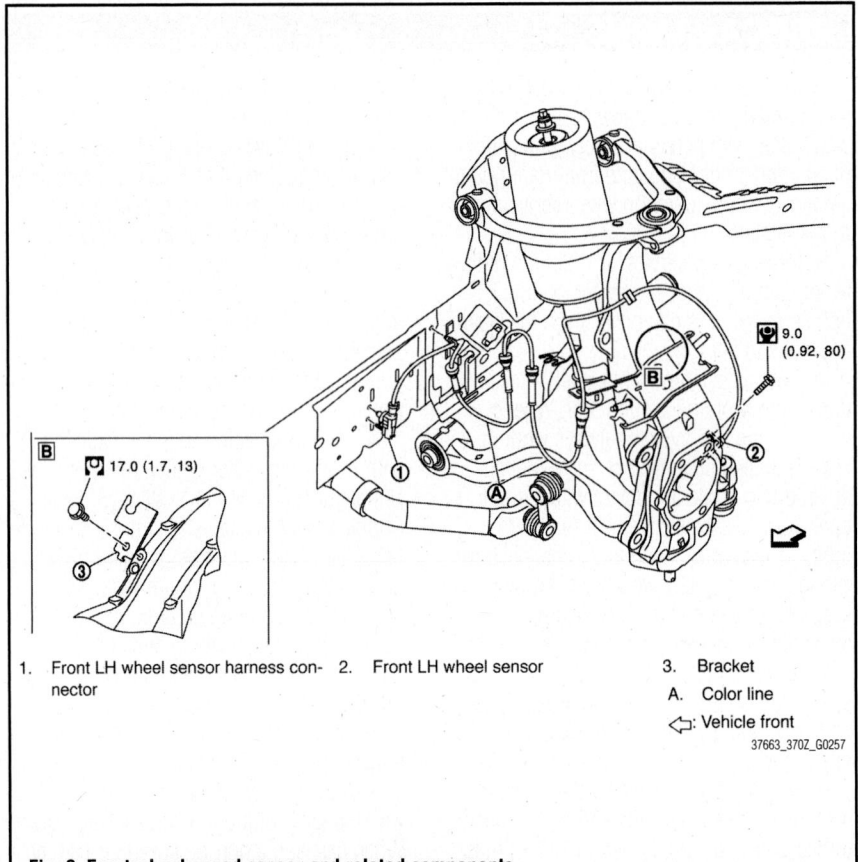

1. Front LH wheel sensor harness connector
2. Front LH wheel sensor
3. Bracket
A. Color line
⬅: Vehicle front

37663_370Z_G0257

Fig. 2 Front wheel speed sensor and related components

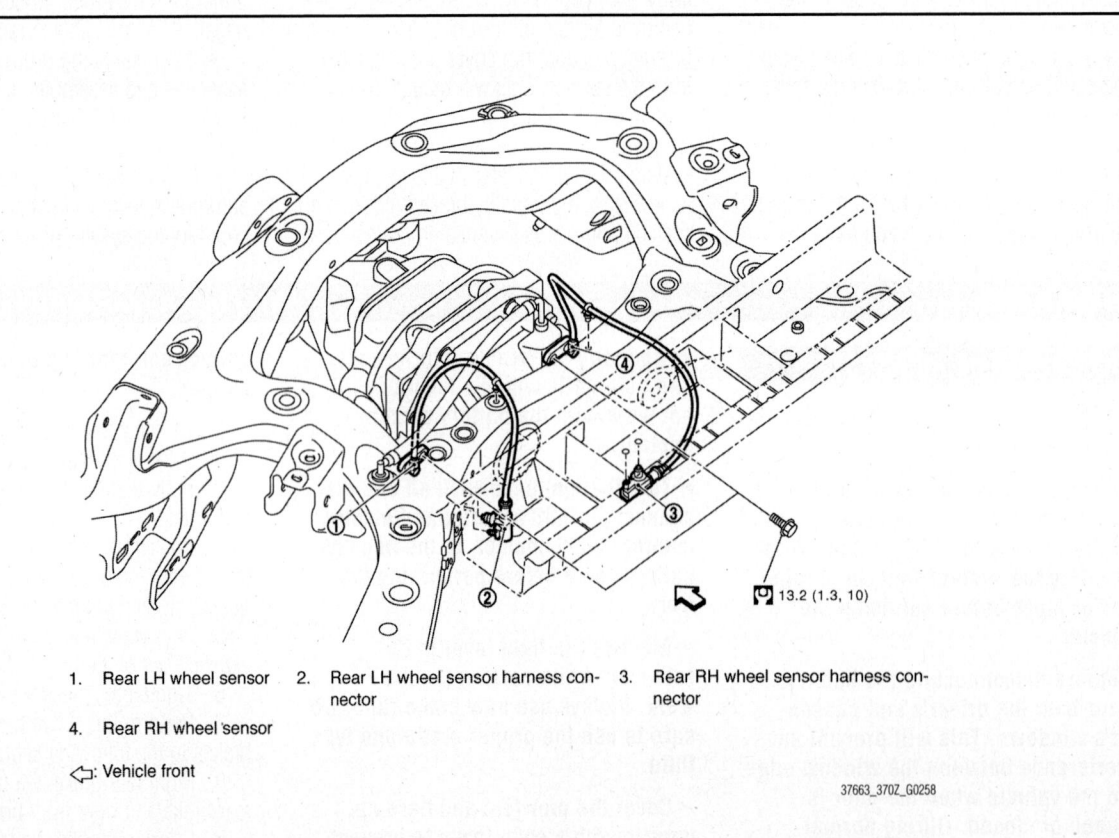

1. Rear LH wheel sensor
2. Rear LH wheel sensor harness connector
3. Rear RH wheel sensor harness connector
4. Rear RH wheel sensor

⬅: Vehicle front

37663_370Z_G0258

Fig. 3 Rear wheel speed sensor and related components

3. Remove the tire and wheel assembly, as required.

4. Never twist or bend sensor harness when removing it.

5. Pull the wheel sensor out without pulling the harness.

6. Be careful not to damage the sensor edges or rotor teeth.

7. Remove the sensor first, before removing the wheel hub and bearing assembly.

To install:

➡Be sure to use new fasteners, as required.

8. Installation is the reverse of the removal procedure.

BRAKES   BLEEDING THE BRAKE SYSTEM

BLEEDING PROCEDURE

1. Before servicing the vehicle, refer to the Precautions Section.

➡If working near and/or around the SRS system and components, be sure to disable the SRS system. After disabling the system wait three minutes or more before servicing the vehicle.

➡Before disconnecting the battery, lower both the driver's and passenger's windows. This will prevent any interference between the window edge and the vehicle when the door is opened or closed. During normal operation the window slightly raises and lowers automatically to prevent any window to vehicle interference. The automatic window function will not work with the battery disconnected.

❊❊ WARNING

Be careful not to splash brake fluid on painted areas; it may cause paint damage. If brake fluid is splashed on painted areas, wash it away with water immediately. All hoses must be free from excessive bending, twisting and pulling.

➡Turn the ignition switch off and disconnect the ABS actuator and electric control unit connector, or the negative battery cable before performing the work.

➡Monitor the fluid level in the reservoir while performing the work. Always use new brake fluid. Be sure to use the proper grade and type fluid.

➡Cover the crowfoot and flare nut wrench with a shop towel to prevent damage to the front four piston type caliper and rear two piston type caliper.

2. Connect a vinyl tube to the bleeder valve of the right rear brake.

3. Fully depress the brake pedal four or five times.

4. Loosen the bleeder valve and bleed the air with the brake pedal depressed, quickly tighten the bleeder valve.

5. Repeat the above step until all air is expelled out of the brake line.

6. Tighten the bleeder valve.

7. Perform the above procedure to the brakes in the following sequence.

8. Right rear brake, left front brake, left rear brake and right front brake.

9. Check and refill the master cylinder, as required.

FLUID FILL PROCEDURE

1. Before servicing the vehicle, refer to the Precautions Section.

➡If working near and/or around the SRS system and components, be sure to disable the SRS system. After disabling the system wait three minutes or more before servicing the vehicle.

➡Before disconnecting the battery, lower both the driver's and passenger's windows. This will prevent any interference between the window edge and the vehicle when the door is opened or closed. During normal operation the window slightly raises and lowers automatically to prevent any window to vehicle interference. The automatic window function will not work with the battery disconnected.

➡Turn the ignition switch off and disconnect the ABS actuator and electric control unit connector, or the negative battery cable before performing the work.

➡Cover the crowfoot and flare nut wrench with a shop towel to prevent damage to the front four piston type caliper and rear two piston type caliper.

2. Check that there is no foreign material in the reservoir tank or around it. Never reuse used fluid.

3. Loosen the bleeder valve.

4. Slowly depress the brake pedal to the full stroke. Release the pedal.

5. Repeat at intervals of two or three seconds until all brake fluid is discharged.

6. Close the bleeder valve with the brake pedal depressed.

7. Repeat the above on each wheel.

8. Bleed the brake system.

BRAKE CALIPERS

REMOVAL & INSTALLATION

See Figures 4 and 5.

1. Before servicing the vehicle, refer to the Precautions Section.

➡**If working near and/or around the SRS system and components, be sure to disable the SRS system. After disabling the system wait three minutes or more before servicing the vehicle.**

➡**Before disconnecting the battery, lower both the driver's and passenger's windows. This will prevent any interference between the window edge and the vehicle when the door is opened or closed. During normal operation the window slightly raises and lowers automatically to prevent any window to vehicle interference. The automatic window function will not work with the battery disconnected.**

2. Raise and safely support the vehicle.
3. Remove the tire and wheel assembly.
4. Hold the rotor, using the wheel nuts.
5. Drain the brake fluid.
6. On two piston caliper, remove the union bolt and copper washer. Discard the washer. Disconnect the brake hose from the caliper. Remove the torque member mounting bolts.
7. On four piston caliper, cover the flare nut wrench with a shop towel. Never scratch the flare nut and the brake tube. Never bend sharply, twist or strongly pull out the brake tube. Over the open end of the brake tube when disconnecting to prevent the entrance of dirt. Remove the caliper mounting bolts.

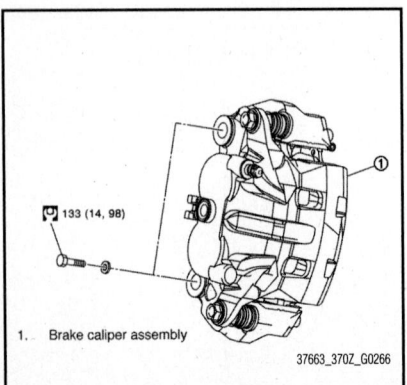

🔧 133 (14, 98)

1. Brake caliper assembly

37663_370Z_G0266

Fig. 4 Front brake caliper and related components—two piston type

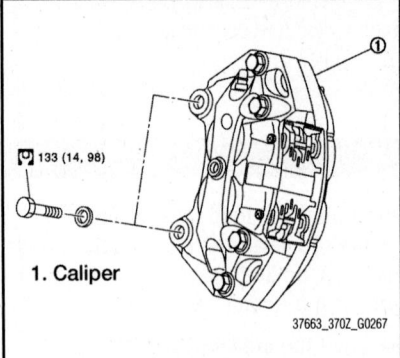

🔧 133 (14, 98)

1. Caliper

37663_370Z_G0267

Fig. 5 Front brake caliper and related components—four piston type

8. Remove the caliper assembly.
9. As required, remove the rotor.

To install:

➡**Be sure to use new fasteners, as required.**

10. Installation is the reverse of the removal procedure.
11. Be sure to use a new copper washer, two piston caliper.
12. Bleed the brake system.

BRAKE PADS

REMOVAL & INSTALLATION

2-Piston Caliper
See Figure 6.

1. Before servicing the vehicle, refer to the Precautions Section.

➡**If working near and/or around the SRS system and components, be sure to disable the SRS system. After disabling the system wait three minutes or more before servicing the vehicle.**

➡**Before disconnecting the battery, lower both the driver's and passenger's windows. This will prevent any interference between the window edge and the vehicle when the door is opened or closed. During normal operation the window slightly raises and lowers automatically to prevent any window to vehicle interference. The automatic window function will not work with the battery disconnected.**

2. Raise and safely support the vehicle.
3. Remove the tire and wheel assembly.
4. Remove the lower sliding pin bolt.
5. Suspend the caliper using mechanics wire.

➡**Do not allow the caliper to hang by the brake line hose.**

6. Remove the pads, shims, shim covers and pad retainers from the torque member.

To install:

➡**Be sure to use new fasteners, as required.**

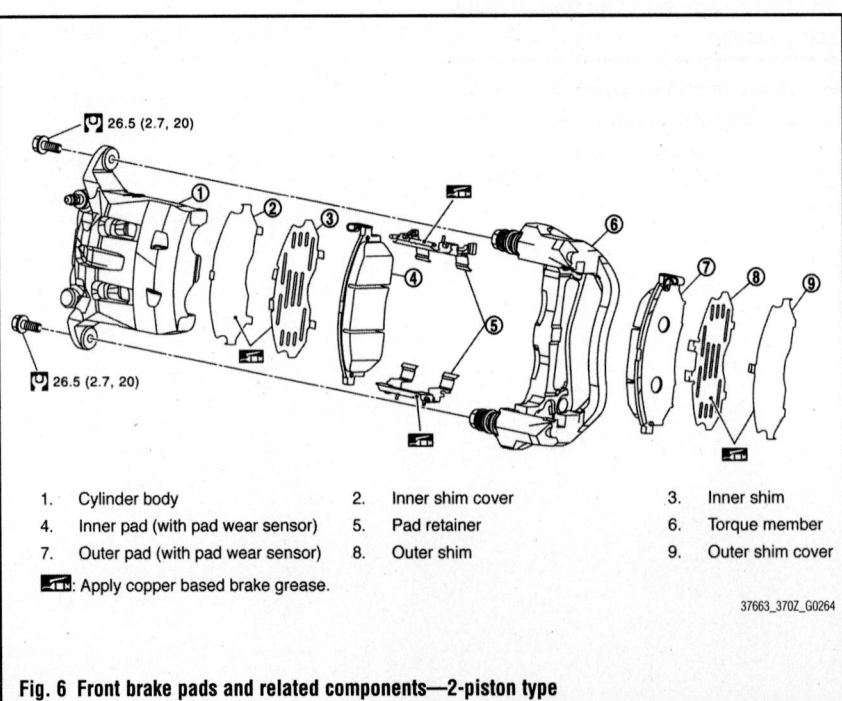

🔧 26.5 (2.7, 20)

🔧 26.5 (2.7, 20)

1.	Cylinder body	2.	Inner shim cover	3.	Inner shim
4.	Inner pad (with pad wear sensor)	5.	Pad retainer	6.	Torque member
7.	Outer pad (with pad wear sensor)	8.	Outer shim	9.	Outer shim cover

🔩: Apply copper based brake grease.

37663_370Z_G0264

Fig. 6 Front brake pads and related components—2-piston type

7. Installation is the reverse of the removal procedure.

8. Apply copper based brake grease to the pad retainers before installing them to the torque member, if the pad retainers were removed.

➡**Both inner and outer pads have a pad return system on the pad retainer. Install the pad return lever securely to the pad wear sensor.**

9. Depress the brake pedal several times to seat the pads and check that no drag feel is present for the disc rotor.

4-Piston Caliper

See Figure 7.

1. Before servicing the vehicle, refer to the Precautions Section.

➡**If working near and/or around the SRS system and components, be sure to disable the SRS system. After disabling the system wait three minutes or more before servicing the vehicle.**

➡**Before disconnecting the battery, lower both the driver's and passenger's windows. This will prevent any interference between the window edge and the vehicle when the door is opened or closed. During normal operation the window slightly raises and lowers automatically to prevent any window to vehicle interference. The automatic window function will not work with the battery disconnected.**

2. Raise and safely support the vehicle.
3. Remove the tire and wheel assembly.
4. Remove the clips from the pad shims.

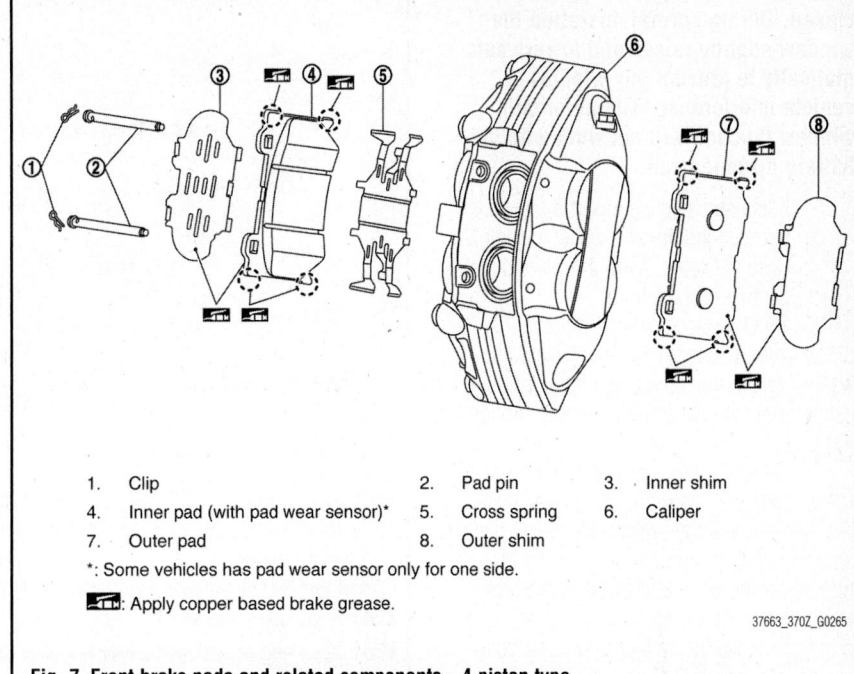

1.	Clip	2.	Pad pin	3.	Inner shim
4.	Inner pad (with pad wear sensor)*	5.	Cross spring	6.	Caliper
7.	Outer pad	8.	Outer shim		

*: Some vehicles has pad wear sensor only for one side.

⬛: Apply copper based brake grease.

37663_370Z_G0265

Fig. 7 Front brake pads and related components—4-piston type

5. Remove the pad pins while holding down the cross spring. Remove the cross spring from the caliper.

6. Using a pliers, remove the brake pads and shims from the caliper.

To install:

➡**Be sure to use new fasteners, as required.**

7. Installation is the reverse of the removal procedure.

8. Apply copper based brake grease to the mating surfaces between the pads and shims. Install the shims to the brake pads.

9. Install the pads to the caliper.

10. Install the upper pad pin from the inner side, then install firmly to the outer side through the hole in the top of the brake pad.

11. Place the top of the cross spring over the upper pad pin, press the cross spring, install the lower pad pin from the inner side to the outer side and secure the cross spring.

12. Install the clips to the pad pins.

➡**If the clip is not fully attached, pad pin or brake pad could fall out while the vehicle is in motion.**

13. Depress the brake pedal several times to seat the pads and check that no drag feel is present for the disc rotor.

BRAKES
REAR DISC BRAKES

BRAKE CALIPERS

REMOVAL & INSTALLATION

See Figures 8 and 9.

1. Before servicing the vehicle, refer to the Precautions Section.

➡**If working near and/or around the SRS system and components, be sure to disable the SRS system. After disabling the system wait three minutes or more before servicing the vehicle.**

➡**Before disconnecting the battery, lower both the driver's and passenger's windows. This will prevent any interference between the window edge and the**

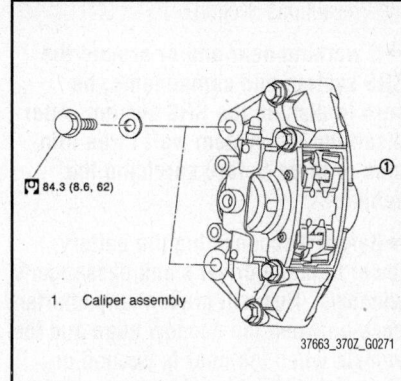

🔧 84.3 (8.6, 62)

1.   Brake caliper assembly

37663_370Z_G0270

Fig. 8 Rear brake caliper and related components—one piston type

🔧 84.3 (8.6, 62)

1.   Caliper assembly

37663_370Z_G0271

Fig. 9 Rear brake caliper and related components—two piston type

vehicle when the door is opened or closed. During normal operation the window slightly raises and lowers automatically to prevent any window to vehicle interference. The automatic window function will not work with the battery disconnected.

2. Raise and safely support the vehicle.

3. Remove the tire and wheel assembly.

4. Hold the rotor, using the wheel nuts.

5. Drain the brake fluid.

6. On two piston caliper, remove the union bolt and copper washer. Discard the washer. Disconnect the brake hose from the caliper. Remove the torque member mounting bolts.

7. On four piston caliper, cover the flare nut wrench with a shop towel. Never scratch the flare nut and the brake tube. Never bend sharply, twist or strongly pull out the brake tube. Over the open end of the brake tube when disconnecting to prevent the entrance of dirt. Remove the caliper mounting bolts.

8. Remove the caliper assembly.

9. As required, remove the rotor.

To install:

➡ Be sure to use new fasteners, as required.

10. Installation is the reverse of the removal procedure.

11. Be sure to use a new cooper washer, two piston caliper.

12. Check for brake drag, correct as required.

13. Bleed the brake system.

BRAKE PADS

REMOVAL & INSTALLATION

1-Piston Caliper

See Figure 10.

1. Before servicing the vehicle, refer to the Precautions Section.

➡ If working near and/or around the SRS system and components, be sure to disable the SRS system. After disabling the system wait three minutes or more before servicing the vehicle.

➡ Before disconnecting the battery, lower both the driver's and passenger's windows. This will prevent any interference between the window edge and the vehicle when the door is opened or closed. During normal operation the window slightly raises and lowers automatically to prevent any window to vehicle interference. The automatic

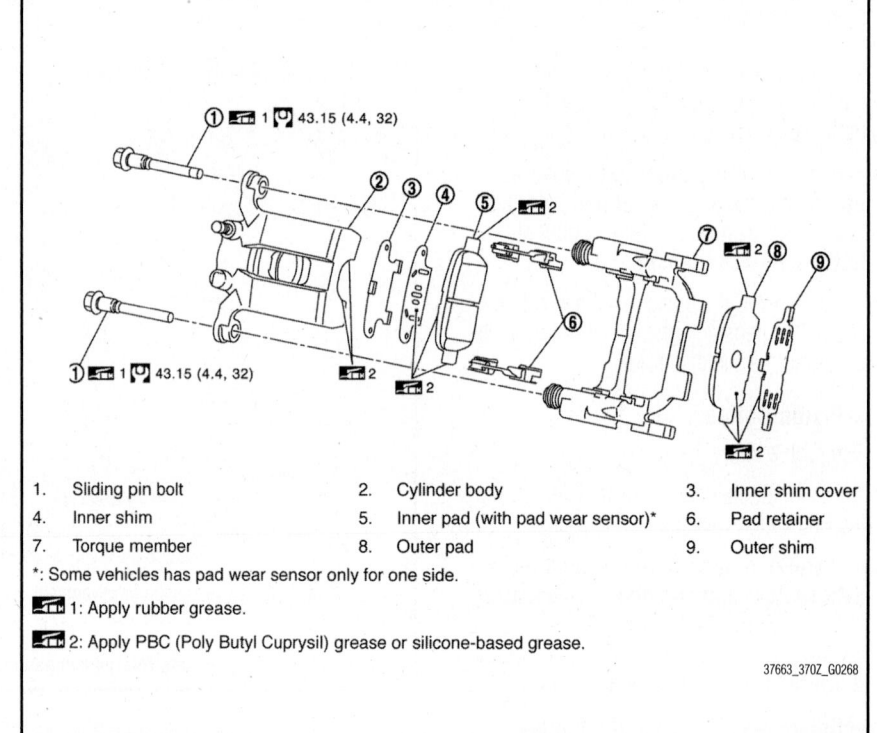

1. Sliding pin bolt
2. Cylinder body
3. Inner shim cover
4. Inner shim
5. Inner pad (with pad wear sensor)*
6. Pad retainer
7. Torque member
8. Outer pad
9. Outer shim

*: Some vehicles has pad wear sensor only for one side.

▣ 1: Apply rubber grease.

▣ 2: Apply PBC (Poly Butyl Cuprysil) grease or silicone-based grease.

37663_370Z_G0268

Fig. 10 Rear brake pads and related components—1-piston type

window function will not work with the battery disconnected.

2. Raise and safely support the vehicle.

3. Remove the tire and wheel assembly.

4. Remove the lower sliding pin bolt.

5. Suspend the caliper using mechanics wire.

➡ Do not allow the caliper to hang by the brake line hose.

6. Remove the pads, shims, shim covers and pad retainers from the torque member.

To install:

➡ Be sure to use new fasteners, as required.

7. Installation is the reverse of the removal procedure.

8. Apply PBC grease to the mating surfaces between the pads and shims.

9. Apply PBC grease to the mating surfaces between the pad retainers and the pads before installing them to the brake pads.

10. Depress the brake pedal several times to seat the pads and check that no drag feel is present for the disc rotor.

2-Piston Caliper

See Figure 11.

1. Before servicing the vehicle, refer to the Precautions Section.

➡ If working near and/or around the SRS system and components, be sure to disable the SRS system. After disabling the system wait three minutes or more before servicing the vehicle.

➡ Before disconnecting the battery, lower both the driver's and passenger's windows. This will prevent any interference between the window edge and the vehicle when the door is opened or closed. During normal operation the window slightly raises and lowers automatically to prevent any window to vehicle interference. The automatic window function will not work with the battery disconnected.

2. Raise and safely support the vehicle.

3. Remove the tire and wheel assembly.

4. Remove the clips from the pad shims.

5. Remove the pad pins while holding down the cross spring. Remove the cross spring from the caliper.

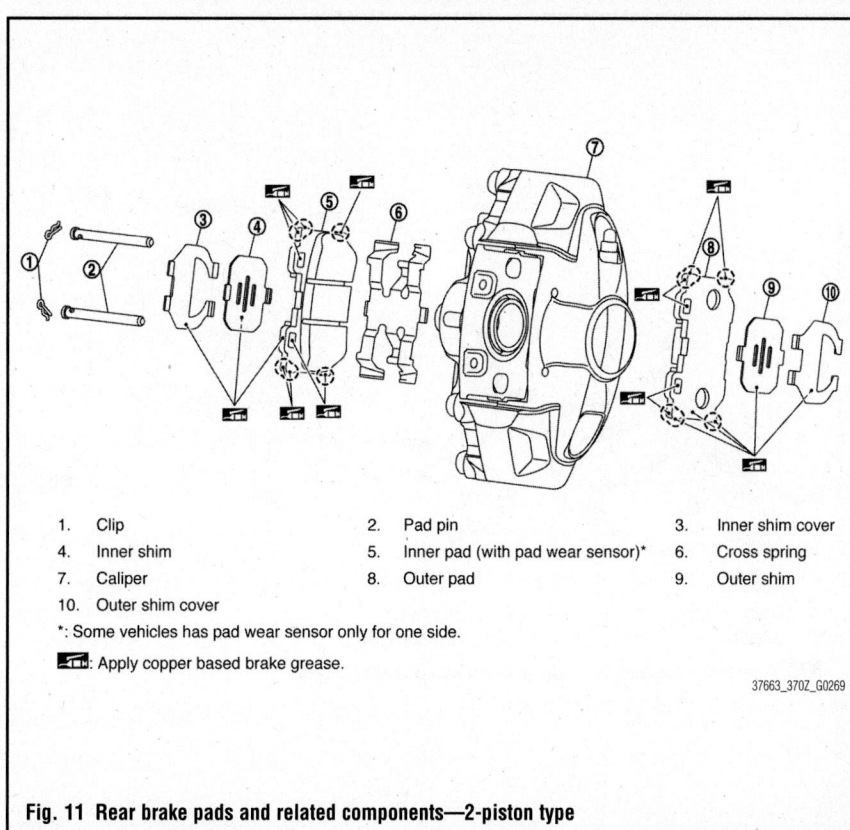

1. Clip
2. Pad pin
3. Inner shim cover
4. Inner shim
5. Inner pad (with pad wear sensor)*
6. Cross spring
7. Caliper
8. Outer pad
9. Outer shim
10. Outer shim cover

*: Some vehicles has pad wear sensor only for one side.

▭: Apply copper based brake grease.

37663_370Z_G0269

Fig. 11 Rear brake pads and related components—2-piston type

6. Using a pliers, remove the brake pads and shims from the caliper.

To install:

➡**Be sure to use new fasteners, as required.**

7. Installation is the reverse of the removal procedure.

8. Apply copper based brake grease to the mating surfaces between the pads and caliper, between the pads and pad pins and between the pads and cross spring.

9. Install the pads to the caliper.

10. Install the upper pad pin from the inner side, then install firmly to the outer side through the hole in the top of the brake pad.

11. Place the top of the cross spring over the upper pad pin, press the cross spring, install the lower pad pin from the inner side to the outer side and secure the cross spring.

12. Install the clips to the pad pins.

➡**If the clip is not fully attached, pad pin or brake pad could fall out while the vehicle is in motion.**

13. Depress the brake pedal several times to seat the pads and check that no drag feel is present for the disc rotor.

BRAKES   PARKING BRAKE

PARKING BRAKE CABLES

ADJUSTMENT

See Figure 12.

1. To perform adjustment operations, remove tire from the vehicle with power tool.

2. Remove the coin pocket. Insert a deep socket wrench to rotate adjusting nut (1) and loosen the cable sufficiently. Then, return the lever.

3. Using wheel nuts, fix the disc rotor to the hub and prevent it from tilting.

4. Remove adjusting hole plug installed on the disc. Using a flat bladed tool, turn the disc in direction "A" as shown in the figure until the disc is locked. After locking, turn the adjuster in the opposite direction by 5 or 6 notches.

5. Rotate the disc to make sure there is no drag. Install the adjusting hole plug.

6. Adjust cable as follows:
   a. Operate lever 10 or more times with a force of 66 ft. lbs. (89 Nm).
   b. Rotate adjusting nut with deep socket to adjust lever stroke.
   c. When parking brake lever is operated with a force of 44 ft. lbs. (60 Nm), check that the stroke is 6 to 7 notches

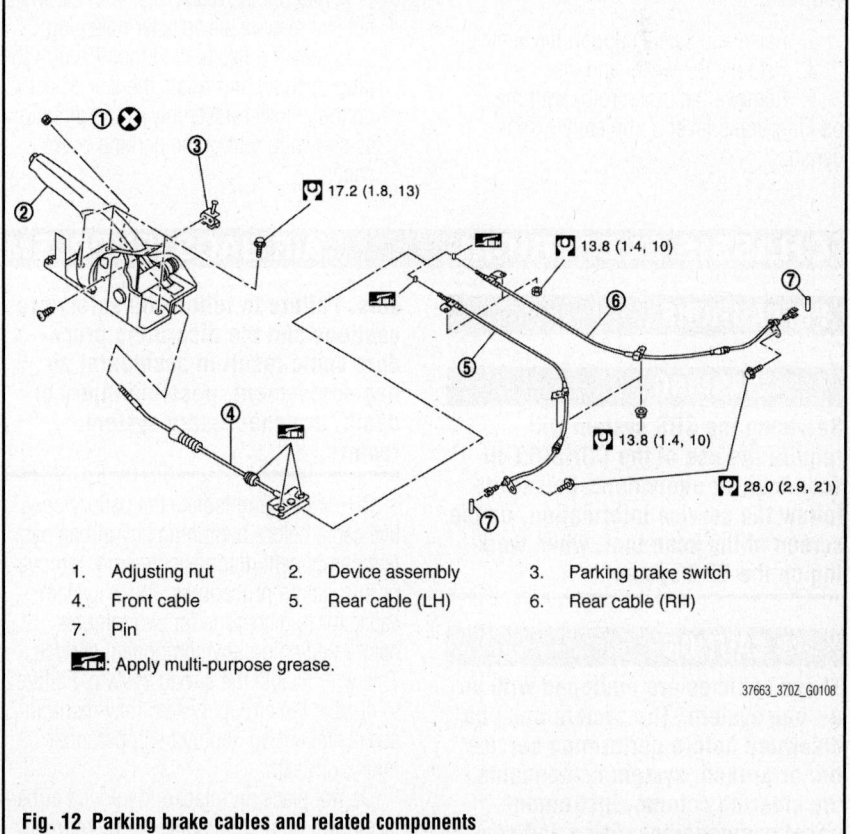

1. Adjusting nut
2. Device assembly
3. Parking brake switch
4. Front cable
5. Rear cable (LH)
6. Rear cable (RH)
7. Pin

▭: Apply multi-purpose grease.

37663_370Z_G0108

Fig. 12 Parking brake cables and related components

(Check it by listening and counting the ratchet clicks).

7. With the lever completely returned, make sure there is no drag on the rear brake.

PARKING BRAKE SHOES

REMOVAL & INSTALLATION

See Figure 13.

1. Before servicing the vehicle, refer to the Precautions Section.

➡️ **If working near and/or around the SRS system and components, be sure to disable the SRS system. After disabling the system wait three minutes or more before servicing the vehicle.**

➡️ **Before disconnecting the battery, lower both the driver's and passenger's windows. This will prevent any interference between the window edge and the vehicle when the door is opened or closed. During normal operation the window slightly raises and lowers automatically to prevent any window to vehicle interference. The automatic window function will not work with the battery disconnected.**

To install:

➡️ **Be sure to use new fasteners, as required.**

2. Raise and safely support the vehicle.
3. Remove the wheel and tire.
4. Remove the brake rotor with the parking brake lever completely disengaged.

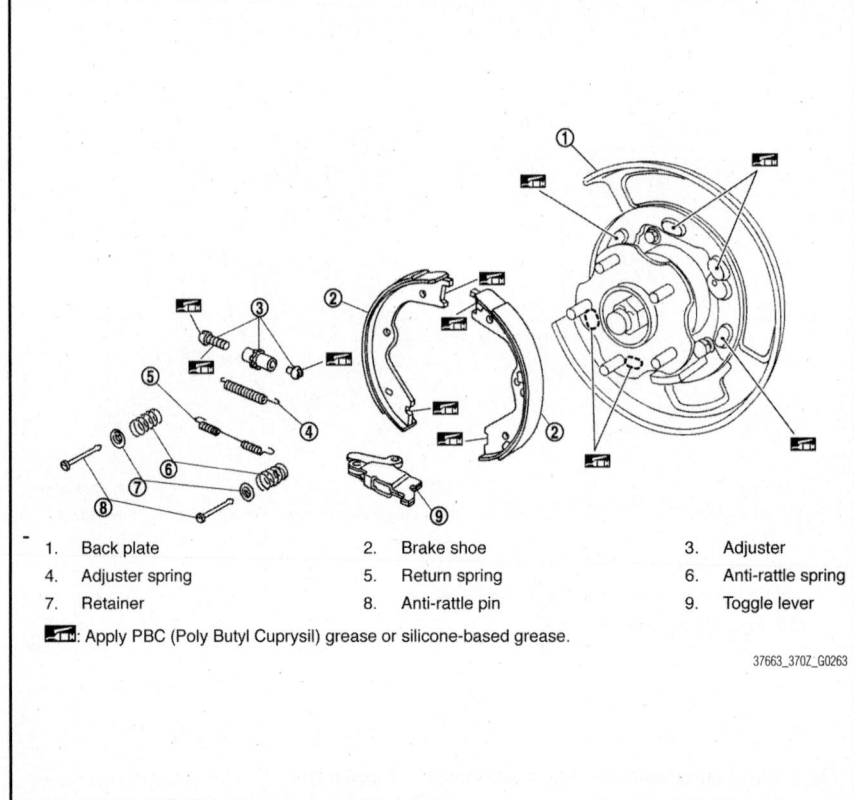

1.	Back plate	2.	Brake shoe	3.	Adjuster
4.	Adjuster spring	5.	Return spring	6.	Anti-rattle spring
7.	Retainer	8.	Anti-rattle pin	9.	Toggle lever

⚙️: Apply PBC (Poly Butyl Cuprysil) grease or silicone-based grease.

37663_370Z_G0263

Fig. 13 Parking brake shoes and related components

5. If the brake rotor cannot be removed, remove as follows:

a. Secure the brake rotor with the wheel nut and remove the adjuster hole plug.

b. Insert a flat-bladed tool through the plug opening and rotate the star wheel on the adjuster assembly in the direction as shown to retract the parking brake shoes.

c. Remove the parking brake shoe springs using a suitable tool.

d. Remove the parking brake shoes and adjuster.

To install:

6. Installation is in the reverse order of removal noting the following:

a. Apply brake grease to the brake shoe contact area.

CHASSIS ELECTRICAL   AIR BAGS (SUPPLEMENTAL RESTRAINT SYSTEM)

PRECAUTIONS

✳️✳️ WARNING

Servicing the SRS system will require the use of the CONSULT-III scan tool, or equivalent. Be sure to follow the service information, on the screen of the scan tool, when working on the SRS system.

✳️✳️ CAUTION

These vehicles are equipped with an air bag system. The system must be disarmed before performing service on, or around, system components, the steering column, instrument panel components, wiring and sen-

sors. Failure to follow the safety precautions and the disarming procedure could result in accidental air bag deployment, possible injury or death, or unnecessary system repairs.

Disconnect and isolate the battery negative cable before beginning any airbag system component diagnosis, testing, removal, or installation procedures. Allow system capacitor to discharge for two minutes before beginning any component service. This will disable the airbag system. Failure to disable the airbag system may result in accidental airbag deployment, personal injury, or death.

Do not place an intact undeployed airbag face down on a solid surface. The airbag

will propel into the air if accidentally deployed and may result in personal injury or death.

When carrying or handling an undeployed airbag, the trim side (face) of the airbag should be pointing towards the body to minimize possibility of injury if accidental deployment occurs. Failure to do this may result in personal injury or death.

Replace airbag system components with OEM replacement parts. Substitute parts may appear interchangeable, but internal differences may result in inferior occupant protection. Failure to do so may result in occupant personal injury or death.

Wear safety glasses, rubber gloves, and long sleeved clothing when cleaning powder residue from vehicle after an airbag deployment. Powder residue emitted from a deployed airbag can cause skin irritation. Flush affected area with cool water if irritation is experienced. If nasal or throat irritation is experienced, exit the vehicle for fresh air until the irritation ceases. If irritation continues, see a physician.

Do not use a replacement airbag that is not in the original packaging. This may result in improper deployment, personal injury, or death.

The factory installed fasteners, screws and bolts used to fasten airbag components have a special coating and are specifically designed for the airbag system. Do not use substitute fasteners. Use only original equipment fasteners listed in the parts catalog when fastener replacement is required.

During, and following, any child restraint anchor service, due to impact event or vehicle repair, carefully inspect all mounting hardware, tether straps, and anchors for proper installation, operation, or damage. If a child restraint anchor is found damaged in any way, the anchor must be replaced. Failure to do this may result in personal injury or death.

Deployed and non-deployed airbags may or may not have live pyrotechnic material within the airbag inflator.

Do not dispose of driver/passenger/curtain airbags or seat belt tensioners unless you are sure of complete deployment. Refer to the Hazardous Substance Control System for proper disposal.

Dispose of deployed airbags and tensioners consistent with state, provincial, local, and federal regulations.

After any airbag component testing or service, do not connect the battery negative cable. Personal injury or death may result if the system test is not performed first.

If the vehicle is equipped with the Occupant Classification System (OCS), do not connect the battery negative cable before performing the OCS Verification Test using the scan tool and the appropriate diagnostic information. Personal injury or death may result if the system test is not performed properly.

Never replace both the Occupant Restraint Controller (ORC) and the Occupant Classification Module (OCM) at the same time. If both require replacement, replace one, then perform the Airbag System test before replacing the other.

Both the ORC and the OCM store Occupant Classification System (OCS) calibration data, which they transfer to one another when one of them is replaced. If both are replaced at the same time, an irreversible fault will be set in both modules and the OCS may malfunction and cause personal injury or death.

If equipped with OCS, the Seat Weight Sensor is a sensitive, calibrated unit and must be handled carefully. Do not drop or handle roughly. If dropped or damaged, replace with another sensor. Failure to do so may result in occupant injury or death.

If equipped with OCS, the front passenger seat must be handled carefully as well. When removing the seat, be careful when setting on floor not to drop. If dropped, the sensor may be inoperative, could result in occupant injury, or possibly death.

If equipped with OCS, when the passenger front seat is on the floor, no one should sit in the front passenger seat. This uneven force may damage the sensing ability of the seat weight sensors. If sat on and damaged, the sensor may be inoperative, could result in occupant injury, or possibly death.

Never use air tools when servicing the SRS system.

DISARMING THE SYSTEM

✲✲ WARNING

Servicing the SRS system will require the use of the CONSULT-III scan tool, or equivalent. Be sure to follow the service information on the screen, of the scan tool, when working on the SRS system.

All SRS electrical wiring harnesses and connectors can be identified with YELLOW and or ORANGE color. Do not use electrical test equipment on any circuit related to the SRS (air bag) sensors. When installing SRS components, always install with the arrow marks facing the front of the vehicle.

To disarm the SRS system turn the ignition switch to **OFF** position. Then, disconnect the both battery cables starting with the negative cable first and wait at least 10 minutes after the cables are disconnected. Be sure to insulate the battery terminal ends.

ARMING THE SYSTEM

To arm the SRS system turn the ignition switch to **OFF** position. Connect the both battery cables starting with the positive cable first.

The SRS or air bag system is equipped with a self-diagnostic operation. After turning the ignition key to the ON or START position, the AIR BAG warning lamp will illuminate for 7 seconds. After 7 seconds, the AIR BAG lamp will extinguish if no malfunction is detected. If the AIR BAG lamp does not extinguish after 7 seconds, check the SRS self-diagnostic system for a malfunction.

DRIVETRAIN

AUTOMATIC TRANSMISSION

DRAIN & REFILL

See Figure 14.

At this time the manufacturer does not provide service information for this component. The following procedure is a guideline and may differ from the vehicle you are servicing.

1. Before servicing the vehicle, refer to the Precautions Section.

➡**If working near and/or around the SRS system and components, be sure to disable the SRS system. After disabling the system wait three minutes or more before servicing the vehicle.**

➡**Before disconnecting the battery, lower both the driver's and passenger's windows. This will prevent any interference between the window edge and the vehicle when the door is opened or closed. During normal operation the window slightly raises and lowers automatically to prevent any window to vehicle interference. The automatic window function will not work with the battery disconnected.**

2. Run the engine until operating temperature is reached.
3. Stop the engine.
4. Loosen the fluid level gauge bolt.
5. Raise and support the vehicle safely.

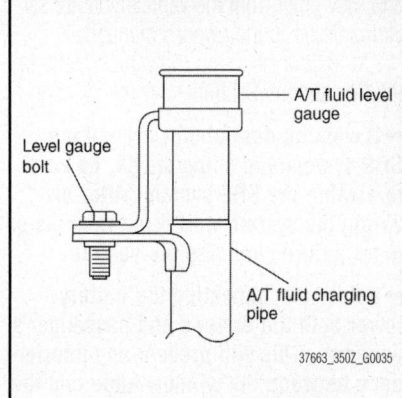

Fig. 14 Automatic transmission fluid level gauge location and related components

6. Loosen the drain plug. Discard the gasket.

➡**When replacing the drain plug use a new gasket and tighten to 25 ft. lbs. (34 Nm).**

7. Drain the fluid from the drain plug and refill with new fluid. Always refill same volume with drained fluid.

8. To replace the fluid, pour in new fluid at the transmission charging pipe with the engine idling and at the same time drain the old fluid from the radiator cooler hose return side.

9. When the color of the fluid coming out is about the same as the color of the new fluid, replacement is complete.

➡**The amount of new fluid to use should be a 30–50 percent increase of the stipulated amount.**

10. The fluid capacity is approximately 10 7/8 quarts (10.3 liters).

➡ **Be sure to use the proper grade and type replacement fluid (genuine NISSAN Matic S ATF). Do not mix with other fluids. Using other than the required fluid will cause deterioration in drivability and transmission damage.**

11. Run the engine at idle speed for about five minutes.

12. Check and correct the fluid level.

13. If the fluid is still dirty, repeat the process.

14. Install the removed fluid level gauge in the fluid charging pipe.

15. Install the fluid level gauge bolt. Tighten to 45 inch lbs. (5.1 Nm).

FLUID LEVEL CHECK

At this time the manufacturer does not provide service information for this component. The following procedure is a guideline and may differ from the vehicle you are servicing. Refer to the owner's manual.

1. Before servicing the vehicle, refer to the Precautions Section.

➡If working near and/or around the SRS system and components, be sure to disable the SRS system. After disabling the system wait three minutes or more before servicing the vehicle.

➡Before disconnecting the battery, lower both the driver's and passenger's windows. This will prevent any interference between the window edge and the vehicle when the door is opened or closed. During normal operation the window slightly raises and lowers auto-

matically to prevent any window to vehicle interference. The automatic window function will not work with the battery disconnected.

2. Run the engine until operating temperature is reached.

3. Check for external fluid leakage.

4. Loosen the level gauge bolt.

5. Before driving, fluid level can be checked at fluid temperatures of 86–122 degrees F using the COLD range on the dipstick.

6. To COLD check the transmission, park the vehicle on a level surface and set the parking brake.

7. Start the engine and move the selector lever through each gear range. Leave the selector lever in the PARK (P) range.

8. Check the fluid level with the engine idling. Always use lint free paper, not a cloth to check the fluid level.

9. Re-insert the gauge. Remove the gauge and note the reading. As required add fluid. Do not overfill.

10. Drive the vehicle for about five minutes in urban areas.

11. Using the CONSULT-III tool, or equivalent, make the fluid temperature about 149 degrees F.

12. Recheck the fluid level, correct as required.

FLUID RECOMMENDATIONS

Genuine NISSAN Matic S ATF is recommended.

MANUAL TRANSMISSION

DRAIN & REFILL

See Figure 15.

1. Before servicing the vehicle, refer to the Precautions Section.

➡If working near and/or around the SRS system and components, be sure to disable the SRS system. After disabling the system wait three minutes or more before servicing the vehicle.

➡Before disconnecting the battery, lower both the driver's and passenger's windows. This will prevent any interference between the window edge and the vehicle when the door is opened or closed. During normal operation the window slightly raises and lowers automatically to prevent any window to vehicle interference. The automatic window function will not work with the battery disconnected.

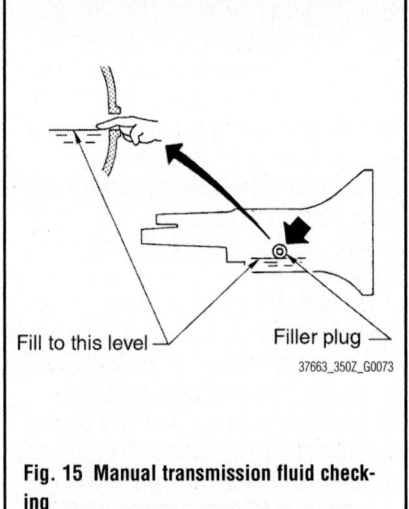
Fill to this level   Filler plug
37663_350Z_G0073

Fig. 15 Manual transmission fluid checking

2. Start the engine and let it run to warm up the transmission.

3. Stop the engine.

4. Raise and safely support the vehicle.

5. Position a drain pan under the transmission.

6. Remove the drain plug. Discard the gasket.

7. Drain the fluid. Be sure to properly dispose of used fluid.

To install:

➡**Be sure to use new fasteners, as required.**

8. Installation is the reverse of the removal procedure.

9. Be sure to use a new gasket.

10. Tighten the drain plug to 25 ft. lbs. (34.5 Nm).

11. The approximate fill capacity is 6 pints.

12. Be sure to use the proper grade and type fluid.

FLUID LEVEL CHECK

1. Remove filler plug and gasket from transmission case.

2. Check the oil level from filler plug mounting hole.

❊❊ WARNING

Never start the engine while checking oil level.

3. Set a gasket on filler plug and then install it to transmission case.

❊❊ WARNING

Never reuse gasket.

4. Tighten filler plug.

FLUID RECOMMENDATIONS

Genuine NISSAN Manual Transmission Fluid (MTF) HQ Multi 75W-85 or API GL-4, Viscosity SAE 75W-85 is recommended.

CLUTCH HYDRAULIC SYSTEM BLEEDING

BLEEDING PROCEDURE

At this time the manufacturer does not provide service information for bleeding the system. The following procedure is a guideline and may differ from the vehicle you are servicing.

Bleeding is required to remove air trapped in the hydraulic system. The bleed screw is located on the clutch slave (operating) cylinder.

Some models are also equipped with a clutch damper mechanism. The clutch damper mechanism is bled in exactly the same manner as the operating cylinder. It should be bled along with the operating cylinder.

1. Before servicing the vehicle, refer to the Precautions Section.
2. Remove the bleed screw dust cap.
3. Attach a transparent vinyl tube to the bleed screw, immersing the free end in a clean container of clean brake fluid.
4. Fill the master cylinder with the proper fluid.
5. Open the bleed screw about ¾ turn.
6. Depress the clutch pedal quickly. Hold it down. Have an assistant tighten the bleed screw. Allow the pedal to return slowly.
7. Repeat the above procedure until no more air bubbles are seen in the fluid container.
8. Remove the bleed tube.
9. Replace the dust cap and refill the master cylinder.
10. Bleed the clutch damper, if equipped.

FLUID FILL PROCEDURE

> ☀☀ **WARNING**
>
> **Keep painted surface on the body or other parts free of clutch fluid. If it spills, wipe up immediately and wash the affected area with water.**

1. Check that there is no foreign material in reservoir tank and then fill with new clutch fluid.

> ☀☀ **WARNING**
>
> **Never reuse drained clutch fluid.**

2. Loosen air bleeder valve, slowly depress clutch pedal to the full stroke and then release clutch pedal.
3. Repeat this operation at intervals of 2 or 3 seconds until new clutch fluid is discharged.
4. Tighten air bleeder valve with the clutch pedal depressed.
5. Perform the air bleeding.

FLUID LEVEL CHECK

1. Check that the fluid level in the reservoir tank is within the specified range (MAX - MIN lines).
2. Visually check for any fluid leakage around the reservoir tank.
3. Check the clutch system for any leakage if the fluid level is extremely low (lower than MIN).

REAR AXLE HOUSING (ASSEMBLY, UNIT)

REMOVAL & INSTALLATION

See Figures 16 through 18.

1. Before servicing the vehicle, refer to the Precautions Section.

➡**If working near and/or around the SRS system and components, be sure to disable the SRS system. After disabling the system wait three minutes or more before servicing the vehicle.**

➡**Before disconnecting the battery, lower both the driver's and passenger's windows. This will prevent any interfer-**

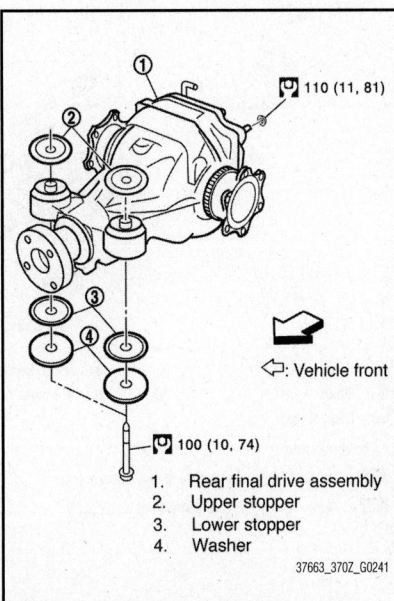

1. Rear final drive assembly
2. Upper stopper
3. Lower stopper
4. Washer

⇦: Vehicle front

37663_370Z_G0241

Fig. 16 Rear drive axle and related components—R200

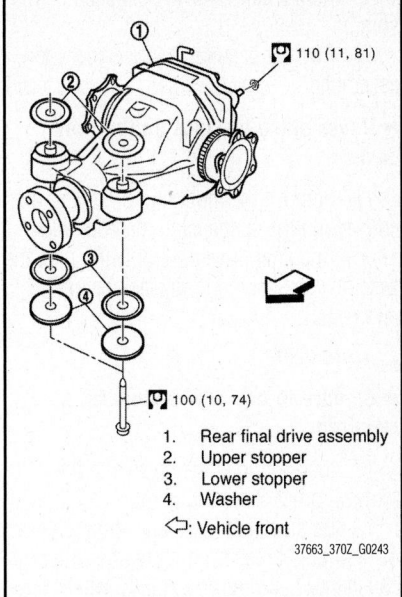

1. Rear final drive assembly
2. Upper stopper
3. Lower stopper
4. Washer

⇦: Vehicle front

37663_370Z_G0243

Fig. 17 Rear drive axle and related components—R200V

ence between the window edge and the vehicle when the door is opened or closed. During normal operation the window slightly raises and lowers automatically to prevent any window to vehicle interference. The automatic window function will not work with the battery disconnected.

2. Raise and support the vehicle safely.
3. Remove the center muffler.
4. Remove the diag brace.
5. Remove the stabilizer bar.
6. Remove the driveshaft.
7. Remove the rear halfshafts from the final drive.
8. Remove the breather hose from the final drive. Discard the hose clamp.

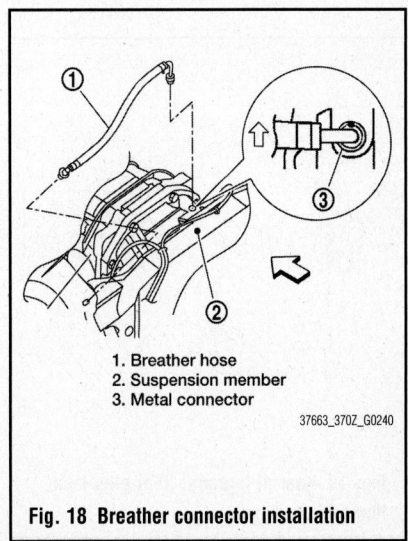

1. Breather hose
2. Suspension member
3. Metal connector

37663_370Z_G0240

Fig. 18 Breather connector installation

9. Remove the rear wheel speed sensors.

10. Position a suitable jack under the assembly.

➡**Never place the jack on the rear cover.**

11. Remove the mounting bolts and nuts connecting the suspension member and remove the final drive assembly. Be sure the assembly is secured in the jack before removing it.

To install:

➡**Be sure to use new fasteners, as required.**

12. Installation is the reverse of the removal procedure.

13. Be sure that there are no pinched or restricted areas on the breather hose caused by bending or winding, when installing it.

14. When installing the new hose clamp, install it at the final drive side with the tab facing downward.

15. If the breather connector was removed, install it as shown in the illustration. Never reuse the breather connector and metal connector.

DRAIN & REFILL

See Figure 19.

1. Before servicing the vehicle, refer to the Precautions Section.

2. Raise and safely support the vehicle.

3. Remove the drain plug and drain the fluid into a suitable container.

4. Discard the gasket.

5. Fill the unit with new gear oil until the oil level reaches the specified level near the filler plug mounting hole.

➡**Be sure to use the proper grade and type gear oil.**

6. Gear oil capacity is 3 pints (1.4 liters).

7. Check the oil level after refilling.

8. Using a new gasket install the drain plug. Tighten to specification.

REAR HALFSHAFT

REMOVAL & INSTALLATION

See Figure 20.

1. Before servicing the vehicle, refer to the Precautions Section.

➡**If working near and/or around the SRS system and components, be sure to disable the SRS system. After disabling the system wait three minutes or more before servicing the vehicle.**

➡**Before disconnecting the battery, lower both the driver's and passenger's windows. This will prevent any interference between the window edge and the vehicle when the door is opened or closed. During normal operation the window slightly raises and lowers automatically to prevent any window to vehicle interference. The automatic window function will not work with the battery disconnected.**

2. Raise and support the vehicle safely.

3. Remove the tire and wheel assemblies.

4. Remove and discard the cotter pin.

5. Loosen the wheel hub locknut.

6. Matchmark the halfshaft and wheel hub and bearing.

7. Remove the diag brace.

8. Remove the center muffler.

9. Patch wheel hub locknut with a piece of wood. Hammer the wood to disengage the wheel hub and bearing assembly from the halfshaft.

➡**Never position the halfshaft at an extreme angle. Do not overextend the slide joint. Properly support the halfshaft, do not allow it to hang unsupported.**

10. Use a suitable puller if the wheel hub and bearing assembly and halfshaft cannot be separated even after performing the above procedure.

11. Remove the wheel locknut.

12. Remove the mounting bolts between the side flange and the halfshaft.

To install:

➡**Be sure to use new fasteners, as required.**

13. Installation is the reverse of the removal procedure.

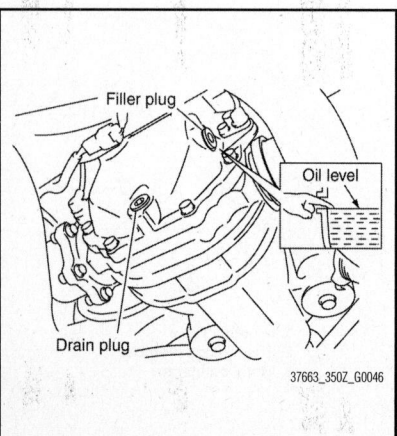

Fig. 19 Rear differential filler plug location and checking

Filler plug

Oil level

Drain plug

37663_350Z_G0046

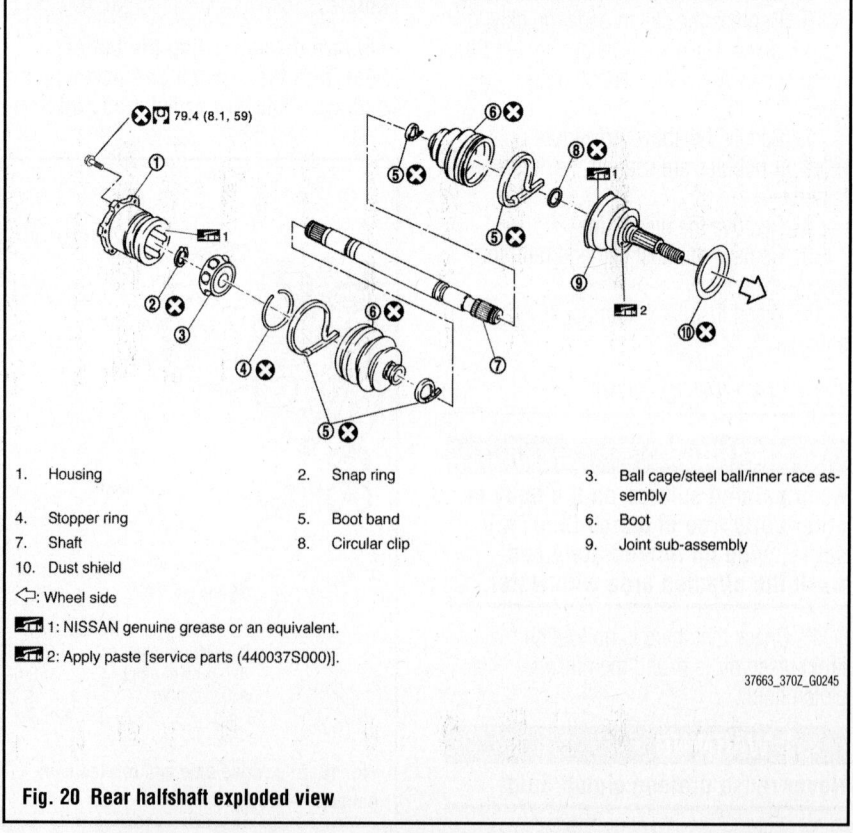

1. Housing
2. Snap ring
3. Ball cage/steel ball/inner race assembly
4. Stopper ring
5. Boot band
6. Boot
7. Shaft
8. Circular clip
9. Joint sub-assembly
10. Dust shield
⟵: Wheel side
🔧 1: NISSAN genuine grease or an equivalent.
🔧 2: Apply paste [service parts (440037S000)].

37663_370Z_G0245

Fig. 20 Rear halfshaft exploded view

14. Clean the matching surface of the half-shaft and wheel hub and bearing assembly.

15. Apply paste part number 440037S000 or equivalent to the surface of the sub joint assembly of the halfshaft.

→**Apply the paste, about 0.04–0.10 ounce) to cover the entire flat surface of the sub joint assembly of the halfshaft.**

16. The wheel hub locknut tightening specification is 136 ft. lbs. (185 Nm).

17. Perform a final tightening of nuts and bolts of each removed component with the vehicle in an unladen position.

ENGINE COOLING

ENGINE COOLANT

DRAIN & REFILL

1. Before servicing the vehicle, refer to the Precautions Section.

→**If working near and/or around the SRS system and components, be sure to disable the SRS system. After disabling the system wait three minutes or more before servicing the vehicle.**

→**Before disconnecting the battery, lower both the driver's and passenger's windows. This will prevent any interference between the window edge and the vehicle when the door is opened or closed. During normal operation the window slightly raises and lowers automatically to prevent any window to vehicle interference. The automatic window function will not work with the battery disconnected.**

❄❄ CAUTION

Never drain the engine coolant when the engine is hot. Wrap a thick cloth around the cap and carefully remove it. First turn the cap a quarter of a turn to release any pressure, and then turn it all the way. Do not allow coolant to come in contact with the drive belts.

2. Raise and safely support the vehicle.
3. Remove the undercover.
4. Open the radiator drain plug and carefully drain the coolant into a suitable container. Be sure to properly dispose of used coolant.
5. Discard the O-ring.
6. Open the water drain plugs on the cylinder block and drain the coolant into a suitable container. Be sure to properly dispose of used coolant.
7. As necessary, remove the reservoir tank and drain the coolant. Clean the tank before reinstalling.
8. If removed install the reservoir tank.
9. Remove the air cleaner assembly.
10. Install the drain plug. Be sure to use a new O-ring.

11. Tighten the plug to 11 inch lbs. (1.2 Nm).
12. Install the engine drain plugs, if removed.
13. Check that all hose clamps are tight.
14. Remove the air relief plug on the heater hose. Discard the O-ring.
15. Fill the radiator and the reservoir with the proper grade and type engine coolant.
16. Specification is approximately 9 quarts (8.5 liters) for the radiator and 7/8ths quart (0.8 liter) for the reservoir at MAX level.
17. When coolant overflows at the air relief hole on the heater hole, install the air relief plug using a new O-ring. Tighten to 11 inch lbs. (1.2 Nm).
18. Install the radiator cap.
19. Warm the engine until the thermostat opens. Warm up time is about ten minutes at 3000 rpm's. Watch the temperature gauge so as not to overheat the engine.
20. Stop the engine and allow cool down to less than 122 degree F.
21. Refill the reservoir tank to the MAX position.
22. Repeat the above steps two or more times with the radiator cap installed until the engine coolant level no longer drops.

23. Check the cooling system for leaks with the engine running.
24. Warm up the engine and check for the sound of coolant flow while running the engine from idle up to 3000 rpm's with the heater temperature set at several positions from COOL to WARM.
25. Repeat the above step three times.
26. If sound is heard bleed air from the cooling system by repeating the above procedure until engine coolant level no longer drops.

FLUID RECOMMENDATIONS

Pre-diluted Genuine NISSAN Long Life Antifreeze/ Coolant (blue) or equivalent.

ELECTRIC ENGINE FAN

REMOVAL & INSTALLATION

See Figure 21.

1. Before servicing the vehicle, refer to the Precautions Section.

→**If working near and/or around the SRS system and components, be sure to disable the SRS system. After disabling the system wait three minutes or more before servicing the vehicle.**

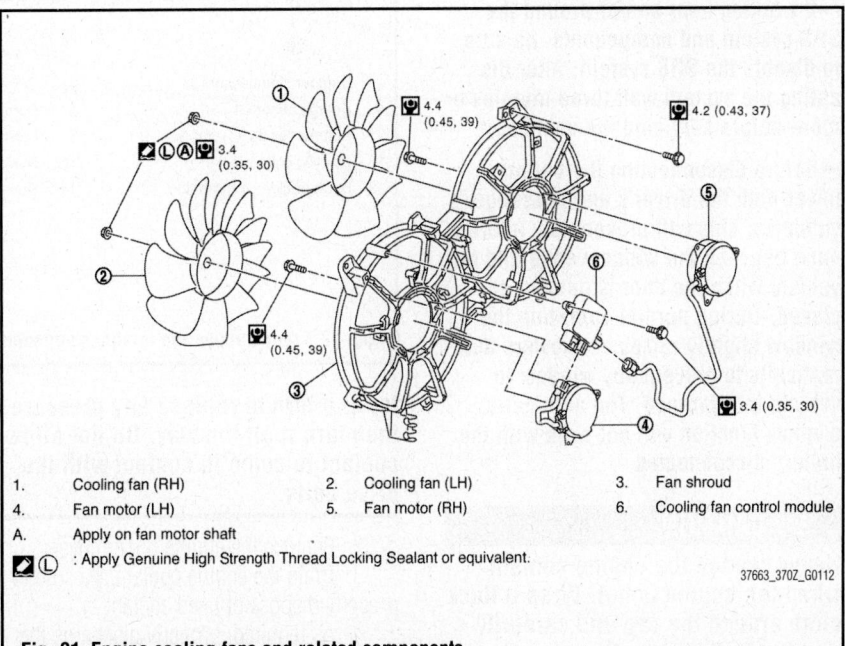

1.	Cooling fan (RH)	2.	Cooling fan (LH)	3.	Fan shroud
4.	Fan motor (LH)	5.	Fan motor (RH)	6.	Cooling fan control module
A.	Apply on fan motor shaft				

🔧Ⓛ : Apply Genuine High Strength Thread Locking Sealant or equivalent.

37663_370Z_G0112

Fig. 21 Engine cooling fans and related components

➡Before disconnecting the battery, lower both the driver's and passenger's windows. This will prevent any interference between the window edge and the vehicle when the door is opened or closed. During normal operation the window slightly raises and lowers automatically to prevent any window to vehicle interference. The automatic window function will not work with the battery disconnected.

2. Remove the reservoir tank.

3. Disconnect the crash zone sensor harness clips from the fan shroud, move the harness to the side.

4. Disconnect the harness connector from the cooling fan control module, move the harness to the side.

5. Remove the under cover.

6. Disconnect and plug the automatic transmission fluid cooler hose from the fan shroud, if equipped.

7. Remove the cooling fan from under the vehicle.

To install:

➡Be sure to use new fasteners, as required.

8. Installation is the reverse of the removal procedure.

RADIATOR

REMOVAL & INSTALLATION

See Figures 22 through 26.

1. Before servicing the vehicle, refer to the Precautions Section.

➡If working near and/or around the SRS system and components, be sure to disable the SRS system. After disabling the system wait three minutes or more before servicing the vehicle.

➡Before disconnecting the battery, lower both the driver's and passenger's windows. This will prevent any interference between the window edge and the vehicle when the door is opened or closed. During normal operation the window slightly raises and lowers automatically to prevent any window to vehicle interference. The automatic window function will not work with the battery disconnected.

✳✳ CAUTION

Never change the engine coolant when the engine is hot. Wrap a thick cloth around the cap and carefully remove it. First turn the cap a quar-

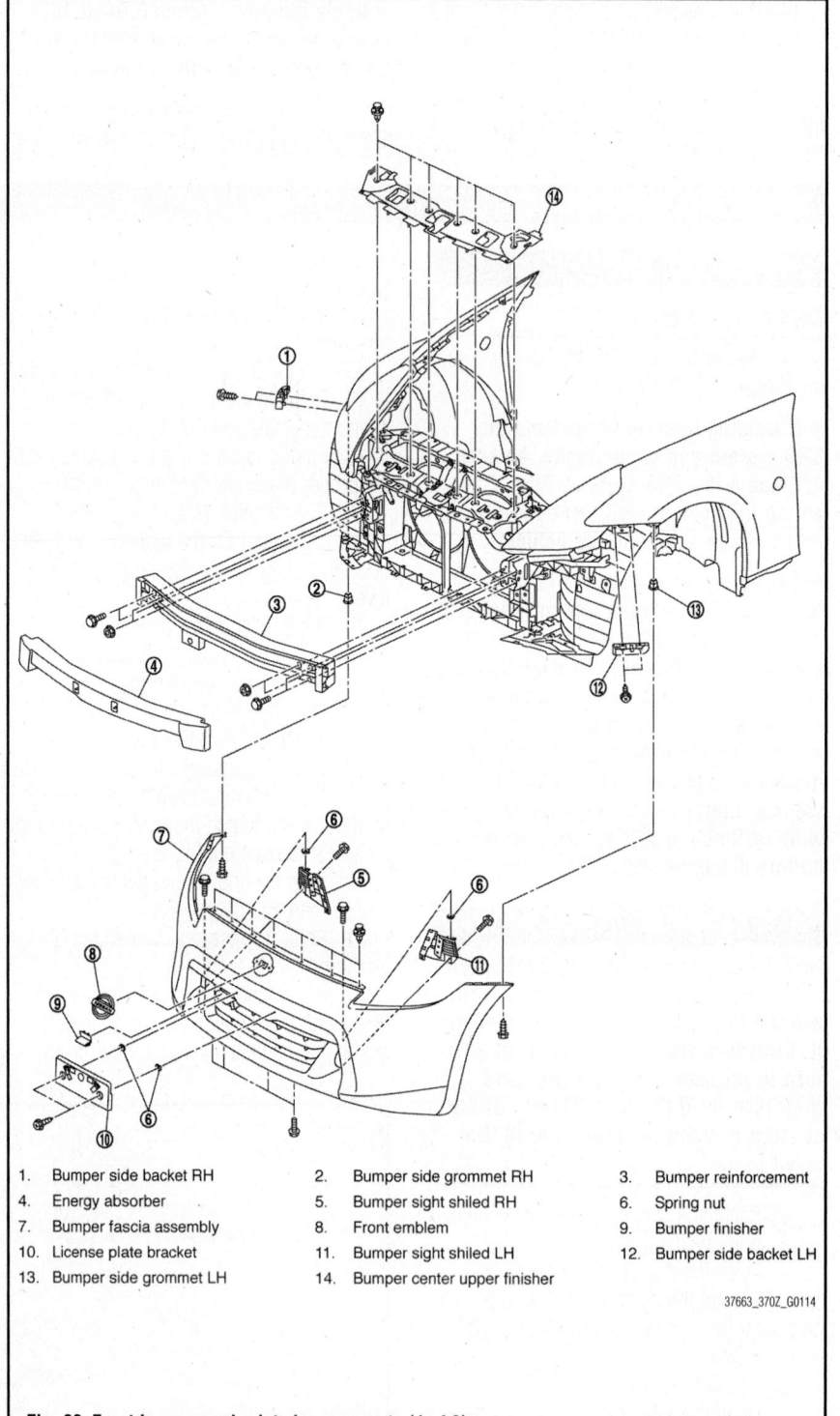

1.	Bumper side backet RH	2.	Bumper side grommet RH	3.	Bumper reinforcement
4.	Energy absorber	5.	Bumper sight shiled RH	6.	Spring nut
7.	Bumper fascia assembly	8.	Front emblem	9.	Bumper finisher
10.	License plate bracket	11.	Bumper sight shiled LH	12.	Bumper side backet LH
13.	Bumper side grommet LH	14.	Bumper center upper finisher		

37663_370Z_G0114

Fig. 22 Front bumper and related components (1 of 2)

ter of a turn to release any pressure, then turn it all the way. Do not allow coolant to come in contact with the drive belts.

2. Remove the engine under cover.

3. Drain the engine coolant. Be sure to properly dispose of used coolant.

4. As required, properly discharge the air conditioning system.

5. Remove the air cleaner assembly.

6. Remove the reservoir tank.

7. Remove the bumper center upper finisher and bumper fascia assembly.

8. Properly support the hood, with a proper support tool.

9. Disconnect the harness clips and

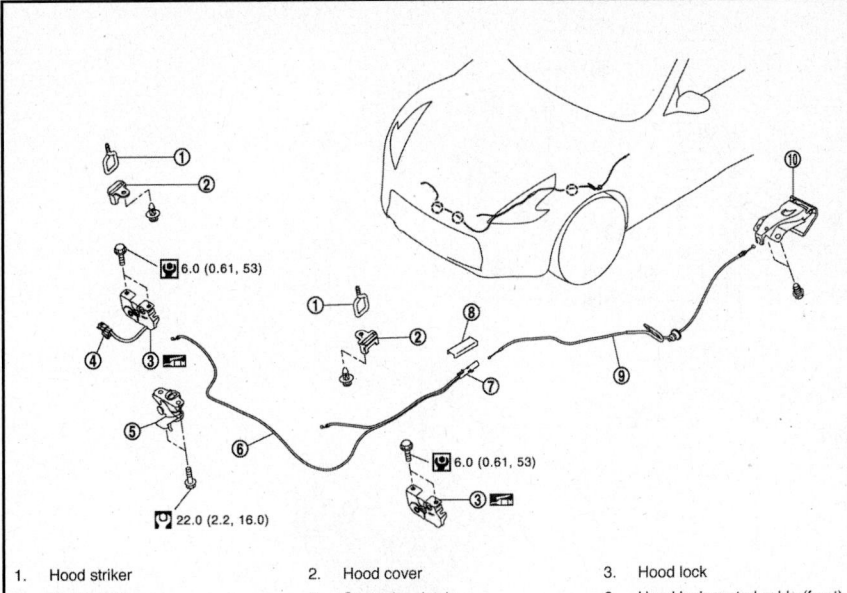

1. Hood striker
2. Hood cover
3. Hood lock
4. Hood switch
5. Secondary latch
6. Hood lock control cable (front)
7. Hood lock control cable protector
8. Hood lock control cable protector cover
9. Hood lock control cable (rear)
10. Hood lock opener

◯ : Clip

37663_370Z_G0115

Fig. 23 Hood latch and related components

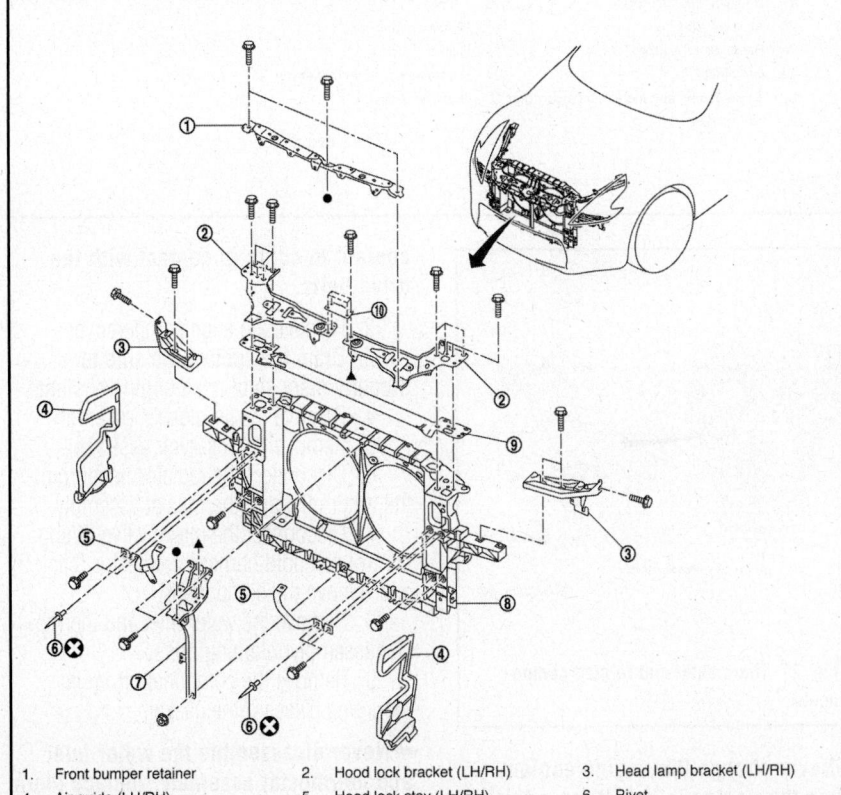

1. Front bumper retainer
2. Hood lock bracket (LH/RH)
3. Head lamp bracket (LH/RH)
4. Air guide (LH/RH)
5. Hood lock stay (LH/RH)
6. Rivet
7. Hood lock stay assembly
8. Radiator core support assembly
9. Radiator core support reinforcement
10. Hood lock bracket (center)

37663_370Z_G0116

Fig. 24 Front bumper and related components (2 of 2)

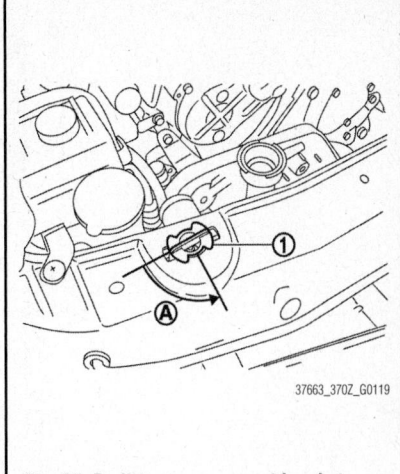

37663_370Z_G0119

Fig. 25 Radiator upper mount bracket removal

hood lock control cable clips from the bumper retainer.

10. Remove the front bumper retainer.

11. Remove the horn. Remove the hood locks.

12. Remove the front combination lamp, left side.

13. Remove the hood lock, right and left brackets. Remove the hood lock stay assembly.

14. Remove the condenser pipe assembly.

15. Remove the hood lock stay mounting bolt.

16. Disconnect and plug the automatic transmission cooler lines at the radiator, if equipped.

17. Remove the upper and lower radiator hoses.

18. Remove the radiator water inlet pipe.

19. Rotate the two upper radiator mount brackets ninety degrees, in the direction shown in the illustration.

20. To remove the radiator and condenser assembly, lift up and pull the assembly forward and then remove the lower mounting rubber from the radiator support.

21. Remove the assembly from the front of the radiator core support.

To install:

➡**Be sure to use new fasteners, as required.**

22. Installation is the reverse of the removal procedure.

23. Be sure to fill the cooling system with the proper grade and type engine coolant.

24. Check that there is no fluid leakage at the automatic transmission lines, if equipped.

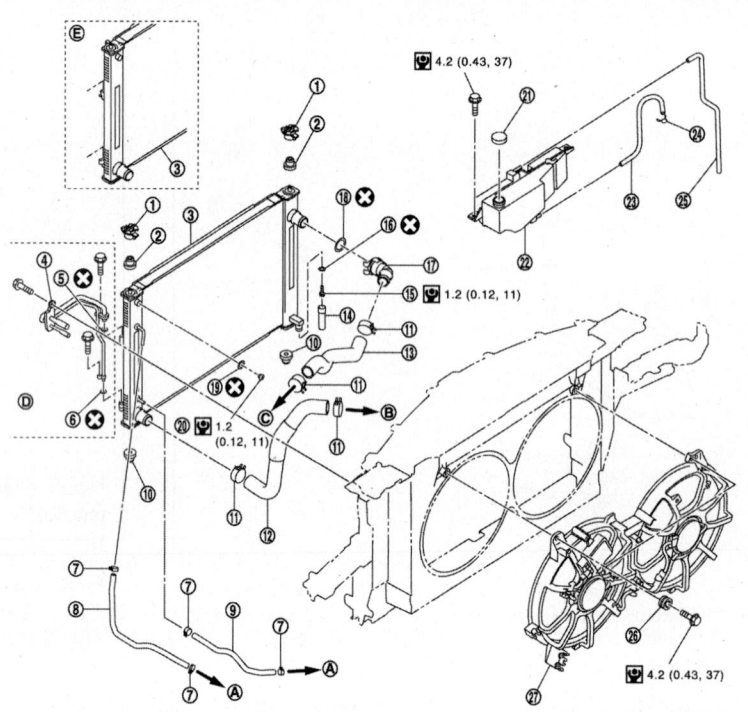

1. Upper mount bracket
2. Mounting rubber (upper)
3. Radiator & condenser assembly
4. Condenser pipe assembly
5. O-ring
6. O-ring
7. Clamp (A/T models)
8. A/T fluid cooler hose (A/T models)
9. A/T fluid cooler hose (A/T models)
10. Mounting rubber (lower)
11. Clamp
12. Radiator hose (lower)
13. Radiator hose (upper)
14. Water drain hose
15. Drain plug
16. O-ring
17. Radiator water inlet pipe
18. O-ring
19. O-ring
20. Air relief plug
21. Reservoir tank cap
22. Reservoir tank
23. Reservoir tank hose
24. Clamp
25. Reservoir tank hose
26. Grommet
27. Radiator cooling fan assembly
A. To transmission (A/T models)
B. To water inlet and thermostat assembly
C. To water outlet
E. M/T models

37663_370Z_G0047

Fig. 26 Radiator and related components

THERMOSTAT

REMOVAL & INSTALLATION

See Figure 27.

1. Before servicing the vehicle, refer to the Precautions Section.

➡️If working near and/or around the SRS system and components, be sure to disable the SRS system. After disabling the system wait three minutes or more before servicing the vehicle.

➡️Before disconnecting the battery, lower both the driver's and passenger's windows. This will prevent any interference between the window edge and the vehicle when the door is opened or closed. During normal operation the window slightly raises and lowers automatically to prevent any window to vehicle interference. The automatic window function will not work with the battery disconnected.

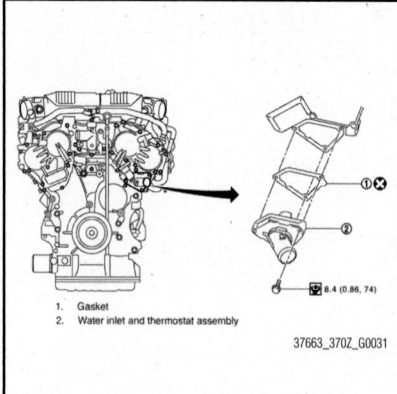

1. Gasket
2. Water inlet and thermostat assembly

37663_370Z_G0031

Fig. 27 Thermostat and related components

➡️Never change the engine coolant when the engine is hot. Wrap a thick cloth around the cap and carefully remove it. First turn the cap a quarter of a turn to release any pressure, then turn it all the way. Do not allow

coolant to come in contact with the drive belts.

2. Remove the engine undercover.
3. Drain the coolant. Be sure to properly dispose of used engine coolant.
4. Remove the air cleaner assembly.
5. Remove the reservoir assembly.
6. Disconnect the radiator hose from the water inlet and thermostat assembly.
7. Disconnect the intake valve timing control solenoid harness connector (bank 2) and remove the component.
8. Remove the water inlet and thermostat assembly retaining bolts.
9. Remove the component from its mounting. Discard the gasket.

➡️Never disassemble the water inlet and thermostat assembly, replace them as a complete unit.

To install:

➡️Be sure to use new fasteners, as required.

10. Installation is the reverse of the removal procedure.

11. Be sure to fill the cooling system with the proper grade and type engine coolant.

WATER PUMP

REMOVAL & INSTALLATION

See Figures 28 through 32.

1. Before servicing the vehicle, refer to the Precautions Section.

➡ **If working near and/or around the SRS system and components, be sure to disable the SRS system. After disabling the system wait three minutes or more before servicing the vehicle.**

➡ **Before disconnecting the battery,**

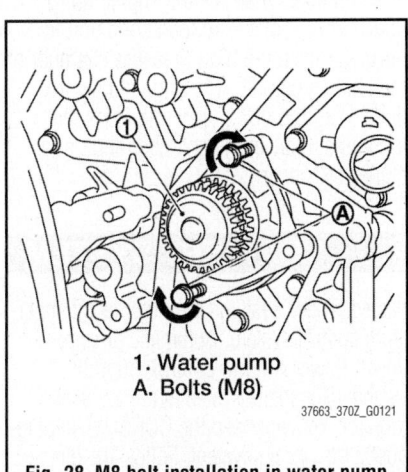

1. Water pump
A. Bolts (M8)

37663_370Z_G0121

Fig. 28 M8 bolt installation in water pump

lower both the driver's and passenger's windows. This will prevent any interference between the window edge and the vehicle when the door is opened or closed. During normal operation the window slightly raises and lowers automatically to prevent any window to vehicle interference. The automatic window function will not work with the battery disconnected.

➡ **Never change the engine coolant when the engine is hot. Wrap a thick cloth around the cap and carefully remove it. First turn the cap a quarter of a turn to release any pressure, then turn it all the way. Do not allow coolant to come in contact with the drive belts.**

2. Remove the engine cover.

3. Properly relieve the fuel system pressure.

4. Disconnect the negative battery cable.

5. Remove the air cleaner assembly.

6. Remove the reservoir tank.

7. Separate the engine harness by removing their brackets from the front timing chain case.

8. Remove the engine undercover.

9. Drain the engine oil. Be sure to properly dispose of used oil.

10. Drain the engine coolant. Be sure to properly dispose of used coolant.

11. Remove the radiator hoses.

12. Remove the cooling fan assembly.

13. Remove the front timing chain case cover.

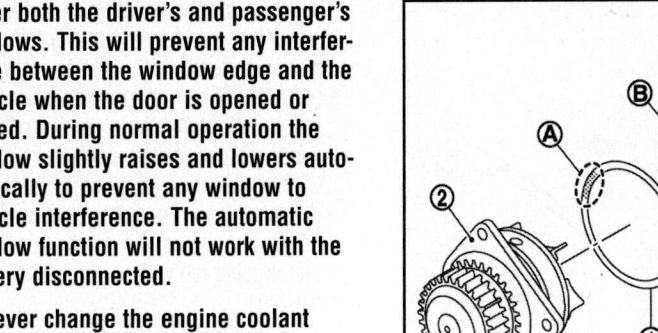

1. O-ring (yellow paint mark)
2. Water pump
3. O-ring (light blue paint mark)
A. Yellow paint mark
B. Light blue paint mark

37663_370Z_G0122

Fig. 30 Water pump and O-ring positioning and color identification

14. To remove the timing chain tensioner (primary), remove the lower mounting bolt. Loosen the upper mounting bolt slowly and then turn the timing chain tensioner (primary) on the upper mounting bolt so that the plunger is fully expanded. Remove the upper mounting bolt and then remove the timing chain tensioner (primary).

➡ **Even if the plunger is fully expanded, it does not drop from the body of the timing chain tensioner (primary).**

15. To remove the water pump, remove the three retaining bolts. Secure a gap between the water pump gear and the timing chain, by turning the crankshaft counterclockwise until the timing chain looseness on the water pump sprocket becomes maximum.

16. Screw M8 bolts (pitch 0.0492 inch) in length approximately 1.97 inch into the water pump upper and lower mounting bolt holes until they reach the timing chain case. Then alternately tighten each bolt for a half a turn, and pull out the water pump.

➡ **Pull the pump straight out preventing the vane from contacting the socket in the installation area. Remove the pump without causing the sprocket to contact the chain.**

17. Remove the M8 bolts and O-rings from the water pump.

To install:

➡ **Be sure to use new fasteners, as required.**

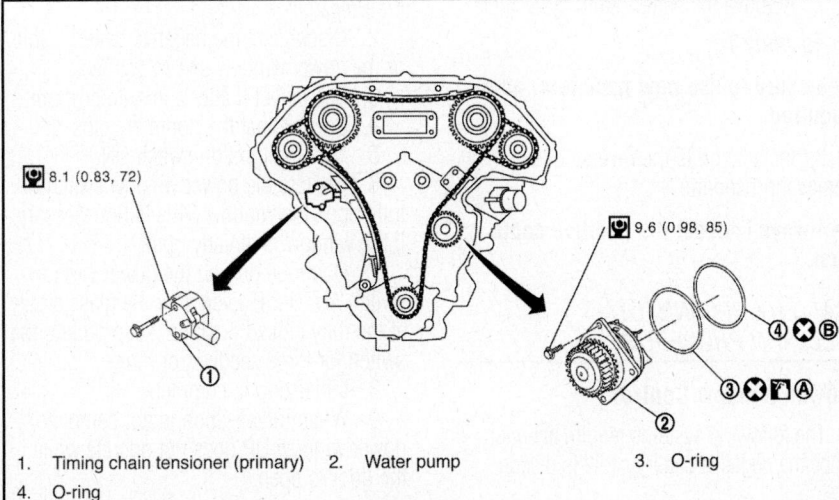

1. Timing chain tensioner (primary)   2. Water pump   3. O-ring
4. O-ring
A. Identify with yellow paint mark   B. Identify with light blue paint mark
                                          Apply engine coolant

37663_370Z_G0120

Fig. 29 Water pump and related components

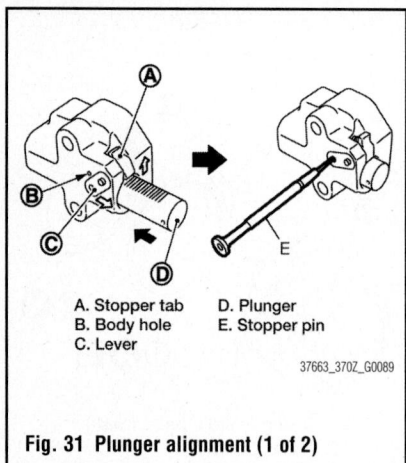

A. Stopper tab   D. Plunger
B. Body hole   E. Stopper pin
C. Lever

37663_370Z_G0089

Fig. 31 Plunger alignment (1 of 2)

18. Apply clean engine oil to the yellow paint marked O-ring. Apply clean coolant to the light blue paint marked O-ring. Install the O-rings on the water pump.

19. Install the water pump. Never allow the cylinder block to nip the O-rings. Check that the chain and pump sprocket are engaged.

20. Tighten the mounting bolts alternately and securely.

21. To install the timing chain tensioner (primary) pull the plunger stopper tab up (or turn lever downward) so as to remove the plunger stopper tab from the ratchet of the plunger. Note that the plunger stopper tab and lever are synchronized.

22. Push the plunger into the inside of the tensioner body. Hold the plunger in the fully compressed position by engaging the plunger stopper tab with the tip of the ratchet.

23. To secure the lever, insert the stopper pin through the hole of the lever into the tensioner body hole. The lever parts and the plunger stopper tab are synchronized, therefore the plunger is secured under this condition.

➡**The illustration shows a suitable tool of 0.047 inch diameter being used as a stopper pin.**

24. Pull out the stopper pin after installing and release the plunger.

25. Check again that the water pump sprocket and timing chain are engaged.

26. Continue the installation in the reverse order of the removal procedure.

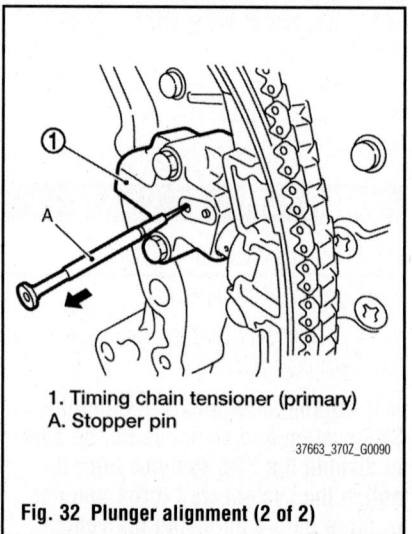

1. Timing chain tensioner (primary)
A. Stopper pin

37663_370Z_G0090

Fig. 32 Plunger alignment (2 of 2)

27. After starting the engine let it idle for three minutes, then rev the engine up to 3000 rpm's under a no load condition to purge air from the high pressure chamber of the chain tensioner. The engine may produce a rattling noise. This indicates that there is still air in the chamber and not a matter of concern.

ENGINE ELECTRICAL

BATTERY

REMOVAL & INSTALLATION

See Figure 33.

1. Before servicing the vehicle, refer to the Precautions Section.

➡**If working near and/or around the SRS system and components, be sure to disable the SRS system. After disabling the system wait three minutes or more before servicing the vehicle.**

➡**Before disconnecting the battery, lower both the driver's and passenger's windows. This will prevent any interference between the window edge and the vehicle when the door is opened or closed. During normal operation the window slightly raises and lowers automatically to prevent any window to vehicle interference. The automatic window function will not work with the battery disconnected.**

2. Remove the battery cover.
3. Remove the cowl top cover.
4. Remove the positive terminal cover.
5. Loosen the cable retaining nuts and remove the cables from the battery cables.

➡**Always remove the negative cable first.**

6. Remove the battery fix frame mounting nuts and battery frame.
7. Remove the battery from the vehicle.

To install:

➡**Be sure to use new fasteners, as required.**

8. Installation is the reverse of the removal procedure.

➡**Always connect the positive cable first.**

BATTERY RECONNECT/ RELEARN PROCEDURE

Power Window Control

The following systems require attention once the negative battery cable is discon-

BATTERY SYSTEM

nected. These systems are Automatic temperature control system, Automatic drive positioner, Power window control, Sunroof system, Sunshade system and Rear view monitor. You will need the CONSULT-III diagnostic tool, or equivalent. Follow the directions on the screen of the tool, as needed.

1. Before servicing the vehicle, refer to the Precautions Section.
2. Disconnect the negative battery cable, or the power window switch connector.
3. Reconnect it after a minute or more.
4. Be sure that the doors are closed.
5. Turn the ignition switch ON.
6. Operate the power window switch to fully open the window. This is unnecessary if the window is already open.
7. Continue pulling the power window switch AUTO-UP. Even after the glass stops at the fully closed position, keep pulling the switch for three seconds or more.
8. Initializing is complete.
9. When initialization is not complete, power window UP does not operate while the door is open.

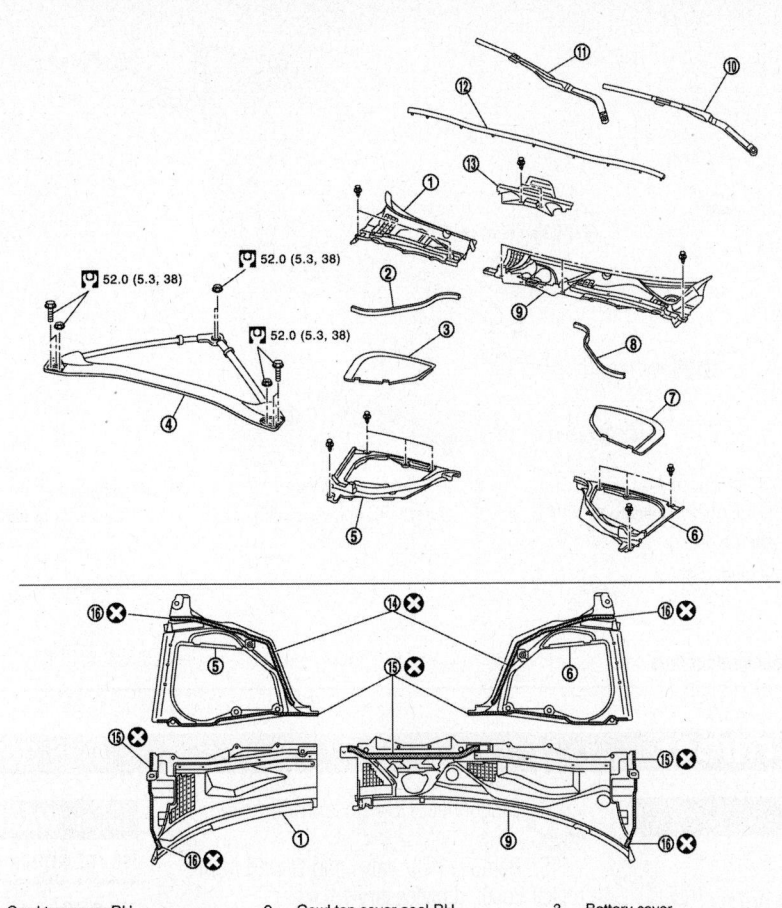

1.	Cowl top cover RH	2.	Cowl top cover seal RH	3.	Battery cover
4.	Front tower bar assembly	5.	Hoodledge cover RH	6.	Hoodledge cover LH
7.	Brake master cylinder cover	8.	Cowl top cover seal LH	9.	Cowl top cover LH
10.	Wiper arm and blade LH	11.	Wiper arm and blade RH	12.	Cowl top cover seal
13.	Cowl top cover center	14.	EPT sealer [t: 5.0 mm (0.197 in)]	15.	EPT sealer [t:3.0 mm (0.118 in)]
16.	EPT sealer [t: 10.0 mm (0.394 in)]				

37663_370Z_G0035

Fig. 33 Cowl top cover and related components

ENGINE ELECTRICAL   CHARGING SYSTEM

ALTERNATOR

REMOVAL & INSTALLATION
See Figure 34.

1. Before servicing the vehicle, refer to the Precautions Section.

➡If working near and/or around the SRS system and components, be sure to disable the SRS system. After disabling the system wait three minutes or more before servicing the vehicle.

➡Before disconnecting the battery, lower both the driver's and passenger's windows. This will prevent any interference between the window edge and the vehicle when the door is opened or closed. During normal operation the window slightly raises and lowers automatically to prevent any window to vehicle interference. The automatic window function will not work with the battery disconnected.

2. Disconnect the negative battery cable.
3. Remove the engine front undercover.
4. Remove the cooling fan assembly.
5. Remove the drive belt.
6. Disconnect the alternator electrical connectors.

7. Remove the harness bracket bolts.
8. Remove the oil pressure switch harness clip. Disconnect the electrical connectors from the oil pressure switch and the oil temperature sensor.
9. Remove the alternator mounting bolts.
10. Remove the alternator from the vehicle, in the downward direction.

To install:

➡Be sure to use new fasteners, as required.

11. Installation is the reverse of the removal procedure.
12. Check belt tension, as required.

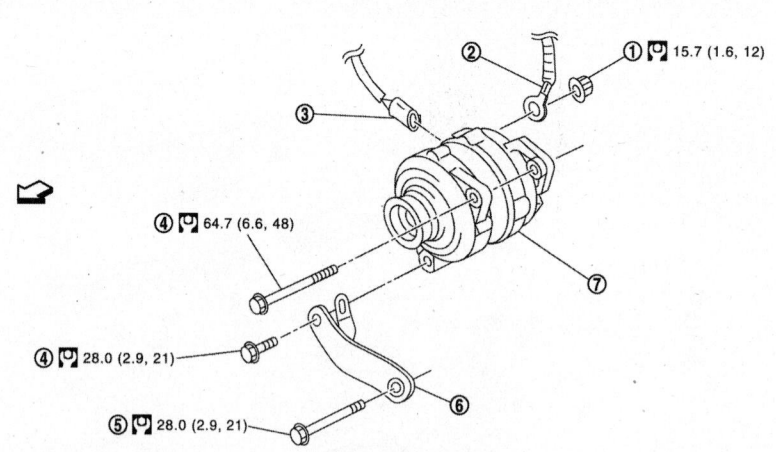

1. "B" terminal nut
4. Alternator mounting bolt
7. Alternator
◁ : Engine front

2. "B" terminal harness
5. Alternator stay mounting bolt

3. Alternator connector
6. Alternator stay

37663_370Z_G0123

Fig. 34 Alternator and related components

ENGINE ELECTRICAL

FIRING ORDERS

See Figure 35.

Firing order: 1–2–3–4–5–6

IGNITION COIL(S)

REMOVAL & INSTALLATION

See Figures 36 through 38.

1. Remove the engine cover.
2. Remove the air cleaner case and air duct (RH and LH).
3. Remove the intake manifold collector.

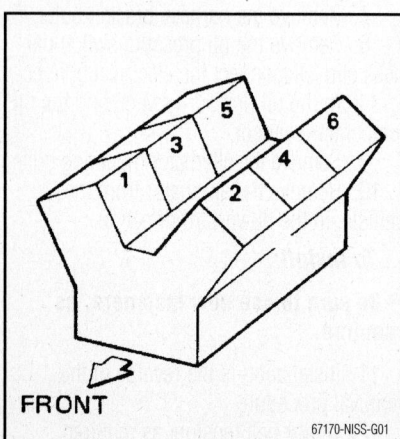

FRONT

67170-NISS-G01

Fig. 35 Distributorless ignition system (one coil on each cylinder)

4. Disconnect PCV hose from rocker cover.
5. Remove PCV valve and O-ring from rocker cover, if necessary.
6. Remove oil filler cap from rocker cover, if necessary.
7. Remove ignition coil.

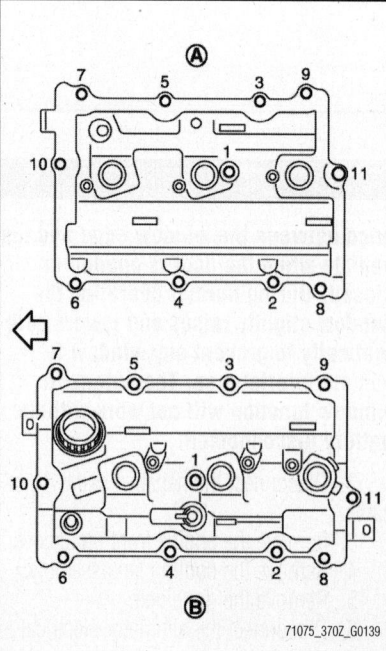

71075_370Z_G0139

Fig. 36 Identifying mounting bolts (A: Bank 1, B: Bank 2, arrow: engine front)

IGNITION SYSTEM

> ❋❋ WARNING
>
> **Never shock it.**

8. Remove harness clips on the rocker cover.
9. Loosen mounting bolts with power toll in reverse order as shown.
10. Remove rocker cover gasket from rocker cover.
11. Use scraper to remove all traces of liquid gasket from cylinder head and VVEL ladder assembly.

> ❋❋ WARNING
>
> **Never scratch or damage the mating surface when cleaning off old liquid gasket.**

To install:

> ❋❋ WARNING
>
> **Do not reuse O-rings.**

12. Apply liquid gasket to the position shown with the
following procedure:
    a. Refer to figure to apply liquid gasket to joint part of VVEL ladder assembly (1) and cylinder head.
    b. Refer to figure to apply liquid gasket in 90 degrees to figure.

➡**Use Genuine RTV Silicone Sealant or equivalent.**

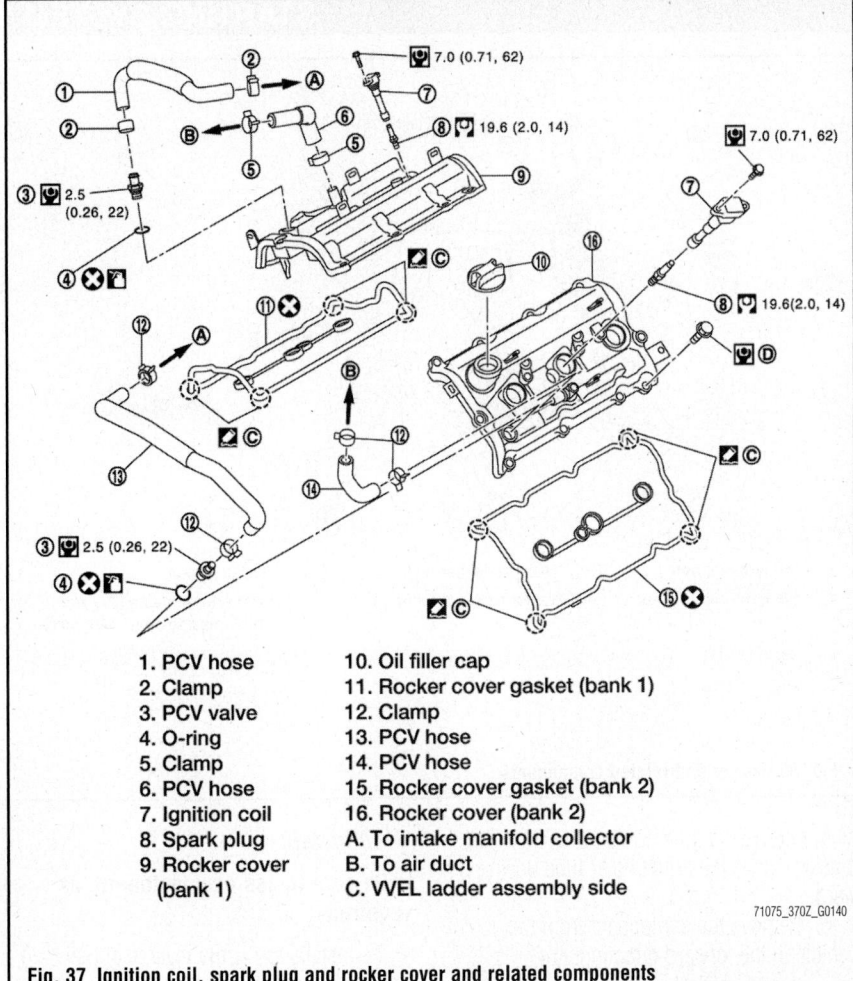

1. PCV hose
2. Clamp
3. PCV valve
4. O-ring
5. Clamp
6. PCV hose
7. Ignition coil
8. Spark plug
9. Rocker cover
   (bank 1)
10. Oil filler cap
11. Rocker cover gasket (bank 1)
12. Clamp
13. PCV hose
14. PCV hose
15. Rocker cover gasket (bank 2)
16. Rocker cover (bank 2)
A. To intake manifold collector
B. To air duct
C. VVEL ladder assembly side

71075_370Z_G0140

Fig. 37 Ignition coil, spark plug and rocker cover and related components

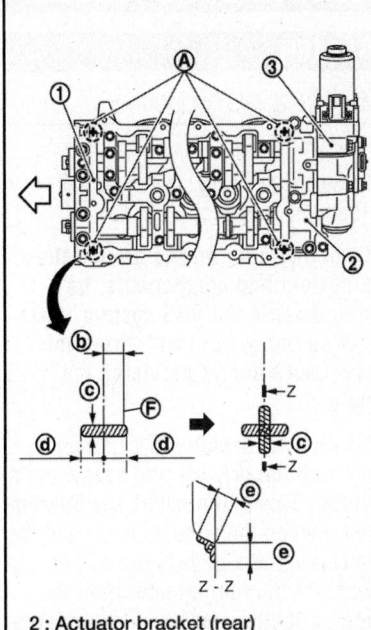

2 : Actuator bracket (rear)
3 : VVEL actuator sub assembly
A : Liquid gasket application point
F : End surface of VVEL ladder assembly
b : 4 mm (0.16 in)
c : 2.5 - 3.5 mm (0.098 - 0.138 in)
d : 5 mm (0.20 in)
e : 10 mm (0.39 in)
Arrow: Engine front

71075_370Z_G0141

Fig. 38 Applying liquid gasket

13. Install rocker cover gasket to rocker cover.

14. Install rocker cover.

a. Check that rocker cover gasket does not drop from the installation groove of rocker cover.

15. Tighten bolts in two steps separately in numerical order as shown.
   • Step 1: 18 inch lbs. (2 Nm)
   • Step 2: 73 inch lbs. (8.3 Nm)

16. Install in the reverse order of removal after this step.

IGNITION TIMING

INSPECTION & ADJUSTMENT

➡The ignition timing is not adjustable. If not within specifications, further diagnostic inspection is required. You will need the CONSULT-III diagnostic tool, or equivalent. Follow the directions on the screen of the tool, as needed.

No timing adjustment is necessary.

SPARK PLUGS

REMOVAL & INSTALLATION

1. Before servicing the vehicle, refer to the Precautions Section.

➡If working near and/or around the SRS system and components, be sure to disable the SRS system. After disabling the system wait three minutes or more before servicing the vehicle.

➡Before disconnecting the battery, lower both the driver's and passenger's windows. This will prevent any interference between the window edge and the vehicle when the door is opened or closed. During normal operation the window slightly raises and lowers automatically to prevent any window to vehicle interference. The automatic window function will not work with the battery disconnected.

2. Remove the engine undercover.

3. Remove the air cleaner assembly.

4. Remove the intake manifold collector.

5. Remove the necessary components in order to gain access to the ignition coils.

6. Disconnect the electrical connectors.

7. Remove the retainer bolt.

8. Remove the component from its mounting.

9. Remove the spark plug from its mounting.

To install:

➡Be sure to use new fasteners, as required.

10. Installation is the reverse of the removal procedure.

STARTER

REMOVAL & INSTALLATION

See Figure 39.

1. Before servicing the vehicle, refer to the Precautions Section.

➡**If working near and/or around the SRS system and components, be sure to disable the SRS system. After disabling the system wait three minutes or more before servicing the vehicle.**

➡**Before disconnecting the battery, lower both the driver's and passenger's windows. This will prevent any interference between the window edge and the vehicle when the door is opened or closed. During normal operation the window slightly raises and lowers automatically to prevent any window to vehicle interference. The automatic window function will not work with the battery disconnected.**

2. Disconnect the negative battery cable. Remove the engine undercover.
3. Disconnect the starter electrical connections.
4. Remove the starter retaining bolts and harness bracket.
5. If equipped with automatic transmis-

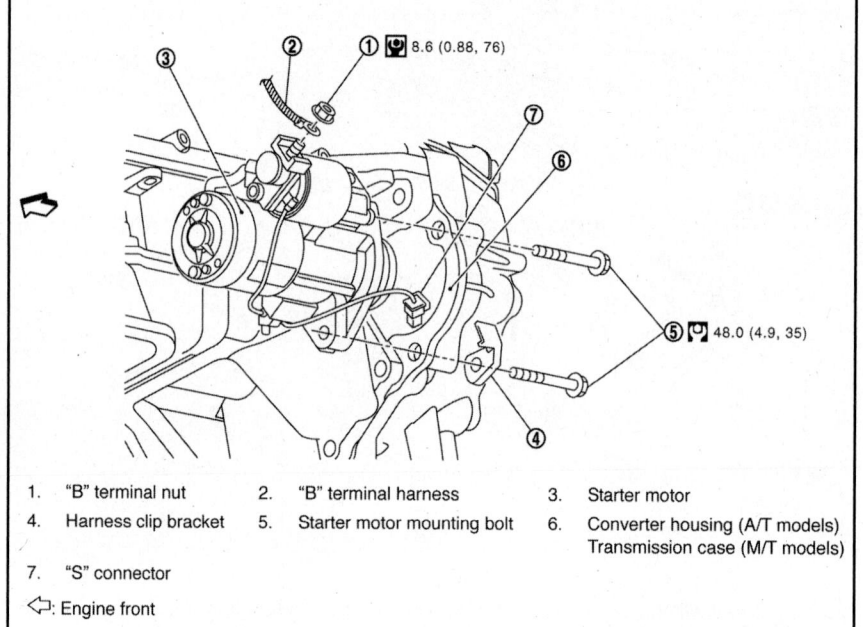

1.	"B" terminal nut	2.	"B" terminal harness	3.	Starter motor
4.	Harness clip bracket	5.	Starter motor mounting bolt	6.	Converter housing (A/T models) Transmission case (M/T models)
7.	"S" connector				

⇦: Engine front

37663_370Z_G0128

Fig. 39 Starter and related components

sion, remove the fluid cooler tube clips and bracket. Move the fluid cooler tube downward.

6. Remove the component from the vehicle in the forward direction.

To install:

➡**Be sure to use new fasteners, as required.**

7. Installation is the reverse of the removal procedure.

ENGINE MECHANICAL

ACCESSORY DRIVE BELT SYSTEM

BELT ROUTINGS

See Figure 40.

INSPECTION

1. Inspect belts for cracks, fraying, wear and oil. If necessary, replace.

REMOVAL & INSTALLATION

Drive Belt

See Figure 41.

1. Before servicing the vehicle, refer to the Precautions Section.

➡**If working near and/or around the SRS system and components, be sure to disable the SRS system. After disabling the system wait three minutes or more before servicing the vehicle.**

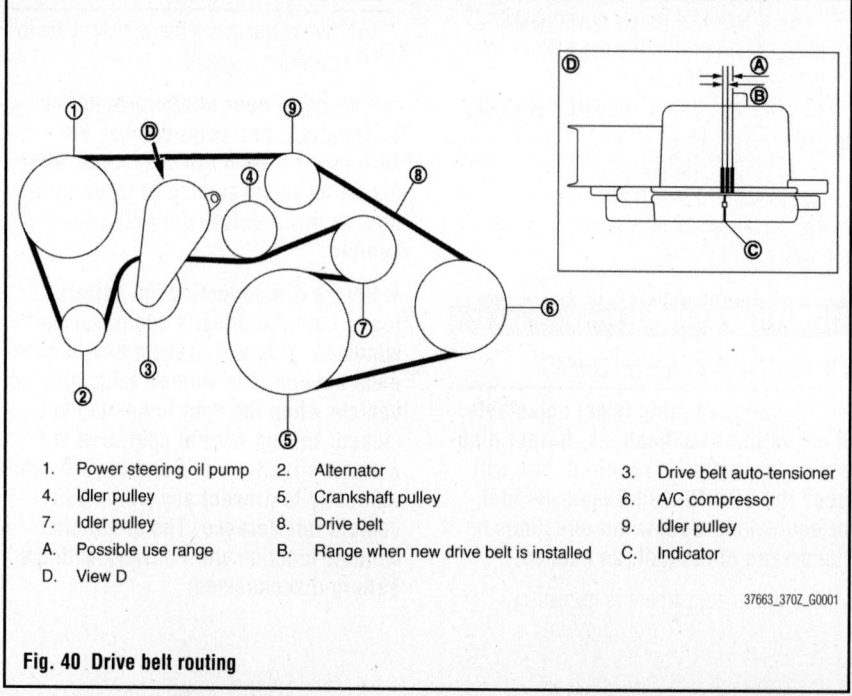

1.	Power steering oil pump	2.	Alternator	3.	Drive belt auto-tensioner
4.	Idler pulley	5.	Crankshaft pulley	6.	A/C compressor
7.	Idler pulley	8.	Drive belt	9.	Idler pulley
A.	Possible use range	B.	Range when new drive belt is installed	C.	Indicator
D.	View D				

37663_370Z_G0001

Fig. 40 Drive belt routing

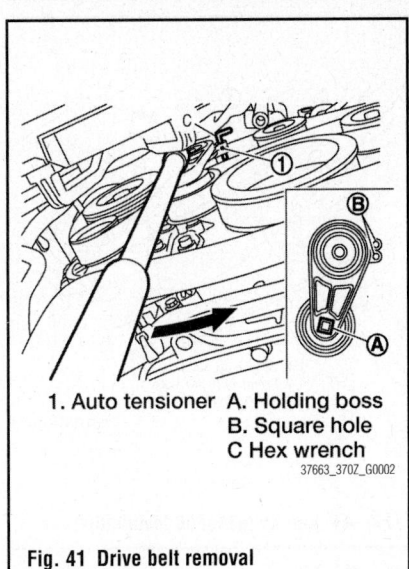

1. Auto tensioner   A. Holding boss
                    B. Square hole
                    C Hex wrench

37663_370Z_G0002

Fig. 41 Drive belt removal

➡Before disconnecting the battery, lower both the driver's and passenger's windows. This will prevent any interference between the window edge and the vehicle when the door is opened or closed. During normal operation the window slightly raises and lowers automatically to prevent any window to vehicle interference. The automatic window function will not work with the battery disconnected.

2. Remove the engine undercover.
3. While securely holding the square hole in the pulley center of the auto tensioner with a spinner handle, move the spinner handle in the direction of the arrow (loosening the drive belt). See illustration.

➡Never place your hand in a location where pinching may occur if the holding tool accidentally comes off.

4. Under the above condition insert a metallic bar about 0.24 inch in diameter through the holding boss to lock the auto tensioner pulley arm.
5. Remove the drive belt.

To install:

➡Be sure to use new fasteners, as required.

6. Installation is the reverse of the removal procedure.
7. Be sure that the belt is securely installed around the pulleys.
8. Be sure that the belt is correctly engaged with the pulley groove.
9. Turn the crankshaft pulley several times to equalize tension between each pulley, then confirm tension of the drive belt at the indicator (notch on fixed side) is within the possible use range.

Drive Belt Idler Pulley & Auto Tensioner

See Figure 42.

1. Remove drive belt.
   a. Keep auto-tensioner pulley arm locked after drive belt is removed.
2. Remove auto-tensioner and idler pulley.
   a. Keep auto-tensioner pulley arm locked to install or remove auto-tensioner.

To install:

3. To install, reverse the removal procedure.

❊❊ WARNING

If there is damage greater than peeled paint, replace drive belt auto-tensioner.

AIR CLEANER

REMOVAL & INSTALLATION

Air Cleaner Assembly

See Figure 43.

1. Before servicing the vehicle, refer to the Precautions Section.

➡If working near and/or around the SRS system and components, be sure to disable the SRS system. After disabling the system wait three minutes or more before servicing the vehicle.

➡Before disconnecting the battery, lower both the driver's and passenger's windows. This will prevent any interference between the window edge and the vehicle when the door is opened or closed. During normal operation the window slightly raises and lowers automatically to prevent any window to vehicle interference. The automatic window function will not work with the battery disconnected.

2. Disconnect the mass air flow sensor harness connector.

➡The sensor is removable under the car-mounted condition.

3. Disconnect the PCV hose.
4. Remove the air cleaner case with the mass air flow sensor and air duct by disconnecting the joints.

➡Matchmark the components, as required.

5. Handle the mass air flow sensor carefully. Never shock the sensor, disassemble the sensor or touch the sensor.

To install:

➡Be sure to use new fasteners, as required.

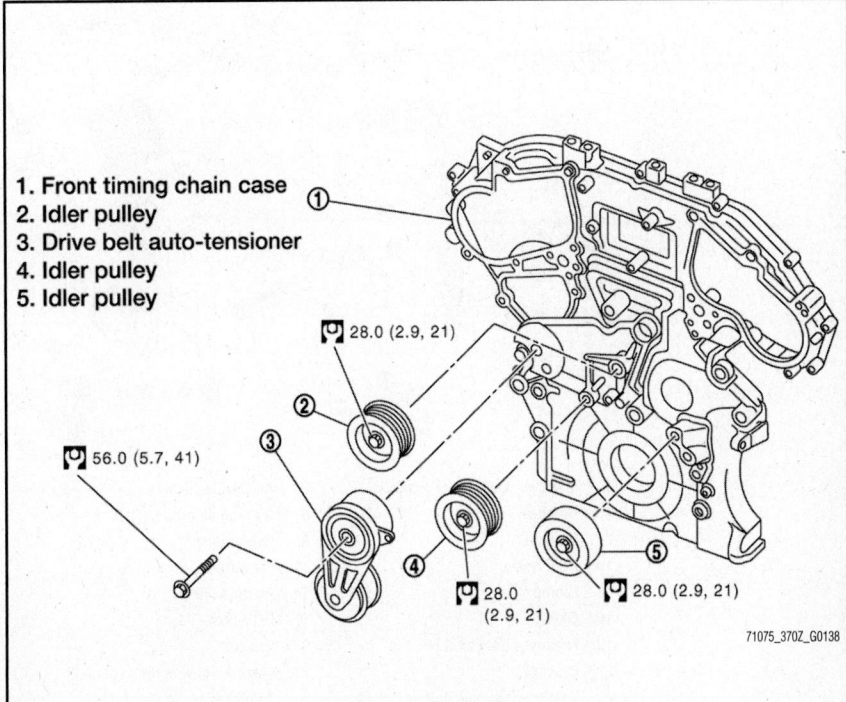

1. Front timing chain case
2. Idler pulley
3. Drive belt auto-tensioner
4. Idler pulley
5. Idler pulley

28.0 (2.9, 21)

56.0 (5.7, 41)

28.0 (2.9, 21)

28.0 (2.9, 21)

71075_370Z_G0138

Fig. 42 Removing and installing the drive belt auto tensioner and idler pulley

6. Installation is the reverse of the removal procedure.

Air Filter Element

See Figure 43

1. Before servicing the vehicle, refer to the Precautions Section.

➡**If working near and/or around the SRS system and components, be sure to disable the SRS system. After disabling the system wait three minutes or more before servicing the vehicle.**

➡**Before disconnecting the battery, lower both the driver's and passenger's windows. This will prevent any interference between the window edge and the vehicle when the door is opened or closed. During normal operation the window slightly raises and lowers automatically to prevent any window to vehicle interference. The automatic window function will not work with the battery disconnected.**

2. Unhook the retaining clips.
3. Remove the holder from the air cleaner case.

4. Remove the air cleaner filter from the holder.

To install:

➡**Be sure to use new fasteners, as required.**

5. Installation is the reverse of the removal procedure.

➡**Install the air cleaner filter by aligning the seal with the notch of the air cleaner case.**

CRANKSHAFT DAMPER

REMOVAL & INSTALLATION

See Figures 44 through 46.

1. Before servicing the vehicle, refer to the Precautions Section.

➡**If working near and/or around the SRS system and components, be sure to disable the SRS system. After disabling the system wait three minutes or more before servicing the vehicle.**

➡**Before disconnecting the battery, lower both the driver's and passenger's windows. This will prevent any**

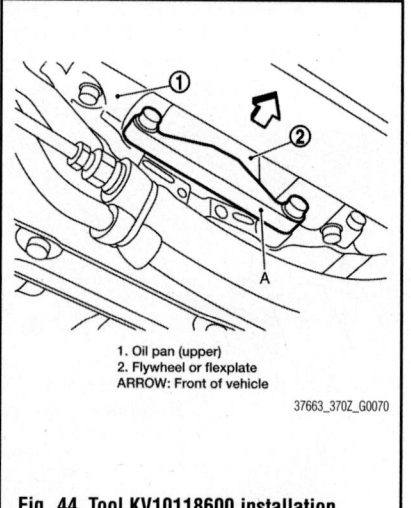

1. Oil pan (upper)
2. Flywheel or flexplate
ARROW: Front of vehicle

37663_370Z_G0070

Fig. 44 Tool KV10118600 installation

interference between the window edge and the vehicle when the door is opened or closed. During normal operation the window slightly raises and lowers automatically to prevent any window to vehicle interference. The automatic window function will not work with the battery disconnected.

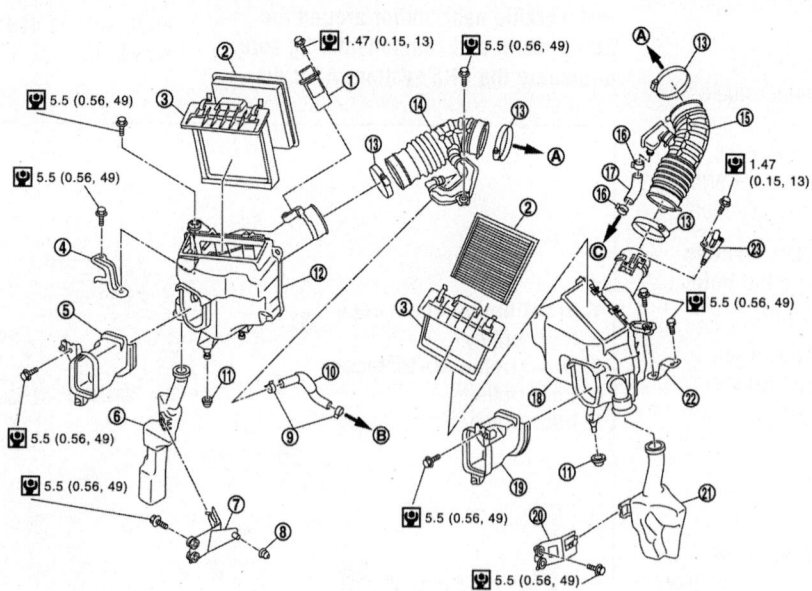

1. Mass air flow sensor (bank 1)	2. Air cleaner filter	3. Holder
4. Bracket	5. Air dust side duct (RH)	6. Resonator (RH)
7. Bracket	8. Grommet	9. Clamp
10. PCV hose	11. Grommet	12. Air cleaner case (RH)
13. Clamp	14. Air duct (RH)	15. Air duct (LH)
16. Clamp	17. PCV hose	18. Air cleaner case (LH)
19. Air dust side duct (LH)	20. Bracket	21. Resonator (LH)
22. Bracket	23. Mass air flow sensor (bank 2)	
A. To electric throttle control actuator	B. To rocker cover (bank 1)	C. To rocker cover (bank 2)

37663_370Z_G0003

Fig. 43 Air cleaner assembly and related components

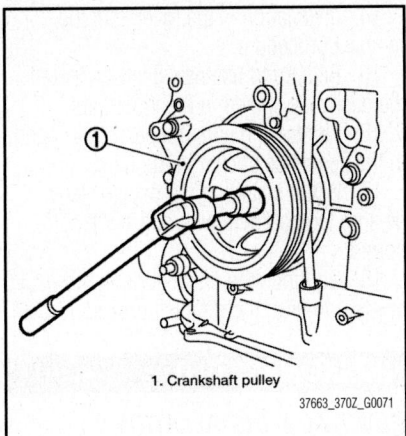

Fig. 45 Crankshaft pulley removal (1 of 2)

Fig. 46 Crankshaft pulley removal (2 of 2)

2. Remove the undercover.
3. Remove the drive belt.
4. Remove the rear cover plate and install the ring gear stopper tool KV10118600 or equivalent.
5. Loosen the pulley bolt and rotate the bolt seating surface at 0.39 inch from its original position.

➡**Never remove the bolt because it is used as a supporting point for a suitable puller.**

6. Position a suitable puller tab on the holes of the pulley and pull the pulley through.

➡**Never put the puller tab on the pulley periphery, because this damages the internal damper.**

7. Remove the crankshaft damper.

To install:

➡**Be sure to use new fasteners, as required**

8. Installation is the reverse of the removal procedure.
9. Be sure to tighten the bolt to specification.

CRANKSHAFT FRONT SEAL

REMOVAL & INSTALLATION

See Figure 47.

1. Before servicing the vehicle, refer to the Precautions Section.

➡**If working near and/or around the SRS system and components, be sure to disable the SRS system. After disabling the system wait three minutes or more before servicing the vehicle.**

➡**Before disconnecting the battery, lower both the driver's and passenger's windows. This will prevent any interference between the window edge and the vehicle when the door is opened or closed. During normal operation the window slightly raises and lowers automatically to prevent any window to vehicle interference. The automatic window function will not work with the battery disconnected.**

2. Remove the undercover.
3. Remove the drive belt.
4. Remove the crankshaft damper.
5. Remove the front seal, using a suitable tool. Be careful not to damage the front timing chain case and crankshaft.
6. Discard the seal.

To install:

➡**Be sure to use new fasteners, as required**

7. Apply clean engine oil to both the seal lip and the dust seal lip.

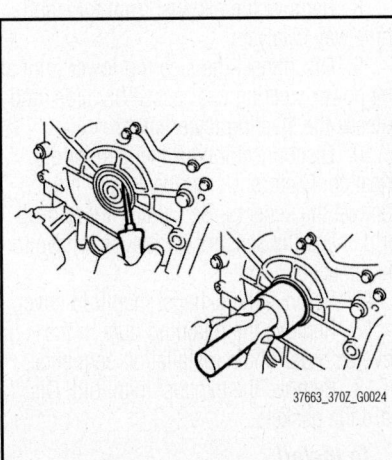

Fig. 47 Front oil seal removal and installation

8. Using a suitable drift, press fit the new seal into position until the height of the seal is level with the mounting surface. Check that the garter spring is in position and the seal lips are not inverted.

➡ **Be careful not to damage the front timing chain case and crankshaft. Press fit straight ahead and avoid causing burrs or tilting the seal.**

9. Continue the installation is the reverse of the removal procedure.

ENGINE COVER

REMOVAL & INSTALLATION

See Figure 48.

1. Loosen mounting bolts of engine cover (front) with power tool.
2. Remove engine cover (front).
3. Loosen mounting bolts and nuts of engine cover (rear) with power tool.
4. Remove engine cover (rear).

To install:

5. To install, reverse the removal procedure.

ENGINE OIL & FILTER

OIL LEVEL CHECK

Check the level on the oil level gauge after parking the vehicle on a level spot and turning off the engine.

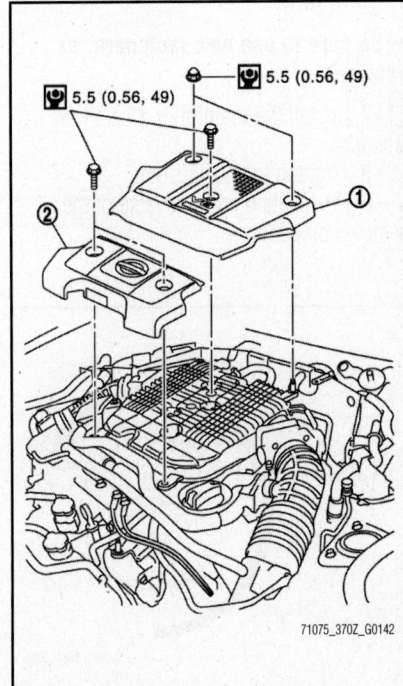

Fig. 48 Removing and installing the engine cover

OIL & FILTER CHANGE

See Figure 49

1. Before servicing the vehicle, refer to the Precautions Section.

➡ **If working near and/or around the SRS system and components, be sure to disable the SRS system. After disabling the system wait three minutes or more before servicing the vehicle.**

➡ **Before disconnecting the battery, lower both the driver's and passenger's windows. This will prevent any interference between the window edge and the vehicle when the door is opened or closed. During normal operation the window slightly raises and lowers automatically to prevent any window to vehicle interference. The automatic window function will not work with the battery disconnected.**

➡ **Be sure that the engine is cold and the engine oil is cold.**

2. Raise and safely support the vehicle.
3. Remove the undercover.
4. Remove the oil drain plug. Discard the washer.
5. Drain the engine oil into a suitable container. Properly dispose of used engine oil.
6. Using oil filter wrench KV10115801, or equivalent remove the oil filter. Discard the filter.

To install:

➡ **Be sure to use new fasteners, as required**

7. Install the drain plug. Use a new washer.
8. Tighten to specification.
9. Installation is the reverse of the removal procedure.

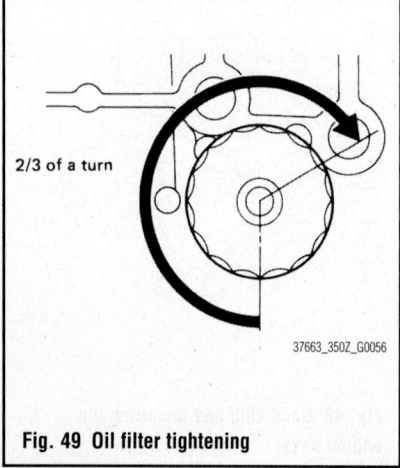

2/3 of a turn

37663_350Z_G0056

Fig. 49 Oil filter tightening

10. Coat the oil filter seal with clean engine oil prior to installation.
11. Do not over tighten the filter to its mounting.
12. Tightening specification should be 13 ft. lbs. (17.7 Nm).
13. Fill the engine with the proper grade and type engine oil.
14. Start the engine and check for leaks, correct as required.

EXHAUST MANIFOLD

REMOVAL & INSTALLATION

See Figures 50 through 52.

1. Before servicing the vehicle, refer to the Precautions Section.

➡ **If working near and/or around the SRS system and components, be sure to disable the SRS system. After disabling the system wait three minutes or more before servicing the vehicle.**

➡ **Before disconnecting the battery, lower both the driver's and passenger's windows. This will prevent any interference between the window edge and the vehicle when the door is opened or closed. During normal operation the window slightly raises and lowers automatically to prevent any window to vehicle interference. The automatic window function will not work with the battery disconnected.**

2. Remove the engine undercover.
3. Drain the engine coolant. Be sure to properly dispose of used coolant.
4. Remove the front tower bar assembly.
5. Remove the engine cover.
6. Remove the air cleaner assembly.
7. Remove the water pipe and heater pipe assemblies.
8. Remove the exhaust front tube and three way catalysts.
9. Disconnect the steering lower joint at the power steering gear assembly side, and release the steering lower shaft.
10. Disconnect the air fuel sensor electrical connectors. Using the proper tool remove the sensors. Be careful not to drop or damage the sensors, or they will have to be replaced.
11. Remove the exhaust manifold cover.
12. Remove the retaining nuts in the reverse order of the installation sequence.
13. Remove the exhaust manifold. Discard the gaskets.

To install:

➡ **Be sure to use new fasteners, as required.**

14. Installation is the reverse of the removal procedure.
15. Be sure to use new gaskets. Install the gasket as shown in the illustration.
16. Tighten the nuts to specification and in the sequence shown in the illustration.
17. When installing the oxygen sensors, be sure to coat them with anti-seize compound.
18. Continue the installation in the reverse order of the removal procedure.

INTAKE MANIFOLD

REMOVAL & INSTALLATION

See Figures 53 through 55.

1. Before servicing the vehicle, refer to the Precautions Section.

➡ **If working near and/or around the SRS system and components, be sure to disable the SRS system. After disabling the system wait three minutes or more before servicing the vehicle.**

➡ **Before disconnecting the battery, lower both the driver's and passenger's windows. This will prevent any interference between the window edge and the vehicle when the door is opened or closed. During normal operation the window slightly raises and lowers automatically to prevent any window to vehicle interference. The automatic window function will not work with the battery disconnected.**

2. Properly relieve the fuel system pressure.
3. Remove the intake manifold collector.
4. Remove the fuel tube and fuel injector assembly.
5. Loosen the intake manifold retaining bolts in the reverse order of the installation.
6. Remove the component from its mounting. Discard the gasket.

➡ **Matchmark the manifold and the cylinder head for proper installation, as these components need to be installed in a specified direction.**

To install:

➡ **Be sure to use new fasteners, as required.**

7. Installation is the reverse of the removal procedure.

➡ **Be sure to use the marks made in the removal for proper alignment and installation.**

8. If the stud bolts were removed, install and tighten to 8 ft. lbs. (10.8 Nm).

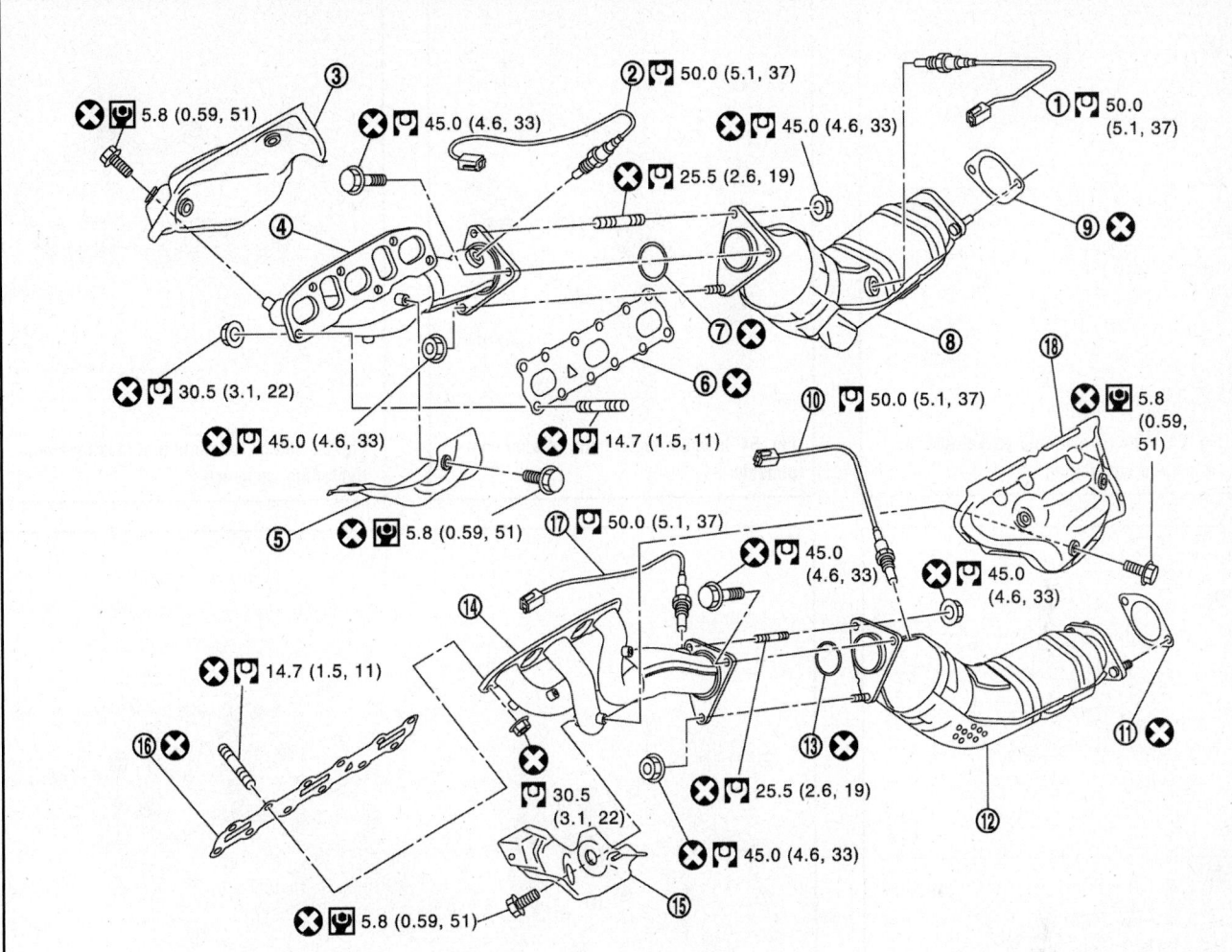

1. Heated oxygen sensor 2 (bank 1)
2. Air fuel ratio sensor 1 (bank 1)
3. Exhaust manifold cover (upper)
4. Exhaust manifold (bank 1)
5. Exhaust manifold cover (lower)
6. Gasket
7. Ring gasket
8. Three way catalyst (bank 1)
9. Gasket
10. Heated oxygen sensor 2 (bank 2)
11. Gasket
12. Three way catalyst (bank 2)
13. Ring gasket
14. Exhaust manifold (bank 2)
15. Exhaust manifold cover (lower)
16. Gasket
17. Air Fuel ratio sensor 1 (bank 2)
18. Exhaust manifold cover (upper)

37663_370Z_G0029

Fig. 50 Exhaust manifold and related components

9. Tighten all mounting bolts to specification and in the proper sequence.

10. Specification is 5 ft. lbs. (7.4 Nm) step one. 19 ft. lbs. (25.5 Nm) step two.

INTAKE MANIFOLD COLLECTOR

REMOVAL & INSTALLATION

See Figures 56 through 58.

1. Before servicing the vehicle, refer to the Precautions Section.

➡**If working near and/or around the SRS system and components, be sure to disable the SRS system. After disabling the system wait three minutes or more before servicing the vehicle.**

➡**Before disconnecting the battery, lower both the driver's and passenger's windows. This will prevent any interference between the window edge and the vehicle when the door is opened or closed. During normal operation the window slightly raises and lowers automatically to prevent any window to vehicle interference. The automatic window function will not work with the battery disconnected.**

2. Remove the engine undercover.

3. Drain the engine coolant. Be sure to properly dispose of used coolant.

4. Remove the front tower bar assembly.

5. Remove the engine cover.

6. Remove the air cleaner assembly.

7. To remove the electric throttle control actuator, Disconnect the water hoses from the component. Disconnect the harness connector.

8. Loosen the mounting bolts in the reverse order of the tightening sequence.

➡**When removing only the intake manifold collector, move the electric throttle**

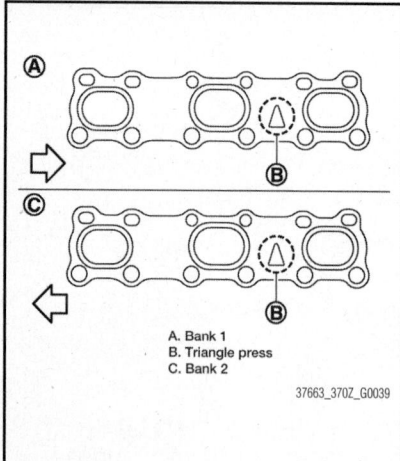

Fig. 51 Exhaust manifold gasket identification and positioning

A. Bank 1
B. Triangle press
C. Bank 2

37663_370Z_G0039

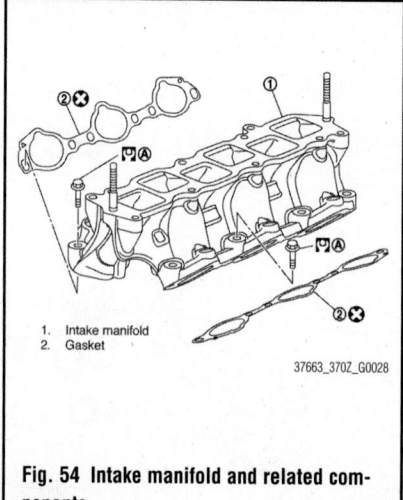

1. Intake manifold
2. Gasket

37663_370Z_G0028

Fig. 54 Intake manifold and related components

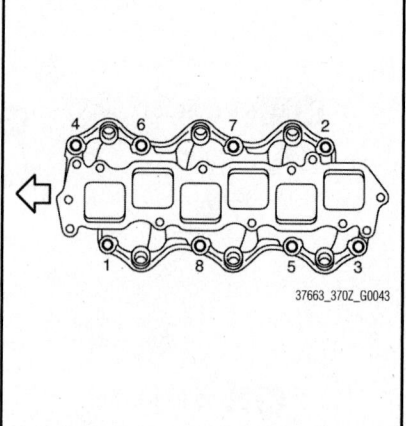

37663_370Z_G0043

Fig. 55 Intake manifold bolt locations and tightening sequence

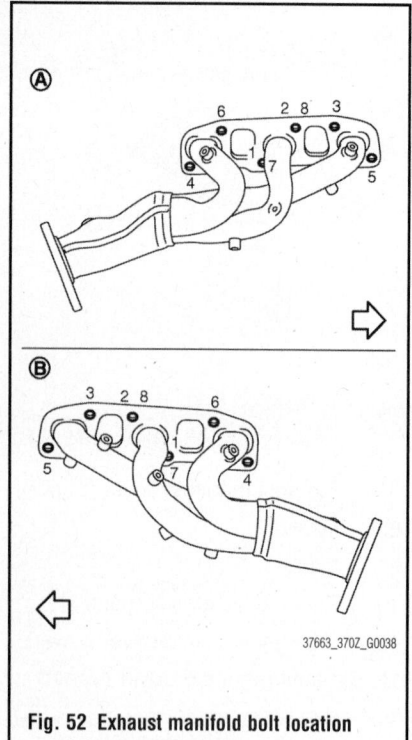

37663_370Z_G0038

Fig. 52 Exhaust manifold bolt location and tightening sequence

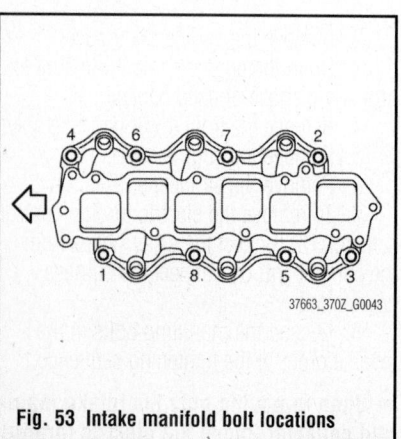

37663_370Z_G0043

Fig. 53 Intake manifold bolt locations

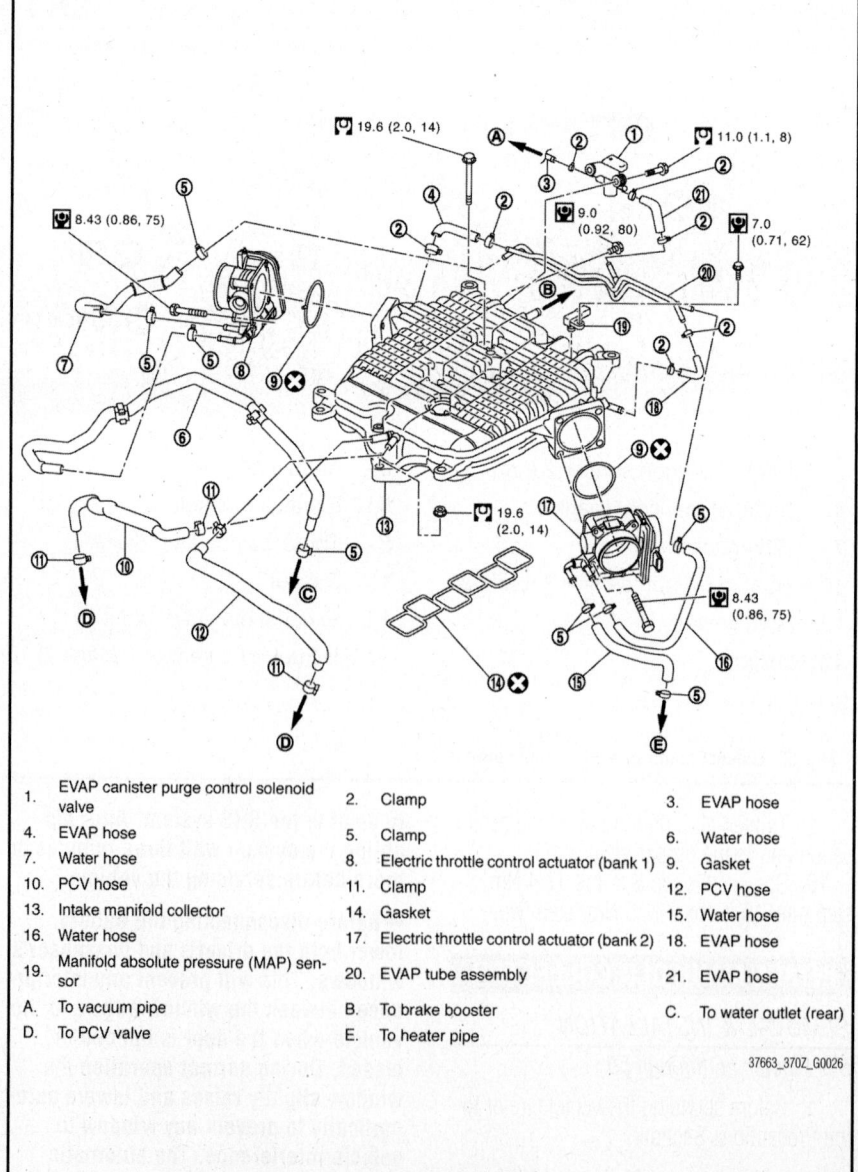

1. EVAP canister purge control solenoid valve	2. Clamp	3. EVAP hose
4. EVAP hose	5. Clamp	6. Water hose
7. Water hose	8. Electric throttle control actuator (bank 1)	9. Gasket
10. PCV hose	11. Clamp	12. PCV hose
13. Intake manifold collector	14. Gasket	15. Water hose
16. Water hose	17. Electric throttle control actuator (bank 2)	18. EVAP hose
19. Manifold absolute pressure (MAP) sensor	20. EVAP tube assembly	21. EVAP hose
A. To vacuum pipe	B. To brake booster	C. To water outlet (rear)
D. To PCV valve	E. To heater pipe	

37663_370Z_G0026

Fig. 56 Intake manifold collector and related components

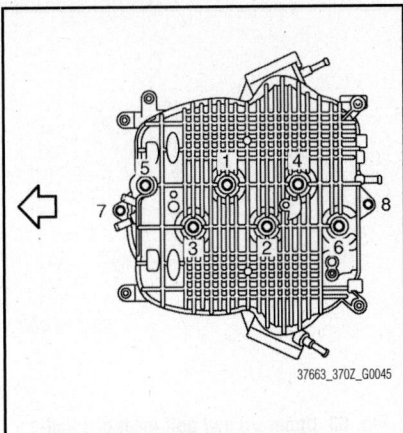

Fig. 57 Intake manifold collector bolt locations and tightening sequence

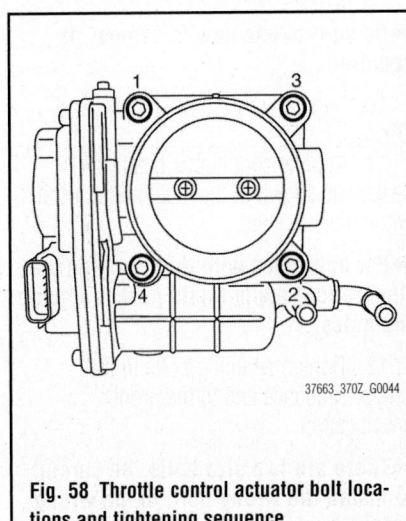

Fig. 58 Throttle control actuator bolt locations and tightening sequence

control actuator without disconnecting the water hose. Handel the component carefully to avoid shock damage.

9. Disconnect the vacuum hose, PCV hose, and EVAP hose from the intake manifold collector.

10. Remove the EVAP canister purge volume control solenoid valve and EVAP tube assembly from the manifold collector.

11. Loosen the retaining bolts and nuts in the reverse order of the tightening sequence and remove the component from its mounting.

To install:

→Be sure to use new fasteners, as required.

12. Installation is the reverse of the removal procedure.

13. Tighten the retaining bolts to specification and in the proper sequence.

14. Specification is 8 ft. lbs. (10.8 Nm).

15. Tighten the electric throttle actuator retaining bolts in the proper sequence.

16. Using the CONSULT-III diagnostic tool or equivalent, perform the throttle valve closed position learning and the idle air volume learning procedures.

OIL PAN

REMOVAL & INSTALLATION

Lower

See Figures 59 through 61.

1. Before servicing the vehicle, refer to the Precautions Section.

→If working near and/or around the SRS system and components, be sure to disable the SRS system. After disabling the system wait three minutes or more before servicing the vehicle.

→Before disconnecting the battery, lower both the driver's and passenger's windows. This will prevent any interference between the window edge and the vehicle when the door is opened or closed. During normal operation the window slightly raises and lowers automatically to prevent any window to vehicle interference. The automatic window function will not work with the battery disconnected.

2. Raise and support the vehicle safely.

3. Remove the engine undercover.

4. Drain the engine oil. Be sure to properly dispose of used oil.

5. Loosen the oil pan retaining bolts in the reverse order of the installation.

6. Insert a seal cutter tool between the upper and lower oil pan sealing surfaces. Never use a screwdriver. Slide the tool, by tapping the side of the tool with a hammer. Remove the lower oil pan.

7. Discard the gasket.

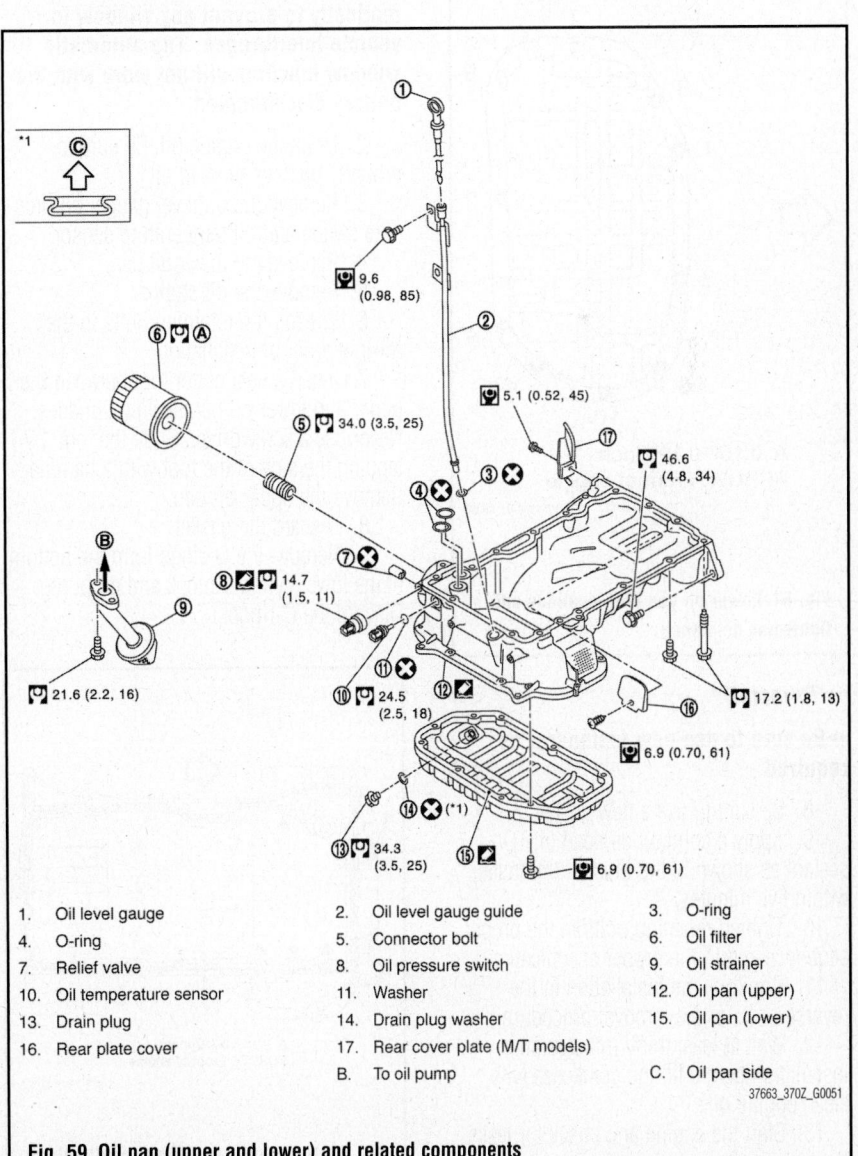

1.	Oil level gauge	2.	Oil level gauge guide	3.	O-ring
4.	O-ring	5.	Connector bolt	6.	Oil filter
7.	Relief valve	8.	Oil pressure switch	9.	Oil strainer
10.	Oil temperature sensor	11.	Washer	12.	Oil pan (upper)
13.	Drain plug	14.	Drain plug washer	15.	Oil pan (lower)
16.	Rear plate cover	17.	Rear cover plate (M/T models)		
B.	To oil pump			C.	Oil pan side

Fig. 59 Oil pan (upper and lower) and related components

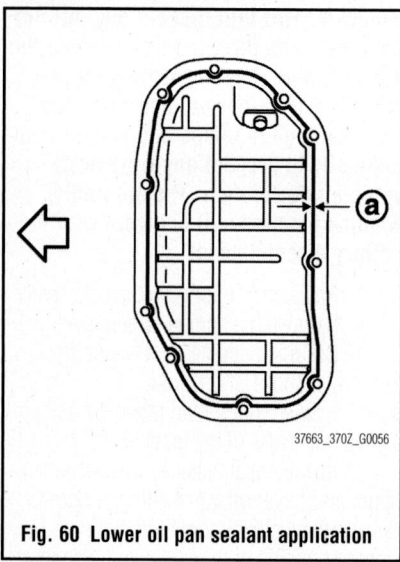

Fig. 60 Lower oil pan sealant application

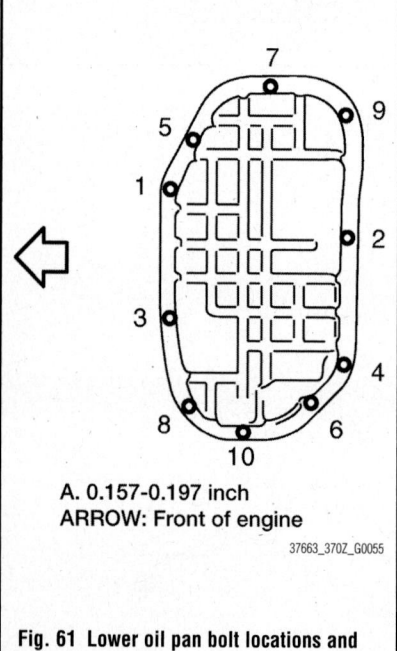

A. 0.157-0.197 inch
ARROW: Front of engine

37663_370Z_G0055

Fig. 61 Lower oil pan bolt locations and tightening sequence

To install:

➡**Be sure to use new fasteners, as required.**

8. Be sure to use a new gasket.

9. Apply a continuous bead of RTV sealant as shown in the illustration. Install within five minutes.

10. Tighten retaining bolts in the proper sequence and to the proper specification.

11. Continue the installation in the reverse order of the removal procedure.

12. Wait at least thirty minutes after installation before fill the crankcase with clean engine oil.

13. Start the engine and check for leaks. Correct as required.

Upper

See Figures 62 and 63.

At this time the manufacturer provides service information for this component with the engine removed from the vehicle and positioned in a suitable holding fixture.

1. Before servicing the vehicle, refer to the Precautions Section.

➡**If working near and/or around the SRS system and components, be sure to disable the SRS system. After disabling the system wait three minutes or more before servicing the vehicle.**

➡**Before disconnecting the battery, lower both the driver's and passenger's windows. This will prevent any interference between the window edge and the vehicle when the door is opened or closed. During normal operation the window slightly raises and lowers automatically to prevent any window to vehicle interference. The automatic window function will not work with the battery disconnected.**

2. Drain the engine oil. Be sure to properly dispose of used oil.

3. Remove the oil level gauge, oil pressure switch and oil temperature sensor.

4. Remove the lower oil pan.

5. Remove the oil strainer.

6. Loosen the retaining bolts in the reverse order of installation.

7. Insert a seal cutter tool between the upper and lower oil pan sealing surfaces. Never use a screwdriver. Slide the tool, by tapping the side of the tool with a hammer. Remove the upper oil pan.

8. Discard the gasket.

9. Remove the O-rings from the bottom of the lower cylinder block and oil pump. Discard the O-rings.

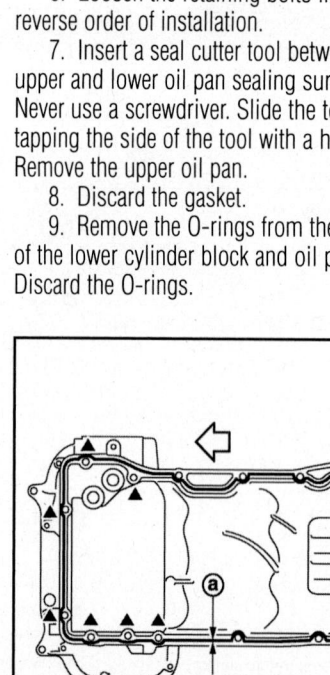

A. 0.157-0.197 inch
ARROW: Front of engine

37663_370Z_G0058

Fig. 62 Upper oil pan sealant application

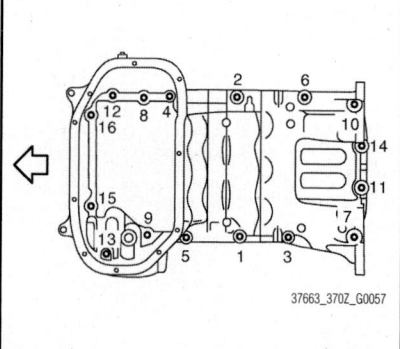

37663_370Z_G0057

Fig. 63 Upper oil pan bolt locations and tightening sequence

To install:

➡**Be sure to use new fasteners, as required.**

10. Be sure to use a new gasket and O-rings.

11. Apply a continuous bead of RTV sealant as shown in the illustration. Install within five minutes.

➡**For bolt holes with the triangle (see illustration) apply liquid gasket outside the holes.**

12. Tighten retaining bolts in the proper sequence and to the proper specification.

➡**There are two size bolts. Be careful to install the wrong bolt, in the wrong hole.**

13. Continue the installation in the reverse order of the removal procedure.

14. Wait at least thirty minutes after installation before fill the crankcase with clean engine oil.

15. Start the engine and check for leaks. Correct as required.

OIL PUMP

REMOVAL & INSTALLATION

See Figure 64.

1. Before servicing the vehicle, refer to the Precautions Section.

➡**If working near and/or around the SRS system and components, be sure to disable the SRS system. After disabling the system wait three minutes or more before servicing the vehicle.**

➡**Before disconnecting the battery, lower both the driver's and passenger's windows. This will prevent any interference between the window edge and the**

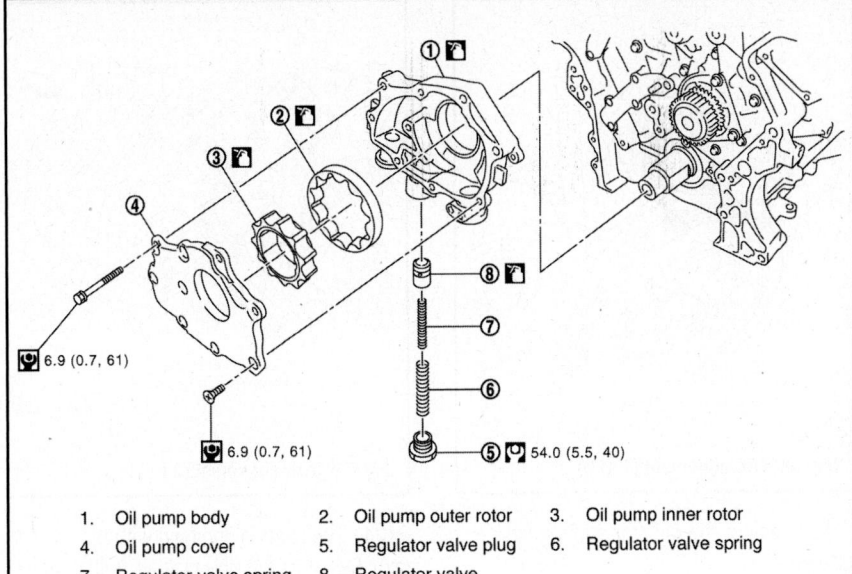

6.9 (0.7, 61)

6.9 (0.7, 61)

8

7

6

5   54.0 (5.5, 40)

1. Oil pump body
2. Oil pump outer rotor
3. Oil pump inner rotor
4. Oil pump cover
5. Regulator valve plug
6. Regulator valve spring
7. Regulator valve spring
8. Regulator valve

37663_370Z_G0060

Fig. 64 Oil pump and related components

vehicle when the door is opened or closed. During normal operation the window slightly raises and lowers automatically to prevent any window to vehicle interference. The automatic window function will not work with the battery disconnected.

2. Remove the upper and lower oil pan assemblies.

3. Remove the timing chain cover and the primary timing chain.

4. Remove the oil pump assembly from its mounting.

To install:

➡**Be sure to use new fasteners, as required.**

5. Installation is the reverse of the removal procedure.

➡**When installing, align the crankshaft flat surfaces with the oil pump inner rotor flat surfaces.**

INSPECTION

1. Clearance between outer rotor and oil pump body: 0.0045–0.0079 in. (0.114–0.200 mm).

2. Tip clearance between inner rotor and outer rotor: Below 0.0071 in. (0.180 mm).

3. Side clearance with a straight-edge between inner rotor and oil pump body: 0.0012–0.0028 in. (0.030–0.070 mm).

4. Side clearance with a straight-edge between outer rotor and oil pump body: 0.0020–0.0043 in. (0.050–0.110 mm).

PISTONS & RINGS

POSITIONING

See Figure 65.

REAR MAIN SEAL

REMOVAL & INSTALLATION

See Figures 66 and 67.

1. Before servicing the vehicle, refer to the Precautions Section.

➡**If working near and/or around the SRS system and components, be sure to disable the SRS system. After disabling the system wait three minutes or more before servicing the vehicle.**

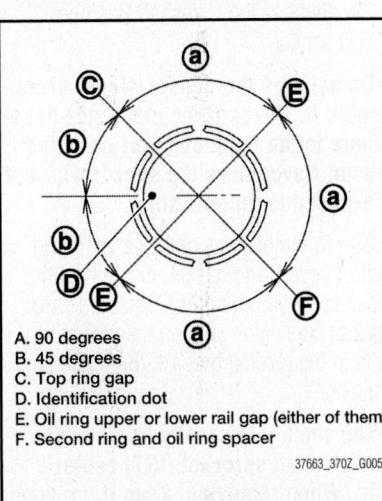

A. 90 degrees
B. 45 degrees
C. Top ring gap
D. Identification dot
E. Oil ring upper or lower rail gap (either of them)
F. Second ring and oil ring spacer

37663_370Z_G0059

Fig. 65 Piston ring positioning

➡Before disconnecting the battery, lower both the driver's and passenger's windows. This will prevent any interference between the window edge and the vehicle when the door is opened or closed. During normal operation the window slightly raises and lowers automatically to prevent any window to vehicle interference. The automatic window function will not work with the battery disconnected.

2. Raise and safely support the vehicle.

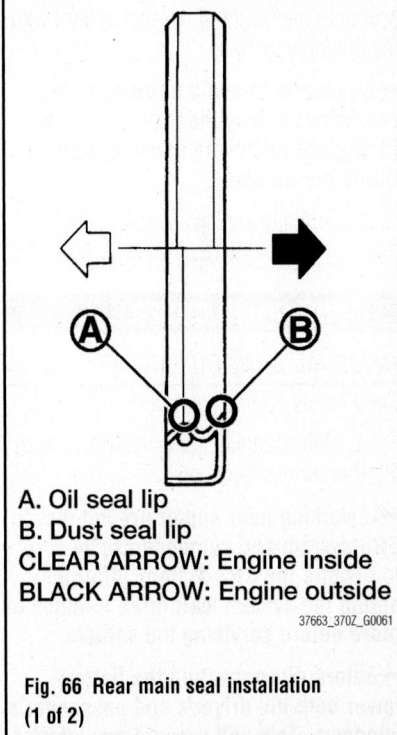

A. Oil seal lip
B. Dust seal lip
CLEAR ARROW: Engine inside
BLACK ARROW: Engine outside

37663_370Z_G0061

Fig. 66 Rear main seal installation (1 of 2)

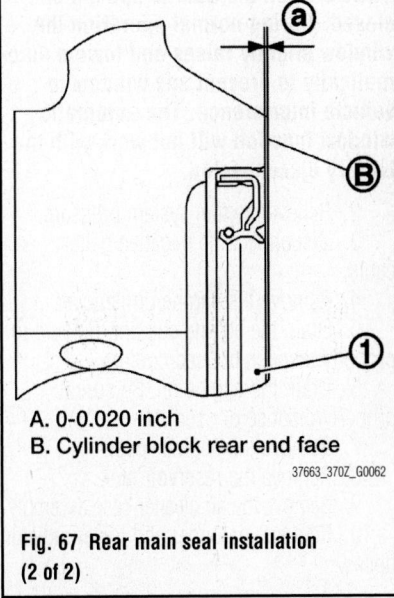

A. 0-0.020 inch
B. Cylinder block rear end face

37663_370Z_G0062

Fig. 67 Rear main seal installation (2 of 2)

3. Remove the transmission from the vehicle.

4. Remove the flexplate or flywheel.

5. Using a suitable tool, carefully remove the seal from its mounting.

To install:

→**Be sure to use new fasteners, as required.**

6. Install the seal so that each seal lip is oriented as shown in the illustration.

7. Press the seal into position as shown in the illustration.

8. Using a suitable drift, press fit the seal until the height of the seal is level with the mounting surface.

→**Be careful to avoid damage to the crankshaft and cylinder block. Press fit straight and avoid causing burrs or tilting the oil seal.**

9. Continue the installation in the reverse order of the removal procedure.

TIMING CHAIN COVER

REMOVAL & INSTALLATION

See Figures 68 through 76.

1. Before servicing the vehicle, refer to the Precautions Section.

→**If working near and/or around the SRS system and components, be sure to disable the SRS system. After disabling the system wait three minutes or more before servicing the vehicle.**

→**Before disconnecting the battery, lower both the driver's and passenger's windows. This will prevent any interference between the window edge and the vehicle when the door is opened or closed. During normal operation the window slightly raises and lowers automatically to prevent any window to vehicle interference. The automatic window function will not work with the battery disconnected.**

2. Relieve the fuel system pressure.

3. Disconnect the negative battery cable.

4. Remove the engine undercover.

5. Drain the engine coolant. Be sure to properly dispose of used coolant.

6. Drain the engine oil. Be sure to properly dispose of used oil.

7. Remove the engine cover.

8. Remove the reservoir tank.

9. Remove the air cleaner case assembly.

10. Remove the upper and lower radiator hoses.

Fig. 68 TDC alignment (1 of 2)

37663_370Z_G0068

11. Remove the radiator cooling fan assembly.

12. Remove the drive belt.

13. Separate the engine harnesses by removing their brackets from the timing chain cover.

14. Remove the intake manifold collector.

15. Remove the fuel sub mounting bolt.

16. Remove the oil level gauge and guide.

17. Remove the air conditioning compressor from the bracket. Secure it to the side. Do not discharge the refrigerant or disconnect the refrigerant hoses.

18. Remove the power steering fluid pump from the bracket with the hoses connected, secure it to the side. Remove the power steering oil pump bracket.

19. Remove the idler pulley, drive belt auto tensioner and bracket.

20. Remove the alternator and alternator bracket.

21. Remove the front water outlet.

22. Remove the Camshaft Position (CMP) sensor.

→**Do not drop the sensor. Never disassemble it. Never allow metal powder to adhere to the magnetic portion of the sensor. Never store the sensor where it is exposed to magnetism.**

23. To remove the intake valve timing control covers and gasket, disconnect the intake valve timing control solenoid valve harness connector. Loosen the mounting bolts in the reverse order of the tightening sequence.

→**The shaft is internally jointed with the camshaft sprocket (INT) center hole. When removing, keep it horizontal until it is completely disconnected.**

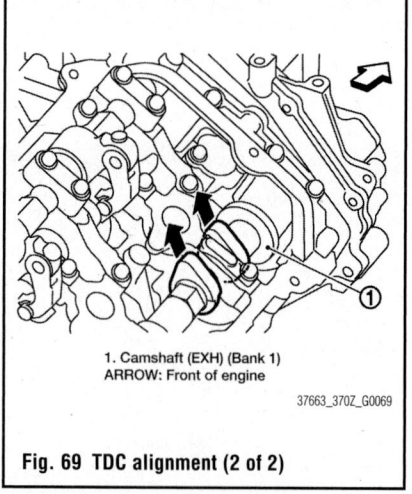

1. Camshaft (EXH) (Bank 1)
ARROW: Front of engine

37663_370Z_G0069

Fig. 69 TDC alignment (2 of 2)

24. The shaft is engaged with the camshaft sprocket (INT) center hole on the inside. Pull straight out so that it does not tilt until the joint is disengaged.

25. Remove the intake valve timing control solenoid valve, if necessary.

→**This valve is not reusable. Never remove it unless required.**

26. Remove the valve covers.

27. To position the engine to TDC on the compression stroke, rotate the crankshaft pulley clockwise to align the timing mark (grooved line without color) with the timing indicator. Check that the exhaust cam noses on No. 1 cylinder (engine front side of bank 1) is located as shown in the illustration. If not turn the crankshaft on complete revolution and align.

28. Remove the crankshaft pulley.

29. Remove the lower oil pan.

30. Loosen the mounting bolts in the front of the upper oil pan in the reverse order of the installation sequence.

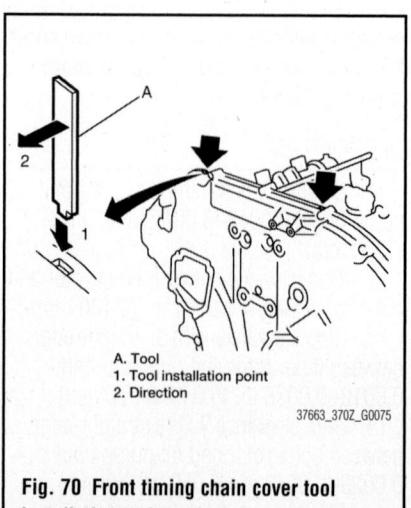

A. Tool
1. Tool installation point
2. Direction

37663_370Z_G0075

Fig. 70 Front timing chain cover tool installation and removal

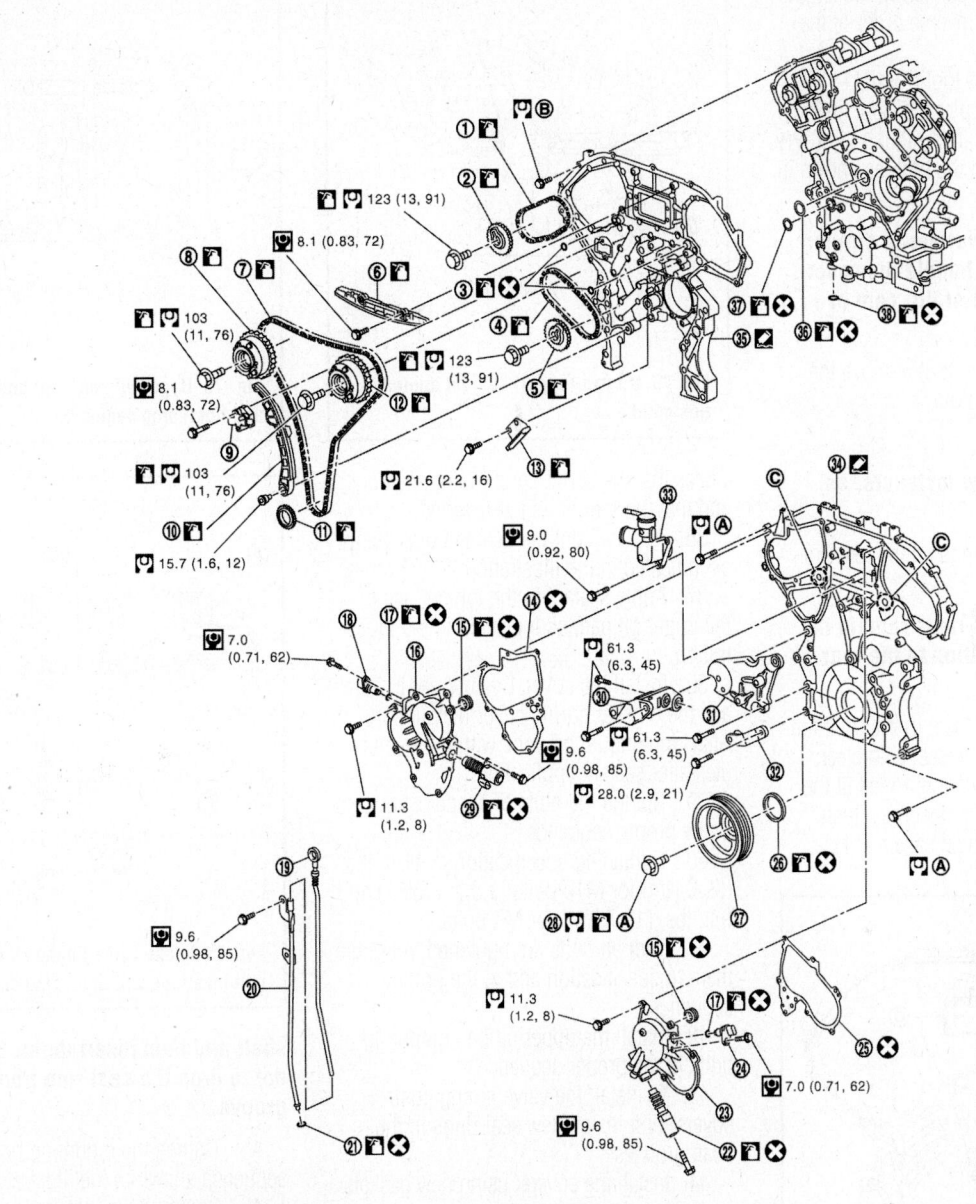

1.	Timing chain (secondary) (bank 1)	2.	Camshaft sprocket (EXH) (bank 1)	3.	O-ring
4.	Timing chain (secondary) (bank 2)	5.	Camshaft sprocket (EXH) (bank 2)	6.	Internal chain guide
7.	Timing chain (primary)	8.	Camshaft sprocket (INT) (bank 1)	9.	Timing chain tensioner (primary)
10.	Slack guide	11.	Crankshaft sprocket	12.	Camshaft sprocket (INT) (bank 2)
13.	Tension guide	14.	Intake valve timing control cover gasket (bank 1)	15.	Seal ring
16.	Intake valve timing control cover (bank 1)	17.	O-ring	18.	Camshaft position sensor (PHASE) (bank 1)
19.	Oil level gauge	20.	Oil level gauge guide	21.	O-ring
22.	Intake valve timing control solenoid valve (bank 2)	23.	Intake valve timing control cover (bank 2)	24.	Camshaft position sensor (PHASE) (bank 2)
25.	Intake valve timing control cover gasket (bank 2)	26.	Front oil seal	27.	Crankshaft pulley
28.	Crankshaft pulley bolt	29.	Intake valve timing control solenoid valve (bank 1)	30.	Power steering oil pump bracket
31.	Idler pulley bracket	32.	Alternator bracket	33.	Water outlet (front)
34.	Front timing chain case	35.	Rear timing chain case	36.	O-ring
37.	O-ring	38.	O-ring	
		C.	Oil filter

37663_370Z_G0010

Fig. 71 Timing chain case and related components

31. To remove the case cover, loosen the mounting bolts in the reverse order of the installation sequence.

32. Insert a suitable tool (KV10111100) into the notch at the top of the front timing chain cover, as shown in the illustration. Pry off the case by moving the tool, as shown in the illustration.

➡**Never use a screwdriver or similar item. After removal handle the cover carefully so it does not tilt, cant or warp under load.**

33. Using the proper tool, remove the front case seal, as required.

To install:

➡**Be sure to use new fasteners, as required.**

34. Install new O-rings on the rear timing chain case cover.

➡**Be sure that the O-rings remain in place during installation to the rear case cover.**

35. Install a new oil seal in the front timing chain cover. Coat the seal with clean engine oil before installation. Press fit the seal into position until it becomes flush with the front timing chain case end face.

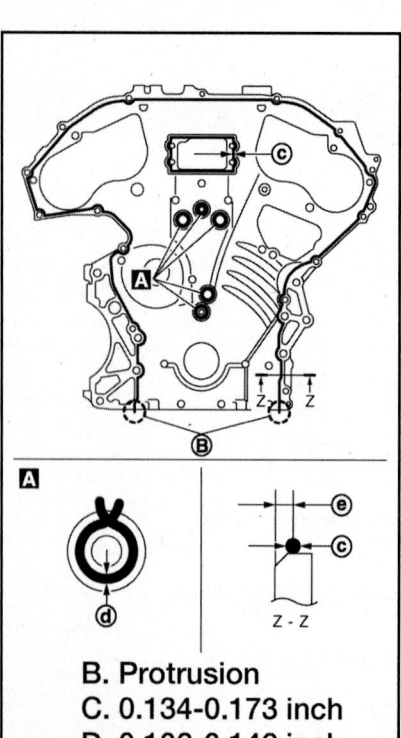

B. Protrusion
C. 0.134-0.173 inch
D. 0.102-0.142 inch
E. 0.157-0.220 inch

37663_370Z_G0091

Fig. 72 Front timing chain cover sealant application points

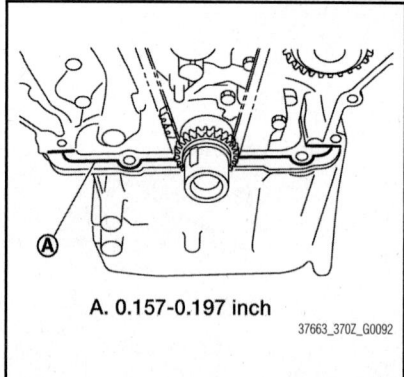

A. 0.157-0.197 inch

37663_370Z_G0092

Fig. 73 Oil pan (upper) sealant application points

Check that the garter spring is in position and that the seal lip is not inverted.

36. Apply a continuous bead of sealant as indicated in the illustration.

37. Apply sealant to the top surface of the upper oil pan, as indicated in the illustration.

38. Install the cover. Be sure not to damage the oil seal during cover installation. Attaching should be done within five minutes after sealant application.

39. Tighten the bolts to specification and in the proper sequence.

40. Tightening specification is 41 ft. lbs. (55.0 Nm) for M10 bolts (1,2,3,4,5,6,7) and 9 ft. lbs. (12.7 NM) for M6 bolts.

41. After all bolts are tightened, retighten them to specification and in the proper sequence.

42. Install the upper oil pan mounting bolts in the proper sequence.

43. To install the valve timing control covers, first install new seal rings in the shaft grooves.

44. Install the covers, using new gaskets.

➡**Align the center of both shaft holes of the camshaft sprocket (INT) and the**

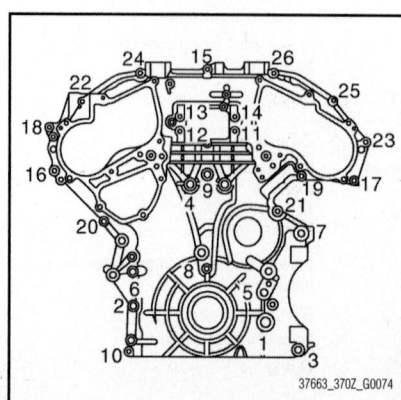

37663_370Z_G0074

Fig. 74 Front timing chain cover bolt locations and tightening sequence

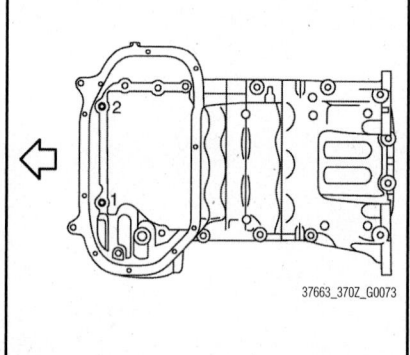

37663_370Z_G0073

Fig. 75 Upper oil pan front bolt locations and tightening sequence

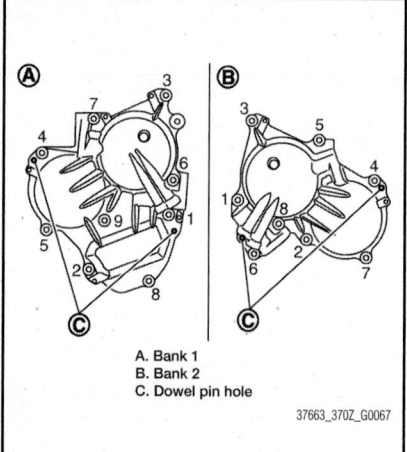

A. Bank 1
B. Bank 2
C. Dowel pin hole

37663_370Z_G0067

Fig. 76 Intake valve timing control cover bolt locations and tightening sequence

shaft and then insert them. Be careful not to drop the seal ring from the shaft groove.

45. Tighten the mounting bolts in the sequence shown in the illustration.

46. Continue the installation in the reverse order of the removal procedure.

TIMING CHAIN, TENSIONER, & SPROCKETS

REMOVAL & INSTALLATION

See Figures 77 through 93.

1. Before servicing the vehicle, refer to the Precautions Section.

➡**If working near and/or around the SRS system and components, be sure to disable the SRS system. After disabling the system wait three minutes or more before servicing the vehicle.**

➡**Before disconnecting the battery, lower both the driver's and passenger's windows. This will prevent any interference between the window edge and the** vehicle when the door is opened or closed. During normal operation the window slightly raises and lowers automatically to prevent any window to vehicle interference. The automatic window function will not work with the battery disconnected.

2. Remove the timing chain cover.

3. Remove the O-ring from the rear timing chain case.

4. To remove the timing chain tensioner (primary), remove the lower mounting bolt. Loosen the upper mounting bolt slowly and then turn the timing chain tensioner (primary) on the upper mounting bolt so that the plunger is fully expanded. Remove the upper mounting bolt and then remove the timing chain tensioner (primary).

➡**Even if the plunger is fully expanded, it does not drop from the body of the timing chain tensioner (primary).**

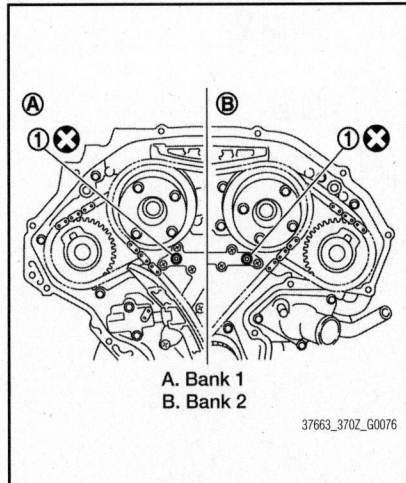

A. Bank 1
B. Bank 2

37663_370Z_G0076

Fig. 77 Rear timing chain case O-ring location

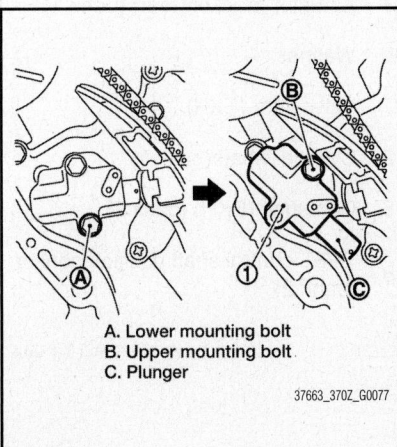

A. Lower mounting bolt
B. Upper mounting bolt
C. Plunger

37663_370Z_G0077

Fig. 78 Timing chain tensioner (primary) location

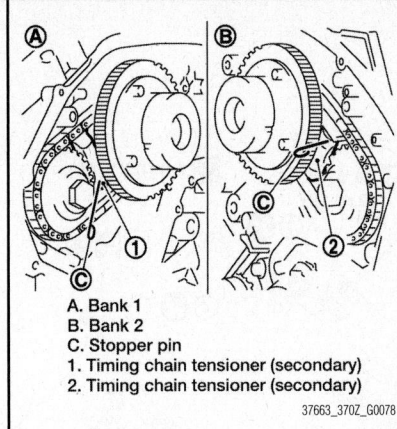

A. Bank 1
B. Bank 2
C. Stopper pin
1. Timing chain tensioner (secondary)
2. Timing chain tensioner (secondary)

37663_370Z_G0078

Fig. 79 Timing chain tensioner (secondary) location

5. Remove the internal chain guide, slack guide and tension guide.

6. Remove the timing chain (primary) and crankshaft sprocket.

➡**After removing the chain (primary), never turn the crankshaft and camshaft separately, or valves may strike the piston heads.**

7. To remove the timing chain (secondary) and camshaft sprockets attach a suitable stopper pin (0.020 inch hard metal) to the chain tensioners (secondary).

8. For removal of the chain tensioners (secondary) refer to the illustration.

➡**Removing VVEL ladder assembly is required.**

9. Remove the camshaft sprocket (EXH) mounting bolt.

➡**Secure the hexagonal portion of the camshaft (EXH) using a wrench to loosen mounting bolt. Never loosen the mounting bolt by securing anything other than camshaft (EXH) hexagonal portion or with tensioning the timing chain.**

10. Remove the camshaft sprocket (INT) mounting bolt.

➡**Secure the hexagonal portion (located between journal No. 1 and journal No. 2) of the driveshaft using a wrench to loosen the mounting bolt. Never loosen the mounting bolt by securing anything other than the driveshaft hexagonal portion or with tensioning the timing chain. When holding the hexagonal part of the driveshaft on the intake side with a wrench, be careful not to allow the wrench to cause interference with other parts.**

➡**Never disassemble the camshaft sprocket (INT). Never loosen bolts (A) as shown in the illustration.**

11. Remove the timing chain (secondary) along with the camshaft sprockets.

12. Remove all traces of gasket material from the front and rear timing chain covers. Be sure to remove all material from the bolt holes and threads. See illustration

To install:

➡**Be sure to use new fasteners, as required.**

➡**See illustration that shows the relationship between the matching mark on each timing chain and that on the corresponding sprocket with components installed.**

13. Check that the dowel pin and crankshaft key are located as shown in the illustration (engine at TDC on the compression stroke).

➡**Though the camshaft does not stop at the position as shown, for placement of cam noses, it is generally accepted that the camshaft is placed in the same direction as that of the illustration.**

➡**Matching marks between the chain and sprockets slip easily. Confirm all matching mark positions repeatedly during the installation process.**

14. To install the timing chains (secondary) and camshaft sprockets, push the plunger of the timing chain tensioner (secondary) and keep it pressed in with a stopper pin.

15. Install the timing chains (secondary) and camshaft sprockets. See illustration.

➡**Illustration shows bank 1 (rear view)**

16. Align the matching marks on the chain (secondary) (orange link) with the ones on the intake and exhaust camshaft sprockets (punched), and install them.

➡**Matching marks for camshaft sprockets (INT) are on the back side of the camshaft sprockets (secondary). There are two types of matching marks, the circle and the oval. They should be used for bank 1 (circle) and bank 2 (oval) respectively.**

17. Shape (orientation of signal plate) of camshaft sprocket (INT) varies depending on the bank position. See illustration.

18. Align dowel pin camshafts with the pin groove on sprockets and install them.

➡**In case that positions of each matching mark and each dowel pin do not fit**

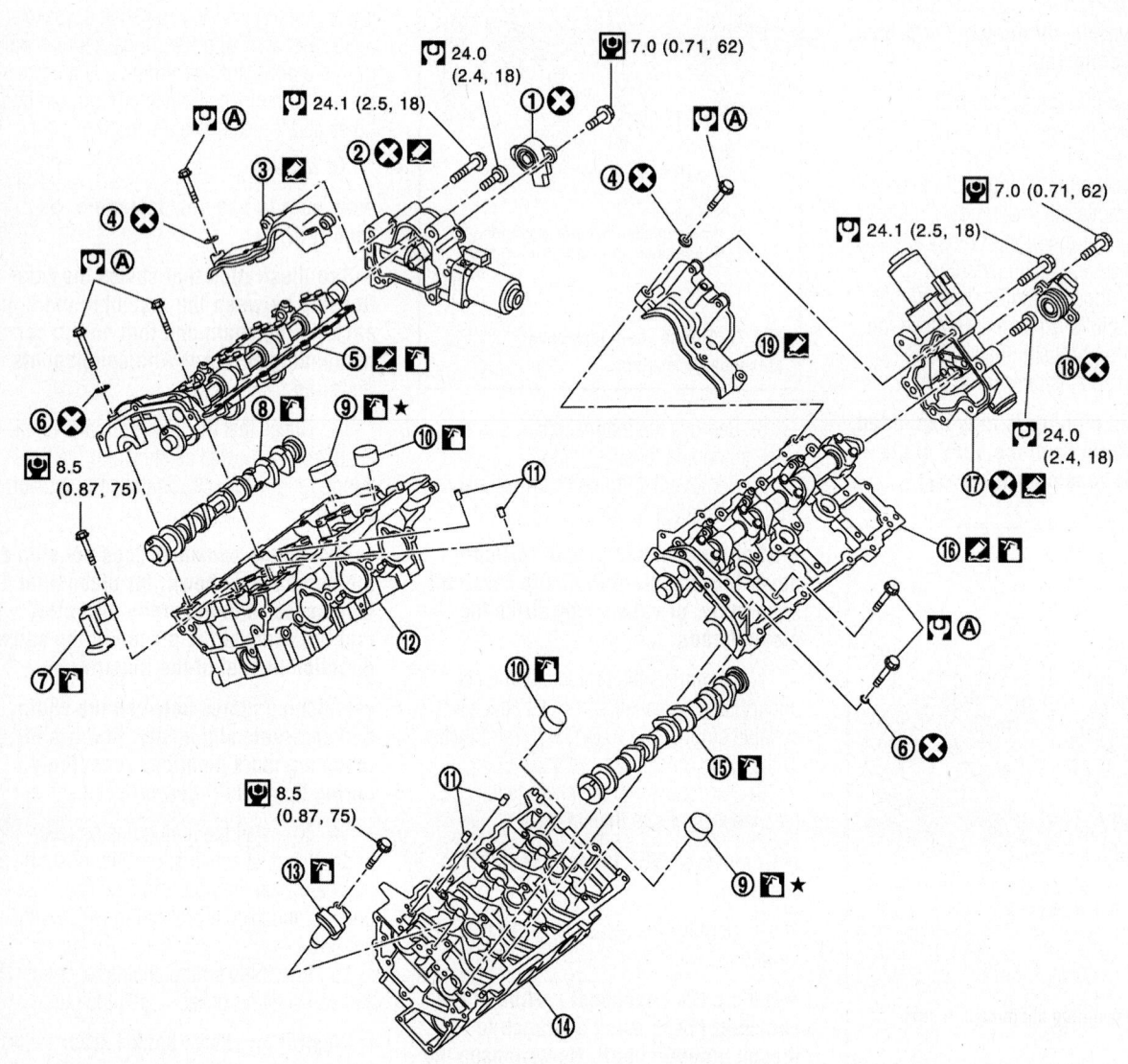

1. VVEL control shaft position sensor (bank 1)
2. VVEL actuator sub assembly (bank 1)
3. Actuator bracket (rear) (bank 1)
4. Washer
5. VVEL ladder assembly (bank 1)
6. Washer
7. Timing chain tensioner (secondary) (bank 1)
8. Camshaft (EXH) (bank 1)
9. Valve lifter (EXH)
10. Valve lifter (INT)
11. Oil filter
12. Cylinder head (bank 1)
13. Timing chain tensioner (secondary) (bank 2)
14. Cylinder head (bank 2)
15. Camshaft (EXH) (bank 2)
16. VVEL ladder assembly (bank 2)
17. VVEL actuator sub assembly (bank 2)
18. VVEL control shaft position sensor (bank 2)
19. Actuator bracket (rear) (bank 2)

37663_370Z_G0004

Fig. 80 Camshaft and related components (VVEL ladder assembly)

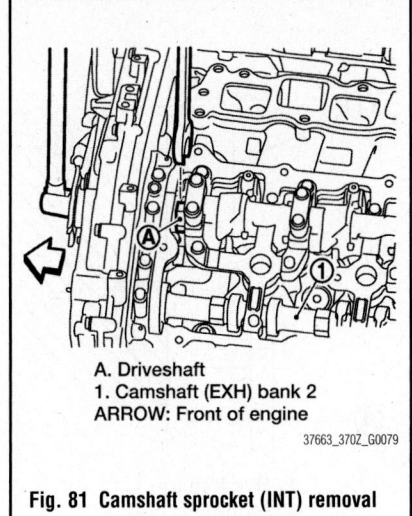

A. Driveshaft
1. Camshaft (EXH) bank 2
ARROW: Front of engine

37663_370Z_G0079

Fig. 81 Camshaft sprocket (INT) removal

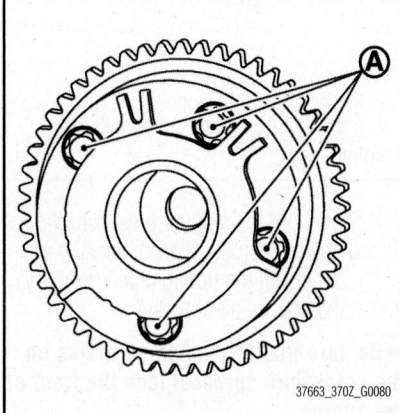

37663_370Z_G0080

Fig. 82 Camshaft sprocket (INT) bolt "A" location

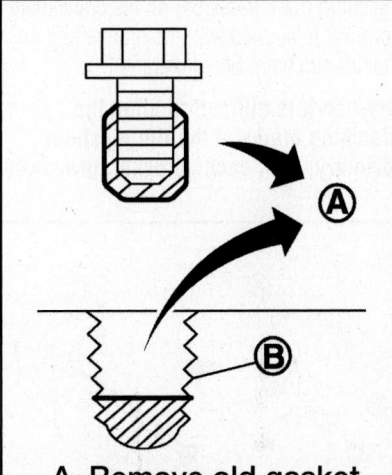

A. Remove old gasket that is stuck
B. Bolt hole

37663_370Z_G0081

Fig. 83 Removing gasket material from bolt hole

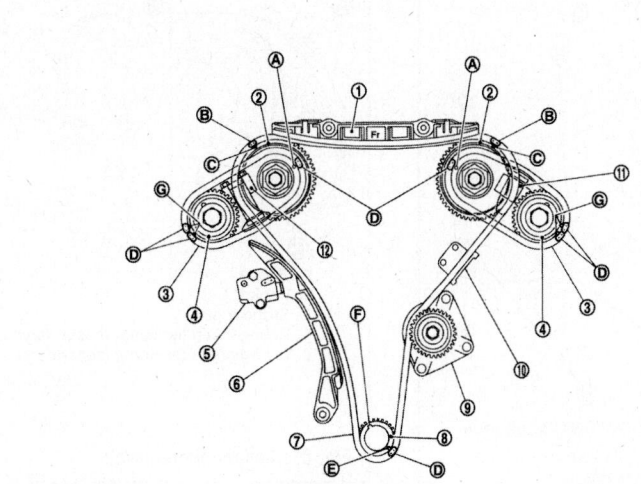

1.	Internal chain guide	2.	Camshaft sprocket (INT)	3.	Timing chain (secondary)
4.	Camshaft sprocket (EXH)	5.	Timing chain tensioner (primary)	6.	Slack guide
7.	Timing chain (primary)	8.	Crankshaft sprocket	9.	Water pump
10.	Tension guide	11.	Timing chain tensioner (secondary) (bank 2)	12.	Timing chain tensioner (secondary) (bank 1)
A.	Matching mark [punched (back side)]	B.	Matching mark (yellow link)	C.	Matching mark (punched)
D.	Matching mark (orange link)	E.	Matching mark (notched)	F.	Crankshaft key

37663_370Z_G0082

Fig. 84 Timing chains and sprockets (relationship) alignment

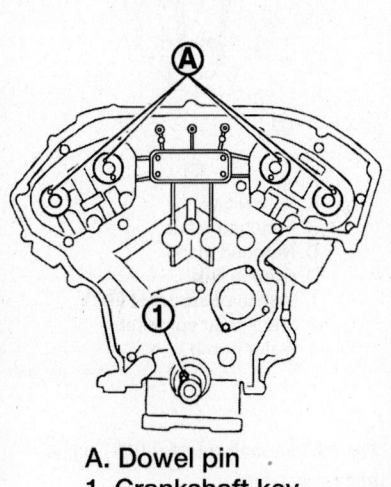

A. Dowel pin
1. Crankshaft key

37663_370Z_G0083

Fig. 85 Camshafts and crankshaft positioning

with matching parts, make fine adjustment to the position holding the hexagonal portion on camshaft (EXH) or driveshaft with wrench or equivalent tool.

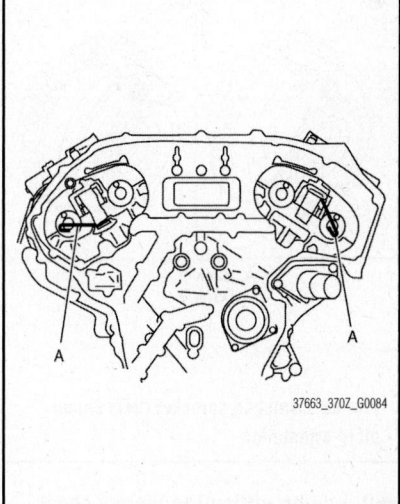

37663_370Z_G0084

Fig. 86 Stopper pin installation and timing chain tensioner (secondary) plunger location

19. Mounting bolts for camshaft sprockets must be tightened in the next step. Tightening by hand is sufficient to prevent the dislocation of the dowel pins.

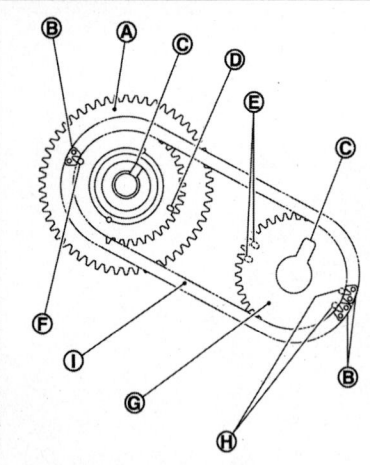

A. Camshaft sprocket (INT) back face
B. Orange link
C. Dowel groove
D. Matching mark (oval)
E. Matching mark (2 oval: on front face)
F. Matching mark (circle)
G. Camshaft sprocket (EXH) back face
H. Matching mark (2 oval: on front face)
I. Timing chain (secondary)

37663_370Z_G0085

Fig. 87 Timing chain tensioner (secondary) and camshaft sprockets—bank 1 rear view

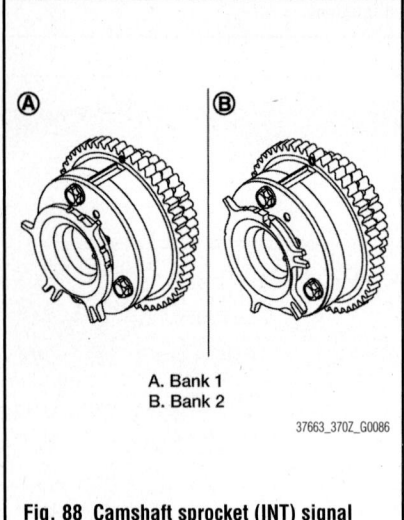

A. Bank 1
B. Bank 2

37663_370Z_G0086

Fig. 88 Camshaft sprocket (INT) signal plate orientation

➡️**It may be difficult to visibly check the dislocation of the matching marks during and after installation. To make the matching easier make a matching mark on the top of the sprocket teeth and its extended line in advance using paint.**

20. Tighten the camshaft sprocket (EXH) mounting bolt.

21. After confirming that the matching marks are aligned, tighten the camshaft

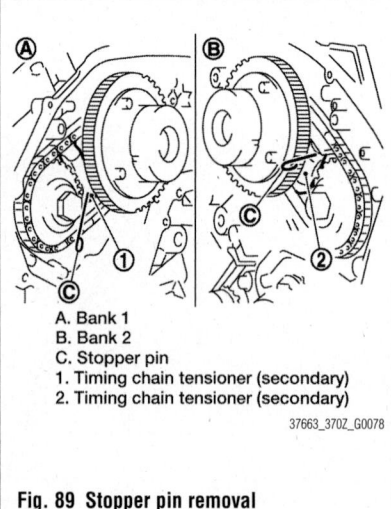

A. Bank 1
B. Bank 2
C. Stopper pin
1. Timing chain tensioner (secondary)
2. Timing chain tensioner (secondary)

37663_370Z_G0078

Fig. 89 Stopper pin removal

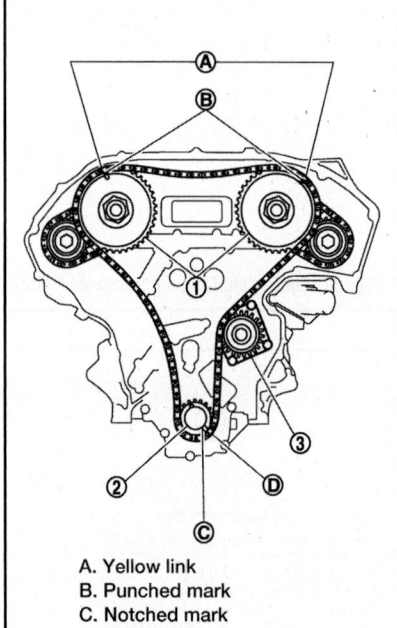

A. Yellow link
B. Punched mark
C. Notched mark
D. Orange link
1. Camshaft sprocket (INT)
2. Crankshaft sprocket
3. Water pump

37663_370Z_G0087

Fig. 90 Camshaft sprocket (INT) signal plate orientation

sprocket (INT) mounting bolt. Secure the hexagonal portion (located between journal No. 1 and Journal No. 2) of the driveshaft using a wrench to tighten the mounting bolt.

➡️ **When holding the hexagonal part of the driveshaft on the intake side with a wrench, be careful not to allow the wrench to cause interference with other parts.**

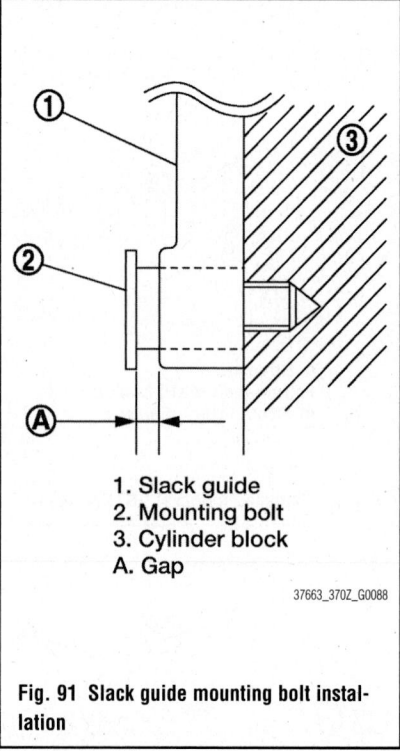

1. Slack guide
2. Mounting bolt
3. Cylinder block
A. Gap

37663_370Z_G0088

Fig. 91 Slack guide mounting bolt installation

22. Pull out the stopper pins from the timing chain tensioners (secondary).

23. To install the timing chain (primary), install the crankshaft sprocket.

➡️**Be sure that the matching marks on the crankshaft sprocket face the front of the engine.**

24. Install the timing chain (primary) so that the matching mark (punched) on the camshaft sprocket (INT) is aligned with the yellow link on the timing chain while the matching mark (notched) on the crankshaft sprocket is aligned with the orange link on the timing chain. See illustration.

➡️**When it is difficult to align the matching marks of the timing chain (primary) with each sprocket, gradually**

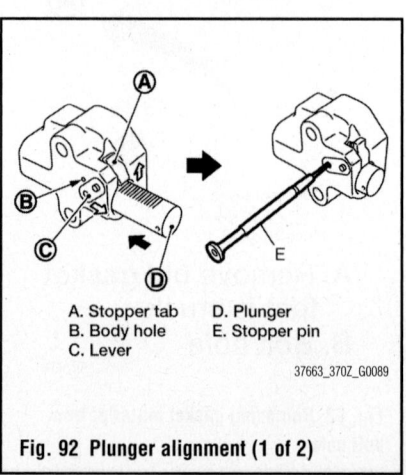

A. Stopper tab   D. Plunger
B. Body hole   E. Stopper pin
C. Lever

37663_370Z_G0089

Fig. 92 Plunger alignment (1 of 2)

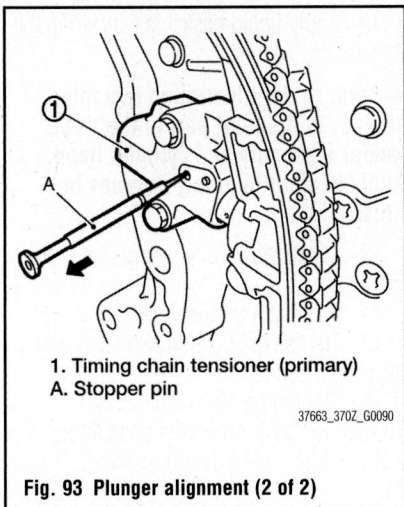

1. Timing chain tensioner (primary)
A. Stopper pin

37663_370Z_G0090

Fig. 93 Plunger alignment (2 of 2)

turn the driveshaft sing a wrench on the hexagonal portion to align it with the matching marks.

25. Install the internal chain guide, slack guide and tension guide.

➡**Never overtighten the slack guide mounting bolt. It is normal for a gap to exist under the bolt seats when mounting bolts are tightened to specification.**

26. To install the timing chain tensioner (primary) pull the plunger stopper tab up (or turn lever downward) so as to remove the plunger stopper tab from the ratchet of the plunger. Note that the plunger stopper tab and lever are synchronized.

27. Push the plunger into the inside of the tensioner body. Hold the plunger in the fully compressed position by engaging the plunger stopper tab with the tip of the ratchet.

28. To secure the lever, insert the stopper pin through the hole of the lever into the tensioner body hole. The lever parts and the plunger stopper tab are synchronized, therefore the plunger is secured under this condition.

➡**The illustration shows a suitable tool of 0.047 inch diameter being used as a stopper pin.**

29. Pull out the stopper pin after installing and release the plunger.

30. Check again that the matching marks on the sprockets and the timing chain have not slipped out of alignment.

31. Install the timing chain cover.

VALVE COVERS

REMOVAL & INSTALLATION
See Figures 94 through 96.

1. Before servicing the vehicle, refer to the Precautions Section.

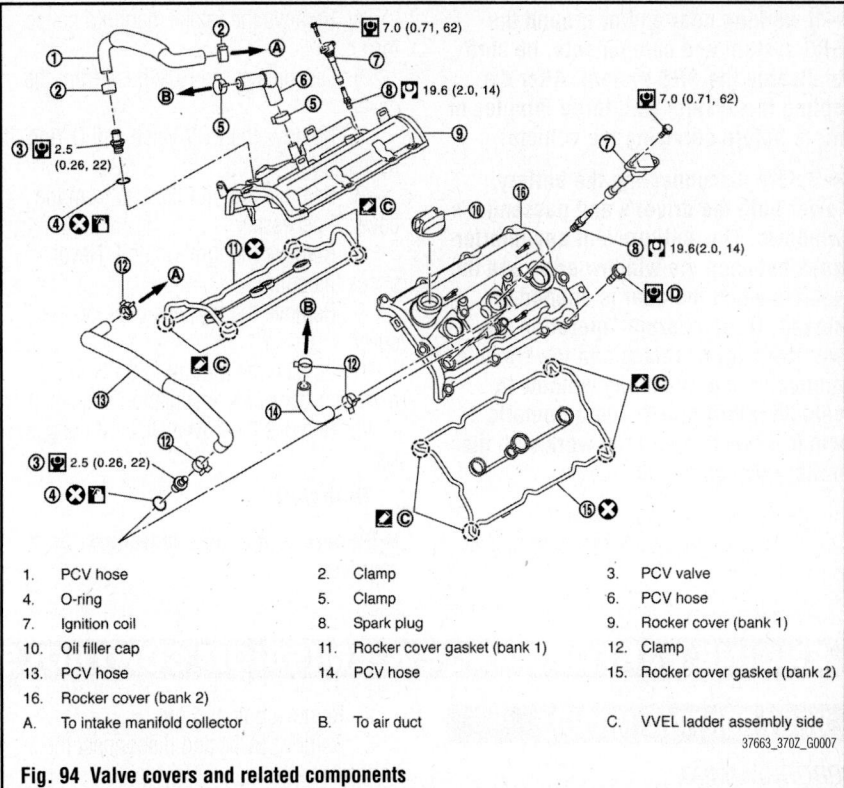

1.	PCV hose	2.	Clamp	3.	PCV valve
4.	O-ring	5.	Clamp	6.	PCV hose
7.	Ignition coil	8.	Spark plug	9.	Rocker cover (bank 1)
10.	Oil filler cap	11.	Rocker cover gasket (bank 1)	12.	Clamp
13.	PCV hose	14.	PCV hose	15.	Rocker cover gasket (bank 2)
16.	Rocker cover (bank 2)				
A.	To intake manifold collector	B.	To air duct	C.	VVEL ladder assembly side

37663_370Z_G0007

Fig. 94 Valve covers and related components

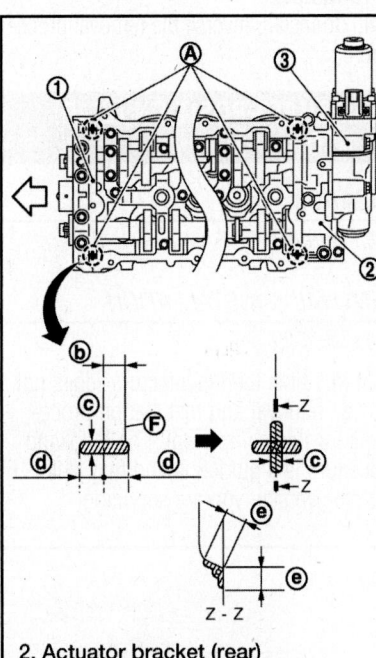

2. Actuator bracket (rear)
3. VVEL actuator sub assembly
A. Sealant application point
F. End surface of VVEL sub assembly
B. 0.16 inch
C. 0.098-0.138 inch
D. 0.20 inch
E 0.39 inch
ARROW: Front of engine

37663_370Z_G0065

Fig. 95 Valve cover sealant application points

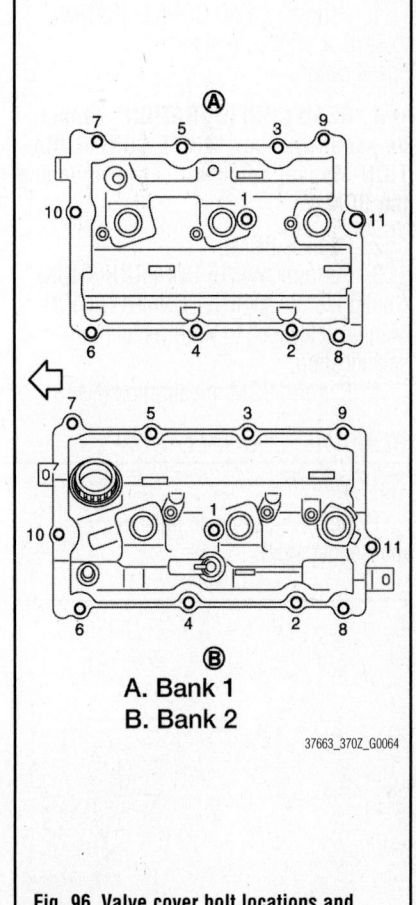

A. Bank 1
B. Bank 2

37663_370Z_G0064

Fig. 96 Valve cover bolt locations and tightening sequence

➡If working near and/or around the SRS system and components, be sure to disable the SRS system. After disabling the system wait three minutes or more before servicing the vehicle.

➡Before disconnecting the battery, lower both the driver's and passenger's windows. This will prevent any interference between the window edge and the vehicle when the door is opened or closed. During normal operation the window slightly raises and lowers automatically to prevent any window to vehicle interference. The automatic window function will not work with the battery disconnected.

2. Remove the engine cover.
3. Remove the air cleaner case and air duct.

4. Remove the intake manifold collector.
5. Disconnect the PCV hose from the cover.
6. Remove the PCV valve and O-ring, if necessary.
7. Remove the oil filler cap from the cover, if necessary.
8. Remove the ignition coil. Never shock the coil.
9. Remove the harness clips on the cover.
10. Loosen the mounting bolts in the reverse order of the installation sequence.
11. Remove the cover. Discard the gasket.

To install:

➡Be sure to use new fasteners, as required.

12. Apply liquid gasket, as shown in the illustration.

➡Refer to the illustration to apply sealant to the joint part of the VVEL ladder assembly and cylinder head. Apply sealant in ninety degrees to illustration.

13. Install the valve cover gasket to the valve cover.
14. Install the rocker cover.
15. Tighten the bolts in two steps and in the proper sequence.
16. Tightening specification is 18 inch lbs. (2.0 Nm), first pass and then 73 inch lbs (8.3 Nm), second pass.
17. Continue the installation in the reverse order of the removal procedure.

ENGINE PERFORMANCE & EMISSION CONTROLS

BODY CONTROL MODULE

PROGRAMMING

CONSULT Configuration

1. Perform "READ CONFIGURATION" to save or print current vehicle specification.

➡If "READ CONFIGURATION" cannot be used, use the "WRITE CONFIGURATION-Manual selection" after replacing the BCM.

2. Replace BCM.
3. Perform "WRITE CONFIGURATION-Config file" or "WRITE CONFIGURATION-Manual selection" to write vehicle specification.
4. Perform BCM initialization (NATS).

REMOVAL & INSTALLATION

See Figure 97.

1. Remove dash side finisher (passenger side).

2. Remove bolt and nut.
3. Remove BCM and disconnect the connector.

To install:
4. To install, reverse the removal procedure.

CAMSHAFT POSITION (CMP) SENSOR

LOCATION
See Figure 98.

REMOVAL & INSTALLATION
See Figure 99.

At this time the manufacturer does not provide removal and installation procedures for this component. The following procedure is a guideline and may differ from the vehicle you are servicing.

1. Before servicing the vehicle, refer to the Precautions Section.

➡If working near and/or around the SRS system and components, be sure to disable the SRS system. After disabling the system wait three minutes or more before servicing the vehicle.

➡Before disconnecting the battery, lower both the driver's and passenger's windows. This will prevent any interference between the window edge and the vehicle when the door is opened or closed. During normal operation the window slightly raises and lowers automatically to prevent any window to vehicle interference. The automatic window function will not work with the battery disconnected.

2. Remove the necessary components in order to gain access to the sensor.

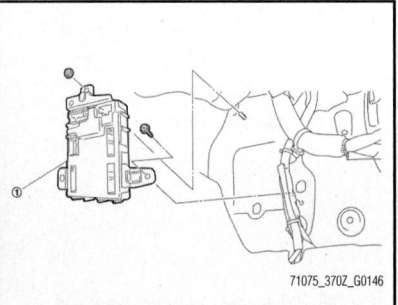

71075_370Z_G0146

Fig. 97 Removing and installing the BCM

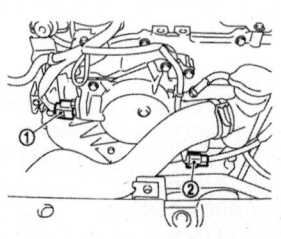 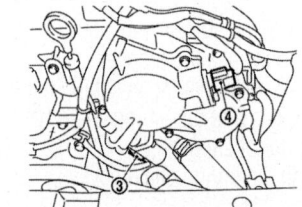

1. Camshaft position sensor (PHASE) (bank 1)
2. Intake valve timing control solenoid valve (bank 1) harness connector
3. Intake valve timing control solenoid valve (bank 2) harness connector
4. Camshaft position sensor (PHASE) (bank 2)

37663_370Z_G0135

Fig. 98 Camshaft Position (CMP) sensor location

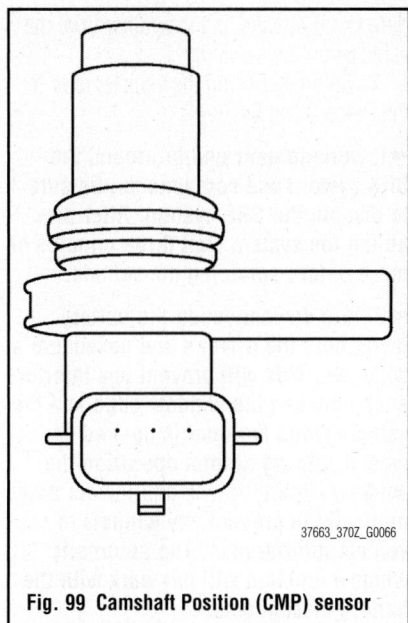

Fig. 99 Camshaft Position (CMP) sensor

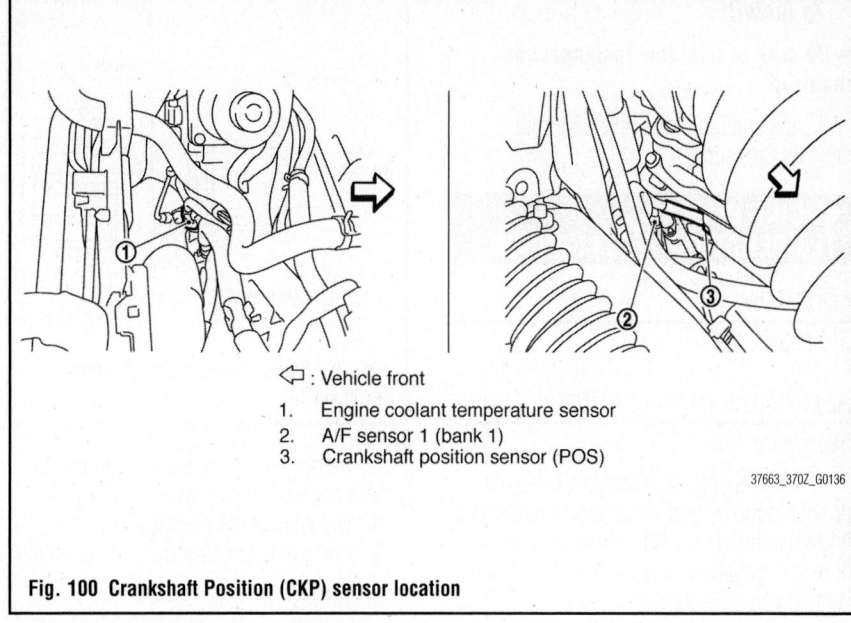

◁ : Vehicle front
1. Engine coolant temperature sensor
2. A/F sensor 1 (bank 1)
3. Crankshaft position sensor (POS)

Fig. 100 Crankshaft Position (CKP) sensor location

3. Disconnect the electrical connector.
4. Remove the retaining bolt.
5. Remove the component from the vehicle.

To install:

➡ **Be sure to use new fasteners, as required.**

6. Installation is the reverse of the removal procedure.

CRANKSHAFT POSITION (CKP) SENSOR

LOCATION
See Figure 100.

This sensor is located on the cylinder block facing the gear teeth on the signal plate.

REMOVAL & INSTALLATION
See Figure 101.

At this time the manufacturer does not provide removal and installation procedures for this component. The following procedure is a guideline and may differ from the vehicle you are servicing.

1. Before servicing the vehicle, refer to the Precautions Section.

➡ **If working near and/or around the SRS system and components, be sure to disable the SRS system. After disabling the system wait three minutes or more before servicing the vehicle.**

➡ **Before disconnecting the battery, lower both the driver's and passenger's windows. This will prevent any interfer-**ence between the window edge and the vehicle when the door is opened or closed. During normal operation the window slightly raises and lowers automatically to prevent any window to vehicle interference. The automatic window function will not work with the battery disconnected.

2. Remove the necessary components in order to gain access to the sensor.
3. Disconnect the electrical connector.
4. Remove the retaining bolt.

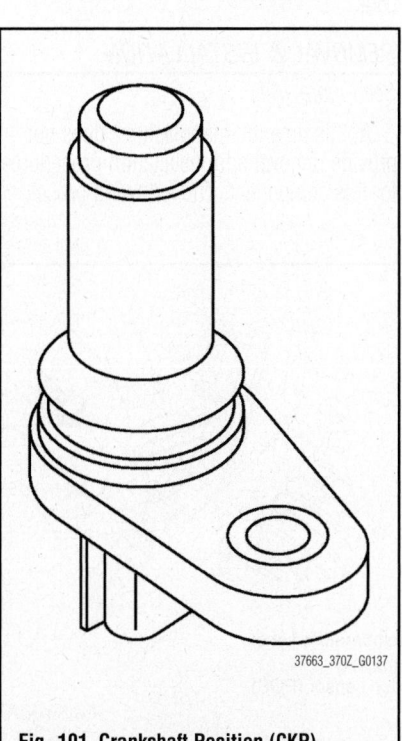

Fig. 101 Crankshaft Position (CKP) sensor

5. Remove the component from the vehicle.

To install:

➡ **Be sure to use new fasteners, as required.**

6. Installation is the reverse of the removal procedure.

ELECTRONIC CONTROL MODULE

LOCATION

The ECM is located on the passenger's side of the vehicle at the kick panel.

REMOVAL & INSTALLATION

1. Before servicing the vehicle, refer to the Precautions Section.

➡ **If working near and/or around the SRS system and components, be sure to disable the SRS system. After disabling the system wait three minutes or more before servicing the vehicle.**

➡ **Before disconnecting the battery, lower both the driver's and passenger's windows. This will prevent any interference between the window edge and the vehicle when the door is opened or closed. During normal operation the window slightly raises and lowers automatically to prevent any window to vehicle interference. The automatic window function will not work with the battery disconnected.**

2. Remove glove box.
3. Disconnect electrical connectors.
4. Remove the component.

To install:

➡ **Be sure to use new fasteners, as required.**

5. Installation is the reverse of the removal procedure.

ENGINE COOLANT TEMPERATURE (ECT) SENSOR

LOCATION

See Figure 102.

REMOVAL & INSTALLATION

See Figure 103.

At this time the manufacturer does not provide removal and installation procedures for this component. The following procedure is a guideline and may differ from the vehicle you are servicing.

1. Before servicing the vehicle, refer to the Precautions Section.

➡ **If working near and/or around the SRS system and components, be sure to disable the SRS system. After disabling the system wait three minutes or more before servicing the vehicle.**

➡ **Before disconnecting the battery, lower both the driver's and passenger's windows. This will prevent any interference between the window edge and the vehicle when the door is opened or closed. During normal operation the window slightly raises and lowers automatically to prevent any window to vehicle interference. The automatic window function will not work with the battery disconnected.**

2. Drain the cooling system. Properly dispose of used engine coolant.

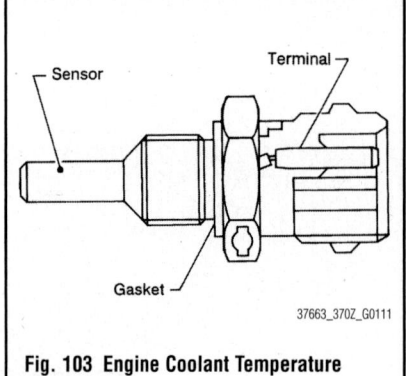

Fig. 103 Engine Coolant Temperature (ECT) sensor

3. Remove the necessary components to gain access to the sensor.
4. Disconnect the electrical sensor.
5. Remove the sensor from its mounting.

To install:

➡ **Be sure to use new fasteners, as required.**

6. Installation is the reverse of the removal procedure.

HEATED OXYGEN SENSOR (HO2S)

LOCATION

This sensor is located in the exhaust stream, after the catalytic converter. Two sensors are used, one for each cylinder bank.

REMOVAL & INSTALLATION

See Figure 104.

At this time the manufacturer does not provide removal and installation procedures for this component. The following proce-

dure is a guideline and may differ from the vehicle you are servicing.

1. Before servicing the vehicle, refer to the Precautions Section.

➡ **If working near and/or around the SRS system and components, be sure to disable the SRS system. After disabling the system wait three minutes or more before servicing the vehicle.**

➡ **Before disconnecting the battery, lower both the driver's and passenger's windows. This will prevent any interference between the window edge and the vehicle when the door is opened or closed. During normal operation the window slightly raises and lowers automatically to prevent any window to vehicle interference. The automatic window function will not work with the battery disconnected.**

2. Raise and safely support the vehicle.
3. Remove the undercover.
4. Disconnect the harness connector.
5. Remove the sensor from its mounting.

➡ **Discard the sensor if it has been dropped.**

To install:

➡ **Be sure to use new fasteners, as required.**

6. Installation is the reverse of the removal procedure.
7. Before installing a new sensor coat the threads with an approved anti-seize lubricant.

INTAKE AIR TEMPERATURE (IAT)/MASS AIR FLOW (MAF) SENSOR

LOCATION

This sensor is placed in the stream of the air intake.

REMOVAL & INSTALLATION

See Figure 105.

At this time the manufacturer does not provide removal and installation procedures for this component. The following procedure is a guideline and may differ from the vehicle you are servicing.

1. Before servicing the vehicle, refer to the Precautions Section.

➡ **If working near and/or around the SRS system and components, be sure to disable the SRS system. After disabling the system wait three minutes or more before servicing the vehicle.**

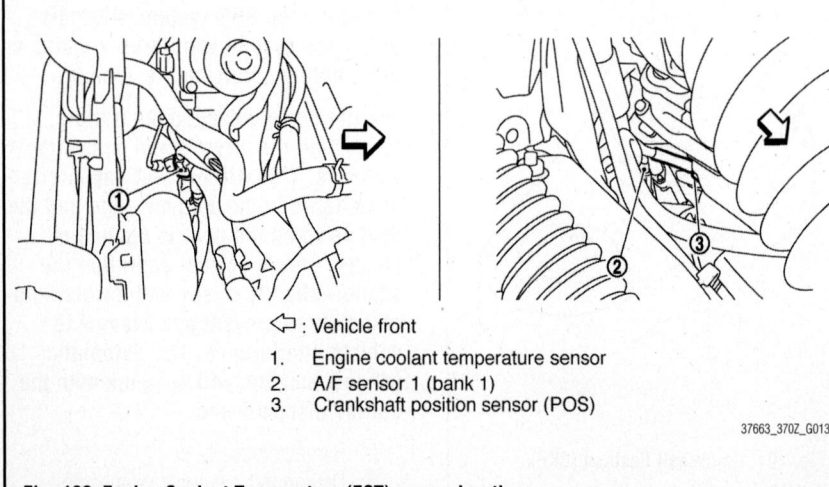

⇦ : Vehicle front

1. Engine coolant temperature sensor
2. A/F sensor 1 (bank 1)
3. Crankshaft position sensor (POS)

Fig. 102 Engine Coolant Temperature (ECT) sensor location

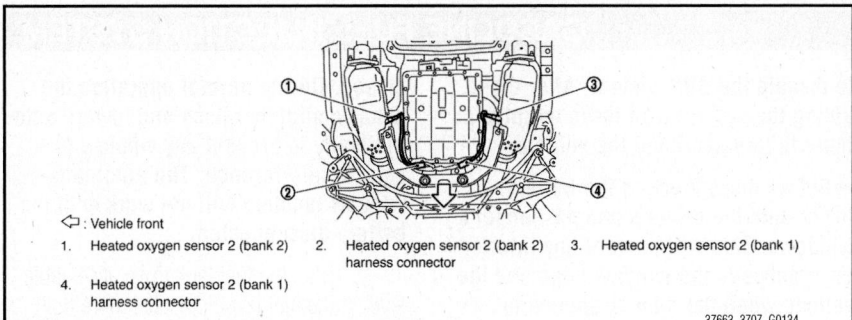

⟵ : Vehicle front
1. Heated oxygen sensor 2 (bank 2) 2. Heated oxygen sensor 2 (bank 2) harness connector 3. Heated oxygen sensor 2 (bank 1)
4. Heated oxygen sensor 2 (bank 1) harness connector

37663_370Z_G0134

Fig. 104 Heated oxygen sensor and related components

➡**Before disconnecting the battery, lower both the driver's and passenger's windows. This will prevent any interference between the window edge and the vehicle when the door is opened or closed. During normal operation the window slightly raises and lowers automatically to prevent any window to vehicle interference. The automatic window function will not work with the battery disconnected.**

2. Disconnect the harness connector from the IAT/MAF sensor.

3. Disconnect the tube clamp at the electric throttle control actuator and at the fresh air intake tube.

4. Remove air cleaner to electric throttle control actuator tube, air cleaner case (upper) with the IAT/MAF sensor attached.

5. Remove IAT/MAF sensor from air cleaner case (upper), as necessary.

6. Remove resonator in the fender, lifting left fender protector, as necessary.

To install:

➡**Be sure to use new fasteners, as required.**

7. Installation is the reverse of the removal procedure.

KNOCK SENSOR (KS)

LOCATION

The knock sensors are located under the intake manifold. There are two of them.

REMOVAL & INSTALLATION

See Figure 106.

At this time the manufacturer does not provide removal and installation procedures for this component, refer to the illustration as required.

MALFUNCTION INDICATOR LIGHT

RESET PROCEDURE

Clearing diagnostic trouble codes resets the MIL.

THROTTLE POSITION SENSOR

LOCATION

This sensor is located in the main hose connecting the intake manifold collector. There are two of these sensors.

REMOVAL & INSTALLATION

See Figures 39 and 107.

At this time the manufacturer does not provide removal and installation procedures for this component, refer to the illustration as required.

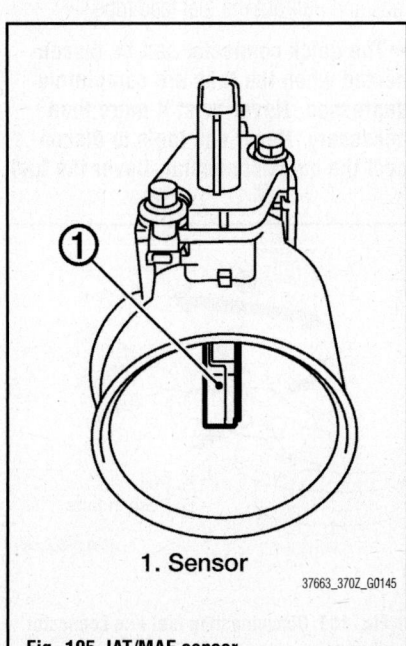

1. **Sensor**

37663_370Z_G0145

Fig. 105 IAT/MAF sensor

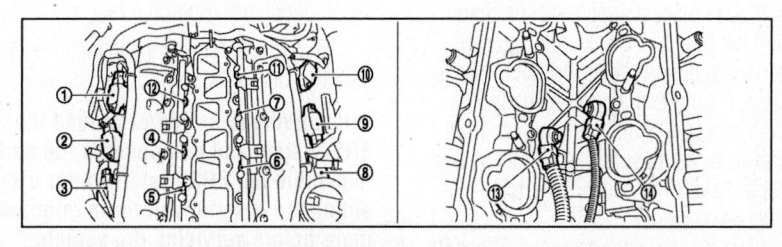

1. Ignition coil No.5 (with power transistor) 2. Ignition coil No.3 (with power transistor) 3. Ignition coil No.1 (with power transistor)
4. Fuel injector No.3 5. Fuel injector No.1 6. Fuel injector No.2
7. Fuel injector No.4 8. Ignition coil No.2 (with power transistor) 9. Ignition coil No.4 (with power transistor)
10. Ignition coil No.6 (with power transistor) 11. Fuel injector No.6 12. Fuel injector No.5
13. Knock sensor (bank 1) 14. Knock sensor (bank 2)

37663_370Z_G0144

Fig. 106 Knock sensors and related components

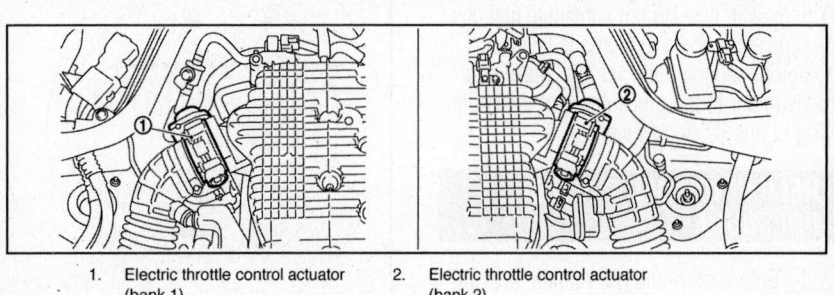

1. Electric throttle control actuator (bank 1) 2. Electric throttle control actuator (bank 2)

37663_370Z_G0148

Fig. 107 TPS sensor (electronic throttle control) location

FUEL SYSTEM SERVICE PRECAUTIONS

Safety is the most important factor when performing not only fuel system maintenance but any type of maintenance. Failure to conduct maintenance and repairs in a safe manner may result in serious personal injury or death. Maintenance and testing of the vehicle's fuel system components can be accomplished safely and effectively by adhering to the following rules and guidelines.

• To avoid the possibility of fire and personal injury, always disconnect the negative battery cable unless the repair or test procedure requires that battery voltage be applied.

• Always relieve the fuel system pressure prior to disconnecting any fuel system component (injector, fuel rail, pressure regulator, etc.), fitting or fuel line connection. Exercise extreme caution whenever relieving fuel system pressure to avoid exposing skin, face and eyes to fuel spray. Please be advised that fuel under pressure may penetrate the skin or any part of the body that it contacts.

• Always place a shop towel or cloth around the fitting or connection prior to loosening to absorb any excess fuel due to spillage. Ensure that all fuel spillage (should it occur) is quickly removed from engine surfaces. Ensure that all fuel soaked cloths or towels are deposited into a suitable waste container.

• Always keep a dry chemical (Class B) fire extinguisher near the work area.

• Do not allow fuel spray or fuel vapors to come into contact with a spark or open flame.

• Always use a back-up wrench when loosening and tightening fuel line connection fittings. This will prevent unnecessary stress and torsion to fuel line piping.

• Always replace worn fuel fitting O-rings with new Do not substitute fuel hose or equivalent where fuel pipe is installed.

Before servicing the vehicle, make sure to also refer to the precautions in the beginning of this section as well.

RELIEVING FUEL SYSTEM PRESSURE

1. Before servicing the vehicle, refer to the Precautions Section.

➡**If working near and/or around the SRS system and components, be sure**

to disable the SRS system. After disabling the system wait three minutes or more before servicing the vehicle.

➡**Before disconnecting the battery, lower both the driver's and passenger's windows. This will prevent any interference between the window edge and the vehicle when the door is opened or closed. During normal operation the window slightly raises and lowers automatically to prevent any window to vehicle interference. The automatic window function will not work with the battery disconnected.**

2. Remove the fuel pump fuse, located in the IPDM E/R.
3. Start the engine.
4. After the engine stalls, crank it two or three times to release all fuel pressure.
5. Turn the ignition switch off.
6. Reinstall the fuel pump fuse after servicing the fuel system.

FUEL PUMP, FUEL LEVEL SENSOR UNIT & FILTER

REMOVAL & INSTALLATION
See Figures 108 through 114.

1. Before servicing the vehicle, refer to the Precautions Section.

➡**If working near and/or around the SRS system and components, be sure to disable the SRS system. After disabling the system wait three minutes or more before servicing the vehicle.**

➡**Before disconnecting the battery, lower both the driver's and passenger's windows. This will prevent any interference between the window edge and the vehicle when the door is opened or**

closed. During normal operation the window slightly raises and lowers automatically to prevent any window to vehicle interference. The automatic window function will not work with the battery disconnected.

2. Drain the fuel tank to an acceptable level. If the fuel level indicates more than the level shown in the illustration, full or almost full, drain the fuel from the tank until the fuel gauge indicates a level as shown in the illustration.

➡**Because fuel will be spilled when removing the main and sub fuel level sensor units for the top of the fuel is above the main and sub fuel level sensor units installed surface. As a guide, fuel level becomes the position as shown in the illustration when approximately 3 3/8 gallons of fuel are removed from the tank.**

3. Properly relieve the fuel system pressure.
4. Remove the fuel tank cap.
5. Remove the rear parcel shelf covers, both left and right.
6. Remove the inspection hole cover.

➡**Right side for main fuel level sensor, fuel filter and fuel pump assembly. Left side for sub fuel level sensor unit.**

7. Disconnect and fuel feed tube. Disconnect the harness connector.
8. Disconnect the quick connector. Hold the sides of the connector, push in the tabs and pull out the fuel feed tube.

➡**The quick connector can be disconnected when the tabs are completely depressed. Never twist it more than necessary. Never use tools to disconnect the quick connector. Cover the fuel**

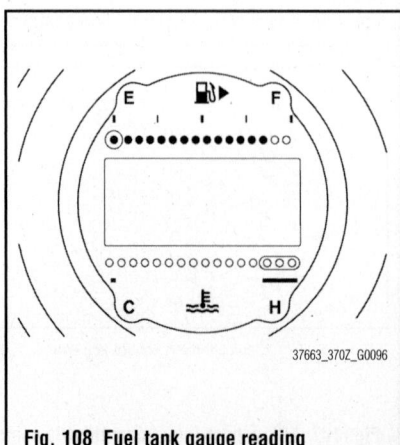

Fig. 108 Fuel tank gauge reading

37663_370Z_G0096

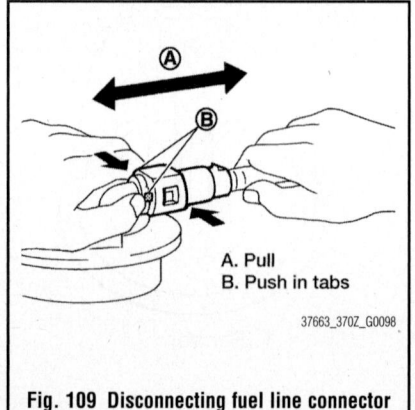

A. Pull
B. Push in tabs

37663_370Z_G0098

Fig. 109 Disconnecting fuel line connector (1 of 2)

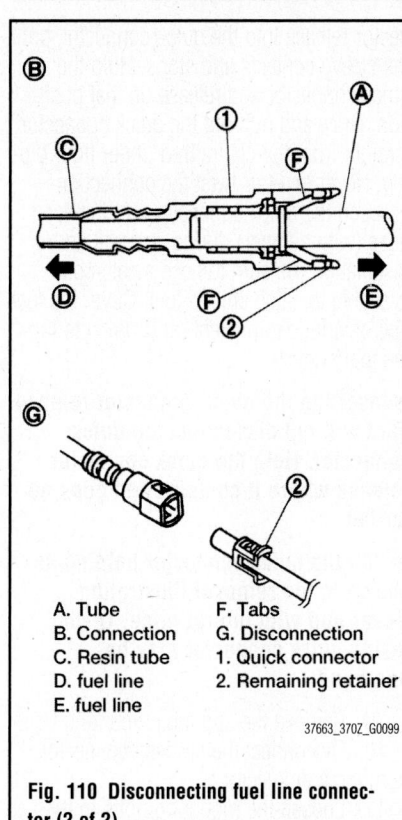

Fig. 110 Disconnecting fuel line connector (2 of 2)

A. Tube
B. Connection
C. Resin tube
D. fuel line
E. fuel line
F. Tabs
G. Disconnection
1. Quick connector
2. Remaining retainer

37663_370Z_G0099

line openings to prevent dirt from entering the fuel system.

9. To remove the main fuel level sensor unit, fuel filter and fuel pump assembly, remove the retainer. Raise the unit and disconnect the quick connector.

10. To remove the sub fuel level sensor, remove the retainer. Raise the component and remove it.

To install:

➡**Be sure to use new fasteners, as required.**

11. Installation is the reverse of the removal procedure.

12. When installing, face the units as shown in the illustration and install them with the knock pun on back aligned with the pin hole on the fuel tank.

13. Install the retainer so that its notch becomes parallel with the notch on the fuel tank. Tighten the retainer bolts evenly.

14. Install the fuel pump fuse, if removed.

15. Turn the ignition switch ON (with the engine stopped), check all connections for fuel leakage, correct as required.

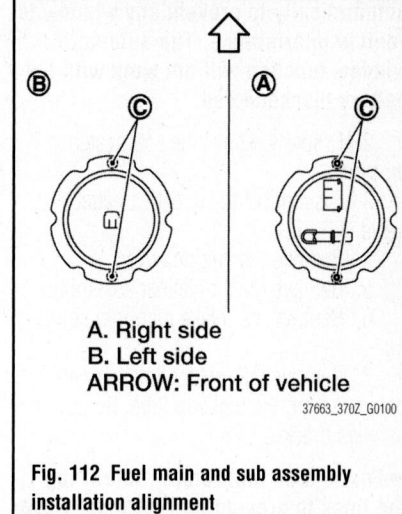

A. Right side
B. Left side
ARROW: Front of vehicle

37663_370Z_G0100

Fig. 112 Fuel main and sub assembly installation alignment

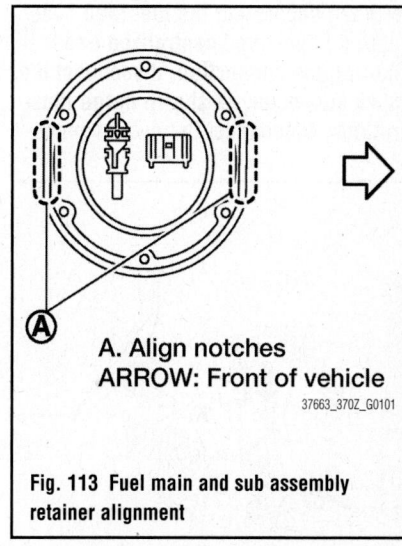

A. Align notches
ARROW: Front of vehicle

37663_370Z_G0101

Fig. 113 Fuel main and sub assembly retainer alignment

16. Start the engine and let it idle, check for fuel leaks. Correct as required.

FUEL RAIL & INJECTORS

REMOVAL & INSTALLATION
See Figures 114 through 119.

1. Before servicing the vehicle, refer to the Precautions Section.

➡**If working near and/or around the SRS system and components, be sure to disable the SRS system. After disabling the system wait three minutes or more before servicing the vehicle.**

➡**Before disconnecting the battery, lower both the driver's and passenger's windows. This will prevent any interference between the window edge and the vehicle when the door is opened or closed. During normal operation the window slightly raises and lowers**

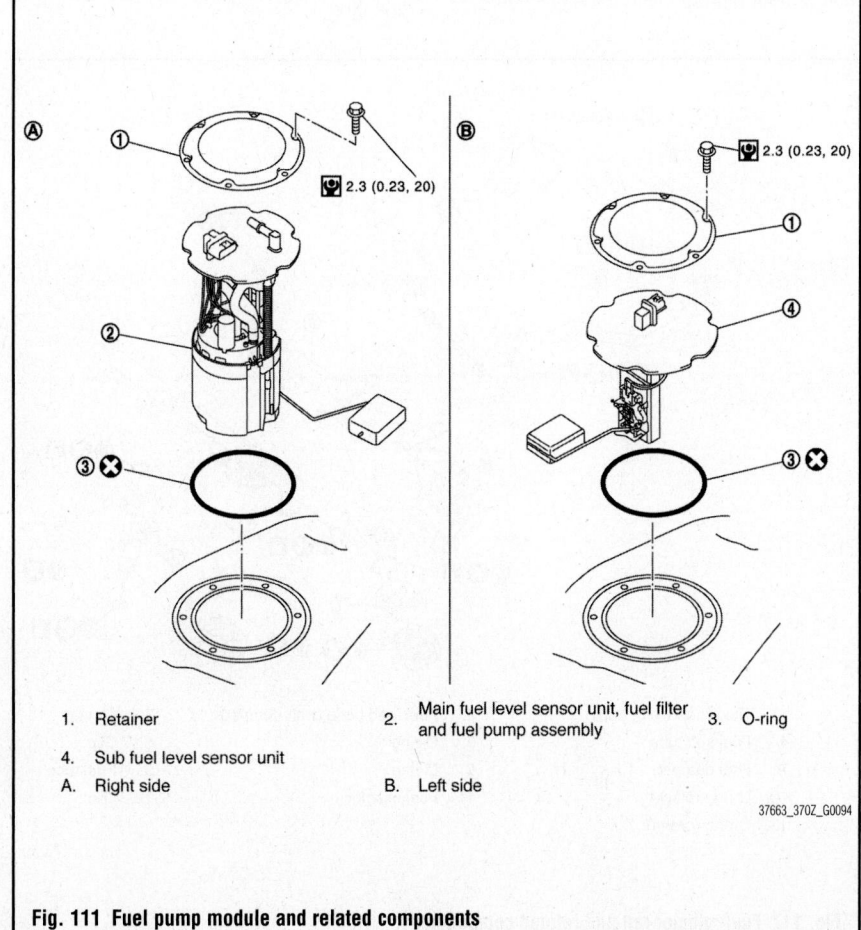

1. Retainer
4. Sub fuel level sensor unit
A. Right side
2. Main fuel level sensor unit, fuel filter and fuel pump assembly
B. Left side
3. O-ring

37663_370Z_G0094

Fig. 111 Fuel pump module and related components

automatically to prevent any window to vehicle interference. The automatic window function will not work with the battery disconnected.

2. Properly relieve the fuel system pressure.

3. Disconnect the negative battery cable.

4. Remove the engine cover.

5. Remove the air cleaner assembly.

6. Remove the intake manifold collector.

7. Remove the fuel feed hose (with damper) from the fuel sub tube. Remove the harness bracket.

➡There is no fuel return route. Plug the lines to prevent fuel leakage. Never separate the damper and the hose.

➡When separating the fuel feed hose (with damper) and centralized under floor piping connection, disconnect the quick connector, as shown in the illustrations. Disconnect the quick connec-

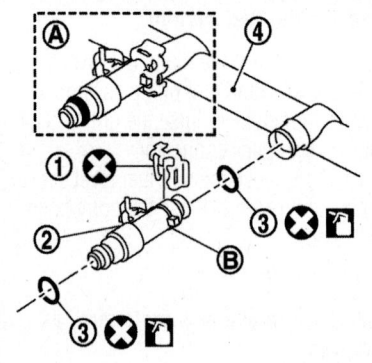

1. Retainer
2. Injector
3. O-ring
4. Rail
A. Installed condition
B. Clip mounting groove

37663_370Z_G0105

Fig. 116 Fuel injector and related components

tor by using a quick connector tool, **J-45488 or equivalent.**

8. To disconnect the quick connector from the centralized under floor piping, with the sleeve side of the quick connector release facing the quick connector, install the quick connector release onto the centralized floor piping. Insert the quick con-

nector release into the quick connector, until the sleeve contacts and stops. Hold the quick connector and release on that position. Draw and pull out the quick connector straight from the centralized under floor piping. Never bend or twist the connection between the quick connector and fuel feed hose (with damper) during removal and/or installation. Be sure to have a catch pan available to catch spilled fuel. Cover the fuel line openings to prevent dirt from entering the fuel system.

➡Inserting the quick connector release hard will not disconnect the quick connector. Hold the quick connector release where it contacts and goes no further.

➡Pull the quick connector holding as shown in the removal illustration. Never pull with lateral force. O-ring inside quick connector may be damaged.

9. Remove the sub tube mounting bolt.

10. Disconnect the harness connector from the fuel injector.

11. Loosen the mounting bolts in the reverse order of the tightening sequence.

12. Remove the assembly from its mounting.

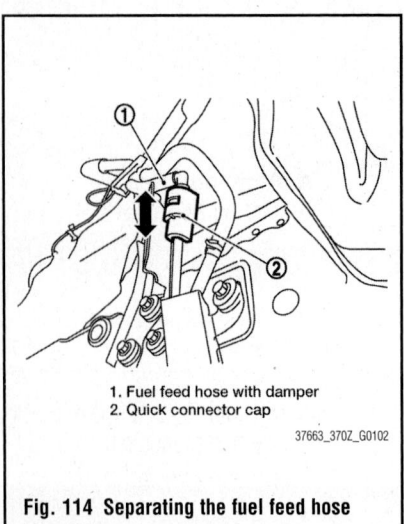

1. Fuel feed hose with damper
2. Quick connector cap

37663_370Z_G0102

Fig. 114 Separating the fuel feed hose (1 of 2)

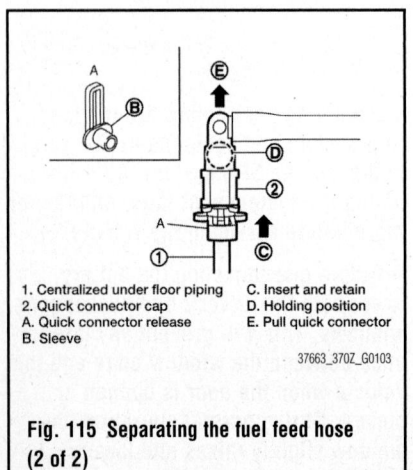

1. Centralized under floor piping
2. Quick connector cap
A. Quick connector release
B. Sleeve
C. Insert and retain
D. Holding position
E. Pull quick connector

37663_370Z_G0103

Fig. 115 Separating the fuel feed hose (2 of 2)

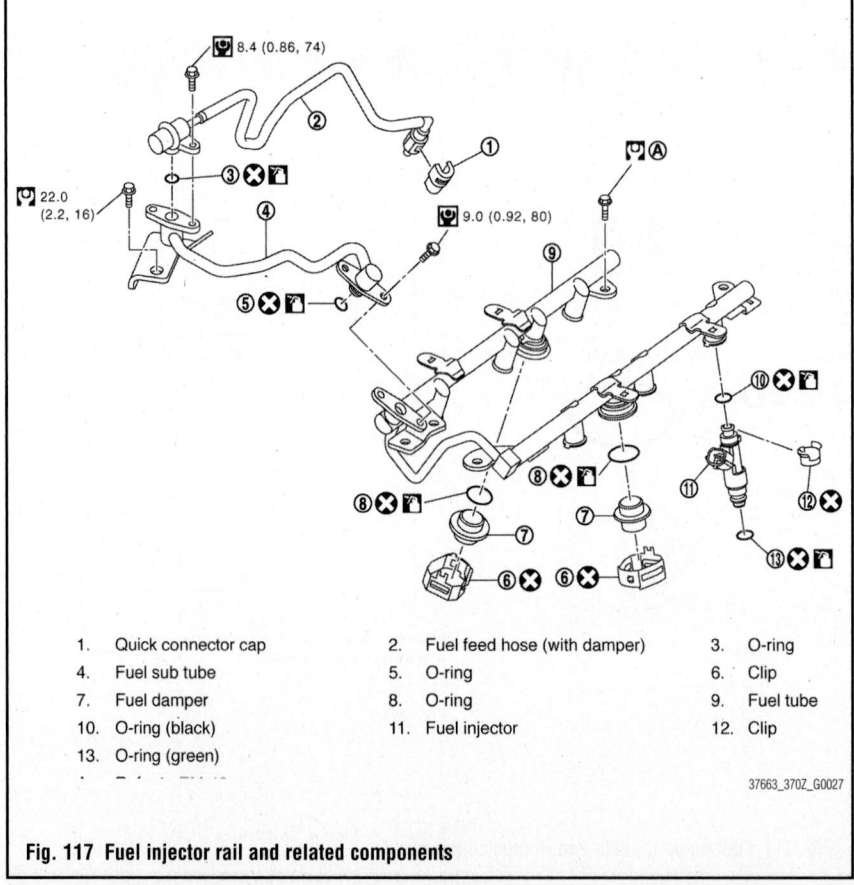

1. Quick connector cap
2. Fuel feed hose (with damper)
3. O-ring
4. Fuel sub tube
5. O-ring
6. Clip
7. Fuel damper
8. O-ring
9. Fuel tube
10. O-ring (black)
11. Fuel injector
12. Clip
13. O-ring (green)

37663_370Z_G0027

Fig. 117 Fuel injector rail and related components

13. Remove the fuel injector from the rail, as required. Discard the O-rings.

14. Do not remove the fuel sub tube and fuel damper.

To install:

➡**Be sure to use new fasteners, as required.**

15. Install new O-rings on the injector, if removed. Lubricate the O-ring with clean engine oil prior to installation.

➡**Fuel tube side O-ring is black, nozzle side is green.**

16. To install the injector, insert a new retaining clip into the mounting groove of the injector. Never reuse the old retaining clip.

17. Do not remove the fuel sub tube and fuel damper. Insert the injector into the fuel tube, with the clip attached.

➡**Insert it while matching it to the axial center. Insert the injector so that the protrusion of the fuel tube matches the cutout of the clip. Check that the tube flange is securely fixed in the flange fixing groove on the clip. Check that the injector does not rotate. See illustration.**

18. Install the fuel injector rail and injectors. Tighten to specification and in the proper sequence.

19. Tightening specification is 7 ft. lbs. (10.1 Nm), first pass. 17 ft. lbs. (23.6 Nm), second pass.

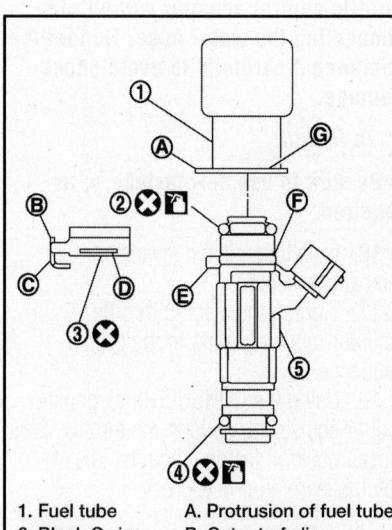

1. Fuel tube
2. Black O-ring
3. Clip
4. Green O-ring
5. Injector

A. Protrusion of fuel tube
B. Cutout of clip
C. Cutout of clip
D. Fixing groove
E. Protrusion of injector
F. Mounting groove
G. Fuel tube flange

37663_370Z_G0106

Fig. 118 Fuel injector installation

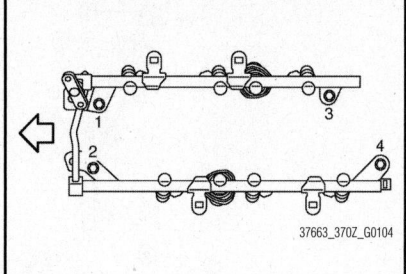

37663_370Z_G0104

Fig. 119 Fuel injector rail bolt locations and tightening sequence

20. Continue the installation in the reverse order of the removal procedure.

FUEL TANK

REMOVAL & INSTALLATION

See Figures 108, 120 and 121.

1. Before servicing the vehicle, refer to the Precautions Section.

➡**If working near and/or around the SRS system and components, be sure to disable the SRS system. After disabling the system wait three minutes or more before servicing the vehicle.**

➡**Before disconnecting the battery, lower both the driver's and passenger's windows. This will prevent any interference between the window edge and the vehicle when the door is opened or closed. During normal operation the window slightly raises and lowers automatically to prevent any window to vehicle interference. The automatic window function will not work with the battery disconnected.**

2. Drain the fuel tank to an acceptable level. If the fuel level indicates more than the level shown in the illustration, full or almost full, drain the fuel from the tank until the fuel gauge indicates a level as shown in the illustration.

➡**Because fuel will be spilled when removing the main and sub fuel level sensor units for the top of the fuel is above the main and sub fuel level sensor units installed surface. As a guide, fuel level becomes the position as shown in the illustration when approximately 3 3/8 gallons of fuel are removed from the tank.**

3. Properly relieve the fuel system pressure.

4. Remove the fuel tank cap.

5. Remove the rear parcel shelf covers, both left and right.

6. Remove the inspection hole cover.

➡**Right side for main fuel level sensor, fuel filter and fuel pump assembly. Left side for sub fuel level sensor unit.**

7. Disconnect and fuel feed tube. Disconnect the harness connector.

8. Disconnect the quick connector. Hold the sides of the connector, push in the tabs and pull out the fuel feed tube.

➡**The quick connector can be disconnected when the tabs are completely depressed. Never twist it more than necessary. Never use tools to disconnect the quick connector. Cover the fuel line openings to prevent dirt from entering the fuel system.**

9. Remove the exhaust front tube, center muffler and main muffler.

10. Remove the driveshaft.

11. Remove the parking brake cables.

12. Remove the rear suspension member assembly.

➡**For this service, halfshaft, final drive and rear suspension member are not required to be separated from one another during removal.**

13. Disconnect the fuel filler hose, vent hose and EVAP hose at the tube side.

14. Properly support the lower part of the fuel tank, with a suitable jack.

➡**Support the position that the fuel tank retaining straps do not engage.**

15. Remove the fuel tank mounting bands.

16. Carefully lower the tank assembly from its mounting. Check that all required hoses and electrical connectors are disconnected before fully lowering the assembly to the ground.

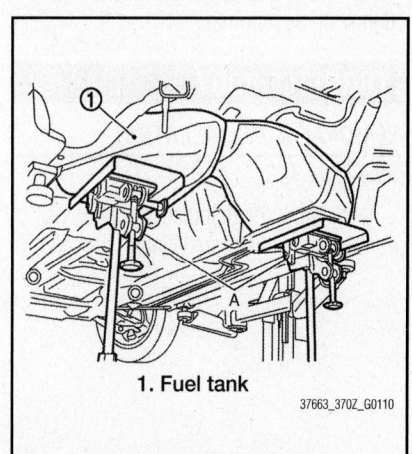

1. Fuel tank

37663_370Z_G0110

Fig. 120 Supporting the fuel tank

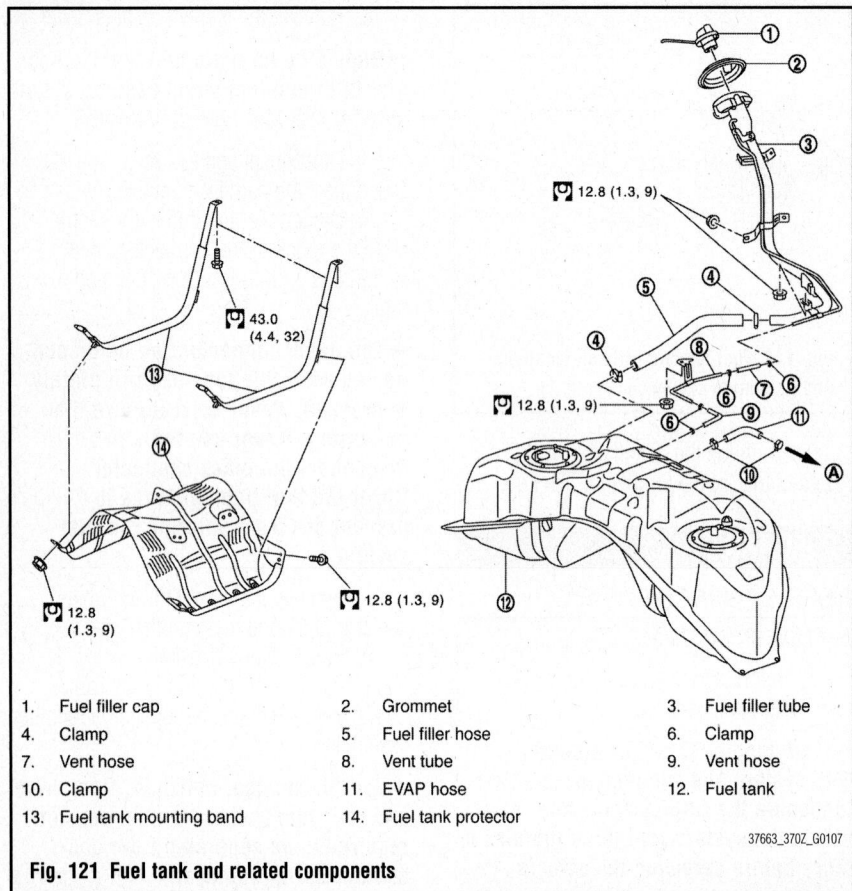

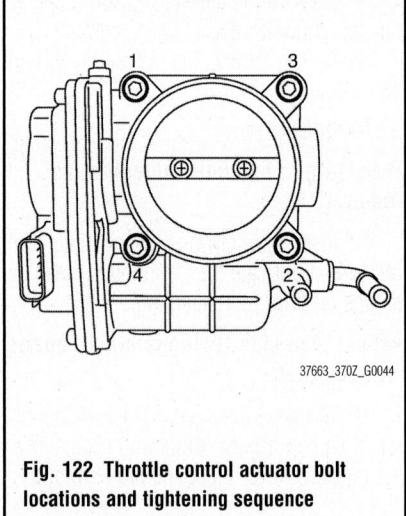

Fig. 122 Throttle control actuator bolt
locations and tightening sequence

1. Fuel filler cap
2. Grommet
3. Fuel filler tube
4. Clamp
5. Fuel filler hose
6. Clamp
7. Vent hose
8. Vent tube
9. Vent hose
10. Clamp
11. EVAP hose
12. Fuel tank
13. Fuel tank mounting band
14. Fuel tank protector

Fig. 121 Fuel tank and related components

To install:

➡**Be sure to use new fasteners, as required.**

17. Installation is the reverse of the removal procedure.

18. Install the fuel pump fuse, if removed.

19. Turn the ignition switch ON (with the engine stopped), check all connections for fuel leakage, correct as required.

20. Start the engine and let it idle, check for fuel leaks. Correct as required.

21. Check and adjust the rear wheel alignment, as required.

THROTTLE BODY

REMOVAL & INSTALLATION

See Figure 122.

At this time the manufacturer does not provide removal and installation procedures for this component. The following procedure is a guideline and may differ from the vehicle you are servicing.

1. Before servicing the vehicle, refer to the Precautions Section.

➡**If working near and/or around the SRS system and components, be sure to disable the SRS system. After disabling the system wait three minutes or more before servicing the vehicle.**

➡**Before disconnecting the battery, lower both the driver's and passenger's windows. This will prevent any interference between the window edge and the vehicle when the door is opened or closed. During normal operation the window slightly raises and lowers automatically to prevent any window to vehicle interference. The automatic window function will not work with the battery disconnected.**

2. Remove the engine undercover.

3. Drain the engine coolant. Be sure to properly dispose of used coolant.

4. Remove the front tower bar assembly.

5. Remove the engine cover.

6. Remove the air cleaner assembly.

7. To remove the electric throttle control actuator, Disconnect the water hoses from the component. Disconnect the harness connector.

8. Loosen the mounting bolts in the reverse order of the tightening sequence.

9. Remove the component from its mounting.

➡**When removing only the intake manifold collector, move the electric throttle control actuator without disconnecting the water hose. Handel the component carefully to avoid shock damage.**

To install:

➡**Be sure to use new fasteners, as required.**

10. Installation is the reverse of the removal procedure.

11. Tighten the electric throttle actuator retaining bolts in the proper sequence.

12. Using the CONSULT-III diagnostic tool or equivalent, perform the throttle valve closed position learning and the idle air volume learning procedures.

HEATING & AIR CONDITIONING SYSTEM

BLOWER MOTOR

REMOVAL & INSTALLATION

See Figure 123.

1. Before servicing the vehicle, refer to the Precautions Section.

➡ **If working near and/or around the SRS system and components, be sure to disable the SRS system. After disabling the system wait three minutes or more before servicing the vehicle.**

➡ **Before disconnecting the battery, lower both the driver's and passenger's windows. This will prevent any interference between the window edge and the vehicle when the door is opened or closed. During normal operation the window slightly raises and lowers automatically to prevent any window to vehicle interference. The automatic window function will not work with the battery disconnected.**

2. Disconnect the blower motor electrical connector.
3. Remove the blower motor retaining screws.
4. Remove the component from its mounting.

To install:

➡ **Be sure to use new fasteners, as required.**

5. Installation is the reverse of the removal procedure.

HEATER CORE

REMOVAL & INSTALLATION

See Figure 124.

1. Before servicing the vehicle, refer to the Precautions Section.

➡ **If working near and/or around the SRS system and components, be sure to disable the SRS system. After disabling the system wait three minutes or more before servicing the vehicle.**

➡ **Before disconnecting the battery, lower both the driver's and passenger's windows. This will prevent any interference between the window edge and the vehicle when the door is opened or closed. During normal operation the window slightly raises and lowers automatically to prevent any window to vehicle interference. The automatic window function will not work with the battery disconnected.**

2. Drain the radiator. Properly dispose of used coolant.
3. Properly discharge the air conditioning system.
4. Remove the heater/cooling unit assembly.
5. Remove the heater piping grommet and heater pipe bracket.
6. Remove the mounting screws and heater pipe cover.
7. Remove the left foot duct.
8. Slide the core leftward and remove.

To install:

➡ **Be sure to use new fasteners, as required.**

9. Installation is the reverse of the removal procedure.
10. Be sure to use new O-rings, coated with clean refrigerant oil prior to installation.
11. Properly refill the cooling system.
12. Properly charge the air conditioning system.

13. Start the engine and check the system for proper operation and refrigerant leakage.

HEATER & COOLING UNIT

REMOVAL & INSTALLATION

See Figures 125 through 128.

1. Before servicing the vehicle, refer to the Precautions Section.

➡ **If working near and/or around the SRS system and components, be sure to disable the SRS system. After disabling the system wait three minutes or more before servicing the vehicle.**

➡ **Before disconnecting the battery, lower both the driver's and passenger's windows. This will prevent any interference between the window edge and the vehicle when the door is opened or closed. During normal operation the window slightly raises and lowers automatically to prevent any window to vehicle interference. The automatic window function will not work with the battery disconnected.**

2. Set the temperature to 60 degrees F.
3. Disconnect the negative battery cable.
4. Properly discharge the air conditioning system
5. Drain the cooling system. Be sure to properly dispose of used coolant.
6. Remove the cowl top cover.
7. Using tool SST: J-45815 disconnect and plug the one touch connectors at the housing.
8. Remove the clamps and disconnect the heater hoses. Plug the hoses.
9. Remove the ventilator duct. Remove the foot grille.
10. Remove the steering column assembly and position it out of the way.
11. Remove the instrument stay, left and right.
12. Remove the drain hose clamp, disconnect the hose.
13. Remove the mounting nuts and remove the ECM.
14. Remove the mounting bolt (B) and mounting screws (C) and remove the heater/cooling unit.
15. Remove the mounting bolts (D), mounting screws (E) and ground bolts (F).

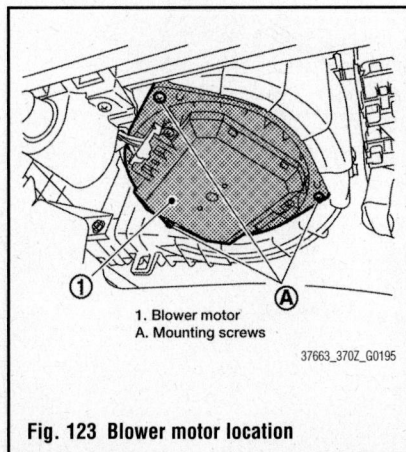

1. Blower motor
A. Mounting screws

37663_370Z_G0195

Fig. 123 Blower motor location

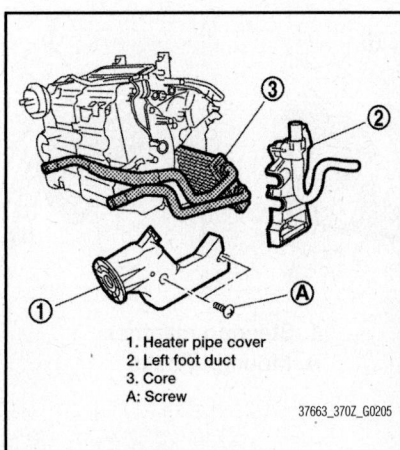

1. Heater pipe cover
2. Left foot duct
3. Core
A: Screw

37663_370Z_G0205

Fig. 124 Heater core and related components

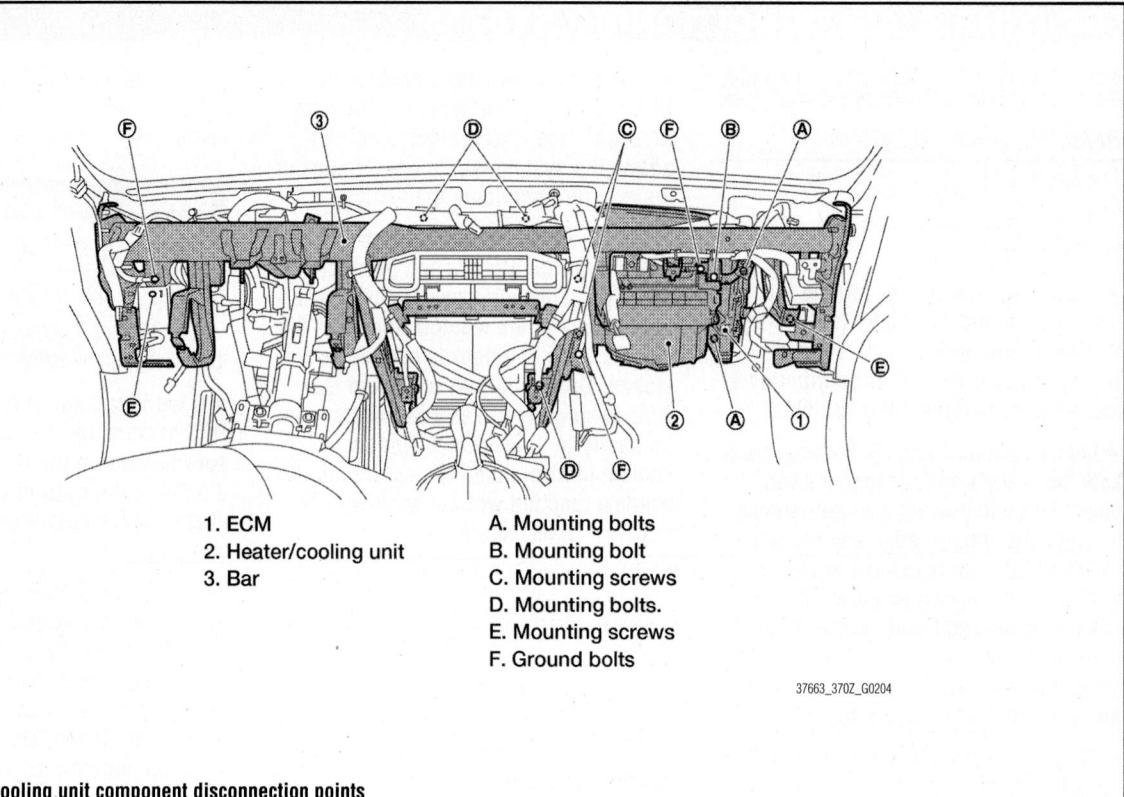

1. ECM
2. Heater/cooling unit
3. Bar

A. Mounting bolts
B. Mounting bolt
C. Mounting screws
D. Mounting bolts.
E. Mounting screws
F. Ground bolts

37663_370Z_G0204

Fig. 125 Heater/cooling unit component disconnection points

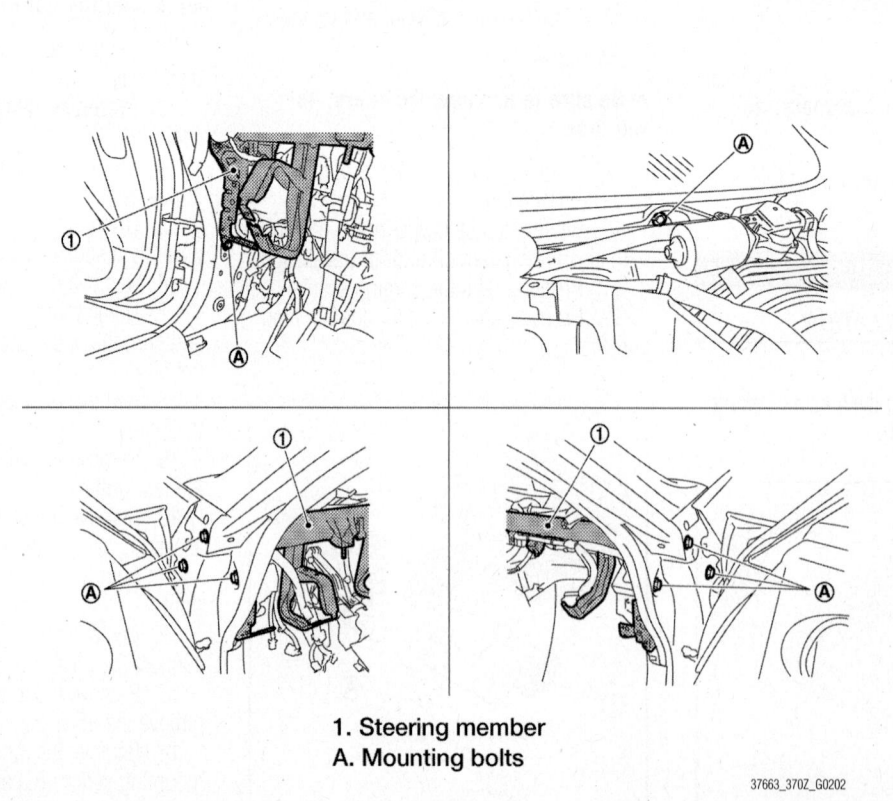

1. Steering member
A. Mounting bolts

37663_370Z_G0202

Fig. 126 Steering member and related components

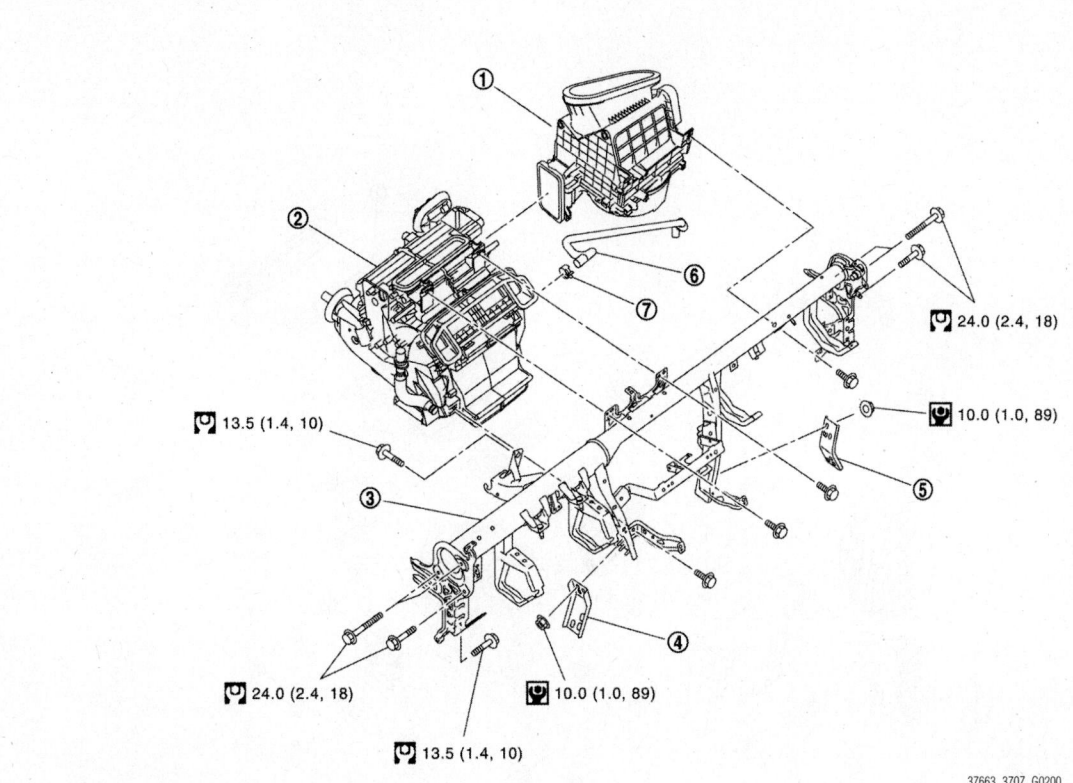

13.5 (1.4, 10)

24.0 (2.4, 18)

10.0 (1.0, 89)

24.0 (2.4, 18)

10.0 (1.0, 89)

13.5 (1.4, 10)

37663_370Z_G0200

Fig. 127 Heater/cooling unit and related components (1 of 2)

16. Disconnect the harness connectors and clips required to remove the steering member.

17. Remove the vehicle harness out of the way.

18. Remove the mounting bolts, and then remove the steering member.

19. Remove the heater/cooling assembly from the vehicle.

To install:

➡**Be sure to use new fasteners, as required.**

20. Installation is the reverse of the removal procedure.

21. Be sure to use new O-rings, coated with clean refrigerant oil prior to installation.

22. Properly charge the air conditioning system.

23. Properly refill the cooling system.

24. Start the engine and check the system for proper operation and refrigerant leakage.

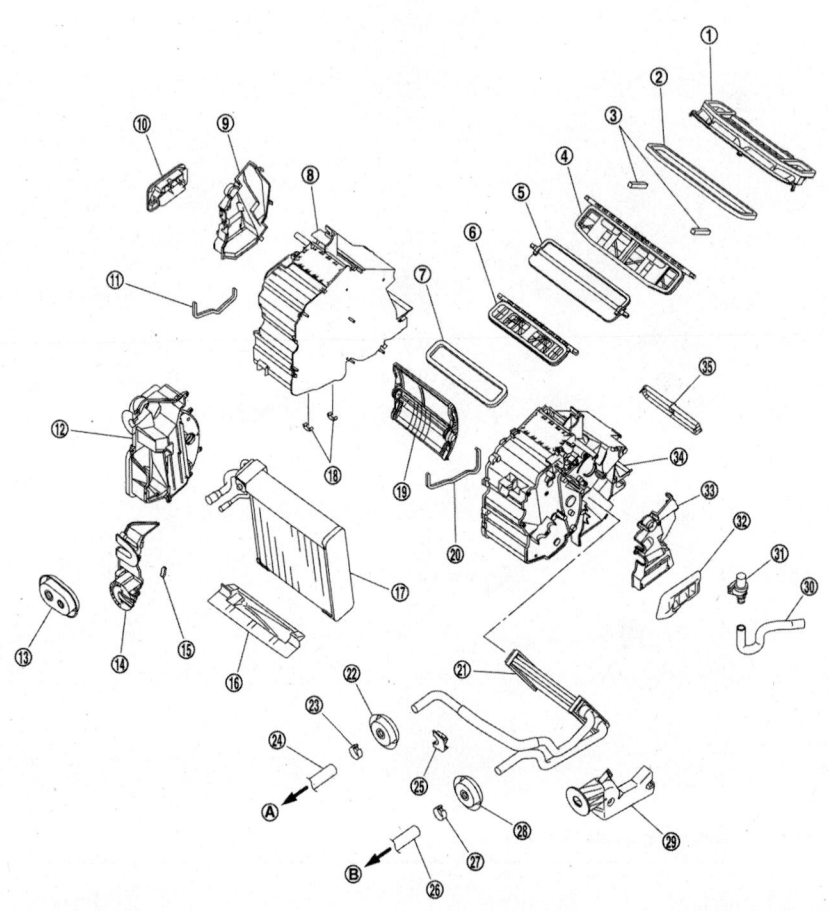

1.	Adapter duct	2.	Ventilator seal	3.	Ventilator seal
4.	Ventilator door	5.	Max. cool door	6.	Defroster door
7.	Defroster seal	8.	Heater & cooling unit case RH	9.	Foot duct RH
10.	Foot grille RH	11.	Case packing	12.	Evaporator cover
13.	Cooler pipe grommet	14.	Evaporator cover adapter	15.	Packing
16.	Insulator	17.	Evaporator assembly	18.	Clip
19.	Air mix door	20.	Case packing	21.	Heater core
22.	Heater pipe grommet	23.	Clamp	24.	Heater hose
25.	Heater pipe bracket	26.	Heater hose	27.	Clamp
28.	Heater pipe grommet	29.	Heater pipe cover	30.	Aspirator hose
31.	Aspirator	32.	Foot grille LH	33.	Foot duct LH
34.	Heater & cooling unit case LH	35.	Cover		
A.	To water outlet (rear)	B.	To heater pipe		

37663_370Z_G0201

Fig. 128 Heater/cooling unit and related components (2 of 2)

STEERING

ELECTRONIC STEERING CONTROL UNIT

REMOVAL & INSTALLATION

See Figure 129.

1. Remove instrument lower panel RH.
2. Disconnect power steering control unit connector.
3. Remove power steering control unit.

To install:

4. To install, reverse the removal procedure.

POWER RACK & PINION STEERING GEAR

REMOVAL & INSTALLATION

See Figures 130 through 134.

1. Before servicing the vehicle, refer to the Precautions Section.

➡ **If working near and/or around the SRS system and components, be sure to disable the SRS system. After disabling the system wait three minutes or more before servicing the vehicle.**

➡ **Before disconnecting the battery, lower both the driver's and passenger's windows. This will prevent any interference between the window edge and the vehicle when the door is opened or closed. During normal operation the** window slightly raises and lowers automatically to prevent any window to vehicle interference. The automatic window function will not work with the battery disconnected.

2. Position the front tires in the straight ahead position.
3. Raise and support the vehicle safely.
4. Remove the tire and wheel assemblies.
5. Remove the front suspension member.
6. Remove the cotter pin and loosen the locknut.
7. Remove the outer steering socket from the steering knuckle, so as not to damage the ball joint boot, using a suitable removal tool.

➡ **Temporarily tighten the nut to prevent damage to the threads and prevent the ball joint remover from sudden drop.**

8. Disconnect and plug the power steering fluid lines. Drain the power steering fluid. Be sure to correctly dispose of used fluid.
9. Remove the power steering solenoid valve harness connector.
10. Remove the rack stay.
11. Remove the lower joint fixing bolts (steering gear side).
12. Separate the lower shaft from the

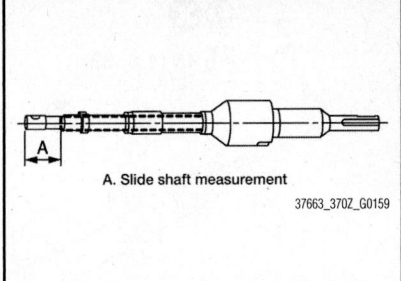

A. Slide shaft measurement

37663_370Z_G0159

Fig. 130 Power steering gear slide shaft measurement

steering gear assembly by sliding the slide shaft. See illustration.

➡ **The spiral cable may be cut if the steering wheel turns while separating the steering column assembly and steering gear assembly. Always lock the steering wheel using string to avoid turning.**

13. Remove the steering gear retaining bolts.
14. Remove the gear assembly from its mounting.

To install:

➡ **Be sure to use new fasteners, as required.**

15. Installation is the reverse of the removal procedure.

➡ **The spiral cable may be cut if the steering wheel turns while separating the steering column assembly and steering gear assembly. Always lock the steering wheel using string to avoid turning.**

16. Tighten the mounting bolts in the order shown in the illustration.
17. Tighten step one, temporary and in step two final.
18. When installing the suction hoses, refer to the illustration.

➡ **Never apply fluid to the hose and the tube. Insert the hose securely until it contacts the spool of the tube. Install the clamp at the hose (0.12–0.31 inch from the edge of the hose.**

19. To install the lower joint to the steering gear, see the illustration. Position the rack of the steering gear in the neutral position.

➡ **To get to the neutral position turn the sub gear assembly and measure the distance of the inner socket, then measure the intermediate position and distance.**

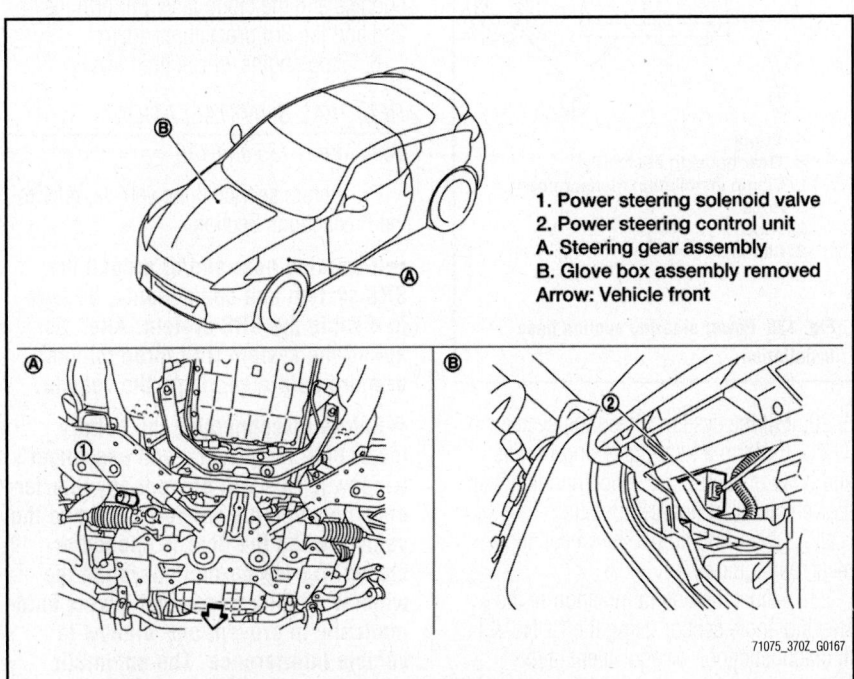

1. Power steering solenoid valve
2. Power steering control unit
A. Steering gear assembly
B. Glove box assembly removed
Arrow: Vehicle front

71075_370Z_G0167

Fig. 129 Electronic power steering system component locations

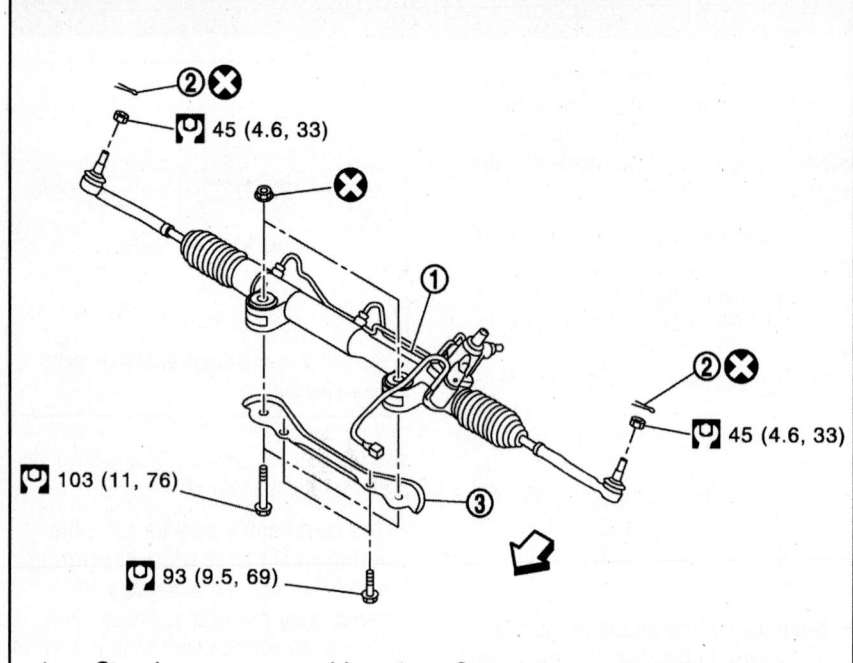

1. Steering gear assembly   2. Cotter pin   3. Rack stay

◁ : Vehicle front

37663_370Z_G0158

Fig. 131 Power steering gear and related components

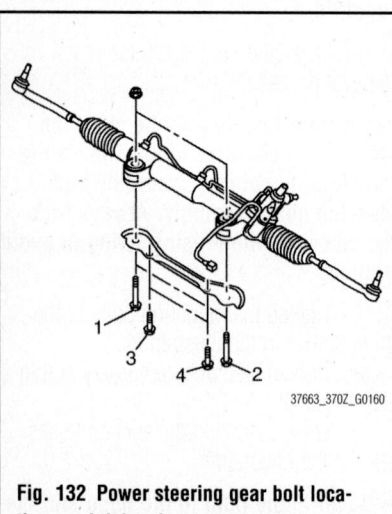

37663_370Z_G0160

Fig. 132 Power steering gear bolt locations and tightening sequence

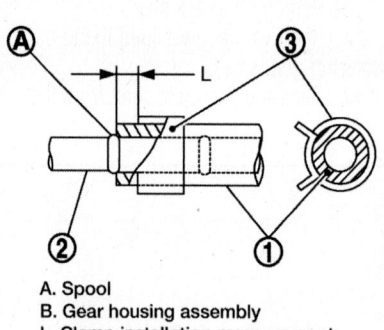

A. Spool
B. Gear housing assembly
L. Clamp installation measurement
1. Hose
2. Tube
3. Clamp

37663_370Z_G0161

Fig. 133 Power steering suction hose installation

Align the rear cover cap projection with the marking position of the gear housing assembly. Install the slip part of the lower joint aligning with the rear cover cap projection. Make sure that the slit part of the lower joint is aligned with the rear cover cap protection and the marking position of the gear housing assembly. Make sure that there is no clearance between the lower joint, gear housing assembly and mounting bolt.

20. Fill the system. Bleed the system.
21. Perform a final tightening of nuts and bolts of each removed component with the vehicle in an unladen position.
22. Check and adjust the wheel alignment, as required.
23. Adjust the neutral position of the steering angle sensor using the CONSULT-III diagnostic tool, or equivalent, after checking the wheel alignment.

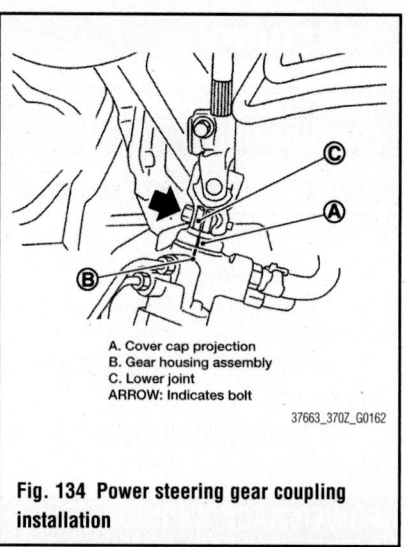

A. Cover cap projection
B. Gear housing assembly
C. Lower joint
ARROW: Indicates bolt

37663_370Z_G0162

Fig. 134 Power steering gear coupling installation

POWER STEERING PUMP

BLEEDING

1. Stop engine, and then turn steering wheel fully to right and left several times.
2. Do not allow steering fluid reservoir tank to go below the low-level line. Check tank frequenter and add fluid as needed.
3. Run engine at idle speed. Turn steering wheel fully to the right and then fully to the left, and keep for about 3 seconds. Then check whether a fluid leakage has occurred.
4. Repeat the 2nd procedure several times at about three seconds interval Check generation of air bubbles and cloud in fluid.
5. If air bubbles and the cloud don't fade, stop engine, hold air bleeding until air bubbles and the cloud fade. Perform the 2nd and the 3rd procedures again.
6. Stop engine, check fluid level.

REMOVAL & INSTALLATION

See Figures 135 and 136.

1. Before servicing the vehicle, refer to the Precautions Section.

➡**If working near and/or around the SRS system and components, be sure to disable the SRS system. After disabling the system wait three minutes or more before servicing the vehicle.**

➡**Before disconnecting the battery, lower both the driver's and passenger's windows. This will prevent any interference between the window edge and the vehicle when the door is opened or closed. During normal operation the window slightly raises and lowers automatically to prevent any window to vehicle interference. The automatic window function will not work with the battery disconnected.**

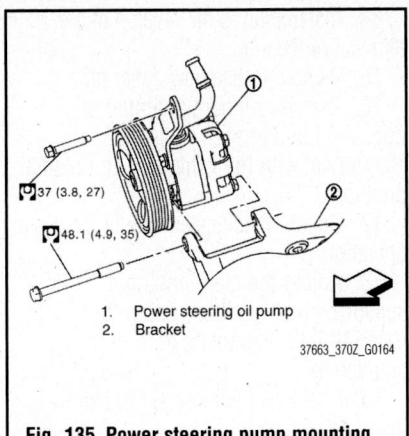

Fig. 135 Power steering pump mounting bolt location

1. Power steering oil pump
2. Bracket

37663_370Z_G0164

2. Drain the power steering fluid from the reservoir. Be sure to properly dispose of used fluid.

3. Remove the air cleaner assembly.

4. Loosen the drive belt.

5. Remove the belt from the steering pump.

6. Remove the pressure sensor connector.

7. Remove the copper washers and eye bolt. Drain fluid from their piping.

8. Remove the suction hose. Drain fluid from their piping.

9. Remove the cooling fan assembly.

10. Remove the pump retaining bolts.

11. Remove the pump from the vehicle.

To install:

➡**Be sure to use new fasteners, as required.**

12. Installation is the reverse of the removal procedure.

13. When installing the suction hoses see note below.

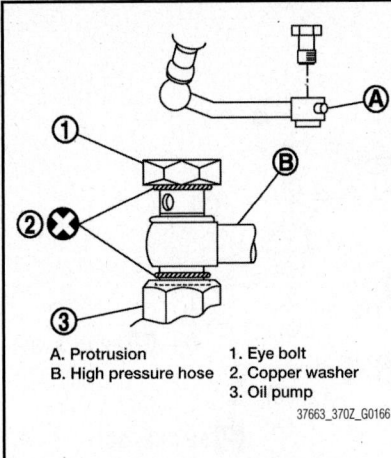

A. Protrusion
B. High pressure hose
1. Eye bolt
2. Copper washer
3. Oil pump

37663_370Z_G0166

Fig. 136 Power steering pump eye bolt and cooper washer installation

➡**Never apply fluid to the hose and the tube. Insert the hose securely until it contacts the spool of the tube. Install the clamp at the hose (0.12–0.31 inch from the edge of the hose.**

14. When installing the eye bolt and copper washer to the pump, see illustration.

➡**Never reuse the cooper washer. Apply clean power steering fluid around the washers, then install the eye bolt. Install the eye bolt with the eye joint (assembled to the high pressure hose) protrusion facing with pump side cutout, and then tighten it to specification.**

15. Adjust the belt tension.

16. Fill the system with the proper grade and type fluid. Bleed the system.

17. Check for fluid leakage, correct as required.

POWER STEERING FLUID

FLUID RECOMMENDATIONS

Genuine NISSAN PSF or equivalent is the recommended power steering fluid. DEXRON® VI type ATF may also be used.

FLUID FILL PROCEDURE

See Figure 137.

1. Stop the engine before performing the fluid level check.

2. Ensure that the fluid level is between the MAX range and the MIN level.

➡**Because fluid level differs within the HOT range and the COLD range, check carefully.**

3. Hot specification temperature is 122–176 degrees F. Cold specification temperature is 32–86 degrees F.

4. Do not overfill the MAX level.

5. Do not reuse old power steering fluid.

6. Be sure to use the proper grade and type replacement fluid.

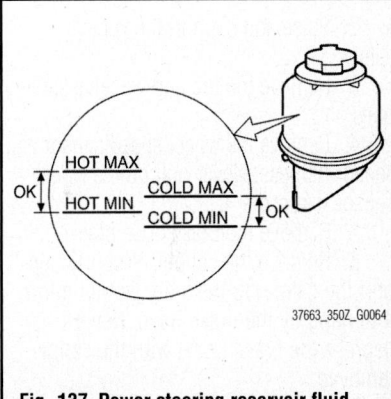

37663_350Z_G0064

Fig. 137 Power steering reservoir fluid markings

SUSPENSION

FRONT SUSPENSION

CROSSMEMBER

REMOVAL & INSTALLATION

See Figure 138.

1. At first, remove engine and transmission assembly with front suspension member downward. Then separate engine, transmission assembly and front suspension member.

2. Remove the following parts.

a. Steering knuckles and wheel hub and bearing assemblies.

b. Steering gear assembly and hydraulic line.

c. Stabilizer bar and stabilizer connecting rods.

d. Transverse links.

e. Remove suspension member stay.

f. Remove suspension member sub stays.

To install:

3. To install, reverse the removal procedure.

➡**Perform final tightening of bolts and nuts at the vehicle installation position (rubber bushing), under unladen condition with tires on level ground.**

KNUCKLE

REMOVAL & INSTALLATION

See Figure 145.

1. Before servicing the vehicle, refer to the Precautions Section.

➡**If working near and/or around the SRS system and components, be sure to disable the SRS system. After disabling the system wait three minutes or more before servicing the vehicle.**

➡**Before disconnecting the battery, lower both the driver's and passenger's windows. This will prevent any interference between the window edge and the vehicle when the door is opened or closed. During normal operation the window slightly raises and lowers automatically to prevent any window to vehicle interference. The automatic window**

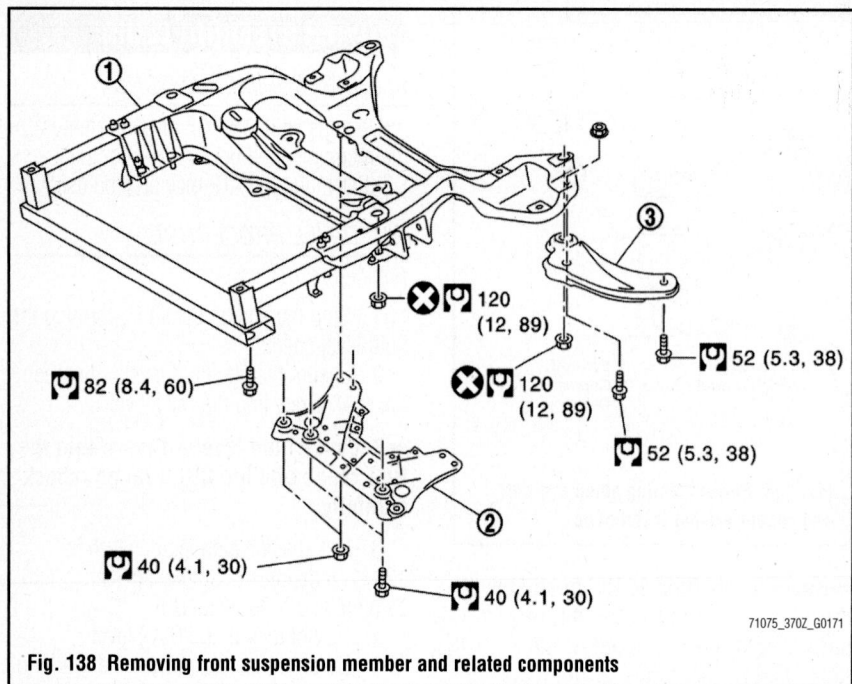

Fig. 138 Removing front suspension member and related components

82 (8.4, 60)

120 (12, 89)

120 (12, 89)

52 (5.3, 38)

52 (5.3, 38)

40 (4.1, 30)

40 (4.1, 30)

71075_370Z_G0171

14. Installation is the reverse of the removal procedure.

15. Be sure to use new cotter pins.

16. Perform a final tightening of nuts and bolts of each removed component with the vehicle in an unladen position.

17. Check wheel speed sensor for proper operation.

18. Adjust the steering angle sensor neutral position, using the CONSULT-III diagnostic tool, or equivalent.

19. Check and adjust the front alignment, as required.

PERFORMANCE DAMPER

REMOVAL & INSTALLATION

See Figure 139.

1. Before servicing the vehicle, refer to the Precautions Section.

function will not work with the battery disconnected.

2. Raise and support the vehicle safely.

3. Remove the tire and wheel assemblies.

4. Remove the wheel speed sensor and sensor harness. Never pull on the wheel sensor harness.

5. Remove the brake hose bracket.

6. Remove the caliper. Properly support the caliper to the side. Do not allow it to hang by the brake hose. Never depress the brake pedal with the caliper removed.

7. Remove the brake rotor.

8. Remove the wheel and hub bearing assembly.

9. As required, remove the splash guard shield.

10. Remove the steering outer socket.

11. Remove the cotter pin of the transverse link and steering knuckle, and then loosen nut.

12. Separate the upper link from the steering knuckle.

13. Separate the transverse link from the steering knuckle, using a ball joint removal tool. Remove the steering knuckle.

➡**Temporarily tighten the nut to prevent damage to the threads and to prevent the removal tool from suddenly coming off.**

To install:

➡**Be sure to use new fasteners, as required.**

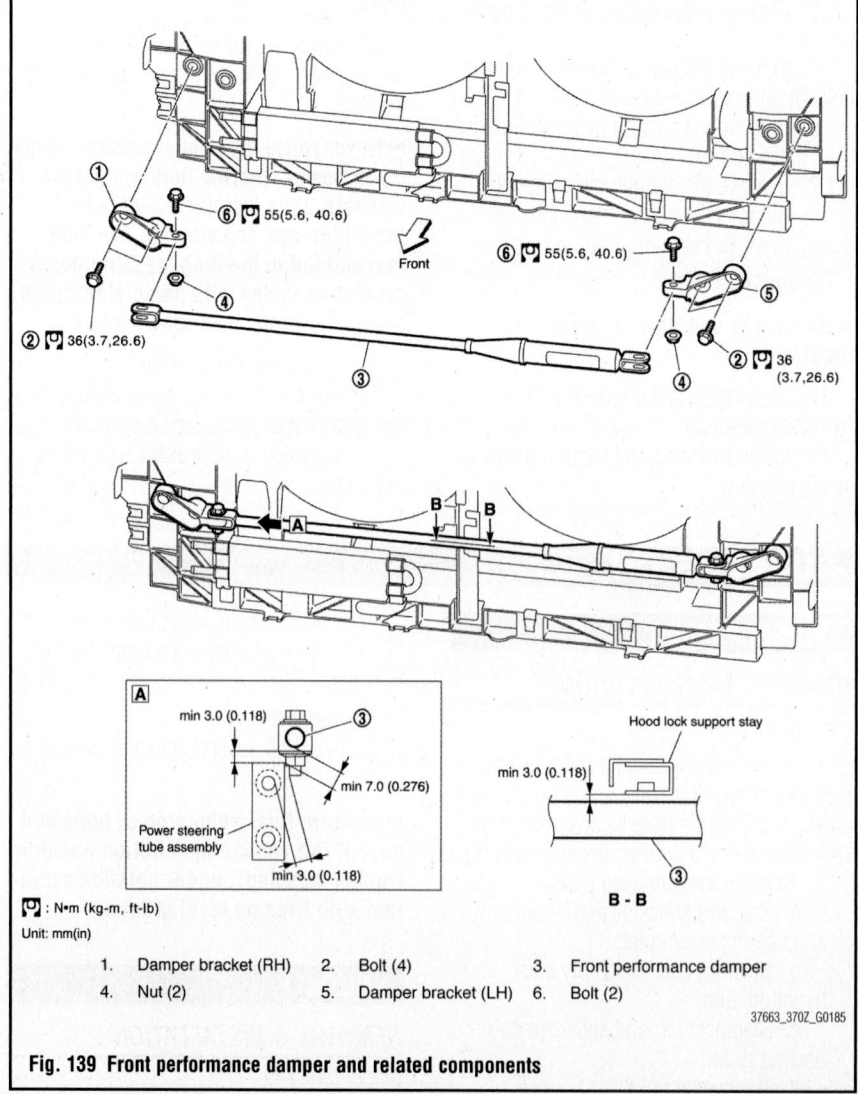

55(5.6, 40.6)

55(5.6, 40.6)

Front

36(3.7,26.6)

36 (3.7,26.6)

A

min 3.0 (0.118)

min 7.0 (0.276)

Power steering tube assembly

min 3.0 (0.118)

Hood lock support stay

min 3.0 (0.118)

B - B

: N•m (kg-m, ft-lb)
Unit: mm(in)

1. Damper bracket (RH)   2. Bolt (4)   3. Front performance damper
4. Nut (2)   5. Damper bracket (LH)   6. Bolt (2)

37663_370Z_G0185

Fig. 139 Front performance damper and related components

➡️ If working near and/or around the SRS system and components, be sure to disable the SRS system. After disabling the system wait three minutes or more before servicing the vehicle.

➡️ Before disconnecting the battery, lower both the driver's and passenger's windows. This will prevent any interference between the window edge and the vehicle when the door is opened or closed. During normal operation the window slightly raises and lowers automatically to prevent any window to vehicle interference. The automatic window function will not work with the battery disconnected.

  2. Remove the front bumper fascia.

  3. Remove the bolts and nuts from the performance damper.

To install:

➡️ Be sure to use new fasteners, as required.

  4. Installation is the reverse of the removal procedure.

  5. When installing, check all clearances to be sure there are no areas of interference.

STABILIZER BAR (SWAY BAR) & LINKS

REMOVAL & INSTALLATION

See Figure 140.

  1. Before servicing the vehicle, refer to the Precautions Section.

➡️ If working near and/or around the SRS system and components, be sure to disable the SRS system. After disabling the system wait three minutes or more before servicing the vehicle.

➡️ Before disconnecting the battery, lower both the driver's and passenger's windows. This will prevent any interference between the window edge and the vehicle when the door is opened or closed. During normal operation the window slightly raises and lowers automatically to prevent any window to vehicle interference. The automatic window function will not work with the battery disconnected.

  2. Raise and safely support the vehicle.

  3. Remove the tire and wheel assemblies.

  4. Remove the engine undercover.

  5. Remove the stabilizer connecting rod. Matchmark to identify for reinstallation.

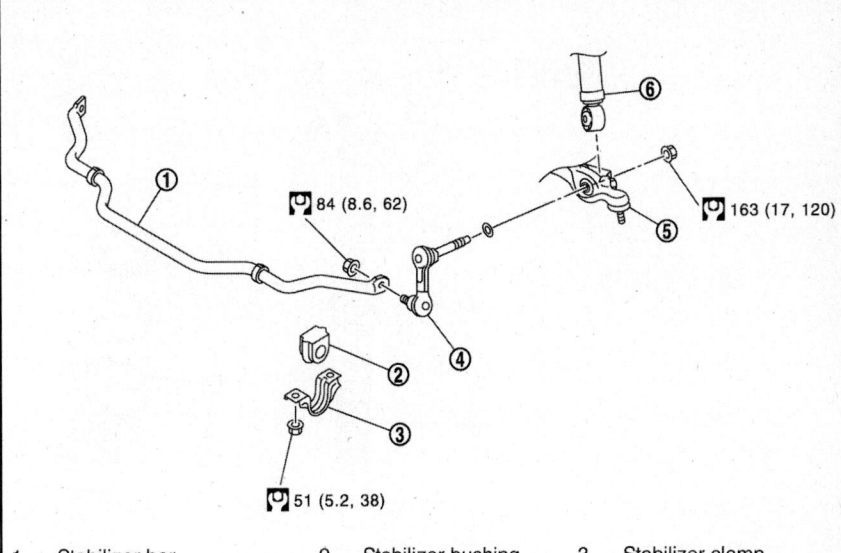

| 1. | Stabilizer bar | 2. | Stabilizer bushing | 3. | Stabilizer clamp |
| 4. | Stabilizer connecting rod | 5. | Transverse link | 6. | Shock absorber |

37663_370Z_G0178

Fig. 140 Front stabilizer bar and related components

  6. Remove the stabilizer clamps and stabilizer bushings.

  7. Remove the stabilizer bar.

To install:

➡️ Be sure to use new fasteners, as required.

  8. Installation is the reverse of the removal procedure.

  9. Be sure to check the matchmarks.

  10. Tighten the mounting nut to specification while holding a hexagonal part of the stabilizer connecting rod side.

  11. Perform a final tightening of nuts and bolts of each removed component with the vehicle in an unladen position.

  12. Check and adjust the front alignment, as required.

STRUTS

REMOVAL & INSTALLATION

See Figure 141.

  1. Before servicing the vehicle, refer to the Precautions Section.

➡️ If working near and/or around the SRS system and components, be sure to disable the SRS system. After disabling the system wait three minutes or more before servicing the vehicle.

➡️ Before disconnecting the battery, lower both the driver's and passenger's windows. This will prevent any interference between the window edge and the

vehicle when the door is opened or closed. During normal operation the window slightly raises and lowers automatically to prevent any window to vehicle interference. The automatic window function will not work with the battery disconnected.

  2. Raise and safely support the vehicle.

  3. Remove the tire and wheel assemblies.

  4. Remove the wheel sensor and harness connector from the shock absorber.

  5. Remove the brake hose bracket.

  6. Remove the stabilizer connecting rod.

  7. Separate the upper link from the steering knuckle.

  8. Remove the shock absorber assembly and gusset.

➡️ If removing the shock is difficult, loosen the upper link mounting nuts (vehicle side).

To install:

➡️ Be sure to use new fasteners, as required.

  9. Installation is the reverse of the removal procedure.

➡️ Never tap on the ball joint cap of the stabilizer connecting rod with a hammer when inserting the stabilizer connecting rod into the transverse link.

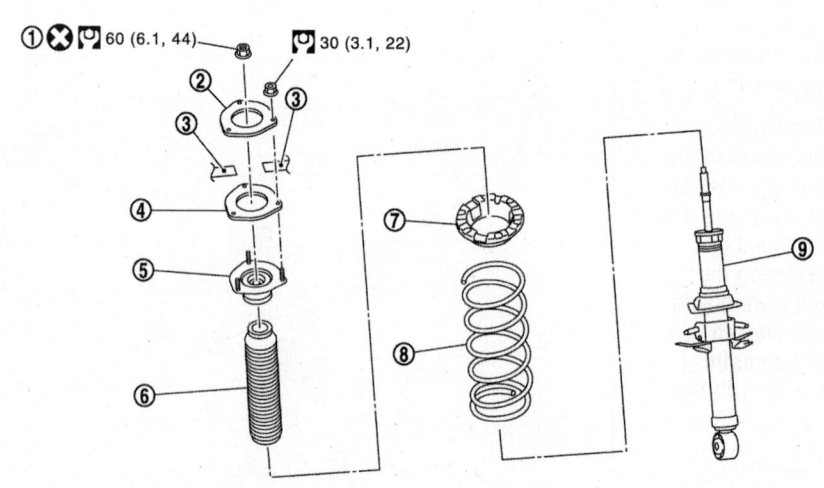

① ✕ ⎕ 60 (6.1, 44)   ⎕ 30 (3.1, 22)

1.	Piston rod lock nut	2.	Gusset	3.	Vehicle body
4.	Mounting seal	5.	Shock absorber mounting bracket	6.	Bound bumper
7.	Rubber seat	8.	Coil spring	9.	Shock absorber

37663_370Z_G0176

Fig. 141 Front strut assembly and related components

10. Perform a final tightening of nuts and bolts of each removed component with the vehicle in an unladen position.

11. Adjust the steering angle sensor neutral position, using the CONSULT-III diagnostic tool, or equivalent.

12. Check and adjust the front alignment, as required.

TOWER BAR

REMOVAL & INSTALLATION

See Figure 142.

1. Before servicing the vehicle, refer to the Precautions Section.

➡️**If working near and/or around the SRS system and components, be sure to disable the SRS system. After disabling the system wait three minutes or more before servicing the vehicle.**

➡️**Before disconnecting the battery, lower both the driver's and passenger's windows. This will prevent any interference between the window edge and the vehicle when the door is opened or closed. During normal operation the window slightly raises and lowers automatically to prevent any window to vehicle interference. The automatic window function will not work with the battery disconnected.**

2. Remove the cowl top cover center.

3. Remove the tower bar retaining nuts.

4. Remove the component from the vehicle.

To install:

➡️**Be sure to use new fasteners, as required.**

5. Installation is the reverse of the removal procedure.

6. Perform a final tightening of nuts and bolts of each removed component with the vehicle in an unladen position.

7. Check and adjust the front alignment, as required.

TRANSVERSE LINK

REMOVAL & INSTALLATION

See Figure 143.

1. Before servicing the vehicle, refer to the Precautions Section.

➡️**If working near and/or around the SRS system and components, be sure to disable the SRS system. After disabling the system wait three minutes or more before servicing the vehicle.**

➡️**Before disconnecting the battery, lower both the driver's and passenger's windows. This will prevent any interference between the window edge and the vehicle when the door is opened or closed. During normal operation the** window slightly raises and lowers automatically to prevent any window to vehicle interference. The automatic window function will not work with the battery disconnected.

2. Raise and safely support the vehicle.

3. Remove the tire and wheel assemblies.

4. Remove the engine undercover.

5. Remove the stabilizer connecting rod.

6. Remove the outer socket from the steering knuckle.

7. Remove the transverse link from the steering knuckle.

8. Position a suitable jack under the transverse link.

9. Remove the transverse link from its mounting.

To install:

➡️**Be sure to use new fasteners, as required.**

10. Installation is the reverse of the removal procedure.

➡️**Never tap on the ball joint cap of the stabilizer connecting rod with a hammer when inserting the stabilizer connecting rod into the transverse link.**

11. Perform a final tightening of nuts and bolts of each removed component with the vehicle in an unladen position.

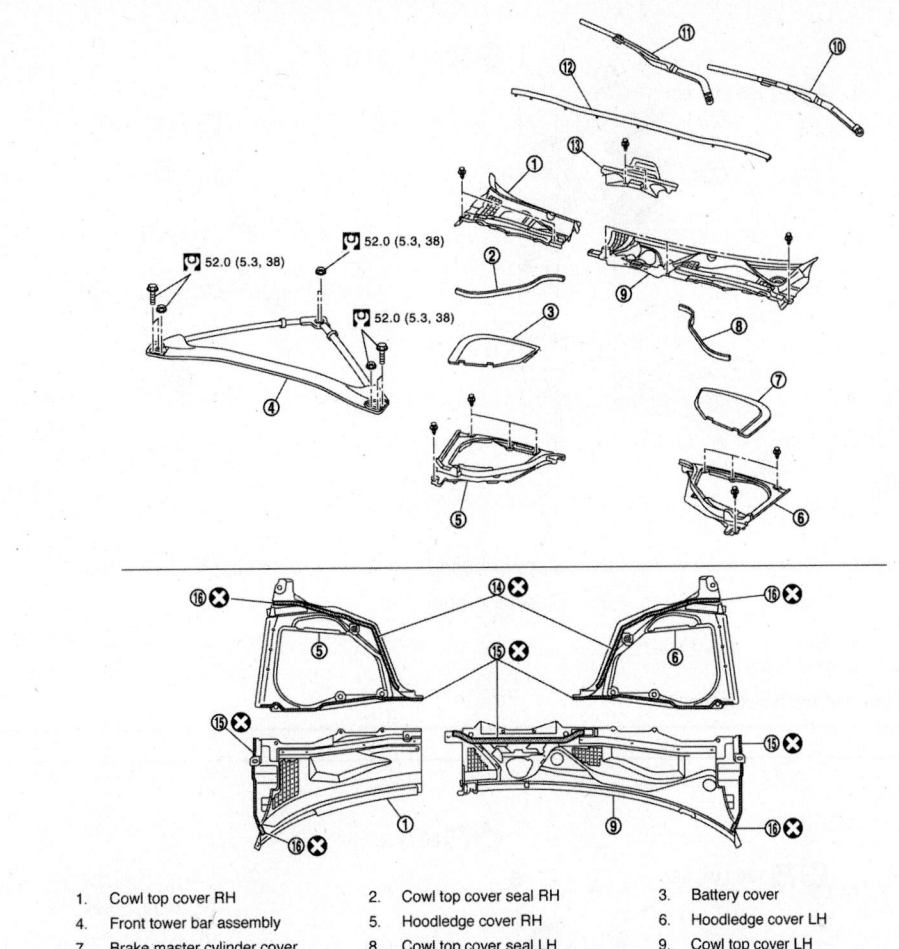

1. Cowl top cover RH
2. Cowl top cover seal RH
3. Battery cover
4. Front tower bar assembly
5. Hoodledge cover RH
6. Hoodledge cover LH
7. Brake master cylinder cover
8. Cowl top cover seal LH
9. Cowl top cover LH
10. Wiper arm and blade LH
11. Wiper arm and blade RH
12. Cowl top cover seal
13. Cowl top cover center
14. EPT sealer [t: 5.0 mm (0.197 in)]
15. EPT sealer [t:3.0 mm (0.118 in)]
16. EPT sealer [t: 10.0 mm (0.394 in)]

37663_370Z_G0035

Fig. 142 Front tower bar and related components

12. Check wheel speed sensor for proper operation.

13. Adjust the steering angle sensor neutral position, using the CONSULT-III diagnostic tool, or equivalent.

14. Check and adjust the front alignment, as required.

UPPER LINK

REMOVAL & INSTALLATION

See Figure 144.

1. Before servicing the vehicle, refer to the Precautions Section.

➡**If working near and/or around the SRS system and components, be sure to disable the SRS system. After disabling the system wait three minutes or more before servicing the vehicle.**

➡**Before disconnecting the battery, lower both the driver's and passenger's windows. This will prevent any interference between the window edge and the vehicle when the door is opened or closed. During normal operation the window slightly raises and lowers automatically to prevent any window to vehicle interference. The automatic window function will not work with the battery disconnected.**

2. Raise and safely support the vehicle.

3. Remove the tire and wheel assemblies.

4. Remove the shock absorber.

5. Remove the upper link from the steering knuckle.

6. Remove the upper link and stopper rubber.

To install:

➡**Be sure to use new fasteners, as required.**

➡**Never tap on the ball joint cap of the stabilizer connecting rod with a hammer when inserting the stabilizer connecting rod into the transverse link.**

7. Perform a final tightening of nuts and bolts of each removed component with the vehicle in an unladen position.

8. Check wheel speed sensor for proper operation.

9. Adjust the steering angle sensor neutral position, using the CONSULT-III diagnostic tool, or equivalent.

10. Check and adjust the front alignment, as required.

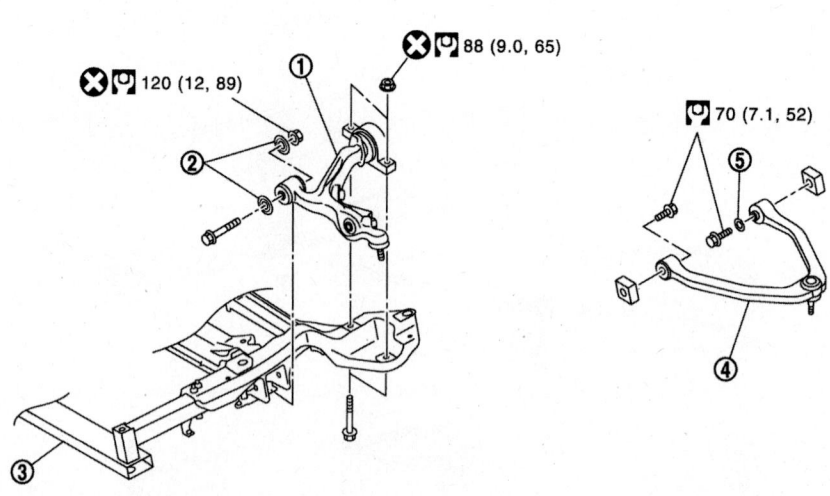

1. Transverse link
2. Stopper bush
3. Front suspension member
4. Upper link
5. Stopper rubber

37663_370Z_G0182

Fig. 143 Front transverse link and related components

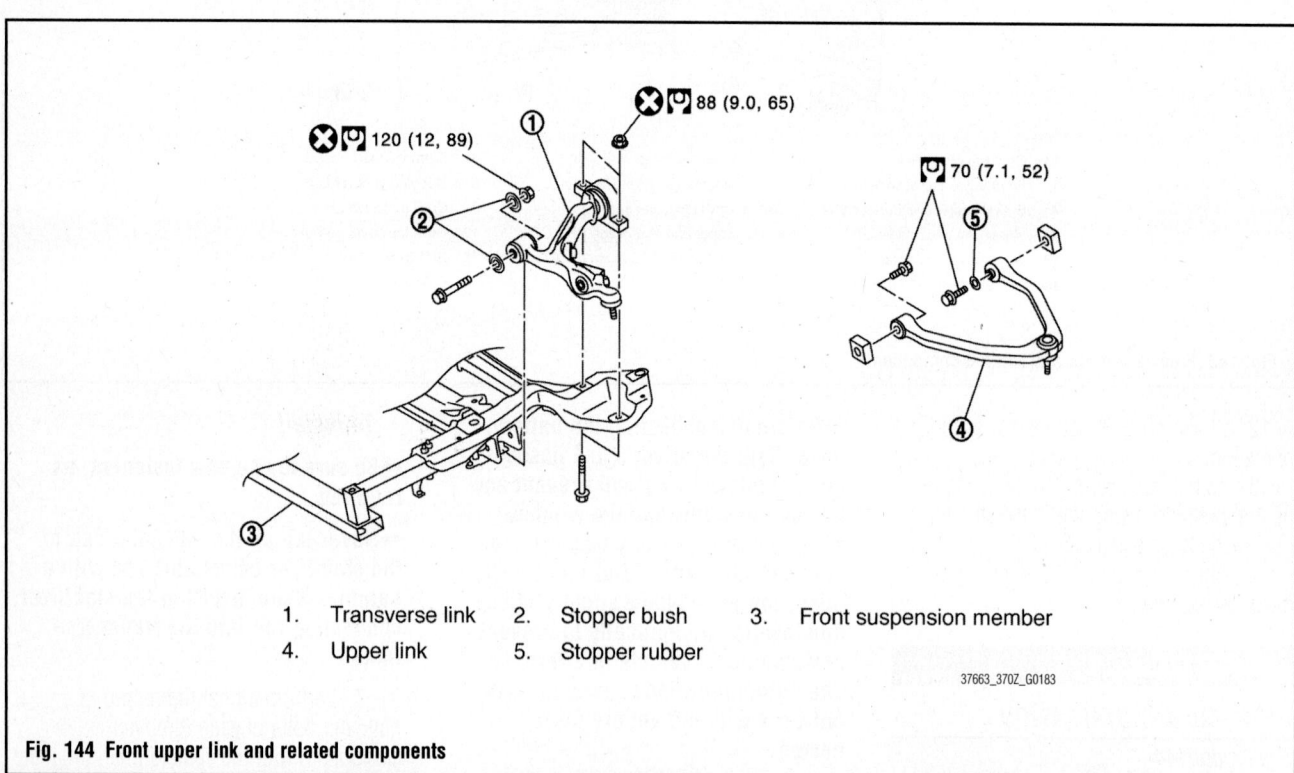

1. Transverse link
2. Stopper bush
3. Front suspension member
4. Upper link
5. Stopper rubber

37663_370Z_G0183

Fig. 144 Front upper link and related components

WHEEL HUBS & BEARINGS

ADJUSTMENT

These bearings are not adjustable. If defective, they must be replaced.

REMOVAL & INSTALLATION

See Figure 145.

1. Before servicing the vehicle, refer to the Precautions Section.

➡**If working near and/or around the SRS system and components, be sure to disable the SRS system. After disabling the system wait three minutes or more before servicing the vehicle.**

➡**Before disconnecting the battery, lower both the driver's and passenger's windows. This will prevent any interference between the window edge and the vehicle when the door is opened or closed. During normal operation the window slightly raises and lowers automatically to prevent any window to**

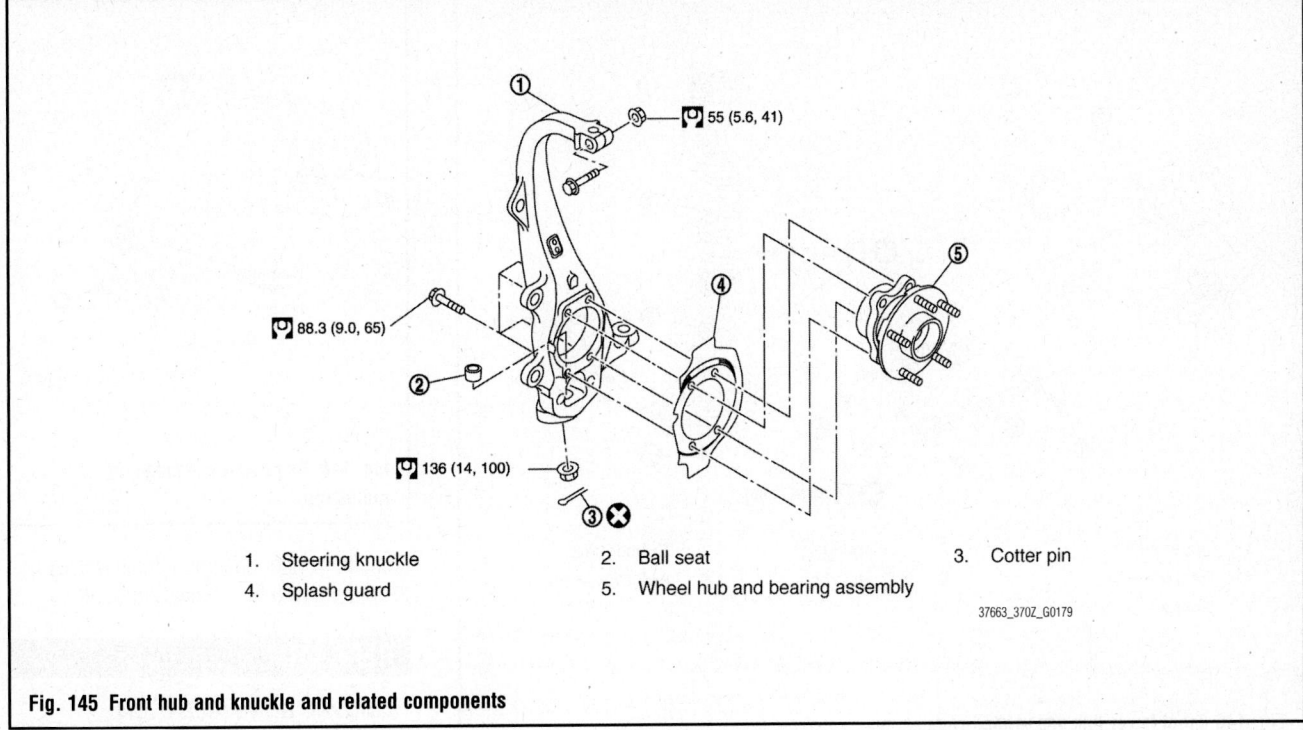

□ 55 (5.6, 41)

□ 88.3 (9.0, 65)

□ 136 (14, 100)

1. Steering knuckle
4. Splash guard
2. Ball seat
5. Wheel hub and bearing assembly
3. Cotter pin

37663_370Z_G0179

Fig. 145 Front hub and knuckle and related components

vehicle interference. The automatic window function will not work with the battery disconnected.

2. Raise and support the vehicle safely.
3. Remove the tire and wheel assemblies.
4. Remove the wheel speed sensor and sensor harness. Never pull on the wheel sensor harness.
5. Remove the brake hose bracket.
6. Remove the caliper. Properly support the caliper to the side. Do not allow it to hang by the brake hose. Never depress the brake pedal with the caliper removed.
7. Remove the brake rotor.
8. Remove the wheel and hub bearing assembly.
9. As required, remove the splash guard shield.

To install:

➡Be sure to use new fasteners, as required.

10. Installation is the reverse of the removal procedure.
11. Perform a final tightening of nuts and bolts of each removed component with the vehicle in an unladen position.
12. Check wheel speed sensor for proper operation.
13. Adjust the steering angle sensor neutral position, using the CONSULT-III diagnostic tool, or equivalent.
14. Check and adjust the front alignment, as required.

SUSPENSION

FRONT LINK

REMOVAL & INSTALLATION

Lower

See Figure 146.

1. Before servicing the vehicle, refer to the Precautions Section.

➡If working near and/or around the SRS system and components, be sure to disable the SRS system. After disabling the system wait three minutes or more before servicing the vehicle.

➡Before disconnecting the battery, lower both the driver's and passenger's windows. This will prevent any interference between the window edge and the vehicle when the door is opened or closed. During normal operation the window slightly raises and lowers automatically to prevent any window to vehicle interference. The automatic window function will not work with the battery disconnected.

2. Raise and support the vehicle safely.
3. Remove the tire and wheel assemblies.
4. Position a suitable jack under the rear axle assembly to relieve tension on the coil spring.
5. Remove the lower link retaining bolts.
6. Remove the component from its mounting.

To install:

➡Be sure to use new fasteners, as required.

7. Installation is the reverse of the removal procedure.

REAR SUSPENSION

8. Perform a final tightening of nuts and bolts of each removed component with the vehicle in an unladen position.
9. Check wheel speed sensor for proper operation.
10. Adjust the steering angle sensor neutral position, using the CONSULT-III diagnostic tool, or equivalent.
11. Check and adjust the front alignment, as required.

RADIUS ARM

REMOVAL & INSTALLATION

See Figure 147.

1. Before servicing the vehicle, refer to the Precautions Section.

➡If working near and/or around the SRS system and components, be sure

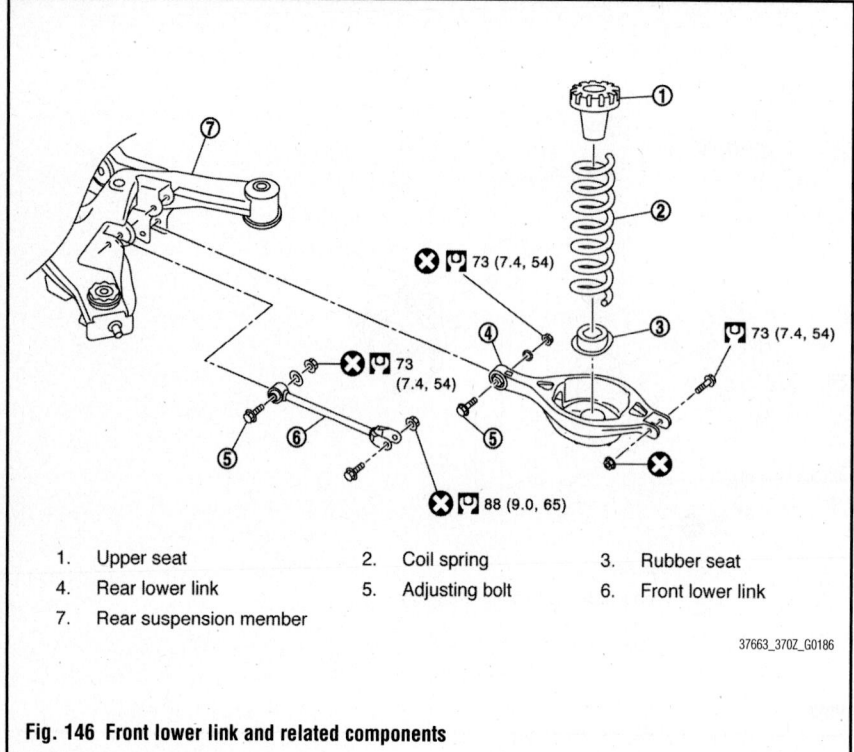

1. Upper seat
2. Coil spring
3. Rubber seat
4. Rear lower link
5. Adjusting bolt
6. Front lower link
7. Rear suspension member

37663_370Z_G0186

Fig. 146 Front lower link and related components

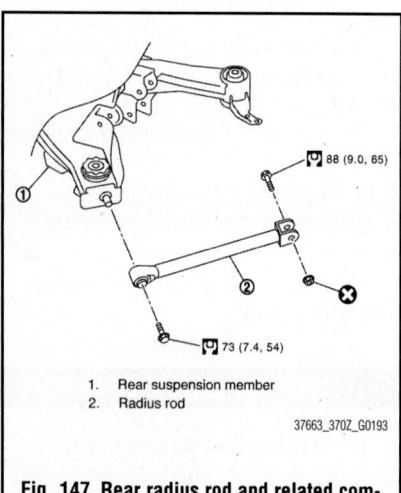

1. Rear suspension member
2. Radius rod

37663_370Z_G0193

Fig. 147 Rear radius rod and related components

to disable the SRS system. After disabling the system wait three minutes or more before servicing the vehicle.

➡Before disconnecting the battery, lower both the driver's and passenger's windows. This will prevent any interference between the window edge and the vehicle when the door is opened or closed. During normal operation the window slightly raises and lowers automatically to prevent any window to vehicle interference. The automatic window function will not work with the battery disconnected.

2. Raise and support the vehicle safely.
3. Remove the tire and wheel assemblies.
4. Remove the radius rod retaining nuts.
5. Remove the component from the vehicle.

To install:

➡Be sure to use new fasteners, as required.

6. Installation is the reverse of the removal procedure.
7. Perform a final tightening of nuts and bolts of each removed component with the vehicle in an unladen position.
8. Check wheel speed sensor for proper operation.
9. Adjust the steering angle sensor neutral position, using the CONSULT-III diagnostic tool, or equivalent.
10. Check and adjust the front alignment, as required.

RADIUS ROD

REMOVAL & INSTALLATION
See Figure 148.

1. Remove tires with power tool.
2. Remove radius rod.

To install:

3. To install, reverse the removal procedure. Perform final tightening or rear suspension member and axle installation

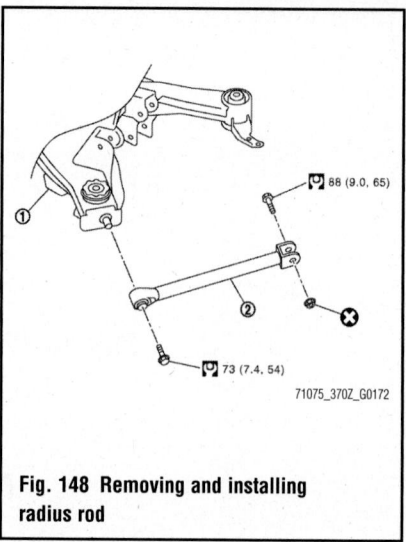

71075_370Z_G0172

Fig. 148 Removing and installing radius rod

position (rubber bushing), under unladen conditions with tires on level ground.

REAR ARM

REMOVAL & INSTALLATION
See Figure 149.

1. Before servicing the vehicle, refer to the Precautions Section.

➡If working near and/or around the SRS system and components, be sure to disable the SRS system. After disabling the system wait three minutes or more before servicing the vehicle.

➡Before disconnecting the battery, lower both the driver's and passenger's windows. This will prevent any interference between the window edge and the vehicle when the door is opened or closed. During normal operation the window slightly raises and lowers automatically to prevent any window to vehicle interference. The automatic window function will not work with the battery disconnected.

2. Raise and support the vehicle safely.
3. Remove the tire and wheel assemblies.
4. Remove the diagonal brace.
5. Remove the stabilizer connecting rod.
6. Remove the halfshaft.
7. Remove the cotter pin of the suspension arm ball joint, and loosen the nut.
8. Remove the suspension arm (rear suspension member side).
9. Use a ball joint removal tool to remove the suspension arm from the axle housing.

➡Be careful not to damage the ball

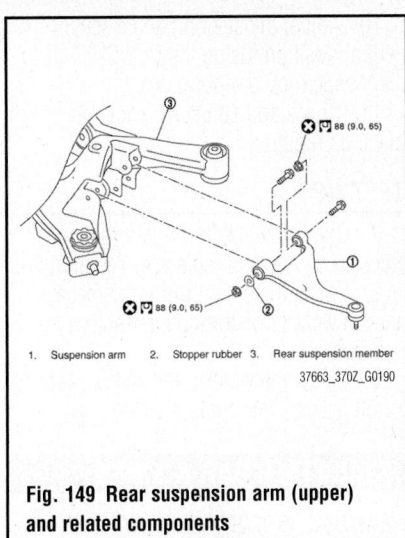

1. Suspension arm 2. Stopper rubber 3. Rear suspension member

37663_370Z_G0190

Fig. 149 Rear suspension arm (upper) and related components

joint boot. Temporarily tighten the nut to prevent damage to the threads and to prevent the tool from coming off.

10. Remove the suspension arm.
11. Remove the stabilizer connecting rod mounting bracket.

To install:

➡️**Be sure to use new fasteners, as required.**

12. Installation is the reverse of the removal procedure.
13. Perform a final tightening of nuts and bolts of each removed component with the vehicle in an unladen position.
14. Check wheel speed sensor for proper operation.
15. Adjust the steering angle sensor neutral position, using the CONSULT-III diagnostic tool, or equivalent.
16. Check and adjust the front alignment, as required.

REAR LOWER LINK & COIL

REMOVAL & INSTALLATION

See Figures 150 through 152.

1. Before servicing the vehicle, refer to the Precautions Section.

➡️**If working near and/or around the SRS system and components, be sure to disable the SRS system. After disabling the system wait three minutes or more before servicing the vehicle.**

➡️**Before disconnecting the battery, lower both the driver's and passenger's windows. This will prevent any interference between the window edge and the vehicle when the door is opened or**

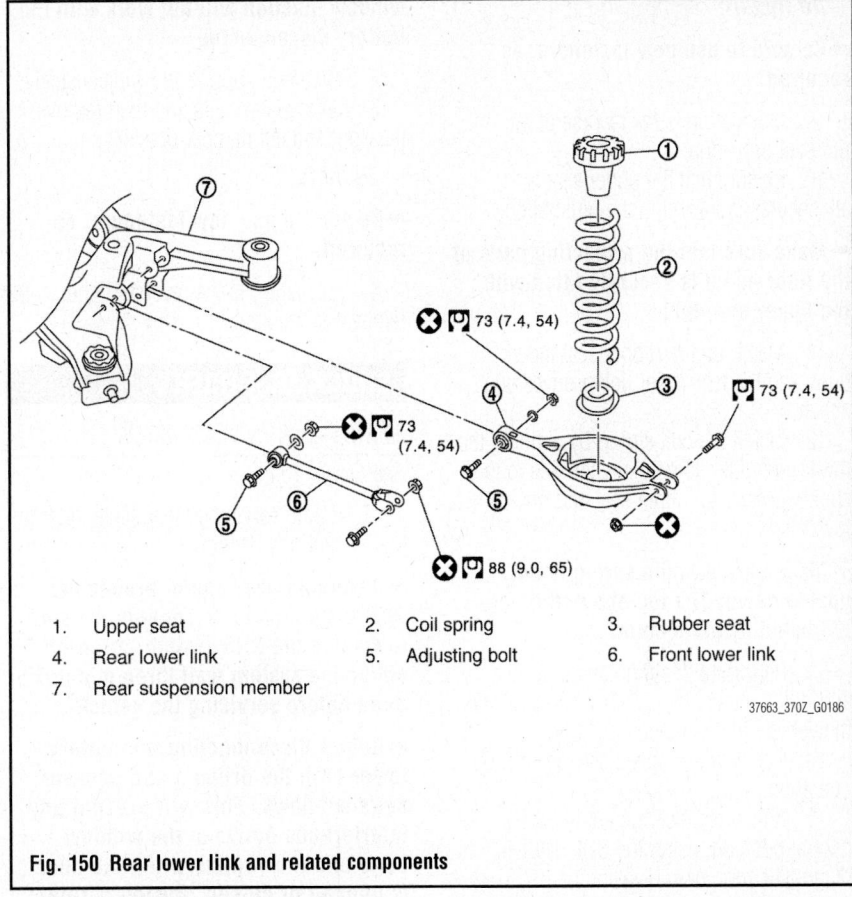

1. Upper seat 2. Coil spring 3. Rubber seat
4. Rear lower link 5. Adjusting bolt 6. Front lower link
7. Rear suspension member

37663_370Z_G0186

Fig. 150 Rear lower link and related components

closed. During normal operation the window slightly raises and lowers automatically to prevent any window to vehicle interference. The automatic window function will not work with the battery disconnected.

2. Raise and support the vehicle safely.
3. Remove the tire and wheel assemblies.

4. Position a suitable jack under the rear lower link to relieve the coil spring tension.
5. Loosen the rear lower link mounting nuts (rear suspension member side).
6. Remove the rear lower link (axle housing side).
7. Slowly lower the jack and remove the upper seat, coil spring and rubber sheet from the rear lower link.
8. Remove the rear lower link.

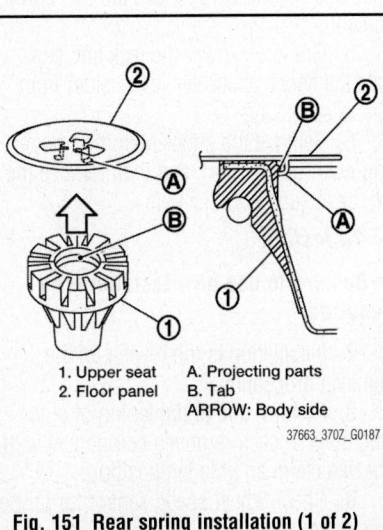

1. Upper seat A. Projecting parts
2. Floor panel B. Tab
ARROW: Body side

37663_370Z_G0187

Fig. 151 Rear spring installation (1 of 2)

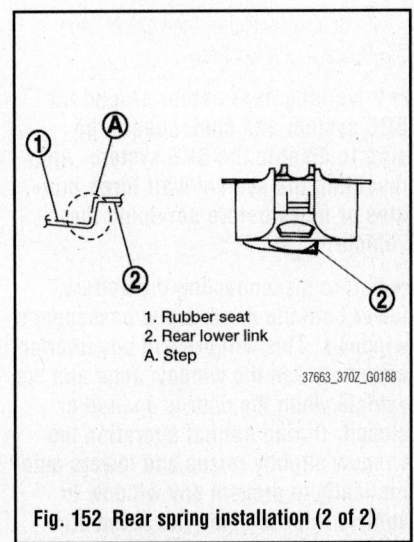

1. Rubber seat
2. Rear lower link
A. Step

37663_370Z_G0188

Fig. 152 Rear spring installation (2 of 2)

To install:

➡**Be sure to use new fasteners, as required.**

9. Installation is the reverse of the removal procedure.

10. Be sure that the upper seat is attached as indicated in the illustration.

➡**Make sure that the projecting parts of the floor panel is securely fitted with the upper seat tab.**

11. Match up the rubber seat indentations and the rear lower link grooves and attach.

12. Install the coil spring by aligning the lower end of the large diameter side to the step between the rubber seat and the rear lower link.

➡**Make sure that the spring is not upside down. The top and bottom are indicated by paint color.**

13. Perform a final tightening of nuts and bolts of each removed component with the vehicle in an unladen position.

14. Check wheel speed sensor for proper operation.

15. Adjust the steering angle sensor neutral position, using the CONSULT-III diagnostic tool, or equivalent.

16. Check and adjust the front alignment, as required.

REAR PERFORMANCE DAMPER

REMOVAL & INSTALLATION

See Figure 153.

➡**Perform this operation in a level place while the vehicle is in the unladen position, in running order. Never tighten bolts while the vehicle is raised or jacked up.**

1. Before servicing the vehicle, refer to the Precautions Section.

➡**If working near and/or around the SRS system and components, be sure to disable the SRS system. After disabling the system wait three minutes or more before servicing the vehicle.**

➡**Before disconnecting the battery, lower both the driver's and passenger's windows. This will prevent any interference between the window edge and the vehicle when the door is opened or closed. During normal operation the window slightly raises and lowers automatically to prevent any window to vehicle interference. The automatic**

window function will not work with the battery disconnected.

2. Raise and support the vehicle safely.

3. Remove the bolts and then remove the right and left damper brackets.

To install:

➡**Be sure to use new fasteners, as required.**

4. Installation is the reverse of the removal procedure.

SHOCK ABSORBERS

REMOVAL & INSTALLATION

See Figure 154.

1. Before servicing the vehicle, refer to the Precautions Section.

➡**If working near and/or around the SRS system and components, be sure to disable the SRS system. After disabling the system wait three minutes or more before servicing the vehicle.**

➡**Before disconnecting the battery, lower both the driver's and passenger's windows. This will prevent any interference between the window edge and the vehicle when the door is opened or closed. During normal operation the window slightly raises and lowers automatically to prevent any window to vehicle interference. The automatic window function will not work with the battery disconnected.**

2. Raise and support the vehicle safely.

3. Remove the tire and wheel assemblies.

4. Position a suitable jack under the rear axle assembly to relieve the coil spring tension.

5. Gradually lower the jack and separate the shock absorber (lower side) from the axle housing.

6. Remove the shock absorber mounting nuts (upper side), and then remove the shock absorber.

To install:

➡**Be sure to use new fasteners, as required.**

7. Installation is the reverse of the removal procedure.

8. Perform a final tightening of nuts and bolts of each removed component with the vehicle in an unladen position.

9. Check wheel speed sensor for proper operation.

10. Adjust the steering angle sensor neutral position, using the CONSULT-III diagnostic tool, or equivalent.

11. Check and adjust the front alignment, as required.

TESTING

1. Check the shock for deformation, cracks, damage and replace as required.

2. Check the piston rod for damage, uneven wear or distortion and replace as required.

3. Check the welded and sealed areas for oil leakage and replace as required.

STABILIZER BAR & LINKS

REMOVAL & INSTALLATION

See Figure 155.

1. Before servicing the vehicle, refer to the Precautions Section.

➡**If working near and/or around the SRS system and components, be sure to disable the SRS system. After disabling the system wait three minutes or more before servicing the vehicle.**

➡**Before disconnecting the battery, lower both the driver's and passenger's windows. This will prevent any interference between the window edge and the vehicle when the door is opened or closed. During normal operation the window slightly raises and lowers automatically to prevent any window to vehicle interference. The automatic window function will not work with the battery disconnected.**

2. Raise and support the vehicle safely.

3. Remove the tire and wheel assemblies.

4. Remove the diagonal brace.

5. Remove the stabilizer connecting rods.

6. Remove the stabilizer clamps. Remove the stabilizer bushings.

7. Remove the stabilizer bar.

8. Remove the stabilizer connecting rod mounting brackets from the suspension arm.

To install:

➡**Be sure to use new fasteners, as required.**

9. Installation is the reverse of the removal procedure.

10. Perform a final tightening of nuts and bolts of each removed component with the vehicle in an unladen position.

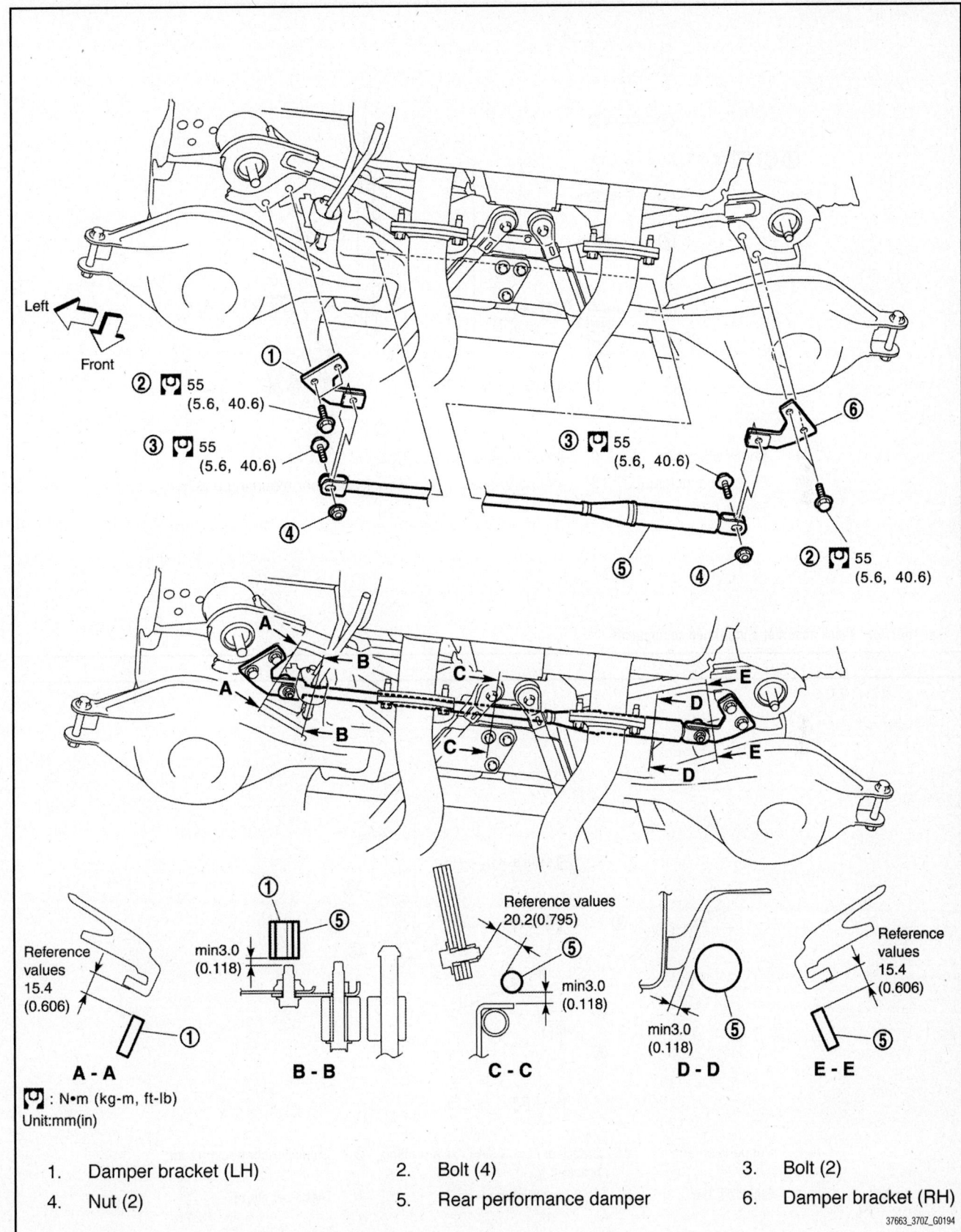

Left

Front

② 🔧 55
(5.6, 40.6)

③ 🔧 55
(5.6, 40.6)

③ 🔧 55
(5.6, 40.6)

② 🔧 55
(5.6, 40.6)

A - A

Reference
values
15.4
(0.606)

B - B

min3.0
(0.118)

C - C

Reference values
20.2(0.795)

min3.0
(0.118)

D - D

min3.0
(0.118)

E - E

Reference
values
15.4
(0.606)

🔧 : N•m (kg-m, ft-lb)
Unit:mm(in)

1. Damper bracket (LH)	2. Bolt (4)	3. Bolt (2)
4. Nut (2)	5. Rear performance damper	6. Damper bracket (RH)

37663_370Z_G0194

Fig. 153 Rear performance damper and related components

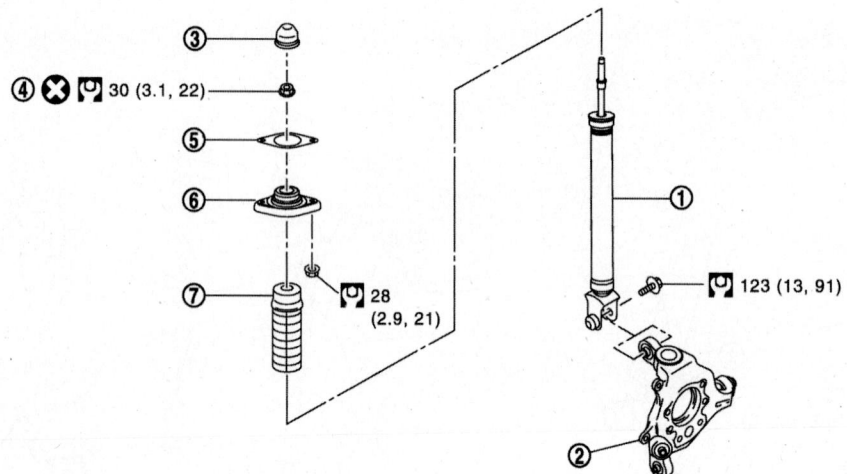

1. Shock absorber
2. Axle housing
3. Cap
4. Piston rod lock nut
5. Mounting seal
6. Shock absorber mounting bracket
7. Bound bumper cover

37663_370Z_G0189

Fig. 154 Rear shock absorber and related components

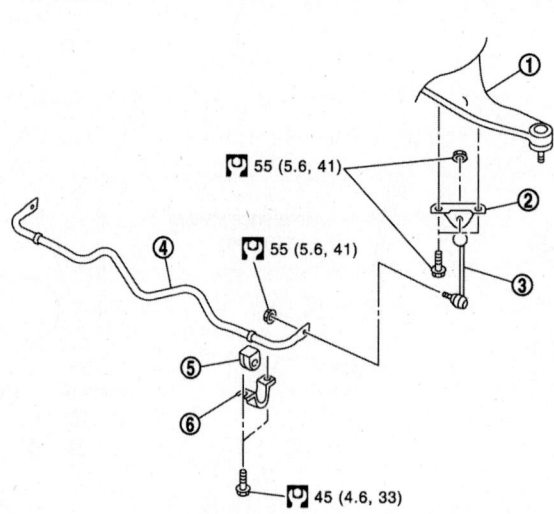

1. Suspension arm
2. Stabilizer connecting rod mounting bracket
3. Stabilizer connecting rod
4. Stabilizer bar
5. Stabilizer bushing
6. Stabilizer clamp

37663_370Z_G0191

Fig. 155 Rear stabilizer bar and related components

11. Check wheel speed sensor for proper operation.

12. Adjust the steering angle sensor neutral position, using the CONSULT-III diagnostic tool, or equivalent.

13. Check and adjust the front alignment, as required.

WHEEL HUBS & BEARINGS

ADJUSTMENT

These bearings are not adjustable. If defective, they must be replaced.

REMOVAL & INSTALLATION

See Figure 156.

1. Before servicing the vehicle, refer to the Precautions Section.

➡**If working near and/or around the SRS system and components, be sure to disable the SRS system. After disabling the system wait three minutes or more before servicing the vehicle.**

➡**Before disconnecting the battery, lower both the driver's and passenger's windows. This will prevent any interference between the window edge and the vehicle when the door is opened or closed. During normal operation the window slightly raises and lowers automatically to prevent any window to vehicle interference. The automatic window function will not work with the battery disconnected.**

2. Raise and support the vehicle safely.

3. Remove the tire and wheel assembly.

4. Remove the caliper. Properly support the caliper to the side. Do not allow it to hang by the brake hose. Never depress the brake pedal with the caliper removed.

5. Remove the brake rotor.

6. Remove the cotter pin. Loosen the wheel hub locknut.

7. matchmark the halfshaft and the wheel hub and bearing assembly.

8. Remove the locknut.

9. Remove the parking brake shoe and parking brake cable.

10. Remove the stabilizer connecting rod (upper side).

11. Remove the coil spring.

12. Properly position a suitable jack under the axle housing.

13. Remove the radius rod.

14. Remove the shock absorber (lower side).

15. Remove the front lower link (axle housing side). Remove the rear lower link (axle housing side)

16. Separate the axle housing from the suspension arm, using the proper tool and remove the axle housing.

➡**Be careful not to damage the ball joint boot. Temporarily tighten the nut to prevent damage to the threads and to prevent the tool from coming off.**

➡**Never place the halfshaft at an extreme angle. Be careful not to overextend the slide joint. Never allow the halfshaft to hang down with proper support.**

17. Remove the wheel hub and bearing assembly.

To install:

➡**Be sure to use new fasteners, as required.**

18. Installation is the reverse of the removal procedure.

19. Use the matchmarks to align removed components.

20. Clean the matching surface of the halfshaft and wheel hub and bearing assembly.

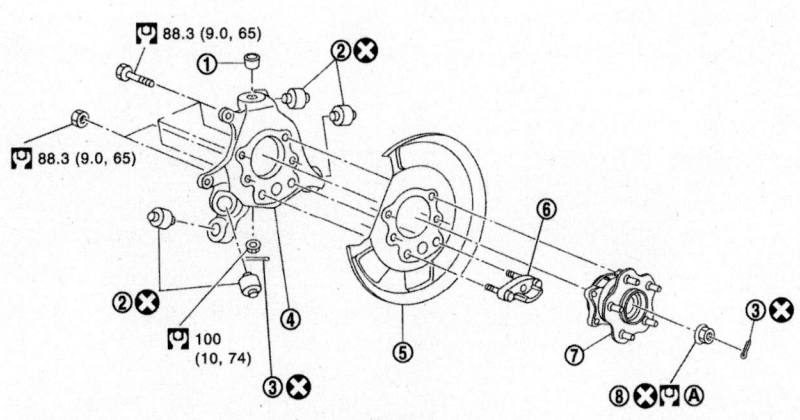

1. Ball seat
2. Bushing
3. Cotter pin
4. Axle housing
5. Back plate
6. Anchor block
7. Wheel hub and bearing assembly
8. Wheel hub lock nut

37663_370Z_G0192

Fig. 156 Rear wheel bearing and hub assembly and related components

21. Apply paste part number 440037S000 or equivalent to the surface of the sub joint assembly of the halfshaft.

➡**Apply the paste, about 0.04–0.10 ounce) to cover the entire flat surface of the sub joint assembly of the halfshaft.**

22. The wheel hub locknut tightening specification is 136 ft. lbs. (185 Nm).

23. Perform a final tightening of nuts and bolts of each removed component with the vehicle in an unladen position.

24. Check wheel speed sensor for proper operation.

25. Adjust the steering angle sensor neutral position, using the CONSULT-III diagnostic tool, or equivalent.

26. Check and adjust the front alignment, as required.

NISSAN

Altima • Altima Hybrid

7

SPECIFICATIONS AND MAINTENANCE CHARTS

ENGINE AND VEHICLE IDENTIFICATION

Engine								Model Year	
Code ①	Liters (cc)	Cu. In.	Cyl.	Fuel Sys.	Engine Type	Eng. Mfg.		Code ②	Year
QR25DE	2.5 (2488)	152	4	MFI	DOHC	Nissan		B	2011
VQ35DE	3.5 (3498)	213	6	MFI	DOHC	Nissan		C	2012

MFI: Multi-port Fuel Injection

DOHC: Double Overhead Camshaft

① The Engine Code is stamped on the engine block near the starter.

② 10th position of the Vehicle Identification Number (VIN)

71075_ALTI_C0001

GENERAL ENGINE SPECIFICATIONS

Year	Model	Engine Displacement Liters	Engine Series ID	Net Horsepower @ rpm	Net Torque @ rpm (ft. lbs.)	Bore x Stroke (in.)	Compression Ratio	Oil Pressure @ rpm
2011	Altima	2.5	QR25DE	175@5600	180@3900	3.50X3.94	9.5:1	43@2000
	Altima	3.5	VQ35DE	270@6000	258@4400	3.76X3.20	10.3:1	43@2000
	Altima Hybrid	2.5	QR25DE	158@6200	162@2800	3.50X3.94	9.5:1	43@2000
2012	Altima	2.5	QR25DE	175@5600	180@3900	3.50X3.94	9.5:1	43@2000
	Altima	3.5	VQ35DE	270@6000	258@4400	3.76X3.20	10.3:1	43@2000
	Altima Hybrid	2.5	QR25DE	158@6200	162@2800	3.50X3.94	9.5:1	43@2000

71075_ALTI_C0002

ENGINE TUNE-UP SPECIFICATIONS

Year	Engine Displacement Liters	Engine ID	Spark Plug Gap (in.)	Ignition Timing (deg.) MT	Ignition Timing (deg.) AT	Fuel Pump (psi) ①	Idle Speed (rpm) MT	Idle Speed (rpm) AT ②	Valve Clearance Intake ③	Valve Clearance Exhaust ③
2011	2.5	QR25DE	0.043	15B	15B	51	650-750	650-750	0.009-0.013	0.010-0.013
	3.5	VQ35DE	0.043	12B	12B	51	550-650	550-650	0.010-0.013	0.011-0.015
2012	2.5	QR25DE	0.043	15B	15B	51	650-750	650-750	0.009-0.013	0.010-0.013
	3.5	VQ35DE	0.043	12B	12B	51	550-650	550-650	0.010-0.013	0.011-0.015

B: Before top dead center

① At idle

② Automatic transmission in neutral

③ Engine cold

71075_ALTI_C0003

CAPACITIES

Year	Model	Engine Displacement Liters	Engine ID	Engine Oil with Filter (qts.)	Transaxle (pts.) 5-Spd	Auto.	Fuel Tank (gal.)	Cooling System (qts.)
2011	Altima	2.5	QR25DE	4.5	7.2	17.5	20.0	8.0
	Altima	3.5	VQ35DE	4.5	3.6	10.8	20.0	8.5
	Altima Hybrid	2.5	QR25DE	4.5	NA	8.66	20.0	①
2012	Altima	2.5	QR25DE	4.5	7.2	17.5	20.0	8.0
	Altima	3.5	VQ35DE	4.5	3.6	10.8	20.0	8.5
	Altima Hybrid	2.5	QR25DE	4.5	NA	8.66	20.0	①

NA: Not Applicable

① Engine coolant, 8 1/8 qts.: Inverter coolant, 3 3/8 qts.

71075_ALTI_C0004

FLUID SPECIFICATIONS

Year	Model	Engine Displacement Liters	Engine Oil	Man. Trans.	Auto. Trans.	Power Steering Fluid	Brake Master Cylinder	Cooling System
2011	Altima	2.5	5W-30	75W-80	Nissan NS-2	Dexron® IV	DOT 3	N-LL
	Altima	3.5	5W-30	75W-80	Nissan NS-2	Dexron® IV	DOT 3	N-LL
	Altima Hybrid	2.5	0W-20 ①	NA	②	NA	DOT 3	N-LL
2012	Altima	2.5	5W-30	75W-80	Nissan NS-2	Dexron® IV	DOT 3	N-LL
	Altima	3.5	5W-30	75W-80	Nissan NS-2	Dexron® IV	DOT 3	N-LL

N-LL: Nissan Long Life coolant

① For warm and hot climates, if 0W-20 is not available, 5W-20 or 5W-30 is applicable.

② Genuine NISSAN Matic W ATF; Using transaxle fluid other than Genuine NISSAN Matic W ATF will damage CVT.

71075_ALTI_C0005

VALVE SPECIFICATIONS

Year	Engine Displacement Liters	Engine ID	Seat Angle (deg.)	Face Angle (deg.)	Spring Test Pressure (lbs. @ in.)	Spring Installed Height (in.)	Stem-to-Guide Clearance (in.) Intake	Exhaust	Stem Diameter (in.) Intake	Exhaust
2011	2.5	QR25DE	45.15-45.45	—	34-39@ 1.39	1.390	0.0008-0.0021	0.0012-0.0025	0.2348-0.2354	0.2344-0.2350
	3.5	VQ35DE	45.15-45.45	—	37-42@ 1.457	1.457	0.0008-0.0021	0.0016-0.0028	0.2348-0.2354	0.2344-0.2350
2012	2.5	QR25DE	45.15-45.45	—	34-39@ 1.39	1.390	0.0008-0.0021	0.0012-0.0025	0.2348-0.2354	0.2344-0.2350
	3.5	VQ35DE	45.15-45.45	—	37-42@ 1.457	1.457	0.0008-0.0021	0.0016-0.0028	0.2348-0.2354	0.2344-0.2350

71075_ALTI_C0006

CAMSHAFT SPECIFICATIONS
All measurements in inches unless noted

Year	Engine Displacement Liters	Engine Code/ID	Journal Dia.	Brg. Oil Clearance	Shaft End-play	Circle Runout	Lobe Height Intake	Lobe Height Exhaust
2011	2.5	QR25DE	①	0.0018-0.0034	0.0045-0.0074	0.0016	1.7644-1.7718	1.7313-1.7388
	3.5	VQ35DE	②	③	0.0045-0.0074	0.0008	1.7904-1.7978	1.7904-1.7978
	2.5 Hybrid	QR25DE	①	0.0018-0.0034	0.0045 0.0074	0.0059	1.7644-1.7718	1.7313-1.7388
2012	2.5	QR25DE	①	0.0018-0.0034	0.0045-0.0074	0.0016	1.7644-1.7718	1.7313-1.7388
	3.5	VQ35DE	②	③	0.0045-0.0074	0.0008	1.7904-1.7978	1.7904-1.7978
	2.5 Hybrid	QR25DE	①	0.0018-0.0034	0.0045 0.0074	0.0059	1.7644-1.7718	1.7313-1.7388

① No. 1: 1.0998-1.1006
  All others: 0.9926-0.9234

② No. 1: 1.0211-1.0218
  All others: 0.9230-0.9238

③ No. 1: 0.0018-0.0034
  All others: 0.0014-0.0030

71075_ALTI_C0007

CRANKSHAFT AND CONNECTING ROD SPECIFICATIONS
All measurements are given in inches.

Year	Engine Displacement Liters	Engine ID	Crankshaft Main Brg. Journal Dia.	Crankshaft Main Brg. Oil Clearance	Crankshaft Shaft End-play	Crankshaft Thrust on No.	Connecting Rod Journal Diameter	Connecting Rod Oil Clearance	Connecting Rod Side Clearance
2011	2.5	QR25DE	2.3206-2.3216	①	0.0039-0.0102	3	1.8898-1.8903	0.0002-0.0007	0.0079-0.0138
	3.5	VQ35DE	2.3603-2.3612	0.0014-0.0018	0.0039-0.0098	3	2.1654-2.1659	0.0002-0.0007	0.0079-0.0138
	2.5 Hybrid	QR25DE	2.1636-2.1645	①	0.0039-0.0102	3	1.8898-1.8903	0.0014-0.0018	0.0079-0.0138
2012	2.5	QR25DE	2.3206-2.3216	①	0.0039-0.0102	3	1.8898-1.8903	0.0002-0.0007	0.0079-0.0138
	3.5	VQ35DE	2.3603-2.3612	0.0014-0.0018	0.0039-0.0098	3	2.1654-2.1659	0.0002-0.0007	0.0079-0.0138
	2.5 Hybrid	QR25DE	2.1636-2.1645	①	0.0039-0.0102	3	1.8898-1.8903	0.0014-0.0018	0.0079-0.0138

① Nos. 1, 3, 5 : 0.0005-0.0009
  Nos. 2, 4 : 0.0007-0.0011

71075_ALTI_C0008

PISTON AND RING SPECIFICATIONS

All measurements are given in inches.

Year	Engine Displacement Liters (ID)	Piston Clearance	Ring Gap			Ring Side Clearance		
			Top Compression	Bottom Compression	Oil Control	Top Compression	Bottom Compression	Oil Control
2011	2.5 (QR25DE)	0.0004-0.0012	0.0083-0.0122	0.0146-0.0205	0.0079-0.0177	0.0016-0.0031	0.0012-0.0028	0.0018-0.0049
	3.5 (VQ35DE)	0.0004-0.0012	0.0091-0.0130	0.0091-0.0130	0.0079-0.0177	0.0018-0.0031	0.0012-0.0028	0.0026-0.0049
2012	2.5 (QR25DE)	0.0004-0.0012	0.0083-0.0122	0.0146-0.0205	0.0079-0.0177	0.0016-0.0031	0.0012-0.0028	0.0018-0.0049
	3.5 (VQ35DE)	0.0004-0.0012	0.0091-0.0130	0.0091-0.0130	0.0079-0.0177	0.0018-0.0031	0.0012-0.0028	0.0026-0.0049

71075_ALTI_C0009

TORQUE SPECIFICATIONS

All readings in ft. lbs.

Year	Engine Displacement Liters	Engine ID	Cylinder Head Bolts	Main Bearing Bolts	Rod Bearing Bolts	Crankshaft Damper Bolts	Flywheel Bolts	Manifold		Spark Plugs	Oil Drain Plug
								Intake	Exhaust		
2011	2.5	QR25DE	①	②	③	NA	76-83	13-15	29-32	18	25
	3.5	VQ35DE	④	⑤	⑥	⑦	61-69	⑧	21-24	18	25
2012	2.5	QR25DE	①	②	③	NA	76-83	13-15	29-32	18	25
	3.5	VQ35DE	④	⑤	⑥	⑦	61-69	⑧	21-24	18	25

NA: Not available

① Step 1: 72 ft. lbs.

Step 2: Loosen completely, then retorque to 29 ft. lbs.

Step 3: Turn each bolt, in sequence, an additional 75 degrees

Step 4: Turn each bolt, in sequence, an additional 75 degrees

② Tighten bolts 11-22 to 19 ft. lbs.

Step 2: Tighten bolts 1-10 to 29 ft. lbs.

Step 3: Tighten bolts 1-10 an additional 60 degrees

③ Step 1: Tighten bolts to 22 ft. lbs.

Step 2: Loosen all bolts

Step 3: tighten bolts to 14 ft. lbs.

Step 4: Tighten bolts an additional 90 degrees

④ Step 1: Tighten to 72 ft. lbs.

Step 2: Loosen bolts completely in reverse order

Step 3: Tighten to 29 ft. lbs.

Step 4: Tighten an additional 103 degrees

Step 5: Repeat Step 4

⑤ Step 1: Shift crankshaft to align bearing beam

Step 2: Tighten all bolts to 24-28 ft. lbs.

Step 3: Tighten an additional 90-95 degrees

⑥ Step 1: Tighten to 14-15 ft. lbs.

Step 2: Tighten an additional 90-95 degrees

⑦ Step 1: Tighten to 32 ft. lbs. an additional 90 degrees

⑧ Step 1: Tighten to 65 inch lbs.

Step 2: tighten to 19 ft. lbs.

71075_ALTI_C0010

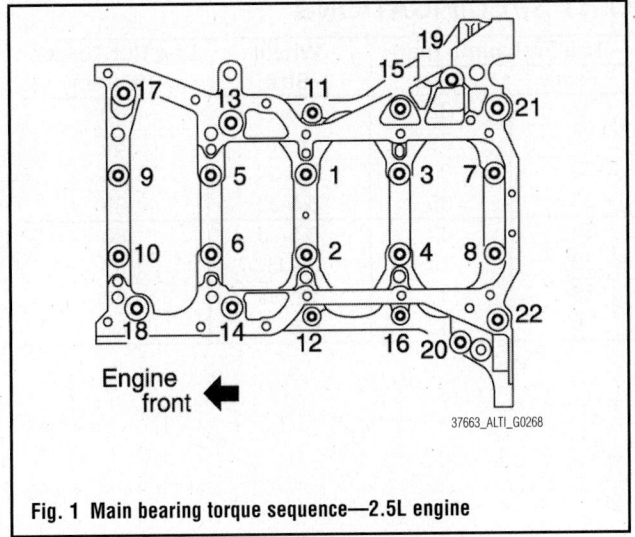

Fig. 1 Main bearing torque sequence—2.5L engine

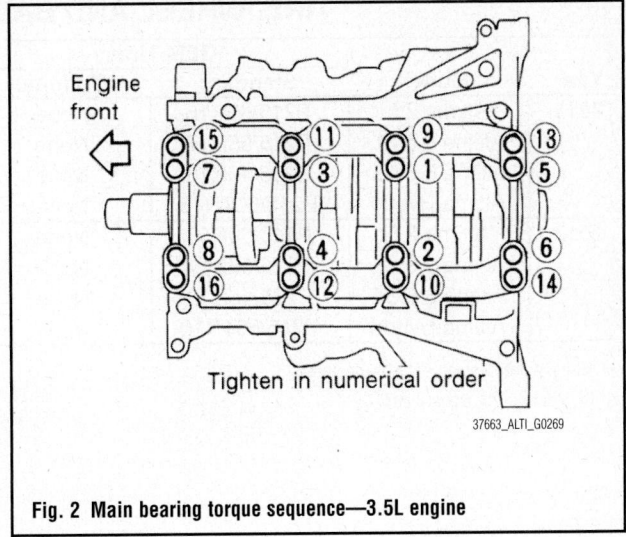

Fig. 2 Main bearing torque sequence—3.5L engine

WHEEL ALIGNMENT

Year	Model		Caster		Camber		Toe-in (in.)
			Range (+/-Deg.)	Preferred Setting (Deg.)	Range (+/-Deg.)	Preferred Setting (Deg.)	
2011	Altima	Front	0.75	4.90	①	②	0.04 +/- 0.04
		Rear	—	—	0.30	-1.25	0.09 +/- 0.06
	Altima	Front	0.55	5.00	0.75	③	0.04 +/- 0.04
	Hybrid	Rear			NA	NA	NA
2012	Altima	Front	0.75	4.90	①	②	0.04 +/- 0.04
		Rear	—	—	0.50	-0.62	0.09 +/- 0.06
	Altima	Front	0.55	5.00	0.75	③	0.04 +/- 0.04
	Hybrid	Rear			NA	NA	NA

① Minus 0.25 degrees, plus 0.75 degrees

② Left, -.050 degrees; Right -0.75 degrees

③ Left, -0.40; Right, -0.65

71075_ALTI_C0011

TIRE, WHEEL AND BALL JOINT SPECIFICATIONS

| Year | Model | OEM Tires | | Tire Pressures (psi) | | Wheel Size | Lug Nut Torque (ft. lbs.) |
		Standard	Optional	Front	Rear		
2011	Altima 2.5	P215/60R16	None	32	32	6.5-JJ	80
	Altima 3.5 SL	P215/55R17	None	33	33	7-JJ	80
	Altima 3.5 SE	P235/45R18	None	33	33	7.5-JJ	80
	Altima Hybrid	P215/60TR16	None	35	35	6.5-JJ	83
2012	Altima 2.5	P215/60R16	None	32	32	6.5-JJ	80
	Altima 3.5 SL	P215/55R17	None	33	33	7-JJ	80
	Altima 3.5 SE	P235/45R18	None	33	33	7.5-JJ	80
	Altima Hybrid	P215/60TR16	None	35	35	6.5-JJ	83

OEM: Original Equipment Manufacturer

PSI: Pounds Per Square Inch

71075_ALTI_C0012

BRAKE SPECIFICATIONS

All measurements in inches unless noted

| Year | Model | | Brake Disc | | | Minimum Lining Thickness | | Brake Caliper | |
			Original Thickness	Minimum Thickness	Maximum Run-out	Front	Rear	Bracket Bolts (ft. lbs.)	Mounting Bolts (ft. lbs.)
2011	Altima	Front	1.024	0.945	0.002	0.079	—	98	20
		Rear	0.354	0.315	0.002	—	0.039	62	32
2012	Altima	Front	1.020	0.945	0.002	0.079	—	98	20
		Rear	0.354	0.315	0.002	—	0.039	62	32

71075_ALTI_C0013

SCHEDULED MAINTENANCE INTERVALS
Nissan—Altima

TO BE SERVICED	TYPE OF SERVICE	VEHICLE MILEAGE INTERVAL (x1000)												
		7.5	15	22.5	30	37.5	45	52.5	60	67.5	75	82.5	90	97.5
Engine oil & filter	R	✓	✓	✓	✓	✓	✓	✓	✓	✓	✓	✓	✓	✓
Brake lines & cables	S/I		✓		✓		✓		✓		✓		✓	
Brake pads, discs, drums & linings	S/I		✓		✓		✓		✓		✓		✓	
Driveshaft boots	S/I		✓		✓		✓		✓		✓		✓	
Exhaust system	S/I				✓				✓				✓	
Transaxle fluid	S/I		✓		✓		✓		✓		✓		✓	
Air cleaner filter	R				✓				✓				✓	
Spark plugs (except platinum)	R				✓				✓				✓	
Spark plugs (iridium and platinum)	R	Replace every 105,000 miles												
Steering gear & linkage, axle & suspension parts	S/I			✓				✓				✓		
Engine coolant	R	Replace every 60,000 miles, then every 30,000 miles												
Inverter coolant	R	Replace every 60,000 miles, then every 30,000 miles												
Drive belts	S/I								✓					
Fuel lines	S/I								✓					
Vapor lines	S/I								✓					
Cabin microfilter	R		✓		✓		✓		✓		✓		✓	
Valve adjustment	S/I	As needed												

R: Replace   S/I: Service or Inspect

FREQUENT OPERATION MAINTENANCE (SEVERE SERVICE)

If a vehicle is operated under any of the following conditions it is considered severe service:

- Extremely dusty areas.

- 50% or more of the vehicle operation is in 32°C (90°F) or higher temperatures, or constant operation in temperatures below 0°C (32°F).

- Prolonged idling (vehicle operation in stop and go traffic).

- Frequent short running periods (engine does not warm to normal operating temperatures).

- Police, taxi, delivery usage or trailer towing usage.

Oil & oil filter: change every 3750 miles.

Brake pads & discs: service or inspect every 7500 miles.

Driveshaft boots: service or inspect every 7500 miles.

Exhaust system: service or inspect every 7500 miles.

Steering gear & linkage, axle & suspension parts: service or inspect every 7500 miles.

Steering linkage ball joints & front suspension ball joints: service or inspect every 7500 miles.

Air cleaner filter: service or inspect every 15,000 miles.

71075_ALTI_C0014

SCHEDULED MAINTENANCE INTERVALS
Nissan—Altima Hybrid

TO BE SERVICED	TYPE OF SERVICE	VEHICLE MILEAGE INTERVAL (x1000)												
		7.5	15	22.5	30	37.5	45	52.5	60	67.5	75	82.5	90	97.5
Engine oil & filter	R	✓	✓	✓	✓	✓	✓	✓	✓	✓	✓	✓	✓	✓
Brake lines & cables	S/I		✓		✓		✓		✓		✓		✓	
Brake pads, discs, drums & linings	S/I		✓		✓		✓		✓		✓		✓	
Driveshaft boots	S/I		✓		✓		✓		✓		✓		✓	
Exhaust system	S/I				✓				✓				✓	
Transaxle fluid	S/I		✓		✓		✓		✓		✓		✓	
Air cleaner filter	R				✓				✓				✓	
Spark plugs (exc. platinum)	R				✓				✓				✓	
Spark plugs (iridium and platinum)	R	Replace every 105,000 miles												
Steering gear & linkage, axle & suspension parts	S/I				✓				✓				✓	
Engine coolant	R	Replace every 60,000 miles, then every 30,000 miles												
Inverter coolant	R	Replace every 60,000 miles, then every 30,000 miles												
Drive belts	S/I								✓					
Fuel lines	S/I								✓					
Vapor lines	S/I								✓					
Cabin microfilter	R		✓		✓		✓		✓		✓		✓	
Valve adjustment	S/I	As needed												

R: Replace   S/I: Service or Inspect

FREQUENT OPERATION MAINTENANCE (SEVERE SERVICE)

If a vehicle is operated under any of the following conditions it is considered severe service:

- Extremely dusty areas.

- 50% or more of the vehicle operation is in 32°C (90°F) or higher temperatures, or constant operation in temperatures below 0°C (32°F).

- Prolonged idling (vehicle operation in stop and go traffic).

- Frequent short running periods (engine does not warm to normal operating temperatures).

- Police, taxi, delivery usage or trailer towing usage.

Oil & oil filter: change every 3750 miles.

Brake pads & discs: service or inspect every 7500 miles.

Driveshaft boots: service or inspect every 7500 miles.

Exhaust system: service or inspect every 7500 miles.

Steering gear & linkage, axle & suspension parts: service or inspect every 7500 miles.

Steering linkage ball joints & front suspension ball joints: service or inspect every 7500 miles.

Air cleaner filter: service or inspect every 15,000 miles.

71075_ALTI_C0015

PRECAUTIONS

Before servicing any vehicle, please be sure to read all of the following precautions, which deal with personal safety, prevention of component damage, and important points to take into consideration when servicing a motor vehicle:

• Never open, service or drain the radiator or cooling system when the engine is hot; serious burns can occur from the steam and hot coolant.

• Observe all applicable safety precautions when working around fuel. Whenever servicing the fuel system, always work in a well-ventilated area. Do not allow fuel spray or vapors to come in contact with a spark, open flame, or excessive heat (a hot drop light, for example). Keep a dry chemical fire extinguisher near the work area. Always keep fuel in a container specifically designed for fuel storage; also, always properly seal fuel containers to avoid the possibility of fire or explosion. Refer to the additional fuel system precautions later in this section.

• Fuel injection systems often remain pressurized, even after the engine has been turned **OFF**. The fuel system pressure must be relieved before disconnecting any fuel lines. Failure to do so may result in fire and/or personal injury.

• Brake fluid often contains polyglycol ethers and polyglycols. Avoid contact with the eyes and wash your hands thoroughly after handling brake fluid. If you do get brake fluid in your eyes, flush your eyes with clean, running water for 15 minutes. If eye irritation persists, or if you have taken brake fluid internally, IMMEDIATELY seek medical assistance.

• The EPA warns that prolonged contact with used engine oil may cause a number of skin disorders, including cancer. You should make every effort to minimize your exposure to used engine oil. Protective gloves should be worn when changing oil. Wash your hands and any other exposed skin areas as soon as possible after exposure to used engine oil. Soap and water, or waterless hand cleaner should be used.

• All new vehicles are now equipped with an air bag system, often referred to as a Supplemental Restraint System (SRS) or Supplemental Inflatable Restraint (SIR) system. The system must be disabled before performing service on or around system components, steering column, instrument panel components, wiring and sensors. Failure to follow safety and disabling procedures could result in accidental air bag deployment, possible personal injury and unnecessary system repairs.

• Always wear safety goggles when working with, or around, the air bag system. When carrying a non-deployed air bag, be sure the bag and trim cover are pointed away from your body. When placing a non-deployed air bag on a work surface, always face the bag and trim cover upward, away from the surface. This will reduce the motion of the module if it is accidentally deployed. Refer to the additional air bag system precautions later in this section.

• Clean, high quality brake fluid from a sealed container is essential to the safe and proper operation of the brake system. You should always buy the correct type of brake fluid for your vehicle. If the brake fluid becomes contaminated, completely flush the system with new fluid. Never reuse any brake fluid. Any brake fluid that is removed from the system should be discarded. Also, do not allow any brake fluid to come in contact with a painted surface; it will damage the paint.

• Never operate the engine without the proper amount and type of engine oil; doing so WILL result in severe engine damage.

• Timing belt maintenance is extremely important. Many models utilize an interference-type, non-freewheeling engine. If the timing belt breaks, the valves in the cylinder head may strike the pistons, causing potentially serious (also time-consuming and expensive) engine damage. Refer to the maintenance interval charts for the recommended replacement interval for the timing belt, and to the timing belt section for belt replacement and inspection.

• Disconnecting the negative battery cable on some vehicles may interfere with the functions of the on-board computer system(s) and may require the computer to undergo a relearning process once the negative battery cable is reconnected.

• When servicing drum brakes, only disassemble and assemble one side at a time, leaving the remaining side intact for reference.

• Only an MVAC-trained, EPA-certified automotive technician should service the air conditioning system or its components.

BRAKES

GENERAL INFORMATION

PRECAUTIONS

• Certain components within the ABS system are not intended to be serviced or repaired individually.

• Do not use rubber hoses or other parts not specifically specified for and ABS system. When using repair kits, replace all parts included in the kit. Partial or incorrect repair may lead to functional problems and require the replacement of components.

• Lubricate rubber parts with clean, fresh brake fluid to ease assembly. Do not use shop air to clean parts; damage to rubber components may result.

• Use only DOT 3 brake fluid from an unopened container.

• If any hydraulic component or line is removed or replaced, it may be necessary to bleed the entire system.

• A clean repair area is essential. Always clean the reservoir and cap thoroughly before removing the cap. The slightest amount of dirt in the fluid may plug an orifice and impair the system function. Perform repairs after components have been thoroughly cleaned; use only denatured alcohol to clean components. Do not allow ABS components to come into contact with any substance containing mineral oil; this includes used shop rags.

• The Anti-Lock control unit is a microprocessor similar to other computer units in the vehicle. Ensure that the ignition switch is **OFF** before removing or installing controller harnesses. Avoid static electricity discharge at or near the controller.

ANTI-LOCK BRAKE SYSTEM (ABS)

• If any arc welding is to be done on the vehicle, the control unit should be unplugged before welding operations begin.

WHEEL SPEED SENSORS

REMOVAL & INSTALLATION

Front Wheel Sensor

See Figure 3.

Note the following:

• Be careful not to damage wheel sensor edge and sensor rotor teeth.

• When removing the front or rear wheel hub, first remove the wheel sensor from the wheel hub. Failure to do so may result in damage to the wheel sensor wires making the sensor inoperative.

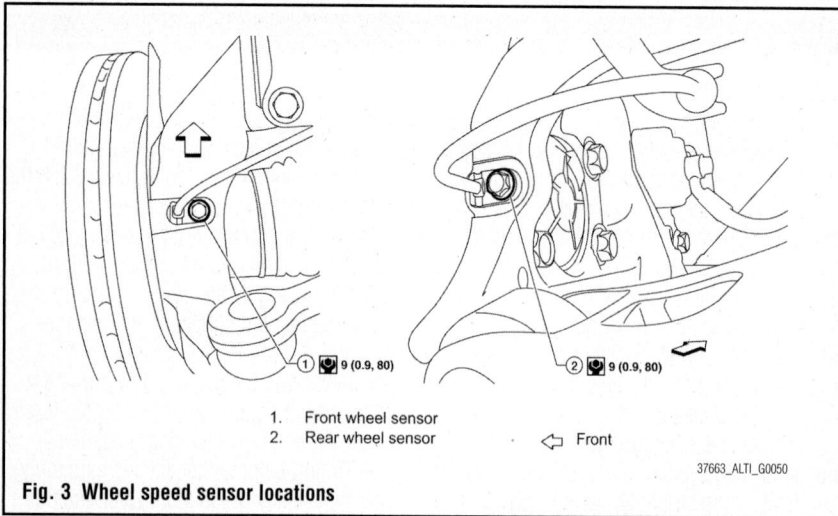

1. Front wheel sensor
2. Rear wheel sensor

⇦ Front

37663_ALTI_G0050

Fig. 3 Wheel speed sensor locations

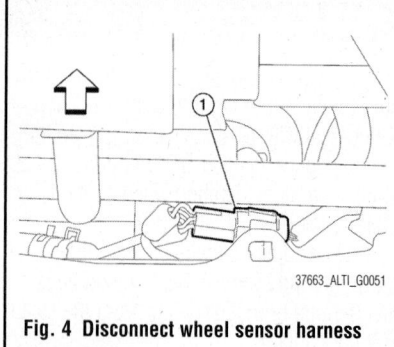

37663_ALTI_G0051

Fig. 4 Disconnect wheel sensor harness connector (1)

• Pull out the wheel sensor, being careful to turn it as little as possible. Do not pull on the wheel sensor harness.

• Before installation, check if foreign objects such as iron fragments are adhered to the pick-up part of the sensor or to the inside of the hole in the wheel hub for the wheel sensor, or if a foreign object is caught in the surface of the mating surface for the sensor rotor. Fix as necessary and then install the wheel sensor.

1. Remove front wheel and tire.
2. Partially front wheel fender protector.
3. Remove wheel sensor bolt and wheel sensor.
4. Remove harness wire from mounts and disconnect wheel sensor harness connector.

To install:

5. Installation is in the reverse order of removal.

Rear Wheel Sensor

See Figures 3 and 4.

➡ **Both rear wheel sensors share one harness and must be replaced as an assembly.**

Note the following:

• Be careful not to damage wheel sensor edge and sensor rotor teeth.

• When removing the front or rear wheel hub, first remove the wheel sensor from the wheel hub. Failure to do so may result in damage to the wheel sensor wires making the sensor inoperative.

• Pull out the wheel sensor, being careful to turn it as little as possible. Do not pull on the wheel sensor harness.

• Before installation, check if foreign objects such as iron fragments are adhered to the pick-up part of the sensor or to the inside of the hole in the wheel hub for the

wheel sensor, or if a foreign object is caught in the surface of the mating surface for the sensor rotor. Fix as necessary and then install the wheel sensor.

1. Remove rear wheel and tire.
2. Remove wheel sensor bolts and wheel sensors from both rear wheel hub and bearing assemblies.
3. Remove harness wire from mounts and harness wire clips from rear suspension member.
4. Disconnect wheel sensor harness connector.

To install:

5. Installation is in the reverse order of removal.

WHEEL SPEED SENSOR RINGS (TOOTHED RINGS)

REMOVAL & INSTALLATION

The front and rear wheel sensor rotors are an integral part of the wheel hubs and cannot be disassembled. When replacing the sensor rotor, replace the wheel hub.

BRAKES   BLEEDING THE BRAKE SYSTEM

BLEEDING PROCEDURE

MANUAL

Standard (Non-Hybrid)

➡ **While bleeding, pay attention to master cylinder fluid level.**

✳✳ CAUTION

Before working, disconnect connectors of ABS actuator and electric unit (control unit) or battery cable from the negative terminal.

1. Connect a vinyl tube to rear right brake caliper bleed valve.
2. Fully depress brake pedal 4 or 5 times.
3. With brake pedal depressed, loosen

bleed valve to bleed air in brake line, and then tighten it immediately.

4. Repeat steps 2 and 3 until all of the air is out of the brake line.
5. Tighten the bleed valve.
6. From step 1 to 5, with master cylinder reservoir tank filled at least half way, bleed air from brake hydraulic line bleed valves in the following order:

• Rear right brake
• Front left brake
• Rear left brake
• Front right brake

Hybrid

✳✳ WARNING

If any DTC is indicated, erase the indicated DTC. After the procedure of

air bleed, perform initialization of linear solenoid valve.

➡ **The brake warning buzzer may be activated during the air bleed procedure. The work can be continued, as it is normal.**

Air Release of Static Pressure System (Front Wheel)

✳✳ WARNING

Monitor the fluid level in the reservoir tank during the air bleeding. Always use new brake fluid for refilling. Never reuse the drained brake fluid.

1. Turn ignition switch OFF.

2. Connect CONSULT-III.

3. Turn ignition switch (READY).

4. When performing air bleed of the static pressure system and suction drain system, remove 2 relays for brake actuator motor beforehand.

5. Connect a vinyl tube to the bleeder valve of the front brake.

6. When performing air bleed, following conditions are required:
- ABS relay No.1 and No.2: ON
- Parking brake: ON
- Shift position: P range
- Vehicle speed: 0 km/h (0 MPH)
- Normal power supply voltage
- Normal communication with HV
- No failure of brake system (except following items):
- Motor relay
- Accumulator
- Fluid level switch
- Calibration for each sensors and linear solenoid
- Test mode diagnostic code

7. Select "AIR REL INHIBIT" in "ACTIVE TEST".

8. Loosen the bleeder valve and bleed air with the brake pedal depressed.

➡**Air bleeding is allowed to start from either right or left.**

9. After a complete air bleeding, tighten bleeder valve to the specified torque.

10. Check that the fluid level is the reservoir tank is within the specified range after air bleeding.

Air Release of Suction Drain System

❊❊ WARNING

Monitor the fluid level in the reservoir tank during the air bleeding. Perform the air bleed procedure within 30 seconds after the transmission of the signal from CONSULT-III. When the air bleed is performed afterward, the re-transmission of the signal from CONSULT-III is needed.

➡**Air bleed from the bleeder valve is not necessary since this operation is to return brake fluid (air).**

1. Turn ignition switch OFF.

2. Connect CONSULT-III.

3. Turn ignition switch (READY).

4. When performing air bleed, following conditions are required:

- ABS relay No.1 and No.2: ON
- Parking brake: ON
- Shift position: P range
- Vehicle speed: 0 km/h (0 MPH)
- Normal power supply voltage
- Normal communication with HV
- No failure of brake system (except following items):
- Motor relay
- Accumulator
- Fluid level switch
- Calibration for each sensors and linear solenoid
- Test mode diagnostic code

5. Select "AIR REL DRAIN" in "ACTIVE TEST".

6. Step on the brake pedal and return brake fluid to reservoir tank.

7. Ensure that no air (bubble) is contained in the brake fluid circulated from reservoir tank.

Air Release of Rear Wheel System

❊❊ WARNING

Monitor the fluid level in the reservoir tank during the air bleeding. Always use new brake fluid for refilling. Never reuse the drained brake fluid.

1. Turn ignition switch OFF.
2. Connect 2 motor relays.
3. Connect CONSULT-III.
4. Turn ignition switch (READY).

➡**If CONSULT-III is frozen, erase the DTC.**

5. Confirm accumulator pressure level by using "DATA MONITOR" in CONSULT-III.

   a. Select "ACC PRESS SEN" in "DATA MONITOR".

   b. Ensure that this voltage is over 3.42 V.

   c. If voltage is under 3.42 V, then step on the brake pedal several time.

6. When performing air bleed, following conditions must be met:
- ABS relay No.1 and No.2: ON
- Parking brake: ON
- Shift position: P range
- Vehicle speed: 0 km/h (0 MPH)
- Normal power supply voltage
- Normal communication with HV
- ABS motor relay No.1 and No.2 are set
- No failure of brake system (except following items):
- Motor relay
- Accumulator
- Fluid level switch

- Calibration for each sensors and linear solenoid
- Test mode diagnostic code

7. Connect a vinyl tube to the bleeder valve of the rear brake.

8. Select "AIR REL INHIBIT" in "ACTIVE TEST".

9. Loosen the bleeder valve and bleed air with the brake pedal depressed.

10. Ensure that there is no air leakage from the bleeder.

11. After a complete air bleeding, tighten bleeder valve to the specified torque.

12. Check that the fluid level is the reservoir tank is within the specified range after air bleeding.

Air Release of Power Supply System

Note the following:
- Monitor the fluid level in the reservoir tank during the air bleeding.
- Always use new brake fluid for refilling. Never reuse the drained brake fluid.
- Perform the air bleed procedure within 10 seconds after the transmission of the signal from CONSULT-III. When the air bleed is performed afterward, the re-transmission of the signal from CONSULT-III is needed.

➡**No need to step on the brake pedal. Air bleeding is necessary for the front left brake only.**

1. Turn ignition switch OFF.
2. Connect CONSULT-III.
3. Turn ignition switch (READY).
4. Connect a vinyl tube to the bleeder valve of the front left brake.
5. When performing air bleed, following conditions must be met:
- ABS relay No.1 and No.2: ON
- Parking brake: ON
- Shift position: P range
- Vehicle speed: 0 km/h (0 MPH)
- Normal power supply voltage
- Normal communication with HV
- ABS motor relay No.1 and No.2 are set
- No failure of brake system (except following items):
- Motor relay
- Accumulator
- Fluid level switch
- Calibration for each sensors and linear solenoid
- Test mode diagnostic code

6. Select "AIR REL PWR SPLY 2" in "ACTIVE TEST".

7. Loosen the bleeder valve.

8. Ensure that there is no air leakage from the bleeder.

9. After a complete air bleeding, tighten bleeder valve to the specified torque.

Air Release of Stroke Simulator System

Air Bleed of Stroke Simulator System 1

✳✳ WARNING

Perform the air bleed procedure within 30 seconds after the transmission of the signal from CONSULTIII. When the air bleed is performed afterward, the re-transmission of the signal from CONSULT-III is needed.

➡**Air bleed from the bleeder is not necessary in this stage. This process is performed to send air contained in the stroke simulator to piping. Pedal operation only and no need of air bleed from the bleeder.**

1. Turn ignition switch OFF.
2. Connect CONSULT-III.
3. Turn ignition switch (READY).
4. When performing air bleed, following conditions must be met:
   - ABS relay No.1 and No.2: ON
   - Parking brake: ON
   - Shift position: P range
   - Vehicle speed: 0 km/h (0 MPH)
   - Normal power supply voltage
   - Normal communication with HV
   - ABS motor relay No.1 and No.2 are set
   - No failure of brake system (except following items):
   - Motor relay
   - Accumulator
   - Fluid level switch
   - Calibration for each sensors and linear solenoid
   - Test mode diagnostic code
5. Select "AIR REL STROKE SIM" in "ACTIVE TEST".
6. Step on the brake pedal 20 times with its stroke fully within continuously 20 to 30 seconds.

Air Release of Stroke Simulator System 2

➡**Air bleeding is necessary for the front left brake only.**

7. Connect a vinyl tube to the bleeder valve of the front left brake.
8. When performing air bleed, following conditions must be met:
   - ABS relay No.1 and No.2: ON
   - Parking brake: ON
   - Shift position: P range

- Vehicle speed: 0 km/h (0 MPH)
- Normal power supply voltage
- Normal communication with HV
- ABS motor relay No.1 and No.2 are set
- No failure of brake system (except following items):
- Motor relay
- Accumulator
- Fluid level switch
- Calibration for each sensors and linear solenoid
- Test mode diagnostic code

9. Select "AIR REL INHIBIT" in "ACTIVE TEST".
10. Loosen the bleeder valve and bleed air with the brake pedal depressed.
11. Ensure that there is no air from the bleeder.
12. Tighten the bleeder valve to the specified torque.
13. Return to previous step "Air Release of Stroke Simulator System 1". Repeat "Air Release of Stroke Simulator System 1" and "Air Release of Stroke Simulator System 2" at least 3 times.

Air Release of High-Pressure Line

✳✳ WARNING

Be careful with fluid level in the reservoir tank because a large amount of brake fluid flows back to the reservoir tank.

➡**Air bleed from the bleeder is not necessary in this stage.**

1. Turn ignition switch OFF.
2. Connect CONSULT-III.
3. Turn ignition switch (READY).
4. When performing air bleed, following conditions must be met:
   - ABS relay No.1 and No.2: ON
   - Parking brake: ON
   - Shift position: P range
   - Vehicle speed: 0 km/h (0 MPH)
   - Normal power supply voltage
   - Normal communication with HV
   - ABS motor relay No.1 and No.2 are set
   - No failure of brake system (except following items):
   - Motor relay
   - Accumulator
   - Fluid level switch
   - Calibration for each sensors and linear solenoid
   - Test mode diagnostic code
5. Select "ACC 0 DOWN" in "ACTIVE TEST".

➡**Return air remaining in the high-pressure line to reservoir tank and open atmosphere.**

6. Repeat 5 times to ensure the circulation of brake fluid since visual judgment of completion is difficult.
7. Fill the brake fluid to the MAX line after completing this operation, with "ACC 0 DOWN" condition.

BLEEDING THE ABS SYSTEM

➡**While bleeding, pay attention to master cylinder fluid level.**

✳✳ CAUTION

Before working, disconnect connectors of ABS actuator and electric unit (control unit) or battery cable from the negative terminal.

1. Connect a vinyl tube to rear right brake caliper bleed valve.
2. Fully depress brake pedal 4 or 5 times.
3. With brake pedal depressed, loosen bleed valve to bleed air in brake line, and then tighten it immediately.
4. Repeat steps 2 and 3 until all of the air is out of the brake line.
5. Tighten the bleed valve.
6. From step 1 to 5, with master cylinder reservoir tank filled at least half way, bleed air from brake hydraulic line bleed valves in the following order:
   - Rear right brake
   - Front left brake
   - Rear left brake
   - Front right brake

✳✳ WARNING

If any DTC is indicated, erase the indicated DTC. After the procedure of air bleed, perform initialization of linear solenoid valve.

➡**The brake warning buzzer may be activated during the air bleed procedure. The work can be continued, as it is normal.**

FLUID FILL PROCEDURE

Standard (Non-Hybrid)

Draining

Note the following precautions:
- Be careful not to splash brake fluid on painted areas; it may cause paint damage. If brake fluid is splashed on painted areas, wash it away with water immediately.
- Before working, disconnect connectors of ABS actuator and electric unit (control

unit) or battery cable from the negative terminal.

1. Connect a vinyl tube to bleed valve.
2. Depress brake pedal, loosen bleed valve, and gradually remove brake fluid.

Refilling

Note the following precautions:
• Refill with new brake fluid "DOT 3".
• Never reuse drained brake fluid.
• Before working, disconnect connectors of ABS actuator and electric unit (control unit) or battery cable from the negative terminal.

1. Make sure there is no foreign material in the reservoir tank, and refill with new brake fluid.

2. Loosen bleed valve, depress brake pedal slowly to full stroke and then release it. Repeat the procedure every 2 or 3 seconds until the new brake fluid comes out, then close the bleed valve while depressing the pedal. Repeat the same work for each wheel.

3. Bleed air.

Hybrid

1. Check that the brake fluid level in reservoir tank is within the specified range between the MAX and MIN lines as shown.

2. Visually check around the reservoir tank for fluid leaks.

3. If the fluid level is excessively low, check the brake system for leaks.

4. Release the parking brake and check if the brake warning lamp goes off. If not, check brake system for fluid leaks.

5. Make sure there is no foreign material in the reservoir tank, and refill with new brake fluid to the proper level.

> **✳✳ WARNING**
>
> **Refill with new brake fluid "DOT 3". Never reuse drained brake fluid.**

BRAKES   FRONT DISC BRAKES

BRAKE CALIPERS

REMOVAL & INSTALLATION

See Figures 5 and 6.

1. Remove front wheel and tires.
2. Drain brake fluid.
3. Remove union bolt and disconnect brake hose from caliper assembly. Discard the copper washers.

> **✳✳ WARNING**
>
> **Do not reuse copper washers.**

4. Remove torque member bolts, and remove brake caliper assembly.

> **✳✳ WARNING**
>
> **Do not drop brake pad.**

5. Remove disc rotor. If reusing the disc rotor apply match marks before removal.

> **✳✳ WARNING**
>
> **If reusing the rotor, make sure to matchmark the installed position of the wheel hub assembly and disc rotor.**

To install:

6. Install disc rotor, align the matching marks if installing the original disc rotor as shown.

> **✳✳ WARNING**
>
> **If reinstalling the old rotor, make sure to align the matchmarks made during removal.**

7. Install brake caliper assembly to vehicle, and tighten torque member bolts to the specified torque.

> **✳✳ WARNING**
>
> **Do not allow oil or any moisture on all contact surfaces between steering knuckle and caliper assembly, bolts, and washer.**

8. Install brake hose to brake caliper assembly with new copper washers. Align the brake hose tab between the protrusions on the caliper assembly as shown. Tighten union bolt to the specified torque.

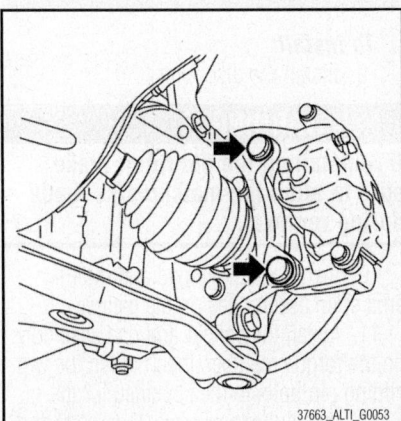

37663_ALTI_G0053

Fig. 5 Remove torque member bolts, and remove brake caliper assembly

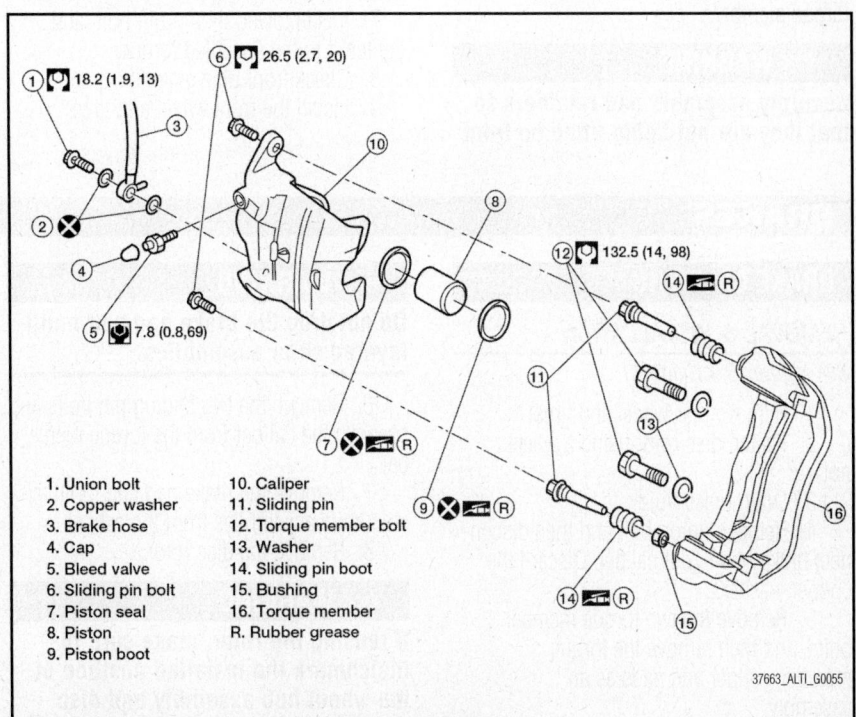

1. Union bolt
2. Copper washer
3. Brake hose
4. Cap
5. Bleed valve
6. Sliding pin bolt
7. Piston seal
8. Piston
9. Piston boot
10. Caliper
11. Sliding pin
12. Torque member bolt
13. Washer
14. Sliding pin boot
15. Bushing
16. Torque member
R. Rubber grease

37663_ALTI_G0055

Fig. 6 Exploded view of front brake caliper assembly

✳✳ WARNING

Do not reuse copper washers.

9. Refill with new brake fluid and bleed air from the brake hydraulic system.
10. Check front disc brakes for drag.
11. Install front wheel and tires.

BRAKE PADS

REMOVAL & INSTALLATION

See Figures 7 and 8.

1. Remove the front wheel and tires.
2. Remove lower sliding pin bolt.
3. Hang caliper with a suitable wire, and remove pads, pad retainers, shims, and shim cover from torque member.

✳✳ WARNING

When removing the pad retainer from the torque member, lift it in the direction indicated by the arrow as shown so that it does not deform.

To install:

4. Apply Molykote M-77 grease or equivalent between the outer shim cover and shim; and the inner multilayered shim and inner pad. Install outer shim, outer shim cover to outer pad, and inner multilayered shim to inner pad.
5. Apply Molykote 7439 grease or equivalent between pad retainers and pad ends. Install pad retainers and pads on torque member.

✳✳ WARNING

Securely assemble pad retainers so that they are not being lifted up from

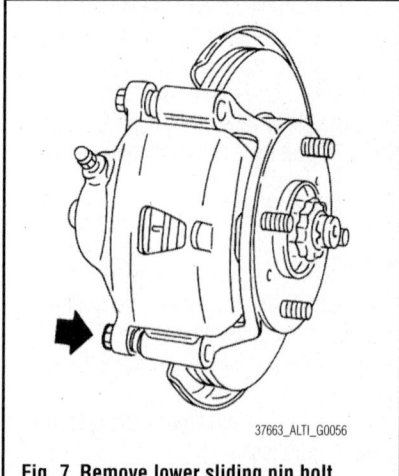

Fig. 7 Remove lower sliding pin bolt

37663_ALTI_G0056

torque member. Both inner and outer pads have a pad return system on the pad retainer. Install pad return lever securely to pad wear sensor.

6. Install caliper over assembled pads on to the torque member.

✳✳ WARNING

When replacing the pads, check brake fluid level in the reservoir tank because brake fluid returns to master cylinder reservoir tank when the piston is compressed.

➡ Use a disc brake piston tool (commercial service tool) to easily press in the piston.

7. Install lower sliding pin bolt, and tighten it to the specified torque.
8. Check front disc brake for drag.
9. Install the front wheel and tires.

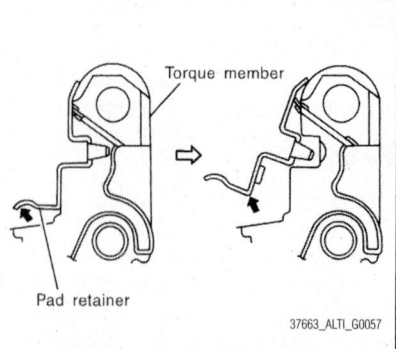

Torque member

Pad retainer

37663_ALTI_G0057

Fig. 8 When removing the pad retainer from the torque member, lift it in the direction indicated by the arrow as shown so that it does not deform

BRAKE BURNISHING PROCEDURE

Burnish contact surfaces between disc rotors and pads according to following procedure after refinishing or replacing rotors, after replacing pads, or if a soft pedal occurs at very low mileage.

✳✳ CAUTION

Be careful of vehicle speed because the brake does not operate easily until pad and disc rotor are securely fitted. Only perform this procedure under safe road and traffic conditions. Use extreme caution.

1. Drive vehicle on straight, flat road.
2. Depress brake pedal with the power to stop vehicle within 3 to 5 seconds until the vehicle stops.
3. Drive without depressing brake for a few minutes to cool the brake.
4. Repeat steps 1 to 3 until pad and disc rotor are securely fitted.

BRAKES

BRAKE CALIPERS

REMOVAL & INSTALLATION

See Figures 9 through 11.

1. Remove rear wheel and tires.
2. Fasten disc rotor using a wheel nut.
3. Drain brake fluid.
4. Remove union bolt and then disconnect brake hose from caliper. Discard the copper washers.
5. Remove the two torque member bolts, and then remove the torque member, caliper and pads as an assembly.

✳✳ WARNING

Do not drop the brake pad and multi-layered shim assemblies.

6. Remove the two sliding pin bolts and separate the caliper from the torque member.
7. Remove the brake pad and multilayered shim assemblies from the caliper.
8. Remove the disc rotor.

✳✳ WARNING

If reusing the rotor, make sure to matchmark the installed position of the wheel hub assembly and disc rotor.

REAR DISC BRAKES

To install:

9. Install the disc rotor.

✳✳ WARNING

If reinstalling the old rotor, make sure to align the matchmarks made during removal.

10. Install the brake pad and multilayered shim assemblies on the caliper.
11. Install the caliper and pad assembly on the torque member, then tighten the two sliding pin bolts to the specified torque.
12. Install the torque member, pads and brake caliper assembly, and tighten the torque member bolts (A) to the specified torque.

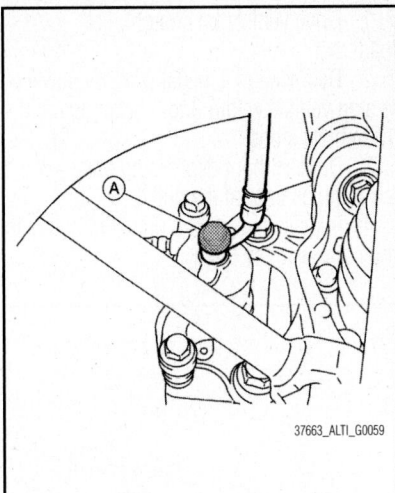

Fig. 9 Remove union bolt (A) and then disconnect brake hose from caliper

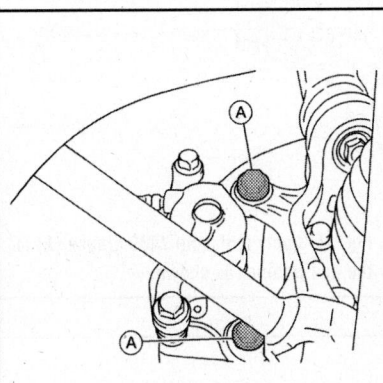

Fig. 10 Remove the two torque member bolts (A), and then remove the torque member, caliper and pads as an assembly

✲✲ WARNING

Before installing, wipe off all oil and moisture on all mating surfaces of rear axle and torque member, threads, bolts and washers.

13. Align the L-shaped pin on the brake hose in the hole in the caliper, then install the brake hose with new copper washers and tighten the union bolt to the specified torque.

✲✲ WARNING

Do not reuse copper washers.

14. Refill with new brake fluid and bleed air.
15. Check rear disc brake for drag.
16. Install rear wheel and tires.

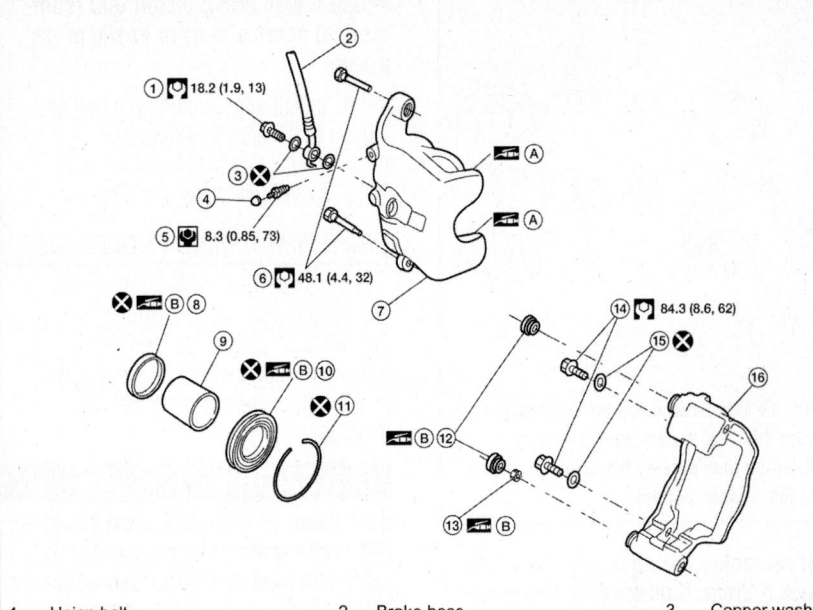

1. Union bolt
2. Brake hose
3. Copper washer
4. Cap
5. Bleed valve
6. Sliding pin bolt
7. Caliper
8. Piston seal
9. Piston
10. Piston boot
11. Retaining ring
12. Sliding pin boot
13. Bushing
14. Torque member bolt
15. Washer
16. Torque member
A. PBC (Poly Butyl Cuprysil) grease or silicone-based grease
B. Rubber grease

Fig. 11 Exploded view of rear disc brake assembly

BRAKE PADS

REMOVAL & INSTALLATION
See Figures 12 through 16.

1. Remove rear wheel and tires.
2. Remove upper sliding pin bolt and swing caliper out supporting it with a suitable wire.

✲✲ WARNING

Do not twist or stretch the brake hose.

3. Remove pads, pad retainers and multilayered shims from torque member.

✲✲ WARNING

When removing the pad retainer from the torque member, lift it in the direction indicated by the arrow as shown so that it does not deform.

To install:
4. Apply Molykote M-77 grease or equivalent to between multilayered shims and brake pads. Install inner multilayered shim to inner pad, and outer multilayered shim to outer pad.
5. Apply Molykote 7439 grease to the pad retainer as shown.

6. Attach pad retainers to torque member, then install brake pads and multilayered shim assemblies.

✲✲ WARNING

When attaching pad retainer, attach it firmly so that it is flush with torque member as shown.

7. Press in piston until pads can be installed, and then install caliper to torque member.

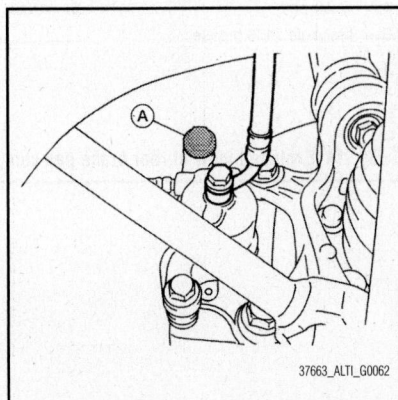

Fig. 12 Remove upper sliding pin bolt (A) and swing caliper out supporting it with a suitable wire

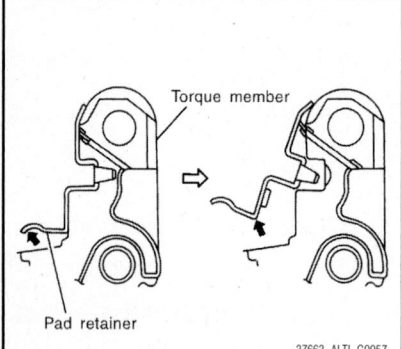

37663_ALTI_G0057

Fig. 13 When removing the pad retainer from the torque member, lift it in the direction indicated by the arrow as shown so that it does not deform

➡️**If replacing the pads with new ones, check a brake fluid level in the reservoir tank because brake fluid returns to master cylinder reservoir tank when you compress the piston.**

➡️**Use a disc brake piston tool (commercial service tool) to easily press piston.**

8. Install upper sliding pin bolt and tighten to the specified torque.
9. Check rear disc brake for drag.
10. Install rear wheel and tires.

BRAKE BURNISHING PROCEDURE

Burnish contact surfaces between disc rotors and pads according to following procedure after refinishing or replacing rotors, after replacing pads, or if a soft pedal occurs at very low mileage.

✳️ CAUTION

Be careful of vehicle speed because the brake does not operate easily until pad and disc rotor are securely fitted. Only perform this procedure under safe road and traffic conditions. Use extreme caution.

1. Drive vehicle on straight, flat road.
2. Depress brake pedal with the power to stop vehicle within 3 to 5 seconds until the vehicle stops.
3. Drive without depressing brake for a few minutes to cool the brake.
4. Repeat steps 1 to 3 until pad and disc rotor are securely fitted.

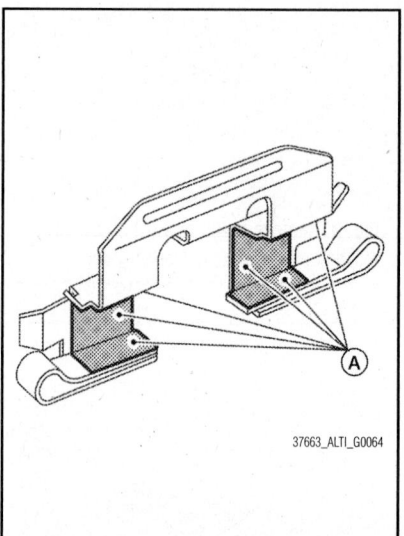

37663_ALTI_G0064

Fig. 15 Apply Molykote 7439 grease (A) to the pad retainer as shown

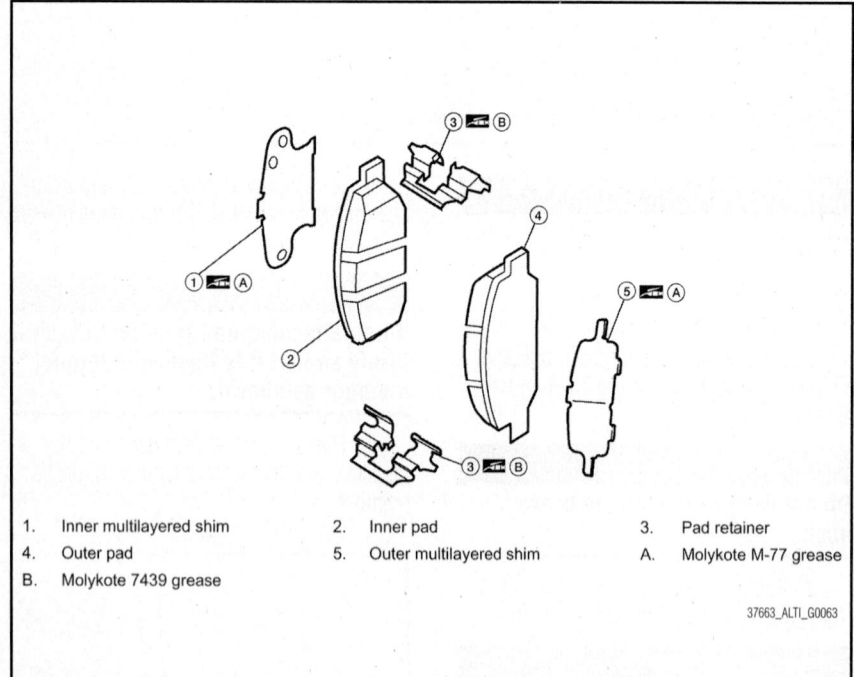

1. Inner multilayered shim
4. Outer pad
B. Molykote 7439 grease

2. Inner pad
5. Outer multilayered shim

3. Pad retainer
A. Molykote M-77 grease

37663_ALTI_G0063

Fig. 14 Exploded view of rear brake pad components

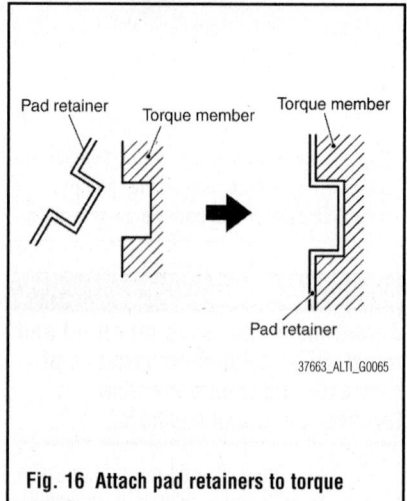

37663_ALTI_G0065

Fig. 16 Attach pad retainers to torque member

BRAKES

ADJUSTMENTS

CABLES

Pedal Type

See Figures 17 through 19.

1. Remove rear wheel and tires.

2. Insert a deep socket wrench onto adjusting nut. Rotate adjusting nut to fully loosen cable, and then release parking brake pedal.

3. Secure disc rotor to hub using wheel nut so as not to tilt disc rotor.

4. Remove adjuster hole plug installed on the disc rotor. Turn the adjuster in direction using a suitable tool or a flat-bladed screwdriver as shown, until disc rotor is locked. Turn the adjuster in the opposite direction by 5 or 6 notches after locking.

5. Rotate disc rotor to make sure that there is no drag. Install the adjuster hole plug.

6. Adjust parking brake cable with the following procedure.

   a. Operate parking brake pedal 10 or more times with a full stroke of 7.6 inches (194.3 mm).

   b. Rotate adjusting nut to adjust parking brake pedal stroke using a deep socket wrench.

   c. Operate parking brake pedal with a force of 66 lbs. (294 N), make sure the pedal stroke is within the specified number (4 to 5) of notches. Check it by listening and counting the ratchet clicks.

   d. Make sure that there is no drag on the parking brake with the parking brake pedal completely released.

Lever Type

See Figures 20 through 22.

1. Fully engage the control lever.

2. Loosen the parking brake cable adjusting nut and fully release the control lever.

3. Adjust clearance of the rear parking brake shoes.

4. Depress the brake pedal fully more than five times.

5. Make sure that no drag exists while rotating the rear wheel and tires.

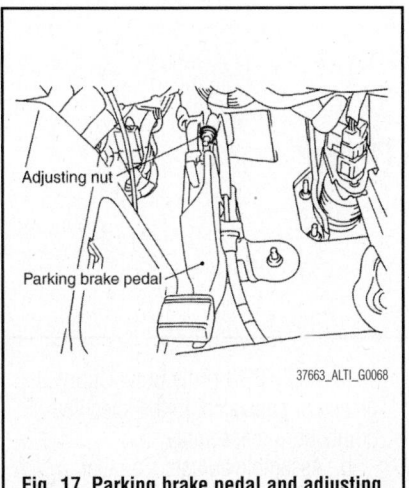

Fig. 17 Parking brake pedal and adjusting nut

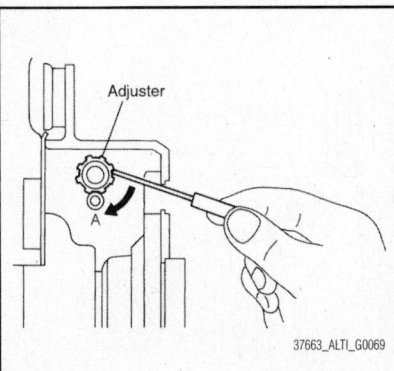

Fig. 18 Turn the adjuster in direction (A) using a suitable tool or a flat-bladed screwdriver as shown, until disc rotor is locked

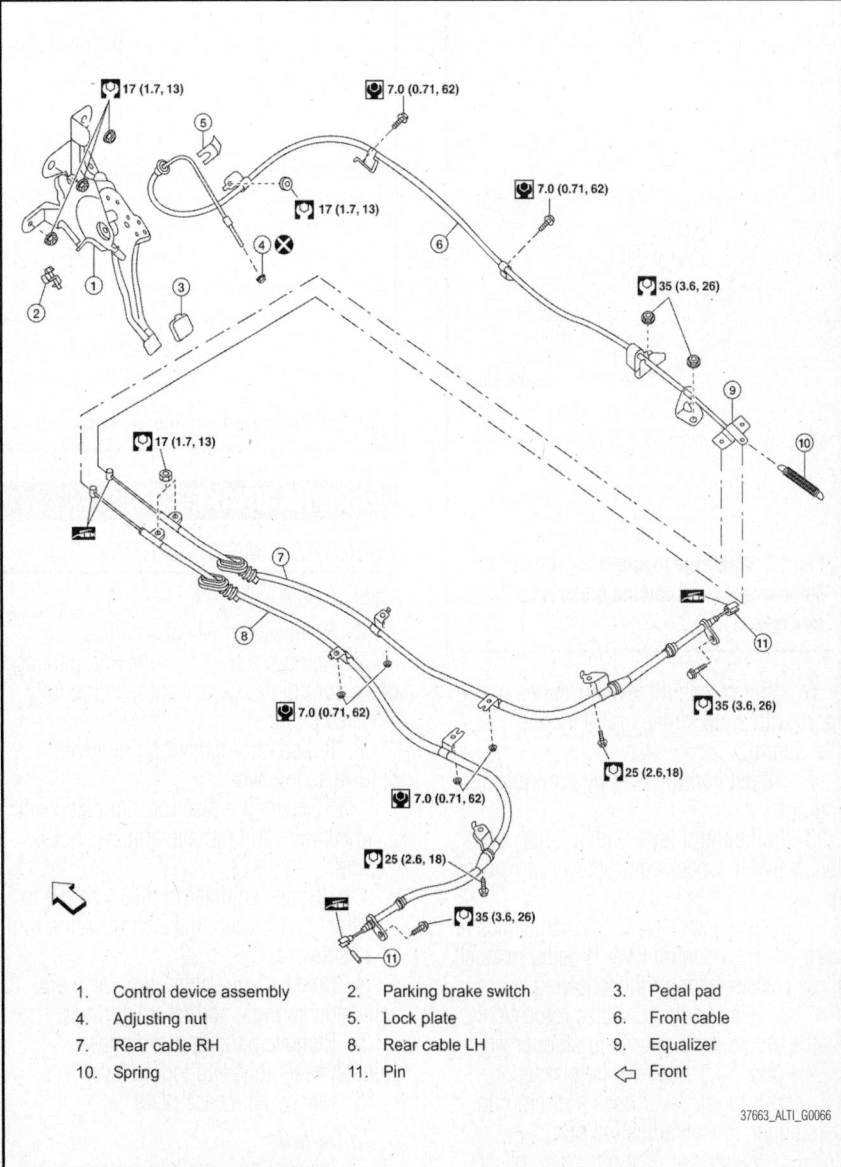

1.	Control device assembly	2.	Parking brake switch	3. Pedal pad
4.	Adjusting nut	5.	Lock plate	6. Front cable
7.	Rear cable RH	8.	Rear cable LH	9. Equalizer
10.	Spring	11.	Pin	Front

Fig. 19 Pedal type parking brake control

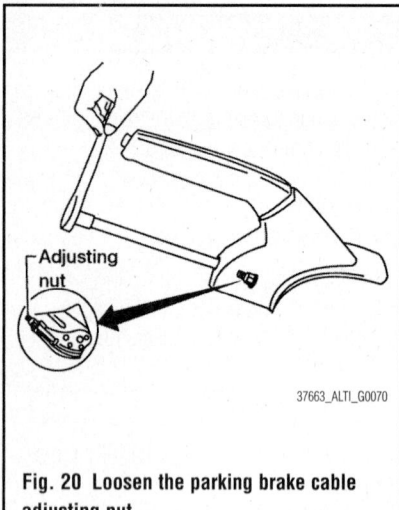

Fig. 20 Loosen the parking brake cable adjusting nut

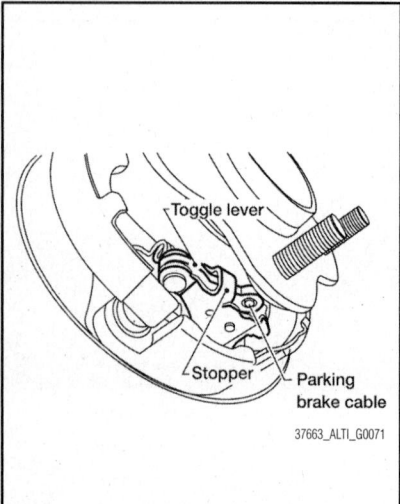

Fig. 21 Verify the toggle lever returns to stopper when the parking brake lever is released

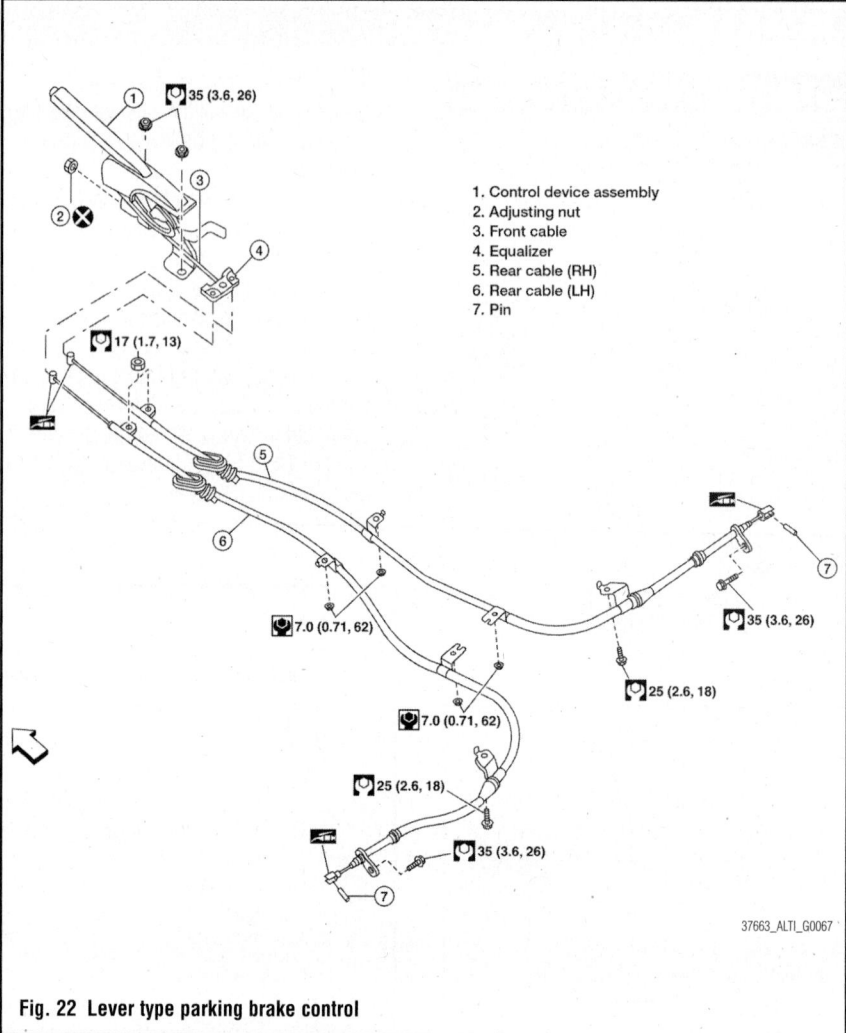

1. Control device assembly
2. Adjusting nut
3. Front cable
4. Equalizer
5. Rear cable (RH)
6. Rear cable (LH)
7. Pin

Fig. 22 Lever type parking brake control

6. Operate control lever 10 times or more with a full stroke of 3.9 inches (99.5 mm).

7. Adjust control lever by turning adjusting nut.

8. Pull control lever with a force of 66 lbs. (294 N). Check control lever stroke and ensure smooth operation.

9. After adjustment, check that there is no drag while the control lever is being released. If drag exists, perform the following:

a. Remove the rear disc rotor. Verify the toggle lever returns to stopper when the parking brake lever is released.

b. If toggle lever does not return to stopper, loosen adjusting nut.

c. Install rear disc rotor and adjust shoe clearance.

PARKING BRAKE SHOES

REMOVAL & INSTALLATION

See Figures 23 and 24.

1. Remove rear wheel and tires.

2. Remove rear disc rotor with parking brake control device assembly in the fully released position.

3. If disc rotor cannot be removed, remove as follows:

a. Secure the disc rotor in place with wheel nuts and remove adjuster hole plug.

b. Rotate adjuster in direction (B) to retract and loosen brake shoe, using tool as shown.

4. Remove anti-rattle pins, retainers, anti-rattle springs, and return springs.

5. Remove parking brake shoes, adjuster assembly, and toggle lever.

6. Remove the back plate.

To install:

7. Installation is in the reverse order of removal. Note the following:

a. Apply PBC (Poly Butyl Cuprysil) grease or equivalent to the specified points during assembly.

b. Assemble adjusters so that threaded part is expanded when rotating it in the direction shown.

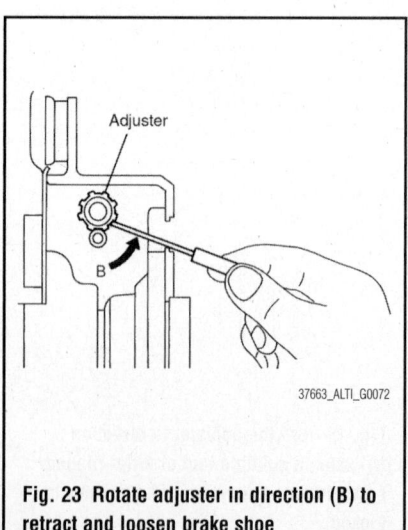

Fig. 23 Rotate adjuster in direction (B) to retract and loosen brake shoe

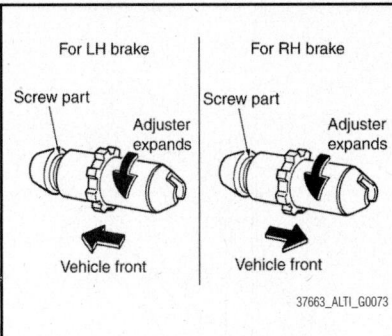

Fig. 24 Shorten adjuster by rotating it as shown

c. Shorten adjuster by rotating it as shown.

d. Check shoe sliding surface and drum inner surface for grease. Wipe it off if it adhere on the surfaces.

e. Perform break-in operation as follows after replacing brake shoes or disc rotors, or if brakes do not function well.

8. Adjust parking brake control device assembly stroke to the specified amount.

9. Perform parking brake break-in (drag run) operation by driving vehicle under the following conditions:

a. Drive the vehicle forward.

b. Maintain vehicle speed at approximately 40 km/h (25 MPH) keeping it constant in forward direction.

c. Apply the parking brake at an operating force of approximately 88 lbs. (400 N) with constant force.

d. Release the parking brake after approximately 10 seconds.

10. Check parking brake control device assembly stroke of parking brake. Readjust as necessary if it is outside the standard specifications.

CHASSIS ELECTRICAL

PRECAUTIONS

The Supplemental Restraint System such as "AIR BAG" and "SEAT BELT PRE-TENSIONER", used along with a front seat belt, helps to reduce the risk or severity of injury to the driver and front passenger for certain types of collision. This system includes seat belt switch inputs and dual stage front air bag modules. The SRS system uses the seat belt switches to determine the front air bag deployment, and may only deploy one front air bag, depending on the severity of a collision and whether the front occupants are belted or unbelted.

Information necessary to service the system safely is included in the SR and SB section of this Service Manual.

Note the following:

• To avoid rendering the SRS inoperative, which could increase the risk of personal injury or death in the event of a

AIR BAGS (SUPPLEMENTAL RESTRAINT SYSTEM)

collision which would result in air bag inflation, all maintenance must be performed by an authorized NISSAN/INFINITI dealer.

• Improper maintenance, including incorrect removal and installation of the SRS, can lead to personal injury caused by unintentional activation of the system. For removal of Spiral Cable and Air Bag Module, see the SR section.

• Do not use electrical test equipment on any circuit related to the SRS unless instructed to in this Service Manual. SRS wiring harnesses can be identified by yellow and/or orange harnesses or harness connectors.

DISARMING THE SYSTEM

✳✳ CAUTION

Before servicing the SRS, turn ignition switch OFF, disconnect both bat-

tery cables and wait at least 3 minutes. For approximately 3 minutes after the cables are removed, it is still possible for the air bag and seat belt pretensioner to deploy. Therefore, do not work on any SRS connectors or wires until at least 3 minutes have passed.

ARMING THE SYSTEM

When the repair work is completed, re-connect both battery cables. With the brake pedal released, turn the push-button ignition switch from ACC position to ON position, then to LOCK position. (The steering wheel will lock when the push-button ignition switch is turned to LOCK position.)

DRIVETRAIN

CONTINUALLY VARIABLE TRANSMISSION (CVT)

DRAIN & REFILL

1. Warm up CVT fluid by driving the vehicle for 10 minutes.

2. Drain CVT fluid from CVT fluid cooler hose (outlet side) and refill with new CVT fluid at CVT fluid charging pipe with the engine running at idle speed.

3. Refill until new CVT fluid comes out from CVT fluid cooler hose (outlet side).

➡**About 30 to 50% extra fluid will be required for this procedure.**

• Use only Genuine NISSAN CVT Fluid NS-2. Do not mix with other fluid.

• Using CVT fluid other than Genuine NISSAN CVT Fluid NS-2 will deteriorate in driveability and CVT durability, and may damage the CVT, which is not covered by the warranty.

• When filling CVT fluid, take care not to scatter heat generating parts such as exhaust.

• Delete CVT fluid deterioration date with CONSULT-III after changing CVT fluid.

4. Check fluid level and condition.

HYBRID VEHICLE TRANSAXLE

DRAIN & REFILL

See Figures 25 and 26.

1. Remove the filler plug and gasket from the transaxle.

2. Remove the drain plug and gasket and drain the fluid from the transaxle.

3. Install the drain plug with a new gasket to the transaxle.

✳✳ WARNING

Do not reuse gasket.

4. Fill the transaxle with specified fluid to the fluid level "A" as shown.

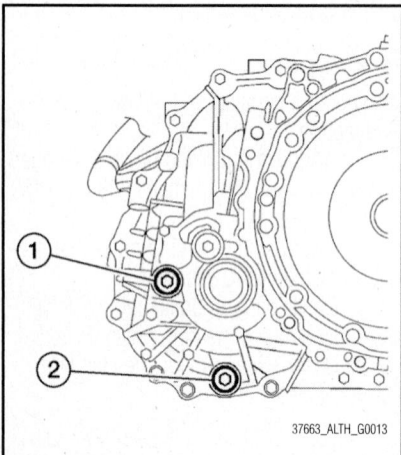

Fig. 25 Filler plug (1) and drain plug (2) locations

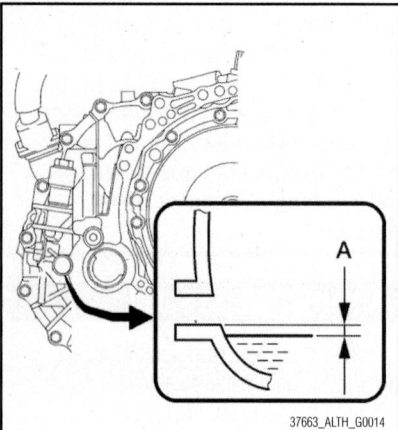

Fig. 26 Fill the transaxle with specified fluid to the fluid level "A" as shown

➡ **"A" maximum is 0.20 inches (5 mm).**

- Use only the specified Genuine Nissan fluid, using fluid other than the Genuine Nissan specified fluid will deteriorate in driveability and durability, and may damage the transaxle, which is not covered by the warranty.
- Do not spill fluid on heat generating parts such as exhaust manifold.
- Do not overfill the transaxle.
- Recheck the fluid level after driving the vehicle to warm up the fluid.

5. Install the filler plug with a new gasket to the transaxle. Tighten to 29 ft. lbs. (39 Nm).

✸✸ WARNING

Do not reuse gasket.

MANUAL TRANSAXLE

DRAIN & REFILL

1. Start engine and let it run to warm up transaxle oil.
2. Stop engine and remove the drain plug to drain the oil.
3. Install the drain plug with a new gasket to the transaxle case. Tighten the drain plug to the specified torque.

✸✸ WARNING

Do not reuse gasket.

4. Remove the filler plug and fill transaxle with new oil.
5. After refilling oil, measure oil level to check if it is within the specification using suitable gauge.

✸✸ CAUTION

Do not start engine while checking oil level. Insert the suitable gauge straight and against the wall of the filler plug hole, then measure the gauge from the top of the filler plug hole to the oil level.

6. Install the filler plug with a new O-ring to the clutch housing.

✸✸ WARNING

Do not reuse O-ring.

7. Tighten filler plug bolt to the specified torque.

FLUID RECOMMENDATIONS

Genuine NISSAN Manual Transmission Fluid (MTF) HQ Multi 75W-85 or equivalent is recommended. If Genuine NISSAN Manual Transmission Fluid (MTF) HQ Multi 75W-85 is hard to obtain, API GL-4, Viscosity SAE 75W-85 may be used as a temporary replacement. However use Genuine NISSAN gear oil as soon as it is available.

CLUTCH HYDRAULIC SYSTEM BLEEDING

BLEEDING PROCEDURE

See Figures 27 through 30.

✸✸ WARNING

Do not spill clutch fluid onto painted surfaces. If it spills, wipe up immediately and wash the affected area with water.

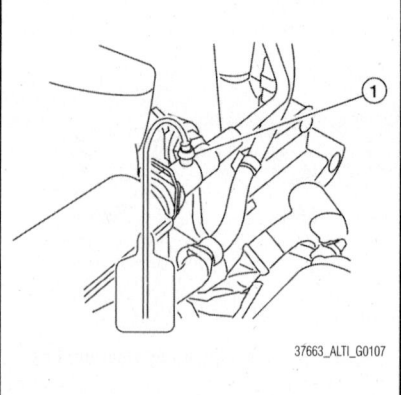

Fig. 27 Connect a transparent vinyl tube and container to the bleeding connector (1) on the CSC

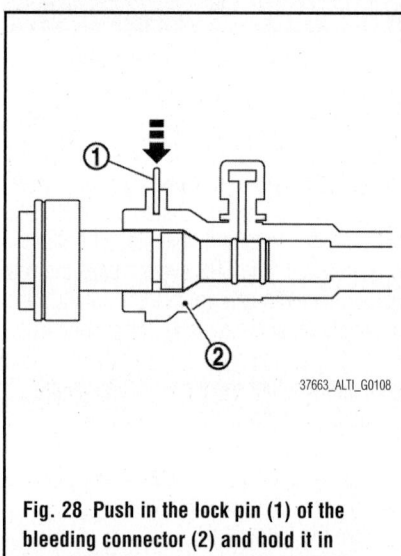

Fig. 28 Push in the lock pin (1) of the bleeding connector (2) and hold it in

Note the following:
- Do not use a vacuum assist or any other type of power bleeder on this system. Use of vacuum assist or power bleeder will not purge all the air from the system.
- Carefully monitor clutch fluid level in reservoir tank during bleeding operation.
- First bleed the air from the bleeding connector on the CSC and then from the air bleed connector valve.

1. Fill master cylinder reservoir tank with new clutch fluid.
2. Connect a transparent vinyl tube and container to the bleeding connector on the CSC.
3. Depress and release the clutch pedal slowly and fully 15 times at an interval of two to three seconds and release the clutch pedal.
4. Bleed the air from the clutch system according to the following:

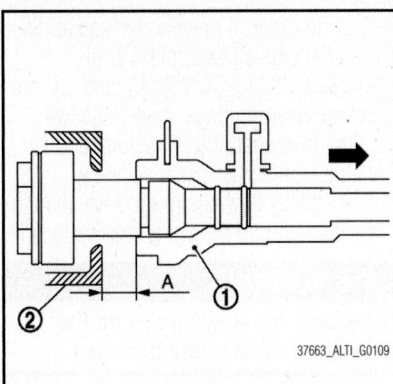

Fig. 29 Slide the bleeding connector (1) away from the transaxle housing (2) to the specified distance (A) of 0.39 inches (10 mm)

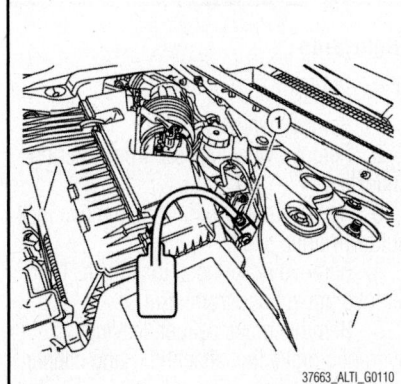

Fig. 30 Connect a transparent vinyl tube and container to the air bleed connector valve (1)

   a. Push in the lock pin of the bleeding connector, and hold it in.

✳✳ CAUTION

Hold the lock pin in to prevent the bleeding connector from separating when fluid pressure is applied.

   b. Slide the bleeding connector away from the transaxle housing to a distance of 0.39 inches (10 mm) to allow air to bleed from the clutch system.
   c. Depress the clutch pedal and hold it down.

➡**Hold the clutch pedal down to prevent air from getting back into the clutch system.**

   5. Return the bleeding connector and lock pin to their original positions.
   6. Release the clutch pedal and wait for five seconds.
   7. Repeat steps 3 through 6 until no air bubbles can be observed in the clutch fluid.

   8. Connect a transparent vinyl tube and container to the air bleed connector valve.
   9. Fully depress the clutch pedal several times.
   10. With clutch pedal depressed, open the air bleed connector valve.
   11. Close the air bleed connector valve.
   12. Release the clutch pedal and wait for five seconds.
   13. Repeat steps 9 through 12 until no air bubble can be observed in the clutch fluid.
   14. Check clutch fluid level in reservoir tank.

FLUID RECOMMENDATIONS

   Genuine NISSAN Super Heavy Duty Brake Fluid or equivalent DOT 3 (US FMVSS No. 116) is recommended and may be found at your local NISSAN dealership.

CLUTCH MASTER CYLINDER

REMOVAL & INSTALLATION

See Figures 31 through 33.

   1. Remove the air cleaner and air duct.
   2. Use one of the following methods to remove hose from master cylinder:

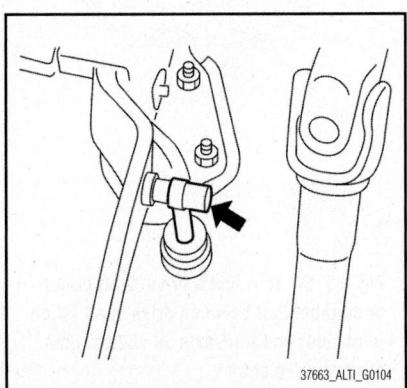

Fig. 31 Remove master cylinder rod end from clutch pedal assembly

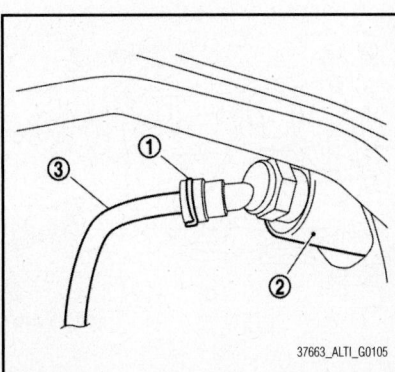

Fig. 32 Remove lock pin (1) from connector of master cylinder (2) and separate clutch tube (3)

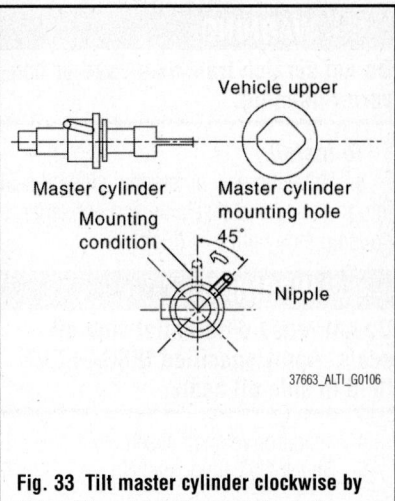

Fig. 33 Tilt master cylinder clockwise by 45° and insert it in the mounting hole

   a. Drain clutch fluid from reservoir tank and remove hose.
   b. Remove hose from master cylinder. Immediately plug hose and reservoir tank to prevent clutch fluid from dripping.

✳✳ WARNING

Do not spill clutch fluid onto painted surfaces. If it spills, wipe up immediately and wash the affected area with water.

   3. Remove master cylinder rod end from clutch pedal assembly.
   4. Remove lock pin from connector of master cylinder and separate clutch tube.
   5. Rotate master cylinder clockwise by 45° and remove from the vehicle.

   To install:
   6. Tilt master cylinder clockwise by 45° and insert it in the mounting hole. Rotate counterclockwise to secure it. At this time, nipple is in the up position.
   7. Install master cylinder rod end to clutch pedal.
   8. Install clutch tube fully into connector of master cylinder.
   9. Install lock pin fully into connector of master cylinder.
   10. Fill with new clutch fluid and bleed clutch hydraulic system.
   11. Inspect clutch pedal operation.
   12. Install the air cleaner and air duct.

FRONT DIFFERENTIAL SIDE OIL SEAL

REMOVAL & INSTALLATION

   1. Remove drive shaft assembly.
   2. Remove the differential side oil seal using suitable tool.

> ✳✳ **WARNING**
>
> **Do not scratch transaxle case or converter housing.**

To install:

3. Drive the new differential side oil seal into the transaxle case side and converter housing side until it is flush.

> ✳✳ **WARNING**
>
> **Do not reuse differential side oil seals. Apply specified NISSAN CVT fluid to side oil seals.**

4. Install drive shaft assembly.
5. Check CVT fluid level.

FRONT HALFSHAFT

REMOVAL & INSTALLATION

Left Side

See Figures 34 through 36.

1. Remove wheel and tire.
2. Remove wheel sensor from steering knuckle.
3. Remove cotter pin. Then remove lock nut from drive shaft.
4. Remove brake hose lock plate. Then remove brake hose from strut.
5. Remove brake caliper, leaving hydraulic brake line attached. Hang caliper aside using wire.
6. Remove front strut to steering knuckle bolts and nuts, then separate steering knuckle front strut.
7. Remove drive shaft from wheel hub and bearing assembly, using a puller or suitable tool.

> ✳✳ **WARNING**
>
> **When removing drive shaft, do not apply an excessive angle to drive shaft joint. Also be careful not to excessively extend slide joint.**

8. Remove the left side drive shaft from the transaxle.
   a. Remove drive shaft from transaxle using Tool and drive shaft puller or suitable tool.
   b. Set Tool KV40107500 and a drive shaft puller or suitable tool between drive shaft (slide joint side) and transaxle as shown, then remove drive shaft.

To install:

9. Installation is in the reverse order of removal. Note the following:

> ✳✳ **WARNING**
>
> **Do not reuse non-reusable parts.**

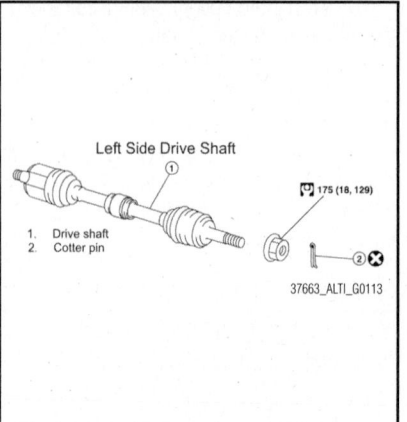

1. Drive shaft
2. Cotter pin

Fig. 34 Left side front drive shaft

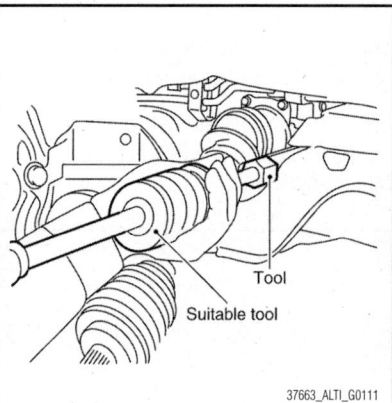

Fig. 35 Set Tool and a drive shaft puller or suitable tool between drive shaft (slide joint side) and transaxle as shown, then remove drive shaft

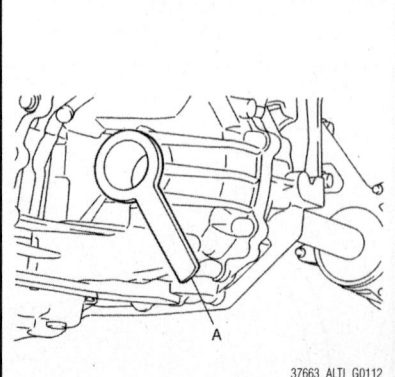

Fig. 36 Place Tool KV38106700 (J-34296) (A) onto oil seal before inserting drive shaft as shown

a. In order to prevent damage to differential side oil seal, place Tool KV38106700 (J-34296) (A) onto oil seal before inserting drive shaft as shown. Slide drive shaft into slide joint and tap with a hammer to install securely.
   b. Install new circlip on drive shaft in the circular clip groove on transaxle side.

> ✳✳ **WARNING**
>
> **Make sure the new circlip on the drive shaft is securely fastened.**

c. After its insertion, try to pull the flange out of the slide joint by hand.

> ✳✳ **WARNING**
>
> **If it pulls out, the circlip is not properly meshed with the transaxle side gear.**

Right Side

See Figures 37 and 38.

1. Remove wheel and tire.
2. Remove wheel sensor from steering knuckle.
3. Remove cotter pin. Then remove lock nut from drive shaft.
4. Remove brake hose lock plate. Then remove brake hose from strut.
5. Remove brake caliper, leaving hydraulic brake line attached. Hang caliper aside using wire.
6. Remove front strut to steering knuckle bolts and nuts, then separate steering knuckle front strut.
7. Remove drive shaft from wheel hub and bearing assembly, using a puller or suitable tool.

> ✳✳ **WARNING**
>
> **When removing drive shaft, do not apply an excessive angle to drive shaft joint. Also be careful not to excessively extend slide joint.**

8. Remove the retaining bracket bolts, and separate drive shaft from transaxle.

To install:

9. Installation is in the reverse order of removal. Note the following:

> ✳✳ **WARNING**
>
> **Do not reuse non-reusable parts.**

a. Tighten retaining bracket bolts and support bearing bracket bolts to specifications.
   b. For QR25DE models, install the retaining bracket with the notch facing up and follow the bolt tightening order.

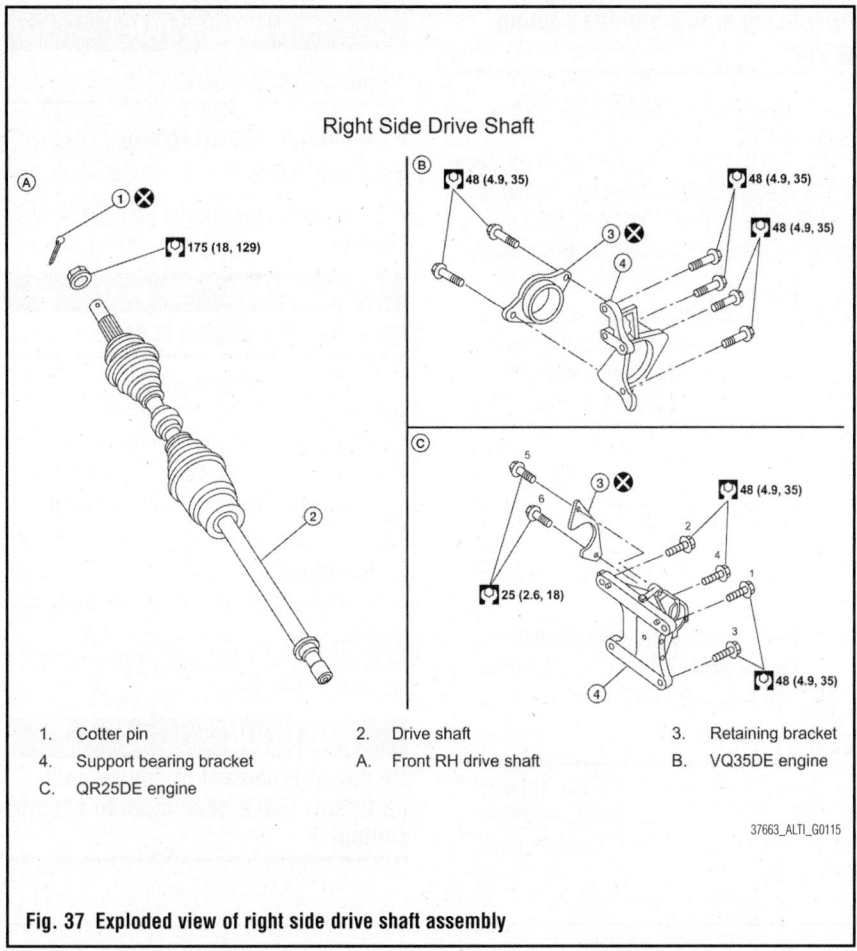

Right Side Drive Shaft

1. Cotter pin
4. Support bearing bracket
C. QR25DE engine
2. Drive shaft
A. Front RH drive shaft
3. Retaining bracket
B. VQ35DE engine

37663_ALTI_G0115

Fig. 37 Exploded view of right side drive shaft assembly

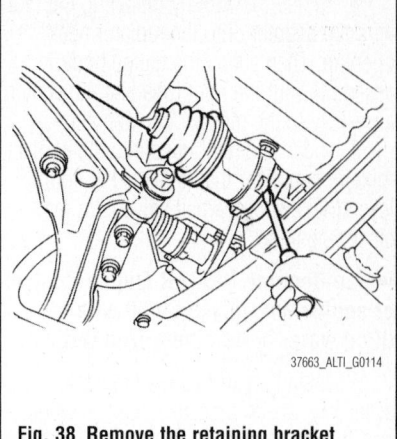

37663_ALTI_G0114

Fig. 38 Remove the retaining bracket bolts, and separate drive shaft from transaxle

c. In order to prevent damage to differential side oil seal, place tool onto oil seal before inserting drive shaft as shown. Slide drive shaft into slide joint and tap with a hammer to install securely.

d. Install new circlip on drive shaft in the circular clip groove on transaxle side.

✳✳ WARNING

Make sure the new circlip on the drive shaft is securely fastened.

e. After its insertion, try to pull the flange out of the slide joint by hand.

✳✳ WARNING

If it pulls out, the circlip is not properly meshed with the transaxle side gear.

ENGINE COOLING

ENGINE COOLANT

DRAIN & REFILL

Draining Engine Coolant

✳✳ CAUTION

To avoid being scalded, never change the coolant when the engine is hot. Wrap a thick cloth around cap and carefully remove the cap. First, turn the cap a quarter of a turn to release built-up pressure. Then push down and turn the cap all the way to remove.

1. Remove the engine undercover.
2. Open the radiator drain plug at the bottom of the radiator and remove the radiator filler cap. This is the only step required when partially draining the cooling system (radiator only).

✳✳ CAUTION

Do not allow the coolant to contact the drive belts.

3. Follow this step for heater core removal/replacement only. Disconnect the upper heater hose at the engine side and apply moderate air pressure into the hose for 30 seconds to blow the excess coolant out of the heater core.
4. When draining all of the coolant in the system, remove the reservoir tank and drain the coolant, then clean the reservoir tank before installation.

✳✳ CAUTION

Do not allow the coolant to contact the drive belts.

5. When draining all of the coolant in the system for engine removal or repair, open the drain plug on the cylinder block.

6. Check the drained coolant for contaminants such as rust, corrosion or discoloration. If the coolant is contaminated, flush the engine cooling system.

Refilling Engine Coolant

1. Install the radiator drain plug. Install the reservoir tank and cylinder block drain plug, if removed for a total system drain or for engine removal or repair.
   a. The radiator must be completely empty of coolant and water.
   b. Apply sealant to the threads of the cylinder block drain plugs. Use Genuine High Performance Thread Sealant or equivalent.
2. If disconnected, reattach the upper radiator hose at the engine side.
3. Set the vehicle heater controls to the full HOT and heater ON position. Turn the vehicle ignition ON with the engine OFF as necessary to activate the heater mode.

4. Install the Tool by installing the radiator cap adapter onto the radiator neck opening. Then attach the gauge body assembly with the refill tube and the venturi assembly to the radiator cap adapter.

5. Insert the refill hose into the coolant mixture container that is placed at floor level. Make sure the ball valve is in the closed position.

➡Use Genuine NISSAN Engine Coolant or equivalent, mixed 50/50 with distilled water or demineralized water.

6. Install an air hose to the venturi assembly, the air pressure must be within specification.

✳✳ CAUTION
The compressed air supply must be equipped with an air dryer.

7. The vacuum gauge will begin to rise and there will be an audible hissing noise. During this process open the ball valve on the refill hose slightly. Coolant will be visible rising in the refill hose. Once the refill hose is full of coolant, close the ball valve. This will purge any air trapped in the refill hose.

8. Continue to draw the vacuum until the gauge reaches 28 inches of vacuum. The gauge may not reach 28 inches in high altitude locations, use the vacuum specifications based on the altitude above sea level.

    a. 0–328 ft. (100 m): 28 inches of vacuum

    b. 984 ft. (300 m): 27 inches of vacuum

    c. 1,641 ft. (500 m): 26 inches of vacuum

    d. 3,281 ft. (1,000 m): 24–25 inches of vacuum

9. When the vacuum gauge has reached the specified amount, disconnect the air hose and wait 20 seconds to see if the system loses any vacuum. If the vacuum level drops, perform any necessary repairs to the system and repeat steps 6 to 8 to bring the vacuum to the specified amount. Recheck for any leaks.

10. Place the coolant container (with the refill hose inserted) at the same level as the top of the radiator. Then open the ball valve on the refill hose so the coolant will be drawn up to fill the cooling system. The cooling system is full when the vacuum gauge reads zero.

✳✳ CAUTION
Do not allow the coolant container to get too low when filling, to avoid air

from being drawn into the cooling system.

11. Remove the Tool from the radiator neck opening.

12. Fill the cooling system reservoir tank to the specified level and install the radiator cap. Run the engine to warm up the cooling system and top up the system as necessary.

FLUSHING

1. Fill the radiator from the filler neck above the radiator upper hose and reservoir tank with clean water and reinstall the radiator filler cap.

2. Run the engine until it reaches normal operating temperature.

3. Rev the engine two or three times under no-load.

4. Stop the engine and wait until it cools down.

5. Drain the water from the system.

6. Repeat steps 1 through 5 until clear water begins to drain from the radiator.

LEVEL CHECK

Check if the reservoir tank coolant level is within MIN to MAX when the engine is cool. Adjust coolant level if it is too much or too little.

ELECTRIC ENGINE FAN

REMOVAL & INSTALLATION

2.5L Engine, Except Hybrid
See Figure 39.

1. Drain engine coolant from the radiator.

✳✳ CAUTION
Perform when engine is cold.

2. Remove air cleaner duct assembly.

3. Disconnect upper radiator hose.

4. Disconnect fan motor connectors.

5. Remove radiator cooling fan assembly.

To install:

6. Installation is in the reverse order of removal.

7. After installation refill engine coolant and check for leaks.

✳✳ CAUTION
Do not spill coolant in engine compartment. Use a shop cloth to absorb coolant.

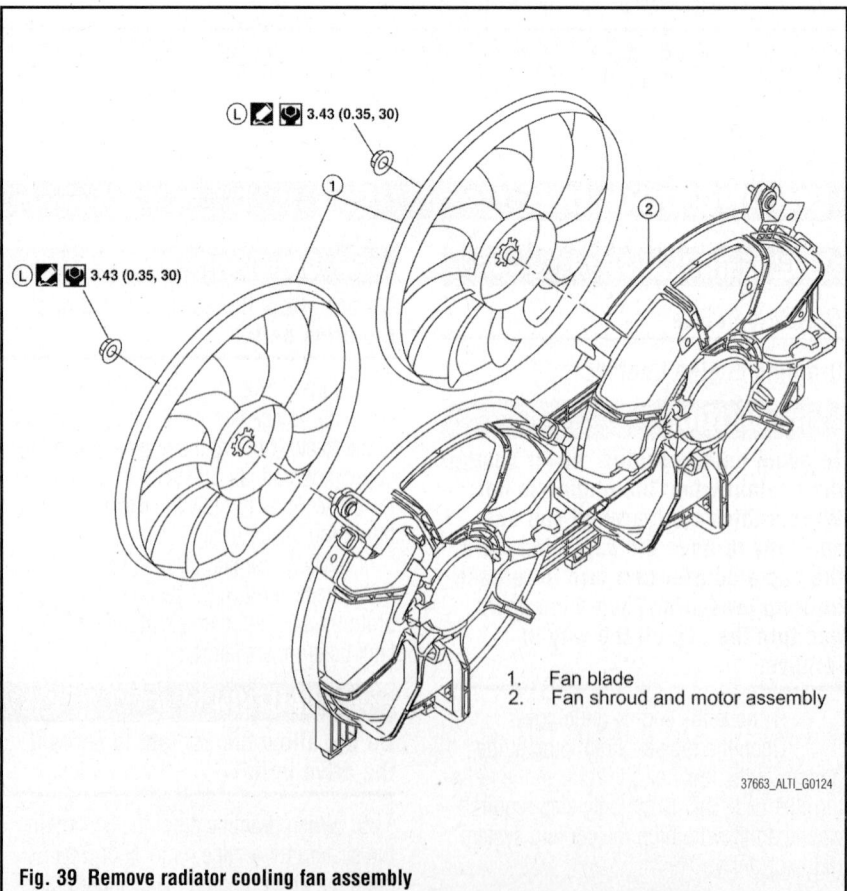

1. Fan blade
2. Fan shroud and motor assembly

37663_ALTI_G0124

Fig. 39 Remove radiator cooling fan assembly

3.5L Engine

See Figure 39.

1. Drain engine coolant from the radiator.

✳✳ CAUTION

Perform when engine is cold.

2. Remove CVT control module (if equipped).
3. Remove battery tray.
4. Disconnect upper radiator hose.
5. Disconnect fan motor connectors.
6. Remove radiator cooling fan assembly.

To install:

7. Installation is in the reverse order of removal.
8. After installation refill engine coolant and check for leaks.

✳✳ CAUTION

Do not spill coolant in engine compartment. Use a shop cloth to absorb coolant.

Hybrid

See Figure 40.

1. Disconnect the 12-volt battery negative terminal.
2. Drain engine coolant from radiator, condenser and liquid tank assembly.

✳✳ WARNING

Perform when engine and inverter are cold.

3. Remove inverter upper bracket.
4. Drain inverter coolant from sub-radiator.
5. Remove air cleaner duct (front).
6. Disconnect radiator upper hose.
7. Remove sub radiator coolant reservoir tank.
8. Disconnect ECM.
9. Remove ECM and bracket assembly.
10. Disconnect radiator cooling fan controller.
11. Remove radiator cooling fan assembly.

To install:

12. Installation is in the reverse order of removal.

➡**Radiator cooling fan is controlled by ECM.**

RADIATOR

REMOVAL & INSTALLATION

Non-Hybrid

See Figure 41.

✳✳ WARNING

Never remove the radiator cap when the engine is hot. Serious burns could occur from high pressure coolant escaping from the radiator. Wrap a thick cloth around the cap. Slowly turn it a quarter turn to allow built-up pressure to escape. Carefully remove the cap by turning it all the way.

1. Drain engine coolant from the radiator.
2. Remove front grille (Sedan only).
3. Remove front bumper fascia (Coupe only).
4. Remove engine undercover.
5. Remove front air duct.
6. Remove A/C condenser.
7. Disconnect upper and lower radiator hoses.
8. Disconnect the CVT oil cooler hoses, if equipped. Plug the hoses to prevent CVT oil loss.
9. Remove radiator.

✳✳ WARNING

Do not damage or scratch the radiator core when removing.

To install:

10. Installation is in the reverse order of removal.

Hybrid

See Figure 42.

✳✳ CAUTION

Never remove the radiator cap when the engine and inverter are hot. Serious burns could occur from high pressure coolant escaping from the radiator, condenser and liquid tank assembly. Wrap a thick cloth around the cap. Slowly turn it a quarter turn to allow built-up pressure to escape. Carefully remove the cap by turning it all the way.

1. Remove engine under cover.
2. Drain engine coolant from radiator, condenser and liquid tank assembly.
3. Remove air cleaner duct (front).
4. Remove radiator, condenser and liquid tank assembly upper hose and lower hose.
5. Remove coolant reservoir hose.
6. Drain inverter coolant from sub-radiator.
7. Remove front bumper reinforcement.
8. Remove sub radiator.

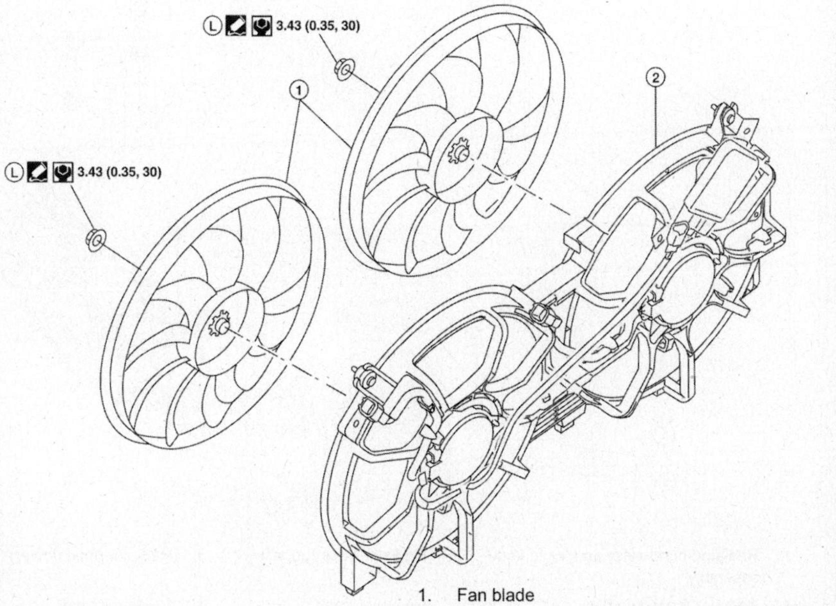

ⓛ 🔧 🔩 3.43 (0.35, 30)

ⓛ 🔧 🔩 3.43 (0.35, 30)

1. Fan blade
2. Radiator cooling fan shroud and motor assembly

37663_ALTH_G0018

Fig. 40 Exploded view of engine fan assembly

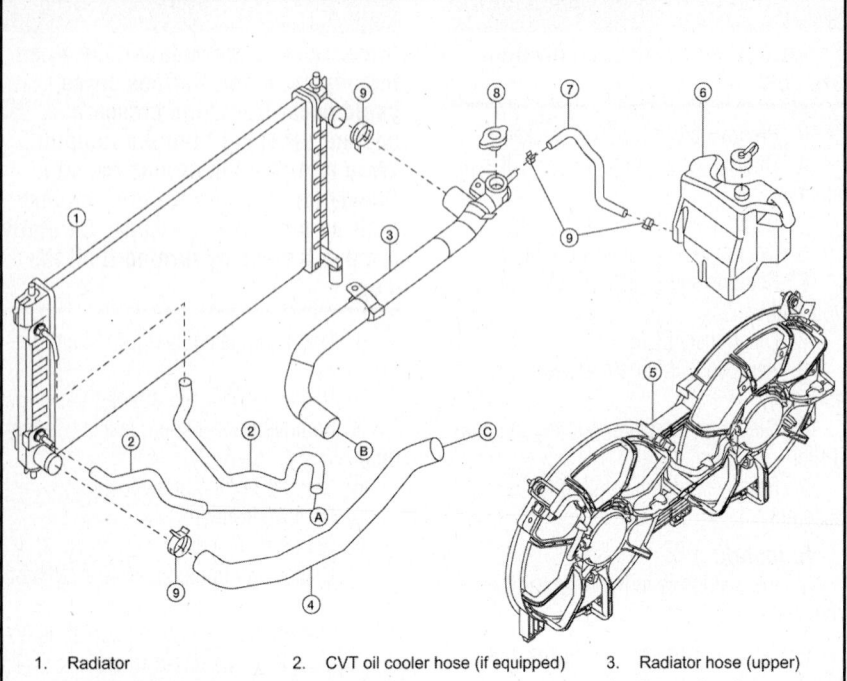

1.	Radiator	2.	CVT oil cooler hose (if equipped)	3.	Radiator hose (upper)
4.	Radiator hose (lower)	5.	Cooling fan	6.	Reservoir tank
7.	Reservoir hose	8.	Radiator filler cap	9.	Clamps
A.	To CVT (if equipped)	B.	To water outlet	C.	To water inlet

37663_ALTI_G0127

Fig. 41 Exploded view of radiator and cooling system assembly

9. Discharge A/C system.

10. Remove high side junction pipe assembly.

11. Remove the refrigerant pressure sensor for installation on new radiator, condenser and liquid tank assembly.

12. Remove both radiator, condenser and liquid tank assembly clips.

13. Remove radiator, condenser and liquid tank assembly.

Note the following:

• Do not damage or scratch the radiator, condenser and liquid tank assembly and sub radiator core when removing.

• When removing refrigerant components from a vehicle, immediately cap (seal) the component to minimize the entry of moisture from the atmosphere.

• When installing refrigerant components to vehicle, never remove the caps (unseal) until just before connecting the components. Connect all refrigerant loop components as quickly as possible to minimize the entry of moisture into system

To install:

14. Installation is in the reverse order of removal.

THERMOSTAT

REMOVAL & INSTALLATION

2.5L Engine, Except Hybrid

See Figure 43.

> ❊❊ **WARNING**
>
> **Never remove the radiator cap when the engine is hot. Serious burns could occur from high pressure coolant escaping from the radiator.**

> ❊❊ **CAUTION**
>
> **Perform when the engine is cold.**

1. Drain engine coolant from the radiator.

2. Remove the air duct.

3. Remove radiator lower hose from the engine coolant inlet side.

4. Remove engine coolant inlet and thermostat.

To install:

5. Installation is in the reverse order of removal.

6. Install the engine coolant temperature sensor.

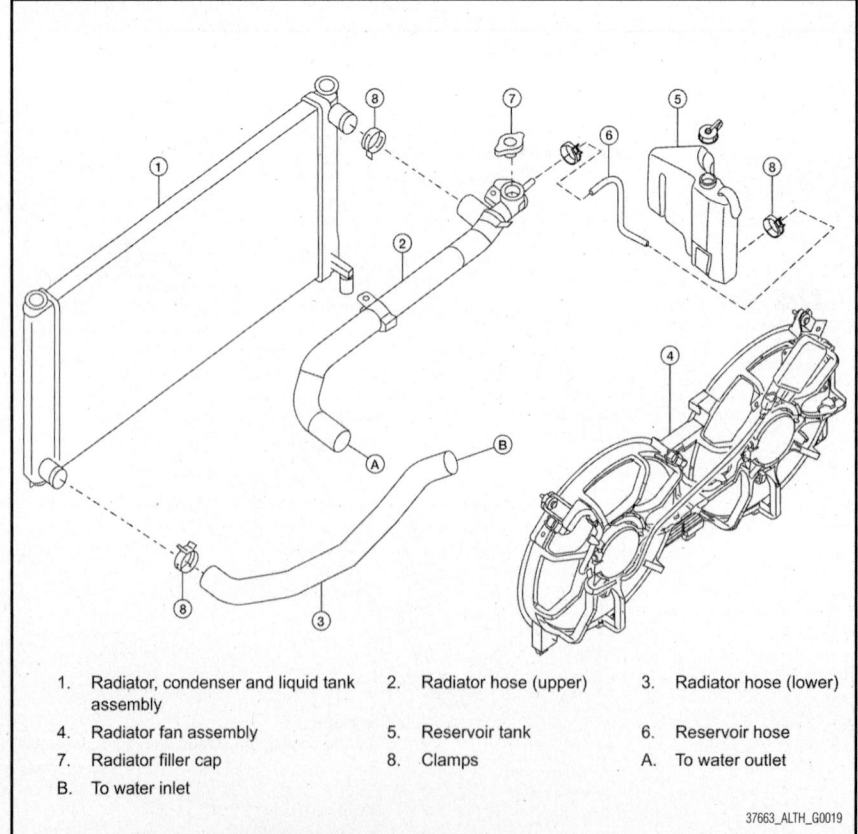

1.	Radiator, condenser and liquid tank assembly	2.	Radiator hose (upper)	3.	Radiator hose (lower)
4.	Radiator fan assembly	5.	Reservoir tank	6.	Reservoir hose
7.	Radiator filler cap	8.	Clamps	A.	To water outlet
B.	To water inlet				

37663_ALTH_G0019

Fig. 42 Exploded view of radiator assembly

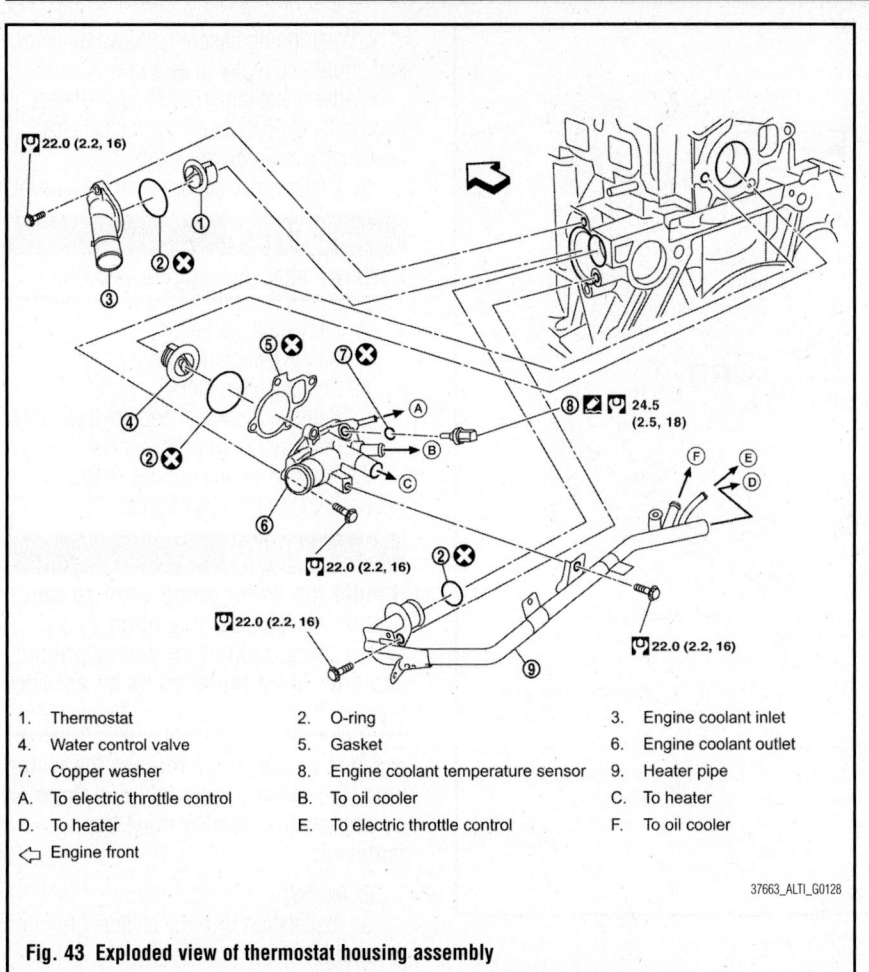

1. Thermostat
2. O-ring
3. Engine coolant inlet
4. Water control valve
5. Gasket
6. Engine coolant outlet
7. Copper washer
8. Engine coolant temperature sensor
9. Heater pipe
A. To electric throttle control
B. To oil cooler
C. To heater
D. To heater
E. To electric throttle control
F. To oil cooler
◁ Engine front

37663_ALTI_G0128

Fig. 43 Exploded view of thermostat housing assembly

➡**Use Genuine RTV Silicone Sealant or equivalent.**

7. Install the thermostat with the whole circumference of the flange part fitting securely inside the rubber ring.

8. Install the thermostat with the jiggle valve facing upwards. The position deviation may be within the range of ±10°.

9. If necessary, to install the heater pipe, first apply a mild detergent to the O-ring and then quickly insert the pipe into the housing.

3.5L Engine

See Figure 44.

✳✳ WARNING

Never remove the radiator cap when the engine is hot. Serious burns could occur from high pressure coolant escaping from the radiator.

✳✳ CAUTION

Perform when engine is cool.

1. Drain engine coolant from the radiator.

2. Remove drive belts.

3. Remove water drain plug on water pump side of the engine.

4. Disconnect lower radiator hose.

5. Remove engine coolant inlet and thermostat assembly.

➡**Do not disassemble engine coolant inlet and thermostat. Replace them as a unit, if necessary.**

To install:

6. Installation is in the reverse order of removal.

7. Install thermostat with jiggle valve facing upward.

8. After installation refill engine coolant and check for leaks.

✳✳ CAUTION

Do not spill coolant in engine compartment. Use a shop cloth to absorb coolant.

Hybrid

See Figure 45.

✳✳ CAUTION

Never remove the radiator cap when the engine is hot. Serious burns could occur from high pressure coolant escaping from the radiator.

✳✳ WARNING

Perform when the engine is cold.

1. Drain engine coolant from radiator, condenser and liquid tank assembly.

2. Remove air cleaner duct (front).

3. Remove radiator lower hose from the engine coolant inlet side.

4. Remove engine coolant inlet and thermostat.

To install:

5. Installation is in the reverse order of removal.

6. Install the engine coolant temperature sensor.

➡**Use Genuine RTV Silicone Sealant or equivalent.**

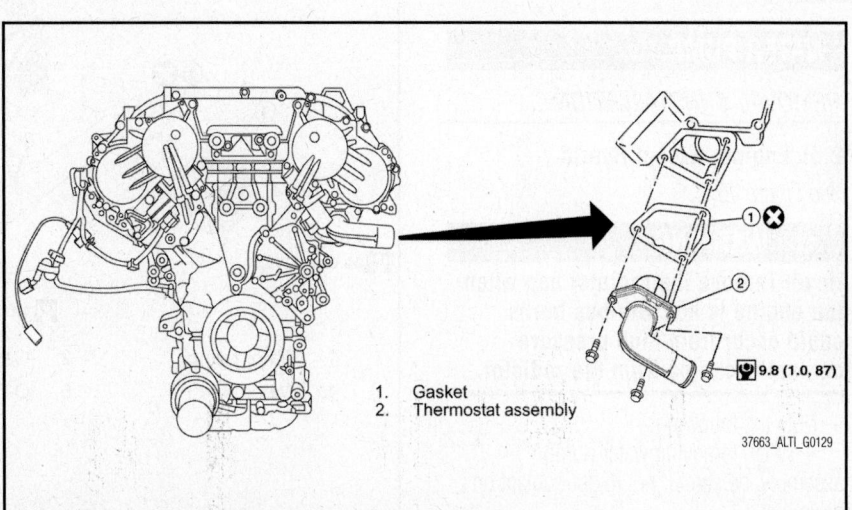

1. Gasket
2. Thermostat assembly

37663_ALTI_G0129

Fig. 44 Exploded view of thermostat housing assembly

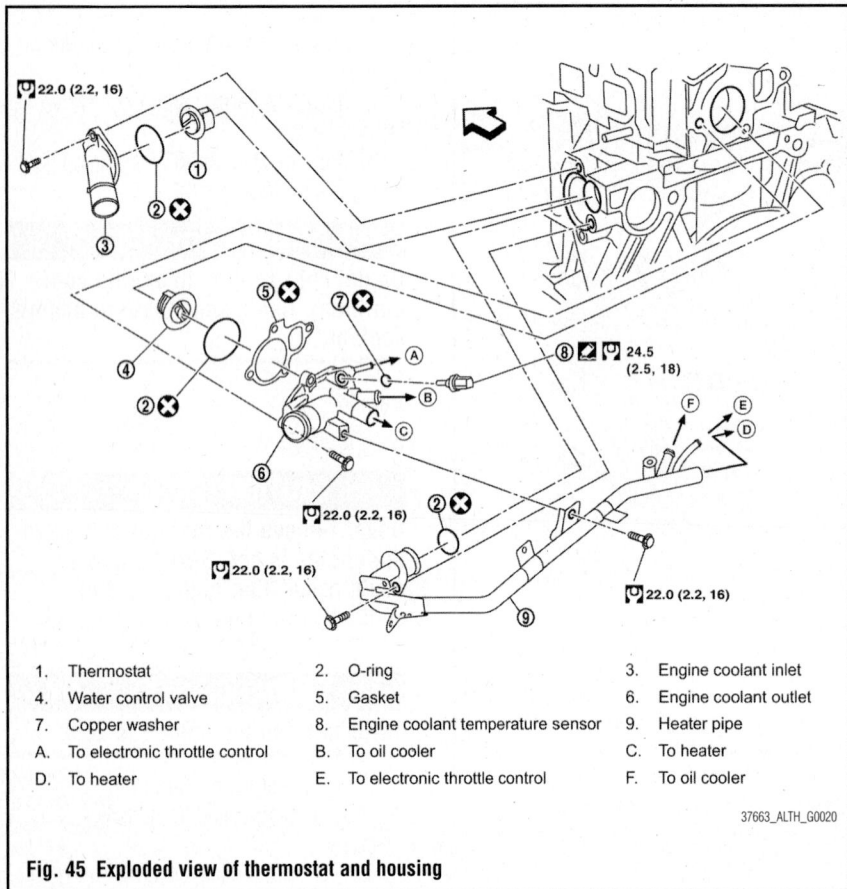

1. Thermostat
2. O-ring
3. Engine coolant inlet
4. Water control valve
5. Gasket
6. Engine coolant outlet
7. Copper washer
8. Engine coolant temperature sensor
9. Heater pipe
A. To electronic throttle control
B. To oil cooler
C. To heater
D. To heater
E. To electronic throttle control
F. To oil cooler

37663_ALTH_G0020

Fig. 45 Exploded view of thermostat and housing

7. Install the thermostat with the whole circumference of the flange part fitting securely inside the rubber ring.

8. Install the thermostat with the jiggle valve facing upwards. The position deviation may be within the range of ±10°.

9. If necessary, to install the heater pipe, first apply a mild detergent to the O-ring and then quickly insert the pipe into the housing.

10. Use a new gasket and O-ring for installation.

WATER PUMP

REMOVAL & INSTALLATION

2.5L Engine, Except Hybrid
See Figure 46.

✳✳ WARNING

Never remove the radiator cap when the engine is hot. Serious burns could occur from high pressure coolant escaping from the radiator.

Note the following:
• When removing water pump assembly, be careful not to get coolant on drive belt.

• Water pump cannot be disassembled and should be replaced as a unit.
• After installing water pump, connect hose and clamp securely, then check for leaks using radiator cap tester.

1. Drain engine coolant from the radiator.

✳✳ CAUTION

Perform when the engine is cold.

2. Remove drive belt.
3. Remove engine cover.
4. Remove alternator.
5. Remove RH wheel and tire assembly.
6. Remove fender protector RH.
7. Remove engine ground strap.
8. Remove the water pump.

✳✳ CAUTION

Handle the water pump vane so that it does not contact any other parts. Water pump cannot be disassembled and should be replaced as an assembly.

➡If it is necessary to remove the water pipe, the exhaust manifold and three way catalyst assembly must be removed.

To install:

9. Installation is in the reverse order of removal.

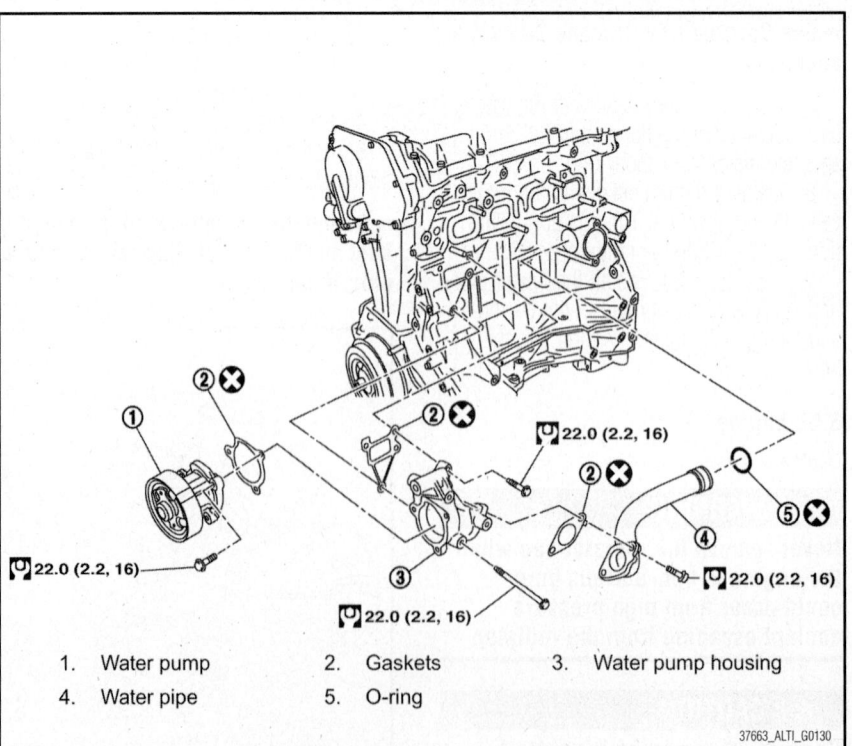

1. Water pump
2. Gaskets
3. Water pump housing
4. Water pipe
5. O-ring

37663_ALTI_G0130

Fig. 46 Exploded view of water pump assembly

10. When inserting water pipe end to cylinder block, apply a neutral detergent to O-ring. Then insert it immediately.

11. After installation refill engine coolant and check for leaks.

3.5L Engine

See Figures 47 through 52.

❊❊ WARNING

Never remove the radiator cap when the engine is hot. Serious burns could occur from high pressure coolant escaping from the radiator.

Note the following:
• When removing water pump assembly, be careful not to get coolant on drive belt.
• Water pump cannot be disassembled and should be replaced as a unit.
• After installing water pump, connect hose and clamp securely, then check for leaks using radiator cap tester.

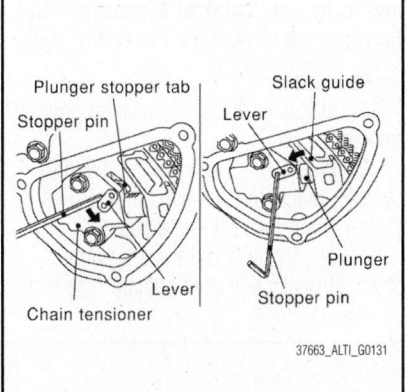

Fig. 47 Insert the stopper pin into the tensioner body hole to hold the lever and keep the plunger stopper tab released

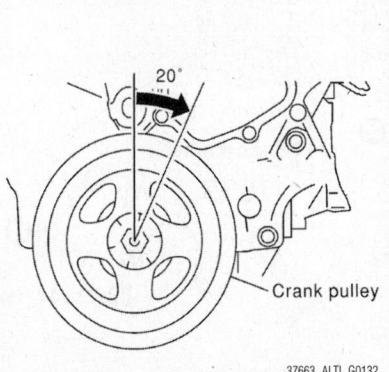

Fig. 48 Make a gap between water pump gear and timing chain, by turning the crankshaft pulley approximately 20° clockwise

1. Drain engine coolant from the radiator.

❊❊ CAUTION

Perform when the engine is cold.

2. Remove engine coolant reservoir tank.

3. Remove RH wheel and tire.

4. Remove the fender protector (RH).

5. Remove drive belts.

6. Remove the drive belt auto tensioner and the idler pulley.

7. Support engine and remove the front engine insulator and bracket.

8. Remove water drain plug on water pump side of cylinder block.

9. Remove IVT control valve cover and water pump cover.

10. Remove the timing chain tensioner assembly.

a. Pull the lever down to release the plunger stopper tab.

b. Insert the stopper pin into the tensioner body hole to hold the lever and keep the plunger stopper tab released.

➡**An Allen wrench is used for a stopper pin as an example.**

c. Insert the plunger stopper tab into the tensioner body by pressing the slack guide.

d. Keep the slack guide pressed and hold the plunger stopper tab in by pushing the stopper pin deeper through the lever and into the chain tensioner body hole.

e. Make a gap between water pump gear and timing chain, by turning the crankshaft pulley approximately 20° clockwise.

11. Remove chain tensioner.

❊❊ WARNING

Be careful not to drop bolts inside chain case.

12. Remove the three water pump bolts. Make a gap between water pump gear and timing chain, by turning crankshaft pulley counterclockwise until timing chain loosens on water pump sprocket.

13. Screw bolts into water pump upper and lower bolt holes until they reach the timing chain case. Then, alternately tighten each bolt for a half turn, and pull out the water pump.

❊❊ WARNING

Pull straight out while preventing vane from contacting socket in installation area. Remove water pump

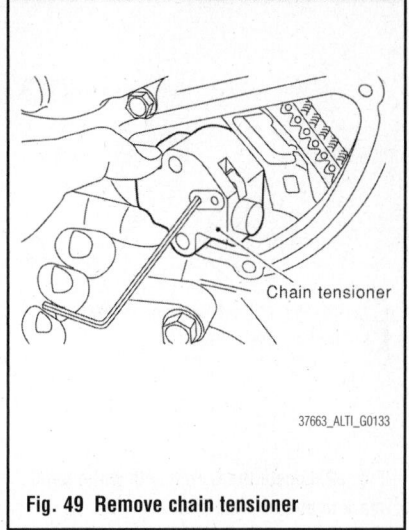

Fig. 49 Remove chain tensioner

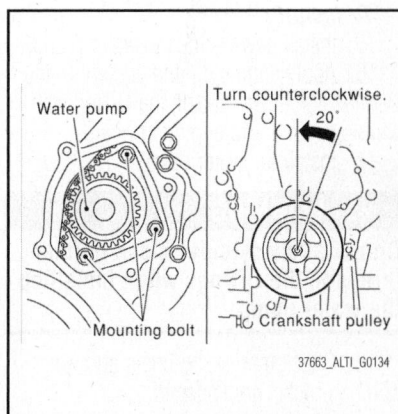

Fig. 50 Remove the three water pump bolts

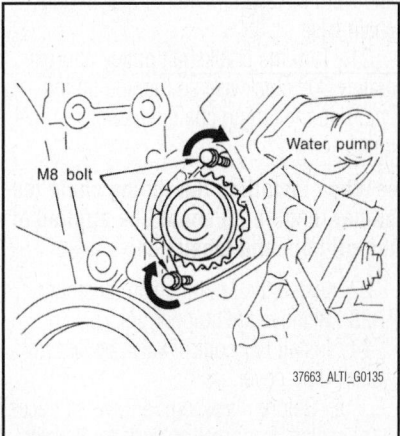

Fig. 51 Screw bolts into water pump upper and lower bolt holes until they reach the timing chain case

without causing sprocket to contact timing chain.

14. Remove bolts and O-rings from water pump.

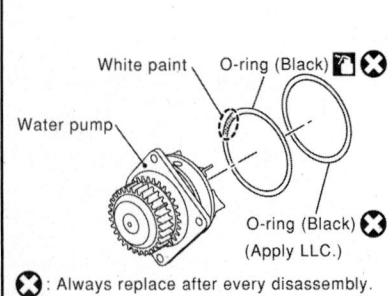

White paint   O-ring (Black)

Water pump

O-ring (Black)
(Apply LLC.)

✖ : Always replace after every disassembly.
⬛ : Lubricate with new engine oil.

37663_ALTI_G0136

Fig. 52 Locate the O-ring with white paint mark to engine front side

To install:

15. Install new O-rings to water pump.
16. Apply engine oil and coolant to the O-rings as shown. Locate the O-ring with white paint mark to engine front side.
17. Install the water pump.

✳✳ WARNING

Do not allow cylinder block to interfere with the O-rings when installing the water pump.

18. Check that timing chain and water pump sprocket are engaged.
19. Insert water pump by tightening bolts alternately and evenly.
20. Remove dust and foreign material completely from backside of chain tensioner and from installation area of rear timing chain case.
21. Turn the crankshaft pulley approximately 20° clockwise so that the timing chain on the timing chain tensioner side is loose.

➡When installing the timing chain tensioner, engine oil should be applied to the oil hole and tensioner.

22. Install the timing chain tensioner.
23. Remove the stopper pin.
24. Install IVT control valve cover and water pump cover.
   a. Before installing, remove all traces of sealant from mating surface of water pump cover and IVT control valve cover using a scraper. Also remove traces of sealant from the mating surface of the front cover.
   b. Apply a continuous bead of RTV Silicone Sealant or equivalent, to mating surface of IVT control valve cover and water pump cover.

25. Install water drain plug on water pump side of cylinder block.
26. Install idler pulley.
27. Installation of remaining components is in the reverse order of removal.
28. After installation refill engine coolant and check for leaks.

✳✳ CAUTION

Do not spill coolant in engine compartment. Use a shop cloth to absorb coolant.

29. After starting engine, let idle for three minutes, then rev engine up to 3,000 rpm under no load to purge air from the high-pressure chamber of the chain tensioner. The engine may produce a rattling noise. This indicates that air still remains in the chamber and is not a matter of concern.

Hybrid
See Figure 53.

✳✳ CAUTION

Never remove the radiator cap when the engine is hot. Serious burns could occur from high pressure coolant escaping from the radiator.

1. Drain engine coolant from radiator, condenser and liquid tank assembly.

✳✳ WARNING

Perform when the engine is cold.

2. Remove drive belt.
3. Remove engine cover.
4. Remove air cleaner duct (front).
5. Remove engine coolant reservoir.
6. Remove RH wheel and tire assembly.
7. Remove fender protector.
8. Remove engine ground strap.
9. Remove idler pulley bracket.
10. Remove the water pump.

✳✳ WARNING

Handle the water pump vane so that it does not contact any other parts.

✳✳ WARNING

Water pump cannot be disassembled, and must be replaced as an assembly.

➡If necessary, the exhaust manifold and three way catalyst assembly must be removed to remove the water pipe.

To install:

11. Installation is in the reverse order of removal.
12. When inserting water pipe end to cylinder block, apply a neutral detergent to O-ring. Then insert it immediately.
13. After installing the water pump, check for leaks using the radiator cap tester.

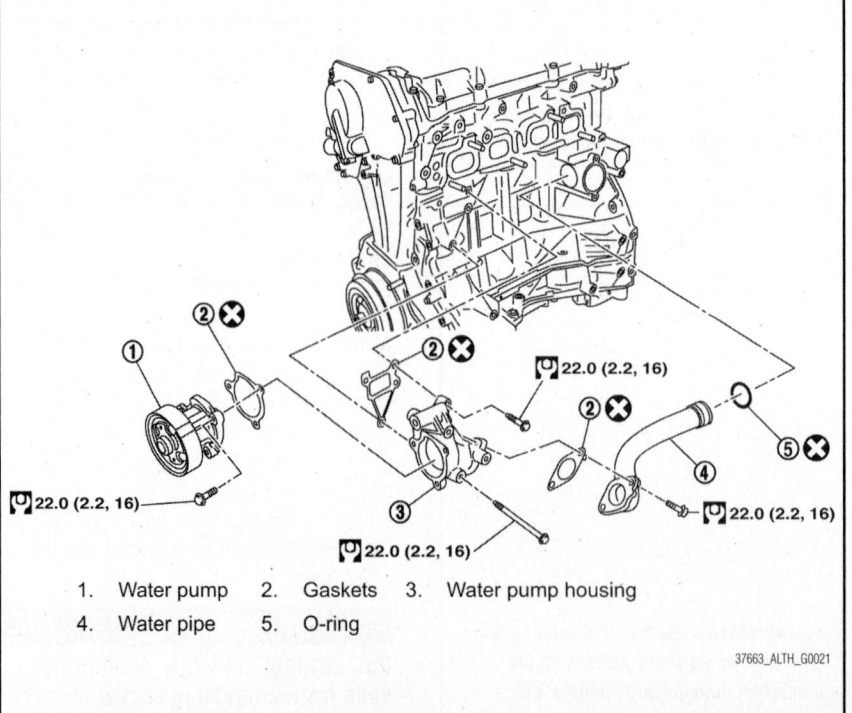

22.0 (2.2, 16)

1. Water pump   2. Gaskets   3. Water pump housing
4. Water pipe   5. O-ring

37663_ALTH_G0021

Fig. 53 Exploded view of water pump assembly

ENGINE ELECTRICAL

Refer to Engine Hybrid System section for information regarding the battery system of hybrid vehicle.

BATTERY

REMOVAL & INSTALLATION

1. Remove air duct (front).
2. Loosen battery terminal nuts, and disconnect both battery terminals.

✳✳ CAUTION

When disconnecting, disconnect the negative terminal first.

3. Remove battery frame nuts and battery frame.
4. Remove battery.

ENGINE ELECTRICAL

Refer to Engine Hybrid System section for information regarding the charging system of hybrid vehicle.

ALTERNATOR

REMOVAL & INSTALLATION

2.5L Engine
See Figure 54.

1. Disconnect the battery negative terminal.
2. Remove engine side undercover.
3. Remove drive belt.
4. Remove "B" terminal nut.
5. Remove air intake duct.
6. Disconnect alternator harness connectors.
7. Remove alternator ground harness bolt.

BATTERY SYSTEM

To install:

5. Installation is in the reverse order of removal.

✳✳ CAUTION

When connecting, connect the positive terminal first.

6. Reset electronic systems as necessary.

CHARGING SYSTEM

8. Remove alternator bolts.
9. Remove alternator assembly upward.

To install:

10. Installation is in the reverse order of removal.

✳✳ WARNING

Be sure to tighten "B" terminal nut carefully.

11. Install alternator and check tension of belt.

➡**For this model, the power generation voltage variable control system that controls the power generation voltage of the alternator has been adopted. Therefore, the power generation voltage variable control system operation inspection should be performed after replacing the alternator, and then make sure that the system operates normally.**

3.5L Engine
See Figure 55.

1. Disconnect the battery negative terminal.
2. Partially drain engine coolant.
3. Remove engine room cover.
4. Remove RH front wheel and tire assembly.
5. Remove engine side undercover.
6. Remove air cleaner and duct assembly.
7. Remove battery tray.
8. Remove cooling fan assembly.
9. Evacuate A/C system.
10. Remove drive belt.
11. Remove the A/C compressor.
12. Remove idler pulley.
13. Remove A/C idler pulley.
14. Disconnect oil pressure switch.
15. Disconnect the alternator harness connectors.
16. Remove the alternator bolt and nuts.
17. Slide the alternator out and remove.

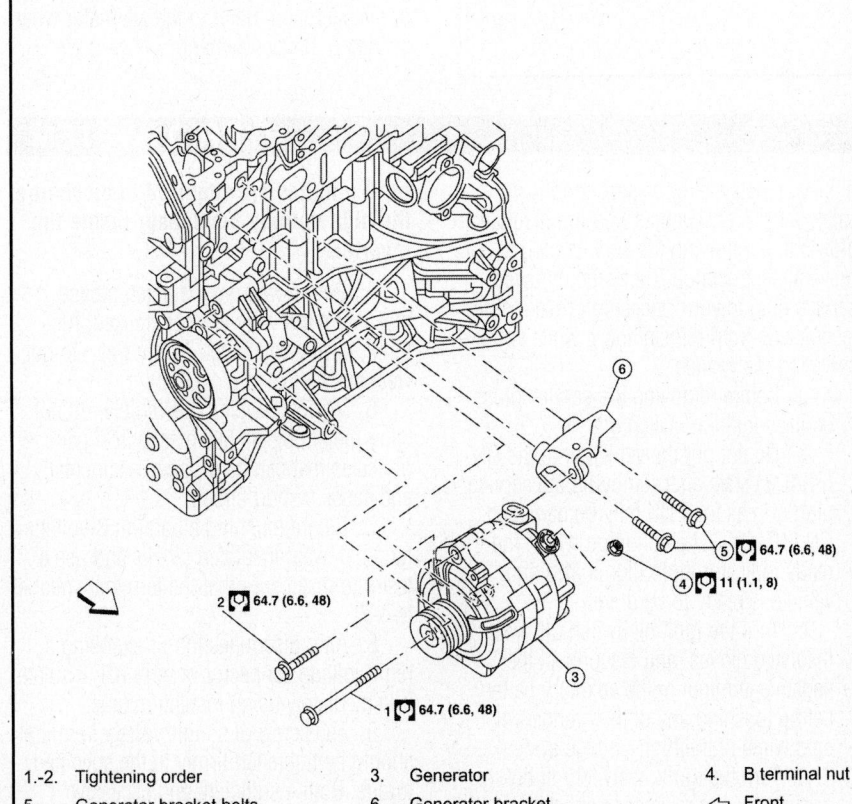

| 1.-2. | Tightening order | 3. | Generator | 4. | B terminal nut |
| 5. | Generator bracket bolts | 6. | Generator bracket | ◁ | Front |

37663_ALTI_G0138

Fig. 54 Exploded view of alternator assembly

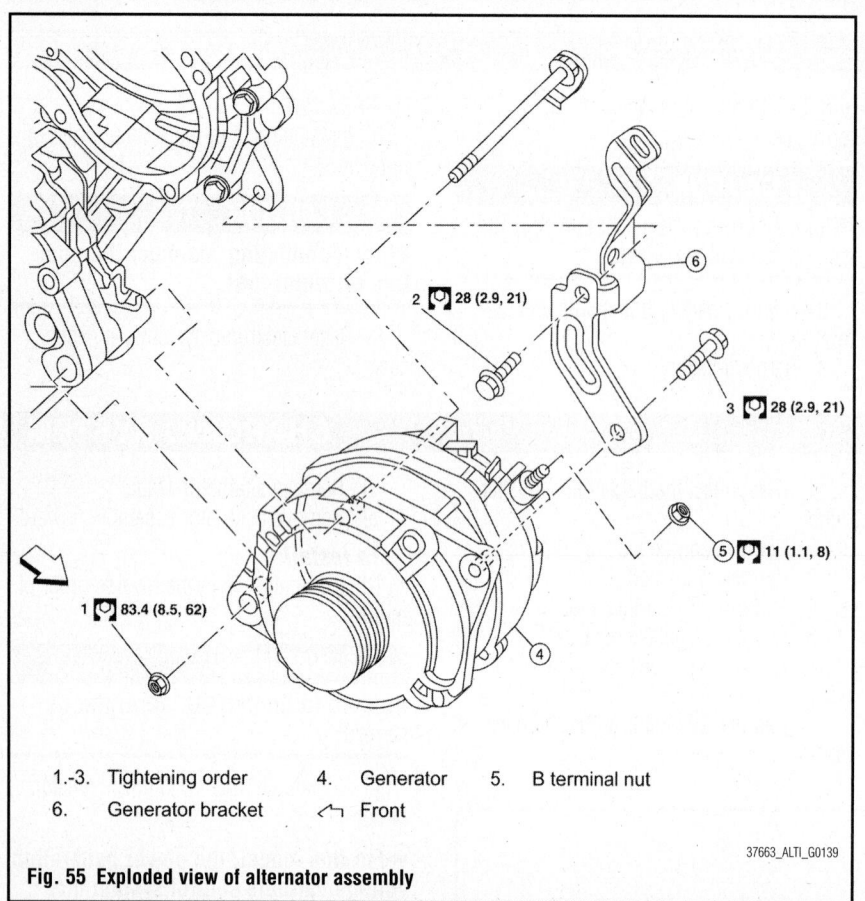

1.-3. Tightening order   4. Generator   5. B terminal nut
6. Generator bracket   ⟵ Front

37663_ALTI_G0139

Fig. 55 Exploded view of alternator assembly

To install:

18. Installation is in the reverse order of removal.

> **※※ WARNING**
> **Be sure to tighten "B" terminal nut carefully.**

19. Install alternator and check tension of belt.

20. For this model, the power generation voltage variable control system that controls the power generation voltage of the alternator has been adopted. Therefore, the power generation voltage variable control system operation inspection should be performed after replacing the alternator, and then make sure that the system operates normally.

21. Make sure that alternator pulley does not rattle.

22. Make sure that alternator pulley nut is tight. Tighten to 87 ft. lbs. (118 Nm).

VOLTAGE REGULATOR

The voltage regulator is an integral part of the alternator. Refer to the alternator when servicing this component.

ENGINE ELECTRICAL

PRECAUTIONS

> **※※ WARNING**
> **Observe the following precautions to ensure safe and proper servicing. These precautions are not described in each individual section.**

➡**The hybrid system contains a 244.8 V high-voltage system with a strong alkali solution of potassium hydroxide.**

> **※※ CAUTION**
> **Be sure to follow the instructions in this manual to handle the system correctly. Failure to do so may result in serious injury or electrocution.**

1. Engineer must undergo special training to be able to perform high-voltage system inspection and servicing.

2. High-voltage cables are colored orange. The HV battery and other high-voltage components have "high voltage" caution labels. Do not carelessly touch these wires and components.

3. Before inspecting or servicing the high-voltage system, be sure to follow safety measures, such as wearing insulated gloves and removing the service plug to prevent electrocution. Carry the removed service plug in your pocket to prevent other technicians from reinstalling it while you are servicing the vehicle.

a. Before removing the service plug, confirm ignition switch off.

b. Do not put the vehicle into the ON (READY) state after removing the service plug grip as the ECU may be damaged. ON (READY): The condition which the ready indicator lamp illuminates and vehicle is ready to be driven.

c. Turn the ignition switch off, wear insulated gloves, and disconnect the negative terminal of the auxiliary battery before touching any of the orange-colored wires of the high-voltage system.

d. Turn the ignition switch off before performing any resistance checks.

e. Turn the ignition switch off before disconnecting or reconnecting any connectors.

4. After removing the service plug, wait 10 minutes before touching any of the high-voltage connectors and terminals.

HYBRID SYSTEM

➡**10 minutes are required to discharge the high-voltage condenser inside the inverter.**

5. Before wearing insulated gloves, make sure that they are not cracked, ruptured, torn, or damaged in any way. Do not wear wet insulated gloves.

6. When servicing the vehicle, do not carry metal objects like mechanical pencils or scales that can be dropped accidentally and cause a short circuit.

7. Before touching a bare high-voltage terminal, wear insulated gloves and use a tester to make sure that the terminal voltage is 0 V.

8. After disconnecting or exposing a high-voltage connector or terminal, insulate it immediately using insulation tape.

9. The screw of a high-voltage terminal should be tightened firmly to the specified torque. Both insufficient and excessive torque can cause failure.

10. Use the "CAUTION: high-voltage. DO NOT TOUCH DURING OPERATION" sign to notify other engineers that a high-voltage system is being inspected and/or repaired.

11. After servicing the high-voltage system and before reinstalling the service plug, check again that you have not left a part or tool inside, that the high-voltage terminal screws are firmly tightened, and that the connectors are correctly connected.

12. Do not place the battery upside down while removing and installing it.

⁂ CAUTION

When engaged in operations such as removal and installation related to high-voltage equipment, use personal protective equipment to avoid death or serious personal injury from electric shock.

AUXILIARY BATTERY

REMOVAL & INSTALLATION

See Figure 56.

1. Remove trunk side finisher (RH).
2. Loosen 12-volt battery terminal nuts, and disconnect both 12-volt battery terminals.

⁂ WARNING

When disconnecting, disconnect the 12-volt battery negative terminal first.

3. Remove the 12-volt battery ventilation tube.
4. Remove 12-volt battery frame nuts and 12-volt battery frame.
5. Remove 12-volt battery.

To install:

6. Installation is the reverse order of removal.

D
B
C
A
A
37663_ALTH_G0022

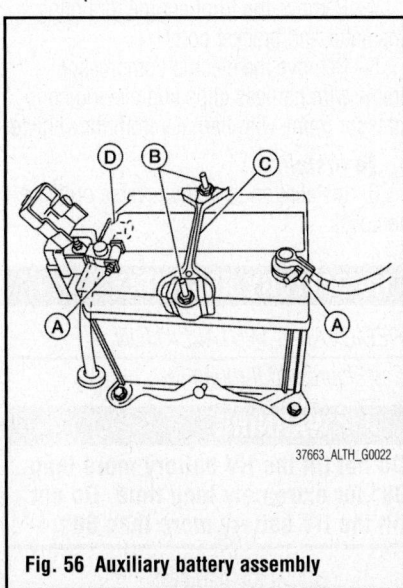

Fig. 56 Auxiliary battery assembly

⁂ WARNING

When connecting, connect the 12-volt battery positive terminal first.

BATTERY SMART UNIT

REMOVAL & INSTALLATION

See Figure 57.

1. Remove the HV relay assembly from the HV battery assembly.
2. Remove the bolt from the battery smart unit.
3. Disconnect the connectors from the battery smart unit and remove it from the HV battery assembly.

To install:

4. Installation is in the reverse order of removal.

9 (0.9, 80)
8.0 (0.8, 71)
21.8 (2.2, 16)
1. Battery shield contact
2. HV wire
3. Filter noise capacitor
4. Ground wire
5. Lock
6. RH cover
7. LH cover
8. Side cover
9. Duct
10. Clip
11. HV vehicle converter
12. Connector
13. Vent hose
14. HV battery assembly
15. Service plug grip
16. Battery smart unit
17. HV relay assembly
A. Refer to installation.
37663_ALTH_G0023

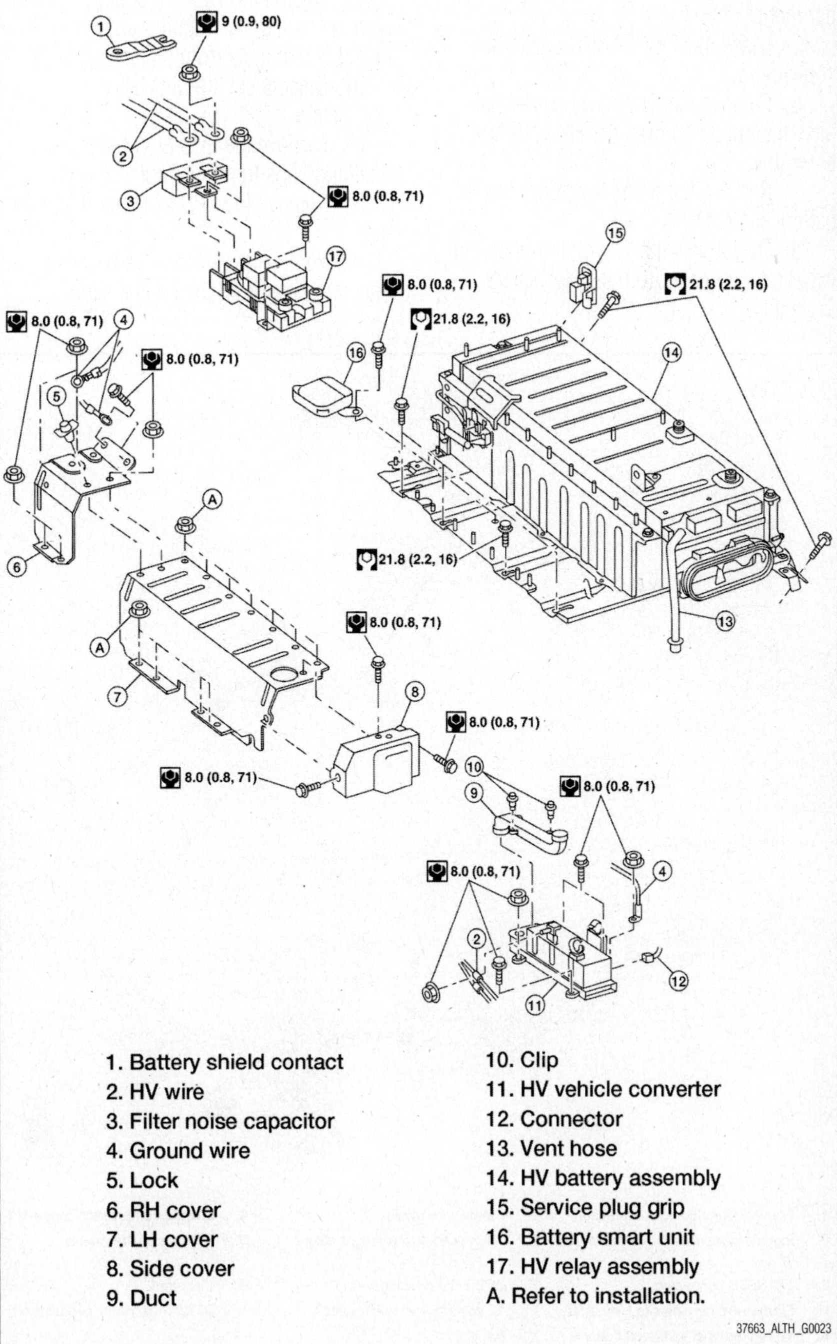

Fig. 57 Exploded view of Battery Smart Unit

FRAME WIRE

REMOVAL & INSTALLATION

Frame Wire (Main)

See Figures 58 and 59.

1. Disconnect the positive 12 volt terminal from the 12 volt battery.
2. Remove the rear seat.
3. Remove the fuel tank.
4. Remove the 12 volt positive battery cable retaining clips from the trunk compartment.
5. Disconnect the DC/DC converter connectors.
6. Remove the DC/DC converter harness retaining clip from the HV battery assembly.
7. Remove the frame wire from the HV battery assembly.
8. Disconnect the 12 volt terminal and cable retaining clip from the HV battery assembly.

9. Remove the frame wire harness retaining clips from the vehicle interior.
10. Remove the air cleaner and air duct.
11. Remove the inverter cover and terminal cover from the inverter.
12. Remove the frame wire inverter connector bolt and disconnect the frame wire inverter connector from the inverter.
13. Remove the HV fuse box cover from the HV fuse box.
14. Remove the HV fuse box terminal cap and nuts from the HV fuse box.
15. Open the HV fuse box side cover and remove the harness retaining clip and HV fuse box terminals from the HV fuse box.
16. Disconnect the EPS ECU connectors.
17. Remove the EPS ECU harness retaining clips from the engine room.
18. Remove the EPS ECU bonding wire bolt.
19. Remove the frame wire harness retaining clips from the engine room.

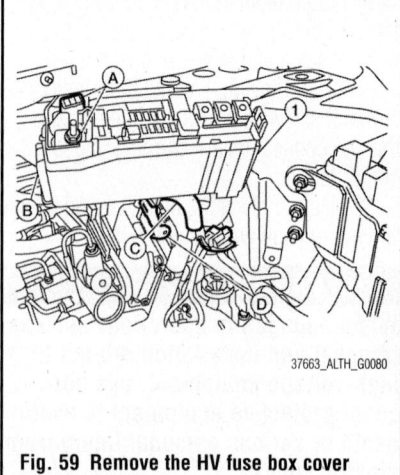

37663_ALTH_G0080

Fig. 59 Remove the HV fuse box cover from the HV fuse box

20. Remove the RH member pin stay.
21. Remove the frame wire retainer nuts and bolts from the underside of vehicle.
22. Remove the frame wire harness assembly with grommet from floor pass through and underside of vehicle.
23. Remove the frame wire harness from the engine room clip and remove the frame wire harness from the engine room.

To install:

24. Installation is in the reverse order of removal.

Frame Wire (Electric Compressor)

1. Remove the air cleaner and air duct.
2. Remove the front terminal cover bolt from the inverter cover and disconnect the electric compressor inverter connector from the inverter.
3. Disconnect the electric compressor connector from the electric compressor.
4. Remove the front engine mounting insulator and bracket bolts.
5. Remove the electric compressor frame wire harness clips and electric compressor frame wire harness from the vehicle.

To install:

6. Installation is in the reverse order of removal.

HV BATTERY ASSEMBLY

REMOVAL & INSTALLATION

See Figures 60 through 68.

❈❈ WARNING

Do not tilt the HV battery more than 30° for extremely long time. Do not tilt the HV battery more than 60°.

1. Remove the rear seat.

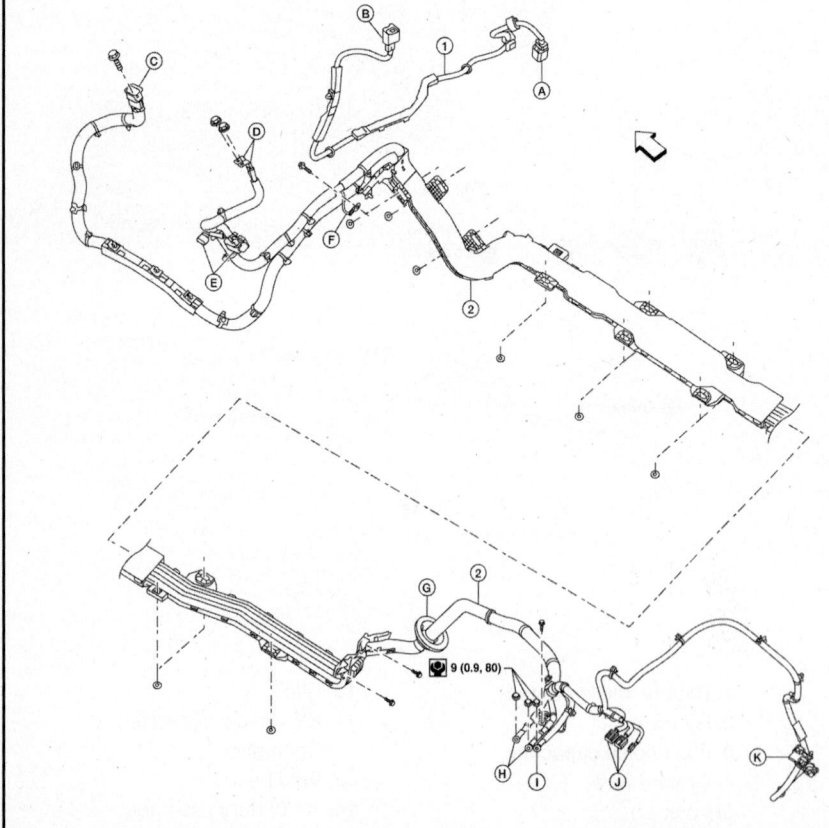

1.	Frame wire (electric compressor)	2.	Frame wire (main)	A.	Electric compressor connector
B.	Electric compressor inverter connector	C.	Frame wire inverter connector	D.	HV fuse box terminals
E.	EPS ECU connectors	F.	EPS ECU bonding wire	G.	Grommet
H.	Frame wire terminals to HV battery	I.	12 volt terminal to HV battery	J.	DC/DC converter connectors
K.	12 volt terminal to 12 volt battery	⇐:	Front		

37663_ALTH_G0079

Fig. 58 Exploded view of frame wire

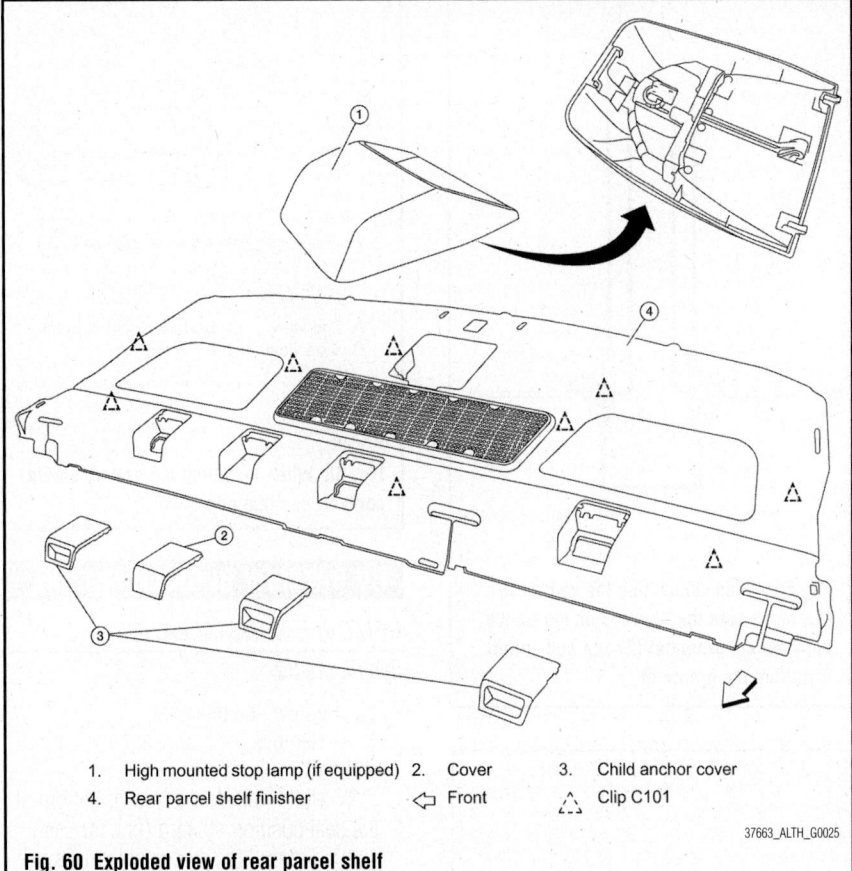

1. High mounted stop lamp (if equipped) 2. Cover 3. Child anchor cover
4. Rear parcel shelf finisher ⟵ Front △ Clip C101

37663_ALTH_G0025

Fig. 60 Exploded view of rear parcel shelf

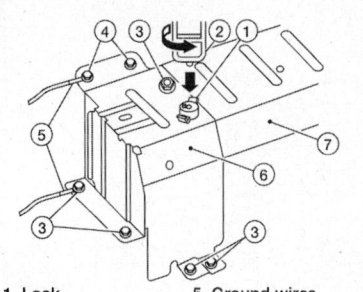

1. Lock 5. Ground wires
2. Service plug grip 6. RH cover
3. Nuts 7. HV battery assembly
4. Bolts

37663_ALTH_G0028

Fig. 62 Remove the lock from the RH cover on the HV battery assembly using the service plug grip

a. Remove the rear seat cushion trim and pad.

b. Pull the lock at the front bottom of the seat cushion forward (one for each side), and pull the seat cushion upward to release the wire from the plastic hook, then pull the seat cushion forward to remove.

c. Remove the seatback anchor bolts.

d. Lift the seatback off the rear parcel panel front hangers and remove the seatback assembly.

2. Remove the rear parcel shelf.

a. Remove high mounted stop lamp (if equipped).

b. Remove rear pillar finisher RH/LH.

c. Thread the rear seat belt RH/LH/Center through vertical opening and release from rear parcel shelf finisher.

d. Remove the clips, then remove rear parcel shelf finisher.

3. Remove the trunk room trim.

a. Remove trunk net rear and trunk net side (if equipped).

b. Remove trunk floor carpet.

c. Release the clips, then remove trunk rear finisher.

d. Release the clips, then remove trunk side finisher (RH/LH).

e. Remove spare tire cover.

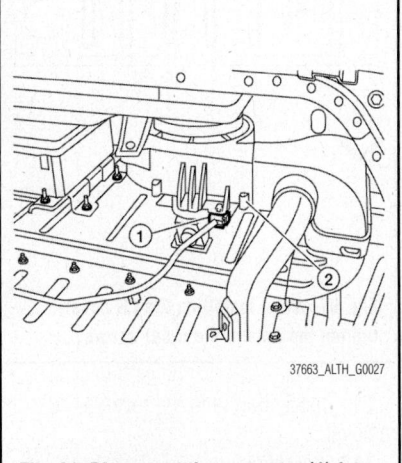

37663_ALTH_G0027

Fig. 61 Disconnect the connector (1) from the HV battery blower motor (2)

4. Remove the inlet and outlet cooling ducts.

5. Disconnect the connector from the HV battery blower motor.

6. Remove the HV battery blower motor harness clips and HV battery blower motor harness from the HV battery.

7. Remove the lock from the RH cover on the HV battery assembly using the service plug grip.

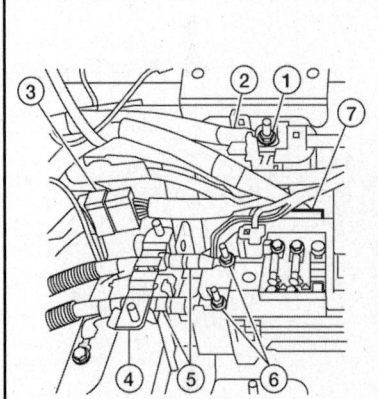

1. 12 volt terminal nut
2. Terminal cable
3. Body harness connector
4. Battery shield contact
5. HV wires
6. HV wire nuts
7. EPS DC/DC converter connector

37663_ALTH_G0029

Fig. 63 Remove the terminal cover and 12 volt terminal nut, then remove the terminal cable and 12 volt harness from the HV battery assembly

8. Remove the nuts, bolts and ground wires from the RH cover.

9. Remove the RH cover from the HV battery assembly.

10. Remove the terminal cover and 12 volt terminal nut, then remove the terminal cable and 12 volt harness from the HV battery assembly.

11. Remove the battery shield contact, HV wire nuts and HV wires from the HV battery assembly.

12. Disconnect the body harness connector from the HV battery assembly.

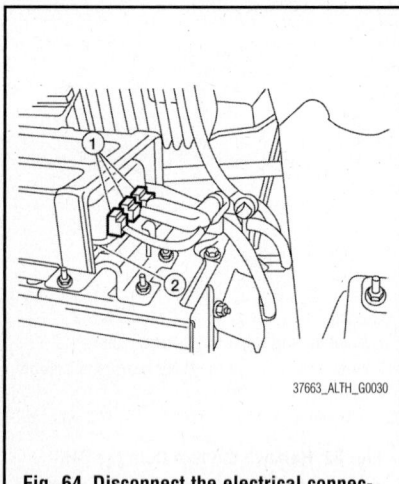

Fig. 64 Disconnect the electrical connectors (1) from the EPS DC/DC converter (2)

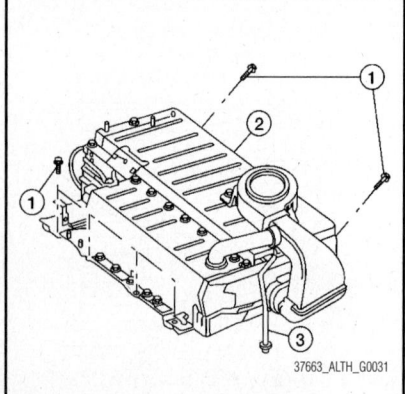

Fig. 65 Disconnect the vent hose (3); remove the HV battery bolts (1) from the HV battery assembly (2)

13. Disconnect the EPS DC/DC converter connector from the HV battery assembly.

14. Remove the harnesses from the HV battery assembly.

15. Disconnect the electrical connectors from the EPS DC/DC converter.

16. Remove the harness clips and harness from the HV battery assembly.

17. Disconnect the vent hose from the vehicle.

18. Remove the HV battery bolts from the HV battery assembly.

19. Remove the HV battery assembly from the vehicle.

20. If necessary, remove the following components from the HV battery assembly:
- The HV battery blower motor and cooling ducts.
- The EPS DC/DC converter.
- The HV relay assembly.
- The battery smart unit.
- The HV vehicle converter.

To install:

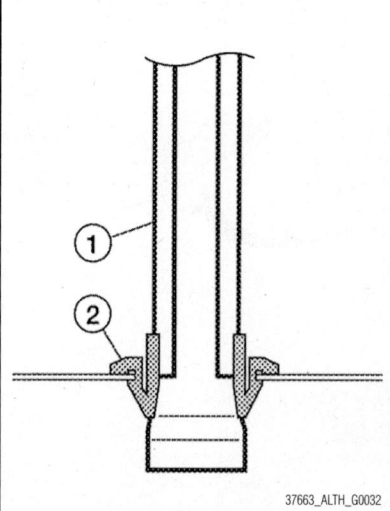

Fig. 66 When connecting the vent hose (1), make sure that there is no clearance between the grommet (2) and body after installing the grommet

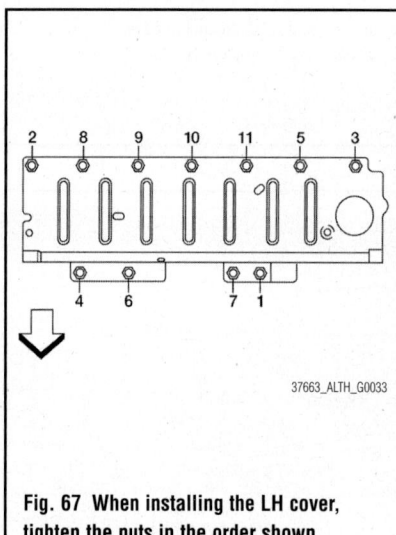

Fig. 67 When installing the LH cover, tighten the nuts in the order shown

21. Installation is in the reverse order of removal.

Note the following:
- When connecting the vent hose, make sure that there is no clearance between the grommet and body after installing the grommet.
- When installing the LH cover, tighten the nuts in the order shown.
- When installing the battery shield contact, position as shown.
- When installing the lock to the RH cover, push the lock into the hole and ensure it is locked.

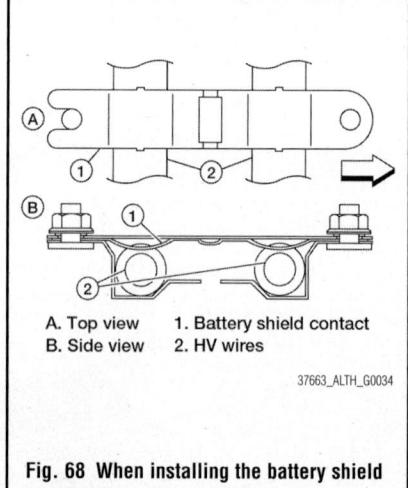

A. Top view   1. Battery shield contact
B. Side view   2. HV wires

Fig. 68 When installing the battery shield contact, position as shown

HV BATTERY BLOWER MOTOR

REMOVAL & INSTALLATION

See Figure 69.

1. Remove the rear seat.
   a. Remove the rear seat cushion trim and pad.
   b. Pull the lock at the front bottom of the seat cushion forward (one for each side), and pull the seat cushion upward to release the wire from the plastic hook, then pull the seat cushion forward to remove.
   c. Remove the seatback anchor bolts.
   d. Lift the seatback off the rear parcel panel front hangers and remove the seatback assembly.

2. Remove the rear parcel shelf finisher.
   a. Remove high mounted stop lamp (if equipped).
   b. Remove rear pillar finisher RH/LH.
   c. Thread the rear seat belt RH/LH/Center through vertical opening and release from rear parcel shelf finisher.
   d. Remove the clips, then remove rear parcel shelf finisher.

3. Remove the trunk room trim.
   a. Remove trunk net rear and trunk net side (if equipped).
   b. Remove trunk floor carpet.
   c. Release the clips, then remove trunk rear finisher.
   d. Release the clips, then remove trunk side finisher (RH/LH).
   e. Remove spare tire cover.

4. Remove the upper and lower inlet duct clips and bolts.

5. Remove the upper and lower inlet duct from the package shelf and HV battery blower motor.

6. Remove the front duct clip and

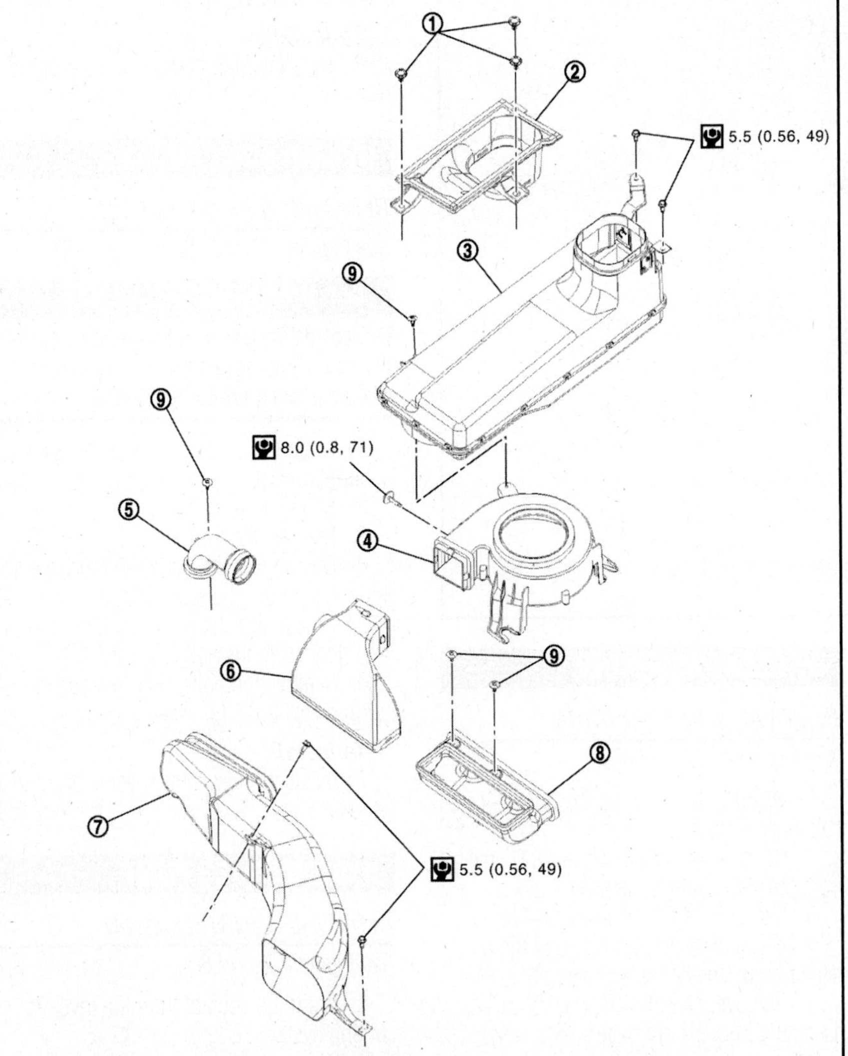

Fig. 69 showing exploded view:

5.5 (0.56, 49)

8.0 (0.8, 71)

5.5 (0.56, 49)

1.	Clip A	2.	Upper inlet duct	3.	Lower inlet duct
4.	HV battery blower motor	5.	Front duct	6.	Rear upper duct
7.	Outlet duct	8.	Rear lower duct	9.	Clip B

37663_ALTH_G0026

Fig. 69 Exploded view showing cooling ducts and HV battery blower motor

remove the front duct from the rear upper duct and HV battery assembly.

7. Separate the rear upper duct from the rear lower duct and remove the rear upper duct from the HV battery blower motor.

8. Disconnect the HV battery blower motor harness connector from the HV battery blower motor.

9. Remove the HV battery blower motor from the HV battery assembly.

To install:

10. Installation is in the reverse order of removal.

HV ECU

REMOVAL & INSTALLATION

Precaution For Replacing Hybrid Vehicle Control ECU

See Figures 70 through 73.

When replacing the hybrid vehicle control ECU, never remove the waterproof sheet.

➡**The hybrid vehicle control ECU is covered with a waterproof sheet. If the**

waterproof sheet is peeled off, the labels on the hybrid vehicle control ECU will be removed together with the waterproof sheet. Consequently important data printed on the label for warranty procedure will be lost.

1. Remove the console side finisher LH.

2. Remove the LH bolts from the HV ECU.

3. Remove the instrument side panel RH.

4. Remove the RH bolts from the HV ECU.

5. Disconnect the drain hose from the heater and cooling unit assembly.

6. Pull out the HV ECU to RH side.

7. Disconnect the HV ECU harness connector from the HV ECU, and remove the HV ECU from the vehicle.

8. If necessary, remove the screws and HV ECU brackets from the HV ECU.

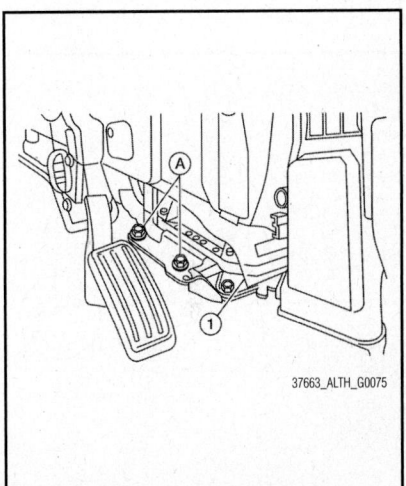

37663_ALTH_G0075

Fig. 70 Remove the LH bolts (A) from the HV ECU (1)

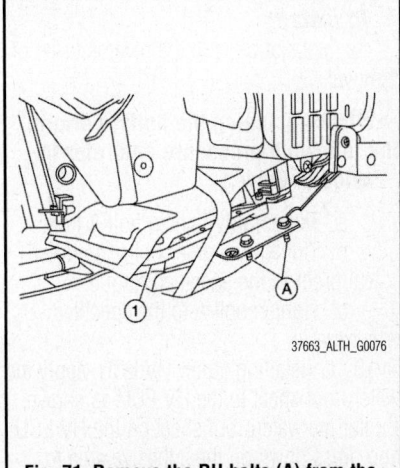

37663_ALTH_G0076

Fig. 71 Remove the RH bolts (A) from the HV ECU (1)

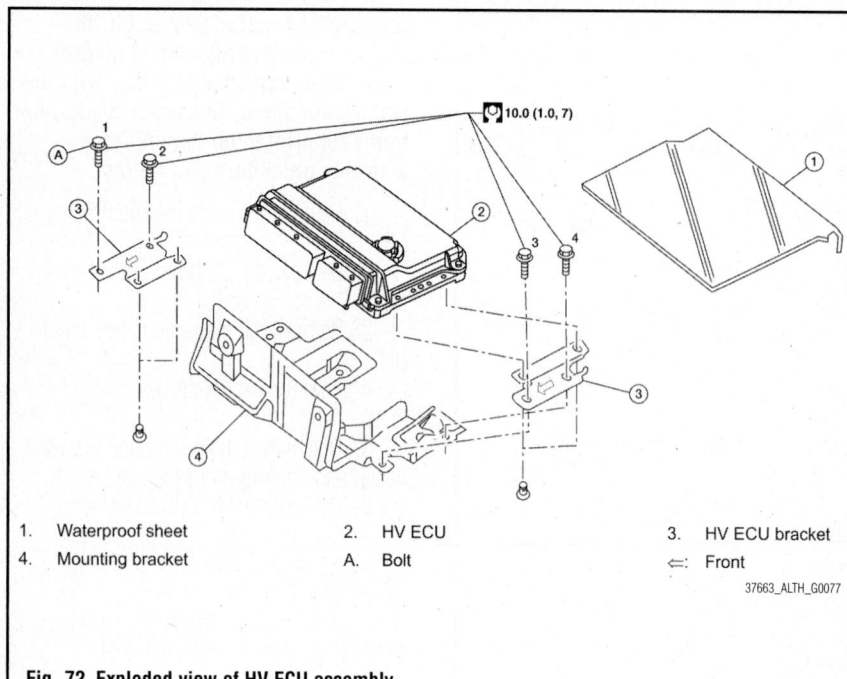

1.	Waterproof sheet	2.	HV ECU
4.	Mounting bracket	A.	Bolt

3.	HV ECU bracket
⇐:	Front

37663_ALTH_G0077

Fig. 72 Exploded view of HV ECU assembly

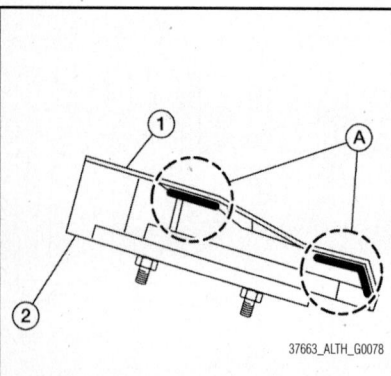

37663_ALTH_G0078

Fig. 73 Apply the waterproof sheet (1) to the HV ECU (2) and press down on the adhesive area (A)

To install:

9. Installation is in the reverse order of removal.

➡**When tightening the bolts, perform the following procedure and refer to "Exploded View".**

a. Temporarily tighten bolt A first.

b. Tighten the other bolts in numerical order to the specified torque.

c. Tighten bolt A to the specified torque.

10. If installing a new HV ECU, apply the waterproof sheet to the HV ECU as shown. Center the waterproof sheet on the HV ECU and press down on the adhesive area to secure the waterproof sheet to the HV ECU.

HV RELAY ASSEMBLY

REMOVAL & INSTALLATION
See Figures 74 through 76.

1. Remove the HV wires from the HV battery assembly.

2. Remove the side cover and LH cover from the HV battery assembly.

3. Remove the filter noise capacitor.

4. Disconnect ground wire and the connectors from the HV relay assembly.

5. Remove the bolts and the HV relay assembly from the HV battery assembly.

To install:

6. Installation is in the reverse order of removal.

7. When installing the LH cover, tighten the nuts in the order shown.

HV VEHICLE CONVERTER

REMOVAL & INSTALLATION
See Figure 74.

1. Disconnect the connectors from the HV relay assembly.

2. Remove the HV wire nut and HV wire from the HV vehicle converter.

3. Remove the ground wire nut and ground wire from the HV vehicle converter.

4. Remove the HV vehicle converter.

a. Remove the HV vehicle converter nut and bolts from the HV vehicle converter.

b. Disconnect the connector from the back of the HV vehicle converter.

c. Remove the HV vehicle converter.

5. Remove the clips and duct from the HV vehicle converter.

To install:

6. Installation is in the reverse order of removal.

HYBRID SUB RADIATOR

REMOVAL & INSTALLATION
See Figures 77 and 78.

✳✳ WARNING

Do not damage or scratch the radiator and condenser assembly and sub radiator core when removing.

1. Drain the coolant from the inverter cooling system.

2. Remove the air duct.

3. Remove the front grille.

4. Disconnect the clamp and the upper outlet hose from the sub radiator.

5. Disconnect the clamp and the lower inlet hose from the sub radiator.

6. Remove the bolts, then remove the sub radiator from the vehicle.

To install:

7. Installation is in the reverse order of removal.

HYBRID WATER PUMP

REMOVAL & INSTALLATION
See Figures 79 and 80.

1. Drain the coolant from the inverter cooling system.

2. Disconnect the water inlet hose and water outlet hose from the water pump with motor and bracket assembly.

3. Remove the bolts from the water pump with motor and bracket assembly and remove from the vehicle.

To install:

4. Installation is in the reverse order of removal.

➡**Do not use the power tool.**

INVERTER WITH CONVERTER ASSEMBLY

REMOVAL & INSTALLATION
See Figures 81 through 86.

1. Drain the coolant from the inverter cooling system.

2. Remove the inverter cover.

3. Remove the air cleaner and air duct.

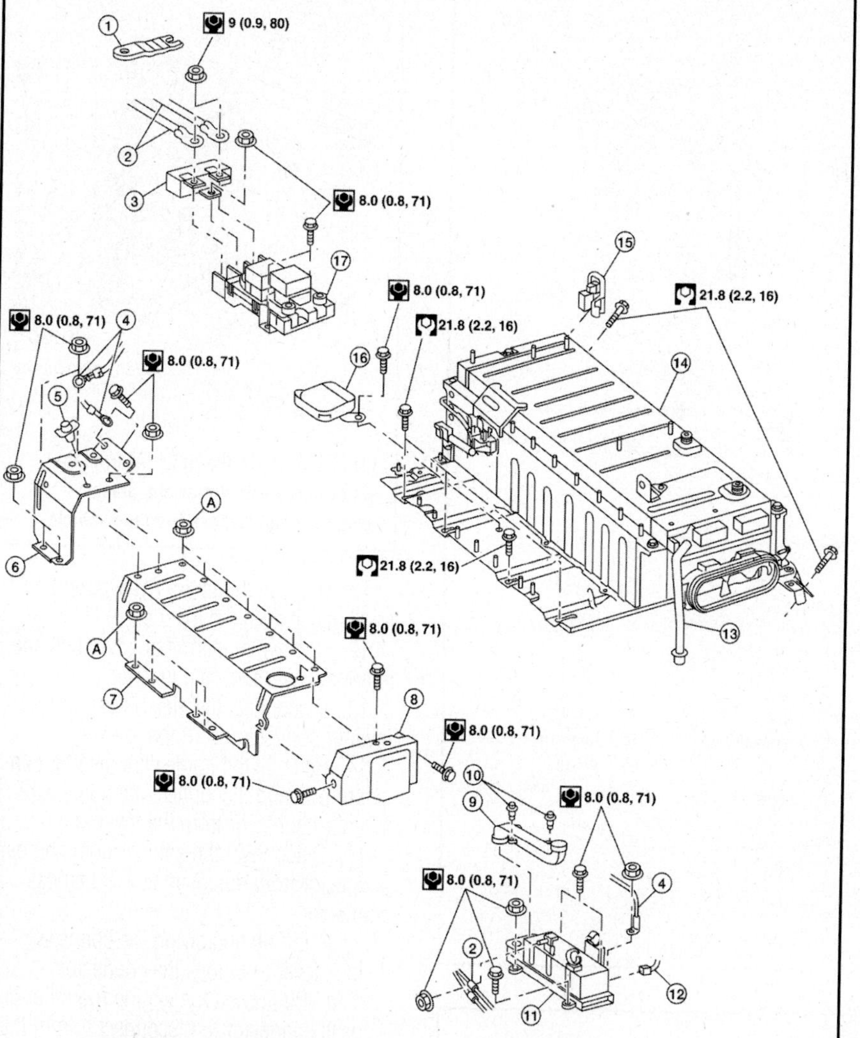

9 (0.9, 80)

8.0 (0.8, 71)

8.0 (0.8, 71)

21.8 (2.2, 16)

8.0 (0.8, 71)

8.0 (0.8, 71)

8.0 (0.8, 71)

21.8 (2.2, 16)

21.8 (2.2, 16)

8.0 (0.8, 71)

8.0 (0.8, 71)

8.0 (0.8, 71)

8.0 (0.8, 71)

8.0 (0.8, 71)

1. Battery shield contact
2. HV wire
3. Filter noise capacitor
4. Ground wire
5. Lock
6. RH cover
7. LH cover
8. Side cover
9. Duct
10. Clip
11. HV vehicle converter
12. Connector
13. Vent hose
14. HV battery assembly
15. Service plug grip
16. Battery smart unit
17. HV relay assembly
A. Refer to installation.

37663_ALTH_G0023

Fig. 74 Exploded view showing the HV relay assembly & converter

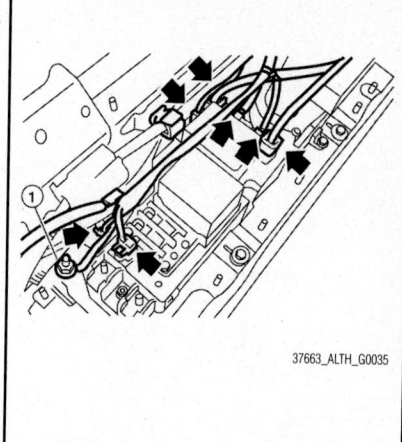

37663_ALTH_G0035

Fig. 75 Disconnect ground wire (1), and the connectors from the HV relay assembly

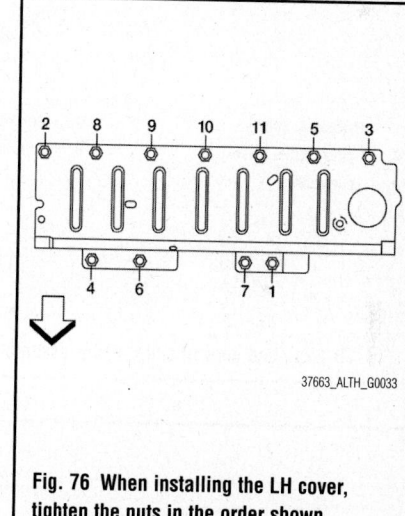

37663_ALTH_G0033

Fig. 76 When installing the LH cover, tighten the nuts in the order shown

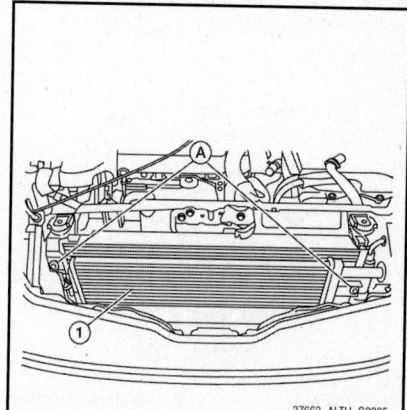

37663_ALTH_G0085

Fig. 77 Remove the bolts (A), then remove the sub radiator (1) from the vehicle

4. Remove the nuts and bolts from the upper center bracket.

5. Remove the inverter upper center bracket.

6. Remove the hoses and bolts from the inverter cooling reservoir tank.

7. Remove the inverter cooling reservoir tank from the vehicle.

8. Disconnect the MG1 and MG2 connectors from the inverter as follows:

a. Remove bolts G, I, J and L as shown.

b. Remove bolts H and K as shown.

c. Disconnect the MG1 and MG2 connectors from the inverter.

9. Remove the MG1 and MG2 harness clips from the bracket and set the MG1 and MG2 harness aside.

10. Remove the coolant hoses from the inverter.

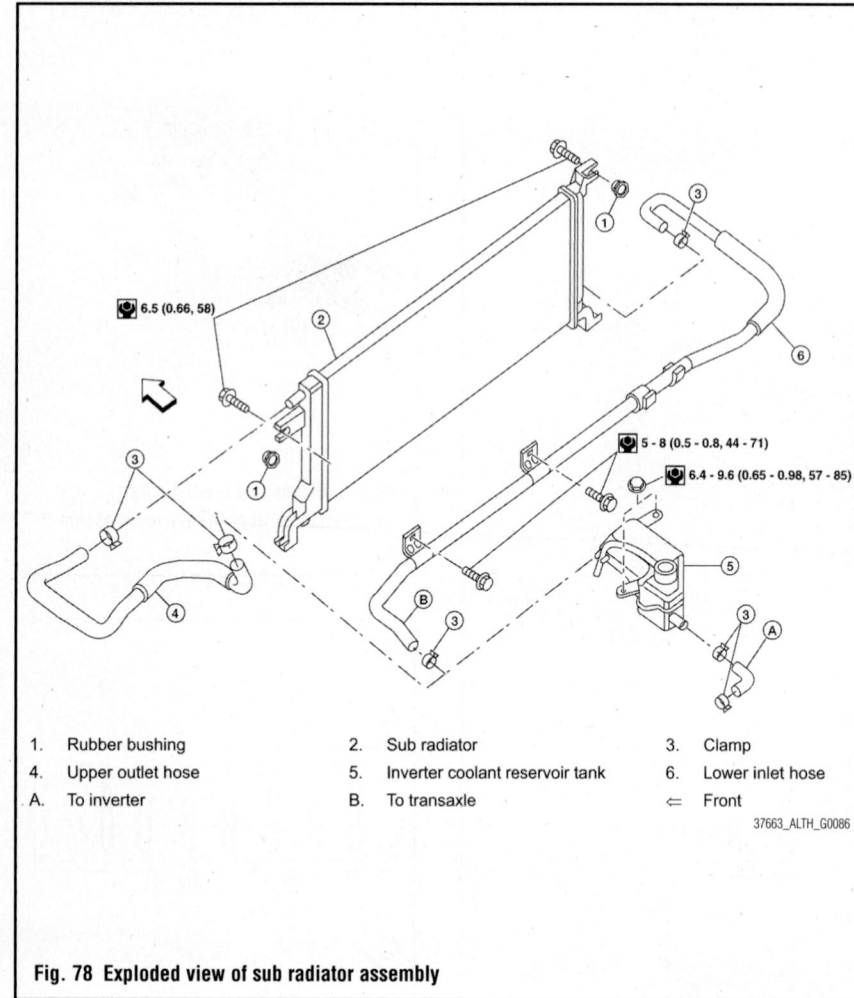

6.5 (0.66, 58)

5 - 8 (0.5 - 0.8, 44 - 71)

6.4 - 9.6 (0.65 - 0.98, 57 - 85)

1.	Rubber bushing	2.	Sub radiator	3.	Clamp
4.	Upper outlet hose	5.	Inverter coolant reservoir tank	6.	Lower inlet hose
A.	To inverter	B.	To transaxle	⇐	Front

37663_ALTH_G0086

Fig. 78 Exploded view of sub radiator assembly

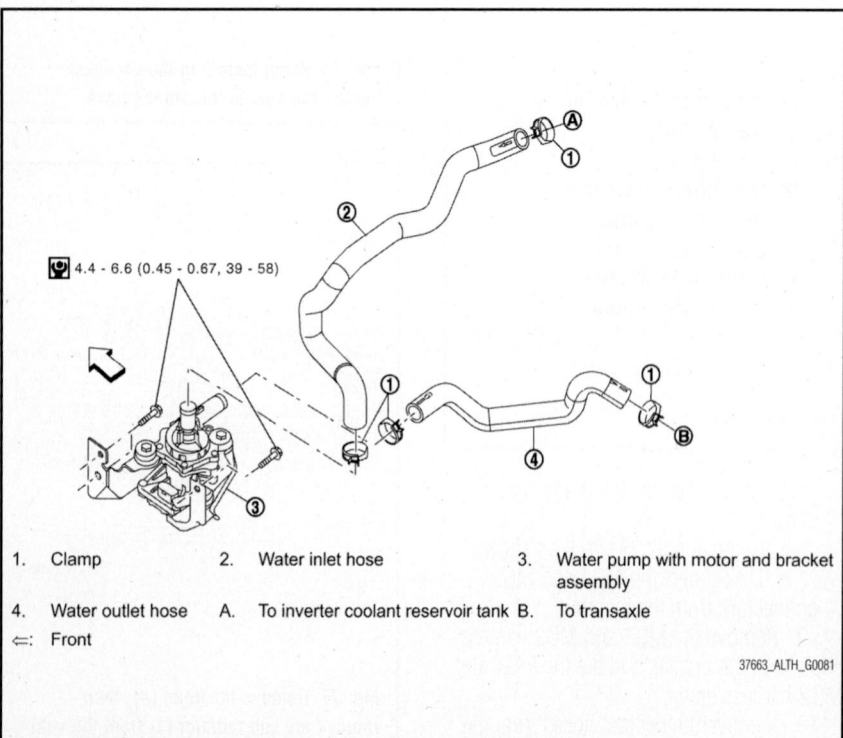

4.4 - 6.6 (0.45 - 0.67, 39 - 58)

1.	Clamp	2.	Water inlet hose	3.	Water pump with motor and bracket assembly
4.	Water outlet hose	A.	To inverter coolant reservoir tank	B.	To transaxle
⇐:	Front				

37663_ALTH_G0081

Fig. 79 Exploded view of Hybrid water pump with motor and bracket assembly

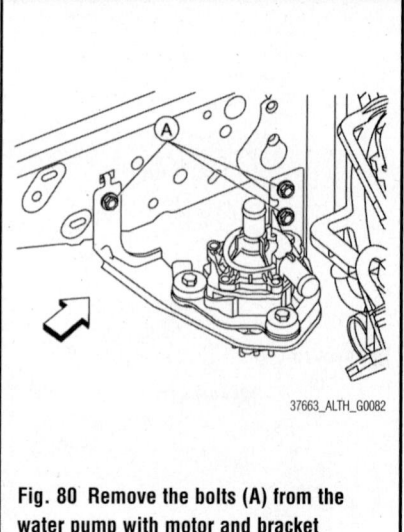

37663_ALTH_G0082

Fig. 80 Remove the bolts (A) from the water pump with motor and bracket assembly and remove from the vehicle

11. Remove the upper RH bracket bolt and bracket from the inverter.

12. Remove the terminal cover bolt and terminal cover from the inverter.

13. Disconnect the electric compressor inverter connector from the inverter.

14. Remove the frame wire inverter connector bolt and disconnect the frame wire inverter connector from the inverter.

15. Disconnect the engine room harness connector from the inverter; EGI harness connector

    a. Lift up and swing the connector lock lever to unlock the connector.

    b. Pull up on the engine room harness connector to disconnect it from the inverter.

16. Remove the engine room harness clip from the bracket and set the engine room harness aside.

17. Disconnect the EGI harness connector from the inverter.

18. Remove the inverter nuts.

19. Remove the inverter from the vehicle.

20. Remove any necessary brackets from the inverter.

To install:

21. Installation is in the reverse order of removal.

22. When installing the inverter, lower RH bracket, lower LH bracket, rear LH bracket, rear RH bracket and upper LH bracket should be attached to the inverter in advance.

23. When lower RH bracket, lower LH bracket, rear LH bracket, rear RH bracket and upper LH bracket are attached to the inverter, they should be touched to anti-rotation at the boss of the inverter.

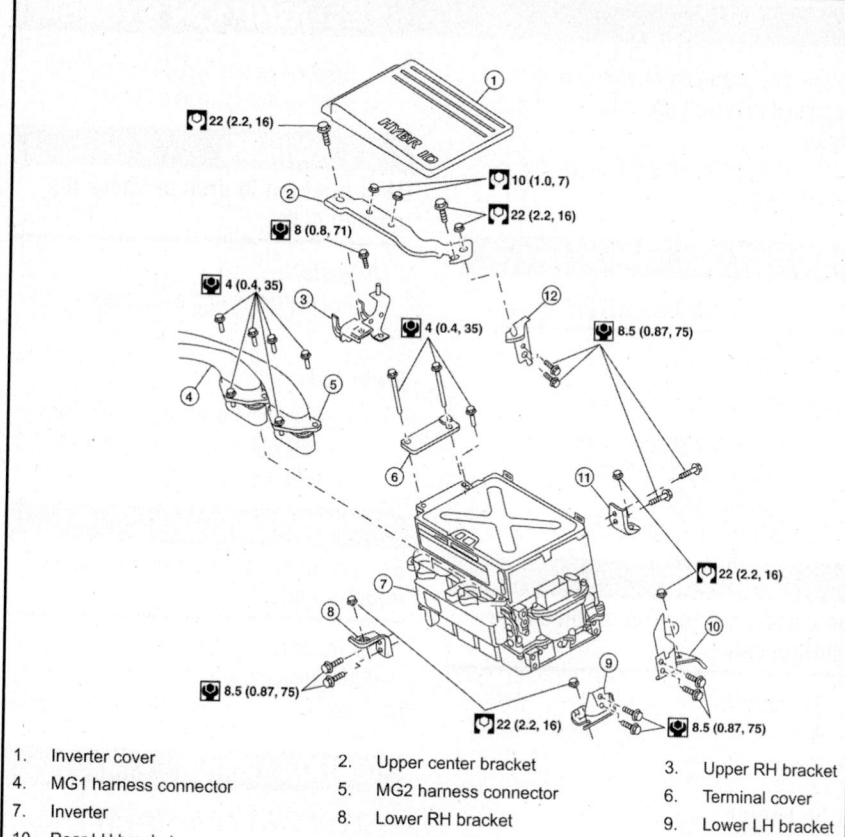

1. Inverter cover
4. MG1 harness connector
7. Inverter
10. Rear LH bracket

2. Upper center bracket
5. MG2 harness connector
8. Lower RH bracket
11. Rear RH bracket

3. Upper RH bracket
6. Terminal cover
9. Lower LH bracket
12. Upper LH bracket

37663_ALTH_G0070

Fig. 81 Exploded view of inverter with converter assembly

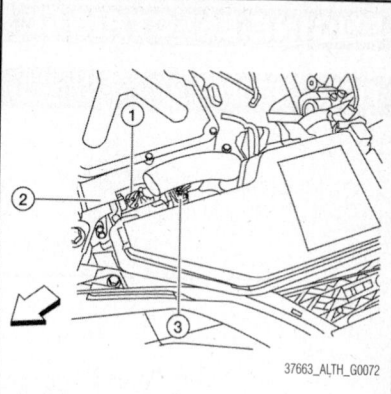

Fig. 84 Disconnect the engine room harness connector (1) from the inverter (2); EGI harness connector (3)

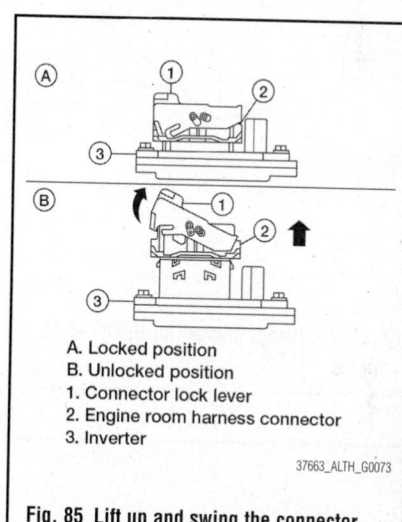

A. Locked position
B. Unlocked position
1. Connector lock lever
2. Engine room harness connector
3. Inverter

37663_ALTH_G0073

Fig. 85 Lift up and swing the connector lock lever to unlock the connector

Fig. 82 Disconnect the MG1 and MG2 connectors from the inverter

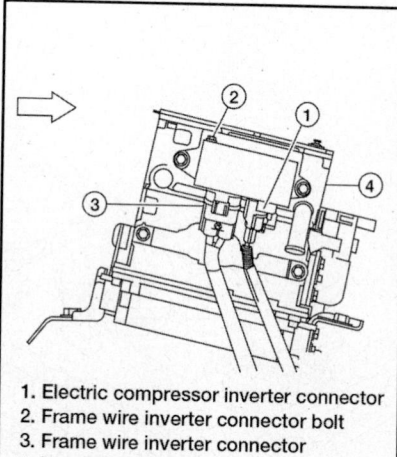

1. Electric compressor inverter connector
2. Frame wire inverter connector bolt
3. Frame wire inverter connector
4. Inverter

37663_ALTH_G0071

Fig. 83 Disconnect the electric compressor inverter connector from the inverter

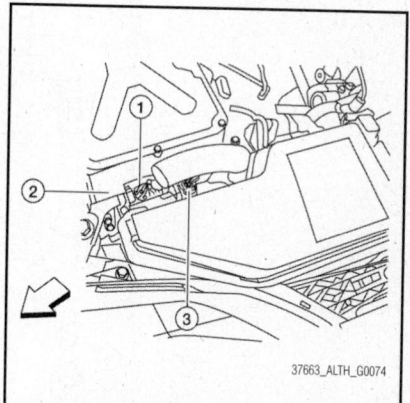

37663_ALTH_G0074

Fig. 86 Disconnect the EGI harness connector (3) from the inverter (2); EGI engine room connector (1)

24. When tightening bolts, perform the following procedure:
  a. Temporarily tighten the bolts A, B, E, F.

  b. Connect MG1 harness connector and MG2 harness connector.
  c. Fully tighten the bolts H, K.

  d. Fully tighten the bolts G, I, J and L.
  e. Fully tighten the bolt F.
  f. Fully tighten the bolts A, B, C, D, E.

ENGINE ELECTRICAL

FIRING ORDERS

See Figure 87.

The firing order for the 2.5L engine is 1–3–4–2. The 2.5L engine is an in-line 4 cylinder engine with cylinders numbered in order front to back of the engine.

The firing order for the 3.5L engine is 1–2–3–4–5–6.

IGNITION COIL

REMOVAL & INSTALLATION

2.5L Engine

See Figure 88.

1. Remove the engine cover.
2. Disconnect the harness connector from the ignition coil.
3. Remove the ignition coil.

✳✳ WARNING

Be careful not to drop or shock the ignition coil.

To install:

4. Installation is in the reverse order of removal.

3.5L Engine

Left Side

See Figure 89.

1. Remove the engine cover.

2. Disconnect ignition coil connector.
3. Remove the ignition coil.

✳✳ WARNING

Be careful not to drop or shock the ignition coil.

To install:

4. Installation is in the reverse order of removal.

Right Side

1. Remove the intake manifold collector.
2. Disconnect ignition coil connector.
3. Remove the ignition coil:

✳✳ WARNING

Be careful not to drop or shock the ignition coil.

To install:

4. Installation is in the reverse order of removal.

IGNITION TIMING

INSPECTION & ADJUSTMENT

The ignition timing is controlled by the ECM. No adjustment is necessary or possible.

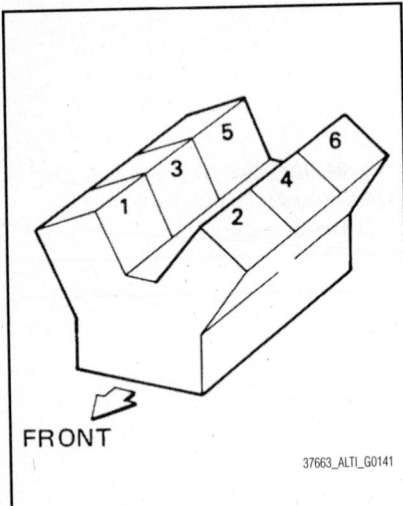

FRONT

37663_ALTI_G0141

Fig. 87 Cylinder number locations—3.5L engine

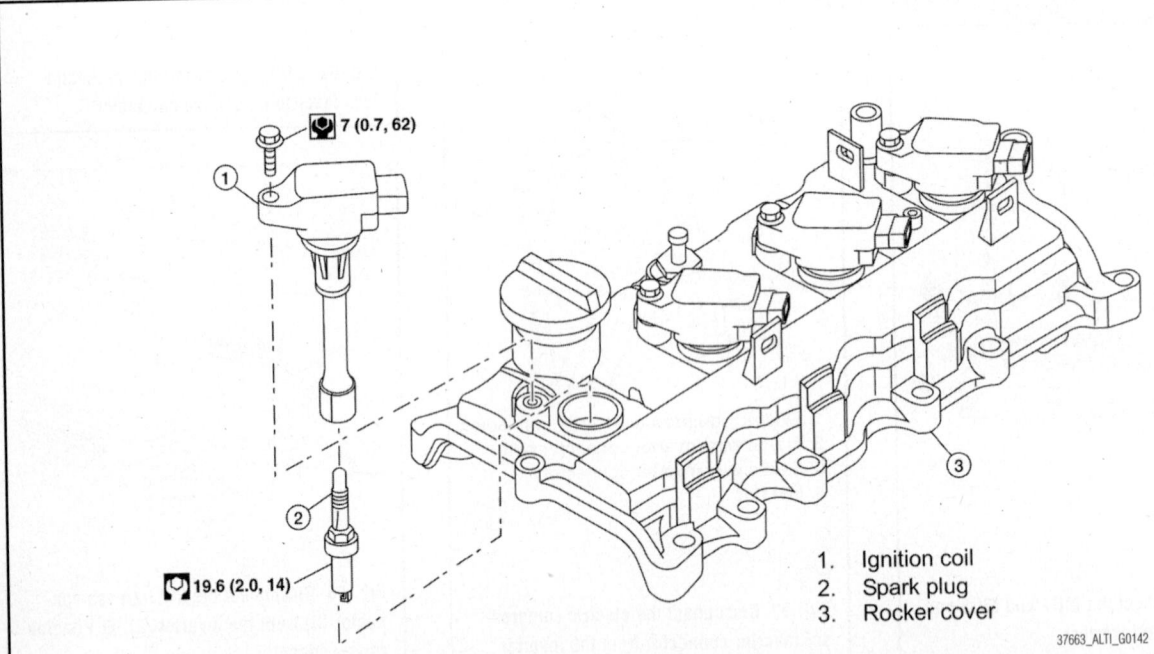

7 (0.7, 62)

19.6 (2.0, 14)

1. Ignition coil
2. Spark plug
3. Rocker cover

37663_ALTI_G0142

Fig. 88 Exploded view of ignition coil & spark plug assemblies

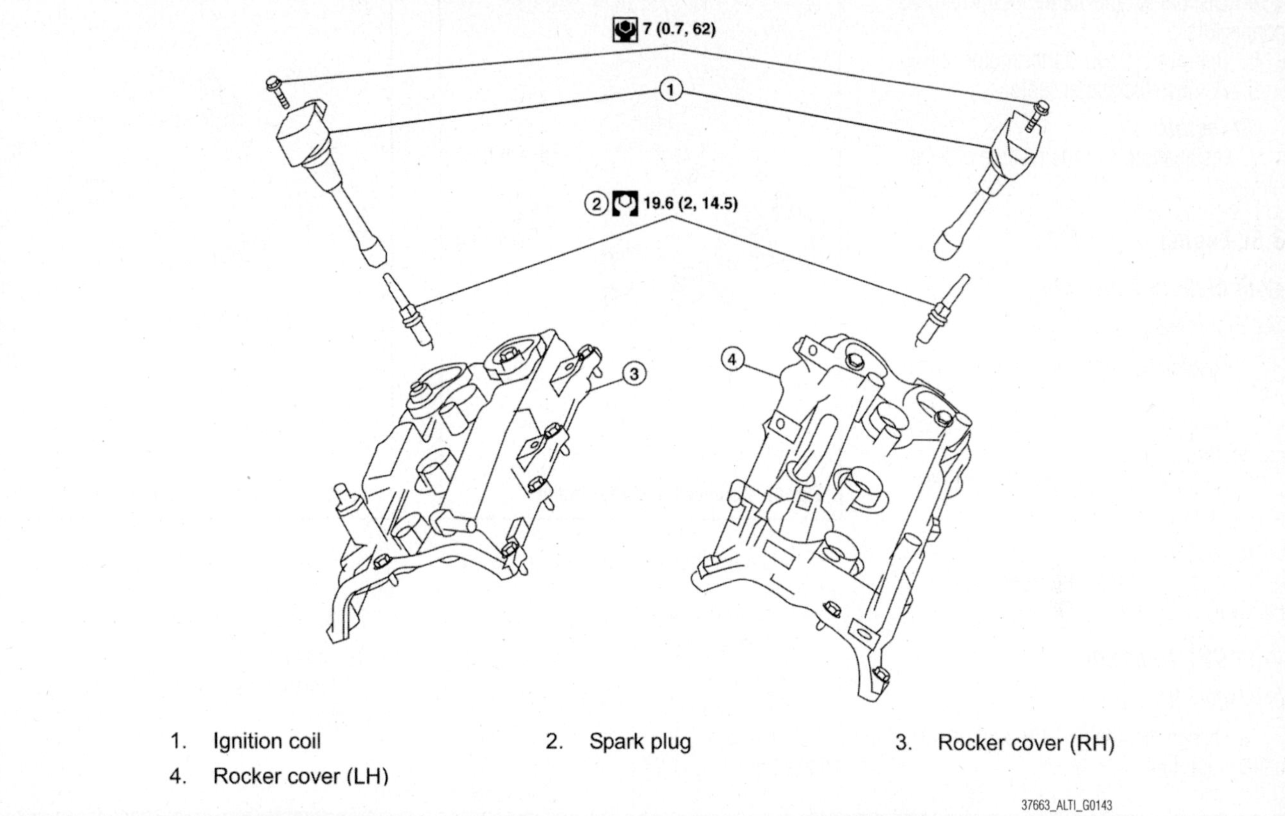

⬛ 7 (0.7, 62)

①

② ⬛ 19.6 (2, 14.5)

③

④

1. Ignition coil
2. Spark plug
3. Rocker cover (RH)
4. Rocker cover (LH)

37663_ALTI_G0143

Fig. 89 Exploded view of ignition coil & spark plug assemblies

SPARK PLUGS

INSPECTION& GAPPING

1. Check spark plugs for burnt or broken tips.

2. Check for cracked insulators.
3. Check spark plug gap and adjust as necessary.

REMOVAL & INSTALLATION
See Figures 88 and 89.

1. Remove the ignition coil.
2. Remove the spark plug with a suitable spark plug wrench.
3. Installation is in the reverse order of removal.

ENGINE ELECTRICAL

Refer to Engine Hybrid System section for information regarding the starting system of hybrid vehicle.

STARTER

REMOVAL & INSTALLATION

2.5L Engine

With Manual Transaxle
See Figure 90.

1. Disconnect the negative battery terminal.
2. Disconnect the starter motor harness connectors.
3. Remove the two starter motor bolts.
4. Remove the starter motor.

To install:
5. Installation is in the reverse order of removal.

⬛ 4 (0.4, 35)

⬛ 67.5 (6.9, 15.2)

37663_ALTI_G0144

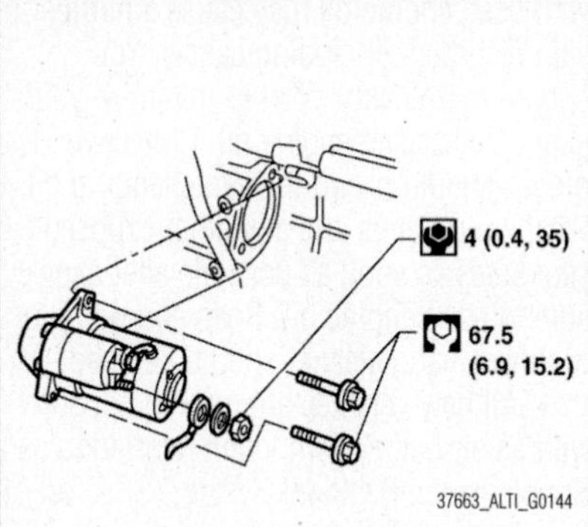

Fig. 90 Remove the starter motor

With CVT Transaxle
See Figure 91.

1. Remove the battery and battery tray bracket.

STARTING SYSTEM

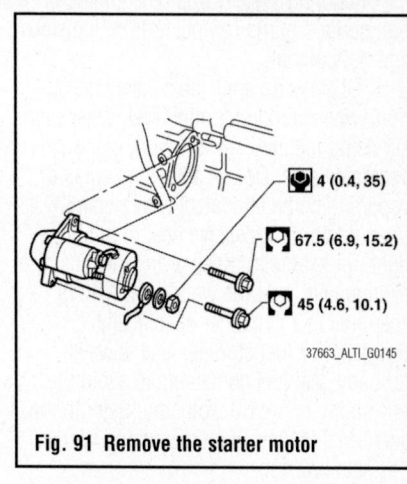

⬛ 4 (0.4, 35)

⬛ 67.5 (6.9, 15.2)

⬛ 45 (4.6, 10.1)

37663_ALTI_G0145

Fig. 91 Remove the starter motor

2. Remove the air cleaner assembly ducts.
3. Disconnect the following:
   • ECM
   • TCM

4. Disconnect the starter motor harness connectors.

5. Remove the two starter motor bolts.

6. Remove the starter motor.

To install:

7. Installation is in the reverse order of removal.

3.5L Engine

With Manual Transaxle

See Figure 92.

1. Disconnect the negative battery terminal.

2. Disconnect the starter motor harness connectors.

3. Remove the two starter motor bolts.

4. Remove the starter motor.

To install:

5. Installation is in the reverse order of removal.

With CVT Transaxle

See Figure 93.

1. Disconnect the negative and positive battery terminal.

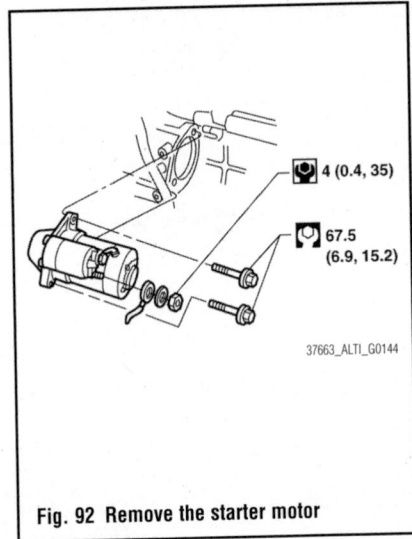

4 (0.4, 35)

67.5 (6.9, 15.2)

37663_ALTI_G0144

Fig. 92 Remove the starter motor

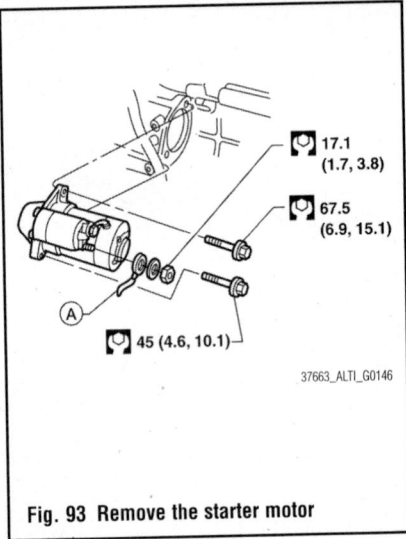

17.1 (1.7, 3.8)

67.5 (6.9, 15.1)

Ⓐ

45 (4.6, 10.1)

37663_ALTI_G0146

Fig. 93 Remove the starter motor

2. Remove the air cleaner assembly and air ducts.

3. Disconnect the following:
   • ECM
   • TCM

4. Remove the battery tray.

5. Disconnect the starter motor harness connectors.

6. Remove the two starter motor bolts.

7. Remove the starter motor.

To install:

8. Installation is in the reverse order of removal.

ENGINE MECHANICAL

PRECAUTIONS

Before servicing any vehicle, please be sure to read all of the following precautions, which deal with personal safety, prevention of component damage, and important points to take into consideration when servicing a motor vehicle:

• Never open, service or drain the radiator or cooling system when the engine is hot; serious burns can occur from the steam and hot coolant.

• Observe all applicable safety precautions when working around fuel. Whenever servicing the item, always work in a well-ventilated area. Do not allow fuel spray or vapors to come in contact with a spark, open flame, or excessive heat (a hot drop light, for example). Keep a dry chemical fire extinguisher near the work area. Always keep fuel in a container specifically designed for fuel storage; also, always properly seal fuel containers to avoid the possibility of fire or explosion. Refer to the additional fuel system precautions later in this section.

• Fuel injection systems often remain pressurized, even after the engine has been turned **OFF**. The fuel system pressure must be relieved before disconnecting any fuel lines. Failure to do so may result in fire and/or personal injury.

• Brake fluid often contains polyglycol ethers and polyglycols. Avoid contact with the eyes and wash your hands thoroughly after handling brake fluid. If you do get brake fluid in your eyes, flush your eyes with clean, running water for 15 minutes. If eye irritation persists, or if you have taken brake fluid internally, IMMEDIATELY seek medical assistance.

• The EPA warns that prolonged contact with used engine oil may cause a number of skin disorders, including cancer. You should make every effort to minimize your exposure to used engine oil. Protective gloves should be worn when changing oil. Wash your hands and any other exposed skin areas as soon as possible after exposure to used engine oil. Soap and water, or waterless hand cleaner should be used.

• All new vehicles are now equipped with an air bag system, often referred to as a Supplemental Restraint System (SRS) or Supplemental Inflatable Restraint (SIR) system. The system must be disabled before performing service on or around system components, steering column, instrument panel components, wiring and sensors. Failure to follow safety and disabling procedures could result in accidental air bag deployment, possible personal injury and unnecessary system repairs.

• Always wear safety goggles when working with, or around, the air bag system. When carrying a non-deployed air bag, be sure the bag and trim cover are pointed away from your body. When placing a non-deployed air bag on a work surface, always face the bag and trim cover upward, away from the surface. This will reduce the motion of the module if it is accidentally deployed. Refer to the additional air bag system precautions later in this section.

• Clean, high quality brake fluid from a sealed container is essential to the safe and proper operation of the brake system. You should always buy the correct type of brake fluid for your vehicle. If the brake fluid becomes contaminated, completely flush the system with new fluid. Never reuse any brake fluid. Any brake fluid that is removed from the system should be discarded. Also, do not allow any brake fluid to come in contact with a painted surface; it will damage the paint.

• Never operate the engine without the proper amount and type of engine oil; doing so WILL result in severe engine damage.

• Timing belt maintenance is extremely important. Many models utilize an interference-type, non-freewheeling engine. If the timing belt breaks, the valves in the cylinder head may strike the pistons, causing potentially serious (also time-consuming and expensive) engine damage. Refer to the maintenance interval charts for the recommended replacement interval for the timing belt, and to the timing belt section for belt replacement and inspection.

- Disconnecting the negative battery cable on some vehicles may interfere with the functions of the on-board computer system(s) and may require the computer to undergo a relearning process once the negative battery cable is reconnected.

- When servicing drum brakes, only disassemble and assemble one side at a time, leaving the remaining side intact for reference.

- Only an MVAC-trained, EPA-certified automotive technician should service the air conditioning system or its components.

ACCESSORY DRIVE BELT SYSTEM

ADJUSTMENT

Belt tension is not manually adjustable, it is automatically adjusted by the drive belt auto-tensioner. If the belt is out of adjustment, replace the accessory drive belt.

BELT ROUTINGS

See Figures 94 through 96.

INSPECTION

⁕⁕ CAUTION

For Hybrid vehicle, inspect the drive belt only when the Hybrid System is off.

1. Make sure that the stamp mark of drive belt auto-tensioner is within the usable range.
2. Check the drive belt auto-tensioner indicator (notch) when the engine is cold.

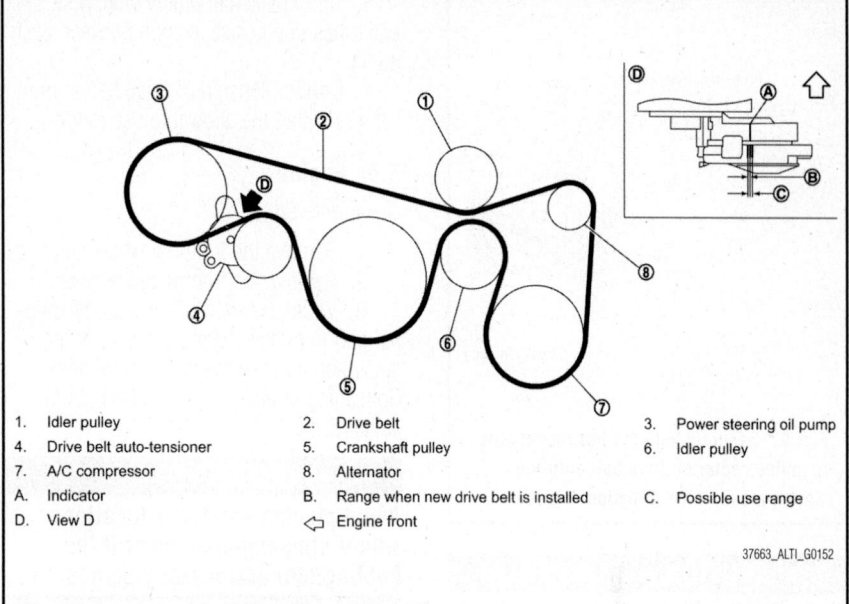

1.	Idler pulley	2.	Drive belt	3.	Power steering oil pump
4.	Drive belt auto-tensioner	5.	Crankshaft pulley	6.	Idler pulley
7.	A/C compressor	8.	Alternator		
A.	Indicator	B.	Range when new drive belt is installed	C.	Possible use range
D.	View D	⟵	Engine front		

37663_ALTI_G0152

Fig. 95 Accessory drive belt routing—3.5L engine

1.	Drive belt auto-tensioner	2.	Crankshaft	3.	Water pump
4.	Idler pulley	A.	Water pump belt working range	B.	Minimum belt length
C.	Nominal position	D.	Maximum belt length	E.	Maximum belt length +0.8%
F.	View F				

37663_ALTH_G0036

Fig. 96 Accessory drive belt routing—Hybrid engine

3. When the new drive belt is installed, check that the belt is in range.
4. Visually check entire belt for wear, damage or cracks.
5. If the indicator is out of allowable use range or belt is damaged, replace the belt.

REMOVAL & INSTALLATION

2.5L Engine

See Figure 97.

1. Remove the fender protector side cover RH.
2. Securely hold the hexagonal part in pulley center of drive belt auto-tensioner, move in the direction of arrow (loosening direction of tensioner) using Tool J-46535.

⁕⁕ WARNING

Avoid placing hand in a location where pinching may occur if the holding tool accidentally comes off.

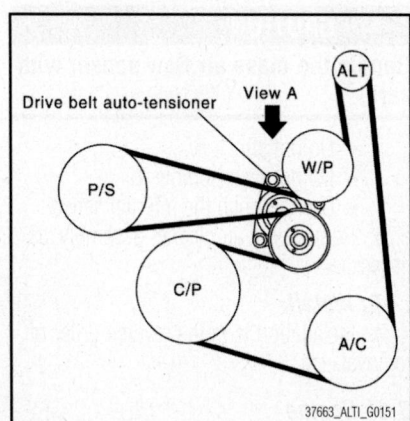

37663_ALTI_G0151

Fig. 94 Accessory drive belt routing— 2.5L engine, except Hybrid

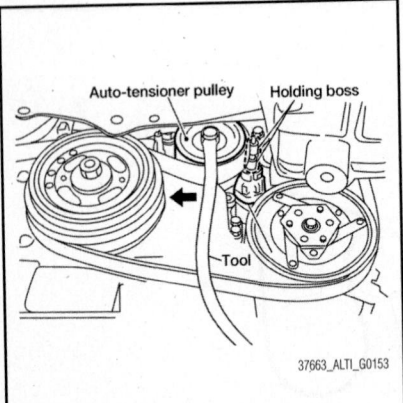

Fig. 97 Securely hold the hexagonal part in pulley center of drive belt auto-tensioner, move in the direction of arrow

❊❊ CAUTION

Do not loosen the auto-tensioner pulley bolt. (Do not turn it counterclockwise.) If turned counterclockwise, the complete auto-tensioner must be replaced as a unit, including pulley.

  3. Insert a rod approximately 0.24 inches (6 mm) in diameter through the rear of tensioner into retaining boss to lock tensioner pulley.
  4. Leave tensioner pulley arm locked until drive belt is installed again.
  5. Loosen drive belt from water pump pulley and then remove it from the other pulleys.

To install:

  6. Install the drive belt onto all of the pulleys except for the water pump pulley. Then install the drive belt onto water pump pulley last.

❊❊ CAUTION

Confirm belts are completely set on the pulleys.

  7. Release tensioner, and apply tension to drive belt.

❊❊ WARNING

Avoid placing hand in a location where pinching may occur if the holding tool accidentally comes off.

❊❊ CAUTION

Do not loosen the auto-tensioner pulley bolt. Don't turn it counterclockwise. If turned counterclockwise, the complete auto-tensioner must be replaced as a unit, including pulley.

  8. Turn crankshaft pulley clockwise several times to equalize tension between each pulley.
  9. Confirm tension of drive belt at indicator is within the allowable use range.

3.5L Engine

See Figure 98.

  1. Remove the front RH wheel and tire.
  2. Remove the front RH side cover.
  3. While securely holding the hexagonal part in pulley center of drive belt auto-tensioner, move in the direction of arrow (loosening direction of tensioner) using suitable tool.

❊❊ WARNING

Avoid placing hand in a location where pinching may occur if the holding tool accidentally comes off.

❊❊ CAUTION

Do not loosen the auto-tensioner pulley bolt. (Do not turn it counterclockwise.) If turned counterclockwise, the complete auto-tensioner must be replaced as a unit, including pulley.

  4. Insert a rod approximately 0.24 inches (6 mm) in diameter through the rear of tensioner into retaining boss to lock tensioner pulley.
  5. Leave tensioner pulley arm locked until belt is installed again.
  6. Loosen drive belt from water pump pulley and then remove it from the other pulleys.

To install:

  7. Install the drive belt onto all of the pulleys.

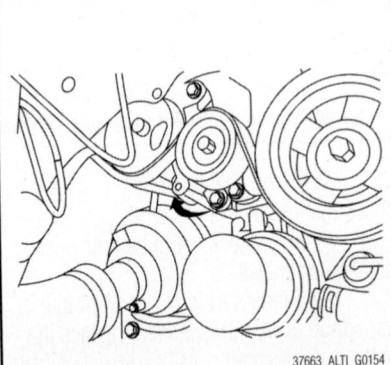

Fig. 98 While securely holding the hexagonal part in pulley center of drive belt auto-tensioner, move in the direction of arrow

❊❊ CAUTION

Confirm belts are completely set on the pulleys.

  8. Release tensioner, and apply tension to belt.

❊❊ WARNING

Avoid placing hand in a location where pinching may occur if the holding tool accidentally comes off.

❊❊ CAUTION

Do not loosen the auto-tensioner pulley bolt. (Don't turn it counterclockwise. If turned counterclockwise, the complete auto-tensioner must be replaced as a unit, including pulley.

  9. Turn crankshaft pulley clockwise several times to equalize tension between each pulley.
  10. Confirm tension of belt at indicator is within the possible use range.

AIR CLEANER

REMOVAL & INSTALLATION

Air Cleaner Assembly

2.5L Engine, Except Hybrid

See Figure 99.

  1. Remove front air duct.
  2. Disconnect the air duct hose clamps at the electric throttle control actuator and the air cleaner assembly.
  3. Remove air duct hose.
  4. Disconnect the mass air flow sensor
  5. Remove mass air flow sensor from air cleaner assembly, as necessary.

❊❊ CAUTION

Handle the mass air flow sensor with care:

  • Do not shock it.
  • Do not disassemble it.
  • Do not touch the internal sensor.
  6. Remove the air cleaner assembly, as necessary.

To install:

  7. Installation is in the reverse order of removal.

3.5L Engine

See Figure 100.

  1. Remove engine room cover.
  2. Remove front air duct.

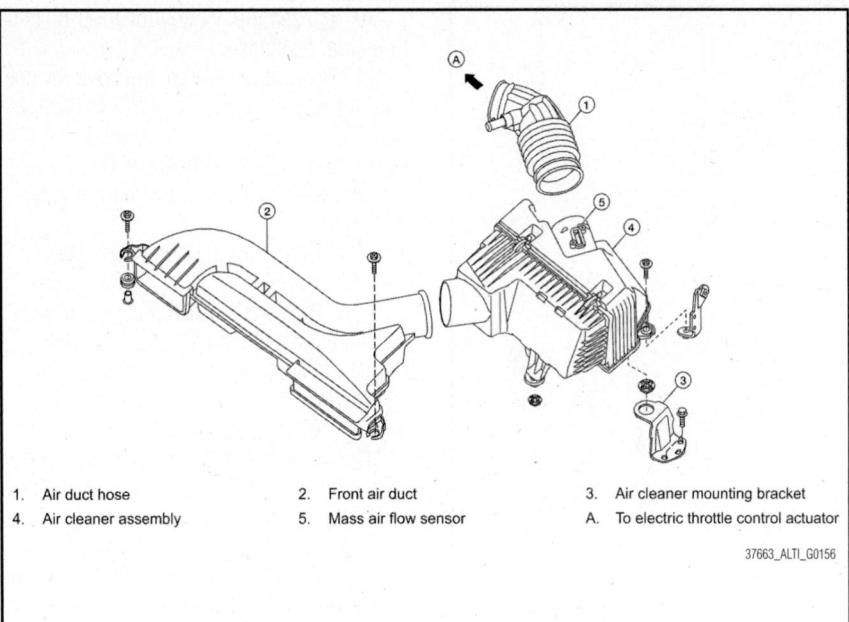

1. Air duct hose
2. Front air duct
3. Air cleaner mounting bracket
4. Air cleaner assembly
5. Mass air flow sensor
A. To electric throttle control actuator

37663_ALTI_G0156

Fig. 99 Exploded view of air cleaner assembly

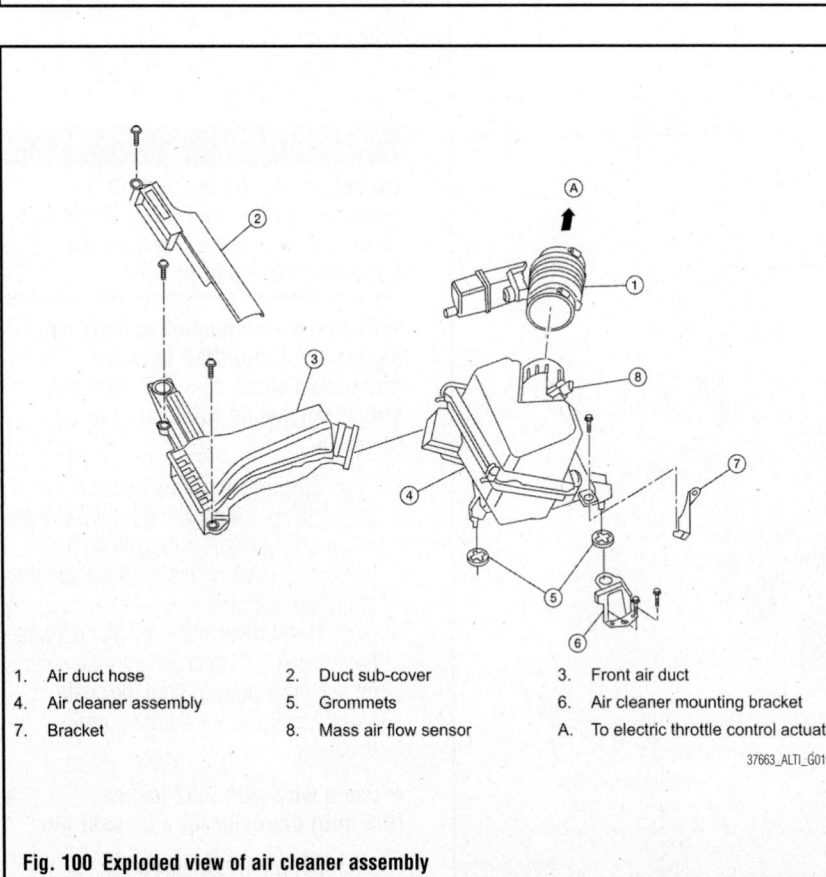

1. Air duct hose
2. Duct sub-cover
3. Front air duct
4. Air cleaner assembly
5. Grommets
6. Air cleaner mounting bracket
7. Bracket
8. Mass air flow sensor
A. To electric throttle control actuator

37663_ALTI_G0157

Fig. 100 Exploded view of air cleaner assembly

3. Disconnect the tube clamp at the electric throttle control actuator and at the air duct hose.

4. Disconnect the blow-by hose.

5. Remove air duct hose.

6. Disconnect mass air flow sensor.

7. Remove mass air flow sensor from air cleaner assembly, as necessary.

✳✳ CAUTION

Handle mass air flow sensor with care:

- Do not shock it.
- Do not disassemble it.
- Do not touch its sensor.

8. Remove air cleaner assembly.

To install:

9. Installation is in the reverse order of removal.

Hybrid

See Figure 101.

1. Remove front air duct.

2. Disconnect the air duct hose clamps at the electric throttle control actuator and the air cleaner assembly.

3. Disconnect the mass air flow sensor

4. Remove mass air flow sensor from air cleaner assembly, as necessary.

✳✳ WARNING

Handle the mass air flow sensor with care:

- Do not shock it.
- Do not disassemble it.
- Do not touch the internal sensor.

5. Remove the air cleaner assembly, as necessary.

To install:

6. Installation is in the reverse order of removal.

Air Filter Element

2.5L Engine

See Figure 102.

➡**It is not necessary to remove the front air duct to replace the air cleaner element.**

1. Unhook the air cleaner case side clips.
2. Remove the air cleaner filter.
3. Install a new air cleaner filter.
4. Lock the air cleaner case side clips.

3.5L Engine

See Figure 103.

➡**It is not necessary to remove the front air duct to replace the air cleaner element.**

1. Unhook the air cleaner case side clips.
2. Remove the air cleaner filter.
3. Install a new air cleaner filter.
4. Lock the air cleaner case side clips.

CAMSHAFT AND VALVE LIFTERS

REMOVAL & INSTALLATION

2.5L Engine

See Figures 104 through 119.

1. Remove the rocker cover.

2. Remove the front right side tire and wheel.

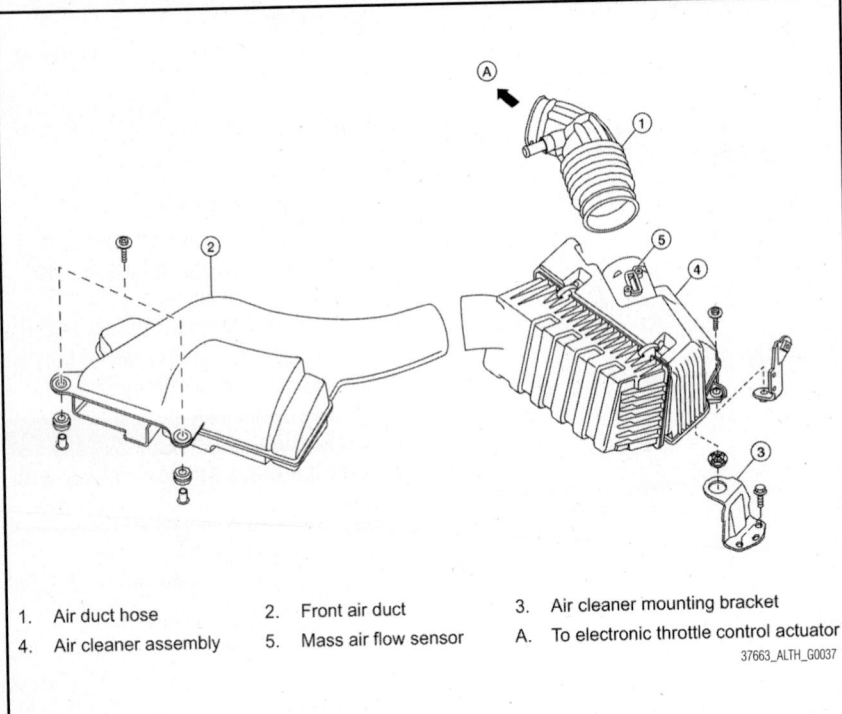

1. Air duct hose
4. Air cleaner assembly
2. Front air duct
5. Mass air flow sensor
3. Air cleaner mounting bracket
A. To electronic throttle control actuator

37663_ALTH_G0037

Fig. 101 Exploded view of air cleaner assembly

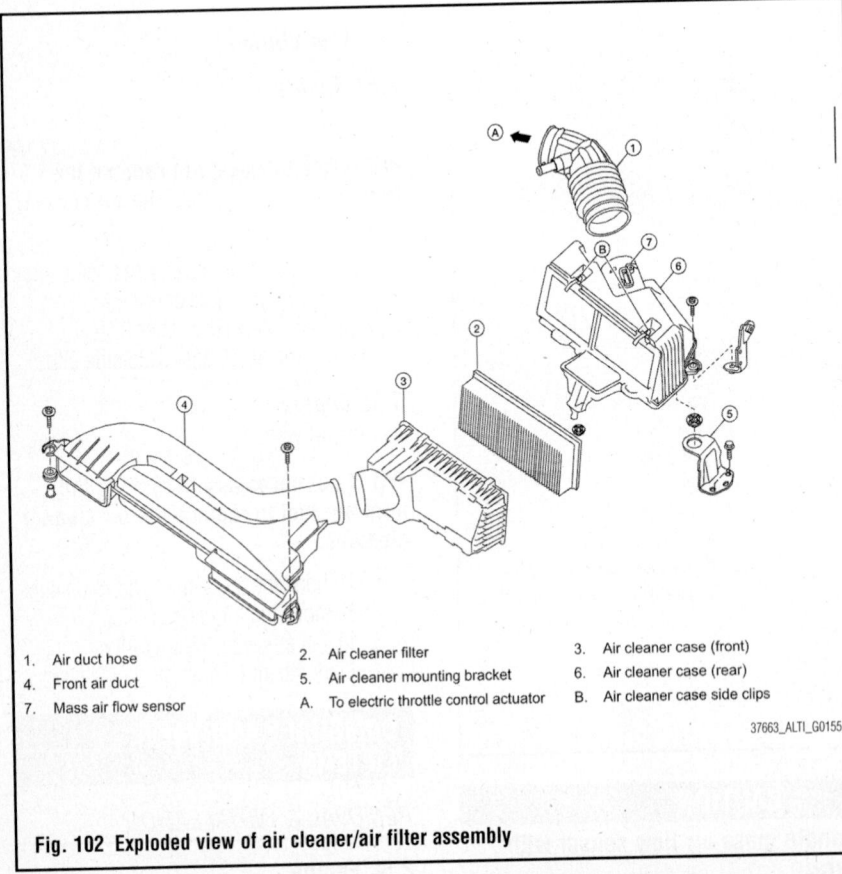

1. Air duct hose
4. Front air duct
7. Mass air flow sensor
2. Air cleaner filter
5. Air cleaner mounting bracket
A. To electric throttle control actuator
3. Air cleaner case (front)
6. Air cleaner case (rear)
B. Air cleaner case side clips

37663_ALTI_G0155

Fig. 102 Exploded view of air cleaner/air filter assembly

3. Remove the RH splash shield.
4. Remove the drive belt.
5. Remove the power steering reservoir.
6. Remove the coolant overflow reservoir tank.
7. Disconnect variable timing control solenoid and camshaft sensor harness connectors.
8. Remove camshaft sensor.
9. Remove camshaft sensor bracket.

10. Loosen the IVT control cover bolts in the order as shown.
11. Remove the IVT control cover by cutting the sealant using Tool KV10111100 (J-37228).
12. Set the No.1 cylinder at TDC on its compression stroke with the following procedure:
   a. Open the splash cover on RH under cover.
   b. Rotate crankshaft pulley clockwise, and align mating marks for TDC with timing indicator on front cover, as shown.
   c. At the same time, make sure that the mating marks on camshaft sprockets are lined up with the yellow links in the timing chain, as shown.
   d. If not, rotate crankshaft pulley one more turn to line up the mating marks to the yellow links, as shown.
13. Pull the timing chain guide out between the camshaft sprockets through front cover.
14. Remove camshaft sprockets with the following procedure.

✳✳ CAUTION

Do not rotate the crankshaft or camshaft while the timing chain is removed. It causes interference between valve and piston.

➡**Chain tension holding work is not necessary. Crankshaft sprocket and timing chain do not disconnect structurally while front cover is attached.**

   a. Line up the mating marks on camshaft sprockets with the yellow links in the timing chain, and paint an indelible mating mark on the sprocket and timing chain link plate.
   b. Push in the tensioner plunger and hold. Insert a stopper pin into the hole on tensioner body to hold the chain tensioner. Remove the timing chain tensioner.

➡**Use a wire with 0.02 inches (0.5 mm) diameter for a stopper pin.**

   c. Secure the hexagonal part of camshaft with a suitable tool.
   d. Loosen the camshaft sprocket mounting bolts and remove the camshaft sprockets.
15. Loosen the camshaft bracket bolts in the order as shown, and remove the camshaft brackets and camshafts.
16. Remove No.1 camshaft bracket by slightly tapping it with a rubber mallet.
17. Remove the valve lifters.

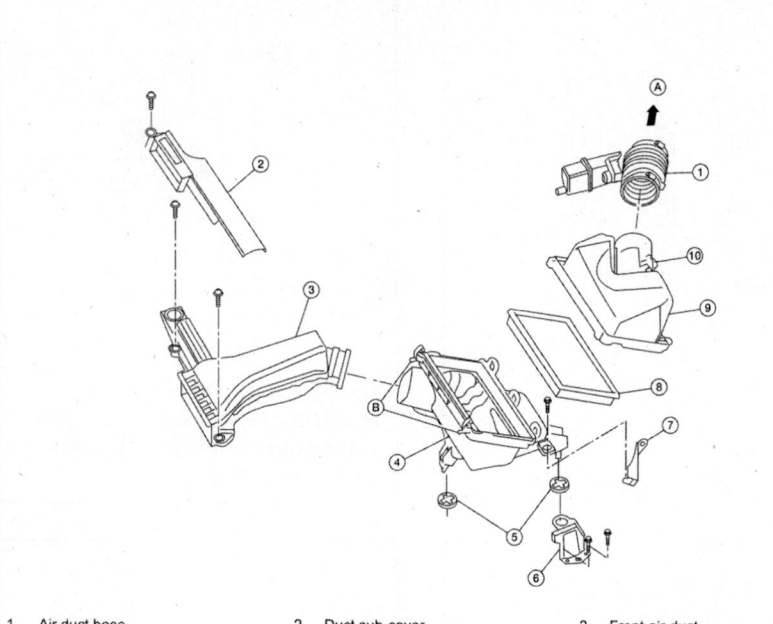

1. Air duct hose
2. Duct sub-cover
3. Front air duct
4. Air cleaner case (lower)
5. Grommets
6. Air cleaner case mounting bracket
7. Bracket
8. Air cleaner filter
9. Air cleaner case (upper)
10. Mass air flow sensor
A. To electric throttle control actuator
B. Air cleaner case side clips

37663_ALTI_G0158

Fig. 103 Exploded view of air cleaner/air filter assembly

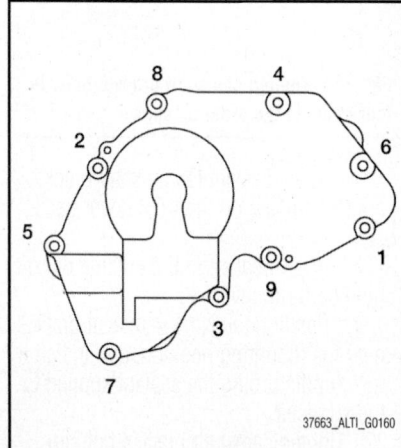

37663_ALTI_G0160

Fig. 104 Loosen the IVT control cover bolts in the order as shown

37663_ALTI_G0161

Fig. 105 Remove the IVT control cover by cutting the sealant

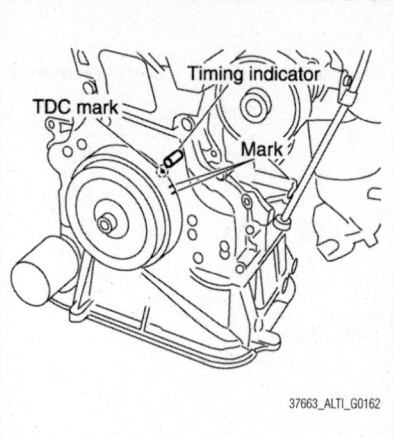

37663_ALTI_G0162

Fig. 106 Set the No.1 cylinder at TDC on its compression stroke

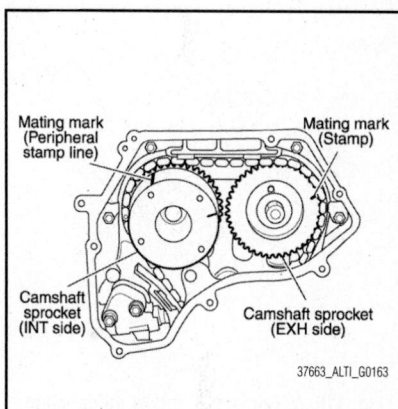

37663_ALTI_G0163

Fig. 107 Make sure that the mating marks on camshaft sprockets are lined up with the yellow links in the timing chain

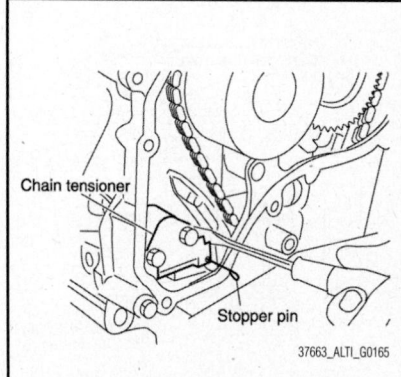

37663_ALTI_G0165

Fig. 108 Push in the tensioner plunger and hold; insert a stopper pin into the hole on tensioner body to hold the chain tensioner

18. Check mounting positions, and set them aside in the order removed.

To install:
19. Install the valve lifter.

➡**Install them in the same position from which they were removed.**

20. Install the camshafts.
   a. The distinction between the intake and exhaust camshafts is in a difference of shapes of the back end:

- A: Exhaust
- B: Intake Signal plate for the Camshaft Position (CMP) sensor (PHASE)

   b. Install camshafts so that the dowel pins on the front side are positioned as shown.
21. Install camshaft brackets.
   a. Install by referring to identification mark on upper surface mark.

   b. Install so that identification mark can be correctly read when viewed from the exhaust side.
   c. Install No. 1 camshaft bracket as follows.

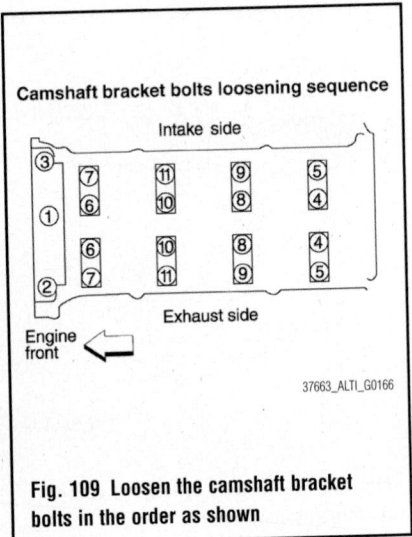

Fig. 109 Loosen the camshaft bracket bolts in the order as shown

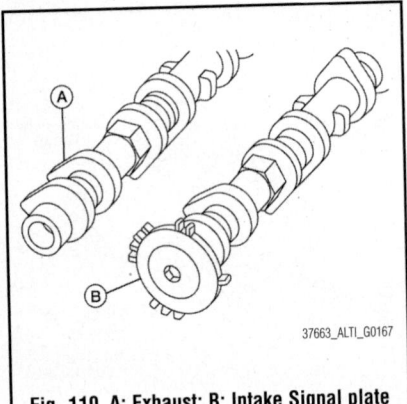

Fig. 110 A: Exhaust; B: Intake Signal plate for the Camshaft Position (CMP) sensor (PHASE)

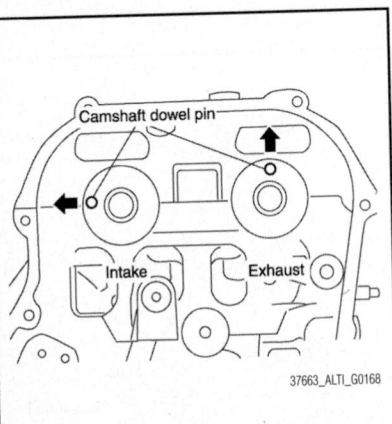

Fig. 111 Install camshafts so that the dowel pins on the front side are positioned as shown

d. Apply sealant to No.1 camshaft bracket as shown.

➡ Use Genuine Silicone RTV Sealant, or equivalent.

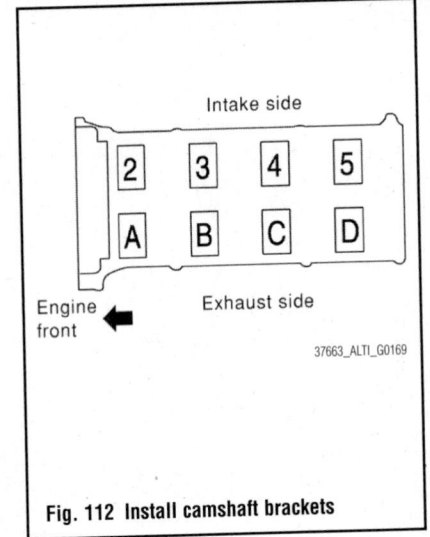

Fig. 112 Install camshaft brackets

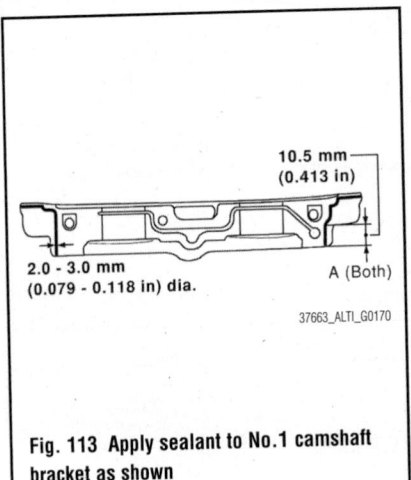

Fig. 113 Apply sealant to No.1 camshaft bracket as shown

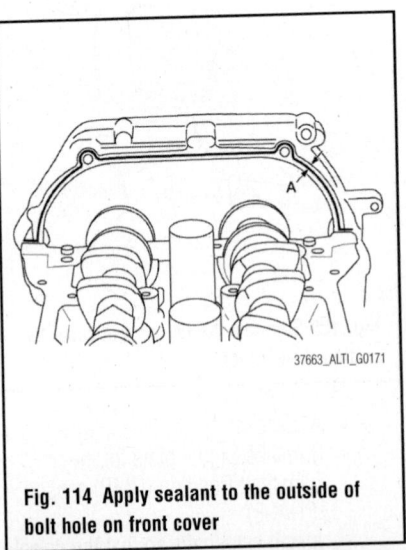

Fig. 114 Apply sealant to the outside of bolt hole on front cover

✳✳ CAUTION

After installation, be sure to wipe off any excessive sealant leaking from part (A) (both on right and left sides).

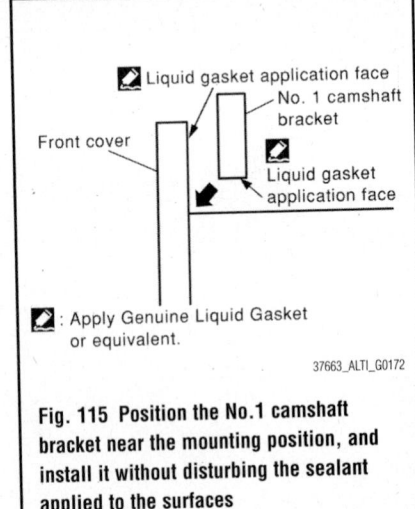

Fig. 115 Position the No.1 camshaft bracket near the mounting position, and install it without disturbing the sealant applied to the surfaces

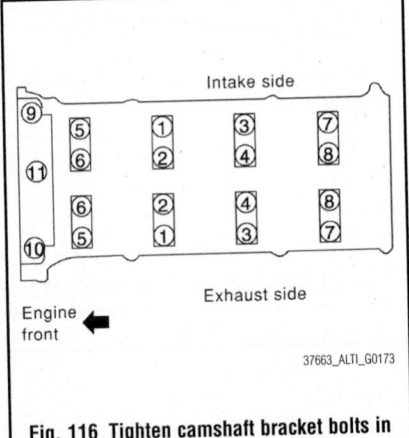

Fig. 116 Tighten camshaft bracket bolts in four steps in the order as shown

e. Apply sealant to camshaft bracket contact surface on the front cover backside.

f. Apply sealant to the outside of bolt hole on front cover.

g. Position the No.1 camshaft bracket near the mounting position, and install it without disturbing the sealant applied to the surfaces.

22. Tighten camshaft bracket bolts in four steps in the order as shown.

a. Step 1: Bolts 9–11: 17 inch lbs. (2. Nm)

b. Step 2: Bolts 1–8: 17 inch lbs. (2. Nm)

c. Step 3: Bolts 1–11: 52 inch lbs. (6 Nm)

d. Step 4: Bolts 1–11: 92 inch lbs. (10 Nm)

✳✳ WARNING

After tightening camshaft bracket bolts, be sure to wipe off excessive sealant from the mating surface of rocker cover and the mating surface of front cover, when installed without the front cover.

23. Install camshaft sprockets.

   a. Install them by lining up the mating marks on each camshaft sprocket with the ones painted on the timing chain during removal.

   b. Before installation of chain tensioner, it is possible to re-match the marks on timing chain with the ones on each sprocket.

✳✳ CAUTION

Aligned mating marks could slip. Therefore, after matching them, hold the timing chain in place by hand. Before and after installing chain tensioner, check again to make sure that mating marks have not slipped.

24. Install chain tensioner.

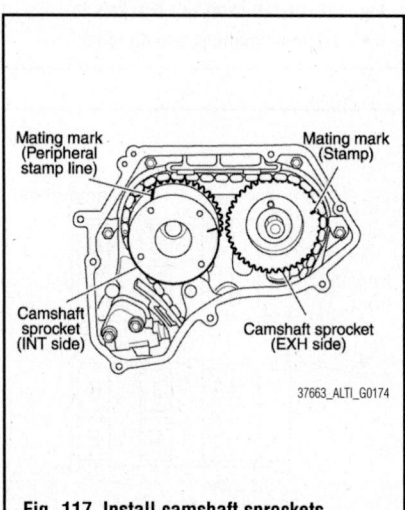

Fig. 117 Install camshaft sprockets

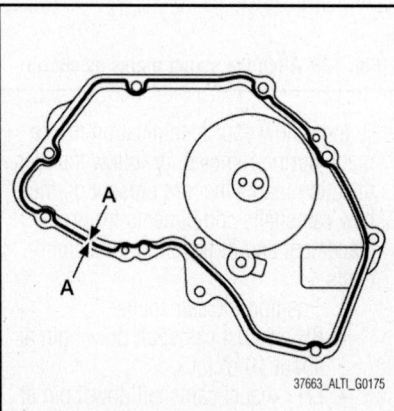

Fig. 118 Apply Genuine Silicone RTV Sealant to the positions as shown

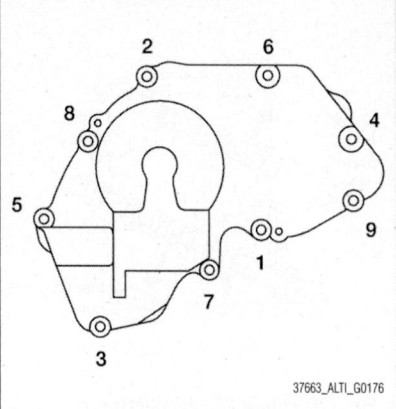

Fig. 119 Tighten the bolts in the numerical order as shown

✳✳ CAUTION

After installation, pull the stopper pin off completely, and make sure that the tensioner is fully released.

25. Install chain guide.
26. Install IVT control cover with the following procedure.

   a. Install IVT control solenoid valve to intake valve timing control cover.

   b. Install O-ring to front cover side.

   c. Apply Genuine Silicone RTV Sealant to the positions as shown.

   d. Install IVT control cover.

   e. Tighten the bolts in the numerical order as shown.

27. Check and adjust valve clearances.

28. Installation of the remaining components is in the reverse order.

3.5L Engine

See Figures 120 through 131.

1. Remove the timing chains.
2. Remove camshaft position brackets (RH shown LH similar).
3. Remove the intake and exhaust camshaft brackets and the camshafts.

   a. Mark the camshafts, camshaft brackets, and bolts so they are placed in the same position and direction for installation.

   b. Equally loosen the camshaft bracket bolts in several steps in the numerical order as shown.

4. Remove valve lifters, if necessary.

➡**Identify installation positions to ensure proper installation.**

5. Remove secondary timing chain tensioner from cylinder head. Remove

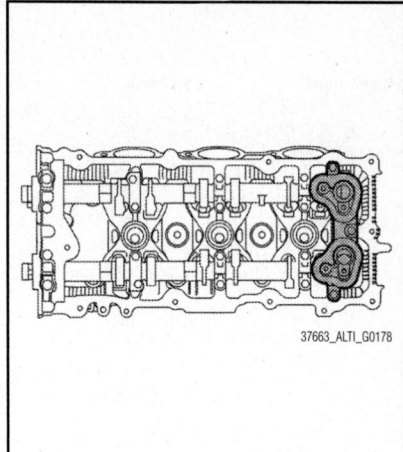

Fig. 120 Remove camshaft position brackets

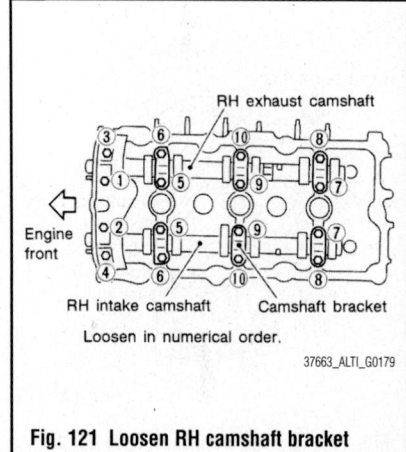

Fig. 121 Loosen RH camshaft bracket bolts in order

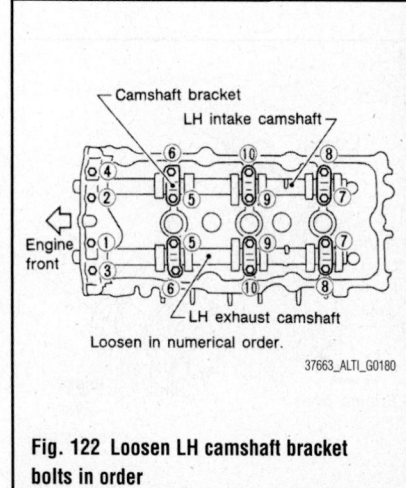

Fig. 122 Loosen LH camshaft bracket bolts in order

secondary tensioner with its stopper pin attached.

➡**Stopper pin was attached when secondary timing chain was removed.**

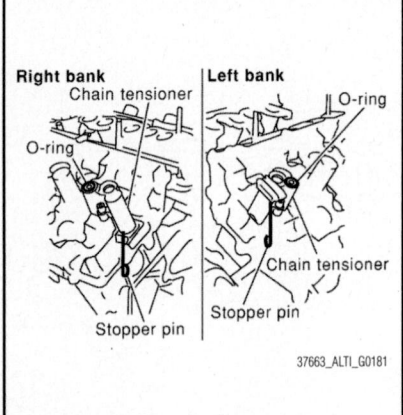

Fig. 123 Remove secondary timing chain tensioner from cylinder head

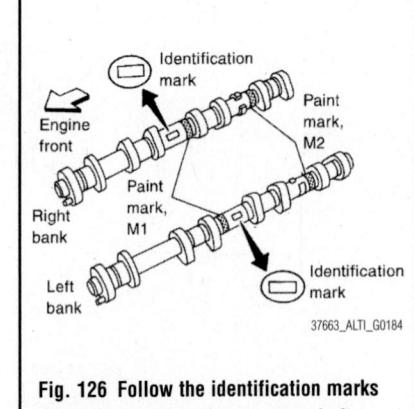

Fig. 126 Follow the identification marks that are present on the new camshafts components for proper placement and direction of the components

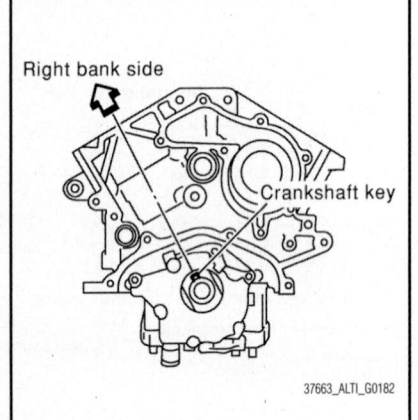

Fig. 124 Turn the crankshaft until No. 1 piston is set at TDC on the compression stroke

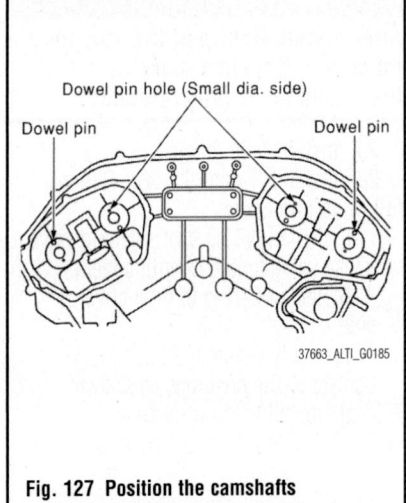

Fig. 127 Position the camshafts

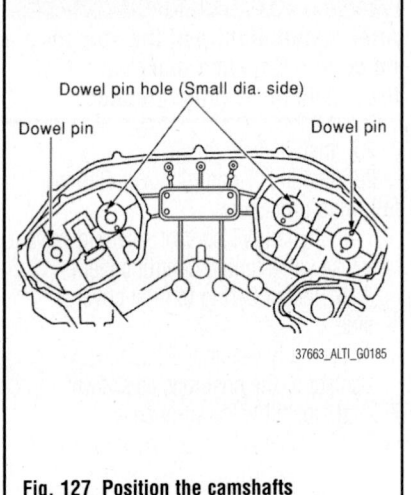

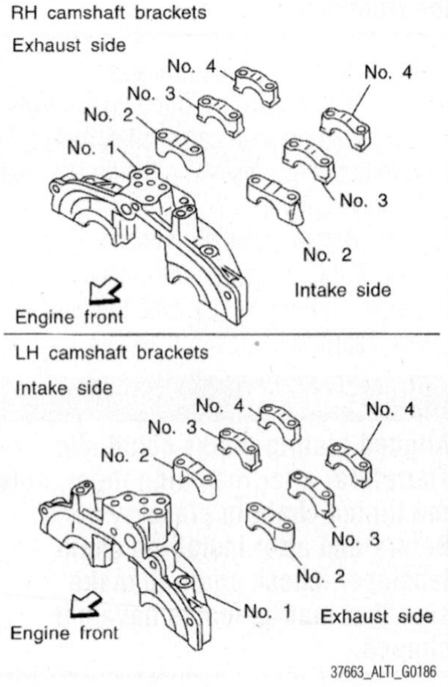

Fig. 128 Install camshaft brackets in their original positions and direction

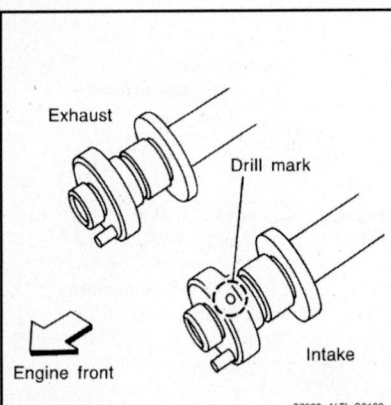

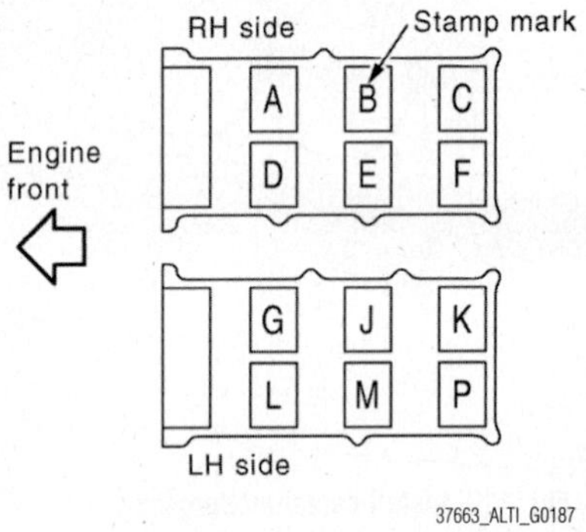

Fig. 125 Intake camshaft has a drill mark on camshaft sprocket mounting flange

Fig. 129 Align the stamp marks as shown

To install:

6. Before installation, remove any old Silicone RTV Sealant from component mating surfaces using a scraper.

a. Remove the old Silicone RTV Sealant from the bolt holes and threads.

b. Do not scratch or damage the mating surfaces.

7. Before installing the front cam bracket, remove the old Silicone RTV Sealant from the mating surface using a scraper. Do not scratch or damage the mating surface.

8. Turn the crankshaft until No. 1 piston is set at TDC on the compression stroke.

➡**The crankshaft key should line up with the right bank cylinder center line as shown.**

9. Install camshaft chain tensioners on both sides of cylinder head.

10. Install valve lifters, if removed.

➡**Install them in original positions.**

11. Install exhaust and intake camshafts and camshaft brackets.

➡**Intake camshaft has a drill mark on camshaft sprocket mounting flange.**

a. Follow your identification marks made during removal, or follow the identification marks that are present on the new camshafts components for proper placement and direction of the components.

b. Position the camshafts:
• RH exhaust camshaft dowel pin at about 10 o'clock.
• LH exhaust camshaft dowel pin at about 2 o'clock.

c. Before installing camshaft brackets, apply sealant to mating surface of No. 1 camshaft bracket.

d. Before installation, wipe off any protruding sealant.

e. Install camshaft brackets in their original positions and direction.

f. Align the stamp marks as shown.

g. If checking and adjusting any part of valve assembly or camshaft, check valve clearance according to the reference data.

h. Tighten the camshaft brackets in the three steps:

- Step 1: Tighten Bolts 7–10, then tighten 1–6 in numerical order to 17 inch lbs. (2 Nm).
- Step 2: Tighten all bolts in numerical order to 52 inch lbs. (6 Nm).
- Step 3: Tighten all bolts in numerical order to 8 ft. lbs. (10 Nm).

12. Measure difference in levels between front end faces of No. 1 camshaft bracket and cylinder head.

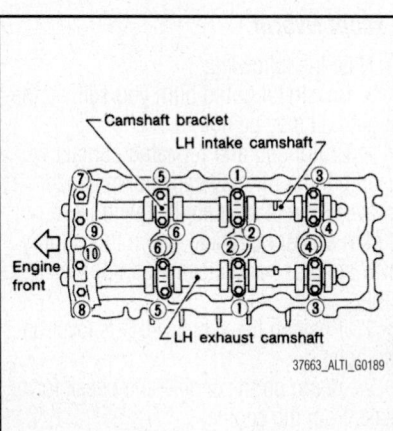

Fig. 130 Tighten the RH camshaft brackets in the three steps, in numerical order

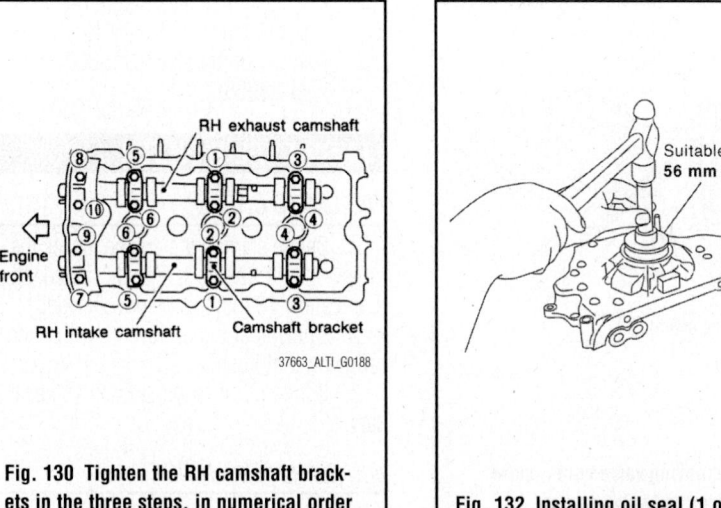

Fig. 131 Tighten the LH camshaft brackets in the three steps, in numerical order

✲✲ WARNING

If measurement is outside the specified range of -0.0055 inches (-0.14 mm), re-install camshaft and camshaft bracket.

13. Install Camshaft Position (CMP) sensor (PHASE) (RH and LH bank.)

14. Install the fuel rail and injectors.

15. Install the timing chains.

CRANKSHAFT FRONT SEAL

REMOVAL & INSTALLATION

See Figures 132 and 133.

1. Remove RH front wheel.
2. Remove fender protector side cover RH.
3. Remove drive belts.
4. Remove crankshaft pulley.
5. Remove front oil seal from front cover.

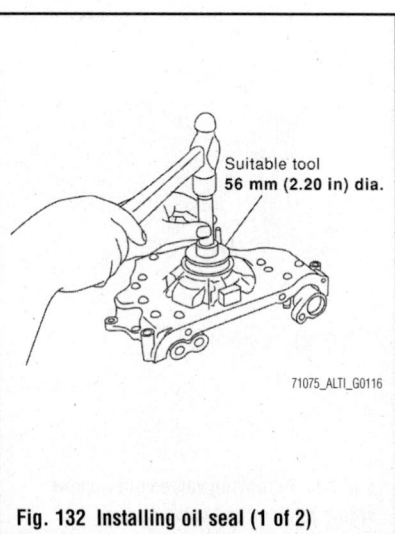

Fig. 132 Installing oil seal (1 of 2)

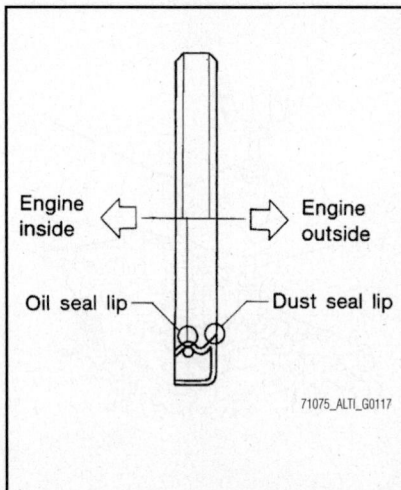

Fig. 133 Installing oil seal (2 of 2)

✲✲ WARNING

Be careful not to scratch front cover.

To install:

6. Install new front oil seal to front cover using suitable tool.

a. Install new oil seal in until it is flush with front end surface of front cover.

✲✲ WARNING

Do not reuse oil seal.

✲✲ WARNING

Be careful not to cause damage to circumference of oil seal.

b. Install new oil seal in the direction shown.

7. Installation of the remaining components is in the reverse order of removal.

CRANKSHAFT PULLEY

REMOVAL & INSTALLATION

2.5L Engine

1. Remove crankshaft pulley with the following procedure:

a. Hold the crankshaft pulley using suitable tool, then loosen the crankshaft pulley bolt, and pull the pulley out about 0.39 inches (10 mm).

b. Attach suitable pulley puller in the M 6 (0.24 inch diameter) thread hole on crankshaft pulley, and remove crankshaft pulley using a suitable puller.

2. Installation is the reverse of removal.

3.5L Engine

See Figures 134 and 135.

1. Remove the following parts:

- Engine undercover.
- Drive belts.

2. Disconnect the battery negative terminal.

3. Remove the crankshaft pulley as follows:

a. Remove the starter motor.

b. Lock the ring gear using Tool KV10117700 (J-44716) attached to the starter bolt hole.

✲✲ WARNING

Do not damage the ring gear teeth, or the signal plate teeth behind the ring gear when setting the stopper.

c. Loosen crankshaft pulley bolt using Tool KV10109300 and locate bolt

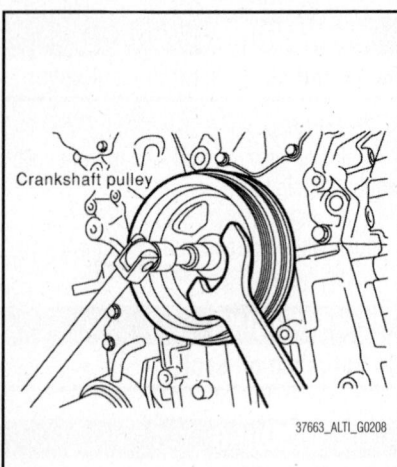

Fig. 134 Loosen crankshaft pulley bolt

Fig. 135 Position a pulley puller at recess hole of crankshaft pulley to remove crankshaft pulley

seating surface at 10 mm (0.39 in) from its original position.

    d. Position a pulley puller at recess hole of crankshaft pulley to remove crankshaft pulley.

☀ WARNING

Do not use a puller claw on crankshaft pulley periphery.

To install:

  4. Install crankshaft pulley and tighten the bolt in two steps.

    a. Lubricate thread and seat surface of the bolt with new engine oil.

    b. For the second step angle tighten using Tool KV10112100 (BT-8653-A).

    c. Step 1: 32 ft. lbs. (44 Nm)

    d. Step 2: Tighten an additional 84–90 degrees

  5. Remove tool attached to the starter bolt hole.

  6. Installation of the remaining components is in reverse order of removal.

CRANKSHAFT REAR COVER & SEAL

REMOVAL & INSTALLATION

See Figures 136 through 138.

  1. Remove camshaft.
  2. Remove valve lifter.
  3. Rotate crankshaft, and set piston whose oil seal is to be removed, to top dead center. This prevents valve from dropping inside cylinder.

☀ WARNING

When rotating crankshaft, be careful to avoid scarring the front cover with the timing chain.

  4. Remove valve collet, valve spring retainer and valve spring using Tool.
  5. Remove valve oil seal using Tool.

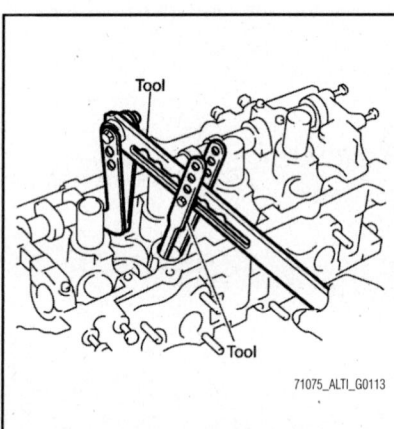

Fig. 136 Removing valve collet, valve spring retainer and valve spring

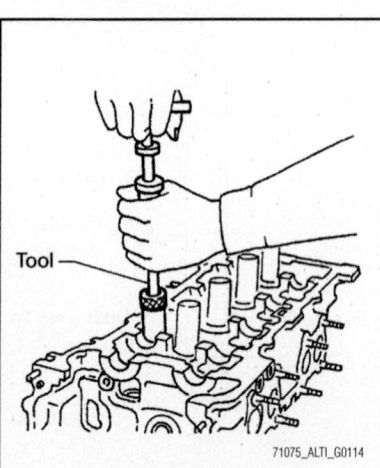

Fig. 137 Removing valve oil seal

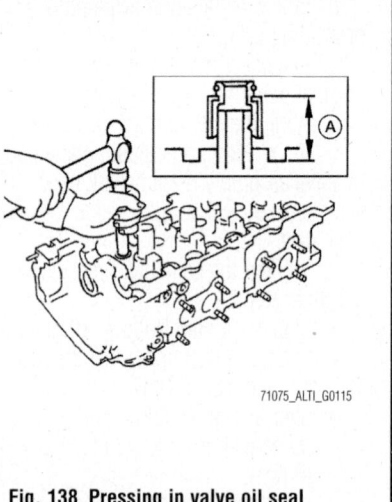

Fig. 138 Pressing in valve oil seal

To install:

  6. Apply new engine oil to new valve oil seal joint surface and seal lip.
  7. Press in valve oil seal to the position shown.

    • Oil seal installed height (A): 0.476 inch
    • Tool number: KV10115600 (J-38958)

ENGINE OIL & FILTER

OIL LEVEL CHECK

  1. Before starting the engine, check the oil level. If the engine is already started, stop it and allow 10 minutes before checking.
  2. Check that the oil level is within the range on the dipstick.
  3. If it is out of range, add oil as necessary.

OIL & FILTER CHANGE

Engine Oil

Except Hybrid

Note the following:
  • Be careful not to burn yourself, as the engine oil may be hot.
  • Prolonged and repeated contact with used engine oil may cause skin cancer: try to avoid direct skin contact with used oil. If skin contact is made, wash thoroughly with soap or hand cleaner as soon as possible.

  1. Position the vehicle so it is level on the hoist.
  2. Warm up the engine and check for oil leaks from the engine.
  3. Stop engine and wait for 10 minutes.
  4. Remove the oil pan drain plug and oil filler cap.
  5. Drain the engine oil.

6. Install the oil pan drain plug with a new washer and refill the engine with new engine oil.

Note the following:

• Be sure to clean the oil pan drain plug and install using a new washer.

• The refill capacity depends on the oil temperature and drain time. Use these specifications for reference only. Always use the dipstick to determine when the proper amount of oil is in the engine.

7. Warm up the engine and check around the drain plug and oil filter for oil leaks.

8. Stop the engine and wait for 10 minutes.

9. Check the oil level using the dipstick.

❈❈ WARNING

Do not overfill the engine oil.

Hybrid

Note the following:

• Be careful not to burn yourself, as the engine oil may be hot.

• Prolonged and repeated contact with used engine oil may cause skin cancer: try to avoid direct skin contact with used oil. If skin contact is made, wash thoroughly with soap or hand cleaner as soon as possible.

1. Position the vehicle so it is level on the hoist.

2. Turn Hybrid System ON and warm up the engine, and check for oil leaks from the engine.

3. Turn Hybrid System OFF and wait for 10 minutes.

4. Remove the oil pan drain plug and oil filler cap.

5. Drain the engine oil.

6. Install the oil pan drain plug with a new washer and refill the engine with new engine oil.

Note the following:

• Be sure to clean the oil pan drain plug and install using a new washer.

• The refill capacity depends on the oil temperature and drain time. Use these specifications for reference only. Always use the dipstick to determine when the proper amount of oil is in the engine.

7. Turn Hybrid System ON and warm up the engine and check the area around the drain plug and oil filter for oil leakage.

8. Turn the Hybrid System OFF and wait for 10 minutes.

9. Check the oil level using the dipstick.

❈❈ WARNING

Do not overfill the engine oil.

Engine Oil Filter

2.5L Engine

See Figure 139.

1. Remove the oil filter using Tool KV10115801 (J-38956).

❈❈ CAUTION

Be careful not to get burned, the engine and engine oil may be hot.

Note the following:

• When removing, prepare a shop cloth to absorb any oil leakage or spillage.

• Do not allow engine oil to adhere to the drive belts.

• Completely wipe off any oil that adheres to the engine and the vehicle.

• The oil filter has a built in pressure relief valve. Use a genuine NISSAN oil filter or equivalent

To install:

2. Remove foreign materials adhering to the oil filter installation surface.

3. Apply clean engine oil to the oil seal contact surface of the new oil filter.

4. Screw the oil filter manually until it touches the installation surface, then tighten it by ⅔ turn. Or tighten to 13 ft. lbs. (18 Nm).

5. Check oil level and add engine oil as necessary.

6. After warming up the engine, check for any engine oil leaks.

3.5L Engine

See Figure 140.

1. Drain engine oil.

2. Remove the engine side under cover.

3. Remove the oil filter using Tool KV10115801 (J-38956).

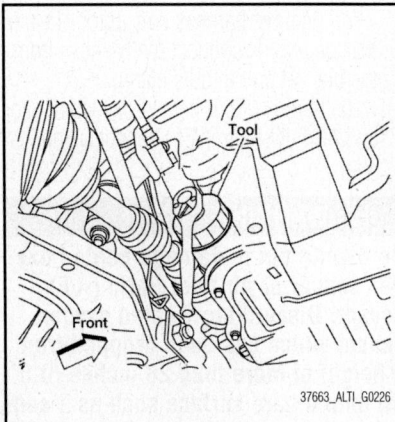

Fig. 139 Remove the oil filter

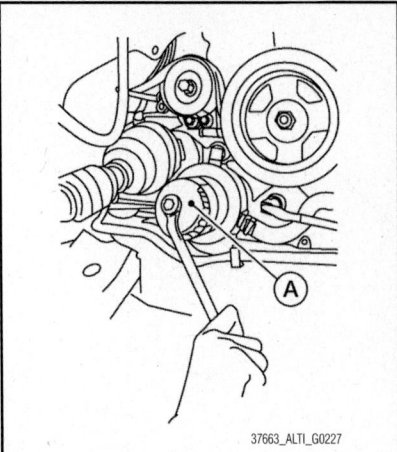

37663_ALTI_G0227

Fig. 140 Remove the oil filter using tool (A) as shown

❈❈ CAUTION

Be careful not to get burned, the engine and engine oil may be hot.

Note the following:

• When removing, prepare a shop cloth to absorb any oil leakage or spillage.

• Do not allow engine oil to adhere to the drive belts.

• Completely wipe off any oil that adheres to the engine and the vehicle.

• The oil filter has a built in pressure relief valve. Use a genuine NISSAN oil filter or equivalent

To install:

4. Remove foreign materials adhering to the oil filter installation surface.

5. Apply clean engine oil to the oil seal contact surface of the new oil filter.

6. Screw the oil filter manually until it touches the installation surface, then tighten it by ⅔ turn. Or tighten to 13 ft. lbs. (18 Nm).

7. Refill the engine with new engine oil.

8. Check the oil level and add engine oil as necessary.

9. After warming up the engine, check for any engine oil leaks.

EXHAUST MANIFOLD

REMOVAL & INSTALLATION

2.5L Engine

See Figures 141 and 142.

1. Remove the engine undercover.

2. Disconnect the electrical connector of Air Fuel Ratio (A/F) sensor 1, and unhook the harness from the bracket and middle clamp on the cover.

3. Remove the Air Fuel Ratio (A/F) sensor 1 using Tool J-44626.

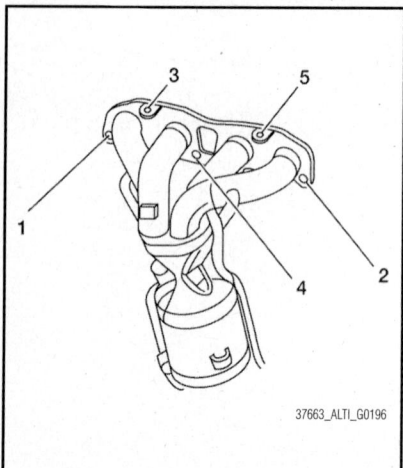

Fig. 141 Loosen the nuts in the reverse order as shown

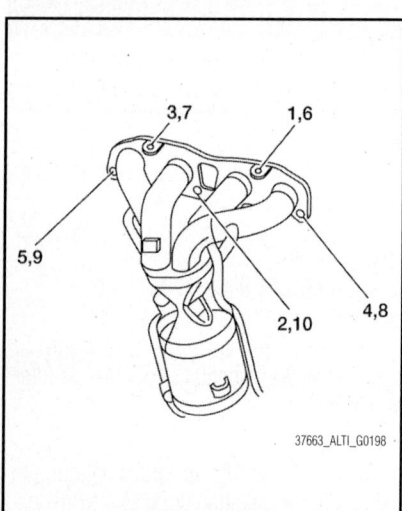

Fig. 142 Tighten the nuts to specification in the numerical order shown

✳✳ WARNING

Be careful not to damage Air Fuel Ratio (A/F) sensor. Discard any Air Fuel Ratio (A/F) sensor which has been dropped from a height of more than 20 inches (0.5 m) onto a hard surface such as a concrete floor; replace with a new one.

4. Remove the exhaust manifold cover (lower).

5. Remove the exhaust front tube.

6. Remove the exhaust manifold cover (upper).

7. Loosen the nuts in the reverse order as shown, on the exhaust manifold and three way catalyst assembly.

8. Remove the exhaust manifold and three way catalyst assembly and gasket. Discard the gasket.

To install:

9. Installation is in the reverse order of removal.

10. Tighten the nuts in two steps to 31 ft. lbs. (42 Nm) in the numerical order shown, on the exhaust manifold and three way catalyst assembly.

11. Clean the Air Fuel Ratio (A/F) sensor 1 threads with the Tool, then apply the anti-seize lubricant to the threads before installing the Air Fuel Ratio (A/F) sensor 1.

✳✳ CAUTION

Do not over-tighten the Air Fuel Ratio (A/F) sensor 1. Doing so may cause damage to the Air Fuel Ratio (A/ F) sensor 1, resulting in a malfunction and the MIL coming on.

3.5L Engine

Right Side

See Figure 143.

✳✳ CAUTION

Perform the work when the exhaust and cooling system have completely cooled down.

1. Remove the front air duct and air cleaner assembly.

2. Disconnect the EVAP vacuum hose and brake booster vacuum hose.

3. Remove cowl top.

4. Remove the front suspension member.

5. Remove the rear engine mounting bracket.

6. Remove the RH three way catalyst support bracket.

7. Remove heated oxygen sensor 2 (bank 1) and Air Fuel Ratio (A/F) sensor 1 (bank 1).

   a. Remove harness connector of each sensor, and disconnect the harness from the bracket and middle clamp.

   b. Remove both heated oxygen sensor and Air Fuel Ratio (A/F) sensor using Tool.

✳✳ WARNING

Be careful not to damage heated oxygen sensor or Air Fuel Ratio (A/F) sensor. Discard any heated oxygen sensor which has been dropped from a height of more than 20 inches (0.5 m) onto a hard surface such as a concrete floor; replace with a new sensor.

8. Remove exhaust manifold and three way catalyst heat shields.

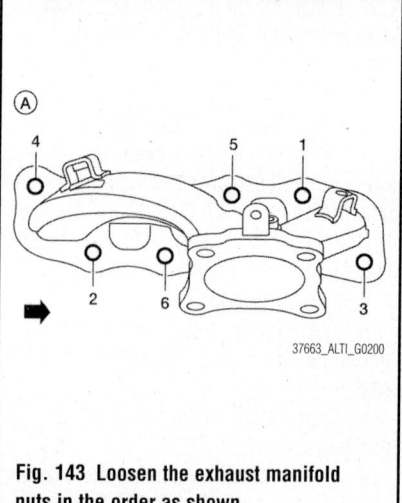

Fig. 143 Loosen the exhaust manifold nuts in the order as shown

9. Remove the three way catalyst (manifold) (bank 1) by loosening the bolts first and then removing the nuts and through bolts.

10. Remove the exhaust manifold RH (A). Loosen the exhaust manifold nuts in the order as shown.

Left Side

See Figures 144 through 147.

✳✳ CAUTION

Perform the work when the exhaust and cooling system have completely cooled down.

1. Remove the front air duct and air cleaner assembly.

2. Remove the cooling fan assembly.

3. Disconnect the heater hoses.

4. Remove the front suspension member.

5. Remove the front engine mounting bracket.

6. Remove the LH three way catalyst support bracket.

7. Remove heated oxygen sensor 2 (bank 2) and Air Fuel Ratio (A/F) sensor 1 (bank 2).

   a. Remove harness connector of each sensor, and disconnect the harness from the bracket and middle clamp.

   b. Remove both heated oxygen sensor and Air Fuel Ratio (A/F) sensor using Tool.

✳✳ WARNING

Be careful not to damage heated oxygen sensor or Air Fuel Ratio (A/F) sensor. Discard any heated oxygen sensor which has been dropped from a height of more than 20 inches (0.5 m) onto a hard surface such as a

concrete floor; replace with a new sensor.

8. Remove exhaust manifold and three way catalyst heat shields.

9. Remove the three way catalyst (manifold) (bank 2) by loosening the bolts first and then removing the nuts and through bolts.

10. Remove the exhaust manifold LH (B). Loosen the exhaust manifold nuts in the order as shown.

To install:

11. Installation is in the reverse order of removal.

12. Install the exhaust manifold nuts in the order as shown RH (A) and LH (B).

13. Install RH and LH three way catalyst support brackets.

14. Hand tighten the three way catalyst support bracket bolts to seat the support brackets.

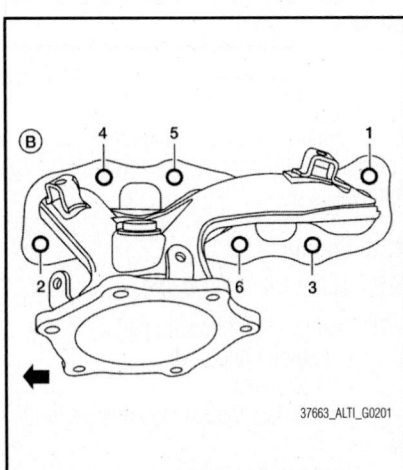

Fig. 144 Loosen the exhaust manifold nuts in the order as shown

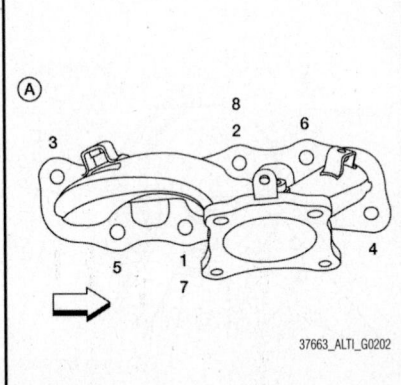

Fig. 145 Install the exhaust manifold nuts in the order as shown RH

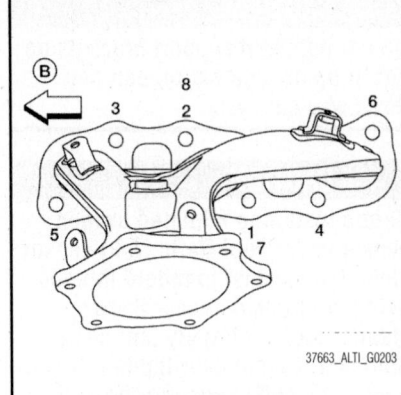

Fig. 146 Install the exhaust manifold nuts in the order as shown LH

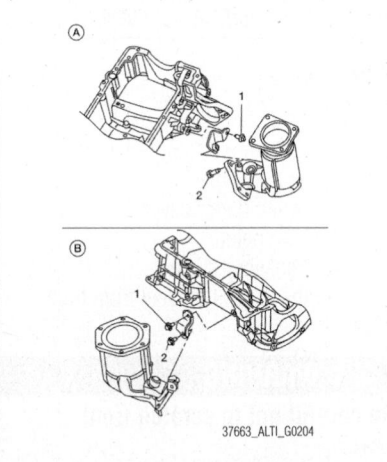

Fig. 147 Install RH (A) and LH (B) three way catalyst support brackets

15. Tighten the bolts to 16 ft. lbs. (22 Nm) in the numerical order as shown.

✸✸ CAUTION

Do not tighten if support brackets do not fit tightly against oil pan and three way catalysts.

✸✸ CAUTION

Before installing a heated oxygen sensor or Air Fuel Ratio (A/F) sensor, clean the exhaust manifold threads using the oxygen sensor thread cleaner tool, and apply anti-seize lubricant. Do not over-tighten the Air Fuel Ratio (A/F) sensor or heated oxygen sensors. Doing so may cause damage.

CVT Models

See Figures 148 through 150.

Note the following:
• Perform the work when the exhaust and cooling system have completely cooled down.

• When removing the front and rear engine mounting through bolts and nuts, lift the engine up slightly for safety.

1. Remove the engine and transaxle assembly.

2. Remove the RH and LH three way catalyst support brackets.

3. Remove heated oxygen sensor 2 (bank 1), heated oxygen sensor 2 (bank 2), Air Fuel Ratio (A/F) sensor 1 (bank 1) and Air Fuel Ratio (A/F) sensor 1 (bank 2).

a. Remove harness connector of each sensor, and disconnect the harness from the bracket and middle clamp.

b. Remove both heated oxygen sensors and air fuel ratio (A/F) sensors using Tool.

✸✸ WARNING

Be careful not to damage heated oxygen sensors or Air Fuel Ratio (A/F) sensors. Discard any heated oxygen sensor which has been dropped from a height of more than 20 inches (0.5 m) onto a hard surface such as a concrete floor; replace with a new sensor.

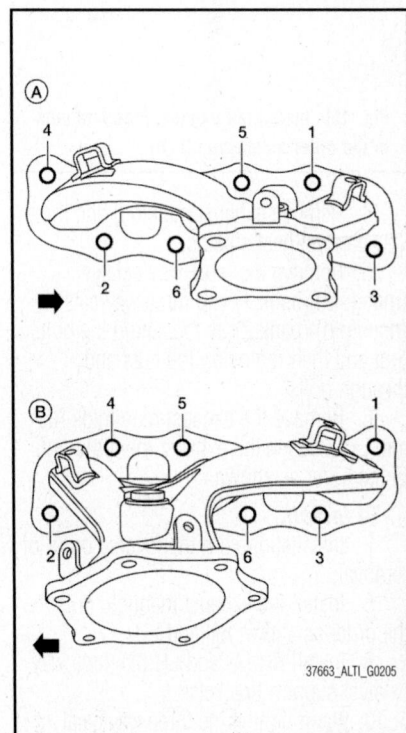

Fig. 148 Remove the exhaust manifolds RH (A) and LH (B); loosen the exhaust manifold nuts in the order as shown

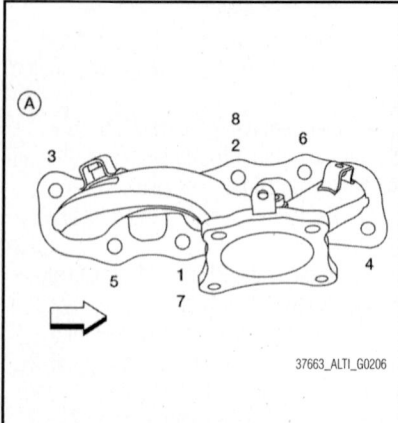

Fig. 149 Install the exhaust manifold nuts in the order as shown RH (A)

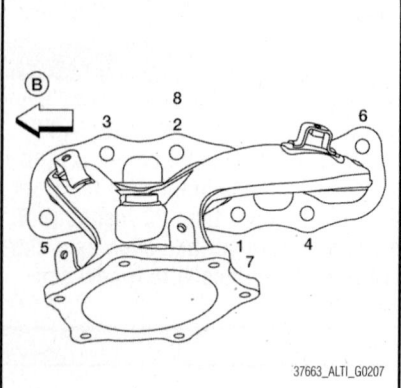

Fig. 150 Install the exhaust manifold nuts in the order as shown LH (B)

4. Remove exhaust manifold and three way catalyst heat shields.

5. Remove the three way catalyst (manifold) (bank 1) and three way catalyst (manifold) (bank 2) by loosening the bolts first and then removing the nuts and through bolts.

6. Remove the exhaust manifolds RH and LH. Loosen the exhaust manifold nuts in the order as shown.

To install:

7. Installation is in the reverse order of removal.

8. Install the exhaust manifold nuts in the order as shown RH and LH.

9. Install RH (A) and LH (B) three way catalyst support brackets.

10. Hand tighten the three way catalyst support bracket bolts to seat the support brackets.

11. Tighten the bolts to 16 ft. lbs. (22 Nm) in the numerical order as shown.

❊❊ CAUTION

Do not tighten if support brackets do not fit tightly against oil pan and three way catalysts.

❊❊ CAUTION

Before installing a heated oxygen sensor or Air Fuel Ratio (A/F) sensor, clean the exhaust manifold threads using the oxygen sensor thread cleaner tool, and apply anti-seize lubricant. Do not over-tighten the Air Fuel Ratio (A/F) sensor or heated oxygen sensors. Doing so may cause damage.

FRONT COVER SEAL

REMOVAL & INSTALLATION

2.5L Engine

See Figures 151 and 152.

1. Remove the following parts:
    - RH front wheel
    - Engine under cover
    - Drive belts
    - Crankshaft pulley
2. Remove front oil seal from front cover.

❊❊ WARNING

Be careful not to scratch front cover.

To install:

3. Install new front oil seal to front cover using suitable tool.

    a. Install new oil seal in until it is flush with front end surface of front cover.

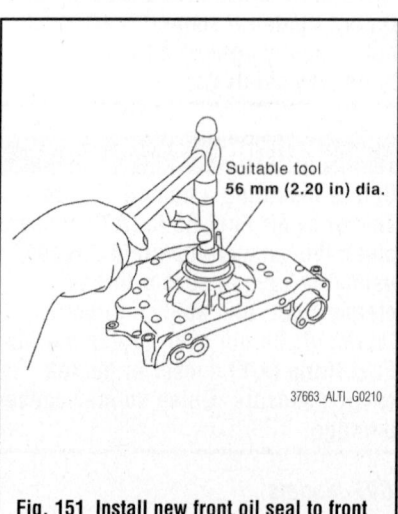

Fig. 151 Install new front oil seal to front cover using suitable tool

---

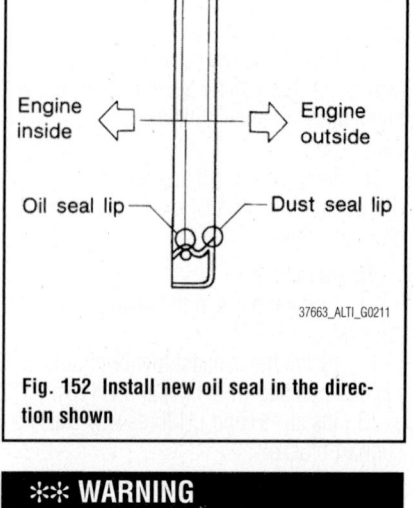

Fig. 152 Install new oil seal in the direction shown

❊❊ WARNING

Do not reuse oil seal. Be careful not to cause damage to circumference of oil seal.

    b. Install new oil seal in the direction shown.

4. Installation of the remaining components is in the reverse order of removal.

3.5L Engine

See Figures 152 through 154.

1. Remove the following parts:
    - Engine undercover.
    - Drive belts.
2. Disconnect the battery negative terminal.

3. Remove the crankshaft pulley as follows:

    a. Remove the starter motor.
    b. Lock the ring gear using Tool

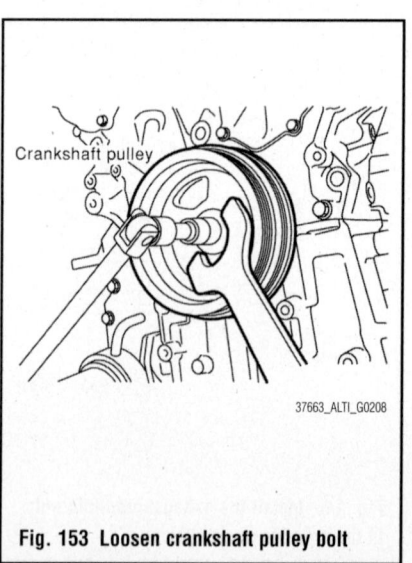

Fig. 153 Loosen crankshaft pulley bolt

Fig. 154 Position a pulley puller at recess hole of crankshaft pulley to remove crankshaft pulley

KV10117700 (J-44716) attached to the starter bolt hole.

✳✳ WARNING

Do not damage the ring gear teeth, or the signal plate teeth behind the ring gear when setting the stopper.

   c. Loosen crankshaft pulley bolt using Tool KV10109300 and locate bolt seating surface at 10 mm (0.39 in) from its original position.

   d. Position a pulley puller at recess hole of crankshaft pulley to remove crankshaft pulley.

✳✳ WARNING

Do not use a puller claw on crankshaft pulley periphery.

  4. Remove front oil seal from front cover.

✳✳ WARNING

Be careful not to damage front cover or crankshaft.

To install:

  5. Apply new engine oil to new oil seal and install.

   a. Install new oil seal in the direction as shown.

✳✳ WARNING

Press fit straight and avoid causing burrs or tilting the oil seal.

   b. Press-fit oil seal until it becomes flush with the timing chain case end face, using suitable tool.

   c. Make sure the garter spring in the oil seal is in position and seal lip is not inverted.

  6. Install crankshaft pulley and tighten the bolt in two steps.

   a. Lubricate thread and seat surface of the bolt with new engine oil.

   b. For the second step angle tighten using Tool KV10112100 (BT-8653-A).

   c. Step 1: 32 ft. lbs. (44 Nm)

   d. Step 2: Tighten an additional 84–90 degrees

  7. Remove tool attached to the starter bolt hole.

  8. Installation of the remaining components is in reverse order of removal.

INTAKE MANIFOLD

REMOVAL & INSTALLATION

2.5L Engine

See Figures 155 through 161.

✳✳ WARNING

To avoid the danger of being scalded, never drain the coolant when the engine is hot.

  1. Release the fuel pressure.

  2. Drain coolant when engine is cooled.

  3. Disconnect the MAF sensor electrical connector.

  4. Remove air cleaner and air duct assembly.

  5. Remove cowl top finisher.

  6. Disconnect the following components at the intake side:

- PCV hose
- EVAP hose and EVAP canister purge volume control solenoid
- Electric throttle control actuator
- Brake booster vacuum hose

  7. Disconnect the fuel quick connector on the engine side.

   a. Remove quick connector cap.

   b. With the sleeve side of tool facing quick connector, install tool onto fuel tube.

   c. Insert tool into quick connector until sleeve contacts and goes no further. Hold the tool on that position.

✳✳ WARNING

Inserting the tool hard will not disconnect quick connector. Hold tool where it contacts and goes no further.

   d. Pull the quick connector straight out from the fuel tube.

- Pull quick connector holding it at the A position, as shown.
- Do not pull with lateral force applied. O-ring inside quick connector may be damaged.

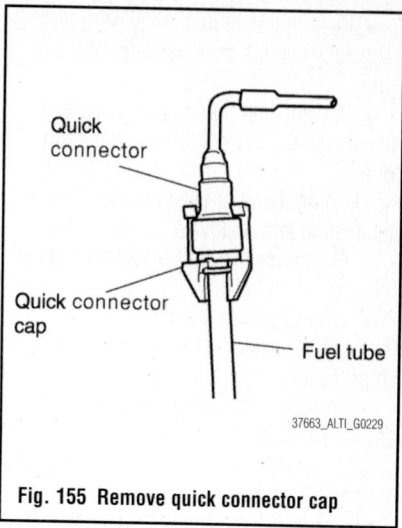

Fig. 155 Remove quick connector cap

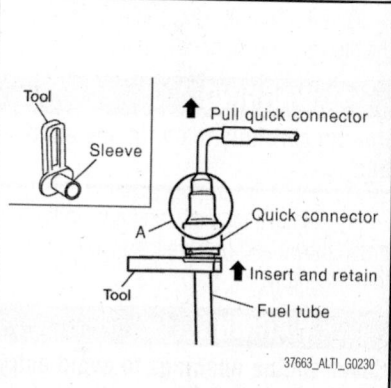

Fig. 156 With the sleeve side of tool facing quick connector, install tool onto fuel tube

- Prepare container and cloth beforehand as fuel will leak out.
- Avoid fire and sparks.
- Be sure to cover openings of disconnected pipes with plug or plastic bag to avoid fuel leakage and entry of foreign materials.

  8. When removing fuel hose quick connector at vehicle piping side, perform as follows.

   a. Remove quick connector cap.

   b. Hold the sides of the connector, push in tabs and pull out the tube. (The figure is shown for reference only.)

   c. If the connector and the tube are stuck together, push and pull several times until they start to move. Then disconnect them by pulling.

Note the following:

- The tube can be removed when the tabs are completely depressed. Do not twist it more than necessary.
- Do not use any tools to remove the quick connector.

• Keep the resin tube away from heat. Be especially careful when welding near the tube.

• Prevent acid liquid such as battery electrolyte etc. from getting on the resin tube.

• Do not bend or twist the tube during installation and removal.

• Do not remove the remaining retainer on tube.

• When the tube is replaced, also replace the retainer with a new one. Retainer color: Green.

• To keep clean the connecting portion and to avoid damage and foreign materials, cover them completely with plastic bags or something similar.

9. Disconnect electric throttle control actuator coolant hoses.

10. Loosen bolts diagonally, and remove the electric throttle control actuator.

✳✳ WARNING

Handle carefully to avoid any damage.

11. Remove the bolts and nuts in the order shown and remove the intake manifold assembly.

✳✳ WARNING

Cover engine openings to avoid entry of foreign materials.

To install:

12. Installation is in the reverse order of removal. Follow the tightening sequences below.

13. Tighten intake manifold bolts and nuts.

  a. Tighten in numerical order as shown.

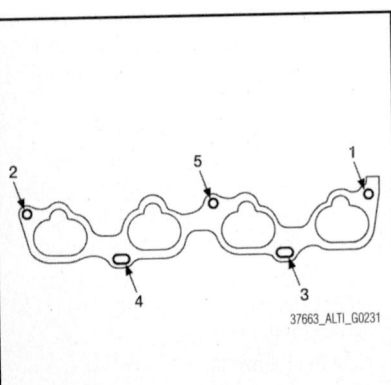

Fig. 157 Remove the bolts and nuts in the order shown

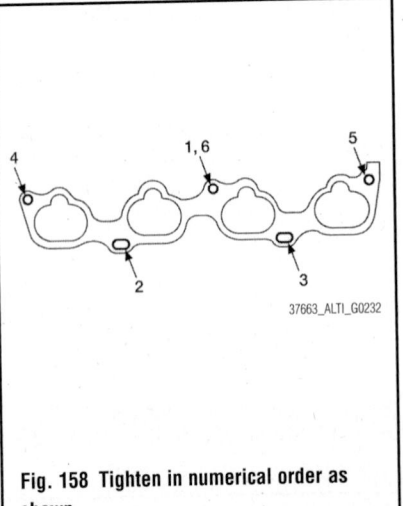

Fig. 158 Tighten in numerical order as shown

✳✳ WARNING

After tightening the five bolts in the order shown, the 1, 6 position designates that the first bolt tightened is to be retightened to specification.

14. Install the Electric Throttle Control Actuator:

  a. Tighten the bolts of electric throttle control actuator equally and diagonally in several steps.

  b. After installation perform procedure in "INSPECTION AFTER INSTALLATION".

15. To connect the quick connector on the fuel hose (engine side):

  a. Make sure no foreign substances are deposited in and around the fuel tube and quick connector, and there is no damage to them.

  b. Thinly apply new engine oil around the fuel tube tip end.

  c. Align center to insert quick connector straight into fuel tube.

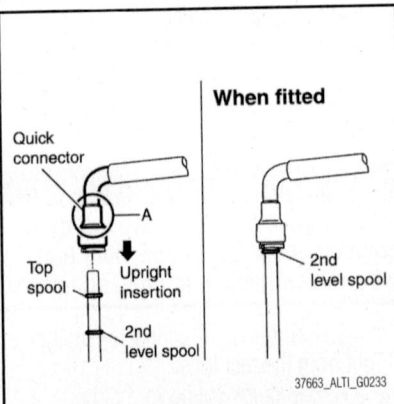

Fig. 159 Hold at position A as shown, when inserting the fuel tube into the quick connector

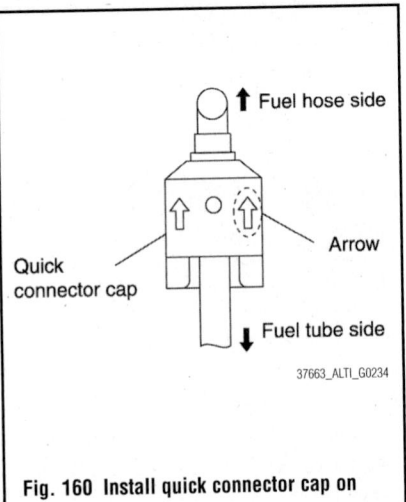

Fig. 160 Install quick connector cap on quick connector joint

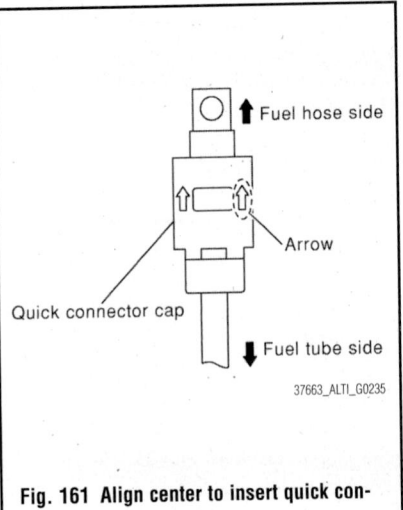

Fig. 161 Align center to insert quick connector straight into fuel tube

  d. Insert fuel tube into quick connector until the top spool on fuel tubes is inserted completely and the second level spool is positioned slightly below the quick connector bottom end.

  e. Hold at position A as shown, when inserting the fuel tube into the quick connector.

  f. Carefully align to center to avoid inclined insertion to prevent damage to the O-ring inside the quick connector.

  g. Insert the fuel tube until you hear a "click" sound and actually feel the engagement.

  h. To avoid misidentification of engagement with a similar sound, be sure to perform the next step.

  i. Before clamping the fuel hose with the hose clamp, pull the quick connector hard by hand, holding at the A position, as shown. Make sure it is completely engaged (connected) so that it does not come off of the fuel tube.

➡**Recommended pulling force is 11 lbs. (50 N).**

   j. Install quick connector cap on quick connector joint.

   16. Direct arrow mark on quick connector cap to upper side (fuel hose side).

   17. Install fuel hose to hose clamp.

   18. To connect the quick connector on the fuel hose (vehicle piping side):

   a. Make sure no foreign substances are deposited in and around the fuel tube and quick connector, and there is no damage to them.

   b. Align center to insert quick connector straight into fuel tube.

   c. Insert fuel tube until a click is heard.

   d. Install quick connector cap on quick connector joint. Direct arrow mark on quick connector cap upper side.

   e. Install fuel hose to hose clamp.

   19. Make sure there is no fuel leakage at connections as follows:

   a. Apply fuel pressure to fuel lines by turning ignition switch ON (with engine stopped). Then check for fuel leaks at connections.

   b. Start the engine and rev it up and check for fuel leaks at connections.

   c. Perform procedures for "Throttle Valve Closed Position Learning" after finishing repairs.

   d. If electric throttle control actuator is replaced, perform procedures for "Idle Air Volume Learning" after finishing repairs.

❊❊ WARNING

Do not touch engine immediately after stopping as engine is extremely hot.

➡**Use mirrors for checking on connections out of the direct line of sight.**

3.5L Engine

See Figures 162 through 166.

❊❊ WARNING

To avoid the danger of being scalded, never drain the coolant when the engine is hot.

   1. Release the fuel pressure.
   2. Disconnect the battery negative terminal.
   3. Remove the cowl top.
   4. Remove the engine cover.
   5. Remove front air duct and air duct hose.
   6. If necessary, remove the electric throttle control actuator bolts in the reverse

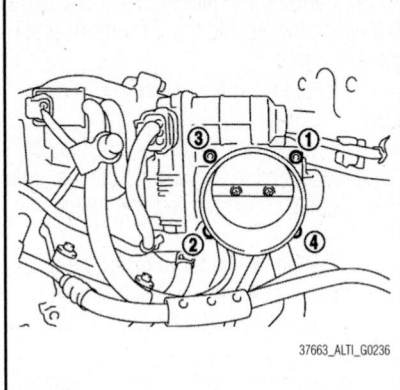

Fig. 162 Remove the electric throttle control actuator bolts in the reverse order as shown

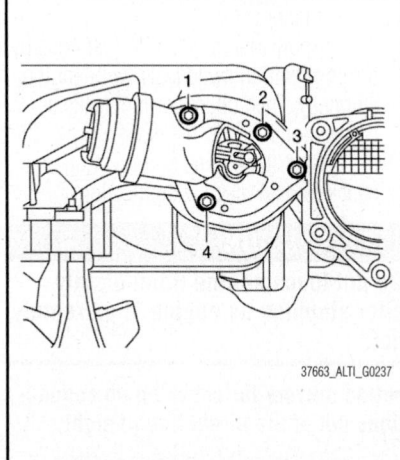

Fig. 163 Remove power valve bolts in the reverse order as shown

order as shown and remove the electric throttle control actuator and position aside.

❊❊ WARNING

Handle carefully to avoid any shock to the electric throttle control actuator. Do not disassemble.

   7. If necessary, remove power valve bolts in the reverse order as shown and remove the power valves.
   8. Disconnect the following:
   • Power brake booster vacuum hose
   • Fuel injector electrical connectors
   • PCV hose
   • Electric throttle control actuator electrical connector
   • EVAP canister purge hose

❊❊ CAUTION

Cover any engine openings to avoid the entry of any foreign material.

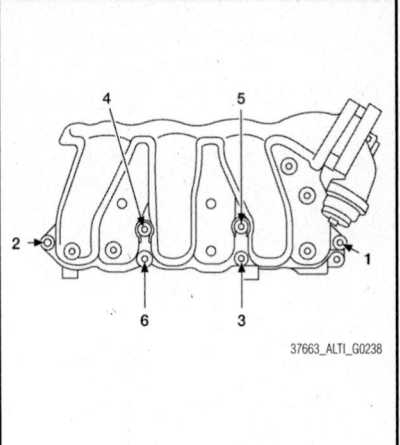

Fig. 164 Loosen the intake manifold collector bolts in the order as shown

   9. Remove the EVAP canister purge volume solenoid valve bracket bolt. Position the valve aside.

   10. Loosen the intake manifold collector bolts in the order as shown, and remove the intake manifold collector and gasket.

   11. If necessary remove the following components:
   • VIAS control solenoid valve
   • EVAP canister purge volume control solenoid valve

   12. Disconnect fuel tube quick connector at vehicle piping side.

   13. To remove the quick connector cap, hold the sides of the connector, push in the tabs and pull out the tube.
   • The tube can be removed when the tabs are completely depressed. Do not twist it more than necessary.
   • Do not use any tools to remove the quick connector.
   • Keep the resin tube away from heat. Be especially careful when welding near the tube.
   • Prevent acid liquids such as battery electrolyte, etc. from getting on the resin tube.
   • Do not bend or twist the tube during removal or installation.
   • Do not remove the remaining retainer on the tube
   • When the tube is replaced, also replace the retainer with a new one.
   • To keep the connecting portion clean and to avoid damage and foreign materials entering, cover the ends of the fuel tubes with plastic bags or something similar.

➡**If the connector and the tube are stuck together, push and pull several times until they start to move. Then disconnect them by pulling.**

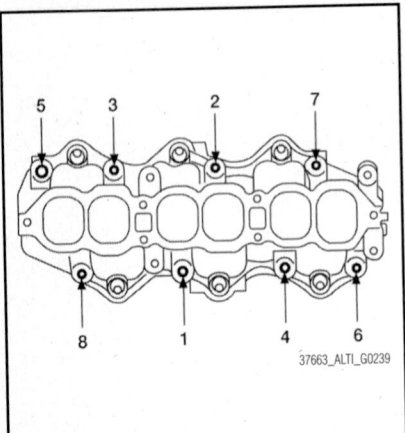

Fig. 165 Loosen the bolts in the order as shown

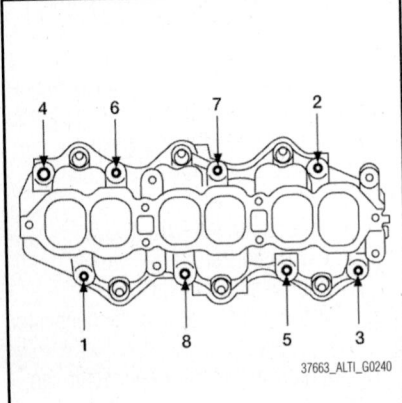

Fig. 166 Install intake manifold bolts in two steps in the numerical order as shown

14. Remove the fuel rail with the fuel injectors attached, from the intake manifold. Remove the fuel injector O-rings and use new O-rings for installation.

15. Loosen the bolts in the order as shown, and remove the intake manifold.

To install:

16. Installation is in the reverse order of removal. Follow the procedure below for specific tightening sequences and procedures.

17. Install intake manifold bolts in two steps in the numerical order as shown.

   a. Step 1: 65 inch lbs. (7 Nm)

   b. Step 2: 19 ft. lbs. (25 Nm)

➡**After installation, it is necessary to re-calibrate the electric throttle control actuator.**

18. Perform the "Throttle Valve Closed Position Learning" when harness connector of the electric throttle control actuator is disconnected.

19. Perform the "Idle Air Volume Learning" when the electric throttle control actuator is replaced.

20. Install the quick connector as follows:

   a. Make sure no foreign substances are deposited in and around the fuel tube and quick connector and that there is no damage.

   b. Align the center to insert the quick connector straight onto the fuel tube.

   c. Insert the fuel tube until a click is heard.

   d. Install the quick connector cap on the quick connector joint. Align the arrow mark on the quick connector cap to the upper side.

   e. Install the fuel hose into the hose clamp.

21. Perform the following inspection after installation:

   a. Apply fuel pressure to fuel lines by turning ignition switch ON (with engine stopped). Then check for fuel leaks at connections.

   b. Start the engine and rev it up and check for fuel leaks at connections.

⁂ **CAUTION**

Do not touch engine immediately after stopping as engine is extremely hot.

➡**Use mirrors for checking on connections out of the direct line of sight.**

OIL PAN

REMOVAL & INSTALLATION

2.5L Engine

See Figures 167 through 172.

⁂ **CAUTION**

To avoid the danger of being scalded, never drain the engine oil when the engine is hot.

1. Drain engine oil.

2. Remove the front exhaust tube.

3. Remove power steering cooler hose bracket from suspension member.

4. Remove the front suspension member for clearance to remove the oil pan.

5. Remove the lower oil pan bolts in the order as shown.

6. Remove the lower oil pan using Tool KV10111100 (J-37228).

➡**Tap gently to cut sealant around the pan; do not damage the mating surface using Tool.**

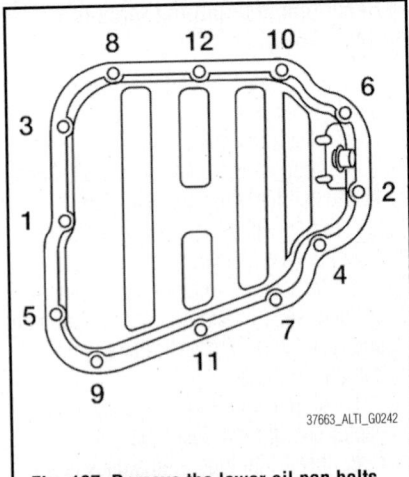

Fig. 167 Remove the lower oil pan bolts in the order

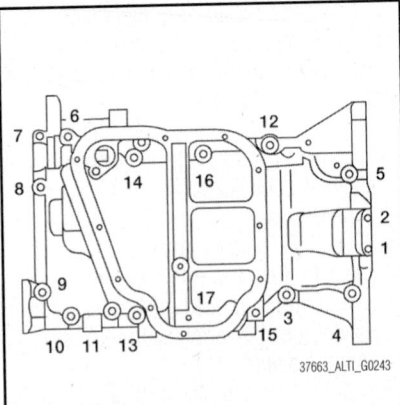

Fig. 168 Loosen the upper oil pan bolts in the order shown

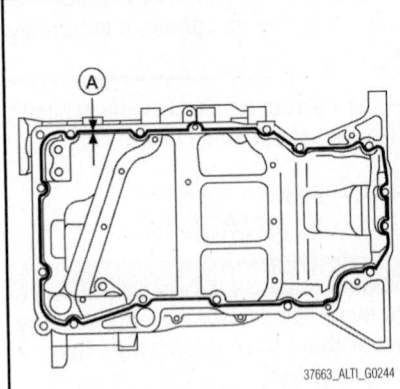

Fig. 169 Apply Genuine Silicone RTV Sealant or equivalent to the upper oil pan (A)

7. Remove the oil strainer.

8. Remove rear plate cover, and four engine-to transaxle bolts.

9. Loosen the upper oil pan bolts in the order shown to remove upper oil pan.

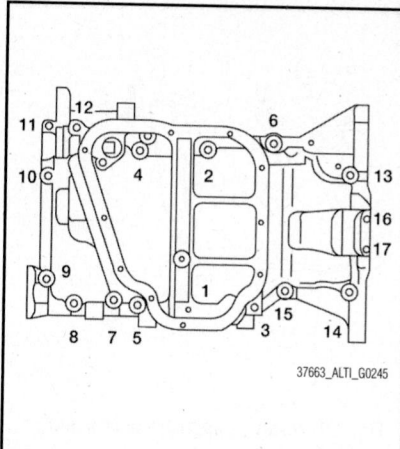

Fig. 170 Tighten the upper oil pan bolts in the order as shown

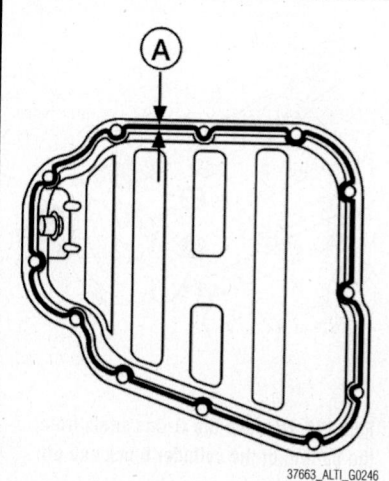

Fig. 171 Apply Genuine Silicone RTV Sealant or equivalent to the lower oil pan (A)

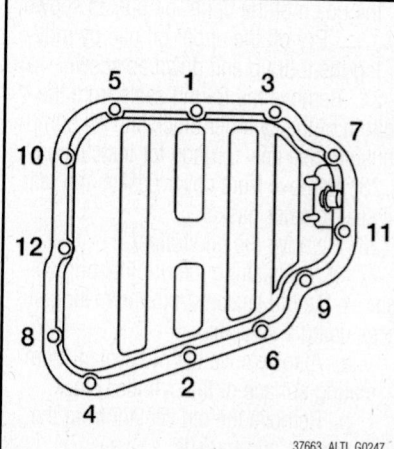

Fig. 172 Tighten the lower oil pan bolts in the numerical order shown

10. Remove upper oil pan using Tool KV10111100 (J-37228).

➡**Tap gently to cut sealant around the pan; do not damage the mating surface using Tool.**

11. Clean the oil strainer screen to remove any foreign material.

To install:

12. Installation is in the reverse order of removal.

13. Apply Genuine Silicone RTV Sealant or equivalent to the upper oil pan as shown.

➡**Install two new O-rings in the upper pan.**

14. Tighten the upper oil pan bolts in the order as shown.

15. Apply Genuine Silicone RTV Sealant or equivalent to the lower oil pan (A) as shown.

16. Tighten the lower oil pan bolts in the numerical order shown.

✳✳ WARNING

Wait at least 30 minutes after the oil pans are installed before filling the engine with oil.

17. Check for any engine oil leaks with the engine at operating temperature and running at idle.

3.5L Engine

Lower Oil Pan

See Figures 173 and 174.

✳✳ WARNING

You should not remove the oil pan until the exhaust system and cooling system have completely cooled off.

1. Drain the engine oil.
2. Loosen the lower oil pan bolts in order as shown.
3. Remove the lower oil pan.
   a. Insert Tool KV10111100 (J-37228) between the lower oil pan and the upper oil pan.

✳✳ WARNING

Be careful not to damage the mating surface. Do not insert a screwdriver, this will damage the mating surfaces.

   b. Slide the Tool by tapping its side with a hammer to remove the lower oil pan from the upper oil pan.
4. If re-installing the original lower oil

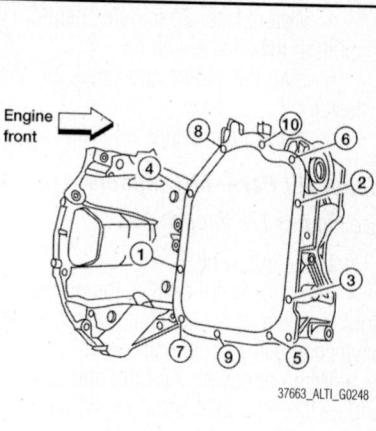

Fig. 173 Loosen the lower oil pan bolts in order as shown

pan, remove the old sealant from the mating surfaces using a scraper.
   a. Also remove the old sealant from mating surface of the upper oil pan.
   b. Remove the old sealant from the bolt holes and threads.

✳✳ WARNING

Do not scratch or damage the mating surfaces when cleaning off the old sealant.

5. Clean oil strainer if any object is attached.

To install:

6. Apply a continuous bead of sealant to the lower oil pan.
   a. Use Genuine Silicone RTV Sealant, or equivalent.
   b. Installation must be done within 5 minutes after applying sealant.
7. Install the lower oil pan. Tighten the lower oil pan bolts in order as shown.

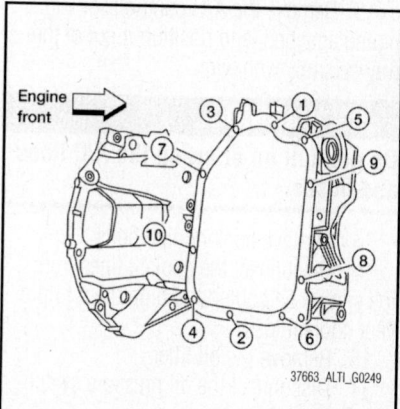

Fig. 174 Tighten the lower oil pan bolts in order as shown

a. Wait at least 30 minutes before refilling the engine with oil.

b. Start the engine and check for leaks.

c. Inspect the engine oil level.

Upper Oil Pan—M/T Models

See Figures 175 through 180.

Note the following:

• You should not remove the oil pan until the exhaust system and cooling system have completely cooled off.

• When removing the front and rear engine through bolts and nuts, lift the engine up slightly for safety.

❄ WARNING

When removing the upper oil pan from the engine, first remove the Crankshaft Position (CKP) sensor (POS). Be careful not to damage sensor edges or signal plate teeth.

1. Drain the engine coolant.

2. Drain engine oil.

3. Disconnect the battery negative terminal.

4. Remove the engine room cover.

5. Remove the front air duct, air duct hose and air cleaner assembly.

6. Remove the cowl top.

7. Remove the intake manifold collector.

8. Remove the cooling fan assembly.

9. Remove the front suspension member.

10. Disconnect the heated oxygen sensors and air flow ratio (A/F) sensors and remove the two catalytic convertors from the exhaust manifolds.

11. Remove the oil gauge and oil gauge guide from the oil pan.

12. Remove the drive belt.

13. Remove the A/C compressor with piping attached, and position it out of the way securely with wire.

❄ WARNING

Do not pull on or crimp the A/C lines and hoses.

14. Remove coolant pipe bolts.

15. Disconnect the coolant lines from the engine oil cooler and plug them to prevent coolant loss.

16. Remove the oil filter.

17. Disconnect the oil pressure switch electrical connector.

18. Remove the oil pressure switch, if necessary and the Crankshaft Position (CKP) sensor (POS) from the upper oil pan.

19. Remove the oil cooler, if necessary.

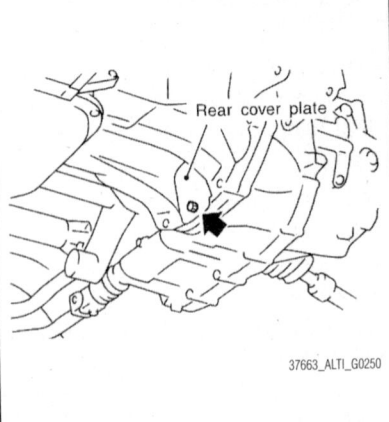

Fig. 175 Remove the rear plate cover from the upper oil pan

Fig. 176 Remove the four upper oil pan to transaxle bolts

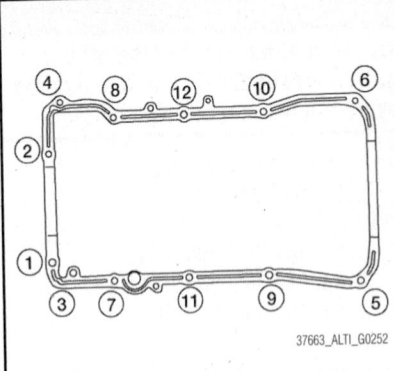

Fig. 177 Loosen the bolts in the order as shown

20. Remove the rear plate cover from the upper oil pan.

21. Remove the lower oil pan.

22. Remove the four upper oil pan to transaxle bolts.

23. Remove the upper oil pan.

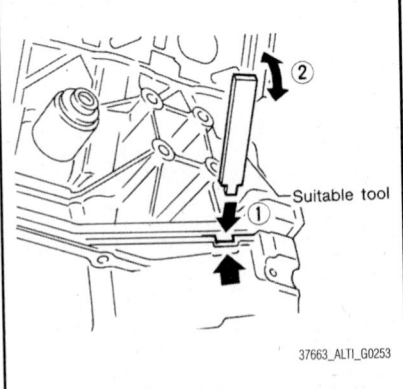

Fig. 178 Insert an appropriate size tool into the notch (1) of the upper oil pan; pry off the upper oil pan by moving the tool up and down (2)

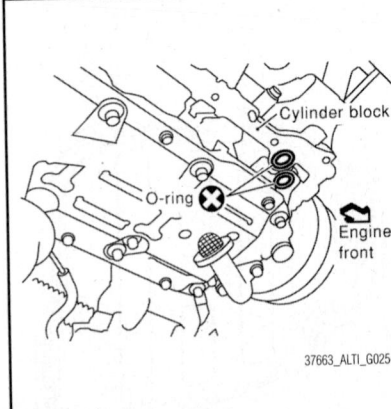

Fig. 179 Remove the O-ring seals from the bottom of the cylinder block and oil pump housing

a. Loosen the bolts in the order as shown.

b. Insert an appropriate size tool into the notch of the upper oil pan as shown.

c. Pry off the upper oil pan by moving the tool up and down as shown.

24. Remove the O-ring seals from the bottom of the cylinder block and oil pump housing, use new O-rings for installation.

25. Remove front cover gasket and rear oil seal retainer gasket.

26. Remove the oil strainer.

27. If re-installing the original oil pan, remove the old sealant from the mating surfaces using a scraper.

a. Also remove the old sealant from mating surface of the cylinder block.

b. Remove the old sealant from the bolt holes and threads.

❄ WARNING

Do not scratch or damage the mating surfaces when cleaning off the old sealant.

28. Clean oil strainer if any object is attached.

To install:

Wait at least 30 minutes before refilling the engine with oil.

29. Install oil strainer and tighten bolt to specified torque.

30. Apply Genuine Silicone RTV Sealant or equivalent, to the front cover gasket and the rear oil seal retainer gasket.

31. Install the front cover gasket and rear oil seal retainer gasket as shown.

32. Apply a bead of sealant to the cylinder block mating surface of the upper oil pan.

   a. Use Genuine Silicone RTV Sealant, or equivalent.

   b. Be sure the sealant is applied to a limited portion as shown.

   c. Attaching should be done within 5 minutes after coating.

33. Install new O-rings on the cylinder block and oil pump body.

34. Install the upper oil pan.

   a. Tighten upper oil pan bolts in the order as shown.

   b. Wait at least 30 minutes before refilling the engine with oil.

35. Install the four upper oil pan to transaxle bolts.

36. Apply a continuous bead of sealant to the lower oil pan.

   a. Use Genuine Silicone RTV Sealant, or equivalent.

   b. Installation must be done within 5 minutes after applying sealant.

37. Install the lower oil pan.

38. Install rear plate cover.

39. Installation of the remaining components is in the reverse order of removal.

   a. Start the engine and check for leaks.

   b. Inspect the engine oil level.

Upper Oil Pan—CVT Models

See Figures 177, 178 and 180.

You should not remove the oil pan until the exhaust system and cooling system have completely cooled off.

When removing the front and rear engine through bolts and nuts, lift the engine up slightly for safety.

When removing the upper oil pan from the engine, first remove the Crankshaft Position (CKP) sensor (POS). Be careful not to damage sensor edges or signal plate teeth.

1. Remove the engine from the vehicle.
2. Drain the engine oil.
3. Remove the oil dipstick.
4. Remove the drive belt.
5. Disconnect the A/C compressor harness connector.
6. Remove the A/C compressor bolts and remove the A/C compressor.
7. Remove coolant pipe bolts.
8. Disconnect the coolant lines from the engine oil cooler.
9. Remove the oil filter and engine oil cooler from the upper oil pan.
10. Remove the oil pressure switch, and the Crankshaft Position (CKP) sensor (POS) from the upper oil pan.
11. Remove the lower oil pan.
12. Remove the upper oil pan.

   a. Loosen the bolts in the order as shown.

   b. Insert an appropriate size tool into the notch of the upper oil pan.

   c. Pry off the upper oil pan by moving the tool up and down.

13. Remove the O-ring seals from the bottom of the cylinder block and oil pump housing. Use new O-rings for installation.

14. Remove front cover gasket and rear oil seal retainer gasket.

15. Remove the oil strainer.

16. If re-installing the original oil pan, remove the old sealant from the mating surfaces using a scraper.

   a. Also remove the old sealant from mating surface of the cylinder block.

   b. Remove the old sealant from the bolt holes and threads.

Do not scratch or damage the mating surfaces when cleaning off the old sealant.

17. Clean oil strainer if any object is attached.

To install:

Wait at least 30 minutes before refilling the engine with oil.

18. Install oil strainer and tighten bolt to specified torque.

19. Apply Genuine Silicone RTV Sealant or equivalent, to the front cover gasket and the rear oil seal retainer gasket as shown.

20. Install the front cover gasket and rear oil seal retainer gasket as shown.

21. Apply a bead of sealant to the cylinder block mating surface of the upper oil pan to a limited portion as shown.

   a. Use Genuine Silicone RTV Sealant, or equivalent.

   b. Attaching should be done within 5 minutes after coating.

22. Install new O-rings on the cylinder block and oil pump body.

23. Install the upper oil pan.

   a. Tighten upper oil pan bolts in the order as shown.

   b. Wait at least 30 minutes before refilling the engine with oil.

24. Install the lower oil pan.

25. Installation of the remaining components is in the reverse order of removal.

   a. Start the engine and check for leaks.

   b. Inspect the engine oil level.

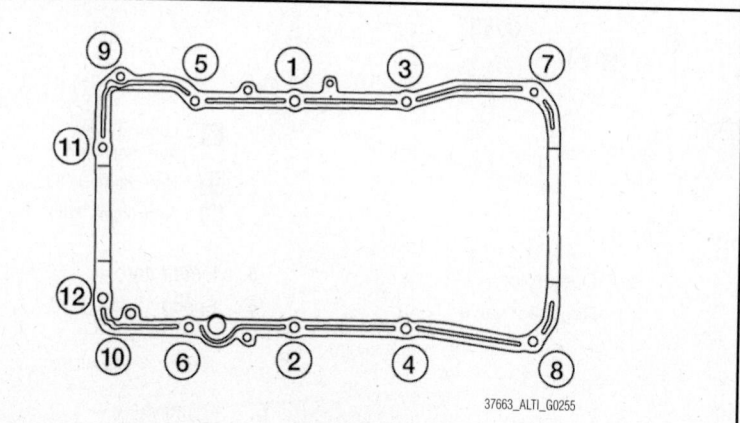

37663_ALTI_G0255

Fig. 180 Tighten upper oil pan bolts in the order as shown

OIL PUMP

REMOVAL & INSTALLATION

2.5L Engine

See Figure 181.

The oil pump is part of the front cover. For removal and installation of the oil pump, it is necessary to remove and install the front cover. Refer to Front Cover section for Removal and Installation.

3.5L Engine

See Figure 182.

1. Remove the engine from the vehicle.
2. Remove the upper oil pan.
3. Remove the timing chain.
4. Remove oil pump assembly.

To install:

5. To install, reverse the removal procedure.

PISTONS & RINGS

POSITIONING

See Figures 183 through 185.

TIMING CHAIN COVER

REMOVAL & INSTALLATION

2.5L Engine

See Figures 186 through 194.

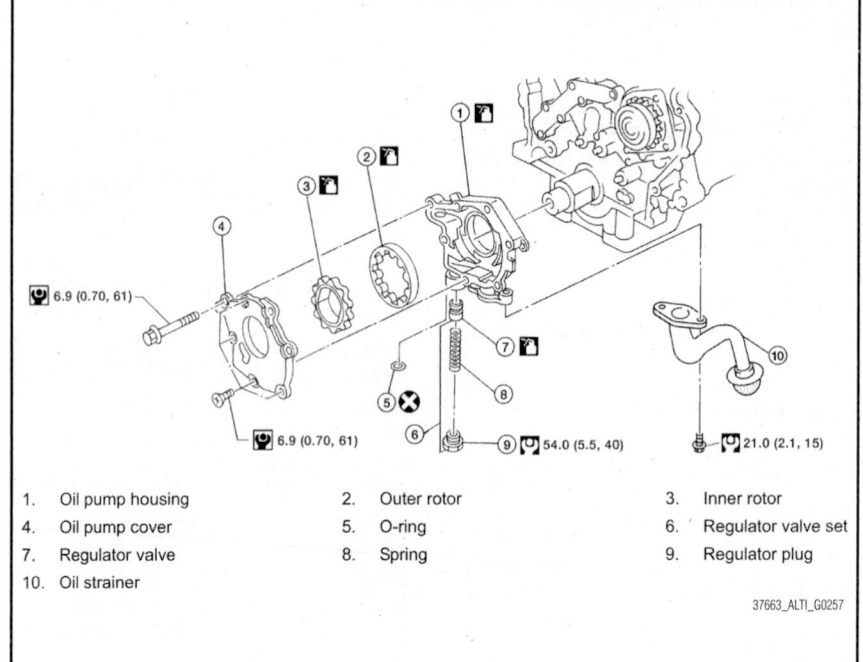

1.	Oil pump housing	2.	Outer rotor	3.	Inner rotor
4.	Oil pump cover	5.	O-ring	6.	Regulator valve set
7.	Regulator valve	8.	Spring	9.	Regulator plug
10.	Oil strainer				

37663_ALTI_G0257

Fig. 182 Exploded view of oil pump assembly

1. Support the engine and transaxle assembly with suitable tools.
2. Remove RH splash shield.
3. Remove the upper and lower oil pan, and oil strainer.
4. Remove the alternator.
5. Remove the engine cover.
6. Disconnect variable timing control solenoid harness connector.
7. Remove the engine ground.
8. Remove the coolant overflow reservoir tank.
9. Position the RH engine compartment fuse and relay box aside.
10. Remove the RH engine mount and bracket.
11. Loosen bolts in the numerical order as shown.

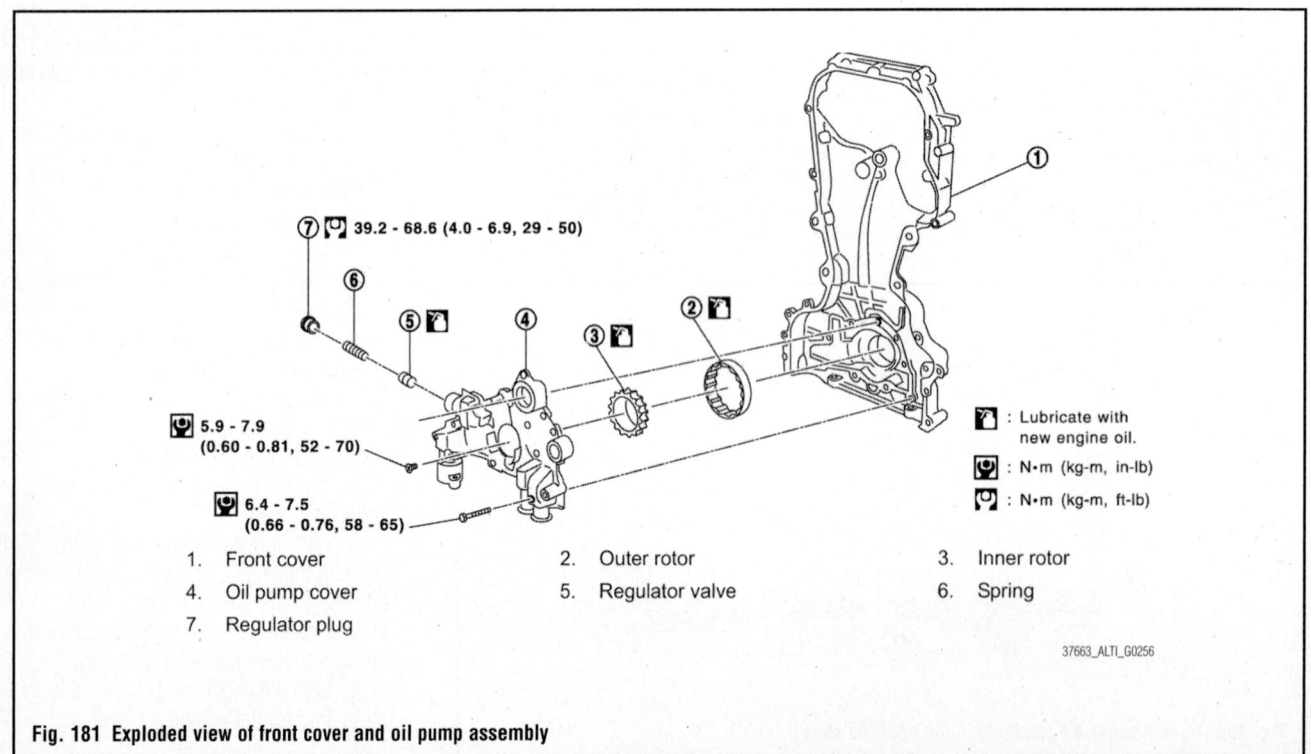

1.	Front cover	2.	Outer rotor	3.	Inner rotor
4.	Oil pump cover	5.	Regulator valve	6.	Spring
7.	Regulator plug				

37663_ALTI_G0256

Fig. 181 Exploded view of front cover and oil pump assembly

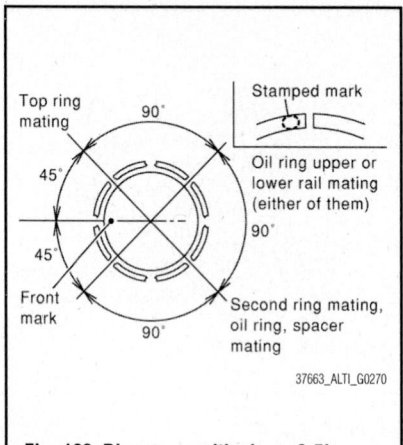

Fig. 183 Ring gap positioning—2.5L Engine

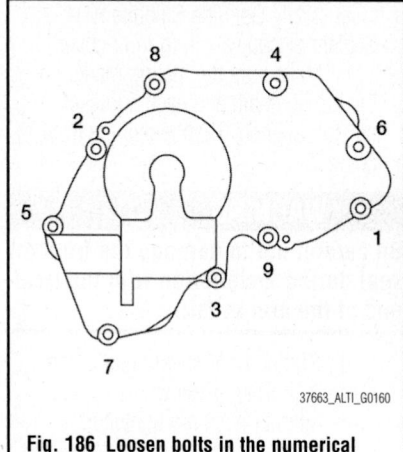

Fig. 186 Loosen bolts in the numerical order as shown

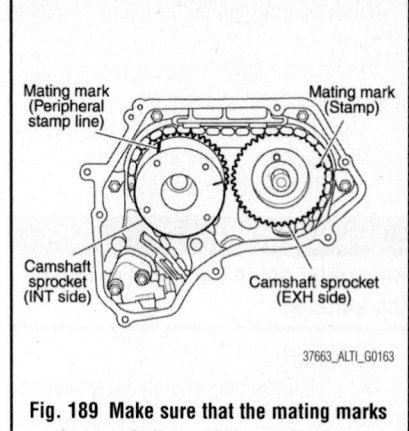

Fig. 189 Make sure that the mating marks on the camshaft sprockets are lined up as shown

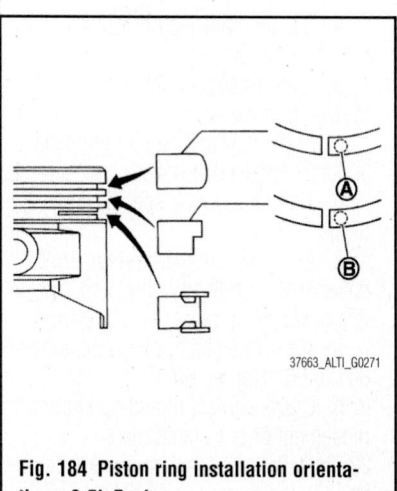

Fig. 184 Piston ring installation orientation—3.5L Engine

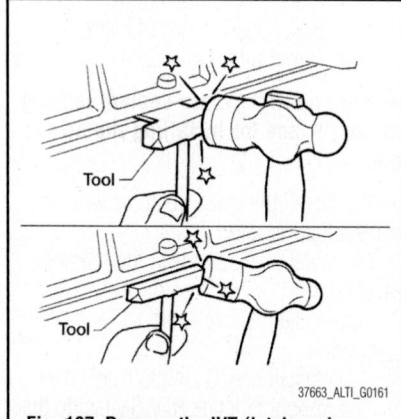

Fig. 187 Remove the IVT (intake valve timing) control cover using Tool KV10111100 (J-37228)

b. At the same time, make sure that the mating marks on the camshaft sprockets are lined up as shown.

c. If not lined up, rotate the crankshaft pulley one more turn to line up the mating marks to the positions as shown.

15. Remove crankshaft pulley with the following procedure:

a. Hold the crankshaft pulley using suitable tool, then loosen the crankshaft pulley bolt, and pull the pulley out about 0.39 inches (10 mm).

b. Attach suitable pulley puller in the thread hole on crankshaft pulley, and

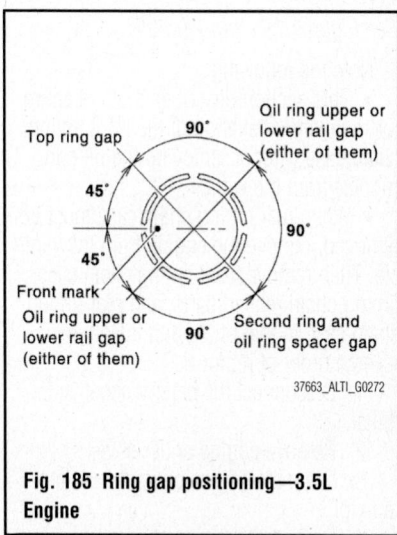

Fig. 185 Ring gap positioning—3.5L Engine

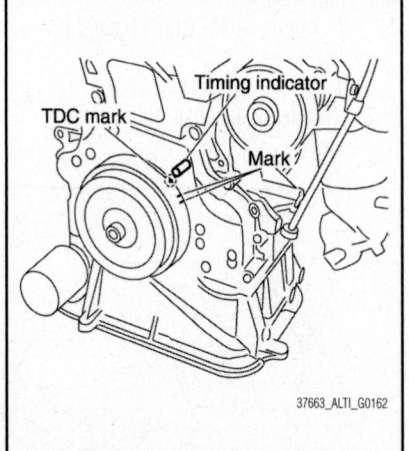

Fig. 188 Set the No.1 cylinder at TDC on the compression stroke

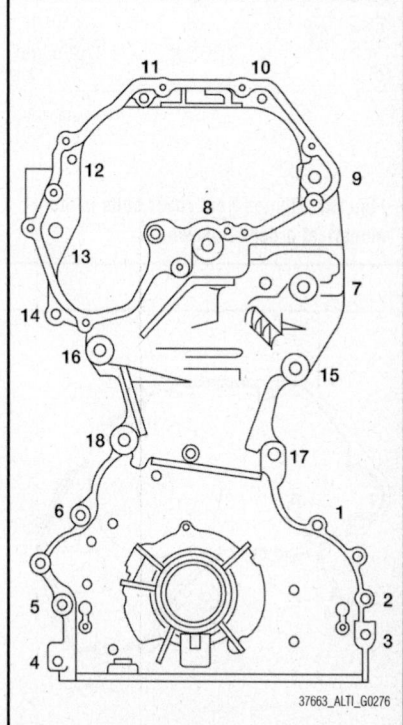

Fig. 190 Loosen the bolts in the numerical order as shown, and remove them

12. Remove the IVT (intake valve timing) control cover using Tool KV10111100 (J-37228).

13. Pull chain guide between camshaft sprockets out through front cover.

14. Set the No.1 cylinder at TDC on the compression stroke with the following procedure:

a. Rotate the crankshaft pulley clockwise and align the mating marks to the timing indicator on the front cover.

remove crankshaft pulley using a suitable puller.

16. Remove the front cover with the following procedure:

a. Loosen the bolts in the numerical order as shown, and remove them.

b. Remove the front cover.

✲✲ WARNING

Be careful not to damage the mounting surface.

To install:

17. Install front cover with the following procedure:

a. Install O-rings to cylinder head and cylinder block.

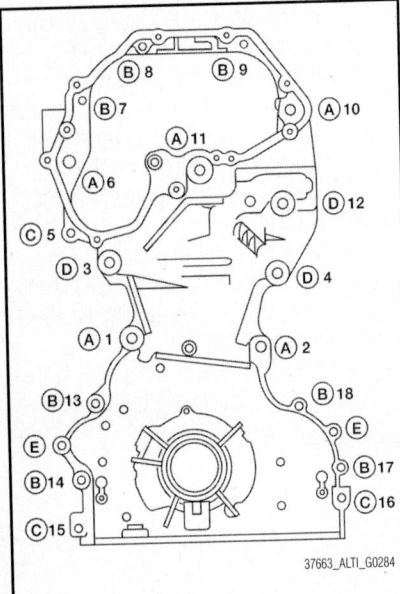

Fig. 191 Tighten front cover bolts in the numerical order as shown

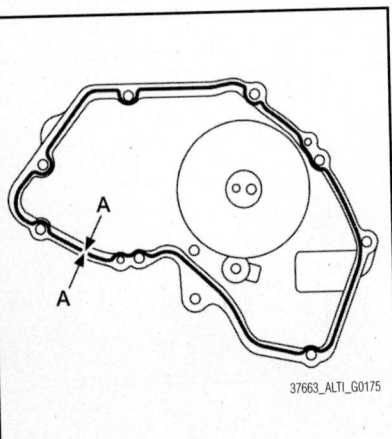

Fig. 192 Apply Silicone RTV Sealant to the IVT cover as shown

b. Apply Genuine Silicone RTV Sealant or equivalent, to front cover.

c. Make sure the mating marks on the timing chain and each sprocket are still aligned. Then install the front cover.

✲✲ WARNING

Be careful not to damage the front oil seal during installation with the front end of the crankshaft.

d. Tighten front cover bolts in the numerical order as shown.

e. After all bolts are tightened, retighten them to the specified torque:

- Bolts A: 36 ft. lbs. (49 Nm)
- Bolts B: 9 ft. lbs. (13 Nm)
- Bolts C: 9 ft. lbs. (13 Nm)
- Bolts D: 36 ft. lbs. (49 Nm)
- Dowel pins

➡ **Wipe off any excess sealant leaking at the surface for installing the oil pan.**

18. Install the chain guide between the camshaft sprockets.

19. Install IVT cover with the following procedure:

a. Install IVT solenoid valve to IVT cover.

b. Install new O-ring to front cover.

c. Apply Silicone RTV Sealant to the IVT cover as shown.

d. Tighten the IVT cover bolts in the numerical order as shown.

20. Insert crankshaft pulley by aligning with crankshaft key.

a. Tap its center with a plastic hammer to insert.

b. Do not tap the belt hook.

21. Tighten crankshaft pulley bolts.

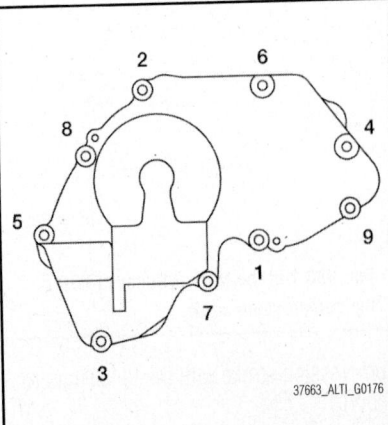

Fig. 193 Tighten the IVT cover bolts in the numerical order as shown

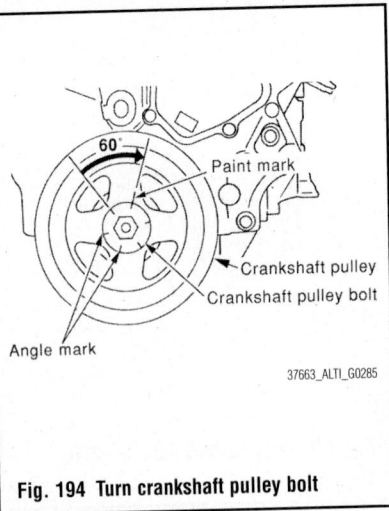

Fig. 194 Turn crankshaft pulley bolt

a. Secure crankshaft pulley with tool to tighten the bolt.

b. Perform angle tightening with the following procedure:

c. Apply new engine oil to threads and seat surfaces of bolts.

d. Tighten to initial specifications: 31 ft. lbs. (42 Nm).

e. Apply a paint mark on the front cover, mating with any one of six easy to recognize stamp marks on bolt flange.

f. Turn crankshaft pulley bolt another 60° to 66° (Target: 60°).

g. Check vertical mounting angle with movement of one stamp mark.

22. Installation of the remaining components is in the reverse order of removal.

3.5L Engine

See Figures 195 through 205.

Note the following:

- This section describes the procedure for removal/installation of the front timing chain case without removing the oil pan (upper) from the vehicle.

- When rear timing chain case must be removed, remove the engine from the vehicle. Then remove front timing chain case, timing chain related parts, and rear timing chain case in this order, and install in reverse order of removal.

1. Disconnect the battery negative terminal.

2. Remove engine under cover.

3. Drain the engine coolant from the radiator.

4. Drain the engine oil.

5. Drain the power steering fluid.

6. Remove engine room cover.

7. Remove front air duct.

8. Remove battery tray.

9. Remove cowl top and cowl top extension.

10. Remove upper radiator hose.

11. Disconnect engine coolant reservoir hose from the radiator and remove engine coolant reservoir.

12. Remove cooling fan assembly.

13. Disconnect lower radiator hose from engine.

14. Recover the A/C system R134a.

15. Remove the starter motor.

16. Disconnect the power steering fluid reservoir tank hose from the power steering pump and fluid cooler and remove the power steering fluid reservoir tank.

17. Remove the front RH wheel and tire.

18. Remove the engine side under cover.

19. Remove the drive belt.

20. Remove the power steering pump.

21. Remove the lower oil pan. Loosen the lower oil pan bolts in order as shown.

   a. Insert Tool between the lower oil pan and the upper oil pan.

   b. Be careful not to damage the mating surface.

✳✳ WARNING

Do not insert a screwdriver, this will damage the mating surfaces.

   c. Slide the Tool by tapping its side with a hammer to remove the lower oil pan from the upper oil pan.

22. Remove upper oil pan bolts in reverse order as shown.

23. Remove the alternator.

24. Disconnect the A/C tubes from the A/C compressor and position aside.

25. Remove the A/C compressor bolts and remove the A/C compressor.

26. Remove the alternator bracket.

27. Support the engine with suitable jack and remove the RH engine insulator, mount and bracket.

28. Remove the rocker covers, if necessary.

➡**Necessary only when removing timing chains.**

29. If removing the timing chains, obtain compression TDC of No. 1 cylinder as follows:

   a. Rotate crankshaft pulley clockwise to align timing mark (grooved line without color) with timing indicator.

   b. Check that intake and exhaust camshaft lobes on No. 1 cylinder (right bank of engine) are located as shown.

   c. If not, turn the crankshaft one revolution (360°) and align as shown.

30. Lock the drive plate using Tool KV10117700 (J-44716) attached to the starter bolt hole.

✳✳ WARNING

Do not damage the ring gear teeth, or the signal plate teeth behind the ring gear, when setting the Tool.

31. Remove the crankshaft pulley as follows:

   a. Loosen crankshaft pulley bolt using suitable tool and locate bolt seating surface at 0.39 inches (10 mm) from its original position.

   b. Position a pulley puller at recess hole of crankshaft pulley to remove crankshaft pulley.

✳✳ WARNING

Do not use a puller claw on crankshaft pulley periphery.

32. Remove engine oil cooler tube bolts and bracket.

33. Disconnect the oil pressure switch harness connector.

34. Disconnect valve timing control harness connector.

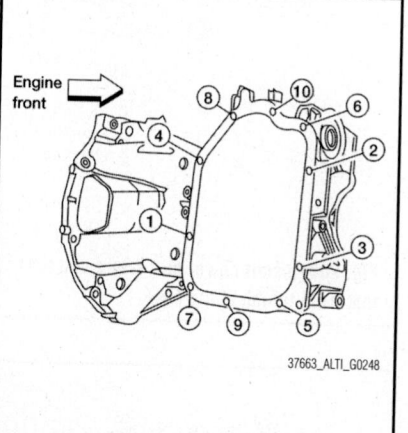

Fig. 195 Loosen the lower oil pan bolts in order as shown

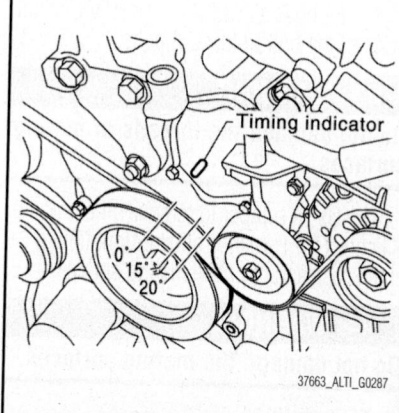

Fig. 197 Rotate crankshaft pulley clockwise to align timing mark (grooved line without color) with timing indicator

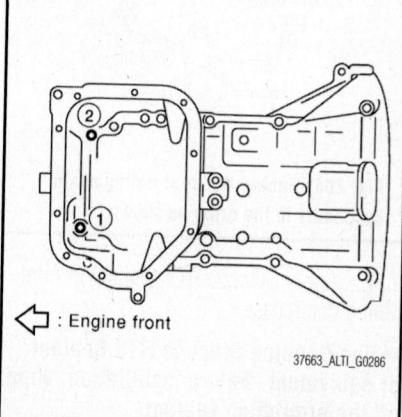

Fig. 196 Remove upper oil pan bolts in reverse order as shown

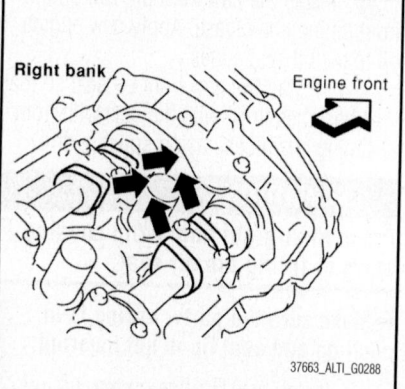

Fig. 198 Check that intake and exhaust camshaft lobes on No. 1 cylinder (right bank of engine) are located as shown

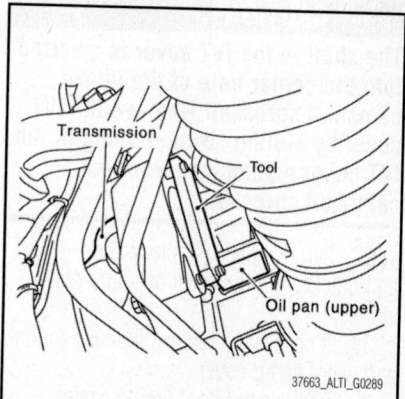

Fig. 199 Lock the drive plate using Tool KV10117700 (J-44716) attached to the starter bolt hole

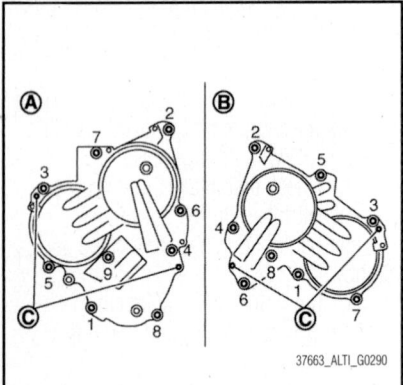

Fig. 200 Remove the Bank 1 (RH) (A) and Bank 2 (LH) (B) IVT covers

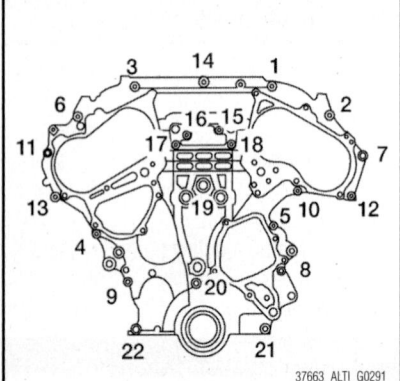

Fig. 201 Loosen the front timing chain case bolts in the order as shown

35. Remove the Bank 1 (RH) and Bank 2 (LH) IVT covers.

36. Loosen the IVT cover bolts in the reverse order as shown.

✳✳ CAUTION

The shaft in the IVT cover is inserted into the center hole of the intake camshaft sprocket. Remove the IVT cover by pulling straight out until the IVT cover disengages from the camshaft sprocket.

37. Remove the A/C idler pulley and bracket and the drive belt auto-tensioner.

38. If necessary, remove the idler pulley and water pump cover.

39. Remove the front timing chain case.

    a. Loosen the front timing chain case bolts in the order as shown.

    b. Insert the appropriate size tool into

the notch at the top of the front timing chain case.

    c. Pry off the case by moving the suitable tool back and forth.

✳✳ WARNING

Do not use a screwdriver or similar tool. After removal, handle carefully so it does not bend, or warp under a load.

40. Remove O-rings from rear timing chain case.

    a. A: Bank 1

    b. B: Bank 2

✳✳ CAUTION

Use new O-rings for installation.

41. Remove the front oil seal from the front timing chain case using a suitable tool.

✳✳ WARNING

Do not damage the front cover.

42. Remove all old Silicone RTV Sealant from all the bolt holes and bolts.

✳✳ WARNING

Do not damage the threads or mating surfaces.

43. Use a scraper to remove all of the old Silicone RTV Sealant from the front timing chain case and opposite mating surfaces.

✳✳ WARNING

Do not damage the mating surfaces.

To install:

44. Install dowel pins (right and left) into front timing chain case up to a point close to taper in order to shorten protrusion length.

45. Install the new front oil seal on the front timing chain case. Apply new engine oil to the oil seal edges.

    a. Install the new front oil seal so that it becomes flush with the face with front timing chain case using suitable drift.

✳✳ WARNING

Press fit straight and avoid causing burrs or tilting the oil seal.

➡**Make sure the garter spring is in position and seal lip is not inverted.**

46. Install new O-rings on rear timing chain case.

➡**Make sure to use new O-rings for installation.**

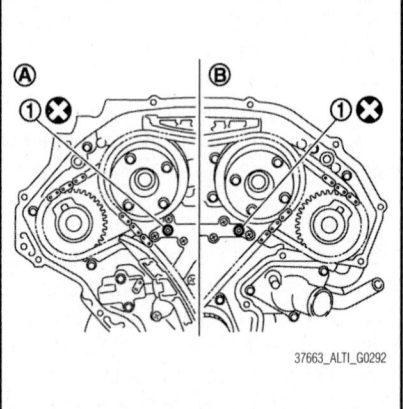

Fig. 202 Remove O-rings (1) from rear timing chain case of Bank 1 (A) and Bank 2 (B)

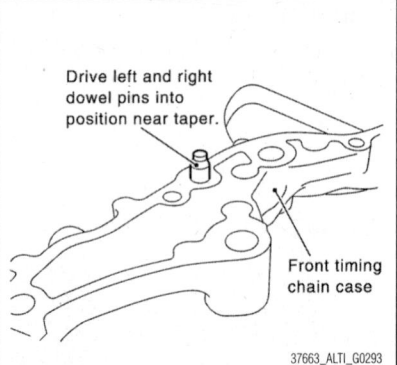

Fig. 203 Install dowel pins (right and left) into front timing chain case

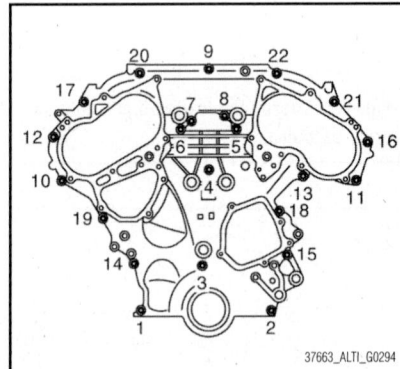

Fig. 204 Tighten the front timing chain case bolts in the order as shown

47. Apply Silicone RTV Sealant to front timing chain case.

➡**Use Genuine Silicone RTV Sealant, or equivalent. Before installation, wipe off the protruding sealant.**

48. Install dowel pin on the rear timing chain case into dowel pin hole in front timing chain case.

49. Loosely install the front timing chain case bolts.

50. Tighten the front timing chain case bolts in the order as shown.

51. Retighten the front timing chain case bolts in the order as shown.

    a. Tighten bolts 1 and 2 to 21 ft. lbs. (28 Nm).

    b. Tighten bolts 3–22 to 9 ft. lbs. (13 Nm).

52. Install upper oil pan bolts.

53. Install lower oil pan.

54. Install IVT control valve covers.

    a. Install new seal rings in shaft grooves.

✳✳ WARNING

When replacing seal rings, replace all rings with new ones on both RH and LH IVT control valve covers.

    b. Install IVT covers with a new gasket to front timing chain case.

    c. Being careful not to move seal ring from the installation groove, align the dowel pins on the front timing chain case with the holes to install valve timing control covers.

    d. Tighten bolts in the numerical order as shown to 8 ft. lbs. (11 Nm).

55. Apply liquid gasket and install the water pump cover, if removed.

➡**Use Genuine Silicone RTV Sealant or equivalent.**

56. Install crankshaft pulley and tighten the bolt in two steps.

    a. Lubricate thread and seat surface of the bolt with new engine oil.

    b. Apply a paint mark for the second step of angle tightening.

    c. Step 1: 32 ft. lbs. (44 Nm)

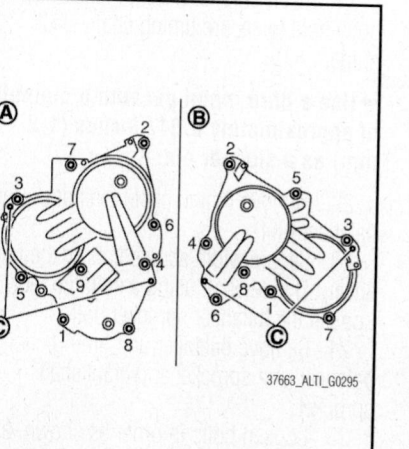

Fig. 205 Tighten bolts in the numerical order as shown

    d. Step 2: Additional 84–90 degrees clockwise

57. Rotate crankshaft pulley in normal direction (clockwise when viewed from front) to confirm it turns smoothly.

58. Installation of the remaining components is in reverse order of removal.

TIMING CHAIN COVER, CHAIN, TENSIONER, & SPROCKETS

REMOVAL & INSTALLATION

2.5L Engine

See Figures 206 through 223.

1. Support the engine and transaxle assembly with suitable tools.

2. Remove RH splash shield.

3. Remove the upper and lower oil pan, and oil strainer.

4. Remove the alternator.

5. Remove the engine cover.

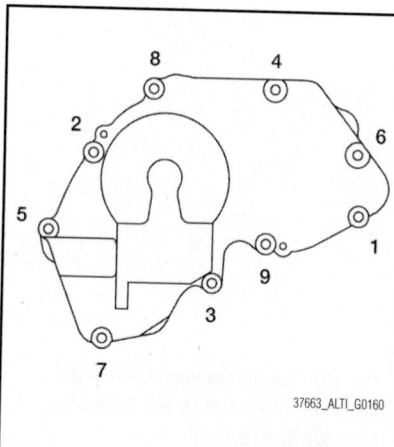

Fig. 206 Loosen bolts in the numerical order as shown

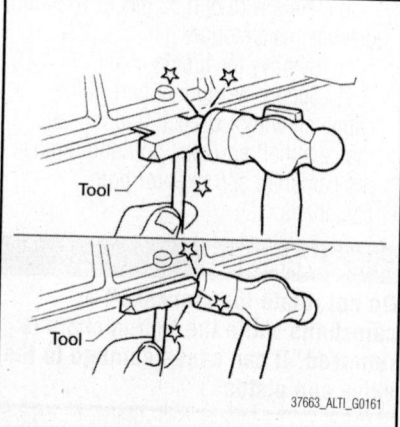

Fig. 207 Remove the IVT (intake valve timing) control cover using Tool KV10111100 (J-37228)

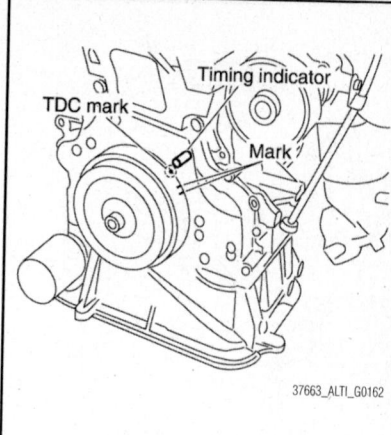

Fig. 208 Set the No.1 cylinder at TDC on the compression stroke

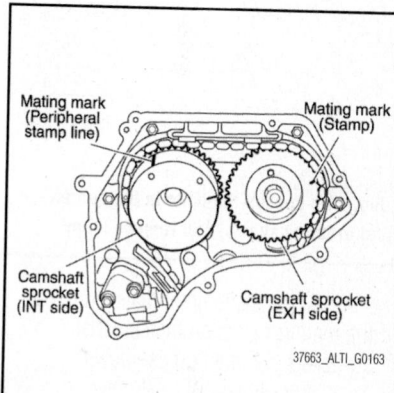

Fig. 209 Make sure that the mating marks on the camshaft sprockets are lined up as shown

6. Disconnect variable timing control solenoid harness connector.

7. Remove the engine ground.

8. Remove the coolant overflow reservoir tank.

9. Position the RH engine compartment fuse and relay box aside.

10. Remove the RH engine mount and bracket.

11. Loosen bolts in the numerical order as shown.

12. Remove the IVT (intake valve timing) control cover using Tool KV10111100 (J-37228).

13. Pull chain guide between camshaft sprockets out through front cover.

14. Set the No.1 cylinder at TDC on the compression stroke with the following procedure:

    a. Rotate the crankshaft pulley clockwise and align the mating marks to the timing indicator on the front cover.

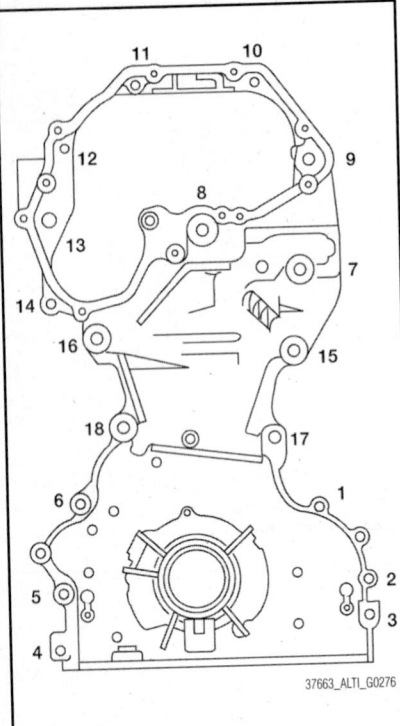

Fig. 210 Loosen the bolts in the numerical order as shown, and remove them

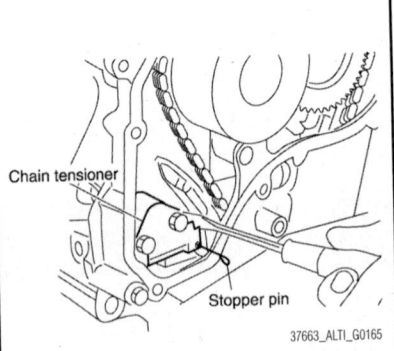

Fig. 211 Insert a stopper pin into the hole on the tensioner body to secure the chain tensioner plunger and remove chain tensioner

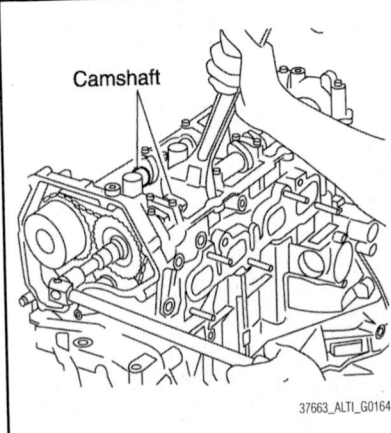

Fig. 212 Secure hexagonal part of the camshaft with a wrench and loosen the camshaft sprocket bolt

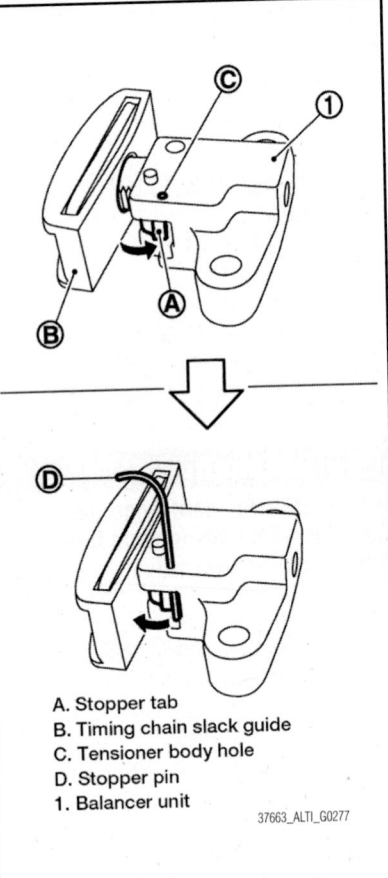

A. Stopper tab
B. Timing chain slack guide
C. Tensioner body hole
D. Stopper pin
1. Balancer unit

Fig. 213 Press stopper tab in the direction shown to push the timing chain slack guide toward timing chain tensioner (for balancer unit)

b. At the same time, make sure that the mating marks on the camshaft sprockets are lined up as shown.

c. If not lined up, rotate the crankshaft pulley one more turn to line up the mating marks to the positions as shown.

15. Remove crankshaft pulley with the following procedure:

a. Hold the crankshaft pulley using suitable tool, then loosen the crankshaft pulley bolt, and pull the pulley out about 0.39 inches (10 mm).

b. Attach suitable pulley puller in the thread hole on crankshaft pulley, and remove crankshaft pulley using a suitable puller.

16. Remove the front cover with the following procedure:

a. Loosen the bolts in the numerical order as shown, and remove them.

b. Remove the front cover.

⁂ WARNING

Be careful not to damage the mounting surface.

17. Remove front oil seal using suitable tool, if necessary.

18. Remove timing chain with the following procedure:

a. Push in the tensioner plunger. Insert a stopper pin into the hole on the tensioner body to secure the chain tensioner plunger and remove chain tensioner.

b. Use a wire of 0.02 inches (0.5 mm) diameter as a stopper pin.

c. Remove the timing chain.

d. Secure hexagonal part of the camshaft with a wrench and loosen the camshaft sprocket bolt and remove the camshaft sprocket for both camshafts.

⁂ WARNING

Do not rotate the crankshaft or camshafts while the timing chain is removed. It can cause damage to the valve and piston.

19. Remove the chain slack guide, tension guide, timing chain, and oil pump drive spacer.

20. Press stopper tab in the direction shown to push the timing chain slack guide toward timing chain tensioner (for balancer unit).

➡The slack guide is released by pressing the stopper tab. As a result, the slack guide can be moved.

21. Insert stopper pin into tensioner body hole to secure timing chain slack guide.

➡Use a hard metal pin with a diameter of approximately 0.047 inches (1.2 mm) as a stopper pin.

22. Remove timing chain tensioner (for balancer unit).

23. Secure width across flats of the balancer LH side shaft using a suitable tool. Loosen the balancer sprocket bolt.

24. Remove balancer unit timing chain, balancer unit sprocket and crankshaft sprocket.

25. Loosen bolts in order as shown, and remove balancer unit.

➡Use Torx® socket size E14.

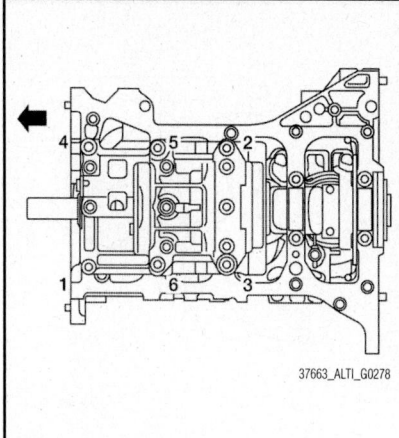

Fig. 214 Loosen bolts in order as shown, and remove balancer unit

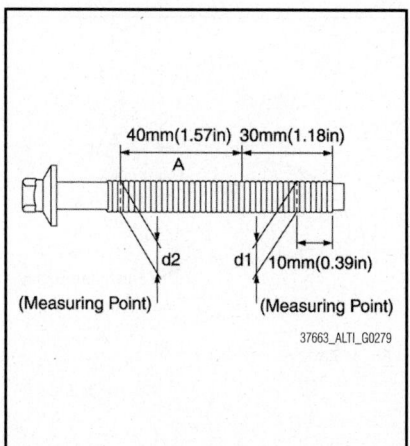

40mm(1.57in) 30mm(1.18in)

A

d2 d1 10mm(0.39in)

(Measuring Point)   (Measuring Point)

Fig. 215 Check the balancer unit bolt outer diameter

※※ **WARNING**

Do not disassemble balancer unit.

26. Check the timing chain for cracks or excessive wear. If a defect is detected, replace it.

27. Check the balancer unit bolt outer diameter.

a. Measure outer diameters (d1, d2) at the two positions as shown.

b. Measure d2 within the range A.

c. If the value difference (d1–d2) exceeds the limit, replace it with a new one.

To install:

➡**There may be two color variations of the link marks (link colors) on the timing chain. There are 26 links between the gold/yellow mating marks on the timing chain; and 64 links between the camshaft sprocket gold/yellow link and the crankshaft sprocket orange/blue**

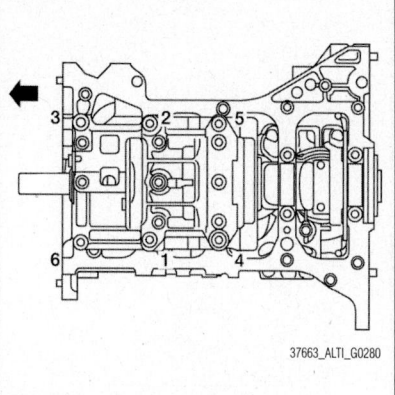

Fig. 216 Install the balancer unit and tighten the bolts in the numerical order as shown

link, on the timing chain side without the tensioner.

28. Make sure the crankshaft key points straight up.

29. Install the balancer unit and tighten the bolts in the numerical order as shown.

a. Step 1: Tighten bolts 1–5 to 31 ft. lbs. (42 Nm). Tighten bolt 6 to 27 ft. lbs. (36 Nm).

b. Step 2: Tighten bolts 1–5 an additional 120°. Tighten bolt 6 an additional 90°.

c. Step 3: Loosen all bolts in reverse order of tightening.

d. Step 4: Tighten bolts 1–5 to 31 ft. lbs. (42 Nm). Tighten bolt 6 to 27 ft. lbs. (36 Nm).

e. Step 5: Tighten bolts 1–5 an additional 120°. Tighten bolt 6 an additional 90°.

Note the following:

• When reusing a bolt, check its outer diameter before installation. Follow the Balancer Unit Bolt Outer Diameter procedure.

• Apply new engine oil to threads and seating surfaces of bolts.

• Check tightening angle with an angle wrench or a protractor.

• Do not make judgment by visual check alone.

30. Install the crankshaft sprocket and timing chain for the balancer unit.

a. Make sure that the crankshaft sprocket is positioned with mating marks on the block and sprocket meeting at the top.

b. Install it by lining up mating marks on each sprocket and timing chain.

31. Install timing chain tensioner (for balancer unit).

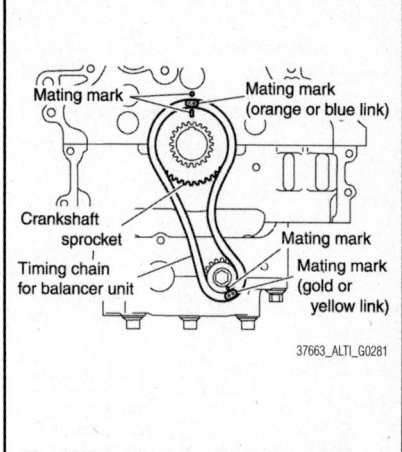

Fig. 217 Install the crankshaft sprocket and timing chain for the balancer unit

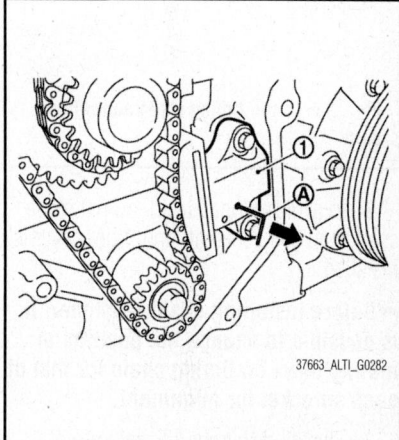

Fig. 218 Install timing chain tensioner (for balancer unit)

a. Fix the plunger at the most compressed position using a stopper pin, and then install it.

b. Securely pull out the stopper pin after installing the timing chain tensioner (for balancer unit).

c. Check matching mark position of balancer unit drive chain and each sprocket again.

32. Install timing chain and related parts.

a. Install by lining up mating marks on each sprocket and timing chain as shown.

b. Before and after installing timing chain tensioner, check again to make sure the mating marks have not slipped.

c. After installing timing chain tensioner, remove the stopper pin, and make sure that the tensioner moves freely.

Note the following:

• For the following note, after the mating marks are aligned, keep them aligned by holding them by hand.

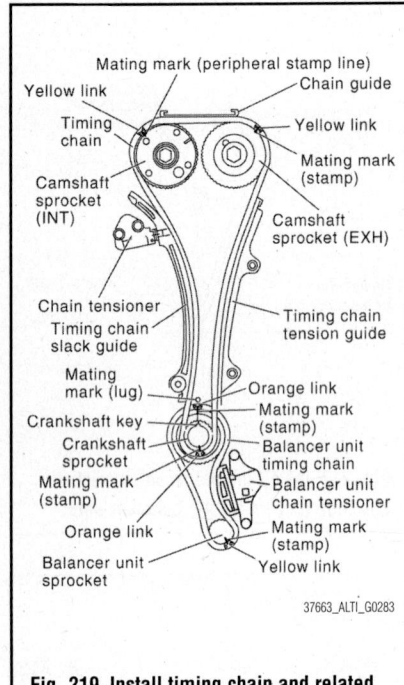

Fig. 219 Install timing chain and related parts

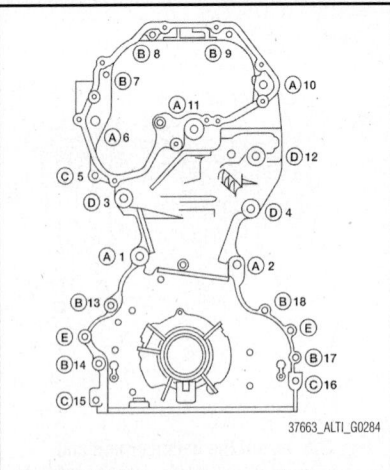

Fig. 220 Tighten front cover bolts in the numerical order as shown

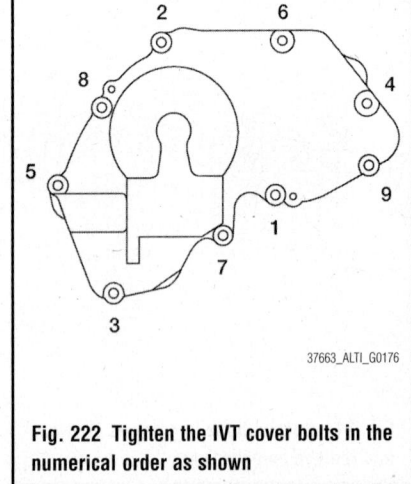

Fig. 222 Tighten the IVT cover bolts in the numerical order as shown

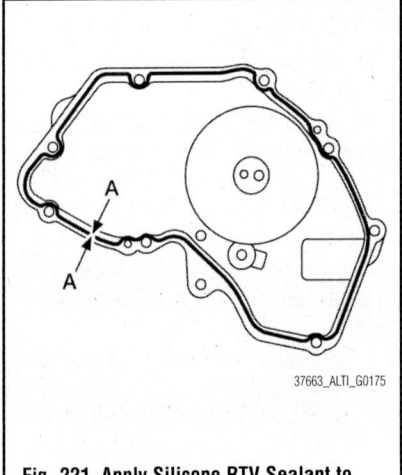

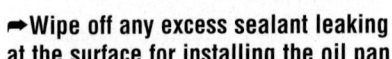

Fig. 221 Apply Silicone RTV Sealant to the IVT cover as shown

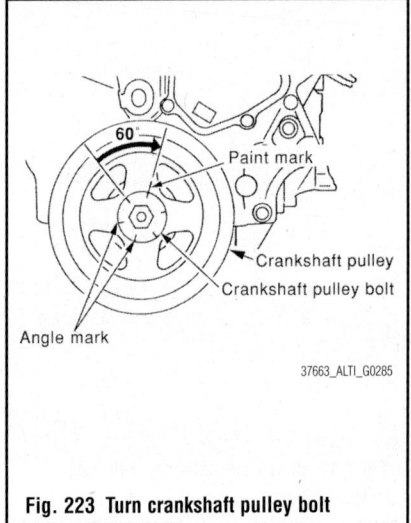

Fig. 223 Turn crankshaft pulley bolt

• To avoid skipped teeth, do not move crankshaft and camshaft until front cover is installed.

➡**Before installing chain tensioner, it is possible to change the position of mating mark on timing chain for that of each sprocket for alignment.**

33. Install new front oil seal to front cover, using suitable tool.
   a. Install new oil seal in until it is flush with front end surface of front cover.

❊❊ WARNING

Do not reuse oil seal. Be careful not to cause damage to circumference of oil seal.

34. Install front cover with the following procedure:
   a. Install O-rings to cylinder head and cylinder block.
   b. Apply Genuine Silicone RTV Sealant or equivalent, to front cover.
   c. Make sure the mating marks on the timing chain and each sprocket are still aligned. Then install the front cover.

❊❊ WARNING

Be careful not to damage the front oil seal during installation with the front end of the crankshaft.

   d. Tighten front cover bolts in the numerical order as shown.
   e. After all bolts are tightened, retighten them to the specified torque:
   • Bolts A: 36 ft. lbs. (49 Nm)
   • Bolts B: 9 ft. lbs. (13 Nm)
   • Bolts C: 9 ft. lbs. (13 Nm)
   • Bolts D: 36 ft. lbs. (49 Nm)
   • Dowel pins

➡**Wipe off any excess sealant leaking at the surface for installing the oil pan.**

35. Install the chain guide between the camshaft sprockets.
36. Install IVT cover with the following procedure:
   a. Install IVT solenoid valve to IVT cover.
   b. Install new O-ring to front cover.
   c. Apply Silicone RTV Sealant to the IVT cover as shown.
   d. Tighten the IVT cover bolts in the numerical order as shown.

37. Insert crankshaft pulley by aligning with crankshaft key.
   a. Tap its center with a plastic hammer to insert.
   b. Do not tap the belt hook.
38. Tighten crankshaft pulley bolts.
   a. Secure crankshaft pulley with tool to tighten the bolt.
   b. Perform angle tightening with the following procedure:
   c. Apply new engine oil to threads and seat surfaces of bolts.
   d. Tighten to initial specifications: 31 ft. lbs. (42 Nm).
   e. Apply a paint mark on the front cover, mating with any one of six easy to recognize stamp marks on bolt flange.
   f. Turn crankshaft pulley bolt another 60° to 66° (Target: 60°).
   g. Check vertical mounting angle with movement of one stamp mark.
39. Installation of the remaining components is in the reverse order of removal.

3.5L Engine

See Figures 224 through 235.

Note the following:

• After removing timing chains, do not turn the crankshaft and camshaft separately, or the valves will strike the pistons.

• When installing camshafts, chain tensioners, oil seals, or other sliding parts, lubricate contacting surfaces with new engine oil.

• Apply new engine oil to bolt threads and seat surfaces when installing camshaft sprockets, camshaft brackets, and crankshaft pulley.

1. Remove front timing chain case.
2. Remove the intake manifold collector.
3. Remove the engine oil dipstick.
4. Place paint marks on the timing chain and sprockets to indicate the correct position of the components for installation.

5. Remove the timing chain tensioner (primary).

   a. Pull lever down and release plunger stopper tab. Plunger stopper tab can be pushed up to release (coaxial structure with lever).

   b. Insert stopper pin into timing chain tensioner (primary) body hole to hold lever, and keep the tab released. An Allen wrench is used for a stopper pin as an example.

   c. Insert plunger into tensioner body by pressing the slack guide.

   d. Keep the slack guide pressed and hold it by pushing the stopper pin through the lever hole and body hole.

   e. Remove the bolts and remove the timing chain tensioner (primary).

6. Remove the internal chain guide and slack guide.

7. Remove timing chain (primary) and crankshaft sprocket.

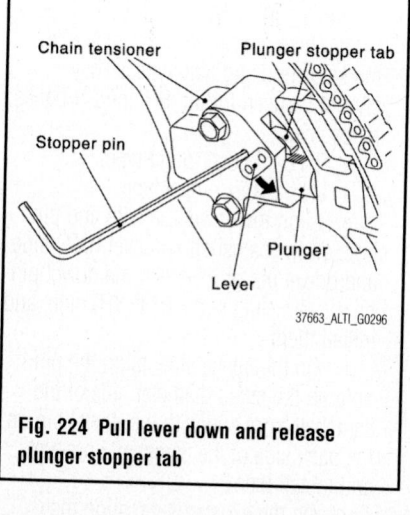

Fig. 224 Pull lever down and release plunger stopper tab

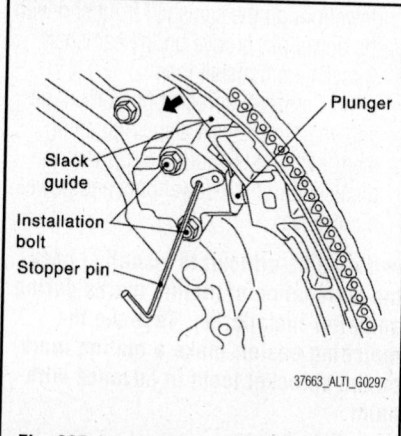

Fig. 225 Insert stopper pin into timing chain tensioner (primary) body hole to hold lever, and keep the tab released

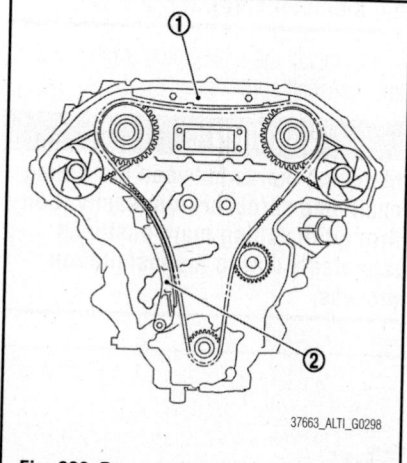

Fig. 226 Remove the internal chain guide (1), and slack guide (2)

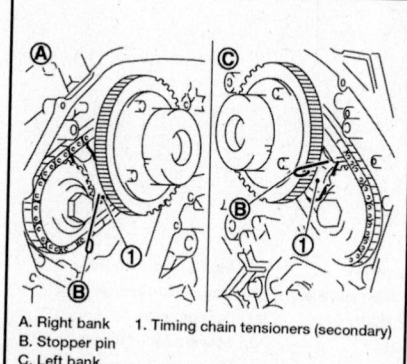

A. Right bank
B. Stopper pin
C. Left bank
1. Timing chain tensioners (secondary)

Fig. 227 Attach a suitable stopper pin to the right and left timing chain tensioners (secondary)

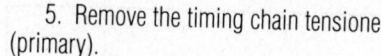

✳✳ WARNING

After removing timing chains, do not turn the crankshaft and camshaft separately, or the valves will strike the pistons.

8. Attach a suitable stopper pin to the right and left timing chain tensioners (secondary).

9. Remove the timing chains (secondary) with camshaft sprockets (INT) and (EXH).

   a. Insert metal or resin plate [0.5 mm (0.020 in)] into guide between timing chain (secondary) and timing chain tensioner (secondary) plunger. Remove camshaft sprocket and timing chain (secondary) with timing chain removed from guide groove.

✳✳ WARNING

Timing chain tensioner plunger can move while stopper pin is inserted in timing chain tensioner. Plunger can come out of tensioner when timing chain is removed. Use caution during removal.

   b. Apply paint to the timing chain and camshaft sprockets for alignment during installation.

   c. Remove the camshaft sprocket (INT) and (EXH) bolts.

   d. Hold the hexagonal portion of the camshaft using a wrench to loosen the bolts.

   e. Handle the sprockets as an assembly.

   f. Remove timing chains (secondary).

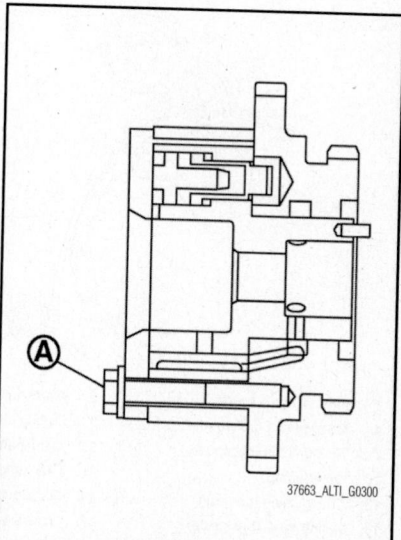

Fig. 228 Do not disassemble the camshaft sprockets (never loosen bolts (A) as shown)

❋❋ WARNING

Avoid impact or dropping the camshaft sprockets. Do not disassemble the camshaft sprockets (never loosen bolts as shown).

10. Remove the tension guide.

11. Check for cracks and any excessive wear of the timing chain. Replace the timing chain as necessary.

To install:

➡ The figure shows the relationship between the mating mark on each timing chain and that on the corresponding sprocket, the components installed.

12. Install the tension guide.

13. Position the crankshaft so No. 1 piston is set at TDC on the compression stroke.

➡ Make sure that the dowel pin hole, dowel pin and crankshaft key are located as shown.

- Camshaft dowel pin hole (intake side): at cylinder head upper face side in each bank.
- Camshaft dowel pin (exhaust side): at cylinder head upper face side in each bank.
- Crankshaft key: at cylinder head side of RH bank.

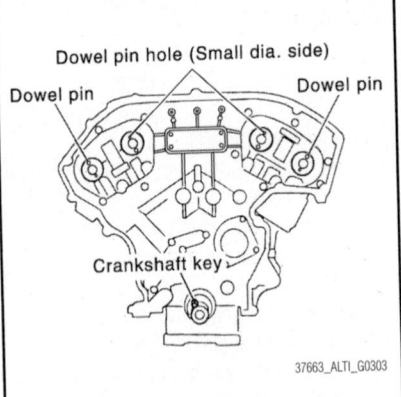

Fig. 230 Position the crankshaft so No. 1 piston is set at TDC on the compression stroke

❋❋ WARNING

Hole on small diameter side must be used for intake camshaft sprocket dowel pin. Do not misidentify (ignore big diameter side).

14. Install the timing chains (secondary) and camshaft sprockets.

❋❋ WARNING

Matching marks between the timing chain and sprockets slip easily. Confirm all matching mark positions repeatedly during the installation process.

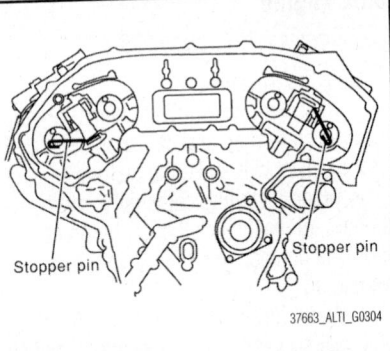

Fig. 231 Push the sleeve of the chain tensioner (secondary) and keep it pressed in with a stopper pin

a. Push the sleeve of the chain tensioner (secondary) and keep it pressed in with a stopper pin.

b. Align the matching marks on the timing chain (secondary), with the ones on the camshaft sprockets (INT) and (EXH) (stamped), and install them.

c. Matching marks for the camshaft sprocket (INT) are on the back side of the secondary sprocket.

d. There are two types of matching marks, round and oval types. They should be used for the RH and LH banks, respectively.
- RH bank: use round type.
- LH bank: use oval type.

e. Align the dowel pin with and pin hole on the camshaft sprocket (INT) side, and dowel pin groove with the dowel pin on the camshaft sprocket (EXH) side, and install them.

f. On the intake side, align the pin hole on the small diameter side of the camshaft front end with the dowel pin on the back side of the camshaft sprocket, and install them.

g. On the exhaust side, align the dowel pin on the camshaft front end with the dowel pin groove on the camshaft sprocket, and install them.

h. Camshaft sprocket bolts must be tightened in the next step. Tightening them by hand is enough to prevent the dislocation of the dowel pins and dowel pin grooves.

➡ It may be difficult to visually check the dislocation of mating marks during and after installation. To make the matching easier, make a mating mark on the sprocket teeth in advance with paint.

15. After confirming the mating marks are aligned, tighten the camshaft sprocket bolts.

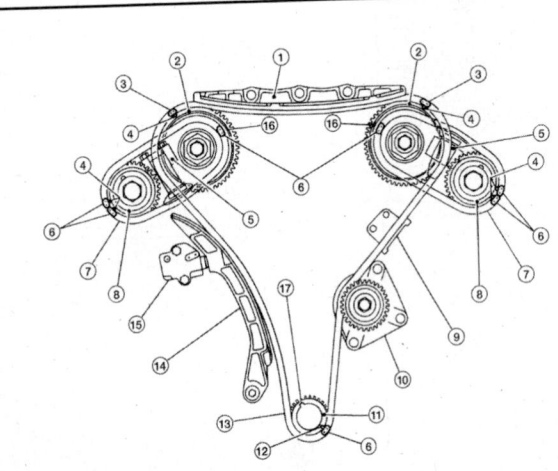

1. Internal chain guide
2. Camshaft sprocket (INT)
3. Mating mark (pink link)
4. Mating mark (punched)
5. Timing chain tensioner (secondary)
6. Mating mark (orange link)
7. Timing chain (secondary)
8. Camshaft sprocket (EXH)
9. Tension guide
10. Water pump
11. Crankshaft sprocket
12. Mating mark (notched)
13. Timing chain (primary)
14. Slack guide
15. Timing chain tensioner (primary)
16. Mating mark (back side)
17. Crankshaft key

Fig. 229 Mating mark locations and relationships to timing chain assembly

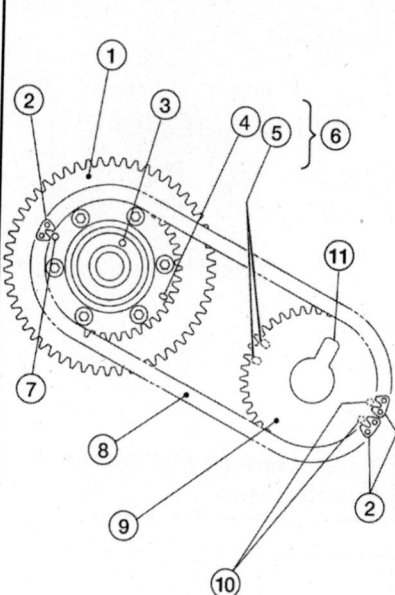

1. Camshaft sprocket (INT) side
2. Secondary timing chain orange link
3. Dowel pin
4. Matching marks
5. Matching marks
6. LH bank: oval type (4 and 5)
7. Matching marks
8. Timing chain (secondary)
9. Camshaft sprocket
10. Matching marks
11. Dowel pin grooves

37663_ALTI_G0305

Fig. 232 Align the matching marks on the timing chain (secondary), with the ones on the camshaft sprockets (INT) and (EXH) (stamped), and install them

➡**Hold the camshaft using a wrench at the hexagonal portion to tighten the bolts.**

16. Pull the stopper pins out from the secondary timing chain tensioners.

17. Install the crankshaft sprocket on the crankshaft.

➡**Make sure the mating marks on the crankshaft sprocket face the front of the engine.**

18. Install the timing chain (primary).

a. Install timing chain (primary) so the mating mark (punched) on camshaft sprocket is aligned with the pink link on the timing chain, while the mating mark (notched) on the crankshaft sprocket is aligned with the orange one on the timing chain, as shown.

b. When it is difficult to align mating marks of the timing chain (primary) with each sprocket, gradually turn the camshaft using a wrench on the hexago-

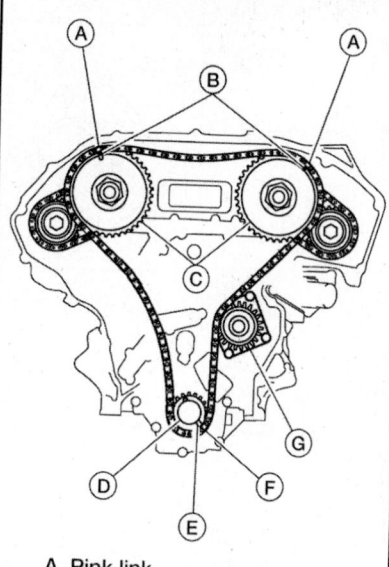

A. Pink link
B. Punched mating mark
C. Camshaft sprockets
D. Crankshaft sprocket
E. Notched mating mark
F. Orange link
G. Water pump

37663_ALTI_G0306

Fig. 233 Install the timing chain (primary)

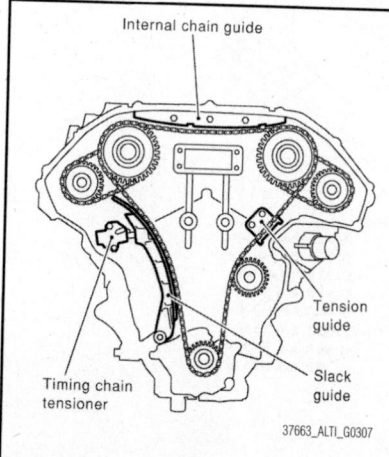

37663_ALTI_G0307

Fig. 234 Install the internal chain guide and slack guide

nal portion to align it with the mating marks.

c. During alignment, be careful to prevent dislocation of mating mark alignments of the secondary timing chains.

19. Install the internal chain guide and slack guide.

⚠ **WARNING**

Do not overtighten the slack guide bolts. It is normal for a gap to exist

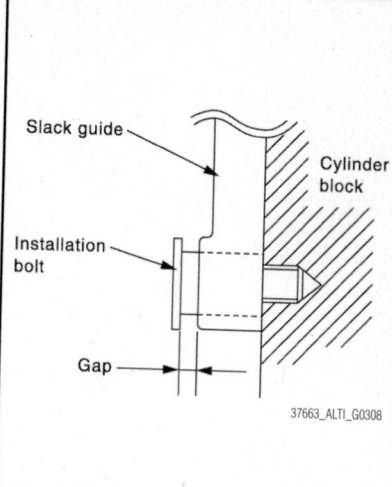

37663_ALTI_G0308

Fig. 235 It is normal for a gap to exist under the bolt seats when the bolts are tightened to specification

under the bolt seats when the bolts are tightened to specification.

20. Install the timing chain tensioner (primary) for the slack guide.

a. When installing the timing chain tensioner (primary), push in the sleeve and keep it pressed in with the stopper pin.

b. Remove any dirt and foreign materials completely from the back and the mounting surfaces of the timing chain tensioner (primary).

c. After installation, pull out the stopper pin while pressing the slack guide.

21. Reconfirm that the matching marks on the sprockets and the timing chain have not slipped out of alignment.

22. Install the front timing chain case.

VALVE COVERS

REMOVAL & INSTALLATION

2.5L Engine

See Figures 236 through 239.

1. Disconnect the battery negative terminal.

2. Remove the engine cover.

3. Remove the front air duct.

4. Remove the blow-by hose.

5. Remove the two brake ECU nuts and set the brake ECU aside.

6. Remove the RH engine mount torque rod.

7. Use a suitable tool to support the engine assembly.

8. Remove the RH engine support bracket.

9. Remove the RH engine mounting bracket.

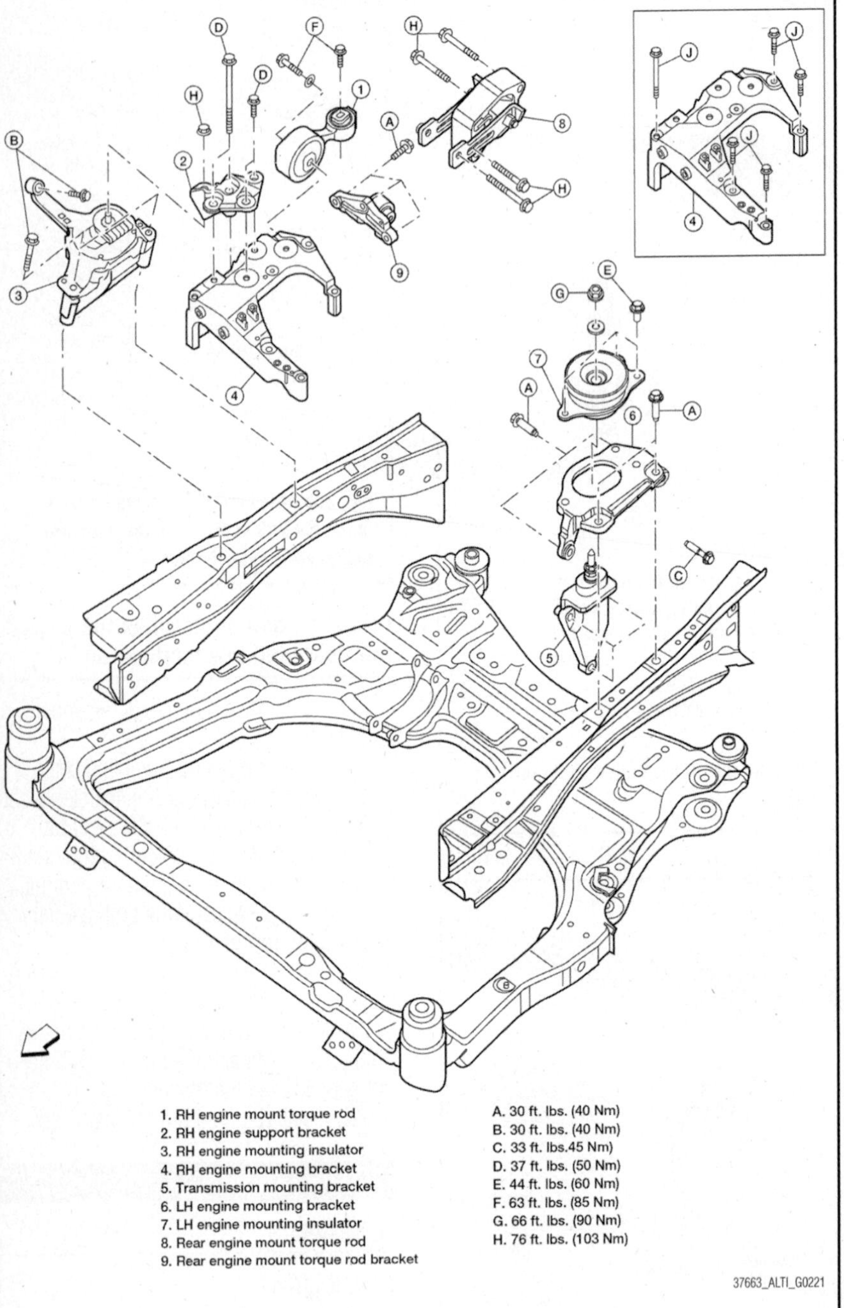

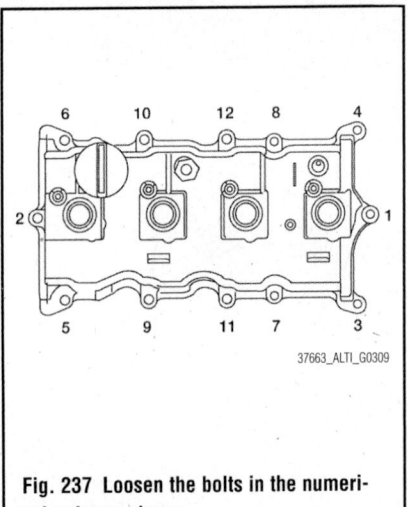

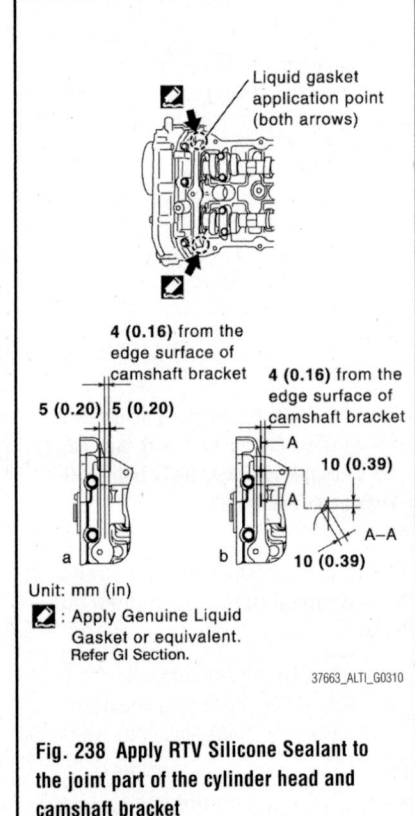

1. RH engine mount torque rod
2. RH engine support bracket
3. RH engine mounting insulator
4. RH engine mounting bracket
5. Transmission mounting bracket
6. LH engine mounting bracket
7. LH engine mounting insulator
8. Rear engine mount torque rod
9. Rear engine mount torque rod bracket
A. 30 ft. lbs. (40 Nm)
B. 30 ft. lbs. (40 Nm)
C. 33 ft. lbs.45 Nm)
D. 37 ft. lbs. (50 Nm)
E. 44 ft. lbs. (60 Nm)
F. 63 ft. lbs. (85 Nm)
G. 66 ft. lbs. (90 Nm)
H. 76 ft. lbs. (103 Nm)
37663_ALTI_G0221

Fig. 236 Exploded view of engine mounts and subframe assembly

6 10 12 8 4
2 1
5 9 11 7 3
37663_ALTI_G0309

Fig. 237 Loosen the bolts in the numerical order as shown

Liquid gasket application point (both arrows)
4 (0.16) from the edge surface of camshaft bracket
5 (0.20)
5 (0.20)
4 (0.16) from the edge surface of camshaft bracket
A
10 (0.39)
A–A
10 (0.39)
a
b
Unit: mm (in)
: Apply Genuine Liquid Gasket or equivalent.
Refer GI Section.
37663_ALTI_G0310

Fig. 238 Apply RTV Silicone Sealant to the joint part of the cylinder head and camshaft bracket

10. Disconnect the PCV hose.
11. Remove the ignition coils.
12. Disconnect the fuel injectors and position the fuel injector harness aside.
13. Loosen the bolts in the numerical order as shown.
14. Remove the rocker cover and the rocker cover gasket. Discard the rocker cover gasket.

WARNING

Do not reuse the rocker cover gasket.

15. Remove the oil filler cap if necessary, to transfer to the new rocker cover.

To install:

16. Apply RTV Silicone Sealant to the joint part of the cylinder head and camshaft bracket using the following steps:
    a. Follow left side illustration to apply sealant to joint part of No.1 camshaft bracket and cylinder head.
    b. Follow right side illustration to apply sealant in a 90 degree angle to the left side illustration.
17. Install the rocker cover and the new rocker cover gasket.

➡The rocker cover gasket must be securely installed in the groove in the rocker cover.

18. Tighten the rocker cover bolts in two steps, in the numerical order as shown.
    a. Step 1: 17 inch lbs. (2 Nm)
    b. Step 2: 74 inch lbs. (8 Nm)
19. Installation of the remaining components is in the reverse order of removal.

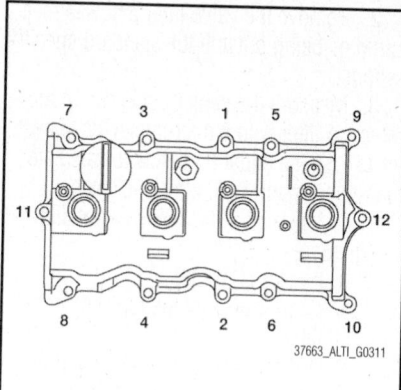

Fig. 239 Tighten the rocker cover bolts in two steps, in numerical order

3.5L Engine

Left Side

See Figures 240 through 242.

1. Remove the engine cover.
2. Remove front air duct.
3. Remove blow by hose from rocker cover.
4. Remove camshaft position sensor. Note the following:
- Handle carefully to avoid dropping and shocks.
- Do not disassemble.
- Do not allow metal powder to adhere to magnetic part at sensor tip.
- Do not place sensors in a location where they are exposed to magnetism.
5. Disconnect the ignition coil connectors.
6. Remove the ignition coils.

✳✳ WARNING

Never shock ignition coils.

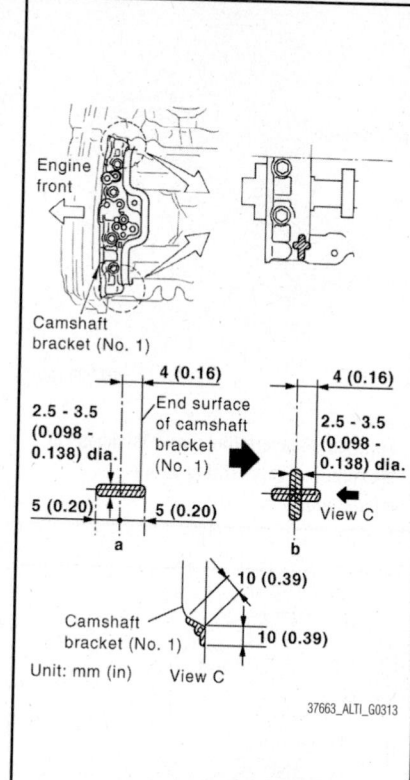

Fig. 241 Apply sealant to the areas on the front corners

7. Remove LH rocker cover bolts from cylinder head as shown.

To install:

8. Installation is in the reverse order of removal.
9. Apply sealant to the areas on front corners.
10. Tighten the rocker cover bolts in two steps in the order as shown.
   a. Step 1: 17 inch lbs. (2 Nm)
   b. Step 2: 74 inch lbs. (8 Nm)

Right Side

See Figures 241, 243 and 244.

1. Remove the engine cover.
2. Remove the front air duct and air duct hose.
3. Remove the intake manifold collector.
4. Remove ignition coils.

✳✳ WARNING

Never shock ignition coils.

5. Remove camshaft position sensor. Note the following:
- Handle carefully to avoid dropping and shocks.
- Do not disassemble.
- Do not allow metal powder to adhere to magnetic part at sensor tip (A).
- Do not place sensors in a location where they are exposed to magnetism.
6. Remove RH rocker cover bolts from cylinder head as shown.

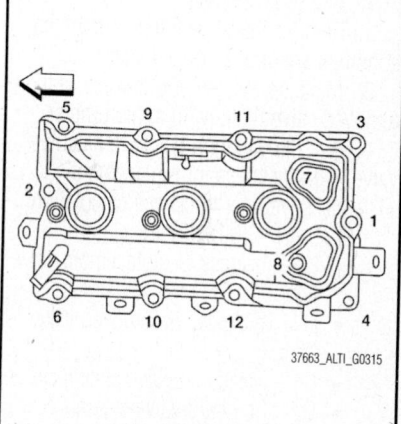

Fig. 243 Remove RH rocker cover bolts from cylinder head as shown

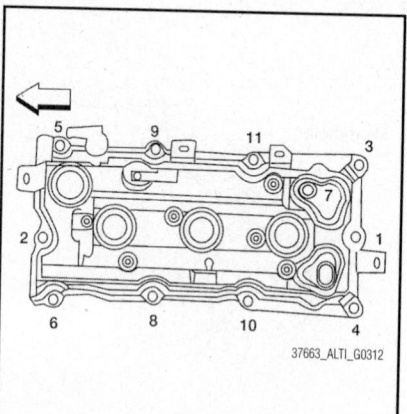

Fig. 240 Remove LH rocker cover bolts from cylinder head

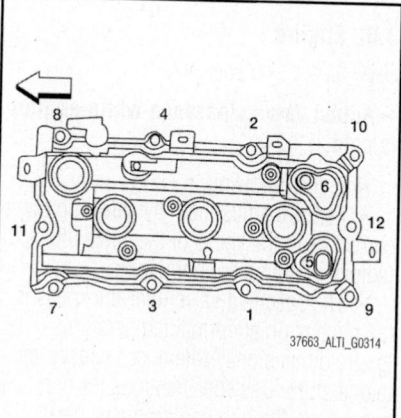

Fig. 242 Tighten the rocker cover bolts in two steps in the order as shown

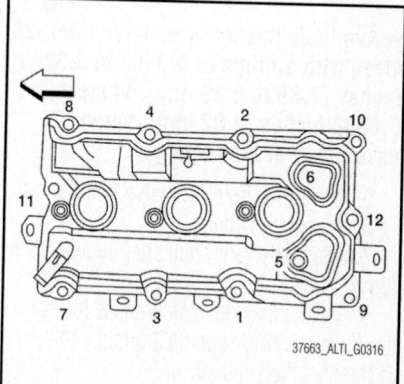

Fig. 244 Tighten the rocker cover bolts in two steps in the order as shown

To install:

7. Installation is in the reverse order of removal.

8. Apply sealant to the areas on the front corners.

9. Tighten the rocker cover bolts in two steps in the order as shown.
   a. Step 1: 17 inch lbs. (2 Nm)
   b. Step 2: 74 inch lbs. (8 Nm)

VALVE LASH (CLEARANCE) ADJUSTMENT

ADJUSTMENT

2.5L Engine

See Figures 245 and 246.

Perform adjustment depending on selected head thickness of valve lifter.

The specified valve lifter thickness is the dimension at normal temperatures. Ignore dimensional differences caused by temperature. Use the specifications for hot engine condition to adjust.

1. Remove camshaft.

2. Remove the valve lifters at the locations that are outside the standard.

3. Measure the center thickness of the removed valve lifters with a micrometer.

4. Use the equation below to calculate valve lifter thickness for replacement.
   a. Valve lifter thickness calculation:
      - $t = t1 + (C1 - C2)$
      - t = Thickness of replacement valve lifter.
      - $t1$ = Thickness of removed valve lifter.
      - $C1$ = Measured valve clearance.
      - $C2$ = Standard valve clearance.
   b. Thickness of a new valve lifter can be identified by stamp marks on the reverse side (inside the cylinder). Stamp mark 696 indicates a thickness of 0.2740 inches (6.96 mm).

➡**Available thickness of valve lifter: 26 sizes with a range of 0.3102 to 0.3299 inches (7.88 to 8.38 mm), in steps of 0.0008 inches (0.02 mm), when assembled at the factory.**

5. Install the selected valve lifter.

6. Install camshaft.

7. Manually turn crankshaft pulley a few turns.

8. Check that valve clearances for cold engine are within specifications, by referring to the specified values.

9. After completing the repair, check valve clearances again with the specifications for warmed engine. Use a feeler gauge to measure the clearance between the valve

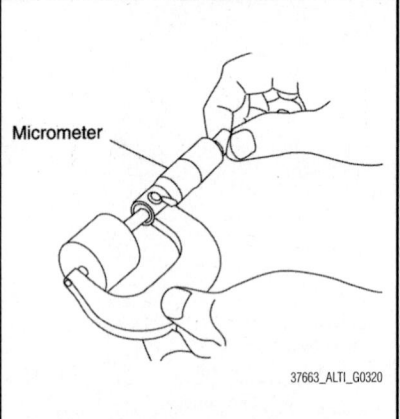

Fig. 245 Measure the center thickness of the removed valve lifters with a micrometer

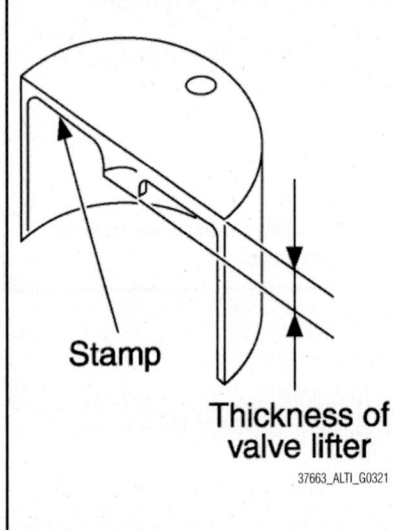

Stamp

Thickness of valve lifter

37663_ALTI_G0321

Fig. 246 Thickness of a new valve lifter can be identified by stamp marks on the reverse side (inside the cylinder)

and camshaft. Make sure the values are within specifications.

3.5L Engine

See Figures 246 and 247.

➡**Adjust valve clearance while engine is cold.**

Note the following:
- Perform adjustment by selecting the correct head thickness of the valve lifter (adjusting shims are not used).
- The specified valve lifter thickness is the dimension at normal temperatures. Ignore dimensional differences caused by temperature. Use specifications for hot engine condition to confirm valve clearances.

1. Remove the camshaft.

2. Remove the valve lifter that was measured as being outside the standard specifications.

3. Measure the center thickness of the removed lifter with a micrometer, as shown.

4. Use the equation below to calculate the replacement valve lifter thickness.
   a. Valve lifter thickness calculation equation:
      - $t = t1 + (C1 - C2)$
      - t = thickness of the replacement lifter
      - $t1$ = thickness of the removed lifter
      - $C1$ = measured valve clearance
      - $C2$ = standard valve clearance
   b. The thickness of the new valve lifter can be identified by the stamp mark on the reverse side (inside the lifter).

➡**Available thickness of the valve lifter (factory setting): 0.3102 –0.3307 inches (7.88–8.40 mm), in 0.0008inches (0.02 mm) increments, in 27 sizes (intake / exhaust).**

5. Install the selected replacement valve lifter.

6. Install the camshaft.

7. Rotate the crankshaft a few turns by hand.

8. Confirm that the valve clearances are within specification.

9. After the engine has been run to full operating temperature, confirm that the valve clearances are within specification.
   a. Intake: Cold: 0.010–0.013 inches (0.26–0.34 mm)
   b. Intake: Hot: 0.012–0.016 inches (0.304–0.416 mm)
   c. Exhaust Cold: 0.011–0.015 inches (0.29–0.37 mm)
   d. Exhaust: Hot: 0.012–0.017 inches (0.308–0.432 mm)

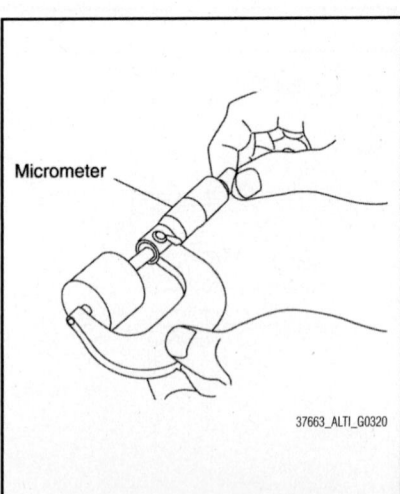

Fig. 247 Measure the center thickness of the removed lifters with a micrometer

INSPECTION

2.5L Engine

See Figures 248 through 250.

Perform this inspection as follows after removal, installation, or replacement of the camshaft or any valve related parts, or if there are any unusual engine conditions due to changes in valve clearance over time (starting, idling, and/or noise).

1. Warm up the engine, then stop it.

2. Remove the fender protector side cover RH.

3. Remove the rocker cover.

4. Turn crankshaft pulley in normal direction (clockwise when viewed from front) to align TDC identification mark (without paint mark) with timing indicator.

5. At this time, check that the both intake and exhaust cam lobes of No. 1 cylinder face outside.

➥**If they do not face outside, turn crankshaft pulley once more.**

6. Measure valve clearances with a feeler gauge at No. 1 INT and EXH, No. 2 INT, and No.3 EXH with No.1 cylinder compression TDC.

   a. Use a feeler gauge to measure the clearance between valve and camshaft.

   b. Valve clearance standards:
   • Cold: Intake: 0.009–0.013 inches (0.24–0.32 mm)
   • Cold: Exhaust: 0.010–0.013 inches (0.26–0.34 mm)
   • Hot: Intake: 0.012–0.016 inches (0.304–0.416 mm)
   • Hot Exhaust: 0.012–0.017 inches (0.308–0.432 mm)

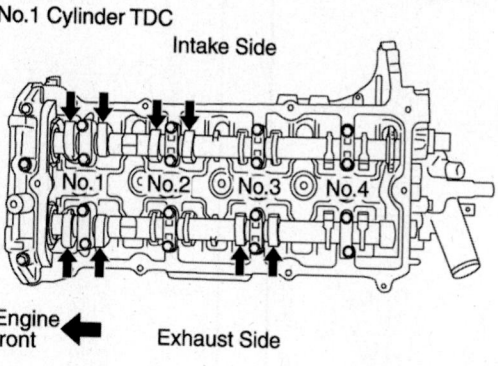

No.1 Cylinder TDC
Intake Side
Engine front — Exhaust Side

37663_ALTI_G0318

Fig. 249 Measure valve clearances at No. 1 INT and EXH, No. 2 INT, and No.3 EXH

7. Turn crankshaft one complete revolution (360°) and align mark on crankshaft pulley with pointer.

8. Measure valve clearances with a feeler gauge at No. 2 EXH, No. 3 INT, and No. 4 INT and EXH with No.4 cylinder compression TDC.

9. If out of specifications, make necessary adjustment.

3.5L Engine

See Figures 251 through 254.

Perform inspection as follows after removal, installation or replacement of

camshaft or valve related parts, or if there is unusual engine conditions regarding valve clearance.

➥**Check valve clearance while engine is cold and not running.**

1. Remove the air duct with air cleaner case, collectors, hoses, wires, harnesses, and connectors.

2. Remove the intake manifold collectors.

3. Remove the ignition coils and spark plugs.

4. Remove the rocker covers.

5. Set No.1 cylinder at TDC on its compression stroke.

   a. Align pointer with TDC mark on crankshaft pulley.

   b. Check that the valve lifters on No.1 cylinder are loose and valve lifters on No.4 are tight. If not, turn the crankshaft

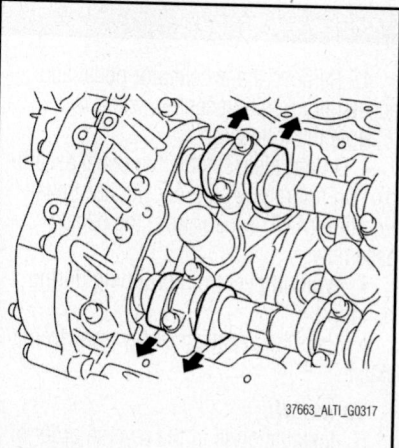

37663_ALTI_G0317

Fig. 248 Check that the both intake and exhaust cam lobes of No. 1 cylinder face outside

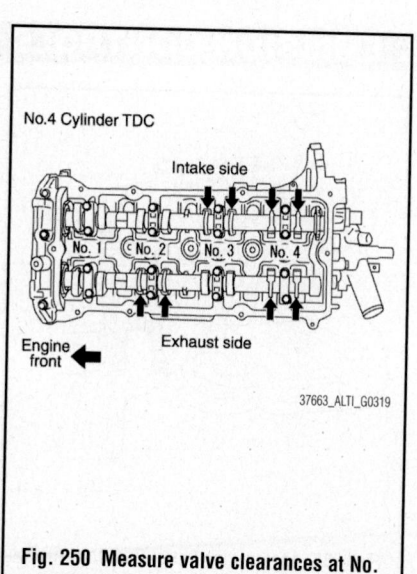

No.4 Cylinder TDC
Intake side
Engine front — Exhaust side

37663_ALTI_G0319

Fig. 250 Measure valve clearances at No. 2 EXH, No. 3 INT, and No. 4 INT and EXH

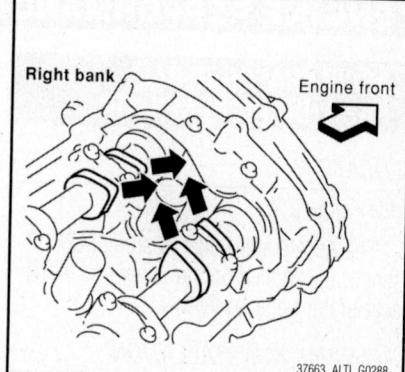

Right bank — Engine front

37663_ALTI_G0288

Fig. 251 Check that the valve lifters on No.1 cylinder are loose and valve lifters on No.4 are tight, if not, turn the crankshaft one full revolution and align as shown

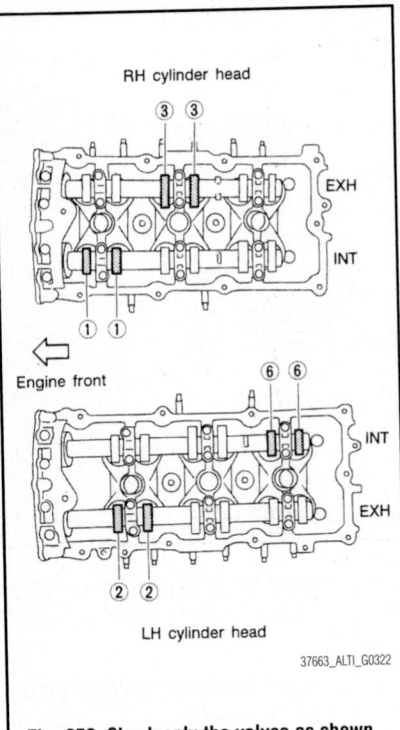

Fig. 252 Check only the valves as shown

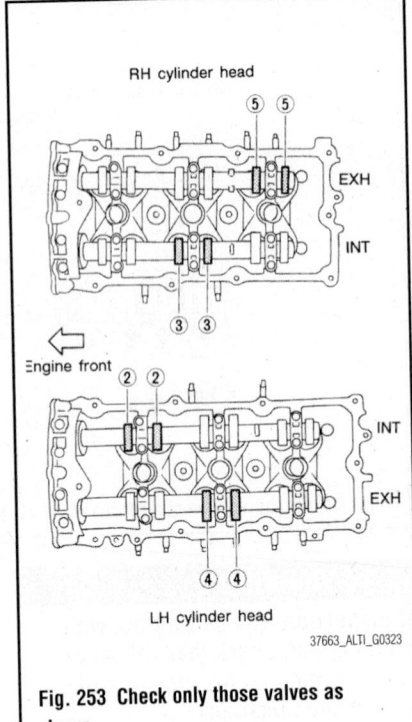

Fig. 253 Check only those valves as shown

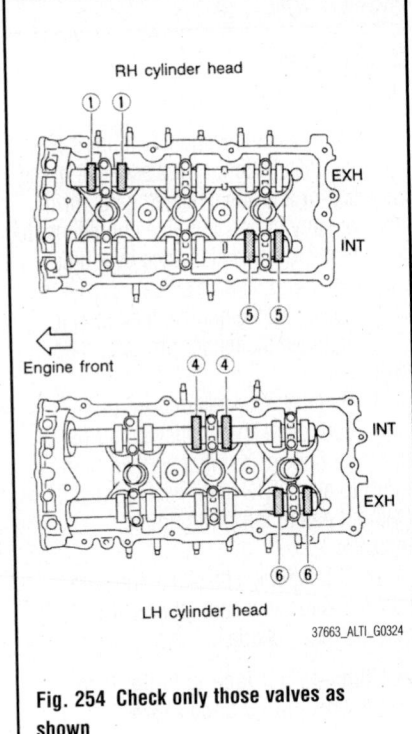

Fig. 254 Check only those valves as shown

one full revolution (360°) and align as shown.

6. Check only the valves as shown.
   - No. 1 Intake
   - No. 2 Exhaust
   - No. 3 Exhaust
   - No. 6 Intake

   a. Using a feeler gauge, measure the clearance between the valve lifter and camshaft.

   b. Record any valve clearance measurements which are out of specification. They will be used later to determine the required replacement lifter size.

7. Turn crankshaft 240°.

8. Set No.3 cylinder at TDC on its compression stroke.

9. Check only those valves as shown.
   - No. 2 Intake
   - No. 3 Intake
   - No. 4 Exhaust
   - No. 5 exhaust

10. Turn the crankshaft 240° and align as above.

11. Set No.5 cylinder at TDC on its compression stroke.

12. Check only those valves as shown.
   - No. 1 Exhaust

   - No. 4 Intake
   - No. 5 Intake
   - No. 6 Exhaust

13. If all valve clearances are within specification, install the following components:
   - Intake manifold collectors
   - Rocker covers
   - All spark plugs
   - All ignition coils

14. If the valve clearances are out of specification, adjust the valve clearances.

ENGINE PERFORMANCE & EMISSION CONTROLS

ACCELERATOR PEDAL POSITION (APP) SENSOR

LOCATION

See Figure 255.

The Accelerator Pedal Position (APP) sensor is installed on the upper end of the accelerator pedal assembly.

REMOVAL & INSTALLATION

1. Disconnect the battery negative terminal.

2. Disconnect the accelerator position sensor electrical connector.

3. Remove the three accelerator pedal nuts.

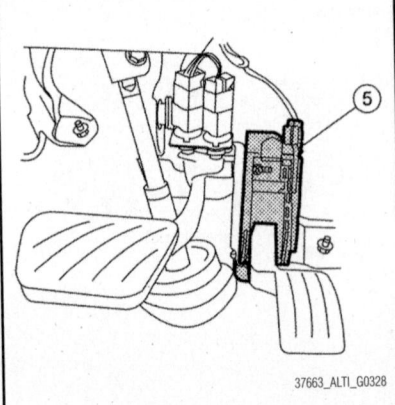

Fig. 255 Accelerator Pedal Position (APP) sensor location

4. Remove the accelerator pedal and accelerator position sensor assembly.

   Note the following:
   - Do not disassemble the pedal assembly. Do not remove the Accelerator Pedal Position (APP) sensor from the pedal assembly.
   - Avoid impact from dropping during handling.
   - Keep the pedal assembly away from water.

   To install:
5. Installation is in the reverse order of removal.

6. Align and install accelerator pedal and accelerator position sensor assembly with locating pins in locating pin holes.

7. Check the accelerator pedal for smooth operation. There should be no binding or sticking when applying or releasing the accelerator pedal.

8. Check that the accelerator pedal moves through the full specified distance of 1.91–2.11inches (49–54 mm) of pedal travel.

✳✳ CAUTION

When the harness connector of the Accelerator Pedal Position (APP) sensor is disconnected, perform the "Accelerator pedal released position learning".

BODY CONTROL MODULE

REMOVAL & INSTALLATION

See Figure 256.

✳✳ WARNING

Before replacing BCM, perform "READ CONFIGURATION" to save or print current vehicle specification.

1. Disconnect the 12-volt battery negative terminal.

2. Remove the combination meter.

3. Remove the BCM screws (A) using a suitable tool, and pull out the BCM (1).

4. Disconnect the BCM connector and remove the BCM (1).

To install:

5. Installation is the reverse of the removal procedure.

6. When replacing BCM, perform "WRITE CONFIGURATION".

7. When replacing BCM, perform the system initialization (NATS). Refer to the CONSULT operation manual for the initialization procedure.

8. When replacing BCM, if new BCM does not come with keyfobs attached, all existing keyfobs must be re-registered. Refer to the CONSULT operation manual for the initialization procedure.

CAMSHAFT POSITION (CMP) SENSOR

REMOVAL & INSTALLATION

1. Turn ignition switch OFF.
2. Loosen the fixing bolt of the sensor.
3. Disconnect Camshaft Position (CMP) sensor (PHASE) harness connector.
4. Remove the sensor.
5. Installation is the reverse of removal.

CRANKSHAFT POSITION (CKP) SENSOR

REMOVAL & INSTALLATION

1. Turn ignition switch OFF.
2. Loosen the fixing bolt of the sensor.
3. Disconnect Crankshaft Position (CKP) sensor (POS) harness connector.
4. Remove the sensor.

To install:

5. Installation is the reverse of removal.

ELECTRONIC CONTROL MODULE (ECM)

LOCATION

See Figure 257.

The ECM is located in the engine compartment, adjacent to the battery.

REMOVAL & INSTALLATION

See Figure 257.

➡The manufacturer does not provide a specific Removal and Installation procedure for this component. Refer to the graphic(s) when servicing this component.

ENGINE COOLANT TEMPERATURE (ECT) SENSOR

LOCATION

See Figures 258 through 260.

REMOVAL & INSTALLATION

1. Turn ignition switch OFF.
2. Disconnect engine coolant temperature sensor harness connector.
3. Remove engine coolant temperature sensor.

To install:

4. Installation is the reverse of removal.

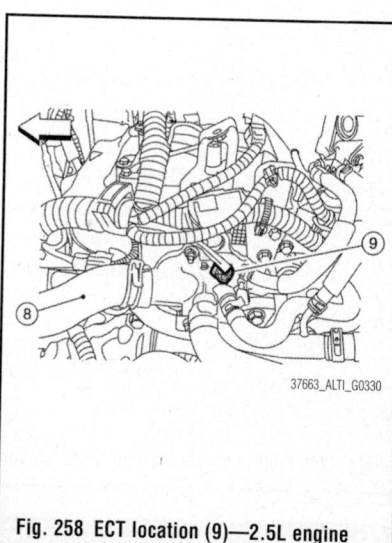

Fig. 258 ECT location (9)—2.5L engine

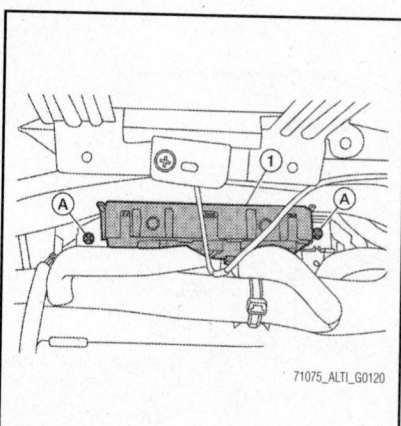

Fig. 256 Removing the screws (A) and BCM (1)

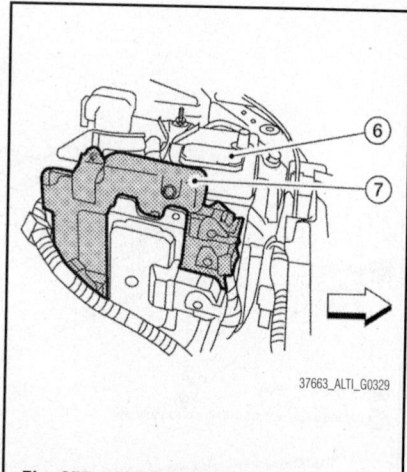

Fig. 257 Location of battery (6) and ECM (7)

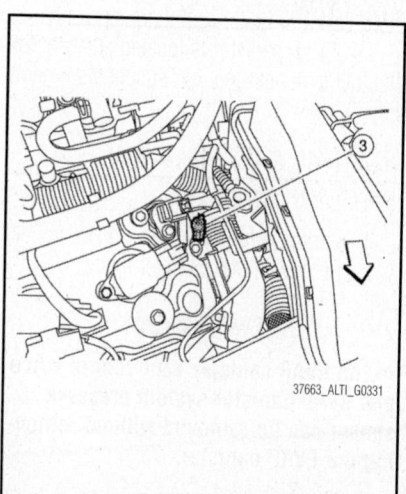

Fig. 259 ECT location (3)—3.5L engine

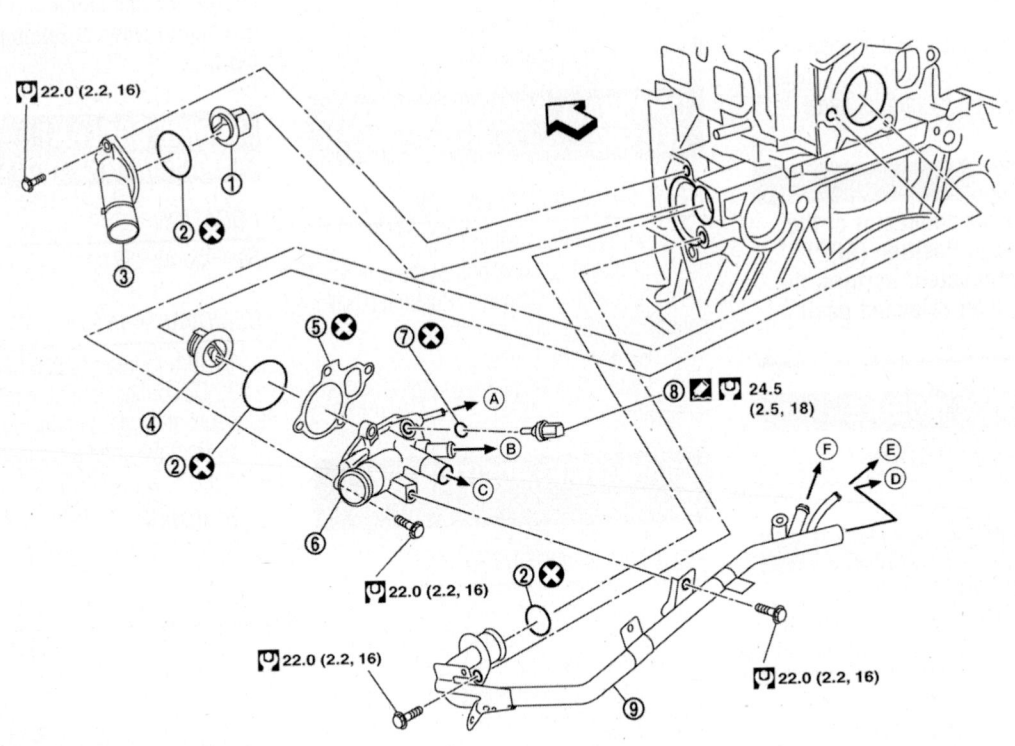

1. Thermostat
4. Water control valve
7. Copper washer
A. To electronic throttle control
D. To heater

2. O-ring
5. Gasket
8. Engine coolant temperature sensor
B. To oil cooler
E. To electronic throttle control

3. Engine coolant inlet
6. Engine coolant outlet
9. Heater pipe
C. To heater
F. To oil cooler

37663_ALTH_G0020

Fig. 260 Engine coolant temperature sensor (8)—Hybrid

EVAP CANISTER

LOCATION

The EVAP canister is located adjacent to the fuel tank near the rear suspension member.

REMOVAL & INSTALLATION

See Figures 261 and 262.

1. Lift up the vehicle on a hoist.
2. Remove EVAP canister fixing bolt.
3. Remove EVAP canister.

➡**The EVAP canister vent control valve and EVAP canister system pressure sensor can be removed without removing the EVAP canister.**

4. Install in the reverse order of removal.

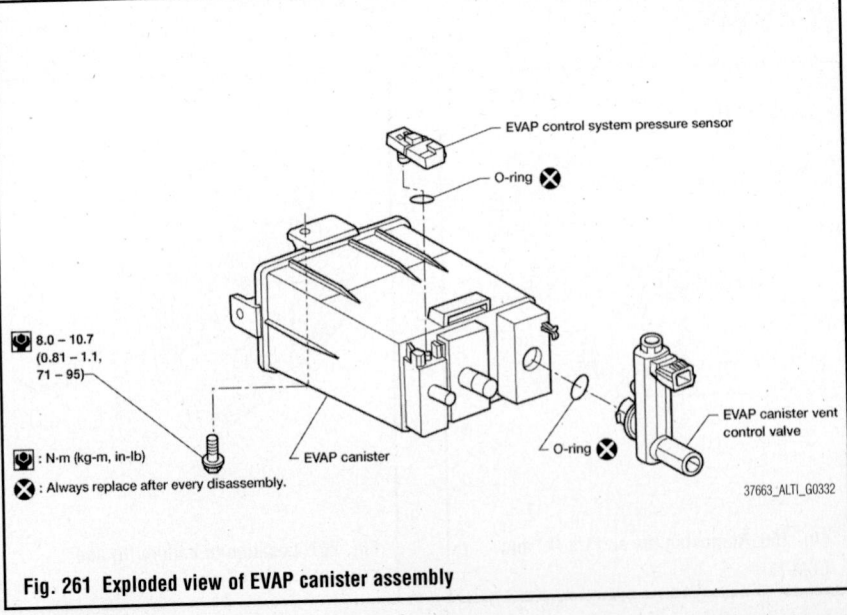

Fig. 261 Exploded view of EVAP canister assembly

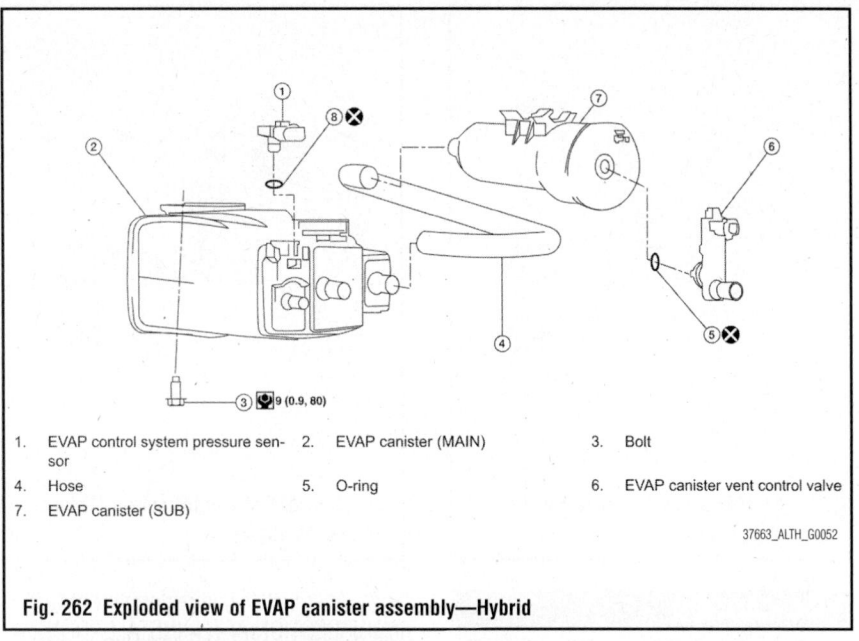

1. EVAP control system pressure sensor
2. EVAP canister (MAIN)
3. Bolt
4. Hose
5. O-ring
6. EVAP canister vent control valve
7. EVAP canister (SUB)

37663_ALTH_G0052

Fig. 262 Exploded view of EVAP canister assembly—Hybrid

HEATED OXYGEN SENSOR (HO2S)

LOCATION

2.5L Engine

See Figure 263.

3.5L Engine

See Figure 264.

Heated oxygen sensor 2 is located downstream of the three way catalyst (manifold).

REMOVAL & INSTALLATION

See Figures 263 and 264.

➡The manufacturer does not provide a specific Removal and Installation procedure for this component. Refer to the graphic(s) when servicing this component.

INTAKE AIR TEMPERATURE (IAT) SENSOR

LOCATION

The Intake Air Temperature (IAT) sensor is an integral part of the Mass Air Flow (MAF) sensor/Intake Air Temperature (IAT) sensor assembly which is mounted on the air intake duct. Refer to the Mass Air Flow section for information regarding servicing this component.

KNOCK SENSOR (KS)

LOCATION

2.5L Engine

See Figure 265.

The knock sensor is attached to the cylinder block.

3.5L Engine

See Figure 266.

REMOVAL & INSTALLATION

See Figures 265 and 266.

➡The manufacturer does not provide a specific Removal and Installation procedure for this component. Refer to the graphic(s) when servicing this component.

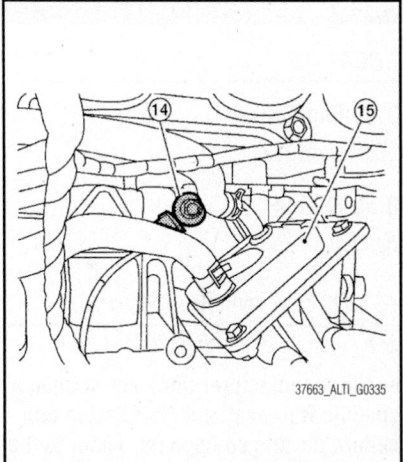

37663_ALTI_G0335

Fig. 265 Knock sensor (14) location near engine oil cooler (15)—2.5L engine

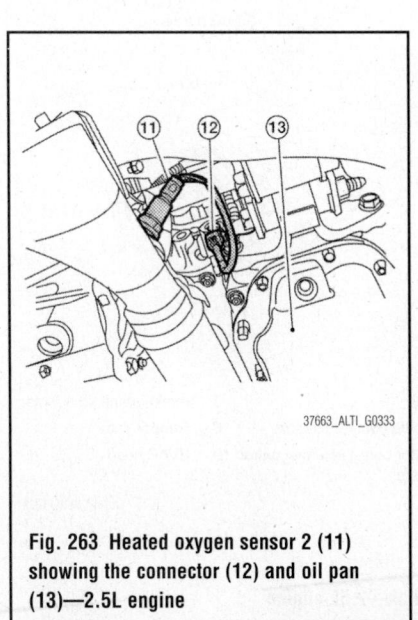

37663_ALTI_G0333

Fig. 263 Heated oxygen sensor 2 (11) showing the connector (12) and oil pan (13)—2.5L engine

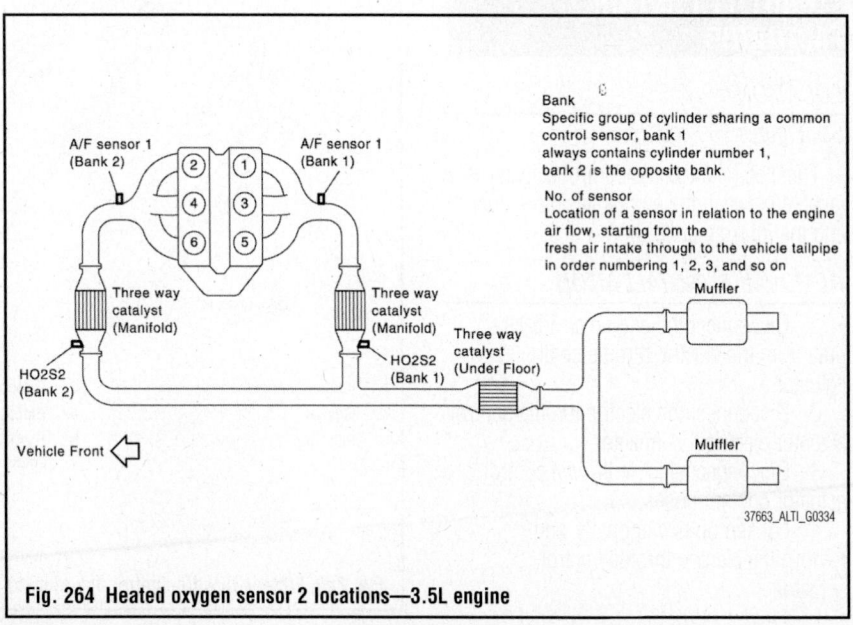

A/F sensor 1 (Bank 2)

A/F sensor 1 (Bank 1)

Three way catalyst (Manifold)

Three way catalyst (Manifold)

HO2S2 (Bank 2)

HO2S2 (Bank 1)

Three way catalyst (Under Floor)

Muffler

Muffler

Vehicle Front

Bank
Specific group of cylinder sharing a common control sensor, bank 1 always contains cylinder number 1, bank 2 is the opposite bank.

No. of sensor
Location of a sensor in relation to the engine air flow, starting from the fresh air intake through to the vehicle tailpipe in order numbering 1, 2, 3, and so on

37663_ALTI_G0334

Fig. 264 Heated oxygen sensor 2 locations—3.5L engine

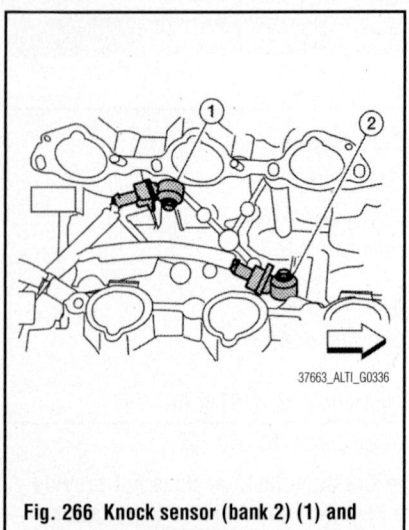

Fig. 266 Knock sensor (bank 2) (1) and knock sensor (bank 1) (2)—3.5L engine

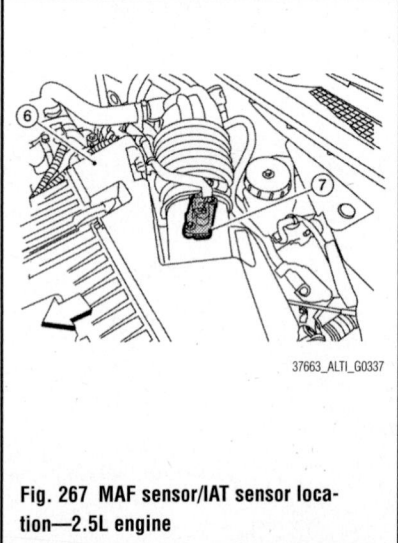

Fig. 267 MAF sensor/IAT sensor location—2.5L engine

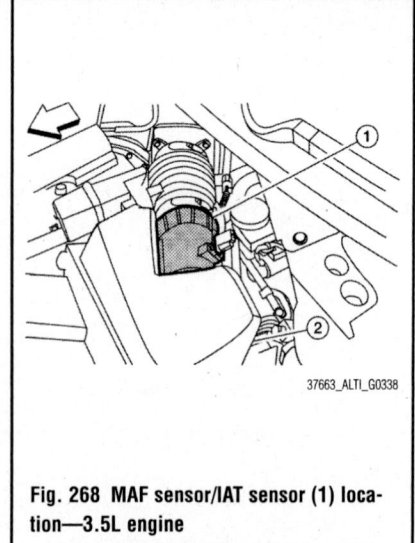

Fig. 268 MAF sensor/IAT sensor (1) location—3.5L engine

MASS AIR FLOW (MAF) SENSOR

LOCATION

2.5L Engine

See Figure 267.

3.5L Engine

See Figure 268.

REMOVAL & INSTALLATION

See Figures 267 through 268.

➡**The manufacturer does not provide a specific Removal and Installation procedure for this component. Refer to the graphic(s) when servicing this component.**

THROTTLE CONTROL ACTUATOR

LOCATION

See Figures 269 and 270.

The electrical throttle control actuator is located between the engine air intake duct and the intake manifold.

REMOVAL & INSTALLATION

1. Disconnect the engine air intake duct from the electric throttle control actuator.
2. Disconnect the electric throttle control actuator electrical connector.
3. Disconnect electric throttle control actuator coolant hoses.
4. Loosen bolts diagonally, and remove the electric throttle control actuator.

✳✳ CAUTION

Handle carefully to avoid any damage.

5. Installation is the reverse of removal.
6. Tighten the bolts of electric throttle control actuator equally and diagonally in several steps.

7. After installation perform procedure in "INSPECTION AFTER INSTALLATION".

Inspection After Installation

1. Make sure there is no fuel leakage at connections as follows:
   a. Apply fuel pressure to fuel lines by turning ignition switch ON (with engine

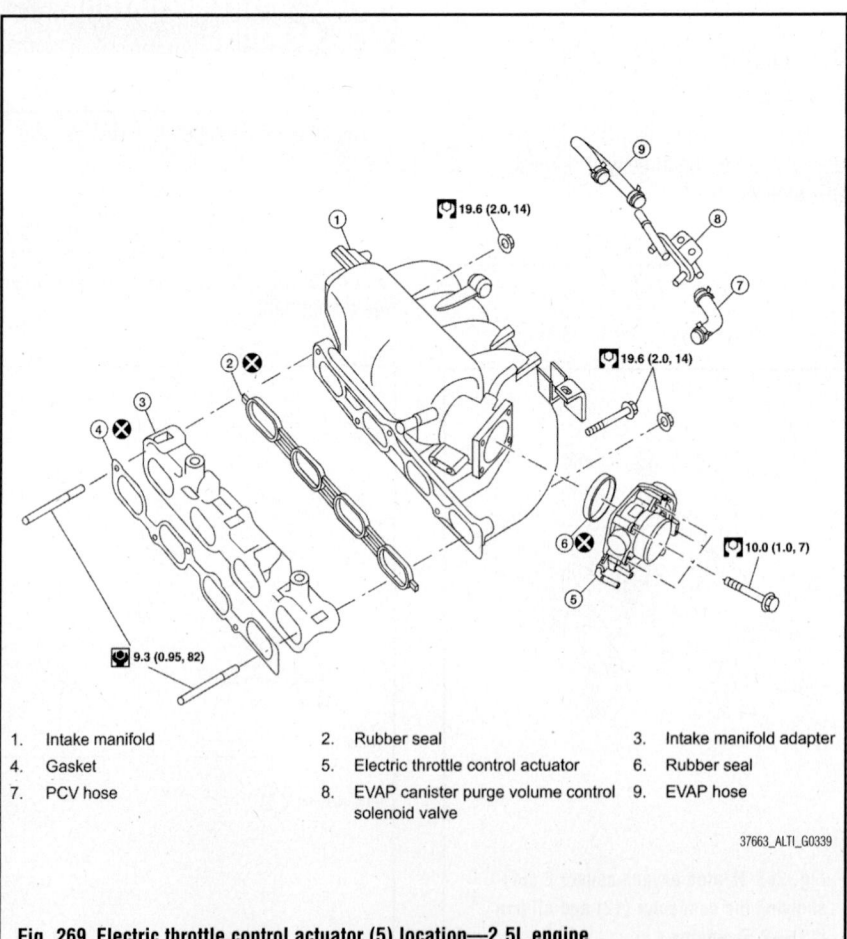

1.	Intake manifold	2.	Rubber seal	3.	Intake manifold adapter
4.	Gasket	5.	Electric throttle control actuator	6.	Rubber seal
7.	PCV hose	8.	EVAP canister purge volume control solenoid valve	9.	EVAP hose

Fig. 269 Electric throttle control actuator (5) location—2.5L engine

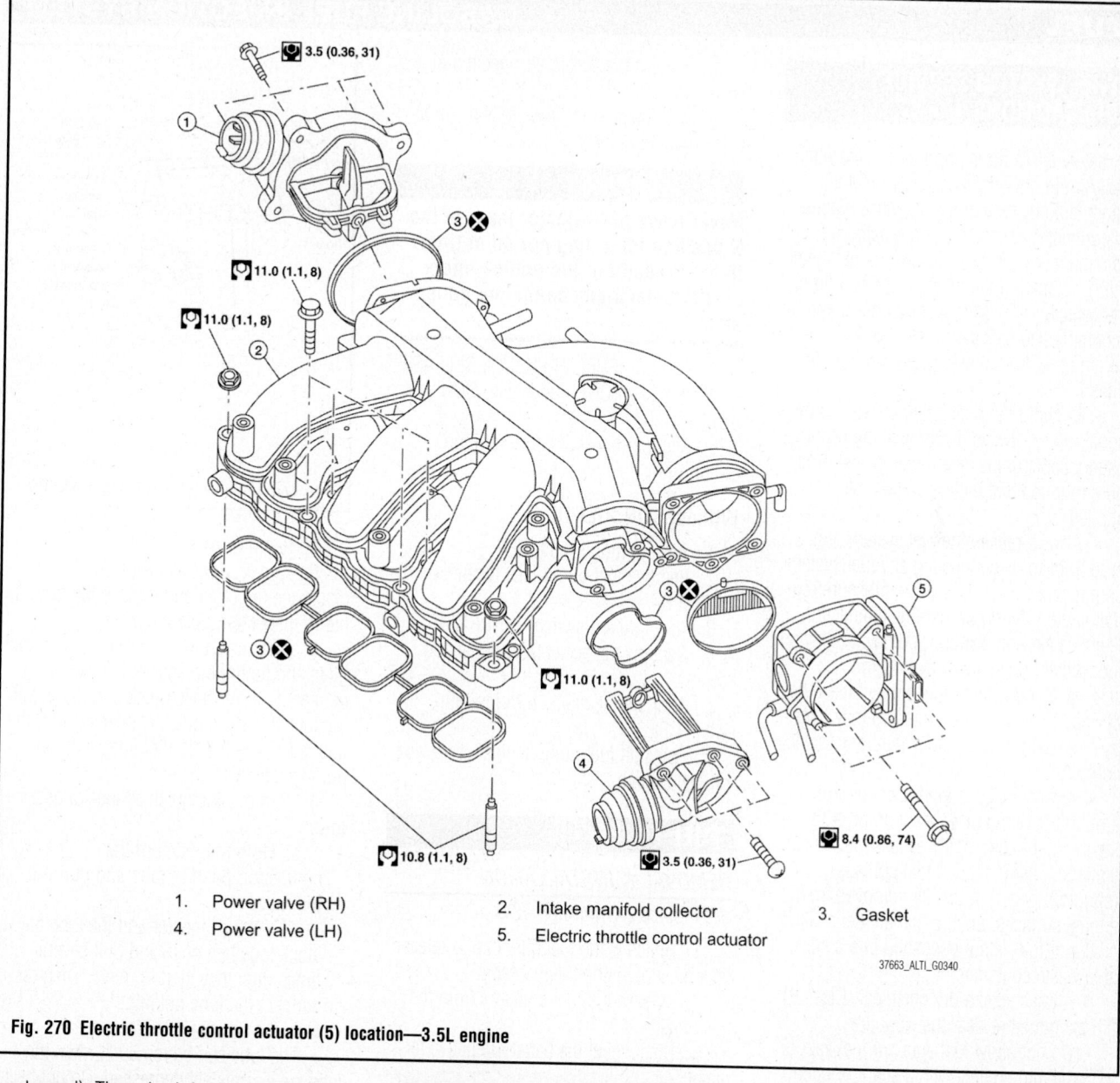

1. Power valve (RH)
4. Power valve (LH)
2. Intake manifold collector
5. Electric throttle control actuator
3. Gasket

37663_ALTI_G0340

Fig. 270 Electric throttle control actuator (5) location—3.5L engine

stopped). Then check for fuel leaks at connections.

b. Start the engine and rev it up and check for fuel leaks at connections.

❈❈ WARNING

Do not touch engine immediately after stopping as engine is extremely hot.

➡**Use mirrors for checking on connections out of the direct line of sight.**

THROTTLE POSITION SENSOR

LOCATION

Electric throttle control actuator consists of throttle control motor, Accelerator Pedal Position (APP) sensor, throttle

position sensor etc. The actuator sends a signal to the ECM, and ECM sends the signal to TCM with CAN communication.

REMOVAL & INSTALLATION

Refer to Throttle Control Actuator section when servicing this component.

FUEL SYSTEM SERVICE PRECAUTIONS

Safety is the most important factor when performing not only fuel system maintenance but any type of maintenance. Failure to conduct maintenance and repairs in a safe manner may result in serious personal injury or death. Maintenance and testing of the vehicle's fuel system components can be accomplished safely and effectively by adhering to the following rules and guidelines.

• To avoid the possibility of fire and personal injury, always disconnect the negative battery cable unless the repair or test procedure requires that battery voltage be applied.

• Always relieve the fuel system pressure prior to disconnecting any fuel system component (injector, fuel rail, pressure regulator, etc.), fitting or fuel line connection. Exercise extreme caution whenever relieving fuel system pressure to avoid exposing skin, face and eyes to fuel spray. Please be advised that fuel under pressure may penetrate the skin or any part of the body that it contacts.

• Always place a shop towel or cloth around the fitting or connection prior to loosening to absorb any excess fuel due to spillage. Ensure that all fuel spillage (should it occur) is quickly removed from engine surfaces. Ensure that all fuel soaked cloths or towels are deposited into a suitable waste container.

• Always keep a dry chemical (Class B) fire extinguisher near the work area.

• Do not allow fuel spray or fuel vapors to come into contact with a spark or open flame.

• Always use a back-up wrench when loosening and tightening fuel line connection fittings. This will prevent unnecessary stress and torsion to fuel line piping.

• Always replace worn fuel fitting O-rings with new Do not substitute fuel hose or equivalent where fuel pipe is installed.

Before servicing the vehicle, make sure to also refer to the precautions in the beginning of this section as well.

RELIEVING FUEL SYSTEM PRESSURE

With CONSULT-III

1. Lift up the vehicle.
2. Turn ignition switch ON (READY).

3. Depress the accelerator pedal and keep it.
4. Shift the selector lever to N position with engine running.

✳✳ WARNING

Never leave the selector lever in the N position for a long period of time. In the N position, the engine operates but electricity cannot be generated.

5. Perform "FUEL PRESSURE RELEASE" in "WORK SUPPORT" mode with CONSULT-III.
6. After engine stalls, turn ignition switch OFF.

Without CONSULT-III

1. Turn ignition switch OFF.
2. Remove fuel pump fuse located in IPDM E/R.
3. Turn ignition switch ON (READY).
4. Depress the accelerator pedal and keep it.
5. After engine stalls, turn ignition switch OFF.
6. Reinstall fuel pump fuse after servicing fuel system.

FUEL FILTER

REMOVAL & INSTALLATION

See Figures 271 through 274.

1. Unscrew the fuel filler cap to release the pressure inside the fuel tank.
2. Release the fuel pressure from the fuel lines.
3. Disconnect the battery negative terminal.

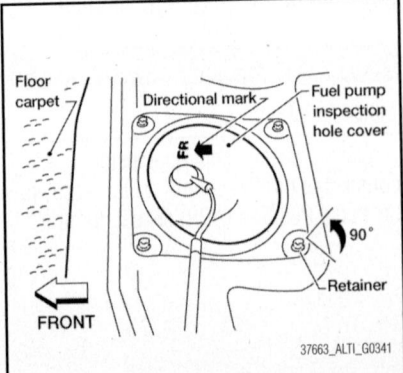

Fig. 271 Turn the four retainers 90° in a clockwise direction and remove the fuel pump inspection hole cover

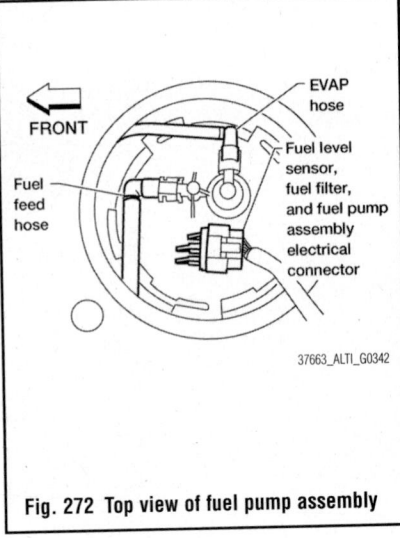

Fig. 272 Top view of fuel pump assembly

4. Remove the rear seat bottom.
5. Turn the four retainers 90° in a clockwise direction and remove the fuel pump inspection hole cover.
6. Disconnect the fuel level sensor, fuel filter, and fuel pump assembly electrical connector, EVAP hose quick connector, and the fuel feed hose quick connector from the fuel level sensor unit, fuel filter, and fuel pump assembly.
7. Remove the quick connector as follows:

   a. Hold the sides of the connector, push in tabs and pull out the tube.

   b. If the connector and the tube are stuck together, push and pull several times until they start to move. Then disconnect them by pulling.

Note the following:

• The tube can be removed when the tabs are completely depressed. Do not twist it more than necessary.

• Do not use any tools to remove the quick connector.

• Keep the resin tube away from heat. Be especially careful when welding near the tube.

• Prevent acid liquid such as battery electrolyte, etc. from getting on the resin tube.

• Do not bend or twist the tube during installation and removal.

• Only when the tube is replaced, remove the remaining retainer on the tube or fuel level sensor, fuel filter, and fuel pump assembly.

• When the tube or fuel level sensor, fuel filter, and fuel pump assembly is replaced, also replace the

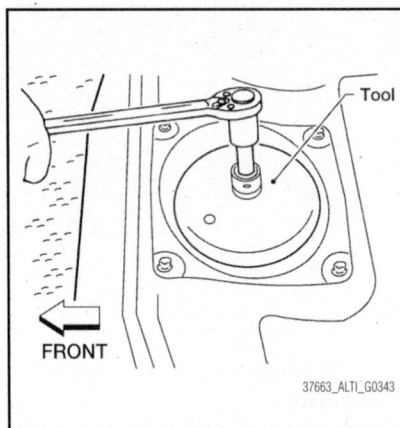

Fig. 273 Remove the lock ring using a socket drive handle and Tool KV991J0090 (J-46214)

retainer with a new one (green colored retainer).

- To keep the connecting portion clean and to avoid damage and foreign materials, cover them completely with plastic bags or something similar.

8. Remove the lock ring using a socket drive handle and Tool KV991J0090 (J-46214) as shown.

> ※※ **WARNING**
>
> **Discard the lock ring, do not reuse the lock ring. Discard the ring seal, do not reuse the ring seal.**

9. Remove the fuel level sensor, fuel filter, and fuel pump assembly.

> ※※ **WARNING**
>
> **Do not bend the float arm during removal. Discard the ring seal, do not reuse the ring seal.**

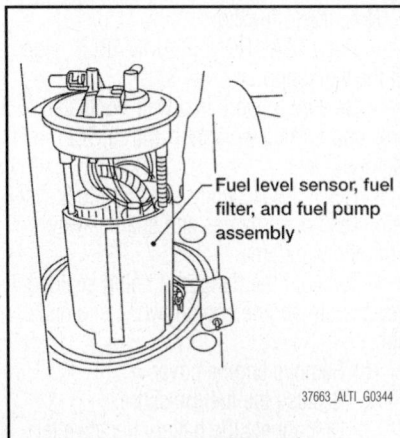

Fuel level sensor, fuel filter, and fuel pump assembly

Fig. 274 Remove the fuel level sensor, fuel filter, and fuel pump assembly

10. Inspect the fuel level sensor, fuel filter, and fuel pump for any defects and foreign materials. Replace as necessary.

To install:

11. Installation is in the reverse order of removal.

12. Install the fuel level sensor, fuel filter, and fuel pump assembly with the fuel feed hose facing the front of the vehicle as shown. Use a new ring seal.

13. Connect the quick connector as follows:

   a. Check the connection for damage or any foreign materials.

   b. Align the connector with the tube, then insert the connector straight into the tube until a click is heard.

14. After the tube is connected, make sure the connection is secure by performing the following checks:

   a. Pull the tube and the connector to make sure they are securely connected.

   b. Visually confirm that the two retainer tabs are connected to the quick connector.

15. Turn the ignition switch to ON (without starting the engine) to apply fuel pressure to the fuel system, then check the connections for fuel leaks.

16. Start the engine and let it idle and check for fuel leaks at the fuel system connections.

FUEL PUMP

REMOVAL & INSTALLATION

See Figures 271 through 274.

1. Unscrew the fuel filler cap to release the pressure inside the fuel tank.

2. Release the fuel pressure from the fuel lines.

3. Disconnect the battery negative terminal.

4. Remove the rear seat bottom.

5. Turn the four retainers 90° in a clockwise direction and remove the fuel pump inspection hole cover.

6. Disconnect the fuel level sensor, fuel filter, and fuel pump assembly electrical connector, EVAP hose quick connector, and the fuel feed hose quick connector from the fuel level sensor unit, fuel filter, and fuel pump assembly.

7. Remove the quick connector as follows:

   a. Hold the sides of the connector, push in tabs and pull out the tube.

   b. If the connector and the tube are stuck together, push and pull several times until they start to move. Then disconnect them by pulling.

Note the following:

- The tube can be removed when the tabs are completely depressed. Do not twist it more than necessary.

- Do not use any tools to remove the quick connector.

- Keep the resin tube away from heat. Be especially careful when welding near the tube.

- Prevent acid liquid such as battery electrolyte, etc. from getting on the resin tube.

- Do not bend or twist the tube during installation and removal.

- Only when the tube is replaced, remove the remaining retainer on the tube or fuel level sensor, fuel filter, and fuel pump assembly.

  - When the tube or fuel level sensor, fuel filter, and fuel pump assembly is replaced, also replace the retainer with a new one (green colored retainer).

  - To keep the connecting portion clean and to avoid damage and foreign materials, cover them completely with plastic bags or something similar.

8. Remove the lock ring using a socket drive handle and Tool KV991J0090 (J-46214) as shown.

> ※※ **WARNING**
>
> **Discard the lock ring, do not reuse the lock ring. Discard the ring seal, do not reuse the ring seal.**

9. Remove the fuel level sensor, fuel filter, and fuel pump assembly.

> ※※ **WARNING**
>
> **Do not bend the float arm during removal. Discard the ring seal, do not reuse the ring seal.**

10. Inspect the fuel level sensor, fuel filter, and fuel pump for any defects and foreign materials. Replace as necessary.

To install:

11. Installation is in the reverse order of removal.

12. Install the fuel level sensor, fuel filter, and fuel pump assembly with the fuel feed hose facing the front of the vehicle as shown. Use a new ring seal.

13. Connect the quick connector as follows:

   a. Check the connection for damage or any foreign materials.

b. Align the connector with the tube, then insert the connector straight into the tube until a click is heard.

14. After the tube is connected, make sure the connection is secure by performing the following checks:

a. Pull the tube and the connector to make sure they are securely connected.

b. Visually confirm that the two retainer tabs are connected to the quick connector.

15. Turn the ignition switch to ON (without starting the engine) to apply fuel pressure to the fuel system, then check the connections for fuel leaks.

16. Start the engine and let it idle and check for fuel leaks at the fuel system connections.

FUEL RAIL & INJECTORS

REMOVAL & INSTALLATION

2.5L Engine

See Figures 275 through 278.

1. Remove engine room cover.
2. Release the fuel pressure.
3. Remove the front air duct.
4. Disconnect the fuel hose quick connector at the fuel tube side.

❋❋ CAUTION

Prepare a container and cloth for catching any spilled fuel. This operation should be performed in a place that is free from any open flames. While hoses are disconnected seal their openings with vinyl bag or similar material to prevent foreign material from entering them.

5. Remove the intake manifold.
6. Disconnect sub-harness for injector at engine front side, and remove it from bracket.
7. Loosen the bolts in the reverse order shown, then remove fuel tube and fuel injectors as an assembly.
8. Remove the fuel injectors from the fuel tube.

a. Release the clip and remove the fuel injector.

b. Pull fuel injector straight out of the fuel tube.

❋❋ WARNING

Be careful not to damage the nozzle. Avoid any impact, such as dropping the fuel injector. Do not disassemble or adjust the fuel injector.

To install:

9. Install new O-rings on the fuel injector, the fuel side black O-ring and the nozzle side green O-ring.

❋❋ CAUTION

Upper and lower O-rings are different. Be careful not to confuse them. Fuel tube side: black O-ring; Nozzle side: green O-ring.

Note the following:
- Lubricate the O-rings lightly with new engine oil.
- Handle O-rings with bare hands only. Do not wear gloves.
- Do not clean O-rings with solvent.
- Make sure that O-ring and its mating part are free of foreign material.
- Be careful not to scratch O-rings during installation.
- Do not twist or stretch the O-ring. If the O-ring was stretched while it is attached, do not insert it into the fuel tube immediately.

10. Install the fuel injector into the fuel tube with the following procedure:

a. Do not reuse the clip, replace it with a new one.

b. Insert the new clip into the clip mounting groove on fuel injector.

c. Insert the clip so that projection of fuel injector matches notch of the clip.

d. Fuel tube side: black O-ring

e. Nozzle side: green O-ring

11. Insert fuel injector into fuel tube with clip attached.

a. Insert it while matching it to the axial center.

b. Insert fuel injector so that projection of fuel tube matches notch of the clip.

c. Make sure that fuel tube flange is securely fixed in flange fixing groove on the clip.

d. Make sure that installation is complete by checking that fuel injector does not rotate or come off.

12. Install fuel tube assembly.

a. Insert the tip of each fuel injector into intake manifold.

b. Tighten the bolts in two steps in the numerical order as shown.

c. Step 1: 7 ft. lbs. (10 Nm)

d. Step 2: 16 ft. lbs. (22 Nm)

➡**After properly connecting fuel tube assembly to injector and fuel hose, check connection for fuel leakage.**

13. Install the intake manifold.

14. Connect the fuel hose quick connector.

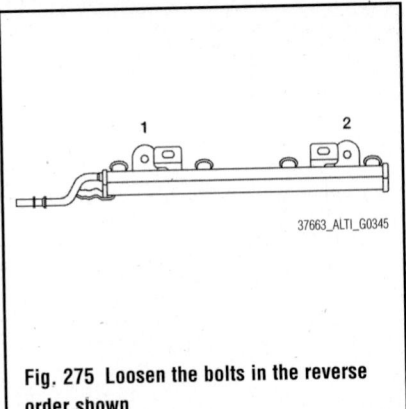

Fig. 275 Loosen the bolts in the reverse order shown

15. Installation of the remaining components is in the reverse order of removal.

16. Make sure there is no fuel leakage at connections as follows:

a. Apply fuel pressure to fuel lines by turning ignition switch ON (with engine stopped). Then check for fuel leaks at connections.

b. Start the engine and rev it up and check for fuel leaks at connections.

17. Perform procedures for "Throttle Valve Closed Position Learning" after finishing repairs.

18. If electric throttle control actuator is replaced, perform procedures for "Idle Air Volume Learning" after finishing repairs.

❋❋ WARNING

Do not touch engine immediately after stopping as engine is extremely hot.

➡**Use mirrors for checking on connections out of the direct line of sight.**

3.5L Engine

See Figures 279 through 284.

Note the following:
- Put a "CAUTION: FLAMMABLE" sign in the workshop.
- Be sure to work in a well-ventilated area and furnish workshop with a CO_2 fire extinguisher.
- Never smoke while servicing fuel system. Keep open flames and sparks away from the work area.
- To avoid the danger of being scalded, never drain engine coolant when engine is hot.

1. Remove engine cover.
2. Release the fuel pressure.
3. Disconnect the battery negative terminal.
4. Remove front wiper arm and extension cowl top.

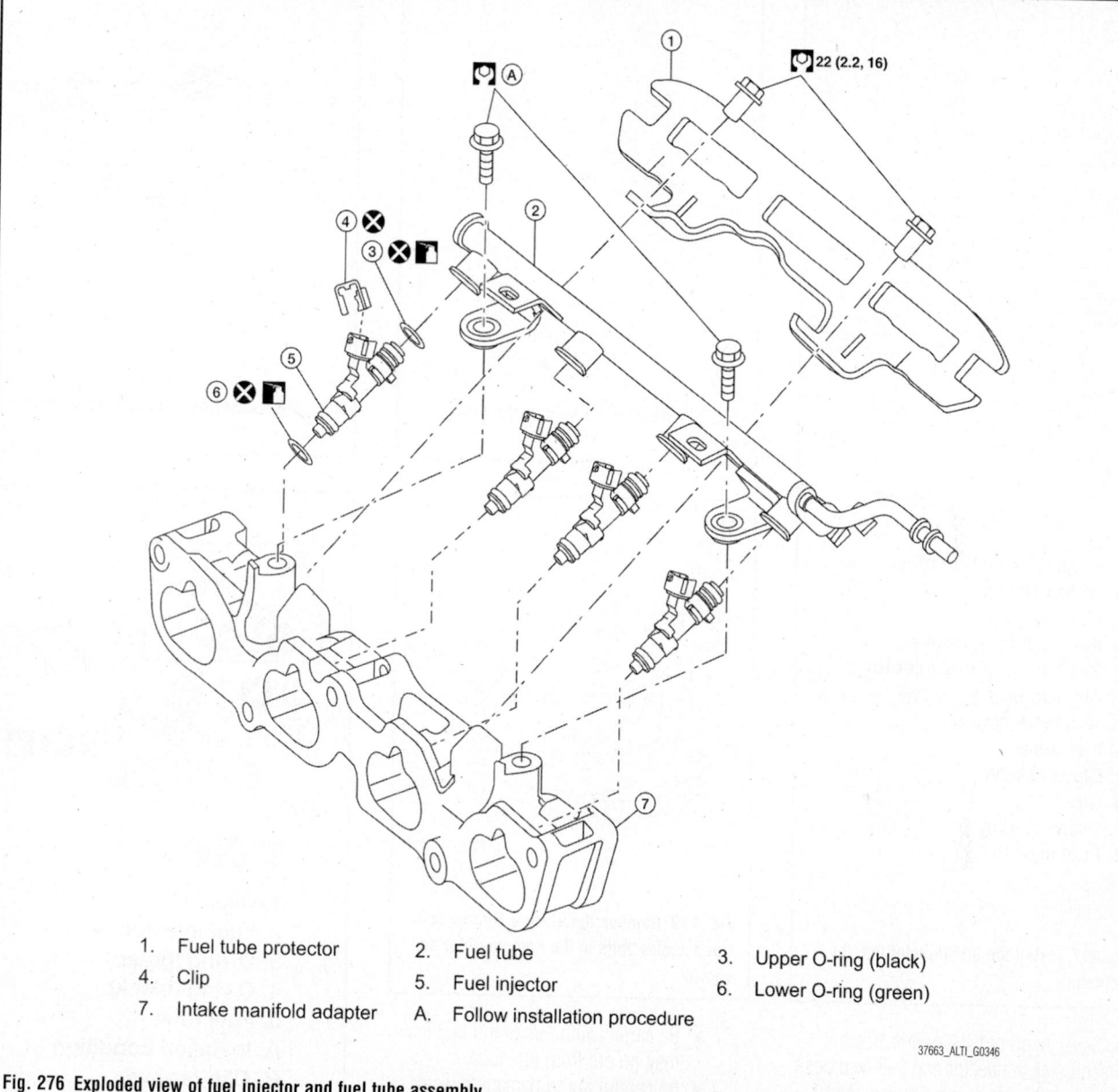

1. Fuel tube protector
2. Fuel tube
3. Upper O-ring (black)
4. Clip
5. Fuel injector
6. Lower O-ring (green)
7. Intake manifold adapter
A. Follow installation procedure

37663_ALTI_G0346

Fig. 276 Exploded view of fuel injector and fuel tube assembly

5. Remove the electric throttle control actuator bolts in the reverse order as shown and remove the electric throttle control actuator.

❋❋ CAUTION

Handle carefully to avoid any shock to the electric throttle control actuator. Do not disassemble.

6. Remove intake manifold collector.
7. When separating fuel feed hose and fuel tube connection, disconnect quick connector as follows:

    a. Remove quick connector cap from quick connector.

    b. Disconnect quick connector from fuel tube as follows:

❋❋ CAUTION

Disconnect quick connector by using the quick connector release (commercial service tool: J-45488, not by picking out retainer tabs.

8. With the sleeve side of quick connector release facing to quick connector, install the quick connector release onto fuel tube.

9. Insert the quick connector release into quick connector until sleeve contacts and goes no further. Hold quick connector release on that position.

❋❋ WARNING

Inserting quick connector release hard will not disconnect quick

connector. Hold quick connector release where it contacts and goes no further.

10. Draw and pull out quick connector straight from fuel tube.
    Note the following:
• Never pull with lateral force applied. O-ring inside quick connector may be damaged.
• Prepare container and cloth beforehand as fuel will leakage out.
• Avoid fire and sparks.
• Keep parts away from heat source. Especially, be careful when welding is performed around them.
• Never expose parts to battery electrolyte or other acids.

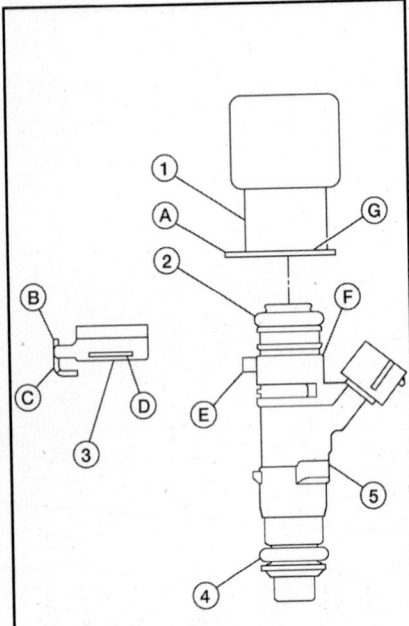

A. Projection of fuel tube
B. Notch of clip
C. Notch of clip
D. Flange fixing groove
E. Projection of fuel injector
F. Clip mounting groove
G. fuel tube flange
1. fuel tube
2. Black O-ring
3. Clip
4. Green O-ring
5. Fuel injector

37663_ALTI_G0347

Fig. 277 Install the fuel injector into the fuel tube

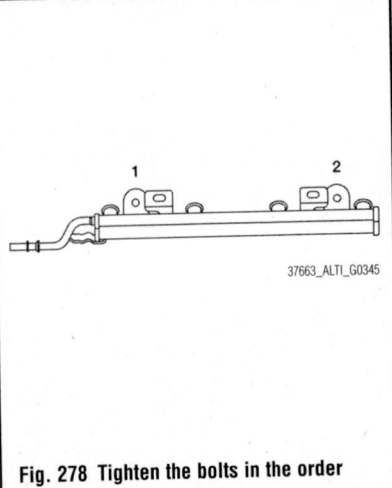

37663_ALTI_G0345

Fig. 278 Tighten the bolts in the order shown

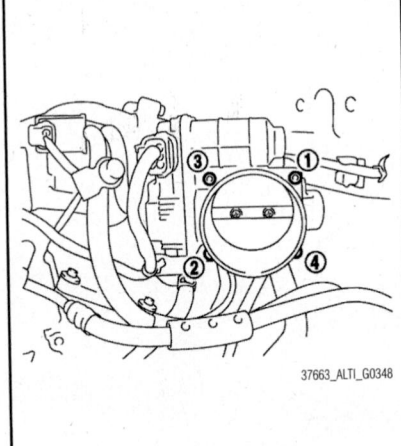

37663_ALTI_G0348

Fig. 279 Remove the electric throttle control actuator bolts in the reverse order as shown

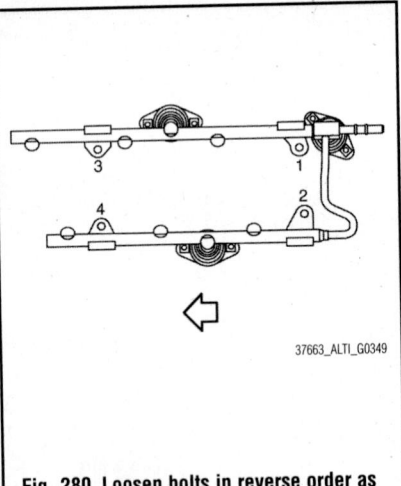

37663_ALTI_G0349

Fig. 280 Loosen bolts in reverse order as shown

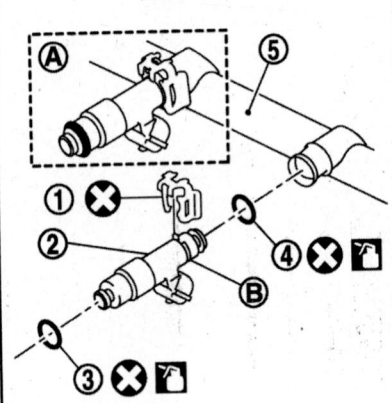

1. Clip
2. Fuel injector
3. O-ring (green)
4. O-ring (black)
5. Fuel tube
A. Installed condition
B. Clip groove

37663_ALTI_G0350

Fig. 281 Remove fuel injector from fuel tube

• Never bend or twist connection between quick connector and fuel feed hose (with damper) during installation/removal.
• To keep clean the connecting portion and to avoid damage and foreign materials, cover them completely with plastic bags or something similar.
11. Disconnect harness connector from fuel injector.
12. Loosen bolts in reverse order as shown, and remove fuel tube and fuel injector assembly.

❈❈ WARNING

Never tilt fuel tube, or remaining fuel in pipes may flow out from pipes.

13. Remove fuel injector from fuel tube as follows:
    a. Open and remove clip.
    b. Remove fuel injector from fuel tube by pulling straight.

• Be careful with remaining fuel that may go out from fuel tube.
• Be careful not to damage injector nozzle during removal.
• Never bump or drop fuel injector.
• Never disassemble fuel injector.
14. Remove fuel damper from fuel tube.

To install:

15. Install fuel damper as follows:
    a. Install new O-ring to fuel tube as shown. When handling new O-ring, be careful of the following caution:
    • Handle O-ring with bare hands. Never wear gloves.
    • Lubricate O-ring with new engine oil.
    • Never clean O-ring with solvent.
    • Check that O-ring and its mating part are free of foreign material.

• When installing O-ring, be careful not to scratch it with tool or fingernails. Also be careful not to twist or stretch O-ring. If O-ring was stretched while it was being attached, never insert it quickly into fuel tube.
• Insert new O-ring straight into fuel tube. Never twist it.
    b. Install spacer to fuel damper.
    c. Insert fuel damper straight into fuel tube.
• Insert straight, checking that the axis is lined up.

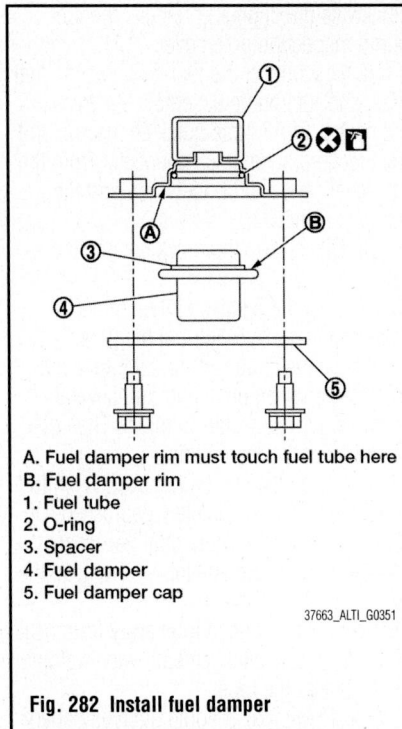

A. Fuel damper rim must touch fuel tube here
B. Fuel damper rim
1. Fuel tube
2. O-ring
3. Spacer
4. Fuel damper
5. Fuel damper cap

37663_ALTI_G0351

Fig. 282 Install fuel damper

- Never pressure-fit with excessive force.
- Insert fuel damper until it is touching the fuel tube.
d. Tighten bolts evenly in turn.
e. After tightening bolts, check that there is no gap between fuel damper cap and fuel tube.
16. Install new O-rings to fuel injector paying attention to the following.
- Upper and lower O-ring are different. Be careful not to confuse them.
- Fuel tube side: Black
- Nozzle side: Green
- Lubricate O-ring with new engine oil.
- Never clean O-ring with solvent.
- Check that O-ring and its mating part are free of foreign material.
- When installing O-ring, be careful not to scratch it with tool or fingernails. Also be careful not to twist or stretch O-ring. If O-ring was stretched while it was being attached, never insert it quickly into fuel tube.
- Insert O-ring straight into fuel injector. Never decenter or twist it.
17. Install fuel injector to fuel tube as follows:
a. Insert clip into clip groove on fuel injector.
b. Insert clip so that protrusion of fuel injector matches cutout of clip.

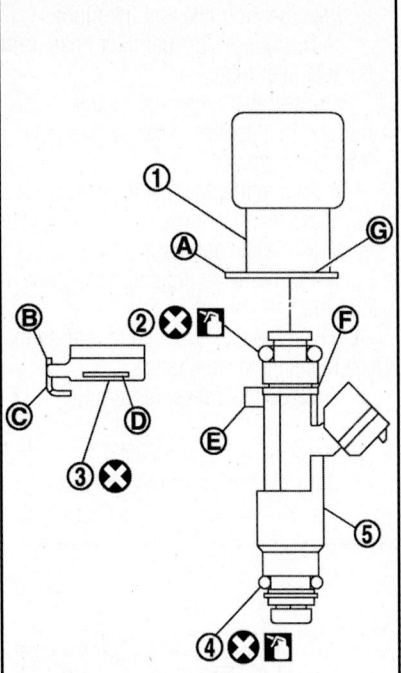

A. Protrusion of fuel tube
B. Cutout of clip
C. Cutout of clip
D. Flange fixing groove on clip
E. Protrusion of fuel injector
F. Clip groove
G. Fuel tube flange
1. Fuel tube
2. O-ring: Black
3. Clip
4. O-ring: Green
5. Fuel injector

37663_ALTI_G0352

Fig. 283 Install fuel injector to fuel tube

❈❈ WARNING

Never reuse clip. Replace it with new one. Be careful to keep clip from interfering with O-ring. If interference occurs, replace O-ring.

c. Insert fuel injector into fuel tube with clip attached.
d. Insert it while matching it to the axial center.
e. Insert fuel injector so that protrusion of fuel tube matches cutout of clip.
f. Check that fuel tube flange is securely fixed in flange fixing groove on clip.
g. Check that installation is complete by checking that fuel injector does not rotate or come off.
h. Check that protrusions of fuel

injectors and fuel tubes are aligned with cutouts of clips after installation.
18. Install fuel tube and fuel injector assembly to intake manifold.

❈❈ WARNING

Be careful not to let tip of injector nozzle come in contact with other parts.

19. Tighten bolts in two steps in numerical order as shown:
a. Step 1: 7 ft. lbs. (10 Nm)
b. Step 2: 16 ft. lbs. (22 Nm)
20. Connect fuel injector harness.
21. Install intake manifold collector.
22. Connect quick connector between fuel feed hose and fuel tube connection with the following procedure:
a. Check no foreign substances are deposited in and around fuel tube and quick connector, and no damage on them.
b. Thinly apply new engine oil around fuel tube from tip end to spool end.
c. Align center to insert quick connector straightly into fuel tube.
d. Insert quick connector to fuel tube until top spool is completely inside quick connector, and 2nd level spool exposes right below quick connector.
Note the following:
- Hold (A) position as shown in the figure when inserting fuel tube into quick connector.
- Carefully align center to avoid inclined insertion to prevent damage to O-ring inside quick connector.
- Insert until you hear a "click" sound and actually feel the engagement.
- To avoid misidentification of engagement with a similar sound, be sure to perform the next step.

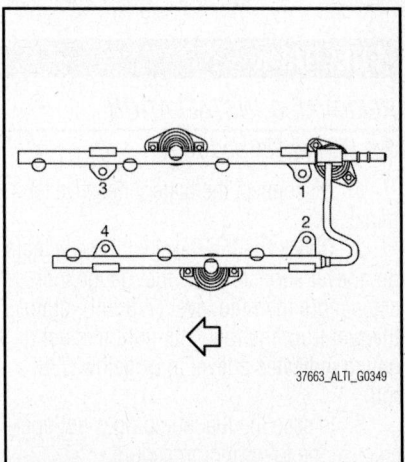

37663_ALTI_G0349

Fig. 284 Tighten bolts in two steps in numerical order

a. Pull quick connector by hand holding position. Check it is completely engaged (connected) so that it does not come out from fuel tube.

b. Install quick connector cap to quick connector.

c. Install quick connector cap with arrow on surface facing in direction of quick connector (fuel feed hose side).

❈❈ CAUTION

If quick connector cap cannot be installed smoothly, quick connector may have not been installed correctly. Check connection again.

d. Secure fuel feed hose to clamp of quick connector cap.

23. Installation is in the reverse order of removal.

24. Make sure there is no fuel leakage at connections as follows:

a. Apply fuel pressure to fuel lines by turning ignition switch ON (with engine stopped). Then check for fuel leaks at connections.

b. Start the engine and rev it up and check for fuel leaks at connections.

❈❈ CAUTION

Do not touch engine immediately after stopping as engine is extremely hot.

➡**Use mirrors for checking on connections out of the direct line of sight.**

25. Perform procedures for "Throttle Valve Closed Position Learning" after finishing repairs.

26. If electric throttle control actuator is replaced, perform procedures for "Idle Air Volume Learning" after finishing repairs.

FUEL TANK

REMOVAL & INSTALLATION

See Figures 285 and 286.

1. Disconnect the battery negative terminal.

2. Check the fuel level with the vehicle on a level surface. If the fuel gauge indicates more than the level (7/8 full), drain the fuel from the fuel tank until the fuel gauge indicates a level at or below (7/8 full).

3. In case the fuel pump does not operate, use the following procedure.

a. Insert fuel tubing of less than 0.98 inches (25 mm) diameter into the fuel

filler tube through the fuel filler opening to drain fuel from the fuel filler tube.

b. Disconnect the fuel filler hose from the fuel filler tube.

c. Insert fuel tubing into the fuel tank through the fuel filler hose to drain fuel from the fuel tank.

d. As a guide, the fuel level reaches or is less than the level on the fuel gauge as shown, when approximately 2 ⅝ US gal. (10L) of fuel is drained from a full fuel tank.

4. Open the fuel filler cap to release the pressure inside the fuel tank.

5. Release fuel pressure from fuel line.

6. Remove rear seat bottom.

7. Turn the four retainers 90° in a

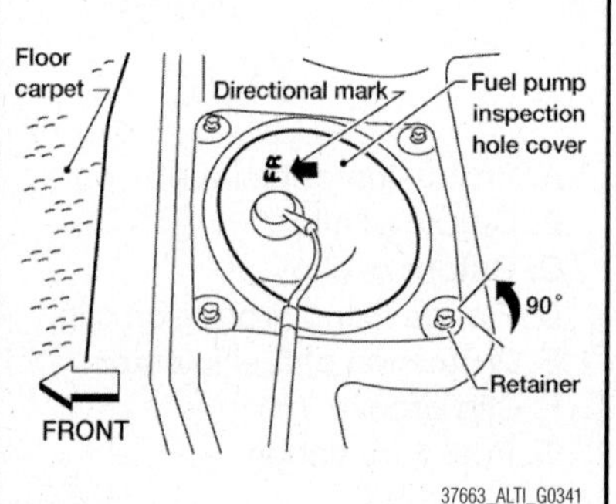

Fig. 285 Turn the four retainers 90° in a clockwise direction and remove the fuel pump inspection hole cover

clockwise direction and remove the fuel pump inspection hole cover.

8. Disconnect the fuel level sensor, fuel filter, and fuel pump assembly electrical connector, EVAP hose quick connector, and the fuel feed hose quick connector from the fuel level sensor unit, fuel filter, and fuel pump assembly.

9. Disconnect the quick connectors as follows:

a. Hold the sides of the connector, push in tabs and pull out the tube.

b. If the connector and the tube are stuck together, push and pull several times until they start to move. Then disconnect them by pulling.

- The tube can be removed when the tabs are completely depressed. Do not twist it more than necessary.
- Do not use any tools to remove the quick connector.
- Keep the resin tube away from heat. Be especially careful when welding near the tube.
- Prevent acid liquid such as battery electrolyte, from getting on the resin tube.
- Do not bend or twist the tube during installation and removal.
- Only when the tube is replaced, remove the remaining retainer on the tube or fuel level sensor, fuel filter, and fuel pump assembly.
- When the tube or fuel level sensor, fuel filter, and fuel pump assembly is replaced, also replace the retainer with a new one (green colored retainer).
- To keep the connecting portion

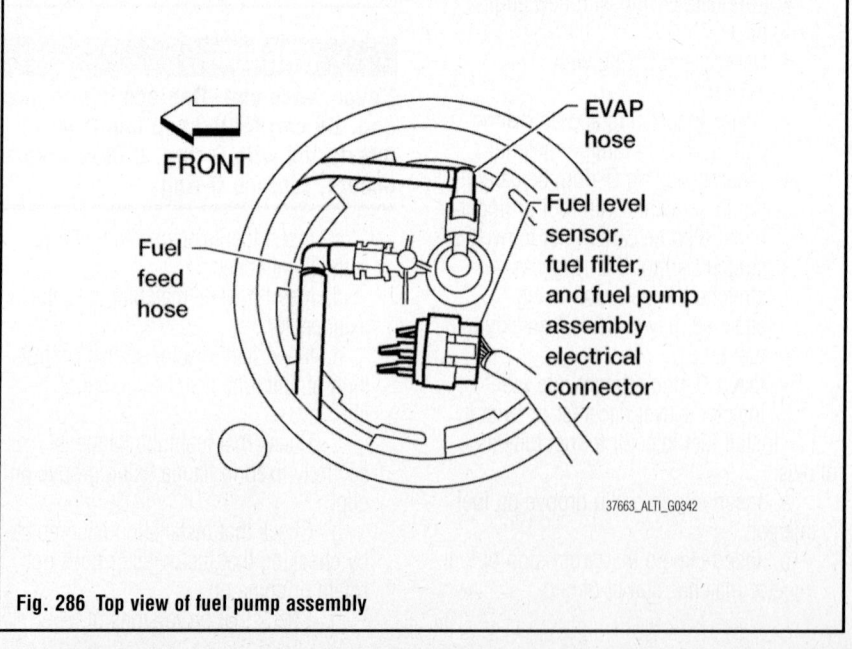

Fig. 286 Top view of fuel pump assembly

clean and to avoid damage and foreign materials, cover them completely with plastic bags or something similar.

10. Remove the center exhaust tube, with muffler(s).

11. Disconnect the fuel filler hose and the recirculation hose at the fuel tank side.

12. Disconnect the three parking brake cable mounting brackets on each cable and position the cables out of the way.

13. Remove the fuel tank protector.

14. Disconnect the fuel tank mounting straps while supporting the fuel tank.

15. Remove the fuel tank.

16. If replacing the fuel tank, remove the fuel level sensor, fuel filter and fuel pump assembly to transfer to the new fuel tank.

To install:

17. Install in the reverse order of removal paying attention to the following.

18. Before tightening the fuel tank mounting straps, temporarily install the filler hose and the recirculation hose.

19. Tighten all fuel tank mounting strap bolts to specification, then tighten the hose clamps.

20. Connect the quick connector as follows:

a. Check the connection for damage or any foreign materials.

b. Align the connector with the tube, then insert the connector straight into the tube until a click is heard.

21. After the tube is connected, make sure the connection is secure by performing the following checks:

a. Pull on the tube and the connector to make sure they are securely connected.

b. Visually confirm that the two retainer tabs are connected to the quick connector.

22. Use the following procedure to check for fuel leaks.

a. Turn the ignition switch ON (without starting the engine). Then check the connections for fuel leaks by applying fuel pressure to the fuel piping.

b. Run the engine and check for fuel leaks at the fuel system tube and hose connections.

HEATING & AIR CONDITIONING SYSTEM

BLOWER MOTOR

REMOVAL & INSTALLATION

See Figure 287.

1. Remove the glove box.

2. Remove the console finisher RH.

3. Disconnect the blower motor connector (1).

4. Remove the screws from the blower unit, then remove the blower unit from the heater and cooling unit assembly.

5. Remove the three blower motor screws (A), then remove the blower motor from the blower unit.

To install:

6. Installation is the reverse of the removal procedure.

HEATER CORE

REMOVAL & INSTALLATION

See Figure 288.

1. Remove the heater and cooling unit assembly.

2. Remove the heater grommet, heater pipe support and heater pipe cover.

3. Remove the heater and cooling unit foot duct LH.

4. Remove the heater core.

To install:

5. Installation is in the reverse order of removal.

➡**Make sure that the aspirator hose is securely attached to the aspirator on the heater and cooling unit foot duct LH.**

HEATER AND COOLING UNIT ASSEMBLY

REMOVAL & INSTALLATION

Non-Hybrid

1. Discharge the refrigerant from the A/C system.

2. Drain the engine coolant from the cooling system.

3. Disconnect the negative battery terminal.

4. Remove the wiper motor and linkage.

5. Remove the upper cowl (for VQ35DE only).

6. Remove the strut tower bar (for VQ35DE only).

7. Remove the lower RH cowl (for VQ35DE only).

8. Disconnect the heater hoses from the heater core pipes.

✳✳ WARNING

Cap or wrap the pipe joint with a suitable material such as vinyl tape to avoid the entry of contaminants into the system.

9. Disconnect the refrigerant lines from the expansion valve.

✳✳ WARNING

Cap or wrap the line joint with a suitable material such as vinyl tape to avoid the entry of contaminants into the system.

10. Remove the instrument panel assembly.

11. Remove the steering column assembly.

12. Disconnect the drain hose.

13. Remove the heater and cooling unit assembly attached to the steering member as one assembly from the vehicle.

14. Remove the blower unit from the heater and cooling unit and steering member assembly.

15. Remove the heater and cooling unit from the steering member.

To install:

16. Installation is in the reverse order of removal.

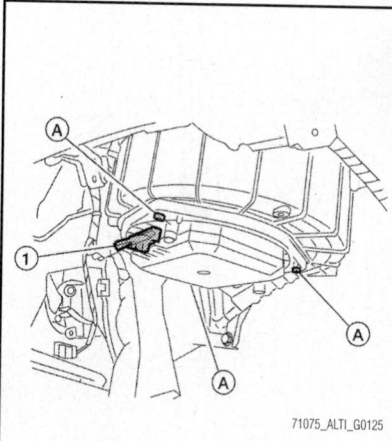

71075_ALTI_G0125

Fig. 287 Disconnecting the connector (1) and removing motor screws (A) and motor

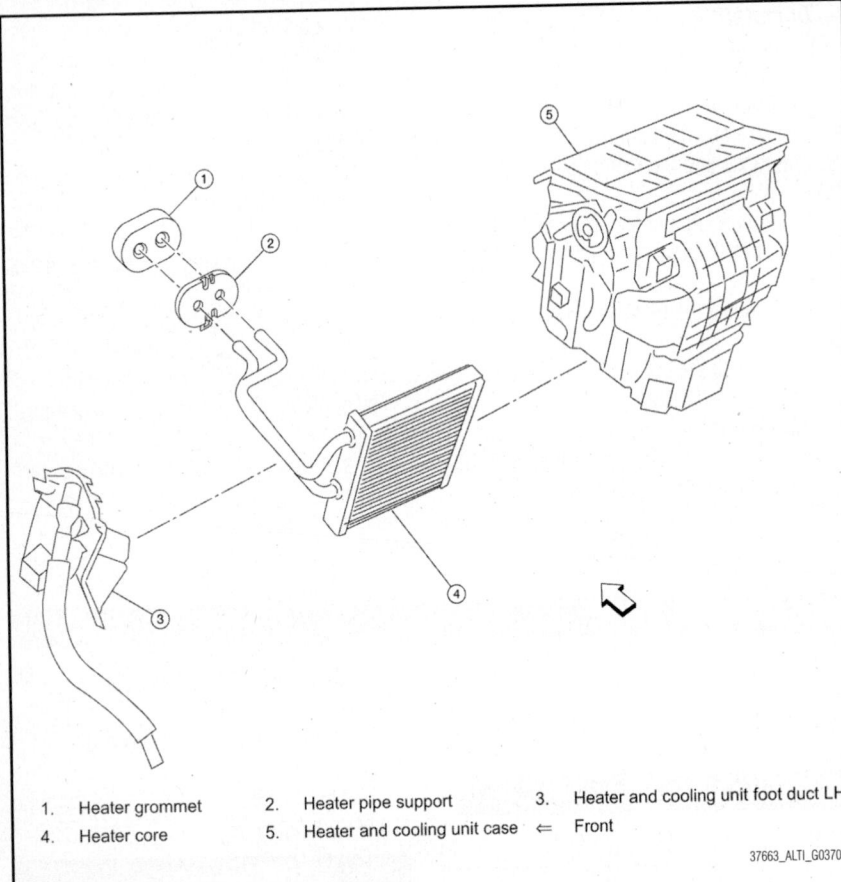

1. Heater grommet
2. Heater pipe support
3. Heater and cooling unit foot duct LH
4. Heater core
5. Heater and cooling unit case ⇐ Front

37663_ALTI_G0370

Fig. 288 Remove the heater core

17. Fill the radiator with the specified water and coolant mixture.
18. Recharge the A/C system.

Hybrid

See Figure 289.

1. Discharge the refrigerant from the A/C system.
2. Drain the engine coolant from the cooling system.
3. Remove the wiper motor and linkage.
4. Disconnect the 12 volt battery negative and positive terminals.
5. Remove the strut tower bar.
6. Remove the lower RH cowl.
7. Disconnect the heater hoses from the heater core pipes.

> ❊❊ **WARNING**
>
> **Cap or wrap the pipe joint with a suitable material such as vinyl tape to avoid the entry of contaminants into the system.**

8. Disconnect the refrigerant lines from the expansion valve.

> ❊❊ **WARNING**
>
> **Cap or wrap the line joint with a suitable material such as vinyl tape to avoid the entry of contaminants into the system.**

28 (2.9, 21)

20 (2, 15)

1. Steering member
2. Heater and cooling unit
3. Blower unit

37663_ALTH_G0058

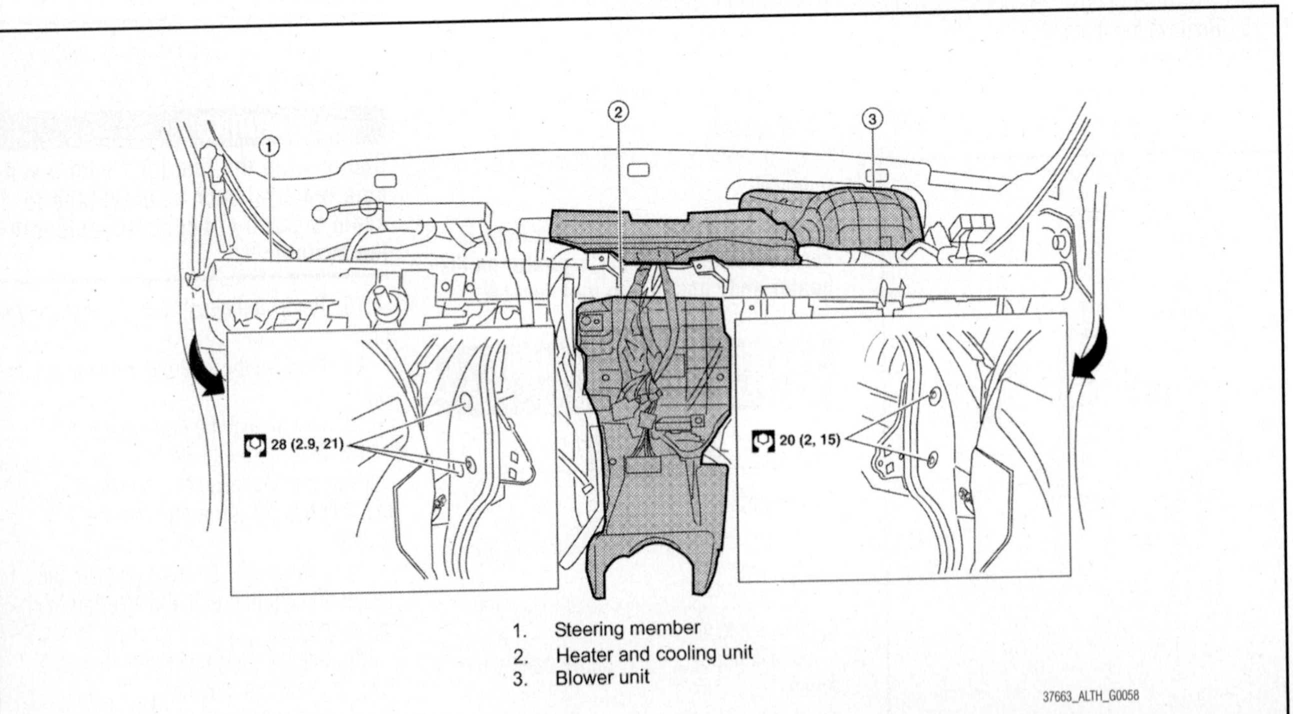

Fig. 289 Heater and cooling unit assembly

9. Remove the instrument panel assembly.

10. Remove the steering column assembly.

11. Disconnect the drain hose.

12. Remove the heater and cooling unit assembly attached to the steering member as one assembly from the vehicle.

13. Remove the blower unit from the heater and cooling unit and steering member assembly.

14. Remove the heater and cooling unit from the steering member.

To install:

15. Installation is in the reverse order of removal.

16. Fill the radiator with the specified water and coolant mixture.

17. Recharge the A/C system.

AUXILIARY HEATING & AIR CONDITIONING

BLOWER MOTOR

REMOVAL & INSTALLATION

See Figure 290.

1. Remove the glove box.

2. Disconnect the blower motor connector.

3. Remove the three blower motor screws and remove the blower motor from the blower unit.

To install:

4. Installation is in the reverse order of removal.

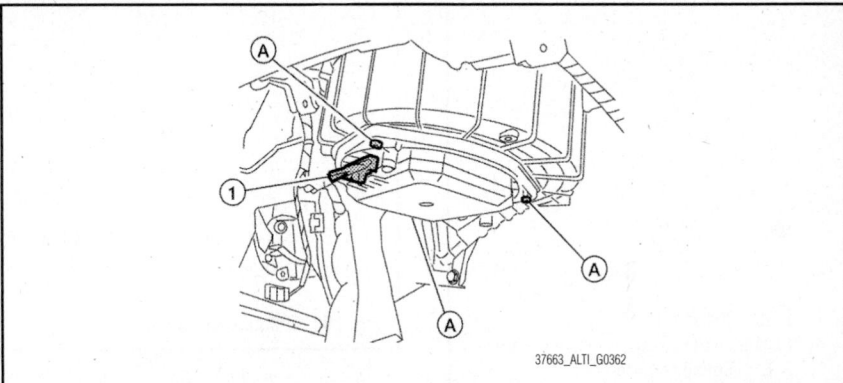

Fig. 290 Disconnect the blower motor connector (1); remove the three blower motor screws (A)

STEERING

ELECTRONIC POWER STEERING (EPS) CONTROL UNIT

REMOVAL & INSTALLATION

See Figures 291 through 293.

1. Remove the engine cover.

2. Remove the front wiper arm cover and wiper arm and blade assembly.

3. Remove the cowl top weatherstrip seal.

4. Remove the cowl top end caps.

5. Remove the cowl top finisher assembly.

6. Disconnect the washer hose.

7. Remove the strut brace.

8. Remove the wiper motor and connecting rod assembly.

9. Disconnect the 12-volt battery negative terminal.

10. Remove the left cowl extension.

11. Disconnect the MAF sensor connector.

12. Remove the air cleaner duct, blow-by hose and air cleaner duct hose.

13. Remove the fuse and fusible link box.

14. Disconnect the harness clips.

15. Disconnect the EPS DC/DC converter connector, EPS sensor connector, and EPS motor connector from the EPS ECU.

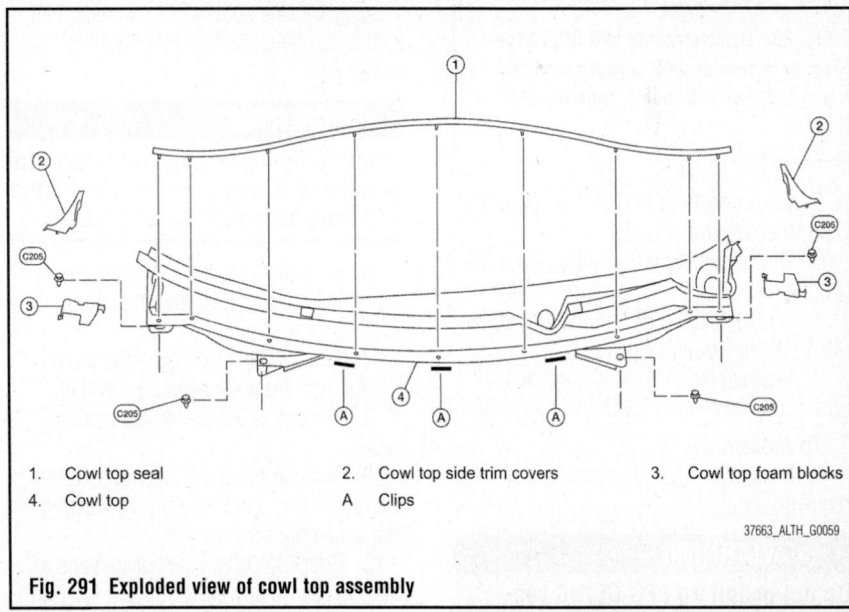

| 1. | Cowl top seal | 2. | Cowl top side trim covers | 3. | Cowl top foam blocks |
| 4. | Cowl top | A | Clips | | |

37663_ALTH_G0059

Fig. 291 Exploded view of cowl top assembly

➡For EPS DC/DC converter connector and EPS motor connector, perform the following:

   a. Pull lock plate up until it stops.
   b. Turn the lock lever until it stops.
   c. Pull the connector to disconnect it.

16. Remove the EPS control unit nut and bolts and EPS control unit.

To install:

17. Installation is in the reverse order of removal.

EPS DC/DC CONVERTER

REMOVAL & INSTALLATION

See Figure 294.

1. Pull the service plug to disconnect the high voltage battery.

2. Remove the rear seat.

3. Remove the EPS DC/DC converter cover nuts and remove the cover.

4. Remove the shield earth nut.

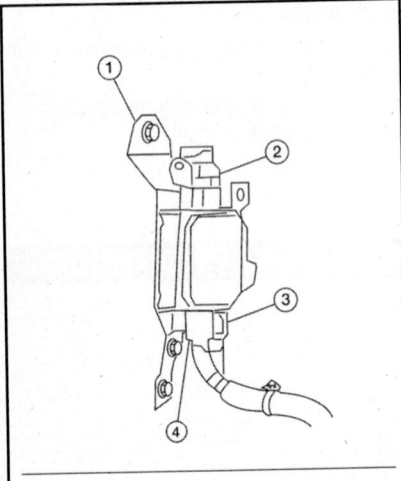

1. EPS ECU
2. EPS DC/DC converter connector
3. EPS sensor connector
4. EPS motor connector
5. Lock plate
6. Lock lever

37663_ALTH_G0060

Fig. 292 Disconnect the EPS DC/DC converter connector, EPS sensor connector, and EPS motor connector from the EPS ECU

5. Disconnect the EPS motor power line (245 V) connector and clip.

6. Disconnect the EPS motor power line ground.

7. Disconnect the EPS motor power line (42 V) connector and clip.

8. Remove the EPS DC/DC converter nuts and remove the EPS DC/DC converter.

To install:

9. Installation is in the reverse order of removal.

✳✳ WARNING

Do not install an EPS DC/DC converter if it has been dropped or shocked, replace with a new one.

POWER RACK & PINION STEERING GEAR & LINKAGE

REMOVAL & INSTALLATION

Non-Hybrid

See Figures 295 through 298.

1. Remove the front tires.

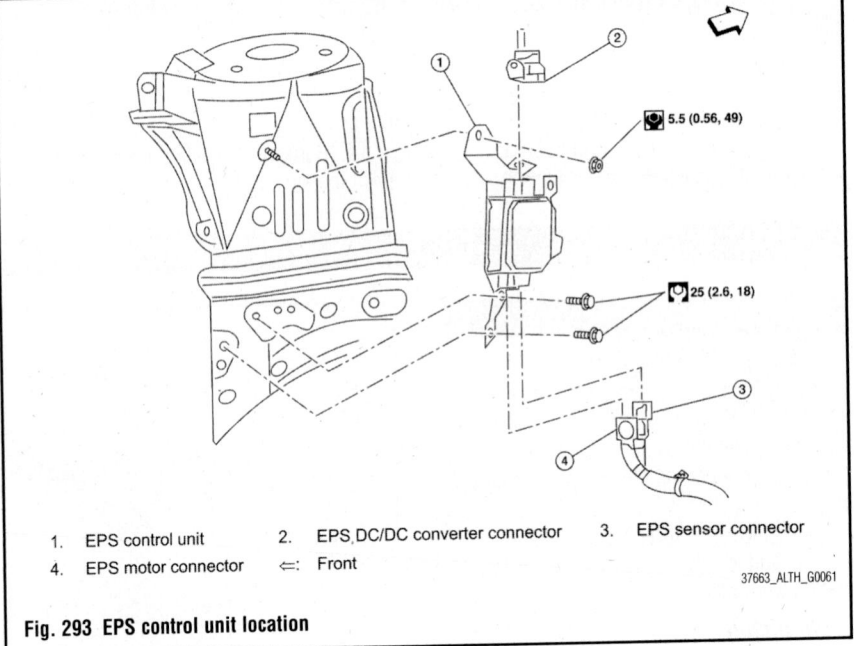

1. EPS control unit
2. EPS DC/DC converter connector
3. EPS sensor connector
4. EPS motor connector
⇐: Front

5.5 (0.56, 49)
25 (2.6, 18)

37663_ALTH_G0061

Fig. 293 EPS control unit location

2. Remove undercover.

3. Remove lower side bolt of lower joint.

4. Remove cotter pin and loosen the nut.

5. Remove steering outer socket from steering knuckle so as not to damage ball joint boot using the Tool HT72520000 (J-25730-A).

✳✳ CAUTION

Temporarily tighten the nut to prevent damage to threads and to prevent the Tool from suddenly coming off.

6. Remove high and low pressure piping of hydraulic piping, and then drain power steering fluid.

7. Remove steering hydraulic piping bracket from front suspension member.

8. Remove SSPS valve harness connector.

9. Remove bolts and nuts of steering gear assembly, and then remove steering gear assembly from vehicle.

10. Check for fluid leaks or damage to steering gear. If any exist, replace steering gear as an assembly.

To install:

11. Installation is in the reverse order of removal.

12. When installing lower joint to steering gear assembly, follow the procedure listed below.

a. Set rack of steering gear in the neutral position.

➡**To get the neutral position of rack, turn gear-sub assembly and measure the distance of inner socket, and then**

measure the intermediate position of the distance.

b. Align rear cover cap projection with the marking position of gear housing assembly.

c. Install slit part of lower joint aligning with the projection of rear cover cap. Make sure that the slit part of lower joint is aligned with both the projection of rear cover cap and the marking position of gear housing assembly.

13. After installation, bleed air from the steering hydraulic system.

14. Perform final tightening of nuts and bolts on each part under unladen conditions with tires on level ground when removing steering gear assembly. Check wheel alignment.

15. Make sure that steering wheel operates smoothly by turning several times from full left stop to full right stop.

Hybrid

See Figures 299 through 301.

1. Remove tires and wheels.

2. Remove engine undercover.

3. Remove stabilizer bar connecting rods from struts and reposition stabilizer bar.

4. Remove front exhaust tube.

5. Remove lower side bolt of lower steering joint.

6. Remove cotter pin and then loosen the nut.

7. Remove steering outer socket from steering knuckle so as not to damage ball joint boot using Tool HT72520000 (J-25730-A) as shown.

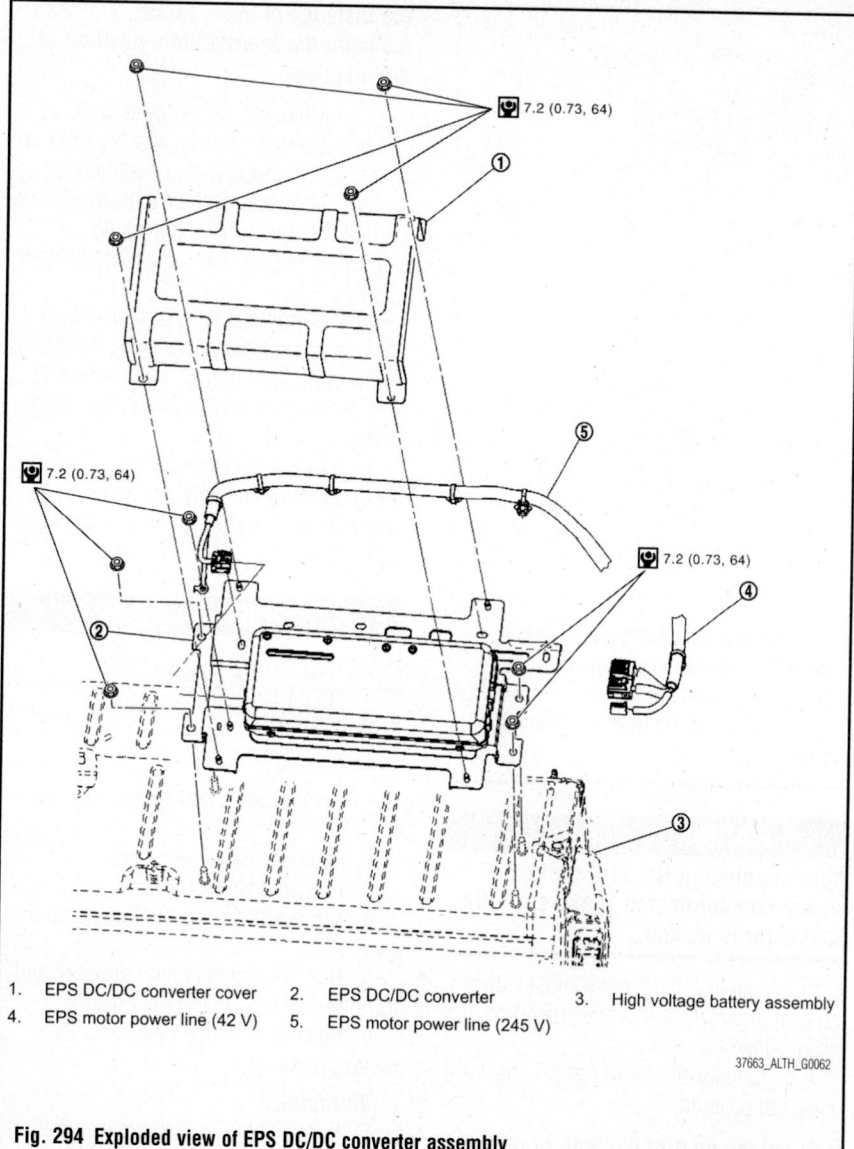

1. EPS DC/DC converter cover
2. EPS DC/DC converter
3. High voltage battery assembly
4. EPS motor power line (42 V)
5. EPS motor power line (245 V)

37663_ALTH_G0062

Fig. 294 Exploded view of EPS DC/DC converter assembly

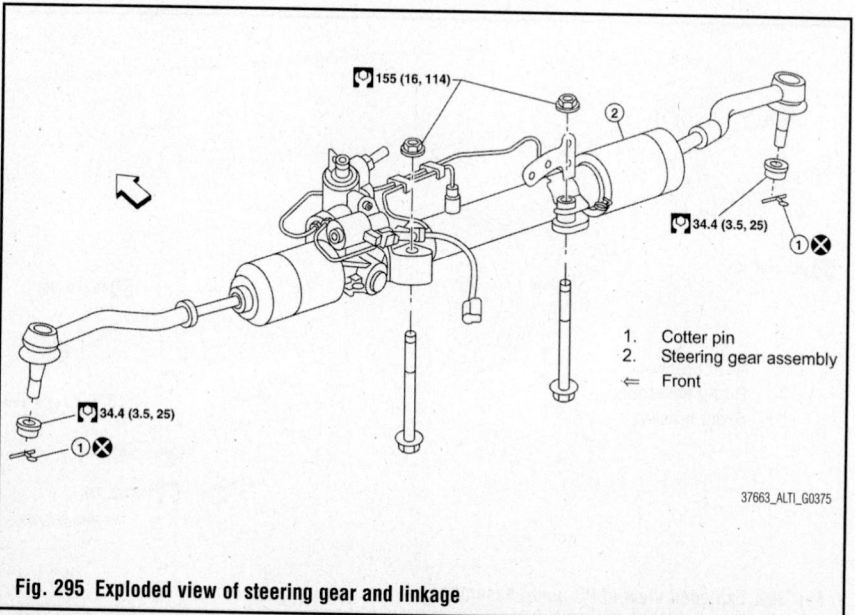

1. Cotter pin
2. Steering gear assembly
⇐ Front

37663_ALTI_G0375

Fig. 295 Exploded view of steering gear and linkage

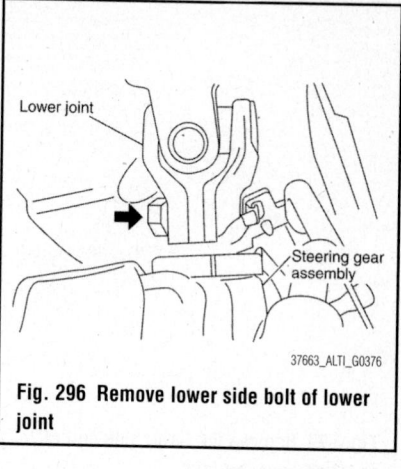

Lower joint

Steering gear assembly

37663_ALTI_G0376

Fig. 296 Remove lower side bolt of lower joint

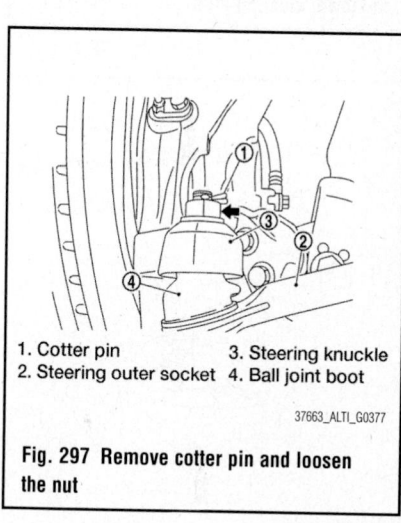

1. Cotter pin   3. Steering knuckle
2. Steering outer socket   4. Ball joint boot

37663_ALTI_G0377

Fig. 297 Remove cotter pin and loosen the nut

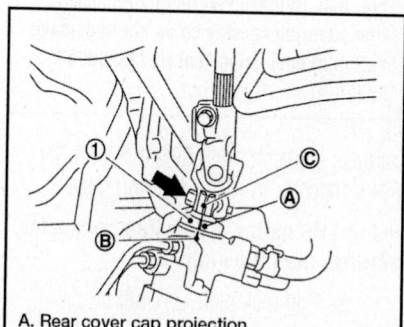

A. Rear cover cap projection
B. Marking position of gear housing assembly
C. slit part of lower joint
1. Rear cover cap

37663_ALTI_G0378

Fig. 298 Align rear cover cap projection with the marking position of gear housing assembly

✳✳ WARNING

Temporarily tighten the nut to prevent damage to threads and to prevent the Tool from suddenly coming off.

8. Disconnect EPS torque sensor harness connector, EPS motor angle sensor

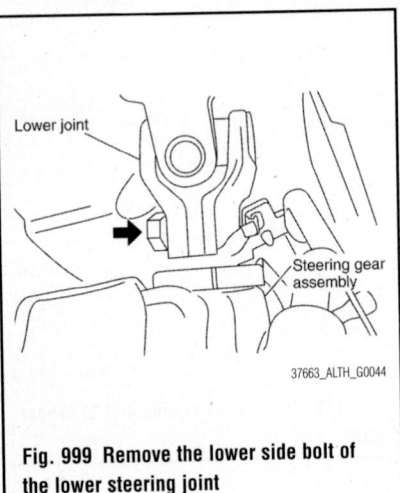

Fig. 999 Remove the lower side bolt of the lower steering joint

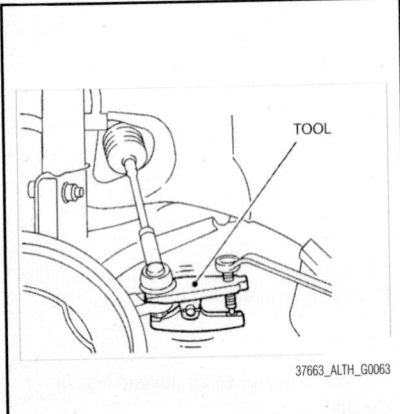

Fig. 300 Remove steering outer socket from steering knuckle so as not to damage ball joint boot using Tool HT72520000 (J-25730-A)

harness connector and EPS motor power line connector from the steering gear.

➡**For EPS motor power line connector, perform the following;**

    a. Pull lock plate up until it stops.
    b. Turn the lock lever until it stops.
    c. Pull EPS motor power line connector to disconnect it.

  9. Remove ground bracket and harness bracket on steering gear.

  10. Remove bolts and nuts of steering gear assembly, and then remove steering gear assembly from vehicle.

  11. Check for damage to steering gear. If any exist, replace steering gear as an assembly.

  To install:
  12. Installation is the reverse order of removal.

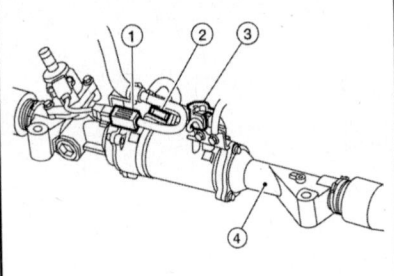

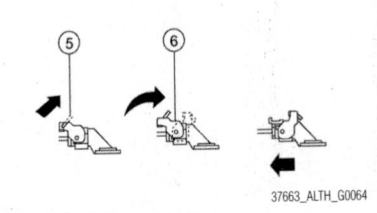

Fig. 301 Disconnect EPS torque sensor harness connector, EPS motor angle sensor harness connector and EPS motor power line connector from the steering gear

❊❊ WARNING

After connecting the EPS motor power line connector, make sure the connector is locked.

  13. When installing lower joint to steering gear assembly, follow the procedure listed below.
    a. Set rack of steering gear in the neutral position.

➡**To get the neutral position of rack, turn gear-sub assembly and measure**

the distance of inner socket, and then measure the intermediate position of the distance.

    b. Install slit part of lower shaft assembly joint aligning with the marking position of steering gear assembly input shaft. Make sure that the slit part of lower joint is aligned with the marking position of gear housing assembly input shaft.

  14. Perform final tightening of nuts and bolts on each part under unladen conditions with tires on level ground when removing steering gear assembly. Check wheel alignment.

  15. Make sure that steering wheel operates smoothly by turning several times from full left stop to full right stop.

POWER STEERING PUMP

REMOVAL & INSTALLATION

2.5L Engine

See Figure 302.

  1. Drain power steering fluid from reservoir tank.
  2. Remove undercover.
  3. Loosen drive belt.
  4. Remove drive belt from oil pump pulley.
  5. Remove piping of high pressure and low pressure (drain fluid from lines).
  6. Remove oil pump bolts, and then remove power steering oil pump.

  To install:
  7. Installation is in the reverse order of removal.

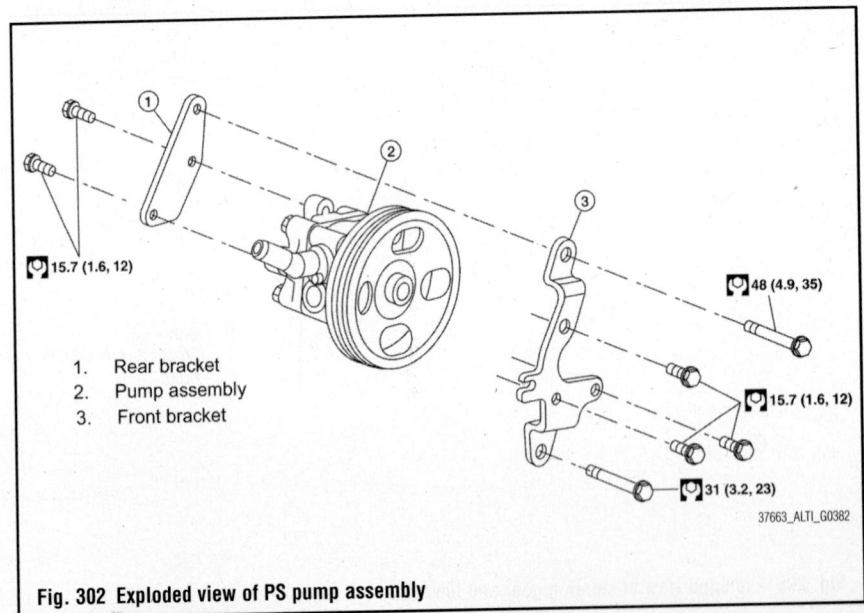

1. Rear bracket
2. Pump assembly
3. Front bracket

15.7 (1.6, 12)
48 (4.9, 35)
15.7 (1.6, 12)
31 (3.2, 23)

Fig. 302 Exploded view of PS pump assembly

3.5L Engine

See Figure 303.

1. Drain power steering fluid from reservoir tank.
2. Remove undercover.
3. Loosen drive belt.
4. Remove drive belt from oil pump pulley.
5. Remove piping of high pressure and low pressure (drain fluid from lines).
6. Remove oil pump bolts, and then remove power steering oil pump.

To install:

7. Installation is in the reverse order of removal.

POWER STEERING FLUID

FLUID RECOMMENDATIONS

The recommended fluid is Genuine Nissan PSF or equivalent.

FLUID LEVEL CHECK

1. Check fluid level with engine stopped.
2. Make sure that fluid level is between MIN and MAX.
3. Fluid levels at HOT (A) and COLD (B) are different. Do not confuse them.
   - HOT (A): fluid temperature 122–176°F
   - COLD (B): fluid temperature 32–86°F

✳✳ WARNING

The fluid level should not exceed the MAX line. Excessive fluid will cause fluid leakage from the cap.

✳✳ WARNING

Do not reuse drained power steering fluid.

FLUID FILL PROCEDURE

1. Fill power steering reservoir while checking fluid level.

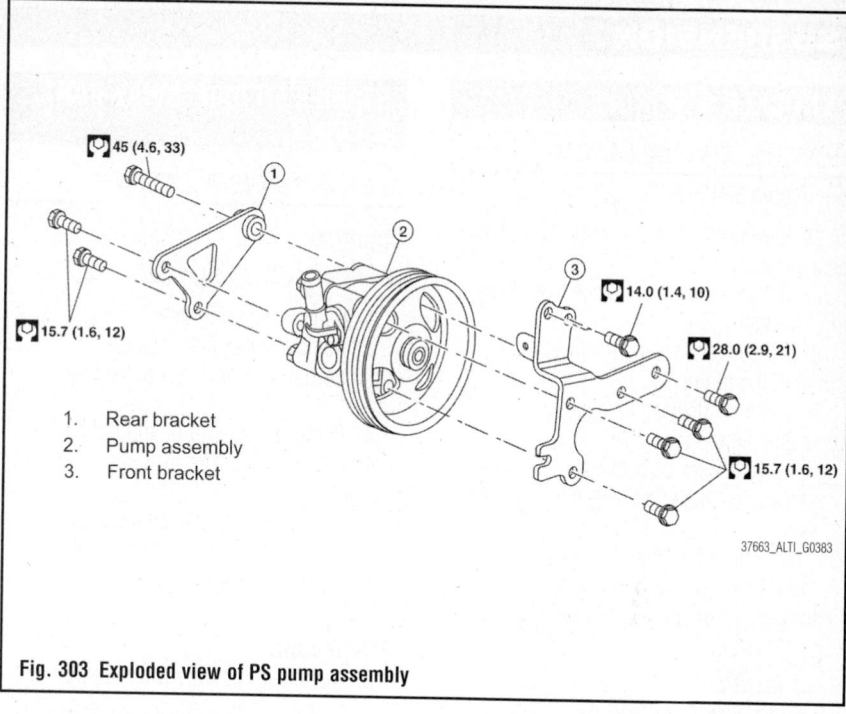

1. Rear bracket
2. Pump assembly
3. Front bracket

37663_ALTI_G0383

Fig. 303 Exploded view of PS pump assembly

2. Bleed air from hydraulic system.
3. Check for fluid leaks.

Air Bleeding Hydraulic System

If air bleeding is not complete, the following symptoms can be observed.

   a. Bubbles are created in reservoir tank.

   b. Clicking noise can be heard from oil pump.

   c. Excessive buzzing in the oil pump.

Fluid noise may occur in the steering gear or oil pump. This does not affect performance or durability of the system.

1. Turn steering wheel several times from full left stop to full right stop with engine off.

✳✳ WARNING

Turn steering wheel while filling reservoir tank with fluid so as not to lower fluid level below the MIN line.

2. Start engine and hold steering wheel at each lock position for 3 seconds at idle to check for fluid leakage.
3. Repeat step 2 above several times at approximately 3 second intervals.

✳✳ WARNING

Do not hold the steering wheel in a locked position for more than 10 seconds. (There is the possibility that oil pump may be damaged.)

4. Check fluid for bubbles and while contamination.
5. Stop engine if bubbles and white contamination do not drain out. Perform step 2 and 3 above after waiting
   until bubbles and white contamination drain out.
6. Stop the engine, and then check fluid level.

SUSPENSION

FRONT SUSPENSION

KNUCKLE & SPINDLE

REMOVAL & INSTALLATION

See Figure 304.

1. Remove front wheel hub and bearing assembly.
2. Remove steering linkage from steering knuckle.
3. Remove the steering knuckle lower pinch bolt and nut.
4. Remove steering knuckle to strut bolts and steering knuckle.
   a. Check for deformity, cracks and damage on each part, replace if necessary.
   b. Check ball joint for boot breakage, axial looseness, and torque of transverse link ball joint and repair as necessary.

To install:

5. Installation is the reverse of the removal procedure.

✳✳ WARNING

Do not reuse non-reusable parts.

STABILIZER BAR (SWAY BAR) & LINKS

REMOVAL & INSTALLATION

Stabilizer

See Figures 305 through 307.

1. Remove steering gear.
2. Remove mounting nuts on upper portion of stabilizer connecting rod.
3. Remove stabilizer clamp bolts.
4. Remove stabilizer from the vehicle.
   a. Check stabilizer, connecting rod, bushing and clamp for deformation, cracks and damage, and replace if necessary.

To install:

5. Installation is the reverse of the removal procedure.
6. When installing stabilizer, make sure that the clamps are facing in the direction shown.
7. Make sure the cut surface of the bushing faces the rear.

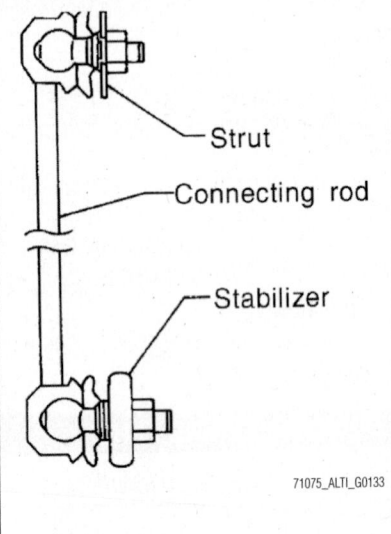

71075_ALTI_G0133

Fig. 305 Removing and installing stabilizer

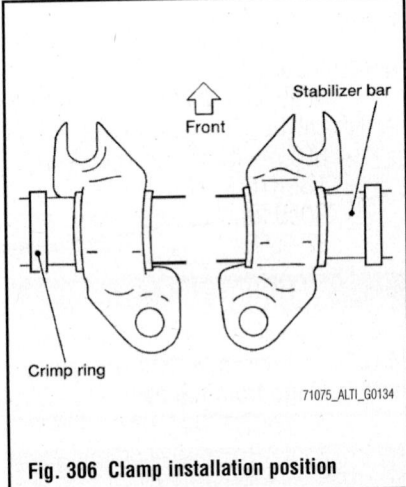

71075_ALTI_G0134

Fig. 306 Clamp installation position

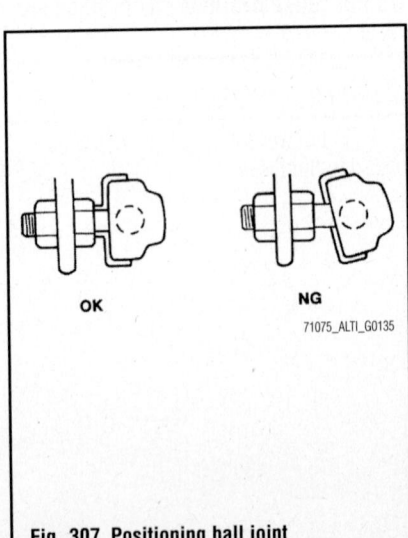

71075_ALTI_G0135

Fig. 307 Positioning ball joint

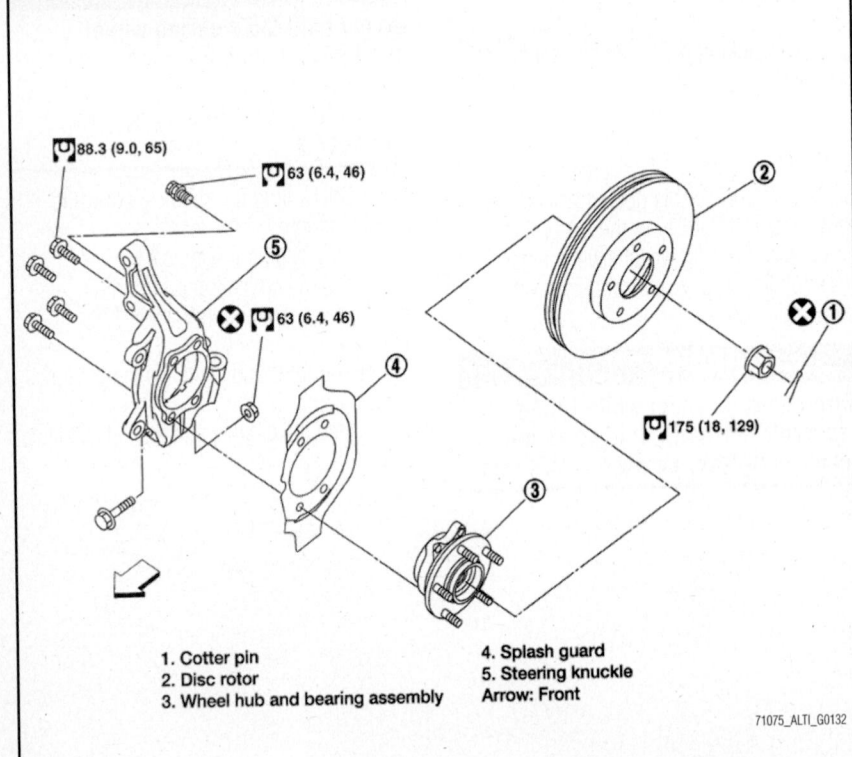

88.3 (9.0, 65)   63 (6.4, 46)

63 (6.4, 46)

175 (18, 129)

1. Cotter pin
2. Disc rotor
3. Wheel hub and bearing assembly
4. Splash guard
5. Steering knuckle
Arrow: Front

71075_ALTI_G0132

Fig. 304 Removing and installing steering knuckle

8. Stabilizer uses pillow ball type connecting rod. Position ball joint with case on pillow ball head parallel to stabilizer.

STRUTS

REMOVAL & INSTALLATION

See Figure 308.

➡**Refer to the exploded view of front suspension while servicing this component.**

1. Remove wheel and tire.
2. Remove brake caliper and reposition aside using wire.

✳✳ CAUTION

Avoid depressing brake pedal with brake caliper removed.

3. Remove wheel sensor electrical harness from strut.
4. Remove brake hose lock plate.
5. Remove steering knuckle to strut bolts and nuts.
6. Remove bolt on strut tower bar then bolts on strut tower and remove strut from vehicle.
7. Check the strut for any oil leakage or other damage and replace as necessary.

To install:

8. Installation is in the reverse order of removal.
9. Refer to "Exploded View" for tightening torque.
10. Be sure arrows on strut mount insulator and spring upper seat are positioned as shown. Also be sure notch in strut spacer is positioned as shown. Then install strut.
11. Assemble upper mounting plate with its notch facing toward the outside.

OVERHAUL

Disassembly

See Figure 309.

1. Install Tool ST35652000 to strut and secure it in a vise.

✳✳ CAUTION

When installing Tool ST35652000, wrap a shop cloth around strut to protect it from damage.

2. Slightly loosen piston rod lock nut.

✳✳ WARNING

Do not remove piston rod lock nut completely. If it is removed completely, the coil spring can jump out and may cause serious damage or injury.

3. Compress coil spring using a commercially available spring compressor.

✳✳ WARNING

Make sure that the pawls of the two spring compressors are firmly hooked on the spring. The spring compressors must be tightened alternately so as not to tilt the spring.

4. Making sure coil spring is free between upper and lower seats, then remove piston rod lock nut.
5. Remove small parts on strut.
   a. Remove strut spacer, strut mount insulator, strut mounting insulator bracket thrust bearing, spring upper seat, and upper rubber seat. Then remove coil spring.
6. Remove bound bumper from spring upper seat.

7. Gradually release spring compressor (commercial service tool), and remove coil spring.

Inspection After Disassembly

Strut

1. Check strut for deformation, cracks, and damage, and replace if necessary.
2. Check piston rod for damage, uneven wear, and distortion, and replace if necessary.
3. Check welded and sealed areas for oil leakage, and replace if necessary.

Insulator and Rubber Parts

1. Check strut mount insulator for cracks and rubber parts for wear. Replace them if necessary.

Coil Spring

1. Check for cracks, wear, and damage, and replace if necessary.

Assembly

See Figures 310 and 311.

1. Compress coil spring using a spring compressor (commercial service tool), and install it onto the strut.

✳✳ WARNING

Face tube side of coil spring downward. Align lower end to spring seat as shown.

✳✳ CAUTION

Be sure spring compressor is securely attached to coil spring. Compress coil spring.

2. Connect bound bumper to spring upper seat.

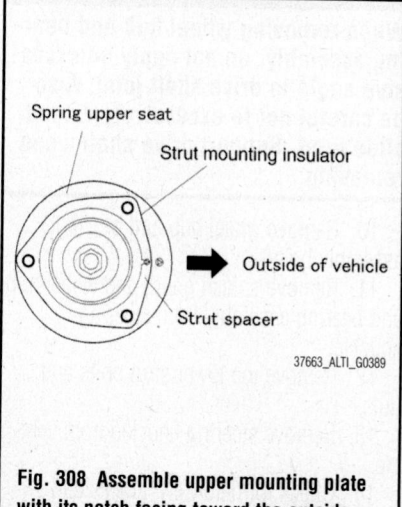

Fig. 308 Assemble upper mounting plate with its notch facing toward the outside

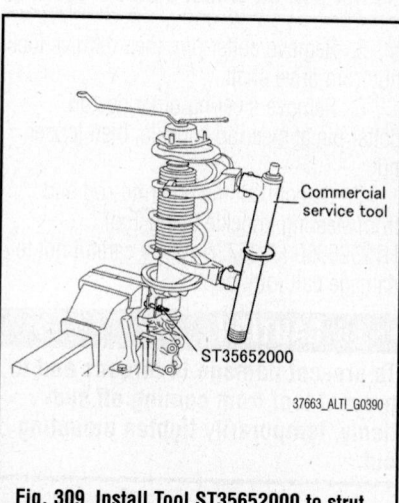

Fig. 309 Install Tool ST35652000 to strut and secure it in a vise

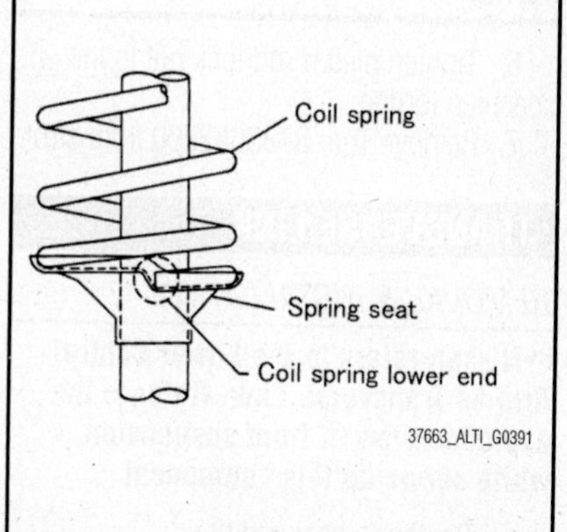

Fig. 310 Align lower end to spring seat as shown

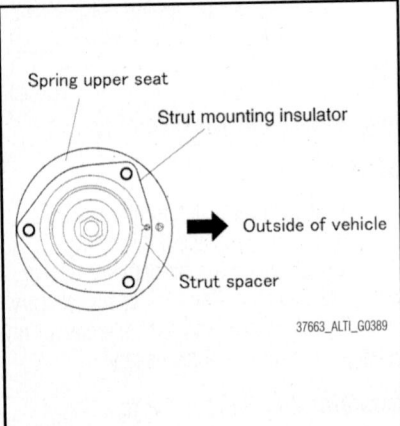

Fig. 311 Assemble upper mounting plate with its notch facing toward the outside

Note the following:
• Be sure to install bound bumper to spring upper seat securely.
• When installing bound bumper, use soapy water. Do not use machine oil or other lubricants.

3. Install small parts to the strut.
• Connect upper rubber seat, spring upper seats, thrust bearing, strut mount insulator, and strut spacer. Temporarily install piston rod lock nut.

> ### ✸✸ WARNING
> **Do not reuse piston rod lock nut.**

4. Be sure arrows on strut mount insulator and spring upper seat are positioned as shown. Also be sure notch in the strut spacer is positioned as shown.

5. Be sure coil spring is properly set in spring rubber seat. Gradually release spring compressor.

> ### ✸✸ WARNING
> **Be sure upper rubber seat is properly aligned to spring upper seat and coil spring.**

6. Tighten piston rod lock nut to the specified torque.

7. Remove Tool ST35652000 from strut.

TRANSVERSE LINK

REMOVAL & INSTALLATION

➡**Nissan refers to the Lower Control Arm as Transverse Link. Refer to the exploded view of front suspension while servicing this component.**

1. Remove wheel and tire.
2. Remove steering knuckle from transverse link.

3. Remove mounting nuts and washers on lower portion of stabilizer connecting rod.
4. Slightly loosen transverse link mounting bolts.
5. Remove transverse link bolts and nuts, and remove transverse link from suspension member.

To install:

6. Installation is in the reverse order of removal.
8. Tighten transverse link bolts with vehicle unladen and all four tires on flat, level ground.
9. After installation, check wheel alignment.

WHEEL HUB & BEARING

ADJUSTMENT

No adjustment is possible. If there is too much play, replace the bearing and/or hub.

REMOVAL & INSTALLATION

See Figures 312 and 313.

1. Remove wheel and tire from vehicle.
2. Remove brake caliper, leaving brake caliper hydraulic lines connected. Reposition brake caliper aside with wire.

➡**Avoid depressing brake pedal while brake caliper is removed.**

3. Put alignment marks on disc rotor and wheel hub and bearing assembly, then remove disc rotor.
4. Remove wheel sensor from steering knuckle.

> ### ✸✸ WARNING
> **Do not pull on wheel sensor harness.**

5. Remove cotter pin, then remove lock nut from drive shaft.
6. Remove steering outer tie-rod cotter pin at steering knuckle, then loosen nut.
7. Disconnect the outer tie-rod end from steering knuckle using Tool HT2520000 (J-25730-A). Be careful not to damage ball joint boot.

> ### ✸✸ CAUTION
> **To prevent damage to threads and to prevent tool from coming off suddenly, temporarily tighten mounting nut.**

8. Remove transverse link and steering knuckle pinch bolt and nut.

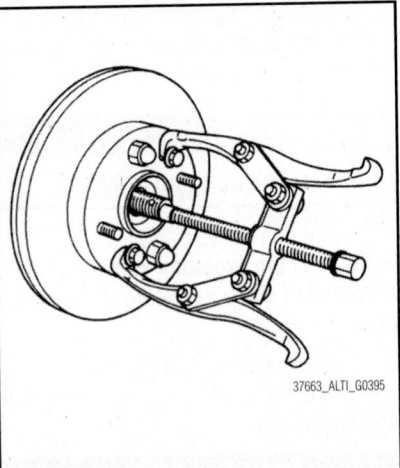

Fig. 312 Remove wheel hub and bearing assembly from drive shaft using a puller

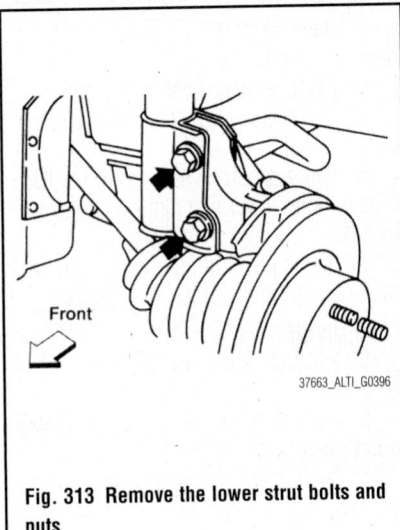

Fig. 313 Remove the lower strut bolts and nuts

9. Remove wheel hub and bearing assembly from drive shaft using a puller or suitable tool.

> ### ✸✸ CAUTION
> **When removing wheel hub and bearing assembly, do not apply an excessive angle to drive shaft joint. Also be careful not to excessively extend slide joint. Support drive shaft when removing.**

10. Remove wheel hub and bearing assembly bolts.
11. Remove splash guard and wheel hub and bearing assembly from steering knuckle.
12. Remove the lower strut bolts and nuts.
13. Remove steering knuckle from vehicle.
14. Check for deformity, cracks and damage on each part, replace if necessary.

15. Check for boot breakage, axial looseness, and torque of transverse link ball joint and repair as necessary.

To install:

16. Installation is in the reverse order of removal.

※※ WARNING

Do not reuse non-reusable parts.

17. When installing wheel hub and bearing assembly to steering knuckle, align

cutout in toner ring cover with wheel sensor mounting hole in steering knuckle.

18. When installing disc rotor on wheel hub and bearing assembly, align the marks.

SUSPENSION

COIL SPRINGS

REMOVAL & INSTALLATION

See Figures 314 and 315.

1. Loosen the rear lower link bolt and nut from the suspension member side.

2. Support the rear lower link by placing a suitable jack under the knuckle.

3. Remove the rear lower link adjusting bolt and nut from the suspension member side.

➡**Do not reuse the adjusting nut, use a new adjusting nut for installation.**

4. Slowly lower the jack to lower the rear lower link and coil spring.

5. Remove the upper rubber seat, coil spring, and lower rubber seat from the rear lower link.

6. Remove rear lower link bolt and nut from the suspension member side.

7. Remove the rear lower link.

To install:

8. Installation is in the reverse order of removal.

➡**Do not reuse the adjusting nut, use a new adjusting nut for installation.**

9. Check that the projecting part inside the upper rubber seat and the bracket flange are attached as shown.

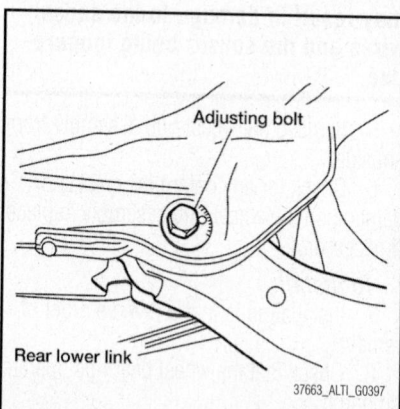

Fig. 314 Remove the rear lower link adjusting bolt and nut from the suspension member side

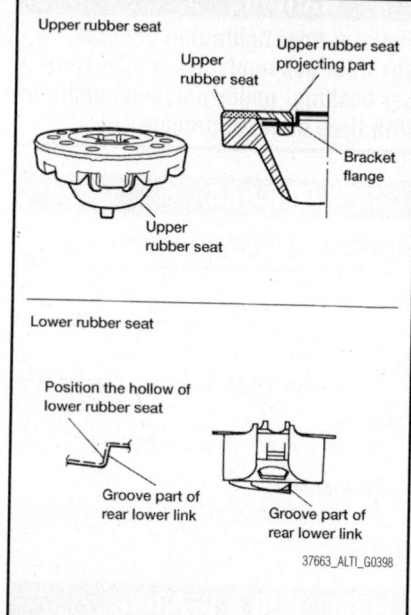

Fig. 315 Proper orientation of coil spring mounting components

10. Check that the projection part outside the upper rubber seat is directed toward the front of the vehicle.

11. Position the hollow of the lower rubber seat with the groove part of the rear lower link.

12. Install the coil spring so that the side with the two paint markers is directed toward the lower side.

13. Check the rear wheel alignment and adjust if necessary.

CONTROL ARMS/LINKS

REMOVAL & INSTALLATION

Front Lower Link

See Figure 316.

1. Remove the front lower link nut and bolt from the knuckle side and the adjusting bolt and nut from the suspension member side.

➡**Do not reuse the adjusting nut, use a new adjusting nut for installation.**

2. Remove the front lower link.

REAR SUSPENSION

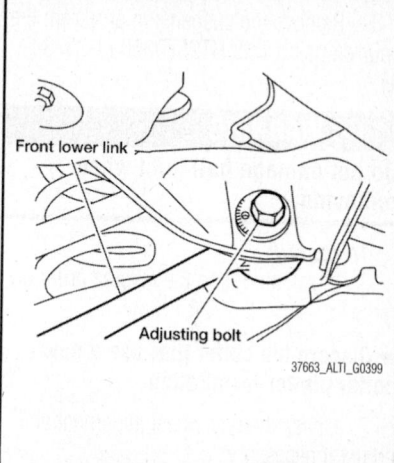

Fig. 316 Remove the front lower link nut and bolt from the knuckle side and the adjusting bolt and nut from the suspension member side

To install:

3. Installation is in the reverse order of removal.

➡**Do not reuse the adjusting nut, use a new adjusting nut for installation.**

4. Check the rear wheel alignment and adjust if necessary.

Radius Rod

➡**Refer to Exploded View of Rear Suspension while servicing this component.**

1. Remove the rear suspension assembly.
2. Remove the radius rod.

To install:

3. Installation is in the reverse order of removal.

4. Check the rear wheel alignment and adjust if necessary.

Rear Lower Link

Refer to Coil Spring Removal and Installation when servicing this component.

Suspension Arm

➡**Refer to Exploded View of Rear Suspension while servicing this component.**

1. Remove the rear suspension assembly.

2. Remove the connecting rod bracket from the suspension arm.

3. Remove the two suspension arm bolts and nuts from the suspension member side of the suspension arm.

4. Remove the ball joint cotter pin and lock nut.

➡**Discard the cotter pin, use a new cotter pin for installation.**

5. Remove the suspension arm from the knuckle using Tool HT2520000 (J-25730-A).

✳✳ WARNING
Do not damage ball joint when removing.

To install:

6. Installation is in the reverse order of removal.

➡**Discard the cotter pin, use a new cotter pin for installation.**

7. Check the rear wheel alignment and adjust if necessary.

RADIUS ROD

REMOVAL & INSTALLATION

1. Remove wheel and tire using power tool.

2. Remove radius rod from knuckle using power tools.

3. Remove radius rod from rear suspension member.

To install:

4. Installation is the reverse of the removal procedure.

a. Check the rear wheel alignment and adjust if necessary.

SHOCK ABSORBERS

REMOVAL & INSTALLATION

1. Remove wheel and tire using power tool.

2. Set a jack under rear lower link to relieve the coil spring tension.

3. Remove shock absorber lower end bolt with a power tool.

4. Gradually lower the jack to remove it from rear lower link.

5. Remove shock absorber assembly upper end nuts with a power tool, and then remove shock absorber assembly from vehicle.

To install:

6. Installation is the reverse of the removal procedure.

✳✳ WARNING
Do not reuse non-reusable parts.

✳✳ WARNING
Tighten the wheel nuts to specification.

✳✳ WARNING
Perform final tightening of shock absorber assembly lower side (rubber bushing) under unladen condition with tires on level ground.

STABILIZER BAR

REMOVAL & INSTALLATION

See Figure 317.

1. Disconnect the stabilizer bar from connecting rod.

2. Remove the stabilizer bar clamps and bushings.

3. Remove the stabilizer bar.

To install:

4. Installation is in the reverse order of removal.

WHEEL HUB & BEARING

ADJUSTMENT

No adjustment is possible. If there is too much play, replace the bearing and/or hub.

REMOVAL & INSTALLATION

See Figure 318.

Wheel hub assembly does not require maintenance. If any of the following symp-

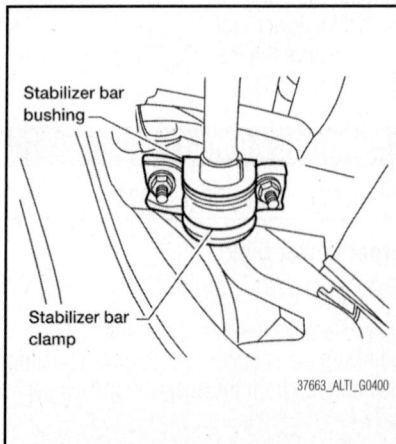

Fig. 317 Remove the stabilizer bar clamps and bushings

Stabilizer bar bushing

Stabilizer bar clamp

37663_ALTI_G0400

toms are noted, replace the wheel hub assembly:

• A growling noise is emitted from the wheel hub assembly while driving.

• The wheel hub assembly drags or turns roughly.

1. Remove the rear wheel and tire.

2. Remove the brake caliper assembly.

a. The brake hose does not need to be disconnected from the brake caliper.

b. Suspend the brake caliper assembly using wire, do not stretch the brake hose.

c. Do not depress the brake pedal, or the caliper piston will pop out.

d. Do not twist the brake hose.

3. Remove the brake rotor.

4. Remove the rear ABS sensor, then move it away from the wheel hub assembly.

✳✳ WARNING
Failure to remove the ABS sensor may result in damage to the sensor wires and the sensor being inoperative.

5. Remove the wheel hub assembly from knuckle.

6. Check for any deformity, cracks, or damage on the wheel hub assembly, replace if necessary.

To install:

7. Installation is in the reverse order of removal.

8. Check that the wheel bearings operate smoothly.

9. Check that the wheel hub bearing axial end play is within specification: 0.004 inches (0.1 mm).

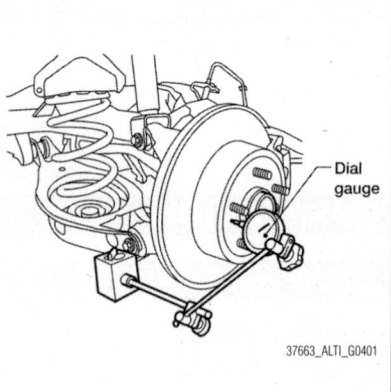

Dial gauge

37663_ALTI_G0401

Fig. 318 Check that the wheel hub bearing axial end play is within specification

NISSAN

Armada

8

SPECIFICATIONS AND MAINTENANCE CHARTS

ENGINE AND VEHICLE IDENTIFICATION

	Engine								Model Year	
									Code ②	Year
Code ①	Liters (cc)	Cu. In.	Cyl.	Fuel Sys.	Engine	Eng. Mfg.			B	2011
VK56DE	5.6 (5552)	338.8	8	MFI	DOHC	Nissan			C	2012

MFI: Multi-port Fuel Injection

DOHC: Double Overhead Camshafts

① Engine VIN: A. Engine VIN: B (FFV= flex fuel vehicle).

② 10th digit of the Vehicle Identification Number (VIN)

71075_ARMA_C0001

GENERAL ENGINE SPECIFICATIONS

Year	Model	Engine Displacement Liters	Engine ID	Net Horsepower @ rpm	Net Torque @ rpm (ft. lbs.)	Bore x Stroke (in.)	Com-pression Ratio	Oil Pressure @ rpm
2011	Armada	5.6	VK56DE	317@5200	385@3400	3.86X3.62	9.8:1	43@2000
2012	Armada	5.6	VK56DE	317@5200	385@3400	3.86X3.62	9.8:1	43@2000

71075_ARMA_C0002

ENGINE TUNE-UP SPECIFICATIONS

Year	Engine Displacement Liters	Engine ID	Spark Plug Gap (in.)	Ignition Timing	Fuel Pump (psi) ①	Idle Speed	Valve Clearance (in.) In.	Valve Clearance (in.) Ex.
2011	5.6	VK56DE	0.043	②	51	600-700	0.010-0.013	0.011-0.015
2012	5.6	VK56DE	0.043	②	51	600-700	0.010-0.013	0.011-0.015

NOTE: The Vehicle Emission Control Information label often reflects specification changes made during production. The label figures must be used if they differ from those in this chart.

① Approximate at idle

② 15 degrees +/- 5 degrees BTDC

71075_ARMA_C0003

CAPACITIES

Year	Model	Engine Displacement Liters	Engine ID	Engine Oil with Filter (qts.)	Transmission (pts.)	Transfer Case (pts.)	Drive Axle Front (pts.)	Drive Axle Rear (pts.)	Fuel Tank (gal.)	Cooling System (qts.)
2011	Armada	5.6	VK56DE	6.75	22.50	6.36	3.380	3.750	28.0	15.25
2012	Armada	5.6	VK56DE	6.75	22.50	6.36	3.380	3.750	28.0	15.25

NOTE: All capacities are approximate. Add fluid gradually and check to be sure a proper fluid level is obtained.

71075_ARMA_C0004

FLUID SPECIFICATIONS

Year	Model	Engine Displ. Liters	Engine Oil	Auto. Trans.	Drive Axle Front	Drive Axle Rear	Transfer Case	Power Steering Fluid	Brake Master Cylinder	Cooling System
2011	Armada	5.6	SAE 5W-30	①	②	②	③	④	⑤	⑥
2012	Armada	5.6	SAE 5W-30	①	②	②	③	④	⑤	⑥

① Genuine Nissan Matic S ATF fluid. If not available genuine Nissan Matic J ATF may be used.

② Front: Nissan differential oil hypoid super GL-5 80W-90 or API GL-5 viscosity SAE 80W-90 gear oil.

  Rear: Nissan differential oil API GL-5 synthetic 75W-90 gear oil.

③ Nissan Matic D ATF

④ Nissan Power Steering Fluid

⑤ Nissan Super Heavy Duty DOT 3

⑥ Nissan Long Life antifreeze (BLUE) or equivalent

71075_ARMA_C0013

VALVE SPECIFICATIONS

Year	Engine Displacement Liters	Engine ID	Seat Angle (deg.)	Face Angle (deg.)	Spring Test Pressure (lbs. @ in.)	Spring Installed Height (in.)	Stem-to-Guide Clearance (in.) Intake	Stem-to-Guide Clearance (in.) Exhaust	Stem Diameter (in.) Intake	Stem Diameter (in.) Exhaust
2011	5.6	VK56DE	45.15-45.45	45	37.0@1.457	1.9913	0.0008-0.0021	0.0012-0.0025	0.2348-0.2354	0.2344-0.2350
2012	5.6	VK56DE	45.15-45.45	45	37.0@1.457	1.9913	0.0008-0.0021	0.0012-0.0025	0.2348-0.2354	0.2344-0.2350

71075_ARMA_C0006

CAMSHAFT SPECIFICATIONS

All measurements are given in inches.

Year	Engine Displ. Liters	Engine ID/VIN	Journal Dia.	Brg. Oil Clearance	Shaft End-play	Runout	Journal Bore	Lobe Height	
								Intake	Exhaust
2011	5.6	VK56DE	1.0217-1.0224	0.0012-0.0028	0.0045-0.0074	0.0008	1.0236-1.0244	1.7663-1.7738	1.7746-1.7821
2012	5.6	VK56DE	1.0278-1.0224	0.0012-0.0028	0.0045-0.0074	0.0008	1.0236-1.0244	1.7663-1.7738	1.7746-1.7821

71075_ARMA_C0007

CRANKSHAFT AND CONNECTING ROD SPECIFICATIONS

All measurements are given in inches.

Year	Engine Displ. Liters	Engine ID	Crankshaft				Connecting Rod		
			Main Brg. Journal Dia.	Main Brg. Oil Clearance	Shaft End-play	Thrust on No.	Journal Diameter	Oil Clearance	Side Clearance
2011	5.6	VK56DE	①	②	0.0039-0.0102	3	③	0.0008-0.0015	0.0079-0.0157
2012	5.6	VK56DE	①	②	0.0039-0.0102	3	③	0.0008-0.0015	0.0079-0.0157

① There are 24 different grades, ranging from 2.5182- 2.5174

② No. 1 and 5: 0.00004- 0.0004

  No. 2, 3 and 4: 0.0003- 0.0007

③ There are 13 different grades, ranging from 2.2441- 2.2446. Specification is for rod bearing housing.

71075_ARMA_C0005

PISTON AND RING SPECIFICATIONS

All measurements are given in inches.

Year	Engine Displacement Liters	Engine ID	Piston Clearance	Ring Gap			Ring Side Clearance		
				Top Comp.	Bottom Comp.	Oil Control	Top Comp.	Bottom Comp.	Oil Control
2011	5.6	VK56DE	0.0004-0.0012	0.0091-0.0130	0.0098-0.0157	0.0079-0.0236	0.0014-0.0033	0.0012-0.0028	0.0006-0.0073
2012	5.6	VK56DE	0.0004-0.0012	0.0091-0.0130	0.0098-0.0157	0.0079-0.0236	0.0014-0.0033	0.0012-0.0028	0.0006-0.0073

71075_ARMA_C0008

TORQUE SPECIFICATIONS

All readings in ft. lbs.

Year	Engine Displacement Liters	Engine ID	Cylinder Head Bolts	Main Bearing Bolts	Rod Bearing Bolts	Crankshaft Damper Bolts	Flywheel Bolts	Manifold		Spark Plugs	Oil Pan Drain Plug
								Intake	Exhaust		
2011	5.6	VK56DE	①	②	③	④	65	NA	25	18	25
2012	5.6	VK56DE	①	②	③	④	65	NA	25	18	25

NA: not available

① Step 1: 33 ft. lbs
   Step 2: +70 degrees clockwise
   Step 3: loosen in reverse order of tightening sequence
   Step 4: 33 ft. lbs.
   Step 5: +60 degrees clockwise
   Step 6: +60 degrees clockwise

② Step 1: cap bolts in order 1-10: 29 ft. lbs.
   Step 2: cap sub bolts in order 11-20: 22 ft. lbs.
   Step 3: cap bolts in order 1-10: +40 degrees
   Step 4: cap sub bolts in order 11-20: +30 degrees
   Step 5: side bolts in order 21-30: 36 ft. lbs.

③ Step 1: 11 ft. lbs.
   Step 2: +90 degrees clockwise

④ Step 1: 69 ft. lbs.
   Step 2: +90 degrees

71075_ARMA_C0009

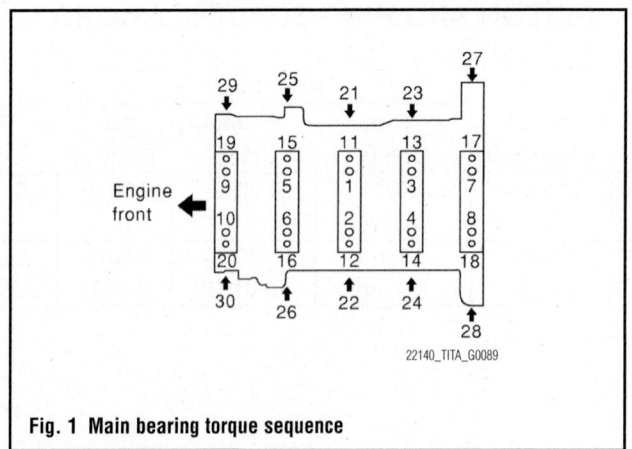

Fig. 1 Main bearing torque sequence

WHEEL ALIGNMENT

Year	Model		Caster Range (+/-Deg.)	Caster Preferred Setting (Deg.)	Camber Range (+/-Deg.)	Camber Preferred Setting (Deg.)	Toe-in (in.)
2011	Armada	2WD	①	①	②	②	NA
		4WD	③	③	④	④	NA
2012	Armada	2WD	①	①	②	②	NA
		4WD	③	③	④	④	NA

NA - Not Available

① Air leveling system: Minimum: 3 degrees 15' (3.25 degrees). Nominal: 4 degrees 0' (4.00 degrees). Maximum: 4 degrees 45' (4.75 degrees).

Except air leveling system: Minimum: 2 degrees 39' (2.65 degrees). Nominal: 3 degrees 24' (3.40 degrees). Maximum: 4 degree 09' (4.15 degrees).

② Minimum: -0 degrees 51' (-0.85 degrees). Nominal: 0 degrees 6' (0.10 degrees). Maximum: 0 degree 39' (0.65 degree).

③ Air leveling system: Minimum: 2 degrees 45' (2.75 degrees). Nominal: 3 degrees 30' (3.50 degrees). Maximum: 4 degrees 15' (4.25 degrees).

Except air leveling system: Minimum: 2 degrees 15' (2.25 degrees). Nominal: 3 degrees 0' (3.00 degrees). Maximum: 3 degrees 45' (3.75 degrees).

④ Minimum: -0 degrees 33' (-0.55 degrees). Nominal: 0 degrees 12' (0.20 degrees). Maximum: 0 degree 57' (0.95 degrees).

71075_ARMA_C0010

TIRE, WHEEL AND BALL JOINT SPECIFICATIONS

Year	Model	OEM Tires		Tire Pressures (psi)		Wheel Size	Ball Joint Inspection	Lug Nut Torque (ft. lbs.)
		Standard	Optional	Front	Rear			
2011	Armada	①	None	35	35	②	③	98
2012	Armada	①	None	35	35	②	③	98

OEM: Original Equipment Manufacturer

PSI: Pounds Per Square Inch

① SV: P265/70R18. SL and Platinum: P275/60R20.

② SV: 18x8JJ. SL and Platinum: 20x8JJ.

③ Axial play

   Upper: 0

71075_ARMA_C0011

BRAKE SPECIFICATIONS

All measurements in inches unless noted

Year	Model		Brake Disc			Minimum Pad Thickness	Brake Caliper	
			Original Thickness	Minimum Thickness	Maximum Runout		Bracket Bolts (ft. lbs.)	Mounting Bolts (ft. lbs.)
2011	Armada	F	1.181	1.102	①	0.039	②	②
		R	0.551	0.472	①	0.039	③	③
2012	Armada	F	1.181	1.102	①	0.039	②	②
		R	0.551	0.472	①	0.039	③	③

① Maximum uneven wear measured at 8 positions: 0.0006. Runout limit, attached to vehicle: Front: 0.001, Rear: 0.002.

② Torque member mounting bolt: 155

   Sliding pin bolt 53

③ Sliding pin bolt: 24

71075_ARMA_C0012

SCHEDULED MAINTENANCE INTERVALS
NISSAN ARMADA

TO BE SERVICED	SERVICE	VEHICLE MILEAGE INTERVAL (x1000)												
		3.75	7.5	15	22.5	30	37.5	45	52.5	60	67.5	75	82.5	90
Engine oil & filter	R	✓	✓	✓	✓	✓	✓	✓	✓	✓	✓	✓	✓	✓
Brake lines & cables	I			✓		✓		✓		✓		✓		✓
Brake pads& rotors	L/I			✓		✓		✓		✓		✓		✓
Driveshaft boots & propeller shaft (4x4)	I					✓				✓				
Automatic transmission, final drive oil & transfer case	I				✓	✓		✓		✓				
LSD gear oil	I			✓		✓		✓		✓		✓		✓
Front wheel bearing grease (4x4)	R					✓				✓				
Air cleaner filter	R					✓				✓				✓
Engine coolant	R									✓				✓
Exhaust system	I						✓	✓	✓	✓				
Spark plugs	R	Replace every 105,000 miles												
Drive belt(s)	I			✓		✓		✓		✓		✓		✓
Cabin air filter	R							✓						✓
Exhaust system	I		✓			✓				✓				✓
Fuel lines	I		✓			✓				✓				
Steering gear (box) & linkage, axle & suspension parts	I					✓				✓				✓
Transfer case	I					✓				✓				✓
Tire rotation			✓	✓	✓	✓	✓	✓	✓	✓	✓	✓	✓	✓
Vapor lines	S/I					✓				✓				✓

R: Replace   S/I: Service or Inspect   L: Lubricate

FREQUENT OPERATION MAINTENANCE (SEVERE SERVICE)

If a vehicle is operated under any of the following conditions it is considered severe service:

- Extremely dusty areas.

- 50% or more of the vehicle operation is in 32°C (90°F) or higher temperatures, constant operation in temp. below 0°C (32°F).

- Prolonged idling (vehicle operation in stop and go traffic).

- Frequent short running periods (engine does not warm to normal operating temperatures).

- Police, taxi, delivery usage or trailer towing usage.

Oil & oil filter: replace every 3750 miles.

Brake pads, discs, drums & linings: service or inspect every 7500 miles.

Driveshaft boots & propeller shaft: service or inspect every 7500 miles.

Exhaust system: service or inspect every 7500 miles.

Final drive oil: Change every 30000 miles if towing a trailer.

Transfer case fluid: Change every 30000 miles if towing a trailer.

Steering gear (box) & linkage, (steering damper-4x4), axle & suspension parts: service or inspect every 7500 miles.

Steering linkage ball joints & front suspension ball joints: service or inspect every 7500 miles.

71075_ARMA_C0014

PRECAUTIONS

Before servicing any vehicle, please be sure to read all of the following precautions, which deal with personal safety, prevention of component damage, and important points to take into consideration when servicing a motor vehicle:

• Never open, service or drain the radiator or cooling system when the engine is hot; serious burns can occur from the steam and hot coolant.

• Observe all applicable safety precautions when working around fuel. Whenever servicing the fuel system, always work in a well-ventilated area. Do not allow fuel spray or vapors to come in contact with a spark, open flame, or excessive heat (a hot drop light, for example). Keep a dry chemical fire extinguisher near the work area. Always keep fuel in a container specifically designed for fuel storage; also, always properly seal fuel containers to avoid the possibility of fire or explosion. Refer to the additional fuel system precautions later in this section.

• Fuel injection systems often remain pressurized, even after the engine has been turned **OFF**. The fuel system pressure must be relieved before disconnecting any fuel lines. Failure to do so may result in fire and/or personal injury.

• Brake fluid often contains polyglycol ethers and polyglycols. Avoid contact with the eyes and wash your hands thoroughly after handling brake fluid. If you do get brake fluid in your eyes, flush your eyes with clean, running water for 15 minutes. If eye irritation persists, or if you have taken brake fluid internally, IMMEDIATELY seek medical assistance.

• The EPA warns that prolonged contact with used engine oil may cause a number of skin disorders, including cancer. You should make every effort to minimize your exposure to used engine oil. Protective gloves should be worn when changing oil. Wash your hands and any other exposed skin areas as soon as possible after exposure to used engine oil. Soap and water, or waterless hand cleaner should be used.

• All new vehicles are now equipped with an air bag system, often referred to as a Supplemental Restraint System (SRS) or Supplemental Inflatable Restraint (SIR) system. The system must be disabled before performing service on or around system components, steering column, instrument panel components, wiring and sensors. Failure to follow safety and disabling procedures could result in accidental air bag deployment, possible personal injury and unnecessary system repairs.

• Always wear safety goggles when working with, or around, the air bag system. When carrying a non-deployed air bag, be sure the bag and trim cover are pointed away from your body. When placing a non-deployed air bag on a work surface, always face the bag and trim cover upward, away from the surface. This will reduce the motion of the module if it is accidentally deployed. Refer to the additional air bag system precautions later in this section.

• Clean, high quality brake fluid from a sealed container is essential to the safe and proper operation of the brake system. You should always buy the correct type of brake fluid for your vehicle. If the brake fluid becomes contaminated, completely flush the system with new fluid. Never reuse any brake fluid. Any brake fluid that is removed from the system should be discarded. Also, do not allow any brake fluid to come in contact with a painted surface; it will damage the paint.

• Never operate the engine without the proper amount and type of engine oil; doing so WILL result in severe engine damage.

• Timing belt maintenance is extremely important. Many models utilize an interference-type, non-freewheeling engine. If the timing belt breaks, the valves in the cylinder head may strike the pistons, causing potentially serious (also time-consuming and expensive) engine damage. Refer to the maintenance interval charts for the recommended replacement interval for the timing belt, and to the timing belt section for belt replacement and inspection.

• Disconnecting the negative battery cable on some vehicles may interfere with the functions of the on-board computer system(s) and may require the computer to undergo a relearning process once the negative battery cable is reconnected.

• When servicing drum brakes, only disassemble and assemble one side at a time, leaving the remaining side intact for reference.

• Only an MVAC-trained, EPA-certified automotive technician should service the air conditioning system or its components.

BRAKES

GENERAL INFORMATION

PRECAUTIONS

• Certain components within the ABS system are not intended to be serviced or repaired individually.

• Do not use rubber hoses or other parts not specifically specified for and ABS system. When using repair kits, replace all parts included in the kit. Partial or incorrect repair may lead to functional problems and require the replacement of components.

• Lubricate rubber parts with clean, fresh brake fluid to ease assembly. Do not use shop air to clean parts; damage to rubber components may result.

• Use only DOT 3 brake fluid from an unopened container.

• If any hydraulic component or line is removed or replaced, it may be necessary to bleed the entire system.

• A clean repair area is essential. Always clean the reservoir and cap thoroughly before removing the cap. The slightest amount of dirt in the fluid may plug an orifice and impair the system function. Perform repairs after components have been thoroughly cleaned; use only denatured alcohol to clean components. Do not allow ABS components to come into contact with any substance containing mineral oil; this includes used shop rags.

• The Anti-Lock control unit is a microprocessor similar to other computer units in the vehicle. Ensure that the ignition switch is **OFF** before removing or installing controller harnesses. Avoid static

ANTI-LOCK BRAKE SYSTEM (ABS)

electricity discharge at or near the controller.

• If any arc welding is to be done on the vehicle, the control unit should be unplugged before welding operations begin.

BLEEDING THE ABS SYSTEM

➡**Be sure that the master cylinder is full of clean fresh brake fluid before starting the bleeding process. Use only the recommended brake fluid when bleeding the system. Do not allow brake fluid to spill on painted surfaces as damage will occur.**

1. Before servicing the vehicle, refer to the Precautions Section.

➥ If working near and/or around the SRS system and components, be sure to disable the SRS system. After disabling the system wait three minutes or more before servicing the vehicle.

➥ Whenever the negative battery cable is disconnected the following components will require resetting. The Idle Air Volume Learning, Steering Angle Sensor Neutral Position, Sunroof Memory Reset/Initialization, Automatic Drive Positioner System, Audio presets and Navigation. Use the CONSULT-III diagnostic tool, or equivalent to perform the required resets.

➥ The automatic back door system must be initialized anytime the battery has been disconnected. Close the back door. Open the back door with the automatic open feature. Do not stop the process until the back door opens completely.

2. Disconnect the negative battery cable.

3. Turn the ignition switch OFF. Disconnect the ABS actuator and electric control unit connector.

4. Connect a vinyl tube to the rear right bleed valve. Be sure to have a catch pan handy to catch excess brake fluid.

5. Fully depress the brake pedal four or five times.

6. With the brake pedal depressed, loosen the bleed valve to let air out, then tighten it immediately.

7. Repeat the above steps until all air is removed from the system. Be sure to keep watch on the brake fluid level and replenish, as necessary.

8. Tighten the bleed valve.

9. Repeat the above steps at each wheel, with the master cylinder reservoir tank filled at least half way.

10. Bleed the remaining components in the following order: front left, rear left and front right.

11. Be sure to perform the reconnect/relearn procedures.

WHEEL SPEED SENSORS

REMOVAL & INSTALLATION

Front

See Figure 2.

1. Before servicing the vehicle, refer to the Precautions Section.

➥ If working near and/or around the SRS system and components, be sure to disable the SRS system. After dis-

abling the system wait three minutes or more before servicing the vehicle.

➥ Whenever the negative battery cable is disconnected the following components will require resetting. The Idle Air Volume Learning, Steering Angle Sensor Neutral Position, Sunroof Memory Reset/Initialization, Automatic Drive Positioner System, Audio presets and Navigation. Use the CONSULT-III diagnostic tool, or equivalent to perform the required resets.

➥ The automatic back door system must be initialized anytime the battery has been disconnected. Close the back door. Open the back door with the automatic open feature. Do not stop the process until the back door opens completely.

2. Disconnect the negative battery cable.

3. Raise and support the vehicle safely.

4. Remove the tire and wheel assembly.

5. Remove the wheel speed sensor mounting screw.

➥ Remove the rotor to gain access to the wheel sensor mounting bolt.

6. Pull the sensor out, being careful to turn it as little as possible. Do not pull on the sensor harness.

7. Disconnect the wheel speed sensor electrical connector.

8. Remove the harness from it mount.

To install:

➥ Be sure to use new fasteners, as required.

9. Inspect the sensor O-ring, replace as required.

10. Before installing the sensor be sure no foreign materials such as iron fragments are adhered to the pick-up part of the sensor or to the inside of the sensor mounting hole or on the rotor mounting surface.

11. Apply a thin coat of a suitable grease to the wheel sensor O-ring and mounting hole.

12. Tighten the sensor retaining bolt to 73 inch lbs.

13. Continue the installation in the reverse order of the removal procedure.

14. Be sure to perform the reconnect/relearn procedures.

Rear

1. Before servicing the vehicle, refer to the Precautions Section.

➥ If working near and/or around the SRS system and components, be sure to disable the SRS system. After disabling the system wait three minutes or more before servicing the vehicle.

➥ Whenever the negative battery cable is disconnected the following components will require resetting. The Idle Air Volume Learning, Steering Angle Sensor Neutral Position, Sunroof Memory Reset/Initialization, Automatic Drive Positioner System, Audio presets and Navigation. Use the CONSULT-III diagnostic tool, or equivalent to perform the required resets.

➥ The automatic back door system must be initialized anytime the battery has been disconnected. Close the back

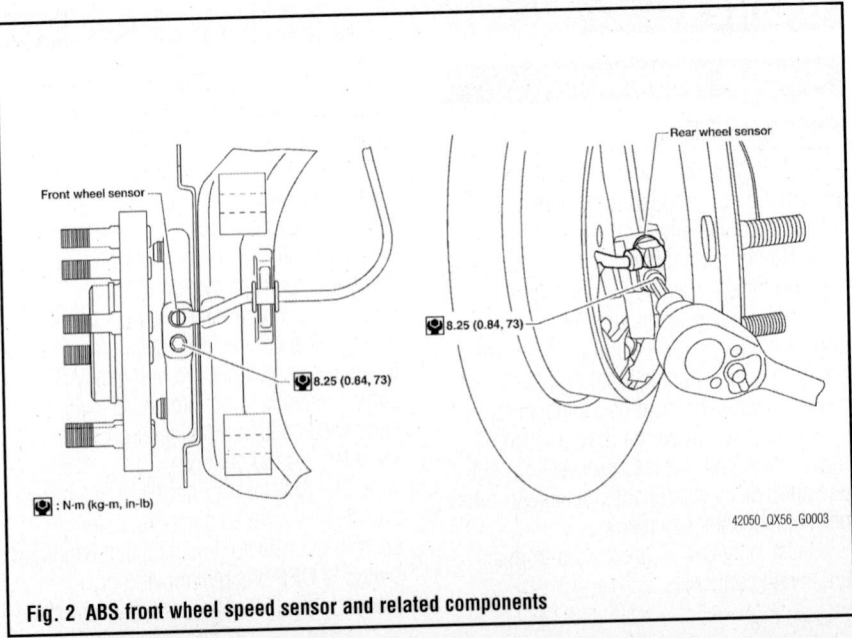

Fig. 2 ABS front wheel speed sensor and related components

door. Open the back door with the automatic open feature. Do not stop the process until the back door opens completely.

2. Disconnect the negative battery cable.

3. Raise and support the vehicle safely.

4. Remove the tire and wheel assembly.

5. Remove the wheel speed sensor mounting screw.

➡**Remove the rear hub and bearing assembly to gain access to the wheel sensor mounting bolt.**

6. Pull the sensor out, being careful to turn it as little as possible. Do not pull on the sensor harness.

7. Disconnect the wheel speed sensor electrical connector.

8. Remove the harness from it mount.

To install:

➡**Be sure to use new fasteners, as required.**

9. Inspect the sensor O-ring, replace as required.

10. Before installing the sensor be sure no foreign materials such as iron fragments are adhered to the pick-up part of the sensor or to the inside of the sensor mounting hole or on the rotor mounting surface.

11. Apply a thin coat of a suitable grease to the wheel sensor O-ring and mounting hole.

12. Tighten the sensor retaining bolt to 73 inch lbs.

13. Continue the installation in the reverse order of the removal procedure.

14. Be sure to perform the reconnect/relearn procedures.

BRAKES

BLEEDING PROCEDURE

MANUAL

See Figures 3 and 4.

➡**Be sure that the master cylinder is full of clean fresh brake fluid before starting the bleeding process. Use only the recommended brake fluid when bleeding the system. Do not allow brake fluid to spill on painted surfaces as damage will occur.**

1. Before servicing the vehicle, refer to the Precautions Section.

➡**If working near and/or around the SRS system and components, be sure to disable the SRS system. After disabling the system wait three minutes or more before servicing the vehicle.**

➡**Whenever the negative battery cable is disconnected the following components will require resetting. The Idle Air Volume Learning, Steering Angle Sensor Neutral Position, Sunroof Memory Reset/Initialization, Automatic Drive Positioner System, Audio presets and Navigation. Use the CONSULT-III**

diagnostic tool, or equivalent to perform the required resets.

➡**The automatic back door system must be initialized anytime the battery has been disconnected. Close the back door. Open the back door with the automatic open feature. Do not stop the process until the back door opens completely.**

2. Disconnect the negative battery cable.

3. Turn the ignition switch OFF. Disconnect the ABS actuator and electric control unit connector.

4. Connect a vinyl tube to the rear right bleed valve. Be sure to have a catch pan handy to catch excess brake fluid.

5. Fully depress the brake pedal four or five times.

6. With the brake pedal depressed, loosen the bleed valve to let air out, then tighten it immediately.

7. Repeat the above steps until all air is removed from the system. Be sure to keep watch on the brake fluid level and replenish, as necessary.

8. Tighten the bleed valve.

9. Repeat the above steps at each wheel, with the master cylinder reservoir tank filled at least half way.

10. Bleed the remaining components in the following order: front left, rear left and front right.

11. Be sure to perform the reconnect/relearn procedures.

FLUID RECOMMENDATIONS

When adding/changing fluid to the brake system be sure to use DOT 3 (US FMVSS No. 116, or equivalent.

LEVEL CHECK

See Figure 5.

1. Before servicing the vehicle, refer to the Precautions Section.

➡**If working near and/or around the SRS system and components, be sure to disable the SRS system. After disabling the system wait three**

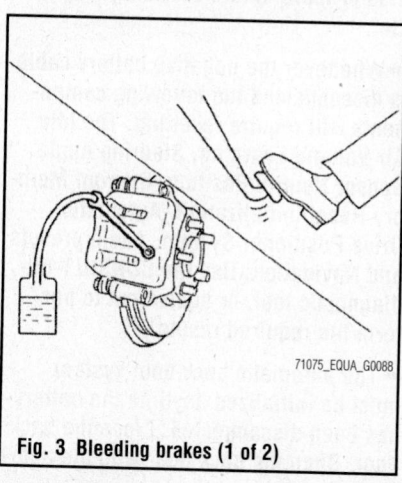

71075_EQUA_G0088

Fig. 3 Bleeding brakes (1 of 2)

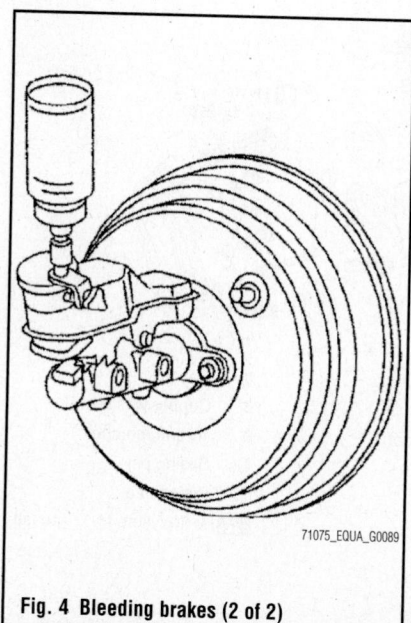

71075_EQUA_G0089

Fig. 4 Bleeding brakes (2 of 2)

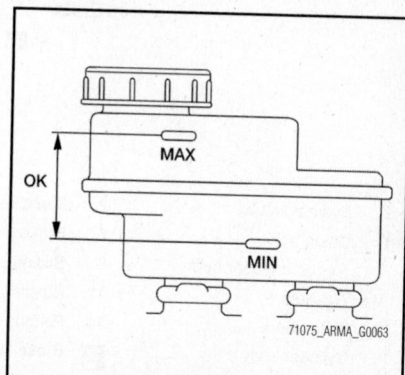

71075_ARMA_G0063

Fig. 5 Fill the brake fluid reservoir tank between the MAX and MIN lines

minutes or more before servicing the vehicle.

➡ Whenever the negative battery cable is disconnected the following components will require resetting. The Idle Air Volume Learning, Steering Angle Sensor Neutral Position, Sunroof Memory Reset/Initialization, Automatic

Drive Positioner System, Audio presets and Navigation. Use the CONSULT-III diagnostic tool, or equivalent to perform the required resets.

➡ The automatic back door system must be initialized anytime the battery has been disconnected. Close the back door. Open the back door with the automatic

open feature. Do not stop the process until the back door opens completely.

2. Check that there is no foreign material in the reservoir tank or around it. Never reuse used fluid.

3. Check that the fluid is within specification.

4. Correct fluid level, as required.

BRAKES

BRAKE CALIPERS

REMOVAL & INSTALLATION

See Figure 6.

1. Before servicing the vehicle, refer to the Precautions Section.

➡ If working near and/or around the SRS system and components, be sure to disable the SRS system. After disabling the system wait three minutes or more before servicing the vehicle.

➡ Whenever the negative battery cable is disconnected the following components will require resetting. The Idle Air Volume Learning, Steering Angle Sensor Neutral Position, Sunroof Memory Reset/Initialization, Automatic

Drive Positioner System, Audio presets and Navigation. Use the CONSULT-III diagnostic tool, or equivalent to perform the required resets.

➡ The automatic back door system must be initialized anytime the battery has been disconnected. Close the back door. Open the back door with the automatic open feature. Do not stop the process until the back door opens completely.

2. Disconnect the negative battery cable.

3. Drain brake fluid as necessary.

4. Raise and safely support the vehicle.

5. Remove or disconnect the following:
- Wheel and tire assembly
- Union bolt

FRONT DISC BRAKES

- Disconnect and plug brake hose. Discard the washer.
- Caliper-to-torque member slide pins, or remove the caliper and torque member as an assembly
- Brake caliper

To install:

➡ Be sure to use new fasteners, as required.

6. Install or connect the following:
- Brake caliper, tighten torque member bolts to specification and the caliper slide pin to specification
- Union bolt and tighten to 13 ft. lbs. (18 Nm)

7. Fill the master cylinder and bleed the brake system.

8. Install the wheel and tire assemblies.

9. Be sure to perform the reconnect/relearn procedures.

BRAKE PADS

REMOVAL & INSTALLATION

See Figures 7 and 8.

1. Before servicing the vehicle, refer to the Precautions Section.

➡ If working near and/or around the SRS system and components, be sure to disable the SRS system. After disabling the system wait three minutes or more before servicing the vehicle.

➡ Whenever the negative battery cable is disconnected the following components will require resetting. The Idle Air Volume Learning, Steering Angle Sensor Neutral Position, Sunroof Memory Reset/Initialization, Automatic Drive Positioner System, Audio presets and Navigation. Use the CONSULT-III diagnostic tool, or equivalent to perform the required resets.

➡ The automatic back door system must be initialized anytime the battery has been disconnected. Close the back door. Open the back door with the auto-

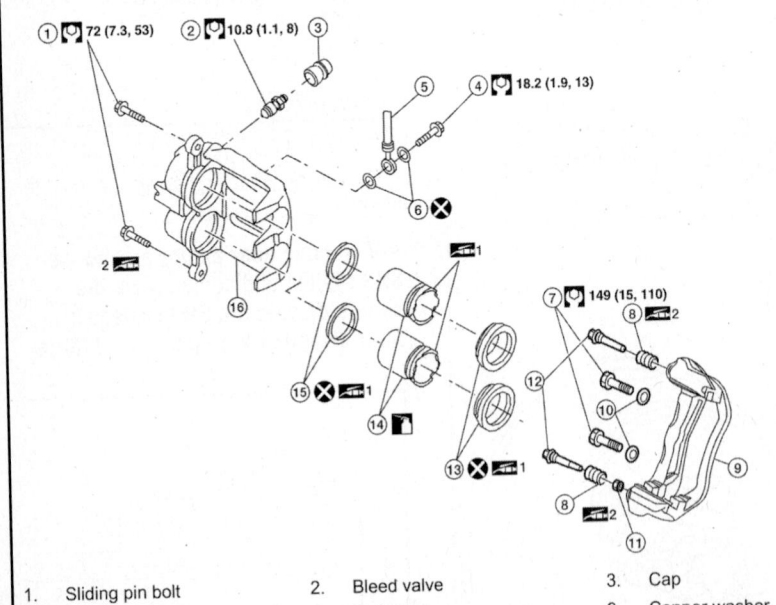

1.	Sliding pin bolt	2.	Bleed valve	3.	Cap
4.	Union bolt	5.	Brake hose	6.	Copper washer
7.	Torque member bolt	8.	Sliding pin boot	9.	Torque member
10.	Washers	11.	Bushing	12.	Sliding pin
13.	Piston boot	14.	Piston	15.	Piston seal
16.	Cylinder body		Brake fluid		1: Molykote M-77 grease

37663_QX56_G0018

Fig. 6 Front brake caliper and related components

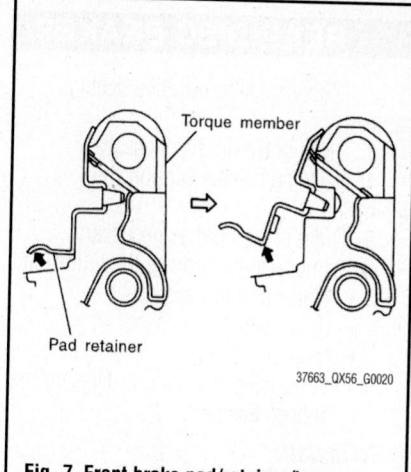

Fig. 7 Front brake pad/retainer/torque member removal

matic open feature. **Do not stop the process until the back door opens completely.**

   2. Disconnect the negative battery cable.

   3. Drain brake fluid as necessary.

   4. Raise and safely support the vehicle.

   5. Remove the wheel and tire assembly.

   6. Remove lower sliding pin bolt.

   7. Suspend brake caliper with a remove and remove brake pad and shim from torque member.

➡**When removing the pad retainer from the torque member, lift it in the direction indicated by the arrow in the illustration.**

To install:

➡**Be sure to use new fasteners, as required.**

   8. Push pistons in so that the pad is firmly installed, using a suitable tool.

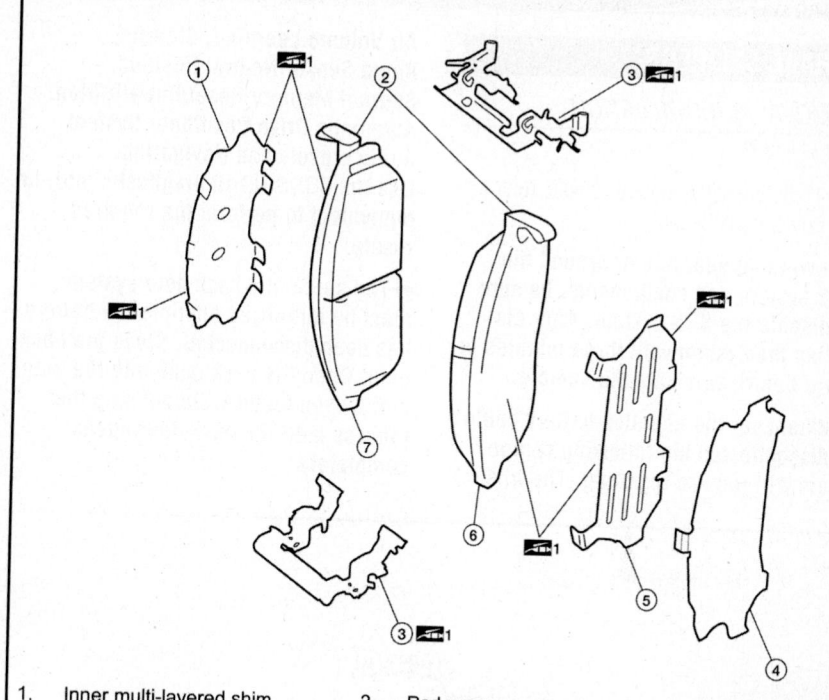

1.	Inner multi-layered shim	2.	Pad wear sensor	3.	Pad retainer
4.	Outer shim cover	5.	Outer shim	6.	Outer pad
7.	Inner pad				

1: Molykote M-77 grease

Fig. 8 Front brake pads and related components

   9. Mount the brake caliper to torque member.

   10. Attach pad retainer to torque member.

➡**Securely assemble the pad retainers so that they are not being lifted up from the torque member. Both inner and outer pads have a pad return system on** the pad retainer. **Install the pad return lever securely to the pad wear sensor.**

   11. Lubricate lower sliding pin bolt with a thin layer of silicone grease and install. Torque to specification.

   12. Install the wheel and tire assembly.

   13. Be sure to perform the reconnect/relearn procedures.

BRAKES

BRAKE CALIPERS

REMOVAL & INSTALLATION
See Figure 9.

1. Before servicing the vehicle, refer to the Precautions Section.

➡ If working near and/or around the SRS system and components, be sure to disable the SRS system. After disabling the system wait three minutes or more before servicing the vehicle.

➡ Whenever the negative battery cable is disconnected the following components will require resetting. The Idle Air Volume Learning, Steering Angle Sensor Neutral Position, Sunroof Memory Reset/Initialization, Automatic Drive Positioner System, Audio presets and Navigation. Use the CONSULT-III diagnostic tool, or equivalent to perform the required resets.

➡ The automatic back door system must be initialized anytime the battery has been disconnected. Close the back door. Open the back door with the automatic open feature. Do not stop the process until the back door opens completely.

2. Disconnect the negative battery cable.
3. Drain brake fluid as necessary.
4. Raise and safely support the vehicle.
5. Drain brake fluid as necessary.
6. Remove or disconnect the following:
   - Wheel and tire assembly
   - Union bolt
   - Mounting bolts
   - Brake caliper assembly. Discard the copper washers.

To install:

➡ Be sure to use new fasteners, as required.

7. Install or connect the following:
   - Brake caliper assembly and tighten mounting bolts to specification
   - Union bolt and tighten to 13 ft. lbs. (18 Nm)
8. Fill the master cylinder and bleed the brake system.
9. Install the wheel and tire assemblies.
10. Be sure to perform the reconnect/relearn procedures.

BRAKE PADS

REMOVAL & INSTALLATION
See Figure 10.

1. Before servicing the vehicle, refer to the Precautions Section.

➡ If working near and/or around the SRS system and components, be sure to disable the SRS system. After disabling the system wait three minutes or more before servicing the vehicle.

➡ Whenever the negative battery cable is disconnected the following components will require resetting. The Idle Air Volume Learning, Steering Angle Sensor Neutral Position, Sunroof Memory Reset/Initialization, Automatic Drive Positioner System, Audio presets and Navigation. Use the CONSULT-III diagnostic tool, or equivalent to perform the required resets.

➡ The automatic back door system must be initialized anytime the battery has been disconnected. Close the back door. Open the back door with the automatic open feature. Do not stop the process until the back door opens completely.

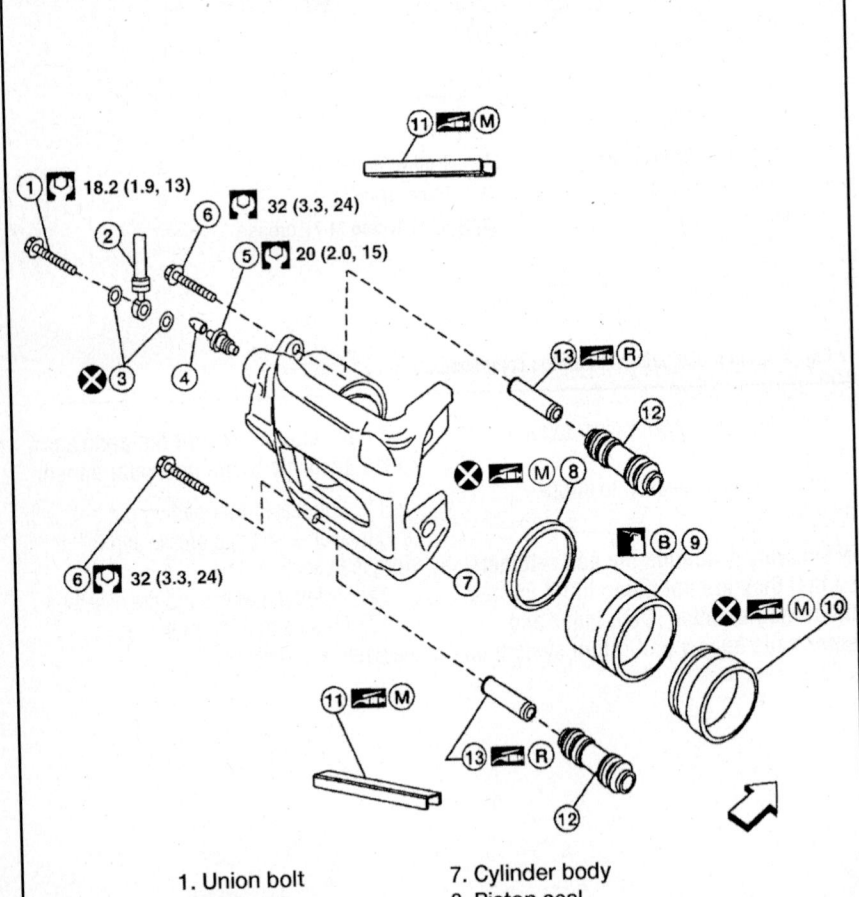

1. Union bolt
2. Brake hose
3. Washer
4. Cap
5. Bleed valve
6. Sliding pin bolt
7. Cylinder body
8. Piston seal
9. Piston
10. Piston boot
11. Knuckle side
12. Sliding sleeve bolt

37663_QX56_G0021

Fig. 9 Rear brake caliper and related components

2. Disconnect the negative battery cable.

3. Drain brake fluid as necessary.

4. Raise and safely support the vehicle.

5. Remove the wheel and tire assembly.

6. Remove the sliding sleeves and pin bolts from the cylinder body

7. Support the cylinder body, with mechanics wire. Remove the pads, shims, cover and retainer.

To install:

8. Push pistons in so that the pad is firmly installed, using a suitable tool.

9. Install pads to the brake caliper.

10. Install top mounting bolt and tighten to specification.

11. Install the wheel and tire assembly.

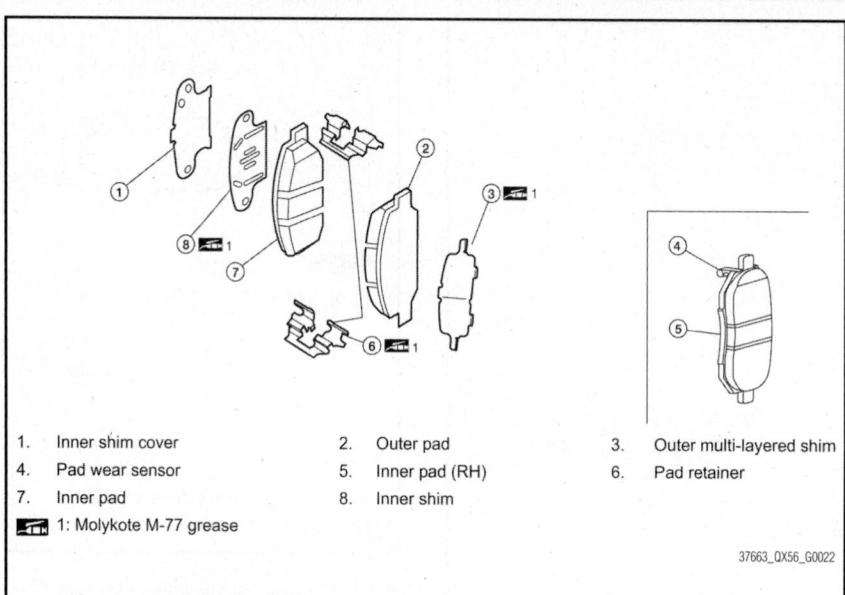

1.	Inner shim cover	2.	Outer pad	3.	Outer multi-layered shim
4.	Pad wear sensor	5.	Inner pad (RH)	6.	Pad retainer
7.	Inner pad	8.	Inner shim		

1: Molykote M-77 grease

37663_QX56_G0022

Fig. 10 Rear brake pads and related components

BRAKES

ADJUSTMENTS

CABLES

1. Before servicing the vehicle, refer to the Precautions Section.

➡ **If working near and/or around the SRS system and components, be sure to disable the SRS system. After disabling the system wait three minutes or more before servicing the vehicle.**

➡ **Whenever the negative battery cable is disconnected the following components will require resetting. The Idle Air Volume Learning, Steering Angle Sensor Neutral Position, Sunroof Memory Reset/Initialization, Automatic Drive Positioner System, Audio presets and Navigation. Use the CONSULT-III diagnostic tool, or equivalent to perform the required resets.**

➡ **The automatic back door system must be initialized anytime the battery has been disconnected. Close the back door. Open the back door with the automatic open feature. Do not stop the process until the back door opens completely.**

2. Disconnect the negative battery cable.

3. Remove the lower instrument panel, driver's side.

4. Partially engage the parking brake pedal to access the adjusting nut.

5. Insert a deep socket wrench to rotate the adjusting nut and loosen the cable sufficiently.

6. Disengage the parking brake pedal.

7. Raise and support the vehicle safely.

8. Remove the tire and wheel assembly.

9. Remove the rotor. Measure the inner diameter at the widest point using tool J-21177A or equivalent.

10. Transfer the recorded measurement less 0.6 mm to the parking brake shoes and adjust accordingly.

11. Using wheel nuts, secure the rotor to the hub to prevent it from tilting.

12. Rotate the rotor to make sure that there is no drag.

13. To adjust the cable operate the pedal ten or more times with a force of 110 lbs.

14. Rotate the adjusting nut with a deep socket to adjust the pedal stroke to specification. Specification is 3–4 notches with a force of 44.1 lbs.

15. With the parking brake pedal completely disengaged, make sure there is no drag on the parking brake.

16. Reassemble and reinstall any removed components.

PARKING BRAKE SHOES

BURNISHING

Perform the parking brake burnishing operation by driving the vehicle forward

PARKING BRAKE

under the following conditions: vehicle speed 25 mph forward direction, parking brake operating force 44.1 lbs set and apply time of 30 seconds. After parking brake burnishing operation, recheck parking brake adjustment, correct as required.

➡ **To prevent the brake lining from getting too hot, allow a cool off period of five minutes between operations. Do not perform excessive break-in operations, because it may cause uneven or early wear of the lining.**

REMOVAL & INSTALLATION

See Figures 11 and 12.

1. Before servicing the vehicle, refer to the Precautions Section.

➡ **If working near and/or around the SRS system and components, be sure to disable the SRS system. After disabling the system wait three minutes or more before servicing the vehicle.**

➡ **Whenever the negative battery cable is disconnected the following components will require resetting. The Idle Air Volume Learning, Steering Angle Sensor Neutral Position, Sunroof Memory Reset/Initialization, Automatic Drive Positioner System, Audio presets and Navigation. Use the CONSULT-III diagnostic tool, or equivalent to perform the required resets.**

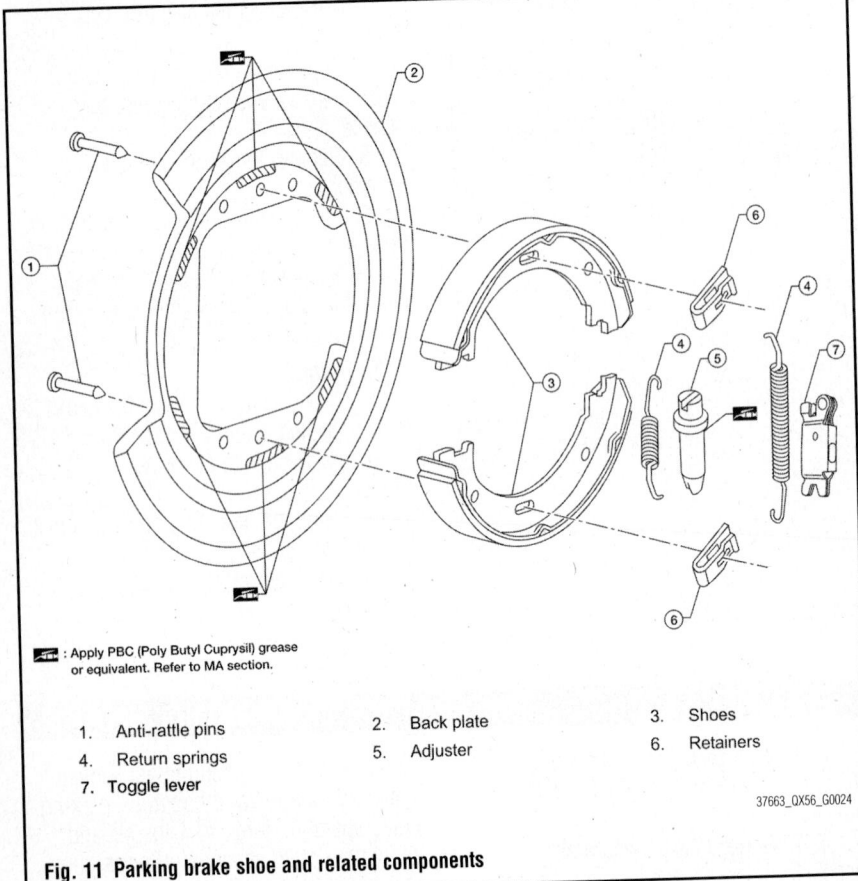

: Apply PBC (Poly Butyl Cuprysil) grease or equivalent. Refer to MA section.

1. Anti-rattle pins
4. Return springs
7. Toggle lever

2. Back plate
5. Adjuster

3. Shoes
6. Retainers

37663_QX56_G0024

Fig. 11 Parking brake shoe and related components

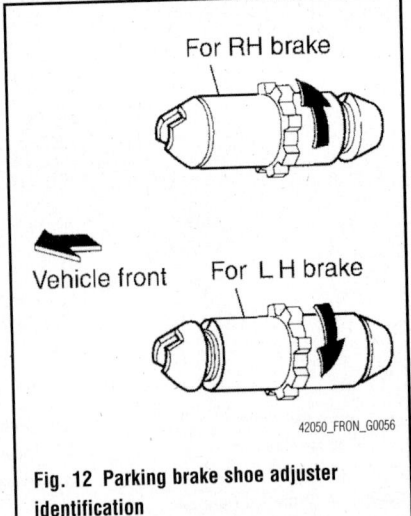

For RH brake

Vehicle front

For L H brake

42050_FRON_G0056

Fig. 12 Parking brake shoe adjuster identification

➡The automatic back door system must be initialized anytime the battery has been disconnected. Close the back door. Open the back door with the automatic open feature. Do not stop the process until the back door opens completely.

2. Disconnect the negative battery cable.

3. Remove the tire and wheel assembly.
4. Be sure that the parking brake lever is in the released position.
5. Remove the rear disc rotor.
6. Remove the return springs.
7. Remove the adjuster.
8. Disconnect the parking brake cable from the toggle lever.
9. Remove the retainers.

10. Remove the anti-rattle pins and shoes.

To install:

➡Be sure to use new fasteners, as required.

11. Apply brake grease to the specified points during reassembly, see illustration for locating points.
12. Assemble the adjuster so that the threaded part expands when rotating it in the direction shown by the arrow. Shorten the adjuster by rotating it in the opposite direction shown by the arrow.
13. Continue the installation in the reverse order of the removal procedure.
14. Adjust the parking brake.
15. Perform the parking brake burnishing operation.
16. Be sure to perform the reconnect/relearn procedures.

CHASSIS ELECTRICAL

AIR BAGS (SUPPLEMENTAL RESTRAINT SYSTEM)

PRECAUTIONS

Disconnect and isolate the battery negative cable before beginning any airbag system component diagnosis, testing, removal, or installation procedures. Allow system capacitor to discharge for two minutes before beginning any component service. This will disable the airbag system. Failure to disable the airbag system may result in accidental airbag deployment, personal injury, or death.

Do not place an intact undeployed airbag face down on a solid surface. The airbag will propel into the air if accidentally deployed and may result in personal injury or death.

When carrying or handling an undeployed airbag, the trim side (face) of the airbag should be pointing towards the body to minimize possibility of injury if accidental deployment occurs. Failure to do this may result in personal injury or death.

Replace airbag system components with OEM replacement parts. Substitute parts may appear interchangeable, but internal differences may result in inferior occupant protection. Failure to do so may result in occupant personal injury or death.

Wear safety glasses, rubber gloves, and long sleeved clothing when cleaning powder residue from vehicle after an airbag deployment. Powder residue emitted from a deployed airbag can cause skin irritation. Flush affected area with cool water if irritation

is experienced. If nasal or throat irritation is experienced, exit the vehicle for fresh air until the irritation ceases. If irritation continues, see a physician.

Do not use a replacement airbag that is not in the original packaging. This may result in improper deployment, personal injury, or death.

The factory installed fasteners, screws and bolts used to fasten airbag components have a special coating and are specifically designed for the airbag system. Do not use substitute fasteners. Use only original equipment fasteners listed in the parts catalog when fastener replacement is required.

During, and following, any child restraint anchor service, due to impact event or vehicle repair, carefully inspect all mounting hardware, tether straps, and anchors for proper installation, operation, or damage. If a child restraint anchor is found damaged in any way, the anchor must be replaced. Failure to do this may result in personal injury or death.

Deployed and non-deployed airbags may or may not have live pyrotechnic material within the airbag inflator.

Do not dispose of driver/passenger/curtain airbags or seat belt tensioners unless you are sure of complete deployment. Refer to the Hazardous Substance Control System for proper disposal.

Dispose of deployed airbags and tensioners consistent with state, provincial, local, and federal regulations.

After any airbag component testing or service, do not connect the battery negative cable. Personal injury or death may result if the system test is not performed first.

If the vehicle is equipped with the Occupant Classification System (OCS), do not connect the battery negative cable before performing the OCS Verification Test using the scan tool and the appropriate diagnostic information. Personal injury or death may result if the system test is not performed properly.

Never replace both the Occupant Restraint Controller (ORC) and the Occupant Classification Module (OCM) at the same time. If both require replacement, replace one, then perform the Airbag System test before replacing the other.

Both the ORC and the OCM store Occupant Classification System (OCS) calibration data, which they transfer to one another when one of them is replaced. If both are replaced at the same time, an irreversible fault will be set in both modules and the OCS may malfunction and cause personal injury or death.

If equipped with OCS, the Seat Weight Sensor is a sensitive, calibrated unit and must be handled carefully. Do not drop or handle roughly. If dropped or damaged, replace with another sensor. Failure to do so may result in occupant injury or death.

If equipped with OCS, the front passenger seat must be handled carefully as well. When removing the seat, be careful when setting on floor not to drop. If dropped, the sensor may be inoperative, could result in occupant injury, or possibly death.

If equipped with OCS, when the passenger front seat is on the floor, no one should sit in the front passenger seat. This uneven force may damage the sensing ability of the seat weight sensors. If sat on and damaged, the sensor may be inoperative, could result in occupant injury, or possibly death.

The Supplemental Restraint System (SRS) such as AIR BAG and SEAT BELT PRE-TENSIONER, used along with a front seat belt, helps to reduce the risk or severity of injury to the driver and front passenger for certain types of collision. This system includes seat belt switch inputs and dual stage front air bag modules. The SRS system uses the seat belt switches to determine the front air bag deployment, and may only deploy one front air bag, depending on the severity of a collision and whether the front occupants are belted or unbelted.

✳✳ CAUTION

Improper maintenance, including incorrect removal and installation of the SRS, can lead to personal injury caused by unintentional activation of the system. Do not use electrical test equipment on any circuit related to the SRS unless instructed. SRS wiring harnesses can be identified by yellow and/or orange harnesses or harness connectors.

✳✳ CAUTION

When working near the Airbag Diagnosis Sensor Unit or other Airbag System sensors with the Ignition ON or engine running, DO NOT use air or electric power tools or strike near the sensor(s) with a hammer. Heavy vibration could activate the sensor(s) and deploy the air bag(s), possibly causing serious injury. When using air or electric power tools or hammers, always switch the Ignition OFF, disconnect the battery, and wait at least 3 minutes before performing any service.

✳✳ CAUTION

Before removing the seat belt pre-tensioner assembly, turn the ignition switch OFF, disconnect both battery terminals and wait at least three minutes. For approximately three minutes after the battery terminals have been removed, it is still possible for the air bag and seat belt pretensioner to deploy. Therefore, do not attempt work on any SRS connectors or wires until at least three minutes have passed. After replacing or reinstalling seat belt pre-tensioner assembly, or reconnecting seat belt pre-tensioner assembly connector, make sure entire SRS operates properly.

➡Do not disassemble buckle or seat belt assembly. Replace anchor bolts if they are deformed or worn out. Never oil tongue and buckle. If any component of seat belt assembly is questionable, do not repair. Replace the whole seat belt assembly. If webbing is cut, frayed, or damaged, replace seat belt assembly. When replacing seat belt assembly, use a genuine Nissan seat belt assembly.

➡After a collision inspect all seat belt assemblies including retractors and attaching hardware after any collision. Nissan recommends that all seat belt assemblies in use during a collision be replaced unless the collision was minor and the belts show no damage and continue to operate properly. Failure to do so could result in serious personal injury in an accident. Seat belt assemblies not in use during a collision should also be replaced if either damage or improper operation is noted. Seat belt pre-tensioner should be replaced even if the seat belts are not in use during a frontal collision in which the air bags are deployed. Replace any seat belt assembly (including anchor bolts) if the seat belt was in use at the time of a collision (except for minor collisions and the belts, retractors and buckles show no damage and continue to operate properly). The seat belt was damaged in an accident. (i.e., torn webbing, bent retractor or guide, etc.). The seat belt attaching point was damaged in an accident. Inspect the seat belt attaching area for damage or distortion and repair as necessary before installing a new seat belt assembly. Anchor bolts are deformed or worn out. The seat belt pre-tensioner should be replaced even if the seat belts are not in use during the collision in which the air bags are deployed.

➡Replace occupant classification system control unit and passenger front seat cushion as an assembly.

➡If the steering wheel was rotated after battery disconnect, perform the following: This Procedure is applied only to models with Intelligent Key system and NATS (NISSAN ANTI-THEFT SYSTEM). Remove and install all control units after disconnecting both battery cables with the ignition knob in the _LOCK_ position. Always use CONSULT to perform self-diagnosis as a part of each function inspection after finishing work. If DTC is detected, perform trouble diagnosis according to self-diagnostic results. For models equipped with the Intelligent Key system and NATS, an electrically controlled steering lock mechanism is adopted on the key cylinder. For this reason, if the battery is disconnected or if the battery is discharged, the steering wheel will lock and steering wheel rotation will become impossible. If steering wheel rotation is required when battery power is interrupted, follow the procedure below before starting the repair operation.

RESET

1. Connect both battery cables.
2. Use the Intelligent Key or mechanical key to turn the ignition switch to the ACC position. At this time, the steering lock will be released.
3. Disconnect both battery cables. The steering lock will remain released and the steering wheel can be rotated.
4. Perform the necessary repair operation.
5. When the repair work is completed, return the ignition switch to the LOCK position before connecting the battery cables. (At this time, the steering lock mechanism will engage).
6. Perform a self-diagnosis check of all control units using CONSULT III diagnostic tool, or equivalent.

DISARMING THE SYSTEM

➡Whenever the negative battery cable is disconnected the following components will require resetting. The Idle Air Volume Learning, Steering Angle Sensor Neutral Position, Sunroof Memory Reset/Initialization, Automatic Drive Positioner System, Audio presets and Navigation. Use the CONSULT-III diagnostic tool, or equivalent to perform the required resets.

➡The automatic back door system must be initialized anytime the battery has been disconnected. Close the back door. Open the back door with the automatic open feature. Do not stop the process until the back door opens completely.

1. Before servicing the vehicle, refer to the Precautions Section.
2. Disconnect the negative battery cable.
3. Disconnect the positive battery cable.
4. Wait at least 3 minutes before working on the vehicle. The air bag system is

designed to retain enough power to deploy the air bag for a short time after the battery has been disconnected.

ARMING THE SYSTEM

➡Once repair work has been completed, return the ignition switch to the LOCK position, before connecting the battery cables. At this time the steering lock mechanism will engage. Install the CONSULT-III diagnostic tool, or equivalent, and follow the directions on the screen of the tool and perform the self-diagnosis check.

CENTERING THE CLOCKSPRING

See Figure 13.

➡The Nissan CONSULT III diagnostic tool, or equivalent is required to perform this procedure.

➡If working near and/or around the SRS system and components, be sure to disable the SRS system. After disabling the system wait three minutes or more before servicing the vehicle.

➡Whenever the negative battery cable is disconnected the following components will require resetting. The Idle Air Volume Learning, Steering Angle Sensor Neutral Position, Sunroof Memory Reset/Initialization, Automatic Drive Positioner System, Audio presets and Navigation. Use the Nissan CONSULT III diagnostic tool, or equivalent to perform the required resets.

➡The automatic back door system must be initialized anytime the battery has been disconnected. Close the back door. Open the back door with the automatic open feature. Do not stop the process until the back door opens completely.

1. Before servicing the vehicle, refer to the Precautions Section.
2. Disconnect the negative battery cable.
3. Refer to the adjustment table to determine if adjustment of the steering angle sensor neutral position is required.

➡To adjust the neutral position of the steering angle sensor, make sure to use the Nissan CONSULT III diagnostic tool, or equivalent. Adjustment cannot be done without a diagnostic tool.

4. Stop the vehicle with the front wheels in the straight-ahead position.
5. On the TOOL screen, touch UTILITY and ST ANG SEN ADJUSTMENT in order.
6. Touch START.

➡Do not touch the steering wheel while adjusting the steering angle sensor.

7. After approximately 10 seconds, touch END.

➡After approximately 60 seconds, the program ends automatically.

8. Turn the ignition switch OFF, then turn it ON again.
9. Run the vehicle with the front wheels in the straight-ahead position, then stop.
10. Select DATA LIST. Then make sure STR ANGLE SIG is within 0 plus or minus 2.5°.
11. Erase the self-diagnosis memory of the ABS actuator and electric unit (control unit) and ECM.
12. After erasing DTC memory, start the engine and drive the vehicle at 19 MPH (30 km/h) or more for approximately 1 minute as the final inspection, and make sure that the ABS warning lamp, VDC OFF indicator lamp, SLIP indicator lamp and brake warning lamp turn OFF.

x: Required −: Not required

Situation	Adjustment of steering angle sensor neutral position
Removing/Installing ABS actuator and electric unit (control unit)	−
Replacing ABS actuator and electric unit (control unit)	x
Removing/Installing steering angle sensor	x
Replacing steering angle sensor	x
Removing/Installing steering components	x
Replacing steering components	x
Removing/Installing suspension components	x
Replacing suspension components	x
Change tires to new ones	−
Tire rotation	−
Adjusting wheel alignment	x
Battery disconnection	x

37671_EQUA_G0032

Fig. 13 Steering angle sensor neutral position adjustment requirement table

DRIVETRAIN

AUTOMATIC TRANSMISSION

DRAIN & REFILL

See Figure 15

1. Before servicing the vehicle, refer to the Precautions Section.

➡ **If working near and/or around the SRS system and components, be sure to disable the SRS system. After disabling the system wait three minutes or more before servicing the vehicle.**

➡ **Whenever the negative battery cable is disconnected the following components will require resetting. The Idle Air Volume Learning, Steering Angle Sensor Neutral Position, Sunroof Memory Reset/Initialization, Automatic Drive Positioner System, Audio presets and Navigation. Use the CONSULT-III diagnostic tool, or equivalent to perform the required resets.**

➡ **The automatic back door system must be initialized anytime the battery has been disconnected. Close the back door. Open the back door with the automatic open feature. Do not stop the process until the back door opens completely.**

2. Disconnect the negative battery cable.
3. Run the engine until operating temperature is reached.
4. Stop the engine.
5. Remove the fluid level gauge.
6. Raise and support the vehicle safely.
7. Loosen the drain plug. Discard the gasket.

➡ **When replacing the drain plug use a new gasket and tighten to 25 ft. lbs. (34 Nm).**

8. Drain the fluid from the drain plug and refill with new fluid. Always refill same volume with drained fluid.

➡ **To flush the old fluid from the oil coolers, pour new fluid into the charging pipe with the engine idling and at the same time drain the old fluid from the auxiliary cooler hose return line. When the color of the fluid coming out is about the same color as new fluid, the procedure is complete. The amount of new fluid used for flushing should be about 30–50 percent increase of the specified capacity.**

9. Install the fluid level gauge.
10. Drive the vehicle until operating temperature is reached.
11. Check and correct the fluid level.
12. If the fluid is still dirty, repeat the process.
13. Install the removed fluid level gauge in the fluid charging pipe.
14. Install the fluid level gauge bolt. Tighten to 45 inch lbs. (5.1 Nm).
15. Be sure to perform the reconnect/relearn procedures.

FLUID LEVEL CHECK

See Figures 14 and 15.

1. Before servicing the vehicle, refer to the Precautions Section.

➡ **If working near and/or around the SRS system and components, be sure to disable the SRS system. After disabling the system wait three minutes or more before servicing the vehicle.**

➡ **Whenever the negative battery cable is disconnected the following components will require resetting. The Idle Air Volume Learning, Steering Angle Sensor Neutral Position, Sunroof Memory Reset/Initialization, Automatic Drive Positioner System, Audio presets and Navigation. Use the CONSULT-III diagnostic tool, or equivalent to perform the required resets.**

➡ **The automatic back door system must be initialized anytime the battery has been disconnected. Close the back door. Open the back door with the**

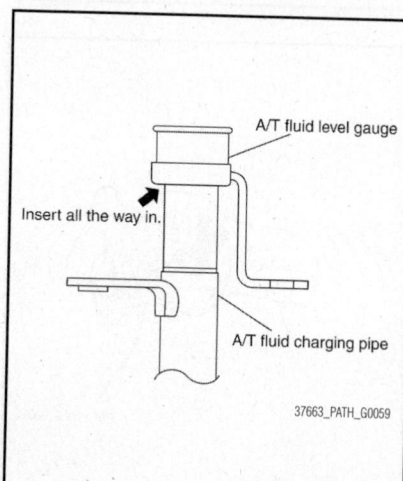

Fig. 14 Automatic transmission fluid level gauge location and related components

automatic open feature. Do not stop the process until the back door opens completely.

2. Run the engine until operating temperature is reached.
3. Check for external fluid leakage.
4. Loosen the level gauge bolt.
5. Before driving, fluid level can be checked at fluid temperatures of 86–122 degrees F using the COLD range on the dipstick.
6. To COLD check the transmission, park the vehicle on a level surface and set the parking brake.
7. Start the engine and move the selector lever through each gear range. Leave the selector lever in the PARK (P) range.
8. Check the fluid level with the engine idling. Always use lint free paper, not a cloth to check the fluid level.
9. Re-insert the gauge. Remove the gauge and note the reading. As required add fluid. Do not overfill.
10. Drive the vehicle for about five minutes in urban areas.
11. Using the CONSULT-III tool, or equivalent, make the fluid temperature about 176 degrees F.
12. Recheck the fluid level, correct as required.

FLUID RECOMMENDATIONS

The automatic transmission uses Nissan approved Matic S ATF transmission fluid. If this fluid is not available Nissan Matic J ATF may be used. If information differs from the owner's manual, use the data in the owner's manual.

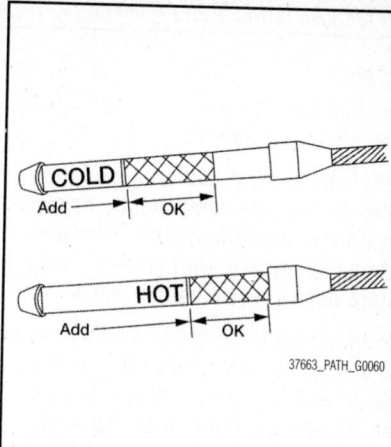

Fig. 15 Automatic transmission fluid dipstick markings

DRIVESHAFT

REMOVAL & INSTALLATION

Rear

1. Before servicing the vehicle, refer to the Precautions Section.

➡ **If working near and/or around the SRS system and components, be sure to disable the SRS system. After disabling the system wait three minutes or more before servicing the vehicle.**

➡ **Whenever the negative battery cable is disconnected the following components will require resetting. The Idle Air Volume Learning, Steering Angle Sensor Neutral Position, Sunroof Memory Reset/Initialization, Automatic Drive Positioner System, Audio presets and Navigation. Use the CONSULT-III diagnostic tool, or equivalent to perform the required resets.**

➡ **The automatic back door system must be initialized anytime the battery has been disconnected. Close the back door. Open the back door with the automatic open feature. Do not stop the process until the back door opens completely.**

2. Disconnect the negative battery cable.

3. Position the selector lever in the N position.

4. Release the parking brake.

5. Raise and support the vehicle safely. Remove the engine under cover, if equipped and as required.

6. Matchmark the driveshaft and companion flange.

7. Remove the driveshaft. Discard the nuts.

To install:

➡ **Be sure to use new fasteners, as required.**

8. Installation is the reverse of the removal procedure.

9. Check for vehicle vibration, correct as required.

10. Be sure to perform the reconnect/relearn procedures.

DRIVESHAFT ANGLE MEASUREMENT

See Figures 16 and 17.

1. Before servicing the vehicle, refer to the Precautions Section.

➡ **If working near and/or around the SRS system and components, be sure**

to disable the SRS system. After disabling the system wait three minutes or more before servicing the vehicle.

➡ **Whenever the negative battery cable is disconnected the following components will require resetting. The Idle Air Volume Learning, Steering Angle Sensor Neutral Position, Sunroof Memory Reset/Initialization, Automatic Drive Positioner System, Audio presets and Navigation. Use the CONSULT-III diagnostic tool, or equivalent to perform the required resets.**

➡ **The automatic back door system must be initialized anytime the battery has been disconnected. Close the back door. Open the back door with the automatic open feature. Do not stop the process until the back door opens completely.**

2. Disconnect the negative battery cable.

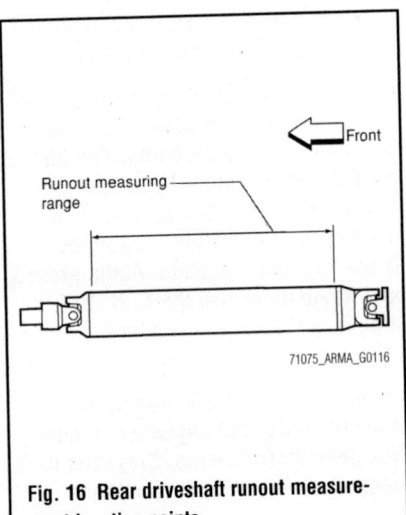

Fig. 16 Rear driveshaft runout measurement location points

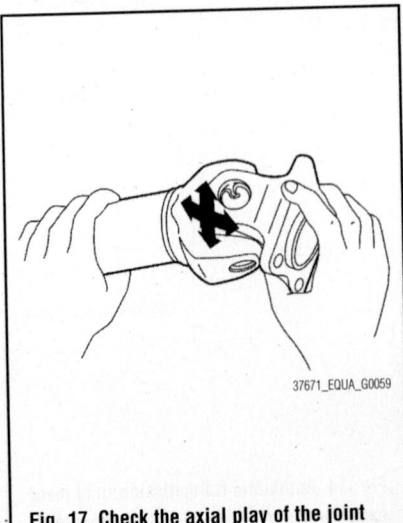

Fig. 17 Check the axial play of the joint

3. Raise and support the vehicle safely.

4. Matchmark the driveshaft and companion flange.

5. Remove the driveshaft. Discard the nuts.

6. Inspect the driveshaft runout, replace the driveshaft as necessary. See illustration. Runout specification should be 0.024 inch (0.60mm).

7. While holding the flange yoke on one side, check the axial play of the joint. See illustration. If the journal axial play exceeds 0.0008 inch (0.02mm), repair or replace the journal parts.

8. Check the driveshaft for dents or cracks. If damage is detected, replace the driveshaft assembly.

To install:

➡ **Be sure to use new fasteners, as required.**

9. Installation is the reverse of the removal procedure.

10. Check for vehicle vibration, correct as required.

❊❊ WARNING

Do not reuse the bolts and nuts. Always install new ones.

11. Be sure to perform the reconnect/relearn procedures.

FRONT AXLE HOUSING (ASSEMBLY, UNIT)

FLUID RECOMMENDATIONS

Be sure to use the proper grade and type fluid when servicing this component. Use API GL-5 viscosity SAE 80W-90 gear oil, or equivalent when servicing/refilling. Approximate capacity is 3 3/8 pints (1.6L).

LEVEL CHECK

See Figure 18.

1. Before servicing the vehicle, refer to the Precautions Section.

➡ **If working near and/or around the SRS system and components, be sure to disable the SRS system. After disabling the system wait three minutes or more before servicing the vehicle.**

➡ **Whenever the negative battery cable is disconnected the following components will require resetting. The Idle Air Volume Learning, Steering Angle Sensor Neutral Position, Sunroof Memory Reset/Initialization, Automatic Drive Positioner System, Audio presets and Navigation. Use the CONSULT-III**

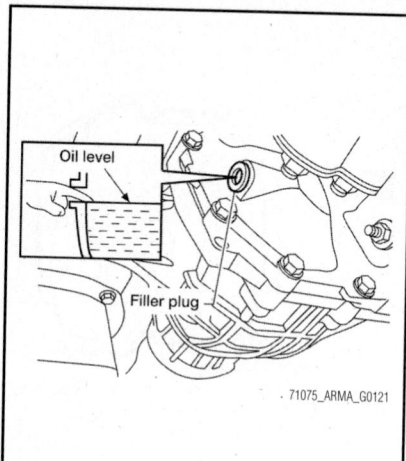

**Fig. 18 Front drive axle plug location—
checking**

diagnostic tool, or equivalent to perform the required resets.

➡The automatic back door system must
be initialized anytime the battery has
been disconnected. Close the back door.
Open the back door with the automatic
open feature. Do not stop the process
until the back door opens completely.

   2. Disconnect the negative battery cable.
   3. Raise and support the vehicle safely,
as required.
   4. Position a catch pan under the
assembly.
   5. Remove the filler plug. Discard the
gasket.
   6. Check for proper fluid level, see illustration.
   7. Correct fluid level, as required.
   8. Install the filler plug, using a new
gasket.
   9. Tighten filler plug to 27 ft. lbs. (36 Nm).

DRAIN & REFILL

   1. Before servicing the vehicle, refer to
the Precautions Section.

➡If working near and/or around the
SRS system and components, be sure
to disable the SRS system. After disabling the system wait three minutes or
more before servicing the vehicle.

➡Whenever the negative battery cable
is disconnected the following components will require resetting. The Idle
Air Volume Learning, Steering Angle
Sensor Neutral Position, Sunroof Memory Reset/Initialization, Automatic
Drive Positioner System, Audio presets
and Navigation. Use the CONSULT-III
diagnostic tool, or equivalent to perform the required resets.

➡The automatic back door system
must be initialized anytime the battery
has been disconnected. Close the back
door. Open the back door with the automatic open feature. Do not stop the
process until the back door opens completely.

   2. Disconnect the negative battery cable.
   3. Raise and support the vehicle safely,
as required.
   4. Position a catch pan under the
assembly.
   5. Remove the drain plug. Discard the
gasket.
   6. Drain the fluid. Be sure to properly
dispose of used fluid.
   7. Fill with the proper grade and type
fluid. Check for proper fluid level.
   8. Correct fluid level, as required.
   9. Install the, using a new gasket.
   10. Tighten checking plug to 27 ft. lbs.
(36 Nm).

REMOVAL & INSTALLATION

See Figures 19 and 20.

   1. Before servicing the vehicle, refer to
the Precautions Section.

➡If working near and/or around the
SRS system and components, be sure
to disable the SRS system. After disabling the system wait three minutes or
more before servicing the vehicle.

➡Whenever the negative battery cable
is disconnected the following components will require resetting. The Idle
Air Volume Learning, Steering Angle
Sensor Neutral Position, Sunroof Memory Reset/Initialization, Automatic
Drive Positioner System, Audio presets
and Navigation. Use the CONSULT-III
diagnostic tool, or equivalent to perform the required resets.

➡The automatic back door system
must be initialized anytime the battery
has been disconnected. Close the back
door. Open the back door with the automatic open feature. Do not stop the
process until the back door opens completely.

   2. Disconnect the negative battery
cable.
   3. Raise and safely support the vehicle.
   4. Remove the under cover, if equipped.
   5. Remove the halfshafts.
   6. Remove the crossmember.
   7. Remove the front halfshaft.
   8. Disconnect the vent hose.
   9. Properly support the assembly, using
the proper service jack.

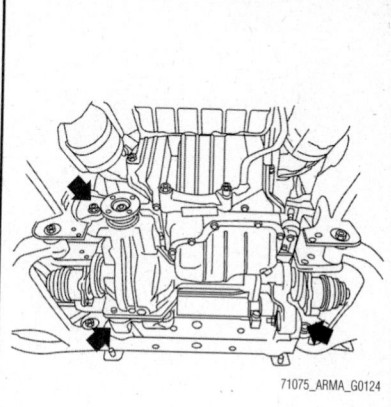

**Fig. 19 Front drive axle retaining bolt
locations**

   10. Remove the retaining bolts and carefully remove the assembly from the
vehicle.

To install:

➡Be sure to use new fasteners, as
required.

   11. Installation is the reverse of the
removal procedure.
   12. Be sure to perform the recon-
nect/relearn procedures.

FRONT DIFFERENTIAL HOUSING COVER

REMOVAL & INSTALLATION

See Figure 21.

   1. Before servicing the vehicle, refer to
the Precautions Section.

➡If working near and/or around the
SRS system and components, be sure
to disable the SRS system. After disabling the system wait three minutes or
more before servicing the vehicle.

➡Whenever the negative battery cable
is disconnected the following components will require resetting. The Idle
Air Volume Learning, Steering Angle
Sensor Neutral Position, Sunroof Memory Reset/Initialization, Automatic
Drive Positioner System, Audio presets
and Navigation. Use the CONSULT-III
diagnostic tool, or equivalent to perform the required resets.

➡The automatic back door system
must be initialized anytime the battery
has been disconnected. Close the back
door. Open the back door with the automatic open feature. Do not stop the

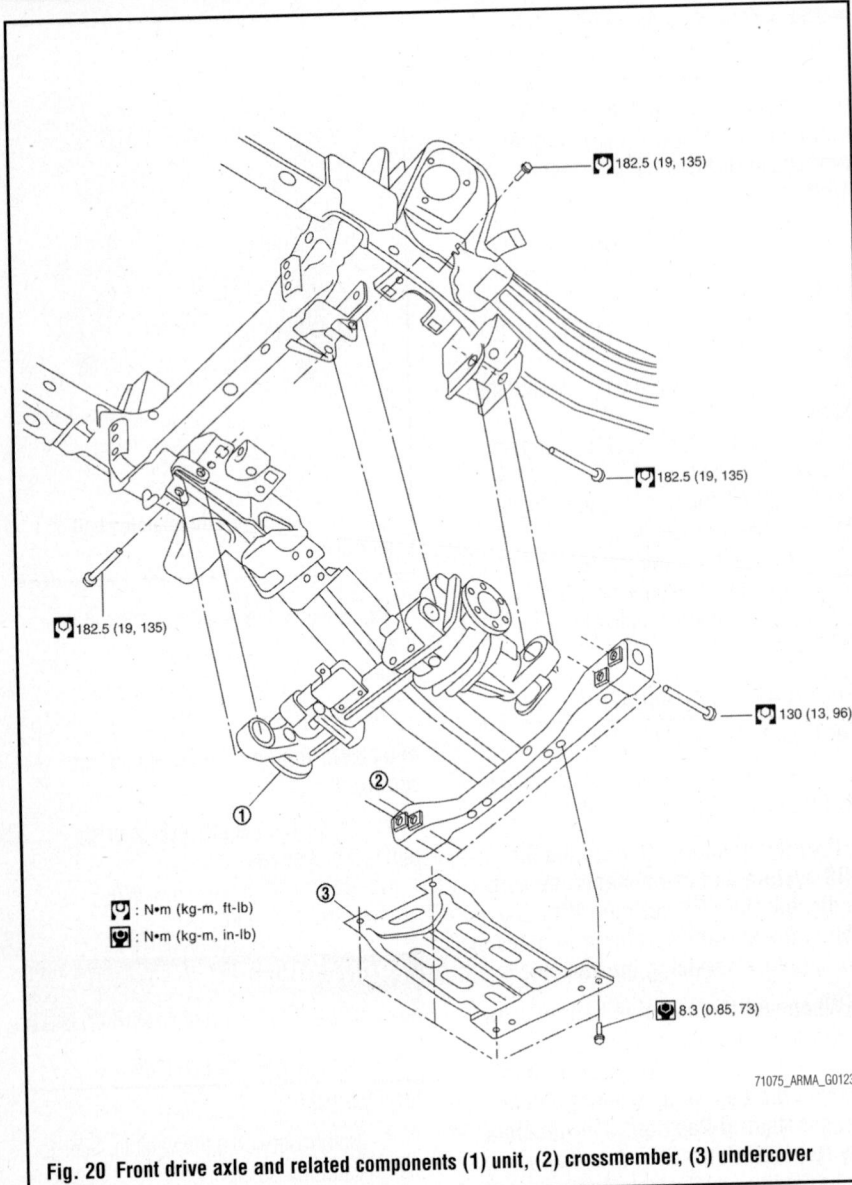

182.5 (19, 135)

182.5 (19, 135)

182.5 (19, 135)

130 (13, 96)

②

①

③

: N•m (kg-m, ft-lb)

: N•m (kg-m, in-lb)

8.3 (0.85, 73)

71075_ARMA_G0123

Fig. 20 Front drive axle and related components (1) unit, (2) crossmember, (3) undercover

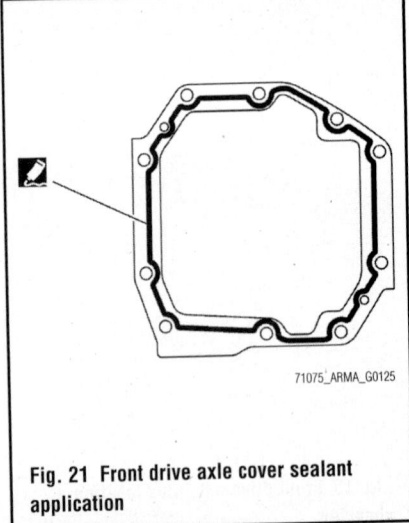

71075_ARMA_G0125

Fig. 21 Front drive axle cover sealant application

process until the back door opens completely.

2. Disconnect the negative battery cable.
3. Raise and support the vehicle safely, as required.
4. Position a catch pan under the assembly.
5. Drain the fluid. Be sure to properly dispose of used fluid.
6. Remove the front axle assembly from the vehicle. Position the unit in a suitable holding fixture.
7. Remove the carrier cover retaining bolts. Remove the cover. Discard the gasket.

To install:

➡**Be sure to use new fasteners, as required.**

8. Installation is the reverse of the removal procedure.

➡**Be sure to properly remove all old RTV sealant on all mating surfaces.**

9. Apply a thin bead (0.12 inch) of genuine RTV sealant, or equivalent, to the mating surface of the carrier cover.
10. Tighten the cover bolts.
11. Be sure to perform the reconnect/relearn procedures.

FRONT PINION OIL SEAL

REMOVAL & INSTALLATION

1. Before servicing the vehicle, refer to the Precautions Section.

➡**If working near and/or around the SRS system and components, be sure to disable the SRS system. After disabling the system wait three minutes or more before servicing the vehicle.**

➡Whenever the negative battery cable is disconnected the following components will require resetting. The Idle Air Volume Learning, Steering Angle Sensor Neutral Position, Sunroof Memory Reset/Initialization, Automatic Drive Positioner System, Audio presets and Navigation. Use the CONSULT-III diagnostic tool, or equivalent to perform the required resets.

➡The automatic back door system must be initialized anytime the battery has been disconnected. Close the back door. Open the back door with the automatic open feature. Do not stop the process until the back door opens completely.

2. Disconnect the negative battery cable.
3. Remove or disconnect the following:
   • Front driveshaft
   • Halfshafts
4. Measure and record the pinion bearing preload using special tool J-25765-A.
5. Loosen the pinion nut while holding the companion flange using special tool J-44195.
6. Remove the companion flange using a suitable tool.
7. Using a punch or drill, place a small hole in the case.
8. Remove the seal using special tool SP8P or equivalent.

To install:

➡**Be sure to use new fasteners, as required.**

9. Press front seal into carrier using a suitable tool.

10. Install companion flange and new pinion nut. Tighten pinion nut until there is no end play and until recorded pinion bearing preload is met plus an additional 5 inch lbs. (0.5 Nm).

11. Install or connect the following:
- Halfshafts
- Front driveshaft

12. Be sure to perform the reconnect/relearn procedures.

FRONT HALFSHAFT

REMOVAL & INSTALLATION

See Figure 22.

1. Before servicing the vehicle, refer to the Precautions Section.

➡**If working near and/or around the SRS system and components, be sure to disable the SRS system. After disabling the system wait three minutes or more before servicing the vehicle.**

➡**Whenever the negative battery cable is disconnected the following components will require resetting. The Idle Air Volume Learning, Steering Angle Sensor Neutral Position, Sunroof Memory Reset/Initialization, Automatic Drive Positioner System, Audio presets and Navigation. Use the CONSULT-III diagnostic tool, or equivalent to perform the required resets.**

➡**The automatic back door system must be initialized anytime the battery has been disconnected. Close the back door. Open the back door with the automatic open feature. Do not stop the process until the back door opens completely.**

2. Disconnect the negative battery cable.
3. Remove or disconnect the following:
- Wheel and tire assembly
- Engine splash guard

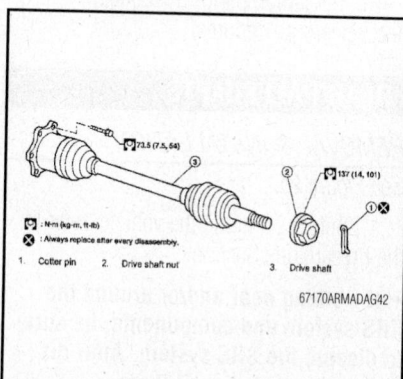

☐ : N·m (kg·m, ft-lb)
☒ : Always replace after every disassembly.
1. Cotter pin   2. Drive shaft nut   3. Drive shaft
67170ARMADAG42

Fig. 22 Front halfshaft and related components

- Wheel speed sensor harness from mount on knuckle, and reposition
- Brake caliper, do not disconnect brake hose or allow caliper to hang by hose
- Coil spring and shock absorber
- Cotter pin and halfshaft nut
- Halfshaft from front differential
- Halfshaft from hub and bearing assembly

To install:

➡**Be sure to use new fasteners, as required.**

4. Install or connect the following:
- Halfshaft into hub
- Halfshaft into front differential
- Halfshaft nut and tighten to 101 ft. lbs. (137 Nm) and replace cotter pin
- Wheel speed sensor harness
- Brake caliper
- Coil spring and shock absorber
- Engine splash guard
- Wheel and tire assembly

5. Be sure to perform the reconnect/relearn procedures.

REAR AXLE HOUSING (ASSEMBLY, UNIT)

REMOVAL & INSTALLATION

1. Before servicing the vehicle, refer to the Precautions Section.

➡**If working near and/or around the SRS system and components, be sure to disable the SRS system. After disabling the system wait three minutes or more before servicing the vehicle.**

➡**Whenever the negative battery cable is disconnected the following components will require resetting. The Idle Air Volume Learning, Steering Angle Sensor Neutral Position, Sunroof Memory Reset/Initialization, Automatic Drive Positioner System, Audio presets and Navigation. Use the CONSULT-III diagnostic tool, or equivalent to perform the required resets.**

➡**The automatic back door system must be initialized anytime the battery has been disconnected. Close the back door. Open the back door with the automatic open feature. Do not stop the process until the back door opens completely.**

2. Disconnect the negative battery cable.
3. Remove the spare tire and wheel assembly.

4. Raise and support the vehicle safely.
5. Drain the gear oil. Be sure to properly dispose of used oil.
6. Remove the tire and wheel assemblies.
7. Remove the driveshaft.
8. Remove the rear stabilizer bar.
9. Disconnect the rear halfshaft from the rear axle assembly. Position it aside using mechanics wire or equivalent.
10. Disconnect the breather hose from the axle cover.
11. Position a suitable jack under the assembly.

➡**Do not position the jack under the aluminum cover.**

12. Remove the rear axle assembly retaining bolts and nuts.
13. Carefully remove the assembly from the vehicle.

To install:

➡**Be sure to use new fasteners, as required.**

14. Installation is the reverse of the removal procedure.
15. Be sure to perform the reconnect/relearn procedures.

FLUID RECOMMENDATIONS

Be sure to use the proper grade and type fluid when servicing this component. Use API GL-5 synthetic gear oil, Viscosity SAE 75W-90 gear oil, or equivalent when servicing/refilling the axle. Approximate capacity is 3.75 pints.

LEVEL CHECK

1. Before servicing the vehicle, refer to the Precautions Section.

➡**If working near and/or around the SRS system and components, be sure to disable the SRS system. After disabling the system wait three minutes or more before servicing the vehicle.**

➡**Whenever the negative battery cable is disconnected the following components will require resetting. The Idle Air Volume Learning, Steering Angle Sensor Neutral Position, Sunroof Memory Reset/Initialization, Automatic Drive Positioner System, Audio presets and Navigation. Use the CONSULT-III diagnostic tool, or equivalent to perform the required resets.**

➡**The automatic back door system must be initialized anytime the battery has been disconnected. Close the back door. Open the back door with the auto-**

matic open feature. Do not stop the process until the back door opens completely.

2. Disconnect the negative battery cable.

3. Raise and safely support the vehicle.

4. Position a catch pan under the assembly.

5. Remove the drain plug.

6. Discard the gasket.

7. Check the oil level, correct as required.

➡**Be sure to use the proper grade and type gear oil.**

8. Check the oil level after refilling.

9. Using a new gasket install the plug. Tighten to specification (25 ft. lbs.).

10. Be sure to perform the reconnect/relearn procedures.

DRAIN & REFILL

See Figure 23.

1. Before servicing the vehicle, refer to the Precautions Section.

➡**If working near and/or around the SRS system and components, be sure to disable the SRS system. After disabling the system wait three minutes or more before servicing the vehicle.**

➡**Whenever the negative battery cable is disconnected the following components will require resetting. The Idle Air Volume Learning, Steering Angle Sensor Neutral Position, Sunroof Memory Reset/Initialization, Automatic Drive Positioner System, Audio presets and Navigation. Use the CONSULT-III diagnostic tool, or equivalent to perform the required resets.**

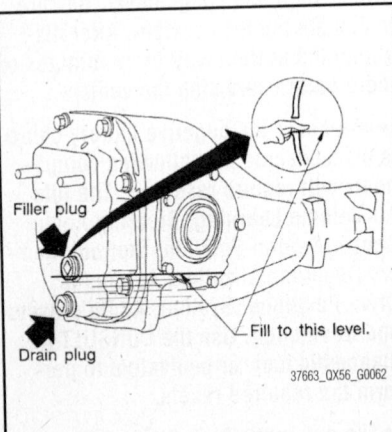

Fig. 23 Rear differential plug location— draining

➡**The automatic back door system must be initialized anytime the battery has been disconnected. Close the back door. Open the back door with the automatic open feature. Do not stop the process until the back door opens completely.**

2. Disconnect the negative battery cable.

3. Raise and safely support the vehicle.

4. Remove the drain plug and drain the fluid into a suitable container.

5. Discard the gasket.

6. Fill the unit with new gear oil until the oil level reaches the specified level near the filler plug mounting hole.

To install:

➡**Be sure to use new fasteners, as required.**

7. Installation is the reverse of the removal procedure.

8. Using a new gasket install the drain plug. Tighten to specification (25 ft. lbs.).

9. Check the oil level after refilling.

➡**Be sure to use the proper grade and type gear oil.**

10. Be sure to perform the reconnect/relearn procedures.

REAR DIFFERENTIAL HOUSING COVER

REMOVAL & INSTALLATION
See Figure 24.

1. Before servicing the vehicle, refer to the Precautions Section.

➡**If working near and/or around the SRS system and components, be sure to disable the SRS system. After disabling the system wait three minutes or more before servicing the vehicle.**

➡**Whenever the negative battery cable is disconnected the following components will require resetting. The Idle Air Volume Learning, Steering Angle Sensor Neutral Position, Sunroof Memory Reset/Initialization, Automatic Drive Positioner System, Audio presets and Navigation. Use the CONSULT-III diagnostic tool, or equivalent to perform the required resets.**

➡**The automatic back door system must be initialized anytime the battery has been disconnected. Close the back door. Open the back door with the automatic open feature. Do not stop the**

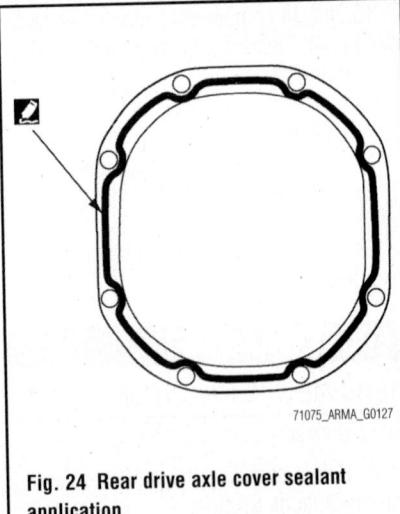

71075_ARMA_G0127

Fig. 24 Rear drive axle cover sealant application

process until the back door opens completely.

2. Disconnect the negative battery cable.

3. Raise and support the vehicle safely, as required.

4. Position a catch pan under the assembly.

5. Drain the fluid. Be sure to properly dispose of used fluid.

6. Remove the rear final drive assembly from the vehicle.

7. Remove the carrier cover retaining bolts. Remove the cover. Discard the gasket.

To install:

➡**Be sure to use new fasteners, as required.**

8. Installation is the reverse of the removal procedure.

➡**Be sure to properly remove all old RTV sealant on all mating surfaces.**

9. Apply a thin bead (0.12 inch) of genuine RTV sealant, or equivalent, to the mating surface of the carrier cover.

10. Be sure to perform the reconnect/relearn procedures.

REAR HALFSHAFTS

REMOVAL & INSTALLATION
See Figure 25.

1. Before servicing the vehicle, refer to the Precautions Section.

➡**If working near and/or around the SRS system and components, be sure to disable the SRS system. After disabling the system wait three minutes or more before servicing the vehicle.**

➡Whenever the negative battery cable is disconnected the following components will require resetting. The Idle Air Volume Learning, Steering Angle Sensor Neutral Position, Sunroof Memory Reset/Initialization, Automatic Drive Positioner System, Audio presets and Navigation. Use the CONSULT-III diagnostic tool, or equivalent to perform the required resets.

➡The automatic back door system must be initialized anytime the battery has been disconnected. Close the back door. Open the back door with the automatic open feature. Do not stop the process until the back door opens completely.

2. Disconnect the negative battery cable.

3. Remove or disconnect the following:
- Wheel and tire assembly
- Stabilizer bar clamp
- Cotter pin and driveshaft nut
- Bolts from the inside flange of the driveshaft

4. Separate the driveshaft from the wheel hub by lightly tapping the end with suitable hammer and wood block.

5. Remove the halfshaft.

✷✷ WARNING

Do not excessively extend the slide joint.

To install:

➡Be sure to use new fasteners, as required.

➡Do not reuse the halfshaft inside flange bolts and cotter pin.

6. Install or connect the following:
- Halfshaft
- Bolts for the inside flange and tighten to 87 ft. lbs. (118 Nm)
- Driveshaft nut and tighten nut to 101 ft. lbs. (137 Nm) and replace cotter pin
- Stabilizer bar clamp
- Wheel and tire assembly

7. Be sure to perform the reconnect/relearn procedures.

PINION OIL SEAL

REMOVAL & INSTALLATION
See Figure 26.

1. Before servicing the vehicle, refer to the Precautions Section.

➡If working near and/or around the SRS system and components, be sure to disable the SRS system. After disabling the system wait three minutes or more before servicing the vehicle.

➡Whenever the negative battery cable is disconnected the following components will require resetting. The Idle Air Volume Learning, Steering Angle Sensor Neutral Position, Sunroof Memory Reset/Initialization, Automatic Drive Positioner System, Audio presets and Navigation. Use the CONSULT-III diagnostic tool, or equivalent to perform the required resets.

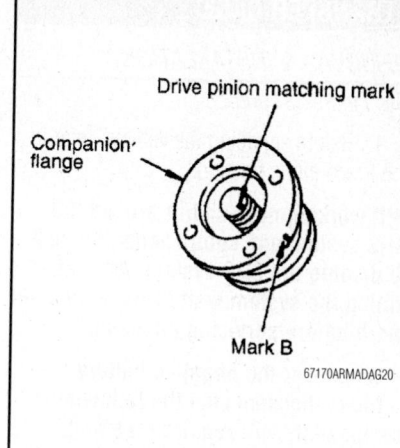

Fig. 26 Companion flange marking

➡The automatic back door system must be initialized anytime the battery has been disconnected. Close the back door. Open the back door with the automatic open feature. Do not stop the process until the back door opens completely.

2. Disconnect the negative battery cable.

3. Raise and safely support the vehicle.

4. Remove the rear driveshaft.

5. Measure and record the total preload.

6. Matchmark the drive pinion to position 'B' on the companion flange.

7. Remove the drive pinion nut using suitable tool.

8. Remove the companion flange using suitable tool.

9. Remove the rear pinion seal using special tool J-34286.

To install:

➡Be sure to use new fasteners, as required.

10. Press the rear pinion seal into the carrier using suitable tool.

11. Align the matchmark on the companion flange to the drive pinion and install the companion flange.

12. Lubricate the drive pinion threads and seating surfaces of the drive pinion nut with grease.

13. Using a new drive pinion nut, tighten to 124–274 ft. lbs. (167–372 Nm).

➡Final torque is determined when adjusting total preload using special tool J-25765-A.

14. Install rear driveshaft.

15. Be sure to perform the reconnect/relearn procedures.

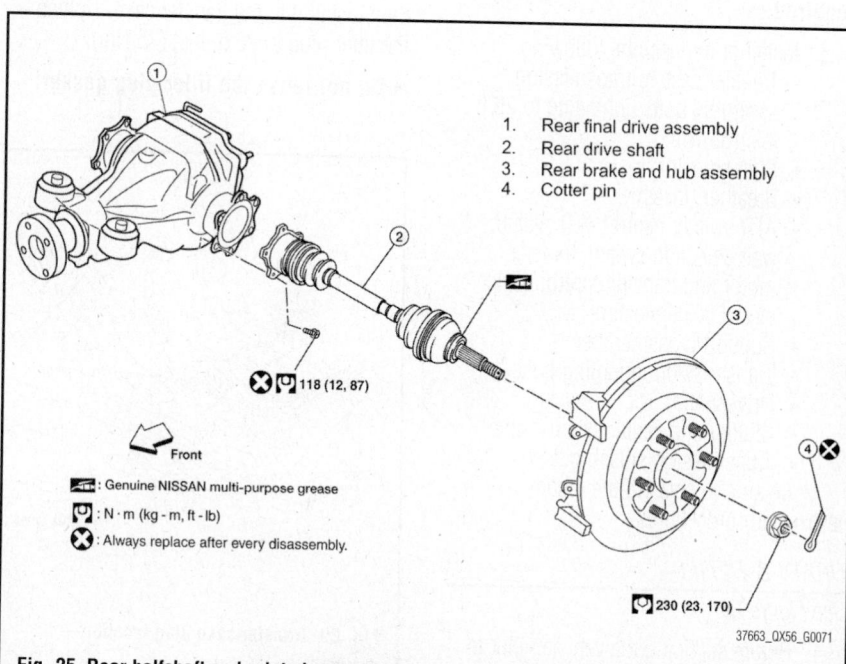

1. Rear final drive assembly
2. Rear drive shaft
3. Rear brake and hub assembly
4. Cotter pin

118 (12, 87)

Front

: Genuine NISSAN multi-purpose grease

: N·m (kg - m, ft - lb)

: Always replace after every disassembly.

230 (23, 170)

37663_QX56_G0071

Fig. 25 Rear halfshaft and related components

TRANSFER CASE

REMOVAL & INSTALLATION

See Figures 27 and 28.

1. Before servicing the vehicle, refer to the Precautions Section.

➡️**If working near and/or around the SRS system and components, be sure to disable the SRS system. After disabling the system wait three minutes or more before servicing the vehicle.**

➡️**Whenever the negative battery cable is disconnected the following components will require resetting. The Idle Air Volume Learning, Steering Angle Sensor Neutral Position, Sunroof Memory Reset/Initialization, Automatic Drive Positioner System, Audio presets and Navigation. Use the CONSULT-III diagnostic tool, or equivalent to perform the required resets.**

➡️**The automatic back door system must be initialized anytime the battery has been disconnected. Close the back door. Open the back door with the automatic open feature. Do not stop the process until the back door opens completely.**

2. Disconnect the negative battery cable.

3. Remove or disconnect the following:
   - Transmission splash guard
   - Center exhaust pipe and muffler
   - Front and rear driveshafts

➡️**Plug rear oil seal after removing rear driveshaft.**

   - Transmission assembly mounting bolts

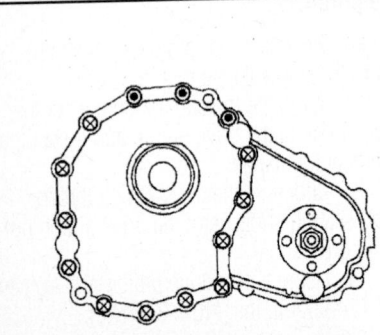

: Transfer ➡️ Automatic transmission
: Automatic transmission ➡️ Transfer

67170ARMADAG41

Fig. 27 Transfer case mounting bolt locations

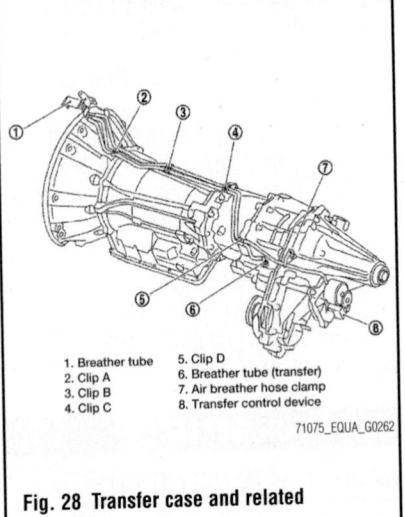

1. Breather tube
2. Clip A
3. Clip B
4. Clip C
5. Clip D
6. Breather tube (transfer)
7. Air breather hose clamp
8. Transfer control device

71075_EQUA_G0262

Fig. 28 Transfer case and related components

4. Support the transmission assembly with a suitable jack and remove the crossmember.

5. Remove or disconnect the following:
   - ATP switch, neutral 4LO switch, wait detection switch, transfer motor and transfer control device electrical connectors
   - Breather hoses
   - Shift actuator from the extension housing
   - Transfer case to transmission assembly bolts
   - Transfer case assembly

To install:

➡️**Be sure to use new fasteners, as required.**

6. Install or connect the following:
   - Transfer case to transmission assembly bolts tightening to 26 ft. lbs. (36 Nm)
   - Shift actuator
   - Breather hoses
   - ATP switch, neutral 4LO switch, wait detection switch, transfer motor and transfer control device electrical connectors
   - Support crossmember
   - Transmission mounting bolts
   - Driveshafts
   - Muffler and center exhaust pipe
   - Transmission splash guard

7. Be sure to perform the reconnect/relearn procedures.

DRAIN & REFILL

See Figure 29.

1. Before servicing the vehicle, refer to the Precautions Section.

➡️**If working near and/or around the SRS system and components, be sure to disable the SRS system. After disabling the system wait three minutes or more before servicing the vehicle.**

➡️**Whenever the negative battery cable is disconnected the following components will require resetting. The Idle Air Volume Learning, Steering Angle Sensor Neutral Position, Sunroof Memory Reset/Initialization, Automatic Drive Positioner System, Audio presets and Navigation. Use the CONSULT-III diagnostic tool, or equivalent to perform the required resets.**

➡️**The automatic back door system must be initialized anytime the battery has been disconnected. Close the back door. Open the back door with the automatic open feature. Do not stop the process until the back door opens completely.**

2. Disconnect the negative battery cable.

3. Raise and safely support the vehicle.

4. Position a catch pan under the component. Be sure to properly dispose of used fluid.

5. Remove the transfer case drain plug. Drain the fluid. Discard the gasket.

6. Reinstall the drain plug using a new gasket.

7. Refill the unit with the proper grade and type fluid.

8. Check oil level and correct as required. Set a new gasket to the filler plug, then install it to the transfer case. Tighten the filler plug to 26 ft. lbs. (35 Nm).

➡️**Do not reuse the filler plug gasket.**

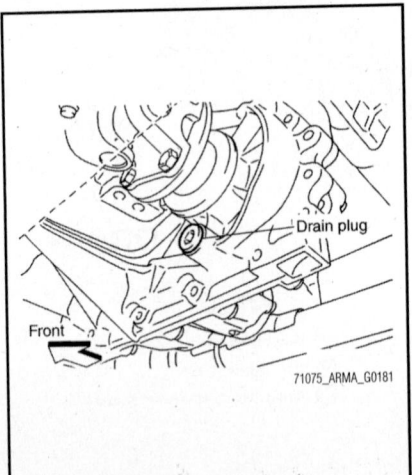

Drain plug

Front

71075_ARMA_G0181

Fig. 29 Transfer case plug location—draining

FLUID LEVEL CHECK

See Figure 30.

1. Before servicing the vehicle, refer to the Precautions Section.

➡️**If working near and/or around the SRS system and components, be sure to disable the SRS system. After disabling the system wait three minutes or more before servicing the vehicle.**

➡️**Whenever the negative battery cable is disconnected the following components will require resetting. The Idle Air Volume Learning, Steering Angle Sensor Neutral Position, Sunroof Memory Reset/Initialization, Automatic Drive Positioner System, Audio presets and Navigation. Use the CONSULT-III diagnostic tool, or equivalent to perform the required resets.**

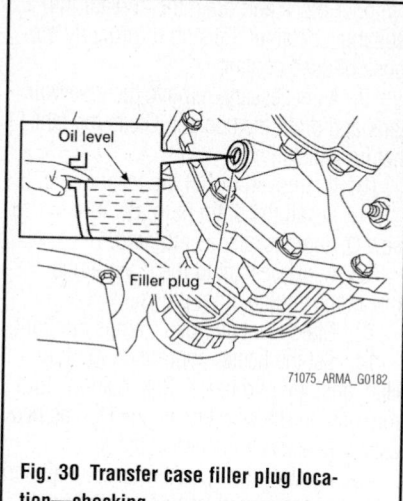

Fig. 30 Transfer case filler plug location—checking

➡️**The automatic back door system must be initialized anytime the battery has been disconnected. Close the back door.**

Open the back door with the automatic open feature. Do not stop the process until the back door opens completely.

2. Disconnect the negative battery cable.

3. Raise and safely support the vehicle.

4. Remove the transfer case checking plug. Discard the gasket.

5. Check oil level and correct as required. Set a new gasket to the plug, then install it to the transfer case. Tighten the filler plug to 26 ft. lbs. (35 Nm).

➡️**Do not reuse the filler plug gasket.**

FLUID RECOMMENDATIONS

The transfer case uses API GL-5 viscosity SAE 80W-90. If information differs from the owner's manual, use the data in the owner's manual.

ENGINE COOLING

BELT-DRIVEN ENGINE FAN

REMOVAL & INSTALLATION

See Figure 31.

➡️**Never remove the radiator cap when the engine is hot. Serious burns could occur from high-pressure engine coolant escaping from the radiator.**

1. Before servicing the vehicle, refer to the Precautions Section.

➡️**If working near and/or around the SRS system and components, be sure to disable the SRS system. After disabling the system wait three minutes or more before servicing the vehicle.**

➡️**Whenever the negative battery cable is disconnected the following components will require resetting. The Idle Air Volume Learning, Steering Angle Sensor Neutral Position, Sunroof Memory Reset/Initialization, Automatic Drive Positioner System, Audio presets and Navigation. Use the CONSULT-III diagnostic tool, or equivalent to perform the required resets.**

➡️**The automatic back door system must be initialized anytime the battery has been disconnected. Close the back door. Open the back door with the automatic open feature. Do not stop the process until the back door opens completely.**

2. Disconnect the negative battery cable.

3. Make sure the engine is cold.

4. Remove the air duct and resonator assembly.

5. Remove the engine front undercover.

6. Remove the lower radiator shroud.

7. Remove the drive belt.

8. Remove the cooling fan retaining bolts.

9. Remove the cooling fan from its mounting.

To install:

➡️**Be sure to use new fasteners, as required.**

10. Installation is the reverse of the removal procedure.

11. Be sure to install the fan with its front mark "F" facing the front of the engine.

12. Be sure to check and refill the cooling using the proper grade and type engine coolant, as required.

13. Start the engine and check for leaks.

14. Start the engine and allow it to reach

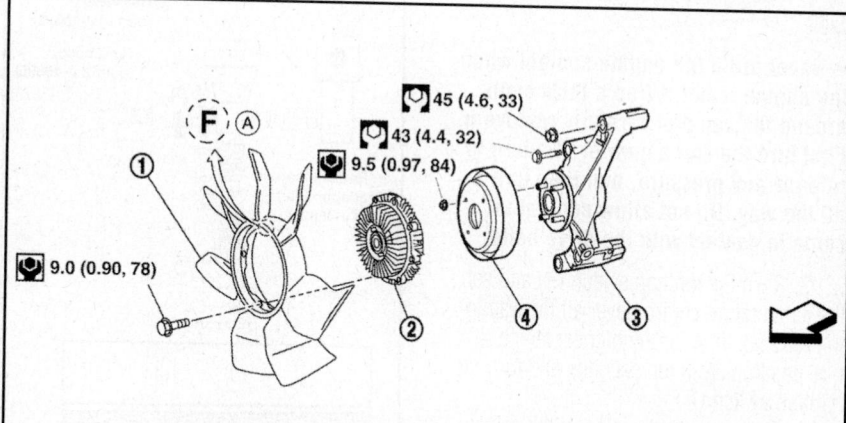

| 1. | Cooling fan | 2. | Fan coupling | 3. | Fan bracket |
| 4. | Cooling fan pulley | A. | Front mark | ⇦ | Engine front |

Fig. 31 Engine cooling fan and related components—crankshaft driven

operation temperature. Recheck the coolant level, fill as required.

15. Be sure to perform the reconnect/relearn procedures.

ENGINE COOLANT

DRAIN & REFILL

See Figure 32.

➡You will need tool KV991J0070, or equivalent, and a compressed air supply to perform this procedure.

1. Before servicing the vehicle, refer to the Precautions Section.

➡If working near and/or around the SRS system and components, be sure to disable the SRS system. After disabling the system wait three minutes or more before servicing the vehicle.

➡Whenever the negative battery cable is disconnected the following components will require resetting. The Idle Air Volume Learning, Steering Angle Sensor Neutral Position, Sunroof Memory Reset/Initialization, Automatic Drive Positioner System, Audio presets and Navigation. Use the CONSULT-III diagnostic tool, or equivalent to perform the required resets.

➡The automatic back door system must be initialized anytime the battery has been disconnected. Close the back door. Open the back door with the automatic open feature. Do not stop the process until the back door opens completely.

2. Disconnect the negative battery cable.

➡Never drain the engine coolant when the engine is hot. Wrap a thick cloth around the cap and carefully remove it. First turn the cap a quarter of a turn to release any pressure, and then turn it all the way. Do not allow coolant to come in contact with the drive belts.

3. Turn the ignition switch ON and set the temperature control lever all the way to the HOT position, or the highest temperature position. Wait ten seconds and turn the ignition switch OFF.

4. Raise and safely support the vehicle.

5. Remove the undercover.

6. Open the radiator drain plug and carefully drain the coolant into a suitable container. Be sure to properly dispose of used coolant.

7. Discard the O-ring.

8. Open the water drain plugs on the cylinder block and drain the coolant into a suitable container. Be sure to properly dispose of used coolant.

9. As necessary, remove the reservoir tank and drain the coolant. Clean the tank before reinstalling.

10. If removed install the reservoir tank.

11. Install the drain plug. Be sure to use new O-ring/gaskets, as required.

12. Apply sealant to the drain plugs. Install the engine drain plugs.

13. Check that all hose clamps are tight.

14. Set the heater controls to the full HOT position and heater ON position. Turn the ignition ON with the engine OFF as necessary to activate the heater mode.

15. Remove the vented reservoir cap and replace it with a non-vented cap before filling the system.

16. Install tool KV991J0070 or equivalent, see illustration.

17. Insert the refill hose into the coolant mixture (placed at floor level). Be sure that the ball valve is in the closed position. Install the air hose to the venture assembly. The air pressure must be within specification (80–120 psi).

➡The compressed air supply must be equipped with an air dryer.

18. The vacuum gauge will begin to rise and there will be an audible hissing noise. During this process open the ball valve on the refill hose, slightly. Coolant will be visi-

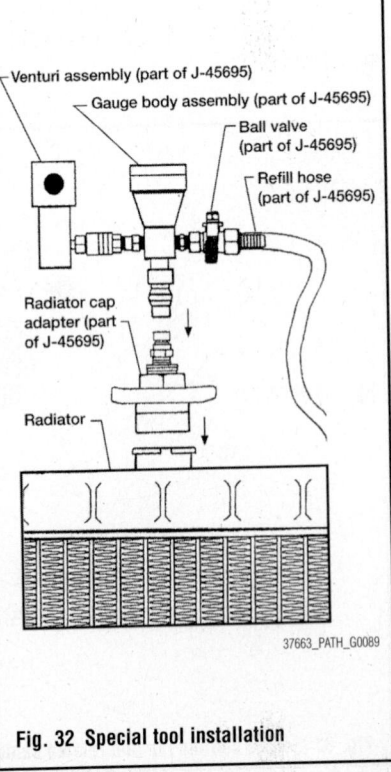

Fig. 32 Special tool installation

Venturi assembly (part of J-45695)

Gauge body assembly (part of J-45695)

Ball valve (part of J-45695)

Refill hose (part of J-45695)

Radiator cap adapter (part of J-45695)

Radiator

37663_PATH_G0089

ble rising in the refill hose. Once the refill hose is full of coolant, close the ball valve. This will purge any trapped air in the refill hose.

19. Continue to draw vacuum until the gauge reaches 28 inches of vacuum.

➡The gauge may not reach specification in high altitude applications. If not see the following specifications. Altitude above sea level: 0-100m (328 ft.) 28 inches of vacuum. Altitude above sea level: 300m (984 ft.) 27 inches of vacuum. Altitude above sea level: 500m (1641 ft.) 26 inches of vacuum. Altitude above sea level: 1000m (3281 ft.) 24-25 inches of vacuum.

20. When the proper specification has been reached, disconnect the air hose and wait 20 seconds to see if the system loses vacuum. If the level drops perform any necessary repairs and repeat the procedure.

21. Place the coolant container (with the refill hose inserted) at the same level as the top of the radiator. Then open the ball valve on the refill hose so the coolant will be drawn up to fill the cooling system. The cooling system is full when the vacuum gauge reads zero.

➡Do not allow the coolant container to get too low when filling, to avoid air from being drawn into the system.

22. Remove the tool from the radiator neck opening and install the radiator cap.

23. Remove the non-vented reservoir cap.

24. Fill the reservoir tank to specification with the proper coolant mixture.

25. Be sure to perform the reconnect/relearn procedures.

FLUID RECOMMENDATIONS

Be sure to use genuine Nissan (blue) coolant when filling/servicing the cooling system.

LEVEL CHECK

1. Before servicing the vehicle, refer to the Precautions Section.

➡If working near and/or around the SRS system and components, be sure to disable the SRS system. After disabling the system wait three minutes or more before servicing the vehicle.

➡Whenever the negative battery cable is disconnected the following components will require resetting. The Idle Air Volume Learning, Steering Angle Sensor Neutral Position, Sunroof Memory Reset/Initialization, Automatic

Drive Positioner System, Audio presets and Navigation. Use the CONSULT-III diagnostic tool, or equivalent to perform the required resets.

→The automatic back door system must be initialized anytime the battery has been disconnected. Close the back door. Open the back door with the automatic open feature. Do not stop the process until the back door opens completely.

Check that the coolant reservoir tank is level and within the MIN and MAX marks. Adjust coolant level, as necessary.

ELECTRIC ENGINE FAN

REMOVAL & INSTALLATION

Motor Driven

See Figure 33.

→Never remove the radiator cap when the engine is hot. Serious burns could occur from high-pressure engine coolant escaping from the radiator.

1. Before servicing the vehicle, refer to the Precautions Section.

→If working near and/or around the SRS system and components, be sure to disable the SRS system. After disabling the system wait three minutes or more before servicing the vehicle.

→Whenever the negative battery cable is disconnected the following components will require resetting. The Idle Air Volume Learning, Steering Angle Sensor Neutral Position, Sunroof Memory Reset/Initialization, Automatic Drive Positioner System, Audio presets and Navigation. Use the CONSULT-III

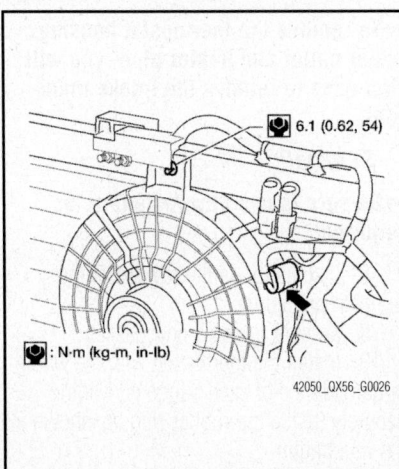

6.1 (0.62, 54)

: N·m (kg-m, in-lb)

42050_QX56_G0026

Fig. 33 Engine cooling fan and related components—electric

diagnostic tool, or equivalent to perform the required resets.

→The automatic back door system must be initialized anytime the battery has been disconnected. Close the back door. Open the back door with the automatic open feature. Do not stop the process until the back door opens completely.

2. Disconnect the negative battery cable.

3. Remove the front bumper fascia.

4. Disconnect the harness connector from the fan motor.

5. Remove the retaining bolt.

6. Remove the fan grille and motor assembly.

To install:

→Be sure to use new fasteners, as required.

7. Installation is the reverse of the removal procedure.

8. Be sure to check and refill the cooling using the proper grade and type engine coolant, as required.

9. Start the engine and check for leaks.

10. Start the engine and allow it to reach operation temperature. Recheck the coolant level, fill as required.

11. Be sure to perform the reconnect/relearn procedures.

RADIATOR

REMOVAL & INSTALLATION
See Figure 34.

✳✳ CAUTION

Never remove the radiator cap when the engine is hot. Serious burns could occur from high-pressure engine coolant escaping from the radiator.

1. Before servicing the vehicle, refer to the Precautions Section.

→If working injury or death may result if the sys and/or around the SRS system and components, be sure to disable the SRS system. After disabling the system wait three minutes or more before servicing the vehicle.

→Whenever the negative battery cable is disconnected the following components will require resetting. The Idle Air Volume Learning, Steering Angle Sensor Neutral Position, Sunroof Memory Reset/Initialization, Automatic

Drive Positioner System, Audio presets and Navigation. Use the CONSULT-III diagnostic tool, or equivalent to perform the required resets.

→The automatic back door system must be initialized anytime the battery has been disconnected. Close the back door. Open the back door with the automatic open feature. Do not stop the process until the back door opens completely.

2. Disconnect the negative battery cable.

3. Make sure the engine is cold before removing the radiator.

4. Remove the engine room cover.

5. Remove the air cleaner and air duct assembly.

6. Drain the engine coolant. Be sure to properly dispose of the used coolant.

7. Disconnect and plug the automatic transmission fluid lines.

8. Disconnect the upper radiator hose. Do not allow coolant to contact the drive belts.

9. Disconnect the lower radiator hose. Do not allow coolant to contact the drive belts.

10. To remove the lower radiator shroud, release the tabs and pull the lower radiator shroud rearwards and down.

11. Remove the radiator shroud upper bolts and remove the upper radiator shroud.

12. Remove the air conditioning condenser bolts and brackets.

→Lift the condenser up and forward to remove it from the radiator.

13. Remove the transmission fluid cooler bolts and the fluid cooler from the radiator. Position it to the side.

14. Lift up and remove the radiator. Be careful not to damage or scratch the air conditioning condenser and radiator core when removing the radiator.

To install:

→Be sure to use new fasteners, as required.

15. Installation is the reverse of the removal procedure.

16. Be sure to refill the cooling using the proper grade and type engine coolant.

17. Start the engine and check for leaks.

18. Start the engine and allow it to reach operation temperature. Recheck the coolant level, fill as required.

19. Be sure to perform the reconnect/relearn procedures.

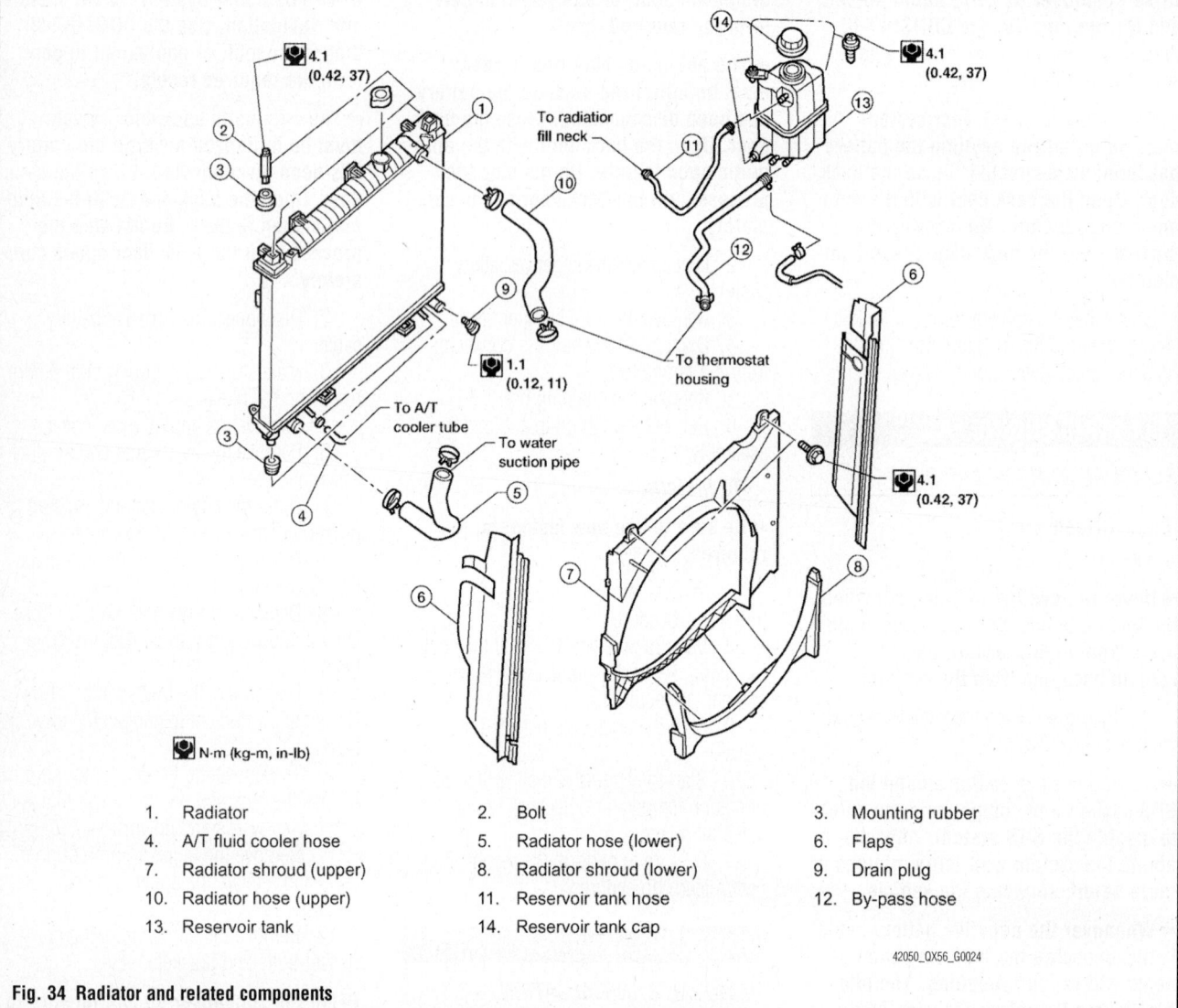

N·m (kg-m, in-lb)

1. Radiator	2. Bolt	3. Mounting rubber
4. A/T fluid cooler hose	5. Radiator hose (lower)	6. Flaps
7. Radiator shroud (upper)	8. Radiator shroud (lower)	9. Drain plug
10. Radiator hose (upper)	11. Reservoir tank hose	12. By-pass hose
13. Reservoir tank	14. Reservoir tank cap	

42050_QX56_G0024

Fig. 34 Radiator and related components

THERMOSTAT

REMOVAL & INSTALLATION

See Figures 35 and 36.

✳✳ CAUTION

Never remove the radiator cap when the engine is hot. Serious burns could occur from high-pressure engine coolant escaping from the radiator.

1. Before servicing the vehicle, refer to the Precautions Section.

➡ If working near and/or around the SRS system and components, be sure to disable the SRS system. After disabling the system wait three minutes or more before servicing the vehicle.

➡ Whenever the negative battery cable is disconnected the following components will require resetting. The Idle Air Volume Learning, Steering Angle Sensor Neutral Position, Sunroof Memory Reset/Initialization, Automatic Drive Positioner System, Audio presets and Navigation. Use the CONSULT-III diagnostic tool, or equivalent to perform the required resets.

➡ The automatic back door system must be initialized anytime the battery has been disconnected. Close the back door. Open the back door with the automatic open feature. Do not stop the process until the back door opens completely.

2. Disconnect the negative battery cable.
3. Make sure the engine is cold.
4. Remove the engine room cover.
5. Remove the air duct and resonator assembly.

6. Disconnect the water suction hose from the water inlet.
7. Remove the water inlet and thermostat.

➡ To remove the thermostat housing, water outlet and heater pipe, you will first have to remove the intake manifold.

To install:

➡ Be sure to use new fasteners, as required.

8. Installation is the reverse of the removal procedure.
9. Be sure to use a new gasket.
10. Install the thermostat with the whole circumference of each flange part fitting securely inside the rubber ring, as shown in the illustration.
11. Be sure to perform the reconnect/relearn procedures.

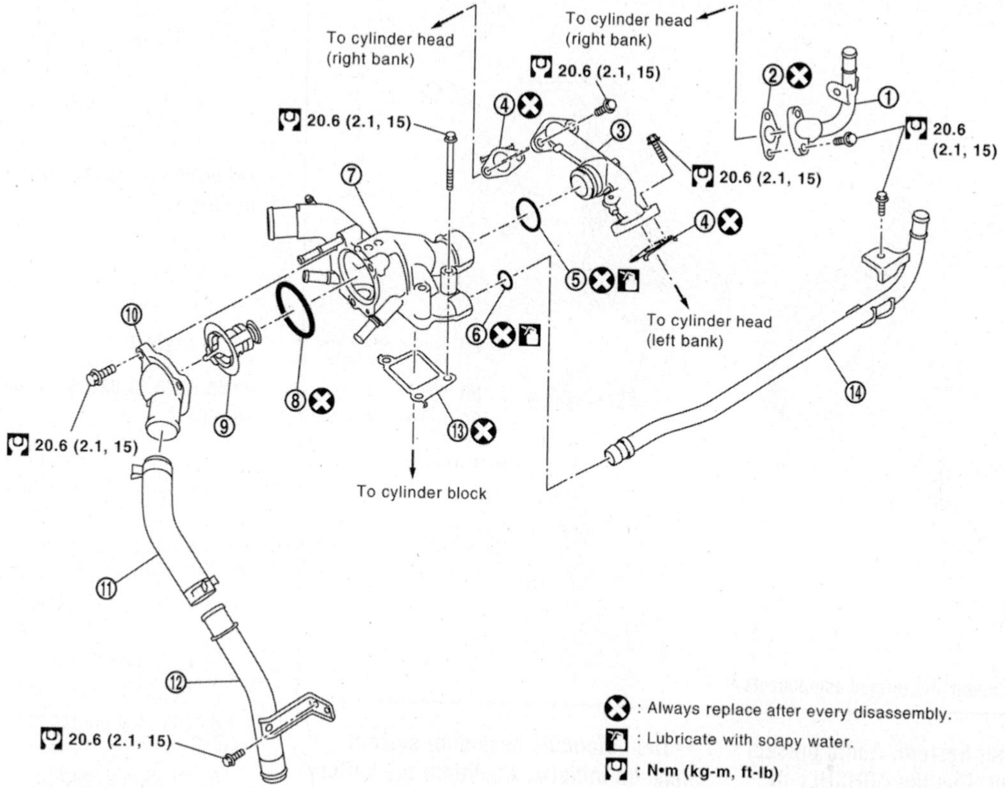

To cylinder head (right bank)
To cylinder head (right bank)
20.6 (2.1, 15)
20.6 (2.1, 15)
20.6 (2.1, 15)
20.6 (2.1, 15)
20.6 (2.1, 15)
20.6 (2.1, 15)
To cylinder head (left bank)
To cylinder block
: Always replace after every disassembly.
: Lubricate with soapy water.
: N•m (kg-m, ft-lb)

1. Heater pipe
2. Gasket
3. Water outlet
4. Gasket
5. O-ring
6. O-ring
7. Thermostat housing
8. Rubber ring
9. Thermostat
10. Water inlet
11. Water suction hose
12. Water suction pipe

37663_QX56_G0064

Fig. 35 Thermostat and related components

Rubber ring
Align projection
Thermostat housing
Thermostat flange
Jiggle valve
Rubber ring
Thermostat
42050_QX56_G0028

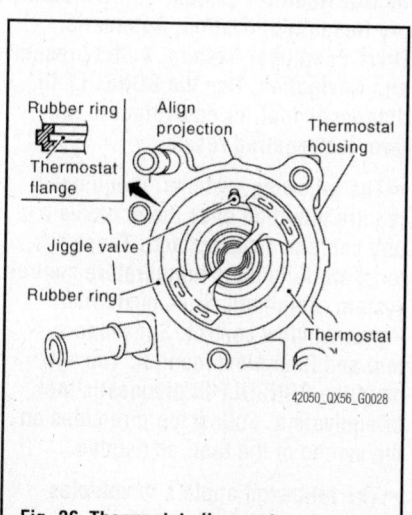

Fig. 36 Thermostat alignment

12. Install the thermostat with the jiggle valve facing upward.
13. Be sure to refill the cooling using the proper grade and type engine coolant.
14. Start the engine and check for leaks.
15. Start the engine and allow it to reach operation temperature. Recheck the coolant level, fill as required.

WATER PUMP

REMOVAL & INSTALLATION

See Figure 37.

✲✲ CAUTION

Never remove the radiator cap when the engine is hot. Serious burns could occur from high-pressure engine coolant escaping from the radiator.

1. Before servicing the vehicle, refer to the Precautions Section.

➡If working near and/or around the SRS system and components, be sure to disable the SRS system. After disabling the system wait three minutes or more before servicing the vehicle.

➡Whenever the negative battery cable is disconnected the following components will require resetting. The Idle Air Volume Learning, Steering Angle Sensor Neutral Position, Sunroof Memory Reset/Initialization, Automatic

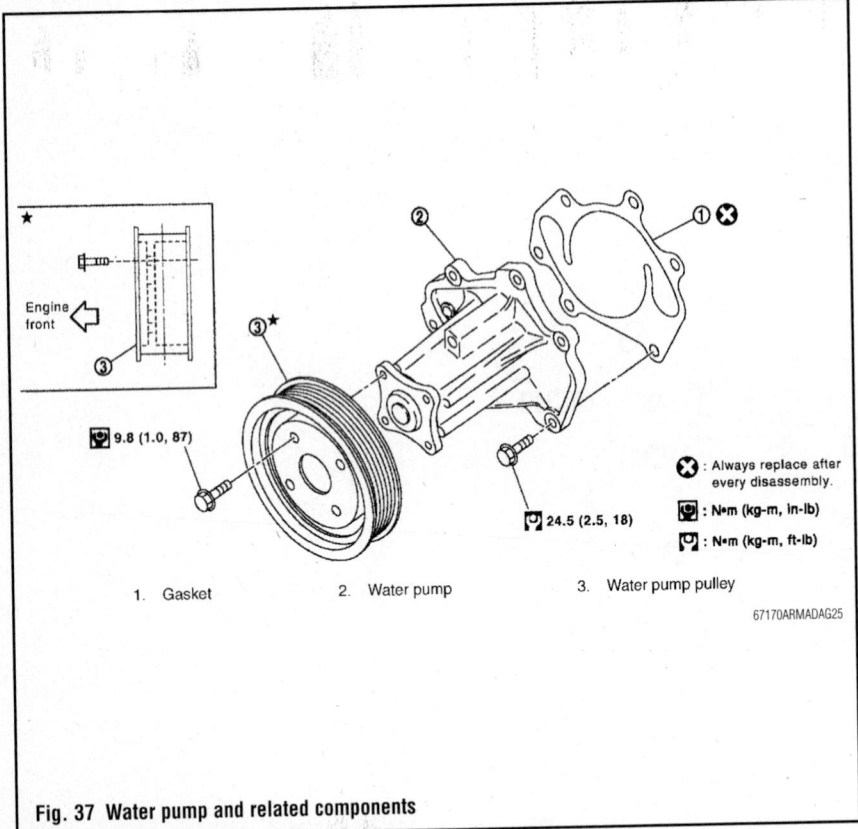

Engine front

9.8 (1.0, 87)

24.5 (2.5, 18)

✕ : Always replace after every disassembly.
⚙ : N•m (kg-m, in-lb)
⚒ : N•m (kg-m, ft-lb)

1. Gasket
2. Water pump
3. Water pump pulley

67170ARMADAG25

Fig. 37 Water pump and related components

Drive Positioner System, Audio presets and Navigation. Use the CONSULT-III diagnostic tool, or equivalent to perform the required resets.

➡The automatic back door system must be initialized anytime the battery has been disconnected. Close the back door. Open the back door with the auto-

matic open feature. Do not stop the process until the back door opens completely.

2. Disconnect the negative battery cable.
3. Make sure the engine is cold.
4. Drain the cooling system.
5. Remove or disconnect the following:
   • Engine room cover
   • Accessory drive belt

➡**Leave tensioner pulley in its fixed position.**

   • Cooling fan
   • Water pump pulley
   • Water pump

To install:

➡**Be sure to use new fasteners, as required.**

6. Install or connect the following:
   • Water pump with a new gasket. Tighten bolts to 18 ft. lbs. (25 Nm)
   • Water pump pulley and tighten bolts to 87 inch lbs. (10 Nm)
   • Accessory drive belt
   • Air intake assembly
   • Engine splash guard
7. Refill the cooling system.
8. Start the engine and check for leaks.
9. Be sure to perform the reconnect/relearn procedures.

ENGINE ELECTRICAL

BATTERY

REMOVAL & INSTALLATION

➡Whenever the negative battery cable is disconnected the following components will require resetting. The Idle Air Volume Learning, Steering Angle Sensor Neutral Position, Sunroof Memory Reset/Initialization, Automatic Drive Positioner System, Audio presets and Navigation. Use the CONSULT-III diagnostic tool, or equivalent to perform the required resets.

1. Before servicing the vehicle, refer to the Precautions Section.

➡If working near and/or around the SRS system and components, be sure to disable the SRS system. After disabling the system wait three minutes or more before servicing the vehicle.

➡The automatic back door system must be initialized anytime the battery has been disconnected. Close the back

door. Open the back door with the automatic open feature. Do not stop the process until the back door opens completely.

2. Disconnect both cables from the battery.

➡Disconnect the negative battery cable, first.

3. Remove the battery cover.
4. Remove the battery clamp nuts and battery clamp.
5. Remove the battery.

To install:
6. Installation is the reverse of the removal procedure.
7. Perform reconnect/relearn procedures.

➡Connect the positive battery cable first.

8. Tighten the battery terminal nut to 31 inch lbs. (3.5 Nm). Tighten the clamp nuts to 11 ft. lbs.3 (3.5 Nm).

BATTERY SYSTEM

BATTERY RECONNECT/RELEARN PROCEDURE

➡Whenever the negative battery cable is disconnected the following components will require resetting. The Idle Air Volume Learning, Steering Angle Sensor Neutral Position, Sunroof Memory Reset/Initialization, Automatic Drive Positioner System, Audio presets and Navigation. Use the CONSULT-III diagnostic tool, or equivalent to perform the required resets.

➡The following systems, if equipped, require attention once the negative battery cable is disconnected. These systems are Automatic temperature control system, Automatic drive positioner, Power window control, Sunshade system and Rear view monitor. You will need the CONSULT-III diagnostic tool, or equivalent. Follow the directions on the screen of the tool, as needed.

➡The following applies to vehicles equipped with Intelligent Key

system and Nissan Anti-Theft System (NATS).

Remove and install all control units after disconnecting both battery cables with the ignition knob in the LOCK position. Always use the CONSULT-III diagnostic tool to perform self-diagnostics as a part of each function inspection after finishing repair work. If DTC's are detected, perform trouble diagnosis according to the self-diagnostic results.

➡**For models equipped with the Intelligent Key system and NATS, an electrically controlled steering lock mechanism is adopted on the key cylinder. For this reason, if the battery is disconnected or discharged the steering wheel will lock and steering wheel rotation will become impossible. If wheel rotation is required when battery power is interrupted, follow the procedure below before starting the repair operation.**

1. Connect both battery cables.

➡**Supply power using jumper cables if the battery is discharged.**

2. Using the Intelligent Key or mechanical key turn the ignition switch to the ACC position. At this time the steering lock will be released.

3. Disconnect both battery cables. The steering lock will remain released and the steering wheel can be rotated.

4. Perform the necessary repair operation.

5. When repair is complete, return the ignition switch to the LOCK position before connecting the battery cables. At this time the steering lock mechanism will engage.

6. Using the CONSULT-III diagnostic tool, perform self-diagnostics.

➡**The following applies to vehicles equipped with Automatic rear door opener/closer system.**

7. Close back door (cargo access door).

8. Open back door with automatic open operation. Switch is located on instrument panel, lower left side.

9. Do not stop the automatic door operation until the door is fully open.

10. Open back door (cargo access door).

Vehicles equipped with engine and transaxle computers may require a relearn procedure after the vehicle battery has been disconnected. Most vehicle computers memorize and store vehicle operational pat-

terns. When the battery is disconnected, the information may be cleared. If the information is cleared, the computer will go into default mode in order to operate the vehicle. The vehicle computer will relearn operational patterns each time the vehicle is restarted. The relearning process may take up to 40 or more key cycles.

When a specific engine component is replaced, a relearn procedure may be required. If the relearn procedure is not performed, the vehicle may exhibit the following:

- Harsh or poor shift quality
- Poor fuel mileage
- Hesitation or stumble
- Unstable idle or stalling
- Lean or rich running conditions

If an accessory component was replaced, a relearn procedure may also be required. The following systems and components may not work properly without a relearn procedure:

- Anti-theft system
- Steering system
- Power window system
- Power sunroof system

➡**If working near and/or around the SRS system and components, be sure to disable the SRS system. After disabling the system wait three minutes or more before servicing the vehicle.**

➡**Whenever the negative battery cable is disconnected the following components will require resetting. The Idle Air Volume Learning, Steering Angle Sensor Neutral Position, Sunroof Memory Reset/Initialization, Automatic Drive Positioner System, Audio presets and Navigation. Use the Nissan CONSULT III diagnostic tool, or equivalent to perform the required resets.**

➡**The automatic back door system must be initialized anytime the battery has been disconnected. Close the back door. Open the back door with the automatic open feature. Do not stop the process until the back door opens completely.**

In most cases a diagnostic tool will be required to perform the required relearn procedures.

Accelerator Pedal Released Position Learning

Accelerator Pedal Released Position Learning is an operation to learn the fully

released position of the accelerator pedal by monitoring the accelerator pedal position sensor output signal. It must be performed each time the harness connector of the accelerator pedal position sensor or ECM is disconnected.

1. Before servicing the vehicle, refer to the Precautions Section.

2. Check that the accelerator pedal is fully released.

3. Turn the ignition switch ON and wait at least 2 seconds.

4. Turn the ignition switch OFF and wait at least 10 seconds.

5. Turn the ignition switch ON and wait at least 2 seconds.

6. Turn the ignition switch OFF and wait at least 10 seconds.

Sunroof Memory Reset/Initialization Learning

Memory Reset Procedure

➡**Do not disconnect the electronic power while the sunroof is operating or within 5 seconds after the sunroof stops (to wipe-out the memory of lid position and operating friction).**

1. Before servicing the vehicle, refer to the Precautions Section.

2. Initialization of the system should be conducted after the following conditions.

- When the battery has been disconnected or discharged
- When the sunroof motor has been disconnected from power
- When the sunroof motor is changed
- When the sunroof does not operate normally (incomplete initialization conditions)

Initialization Procedure

If the sunroof does not close or open automatically, use the following procedure to return sunroof operation to normal.

3. Turn the ignition switch ON.

4. Push and hold the sunroof tilt switch in the forward (DOWN) position until the sunroof is fully closed.

5. After the sunroof has closed all the way, push and hold the tilt switch forward (DOWN) again for more than 2 seconds to relearn the motor position.

6. Initialization is complete if the sunroof operates normally.

Throttle Valve Closed Position Learning

Throttle Valve Closed Position Learning is an operation to learn the fully closed position of the throttle valve by monitoring the throttle position sensor output signal. It must be performed each time the harness connector of electric throttle control actuator or ECM is disconnected.

1. Before servicing the vehicle, refer to the Precautions Section.
2. Check that the accelerator pedal is fully released.
3. Turn the ignition switch ON.

4. Turn the ignition switch OFF and wait at least 10 seconds.
5. Check that throttle valve moves during the above 10 seconds by confirming the operating sound.

ENGINE ELECTRICAL   CHARGING SYSTEM

ALTERNATOR

REMOVAL & INSTALLATION
See Figure 38.

1. Before servicing the vehicle, refer to the Precautions Section.

➡**If working near and/or around the SRS system and components, be sure to disable the SRS system. After disabling the system wait three minutes or more before servicing the vehicle.**

➡**Whenever the negative battery cable is disconnected the following components will require resetting. The Idle Air Volume Learning, Steering Angle Sensor Neutral Position, Sunroof Memory Reset/Initialization, Automatic Drive Positioner System, Audio presets and Navigation. Use the CONSULT-III diagnostic tool, or equivalent to perform the required resets.**

➡**Whenever the negative battery cable is disconnected the following components will require resetting. The Idle Air Volume Learning, Steering Angle Sensor Neutral Position, Sunroof Memory Reset/Initialization, Automatic Drive Positioner System, Audio presets and Navigation. Use the CONSULT-III diagnostic tool, or equivalent to perform the required resets.**

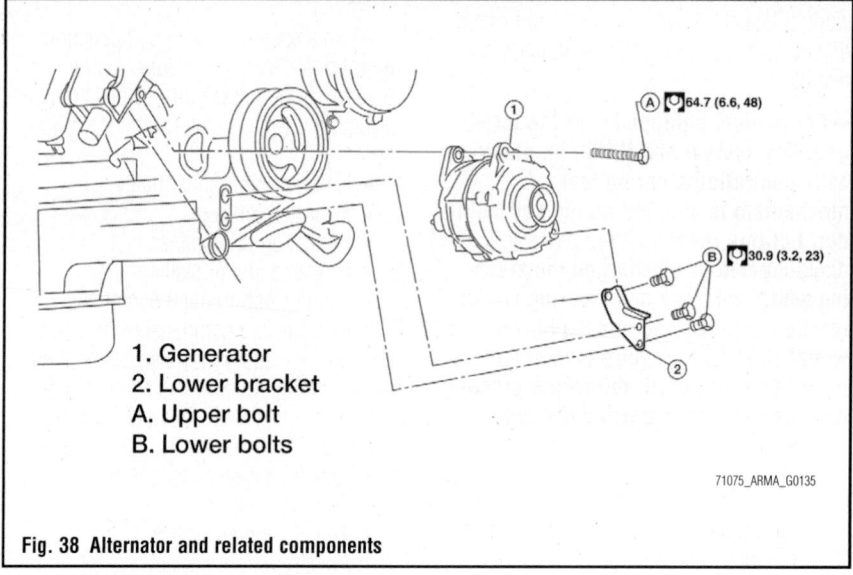

1. Generator
2. Lower bracket
A. Upper bolt
B. Lower bolts

71075_ARMA_G0135

Fig. 38 Alternator and related components

2. Disconnect the negative battery cable.
3. Remove or disconnect the following:
   - Fan shroud
   - Drive belt
   - Lower alternator bracket
   - Alternator upper bolt
   - Alternator harness connectors
   - Alternator

   To install:

➡**Be sure to use new fasteners, as required.**

4. Install or connect the following:
   - Alternator
   - Alternator harness connectors
   - Upper bolt, tighten to 48 ft. lbs. (65 Nm)
   - Lower bracket, tighten to 23 ft. lbs (31 Nm)
   - Drive belt
   - Fan shroud
   - Negative battery cable
5. Be sure to perform the reconnect/relearn procedures.

ENGINE ELECTRICAL   IGNITION SYSTEM

IGNITION COIL

REMOVAL & INSTALLATION
See Figure 39.

1. Before servicing the vehicle, refer to the Precautions Section.

➡**If working near and/or around the SRS system and components, be sure to disable the SRS system. After disabling the system wait three minutes or more before servicing the vehicle.**

➡**Whenever the negative battery cable is disconnected the following components will require resetting. The Idle Air Volume Learning, Steering Angle**

Sensor Neutral Position, Sunroof Memory Reset/Initialization, Automatic Drive Positioner System, Audio presets and Navigation. Use the CONSULT-III diagnostic tool, or equivalent to perform the required resets.

➡**The automatic back door system must be initialized anytime the battery has been disconnected. Close the back door. Open the back door with the automatic open feature. Do not stop the process until the back door opens completely.**

2. Disconnect the negative battery cable.
3. Remove the engine room cover.

Remove the air cleaner assembly, as required.

4. Disconnect the harness connector from the ignition coil.
5. Remove the ignition coil retaining bolt.
6. Remove the ignition coil.

To install:

➡**Be sure to use new fasteners, as required.**

7. Install the ignition coil, torque the retaining bolt to 85 inch lbs. (10 Nm).
8. Connect the harness coil.
9. Connect the negative battery cable.
10. Be sure to perform the reconnect/relearn procedures.

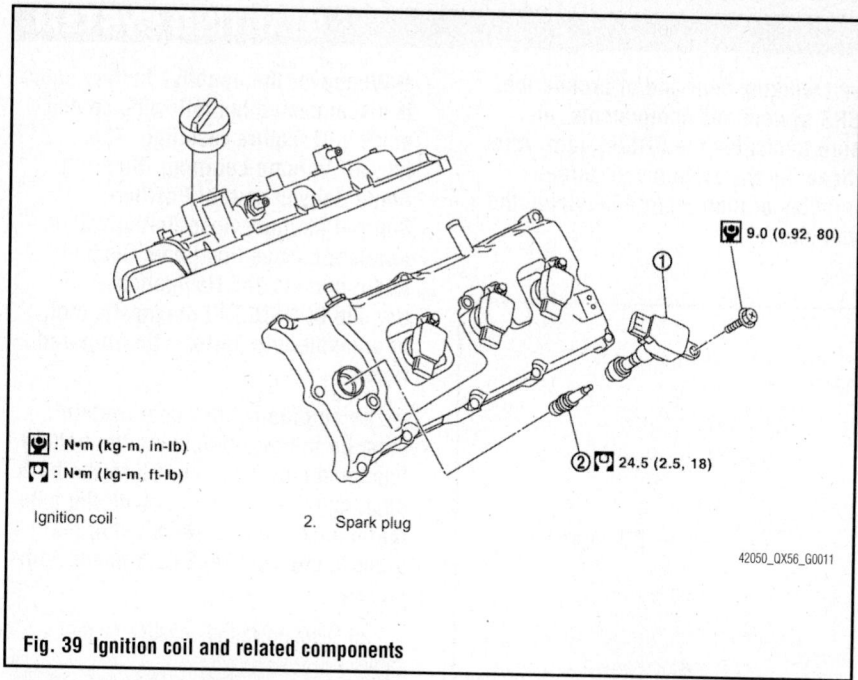

9.0 (0.92, 80)

①

② 24.5 (2.5, 18)

: N•m (kg-m, in-lb)

: N•m (kg-m, ft-lb)

Ignition coil   2.   Spark plug

42050_QX56_G0011

Fig. 39 Ignition coil and related components

FIRING ORDERS

1–8–7–3–6–5–4–2

IGNITION TIMING

INSPECTION & ADJUSTMENT

The ignition timing is controlled by the Powertrain Control Module (PCM). No adjustment is necessary or possible.

1. Before servicing the vehicle, refer to the Precautions Section.

➡**If working near and/or around the SRS system and components, be sure to disable the SRS system. After disabling the system wait three minutes or more before servicing the vehicle.**

➡**Whenever the negative battery cable is disconnected the following components will require resetting. The Idle Air Volume Learning, Steering Angle**

Sensor Neutral Position, Sunroof Memory Reset/Initialization, Automatic Drive Positioner System, Audio presets and Navigation. Use the CONSULT-III diagnostic tool, or equivalent to perform the required resets.

➡**The automatic back door system must be initialized anytime the battery has been disconnected. Close the back door. Open the back door with the automatic open feature. Do not stop the process until the back door opens completely.**

2. Remove the number one ignition coil.

3. Connect the number one ignition coil and spark plug with a suitable high tension wire.

4. Attach the timing light clamp to the wire.

5. Check the ignition timing.

SPARK PLUGS

REMOVAL & INSTALLATION

1. Before servicing the vehicle, refer to the Precautions Section.

➡**If working near and/or around the SRS system and components, be sure to disable the SRS system. After disabling the system wait three minutes or more before servicing the vehicle.**

➡**Whenever the negative battery cable is disconnected the following components will require resetting. The Idle Air Volume Learning, Steering Angle Sensor Neutral Position, Sunroof Memory Reset/Initialization, Automatic Drive Positioner System, Audio presets and Navigation. Use the CONSULT-III diagnostic tool, or equivalent to perform the required resets.**

2. Disconnect the negative battery cable.

3. Disconnect the harness connector from the ignition coil.

4. Remove the ignition coil retaining bolt.

5. Remove the ignition coil.

6. Remove the spark plug using a spark plug socket and wrench.

To install:

➡**Be sure to use new fasteners, as required.**

7. Be sure the spark plug gap is to specification (0.043 in.).

8. Carefully install the spark plug and torque to specification, 18 ft. lbs. (25 Nm).

9. Install the ignition coil, torque the retaining bolt to 80 inch lbs. (9 Nm).

10. Connect the harness coil.

11. Connect the negative battery cable.

12. Be sure to perform the reconnect/relearn procedures.

ENGINE ELECTRICAL

STARTING SYSTEM

STARTER

REMOVAL & INSTALLATION

See Figure 40.

1. Before servicing the vehicle, refer to the Precautions Section.

➡If working near and/or around the SRS system and components, be sure to disable the SRS system. After disabling the system wait three minutes or more before servicing the vehicle.

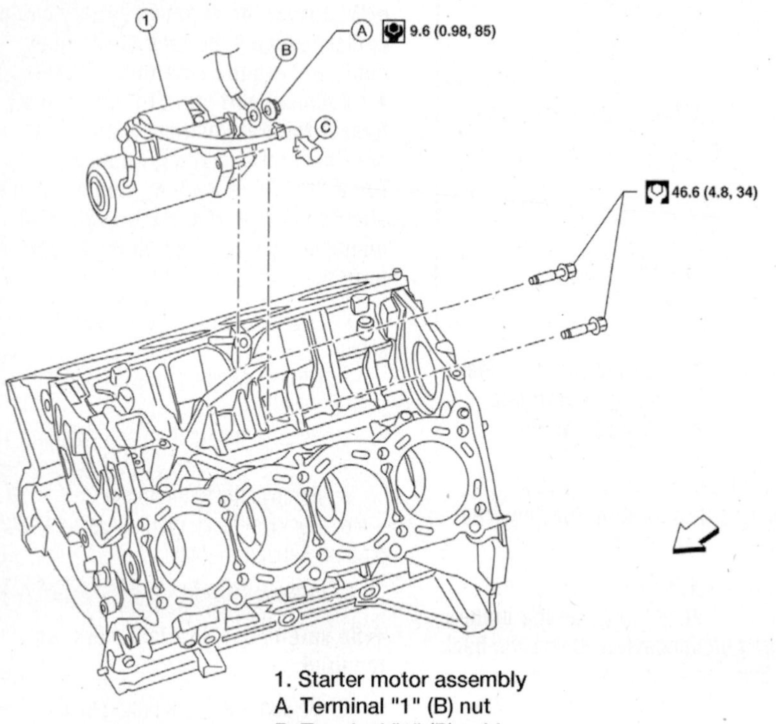

1. Starter motor assembly
A. Terminal "1" (B) nut
B. Terminal "1" (B) cable
C. Terminal "2" (S) connector Vehicle front

9.6 (0.98, 85)

46.6 (4.8, 34)

71075_ARMA_G0137

Fig. 40 Starter and related components

➡Whenever the negative battery cable is disconnected the following components will require resetting. The Idle Air Volume Learning, Steering Angle Sensor Neutral Position, Sunroof Memory Reset/Initialization, Automatic Drive Positioner System, Audio presets and Navigation. Use the CONSULT-III diagnostic tool, or equivalent to perform the required resets.

➡The automatic back door system must be initialized anytime the battery has been disconnected. Close the back door. Open the back door with the automatic open feature. Do not stop the process until the back door opens completely.

2. Disconnect the negative battery cable.
3. Remove the intake manifold.
4. Remove the starter harness connectors.
5. Remove the starter retaining bolts.
6. Remove the starter from its mounting.

To install:

➡Be sure to use new fasteners, as required.

7. Installation is the reverse of the removal procedure.
8. Tighten the retaining bolts to 34 ft. lbs. (46 Nm).
9. Tighten the terminal nut to 8 ft. lbs. (10.8 Nm).
10. Be sure to perform the reconnect/relearn procedures.

ENGINE MECHANICAL

ACCESSORY DRIVE BELT SYSTEM

ADJUSTMENT

Drive belt tension is not necessary, as it is automatically adjusted by the auto tensioner.

BELT ROUTING

See Figure 41.

INSPECTION

Inspect the drive belt for signs of glazing or cracking. A glazed belt will be perfectly smooth from slippage, while a good belt will have a slight texture of fabric visible. Cracks will usually start at the inner edge of the belt and run outward. All worn or damaged drive belts should be replaced immediately.

REMOVAL & INSTALLATION

1. Before servicing the vehicle, refer to the Precautions Section.

➡If working near and/or around the SRS system and components, be sure to disable the SRS system. After disabling the system wait three minutes or more before servicing the vehicle.

➡Whenever the negative battery cable is disconnected the following components will require resetting. The Idle Air Volume Learning, Steering Angle Sensor Neutral Position, Sunroof Memory Reset/Initialization, Automatic Drive Positioner System, Audio presets and Navigation. Use the CONSULT-III diagnostic tool, or equivalent to perform the required resets.

➡The automatic back door system must be initialized anytime the battery

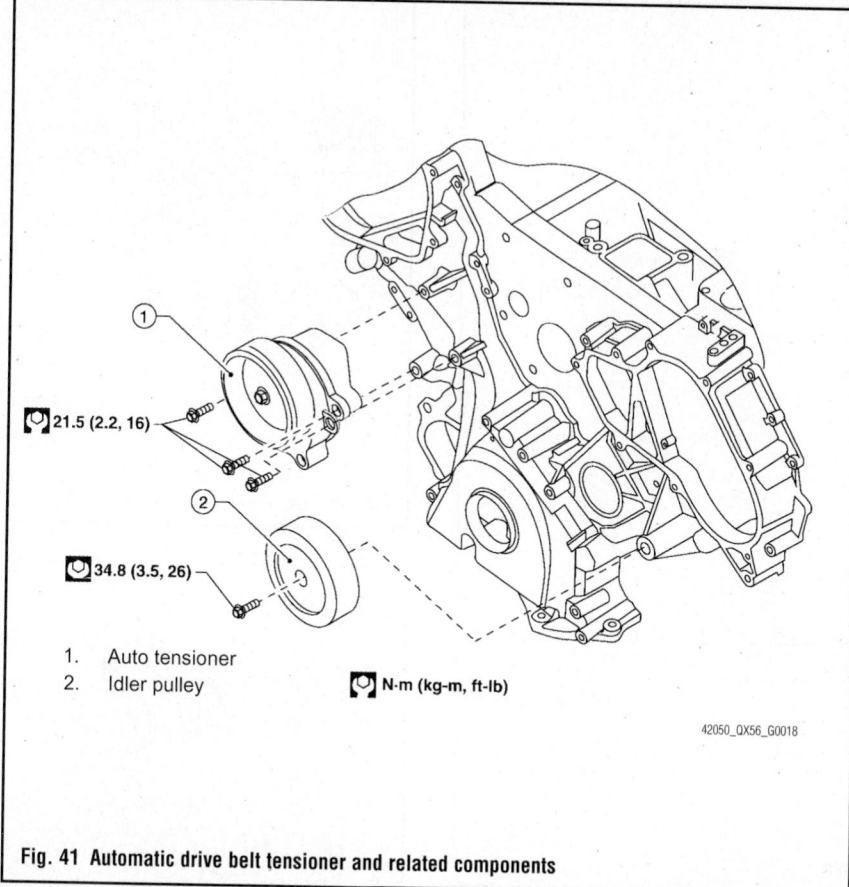

21.5 (2.2, 16)

34.8 (3.5, 26)

1. Auto tensioner
2. Idler pulley

N·m (kg-m, ft-lb)

42050_QX56_G0018

Fig. 41 Automatic drive belt tensioner and related components

5.5 (0.56, 49)

37663_QX56_G0049

Fig. 42 Engine room cover and related components

has been disconnected. Close the back door. Open the back door with the automatic open feature. Do not stop the process until the back door opens completely.

2. Disconnect the negative battery cable.

3. Remove the engine room cover.

4. Remove the air duct and resonator assembly.

5. Install special tool J-46535, or equivalent on the auto tensioner pulley bolt and move it upward.

➡**Avoid placing your hand in a location where pinching may occur if the holding tool accidentally comes off.**

6. Remove the drive belt from the vehicle.

To install:

➡**Be sure to use new fasteners, as required.**

7. Installation is the reverse of the removal procedure.

8. Be sure that the belt is securely installed around all pulleys.

9. Rotate the crankshaft several times clockwise to equalize belt tension between the pulleys.

10. Make sure that the belt tension is within the allowable working range, using the indicator notch on the auto tensioner.

11. Be sure to perform the reconnect/relearn procedures.

AIR CLEANER

REMOVAL & INSTALLATION

Air Cleaner Assembly
See Figures 42 and 43.

1. Before servicing the vehicle, refer to the Precautions Section.

➡**If working near and/or around the SRS system and components, be sure to disable the SRS system. After disabling the system wait three minutes or more before servicing the vehicle.**

➡**Whenever the negative battery cable is disconnected the following components will require resetting. The Idle Air Volume Learning, Steering Angle Sensor Neutral Position, Sunroof Memory Reset/Initialization, Automatic Drive Positioner System, Audio presets and Navigation. Use the CONSULT-III**

diagnostic tool, or equivalent to perform the required resets.

➡**The automatic back door system must be initialized anytime the battery has been disconnected. Close the back door. Open the back door with the automatic open feature. Do not stop the process until the back door opens completely.**

2. Disconnect the negative battery cable.

3. Remove the engine room cover.

4. Disconnect the harness connector from the upper air cleaner case.

5. Remove the air duct/resonator assembly and air cleaner case.

6. Remove the air cleaner filter.

7. Remove the lower air cleaner case.

To install:

➡**Be sure to use new fasteners, as required.**

8. Installation is the reverse of removal procedure.

9. Be sure to perform the reconnect/relearn procedures.

Air Filter Element
See Figures 42 and 43.

1. Before servicing the vehicle, refer to the Precautions Section.

➡**If working near and/or around the SRS system and components, be sure to disable the SRS system. After disabling the system wait three minutes or more before servicing the vehicle.**

➡**Whenever the negative battery cable is disconnected the following components will require resetting. The Idle Air Volume Learning, Steering Angle**

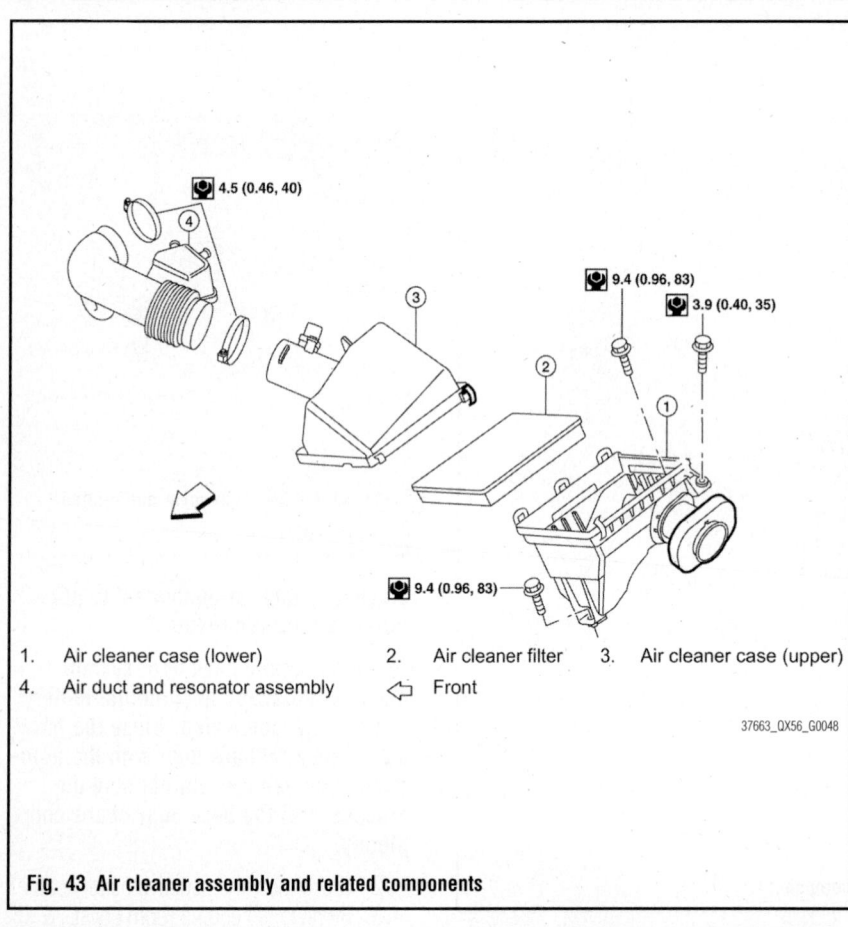

4.5 (0.46, 40)

9.4 (0.96, 83)

3.9 (0.40, 35)

9.4 (0.96, 83)

1. Air cleaner case (lower)
4. Air duct and resonator assembly
2. Air cleaner filter
3. Air cleaner case (upper)
⟵ Front

37663_QX56_G0048

Fig. 43 Air cleaner assembly and related components

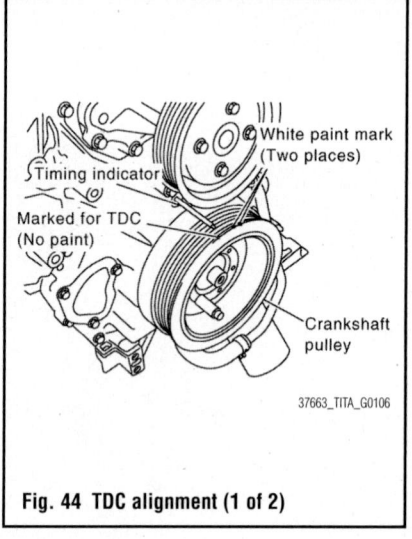

White paint mark (Two places)

Timing indicator

Marked for TDC (No paint)

Crankshaft pulley

37663_TITA_G0106

Fig. 44 TDC alignment (1 of 2)

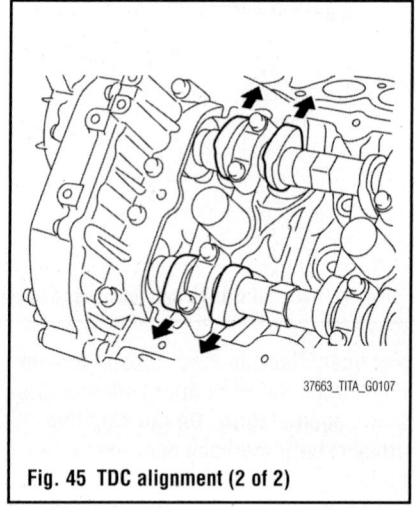

37663_TITA_G0107

Fig. 45 TDC alignment (2 of 2)

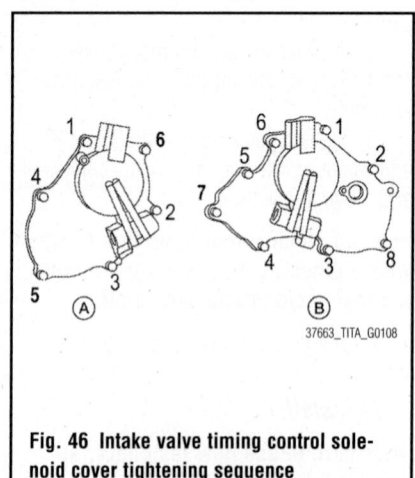

37663_TITA_G0108

Fig. 46 Intake valve timing control solenoid cover tightening sequence

Sensor Neutral Position, Sunroof Memory Reset/Initialization, Automatic Drive Positioner System, Audio presets and Navigation. Use the CONSULT-III diagnostic tool, or equivalent to perform the required resets.

➡The automatic back door system must be initialized anytime the battery has been disconnected. Close the back door. Open the back door with the automatic open feature. Do not stop the process until the back door opens completely.

2. Disconnect the negative battery cable.

3. Remove the engine room cover.

4. Disconnect the harness connector from the upper air cleaner case.

5. Remove the air duct/resonator assembly and air cleaner case.

6. Remove the air cleaner filter.

To install:

➡Be sure to use new fasteners, as required.

7. Installation is the reverse of the removal procedure.

8. Be sure to perform the reconnect/relearn procedures.

CAMSHAFT & BEARINGS

REMOVAL & INSTALLATION

See Figures 44 through 60.

1. Before servicing the vehicle, refer to the Precautions Section.

➡If working near and/or around the SRS system and components, be sure to disable the SRS system. After disabling the system wait three minutes or more before servicing the vehicle.

➡Whenever the negative battery cable is disconnected the following components will require resetting. The Idle Air Volume Learning, Steering Angle Sensor Neutral Position, Sunroof Memory Reset/Initialization, Automatic Drive Positioner System, Audio presets and Navigation. Use the CONSULT-III diagnostic tool, or equivalent to perform the required resets.

➡The automatic back door system must be initialized anytime the battery has been disconnected. Close the back door. Open the back door with the automatic open feature. Do not stop the process until the back door opens completely.

2. Disconnect the negative battery cable.

3. Remove the engine cover.

4. Remove the air cleaner assembly.

5. Remove the power steering reservoir tank bolts. Position the unit to the side.

6. Remove the valve covers.

7. Remove the spark plugs.

8. Remove the drive belt.

9. Be sure that the number one cylinder is at TDC on the compression stroke.

➡**Turn the crankshaft pulley clockwise to align the TDC identification notch (without paint mark) with the timing indicator on the front cover. At this time make sure that the intake and exhaust cam lobes of the number one cylinder (top front on left bank) point outside. If not turn the crankshaft pulley once more. See illustration**

10. Remove the CMP sensor.

11. Remove the intake valve timing control position sensors (right and left).

12. Remove the intake valve timing control solenoid valves (right and left).

13. Loosen and remove the intake valve timing control valve cover (right and left)

bolts in the reverse order of the tightening sequence.

14. Paint alignment marks on the right bank (A) timing chain links (C) and left bank (B) timing chain links (D) and align with the camshaft sprocket alignment marks (E) and (F). See illustration.

15. To remove the left tensioner, squeeze the return proof clip ends using a suitable tool and push the plunger into the tensioner body. Secure the plunger using a stopper pin (hard wire 0.04 inch in diameter). Remove the bolts and the tensioner.

➡**The plunger, spring and spring seat pop out when squeezing the return proof clip without holding the plunger head. It may cause serious injuries. Always hold the plunger head when removing.**

➡**Stop the plunger in the fully extended position using the return**

proof clip (1) if the stopper pin is removed. Push the plunger (2) into the chain tensioner body while squeezing the return proof clip (1). Secure it using a stopper pin (3). See illustration.

16. Remove the chain tensioner cover from the front cover, using tool KV10111100, or equivalent. Do not damage the mating surfaces.

17. To remove the left tensioner, squeeze the return proof clip ends using a suitable tool and push the plunger into the tensioner body. Secure the plunger using a stopper pin (hard wire 0.04 inch in diameter). Remove the bolts and the tensioner.

➡**The plunger, spring and spring seat pop out when squeezing the return proof clip without holding the plunger head. It may cause serious injuries. Always hold the plunger head when removing.**

➡**If it is difficult to push the plunger on the tensioner, remove the plunger under the extended condition.**

18. Loosen the camshaft sprocket bolts and remove the sprockets.

➡**To avoid interference between the valves and pistons, do not turn the crankshaft or camshaft with the timing chain disconnected.**

19. Remove the front cover bolts. See illustration for location (arrow).

20. Remove the camshaft bracket bolts in the reverse order of the tightening sequence. Remove the number one camshaft bracket. The bottom of the front surface of the bracket will be stuck because of liquid gasket.

21. Remove the camshaft. Remove the lifters, as necessary.

To install:

➡**Be sure to use new fasteners, as required.**

22. Install the camshafts. Be sure that the camshafts are properly identified. See illustrations.

23. Install the dowel pins at the front of the camshaft. See illustration for proper direction.

24. Install the camshaft brackets.

➡**Install by referring to the illustration location mark on the upper surface. Install so that the installation mark can be correctly read when viewed from the intake manifold side.**

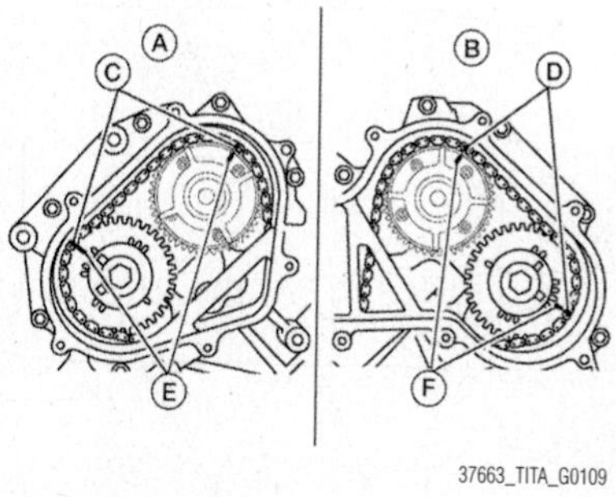

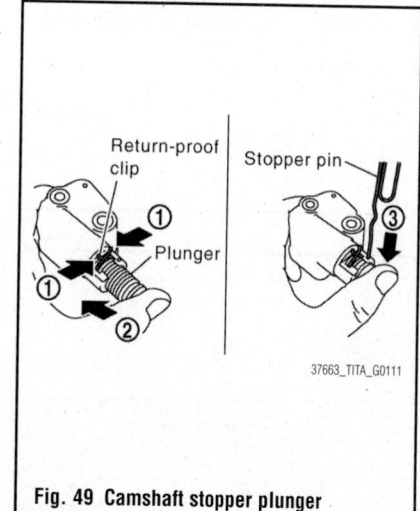

Fig. 47 Camshaft sprocket/chain link alignment

37663_TITA_G0109

Fig. 49 Camshaft stopper plunger retention

37663_TITA_G0111

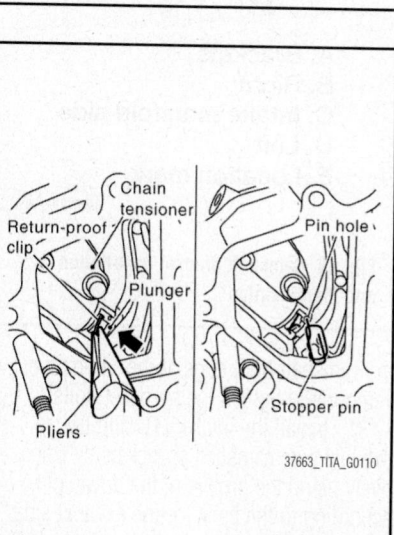

Fig. 48 Camshaft stopper pin installation

37663_TITA_G0110

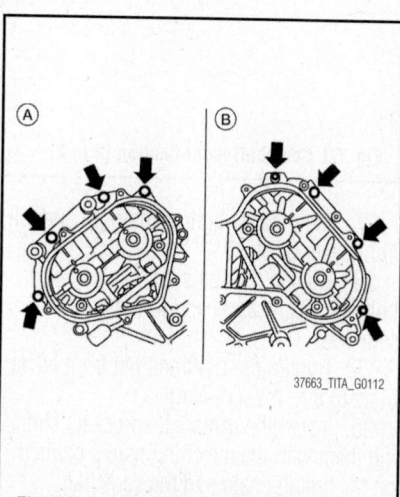

Fig. 50 Camshaft front cover bolt location (arrow)

37663_TITA_G0112

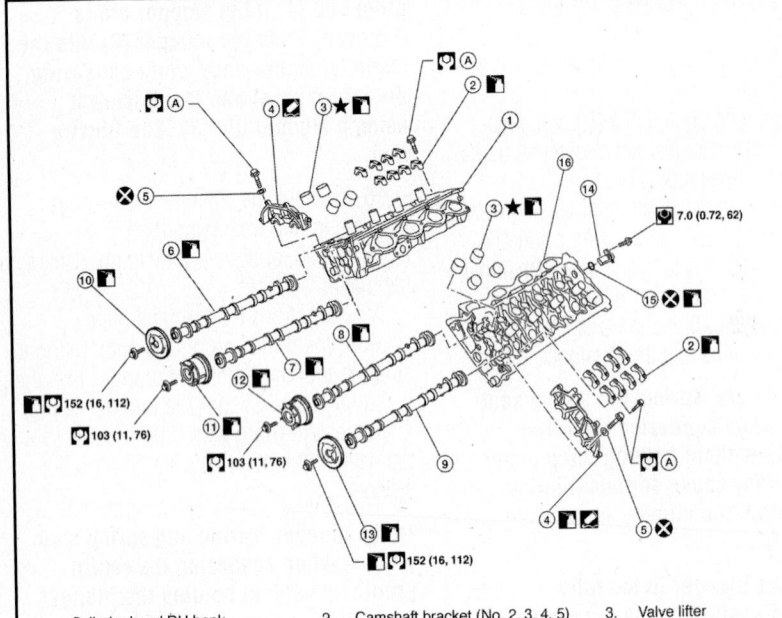

Fig. 51 **Camshafts and related components**

1. Cylinder head RH bank
2. Camshaft bracket (No. 2, 3, 4, 5)
3. Valve lifter
4. Camshaft bracket (No. 1)
5. Seal washer
6. Camshaft RH bank EXH
7. Camshaft RH bank INT
8. Camshaft LH bank INT
9. Camshaft LH bank EXH
10. Camshaft sprocket RH bank EXH
11. Camshaft sprocket RH bank INT (VTC)
12. Camshaft sprocket LH bank INT (VTC)
13. Camshaft sprocket LH bank EXH
14. Camshaft position sensor (PHASE)
15. O-ring
16. Cylinder head LH bank

37663_TITA_G0105

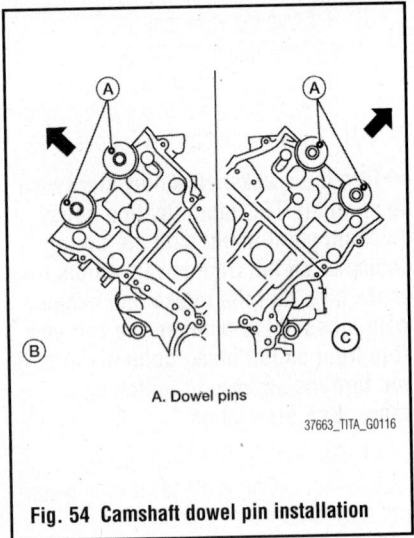

A. Dowel pins

37663_TITA_G0116

Fig. 54 **Camshaft dowel pin installation**

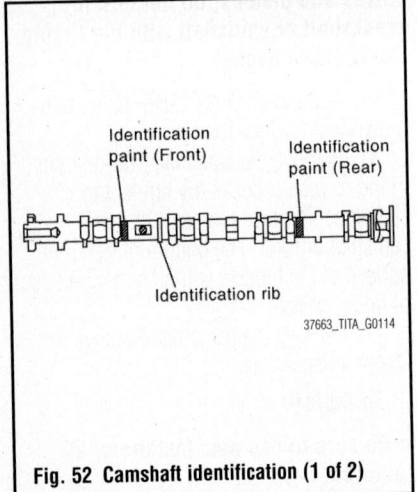

Fig. 52 **Camshaft identification (1 of 2)**

Identification paint (Front)

Identification paint (Rear)

Identification rib

37663_TITA_G0114

Bank	INT EXH		Identification paint (front)	Identification paint (rear)	Identification rib
RH	INT		Pink	—	Yes
	EXH		—	Orange	Yes
LH	INT		Pink	—	No
	EXH		—	Orange	No

37663_TITA_G0115

Fig. 53 **Camshaft identification (2 of 2)**

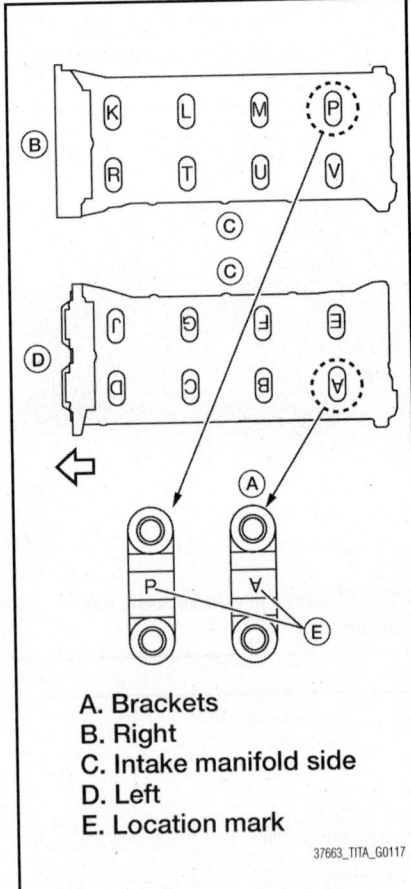

A. Brackets
B. Right
C. Intake manifold side
D. Left
E. Location mark

37663_TITA_G0117

Fig. 55 **Camshaft bracket installation and identification**

25. To install the number one camshaft bracket, apply liquid gasket as shown in the illustration. Be sure to wipe off any excessive gasket after installation.

26. Apply liquid gasket to the back side of the left front cover and the right front cover. Bead diameter should be 0.102–0.142 inch. Position the number one camshaft bracket close to the mounting position and then install it to prevent from touching gasket applied to each surface.

27. Temporarily tighten the right and left front cover bolts.

28. Tighten the camshaft bracket bolts to specification and in the proper sequence.

29. Tighten the right and left front cover bolts to 8 ft. lbs. (11 Nm).

30. Install the camshaft sprockets aligning them with the matching marks painted on the timing chain and the camshaft sprockets, before removal. Align the sprocket key groove with the dowel pin on

the camshaft front edge at the same time. Temporarily tighten the sprocket bolts.

31. Install the intake VTC and the exhaust side camshaft sprockets by selectively using the groove of the dowel pin according to the bank for the exhaust side camshaft sprockets, (common part used for both exhaust banks).

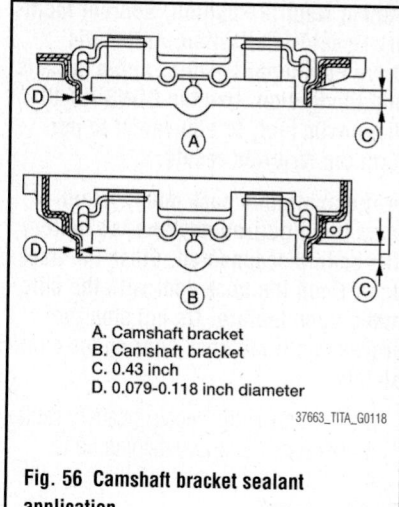

A. Camshaft bracket
B. Camshaft bracket
C. 0.43 inch
D. 0.079-0.118 inch diameter

37663_TITA_G0118

Fig. 56 Camshaft bracket sealant application

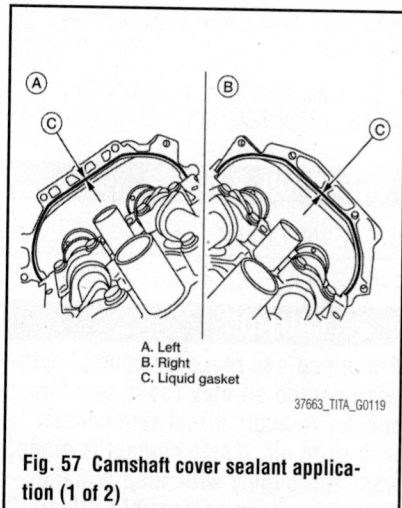

A. Left
B. Right
C. Liquid gasket

37663_TITA_G0119

Fig. 57 Camshaft cover sealant application (1 of 2)

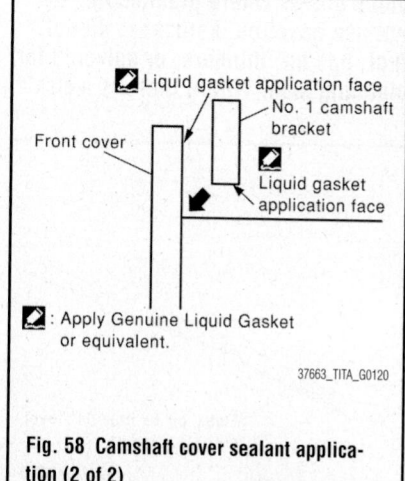

Liquid gasket application face

No. 1 camshaft bracket

Front cover

Liquid gasket application face

: Apply Genuine Liquid Gasket or equivalent.

37663_TITA_G0120

Fig. 58 Camshaft cover sealant application (2 of 2)

➡**Use the groove marked "R" for right bank and "L" for left bank.**

32. Lock the hex part of the camshaft in the same way as for removal. Tighten the sprocket bolts.

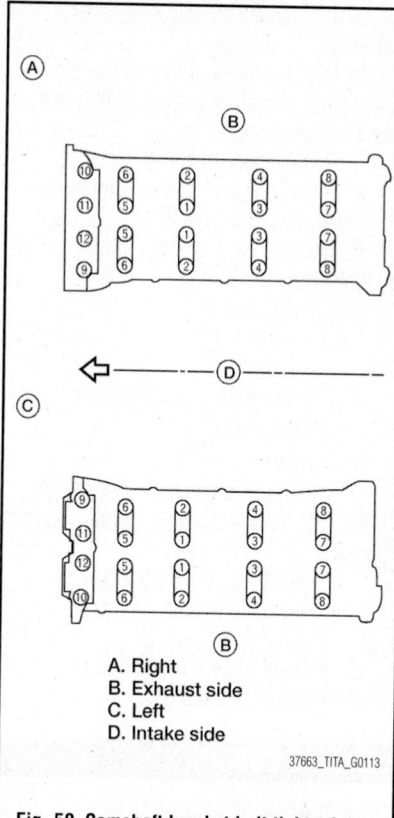

A. Right
B. Exhaust side
C. Left
D. Intake side

37663_TITA_G0113

Fig. 59 Camshaft bracket bolt tightening sequence

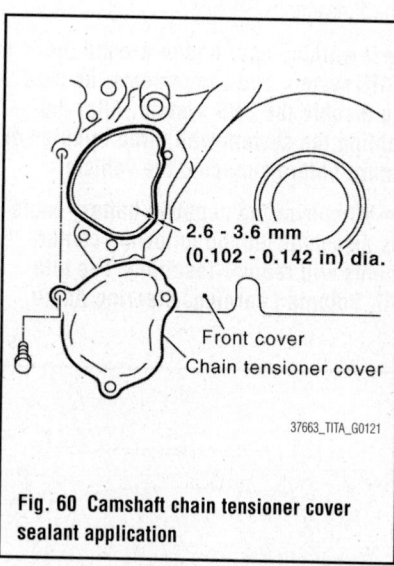

2.6 - 3.6 mm
(0.102 - 0.142 in) dia.

Front cover
Chain tensioner cover

37663_TITA_G0121

Fig. 60 Camshaft chain tensioner cover sealant application

33. Check that the timing marks are properly aligned.

34. To install the chain tensioner, compress the plunger and hold it using a stopper pin. Loosen the slack guide timing chain by rotating the camshaft hex part if mounting space is small. Tighten the tensioner bolts to 61 inch lbs. (7 Nm).

35. Remove the stopper pin and release the plunger, then apply tension to the chain.

36. Install the chain tensioner cover onto the front cover. Apply liquid gasket. See illustration. Tighten the bolts to 80 inch lbs. (9 Nm).

37. Check and adjust valve clearances.

38. Continue the installation in the reverse order of the removal procedure.

39. Be sure to perform the reconnect/relearn procedures.

CRANKSHAFT FRONT SEAL

REMOVAL & INSTALLATION

See Figure 61.

1. Before servicing the vehicle, refer to the Precautions Section.

➡**If working near and/or around the SRS system and components, be sure to disable the SRS system. After disabling the system wait three minutes or more before servicing the vehicle.**

➡**Whenever the negative battery cable is disconnected the following components will require resetting. The Idle Air Volume Learning, Steering Angle Sensor Neutral Position, Sunroof Memory Reset/Initialization, Automatic Drive Positioner System, Audio presets and Navigation. Use the CONSULT-III diagnostic tool, or equivalent to perform the required resets.**

➡**The automatic back door system must be initialized anytime the battery has been disconnected. Close the back door. Open the back door with the automatic open feature. Do not stop the process until the back door opens completely.**

2. Disconnect the negative battery cable.

3. Remove the engine cover, engine undercover and air cleaner assembly, as required for access.

4. Remove the drive belt.

5. Remove the radiator.

6. Remove the necessary components to gain access to the crankshaft pulley/damper.

7. Remove the crankshaft pulley/damper.

8. Remove the oil seal using a suitable tool.

To install:

9. Apply new engine oil to both the oil seal lip and dust seal lip of the new front oil seal.

10. Install the front oil seal so that each seal lip is oriented as shown.

11. Install the crankshaft damper pulley.

12. Tighten the crankshaft pulley bolt as follows:

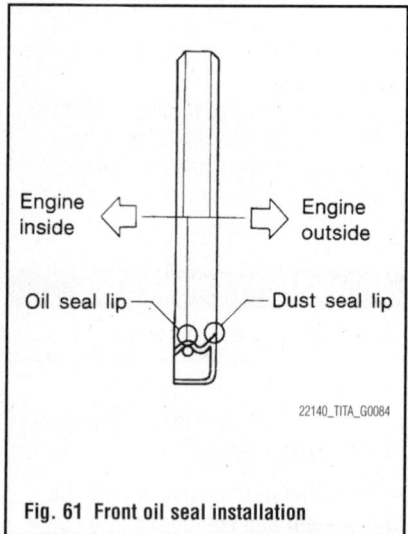

Fig. 61 Front oil seal installation

- Step 1: 69 ft. lbs. (93 Nm)
- Step 2: Additional 90°(angle tightening)

13. Be sure to fill the cooling system with the proper grade and type engine coolant.

14. Be sure to perform the reconnect/relearn procedures.

CRANKSHAFT PULLEY

REMOVAL & INSTALLATION

At this time the manufacturer does not provide removal and installation procedures for this component. The following procedure is a guideline and may differ from the vehicle you are servicing.

1. Before servicing the vehicle, refer to the Precautions Section.

➡If working near and/or around the SRS system and components, be sure to disable the SRS system. After disabling the system wait three minutes or more before servicing the vehicle.

➡Whenever the negative battery cable is disconnected the following components will require resetting. The Idle Air Volume Learning, Steering Angle Sensor Neutral Position, Sunroof Memory Reset/Initialization, Automatic Drive Positioner System, Audio presets and Navigation. Use the CONSULT-III diagnostic tool, or equivalent to perform the required resets.

➡The automatic back door system must be initialized anytime the battery has been disconnected. Close the back door. Open the back door with the automatic open feature. Do not stop the process until the back door opens completely.

2. Disconnect the negative battery cable.

3. Remove the engine cover, engine undercover and air cleaner assembly, as required for access.

4. Remove the drive belt.

5. Remove the necessary components to gain access to the crankshaft damper.

6. Remove the crankshaft pulley using suitable tool.

7. Set the bolts in the two bolt holes 0.04 inch (M6 × 1.0 mm) on the front surface.

8. Remove the crankshaft pulley from the crankshaft using tool.

To install:

9. Install the crankshaft damper pulley.

10. Tighten the crankshaft pulley bolt as follows:

- Step 1: 69 ft. lbs. (93 Nm)
- Step 2: Additional 90°(angle tightening)

11. Be sure to perform the reconnect/relearn procedures.

ENGINE COVER

REMOVAL & INSTALLATION
See Figure 62.

1. Before servicing the vehicle, refer to the Precautions Section.

➡If working near and/or around the SRS system and components, be sure to disable the SRS system. After disabling the system wait three minutes or more before servicing the vehicle.

➡Whenever the negative battery cable is disconnected the following components will require resetting. The Idle Air Volume Learning, Steering Angle Sensor Neutral Position, Sunroof Memory Reset/Initialization, Automatic Drive Positioner System, Audio presets and Navigation. Use the CONSULT-III diagnostic tool, or equivalent to perform the required resets.

➡The automatic back door system must be initialized anytime the battery has been disconnected. Close the back door. Open the back door with the automatic open feature. Do not stop the process until the back door opens completely.

2. Disconnect the negative battery cable.

3. Remove the cover retaining bolts.

4. Lift up on the cover to disengage the snap fit mounts.

To install:

➡Be sure to use new fasteners, as required.

5. Installation is the reverse of the removal procedure.

ENGINE OIL & FILTER

OIL LEVEL CHECK
See Figure 63.

✳✳ CAUTION

Prolonged and repeated contact with used engine oil may cause skin cancer. Try to avoid direct skin contact with used oil. If skin contact is made, wash thoroughly with soap or hand cleaner as soon as possible. Wear protective clothing, including impervious gloves where practicable. Do not use gasoline, kerosene, diesel fuel, gas oil, thinners, or solvents for cleaning skin. Where there is a risk

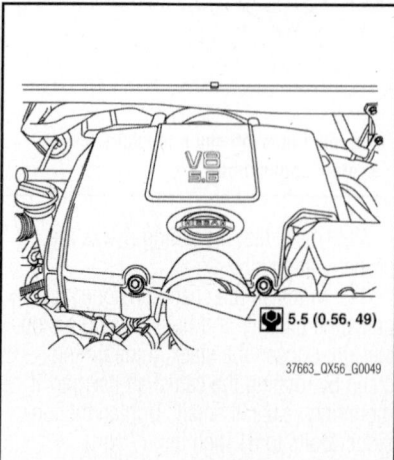

Fig. 62 Engine room cover and related components

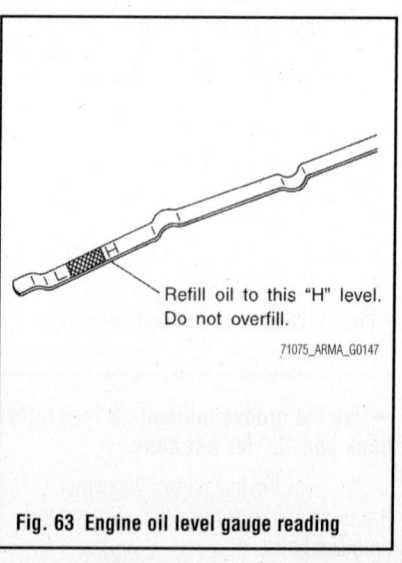

Fig. 63 Engine oil level gauge reading

of eye contact, eye protection should be worn, for example, chemical goggles or face shields; in addition an eye wash facility should be provided.

➥Be sure to check the engine oil with the vehicle engine OFF and parked on a level surface. If the engine was running prior to checking the oil, turn it off and wait 10 minutes.

1. Before servicing the vehicle, refer to the Precautions Section.
2. Remove the oil level gauge and wipe it clean with a shop towel. Be sure to properly dispose of the used shop towel.
3. Insert the gauge into its mounting.
4. Remove the oil level gauge and check the reading. See illustration.
5. Correct, oil level as required. Be sure to use the proper grade and type engine oil.

OIL & FILTER CHANGE

See Figure 64.

1. Before servicing the vehicle, refer to the Precautions Section.

➥If working near and/or around the SRS system and components, be sure to disable the SRS system. After disabling the system wait three minutes or more before servicing the vehicle.

➥Whenever the negative battery cable is disconnected the following components will require resetting. The Idle Air Volume Learning, Steering Angle Sensor Neutral Position, Sunroof Memory Reset/Initialization, Automatic Drive Positioner System, Audio presets and Navigation. Use the CONSULT-III diagnostic tool, or equivalent to perform the required resets.

2. Disconnect the negative battery cable.

➥Be sure that the engine is cold and the engine oil is cold.

➥The automatic back door system must be initialized anytime the battery has been disconnected. Close the back door. Open the back door with the automatic open feature. Do not stop the process until the back door opens completely.

3. Raise and safely support the vehicle.
4. Remove the undercover, if equipped.
5. Remove the oil drain plug. Discard the washer.
6. Drain the engine oil into a suitable container. Properly dispose of used engine oil.

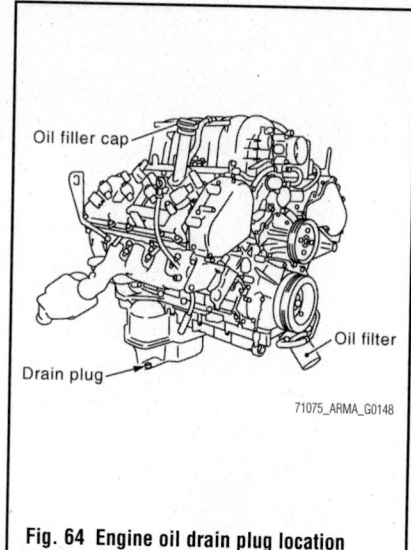

Oil filler cap

Oil filter

Drain plug

71075_ARMA_G0148

Fig. 64 Engine oil drain plug location

7. Using oil filter wrench remove the oil filter. Discard the filter.

To install:

➥Be sure to use new fasteners, as required

8. Install the drain plug. Use a new washer.
9. Tighten to specification, 25 ft. lbs. (34 Nm).
10. Coat the oil filter seal with clean engine oil prior to installation.
11. Do not over tighten the filter to its mounting.
12. Tightening specification should be 13 ft. lbs. (17.7 Nm).
13. Fill the engine with the proper grade and type engine oil.
14. Start the engine and check for leaks, correct as required.

ENGINE MOUNTS

REMOVAL & INSTALLATION

See Figure 65.

At this time the manufacturer does not provide removal and installation procedures for this component, refer to the illustration as required.

Before servicing the vehicle, refer to the Precautions Section.

➥If working near and/or around the SRS system and components, be sure to disable the SRS system. After disabling the system wait three minutes or more before servicing the vehicle.

➥Whenever the negative battery cable is disconnected the following components will require resetting. The Idle Air Volume Learning, Steering Angle

Sensor Neutral Position, Sunroof Memory Reset/Initialization, Automatic Drive Positioner System, Audio presets and Navigation. Use the CONSULT-III diagnostic tool, or equivalent to perform the required resets.

➥The automatic back door system must be initialized anytime the battery has been disconnected. Close the back door. Open the back door with the automatic open feature. Do not stop the process until the back door opens completely.

EXHAUST MANIFOLD

REMOVAL & INSTALLATION

See Figures 66 and 67.

1. Before servicing the vehicle, refer to the Precautions Section.

➥If working near and/or around the SRS system and components, be sure to disable the SRS system. After disabling the system wait three minutes or more before servicing the vehicle.

➥Whenever the negative battery cable is disconnected the following components will require resetting. The Idle Air Volume Learning, Steering Angle Sensor Neutral Position, Sunroof Memory Reset/Initialization, Automatic Drive Positioner System, Audio presets and Navigation. Use the CONSULT-III diagnostic tool, or equivalent to perform the required resets.

➥The automatic back door system must be initialized anytime the battery has been disconnected. Close the back door. Open the back door with the automatic open feature. Do not stop the process until the back door opens completely.

2. Disconnect the negative battery cable.
3. Raise and support the vehicle safely.
4. Remove the engine under cover, if equipped.
5. Remove the front final drive assembly, if equipped.
6. Remove the main muffler and center exhaust tube.
7. Remove the front exhaust tubes.
8. Remove the tire and wheel assemblies.
9. Remove the fender protectors.
10. Remove the A/F sensors. Do not drop the sensors. If the sensor is dropped it must be replaced.

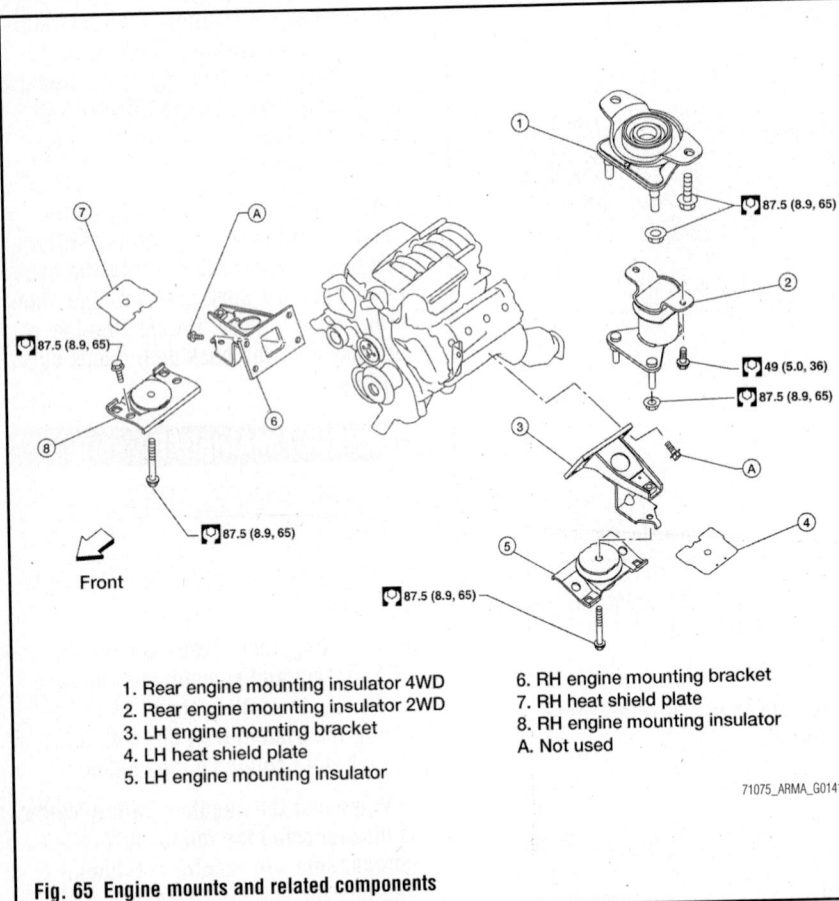

1. Rear engine mounting insulator 4WD
2. Rear engine mounting insulator 2WD
3. LH engine mounting bracket
4. LH heat shield plate
5. LH engine mounting insulator
6. RH engine mounting bracket
7. RH heat shield plate
8. RH engine mounting insulator
A. Not used

71075_ARMA_G0141

Fig. 65 Engine mounts and related components

HYDRAULIC LASH ADJUSTERS

REMOVAL & INSTALLATION

See Figure 68.

At this time the manufacturer does not provide removal and installation procedures for this component, refer to the illustration as required.

Before servicing the vehicle, refer to the Precautions Section.

➡ **If working near and/or around the SRS system and components, be sure to disable the SRS system. After disabling the system wait three minutes or more before servicing the vehicle.**

➡ **Whenever the negative battery cable is disconnected the following components will require resetting. The Idle Air Volume Learning, Steering Angle Sensor Neutral Position, Sunroof Memory Reset/Initialization, Automatic Drive Positioner System, Audio presets and Navigation. Use the CONSULT-III diagnostic tool, or equivalent to perform the required resets.**

➡ **The automatic back door system must be initialized anytime the battery has been disconnected. Close the back door. Open the back door with the auto-**

11. Properly support the engine.
12. Remove the engine mounting insulator.
13. Remove the exhaust manifold cover.
14. Remove the engine mounting bracket.
15. On right side remove the oil level dipstick.
16. Remove the left exhaust manifold nuts/bolts in the reverse order of the installation sequence.
17. Remove the exhaust manifold. Discard the gaskets.

To install:

➡ **Be sure to use new fasteners, as required.**

18. Installation is the reverse of the removal procedure.
19. Install new gaskets with the top of the triangular UP mark on it facing up and its coated face (gray side) toward the exhaust manifold side.
20. Tighten the retaining nuts/bolts to specification and in the proper sequence.
21. Be sure to perform the reconnect/relearn procedures.

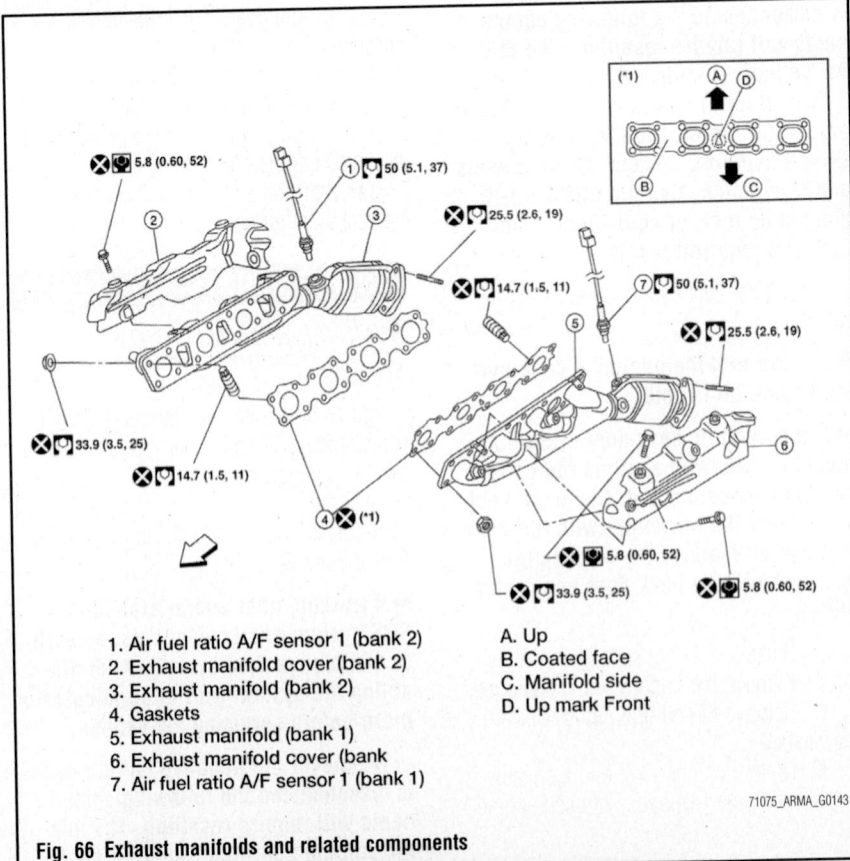

1. Air fuel ratio A/F sensor 1 (bank 2)
2. Exhaust manifold cover (bank 2)
3. Exhaust manifold (bank 2)
4. Gaskets
5. Exhaust manifold (bank 1)
6. Exhaust manifold cover (bank 1)
7. Air fuel ratio A/F sensor 1 (bank 1)

A. Up
B. Coated face
C. Manifold side
D. Up mark Front

71075_ARMA_G0143

Fig. 66 Exhaust manifolds and related components

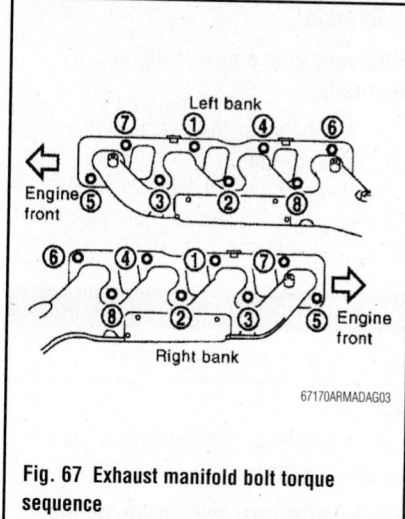

Fig. 67 Exhaust manifold bolt torque sequence

matic open feature. Do not stop the process until the back door opens completely.

INTAKE MANIFOLD

REMOVAL & INSTALLATION

Lower Manifold

See Figures 69 and 70.

1. Before servicing the vehicle, refer to the Precautions Section.

➡ If working near and/or around the SRS system and components, be sure to disable the SRS system. After disabling the system wait three minutes or more before servicing the vehicle.

➡ Whenever the negative battery cable is disconnected the following components will require resetting. The Idle Air Volume Learning, Steering Angle Sensor Neutral Position, Sunroof Memory Reset/Initialization, Automatic Drive Positioner System, Audio presets and Navigation. Use the CONSULT-III diagnostic tool, or equivalent to perform the required resets.

➡ The automatic back door system must be initialized anytime the battery has been disconnected. Close the back door. Open the back door with the automatic open feature. Do not stop the process until the back door opens completely.

2. Disconnect the negative battery cable.
3. Drain the cooling system.
4. Relieve the fuel system pressure.
5. Remove or disconnect the following:

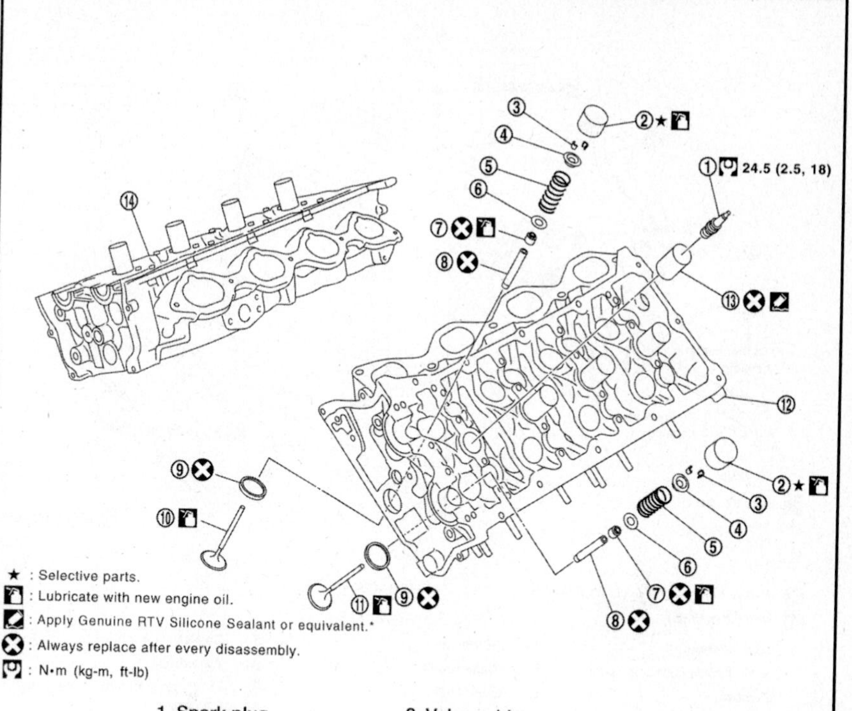

★ : Selective parts.
▨ : Lubricate with new engine oil.
▧ : Apply Genuine RTV Silicone Sealant or equivalent.*
⊗ : Always replace after every disassembly.
▨ : N·m (kg-m, ft-lb)

1. Spark plug
2. Valve lifter
3. Valve collet
4. Valve spring retainer
5. Valve spring
6. Valve spring seat
7. Valve oil seal
8. Valve guide
9. Valve seat
10. Valve (INT)
11. Valve (EXH)
12. Cylinder head (LH bank)
13. Spark plug tube
14. Cylinder head (RH bank)

71075_ARMA_G0140

Fig. 68 Valve lifters and related components

- Engine room cover
- Air intake assembly
- Fuel tube quick connector using special tool J-45488
- Wiring harnesses and brackets from manifold
- Vacuum hoses
- PCV hose and tube
- Electric throttle control actuator, loosening bolts diagonally
- Fuel injectors
- Fuel tube assembly
- Intake manifold, removing bolts in reverse order of installation

To install:

➡ Be sure to use new fasteners, as required.

6. Install the intake manifold with new gaskets. Tighten the bolts in order as shown.

7. Install or connect the following:
- Fuel tube assembly
- Fuel injectors
- Electronic throttle control actuator,

tightening the bolts in several steps
- PCV hose
- Vacuum hoses
- Wiring harnesses

8. Connect the fuel tube as follows:
   a. Apply a thin layer of engine oil on the tube from tip end to spool end.
   b. Insert tube into quick connector past the white identification mark.
   c. Insert tube into quick connector until top spool is completely inside the connector and 2nd level spool is exposed right below the connector.
   d. Pull slightly on the quick connector to ensure it is fully engaged.
   e. Install quick connector cap on quick connector joint.

9. Install or connect the following:
- Air intake assembly
- Engine cover

10. Refill the cooling system.
11. Start engine and check for leaks.
12. Be sure to perform the reconnect/relearn procedures.

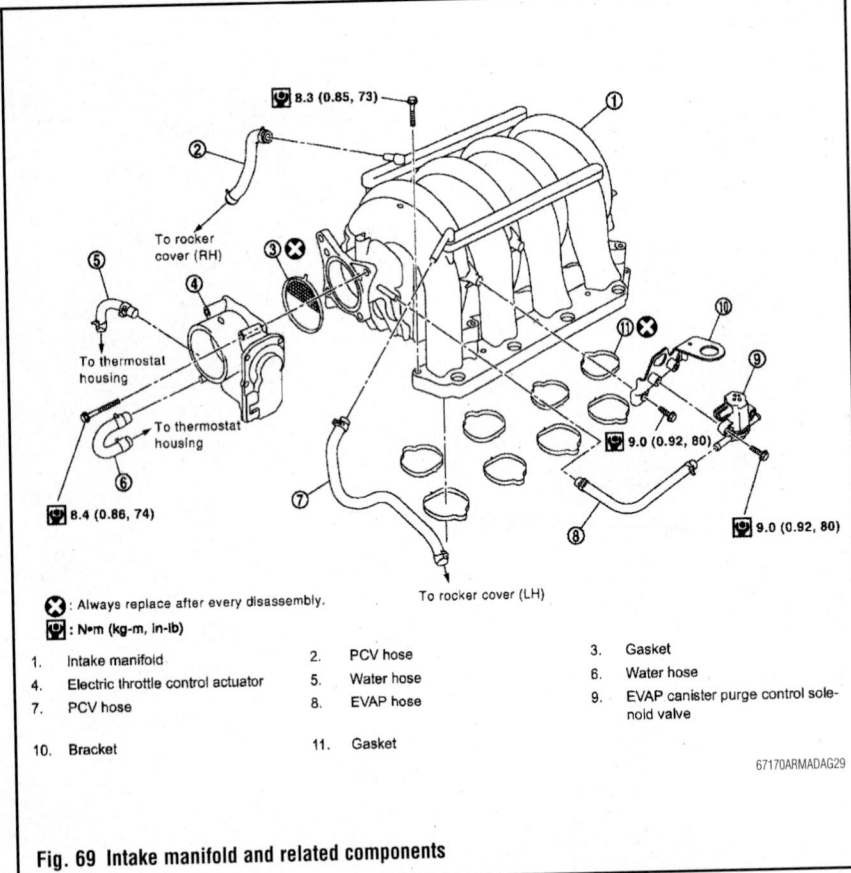

8.3 (0.85, 73)

To rocker cover (RH)

To thermostat housing

To thermostat housing

8.4 (0.86, 74)

9.0 (0.92, 80)

9.0 (0.92, 80)

To rocker cover (LH)

⊗ : Always replace after every disassembly.

▣ : N•m (kg-m, in-lb)

1. Intake manifold
4. Electric throttle control actuator
7. PCV hose
10. Bracket

2. PCV hose
5. Water hose
8. EVAP hose
11. Gasket

3. Gasket
6. Water hose
9. EVAP canister purge control solenoid valve

67170ARMADAG29

Fig. 69 Intake manifold and related components

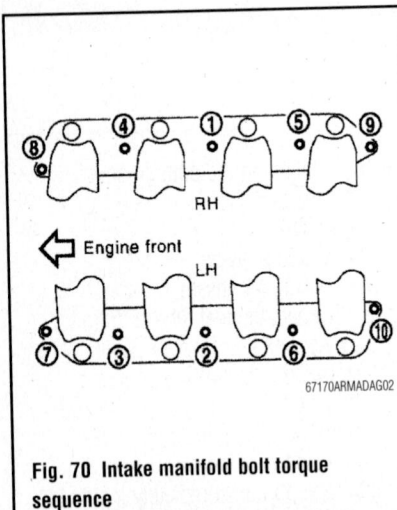

RH

Engine front

LH

67170ARMADAG02

Fig. 70 Intake manifold bolt torque sequence

OIL COOLER

REMOVAL & INSTALLATION

See Figure 71.

1. Before servicing the vehicle, refer to the Precautions Section.

➡ If working near and/or around the SRS system and components, be sure to disable the SRS system. After disabling the system wait three minutes or more before servicing the vehicle.

➡ Whenever the negative battery cable is disconnected the following components will require resetting. The Idle Air Volume Learning, Steering Angle Sensor Neutral Position, Sunroof Memory Reset/Initialization, Automatic Drive Positioner System, Audio presets and Navigation. Use the CONSULT-III diagnostic tool, or equivalent to perform the required resets.

➡ The automatic back door system must be initialized anytime the battery has been disconnected. Close the back door. Open the back door with the automatic open feature. Do not stop the process until the back door opens completely.

2. Disconnect the negative battery cable.
3. Remove the engine undercover.
4. Drain the engine coolant, when removing the water hoses. Be sure to properly dispose of used coolant.
5. Disconnect the water hoses from the oil cooler.
6. Drain the engine oil. Remove the oil filter.
7. Remove the connector bolt, oil cooler and O-ring. Discard the O-ring.

To install:

➡ Be sure to use new fasteners, as required.

8. Installation is the reverse of the removal procedure.
9. If replacement of the relief valve is necessary, install a new one. Do not attempt to repair the old one. Discard it!

OIL PUMP

REMOVAL & INSTALLATION

See Figures 72 and 73.

1. Before servicing the vehicle, refer to the Precautions Section.

➡ If working near and/or around the SRS system and components, be sure to disable the SRS system. After disabling the system wait three minutes or more before servicing the vehicle.

➡ Whenever the negative battery cable is disconnected the following components will require resetting. The Idle Air Volume Learning, Steering Angle Sensor Neutral Position, Sunroof Memory Reset/Initialization, Automatic Drive Positioner System, Audio presets and Navigation. Use the CONSULT-III diagnostic tool, or equivalent to perform the required resets.

➡ The automatic back door system must be initialized anytime the battery has been disconnected. Close the back door. Open the back door with the automatic open feature. Do not stop the process until the back door opens completely.

2. Disconnect the negative battery cable.
3. Remove or disconnect the following:
   • Timing chain cover
   • Oil pump drive spacer
   • Oil pump

To install:

➡ Be sure to use new fasteners, as required.

4. Install or connect the following:
   • Oil pump
   • Oil pump drive spacer
   • Timing chain cover

➡ When inserting the oil pump drive spacer, align the crankshaft key and the flat face of the inner rotor. If they are not aligned rotate the pump inner rotor by hand. Make sure that each part is aligned and tap lightly until it reaches the end.

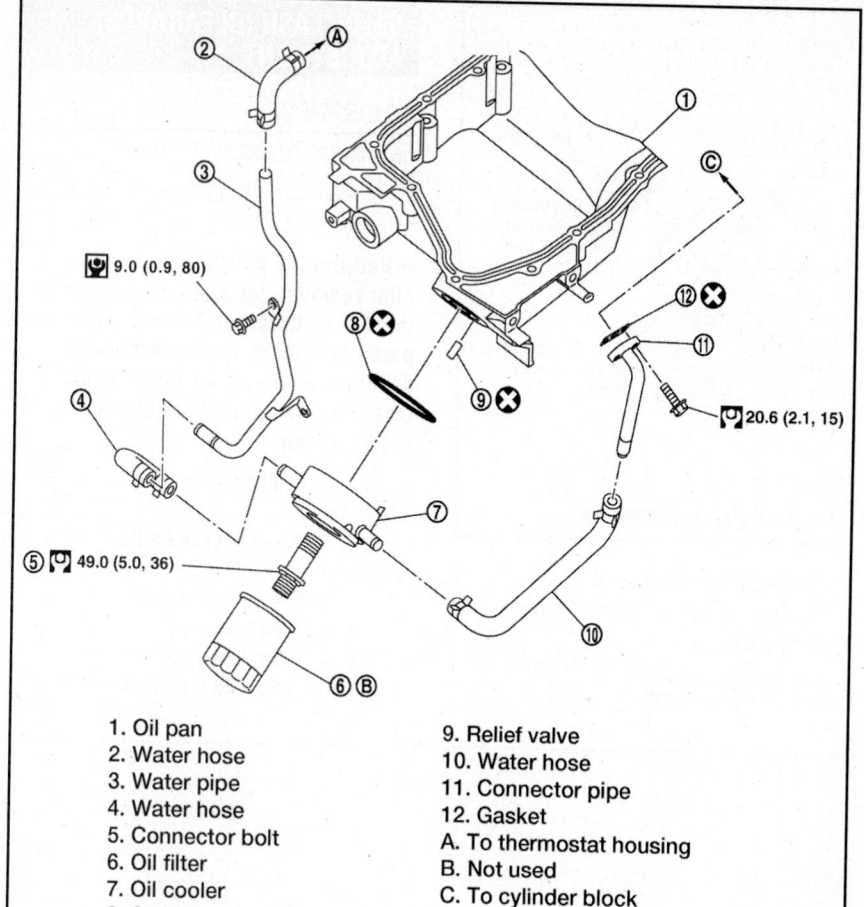

Fig. 71 Engine oil cooler and related components

1. Oil pan
2. Water hose
3. Water pipe
4. Water hose
5. Connector bolt
6. Oil filter
7. Oil cooler
8. O-ring
9. Relief valve
10. Water hose
11. Connector pipe
12. Gasket
A. To thermostat housing
B. Not used
C. To cylinder block

71075_ARMA_G0144

5. Be sure to perform the reconnect/relearn procedures.

PISTONS & RINGS

POSITIONING

See Figures 74 and 75.

ROCKER COVER

REMOVAL & INSTALLATION

See Figures 76 through 78.

1. Before servicing the vehicle, refer to the Precautions Section.

➡ If working near and/or around the SRS system and components, be sure to disable the SRS system. After disabling the system wait three minutes or more before servicing the vehicle.

➡ Whenever the negative battery cable is disconnected the following components will require resetting. The Idle Air Volume Learning, Steering Angle Sensor Neutral Position, Sunroof Mem-

ory Reset/Initialization, Automatic Drive Positioner System, Audio presets and Navigation. Use the CONSULT-III diagnostic tool, or equivalent to perform the required resets.

➡ The automatic back door system must be initialized anytime the battery has been disconnected. Close the back door. Open the back door with the automatic open feature. Do not stop the process until the back door opens completely.

2. Disconnect the negative battery cable.

3. Remove or disconnect the following:
   • Engine room cover
   • Air duct and resonator assembly, as required
   • Harness on the upper rocker cover and its peripheral and aside
   • Electric throttle control actuator, loosening the bolts diagonally, as required
   • Ignition coils
   • PCV hose from the PCV control valves

❊❊ CAUTION

Do not handle valve cover (B) by oil filler neck.

   • Bolts in reverse order shown in tightening sequence
   • Valve cover
   • discard the gaskets

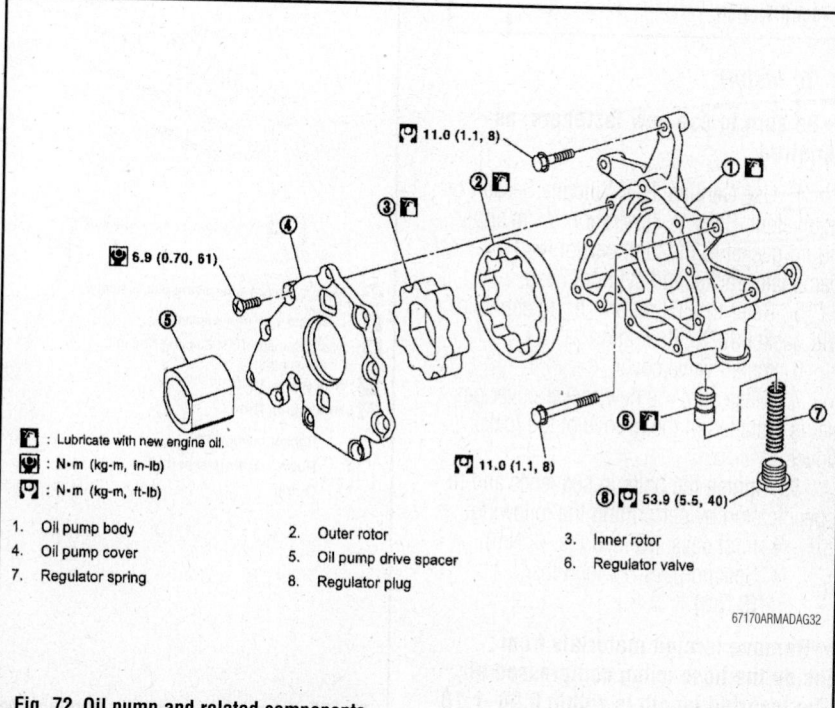

1. Oil pump body
2. Outer rotor
3. Inner rotor
4. Oil pump cover
5. Oil pump drive spacer
6. Regulator valve
7. Regulator spring
8. Regulator plug

67170ARMADAG32

Fig. 72 Oil pump and related components

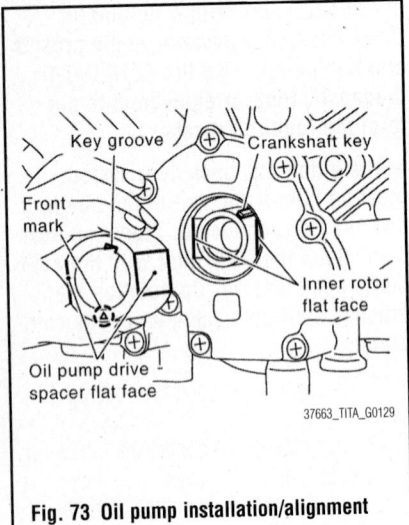

Fig. 73 Oil pump installation/alignment

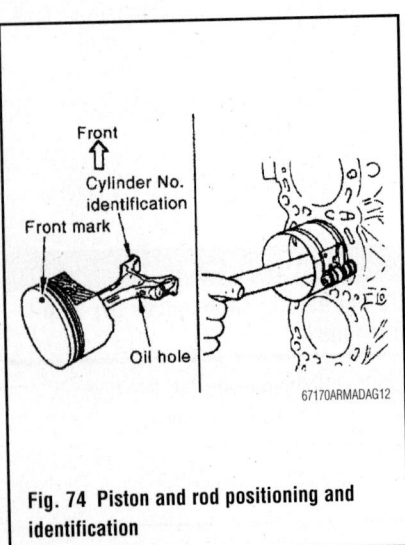

Fig. 74 Piston and rod positioning and identification

To install:

➡ **Be sure to use new fasteners, as required.**

4. Use Genuine RTV Silicone Sealant or equivalent. Refer to illustration "a" to apply liquid gasket to the joint part of No.1 camshaft bracket and cylinder head.

5. Refer to illustration _b_ to apply liquid gasket 90°to illustration "a".

6. Install valve cover.

7. Make sure the new rocker cover gasket is installed in the groove of the rocker cover.

8. Tighten the bolts in sequence and to specification by performing the following:
- First pass: 18 inch lbs. (2 Nm)
- Second pass: 73 inch lbs. (8 Nm)

➡ **Remove foreign materials from inside the hose using compressed air. The inserted length is within 0.98–1.18 inches (25–30 mm).**

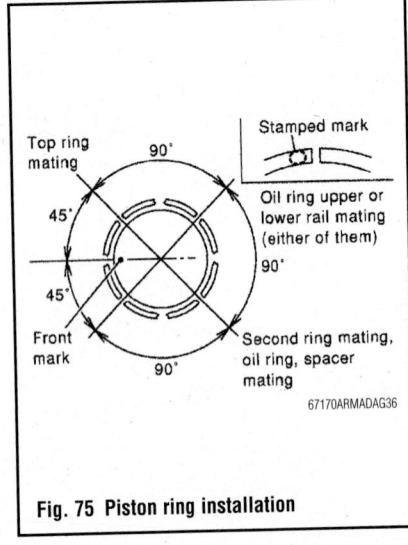

Fig. 75 Piston ring installation

9. Install the PCV hoses

10. To complete installation, reverse remaining removal procedures.

11. Be sure to perform the reconnect/relearn procedures.

VALVE LASH (CLEARANCE) ADJUSTMENT

ADJUSTMENT

See Figures 79 and 80.

1. Before servicing the vehicle, refer to the Precautions Section.

➡ **Perform the following inspection after removal, installation or replacement of camshaft or valve-related parts, or if there are unusual engine conditions due to changes in valve clearance over time (starting, idling, and/or noise).**

2. Run engine to operating temperature.

3. Remove or disconnect the following:
- Battery cover, if equipped
- Engine room cover
- Air intake assembly
- Left and right rocker covers

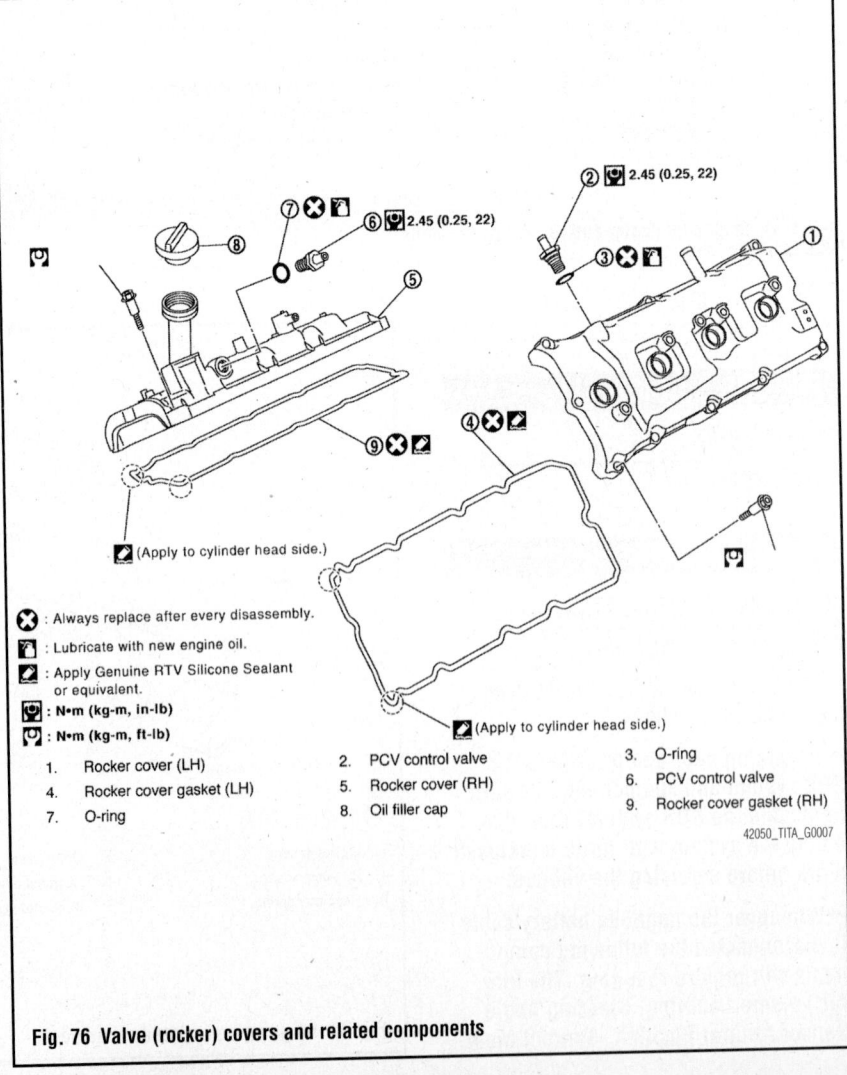

Fig. 76 Valve (rocker) covers and related components

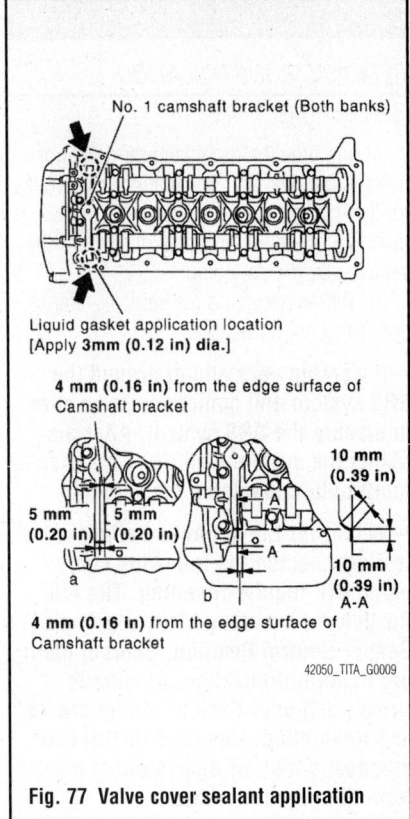

Fig. 77 Valve cover sealant application

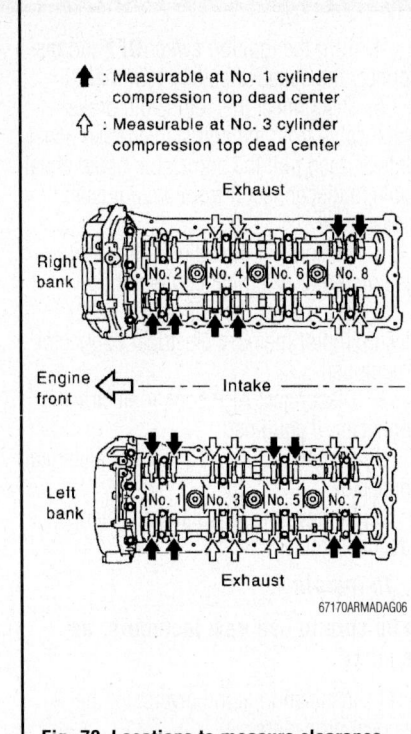

Fig. 79 Locations to measure clearance with No. 1 cylinder at TDC

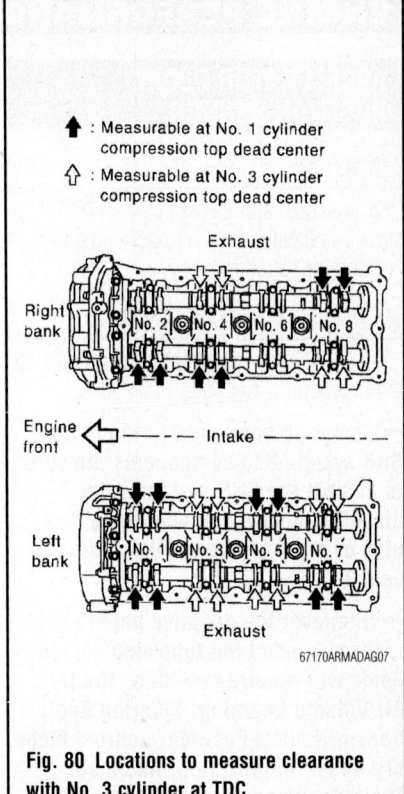

Fig. 80 Locations to measure clearance with No. 3 cylinder at TDC

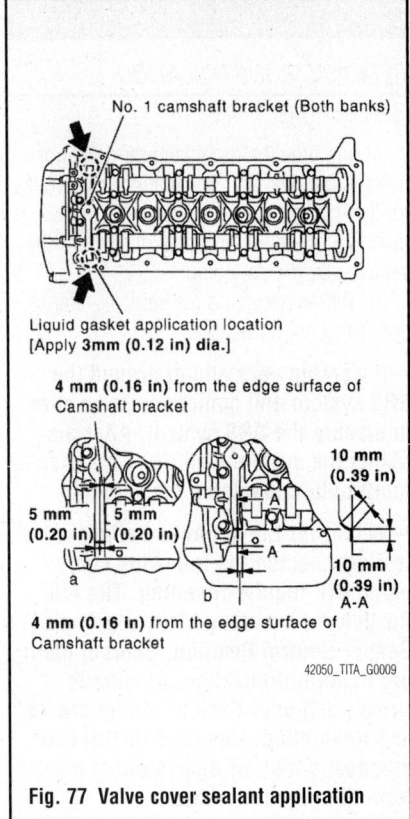

Fig. 78 Valve cover bolt tightening sequence

4. Turn the crankshaft pulley clockwise to Top Dead Center (TDC) identification notch with timing indicator.

5. Ensure that both the intake and exhaust cam noses of the No. 1 cylinder face outside.

6. Measure the valve clearances at locations shown in figure.

7. Turn the crankshaft pulley clockwise 270 degrees from the position of No. 1 cylinder compression to obtain No. 3 cylinder compression TDC.

8. Measure the valve clearances at locations shown in the figure.

9. Turn crankshaft pulley clockwise 90 degrees and measure the intake and exhaust valve clearance of No. 6 cylinder and exhaust valve clearance of No. 2 cylinder.

10. To adjust the valves, remove camshaft and valve lifter(s) out of specification.

11. Install replacement valve lifter(s).

12. Install the camshaft.

13. Manually turn the crankshaft pulley several turns.

14. Recheck valve clearances with engine at operating temperature.

INSPECTION

1. Before servicing the vehicle, refer to the Precautions Section.

2. Remove camshaft and valve lifter(s) out of specification.

3. Install replacement valve lifter(s).

4. Install the camshaft.

5. Manually turn the crankshaft pulley several turns.

6. Recheck valve clearances with engine at operating temperature.

ENGINE PERFORMANCE & EMISSION CONTROLS

ACCELERATOR PEDAL POSITION SENSOR

LOCATION

The Accelerator Pedal Position (APP) sensor is located on the upper end of the accelerator pedal assembly.

REMOVAL & INSTALLATION

1. Before servicing the vehicle, refer to the Precautions Section.

➡ If working near and/or around the SRS system and components, be sure to disable the SRS system. After disabling the system wait three minutes or more before servicing the vehicle.

➡ Whenever the negative battery cable is disconnected the following components will require resetting. The Idle Air Volume Learning, Steering Angle Sensor Neutral Position, Sunroof Memory Reset/Initialization, Automatic Drive Positioner System, Audio presets and Navigation. Use the CONSULT-III diagnostic tool, or equivalent to perform the required resets.

➡ The automatic back door system must be initialized anytime the battery has been disconnected. Close the back door. Open the back door with the automatic open feature. Do not stop the process until the back door opens completely.

2. Disconnect the negative battery cable.

✳✳ CAUTION

Do not disassemble the accelerator pedal adjusting mechanism. Before removal and installation, the accelerator and brake pedals must be in the front most position. This is to align the base position of the accelerator and brake pedals. Do not disassemble the accelerator pedal assembly. Do not remove the Accelerator Pedal Position (APP) sensor from the accelerator pedal bracket. Avoid damage from dropping the accelerator pedal assembly during handling. Keep the accelerator pedal assembly away from water.

3. Move the accelerator and brake pedals to the front most position.

4. Turn the ignition switch **OFF** and disconnect the negative battery terminal.

5. Disconnect the adjustable brake pedal cable from the adjustable brake pedal. Unlock, then pull the adjustable brake pedal cable to disconnect it from the adjustable brake pedal.

6. Disconnect the adjustable pedal electric motor electrical connector.

7. Disconnect the adjustable pedal electric motor memory electrical connector, if equipped.

8. Disconnect APP sensor electrical connector, if equipped.

9. Disconnect the APP sensor electrical connector.

10. Remove the adjustable accelerator pedal assembly.

To install:

➡ Be sure to use new fasteners, as required.

11. Installation is the reverse of the removal procedure.

12. Be sure to perform the reconnect/relearn procedures.

➡ Check that the accelerator pedal moves freely within the specified ranges, if equipped with adjustable pedals. Specification is 1.90 inch for total pedal applied stroke and 5.41 +/- 0.39 inch for total pedal height.

ADJUSTMENT

The accelerator pedal released position learning is an operation to learn the fully released position of the accelerator pedal by monitoring the pedal position sensor output. I must be performed each time the harness connector of the pedal or ECM is disconnected.

1. Be sure the pedal is fully released.

2. Turn the ignition switch ON and wait 2 seconds.

3. Turn the ignition switch OFF and wait 10 seconds.

4. Turn the ignition switch ON and wait 2 seconds.

5. Turn the ignition switch OFF and wait 10 seconds.

AIR-FUEL RATIO SENSOR

LOCATION

The Air Fuel Ratio (AF) sensors are located in the left and right exhaust manifold assembly.

REMOVAL & INSTALLATION

See Figure 66.

At this time the manufacturer does not provide removal and installation procedures for this component. The following procedure is a guideline and may differ from the vehicle you are servicing.

1. Before servicing the vehicle, refer to the Precautions Section.

➡ If working near and/or around the SRS system and components, be sure to disable the SRS system. After disabling the system wait three minutes or more before servicing the vehicle.

➡ Whenever the negative battery cable is disconnected the following components will require resetting. The Idle Air Volume Learning, Steering Angle Sensor Neutral Position, Sunroof Memory Reset/Initialization, Automatic Drive Positioner System, Audio presets and Navigation. Use the CONSULT-III diagnostic tool, or equivalent to perform the required resets.

➡ The automatic back door system must be initialized anytime the battery has been disconnected. Close the back door. Open the back door with the automatic open feature. Do not stop the process until the back door opens completely.

2. Disconnect the negative battery cable.

3. Remove the air cleaner assembly, as required.

➡ Raise and support the vehicle and remove the engine undercover, as required.

4. Disconnect the sensor electrical connector.

5. Carefully remove the sensor from its mounting.

To install:

➡ Be sure to use new fasteners, as required.

6. Installation is the reverse of the removal procedure.

7. Be sure to perform the reconnect/relearn procedures.

BODY CONTROL MODULE

PROGRAMMING

➡ The Nissan CONSULT III diagnostic tool, or equivalent is required to perform diagnostic testing and programming of this component.

REMOVAL & INSTALLATION
See Figure 81.

1. Before servicing the vehicle, refer to the Precautions Section.

➡If working near and/or around the SRS system and components, be sure to disable the SRS system. After disabling the system wait three minutes or more before servicing the vehicle.

➡Whenever the negative battery cable is disconnected the following components will require resetting. The Idle Air Volume Learning, Steering Angle Sensor Neutral Position, Sunroof Memory Reset/Initialization, Automatic Drive Positioner System, Audio presets and Navigation. Use the CONSULT-III diagnostic tool, or equivalent to perform the required resets.

➡The automatic back door system must be initialized anytime the battery has been disconnected. Close the back door. Open the back door with the automatic open feature. Do not stop the process until the back door opens completely.

2. Disconnect the negative battery cable.
3. Remove the lower trim panel.
4. Remove the component retaining screws.
5. Disconnect the electrical connector.
6. Remove the component from its mounting.

To install:

➡Be sure to use new fasteners, as required.

7. Installation is the reverse of the removal procedure.
8. Be sure to reconfigure this component, using the CONSULT III diagnostic tool, or equivalent..

CAMSHAFT POSITION SENSOR

LOCATION

The Camshaft Position (CMP) sensors are located on the timing cover, facing the engine.

REMOVAL & INSTALLATION
See Figure 82.

1. Before servicing the vehicle, refer to the Precautions Section.

➡If working near and/or around the SRS system and components, be sure to disable the SRS system. After disabling the system wait three minutes or more before servicing the vehicle.

➡Whenever the negative battery cable is disconnected the following components will require resetting. The Idle Air Volume Learning, Steering Angle Sensor Neutral Position, Sunroof Memory Reset/Initialization, Automatic Drive Positioner System, Audio presets and Navigation. Use the CONSULT-III diagnostic tool, or equivalent to perform the required resets.

➡The automatic back door system must be initialized anytime the battery has been disconnected. Close the back door. Open the back door with the automatic open feature. Do not stop the process until the back door opens completely.

2. Disconnect the negative battery cable.
3. Remove the engine cover.
4. Remove air intake duct.
5. Disconnect the camshaft position sensor.
6. Remove the bolt and the Camshaft Position (CMP) sensor.

To install:
7. Install the CMP sensor and tighten the bolt.
8. Reconnect the camshaft electrical sensor.
9. Install the air intake duct.
10. Install the engine cover.
11. Be sure to perform the reconnect/relearn procedures.

CRANKSHAFT POSITION SENSOR

LOCATION

The Crankshaft Position (CKP) sensor is located on the transmission assembly facing the gear teeth (cogs) of the signal plate.

REMOVAL & INSTALLATION
See Figure 83.

1. Before servicing the vehicle, refer to the Precautions Section.

➡If working near and/or around the SRS system and components, be sure to disable the SRS system. After disabling the system wait three minutes or more before servicing the vehicle.

➡Whenever the negative battery cable is disconnected the following components will require resetting. The Idle Air Volume Learning, Steering Angle Sensor Neutral Position, Sunroof Mem-

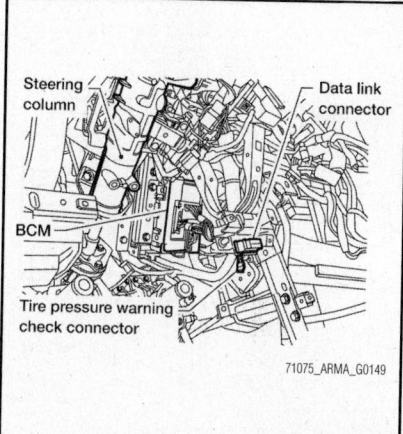

71075_ARMA_G0149

Fig. 81 Body control module and related components

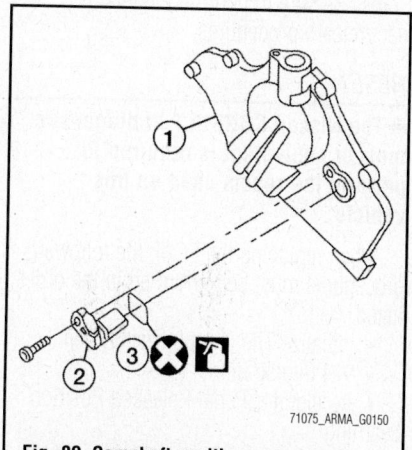

71075_ARMA_G0150

Fig. 82 Camshaft position sensor and related components (1) VTC valve cover, (2) CPS, (3) O-ring

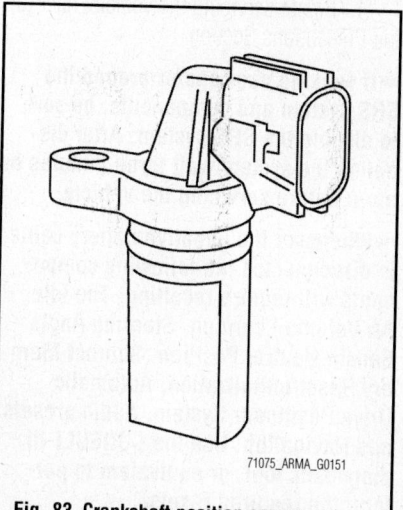

71075_ARMA_G0151

Fig. 83 Crankshaft position sensor

ory Reset/Initialization, Automatic Drive Positioner System, Audio presets and Navigation. Use the CONSULT-III diagnostic tool, or equivalent to perform the required resets.

➡The automatic back door system must be initialized anytime the battery has been disconnected. Close the back door. Open the back door with the automatic open feature. Do not stop the process until the back door opens completely.

2. Disconnect the negative battery cable.
3. Raise and support the vehicle safely.
4. Disconnect the Crankshaft Position (CKP) sensor connector.
5. Remove the mounting bolt and CKP sensor.

To install:
6. Install the CKP sensor and tighten the mounting bolt.
7. Reconnect the CKP sensor connector.
8. Lower the vehicle.
9. Be sure to perform the reconnect/relearn procedures.

ELECTRONIC CONTROL MODULE

LOCATION

The Electronic Control Module (ECM) is located in the engine room passenger side behind battery.

REMOVAL & INSTALLATION
See Figure 84.

At this time the manufacturer does not provide removal and installation procedures for this component. The following procedure is a guideline and may differ from the vehicle you are servicing.
1. Before servicing the vehicle, refer to the Precautions Section.

➡If working near and/or around the SRS system and components, be sure to disable the SRS system. After disabling the system wait three minutes or more before servicing the vehicle.

➡Whenever the negative battery cable is disconnected the following components will require resetting. The Idle Air Volume Learning, Steering Angle Sensor Neutral Position, Sunroof Memory Reset/Initialization, Automatic Drive Positioner System, Audio presets and Navigation. Use the CONSULT-III diagnostic tool, or equivalent to perform the required resets.

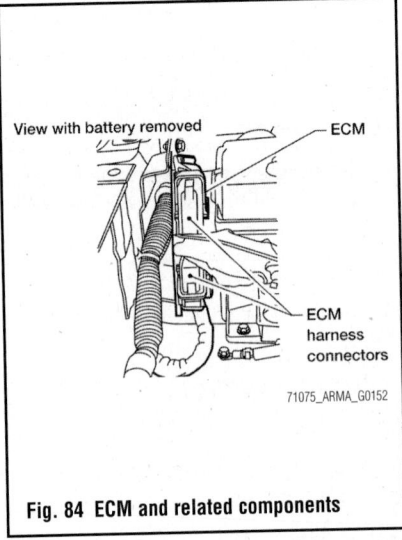

View with battery removed — ECM

ECM harness connectors

71075_ARMA_G0152

Fig. 84 ECM and related components

➡The automatic back door system must be initialized anytime the battery has been disconnected. Close the back door. Open the back door with the automatic open feature. Do not stop the process until the back door opens completely.

2. Disconnect the negative battery cable.
3. Disconnect the positive battery cable and remove the battery, as required..
4. Carefully remove the Electronic Control Module (ECM) harness connectors.
5. Remove the ECM mounting bolts and the ECM.

To install:
6. Install the ECM and mounting bolts and tighten to 62 inch lbs. (7 Nm).
7. Carefully install the ECM harness connectors.
8. Install the battery.
9. Reconnect the battery cables.
10. Be sure to perform the reconnect/relearn procedures.

RESET

➡The Nissan CONSULT III diagnostic tool, or equivalent is required to perform the resets used on this vehicle.

When replacing the ECM, the following procedures must be performed in the order listed.
- Initialize The Immobilizer System
- VIN Registration
- Accelerator Pedal Released Position Learning
- Throttle Valve Closed Position Learning
- Idle Air Volume Learning

ENGINE COOLANT TEMPERATURE SENSOR

LOCATION

The Engine Coolant Temperature (ECT) sensor is mounted in the front of the intake manifold. It is just to the right of the throttle body.

REMOVAL & INSTALLATION
See Figure 85.

✳✳ CAUTION

Never open, service or drain the radiator or cooling system when hot; serious burns can occur from the steam and hot coolant. Also, when draining engine coolant, keep in mind that cats and dogs are attracted to ethylene glycol antifreeze and could drink any that is left in an uncovered container or in puddles on the ground. This will prove fatal in sufficient quantities. Always drain coolant into a sealable container. Coolant should be reused unless it is contaminated or is several years old.

1. Before servicing the vehicle, refer to the Precautions Section.

➡If working near and/or around the SRS system and components, be sure to disable the SRS system. After disabling the system wait three minutes or more before servicing the vehicle.

➡Whenever the negative battery cable is disconnected the following components will require resetting. The Idle Air Volume Learning, Steering Angle Sensor Neutral Position, Sunroof Memory Reset/Initialization, Automatic

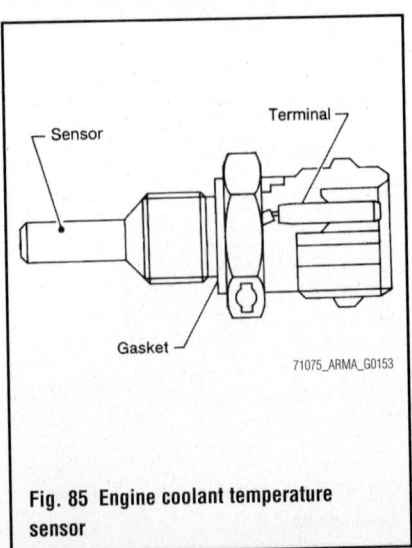

Sensor — Terminal

Gasket

71075_ARMA_G0153

Fig. 85 Engine coolant temperature sensor

Drive Positioner System, Audio presets and Navigation. Use the CONSULT-III diagnostic tool, or equivalent to perform the required resets.

2. Disconnect the negative battery cable.
3. Remove the engine cover.
4. Remove the intake air duct.
5. Partially drain the cooling system.
6. Disconnect the harness connector.
7. Remove the Engine Coolant Temperature (ECT) sensor.

To install:

➡Be sure to use new fasteners, as required.

8. Install the ECT sensor and carefully tighten.
9. Reconnect the harness connector.
10. Install the intake air duct.
11. Install the engine cover.
12. Refill the engine coolant.
13. Be sure to perform the reconnect/relearn procedures.

EVAP CANISTER

LOCATION

This component is located under the vehicle near the fuel tank.

REMOVAL & INSTALLATION

At this time the manufacturer does not provide removal and installation procedures for this component. The following procedure is a guideline and may differ from the vehicle you are servicing.

1. Before servicing the vehicle, refer to the Precautions Section.

➡If working near and/or around the SRS system and components, be sure to disable the SRS system. After disabling the system wait three minutes or more before servicing the vehicle.

➡Whenever the negative battery cable is disconnected the following components will require resetting. The Idle Air Volume Learning, Steering Angle Sensor Neutral Position, Sunroof Memory Reset/Initialization, Automatic Drive Positioner System, Audio presets and Navigation. Use the CONSULT-III diagnostic tool, or equivalent to perform the required resets.

➡The automatic back door system must be initialized anytime the battery has been disconnected. Close the back door. Open the back door with the automatic open feature. Do not stop the

process until the back door opens completely.

2. Disconnect the negative battery cable.
3. Raise and support the vehicle safely.
4. Remove the left rear tire and wheel assembly.
5. Remove the left rear seat belt anchor which is located next to the EVAP canister.
6. Remove the fuel tank shield, as required.
7. Disconnect the component electrical connectors.
8. Disconnect the hoses.
9. Remove the mounting clips, nuts and or screws.
10. Remove the component from the vehicle.

To install:

➡Be sure to use new fasteners, as required.

11. Installation is the reverse of the removal procedure.
12. Be sure to perform the reconnect/relearn procedures.

HEATED OXYGEN SENSORS (HO2S)

LOCATION

The Heated Oxygen sensors (HO2S) are located after the exhaust manifold converter assembly, in the lower part of the exhaust system.

REMOVAL & INSTALLATION

See Figure 86.

1. Before servicing the vehicle, refer to the Precautions Section.

➡If working near and/or around the SRS system and components, be sure to disable the SRS system. After disabling the system wait three minutes or more before servicing the vehicle.

➡Whenever the negative battery cable is disconnected the following components will require resetting. The Idle Air Volume Learning, Steering Angle Sensor Neutral Position, Sunroof Memory Reset/Initialization, Automatic Drive Positioner System, Audio presets

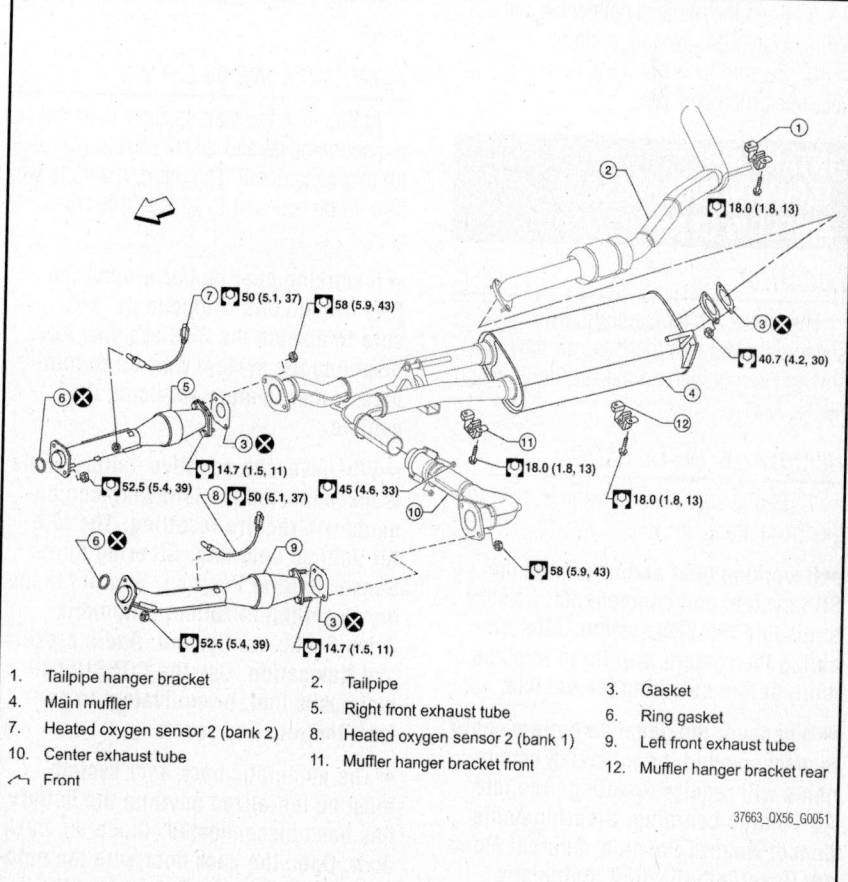

1. Tailpipe hanger bracket
2. Tailpipe
3. Gasket
4. Main muffler
5. Right front exhaust tube
6. Ring gasket
7. Heated oxygen sensor 2 (bank 2)
8. Heated oxygen sensor 2 (bank 1)
9. Left front exhaust tube
10. Center exhaust tube
11. Muffler hanger bracket front
12. Muffler hanger bracket rear
⬑ Front

37663_QX56_G0051

Fig. 86 Heated oxygen sensors and related components

and Navigation. Use the CONSULT-III diagnostic tool, or equivalent to perform the required resets.

➡The automatic back door system must be initialized anytime the battery has been disconnected. Close the back door. Open the back door with the automatic open feature. Do not stop the process until the back door opens completely.

   2. Disconnect the negative battery cable.

   3. Raise and safely support the vehicle.

   4. Remove the engine undercover, as needed.

   5. Unplug the Heated Oxygen (HO2S) sensor harness.

   6. Using an O2 wrench remove the HO2S sensor.

➡Lower the exhaust in needed.

To install:

➡Be sure to use new fasteners, as required.

   7. Install the HO2S sensor and tighten to 37 ft. lbs. (50 Nm).

   8. Install the harness connector.

   9. Keep the harness connector and wiring away from exhaust system.

   10. Be sure to perform the reconnect/relearn procedures.

INTAKE AIR TEMPERATURE/MASS AIRFLOW SENSOR

LOCATION

   The Intake Air Temperature (IAT) sensor is integral to the mass air flow sensor, and is mounted on the air filter housing lid.

REMOVAL & INSTALLATION

   1. Before servicing the vehicle, refer to the Precautions Section.

➡If working near and/or around the SRS system and components, be sure to disable the SRS system. After disabling the system wait three minutes or more before servicing the vehicle.

➡Whenever the negative battery cable is disconnected the following components will require resetting. The Idle Air Volume Learning, Steering Angle Sensor Neutral Position, Sunroof Memory Reset/Initialization, Automatic Drive Positioner System, Audio presets and Navigation. Use the CONSULT-III

diagnostic tool, or equivalent to perform the required resets.

➡The automatic back door system must be initialized anytime the battery has been disconnected. Close the back door. Open the back door with the automatic open feature. Do not stop the process until the back door opens completely.

   2. Disconnect the negative battery cable.

   3. Remove the engine room cover.

   4. Remove the Intake Air Temperature (IAT/MAF) sensor harness.

   5. Remove the mounting screws and the IAT/MAF sensor.

To install:

   6. Install the IAT/MAF sensor.

   7. Install the harness connector.

   8. Install the engine room cover.

   9. Be sure to perform the reconnect/relearn procedures.

KNOCK SENSOR (KS)

LOCATION

   The Knock (KS) sensors are mounted under the intake manifold on the cylinder block.

REMOVAL & INSTALLATION

   At this time the manufacturer does not provide removal and installation procedures for this component. The intake manifold will have to be removed to service this component.

➡If working near and/or around the SRS system and components, be sure to disable the SRS system. After disabling the system wait three minutes or more before servicing the vehicle.

➡Whenever the negative battery cable is disconnected the following components will require resetting. The Idle Air Volume Learning, Steering Angle Sensor Neutral Position, Sunroof Memory Reset/Initialization, Automatic Drive Positioner System, Audio presets and Navigation. Use the CONSULT-III diagnostic tool, or equivalent to perform the required resets.

➡The automatic back door system must be initialized anytime the battery has been disconnected. Close the back door. Open the back door with the automatic open feature. Do not stop the process until the back door opens completely.

MALFUNCTION INDICATOR LIGHT

RESET PROCEDURE

   Clearing diagnostic trouble codes resets MIL.

   Proper operation of the Malfunction Indicator Light (MIL):

• The MIL will illumine with the ignition switch ON and the engine OFF

• The MIL will turn OFF when the engine is started

• The MIL will remain ON if the self-diagnostic system has detected a malfunction

• The MIL may turn OFF if the malfunction is no longer present

• If the MIL is illuminated and then the engine stalls, the MIL will remain illuminated as long as the ignition switch is ON

• If the MIL is not illuminated and the engine stalls, the MIL will not illuminate until the ignition switch is cycled OFF, then ON

   1. Before servicing the vehicle, refer to the Precautions Section.

   2. Resetting the MIL:

• The control module turns OFF the MIL after 3 consecutive ignition cycles that the diagnostic system runs and does not fail

• The control module turns OFF the MIL after a current Diagnostic Trouble Code (DTC) clears when the diagnostic cycle runs and passes

• There may still be a history of DTC's stored in the system. These will clear after 40 consecutive warm-up cycles, if no failures are reported by any other related diagnostic system

• Manual resetting of the MIL and any DTC stored in the system, requires the use of an OBD2 scan tool connected to the Data Link Connector (DLC) for communication with the vehicle. Follow the instructions of the scan tool for both retrieval and resetting of DTC's. The scan tool can be used to command the MIL off.

➡If the error symptoms causing the MIL to illuminate have been corrected, the MIL will return to normal operation.

   3. If a DTC is present, record the code and troubleshoot the fault.

MASS AIR FLOW SENSOR

Refer to Intake Air Temperature (IAT) Sensor.

PCV VALVE

LOCATION

The PCV valve is located on top of the engine, in one of the valve rocker covers.

REMOVAL & INSTALLATION

At this time the manufacturer does not provide removal and installation procedures for this component. The following procedure is a guideline and may differ from the vehicle you are servicing.

1. Before servicing the vehicle, refer to the Precautions Section.

➡If working near and/or around the SRS system and components, be sure to disable the SRS system. After disabling the system wait three minutes or more before servicing the vehicle.

➡Whenever the negative battery cable is disconnected the following components will require resetting. The Idle Air Volume Learning, Steering Angle Sensor Neutral Position, Sunroof Memory Reset/Initialization, Automatic Drive Positioner System, Audio presets and Navigation. Use the CONSULT-III diagnostic tool, or equivalent to perform the required resets.

➡The automatic back door system must be initialized anytime the battery has been disconnected. Close the back door. Open the back door with the automatic open feature. Do not stop the process until the back door opens completely.

2. Disconnect the negative battery cable.
3. Remove the necessary components in order to gain access to the component.
4. Disconnect the PCV hose.
5. Remove the valve from its mounting.

To install:

➡Be sure to use new fasteners, as required.

6. Installation is the reverse of the removal procedure.

7. Be sure to perform the reconnect/relearn procedures.

THROTTLE POSITION SENSOR

LOCATION

See Figure 87.

The Throttle Position(TPS) sensor is integral to the electric Throttle Control actuator. The Throttle Control actuator is mounted at the front of the intake manifold.

REMOVAL & INSTALLATION

See Figure 88.

At this time the manufacturer does not provide removal and installation procedures for this component. The following procedure is a guideline and may differ from the vehicle you are servicing.

1. Before servicing the vehicle, refer to the Precautions Section.

➡If working near and/or around the SRS system and components, be sure to disable the SRS system. After disabling the system wait three minutes or more before servicing the vehicle.

➡Whenever the negative battery cable is disconnected the following components will require resetting. The Idle Air Volume Learning, Steering Angle Sensor Neutral Position, Sunroof Memory Reset/Initialization, Automatic Drive Positioner System, Audio presets and Navigation. Use the CONSULT-III diagnostic tool, or equivalent to perform the required resets.

➡The automatic back door system must be initialized anytime the battery has been disconnected. Close the back door. Open the back door with the automatic open feature. Do not stop the process until the back door opens completely.

2. Disconnect the negative battery cable.
3. Drain the cooling system, as required. Be sure to properly dispose of used engine coolant.
4. Remove the engine room cover. Remove the air intake duct.
5. Disconnect harness connector.
6. Disconnect water hoses.
7. Loosen the throttle body assembly mounting bolts in reverse order of the tightening sequence.

To install:

➡Be sure to use new fasteners, as required.

8. Install the throttle body assembly with a new gasket.
9. Tighten the mounting bolts in sequence to 74 inch lbs. (8.4 Nm).
10. Reconnect the water hose.
11. Reconnect the harness connector.
12. Reconnect the air intake duct.
13. Fill the cooling system with the proper grade and type engine coolant.
14. Be sure to perform the reconnect/relearn procedures.

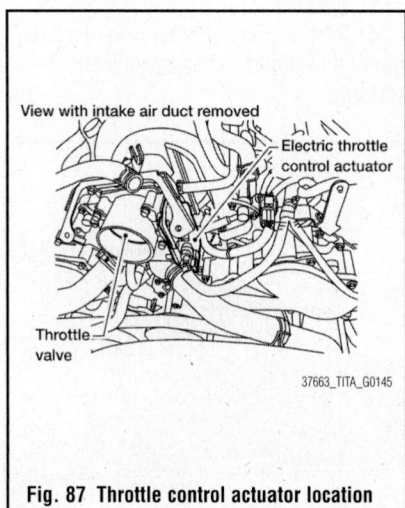

View with intake air duct removed

Electric throttle control actuator

Throttle valve

37663_TITA_G0145

Fig. 87 Throttle control actuator location

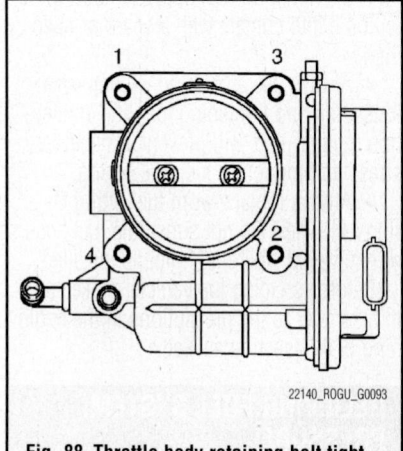

22140_ROGU_G0093

Fig. 88 Throttle body retaining bolt tightening sequence

FUEL SYSTEM SERVICE PRECAUTIONS

Safety is the most important factor when performing not only fuel system maintenance but any type of maintenance. Failure to conduct maintenance and repairs in a safe manner may result in serious personal injury or death. Maintenance and testing of the vehicle's fuel system components can be accomplished safely and effectively by adhering to the following rules and guidelines.

• To avoid the possibility of fire and personal injury, always disconnect the negative battery cable unless the repair or test procedure requires that battery voltage be applied.

• Always relieve the fuel system pressure prior to disconnecting any fuel system component (injector, fuel rail, pressure regulator, etc.), fitting or fuel line connection. Exercise extreme caution whenever relieving fuel system pressure to avoid exposing skin, face and eyes to fuel spray. Please be advised that fuel under pressure may penetrate the skin or any part of the body that it contacts.

• Always place a shop towel or cloth around the fitting or connection prior to loosening to absorb any excess fuel due to spillage. Ensure that all fuel spillage (should it occur) is quickly removed from engine surfaces. Ensure that all fuel soaked cloths or towels are deposited into a suitable waste container.

• Always keep a dry chemical (Class B) fire extinguisher near the work area.

• Do not allow fuel spray or fuel vapors to come into contact with a spark or open flame.

• Always use a back-up wrench when loosening and tightening fuel line connection fittings. This will prevent unnecessary stress and torsion to fuel line piping.

• Always replace worn fuel fitting O-rings with new Do not substitute fuel hose or equivalent where fuel pipe is installed.

Before servicing the vehicle, make sure to also refer to the precautions in the beginning of this section as well.

RELIEVING FUEL SYSTEM PRESSURE

With CONSULT-III®

1. Turn ignition switch **ON**.
2. Perform "FUEL PRESSURE RELEASE" in "WORK SUPPORT" mode with CONSULT-III®.

3. Start engine.
4. After engine stalls, turn over the engine two or three times to release all fuel pressure.
5. Turn ignition switch **OFF**.

Without CONSULT-III®

See Figure 89.

1. Before servicing the vehicle, refer to the Precautions Section.

➡ If working near and/or around the SRS system and components, be sure to disable the SRS system. After disabling the system wait three minutes or more before servicing the vehicle.

➡ Whenever the negative battery cable is disconnected the following components will require resetting. The Idle Air Volume Learning, Steering Angle Sensor Neutral Position, Sunroof Memory Reset/Initialization, Automatic Drive Positioner System, Audio presets and Navigation. Use the CONSULT-III diagnostic tool, or equivalent to perform the required resets.

➡ The automatic back door system must be initialized anytime the battery has been disconnected. Close the back door. Open the back door with the automatic open feature. Do not stop the process until the back door opens completely.

2. Remove fuel pump fuse located in IPDM E/R.
3. Start engine.
4. After engine stalls, turn over engine two or three times to release all fuel pressure.

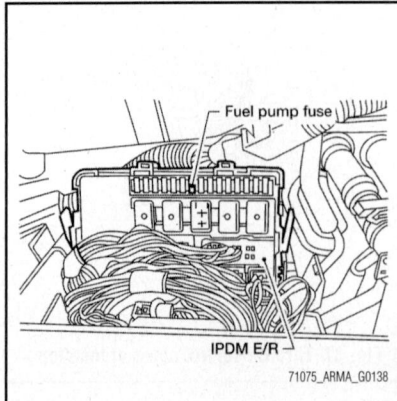

Fig. 89 IPDM E/R fuse and fuel pump fuse location

5. Turn ignition switch **OFF**.
6. Disconnect the negative battery cable.
7. Reinstall fuel pump fuse after servicing fuel system.
8. Be sure to perform the reconnect/relearn procedures.

FUEL FILTER

REMOVAL & INSTALLATION

➡ The fuel filter is part of the fuel pump assembly.

➡ This vehicle uses quick connect fittings. Be sure to properly relieve the fuel system pressure before disconnecting any of these fittings. Always replace O-rings and clamps with new ones. Do not bend, twist or kink hoses when they are being removed or installed. Be sure that the clamp screw does not contact adjacent parts. When tightening the high pressure rubber hose clamp make sure the clamp end is 0.12 inch from the hose end. After connecting these fittings make sure that the connectors are secure. Check for fuel leakage at these connections turn the ignition key to the ON position (do not start the engine), correct as required. Start the engine, raise the idle, and verify that there are no fuel leaks, correct as required.

FUEL PUMP MODULE

REMOVAL & INSTALLATION

See Figure 90.

1. Before servicing the vehicle, refer to the Precautions Section.

➡ If working near and/or around the SRS system and components, be sure to disable the SRS system. After disabling the system wait three minutes or more before servicing the vehicle.

➡ Whenever the negative battery cable is disconnected the following components will require resetting. The Idle Air Volume Learning, Steering Angle Sensor Neutral Position, Sunroof Memory Reset/Initialization, Automatic Drive Positioner System, Audio presets and Navigation. Use the CONSULT-III diagnostic tool, or equivalent to perform the required resets.

➡ The automatic back door system must be initialized anytime the battery has been disconnected. Close the back door. Open the back door with the automatic open feature. Do not stop the process until the back door opens completely.

➡Be sure to check the fuel gauge indicator. Make sure that it reads not more than half a tank. If not, properly drain fuel until the gauge reads half a tank or less.

➡This vehicle uses quick connect fittings. Be sure to properly relieve the fuel system pressure before disconnecting any of these fittings. Always replace O-rings and clamps with new ones. Do not bend, twist or kink hoses when they are being removed or installed. Be sure that the clamp screw does not contact adjacent parts. When tightening the high pressure rubber hose clamp make sure the clamp end is 0.12 inch from the hose end. After connecting these fittings make sure that the connectors are secure. Check for fuel leakage at these connections turn the ignition key to the ON position (do not start the engine), correct as required. Start the engine, raise the idle, and verify that there are no fuel leaks, correct as required.

2. Disconnect the negative battery cable.
3. Relieve the fuel system pressure.
4. Drain the fuel tank to an acceptable level, as necessary.
5. Remove fuel filler cap to release pressure from inside tank.
6. Remove left hand rear inner fender liner.
7. Disconnect fuel filler hose from fuel filler pipe.
8. Drain fuel tank through the fuel filler hose using a suitable hose.
9. Remove or disconnect the following:
   - Second row left hand seat
   - Third row seat
   - Second and third row seat belt buckles mounted on floor
   - Left hand center pillar trim
   - Left hand rear trim panel
   - Left hand rear side door kick plate and weather stripping
   - Second row rear center console and base, if equipped
   - Inspection hole cover under carpet by turning retainers 90 degrees
   - Electrical connectors
   - EVAP hose
   - Fuel supply hose
   - Lockring using special tool J-46214

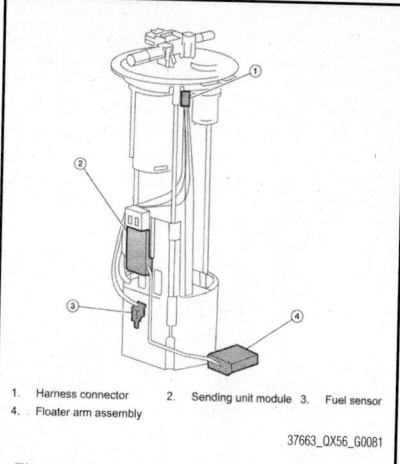

1. Harness connector   2. Sending unit module   3. Fuel sensor
4. Floater arm assembly

37663_QX56_G0081

Fig. 90 Fuel pump module and related components

   - Fuel level sensor
   - Fuel filter
   - Fuel pump assembly

To install:

➡Be sure to use new fasteners, as required.

10. Install or connect the following:
    - Fuel pump assembly
    - Fuel filter
    - Fuel level sensor
    - Lockring using special tool J-46214
    - Fuel supply hose
    - EVAP hose
    - Electrical connectors
    - Inspection hole cover
    - Second row rear center console and base, if equipped
    - Left hand rear side door kick plate and weather stripping
    - Left hand rear trim panel
    - Left hand center pillar trim
    - Second and third row seat belt buckles
    - Third row seat
    - Second row left hand seat
    - Fuel filler hose to fuel filler pipe
    - Left hand rear inner fender liner
11. Start the engine and check for leaks.
12. Be sure to perform the reconnect/relearn procedures.

FUEL RAIL & INJECTORS

REMOVAL & INSTALLATION
See Figure 91.

1. Before servicing the vehicle, refer to the Precautions Section.

➡If working near and/or around the

SRS system and components, be sure to disable the SRS system. After disabling the system wait three minutes or more before servicing the vehicle.

➡Whenever the negative battery cable is disconnected the following components will require resetting. The Idle Air Volume Learning, Steering Angle Sensor Neutral Position, Sunroof Memory Reset/Initialization, Automatic Drive Positioner System, Audio presets and Navigation. Use the CONSULT-III diagnostic tool, or equivalent to perform the required resets.

➡The automatic back door system must be initialized anytime the battery has been disconnected. Close the back door. Open the back door with the automatic open feature. Do not stop the process until the back door opens completely.

➡This vehicle uses quick connect fittings. Be sure to properly relieve the fuel system pressure before disconnecting any of these fittings. Always replace O-rings and clamps with new ones. Do not bend, twist or kink hoses when they are being removed or installed. Be sure that the clamp screw does not contact adjacent parts. When tightening the high pressure rubber hose clamp make sure the clamp end is 0.12 inch from the hose end. After connecting these fittings make sure that the connectors are secure. Check for fuel leakage at these connections turn the ignition key to the ON position (do not start the engine), correct as required. Start the engine, raise the idle, and verify that there are no fuel leaks, correct as required.

2. Disconnect the negative battery cable.
3. Remove engine cover. Remove the air cleaner assembly.
4. Relieve fuel system pressure.
5. Remove or disconnect the following:
   - Fuel injector harness connectors
   - Fuel hose assembly from right and left fuel rails
   - Fuel injectors with fuel rail as an assembly
   - Fuel injector from fuel rail

To install:

➡Be sure to use new fasteners, as required.

6. Install or connect the following:

➡Always use a new O-ring when reinstalling the fuel injector to the fuel rail.

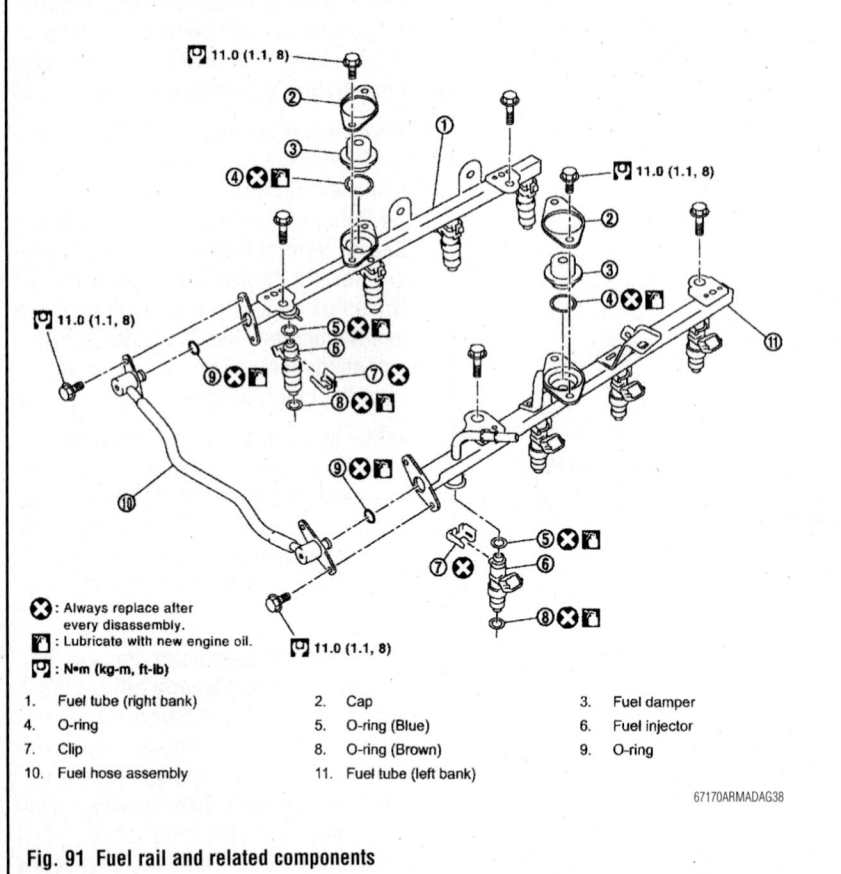

: Always replace after every disassembly.
: Lubricate with new engine oil.
: N•m (kg-m, ft-lb)

1. Fuel tube (right bank)
2. Cap
3. Fuel damper
4. O-ring
5. O-ring (Blue)
6. Fuel injector
7. Clip
8. O-ring (Brown)
9. O-ring
10. Fuel hose assembly
11. Fuel tube (left bank)

67170ARMADAG38

Fig. 91 Fuel rail and related components

- New clip onto the fuel injector
- Fuel injector to fuel rail
- Fuel injectors and fuel rail as an assembly to the intake manifold. Tighten the bolts to 8 ft. lbs. (11 Nm).
- Fuel hose assembly
- Fuel injector harness connectors
- Negative battery cable
- Engine cover
7. Start engine and check for leaks.
8. Be sure to perform the reconnect/relearn procedures.

FUEL TANK

DRAINING

See Figure 92.

1. Before servicing the vehicle, refer to the Precautions Section.

➡If working near and/or around the SRS system and components, be sure to disable the SRS system. After disabling the system wait three minutes or more before servicing the vehicle.

➡Whenever the negative battery cable is disconnected the following components will require resetting. The Idle Air Volume Learning, Steering Angle Sensor Neutral Position, Sunroof Memory Reset/Initialization, Automatic Drive Positioner System, Audio presets and Navigation. Use the CONSULT-III diagnostic tool, or equivalent to perform the required resets.

➡The automatic back door system must be initialized anytime the battery has been disconnected. Close the back door. Open the back door with the automatic open feature. Do not stop the process until the back door opens completely.

➡This vehicle uses quick connect fittings. Be sure to properly relieve the fuel system pressure before disconnecting any of these fittings. Always replace O-rings and clamps with new ones. Do not bend, twist or kink hoses when they are being removed or installed. Be sure that the clamp screw does not contact adjacent parts. When tightening the high pressure rubber hose clamp make sure the clamp end is 0.12 inch from the hose end. After connecting these fittings make sure that the connectors are secure. Check for fuel leakage at these connections turn the ignition key to the ON position (do not start the engine), correct as required. Start the engine, raise the idle, and verify that there are no fuel leaks, correct as required.

2. Disconnect the negative battery cable.
3. Remove the fuel filler cap to release the pressure from inside the fuel tank.
4. raise and safely support the vehicle.
5. Remove the LH rear wheel and tire.
6. Check the fuel level on level gauge. If the fuel gauge indicates more than the level (full or almost full), drain the fuel from the fuel tank until the fuel gauge indicates an acceptable fuel level.

➡Fuel will be spilled when removing the fuel level sensor, fuel filter, and fuel pump assembly for the fuel level is above the fuel level sensor, fuel filter, and fuel pump assembly fuel tank opening.

- As a guide, the fuel level reaches the fuel gauge position as shown, or less, when approximately 4 US gal (15L) of fuel are drained from the fuel tank.

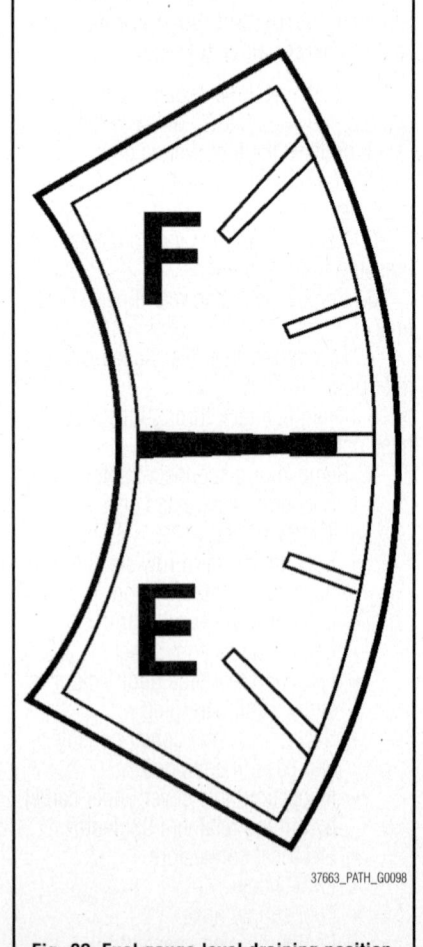

37663_PATH_G0098

Fig. 92 Fuel gauge level draining position

- If the fuel pump does not operate, use the following procedure to drain the fuel to the specified level.

  a. Insert a suitable hose of less than 15 mm (0.59 in.) diameter into the fuel filler pipe through the fuel filler opening to drain the fuel from fuel filler pipe.

  b. Remove the fuel filler pipe shield.

  c. Disconnect the fuel filler hose from the fuel filler pipe.

  d. Insert a suitable hose into the fuel tank through the fuel filler hose to drain the fuel from the fuel tank.

7. Release the fuel pressure from the fuel lines.

8. Be sure to perform the reconnect/relearn procedures.

REMOVAL & INSTALLATION

See Figure 93.

1. Before servicing the vehicle, refer to the Precautions Section.

➡️**If working near and/or around the SRS system and components, be sure to disable the SRS system. After disabling the system wait three minutes or more before servicing the vehicle.**

➡️**Whenever the negative battery cable is disconnected the following components will require resetting. The Idle Air Volume Learning, Steering Angle Sensor Neutral Position, Sunroof Memory Reset/Initialization, Automatic Drive Positioner System, Audio presets and Navigation. Use the CONSULT-III diagnostic tool, or equivalent to perform the required resets.**

➡️**The automatic back door system must be initialized anytime the battery has been disconnected. Close the back door. Open the back door with the automatic open feature. Do not stop the process until the back door opens completely.**

➡️**This vehicle uses quick connect fittings. Be sure to properly relieve the fuel system pressure before disconnecting any of these fittings. Always replace O-rings and clamps with new ones. Do not bend, twist or kink hoses when they are being removed or installed. Be sure that the clamp screw does not contact adjacent parts. When tightening the high pressure rubber hose clamp make sure the clamp end is 0.12 inch from the hose end. After connecting these fittings make sure that the connectors are secure. Check for fuel leakage at these connections**

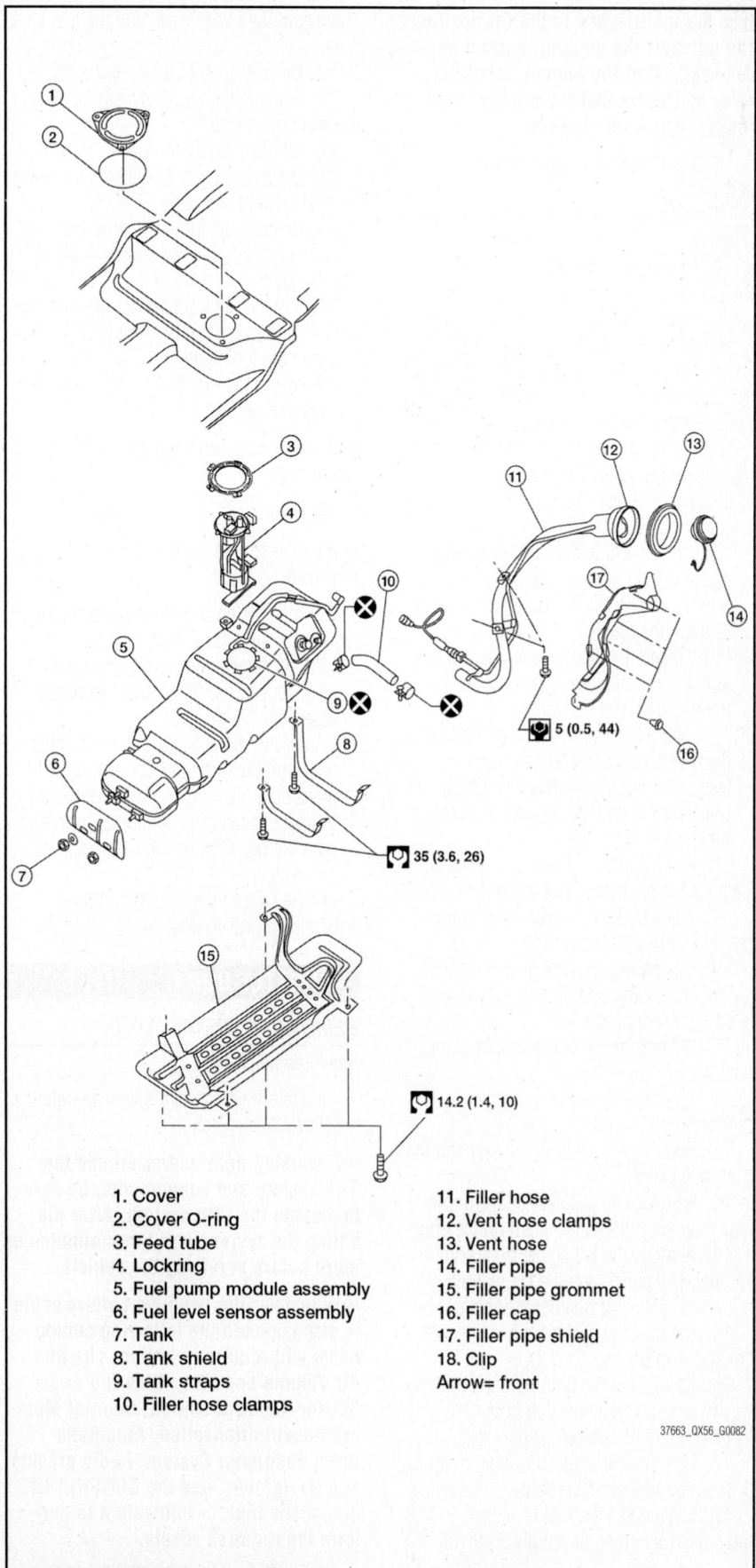

1. Cover
2. Cover O-ring
3. Feed tube
4. Lockring
5. Fuel pump module assembly
6. Fuel level sensor assembly
7. Tank
8. Tank shield
9. Tank straps
10. Filler hose clamps
11. Filler hose
12. Vent hose clamps
13. Vent hose
14. Filler pipe
15. Filler pipe grommet
16. Filler cap
17. Filler pipe shield
18. Clip
Arrow= front

37663_QX56_G0082

Fig. 93 Fuel tank and related components

turn the ignition key to the ON position (do not start the engine), correct as required. Start the engine, raise the idle, and verify that there are no fuel leaks, correct as required.

2. Disconnect the negative battery cable.

3. Drain the fuel from the fuel tank, if necessary.

4. Remove the fuel filler cap to release the pressure from inside the fuel tank.

5. Check the fuel level on level gauge. If the fuel gauge indicates more than the level as shown (full or almost full), drain the fuel from the fuel tank until the fuel gauge indicates the level as shown, or less.

6. If the fuel pump does not operate, use the following procedure to drain the fuel to the specified level after disconnecting the fuel filler hose from the fuel filler pipe:

   a. Insert a suitable hose of less than 15 mm (0.59 in.) diameter into the fuel filler pipe through the fuel filler opening to drain the fuel from fuel filler pipe.

   b. Insert a suitable hose into the fuel tank through the fuel filler hose to drain the fuel from the fuel tank.

   c. As a guide, the fuel level reaches the fuel gauge position as shown, or less, when approximately 3¾US gallons (14 liters) of fuel are drained from the fuel tank.

7. Raise and safely support the vehicle. Remove the LH rear wheel and tire.

8. Remove the four clips and remove the rear fender protector, front.

9. Disconnect the fuel filler hose from the fuel filler pipe and disconnect the vent hose quick connector.

10. Release the fuel pressure from the fuel lines.

11. Disconnect the battery negative terminal.

12. Remove the second row seat and the third row LH seat.

13. Remove the second and third row rear seat belt buckles mounted on the floor.

14. Remove the LH center pillar trim, the LH rear trim panel, and the LH rear side door kick plate and weather stripping.

15. Remove the second row rear center console and base.

16. Reposition the floor carpet out of the way to access the inspection hole cover, located under the center LH rear seat.

17. Remove the inspection hole cover by turning the retainers 90 degrees clockwise.

18. Disconnect the fuel level sensor, fuel filter, and fuel pump assembly electrical connector, the EVAP hose, and the fuel feed hose.

19. Disconnect the quick connector

20. Remove the four bolts and remove the fuel tank shield.

21. Remove the driveshaft.

22. Disconnect fuel filler hose, and vent hose at the fuel tank side.

23. Remove the fuel tank strap bolts while supporting the fuel tank with a suitable lift jack.

24. Disconnect the EVAP hose from the molded clip in the top of the fuel tank while lowering the fuel tank.

25. Lower the fuel tank using a suitable lift jack and remove it.

➡ If necessary, remove the lockring using tool.

To install:

➡ Be sure to use new fasteners, as required.

26. Installation is in the reverse order of removal, noting the following:

   a. For installation, use a new fuel level sensor, fuel filter, and fuel pump assembly O-ring.

   b. After installing the quick connectors, pull the tube and the connector to make sure they are securely connected. Visually inspect the connector to make sure the two retainer tabs are securely connected.

27. Be sure to perform the reconnect/relearn procedures.

THROTTLE BODY

REMOVAL & INSTALLATION

See Figure 94.

1. Before servicing the vehicle, refer to the Precautions Section.

➡ If working near and/or around the SRS system and components, be sure to disable the SRS system. After disabling the system wait three minutes or more before servicing the vehicle.

➡ Whenever the negative battery cable is disconnected the following components will require resetting. The Idle Air Volume Learning, Steering Angle Sensor Neutral Position, Sunroof Memory Reset/Initialization, Automatic Drive Positioner System, Audio presets and Navigation. Use the CONSULT-III diagnostic tool, or equivalent to perform the required resets.

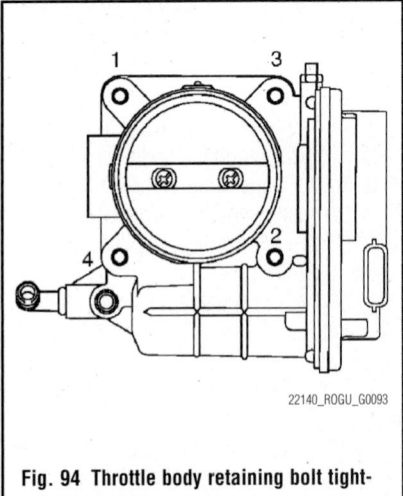

22140_ROGU_G0093

Fig. 94 Throttle body retaining bolt tightening sequence

➡ The automatic back door system must be initialized anytime the battery has been disconnected. Close the back door. Open the back door with the automatic open feature. Do not stop the process until the back door opens completely.

2. Disconnect the negative battery cable.

3. Partially drain the engine coolant.

4. Remove the engine room cover.

5. Remove the air duct and resonator assembly.

6. Drain the engine coolant. Be sure to properly dispose of used coolant.

7. Disconnect the hoses from the unit.

8. Remove the 4 mounting bolts.

9. Remove electric throttle control actuator by loosening bolts diagonally.

10. Remove the old gasket and discard it.

To install:

➡ Be sure to use new fasteners, as required.

11. Install a new gasket and the throttle body.

12. Install the 4 mounting bolts in alternate sequence and tighten to 74 inch lbs. (8.4 Nm).

13. Reconnect the hoses to the throttle body.

14. Reconnect the air duct and resonator assembly.

15. As required, fill the cooling system.

16. Install the engine cover.

17. Be sure to perform the reconnect/relearn procedures.

HEATING & AIR CONDITIONING SYSTEM

BLOWER MOTOR

REMOVAL & INSTALLATION

See Figure 95.

1. Before servicing the vehicle, refer to the Precautions Section.

➡**If working near and/or around the SRS system and components, be sure to disable the SRS system. After disabling the system wait three minutes or more before servicing the vehicle.**

➡**Whenever the negative battery cable is disconnected the following components will require resetting. The Idle Air Volume Learning, Steering Angle Sensor Neutral Position, Sunroof Memory Reset/Initialization, Automatic Drive Positioner System, Audio presets and Navigation. Use the CONSULT-III diagnostic tool, or equivalent to perform the required resets.**

➡**The automatic back door system must be initialized anytime the battery has been disconnected. Close the back door. Open the back door with the automatic open feature. Do not stop the process until the back door opens completely.**

2. Remove the glove box assembly.
3. Disconnect the front blower motor electrical connector.
4. Remove the blower retaining screws.
5. Remove the blower motor from its mounting.

To install:

➡**Be sure to use new fasteners, as required.**

6. Installation is the reverse of the removal procedure.
7. Be sure to perform the reconnect/relearn procedures.

HEATER/COOLING UNIT

REMOVAL & INSTALLATION

See Figures 96 through 98.

1. Before servicing the vehicle, refer to the Precautions Section.

➡**If working near and/or around the SRS system and components, be sure to disable the SRS system. After disabling the system wait three minutes or more before servicing the vehicle.**

➡**Whenever the negative battery cable is disconnected the following components will require resetting. The Idle Air Volume Learning, Steering Angle Sensor Neutral Position, Sunroof Memory Reset/Initialization, Automatic Drive Positioner System, Audio presets and Navigation. Use the CONSULT-III diagnostic tool, or equivalent to perform the required resets.**

➡**The automatic back door system must be initialized anytime the battery has been disconnected. Close the back door. Open the back door with the automatic open feature. Do not stop the process until the back door opens completely.**

2. Position the front seats in the rearmost position.
3. Disconnect the negative battery cable.
4. Properly discharge the A/C system.
5. Drain the cooling system. Be sure to properly dispose of used coolant.
6. Disconnect the heater hoses at the heater core.
7. Disconnect and plug the refrigerant lines at the evaporator core.
8. Remove the instrument panel.
9. Remove the center console.
10. Disconnect the instrument panel wire harness at the right and left in-line connector brackets, and the fuse block (JB) electrical connectors.
11. Disconnect the steering member from each side of the vehicle body.
12. Remove the heater/cooling unit with it attached to the steering member from the vehicle.

➡**Use care not to damage the seats or interior trim panels.**

13. Remove the heater/cooling unit from the steering member.

To install:

➡**Be sure to use new fasteners, as required.**

14. Installation is the reverse of the removal procedure.
15. Be sure to use new O-rings coated with clean refrigerant oil, as required.
16. Fill the cooling system with the proper grade and the coolant.
17. Properly recharge the A/C system.
18. Start the engine and check for leaks, correct as required.
19. Be sure to perform the reconnect/relearn procedures.

HEATER CORE

REMOVAL & INSTALLATION

See Figure 98.

1. Before servicing the vehicle, refer to the Precautions Section.

➡**If working near and/or around the SRS system and components, be sure to disable the SRS system. After disabling the system wait three minutes or more before servicing the vehicle.**

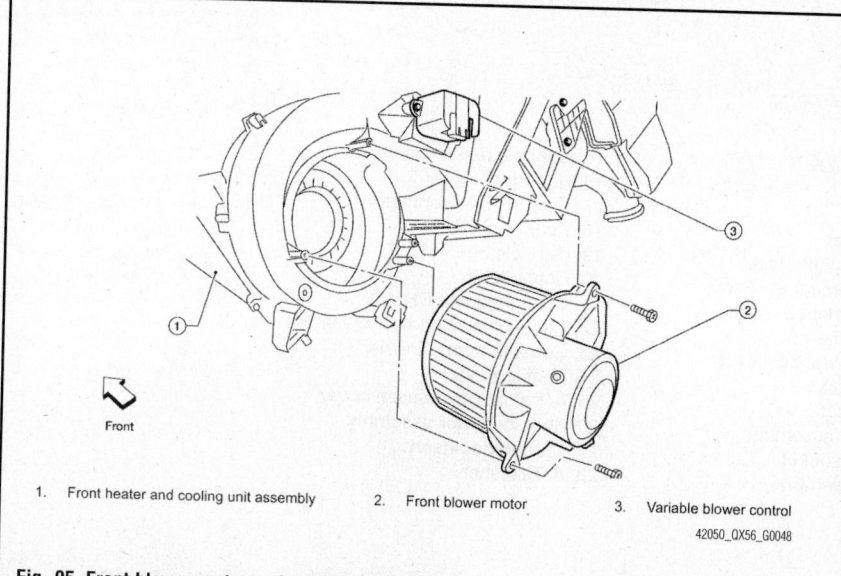

1. Front heater and cooling unit assembly
2. Front blower motor
3. Variable blower control

42050_QX56_G0048

Fig. 95 Front blower motor and related components

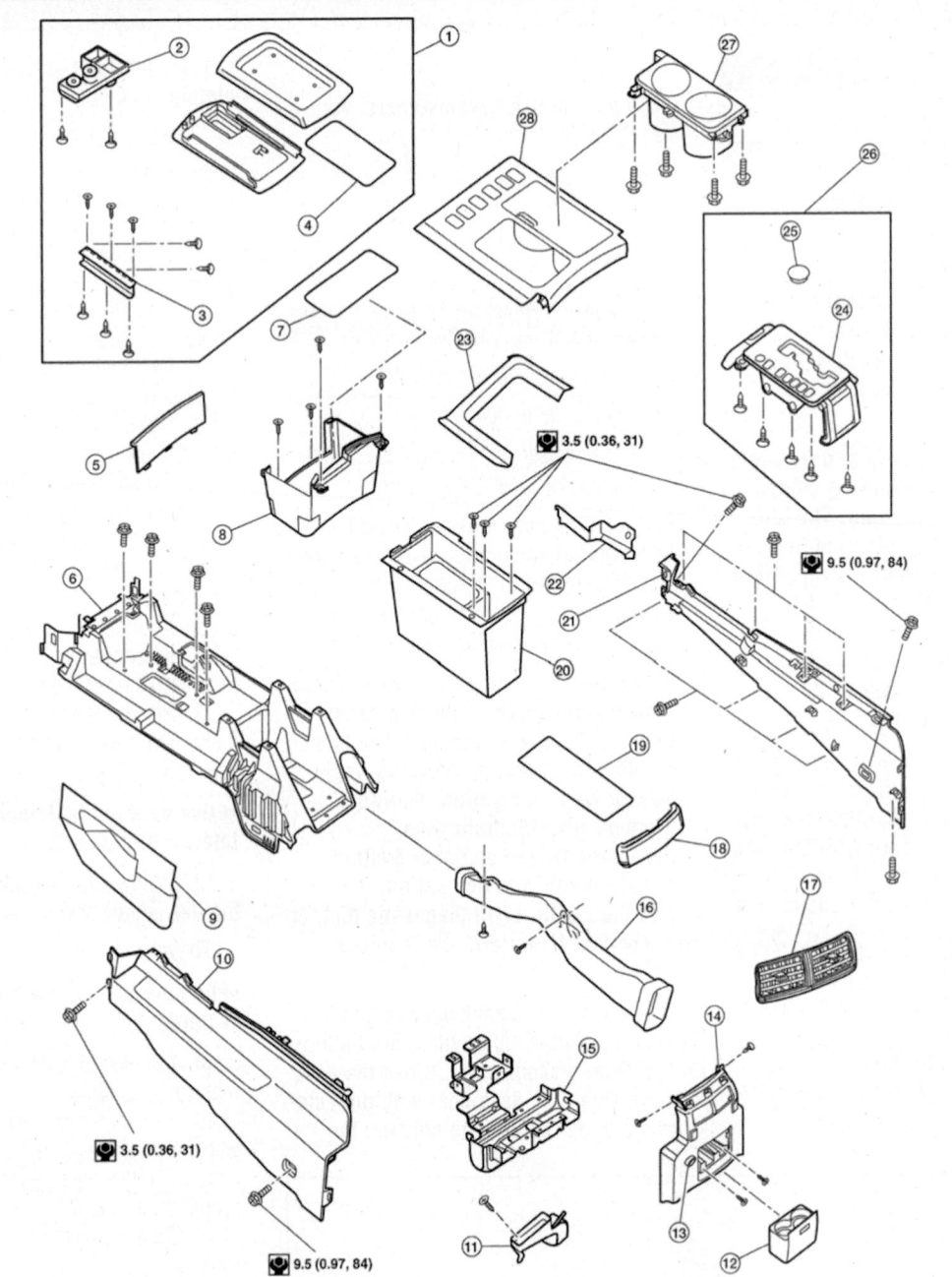

3.5 (0.36, 31)

9.5 (0.97, 84)

3.5 (0.36, 31)

9.5 (0.97, 84)

1. Console lid assembly
2. Console lid latch
3. Console lid hinge
4. Console lid bin mat
5. Storage compartment mask
6. Console reinforcement assembly
7. Storage compartment bin mat
8. Storage compartment
9. Center console lower cover LH
10. Console cover LH
11. Rear console duct
12. Rear cup holder assembly
13. Console power socket
14. Rear finisher assembly
15. Console rear bracket
16. Heat duct
17. Rear finisher ventilator
18. Rear upper finisher
19. Console bin mat
20. Console bin
21. Console cover RH
22. Center console lower cover RH
23. Console upper finisher
24. Shift indicator
25. A/T shift lock release cover
26. Shift indicator assembly
27. Cup holder insert
28. A/T finisher

71075_ARMA_G0032

Fig. 96 Front console and related components

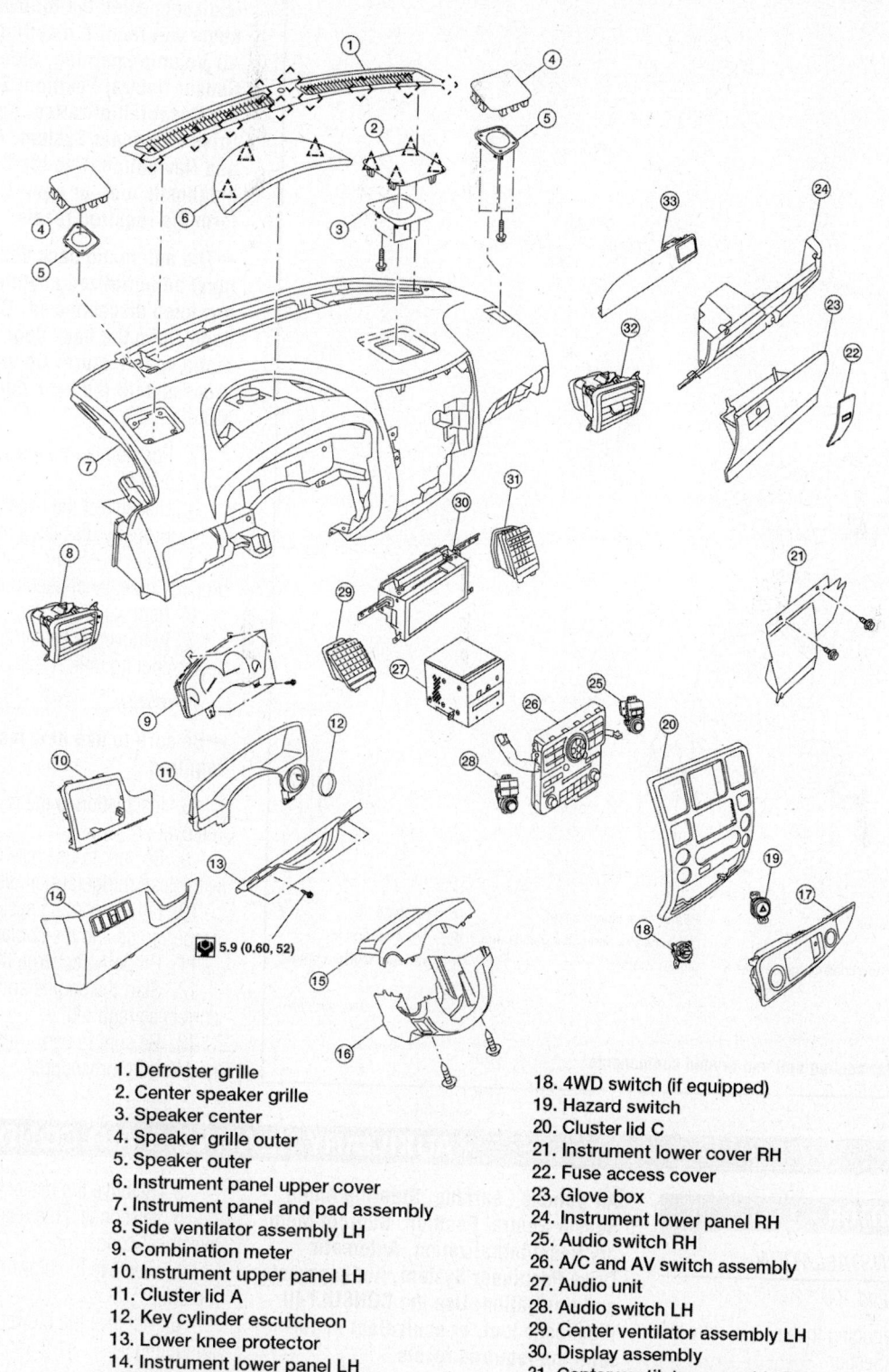

1. Defroster grille
2. Center speaker grille
3. Speaker center
4. Speaker grille outer
5. Speaker outer
6. Instrument panel upper cover
7. Instrument panel and pad assembly
8. Side ventilator assembly LH
9. Combination meter
10. Instrument upper panel LH
11. Cluster lid A
12. Key cylinder escutcheon
13. Lower knee protector
14. Instrument lower panel LH
15. Steering column cover upper
16. Steering column cover lower
17. Cluster lid C lower

18. 4WD switch (if equipped)
19. Hazard switch
20. Cluster lid C
21. Instrument lower cover RH
22. Fuse access cover
23. Glove box
24. Instrument lower panel RH
25. Audio switch RH
26. A/C and AV switch assembly
27. Audio unit
28. Audio switch LH
29. Center ventilator assembly LH
30. Display assembly
31. Center ventilator assembly RH
32. Side ventilator assembly RH
33. Instrument upper panel RH
Metal clip

71075_ARMA_G0033

Fig. 97 Instrument panel and related components

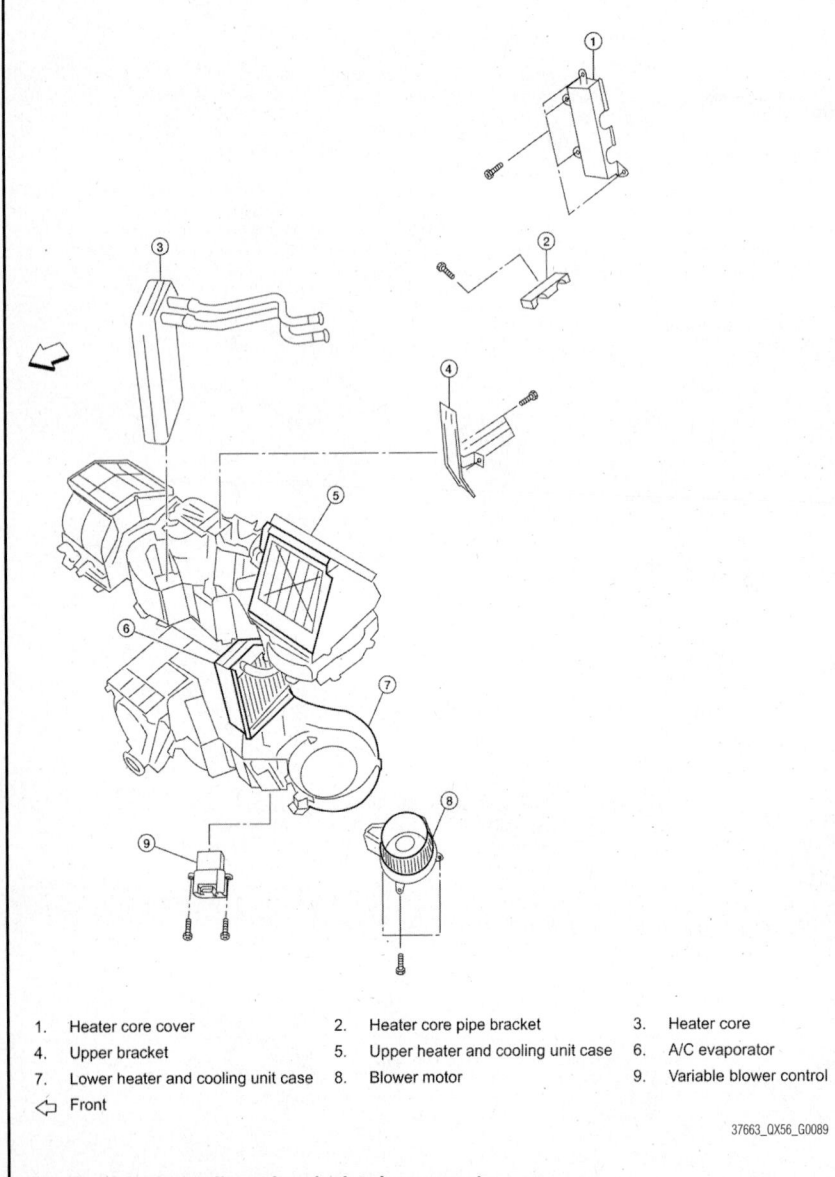

1. Heater core cover
2. Heater core pipe bracket
3. Heater core
4. Upper bracket
5. Upper heater and cooling unit case
6. A/C evaporator
7. Lower heater and cooling unit case
8. Blower motor
9. Variable blower control
⇦ Front

37663_QX56_G0089

Fig. 98 AC heater/cooling unit and related components

➡Whenever the negative battery cable is disconnected the following components will require resetting. The Idle Air Volume Learning, Steering Angle Sensor Neutral Position, Sunroof Memory Reset/Initialization, Automatic Drive Positioner System, Audio presets and Navigation. Use the CONSULT-III diagnostic tool, or equivalent to perform the required resets.

➡The automatic back door system must be initialized anytime the battery has been disconnected. Close the back door. Open the back door with the automatic open feature. Do not stop the process until the back door opens completely.

2. Position the front seats in the rear-most position.
3. Disconnect the negative battery cable.
4. Properly discharge the A/C system.
5. Drain the cooling system. Be sure to properly dispose of used coolant.
6. Remove the heater/cooling unit.
7. Remove the heater core from the heater/cooling unit.

To install:

➡Be sure to use new fasteners, as required.

8. Installation is the reverse of the removal procedure.
9. Be sure to use new O-rings coated with clean refrigerant oil, as required.
10. Fill the cooling system with the proper grade and the coolant.
11. Properly recharge the A/C system.
12. Start the engine and check for leaks, correct as required.
13. Be sure to perform the reconnect/relearn procedures.

AUXILIARY HEATING & AIR CONDITIONING

BLOWER MOTOR

REMOVAL & INSTALLATION

See Figures 99 and 100.

1. Before servicing the vehicle, refer to the Precautions Section.

➡If working near and/or around the SRS system and components, be sure to disable the SRS system. After disabling the system wait three minutes or more before servicing the vehicle.

➡Whenever the negative battery cable is disconnected the following components will require resetting. The Idle Air Volume Learning, Steering Angle Sensor Neutral Position, Sunroof Memory Reset/Initialization, Automatic Drive Positioner System, Audio presets and Navigation. Use the CONSULT-III diagnostic tool, or equivalent to perform the required resets.

➡The automatic back door system must be initialized anytime the battery has been disconnected. Close the back door. Open the back door with the automatic open feature. Do not stop the process until the back door opens completely.

2. Disconnect the negative battery cable.

3. Remove the heater/cooling unit.
4. Disconnect the rear blower motor electrical connector.
5. Remove the blower retaining screws.
6. Remove the blower motor from its mounting.

To install:

➡Be sure to use new fasteners, as required.

7. Installation is the reverse of the removal procedure.
8. Be sure to use new O-rings coated with clean refrigerant oil, as required.

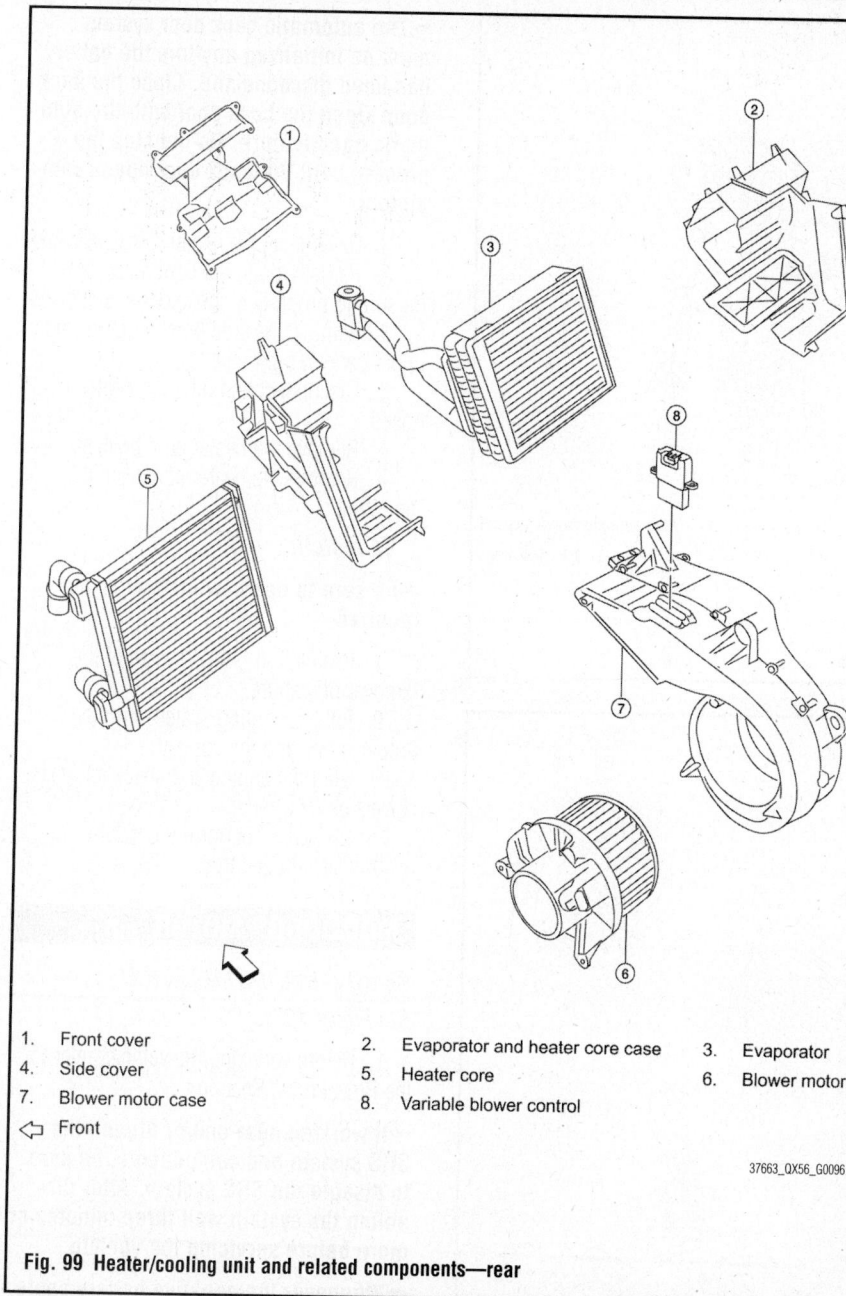

1. Front cover
4. Side cover
7. Blower motor case
⬅ Front

2. Evaporator and heater core case
5. Heater core
8. Variable blower control

3. Evaporator
6. Blower motor

37663_QX56_G0096

Fig. 99 Heater/cooling unit and related components—rear

9. Fill the cooling system with the proper grade and the coolant.
10. Properly recharge the A/C system.
11. Start the engine and check for leaks, correct as required.
12. Be sure to perform the reconnect/relearn procedures.

HEATER/COOLING UNIT

REMOVAL & INSTALLATION

See Figure 101.

1. Before servicing the vehicle, refer to the Precautions Section.

➡If working near and/or around the SRS system and components, be sure to disable the SRS system. After disabling the system wait three minutes or more before servicing the vehicle.

➡Whenever the negative battery cable is disconnected the following components will require resetting. The Idle Air Volume Learning, Steering Angle Sensor Neutral Position, Sunroof Memory Reset/Initialization, Automatic Drive Positioner System, Audio presets and Navigation. Use the CONSULT-III diagnostic tool, or equivalent to perform the required resets.

➡The automatic back door system must be initialized anytime the battery has been disconnected. Close the back door. Open the back door with the automatic open feature. Do not stop the process until the back door opens completely.

2. Disconnect the negative battery cable.
3. Properly discharge the A/C system.
4. Drain the cooling system. Be sure to properly dispose of used coolant.
5. Remove the side finisher lower and upper panels, right side.
6. Disconnect and plug the heater hoses.
7. Disconnect and plug the refrigerant lines.
8. Disconnect the electrical connectors.
9. Disconnect the ducts and drain tubes.
10. Remove the heater/cooling unit from the vehicle.

To install:

➡Be sure to use new fasteners, as required.

11. Installation is the reverse of the removal procedure.
12. Be sure to use new O-rings coated with clean refrigerant oil, as required.
13. Fill the cooling system with the proper grade and the coolant.
14. Properly recharge the A/C system.
15. Start the engine and check for leaks, correct as required.
16. Be sure to perform the reconnect/relearn procedures.

HEATER CORE

REMOVAL & INSTALLATION

See Figure 102.

1. Before servicing the vehicle, refer to the Precautions Section.

➡If working near and/or around the SRS system and components, be sure to disable the SRS system. After disabling the system wait three minutes or more before servicing the vehicle.

➡Whenever the negative battery cable is disconnected the following components will require resetting. The Idle Air Volume Learning, Steering Angle Sensor Neutral Position, Sunroof Memory Reset/Initialization, Automatic Drive Positioner System, Audio presets and Navigation. Use the CONSULT-III diagnostic tool, or equivalent to perform the required resets.

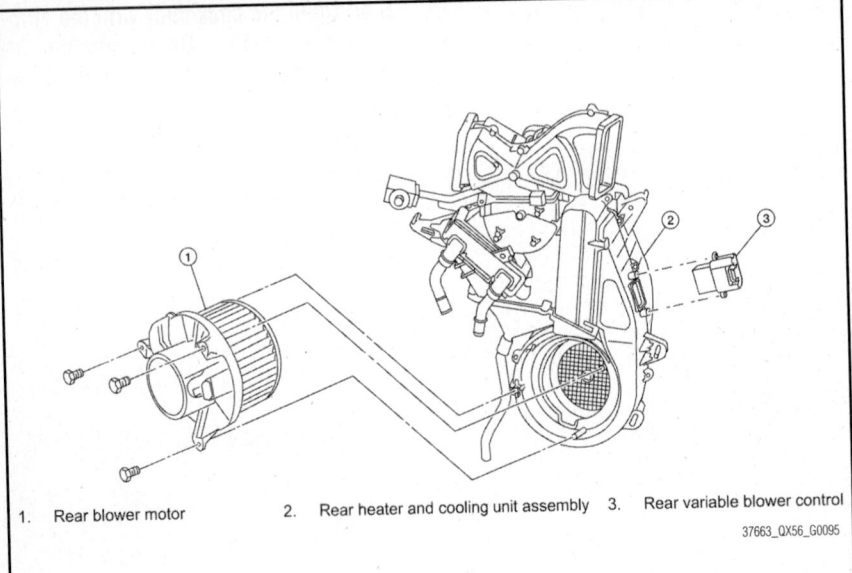

1. Rear blower motor   2. Rear heater and cooling unit assembly   3. Rear variable blower control

37663_QX56_G0095

Fig. 100 Blower motor and related components—rear

1. Front cover   2. Evaporator and heater core case   3. Evaporator
4. Side cover   5. Heater core   6. Blower motor
7. Blower motor case   8. Variable blower control
◁ Front

37663_QX56_G0096

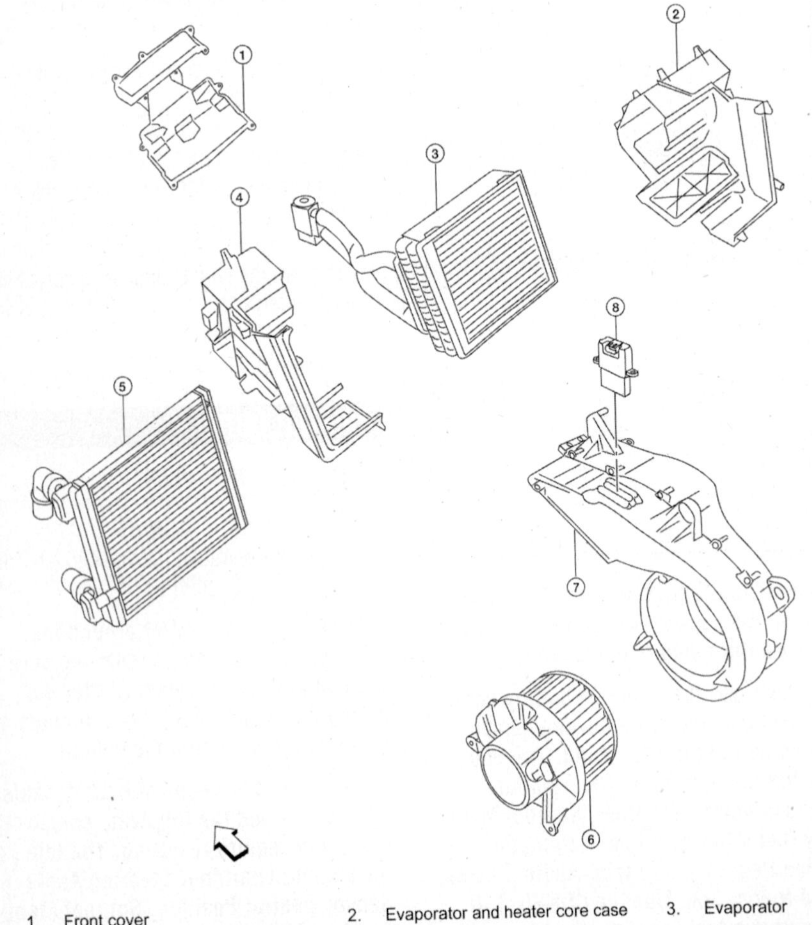

Fig. 101 Heater/cooling unit and related components—rear

➡ The automatic back door system must be initialized anytime the battery has been disconnected. Close the back door. Open the back door with the automatic open feature. Do not stop the process until the back door opens completely.

2. Disconnect the negative battery cable.
3. Partially drain the cooling system. Be sure to properly dispose of used coolant.
4. Remove the side finisher lower and upper panels, right side.
5. Disconnect and plug the heater hoses.
6. Remove the heater core bracket.
7. Remove the heater core from its mounting.

To install:

➡ Be sure to use new fasteners, as required.

8. Installation is the reverse of the removal procedure.
9. Fill the cooling system with the proper grade and the coolant.
10. Start the engine and check for leaks, correct as required.
11. Be sure to perform the reconnect/relearn procedures.

MODE DOOR MOTOR

REMOVAL & INSTALLATION

See Figure 103.

1. Before servicing the vehicle, refer to the Precautions Section.

➡ If working near and/or around the SRS system and components, be sure to disable the SRS system. After disabling the system wait three minutes or more before servicing the vehicle.

➡ Whenever the negative battery cable is disconnected the following components will require resetting. The Idle Air Volume Learning, Steering Angle Sensor Neutral Position, Sunroof Memory Reset/Initialization, Automatic Drive Positioner System, Audio presets and Navigation. Use the CONSULT-III diagnostic tool, or equivalent to perform the required resets.

➡ The automatic back door system must be initialized anytime the battery has been disconnected. Close the back door. Open the back door with the automatic open feature. Do not stop the process until the back door opens completely.

2. Disconnect the negative battery cable.

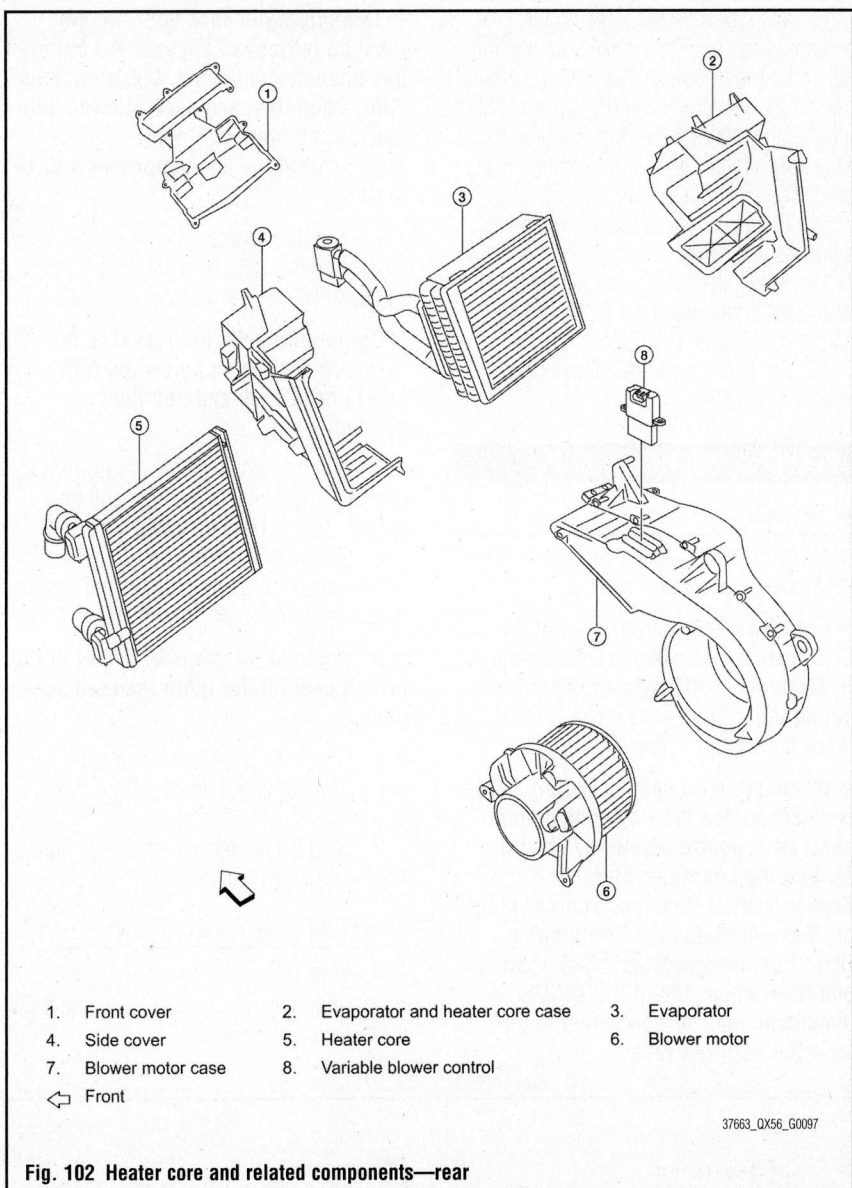

1. Front cover
4. Side cover
7. Blower motor case
⬅ Front

2. Evaporator and heater core case
5. Heater core
8. Variable blower control

3. Evaporator
6. Blower motor

37663_QX56_G0097

Fig. 102 Heater core and related components—rear

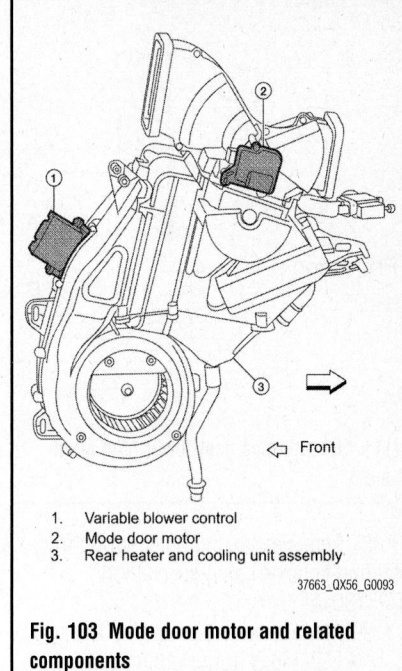

1. Variable blower control
2. Mode door motor
3. Rear heater and cooling unit assembly

37663_QX56_G0093

Fig. 103 Mode door motor and related components

3. Remove the luggage side finisher, right side.

4. Remove the rear heater/cooling unit assembly.

5. Disconnect the electrical connector.

6. Remove the retaining screws.

7. Remove the component from its mounting.

To install:

➡**Be sure to use new fasteners, as required.**

8. Installation is the reverse of the removal procedure.

9. Be sure to perform the recon-nect/relearn procedures.

STEERING

POWER RACK & PINION STEERING GEAR

REMOVAL & INSTALLATION

See Figures 104 and 105.

1. Before servicing the vehicle, refer to the Precautions Section.

➡**If working near and/or around the SRS system and components, be sure to disable the SRS system. After dis-abling the system wait three minutes or more before servicing the vehicle.**

➡**Whenever the negative battery cable is disconnected the following compo-nents will require resetting. The Idle**

Air Volume Learning, Steering Angle Sensor Neutral Position, Sunroof Mem-ory Reset/Initialization, Automatic Drive Positioner System, Audio presets and Navigation. Use the CONSULT-III diagnostic tool, or equivalent to per-form the required resets.

➡**The automatic back door system must be initialized anytime the battery has been disconnected. Close the back door. Open the back door with the auto-matic open feature. Do not stop the process until the back door opens com-pletely.**

2. Ensure the wheels are in the straight-ahead position.

3. Disconnect the negative battery cable.

4. Remove or disconnect the following:
   - Wheels and tires
   - Engine splash guard

5. On 4WD, remove front final drive and support the halfshafts.

6. Remove cotter pin at steering outer socket and loosen mounting nut.

7. With the steering wheel in the straight ahead position, make sure that the slit of the lower joint (A) fits with the projec-tion on the rear cover cap (B), while check-ing that the mark on the steering gear assembly aligns with the mark on the rear cover cap. See illustration.

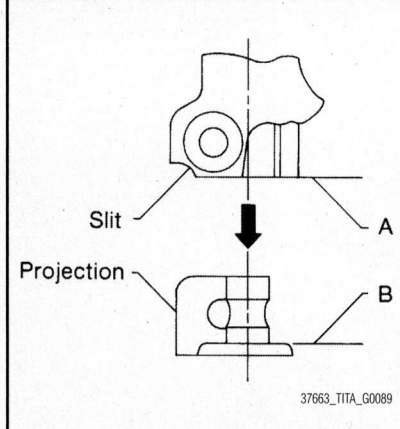

Fig. 104 Steering gear lower joint alignment

8. Remove steering outer socket from steering knuckle using special tool J-25730-A.

9. Remove or disconnect the following:
- Oil pipes from steering gear assembly
- Lower joint mounting bolt from lower shaft
- Mounting bolts and nuts from steering gear assembly
- Steering gear assembly

To install:

➡**Be sure to use new fasteners, as required.**

10. Installation is the reverse of the removal procedure.

11. With the steering wheel in the straight ahead position, make sure that the slit of the lower joint (A) fits with the projection on the rear cover cap (B), while checking that the mark on the steering gear assembly aligns with the mark on the rear cover cap. See illustration.

12. Check the wheel alignment and adjust as necessary.

13. Adjust the steering angle sensor neutral position, using the CONSULT-III diagnostic tool, or equivalent.

14. Be sure to perform the reconnect/relearn procedures.

POWER STEERING PUMP

BLEEDING

1. Before servicing the vehicle, refer to the Precautions Section.

➡**If working near and/or around the SRS system and components, be sure to disable the SRS system. After disabling the system wait three minutes or more before servicing the vehicle.**

➡**Whenever the negative battery cable is disconnected the following components will require resetting. The Idle Air Volume Learning, Steering Angle Sensor Neutral Position, Sunroof Memory Reset/Initialization, Automatic Drive Positioner System, Audio presets and Navigation. Use the CONSULT-III diagnostic tool, or equivalent to perform the required resets.**

➡**The automatic back door system must be initialized anytime the battery has been disconnected. Close the back door. Open the back door with the automatic open feature. Do not stop the process until the back door opens completely.**

2. Stop the engine.

3. Turn the steering wheel fully to the right and left several times.

➡**Do not allow the fluid level in the reservoir tank to go below the MIN level line. Check and add fluid as needed.**

4. Run the engine at idle speed. Turn the steering wheel fully to the right and then fully to the left. Hold for about three seconds. Check for fluid leakage.

5. Repeat the above step several times at three second intervals.

➡**Do not hold the steering wheel in the locked position for more than ten seconds.**

6. Check for air bubbles or cloudy fluid. If found, repeat the bleeding procedure.

7. Stop the engine and check the fluid level. Correct as required.

REMOVAL & INSTALLATION
See Figure 106.

1. Before servicing the vehicle, refer to the Precautions Section.

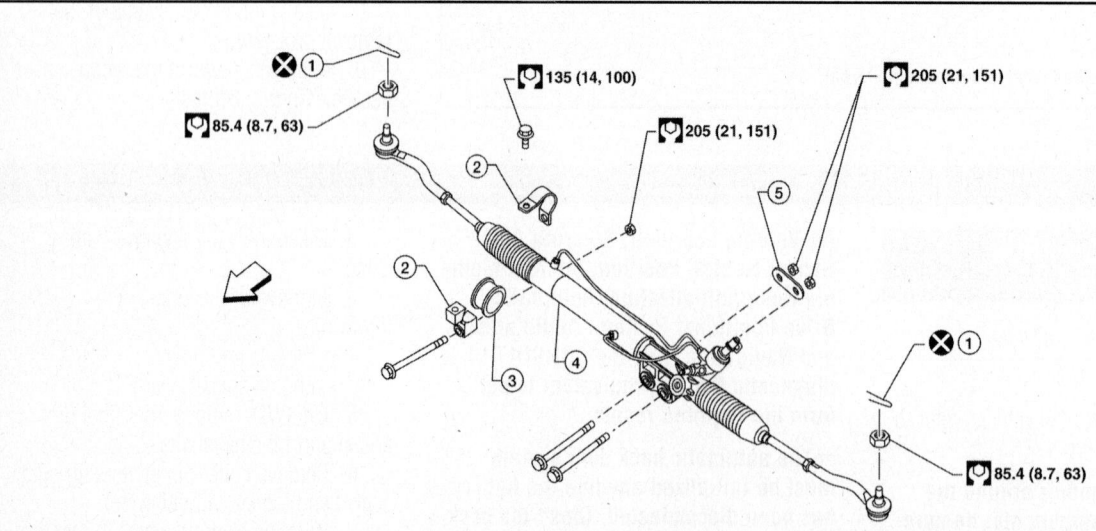

1. Cotter pin
2. Steering gear bracket
3. Steering gear insulator
4. Steering gear assembly
5. Washer

Fig. 105 Power steering gear and related components

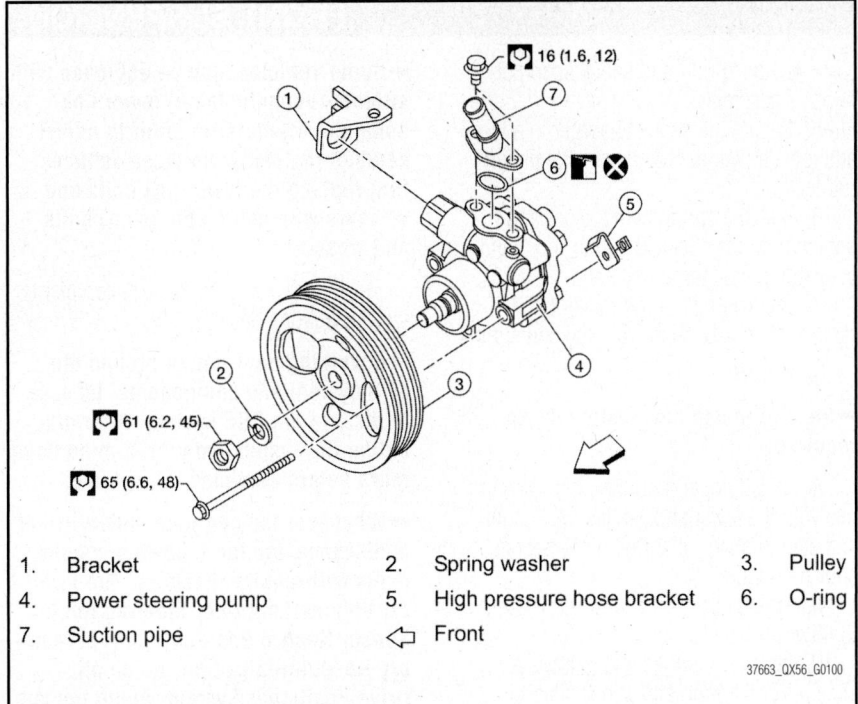

16 (1.6, 12)

61 (6.2, 45)

65 (6.6, 48)

1. Bracket
2. Spring washer
3. Pulley
4. Power steering pump
5. High pressure hose bracket
6. O-ring
7. Suction pipe
⟨ Front

37663_QX56_G0100

Fig. 106 Power steering pump and related components

➡ **If working near and/or around the SRS system and components, be sure to disable the SRS system. After disabling the system wait three minutes or more before servicing the vehicle.**

➡ **Whenever the negative battery cable is disconnected the following components will require resetting. The Idle Air Volume Learning, Steering Angle Sensor Neutral Position, Sunroof Memory Reset/Initialization, Automatic Drive Positioner System, Audio presets and Navigation. Use the CONSULT-III diagnostic tool, or equivalent to perform the required resets.**

➡ **The automatic back door system must be initialized anytime the battery has been disconnected. Close the back door. Open the back door with the automatic open feature. Do not stop the process until the back door opens completely.**

2. Disconnect the negative battery cable.

3. Drain the power steering fluid into a suitable container. Properly discard the used fluid.

4. Remove the engine room cover.

5. Remove the air duct assembly.

6. Remove the power steering reservoir tank.

7. Remove the drive belt.

8. Disconnect the pressure sensor electrical connector.

9. Remove the high pressure and the low pressure lines from the power steering fluid pump.

10. Remove the pump mounting bolts.

11. Remove the pump from the vehicle.

To install:

➡ **Be sure to use new fasteners, as required.**

12. Installation is the reverse of removal procedure.

13. Bleed the power steering system.

➡ **The drive belt tension is automatic and requires no adjustment.**

14. Be sure to perform the reconnect/relearn procedures.

POWER STEERING FLUID

FLUID RECOMMENDATIONS

Use genuine Nissan power steering fluid (PSF), or equivalent when servicing the power steering system. The system capacity is 2 1/8 pints.

FLUID FILL PROCEDURE

See Figure 107.

✳✳ CAUTION

Used fluid is considerably more dangerous than new fluid. Avoid skin contact with used fluid.

1. Before servicing the vehicle, refer to the Precautions Section.

➡ **If working near and/or around the SRS system and components, be sure to disable the SRS system. After disabling the system wait three minutes or more before servicing the vehicle.**

➡ **Whenever the negative battery cable is disconnected the following components will require resetting. The Idle Air Volume Learning, Steering Angle Sensor Neutral Position, Sunroof Memory Reset/Initialization, Automatic Drive Positioner System, Audio presets and Navigation. Use the CONSULT-III diagnostic tool, or equivalent to perform the required resets.**

➡ **The automatic back door system must be initialized anytime the battery has been disconnected. Close the back door. Open the back door with the automatic open feature. Do not stop the process until the back door opens completely.**

2. Inspect the power steering fluid level in the power steering reservoir. Do allow the fluid to drop below the MIN marking.

3. Remove the power steering reservoir cap.

4. Add fluid, as necessary, referring to the scale on the reservoir tank.
   - HOT range for fluid temperatures: 122–176°F (50–80°C)
   - COLD range for fluid temperatures: 32–86°F (0–30°C)

➡ **Do not overfill the fluid. Do not reuse any used power steering fluid.**

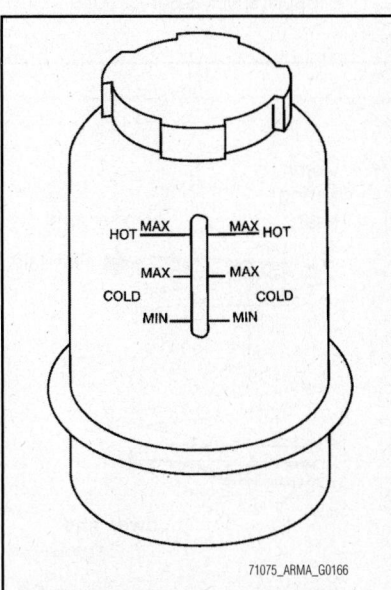

71075_ARMA_G0166

Fig. 107 Power steering reservoir tank fill markings

COIL SPRINGS

REMOVAL & INSTALLATION

See Figure 108.

1. Before servicing the vehicle, refer to the Precautions Section.

➡ If working near and/or around the SRS system and components, be sure to disable the SRS system. After disabling the system wait three minutes or more before servicing the vehicle.

➡ Whenever the negative battery cable is disconnected the following components will require resetting. The Idle Air Volume Learning, Steering Angle Sensor Neutral Position, Sunroof Memory Reset/Initialization, Automatic Drive Positioner System, Audio presets and Navigation. Use the CONSULT-III diagnostic tool, or equivalent to perform the required resets.

➡ The automatic back door system must be initialized anytime the battery has been disconnected. Close the back door. Open the back door with the automatic open feature. Do not stop the process until the back door opens completely.

2. Disconnect the negative battery cable.
3. Raise and safely support the vehicle.
4. Remove or disconnect the following:
   - Wheel and tire assembly
   - Lower shock absorber bolt
   - Upper shock absorber bolts

- Coil spring and shock absorber assembly

5. Secure the shock absorber in a vice and loosen (without removing) the piston rod locknut.
6. Install a spring compressor and tighten until the shock absorber mounting insulator can be turned by hand.
7. Remove piston rod locknut and remove shock absorber from the coil spring.

To install:

➡ Be sure to use new fasteners, as required.

8. Install upper mounting insulator in line with the lower shock absorber mount and step in shock absorber lower seat as shown in figure.
9. Tighten the new piston rod locknut to 40 ft. lbs. (54 Nm).
10. Install or connect the following:
    - Coil spring and shock absorber assembly
    - Upper shock absorber bolts and tighten to 22 ft. lbs (30 Nm)
    - Lower shock absorber bolt and tighten to 99 ft. lbs. (134 Nm)
    - Wheel and tire assembly
11. Check wheel alignment and adjust as necessary.
12. Be sure to perform the reconnect/relearn procedures.

LOWER BALL JOINTS

REMOVAL & INSTALLATION

At this time the manufacturer does not provide removal and installation procedures for this component. The lower ball joint is part of the lower control arm (lower link) assembly.

➡ Some vehicles may be equipped with straight (nonadjustable) lower link bolts and washers. In order to adjust camber and caster on these vehicles, first replace the lower link bolts and washers with adjustable (cam) bolts and washers.

➡ Nissan/Infiniti refers to the lower control arm as a lower link.

LOWER CONTROL ARMS

REMOVAL & INSTALLATION

See Figure 109.

➡ Nissan/Infiniti refers to the lower control arm as a lower link.

➡ Some vehicles may be equipped with straight (nonadjustable) lower link bolts and washers. In order to adjust camber and caster on these vehicles, first replace the lower link bolts and washers with adjustable (cam) bolts and washers.

1. Before servicing the vehicle, refer to the Precautions Section.

➡ If working near and/or around the SRS system and components, be sure to disable the SRS system. After disabling the system wait three minutes or more before servicing the vehicle.

➡ Whenever the negative battery cable is disconnected the following components will require resetting. The Idle Air Volume Learning, Steering Angle Sensor Neutral Position, Sunroof Memory Reset/Initialization, Automatic Drive Positioner System, Audio presets and Navigation. Use the CONSULT-III diagnostic tool, or equivalent to perform the required resets.

➡ The automatic back door system must be initialized anytime the battery has been disconnected. Close the back door. Open the back door with the automatic open feature. Do not stop the process until the back door opens completely.

2. Disconnect the negative battery cable.
3. Raise and safely support the vehicle.
4. Remove the tire and wheel assembly.

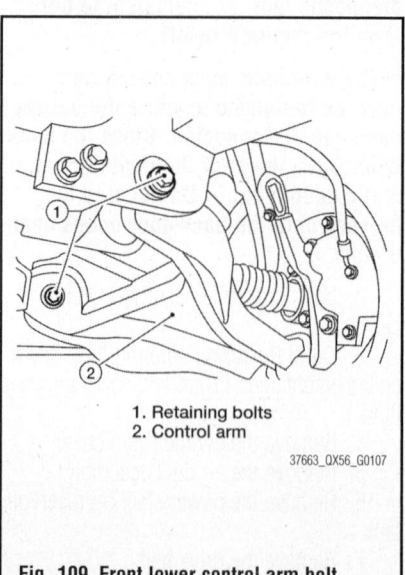

1. Retaining bolts
2. Control arm

37663_QX56_G0107

Fig. 109 Front lower control arm bolt locations

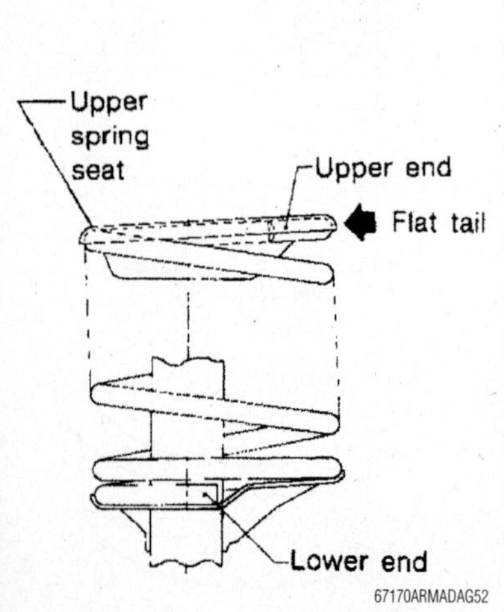

Fig. 108 Front coil spring positioning

67170ARMADAG52

5. Remove the lower shock absorber retaining bolt.

6. Remove the stabilizer bar connecting rod lower nut.

7. Remove the pinch bolt from the steering knuckle, than separate the lower link ball joint from the steering knuckle.

8. Remove the lower link bolts and nuts.

9. Remove the lower link.

To install:

➡ **Be sure to use new fasteners, as required.**

10. Installation is the reverse of the removal procedure.

11. Check and adjust alignment, as required.

12. Be sure to perform the reconnect/relearn procedures.

SHOCK ABSORBERS

REMOVAL & INSTALLATION

See Figures 110 and 111.

1. Before servicing the vehicle, refer to the Precautions Section.

➡ **If working near and/or around the SRS system and components, be sure to disable the SRS system. After disabling the system wait three minutes or more before servicing the vehicle.**

➡ **Whenever the negative battery cable is disconnected the following components will require resetting. The Idle Air Volume Learning, Steering Angle Sensor Neutral Position, Sunroof Memory Reset/Initialization, Automatic Drive Positioner System, Audio presets and Navigation. Use the CONSULT-III diagnostic tool, or equivalent to perform the required resets.**

➡ **The automatic back door system must be initialized anytime the battery has been disconnected. Close the back door. Open the back door with the automatic open feature. Do not stop the process until the back door opens completely.**

2. Disconnect the negative battery cable.

3. Raise and safely support the vehicle.

4. Remove or disconnect the following:
   • Wheel and tire assembly
   • Lower shock absorber bolt
   • Upper shock absorber bolts
   • Coil spring and shock absorber assembly

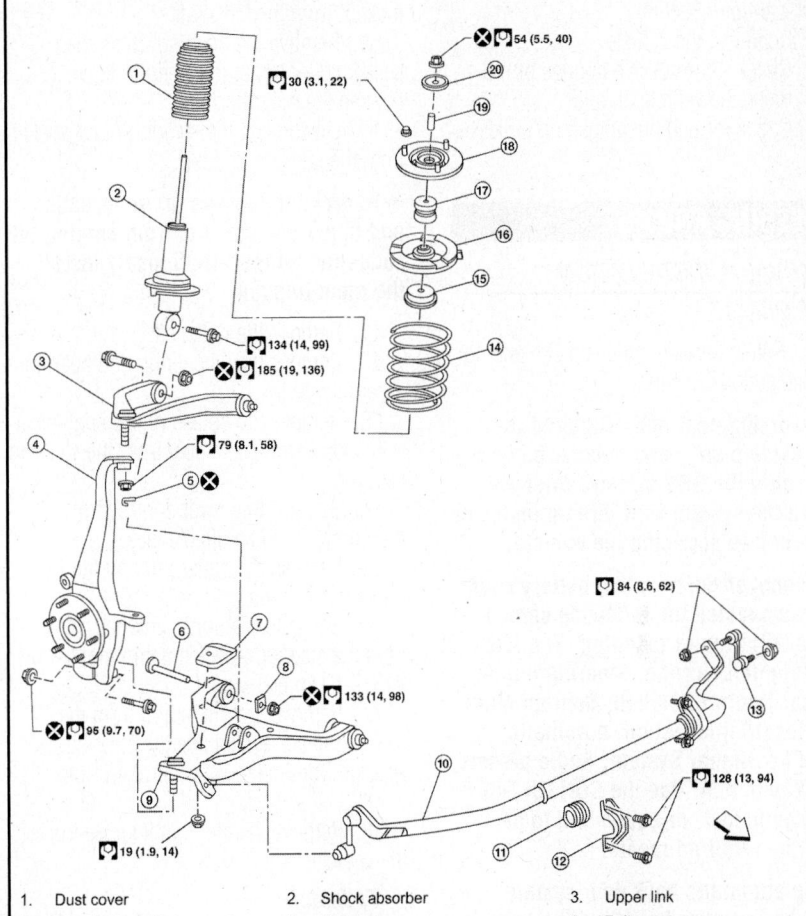

1.	Dust cover	2.	Shock absorber	3.	Upper link
4.	Steering knuckle	5.	Cotter pin	6.	Bolt
7.	Jounce bumper	8.	Washer	9.	Lower link
10.	Stabilizer bar	11.	Stabilizer bar bushing	12.	Stabilizer bar mounting bracket
13.	Connecting rod	14.	Coil spring	15.	Upper seat
16.	Upper spring seat	17.	Shock absorber bushing	18.	Shock absorber mounting insulator
19.	Spacer	20.	Washer	⇦	Front

37663_QX56_G0104

Fig. 110 Front shock absorber and related components

To install:

➡ **Be sure to use new fasteners, as required.**

5. Installation is the reverse of the removal procedure.

6. Install upper mounting insulator in line with the lower shock absorber mount and step in shock absorber lower seat as shown in figure.

7. Check wheel alignment and adjust as necessary.

8. Be sure to perform the reconnect/relearn procedures.

TESTING

➡ **If the vehicle is equipped with automatic load leveling system, be sure to check for air leaks and loose or worn connection points.**

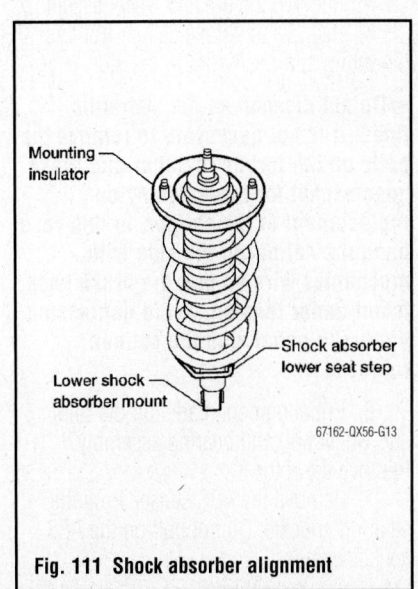

67162-QX56-G13

Fig. 111 Shock absorber alignment

1. Before servicing the vehicle, refer to the Precautions Section.
2. Road test the vehicle.
3. Check for excessive bounce or roll.
4. Raise the vehicle on a lift.
5. Check for bad bushings and oil leakage.

STEERING KNUCKLE

REMOVAL & INSTALLATION

See Figure 112.

1. Before servicing the vehicle, refer to the Precautions Section.

→**If working near and/or around the SRS system and components, be sure to disable the SRS system. After disabling the system wait three minutes or more before servicing the vehicle.**

→**Whenever the negative battery cable is disconnected the following components will require resetting. The Idle Air Volume Learning, Steering Angle Sensor Neutral Position, Sunroof Memory Reset/Initialization, Automatic Drive Positioner System, Audio presets and Navigation. Use the CONSULT-III diagnostic tool, or equivalent to perform the required resets.**

→**The automatic back door system must be initialized anytime the battery has been disconnected. Close the back door. Open the back door with the automatic open feature. Do not stop the process until the back door opens completely.**

2. Disconnect the negative battery cable.
3. Raise and support the vehicle safely.
4. Remove the tire and wheel assembly.
5. Remove the brake caliper from its mounting and position it to the side.

→**Do not disconnect the hydraulic lines. It is not necessary to remove the bolts on the torque member and brake hose except for disassembly or replacement of the caliper. In this case hang the caliper to the side with mechanics wire so that the brake hose is not under tension. Avoid depressing the brake pedal with the caliper removed.**

6. Put alignment marks on the rotor and wheel hub and bearing assembly. Remove the rotor.
7. Remove the ABS sensor from the steering knuckle. Do not pull on the ABS sensor harness.

8. Remove the cotter pin. Remove the locknut from the halfshaft.
9. Remove the steering outer shaft socket cotter pin at the steering knuckle. Loosen the mounting nut.
10. Disconnect the steering outer socket from the steering knuckle.

→**To prevent damage to the threads and to prevent the tool from coming off suddenly, temporarily loosely install the mounting nut.**

11. Remove the halfshaft.
12. Remove the wheel hub and bearing assembly bolts.
13. Remove the splash guard and wheel hub and bearing assembly from the steering knuckle.
14. Support the lower control arm assembly, using a suitable jack.
15. Remove the cotter pin and nut from the upper ball joint.
16. Separate the upper link ball joint from the steering knuckle using tool J-24319-01 or equivalent.
17. Remove the pinch bolt from the steering knuckle. Remove the steering knuckle from the lower control arm ball joint.
18. Remove the steering knuckle from the vehicle.

To install:

→**Be sure to use new fasteners, as required.**

19. Installation is the reverse of the removal procedure.

20. Be sure to use the alignment marks made during the removal procedure when reinstalling removed components.
21. Check and adjust the front end alignment, as required.
22. Be sure to perform the reconnect/relearn procedures.

STABILIZER BAR

REMOVAL & INSTALLATION

See Figure 113.

1. Before servicing the vehicle, refer to the Precautions Section.

→**If working near and/or around the SRS system and components, be sure to disable the SRS system. After disabling the system wait three minutes or more before servicing the vehicle.**

→**Whenever the negative battery cable is disconnected the following components will require resetting. The Idle Air Volume Learning, Steering Angle Sensor Neutral Position, Sunroof Memory Reset/Initialization, Automatic Drive Positioner System, Audio presets and Navigation. Use the CONSULT-III diagnostic tool, or equivalent to perform the required resets.**

→**The automatic back door system must be initialized anytime the battery has been disconnected. Close the back door. Open the back door with the automatic open feature. Do not stop the**

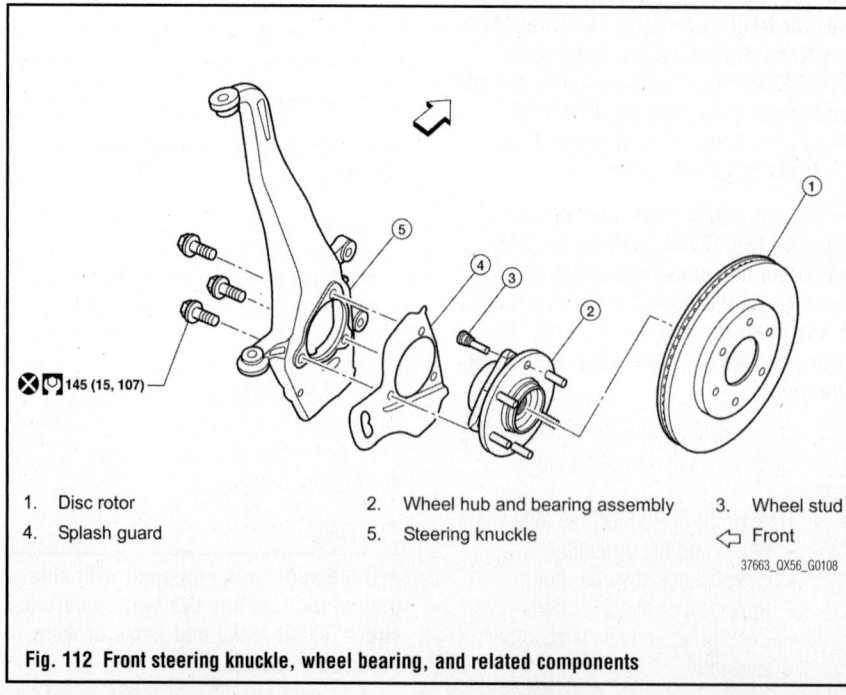

145 (15, 107)

| 1. Disc rotor | 2. Wheel hub and bearing assembly | 3. Wheel stud |
| 4. Splash guard | 5. Steering knuckle | ⇦ Front |

37663_QX56_G0108

Fig. 112 Front steering knuckle, wheel bearing, and related components

process until the back door opens completely.

2. Disconnect the negative battery cable.

3. Raise and safely support the vehicle.

4. Remove the tire and wheel assembly.

5. Remove the engine under cover.

6. Remove the stabilizer bar mounting bracket retaining bolts and rubber bushings.

7. Remove the connecting rod nuts.

8. Remove the stabilizer bar from the vehicle.

To install:

➡Be sure to use new fasteners, as required.

9. Installation is the reverse of the removal procedure.

10. Be sure to perform the reconnect/relearn procedures.

UPPER CONTROL ARMS

REMOVAL & INSTALLATION

See Figures 114 and 115.

➡Nissan/Infiniti refers to the upper control arm as a upper link.

1. Before servicing the vehicle, refer to the Precautions Section.

➡If working near and/or around the SRS system and components, be sure to disable the SRS system. After disabling the system wait three minutes or more before servicing the vehicle.

➡Whenever the negative battery cable is disconnected the following components will require resetting. The Idle Air Volume Learning, Steering Angle Sensor Neutral Position, Sunroof Memory Reset/Initialization, Automatic Drive Positioner System, Audio presets and Navigation. Use the CONSULT-III diagnostic tool, or equivalent to perform the required resets.

➡The automatic back door system must be initialized anytime the battery has been disconnected. Close the back door. Open the back door with the automatic open feature. Do not stop the process until the back door opens completely.

2. Disconnect the negative battery cable.

3. Raise and safely support the vehicle.

4. Remove the tire and wheel assembly.

➡Remove the fender protector to access the upper control arm.

5. Remove or disconnect the following:
- Cotter pin and nut from upper ball joint

6. Separate upper ball joint stud from steering knuckle using special tool J-24319-01.

7. Remove the following:
- Upper control arm mounting bolts. See illustration for bolt locations.
- Upper control arm

To install:

➡Be sure to use new fasteners, as required.

8. Installation is the reverse of the removal procedure.

9. Check and adjust alignment, as required.

10. Be sure to perform the reconnect/relearn procedures.

WHEEL HUB & BEARING (SEALED UNIT)

REMOVAL & INSTALLATION

See Figure 112.

1. Before servicing the vehicle, refer to the Precautions Section.

➡If working near and/or around the SRS system and components, be sure

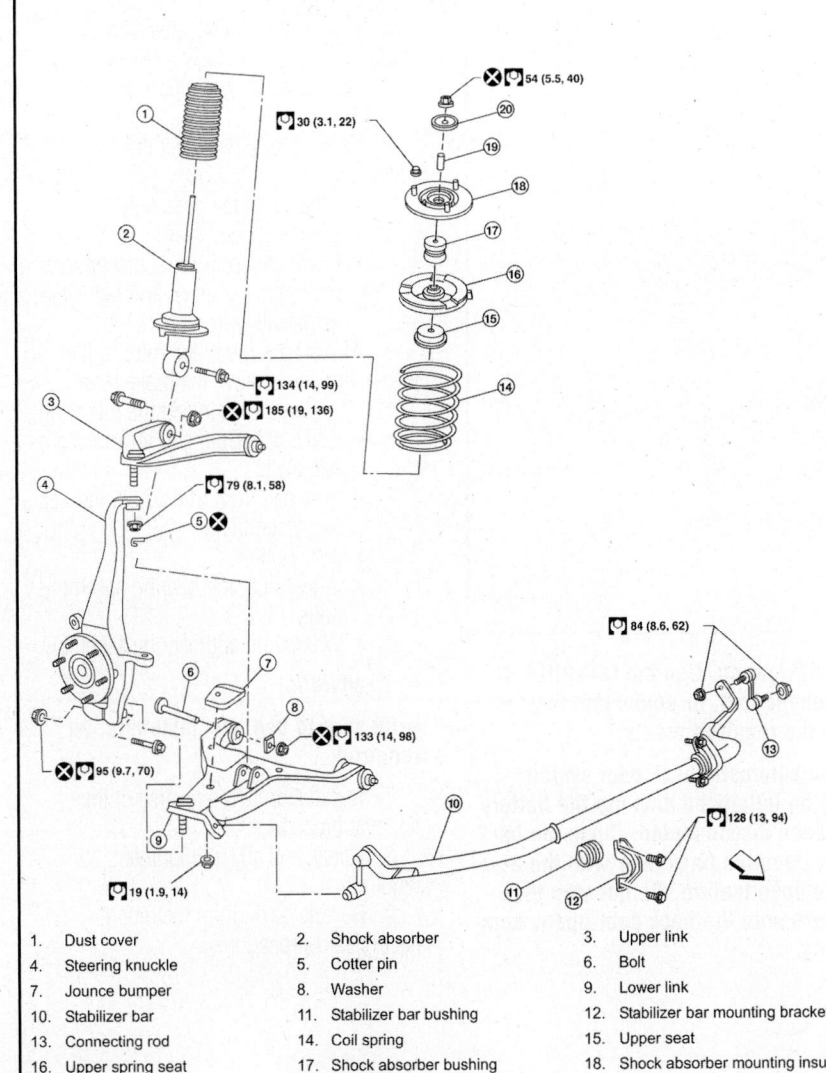

1.	Dust cover	2.	Shock absorber	3.	Upper link
4.	Steering knuckle	5.	Cotter pin	6.	Bolt
7.	Jounce bumper	8.	Washer	9.	Lower link
10.	Stabilizer bar	11.	Stabilizer bar bushing	12.	Stabilizer bar mounting bracket
13.	Connecting rod	14.	Coil spring	15.	Upper seat
16.	Upper spring seat	17.	Shock absorber bushing	18.	Shock absorber mounting insulator
19.	Spacer	20.	Washer	⟵	Front

37663_QX56_G0104

Fig. 113 Front stabilizer bar and related components

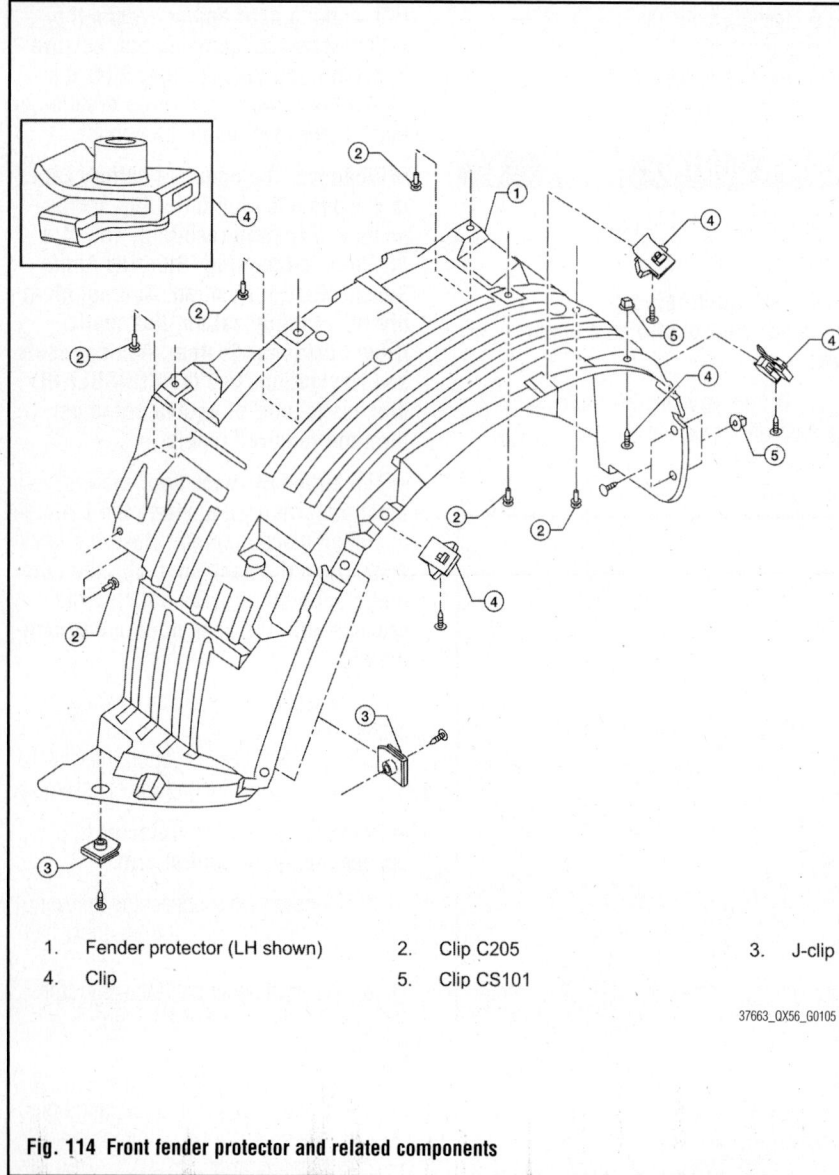

1. Fender protector (LH shown) 2. Clip C205 3. J-clip
4. Clip 5. Clip CS101

37663_QX56_G0105

Fig. 114 Front fender protector and related components

Upper link

37663_QX56_G0106

Fig. 115 Front upper control arm (upper link) bolt locations

2. Disconnect the negative battery cable.

3. raise and safely support the vehicle.

4. Remove or disconnect the following:
- Wheel and tire assembly
- Engine splash guard
- Brake caliper without disconnecting the hydraulic lines, and reposition aside with wire

5. Matchmark the brake rotor to the wheel hub and remove the brake rotor.

6. Remove or disconnect the following:
- 4WD, cotter pin and locknut from halfshaft
- Halfshaft from wheel hub and bearing assembly
- ABS sensor
- Wheel hub and bearing assembly bolts
- Wheel hub and bearing assembly

To install:

➡**Be sure to use new fasteners, as required.**

7. Installation is the reverse of the removal procedure.

8. Check and adjust alignment, as required.

9. Be sure to perform the reconnect/relearn procedures.

to disable the SRS system. After disabling the system wait three minutes or more before servicing the vehicle.

➡**Whenever the negative battery cable is disconnected the following components will require resetting. The Idle Air Volume Learning, Steering Angle Sensor Neutral Position, Sunroof Memory Reset/Initialization, Automatic Drive Positioner System, Audio presets** and Navigation. Use the CONSULT-III diagnostic tool, or equivalent to perform the required resets.

➡**The automatic back door system must be initialized anytime the battery has been disconnected. Close the back door. Open the back door with the automatic open feature. Do not stop the process until the back door opens completely.**

SUSPENSION   **REAR SUSPENSION**

COIL SPRINGS

REMOVAL & INSTALLATION

See Figures 116 and 117.

1. Before servicing the vehicle, refer to the Precautions Section.

➡ **If working near and/or around the SRS system and components, be sure to disable the SRS system. After disabling the system wait three minutes or more before servicing the vehicle.**

➡ **Whenever the negative battery cable is disconnected the following components will require resetting. The Idle Air Volume Learning, Steering Angle Sensor Neutral Position, Sunroof Memory Reset/Initialization, Automatic Drive Positioner System, Audio presets and Navigation. Use the CONSULT-III diagnostic tool, or equivalent to perform the required resets.**

➡ **The automatic back door system must be initialized anytime the battery has been disconnected. Close the back door. Open the back door with the automatic open feature. Do not stop the process until the back door opens completely.**

2. Disconnect the negative battery cable.

3. Raise and safely support the vehicle.

4. Remove the tire and wheel assembly.

5. Release the air pressure from the rear load leveling air suspension system using the CONSULT-III® "EXHAUST SOLENOID" active test.

6. Remove the height sensor arm bracket bolt from the left-hand rear lower link.

7. Place a suitable jack under the rear lower link and relieve the coil spring tension.

8. Loosen the rear lower link adjusting bolt and nut connected to the rear suspension member.

9. Remove the rear lower link bolt and nut from the knuckle.

10. Slowly lower the jack to relieve the coil spring tension.

11. Remove the coil spring.

To install:

➡ **Be sure to use new fasteners, as required.**

12. Installation is the reverse of the removal procedure.

➡ **When installing the rubber seats for the coil spring, ensure the embossed arrow points outward toward the wheel.**

13. Be sure to perform the reconnect/relearn procedures.

REAR SUSPENSION ARM

REMOVAL & INSTALLATION

See Figure 116.

➡ **Nissan/Infiniti refers to the upper control arm as rear suspension arm.**

1. Before servicing the vehicle, refer to the Precautions Section.

➡ **If working near and/or around the SRS system and components, be sure to disable the SRS system. After disabling the system wait three minutes or more before servicing the vehicle.**

➡ **Whenever the negative battery cable is disconnected the following components will require resetting. The Idle Air Volume Learning, Steering Angle Sensor Neutral Position, Sunroof Memory Reset/Initialization, Automatic Drive Positioner System, Audio presets and Navigation. Use the CONSULT-III diagnostic tool, or equivalent to perform the required resets.**

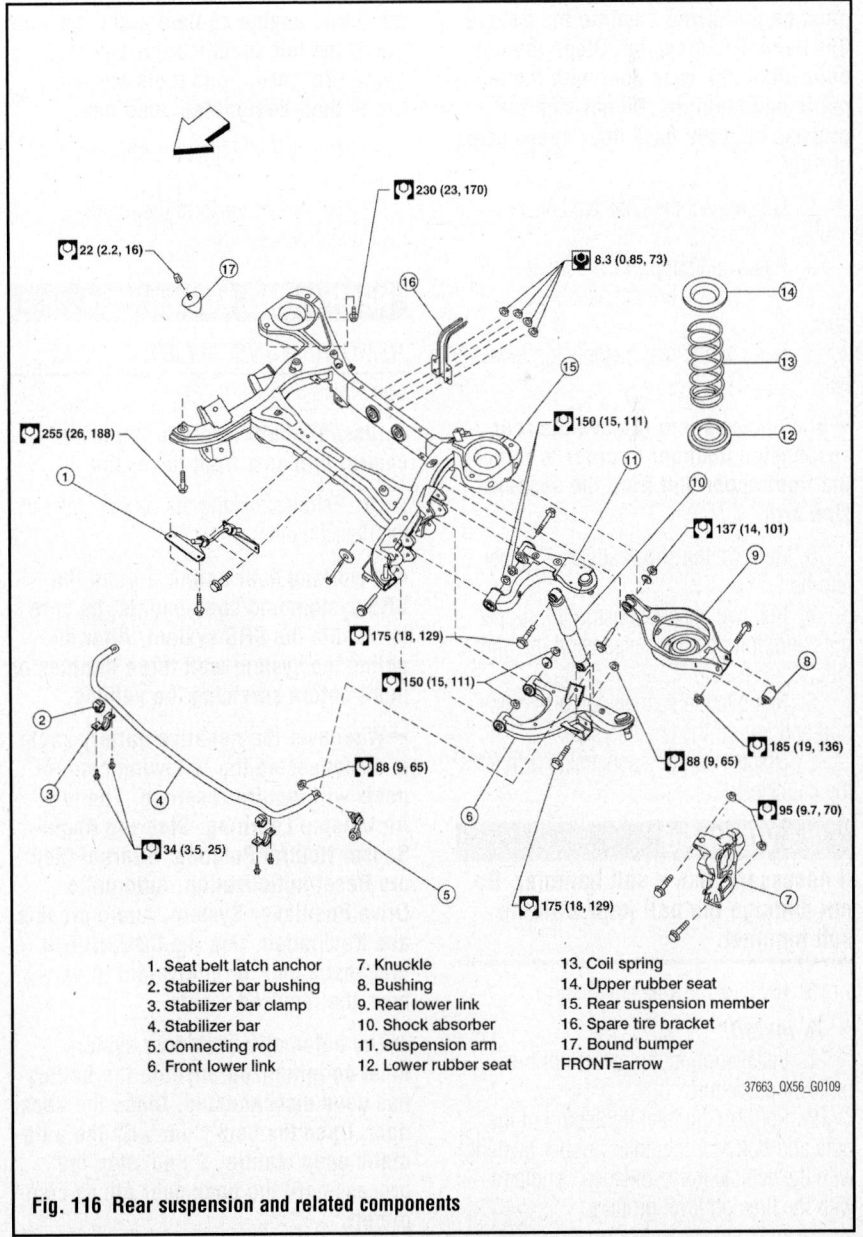

1. Seat belt latch anchor	7. Knuckle	13. Coil spring
2. Stabilizer bar bushing	8. Bushing	14. Upper rubber seat
3. Stabilizer bar clamp	9. Rear lower link	15. Rear suspension member
4. Stabilizer bar	10. Shock absorber	16. Spare tire bracket
5. Connecting rod	11. Suspension arm	17. Bound bumper
6. Front lower link	12. Lower rubber seat	FRONT=arrow

37663_QX56_G0109

Fig. 116 Rear suspension and related components

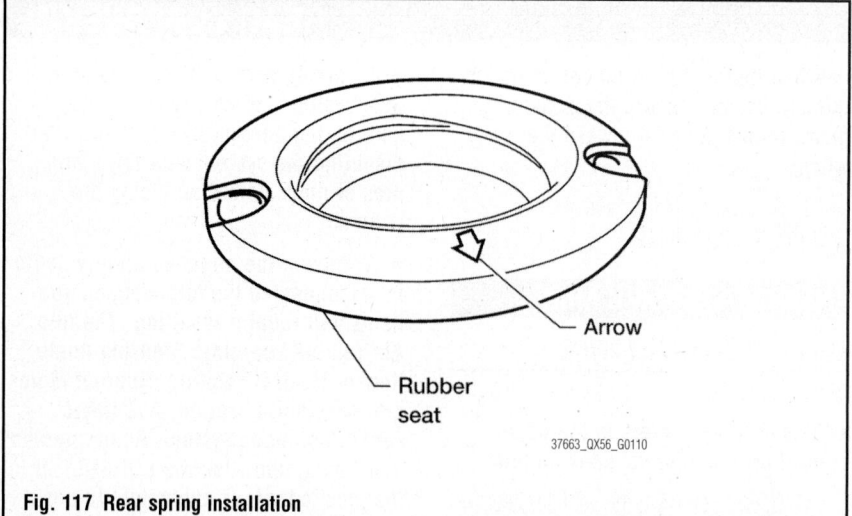

Fig. 117 Rear spring installation

➡The automatic back door system must be initialized anytime the battery has been disconnected. Close the back door. Open the back door with the automatic open feature. Do not stop the process until the back door opens completely.

2. Disconnect the negative battery cable.

3. Raise and support the vehicle safely.

4. Remove the tire and wheel assemblies.

5. Remove the rear suspension member.

➡It is necessary to remove the rear suspension member in order to remove the front upper bolt from the suspension arm.

6. Remove the shock absorber upper end bolt.

7. Remove the suspension arm upper nuts and bolts on the suspension member side.

8. Remove the suspension arm pinch bolt and nut on the knuckle side.

9. Disconnect the suspension arm from the knuckle.

✳✳ WARNING

If necessary, use a soft hammer. Do not damage the ball joint with the soft hammer.

10. Remove the suspension arm.

To install:

11. Installation is the reverse of the removal procedure.

12. Perform the final tightening of the nuts and bolts for the links (rubber bushing) with the vehicle in the unladen condition with the tires on level ground.

➡Unladen condition means that the fuel tank, engine coolant and lubricants are at the full specification and the spare tire, jack, hand tools and mats are in their designated positions.

13. Check and adjust the alignment, as required.

14. Be sure to perform the reconnect/relearn procedures.

REAR FRONT LOWER LINK

REMOVAL & INSTALLATION

See Figure 116.

➡Nissan/Infiniti refers to the lower control arm as a front lower link.

1. Before servicing the vehicle, refer to the Precautions Section.

➡If working near and/or around the SRS system and components, be sure to disable the SRS system. After disabling the system wait three minutes or more before servicing the vehicle.

➡Whenever the negative battery cable is disconnected the following components will require resetting. The Idle Air Volume Learning, Steering Angle Sensor Neutral Position, Sunroof Memory Reset/Initialization, Automatic Drive Positioner System, Audio presets and Navigation. Use the CONSULT-III diagnostic tool, or equivalent to perform the required resets.

➡The automatic back door system must be initialized anytime the battery has been disconnected. Close the back door. Open the back door with the automatic open feature. Do not stop the process until the back door opens completely.

2. Disconnect the negative battery cable.

3. Raise and support the vehicle safely.

4. Remove the tire and wheel assemblies.

5. Release the air pressure from the rear load leveling air suspension system using the CONSULT-III® "EXHAUST SOLENOID" active test.

6. Remove the shock absorber lower end bolt.

7. Remove the adjusting bolt and nut, and the bolt and nut from the front lower link and rear suspension member.

8. Remove the front lower link pinch bolt and nut on the knuckle side.

9. Disconnect the front lower link from the knuckle.

✳✳ WARNING

If necessary, use a soft hammer. Do not damage the ball joint with the soft hammer.

10. Remove the front lower link.

To install:

11. Installation is the reverse of the removal procedure.

12. Perform the final tightening of the nuts and bolts for the links (rubber bushing) with the vehicle in the unladen condition with the tires on level ground.

➡Unladen condition means that the fuel tank, engine coolant and lubricants are at the full specification and the spare tire, jack, hand tools and mats are in their designated positions.

13. Check and adjust the alignment, as required.

14. Be sure to perform the reconnect/relearn procedures.

REAR LOWER LINK AND SPRING

REMOVAL & INSTALLATION

See Figure 116.

➡Nissan/Infiniti refers to the lower control arm as a rear lower link.

1. Before servicing the vehicle, refer to the Precautions Section.

➡If working near and/or around the SRS system and components, be sure to disable the SRS system. After disabling the system wait three minutes or more before servicing the vehicle.

➡Whenever the negative battery cable is disconnected the following components will require resetting. The Idle Air Volume Learning, Steering Angle Sensor Neutral Position, Sunroof Memory Reset/Initialization, Automatic Drive Positioner System, Audio presets and Navigation. Use the CONSULT-III diagnostic tool, or equivalent to perform the required resets.

➡The automatic back door system must be initialized anytime the battery has been disconnected. Close the back door. Open the back door with the automatic open feature. Do not stop the process until the back door opens completely.

2. Disconnect the negative battery cable.

3. Raise and support the vehicle safely.

4. Remove the tire and wheel assemblies.

5. Release the air pressure from the rear load leveling air suspension system using the CONSULT-III® "EXHAUST SOLENOID" active test.

6. Remove the height sensor arm bracket bolt from the left-hand rear lower link.

7. Place a suitable jack under the rear lower link and relieve the coil spring tension.

8. Loosen the rear lower link adjusting bolt and nut connected to the rear suspension member.

9. Remove the rear lower link bolt and nut from the knuckle.

10. Slowly lower the jack to relieve the coil spring tension.

11. Remove the coil spring.

12. Remove the upper rubber seat, coil spring and lower rubber seat from the rear lower link.

13. Remove the rear lower link adjusting bolt and nut from the rear suspension member.

14. Remove the rear lower link from its mounting.

To install:
15. Installation is the reverse of the removal procedure.

16. When installing the upper and lower rubber seats for the rear coil springs, the arrow embossed on the rubber seats must point out toward the wheel and tire assembly.

17. Tighten the rear lower link bolt to knuckle to 70 ft. lbs. (95 Nm).

18. Tighten the rear lower link adjusting bolt to rear suspension member to 101 ft. lbs. (137 Nm).

19. Tighten the height sensor arm bracket bolt to left-head rear lower link to 9 ft. lbs. (12 Nm).

20. Perform the final tightening of the nuts and bolts for the links (rubber bushing) with the vehicle in the unladen condition with the tires on level ground.

➡Unladen condition means that the fuel tank, engine coolant and lubricants are at the full specification and the spare tire, jack, hand tools and mats are in their designated positions.

21. Check and adjust the alignment, as required.

SHOCK ABSORBERS

REMOVAL & INSTALLATION
See Figure 118.

1. Before servicing the vehicle, refer to the Precautions Section.

➡If working near and/or around the SRS system and components, be sure to disable the SRS system. After disabling the system wait three minutes or more before servicing the vehicle.

➡Whenever the negative battery cable is disconnected the following components will require resetting. The Idle Air Volume Learning, Steering Angle Sensor Neutral Position, Sunroof Memory Reset/Initialization, Automatic

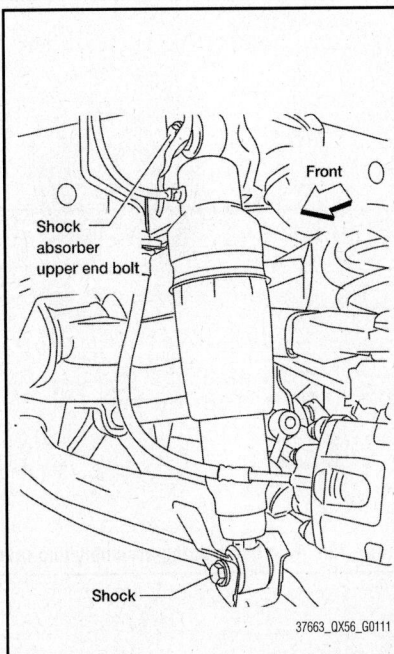

Shock absorber upper end bolt

Front

Shock

37663_QX56_G0111

Fig. 118 Rear shock absorber and related components

Drive Positioner System, Audio presets and Navigation. Use the CONSULT-III diagnostic tool, or equivalent to perform the required resets.

➡The automatic back door system must be initialized anytime the battery has been disconnected. Close the back door. Open the back door with the automatic open feature. Do not stop the process until the back door opens completely.

2. Disconnect the negative battery cable.

3. Raise and safely support the vehicle.

4. Remove the tire and wheel assemblies.

5. Release the air pressure from the rear load leveling air suspension system using the CONSULT-III® "EXHAUST SOLENOID" active test.

6. Remove or disconnect the following:
- Rear fender protector
- Rear load leveling air suspension hose from the shock absorber
- Shock absorber upper and lower end bolts
- Shock absorber

To install:

➡Be sure to use new fasteners, as required.

7. Installation is the reverse of the removal procedure.

8. Be sure to perform the reconnect/relearn procedures.

TESTING

➡If the vehicle is equipped with automatic load leveling system, be sure to check for air leaks and loose or worn connection points.

1. Before servicing the vehicle, refer to the Precautions Section.

2. Road test the vehicle.

3. Check for excessive bounce or roll.

4. Raise the vehicle on a lift.

5. Check for bad bushings and oil leakage.

STABILIZER BAR (SWAY BAR) & LINKS

REMOVAL & INSTALLATION
See Figure 119.

1. Before servicing the vehicle, refer to the Precautions Section.

➡If working near and/or around the SRS system and components, be sure to disable the SRS system. After

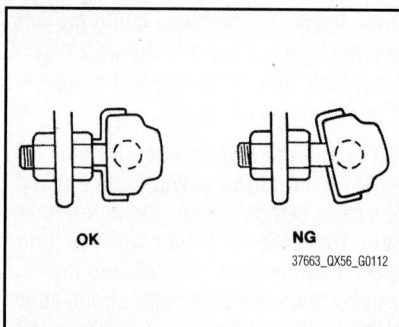

Fig. 119 Rear stabilizer bar bushing positioning

disabling the system wait three minutes or more before servicing the vehicle.

→Whenever the negative battery cable is disconnected the following components will require resetting. The Idle Air Volume Learning, Steering Angle Sensor Neutral Position, Sunroof Memory Reset/Initialization, Automatic Drive Positioner System, Audio presets and Navigation. Use the CONSULT-III diagnostic tool, or equivalent to perform the required resets.

→The automatic back door system must be initialized anytime the battery has been disconnected. Close the back door. Open the back door with the automatic open feature. Do not stop the process until the back door opens completely.

2. Disconnect the negative battery cable.

3. Raise and safely support the vehicle.

4. Disconnect the bar ends from the connecting rods.

5. Remove the bar clamps. Remove the bar bushings.

6. Remove the stabilizer bar.

To install:

→Be sure to use new fasteners, as required.

7. Installation is the reverse of the removal procedure.

→Install the stabilizer bar with the ball joint sockets properly aligned. Install the bar bushing and clamp so that they are positioned inside of the sideslip prevention clamp on the bar. See illustration

8. Be sure to perform the reconnect/relearn procedures.

WHEEL HUB & BEARING (SEALED UNIT)

REMOVAL & INSTALLATION

See Figure 120.

1. Before servicing the vehicle, refer to the Precautions Section.

→If working near and/or around the SRS system and components, be sure to disable the SRS system. After disabling the system wait three minutes or more before servicing the vehicle.

→Whenever the negative battery cable is disconnected the following components will require resetting. The Idle Air Volume Learning, Steering Angle Sensor Neutral Position, Sunroof Memory Reset/Initialization, Automatic Drive Positioner System, Audio presets and Navigation. Use the CONSULT-III diagnostic tool, or equivalent to perform the required resets.

→The automatic back door system must be initialized anytime the battery has been disconnected. Close the back door. Open the back door with the automatic open feature. Do not stop the process until the back door opens completely.

2. Disconnect the negative battery cable.

3. Raise and support the vehicle safely.

4. Remove or disconnect the following:
   - Wheel and tire assembly
   - Brake caliper without disconnecting the hydraulic lines, and reposition aside with wire
   - Brake rotor
   - Cotter pin and nut from driveshaft
   - Driveshaft
   - Wheel hub and bearing assembly bolts

5. Pulling out the wheel hub and bearing assembly slightly, remove the ABS sensor.

6. Remove the wheel hub and bearing assembly.

To install:

→Be sure to use new fasteners, as required.

7. Install or connect the following:
   - ABS sensor
   - Wheel hub and bearing assembly, using new bolts
   - Driveshaft
   - Lock nut and new cotter pin
   - Brake rotor
   - Brake caliper
   - Wheel and tire assembly

8. Be sure to perform the reconnect/relearn procedures.

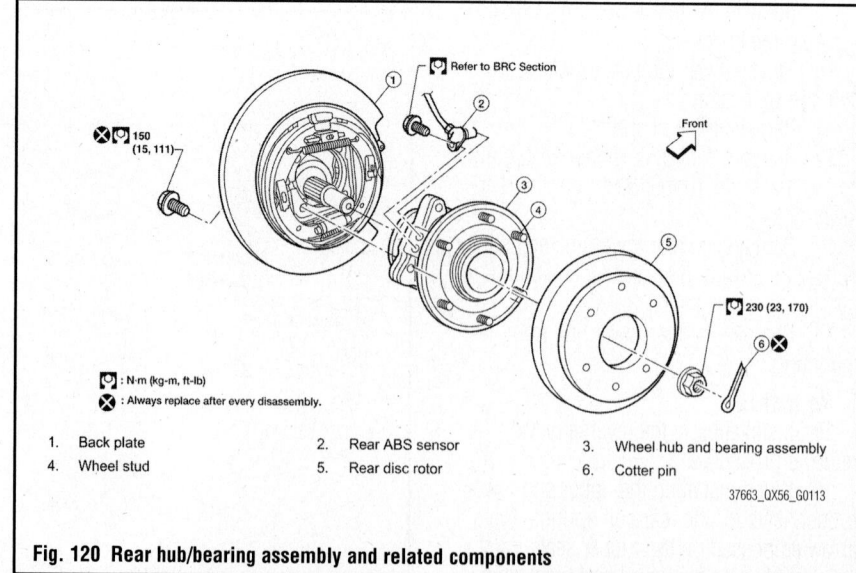

: N·m (kg-m, ft-lb)

: Always replace after every disassembly.

| 1. | Back plate | 2. | Rear ABS sensor | 3. | Wheel hub and bearing assembly |
| 4. | Wheel stud | 5. | Rear disc rotor | 6. | Cotter pin |

Fig. 120 Rear hub/bearing assembly and related components

NISSAN

Cube

9

SPECIFICATIONS AND MAINTENANCE CHARTS

ENGINE AND VEHICLE IDENTIFICATION

			Engine				Model Year	
Code ①	Liters (cc)	Cu. In.	Cyl.	Fuel Sys.	Engine Type	Eng. Mfg.	Code ②	Year
MR18DE	1.8 (1797)	110	4	MFI	DOHC	Nissan	B	2011
							C	2012

MFI: Multi-port Fuel Injection

DOHC: Double Overhead Camshaft

① The Engine Code is stamped on the engine block near the starter.

② 10th position of the Vehicle Identification Number (VIN)

71075_CUBE_C0001

GENERAL ENGINE SPECIFICATIONS

Year	Model	Engine Displacement Liters	Engine Series ID	Net Horsepower @ rpm	Net Torque @ rpm (ft. lbs.)	Bore x Stroke (in.)	Com-pression Ratio	Oil Pressure @ rpm
2011	Cube	1.8	MR18DE	122@5200	127@4800	3.31X3.19	9.9:1	29@2000
2012	Cube	1.8	MR18DE	122@5200	127@4800	3.31X3.19	9.9:1	29@2000

71075_CUBE_C0002

ENGINE TUNE-UP SPECIFICATIONS

Year	Engine Displacement Liters	Engine ID	Spark Plug Gap (in.)	Ignition Timing (deg.) MT	Ignition Timing (deg.) AT	Fuel Pump (psi) ①	Idle Speed (rpm) MT	Idle Speed (rpm) AT ②	Valve Clearance Intake ③	Valve Clearance Exhaust ③
2011	1.8	MR18DE	0.043	13B	13B	51	650-750	650-750	0.010-0.013	0.011-0.015
2012	1.8	MR18DE	0.043	13B	13B	51	650-750	650-750	0.010-0.013	0.011-0.015

B: Before top dead center

① At idle

② Automatic transmission in neutral

③ Engine cold

71075_CUBE_C0003

CAPACITIES

Year	Model	Engine Displacement Liters	Engine ID	Engine Oil with Filter (qts.)	Transmission (pts.)		Fuel Tank (gal.)	Cooling System (qts.)
					5-Spd	Auto.		
2011	Cube	1.8	MR18DE	4.25	4.25	15.75	13.3	①
2012	Cube	1.8	MR18DE	4.25	4.25	15.75	13.3	①

① CVT models: 7.5 qts.
  M/T models: 7.25 qts.

71075_CUBE_C0004

FLUID SPECIFICATIONS

Year	Model	Engine Displ. Liters	Engine Oil	Man. Trans.	Auto. Trans.	Power Steering Fluid	Brake Master Cylinder	Cooling System
2011	Cube	1.8	5W-30	75W-80	Nissan NS-2	NA	DOT 3	N-LL
2012	Cube	1.8	5W-30	75W-80	Nissan NS-2	NA	DOT 3	N-LL

NA: Not Applicable

N-LL: Nissan Long Life coolant

71075_CUBE_C0005

VALVE SPECIFICATIONS

Year	Engine ID	Engine Displacement Liters	Seat Angle (deg.)	Face Angle (deg.)	Spring Test Pressure (lbs. @ in.)	Spring Installed Height (in.)	Stem-to-Guide Clearance (in.)		Stem Diameter (in.)	
							Intake	Exhaust	Intake	Exhaust
2011	MR18DE	1.8	45.15-45.45	—	①	1.390	0.0008-0.0021	0.0012-0.0025	0.2152-0.2157	0.2148-0.2154
2012	MR18DE	1.8	45.15-45.45	—	①	1.390	0.0008-0.0021	0.0012-0.0025	0.2152-0.2157	0.2148-0.2154

① Intake: 34-39 lbs. @ 1.39 inches
  Exhaust: 31-36 lbs. @ 1.39 inches

71075_CUBE_C0006

CAMSHAFT SPECIFICATIONS
All measurements in inches unless noted

Year	Engine Displacement Liters	Engine ID	Journal Dia.	Brg. Oil Clearance	Shaft End-play	Circle Runout	Lobe Height	
							Intake	Exhaust
2011	1.8	MR18DE	①	②	0.0030-0.0060	0.0008	1.7561-1.7636	1.6998-1.7073
2012	1.8	MR18DE	①	②	0.0030-0.0060	0.0008	1.7561-1.7636	1.6998-1.7073

① No. 1: 1.0998-1.1006
   All others: 0.98236-0.9831

② No. 1: 0.0018-0.0034 inches
   All others: 0.0012-0.0028 inches

71075_CUBE_C0010

CRANKSHAFT AND CONNECTING ROD SPECIFICATIONS
All measurements are given in inches.

Year	Engine Displacement Liters	Engine ID	Crankshaft				Connecting Rod		
			Main Brg. Journal Dia.	Main Brg. Oil Clearance	Shaft End-play	Thrust on No.	Journal Diameter	Oil Clearance	Side Clearance
2011	1.8	MR18DE	2.0457-2.0464	①	0.0039-0.0102	3	1.8504-1.8509	0.0002-0.0009	0.0079-0.0138
2012	1.8	MR18DE	2.0457-2.0464	①	0.0039-0.0102	3	1.8504-1.8509	0.0002-0.0009	0.0079-0.0138

① Nos. 1, 4, 5 : 0.0009-0.0013
   Nos. 2, 3 : 0.0005-0.0009

71075_CUBE_C0009

GENERAL ENGINE SPECIFICATIONS

Year	Model	Engine Displacement Liters	Engine Series ID	Net Horsepower @ rpm	Net Torque @ rpm (ft. lbs.)	Bore x Stroke (in.)	Compression Ratio	Oil Pressure @ rpm
2011	Cube	1.8	MR18DE	122@5200	127@4800	3.31X3.19	9.9:1	29@2000
2012	Cube	1.8	MR18DE	122@5200	127@4800	3.31X3.19	9.9:1	29@2000

71075_CUBE_C0002

TORQUE SPECIFICATIONS

All readings in ft. lbs.

Year	Engine Displacement Liters	Engine ID	Cylinder Head Bolts	Main Bearing Bolts	Rod Bearing Bolts	Crankshaft Damper Bolts	Flywheel Bolts	Manifold		Spark Plugs	Oil Drain Plug
									Intake	Exhaust		
2011	1.8	MR18DE	①	②	③	④	80	20	25	14	25
2012	1.8	MR18DE	①	②	③	④	80	20	25	14	25

① Step 1: 30 ft. lbs.

  Step 2: Turn each bolt, in sequence, an additional 100 degrees

  Step3: Loosen completely, then retorque to 30 ft. lbs.

  Step 4: Turn each bolt, in sequence, an additional 100 degrees

  Step 5: Turn each bolt, in sequence, an additional 100 degrees

② Step 1: Tighten bolts 25 ft. lbs.

  Step 2: Tighten bolts an additional 60 degrees

③ Step 1: Tighten bolts to 20 ft. lbs.

  Step 2: Loosen all bolts

  Step 3: tighten bolts to 14 ft. lbs.

  Step 4: Tighten bolts an additional 60 degrees

④ Step 1: Tighten to 51 ft. lbs.

  Step 2: Loosen bolt completely

  Step 3: Tighten to 22 ft. lbs.

  Step 4: Tighten an additional 60 degrees

71075_CUBE_C0007

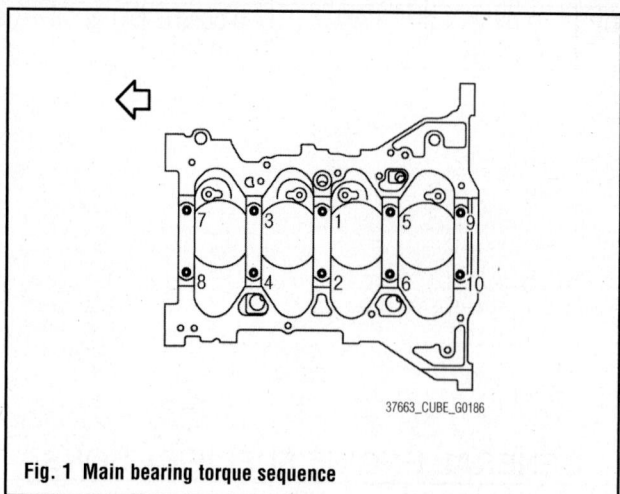

37663_CUBE_G0186

Fig. 1 Main bearing torque sequence

WHEEL ALIGNMENT

Year	Model		Caster		Camber		Toe-in (in.)
			Range (+/-Deg.)	Preferred Setting (Deg.)	Range (+/-Deg.)	Preferred Setting (Deg.)	
2011	Cube	Front	0.75	4.67	0.75	-0.17	0.05 +0.02/- 0.05
		Rear	—	—	0.50	-1.51	0.118 +0.157/-0.079
2012	Cube	Front	0.75	4.67	0.75	-0.17	0.05 +0.02/- 0.05
		Rear	—	—	0.50	-1.51	0.118 +0.157/-0.079

71075_CUBE_C0011

TIRE, WHEEL AND BALL JOINT SPECIFICATIONS

Year	Model	OEM Tires		Tire Pressures (psi)		Wheel Size	Lug Nut Torque (ft. lbs.)
		Standard	Optional	Front	Rear		
2011	Cube	P195/60R15	P195/55R16	33	33	6.5-JJ	80
2012	Cube	P195/60R15	P195/55R16	33	33	6.5-JJ	80

OEM: Original Equipment Manufacturer

PSI: Pounds Per Square Inch

71075_CUBE_C0012

BRAKE SPECIFICATIONS

All measurements in inches unless noted

Year	Model		Brake Disc			Brake Drum Diameter			Minimum Lining Thickness		Brake Caliper	
			Original Thickness	Minimum Thickness	Maximum Run-out	Original Inside Diameter	Max. Wear Limit	Maximum Machine Diameter	Front	Rear	Bracket Bolts (ft. lbs.)	Mounting Bolts (ft. lbs.)
2011	Cube	F	0.945	0.866	0.001	—	—	—	0.079	—	62	20
		R	0.354	0.315	0.002	9.00	9.06	—	—	0.079	—	—
2012	Cube	F	0.945	0.866	0.001	—	—	—	0.079	—	62	20
		R	0.354	0.315	0.002	9.00	9.06	—	—	0.079	—	—

71075_CUBE_C0013

SCHEDULED MAINTENANCE INTERVALS
Nissan—Cube

TO BE SERVICED	TYPE OF SERVICE	VEHICLE MILEAGE INTERVAL (x1000)												
		7.5	15	22.5	30	37.5	45	52.5	60	67.5	75	82.5	90	97.5
Engine oil & filter	R	✓	✓	✓	✓	✓	✓	✓	✓	✓	✓	✓	✓	✓
Brake lines & cables	S/I		✓		✓		✓		✓		✓		✓	
Brake pads, discs, drums & linings	S/I		✓		✓		✓		✓		✓		✓	
Driveshaft boots	S/I		✓		✓		✓		✓		✓		✓	
Exhaust system	S/I				✓				✓				✓	
Transaxle fluid	S/I		✓		✓		✓		✓		✓		✓	
Air cleaner filter	R				✓				✓				✓	
Spark plugs (except platinum)	R				✓				✓				✓	
Spark plugs (iridium and platinum)	R	Replace every 105,000 miles												
Steering gear & linkage, axle & suspension parts	S/I				✓				✓				✓	
Engine coolant	R	Replace every 60,000 miles, then every 30,000 miles												
Inverter coolant	R	Replace every 60,000 miles, then every 30,000 miles												
Drive belts	S/I								✓					
Fuel lines	S/I								✓					
Vapor lines	S/I								✓					
Cabin microfilter	R		✓		✓		✓		✓		✓		✓	
Valve adjustment	S/I	As needed												

R: Replace   S/I: Service or Inspect

FREQUENT OPERATION MAINTENANCE (SEVERE SERVICE)

If a vehicle is operated under any of the following conditions it is considered severe service:

- Extremely dusty areas.

- 50% or more of the vehicle operation is in 32°C (90°F) or higher temperatures, or constant operation in temperatures below 0°C (32°F).

- Prolonged idling (vehicle operation in stop and go traffic).

- Frequent short running periods (engine does not warm to normal operating temperatures).

- Police, taxi, delivery usage or trailer towing usage.

Oil & oil filter: change every 3750 miles.

Brake pads & discs: service or inspect every 7500 miles.

Driveshaft boots: service or inspect every 7500 miles.

Exhaust system: service or inspect every 7500 miles.

Steering gear & linkage, axle & suspension parts: service or inspect every 7500 miles.

Steering linkage ball joints & front suspension ball joints: service or inspect every 7500 miles.

Air cleaner filter: service or inspect every 15,000 miles.

71075_CUBE_C0014

PRECAUTIONS

Before servicing any vehicle, please be sure to read all of the following precautions, which deal with personal safety, prevention of component damage, and important points to take into consideration when servicing a motor vehicle:

• Never open, service or drain the radiator or cooling system when the engine is hot; serious burns can occur from the steam and hot coolant.

• Observe all applicable safety precautions when working around fuel. Whenever servicing the fuel system, always work in a well-ventilated area. Do not allow fuel spray or vapors to come in contact with a spark, open flame, or excessive heat (a hot drop light, for example). Keep a dry chemical fire extinguisher near the work area. Always keep fuel in a container specifically designed for fuel storage; also, always properly seal fuel containers to avoid the possibility of fire or explosion. Refer to the additional fuel system precautions later in this section.

• Fuel injection systems often remain pressurized, even after the engine has been turned **OFF**. The fuel system pressure must be relieved before disconnecting any fuel lines. Failure to do so may result in fire and/or personal injury.

• Brake fluid often contains polyglycol ethers and polyglycols. Avoid contact with the eyes and wash your hands thoroughly after handling brake fluid. If you do get brake fluid in your eyes, flush your eyes with clean, running water for 15 minutes. If eye irritation persists, or if you have taken brake fluid internally, IMMEDIATELY seek medical assistance.

• The EPA warns that prolonged contact with used engine oil may cause a number of skin disorders, including cancer. You should make every effort to minimize your exposure to used engine oil. Protective gloves should be worn when changing oil. Wash your hands and any other exposed skin areas as soon as possible after exposure to used engine oil. Soap and water, or waterless hand cleaner should be used.

• All new vehicles are now equipped with an air bag system, often referred to as a Supplemental Restraint System (SRS) or Supplemental Inflatable Restraint (SIR) system. The system must be disabled before performing service on or around system components, steering column, instrument panel components, wiring and sensors. Failure to follow safety and disabling procedures could result in accidental air bag deployment, possible personal injury and unnecessary system repairs.

• Always wear safety goggles when working with, or around, the air bag system. When carrying a non-deployed air bag, be sure the bag and trim cover are pointed away from your body. When placing a non-deployed air bag on a work surface, always face the bag and trim cover upward, away from the surface. This will reduce the motion of the module if it is accidentally deployed. Refer to the additional air bag system precautions later in this section.

• Clean, high quality brake fluid from a sealed container is essential to the safe and proper operation of the brake system. You should always buy the correct type of brake fluid for your vehicle. If the brake fluid becomes contaminated, completely flush the system with new fluid. Never reuse any brake fluid. Any brake fluid that is removed from the system should be discarded. Also, do not allow any brake fluid to come in contact with a painted surface; it will damage the paint.

• Never operate the engine without the proper amount and type of engine oil; doing so WILL result in severe engine damage.

• Timing belt maintenance is extremely important. Many models utilize an interference-type, non-freewheeling engine. If the timing belt breaks, the valves in the cylinder head may strike the pistons, causing potentially serious (also time-consuming and expensive) engine damage. Refer to the maintenance interval charts for the recommended replacement interval for the timing belt, and to the timing belt section for belt replacement and inspection.

• Disconnecting the negative battery cable on some vehicles may interfere with the functions of the on-board computer system(s) and may require the computer to undergo a relearning process once the negative battery cable is reconnected.

• When servicing drum brakes, only disassemble and assemble one side at a time, leaving the remaining side intact for reference.

• Only an MVAC-trained, EPA-certified automotive technician should service the air conditioning system or its components.

BRAKES

PRECAUTIONS

✳✳ CAUTION

Clean any dust from the front brake and rear brake with a vacuum dust collector. Never blow with compressed air.

Note the following:
• Only use "DOT 3" brake fluid.
• Never reuse drained brake fluid.
• Never spill or splash brake fluid on painted surfaces. Brake fluid may seriously damage paint. Wipe it off immediately and wash with water if it gets on a painted surface.
• After pressing the brake pedal more deeply or harder than normal driving, such as air bleeding, check each item of brake pedal. Adjust brake pedal if it is outside the standard value.

• Always clean with new brake fluid when cleaning the master cylinder, brake caliper and other components.

• Never use mineral oils such as gasoline or light oil to clean. They may damage rubber parts and cause improper operation.

• Always loosen the brake tube flare nut with a flare nut wrench.

• Tighten the brake tube flare nut to the specified torque with a crowfoot and torque wrench.

• Always confirm the specified tightening torque when installing the brake pipes.

• Brake system is an important safety part. If a brake fluid leak is detected, always disassemble the affected part. If a malfunc-

ANTI-LOCK BRAKE SYSTEM

tion is detected, replace part with a new one.

• Turn the ignition switch OFF and disconnect the ABS actuator and electric unit (control unit) connector or the battery negative terminal before performing the work.

• Check that no brake fluid leakage is present after replacing the parts.

• Burnish the brake contact surfaces after refinishing or replacing rotors, after replacing pads, or if a soft pedal occurs at very low mileage.

• Slight vibrations are felt on the brake pedal and the operation noises occur, when VDC, TCS or ABS is activated.

• When starting engine or when starting vehicle just after starting engine, brake pedal may vibrate or motor operating noise may be heard from engine compartment. This is normal condition.

• Stopping distance is longer than that of vehicles without ABS when the vehicle drives on rough, gravel, or snow-covered (fresh, deep snow) roads.

• When an error is indicated by ABS or another warning lamp, collect all necessary information from customer (what symptoms are present under what conditions) and check for estimate causes before starting diagnostic servicing. Besides electrical system inspection, check brake booster operation, brake fluid level, and oil leaks.

• If tire size and type are used in an improper combination, or brake pads are not Genuine NISSAN parts, stopping distance or steering stability may deteriorate.

• ABS might be out of order or malfunctions by putting a radio (wiring inclusive), an antenna and a lead-in wire near the control unit.

• If aftermarket parts (car stereo, CD player, etc.) have been installed, check for incidents such as harness pinches, open circuits, and improper wiring.

• VDC system may not operate normally or a VDC OFF indicator lamp or SLIP indicator lamp may light.

• When replacing the following parts with parts other than genuine parts or making modifications: Suspension related parts (shock absorber, spring, bushing, etc.), tires, wheels (other than specified sizes), brake-related parts (pad, rotor, caliper, etc.), engine-related parts (muffler, ECM, etc.) and body reinforcement-related parts (roll bar, tower bar, etc.).

• When driving with worn or deteriorated suspension, tires and brake-related parts.

BLEEDING THE ABS SYSTEM

✳✳ WARNING
Turn the ignition switch OFF and disconnect the ABS actuator and electric unit (control unit) connector or the battery negative terminal before performing the work. Monitor the fluid level in the reservoir tank while performing the air bleeding. Always use new brake fluid for refilling. Never reuse the drained brake fluid.

1. Connect a vinyl tube to the bleeder valve of the rear right brake.
2. Fully depress the brake pedal 4 to 5 times.

3. Loosen the bleeder valve and bleed air with the brake pedal depressed, and then quickly tighten the bleeder valve.
4. Repeat steps 2 and 3 until all of the air is out of the brake line.
5. Tighten the bleeder valve.
6. Perform steps 1 to 5 for the rear right brake, front left brake, rear left brake, and front right brake in that order.
7. Check that the fluid level in the reservoir tank is within the specified range after air bleeding.
8. Check each item of brake pedal. Adjust it if the measurement value is not the standard.

WHEEL SPEED SENSORS

REMOVAL & INSTALLATION

Front
See Figure 2.

1. Remove the fender protector.
2. Remove the wheel sensor from steering knuckle.

✳✳ WARNING
Never twist sensor harness as much as possible, when removing it. Pull wheel sensors out without pulling sensor harness.

3. Remove the wheel sensor harness from vehicle.

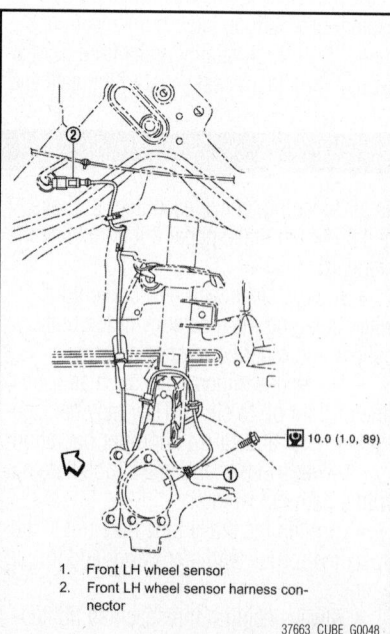

1. Front LH wheel sensor
2. Front LH wheel sensor harness connector

37663_CUBE_G0048

Fig. 2 Front wheel speed sensor assembly

✳✳ WARNING
Never twist sensor harness as much as possible, when removing it. Pull wheel sensors out without pulling sensor harness.

To install:
4. Note the following, and install in the reverse order of the removal.

a. Make sure there is no foreign material such as iron chips on and in the mounting hole of the wheel sensor.

b. Make sure no foreign material has been caught in the sensor rotor. Remove any foreign material and clean the mount. Replace the wheel sensor if necessary.

c. When installing wheel sensor, be sure to press rubber grommets in until they lock at locations shown above in the figure. When installed, harness must not be twisted.

Rear
See Figure 3.

1. Remove wheel sensor from wheel hub and bearing assembly.

✳✳ WARNING
Never twist sensor harness as much as possible, when removing it. Pull wheel sensors out without pulling sensor harness.

2. Remove wheel sensor harness from vehicle.

✳✳ WARNING
Never twist sensor harness as much as possible, when removing it. Pull wheel sensors out without pulling sensor harness.

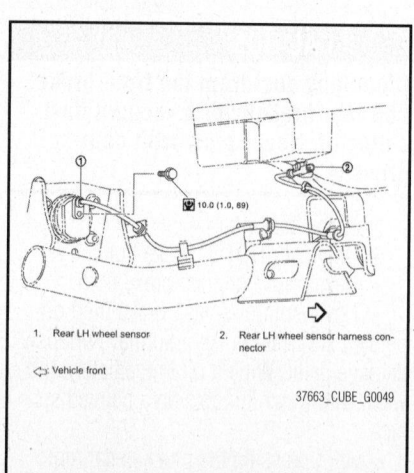

1. Rear LH wheel sensor
2. Rear LH wheel sensor harness connector

⟵ Vehicle front

37663_CUBE_G0049

Fig. 3 Rear wheel speed sensor assembly

To install:

3. Note the following, and install in the reverse order of the removal.

a. Make sure there is no foreign material such as iron chips on and in the mounting hole of the wheel sensor.

b. Make sure no foreign material has been caught in the sensor rotor. Remove any foreign material and clean the mount. Replace the wheel sensor if necessary.

c. When installing wheel sensor, be sure to press rubber grommets in until they lock at locations shown above in the figure. When installed, harness must not be twisted.

BRAKES   BLEEDING THE BRAKE SYSTEM

BLEEDING PROCEDURE

BLEEDING THE MASTER CYLINDER

In-Vehicle

See Figure 4.

> ※ **WARNING**
>
> **Turn the ignition switch OFF and disconnect the ABS actuator and electric unit (control unit) connector or the battery negative terminal before performing the work. Monitor the fluid level in the reservoir tank while performing the air bleeding. Always use new brake fluid for refilling. Never reuse the drained brake fluid.**

1. Connect a vinyl tube to the bleeder valve of the rear right brake.

2. Fully depress the brake pedal 4 to 5 times.

3. Loosen the bleeder valve and bleed air with the brake pedal depressed, and then quickly tighten the bleeder valve.

4. Repeat steps 2 and 3 until all of the air is out of the brake line.

5. Tighten the bleeder valve.

6. Perform steps 1 to 5 for the rear right brake, front left brake, rear left brake, and front right brake in that order.

7. Check that the fluid level in the reservoir tank is within the specified range after air bleeding.

8. Check each item of brake pedal. Adjust it if the measurement value is not the standard.

FLUID FILL PROCEDURE

See Figure 5.

> ※ **WARNING**
>
> **Turn the ignition switch OFF and disconnect the ABS actuator and electric unit (control unit) connector or the battery negative terminal before performing work.**

1. Check that there is no foreign material in the reservoir tank, and refill with new brake fluid.

> ※ **WARNING**
>
> **Never reuse drained brake fluid.**

2. Loosen the bleeder valve, slowly depress the brake pedal to the full stroke, and then release the pedal. Repeat this operation at intervals of 2 or 3 seconds until all brake fluid is

3. discharged.

4. Then close the bleeder valve with the brake pedal depressed.

5. Repeat the same work on each wheel.

6. Perform the air bleeding.

7. Check for brake fluid leakage from the master cylinder mounting face, reservoir tank mounting face and brake tube connections.

BRAKE FLUID

FLUID RECOMMENDATIONS

Be sure to use particular brake fluid either as indicated on reservoir cap of that vehicle or recommended in owner's manual which comes along with that vehicle. Use of any other fluid is strictly prohibited.

LEVEL CHECK

See Figure 6.

> ※ **CAUTION**
>
> **Brake fluid contains polyglycol ethers and polyglycols. Avoid contact with the eyes and wash your hands thoroughly after handling brake fluid. If you do get brake fluid in your eyes, flush your eyes with clean, running water for 15 minutes. If eye irritation persists, or if you have taken brake fluid internally, IMMEDIATELY seek medical assistance.**

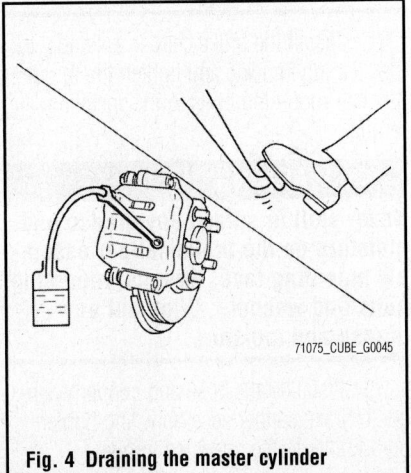

Fig. 4 Draining the master cylinder

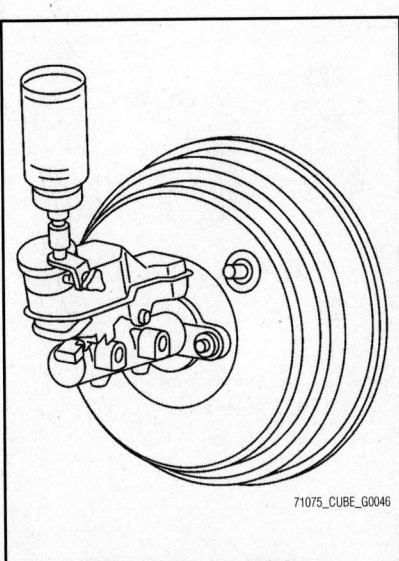

Fig. 5 Filling the master cylinder

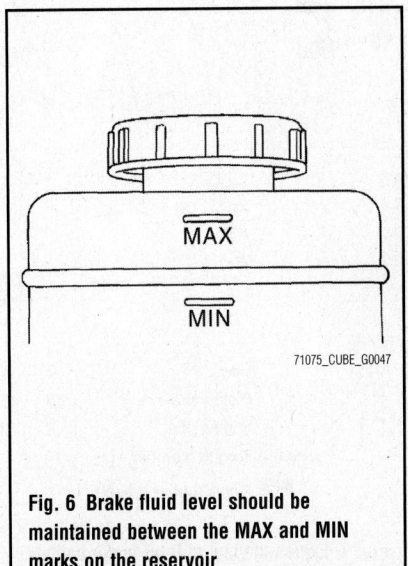

Fig. 6 Brake fluid level should be maintained between the MAX and MIN marks on the reservoir

> ※※ **WARNING**
>
> Clean, high quality brake fluid is essential to the safe and proper operation of the brake system. You should always buy the highest quality brake fluid that is available. If the brake fluid becomes contaminated, drain and flush the system, then refill the master cylinder with new fluid. Never reuse any brake fluid. Any brake fluid that is removed from the system

should be discarded. Also, do not allow any brake fluid to come in contact with a painted surface; it will damage the paint.

1. Before servicing the vehicle, refer to the Precautions Section.
   - Check that the fluid level in the reservoir tank is within the specified range (MAX – MIN lines).
   - Visually check for any brake fluid leakage around the reservoir tank.

- Check the brake system for any leakage if the fluid level is extremely low (lower than MIN).
- Check the brake system for fluid leakage if the warning lamp remains illuminated even after the parking brake is released.
- Check the reservoir tank for the mixing of foreign matter (e.g. dust) and oils other than brake fluid.

BRAKES

BRAKE CALIPERS

REMOVAL & INSTALLATION

See Figure 7.

> ※※ **CAUTION**
>
> Clean any dust from the brake caliper and brake pads with a vacuum dust collector. Never blow with compressed air.

> ※※ **WARNING**
>
> Never spill or splash brake fluid on painted surfaces. Brake fluid may seriously damage paint. Wipe it out immediately and wash with water if it gets on a protect surface.

1. Remove tires and wheels.
2. Fix the disc rotor using wheel nuts.
3. Drain brake fluid.

> ※※ **WARNING**
>
> Never spill or splash brake fluid on the disc rotor.

4. Remove union bolt and copper washer, and disconnect brake hose from caliper assembly.

> ※※ **WARNING**
>
> Never depress the brake pedal. Brake fluid may splash while removing the brake hose.

5. Remove torque member mounting bolts, and remove brake caliper assembly.

> ※※ **WARNING**
>
> Never drop brake pad and caliper assembly.

6. Remove disc rotor.

FRONT DISC BRAKES

> ※※ **WARNING**
>
> Put matching marks on the wheel hub and bearing assembly and the disc rotor before removing the disc rotor. Never drop disc rotor.

To install:

> ※※ **CAUTION**
>
> Clean any dust from the brake caliper and brake pads with a vacuum dust collector. Never blow with compressed air.

> ※※ **WARNING**
>
> Never depress the brake pedal. Brake fluid may splash while removing the brake hose. Never spill or splash brake fluid on painted surfaces. Brake fluid may seriously damage paint. Wipe it out immediately and wash with water if it gets on a protect surface.

7. Install disc rotor.

> ※※ **WARNING**
>
> Align the matching marks that have been made during removal when reusing the disc rotor.

8. Install the brake caliper assembly to the steering knuckle and tighten the torque member mounting bolts to the specified torque.

> ※※ **WARNING**
>
> Never spill or splash any grease and moisture on the brake caliper assembly mounting face, threads, mounting bolts and washers. Wipe out any grease and moisture.

9. Install brake hose and copper washers to brake caliper assembly, and tighten union bolts to the specified torque.

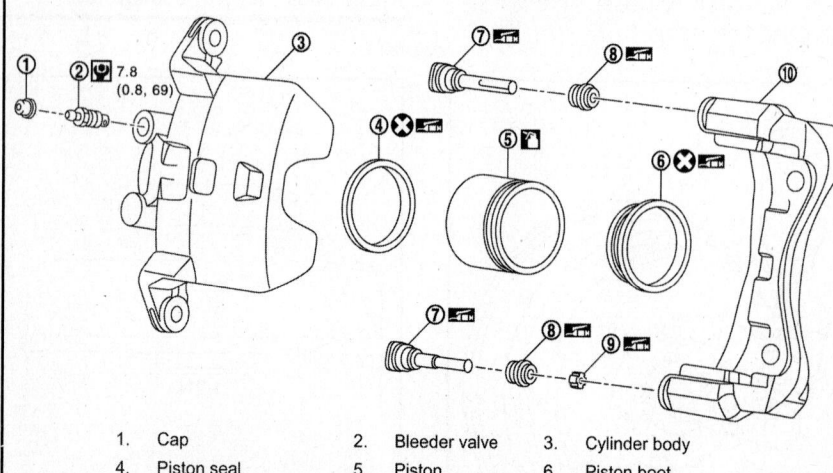

1. Cap
2. Bleeder valve
3. Cylinder body
4. Piston seal
5. Piston
6. Piston boot
7. Sliding pin
8. Sliding pin boot
9. Bushing
10. Torque member

: Apply rubber grease.

37663_CUBE_G0050

Fig. 7 Exploded view of front brake caliper

❋❋ WARNING

Never reuse copper washer.

10. Refill with new brake fluid and perform the air bleeding.

❋❋ WARNING

Never reuse drained brake fluid. Never spill or splash brake fluid on the disc rotor.

11. Check a drag of front disc brake. If any drag is found, inspect the installation.
12. Install tires.

BRAKE PADS

REMOVAL & INSTALLATION

See Figure 8.

❋❋ CAUTION

Clean any dust from the brake caliper and brake pads with a vacuum dust collector. Never blow with compressed air.

❋❋ WARNING

Never depress the brake pedal while removing the brake pads because the piston may pop out. Never spill or splash brake fluid on the disc rotor.

1. Remove tires and wheels.
2. Remove lower sliding pin bolt.
3. Suspend the cylinder body with suitable wire so that the brake hose will not stretch. Then remove the brake pads, shims, shim covers and pad retainers from the torque member.
Note the following:
• Never deform the pad retainer when removing the pad retainer from the torque member.
• Never damage the piston boot.
• Never drop the brake pads, shims, and the shim covers.
• Remember each position of the removed brake pads.

To install:

❋❋ CAUTION

Clean any dust from the brake caliper and brake pads with a vacuum dust collector. Never blow with compressed air.

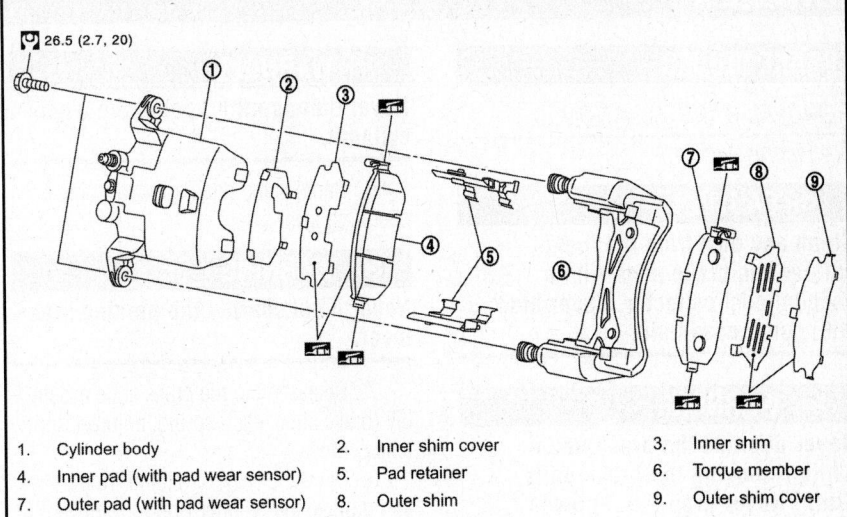

⊡ 26.5 (2.7, 20)

1. Cylinder body	2. Inner shim cover	3. Inner shim
4. Inner pad (with pad wear sensor)	5. Pad retainer	6. Torque member
7. Outer pad (with pad wear sensor)	8. Outer shim	9. Outer shim cover

◩ : Apply PBC (Poly Butyl Cuprysil) grease or silicone-based grease.

37663_CUBE_G0051

Fig. 8 Exploded view of front brake pads

❋❋ WARNING

Never depress the brake pedal while removing the brake pads or the cylinder body because the piston may pop out. Never spill or splash brake fluid on the disc rotor.

4. Install the pad retainers to the torque member if the pad retainers has been removed.

❋❋ WARNING

Securely assemble the pad retainers so that it will not be lifted up from the torque member. Never deform the pad retainers.

5. Apply PBC (Poly Butyl Cuprysil) grease or silicone-based grease to the mating faces between the inner shim and the inner pad, and install them to the inner pad.

❋❋ WARNING

Always replace the shim together with the shim cover when replacing the brake pad.

6. Apply PBC (Poly Butyl Cuprysil) grease or silicone-based grease to the mating faces between the outer shim and the outer shim cover, and install them to the outer pad.

❋❋ WARNING

Always replace the shim together with the shim cover when replacing the brake pad.

7. Apply PBC (Poly Butyl Cuprysil) grease or silicone-based grease to the mating faces between the brake pads and the pad retainers, and Install the brake pads to the torque member.
8. Install cylinder body to torque member.

❋❋ WARNING

Never damage the piston boot. When replacing brake pad with new one, check a brake fluid level in the reservoir tank because brake fluid returns to master cylinder reservoir tank when pressing piston in.

➡Use a disc brake piston tool to easily press piston.

9. Install the lower sliding pin bolt and tighten it to the specified torque.
10. Depress the brake pedal several times to check that no drag feel is present for the front disc brake.
11. Install tires.

BRAKES

REAR DRUM BRAKES

BRAKE DRUMS

REMOVAL & INSTALLATION

See Figures 9 and 10.

✳✳ CAUTION

Clean any dust from the brake caliper and brake pads with a vacuum dust collector. Never blow with compressed air.

✳✳ WARNING

Never depress the brake pedal while removing the brake pads drum. Never drop the removed parts.

1. Remove tires and wheels.
2. Perform drain the brake fluid remove or disassemble the wheel cylinder.
3. Remove the brake drum with the parking brake lever. If brake drum is difficult to the brake drum, remove according to the following procedure.
   a. Remove the plug from brake drum.
   b. Pull the adjuster lever from the plug hole of brake drum using a suitable wire, rotate the adjuster in the direction of the arrow using a suitable tool, and then compress the expanded brake shoe.
4. Press and rotate the retainer, and then remove the retainer, spring and shoe hold pin.
5. Remove the brake shoe assembly (brake shoe, each spring, adjuster and adjuster lever).

✳✳ WARNING

Never damage the boot of the wheel cylinder.

6. Remove the parking brake cable from operating lever.

✳✳ WARNING

Never bend sharply the parking brake lever.

7. Disassemble the brake shoe assembly (brake shoe, each spring, adjuster and adjuster lever).
8. Open the joint of retainer ring and remove the retainer ring to remove the operating lever from the brake shoe pin.
9. Separate the brake tube from the wheel cylinder.
10. Remove the wheel cylinder from back plate.

To install:

✳✳ CAUTION

Clean any dust from the brake caliper and brake pads with a vacuum dust collector. Never blow with compressed air.

✳✳ WARNING

Never depress the brake pedal while removing the brake drum. Never spill or splash brake fluid on the brake drum.

11. Note the following, and install in the reverse of removal.

   a. After installing the retainer ring, close the joint of retainer ring until securely closed.
   b. When disassembled adjuster, confirm the difference between left and right wheel for assemble.
   c. Apply PBC (Poly Butyl Cuprysil) silicone-based grease to the adjuster screw.
   d. Apply PBC (Poly Butyl Cuprysil) silicone-based grease to the mating faces between the adjuster and brake shoe.
   e. Apply PBC (Poly Butyl Cuprysil) silicone-based grease to the mating faces between the back plate and brake shoe.
   f. Shorten adjuster by rotating it.
   g. Install the brake shoe assembly so that it does damage the wheel cylinder.
   h. Check the component parts of brake shoe assembly are installed properly.
   i. Check the brake shoe sliding surface and brake drum inner surface for grease. Wipe it out any adheres to the surfaces.
   j. Perform the air bleeding when removed or disassembled the wheel cylinder.
   k. Adjust the brake shoe (parking brake lever stroke) after install and air bleeding.

BRAKE SHOES

REMOVAL & INSTALLATION

See Figures 9 and 10.

✳✳ CAUTION

Clean any dust from the brake caliper and brake pads with a vacuum dust collector. Never blow with compressed air.

✳✳ WARNING

Never depress the brake pedal while removing the brake pads drum. Never drop the removed parts.

1. Remove tires and wheels.
2. Perform drain the brake fluid when remove or disassemble the wheel cylinder.
3. Remove the brake drum with the parking brake lever. If brake drum is difficult to the brake drum, remove according to the following procedure.
   a. Remove the plug from brake drum.
   b. Pull the adjuster lever from the plug hole of brake drum using a suitable wire, rotate the adjuster in the direction

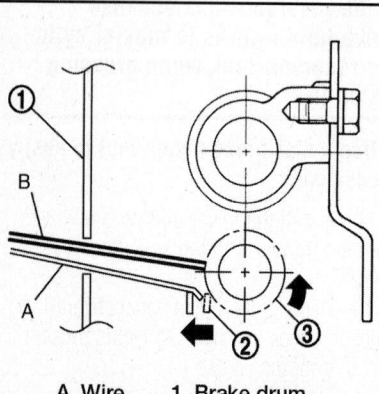

A. Wire   1. Brake drum
B. Tool   2. Adjuster lever

37663_CUBE_G0052

Fig. 9 Rotate the adjuster in the direction of the arrow using a suitable tool

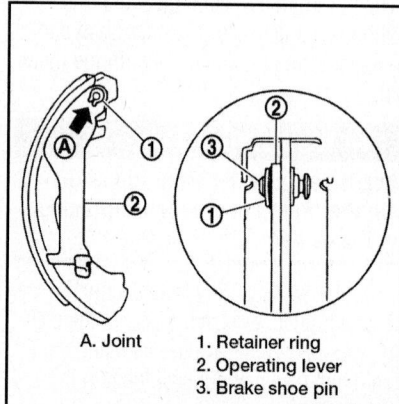

A. Joint   1. Retainer ring
           2. Operating lever
           3. Brake shoe pin

37663_CUBE_G0053

Fig. 10 Open the joint of retainer ring and remove the retainer ring to remove the operating lever from the brake shoe pin

of the arrow using a suitable tool, and then compress the expanded brake shoe.

4. Press and rotate the retainer, and then remove the retainer, spring and shoe hold pin.

5. Remove the brake shoe assembly (brake shoe, each spring, adjuster and adjuster lever).

✳✳ WARNING

Never damage the boot of the wheel cylinder.

6. Remove the parking brake cable from operating lever.

✳✳ WARNING

Never bend sharply the parking brake lever.

7. Disassemble the brake shoe assembly (brake shoe, each spring, adjuster and adjuster lever).

8. Open the joint of retainer ring and remove the retainer ring to remove the operating lever from the brake shoe pin.

9. Separate the brake tube from the wheel cylinder.

10. Remove the wheel cylinder from back plate.

To install:

✳✳ CAUTION

Clean any dust from the brake caliper and brake pads with a vacuum dust collector. Never blow with compressed air.

✳✳ WARNING

Never depress the brake pedal while removing the brake drum. Never spill or splash brake fluid on the brake drum.

11. Note the following, and install in the reverse of removal.

a. After installing the retainer ring, close the joint of retainer ring until securely closed.

b. When disassembled adjuster, confirm the difference between left and right wheel for assemble.

c. Apply PBC (Poly Butyl Cuprysil) silicone-based grease to the adjuster screw.

d. Apply PBC (Poly Butyl Cuprysil) silicone-based grease to the mating faces between the adjuster and brake shoe.

e. Apply PBC (Poly Butyl Cuprysil) silicone-based grease to the mating faces between the back plate and brake shoe.

f. Shorten adjuster by rotating it.

g. Install the brake shoe assembly so that it does damage the wheel cylinder.

h. Check the component parts of brake shoe assembly are installed properly.

i. Check the brake shoe sliding surface and brake drum inner surface for grease. Wipe it out any adheres to the surfaces.

j. Perform the air bleeding when removed or disassembled the wheel cylinder.

k. Adjust the brake shoe (parking brake lever stroke) after install and air bleeding.

BRAKES

ADJUSTMENTS

Lever Stroke

1. Operate the parking brake lever with a force of 44 lbs. (196 N). Check that the lever stroke is within the specified number of notches. (Check it by listening to the clicks of the ratchet.)

2. When parking brake warning lamp turns ON, check that the lever stroke is within the specified number of notches. (Check it by listening to the clicks of the ratchet.)

Inspect Components

1. Check each component for installation condition such as looseness.

2. Check the parking brake lever assembly for bend, damage and cracks. Replace if necessary.

3. Check the cables and equalizer for wear, damage and cracks. Replace if necessary.

4. Check the parking brake switch, and replace it if necessary.

Adjustment

See Figure 11.

1. Remove the console mask.

2. Pull parking brake lever until a socket wrench can be inserted to adjusting nut.

3. Release the parking brake lever by turning the adjusting nut with a socket wrench and loosening the cable.

4. Depress the brake pedal with a force of 44 lbs. (196 N) about 10 times and adjust the brake shoe clearance.

✳✳ WARNING

Make sure to securely depress the brake pedal.

5. Check a drag of rear drum brake.

6. Adjust the cable with the following procedure.

a. When replace parking brake cable, operate the parking brake lever with a force of 110 lbs. (490 N) for 10 strokes or more.

b. Adjust the parking brake lever stroke by turning the adjusting nut with a deep socket wrench.

✳✳ WARNING

Never reuse the adjusting nut if the nut is removed.

PARKING BRAKE

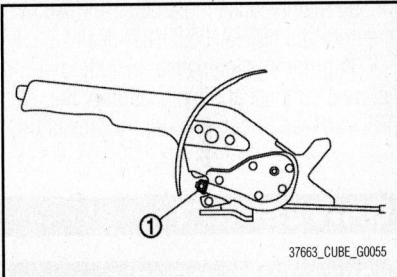

37663_CUBE_G0055

Fig. 11 Pull parking brake lever until a socket wrench can be inserted to adjusting nut (1)

c. Operate the parking brake lever with a force of 44 lbs. (196 N). Check that the lever stroke is within the specified number of notches. (Check it by listening to the clicks of the ratchet.)

d. Rotate the brake drum with the parking brake lever released and check that there is no drag.

PARKING BRAKE SHOES

REMOVAL & INSTALLATION

See Rear Drum Brake.

CHASSIS ELECTRICAL AIR BAGS (SUPPLEMENTAL RESTRAINT SYSTEM)

PRECAUTIONS

Precaution for Supplemental Restraint System (SRS) "Air Bag" And "Seat Belt Pre-Tensioner"

The Supplemental Restraint System such as "AIR BAG" and "SEAT BELT PRE-TENSIONER", used along with a front seat belt, helps to reduce the risk or severity of injury to the driver and front passenger for certain types of collision. This system includes seat belt switch inputs and dual stage front air bag modules. The SRS system uses the seat belt switches to determine the front air bag deployment, and may only deploy one front air bag, depending on the severity of a collision and whether the front occupants are belted or unbelted.

Information necessary to service the system safely is included in the "SRS AIR BAG" and "SEAT BELT" of this Service Manual.

Note the following:

• To avoid rendering the SRS inoperative, which could increase the risk of personal injury or death in the event of a collision which would result in air bag inflation, all maintenance must be performed by an authorized NISSAN/INFINITI dealer.

• Improper maintenance, including incorrect removal and installation of the SRS, can lead to personal injury caused by

unintentional activation of the system. For removal of Spiral Cable and Air Bag Module, see the "SRS AIR BAG".

• Do not use electrical test equipment on any circuit related to the SRS unless instructed to in this Service Manual. SRS wiring harnesses can be identified by yellow and/or orange harnesses or harness connectors.

Precautions When Using Power Tools (Air Or Electric) And Hammers

Note the following:

• When working near the Air Bag Diagnosis Sensor Unit or other Air Bag System sensors with the ignition ON or engine running, DO NOT use air or electric power tools or strike near the sensor(s) with a hammer. Heavy vibration could activate the sensor(s) and deploy the air bag(s), possibly causing serious injury.

• When using air or electric power tools or hammers, always switch the ignition OFF, disconnect the battery, and wait at least 3 minutes before performing any service.

Precaution Necessary for Steering Wheel Rotation after Battery Disconnect

Note the following:

• Before removing and installing any control units, first turn the push-button

ignition switch to the LOCK position, then disconnect both battery cables.

• After finishing work, confirm that all control unit connectors are connected properly, then re-connect both battery cables.

• Always use CONSULT-III to perform self-diagnosis as a part of each function inspection after finishing work.

• If a DTC is detected, perform trouble diagnosis according to self-diagnosis results.

• For vehicle with steering lock unit, if the battery is disconnected or discharged, the steering wheel will lock and cannot be turned.

• If turning the steering wheel is required with the battery disconnected or discharged, follow the operation procedure below before starting the repair operation.

DISARMING THE SYSTEM

Before servicing, turn ignition switch OFF, disconnect battery negative terminal and wait 3 minutes or more.

ARMING THE SYSTEM

Reconnect the battery cables. Wait several seconds to ensure that no system malfunction is detected by the air bag warning lamp.

DRIVESHAFT

CONTINUALLY VARIABLE TRANSMISSION (CVT)

DRAIN AND REFILL

1. Remove drain plug from oil pan and then the CVT fluid.
2. Remove drain plug gasket from drain plug.
3. Install drain plug gasket to drain plug.

✳✳ WARNING

Never reuse drain plug gasket.

4. Install drain plug to oil pan.
5. Fill CVT fluid from CVT fluid charging pipe to the specified level.

• Always use the specified fluid. If use, misuse, or mixing of fluid other than the specified fluid occurs, original performance can-

not be obtained or it may cause serious malfunctions.

• CVT fluid is not reusable. Never reuse CVT fluid.
• Always use shop paper. Never use shop cloth.
• After replacement, always perform CVT fluid leakage check.
• Delete CVT fluid deterioration date with CONSULT-III after changing CVT fluid.

6. After engine warms up, drive the vehicle in an urban area for approximately 10 minutes.

➡**When ambient temperature is 68°F (20°C), it takes about 10 minutes for the CVT fluid to warm up to 122 to 176°F (50 to 80°C).**

7. Check CVT fluid level and condition.
8. Repeat steps 1 to 6 if CVT fluid has been contaminated.

MANUAL TRANSAXLE

DRAIN & REFILL

Draining
See Figure 12.

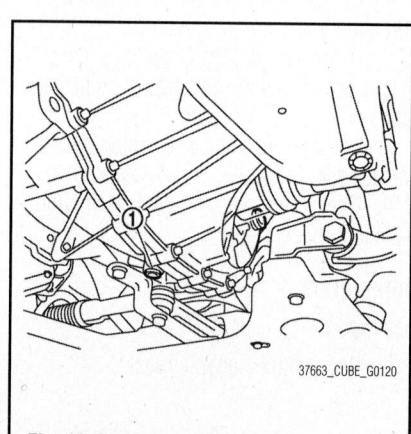

37663_CUBE_G0120

Fig. 12 Remove drain plug (1)

1. Start engine and let it run to warm up transaxle.

2. Stop engine. Remove drain plug and gasket, using a socket and then drain gear oil.

3. Set a gasket on drain plug and install it to clutch housing, using a socket.

✳✳ WARNING

Never reuse gasket.

4. Tighten drain plug to the specified torque.

Refilling

See Figure 13.

1. Remove filler plug and gasket from transaxle case.

2. Fill with new gear oil until oil level reaches the specified limit at filler plug mounting hole as shown.

3. After refilling gear oil, check the oil level.

4. Set a gasket on filler plug and then install it to transaxle case.

✳✳ WARNING

Never reuse gasket.

5. Tighten filler plug to the specified torque.

FLUID LEVEL CHECK

See Figure 14.

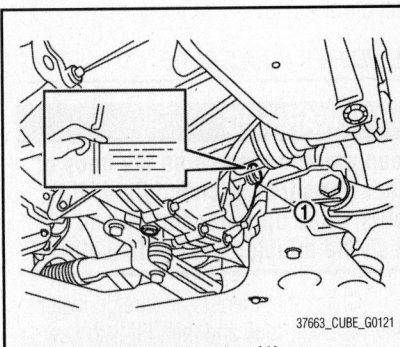

Fig. 13 Remove filler plug (1)

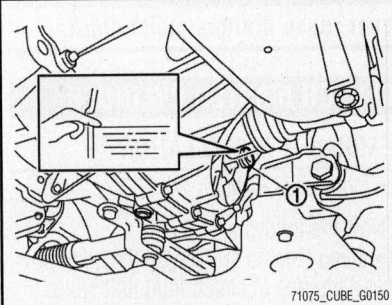

Fig. 14 Checking the fluid level and locating filler plug (1)—M/T

FLUID RECOMMENDATIONS

Genuine NISSAN gear Oil (Chevron Texaco ETL 8997B) 75W-80, or equivalent. If Genuine NISSAN gear oil is not available, API GL-4, Viscosity SAE 75W-80 may be used as a temporary replacement. However, use Genuine NISSAN gear oil as soon as it is available.

CLUTCH

PEDAL HEIGHT ADJUSTMENT

See Figures 15 through 17.

The Height of Clutch Pedal
1. Turn the floor carpet.
2. Check that the clutch pedal height "H1" from the dash lower panel is within the reference value (6.83–7.23 inches).
3. Replace clutch pedal if the height is outside the reference value.

Clutch Pedal Height When Disengaging The Clutch

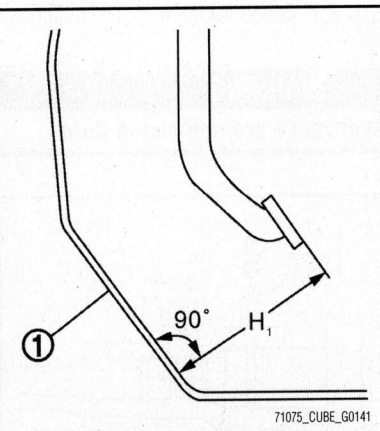

Fig. 15 Check that the clutch pedal height "H1" from the dash lower panel (1) is within the reference value

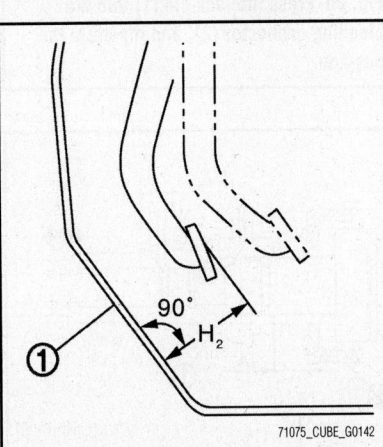

Fig. 16 check that the clutch pedal height "H2" from the dash lower panel (1) is within the reference value

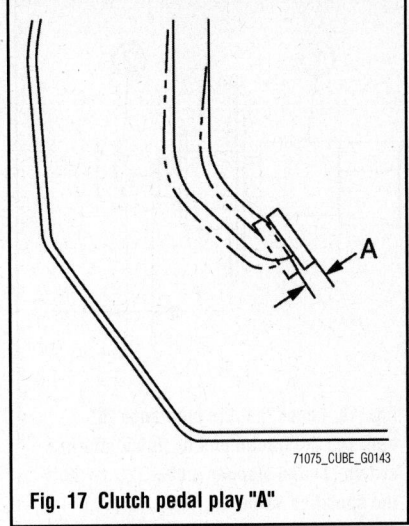

Fig. 17 Clutch pedal play "A"

4. Turn the floor carpet.
5. Start the engine and run at idle.
6. Apply the parking brake.
7. Depress the brake pedal.
8. Fully depress clutch pedal and shift to the 1st gear.
9. Gradually release the clutch pedal and check that the clutch pedal height "H2" from the dash lower panel is within the reference value (3.15 inches or more) with a scale immediately before the clutch is engaged.

➡**Although the clutch pedal height differs according to whether the clutch gets disengaged or engaged, clutch-engaged case is regarded as clutch-disengaged case for easier inspection.**

10. Replace clutch pedal if the height is outside the reference value.
Clutch Pedal Play
11. Push the pedal pad by hand until a resistance can be felt and check that the play "A" on the upper surface of the pedal pad is within the reference value.
   a. Clutch pedal play "A" [Looseness at clutch pedal pin]: 0.08 – 0.31 in [0 – 0.051 in].
12. Replace clutch pedal if the play is outside the reference value.

SHIFT INTERLOCK SWITCH ADJUSTMENT

See Figures 18 and 19.

Position of Clutch Interlock Switch
1. Check that the clearance "C" between the thread end of clutch interlock switch and stopper rubber is within the specified value (0.0291 – 0.0772 inches) while clutch pedal is fully depressed.
Position of ASCD Clutch Switch (With ASCD)
2. Check that the clearance "A" between

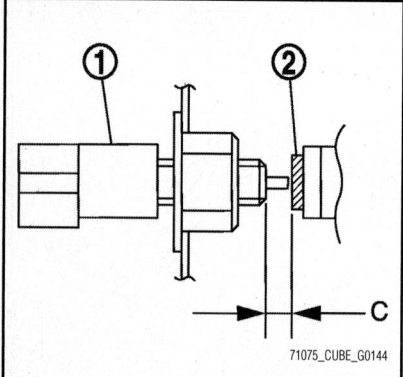

Fig. 18 Check that the clearance "C" between the thread end of clutch interlock switch (1) and stopper rubber (2) is within the specified value

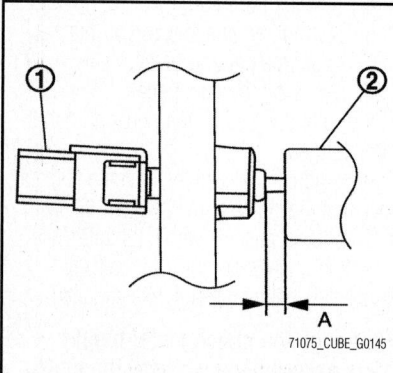

Fig. 19 Check that the clearance "A" between the thread end of ASCD clutch switch (1) and clutch pedal (2) is within the specified value

the thread end of ASCD clutch switch (1) and clutch pedal (2) is within the specified value (0.0291 – 0.0772 inches) while clutch pedal is fully released.

Adjustment
Position of Clutch Interlock Switch

3. Loosen clutch interlock switch by turning it 45 degrees counterclockwise.

4. Press the clutch interlock switch to bring clearance between the thread end of clutch interlock switch and stopper rubber to the specified value (0.0291–0.0772 inches) while clutch pedal is fully depressed.

5. Fasten clutch interlock switch by turning it 45 degrees clockwise.
Position of ASCD Clutch Switch (With ASCD)

6. Loosen ASCD clutch switch by turning it 45 degrees counterclockwise.

7. Press the ASCD clutch switch to bring clearance between the thread end of ASCD clutch switch and clutch pedal to the

specified value (0.0291 – 0.0772 inches) while clutch pedal is fully released.

8. Fasten ASCD clutch switch by turning it 45 degrees clockwise.

HYDRAULIC SYSTEM BLEEDING

BLEEDING PROCEDURE
See Figures 20 and 21.

Note the following:
• Monitor clutch fluid level in reservoir tank so as not to empty the tank.
• Keep painted surface on the body or other parts free of clutch fluid. If it spills, wipe up immediately and wash the affected area with water.

➡Do not use a vacuum assist or any other type of power bleeder on this system. Use of a vacuum assist or power bleeder will not purge all the air from the system.

1. Fill reservoir tank with new clutch fluid.

❊❊ WARNING
Never reuse drained clutch fluid.

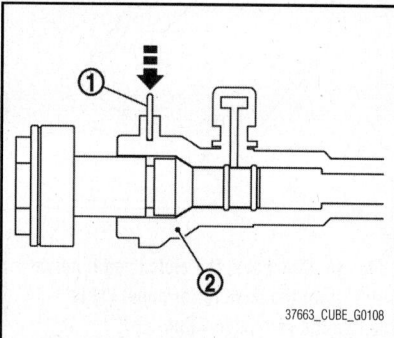

Fig. 20 Press the lock pin (1) into the bleeding connector (2), and maintain the position

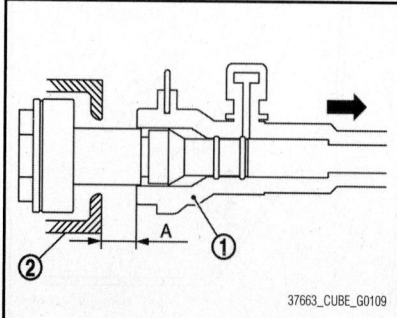

Fig. 21 Slide bleeding connector (1) in the direction of the arrow as shown

2. Connect a transparent vinyl hose to air bleeder of bleeding connector.

3. "Depress" and "release" the clutch pedal slowly and fully 15 times at an interval of 2 to 3 seconds and release the clutch pedal.

4. Press the lock pin into the bleeding connector and maintain the position.

❊❊ WARNING
Since the inside of clutch tube is under hydraulic pressure, hold the tube to prevent it from getting disconnected.

5. Slide bleeding connector in the direction of the arrow as shown.

6. Depress the clutch pedal soon and hold it, and then bleed the air from the piping.

❊❊ WARNING
Since the inside of clutch tube is under hydraulic pressure, hold the tube to prevent it from getting disconnected.

7. Return clutch tube and lock pin in their original positions.

8. Release clutch pedal and wait for 5 seconds.

9. Repeat steps 3 to 8 until no bubbles are observed in the clutch fluid.

10. Check that the fluid level in the reservoir tank is within the specified range after air bleeding.

FLUID FILL PROCEDURE

❊❊ WARNING
Keep painted surface on the body or other parts free of clutch fluid. If it spills, wipe up immediately and wash the affected area with water.

1. Check that there is no foreign material in reservoir tank and then fill with new clutch fluid.

❊❊ WARNING
Never reuse drained clutch fluid.

CLUTCH MASTER CYLINDER

REMOVAL & INSTALLATION
See Figures 22 through 24.

Note the following:
• Keep painted surface on the body or other parts free of clutch fluid. If it spills, wipe up immediately and wash the affected area with water.
• Never disassemble master cylinder.

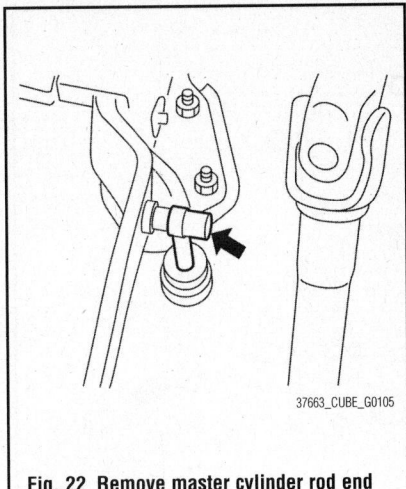

Fig. 22 Remove master cylinder rod end from clutch pedal

1. Drain clutch fluid.
2. Remove air duct (inlet).
3. Remove battery.
4. Remove air cleaner case and air ducts.
5. Remove hose from reservoir tank assembly and master cylinder.
6. Remove master cylinder rod end from clutch pedal.
7. Remove lock pin from connector of master cylinder and separate clutch tube.
8. Rotate master cylinder clockwise by 45 degrees and then remove master cylinder from the vehicle.

To install:

❈❈ WARNING

Keep painted surface on the body or other parts free of clutch fluid. If it spills, wipe up immediately and wash the affected area with water.

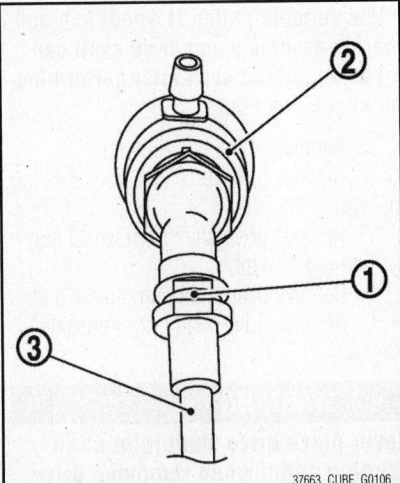

Fig. 23 Remove lock pin (1) from connector of master cylinder (2) and separate clutch tube (3)

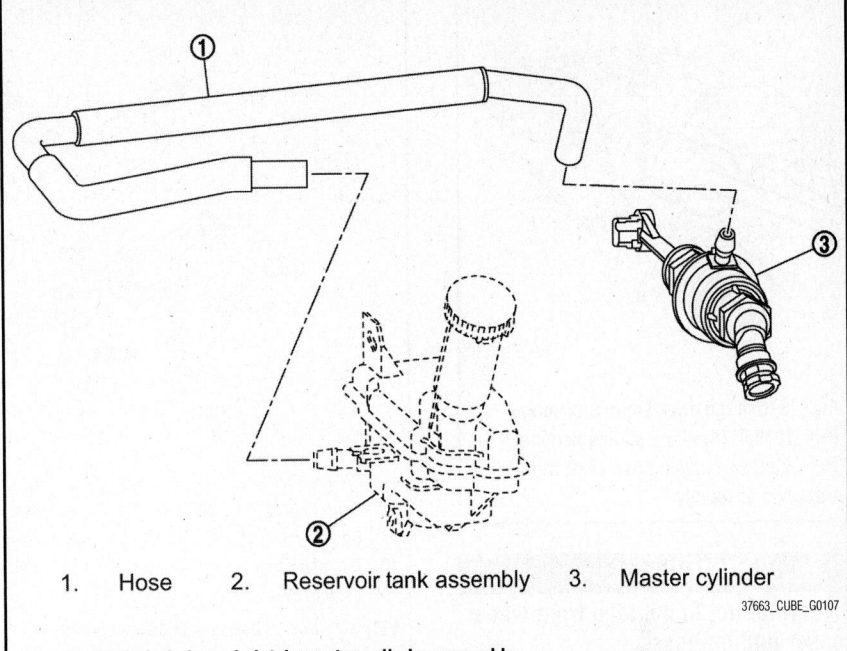

| 1. | Hose | 2. | Reservoir tank assembly | 3. | Master cylinder |

Fig. 24 Exploded view of clutch master cylinder assembly

9. Tilt master cylinder clockwise by 45 degrees and insert it to the mounting hole. Rotate counterclockwise and secure it with the nipple in an upward position.
10. Install master cylinder rod end to clutch pedal.

❈❈ WARNING

Press master cylinder rod end into clutch pedal until it stops.

11. Install hose to reservoir tank assembly and master cylinder.

❈❈ WARNING

Set hose with painted mark facing upward.

12. Install lock pin into connector of master cylinder until it stops.
13. Install clutch tube into connector of master cylinder until it stops.
14. Install air cleaner case and air ducts.
15. Install battery.
16. Install air duct (inlet).
17. Fill with clutch fluid.
18. Check the clutch fluid level and clutch fluid leakage.
19. Check the clutch pedal and clutch interlock switch position.
20. Check the ASCD clutch switch position. (With ASCD)
21. Adjust the clutch interlock switch position.

22. Adjust the ASCD clutch switch position. (With ASCD)
23. Bleed air from the clutch hydraulic system.

FRONT HALFSHAFT

REMOVAL & INSTALLATION

Left Side
See Figures 25 through 27.

1. Remove tires and wheels.
2. Remove cotter pin, and then loosen wheel hub lock nut.
3. Patch wheel hub lock nut with a piece of wood. Hammer the wood to disengage wheel hub and bearing assembly from drive shaft.

❈❈ WARNING

Never place drive shaft joint at an extreme angle. Also be careful not to overextend slide joint. Never allow drive shaft to hang down without support for joint sub-assembly, shaft and the other parts.

➡Use suitable puller, if wheel hub and bearing assembly and drive shaft cannot be separated even after performing the above procedure.

4. Remove wheel hub lock nut.
5. Remove transverse link from steering knuckle.
6. Remove shaft assembly from wheel hub and bearing assembly.

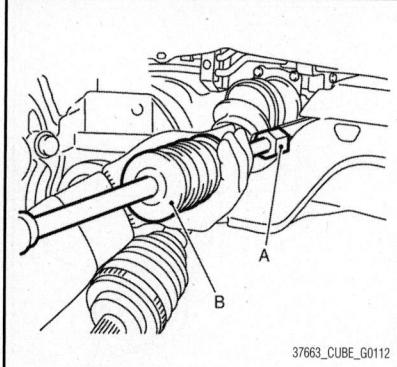

Fig. 25 Use the drive shaft attachment KV40107500 (A) and a sliding hammer (B), and then remove drive shaft from transaxle assembly

> ※※ **WARNING**
>
> **Be careful not to damage front wheel sensor and harness.**

7. Use the drive shaft attachment KV40107500 and a sliding hammer while inserting tip of the drive shaft attachment between shaft and transaxle assembly, and then remove drive shaft from transaxle assembly.

> ※※ **WARNING**
>
> **Never place drive shaft joint at an extreme angle when removing drive shaft. Also be careful not to over-extend slide joint. Confirm that the circular clip is attached to the drive shaft.**

To install:

8. Note the following for the transaxle side, and install in the reverse order of removal.

9. Always replace differential side oil seal with new one when installing drive shaft.

10. Place the protector KV38107900 onto transaxle assembly to prevent damage to the oil seal while inserting drive shaft. Slide drive shaft sliding joint and tap with a hammer to install securely.

> ※※ **WARNING**
>
> **Check that circular clip is completely engaged.**

11. Clean the matching surface of wheel hub lock nut and wheel hub and bearing assembly.

> ※※ **WARNING**
>
> **Never apply lubricating oil to these matching surface.**

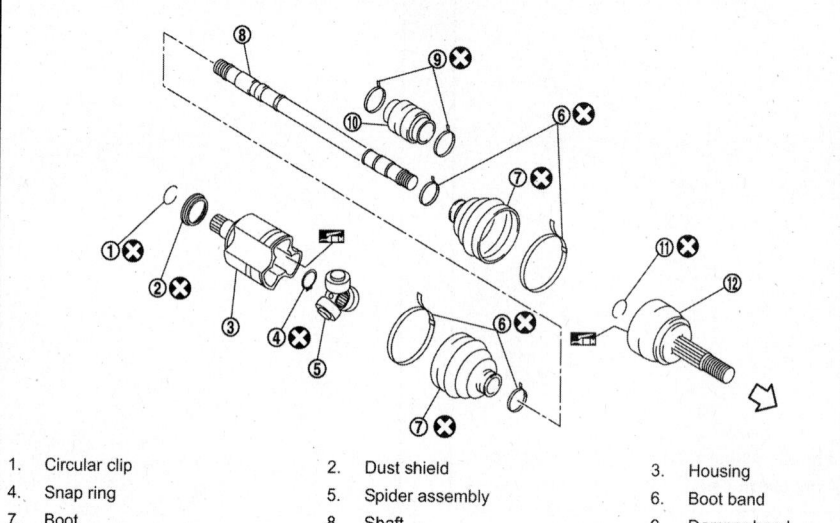

1.	Circular clip	2.	Dust shield	3.	Housing
4.	Snap ring	5.	Spider assembly	6.	Boot band
7.	Boot	8.	Shaft	9.	Damper band
10.	Dynamic damper	11.	Circular clip	12.	Joint sub-assembly

⟵ : Wheel side

▥ : Fill NISSAN Genuine grease or equivalent.

37663_CUBE_G0110

Fig. 26 Exploded view of CV joint and boot assembly—LH side

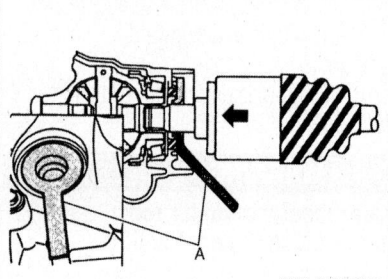

37663_CUBE_G0113

Fig. 27 Place the protector KV38107900 (A) onto transaxle assembly to prevent damage to the oil seal while inserting drive shaft

12. Tighten the wheel hub lock nut to the specified torque.

> ※※ **WARNING**
>
> **Never use a power tool to tighten the wheel hub lock nut.**

13. Perform the final tightening of each of parts under unladen conditions, which were removed when removing wheel hub and bearing assembly and axle housing.

14. Never reuse cotter pin.

Right Side

See Figures 27 through 29.

1. Remove tires and wheels.

2. Remove wheel sensor and sensor harness if necessary.

3. Remove cotter pin, and then loosen wheel hub lock nut.

4. Patch wheel hub lock nut with a piece of wood. Hammer the wood to disengage wheel hub and bearing assembly from drive shaft.

> ※※ **WARNING**
>
> **Never place drive shaft joint at an extreme angle. Also be careful not to overextend slide joint. Never allow drive shaft to hang down without support for joint sub-assembly, shaft and the other parts.**

➡**Use suitable puller, if wheel hub and bearing assembly and drive shaft cannot be separated even after performing the above procedure.**

5. Remove wheel hub lock nut.

6. Remove transverse link from steering knuckle.

7. Remove drive shaft from wheel hub and bearing assembly.

8. Remove bearing housing plate bolts.

9. Remove drive shaft from transaxle assembly.

> ※※ **WARNING**
>
> **Never place drive shaft joint at an extreme angle when removing drive shaft. Also be careful not to over-extend slide joint.**

10. Remove support bearing bracket.

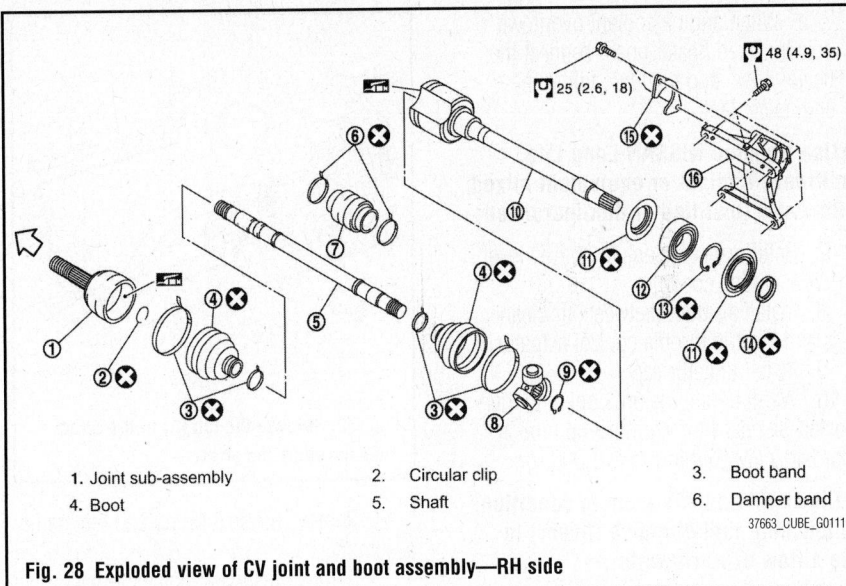

1. Joint sub-assembly
2. Circular clip
3. Boot band
4. Boot
5. Shaft
6. Damper band

37663_CUBE_G0111

Fig. 28 Exploded view of CV joint and boot assembly—RH side

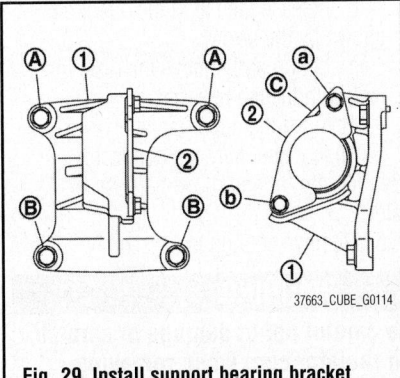

37663_CUBE_G0114

Fig. 29 Install support bearing bracket

To install:

11. Note the following for the transaxle side, and install in the reverse order of removal.

12. Always replace differential side oil seal with new one when installing drive shaft.

13. Install support bearing bracket in following procedure:

    a. Temporarily tighten mounting bolts, then tighten them to specified torque.

    b. Set plate so that notch faces the upper side.

    c. Temporarily tighten mounting bolts, then tighten them.

✳✳ WARNING
Never reuse plate.

14. Place the protector KV38107900 onto transaxle assembly to prevent damage to the oil seal while inserting drive shaft. Slide drive shaft sliding joint and tap with a hammer to install securely.

WHEEL SIDE

1. Clean the matching surface of wheel hub lock nut and wheel hub and bearing assembly.

✳✳ WARNING
Never apply lubricating oil to these matching surface.

2. Tighten the wheel hub lock nut to the specified torque.

✳✳ WARNING
Never use a power tool to tighten the wheel hub lock nut.

3. Perform the final tightening of each of the parts, which were removed when removing wheel hub and bearing assembly and axle housing, under unladen conditions.

✳✳ WARNING
Never reuse cotter pin.

ENGINE COOLING

ENGINE COOLANT

BLEEDING

1. Install reservoir tank if removed and radiator drain plug.

✳✳ WARNING
Be sure to clean drain plug and install with new O-ring.

➡ If water drain plugs on cylinder block are removed, close and install them.

2. Remove air duct (between air cleaner case and electric throttle control actuator).

3. Disconnect heater hose at the firewall.

4. Fill radiator and reservoir tank with water and reinstall radiator cap.

➡ When engine coolant over flows disconnected heater hose, connect heater hose, and continue filling the engine coolant.

5. Install air duct (between air cleaner case and electric throttle control actuator).

6. Run the engine and warm it up to normal operating temperature.

7. Rev the engine two or three times under no-load.

8. Stop the engine and wait until it cools down.

9. Drain water from the system.

10. Repeat steps 1 through 9 until clear water begins to drain from radiator.

DRAIN & REFILL

Draining
See Figure 30.

✳✳ CAUTION
Never remove radiator cap when engine is hot. Serious burns may occur from high-pressure engine coolant escaping from radiator.

✳✳ WARNING
Wrap a thick cloth around the radiator cap. Slowly turn it a quarter of a turn to release built-up pressure. Then open it all the way.

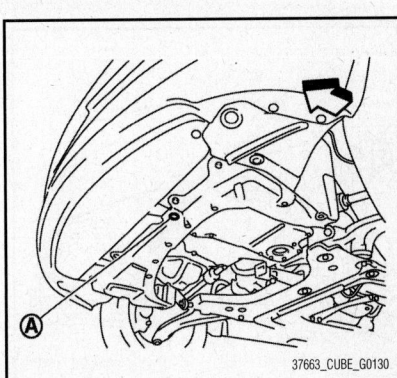

37663_CUBE_G0130

Fig. 30 Open radiator drain plug (A) at the bottom of radiator

1. Remove engine under cover.
2. Open radiator drain plug at the bottom of radiator, and then remove radiator cap.

✳✳ WARNING

Perform this step when engine is cold.

➡**When draining all of engine coolant in the system, open water drain plugs on cylinder block.**

3. Remove reservoir tank if necessary, and drain engine coolant and clean reservoir tank before installing.
4. Check drained engine coolant for contaminants such as rust, corrosion or discoloration. If contaminated, flush the engine cooling system.

Refilling

See Figure 31.

1. Install reservoir tank if removed and radiator drain plug.

✳✳ WARNING

Be sure to clean drain plug and install with new O-ring.

2. If water drain plugs on cylinder block are removed, close and install them.
3. Check that each hose clamp has been firmly tightened.
4. Remove air duct (between air cleaner case and electric throttle control actuator).
5. Disconnect heater hose at the firewall.
6. Fill radiator to overflow opening on radiator neck.

✳✳ WARNING

Never spill the engine coolant on to electronic equipment (alternator etc.).

a. Pour coolant slowly of less than 2⅛ (2 L) a minute to allow air in system to escape.

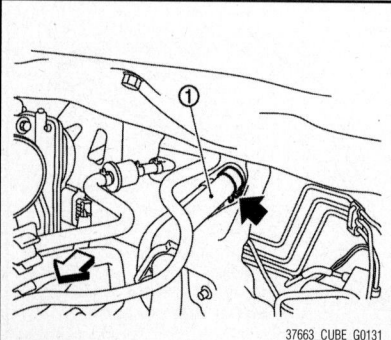

37663_CUBE_G0131

Fig. 31 Disconnect heater hose (1) at the firewall

b. When engine coolant overflows disconnected heater hose, connect the heater hose, and continue filling the engine coolant.

➡**Use Genuine NISSAN Long Life Antifreeze/Coolant or equivalent mixed with water (distilled or demineralized).**

7. Refill reservoir tank to "MAX" level line with engine coolant.
8. Install air duct (between air cleaner case and electric throttle control actuator).
9. Install radiator cap.
10. Warm up engine until opening thermostat. Standard for warming-up time is approximately 10 minutes at 3,000 rpm.

➡**Check thermostat opening condition by touching radiator hose (lower) to see a flow of warm water.**

✳✳ WARNING

Watch water temperature gauge so as not to overheat engine.

11. Stop the engine and cool down to less than approximately 122°F (50°C).
a. Cool down using fan to reduce the time.
b. If necessary, refill radiator up to filler neck with engine coolant.
12. Refill reservoir tank to "MAX" level line with engine coolant.
13. Repeat steps 5 through 10 two or more times with radiator cap installed until engine coolant level no longer drops.
14. Check cooling system for leakage with engine running.
15. Warm up the engine, and check for sound of engine coolant flow while running engine from idle up to 3,000 rpm with heater temperature controller set at several position between "COOL" and "WARM".

➡**Sound may be noticeable at heater unit.**

16. Repeat step 14 three times.
17. If sound is heard, bleed air from cooling system by repeating step 5 through 10 until reservoir tank level no longer drops.

ELECTRIC ENGINE FAN

REMOVAL & INSTALLATION

See Figures 32 and 33.

1. Drain engine coolant from radiator.

✳✳ WARNING

Perform this step engine is cold. Never spill engine coolant on drive belt.

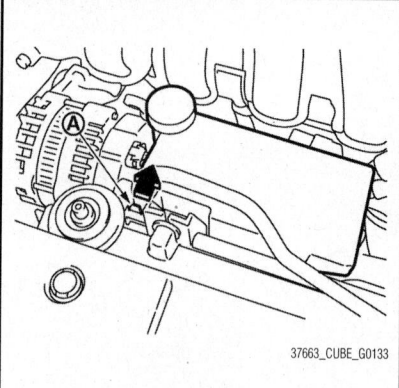

37663_CUBE_G0133

Fig. 32 Release the tab (A) in the direction shown by the arrow

2. Remove air duct (inlet) and resonator assembly.
3. Remove reservoir tank as follows:
a. Disconnect reservoir tank hose.
b. Release the tab in the direction shown by the arrow.
c. Lift up and remove the reservoir tank with the tab released.
4. Remove the upper radiator hose.
5. Disconnect harness connector from fan motor, and move harness to aside.
6. Remove cooling fan assembly.

✳✳ WARNING

Be careful not to damage or scratch on radiator core when removing.

To install:

7. Note the following, and install in the reverse order of removal.

✳✳ WARNING

Only use genuine parts for fan shroud mounting bolt and observe the specified torque (to prevent radiator from being damaged).

➡**Cooling fan is controlled by ECM.**

RADIATOR

REMOVAL & INSTALLATION

See Figures 32, 34 through 37.

✳✳ CAUTION

Never remove radiator cap when engine is hot. Serious burns may occur from high-pressure engine coolant escaping from radiator. Wrap a thick cloth around the radiator cap. Slowly turn it a quarter of a turn to release built-up pressure. Then turn it all the way.

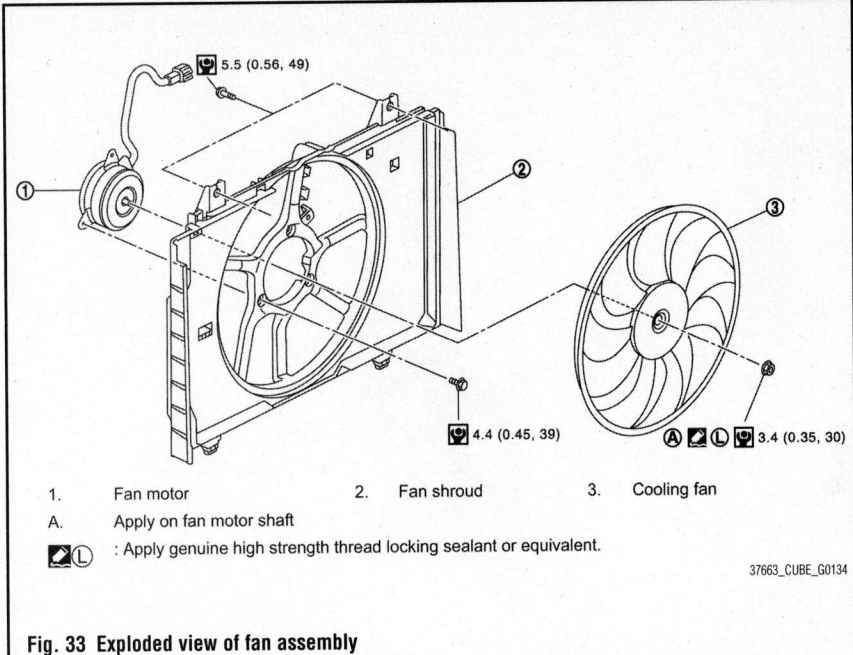

1. Fan motor   2. Fan shroud   3. Cooling fan
A. Apply on fan motor shaft
🔧Ⓛ : Apply genuine high strength thread locking sealant or equivalent.

37663_CUBE_G0134

Fig. 33 Exploded view of fan assembly

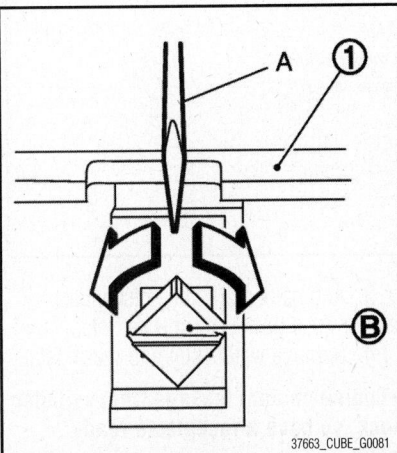

37663_CUBE_G0081

Fig. 34 Disengage front grille mounting clips (B) by rotating 45° using a flat-bladed screwdriver (A) through access hole of front grille (1) upper

1. Drain engine coolant from radiator.

※※ WARNING

Perform this step when the engine is cold. Never spill engine coolant on drive belt.

2. Remove air duct (inlet) and resonator assembly.
3. Remove reservoir tank as follows:
   a. Disconnect reservoir tank hose.
   b. Release the tab in the direction shown by the arrow.
   c. Lift up and remove the reservoir tank with tab released.
4. Remove radiator hose (upper and lower).

5. Disconnect harness connector from fan motor, and move harness aside.
6. Remove cooling fan assembly.

※※ WARNING

Be careful not to damage or scratch the radiator core.

7. Remove the front grille assembly.

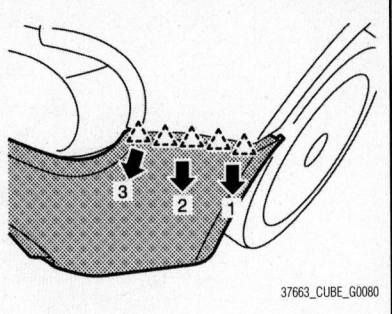

37663_CUBE_G0080

Fig. 35 Pull bumper fascia assembly side toward the vehicle side as shown by the arrows

   a. Remove front grille upper side fixing clips.
   b. Disengage front grille mounting clips by rotating 45°using a flat-bladed screwdriver through access hole of front grille upper while pulling front grill toward vehicle front.
8. Remove the front bumper fascia assembly.

※※ WARNING

Bumper fascia is made of resin. Never apply strong force to it, and be careful to prevent contact with oil.

   a. Fully open hood assembly.
   b. Remove bumper fascia upper fixing clips.

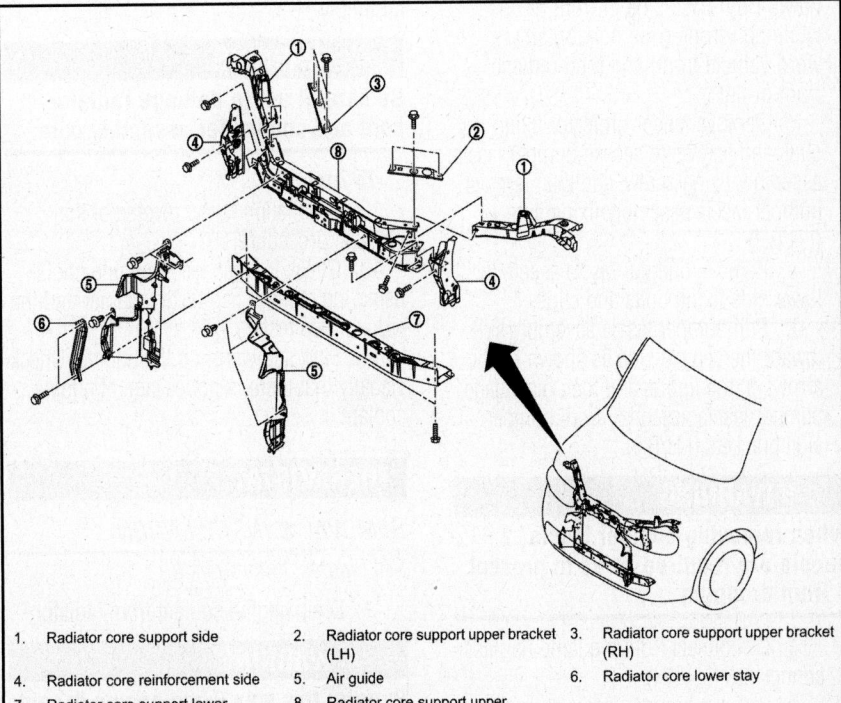

1.	Radiator core support side	2.	Radiator core support upper bracket (LH)	3.	Radiator core support upper bracket (RH)
4.	Radiator core reinforcement side	5.	Air guide	6.	Radiator core lower stay
7.	Radiator core support lower	8.	Radiator core support upper		

37663_CUBE_G0135

Fig. 36 Remove radiator core support (upper)

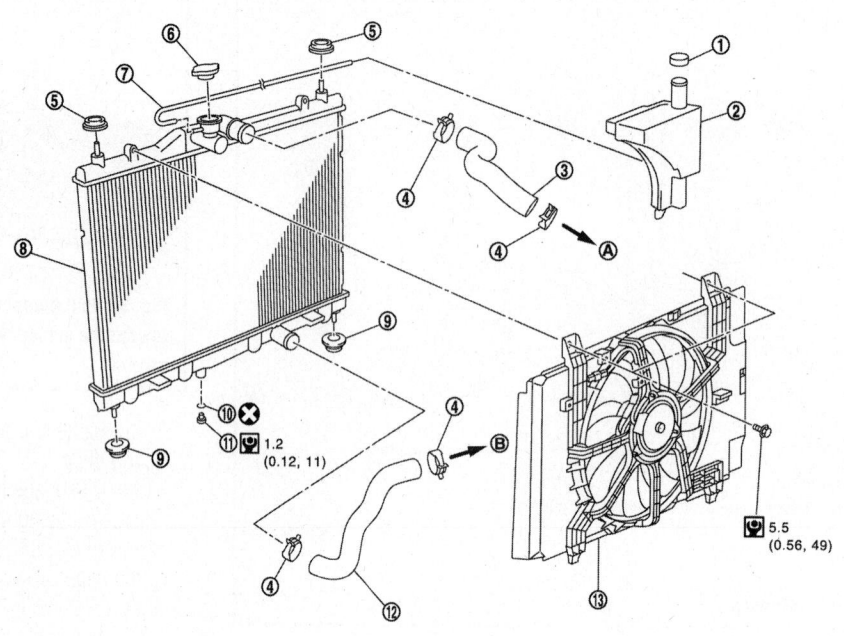

1. Reservoir tank cap
2. Reservoir tank
3. Radiator hose (upper)
4. Clamp
5. Mounting rubber (upper)
6. Radiator cap
7. Reservoir tank hose
8. Radiator
9. Mounting rubber (lower)
10. O-ring
11. Drain plug
12. Radiator hose (lower)
13. Cooling fan assembly
A. To water outlet
B. To water inlet

37663_CUBE_G0136

Fig. 37 Exploded view of radiator assembly

c. Disengage front grille fixing pawls from back side of front grille while pull front grille horizontally to word vehicle front, and then remove front grille.

d. Remove fender protector fixing clips and screws to access bumper fascia assembly fixing screw, and then remove bumper fascia assembly fixing screws (LH/RH).

e. Remove bumper fascia assembly lower side fixing bolts and clips.

f. Pull bumper fascia assembly side toward the vehicle side as shown by the arrows in the figure, and then disengage bumper fascia assembly from bumper side brackets (LH/RH).

❋❋ WARNING

When removing bumper fascia, 2 people are required so as to prevent it from dropping.

g. Disconnect front fog lamp harness connectors (LH/RH).

h. Remove bumper fascia assembly.

9. Remove the front combination lamp assembly (RH and LH).

10. Remove radiator core support (upper).

11. Pull up and remove the radiator assembly.

❋❋ WARNING

Be careful not to damage radiator core and condenser assembly core.

To install:

12. Installation is the reverse of the removal procedure.

13. Check for leakage of engine coolant using the radiator cap tester adapter and the radiator cap tester.

14. Start and warm up the engine. Check visually that there is no leakage of engine coolant.

THERMOSTAT

REMOVAL & INSTALLATION
See Figures 38 through 40.

1. Drain engine coolant from radiator.

❋❋ WARNING

Perform this step when engine is cold.

2. Remove air duct (inlet) and resonator assembly.

3. Add paint mark, then disconnect radiator hose (lower) from water inlet.

4. Remove water inlet and thermostat.

➡**Engine coolant leakage from cylinder block, so have a receptacle ready below.**

5. Remove thermostat housing with the following procedure:

a. Remove A/C compressor with A/C piping connected, and temporarily fasten it on vehicle with a rope.

b. Remove water pump.

c. Remove alternator.

To install:

6. Note the following, and install in the reverse order of removal.

7. Install thermostat with making rubber ring groove fit to thermostat flange with the whole circumference.

8. Install thermostat with jiggle valve facing upwards.

9. Check for leakage of engine coolant using the radiator cap tester adapter and the radiator cap tester.

10. Start and warm up the engine. Check visually that there is no leakage of engine coolant.

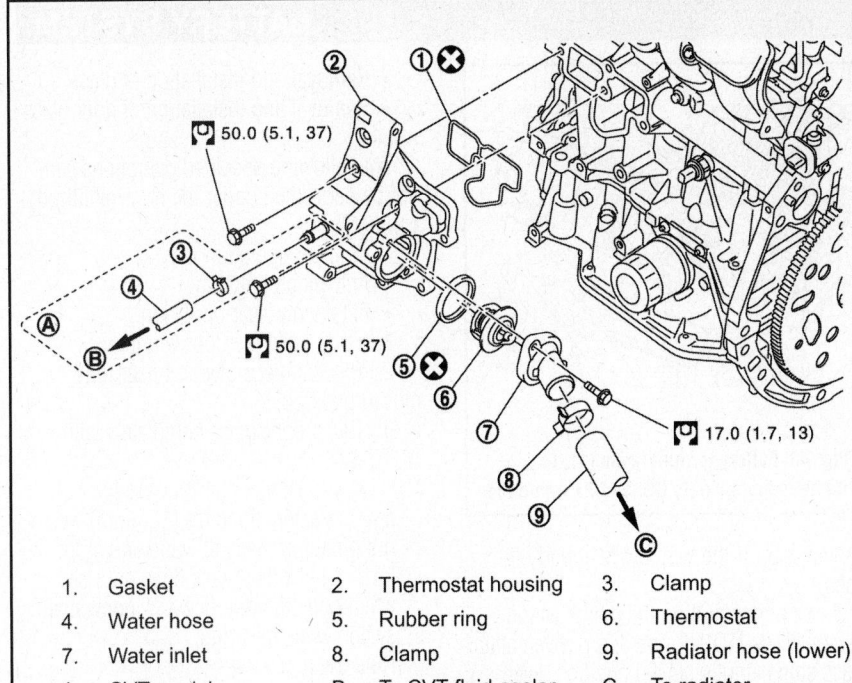

1. Gasket
2. Thermostat housing
3. Clamp
4. Water hose
5. Rubber ring
6. Thermostat
7. Water inlet
8. Clamp
9. Radiator hose (lower)
A. CVT models
B. To CVT fluid cooler
C. To radiator

37663_CUBE_G0137

Fig. 38 Exploded view of thermostat housing assembly

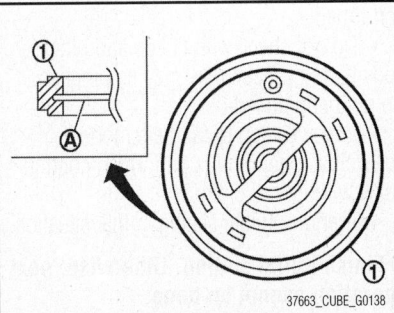

37663_CUBE_G0138

Fig. 39 Install thermostat with making rubber ring (1) groove fit to thermostat flange (A) with the whole circumference

WATER PUMP

REMOVAL & INSTALLATION

See Figures 41 and 42.

1. Drain engine coolant from radiator.

❋❋ WARNING

Perform this step when the engine is cold. Never spill engine coolant on drive belt.

2. Remove front fender protector (RH).
3. Remove drive belt.
4. Remove water pump.

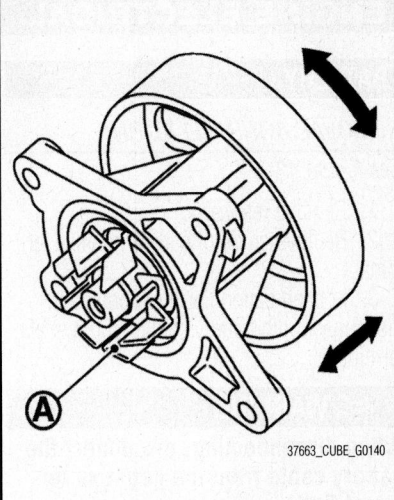

37663_CUBE_G0140

Fig. 41 Check visually that there is no significant dirt or rusting on water pump body and vane (A)

➡Engine coolant leakage from cylinder block, so have a receptacle ready below.

❋❋ WARNING

Handle water pump vane so that it does not contact any other parts. Water pump cannot be disassembled and should be replaced as a unit.

5. Check visually that there is no significant dirt or rusting on water pump body and vane.
6. Check that there is no looseness in vane shaft, and that it turns smoothly when rotated by hand.
7. Replace water pump, if necessary.

To install:

8. Installation is the reverse of the removal procedure.

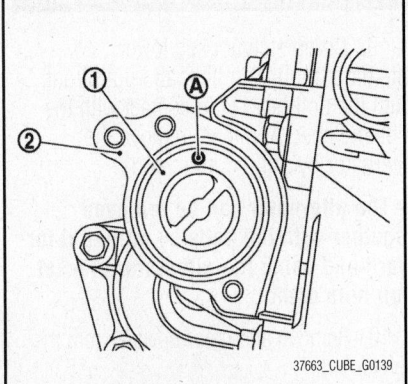

37663_CUBE_G0139

Fig. 40 Install thermostat (1) with jiggle valve (A) facing upwards; thermostat housing (2)

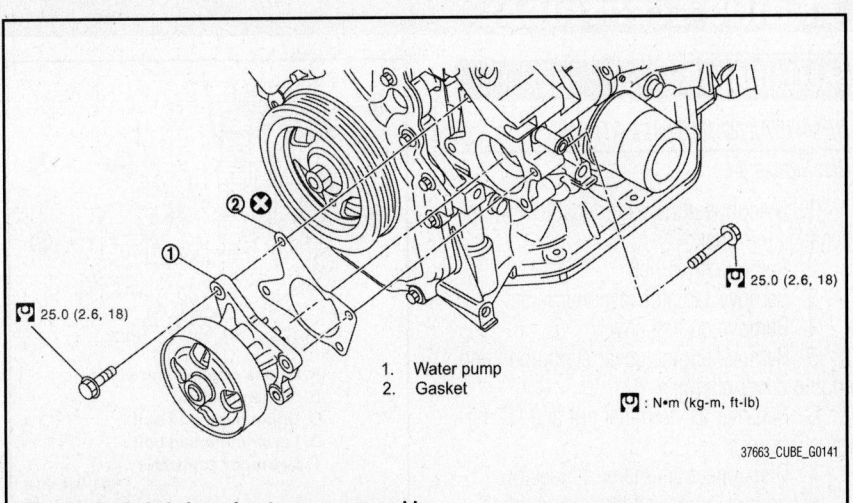

1. Water pump
2. Gasket

: N•m (kg-m, ft-lb)

37663_CUBE_G0141

Fig. 42 Exploded view of water pump assembly

ENGINE ELECTRICAL

BATTERY SYSTEM

BATTERY

REMOVAL & INSTALLATION

See Figure 43.

1. Remove resonator.
2. Remove cover of battery positive terminal.
3. Loosen battery terminal nuts, and disconnect both battery cables from battery terminals.

✳✳ WARNING

When disconnecting, disconnect the battery cable from the negative terminal first.

4. Remove battery fix frame mounting nuts and battery fix frame.
5. Remove battery.

To install:

6. Installation is the reverse of the removal procedure.

✳✳ WARNING

When connecting, connect the battery cable to the positive terminal first.

7. Reset electronic systems as necessary.

BATTERY RECONNECT/ RELEARN PROCEDURE

Initialization Procedure

1. Disconnect battery minus terminal or power window main switch connector. Reconnect it after a minute or more.
2. Turn ignition switch ON.
3. Operate power window switch to fully open the window. (This operation is

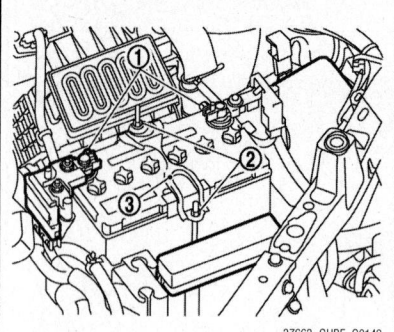

Fig. 43 Battery terminal nuts (1), fix frame mounting nuts (2), and fix frame (3)

37663_CUBE_G0142

unnecessary if the window is already fully open.)

4. Continue pulling the power window switch UP (AUTO-UP operation). Even after glass stops at fully closed position, keep pulling the switch for 2 seconds or more.
5. Initializing procedure is completely.
6. Inspect anti-pinch function.

Check Anti-Pinch Function

If any of the following work has been done Initial setting is necessary.
- Power supply to the power window main switch or power window motor is cut off by the removal of battery terminal or the battery fuse is blown.
- Disconnection and connection of power window main switch harness connector.
- Removal and installation of motor from regulator assembly.
- Operation of regulator assembly as an independent unit.

- Removal and installation of glass.
- Removal and installation of door glass run.

The following specified operations cannot be performed under the non-initialized condition:
- Auto-up operation
- Anti-pinch function
- Retained power operation
1. Fully open the door window.
2. Place a piece of wood near fully closed position.
3. Close door glass completely with AUTO-UP.
   a. Check that glass lowers for approximately 6 inches (150 mm) without pinching piece of wood and stops.
   b. Check that glass does not rise when operating the power window main switch while lowering.
Note the following:
- Perform initial setting when auto-up operation or anti-pinch function does not operate normally.
- Check that AUTO-UP operates before inspection when system initialization is performed.
- Do not check with hands and other body parts because they may be pinched. Do not get pinched.
- It may switch to fail-safe mode if open/close operation is performed continuously without full close.
4. Perform initial setting in that situation.

➡**Finish initial setting. Otherwise, next operation cannot be done;**

- Auto-up operation
- Anti-pinch function
- Retained power operation

ENGINE ELECTRICAL

CHARGING SYSTEM

ALTERNATOR

REMOVAL & INSTALLATION

See Figure 44.

1. Disconnect the battery cable from the negative terminal.
2. Remove drive belt.
3. Remove radiator reservoir tank.
4. Remove engine cover.
5. Remove engine cover clamp bolt and engine cover clamp.
6. Remove "B" terminal nut and "B" terminal harness.
7. Disconnect alternator connector.
8. Remove upper alternator mounting bolt.

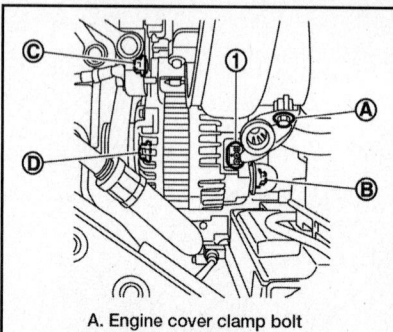

A. Engine cover clamp bolt
B. "B" terminal nut
C. Upper mounting bolt
D. Lower mounting bolt
1. Alternator connector

37663_CUBE_G0143

Fig. 44 Removing the alternator

9. Completely loosen lower alternator mounting bolt, and pull it out until the bolt head is in contact with the side member. And then, remove the alternator by pulling it forward.

➡**The alternator can be removed together with the bolts by pulling it forward and using the alternator bracket bolt hole cutout.**

10. Remove alternator upward from the vehicle.

To install:

11. Installation is the reverse of the removal procedure.

✳✳ WARNING
Be sure to tighten "B" terminal nut carefully.

12. Temporarily tighten the alternator bolts in order from the lower to the upper, and then tighten them in order from the upper to the lower.

✳✳ WARNING
For the alternator, the front side (pulley side) surface is the reference surface. Fit the reference surface to the alternator mounting part, and then tighten the bolts.

13. Check tension of the accessory drive belt.

ENGINE ELECTRICAL

FIRING ORDERS

The firing order for the MR18DE engine is 1–3–4–2.

IGNITION COIL

REMOVAL & INSTALLATION
See Figure 45.

1. Remove intake manifold.
2. Remove ignition coil.

✳✳ WARNING
Never drop or shock ignition coil.
Never disassemble ignition coil.

3. Installation is the reverse of removal.

IGNITION TIMING

INSPECTION

The ignition timing is $13° +/- 5°$BTDC at idle speed.

ADJUSTMENT

The ignition timing is controlled by the ECM. No adjustment is necessary or possible.

IGNITION SYSTEM

SPARK PLUGS

REMOVAL & INSTALLATION
See Figure 45.

1. Remove ignition coil.
2. Remove spark plug with a spark plug wrench.

✳✳ WARNING
Never drop or shock spark plug.

To install:
3. Installation is the reverse of the removal procedure.

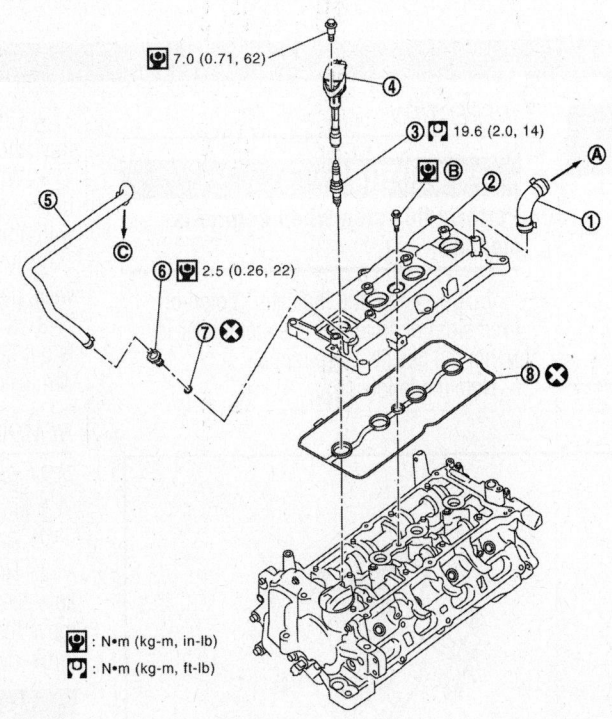

☑ 7.0 (0.71, 62)

④

③☑ 19.6 (2.0, 14)

☑Ⓑ ②

Ⓐ

①

⑤

Ⓒ

⑥☑ 2.5 (0.26, 22)

⑦✗

⑧✗

☑ : N•m (kg-m, in-lb)
☑ : N•m (kg-m, ft-lb)

1. PCV hose	2. Rocker cover	3. Spark plug
4. Ignition coil	5. PCV hose	6. PCV valve
7. O-ring	8. Rocker cover gasket	
	Tightening must be done following	
A. To air duct assembly	B. the installation procedure.	C. To intake manifold

37663_CUBE_G0145

Fig. 45 Exploded view of ignition coil, spark plug, and rocker cover assembly

STARTER

REMOVAL & INSTALLATION

See Figure 46.

1. Disconnect the battery cable from the negative terminal.
2. Remove air duct (inlet).
3. Remove radiator reservoir tank.
4. Disconnect oil pressure switch connector.
5. Remove "B" terminal nut and "B" terminal harness.
6. Remove "S" terminal nut and "S" terminal harness.
7. Remove starter motor mounting bolts.
8. Remove starter motor upward from the vehicle.

To install:

9. Installation is the reverse of the removal procedure.

✳✳ WARNING

Be sure to tighten "B" terminal nut carefully.

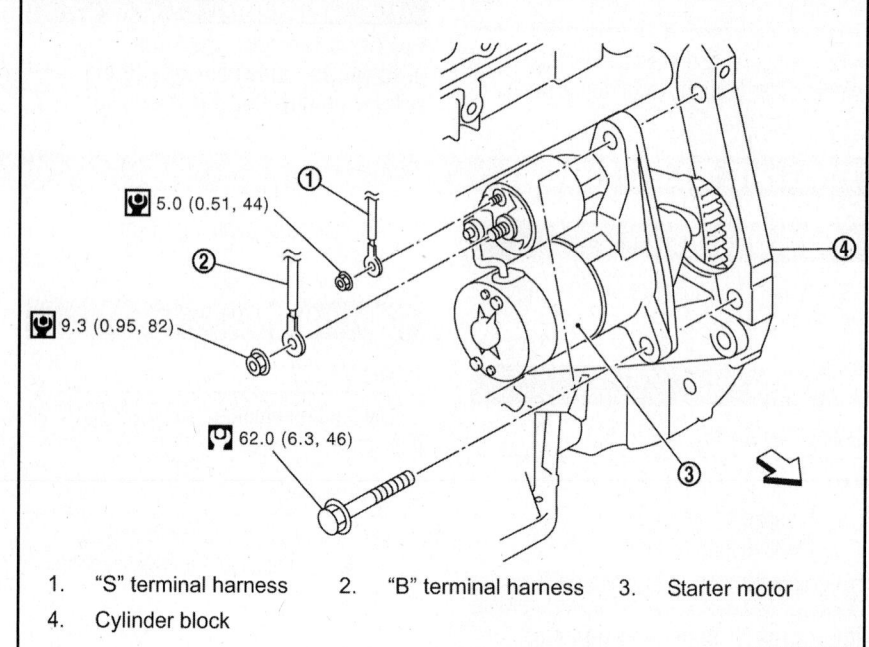

1. "S" terminal harness   2. "B" terminal harness   3. Starter motor
4. Cylinder block

37663_CUBE_G0146

Fig. 46 Exploded view of starter removal

ENGINE MECHANICAL

ACCESSORY DRIVE BELT SYSTEM

ADJUSTMENT

Belt tension is not necessary, as it is automatically adjusted by drive belt auto-tensioner.

BELT ROUTINGS

See Figure 47.

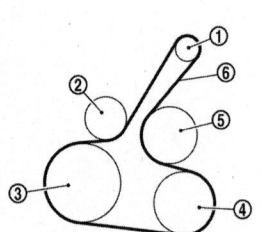

1. Alternator
4. A/C compressor (models with A/C) Idler pulley (models without A/C)
A. Possible use range

2. Drive belt auto-tensioner
5. Water pump
B. Range when new drive belt is installed

3. Crankshaft pulley
6. Drive belt
C. Indicator

37663_CUBE_G0148

Fig. 47 Accessory drive belt routing

INSPECTION

✳✳ CAUTION

Perform this step when engine is stopped.

1. Check that the indicator (notch on fixed side) of drive belt auto-tensioner is within the possible use range.
   Note the following:

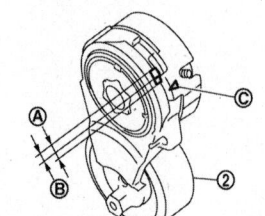

- Check the drive belt auto-tensioner indication when the engine is cold.
- When new drive belt is installed, the indicator (notch on fixed side) should be within the range.

2. Visually check entire drive belt for wear, damage or cracks.
3. If the indicator (notch on fixed side) is out of the possible use range or belt is damaged, replace drive belt.

REMOVAL & INSTALLATION

See Figure 48.

1. Remove front wheel and tire (RH).
2. Remove front fender protector (RH).
3. Hold the hexagonal part of drive belt auto-tensioner with a wrench securely. Then move the wrench handle in the direction of arrow (loosening direction of tensioner).

✳✳ WARNING

Avoid placing hand in a location where pinching may occur if the holding tool accidentally comes off.

4. Insert a rod approximately 0.24 inches (6 mm) in diameter such as short-length screwdriver into the hole of the retaining boss to fix drive belt auto-tensioner.

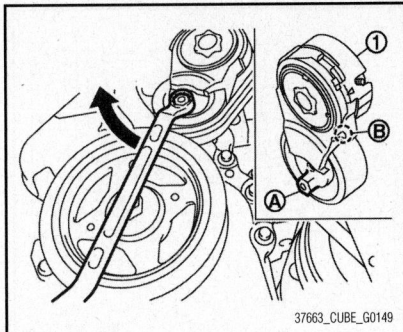

Fig. 48 Hold the hexagonal part (A) of drive belt auto-tensioner (1) with a wrench securely; then move the wrench handle in the direction of arrow; insert a rod into hole (B)

➡**Keep drive belt auto-tensioner pulley arm locked after drive belt is removed.**

5. Remove drive belt.

To install:

6. Install drive belt.

✳✳ WARNING

Confirm drive belt is completely set to pulleys. Check for engine oil, working fluid and engine coolant are not adhered to drive belt and each pulley groove.

7. Release drive belt auto-tensioner, and apply tension to drive belt.

8. Turn crankshaft pulley clockwise several times to equalize tension between each pulley.

9. Confirm tension of drive belt at indicator (notch on fixed side) is within the possible use range.

AIR CLEANER

REMOVAL & INSTALLATION

Air Cleaner Assembly

See Figure 49.

1. Remove air duct (inlet) and resonator assembly.

2. Remove engine cover.

3. Remove the air cleaner filter from the air cleaner case.

4. Remove air duct [between air duct (inlet) and air cleaner case] from the air cleaner case.

5. Disconnect PCV hose.

6. Remove the air duct (between air cleaner case and electric throttle control actuator).

➡**Add matching marks if necessary for easier installation.**

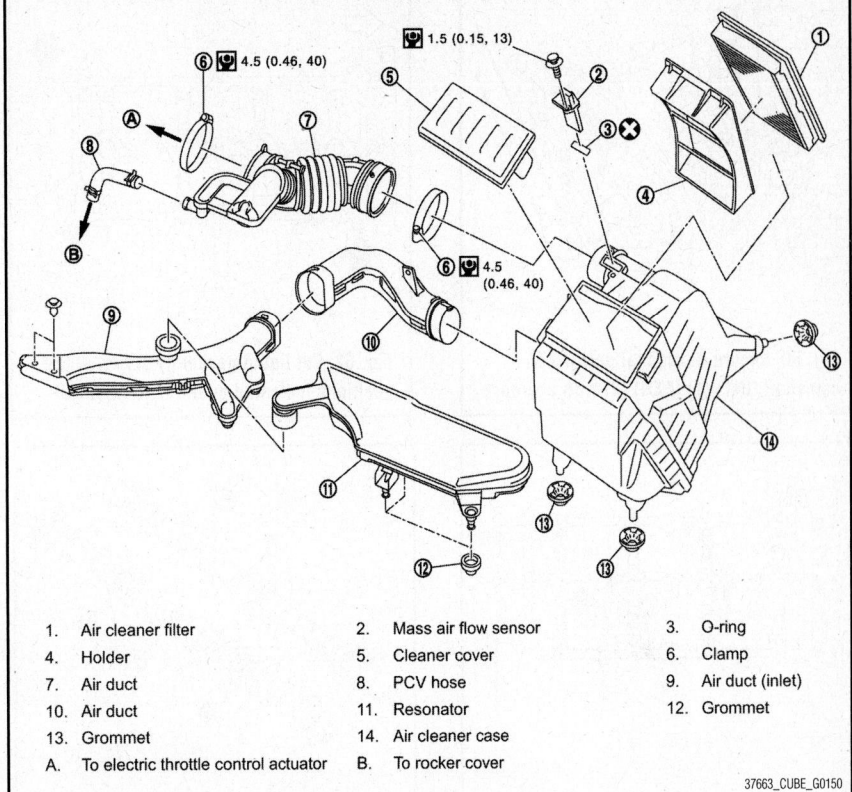

1. Air cleaner filter	2. Mass air flow sensor	3. O-ring
4. Holder	5. Cleaner cover	6. Clamp
7. Air duct	8. PCV hose	9. Air duct (inlet)
10. Air duct	11. Resonator	12. Grommet
13. Grommet	14. Air cleaner case	
A. To electric throttle control actuator	B. To rocker cover	

37663_CUBE_G0150

Fig. 49 Exploded view of air cleaner and air duct assembly

7. Remove air cleaner case with the following procedure.

    a. Remove battery.

    b. Disconnect Mass Air Flow (MAF) sensor harness connector.

    c. Remove the air cleaner case.

8. Remove Mass Air Flow (MAF) sensor from air cleaner case, if necessary.

✳✳ WARNING

Handle the Mass Air Flow (MAF) sensor with following cares:

- Never shock the Mass Air Flow (MAF) sensor.
- Never disassemble the Mass Air Flow (MAF) sensor.
- Never touch the sensor of the Mass Air Flow (MAF) sensor.

To install:

9. Note the following, and install in the reverse order of removal.

10. Align marks. Attach each joint. Screw clamps firmly.

Air Filter Element

1. Push in the tabs at both ends of the air cleaner cover.

2. Pull up the air cleaner cover and remove it.

3. Remove the air cleaner filter and holder assembly from the air cleaner case.

4. Remove the air cleaner filter from the holder.

To install:

5. Installation is the reverse of the removal procedure.

CAMSHAFT, SPROCKET, BEARINGS & VALVE LIFTERS

REMOVAL & INSTALLATION

See Figures 50 through 56.

✳✳ WARNING

The rotating direction in the text indicates all directions seen from the engine front.

1. Remove the following parts:
- Intake manifold
- Rocker cover
- Front cover and timing chain related parts

➡**Removal of oil pump drive related part is not necessary.**

2. Remove Camshaft Position (CMP) sensor (PHASE) from camshaft bracket. Note the following:

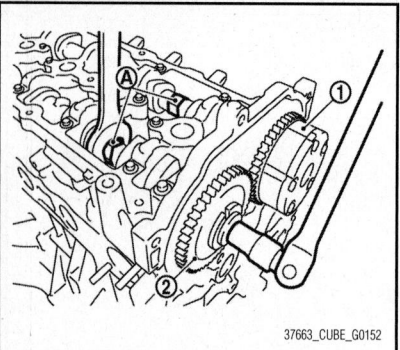

Fig. 50 Secure hexagonal part (A) of camshaft (INT) (1) (EXH) (2) with a wrench

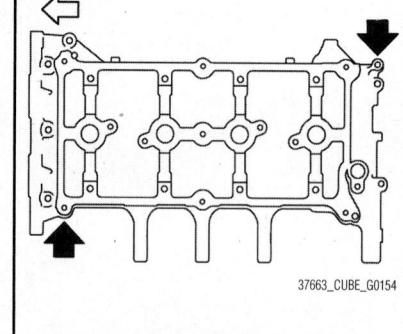

Fig. 52 Cut liquid gasket by prying at the positions indicated by the black arrows

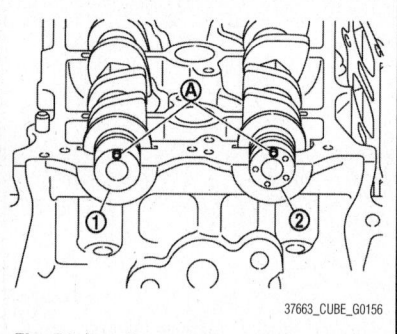

Fig. 54 Install camshafts so that camshaft dowel pins (A) on the front side are positioned as shown

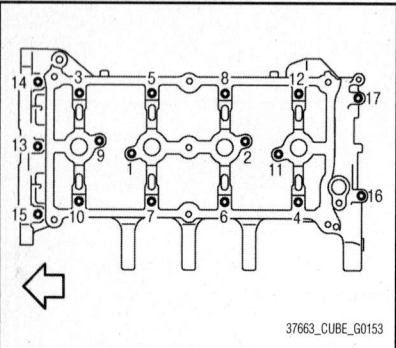

Fig. 51 Loosen mounting bolts in reverse order as shown

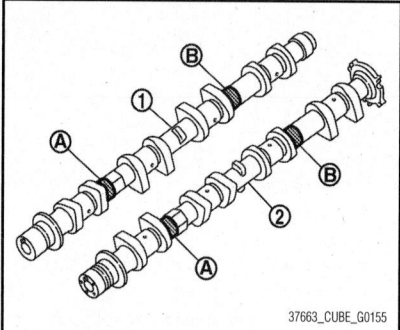

Fig. 53 Identifying intake and exhaust camshafts

- Handle Camshaft Position (CMP) sensor (PHASE) carefully and avoid impacts.
- Never disassemble Camshaft Position (CMP) sensor (PHASE).
- Never place sensor where it is exposed to magnetism.

3. Put a matching mark on the camshaft sprocket (INT) and the camshaft bracket.

➡️It prevents the knock pin of the camshaft (INT) from engaging with the incorrect pin hole when installing the camshaft sprocket (INT).

4. Remove camshaft sprockets.
  a. Secure hexagonal part of camshaft with a wrench.
  b. Loosen camshaft sprocket mounting bolts and remove camshaft sprocket.

⁕⁕ WARNING

Never rotate crankshaft or camshaft while timing chain is removed. It causes interference between valve and piston.

5. Remove camshaft bracket with the following procedure:
  a. Loosen mounting bolts in reverse order as shown.
  b. Cut liquid gasket by prying at the

positions indicated by the black arrows, and then remove the camshaft bracket.
6. Remove camshafts.
7. Remove valve lifters.

⁕⁕ WARNING

Identify installation positions, and store them in order.

8. Remove signal plate from camshaft (INT), if necessary.

To install:
9. Install valve lifters.

⁕⁕ WARNING

Install them in their original positions.

10. Install camshafts.
  a. Clean camshaft journal to remove any foreign material.
  b. Distinguish between the intake and the exhaust by looking at the different shapes of the front and rear ends of the camshaft or using the identification colors.
  - Camshaft (INT) A: Yellow
  - Camshaft (EXH) B: Yellow
  c. Install camshafts so that camshaft dowel pins on the front side are positioned as shown.

11. Install camshaft bracket with the following procedure:
  a. Remove foreign material completely from camshaft bracket backside and from cylinder head installation face.
  b. Apply liquid gasket to camshaft bracket as shown.
  c. Tighten mounting bolts of camshaft brackets in three steps, in numerical order as shown.
  - Step 1: 17 inch lbs. (2 Nm)
  - Step 2: 52 inch lbs. (6 Nm)
  - Step 3: 84 inch lbs. (9.5 Nm)

➡️There are two types of mounting bolts. All bolt thread lengths, except numbers 13, 14, and 15, are 1.378 inches (35 mm) long. Bolt numbers 13, 14, and 15 are 2.264 inches (57.5 mm) long.

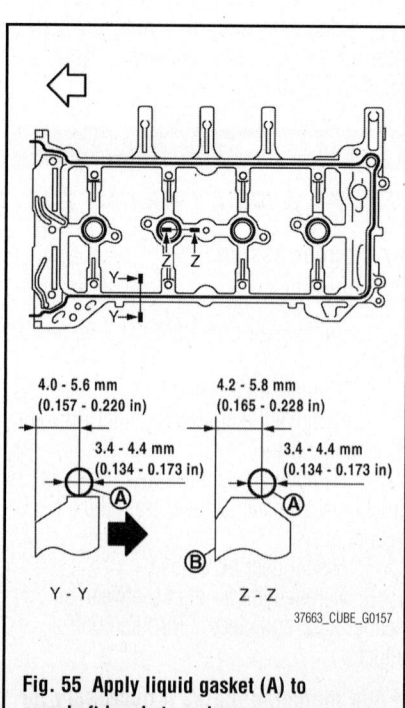

Fig. 55 Apply liquid gasket (A) to camshaft bracket as shown

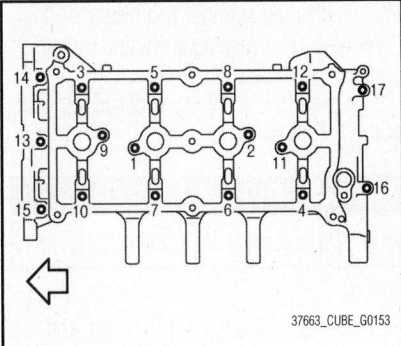

Fig. 56 Tighten mounting bolts in numerical order as shown

✳✳ WARNING

After tightening mounting bolts of camshaft brackets, be sure to wipe off excessive liquid gasket from the mating surface of cylinder head.

12. Install the camshaft sprocket (INT) to the camshaft (INT) with the following procedure.

a. Refer to the paint mark made during removal. Securely align the knock pin and the pin hole, and then install them.

b. Hold the hexagonal part of camshaft (INT) using wrench to tighten mounting bolt.

c. Tighten camshaft (INT) mounting bolt to 26 ft. lbs. (35 Nm).

d. Tighten an additional 67 degrees clockwise (angle tightening).

✳✳ WARNING

Check the tightening angle by using an angle wrench KV10112100 (BT8653-A) or protractor. Never judge by visual inspection without an angle wrench.

13. Install camshaft sprocket (EXH).

➡Secure the hexagonal part of camshaft (EXH) using wrench to tighten mounting bolt to 65 ft. lbs. (88 Nm).

14. Install timing chain and related parts.
15. Inspect and adjust valve clearance.
16. Install remaining parts in the reverse order of removal.

CRANKSHAFT FRONT SEAL

REMOVAL & INSTALLATION

See Figure 57.

1. Remove the following parts:
- Front fender protector (RH)
- Drive belt
- Crankshaft pulley

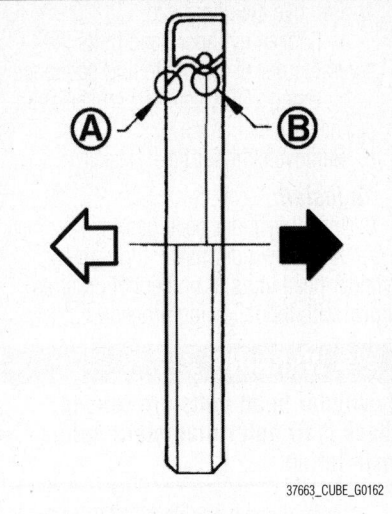

Fig. 57 Install front oil seal so that each seal lip is oriented as shown

2. Remove front oil seal with a suitable tool.

✳✳ WARNING

Be careful not to damage front cover and crankshaft.

To install:

3. Apply new engine oil to new front oil seal joint surface and seal lip.

4. Install front oil seal so that each lip is oriented as shown.

5. Press-fit front oil seal using a suitable drift with outer diameter 2.24 inches (57 mm) and inner diameter 1.77 inches (45 mm).

✳✳ WARNING

Press-fit oil seal straight to avoid causing burrs or tilting.

6. Install in the reverse order of removal, for the rest of parts.

CRANKSHAFT PULLEY

REMOVAL & INSTALLATION

See Figures 58 through 60.

1. Remove crankshaft pulley with the following procedure:

a. Fix crankshaft pulley with a pulley holder, loosen crankshaft pulley bolt, and locate bolt seating surface at 0.39 inches (10 mm) from its original position.

✳✳ WARNING

Never remove the crankshaft pulley bolt as they will be used as a supporting point for the pulley remover.

Fig. 58 Fix crankshaft pulley (1) with a pulley holder (A)

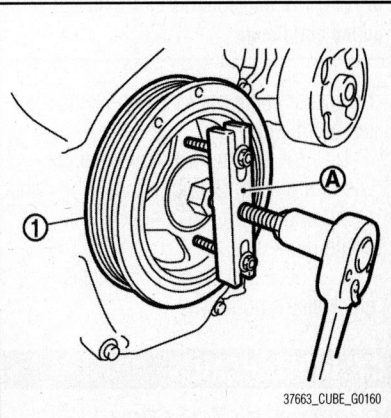

Fig. 59 Attach a pulley puller KV11103000 in the M6 thread hole on crankshaft pulley (1)

b. Attach a pulley puller KV11103000 in the M6 thread hole on crankshaft pulley and remove crankshaft pulley.

To install:

2. Install crankshaft pulley with the following procedure:

a. When inserting crankshaft pulley with a plastic hammer, tap on its center portion (not circumference).

✳✳ WARNING

Never damage front oil seal lip section.

b. Secure crankshaft pulley with a pulley holder.

c. Apply new engine oil to thread and seat surfaces of crankshaft pulley bolt.

d. Tighten crankshaft pulley bolt to 51 ft. lbs. (69 Nm).

e. Completely loosen the bolt.

f. Tighten crankshaft pulley bolt 22 ft. lbs. (29 Nm).

g. Put a paint mark on crankshaft pulley, matching with any one of six easy to

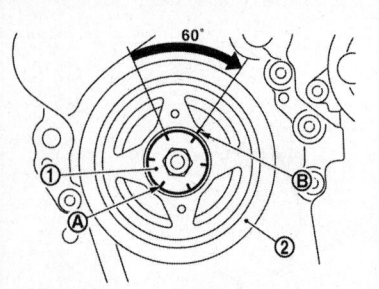

A. Angle marks
B. Paint mark
1. Crankshaft pulley bolt flange
2. Crankshaft pulley

37663_CUBE_G0161

Fig. 60 Put a paint mark on crankshaft pulley, matching with any one of six easy to recognize angle marks on crankshaft pulley bolt flange

recognize angle marks on crankshaft pulley bolt flange.

h. Turn another 60 degrees clockwise (angle tightening). Check the tightening angle with movement of one angle mark.

i. Check that crankshaft rotates clockwise smoothly.

CYLINDER HEAD

REMOVAL & INSTALLATION

See Figures 61 and 62.

1. Release fuel pressure.
2. Drain engine coolant and engine oil.
3. Remove the following components and related parts:
   - Exhaust manifold
   - Intake manifold
   - Fuel injector and fuel tube assembly
   - Water outlet
   - Rocker cover
   - Front cover, timing chain
   - Camshaft

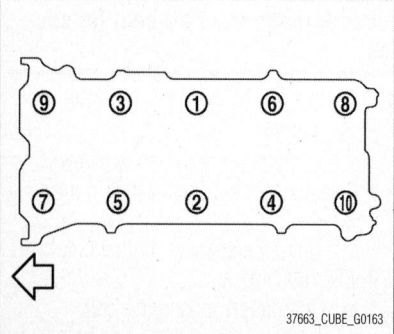

37663_CUBE_G0163

Fig. 61 Tighten cylinder head bolts in numerical order as shown

4. Remove cylinder head.
   a. Loosen cylinder head bolts in the reverse order of the tightening sequence.
   b. Using TORX® socket, loosen cylinder head bolts.
5. Remove cylinder head gasket.

To install:

6. Install cylinder head gasket.
7. Install cylinder head, and tighten cylinder head bolts in numerical order as shown with the following procedure.

✳✳ WARNING

If cylinder head bolts are reused, check their outer diameters before installation.

   a. Apply new engine oil to threads and seating surface of mounting bolts.
   b. Tighten all cylinder head bolts to 30 ft. lbs. (40 Nm).
   c. Turn all cylinder head bolts 100 degrees clockwise (angle tightening).

✳✳ WARNING

Check and confirm the tightening angle by using an angle wrench KV10112100 (BT8653-A) or protractor. Never judge by visual inspection without the tool.

   d. Completely loosen.

✳✳ WARNING

In this step, loosen cylinder head bolts in reverse order of that indicated in the figure.

   e. Tighten all cylinder head bolts to 30 ft. lbs. (40 Nm).
   f. Turn all cylinder head bolts 100 degrees clockwise (angle tightening).

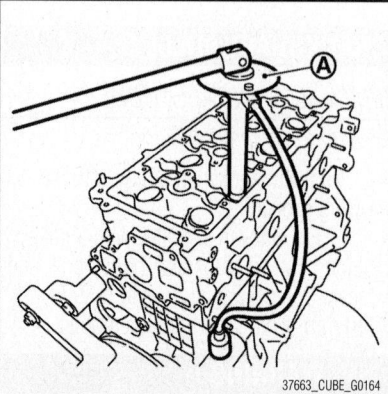

37663_CUBE_G0164

Fig. 62 Check and confirm the tightening angle by using an angle wrench (A) or protractor

   g. Turn all cylinder head bolts 100 degrees clockwise again (angle tightening).
8. Install in the reverse order of removal, for the rest of parts.

ENGINE COVER

REMOVAL & INSTALLATION
See Figure 63.

Refer to the illustration to remove and install the engine cover.

ENGINE OIL & FILTER

REMOVAL & INSTALLATION

Draining

Note the following:
- Be careful not to get burned, as engine oil may be hot.
- Prolonged and repeated contact with used engine oil may cause skin cancer. Try to avoid direct skin contact with used engine oil. If skin contact is made, wash thoroughly with soap or hand cleaner as soon as possible.

1. Warm up the engine, and check for engine oil leakage from engine components.
2. Stop the engine and wait for 10 minutes.
3. Loosen oil filler cap.
4. Remove drain plug and then drain engine oil.

Refilling

1. Install drain plug with new drain plug washer.

✳✳ WARNING

Be sure to clean drain plug and install with new drain plug washer.

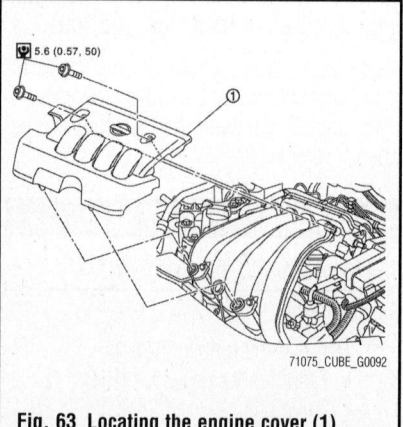

71075_CUBE_G0092

Fig. 63 Locating the engine cover (1)

2. Refill with new engine oil.

3. Warm up engine and check area around drain plug and oil filter for engine oil leakage.

4. Stop engine and wait for 10 minutes.

5. Check the engine oil level.

Oil Filter

See Figure 64.

1. Remove engine under cover.

2. Using oil filter wrench KV10115801 (J-38956), remove oil filter. Note the following:

• Oil filter is provided with relief valve. Use genuine NISSAN oil filter or equivalent.

• Be careful not to get burned when engine and engine oil may be hot.

• When removing, prepare a shop cloth to absorb any engine oil leakage or spillage.

• Completely wipe off any engine oil that adheres to engine and vehicle.

To install:

3. Remove foreign materials adhering to the oil filter installation surface.

4. Apply new engine oil to the oil seal contact surface of new oil filter.

5. Screw oil filter manually until it touches the installation surface, then tighten it by ⅔ turn (A). Or tighten to 13 ft. lbs. (18 Nm).

6. Check the engine oil level.

7. Start the engine, and check that there is no leakage of engine oil.

8. Stop the engine and wait for 10 minutes.

9. Check the engine oil level, and adjust the level.

EXHAUST MANIFOLD

REMOVAL & INSTALLATION

See Figures 65 through 67.

1. Remove exhaust front tube.

2. Remove exhaust manifold cover.

3. Remove the air fuel ratio sensor 1.

   a. Using heated oxygen sensor wrench KV10117100 (J-3647-A), remove air fuel ratio sensor 1.

✳✳ WARNING

Handle air fuel ratio sensor 1 carefully and avoid impacts.

➡**The exhaust manifold can be removed and installed without removing the air fuel ratio sensor 1 (Disassembly of harness connector is necessary).**

4. Remove exhaust manifold stay.

5. Remove exhaust manifold.

6. Loosen nuts in reverse order of the tightening sequence.

➡**Disregard Nos. 6 to 8 when loosening.**

7. Remove gasket.

✳✳ WARNING

Cover engine openings to avoid entry of foreign materials.

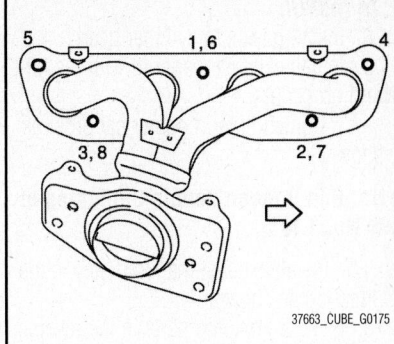

37663_CUBE_G0175

Fig. 66 Tighten nuts in numerical order as shown

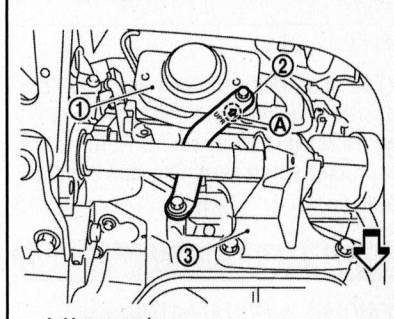

A. Upper mark
1. Exhaust manifold
2. Exhaust manifold stay
3. Drive shaft support bearing bracket

37663_CUBE_G0176

Fig. 67 Install exhaust manifold stay in the direction as shown

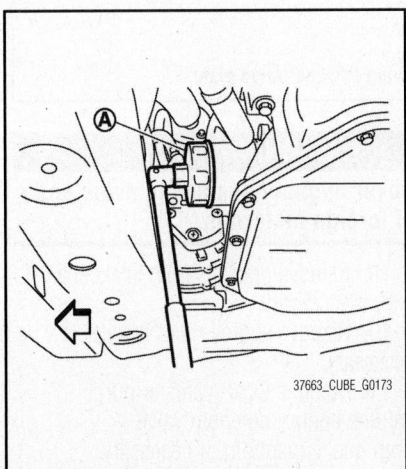

37663_CUBE_G0173

Fig. 64 Using oil filter wrench (A), remove oil filter

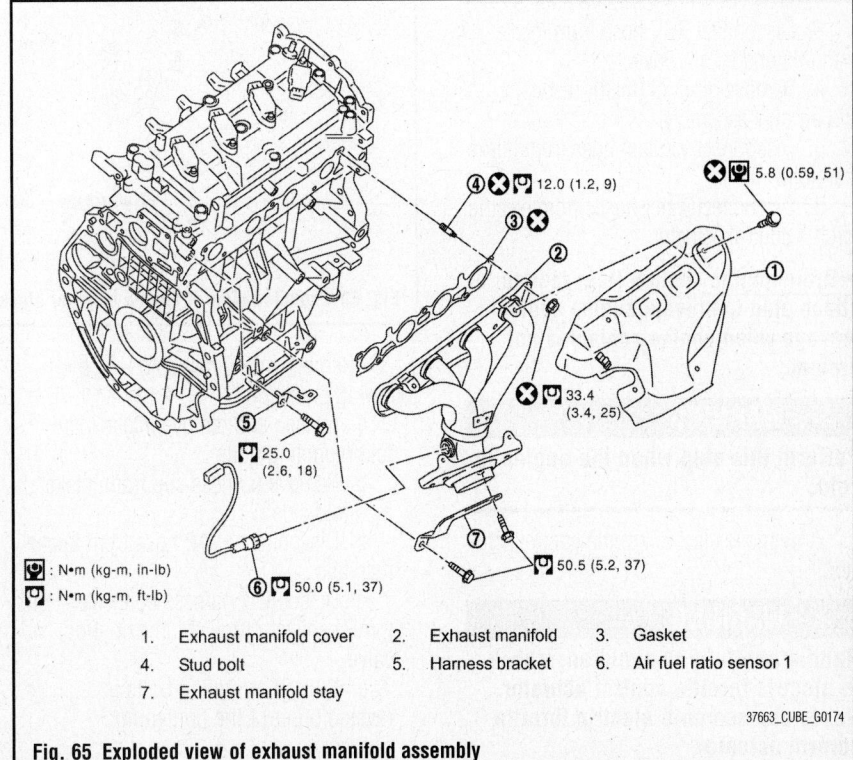

: N•m (kg-m, in-lb)
: N•m (kg-m, ft-lb)

④ ❌ 🔧 12.0 (1.2, 9)
③ ❌
②
① ❌ 🔧 5.8 (0.59, 51)
❌ 🔧 33.4 (3.4, 25)
⑤ 🔧 25.0 (2.6, 18)
⑦ 🔧 50.5 (5.2, 37)
⑥ 🔧 50.0 (5.1, 37)

1. Exhaust manifold cover	2. Exhaust manifold	3. Gasket
4. Stud bolt	5. Harness bracket	6. Air fuel ratio sensor 1
7. Exhaust manifold stay		

37663_CUBE_G0174

Fig. 65 Exploded view of exhaust manifold assembly

To install:

8. Install gasket to cylinder head.

9. Install exhaust manifold with the following procedure:

a. Tighten nuts in numerical order as shown.

➡ **No. 6 to 8 mean double tightening of nuts No. 1 to 3.**

b. Install exhaust manifold stay in the direction as shown.

10. Install remaining parts in the reverse order of removal.

FLEXPLATE (DRIVEPLATE)

REMOVAL & INSTALLATION

See Figure 68.

➡ **The manufacturer does not provide a specific Removal and Installation procedure for this component. Refer to the graphic(s) when servicing this component.**

INTAKE MANIFOLD

REMOVAL & INSTALLATION

See Figures 69 through 71.

1. Remove engine cover.
2. Pull out oil level gauge.

✳ WARNING

Cover the oil level gauge guide openings to avoid entry of foreign materials.

3. Disconnect PCV hose from intake manifold and rocker cover.

4. Remove air duct (inlet), resonator and air duct assembly.

5. Disconnect vacuum hose from intake manifold.

6. Disconnect water hoses from electric throttle control actuator.

➡ **Drain engine coolant from radiator or attach plug to prevent engine coolant leakage when engine coolant is not drained.**

✳ WARNING

Perform this step when the engine is cold.

7. Remove electric throttle control actuator.

✳ WARNING

Handle carefully to avoid any shock to electric throttle control actuator. Never disassemble electric throttle control actuator.

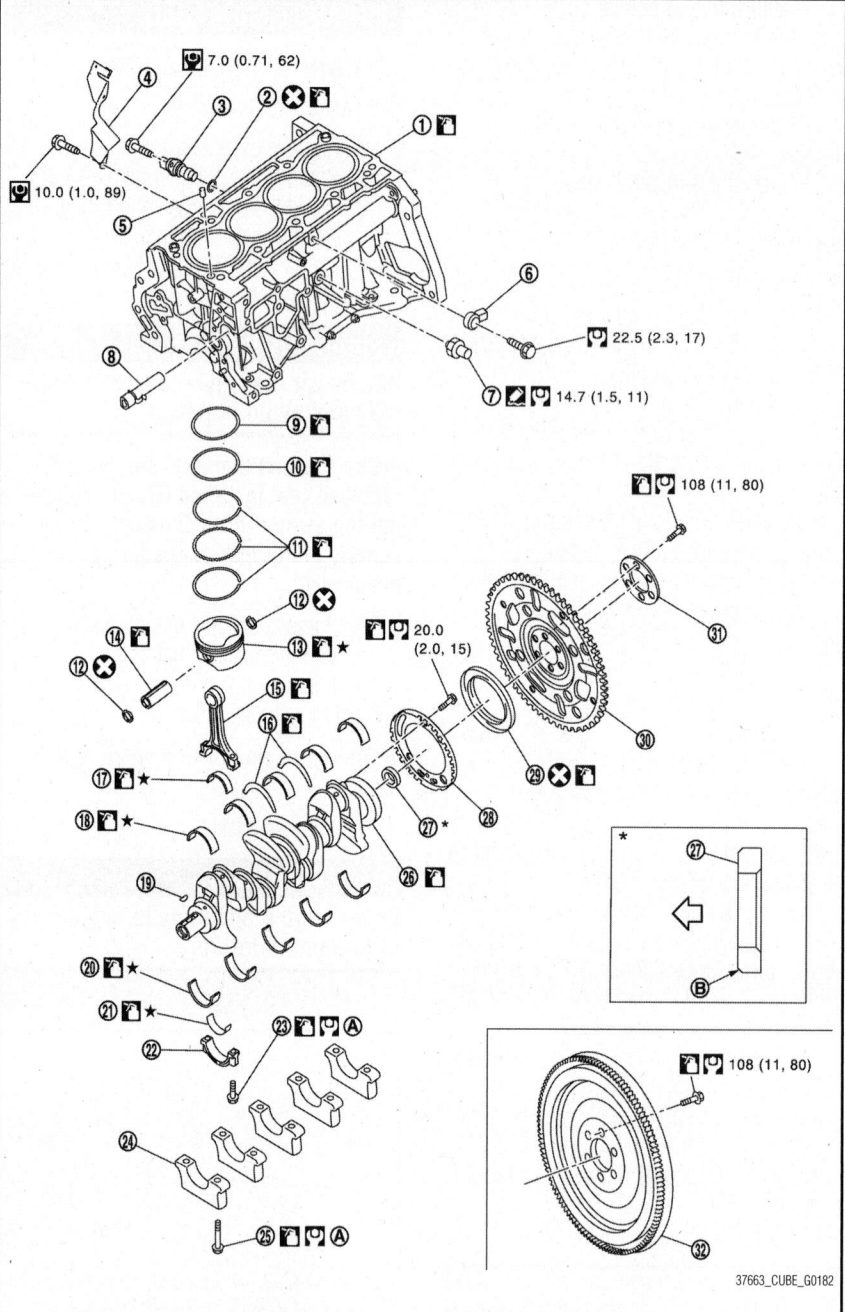

Fig. 68 Exploded view of engine cylinder block showing flywheel/drive plate

8. Remove intake manifold with the following procedure:

a. Loosen and remove intake manifold mounting bolts.

b. Remove harness clip from intake manifold side.

c. Disconnect EVAP hose from intake manifold.

d. Disconnect harness connector from EVAP canister purge volume control valve.

e. Loosen mounting bolts in reverse order of the tightening sequence.

✳ WARNING

Cover engine openings to avoid entry of foreign materials.

9. Remove brackets from intake manifold, if necessary.

10. Remove engine cover bracket, if necessary.

11. Remove EVAP canister purge volume control solenoid valve from intake manifold, if necessary.

To install:

12. Note the following, and install in the reverse order of removal.

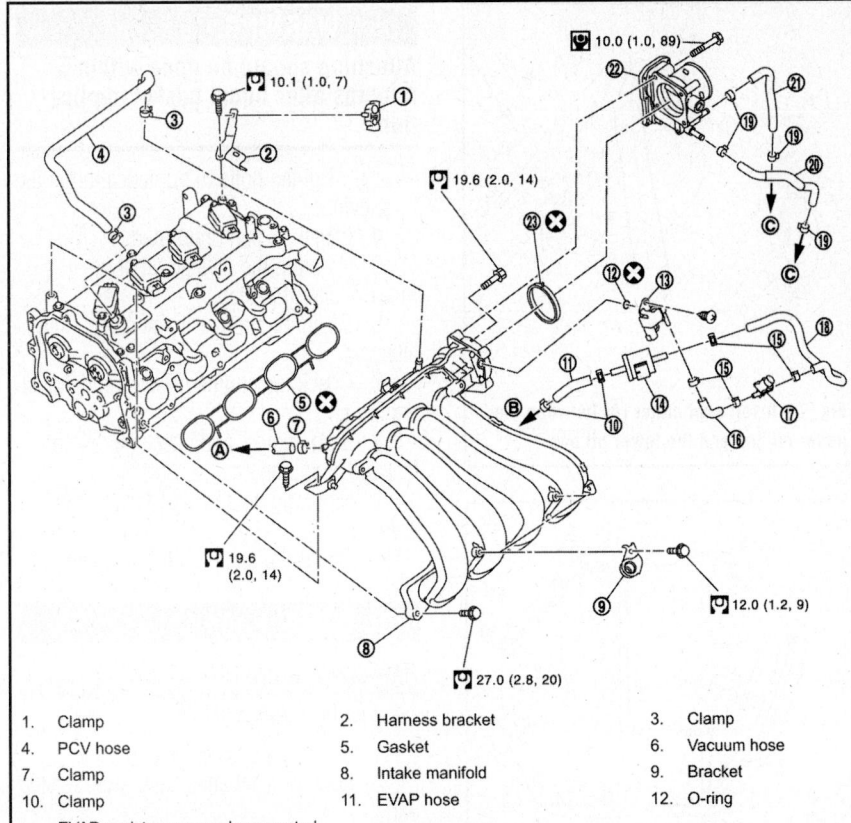

1. Clamp
2. Harness bracket
3. Clamp
4. PCV hose
5. Gasket
6. Vacuum hose
7. Clamp
8. Intake manifold
9. Bracket
10. Clamp
11. EVAP hose
12. O-ring
13. EVAP canister purge volume control solenoid valve
14. EVAP vacuum tank
15. Clamp
16. EVAP hose
17. EVAP service port
18. EVAP hose
19. Clamp
20. Water hose
21. Water hose
22. Electric throttle control actuator
23. Gasket
A. To brake booster
B. To centralized under-floor piping
C. To water outlet

37663_CUBE_G0177

Fig. 69 Exploded view of intake manifold assembly

c. Tighten intake manifold mounting bolt A. Then tighten intake manifold mounting bolt B.

15. Tighten bolts of electric throttle control actuator equally and diagonally in several steps.

16. Perform "Throttle Valve Closed Position Learning" after repair when removing harness connector of the electric throttle control actuator.

17. Perform "Throttle Valve Closed Position Learning" and "Idle Air Volume Learning" after repair when replacing electric throttle control actuator.

THROTTLE VALVE CLOSED POSITION LEARNING

1. Make sure that accelerator pedal is fully released.
2. Turn ignition switch ON.
3. Turn ignition switch OFF and wait at least 10 seconds.

➡**Make sure that throttle valve moves during above 10 seconds by confirming the operating sound.**

IDLE AIR VOLUME LEARNING

Make sure that all of the following conditions are satisfied.

➡**Learning will be cancelled if any of the following conditions are missed for even a moment.**

- Battery voltage: More than 12.9 V (At idle)
- Engine coolant temperature: 158–212°F (70–100°C)
- Selector lever: P or N (CVT), Neutral (M/T)
- Electric load switch: OFF (Air conditioner, headlamp, rear window defogger)

➡**On vehicles equipped with daytime light systems, if the parking brake is applied before the engine is started the headlamp will not be illuminated.**

- Steering wheel: Neutral (Straight-ahead position)
- Vehicle speed: Stopped
- Transmission: Warmed-up
- CVT models: With CONSULT-III: Drive vehicle until "FLUID TEMP SE" in "DATA MONITOR" mode of "TRANSMISSION" system indicates less than 0.9 V.
- CVT models: Without CONSULT-III: Drive vehicle for 10 minutes.
- M/T models: Drive vehicle for 10 minutes.

13. Check if gasket is not dropped from the installation groove of intake manifold.

14. Install intake manifold with the following procedure:

a. Tighten intake manifold mounting bolts in numerical order as shown.

b. Tighten No. 1 bolt again.

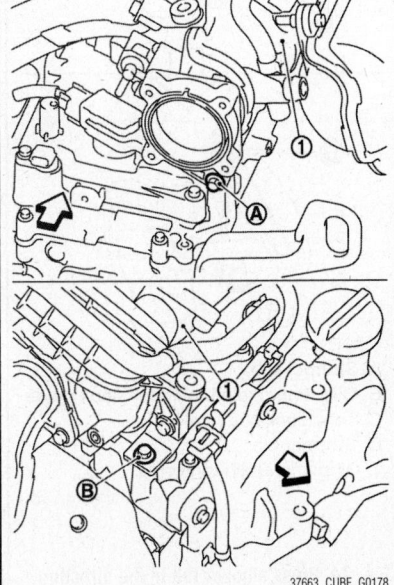

37663_CUBE_G0178

Fig. 70 Loosen and remove intake manifold (1) mounting bolts (A) and (B)

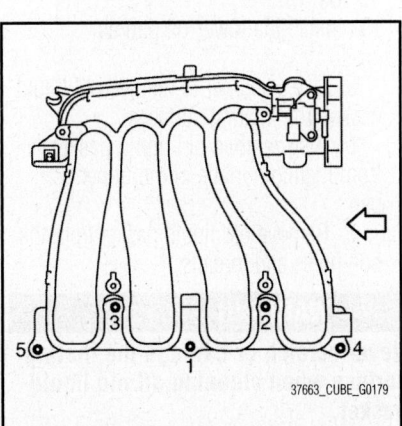

37663_CUBE_G0179

Fig. 71 Tighten intake manifold mounting bolts in numerical order

With CONSULT-III

1. Perform Accelerator Pedal Released Position Learning.

2. Perform Throttle Valve Closed Position Learning.

3. Start engine and warm it up to normal operating temperature.

4. Select "IDLE AIR VOL LEARN" in "WORK SUPPORT" mode.

5. Touch "START" and wait 20 seconds.

Without CONSULT-III

Note the following:

• It is better to count the time accurately with a clock.

• It is impossible to switch the diagnostic mode when an accelerator pedal position sensor circuit has a malfunction.

1. Perform Accelerator Pedal Released Position Learning.

2. Perform Throttle Valve Closed Position Learning.

3. Start engine and warm it up to normal operating temperature.

4. Turn ignition switch OFF and wait at least 10 seconds.

5. Confirm that accelerator pedal is fully released, turn ignition switch ON and wait 3 seconds.

6. Repeat the following procedure quickly five times within 5 seconds.

   a. Fully depress the accelerator pedal.

   b. Fully release the accelerator pedal.

7. Wait 7 seconds, fully depress the accelerator pedal and keep it for approx. 20 seconds until the MIL stops blinking and turned ON.

8. Fully release the accelerator pedal within 3 seconds after the MIL turned ON.

9. Start engine and let it idle.

OIL PAN

REMOVAL & INSTALLATION

Lower Oil Pan

See Figures 72 and 73.

1. Drain engine oil.

2. Remove the lower oil pan with the following procedure:

   a. Loosen the lower oil pan mounting bolts in reverse order of the tightening sequence.

   b. Insert seal cutter KV10111100 (J-37228) between the upper oil pan and the lower oil pan.

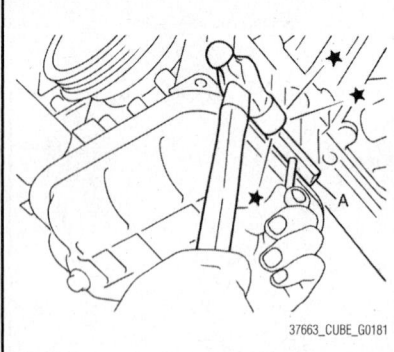

Fig. 72 Insert seal cutter (A) between the upper oil pan and the lower oil pan

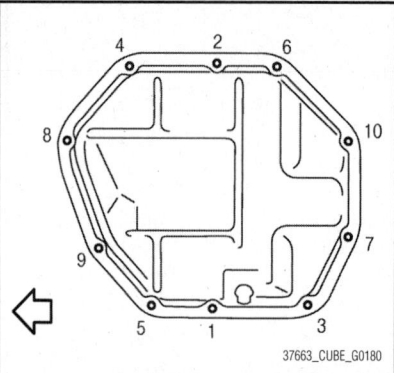

Fig. 73 Tighten bolts in numerical order

✳✳ WARNING

Be careful not to damage the mating surface. Never insert a screwdriver. This damages the mating surfaces.

   c. Slide the seal cutter KV10111100 (J-37228) by tapping on the side of tool with a hammer.

   d. Remove the lower oil pan.

To install:

3. Install the lower oil pan as follows:

   a. Use a scraper to remove old liquid gasket from mating surfaces.

   b. Also remove old liquid gasket from mating surface of the upper oil pan.

   c. Remove old liquid gasket from the bolt holes and threads.

✳✳ WARNING

Never scratch or damage the mating surface when cleaning off old liquid gasket.

   d. Apply a continuous bead of liquid gasket with a tube presser.

✳✳ WARNING

Attaching should be done within 5 minutes after liquid gasket application.

   e. Tighten bolts in numerical order as shown.

4. Install oil pan drain plug.

5. Install in the reverse order of removal after this step.

6. Clean oil strainer if any object attached.

7. Check the engine oil level and adjust engine oil.

8. Start engine, and check there is no leakage of engine oil.

9. Stop engine and wait for 10 minutes.

10. Check the engine oil level again.

OIL PUMP

REMOVAL & INSTALLATION

See Figures 74 through 76.

1. Remove the lower oil pan.

2. Remove front cover, and other related parts.

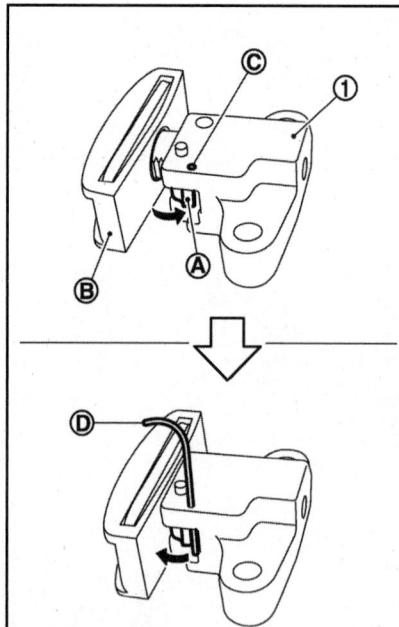

A. Stopper tab
B. Oil pump chain tensioner slack guide
C. Tensioner body hole
D. Stopper pin
1. Oil pump chain tensioner

37663_CUBE_G0183

Fig. 74 Press stopper tab in the direction shown to push the oil pump chain tensioner slack guide toward oil pump chain tensioner

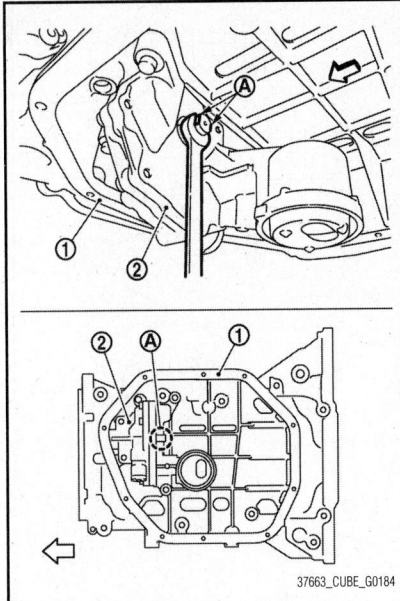

Fig. 75 WAF part of oil pump shaft (A), upper oil pan (1), and oil pump (2)

3. Remove oil pump sprocket with the following procedure:

➡**Add matching mark if necessary for easier installation.**

a. Press stopper tab in the direction shown to push the oil pump chain tensioner slack guide toward oil pump chain tensioner.

➡**The oil pump chain tensioner slack guide is released by pressing the stopper tab. As the result, the oil pump chain tensioner slack guide can be moved.**

b. Insert a stopper pin into tensioner body hole to secure the oil pump chain tensioner slack guide.

➡**Use a hard metal pin with the diameter of approximately 0.47 inches (1.2 mm) as a stopper pin.**

c. Remove oil pump chain tensioner.

➡**When the holes on lever and tensioner body cannot be aligned, align these holes by slightly moving the oil pump chain tensioner slack guide.**

d. Hold the WAF part of oil pump shaft, and then loosen the oil pump sprocket bolt and remove it.

❉❉ **WARNING**

Secure the oil pump shaft with the WAF part. Never loosen the oil pump sprocket bolt by tightening the oil pump drive chain.

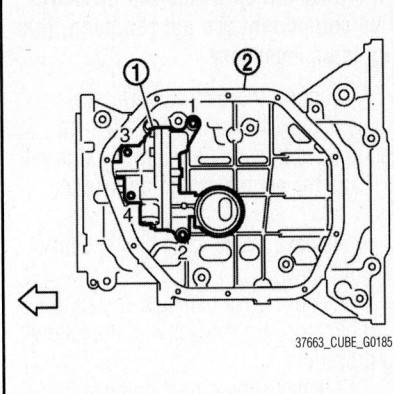

Fig. 76 Tighten bolts in numerical order

e. Remove oil pump sprocket.

4. Remove oil pump. Loosen bolts in reverse order of the tightening sequence.

To install:

5. Note the following, and install in the reverse order of removal.

6. Tighten bolts in numerical order as shown.

7. Check the engine oil level.

8. Start the engine, and check that there is no leakage of engine oil.

9. Stop the engine and wait for 10 minutes.

10. Check the engine oil level, and adjust the level.

PISTONS & RINGS

POSITIONING

See Figure 77.

TIMING CHAIN COVER

REMOVAL & INSTALLATION

See Figures 78 through 84.

➡**The rotating direction in the text indicates all directions viewed from the engine front.**

1. Remove front fender protector (RH).
2. Drain engine oil.

❉❉ **WARNING**

Perform this step when engine is cold.

3. Remove the following parts:
- Intake manifold
- Rocker cover
- Drive belt
- Ground cable (between front cover and radiator core support)

4. Set No. 1 cylinder at TDC on its compression stroke with the following procedure:

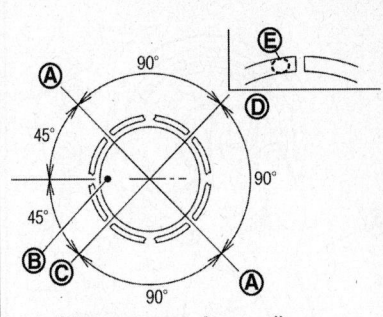

A. Oil ring upper or lower rail gap
B. Front mark
C. Second ring and oil ring spacer gap
D. Top ring gap
E. Stamped mark

Fig. 77 Piston ring gap locations

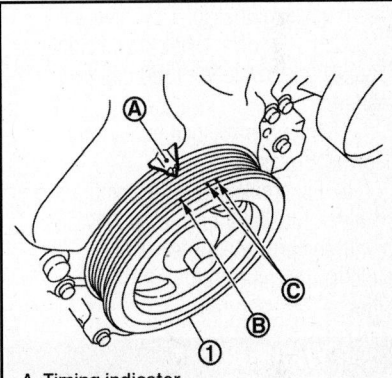

A. Timing indicator
B. TDC mark (no paint)
C. White paint mark (not used for service)
1. Crankshaft pulley

Fig. 78 Rotate crankshaft pulley clockwise and align TDC mark (no paint) to timing indicator on front cover

Fig. 79 Check that the cam noses of the No. 1 cylinder are located as shown

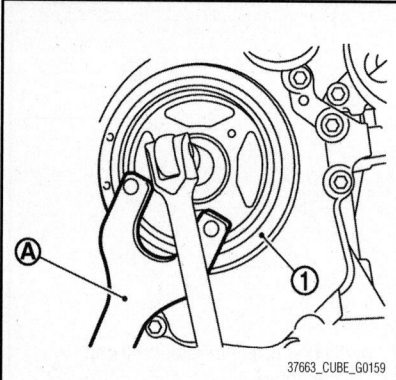

Fig. 80 Fix crankshaft pulley (1) with a pulley holder (A)

a. Rotate crankshaft pulley clockwise and align TDC mark (no paint) to timing indicator on front cover.

b. At the same time, check that the cam noses of the No. 1 cylinder are located as shown. If not, rotate crankshaft pulley one revolution (360 degrees) and align.

5. Remove crankshaft pulley with the following procedure:

a. Fix crankshaft pulley with a pulley holder, loosen crankshaft pulley bolt until the bolt seating surface is 0.39 inches (10 mm) from its original position.

✳✳ WARNING

Do not remove the crankshaft pulley bolt as it will be used as a supporting point for the pulley puller.

b. Attach a pulley puller KV11103000 in the M6 thread hole on crankshaft pulley, and remove crankshaft pulley.

6. Remove the lower oil pan.

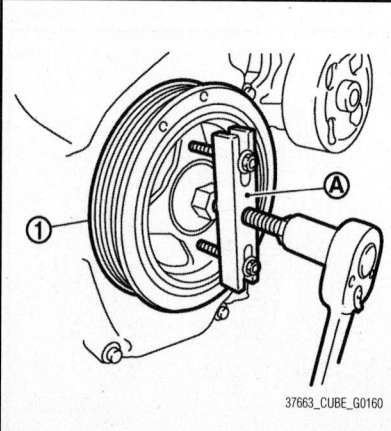

Fig. 81 Attach a pulley puller KV11103000 in the M6 thread hole on crankshaft pulley (1)

➡If crankshaft sprocket and oil pump drive component are not removed, this step is unnecessary.

7. Support the bottom surface of engine using a transmission jack, and then remove the engine mounting bracket (RH) and the engine mounting insulator (RH).

8. Remove intake valve timing control solenoid valve.

9. Remove drive belt auto-tensioner.

10. Remove front cover with the following procedure:

a. Remove the mounting bolts in reverse order of the tightening sequence shown.

b. Cut liquid gasket by prying at the positions shown, and then remove the front cover.

✳✳ WARNING

Be careful not to damage the mating surface.

11. Remove front oil seal from front cover.

✳✳ WARNING

Be careful not to damage front cover.

To install:

12. Install front cover with the following procedure:

a. Install new O-ring to cylinder block.

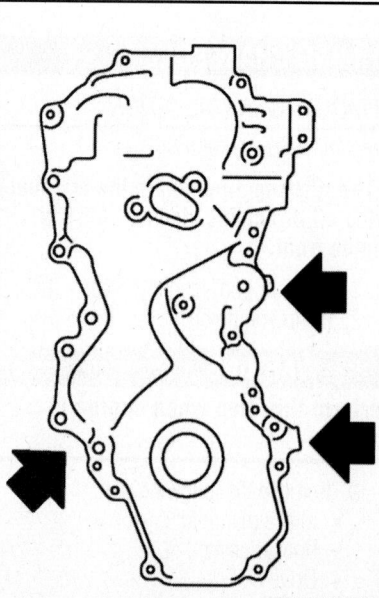

Fig. 82 Cut liquid gasket by prying at the positions shown

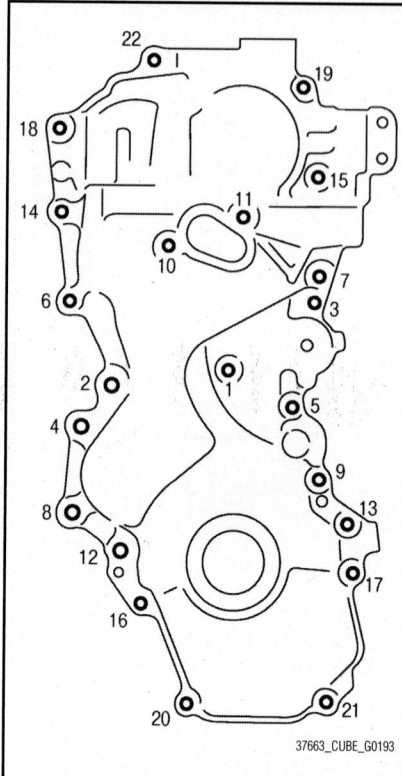

Fig. 83 Tighten mounting bolts in numerical order

✳✳ WARNING

Do not misalign O-ring.

b. Apply a continuous bead of liquid gasket to front cover.

c. Check that matching marks of timing chain and each sprockets are still aligned. Then install front cover.

✳✳ WARNING

Check O-ring on cylinder block is correctly installed. Be careful not to damage front oil seal by contact with front end of crankshaft.

d. Install front cover, and tighten mounting bolts in numerical order as shown.

✳✳ WARNING

Attaching should be done within 5 minutes after liquid gasket application.

e. After all bolts are tightened, retighten them to specified torque in numerical order as shown.

✳✳ WARNING

Be sure to wipe off any excessive liquid gasket leaking.

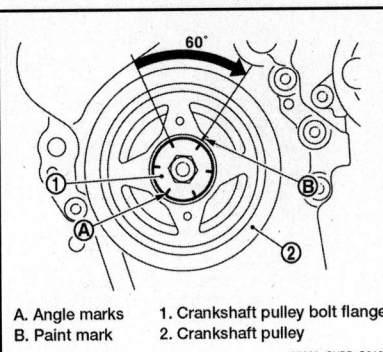

A. Angle marks
B. Paint mark
1. Crankshaft pulley bolt flange
2. Crankshaft pulley

37663_CUBE_G0161

Fig. 84 Put a paint mark on crankshaft pulley, matching with any one of six easy to recognize angle marks on crankshaft pulley bolt flange

13. Install crankshaft pulley with the following procedure:

a. When inserting crankshaft pulley with a plastic hammer, tap on its center portion (not circumference).

❋❋ WARNING

Never damage front oil seal lip section.

b. Secure crankshaft pulley with a pulley holder.

c. Apply new engine oil to thread and seat surfaces of crankshaft pulley bolt.

d. Tighten crankshaft pulley bolt.

e. Completely loosen.

f. Tighten crankshaft pulley bolt.

g. Put a paint mark on crankshaft pulley, matching with any one of six easy to recognize angle marks on crankshaft pulley bolt flange.

h. Turn another 60 degrees clockwise (angle tightening). Check the tightening angle with movement of one angle mark.

i. Check that crankshaft rotates clockwise smoothly.

14. Install remaining parts in the reverse order of removal.

15. Before starting engine, check oil/fluid levels including engine coolant and engine oil. If less than required quantity, fill to the specified level.

16. Use procedure below to check for fuel leakage.

a. Turn ignition switch "ON" (with engine stopped). With fuel pressure applied to fuel piping, check for fuel leakage at connection points.

b. Start engine. With engine speed increased, check again for fuel leakage at connection points.

17. Run engine to check for unusual noise and vibration.

➡ If hydraulic pressure inside chain tensioner drops after removal/installation, slack in guide may generate a pounding noise during and just after the engine start. However, this does not indicate an unusualness. Noise will stop after hydraulic pressure rises.

18. Warm up engine thoroughly to check there is no leakage of fuel, or any oil/fluids including engine oil and engine coolant.

19. Bleed air from lines and hoses of applicable lines, such as in cooling system.

20. After cooling down engine, again check oil/fluid levels including engine oil and engine coolant. Refill to the specified level, if necessary.

TIMING CHAIN, TENSIONER, & SPROCKETS

REMOVAL & INSTALLATION

See Figures 78 through 84 and 85 through 95.

➡ **The rotating direction in the text indicates all directions viewed from the engine front.**

1. Remove front fender protector (RH).
2. Drain engine oil.

❋❋ WARNING

Perform this step when engine is cold.

3. Remove the following parts:
   • Intake manifold
   • Rocker cover
   • Drive belt
   • Ground cable (between front cover and radiator core support)
4. Set No. 1 cylinder at TDC on its compression stroke with the following procedure:

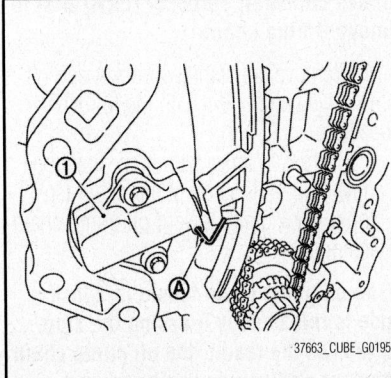

37663_CUBE_G0195

Fig. 85 Insert a stopper pin (A) into the body hole, and then fix it with the plunger pushed in; remove timing chain tensioner (1)

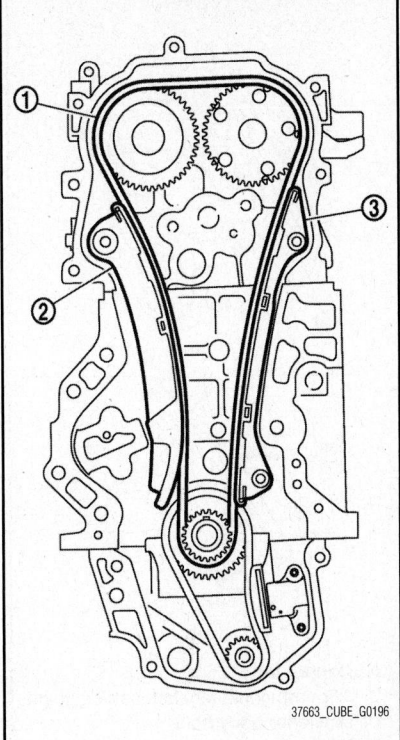

37663_CUBE_G0196

Fig. 86 Remove slack guide (2), tension guide (3) and timing chain (1)

a. Rotate crankshaft pulley clockwise and align TDC mark (no paint) to timing indicator on front cover.

b. At the same time, check that the cam noses of the No. 1 cylinder are located as shown. If not, rotate crankshaft pulley one revolution (360 degrees) and align.

5. Remove crankshaft pulley with the following procedure:

a. Fix crankshaft pulley with a pulley holder, loosen crankshaft pulley bolt until the bolt seating surface is 0.39 inches (10 mm) from its original position.

❋❋ WARNING

Do not remove the crankshaft pulley bolt as it will be used as a supporting point for the pulley puller.

b. Attach a pulley puller KV11103000 in the M6 thread hole on crankshaft pulley, and remove crankshaft pulley.

6. Remove the lower oil pan.

➡ **If crankshaft sprocket and oil pump drive component are not removed, this step is unnecessary.**

7. Support the bottom surface of engine using a transmission jack, and then remove the engine mounting bracket (RH) and the engine mounting insulator (RH).

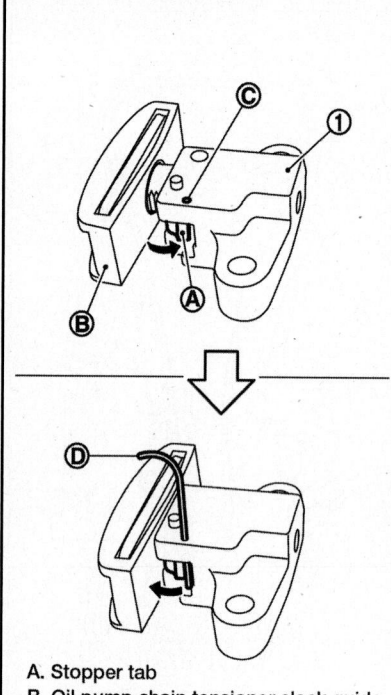

A. Stopper tab
B. Oil pump chain tensioner slack guide
C. Tensioner body hole
D. Stopper pin
1. Oil pump chain tensioner

37663_CUBE_G0183

Fig. 87 Press stopper tab in the direction shown to push the oil pump chain tensioner slack guide toward oil pump chain tensioner

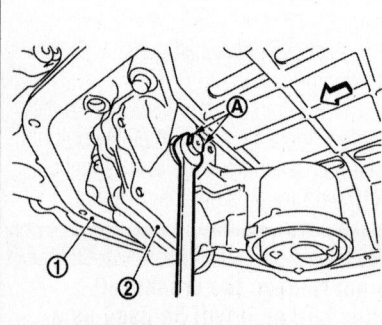

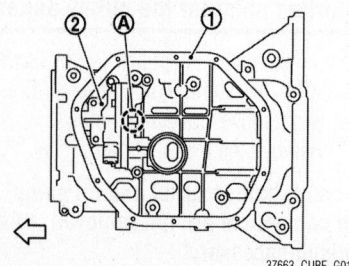

37663_CUBE_G0184

Fig. 88 WAF part of oil pump shaft (A), upper oil pan (1), and oil pump (2)

8. Remove intake valve timing control solenoid valve.

9. Remove drive belt auto-tensioner.

10. Remove front cover with the following procedure:

   a. Remove the mounting bolts in reverse order as shown.

   b. Cut liquid gasket by prying at the positions shown, and then remove the front cover.

✳✳ WARNING
Be careful not to damage the mating surface.

11. Remove front oil seal from front cover.

✳✳ WARNING
Be careful not to damage front cover.

12. Remove timing chain tensioner with the following procedure:

   a. Push in timing chain tensioner plunger.

   b. Insert a stopper pin into the body hole, and then fix it with the plunger pushed in.

➡ **Use approximately 0.059 inches (1.5 mm) diameter. hard metal pin as a stopper pin.**

   c. Remove timing chain tensioner.

13. Remove slack guide, tension guide, and timing chain.

✳✳ WARNING
Never rotate each crankshaft and camshaft individually while timing chain is removed. It causes interference between valve and piston.

➡ **If timing chain is difficult to remove, remove camshaft sprocket (EXH) first to remove timing chain.**

14. Remove crankshaft sprocket and oil pump drive component with the following procedure:

   a. Press stopper tab in the direction shown to push the oil pump chain tensioner slack guide toward oil pump chain tensioner.

➡ **The oil pump chain tensioner slack guide is released by pressing the stopper tab. As the result, the oil pump chain tensioner slack guide can be moved.**

   b. Insert a stopper pin into tensioner body hole to secure the oil pump chain tensioner slack guide.

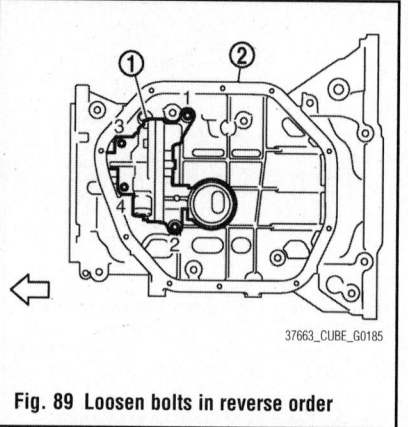

37663_CUBE_G0185

Fig. 89 Loosen bolts in reverse order

➡ **Use a hard metal pin with the diameter of approximately 0.47 inches (1.2 mm) as a stopper pin.**

   c. Remove oil pump chain tensioner.

➡ **When the holes on lever and tensioner body cannot be aligned, align these holes by slightly moving the oil pump chain tensioner slack guide.**

   d. Hold the WAF part of oil pump shaft, and then loosen the oil pump sprocket bolt and remove it.

✳✳ WARNING
Secure the oil pump shaft with the WAF part. Never loosen the oil pump sprocket bolt by tightening the oil pump drive chain.

   e. Remove oil pump sprocket.

15. Remove oil pump. Loosen bolts in reverse order as shown.

✳✳ WARNING
Secure the oil pump shaft with the WAF part. Never loosen the oil pump sprocket bolt by tightening the oil pump drive chain.

   a. Remove crankshaft sprocket, oil pump sprocket, and oil pump drive chain as a set.

16. Remove tension guide (front cover side) from front cover, if necessary.

To install:
The figure shows the relationship between the matching mark on each timing chain and that on the corresponding sprocket, with the components installed.

17. Check that crankshaft key points straight up.

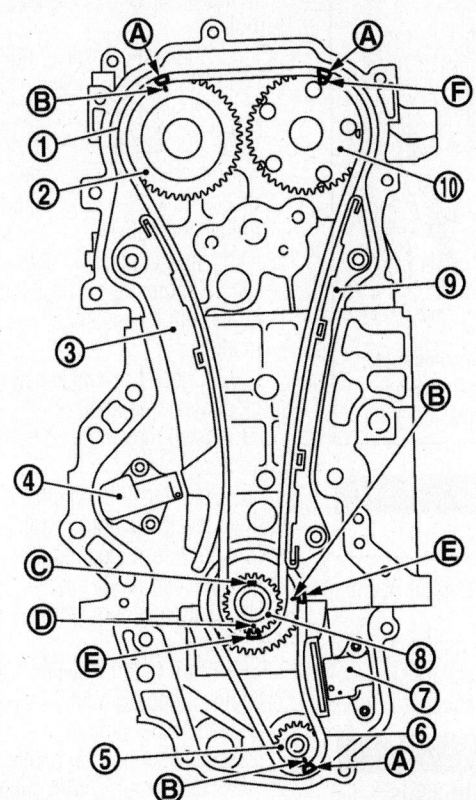

1. Timing chain
2. Camshaft sprocket (EXH)
3. Slack guide
4. Timing chain tensioner
5. Oil pump sprocket
6. Oil pump drive chain
7. Oil pump chain tensioner
8. Crankshaft sprocket
9. Tension guide
10. Camshaft sprocket (INT)

A. Matching mark (dark blue link)
B. Matching mark (stamping)
C. Crankshaft key position (straight up)
D. Matching mark (stamping)
E. Matching mark (orange link)
F. Matching mark (outer groove*)

*: There are two outer grooves in camshaft sprocket (INT).
   The wider one is a matching mark.

37663_CUBE_G0197

Fig. 90 Timing chain assembly

18. If the tension guide (front cover side) is removed, install it to the front cover.

❋❋ WARNING

Check the joint condition by sound or feeling.

19. Install crankshaft sprocket, oil pump sprocket, and oil pump drive chain.
   a. Install it by aligning matching marks on each sprockets and oil pump drive chain.
   b. If these matching marks are not aligned, rotate the oil pump shaft slightly to correct the position.

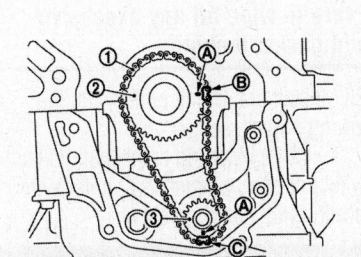

A. Matching mark (stamping)
B. Matching mark (orange link)
C. Matching mark (dark blue link)

1. Oil pump drive chain
2. Crankshaft sprocket
3. Oil pump sprocket

37663_CUBE_G0198

Fig. 91 Install crankshaft sprocket, oil pump sprocket, and oil pump drive chain

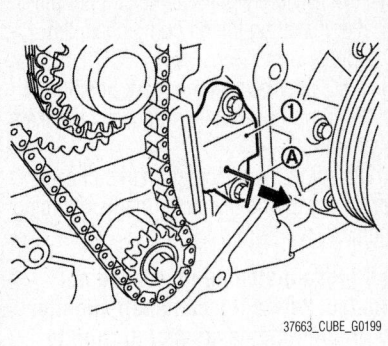

37663_CUBE_G0199

Fig. 92 Install oil pump chain tensioner (1); fix the plunger at the most compressed position using a stopper pin (A)

❋❋ WARNING

Check matching mark position of each sprockets after installing the oil pump drive chain.

20. Hold the WAF part of oil pump shaft (A), and then tighten the oil pump shaft sprocket bolt.

❋❋ WARNING

Secure the oil pump shaft with the WAF part. Never loosen the oil pump shaft sprocket bolt by tightening the oil pump drive chain.

21. Install oil pump chain tensioner.
   a. Fix the plunger at the most compressed position using a stopper pin, and then install it.

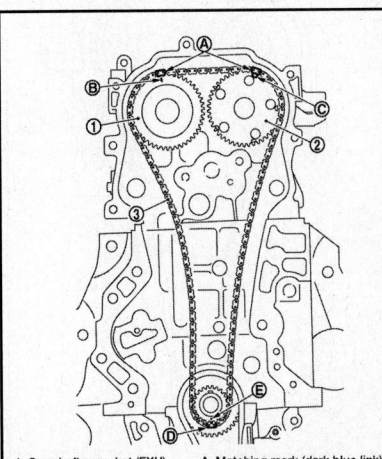

1. Camshaft sprocket (EXH)
2. Camshaft sprocket (INT)
3. Timing chain

A. Matching mark (dark blue link)
B. Matching mark (stamping)
C. Matching mark (outer groove*)
D. Matching mark (orange link)
E. Matching mark (stamping)

*: There are 2 outer grooves in camshaft sprocket (INT).
   The wider one is a matching mark.

37663_CUBE_G0200

Fig. 93 Align the matching marks of each sprocket with the matching marks of timing chain

b. Securely pull out the stopper pin after installing the oil pump chain tensioner.

c. Check matching mark position of oil pump drive chain and each sprockets again.

22. Align the matching marks of each sprocket with the matching marks of timing chain.

➡ **If these matching marks are not aligned, rotate the camshaft slightly by holding the hexagonal portion to correct the position.**

✵✵ WARNING

Check matching mark position of each sprocket and timing chain again after installing the timing chain.

23. Install the slack guide and the tension guide.

24. Install timing chain tensioner.

a. Fix the plunger at the most compressed position using a stopper pin, and then install it.

b. Securely pull out the stopper pin after installing the timing chain tensioner.

25. Check matching mark position of timing chain and each sprockets again.

26. Install front oil seal.

27. Install front cover with the following procedure:

a. Install new O-ring to cylinder block.

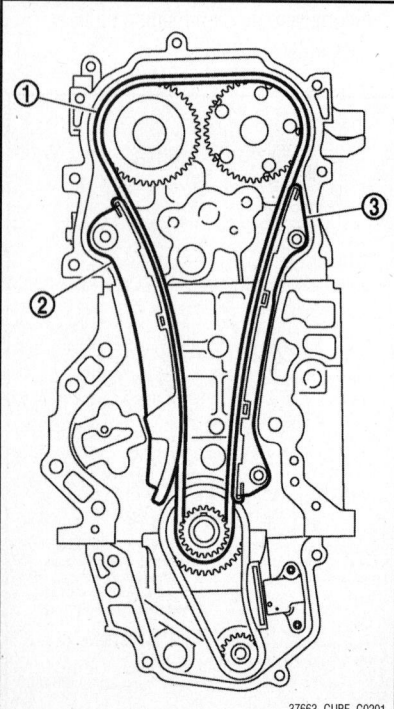

Fig. 94 Install the slack guide (2) and the tension guide (3)

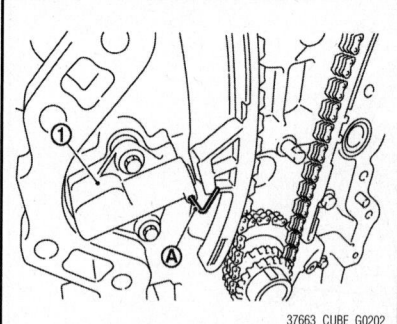

Fig. 95 Install timing chain tensioner (1); fix the plunger at the most compressed position using a stopper pin (A)

✵✵ WARNING

Do not misalign O-ring.

b. Apply a continuous bead of liquid gasket to front cover.

c. Check that matching marks of timing chain and each sprockets are still aligned. Then install front cover.

✵✵ WARNING

Check O-ring on cylinder block is correctly installed. Be careful not to damage front oil seal by contact with front end of crankshaft.

d. Install front cover, and tighten mounting bolts in numerical order as shown.

✵✵ WARNING

Attaching should be done within 5 minutes after liquid gasket application.

e. After all bolts are tightened, retighten them to specified torque in numerical order as shown.

✵✵ WARNING

Be sure to wipe off any excessive liquid gasket leaking.

28. Install crankshaft pulley with the following procedure:

a. When inserting crankshaft pulley with a plastic hammer, tap on its center portion (not circumference).

✵✵ WARNING

Never damage front oil seal lip section.

b. Secure crankshaft pulley with a pulley holder.

c. Apply new engine oil to thread

and seat surfaces of crankshaft pulley bolt.

d. Tighten crankshaft pulley bolt.

e. Completely loosen.

f. Tighten crankshaft pulley bolt.

g. Put a paint mark on crankshaft pulley, matching with any one of six easy to recognize angle marks on crankshaft pulley bolt flange.

h. Turn another 60 degrees clockwise (angle tightening). Check the tightening angle with movement of one angle mark.

i. Check that crankshaft rotates clockwise smoothly.

29. Install remaining parts in the reverse order of removal.

30. Before starting engine, check oil/fluid levels including engine coolant and engine oil. If less than required quantity, fill to the specified level.

31. Use procedure below to check for fuel leakage.

a. Turn ignition switch "ON" (with engine stopped). With fuel pressure applied to fuel piping, check for fuel leakage at connection points.

b. Start engine. With engine speed increased, check again for fuel leakage at connection points.

32. Run engine to check for unusual noise and vibration.

➡ **If hydraulic pressure inside chain tensioner drops after removal/installation, slack in guide may generate a pounding noise during and just after the engine start. However, this does not indicate an unusualness. Noise will stop after hydraulic pressure rises.**

33. Warm up engine thoroughly to check there is no leakage of fuel, or any oil/fluids including engine oil and engine coolant.

34. Bleed air from lines and hoses of applicable lines, such as in cooling system.

35. After cooling down engine, again check oil/fluid levels including engine oil and engine coolant. Refill to the specified level, if necessary.

VALVE (ROCKER ARM) COVER

REMOVAL & INSTALLATION

See Figure 96.

1. Remove intake manifold.
2. Remove ignition coil.

✵✵ WARNING

Never drop or shock ignition coil. Never disassemble ignition coil.

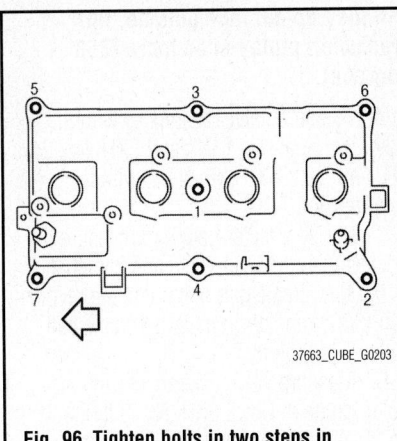

Fig. 96 Tighten bolts in two steps in numerical order as shown

3. Remove rocker cover.

4. Remove the bolts in reverse order of the tightening sequence.

5. Remove PCV valve and PCV hose, if necessary.

6. Remove rocker cover gasket from rocker cover.

7. Use scraper to remove all traces of liquid gasket from cylinder head and front cover.

✲✲ WARNING

Never scratch or damage the mating surface when cleaning off old liquid gasket.

To install:

8. Install the rocker cover gasket to rocker cover.

✲✲ WARNING

Check the gasket is not dropped.

9. Install rocker cover.

10. Tighten bolts in two steps in numerical order as shown.
   a. Step 1: 17 inch lbs. (2 Nm)
   b. Step 2: 74 inch lbs. (8 Nm)

11. Install in the reverse order of removal, for the rest of parts.

VALVE LASH (CLEARANCE) ADJUSTMENT

ADJUSTMENT

See Figures 97 and 98.

Perform adjustment depending on selected head thickness of valve lifter.

1. Remove camshaft.

2. Remove the valve lifters at the locations that are out of the standard.

3. Measure the center thickness of the removed valve lifters with a micrometer.

4. Use the equation below to calculate valve lifter thickness for replacement.

a. Valve lifter thickness calculation:
- $t = t1 + (C1 - C2)$
- t = Thickness of replacement valve lifter.
- t1 = Thickness of removed valve lifter.
- C1 = Measured valve clearance.
- C2 = Standard valve clearance.

b. Thickness of a new valve lifter can be identified by stamp marks on the reverse side (inside the cylinder). Stamp mark 302 indicates a thickness of 0.1189 inches (3.02 mm).

➡**Available thickness of valve lifter: 26 sizes with a range of 0.1181 to 0.1378 inches (3.00 to 3.50 mm), in steps of 0.0008 inches (0.02 mm), when assembled at the factory.**

5. Install the selected valve lifter.

6. Install camshaft.

7. Manually turn crankshaft pulley a few turns.

8. Check that valve clearances for cold

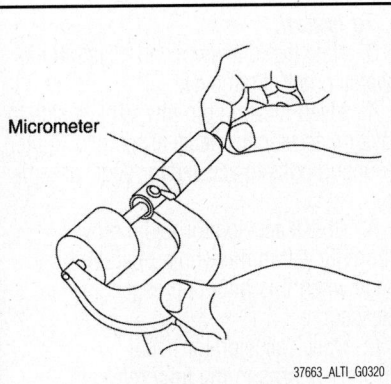

Fig. 97 Measure the center thickness of the removed valve lifters with a micrometer

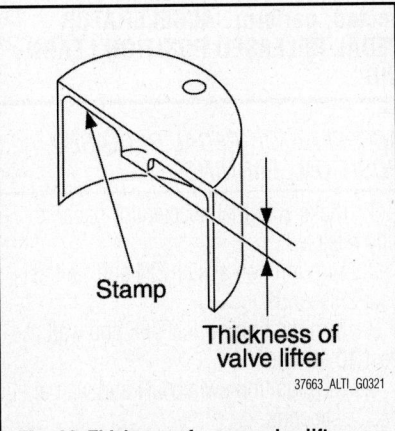

Fig. 98 Thickness of a new valve lifter can be identified by stamp marks on the reverse side (inside the cylinder)

engine are within specifications, by referring to the specified values.

9. Install remaining parts in the reverse order of removal.

10. Warm up the engine, and check for unusual noise and vibration.

INSPECTION

See Figures 99 through 101.

Perform this inspection as follows after removal, installation, or replacement of the camshaft or any valve related parts, or if there are any unusual engine conditions due to changes in valve clearance.

1. Remove the rocker cover.

2. Turn crankshaft pulley in normal direction (clockwise when viewed from front) to align TDC identification mark (without paint mark) with timing indicator.

3. At this time, check that the both intake and exhaust cam lobes of No. 1 cylinder face inside.

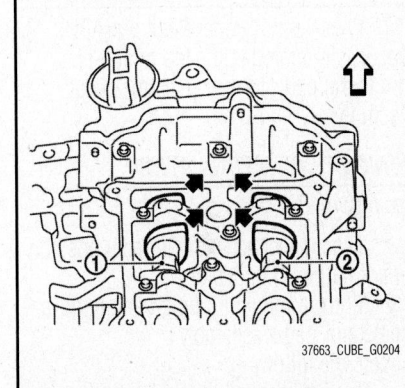

Fig. 99 Check that the both intake and exhaust cam lobes of No. 1 cylinder face inside

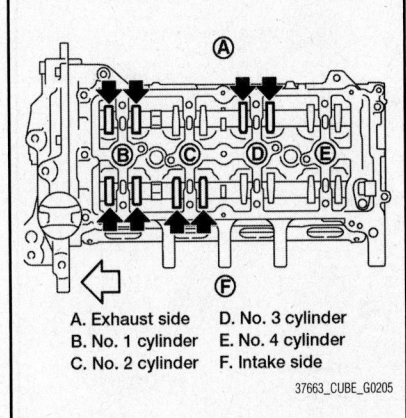

A. Exhaust side   D. No. 3 cylinder
B. No. 1 cylinder E. No. 4 cylinder
C. No. 2 cylinder F. Intake side

Fig. 100 Measure valve clearances at No. 1 INT and EXH, No. 2 INT, and No.3 EXH

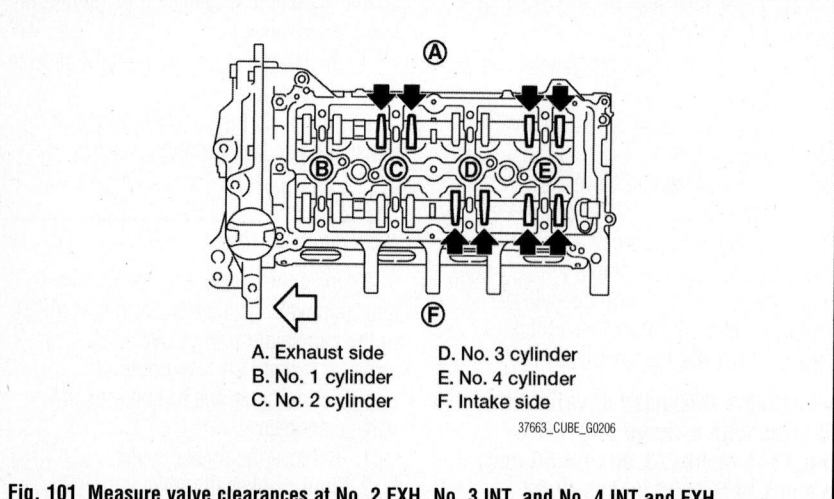

A. Exhaust side
B. No. 1 cylinder
C. No. 2 cylinder
D. No. 3 cylinder
E. No. 4 cylinder
F. Intake side

37663_CUBE_G0206

Fig. 101 Measure valve clearances at No. 2 EXH, No. 3 INT, and No. 4 INT and EXH

➡**If they do not face outside, turn crankshaft pulley once more (360 degrees).**

4. Measure valve clearances with a feeler gauge at No. 1 INT and EXH, No. 2 INT, and No.3 EXH with No.1 cylinder compression TDC.

   a. Use a feeler gauge to measure the clearance between valve and camshaft.

5. Turn crankshaft one complete revolution (360°) and align mark on crankshaft pulley with pointer.

6. Measure valve clearances with a feeler gauge at No. 2 EXH, No. 3 INT, and No. 4 INT and EXH with No.4 cylinder compression TDC.

7. If out of specifications, make necessary adjustment.

ENGINE PERFORMANCE & EMISSION CONTROLS

ACCELERATOR PEDAL POSITION SENSOR

LOCATION

The Accelerator Pedal Position (APP) sensor is mounted at the top of and is an integral part of the accelerator pedal assembly.

REMOVAL & INSTALLATION

See Figure 102.

1. Disconnect accelerator pedal position sensor harness connector.
2. Loosen mounting bolts, and remove accelerator pedal assembly.
   Note the following:
   • Never disassemble accelerator lever. Never remove accelerator pedal position sensor from accelerator lever.

• Avoid impact from dropping etc. during handling.
• Be careful to keep accelerator lever away from water.

To install:

3. Note the following, and install in the reverse order of removal.

4. Insert locating pin into vehicle side to position accelerator pedal assembly. Tighten mounting bolts to accelerator pedal assembly.

5. Check accelerator pedal moves smoothly within the whole operation range when it is fully depressed and released.

6. Check accelerator pedal securely returns to the fully released position.

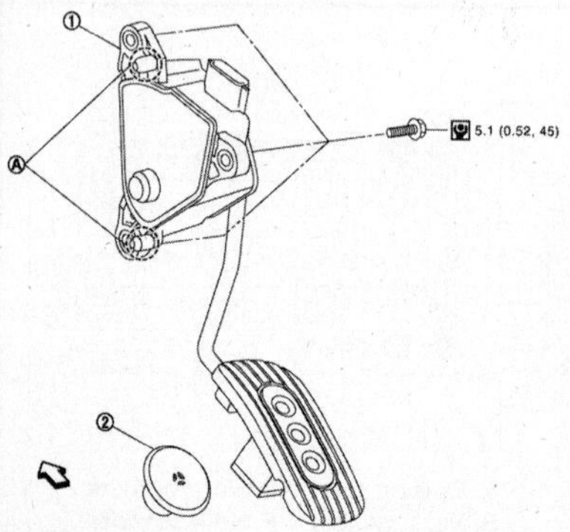

✳✳ WARNING

When harness connector of accelerator pedal position sensor is disconnected, perform "ACCELERATOR PEDAL RELEASED POSITION LEARNING".

ACCELERATOR PEDAL RELEASED POSITION LEARNING

1. Make sure that accelerator pedal is fully released.
2. Turn ignition switch ON and wait at least 2 seconds.
3. Turn ignition switch OFF and wait at least 10 seconds.
4. Turn ignition switch ON and wait at least 2 seconds.
5. Turn ignition switch OFF and wait at least 10 seconds.

AIR-FUEL RATIO SENSOR

LOCATION

See Figure 103.

REMOVAL & INSTALLATION

➡**The manufacturer does not provide a specific Removal and Installation procedure for this component. Refer to the graphic(s) when servicing this component.**

CAMSHAFT POSITION (CMP) SENSOR

LOCATION

See Figure 104.

Refer to the accompanying illustration.

REMOVAL & INSTALLATION

See Figure 105.

1. Turn ignition switch OFF.
2. Loosen the fixing bolt of the sensor.
3. Disconnect Camshaft Position (CMP) sensor (PHASE) harness connector.
4. Remove the sensor.
5. Installation is the reverse of removal.

CRANKSHAFT POSITION (CKP) SENSOR

LOCATION

The Crankshaft Position (CKP) sensor (POS) is located on the oil pan facing the gear teeth (cogs) of the signal plate. It

1. Accelerator pedal assembly   2. Accelerator pedal stopper (Under floor carpet)
A. Locating pin

37663_CUBE_G0208

Fig. 102 Accelerator pedal assembly

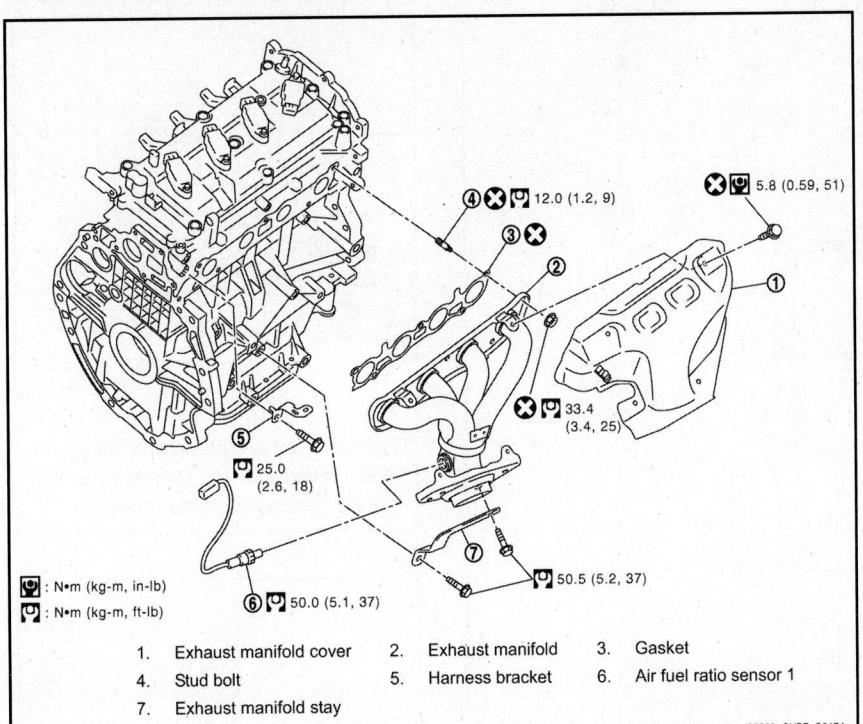

: N•m (kg-m, in-lb)

: N•m (kg-m, ft-lb)

1.	Exhaust manifold cover	2.	Exhaust manifold	3.	Gasket
4.	Stud bolt	5.	Harness bracket	6.	Air fuel ratio sensor 1
7.	Exhaust manifold stay				

37663_CUBE_G0174

Fig. 103 Exploded view of exhaust manifold assembly showing air fuel ratio sensor 1 location

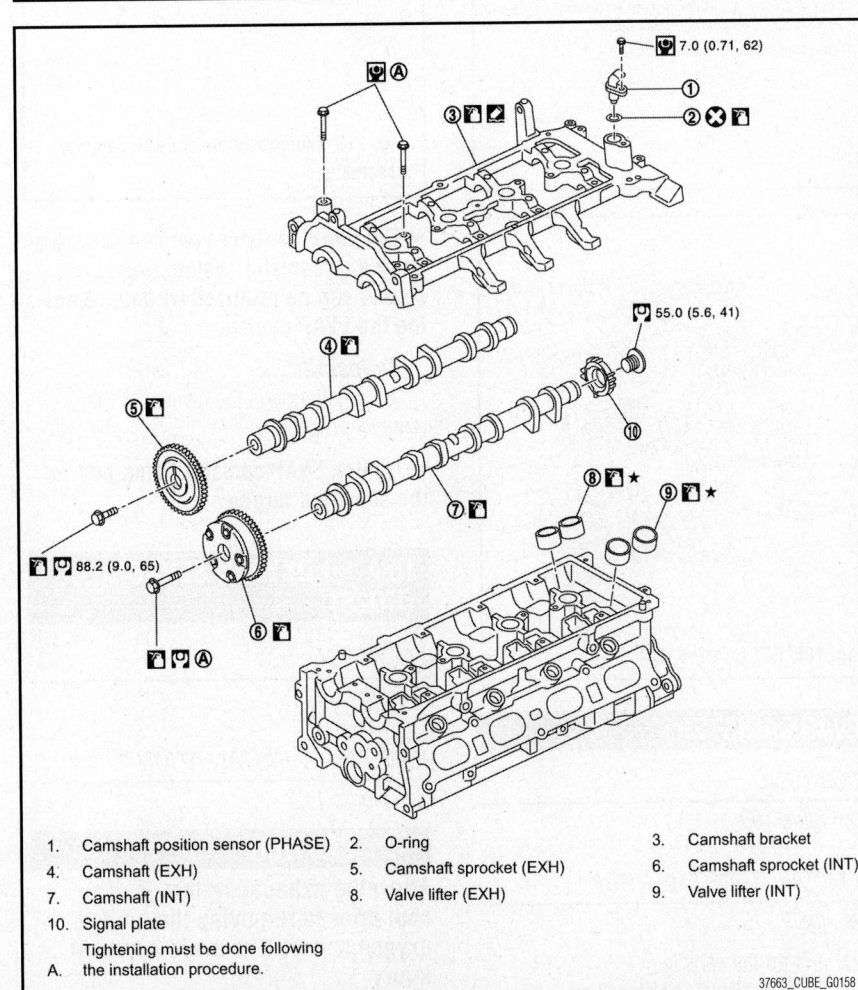

1.	Camshaft position sensor (PHASE)	2.	O-ring	3.	Camshaft bracket
4.	Camshaft (EXH)	5.	Camshaft sprocket (EXH)	6.	Camshaft sprocket (INT)
7.	Camshaft (INT)	8.	Valve lifter (EXH)	9.	Valve lifter (INT)
10.	Signal plate				
	Tightening must be done following				
A.	the installation procedure.				

37663_CUBE_G0158

Fig. 104 Exploded view of camshaft assembly showing Camshaft Position (CMP) sensor

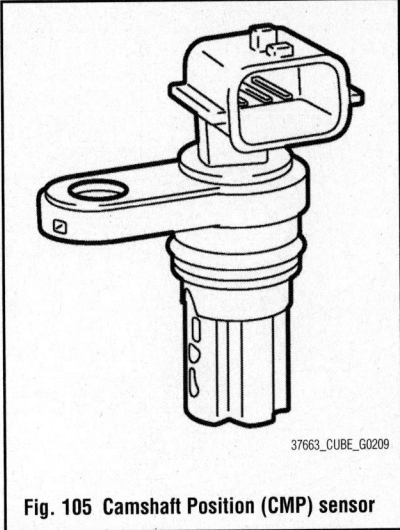

37663_CUBE_G0209

Fig. 105 Camshaft Position (CMP) sensor

detects the fluctuation of the engine revolution.

REMOVAL & INSTALLATION

See Figure 106.

1. Turn ignition switch OFF.
2. Loosen the fixing bolt of the sensor.
3. Disconnect Crankshaft Position (CKP) sensor (POS) harness connector.
4. Remove the sensor.
5. Installation is the reverse of removal.

ELECTRONIC CONTROL MODULE

LOCATION

See Figure 107.

Refer to the accompanying illustration.

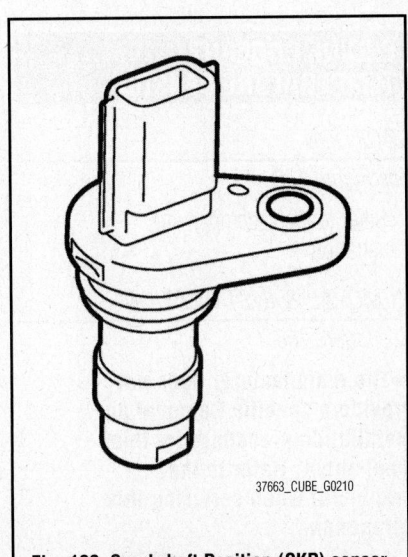

37663_CUBE_G0210

Fig. 106 Crankshaft Position (CKP) sensor

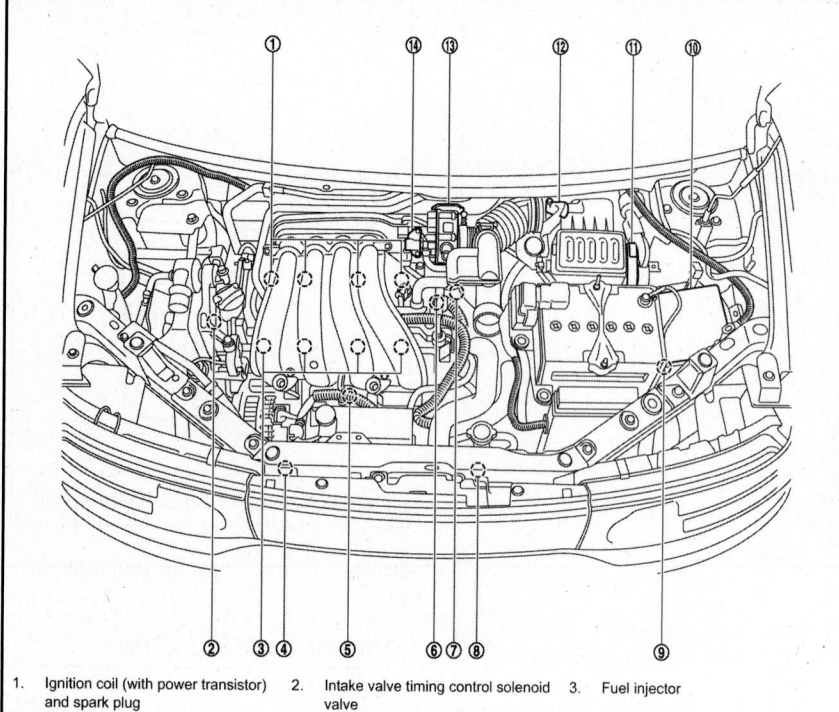

1. Ignition coil (with power transistor) and spark plug
2. Intake valve timing control solenoid valve
3. Fuel injector
4. Refrigerant pressure sensor
5. Knock sensor
6. Camshaft position sensor (PHASE)
7. Engine coolant temperature sensor
8. Cooling fan motor
9. Battery current sensor
10. IPDM E/R
11. ECM
12. Mass air flow sensor (with intake air temperature sensor)
13. Electric throttle control actuator (with built in throttle position sensor and throttle control motor)
14. EVAP canister purge volume control solenoid valve

37663_CUBE_G0207

Fig. 107 Engine control component locations

REMOVAL & INSTALLATION

See Figure 107.

➡The manufacturer does not provide a specific Removal and Installation procedure for this component. Refer to the graphic(s) when servicing this component.

ENGINE COOLANT TEMPERATURE SENSOR

LOCATION

See Figure 108.

Refer to the accompanying illustration.

REMOVAL & INSTALLATION

See Figure 108.

➡The manufacturer does not provide a specific Removal and Installation procedure for this component. Refer to the graphic(s) when servicing this component.

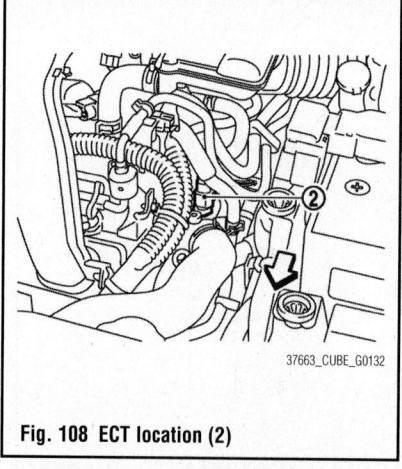

37663_CUBE_G0132

Fig. 108 ECT location (2)

EVAP CANISTER

LOCATION

See Figure 109.

REMOVAL & INSTALLATION

See Figure 110.

1. Lift up the vehicle.
2. Remove EVAP canister fixing bolt.
3. Remove EVAP canister.

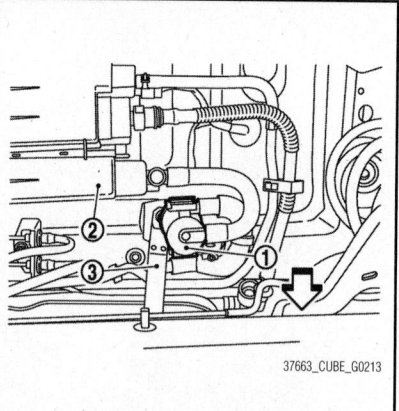

37663_CUBE_G0213

Fig. 109 EVAP canister location showing EVAP canister filter (1), EVAP canister (2), and fuel tank mounting band (RH) (3)

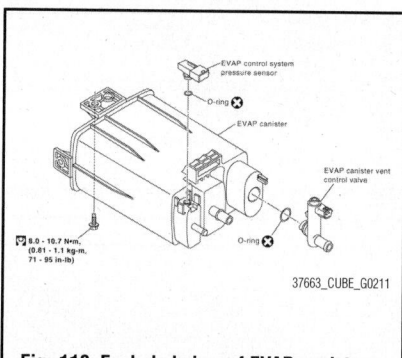

37663_CUBE_G0211

Fig. 110 Exploded view of EVAP canister assembly

➡The EVAP canister vent control valve and EVAP canister system pressure sensor can be removed without removing the EVAP canister.

To install:

4. Installation is the reverse of the removal procedure.

➡Tighten EVAP canister fixing bolt to the specified torque.

HEATED OXYGEN SENSOR (HO2S)

LOCATION

See Figure 111.

REMOVAL & INSTALLATION

See Figure 112.

✳✳ CAUTION

Allow the exhaust system to cool prior to removing the heated oxygen sensor 2 to avoid personal injury.

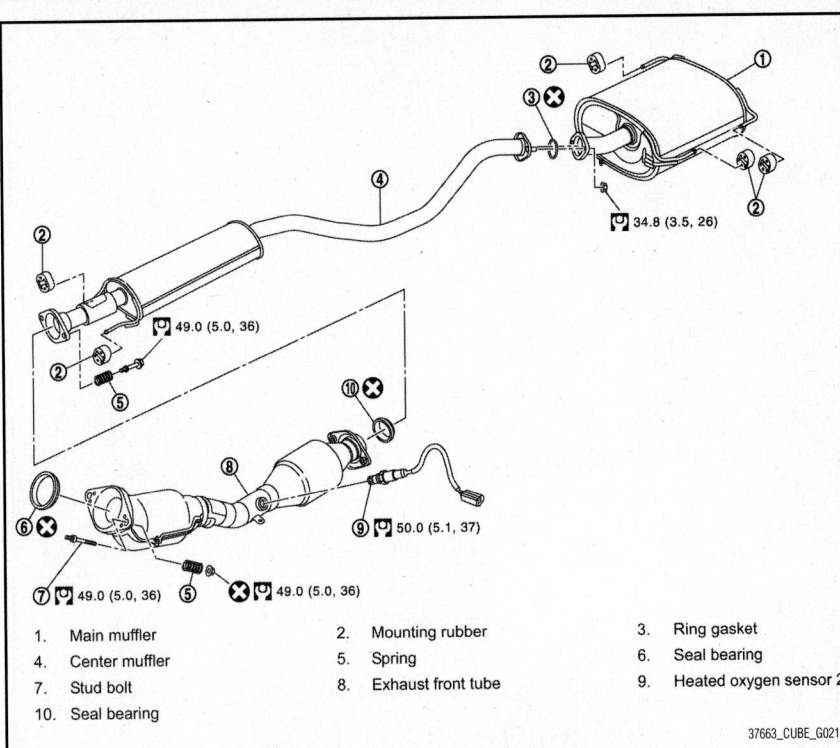

1. Main muffler
2. Mounting rubber
3. Ring gasket
4. Center muffler
5. Spring
6. Seal bearing
7. Stud bolt
8. Exhaust front tube
9. Heated oxygen sensor 2
10. Seal bearing

37663_CUBE_G0215

Fig. 111 Exploded view of exhaust system showing heated oxygen sensor 2 location

1. Remove heated oxygen sensor 2 with following procedure:

a. Using heated oxygen sensor wrench KV10114400 (J-38365), remove the heated oxygen sensor 2.

✳✳ WARNING

Be careful not to damage heated oxygen sensor 2.

To install:

2. Note the following, and install in the reverse order of removal.

Note the following:

• Discard any heated oxygen sensor 2

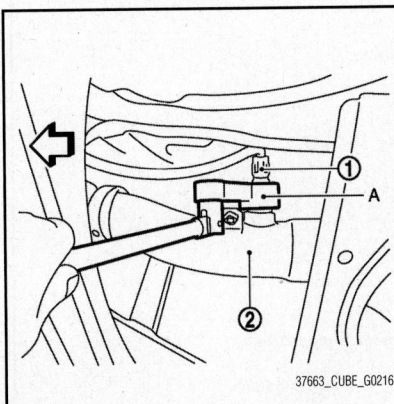

37663_CUBE_G0216

Fig. 112 Using heated oxygen sensor wrench (A), remove the heated oxygen sensor 2 (1)

which has been dropped onto a hard surface such as a concrete floor. Use a new one.

• Before installing a new heated oxygen sensor 2, clean exhaust system threads using the heated oxygen sensor thread cleaner (J-43897-18 or J-43897-12) and apply anti-seize lubricant.

• Never over torque heated oxygen sensor 2. Doing so may cause damage to the heated oxygen sensor 2, resulting in the "MIL" coming on.

INTAKE AIR TEMPERATURE SENSOR

LOCATION

The intake air temperature sensor is built-into Mass Air Flow (MAF) sensor. The sensor detects intake air temperature and transmits a signal to the ECM.

The temperature sensing unit uses a thermistor which is sensitive to the change in temperature. Electrical resistance of the thermistor decreases in response to the temperature rise.

REMOVAL & INSTALLATION

The Intake Air Temperature (IAT) sensor is an integral part of the Mass Air Flow (MAF) sensor/Intake Air Temperature (IAT) sensor assembly which is mounted on the air intake duct. Refer to the Mass Air Flow

section for information regarding servicing this component.

KNOCK SENSOR (KS)

LOCATION

See Figure 113.

The knock sensor is attached to the cylinder block.

REMOVAL & INSTALLATION

See Figure 113.

➡**The manufacturer does not provide a specific Removal and Installation procedure for this component. Refer to the graphic(s) when servicing this component.**

MALFUNCTION INDICATOR LIGHT

RESET PROCEDURE

The Malfunction Indicator Lamp (MIL) is located on the combination meter. The MIL will light up when the ignition switch is turned ON without the engine running. This is a bulb check.

When the engine is started, the MIL should go off. If the MIL remains on, the on board diagnostic system has detected an engine system malfunction.

MASS AIR FLOW (MAF) SENSOR

LOCATION

The Mass Air Flow (MAF) sensor is placed in the stream of intake air. It measures the intake flow rate by measuring a part of the entire intake flow. The Mass Air Flow (MAF) sensor controls the temperature of the hot wire to a certain amount. The heat generated by the hot wire is reduced as the intake air flows around it. The more air, the greater the heat loss.

Therefore, the electric current supplied to hot wire is changed to maintain the temperature of the hot wire as air flow increases. The ECM detects the air flow by means of this current change.

REMOVAL & INSTALLATION

See Figure 114.

➡**The manufacturer does not provide a specific Removal and Installation procedure for this component. Refer to the graphic(s) when servicing this component.**

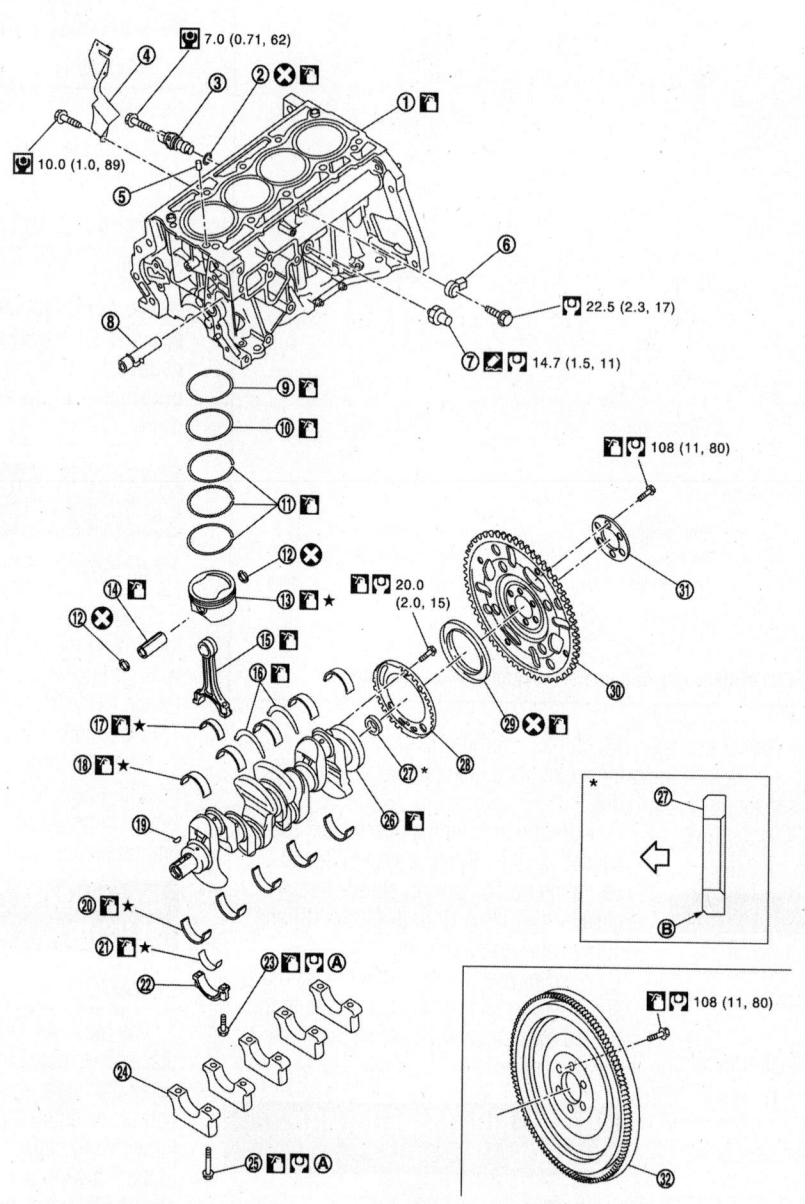

1. Cylinder block
2. O-ring
3. Crankshaft position sensor (POS)
4. Crankshaft position sensor (POS) cover
5. Oil filter (for intake valve timing control)
6. Knock sensor
7. Oil pressure switch
8. Cylinder block heater (for Canada)
9. Top ring
10. Second ring
11. Oil ring
12. Snap ring
13. Piston
14. Piston pin
15. Connecting rod
16. Thrust bearing
17. Connecting rod bearing (upper)
18. Main bearing (upper)
19. Crankshaft key
20. Main bearing (lower)
21. Connecting rod bearing (lower)
22. Connecting rod cap
23. Connecting rod cap bolt
24. Main bearing cap
25. Main bearing cap bolt
26. Crankshaft
27. Pilot converter (CVT models)
28. Signal plate
29. Rear oil seal
30. Drive plate (CVT models)
31. Reinforcement plate (CVT models)
32. Flywheel (M/T models)
A. Tightening must be done following the assembly procedure.
B. Chamfered

37663_CUBE_G0217

Fig. 113 Exploded view of cylinder block assembly showing knock sensor location

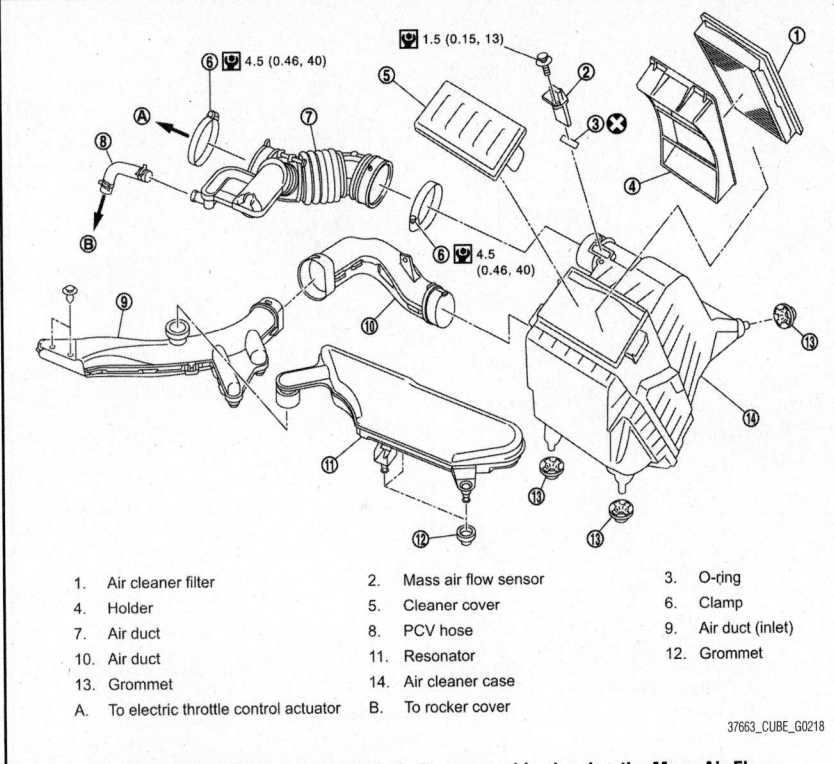

1. Air cleaner filter
2. Mass air flow sensor
3. O-ring
4. Holder
5. Cleaner cover
6. Clamp
7. Air duct
8. PCV hose
9. Air duct (inlet)
10. Air duct
11. Resonator
12. Grommet
13. Grommet
14. Air cleaner case
A. To electric throttle control actuator
B. To rocker cover

37663_CUBE_G0218

Fig. 114 Exploded view of air cleaner and air duct assembly showing the Mass Air Flow (MAF) sensor location

THROTTLE POSITION SENSOR (TPS)

LOCATION

The Throttle Position Sensor (TPS) is an integral part of the electric throttle control actuator.

REMOVAL & INSTALLATION

Refer to the electric throttle control actuator section when servicing this component.

VEHICLE SPEED SENSOR

LOCATION

See Figure 117.

REMOVAL & INSTALLATION

1. Remove battery.
2. Remove air duct (inlet), air duct and air cleaner case.
3. Remove battery bracket.
4. Remove control cable from manual lever.

POSITIVE CRANKCASE VENTILATION (PCV) VALVE

LOCATION

See Figure 115.

REMOVAL & INSTALLATION

See Figure 115.

➡The manufacturer does not provide a specific Removal and Installation procedure for this component. Refer to the graphic(s) when servicing this component.

THROTTLE CONTROL ACTUATOR

LOCATION

See Figure 116.

REMOVAL & INSTALLATION

See Figure 116.

➡The manufacturer does not provide a specific Removal and Installation procedure for this component. Refer to the graphic(s) when servicing this component.

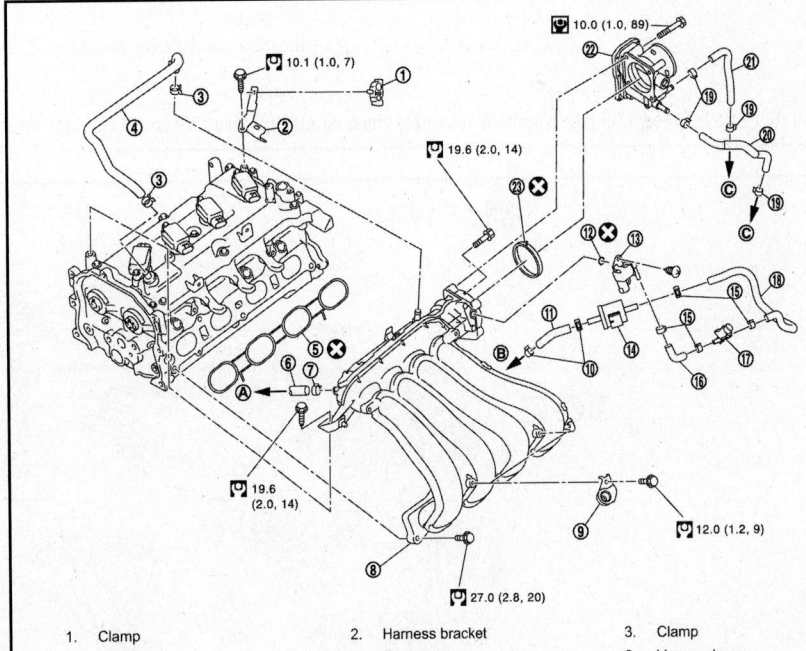

1. Clamp
2. Harness bracket
3. Clamp
4. PCV hose
5. Gasket
6. Vacuum hose
7. Clamp
8. Intake manifold
9. Bracket
10. Clamp
11. EVAP hose
12. O-ring
13. EVAP canister purge volume control solenoid valve
14. EVAP vacuum tank
15. Clamp
16. EVAP hose
17. EVAP service port
18. EVAP hose
19. Clamp
20. Water hose
21. Water hose
22. Electric throttle control actuator
23. Gasket
A. To brake booster
B. To centralized under-floor piping
C. To water outlet

37663_CUBE_G0177

Fig. 115 Exploded view of intake manifold assembly showing PCV valve location

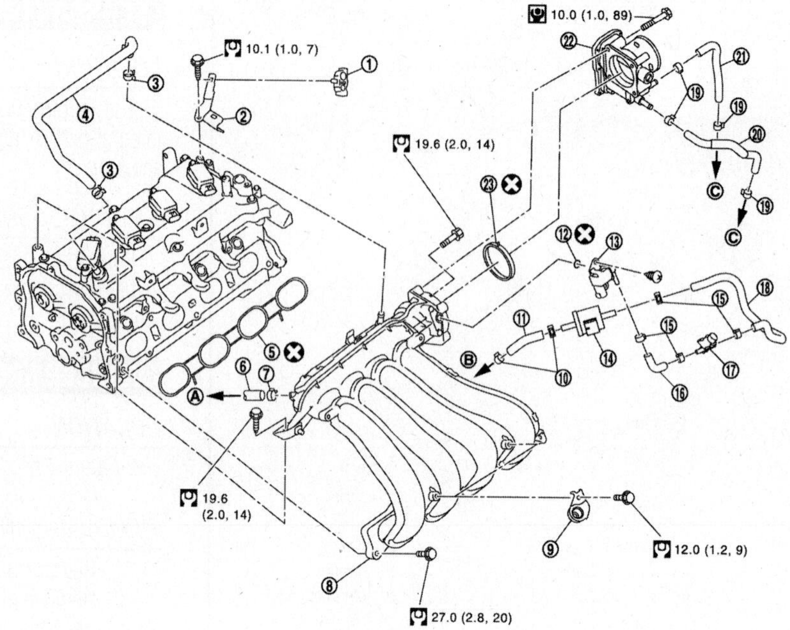

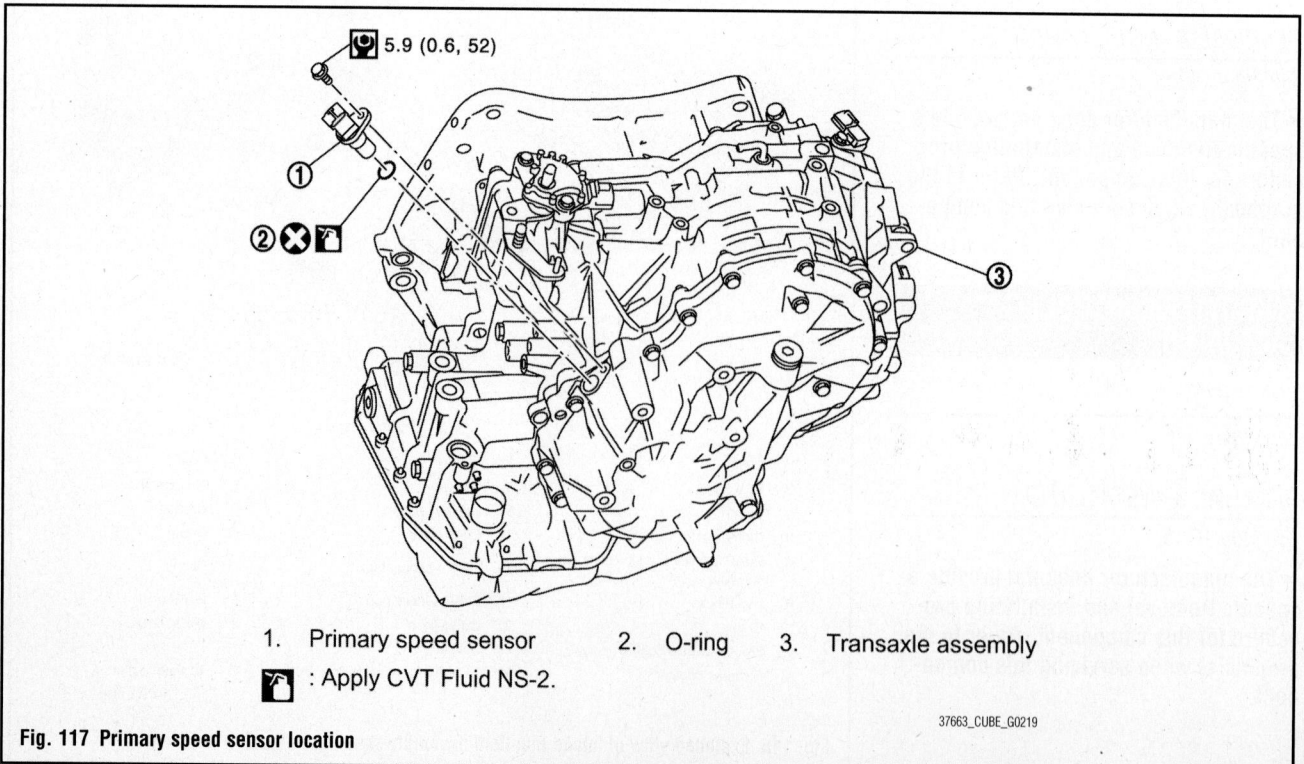

Fig. 116 Exploded view of intake manifold assembly showing electric throttle control actuator location

1. Clamp	2. Harness bracket	3. Clamp
4. PCV hose.	5. Gasket	6. Vacuum hose
7. Clamp	8. Intake manifold	9. Bracket
10. Clamp	11. EVAP hose	12. O-ring
13. EVAP canister purge volume control solenoid valve	14. EVAP vacuum tank	15. Clamp
16. EVAP hose	17. EVAP service port	18. EVAP hose
19. Clamp	20. Water hose	21. Water hose
22. Electric throttle control actuator	23. Gasket	
A. To brake booster	B. To centralized under-floor piping	C. To water outlet

37663_CUBE_G0177

1. Primary speed sensor   2. O-ring   3. Transaxle assembly

: Apply CVT Fluid NS-2.

37663_CUBE_G0219

Fig. 117 Primary speed sensor location

5. Place manual lever to "L" position.
6. Disconnect primary speed sensor connector.
7. Remove primary speed sensor.
8. Remove O-ring from primary speed sensor.

To install:

9. Note the following, and install in the reverse order of removal.

✳✳ WARNING

Never reuse O-ring. Apply CVT fluid to O-ring.

10. Check for CVT fluid leakage and check CVT fluid level.
11. Check the CVT position.

FUEL   GASOLINE FUEL INJECTION SYSTEM

FUEL SYSTEM SERVICE PRECAUTIONS

Safety is the most important factor when performing not only fuel system maintenance but any type of maintenance. Failure to conduct maintenance and repairs in a safe manner may result in serious personal injury or death. Maintenance and testing of the vehicle's fuel system components can be accomplished safely and effectively by adhering to the following rules and guidelines.

• To avoid the possibility of fire and personal injury, always disconnect the negative battery cable unless the repair or test procedure requires that battery voltage be applied.

• Always relieve the fuel system pressure prior to disconnecting any fuel system component (injector, fuel rail, pressure regulator, etc.), fitting or fuel line connection. Exercise extreme caution whenever relieving fuel system pressure to avoid exposing skin, face and eyes to fuel spray. Please be advised that fuel under pressure may penetrate the skin or any part of the body that it contacts.

• Always place a shop towel or cloth around the fitting or connection prior to loosening to absorb any excess fuel due to spillage. Ensure that all fuel spillage (should it occur) is quickly removed from engine surfaces. Ensure that all fuel soaked cloths or towels are deposited into a suitable waste container.

• Always keep a dry chemical (Class B) fire extinguisher near the work area.

• Do not allow fuel spray or fuel vapors to come into contact with a spark or open flame.

• Always use a back-up wrench when loosening and tightening fuel line connection fittings. This will prevent unnecessary stress and torsion to fuel line piping.

• Always replace worn fuel fitting O-rings with new Do not substitute fuel hose or equivalent where fuel pipe is installed.

Before servicing the vehicle, make sure to also refer to the precautions in the beginning of this section as well.

RELIEVING FUEL SYSTEM PRESSURE

RELIEVING

With CONSULT-III

1. Turn ignition switch ON.
2. Perform "FUEL PRESSURE RELEASE" in "WORK SUPPORT" mode with CONSULT-III.
3. Start engine.
4. After engine stalls, crank it two or three times to release all fuel pressure.
5. Turn ignition switch OFF.

Without CONSULT-III

1. Remove fuel pump fuse located in IPDM E/R.
2. Start engine.
3. After engine stalls, crank it two or three times to release all fuel pressure.
4. Turn ignition switch OFF.
5. Reinstall fuel pump fuse after servicing fuel system.

FUEL PUMP, FUEL FILTER & SENSOR UNIT

REMOVAL & INSTALLATION
See Figures 118 through 122.

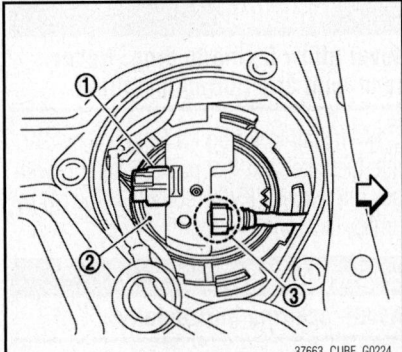

Fig. 118 Disconnect harness connector (1) and quick connector (3) from the fuel level sensor unit, fuel filter and fuel pump assembly (2)

1. Release the fuel pressure from the fuel lines.
2. Check fuel level on fuel gauge. If fuel gauge indicates more than the level as shown (full or almost full), drain fuel from fuel tank until fuel gauge indicates level as shown or below.

➡**Because fuel will be spilled when removing fuel level sensor units for the top of the fuel is above the fuel level sensor units installation surface.**

a. As a guide, drain approximately 2⅝ gal. (10L) of fuel.

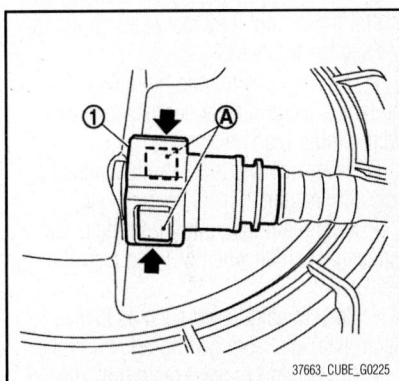

Fig. 119 Pinch quick connector square-part (A) with your fingers, and pull out the quick connector (1)

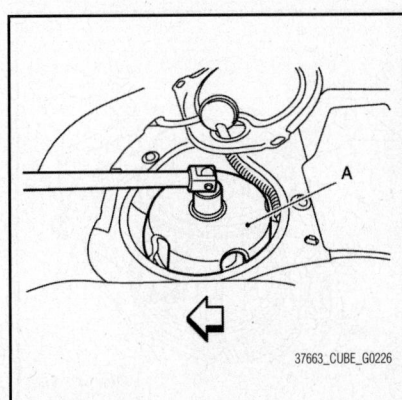

Fig. 120 Remove lock ring for fuel level sensor unit, fuel filter and fuel pump assembly with fuel tank lock ring wrench (A) by turning counterclockwise

b. In a case that fuel pump does not operate, perform the following procedure.

c. Insert hose of less than 1 inch (25 mm) in diameter into fuel filler tube through fuel filler opening to draw fuel from fuel filler tube.

d. Disconnect fuel filler hose from fuel filler tube.

e. Insert hose into fuel tank through fuel filler hose to draw fuel from fuel tank.

3. Open fuel filler lid.

4. Open filler cap and release the pressure inside fuel tank.

5. Remove rear seat.

6. Remove inspection hole cover. Using a screwdriver, remove it by turning clips clockwise by 90 degrees.

7. Disconnect harness connector and quick connector.

8. Remove quick connector in the following procedures.

   a. Pinch quick connector square-part with your fingers, and pull out the quick connector by hand.

   b. If quick connector and tube on sender unit are stuck, push and pull several times until they move, and pull out. Note the following:

• Quick connector can be removed when the tabs are completely depressed. Never twist it more than necessary.

• Never use any tools to disconnected quick connector.

• Keep resin tube away from heat. Be especially careful when welding near the resin tube.

• Prevent acid liquid such as battery electrolyte, etc. from getting on resin tube.

• Never bend or twist resin tube during installation and disconnection.

• To keep the connecting portion clean and to avoid damage and foreign materials,

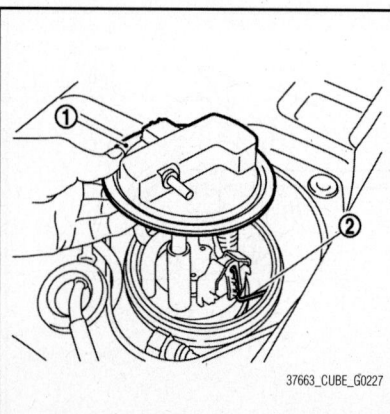

Fig. 121 Remove fuel level sensor unit, fuel filter and fuel pump assembly (1); float arm (2)

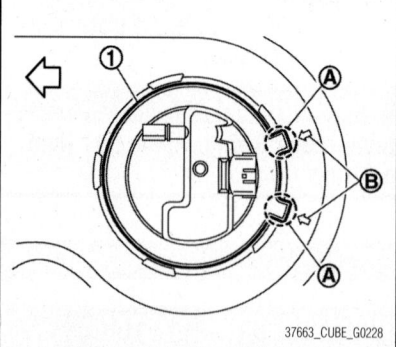

Fig. 122 Align (A) with (B) as shown; install fuel level sensor unit, fuel filter and fuel pump assembly (1)

cover them completely with plastic bags or something similar.

• Never insert plug, preventing damage on O-ring in quick connector.

9. Remove lock ring for fuel level sensor unit, fuel filter and fuel pump assembly with fuel tank lock ring wrench KV991J0090 (J-46214) by turning counterclockwise.

10. Remove fuel level sensor unit, fuel filter and fuel pump assembly.

Note the following:

• Never bend float arm during removal.

• Never pollute the inside by residue fuel. Draw out avoiding inclination by supporting with a cloth.

• Never cause impacts such by dropping when handling components.

To install:

11. Note to the following, and install in the reverse order of removal.

12. Install new O-ring to fuel tank without any twist.

13. Align A with B as shown. Install fuel level sensor unit, fuel filter and fuel pump assembly to fuel tank.

✲✲ WARNING

Never allow O-ring to drop. Never bend float arm during installing.

14. Install lock ring for fuel level sensor unit, fuel filter and fuel pump assembly with lock ring wrench KV991J0090 (J-46214) by turning clockwise.

✲✲ WARNING

Install lock ring horizontally.

15. Connect quick connector as follows:
   a. Check the connection for damage or any foreign materials.
   b. Align the connector with the tube, then insert the connector straight into the tube until a click sound is heard.

---

   c. After connecting, check that the connection is secure by following method.

   d. Visually confirm that the two tabs are connected to the connector.

   e. Pull (A) the tube and the connector to check they are securely connected.

16. Turn ignition switch "ON" (with engine stopped), then check connections for leakage by applying fuel pressure to fuel piping.

17. Start engine and let it idle and check there are no fuel leakage at the fuel system connections.

FUEL RAIL & INJECTORS

REMOVAL & INSTALLATION

See Figures 123 through 127.

✲✲ WARNING

Never remove or disassemble parts unless instructed.

Note the following:

• Be sure to work in a well-ventilated area and furnish workshop with a CO_2 fire extinguisher.

• Never smoke while servicing fuel system. Keep open flames and sparks away from the work area.

1. Release the fuel pressure.

2. Remove intake manifold.

3. Disconnect quick connector with the following procedure.

   a. Disconnect fuel feed tube from fuel tube.

➡**There is no fuel return path.**

   b. Remove quick connector cap (engine side) from quick connector connection.

Fig. 123 Disconnect fuel feed tube (1) from fuel tube (3); remove quick connector cap (engine side) (2)

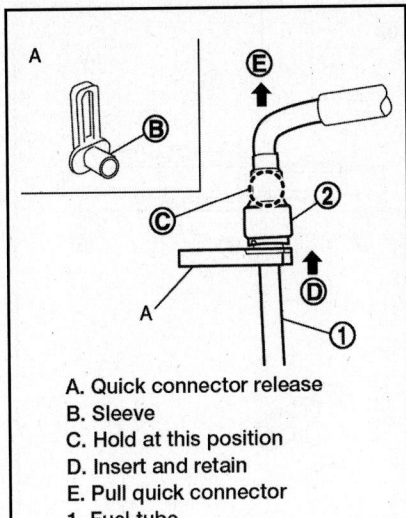

A. Quick connector release
B. Sleeve
C. Hold at this position
D. Insert and retain
E. Pull quick connector
1. Fuel tube
2. Quick connector

37663_CUBE_G0230

Fig. 124 Insert quick connector release into quick connector until sleeve contacts and goes no further

c. With the sleeve side of quick connector release facing quick connector, install quick connector release onto fuel tube.

d. Insert quick connector release into quick connector until sleeve contacts and goes no further. Hold quick connector release on that position.

✳✳ WARNING

Inserting quick connector release hard will not disconnect quick connector. Hold quick connector release where it contacts and goes no further.

e. Draw and pull out quick connector straight from fuel tube.
Note the following:
• Pull quick connector at holding position.
• Never pull with lateral force applied.

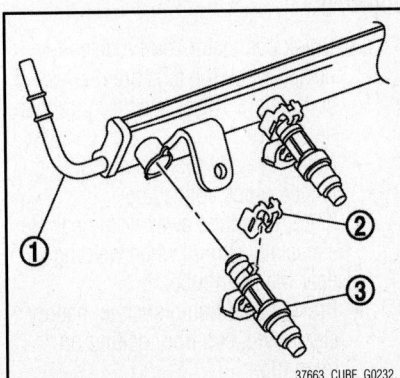

Fig. 125 Open and remove clip (2); remove fuel injector (3) from fuel tube (1) by pulling straight

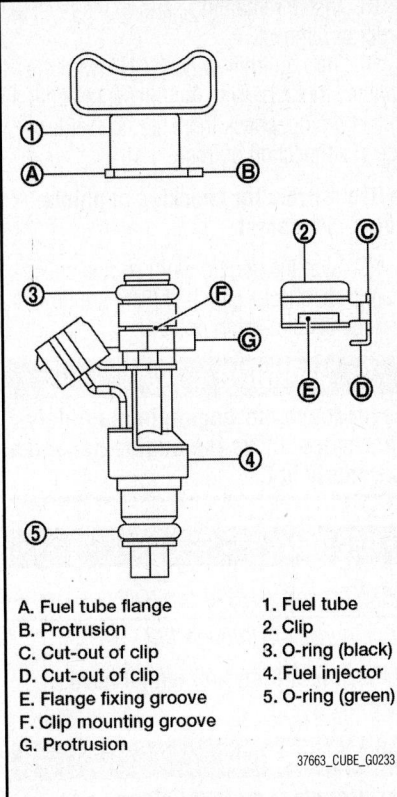

A. Fuel tube flange
B. Protrusion
C. Cut-out of clip
D. Cut-out of clip
E. Flange fixing groove
F. Clip mounting groove
G. Protrusion
1. Fuel tube
2. Clip
3. O-ring (black)
4. Fuel injector
5. O-ring (green)

37663_CUBE_G0233

Fig. 126 Install fuel injector to fuel tube

O-ring inside quick connector may be damaged.
• Prepare container and cloth beforehand as fuel will leakage out.
• Avoid fire and sparks.
• Keep parts away from heat source. Especially, be careful when welding is performed around them.
• Never expose parts to battery electrolyte or other acids.
• Never bend or twist connection between quick connector and fuel feed tube during installation/removal.
• To keep clean the connecting portion and to avoid damage and foreign materials, cover them completely with plastic bags, etc. or something similar.
4. Disconnect harness connector from fuel injector.
5. Remove fuel tube and fuel injector assembly.
6. Loosen mounting bolts in reverse order of the tightening sequence.
• When removing, be careful to avoid any interference with fuel injector.
• Use a shop cloth to absorb any fuel leakage from fuel tube.
7. Remove fuel injector from fuel tube with the following procedure:
a. Open and remove clip.
b. Remove fuel injector from fuel tube by pulling straight.

Note the following:
• Be careful with remaining fuel that may go out from fuel tube.
• Be careful not to damage fuel injector nozzle during removal.
• Never bump or drop fuel injector.
• Never disassemble fuel injector.

To install:
8. Note the following, and install O-rings to fuel injector.
• Upper and lower O-rings are different. Be careful not to confuse them.
• Handle O-ring with bare hands. Never wear gloves.
• Lubricate O-ring with new engine oil.
• Never clean O-ring with solvent.
• Check that O-ring and its mating part are free of foreign material.
• When installing O-ring, be careful not to scratch it with tool or fingernails. Also be careful not to twist or stretch O-ring. If O-ring is stretched while installing, never insert it quickly into fuel tube.
• Insert O-ring straight into fuel tube. Never decenter or twist it.
9. Install fuel injector to fuel tube with the following procedure:
a. Insert clip into clip mounting groove on fuel injector.
b. Insert clip so that protrusion of fuel injector matches cutout of clip.

✳✳ WARNING

Never reuse clip. Replace it with a new one. Be careful to keep clip from interfering with O-ring. If interference occurs, replace O-ring.

c. Insert fuel injector into fuel tube with clip attached.
d. Insert it while matching it to the axial center.
e. Insert fuel injector so that protrusion of fuel tube matches cut-out of clip.

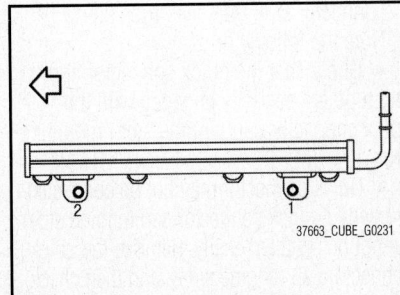

37663_CUBE_G0231

Fig. 127 Tighten mounting bolts in numerical order as shown

f. Check that fuel tube flange is securely fixed in flange fixing groove on clip.

g. Check that installation is complete by checking that fuel injector does not rotate or come off.

10. Set fuel tube and fuel injector assembly at its position for installation on cylinder head.

✷✷ WARNING

For installation, be careful not to interfere with fuel injector nozzle.

11. Install fuel tube and injector assembly onto cylinder.

12. Tighten mounting bolts in numerical order as shown.

13. Connect harness connector to fuel injector.

14. Connect fuel feed tube with the following procedure.

a. Check for damage or foreign material on the fuel tube and quick connector.

b. Apply new engine oil lightly to area around the top of fuel tube.

c. Align center to insert quick connector straightly into fuel tube.

d. Insert quick connector to fuel tube until the top spool on fuel tube is inserted completely and the 2nd level spool is positioned slightly below quick connector bottom end.

- Carefully align center to avoid inclined insertion to prevent damage to O-ring inside quick connector.
- Insert until you hear a "click" sound and actually feel the engagement.
- To avoid misidentification of engagement with a similar sound, be sure to perform the next step.

e. Pull quick connector hard by hand holding position. Check it is completely engaged (connected) so that it does not come out from fuel tube.

f. Install quick connector cap (engine side) to quick connector connection.

g. Install quick connector cap (engine side) with the side arrow facing quick connector side (fuel feed tube side). Note the following:

- Check that the quick connector and fuel tube are securely engaged with the quick connector cap (engine side) mounting groove.
- Quick connector may not be connected correctly if quick connector cap (engine side) cannot be installed easily. Remove the quick connector cap (engine side), and then check the connection of quick connector again.

a. Install fuel feed hose to hose clamp.

15. Install remaining parts in the reverse order of removal.

16. Turn ignition switch "ON" (with the engine stopped). With fuel pressure applied to fuel piping, check there are no fuel leakage at connection points.

➡**Use mirrors for checking at points out of clear sight.**

17. Start the engine. With engine speed increased, check again that there are no fuel leakage at connection points.

✷✷ WARNING

Never touch the engine immediately after stopped, as the engine becomes extremely hot.

FUEL TANK

REMOVAL & INSTALLATION

See Figures 128 through 133.

➡**Drain fuel from fuel tank if necessary.**

✷✷ WARNING

Perform work on level place.

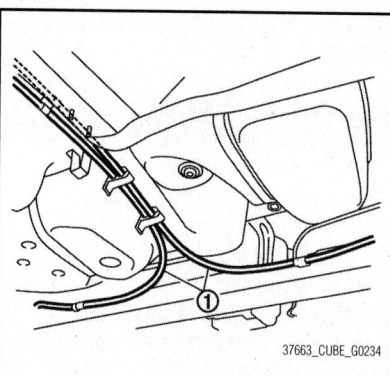

Fig. 128 Move parking brake cable (1) from the lower face of fuel tank

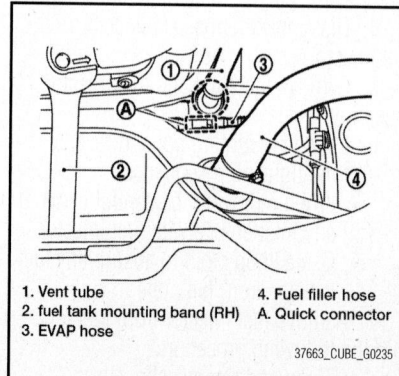

1. Vent tube
2. fuel tank mounting band (RH)
3. EVAP hose
4. Fuel filler hose
A. Quick connector

37663_CUBE_G0235

Fig. 129 Disconnect fuel filler hose at fuel tank side

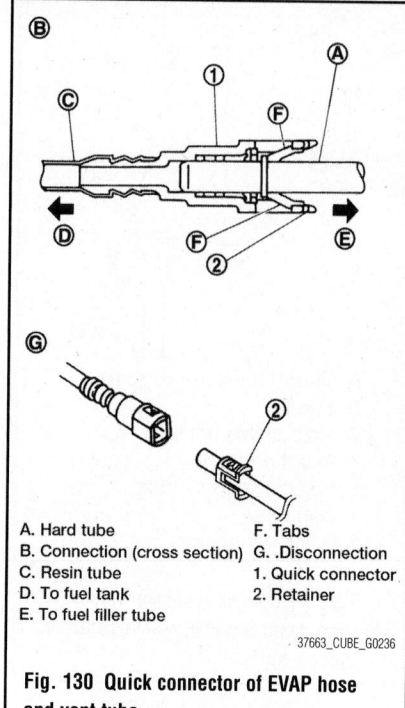

A. Hard tube
B. Connection (cross section)
C. Resin tube
D. To fuel tank
E. To fuel filler tube
F. Tabs
G. .Disconnection
1. Quick connector
2. Retainer

37663_CUBE_G0236

Fig. 130 Quick connector of EVAP hose and vent tube

1. Perform steps 2 to 7 of "Removal" in "Fuel Level Sensor Unit, Fuel Filter And Fuel Pump Assembly" on fuel level sensor unit, fuel filter and fuel pump assembly.

2. Remove center muffler.

3. Remove insulator on vehicle side located above center and main mufflers.

4. Move parking brake cable from the lower face of fuel tank. Then remove clips for parking brake cable.

5. Remove brake tube protector.

6. Disconnect fuel filler hose at fuel tank side.

7. Disconnect EVAP hose and vent tube at the position shown.

➡**Instruction for quick connector of EVAP hose and vent tube, refer to the following:**

- Quick connector can be disconnected when the tabs are depressed completely. Never twist it more than necessary.
- Never use any tools to disconnected quick connector.
- Keep resin tube away from heat. Be especially careful when welding near the resin tube.
- Prevent acid liquid such as battery electrolyte, etc. from getting on resin tube.
- Never bend or twist resin tube during installation and disconnection.
- Never remove the remaining retainer on hard tube (or the equiv-

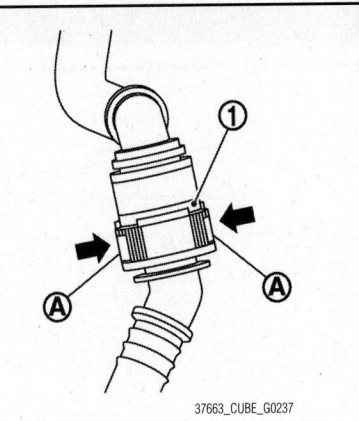

Fig. 131 Pinch quick connector square-part (A) with your fingers, and pull out the quick connector (1) by hand

alent) except when resin tube or retainer is replaced.

- When resin tube or hard tube (or the equivalent) is replaced, also replace retainer with new green one.
- To keep the connecting portion clean and to avoid damage and foreign materials, cover them completely with plastic bags or something similar.
- Never insert plug, preventing

damage on O-ring in quick connector.

8. Remove quick connector in the following procedures.

a. Pinch quick connector square-part with your fingers, and pull out the quick connector by hand.

b. If quick connector and tube on vehicle are stuck, push and pull several times until they move, and pull out.

- The tube can be removed when the tabs are completely depressed. Never twist it more than necessary.
- Never use any tools to disconnect quick connector.
- Keep the resin tube away from heat. Be especially careful when welding near the tube.
- Prevent acid liquid such as battery electrolyte, etc. from getting on the resin tube.
- Never bend or twist resin tube during installation and disconnection.
- To keep the connecting portion clean and to avoid damage and foreign materials, cover them completely with plastic bags or something similar.

- Never insert plug, preventing damage on O-ring in quick connector.

9. Support the center part of fuel tank with transmission jack.

✳✳ WARNING

Securely support the fuel tank with a piece of wood.

10. Remove EVAP canister filter.
11. Remove fuel tank mounting bands (RH and LH).
12. Lower transmission jack carefully to remove fuel tank while holding it by hand.

✳✳ WARNING

Fuel tank may be in an unstable condition because of the shape of fuel tank bottom. Never rely on jack too much. Be sure to hold tank securely.

To install:

13. Note the following, and install in the reverse order of removal.
14. Temporarily tighten bolts except #4 in numerical order as shown.
15. Tighten bolt #4 to 18 ft. lbs. (25 Nm), pressing fuel tank in the direction shown by black arrow.
16. Tighten bolts 1, 2, and 3 to 18 ft. lbs. (25 Nm) in the reverse order as shown.
17. Insert the fuel filler tube into the fuel hose to the length of 1.38 inches (35 mm).
18. Securely clamp the hose and tube. Be sure hose clamp is not placed on swelled area of fuel filler tube.
19. Check connections for damage or foreign material.
20. Align the matching side connection part with the center of shaft, and insert connector straight until it clicks.
21. Turn ignition switch "ON" (with engine stopped), and check connections for leakage by applying fuel pressure to fuel piping.
22. Start engine and rev it up and check there are no fuel leakage at the fuel system tube and hose connections.

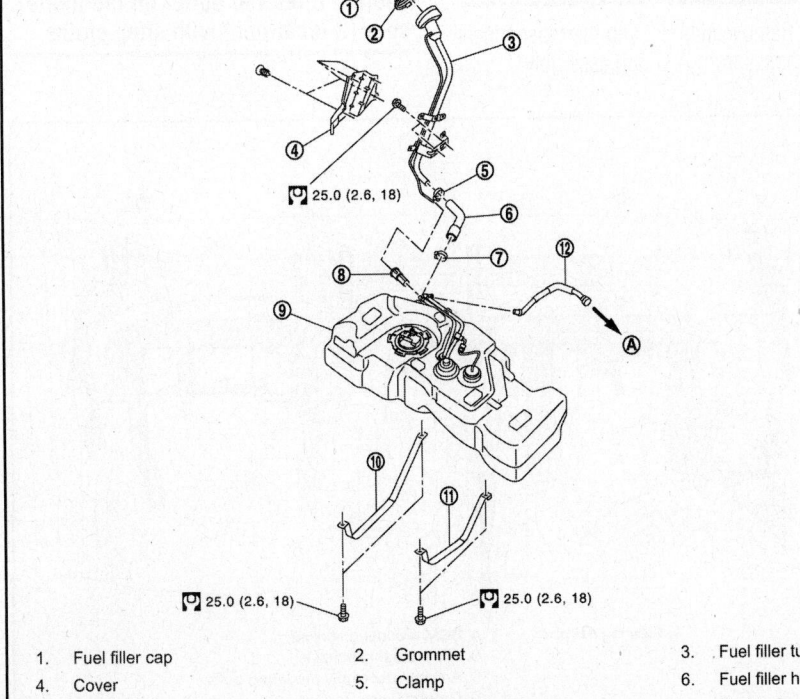

1. Fuel filler cap	2. Grommet	3. Fuel filler tube
4. Cover	5. Clamp	6. Fuel filler hose
7. Clamp	8. EVAP hose protector	9. Fuel tank
10. Fuel tank mounting band (RH)	11. Fuel tank mounting band (LH)	12. Vent tube
A. To EVAP canister		

Fig. 132 Exploded view of the fuel tank assembly

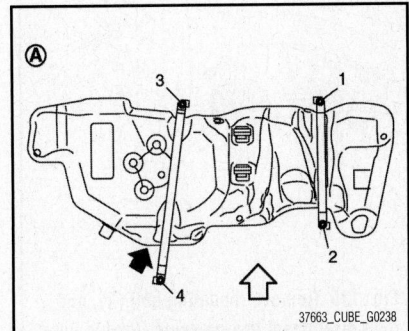

Fig. 133 Temporarily tighten bolts except #4 in numerical order as shown

HEATING, VENTILATION, & AIR CONDITIONING

A/C UNIT

REMOVAL & INSTALLATION

See Figures 134 through 139.

> ❊❊ **WARNING**
>
> **Perform lubricant return operation before each refrigeration system disassembly. However, if a large amount of refrigerant or lubricant is detected, never perform lubricant return operation.**

1. Use a refrigerant collecting equipment (for HFC-134a) to discharge the refrigerant.
2. Drain engine coolant from cooling system.
3. Remove cowl top extension.
4. Remove mounting nut and lower

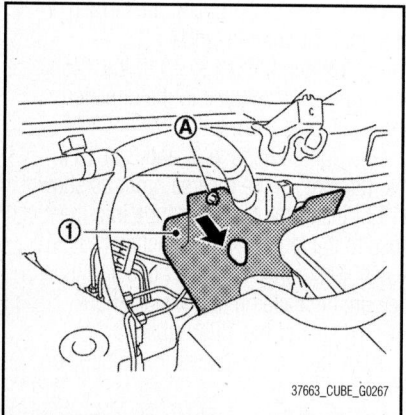

Fig. 134 Remove mounting nut (A), and lower dash insulator (1)

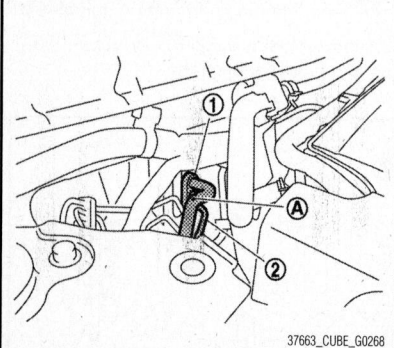

Fig. 135 Remove mounting bolt (A), and then disconnect low-pressure flexible hose (1) and high-pressure pipe (2) from expansion valve

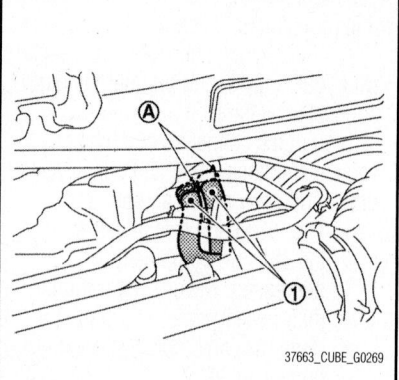

Fig. 136 Remove clamps (A), and then disconnect heater hoses (1) from A/C unit assembly

dash insulator; position without the hindrance for work. (If equipped)

5. Remove mounting bolt, and then disconnect low-pressure flexible hose and high-pressure pipe from expansion valve. (If equipped)

> ❊❊ **WARNING**
>
> **Cap or wrap the joint of the A/C piping and expansion valve with suitable material such as vinyl tape to avoid the entry of air.**

6. Remove clamps, and then disconnect heater hoses from A/C unit assembly.

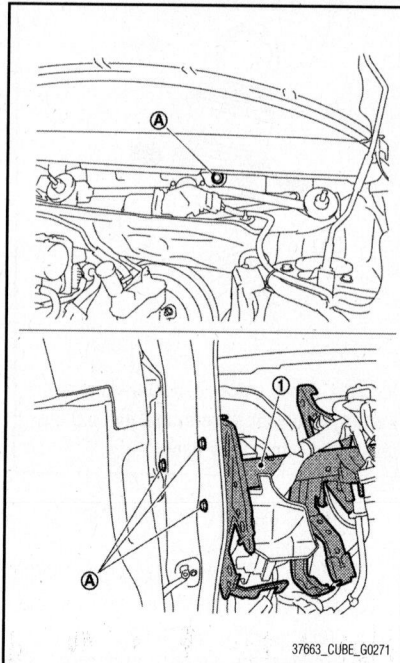

Fig. 138 Remove mounting bolts (A), and then remove steering member (1) from the vehicle

> ❊❊ **WARNING**
>
> **Some coolant may spill when heater hoses are disconnected. Close off the coolant inlet and outlet on the heater core (2 locations) with shop cloths.**

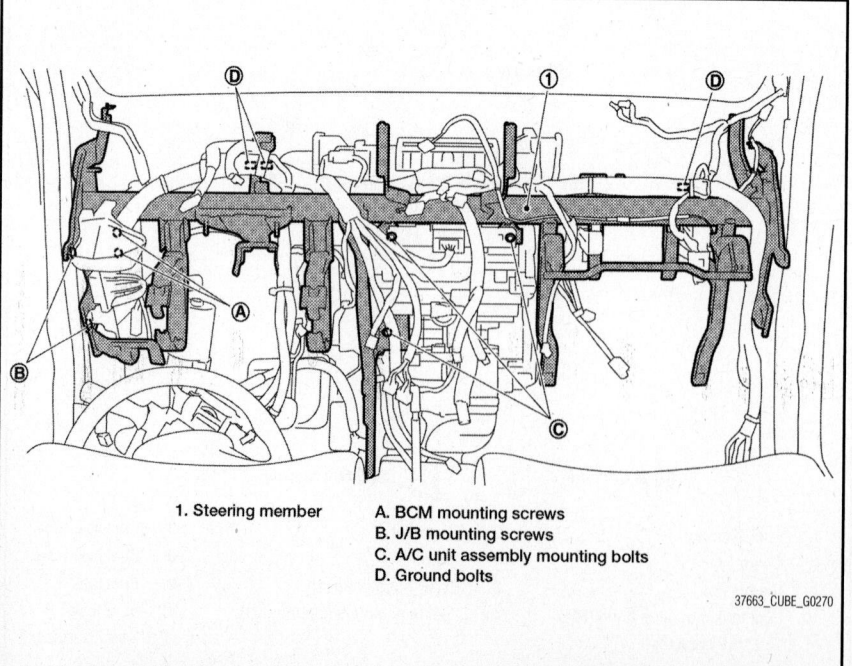

1. Steering member

A. BCM mounting screws
B. J/B mounting screws
C. A/C unit assembly mounting bolts
D. Ground bolts

Fig. 137 Disconnect the harness connectors and clips required to remove the steering member

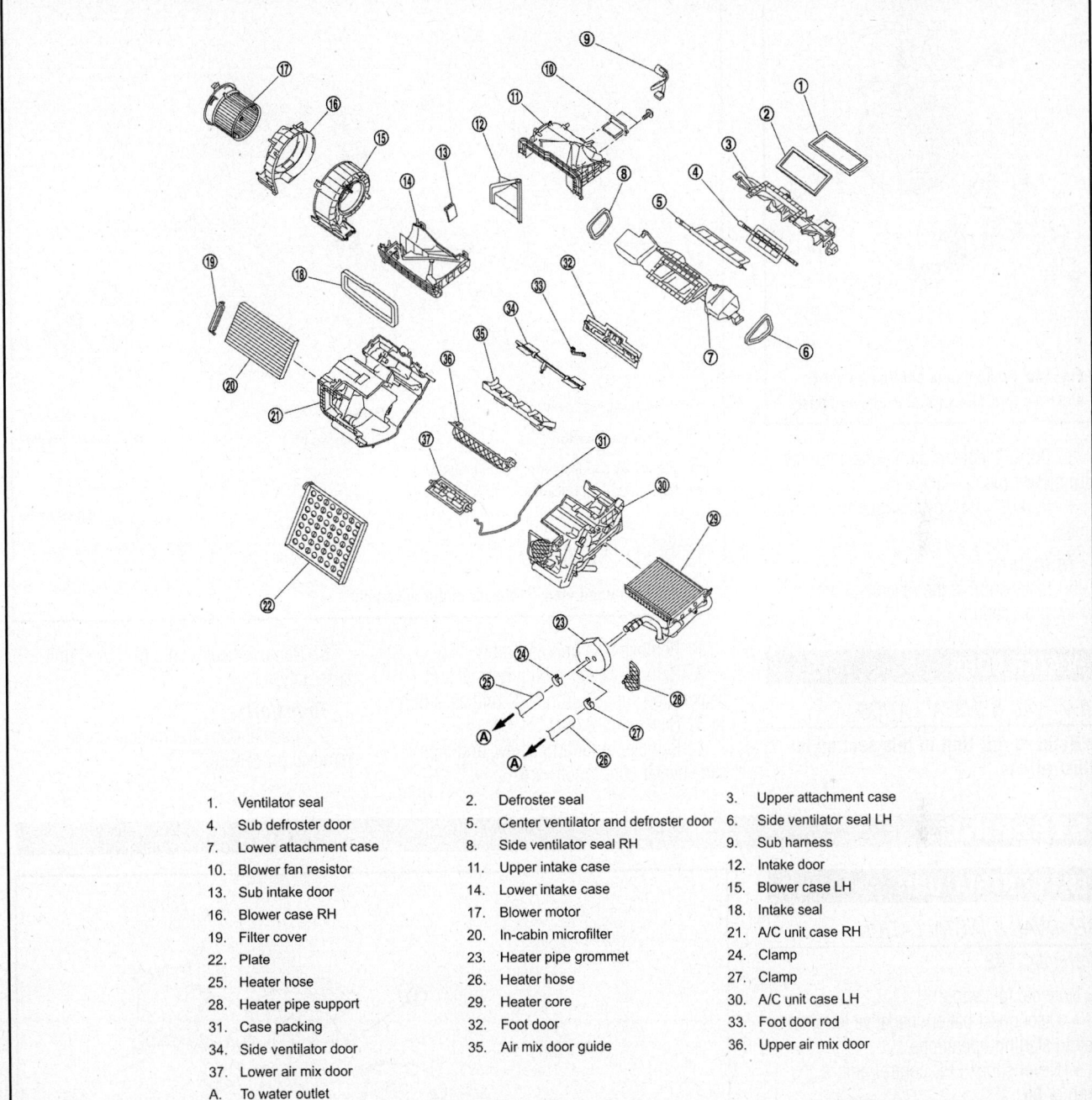

1. Ventilator seal
2. Defroster seal
3. Upper attachment case
4. Sub defroster door
5. Center ventilator and defroster door
6. Side ventilator seal LH
7. Lower attachment case
8. Side ventilator seal RH
9. Sub harness
10. Blower fan resistor
11. Upper intake case
12. Intake door
13. Sub intake door
14. Lower intake case
15. Blower case LH
16. Blower case RH
17. Blower motor
18. Intake seal
19. Filter cover
20. In-cabin microfilter
21. A/C unit case RH
22. Plate
23. Heater pipe grommet
24. Clamp
25. Heater hose
26. Heater hose
27. Clamp
28. Heater pipe support
29. Heater core
30. A/C unit case LH
31. Case packing
32. Foot door
33. Foot door rod
34. Side ventilator door
35. Air mix door guide
36. Upper air mix door
37. Lower air mix door
A. To water outlet

37663_CUBE_G0272

Fig. 139 Exploded view of A/C unit assembly

7. Remove side ventilator duct.
8. Move steering column assembly to a position where it does not inhibit work.
9. Remove instrument stay.
10. Disconnect drain hose from A/C unit assembly.
11. Disconnect the harness connectors and clips required to remove the steering member, and then move the vehicle harness to the position without hindrance for work.
12. Remove the BCM mounting screws.
13. Remove the J/B mounting screws.
14. Remove the A/C unit assembly mounting bolts.
15. Remove ground bolts.
16. Remove mounting bolts, and then remove steering member from the vehicle.
17. Remove A/C unit assembly from the vehicle.

To install:

18. Installation is the reverse of the removal procedure.
19. Replace O-rings with new ones.

Apply compressor oil to them when installing.
20. Check for leakages when recharging refrigerant.

BLOWER MOTOR

REMOVAL & INSTALLATION

See Figures 140 and 141.

1. Remove remote keyless entry receiver.
2. Disconnect blower motor connector.

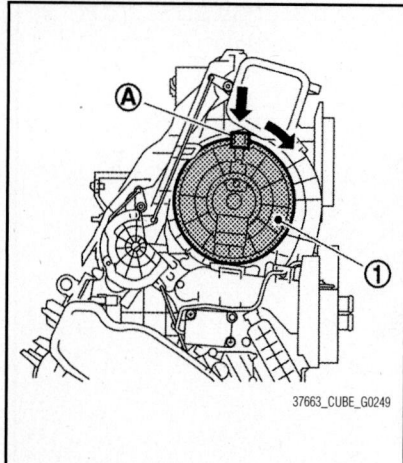

Fig. 140 Press flange holding hook (A), and then turn blower motor (1) clockwise

3. Press flange holding hook, and then turn blower motor clockwise.

4. Pull outside, and then remove blower motor.

To install:

5. Installation is the reverse of the removal procedure.

HEATER CORE

REMOVAL & INSTALLATION

➡ **Refer to A/C Unit in this section for illustrations.**

STEERING

EPS CONTROL UNIT

REMOVAL & INSTALLATION

See Figure 142.

Note the following:
- Disconnect battery negative terminal before starting operations.
- Never shock EPS control unit, e.g. drop or hit.
- Never get EPS control unit wet with water or other liquid. Also, do not give EPS control unit a radical temperature change to avoid getting water drops.
- Never disassemble or remodel EPS control unit, EPS motor, torque sensor, harness and connectors.

1. Remove instrument lower panel LH.
2. Remove knee protector.
3. Disconnect EPS control unit connectors.

✳✳ WARNING

Hold and pull the connector housing, not pulling harness, when discon-

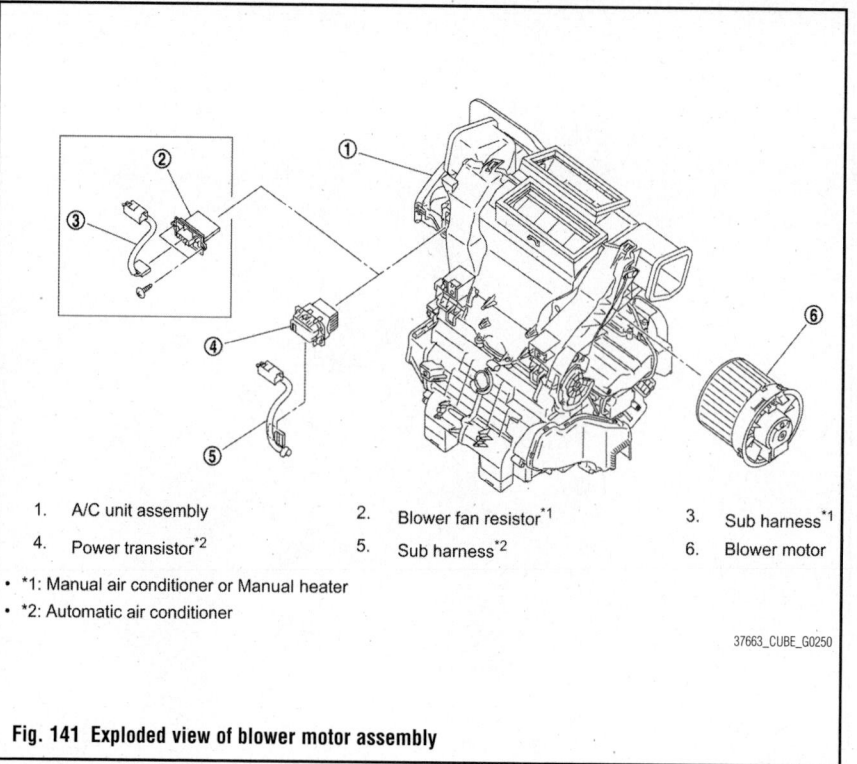

1. A/C unit assembly	2. Blower fan resistor*1	3. Sub harness*1
4. Power transistor*2	5. Sub harness*2	6. Blower motor

- *1: Manual air conditioner or Manual heater
- *2: Automatic air conditioner

Fig. 141 Exploded view of blower motor assembly

1. Remove A/C unit assembly.
2. Remove heater pipe grommet and heater pipe support from A/C unit assembly.
3. Remove foot duct LH.
4. Remove mounting screw, and then slide heater core to leftward.

5. Remove heater core from A/C unit assembly.

To install:

6. Installation is the reverse of the removal procedure.

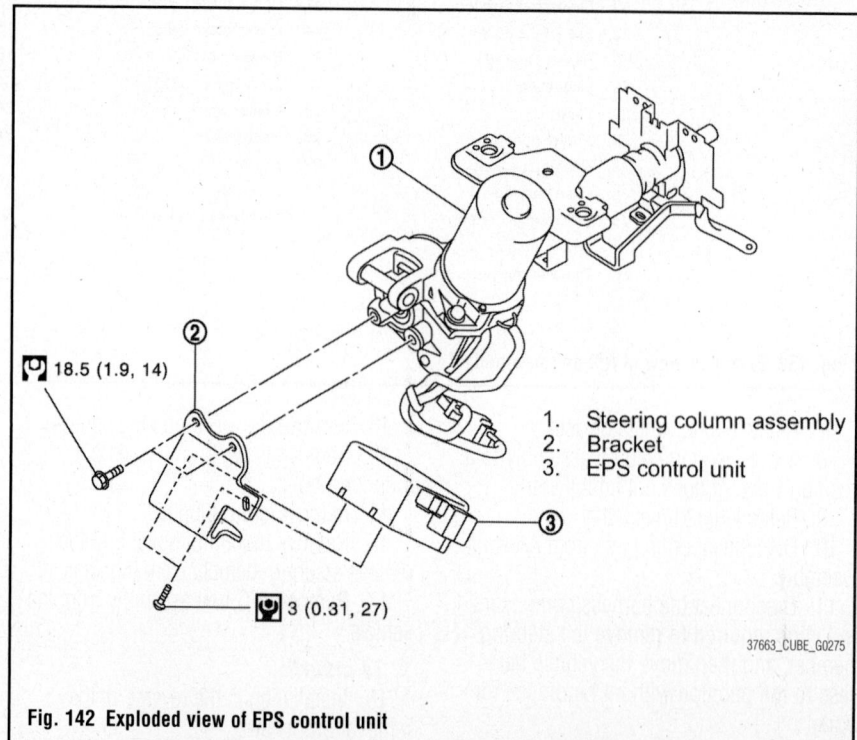

18.5 (1.9, 14)

3 (0.31, 27)

1. Steering column assembly
2. Bracket
3. EPS control unit

Fig. 142 Exploded view of EPS control unit

necting connectors. **Also, do not grip, collapse or apply excessive force to the connector.**

4. Remove EPS control unit from steering column assembly.

To install:

5. Note the following, and install in the reverse order of removal.

6. Check that harness is not damaged when installing EPS control unit. Also, check that EPS control unit is installed without trapping harness of foreign materials.

7. After installing steering column assembly, perform self-diagnosis with CONSULT-III to ensure correct operation.

POWER RACK & PINION STEERING GEAR

REMOVAL & INSTALLATION

See Figure 143.

1. Set vehicle to the straight-ahead position.

2. Remove bolt intermediate shaft (steering gear assembly side).

✳✳ WARNING

Spiral cable may be cut if steering wheel turns while separating steering column assembly and steering gear assembly. Always fix the steering wheel using string to avoid.

3. Remove tires and wheels.

4. Remove steering outer socket from steering knuckle so as not to damage ball joint boot using suitable ball joint remover.

✳✳ WARNING

Temporarily tighten the nut to prevent damage to threads and to prevent the ball joint remover from sudden drop turning.

5. Remove front stabilizer connecting rod.

6. Support front suspension member with a suitable jack.

7. Remove rear torque and engine mounting bracket mounting bolts.

8. Remove the mounting bolts and nuts of steering gear assembly.

9. Remove member stay, front suspension member fixing bolts.

10. Lower the suitable jack for the front suspension member to the steering gear assembly can be removed.

To install:

11. Note the following, and install in the reverse order of removal.

✳✳ WARNING

Spiral cable may be cut if steering wheel turns while separating steering column assembly and steering gear assembly. Always fix the steering wheel using string to avoid turning.

12. Clean mounting surface on the body side of fire wall seal when installing steering gear assembly.

13. Perform final tightening of nuts and bolts on each part under unladen conditions with tires on level ground when removing steering gear assembly. Check wheel alignment.

14. Rotate steering wheel to check for de-centered condition, binding, noise or excessive steering effort.

15. Adjust neutral position of steering angle sensor after checking wheel alignment.

ADJUSTMENT OF STEERING ANGLE SENSOR NEUTRAL POSITION

✳✳ WARNING

To adjust neutral position of steering angle sensor, make sure to use CONSULT-III. (Adjustment cannot be done without CONSULT-III.)

1. Stop vehicle with front wheels in straight-ahead position.

2. On the CONSULT-III screen, touch "WORK SUPPORT" and "ST ANGLE SENSOR ADJUSTMENT" in order.

3. Touch "START".

✳✳ WARNING

Never touch steering wheel while adjusting steering angle sensor.

4. After approximately 10 seconds, touch "END".

➡**After approximately 60 seconds, it ends automatically.**

5. Turn the ignition switch OFF, then turn it ON again.

✳✳ WARNING

Be sure to perform above operation.

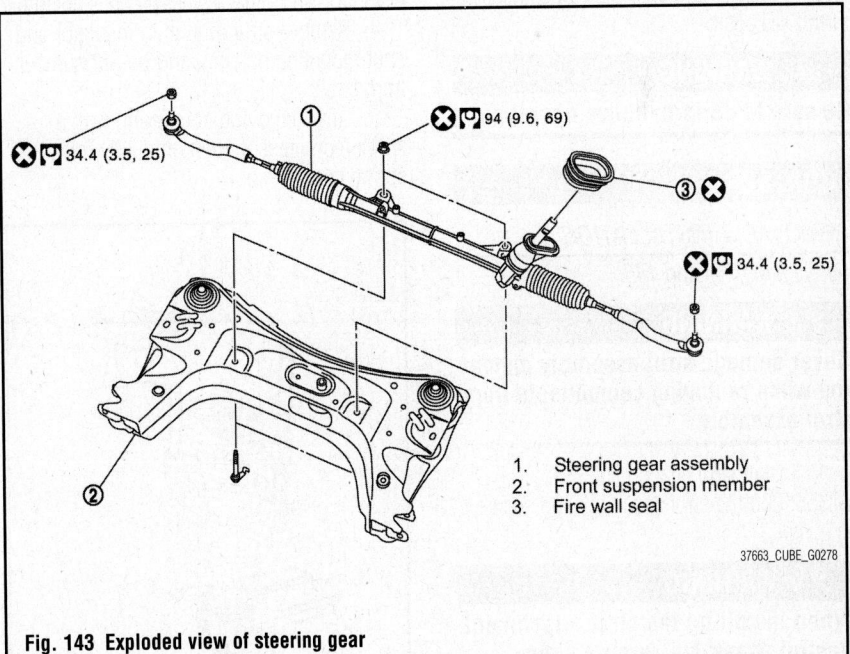

1. Steering gear assembly
2. Front suspension member
3. Fire wall seal

37663_CUBE_G0278

Fig. 143 Exploded view of steering gear

ADJUSTMENT OF STEERING ANGLE SENSOR NEUTRAL POSITION

PROCEDURE

※※ WARNING

To adjust neutral position of steering angle sensor, make sure to use CONSULT-III. (Adjustment cannot be done without CONSULT-III.)

1. Stop vehicle with front wheels in straight-ahead position.
2. On the CONSULT-III screen, touch "WORK SUPPORT" and "ST ANGLE SENSOR ADJUSTMENT" in order.
3. Touch "START".

※※ WARNING

Never touch steering wheel while adjusting steering angle sensor.

4. After approximately 10 seconds, touch "END".

➡**After approximately 60 seconds, it ends automatically.**

5. Turn the ignition switch OFF, then turn it ON again.

※※ WARNING

Be sure to perform above operation.

COIL SPRINGS

REMOVAL & INSTALLATION

See Figures 144 and 145.

※※ WARNING

Never damage strut assembly piston rod when removing components from strut assembly.

1. Install strut attachment SST: ST35652000 to strut assembly and secure it in a vise.

※※ WARNING

When installing the strut attachment to strut assembly, wrap a shop cloth around strut to protect from damage.

2. Using a spring compressor, compress coil spring between strut mounting bearing and lower seat (strut assembly) until coil spring with a spring compressor is free.

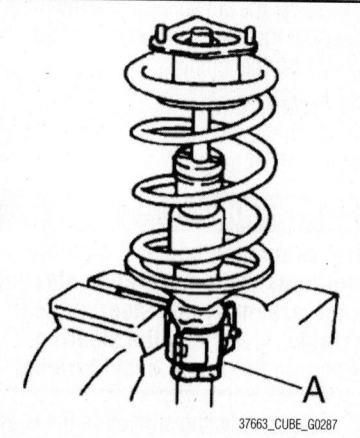

37663_CUBE_G0287

Fig. 144 Install strut attachment (A) to strut assembly and secure it in a vise

※※ WARNING

Be sure a spring compressor is securely attached to coil spring. Compress coil spring.

3. Check coil spring with a spring compressor between strut mounting bearing and lower seat (strut assembly) is free. And then remove piston rod lock nut while securing the piston rod tip so that piston rod does not turn.
4. Remove strut mounting insulator and strut mounting bearing, and bound bumper from strut.
5. After removing coil spring with a spring compressor, then gradually release a spring compressor.

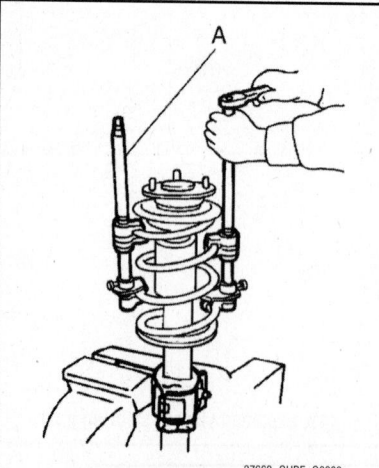

37663_CUBE_G0288

Fig. 145 Using a spring compressor (A), compress coil spring between strut mounting bearing and lower seat (strut assembly) until coil spring with a spring compressor is free

※※ WARNING

Loosen while making sure coil spring attachment position does not move.

6. Remove the strut attachment SST: ST35652000 from strut.
7. Check coil spring for cracks, wear or damage. Replace it if necessary.

KNUCKLE & SPINDLE

REMOVAL & INSTALLATION

See Figure 146.

1. Remove tires and wheels.
2. Remove wheel sensor and sensor harness.
3. Remove lock plate from strut assembly.
4. Remove caliper assembly. Hang caliper assembly not to interfere with work.

※※ WARNING

Never depress brake pedal while brake caliper is removed.

5. Remove disc rotor.
6. Remove cotter pin, and then loosen wheel hub lock nut.
7. Patch wheel hub lock nut with a piece of wood. Hammer the wood to disengage wheel hub and bearing assembly from drive shaft.

- Never place drive shaft joint at an extreme angle. Also be careful not to overextend slide joint.
- Never allow drive shaft to hang down without support for joint subassembly, shaft and the other parts.

➡**Use suitable puller, if wheel hub and bearing assembly and drive shaft cannot be separated even after performing the above procedure.**

8. Remove wheel hub lock nut.
9. Remove wheel hub and bearing assembly, and then remove splash guard.
10. Suspend the drive shaft with suitable wire.
11. Remove steering outer socket from steering knuckle.
12. Remove steering knuckle from transverse link.
13. Remove steering knuckle from strut assembly.

To install:

14. Note the following, and install in the reverse order of the removal.

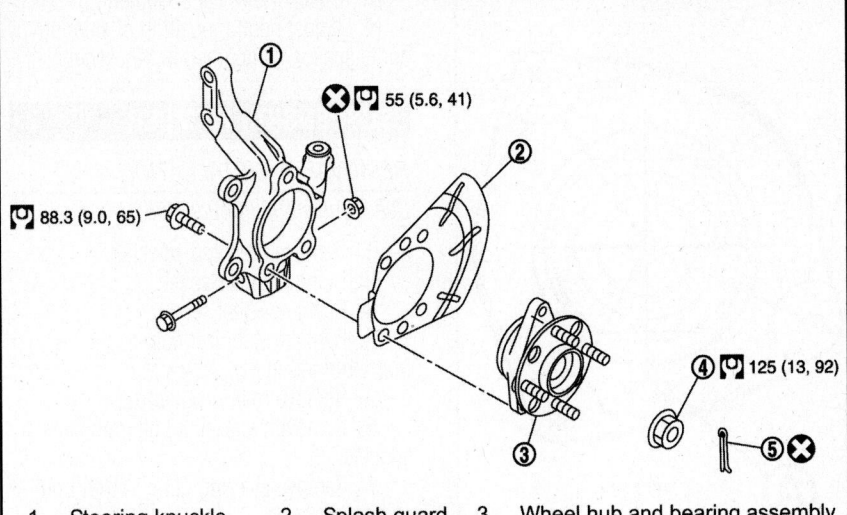

1. Steering knuckle
2. Splash guard
3. Wheel hub and bearing assembly
4. Wheel hub lock nut
5. Cotter pin

37663_CUBE_G0283

Fig. 146 Exploded view of steering knuckle assembly

a. Clean the matching surface of wheel hub lock nut and wheel hub and bearing assembly.

✳✳ WARNING

Never apply lubricating oil to these matching surface.

b. Tighten the wheel hub lock nut to 92 ft. lbs. (125 Nm).

✳✳ WARNING

Never use a power tool to tighten the wheel hub lock nut.

c. Perform the final tightening of each of parts under unladen conditions, which were removed when removing wheel hub and bearing assembly and axle housing.

d. Never reuse cotter pin.

15. Check wheel sensor harness for proper connection.

16. Check the wheel alignment.

17. Adjust neutral position of steering angle sensor, as outlined in this section.

LOWER TRANSVERSE LINK

REMOVAL & INSTALLATION

See Figure 148.

1. Remove tires and wheels.
2. Remove under cover.
3. Remove transverse link from steering knuckle.
4. Remove transverse link from suspension member.

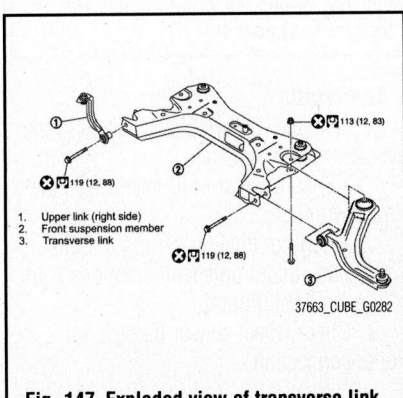

1. Upper link (right side)
2. Front suspension member
3. Transverse link

37663_CUBE_G0282

Fig. 147 Exploded view of transverse link

➡**Support the point around from upper link (right side) with a jack when removing transverse link (right side).**

To install:

5. Note the following, and install in the reverse order of removal.

6. Perform final tightening of bolts and nuts at the front suspension member, under unladen conditions with tires on level ground.

7. Check wheel sensor harness for proper connector.

8. Check wheel alignment.

9. Adjust neutral position of steering angle sensor, as outlined in this section.

STABILIZER BAR

REMOVAL & INSTALLATION

See Figures 148 and 149.

1. Remove tires and wheels.

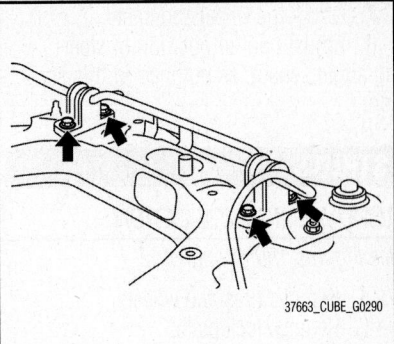

37663_CUBE_G0290

Fig. 148 Remove mounting bolts from the stabilizer clamps

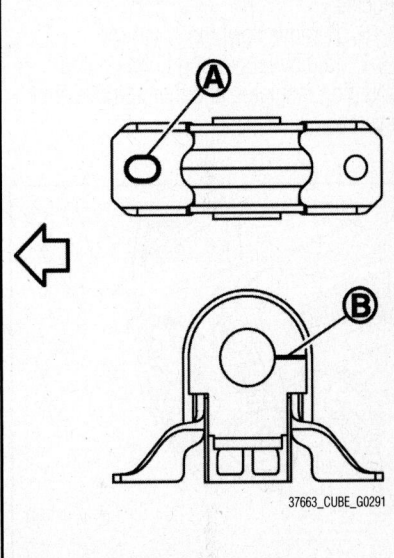

37663_CUBE_G0291

Fig. 149 Proper installation showing elongated hole (A) in clamp and slit (B) in bushing

2. Remove front suspension member.

3. Remove stabilizer connecting rod.

4. Remove mounting bolts from the stabilizer clamps, and then remove stabilizer clamps and stabilizer bushings from front suspension member.

5. Remove stabilizer bar.

To install:

6. Note the following, and install in the reverse order of removal.

a. Install stabilizer clamps so that the elongated hole faces toward the vehicle front.

b. Install stabilizer bushings so that the slit faces the vehicle rear side.

c. Tighten the mounting nut to the specified torque while holding a hexagonal part of stabilizer connecting rod side.

d. Perform final tightening of bolts and nuts at the vehicle installation position (rubber bushing), under unladen conditions with tires on level ground.

7. Check the wheel alignment.

8. Adjust neutral position of steering angle sensor, as outlined in this section.

STRUTS

REMOVAL & INSTALLATION

See Figures 150 through 152.

1. Remove tires and wheels.

2. Remove lock plate.

3. Remove wheel sensor.

4. Remove stabilizer connecting rod from strut assembly.

5. Remove strut assembly from steering knuckle.

6. Remove cowl top cover rod.

7. Remove mounting bolts of strut mounting insulator, and then remove strut assembly.

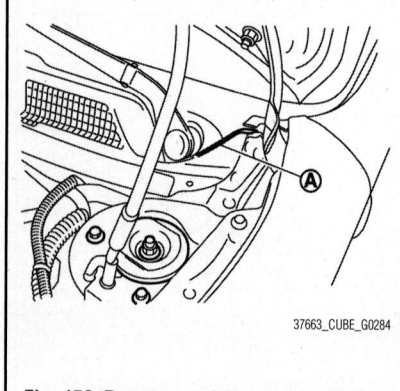

Fig. 150 Remove cowl top cover rod (A)

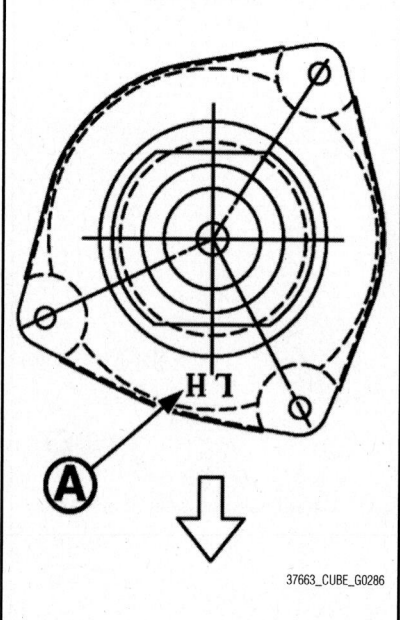

37663_CUBE_G0286

Fig. 152 Ensure the strut ID letters (A) face the body front side

To install:

8. Note the following, and install in the reverse order of removal.

 a. Ensure the strut ID letters face the body front side.

 b. Perform final tightening of bolts and nuts, under unladen conditions with tires on level ground.

9. Check wheel sensor harness for proper connection.

10. Check the wheel alignment.

11. Adjust neutral position of steering angle sensor, as outlined in this section.

SUSPENSION MEMBER

REMOVAL & INSTALLATION

See Figures 153 through 156.

1. Remove tires and wheels.

2. Remove under cover.

3. Remove wheel sensor.

4. Remove stabilizer connecting rod from strut assembly.

5. Remove rear torque rod.

6. Remove transverse link from steering knuckle.

7. Remove steering outer socket from steering knuckle.

8. Remove intermediate shaft from steering gear assembly.

9. Set suitable jack under front suspension member.

✳✳ WARNING

Check the stable condition when using a jack.

10. Remove suspension member stay rear mounting bolts.

11. Remove upper link mounting bolts (right side: vehicle body side, left side: suspension member side), and suspension member mounting bolts.

12. Gradually lower the jack to remove front suspension member from vehicle body.

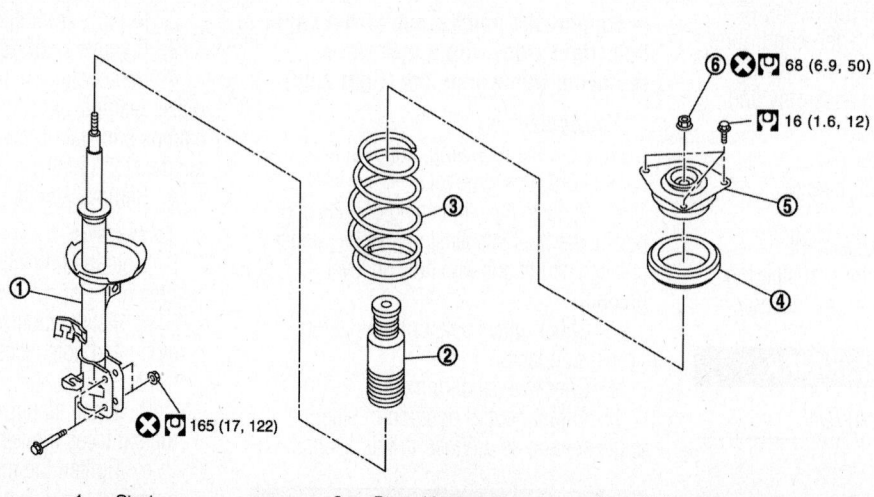

1. Strut
2. Bound bumper
3. Coil spring
4. Strut mounting bearing
5. Strut mounting insulator
6. Piston rod lock nut

37663_CUBE_G0285

Fig. 151 Exploded view of front coil spring and strut assembly

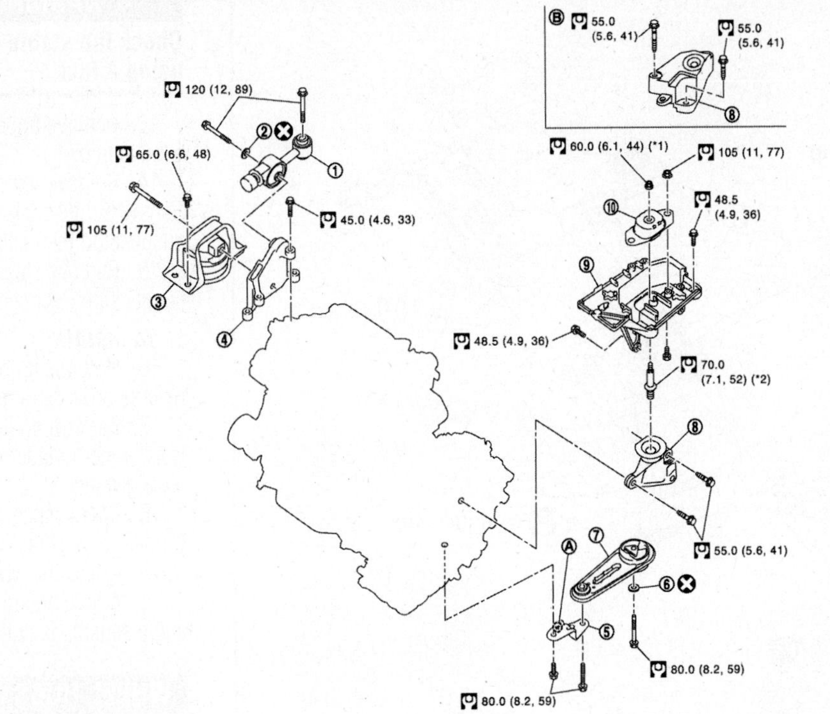

1. Upper torque rod (RH)
2. Washer
3. Engine mounting insulator (RH)
4. Engine mounting bracket (RH)
5. Rear engine mounting bracket
6. Washer
7. Rear torque rod
8. Engine mounting bracket (LH)
9. Engine mounting bracket support (LH)
10. Engine mounting insulator (LH)
A. Front mark
B. M/T models

37663_CUBE_G0165

Fig. 153 Exploded view of engine mount assemblies

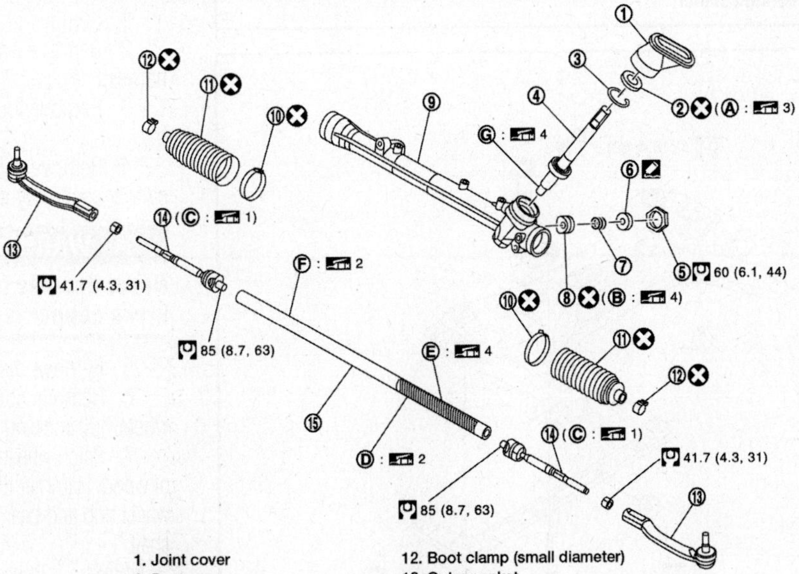

1. Joint cover
2. Dust cover
3. Snap ring
4. Pinion assembly
5. Lock nut
6. Adjusting screw
7. Spring
8. Retainer
9. Gear housing assembly
10. Boot clamp (large diameter)
11. Boot
12. Boot clamp (small diameter)
13. Outer socket
14. Inner socket
15. Rack assembly
A. Oil seal lip
B. Around retainer
C. Inner socket
D. Teeth part of rack assembly
E. back of rack assembly
F. Bush part of rack assembly
G. Needle bearing

37663_CUBE_G0279

Fig. 154 Exploded view of steering gear and linkage

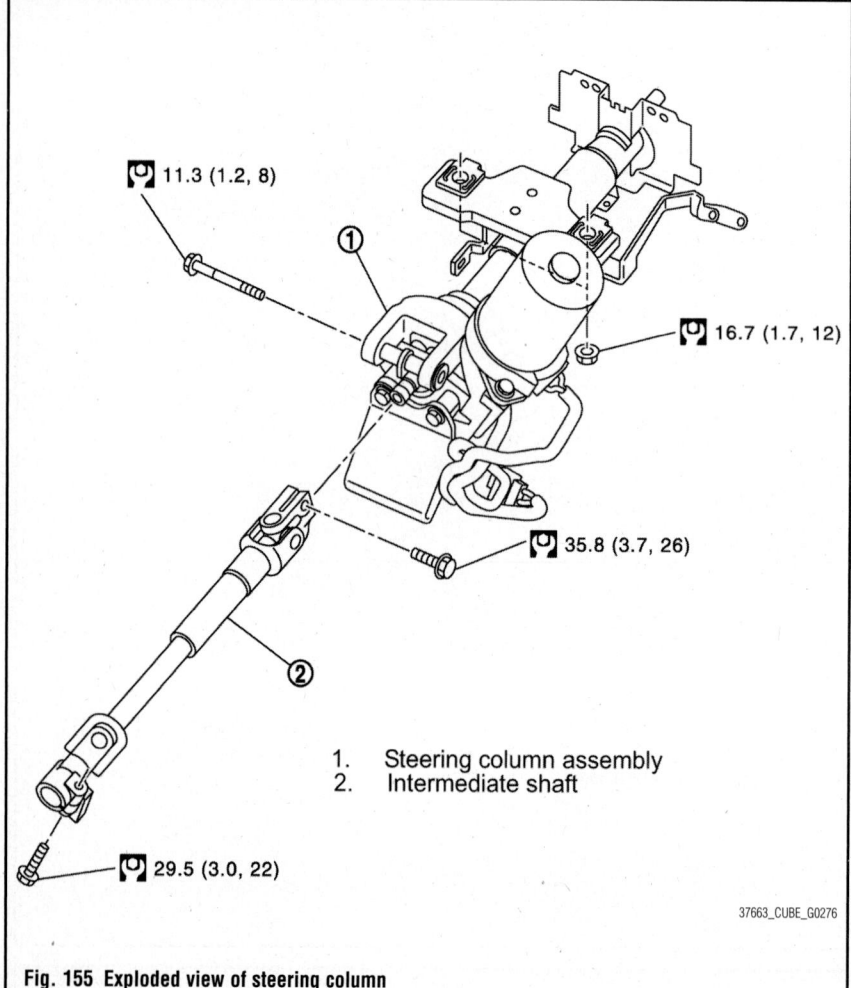

⬚ 11.3 (1.2, 8)

⬚ 16.7 (1.7, 12)

⬚ 35.8 (3.7, 26)

⬚ 29.5 (3.0, 22)

1. Steering column assembly
2. Intermediate shaft

37663_CUBE_G0276

Fig. 155 Exploded view of steering column

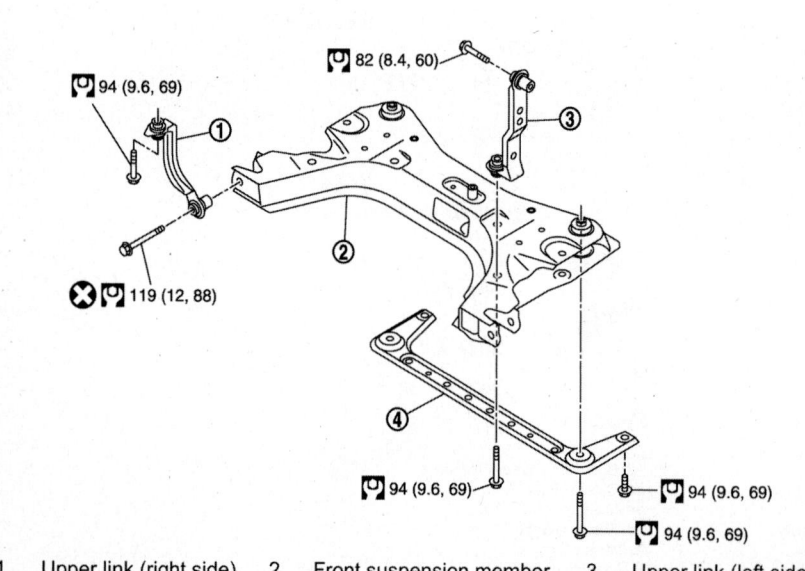

⬚ 94 (9.6, 69)

⬚ 82 (8.4, 60)

✗ ⬚ 119 (12, 88)

⬚ 94 (9.6, 69)

⬚ 94 (9.6, 69)

⬚ 94 (9.6, 69)

1. Upper link (right side) 2. Front suspension member 3. Upper link (left side)
4. Member stay

37663_CUBE_G0293

Fig. 156 Exploded view of front suspension member assembly

Check the stable condition when using a jack.

13. Remove upper link (left side) from vehicle body.

14. Remove upper link (right side), transverse links, stabilizer assembly from suspension member.

15. Remove steering gear assembly from suspension member.

To install:

16. Note the following, and install in the reverse order of removal.

17. Perform final tightening of bolts and nuts, under unladen conditions with tires on level ground.

18. Check wheel sensor harness for proper connection.

19. Check the wheel alignment.

20. Adjust neutral position of steering angle sensor, as outlined in this section.

WHEEL HUBS & BEARINGS

ADJUSTMENT

No adjustments are possible. If the axle end play is greater than 0.002 inches (0.05mm), replace the wheel bearing.

REMOVAL & INSTALLATION

See Figure 157.

1. Remove tires and wheels.

2. Remove wheel sensor and sensor harness.

3. Remove lock plate from strut assembly.

4. Remove caliper assembly. Hang caliper assembly not to interfere with work.

Never depress brake pedal while brake caliper is removed.

5. Remove disc rotor.

6. Remove cotter pin, and then loosen wheel hub lock nut.

7. Patch wheel hub lock nut with a piece of wood. Hammer the wood to disengage wheel hub and bearing assembly from drive shaft.

Note the following:

• Never place drive shaft joint at an extreme angle. Also be careful not to overextend slide joint.

• Never allow drive shaft to hang down without support for joint sub-assembly, shaft and the other parts.

➡Use suitable puller, if wheel hub and bearing assembly and drive shaft can-

not be separated even after performing
the above procedure.

8. Remove wheel hub lock nut.

9. Remove wheel hub and bearing
assembly, and then remove splash guard.

10. Suspend the drive shaft with suitable
wire.

11. Remove steering outer socket from
steering knuckle.

12. Remove steering knuckle from transverse link.

13. Remove steering knuckle from strut
assembly.

To install:

14. Note the following, and install in the
reverse order of the removal.

a. Clean the matching surface of
wheel hub lock nut and wheel hub and
bearing assembly.

❊❊ WARNING

**Never apply lubricating oil to these
matching surface.**

b. Tighten the wheel hub lock nut to
92 ft. lbs. (125 Nm).

❊❊ WARNING

**Never use a power tool to tighten the
wheel hub lock nut.**

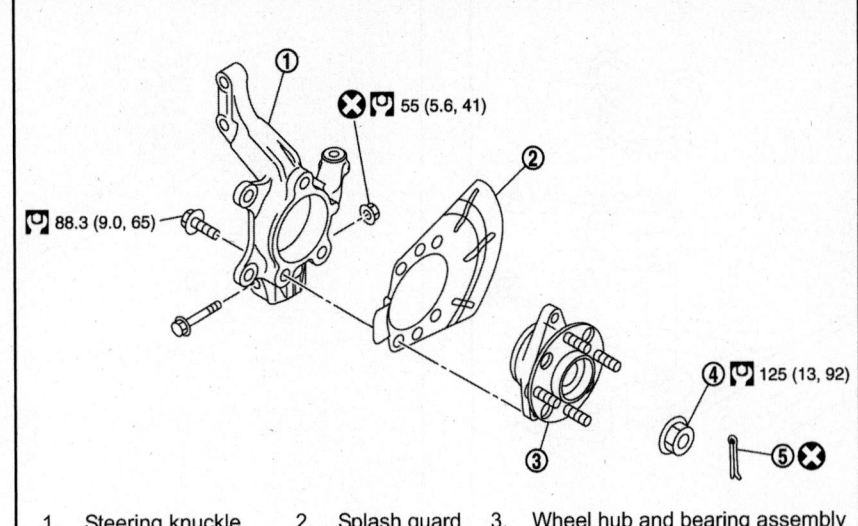

1. Steering knuckle
2. Splash guard
3. Wheel hub and bearing assembly
4. Wheel hub lock nut
5. Cotter pin

37663_CUBE_G0283

Fig. 157 Exploded view of steering knuckle assembly

c. Perform the final tightening of
each of parts under unladen conditions,
which were removed when removing
wheel hub and bearing assembly and
axle housing.

d. Never reuse cotter pin.

15. Check wheel sensor harness for
proper connection.

16. Check the wheel alignment.

17. Adjust neutral position of steering angle sensor, as outlined in this section.

SUSPENSION

ADJUSTMENT OF STEERING ANGLE SENSOR NEUTRAL POSITION

PROCEDURE

❊❊ WARNING

**To adjust neutral position of steering
angle sensor, make sure to use
CONSULT-III. (Adjustment cannot be
done without CONSULT-III.)**

1. Stop vehicle with front wheels in
straight-ahead position.

2. On the CONSULT-III screen,
touch "WORK SUPPORT" and
"ST ANGLE SENSOR ADJUSTMENT"
in order.

3. Touch "START".

❊❊ WARNING

**Never touch steering wheel while
adjusting steering angle sensor.**

4. After approximately 10 seconds,
touch "END".

➤After approximately 60 seconds,
it ends automatically.

5. Turn the ignition switch OFF, then
turn it ON again.

❊❊ WARNING

Be sure to perform above operation.

COIL SPRINGS

REMOVAL & INSTALLATION
See Figures 158 and 159.

1. Remove tires and wheels.

2. Set jack under rear suspension
beam.

❊❊ WARNING

**Check the stable condition when
using a jack.**

3. Remove right and left rear shock
absorber mounting bolts (lower side).

4. Slowly lower jack, then remove upper
rubber seat, coil spring and lower rubber
seat from rear suspension beam.

REAR SUSPENSION

❊❊ WARNING

**Check the stable condition when
using a jack.**

5. Check lubber seat and coil spring for
deformation, crack, and damage. Replace it
if necessary.

To install:

6. Note the following, and install in the
reverse order of removal.

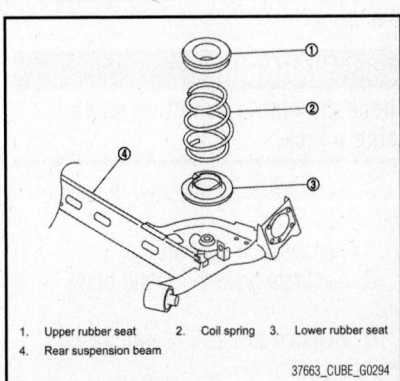

1. Upper rubber seat
2. Coil spring
3. Lower rubber seat
4. Rear suspension beam

37663_CUBE_G0294

**Fig. 158 Exploded view of rear coil spring
assembly**

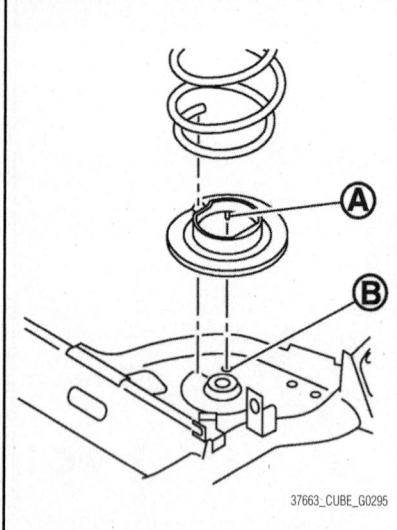

Fig. 159 Install the lower rubber seat with projection (A) as shown attached to rear suspension beam mounting hole (B)

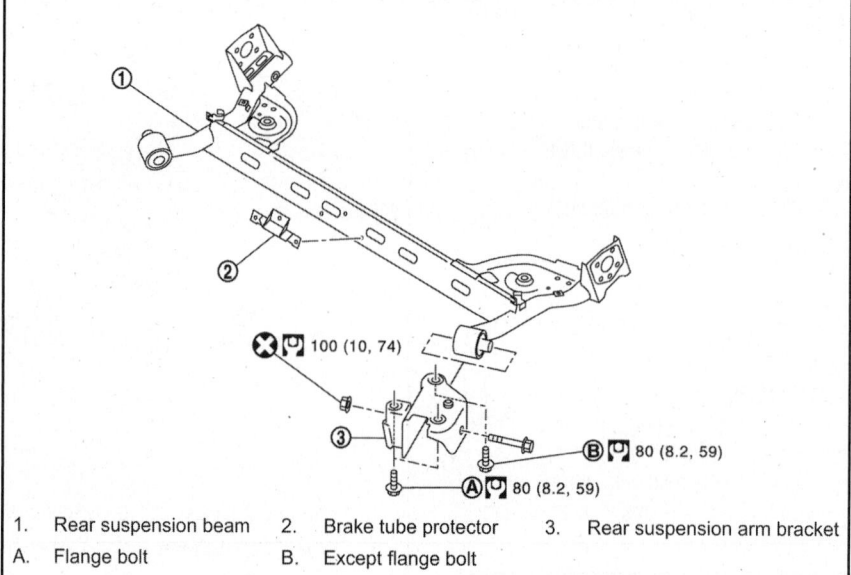

1. Rear suspension beam
2. Brake tube protector
3. Rear suspension arm bracket
A. Flange bolt
B. Except flange bolt

⊗ 🔧 100 (10, 74)

Ⓑ 🔧 80 (8.2, 59)

Ⓐ 🔧 80 (8.2, 59)

37663_CUBE_G0296

Fig. 160 Exploded view of rear suspension beam assembly

a. Install the lower rubber seat with projection as shown attached to rear suspension beam mounting hole.

b. Match up lower rubber seat indentions and rear suspension beam grooves and attach.

7. Check wheel alignment.

CROSSMEMBER

REMOVAL & INSTALLATION

See Figure 160.

1. Remove tires and wheels.
2. Drain brake fluid.
3. Remove parking brake cable and brake drum from rear suspension beam.
4. Remove wheel sensor and sensor harness.
5. Set suitable jack under rear suspension beam.

✳✳ WARNING

Check the stable condition when using a jack.

6. Remove shock absorber mounting bolts (lower side).
7. Remove coil springs.
8. Separate brake hose and brake tube.
9. Remove suspension arm bracket mounting bolts.
10. Slowly lower jack, remove suspension arm bracket and rear suspension beam from vehicle body.

✳✳ WARNING

Check the stable condition when using a jack.

11. Remove wheel hub and bearing assembly.
12. Remove drum brake assembly.
13. Remove suspension arm bracket from rear suspension beam.
14. Remove brake tube protector from rear suspension beam.
15. Check rear suspension beam and rear suspension beam bracket for deformation, cracks or damage. Replace the part if necessary.

To install:

16. Note the following, and install in the reverse order of removal.

a. Perform final tightening of rear suspension beam installation position (rubber bushing), under unladen conditions with tires on level ground.

b. Refill with new brake fluid and perform the air bleeding.

17. Check wheel sensor harness for proper connection.
18. Adjust parking brake.
19. Check the wheel alignment.
20. Adjust neutral position of steering angle sensor, as outlined in this section.

SHOCK ABSORBERS

REMOVAL & INSTALLATION

See Figures 161 through 163.

1. Remove tires and wheels.

2. Set suitable jack under rear suspension beam.

✳✳ WARNING

Check the stable condition when using a jack.

3. Remove shock absorber mounting bolt (lower side).
4. Remove luggage lid from luggage side finisher.
5. Remove shock absorber mounting nut (upper side), and then remove washer and bushing.
6. Remove shock absorber assembly.
7. Check the following items, and replace the part if necessary.

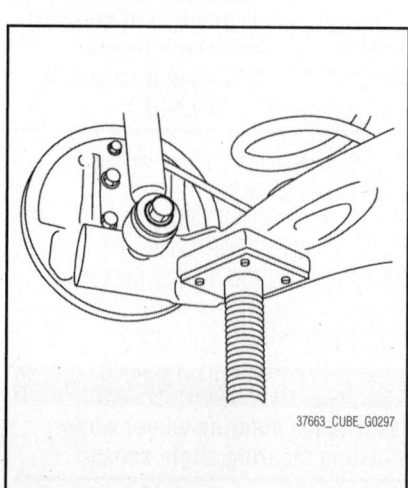

37663_CUBE_G0297

Fig. 161 Set suitable jack under rear suspension beam

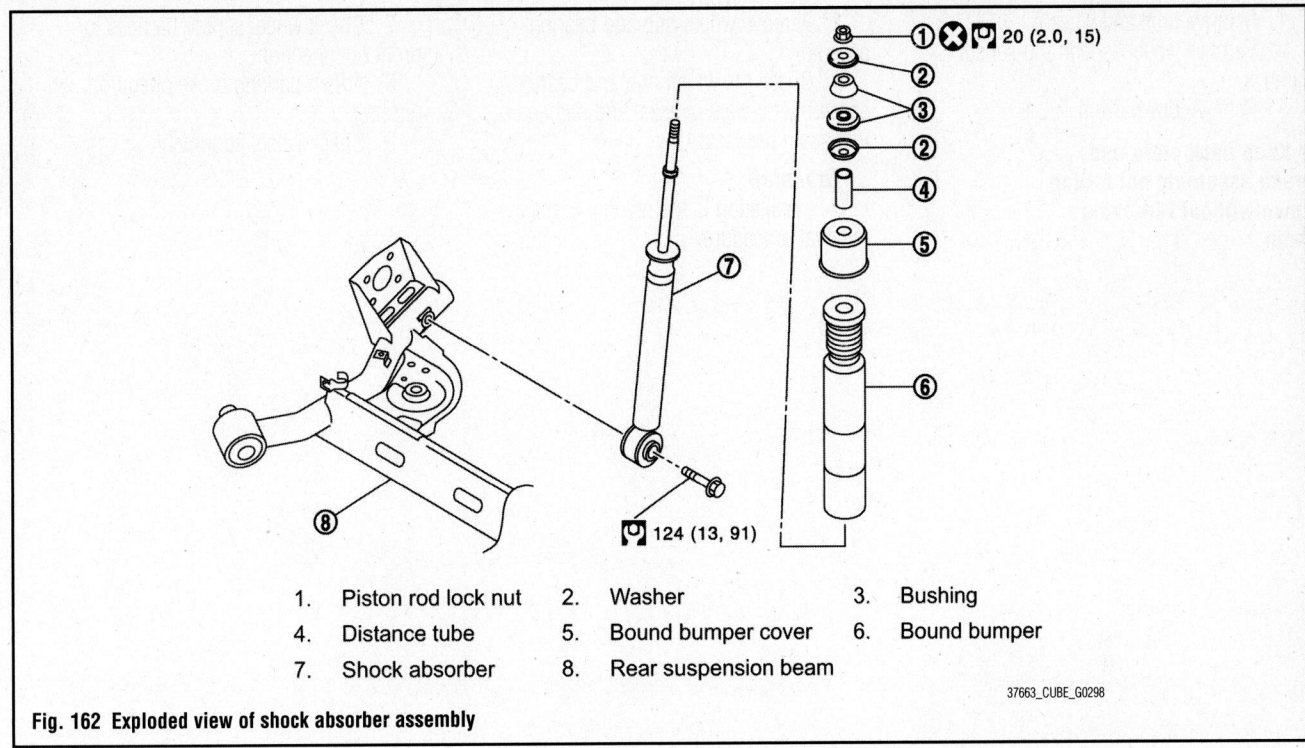

1. Piston rod lock nut
2. Washer
3. Bushing
4. Distance tube
5. Bound bumper cover
6. Bound bumper
7. Shock absorber
8. Rear suspension beam

37663_CUBE_G0298

Fig. 162 Exploded view of shock absorber assembly

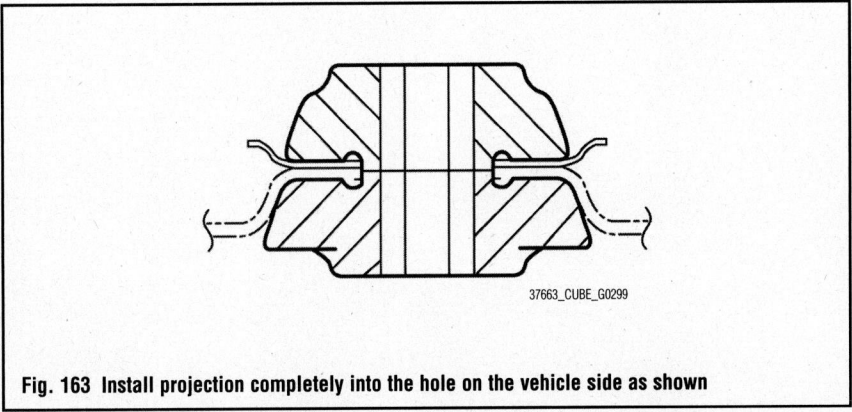

37663_CUBE_G0299

Fig. 163 Install projection completely into the hole on the vehicle side as shown

- Shock absorber for deformation, cracks, and other damage.
- Piston rod for damage, uneven wear, and distortion.

8. Check for cracks and damage. Replace it if necessary.

9. Check for cracks and damage. Replace it if necessary.

To install:

10. Note the following, and install in the reverse order of removal.

a. Perform final tightening of bolts and nuts at the shock absorber lower side (rubber bushing), under unladen conditions with tires on level ground.

b. Install projection completely into the hole on the vehicle side as shown, when installing bushing.

c. Hold the head of shock absorber

piston rod to prevent it from rotating, then tighten piston rod lock nut with a standard tightening torque value.

WHEEL HUBS & BEARINGS

ADJUSTMENT

No adjustments are possible. If the axle end play is greater than 0.002 inches (0.05mm), replace the wheel bearing.

REMOVAL & INSTALLATION

See Figure 164.

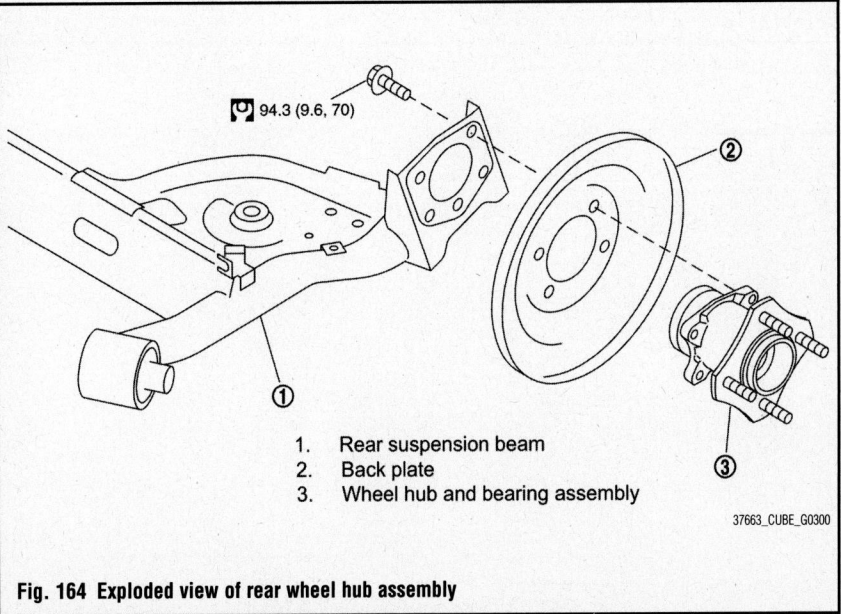

1. Rear suspension beam
2. Back plate
3. Wheel hub and bearing assembly

37663_CUBE_G0300

Fig. 164 Exploded view of rear wheel hub assembly

1. Remove tires and wheels.
2. Remove wheel sensor and sensor harness.
3. Remove brake drum.

➡**Keep back plate and brake assembly not falling down without removing them.**

4. Remove wheel hub and bearing assembly.
5. Check the wheel hub and bearing assembly for wear, cracks, and damage. Replace if necessary.

To install:

6. Installation is the reverse of the removal procedure.

7. Check wheel sensor harness for proper connection.
8. Adjust parking brake operation (stroke).
9. Check wheel alignment.

SPECIFICATIONS AND MAINTENANCE CHARTS

ENGINE AND VEHICLE IDENTIFICATION

| | Engine | | | | | | Model Year | |
ID/Code	Liters (cc)	Cu. In.	Cyl.	Fuel Sys.	Engine Type	Eng. Mfg.	Code ①	Year
QR25DE	2.5 (2,488)	152	4	MFI	DOHC	Nissan	B	2011
VQ40DE	4.0 (3,954)	241	6	MFI	DOHC	Nissan	C	2012

MFI: Multi-port Fuel Injection

DOHC: Double Overhead Camshafts

① 10th digit of the Vehicle Identification Number (VIN)

71075_FRON_C0001

GENERAL ENGINE SPECIFICATIONS

All measurements are given in inches.

Year	Model	Engine Displacement Liters	Engine Series ID	Net Horsepower @ rpm	Net Torque @ rpm (ft. lbs.)	Bore x Stroke (in.)	Compression Ratio	Oil Pressure @ rpm
2011	Frontier	2.5	QR25DE	152@5200	171@4400	3.50 x 3.94	9.5:1	43@2000
		4.0	VQ40DE	261@5600	281@4000	3.76 x 3.62	9.7:1	43@2000
2012	Frontier	2.5	QR25DE	152@5200	171@4400	3.50 x 3.94	9.5:1	43@2000
		4.0	VQ40DE	261@5600	281@4000	3.76 x 3.62	9.7:1	43@2000

71075_FRON_C0002

GASOLINE ENGINE TUNE-UP SPECIFICATIONS

Year	Engine Displacement Liters	Engine ID	Spark Plug Gap (in.)	Ignition Timing (deg.) MT	Ignition Timing (deg.) AT	Fuel Pump (psi)	Idle Speed (RPM) MT	Idle Speed (RPM) AT ①	Valve Clearance (in.) Intake	Valve Clearance (in.) Exhaust
2011	2.5	QR25DE	0.043	10-20B	10-20B	51 ②	575-675	650-750	③	④
	4.0	VQ40DE	0.043	10-20B	10-20B	51 ②	575-675	650-750	⑤	⑥
2012	2.5	QR25DE	0.043	10-20B	10-20B	51 ②	575-675	650-750	③	④
	4.0	VQ40DE	0.043	10-20B	10-20B	51 ②	575-675	650-750	⑤	⑥

NOTE: The Vehicle Emission Control Information label often reflects specification changes made during production.

The label figures must be used if they differ from those in this chart.

B: Before top dead center

① Automatic Transmission (AT) in Neutral

② At idle

③ 0.009-0.013 inch (cold)

0.012-0.016 inch (hot)

④ 0.010-0.013 inch (cold)

0.012-0.017 inch (hot)

⑤ 0.010-0.013 inch (cold)

0.012-0.016 inch (hot)

⑥ 0.011-0.015 inch (cold)

0.012-0.017 inch (hot)

71075_FRON_C0003

CAPACITIES

Year	Model	Engine Displacement Liters	Engine ID	Engine Oil with Filter (qts.)	Transmission (pts.)		Transfer Case (pts.)	Drive Axle		Fuel Tank (gal.)	Cooling System (qts.)
					Manual	Auto. ①		Front (pts.)	Rear (pts.)		
2011	Frontier	2.5	QR25DE	4.9	6.1	21.8	NA	NA	3.4	21.1	10.0
		4.0	VQ40DE	5.4	NA	21.8	4.25	1.75	②	21.1	10.8
2012	Frontier	2.5	QR25DE	4.9	6.1	21.8	NA	NA	3.4	21.1	10.0
		4.0	VQ40DE	5.4	NA	21.8	4.25	1.75	②	21.1	10.8

NA: Not Applicable

NOTE: All capacities are approximate. Add fluid gradually and check to be sure a proper fluid level is obtained.

① Drain and refill

71075_FRON_C0004

FLUID SPECIFICATIONS

Year	Model	Engine Displacement Liters (ID)	Engine Oil	Manual Trans.	Auto. Trans.	Transfer Case	Drive Axle		Power Steering Fluid	Brake Master Cylinder	Cooling System
							Front	Rear			
2011	Frontier	2.5 (QR25DE)	5W-30	①	②	NA	NA	③	NS	DOT 3	④
		4.0 (VQ40DE)	5W-30	NA	②	⑤	⑥	⑦	NS	DOT 3	④
2012	Frontier	2.5 (QR25DE)	5W-30	①	②	NA	NA	③	NS	DOT 3	④
		4.0 (VQ40DE)	5W-30	NA	②	⑤	⑥	⑦	NS	DOT 3	④

NA: Not Applicable

NS: Not specified by manufacturer at date of publication

DOT: Department Of Transportation

① Manual Transmission Fluid (MTF) API GL-4, Viscosity SAE 75W-85

② Nissan approved Matic S ATF. If Matic S ATF is not available, NISSAN approved Matic J ATF may also be used.

③ API GL-5 synthetic gear oil, Viscosity SAE 75W-90

④ Nissan genuine coolant (blue), or equivalent

⑤ Nissan approved Matic D AT

⑥ API GL-5 Viscosity SAE 80W-90. For hot climates, viscosity SAE 90 is suitable for ambient temperatures above 32 degrees F (0 degrees C).

⑦ C200 axle: API GL-5 synthetic gear oil, Viscosity SAE 75W-90

M226 axle: API GL-5 synthetic gear oil, Viscosity SAE 75W-140

71075_FRON_C0005

VALVE SPECIFICATIONS

Year	Engine Displacement Liters	Engine ID	Seat Angle (deg.)	Face Angle (deg.)	Spring Test Pressure (lbs. @ in.)	Spring Installed Height (in.)	Stem-to-Guide Clearance (in.)		Stem Diameter (in.)	
							Intake	Exhaust	Intake	Exhaust
2011	2.5	QR25DE	44.15-45.45	45.00	①	1.390	0.0008-0.0021	0.0012-0.0025	0.2348-0.2354	0.2344-0.2350
	4.0	VQ40DE	44.15-45.45	45.00	②	1.457	0.0008-0.0021	0.0012-0.0025	0.2348-0.2354	0.2344-0.2350
2012	2.5	QR25DE	44.15-45.45	45.00	①	1.390	0.0008-0.0021	0.0012-0.0025	0.2348-0.2354	0.2344-0.2350
	4.0	VQ40DE	44.15-45.45	45.00	②	1.457	0.0008-0.0021	0.0012-0.0025	0.2348-0.2354	0.2344-0.2350

① Intake valve open: 79-89@0.996

Exhaust valve open: 71-81@1.054

② Installation: 37-42@1.457

Valve open: 84-95@1.071

71075_FRON_C0006

CAMSHAFT AND BEARING SPECIFICATIONS CHART

All measurements are given in inches.

Year	Engine Displ. Liters	Engine ID	Journal Dia.	Brg. Oil Clearance	Shaft End-play	Runout	Journal Bore	Lobe Lift	
								Intake	Exhaust
2011	2.5	QR25DE	①	0.0018-0.0034	0.0045-0.0074	0.0008	②	1.7722-1.7797	1.7313-1.7388
	4.0	VQ40DE	③	④	0.0045-0.0074	0.0008	⑤	1.7900-1.7974	1.7746-1.7821
2012	2.5	QR25DE	①	0.0018-0.0034	0.0045-0.0074	0.0008	②	1.7722-1.7797	1.7313-1.7388
	4.0	VQ40DE	③	④	0.0045-0.0074	0.0008	⑤	1.7900-1.7974	1.7746-1.7821

① Number 1: 1.0998-1.1006 inches

Numbers 2, 3, 4, 5: 0.9226-0.9234 inch

② Number 1: 1.1024-1.1032 inches

Numbers 2, 3, 4, 5: 0.9252-0.9260 inch

③ Number 1: 1.0211-1.0218 inches

Numbers 2, 3, 4: 0.9230-0.9238 inch

④ Number 1: 0.0018-0.0034 inch

Numbers 2, 3, 4: 0.0014-0.0030 inch

⑤ Number 1: 1.0236-1.0244 inches

71075_FRON_C0007

CRANKSHAFT AND CONNECTING ROD SPECIFICATIONS

All measurements are given in inches.

Year	Engine Displacement Liters	Engine ID	Crankshaft				Connecting Rod		
			Main Brg. Journal Dia.	Main Brg. Oil Clearance	Shaft End-play	Thrust on No.	Journal Diameter	Oil Clearance	Side Clearance
2011	2.5	QR25DE	①	②	0.0039-0.0102	3	③	0.0014-0.0018	0.0079-0.0138
	4.0	VQ40DE	④	0.0014-0.0018	0.0039-0.0098	3	2.2441-2.2446	0.0013-0.0023	0.0079-0.0138
2012	2.5	QR25DE	①	②	0.0039-0.0102	3	③	0.0014-0.0018	0.0079-0.0138
	4.0	VQ40DE	④	0.0014-0.0018	0.0039-0.0098	3	2.2441-2.2446	0.0013-0.0023	0.0079-0.0138

NS: Not specified by manufacturer at date of publication

① There are 24 different grades, ranging from A (2.1645 inches) to 7 (2.1636 inches)

② Numbers 1, 3, 5: 0.0011-0.0017 inch

Numbers 2, 4: 0.0016-0.0022 inch

③ There are 13 different grades, ranging from 0 (1.8898 inches) to C (1.8903 inches)

④ There are 24 different grades, ranging from A (2.7549 inches) to 7 (2.7540 inches)

71075_FRON_C0008

PISTON AND RING SPECIFICATIONS

All measurements are given in inches.

Year	Engine Displ. Liters	Engine ID	Piston Clearance	Ring Gap			Ring Side Clearance		
				Top Compression	Bottom Compression	Oil Control	Top Compression	Bottom Compression	Oil Control
2011	2.5	QR25DE	0.0004-0.0012	0.0083-0.0122	0.0126-0.0185	0.0079-0.0236	0.0018-0.0031	0.0012-0.0028	0.0026-0.0053
	4.0	VQ40DE	0.0004-0.0012	0.0091-0.0130	0.0130-0.0189	0.0079-0.0197	0.0018-0.0031	0.0012-0.0028	0.0026-0.0053
2012	2.5	QR25DE	0.0004-0.0012	0.0083-0.0122	0.0126-0.0185	0.0079-0.0236	0.0018-0.0031	0.0012-0.0028	0.0026-0.0053
	4.0	VQ40DE	0.0004-0.0012	0.0091-0.0130	0.0130-0.0189	0.0079-0.0197	0.0018-0.0031	0.0012-0.0028	0.0026-0.0053

71075_FRON_C0009

TORQUE SPECIFICATIONS

All readings in ft. lbs.

Year	Engine Displacement Liters	Engine ID	Cylinder Head Bolts	Main Bearing Bolts	Rod Bearing Bolts	Crankshaft Damper Bolts	Flywheel Bolts	Manifold		Spark Plugs	Oil Pan Drain Plug
								Intake	Exhaust		
2011	2.5	QR25DE	①	②	③	④	80	⑤	⑥	14	25
	4.0	VQ40DE	⑦	⑧	14	⑨	65	⑩	⑥	18	25
2012	2.5	QR25DE	①	②	③	④	80	⑤	⑥	14	25
	4.0	VQ40DE	⑦	⑧	14	⑨	65	⑩	⑥	18	25

① Step 1: 37 ft. lbs.
Step 2: Plus 60 degrees clockwise
Step 3: Loosen completely to 0 ft. lbs.
Step 4: 29 ft. lbs.
Step 5: Plus 75 degrees clockwise
Step 6: Plus 75 degrees clockwise

② Step 1: bolts 11-22 to 19 ft. lbs.
Step 2: bolts 1-10 to 29 ft. lbs.
Step 3: bolts 1-10, plus 60-65 degrees

③ Step 1: 20 ft. lbs.
Step 2: Loosen to 0 ft. lbs.
Step 3: 14 ft. lbs.
Step 4: Plus 85-95 degrees

④ Step 1: 31 ft. lbs.
Step 2: Plus 60 degrees

⑤ 83 inch lbs.

⑥ Stud bolts: 11 ft. lbs.
Nuts: 22 ft. lbs.

⑦ Step 1: 72 ft. lbs.
Step 2: Loosen completely to 0 ft. lbs.
Step 3: 29 ft. lbs.
Step 4: Plus 90 degrees clockwise
Step 5: Plus 90 degrees clockwise

⑧ Bolts: 17-24 (M8) to 16 ft. lbs.
Install rear main seal
Bolts: 1-16 (M10) to 26 ft. lbs.
Bolts: 1-16 (M10), plus 90 degrees clockwise

⑨ Step 1: 33 ft. lbs.
Step 2: Plus 84-90 degrees clockwise

⑩ Intake manifold collector:
Bolts and nuts: 96 inch lbs.
Stud bolts: 61 inch lbs.
Intake manifold:
Bolts and nuts: 65 inch lbs. and then to 21 ft. lbs.
Studs: 96 inch lbs.

71075_FRON_C0010

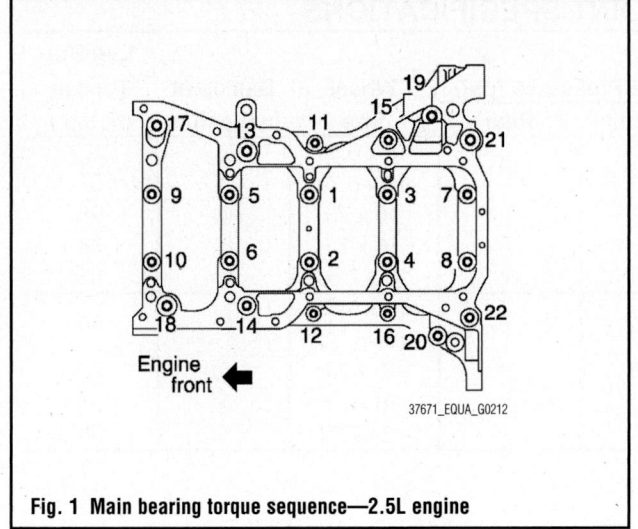

Fig. 1 Main bearing torque sequence—2.5L engine

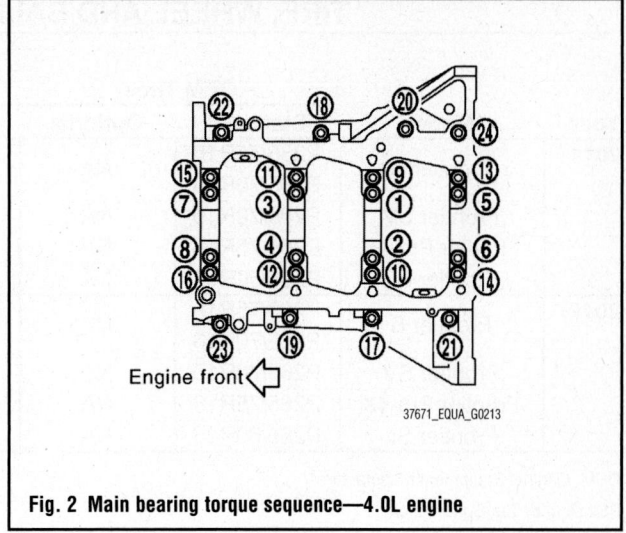

Fig. 2 Main bearing torque sequence—4.0L engine

WHEEL ALIGNMENT

Year	Model		Caster		Camber		Toe-in (Deg.)
			Range (+/-Deg.)	Preferred Setting (Deg.)	Range (+/-Deg.)	Preferred Setting (Deg.)	
2011	Frontier	2WD	0.75	3.00	0.75	0.25	0.12+/-0.04
		4WD	0.75	2.75	0.75	0.50	0.12+/-0.04
2012	Frontier	2WD	0.75	3.00	0.75	0.25	0.12+/-0.04
		4WD	0.75	2.75	0.75	0.50	0.12+/-0.04

NOTE: Measurements given for an unladen vehicle with fuel, coolant, and engine oil must be full; spare tire, jack, hand tools, and mats in designated positions. Some vehicles may be equipped with straight (non-adjustable) lower link bolts and washers. In order to adjust camber and caster on these vehicles, first replace the lower link bolts and washers with adjustable (cam) bolts and washers.

71075_FRON_C0011

TIRE, WHEEL AND BALL JOINT SPECIFICATIONS

Year	Model	OEM Tires Standard	OEM Tires Optional	Tire Pressures (psi) Front	Tire Pressures (psi) Rear	Wheel Size	Ball Joint Inspection	Lug Nut Torque (ft. lbs.)
2011	Frontier S	P235/75R15 P265/70R16	NA	①	①	15 x 6.5J 16 x 6.7J	②	98
	Frontier SV	P265/70R16	NA	①	①	16 x 7J	②	98
	Frontier Pro-4X	P265/75R16	NA	①	①	16 x 7J	②	98
	Frontier SL	P265/60R18	NA	①	①	18 x 7.5J	②	98
2012	Frontier S	P235/75R15 P265/70R16	NA	①	①	15 x 6.5J 16 x6. 7J	②	98
	Frontier SV	P265/70R16	NA	①	①	16 x 7J	②	98
	Frontier Pro-4X	P265/75R16	NA	①	①	16 x 7J	②	98
	Frontier SL	P265/60R18	NA	①	①	18 x7.5J	②	98

OEM: Original Equipment Manufacturer

PSI: Pounds Per Square Inch

NA: Not Applicable

① 35 psi. If specification differs from tire placard on the vehicle, use tire placard specification.

② Swinging force: Lower (at groove) 3-33 ft. lbs.; Upper (at hole) 2-18 ft. lbs.

Turning force: Lower 5-56 inch lbs.; Upper 5-43 inch lbs.

Vertical end play: 0 inch

71075_FRON_C0012

BRAKE SPECIFICATIONS
All measurements in inches unless noted

Year	Model		Brake Disc Original Thickness	Brake Disc Minimum Thickness	Brake Disc Maximum Runout	Minimum Lining Thickness	Brake Caliper Bracket Bolts (ft. lbs.)	Brake Caliper Guide Pin Bolts (ft. lbs.)
2011	Frontier	Front	1.102	1.024	0.002	0.079	136	20
		Rear	0.709	0.630	0.002	0.079	63	20
2012	Frontier	Front	1.102	1.024	0.002	0.079	136	20
		Rear	0.709	0.630	0.002	0.079	63	20

F: Front

R: Rear

71075_FRON_C0013

SCHEDULED MAINTENANCE INTERVALS
Nissan Frontier

TO BE SERVICED	TYPE OF SERVICE	VEHICLE MILEAGE INTERVAL (x1000)												
		7.5	15	22.5	30	37.5	45	52.5	60	67.5	75	82.5	90	97.5
Engine oil & filter	R	✓	✓	✓	✓	✓	✓	✓	✓	✓	✓	✓	✓	✓
Brake lines & cables	S/I		✓		✓		✓		✓		✓		✓	
Brake pads, discs, drums & linings	S/I		✓		✓		✓		✓		✓		✓	
Driveshaft boots & propeller shaft	S/I				✓				✓				✓	
Front wheel bearings (4X2)	S/I				✓				✓				✓	
Front wheel bearings (4X4)	S/I				✓				✓				✓	
Automatic & manual transmission, transfer & differential gear oil ①	S/I		✓		✓		✓		✓		✓		✓	
Air cleaner filter	R				✓				✓				✓	
Engine coolant	R								✓				✓	
Spark plugs (platinum)	R	replace every 105,000 miles												
Drive belt(s)	S/I				✓				✓				✓	
Exhaust system	S/I				✓				✓				✓	
Fuel lines	S/I				✓				✓				✓	
Steering gear (box) & linkage, axle & suspension parts	S/I				✓				✓				✓	
Vapor lines	S/I				✓				✓				✓	
Tires (rotate)	S/I	✓	✓	✓	✓	✓	✓	✓	✓	✓	✓	✓	✓	✓
Timing belt ②	R													

R: Replace   S/I: Service or Inspect

① Differential (w/limited-slip differential) oil: replace oil every 30,000 miles, 2007-2008 vehicles.

② Timing belt: replace at 105,000 miles.

FREQUENT OPERATION MAINTENANCE (SEVERE SERVICE)

If a vehicle is operated under any of the following conditions it is considered severe service:

- Extremely dusty areas.

- 50% or more of the vehicle operation is in 32°C (90°F) or higher temperatures, or constant operation in temperatures below 0°C (32°F).

- Prolonged idling (vehicle operation in stop and go traffic).

- Frequent short running periods (engine does not warm to normal operating temperatures).

- Police, taxi, delivery usage or trailer towing usage.

Oil & oil filter: replace every 3750 miles.

Brake pads, discs, drums & linings: service or inspect every 7500 miles.

Driveshaft boots & propeller shaft: service or inspect every 7500 miles.

Exhaust system: service or inspect every 7500 miles.

Steering gear (box) & linkage, (steering damper-4X4), axle & suspension parts: service or inspect every 7500 miles.

Steering linkage ball joints & front suspension ball joints: service or inspect every 7500 miles.

71075_FRON_C0014

PRECAUTIONS

Before servicing any vehicle, please be sure to read all of the following precautions, which deal with personal safety, prevention of component damage, and important points to take into consideration when servicing a motor vehicle:

• Never open, service or drain the radiator or cooling system when the engine is hot; serious burns can occur from the steam and hot coolant.

• Observe all applicable safety precautions when working around fuel. Whenever servicing the fuel system, always work in a well-ventilated area. Do not allow fuel spray or vapors to come in contact with a spark, open flame, or excessive heat (a hot drop light, for example). Keep a dry chemical fire extinguisher near the work area. Always keep fuel in a container specifically designed for fuel storage; also, always properly seal fuel containers to avoid the possibility of fire or explosion. Refer to the additional fuel system precautions later in this section.

• Fuel injection systems often remain pressurized, even after the engine has been turned **OFF**. The fuel system pressure must be relieved before disconnecting any fuel lines. Failure to do so may result in fire and/or personal injury.

• Brake fluid often contains polyglycol ethers and polyglycols. Avoid contact with the eyes and wash your hands thoroughly after handling brake fluid. If you do get brake fluid in your eyes, flush your eyes with clean, running water for 15 minutes. If eye irritation persists, or if you have taken brake fluid internally, IMMEDIATELY seek medical assistance.

• The EPA warns that prolonged contact with used engine oil may cause a number of skin disorders, including cancer. You should make every effort to minimize your exposure to used engine oil. Protective gloves should be worn when changing oil. Wash your hands and any other exposed skin areas as soon as possible after exposure to used engine oil. Soap and water, or waterless hand cleaner should be used.

• All new vehicles are now equipped with an air bag system, often referred to as a Supplemental Restraint System (SRS) or Supplemental Inflatable Restraint (SIR) system. The system must be disabled before performing service on or around system components, steering column, instrument panel components, wiring and sensors. Failure to follow safety and disabling procedures could result in accidental air bag deployment, possible personal injury and unnecessary system repairs.

• Always wear safety goggles when working with, or around, the air bag system. When carrying a non-deployed air bag, be sure the bag and trim cover are pointed away from your body. When placing a non-deployed air bag on a work surface, always face the bag and trim cover upward, away from the surface. This will reduce the motion of the module if it is accidentally deployed. Refer to the additional air bag system precautions later in this section.

• Clean, high quality brake fluid from a sealed container is essential to the safe and proper operation of the brake system. You should always buy the correct type of brake fluid for your vehicle. If the brake fluid becomes contaminated, completely flush the system with new fluid. Never reuse any brake fluid. Any brake fluid that is removed from the system should be discarded. Also, do not allow any brake fluid to come in contact with a painted surface; it will damage the paint.

• Never operate the engine without the proper amount and type of engine oil; doing so WILL result in severe engine damage.

• Timing belt maintenance is extremely important. Many models utilize an interference-type, non-freewheeling engine. If the timing belt breaks, the valves in the cylinder head may strike the pistons, causing potentially serious (also time-consuming and expensive) engine damage. Refer to the maintenance interval charts for the recommended replacement interval for the timing belt, and to the timing belt section for belt replacement and inspection.

• Disconnecting the negative battery cable on some vehicles may interfere with the functions of the on-board computer system(s) and may require the computer to undergo a relearning process once the negative battery cable is reconnected.

• When servicing drum brakes, only disassemble and assemble one side at a time, leaving the remaining side intact for reference.

• Only an MVAC-trained, EPA-certified automotive technician should service the air conditioning system or its components.

BRAKES

GENERAL INFORMATION

PRECAUTIONS

• Certain components within the ABS system are not intended to be serviced or repaired individually.

• Do not use rubber hoses or other parts not specifically specified for and ABS system. When using repair kits, replace all parts included in the kit. Partial or incorrect repair may lead to functional problems and require the replacement of components.

• Lubricate rubber parts with clean, fresh brake fluid to ease assembly. Do not use shop air to clean parts; damage to rubber components may result.

• Use only DOT 3 brake fluid from an unopened container.

• If any hydraulic component or line is removed or replaced, it may be necessary to bleed the entire system.

• A clean repair area is essential. Always clean the reservoir and cap thoroughly before removing the cap. The slightest amount of dirt in the fluid may plug an orifice and impair the system function. Perform repairs after components have been thoroughly cleaned; use only denatured alcohol to clean components. Do not allow ABS components to come into contact with any substance containing mineral oil; this includes used shop rags.

• The Anti-Lock control unit is a microprocessor similar to other computer units in the vehicle. Ensure that the ignition switch is **OFF** before removing or installing controller harnesses. Avoid static electricity discharge at or near the controller.

ANTI-LOCK BRAKE SYSTEM (ABS)

• If any arc welding is to be done on the vehicle, the control unit should be unplugged before welding operations begin.

WHEEL SPEED SENSORS

REMOVAL & INSTALLATION

See Figure 3.

➡**If working near and/or around the SRS system and components, be sure to disable the SRS system. After disabling the system wait three minutes or more before servicing the vehicle.**

➡**Whenever the negative battery cable is disconnected the following components will require resetting. The Idle Air Volume Learning, Steering Angle Sensor Neutral Position, Sunroof Memory Reset/Initialization, Automatic**

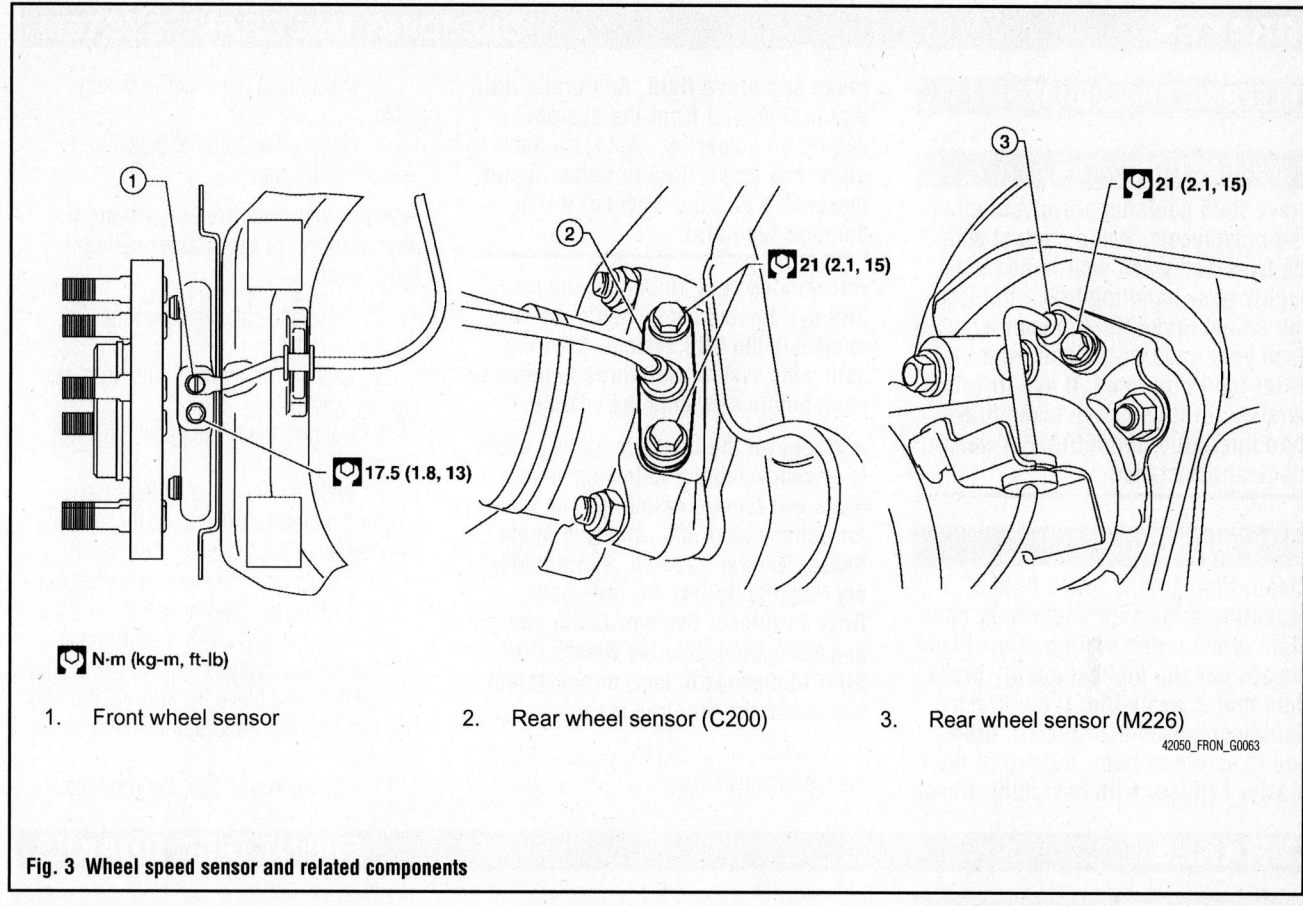

N·m (kg-m, ft-lb)

17.5 (1.8, 13)

21 (2.1, 15)

21 (2.1, 15)

1. Front wheel sensor
2. Rear wheel sensor (C200)
3. Rear wheel sensor (M226)

42050_FRON_G0063

Fig. 3 Wheel speed sensor and related components

Drive Positioner System, Audio presets and Navigation. Use the Nissan CONSULT III diagnostic tool, or equivalent to perform the required resets.

1. Before servicing the vehicle, refer to the Precautions Section.
2. Disconnect the negative battery cable.
3. Raise and safely support the vehicle.
4. Remove the tire and wheel assembly.
5. Remove the brake rotor, for front wheel speed sensor access.

6. Remove the wheel speed sensor mounting bolts and clip.
7. Pull the sensor out, being careful to turn it as little as possible. Do not pull on the sensor harness.

To install:

8. Before installing the sensor, be sure no foreign materials such as iron fragments adhere to the pick-up part of the sensor or to the inside of the sensor mounting hole or on the rotor mounting surface.

9. Inspect and replace the wheel sensor if damaged.
10. Clean the wheel sensor hole and mating surface with brake cleaner and a lint-free cloth. Be careful that dirt and debris do not enter the hub and bearing assembly or the rear axle.
11. Be sure the harness is not twisted when installed.
12. Continue the installation in the reverse order of the removal procedure.

BRAKES   BLEEDING THE BRAKE SYSTEM

BLEEDING PROCEDURE

✷✷ CAUTION

Brake fluid contains polyglycol ethers and polyglycols. Avoid contact with the eyes and wash your hands thoroughly after handling brake fluid. If you do get brake fluid in your eyes, flush your eyes with clean, running water for 15 minutes. If eye irritation persists, or if you have taken brake fluid internally, IMMEDIATELY seek medical assistance.

✷✷ WARNING

Clean, high quality brake fluid is essential to the safe and proper operation of the brake system. You should always buy the highest quality brake fluid that is available. If the brake fluid becomes contaminated, drain and flush the system, then refill the master cylinder with new fluid. Never reuse any brake fluid. Any brake fluid that is removed from the system should be discarded. Also, do not allow any brake fluid to come in contact with a painted surface; it will damage the paint.

➡ If working near and/or around the SRS system and components, be sure to disable the SRS system. After disabling the system wait three minutes or more before servicing the vehicle.

➡ Whenever the negative battery cable is disconnected the following components will require resetting. The Idle Air Volume Learning, Steering Angle Sensor Neutral Position, Sunroof Memory Reset/Initialization, Automatic Drive Positioner System, Audio presets and Navigation. Use the Nissan CONSULT III diagnostic tool, or equivalent to perform the required resets.

1. Before servicing the vehicle, refer to the Precautions Section.

2. Disconnect the negative battery cable.

3. Clean all around the brake fluid reservoir filler cap.

➡ While bleeding the brake system, pay attention to the master cylinder fluid level.

4. Raise and safely support the vehicle.

5. Attach a vinyl tube to the right, rear bleeder valve.

6. Depress the brake pedal fully 4 or 5 times.

7. With the brake pedal depressed, loosen the bleeder valve to let the air out, then tighten it immediately.

8. Repeat until no more air comes out.

9. Tighten the bleeder valve.

10. Perform steps 4–8 at each wheel, with the master cylinder reservoir tank filled at least half way, bleed the air from the front left, rear left, and front right bleed valve, in that order.

11. Fill the master cylinder reservoir.

BRAKES   FRONT DISC BRAKES

BRAKE CALIPERS

REMOVAL & INSTALLATION

See Figure 4.

➡ If working near and/or around the SRS system and components, be sure to disable the SRS system. After disabling the system wait three minutes or more before servicing the vehicle.

➡ Whenever the negative battery cable is disconnected the following components will require resetting. The Idle Air Volume Learning, Steering Angle Sensor Neutral Position, Sunroof Memory Reset/Initialization, Automatic Drive Positioner System, Audio presets and Navigation. Use the Nissan CONSULT III diagnostic tool, or equivalent to perform the required resets.

1. Before servicing the vehicle, refer to the Precautions Section.

2. Disconnect the negative battery cable.

3. Drain the brake fluid, as necessary.

4. Raise and safely support the vehicle.

5. Remove the tire and wheel assembly.

6. Remove the bolt attaching the brake hose to the caliper. Plug the brake hose to prevent brake fluid loss.

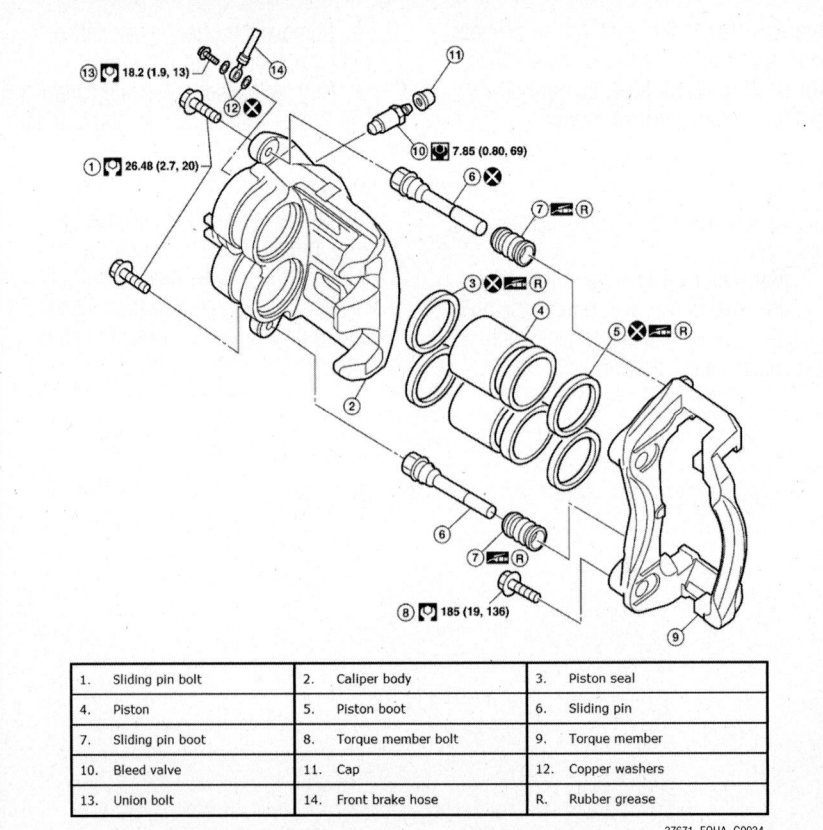

1.	Sliding pin bolt	2.	Caliper body	3.	Piston seal
4.	Piston	5.	Piston boot	6.	Sliding pin
7.	Sliding pin boot	8.	Torque member bolt	9.	Torque member
10.	Bleed valve	11.	Cap	12.	Copper washers
13.	Union bolt	14.	Front brake hose	R.	Rubber grease

37671_EQUA_G0034

Fig. 4 Front disc brake caliper and related components

7. Remove the caliper support mounting bolts and lift the caliper assembly from the knuckle.

To install

8. Position the caliper assembly onto the knuckle and install the bolts. Make sure the rotor fits between the brake pads. Torque the bolts to specification.

9. Using new copper washers, connect the brake hose to the caliper. Torque the brake hose attaching bolt to specification.

10. Bleed the brake system.

11. Apply the brake pedal and inspect the system. Ensure proper operation and no leakage.

12. Install the tire and wheel assembly. Tighten the wheel nuts to 98 ft. lbs. (133 Nm).

13. Lower the vehicle and road test.

BRAKE PADS

REMOVAL & INSTALLATION

See Figure 5.

➡ **If working near and/or around the SRS system and components, be sure to disable the SRS system. After disabling the system wait three minutes or more before servicing the vehicle.**

➡ **Whenever the negative battery cable is disconnected the following components will require resetting. The Idle Air Volume Learning, Steering Angle Sensor Neutral Position, Sunroof Memory Reset/Initialization, Automatic Drive Positioner System, Audio presets and Navigation. Use the Nissan CONSULT III diagnostic tool, or equivalent to perform the required resets.**

1. Before servicing the vehicle, refer to the Precautions Section.

2. Disconnect the negative battery cable.

3. Remove the master cylinder reservoir cap.

4. Partially drain the brake fluid.

5. Raise and safely support the vehicle.

6. Remove the tire and wheel assembly.

7. Remove the caliper sliding pin bolts.

8. Support the caliper body with a suitable wire to avoid pulling on the front brake hose.

9. Remove the front inner and outer brake pads, shims, shim cover, pad return spring, and retainers from the torque member.

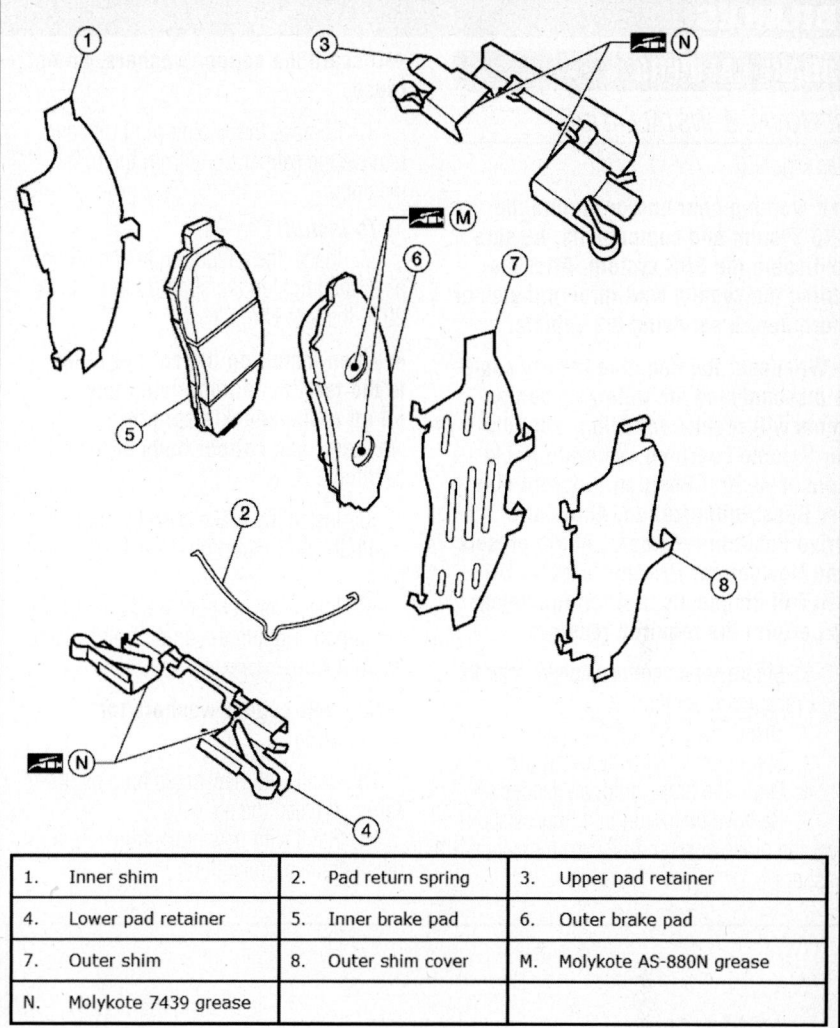

1.	Inner shim	2.	Pad return spring	3.	Upper pad retainer
4.	Lower pad retainer	5.	Inner brake pad	6.	Outer brake pad
7.	Outer shim	8.	Outer shim cover	M.	Molykote AS-880N grease
N.	Molykote 7439 grease				

37671_EQUA_G0036

Fig. 5 Front disc brake pads and related components

To install:

10. Apply Molykote® AS-880N grease, or equivalent, between the outer brake pad backing and the shims, then attach the shim and shim cover to the brake pads.

11. Apply Molykote® 7439 grease to the upper pad retainer.

12. Attach the pad retainer to the torque member, then install the brake pad, shims, and shim covers to the torque member.

➡ **When attaching the pad retainer, attach it firmly so that it is flush with the torque member.**

13. Using a suitable tool, push the pistons into the caliper body.

➡ **When pushing in the piston, brake fluid returns to the master cylinder reservoir tank. Watch the level of the brake fluid in the reservoir tank.**

14. Install the pad return spring to the bottom edge of the brake pads in the holes provided.

15. Install the caliper body.

16. Install the sliding pin bolts and tighten to 20 ft. lbs. (27 Nm).

17. Check the brakes for drag.

18. Inspect the brake fluid level, then install the master cylinder reservoir cap.

19. Install the front wheel and tire. Tighten the wheel nuts to 98 ft. lbs. (133 Nm).

BRAKE CALIPERS

REMOVAL & INSTALLATION

See Figure 6.

➡ **If working near and/or around the SRS system and components, be sure to disable the SRS system. After disabling the system wait three minutes or more before servicing the vehicle.**

➡ **Whenever the negative battery cable is disconnected the following components will require resetting. The Idle Air Volume Learning, Steering Angle Sensor Neutral Position, Sunroof Memory Reset/Initialization, Automatic Drive Positioner System, Audio presets and Navigation. Use the Nissan CONSULT III diagnostic tool, or equivalent to perform the required resets.**

1. Before servicing the vehicle, refer to the Precautions Section.
2. Disconnect the negative battery cable.
3. Remove the rear wheel and tire.
4. Drain the brake fluid, as needed.
5. Remove the union bolt, then disconnect the brake hose and discard the copper washers.

➡ **Discard the copper washers, do not reuse.**

6. Remove the sliding pin bolts and remove the caliper body from the torque member.

To install:

7. Install the caliper body and sliding pins, then tighten the sliding pin bolts to 20 ft. lbs. (27 Nm).

➡ **When installing the caliper body to the torque member, wipe any oil off of the knuckle spindle, washers, and caliper body attachment surfaces.**

8. Install the brake hose by aligning it with the protrusion on the caliper body.
9. Install new copper washers and the union bolt. Tighten the union bolt to 13 ft. lbs. (18 Nm).

➡ **Use new copper washers for installation.**

10. Refill with new brake fluid as necessary and bleed the air.
11. Refill with new brake fluid. Do not reuse drained brake fluid.

12. Install the rear wheel and tire. Tighten the wheel nuts to 98 ft. lbs. (133 Nm).
13. Lower the vehicle and road test.

BRAKE PADS

REMOVAL & INSTALLATION

See Figure 7.

➡ **If working near and/or around the SRS system and components, be sure to disable the SRS system. After disabling the system wait three minutes or more before servicing the vehicle.**

➡ **Whenever the negative battery cable is disconnected the following components will require resetting. The Idle Air Volume Learning, Steering Angle Sensor Neutral Position, Sunroof Memory Reset/Initialization, Automatic Drive Positioner System, Audio presets and Navigation. Use the Nissan CONSULT III diagnostic tool, or equivalent to perform the required resets.**

1. Before servicing the vehicle, refer to the Precautions Section.
2. Disconnect the negative battery cable.
3. Remove the rear wheel and tire.
4. Drain the brake fluid, as needed.
5. Remove caliper sliding pin bolts.
6. Support the caliper body with a suitable wire to avoid pulling on the rear brake hose.
7. Remove the rear inner and outer brake pads, shims, and retainers from the torque member.

To install:

8. Apply Molykote® AS-880N grease between the brake pad back plates and shims, then attach the shims to the brake pads.
9. Apply Molykote® 7439 grease to the pad retainers.
10. Attach the pad retainer to the torque member, then install the brake pad and shim assemblies.

➡ **When attaching the pad retainer, attach it firmly so that it is flush with the torque member.**

11. Using a suitable tool push the piston into the caliper body.

➡ **By pushing in the piston, brake fluid returns to the master cylinder reservoir tank. Watch the level of fluid in the reservoir tank.**

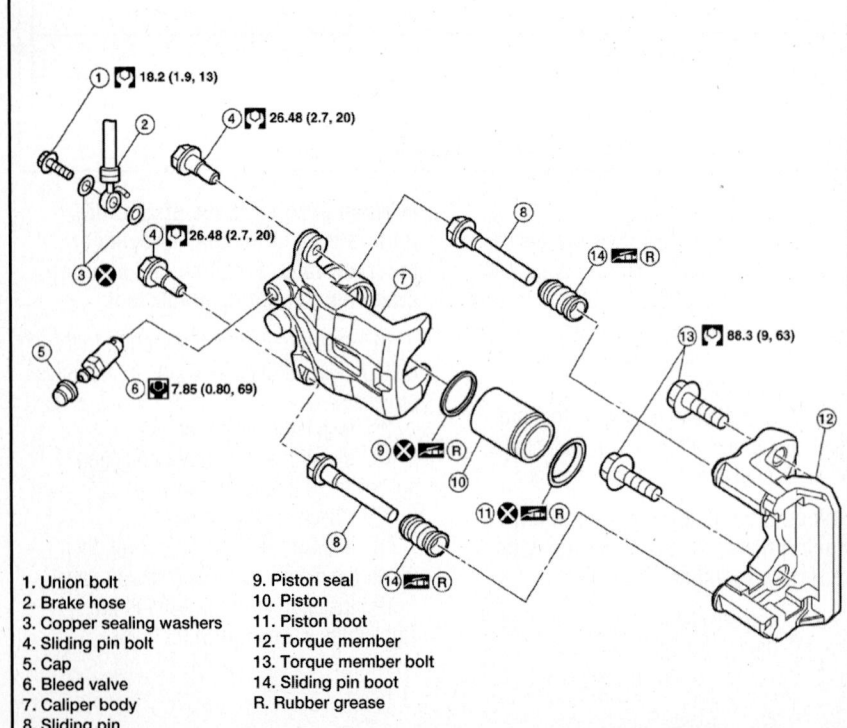

1. Union bolt
2. Brake hose
3. Copper sealing washers
4. Sliding pin bolt
5. Cap
6. Bleed valve
7. Caliper body
8. Sliding pin
9. Piston seal
10. Piston
11. Piston boot
12. Torque member
13. Torque member bolt
14. Sliding pin boot
R. Rubber grease

71082_EQUA_G0097

Fig. 6 Rear disc brake caliper and related components

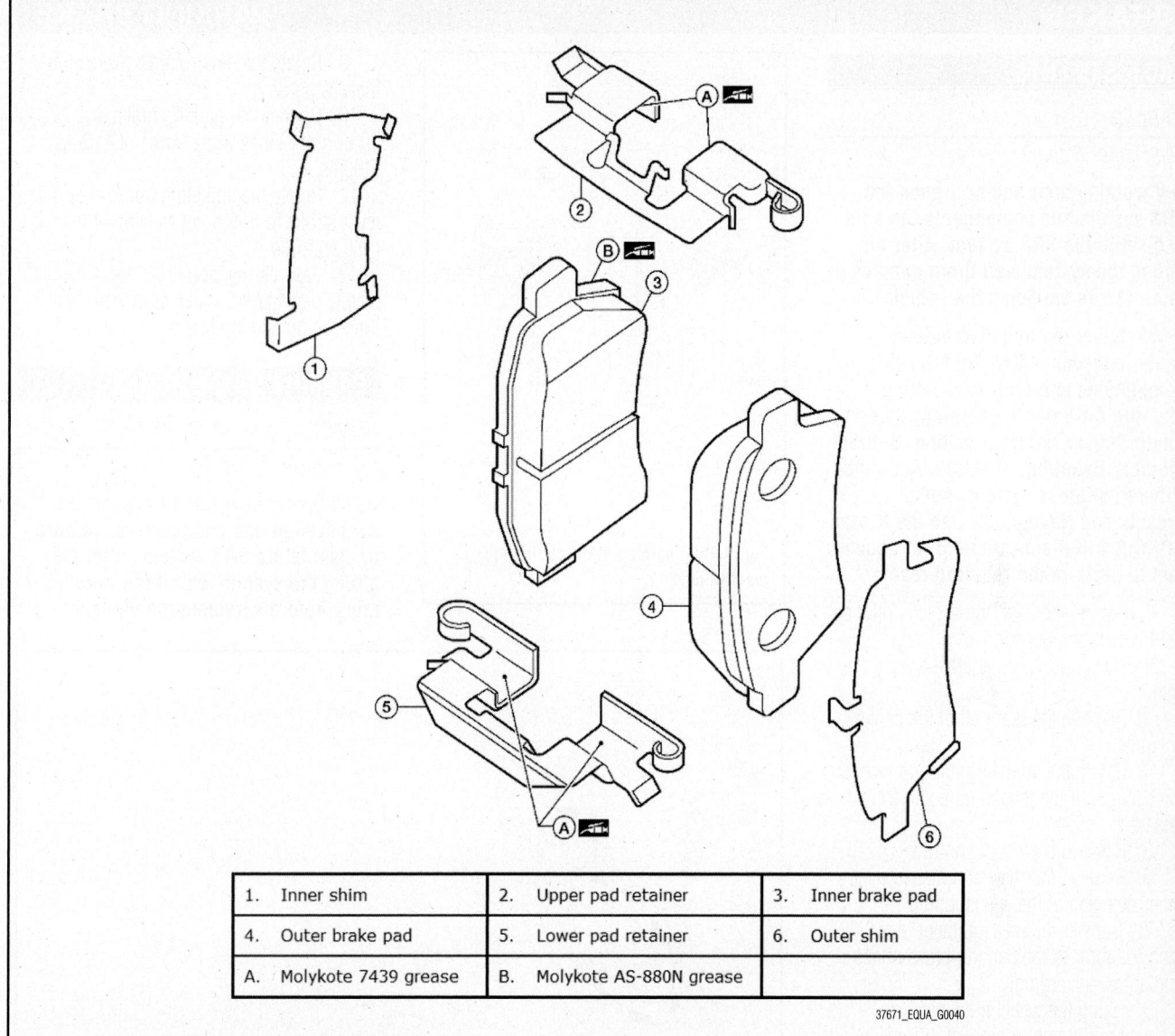

1. Inner shim	2. Upper pad retainer	3. Inner brake pad
4. Outer brake pad	5. Lower pad retainer	6. Outer shim
A. Molykote 7439 grease	B. Molykote AS-880N grease	

37671_EQUA_G0040

Fig. 7 Rear disc brake pads and related components

12. Install the caliper body.

13. Install the sliding pin bolts and tighten to 20 ft. lbs. (27 Nm).

14. Check the brakes for drag.

15. Inspect the brake fluid level, then install the master cylinder reservoir cap.

16. Install the rear wheel and tire. Tighten the wheel nuts to 98 ft. lbs. (133 Nm).

BRAKES

ADJUSTMENTS

CABLES

See Figures 8 and 9.

➡ If working near and/or around the SRS system and components, be sure to disable the SRS system. After disabling the system wait three minutes or more before servicing the vehicle.

➡ Whenever the negative battery cable is disconnected the following components will require resetting. The Idle Air Volume Learning, Steering Angle Sensor Neutral Position, Sunroof Memory Reset/Initialization, Automatic Drive Positioner System, Audio presets and Navigation. Use the Nissan CONSULT III diagnostic tool, or equivalent to perform the required resets.

1. Before servicing the vehicle, refer to the Precautions Section.
2. Disconnect the negative battery cable.
3. Remove the rear half of the center console.
4. Rotate the adjusting nut and loosen the cable until the tension is sufficiently released.
5. Remove the wheel and tire.
6. Remove the rotor and measure the inner diameter at the widest point.
7. Transfer the measurement less 0.02 inch (0.6mm) to the parking brake shoes and adjust accordingly.
8. Using the wheel nuts, secure the disc to the hub to prevent it from tilting.

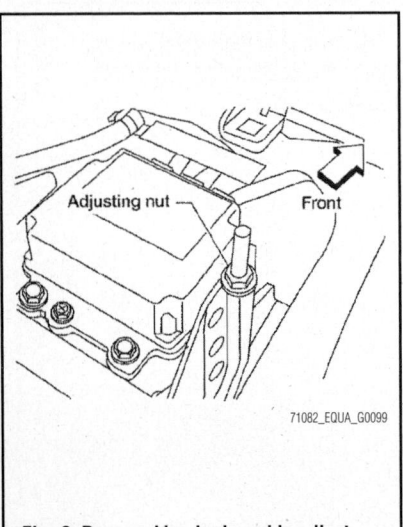

Adjusting nut — Front

71082_EQUA_G0099

Fig. 8 Rear parking brake cable adjustment (1 of 2)

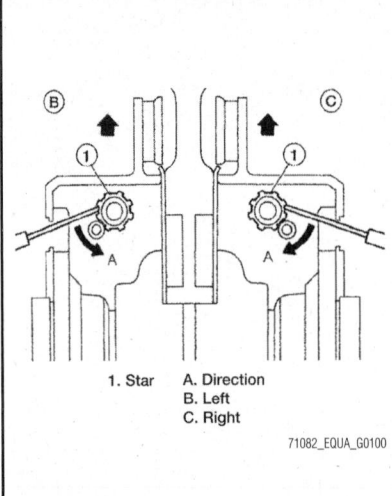

1. Star
A. Direction
B. Left
C. Right

71082_EQUA_G0100

Fig. 9 Rear parking brake cable adjustment (2 of 2)

9. Rotate the disc rotor to make sure there is no drag.
10. Operate the parking brake lever 10 or more times with a force of 110 lbs. (490 N).
11. Rotate the adjusting nut to adjust the lever stroke to 6–8 notches with 44 lbs. (196 N) force.
12. With the parking brake lever completely disengaged, make sure there is no drag on the parking brake.

PARKING BRAKE SHOES

REMOVAL & INSTALLATION

See Figures 10 and 11.

➡ If working near and/or around the SRS system and components, be sure to disable the SRS system. After disabling the system wait three minutes or more before servicing the vehicle.

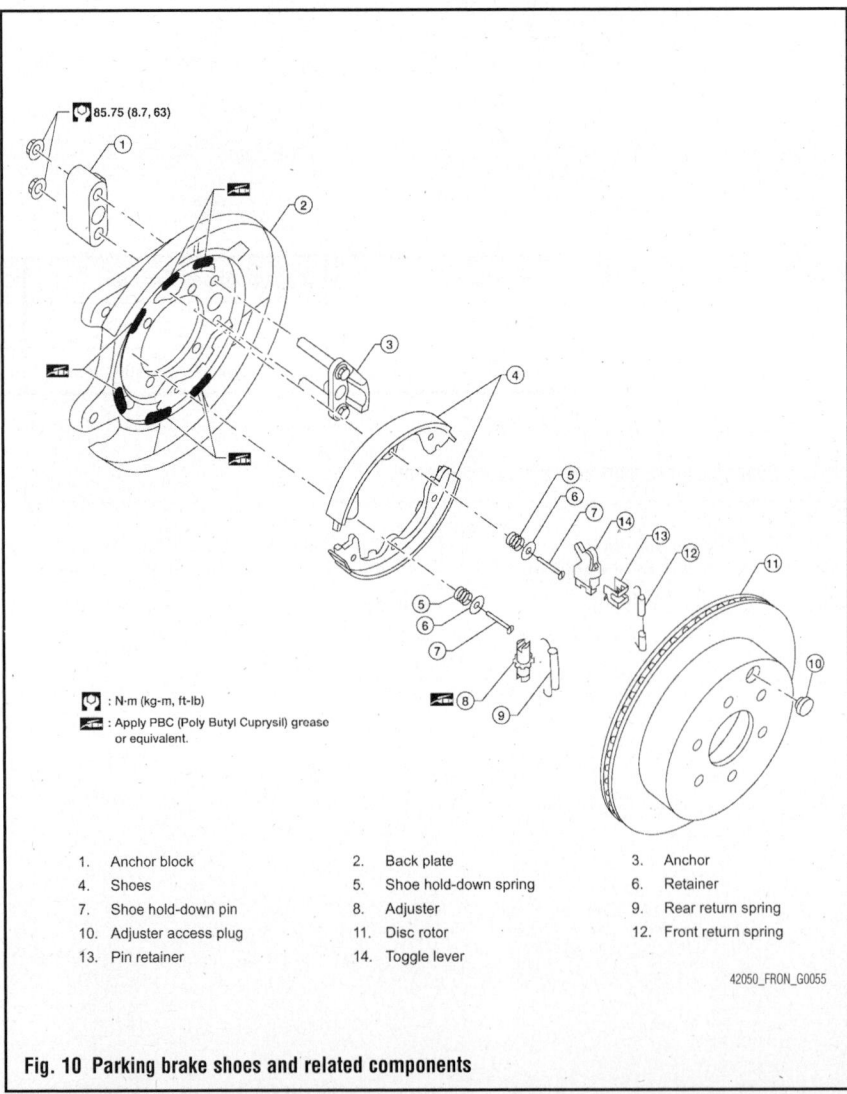

85.75 (8.7, 63)

🔧 : N-m (kg-m, ft-lb)
⬅ : Apply PBC (Poly Butyl Cuprysil) grease or equivalent.

1.	Anchor block	2.	Back plate	3.	Anchor
4.	Shoes	5.	Shoe hold-down spring	6.	Retainer
7.	Shoe hold-down pin	8.	Adjuster	9.	Rear return spring
10.	Adjuster access plug	11.	Disc rotor	12.	Front return spring
13.	Pin retainer	14.	Toggle lever		

42050_FRON_G0055

Fig. 10 Parking brake shoes and related components

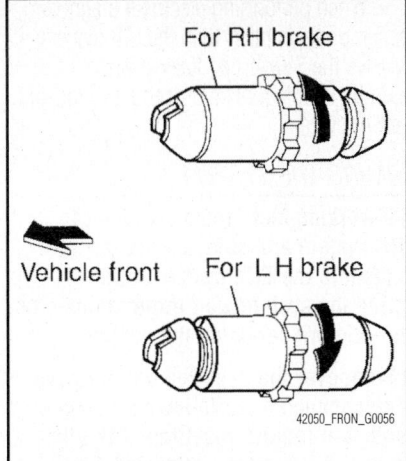

Fig. 11 Parking brake shoe adjuster identification

➡Whenever the negative battery cable is disconnected the following components will require resetting. The Idle Air Volume Learning, Steering Angle Sensor Neutral Position, Sunroof Memory Reset/Initialization, Automatic Drive Positioner System, Audio presets and Navigation. Use the Nissan CONSULT III diagnostic tool, or equivalent to perform the required resets.

1. Before servicing the vehicle, refer to the Precautions Section.
2. Disconnect the negative battery cable.

✳✳ CAUTION

Clean the brakes with a vacuum dust collector to minimize the hazard of airborne particles or other materials.

➡Remove the disc rotor with the parking brake completely disengaged.

3. Raise and safely support the vehicle.
4. Release the parking brake.
5. Remove the rear wheels.
6. Remove the rotor.
7. Remove the return springs.
8. Remove the adjuster.
9. Remove the retainers, anti-rattle pins, and shoes.
10. Remove the pin retainer. Disconnect the parking brake cable from the toggle lever.
11. Remove the back plate.

To install:

12. Installation is the reverse of the removal procedure.
13. Assemble the adjuster so that the threaded part expands when rotating it in the direction shown by the arrow. Shorten the adjuster by rotating it in the opposite direction shown by the arrow.
14. Perform the parking brake break-in operation as follows: Safely drive forward at approximately 25 mph (40 km/h) with the parking brake set with a force of approx. 45 lbs. (200 N) for about 30 seconds.
15. After the break-in operation, check the pedal stroke of the parking brake. Readjust, if necessary.

➡To prevent the lining from getting too hot, allow a cool off period of approximately 5 minutes after every break-in operation.

16. Check and adjust the parking brake pedal stroke. Correct as required.

CHASSIS ELECTRICAL — AIR BAGS (SUPPLEMENTAL RESTRAINT SYSTEM)

✳✳ CAUTION

These vehicles are equipped with an air bag system. The system must be disarmed before performing service on, or around, system components, the steering column, instrument panel components, wiring and sensors. Failure to follow the safety precautions and the disarming procedure could result in accidental air bag deployment, possible injury and unnecessary system repairs.

SERVICE PRECAUTIONS

Disconnect and isolate the battery negative cable before beginning any airbag system component diagnosis, testing, removal, or installation procedures. Allow system capacitor to discharge for two minutes before beginning any component service. This will disable the airbag system. Failure to disable the airbag system may result in accidental airbag deployment, personal injury, or death.

Do not place an intact undeployed airbag face down on a solid surface. The airbag will propel into the air if accidentally deployed and may result in personal injury or death.

When carrying or handling an undeployed airbag, the trim side (face) of the airbag should be pointing away from the body to minimize possibility of injury if accidental deployment occurs. Failure to do this may result in personal injury or death.

Replace airbag system components with OEM replacement parts. Substitute parts may appear interchangeable, but internal differences may result in inferior occupant protection. Failure to do so may result in occupant personal injury or death.

Wear safety glasses, rubber gloves, and long sleeved clothing when cleaning powder residue from vehicle after an airbag deployment. Powder residue emitted from a deployed airbag can cause skin irritation. Flush affected area with cool water if irritation is experienced. If nasal or throat irritation is experienced, exit the vehicle for fresh air until the irritation ceases. If irritation continues, see a physician.

Do not use a replacement airbag that is not in the original packaging. This may result in improper deployment, personal injury, or death.

The factory installed fasteners, screws and bolts used to fasten airbag components have a special coating and are specifically designed for the airbag system. Do not use substitute fasteners. Use only original equipment fasteners listed in the parts catalog when fastener replacement is required.

During, and following, any child restraint anchor service, due to impact event or vehicle repair, carefully inspect all mounting hardware, tether straps, and anchors for proper installation, operation, or damage. If a child restraint anchor is found damaged in any way, the anchor must be replaced. Failure to do this may result in personal injury or death.

Deployed and non-deployed airbags may or may not have live pyrotechnic material within the airbag inflator.

Do not dispose of driver/passenger/curtain airbags or seat belt tensioners unless you are sure of complete deployment. Refer to the Hazardous Substance Control System for proper disposal.

Dispose of deployed airbags and tensioners consistent with state, provincial, local, and federal regulations.

After any airbag component testing or service, do not connect the battery negative cable. Personal injury or death may result if the system test is not performed first.

If the vehicle is equipped with the Occupant Classification System (OCS), do not connect the battery negative cable before performing the OCS Verification Test using the scan tool and the appropriate diagnostic information. Personal injury or death may result if the system test is not performed properly.

Never replace both the Occupant

Restraint Controller (ORC) and the Occupant Classification Module (OCM) at the same time. If both require replacement, replace one, then perform the Airbag System test before replacing the other.

Both the ORC and the OCM store Occupant Classification System (OCS) calibration data, which they transfer to one another when one of them is replaced. If both are replaced at the same time, an irreversible fault will be set in both modules and the OCS may malfunction and cause personal injury or death.

If equipped with OCS, the Seat Weight Sensor is a sensitive, calibrated unit and must be handled carefully. Do not drop or handle roughly. If dropped or damaged, replace with another sensor. Failure to do so may result in occupant injury or death.

If equipped with OCS, the front passenger seat must be handled carefully as well. When removing the seat, be careful when setting on floor not to drop. If dropped, the sensor may be inoperative, could result in occupant injury, or possibly death.

If equipped with OCS, when the passenger front seat is on the floor, no one should sit in the front passenger seat. This uneven force may damage the sensing ability of the seat weight sensors. If sat on and damaged, the sensor may be inoperative, could result in occupant injury, or possibly death.

DISARMING THE SYSTEM

➡**If working near and/or around the SRS system and components, be sure to disable the SRS system. After disabling the system wait three minutes or more before servicing the vehicle.**

➡**Whenever the negative battery cable is disconnected the following components will require resetting. The Idle Air Volume Learning, Steering Angle Sensor Neutral Position, Sunroof Memory Reset/Initialization, Automatic**

Drive Positioner System, Audio presets and Navigation. Use the Nissan CONSULT III diagnostic tool, or equivalent to perform the required resets.

1. Before servicing the vehicle, refer to the Precautions Section.
2. Disconnect the negative battery cable.
3. Turn the ignition switch to **OFF**.
4. Isolate the negative battery cable and isolate it from accidental reconnection. Insulate the cable end with high-quality electrical tape or a similar non-conductive wrapping.
5. Disconnect the positive battery cable.
6. Wait at least 3 minutes for the system capacitor to discharge before performing any service. The airbag system is designed to retain enough voltage to deploy the airbag for a short period of time after the battery has been disconnected.

➡**DTC's will be lost when the negative battery cable is disconnected.**

There are several reasons for disabling the SIR system, such as repairs to the SIR system or servicing a component near or attached to an SIR component. There are several ways to disable the SIR system depending on what type of service is being performed.

• If the vehicle was involved in an accident with an air bag deployment: Disconnect the negative battery cable
• When performing SIR diagnostics: Follow the appropriate SIR service diagnostic procedure(s)
• When removing or replacing an SIR component or a component attached to an SIR component: Disconnect the negative battery cable
• If the vehicle is suspected of having shorted electrical wires: Disconnect the negative battery cable

• When performing electrical diagnosis on components other than the SIR system: Remove the SIR/Airbag fuse(s) when indicated by the diagnostic procedure to disable the SIR system

ARMING THE SYSTEM

➡**If working near and/or around the SRS system and components, be sure to disable the SRS system. After disabling the system wait three minutes or more before servicing the vehicle.**

➡**Whenever the negative battery cable is disconnected the following components will require resetting. The Idle Air Volume Learning, Steering Angle Sensor Neutral Position, Sunroof Memory Reset/Initialization, Automatic Drive Positioner System, Audio presets and Navigation. Use the Nissan CONSULT III diagnostic tool, or equivalent to perform the required resets.**

1. Before servicing the vehicle, refer to the Precautions Section.
2. Be sure the ignition switch is in the **OFF** position.
3. Install the fuses, if removed.
4. Connect the positive battery cable.
5. Connect the negative battery cable.

❉❉ CAUTION

As an added precaution, make sure no one is in the vehicle when reconnecting the negative battery cable.

6. To confirm proper system operation, turn the ignition switch to the **ON** position. The SRS indicator light should light for at least 7 seconds and then go off.
7. If the AIR BAG warning indicator does not operate as described, perform a diagnostic system check, using the Nissan CONSULT III diagnostic tool, or equivalent..

DRIVETRAIN

AUTOMATIC TRANSMISSION

DRAIN & REFILL

See Figure 12.

➡If working near and/or around the SRS system and components, be sure to disable the SRS system. After disabling the system wait three minutes or more before servicing the vehicle.

➡Whenever the negative battery cable is disconnected the following components will require resetting. The Idle Air Volume Learning, Steering Angle Sensor Neutral Position, Sunroof Memory Reset/Initialization, Automatic Drive Positioner System, Audio presets and Navigation. Use the Nissan CONSULT III diagnostic tool, or equivalent to perform the required resets.

1. Before servicing the vehicle, refer to the Precautions Section.
2. Drive the vehicle to warm up the A/T fluid to approximately 176° F (80° C).
3. Stop the engine.
4. Disconnect the negative battery cable.
5. Raise and safely support the vehicle.
6. Remove the A/T fluid level gauge (dipstick).
7. Raise and safely support the vehicle.
8. Drain the A/T fluid from the drain plug hole.

To install:

9. Install the drain plug with a new gasket and tighten the drain plug to 25 ft. lbs. (34 Nm).
10. Refill the transmission with the proper amount and type of new A/T fluid.

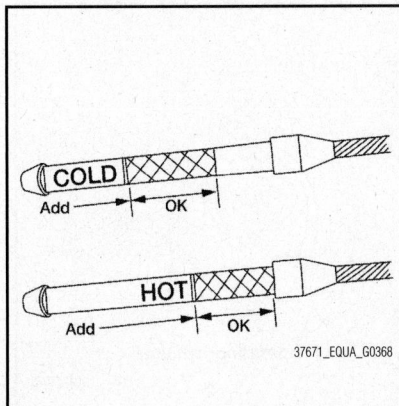

Fig. 12 Automatic transmission dipstick markings

Refill with the same volume as the drained A/T fluid. Use the A/T fluid level gauge to check the A/T fluid level. Add A/T fluid as necessary.

11. To flush out the old A/T fluid from the transmission oil coolers, pour new A/T fluid into the A/T fluid charging pipe with the engine idling and at the same time drain the old A/T fluid from the auxiliary transmission oil cooler hose return line.

12. When the color of the A/T fluid coming out of the auxiliary transmission oil cooler hose return line is about the same as the color of the new A/T fluid, flushing out the old A/T fluid is complete. The amount of new A/T fluid used for flushing should be a 30–50 percent increase of the specified capacity.

✳✳ WARNING

If NISSAN approved Matic S ATF is not available, NISSAN approved Matic J ATF may also be used. Using automatic transmission fluid other than NISSAN approved Matic S ATF or Matic J ATF will cause deterioration in driveability and automatic transmission durability, and may damage the automatic transmission.

➡When filling the transmission with A/T fluid, do not spill the A/T fluid on any heat generating parts such as the exhaust manifold. Do not reuse the drain plug gasket.

13. Install the A/T fluid level gauge and tighten the A/T fluid level gauge bolt to 45 inch lbs. (5 Nm).
14. Drive the vehicle to warm up the A/T fluid to approximately 176° F (80° C).
15. Check the fluid level and condition. If the A/T fluid is still dirty, repeat the drain and refill procedure.
16. Install the A/T fluid level gauge in the A/T fluid charging pipe and install the A/T fluid level gauge bolt to 45 inch lbs. (5 Nm).

FLUID LEVEL CHECK

See Figure 13.

➡If working near and/or around the SRS system and components, be sure to disable the SRS system. After disabling the system wait three minutes or more before servicing the vehicle.

➡Whenever the negative battery cable is disconnected the following components will require resetting. The Idle Air Volume Learning, Steering Angle

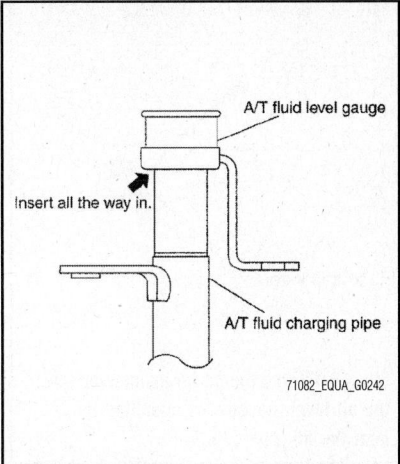

Fig. 13 Automatic transmission dipstick location

Sensor Neutral Position, Sunroof Memory Reset/Initialization, Automatic Drive Positioner System, Audio presets and Navigation. Use the Nissan CONSULT III diagnostic tool, or equivalent to perform the required resets.

1. Before servicing the vehicle, refer to the Precautions Section.
2. Drive the vehicle to warm up the A/T fluid to approximately 176° F (80° C).
3. Stop the engine.
4. Remove the A/T fluid level gauge (dipstick).
5. Wipe the dipstick and reinsert it.
6. Remove the A/T fluid level gauge (dipstick) and check the reading..
7. Correct fluid level, as required.

FLUID RECOMMENDATIONS

The automatic transmission uses Nissan approved Matic S ATF transmission fluid. If this fluid is not available Nissan Matic J ATF may be used. If information differs from the owner's manual, use the data in the owner's manual.

MANUAL TRANSMISSION

DRAIN & REFILL

See Figure 14.

The manual transmission uses API GL-4, viscosity SAE 75W-85 MTF fluid. If information differs from the owner's manual, use the data in the owner's manual.

➡If working near and/or around the SRS system and components, be sure to disable the SRS system. After disabling the system wait three minutes or more before servicing the vehicle.

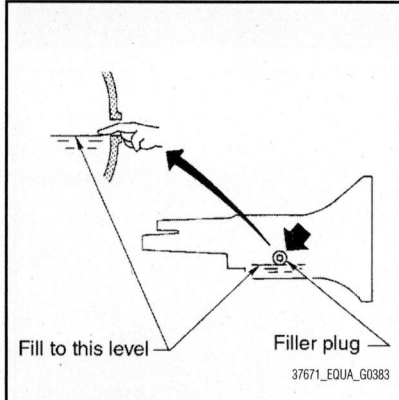

Fig. 14 Fill the manual transmission until the oil level reaches the specified limit near the filler plug hole

➡Whenever the negative battery cable is disconnected the following components will require resetting. The Idle Air Volume Learning, Steering Angle Sensor Neutral Position, Sunroof Memory Reset/Initialization, Automatic Drive Positioner System, Audio presets and Navigation. Use the Nissan CONSULT III diagnostic tool, or equivalent to perform the required resets.

1. Before servicing the vehicle, refer to the Precautions Section.
2. Start the engine and let it run to warm up the transmission.
3. Disconnect the negative battery cable.
4. Stop the engine. Raise and safely support the vehicle.
5. Remove the transmission drain plug and drain the oil.

To install:

6. Set a new gasket on the drain plug and install it to the transmission. Tighten the drain plug to 23–28 ft. lbs. (30–39 Nm).

➡Do not reuse the drain plug gasket.

7. Remove the filler plug. Fill with the proper type of new oil until the oil level reaches the specified limit near the filler plug hole.
8. After refilling the oil, check oil level. Set a new gasket to the filler plug, then install it to the transmission. Tighten the filler plug to 23–28 ft. lbs. (30–39 Nm).

➡Do not reuse the filler plug gasket.

FLUID LEVEL CHECK

➡If working near and/or around the SRS system and components, be sure to disable the SRS system. After disabling the system wait three minutes or more before servicing the vehicle.

➡Whenever the negative battery cable is disconnected the following components will require resetting. The Idle Air Volume Learning, Steering Angle Sensor Neutral Position, Sunroof Memory Reset/Initialization, Automatic Drive Positioner System, Audio presets and Navigation. Use the Nissan CONSULT III diagnostic tool, or equivalent to perform the required resets.

1. Before servicing the vehicle, refer to the Precautions Section.
2. Start the engine and let it run to warm up the transmission.
3. Disconnect the negative battery cable.
4. Stop the engine. Raise and safely support the vehicle.
5. Remove the transmission drain plug. Discard the gasket.
6. Check oil level and correct as required. Set a new gasket to the filler plug, then install it to the transmission. Tighten the filler plug to 23–28 ft. lbs. (30–39 Nm).

➡Do not reuse the filler plug gasket.

FLUID RECOMMENDATIONS

The manual transmission uses API GL-4, viscosity SAE 75W-85 MTF fluid. If information differs from the owner's manual, use the data in the owner's manual.

CLUTCH HYDRAULIC SYSTEM BLEEDING

BLEEDING PROCEDURE

See Figure 15.

✳✳ WARNING

Do not spill brake fluid on painted surfaces. If it spills, wipe up any brake fluid immediately and wash the affected area with water.

➡Do not use a vacuum assist or any other type of power bleeder on this system. Use of a vacuum assist or power bleeder will not purge all of the air from the system. Monitor the fluid level in the reservoir tank to make sure it does not empty.

➡If working near and/or around the SRS system and components, be sure to disable the SRS system. After disabling the system wait three minutes or more before servicing the vehicle.

➡Whenever the negative battery cable is disconnected the following components will require resetting. The Idle Air Volume Learning, Steering Angle

Sensor Neutral Position, Sunroof Memory Reset/Initialization, Automatic Drive Positioner System, Audio presets and Navigation. Use the Nissan CONSULT III diagnostic tool, or equivalent to perform the required resets.

1. Before servicing the vehicle, refer to the Precautions Section.
2. Top off the reservoir with new brake fluid.
3. Connect a transparent vinyl tube and container to the air bleeder valve on the clutch operating cylinder.
4. Fully depress the clutch pedal several times.
5. With the clutch pedal depressed, open the bleeder valve to release the air.
6. Close the bleeder valve.
7. Repeat steps 4 to 6 until clear brake fluid comes out of the air bleeder valve.
8. Tighten the air bleeder to 70 inch lbs. (8 Nm).

FLUID FILL PROCEDURE

✳✳ WARNING

Do not spill brake fluid on painted surfaces. If it spills, wipe up any brake fluid immediately and wash the affected area with water.

➡If working near and/or around the SRS system and components, be sure to disable the SRS system. After disabling the system wait three minutes or more before servicing the vehicle.

➡Whenever the negative battery cable is disconnected the following components will require resetting. The Idle Air Volume Learning, Steering Angle Sensor Neutral Position, Sunroof Memory Reset/Initialization, Automatic

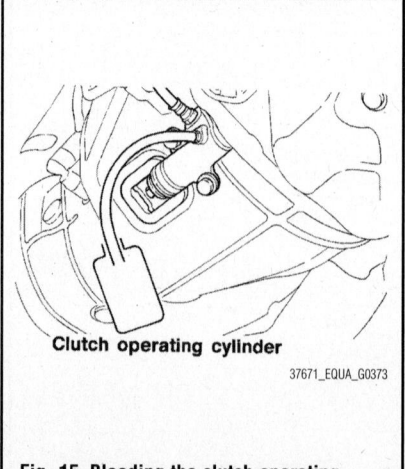

Fig. 15 Bleeding the clutch operating (slave) cylinder

Drive Positioner System, Audio presets and Navigation. Use the Nissan CON-SULT III diagnostic tool, or equivalent to perform the required resets.

1. Before servicing the vehicle, refer to the Precautions Section.
2. Disconnect the negative battery cable.
3. Top off the reservoir with new brake fluid.

DRIVESHAFT

REMOVAL & INSTALLATION

3S1310 Type

See Figure 16.

➡If working near and/or around the SRS system and components, be sure to disable the SRS system. After disabling the system wait three minutes or more before servicing the vehicle.

➡Whenever the negative battery cable is disconnected the following components will require resetting. The Idle Air Volume Learning, Steering Angle Sensor Neutral Position, Sunroof Memory Reset/Initialization, Automatic Drive Positioner System, Audio presets and Navigation. Use the Nissan CON-

SULT III diagnostic tool, or equivalent to perform the required resets.

1. Before servicing the vehicle, refer to the Precautions Section.
2. Disconnect the negative battery cable.
3. Put the transmission in neutral and release the parking brake.
4. Put matching marks on the rear driveshaft flange yoke and the rear final drive companion flange.

> ✳✳ WARNING
>
> **For matching marks, use paint. Never damage or mar the flange yoke and companion flange of the front final drive.**

5. Remove the bolts, then remove the driveshaft from the rear final drive and transmission.
6. Remove the driveshaft from the vehicle.

To install:

7. Installation is the reverse of the removal procedure.
8. Inspect the driveshaft runout. If the runout exceeds 0.024 inch (0.6mm), replace the driveshaft assembly.

9. While holding the flange yoke on one side, check the axial play of the joint. If the journal axial play exceeds 0.0008 inch (0.02mm), repair or replace the journal parts.
10. Check the driveshaft for dents or cracks. If damage is detected, replace the driveshaft assembly.
11. After installation, check for vibration by driving the vehicle.

> ✳✳ WARNING
>
> **Do not reuse the bolts and nuts. Always install new ones.**

3S1330 Type

See Figure 17.

➡If working near and/or around the SRS system and components, be sure to disable the SRS system. After disabling the system wait three minutes or more before servicing the vehicle.

➡Whenever the negative battery cable is disconnected the following components will require resetting. The Idle Air Volume Learning, Steering Angle Sensor Neutral Position, Sunroof Memory Reset/Initialization, Automatic Drive Positioner System, Audio presets and Navigation. Use the Nissan CON-SULT III diagnostic tool, or equivalent to perform the required resets.

1. Before servicing the vehicle, refer to the Precautions Section.
2. Disconnect the negative battery cable.
3. Put the transmission in neutral and release the parking brake.
4. Put matching marks on the rear driveshaft flange yoke and the rear final drive companion flange.

> ✳✳ WARNING
>
> **For matching marks, use paint. Never damage or mar the flange yoke and companion flange of the front final drive.**

5. Remove the bolts, then remove the driveshaft from the rear final drive and transmission.
6. Remove the driveshaft from the vehicle.

To install:

7. Installation is the reverse of the removal procedure.
8. Inspect the driveshaft runout. If the runout exceeds 0.024 inch (0.6mm), replace the driveshaft assembly.

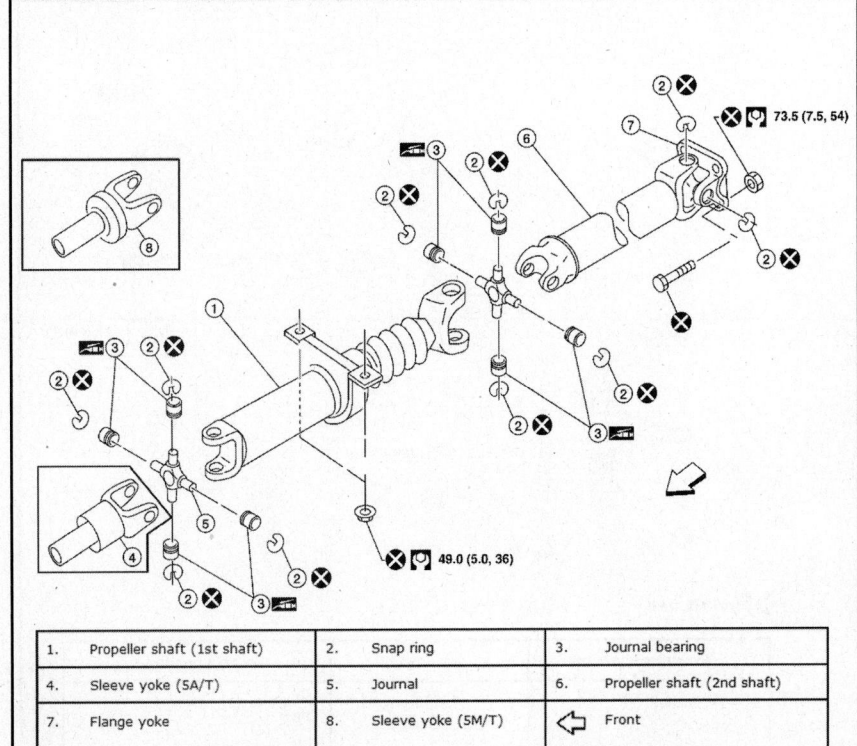

73.5 (7.5, 54)

49.0 (5.0, 36)

1.	Propeller shaft (1st shaft)	2.	Snap ring	3.	Journal bearing
4.	Sleeve yoke (5A/T)	5.	Journal	6.	Propeller shaft (2nd shaft)
7.	Flange yoke	8.	Sleeve yoke (5M/T)	⇦	Front

37671_EQUA_G0066

Fig. 16 Driveshaft and related components—3S1310 type

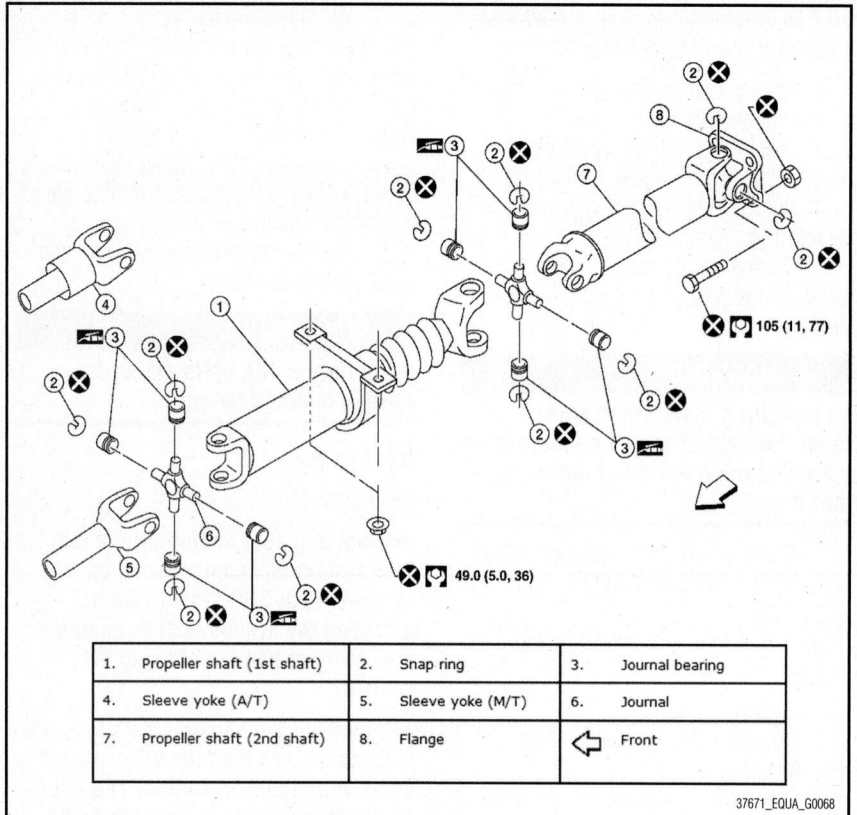

1.	Propeller shaft (1st shaft)	2.	Snap ring	3.	Journal bearing
4.	Sleeve yoke (A/T)	5.	Sleeve yoke (M/T)	6.	Journal
7.	Propeller shaft (2nd shaft)	8.	Flange	⇦	Front

37671_EQUA_G0068

Fig. 17 Driveshaft and related components—3S1330 type

Drive Positioner System, Audio presets and Navigation. Use the Nissan CONSULT III diagnostic tool, or equivalent to perform the required resets.

1. Before servicing the vehicle, refer to the Precautions Section.
2. Disconnect the negative battery cable.
3. Put matching marks on the front driveshaft flange yoke and the front final drive companion flange.

❊❊ WARNING

For matching marks, use paint. Never damage or mar the flange yoke and companion flange of the front final drive.

4. Put matching marks on the front driveshaft flange yoke and the transfer companion flange.
5. Remove the bolts and then remove the front driveshaft from the front final drive and transfer.
6. Remove the driveshaft from the vehicle.

To install:

7. Installation is the reverse of the removal procedure.

9. While holding the flange yoke on one side, check the axial play of the joint. If the journal axial play exceeds 0.0008 inch (0.02mm), repair or replace the journal parts.

10. Check the driveshaft for dents or cracks. If damage is detected, replace the driveshaft assembly.

11. After installation, check for vibration by driving the vehicle.

❊❊ WARNING

Do not reuse the bolts and nuts. Always install new ones.

2F1310 Type

See Figure 18.

➡If working near and/or around the SRS system and components, be sure to disable the SRS system. After disabling the system wait three minutes or more before servicing the vehicle.

➡Whenever the negative battery cable is disconnected the following components will require resetting. The Idle Air Volume Learning, Steering Angle Sensor Neutral Position, Sunroof Memory Reset/Initialization, Automatic

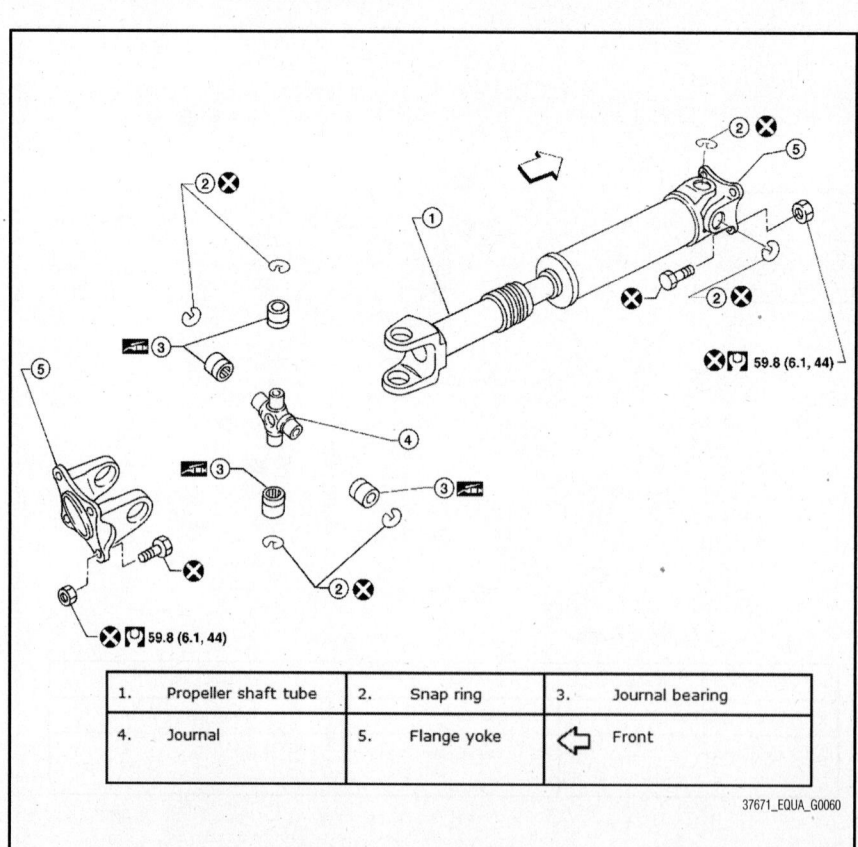

1.	Propeller shaft tube	2.	Snap ring	3.	Journal bearing
4.	Journal	5.	Flange yoke	⇦	Front

37671_EQUA_G0060

Fig. 18 Driveshaft and related components—2F1310 type

8. Inspect the driveshaft runout. If the runout exceeds 0.024 inch (0.6mm), replace the driveshaft assembly.

9. While holding the flange yoke on one side, check the axial play of the joint. If the journal axial play exceeds 0.0008 inch (0.02mm), repair or replace the journal parts.

10. Check the driveshaft tube surface for dents or cracks. If damage is detected, replace the driveshaft assembly.

11. After installation, check for vibration by driving the vehicle.

�֍✶ WARNING

Do not reuse the bolts and nuts. Always install new ones.

2S1330 Type

See Figure 19.

➡If working near and/or around the SRS system and components, be sure to disable the SRS system. After disabling the system wait three minutes or more before servicing the vehicle.

➡Whenever the negative battery cable is disconnected the following compo-

nents will require resetting. The Idle Air Volume Learning, Steering Angle Sensor Neutral Position, Sunroof Memory Reset/Initialization, Automatic Drive Positioner System, Audio presets and Navigation. Use the Nissan CONSULT III diagnostic tool, or equivalent to perform the required resets.

1. Before servicing the vehicle, refer to the Precautions Section.
2. Disconnect the negative battery cable.
3. Move the M/T or A/T shift selector lever to the N position and release the parking brake.
4. Put matching marks on the front driveshaft flange yoke and the front final drive companion flange.

✶✶ WARNING

For matching marks, use paint. Never damage or mar the flange yoke and companion flange of the front final drive.

5. Remove the bolts, then remove the driveshaft from the rear final drive and transfer.
6. Remove the driveshaft from the vehicle.

To install:

7. Installation is the reverse of the removal procedure.

8. Inspect the driveshaft runout. If the runout exceeds 0.04 inch (1.02mm), replace the driveshaft assembly.

9. While holding the flange yoke on one side, check the axial play of the joint. If the journal axial play exceeds 0.0008 inch (0.02mm), repair or replace the journal parts.

10. Check the driveshaft tube for dents or cracks. If damage is detected, replace the driveshaft assembly.

11. After installation, check for vibration by driving the vehicle.

12. If driveshaft assembly or final drive assembly has been replaced, connect them as follows:

a. Face the companion flange mark of the final drive upward. With the mark facing upward, couple the driveshaft and the final drive so that the matching mark of the driveshaft can be positioned as close as possible with the matching mark of the final drive companion flange.

b. Tighten the driveshaft and final drive bolts and nuts to 77 ft. lbs. (105 Nm).

✶✶ WARNING

Do not reuse the bolts and nuts. Always install new ones.

3S1330-2BJ100 Type

See Figure 20.

➡If working near and/or around the SRS system and components, be sure to disable the SRS system. After disabling the system wait three minutes or more before servicing the vehicle.

➡Whenever the negative battery cable is disconnected the following components will require resetting. The Idle Air Volume Learning, Steering Angle Sensor Neutral Position, Sunroof Memory Reset/Initialization, Automatic Drive Positioner System, Audio presets and Navigation. Use the Nissan CONSULT III diagnostic tool, or equivalent to perform the required resets.

1. Before servicing the vehicle, refer to the Precautions Section.
2. Disconnect the negative battery cable.
3. Put the transmission in neutral and release the parking brake.
4. Put matching marks on the rear driveshaft flange yoke and the rear final drive companion flange.

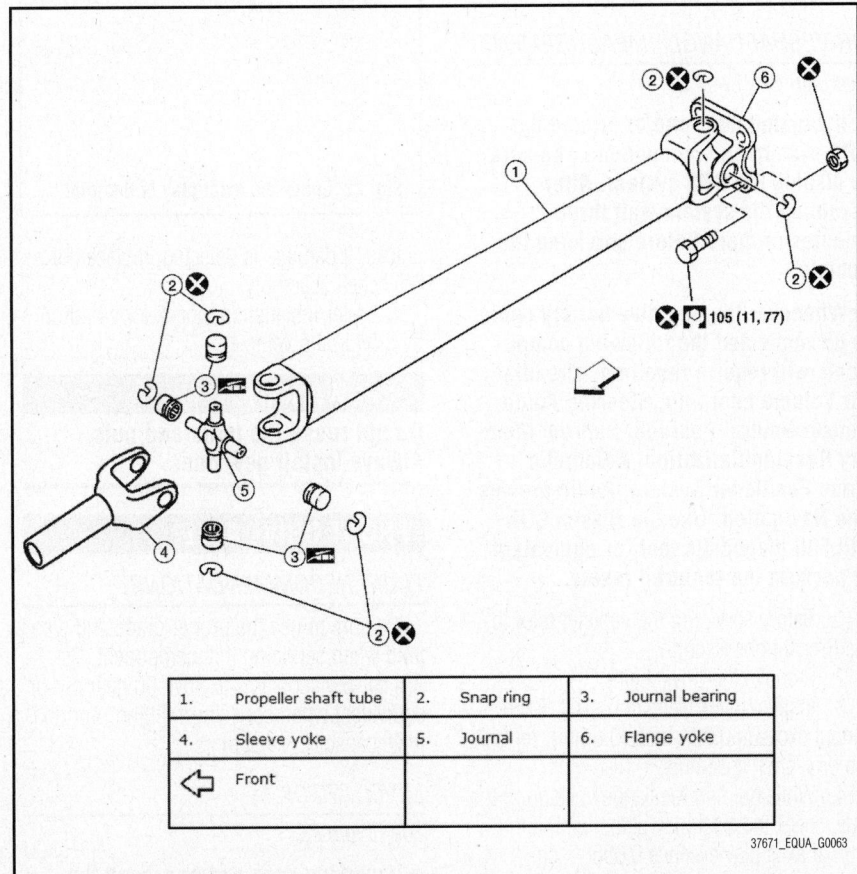

105 (11, 77)

1.	Propeller shaft tube	2.	Snap ring	3.	Journal bearing
4.	Sleeve yoke	5.	Journal	6.	Flange yoke
⇦	Front				

37671_EQUA_G0063

Fig. 19 Driveshaft and related components—2S1330 type

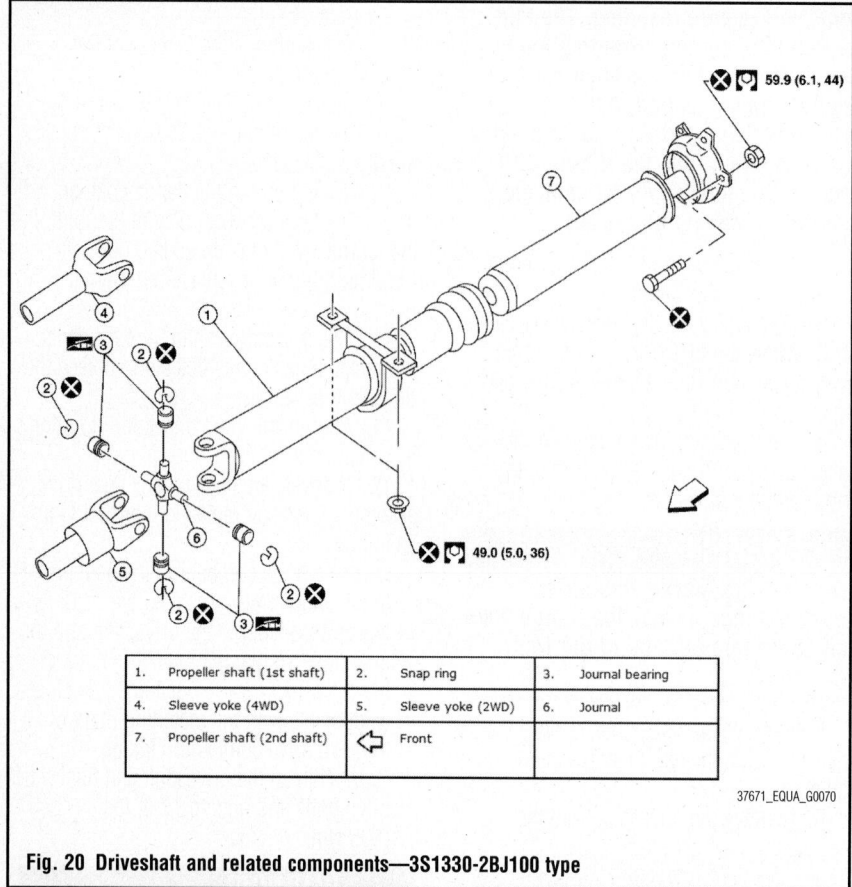

1.	Propeller shaft (1st shaft)	2.	Snap ring	3.	Journal bearing
4.	Sleeve yoke (4WD)	5.	Sleeve yoke (2WD)	6.	Journal
7.	Propeller shaft (2nd shaft)	⇐	Front		

37671_EQUA_G0070

Fig. 20 Driveshaft and related components—3S1330-2BJ100 type

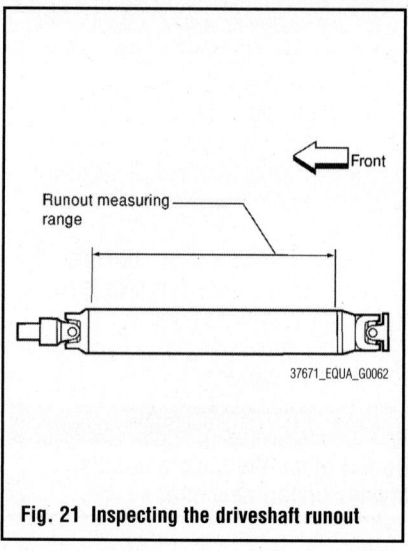

37671_EQUA_G0062

Fig. 21 Inspecting the driveshaft runout

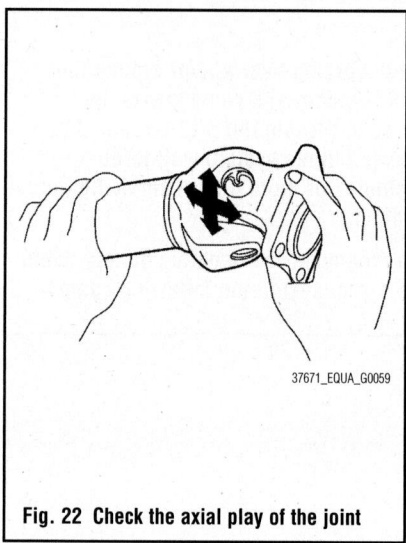

37671_EQUA_G0059

Fig. 22 Check the axial play of the joint

✳✳ WARNING

For matching marks, use paint. Never damage or mar the flange yoke and companion flange of the front final drive.

5. Remove the bolts, then remove the driveshaft from the rear final drive and transmission or transfer.

6. Remove the driveshaft from the vehicle.

To install:

7. Installation is the reverse of the removal procedure.

8. Inspect the driveshaft runout. If the runout exceeds 0.024 inch (0.6mm), replace the driveshaft assembly.

9. While holding the flange yoke on one side, check the axial play of the joint. If the journal axial play exceeds 0.0008 inch (0.02mm), repair or replace the journal parts.

10. Check the driveshaft for dents or cracks. If damage is detected, replace the driveshaft assembly.

11. After installation, check for vibration by driving the vehicle.

✳✳ WARNING

Do not reuse the bolts and nuts. Always install new ones.

DRIVESHAFT ANGLE MEASUREMENT

See Figures 21 and 22.

➡ If working near and/or around the SRS system and components, be sure to disable the SRS system. After disabling the system wait three minutes or more before servicing the vehicle.

➡ Whenever the negative battery cable is disconnected the following components will require resetting. The Idle Air Volume Learning, Steering Angle Sensor Neutral Position, Sunroof Memory Reset/Initialization, Automatic Drive Positioner System, Audio presets and Navigation. Use the Nissan CONSULT III diagnostic tool, or equivalent to perform the required resets.

1. Before servicing the vehicle, refer to the Precautions Section.

2. Remove the driveshaft.

3. Inspect the driveshaft runout. If the runout exceeds 0.024 inch (0.6mm), replace the driveshaft assembly.

4. While holding the flange yoke on one side, check the axial play of the joint. If the journal axial play exceeds 0.0008 inch (0.02mm), repair or replace the journal parts.

5. Check the driveshaft for dents or cracks. If damage is detected, replace the driveshaft assembly.

6. After installation, check for vibration by driving the vehicle.

✳✳ WARNING

Do not reuse the bolts and nuts. Always install new ones.

FRONT DRIVE AXLE

FLUID RECOMMENDATIONS

Be sure to use the proper grade and type fluid when servicing this component. Use API GL-5 viscosity SAE 80W-90 gear oil, or equivalent when servicing/refilling. Approximate capacity is 1.5 pints.

LEVEL CHECK

See Figure 23.

➡ If working near and/or around the SRS system and components, be sure

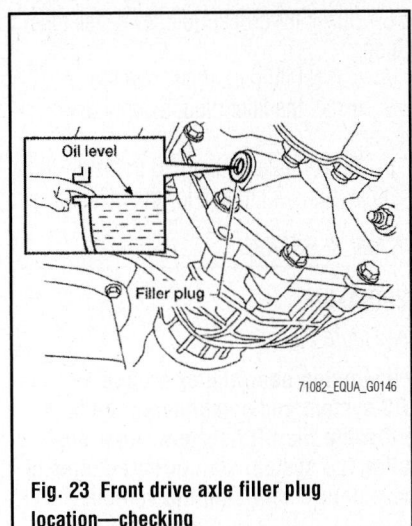

Fig. 23 Front drive axle filler plug location—checking

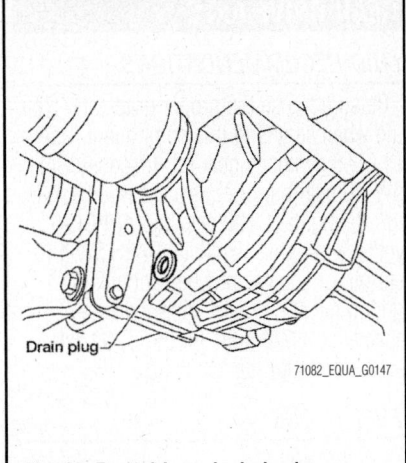

Fig. 24 Front drive axle drain plug location—draining

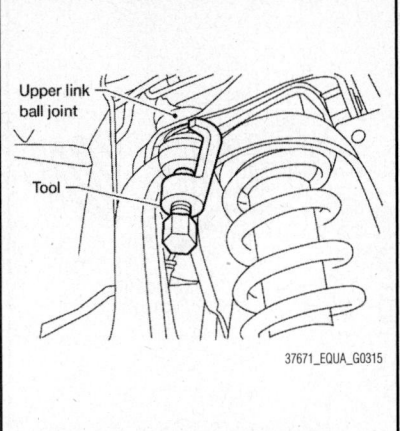

Fig. 25 Using special tool ST29020001 (J-24319-01) to separate the upper arm (link) ball joint from the steering knuckle

to disable the SRS system. After disabling the system wait three minutes or more before servicing the vehicle.

➡Whenever the negative battery cable is disconnected the following components will require resetting. The Idle Air Volume Learning, Steering Angle Sensor Neutral Position, Sunroof Memory Reset/Initialization, Automatic Drive Positioner System, Audio presets and Navigation. Use the Nissan CONSULT III diagnostic tool, or equivalent to perform the required resets.

1. Before servicing the vehicle, refer to the Precautions Section.
2. Disconnect the negative battery cable.
3. Raise and support the vehicle safely, as required.
4. Position a catch pan under the assembly.
5. Remove the filler plug. Discard the gasket.
6. Check for proper fluid level, see illustration.
7. Correct fluid level, as required.
8. Install the filler plug, using a new gasket.
9. Tighten filler plug to 25 ft. lbs.

DRAIN & REFILL

See Figure 24.

➡If working near and/or around the SRS system and components, be sure to disable the SRS system. After disabling the system wait three minutes or more before servicing the vehicle.

➡Whenever the negative battery cable is disconnected the following components will require resetting. The Idle Air Volume Learning, Steering Angle Sensor Neutral Position, Sunroof

Memory Reset/Initialization, Automatic Drive Positioner System, Audio presets and Navigation. Use the Nissan CONSULT III diagnostic tool, or equivalent to perform the required resets.

1. Before servicing the vehicle, refer to the Precautions Section.
2. Disconnect the negative battery cable.
3. Raise and support the vehicle safely, as required.
4. Position a catch pan under the assembly.
5. Remove the drain plug. Discard the gasket.
6. Drain the fluid. Be sure to properly dispose of used fluid.
7. Install the drain plug, using a new gasket.
8. Tighten drain plug to 25 ft. lbs.
9. Fill with the proper grade and type fluid. Check for proper fluid level.
10. Correct fluid level, as required.
11. Install the filler plug, using a new gasket.
12. Tighten filler plug to 25 ft. lbs.

HALFSHAFT

REMOVAL & INSTALLATION

See Figures 25 through 27.

➡If working near and/or around the SRS system and components, be sure to disable the SRS system. After disabling the system wait three minutes or more before servicing the vehicle.

➡Whenever the negative battery cable is disconnected the following components will require resetting. The Idle

Air Volume Learning, Steering Angle Sensor Neutral Position, Sunroof Memory Reset/Initialization, Automatic Drive Positioner System, Audio presets and Navigation. Use the Nissan CONSULT III diagnostic tool, or equivalent to perform the required resets.

1. Before servicing the vehicle, refer to the Precautions Section.
2. Disconnect the negative battery cable.
3. Remove the wheel and tire assembly.
4. Remove the rear engine under cover.
5. Remove the wheel sensor harness from the mount on the knuckle, then disconnect the wheel sensor harness connector.

❋❋ WARNING

Do not pull on the wheel sensor harness.

6. Remove the wheel hub and bearing assembly.

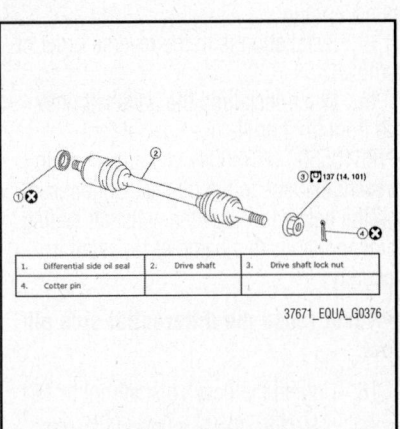

| 1. | Differential side oil seal | 2. | Drive shaft | 3. | Drive shaft lock nut |
| 4. | Cotter pin | | | | |

Fig. 26 Front halfshaft and related components

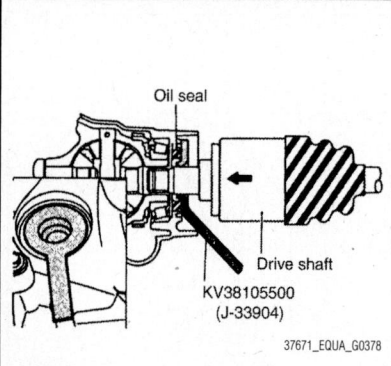

Fig. 27 Using tool KV38105500 (J-33904) to prevent damage to the oil seal while inserting the halfshaft

➡️It is not necessary to remove the wheel sensor from the wheel hub when the wheel hub is not being replaced. Carefully feed the wheel sensor harness through the hole in the splash shield.

7. Separate the upper control arm (link) ball joint stud from the steering knuckle using special tool ST29020001 (J-24319-01), or equivalent.

8. Support the lower control arm with a jack.

9. Pry the halfshaft front the final drive using a suitable tool.

10. Remove the differential side oil seal.

To install:

11. Move the joint up, down, left, right, and in an axial direction. Check for any rough movement or significant looseness.

12. Check the boot for cracks or other damage, and for grease leakage.

13. If damaged, disassemble the half-shaft to verify damage, and repair or replace as necessary.

14. Installation is in the reverse order of removal.

15. When installing the halfshaft onto the front final drive, use special tool KV38105500 (J-33904), or equivalent, to prevent damage to the oil seal while inserting the halfshaft. Slide the halfshaft sliding joint and tap with a hammer to install it securely.

➡️**Never reuse the differential side oil seal.**

16. Tighten the new halfshaft nut to 101 ft. lbs. (137 Nm). Insert a new cotter pin.

17. Tighten the wheel nuts to 98 ft. lbs. (133 Nm).

REAR DRIVE AXLE

FLUID RECOMMENDATIONS

Be sure to use the proper grade and type fluid when servicing this component. Use API GL-5 synthetic gear oil, Viscosity SAE 75W-90 gear oil, or equivalent when servicing/refilling the C200 axle and API GL-5 synthetic gear oil, Viscosity SAE 75W-140 gear oil, or equivalent when servicing/refilling the M226 axle . Approximate capacity is 2. 7/8 pints for the C200 axle and 3.5 pints for the M226 axle.

LEVEL CHECK

See Figure 28.

➡️**If working near and/or around the SRS system and components, be sure to disable the SRS system. After disabling the system wait three minutes or more before servicing the vehicle.**

➡️**Whenever the negative battery cable is disconnected the following components will require resetting. The Idle Air Volume Learning, Steering Angle Sensor Neutral Position, Sunroof Memory Reset/Initialization, Automatic Drive Positioner System, Audio presets and Navigation. Use the Nissan CONSULT III diagnostic tool, or equivalent to perform the required resets.**

1. Before servicing the vehicle, refer to the Precautions Section.

2. Disconnect the negative battery cable.

3. Raise and support the vehicle safely, as required.

4. Position a catch pan under the assembly.

5. Remove the filler plug. Discard the gasket.

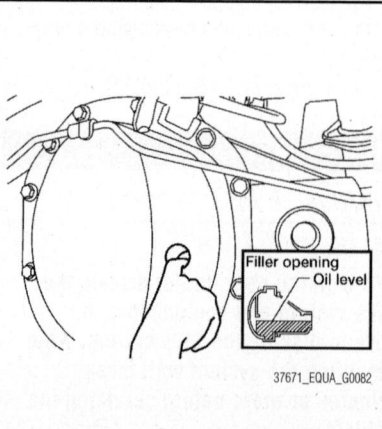

Fig. 28 Rear drive axle filler plug location—checking

6. Check for proper fluid level, see illustration.

7. Correct fluid level, as required.

8. Install the filler plug, using a new gasket.

9. Tighten filler plug to 36 ft. lbs. for the C200 axle and 32 ft. lbs. for the M226 axle

DRAIN & REFILL

C200 Axle

See Figure 29.

➡️**If working near and/or around the SRS system and components, be sure to disable the SRS system. After disabling the system wait three minutes or more before servicing the vehicle.**

➡️**Whenever the negative battery cable is disconnected the following components will require resetting. The Idle Air Volume Learning, Steering Angle Sensor Neutral Position, Sunroof Memory Reset/Initialization, Automatic Drive Positioner System, Audio presets and Navigation. Use the Nissan CONSULT III diagnostic tool, or equivalent to perform the required resets.**

1. Before servicing the vehicle, refer to the Precautions Section.

2. Disconnect the negative battery cable.

3. Raise and support the vehicle safely, as required.

4. Position a catch pan under the assembly.

5. Remove the drain plug from the rear final drive assembly to drain the differential gear oil.

6. Install the drain plug with a new gasket to the rear final drive assembly. Tighten to 36 ft. lbs. (49 Nm).

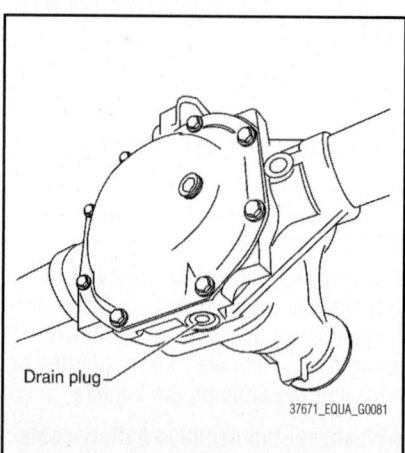

Fig. 29 Remove the drain plug from the rear final drive assembly to drain the differential gear oil—C200 axle

➡ **Do not reuse the drain plug gasket.**

To install:

7. Remove the filler plug from the rear final drive assembly.

8. Fill the rear final drive assembly with new differential gear oil of the proper type until the level reaches the specified level near the filler plug hole. Use API GL-5 synthetic gear oil, Viscosity SAE 75W-90.

9. Install the filler plug to the rear final drive assembly, with sealant applied on the threads (use high performance thread sealant, or equivalent). Tighten to 36 ft. lbs. (49 Nm).

M226 Axle

See Figure 30.

➡ **If working near and/or around the SRS system and components, be sure to disable the SRS system. After disabling the system wait three minutes or more before servicing the vehicle.**

➡ **Whenever the negative battery cable is disconnected the following components will require resetting. The Idle Air Volume Learning, Steering Angle Sensor Neutral Position, Sunroof Memory Reset/Initialization, Automatic Drive Positioner System, Audio presets and Navigation. Use the Nissan CONSULT III diagnostic tool, or equivalent to perform the required resets.**

1. Before servicing the vehicle, refer to the Precautions Section.

2. Disconnect the negative battery cable.

3. Raise and support the vehicle safely, as required.

4. Position a catch pan under the assembly.

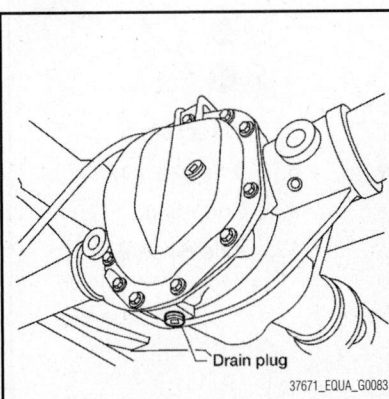

Drain plug

37671_EQUA_G0083

Fig. 30 Remove the drain plug from the rear final drive assembly to drain the differential gear oil—M226 axle

5. Remove the drain plug from the rear final drive assembly to drain the differential gear oil.

6. Install the drain plug with sealant applied on the threads to the rear final drive assembly. Tighten to 32 ft. lbs. (43 Nm).

➡ **Use High Performance Thread Sealant or equivalent.**

To install:

7. Remove the filler plug from the rear final drive assembly.

8. Fill the rear final drive assembly with new differential gear oil until the level reaches the specified level near the filler plug hole. Use API GL-5 synthetic gear oil, Viscosity SAE 75W-140.

9. Install the filler plug to the rear final drive assembly, with sealant applied on the threads (use high performance thread sealant, or equivalent). Tighten to 32 ft. lbs. (43 Nm).

REAR AXLE SHAFT, BEARING & SEAL

REMOVAL & INSTALLATION

See Figures 31 through 33.

➡ **If working near and/or around the SRS system and components, be sure to disable the SRS system. After disabling the system wait three minutes or more before servicing the vehicle.**

➡ **Whenever the negative battery cable is disconnected the following components will require resetting. The Idle Air Volume Learning, Steering Angle Sensor Neutral Position, Sunroof Memory Reset/Initialization, Automatic Drive Positioner System, Audio presets and Navigation. Use the Nissan CONSULT III diagnostic tool, or equivalent to perform the required resets.**

1. Before servicing the vehicle, refer to the Precautions Section.

2. Disconnect the negative battery cable.

3. Remove or disconnect the following:
- Rear wheel and tire assembly
- Wheel speed sensor
- Brake rotor
- Brake caliper assembly
- Parking brake cable
- Brake fluid line
- Bearing cage and backing plate bolts
- Axle shaft assembly
- Axle seal
- Wheel speed sensor rotor
- Snapring and shim washer

- Bearing ring retainer
- Back plate and torque member
- Axle bearing studs
- Wheel bearing
- Grease catcher

To install:

➡ **Use new seals, bearings, circlips, and snaprings for assembly.**

4. Install the grease catcher.

5. Install the wheel studs through the grease catcher into the axle shaft using a suitable press.

➡ **All wheel studs must be pressed in at the same time and should be flush with the grease catcher when installed.**

6. Position the axle bearing on the back plate and torque member.

7. Install the axle bearing studs using a suitable press to attach the axle bearing to the back plate and torque member.

➡ **Always replace the axle bearing with a new one.**

8. Install the back plate and torque member, new axle bearing, and new bearing ring retainer on the axle shaft using a suitable press. Do not exceed 11 tons of force.

9. Press the new bearing ring retainer on the axle shaft with the taper side positioned toward the press.

➡ **Always replace the bearing ring retainer with a new one.**

10. Select the correct size shim washer. Select the size of shim washer so that the installed snapring-to-shim washer clearance is 0.008 inch (0.20mm) or less.

11. Install a new snapring on the axle shaft.

12. Do not over spread the snapring when installing, measure the outer diameter of the snapring after installation and replace it if the snapring outer diameter exceeds 1.87 inch (47.50mm) maximum.

13. Check the snapring to shim washer clearance. Repeat the previous steps, as necessary.

14. Perform a break-in rotation of the wheel bearing.
- Rotate the wheel bearing in the forward direction for a minimum of 10 revolutions at 50–70 RPM
- Rotate the wheel bearing in the reverse direction for a minimum of 10 revolutions at 50–70 RPM

15. Measure the rotational torque of the wheel bearing. The rotational torque should be 16 inch lbs. (2 Nm) at 8–12 RPM.

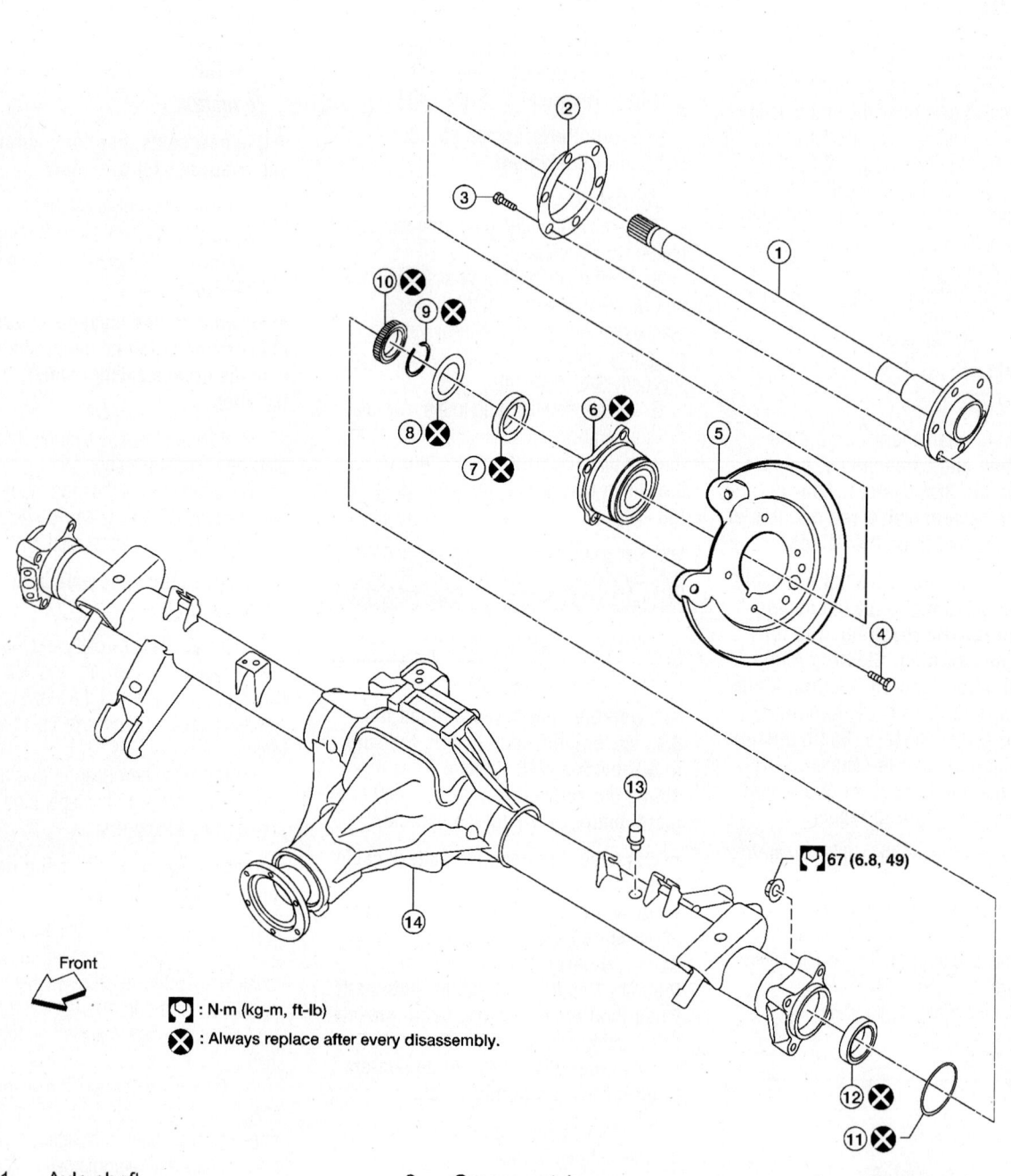

Front

⊡ : N·m (kg-m, ft-lb)

⊗ : Always replace after every disassembly.

1. Axle shaft	2. Grease catcher	3. Wheel stud
4. Axle bearing stud	5. Back plate and torque member	6. Axle bearing
7. Bearing ring retainer	8. Shim washer	9. Snap ring
10. ABS sensor rotor	11. O-ring	12. Axle oil seal
13. Breather	14. Rear final drive	

22140_FRON_G0001

Fig. 31 Rear axle shaft, bearing and seal—C200

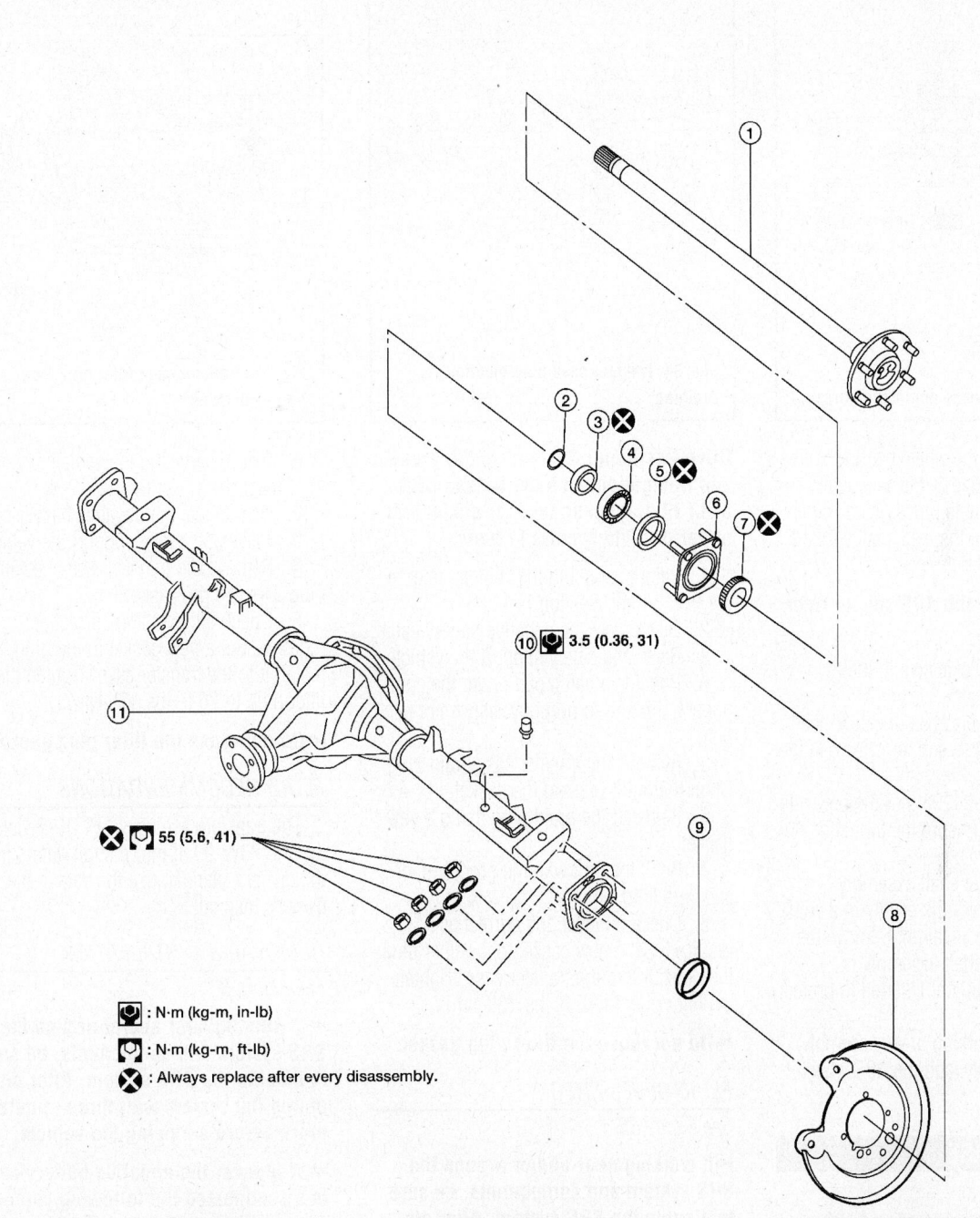

10 ⬚ 3.5 (0.36, 31)

✖ ⬚ 55 (5.6, 41)

⬚ : N·m (kg-m, in-lb)

⬚ : N·m (kg-m, ft-lb)

✖ : Always replace after every disassembly.

1.	Axle shaft	2.	Snap ring	3.	Bearing ring retainer
4.	Axle shaft bearing	5.	Axle oil seal	6.	Axle shaft bearing cage
7.	ABS sensor rotor	8.	Back plate and torque member	9.	Axle shaft bearing cup
10.	Breather	11.	Rear final drive		

22140_FRON_G0002

Fig. 32 Rear axle shaft, bearing and seal—M226

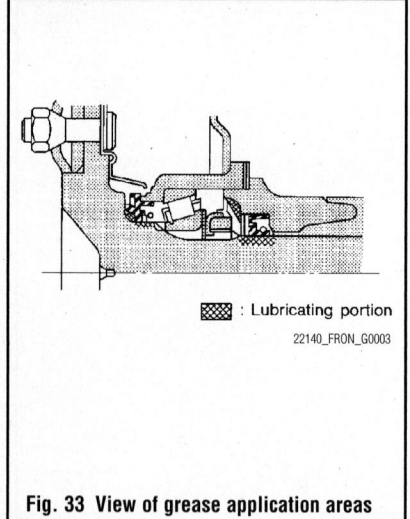

Fig. 33 View of grease application areas

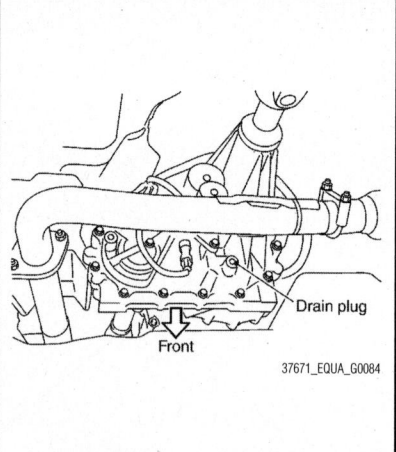

Fig. 34 Transfer case plug location—draining

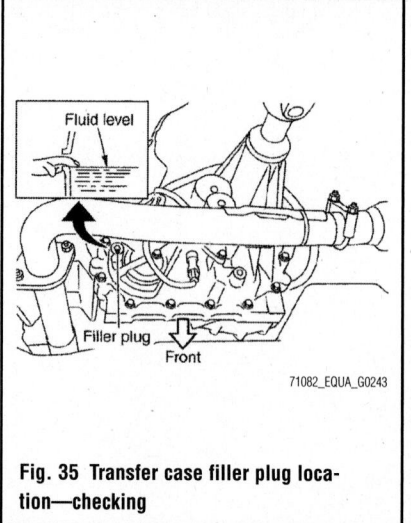

Fig. 35 Transfer case filler plug location—checking

16. Inspect that the wheel bearing is free from axial play relative to the axle shaft.

17. Using a suitable press, install a new ABS sensor rotor on the axle shaft with the notch side away from the press.

➡**Always replace the ABS sensor rotor with a new one.**

18. Install a new axle seal in the housing.

19. Apply multi-purpose grease to the recess of the axle case end, as shown in the illustration.

20. Insert tool J-34296 into the new axle oil seal as a guide. Ensure the tool ends do not overlap.

21. Insert the axle shaft assembly. Tighten the axle shaft nuts evenly in a criss-cross pattern to specification. Remove the tool when the axle shaft assembly is approximately 90 percent inserted to protect the new axle oil seal.

22. Install the parking brake assembly, rear caliper assembly, and the ABS wheel sensor.

TRANSFER CASE

DRAIN & REFILL
See Figure 34.

➡**If working near and/or around the SRS system and components, be sure to disable the SRS system. After disabling the system wait three minutes or more before servicing the vehicle.**

➡**Whenever the negative battery cable is disconnected the following components will require resetting. The Idle Air Volume Learning, Steering Angle Sensor Neutral Position, Sunroof Memory Reset/Initialization, Automatic**

Drive Positioner System, Audio presets and Navigation. Use the Nissan CONSULT III diagnostic tool, or equivalent to perform the required resets.

1. Before servicing the vehicle, refer to the Precautions Section.
2. Disconnect the negative battery cable.
3. Raise and safely support the vehicle.
4. Position a catch pan under the component. Be sure to properly dispose of used fluid.
5. Remove the transfer case drain plug. Drain the fluid. Discard the gasket.
6. Reinstall the drain plug using a new gasket.
7. Refill the unit with the proper grade and type fluid.
8. Check oil level and correct as required. Set a new gasket to the filler plug, then install it to the transfer case. Tighten the filler plug to 26 ft. lbs. (35 Nm).

➡**Do not reuse the filler plug gasket.**

FLUID LEVEL CHECK
See Figure 35.

➡**If working near and/or around the SRS system and components, be sure to disable the SRS system. After disabling the system wait three minutes or more before servicing the vehicle.**

➡**Whenever the negative battery cable is disconnected the following components will require resetting. The Idle Air Volume Learning, Steering Angle Sensor Neutral Position, Sunroof Memory Reset/Initialization, Automatic Drive Positioner System, Audio presets and Navigation. Use the Nissan CONSULT III diagnostic tool, or equivalent to perform the required resets.**

1. Before servicing the vehicle, refer to the Precautions Section.
2. Disconnect the negative battery cable.
3. Raise and safely support the vehicle.
4. Remove the transfer case checking plug. Discard the gasket.
5. Check oil level and correct as required. Set a new gasket to the plug, then install it to the transfer case. Tighten the filler plug to 26 ft. lbs. (35 Nm).

➡**Do not reuse the filler plug gasket.**

FLUID RECOMMENDATIONS

The transfer case uses API GL-5 viscosity SAE 80W-90. If information differs from the owner's manual, use the data in the owner's manual.

REMOVAL & INSTALLATION
See Figure 36.

➡**If working near and/or around the SRS system and components, be sure to disable the SRS system. After disabling the system wait three minutes or more before servicing the vehicle.**

➡**Whenever the negative battery cable is disconnected the following components will require resetting. The Idle Air Volume Learning, Steering Angle Sensor Neutral Position, Sunroof Memory Reset/Initialization, Automatic Drive Positioner System, Audio presets and Navigation. Use the Nissan CONSULT III diagnostic tool, or equivalent to perform the required resets.**

1. Before servicing the vehicle, refer to the Precautions Section.
2. Disconnect the negative battery cable.
3. Switch the 4WD shift switch to 2WD. Set the transfer assembly to 2WD.

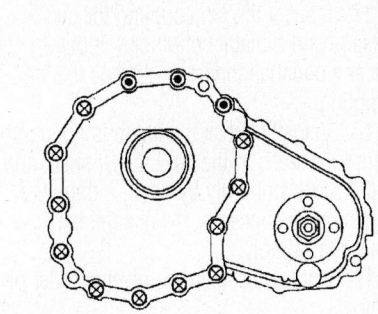

**: Transfer → Transmission

⊗ : Transmission → Transfer

09482_FRON_G0111

Fig. 36 Transfer case bolt tightening sequence

4. Raise and safely support the vehicle.

5. Partially drain the transfer fluid.

6. Remove the transmission under-cover.

7. Remove the center exhaust tube and main muffler.

8. Remove the front and rear drive-shafts.

✳✳ WARNING

Do not damage the spline, sleeve yoke, or rear oil seal when removing the rear driveshaft.

9. Insert a plug into the rear oil seal after removing the rear driveshaft.

10. Remove transmission-to-crossmember bolts.

11. Position 2 suitable jacks under the transmission and transfer assembly.

12. Remove the transmission crossmember.

➡ **Support the transmission and transfer assembly using 2 suitable jacks while removing the transmission crossmember.**

13. Disconnect the electrical connectors from the following:
- The ATP switch
- The 4LO switch
- The wait detection switch
- The transfer control device

14. Remove the wire harness from the retainers.

15. Disconnect each air breather hose from the transfer control device and the breather tube (transfer).

16. Remove the transfer-to-transmission and transmission-to-transfer bolts.

✳✳ WARNING

Support the transfer assembly with a suitable jack while removing it.

17. Remove the transfer assembly from the vehicle.

✳✳ WARNING

Do not damage automatic transmission rear oil seal.

To install:

18. Installation is the reverse of the removal procedure.

19. Tighten the transfer case-to-transmission retaining bolts to 27 ft. lbs. (36 Nm).

20. Fill the transfer with new fluid to the proper level.

21. Check the transfer fluid.

22. Start and run the engine for 1 minute. Stop the engine and recheck the transfer fluid.

23. After the installation, check the 4WD shift indicator pattern. If it is not right, adjust the position between the transfer assembly and transfer control unit.

ENGINE COOLING

BELT-DRIVEN ENGINE FAN

REMOVAL & INSTALLATION

2.5L Engine

See Figure 37.

✳✳ CAUTION

Never open, service or drain the radiator or cooling system when hot; serious burns can occur from the steam and hot coolant. Also, when draining engine coolant, keep in mind that cats and dogs are attracted to ethylene glycol antifreeze and could drink any that is left in an uncovered container or in puddles on the ground. This will prove fatal in sufficient quantities. Always drain coolant into a sealable container. Coolant should be reused unless it is contaminated or is several years old.

➡ **If working near and/or around the SRS system and components, be sure to disable the SRS system. After disabling the system wait three minutes or more before servicing the vehicle.**

➡ **Whenever the negative battery cable is disconnected the following components will require resetting. The Idle Air Volume Learning, Steering Angle Sensor Neutral Position, Sunroof Memory Reset/Initialization, Automatic Drive Positioner System, Audio presets and Navigation. Use the Nissan CONSULT III diagnostic tool, or equivalent to perform the required resets.**

1. Before servicing the vehicle, refer to the Precautions Section.

2. Disconnect the negative battery cable.

3. Remove the engine front under cover.

4. Partially drain the engine coolant.

5. Remove the air duct and resonator assembly and air duct mounting brackets.

6. Remove the upper radiator hose.

7. Disconnect the reservoir tank hose from the upper shroud and radiator.

8. Remove the upper and lower fan shrouds.

9. Remove the drive belt.

10. Remove the cooling fan.

11. Remove the fan coupling, if necessary.

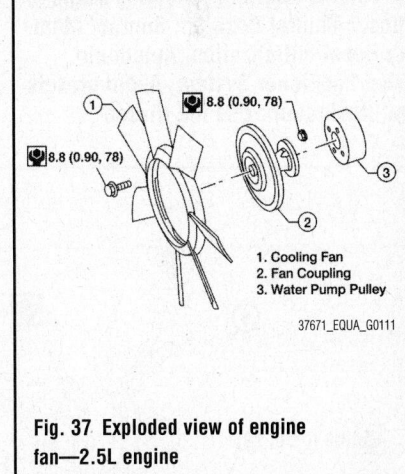

1. Cooling Fan
2. Fan Coupling
3. Water Pump Pulley

8.8 (0.90, 78)

8.8 (0.90, 78)

37671_EQUA_G0111

Fig. 37 Exploded view of engine fan—2.5L engine

12. Remove the water pump pulley, if necessary.

To install:

13. Inspect the fan coupling for oil leakage and bimetal conditions. If there are any unusual concerns, replace the fan coupling.

14. Inspect the cooling fan for cracks or unusual bends. If there are any unusual concerns, replace the cooling fan.

15. Install the cooling fan with its front mark "F" facing the front of the engine.

16. Start and warm up the engine. Visually inspect for leaks of the engine coolant.

4.0L Engine

See Figure 38.

> ❊❊ **CAUTION**
>
> **Never open, service or drain the radiator or cooling system when hot; serious burns can occur from the steam and hot coolant. Also, when draining engine coolant, keep in mind that cats and dogs are attracted to ethylene glycol antifreeze and could drink any that is left in an uncovered container or in puddles on the ground. This will prove fatal in sufficient quantities. Always drain coolant into a sealable container. Coolant should be reused unless it is contaminated or is several years old.**

➡If working near and/or around the SRS system and components, be sure to disable the SRS system. After disabling the system wait three minutes or more before servicing the vehicle.

➡Whenever the negative battery cable is disconnected the following components will require resetting. The Idle Air Volume Learning, Steering Angle Sensor Neutral Position, Sunroof Memory Reset/Initialization, Automatic Drive Positioner System, Audio presets and Navigation. Use the Nissan

CONSULT III diagnostic tool, or equivalent to perform the required resets.

1. Before servicing the vehicle, refer to the Precautions Section.

2. Disconnect the negative battery cable.

3. Remove engine under cover.

4. Partially drain the engine coolant from the radiator.
   - Perform this step when engine is cold
   - Do not spill engine coolant on the drive belts

5. Remove the engine room cover.

6. Remove air duct and resonator assembly.

7. Remove reservoir tank hose from the shroud.

8. Removal the radiator hose (upper) from the radiator.

9. Release the radiator shroud (lower) from the radiator shroud (upper) and position aside. Release the tabs, pull the radiator shroud (lower) rearwards and down.

10. Remove the radiator shroud (upper) bolts and remove the radiator shroud (upper).

11. Remove the drive belt.

12. Remove the engine cooling fan.

13. Remove the fan coupling, if necessary.

14. Remove the cooling fan pulley, if necessary.

15. Remove the drive belt auto-tensioner, if necessary.

16. Remove the fan bracket, if necessary.

To install:

17. Inspect the fan coupling for oil leakage and bimetal conditions. If there are any unusual concerns, replace the fan coupling.

18. Visually check that there is no significant looseness in the fan bracket shaft, and that it turns smoothly by hand. If there are any unusual concerns, replace the fan bracket assembly.

19. Installation is in the reverse order of removal.

20. Install the cooling fan with its front mark "F" facing the front of the engine.

21. Start and warm up the engine. Visually inspect for leaks of the engine coolant.

ENGINE COOLANT

BLEEDING

> ❊❊ **CAUTION**
>
> **Never open, service, or drain the radiator or cooling system when hot; serious burns can occur from the steam and hot coolant. When draining engine coolant, keep in mind that cats and dogs are attracted to ethylene glycol antifreeze and could drink any that is left in an uncovered container or in puddles on the ground. This may prove fatal in sufficient quantities. Always drain coolant into a sealable container. Coolant should be reused unless it is contaminated or is several years old.**

> ❊❊ **WARNING**
>
> **Use engine coolant at a concentration that meets the environmental conditions in which the vehicle is driven, otherwise engine damage could occur. The engine has aluminum parts and must be protected by an ethylene-glycol-based coolant to prevent corrosion and freezing.**

1. Before servicing the vehicle, refer to the Precautions Section.

2. Drain the engine coolant from the engine cooling system.

3. Fill the radiator and the reservoir tank (to the "MAX" line) with water. Reinstall the radiator cap and leave the vented reservoir cap off.

4. Run the engine until it reaches the normal operating temperature.

5. Press the engine accelerator 2–3 times under no-load.

6. Stop the engine and wait until it cools down.

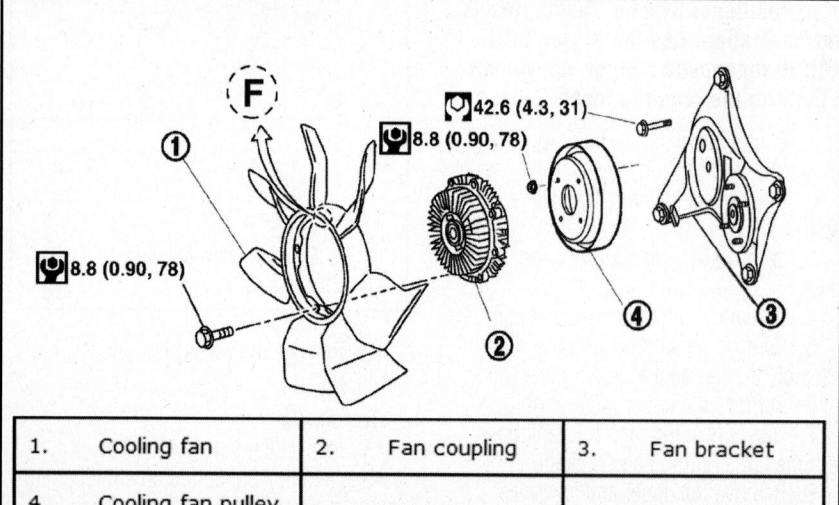

1.	Cooling fan	2.	Fan coupling	3.	Fan bracket
4.	Cooling fan pulley				

37671_EQUA_G0116

Fig. 38 Exploded view of engine cooling fan—4.0L engine

7. Drain the water from the engine cooling system.

8. Repeat steps 3 through 7 until clear water begins to drain from the radiator.

DRAIN & REFILL

See Figures 39 and 40.

✳✳ CAUTION

Never open, service, or drain the radiator or cooling system when hot; serious burns can occur from the steam and hot coolant. When draining engine coolant, keep in mind that cats and dogs are attracted to ethylene glycol antifreeze and could drink any that is left in an uncovered container or in puddles on the ground. This may prove fatal in sufficient quantities. Always drain coolant into a sealable container. Coolant should be reused unless it is contaminated or is several years old.

✳✳ WARNING

Use engine coolant at a concentration that meets the environmental conditions in which the vehicle is driven, otherwise engine damage could occur. The engine has aluminum parts and must be protected by an ethylene-glycol-based coolant to prevent corrosion and freezing.

➡ **If working near and/or around the SRS system and components, be sure to disable the SRS system. After disabling the system wait three minutes or more before servicing the vehicle.**

➡ **Whenever the negative battery cable is disconnected the following components will require resetting. The Idle Air Volume Learning, Steering Angle Sensor Neutral Position, Sunroof Memory Reset/Initialization, Automatic Drive Positioner System, Audio presets and Navigation. Use the Nissan CONSULT III diagnostic tool, or equivalent to perform the required resets.**

1. Before servicing the vehicle, refer to the Precautions Section.

2. Disconnect the negative battery cable.

3. Wrap a thick cloth around the radiator cap to carefully remove the cap. First, turn the cap a quarter of a turn to release any built-up pressure, then push down and turn the cap all the way to remove it.

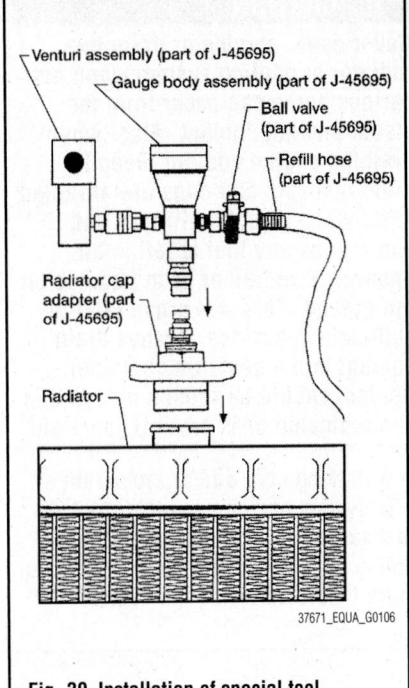

Fig. 39 Installation of special tool KV991J0070 (J-45695) onto the radiator

4. Turn the ignition switch ON and set the temperature control lever all the way to the HOT position or the highest temperature position. Wait 10 seconds and turn the ignition switch OFF.

5. Remove the engine front under cover.

6. Open the radiator drain plug at the bottom of the radiator, and remove the reservoir cap. This is the only step required when partially draining the cooling system (radiator only).

➡ **Do not allow the coolant to contact the drive belts.**

7. To further drain the system, follow the next steps.

8. For heater core removal/replacement only, disconnect the upper heater hose at the engine side and apply moderate air pressure (15 psi maximum pressure) into the hose for 30 seconds to blow the excess coolant out of the heater core.

9. When draining all of the coolant in the system for engine removal or repair, it is necessary to drain the cylinder block. Remove the cylinder block drain plug to drain the cylinder block.

10. Remove the reservoir tank to drain the engine coolant, then clean the reservoir tank before installing it.

11. Check the drained coolant for contaminants such as rust, corrosion, or discoloration. If the coolant is contaminated, flush the engine cooling system.

To install:

12. Close the radiator drain plug. Install the reservoir tank and cylinder block drain plug, if removed for a total system drain or for engine removal or repair. Apply sealant to the threads of the cylinder block drain plugs. Use Genuine High Performance Thread Sealant, or equivalent. The radiator must be completely empty of coolant and water.

13. Set the vehicle heater controls to the full HOT and heater ON position. Turn the vehicle ignition ON with the engine OFF as necessary to activate the heater mode.

14. Remove the vented reservoir cap and replace it with a non-vented reservoir cap before filling the cooling system.

15. Install the special tool KV991J0070 (J-45695) by installing the radiator cap adapter onto the radiator neck opening. Then attach the gauge body assembly with the refill tube and the venturi assembly to the radiator cap adapter.

16. Insert the refill hose into the coolant mixture container that is placed at floor level. Make sure the ball valve is in the closed position. Use recommended coolant.

17. Install an air hose to the venturi assembly, the air pressure must be within

Altitude above sea level	Vacuum gauge reading
328 ft. (0-100m)	28 inches of vacuum
984 ft. (300m)	27 inches of vacuum
1,641 ft. (500m)	26 inches of vacuum
3,281 ft. (1,000m)	24-25 inches of vacuum

37671_EQUA_G0107

Fig. 40 Table of vacuum specifications per altitude

specification. Compressed air supply pressure: 80–119 psi (549–824 kPa).

➡️**The compressed air supply must be equipped with an air dryer.**

18. The vacuum gauge will begin to rise and there will be an audible hissing noise. During this process open the ball valve on the refill hose slightly. Coolant will be visible rising in the refill hose. Once the refill hose is full of coolant, close the ball valve. This will purge any air trapped in the refill hose.

19. Continue to draw the vacuum until the gauge reaches 28 inches of vacuum. The gauge may not reach 28 inches in high altitude locations, use the vacuum specifications below based on the altitude above sea level.

20. When the vacuum gauge has reached the specified amount, disconnect the air hose and wait 20 seconds to see if the system loses any vacuum. If the vacuum level drops, perform any necessary repairs to the system and repeat the previous steps to bring the vacuum to the specified amount. Recheck for any leaks.

21. Place the coolant container (with the refill hose inserted) at the same level as the top of the radiator. Then open the ball valve on the refill hose so the coolant will be drawn up to fill the cooling system. The cooling system is full when the vacuum gauge reads zero.

➡️**To avoid air from being drawn into the cooling system, do not allow the coolant container to get too low when filling.**

22. Remove the special tool from the radiator neck opening and install the radiator cap.

23. Remove the non-vented reservoir cap.

24. Fill the cooling system reservoir tank to the specified level. Run the engine to warm up the cooling system and top up the system as necessary before installing the vented reservoir cap.

FLUID RECOMMENDATIONS

Be sure to use genuine Nissan (blue) coolant when filling/servicing the cooling system.

RADIATOR

REMOVAL & INSTALLATION

2.5L Engine
See Figure 41.

➡️**If working near and/or around the SRS system and components, be sure to disable the SRS system. After disabling the system wait three minutes or more before servicing the vehicle.**

➡️**Whenever the negative battery cable is disconnected the following components will require resetting. The Idle Air Volume Learning, Steering Angle Sensor Neutral Position, Sunroof Memory Reset/Initialization, Automatic Drive Positioner System, Audio presets and Navigation. Use the Nissan CONSULT III diagnostic tool, or equivalent to perform the required resets.**

1. Before servicing the vehicle, refer to the Precautions Section.

2. Disconnect the negative battery cable.

3. Do not remove the radiator cap when the engine is hot. Serious burns could occur from the high-pressure engine coolant escaping from the radiator. Wrap a thick cloth around the cap. Slowly turn it a quarter of a turn to release built-up pressure. Carefully remove radiator cap by turning it all the way.

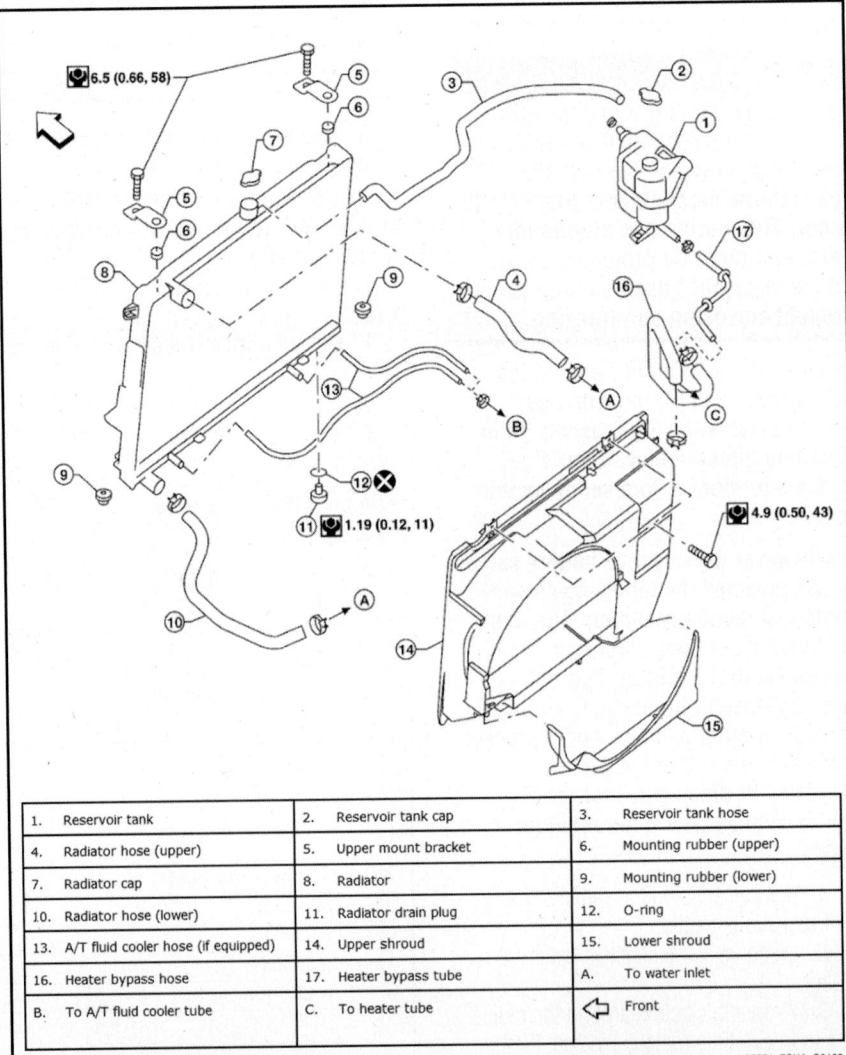

1.	Reservoir tank	2.	Reservoir tank cap	3.	Reservoir tank hose
4.	Radiator hose (upper)	5.	Upper mount bracket	6.	Mounting rubber (upper)
7.	Radiator cap	8.	Radiator	9.	Mounting rubber (lower)
10.	Radiator hose (lower)	11.	Radiator drain plug	12.	O-ring
13.	A/T fluid cooler hose (if equipped)	14.	Upper shroud	15.	Lower shroud
16.	Heater bypass hose	17.	Heater bypass tube	A.	To water inlet
B.	To A/T fluid cooler tube	C.	To heater tube	⬅️	Front

37671_EQUA_G0109

Fig. 41 Radiator and related components—2.5L engine

4. Remove engine under cover.

5. Remove the air cleaner assembly.

6. Drain engine coolant from radiator.

- Perform this step when the engine is cold
- Do not spill engine coolant on the drive belts

7. Remove the air duct and resonator assembly and air duct brackets.

8. Remove the reservoir tank hose.

9. Remove the upper and lower radiator hoses.

10. Disconnect Automatic Transmission (A/T) fluid cooler hoses, if equipped. Install blind plugs to avoid leakage of A/T fluid.

11. Remove the lower and upper shrouds.

12. Remove the front grille.

13. Remove the upper radiator mounting bracket bolts.

14. Remove the 2 A/C condenser bolts.

15. Remove the radiator as follows:

✳✳ WARNING

Do not damage or scratch the A/C condenser and radiator core when removing.

a. While lifting and pulling the radiator in a rearward direction, disassemble the mounting rubber (lower) from the radiator core support center.

➡**Because the A/C condenser is attached to the front-lower portion of the radiator, moving it in the rearward direction should be at a minimum.**

b. Lift the A/C condenser up and remove the radiator after disengaging the fitting at the front-bottom surface.

✳✳ WARNING

Lifting the A/C condenser should be minimum to prevent a load to the A/C piping.

c. After removing the radiator, put the A/C condenser on the radiator core support center to prevent a load to the A/C piping, and temporarily secure it with a rope or by similar means.

To install:

16. Installation is in the reverse order of removal.

17. Fill the engine coolant.

18. Start and warm up the engine. Visually inspect for leaks of the engine coolant.

4.0L Engine

See Figure 42.

✳✳ CAUTION

Never open, service or drain the radiator or cooling system when hot; serious burns can occur from the steam and hot coolant. Also, when draining engine coolant, keep in mind that cats and dogs are attracted to ethylene glycol antifreeze and could drink any that is left in an uncovered container or in puddles on the ground. This will prove fatal in sufficient quantities. Always drain coolant into a sealable container. Coolant should be reused unless it is contaminated or is several years old.

➡**If working near and/or around the SRS system and components, be sure to disable the SRS system. After disabling the system wait three minutes or more before servicing the vehicle.**

➡**Whenever the negative battery cable is disconnected the following components will require resetting. The Idle Air Volume Learning, Steering Angle Sensor Neutral Position, Sunroof Memory Reset/Initialization, Automatic Drive Positioner System, Audio presets and Navigation. Use the Nissan CONSULT III diagnostic tool, or equivalent to perform the required resets.**

1. Before servicing the vehicle, refer to the Precautions Section.

2. Disconnect the negative battery cable.

3. Do not remove the radiator cap when the engine is hot. Serious burns could occur from the high-pressure engine coolant escaping from the radiator. Wrap a thick cloth around the cap. Slowly turn it a quarter of a turn to release built-up pressure. Carefully remove radiator cap by turning it all the way.

4. Remove engine under cover.

5. Remove the air cleaner assembly.

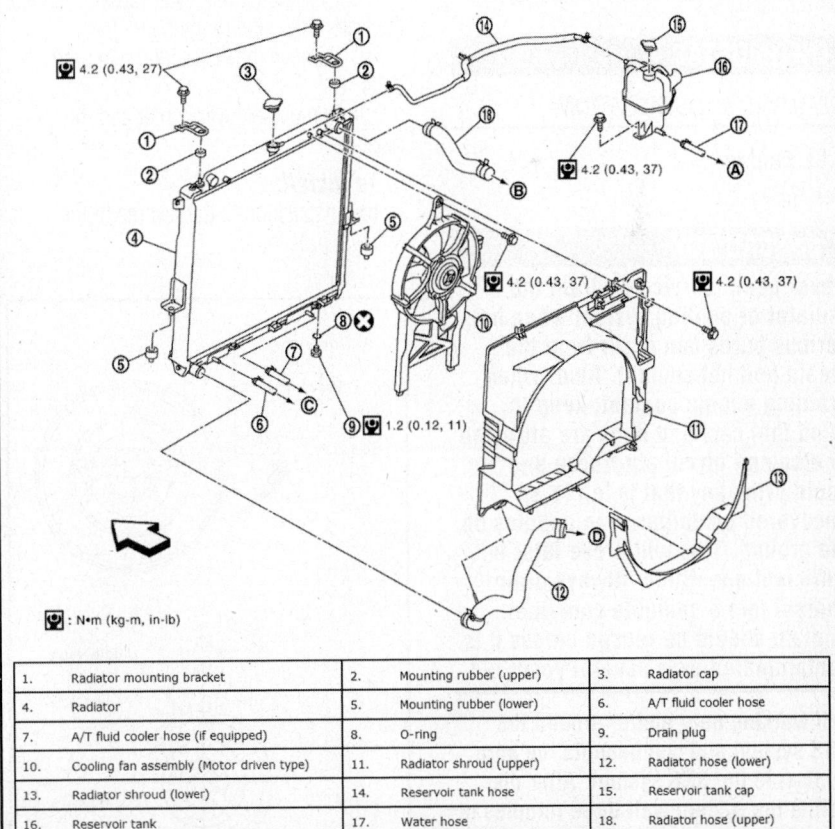

1.	Radiator mounting bracket	2.	Mounting rubber (upper)	3.	Radiator cap
4.	Radiator	5.	Mounting rubber (lower)	6.	A/T fluid cooler hose
7.	A/T fluid cooler hose (if equipped)	8.	O-ring	9.	Drain plug
10.	Cooling fan assembly (Motor driven type)	11.	Radiator shroud (upper)	12.	Radiator hose (lower)
13.	Radiator shroud (lower)	14.	Reservoir tank hose	15.	Reservoir tank cap
16.	Reservoir tank	17.	Water hose	18.	Radiator hose (upper)
A.	To heater return tube	B.	To water pipe	C.	To A/T cooler tube
D.	To water inlet and thermostat assembly	⇦	Vehicle front		

37671_EQUA_G0114

Fig. 42 Radiator and related components—4.0L engine

6. Position a drain pan under the radiator drain cock. Drain the radiator. Be sure to properly dispose of used coolant.

7. Remove the reservoir tank hose.

8. Remove the PCV hose.

9. Remove the upper and lower radiator hoses.

10. Disconnect and plug the transmission fluid lines, if equipped.

11. Remove the engine cooling fan motor.

12. Remove the front grille.

13. Remove the upper radiator mounting bracket bolts.

14. Remove the two air conditioning condenser bolts.

15. Lift and pull the radiator rearward to disengage the mounting rubber (lower) from the core support center.

16. Carefully remove the radiator assembly from the vehicle.

To install:

17. Installation is in the reverse order of removal.

18. Fill the engine coolant.

19. Start and warm up the engine. Visually inspect for leaks of the engine coolant.

THERMOSTAT

REMOVAL & INSTALLATION

2.5L Engine

See Figure 43.

✸✸ CAUTION

Never open, service or drain the radiator or cooling system when hot; serious burns can occur from the steam and hot coolant. Also, when draining engine coolant, keep in mind that cats and dogs are attracted to ethylene glycol antifreeze and could drink any that is left in an uncovered container or in puddles on the ground. This will prove fatal in sufficient quantities. Always drain coolant into a sealable container. Coolant should be reused unless it is contaminated or is several years old.

➡If working near and/or around the SRS system and components, be sure to disable the SRS system. After disabling the system wait three minutes or more before servicing the vehicle.

➡Whenever the negative battery cable is disconnected the following components will require resetting. The Idle Air Volume Learning, Steering Angle Sensor Neutral Position, Sunroof Memory Reset/Initialization, Automatic Drive Positioner System, Audio presets and Navigation. Use the Nissan CONSULT III diagnostic tool, or equivalent to perform the required resets.

1. Before servicing the vehicle, refer to the Precautions Section.

2. Disconnect the negative battery cable.

3. Do not remove the radiator cap when the engine is hot. Serious burns could occur from the high-pressure engine coolant escaping from the radiator. Wrap a thick cloth around the cap. Slowly turn it a quarter of a turn to release built-up pressure. Carefully remove radiator cap by turning it all the way.

4. Remove engine under cover.

5. Drain engine coolant from the radiator.

- Perform this step when the engine is cold
- Do not spill engine coolant on the drive belts

6. Remove the air duct and resonator assembly and air duct brackets.

7. Disconnect the lower radiator hose at the water inlet side (engine side).

8. Remove the water inlet retaining bolts.

9. Remove the water inlet and the thermostat.

To install:

10. Installation is the reverse of the removal procedure.

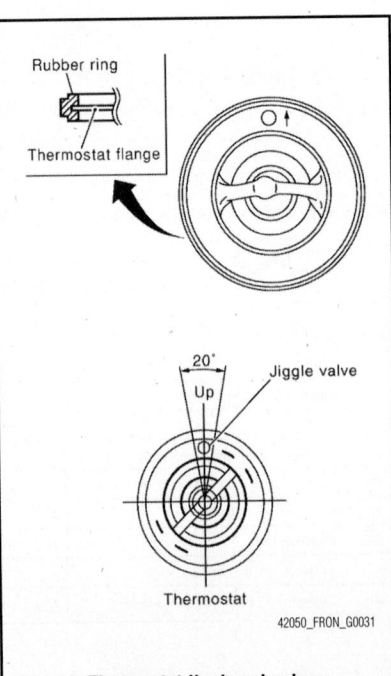

Fig. 43 Thermostat jiggle valve location—2.5L engine

11. Be sure to apply a continuous bead of the proper grade and type RTV sealant to the housing.

12. Install the thermostat with the rubber ring groove positioned to fit the thermostat flange (the whole circumference).

13. Install the thermostat with the jiggle valve facing upward.

➡**The position may deviate within a range of 20 degrees.**

14. Be sure to refill the cooling system using the proper grade and type of engine coolant.

15. Start and warm up the engine. Visually inspect for leaks of the engine coolant.

16. Recheck the coolant level, fill as required.

4.0L Engine

See Figure 44.

✸✸ CAUTION

Never open, service or drain the radiator or cooling system when hot; serious burns can occur from the steam and hot coolant. Also, when draining engine coolant, keep in mind that cats and dogs are attracted to ethylene glycol antifreeze and could drink any that is left in an uncovered container or in puddles on the ground. This will prove fatal in sufficient quantities. Always drain coolant into a sealable container. Coolant should be reused unless it is contaminated or is several years old.

➡If working near and/or around the SRS system and components, be sure to disable the SRS system. After disabling the system wait three minutes or more before servicing the vehicle.

➡Whenever the negative battery cable is disconnected the following components will require resetting. The Idle Air Volume Learning, Steering Angle Sensor Neutral Position, Sunroof Memory Reset/Initialization, Automatic Drive Positioner System, Audio presets and Navigation. Use the Nissan CONSULT III diagnostic tool, or equivalent to perform the required resets.

1. Before servicing the vehicle, refer to the Precautions Section.

2. Disconnect the negative battery cable.

3. Do not remove the radiator cap when the engine is hot. Serious burns could occur from the high-pressure engine coolant escaping from the radiator. Wrap a

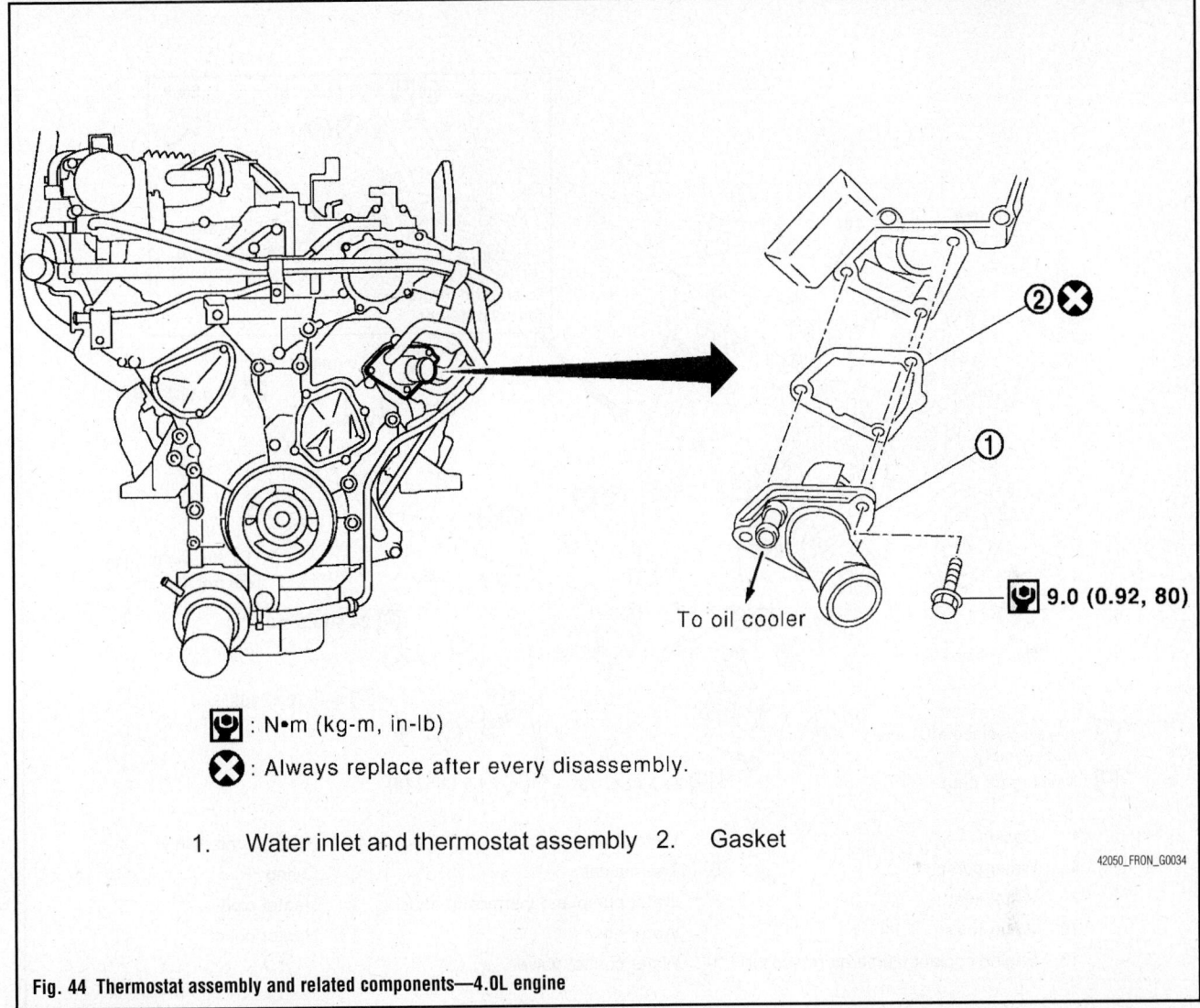

: N•m (kg-m, in-lb)

: Always replace after every disassembly.

1.   Water inlet and thermostat assembly   2.   Gasket

42050_FRON_G0034

Fig. 44 Thermostat assembly and related components—4.0L engine

thick cloth around the cap. Slowly turn it a quarter of a turn to release built-up pressure. Carefully remove radiator cap by turning it all the way.

4. Remove engine under cover.

5. Drain engine coolant from the radiator.

- Perform this step when the engine is cold
- Do not spill engine coolant on the drive belts

6. Remove the air duct and resonator assembly and air cleaner case.

7. Disconnect radiator hose (lower) and oil cooler hose from water inlet and thermostat assembly.

8. Remove the water inlet retaining bolts.

9. Remove the water inlet and the thermostat assembly.

➥Do not disassemble the water inlet and thermostat assembly. Replace as a unit, if required.

To install:

10. Installation is the reverse of the removal procedure.

11. Be sure to refill the cooling using the proper grade and type engine coolant.

12. Start and warm up the engine. Visually inspect for leaks of the engine coolant.

13. Recheck the coolant level, fill as required.

WATER PUMP

REMOVAL & INSTALLATION

2.5L Engine

See Figure 45.

❊❊ CAUTION

Never open, service or drain the radiator or cooling system when hot; serious burns can occur from the

steam and hot coolant. Also, when draining engine coolant, keep in mind that cats and dogs are attracted to ethylene glycol antifreeze and could drink any that is left in an uncovered container or in puddles on the ground. This will prove fatal in sufficient quantities. Always drain coolant into a sealable container. Coolant should be reused unless it is contaminated or is several years old.

➥If working near and/or around the SRS system and components, be sure to disable the SRS system. After disabling the system wait three minutes or more before servicing the vehicle.

➥Whenever the negative battery cable is disconnected the following components will require resetting. The Idle Air Volume Learning, Steering Angle Sensor Neutral Position, Sunroof Memory Reset/Initialization, Automatic

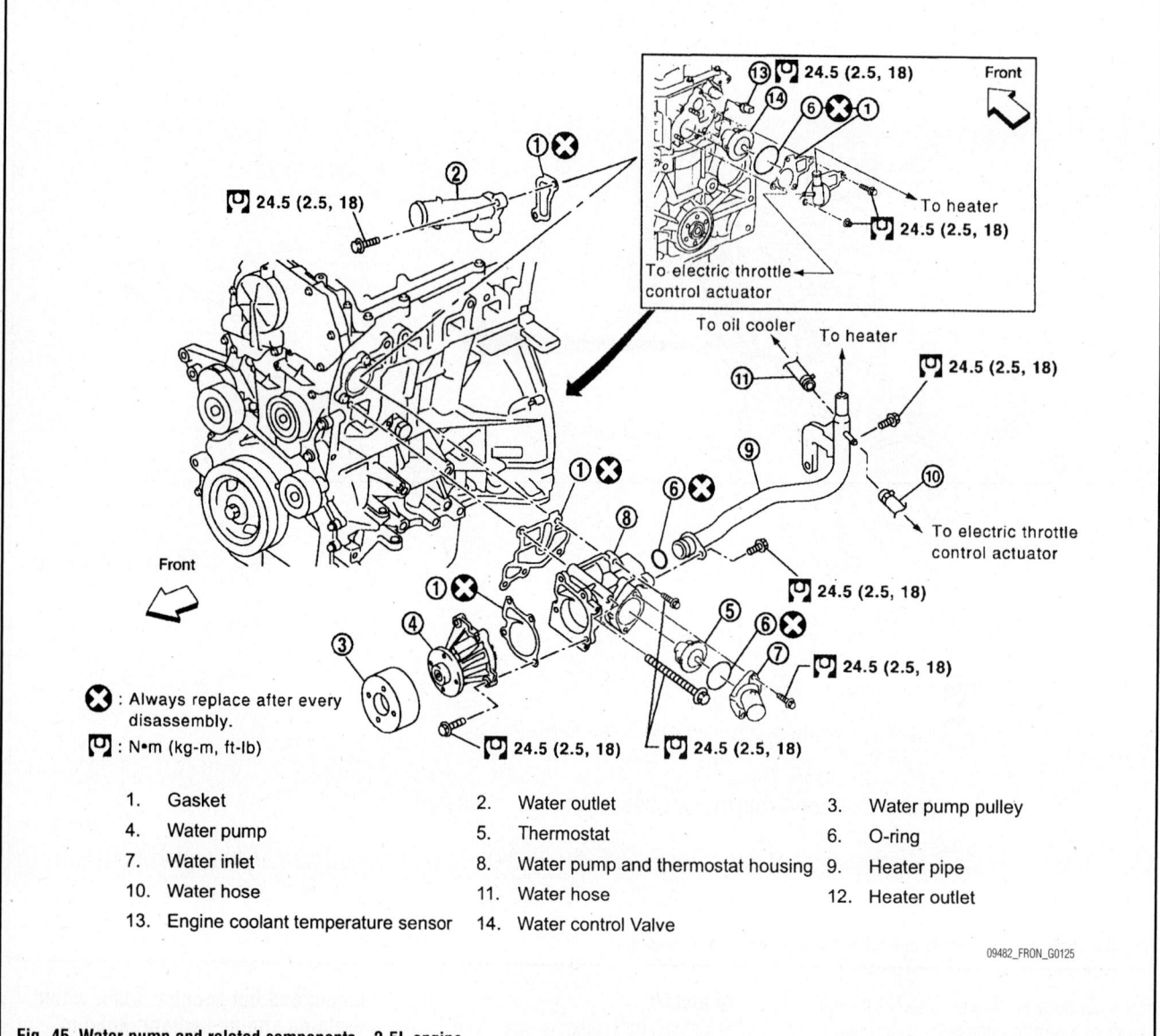

1. Gasket
4. Water pump
7. Water inlet
10. Water hose
13. Engine coolant temperature sensor

2. Water outlet
5. Thermostat
8. Water pump and thermostat housing
11. Water hose
14. Water control Valve

3. Water pump pulley
6. O-ring
9. Heater pipe
12. Heater outlet

09482_FRON_G0125

Fig. 45 Water pump and related components—2.5L engine

Drive Positioner System, Audio presets and Navigation. Use the Nissan CONSULT III diagnostic tool, or equivalent to perform the required resets.

1. Before servicing the vehicle, refer to the Precautions Section.

2. Disconnect the negative battery cable.

3. Do not remove the radiator cap when the engine is hot. Serious burns could occur from the high-pressure engine coolant escaping from the radiator. Wrap a thick cloth around the cap. Slowly turn it a quarter of a turn to release built-up pressure. Carefully remove radiator cap by turning it all the way.

4. Remove engine under cover.

5. Drain engine coolant from the radiator.

- Perform this step when the engine is cold
- Do not spill engine coolant on the drive belts

6. Remove the cooling fan and water pump pulley.

7. Remove the water pump bolts retaining bolts and remove the water pump from the engine.

- Handle the water pump vane so that it does not contact any other parts
- The water pump cannot be disassembled and should be replaced as a unit

➡**Engine coolant will leak from the cylinder block; have a receptacle ready.**

To install:

8. Visually check for dirt or rusting on the water pump body and vane.

9. Ensure that the vane shaft is not loose and that it turns smoothly when rotated by hand.

10. Replace the water pump, if necessary.

11. Installation is in the reverse order of removal.

12. When inserting the heater pipe end into the water pump and thermostat housing, apply a neutral detergent to the O-ring. Then insert it immediately.

13. Be sure to fill the cooling system with the proper grade and type engine coolant.

14. Start and warm up the engine. Visually check for leaks of engine coolant.

4.0L Engine

See Figures 46 through 52.

❊❊ CAUTION

Never open, service or drain the radiator or cooling system when hot; serious burns can occur from the steam and hot coolant. Also, when draining engine coolant, keep in mind that cats and dogs are attracted to ethylene glycol antifreeze and could drink any that is left in an uncovered container or in puddles on the ground. This will prove fatal in sufficient quantities. Always drain coolant into a sealable container. Coolant should be reused unless it is contaminated or is several years old.

➡ If working near and/or around the SRS system and components, be sure to disable the SRS system. After disabling the system wait three minutes or more before servicing the vehicle.

➡ Whenever the negative battery cable is disconnected the following components will require resetting. The Idle Air Volume Learning, Steering Angle Sensor Neutral Position, Sunroof Memory Reset/Initialization, Automatic Drive Positioner System, Audio presets and Navigation. Use the Nissan CONSULT III diagnostic tool, or equivalent to perform the required resets.

1. Before servicing the vehicle, refer to the Precautions Section.

2. Disconnect the negative battery cable.

3. Do not remove the radiator cap when the engine is hot. Serious burns could occur from the high-pressure engine coolant escaping from the radiator. Wrap a thick cloth around the cap. Slowly turn it a quarter of a turn to release built-up pressure. Carefully remove radiator cap by turning it all the way.

4. Remove engine under cover.

5. Drain engine coolant from the radiator.

- Perform this step when the engine is cold
- Do not spill engine coolant on the drive belts

➡ When removing the water pump assembly, be careful not to get engine coolant on the timing chain. The water pump cannot be disassembled and should be replaced as a unit.

6. Remove the engine room cover.

7. Remove the air duct and resonator assembly.

8. Remove the drive belt.

9. Remove the radiator hose (upper).

10. Remove the coolant reservoir hose from the radiator.

11. Remove the engine cooling fan.

12. Remove the chain tensioner cover and the water pump cover from the front timing chain case, using special tool KV10111100 (J-37228), or equivalent.

13. Remove the timing chain tensioner (primary) as follows:

a. Loosen the clip of the timing chain tensioner and release the plunger stopper.

b. Insert the plunger into the tensioner body by pressing the slack guide.

c. Keep the slack guide pressed and hold the plunger in by pushing the stopper pin through the tensioner body hole and the plunger groove.

d. Turn the crankshaft pulley clockwise so that the timing chain on the timing chain tensioner side is loose.

e. Remove the bolts and remove the timing chain tensioner (primary).

❊❊ WARNING

Be careful not to drop the bolts inside the timing chain case.

14. Remove the 3 water pump bolts. Secure a gap between the water pump gear and the timing chain, by turning the crankshaft pulley counterclockwise until the timing chain is loose on the water pump sprocket.

15. Screw M8 bolts 1.97 inches (50mm) in length with a pitch of 0.049 inch (1.25mm) into the water pumps upper and lower bolt holes until they reach the timing chain case. Then, alternately tighten each bolt for a half turn, and pull out the water pump.

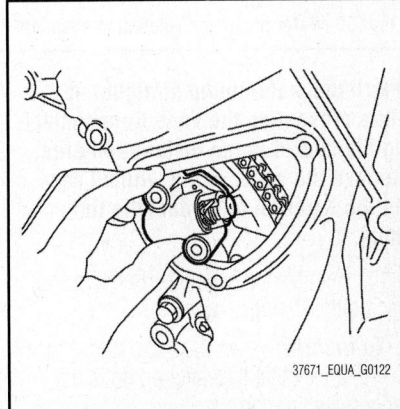

Fig. 48 Remove the bolts and remove the primary timing chain tensioner—4.0L engine

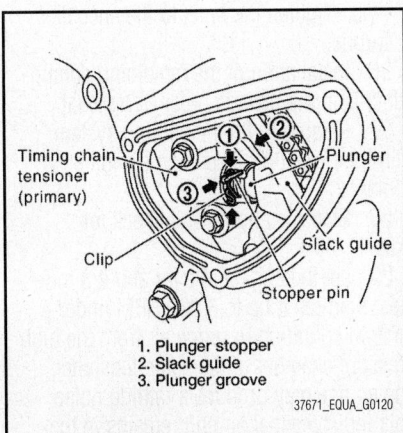

1. Plunger stopper
2. Slack guide
3. Plunger groove

Fig. 46 Loosen the clip of the timing chain tensioner and release the plunger stopper—4.0L engine

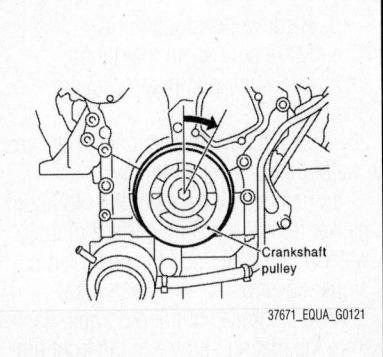

Fig. 47 Turn the crankshaft pulley clockwise so that the timing chain on the timing chain tensioner side is loose—4.0L engine

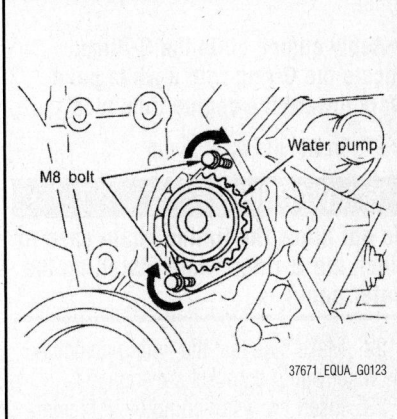

Fig. 49 Removing the water pump using M8 bolts—4.0L engine

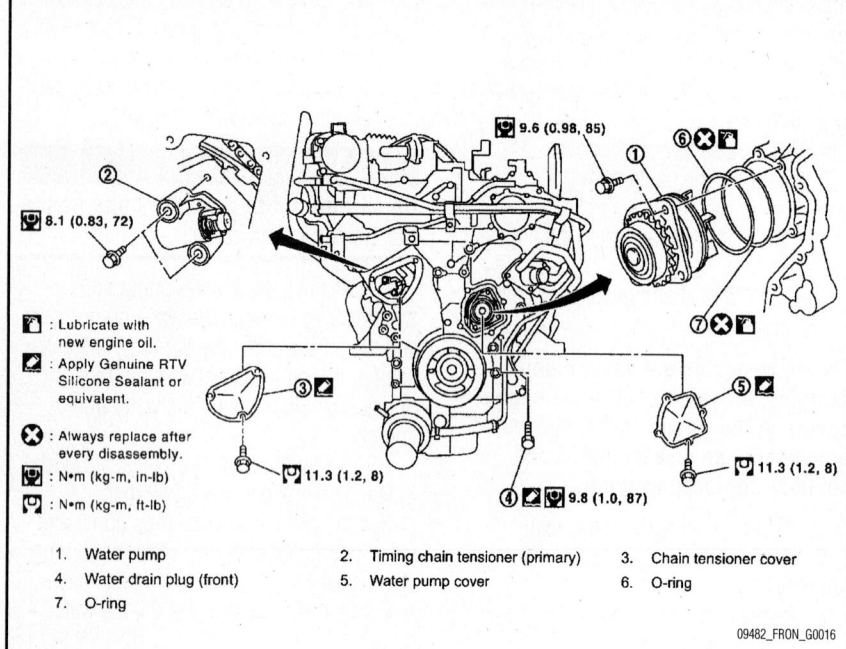

Fig. 50 Water pump and related components—4.0L engine

The following legend appears within the figure:

9.6 (0.98, 85)

8.1 (0.83, 72)

🛢 : Lubricate with new engine oil.

▨ : Apply Genuine RTV Silicone Sealant or equivalent.

❌ : Always replace after every disassembly.

⊡ : N•m (kg-m, in-lb)

⊡ : N•m (kg-m, ft-lb)

11.3 (1.2, 8)

9.8 (1.0, 87)

11.3 (1.2, 8)

1. Water pump
2. Timing chain tensioner (primary)
3. Chain tensioner cover
4. Water drain plug (front)
5. Water pump cover
6. O-ring
7. O-ring

09482_FRON_G0016

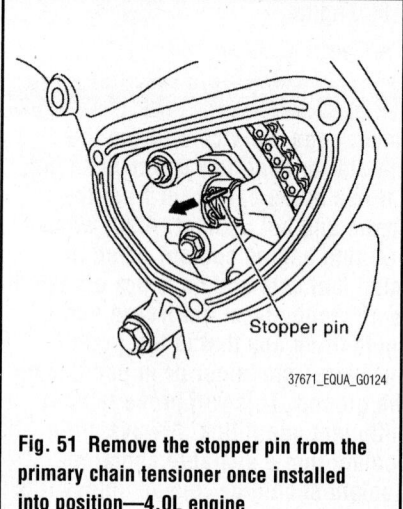

Fig. 51 Remove the stopper pin from the primary chain tensioner once installed into position—4.0L engine

Stopper pin

37671_EQUA_G0124

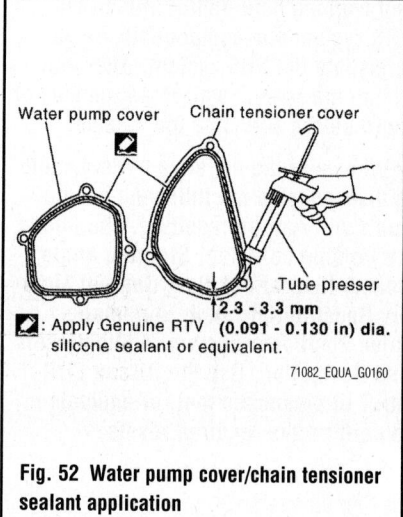

Fig. 52 Water pump cover/chain tensioner sealant application

Water pump cover   Chain tensioner cover

Tube presser
2.3 - 3.3 mm

▨ : Apply Genuine RTV (0.091 - 0.130 in) dia. silicone sealant or equivalent.

71082_EQUA_G0160

➡**Pull the water pump straight out while preventing the vane from contacting the socket in the installation area. Remove the water pump without causing the sprocket to contact the timing chain.**

16. Remove the M8 bolts. Remove and discard the O-rings.

To install:

17. Check that the water pump is not badly rusted or corroded.

18. Check for rough operation due to excessive endplay.

19. Replace the water pump, if necessary.

20. Install new O-rings to the water pump.

➡**Apply engine oil to the O-rings. Locate the O-ring with a white paint mark toward the engine front side.**

21. Install the water pump.

❋❋ **WARNING**

Do not allow the timing chain case to pinch the O-rings when installing the water pump.

22. Make sure that the timing chain and the water pump sprocket are engaged.

23. Insert the water pump by tightening the bolts alternately and evenly.

24. Install the timing chain tensioner (primary) as follows:

a. Remove any dust and foreign material completely from the backside of the timing chain tensioner (primary) and from the installation area of the rear timing chain case.

b. Turn the crankshaft pulley clockwise so that the timing chain on the timing chain tensioner (primary) side is loose.

c. Install the timing chain tensioner (primary) with its stopper pin attached.

❋❋ **WARNING**

Be careful not to drop the bolts inside the timing chain case.

d. Remove the stopper pin.

e. Make sure again that timing chain and water pump sprocket are engaged.

25. Install the chain tensioner cover and the water pump cover as follows:

a. Remove all traces of the old liquid gasket from the mating surface of the water pump cover and the chain tensioner cover using a scraper. Also, remove the traces of the old liquid gasket from the mating surface of the front timing chain case.

b. Apply a continuous bead of liquid gasket, to the mating surface of the chain

tensioner and the water pump cover. Use Genuine RTV Silicone Sealant, or equivalent.

c. Attach the mating surfaces within 5 minutes after applying liquid gasket.

d. Tighten the bolts to the specified torque.

26. Installation of the remaining components is in the reverse order of removal.

27. Be sure to fill the cooling system with the proper grade and type engine coolant.

28. Start the engine and check for leaks.

29. Let the engine idle for about 3 minutes, then rev it up to 3,000 RPM under a no load condition to purge air from the high pressure chamber of the chain tensioner. The engine may produce a rattling noise. This indicates that air still remains in the chamber and is not a matter of concern.

BATTERY

REMOVAL & INSTALLATION

See Figure 53.

Batteries produce explosive gases, contain corrosive acids, and supply levels of electrical current high enough to cause burns. In order to reduce the risk of personal injury while working near a battery, observe the following:

• Protect eyes from battery acid, a suitable pair of industrial grade safety glasses should be worn when removing or servicing a battery

• Avoid leaning over the battery whenever possible

• Do not expose the battery to open flames or sparks

• Do not allow battery acid to come in contact with eyes or skin. Flush any contacted area with clean water immediately. Get medical help.

• Remove metallic jewelry to avoid injury by accidental arcing of the battery current

➡ **Note the location of the positive and negative cables prior to service of the battery or related components.**

✳✳ CAUTION

The Supplemental Restraint System (SRS) is active for a certain length of time after the power supply has been disconnected. Wait for a minimum of 3 minutes before disconnecting or removing any SRS components.

✳✳ CAUTION

For vehicles with an auxiliary battery, make sure that the vehicle's electrical system is fully depowered and no other power source is connected.

➡ **If working near and/or around the SRS system and components, be sure to disable the SRS system. After disabling the system wait three minutes or more before servicing the vehicle.**

➡ **Whenever the negative battery cable is disconnected the following components will require resetting. The Idle**

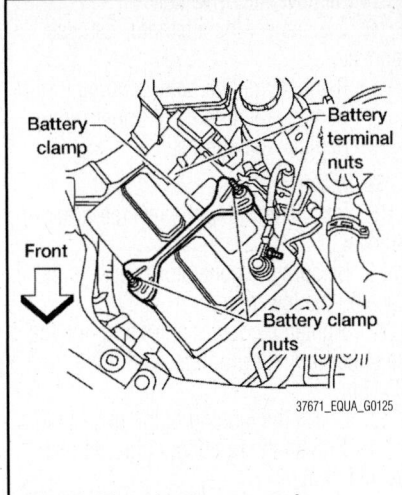

Fig. 53 View of battery removal

Air Volume Learning, Steering Angle Sensor Neutral Position, Sunroof Memory Reset/Initialization, Automatic Drive Positioner System, Audio presets and Navigation. Use the Nissan CONSULT III diagnostic tool, or equivalent to perform the required resets.

1. Before servicing the vehicle, refer to the Precautions Section.
2. Disconnect the negative battery cable.
3. Disconnect the positive battery cable.
4. Remove the battery clamp nuts and the battery clamp.
5. Remove the battery.

To install:
6. Installation is in the reverse order of removal.
7. Connect the positive battery terminal first.
8. Tighten the battery clamp nuts to 35 inch lbs. (4 Nm).
9. Tighten the battery terminal nuts to 30 inch lbs. (3 Nm).
10. Reset electronic systems as necessary.

BATTERY RECONNECT/ RELEARN PROCEDURE

Vehicles equipped with engine and transaxle computers may require a relearn

procedure after the vehicle battery has been disconnected. Most vehicle computers memorize and store vehicle operational patterns. When the battery is disconnected, the information may be cleared. If the information is cleared, the computer will go into default mode in order to operate the vehicle. The vehicle computer will relearn operational patterns each time the vehicle is restarted. The relearning process may take up to 40 or more key cycles.

When a specific engine component is replaced, a relearn procedure may be required. If the relearn procedure is not performed, the vehicle may exhibit the following:

• Harsh or poor shift quality
• Poor fuel mileage
• Hesitation or stumble
• Unstable idle or stalling
• Lean or rich running conditions

If an accessory component was replaced, a relearn procedure may also be required. The following systems and components may not work properly without a relearn procedure:

• Anti-theft system
• Steering system
• Power window system
• Power sunroof system

➡ **If working near and/or around the SRS system and components, be sure to disable the SRS system. After disabling the system wait three minutes or more before servicing the vehicle.**

➡ **Whenever the negative battery cable is disconnected the following components will require resetting. The Idle Air Volume Learning, Steering Angle Sensor Neutral Position, Sunroof Memory Reset/Initialization, Automatic Drive Positioner System, Audio presets and Navigation. Use the Nissan CONSULT III diagnostic tool, or equivalent to perform the required resets.**

In most cases a diagnostic tool will be required to perform the required relearn procedures.

ENGINE ELECTRICAL

CHARGING SYSTEM

ALTERNATOR

REMOVAL & INSTALLATION

2.5L Engine

See Figure 54.

➡If working near and/or around the SRS system and components, be sure to disable the SRS system. After disabling the system wait three minutes or more before servicing the vehicle.

➡Whenever the negative battery cable is disconnected the following components will require resetting. The Idle Air Volume Learning, Steering Angle Sensor Neutral Position, Sunroof Memory Reset/Initialization, Automatic Drive Positioner System, Audio presets and Navigation. Use the Nissan CONSULT III diagnostic tool, or equivalent to perform the required resets.

1. Before servicing the vehicle, refer to the Precautions Section.
2. Disconnect the negative battery cable.
3. Remove the engine under cover.
4. Remove front RH fender protector.

5. Remove the drive belt.
6. Disconnect the alternator harness connectors.
7. Remove the alternator mounting nut.
8. Remove the alternator upper bolt.
9. Remove the alternator.

To install:

10. Installation is in the reverse order of removal.
11. Be sure the generator spacer is in place on the lower stud.
12. Tighten the lower mounting nut and the upper mounting bolt to 48 ft. lbs. (65 Nm).
13. Check the tension of the drive belt.
14. Tighten the terminal nut to 96 inch lbs. (11 Nm).

➡Be sure to tighten the terminal nut carefully.

15. For this model, the power generation voltage variable control system that controls the power generation voltage of the generator has been adopted. Therefore, the power generation voltage variable control system operation inspection should be performed after replacing the generator, and then make sure that the system operates normally.

4.0L Engine

See Figure 55.

➡If working near and/or around the SRS system and components, be sure to disable the SRS system. After disabling the system wait three minutes or more before servicing the vehicle.

➡Whenever the negative battery cable is disconnected the following components will require resetting. The Idle Air Volume Learning, Steering Angle Sensor Neutral Position, Sunroof Memory Reset/Initialization, Automatic Drive Positioner System, Audio presets and Navigation. Use the Nissan CONSULT III diagnostic tool, or equivalent to perform the required resets.

1. Before servicing the vehicle, refer to the Precautions Section.
2. Disconnect the negative battery cable.
3. Partially drain the engine coolant.
4. Remove the engine room cover.
5. Remove the air duct and resonator assembly.
6. Remove the upper radiator hose.
7. Disconnect the coolant reservoir hose from the radiator.
8. Remove the fan shroud.
9. Remove the engine cooling fan (motor driven type).
10. Remove the drive belt.
11. Remove alternator stay.
12. Remove the alternator upper bolt.
13. Disconnect the alternator harness connectors.
14. Remove the alternator from the vehicle.

To install:

15. Installation is in the reverse order of removal.
16. Tighten the upper mounting bolt to 48 ft. lbs. (65 Nm).
17. Tighten the alternator stay bolts to 21 ft. lbs. (28 Nm).
18. Check the tension of the drive belt.
19. Tighten the terminal nut to 96 inch lbs. (11 Nm).

➡Be sure to tighten the terminal nut carefully.

20. For this model, the power generation voltage variable control system that controls the power generation voltage of the generator has been adopted. Therefore, the power generation voltage variable control system operation inspection should be performed after replacing the alternator, and then make sure that the system operates normally.

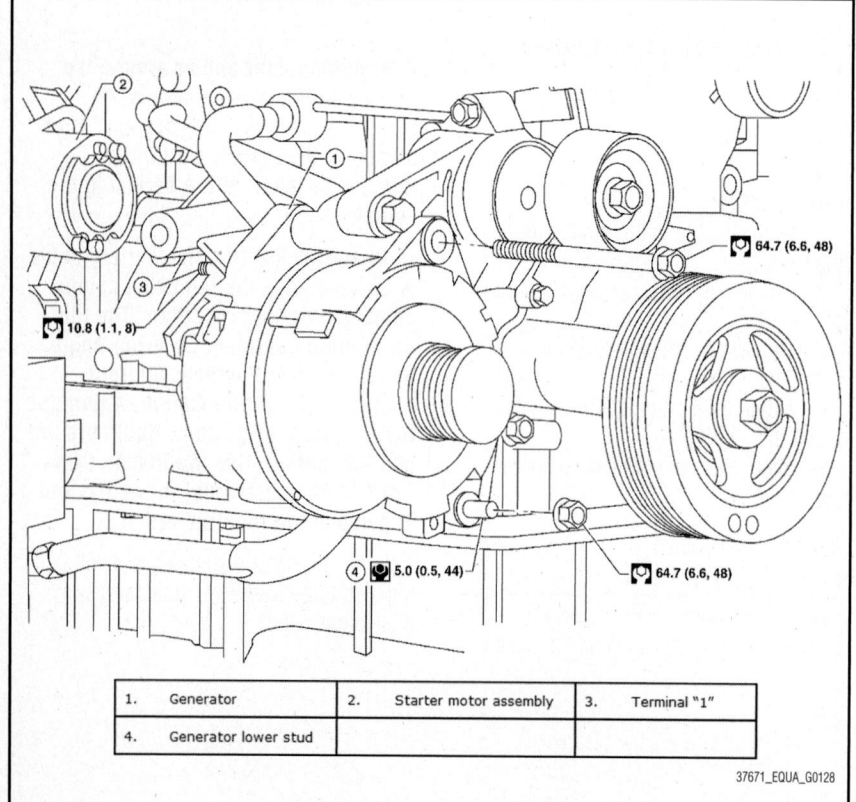

| 1. | Generator | 2. | Starter motor assembly | 3. | Terminal "1" |
| 4. | Generator lower stud | | | | |

37671_EQUA_G0128

Fig. 54 Alternator and related components—2.5L engine

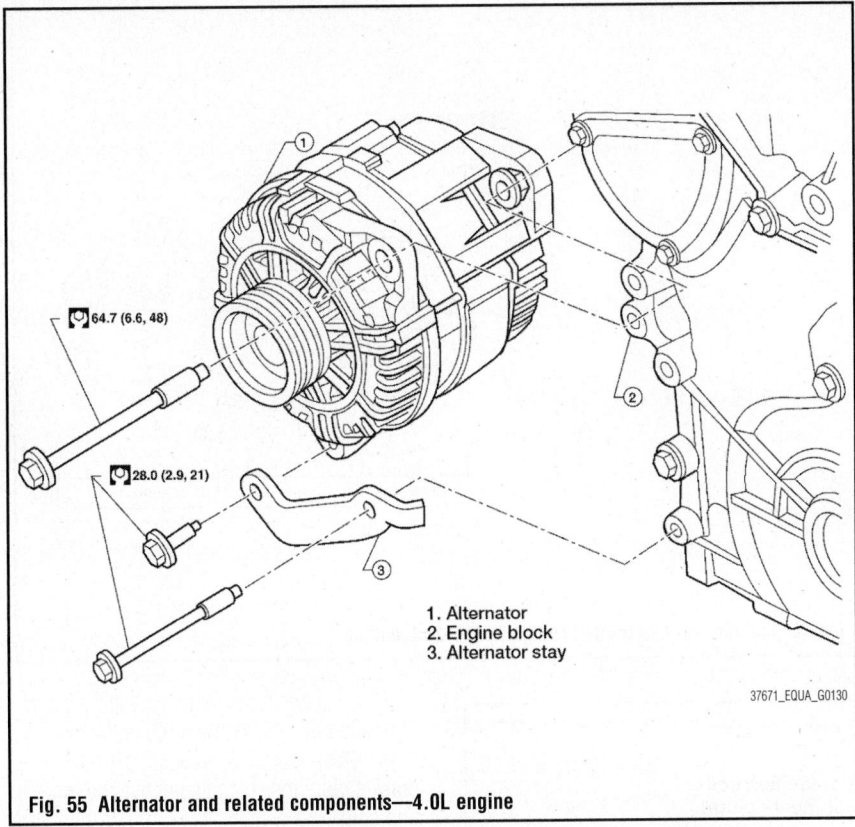

Fig. 55 Alternator and related components—4.0L engine

1. Alternator
2. Engine block
3. Alternator stay

37671_EQUA_G0130

23. Ensure that the transmission is in "P" or "N" position and all of the electric loads and A/C are turned OFF.

24. Select "ALTERNATOR DUTY" in "Active Test" of "Engine / Powertrain", and then check the value of "BATTERY VOLT" monitor when the DUTY value of "ALTERNATOR DUTY" is set to 40.0 percent.

- 2 seconds after setting the DUTY to 40.0 percent, the voltage should be 12–13.6 volts.

25. Check the value of "BATTERY VOLT" monitor when the DUTY value of "ALTERNATOR DUTY" is set to 80.0 percent.

- 20 seconds after setting the DUTY value of "ALTERNATOR DUTY" to 80.0 percent, the voltage should be 0.5 volts greater than that of the 40.0 percent DUTY value.
- If the values are not according to specification, check for a wiring or component failure.

26. For this model, the power generation voltage variable control system that controls the power generation voltage of the generator has been adopted. Therefore, the power generation voltage variable control system operation inspection should be performed after replacing the generator, and then make sure that the system operates normally.

➡When performing this inspection, always use a charged battery. When the charging rate of the battery is low, the response speed of the voltage change will become slow.

This can cause an incorrect inspection.

21. Perform a DTC check with the Tool.
22. Start the engine with TOOL connected.

ENGINE ELECTRICAL

FIRING ORDERS

See Figures 56 and 57.

IGNITION COILS

REMOVAL & INSTALLATION

2.5L Engine

See Figure 58.

➡If working near and/or around the SRS system and components, be sure to disable the SRS system. After disabling the system wait three minutes or more before servicing the vehicle.

➡Whenever the negative battery cable is disconnected the following components will require resetting. The Idle Air Volume Learning, Steering Angle Sensor Neutral Position, Sunroof Memory Reset/Initialization, Automatic Drive Positioner System, Audio presets and Navigation. Use the Nissan CON-

SULT III diagnostic tool, or equivalent to perform the required resets.

1. Before servicing the vehicle, refer to the Precautions Section.
2. Disconnect the negative battery cable.
3. Remove the intake manifold.
4. Disconnect the harness connector from the ignition coil.
5. Remove the ignition coil retaining bolt.
6. Remove the ignition coil from the vehicle.

To install:

7. Installation is the reverse of the removal procedure.
8. Tighten the ignition coil retaining bolt to 62 inch lbs. (7 Nm).

4.0L Engine

Left Bank

See Figure 59.

➡If working near and/or around the SRS system and components, be sure

IGNITION SYSTEM

to disable the SRS system. After disabling the system wait three minutes or more before servicing the vehicle.

➡Whenever the negative battery cable is disconnected the following components will require resetting. The Idle Air Volume Learning, Steering Angle Sensor Neutral Position, Sunroof Memory Reset/Initialization, Automatic Drive Positioner System, Audio presets and Navigation. Use the Nissan CONSULT III diagnostic tool, or equivalent to perform the required resets.

1. Before servicing the vehicle, refer to the Precautions Section.
2. Disconnect the negative battery cable.
3. Remove engine room cover.
4. Move aside the harness, harness bracket, and hoses located above the ignition coil.
5. Disconnect the harness connector from the ignition coil.
6. Remove the ignition coil from the vehicle.

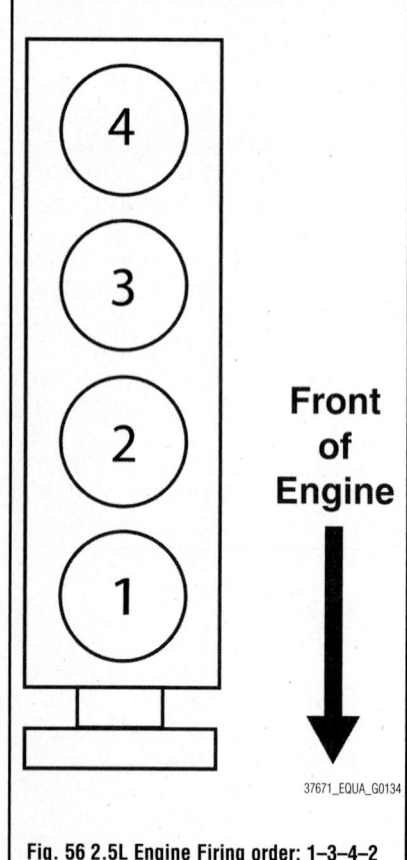

Fig. 56 2.5L Engine Firing order: 1–3–4–2

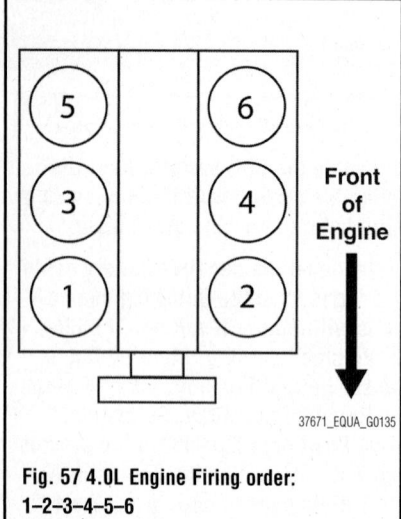

Fig. 57 4.0L Engine Firing order: 1–2–3–4–5–6

Front of Engine

To install:

7. Installation is the reverse of the removal procedure.

8. Tighten the ignition coil retaining bolt to 62 inch lbs. (7 Nm).

Right Bank

➡If working near and/or around the SRS system and components, be sure to disable the SRS system. After disabling the system wait three

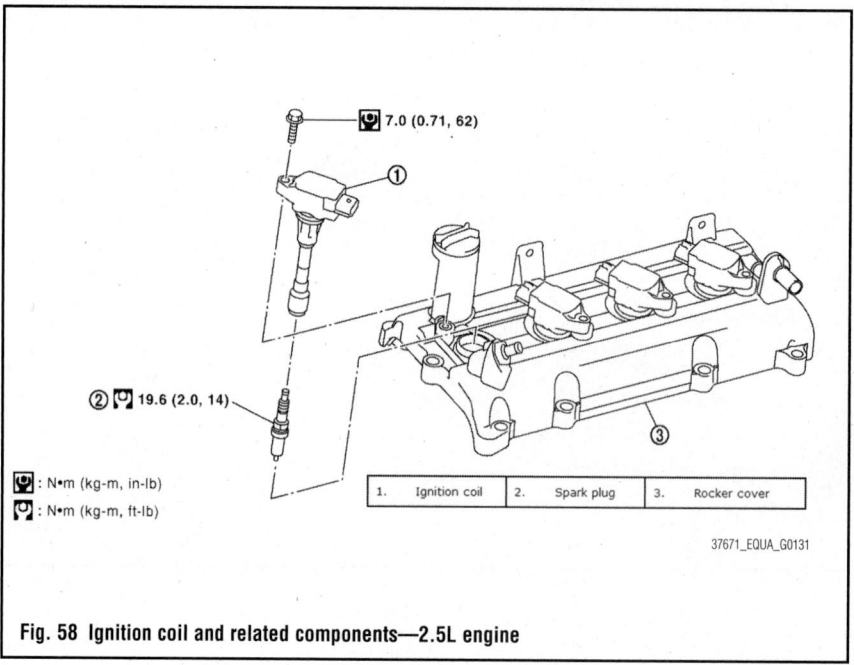

7.0 (0.71, 62)

19.6 (2.0, 14)

: N•m (kg-m, in-lb)

: N•m (kg-m, ft-lb)

| 1. | Ignition coil | 2. | Spark plug | 3. | Rocker cover |

37671_EQUA_G0131

Fig. 58 Ignition coil and related components—2.5L engine

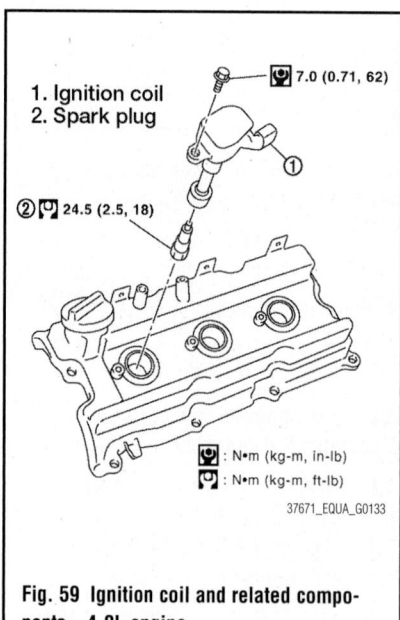

1. Ignition coil
2. Spark plug

7.0 (0.71, 62)

24.5 (2.5, 18)

: N•m (kg-m, in-lb)

: N•m (kg-m, ft-lb)

37671_EQUA_G0133

Fig. 59 Ignition coil and related components—4.0L engine

minutes or more before servicing the vehicle.

➡Whenever the negative battery cable is disconnected the following components will require resetting. The Idle Air Volume Learning, Steering Angle Sensor Neutral Position, Sunroof Memory Reset/Initialization, Automatic Drive Positioner System, Audio presets and Navigation. Use the Nissan CONSULT III diagnostic tool, or equivalent to perform the required resets.

1. Before servicing the vehicle, refer to the Precautions Section.

2. Disconnect the negative battery cable.

3. Remove intake manifold collector.

4. Move aside the harness, harness bracket, and hoses located above the ignition coil.

5. Disconnect the harness connector from the ignition coil.

6. Remove the ignition coil from the vehicle.

To install:

7. Installation is the reverse of the removal procedure.

8. Tighten the ignition coil retaining bolt to 62 inch lbs. (7 Nm).

IGNITION TIMING

INSPECTION & ADJUSTMENT

All engines use a fixed ignition timing system. Basic ignition timing is not adjustable. All spark advance is determined by the Engine Control Module (ECM).

➡If working near and/or around the SRS system and components, be sure to disable the SRS system. After disabling the system wait three minutes or more before servicing the vehicle.

➡Whenever the negative battery cable is disconnected the following components will require resetting. The Idle Air Volume Learning, Steering Angle Sensor Neutral Position, Sunroof Memory Reset/Initialization, Automatic Drive Positioner System, Audio presets and Navigation. Use the Nissan CONSULT III diagnostic tool, or equivalent to perform the required resets.

1. Before servicing the vehicle, refer to the Precautions Section.

2. Remove the number one ignition coil.

3. Connect the number one ignition coil and spark plug with a suitable high tension wire.

4. Attach the timing light clamp to the wire.

5. Check the ignition timing.

The ignition timing is controlled by the Engine Control Module (ECM). No adjustment is necessary.

SPARK PLUGS

REMOVAL & INSTALLATION

2.5L Engine

➡If working near and/or around the SRS system and components, be sure to disable the SRS system. After disabling the system wait three minutes or more before servicing the vehicle.

➡Whenever the negative battery cable is disconnected the following components will require resetting. The Idle Air Volume Learning, Steering Angle Sensor Neutral Position, Sunroof Memory Reset/Initialization, Automatic Drive Positioner System, Audio presets and Navigation. Use the Nissan CONSULT III diagnostic tool, or equivalent to perform the required resets.

1. Before servicing the vehicle, refer to the Precautions Section.

2. Disconnect the negative battery cable.

3. Remove the intake manifold.

➡If removing the number one spark plug only, it is not necessary to remove the intake manifold.

4. Remove the ignition coil.

5. Remove the spark plug using a spark plug socket and wrench.

To install:

6. Installation is the reverse of the removal procedure.

7. Ensure the spark plug gap is to specification: 0.043 inch (1.1mm).

8. Tighten the spark plug to 14 ft. lbs. (20 Nm).

4.0L Engine

➡If working near and/or around the SRS system and components, be sure to disable the SRS system. After disabling the system wait three minutes or more before servicing the vehicle.

➡Whenever the negative battery cable is disconnected the following components will require resetting. The Idle Air Volume Learning, Steering Angle Sensor Neutral Position, Sunroof Memory Reset/Initialization, Automatic Drive Positioner System, Audio presets and Navigation. Use the Nissan CONSULT III diagnostic tool, or equivalent to perform the required resets.

1. Before servicing the vehicle, refer to the Precautions Section.

2. Disconnect the negative battery cable.

3. Remove the engine room cover.

4. Remove the ignition coil.

➡On right side it may be necessary to remove the intake collector.

5. Remove the spark plug using a spark plug socket and wrench.

To install:

6. Installation is the reverse of the removal procedure.

7. Ensure the spark plug gap is to specification: 0.043 inch (1.1mm).

8. Tighten the spark plug to 18 ft. lbs. (25 Nm).

ENGINE ELECTRICAL

STARTING SYSTEM

STARTER

REMOVAL & INSTALLATION

2.5L Engine

See Figure 60.

➡If working near and/or around the SRS system and components, be sure to disable the SRS system. After disabling the system wait three minutes or more before servicing the vehicle.

➡Whenever the negative battery cable is disconnected the following components will require resetting. The Idle Air Volume Learning, Steering Angle Sensor Neutral Position, Sunroof Memory Reset/Initialization, Automatic Drive Positioner System, Audio presets and Navigation. Use the Nissan CONSULT III diagnostic tool, or equivalent to perform the required resets.

1. Before servicing the vehicle, refer to the Precautions Section.

2. Disconnect the negative battery cable.

3. Disconnect the negative battery cable.

4. Remove the air cleaner cover and the air cleaner to intake manifold collector duct.

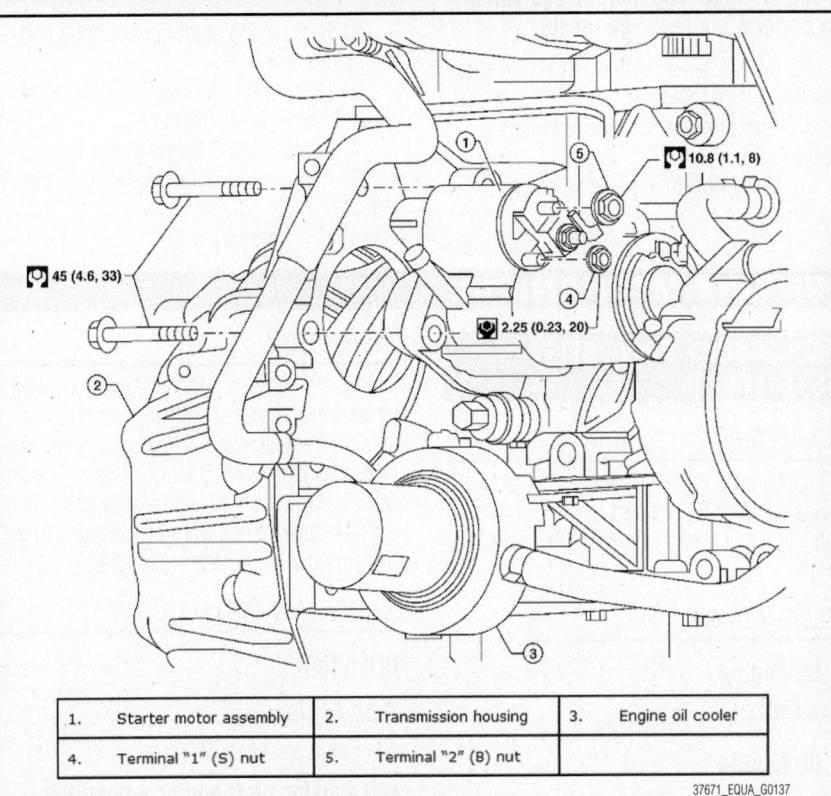

| 1. | Starter motor assembly | 2. | Transmission housing | 3. | Engine oil cooler |
| 4. | Terminal "1" (S) nut | 5. | Terminal "2" (B) nut | | |

37671_EQUA_G0137

Fig. 60 Starter and related components—2.5L engine

5. Remove the "S" terminal and terminal "B" nuts.

6. Remove the 2 starter motor bolts.

7. Remove the starter motor from the vehicle.

To install:

8. Installation is in the reverse order of removal.

❋❋ WARNING

Be sure to tighten the terminal nuts carefully.

9. Use the following illustration for torque values.

4.0L Engine

See Figure 61.

➡️ If working near and/or around the SRS system and components, be sure to disable the SRS system. After disabling the system wait three minutes or more before servicing the vehicle.

➡️ Whenever the negative battery cable is disconnected the following components will require resetting. The Idle Air Volume Learning, Steering Angle Sensor Neutral Position, Sunroof Memory Reset/Initialization, Automatic Drive Positioner System, Audio presets and Navigation. Use the Nissan CONSULT III diagnostic tool, or equivalent to perform the required resets.

1. Before servicing the vehicle, refer to the Precautions Section.

2. Disconnect the negative battery cable.

3. Disconnect the negative battery cable.

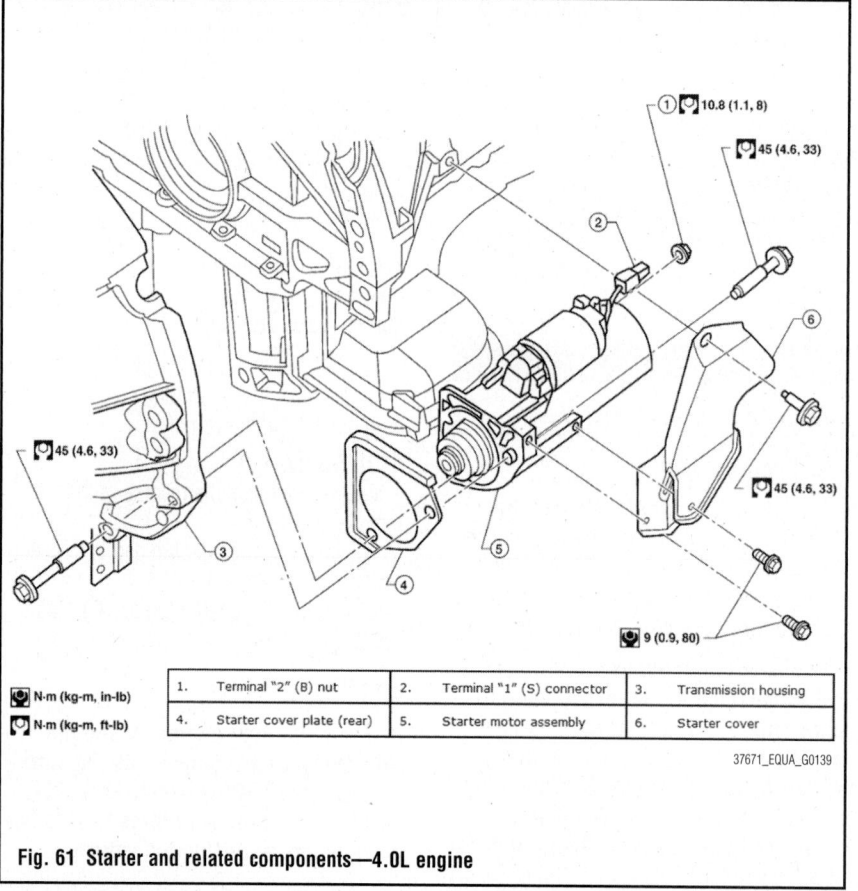

⟳ N·m (kg-m, in-lb)				
⟲ N·m (kg-m, ft-lb)				

1.	Terminal "2" (B) nut	2.	Terminal "1" (S) connector	3.	Transmission housing
4.	Starter cover plate (rear)	5.	Starter motor assembly	6.	Starter cover

37671_EQUA_G0139

Fig. 61 Starter and related components—4.0L engine

4. Raise and support the vehicle safely, as required.

5. Remove the starter cover bolts and the starter cover.

6. Disconnect terminal "1" (S) connector and terminal "2" (B) nut.

7. Remove the 2 starter motor bolts.

8. Remove the starter motor from the vehicle.

To install:

9. Installation is in the reverse order of removal.

❋❋ WARNING

Be sure to tighten the terminal nuts carefully.

10. Use the following illustration for torque values.

ENGINE MECHANICAL

ACCESSORY DRIVE BELT SYSTEM

ADJUSTMENT

There is no manual drive belt tension adjustment. The drive belt tension is automatically adjusted by the drive belt auto tensioner.

BELT ROUTING

2.5L Engine

See Figure 62.

4.0L Engine

See Figure 63.

INSPECTION

Inspect the drive belt for signs of glazing or cracking. A glazed belt will be perfectly smooth from slippage, while a good belt will have a slight texture of fabric visible. Cracks will usually start at the inner edge of the belt and run outward. All worn or damaged drive belts should be replaced immediately.

REMOVAL & INSTALLATION

Drive Belt

2.5L Engine

See Figures 64 and 65.

➡️ If working near and/or around the SRS system and components, be sure to disable the SRS system. After disabling the system wait three minutes or more before servicing the vehicle.

➡️ Whenever the negative battery cable is disconnected the following components will require resetting. The Idle Air Volume Learning, Steering Angle Sensor Neutral Position, Sunroof Memory Reset/Initialization, Automatic Drive Positioner System, Audio presets and Navigation. Use the Nissan CONSULT III diagnostic tool, or equivalent to perform the required resets.

1. Before servicing the vehicle, refer to the Precautions Section.

2. Disconnect the negative battery cable.

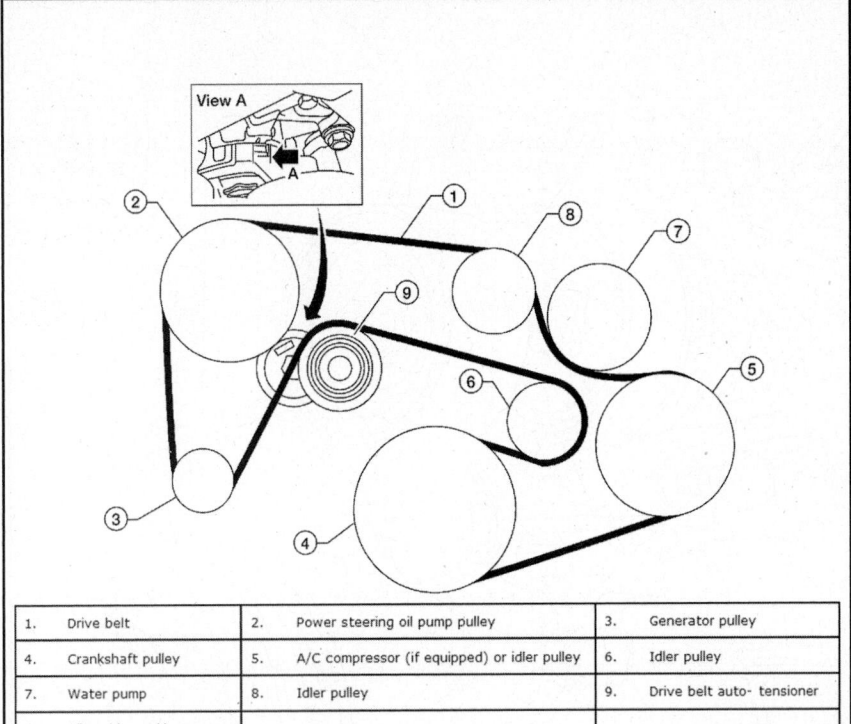

Fig. 62 Drive belt routing—2.5L engine

1.	Drive belt	2.	Power steering oil pump pulley	3.	Generator pulley
4.	Crankshaft pulley	5.	A/C compressor (if equipped) or idler pulley	6.	Idler pulley
7.	Water pump	8.	Idler pulley	9.	Drive belt auto- tensioner
A.	Allowable working range				

37671_EQUA_G0141

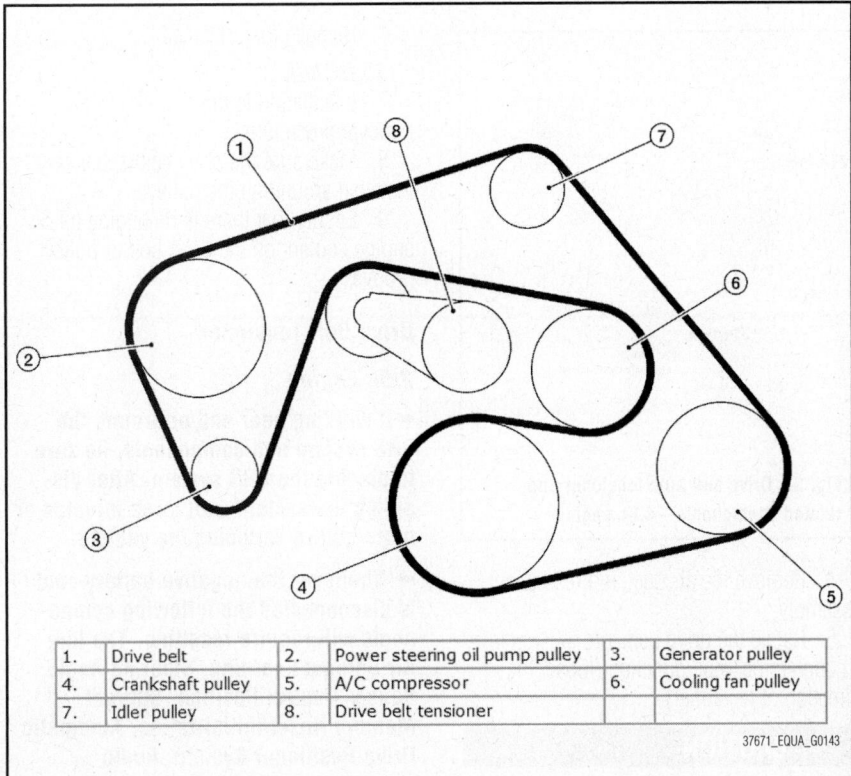

Fig. 63 Drive belt routing—4.0L engine

1.	Drive belt	2.	Power steering oil pump pulley	3.	Generator pulley
4.	Crankshaft pulley	5.	A/C compressor	6.	Cooling fan pulley
7.	Idler pulley	8.	Drive belt tensioner		

37671_EQUA_G0143

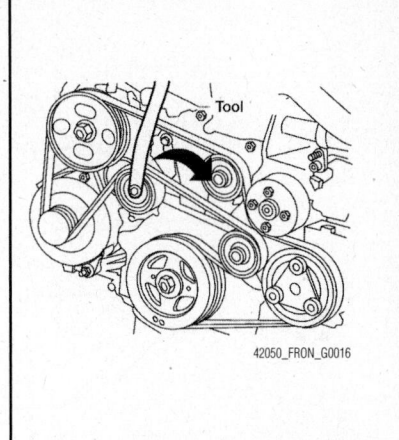

42050_FRON_G0016

Fig. 64 Drive belt tension tool installation and removal direction—2.5L engine

3. Install special tool J-46535, or equivalent, on the auto tensioner pulley bolt and move in a clockwise direction (loosening direction of tensioner).

❊❊ CAUTION

Avoid placing your hand in a location where pinching may occur if the holding tool accidentally comes off.

➡**Do not loosen the auto-tensioner pulley bolt. (Do not turn it counterclockwise.) If it is turned counterclockwise, the complete auto-tensioner must be replaced as a unit, including the pulley.**

4. Remove the drive belt.

To install:

5. Installation is the reverse of the removal procedure.

6. Confirm the belts are completely set on the pulleys.

7. Ensure that there is no engine oil or engine coolant on the drive belt or pulley grooves.

8. Turn the crankshaft pulley clockwise several times to equalize the tension between each pulley.

9. Confirm the tension of the drive belt indicator (fixed side) is within the allowable working range.

4.0L Engine

See Figures 66 and 67.

➡**If working near and/or around the SRS system and components, be sure to disable the SRS system. After disabling the system wait three minutes or more before servicing the vehicle.**

➡**Whenever the negative battery cable is disconnected the following components will require resetting. The Idle**

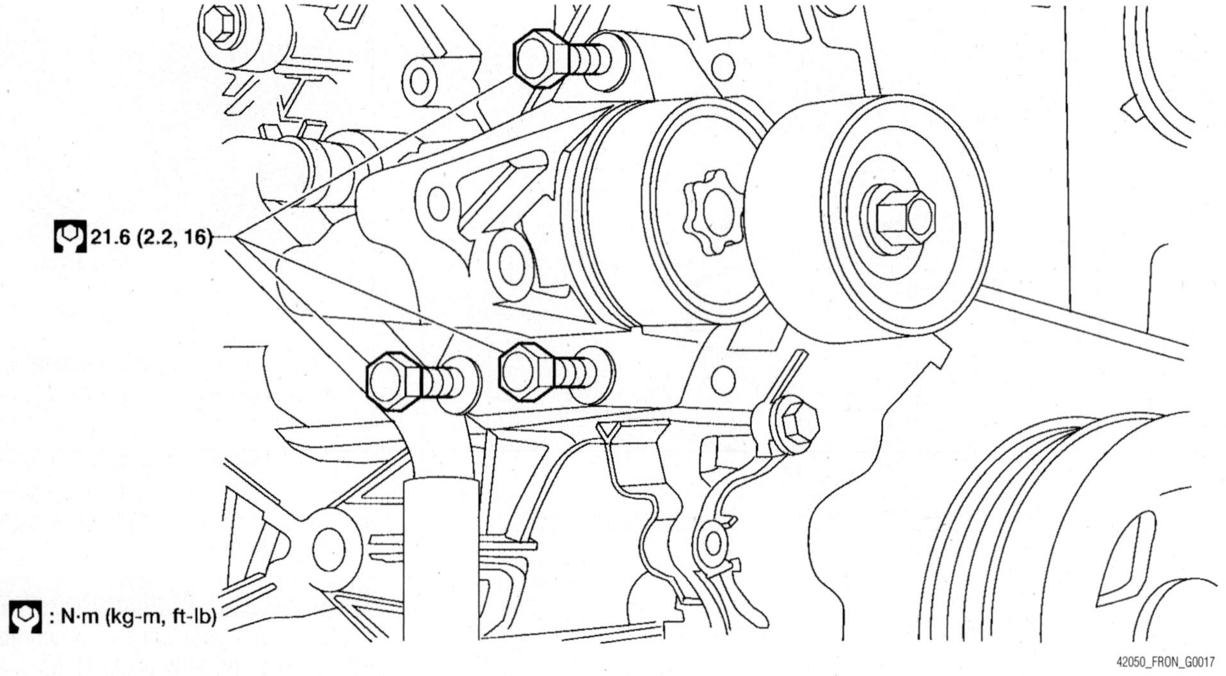

21.6 (2.2, 16)

: N·m (kg-m, ft-lb)

42050_FRON_G0017

Fig. 65 Drive belt auto tensioner and related components—2.5L engine

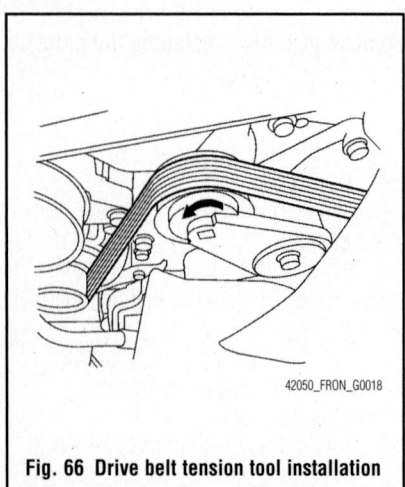

42050_FRON_G0018

Fig. 66 Drive belt tension tool installation and removal direction—4.0L engine

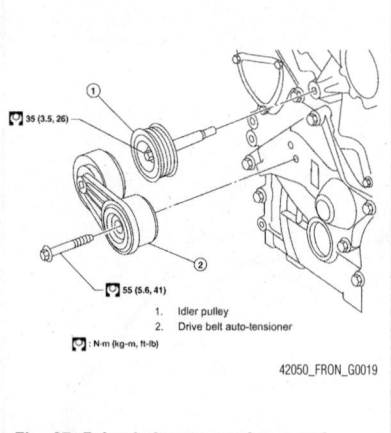

35 (3.5, 26)

55 (5.6, 41)

1. Idler pulley
2. Drive belt auto-tensioner

: N·m (kg-m, ft-lb)

42050_FRON_G0019

Fig. 67 Drive belt auto tensioner and related components—4.0L engine

Air Volume Learning, Steering Angle Sensor Neutral Position, Sunroof Memory Reset/Initialization, Automatic Drive Positioner System, Audio presets and Navigation. Use the Nissan CONSULT III diagnostic tool, or equivalent to perform the required resets.

1. Before servicing the vehicle, refer to the Precautions Section.
2. Disconnect the negative battery cable.
3. Remove the engine room cover.

4. Remove the air duct and resonator assembly.
5. Rotate the drive belt auto tensioner in a counterclockwise direction (loosening direction of tensioner).

※※ CAUTION

Avoid placing your hand in a location where pinching may occur if the tool accidentally comes off.

6. Remove the drive belt.

To install:

7. Installation is the reverse of the removal procedure.
8. Make sure the drive belt is securely installed around all the pulleys.
9. Ensure that there is no engine oil or engine coolant on the drive belt or pulley grooves.

Drive Belt Tensioner

2.5L Engine

➡**If working near and/or around the SRS system and components, be sure to disable the SRS system. After disabling the system wait three minutes or more before servicing the vehicle.**

➡**Whenever the negative battery cable is disconnected the following components will require resetting. The Idle Air Volume Learning, Steering Angle Sensor Neutral Position, Sunroof Memory Reset/Initialization, Automatic Drive Positioner System, Audio presets and Navigation. Use the Nissan CONSULT III diagnostic tool, or equivalent to perform the required resets.**

1. Before servicing the vehicle, refer to the Precautions Section.

2. Disconnect the negative battery cable.

3. Partially drain the engine coolant. Be sure to dispose of used coolant properly.

4. Remove the air cleaner assembly.

5. Remove the drive belt.

6. Disconnect the upper radiator hose at the radiator. Disconnect the reservoir hose at the radiator.

7. Remove the lower and upper radiator shrouds.

8. Remove the power steering pump and position it to the side.

9. Remove the alternator.

10. Remove the tensioner.

To install:

➡**Be sure to use new fasteners, as required.**

11. Installation is the reverse of the removal procedure.

4.0L Engine

➡**If working near and/or around the SRS system and components, be sure to disable the SRS system. After disabling the system wait three minutes or more before servicing the vehicle.**

➡**Whenever the negative battery cable is disconnected the following components will require resetting. The Idle Air Volume Learning, Steering Angle Sensor Neutral Position, Sunroof Memory Reset/Initialization, Automatic Drive Positioner System, Audio presets and Navigation. Use the Nissan CONSULT III diagnostic tool, or equivalent to perform the required resets.**

1. Before servicing the vehicle, refer to the Precautions Section.

2. Disconnect the negative battery cable.

3. Partially drain the engine coolant. Be sure to dispose of used coolant properly.

4. Remove the air cleaner assembly.

5. Remove the drive belt.

6. Remove the tensioner.

To install:

➡**Be sure to use new fasteners, as required.**

7. Installation is the reverse of the removal procedure.

AIR CLEANER

REMOVAL & INSTALLATION

Air Cleaner Assembly

2.5L Engine

See Figure 68.

➡**If working near and/or around the SRS system and components, be sure to disable the SRS system. After disabling the system wait three minutes or more before servicing the vehicle.**

➡**Whenever the negative battery cable is disconnected the following components will require resetting. The Idle Air Volume Learning, Steering Angle Sensor Neutral Position, Sunroof Memory Reset/Initialization, Automatic Drive Positioner System, Audio presets and Navigation. Use the Nissan CONSULT III diagnostic tool, or equivalent to perform the required resets.**

1. Before servicing the vehicle, refer to the Precautions Section.

2. Disconnect the negative battery cable.

➡**Add mating marks as necessary for easier installation.**

3. Remove the breather hose from the air duct.

4. Disconnect the Mass Air Flow (MAF) sensor.

5. Loosen the air duct clamps and remove the air duct.

6. Remove the air duct and resonator assembly bolts and remove the air duct and resonator assembly. Remove the resonator in the fender by lifting the left fender protector, as necessary.

7. Remove air cleaner case.

8. Remove the MAF sensor, if necessary.

To install:

9. Inspect the air duct and resonator assembly for cracks or tears.

10. Replace the air duct and resonator assembly, if necessary.

11. Installation is in the reverse order of removal.

12. Align the marks. Attach each joint. Tighten the clamps firmly.

13. Install the air duct and resonator assembly to the air cleaner case.

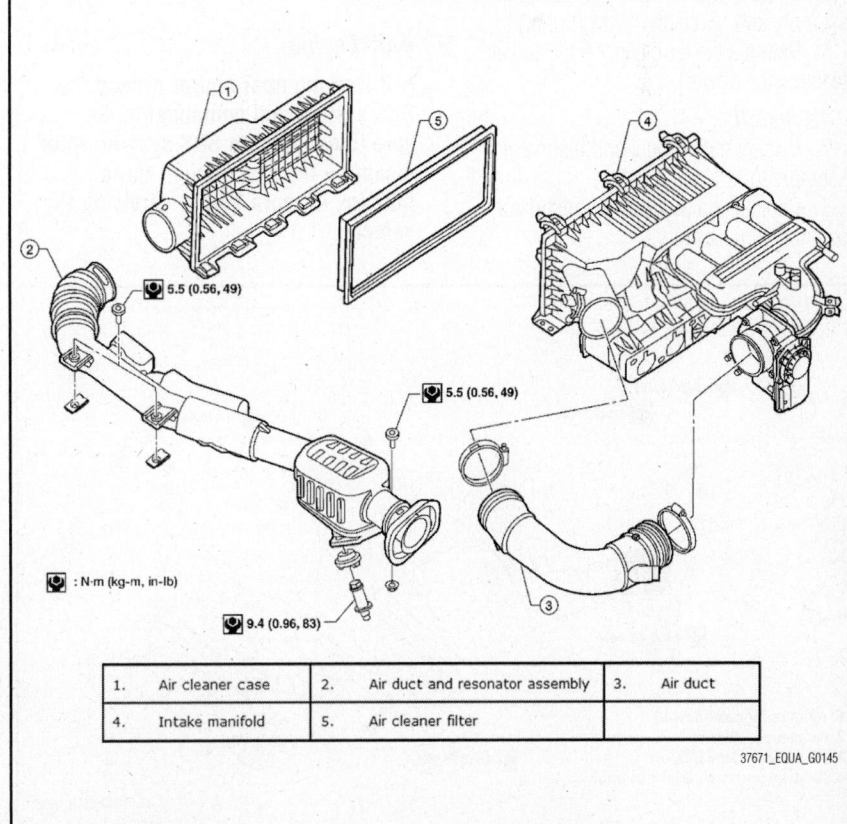

| 1. | Air cleaner case | 2. | Air duct and resonator assembly | 3. | Air duct |
| 4. | Intake manifold | 5. | Air cleaner filter | | |

37671_EQUA_G0145

Fig. 68 Air cleaner assembly and related components—2.5L engine

4.0L Engine

See Figure 69.

➡️If working near and/or around the SRS system and components, be sure to disable the SRS system. After disabling the system wait three minutes or more before servicing the vehicle.

➡️Whenever the negative battery cable is disconnected the following components will require resetting. The Idle Air Volume Learning, Steering Angle Sensor Neutral Position, Sunroof Memory Reset/Initialization, Automatic Drive Positioner System, Audio presets and Navigation. Use the Nissan CONSULT III diagnostic tool, or equivalent to perform the required resets.

1. Before servicing the vehicle, refer to the Precautions Section.
2. Disconnect the negative battery cable.

➡️Handle the Mass Air Flow (MAF) sensor with care. Do not shock, disassemble, or touch its sensor.

➡️Add marks on components, as necessary, for easier installation.

3. Remove the engine room cover.
4. Disconnect the harness connector from the air cleaner case (upper).
5. Remove the air duct and resonator assembly and air cleaner case (upper).
6. Remove the air cleaner filter and air cleaner case (lower).

To install:

7. Inspect the air duct and resonator assembly for cracks or tears.
8. Replace the air duct and resonator assembly, if necessary.

9. Installation is in the reverse order of removal.

Air Filter Element

2.5L Engine

See Figure 70.

➡️If working near and/or around the SRS system and components, be sure to disable the SRS system. After disabling the system wait three minutes or more before servicing the vehicle.

➡️Whenever the negative battery cable is disconnected the following components will require resetting. The Idle Air Volume Learning, Steering Angle Sensor Neutral Position, Sunroof Memory Reset/Initialization, Automatic Drive Positioner System, Audio presets and Navigation. Use the Nissan CONSULT III diagnostic tool, or equivalent to perform the required resets.

1. Before servicing the vehicle, refer to the Precautions Section.
2. Disconnect the negative battery cable.
3. Unfasten the clips and lift up the air cleaner case (upper).
4. Remove the air cleaner filter.
5. Installation is in the reverse order of removal.

4.0L Engine

➡️If working near and/or around the SRS system and components, be sure to disable the SRS system. After disabling the system wait three minutes or more before servicing the vehicle.

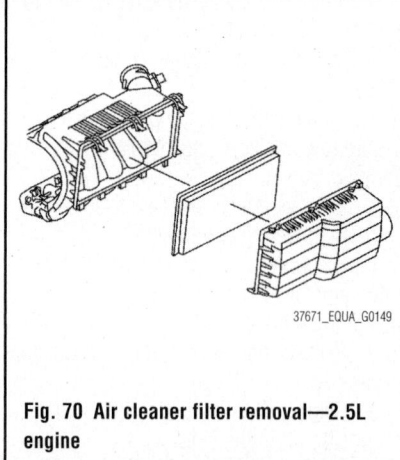

37671_EQUA_G0149

Fig. 70 Air cleaner filter removal—2.5L engine

➡️Whenever the negative battery cable is disconnected the following components will require resetting. The Idle Air Volume Learning, Steering Angle Sensor Neutral Position, Sunroof Memory Reset/Initialization, Automatic Drive Positioner System, Audio presets and Navigation. Use the Nissan CONSULT III diagnostic tool, or equivalent to perform the required resets.

1. Before servicing the vehicle, refer to the Precautions Section.
2. Disconnect the negative battery cable.
3. Unhook the clips, and lift the air cleaner case (upper).
4. Remove the air cleaner filter.
5. Installation is in the reverse order of removal.

CAMSHAFT & BEARINGS

REMOVAL & INSTALLATION

2.5L Engine

See Figures 71 through 81.

➡️The procedure below describes removal and installation of the camshaft without removing the front cover.

➡️If working near and/or around the SRS system and components, be sure to disable the SRS system. After disabling the system wait three minutes or more before servicing the vehicle.

➡️Whenever the negative battery cable is disconnected the following components will require resetting. The Idle Air Volume Learning, Steering Angle Sensor Neutral Position, Sunroof Memory Reset/Initialization, Automatic Drive Positioner System, Audio presets and Navigation. Use the Nissan

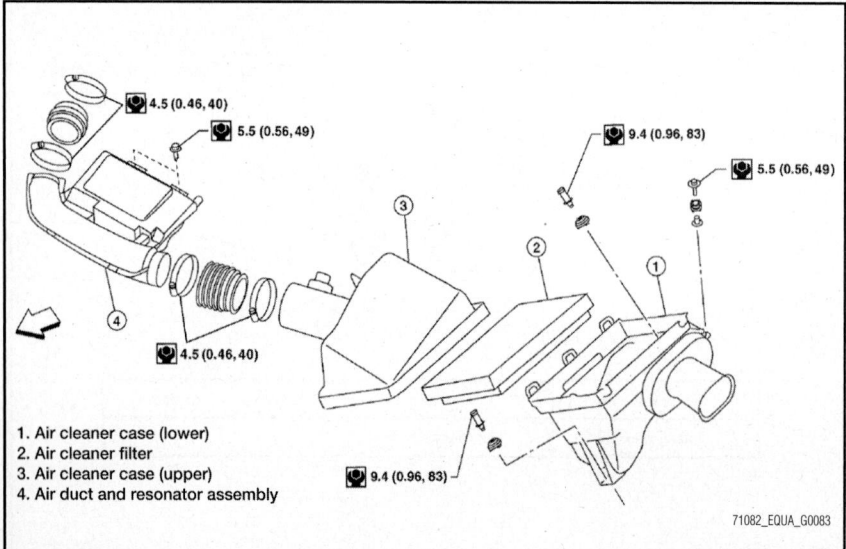

4.5 (0.46, 40)
5.5 (0.56, 49)
9.4 (0.96, 83)
5.5 (0.56, 49)
4.5 (0.46, 40)
9.4 (0.96, 83)

1. Air cleaner case (lower)
2. Air cleaner filter
3. Air cleaner case (upper)
4. Air duct and resonator assembly

71082_EQUA_G0083

Fig. 69 Air cleaner assembly and related components—4.0L engine

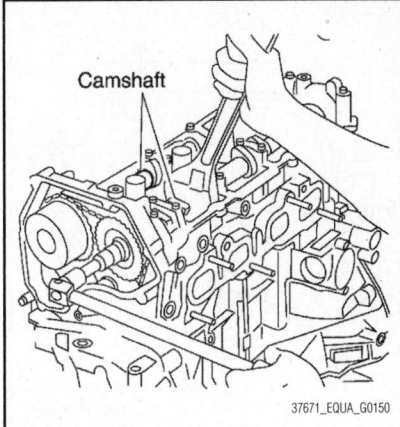

Fig. 71 Secure the camshaft with a suitable tool and loosen the camshaft sprocket bolts—2.5L engine

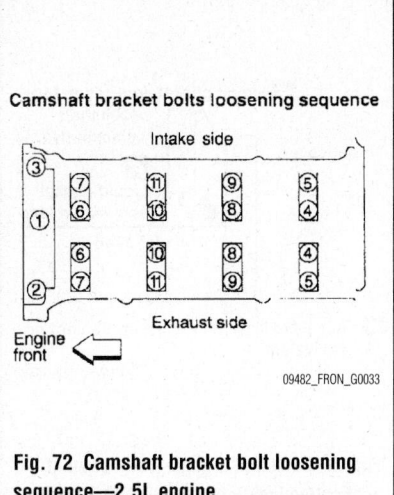

Fig. 72 Camshaft bracket bolt loosening sequence—2.5L engine

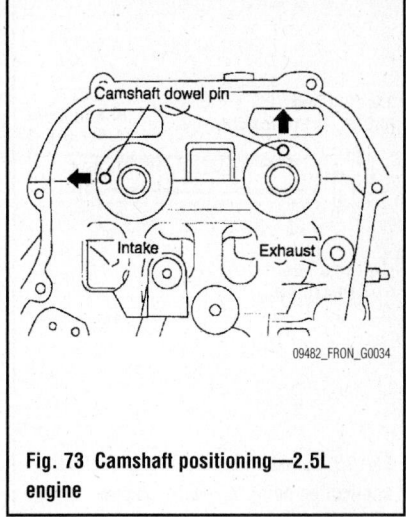

Fig. 73 Camshaft positioning—2.5L engine

CONSULT III diagnostic tool, or equivalent to perform the required resets.

1. Before servicing the vehicle, refer to the Precautions Section.
2. Disconnect the negative battery cable.
3. Properly relieve the fuel system pressure.
4. Remove the rocker cover.
5. Remove the drive belt.
6. Disconnect and remove the camshaft position sensor (PHASE).
7. Disconnect the Intake Valve Timing (IVT) control solenoid electrical connector.
8. Disconnect the ground electrical connectors from the front cover.
9. Remove the IVT control solenoid retaining bolts. Be sure to remove the bolts by reversing the order of the tightening torque sequence.
10. Remove the cover by cutting the sealant using tool KV10111100 (J-37228), or equivalent.
11. Position the number one cylinder on its compression stroke by rotating the crankshaft pulley clockwise. Align the mating marks for Top Dead Center (TDC) with the timing indicator on the front cover.
12. At the same time, make sure that the mating marks on the camshaft sprockets are lined up with the yellow links in the timing chain. If not, rotate the crankshaft one more turn to line up the mating marks to the yellow links.
13. Pull the timing chain guide out between the camshaft sprockets through the front cover.
14. Line up the mating marks on the camshaft sprockets with the yellow links in the timing chain and paint an indelible mating mark on the sprocket and timing chain link plate.

✳✳ WARNING

Do not rotate the crankshaft or the camshaft while the timing chain is removed. Interference between valve and piston may result.

➡**Maintaining chain tension is not necessary. The crankshaft sprocket and timing chain do not disconnect structurally while the front cover is attached.**

15. Push in the tensioner plunger and hold. Insert a stopper pin into the hole on the tensioner body to hold the chain tensioner. Remove the timing chain tensioner.

➡**Use a wire with 0.02 inch (0.5mm) diameter for a stopper pin.**

16. Secure the hexagonal part of the camshaft with a suitable tool. Loosen the camshaft sprocket bolts and remove the camshaft sprockets.
17. Loosen the camshaft bracket bolts. Be sure to remove the bolts by following the bolt removal sequence.
18. Remove the camshafts and brackets from the engine.
19. Remove the number one camshaft bracket by tapping lightly with a rubber mallet. Note the positions for installation.
20. As necessary, remove the valve lifters. Be sure to keep them in the proper order for installation.

To install:
21. Inspect the camshafts, replace as required.
22. Install the camshafts so that the camshaft dowel pins on the front side are positioned as indicated in the illustration.

23. Remove any foreign material from the camshaft bracket backside and from the cylinder head face.
24. Install the camshaft brackets (No. 2–No. 5) aligning the identification marks on the upper surface as indicated in the illustration.

➡**Install so that the identification mark can be correctly read when viewed from the exhaust side.**

25. To install camshaft bracket No. 1, apply liquid gasket to the bracket.

➡**After installation, be sure to wipe excessive gasket material from part "A", as indicated in the illustration. Be sure to use genuine RTV silicone sealant, or equivalent.**

26. Apply liquid gasket to camshaft bracket No. 1 contact surface on the front cover backside. Apply liquid gasket to the outside bolt hole on the front cover. Be sure

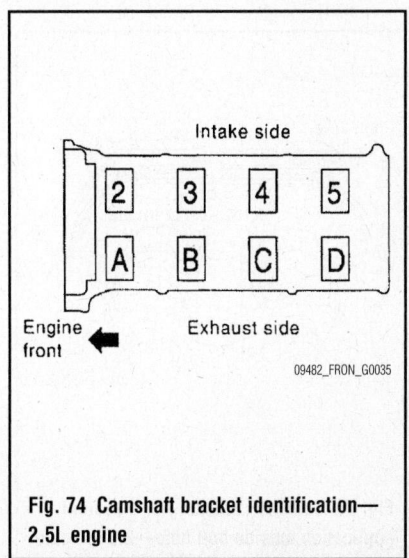

Fig. 74 Camshaft bracket identification—2.5L engine

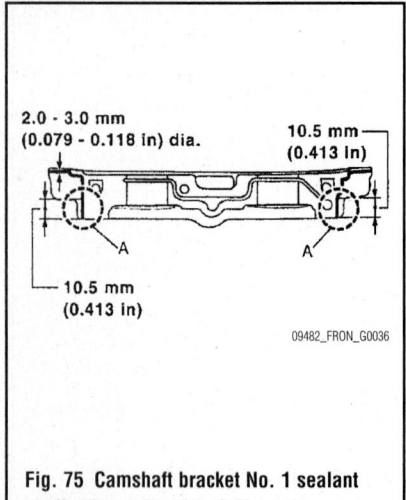

Fig. 75 Camshaft bracket No. 1 sealant application point "A"—2.5L engine

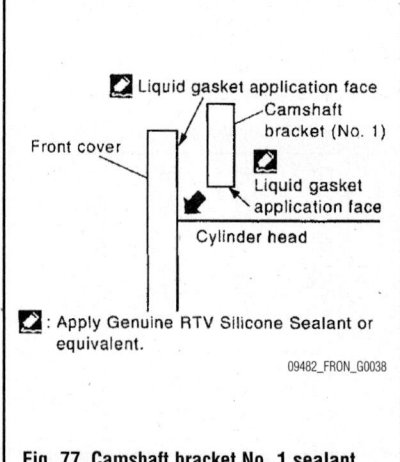

Fig. 77 Camshaft bracket No. 1 sealant application locating points—2.5L engine

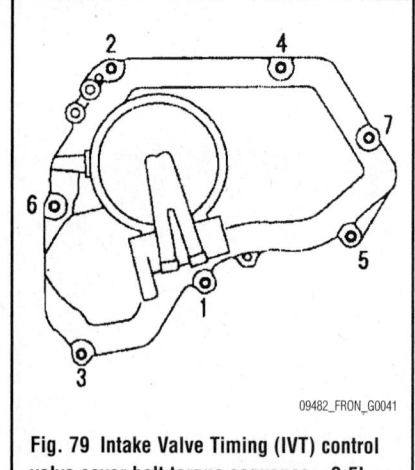

Fig. 79 Intake Valve Timing (IVT) control valve cover bolt torque sequence—2.5L engine

to use genuine RTV silicone sealant or equivalent.

27. Locate camshaft bracket No. 1 near installation position and install it without disturbing the liquid gasket applied to the surfaces. Be sure to use genuine RTV silicone sealant, or equivalent.

28. Tighten the camshaft bracket bolts in the proper sequence and to specification:

   a. Bolts 9–11 to 17 inch lbs. (2 Nm).
   b. Bolts 1–8 to 17 inch lbs. (2 Nm).
   c. Bolts 1–11 to 52 inch lbs. (6 Nm).
   d. Bolts 1–11 to 92 inch lbs.
   (10 Nm).

➡**After tightening the bolts be sure to wipe off any excessive liquid gasket. Be sure to use genuine RTV silicone sealant, or equivalent.**

29. Install the camshaft position sensor. Install the camshaft sprockets by aligning the mating marks on each camshaft sprocket with the paint marks on the timing

chain link plates, which were made during removal.

➡**Aligned mating marks could slip. Therefore, after matching them, hold the timing chain in place by hand. Before and after installing the chain tensioner, make sure that the mating marks have not slipped.**

30. Install the chain tensioner. After installation, pull the stopper pin off completely, and make sure that the chain tensioner plunger is released.

➡**Before installation of the chain tensioner, it is possible to rematch the marks on the timing chain with new ones on each sprocket.**

31. Install the chain guide. Install the oil rings to the camshaft sprocket (INT) insertion points on the backside of the IVT control cover. Install the O-ring to the front cover.

32. Apply a 0.083–0.122 inch (2.1–3.1mm) diameter bead of liquid gasket to the IVT control cover. Be sure to use genuine RTV silicone sealant, or equivalent.

33. Install the cover. Tighten the bolts in the proper sequence. Connect the ground cables and install the harness clip.

34. Check and adjust the valve clearance, as required.

➡**If hydraulic pressure inside the timing chain tensioner drops after removal/installation, slack in the guide may generate a pounding noise during and just after engine start. This is normal. The noise will stop after the hydraulic pressure rises.**

35. Continue the installation in the reverse order of the removal procedure.

36. Apply liquid gasket, be sure to use genuine RTV silicone sealant, or equivalent, to the positions shown in the illustration.

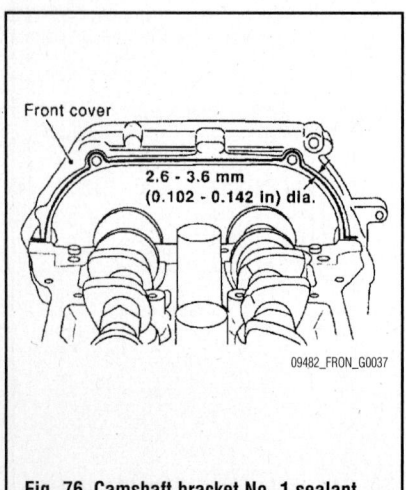

Fig. 76 Camshaft bracket No. 1 sealant application outside bolt hole—2.5L engine

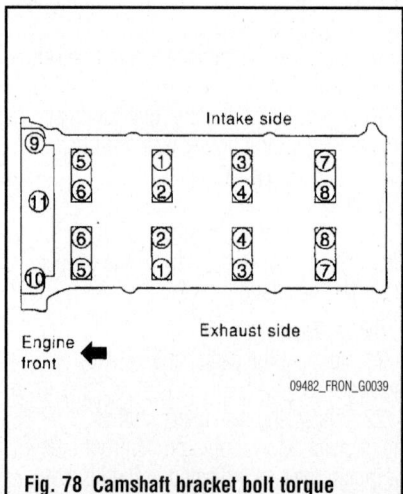

Fig. 78 Camshaft bracket bolt torque sequence—2.5L engine

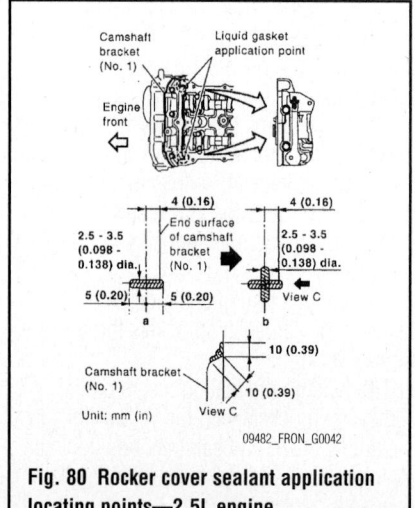

Fig. 80 Rocker cover sealant application locating points—2.5L engine

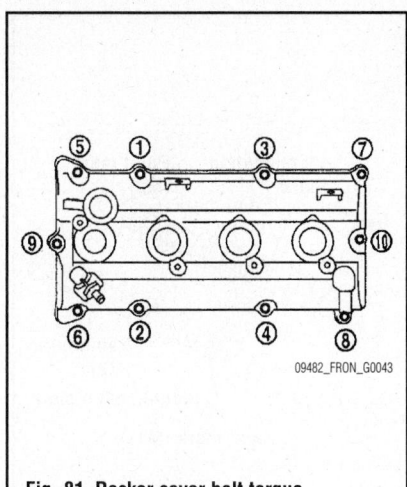

Fig. 81 Rocker cover bolt torque sequence—2.5L engine

37. Install the rocker cover. Torque the retaining bolts to 18 inch lbs. (2 Nm) and then to 73 inch lbs. (8 Nm), in the proper sequence.

38. Inspect the camshaft sprocket (INT) oil groove.

➡**Perform this inspection only when DTC P0011 or DTC P0021 are detected in self-diagnostic results.**

39. Be sure the engine is cold. Check and adjust the oil level, as required.

40. Properly release the fuel system pressure. Disconnect the ignition coil and injector harness connectors.

➡**This is being done to prevent the engine from unintentionally being started while checking.**

41. Remove the intake valve timing control solenoid valve.

42. Crank the engine, and then make sure that engine oil comes out from the camshaft bracket (No. 1) oil hole.

✳✳ CAUTION

Be careful not to touch rotating parts (drive belt, idler pulley, crankshaft pulley, etc.), as injury could result.

➡**Oil may squirt from the IVT control solenoid valve installation hole during engine cranking. Use a shop towel to prevent oil from squirting on engine components.**

43. Clean the oil groove between the oil strainer and the intake timing control solenoid valve if engine oil does not come out from the camshaft bracket (No. 1) oil hole.

44. Remove the components between the IVT control solenoid valve and the camshaft sprocket (INT). Check each oil groove for clogging.

45. After inspection, install any removed components.

4.0L Engine

See Figures 82 through 91.

➡**If working near and/or around the SRS system and components, be sure to disable the SRS system. After disabling the system wait three minutes or more before servicing the vehicle.**

➡**Whenever the negative battery cable is disconnected the following components will require resetting. The Idle Air Volume Learning, Steering Angle Sensor Neutral Position, Sunroof Memory Reset/Initialization, Automatic Drive Positioner System, Audio presets and Navigation. Use the Nissan CONSULT III diagnostic tool, or equivalent to perform the required resets.**

1. Before servicing the vehicle, refer to the Precautions Section.
2. Disconnect the negative battery cable.
3. Properly relieve the fuel system pressure.
4. Disconnect the negative battery cable.
5. Remove the front wheels and tires.
6. Remove the front fender protectors.
7. Remove the intake manifold collector.
8. Remove the front timing chain case, camshaft sprocket, timing chains, and rear timing chain case.
9. Remove the camshaft position sensor (PHASE) (right and left banks) from the cylinder head back side.

➡**Handle the camshaft position sensor carefully to avoid dropping and shocks.**

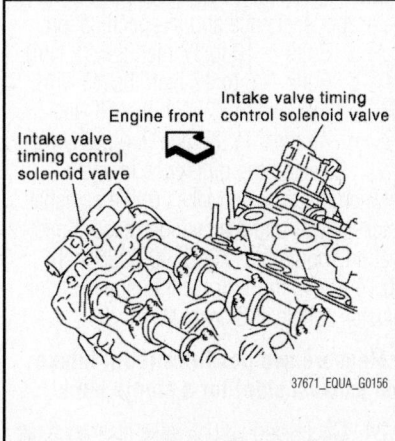

Fig. 82 Remove the Intake Valve Timing (IVT) control solenoid valves—4.0L engine

Do not disassemble. Do not place in a location where the sensor can be exposed to magnetism.

10. Remove the Intake Valve Timing (IVT) control solenoid valves. Discard IVT control solenoid valve gaskets and use new gaskets for installation.

11. Remove the camshaft brackets. Equally loosen the camshaft bracket bolts in several steps in reverse order of the torque sequence.

➡**Mark the camshafts, camshaft brackets, and bolts so they are placed in the same position and direction for installation.**

12. Remove the camshafts.
13. Remove the valve lifters, if necessary. Identify the installation positions and store them.
14. Remove the timing chain tensioner (secondary) from the cylinder head. Remove the timing chain tensioner (secondary) with its stopper pin attached.

➡**The stopper pin was attached when the timing chain (secondary) was removed.**

To install:

15. Inspect the camshafts, replace as required.
16. Install the timing chain tensioners (secondary) on both sides of the cylinder head. Be sure to use new O-rings.

➡**Install the tensioner with its stopper pin attached. Install the tensioner with the sliding part facing downward on the right cylinder head and with the sliding part facing upward on the left cylinder head.**

17. Install the valve lifters, in their original bores.
18. Install the camshafts, with the dowel pin attached to its front end face on the exhaust side.

➡**Follow the identification marks for proper placement and direction.**

19. Install the camshaft so that the dowel pin hole and dowel pin on the front end face are positioned as shown in the illustration (No. 1 piston at TDC on its compression stroke).

➡**Large and small pin holes are located on the front end face of the camshaft (INT), at intervals of 180 degrees. Face the small diameter side pin hole upward (in the cylinder head upper face direction).**

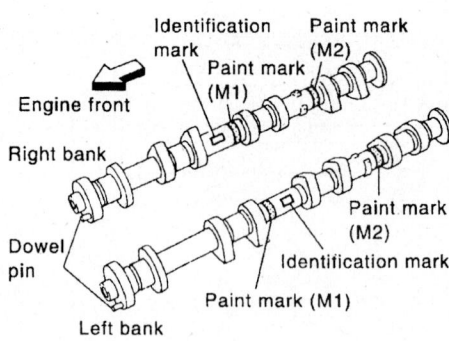

Bank	INT/EXH	Dowel pin	Paint marks		Identification mark
			M1	M2	
RH	INT	No	Green	No	RE
	EXH	Yes	No	White	RE
LH	INT	No	Green	No	LH
	EXH	Yes	No	White	LH

09482_FRON_G0044

Fig. 83 Camshaft identification—4.0L engine

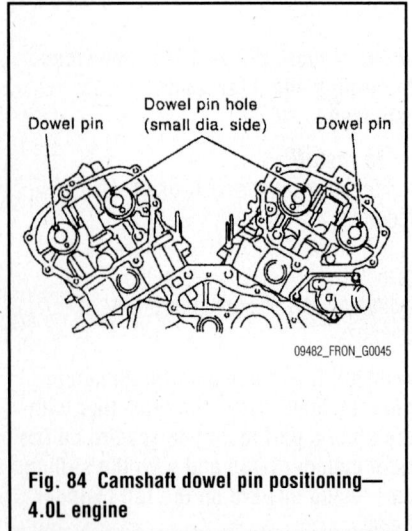

09482_FRON_G0045

Fig. 84 Camshaft dowel pin positioning—4.0L engine

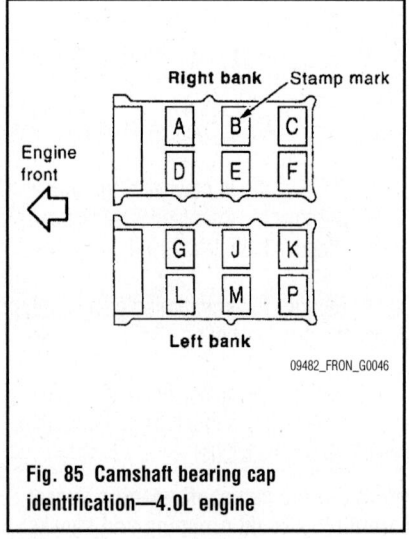

09482_FRON_G0046

Fig. 85 Camshaft bearing cap identification—4.0L engine

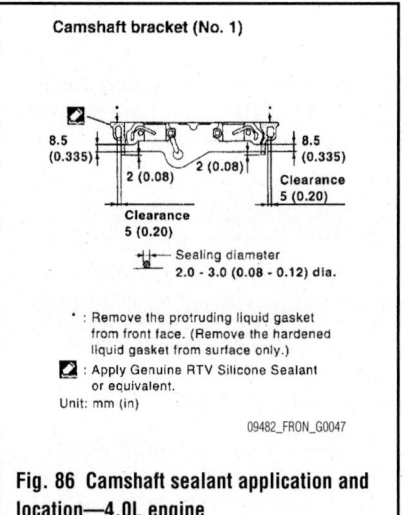

09482_FRON_G0047

Fig. 86 Camshaft sealant application and location—4.0L engine

➡**Though the camshaft does not stop at the portion as shown, for placement of the cam nose, it is generally accepted that the camshaft is placed for the same direction as shown.**

20. Install the camshaft brackets in the same position that they were removed. Install brackets No. 2–No. 4 aligning the stamp marks as indicated in the illustration.

➡**There are no identification marks indicating left or right for camshaft bracket No. 1.**

21. Apply liquid gasket to the mating surfaces of camshaft bracket No. 1 as shown in the illustration on both the left and right cylinder heads. Be sure to use genuine RTV sealant, or equivalent.

22. Tighten the camshaft bracket bolts in the proper sequence and to specification:
    a. Bolts 7–10 to 17 inch lbs. (2 Nm).
    b. Bolts 1–6 to 17 inch lbs. (2 Nm).
    c. All bolts to 52 inch lbs. (6 Nm).
    d. All bolts to 92 inch lbs. (10 Nm).
23. Measure the difference in levels between the front end faces of the camshaft bracket No. 1 and the cylinder head. Specification should be -0.0055–0.0055 inch (-0.14–0.14mm). If not within specification, reinstall camshaft bracket No. 1.

➡**Measure two positions (both intake and exhaust side) for a single bank.**

24. Check and adjust valve clearance, as required.
25. Apply liquid gasket, be sure to use genuine RTV silicone sealant, or equivalent, to the positions shown in the illustration.

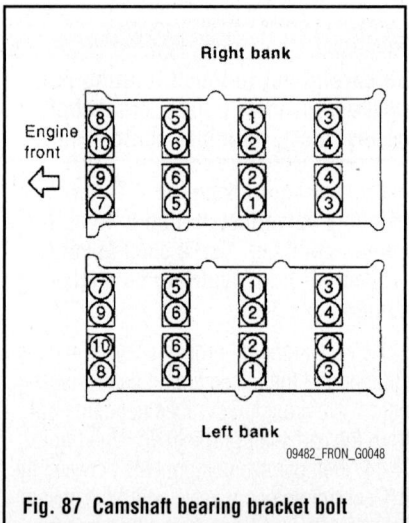

09482_FRON_G0048

Fig. 87 Camshaft bearing bracket bolt torque sequence—4.0L engine

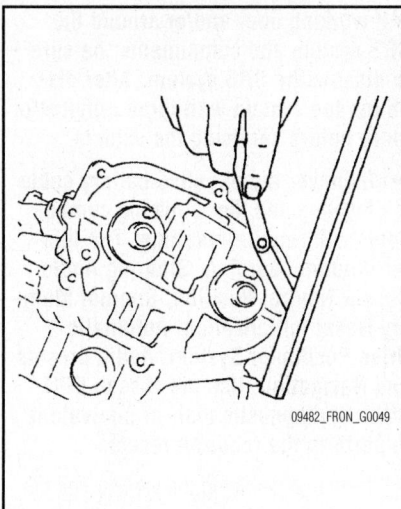

Fig. 88 Camshaft bracket and cylinder head measurement—4.0L engine

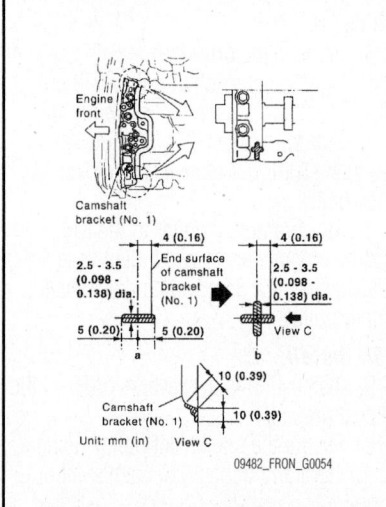

Fig. 89 Rocker cover sealant application locating points—4.0L engine

26. Install the rocker cover.

27. Continue the installation in the reverse order of the removal procedure.

28. Inspect the camshaft sprocket (INT) oil groove.

➡**Perform this inspection only when DTC P0011 or DTC P0021 is detected in self-diagnostic results.**

29. Be sure the engine is cold. Check and adjust oil level, as required.

30. Properly release the fuel system pressure. Disconnect the ignition coil and injector harness connectors.

➡**This is being done to prevent the engine from unintentionally being started while checking.**

31. Remove the intake valve timing control solenoid valve.

32. Crank the engine, and then make sure that engine oil comes out from the camshaft bracket (No. 1) oil hole.

✷✷ CAUTION

Be careful not to touch rotating parts, (drive belt, idler pulley, crankshaft pulley, etc.) as injury could result.

➡**Oil may squirt from the intake valve timing control solenoid valve installation hole during engine cranking. Use a shop towel to prevent oil from squirting on engine components.**

33. Clean the oil groove between the oil strainer and the intake timing control solenoid valve if engine oil does not come out from the camshaft bracket (No. 1) oil hole.

34. Remove the components between the intake valve timing control solenoid valve

and the camshaft sprocket (INT). Check each oil groove for clogging.

35. After inspection, install any removed components.

CRANKSHAFT FRONT SEAL

REMOVAL & INSTALLATION

2.5L Engine
See Figure 92.

➡**If working near and/or around the SRS system and components, be sure to disable the SRS system. After disabling the system wait three minutes or more before servicing the vehicle.**

➡**Whenever the negative battery cable is disconnected the following components will require resetting. The Idle Air Volume Learning, Steering Angle Sensor Neutral Position, Sunroof Memory Reset/Initialization, Automatic Drive Positioner System, Audio presets and Navigation. Use the Nissan CONSULT III diagnostic tool, or equivalent to perform the required resets.**

1. Before servicing the vehicle, refer to the Precautions Section.

2. Disconnect the negative battery cable.

3. Remove the engine undercover.

4. Remove the fan shroud. Remove the cooling fan.

5. Remove the drive belt.

6. Remove the crankshaft pulley.

7. Using a seal removal tool, remove the oil seal from its mounting.

➡**Be careful not to damage the front cover and/or the crankshaft.**

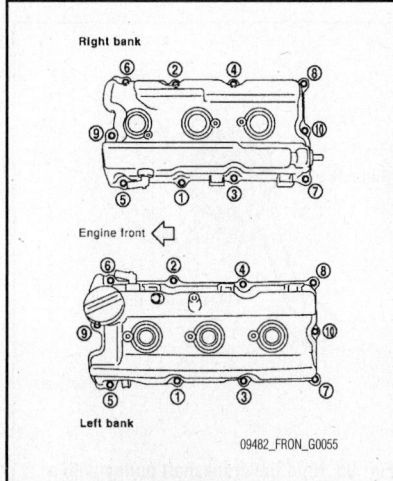

Fig. 90 Rocker cover bolt torque sequence—4.0L engine

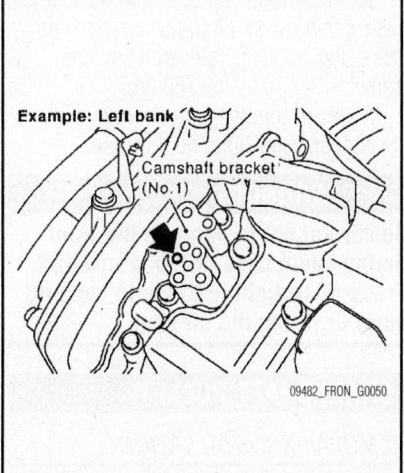

Fig. 91 Camshaft bracket (No. 1) oil hole location

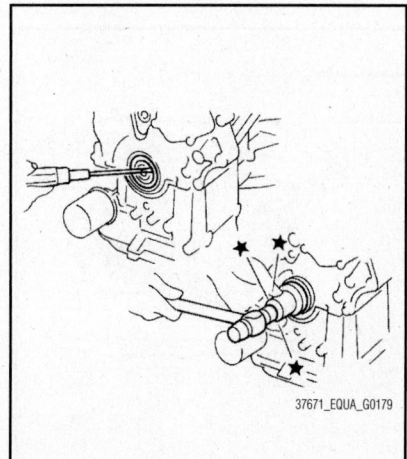

Fig. 92 View of removal and installation of the crankshaft front seal—2.5L engine

To install:

8. Installation is the reverse order of the removal procedure.

9. Press fit the seal until it is flush with the front end surface of the front cover, using the proper tools.

4.0L Engine

See Figures 93 through 95.

➡If working near and/or around the SRS system and components, be sure to disable the SRS system. After disabling the system wait three minutes or more before servicing the vehicle.

➡Whenever the negative battery cable is disconnected the following components will require resetting. The Idle Air Volume Learning, Steering Angle Sensor Neutral Position, Sunroof Memory Reset/Initialization, Automatic Drive Positioner System, Audio presets and Navigation. Use the Nissan CONSULT III diagnostic tool, or equivalent to perform the required resets.

1. Before servicing the vehicle, refer to the Precautions Section.
2. Disconnect the negative battery cable.
3. Remove the engine undercover.
4. Remove the drive belts.
5. Remove the cooling fan.
6. Remove the crankshaft pulley.
7. Using a seal removal tool, remove the oil seal from its mounting.

➡Be careful not to damage the front cover and/or the crankshaft.

To install:

8. Installation is the reverse order of the removal procedure.

9. Install a new front oil seal on the front timing chain case.

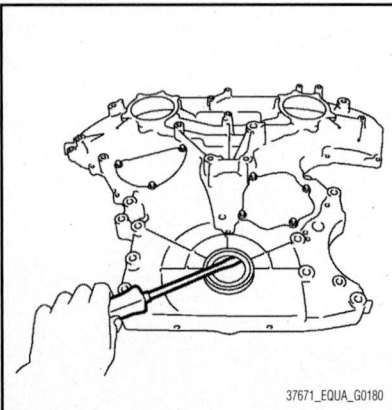

Fig. 93 View of crankshaft front seal removal—4.0L engine

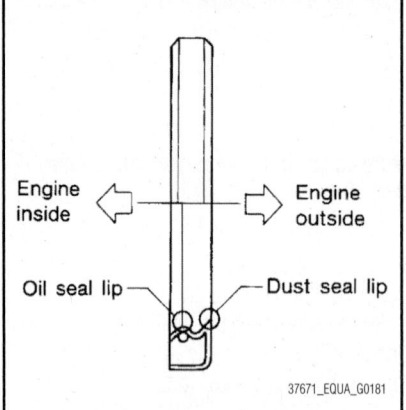

Fig. 94 Install the front oil seal so that each seal lip is oriented as shown—4.0L engine

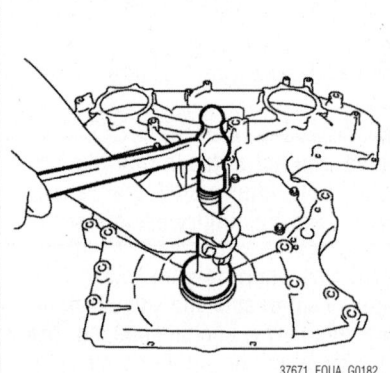

Fig. 95 Press-fit the oil seal until it becomes flush with the front timing chain case end face using a suitable drift—4.0L engine

10. Apply new engine oil to both the oil seal lip and the dust seal lip.

11. Install the front oil seal so that each seal lip is oriented properly.

12. Press-fit the oil seal until it becomes flush with the front timing chain case end face using a suitable drift with an outer diameter of 2.36 inches (60mm).

13. Make sure the garter spring is in position and seal lip is not inverted.

✳✳ WARNING

Be careful not to damage the front timing chain case or the crankshaft. Press-fit straight in to avoid causing burrs or tilting the oil seal.

CRANKSHAFT PULLEY

REMOVAL & INSTALLATION

2.5L Engine

See Figures 96 and 97.

➡If working near and/or around the SRS system and components, be sure to disable the SRS system. After disabling the system wait three minutes or more before servicing the vehicle.

➡Whenever the negative battery cable is disconnected the following components will require resetting. The Idle Air Volume Learning, Steering Angle Sensor Neutral Position, Sunroof Memory Reset/Initialization, Automatic Drive Positioner System, Audio presets and Navigation. Use the Nissan CONSULT III diagnostic tool, or equivalent to perform the required resets.

1. Before servicing the vehicle, refer to the Precautions Section.
2. Disconnect the negative battery cable.
3. Remove the engine undercover.
4. Remove the fan shroud. Remove the cooling fan.
5. Remove the drive belt.
6. Hold the crankshaft pulley with a suitable tool and loosen the retaining bolt.
7. Pull the crankshaft pulley out about 0.39 inch (10mm). Remove the crankshaft pulley bolt.
8. Attach a pulley puller in the M6 (0.24 inch diameter) thread hole on the crankshaft pulley. Remove the crankshaft damper pulley.

To install:

9. Installation is the reverse order of the removal procedure.

10. Secure the crankshaft pulley using a suitable tool and tighten the crankshaft pulley bolt.

   a. Apply new engine oil to the thread and seat surfaces of the crankshaft pulley bolt.

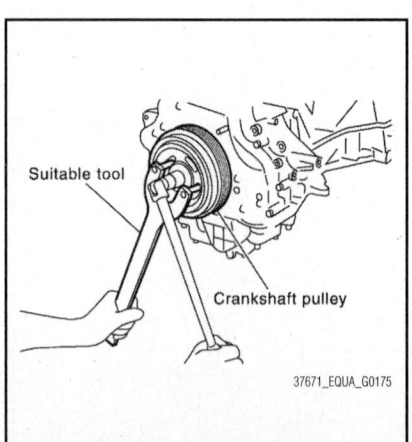

Fig. 96 Hold the crankshaft pulley with a suitable tool and loosen the crankshaft pulley retaining bolt—2.5L engine

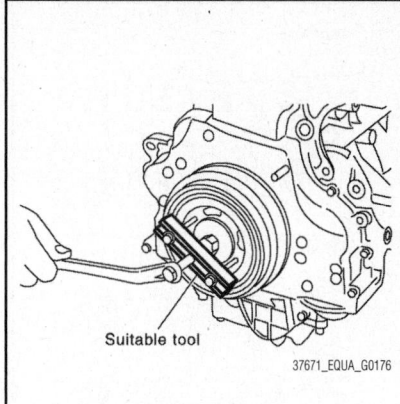

Fig. 97 Attach a pulley puller in the M6 thread hole on the crankshaft pulley for removal—2.5L engine

b. Tighten the crankshaft pulley bolt to 31 ft. lbs. (42 Nm).

c. Put a paint mark on crankshaft pulley, mating with any one of six easy to recognize angle marks on the bolt flange.

d. Turn another 60° clockwise (angle tightening).

4.0L Engine

See Figures 98 and 99.

➡If working near and/or around the SRS system and components, be sure to disable the SRS system. After disabling the system wait three minutes or more before servicing the vehicle.

➡Whenever the negative battery cable is disconnected the following components will require resetting. The Idle Air Volume Learning, Steering Angle Sensor Neutral Position, Sunroof Memory Reset/Initialization, Automatic Drive Positioner System, Audio presets and Navigation. Use the Nissan CONSULT III diagnostic tool, or equivalent to perform the required resets.

1. Before servicing the vehicle, refer to the Precautions Section.
2. Disconnect the negative battery cable.
3. Remove the engine undercover.
4. Remove the drive belts.
5. Remove the cooling fan.
6. Remove the starter motor.
7. Install and set the holding special tool J-48761, or equivalent.
8. Loosen the crankshaft pulley retaining bolt and locate the bolt seating surface, which is about 0.39 inch (10mm) from its original position.

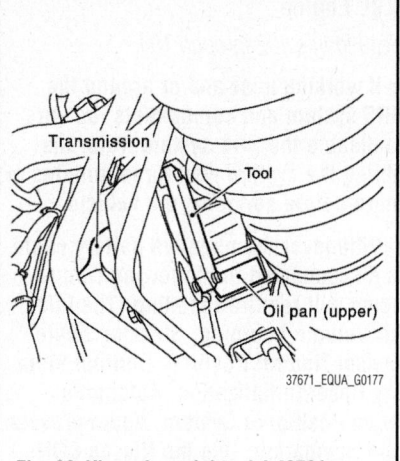

Fig. 98 View of special tool J-48761 installed—4.0L engine

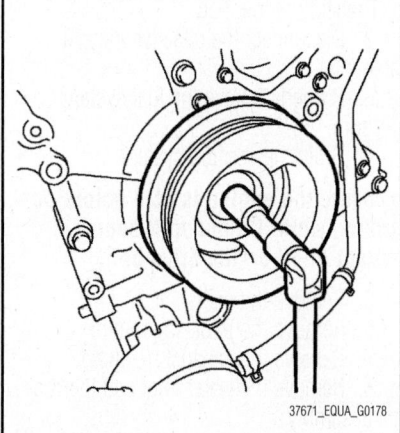

Fig. 99 Removing the crankshaft pulley bolt—4.0L engine

➡Do not remove the crankshaft pulley bolt. Keep the loosened pulley bolt in place to protect the removed crankshaft pulley from dropping.

9. Pull the pulley with both hands and remove it from its mounting. Remove the bolt and pulley from the engine.

To install:

10. Installation is the reverse order of the removal procedure.

11. Install the crankshaft pulley, taking care not to damage the front oil seal.

12. When press-fitting the crankshaft pulley with a plastic hammer, tap on its center portion (not the circumference).

13. Tighten the crankshaft pulley bolt in two steps:

a. Step 1: 33 ft. lbs. (44 Nm).
b. Step 2: 84–90° clockwise

14. Rotate the crankshaft pulley in the normal direction (clockwise when viewed from front) to confirm it turns smoothly.

CYLINDER HEAD

REMOVAL & INSTALLATION

2.5L Engine

See Figures 100 and 101.

➡If working near and/or around the SRS system and components, be sure to disable the SRS system. After disabling the system wait three minutes or more before servicing the vehicle.

➡Whenever the negative battery cable is disconnected the following components will require resetting. The Idle Air Volume Learning, Steering Angle Sensor Neutral Position, Sunroof Memory Reset/Initialization, Automatic Drive Positioner System, Audio presets and Navigation. Use the Nissan CONSULT III diagnostic tool, or equivalent to perform the required resets.

1. Before servicing the vehicle, refer to the Precautions Section.
2. Disconnect the negative battery cable.
3. Properly relieve the fuel system pressure.
4. Drain the cooling system.

➡Ensure the engine is cold before performing work. Do not spill engine coolant or oil on the drive belt.

5. Drain the engine oil.
6. Remove the intake manifold and fuel tube assembly.
7. Remove the fuel injector and fuel tube assembly.
8. Remove the exhaust manifold and the three way catalyst.
9. Remove the water outlet. Remove the heater outlet.
10. Remove the front cover and the timing chain.

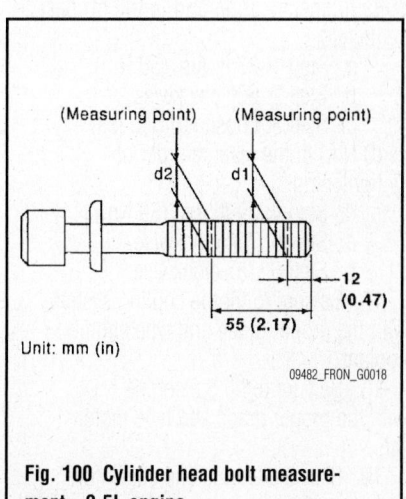

Fig. 100 Cylinder head bolt measurement—2.5L engine

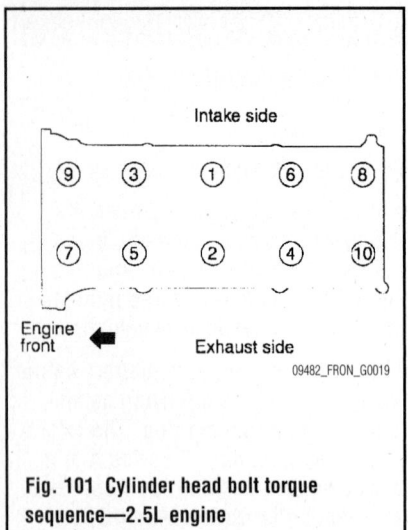

Fig. 101 Cylinder head bolt torque sequence—2.5L engine

11. Remove the camshafts.

12. Remove the cylinder head retaining bolts. Be sure to remove the bolts by reversing the order of the tightening torque sequence.

13. Remove the cylinder head from the engine. Discard the gasket.

To install:

14. Installation is the reverse of the removal procedure.

15. Be sure to inspect the cylinder head bolts. Replace as required.
- Cylinder head bolts are tightened by plastic zone tightening method. Whenever the size difference between (d1) and (d2) exceeds the limit, replace the bolt with a new one
- Limit difference: 0.0091 inch (0.23mm)

16. Install the new cylinder head gasket. Apply engine oil to the cylinder head bolt threads and seating surfaces. Torque the cylinder head bolts to specification and in the proper sequence.
- a. Step 1: 37 ft. lbs. (50 Nm).
- b. Step 2: 60° clockwise.
- c. Step 3: Loosen to 0 ft. lbs. (0 Nm) in the reverse order of tightening.
- d. Step 4: 29 ft. lbs. (39 Nm).
- e. Step 5: 75° clockwise.
- f. Step 6: 75° clockwise.

17. Be sure to fill the cooling system with the proper grade and type engine coolant.

18. Be sure to fill the engine with the proper grade and type motor oil.

19. Start the engine and check for leaks.

4.0L Engine

See Figures 102 through 104.

➡If working near and/or around the SRS system and components, be sure to disable the SRS system. After disabling the system wait three minutes or more before servicing the vehicle.

➡Whenever the negative battery cable is disconnected the following components will require resetting. The Idle Air Volume Learning, Steering Angle Sensor Neutral Position, Sunroof Memory Reset/Initialization, Automatic Drive Positioner System, Audio presets and Navigation. Use the Nissan CONSULT III diagnostic tool, or equivalent to perform the required resets.

1. Before servicing the vehicle, refer to the Precautions Section.

2. Disconnect the negative battery cable.

3. Properly relieve the fuel system pressure.

4. Drain the cooling system.

➡Ensure the engine is cold before performing work. Do not spill engine coolant or oil on the drive belt.

5. Remove the camshaft.

6. Remove the intake manifold.

7. Remove the exhaust manifold.

8. Remove the water inlet and thermostat assembly.

9. Remove the water outlet, water pipe and heater pipe.

10. Remove the cylinder head retaining bolts. Be sure to remove the bolts by reversing the order of the tightening torque sequence.

11. Remove the cylinder head from the engine. Discard the gasket.

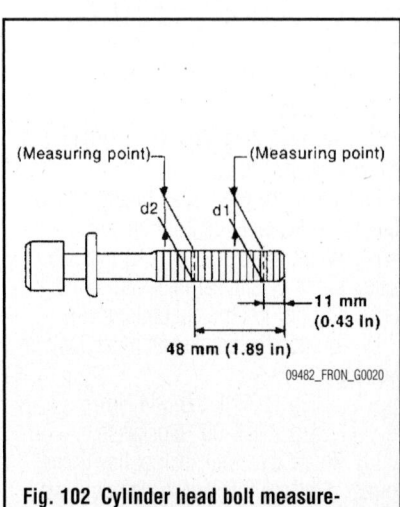

Fig. 102 Cylinder head bolt measurement—4.0L engine

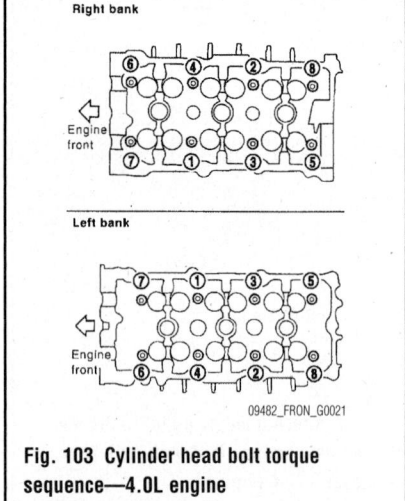

Fig. 103 Cylinder head bolt torque sequence—4.0L engine

To install:

12. Installation is the reverse of the removal procedure.

13. Be sure to inspect the cylinder head bolts. Replace as required.
- Cylinder head bolts are tightened by plastic zone tightening method. Whenever the size difference between (d1) and (d2) exceeds the limit, replace the bolt with a new one
- Limit difference: 0.0043 inch (0.11mm)

14. Install the new cylinder head gasket. Turn the crankshaft until the number one piston is at TDC.

➡The crankshaft key should line up with the right bank center line, see illustration.

15. Torque the cylinder head bolts to specification and in the proper sequence.
- a. Step 1: 72 ft. lbs. (98 Nm).

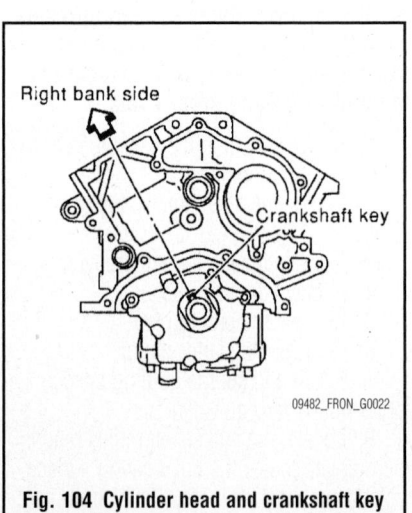

Fig. 104 Cylinder head and crankshaft key alignment—4.0L engine

b. Step 2: Loosen to 0 ft. lbs. (0 Nm) in the reverse order of tightening.

c. Step 3: 29 ft. lbs. (39 Nm).

d. Step 4: 90° clockwise.

e. Step 5: 90° clockwise.

16. Measure the distance between the front end faces of the cylinder block and the cylinder head on both the left and right banks. If the measured value is not within specification reinstall the cylinder head. Specification is 0.555–0.587 inch (14.1–14.9mm).

17. Be sure to fill the cooling system with the proper grade and type engine coolant.

18. Start the engine and check for leaks.

➡ **If the hydraulic pressure inside the timing chain tensioner drops after removal and installation, slack in the guide may generate a pounding noise during and just after engine start. This is normal. The noise will stop after the hydraulic pressure rises.**

ENGINE COVER

REMOVAL & INSTALLATION

4.0L Engine

See Figure 105.

➡ **If working near and/or around the SRS system and components, be sure to disable the SRS system. After disabling the system wait three minutes or more before servicing the vehicle.**

➡ **Whenever the negative battery cable is disconnected the following components will require resetting. The Idle Air Volume Learning, Steering Angle Sensor Neutral Position, Sunroof Mem-**ory Reset/Initialization, Automatic Drive Positioner System, Audio presets and Navigation. Use the Nissan CONSULT III diagnostic tool, or equivalent to perform the required resets.

1. Before servicing the vehicle, refer to the Precautions Section.

2. Disconnect the negative battery cable.

3. Remove the cover retaining bolts.

4. Lift up on the cover to disengage the snap fit mounts.

To install:

➡ **Be sure to use new fasteners, as required.**

5. Installation is the reverse of the removal procedure.

ENGINE OIL & FILTER

OIL LEVEL CHECK

See Figure 106.

❋❋ CAUTION

Prolonged and repeated contact with used engine oil may cause skin cancer. Try to avoid direct skin contact with used oil. If skin contact is made, wash thoroughly with soap or hand cleaner as soon as possible. Wear protective clothing, including impervious gloves where practicable. Do not use gasoline, kerosene, diesel fuel, gas oil, thinners, or solvents for cleaning skin. Where there is a risk of eye contact, eye protection should be worn, for example, chemical goggles or face shields; in addition an eye wash facility should be provided.

➡ **Be sure to check the engine oil with the vehicle engine OFF and parked on a level surface. If the engine was running prior to checking the oil, turn it off and wait 10 minutes.**

1. Before servicing the vehicle, refer to the Precautions Section.

2. Remove the oil level gauge and wipe it clean with a shop towel. Be sure to properly dispose of the used shop towel.

3. Insert the gauge into its mounting.

4. Remove the oil level gauge and check the reading. See illustration.

5. Correct, oil level as required. Be sure to use the proper grade and type engine oil.

OIL & FILTER CHANGE

2.5L Engine

See Figure 107.

❋❋ CAUTION

Prolonged and repeated contact with used engine oil may cause skin cancer. Try to avoid direct skin contact with used oil. If skin contact is made, wash thoroughly with soap or hand cleaner as soon as possible. Wear protective clothing, including impervious gloves where practicable. Do not use gasoline, kerosene, diesel fuel, gas oil, thinners, or solvents for cleaning skin. Where there is a risk of eye contact, eye protection should be worn, for example, chemical goggles or face shields; in addition an eye wash facility should be provided.

➡ **If working near and/or around the SRS system and components, be sure to disable the SRS system. After dis-**

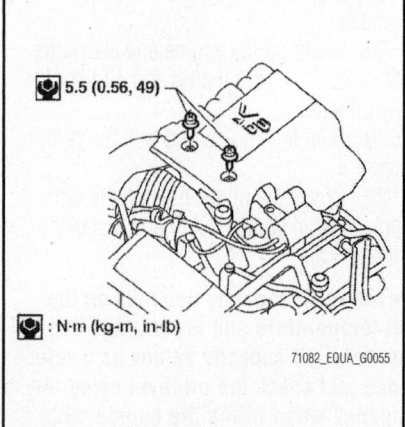

Fig. 105 Engine cover and related components—4.0L engine

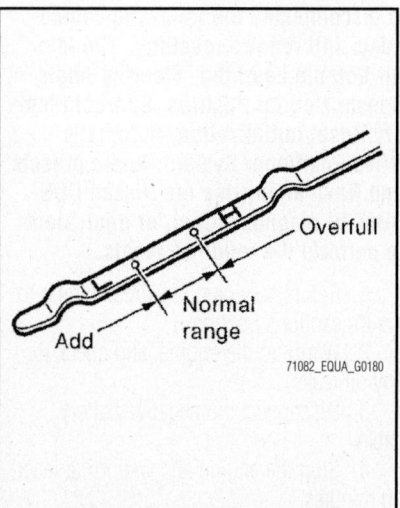

Fig. 106 Engine oil level gauge reading

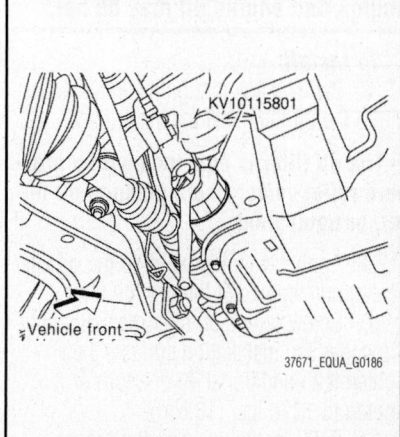

Fig. 107 Removing the oil filter using special tool KV10115801 (J-38956)—2.5L engine

abling the system wait three minutes or more before servicing the vehicle.

➡Whenever the negative battery cable is disconnected the following components will require resetting. The Idle Air Volume Learning, Steering Angle Sensor Neutral Position, Sunroof Memory Reset/Initialization, Automatic Drive Positioner System, Audio presets and Navigation. Use the Nissan CONSULT III diagnostic tool, or equivalent to perform the required resets.

1. Before servicing the vehicle, refer to the Precautions Section.
2. Warm up the engine and check for any oil leaks.
3. Disconnect the negative battery cable.
4. Stop the engine and wait for at least 10 minutes.
5. Remove the oil drain plug and oil filler cap to drain the old oil. Discard the washer.
6. Install a new washer on the oil drain plug, then install the oil drain plug in the oil pan. Tighten the drain plug to 25 ft. lbs. (34 Nm).

➡Clean the drain plug before installing a new washer.

7. Remove the oil filter using special tool KV10115801 (J-38956), or equivalent.
- When removing the oil filter, position a shop cloth to absorb any engine oil leaks or spills
- Do not allow engine oil to adhere to the drive belts
- Completely wipe off any engine oil that adheres to the engine and the vehicle

❊❊ CAUTION

Be careful not to burn yourself, as the engine and engine oil may be hot.

To install:
8. Remove any foreign materials adhering to the oil filter seal mating surface.

➡The oil filter is equipped with a pressure relief valve. Use a genuine oil filter, or equivalent.

9. Apply clean engine oil to the oil filter seal circumference of the new oil filter.
10. Screw on the oil filter manually until it touches the installation surface, then tighten it an additional ⅔ of a turn or tighten to 13 ft. lbs. (18 Nm).
11. Refill the engine with the proper grade and amount of new engine oil.
12. Inspect the engine for oil leaks.
13. Check the engine oil level.

14. Warm up the engine and check the area around the drain plug and oil filter for any oil leaks.
15. Stop the engine and wait for 10 minutes.
16. Check the oil level using the dipstick. Add oil as necessary and install the oil filler cap.

➡The refill capacity depends on the oil temperature and drain time. Use the refill oil capacity values as a reference and check the oil level using the dipstick when filling the engine with oil.

➡Do not overfill the engine with oil.

4.0L Engine

❊❊ CAUTION

Prolonged and repeated contact with used engine oil may cause skin cancer. Try to avoid direct skin contact with used oil. If skin contact is made, wash thoroughly with soap or hand cleaner as soon as possible. Wear protective clothing, including impervious gloves where practicable. Do not use gasoline, kerosene, diesel fuel, gas oil, thinners, or solvents for cleaning skin. Where there is a risk of eye contact, eye protection should be worn, for example, chemical goggles or face shields; in addition an eye wash facility should be provided. Hot oil can scald.

➡If working near and/or around the SRS system and components, be sure to disable the SRS system. After disabling the system wait three minutes or more before servicing the vehicle.

➡Whenever the negative battery cable is disconnected the following components will require resetting. The Idle Air Volume Learning, Steering Angle Sensor Neutral Position, Sunroof Memory Reset/Initialization, Automatic Drive Positioner System, Audio presets and Navigation. Use the Nissan CONSULT III diagnostic tool, or equivalent to perform the required resets.

1. Before servicing the vehicle, refer to the Precautions Section.
2. Warm up the engine, and check for any oil leaks.
3. Disconnect the negative battery cable.
4. Stop the engine and wait for at least 10 minutes.
5. Remove the engine front under cover.

6. Remove the oil drain plug and oil filler cap to drain the old oil.
7. Install a new washer on the oil drain plug, then install the oil drain plug in the oil pan. Tighten the oil drain plug to 25 ft. lbs. (34 Nm).

➡Clean the drain plug before installing a new washer.

8. Remove the oil filter using special tool KV10115801 (J-38956), or equivalent.
- When removing the oil filter, position a shop cloth to absorb any engine oil leaks or spills
- Do not allow engine oil to adhere to the drive belts
- Completely wipe off any engine oil that adheres to the engine and the vehicle

❊❊ CAUTION

Be careful not to burn yourself, as the engine and engine oil may be hot.

To install:
9. Remove any foreign materials adhering to the oil filter seal mating surface.
10. Apply clean engine oil to the oil filter seal circumference of the new oil filter.
11. Screw on the oil filter manually until it touches the installation surface, then tighten it ⅔ of a turn or tighten to 13 ft. lbs. (18 Nm).
12. Refill the engine with the proper grade and amount of new engine oil.
13. Inspect the engine for oil leaks.
14. Install the engine front under cover.
15. Check the engine oil level.
16. Start the engine and check for engine oil leaks.
17. Stop the engine and wait for 10 minutes.
18. Warm up the engine and check the area around the drain plug and oil filter for any oil leaks.
19. Stop the engine and wait for 10 minutes.
20. Check the oil level using the dipstick. Add oil as necessary and install the oil filler cap.

➡The refill capacity depends on the oil temperature and drain time. Use the refill oil capacity values as a reference and check the oil level using the dipstick when filling the engine with oil.

➡Do not overfill the engine with oil.

EXHAUST MANIFOLD

REMOVAL & INSTALLATION

2.5L Engine

See Figures 108 and 109.

Catalytic converter precautions:

• If a large amount of unburned fuel flows into the catalyst, the catalyst temperature will be excessively high.

• Use unleaded gasoline only. Leaded gasoline will seriously damage the three way catalyst.

• When checking for ignition spark or measuring engine compression, make tests quickly and only when necessary.

• Do not run the engine when the fuel tank level is low, otherwise the engine may misfire, causing damage to the catalyst.

• Do not place the vehicle over flammable material. Keep flammable material off the exhaust pipe and the three way catalyst.

✳✳ CAUTION

In order to avoid being burned, do not service the exhaust system while it is still hot. Service the system when it is cool.

✳✳ CAUTION

Always wear protective goggles and gloves when removing exhaust parts as falling rust and sharp edges from worn exhaust components could result in serious personal injury.

The exhaust manifold is removed with the three way catalyst as an assembly.

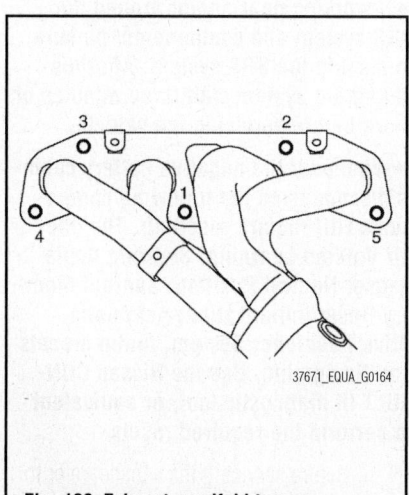

Fig. 108 Exhaust manifold torque sequence—2.5L engine

37671_EQUA_G0164

→If working near and/or around the SRS system and components, be sure to disable the SRS system. After disabling the system wait three minutes or more before servicing the vehicle.

→Whenever the negative battery cable is disconnected the following components will require resetting. The Idle Air Volume Learning, Steering Angle Sensor Neutral Position, Sunroof Memory Reset/Initialization, Automatic Drive Positioner System, Audio presets and Navigation. Use the Nissan CONSULT III diagnostic tool, or equivalent to perform the required resets.

1. Before servicing the vehicle, refer to the Precautions Section.

2. Disconnect the negative battery cable.

3. Disconnect the harness connector of the air fuel ratio sensor 1 and the harness from the bracket and the middle clamp.

4. Remove the air fuel ratio sensor 1 using special tool J-44626, or equivalent.

✳✳ WARNING

Be careful not to damage the air fuel ratio sensor 1. Discard any air fuel ratio sensor 1 which has been

dropped from a height of more than 20 inches (0.5m) onto a hard surface such as a concrete floor. Replace as needed.

5. Remove the exhaust front tube.

6. Remove the exhaust manifold cover.

7. Remove the bracket between the exhaust manifold/three way catalyst assembly and the transmission assembly.

8. Loosen the nuts in the reverse order of the torque sequence to remove the exhaust manifold/three way catalyst assembly.

9. Remove the exhaust manifold gasket.

✳✳ WARNING

Cover the engine openings to avoid entry of foreign materials.

To install:

10. Check the surface distortion of the exhaust manifold/three way catalyst assembly mating surface with a straightedge and a feeler gauge.

• Limit: 0.012 inch (0.3mm)

• If the limit is exceeded, replace the exhaust manifold/three way catalyst assembly.

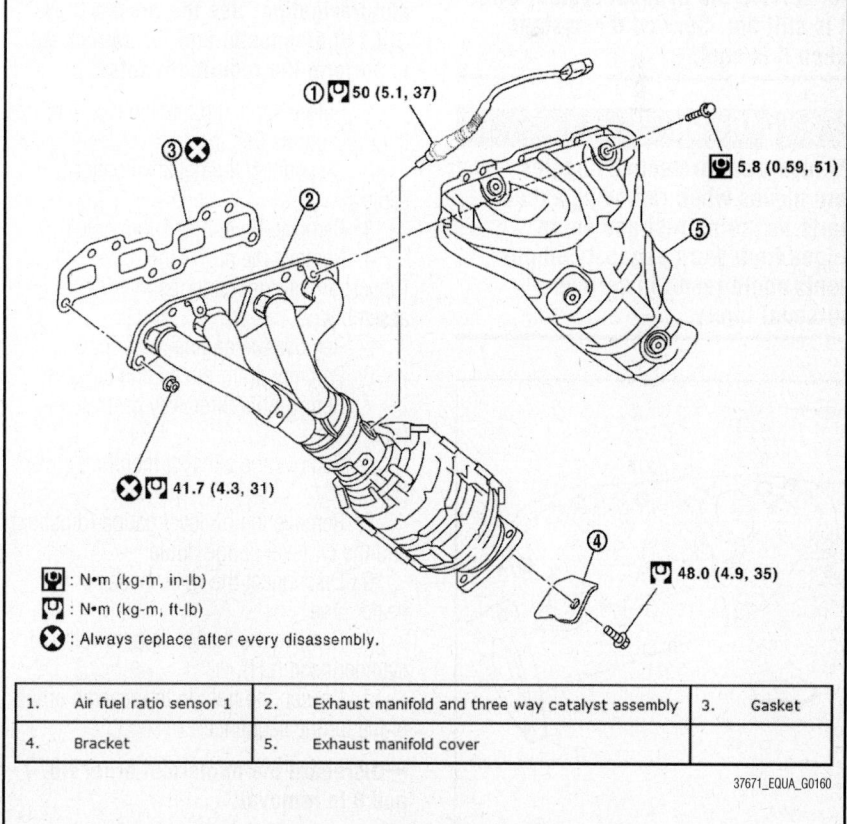

| 1. | Air fuel ratio sensor 1 | 2. | Exhaust manifold and three way catalyst assembly | 3. | Gasket |
| 4. | Bracket | 5. | Exhaust manifold cover | | |

37671_EQUA_G0160

Fig. 109 Exhaust manifold and 3-way catalyst (exploded view)—2.5L engine

11. Installation is in the reverse order of removal.

12. If the stud bolts were removed, install them and tighten to 11 ft. lbs. (15 Nm).

13. Tighten the nuts in the torque sequence according to the illustrations shown.

14. Tighten the nuts in numerical order as shown again.

✳✳ WARNING

Do not over-tighten the air fuel ratio sensor 1. Doing so may cause damage to the air fuel ratio sensor 1, resulting in the MIL coming on.

15. Before installing a new air fuel ratio sensor 1, clean the exhaust system threads using oxygen sensor thread cleaner and apply anti-seize lubricant. Use cleaner J-43897-12 or J-43897 18, or equivalent.

4.0L Engine

Left Side Manifold

See Figures 110 and 111.

✳✳ CAUTION

In order to avoid being burned, do not service the exhaust system while it is still hot. Service the system when it is cool.

✳✳ CAUTION

Always wear protective goggles and gloves when removing exhaust parts as falling rust and sharp edges from worn exhaust components could result in serious personal injury.

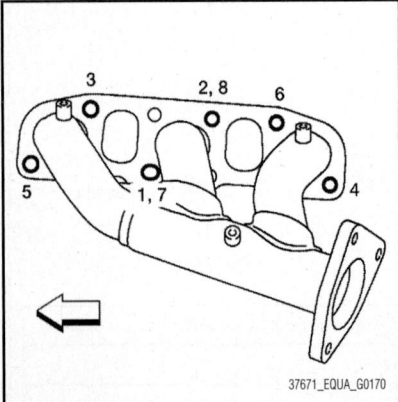

37671_EQUA_G0170

Fig. 110 Exhaust manifold torque sequence (left side)—4.0L engine

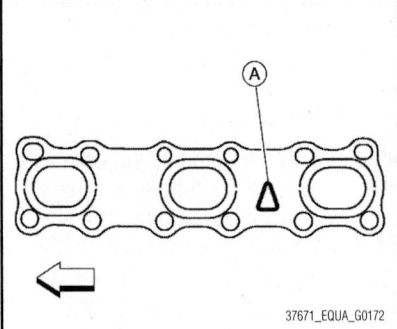

37671_EQUA_G0172

Fig. 111 Install the exhaust manifold gasket with the identification hole (A) pointing as shown—4.0L engine

➡**If working near and/or around the SRS system and components, be sure to disable the SRS system. After disabling the system wait three minutes or more before servicing the vehicle.**

➡**Whenever the negative battery cable is disconnected the following components will require resetting. The Idle Air Volume Learning, Steering Angle Sensor Neutral Position, Sunroof Memory Reset/Initialization, Automatic Drive Positioner System, Audio presets and Navigation. Use the Nissan CONSULT III diagnostic tool, or equivalent to perform the required resets.**

1. Before servicing the vehicle, refer to the Precautions Section.
2. Disconnect the negative battery cable.
3. Remove the engine room cover.
4. Remove the air cleaner case (upper) and air duct and resonator assembly.
5. Remove the engine under cover.
6. Partially drain the engine coolant.
7. Remove the three way catalyst (LH).
8. Remove the exhaust manifold cover (LH).
9. Remove the oil level gauge (dipstick) and the oil level gauge guide.
10. Disconnect the water hoses at the heater pipe.
11. Remove the heater pipe from the cylinder head (LH).
12. Loosen the nuts in the reverse order of the torque sequence.

➡**Disregard the numerical order No. 7 and 8 in removal.**

13. Remove the exhaust manifold (LH).
14. Remove the gaskets.

✳✳ WARNING

Cover the engine openings to avoid entry of foreign materials.

To install:

15. Check the surface distortion of the exhaust manifold mating surface with a straightedge and feeler gauge.
- Limit: 0.012 inch (0.3mm)
- If the limit is exceeded, replace the exhaust manifold.

16. Install the exhaust manifold gasket with the identification hole as shown.

17. If the exhaust manifold studs were removed, install and tighten to 11 ft. lbs. (15 Nm).

18. Install the exhaust manifold and tighten the nuts in numerical order shown.

➡**Use new exhaust manifold nuts for installation.**

19. Tighten nuts No. 1 and 2 in two steps. The numerical order No. 7 and 8 show the second step.

Right Side Manifold

See Figure 112.

✳✳ CAUTION

In order to avoid being burned, do not service the exhaust system while it is still hot. Service the system when it is cool.

✳✳ CAUTION

Always wear protective goggles and gloves when removing exhaust parts as falling rust and sharp edges from worn exhaust components could result in serious personal injury.

➡**If working near and/or around the SRS system and components, be sure to disable the SRS system. After disabling the system wait three minutes or more before servicing the vehicle.**

➡**Whenever the negative battery cable is disconnected the following components will require resetting. The Idle Air Volume Learning, Steering Angle Sensor Neutral Position, Sunroof Memory Reset/Initialization, Automatic Drive Positioner System, Audio presets and Navigation. Use the Nissan CONSULT III diagnostic tool, or equivalent to perform the required resets.**

1. Before servicing the vehicle, refer to the Precautions Section.
2. Disconnect the negative battery cable.

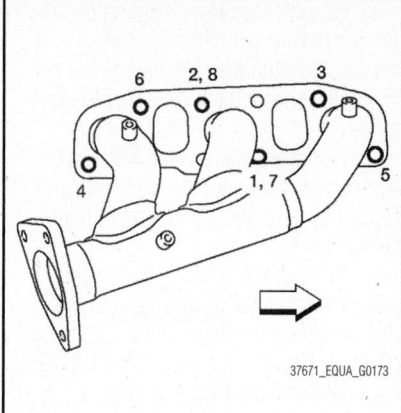

Fig. 112 Exhaust manifold torque sequence (right side)—4.0L engine

3. Remove the three way catalyst (RH).

4. Remove the heat shield from the lower dash panel.

5. Remove the support bolts from the A/T fluid charging pipe (automatic transmission models).

6. Loosen the nuts in the reverse order of the torque sequence.

➡Disregard the numerical order No. 7 and 8 in removal.

7. Remove the exhaust manifold (RH) and the exhaust manifold cover (RH) together.

8. Remove the gaskets.

✳✳ WARNING

Cover the engine openings to avoid entry of foreign materials.

To install:

9. Check the surface distortion of the exhaust manifold mating surface with a straightedge and feeler gauge.

- Limit: 0.012 inch (0.3mm)
- If the limit is exceeded, replace the exhaust manifold.

➡**If necessary, a crowfoot wrench may be used to tighten the exhaust manifold nuts.**

10. Install the exhaust manifold gasket with the identification hole as shown.

11. If the exhaust manifold studs were removed, install and tighten to 11 ft. lbs. (15 Nm).

12. Install the exhaust manifold and tighten the nuts in numerical order shown.

➡**Use new exhaust manifold nuts for installation.**

13. Tighten nuts No. 1 and 2 in two steps. The numerical order No. 7 and 8 show the second step.

FRONT COVER & SEAL

REMOVAL & INSTALLATION

2.5L Engine

See Figures 113 through 115.

At this time the manufacturer does not provide removal and installation procedures for this component. The following procedure is a guideline and may differ from the vehicle you are servicing.

➡**If working near and/or around the SRS system and components, be sure to disable the SRS system. After disabling the system wait three minutes or more before servicing the vehicle.**

➡**Whenever the negative battery cable is disconnected the following components will require resetting. The Idle Air Volume Learning, Steering Angle Sensor Neutral Position, Sunroof Memory Reset/Initialization, Automatic Drive Positioner System, Audio presets and Navigation. Use the Nissan CONSULT III diagnostic tool, or equivalent to perform the required resets.**

1. Before servicing the vehicle, refer to the Precautions Section.

2. Disconnect the negative battery cable.

3. Properly relieve the fuel system pressure.

4. Remove the engine undercover.

5. Remove the air cleaner and the air duct assembly.

6. Remove the spark plugs.

7. Disconnect the PCV hose from the rocker cover. Remove the ignition coil.

8. Remove the PCV valve and O-ring from the rocker cover, if necessary.

9. Remove the oil filler cap from the rocker cover, if necessary.

10. Remove the rocker cover retaining bolts. Be sure to remove the bolts by reversing the order of the tightening torque sequence.

11. Remove the rocker cover. Discard the gasket.

12. Remove the coolant reservoir tank. Remove the auxiliary drive belt autotensioner.

13. Remove the alternator. Remove the strut tower brace.

14. Remove the air conditioning compressor and position it to the side. Do not disconnect the refrigerant lines.

15. Remove the power steering pump and reservoir tank; position the assembly to the side. Do not disconnect the fluid lines.

16. Remove the oil pan. Remove the strainer.

17. Remove the Intake Valve Timing (IVT) control cover bolts. Be sure to remove the bolts by reversing the order of the tightening sequence.

18. Remove the cover by cutting the sealant using tool KV10111100 (J-37228), or equivalent.

19. Remove and discard the oil seal.

To install:

20. Install the front cover oil seal. Install O-rings to the cylinder head and the cylinder block.

21. Apply a continuous bead of liquid gasket to the front cover. Be sure to use genuine RTV sealant, or equivalent.

➡**Sealant application instructions differ depending on position, refer to the illustration for positioning. Detail "A", cross over the start of the application and the end. Detail "B", apply liquid**

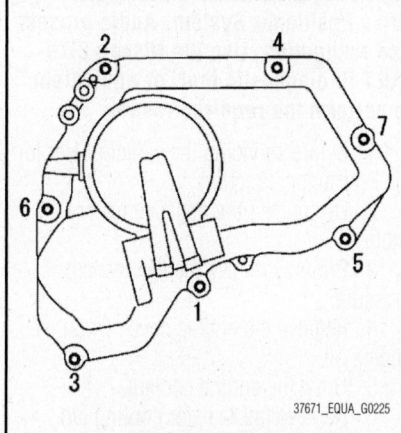

Fig. 113 Intake Valve Timing (IVT) control cover bolt tightening sequence—2.5L engine

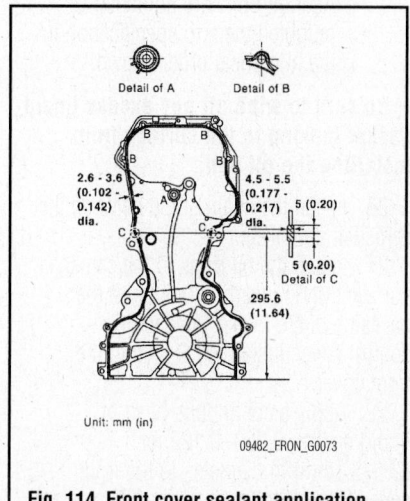

Fig. 114 Front cover sealant application points—2.5L engine

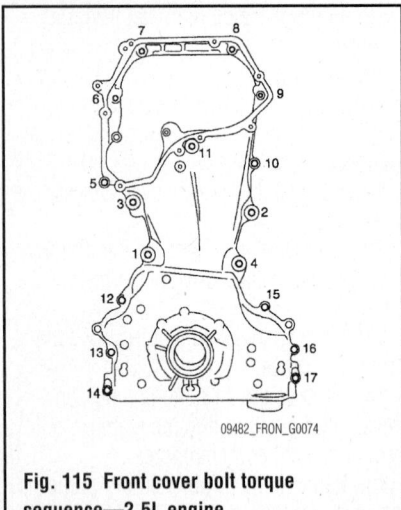

Fig. 115 Front cover bolt torque sequence—2.5L engine

gasket outside of the bolt holes. For all bolt holes other than "B", apply to the inside. Detail "C", between here only, apply a bead of sealant 0.177–0.217 inch (4.5–5.5mm) diameter.

22. Make sure that the mating marks of the chain and each sprocket are still aligned.

23. Install the front cover. Torque the retaining bolts to specification and in the proper sequence.

24. Use the following for locating the M6 bolts:
  - Bolt position 5, 10, 14 and 17—Bolt length: 1.77 inches (45mm)
  - Bolts except the above (except 1 to 4)—Bolt length: 0.79 inch (20mm)

25. Tighten the bolts to the specified torque:
  - Bolt position 5–17: 9 ft. lbs. (13 Nm)
  - Bolt position 1–4: 36 ft. lbs. (49 Nm)
  - After all bolts are tightened, retighten them to specification in the numerical order shown

➡**Be sure to wipe off any excess liquid gasket leaking to the surface from installing the oil pan.**

26. Install the chain guide between the camshaft sprockets.

27. Install the oil rings to the camshaft sprocket (INT) insertion points on the backside of the intake valve timing control cover. Install the O-ring to the front cover.

28. Apply a continuous bead of liquid gasket, 0.083–0.122 inch (2.1–3.1mm) in diameter, to the front cover. Be sure to use genuine RTV sealant, or equivalent.

29. Install the intake valve timing control cover. Tighten the bolts in the proper sequence to specification.

30. Install the intake valve timing control solenoid valve to the intake valve timing control cover, if removed.

31. Connect the ground cables, and install the harness clip.

32. Install the crankshaft pulley.

33. When installing the rocker cover, apply liquid gasket, be sure to use genuine RTV silicone sealant, or equivalent, to the positions shown in the illustration. Refer to figure "a" to apply liquid gasket to joint part of camshaft bracket No. 1 and cylinder head. Refer to figure "b" to apply liquid gasket in 90 degrees to figure "b".

34. Install the rocker cover.

35. Continue the installation in the reverse order of the removal procedure.

➡**If hydraulic pressure inside the timing chain tensioner drops after removal/installation, slack in the guide may generate a pounding noise during and just after engine start. This is normal. The noise should stop after the hydraulic pressure rises.**

4.0L Engine

See Figures 116 through 119.

➡**If working near and/or around the SRS system and components, be sure to disable the SRS system. After disabling the system wait three minutes or more before servicing the vehicle.**

➡**Whenever the negative battery cable is disconnected the following components will require resetting. The Idle Air Volume Learning, Steering Angle Sensor Neutral Position, Sunroof Memory Reset/Initialization, Automatic Drive Positioner System, Audio presets and Navigation. Use the Nissan CONSULT III diagnostic tool, or equivalent to perform the required resets.**

1. Before servicing the vehicle, refer to the Precautions Section.

2. Disconnect the negative battery cable.

3. Properly relieve the fuel system pressure.

4. Remove the engine cover. Drain the engine oil.

5. Drain the engine coolant.

6. Remove the radiator cooling fan assembly. Remove the drive belts.

7. Separate the engine wiring harnesses by removing their brackets from the front timing chain case.

8. Remove the power steering pump from the bracket with the fluid hoses attached. Position the assembly to the side. Do not disconnect the hoses. Remove the bracket.

9. Remove the alternator. Remove the water bypass hose, water hose clamp and idler pulley bracket from the front timing chain case.

10. Remove the left and right intake valve timing control covers. Loosen the bolts in the reverse order of the tightening sequence. Use tool KV10111100 (J-37228), or equivalent to cut the liquid gasket seal.

➡**The shaft is internally jointed with the camshaft sprocket (INT) center hole. When removing, keep it horizontal until it is completely disconnected.**

11. Remove the collared O-rings from the front timing chain case on both the left and right side.

12. Remove the A/C compressor bolts. Secure the component to the side.

13. Loosen the crankshaft pulley retaining bolt and locate the bolt seating surface, which is about 0.39 inch (10mm) from its original position.

➡**Do not remove the crankshaft pulley bolt. Keep the loosened pulley bolt in place to protect the removed crankshaft pulley from dropping.**

14. Pull the pulley with both hands and remove it from its mounting. Remove the bolt and pulley from the engine.

15. Loosen and remove the 2 bolts of the upper oil pan.

16. Loosen the front timing chain cover retaining bolts in the reverse order of the tightening sequence.

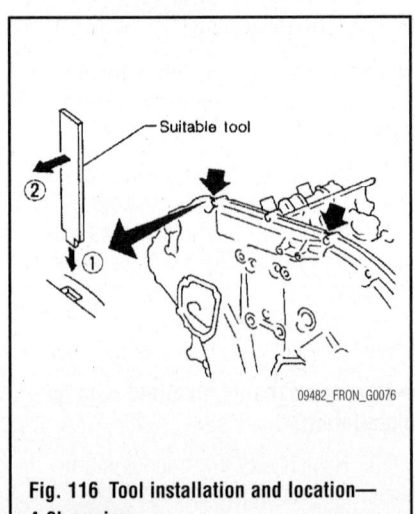

Fig. 116 Tool installation and location—4.0L engine

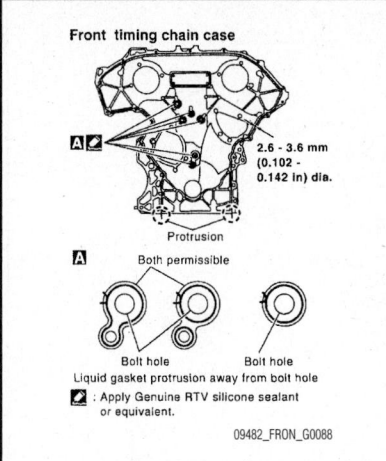

Fig. 117 Front timing chain cover sealant application—4.0L engine

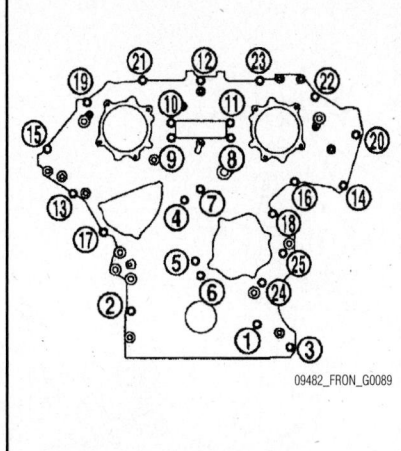

Fig. 118 Front cover bolt torque sequence—4.0L engine

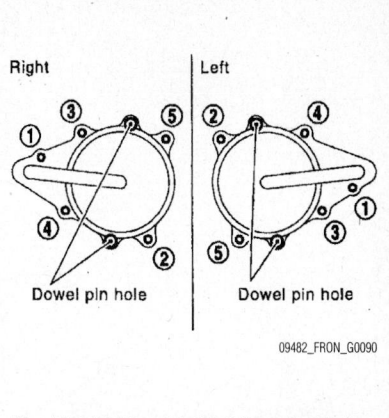

Fig. 119 Right and left intake valve timing control cover bolt torque sequence—4.0L engine

17. Insert a suitable tool in the notch at the top of the front timing chain case and pry off the case by moving the tool as shown in the illustration. Use tool KV10111100 (J-37228), or equivalent to cut the liquid gasket seal.

➡️Do not use a screwdriver or similar tool. After removal, handle the front timing chain cover case carefully so it does not tilt or warp under a load.

18. Remove the water pump cover and chain tensioner cover from the front timing chain case cover, as required.

To install:

19. Install a new front seal in the front timing chain case cover.

20. Install the water pump cover and chain tensioner cover to the front timing chain case cover. Apply a continuous bead of liquid gasket 0.091–0.130 inch (2.31–3.30mm) in diameter to the front timing chain case cover before installing the water pump cover and chain tensioner cover. Be sure to use genuine RTV sealant, or equivalent.

21. Before installing the front timing chain case cover apply a continuous bead of liquid gasket 0.102–0.142 inch (2.6–3.6mm) in diameter to the front timing chain case back side, as shown in the illustration. Be sure to use genuine RTV sealant, or equivalent.

22. Install new O-rings on the rear timing chain case. To assemble the front timing chain case cover, fit the lower end of the front timing chain case tightly onto the top face of the oil pan (upper). From the fitting point, make the entire front timing chain case contact the rear timing chain case completely.

➡️Since the front timing chain case cover is offset for the difference of bolt holes, tighten the bolts temporarily while holding the front timing chain case cover from the front and the top. Then, insert a dowel pin while holding the front timing chain case cover from the front and the top.

23. Once the cover is installed, torque the retaining bolts to specification and in the proper sequence. There are 4 different types of bolts:

- Bolt diameter 0.39 inch (10mm)—Bolt position 1–5: Tighten to 41 ft. lbs. (55 Nm)
- Bolt diameter 0.24 inch (6mm)—Bolt position 6–25: Tighten to 9 ft. lbs. (13 Nm)
- After all bolts are tightened, retighten them to specification and in the proper sequence

24. Install the 2 bolts in the oil pan (upper). Torque to 16 ft. lbs. (22 Nm).

25. Install new seal rings in the shaft grooves of the right and left intake valve timing control covers.

26. Apply a continuous bead of liquid gasket 0.083–0.122 inch (2.1–3.1mm) in diameter to the covers. Be sure to use genuine RTV sealant, or equivalent.

27. Install new collared O-rings in the front timing chain case oil hole (left and right sides). Be careful not to move the seal ring from the installation groove, align the dowel pins on the front timing chain case with the holes to install the intake valve timing control covers.

28. Tighten the bolts in sequence and to specification.

29. Continue the installation in the reverse order of the removal procedure.

➡️If hydraulic pressure inside the timing chain tensioner drops after removal/installation, slack in the guide may generate a pounding noise during and just after engine start. This is normal. The noise should stop after the hydraulic pressure rises.

INTAKE MANIFOLD

REMOVAL & INSTALLATION

2.5L Engine

See Figures 120 through 122.

➡️If working near and/or around the SRS system and components, be sure to disable the SRS system. After disabling the system wait three minutes or more before servicing the vehicle.

➡️Whenever the negative battery cable is disconnected the following components will require resetting. The Idle Air Volume Learning, Steering Angle Sensor Neutral Position, Sunroof Memory Reset/Initialization, Automatic Drive Positioner System, Audio presets and Navigation. Use the Nissan CONSULT III diagnostic tool, or equivalent to perform the required resets.

1. Before servicing the vehicle, refer to the Precautions Section.

2. Disconnect the negative battery cable.

3. Properly relieve the fuel system pressure.

4. Disconnect the negative battery cable. Drain the cooling system.

5. Remove the air cleaner case, air cleaner and air duct.

6. Disconnect and plug the water hoses from the electric throttle control actuator.

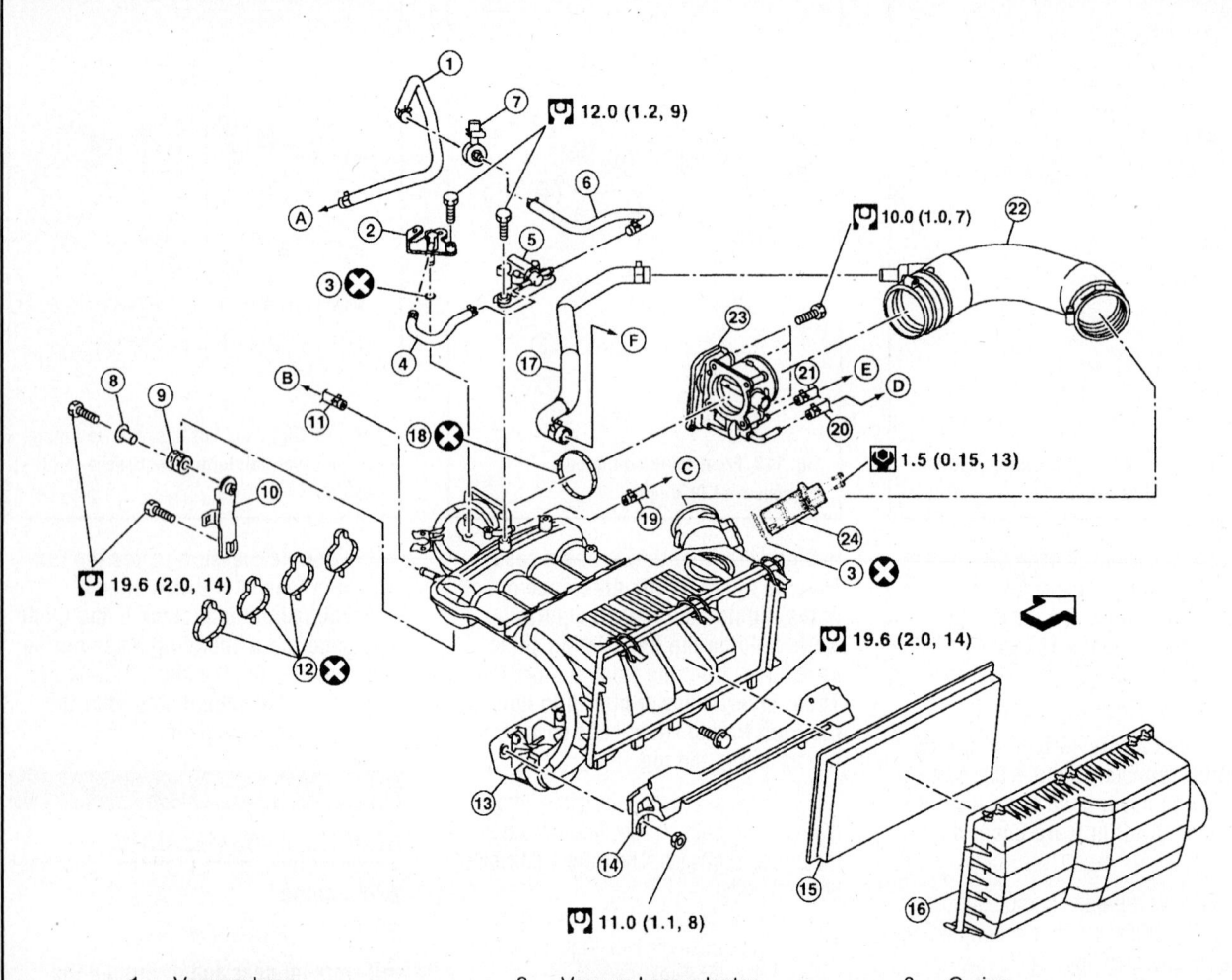

1. Vacuum hose
2. Vacuum hose adapter
3. O–ring
4. Vacuum hose
5. EVAP canister purge volume control solenoid valve
6. Vacuum hose
7. Service port
8. Collar
9. Grommet
10. Intake manifold support
11. Vacuum hose
12. Gasket
13. Intake manifold
14. Fuel tube protector
15. Air cleaner
16. Air cleaner case
17. PCV hose
18. Gasket
19. PCV hose
20. Water hose
21. Water hose
22. Air duct
23. Electric throttle control actuator
24. Mass air flow sensor
A. To vacuum pipe (EVAP canister)
B. To brake booster
C. To PCV valve
D. To heater outlet
E. To heater pipe
F. To rocker cover
→ Engine front

09482_FRON_G0027

Fig. 120 Intake manifold and related components—2.5L engine

7. Remove the mass air flow sensor from the intake manifold. Remove the quick connector cap and disconnect the quick connector at the engine side.

8. Remove the electric throttle control actuator harness connector and retaining bolts. Be sure to remove the bolts by reversing the order of the tightening torque sequence.

9. Remove the electric throttle control actuator and gasket.

10. Disconnect the harness, vacuum hoses, and PCV hoses from the intake manifold and position them to the side.

11. Remove intake manifold support.

12. Remove the intake manifold retaining bolts. Be sure to remove the bolts by reversing the order of the tightening torque sequence.

13. Remove the intake manifold, fuel tube protector, and gasket from the engine.

14. As necessary, remove the EVAP canister purge volume solenoid valve and

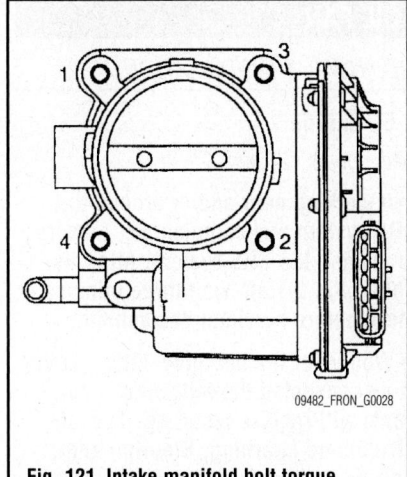

Fig. 121 Intake manifold bolt torque sequence—2.5L engine

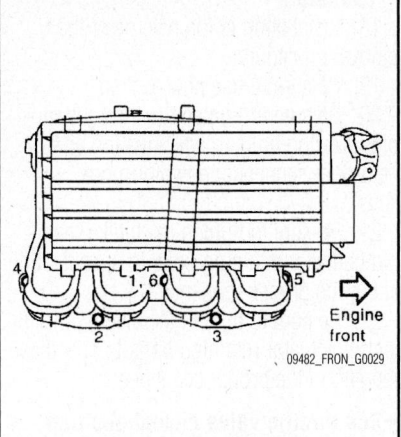

Fig. 122 Electric throttle control actuator bolt torque sequence—2.5L engine

vacuum hose adapter from the intake manifold.

15. Disconnect the sub frame harness from the fuel injectors. Remove the fuel tube and fuel injector assembly from the intake manifold, if required.

✳✳ WARNING

Cover engine openings to avoid entry of foreign materials. Do not disassemble the intake manifold.

To install:

16. Installation is the reverse of the removal procedure.

17. Be sure to use new gaskets.

18. If the stud bolts were removed, install and tighten to 83 inch lbs. (9 Nm).

19. Be sure to tighten the intake manifold retaining bolts to 83 inch lbs. (9 Nm) in the proper sequence.

➡**Refer to the torque sequence illustration, No. 6 means double tightening of bolt No. 1.**

20. Use the following for locating bolts and nuts:

- M8 x 1.50 inches (38mm) are green in color: No. 1, 6
- M8 x 1.38 inches (35mm): No. 2, 3
- Nuts: No. 4, 5

21. Be sure to tighten the electric throttle control actuator retaining bolts to specification and in the proper sequence.

➡**See the Throttle Valve Closed Position learning and Idle Air Volume Learning procedures, for relearning information.**

22. Be sure to fill the cooling system with the proper grade and type engine coolant.

23. Start the engine and check for leaks.

4.0L Engine

See Figures 123 through 127.

➡**The upper intake manifold is also referred to as the intake manifold collector and is removed in conjunction with the lower intake manifold.**

➡**If working near and/or around the SRS system and components, be sure to disable the SRS system. After disabling the system wait three minutes or more before servicing the vehicle.**

➡Whenever the negative battery cable is disconnected the following components will require resetting. The Idle Air Volume Learning, Steering Angle Sensor Neutral Position, Sunroof Memory Reset/Initialization, Automatic Drive Positioner System, Audio presets and Navigation. Use the Nissan CONSULT III diagnostic tool, or equivalent to perform the required resets.

1. Before servicing the vehicle, refer to the Precautions Section.

2. Disconnect the negative battery cable.

3. Properly relieve the fuel system pressure.

4. Disconnect the negative battery cable.

5. Drain the cooling system.

6. Remove the engine cover. Remove the air cleaner case (upper) with the mass air flow sensor and air duct assembly.

7. Disconnect the water hoses from the electric throttle control actuator. Disconnect the harness connector.

8. Remove the electric throttle control actuator retaining bolts. Be sure to remove the bolts by reversing the order of the tightening torque sequence.

9. Remove the electric throttle control actuator.

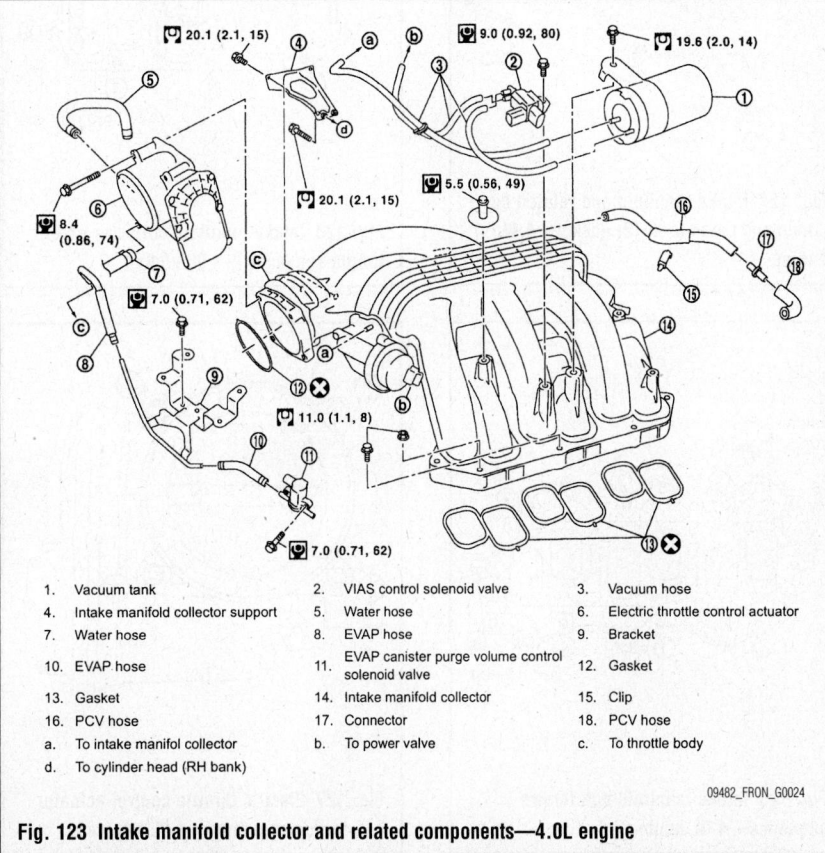

1.	Vacuum tank	2.	VIAS control solenoid valve	3.	Vacuum hose
4.	Intake manifold collector support	5.	Water hose	6.	Electric throttle control actuator
7.	Water hose	8.	EVAP hose	9.	Bracket
10.	EVAP hose	11.	EVAP canister purge volume control solenoid valve	12.	Gasket
13.	Gasket	14.	Intake manifold collector	15.	Clip
16.	PCV hose	17.	Connector	18.	PCV hose
a.	To intake manifol collector	b.	To power valve	c.	To throttle body
d.	To cylinder head (RH bank)				

Fig. 123 Intake manifold collector and related components—4.0L engine

10. Remove the brake booster vacuum hose and the PCV hose. Remove the intake manifold collector support.

11. Disconnect the EVAP hoses and harness connector from the EVAP canister purge volume control solenoid valve. Remove the EVAP canister purge volume control solenoid valve.

12. Remove the VIAS control solenoid valve and vacuum tank.

13. Remove the intake manifold collector retaining bolts. Be sure to remove the bolts by reversing the order of the tightening torque sequence.

14. Remove the intake manifold collector from the engine.

15. Remove the fuel tube and fuel injector assembly.

16. Remove the intake manifold retaining bolts. Be sure to remove the bolts by reversing the order of the tightening torque sequence.

17. Remove the intake manifold from the engine.

To install:

18. Installation is the reverse of the removal procedure.

19. Be sure to use new gaskets.

20. Be sure to tighten the intake manifold retaining bolts to specification and in the proper sequence in two or more steps.

21. Be sure to tighten the intake manifold collector retaining bolts to specification and in the proper sequence.

22. Be sure to tighten the electric throttle control actuator retaining bolts to specification and in the proper sequence.

➡See throttle valve closed position learning and idle air volume learning procedures, for relearning information.

23. Be sure to fill the cooling system with the proper grade and type engine coolant.

24. Start the engine and check for leaks.

OIL PAN

REMOVAL & INSTALLATION

2.5L Engine

See Figures 128 and 129.

➡If working near and/or around the SRS system and components, be sure to disable the SRS system. After disabling the system wait three minutes or more before servicing the vehicle.

➡Whenever the negative battery cable is disconnected the following components will require resetting. The Idle Air Volume Learning, Steering Angle Sensor Neutral Position, Sunroof Memory Reset/Initialization, Automatic Drive Positioner System, Audio presets and Navigation. Use the Nissan CONSULT III diagnostic tool, or equivalent to perform the required resets.

1. Before servicing the vehicle, refer to the Precautions Section.

2. Disconnect the negative battery cable.

3. Raise and safely support the vehicle.

4. Remove the engine under cover. Drain the engine oil.

5. Remove the third crossmember assembly.

6. Remove the lower joint shaft pinch bolt at the steering gear.

7. Remove the steering gear bolts.

8. If equipped with an automatic transmission, remove the fluid cooler tube.

9. Loosen the oil pan retaining bolts, in the reverse order of the installation sequence.

10. Insert a seal cutter tool between the oil pan and the cylinder block, and slide it by tapping on the side of the tool with a hammer.

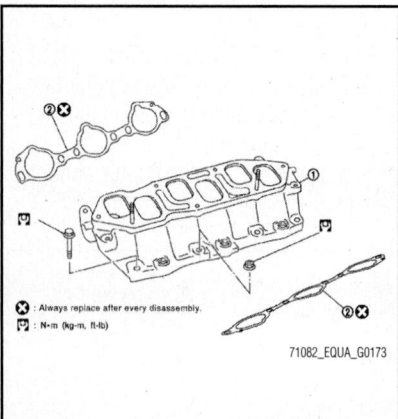

Fig. 124 Intake manifold and related components (1) manifold, (2) gasket—4.0L engine

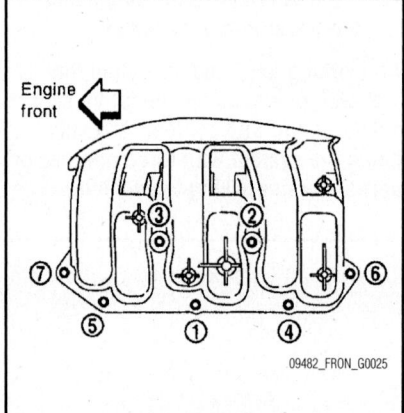

Fig. 126 Intake manifold collector bolt torque sequence—4.0L engine

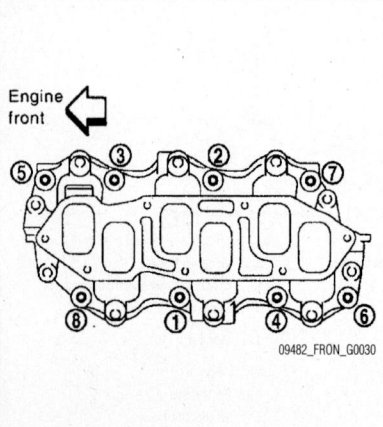

Fig. 125 Intake manifold bolt torque sequence—4.0L engine

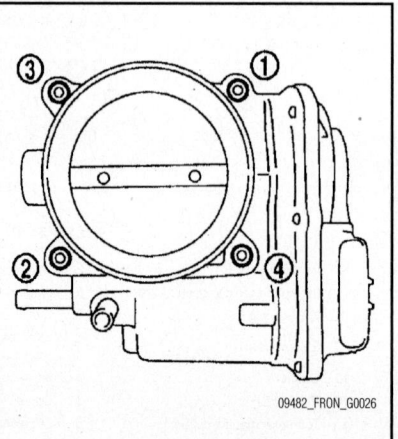

Fig. 127 Electric throttle control actuator bolt torque sequence—4.0L engine

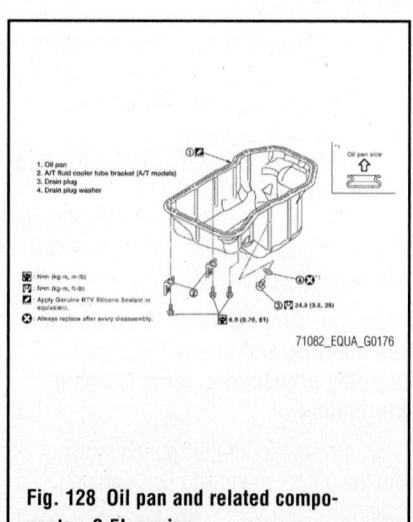

Fig. 128 Oil pan and related components—2.5L engine

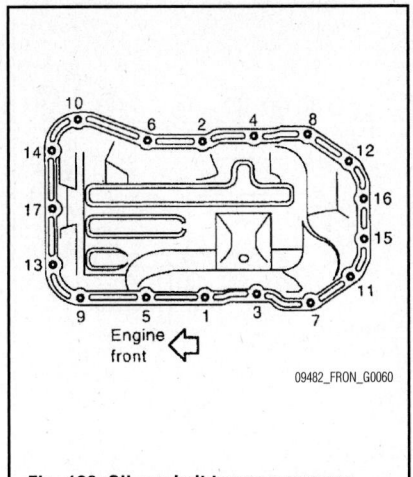

Fig. 129 Oil pan bolt torque sequence—2.5L engine

11. Remove the oil pan from the engine.

To install:

12. Be sure to clean all the oil gasket material from both the oil pan and the cylinder block surfaces, using the proper tools.

✷✷ WARNING

Do not scratch or damage the mating surfaces when cleaning off the old liquid gasket.

13. Apply a continuous bead of sealant 0.138–0.177 inches (3.5–4.5mm) to the oil pan mating surface.

14. Install the oil pan to the cylinder block. This must be done within 5 minutes after applying the liquid gasket.

15. Torque the oil pan bolts to 61 inch lbs. (7 Nm) in the proper sequence.

16. Continue the installation in the reverse order of the removal procedure.

➡**Wait 30 minutes after installation of the oil pan to allow the sealant to cure before adding oil.**

17. Fill the crankcase with the proper grade and amount of oil.

18. Start the engine and check for leaks.

4.0L Engine

Lower

See Figures 130 through 132.

➡**If working near and/or around the SRS system and components, be sure to disable the SRS system. After disabling the system wait three minutes or more before servicing the vehicle.**

➡**Whenever the negative battery cable is disconnected the following components will require resetting. The Idle Air Volume Learning, Steering Angle**

Sensor Neutral Position, Sunroof Memory Reset/Initialization, Automatic Drive Positioner System, Audio presets and Navigation. Use the Nissan CONSULT III diagnostic tool, or equivalent to perform the required resets.

1. Before servicing the vehicle, refer to the Precautions Section.
2. Disconnect the negative battery cable.
3. Raise and safely support the vehicle.
4. Remove the engine under cover. Drain the engine oil.
5. Loosen the oil pan retaining bolts, in the reverse order of the installation sequence.
6. Insert a seal cutter tool between the oil pan and the cylinder block, and slide it by tapping on the side of the tool with a hammer.
7. Remove the oil pan from the engine.

To install:

8. Be sure to clean all the oil gasket material from both the oil pan and the cylinder block surfaces, using the proper tools.
9. Apply a continuous bead of sealant 0.138–0.177 inches (3.5–4.5mm) to the oil pan mating surface. Be sure to use genuine RTV sealant, or equivalent.

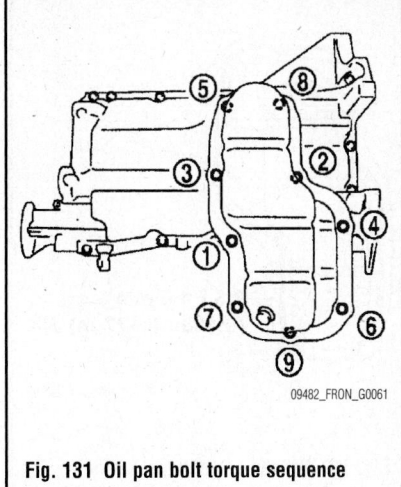

Fig. 131 Oil pan bolt torque sequence (lower)—4.0L engine

10. Install the oil pan to the cylinder block. This must be done within 5 minutes after applying the liquid gasket.

11. Torque the bolts to 80 inch lbs. (9 Nm) in the proper sequence.

12. Continue the installation in the reverse order of the removal procedure.

➡**Wait 30 minutes after installation of the oil pan to allow the sealant to cure before adding oil.**

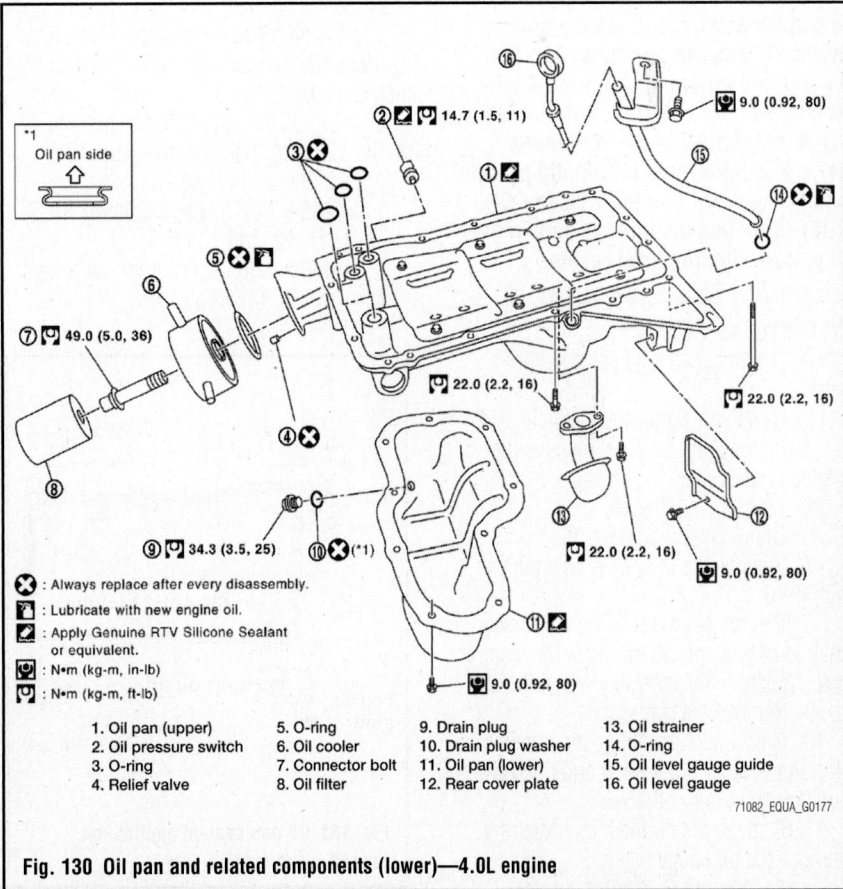

1. Oil pan (upper)
2. Oil pressure switch
3. O-ring
4. Relief valve
5. O-ring
6. Oil cooler
7. Connector bolt
8. Oil filter
9. Drain plug
10. Drain plug washer
11. Oil pan (lower)
12. Rear cover plate
13. Oil strainer
14. O-ring
15. Oil level gauge guide
16. Oil level gauge

Fig. 130 Oil pan and related components (lower)—4.0L engine

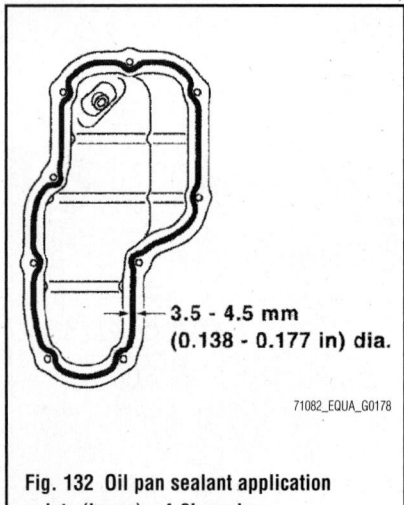

Fig. 132 Oil pan sealant application points (lower)—4.0L engine

13. Fill the crankcase with the proper grade and amount of oil.

14. Start the engine and check for leaks.

Upper

See Figures 133 and 134.

➡️ If working near and/or around the SRS system and components, be sure to disable the SRS system. After disabling the system wait three minutes or more before servicing the vehicle.

➡️ Whenever the negative battery cable is disconnected the following components will require resetting. The Idle Air Volume Learning, Steering Angle Sensor Neutral Position, Sunroof Memory Reset/Initialization, Automatic Drive Positioner System, Audio presets and Navigation. Use the Nissan CONSULT III diagnostic tool, or equivalent to perform the required resets.

1. Before servicing the vehicle, refer to the Precautions Section.

2. Disconnect the negative battery cable.

3. Raise and support the vehicle safely.

4. Remove the air duct. Remove the engine under cover.

5. Drain the engine oil.

6. Drain the engine coolant.

7. Remove the final drive, if equipped with 4WD.

8. Disconnect the steering gear lower shaft joint bolt and steering gear nuts and bolts, position the assembly out of the way.

9. Remove the starter.

10. Disconnect the automatic transmission fluid cooler brackets, if equipped and position them out of the way.

11. Remove the oil filter, as necessary. Remove the oil cooler.

12. Remove the lower oil pan. Remove the oil strainer.

13. Remove the transmission joint bolts which pierce the oil pan.

14. Remove the rear cover plate.

15. Loosen the upper oil pan retaining bolts, in the reverse order of the installation sequence.

16. Insert a seal cutter tool between the oil pan and the cylinder block, and slide it by tapping on the side of the tool with a hammer.

17. Remove the oil pan from the engine. Remove the O-rings from the bottom lower cylinder block and oil pump.

To install:

18. Be sure to clean all the oil gasket material from both the oil pan and the cylinder block surfaces, using the proper tools.

19. Install new O-rings on the bottom lower cylinder block and oil pump.

20. Apply a continuous bead of sealant 0.138–0.177 inches (3.5–4.5mm) to the lower cylinder block mating surfaces of the upper oil pan. Be sure to use genuine RTV sealant, or equivalent.

➡️ For bolt holes marked with a solid black triangle, apply liquid gasket outside the hole. Apply a bead of sealant 0.177–0.217 inch (4.5–5.5mm) in diameter to the "A" section.

21. Install the upper oil pan. This must be done within 5 minutes after applying the liquid gasket.

22. Torque the upper oil pan bolts to 16 ft. lbs. (22 Nm) in the proper sequence. There are 2 types of bolts:
- M8 x 3.97 inches (100mm) No. 7, 11, 12, and 13
- M8 x 0.98 inch (25mm) all except No. 7, 11, 12 and 13

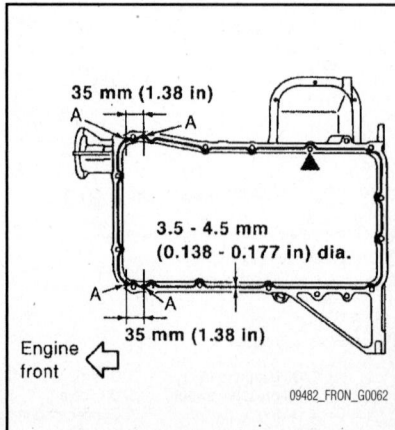

Fig. 133 Oil pan sealant application (upper)—4.0L engine

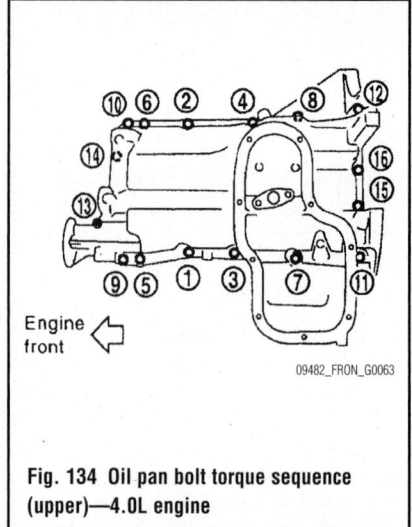

Fig. 134 Oil pan bolt torque sequence (upper)—4.0L engine

23. Tighten the transmission joint bolts.

24. Install the oil strainer to the upper oil pan.

25. Continue the installation in the reverse order of the removal procedure.

➡️ Wait 30 minutes after installation of the oil pan to allow the sealant to cure before adding oil.

26. Fill the crankcase with the proper grade and amount of oil.

27. Start the engine and check for leaks.

OIL PUMP

REMOVAL & INSTALLATION

2.5L Engine

See Figures 135 through 140.

The oil pump is integral to the balancer unit and is serviced as an assembly. Do not disassemble the balancer unit.

➡️ If working near and/or around the SRS system and components, be sure to disable the SRS system. After disabling the system wait three minutes or more before servicing the vehicle.

➡️ Whenever the negative battery cable is disconnected the following components will require resetting. The Idle Air Volume Learning, Steering Angle Sensor Neutral Position, Sunroof Memory Reset/Initialization, Automatic Drive Positioner System, Audio presets and Navigation. Use the Nissan CONSULT III diagnostic tool, or equivalent to perform the required resets.

1. Before servicing the vehicle, refer to the Precautions Section.

2. Disconnect the negative battery cable.

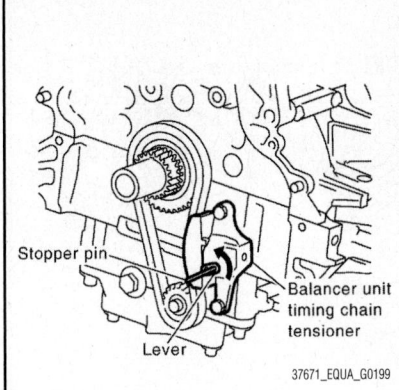

Fig. 135 Balancer unit timing chain tensioner release—2.5L engine

3. Remove the timing chain.

4. Remove the balancer unit timing chain tensioner with the following procedure:

    a. Lift the lever up and release the ratchet claw for return proof.

    b. Push the tensioner sleeve in and hold it.

    c. Matching the hole on the lever with the one on the body, insert a stopper pin to secure the tensioner sleeve.

➡**Use a hard metal pin as a stopper pin—approximately 0.04 inch (1mm) in diameter.**

    d. Remove the balancer unit timing chain tensioner.

5. Secure the hexagonal portion of the balancer shaft using a suitable tool. Loosen the balancer unit sprocket bolt.

6. Remove the balancer unit timing chain, balancer unit sprocket, and crankshaft sprocket.

➡**When removing the balancer unit timing chain, remove the crankshaft sprocket and balancer unit sprocket at the same time.**

7. Loosen the bolts of the balancer unit in reverse order of torque sequence.

❋❋ **WARNING**

Do not disassemble the balancer unit.

➡**Use TORX® socket size E14 for bolts No. 1 to 4.**

To install:

8. Check the balancer unit timing chain for cracks and any excessive wear at the roller links of the chain. Replace the timing chain, if necessary.

9. Measure the balancer unit bolt outer diameters at 2 positions.

- If a reduction appears in (A) range, regard it as (d2)
- Limit of (d1) minus (d2): 0.0059 inch (0.15mm)
- If the limit is exceeded, replace the balancer unit bolt with a new one.

10. Measure the balancer unit bolt length.

- If it exceeds the limit, replace the balancer unit bolt with a new one.
- Limit: 6.77 inches (172mm)

11. Make sure that the crankshaft key points straight up.

12. Install the O-ring to the balancer unit.

13. Tighten the bolts in numerical order with the following procedure to install the balancer unit, using special tool KV10112100 (BT8653-A), or equivalent. Apply new engine oil to the threads and seat surfaces of the bolts.

❋❋ **WARNING**

If the bolts are to be re-used, check their outer diameter before installation.

❋❋ **WARNING**

Check tightening angle using the special tool or a protractor. Do not make a judgment by a visual check alone.

    a. Step 1—Bolts 1–4: 35 ft. lbs. (48 Nm).

    b. Step 2—Bolts 1–4: 100° clockwise.

    c. Step 3—Bolts 1–4: 0 ft. lbs. (0 Nm) in reverse order of the tightening sequence.

    d. Step 4—Bolts 1–4: 35 ft. lbs. (48 Nm).

    e. Step 5—Bolts 1–4: 100° clockwise.

    f. Step 6—Bolts 5–6: 22 ft. lbs. (30 Nm).

14. Install the crankshaft sprocket, balancer unit sprocket, and balancer unit timing chain.

    a. Make sure that the crankshaft sprocket is positioned with the mating marks on the cylinder block and crankshaft sprocket meeting at the top.

    b. Install by aligning the mating marks on each sprocket and the balancer unit timing chain.

15. Secure the hexagonal portion of the balancer shaft using a suitable tool. Tighten the balancer unit sprocket bolt to 48 ft. lbs. (65 Nm).

➡**Install the crankshaft sprocket, balancer unit sprocket, and balancer unit timing chain at the same time.**

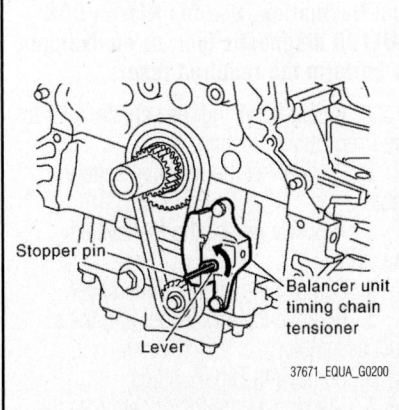

Fig. 136 Balancer unit bolt torque sequence shown—2.5L engine

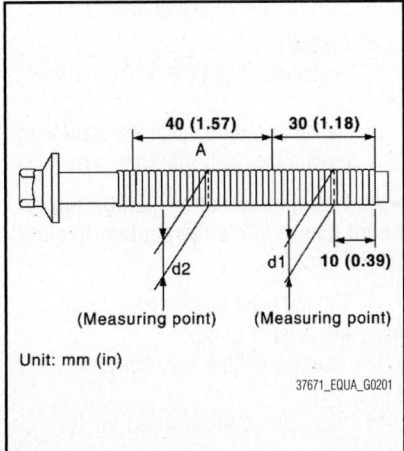

Fig. 137 Measuring balancer unit bolt diameters—2.5L engine

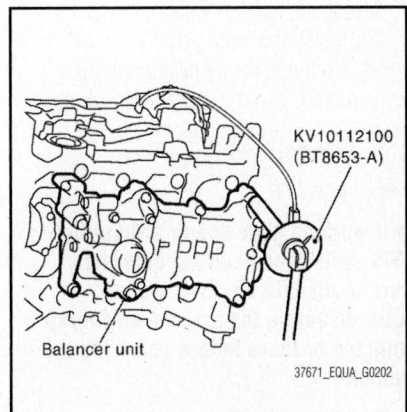

Fig. 138 Special tool KV10112100 (BT8653-A) shown tightening balancer unit bolts—2.5L engine

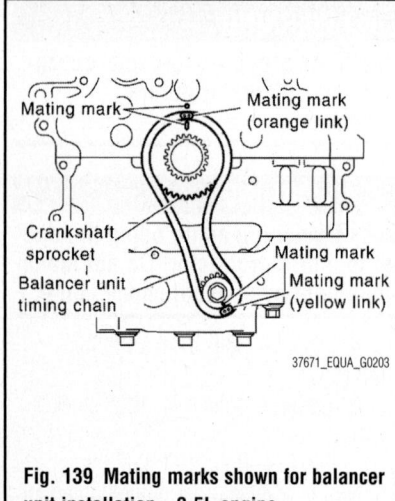

Fig. 139 Mating marks shown for balancer unit installation—2.5L engine

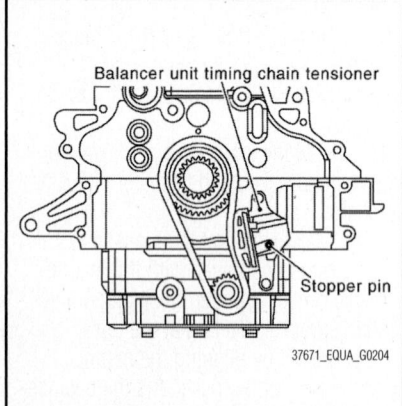

Fig. 140 Install the balancer unit timing chain tensioner and remove the stopper pin—2.5L engine

16. Install the balancer unit timing chain tensioner.

a. After installation, make sure the mating marks have not slipped.

b. Remove the stopper pin and release the tensioner sleeve.

17. Install the timing chain.

18. Continue the installation in the reverse order of the removal procedure.

4.0L Engine

See Figure 141.

➡If working near and/or around the SRS system and components, be sure to disable the SRS system. After disabling the system wait three minutes or more before servicing the vehicle.

➡Whenever the negative battery cable is disconnected the following components will require resetting. The Idle Air Volume Learning, Steering Angle

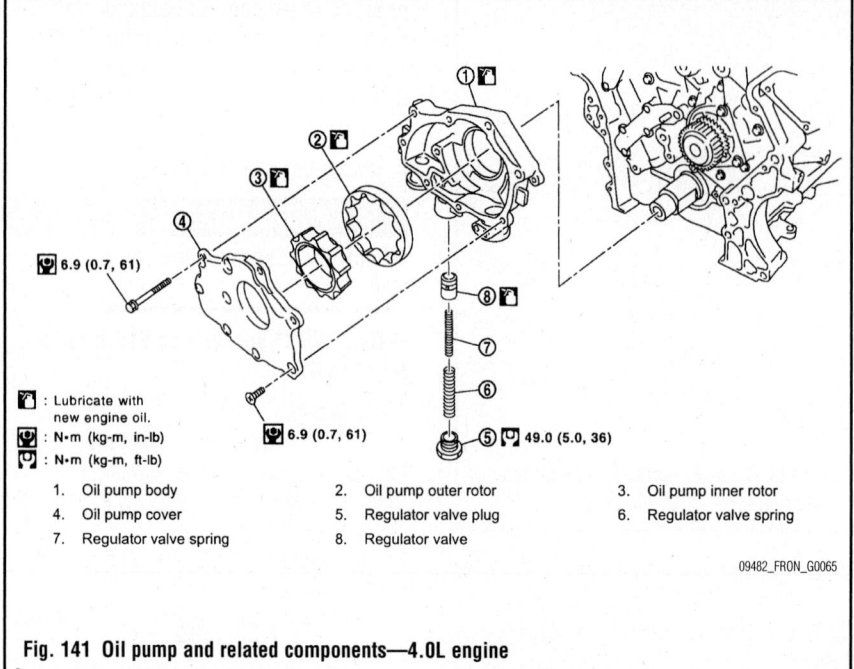

Fig. 141 Oil pump and related components—4.0L engine

Sensor Neutral Position, Sunroof Memory Reset/Initialization, Automatic Drive Positioner System, Audio presets and Navigation. Use the Nissan CONSULT III diagnostic tool, or equivalent to perform the required resets.

1. Before servicing the vehicle, refer to the Precautions Section.

2. Disconnect the negative battery cable.

3. Remove the front wheels and tires.

4. Remove the front fender protectors.

5. Drain the engine oil.

6. Drain the engine coolant.

7. Remove the air duct and resonator assembly and the air cleaner case (upper).

8. Remove the timing chain (primary) only.

9. Remove the oil pump assembly.

To install:

10. Installation is in the reverse order of removal.

11. When installing, align the crankshaft flat faces with the inner rotor flat faces.

➡Wait 30 minutes after installation of the oil pan to allow the sealant to cure before adding oil.

12. Fill the crankcase with the proper grade and amount of oil.

13. Start the engine and check for leaks.

14. Stop the engine and wait for 10 minutes.

15. Check the engine oil level and adjust as required.

TIMING CHAIN COVER, CHAIN, TENSIONER, & SPROCKETS

REMOVAL & INSTALLATION

2.5L Engine

See Figures 142 through 151.

➡If working near and/or around the SRS system and components, be sure to disable the SRS system. After disabling the system wait three minutes or more before servicing the vehicle.

➡Whenever the negative battery cable is disconnected the following components will require resetting. The Idle Air Volume Learning, Steering Angle Sensor Neutral Position, Sunroof Memory Reset/Initialization, Automatic Drive Positioner System, Audio presets and Navigation. Use the Nissan CONSULT III diagnostic tool, or equivalent to perform the required resets.

1. Before servicing the vehicle, refer to the Precautions Section.

2. Disconnect the negative battery cable.

3. Properly relieve the fuel system pressure.

4. Remove the engine undercover.

5. Remove the air cleaner and the air duct assembly.

6. Remove the spark plugs.

7. Disconnect the PCV hose from the rocker cover. Remove the ignition coil.

8. Remove the PCV valve and O-ring from the rocker cover, if necessary.

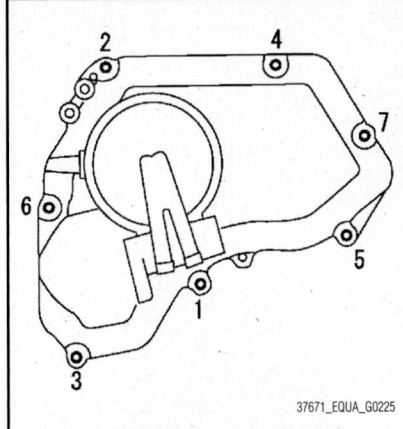

Fig. 142 Intake Valve Timing (IVT) control cover bolt tightening sequence—2.5L engine

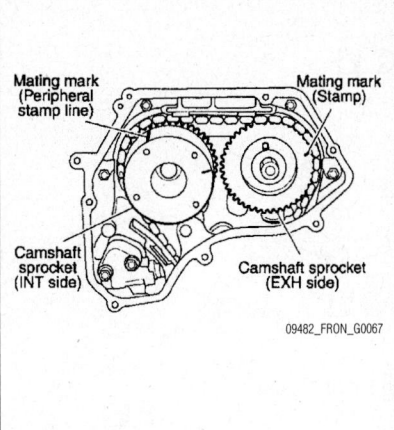

Fig. 143 Timing chain alignment marks—2.5L engine

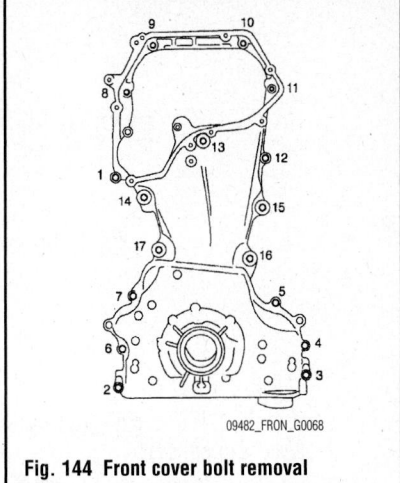

Fig. 144 Front cover bolt removal sequence—2.5L engine

9. Remove the oil filler cap from the rocker cover, if necessary.

10. Remove the rocker cover retaining bolts. Be sure to remove the bolts by reversing the order of the tightening torque sequence.

11. Remove the rocker cover. Discard the gasket.

12. Remove the coolant reservoir tank. Remove the auxiliary drive belt auto-tensioner.

13. Remove the alternator. Remove the strut tower brace.

14. Remove the air conditioning compressor and position it to the side. Do not disconnect the refrigerant lines.

15. Remove the power steering pump and reservoir tank; position the assembly to the side. Do not disconnect the fluid lines.

16. Remove the oil pan. Remove the strainer.

17. Remove the Intake Valve Timing (IVT) control cover bolts. Be sure to remove the bolts by reversing the order of the tightening sequence.

18. Remove the cover by cutting the sealant using tool KV10111100 (J-37228), or equivalent.

19. Position the number one cylinder on its compression stroke by rotating the crankshaft pulley clockwise. Align the mating marks for Top Dead Center (TDC) with the timing indicator on the front cover.

20. At the same time, make sure that the mating marks on the camshaft sprockets are lined up as indicated in the illustration. If not, rotate the crankshaft one more turn to line up the mating marks.

21. Remove the crankshaft pulley.

22. Loosen the front cover retaining bolts, in the order indicated in the bolt loosening sequence illustration. Remove the front cover. Be careful not to damage the mating surfaces.

23. Using a seal removal tool, remove the oil seal, as required.

24. To remove the timing chain, push in on the chain tensioner plunger. Insert a stopper pin into the hole on the chain tensioner body to secure the chain tensioner plunger. Remove the chain tensioner. Remove the timing chain.

➡Use a metal pin as a stopper pin approximately 0.02 inch (0.5mm) diameter.

✳✳ WARNING

Do not rotate the crankshaft or camshafts with the chain removed. It causes interference between the valves and the pistons.

25. Remove the camshaft sprockets.

26. Remove the timing chain slack guide, timing chain tensioner guide and spacer.

27. Remove the balancer unit timing chain tensioner by lifting the lever up and releasing the ratchet claw for return proof. Push the tensioner sleeve in and hold it. Matching the hole on the lever with the one on the body, insert a stopper pin to secure the tensioner sleeve. Remove the balancer unit timing chain tensioner.

➡Use a metal pin as a stopper pin approximately 0.04 inch (1mm) diameter.

28. Secure the hexagonal portion of the balancer shaft using a suitable tool. Loosen the balancer unit sprocket bolt.

29. Remove the balancer unit timing chain, balancer unit sprocket and crankshaft sprocket.

➡**When removing the balancer unit timing chain, remove the crankshaft sprocket and balancer unit sprocket at the same time.**

30. Loosen the balancer unit mounting bolts, in the reverse order of the tightening sequence. Remove the balancer unit. Do not disassemble the balancer unit. Bolts No. 1–4 use Torx® head socket size E14.

To install:

31. Check the chain for cracks and excessive wear, replace as required.

32. Measure the balancer unit bolt outer diameters ("d1" and "d2") at two positions, as shown in the illustration. If reduction appears in the "A" range, regard it as "d2". Specification is as follows: ("d1" - "d2"): 0.0059 inch (0.15mm). If it exceeds the specification (large difference in dimensions) replace the balancer unit bolt with a new one.

33. Measure the balancer bolt unit length. If it exceeds the specification replace the balancer unit bolt with a new one. Specification is 6.77 inches (172mm).

34. Make sure that the crankshaft key is pointing straight up. Install the O-ring to the balancer unit.

35. Install the balancer unit. Apply engine oil to the bolt threads and sealing surfaces. Torque the bolts to specification and in the proper sequence using tool KV10112100 (BT8653-A), or equivalent.

✳✳ WARNING

If the bolts are to be re-used, check their outer diameter before installation.

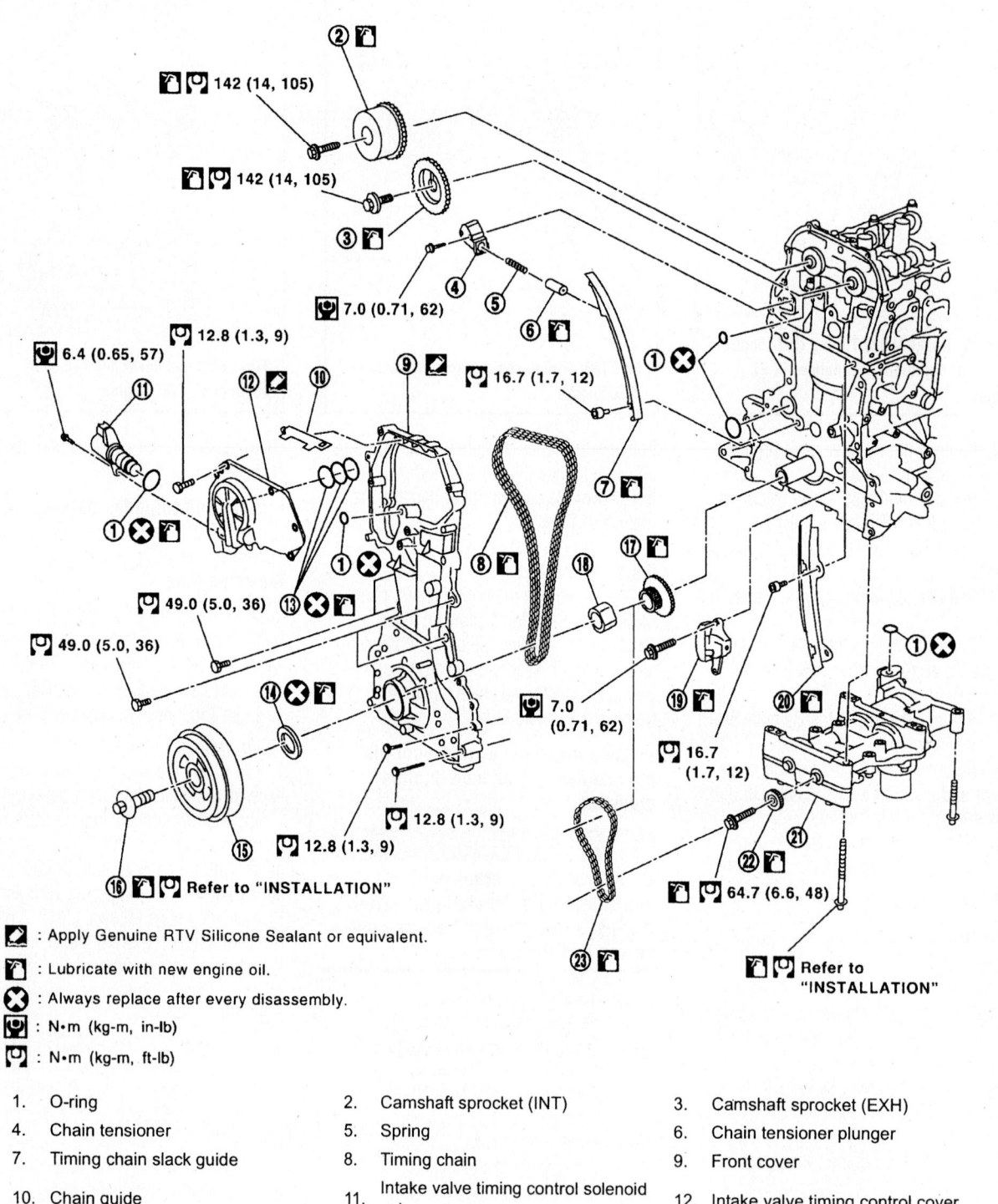

Fig. 145 Timing chain and related components—2.5L engine

142 (14, 105)

142 (14, 105)

7.0 (0.71, 62)

6.4 (0.65, 57)

12.8 (1.3, 9)

16.7 (1.7, 12)

49.0 (5.0, 36)

49.0 (5.0, 36)

7.0 (0.71, 62)

16.7 (1.7, 12)

12.8 (1.3, 9)

12.8 (1.3, 9)

64.7 (6.6, 48)

Refer to "INSTALLATION"

Refer to "INSTALLATION"

: Apply Genuine RTV Silicone Sealant or equivalent.

: Lubricate with new engine oil.

: Always replace after every disassembly.

: N•m (kg-m, in-lb)

: N•m (kg-m, ft-lb)

1. O-ring	2. Camshaft sprocket (INT)	3. Camshaft sprocket (EXH)
4. Chain tensioner	5. Spring	6. Chain tensioner plunger
7. Timing chain slack guide	8. Timing chain	9. Front cover
10. Chain guide	11. Intake valve timing control solenoid valve	12. Intake valve timing control cover
13. Oil ring	14. Front oil seal	15. Crankshaft pulley
16. Crankshaft pulley bolt	17. Crankshaft sprocket	18. Spacer
19. Balancer unit timing chain tensioner	20. Timing chain tension guide	21. Balancer unit
22. Balancer unit sprocket	23. Balancer unit timing chain	

09482_FRON_G0066

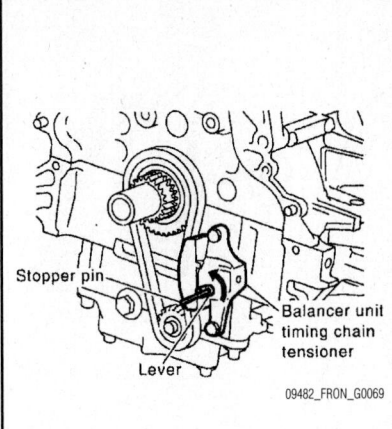

Fig. 146 Balance unit stopper pin installation—2.5L engine

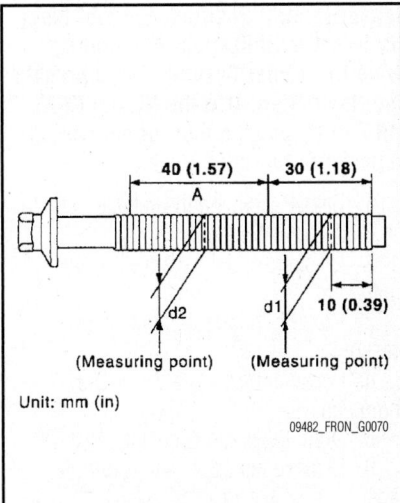

Unit: mm (in)

Fig. 147 Balance unit bolt measurement—2.5L engine

✹✹ WARNING

Check the tightening angle using the special tool or a protractor. Do not make a judgment by a visual check alone.

   a. Step 1—Bolts 1–4: 35 ft. lbs. (48 Nm).
   b. Step 2—Bolts 1–4: 100° clockwise.
   c. Step 3—Bolts 1–4: 0 ft. lbs. (0 Nm) in reverse order of the tightening sequence.
   d. Step 4—Bolts 1–4: 35 ft. lbs. (48 Nm).
   e. Step 5—Bolts 1–4: 100° clockwise.
   f. Step 6—Bolts 5–6: 22 ft. lbs. (30 Nm).

36. Install the crankshaft sprocket, balancer unit sprocket, and balancer timing chain.

37. Make sure that the crankshaft sprocket is positioned with the mating marks on the cylinder block and crankshaft sprocket meeting at the top.

38. Install it by aligning the mating marks on each sprocket and balancer unit timing chain.

39. Secure the hexagonal portion of the balancer shaft using a suitable tool. Tighten the balancer unit sprocket bolt to specification.

➡**Install the crankshaft sprocket, balancer unit sprocket, and balancer unit timing chain at the same time.**

40. Install the balancer unit timing chain tensioner.

➡**After installation, make sure that the mating marks have not slipped. Remove the stopper pin and release the tensioner sleeve.**

41. Align the mating marks on each sprocket and timing chain. Install the timing chain and related parts.

42. Before and after installing the chain tensioner, check again to be sure that the mating marks have not slipped.

43. After installing the chain tensioner, remove the stopper pin. Make sure that the tensioner moves freely.

➡**After the mating marks are aligned, keep them aligned by holding them with your hand. To avoid skipped teeth, do not rotate the crankshaft and camshaft until the cover is installed.**

➡**Before installing the chain tensioner, it is possible to change the position of the mating mark on the timing chain for that on each sprocket for alignment.**

44. Install the front cover oil seal. Install O-rings to the cylinder head and the cylinder block.

45. Apply a continuous bead of liquid gasket to the front cover. Be sure to use genuine RTV sealant, or equivalent.

➡**Sealant application instructions differ depending on position, refer to the illustration for positioning. Detail "A", cross over the start of the application and the end. Detail "B", apply liquid gasket outside of the bolt holes. For all bolt holes other than "B", apply to the inside. Detail "C", between here only, apply a bead of sealant 0.177–0.217 inch (4.5–5.5mm) diameter.**

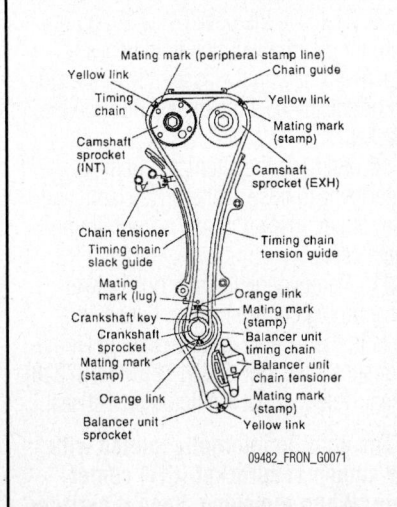

Fig. 148 Timing chain alignment—2.5L engine

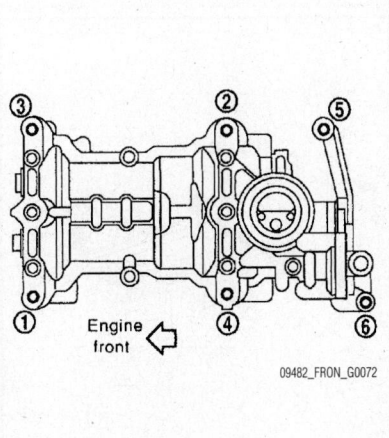

Fig. 149 Balance unit bolt torque sequence—2.5L engine

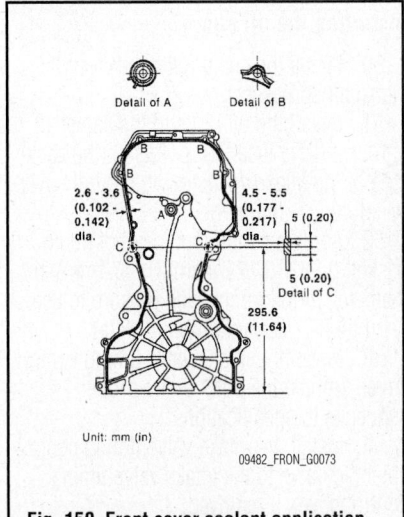

Unit: mm (in)

Fig. 150 Front cover sealant application with respect to positioning—2.5L engine

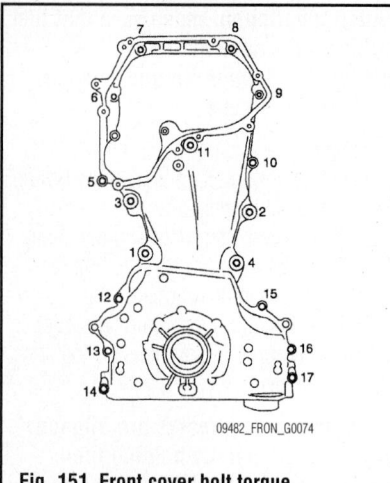

Fig. 151 Front cover bolt torque sequence—2.5L engine

46. Make sure that the mating marks of the chain and each sprocket are still aligned.

47. Install the front cover. Torque the retaining bolts to specification and in the proper sequence.

48. Use the following for locating the M6 bolts:

- Bolt position 5, 10, 14 and 17— Bolt length: 1.77 inches (45mm)
- Bolts except the above (except 1 to 4)—Bolt length: 0.79 inch (20mm)

49. Tighten the bolts to the specified torque:

- Bolt position 5–17: 9 ft. lbs. (13 Nm)
- Bolt position 1–4: 36 ft. lbs. (49 Nm)
- After all bolts are tightened, retighten them to specification in the numerical order shown

➡**Be sure to wipe off any excess liquid gasket leaking to the surface from installing the oil pan.**

50. Install the chain guide between the camshaft sprockets.

51. Install the oil rings to the camshaft sprocket (INT) insertion points on the backside of the intake valve timing control cover. Install the O-ring to the front cover.

52. Apply a continuous bead of liquid gasket, 0.083–0.122 inch (2.1–3.1mm) in diameter, to the front cover. Be sure to use genuine RTV sealant, or equivalent.

53. Install the intake valve timing control cover. Tighten the bolts in the proper sequence to specification.

54. Install the intake valve timing control solenoid valve to the intake valve timing control cover, if removed.

55. Connect the ground cables, and install the harness clip.

56. Install the crankshaft pulley.

57. When installing the rocker cover, apply liquid gasket, be sure to use genuine RTV silicone sealant, or equivalent, to the positions shown in the illustration. Refer to figure "a" to apply liquid gasket to joint part of camshaft bracket No. 1 and cylinder head. Refer to figure "b" to apply liquid gasket in 90 degrees to figure "b".

58. Install the rocker cover.

59. Continue the installation in the reverse order of the removal procedure.

➡**If hydraulic pressure inside the timing chain tensioner drops after removal/installation, slack in the guide may generate a pounding noise during and just after engine start. This is normal. The noise should stop after the hydraulic pressure rises.**

4.0L Engine

See Figures 152 through 167.

➡**The procedure below describes the removal and installation of the front timing case and timing chain related parts and rear timing chain case, when the upper oil pan needs to be removed or installed. When only the timing chain (primary) is being removed it is not necessary to remove the rocker covers.**

➡**If working near and/or around the SRS system and components, be sure to disable the SRS system. After disabling the system wait three minutes or more before servicing the vehicle.**

➡**Whenever the negative battery cable is disconnected the following components will require resetting. The Idle**

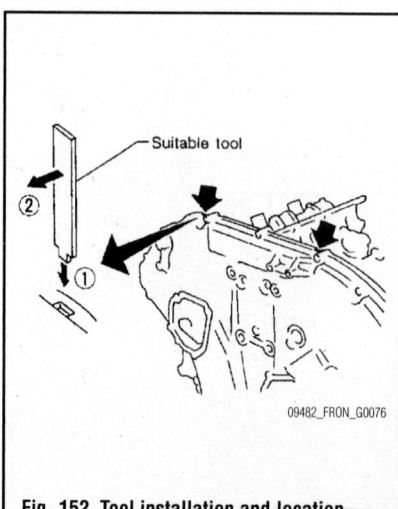

Fig. 152 Tool installation and location— 4.0L engine

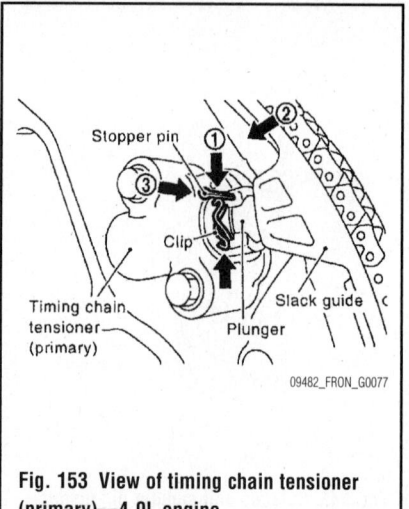

Fig. 153 View of timing chain tensioner (primary)—4.0L engine

Air Volume Learning, Steering Angle Sensor Neutral Position, Sunroof Memory Reset/Initialization, Automatic Drive Positioner System, Audio presets and Navigation. Use the Nissan CONSULT III diagnostic tool, or equivalent to perform the required resets.

1. Before servicing the vehicle, refer to the Precautions Section.

2. Disconnect the negative battery cable.

3. Properly relieve the fuel system pressure.

4. Remove the engine cover. Drain the engine oil.

5. Drain the engine coolant.

6. Remove the upper and lower oil pans.

7. Remove the radiator cooling fan assembly. Remove the drive belts.

8. Separate the engine wiring harnesses by removing their brackets from the front timing chain case.

9. Remove the power steering pump from the bracket with the fluid hoses attached. Position the assembly to the side. Do not disconnect the hoses. Remove the bracket.

10. Remove the alternator. Remove the water bypass hose, water hose clamp and idler pulley bracket from the front timing chain case.

11. Remove the left and right intake valve timing control covers. Loosen the bolts in the reverse order of the tightening sequence. Use tool KV10111100 (J-37228), or equivalent to cut the liquid gasket seal.

➡**The shaft is internally jointed with the camshaft sprocket (INT) center hole. When removing, keep it horizontal until it is completely disconnected.**

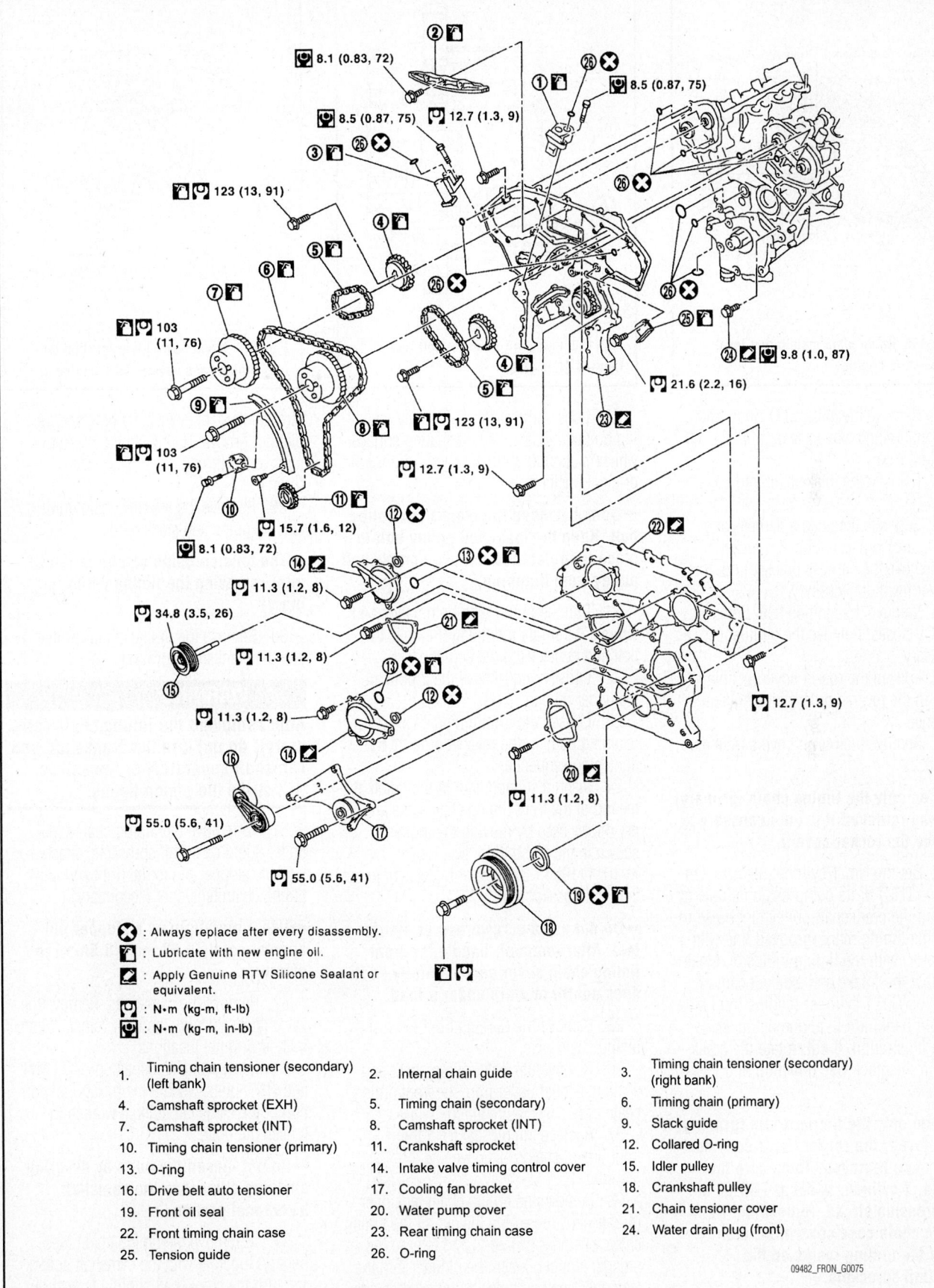

8.1 (0.83, 72)

8.5 (0.87, 75)

8.5 (0.87, 75)   12.7 (1.3, 9)

9.8 (0.87, 75)

123 (13, 91)

103 (11, 76)

103 (11, 76)

9.8 (1.0, 87)

21.6 (2.2, 16)

123 (13, 91)

12.7 (1.3, 9)

15.7 (1.6, 12)

8.1 (0.83, 72)

34.8 (3.5, 26)

11.3 (1.2, 8)

11.3 (1.2, 8)

11.3 (1.2, 8)

12.7 (1.3, 9)

11.3 (1.2, 8)

55.0 (5.6, 41)

55.0 (5.6, 41)

⊗ : Always replace after every disassembly.

🔧 : Lubricate with new engine oil.

◩ : Apply Genuine RTV Silicone Sealant or
    equivalent.

🔩 : N·m (kg-m, ft-lb)

🔩 : N·m (kg-m, in-lb)

1. Timing chain tensioner (secondary)
    (left bank)
2. Internal chain guide
3. Timing chain tensioner (secondary)
    (right bank)
4. Camshaft sprocket (EXH)
5. Timing chain (secondary)
6. Timing chain (primary)
7. Camshaft sprocket (INT)
8. Camshaft sprocket (INT)
9. Slack guide
10. Timing chain tensioner (primary)
11. Crankshaft sprocket
12. Collared O-ring
13. O-ring
14. Intake valve timing control cover
15. Idler pulley
16. Drive belt auto tensioner
17. Cooling fan bracket
18. Crankshaft pulley
19. Front oil seal
20. Water pump cover
21. Chain tensioner cover
22. Front timing chain case
23. Rear timing chain case
24. Water drain plug (front)
25. Tension guide
26. O-ring

09482_FRON_G0075

Fig. 154 Timing chain and related components—4.0L engine

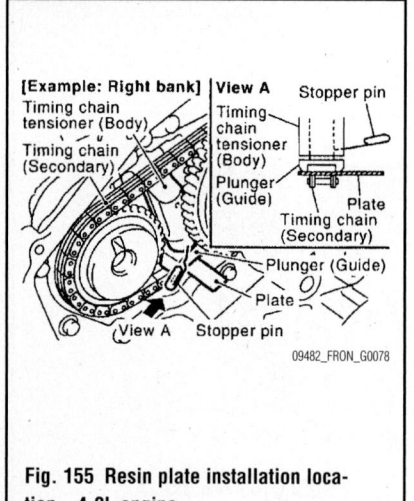

Fig. 155 Resin plate installation location—4.0L engine

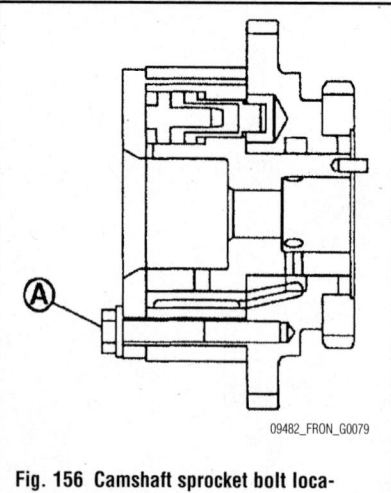

Fig. 156 Camshaft sprocket bolt location—4.0L engine

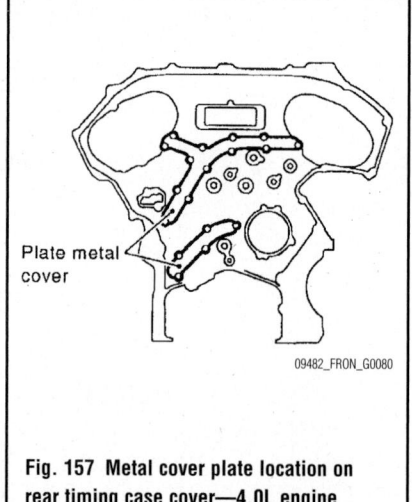

Fig. 157 Metal cover plate location on rear timing case cover—4.0L engine

12. Remove the collared O-rings from the front timing chain case on both the left and right side.

13. Remove the intake manifold collector.

14. Separate the engine harness and remove their brackets from the rocker covers. Remove the harness bracket from the cylinder head, if necessary.

15. Remove the ignition coil. Remove the PCV hoses. Remove the oil filler cap, if necessary.

16. Loosen the rocker cover retaining bolts, in the reverse order of the tightening sequence.

17. Remove the rocker covers from the engine.

➡**When only the timing chain (primary) is being removed it is not necessary to remove the rocker covers.**

18. Set the No. 1 cylinder at Top Dead Center (TDC) of its compression stroke by rotating the crankshaft pulley clockwise to align the timing mark (grooved line without color) with the timing indicator. Make sure that the intake and exhaust cam noses on No. 1 cylinder (engine front side on right bank) are in alignment as shown in the illustration. If not, rotate the crankshaft in the clockwise direction 360 degrees.

➡**When only the timing chain (primary) is removed, the rocker cover does not need to be removed. To be sure that the No. 1 cylinder is set at TDC on the compression stroke, remove the front timing chain case cover first, then check the mating marks on the camshaft sprockets.**

19. Remove the starter. Position tool KV10117700 (J-44716), or equivalent.

20. Loosen the crankshaft pulley retaining bolt and locate the bolt seating surface, which is about 0.39 inch (10mm) from its original position.

➡**Do not remove the crankshaft pulley bolt. Keep the loosened pulley bolt in place to protect the removed crankshaft pulley from dropping.**

21. Pull the pulley with both hands and remove it from its mounting. Remove the bolt and pulley from the engine.

22. Loosen and remove the 2 bolts of the upper oil pan.

23. Loosen the front timing chain cover retaining bolts in the reverse order of the tightening sequence.

24. Insert a suitable tool in the notch at the top of the front timing chain case and pry off the case by moving the tool as shown in the illustration. Use tool KV10111100 (J-37228), or equivalent to cut the liquid gasket seal.

➡**Do not use a screwdriver or similar tool. After removal, handle the front timing chain cover case carefully so it does not tilt or warp under a load.**

25. Remove the O-rings from the rear timing chain case.

26. Remove the water pump cover and chain tensioner cover from the front timing chain case cover, as required.

27. Remove the oil seal from the front timing chain case cover, as required.

28. Remove the timing chain tensioner (primary) by loosening the clip of the timing chain tensioner (primary) and release the plunger stopper. Insert the plunger into the tensioner body by pressing the slack guide. Keep the slack guide pressed and hold the plunger in by pushing the stopper pin

through the tensioner body hole and the plunger groove. Remove the bolts and remove the timing chain tensioner (primary).

29. Remove the internal chain guide, tension guide and slack guide.

➡**The tension guide can be removed after removing the timing chain (primary).**

30. Remove the timing chain (primary) and the crankshaft sprocket.

✷✷ WARNING

After removing the timing chain (primary), do not turn the crankshaft and camshaft separately or the valves will strike the piston heads.

31. To remove the timing chain (secondary) and camshaft sprockets, attach a suitable stopper pin to the right and left timing chain tensioner (secondary).

➡**Use a metal pin as a stopper pin approximately 0.02 inch (0.5mm) in diameter.**

32. Remove the camshafts. Remove the valve lifters. Identify them for reinstallation in their original locations.

33. Remove the camshaft sprocket (INT and EXH) bolts. Secure the hexagonal portion of the camshaft using a wrench to loosen the bolts.

➡**Do not loosen the bolts by securing anything other than the camshaft hexagonal portion.**

34. To remove the timing chain (secondary) together with the camshaft sprockets, turn the crankshaft slightly to secure slackness of the timing chain on the timing chain tensioner (secondary) side.

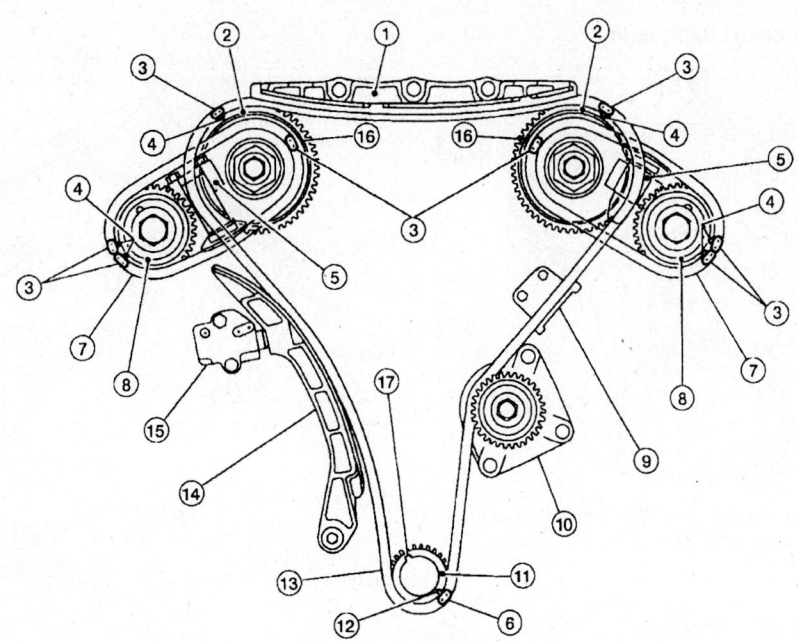

1. Internal chain guide
2. Camshaft sprocket (intake)
3. Mating mark (copper link)
4. Mating mark (punched)
5. Secondary timing chain tensioner
6. Mating mark (yellow link)
7. Secondary timing chain
8. Camshaft sprocket (exhaust)
9. Tensioner guide
10. Water pump
11. Crankshaft sprocket
12. Mating mark (notched)
13. Primary timing chain
14. Slack guide
15. Primary timing chain tensioner
16. Mating mark (back side)
17. Crankshaft key

09482_FRON_G0081

Fig. 158 Timing chain alignment—4.0L engine

35. Insert a 0.02 inch (0.5mm) thick metal or resin plate between the timing chain and timing chain plunger (guide). Remove the timing chain (secondary) together with the camshaft sprockets with the timing chain loose from the guide groove.

✳ CAUTION

Be careful of the plunger coming off when removing the timing chain (secondary). This is because the plunger of the timing chain tensioner (secondary) moves during operation, leading it to coming off its fixed stopper pin.

➡ **The camshaft sprocket (INT) is a one piece integrated design sprocket for the timing chain (primary) and for the timing chain (secondary). When handling the sprocket, avoid shock to the sprocket. Do not disassemble or loosen bolt "A", as shown in the illustration.**

36. Remove the water pump.
37. Remove the rear timing chain case cover bolts, in the reverse order of the tightening sequence. Using the proper tool, cut the liquid gasket sealant seal. Remove the cover.

➡ **Do not remove the metal cover of the oil passage. After removal, handle the case carefully so it does not tilt or warp under a load.**

38. Remove the O-rings from the cylinder head and No. 1 camshaft bracket. Remove the O-rings from the cylinder block.
39. If necessary, remove the timing chain tensioners (secondary) from the cylinder head by first removing the No. 1 camshaft bracket. Remove the timing chain tensioners (secondary) with the stopper pin attached.

To install:

40. Check the chain for cracks and excessive wear, replace as required.
41. Be sure to remove all old gasket material from bolts and bolt holes.
42. If removed, install the timing chain tensioners (secondary) to the cylinder head.

43. Install camshaft bracket No. 1.
44. To install the rear timing chain case cover, first install new O-rings to the cylinder block. Install new O-rings to the cylinder head and camshaft bracket No. 1.
45. Apply liquid gasket sealant to the rear timing chain case back side, as shown in the illustration. Be sure to use genuine RTV sealant, or equivalent.

➡ **For "A" in the figure, completely wipe out excessive liquid gasket extended on a portion touching at engine coolant. Apply liquid gasket on the installation position of the water pump and cylinder head very completely.**

46. Align the rear timing case with dowel pins (right and left) on the cylinder block. Install the rear timing chain case. Make sure that the O-rings stay in place during installation to the cylinder block, cylinder head and camshaft bracket No. 1.
47. Tighten the bolts to specification and in the proper sequence.

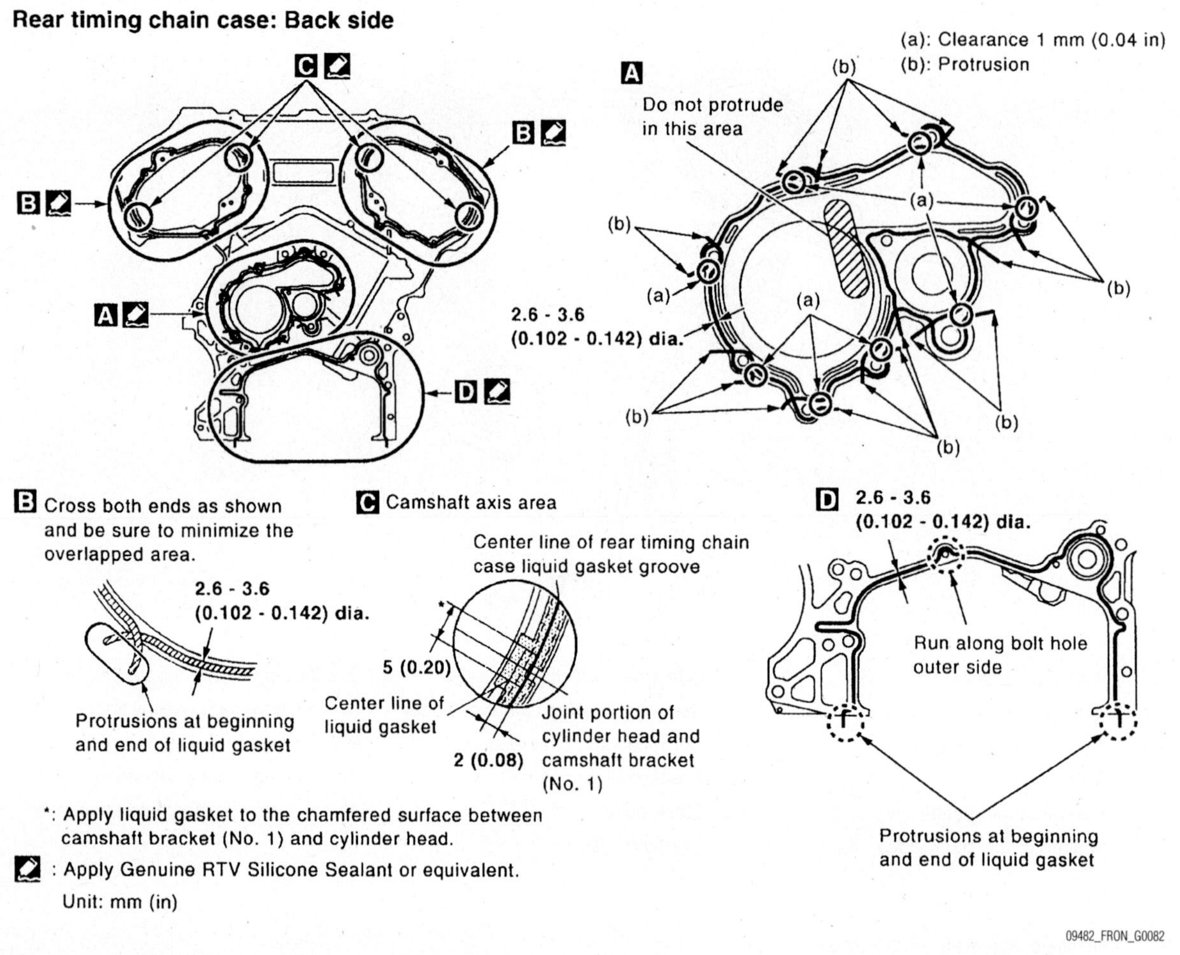

Rear timing chain case: Back side

(a): Clearance 1 mm (0.04 in)
(b): Protrusion

Do not protrude in this area

2.6 - 3.6 (0.102 - 0.142) dia.

B Cross both ends as shown and be sure to minimize the overlapped area.

2.6 - 3.6 (0.102 - 0.142) dia.

Protrusions at beginning and end of liquid gasket

C Camshaft axis area

Center line of rear timing chain case liquid gasket groove

5 (0.20)

Center line of liquid gasket

2 (0.08)

Joint portion of cylinder head and camshaft bracket (No. 1)

D 2.6 - 3.6 (0.102 - 0.142) dia.

Run along bolt hole outer side

Protrusions at beginning and end of liquid gasket

*: Apply liquid gasket to the chamfered surface between camshaft bracket (No. 1) and cylinder head.

: Apply Genuine RTV Silicone Sealant or equivalent.

Unit: mm (in)

09482_FRON_G0082

Fig. 159 Rear timing chain cover sealant application—4.0L engine

- Bolt length 0.79 inch (20mm)— Bolt position 1, 2, 3, 6, 7, 8, 9, and 10: Tighten to 9 ft. lbs. (13 Nm)
- Bolt length 0.63 inch (16mm)— Bolt position 4, 5, and 11: Tighten to 9 ft. lbs. (13 Nm)
- Bolt length 0.63 inch (16mm)— Bolt position 12–26: Tighten to 11 ft. lbs. (15 Nm)

48. After all bolts are tightened, retighten them to specification and in the proper sequence.

➡**Be sure to wipe off any excess liquid gasket leaking to the surface for installing the oil pan.**

49. After installing the rear timing case, check the surface height deference between the rear timing chain case and the lower cylinder block. Specification should be −0.0094–0.0055 inch (−0.24–0.14mm). If not within specification, repeat the installation procedure.

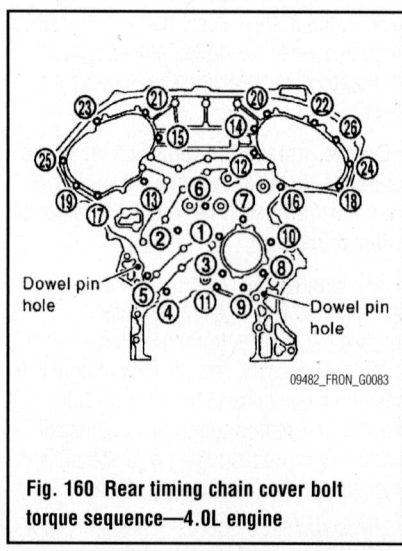

Dowel pin hole

Dowel pin hole

09482_FRON_G0083

Fig. 160 Rear timing chain cover bolt torque sequence—4.0L engine

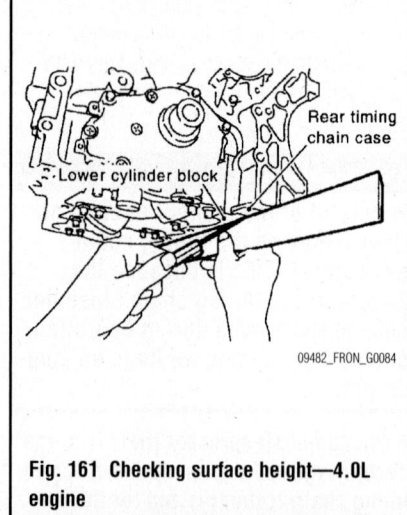

Rear timing chain case

Lower cylinder block

09482_FRON_G0084

Fig. 161 Checking surface height—4.0L engine

50. Install the water pump, using new O-rings.

51. Make sure that the dowel pin hole, dowel pin of the camshaft, and the crank-

shaft key are located with the number one piston at TDC on the compression stroke.

➡**Though the camshaft does not stop at the position, as shown in the illustra-**

tion, for placement of the cam nose it is generally accepted that the camshaft is placed in the same direction as the illustration. The camshaft dowel pin hole (intake side): at the cylinder head upper face side in each bank. Camshaft dowel pin hole (exhaust side): at the cylinder head upper face side in each bank. Crankshaft key: at the cylinder head side of the right bank. The hole on the small diameter side must be used for the intake side dowel pin hole.

52. To install the timing chains (secondary) and camshaft sprockets, push the plunger of the timing chain tensioner (secondary) and keep it pressed in with the stopper pin.

✳✳ WARNING

The mating surfaces between the timing chain and sprockets slip easily. Confirm all mating mark positions repeatedly during the installation process.

53. Install the timing chains (secondary) and camshaft sprockets (INT and EXH).

54. Align the mating marks on the timing chain (secondary) cooper color link, with the ones on the camshaft sprockets (INT and EXH) punched and install them.

➡**Mating marks for the camshaft sprocket (INT) are on the back side of the camshaft sprocket (secondary). There are 2 types of mating marks, circle and oval. They should be used for the right and the left banks, respectively. Right bank: circle type. Left bank: oval type.**

55. Align the dowel pin and pin hole on the camshafts with the groove and the dowel pin on the sprockets, and install them.

56. On the exhaust side, align the pin hole on the small diameter side of the camshaft front end with the dowel pin on the back side of the camshaft sprocket, and install them.

57. On the exhaust side, align the dowel pin on the camshaft front end with the pin groove on the camshaft sprocket, and install them.

➡**In case that the positions of each mating mark and each dowel pin will not fit on the mating marks, make a fine adjustment to the position holding the hexagonal portion on the camshaft with a wrench, or equivalent.**

➡**Bolts for the camshaft sprockets must be tightened. Tightening them by hand is enough to prevent the dislocation of the dowel pins. It may be difficult to visually check the dislocation of mating marks during and after installation. To make the matching easier, make a mating mark on the top of the sprocket teeth and its extended line in advance with paint.**

58. After confirming that the mating marks are aligned, tighten the camshaft sprocket bolts.

59. Pull the stopper pins out from the timing chain tensioners (secondary). Install the tension guide.

60. To install the timing chain (primary), install the crankshaft sprocket. Be sure that the mating marks on the crankshaft sprocket face the front of the engine.

61. Install the timing chain (primary).

➡**Install the timing chain (primary) so that the mating mark punched (B) on** the camshaft sprocket is aligned with the copper link (A) on the timing chain, while the mating mark notched (E) on the crankshaft sprocket (D) is aligned with the yellow link (F) on the timing chain, as shown in the illustration. If it is difficult to align mating marks (A) with (B) and (E) with (F) of the timing chain (primary) with each sprocket, gradually turn the camshaft using a wrench on the hexagonal portion to align it with the timing marks. During alignment be careful to prevent dislocation of the mating marks alignments of the timing chains (secondary). Note (G) indicates the water pump.

62. Install the internal chain guide, slack guide and timing chain tensioner (primary).

➡**Do not over tighten the slack guide bolts. It is normal for a gap to exist under the bolt seats when the bolts are tightened to specification.**

63. When installing the timing chain tensioner (primary), push in the plunger and keep it pressed in with the stopper pin. Remove any dirt on the surfaces. After installation, pull out the stopper pin by pressing the slack guide.

64. Make sure, again, that the mating marks on the camshaft sprockets and timing chain have not slipped out of alignment. Install new O-rings on the rear timing chain case.

65. Install a new front seal in the front timing chain case cover.

66. Install the water pump cover and chain tensioner cover to the front timing chain case cover. Apply a continuous bead of liquid gasket 0.091–0.130 inch (2.31–3.30mm) in diameter to the front timing chain case cover before installing the

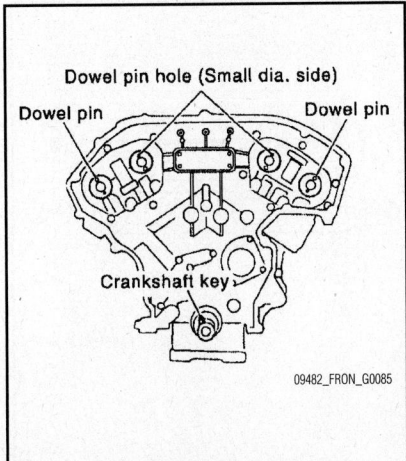

Fig. 162 Dowel pin and crankshaft key alignment—4.0L engine

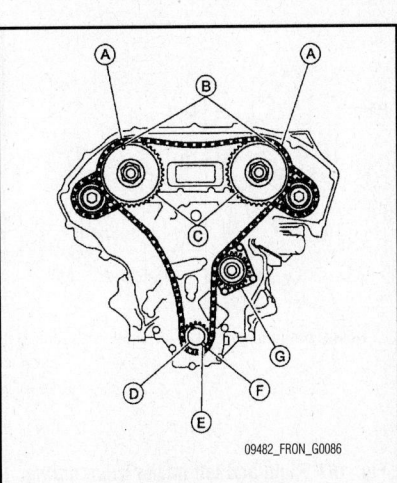

Fig. 163 Timing chain (primary) alignment—4.0L engine

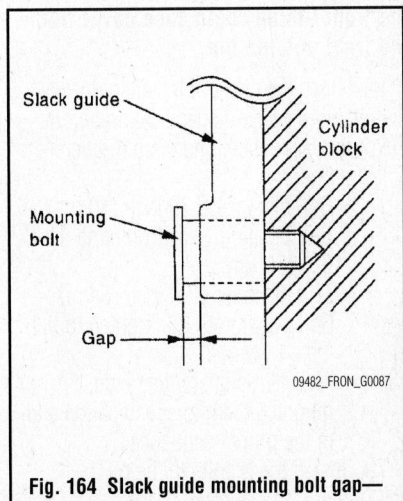

Fig. 164 Slack guide mounting bolt gap—4.0L engine

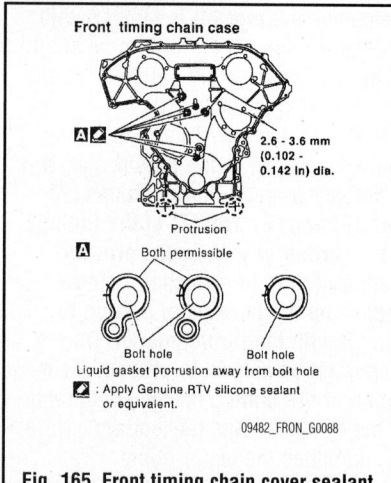

Fig. 165 Front timing chain cover sealant application—4.0L engine

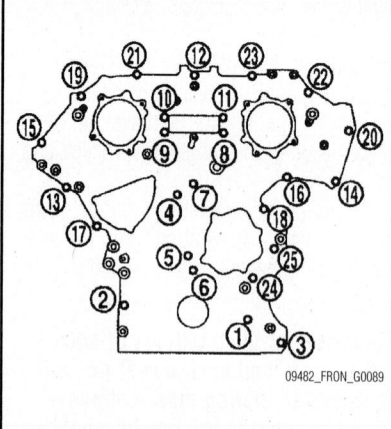

Fig. 166 Front timing chain cover bolt torque sequence—4.0L engine

water pump cover and chain tensioner cover. Be sure to use genuine RTV sealant, or equivalent.

67. Before installing the front timing chain case cover apply a continuous bead of liquid gasket 0.102–0.142 inch (2.6–3.6mm) in diameter to the front timing chain case back side, as shown in the illustration. Be sure to use genuine RTV sealant, or equivalent.

68. Install new O-rings on the rear timing chain case. To assemble the front timing chain case cover, fit the lower end of the front timing chain case tightly onto the top face of the oil pan (upper). From the fitting point, make the entire front timing chain case contact the rear timing chain case completely.

➡**Since the front timing chain case cover is offset for the difference of bolt holes, tighten the bolts temporarily while holding the front timing chain case cover from the front and the top. Then, insert a dowel pin while holding the front timing chain case cover from the front and the top.**

69. Once the cover is installed, torque the retaining bolts to specification and in the proper sequence. There are 4 different types of bolts:
- Bolt diameter 0.39 inch (10mm)—Bolt position 1–5: Tighten to 41 ft. lbs. (55 Nm)
- Bolt diameter 0.24 inch (6mm)—Bolt position 6–25: Tighten to 9 ft. lbs. (13 Nm)
- After all bolts are tightened, retighten them to specification and in the proper sequence

70. Install the 2 bolts in the oil pan (upper). Torque to 16 ft. lbs. (22 Nm).

71. Install new seal rings in the shaft grooves of the right and left intake valve timing control covers.

72. Apply a continuous bead of liquid gasket 0.083–0.122 inch (2.1–3.1mm) in diameter to the covers. Be sure to use genuine RTV sealant, or equivalent.

73. Install new collared O-rings in the front timing chain case oil hole (left and right sides). Be careful not to move the seal ring from the installation groove, align the dowel pins on the front timing chain case with the holes to install the intake valve timing control covers.

74. Tighten the bolts in sequence and to specification.

75. Install the crankshaft pulley.

76. Install the upper and lower oil pans.

77. Install the intake manifold collector.

78. Before installing the rocker cover, apply liquid gasket, be sure to use genuine

RTV silicone sealant or equivalent, to the positions shown in the illustration. Refer to figure "a" to apply liquid gasket to joint part of camshaft bracket No. 1 and cylinder head. Refer to figure "b" to apply liquid gasket to the figure "a" squarely.

79. Install the rocker cover. Torque the retaining bolts to 17 inch lbs. (2 Nm) and then to 74 inch lbs. (8 Nm), in the proper sequence.

80. Continue the installation in the reverse order of the removal procedure.

➡**If hydraulic pressure inside the timing chain tensioner drops after removal/installation, slack in the guide may generate a pounding noise during and just after engine start. This is normal. The noise should stop after the hydraulic pressure rises.**

VALVE (ROCKER) COVERS

REMOVAL & INSTALLATION

2.5L Engine

See Figures 168 and 169.

➡**If working near and/or around the SRS system and components, be sure to disable the SRS system. After disabling the system wait three minutes or more before servicing the vehicle.**

➡**Whenever the negative battery cable is disconnected the following components will require resetting. The Idle Air Volume Learning, Steering Angle Sensor Neutral Position, Sunroof Memory Reset/Initialization, Automatic Drive Positioner System, Audio presets and Navigation. Use the Nissan CONSULT III diagnostic tool, or equivalent to perform the required resets.**

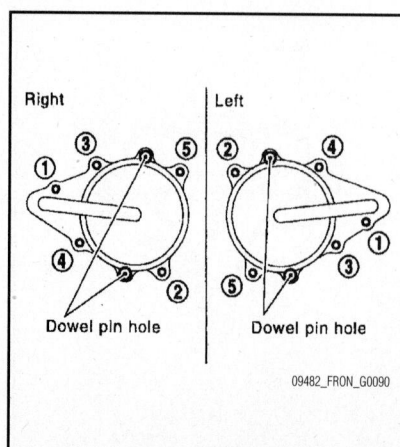

Fig. 167 Right and left intake valve timing control cover bolt torque sequence—4.0L engine

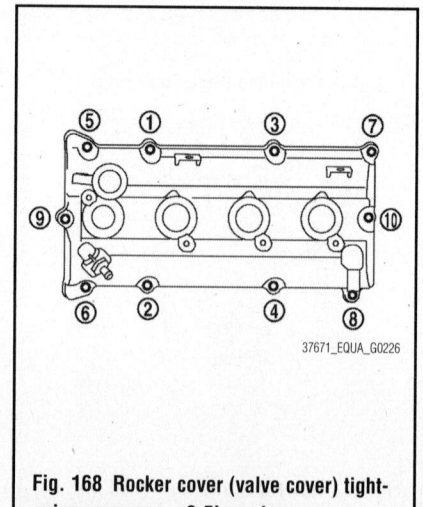

Fig. 168 Rocker cover (valve cover) tightening sequence—2.5L engine

1. Before servicing the vehicle, refer to the Precautions Section.

2. Disconnect the negative battery cable.

3. Remove the intake manifold.

4. Disconnect the PCV hose from rocker cover (valve cover).

5. Remove ignition coils.

6. Remove the PCV valve and O-ring from the rocker cover, if necessary.

7. Remove the oil filler cap from the rocker cover, if necessary.

8. Loosen the bolts in the reverse order of the tightening sequence.

9. Remove the rocker cover gasket from the rocker cover.

10. Use a scraper to remove all the traces of liquid gasket from the cylinder head and the camshaft bracket (No. 1).

✳✳ WARNING

Do not scratch or damage the mating surface when cleaning off old liquid gasket.

To install:

11. Apply liquid gasket using special tool WS39930000, or equivalent, to the joint of the rocker cover, cylinder head, and camshaft bracket (No. 1).

➡**Use Genuine RTV Silicone Sealant, or equivalent.**

12. Install a new rocker cover gasket to the rocker cover.

13. Install the rocker cover.

➡**Check to be sure the rocker cover gasket has not dropped from the installation groove of the rocker cover.**

14. Tighten the bolts in 2 steps in numerical order, as shown:
    a. Step 1: 18 inch lbs. (2 Nm).
    b. Step 2: 73 inch lbs. (8 Nm).

15. Installation of the remaining components is in the reverse order of removal.

4.0L Engine

Left Bank

See Figures 170 and 171.

➡**If working near and/or around the SRS system and components, be sure to disable the SRS system. After disabling the system wait three minutes or more before servicing the vehicle.**

➡**Whenever the negative battery cable is disconnected the following components will require resetting. The Idle Air Volume Learning, Steering Angle Sensor Neutral Position, Sunroof Memory Reset/Initialization, Automatic Drive Positioner System, Audio presets and Navigation. Use the Nissan CONSULT III diagnostic tool, or equivalent to perform the required resets.**

1. Before servicing the vehicle, refer to the Precautions Section.

2. Disconnect the negative battery cable.

3. Remove the engine room cover.

4. Remove the air duct and resonator assembly.

5. Separate the engine harness by removing the brackets from the rocker covers (valve covers).

6. Remove the harness bracket from the cylinder head, if necessary.

7. Disconnect and remove the intake valve timing control solenoid valve (LH bank).

8. Remove the ignition coils.

9. Remove the PCV hoses from the rocker covers.

10. Remove the oil filler cap from the rocker cover (LH), if necessary.

11. Remove the rocker cover bolts in reverse order of the tightening sequence.

12. Remove the rocker cover gaskets from the rocker covers.

13. Use a scraper to remove all traces of liquid gasket from the cylinder head and the camshaft bracket (No. 1).

✳✳ WARNING

Do not scratch or damage the mating surface when cleaning off the old liquid gasket.

To install:

14. Apply liquid gasket using tool WS39930000, or equivalent, to the joint of the rocker cover, the cylinder head, and the camshaft bracket (No. 1).

➡**Use Genuine RTV Silicone Sealant, or equivalent.**

15. Install a new rocker cover gasket to the rocker cover.

16. Install the rocker cover.

➡**Check to be sure the rocker cover gasket has not dropped from the installation groove of the rocker cover.**

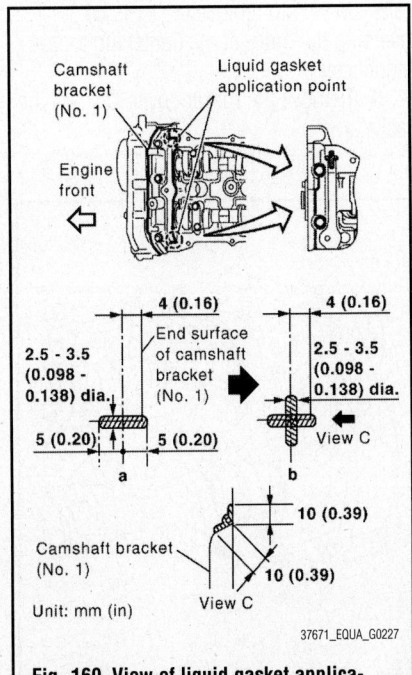

Fig. 169 View of liquid gasket application points—2.5L engine

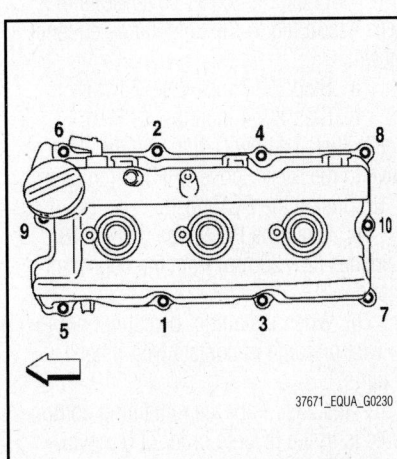

Fig. 170 Rocker cover (valve cover) tightening sequence (left bank)—4.0L engine

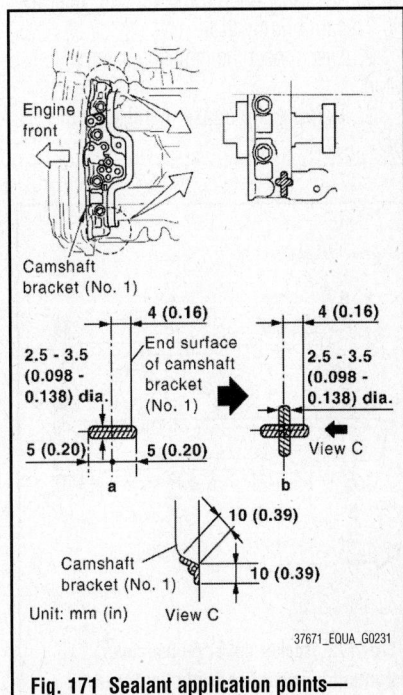

Fig. 171 Sealant application points—4.0L engine

17. Tighten the rocker cover bolts in 2 steps in the tightening sequence shown.
   a. Step 1: 17 inch lbs. (2 Nm).
   b. Step 2: 74 inch lbs. (8 Nm).
18. Install the oil filler cap to the rocker cover (LH), if removed.
19. Install the PCV hose.
   a. Insert the PCV hose 0.98–1.18 inches (25–30mm) from the connector end.
   b. When installing, be careful not to twist or come in contact with other parts.
20. Installation of the remaining components is in the reverse order of removal.
21. Check engine oil level and adjust as necessary.

Right Bank

See Figures 171 and 172.

➡ If working near and/or around the SRS system and components, be sure to disable the SRS system. After disabling the system wait three minutes or more before servicing the vehicle.

➡ Whenever the negative battery cable is disconnected the following components will require resetting. The Idle Air Volume Learning, Steering Angle Sensor Neutral Position, Sunroof Memory Reset/Initialization, Automatic Drive Positioner System, Audio presets and Navigation. Use the Nissan CONSULT III diagnostic tool, or equivalent to perform the required resets.

1. Before servicing the vehicle, refer to the Precautions Section.
2. Disconnect the negative battery cable.
3. Remove the intake manifold collector.

4. Separate the engine harness by removing the brackets from the rocker covers.
5. Remove the harness bracket from the cylinder head (RH).
6. Disconnect and remove the intake valve timing control solenoid valve (RH bank).
7. Remove the ignition coils.
8. Remove the PCV hoses from the rocker cover.
9. Remove the PCV valve and O-ring from the rocker cover (RH), if necessary.
10. Remove the rocker cover bolts in the reverse order of the tightening sequence.
11. Remove the rocker cover gaskets from the rocker covers.
12. Use a scraper to remove all the traces of liquid gasket from the cylinder head and the camshaft bracket (No. 1).

✳✳ WARNING

Do not scratch or damage the mating surface when cleaning off the old liquid gasket.

To install:

13. Apply liquid gasket using special tool WS39930000, or equivalent, to the joint part among the rocker cover, the cylinder head, and the camshaft bracket (No. 1).

➡ Use Genuine RTV Silicone Sealant, or equivalent.

14. Install a new rocker cover gasket to the rocker cover.
15. Install the rocker cover.

➡ Check to be sure the rocker cover gasket has not dropped from the installation groove of the rocker cover.

16. Tighten the rocker cover bolts in 2 steps according to the tightening sequence shown.
   a. Step 1: 17 inch lbs. (2 Nm).
   b. Step 2: 74 inch lbs. (8 Nm).
17. Install a new O-ring and the PCV valve to the rocker cover (RH), if removed.
18. Install the PCV hose.
   a. Insert the PCV hose 0.98–1.18 inches (25–30mm) from the connector end.
   b. When installing, be careful not to twist or come in contact with other parts.
19. Installation of the remaining components is in the reverse order of removal.
20. Check engine oil level and adjust as necessary.

VALVE LASH (CLEARANCE) ADJUSTMENT

ADJUSTMENT

2.5L Engine

See Figures 173 through 175.

➡ If working near and/or around the SRS system and components, be sure to disable the SRS system. After disabling the system wait three minutes or more before servicing the vehicle.

➡ Whenever the negative battery cable is disconnected the following components will require resetting. The Idle Air Volume Learning, Steering Angle Sensor Neutral Position, Sunroof Memory Reset/Initialization, Automatic Drive Positioner System, Audio presets and Navigation. Use the Nissan CONSULT III diagnostic tool, or equivalent to perform the required resets.

1. Before servicing the vehicle, refer to the Precautions Section.
2. Disconnect the negative battery cable.
3. Drain the cooling system.
4. Remove the intake manifold.
5. Disconnect the PCV hose from the rocker cover. Remove the ignition coil.
6. Remove the PCV valve and O-ring from the rocker cover, if necessary.
7. Remove the oil filler cap from the rocker cover, if necessary.
8. Remove the rocker cover retaining bolts. Be sure to remove the bolts by reversing the order of the tightening torque sequence.
9. Remove the rocker cover. Discard the gasket.

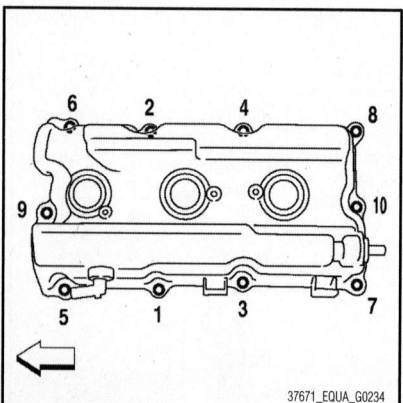

Fig. 172 Rocker cover (valve cover) tightening sequence (right bank)—4.0L engine

37671_EQUA_G0234

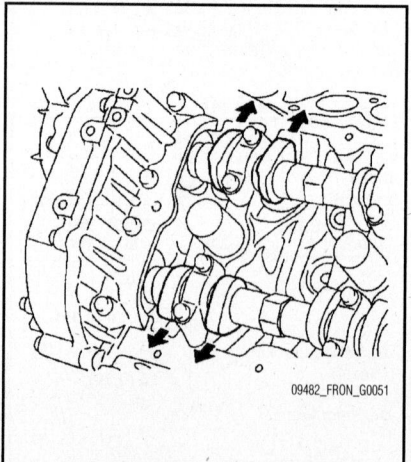

Fig. 173 View of No. 1 cylinder at TDC (compression stroke)—2.5L engine

09482_FRON_G0051

Measuring position		No. 1 CYL.	No. 2 CYL.	No. 3 CYL.	No. 4 CYL.
No. 1 cylinder at compression TDC	INT	×	×		
	EXH	×		×	

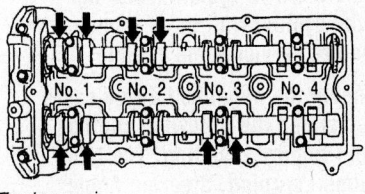

No. 1 cylinder compression TDC
Intake side
Engine front
Exhaust side
09482_FRON_G0052

Fig. 174 Valve adjustment measurement No. 1 cylinder at TDC (compression stroke)—2.5L engine

Measuring position		No. 1 CYL.	No. 2 CYL.	No. 3 CYL.	No. 4 CYL.
No. 4 cylinder at compression TDC	INT			×	×
	EXH		×		×

No. 4 cylinder compression TDC
Intake side
Engine front
Exhaust side
09482_FRON_G0053

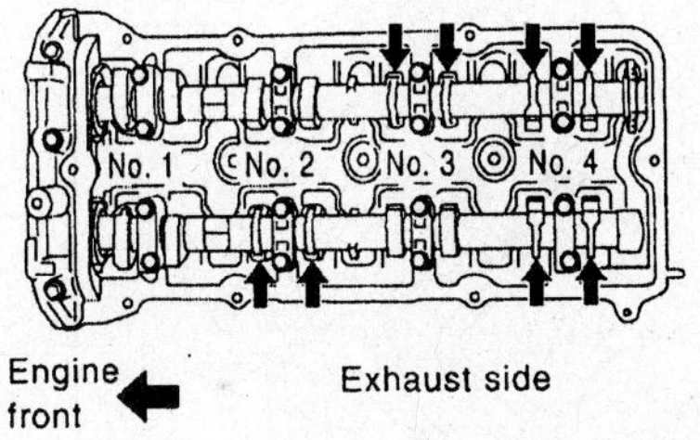

Fig. 175 Valve adjustment measurement No. 4 cylinder at TDC (compression stroke)—2.5L engine

10. Remove the undercover. Remove the lower radiator shroud.

11. Set the No. 1 cylinder at Top Dead Center (TDC) of its compression stroke by rotating the crankshaft pulley clockwise to align the TDC mark to the timing indicator on the front cover. At the same time make sure that both the intake and exhaust cam noses of the No. 1 cylinder face outward, as indicated by the arrows in the illustration. If not, rotate the crankshaft in the clockwise direction 360 degrees.

12. Use a feeler gauge and measure the clearance between the valve lifter and the camshaft.

13. With the No. 1 piston at TDC, refer to the illustration and measure the valve clearances at the locations marked with an "X". The "X" locations are indicated in the illustration with an arrow.

14. Rotate the crankshaft pulley clockwise 360 degrees and align the TDC

mark to the timing indicator on the front cover.

15. With the No. 4 piston at TDC, refer to the illustration and measure the valve clearances at the locations marked with an "X". The "X" locations are indicated in the illustration with an arrow.

16. If measurements are not within specification, proceed to the next step.

17. Remove the camshaft. Remove the valve lifters that are not within specification.

18. Measure the center thickness of the removed lifters, using a micrometer.

19. Use the equation (t = t1+(C1-C2)) to calculate the valve lifter thickness for replacement.

➡**t = valve of lifter thickness to be replaced. t1 = removed valve lifter thickness. C1 = measured valve clearance. C2 = standard valve clearance.**

20. The thickness of the new valve lifter can be identified by the stamp mark on the reverse side (inside the cylinder). The stamp mark "696" indicates 0.2740 inch (6.96mm) thickness.

➡**Available thickness of a valve lifter ranges from 0.2740–0.2937 inch (6.96–7.46mm) in steps of 0.0008 inch (0.02mm). There are 26 different sizes.**

21. Install the selected valve lifters.

22. Install the camshaft.

23. Manually rotate the crankshaft pulley in the clockwise direction a few rotations.

24. Check the valve clearance and be sure it is within specification.

25. When installing the rocker cover, apply liquid gasket, be sure to use genuine RTV silicone sealant, or equivalent.

26. Install the rocker cover.

27. Continue the installation in the reverse of the removal procedure.

4.0L Engine

See Figures 176 through 179.

➡If working near and/or around the SRS system and components, be sure to disable the SRS system. After disabling the system wait three minutes or more before servicing the vehicle.

➡Whenever the negative battery cable is disconnected the following components will require resetting. The Idle Air Volume Learning, Steering Angle Sensor Neutral Position, Sunroof Memory Reset/Initialization, Automatic Drive Positioner System, Audio presets and Navigation. Use the Nissan CONSULT III diagnostic tool, or equivalent to perform the required resets.

1. Before servicing the vehicle, refer to the Precautions Section.

2. Disconnect the negative battery cable.

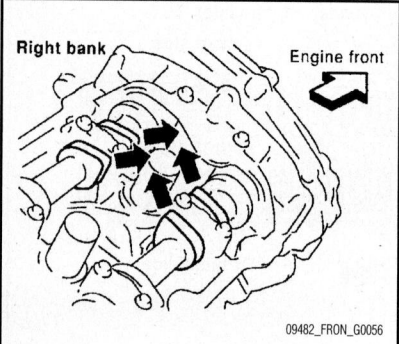

Fig. 176 No. 1 cylinder at TDC (compression stroke)—4.0L engine

3. Remove the engine under cover.

4. Remove the intake manifold collector.

5. Separate the engine harness and remove the brackets from the rocker covers. Remove the harness bracket from the cylinder head, if necessary.

6. Remove the ignition coil. Remove the PCV hoses. Remove the oil filler cap, if necessary.

7. Remove the rocker covers from the engine.

8. Set the No. 1 cylinder at Top Dead Center (TDC) of its compression stroke by rotating the crankshaft pulley clockwise to align the timing mark (grooved line without color) with the timing indicator. Make sure that the intake and exhaust cam noses on No. 1 cylinder (engine front side on right bank) are in alignment as shown in the illustration. If not, rotate the crankshaft in the clockwise direction 360 degrees.

9. Use a feeler gauge and measure the clearance between the valve lifter and the camshaft.

10. With the No. 1 piston at TDC, refer to the illustration and measure the valve clearances at the locations marked with an "X". The "X" locations are indicated in the illustration with an arrow.

11. Rotate the crankshaft pulley clockwise 240 degrees (when viewed from the engine front) to align No. 3 cylinder at TDC on the compression stroke.

➡The crankshaft pulley bolt flange has a stamped line every 60 degrees, which can be used as a guide to rotation angle.

12. With the No. 3 piston at TDC, refer to the illustration and measure the valve clearances at the locations marked with an "X". The "X" locations are indicated in the illustration with an arrow.

13. Rotate the crankshaft pulley clockwise 240 degrees (when viewed from the engine front) to align No. 5 cylinder at TDC on the compression stroke.

➡The crankshaft pulley bolt flange has a stamped line every 60 degrees, which can be used as a guide to rotation angle.

14. With the No. 5 piston at TDC, refer to the illustration and measure the valve clearances at the locations marked with an "X". The "X" locations are indicated in the illustration with an arrow.

15. If measurements are not within specification, proceed to the next step.

16. Remove the camshaft. Remove the valve lifters that are not within specification.

17. Measure the center thickness of the removed lifters, using a micrometer.

18. Use the equation ($t = t_1+(C_1-C_2)$) to calculate the valve lifter thickness for replacement.

➡t = valve lifter thickness to be replaced. t1 = removed valve lifter thickness. C1 = measured valve clearance. C2 = standard valve clearance.

19. The intake valve lifter thickness of the new valve lifter can be identified by the stamp mark on the reverse side (inside the cylinder). The stamp mark "788U" indicates 0.3102 inch (7.88mm) thickness.

Measuring position (right bank)		No. 1 CYL.	No. 3 CYL.	No. 5 CYL.
No. 1 cylinder at compression TDC	EXH		×	
	INT	×		
Measuring position (left bank)		No. 2 CYL.	No. 4 CYL.	No. 6 CYL.
No. 1 cylinder at compression TDC	INT			×
	EXH	×		

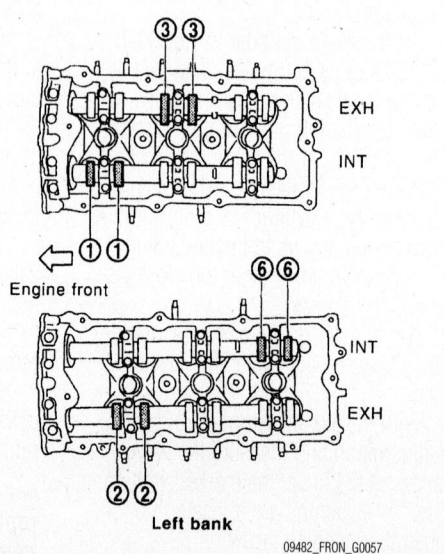

Fig. 177 Valve adjustment measurement No. 1 cylinder at TDC (compression stroke)—4.0L engine

Measuring position (right bank)		No. 1 CYL.	No. 3 CYL.	No. 5 CYL.
No. 3 cylinder at compression TDC	EXH			×
	INT		×	
Measuring position (left bank)		No. 2 CYL.	No. 4 CYL.	No. 6 CYL.
No. 3 cylinder at compression TDC	INT	×		
	EXH		×	

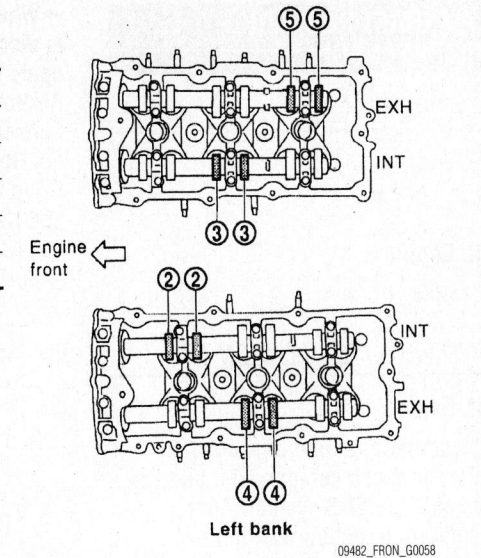

Fig. 178 Valve adjustment measurement No. 3 cylinder at TDC (compression stroke)—4.0L engine

Measuring position (right bank)		No. 1 CYL.	No. 3 CYL.	No. 5 CYL.
No. 5 cylinder at compression TDC	EXH	×		
	INT			×
Measuring position (left bank)		No. 2 CYL.	No. 4 CYL.	No. 6 CYL.
No. 5 cylinder at compression TDC	INT		×	
	EXH			×

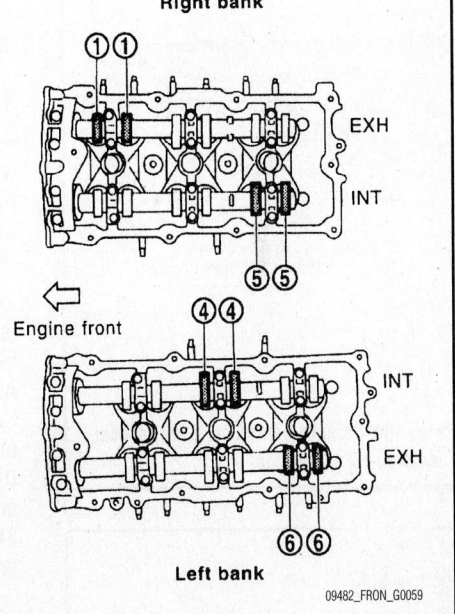

Fig. 179 Valve adjustment measurement No. 5 cylinder at TDC (compression stroke)—4.0L engine

➡**Available thickness of a valve lifter ranges from 0.3102–0.3307 inch (7.88–8.40mm) in steps of 0.0008 inch (0.02mm). There are 27 different sizes.**

20. Exhaust valve lifter thickness of the new valve lifter can be identified by the stamp mark on the reverse side (inside the cylinder). The stamp mark "N788" indicates 0.3102 inch (7.88mm) thickness.

➡**Available thickness of a valve lifter ranges from 0.3102–0.3291 inch (7.88–8.36mm) in steps of 0.0008 inch (0.02mm). There are 25 different sizes.**

21. Install the selected valve lifters.
22. Install the camshaft.
23. Manually rotate the crankshaft pulley in the clockwise direction a few rotations.

24. Check the valve clearance and be sure it is within specification.
25. When installing the rocker cover, apply liquid gasket, be sure to use genuine RTV silicone sealant, or equivalent.
26. Install the rocker cover.
27. Continue the installation in the reverse of the removal procedure.

ENGINE PERFORMANCE & EMISSION CONTROLS

CAMSHAFT POSITION (CMP) SENSOR

LOCATION

2.5L Engine

See Figure 180.

4.0L Engine

See Figure 181.

REMOVAL & INSTALLATION

2.5L Engine

→If working near and/or around the SRS system and components, be sure to disable the SRS system. After disabling the system wait three minutes or more before servicing the vehicle.

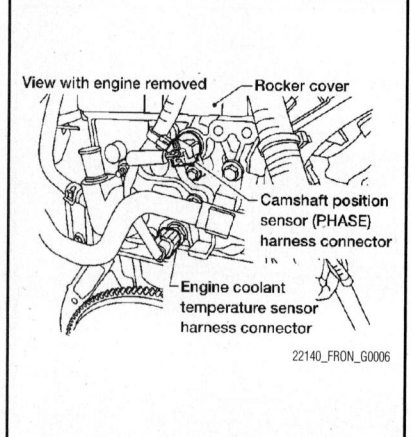

Fig. 180 Camshaft Position (CMP) sensor location—2.5L engine

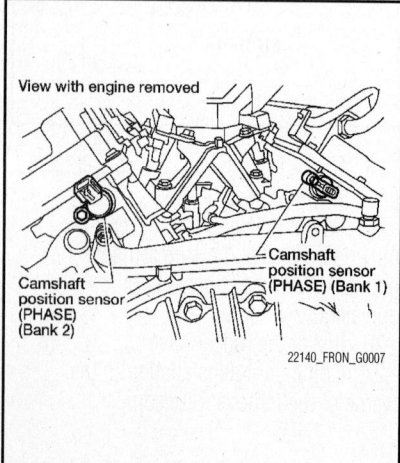

Fig. 181 Camshaft Position (CMP) sensor(s) location—4.0L engine

→Whenever the negative battery cable is disconnected the following components will require resetting. The Idle Air Volume Learning, Steering Angle Sensor Neutral Position, Sunroof Memory Reset/Initialization, Automatic Drive Positioner System, Audio presets and Navigation. Use the Nissan CONSULT III diagnostic tool, or equivalent to perform the required resets.

1. Before servicing the vehicle, refer to the Precautions Section.
2. Disconnect the negative battery cable.
3. Loosen the fixing bolt of the sensor.
4. Disconnect the CMP sensor (PHASE) harness connector.
5. Remove the CMP sensor.

To install:

6. Installation is the reverse of the removal procedure.
7. Visually check the sensor for chipping.
8. Tighten the CMP sensor (PHASE) fixing bolt to 62 inch lbs. (7 Nm).

4.0L Engine

→If working near and/or around the SRS system and components, be sure to disable the SRS system. After disabling the system wait three minutes or more before servicing the vehicle.

→Whenever the negative battery cable is disconnected the following components will require resetting. The Idle Air Volume Learning, Steering Angle Sensor Neutral Position, Sunroof Memory Reset/Initialization, Automatic Drive Positioner System, Audio presets and Navigation. Use the Nissan CONSULT III diagnostic tool, or equivalent to perform the required resets.

1. Before servicing the vehicle, refer to the Precautions Section.
2. Disconnect the negative battery cable.
3. Loosen the fixing bolt of the sensor.
4. Disconnect the CMP sensor (PHASE) harness connector.
5. Remove the CMP sensor.

To install:

6. Installation is the reverse of the removal procedure.
7. Visually check the sensor for chipping.
8. Tighten the CMP sensor (PHASE) fixing bolt to 86 inch lbs. (10 Nm).

CRANKSHAFT POSITION (CKP) SENSOR

LOCATION

2.5L Engine

See Figure 182.

4.0L Engine

See Figure 183.

REMOVAL & INSTALLATION

2.5L Engine

→If working near and/or around the SRS system and components, be sure to disable the SRS system. After disabling the system wait three minutes or more before servicing the vehicle.

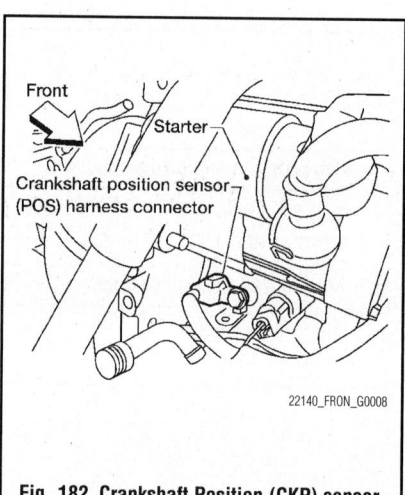

Fig. 182 Crankshaft Position (CKP) sensor location—2.5L engine

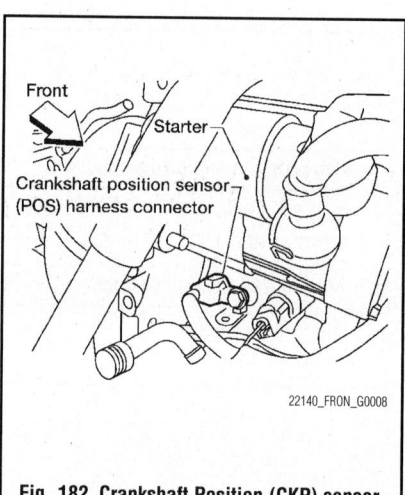

Fig. 183 Crankshaft Position (CKP) sensor location—4.0L engine

➡Whenever the negative battery cable is disconnected the following components will require resetting. The Idle Air Volume Learning, Steering Angle Sensor Neutral Position, Sunroof Memory Reset/Initialization, Automatic Drive Positioner System, Audio presets and Navigation. Use the Nissan CONSULT III diagnostic tool, or equivalent to perform the required resets.

1. Before servicing the vehicle, refer to the Precautions Section.
2. Disconnect the negative battery cable.
3. Loosen the fixing bolt of the sensor.
4. Disconnect the CKP sensor (POS) harness connector.
5. Remove the CKP sensor.

To install:
6. Installation is the reverse of removal procedure.
7. Visually check the sensor for chipping.
8. Tighten the CKP sensor (POS) fixing bolt to 62 inch lbs. (7 Nm).

4.0L Engine

➡If working near and/or around the SRS system and components, be sure to disable the SRS system. After disabling the system wait three minutes or more before servicing the vehicle.

➡Whenever the negative battery cable is disconnected the following components will require resetting. The Idle Air Volume Learning, Steering Angle Sensor Neutral Position, Sunroof Memory Reset/Initialization, Automatic Drive Positioner System, Audio presets and Navigation. Use the Nissan CONSULT III diagnostic tool, or equivalent to perform the required resets.

1. Before servicing the vehicle, refer to the Precautions Section.
2. Disconnect the negative battery cable.
3. Loosen the fixing bolt of the sensor.
4. Disconnect the CKP sensor (POS) harness connector.
5. Remove the CKP sensor.

To install:
6. Installation is the reverse of removal procedure.
7. Visually check the sensor for chipping.

ENGINE CONTROL MODULE (ECM)

LOCATION

The Engine Control Module (ECM) is located in the engine room passenger side behind the engine coolant reservoir tank.

REMOVAL & INSTALLATION
See Figure 184.

At this time the manufacturer does not provide removal and installation procedures for this component. The following procedure is a guideline and may differ from the vehicle you are servicing.

➡If working near and/or around the SRS system and components, be sure to disable the SRS system. After disabling the system wait three minutes or more before servicing the vehicle.

➡Whenever the negative battery cable is disconnected the following components will require resetting. The Idle Air Volume Learning, Steering Angle Sensor Neutral Position, Sunroof Memory Reset/Initialization, Automatic Drive Positioner System, Audio presets and Navigation. Use the Nissan CONSULT III diagnostic tool, or equivalent to perform the required resets.

➡The Nissan CONSULT III diagnostic tool, or equivalent is required to perform the resets used on this vehicle.

1. Before servicing the vehicle, refer to the Precautions Section.
2. Disconnect the negative battery cable.
3. Remove the air cleaner assembly, as required.
4. Disconnect the electrical connectors.
5. Remove the component retaining screws and/or clips.
6. Remove the component from its mounting.

To install:

➡Be sure to use new fasteners, as required.

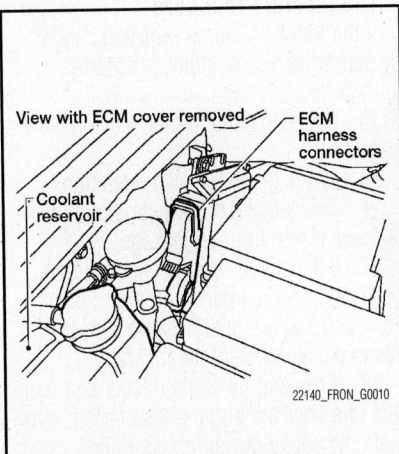

Fig. 184 Engine Control Module (ECM) and related components

7. Installation is the reverse of the removal procedure.
8. Reset the required components.

RESET

➡The Nissan CONSULT III diagnostic tool, or equivalent is required to perform the resets used on this vehicle.

When replacing the ECM, the following procedures must be performed in the order listed.
• Initialize The Immobilizer System
• VIN Registration
• Accelerator Pedal Released Position Learning
• Throttle Valve Closed Position Learning
• Idle Air Volume Learning

Accelerator Pedal Released Position Learning

Accelerator Pedal Released Position Learning is an operation to learn the fully released position of the accelerator pedal by monitoring the accelerator pedal position sensor output signal. It must be performed each time the harness connector of the accelerator pedal position sensor or ECM is disconnected.

1. Before servicing the vehicle, refer to the Precautions Section.
2. Check that the accelerator pedal is fully released.
3. Turn the ignition switch ON and wait at least 2 seconds.
4. Turn the ignition switch OFF and wait at least 10 seconds.
5. Turn the ignition switch ON and wait at least 2 seconds.
6. Turn the ignition switch OFF and wait at least 10 seconds.

Idle Air Volume Learning

Idle Air Volume Learning is an operation to learn the idle air volume that keeps each engine within the specific range. It must be performed under the following conditions:
• Each time the electric throttle control actuator or ECM is replaced
• The idle speed or ignition timing is out of specification

1. Before servicing the vehicle, refer to the Precautions Section.

Preparation
2. Before performing the Idle Air Volume Learning procedure, check that all of the following conditions are satisfied. Learning will be cancelled if any of the following conditions are missed for even a moment:

- Battery voltage: More than 12.9 V (at idle)
- Engine coolant temperature: 158–212° F (70–100° C)
- PNP switch (M/T): ON
- Selector lever (A/T): P or N
- Electric load switch: OFF (air conditioner, headlamp, rear window defogger). On vehicles equipped with daytime light systems, if the parking brake is applied before the engine is started, the headlamp will not illuminate
- Steering wheel: Neutral (straight-ahead position)
- Vehicle speed: Stopped
- Transmission: Warmed-up

➡**With TOOL: Drive the vehicle until "ATF TEMP SE" in "Data List" mode of the "A/T" system indicates less than 0.9 V. Without TOOL: Drive the vehicle for 10 minutes.**

Operations Procedure—with TOOL

3. Perform the Accelerator Pedal Released Position Learning procedure.
4. Perform the Throttle Valve Closed Position Learning procedure.
5. Start the engine and warm it up to a normal operating temperature.
6. Check that all items listed under the topic PREPARATION are in good order.
7. Select "IDLE AIR VOL LEARN" in "Utility" mode.
8. Touch "START" and wait 20 seconds.
9. Check that "CMPLT" is displayed on the Tool screen. If "CMPLT" is not displayed, Idle Air Volume Learning will not be carried out successfully. In this case, find the cause of the incident by referring to the Diagnostic Procedure.
10. Rev up the engine 2–3 times and check that the idle speed and the ignition timing are within specifications.
   a. Idle speed M/T: 625 plus or minus 50 RPM (in Neutral position).
   b. Idle speed A/T: 625 plus or minus 50 RPM (in P or N position).
   c. Ignition timing M/T: 15 plus or minus 5° BTDC (in Neutral position).
   d. Ignition timing A/T: 15 plus or minus 5° BTDC (in P or N position).

Operations Procedure—without TOOL

➡**It is better to count the time accurately with a clock. It is impossible to switch the diagnostic mode when an accelerator pedal position sensor circuit has a malfunction.**

11. Perform the Accelerator Pedal Released Position Learning procedure.
12. Perform the Throttle Valve Closed Position Learning procedure.
13. Start the engine and warm it up to a normal operating temperature.
14. Check that all items listed under the topic PREPARATION are in good order.
15. Turn ignition switch OFF and wait at least 10 seconds.
16. Confirm that accelerator pedal is fully released, then turn the ignition switch ON and wait 3 seconds.
17. Repeat the following procedure quickly 5 times within 5 seconds.
   a. Fully depress the accelerator pedal.
   b. Fully release the accelerator pedal.
18. Wait 7 seconds, fully depress the accelerator pedal for approx. 20 seconds until the MIL stops blinking and turns ON.
19. Fully release the accelerator pedal within 3 seconds after the MIL turns ON.
20. Start the engine and let it idle.
21. Wait 20 seconds.
22. Rev up the engine 2–3 times and check that idle speed and ignition timing are within specifications.
   a. Idle speed M/T: 625 plus or minus 50 RPM (in Neutral position).
   b. Idle speed A/T: 625 plus or minus 50 RPM (in P or N position).
   c. Ignition timing M/T: 15 plus or minus 5° BTDC (in Neutral position).
   d. Ignition timing A/T: 15 plus or minus 5° BTDC (in P or N position).
23. If the idle speed and the ignition timing are not within specification, Idle Air Volume Learning will not be carried out successfully. In this case, find the cause of the incident by referring to the DIAGNOSTIC PROCEDURE.

Diagnostic Procedure

If the Idle Air Volume Learning cannot be performed successfully, proceed as follows:

24. Check that the throttle valve is fully closed.
25. Check the PCV valve operation.
26. Check that the downstream of throttle valve is free from air leakage.
27. If the above 3 items check out OK, engine component parts and their installation condition are questionable. Check and eliminate the cause of the incident.
28. If the engine stalls or has an incorrect idle after the engine has started, eliminate the cause of the incident and perform the Idle Air Volume Learning again.

Initialize the Immobilizer System

1. Before servicing the vehicle, refer to the Precautions Section.
2. Perform an initialization of the immobilizer system and registration of all immobilizer system ignition key IDs.
3. Install the ECM into the vehicle.
4. Using a registered key, turn the ignition switch to ON.

➡**Use the key that has been used before performing the ECM replacement.**

5. Maintain the ignition switch in the ON position for at least 5 seconds.
6. Turn the ignition switch to OFF.
7. Start the engine.
   - If the engine starts, the procedure is completed.
   - If the engine fails to start, initialize the control unit using the Tool and follow the operation manual.

Throttle Valve Closed Position Learning

Throttle Valve Closed Position Learning is an operation to learn the fully closed position of the throttle valve by monitoring the throttle position sensor output signal. It must be performed each time the harness connector of electric throttle control actuator or ECM is disconnected.

1. Before servicing the vehicle, refer to the Precautions Section.
2. Check that the accelerator pedal is fully released.
3. Turn the ignition switch ON.
4. Turn the ignition switch OFF and wait at least 10 seconds.
5. Check that throttle valve moves during the above 10 seconds by confirming the operating sound.

VIN Registration

VIN Registration is an operation registering the VIN in the ECM. It must be performed each time the ECM is replaced.

1. Perform the VIN Registration procedure.
2. Before servicing the vehicle, refer to the Precautions Section.
3. Check the VIN of the vehicle and note it.
4. Turn ignition switch ON and engine stopped.
5. Select "VIN REGISTRATION" on the Tool in "Utility" mode.
6. Follow the instructions of the TOOL display.

ENGINE COOLANT TEMPERATURE (ECT) SENSOR

LOCATION

2.5L Engine

See Figure 185.

4.0L Engine

See Figure 186.

REMOVAL & INSTALLATION

2.5L Engine

> **✷✷ CAUTION**
>
> Never open, service or drain the radiator or cooling system when hot; serious burns can occur from the steam and hot coolant. Also, when draining engine coolant, keep in mind that cats and dogs are attracted

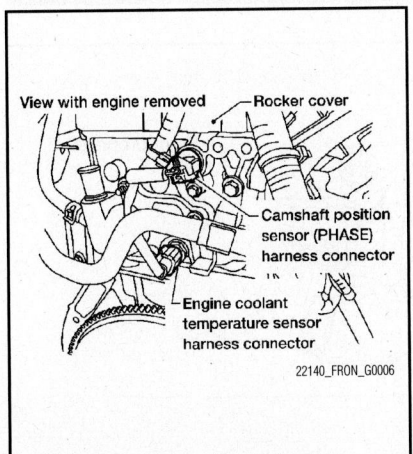

Fig. 185 Engine Coolant Temperature (ECT) sensor location—2.5L engine

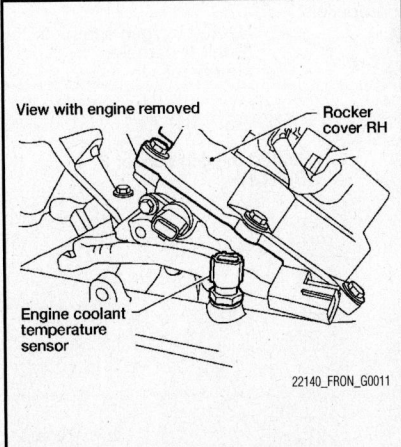

Fig. 186 Engine Coolant Temperature (ECT) sensor location—4.0L engine

to ethylene glycol antifreeze and could drink any that is left in an uncovered container or in puddles on the ground. This will prove fatal in sufficient quantities. Always drain coolant into a sealable container. Coolant should be reused unless it is contaminated or is several years old.

➡ If working near and/or around the SRS system and components, be sure to disable the SRS system. After disabling the system wait three minutes or more before servicing the vehicle.

➡ Whenever the negative battery cable is disconnected the following components will require resetting. The Idle Air Volume Learning, Steering Angle Sensor Neutral Position, Sunroof Memory Reset/Initialization, Automatic Drive Positioner System, Audio presets and Navigation. Use the Nissan CONSULT III diagnostic tool, or equivalent to perform the required resets.

1. Before servicing the vehicle, refer to the Precautions Section.
2. Disconnect the negative battery cable.
3. Loosen the fixing bolt of the Engine Coolant Temperature (ECT) sensor.
4. Disconnect the ECT sensor harness connector.
5. Remove the ECT sensor.

To install:

6. Installation is the reverse of the removal procedure.
7. Tighten the ECT sensor fixing bolt to 18 ft. lbs. (25 Nm).
8. Fill the cooling system and check for leaks.

4.0L Engine

> **✷✷ CAUTION**
>
> Never open, service or drain the radiator or cooling system when hot; serious burns can occur from the steam and hot coolant. Also, when draining engine coolant, keep in mind that cats and dogs are attracted to ethylene glycol antifreeze and could drink any that is left in an uncovered container or in puddles on the ground. This will prove fatal in sufficient quantities. Always drain coolant into a sealable container. Coolant should be reused unless it is contaminated or is several years old.

➡ If working near and/or around the SRS system and components, be sure to disable the SRS system. After dis-

abling the system wait three minutes or more before servicing the vehicle.

➡ Whenever the negative battery cable is disconnected the following components will require resetting. The Idle Air Volume Learning, Steering Angle Sensor Neutral Position, Sunroof Memory Reset/Initialization, Automatic Drive Positioner System, Audio presets and Navigation. Use the Nissan CONSULT III diagnostic tool, or equivalent to perform the required resets.

1. Before servicing the vehicle, refer to the Precautions Section.
2. Disconnect the negative battery cable.
3. Loosen the fixing bolt of the Engine Coolant Temperature (ECT) sensor.
4. Disconnect the ECT sensor harness connector.
5. Remove the ECT sensor.

To install:

6. Installation is the reverse of the removal procedure.
7. Tighten the ECT sensor fixing bolt to 18 ft. lbs. (25 Nm).
8. Fill the cooling system and check for leaks.

HEATED OXYGEN SENSOR (HO2S)

LOCATION

2.5L Engine

This sensor is located under the vehicle before the catalytic converter (under floor converter)

4.0L Engine

This sensor(s) is located after the catalytic converter (manifold converter(s).

REMOVAL & INSTALLATION

2.5L Engine

See Figure 187.

> **✷✷ CAUTION**
>
> Perform the operation with the exhaust system fully cooled. The system will be hot just after the engine stops.

➡ If working near and/or around the SRS system and components, be sure to disable the SRS system. After disabling the system wait three minutes or more before servicing the vehicle.

➡Whenever the negative battery cable is disconnected the following components will require resetting. The Idle Air Volume Learning, Steering Angle Sensor Neutral Position, Sunroof

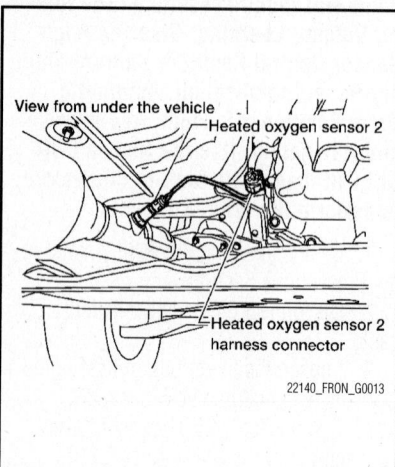

Fig. 187 Heated Oxygen Sensor (HO2S) and related components—2.5L engine

Memory Reset/Initialization, Automatic Drive Positioner System, Audio presets and Navigation. Use the Nissan CONSULT III diagnostic tool, or equivalent to perform the required resets.

1. Before servicing the vehicle, refer to the Precautions Section.
2. Disconnect the negative battery cable.
3. Raise and safely support the vehicle, as required.
4. Disconnect the sensor harness connector.
5. Remove the sensor using the heated oxygen sensor wrench KV10114400 (J-38365), or equivalent.

✳✳ WARNING

Be careful not to damage the sensor. Discard any sensor which has been dropped from a height of more than 19.7 inches (0.5m) onto a hard surface such as a concrete floor. Replace with a new one.

To install:

✳✳ WARNING

Do not over-tighten the sensor. Doing so may cause damage to the sensor, resulting in the MIL coming on.

6. Before installing a new sensor, clean the exhaust system threads using a suitable tool and apply antiseize lubricant. Use oxygen sensor thread cleaner J-43897-12 or J-43897-18, or equivalent.
7. Install the sensor using special tool KV10114400 (J-38365), or equivalent. Tighten to 37 ft. lbs. (50 Nm).

4.0L Engine

See Figure 188.

✳✳ CAUTION

Perform the operation with the exhaust system fully cooled. The system will be hot just after the engine stops.

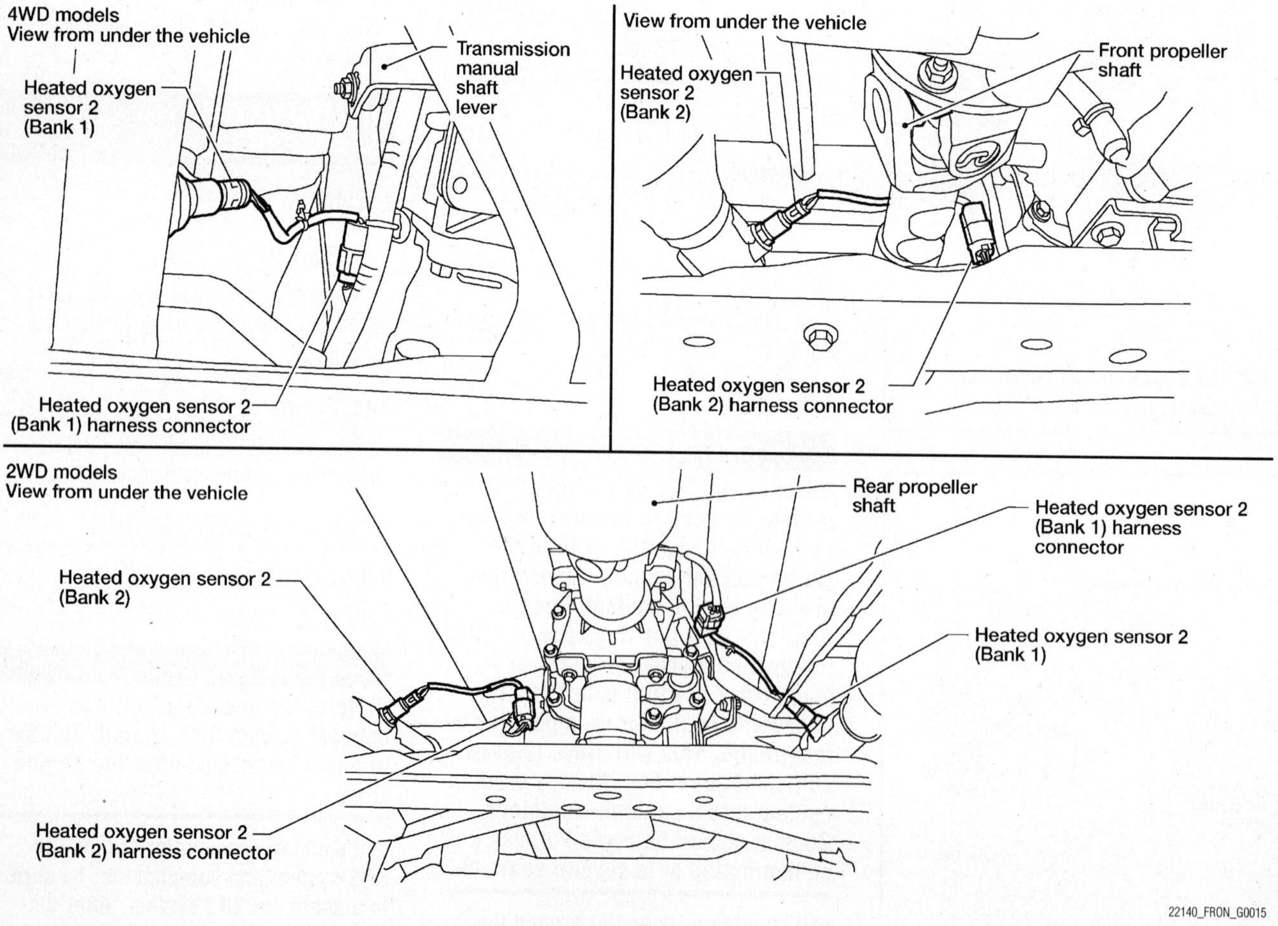

4WD models
View from under the vehicle

Heated oxygen sensor 2 (Bank 1)

Transmission manual shaft lever

Heated oxygen sensor 2 (Bank 1) harness connector

View from under the vehicle

Heated oxygen sensor 2 (Bank 2)

Front propeller shaft

Heated oxygen sensor 2 (Bank 2) harness connector

2WD models
View from under the vehicle

Heated oxygen sensor 2 (Bank 2)

Rear propeller shaft

Heated oxygen sensor 2 (Bank 1) harness connector

Heated oxygen sensor 2 (Bank 1)

Heated oxygen sensor 2 (Bank 2) harness connector

Fig. 188 Heated Oxygen Sensor (HO2S) and related components—4.0L engine

➡If working near and/or around the SRS system and components, be sure to disable the SRS system. After disabling the system wait three minutes or more before servicing the vehicle.

➡Whenever the negative battery cable is disconnected the following components will require resetting. The Idle Air Volume Learning, Steering Angle Sensor Neutral Position, Sunroof Memory Reset/Initialization, Automatic Drive Positioner System, Audio presets and Navigation. Use the Nissan CONSULT III diagnostic tool, or equivalent to perform the required resets.

1. Before servicing the vehicle, refer to the Precautions Section.
2. Disconnect the negative battery cable.
3. Raise and safely support the vehicle, as required.
4. Disconnect the sensor harness connector.
5. Remove the sensor using the heated oxygen sensor wrench KV10114400 (J-38365), or equivalent.

✳✳ WARNING

Be careful not to damage the sensor. Discard any sensor which has been dropped from a height of more than 19.7 inches (0.5m) onto a hard surface such as a concrete floor. Replace with a new one.

To install:

✳✳ WARNING

Do not over-tighten the sensor. Doing so may cause damage to the sensor, resulting in the MIL coming on.

6. Before installing a new sensor, clean the exhaust system threads using a suitable tool and apply antiseize lubricant. Use oxygen sensor thread cleaner J-43897-12 or J-43897-18, or equivalent.
7. Install the sensor using special tool KV10114400 (J-38365), or equivalent. Tighten to 37 ft. lbs. (50 Nm).

INTAKE AIR TEMPERATURE (IAT)/MASS AIR FLOW (MAF) SENSOR

LOCATION

2.5L Engine

See Figure 189.

The Intake Air Temperature (IAT) sensor is built into the Mass Air Flow (MAF) sensor.

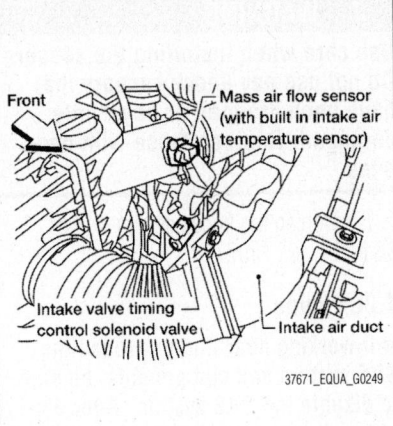

Fig. 189 Mass Air Flow (MAF)/Idle Air Control (IAC) valve location—2.5L engine

4.0L Engine

See Figure 190.

The Intake Air Temperature (IAT) sensor is built into the Mass Air Flow (MAF) sensor.

REMOVAL & INSTALLATION

2.5L Engine

➡If working near and/or around the SRS system and components, be sure to disable the SRS system. After disabling the system wait three minutes or more before servicing the vehicle.

➡Whenever the negative battery cable is disconnected the following components will require resetting. The Idle Air Volume Learning, Steering Angle Sensor Neutral Position, Sunroof Memory Reset/Initialization, Automatic Drive Positioner System, Audio presets and Navigation. Use the Nissan

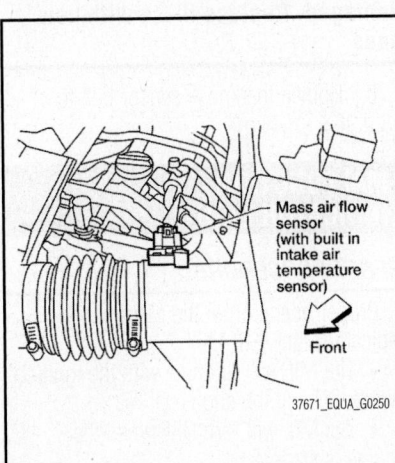

Fig. 190 Mass Air Flow (MAF)/Idle Air Control (IAC) valve location—4.0L engine

CONSULT III diagnostic tool, or equivalent to perform the required resets.

1. Before servicing the vehicle, refer to the Precautions Section.
2. Disconnect the negative battery cable.
3. Remove the Mass Air Flow (MAF) sensor harness.
4. Remove the MAF sensor.

To install:
5. Install the MAF sensor tighten the MAF attaching screws to 13 inch lbs. (2 Nm).
6. Install the MAF sensor harness.

4.0L Engine

➡If working near and/or around the SRS system and components, be sure to disable the SRS system. After disabling the system wait three minutes or more before servicing the vehicle.

➡Whenever the negative battery cable is disconnected the following components will require resetting. The Idle Air Volume Learning, Steering Angle Sensor Neutral Position, Sunroof Memory Reset/Initialization, Automatic Drive Positioner System, Audio presets and Navigation. Use the Nissan CONSULT III diagnostic tool, or equivalent to perform the required resets.

1. Before servicing the vehicle, refer to the Precautions Section.
2. Disconnect the negative battery cable.
3. Remove the Mass Air Flow (MAF) sensor harness.
4. Remove the MAF sensor.

To install:
5. Install the MAF sensor tighten the MAF attaching screws.
6. Install the MAF sensor harness.

KNOCK SENSOR (KS)

LOCATION

2.5L Engine

See Figure 191.

4.0L Engine

See Figure 192.

REMOVAL & INSTALLATION

2.5L Engine

➡If working near and/or around the SRS system and components, be sure to disable the SRS system. After disabling the system wait three minutes or more before servicing the vehicle.

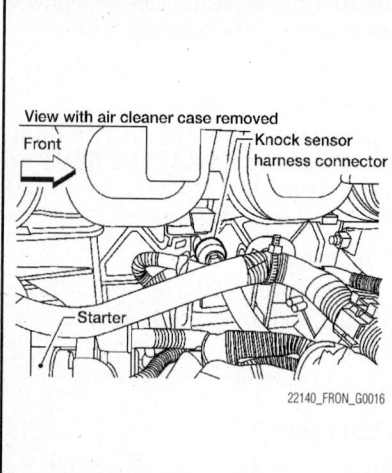

Fig. 191 Knock Sensor (KS) location—2.5L engine

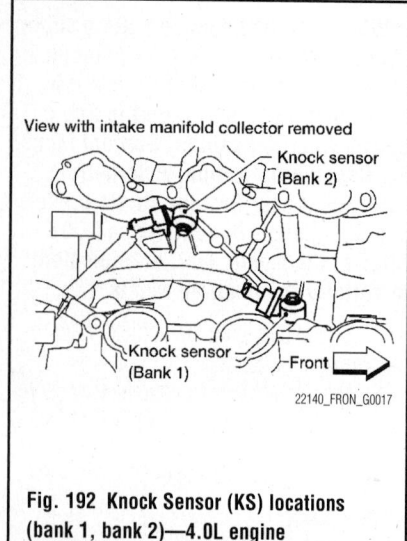

Fig. 192 Knock Sensor (KS) locations (bank 1, bank 2)—4.0L engine

➡Whenever the negative battery cable is disconnected the following components will require resetting. The Idle Air Volume Learning, Steering Angle Sensor Neutral Position, Sunroof Memory Reset/Initialization, Automatic Drive Positioner System, Audio presets and Navigation. Use the Nissan CONSULT III diagnostic tool, or equivalent to perform the required resets.

1. Before servicing the vehicle, refer to the Precautions Section.
2. Disconnect the negative battery cable.
3. Remove the sensor harness. Remove the knock sensor.

To install:
4. Installation is the reverse of the removal procedure.

✳✳ WARNING

Use care when installing the sensor. Do not use any knock sensors that have been dropped or physically damaged. Replace these with new ones.

5. Tighten the knock sensor bolt to 16 ft. lbs. (21 Nm).

4.0L Engine

➡If working near and/or around the SRS system and components, be sure to disable the SRS system. After disabling the system wait three minutes or more before servicing the vehicle.

➡Whenever the negative battery cable is disconnected the following components will require resetting. The Idle Air Volume Learning, Steering Angle Sensor Neutral Position, Sunroof Memory Reset/Initialization, Automatic Drive Positioner System, Audio presets and Navigation. Use the Nissan CONSULT III diagnostic tool, or equivalent to perform the required resets.

1. Before servicing the vehicle, refer to the Precautions Section.
2. Disconnect the negative battery cable.
3. Remove the intake collector to gain access to the knock sensors.
4. Remove the sensor harness. Remove the sensor.

To install:
5. Installation is the reverse of the removal procedure.

✳✳ WARNING

Use care when installing the sensor. Do not use any knock sensors that have been dropped or physically damaged. Replace these with new ones.

6. Tighten the knock sensor bolt to 13 ft. lbs. (18 Nm).

MALFUNCTION INDICATOR LIGHT

RESET PROCEDURE

Proper operation of the Malfunction Indicator Light (MIL):
• The MIL will illumine with the ignition switch ON and the engine OFF
• The MIL will turn OFF when the engine is started
• The MIL will remain ON if the self-diagnostic system has detected a malfunction

• The MIL may turn OFF if the malfunction is no longer present
• If the MIL is illuminated and then the engine stalls, the MIL will remain illuminated as long as the ignition switch is ON
• If the MIL is not illuminated and the engine stalls, the MIL will not illuminate until the ignition switch is cycled OFF, then ON

1. Before servicing the vehicle, refer to the Precautions Section.
2. Resetting the MIL:
• The control module turns OFF the MIL after 3 consecutive ignition cycles that the diagnostic system runs and does not fail
• The control module turns OFF the MIL after a current Diagnostic Trouble Code (DTC) clears when the diagnostic cycle runs and passes
• There may still be a history of DTC's stored in the system. These will clear after 40 consecutive warm-up cycles, if no failures are reported by any other related diagnostic system
• Manual resetting of the MIL and any DTC stored in the system, requires the use of an OBD2 scan tool connected to the Data Link Connector (DLC) for communication with the vehicle. Follow the instructions of the scan tool for both retrieval and resetting of DTC's. The scan tool can be used to command the MIL off.

➡If the error symptoms causing the MIL to illuminate have been corrected, the MIL will return to normal operation.

3. If a DTC is present, record the code and troubleshoot the fault.

THROTTLE CONTROL ACTUATOR

LOCATION

2.5L Engine
See Figure 193.

4.0L Engine
See Figure 194.

REMOVAL & INSTALLATION

2.5L Engine

➡If working near and/or around the SRS system and components, be sure

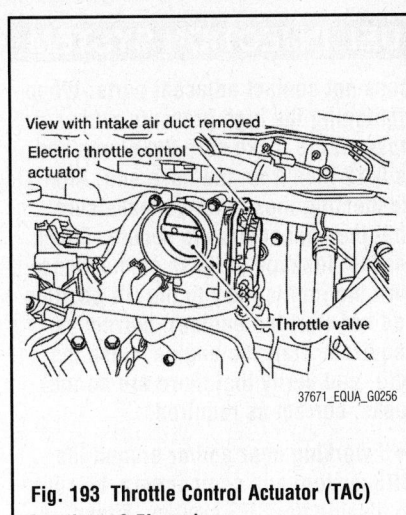

Fig. 193 Throttle Control Actuator (TAC) location—2.5L engine

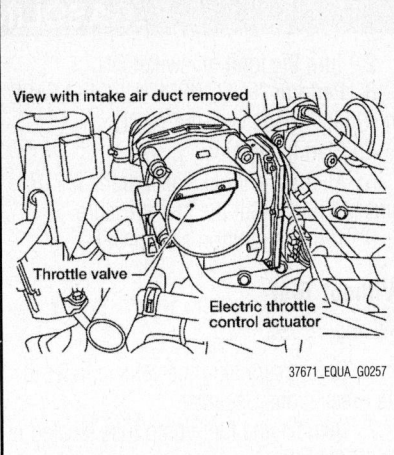

Fig. 194 Throttle Control Actuator (TAC) location—4.0L engine

to disable the SRS system. After disabling the system wait three minutes or more before servicing the vehicle.

➡Whenever the negative battery cable is disconnected the following components will require resetting. The Idle Air Volume Learning, Steering Angle Sensor Neutral Position, Sunroof Memory Reset/Initialization, Automatic Drive Positioner System, Audio presets and Navigation. Use the Nissan CONSULT III diagnostic tool, or equivalent to perform the required resets.

1. Before servicing the vehicle, refer to the Precautions Section.
2. Disconnect the negative battery cable.
3. Remove the air cleaner ductwork.
4. Disconnect the electric throttle control actuator harness connector.
5. Remove the electric throttle control actuator bolts and unit.

To install:
6. Installation is the reverse of the removal procedure.
7. Tighten the electric throttle control actuator mounting bolts to 89 inch lbs. (10 Nm).
8. Perform the throttle position learning.
9. Perform the idle air volume learning.

4.0L Engine

➡If working near and/or around the SRS system and components, be sure to disable the SRS system. After disabling the system wait three minutes or more before servicing the vehicle.

➡Whenever the negative battery cable is disconnected the following components will require resetting. The Idle Air Volume Learning, Steering Angle Sensor Neutral Position, Sunroof Memory Reset/Initialization, Automatic Drive Positioner System, Audio presets and Navigation. Use the Nissan CON-

SULT III diagnostic tool, or equivalent to perform the required resets.

1. Before servicing the vehicle, refer to the Precautions Section.
2. Disconnect the negative battery cable.
3. Remove the air cleaner ductwork.
4. Disconnect the electric throttle control actuator harness connector.
5. Remove the electric throttle control actuator bolts and unit.

To install:
6. Installation is the reverse of the removal procedure.
7. Tighten the electric throttle control actuator mounting bolts to 74 inch lbs. (8 Nm).
8. Perform the throttle position learning.
9. Perform the idle air volume learning.

THROTTLE POSITION SENSOR

LOCATION

2.5L Engine

The electric throttle control actuator includes a throttle control motor and Throttle Position Sensor (TPS). The TPS responds to the throttle valve movement. The TPS has two sensors.

4.0L Engine

The electric throttle control actuator includes a throttle control motor and Throttle Position Sensor (TPS). The TPS responds to the throttle valve movement. The TPS has two sensors.

REMOVAL & INSTALLATION

➡The Throttle Position Sensor (TPS) is replaced with the electric throttle control actuator

FUEL SYSTEM SERVICE PRECAUTIONS

Safety is the most important factor when performing not only fuel system maintenance but any type of maintenance. Failure to conduct maintenance and repairs in a safe manner may result in serious personal injury or death. Maintenance and testing of the vehicle's fuel system components can be accomplished safely and effectively by adhering to the following rules and guidelines.

- To avoid the possibility of fire and personal injury, always disconnect the negative battery cable unless the repair or test procedure requires that battery voltage be applied.

- Always relieve the fuel system pressure prior to disconnecting any fuel system component (injector, fuel rail, pressure regulator, etc.), fitting or fuel line connection. Exercise extreme caution whenever relieving fuel system pressure to avoid exposing skin, face and eyes to fuel spray. Please be advised that fuel under pressure may penetrate the skin or any part of the body that it contacts.

- Always place a shop towel or cloth around the fitting or connection prior to loosening to absorb any excess fuel due to spillage. Ensure that all fuel spillage (should it occur) is quickly removed from engine surfaces. Ensure that all fuel soaked cloths or towels are deposited into a suitable waste container.

- Always keep a dry chemical (Class B) fire extinguisher near the work area.

- Do not allow fuel spray or fuel vapors to come into contact with a spark or open flame.

- Always use a back-up wrench when loosening and tightening fuel line connection fittings. This will prevent unnecessary stress and torsion to fuel line piping.

- Always replace worn fuel fitting O-rings with new Do not substitute fuel hose or equivalent where fuel pipe is installed.

Before servicing the vehicle, make sure to also refer to the precautions in the beginning of this section as well.

RELIEVING FUEL SYSTEM PRESSURE

With Tool

1. Before servicing the vehicle, refer to the Precautions Section.

2. Turn the ignition switch ON.
3. Perform "FUEL PRESSURE RELEASE" in "Utility" mode with the TOOL.
4. Start the engine.
5. After the engine stalls, crank it 2–3 times to release all the fuel pressure.
6. Turn the ignition switch OFF.

Without Tool

See Figure 195.

1. Before servicing the vehicle, refer to the Precautions Section.
2. Remove the fuel pump fuse located in the IPDM E/R.
3. Start the engine.
4. After the engine stalls, crank it 2–3 times to release all the fuel pressure.
5. Turn the ignition switch OFF.
6. When repairs are complete, replace the fuel pump fuse and connect the negative battery cable.

FUEL PUMP MODULE

REMOVAL & INSTALLATION

See Figure 196.

➡Be sure to check the fuel gauge indicator. Make sure that it reads not more than half a tank. If not, properly drain fuel until the gauge reads half a tank or less.

➡This vehicle uses quick connect fittings. Be sure to properly relieve the fuel system pressure before disconnecting any of these fittings. Always replace O-rings and clamps with new ones. Do not bend, twist or kink hoses when they are being removed or installed. Be sure that the clamp screw

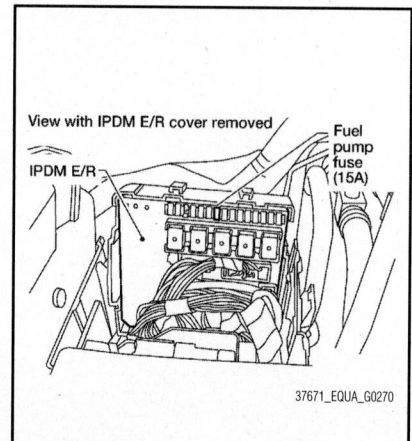

View with IPDM E/R cover removed

IPDM E/R

Fuel pump fuse (15A)

37671_EQUA_G0270

Fig. 195 Remove the fuel pump fuse located in the IPDM E/R

does not contact adjacent parts. When tightening the high pressure rubber hose clamp make sure the clamp end is 0.12 inch from the hose end. After connecting these fittings make sure that the connectors are secure. Check for fuel leakage at these connections turn the ignition key to the ON position (do not start the engine), correct as required. Start the engine, raise the idle, and verify that there are no fuel leaks, correct as required.

➡If working near and/or around the SRS system and components, be sure to disable the SRS system. After disabling the system wait three minutes or more before servicing the vehicle.

➡Whenever the negative battery cable is disconnected the following components will require resetting. The Idle Air Volume Learning, Steering Angle Sensor Neutral Position, Sunroof Memory Reset/Initialization, Automatic Drive Positioner System, Audio presets and Navigation. Use the Nissan CONSULT III diagnostic tool, or equivalent to perform the required resets.

1. Before servicing the vehicle, refer to the Precautions Section.
2. Disconnect the negative battery cable.
3. Properly relieve the fuel system pressure.
4. Remove the fuel filler cap to release the pressure from inside the fuel tank.
5. Raise and safely support the vehicle.
6. Remove the left rear wheel and tire.

➡Fuel will be spilled when removing the fuel pump module for the fuel level is above the fuel level sensor, fuel filter, and fuel pump assembly fuel tank opening.

7. Remove the fuel tank shield.
8. Remove the fuel tank strap bolts while supporting the fuel tank with a suitable lift jack.
9. Lower the fuel tank using a suitable lift jack to access the top of the fuel pump module.
10. Disconnect the fuel pump module electrical connector, EVAP hose, and the fuel feed hose. Disconnect the fuel feed hose from the molded clip in the side of the fuel tank.
11. Disconnect the quick connector as follows:

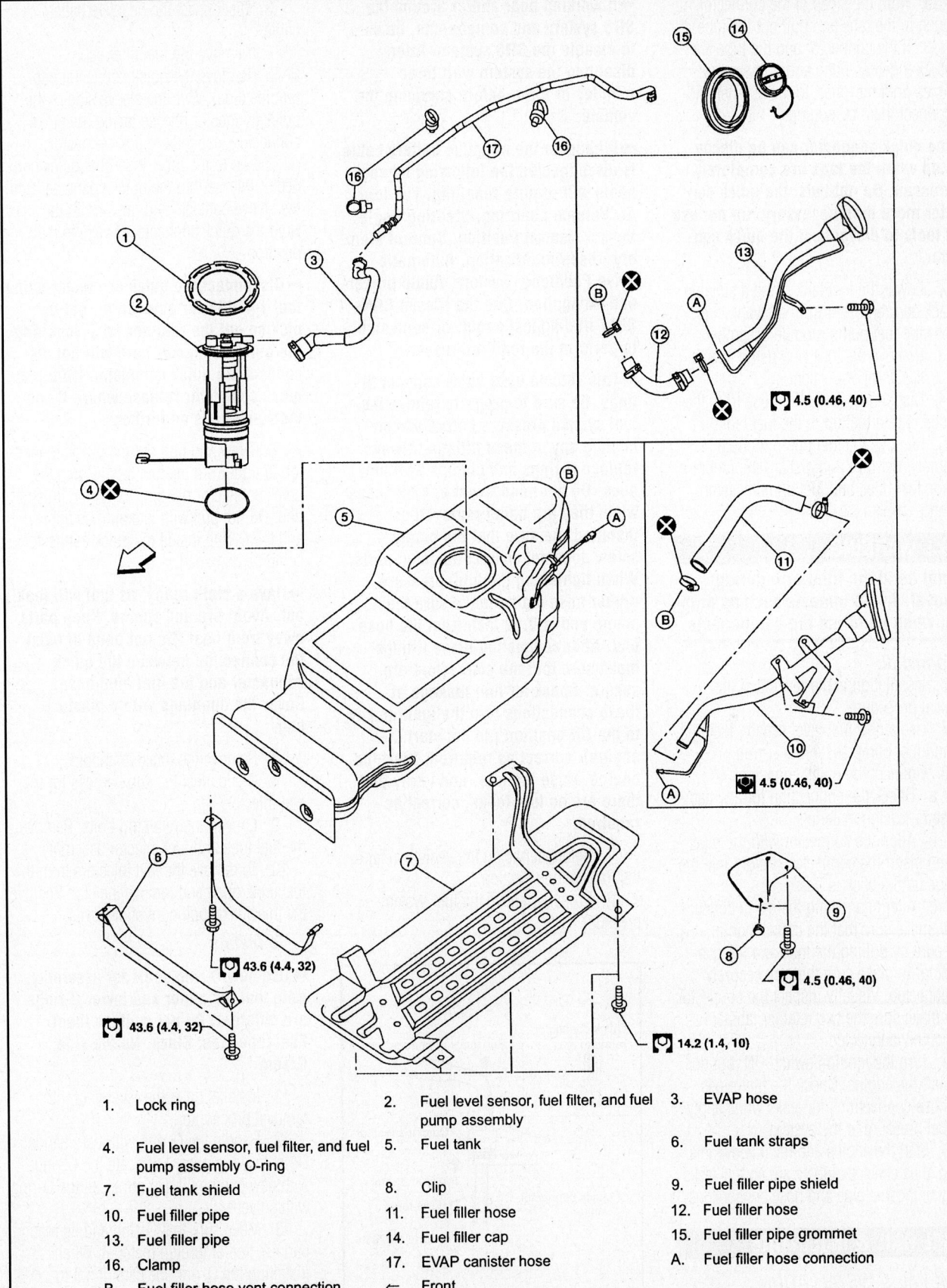

1.	Lock ring	
4.	Fuel level sensor, fuel filter, and fuel pump assembly O-ring	
7.	Fuel tank shield	
10.	Fuel filler pipe	
13.	Fuel filler pipe	
16.	Clamp	
B.	Fuel filler hose vent connection	

1. Lock ring
2. Fuel level sensor, fuel filter, and fuel pump assembly
3. EVAP hose
4. Fuel level sensor, fuel filter, and fuel pump assembly O-ring
5. Fuel tank
6. Fuel tank straps
7. Fuel tank shield
8. Clip
9. Fuel filler pipe shield
10. Fuel filler pipe
11. Fuel filler hose
12. Fuel filler hose
13. Fuel filler pipe
14. Fuel filler cap
15. Fuel filler pipe grommet
16. Clamp
17. EVAP canister hose
A. Fuel filler hose connection
B. Fuel filler hose vent connection
⇐ Front

09482_FRON_G0093

Fig. 196 Fuel pump module unit and related components

a. Hold the sides of the connector, push in the tabs and pull out the tube.

b. If the connector and the tube are stuck together, push and pull several times until they start to move. Then disconnect them by pulling.

➡ **The quick connector can be disconnected when the tabs are completely depressed. Do not twist the quick connector more than necessary. Do not use any tools to disconnect the quick connector.**

12. Lower the fuel tank using a suitable lift jack and remove it from the vehicle to access the fuel pump module assembly.

13. Remove the lock ring using special tool J-45722, or equivalent.

14. Disconnect the EVAP hose from the molded clip in the top of the fuel tank.

15. Remove the fuel pump module assembly. Remove and discard the fuel level sensor, fuel filter, and fuel pump assembly O-ring.

❊❊ WARNING

Do not bend the float arm during removal. Avoid impacts such as dropping when handling the components.

To install:

16. Installation is the reverse of the removal procedure.

17. Use a new fuel level sensor, fuel filter, and fuel pump assembly O-ring.

18. Connect the quick connector.

a. Check the connection for any damage or foreign materials.

b. Align the connector with the pipe, then insert the connector straight into the pipe until a click is heard.

c. After connecting the quick connector, make sure that the connection is secure by pulling the tube and the connector to make sure they are securely connected. Visually inspect the connector to make sure the two retainer tabs are securely connected.

19. Turn the ignition switch ON, but do not start the engine. Check the fuel pipes and hose connections for leaks while applying fuel pressure to the system.

20. Start the engine and rev it above idle speed, then check that there are no fuel leaks at any of the fuel pipe and hose connections.

FUEL RAIL & INJECTORS

REMOVAL & INSTALLATION

2.5L Engine

See Figures 197 through 199.

➡ **If working near and/or around the SRS system and components, be sure to disable the SRS system. After disabling the system wait three minutes or more before servicing the vehicle.**

➡ **Whenever the negative battery cable is disconnected the following components will require resetting. The Idle Air Volume Learning, Steering Angle Sensor Neutral Position, Sunroof Memory Reset/Initialization, Automatic Drive Positioner System, Audio presets and Navigation. Use the Nissan CONSULT III diagnostic tool, or equivalent to perform the required resets.**

➡ **This vehicle uses quick connect fittings. Be sure to properly relieve the fuel system pressure before disconnecting any of these fittings. Always replace O-rings and clamps with new ones. Do not bend, twist or kink hoses when they are being removed or installed. Be sure that the clamp screw does not contact adjacent parts. When tightening the high pressure rubber hose clamp make sure the clamp end is 0.12 inch from the hose end. After connecting these fittings make sure that the connectors are secure. Check for fuel leakage at these connections turn the ignition key to the ON position (do not start the engine), correct as required. Start the engine, raise the idle, and verify that there are no fuel leaks, correct as required.**

1. Before servicing the vehicle, refer to the Precautions Section.

2. Properly relieve the fuel system pressure.

Quick connector release

Sleeve

Pull quick connector.

Quick connector

A

Insert and retain.

Quick connector release

Fuel tube

09482_FRON_G0096

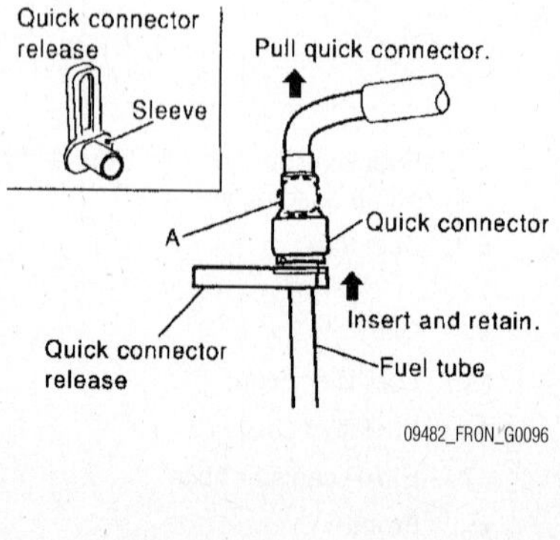

Fig. 197 Quick connector release location "A"—2.5L engine

3. Disconnect the negative battery cable.

4. Remove the fuel filler cap.

5. Remove the quick connector cap (engine side). With the sleeve side of the quick connector release facing the quick connector, install the quick connector release on to the tube. Insert the quick connector release into the quick connector until the sleeve contacts and goes no further. Hold the quick connector release in that position.

➡ **Disconnect the quick connector using tool J-45488, or equivalent, not by picking out the retainer tabs. Inserting the quick connector hard will not disconnect the quick connector. Hold the quick connector release where it contacts and goes no further.**

6. Draw and pull out the quick connector straight from the fuel tube. Grasp the quick connector holding "A" in the illustration. Do not pull with lateral force applied and the O-ring inside the quick connector could be damaged.

➡ **Have a cloth ready, as fuel will leak out. Avoid fire and sparks. Keep parts away from heat. Do not bend or twist the connection between the quick connector and the fuel feed hose. Cover the openings with a plastic bag.**

7. Remove the intake manifold.

8. Disconnect the sub harness for the fuel injector.

9. Loosen the retaining bolts. Remove the fuel tube and fuel injector assembly.

10. To remove the fuel injectors from the fuel tube, open and remove the clip. Remove the injector by pulling it straight out.

To install:

➡ **Use new O-ring seals for assembly. Note that the upper and lower O-rings are different. Do not confuse them. Fuel tube side: Black. Nozzle side: Green.**

11. Installation is the reverse of the removal procedure.

12. Handle the O-ring with bare hands. Do not wear gloves. Lubricate the O-ring with new engine oil. Do not clean the O-ring with solvent.

13. Make sure that O-ring and its mating part are free of foreign material. When installing the O-ring, be careful not to scratch it with a tool or fingernails. Do not twist or stretch the O-ring. If the O-ring was stretched while it was being attached, allow it to retract before inserting it into the fuel

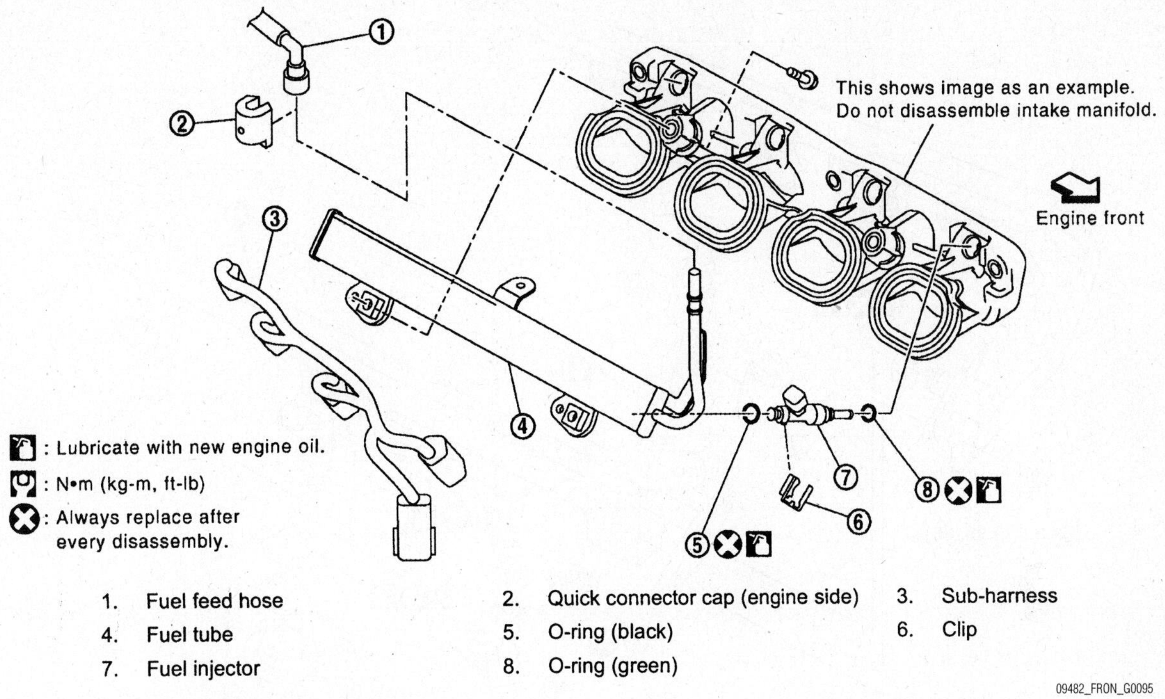

This shows image as an example.
Do not disassemble intake manifold.

Engine front

⬡ : Lubricate with new engine oil.

⬡ : N•m (kg-m, ft-lb)

✖ : Always replace after every disassembly.

1. Fuel feed hose
4. Fuel tube
7. Fuel injector

2. Quick connector cap (engine side)
5. O-ring (black)
8. O-ring (green)

3. Sub-harness
6. Clip

09482_FRON_G0095

Fig. 198 Fuel injector tube and related components—2.5L engine

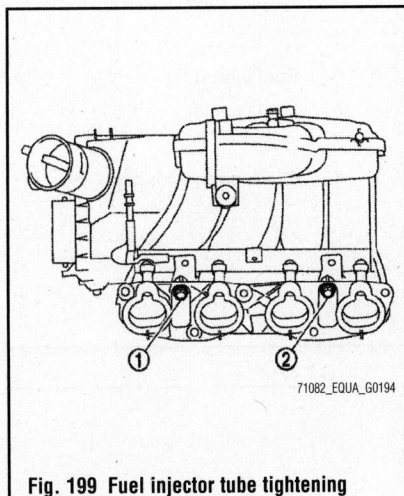

71082_EQUA_G0194

Fig. 199 Fuel injector tube tightening sequence—2.5L engine

tube. Insert a new O-ring straight into the fuel tube. Do not angle or twist it.

14. When installing the fuel feed tube be sure to torque the retaining bolts to 9 ft. lbs. (12 Nm) and then to 21 ft. lbs. (28 Nm) in an alternating order.

15. Turn the ignition switch ON, but do not start the engine. Check the fuel lines and hose connections for leaks while applying fuel pressure to the system.

16. Start the engine and check for fuel leaks, correct as required.

4.0L Engine

See Figures 200 and 201.

➡If working near and/or around the SRS system and components, be sure to disable the SRS system. After disabling the system wait three minutes or more before servicing the vehicle.

➡Whenever the negative battery cable is disconnected the following components will require resetting. The Idle Air Volume Learning, Steering Angle Sensor Neutral Position, Sunroof Memory Reset/Initialization, Automatic Drive Positioner System, Audio presets and Navigation. Use the Nissan CONSULT III diagnostic tool, or equivalent to perform the required resets.

➡This vehicle uses quick connect fittings. Be sure to properly relieve the fuel system pressure before disconnecting any of these fittings. Always replace O-rings and clamps with new ones. Do not bend, twist or kink hoses when they are being removed or installed. Be sure that the clamp screw does not contact adjacent parts. When tightening the high pressure rubber hose clamp make sure the clamp end is 0.12 inch from the hose end. After connecting these fittings make sure

that the connectors are secure. Check for fuel leakage at these connections turn the ignition key to the ON position (do not start the engine), correct as required. Start the engine, raise the idle, and verify that there are no fuel leaks, correct as required.

1. Before servicing the vehicle, refer to the Precautions Section.
2. Properly relieve the fuel system pressure.
3. Disconnect the negative battery cable.
4. Remove the fuel filler cap.
5. Remove the intake manifold collector.
6. Remove the quick connector cap (engine side). With the sleeve side of the quick connector release facing the quick connector, install the quick connector release on to the tube. Insert the quick connector release into the quick connector until the sleeve contacts and goes no further. Hold the quick connector release in that position.

➡Disconnect the quick connector using tool J-45488, or equivalent, not by picking out the retainer tabs. Inserting the quick connector hard will not disconnect the quick connector. Hold the quick connector release where it contacts and goes no further.

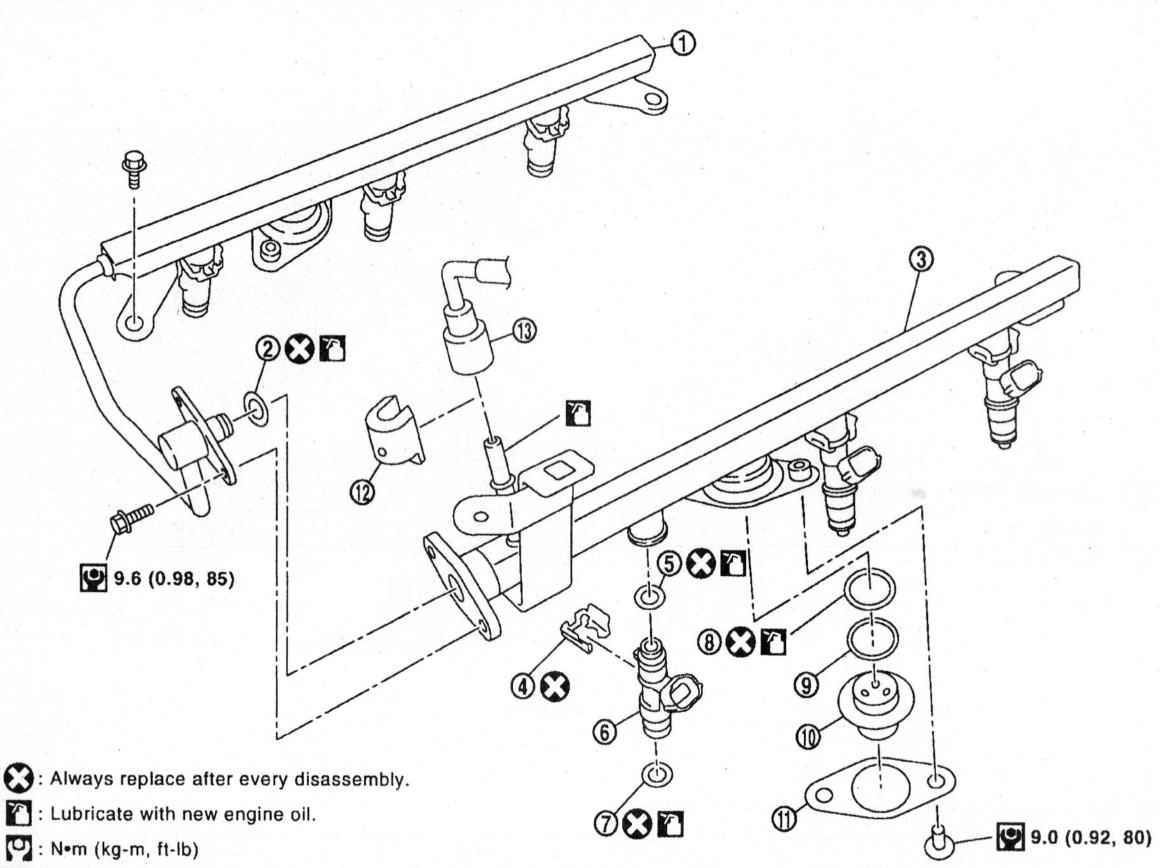

⊗ : Always replace after every disassembly.

⬛ : Lubricate with new engine oil.

🔧 : N•m (kg-m, ft-lb)

🔧 : N•m (kg-m, in-lb)

🔧 9.6 (0.98, 85)

🔧 9.0 (0.92, 80)

1.	Fuel tube (RH)	2.	O-ring	3.	Fuel tube (LH)
4.	Clip	5.	O-ring (blue)	6.	Fuel injector
7.	O-ring (brown)	8.	O-ring	9.	Spacer
10.	Fuel damper	11.	Fuel damper cap	12.	Quick connector cap
13.	Fuel feed hose				

09482_FRON_G0097

Fig. 200 Fuel injector tube and related components—4.0L engine

7. Draw and pull out the quick connector straight from the fuel tube. Grasp the quick connector holding "A" in the illustration. Do not pull with lateral force applied and the O-ring inside the quick connector could be damaged.

➡**Have a cloth ready, as fuel will leak out. Avoid fire and sparks. Keep parts away from heat. Do not bend or twist the connection between the quick connector and the fuel feed hose. Cover the openings with a plastic bag.**

8. Remove the PCV hose between the rocker covers.

9. Disconnect the harness for the fuel injector.

10. Loosen the retaining bolts. Remove the fuel tube and fuel injector assembly.

Remove the bolts which connect the left and right fuel tubes.

11. To remove the fuel injectors from the fuel tube, open and remove the clip. Remove the injector by pulling it straight out.

12. Disconnect the right fuel tube from the left fuel tube. Loosen the bolts, to remove the fuel damper cap and fuel damper, if necessary.

To install:

➡**Use new O-ring seals for assembly. Note that the upper and lower O-rings are different. Do not confuse them. Fuel tube side: Blue. Nozzle side: Brown.**

13. Installation is the reverse of the removal procedure.

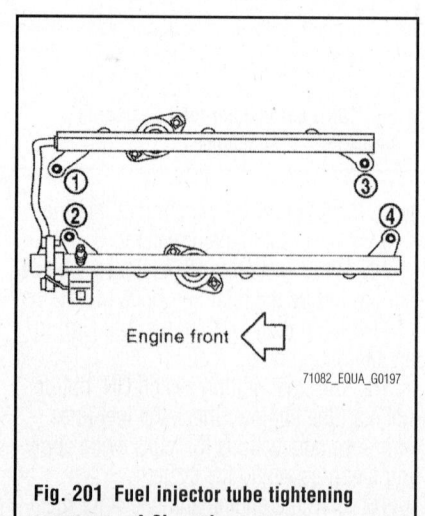

Engine front

71082_EQUA_G0197

Fig. 201 Fuel injector tube tightening sequence—4.0L engine

14. Handle the O-ring with bare hands. Do not wear gloves. Lubricate the O-ring with new engine oil. Do not clean the O-ring with solvent.

15. Make sure that O-ring and its mating part are free of foreign material. When installing the O-ring, be careful not to scratch it with a tool or fingernails. Do not twist or stretch the O-ring. If the O-ring was stretched while it was being attached, allow it to retract before inserting it into the fuel tube. Insert a new O-ring straight into the fuel tube. Do not angle or twist it.

16. When installing the fuel feed tube be sure to torque the retaining bolts to 7 ft. lbs. (10 Nm) and then to 16 ft. lbs. (22 Nm) in an alternating order.

17. Turn the ignition switch ON, but do not start the engine. Check the fuel lines and hose connections for leaks while applying fuel pressure to the system.

18. Start the engine and check for fuel leaks, correct as required.

FUEL TANK

DRAINING

Observe the following precautions:
• Observe all applicable safety precautions when working around the fuel system. Whenever servicing the fuel system, always work in a well-ventilated area. Do not allow fuel spray or vapors to come in contact with a spark or open flame
• Keep a dry chemical fire extinguisher near the work area
• Always keep fuel in a container specifically designed for fuel storage. Properly seal fuel containers to avoid the possibility of fire or explosion
• Use of a fresh air mask is highly recommended
• Always follow the safety recommendations of the fuel extractor you are using to avoid the potential of fire or explosion

➡**If working near and/or around the SRS system and components, be sure to disable the SRS system. After disabling the system wait three minutes or more before servicing the vehicle.**

➡**Whenever the negative battery cable is disconnected the following components will require resetting. The Idle Air Volume Learning, Steering Angle Sensor Neutral Position, Sunroof Memory Reset/Initialization, Automatic Drive Positioner System, Audio presets and Navigation. Use the Nissan**

CONSULT III diagnostic tool, or equivalent to perform the required resets.

1. Before servicing the vehicle, refer to the Precautions Section.
2. Properly relieve the fuel system pressure.
3. Disconnect the negative battery cable.
4. Remove the fuel filler cap.
5. Insert a suitable hose of less than 0.59 inch (15mm) diameter into the fuel filler pipe through the fuel filler opening to drain the fuel from fuel filler pipe.
6. Remove the fuel filler pipe shield.
7. Disconnect the fuel filler hose from the fuel filler pipe.
8. Insert a suitable hose into the fuel tank through the fuel filler hose to drain the fuel from the fuel tank.

REMOVAL & INSTALLATION

➡**Be sure to check the fuel gauge indicator. Make sure that it reads not more than half a tank. If not, properly drain fuel until the gauge reads half a tank or less.**

➡**This vehicle uses quick connect fittings. Be sure to properly relieve the fuel system pressure before disconnecting any of these fittings. Always replace O-rings and clamps with new ones. Do not bend, twist or kink hoses when they are being removed or installed. Be sure that the clamp screw does not contact adjacent parts. When tightening the high pressure rubber hose clamp make sure the clamp end is 0.12 inch from the hose end. After connecting these fittings make sure that the connectors are secure. Check for fuel leakage at these connections turn the ignition key to the ON position (do not start the engine), correct as required. Start the engine, raise the idle, and verify that there are no fuel leaks, correct as required.**

➡**If working near and/or around the SRS system and components, be sure to disable the SRS system. After disabling the system wait three minutes or more before servicing the vehicle.**

➡**Whenever the negative battery cable is disconnected the following components will require resetting. The Idle Air Volume Learning, Steering Angle Sensor Neutral Position, Sunroof Memory Reset/Initialization, Automatic Drive Positioner System, Audio presets**

and Navigation. Use the Nissan CONSULT III diagnostic tool, or equivalent to perform the required resets.

1. Before servicing the vehicle, refer to the Precautions Section.
2. Disconnect the negative battery cable.
3. Properly relieve the fuel system pressure.
4. Remove the fuel filler cap to release the pressure from inside the fuel tank.
5. Raise and support the vehicle safely.
6. Remove the left rear wheel and tire.

➡**Fuel will be spilled when removing the fuel pump module for the fuel level is above the fuel level sensor, fuel filter, and fuel pump assembly fuel tank opening.**

7. Remove the fuel tank shield.
8. Remove the fuel tank strap bolts while supporting the fuel tank with a suitable lift jack.
9. Lower the fuel tank using a suitable lift jack to access the top of the fuel pump module.
10. Disconnect the fuel pump module electrical connector, EVAP hose, and the fuel feed hose. Disconnect the fuel feed hose from the molded clip in the side of the fuel tank.
11. Disconnect the quick connector as follows:
   a. Hold the sides of the connector, push in the tabs and pull out the tube.
   b. If the connector and the tube are stuck together, push and pull several times until they start to move. Then disconnect them by pulling.

➡**The quick connector can be disconnected when the tabs are completely depressed. Do not twist the quick connector more than necessary. Do not use any tools to disconnect the quick connector.**

12. Lower the fuel tank using a suitable lift jack and remove it from the vehicle to access the fuel pump module assembly.
13. Remove the lock ring using special tool J-45722, or equivalent.
14. Disconnect the EVAP hose from the molded clip in the top of the fuel tank.
15. Remove the fuel pump module assembly, as necessary. Remove and discard the fuel level sensor, fuel filter, and fuel pump assembly O-ring.

❋❋ WARNING

Do not bend the float arm during removal. Avoid impacts such as dropping when handling the components.

16. Remove the fuel tank from the vehicle.

To install:

17. Installation is the reverse of the removal procedure.

18. Use a new fuel level sensor, fuel filter, and fuel pump assembly O-ring.

19. Connect the quick connector.

   a. Check the connection for any damage or foreign materials.

   b. Align the connector with the pipe, then insert the connector straight into the pipe until a click is heard.

   c. After connecting the quick connector, make sure that the connection is secure by pulling the tube and the connector to make sure they are securely connected. Visually inspect the connector to make sure the two retainer tabs are securely connected.

20. Turn the ignition switch ON, but do not start the engine. Check the fuel pipes and hose connections for leaks while applying fuel pressure to the system.

21. Start the engine and rev it above idle speed, then check that there are no fuel leaks at any of the fuel pipe and hose connections.

THROTTLE BODY

REMOVAL & INSTALLATION

2.5L Engine

➡ If working near and/or around the SRS system and components, be sure to disable the SRS system. After disabling the system wait three minutes or more before servicing the vehicle.

➡ Whenever the negative battery cable is disconnected the following components will require resetting. The Idle Air Volume Learning, Steering Angle Sensor Neutral Position, Sunroof Memory Reset/Initialization, Automatic Drive Positioner System, Audio presets and Navigation. Use the Nissan CONSULT III diagnostic tool, or equivalent to perform the required resets.

1. Before servicing the vehicle, refer to the Precautions Section.

2. Disconnect the negative battery cable.

3. Remove the air cleaner ductwork.

4. Disconnect the electric throttle control actuator harness connector.

5. Remove the electric throttle control actuator bolts and unit.

To install:

6. Installation is the reverse of the removal procedure.

7. Tighten the electric throttle control actuator mounting bolts to 89 inch lbs. (10 Nm).

8. Perform the throttle position learning.

9. Perform the idle air volume learning.

4.0L Engine

➡ If working near and/or around the SRS system and components, be sure to disable the SRS system. After disabling the system wait three minutes or more before servicing the vehicle.

➡ Whenever the negative battery cable is disconnected the following components will require resetting. The Idle Air Volume Learning, Steering Angle Sensor Neutral Position, Sunroof Memory Reset/Initialization, Automatic Drive Positioner System, Audio presets and Navigation. Use the Nissan CONSULT III diagnostic tool, or equivalent to perform the required resets.

1. Before servicing the vehicle, refer to the Precautions Section.

2. Disconnect the negative battery cable.

3. Remove the air cleaner ductwork.

4. Disconnect the electric throttle control actuator harness connector.

5. Remove the electric throttle control actuator bolts and unit.

To install:

6. Installation is the reverse of the removal procedure.

7. Tighten the electric throttle control actuator mounting bolts to 74 inch lbs. (8 Nm).

8. Perform the throttle position learning.

9. Perform the idle air volume learning.

HEATING & AIR CONDITIONING

BLOWER MOTOR

REMOVAL & INSTALLATION
See Figure 202.

➡ If working near and/or around the SRS system and components, be sure to disable the SRS system. After disabling the system wait three minutes or more before servicing the vehicle.

➡ Whenever the negative battery cable is disconnected the following components will require resetting. The Idle Air Volume Learning, Steering Angle Sensor Neutral Position, Sunroof Memory Reset/Initialization, Automatic Drive Positioner System, Audio presets and Navigation. Use the Nissan CONSULT III diagnostic tool, or equivalent to perform the required resets.

1. Before servicing the vehicle, refer to the Precautions Section.

2. Disconnect the negative battery cable.

3. Remove the lower glove box assembly.

4. Disconnect the blower motor electrical connector.

5. Remove the blower motor retaining screws.

6. Remove the blower motor from its mounting.

To install:

7. Installation is the reverse of the removal procedure.

8. Connect the battery positive cable and then the negative terminal. Follow the relearn procedures for battery reconnection.

HEATER CORE

REMOVAL & INSTALLATION
See Figures 203 through 208.

➡ If working near and/or around the SRS system and components, be sure to disable the SRS system. After disabling the system wait three minutes or more before servicing the vehicle.

➡ Whenever the negative battery cable is disconnected the following components will require resetting. The Idle Air Volume Learning, Steering Angle Sensor Neutral Position, Sunroof Memory Reset/Initialization, Automatic Drive Positioner System, Audio presets and Navigation. Use the Nissan CONSULT III diagnostic tool, or equivalent to perform the required resets.

1. Before servicing the vehicle, refer to the Precautions Section.

2. Disconnect the negative battery cable.

3. Position the front wheels in the straight ahead direction.

4. Disconnect the negative battery cable. Disconnect the positive battery cable.

5. Drain the engine coolant.

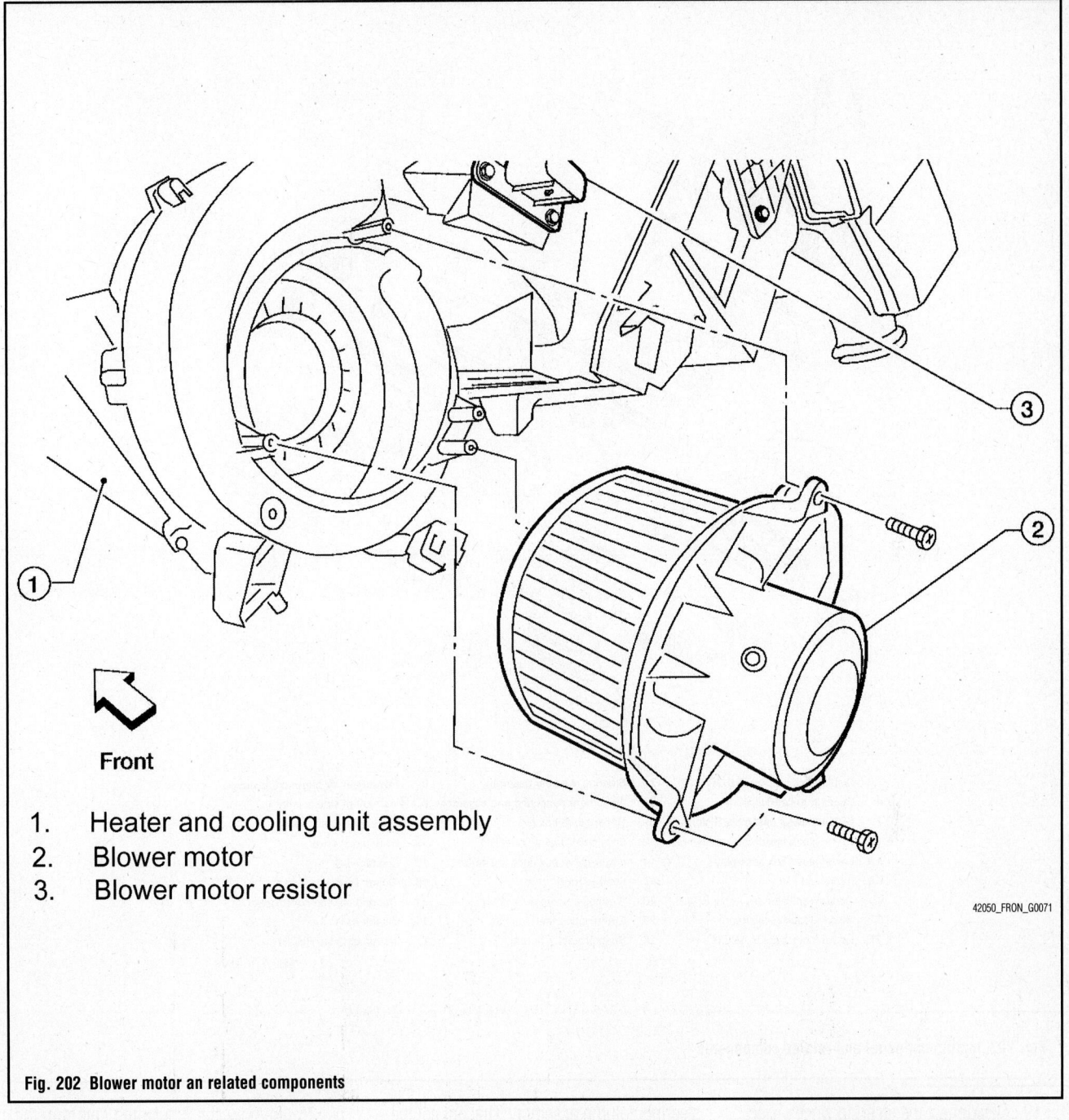

Front

1. Heater and cooling unit assembly
2. Blower motor
3. Blower motor resistor

42050_FRON_G0071

Fig. 202 Blower motor an related components

6. Properly discharge the air conditioning system.

7. If equipped with the 4.0L engine, remove the right side heater core pipe nuts.

8. Disconnect the heater core hoses from the heater core.

9. Disconnect the air conditioning refrigerant lines from the expansion valve.

10. Position the front seats in the rearmost position on the seat tracks.

11. Remove the upper front pillar trim panel. Remove the steering lock escutcheon. Remove the cluster lid "A". Remove the combination meter. Disconnect the electrical connections.

12. Remove the optical sensor. Remove the audio unit. Remove the cluster lid "D".

13. Remove the glove box. Remove the 2 bolts, through the glove box opening, retaining the front passenger's side air bag module to the steering member. Disconnect the air bag module connectors.

14. Remove the instrument stay right side and left side bolts. Remove the instrument panel.

15. Remove the 2 front floor ducts.

16. To remove the driver's side air bag module, locate the retaining clip access hole under the steering wheel. Insert a suitable blunt tool (4–6mm in size).

→Do not use sharp edged objects, such as a screwdriver, to release the driver's side airbag module from the steering wheel as SRS components may be unintentionally damaged.

17. Press upward, toward the center of the steering wheel, on the retaining clip until the air bag module is released from the steering wheel.

18. Lift the air bag module from the steering wheel. Disconnect the electrical connectors. Remove the air bag module.

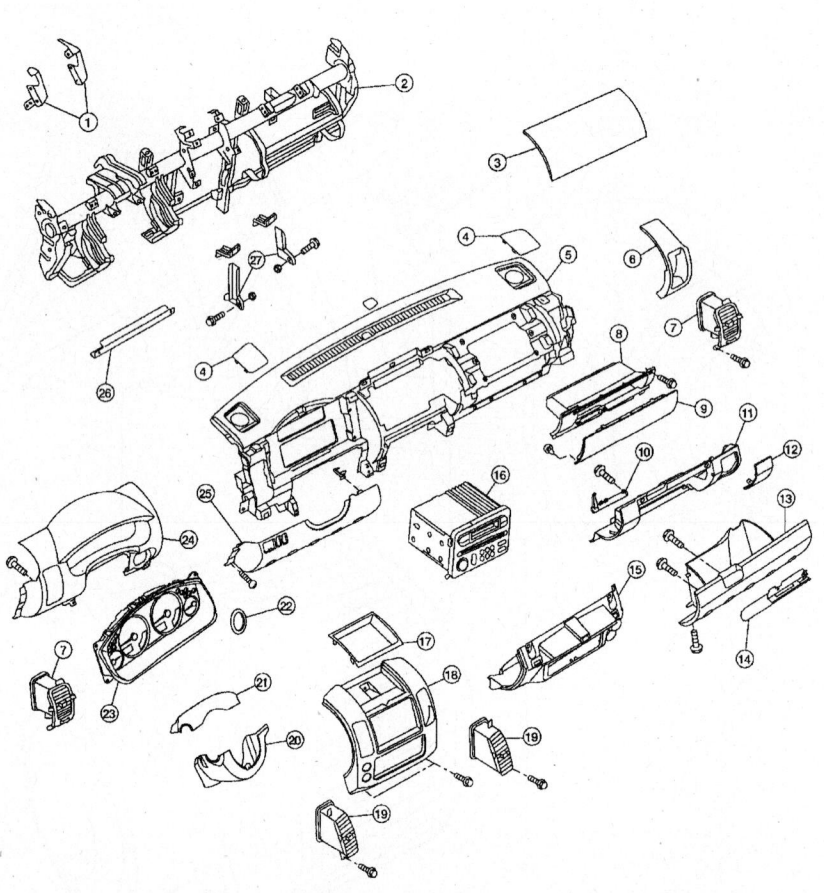

1. Display unit bracket RH/LH
2. Steering member assembly
3. Passenger air bag module cover
4. Speaker grille RH/LH
5. Instrument panel and pad assembly
6. Instrument side finisher
7. Side ventilator assembly RH/LH
8. Upper glove box bin
9. Upper glove box door
10. Lower glove box damper assembly
11. Lower instrument panel RH
12. Fuse block cover
13. Lower glove box assembly
14. Lower glove box latch assembly
15. Cluster lid D
16. Audio unit
17. Storage tray
18. Cluster lid C
19. Center ventilator assembly RH/LH
20. Steering column cover lower
21. Steering column cover upper
22. Steering lock escutcheon
23. Combination meter
24. Cluster lid A
25. Lower instrument panel LH
26. Knee protector brace
27. Instrument stay RH/LH

09482_FRON_G0009

Fig. 203 Instrument panel and related components

19. Disconnect the steering wheel switches. Remove the steering wheel center nut. Using a steering wheel removal tool, remove the steering wheel.

20. Remove the steering column upper and lower covers. Disconnect the wiper and washer switch connector. While pressing the tabs, pull the wiper and washer switch away from the spiral cable to remove it.

21. Disconnect the light and turn signal switch connector. While pressing the tabs, pull the light and turn signal switch toward the driver's door to remove it.

22. Remove the screws. While pressing the tab, pull the spiral cable away from the

steering column assembly. Disconnect the electrical connectors.

✳✳ WARNING

With the steering linkage disconnected, the spiral cable may snap by turning the steering wheel beyond the limited number of turns. The spiral cable can be turned counterclockwise about 2.5 turns from the neutral position.

23. Remove the lower knee protector.
24. Remove the locknut and bolt from the upper joint and then separate the upper joint from the upper shaft.

25. Remove the 3 nuts and bolt from the steering column and then remove the steering column assembly from the steering member.

26. Remove the hole cover seal and clamp. Remove the hole cover nuts, remove the hole cover from the dash panel.

27. Remove the bolt from the lower joint of the lower joint shaft and remove the lower joint shaft from the vehicle.

28. Disconnect the instrument panel wire harness at the right and left in-line connector brackets, and the fuse block (SMJ) electrical connectors.

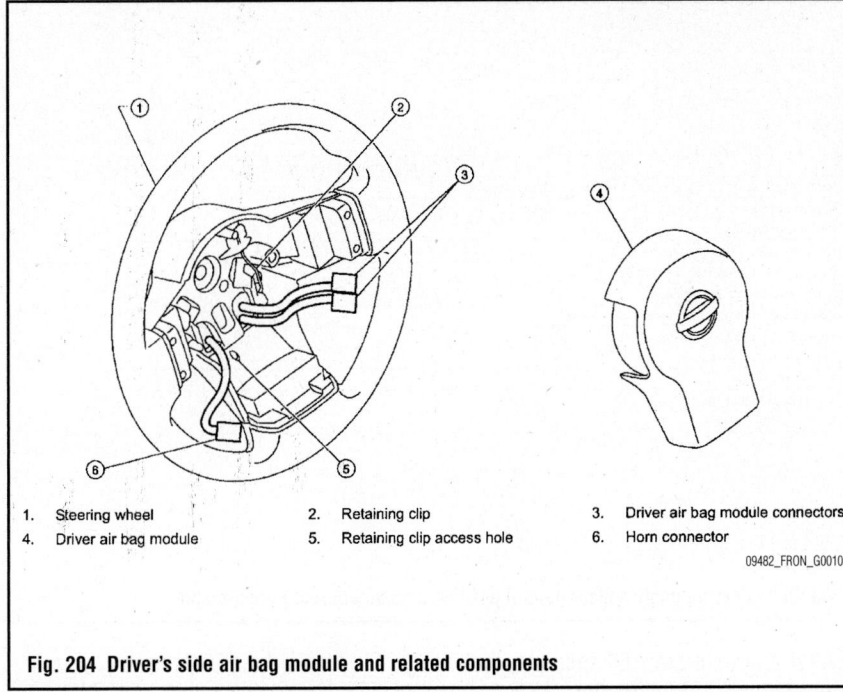

1. Steering wheel
2. Retaining clip
3. Driver air bag module connectors
4. Driver air bag module
5. Retaining clip access hole
6. Horn connector

09482_FRON_G0010

Fig. 204 Driver's side air bag module and related components

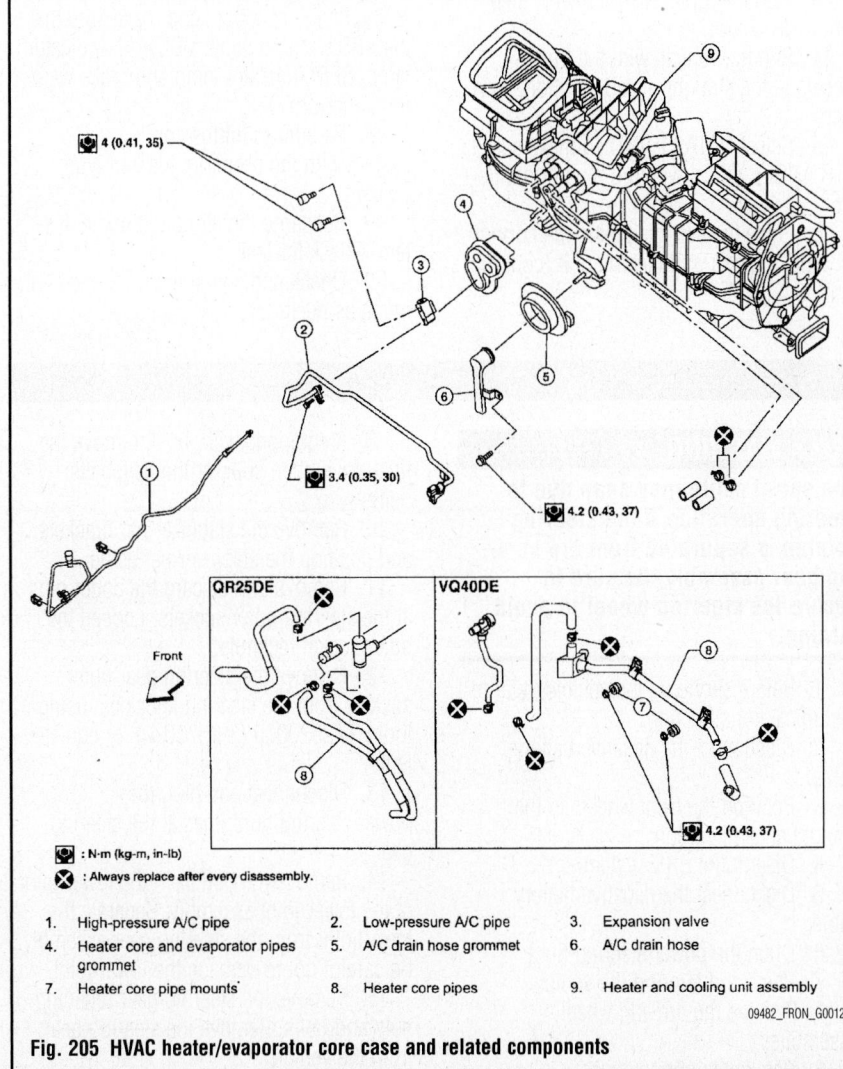

4 (0.41, 35)

3.4 (0.35, 30)

4.2 (0.43, 37)

QR25DE | VQ40DE

Front

4.2 (0.43, 37)

🔲 : N·m (kg-m, in-lb)

✖ : Always replace after every disassembly.

1. High-pressure A/C pipe
2. Low-pressure A/C pipe
3. Expansion valve
4. Heater core and evaporator pipes grommet
5. A/C drain hose grommet
6. A/C drain hose
7. Heater core pipe mounts
8. Heater core pipes
9. Heater and cooling unit assembly

09482_FRON_G0012

Fig. 205 HVAC heater/evaporator core case and related components

29. Remove the covers and then remove the three steering member bolts from each side to disconnect the steering member from the vehicle body.

30. Remove the heater/evaporator case assembly with it attached to the steering member from the vehicle.

31. Separate the steering member from the heater/evaporator unit.

32. Remove the heater cover retaining screws. Remove the cover.

33. Remove the heater core and the evaporator pipe bracket. Remove the heater core.

To install:

34. Installation is the reverse of the removal procedure.

➡**If the in-cabin microfilters are contaminated with coolant, replace them.**

35. Be sure to use new steering column retaining bolts and pinch bolt, as required.

➡**When installing the steering column, finger tighten all of the lower bracket and joint bolts and then tighten them to specification. Do not apply undue stress to the steering column.**

36. With the wheels in the straight ahead position align the slit of the lower joint with the projection on the dust cover. Insert the joint until surface "A" contacts surface "B".

37. Be sure to align the spiral cable correctly when installing the steering wheel. Make sure that the cable is in the neutral position. The neutral position is detected by turning left 2.6 revolutions from the right end position and ending with the locating pin at the top.

38. Refer to the illustration to determine if adjustment of the steering angle sensor neutral position is required.

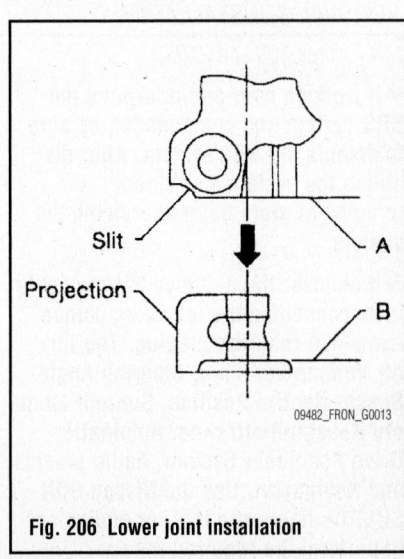

Slit

Projection

A

B

09482_FRON_G0013

Fig. 206 Lower joint installation

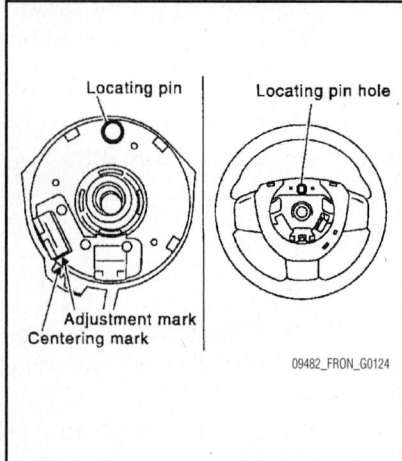

Fig. 207 Spiral cable installation and locating point

Situation	Adjustment of steering angle sensor neutral position
Removing/Installing ABS actuator and electric unit (control unit)	—
Replacing ABS actuator and electric unit (control unit)	×
Removing/Installing steering angle sensor	×
Replacing steering angle sensor	×
Removing/Installing steering components	×
Replacing steering components	×
Removing/Installing suspension components	×
Replacing suspension components	×
Change tires to new ones	—
Tire rotation	—
Adjusting wheel alignment	×
Battery disconnection	×

×: Required –: Not required

37671_EQUA_G0032

Fig. 208 Steering angle sensor neutral position adjustment requirement table

➡To adjust the neutral position of the steering angle sensor, make sure to use the Nissan CONSULT III diagnostic tool, or equivalent. Adjustment cannot be done without a diagnostic tool

39. Stop the vehicle with the front wheels in the straight-ahead position.

40. On the TOOL screen, touch UTILITY and ST ANG SEN ADJUSTMENT in order.

41. Touch START.

➡Do not touch the steering wheel while adjusting the steering angle sensor.

42. After approximately 10 seconds, touch END.

➡After approximately 60 seconds, the program ends automatically.

43. Turn the ignition switch OFF, then turn it ON again.

44. Run the vehicle with the front wheels in the straight-ahead position, then stop.

45. Select DATA LIST. Then make sure STR ANGLE SIG is within 0 plus or minus 2.5°.

46. Erase the self-diagnosis memory of the ABS actuator and electric unit (control unit) and ECM.

47. After erasing DTC memory, start the engine and drive the vehicle at 19 MPH (30 km/h) or more for approximately 1 minute as the final inspection, and make sure that the ABS warning lamp, VDC OFF indicator lamp, SLIP indicator lamp and brake warning lamp turn OFF.

48. Be sure to fill the cooling system with the proper grade and type coolant.

49. Recharge the air conditioning system. Check for leaks.

50. Check and adjust the front end alignment, as necessary.

STEERING

POWER RACK & PINION STEERING GEAR

REMOVAL & INSTALLATION

See Figures 209 and 210.

➡If working near and/or around the SRS system and components, be sure to disable the SRS system. After disabling the system wait three minutes or more before servicing the vehicle.

➡Whenever the negative battery cable is disconnected the following components will require resetting. The Idle Air Volume Learning, Steering Angle Sensor Neutral Position, Sunroof Memory Reset/Initialization, Automatic Drive Positioner System, Audio presets and Navigation. Use the Nissan CONSULT III diagnostic tool, or equivalent to perform the required resets.

❊❊ WARNING

The spiral cable may snap due to steering operation if the steering column is separated from the steering gear assembly. Be sure to secure the steering wheel to avoid turning.

1. Before servicing the vehicle, refer to the Precautions Section.

2. Disconnect the negative battery cable.

3. Position the front wheels in the straight ahead position.

4. Disarm the SRS system.

5. Disconnect the negative battery cable.

6. Drain the power steering fluid.

7. Raise and support the vehicle safely. Remove the tire and wheel assemblies.

8. Remove the undercover.

9. If equipped with 4WD, remove the final drive, then support the halfshafts, using wire.

10. Remove the stabilizer bar brackets, and position the stabilizer bar aside.

11. Remove and discard the cotter pins at the steering outer sockets. Loosen the outer socket locknuts.

12. Remove the steering gear outer sockets from the steering knuckles, using tool HT72520000 (J-25730-A), or equivalent.

13. Disconnect and plug the power steering fluid lines at the steering gear.

14. Remove the bolt from the lower joint of the lower joint assembly. Separate the lower joint from the steering gear assembly. Be careful not to damage the lower joint.

15. Remove the steering gear retaining nuts and bolts. Remove the steering gear from the vehicle.

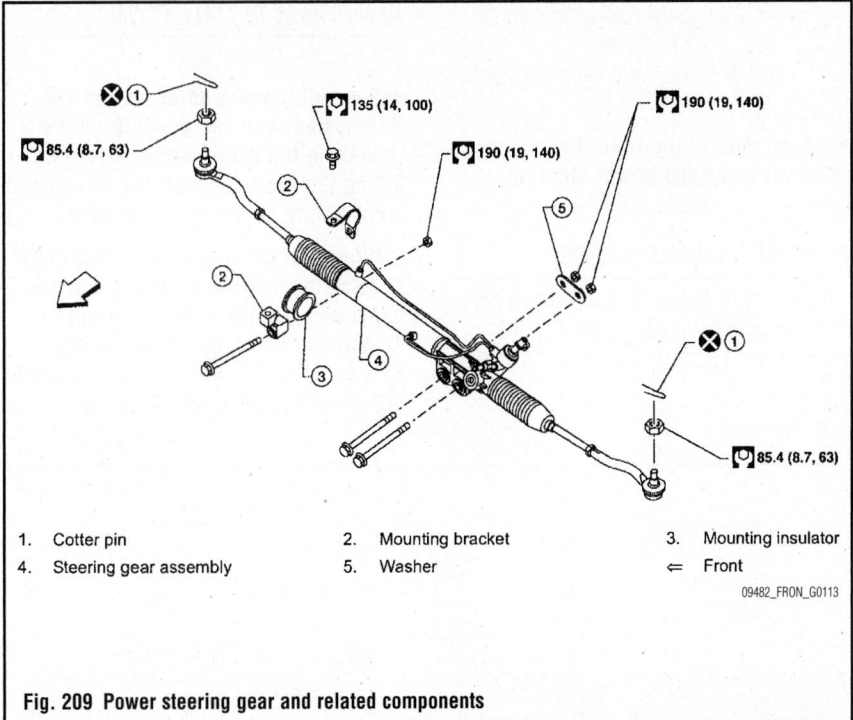

1. Cotter pin
2. Mounting bracket
3. Mounting insulator
4. Steering gear assembly
5. Washer
⇐ Front

09482_FRON_G0113

Fig. 209 Power steering gear and related components

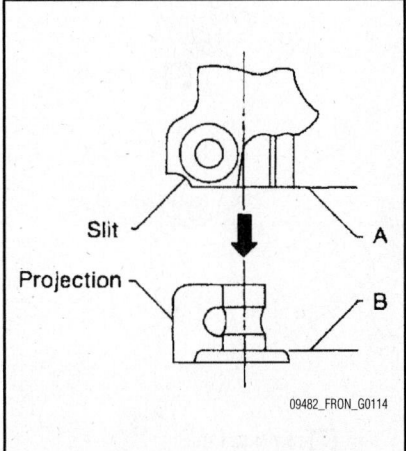

09482_FRON_G0114

Fig. 210 Power steering gear lower joint installation alignment

To install:

16. With the steering wheel in the straight ahead position, align the slit of the lower joint with the projection on the dust cover. Insert the joint until both surfaces contact each other.

17. Continue the installation in the reverse order of the removal procedure.

18. Check and adjust the front alignment, as required.

19. Bleed the power steering system.

20. Fill the power steering pump with the proper grade and type of fluid.

21. Perform the steering angle neutral position adjustment.

POWER STEERING PUMP

BLEEDING

➡ If working near and/or around the SRS system and components, be sure to disable the SRS system. After disabling the system wait three minutes or more before servicing the vehicle.

➡ Whenever the negative battery cable is disconnected the following components will require resetting. The Idle Air Volume Learning, Steering Angle Sensor Neutral Position, Sunroof Memory Reset/Initialization, Automatic Drive Positioner System, Audio presets and Navigation. Use the Nissan CONSULT III diagnostic tool, or equivalent to perform the required resets.

1. Before servicing the vehicle, refer to the Precautions Section.

2. Fill the power steering system with the proper grade and type of steering fluid.

➡ Do not allow the fluid level in the reservoir tank to go below the MIN level line. Check and add fluid as needed.

3. Raise and safely support the vehicle.

4. Quickly turn the steering wheel to the full right and left detents and lightly touch the steering stoppers.

➡ Do not hold the steering wheel in the locked position for more than 10 seconds.

5. Repeat this operation until the fluid level no longer decreases.

6. Start the engine.

7. Quickly turn the steering wheel to the full right and left detents and lightly touch the steering stoppers.

➡ Do not hold the steering wheel in the locked position for more than 10 seconds.

8. Check for air bubbles or cloudy fluid. If found, repeat the bleeding procedure.

9. Stop the engine and check the fluid level. Correct as required.

FLUID FILL PROCEDURE

See Figure 211.

✳✳ CAUTION

Used fluid is considerably more dangerous than new fluid. Avoid skin contact with used fluid.

➡ If working near and/or around the SRS system and components, be sure to disable the SRS system. After disabling the system wait three minutes or more before servicing the vehicle.

➡ Whenever the negative battery cable is disconnected the following components will require resetting. The Idle Air Volume Learning, Steering Angle Sensor Neutral Position, Sunroof

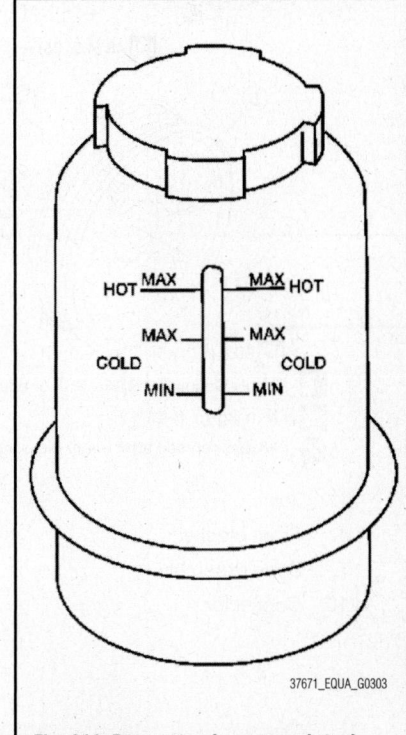

37671_EQUA_G0303

Fig. 211 Power steering reservoir tank fill markings

Memory Reset/Initialization, Automatic Drive Positioner System, Audio presets and Navigation. Use the Nissan CONSULT III diagnostic tool, or equivalent to perform the required resets.

1. Before servicing the vehicle, refer to the Precautions Section.

2. Inspect the power steering fluid level in the power steering reservoir. Do allow the fluid to drop below the MIN marking.

3. Remove the power steering reservoir cap.

4. Add fluid, as necessary, referring to the scale on the reservoir tank.

- HOT range for fluid temperatures: 122–176° F (50–80° C)
- COLD range for fluid temperatures: 32–86° F (0–30° C)

➡Do not overfill the fluid. Do not reuse any used power steering fluid.

FLUID RECOMMENDATIONS

Use genuine Nissan power steering fluid (PSF), or equivalent when servicing the power steering system. The system capacity is 2 1/8 pints.

REMOVAL & INSTALLATION

See Figures 212 and 213.

➡If working near and/or around the SRS system and components, be sure to disable the SRS system. After disabling the system wait three minutes or more before servicing the vehicle.

➡Whenever the negative battery cable is disconnected the following components will require resetting. The Idle Air Volume Learning, Steering Angle Sensor Neutral Position, Sunroof Memory Reset/Initialization, Automatic

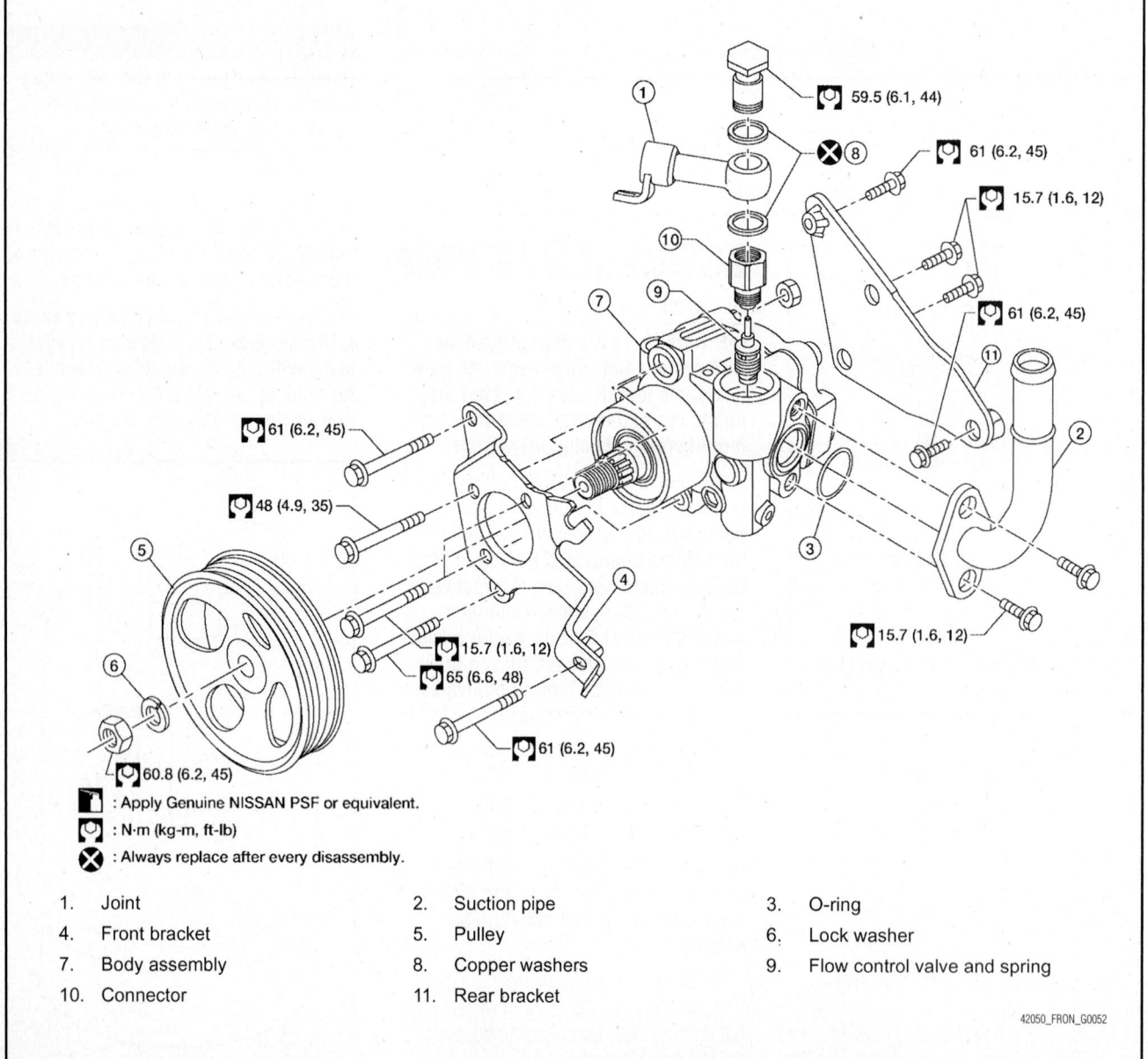

: Apply Genuine NISSAN PSF or equivalent.

: N·m (kg-m, ft-lb)

: Always replace after every disassembly.

1. Joint	2. Suction pipe	3. O-ring
4. Front bracket	5. Pulley	6. Lock washer
7. Body assembly	8. Copper washers	9. Flow control valve and spring
10. Connector	11. Rear bracket	

42050_FRON_G0052

Fig. 212 Power steering pump and related components—2.5L engine

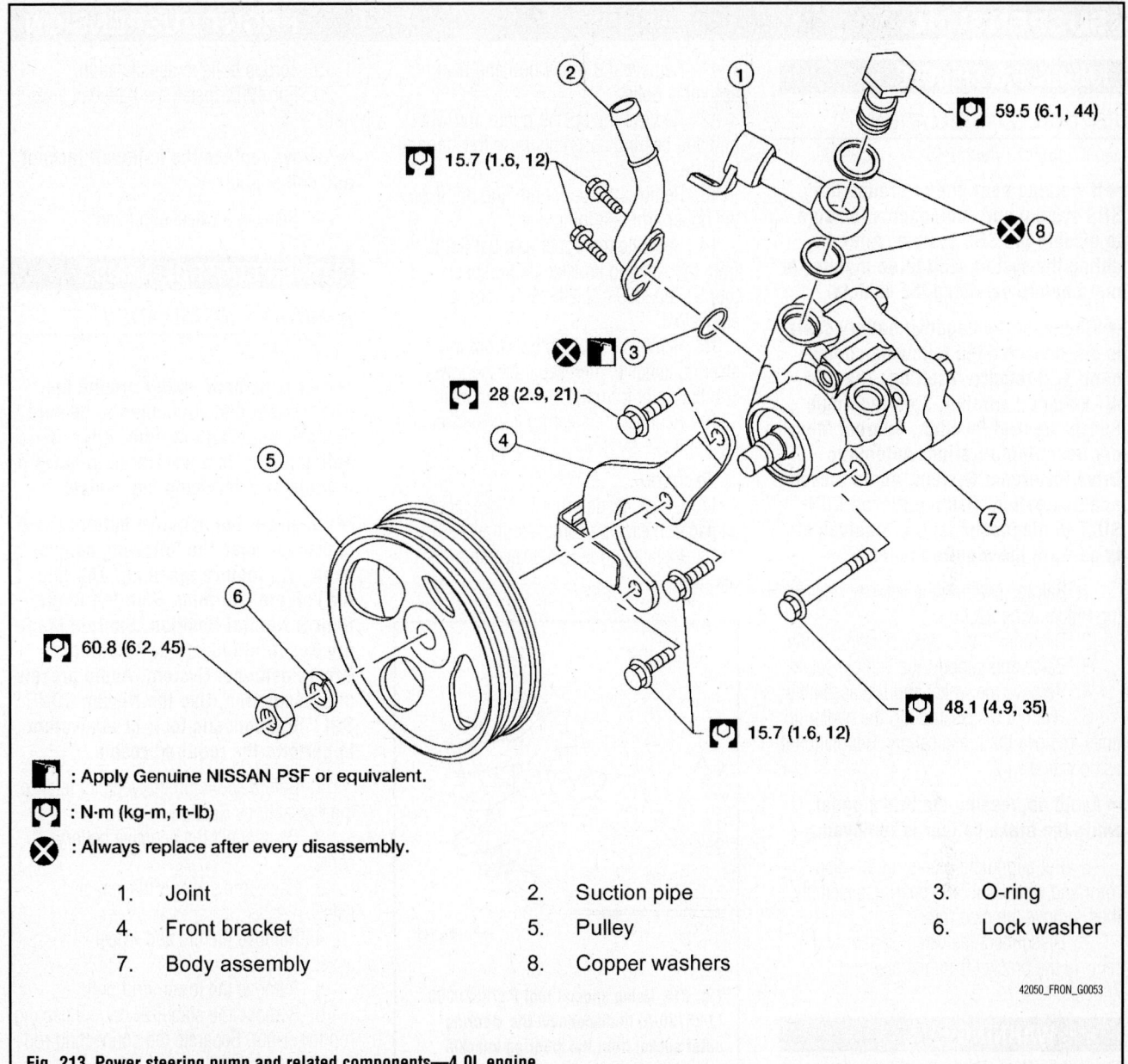

Fig. 213 Power steering pump and related components—4.0L engine

[box] 59.5 (6.1, 44)

[box] 15.7 (1.6, 12)

⊗ 🖐 3

[box] 28 (2.9, 21)

⊗ 8

[box] 60.8 (6.2, 45)

[box] 15.7 (1.6, 12)

[box] 48.1 (4.9, 35)

🖐 : Apply Genuine NISSAN PSF or equivalent.

[box] : N·m (kg-m, ft-lb)

⊗ : Always replace after every disassembly.

1. Joint	2. Suction pipe	3. O-ring
4. Front bracket	5. Pulley	6. Lock washer
7. Body assembly	8. Copper washers	

42050_FRON_G0053

Drive Positioner System, Audio presets and Navigation. Use the Nissan CONSULT III diagnostic tool, or equivalent to perform the required resets.

1. Before servicing the vehicle, refer to the Precautions Section.

2. Disconnect the negative battery cable.

3. Drain the power steering fluid from the reservoir tank. Properly dispose of used fluid.

4. On the 4.0L engine, remove the engine cover.

5. Remove the air duct assembly.

6. Remove the drive belt.

7. Disconnect the pressure sensor electrical connector.

8. Disconnect and plug the fluid lines.

9. Remove the pump retaining bolts.

10. Remove the pump from the vehicle.

To install:

11. Installation is the reverse of the removal procedure.

12. Bleed the power steering system.

SUSPENSION

FRONT SUSPENSION

KNUCKLE & SPINDLE

REMOVAL & INSTALLATION

See Figures 214 and 215.

➡If working near and/or around the SRS system and components, be sure to disable the SRS system. After disabling the system wait three minutes or more before servicing the vehicle.

➡Whenever the negative battery cable is disconnected the following components will require resetting. The Idle Air Volume Learning, Steering Angle Sensor Neutral Position, Sunroof Memory Reset/Initialization, Automatic Drive Positioner System, Audio presets and Navigation. Use the Nissan CONSULT III diagnostic tool, or equivalent to perform the required resets.

1. Before servicing the vehicle, refer to the Precautions Section.
2. Disconnect the negative battery cable.
3. Raise and support the vehicle safely.
4. Remove the wheel and tire assembly.
5. Without disassembling the hydraulic lines, remove the brake caliper. Reposition it aside with a wire.

➡Avoid depressing the brake pedal while the brake caliper is removed.

6. Put alignment marks on the disc rotor and wheel hub and bearing assembly, then remove the disc rotor.
7. Disconnect the wheel sensor and remove the bracket from the steering knuckle.

❊❊ WARNING

Do not pull on the wheel sensor harness.

8. On 4WD models, remove the cotter pin, then remove the locknut from the halfshaft.
9. Remove the steering outer socket cotter pin at the steering knuckle, then loosen the mounting nut.
10. Disconnect the steering outer socket from the steering knuckle using special tool HT72520000 (J-25730-A), or equivalent.

❊❊ WARNING

Be careful not to damage ball joint boot.

➡To prevent damage to the threads and to prevent the tool from coming off suddenly, temporarily tighten the nut.

11. Remove the wheel hub and bearing assembly bolts.
12. Remove the splash guard and wheel hub and bearing assembly from the steering knuckle.
13. Remove the cotter pin and nut from the upper arm ball joint.
14. Separate the upper arm ball joint from the steering knuckle using special tool ST29020001 (J-24319-01), or equivalent.
15. Remove the pinch bolt from the steering knuckle, then separate the lower arm ball joint from the steering knuckle.
16. Remove the steering knuckle from the vehicle.

To install:

17. Check for deformity, cracks, and damage on each part, replace if necessary.
18. Installation is in the reverse order of removal.

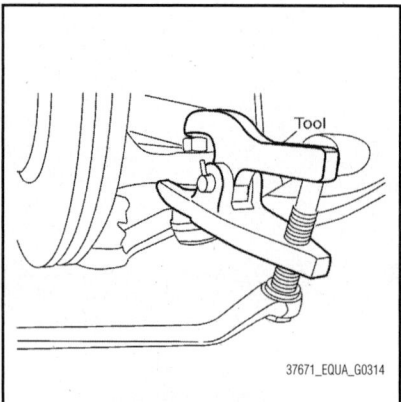

37671_EQUA_G0314

Fig. 214 Using special tool HT72520000 (J-25730-A) to disconnect the steering outer socket from the steering knuckle

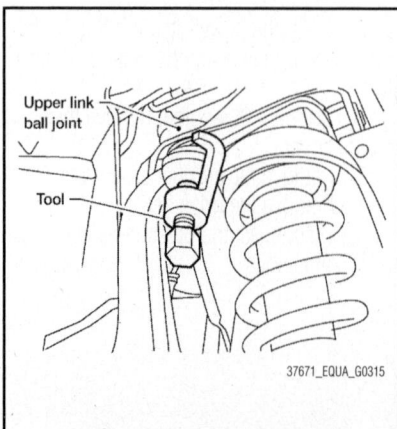

Upper link ball joint

Tool

37671_EQUA_G0315

Fig. 215 Using special tool ST29020001 (J-24319-01) to separate the upper arm (link) ball joint from the steering knuckle

19. Torque bolts to specification.
20. For 4WD, install the halfshaft locknut.

➡Always replace the halfshaft locknut and cotter pin.

21. Perform a wheel alignment.

LOWER CONTROL ARMS

REMOVAL & INSTALLATION

See Figure 216.

➡If working near and/or around the SRS system and components, be sure to disable the SRS system. After disabling the system wait three minutes or more before servicing the vehicle.

➡Whenever the negative battery cable is disconnected the following components will require resetting. The Idle Air Volume Learning, Steering Angle Sensor Neutral Position, Sunroof Memory Reset/Initialization, Automatic Drive Positioner System, Audio presets and Navigation. Use the Nissan CONSULT III diagnostic tool, or equivalent to perform the required resets.

1. Before servicing the vehicle, refer to the Precautions Section.
2. Disconnect the negative battery cable.
3. Raise and support the vehicle safely.
4. Remove the tire and wheel assembly.
5. Remove the lower strut bolt.
6. Remove the stabilizer bar connecting rod lower nut. Separate the connecting rod from the lower control arm.
7. If equipped with 4WD, remove the halfshaft.
8. Remove the pinch bolt from the steering knuckle. Separate the lower control arm ball joint stud from the steering knuckle, using the proper tool.
9. Remove the lower control arm adjusting bolts and nuts. Lower the control arm and remove it from the vehicle.
10. Remove the jounce bumper from the lower control arm.

➡Some vehicles may be equipped with straight (non-adjustable) lower link bolts and washers. In order to adjust camber and caster on these vehicles, first replace the lower link bolts and washers with adjustable (cam) bolts and washers.

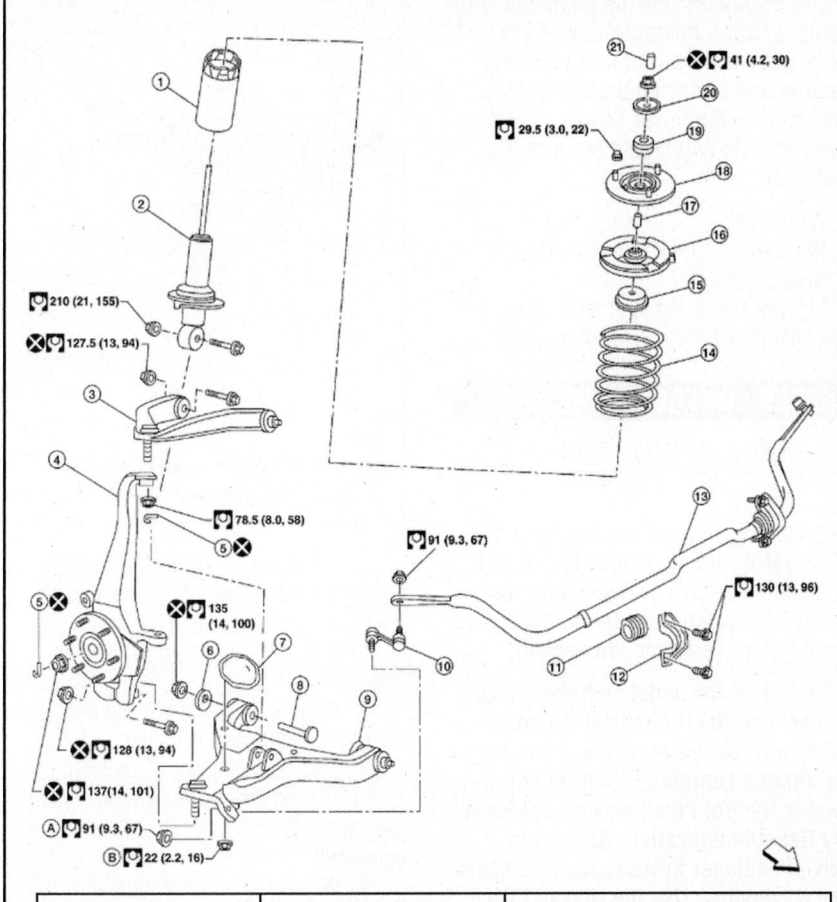

Fig. 216 Front suspension component locations

1.	Dust cover	2.	Shock absorber	3.	Upper link
4.	Steering knuckle	5.	Cotter pin	6.	Washer
7.	Jounce bumper	8.	Bolt	9.	Lower link
10.	Connecting rod	11.	Stabilizer bar bushing	12.	Stabilizer bar mounting bracket
13.	Stabilizer bar	14.	Coil spring	15.	Dust cover cap
16.	Upper spring seat	17.	Spacer	18.	Shock absorber mounting insulator
19.	Spacer	20.	Washer	21.	Cap
A.	To connecting rod	B.	To jounce bumper		Vehicle front

37671_EQUA_G0310

To install:

11. Installation is the reverse of the removal procedure.

12. Be sure to replace all wearable components, as required.

13. Check and adjust the front alignment, as required.

➡**After removing/installing or replacing steering and suspension components which effect wheel alignment, be sure to adjust the neutral position of the steering angle sensor before running the vehicle.**

14. If equipped with VDC, adjust the steering angle sensor.

STABILIZER BAR (SWAY BAR) & LINKS

REMOVAL & INSTALLATION

Stabilizer Bar

See Figures 216 and 217.

➡**If working near and/or around the SRS system and components, be sure to disable the SRS system. After disabling the system wait three minutes or more before servicing the vehicle.**

➡**Whenever the negative battery cable is disconnected the following components will require resetting. The Idle Air Volume Learning, Steering Angle Sensor Neutral Position, Sunroof Memory Reset/Initialization, Automatic Drive Positioner System, Audio presets and Navigation. Use the Nissan CONSULT III diagnostic tool, or equivalent to perform the required resets.**

1. Before servicing the vehicle, refer to the Precautions Section.

2. Disconnect the negative battery cable.

3. Remove the front valance center.

4. Raise and support the vehicle safely.

5. Remove the engine undercover.

6. Remove the connecting rod nuts.

7. Loosen the top bolts for the stabilizer bar mounting brackets. Remove the lower bolts from the mounting brackets.

8. Remove the stabilizer bar from the vehicle.

9. Remove the bushings from the stabilizer bar.

➡**Some vehicles may be equipped with straight (non-adjustable) lower link bolts and washers. In order to adjust camber and caster on these vehicles, first replace the lower link bolts and washers with adjustable (cam) bolts and washers.**

To install:

10. Installation is the reverse of the removal procedure.

11. Check the stabilizer bar for twist and deformation. Replace if necessary.

12. Check rubber bushing for cracks, wear, and deterioration. Replace if necessary.

13. Tighten all nuts and bolts to specification.

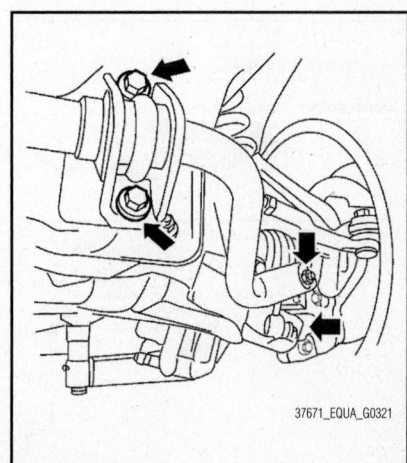

37671_EQUA_G0321

Fig. 217 Remove the stabilizer mounting brackets

STRUTS

REMOVAL & INSTALLATION

See Figures 216 and 218.

➡If working near and/or around the SRS system and components, be sure to disable the SRS system. After disabling the system wait three minutes or more before servicing the vehicle.

➡Whenever the negative battery cable is disconnected the following components will require resetting. The Idle Air Volume Learning, Steering Angle Sensor Neutral Position, Sunroof Memory Reset/Initialization, Automatic Drive Positioner System, Audio presets and Navigation. Use the Nissan CONSULT III diagnostic tool, or equivalent to perform the required resets.

1. Before servicing the vehicle, refer to the Precautions Section.
2. Disconnect the negative battery cable.
3. Raise and support the vehicle safely.
4. Remove the wheel and tire assembly.
5. Support the lower control arm (link) using a suitable jack.
6. Remove the connecting rod upper joints from the stabilizer bar. Swing the stabilizer bar down, repositioning it out of the way, to access the shock absorber lower mount.
7. Remove the shock absorber lower bolt and nut.
8. Remove the 3 shock absorber upper mounting nuts.
9. Remove the coil spring and shock absorber assembly. Turn the steering knuckle out to gain enough clearance for removal.

➡Some vehicles may be equipped with straight (non-adjustable) lower link bolts and washers. In order to adjust camber and caster on these vehicles, first replace the lower link bolts and washers with adjustable (cam) bolts and washers.

To install:

10. Installation is the reverse of the removal procedure.
11. The step in the strut assembly lower seat should face the outside of the vehicle.

UPPER CONTROL ARMS

REMOVAL & INSTALLATION

See Figures 216, 219 and 220.

➡If working near and/or around the SRS system and components, be sure to disable the SRS system. After disabling the system wait three minutes or more before servicing the vehicle.

➡Whenever the negative battery cable is disconnected the following components will require resetting. The Idle Air Volume Learning, Steering Angle Sensor Neutral Position, Sunroof Memory Reset/Initialization, Automatic Drive Positioner System, Audio presets and Navigation. Use the Nissan CONSULT III diagnostic tool, or equivalent to perform the required resets.

1. Before servicing the vehicle, refer to the Precautions Section.
2. Disconnect the negative battery cable.
3. Raise and support the vehicle safely.
4. Remove the tire and wheel assembly.
5. Using a suitable jack, support the lower control arm.

Fig. 220 Remove the upper control arm (link) retaining bolts and nuts

6. If working on the left side, remove the bolt from the lower joint of the lower joint shaft, then reposition the lower joint shaft out of the way. Do not damage the lower joint.
7. Remove the cotter pin and nut from the upper control arm ball joint.
8. Separate the upper control arm ball joint stud from the steering knuckle, using special tool ST29020001 (J-24319-01), or equivalent.
9. Remove the upper control arm retaining bolts and nuts.
10. Remove the upper control arm from the vehicle.

➡Some vehicles may be equipped with straight (non-adjustable) lower link bolts and washers. In order to adjust camber and caster on these vehicles, first replace the lower link bolts and washers with adjustable (cam) bolts and washers.

To install:

11. Installation is the reverse of the removal procedure.
12. Be sure to replace all wearable components, as required.
13. Check and adjust the front alignment, as required.

WHEEL HUBS & BEARINGS

ADJUSTMENT

The front wheel bearings are not adjustable. If the lateral run-out on the hub with the disc removed exceeds specification, the hub must be replaced.

1. Move the wheel hub in the axial direction by hand. Make sure there is no looseness of the wheel bearing. Axial end-play limit: 0.002 inch (0.05mm) or less. If

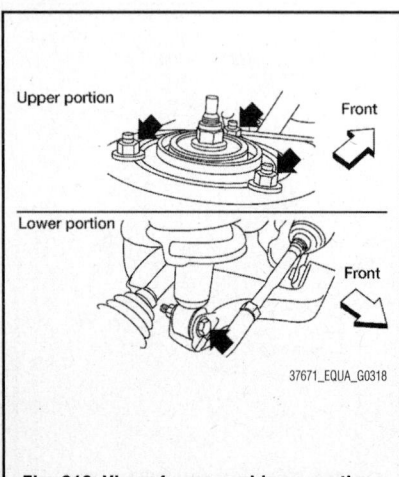

Fig. 218 View of upper and lower portion of the strut assembly

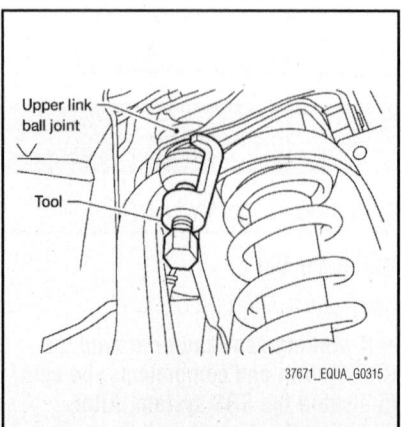

Fig. 219 Using special tool ST29020001 (J-24319-01) to separate the upper arm (link) ball joint from the steering knuckle

out of specification, replace the wheel hub and bearing assembly.

2. Rotate the wheel hub and make sure there is no unusual noise or other irregular conditions. If there are any irregular conditions, replace the wheel hub and bearing assembly.

REMOVAL & INSTALLATION

See Figure 221.

➡ **If working near and/or around the SRS system and components, be sure to disable the SRS system. After disabling the system wait three minutes or more before servicing the vehicle.**

➡ **Whenever the negative battery cable is disconnected the following components will require resetting. The Idle Air Volume Learning, Steering Angle Sensor Neutral Position, Sunroof Memory Reset/Initialization, Automatic Drive Positioner System, Audio presets and Navigation. Use the Nissan CONSULT III diagnostic tool, or equivalent to perform the required resets.**

1. Before servicing the vehicle, refer to the Precautions Section.

2. Disconnect the negative battery cable.

3. Raise and support the vehicle safely.

4. Remove the tire and wheel assembly.

5. Without disassembling the hydraulic lines, remove caliper torque member bolts. Then, reposition the brake caliper aside with wire.

➡ **Do not press the brake pedal while the brake caliper is removed.**

6. Matchmark the disc rotor and wheel hub and bearing assembly, then remove the disc rotor.

7. On models with 4WD, remove the

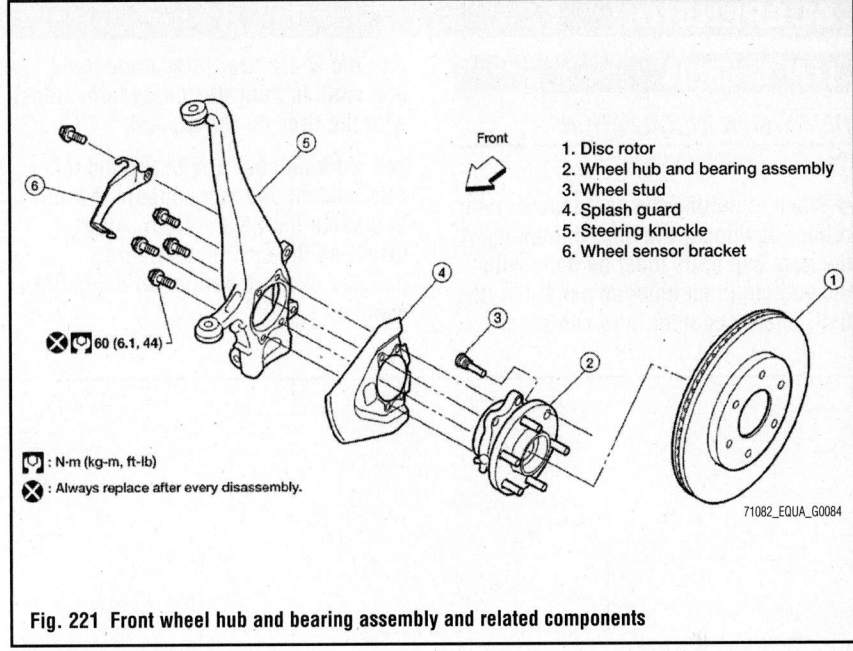

1. Disc rotor
2. Wheel hub and bearing assembly
3. Wheel stud
4. Splash guard
5. Steering knuckle
6. Wheel sensor bracket

⊗ 🔧 60 (6.1, 44)

🔧 : N·m (kg-m, ft-lb)

⊗ : Always replace after every disassembly.

71082_EQUA_G0084

Fig. 221 Front wheel hub and bearing assembly and related components

cotter pin, then remove the locknut from the halfshaft.

8. Remove the wheel sensor from the wheel hub and bearing assembly.

- Inspect the wheel sensor O-ring, replace the wheel sensor assembly if damaged
- Clean the wheel sensor hole and mounting surface with a suitable brake cleaner and clean lint-free shop rag. Be careful that dirt and debris does not enter the axle bearing area
- Apply a coat of suitable grease to the wheel sensor O-ring and mounting hole

✳✳ WARNING

Do not pull on the wheel sensor harness.

9. On models with 4WD, separate the

halfshaft from the wheel hub and bearing assembly.

10. Remove the wheel hub and bearing assembly bolts.

11. Remove the splash guard and wheel hub and bearing assembly from the steering knuckle.

12. Carefully remove the wheel sensor and harness through the hole in the splash guard.

To install:

13. Check for deformity, cracks, and damage on each part and replace if necessary.

14. Installation is in the reverse order of removal.

15. Use new bolts when installing the wheel hub and bearing assembly.

16. When installing the disc rotor on the wheel hub and bearing assembly, position the disc rotor according to the alignment matchmark.

SUSPENSION

LEAF SPRINGS

REMOVAL & INSTALLATION

See Figures 222 and 223.

➡When installing the components with rubber bushings, the final tightening of the nuts and bolts must be done with the vehicle in an unladen condition (the fuel, engine coolant, and engine oil full; the spare tire, jack, hand tools, and mats in their designated positions) with the tires on the ground.

➡If working near and/or around the SRS system and components, be sure to disable the SRS system. After disabling the system wait three minutes or more before servicing the vehicle.

➡Whenever the negative battery cable is disconnected the following components will require resetting. The Idle Air Volume Learning, Steering Angle Sensor Neutral Position, Sunroof Memory Reset/Initialization, Automatic Drive Positioner System, Audio presets and Navigation. Use the Nissan CONSULT III diagnostic tool, or equivalent to perform the required resets.

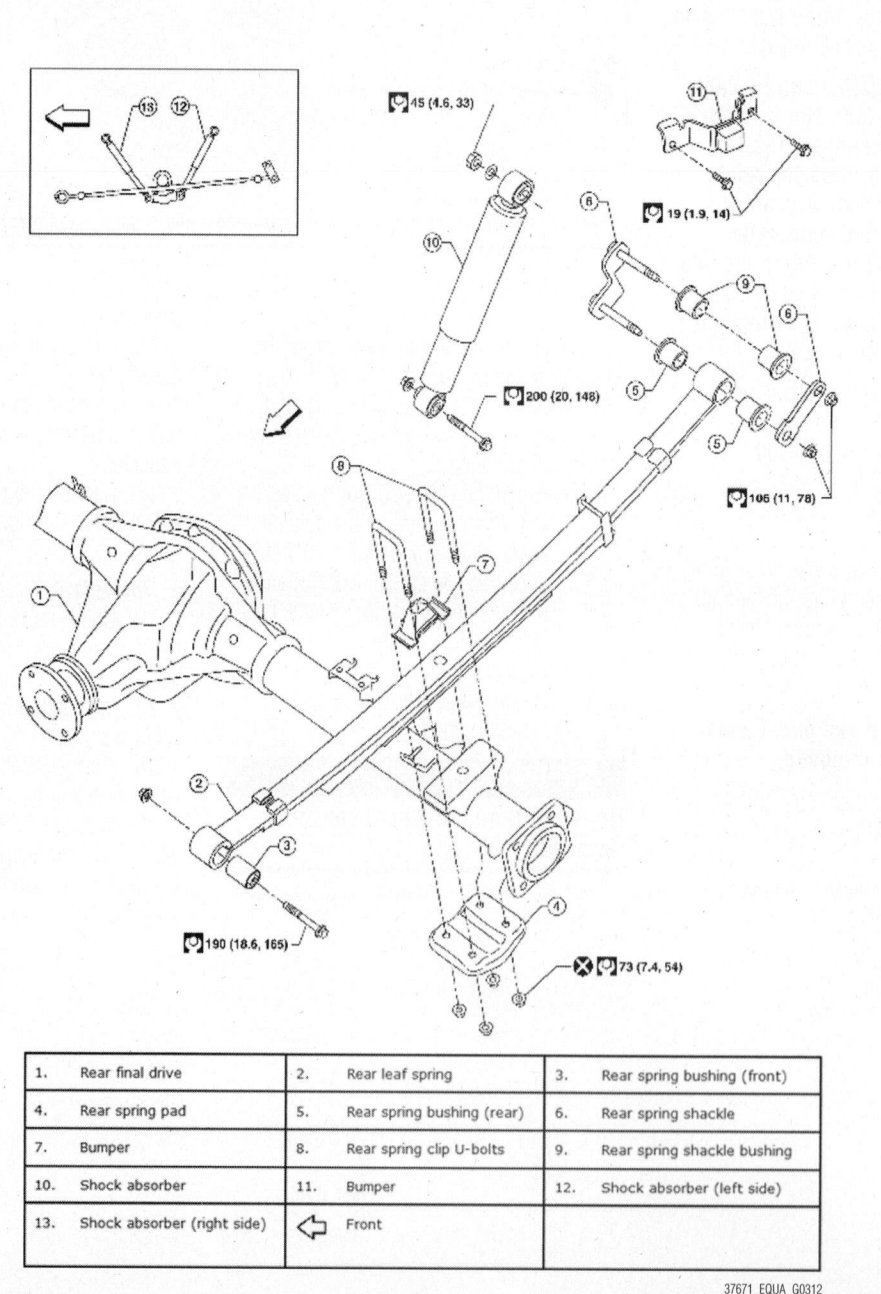

1.	Rear final drive	2.	Rear leaf spring	3.	Rear spring bushing (front)
4.	Rear spring pad	5.	Rear spring bushing (rear)	6.	Rear spring shackle
7.	Bumper	8.	Rear spring clip U-bolts	9.	Rear spring shackle bushing
10.	Shock absorber	11.	Bumper	12.	Shock absorber (left side)
13.	Shock absorber (right side)	⬅	Front		

37671_EQUA_G0312

Fig. 222 Rear suspension components

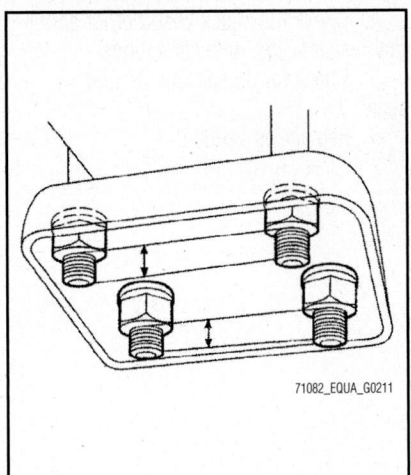

Fig. 223 Rear spring retaining bolts (arrow indicates tolerance specification)

1. Before servicing the vehicle, refer to the Precautions Section.

2. Disconnect the negative battery cable.

3. Raise and safely support the vehicle.

4. Remove the tire and wheel assembly.

5. Support the rear final drive assembly with a suitable jack to relieve the tension from the rear leaf spring.

➡**The axle weight should be supported, but there should be no compression in the rear leaf spring.**

6. Remove the 4 rear spring clip U-bolt nuts, then remove the rear spring pad and bumper.

7. Remove the rear spring shackle and bushings.

8. Remove the rear leaf spring front nut and bolt.

9. Remove the rear leaf spring.

To install:

10. Check the rear leaf spring for any cracks or damage. Replace the rear leaf spring if necessary.

11. Check the rear spring shackle, rear spring clip U-bolts, bumper, and rear spring pad for excessive wear, cracks, straightness, and damage. Replace any components if necessary.

12. Check all bushings for deformation and cracks. Replace any bushings if necessary.

13. Apply soapsuds to all of the rubber bushings.

14. Install the rear spring shackle and rear leaf spring front nut and bolt. Finger-tighten the nuts.

15. Install the rear spring clip U-bolts

and bumper on top of the rear leaf spring.

16. Install the bumper and rear spring pad, then finger-tighten the nuts under the axle case.

17. Tighten the rear spring clip U-bolt nuts diagonally and evenly to 54 ft. lbs. (73 Nm).

➡**Tighten the rear spring clip U-bolt nuts so the lengths of all the exposed rear spring clip U-bolt threads under the spring pad are equal in length within a tolerance of 0.12 inch (3mm).**

18. Remove the jack supporting the rear final drive assembly and bounce the rear of the vehicle to stabilize the suspension.

19. Tighten the rear spring shackle nuts, rear leaf spring front nut, and shock absorber nuts to specification.

    a. Lower shock absorber bolt/nut to 148 ft. lbs. (200 Nm).

    b. Front spring nut and bolt to 165 ft. lbs. (190 Nm).

    c. Rear spring shackle nuts to 78 ft. lbs. (106 Nm).

➡**When installing the components with rubber bushings, the final nut tightening must be carried out under unladen conditions with the tires on level ground.**

SHOCK ABSORBERS

REMOVAL & INSTALLATION
See Figures 222 and 224.

➡**If working near and/or around the SRS system and components, be sure to disable the SRS system. After disabling the system wait three minutes or more before servicing the vehicle.**

➡**Whenever the negative battery cable is disconnected the following components will require resetting. The Idle Air Volume Learning, Steering Angle Sensor Neutral Position, Sunroof Memory Reset/Initialization, Automatic Drive Positioner System, Audio presets and Navigation. Use the Nissan CONSULT III diagnostic tool, or equivalent to perform the required resets.**

1. Before servicing the vehicle, refer to the Precautions Section.

2. Disconnect the negative battery cable.

3. Raise and safely support the vehicle.

4. Support the rear final drive and suspension assembly using a suitable jack.

5. Remove the shock absorber upper and lower nuts and bolts.

6. Remove the shock absorber from the vehicle.

To install:

7. Inspect the shock absorber for any oil leaks, cracks, or deformations. Replace the shock absorber as necessary.

8. Installation is in the reverse order of removal.

9. Tighten the upper shock absorber nut to 33 ft. lbs. (45 Nm).

10. Tighten the lower shock absorber bolt/nut to 148 ft. lbs. (200 Nm).

TESTING

The easiest test of the shock absorber is to simply push down on one corner of the unladen vehicle and release it. Observe the motion of the body as it is released. In most cases, it will come up beyond its original rest position, dip back below it, and settle quickly to rest. This shows that the damper is controlling the spring action. Any tendency to excessive pitch (up-and-down) motion or failure to return to rest within 2–3 cycles, is a sign of poor function within the shock absorber.

Oil-filled shocks may have a light film of oil around the seal, resulting from normal breathing and air exchange. This should NOT be taken as a sign of failure, but any sign of thick or running oil indicates failure. Gas-filled shocks may also show some film at the shaft; if the gas has leaked out, the shock will have almost no resistance to motion.

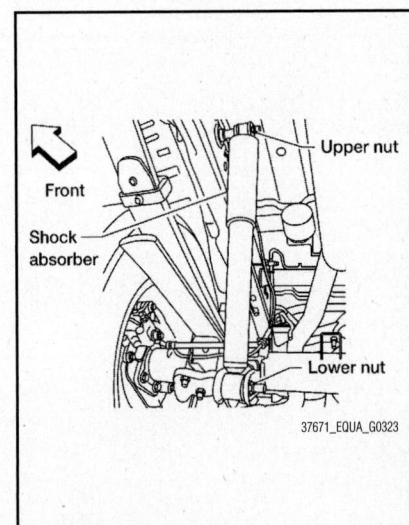

Fig. 224 Rear shock absorber and mounting bolts

While each shock absorber can be replaced individually, it is recommended that they be changed as a pair (both front or both rear) to maintain equal response on both sides of the vehicle. If one side has failed, its mate may also be weak.

1. Before servicing the vehicle, refer to the Precautions Section.

2. Check the rubber parts for damage or deterioration.

3. Check the spring for correct height, deformation, deterioration, or damage.

4. Check the shock absorber for abnormal resistance or unusual sounds.

5. Check for oil seepage around seals.

6. Replace as needed.

NISSAN

JUKE

SPECIFICATIONS AND MAINTENANCE CHARTS

ENGINE AND VEHICLE IDENTIFICATION

Engine							Model Year	
Code (VIN) ①	Liters	cc	Cyl.	Fuel Sys.	Engine Type	Eng. Mfg.	Code ②	Year
MR16DDT (A)	1.6	1618	4	MFI	DOHC	Nissan	B	2011
							C	2012

① 8th position of VIN

② 10th position of VIN

71075_JUKE_C0001

GENERAL ENGINE SPECIFICATIONS

All measurements are given in inches.

Year	Model	Engine Displacement Liters (cc)	Engine ID/VIN	Fuel System Type	Net Horsepower @ rpm	Net Torque @ rpm (ft. lbs.)	Bore x Stroke (in.)	Compression Ratio	Oil Pressure @ rpm
2011	Juke	1.6 (1618)	MR16DDT/A	MFI	188 @ 5600	177 @ 2000	3.138 x 3.193	9.5:1	37.7 @ 2000
2012	Juke	1.6 (1618)	MR16DDT/A	MFI	188 @ 5600	177 @ 2000	3.138 x 3.193	9.5:1	37.7 @ 2000

71075_JUKE_C0002

ENGINE TUNE-UP SPECIFICATIONS

Year	Engine Displacement Liters	Engine ID/VIN	Spark Plug Gap (in.)	Ignition Timing (deg.) MT	Ignition Timing (deg.) AT	Fuel Pump (psi)	Idle Speed (rpm) MT	Idle Speed (rpm) AT	Valve Clearance Intake	Valve Clearance Exhaust
2011	1.6	MR16DDT/A	0.035	6-10	4-8	73	550-650	600-700	0.012-0.016	0.012-0.017
2012	1.6	MR16DDT/A	0.035	6-10	4-8	73	550-650	600-700	0.012-0.016	0.012-0.017

71075_JUKE_C0003

CAPACITIES

Year	Model	Engine Disp. Liters	Engine ID/VIN	Engine Oil with Filter	Transaxle (pts.)		Drive Axle (pts.)		Transfer Case (pts.)	Fuel Tank (gal.)	Cooling System (qts.)
					Auto.	Manual	Front	Rear			
2011	Juke 2WD	1.6	MR16DDT/A	4.75	17.66	8.5	0.875	0.875	1.50	13.3	8.5
	Juke AWD	1.6	MR16DDT/A	4.75	18.25	8.5	0.875	0.875	1.50	11.9	8.5
2012	Juke 2WD	1.6	MR16DDT/A	4.75	17.66	8.5	0.875	0.875	1.50	13.3	8.5
	Juke AWD	1.6	MR16DDT/A	4.75	18.25	8.5	0.875	0.875	1.50	11.9	8.5

NOTE: All capacities are approximate. Add fluid gradually and ensure a proper fluid level is obtained.

71075_JUKE_C0004

FLUID SPECIFICATIONS

Year	Model	Engine Disp. Liters	Engine Oil	Manual Trans.	Auto. Trans.	Transfer Case	Power Steering Fluid	Brake Master Cylinder	Cooling System
2011	Juke	1.6	①	②	③	④	⑥	DOT 3	⑤
2012	Juke	1.6	①	②	③	④	⑥	DOT 3	⑤

DOT: Department Of Transpotation

① Engine oil with API Certification Mark and Viscosity SAE 5W-30

② Genuine NISSAN gear oil (Chevron Texaco ETL8997B) 75W-80 or equivalent (If Genuine Nissan Gear oil is not available, API G/L-4, Viscosity
SAE 75W-80 may be used as a temporary replacement. However use Genuine NISSAN gear oil as soon as it is available.)

③ Genuine NISSAN CVT Fluid NS-2 (Use only Genuine Nissan CVT fluid NS-2. Using transmission fluid other than Genuine NISSAN CVT Fluid
NS-2 will damage the CVT, whish is not covered by the NISSAN new vehicle limited warranty.

④ Genuine NISSAN Differential oil Hypoid Super GL-5 80W-90 or API GL-5 Viscosity SAE-80W-90

⑤ Pre-diluted Genuine NISSAN Long Life Anti-Freeze/Coolant (blue) or equivalent

⑥ Nissan PSF or equivalent. DEXRON VI type ATF may be used

71075_JUKE_C0005

VALVE SPECIFICATIONS

Year	Engine Displacement Liters	Engine ID/VIN	Seat Angle (deg.)	Face Angle (deg.)	Spring Test Pressure (lbs. @ in.)	Spring Free-Length (in.)	Spring Installed Height (in.)	Stem-to-Guide Clearance (in.)		Stem Diameter (in.)	
								Intake	Exhaust	Intake	Exhaust
2011	1.6	MR16DDT/A	45	NA	①	②	1.514	0.0008-0.0021	0.0012-0.0025	0.2152-0.2157	0.2148-0.2154
2012	1.6	MR16DDT/A	45	NA	①	②	1.514	0.0008-0.0021	0.0012-0.0025	0.2152-0.2157	0.2148-0.2154

① Intake: 77.3-88.1 @ 1.1362

Exhaust: 101.2-112.9 @ 1.1823

② Intake: 1.945-1.953

71075_JUKE_C0006

CAMSHAFT SPECIFICATIONS
All measurements in inches unless noted

Year	Engine Displacement Liters	Engine Code/VIN	Journal Diameter	Brg. Oil Clearance	Shaft End-play	Runout	Journal Bore	Lobe Height Intake	Exhaust
2011	1.6	MR16DDT/A	①	②	0.0030-0.006	Less than 0.0008	NA	1.7661-1.7636	1.6998-1.7073
2012	1.6	MR16DDT/A	①	②	0.0030-0.006	Less than 0.0008	NA	1.7661-1.7636	1.6998-1.7073

① No. 1: 1.0998-1.1006
No. 2, 3, 4, 5: 0.9823-0.9831
② No. 1: 0.0018-0.0034
No. 2, 3, 4, 5: 0.0012-0.0028

71075_JUKE_C0007

CRANKSHAFT AND CONNECTING ROD SPECIFICATIONS
All measurements are given in inches.

Year	Engine Disp. Liters	Engine ID/VIN	Main Brg. Journal Dia.	Main Brg. Oil Clearance	Shaft End-play	Thrust on No.	Journal Diameter	Oil Clearance	Side Clearance
2011	1.6	MR16DDT/A	①	0.0026	0.0118	3	②	0.0028	NA
2012	1.6	MR16DDT/A	①	0.0026	0.0118	3	②	0.0028	NA

① Grade No. A: 2.0464-2.0464
Grade No. B: 2.0463-2.0464
Grade No. C: 2.0463-2.0463
Grade No. D: 2.0463-2.0463
Grade No. E: 2.0462-2.0463
Grade No. F: 2.0462-2.0462
Grade No. G: 2.0461-2.0462
Grade No. H: 2.0461-2.0461
Grade No. J: 2.0461-2.0461
Grade No. K: 2.0460-2.0461
Grade No. L: 2.0460-2.0460
Grade No. M: 2.0459-2.0460
Grade No. N: 2.0459-2.0459
Grade No. P: 2.0459-2.0459
Grade No. R: 2.0458-2.0459
Grade No. S: 2.0458-2.0458
Grade No. T: 2.0457-2.0458
Grade No. U: 2.0457-2.0457
Grade No. V: 2.0457-2.0457
Grade No. W: 2.0456-2.0457

② Grade No. A: 1.7311-1.7311
Grade No. B: 1.7311-1.7311
Grade No. C: 1.7310-1.7311
Grade No. D: 1.7310-1.7310
Grade No. E: 1.7309-1.7310
Grade No. F: 1.7309-1.7309
Grade No. G: 1.7309-1.7309
Grade No. H: 1.7308-1.7309
Grade No. J: 1.7308-1.7308
Grade No. K: 1.7307-1.7308
Grade No. L: 1.7307-1.7307
Grade No. M: 1.7307-1.7307
Grade No. N: 1.7306-1.7307
Grade No. P: 1.7306-1.7306
Grade No. R: 1.7305-1.7306
Grade No. S: 1.7305-1.7305
Grade No. T: 1.7305-1.7305
Grade No. U: 1.7304-1.7305

71075_JUKE_C0008

PISTON AND RING SPECIFICATIONS
All measurements are given in inches.

Year	Engine Disp. Liters	Engine ID/VIN	Piston Clearance	Ring Gap			Ring Side Clearance		
				Top Compression	Bottom Compression	Oil Control	Top Compression	Bottom Compression	Oil Control
2011	1.6	MR16DDT/A	0.0031	0.0075-0.0114	0.0114-0.0173	0.0059-0.0177	0.0016-0.0031	0.0012-0.0028	0.0022-0.0061
2012	1.6	MR16DDT/A	0.0031	0.0075-0.0114	0.0114-0.0173	0.0059-0.0177	0.0016-0.0031	0.0012-0.0028	0.0022-0.0061

71075_JUKE_C0009

TORQUE SPECIFICATIONS
All readings in ft. lbs.

Year	Engine Disp. Liters	Engine ID/VIN	Cylinder Head Bolts	Main Bearing Bolts	Rod Bearing Bolts	Crankshaft Damper Bolts	Flywheel Bolts	Manifold		Spark Plugs	Oil Pan Drain Plug
								Intake	Exhaust		
2011	1.6	MR16DDT/A	①	②	③	NA	80	20	51	14	25
2012	1.6	MR16DDT/A	①	②	③	NA	80	20	51	14	25

① Step 1: Tighten to 30 ft. lbs. (40 Nm)

   Step 2: Turn 100 degrees clockwise

   Step 3: Completely loosen

   Step 4: Tighten to 30 ft. lbs. (40 Nm)

   Step 5: Turn 95 degrees clockwise

   Step 6: Turn 95 degrees clockwise

② Step 1: Tighten to 25 ft. lbs. (34.3 Nm)

   Step 2: Turn 70 degrees clockwise in order from No. 1-10

③ Step 1: Tighten to 20 ft. lbs. (27.4 Nm)

   Step 2: Completely loosen

   Step 3: Tighten to 14 ft. lbs. (19.6 Nm)

   Step 4: Turn 60 degrees clockwise

71075_JUKE_C0010

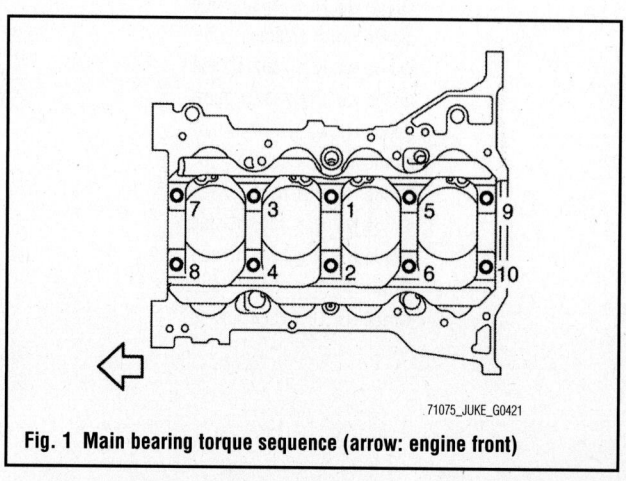

71075_JUKE_G0421

Fig. 1 Main bearing torque sequence (arrow: engine front)

WHEEL ALIGNMENT

Year	Model		Caster Range (+/-Deg.)	Caster Preferred Setting (Deg.)	Camber Range (+/-Deg.)	Camber Preferred Setting (Deg.)	Toe-in (in.)
2011	Juke	F	①	①	②	②	③
		R	NA	NA	④ ⑤	④ ⑤	⑥ ⑦
2012	Juke	F	①	①	②	②	③
		R	NA	NA	④ ⑤	④ ⑤	⑥ ⑦

① 2WD Minimum: 3 degrees 45' (3.75 degrees)
  2WD Nominal: 4 degrees 30' (4.50 degrees)
  2WD Maximum: 5 degrees 15' (5.25 degrees)
  AWD Minimum: 3 degrees 50' (3.84 degrees)
  AWD Nominal: 4 degrees 35' (4.58 degrees)
  AWD Maximum: 5 degrees 20' (5.33 degrees)
② Minimum: -1 degree 10' (-1.16 degree)
  Nominal: -0 degree 25' (-0.42 degree)
  Maximum: 0 degree 20' (0.33 degree)
③ Minimum: 0.04 inch
  Nominal: 0.08 inch
  Maximum: 0.12 inch

④ 2WD Minimum: -2 degree 01' (-2.01 degree)
  2WD Nominal: -1 degree 31' (-1.52 degree)
  2WD Maximum: -1 degree 01' (-1.02 degree)
⑤ AWD Minimum: -0 degree 45' (-0.75 degree)
  AWD Nominal: 0 degree 00' (0.00 degree)
  AWD Maximum: 0 degree 45' (0.75 degree)
⑥ 2WD Minimum: 0.004 in
  2WD Nominal: 0.161 inch
  2WD Maximum: 0.319 inch
⑦ AWD Minimum: 0.043 inch
  AWD Nominal: 0.122 inch
  AWD Maximum: 0.201 inch

71075_JUKE_C0012

TIRE, WHEEL AND BALL JOINT SPECIFICATIONS

Year	Model	OEM Tires Standard	OEM Tires Optional	Tire Pressures (psi) Front	Tire Pressures (psi) Rear	Wheel Size	Ball Joint Inspection	Lug Nut (ft. lbs.)
2011	Juke	P215/55R17	—	33-36	33-36	93V	NA	80
		—	T135/80(90)D16	60	60	101/102M	NA	80
2012	Juke	P215/55R17	—	33-36	33-36	93V	NA	80
		—	T135/80(90)D16	60	60	101/102M	NA	80

OEM: Original Equipment Manufacturer

PSI: Pounds Per Square Inch

NA: Information not available

71075_JUKE_C0013

BRAKE SPECIFICATIONS
All measurements in inches unless noted

Year	Model		Brake Disc Original Thickness	Brake Disc Minimum Thickness	Brake Disc Max. Runout	Brake Drum Diameter Original Inside Diameter	Brake Drum Diameter Max. Wear Limit	Brake Drum Diameter Maximum Machine Diamter	Minimum Pad/Lining Thickness Front	Minimum Pad/Lining Thickness Rear	Brake Caliper Bracket Bolts (ft. lbs.)	Brake Caliper Mounting Bolts (ft. lbs.)
2011	Juke	F	NA	0.945	0.0014	NA	NA	NA	0.079	0.079	122	NA
		R	NA	0.315	0.0040	NA	NA	NA	0.079	0.079	62	NA
2012	Juke	F	NA	0.945	0.0014	NA	NA	NA	0.079	0.079	122	NA
		R	NA	0.315	0.0040	NA	NA	NA	0.079	0.079	62	NA

F: Front

R: Rear

NA: Information not available

71075_JUKE_C0014

SCHEDULED MAINTENANCE INTERVALS
Nissan Juke

TO BE SERVICED	TYPE OF SERVICE	VEHICLE MILEAGE INTERVAL (x1000)							
		7.5	15	22.5	30	37.5	45	52.5	60
Brake lines & cables	Inspect		✓		✓		✓		✓
Brake pads & rotors	Inspect		✓		✓		✓		✓
Brake fluid	Replace				✓				✓
Tires (rotate, inspect the tire tread for wear, and adjust air pressure)	Rotate/ Inspect	✓	✓	✓	✓	✓	✓	✓	✓
Fluid levels	Inspect, add as necessary	✓	✓	✓	✓	✓	✓	✓	✓
Windshield washer spray, wiper operation, clean all wiper blades	Inspect and Service, as necessary	✓	✓	✓	✓	✓	✓	✓	✓
Lights, warning indicators & horn	Inspect	✓	✓	✓	✓	✓	✓	✓	✓
Battery fluid level	Inspect, add as necessary	✓	✓	✓	✓	✓	✓	✓	✓
Air cleaner	Replace				✓				✓
Drive belt	Inspect and Service, as necessary								✓
Spark plugs (Iridium)	Replace	Every 105,000 miles							
Engine coolant	Replace	Every 105,000 miles or 84 months, than every 75,000 miles or 60 months							
EVAP vapor lines	Inspect				✓				✓
Exhaust system	Inspect				✓				✓
Fuel lines	Inspect				✓				✓
Parking brake	Inspect	✓	✓	✓	✓	✓	✓	✓	✓
CVT fluid	Inspect		✓		✓		✓		✓
Manual transmission fluid	Inspect		✓		✓		✓		✓
Steering gear & linkage, axle & suspension parts	Inspect				✓				✓
In-cabin microfilter	Replace		✓		✓		✓		✓
Engine oil & filter	Replace	✓	✓	✓	✓	✓	✓	✓	✓
Drive shaft boots	Inspect		✓		✓		✓		✓
Transfer case fluid	Inspect		✓		✓		✓		✓
Differential gear oil	Inspect		✓		✓		✓		✓
Underbody	Clean	✓	✓	✓	✓	✓	✓	✓	✓
Propeller shaft	Inspect		✓		✓		✓		✓

FREQUENT OPERATION MAINTENANCE (SEVERE SERVICE)

If a vehicle is operated under any of the following conditions it is considered severe service:

Repeated short trips of 5 miles or less.

Repeated short trips of 10 miles or less with the outside temperature below freezing.

Extremely dusty areas

Prolonged idling (vehicle operation in stop and go traffic).

Frequent short running periods (engine does not warm to normal operating temperatures).

Police, taxi, delivery usage or trailer towing usage.

PRECAUTIONS

Before servicing any vehicle, please be sure to read all of the following precautions, which deal with personal safety, prevention of component damage, and important points to take into consideration when servicing a motor vehicle:

• Never open, service or drain the radiator or cooling system when the engine is hot; serious burns can occur from the steam and hot coolant.

• Observe all applicable safety precautions when working around fuel. Whenever servicing the fuel system, always work in a well-ventilated area. Do not allow fuel spray or vapors to come in contact with a spark, open flame, or excessive heat (a hot drop light, for example). Keep a dry chemical fire extinguisher near the work area. Always keep fuel in a container specifically designed for fuel storage; also, always properly seal fuel containers to avoid the possibility of fire or explosion. Refer to the additional fuel system precautions later in this section.

• Fuel injection systems often remain pressurized, even after the engine has been turned **OFF**. The fuel system pressure must be relieved before disconnecting any fuel lines. Failure to do so may result in fire and/or personal injury.

• Brake fluid often contains polyglycol ethers and polyglycols. Avoid contact with the eyes and wash your hands thoroughly after handling brake fluid. If you do get brake fluid in your eyes, flush your eyes with clean, running water for 15 minutes. If eye irritation persists, or if you have taken

brake fluid internally, IMMEDIATELY seek medical assistance.

• The EPA warns that prolonged contact with used engine oil may cause a number of skin disorders, including cancer. You should make every effort to minimize your exposure to used engine oil. Protective gloves should be worn when changing oil. Wash your hands and any other exposed skin areas as soon as possible after exposure to used engine oil. Soap and water, or waterless hand cleaner should be used.

• All new vehicles are now equipped with an air bag system, often referred to as a Supplemental Restraint System (SRS) or Supplemental Inflatable Restraint (SIR) system. The system must be disabled before performing service on or around system components, steering column, instrument panel components, wiring and sensors. Failure to follow safety and disabling procedures could result in accidental air bag deployment, possible personal injury and unnecessary system repairs.

• Always wear safety goggles when working with, or around, the air bag system. When carrying a non-deployed air bag, be sure the bag and trim cover are pointed away from your body. When placing a non-deployed air bag on a work surface, always face the bag and trim cover upward, away from the surface. This will reduce the motion of the module if it is accidentally deployed. Refer to the additional air bag system precautions later in this section.

• Clean, high quality brake fluid from a sealed container is essential to the safe and

proper operation of the brake system. You should always buy the correct type of brake fluid for your vehicle. If the brake fluid becomes contaminated, completely flush the system with new fluid. Never reuse any brake fluid. Any brake fluid that is removed from the system should be discarded. Also, do not allow any brake fluid to come in contact with a painted surface; it will damage the paint.

• Never operate the engine without the proper amount and type of engine oil; doing so WILL result in severe engine damage.

• Timing belt maintenance is extremely important. Many models utilize an interference-type, non-freewheeling engine. If the timing belt breaks, the valves in the cylinder head may strike the pistons, causing potentially serious (also time-consuming and expensive) engine damage. Refer to the maintenance interval charts for the recommended replacement interval for the timing belt, and to the timing belt section for belt replacement and inspection.

• Disconnecting the negative battery cable on some vehicles may interfere with the functions of the on-board computer system(s) and may require the computer to undergo a relearning process once the negative battery cable is reconnected.

• When servicing drum brakes, only disassemble and assemble one side at a time, leaving the remaining side intact for reference.

• Only an MVAC-trained, EPA-certified automotive technician should service the air conditioning system or its components.

BRAKES

GENERAL INFORMATION

PRECAUTIONS

• Certain components within the ABS system are not intended to be serviced or repaired individually.

• Do not use rubber hoses or other parts not specifically specified for and ABS system. When using repair kits, replace all parts included in the kit. Partial or incorrect repair may lead to functional problems and require the replacement of components.

• Lubricate rubber parts with clean, fresh brake fluid to ease assembly. Do not use shop air to clean parts; damage to rubber components may result.

• Use only DOT 3 brake fluid from an unopened container.

• If any hydraulic component or line is

removed or replaced, it may be necessary to bleed the entire system.

• A clean repair area is essential. Always clean the reservoir and cap thoroughly before removing the cap. The slightest amount of dirt in the fluid may plug an orifice and impair the system function. Perform repairs after components have been thoroughly cleaned; use only denatured alcohol to clean components. Do not allow ABS components to come into contact with any substance containing mineral oil; this includes used shop rags.

• The Anti-Lock control unit is a microprocessor similar to other computer units in the vehicle. Ensure that the ignition switch is **OFF** before removing or installing controller harnesses. Avoid static electricity discharge at or near the controller.

• If any arc welding is to be done on the

ANTI-LOCK BRAKE SYSTEM (ABS)

vehicle, the control unit should be unplugged before welding operations begin.

WHEEL SPEED SENSORS

REMOVAL & INSTALLATION

Front

See Figure 2.

1. Remove tires.
2. Remove the fender protector (front).
3. Remove front wheel sensor from steering knuckle.

✳✳ WARNING

Never rotate and never pull front wheel sensor as much as possible, when pulling out.

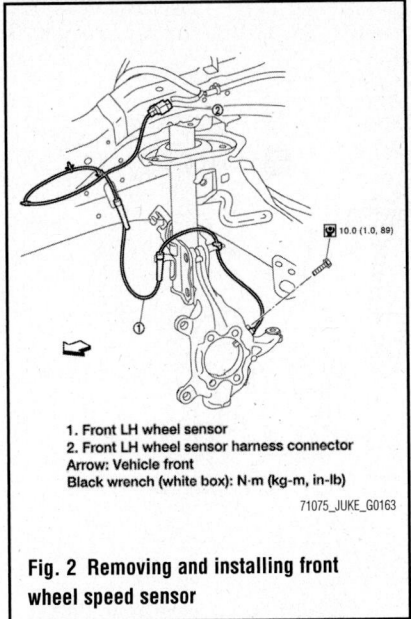

1. Front LH wheel sensor
2. Front LH wheel sensor harness connector
Arrow: Vehicle front
Black wrench (white box): N·m (kg-m, in-lb)

71075_JUKE_G0163

Fig. 2 Removing and installing front wheel speed sensor

4. Remove front wheel sensor harness from the vehicle.

✲✲ WARNING

Do not twist or pull front wheel sensor harness when removing.

To install:

5. Note the following, and install in the reverse order of the removal:

a. Check that there is no foreign material like iron powder or damage on inner surface of front wheel sensor mounting hole of steering knuckle and sensor rotor. Install after cleaning when there are foreign material like iron powder, or replace when there is a malfunction.

b. Never twist front wheel sensor harness when installing front wheel sensor. Check that grommet is fully inserted to bracket. Check that front wheel sensor harness is not twisted after installation.

Rear

See Figures 3 through 5

➡**Rear RH wheel sensor is symmetrically opposite of LH.**

1. Remove rear wheel sensor from wheel hub and bearing assembly (2WD).

✲✲ WARNING

Never rotate or pull rear wheel sensor as much as possible, when pulling out.

2. Remove rear wheel sensor from axle housing (AWD).

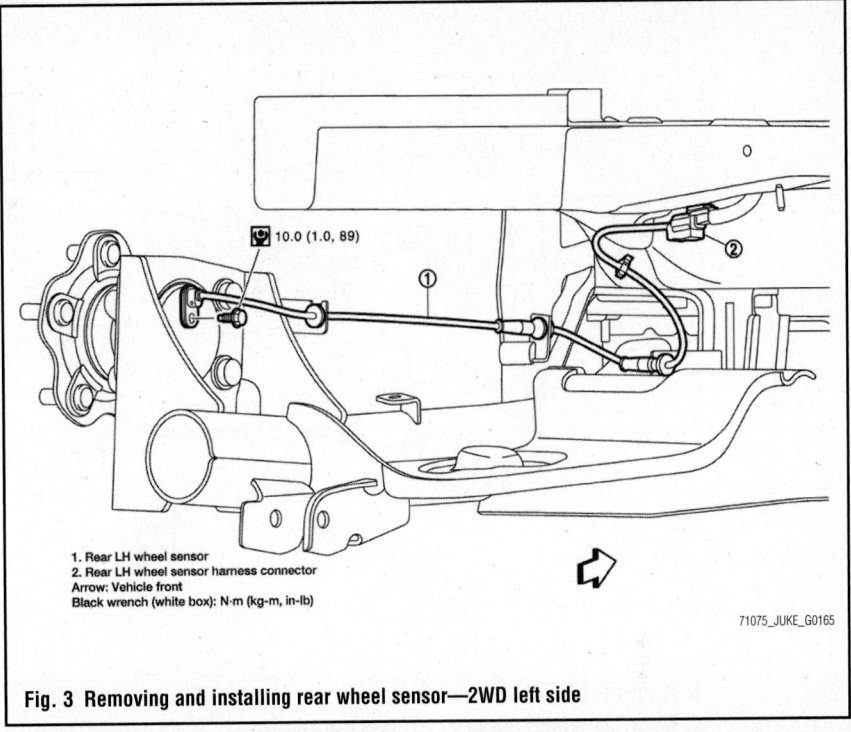

1. Rear LH wheel sensor
2. Rear LH wheel sensor harness connector
Arrow: Vehicle front
Black wrench (white box): N·m (kg-m, in-lb)

71075_JUKE_G0165

Fig. 3 Removing and installing rear wheel sensor—2WD left side

✲✲ WARNING

Never rotate or pull rear wheel sensor as much as possible, when pulling out.

3. Remove rear wheel sensor harness from the vehicle.

✲✲ WARNING

Do not twist or pull rear wheel sensor harness when removing.

To install:

4. Note the following, and install in the reverse order of the removal:

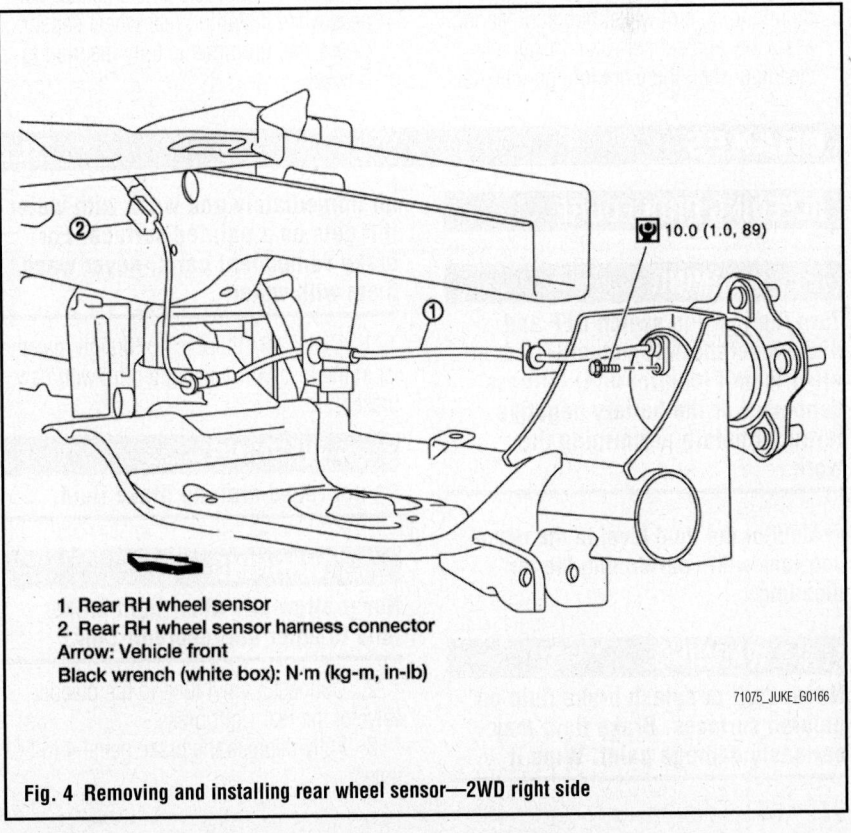

1. Rear RH wheel sensor
2. Rear RH wheel sensor harness connector
Arrow: Vehicle front
Black wrench (white box): N·m (kg-m, in-lb)

71075_JUKE_G0166

Fig. 4 Removing and installing rear wheel sensor—2WD right side

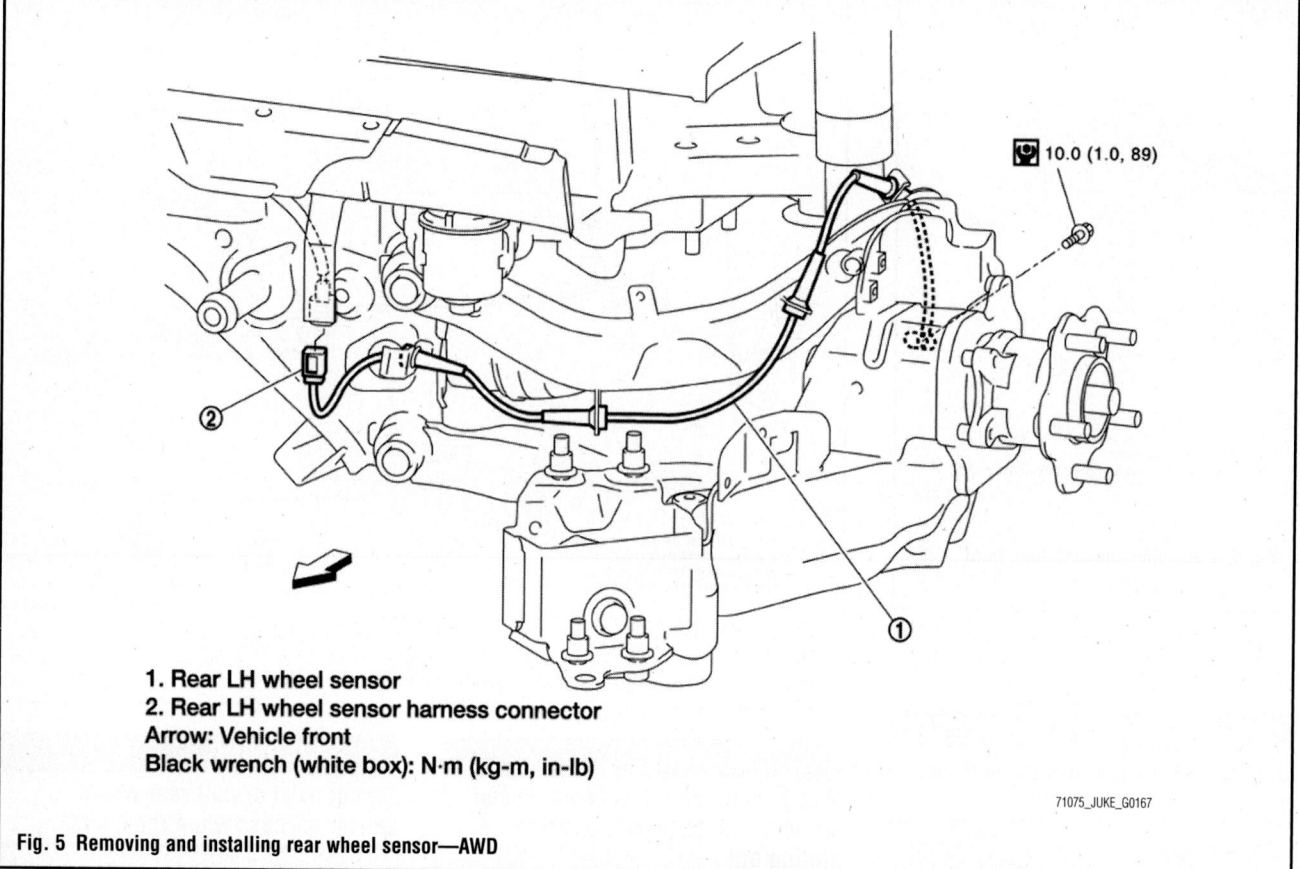

10.0 (1.0, 89)

1. Rear LH wheel sensor
2. Rear LH wheel sensor harness connector
Arrow: Vehicle front
Black wrench (white box): N·m (kg-m, in-lb)

71075_JUKE_G0167

Fig. 5 Removing and installing rear wheel sensor—AWD

a. Check that there is no foreign material like iron powder or damage on inner surface of rear wheel sensor mounting hole of wheel hub and bearing assembly and sensor rotor. Install after cleaning when there are foreign material like iron powder, or replace when there is a malfunction.

b. Never twist rear wheel sensor harness when installing rear wheel sensor. Check that grommet is fully inserted to bracket.

c. Check that rear wheel sensor harness is not twisted after installation.

d. Check that the identification line of the rear wheel sensor is faced upward.

BRAKES   BLEEDING THE BRAKE SYSTEM

BLEEDING PROCEDURE

❊❊ WARNING

Turn the ignition switch OFF and disconnect the ABS actuator and electric unit (control unit) harness connector or the battery negative terminal before performing the work.

➡Monitor the fluid level in the reservoir tank while performing the air bleeding

❊❊ WARNING

Never spill or splash brake fluid on painted surfaces. Brake fluid may seriously damage paint. Wipe it

off immediately and wash with water if it gets on a painted surface. For brake component parts, never wash them with water.

1. Check that there is no foreign material in the reservoir tank, and refill with new brake fluid.

❊❊ WARNING

Do not reuse drained brake fluid.

❊❊ WARNING

Never allow oils other than brake fluid to enter the reservoir tank.

2. Connect a vinyl tube to the bleeder valve of the rear right brake.
3. Fully depress the brake pedal 4 to 5 times.

4. Loosen the bleeder valve and bleed air with the brake pedal depressed, and then quickly tighten the bleeder valve.
5. Repeat steps 3 and 4 until all of the air is out of the brake line.
6. Tighten the bleeder valve to the specified torque.
7. Perform steps 2 to 6. Occasionally fill with the brake fluid in order to keep it in the reservoir tank at least half of MAX line. Bleed air in the following order: rear right brake, front left brake, rear left brake and front right brake in order.
8. Check that the fluid level in the reservoir tank is within the specified range after air bleeding.
9. Check each item of brake pedal. Adjust it if the measurement value is not the standard.

FLUID FILL PROCEDURE

✳✳ WARNING

Turn the ignition switch OFF and disconnect the ABS actuator and electric unit (control unit) harness connector or the battery negative terminal before performing work.

✳✳ WARNING

Never spill or splash brake fluid on painted surfaces. Brake fluid may seriously damage paint. Wipe it off immediately and wash with water if it gets on a painted surface. For brake component parts, never wash them with water.

1. Check that there is no foreign material in the reservoir tank, and refill with new brake fluid.

✳✳ WARNING

Do not reuse drained brake fluid.

✳✳ WARNING

Never allow foreign matter (e.g. dust) and oils other than brake fluid to enter the reservoir tank.

2. Loosen the bleeder valve, slowly depress the brake pedal to the full stroke, and then release the pedal. Repeat this operation at intervals of 2 or 3 seconds until new brake fluid is discharged. Then close the bleeder valve with the brake pedal depressed. Repeat the same work on each wheel.
3. Perform the air bleeding procedure.

FLUID RECOMMENDATIONS

Genuine NISSAN Super Heavy Duty Brake Fluid or equivalent DOT 3 (US FMVSS No. 116).

LEVEL CHECK

1. Check that the fluid level in the reservoir tank is within the specified range (MAX - MIN lines).
2. Visually check for any brake fluid leakage around the reservoir tank.
3. Check the brake system for any leakage if the fluid level is extremely low (lower than MIN).
4. Check the brake system for fluid leakage if the warning lamp remains illuminated even after the parking brake is released.
5. Check the reservoir tank for the mixing of foreign matter (e.g. dust) and oils other than brake fluid.

BRAKES

BRAKE CALIPERS

REMOVAL & INSTALLATION
See Figures 6 and 7.

✳✳ WARNING

Never spill or splash brake fluid on painted surfaces. Brake fluid may seriously damage paint. Wipe it out immediately and wash with water if it gets on a protect surface. For brake component parts, never wash them with water.

✳✳ WARNING

Do not depress the brake pedal while removing the brake hose. If this is not complied with, brake fluid may splash.

✳✳ WARNING

If the brake fluid or grease adheres to the disc rotor, quickly wipe it off.

1. Remove tires with power tool.
2. Fix the disc rotor using wheel nuts.
3. Drain brake fluid.
4. Separate brake hose from caliper assembly.
5. Remove torque member mounting bolts, and remove brake caliper assembly.

✳✳ WARNING

Never drop brake pad and caliper assembly or disc rotor.

FRONT DISC BRAKES

6. When removing disc rotor. Put matching marks on the wheel hub assembly and the disc rotor before removing the disc rotor.

   To install:
7. Install disc rotor.
8. Install the brake caliper assembly to the steering knuckle and tighten the torque member mounting bolts.

✳✳ WARNING

Never spill or splash any grease and moisture on the brake caliper assembly mounting face, threads, mounting bolts and washers. Wipe out any grease and moisture.

9. Install brake hose.
10. Perform the air bleeding.

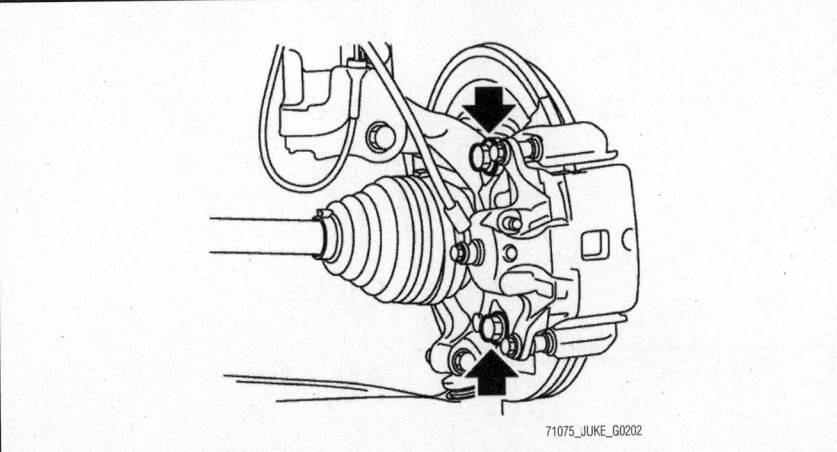

Fig. 6 Removing torque member mounting bolts and caliper assembly

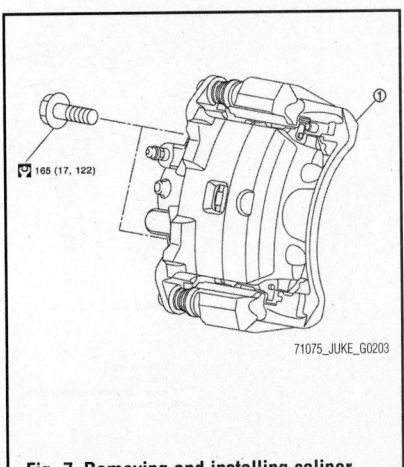

Fig. 7 Removing and installing caliper assembly (1)

11. Check a drag of front disc brake.
12. Install tires.
13. Perform inspection after installation.

BRAKE PADS

REMOVAL & INSTALLATION

See Figures 8 through 11.

> ❊❊ **WARNING**
>
> **Never depress the brake pedal while removing the brake pads because the piston may pop out.**

> ❊❊ **WARNING**
>
> **If the brake fluid or grease adheres to the disc rotor, quickly wipe it off.**

1. Remove tires with power tool.
2. Remove lower sliding pin bolt.
3. Suspend the cylinder body with suitable wire so that the brake hose will not stretch.
4. Remove the brake pads, shims, shim covers and pad retainers from the torque member.

> ❊❊ **WARNING**
>
> **Never deform the pad retainer (2) when removing the pad retainer from the torque member (1).**

> ❊❊ **WARNING**
>
> **Never damage the piston boot.**

> ❊❊ **WARNING**
>
> **Never drop the brake pads, shims, and the shim covers.**

> ❊❊ **WARNING**
>
> **Remember each position of the removed brake pads.**

5. Perform inspection after removal.

To install:

6. Install the pad retainers (1) to the torque member (2) if the pad retainers has been removed.

> ❊❊ **WARNING**
>
> **Securely assemble the pad retainers so that it will not be lifted up from the torque member.**

> ❊❊ **WARNING**
>
> **Never deform the pad retainers.**

7. Apply MOLYKOTE® AS880N or silicone-based grease to the mating faces (A) between the inner pad and the inner shim, and install the inner shims to the inner pad.

> ❊❊ **WARNING**
>
> **Always replace the shim together when replacing the brake pad.**

8. Apply MOLYKOTE® 7439 or equivalent to the mating faces between the inner pad and the pad retainers.
9. Apply MOLYKOTE® AS880N or silicone-based grease to the mating faces between the outer shim cover and the outer shim, and install the outer shim and outer shim cover to the outer pad.

> ❊❊ **WARNING**
>
> **Always replace the shim together with the shim cover when replacing the brake pad.**

10. Apply MOLYKOTE® 7439 or equivalent to the mating faces between the outer pad and the pad retainers.
11. Install the brake pads to the torque member.

> ❊❊ **WARNING**
>
> **Both inner and outer pads have a pad return system on the pad retainer.**

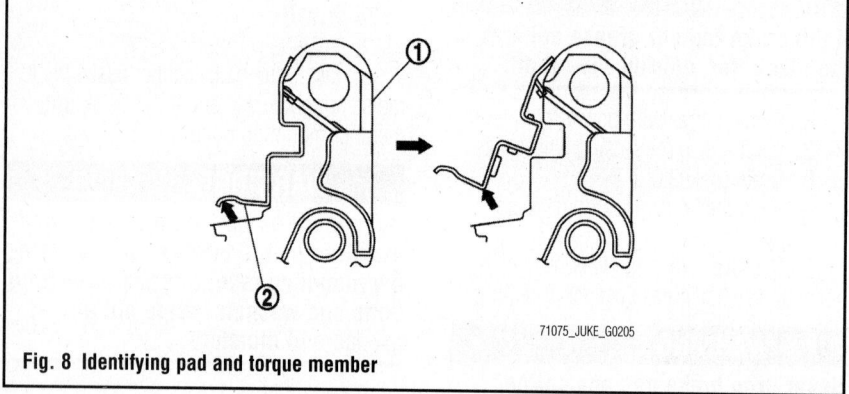

71075_JUKE_G0205

Fig. 8 Identifying pad and torque member

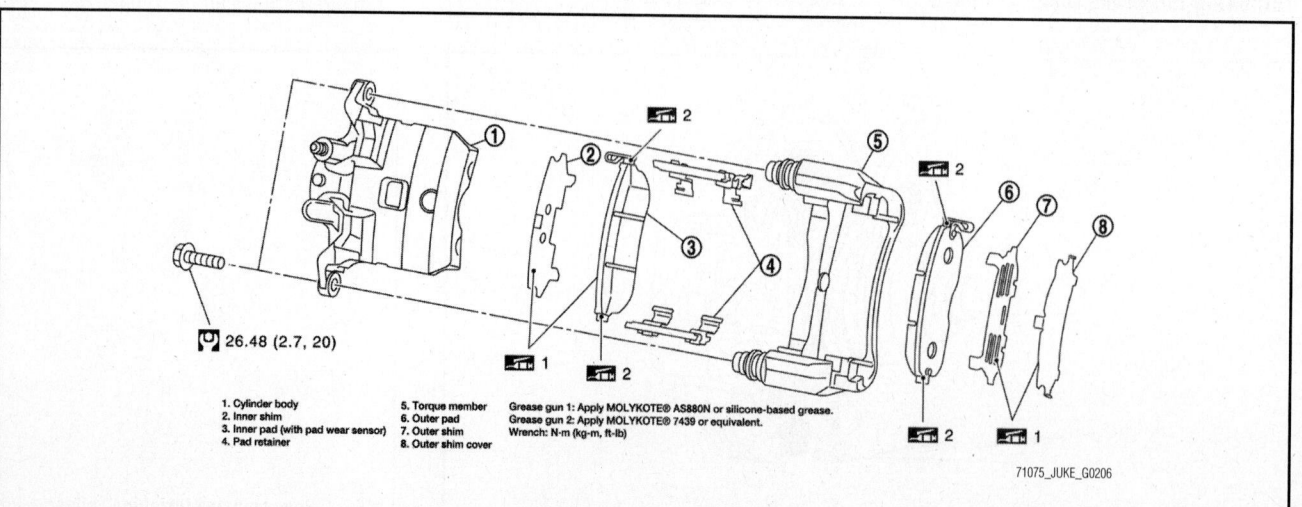

1. Cylinder body
2. Inner shim
3. Inner pad (with pad wear sensor)
4. Pad retainer
5. Torque member
6. Outer pad
7. Outer shim
8. Outer shim cover

Grease gun 1: Apply MOLYKOTE® AS880N or silicone-based grease.
Grease gun 2: Apply MOLYKOTE® 7439 or equivalent.
Wrench: N·m (kg-m, ft-lb)

🔧 26.48 (2.7, 20)

71075_JUKE_G0206

Fig. 9 Removing and installing brake pad

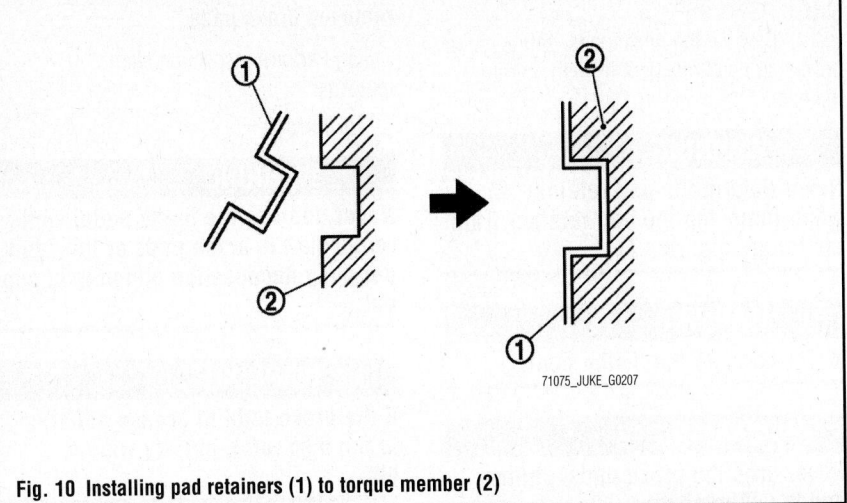

Fig. 10 Installing pad retainers (1) to torque member (2)

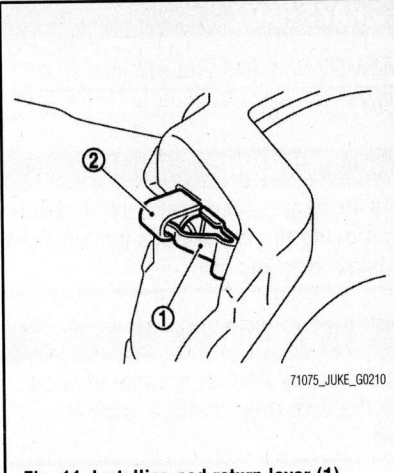

Fig. 11 Installing pad return lever (1) securely to pad retainer (2)

Install pad return lever (1) securely to pad retainer (2).

❊❊ WARNING

Never deform the pad retainers.

12. Install cylinder body to torque member.

❊❊ WARNING

Never damage the piston boot.

13. When replacing brake pad with new one, check a brake fluid level in the reservoir tank because brake fluid returns to master cylinder reservoir tank when pressing piston in.

➡Use a disc brake piston tool to easily press piston.

14. Install the lower sliding pin bolt and tighten.

15. Depress the brake pedal several times to check that no drag feel is present for the front disc brake.

16. Install tires.

BRAKES

BRAKE CALIPERS

REMOVAL & INSTALLATION
See Figure 12.

❊❊ WARNING

Never spill or splash brake fluid on painted surfaces. Brake fluid may seriously damage paint. Wipe it out immediately and wash with water if it gets on a protect surface. For brake component parts, never wash them with water.

❊❊ WARNING

Never depress the brake pedal while removing the brake hose. If this is not complied with, brake fluid may splash.

❊❊ WARNING

If the brake fluid or grease adheres to the disc rotor, quickly wipe it off.

1. Remove tires with power tool.
2. Fix the disc rotor using wheel nuts.
3. Drain brake fluid.

4. Separate brake hose from caliper assembly.

5. Remove torque member mounting bolts, and remove brake caliper assembly.

❊❊ WARNING

Never drop brake pad and caliper assembly.

To install:

❊❊ WARNING

Never depress the brake pedal while removing the brake hose. If this is not complied with, brake fluid may splash.

❊❊ WARNING

If the brake fluid or grease adheres to the disc rotor, quickly wipe it off.

6. Install the brake caliper assembly to the axle housing and tighten the torque member mounting bolts to 62 ft. lbs. (84.3 Nm).

❊❊ WARNING

Never spill or splash any grease and moisture on the brake caliper assem-

REAR DISC BRAKES

bly mounting face, threads, mounting bolts and washers. Wipe out any grease and moisture.

7. Install brake hose.
8. Perform the air bleeding.
9. Check a drag of rear disc brake.
10. Install tires.
11. Perform inspection after installation.

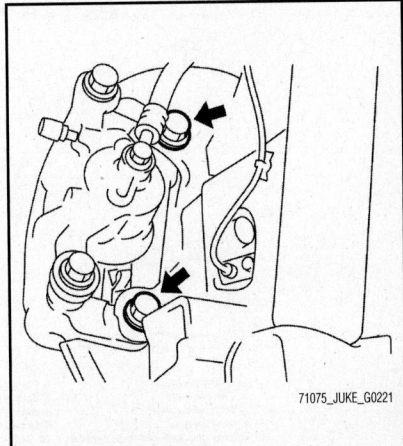

Fig. 12 Removing torque member mounting bolts and brake caliper assembly

BRAKE PADS

REMOVAL & INSTALLATION
See Figures 13 through 15.

> ❄❄ **WARNING**
>
> **Never depress the brake pedal while removing the brake pads because the piston may pop out.**

> ❄❄ **WARNING**
>
> **If the brake fluid or grease adheres to the disc rotor, quickly wipe it off.**

1. Remove tires with power tool.
2. Remove lower sliding pin bolt.
3. Suspend the cylinder body with suitable wire so that the brake hose will not stretch.
4. Remove the brake pads, shims, shim covers and pad retainers from the torque member.

> ❄❄ **WARNING**
>
> **Never deform the pad retainer (2) when removing the pad retainer from the torque member (1).**

> ❄❄ **WARNING**
>
> **Never damage the piston boot.**

> ❄❄ **WARNING**
>
> **Never drop the brake pads, shims, and the shim covers.**

➡ **Remember each position of the removed brake pads.**

5. Perform inspection after removal.

To install:

> ❄❄ **WARNING**
>
> **Never depress the brake pedal while removing the brake pads or the cylinder body because the piston may pop out.**

> ❄❄ **WARNING**
>
> **If the brake fluid or grease adheres to the disc rotor, quickly wipe it off.**

6. Install the pad retainers to the torque member if the pad retainers has been removed.

> ❄❄ **WARNING**
>
> **Securely assemble the pad retainers so that it will not be lifted up from the torque member.**

> ❄❄ **WARNING**
>
> **Never deform the pad retainers.**

7. Apply MOLYKOTE® AS880N or silicone-based grease to the mating faces between the brake pads and the shims, and install the shims to the brake pad.

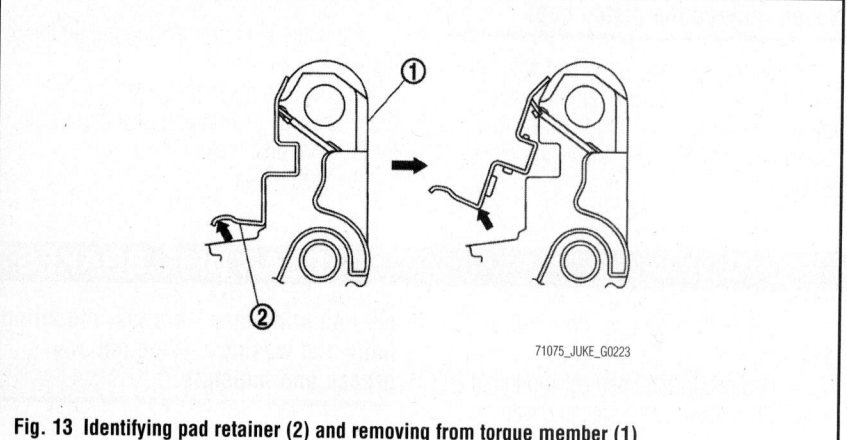

71075_JUKE_G0223

Fig. 13 Identifying pad retainer (2) and removing from torque member (1)

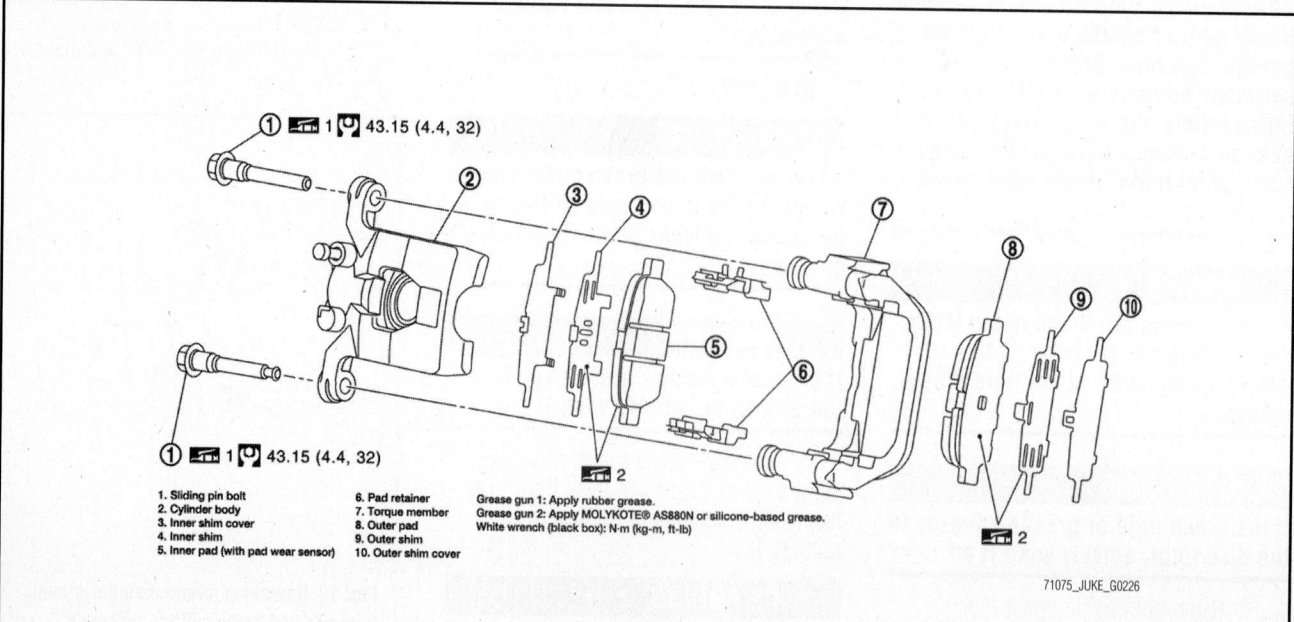

1 ⬛ 1 43.15 (4.4, 32)

1 ⬛ 1 43.15 (4.4, 32)

1. Sliding pin bolt	6. Pad retainer
2. Cylinder body	7. Torque member
3. Inner shim cover	8. Outer pad
4. Inner shim	9. Outer shim
5. Inner pad (with pad wear sensor)	10. Outer shim cover

Grease gun 1: Apply rubber grease.
Grease gun 2: Apply MOLYKOTE® AS880N or silicone-based grease.
White wrench (black box): N·m (kg-m, ft-lb)

71075_JUKE_G0226

Fig. 14 Removing and installing rear brake pad

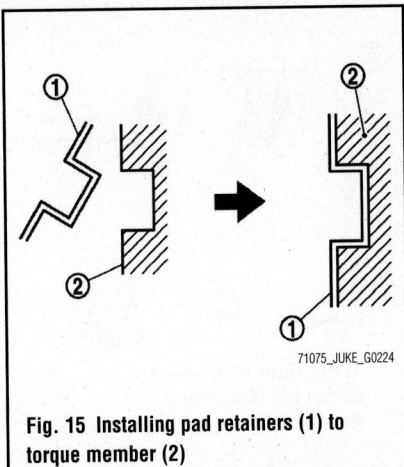

Fig. 15 Installing pad retainers (1) to torque member (2)

✳✳ WARNING

Always replace the shim together with the shim cover when replacing the brake pad.

8. Install the brake pads to the torque member.

9. Install cylinder body to torque member.

✳✳ WARNING

Never damage the piston boot.

✳✳ WARNING

When replacing brake pad with

new one, check a brake fluid level in the reservoir tank because brake fluid returns to master cylinder reservoir tank when pressing piston in.

➡ **Use a disc brake piston tool to easily press piston.**

10. Install the lower sliding pin bolt and tighten it to 32 ft. lbs. (43 Nm).

11. Depress the brake pedal several times to check that no drag feel is present for the rear disc brake.

12. Install tires.

BRAKES

PARKING BRAKE

ADJUSTMENTS

CABLES

See Figures 16 and 17.

1. Remove rear tires with power tool.
2. Fix the disc rotor using wheel nut.
3. Remove the console mask.
4. Remove the adjusting hole plug from the disc rotor. Turn the adjuster (1) in the direction (A) as shown in the figure using a suitable tool until the disc rotor is locked.
5. Turn back the adjuster 7 notches from the locked position.
6. Rotate the disc rotor to check that there is no drag. Install the adjuster hole plug.
7. Adjust the cable with the following procedure.

    a. Operate and maintain the parking brake lever with a force of 90 lbs. (400 N) for 15 minutes or more.

    b. Adjust the parking brake lever

stroke by turning the adjusting nut (1) with a socket wrench.

✳✳ WARNING

Make sure to securely depress the brake pedal.

    c. Operate the parking brake lever with a force of 44 lbs. (196 N). Check that the lever stroke is within the specified number of notches. (Check it by listening to the clicks of the ratchet.)

    d. Rotate the disc rotor to check that there is no drag.

CONTROL ASSEMBLY

1. Operate the parking brake lever with a force of 44 lbs. (196 N). Check that the lever stroke is within the 9-10 notches. (Check it by listening to the clicks of the ratchet.)

2. When brake warning lamp turns ON, check that the lever stroke is 1 notch.

PARKING BRAKE SHOES

REMOVAL & INSTALLATION

See Figures 18 through 22.

1. Remove rear tires with power tool.
2. Remove disc rotor.
3. Remove return spring (1) of the upper side.
4. Remove return spring (1) of lower side.
5. Remove spring (1).

✳✳ WARNING

Never drop the removed parts.

6. Remove parking brake shoes, adjuster, brake strut and toggle lever.

✳✳ WARNING

The parking brake shoes for the front wheels are made of different materials from those for the rear wheels. Never misidentify them when removing.

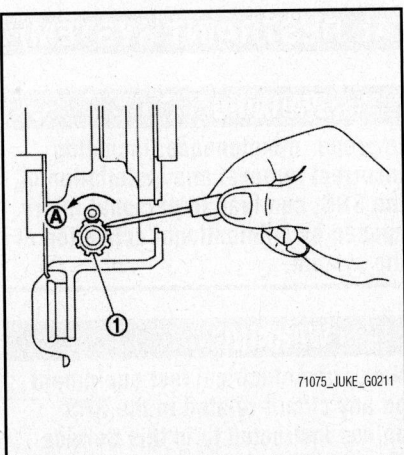

Fig. 16 Turning the adjuster (1) in the direction (A) as shown

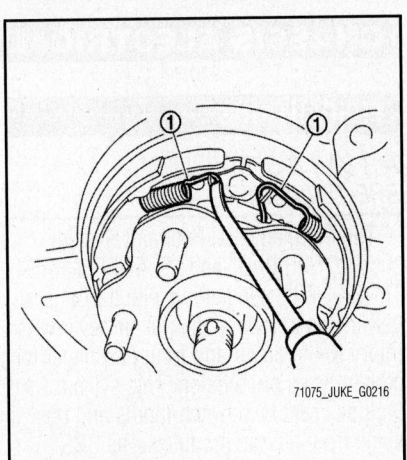

Fig. 18 Removing return spring (1) of upper side

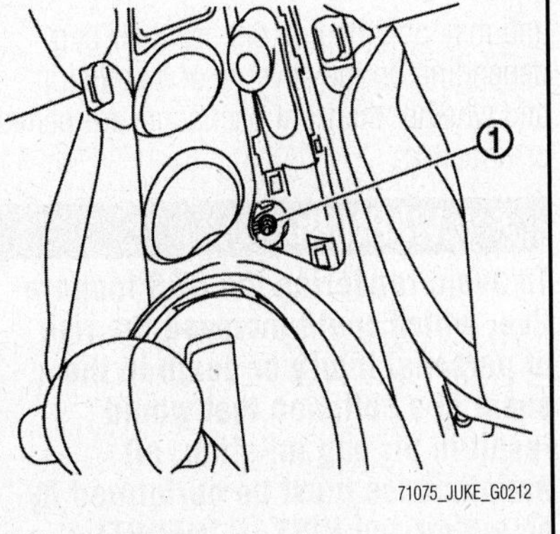

Fig. 17 Adjusting the parking brake lever stroke by turning the adjusting nut (1)

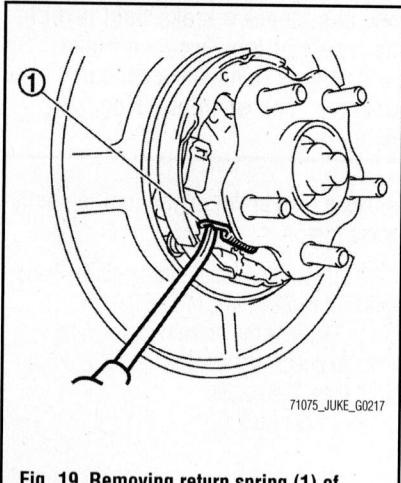

Fig. 19 Removing return spring (1) of lower side

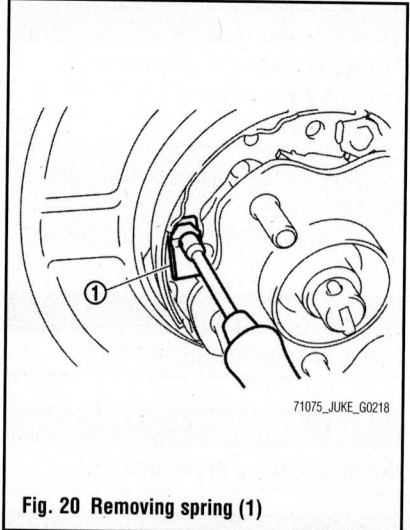

Fig. 20 Removing spring (1)

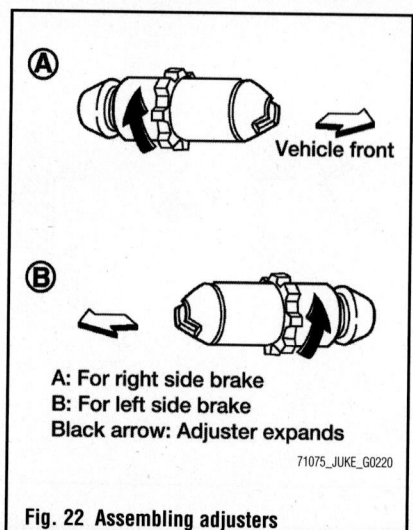

A: For right side brake
B: For left side brake
Black arrow: Adjuster expands

Fig. 22 Assembling adjusters

7. Press the rear cable spring (1) against spring tension to remove rear cable (3) from the clamp (A) of toggle lever (2).

❊❊ WARNING

Never bend rear cable.

8. Remove the back plate.

To install:

9. Note the following, and install in the reverse order of the removal procedure:

    a. Apply PBC (Poly Butyl Cuprysil) grease of silicone-based grease to the back plate and brake shoe.

    b. The parking brake shoes for the front wheels are made of different materials from those for the rear wheels. Never misidentify them when removing and replacing.

    c. Assemble adjusters so the threaded part is expanded when rotating it in the direction shown.

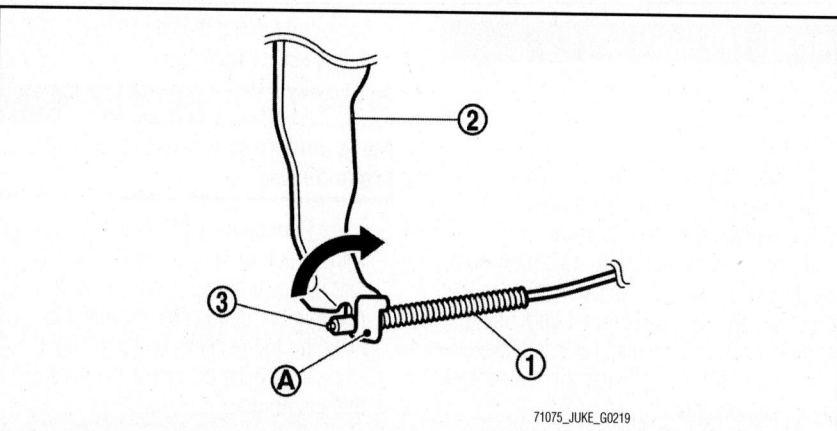

Fig. 21 Pressing rear cable spring (1) against spring tension to remove rear cable (3) from the clamp (A) of toggle lever (2)

    d. Shorten adjuster by rotating it.

    e. When disassembling apply PBC (Poly Butyl Cuprysil) grease or silicone-based grease to threads.

    f. Check that the component parts of the parking brake shoe are properly installed.

    g. Check brake shoe sliding surface and drum inner surface for grease. Wipe is off if it adhered on the surfaces.

CHASSIS ELECTRICAL   AIR BAGS (SUPPLEMENTAL RESTRAINT SYSTEM)

PRECAUTIONS

AIR BAG & SEAT BELT PRE-TENSIONER

The Supplemental Restraint System such as "AIR BAG" and "SEAT BELT PRE-TENSIONER", used along with a front seat belt, helps to reduce the risk or severity of injury to the driver and front passenger for certain types of collision. This system includes seat belt switch inputs and dual stage front air bag modules. The SRS system uses the seat belt switches to determine the front air bag deployment, and may only deploy one front air bag, depending on the severity of a collision and whether the front occupants are belted or unbelted.

❊❊ WARNING

To avoid rendering the SRS inoperative, which could increase the risk of personal injury or death in the event of a collision that would result in air bag inflation, all maintenance must be performed by an authorized NISSAN/INFINITI dealer.

❊❊ WARNING

Improper maintenance, including incorrect removal and installation of the SRS, can lead to personal injury caused by unintentional activation of the system.

❊❊ WARNING

Never use electrical test equipment on any circuit related to the SRS unless instructed to in this Service Manual. SRS wiring harnesses can be identified by yellow and/or

orange harnesses or harness connectors.

USING POWER TOOLE (AIR OR ELECTRIC) AND HAMMERS

✳✳ CAUTION

When working near the Air Bag Diagnosis Sensor Unit or other Air Bag System sensors with the ignition ON or engine running, never use air or electric power tools or strike near the sensor(s) with a hammer. Heavy vibration could activate the sensor(s) and deploy the air bag(s), possibly causing serious injury.

✳✳ CAUTION

When using air or electric power tools or hammers, always switch the ignition OFF, disconnect the battery, and wait at least 3 minutes before performing any service.

SERVICE

✳✳ WARNING

Never use electrical test equipment to check SRS circuits unless

instructed to in this Service Manual.

1. Before servicing the SRS, turn ignition switch OFF, disconnect battery negative terminal and wait 3 minutes or more.

✳✳ WARNING

For approximately 3 minutes after the cables are removed, it is still possible for the air bag and seat belt pretensioner to deploy. Therefore, never work on any SRS connectors or wires until at least 3 minutes have passed.

✳✳ WARNING

Diagnosis sensor unit must always be installed with their arrow marks "⇐" pointing towards the front of the vehicle for proper operation. Also check diagnosis sensor unit for cracks, deformities or rust before installation and replace as required.

✳✳ WARNING

The spiral cable must be aligned with the neutral position since its rotations are limited. Never turn steering

wheel and column after removal of steering gear.

✳✳ CAUTION

Handle air bag module carefully. Always place driver and front passenger air bag modules with the pad side facing upward and seat mounted front side air bag module standing with the stud bolt side facing down.

✳✳ WARNING

Conduct self-diagnosis to check entire SRS for proper function after replacing any components.

✳✳ WARNING

After air bag inflates, the front instrument panel assembly should be replaced if damaged.

✳✳ WARNING

Always replace instrument panel pad following front passenger air bag deployment.

DRIVETRAIN

CONTINUALLY VARIABLE TRANSMISSION (CVT)

DRAIN & REFILL

✳✳ WARNING

Replace drain plug gasket with new ones at the final stage of the operation when installing.

1. Remove drain plug from oil pan.
2. Remove drain plug gasket from drain plug.
3. Install drain plug gasket to drain plug.

✳✳ WARNING

Never reuse drain plug gasket.

4. Install drain plug to oil pan. Tighten to 25 ft. lbs. (34 Nm).
5. Fill CVT fluid from CVT fluid charging pipe to the specified level.

✳✳ WARNING

Use only Genuine NISSAN CVT Fluid NS-2. Never mix with other fluid.

✳✳ WARNING

Using CVT fluid other than Genuine NISSAN CVT Fluid NS-2 will deteriorate in driveability and CVT durability, and may damage the CVT, which is not covered by the NISSAN new vehicle limited warranty.

✳✳ WARNING

When filling CVT fluid, take care not to scatter heat generating parts such as exhaust.

➡Sufficiently shake the container of CVT fluid before using.

✳✳ WARNING

Delete CVT fluid deterioration date with CONSULT after changing CVT fluid.

a. CVT fluid: NISSAN CVT Fluid NS-2

✳✳ WARNING

Use only Genuine NISSAN CVT Fluid NS-2. Using transmission fluid other than Genuine NISSAN CVT Fluid NS-2 will damage the CVT, which is not covered by the NISSAN new vehicle limited warranty.

b. Fluid capacity: 2WD: 8⅝ (8.2L), AWD: 9⅛ (8.6L)
6. With the engine warmed up, drive the vehicle in an urban area.

➡When ambient temperature is 68°F (20°C), it takes about 10 minutes for the CVT fluid to warm up to 122 to 176°F (50 to 80°C).

7. Check CVT fluid level and condition.
    a. Repeat steps 1 to 5 if CVT fluid has been contaminated.
8. Select "Data Monitor" in "TRANSMISSION" using CONSULT.
9. Select "CONFORM CVTF DETERIORTN".
10. Select "Erase".

MANUAL TRANSAXLE

DRAIN & REFILL

Draining

1. Start engine and let it run to warm up transaxle.

2. Stop engine. Remove drain plug and gasket, using a socket [Commercial service tool] and then drain gear oil.

3. Set a gasket on drain plug and install it to clutch housing, using a socket [Commercial service tool].

➡**Never reuse gasket.**

4. Tighten drain plug to 18 ft. lbs. (24 Nm).

Refilling

See Figure 23.

1. Remove filler plug (1) and gasket from transaxle case.

2. Fill with new gear oil until oil level reaches the specified limit at filler plug mounting hole as shown in the figure.

   a. Oil grade: API GL-4, Viscosity SAE 75W-80

   b. Capacity (Approx.): 4¼pt. (2 L)

3. After refilling gear oil, check the oil level.

4. Set a gasket on filler plug and then install it to transaxle case.

➡**Never reuse gasket.**

5. Tighten filler plug to 1.8 ft. lbs. (2 Nm).

FLUID LEVEL CHECK

1. Remove filler plug and gasket from transaxle case.

2. Check the oil level from filler plug mounting hole.

❋❋ WARNING

Never start engine while checking oil level.

3. Set a gasket on filler plug and then install it to transaxle case.

➡**Never reuse gasket.**

4. Tighten filler plug to 1.8 ft. lbs. (2 Nm).

FLUID RECOMMENDATIONS

   a. Oil grade: API GL-4, Viscosity SAE 75W-80

CLUTCH

FLUID LEVEL CHECK

1. Check that the fluid level in the reservoir tank is within the specified range (MAX – MIN lines).

2. Visually check for any clutch fluid leakage around the reservoir tank.

3. Check the clutch system for any leakage if the fluid level is extremely low (lower than MIN).

FLUID RECOMMENDATIONS

Genuine NISSAN Super Heavy Duty Brake Fluid or equivalent DOT 3 (US FMVSS No. 116).

CLUTCH MASTER CYLINDER

BENCH BLEEDING PROCEDURE

See Figures 24 and 25.

1. Monitor clutch fluid level in reservoir tank so as not to empty the tank.

❋❋ WARNING

Keep painted surface on the body or other parts free of clutch fluid. If it
spills, wipe up immediately and wash the affected area with water.

➡**Do not use a vacuum assist or any other type of power bleeder on this system. Use of a vacuum assist or power bleeder will not purge all the air from the system.**

2. Fill reservoir tank with new clutch fluid.

❋❋ WARNING

Never reuse drained clutch fluid.

3. Connect a transparent vinyl hose to air bleeder of bleeding connector.

4. "Depress" and "release" the clutch pedal slowly and fully 15 times at an interval of 2 to 3 seconds and release the clutch pedal.

5. Press the lock pin (1) into the bleeding connector (2), and maintain the position.

6. Slide bleeding connector (1) in the direction of the arrow as shown in the figure.

7. Depress the clutch pedal soon and hold it, and then bleed the air from the piping.

❋❋ WARNING

Since the inside of clutch tube is under hydraulic pressure, hold the tube to prevent it from getting disconnected.

8. Return clutch tube and lock pin in their original positions.

9. Release clutch pedal and wait for 5 seconds.

10. Repeat steps 3 to 8 until no bubbles are observed in the clutch fluid.

11. Check that the fluid level in the reservoir tank is within the specified range after air bleeding.

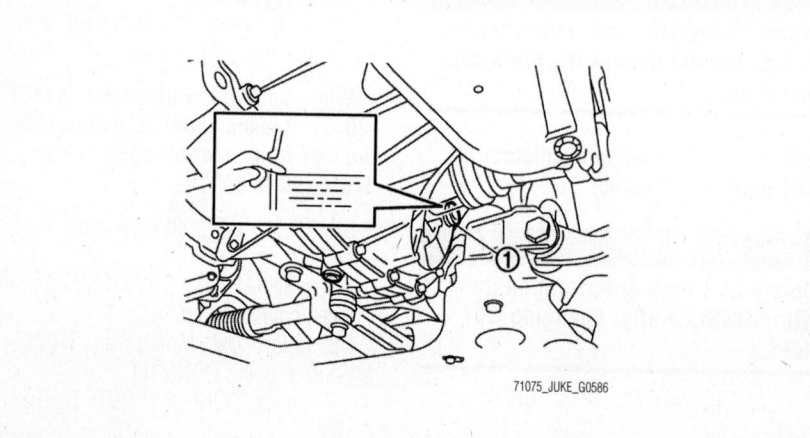

71075_JUKE_G0586

Fig. 23 Locating filler plug (1)

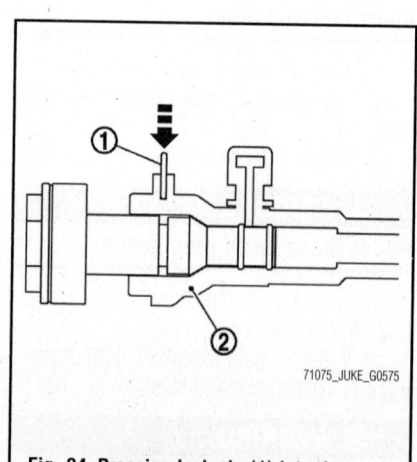

71075_JUKE_G0575

Fig. 24 Pressing lock pin (1) into the bleeding connector (2)

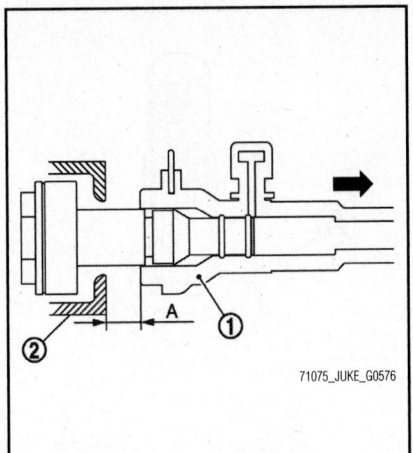

Fig. 25 Sliding bleeding connector (1) away from clutch housing (2)

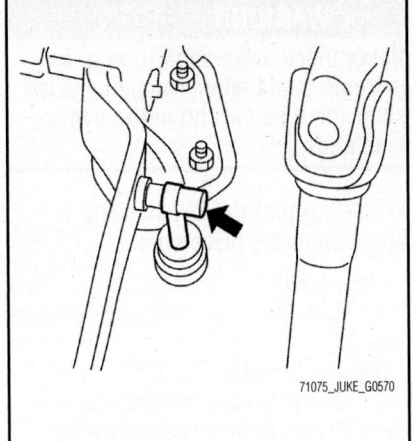

Fig. 26 Removing master cylinder rod end (arrow) from clutch pedal

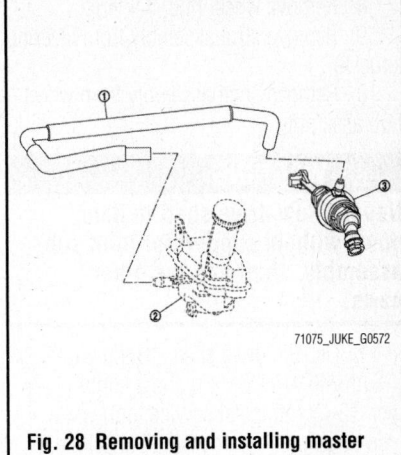

Fig. 28 Removing and installing master cylinder (3), reservoir hose (1) and tank (2)

REMOVAL & INSTALLATION

See Figures 26 through 29.

✳✳ WARNING

Keep painted surface on the body or other parts free of clutch fluid. If it spills, wipe up immediately and wash the affected area with water.

➡**Never disassemble clutch master cylinder.**

1. Drain clutch fluid.
2. Remove air cleaner case.
3. Remove reservoir hose from reservoir tank and master cylinder.
4. Remove master cylinder rod end (arrow) from clutch pedal.
5. Pull up the lock pin (1) from connector of master cylinder (2) and separate clutch tube (3).
6. Rotate master cylinder clockwise by 45 degrees, and then remove master cylinder from the vehicle.

To install:

7. Tilt master cylinder clockwise by 45 degrees and insert it to the mounting hole. Rotate counterclockwise and secure it. At this time, nipple (1) is upward of the vehicle.
8. Install master cylinder rod end to clutch pedal.

✳✳ WARNING

Press master cylinder rod end into clutch pedal until it stops.

9. Install reservoir hose to reservoir tank and master cylinder.

✳✳ WARNING

Set reservoir hose with painted mark facing upward.

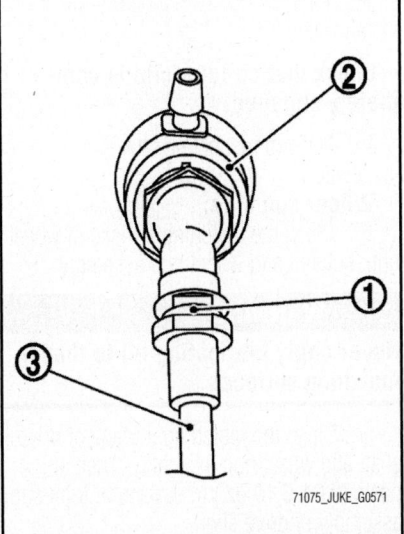

Fig. 27 Pulling lock pin (1) from connector of master cylinder (2) and separating clutch tube (3)

10. Press down the lock pin into connector of master cylinder until it stops.
11. Install clutch tube into connector of master cylinder until it stops.
12. For the next step and after, install in the reverse order of removal.

FRONT DRIVE SHAFT

REMOVAL & INSTALLATION

Left Side

See Figures 30 through 32.

1. Remove tires with power tool.
2. Remove wheel sensor and sensor harness.
3. Remove lock plate from strut assembly.

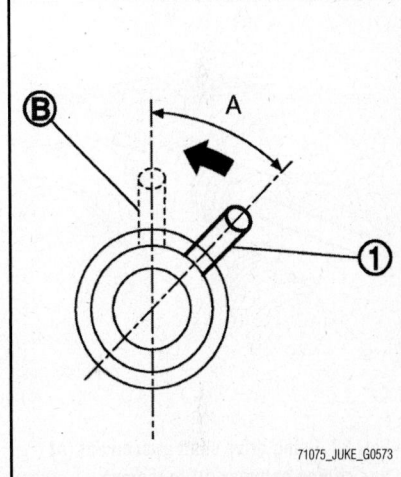

Fig. 29 Tilting master cylinder clockwise 45°(A) and inserting it to the mounting hole. Rotate clockwise and secure it. At this time, nipple (1) is upward of the vehicle (B: mounting condition)

4. Remove caliper assembly. Hang caliper assembly not to interfere with work.

✳✳ WARNING

Never depress brake pedal while brake caliper is removed.

5. Remove disc rotor.
6. Remove cotter pin, and adjusting cap, and then loosen wheel hub lock nut with power tool.
7. Patch wheel hub lock nut with a piece of wood. Hammer the wood to disengage wheel hub assembly from drive shaft.

➡**Use suitable puller, if wheel hub assembly and drive shaft cannot be separated even after performing the above procedure.**

8. Remove wheel hub lock nut.

9. Remove strut assembly from steering knuckle.

10. Remove shaft assembly from wheel hub assembly.

❊❊ WARNING

Never allow drive shaft to hang down without support for joint sub-assembly, shaft and the other parts.

11. Use the drive shaft attachment (A) [SST:KV40107500 (-)] and a sliding hammer (B) (commercial service tool) while inserting tip of the drive shaft attachment between shaft and transaxle assembly, and then remove drive shaft from transaxle assembly.

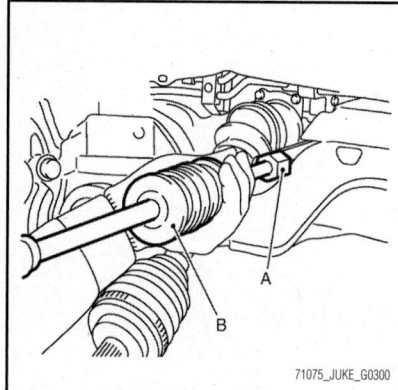

Fig. 30 Using drive shaft attachment (A) and sliding hammer (B) to remove transaxle assembly

❊❊ WARNING

Never place drive shaft joint at an extreme angle when removing drive shaft. Also be careful not to overextend slide joint.

➡ Confirm that the circular clip is attached to the drive shaft.

To install:

12. To install, reverse the removal procedure.

Transaxle side:

13. Always replace differential side oil seal with new one when installing drive shaft.

14. Place the protector onto transaxle assembly to prevent damage to the oil seal while inserting drive shaft. Slide drive shaft sliding joint and tap with a hammer to install securely.

➡ Check that circular clip is completely engaged.

15. Perform inspection after installation.

Wheel hub side:

16. Clean the matching surface of wheel hub lock nut and wheel hub assembly.

❊❊ WARNING

Never apply lubricating oil to these matching surfaces.

17. Clean the matching surface of drive shaft and wheel hub assembly. Then apply paste (0.04-0.10 oz.) to surface of joint sub assembly of drive shaft.

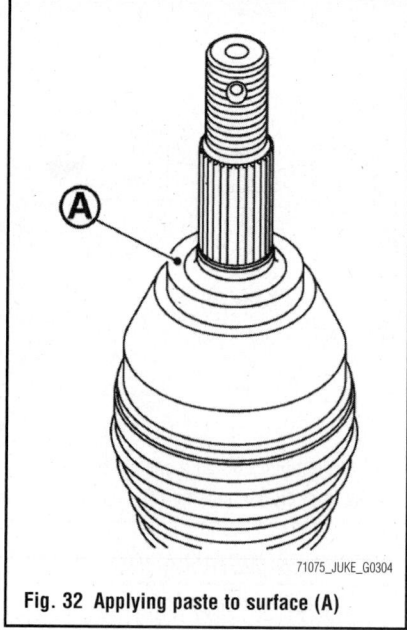

Fig. 32 Applying paste to surface (A)

18. Torque wheel hub lock nut to 133-136 ft. lbs. (180-185 Nm).

❊❊ WARNING

Since the drive shaft is assembled by press-fitting, use the tightening torque range for the wheel hub lock nut.

❊❊ WARNING

Be sure to use torque wrench to tighten the NEW wheel hub lock nut. Never use a power tool. Never reuse the hub lock nut.

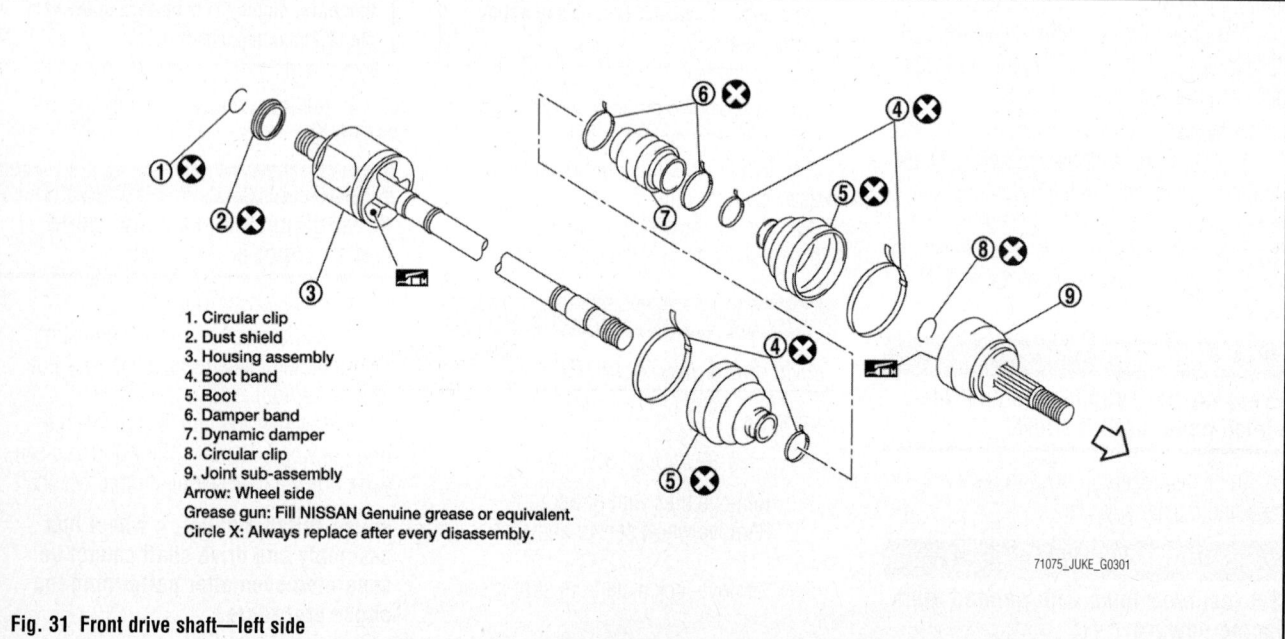

1. Circular clip
2. Dust shield
3. Housing assembly
4. Boot band
5. Boot
6. Damper band
7. Dynamic damper
8. Circular clip
9. Joint sub-assembly
Arrow: Wheel side
Grease gun: Fill NISSAN Genuine grease or equivalent.
Circle X: Always replace after every disassembly.

Fig. 31 Front drive shaft—left side

➡**Wheel hub lock nut tightening torque does not over torque for avoiding axle noise, and does not less than torque for avoiding looseness.**

19. Align the matching marks that have been made during removal when reusing the disc rotor.

20. When installing a cotter pin and adjusting cap, securely bend the basal portion to prevent rattles.

✳✳ WARNING
Never reuse cotter pin.

21. Perform the final tightening of each of parts under unladen conditions,
   which were removed when removing wheel hub assembly and steering knuckle.

22. Perform inspection after installation.

Right Side

See Figures 33 through 36.

1. Remove tires with power tool.
2. Remove wheel sensor and sensor harness if necessary.
3. Remove lock plate from strut assembly.
4. Remove caliper assembly. Hang caliper assembly not to interfere with work.

✳✳ WARNING
Never depress brake pedal while brake caliper is removed.

5. Remove disc rotor.
6. Remove cotter pin, and adjusting cap, and then loosen wheel hub lock nut with power tool.
7. Patch wheel hub lock nut with a piece of wood. Hammer the wood to disengage wheel hub assembly from drive shaft.

➡**Use suitable puller, if wheel hub assembly and drive shaft cannot be separated even after performing the above procedure.**

8. Remove wheel hub lock nut.
9. Remove strut assembly from steering knuckle.
10. Remove drive shaft from wheel hub assembly.

✳✳ WARNING
Never allow drive shaft to hang down without support for joint sub-assembly, shaft and the other parts.

11. Remove bearing housing plate bolts (2WD) or bearing housing bolts (AWD).

12. Remove drive shaft assembly from transaxle assembly.

✳✳ WARNING
Never place drive shaft joint at an extreme angle when removing drive shaft. Also be careful not to overextend slide joint.

13. Remove support bearing bracket. (2WD)

14. Perform inspection after removal.

To install:
15. Note the following, and install in the reverse order of the removal procedure:

Transaxle Side (2WD):
16. Always replace differential side oil seal with new one when installing drive shaft.

17. Install support bearing bracket (1) in following procedure.
   a. Temporarily tighten mounting bolts (A), (B), (C), then tighten them to 32 ft. lbs. (44 Nm).
   b. Set plate (2) so that notch (D) becomes upper side. Temporarily tighten mounting bolts (E), (F), then tighten them to 18 ft. lbs. (25 Nm).

✳✳ WARNING
Never reuse plate.

18. Place the protector (A) [SST:KV38107900 (-)] onto transaxle assembly to prevent damage to the oil seal while inserting drive shaft. Slide drive shaft sliding joint and tap with a hammer to install
securely.

19. To install mounting nuts of the heat insulator (1), temporarily tighten them in numerical order shown in the figure and tighten them to the 14 ft. lbs. (19 Nm).

20. Perform inspection after removal.

Transaxle side (AWD):
21. Always replace differential side oil seal with new one when installing drive shaft.

22. Place the protector (A) [SST:KV38107900 (-)] onto transaxle assembly to prevent damage to the oil seal while inserting drive shaft. Slide drive shaft sliding joint and tap with a hammer to install
securely.

23. Tighten the bearing housing bolt to 35 ft. lbs. (47 Nm).

24. Perform inspection after removal.

Wheel hub side:
25. Clean the matching surface of wheel hub lock nut and wheel hub assembly.

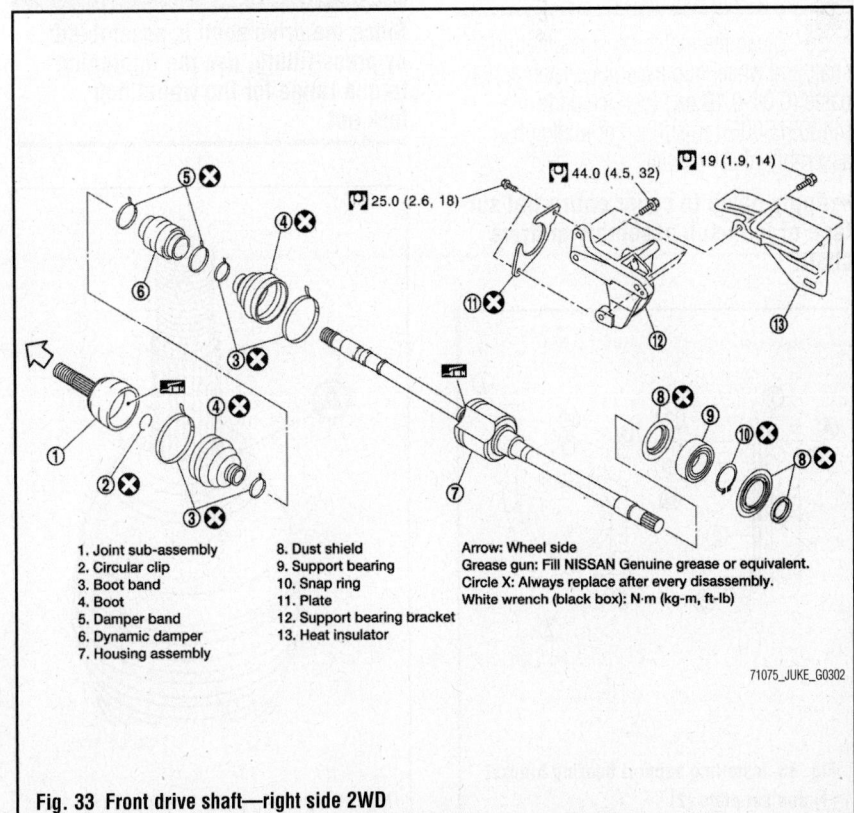

1. Joint sub-assembly
2. Circular clip
3. Boot band
4. Boot
5. Damper band
6. Dynamic damper
7. Housing assembly
8. Dust shield
9. Support bearing
10. Snap ring
11. Plate
12. Support bearing bracket
13. Heat insulator

Arrow: Wheel side
Grease gun: Fill NISSAN Genuine grease or equivalent.
Circle X: Always replace after every disassembly.
White wrench (black box): N·m (kg-m, ft-lb)

71075_JUKE_G0302

Fig. 33 Front drive shaft—right side 2WD

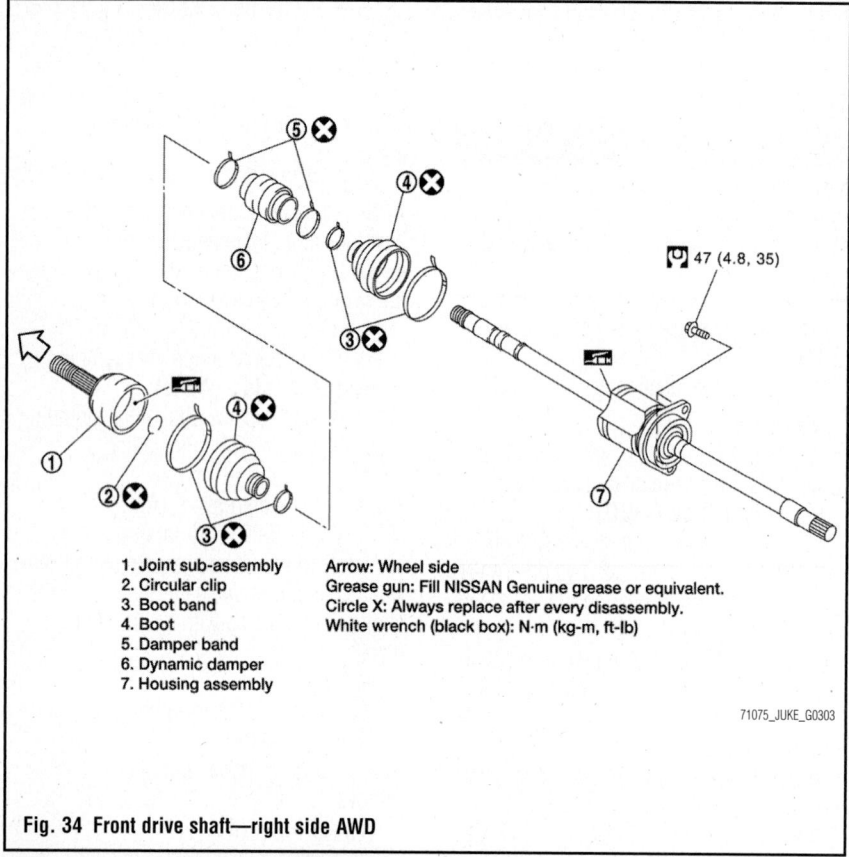

1. Joint sub-assembly
2. Circular clip
3. Boot band
4. Boot
5. Damper band
6. Dynamic damper
7. Housing assembly

Arrow: Wheel side
Grease gun: Fill NISSAN Genuine grease or equivalent.
Circle X: Always replace after every disassembly.
White wrench (black box): N·m (kg-m, ft-lb)

71075_JUKE_G0303

Fig. 34 Front drive shaft—right side AWD

✳✳ WARNING

Do not apply lubricating oil to these matching surfaces.

26. Clean the matching surface of drive shaft and wheel hub assembly. Then apply paste (0.04-0.10 oz.) [service parts (440037S000)] to surface of joint sub assembly of drive shaft.

➡**Apply paste to cover entire flat surface of joint sub assembly of drive shaft.**

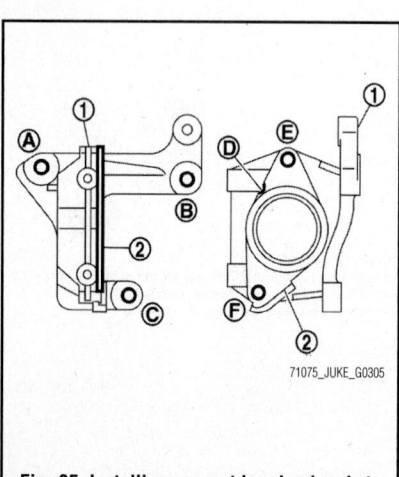

71075_JUKE_G0305

Fig. 35 Installing support bearing bracket (1) and set plate (2)

27. Torque wheel hub lock nut to 133-136 ft. lbs. (180-185 Nm).

✳✳ WARNING

Since the drive shaft is assembled by press-fitting, use the tightening torque range for the wheel hub lock nut.

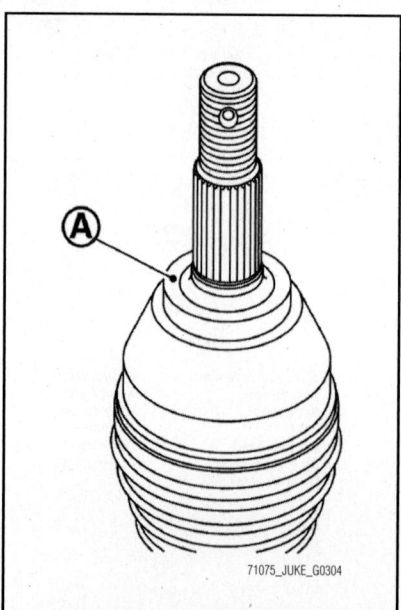

71075_JUKE_G0304

Fig. 36 Applying paste to surface (A)

✳✳ WARNING

Be sure to use torque wrench to tighten the wheel hub lock nut. Never use a power tool.

✳✳ WARNING

Never reuse wheel hub lock nut or cotter pins.

➡**Wheel hub lock nut tightening torque does not over torque for avoiding axle noise, and does not less than torque for avoiding looseness.**

28. Align the matching marks that have been made during removal when reusing the disc rotor.

29. When installing a cotter pin and adjusting cap, securely bend the basal portion to prevent rattles.

30. Perform the final tightening of each of parts under unladen conditions, which were removed when removing wheel hub assembly and steering knuckle.

31. Perform inspection after installation.

FRONT PINION OIL SEAL

REMOVAL & INSTALLATION

Front

See Figures 37 through 43.

Verify identification stamp of replacement frequency put in the lower part of gear carrier to determine replacement for collapsible spacer when replacing front oil seal. If collapsible spacer replacement is necessary, remove final drive assemble and disassemble it to replace front oil seal and collapsible spacer.

➡**The reuse of collapsible spacer is prohibited in principle. However, it is reusable on a one-time basis only in cases when replacing front oil seal.**

Identification stamp of replacement frequency of front oil seal

a. The diagonally shaded area in the figure shows stamping point for replacement frequency of front oil seal.

b. The following shows if collapsible spacer replacement is needed before replacing front oil seal.

When collapsible spacer replacement is required, disassemble final drive assembly to replace collapsible spacer and front oil seal.

• Stamp: Collapsible spacer replacement

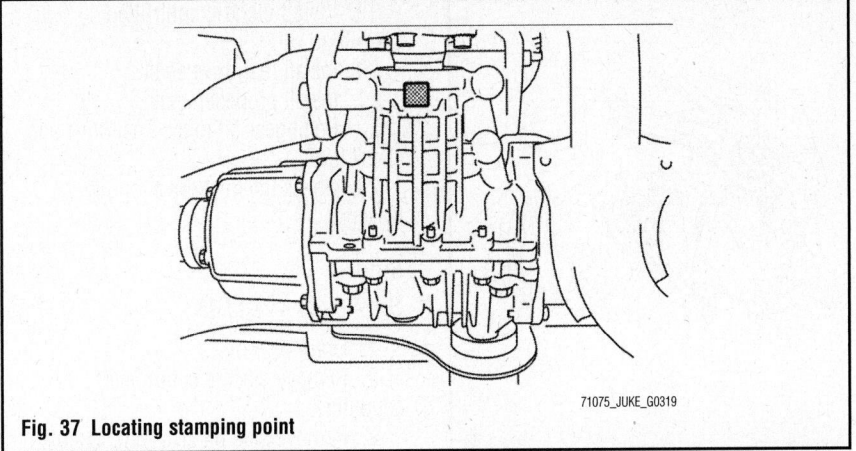

Fig. 37 Locating stamping point

- No stamp: Not required
- "0" or "0" on the far right of stamp: Required
- "01" or "1" on the far right of stamp: Not required

➡**After replacing front oil seal, make a stamping on the stamping point in accordance with the table below in order to identify replacement frequency. Make the stamping from left to right.**

1. Drain gear oil.
2. Make a judgment if a collapsible spacer replacement is required.
3. Remove propeller shaft.
4. Remove rear drive shaft.
5. Remove electric controlled couplings.
6. Measure the total preload with the preload gauge (A) [SST: ST3127S000 (J-25765-A)].

➡**Record the preload measurement.**

7. Put matching mark (A) on the end of the drive pinion. The matching mark should be in line with the matching mark (B) on companion flange (1).

❋❋ WARNING

For matching mark, use paint. Never damage companion flange and drive pinion.

8. Remove companion flange lock nut using the flange wrench (A) (Commercial service tool).
9. Remove companion flange using the puller (A) (Commercial service tool).
10. Remove front oil seal using the suitable oil seal remover. Discard the oil seal. Never reuse an oil seal.

❋❋ WARNING

Never damage gear carrier and drive pinion.

To install:

11. Apply multi-purpose grease to front oil seal lips.
12. Using the drift (Commercial service tool), drive front oil seal until it becomes flush with the gear carrier end.

❋❋ WARNING

Never incline oil seal when installing.

13. Align the matching mark of drive pinion with the matching mark of companion flange (1), and then install the companion flange.
14. Temporarily tighten drive pinion lock nut to drive pinion, using the flange wrench (Commercial service tool).

❋❋ WARNING

Never reuse drive pinion lock nut.

❋❋ WARNING

Apply anti-corrosion oil to the thread and seat of new drive pinion lock nut.

15. Tighten drive pinion lock nut within the limits of torque so as to keep the pinion

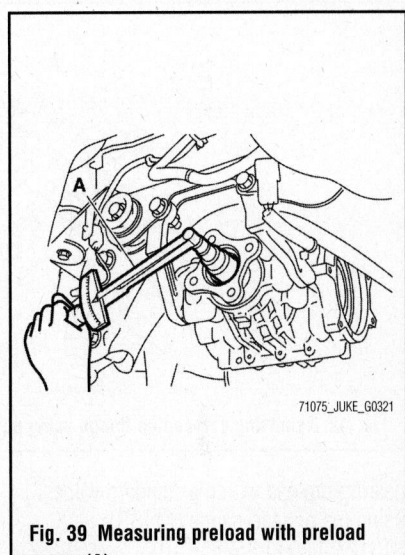

Fig. 39 Measuring preload with preload gauge (A)

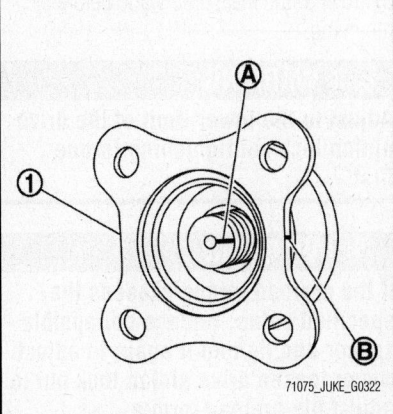

Fig. 40 Putting matching mark (A) on the end of the drive pinion and lining up with matching mark (B) on companion flange (1)

Stamp before stamping	Stamping on the far right	Stamping
No stamp	0	0
"0" (Front oil seal was replaced once.)	1	01
"01" (Collapsible spacer and front oil seal were replaced last time.)	0	010
"0" is on the far right. (Only front oil seal was replaced last time.)	1	...01
"1" is on the far right. (Collapsible spacer and front oil seal were replaced last time.)	0	...010

71075_JUKE_G0320

Fig. 38 Stamping

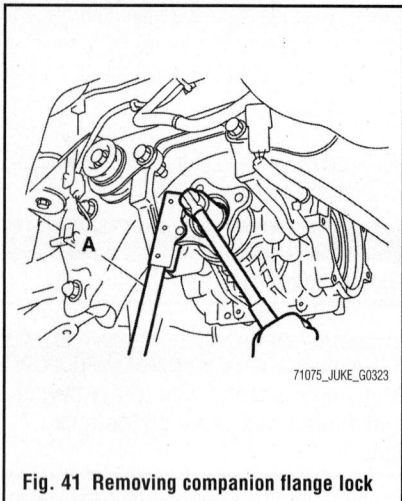

Fig. 41 Removing companion flange lock nut with flange wrench (A)

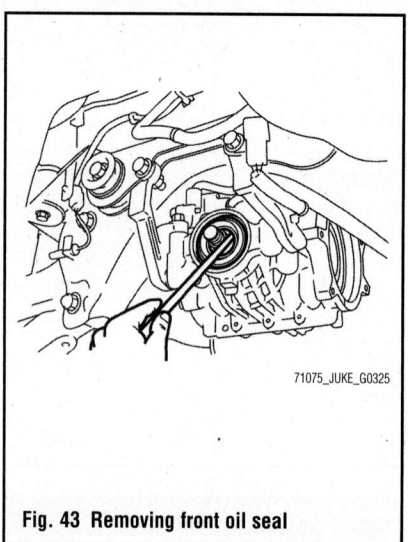

Fig. 43 Removing front oil seal

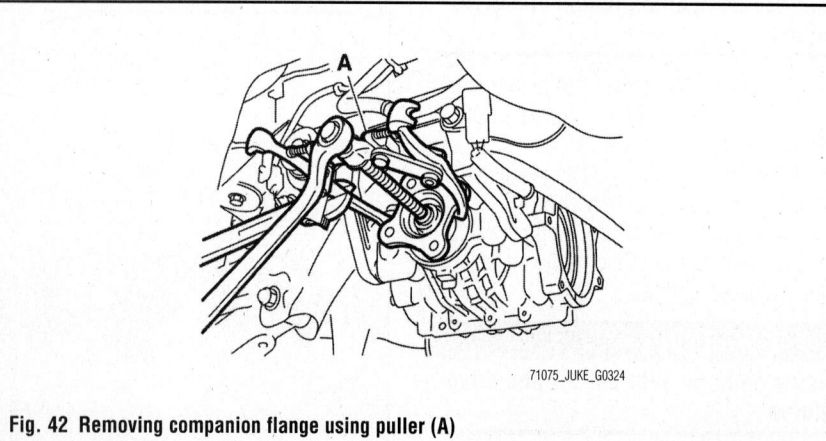

Fig. 42 Removing companion flange using puller (A)

bearing preload within a standard values, using the preload gauge (A) [SST: ST3127S000 (J-25765- A)].

Total preload torque:
A value that ass 1.0-3.0 inch lbs. (0.1-0.4 Nm) to the measured value before removing.

> ✱✱ **WARNING**
>
> **Adjust to the lower limit of the drive pinion lock nut tightening torque first.**

> ✱✱ **WARNING**
>
> **If the preload torque exceeds the specified value, replace collapsible spacer and tighten it again to adjust. Never loosen drive pinion lock nut to adjust the preload torque.**

16. Check for companion flange runout as follows:
  a. For companion flange face, fit a dial indicator onto the companion flange face (inner side of the propeller shaft mounting bolt holes). For inner side of the companion flange, fit a test indicator to the inner side of companion flange (socket diameter).
  b. Rotate companion flange to check for runout.
17. If the runout value is outside the runout limit 0.0020 inch (0.05 mm), follow the procedure below to adjust.
  a. Check for runout while changing the phase between companion flange and drive pinion by 90° step, and search for the position where the runout is the minimum.
  b. If the runout value is still outside of the limit after the phase has been changed, replace companion flange.
  c. If the runout value is still outside of the limit after companion flange has been replaced, possible cause will be an assembly malfunction of drive pinion.
18. Make a stamping for identification of front oil seal replacement frequency.

19. Install electric controlled couplings.
20. Install rear drive shaft.
21. Install propeller shaft.
22. Refill gear oil to the final drive and check level.
23. Check the final drive for oil leakage.

Side

See Figures 44 and 45.

1. Drain gear oil.
2. Remove electric controlled couplings.
3. Remove side oil seal with suitable oil seal remover. Discard the seal and use a new one for installation.

> ✱✱ **WARNING**
>
> **Never damage gear carrier and rear cover.**

> *To install:*

4. Install side oil seal (right) until it becomes flush with the gear carrier end, using the drift [SST: ST33400001 (J-26082)].

> ✱✱ **WARNING**
>
> **When installing, never incline oil seal.**

> ✱✱ **WARNING**
>
> **Apply multi-purpose grease onto oil seal lip, and gear oil onto the circumference of oil seal.**

5. Install a NEW side oil seal (left) until it becomes flush with the gear carrier end, using the drift [SST: KV38100500 (—)].

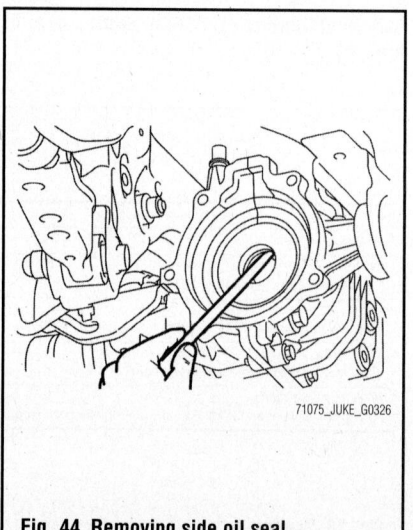

Fig. 44 Removing side oil seal

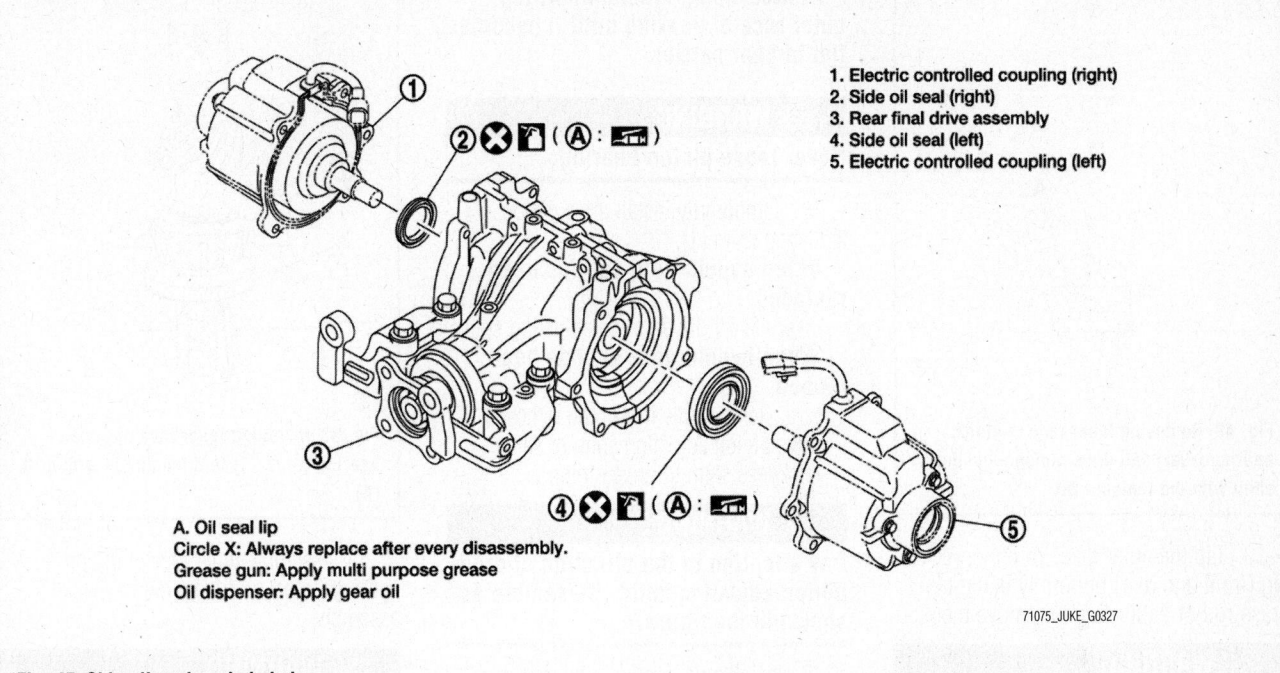

1. Electric controlled coupling (right)
2. Side oil seal (right)
3. Rear final drive assembly
4. Side oil seal (left)
5. Electric controlled coupling (left)

A. Oil seal lip
Circle X: Always replace after every disassembly.
Grease gun: Apply multi purpose grease
Oil dispenser: Apply gear oil

71075_JUKE_G0327

Fig. 45 Side oil seal exploded view

※※ WARNING

When installing, never incline oil seal.

➡ **Apply multi-purpose grease onto oil seal lip, and gear oil onto the circumference of oil seal.**

6. Install electric controlled couplings.
7. Refill gear oil to the final drive and check level.
8. When oil leaks while removing, check oil level after the installation.

RING & PINION

REMOVAL & INSTALLATION

See Figures 46 through 54.

1. Remove electric controlled couplings.
2. Remove center stem assembly.
3. Remove drive pinion lock nut with flange wrench (A) (commercial service tool).
4. Using paint, put matching mark (A) on the end of drive pinion. The matching mark should be in line with the matching mark (B) on companion flange (1).

➡ **The matching mark on the final drive companion flange indicates the maximum vertical runout position.**

➡ **When replacing companion flange, matching mark is not necessary.**

5. Remove companion flange using the puller (Commercial service tool).
6. Press drive pinion assembly out of gear carrier.

※※ WARNING

Never drop drive pinion assembly.

7. Remove front oil seal.
8. Remove inner race of pinion bearing (front).
9. Remove collapsible spacer.
10. Remove inner race of pinion bearing (rear) and drive pinion adjusting shim with the replacer (A) (Commercial service tool).
11. Remove drive pinion adjusting shim.

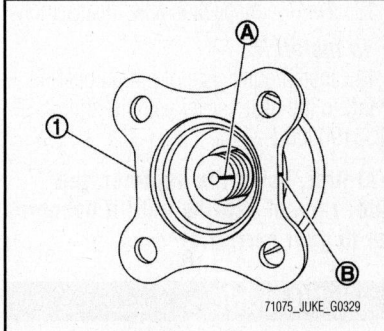

71075_JUKE_G0329

Fig. 47 Putting matching mark (A) on the end of drive pinion and aligning with matching mark (B) on companion flange (1)

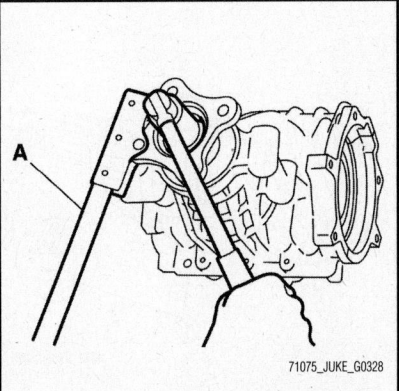

71075_JUKE_G0328

Fig. 46 Removing drive pinion lock nut with flange wrench (A)

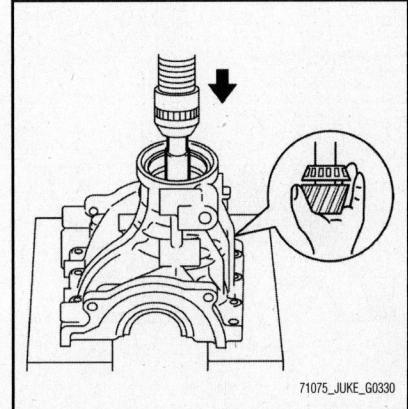

71075_JUKE_G0330

Fig. 48 Pressing drive pinion assembly out of gear carrier

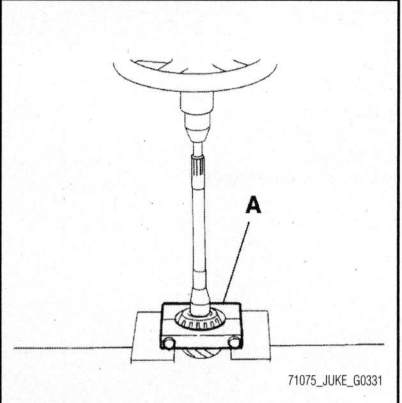

Fig. 49 Removing inner race of pinion bearing (rear) and drive pinion adjusting shim with the replacer (A)

12. Tap the outer races of pinion bearing (front and rear) uniformly using the brass rod or equivalent to remove them.

✳✳ WARNING

Never damage gear carrier.

13. Perform inspection after disassembly.
To install:
14. Install outer race of pinion bearing (front) to the gear carrier with the drift [SST: ST33190000 (—)].

➡**At first, using the hammer, tap outer race of bearing until it becomes flat to gear carrier.**

✳✳ WARNING

Never reuse pinion bearing.

15. Install outer race of pinion bearing (rear) to the gear carrier with the drift [SST: ST37830000 (—)].

➡**At first, using the hammer, tap outer race of bearing until it becomes flat to gear carrier.**

✳✳ WARNING

Never reuse pinion bearing.

16. Temporarily install drive pinion adjusting shim (1).
   When hypoid gear set has been replaced
   a. Select drive pinion adjusting shim.
   When hypoid gear set has been reused
   a. Temporarily install the removed drive pinion adjusting shim or same thickness shim to drive pinion.

✳✳ WARNING

Pay attention to the direction of drive pinion adjusting shim. (Assemble as shown in the figure.)

17. Install inner race of pinion bearing (rear) (1) to drive pinion with the drift (A) [SST: ST33032000 (—)].

➡**Never reuse pinion bearing.**

18. Assemble collapsible spacer (1) to drive pinion (2).

✳✳ WARNING

Be careful of the mounting direction of collapsible spacer.

➡**Never reuse collapsible spacer.**

19. Assemble drive pinion into gear carrier.

✳✳ WARNING

Apply gear oil to pinion bearing.

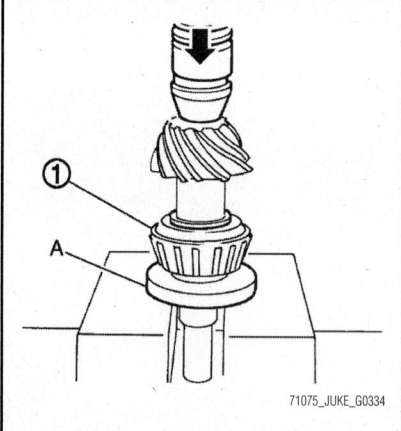

Fig. 52 Installing inner race of pinion bearing (rear) (1) to drive pinion with drift (A)

20. Assemble inner race of pinion bearing (front) to drive pinion assembly.

✳✳ WARNING

Never reuse pinion bearing.

➡**Apply gear oil to pinion bearing.**

21. Using the drift (A)[SST: KV37710000 (—)], press the inner race of pinion bearing (front) to drive pinion as far as drive pinion nut can be tightened.

22. Using the drift (Commercial service tool), drive front oil seal until it becomes flush with the gear carrier end.

➡**Never reuse oil seal.**

✳✳ WARNING

When installing, never incline oil seal.

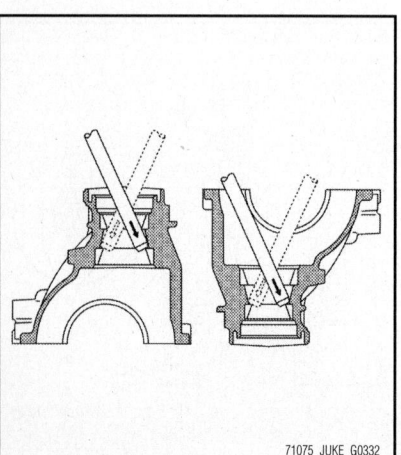

Fig. 50 Tapping outer races of pinion bearing (front and rear)

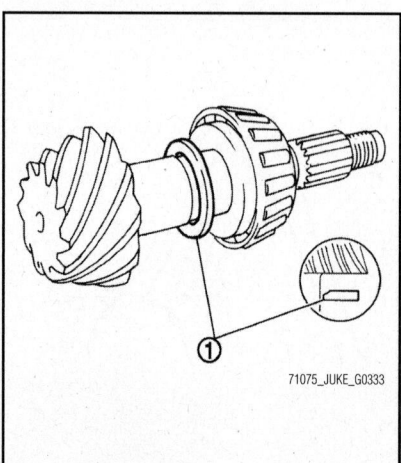

Fig. 51 Temporarily installing drive pinion adjusting shim (1)

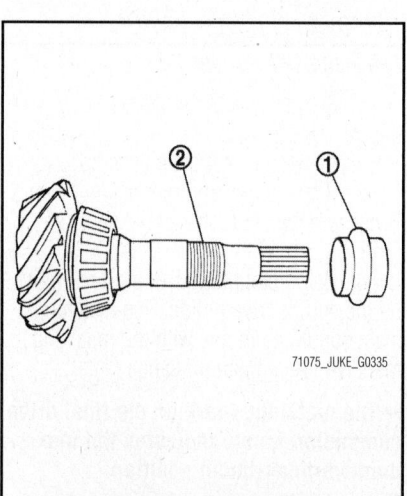

Fig. 53 Assembling collapsible spacer (1) to drive pinion (2)

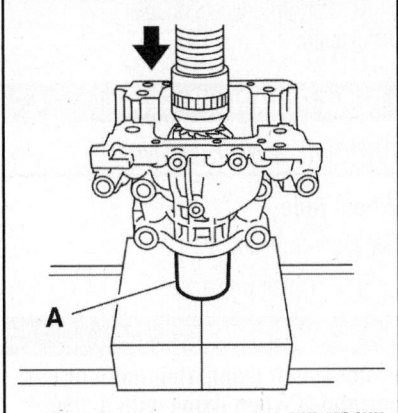

Fig. 54 Using drift (A) to press inner race of pinion bearing (front) to drive pinion as far as drive pinion nut can be tightened

✳✳ WARNING

Apply multi-purpose grease onto oil seal lips, and gear oil onto the circumference of oil seal.

23. Install companion flange.

➡**When reusing drive pinion, align the matching mark of drive pinion with the matching mark of companion flange, and then install companion flange.**

24. Apply anti-corrosion oil to the thread and seat of a NEW drive pinion lock nut, and temporarily tighten drive pinion lock nut to drive pinion, using the flange wrench (Commercial service tool).

25. Adjust to the drive pinion lock nut tightening torque and pinion bearing preload torque, using the preload gauge (A) [SST: ST3127S000 (J-25765-A)].

✳✳ WARNING

Adjust to the lower limit of the drive pinion lock nut tightening torque first. Torque is 99-216 ft. lbs. (134-203 Nm).

✳✳ WARNING

If the preload torque exceeds the specified value, replace collapsible spacer and tighten it again to adjust. Never loosen drive pinion lock nut to adjust the preload torque.

✳✳ WARNING

After adjustment, rotate drive pinion back and forth 2 to 3 times to check for unusual noise, rotation malfunction, and other malfunctions.

26. Install center stem assembly.
27. Check and adjust drive gear runout, tooth contact, drive gear to drive pinion backlash, and companion flange runout.
28. Check total preload torque.
29. Install rear cover.
30. Install electric controlled couplings.

REAR DRIVE AXLE

FLUID RECOMMENDATIONS

Genuine NISSAN Differential oil Hypoid Super GL-5 80W-90 or API GL-5, Viscosity SAE 80W-90.

➡**For hot climates, viscosity SAE 90 is suitable for ambient temperatures above 32°F (0°C).**

LEVEL CHECK

See Figure 55.

1. Remove filler plug (1) and check oil level from filler plug mounting hole.
2. Set a new gasket on filler plug and install it on final drive assembly.

✳✳ WARNING

Do not reuse gasket.

DRAIN & REFILL

See Figures 56 and 57.

1. Stop engine.
2. Remove drain plug (1) and drain gear oil.
3. Set a new gasket on drain plug and install it to final drive assembly and tighten to 26 ft. lbs. (35 Nm).

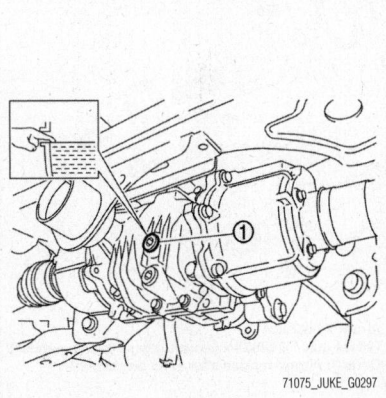

Fig. 55 Removing filler plug and checking oil level

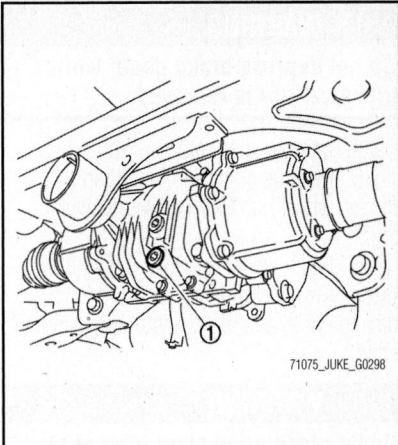

Fig. 56 Removing drain plug (1) and draining gear oil

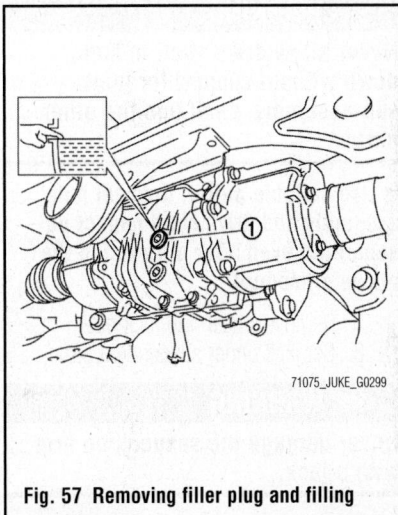

Fig. 57 Removing filler plug and filling with gear oil

✳✳ WARNING

Do not reuse gasket.

4. Remove filler plug (1). Fill with new gear oil until oil level reaches the specified level near filler plug mounting hole.
5. After installing oil, check oil level. Set a new gasket to filler plug, then install it to final drive assembly.

REAR DRIVE SHAFT, BEARING & SEAL

REMOVAL & INSTALLATION

See Figure 58.

1. Remove tires with power tool.
2. Remove wheel sensor.
3. Remove caliper assembly. Hang caliper assembly in a place where it will not interfere with work.

※※ WARNING

Do not depress brake pedal while brake caliper is removed.

4. Remove disc rotor.

5. Remove cotter pin, and then loosen wheel hub lock nut with power tool.

6. Patch wheel hub lock nut with a piece of wood. Hammer the wood to disengage wheel hub assembly from drive shaft.

※※ WARNING

Never place drive shaft joint at an extreme angle. Also be careful not to overextend slide joint.

※※ WARNING

Never allow drive shaft to hang down without support for joint sub-assembly, shaft and the other parts.

→Use suitable puller, if wheel hub assembly and drive shaft cannot be separated even after performing the above procedure.

7. Remove wheel hub lock nut.
8. Set jack under suspension arm.

※※ WARNING

Never damage the suspension arm with a jack.

※※ WARNING

Check the stable condition when using a jack.

9. Remove stabilizer link.
10. Remove shock absorber from suspension arm.
11. Remove upper link from suspension arm.
12. Remove lower link from suspension arm.
13. Remove drive shaft from final drive assembly.

→**Confirm that the circular clip is attached to the drive shaft.**

14. Perform inspection after removal.

To install:

15. Note the following, and install in the reverse order of the removal procedure.

16. Place the protector [SST:KV38107900 (—)] onto final drive assembly to prevent damage to the oil seal while inserting drive shaft. Slide drive shaft sliding joint and tap with a hammer to install securely.

※※ WARNING

Check that circular clip is completely engaged.

17. Perform final tightening of bolts and nuts at suspension arm (rubber bushing), under unladen conditions with tires on level ground.

18. Perform inspection after installation.

REAR PINION OIL SEAL

REMOVAL & INSTALLATION

Wheel Side
See Figures 59 and 60.

1. Fix shaft with a vise.

※※ WARNING

Protect shaft using aluminum or copper plates when fixing with a vise.

2. If sensor rotor needs to be removed, use a bearing replacer (A) and puller (B).

3. Remove boot bands. Then remove boot from joint sub-assembly.

4. Screw drive shaft puller (A) (commercial service tool) into joint sub-assembly screw part to a length of 30 mm (1.18 in) or more. Support drive shaft with one hand and pull out joint sub-assembly from shaft.

→**Align drive shaft puller and drive shaft and remove them by pulling firmly and uniformly.**

※※ WARNING

If joint sub-assembly cannot be removed after five or more unsuccessful attempts, replace shaft and joint sub assembly as a set.

5. Remove circular clip from shaft.
6. Remove boot from shaft.
7. Perform inspection after disassembly.

To install:

8. If sensor rotor is removed, use a drift [SST:KV38100500 (—)] and color (B)

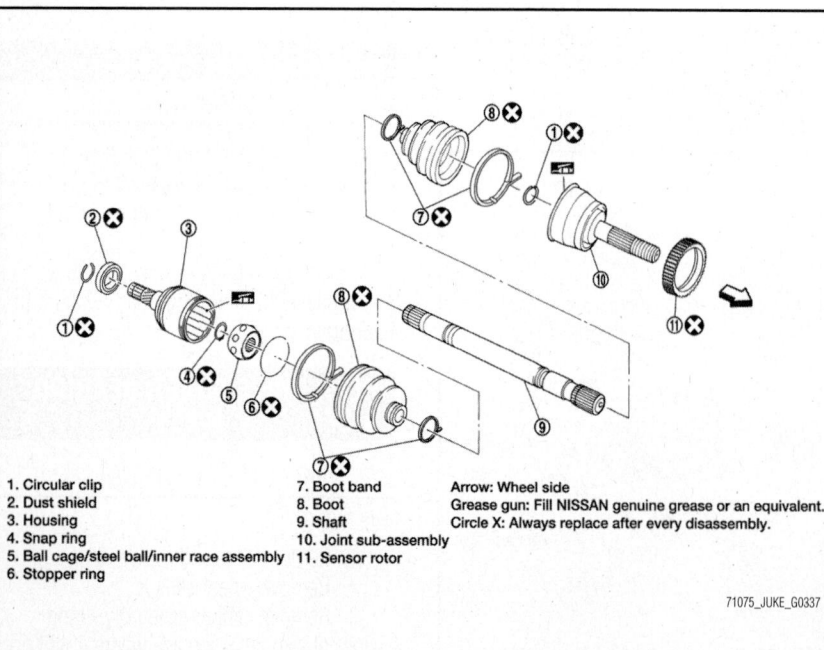

1. Circular clip
2. Dust shield
3. Housing
4. Snap ring
5. Ball cage/steel ball/inner race assembly
6. Stopper ring
7. Boot band
8. Boot
9. Shaft
10. Joint sub-assembly
11. Sensor rotor

Arrow: Wheel side
Grease gun: Fill NISSAN genuine grease or an equivalent.
Circle X: Always replace after every disassembly.

71075_JUKE_G0337

Fig. 58 Removing and installing rear drive shaft

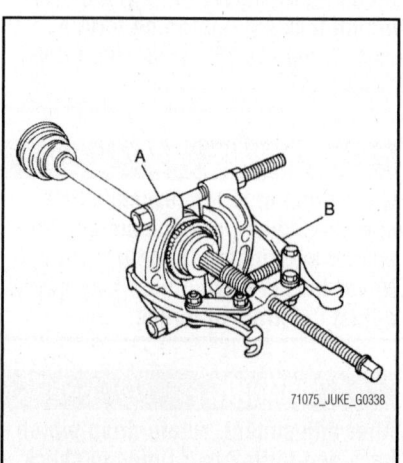

71075_JUKE_G0338

Fig. 59 Removing sensor rotor with bearing replacer (A) and puller (B)

[SST:KV40101840 (—)] to press in a new one.

✳✳ WARNING

Never reuse sensor rotor.

9. Clean the old grease on joint sub-assembly with paper waste.

10. Fill serration slot joint sub-assembly with NISSAN genuine grease or equivalent until the serration slot and ball groove become full to the brim.

➡ **After applying grease, use a shop cloth to wipe off old grease that has oozed out.**

11. Install boot and boot bands to shaft.

✳✳ WARNING

Wrap serration on shaft with tape to protect the boot from damage.

12. Never reuse boot and boot band.

13. Remove the tape wrapped around the serrated on shaft.

14. Position the circular clip on groove at the shaft edge.

✳✳ WARNING

Never reuse circular clip.

➡ **Drive joint inserter is recommended when installing circular clip.**

15. Align both center axles of the shaft edge and joint sub-assembly. Then assemble shaft with joint sub-assembly holding circular clip.

16. Install joint sub-assembly to shaft using plastic hammer.

➡ **Check circular clip is properly positioned on groove of the joint sub-assembly.**

✳✳ WARNING

Confirm that joint sub-assembly is correctly engaged while rotating drive shaft.

17. Apply the specified amount of grease into the boot inside from large diameter side of boot.

18. Install the boot securely into grooves (indicated by "*" marks) shown in the figure.

✳✳ WARNING

If grease adheres to the boot mounting surface (indicated by "*" mark) on the shaft or joint sub-assembly,

boot may be removed. Remove all grease from the boot mounting surface.

19. To prevent the deformation of the boot, adjust the boot installation length (L) to the specified value shown below by inserting the suitable tool into inside of the boot from the large diameter side of the boot and discharging the inside air.

✳✳ WARNING

If the boot installation length exceeds the standard, it may cause breakage of the boot.

✳✳ WARNING

Be careful not to touch the inside of the boot with a tip of tool.

➡ **Install boot bands securely.**

20. Secure housing and shaft, and then make sure that they are in the correct position when rotating boot. Reinstall them using boot bands when boot installation positions become incorrect.

✳✳ WARNING

Do not reuse boot band.

Final Drive Side

1. Fix shaft with a vise.

✳✳ WARNING

Protect shaft using aluminum or copper plates when fixing with a vise.

2. Remove circular clip from housing.

3. Remove dust shield from housing.

4. Remove boot bands, and then remove boot from housing.

5. Put matching marks on housing and shaft, using paint or equivalent. Never scratch the surfaces when making match-marks.

6. Remove stopper ring with suitable tool, and pull out housing.

7. Put matching marks on ball cage/steel ball/inner race assembly and shaft, using paint or equivalent. Do not scratch the surface.

8. Remove snap ring, then remove ball cage/steel ball/inner race assembly from shaft.

9. Remove boot from shaft.

10. Perform inspection after disassembly.

To install:

11. Clean the oil grease on housing with paper waste.

12. Install boot and boot bands to shaft.

➡ **Wrap serration on shaft with tape to protect boot from damage.**

✳✳ WARNING

Never reuse boot and boot bands.

13. Remove the tape wrapped around the serration on shaft.

14. Install ball cage/steel ball/inner race assembly to shaft, and secure them tightly with a snap ring.

✳✳ WARNING

Never reuse snap ring.

15. Apply the appropriate amount of grease onto housing and slide surface.

16. Install housing.

17. Install stopper ring to housing.

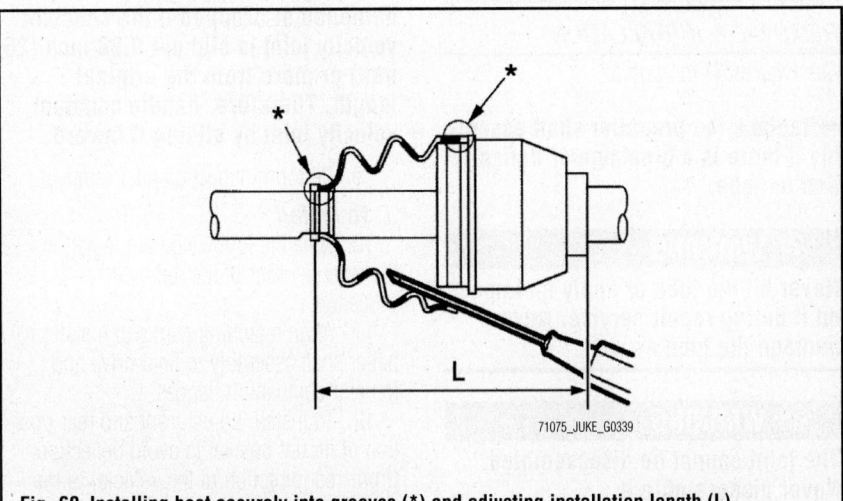

71075_JUKE_G0339

Fig. 60 Installing boot securely into grooves (*) and adjusting installation length (L)

※ **WARNING**

Never reuse stopper ring.

18. After installed, pull shaft to check engagement between housing and stopper ring.

19. Install boot securely into grooves.

※ **WARNING**

If grease adheres to the boot mounting surface on shaft or housing, boot may be removed. Remove all grease from the surface.

20. To prevent the deformation of the boot, adjust the boot installation length by inserting the suitable tool into the inside of the boot from the large diameter side of boot and discharging inside air.

※ **WARNING**

If the boot installation length exceeds the standard, it may cause brakeage of the boot.

➡Be careful not to touch the inside of the boot with the tip of tool

21. Install boot bands securely.
22. Never reuse boot band.
23. Secure housing and shaft, and then make sure that they are in the correct position when rotating boot. Reinstall them using new boot bands when boot installation positions become incorrect.
24. Install new dust shield to housing.
25. Install circular clip to housing.

※ **WARNING**

Never reuse circular clip.

PROPELLER SHAFT

REMOVAL & INSTALLATION

See Figures 61 through 67.

➡Replace the propeller shaft assembly if there is a breakage or deflection on tube.

※ **WARNING**

Never hit the tube or apply an impact on it during repair service. Never damage the tube as well.

※ **WARNING**

The joint cannot be disassembled. Never disassemble it.

※ **WARNING**

If constant velocity joint was bent during propeller shaft assembly removal, installation, or transportation, its boot may be damaged. Wrap boot interference area to metal part with shop cloth or rubber to protect boot from breakage.

1. Shift the transaxle to the neutral position, and then release the parking brake.

2. Put matching marks on propeller shaft flange yoke and final drive companion flanges using paint.

3. Put matching marks on propeller shaft flange yoke and transfer companion flanges using paint.

4. Loosen mounting nuts of center bearing mounting bracket.

➡Tighten mounting nuts temporarily.

5. Remove propeller shaft assembly fixing bolts and nuts.

6. Remove center bearing mounting bracket fixing nuts.

7. Remove propeller shaft assembly.

※ **CAUTION**

This procedure requires 2 workers. Constant velocity joint must be handled with care.

8. If constant velocity joint was bent during propeller shaft assembly removal, installation, or transportation, its boot (1) may be damaged. Wrap boot interference area to metal part (2) with shop cloth or rubber to protect boot from breakage.

➡Since no retaining pin is included in sliding direction, the boot may be damaged or dropped if the constant velocity joint is slid out 0.98 inch (25 mm) or more from the original length. Therefore, handle constant velocity joint by sliding it inward.

9. Perform inspection after removal.

To install:
10. Note the following, and install in the reverse order of the removal procedure.

11. Align matching marks to install propeller shaft assembly to final drive and transfer companion flanges.

12. To install, adjust front and rear position of mount bracket to avoid deflection (front-rear direction of the vehicle) to the center bearing insulator.

13. Perform inspection after installation.

14. After tightening the bolts and nuts to the specification torque, check that the bolts (3) on the flange side is tightened as shown.

15. If propeller shaft assembly or final drive assembly has been replaced, connect them as follows:

※ **WARNING**

Constant velocity joint of a new propeller shaft has a preinstalled protector (1). Protector must be removed after installing propeller shaft.

16. Install propeller shaft (1) while aligning its matching mark (A) of propeller shaft with the matching mark (B) of final drive (2) on the joint as close as possible.

17. Temporary tighten bolts and nuts.

18. Press down propeller shaft (1) with matching mark (C) of final drive (2) facing upward. Then tighten fixing bolts and nuts to the specified torque.

19. After assembly, perform a driving test to check propeller shaft vibration. If

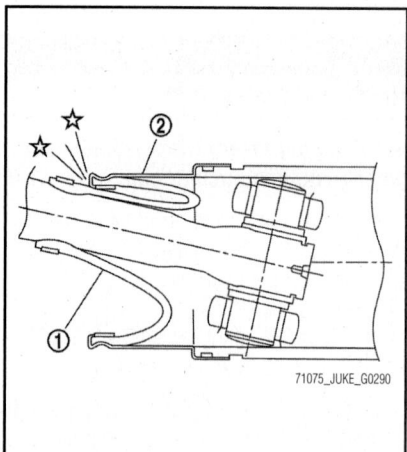

71075_JUKE_G0290

Fig. 61 Protect boot (1) interference area to metal part (2) with shop towel

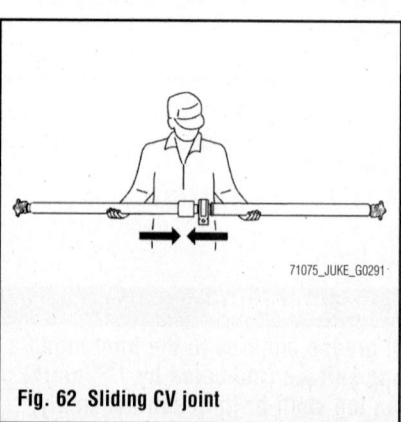

71075_JUKE_G0291

Fig. 62 Sliding CV joint

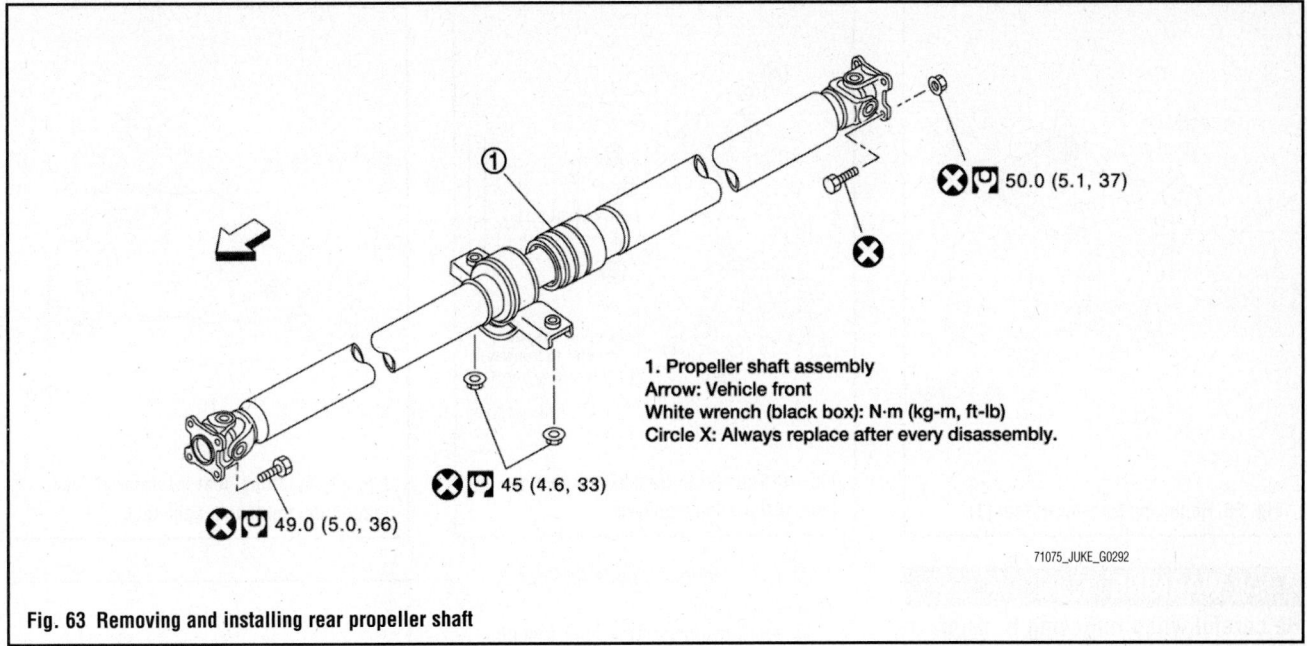

1. Propeller shaft assembly
Arrow: Vehicle front
White wrench (black box): N·m (kg-m, ft-lb)
Circle X: Always replace after every disassembly.

❌ 🔧 50.0 (5.1, 37)

❌ 🔧 45 (4.6, 33)

❌ 🔧 49.0 (5.0, 36)

71075_JUKE_G0292

Fig. 63 Removing and installing rear propeller shaft

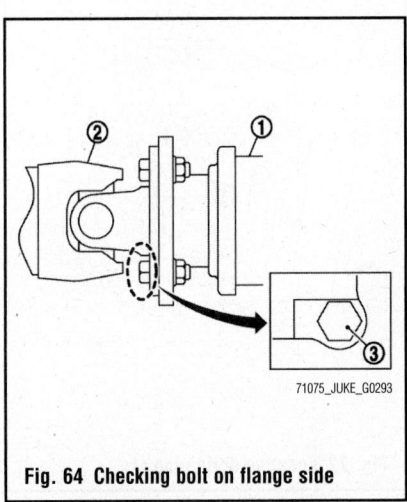

71075_JUKE_G0293

Fig. 64 Checking bolt on flange side

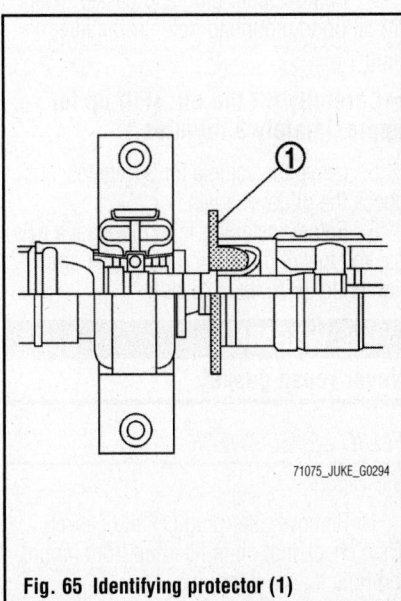

71075_JUKE_G0294

Fig. 65 Identifying protector (1)

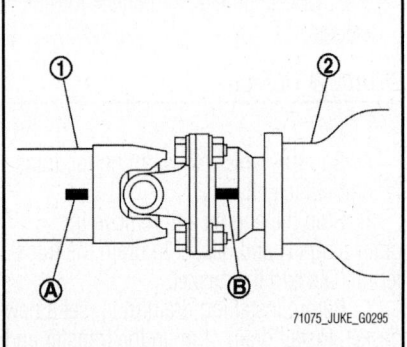

71075_JUKE_G0295

Fig. 66 Installing propeller shaft (1) while aligning its matching mark (A) with matching mark (B) of final drive (2)

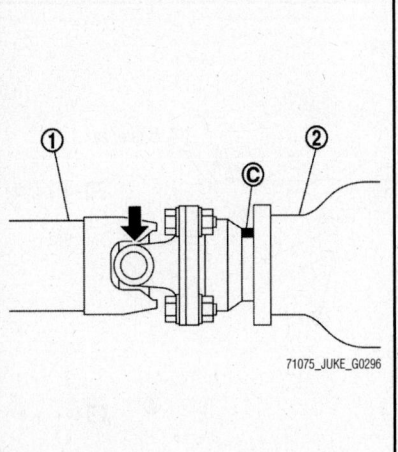

71075_JUKE_G0296

Fig. 67 Pressing propeller shaft (1) with matching mark (C) of final drive (2) facing upward

vibration occurred, separate propeller shaft from final drive or transfer. Reinstall companion flange by changing the phase between companion flange and propeller shaft by the one bolt hole at a time. Then perform driving test and check propeller shaft vibration again at each point.

TRANSFER CASE

REMOVAL & INSTALLATION

See Figures 68 through 71.

1. Separate the propeller shaft from transfer. Then suspend it by wire, etc.

❄ WARNING

Constant velocity joint must be handled with care.

2. Remove right side drive shaft.
3. Remove catalyst convertor support bracket (RH).
4. Remove heat insulator.
5. Remove transfer gusset (1).
6. Remove catalyst convertor support bracket rear.
7. Remove rear torque rod and rear torque rod bracket.
8. Remove bolts fixing transaxle assembly and transfer assembly.
9. Remove transfer assembly from the vehicle.

❄ WARNING

Never damage ring gear shaft.

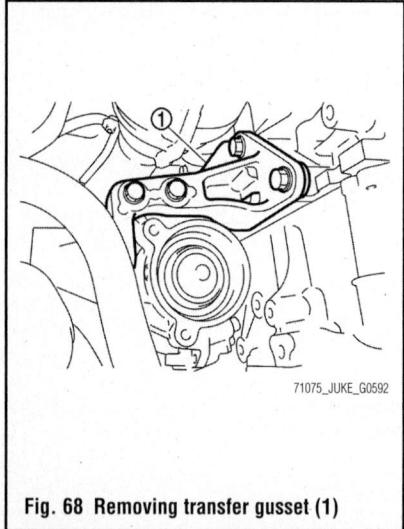

Fig. 68 Removing transfer gusset (1)

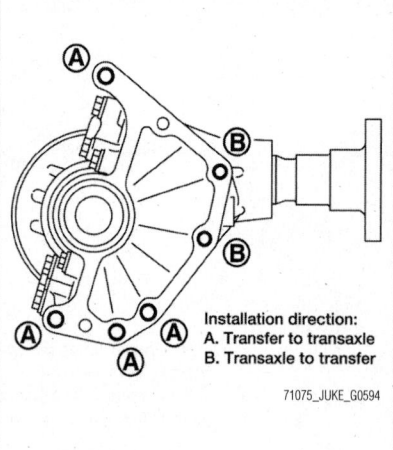

Installation direction:
A. Transfer to transaxle
B. Transaxle to transfer

Fig. 70 Transfer to transaxle mounting
bolts installation standard

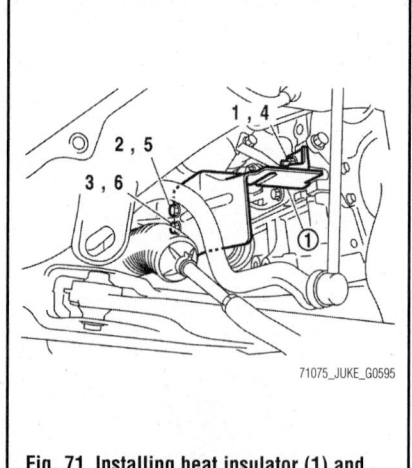

Fig. 71 Installing heat insulator (1) and
identifying tightening sequence

※※ **WARNING**

**Be careful when removing transfer
assembly form the vehicle because it
is heavy.**

To install:

10. Note the following, and install
in the reverse order of the removal procedure:

a. When installing the transfer assembly to the transaxle assembly, install the
mounting bolts following the standard
below.

b. When installing transfer assembly
to transaxle assembly, replace differential
side oil seal (Converter housing side) of
transaxle.

c. Never damage differential side oil
seal (Converter housing side) of
transaxle.

d. When installing heat insulator (1),

tighten the mounting bolts and nut in
numerical order.

e. Perform inspection after installation.

f. Check oil level and check for oil
leakage.

DRAIN & REFILL

See Figure 72.

1. Run the vehicle to warm up the transfer unit sufficiently.

2. Stop the engine and remove the
drain plug (1) and gasket to drain the transfer oil. Discard the gasket.

3. Before installing drain plug, set a new
gasket. Install drain plug on the transfer and
tighten to 26 ft. lbs. (35 Nm).

※※ **WARNING**

Never reuse gasket.

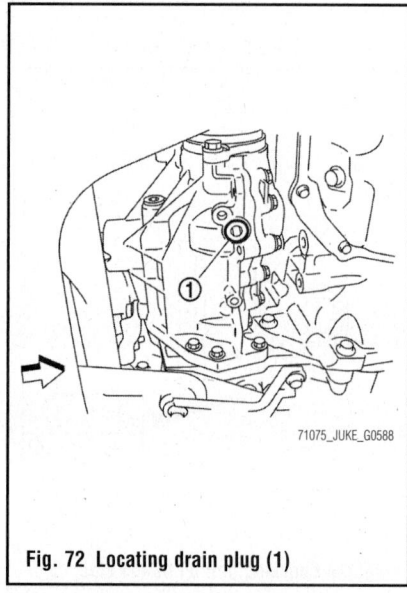

Fig. 72 Locating drain plug (1)

4. Remove filler plug and gasket. Then
fill oil up to mounting hole for the filler
plug.

➡**Carefully fill the oil. (Fill up for
approximately 3 minutes.)**

5. Leave the vehicle for 3 minutes, and
check the oil level again.

6. Before installing filler plug, set a new
gasket. Install filler plug on transfer and
tighten to 26 ft. lbs. (35 Nm)

※※ **WARNING**

Never reuse gasket.

FLUID LEVEL CHECK

See Figure 73.

1. Remove filler plug (1) and gasket.
Then check that oil is filled up from mounting hole for the filler plug.

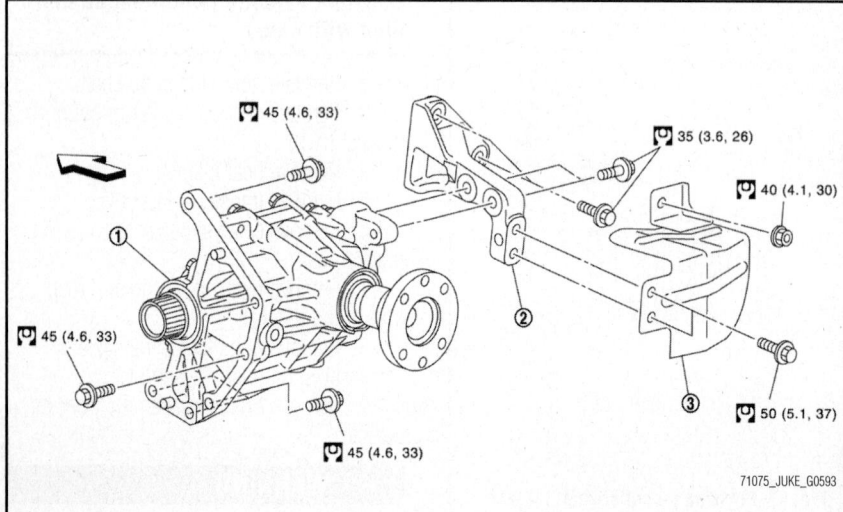

Fig. 69 Removing and installing transfer case

⁘ WARNING

Do not start engine while checking oil level.

2. Before installing filler plug, set a new gasket. Install filler plug on transfer and tighten to 26 ft. lbs. (35 Nm).

⁘ WARNING

Never reuse gasket.

FLUID RECOMMENDATIONS

Genuine NISSAN Differential oil Hypoid Super GL-5 80W-90 or API GL-5, Viscosity SAE 80W-90.

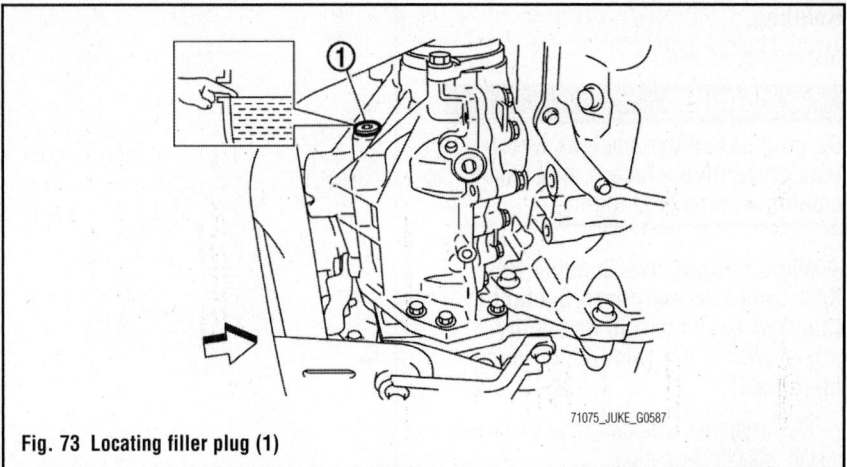

Fig. 73 Locating filler plug (1)

ENGINE COOLING

ENGINE COOLANT

BLEEDING

See Figure 74.
1. Install radiator drain plug.

⁘ WARNING

Be sure to clean drain plug and install with new O-ring.

   a. If water drain plugs on cylinder block are removed, close and tighten them.
2. Remove air duct (suction side), air cleaner cover assembly and air cleaner body assembly.
3. Disconnect vacuum hose break booster side, and remove vacuum tube from clamp.
4. Disconnect heater hose (1) at position (arrow) in the figure.
   a. Enhance heater as high as possible.
5. Fill radiator and reservoir tank with water and reinstall radiator cap.
   a. When engine coolant over flows disconnected heater hose, connect heater hose, and continue filling the engine coolant.
6. Connect vacuum hose, and install vacuum tube.
7. Install air duct (suction side), air cleaner cover assembly and air cleaner body assembly.
8. Run the engine and warm it up to normal operating temperature.
9. Rev the engine two or three times under no-load.
10. Stop the engine and wait until it cools down.
11. Drain water from the system.

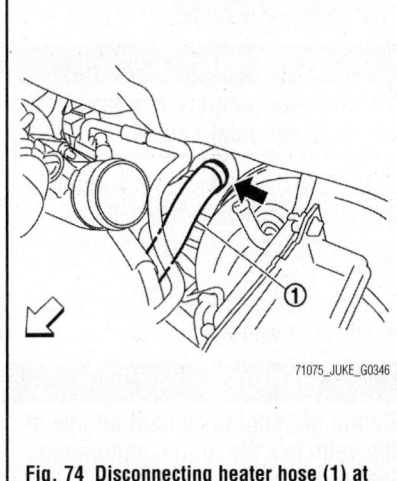

Fig. 74 Disconnecting heater hose (1) at position (arrow)

12. Repeat steps 1 through 9 until clear water begins to drain from radiator.

DRAIN & REFILL

Draining
See Figure 75.

⁘ CAUTION

Never remove radiator cap when engine is hot. Serious burns may occur from high-pressure engine coolant escaping from radiator.

⁘ WARNING

Wrap a thick cloth around the radiator cap. Slowly turn it a quarter of a turn to release built-up pressure. Then turn it all the way.

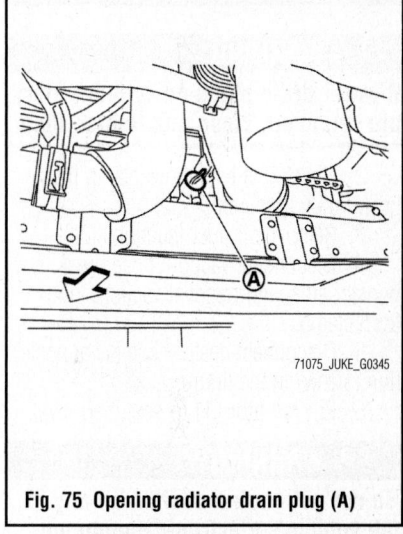

Fig. 75 Opening radiator drain plug (A)

1. Connect drain hose.
   a. Use a genera-purpose hose (0.31 inch wide x 11.81 inch long).
2. Open radiator drain plug (A) at the bottom of radiator, and then remove radiator cap.

⁘ WARNING

Perform this step when engine is cold.

3. When draining all of engine coolant in the system, open water drain plugs on cylinder block.
4. Remove reservoir tank if necessary, and drain engine coolant and clean reservoir tank before installing.
5. Check drained engine coolant for contaminants such as rust, corrosion or discoloration. If contaminated, flush the engine cooling system.
6. Disconnect drain hose.

Refilling

See Figures 76 and 77.

❊❊ WARNING

Do not put additive such as water leak preventive, since it may cause cooling waterway clogging.

➡**When refilling use Genuine NISSAN Long Life Antifreeze/Coolant (blue) or equivalent in its quality mixed with water (distilled or demineralized).**

1. Install reservoir tank if removed, and install radiator drain plug.

❊❊ WARNING

Be sure to clean drain plug and install with new O-ring.

❊❊ WARNING

If water drain plugs on cylinder block are removed, close and tighten them.

2. Check that each hose clamp has been firmly tightened.
3. Remove air duct (suction side).
4. Disconnect vacuum hose break booster side, and removal vacuum tube from clamp.
5. Disconnect heater hose (1) at position (arrow) in the figure.
6. Fill radiator (1) to specified level.

❊❊ WARNING

Do not get engine coolant on any of the vehicle's electronic equipment (alternator etc.).

❊❊ WARNING

Pour coolant slowly of less than 2 (2-1/8 US qts., 1-3/4 Imp qts.) a minute to allow air in system to escape.

❊❊ WARNING

When engine coolant overflows disconnected heater hose, connect heater hose, and continue filling the engine coolant.

7. Refill reservoir tank to "MAX" level line with engine coolant.
8. Install air duct (suction side).
9. Install radiator cap.
10. Warm up engine until opening thermostat. Standard for warming-up time is approximately 10 minutes at 3,000 rpm.
   a. Check thermostat opening condi-

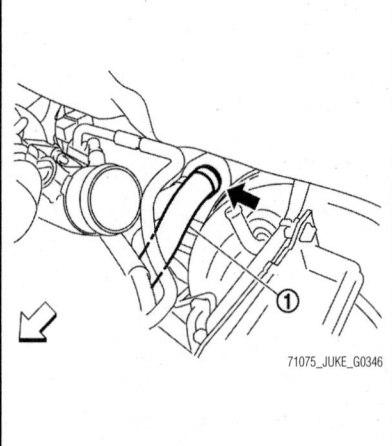

Fig. 76 Disconnecting heater hose (1) at position (arrow)

71075_JUKE_G0346

tion by touching radiator hose (lower) to see a flow of warm water.

❊❊ WARNING

Watch water temperature gauge so as not to overheat engine.

11. Stop the engine and cool down to less than approximately 122°F (50°C).
   a. Cool down using fan to reduce the time.
   b. If necessary, refill radiator up to filler neck with engine coolant.

❊❊ WARNING

Do not get engine coolant on any of the vehicle's electronic equipment (alternator etc.).

12. Refill reservoir tank to "MAX" level line with engine coolant.
13. Repeat steps 6 through 11 two or more times with radiator cap installed until engine coolant level no longer drops.
14. Check cooling system for leakage with engine running.
15. Warm up the engine, and check for sound of engine coolant flow while running engine from idle up to 3,000 rpm with heater temperature controller set at several position between "COOL" and "WARM".
   a. Sound may be noticeable at heater unit.
16. Repeat step 15 three times.
17. If sound is heard, bleed air from cooling system by repeating step 6 through 11 until reservoir tank level no longer drops.

FLUID RECOMMENDATIONS

When refilling use Genuine NISSAN Long Life Antifreeze/Coolant (blue) or

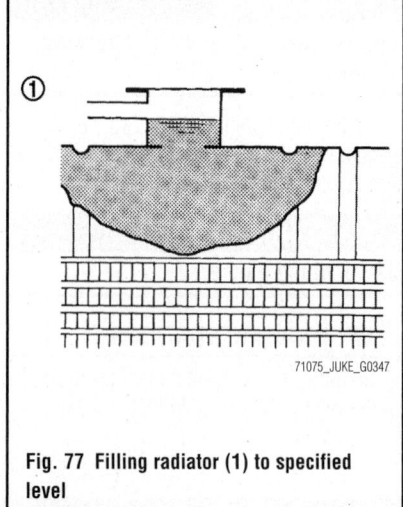

Fig. 77 Filling radiator (1) to specified level

71075_JUKE_G0347

equivalent in its quality mixed with water (distilled or demineralized).

LEVEL CHECK

1. Check that the reservoir tank engine coolant level is within the "MIN" to "MAX" when the engine is cool.

ELECTRIC ENGINE FAN

REMOVAL & INSTALLATION

See Figure 78.

1. Drain engine coolant.

❊❊ WARNING

Perform this step engine is cold.

❊❊ WARNING

Never spill engine coolant on drive belt.

2. Remove engine cover.
3. Remove front bumper.
4. Remove radiator core support upper.
5. Disconnect cooling fan harness connector.
6. Remove reservoir tank.
7. Remove radiator hose (upper).
8. Remove cooling fan assembly.

❊❊ WARNING

Be careful not to damage or scratch on radiator core when removing.

To install:
9. Note the following, and install in the reverse order of the removal procedure:
   a. Only use genuine parts for fan shroud mounting bolt and observe the specified torque (to prevent radiator from being damaged).

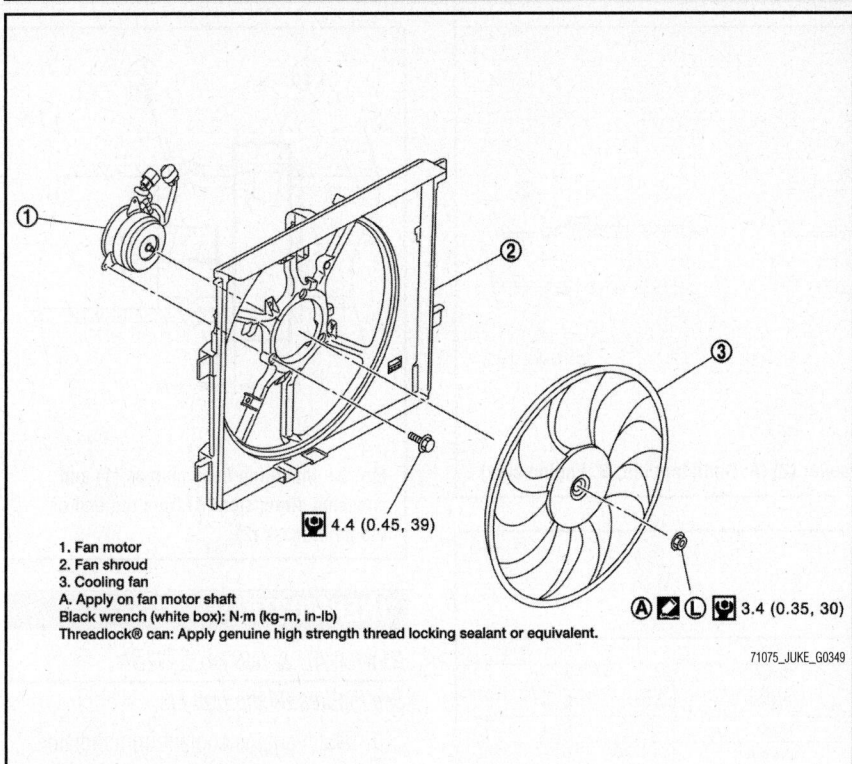

1. Fan motor
2. Fan shroud
3. Cooling fan
A. Apply on fan motor shaft
Black wrench (white box): N·m (kg-m, in-lb)
Threadlock® can: Apply genuine high strength thread locking sealant or equivalent.

4.4 (0.45, 39)

Ⓐ ✎ Ⓛ 3.4 (0.35, 30)

71075_JUKE_G0349

Fig. 78 Removing and installing cooling fan

RADIATOR

REMOVAL & INSTALLATION

See Figures 79 through 83.

❊ CAUTION

Never remove radiator cap when engine is hot. Serious burns may occur from high-pressure engine coolant escaping from engine cooling system.

❊ WARNING

Wrap a thick cloth around the radiator cap. Slowly turn it a quarter of a turn to release built-up pressure. Then turn it all the way.

1. Drain engine coolant from radiator.

❊ WARNING

Perform this step when the engine is cold.

❊ WARNING

Never spill engine coolant on drive belt.

2. Remove engine cover.
3. Remove engine under cover.
4. Remove radiator hose (upper and lower).

5. Remove front bumper.
6. Remove radiator core support upper.
7. Disconnect cooling fan harness connector.
8. Remove reservoir tank.
9. Remove cooling fan assembly.

❊ WARNING

Never damage or scratch the radiator core.

10. Remove condenser from radiator and temporarily fasten it on vehicle with a rope.
11. Pull up and remove the radiator assembly.

❊ WARNING

Never damage radiator core and condenser assembly core.

To install:

❊ WARNING

Do not reuse O-rings.

12. Note the following, and install in the reverse order of the removal procedure:

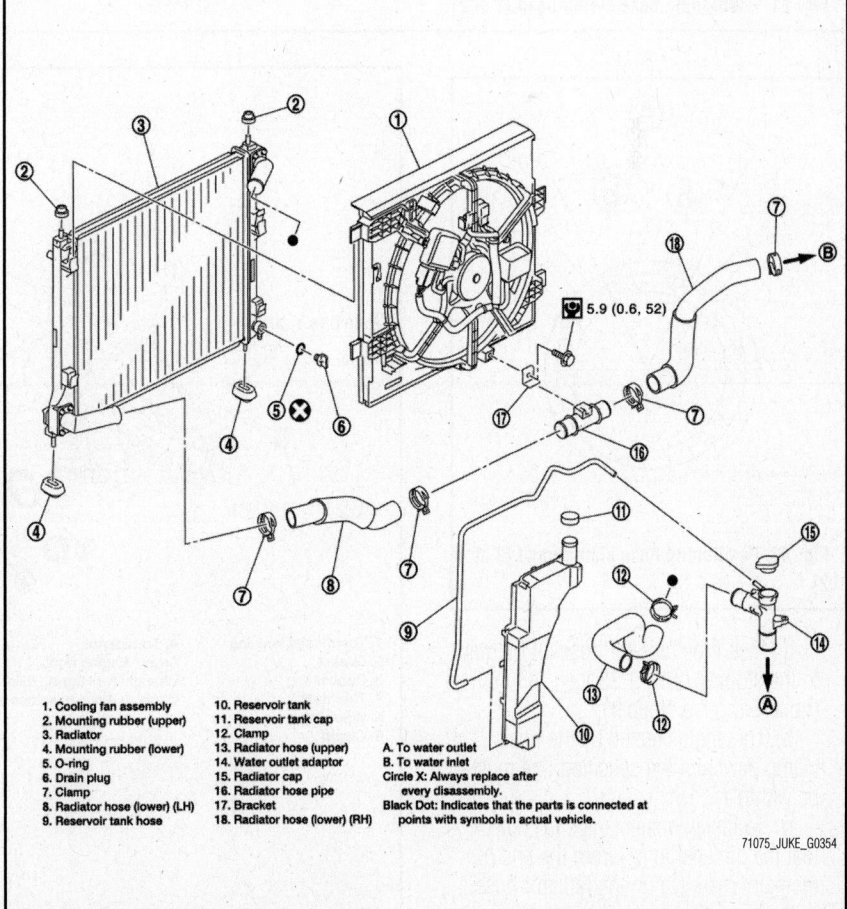

5.9 (0.6, 52)

1. Cooling fan assembly
2. Mounting rubber (upper)
3. Radiator
4. Mounting rubber (lower)
5. O-ring
6. Drain plug
7. Clamp
8. Radiator hose (lower) (LH)
9. Reservoir tank hose
10. Reservoir tank
11. Reservoir tank cap
12. Clamp
13. Radiator hose (upper)
14. Water outlet adaptor
15. Radiator cap
16. Radiator hose pipe
17. Bracket
18. Radiator hose (lower) (RH)

A. To water outlet
B. To water inlet
Circle X: Always replace after every disassembly.
Black Dot: Indicates that the parts is connected at points with symbols in actual vehicle.

71075_JUKE_G0354

Fig. 79 Removing and installing radiator

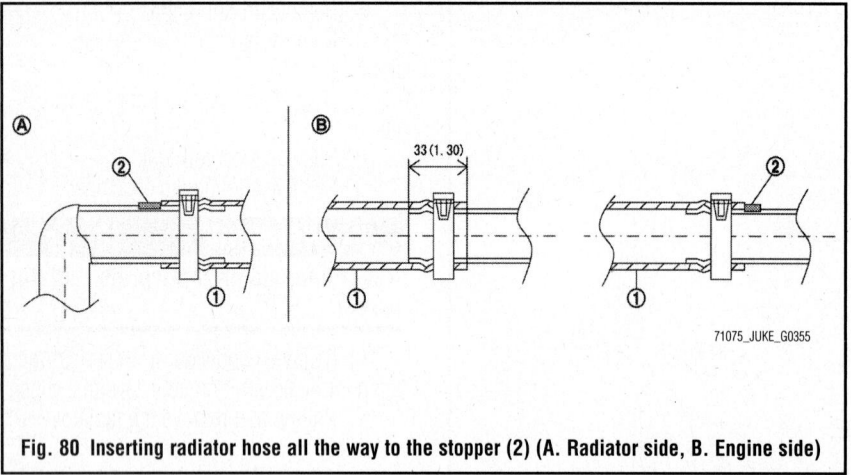

Fig. 80 Inserting radiator hose all the way to the stopper (2) (A. Radiator side, B. Engine side)

Radiator hose	Hose end	Paint mark	Position of hose clamp*
Radiator hose (upper)	Radiator side	Upper	A
	Engine side	Upper	B
Radiator hose (lower) (RH)	Radiator side	Upper	C
	Engine side	Front side	D

71075_JUKE_G0356

Fig. 81 Positioning hose clamp pawl (1 of 2)

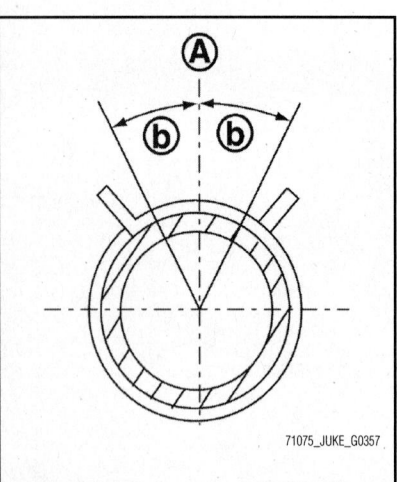

Fig. 82 Positioning hose clamp pawl (2 of 2)

a. Insert the radiator hose all the way to the stopper or by 1.30 inch (33 mm) (hose without a stopper).

b. The angle created by the hose clamp pawl and the specified line must be within +/- 15°.

c. To install hose clamps (1), check that the dimension (A) from the end of the paint mark (2) on the radiator hose to the hose clamp is within the reference value of 0.12 inch (3 mm).

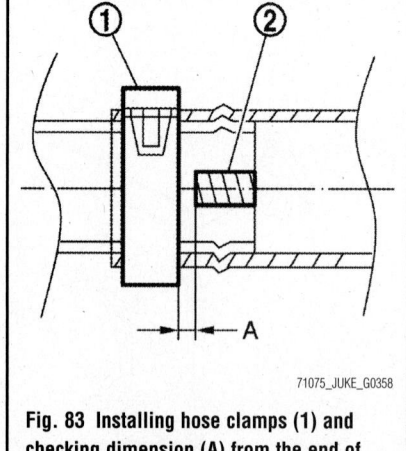

Fig. 83 Installing hose clamps (1) and checking dimension (A) from the end of the paint mark (2)

THERMOSTAT

REMOVAL & INSTALLATION
See Figures 84 through 86.

1. Drain engine coolant from radiator.

✳✳ WARNING
Perform this step when engine is cold.

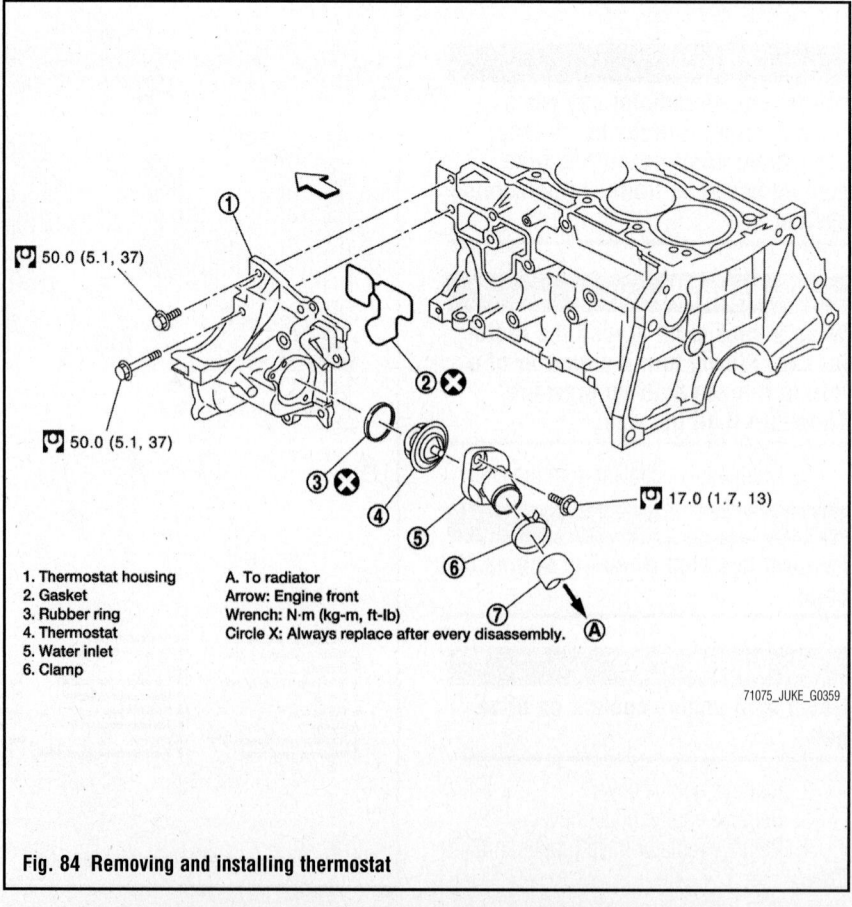

1. Thermostat housing
2. Gasket
3. Rubber ring
4. Thermostat
5. Water inlet
6. Clamp

A. To radiator
Arrow: Engine front
Wrench: N·m (kg-m, ft-lb)
Circle X: Always replace after every disassembly.

71075_JUKE_G0359

Fig. 84 Removing and installing thermostat

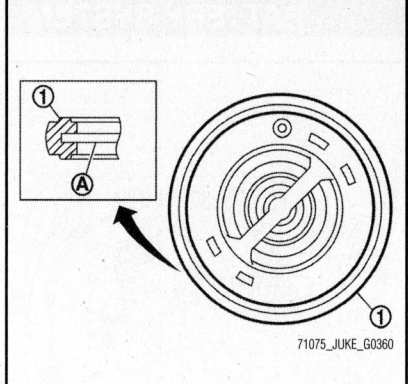

Fig. 85 Installing thermostat with making rubber ring (1) groove fit to thermostat flange (A) with the whole circumference

2. Remove intake manifold.
3. Disconnect radiator hose (lower) (RH) from water inlet.
4. Remove water inlet and thermostat.

➡**Engine coolant leakage from cylinder block, so have a receptacle ready below.**

Thermostat housing
5. Drain engine coolant.
6. Remove alternator.
7. Remove water pump.
8. Disconnect water hose, and then remove thermostat housing.

To install:
9. Note the following, and install in the reverse order of the removal procedure:

a. Install thermostat with making rubber ring (1) groove fit to thermostat flange (A) with the whole circumference.

b. Install thermostat (2) with jiggle valve (A) facing upwards.

WATER PUMP

REMOVAL & INSTALLATION
See Figure 87.

1. Drain engine coolant from radiator.

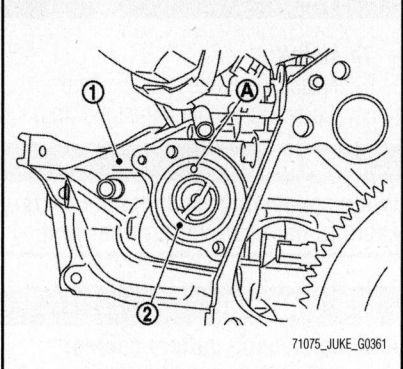

Fig. 86 Installing thermostat (2) with jiggle valve (A) facing upwards

✳✳ **WARNING**

Perform this step when the engine is cold.

✳✳ **WARNING**

Never spill engine coolant on drive belt.

2. Steer front wheel to the right.
3. Remove front fender protector (RH)
4. Remove drive belt.
5. Remove water pump.

➡**Engine coolant will leak from cylinder block, so have a receptacle ready below.**

✳✳ **WARNING**

Handle water pump vane so that it does not contact any other parts.

➡**Water pump cannot be disassembled and should be replaced as a unit.**

To install:
6. Installation is the reverse of the removal procedure.

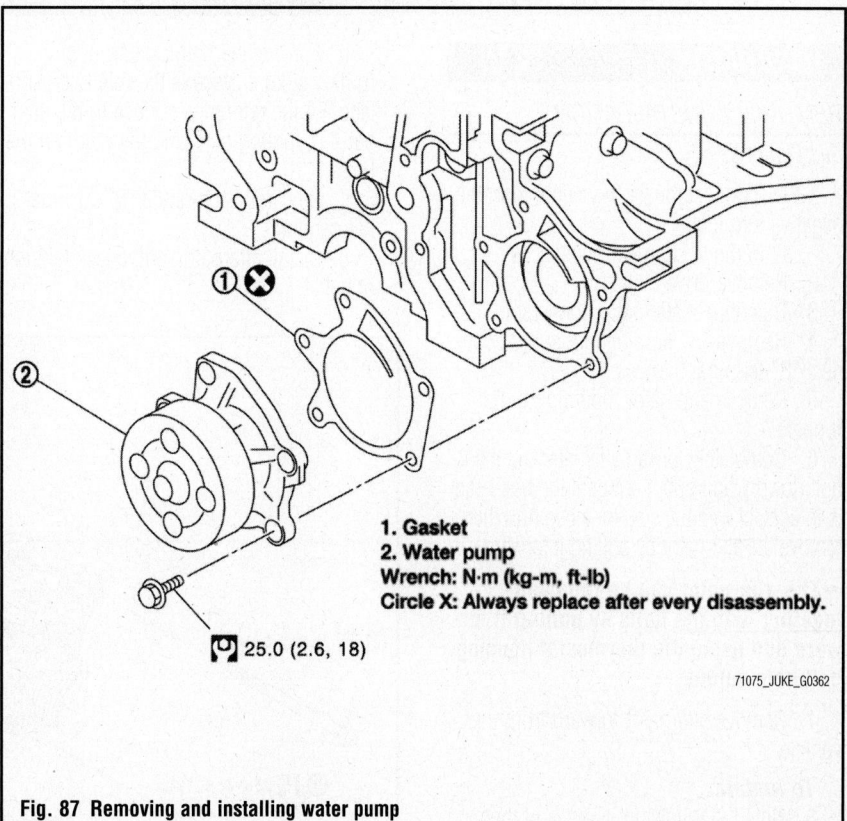

1. Gasket
2. Water pump
Wrench: N·m (kg-m, ft-lb)
Circle X: Always replace after every disassembly.

🔧 25.0 (2.6, 18)

Fig. 87 Removing and installing water pump

ENGINE ELECTRICAL   BATTERY SYSTEM

BATTERY

REMOVAL & INSTALLATION

See Figure 88.

1. Disconnect the battery cable from the negative terminal.

> ※※ **WARNING**
>
> **When disconnecting, disconnect the battery cable from the negative terminal first.**

2. Remove cover of battery positive terminal.
3. Disconnect the battery cable from the positive terminal.
4. Remove battery fix frame mounting nuts and battery fix frame.
5. Remove battery.

To install:

6. Installation is the reverse of the removal procedure, noting the following:

> ※※ **WARNING**
>
> **When connecting, connect the battery cable to the positive terminal first.**

> ※※ **WARNING**
>
> **After connecting battery cables, ensure that they are tightly clamped to battery terminals for good contact.**

> ※※ **WARNING**
>
> **Check battery terminal for poor connection caused by corrosion.**

7. Reset electronic systems as necessary.

5.4 (0.55, 48)

3.9 (0.40, 35)

71075_JUKE_G0368

Fig. 88 Removing and installing battery

ENGINE ELECTRICAL   CHARGING SYSTEM

ALTERNATOR

REMOVAL & INSTALLATION

See Figure 89.

1. Disconnect the battery cable from the negative terminal.
   a. Remove charge air cooler.
2. Remove drive belt.
3. Disconnect alternator connector.
4. Remove "B" terminal nut and disconnect "B" terminal harness.
5. Remove alternator mounting bolt (upper).
6. Completely loosen alternator mounting bolt (lower), and pull it out until the bolt head is in contact with the side member. And then, remove the alternator by pulling it forward.

➡**The alternator can be removed together with the bolts by pulling it forward and using the thermostat housing bolt hole cutout.**

7. Remove alternator forward from the vehicle.

To install:

8. Note the following items, and then install in the reverse order of the removal procedure:
   a. Temporarily tighten the alternator bolts in order from the lower to the upper, and then tighten them in order from the upper to the lower.
   b. For the alternator, the front side (pulley side) surface is the reference surface. Fit the reference surface to the alternator mounting part, and then tighten the bolts.
   c. Be careful to tighten "B" terminal nut carefully.
   d. Install alternator, and check tension of belt.

➡**For this model, the power generation voltage variable control system that controls the power generation voltage of the alternator has been adopted. Therefore, the power generation voltage variable control system operation inspection should be performed after replacing the alternator, and then make sure that the system operates normally.**

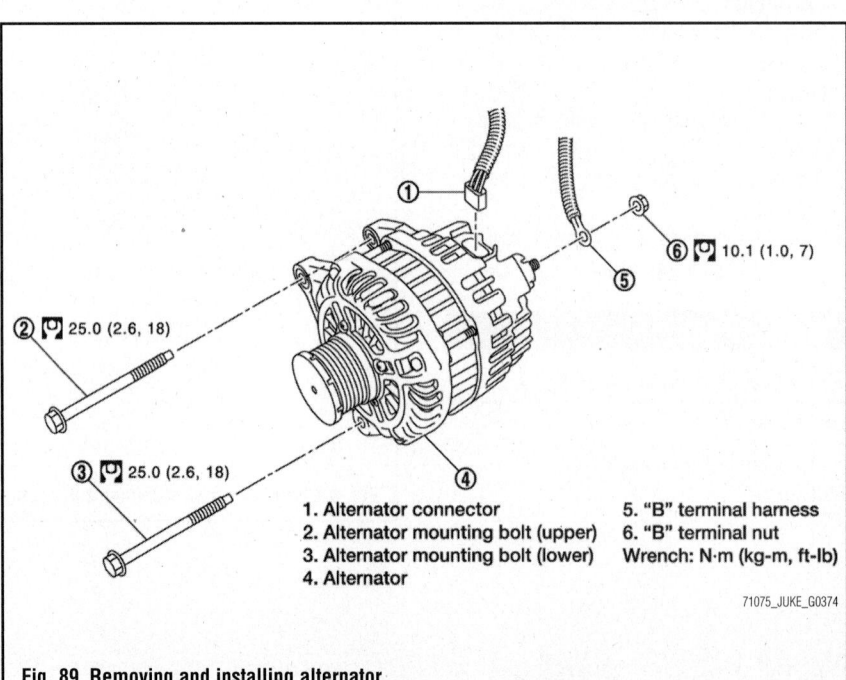

1. Alternator connector
2. Alternator mounting bolt (upper)
3. Alternator mounting bolt (lower)
4. Alternator
5. "B" terminal harness
6. "B" terminal nut
Wrench: N·m (kg-m, ft-lb)

② 25.0 (2.6, 18)
③ 25.0 (2.6, 18)
⑥ 10.1 (1.0, 7)

71075_JUKE_G0374

Fig. 89 Removing and installing alternator

ENGINE ELECTRICAL   IGNITION SYSTEM

IGNITION COIL

REMOVAL & INSTALLATION

See Figure 90.

1. Drain engine coolant.
2. Remove engine cover.
3. Remove air inlet tube assembly.
4. Remove PCV hose.
5. Remove rocker cover protector.
6. Disconnect ignition coil harness connector, and then remove ignition coil.

> ※※ **WARNING**
>
> **Never drop or shock ignition coil. Do not disassemble ignition coil.**

7. Move ignition harness.
8. Remove rocker cover.
   a. Loosen bolts in reverse order shown in the figure.
9. Remove PCV valve and PCV hose, if necessary.
10. Remove rocker cover gasket from rocker cover.

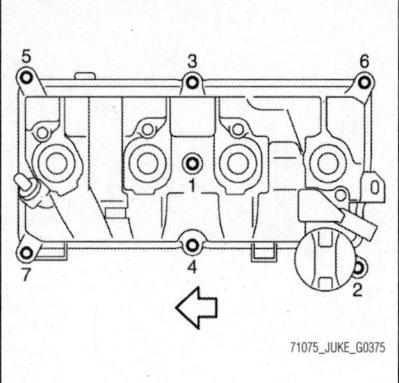

Fig. 90 Bolt removal and installation (arrow: engine front)

To install:

> ※※ **WARNING**
>
> **Do not reuse O-rings.**

11. Install the rocker cover gasket to rocker cover.

> ※※ **WARNING**
>
> **Check the gasket is not dropped.**

12. Install rocker cover.
13. Tighten bolts in two steps separately in numerical order as shown in the figure.
   a. Step 1: 17 inch lbs. (1.96 Nm)
   b. Step 2: 74 inch lbs. (8.33 Nm)
14. Install in the reverse order of removal, for the rest of parts.

SPARK PLUGS

REMOVAL & INSTALLATION

1. Remove engine cover.
2. Remove air inlet tube assembly.
3. Remove ignition coil.
4. Remove spark plug with a spark plug wrench (commercial service tool: 14 mm).

> ※※ **WARNING**
>
> **Do not drop or shock spark plug.**

To install:
5. To install, reverse the removal procedure.

ENGINE ELECTRICAL   STARTING SYSTEM

STARTER

REMOVAL & INSTALLATION

See Figures 91 through 93.

1. Disconnect the battery cable from the negative terminal.
2. Drain engine coolant from radiator.
3. Remove charge air cooler.
4. Remove CVT water hose on thermostat housing side (CVT models).
   a. Remove radiator hose (lower) on water inlet side.
5. Move CVT water hose A and radiator hose (lower) to a location where they do not inhibit work.
6. Open "B" terminal cover, in the direction indicated by an arrow, as shown in the figure.
7. Remove "B" terminal nut and "B" terminal harness.
8. Remove "S" terminal nut and "S" terminal harness.
9. Disconnect harness connector (1) from crankshaft position sensor.
10. Remove harness fixing clip (A) from oil pan (upper), and then move harness (2)

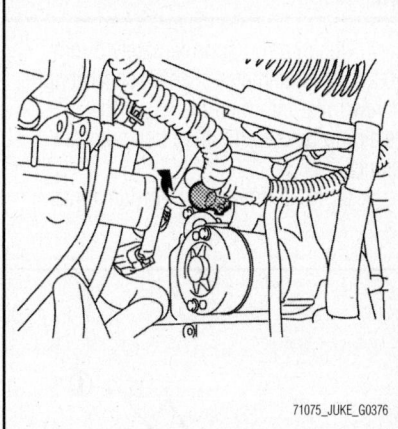

Fig. 91 Opening "B" terminal cover in direction of arrow

to a location where they do not inhibit work.
11. Remove starter motor mounting bolts.
12. Remove starter motor forward from the vehicle.

To install:
13. Note the following items, and install in the reverse order of removal.

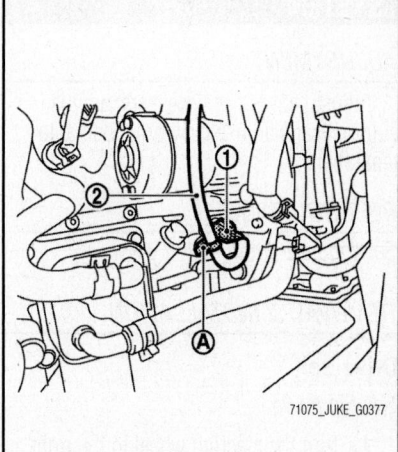

Fig. 92 Disconnecting harness connector (1) from crankshaft position sensor and removing harness fixing clip (A) from oil pan (upper) and removing harness (2)

   a. Be careful to tighten "B" terminal nut to 44 inch lbs. (5 Nm).
   b. After work is complete fill engine coolant.

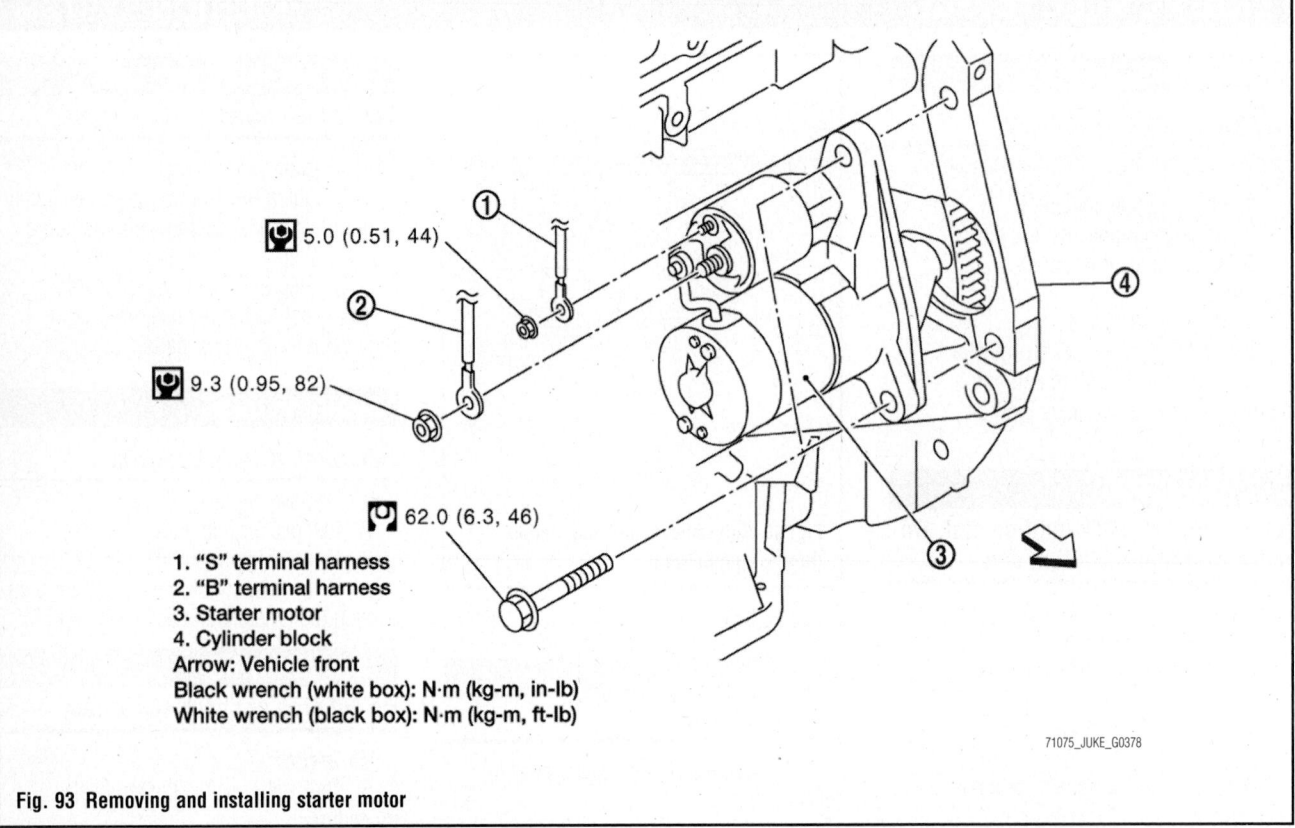

1. "S" terminal harness
2. "B" terminal harness
3. Starter motor
4. Cylinder block
Arrow: Vehicle front
Black wrench (white box): N·m (kg-m, in-lb)
White wrench (black box): N·m (kg-m, ft-lb)

71075_JUKE_G0378

Fig. 93 Removing and installing starter motor

ENGINE MECHANICAL

ACCESSORY DRIVE BELT SYSTEM

ADJUSTMENT

Adjustment is not necessary, as it is automatically adjusted by drive belt auto tensioner.

BELT ROUTINGS

See Figure 94.

REMOVAL & INSTALLATION

Drive Belt

See Figure 95.

1. Turn the steering wheel to the right.
2. Remove the front fender protector (RH) front side bolts and clips. And keep a service area.
3. Hold the hexagonal part (A) of drive belt auto-tensioner (1) with a wrench securely. Then move the wrench handle in the direction of arrow (loosening direction of tensioner).

✳✳ CAUTION

Avoid placing hand in a location where it may get pinched if the

holding tool accidently comes off.

4. Insert a rod approximately 6 mm (0.24 in) in diameter such as short-length screwdriver into the hole (B) of the retaining boss to fix drive belt auto-tensioner.

 a. Keep drive belt auto-tensioner

pulley arm locked after drive belt is removed.

5. Remove drive belt.

To install:

6. Install drive belt.

✳✳ WARNING

Confirm drive belt is completely set to pulleys.

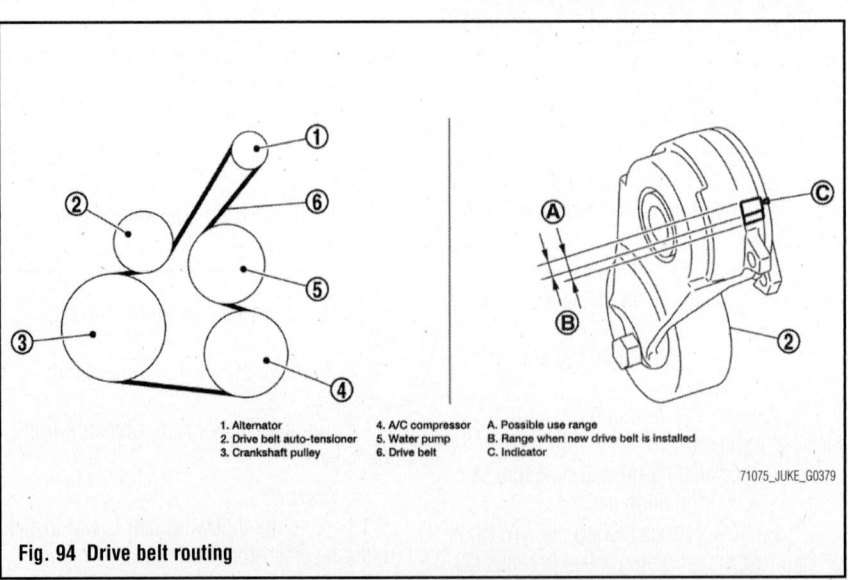

1. Alternator
2. Drive belt auto-tensioner
3. Crankshaft pulley
4. A/C compressor
5. Water pump
6. Drive belt
A. Possible use range
B. Range when new drive belt is installed
C. Indicator

71075_JUKE_G0379

Fig. 94 Drive belt routing

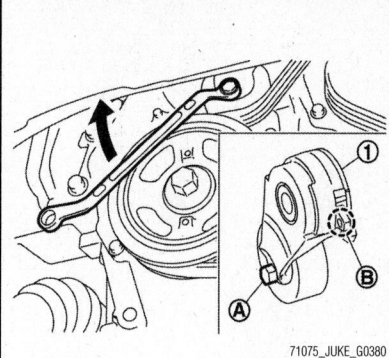

Fig. 95 Moving hexagonal part (A) of drive belt auto tensioner (1) in direction of arrow and inserting screwdriver into hole (B) to fix drive belt auto tensioner

✳✳ WARNING

Make sure there is no engine oil, engine coolant or other fluids on the drive belt or pulley grooves.

7. Release drive belt auto-tensioner, and apply tension to drive belt.

8. Turn crankshaft pulley clockwise several times to equalize tension between each pulley.

9. Confirm tension of drive belt at indicator (notch on fixed side) is within the possible use range.

Drive Belt Auto Tensioner & Idler Pulley

See Figure 96.

1. Loosen mounting bolt and remove drive belt auto-tensioner.

To install:

2. Installation is the reverse of the removal procedure.

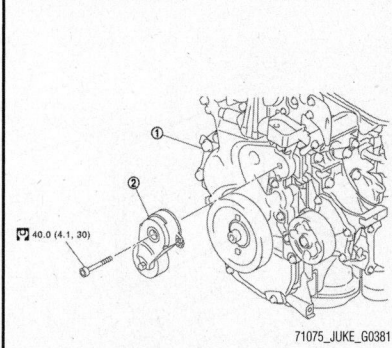

Fig. 96 Removing and installing drive belt auto tensioner (2) and front cover (1) (White wrench (black box): Nm (kg-m, ft-lb))

✳✳ WARNING

When installing drive belt auto-tensioner, be careful not to interfere with water pump pulley.

AIR CLEANER

REMOVAL & INSTALLATION

Air Cleaner Assembly & Air Duct

See Figure 97.

➡**Mass air flow sensor is removable under the car-mounted condition.**

1. Remove engine cover.
   a. Remove air duct inlet (upper).
2. Remove the air cleaner filter from the air cleaner case.
3. Disconnect mass air flow sensor harness connector, and remove harness clamp from air cleaner body.
4. Remove air cleaner body assembly.
5. Remove the air duct (suction side).
   a. Add matching marks if necessary for easier installation.
6. Remove air cleaner cover assembly.
7. Remove mass air flow sensor from air cleaner cover, if necessary.

✳✳ WARNING

Handle the mass air flow sensor with following cares.

   a. Never shock the mass air flow sensor.
   b. Never disassemble the mass air flow sensor.
   c. Never touch the sensor of the mass air flow sensor.
8. Remove air duct inlet (lower) with the following procedure.
   a. Remove fender protector (LH).
   b. Remove air duct inlet (lower).

To install:

9. Note the following, and install in the reverse order of the removal procedure:
   a. Align marks. Attach each joint. Screw clamps firmly.
   b. Fixing clips shall be fixed after inserting air cleaner body assembly protrusion to air cleaner case botch hole.
   c. Make sure whether air cleaner body has been firmly installed by shaking it.

Air Filter Element

See Figures 98 and 99.

1. Remove air duct inlet (upper) (1).
2. Unhook the tabs (A) of both ends of the air cleaner cover.
3. Remove the air cleaner filter (1) and air cleaner body (2) from the air cleaner case.
4. Remove the air cleaner filter from the air cleaner body.

To install:

5. Note the following, and install in the reverse order of the removal procedure:
   a. Fixing clips shall be fixed after inserting air cleaner body protrusion to air cleaner case notch hole.
   b. Make sure that whether air cleaner body has been firmly installed by shaking it.

CAMSHAFT

REMOVAL & INSTALLATION

See Figures 100 through 107.

✳✳ WARNING

The rotating direction in the text indicates all directions seen from the engine front.

1. Remove the following parts.
   • Intake manifold
   • Rocker cover
   • Front cover and timing chain related parts

➡**Removal of oil pump drive related part is not necessary.**

2. Remove camshaft position sensor (PHASE) and exhaust valve timing control position sensor from camshaft bracket.

✳✳ WARNING

Handle camshaft position sensor (PHASE) and exhaust valve timing control position sensor carefully and avoid impacts. Never disassembly the camshaft position sensor (PHASE) and exhaust valve timing control position sensor.

✳✳ WARNING

Never place sensor where it is exposed to magnetism.

3. Put the matching mark (A) on the camshaft sprocket (INT) (2), camshaft sprocket (EXH) (3) and the camshaft bracket (1) as shown in the figure.

➡**It prevents the knock pin of the camshaft (INT) from engaging with the**

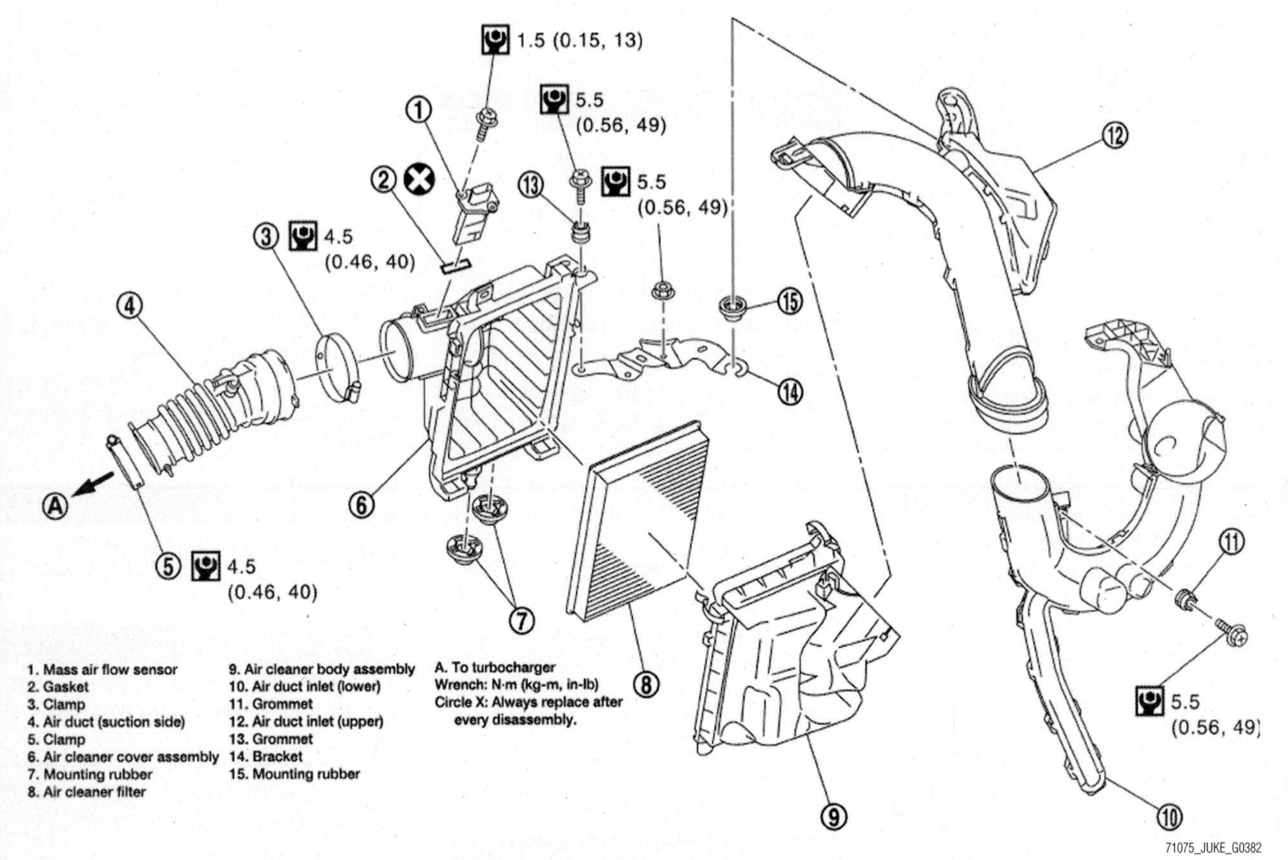

1.5 (0.15, 13)

5.5 (0.56, 49)

5.5 (0.56, 49)

4.5 (0.46, 40)

4.5 (0.46, 40)

5.5 (0.56, 49)

1. Mass air flow sensor
2. Gasket
3. Clamp
4. Air duct (suction side)
5. Clamp
6. Air cleaner cover assembly
7. Mounting rubber
8. Air cleaner filter

9. Air cleaner body assembly
10. Air duct inlet (lower)
11. Grommet
12. Air duct inlet (upper)
13. Grommet
14. Bracket
15. Mounting rubber

A. To turbocharger
Wrench: N·m (kg-m, in-lb)
Circle X: Always replace after
every disassembly.

71075_JUKE_G0382

Fig. 97 Removing and installing air cleaner and air duct

incorrect pin hole when installing the camshaft sprocket (INT).

4. Remove camshaft sprockets (INT and EXH).

a. Secure hexagonal part (A) of camshaft with a wrench. Loosen camshaft sprocket mounting bolts and remove camshaft sprocket.

※ WARNING

Do not rotate crankshaft or camshaft while timing chain is removed. It causes interference between valve and piston.

※ WARNING

Never loosen the mounting bolts with securing anything other than the camshaft hexagonal part or with tensioning the timing chain.

5. Remove camshaft bracket with the following procedure:

a. Loosen mounting bolts in reverse order as shown in the figure.

b. Cut liquid gasket by prying the position (arrow) shown in the figure, and then remove the camshaft bracket.

※ WARNING

Do not damage the mating surface.

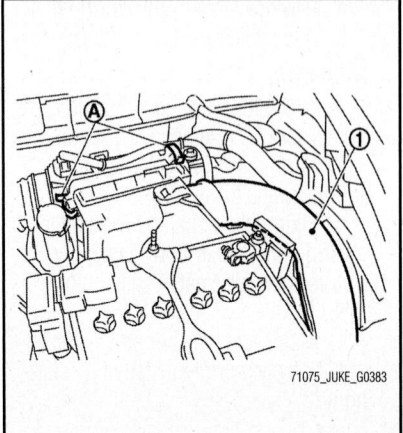

71075_JUKE_G0383

Fig. 98 Removing air duct inlet (upper) (1) and unhooking tabs (A)

※ WARNING

A more adhesive liquid gasket is applied compared to previous types when shipped, so it should not be forced off the position not specified.

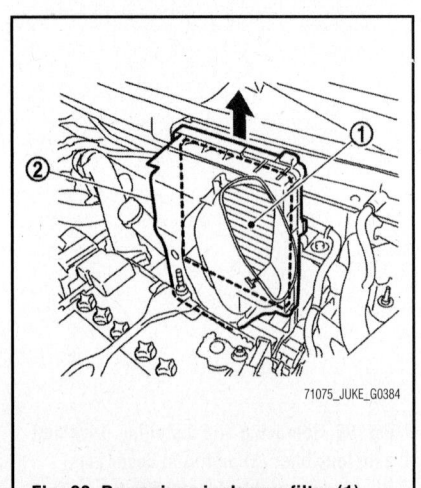

71075_JUKE_G0384

Fig. 99 Removing air cleaner filter (1) and air cleaner body (2)

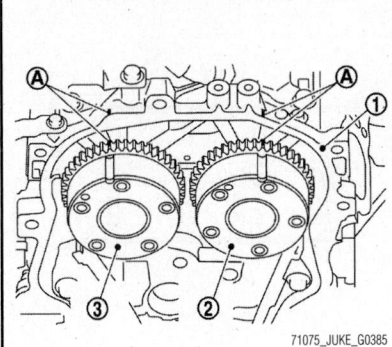

Fig. 100 Putting matching mark (A) on camshaft sprocket (INT) (2), camshaft sprocket (EXH) (3) and camshaft bracket (1)

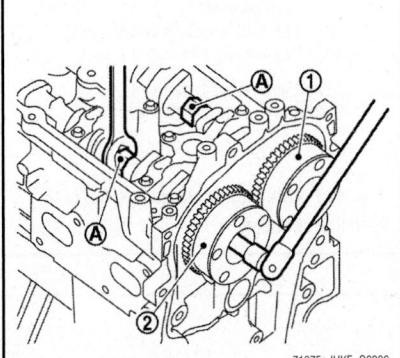

Fig. 101 Securing hexagonal part (A) of camshaft with wrench and removing camshaft sprocket (1: camshaft sprocket (INT), 2: camshaft sprocket (EXH))

6. Remove camshafts.
7. Remove valve lifters.
   a. Identify installation positions, and store them without mixing them up.
8. Remove signal plate from camshaft, if necessary.

To install:

❋❋ WARNING
Do not reuse O-rings.

9. Install valve lifters.
   a. Install them in the original positions.
10. Install camshafts.
   a. Clean camshaft journal to remove any foreign material.
   b. Distinguish between the intake and the exhaust by looking at the different shapes of the front and rear ends of the camshaft or using the identification colors (A) and (B).

c. Install camshafts so that camshaft dowel pins (A) on the front side are positioned as shown.
   d. Apply liquid gasket (A) to camshaft bracket as shown in the figure.

➡**Use Genuine RTV silicon sealant or equivalent.**

e. Tighten mounting bolts of camshaft brackets in the following steps, in numerical order as shown in the figure.
   f. There are two types of mounting bolts. Refer to the following for locating bolts.
   • M6 bolts [thread length: 2.264 inch (57.5 mm)]: 13, 14, and 15 in the figure
   • M6 bolts [thread length: 1.378 inch (35.0 mm)]: except the above

Camshaft bolt tightening sequence:
   • Step 1: tighten in order to 17 inch lbs. (1.96 Nm)

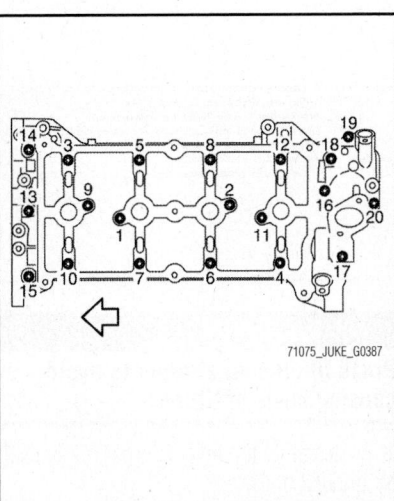

Fig. 102 Mounting bolts installation order

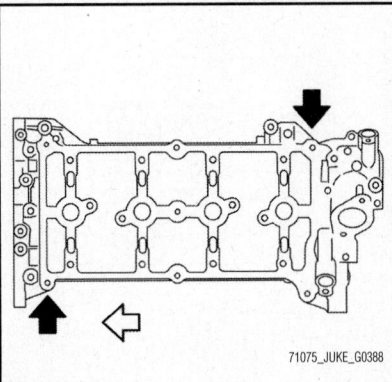

Fig. 103 Identifying prying position for gasket removal

• Step 2: tighten in order to 52 inch lbs. (5.88 Nm)
   • Step 3: tighten in order to 84 inch lbs. (9.5 Nm)

❋❋ WARNING
After tightening mounting bolts of camshaft brackets, be sure to wipe off excessive liquid gasket from the mating surface of cylinder head.

11. Install the camshaft sprocket to the camshaft with the following procedure.
   a. When the camshaft sprocket (INT) and camshaft sprocket (EXH) is removed, refer to the paint mark (A) put according to step "3". Securely align the knock pin and the pin hole, and then install them.
   b. Tighten bolts in the following steps. Secure the hexagonal part of camshaft using wrench to tighten mounting bolt.

Bolt tightening sequence:
   • Tighten camshaft mounting bolt
   • Turn 30 degrees clockwise (angle tightening)

❋❋ WARNING
Check the tightening angle by using an angle wrench or protractor. Never judge by visual inspection without an angle wrench.

12. Install timing chain and related parts.
13. Inspect and adjust valve clearance.
14. Install remaining parts in the reverse order of removal.

CRANKSHAFT FRONT SEAL

REMOVAL & INSTALLATION
See Figure 109.

1. Remove the front fender, drive belt and crankshaft pulley.
2. Remove front oil seal with a suitable tool.

❋❋ WARNING
Be careful not to damage front cover and crankshaft.

To install:
3. Apply new engine oil to new front oil seal joint surface and seal lip.
4. Install front oil seal so that each seal lip is oriented as shown.
   a. Press fit front oil seal using a suitable drift with outer diameter 2.24 inch (55 mm) and inner diameter 1.77 inch (45 mm).

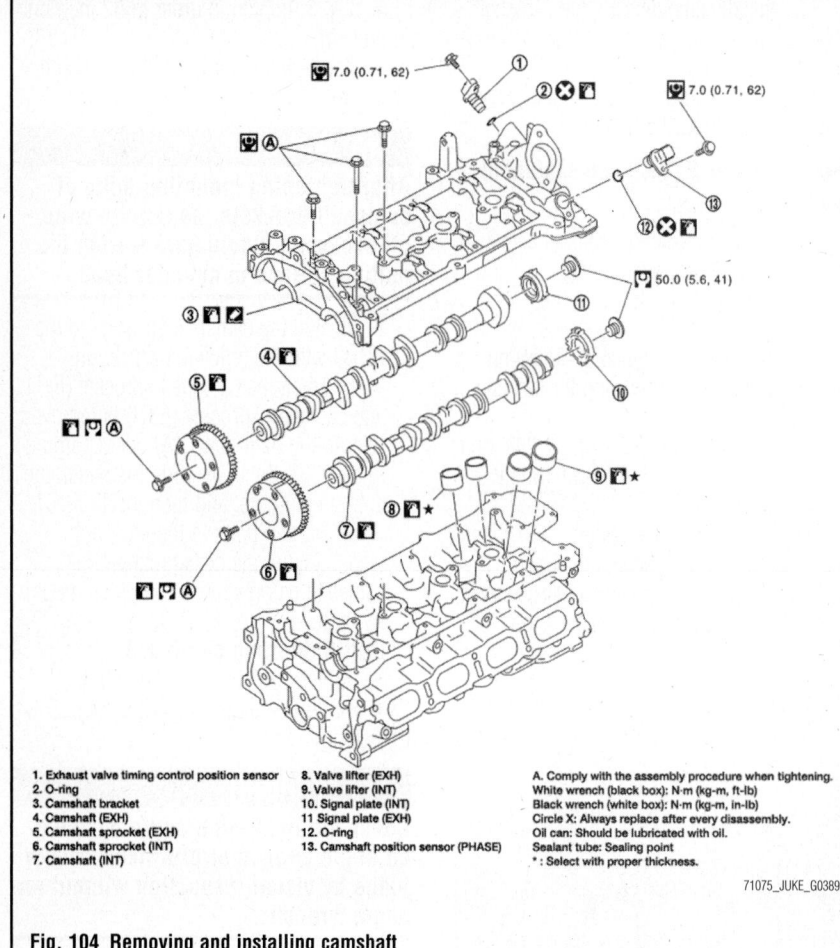

1. Exhaust valve timing control position sensor
2. O-ring
3. Camshaft bracket
4. Camshaft (EXH)
5. Camshaft sprocket (EXH)
6. Camshaft sprocket (INT)
7. Camshaft (INT)
8. Valve lifter (EXH)
9. Valve lifter (INT)
10. Signal plate (INT)
11 Signal plate (EXH)
12. O-ring
13. Camshaft position sensor (PHASE)

A. Comply with the assembly procedure when tightening.
White wrench (black box): N·m (kg-m, ft-lb)
Black wrench (white box): N·m (kg-m, in-lb)
Circle X: Always replace after every disassembly.
Oil can: Should be lubricated with oil.
Sealant tube: Sealing point
* : Select with proper thickness.

71075_JUKE_G0389

Fig. 104 Removing and installing camshaft

b. Within 0.012 inch (0.3 mm) toward engine front (crankshaft pulley side).
c. Within 0.020 inch (0.5 mm) toward engine rear (crankshaft sprocket side).

❋❋ WARNING
Be careful not to damage front cover and crankshaft.

❋❋ WARNING
Press fit oil seal straight to avoid causing burrs of tilting.

5. Install in the reverse order of removal, for the rest of parts.

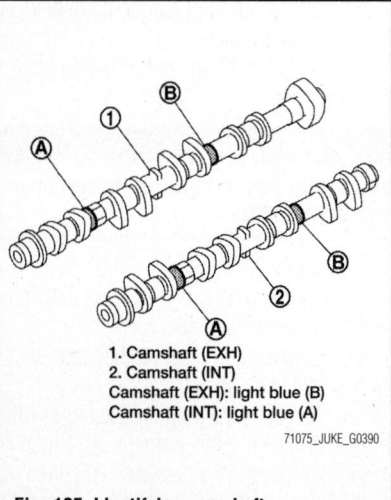

1. Camshaft (EXH)
2. Camshaft (INT)
Camshaft (EXH): light blue (B)
Camshaft (INT): light blue (A)

71075_JUKE_G0390

Fig. 105 Identifying camshafts

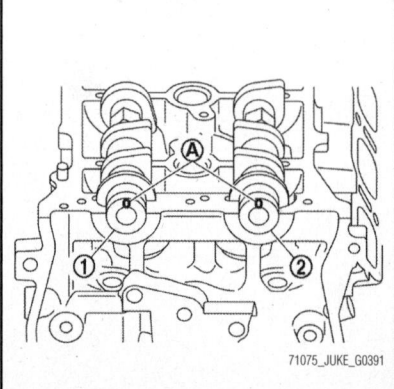

71075_JUKE_G0391

Fig. 106 Positioning camshafts (1: EXH, 2: INT) and dowel pins (A)

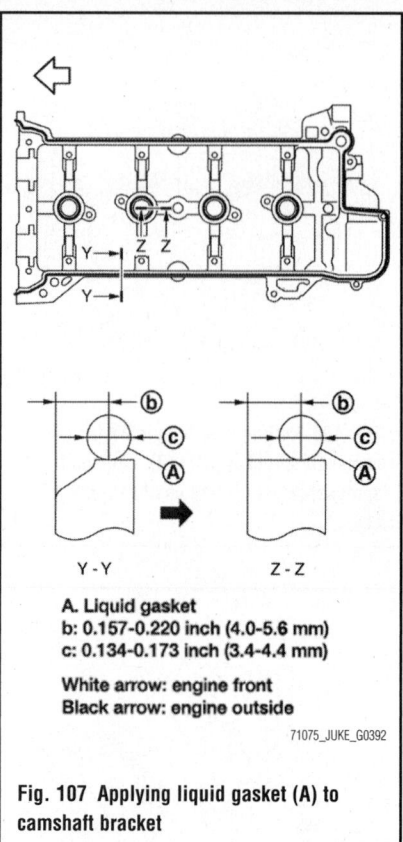

A. Liquid gasket
b: 0.157-0.220 inch (4.0-5.6 mm)
c: 0.134-0.173 inch (3.4-4.4 mm)

White arrow: engine front
Black arrow: engine outside

71075_JUKE_G0392

Fig. 107 Applying liquid gasket (A) to camshaft bracket

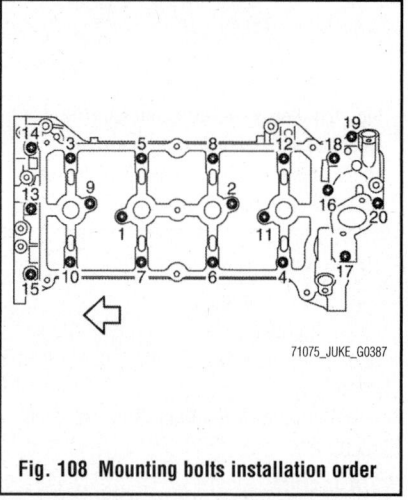

71075_JUKE_G0387

Fig. 108 Mounting bolts installation order

CRANKSHAFT REAR COVER & SEAL

REMOVAL & INSTALLATION
See Figures 109 and 110.

1. Remove transaxle assembly.
2. Remove clutch cover and clutch disc (M/T models).
3. Remove drive plate (CVT models) or flywheel (M/T models)
4. Remove rear oil seal with a suitable tool.

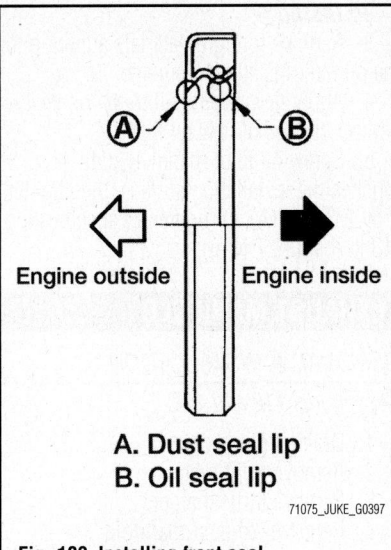

A. Dust seal lip
B. Oil seal lip

71075_JUKE_G0397

Fig. 109 Installing front seal

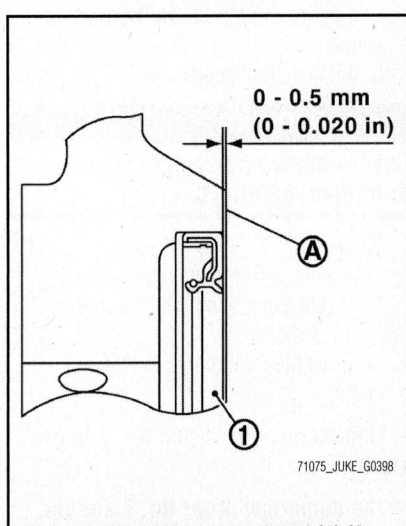

0 - 0.5 mm
(0 - 0.020 in)

Ⓐ

①

71075_JUKE_G0398

Fig. 110 Pressing in rear oil seal (1) (A. rear end surface of cylinder block)

✱✱ WARNING

Be careful not to damage crankshaft and cylinder block.

To install:
5. Apply the liquid gasket lightly to entire outside area of new rear oil seal. Use Genuine RTV silicon sealant or equivalent.
6. Install rear oil seal so that each seal lip is oriented as shown.
   a. Press-fit rear oil seal with a suitable drift (A) outer diameter 4.53 inch (115 mm) and inner diameter 3.54 inch (90 mm).

✱✱ WARNING

Be careful not to damage crankshaft and cylinder block.

---

b. Press-fit oil seal straight to avoid causing burrs or tilting.

✱✱ WARNING

Never touch grease applied onto oil seal lip.

   c. Press in rear oil seal (1) to the position as shown in the figure.
7. Install in the reverse order of removal, for the rest of parts.

CYLINDER HEAD

REMOVAL & INSTALLATION
See Figures 111 and 112.

✱✱ WARNING

Handling and disposal of sodium-filled exhaust valves requires special care and consideration. Under conditions such as breakage with subsequent contact with water, metal sodium which lines the inner portion of exhaust valve will react violently, forming sodium hydroxide and hydrogen which may result in an explosion. Sodium-filled exhaust valve is identified on the top of its stem as shown in illustration.

1. Remove the following components and related parts:
   • Exhaust manifold
   • Intake manifold
   • Fuel injector
   • Water outlet
   • Rocker cover
   • Front cover, timing chain
   • Camshaft
2. Remove cylinder head.
   a. Using TORX® wrench loosen cylinder head bolts in reverse order as shown.

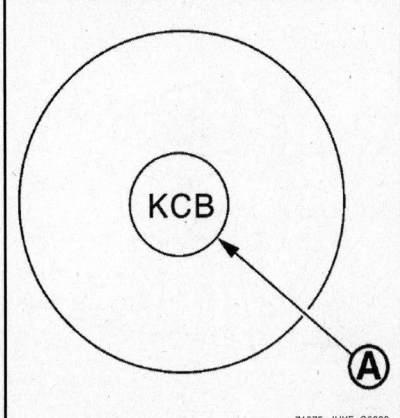

KCB

Ⓐ

71075_JUKE_G0399

Fig. 111 Sodium filled exhaust valve mark (A)

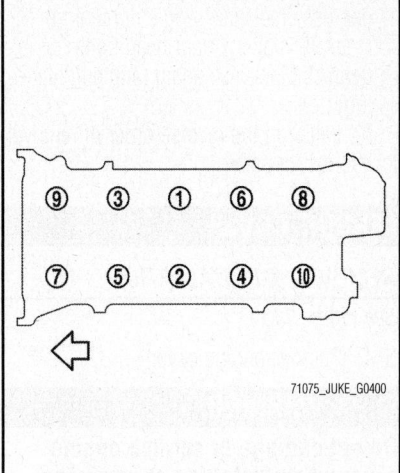

⑨ ③ ① ⑥ ⑧

⑦ ⑤ ② ④ ⑩

71075_JUKE_G0400

Fig. 112 Cylinder head bolt tightening sequence (arrow: engine front)

3. Remove the cylinder head gasket.
To install:
4. Install cylinder head gasket.
5. Install cylinder head, and tighten cylinder head bolts in numerical order as shown in the figure with the following procedure.

✱✱ WARNING

If cylinder head bolts are reused, check their outer diameters before installation.

   a. Apply new engine oil to threads and seating surface of mounting bolts.
   b. Tighten all cylinder head bolts to 30 ft. lbs. (40 Nm).
   c. Turn all cylinder head bolts 100 degrees clockwise (angle tightening).

✱✱ WARNING

Check and confirm the tightening angle by using an angle wrench [SST: KV10112100 (BT8653-A)] or protractor.

✱✱ WARNING

Never judge by visual inspection without the tool.

   d. Completely loosen.

✱✱ WARNING

In this step, loosen cylinder head bolts in reverse order that indicated in the figure.

   e. Tighten all cylinder head bolts to 30 ft. lbs. (40 Nm).

f. Turn all cylinder head bolts 95 degrees clockwise (angle tightening).

g. Turn all cylinder head bolts 95 degrees clockwise again (angle tightening).

6. Install in the reverse order of removal, for the rest of parts.

ENGINE COVER

REMOVAL & INSTALLATION

See Figure 113.

1. Remove engine cover.

❊❊ WARNING

Do not damage or scratch engine cover when installing or removing.

To install:

2. Install in reverse order of removal.

ENGINE OIL & FILTER

OIL LEVEL CHECK

See Figure 114.

Before starting engine, put vehicle horizontally and check the engine oil level. If engine is already started, stop it and allow 10 minutes before checking.

1. Pull out oil level gauge and wipe it clean.

2. Insert oil level gauge and check that the engine oil level is within

the range shown in the figure.

3. If it is out of range, adjust it.

OIL & FILTER CHANGE

See Figure 115.

1. Remove engine under cover.

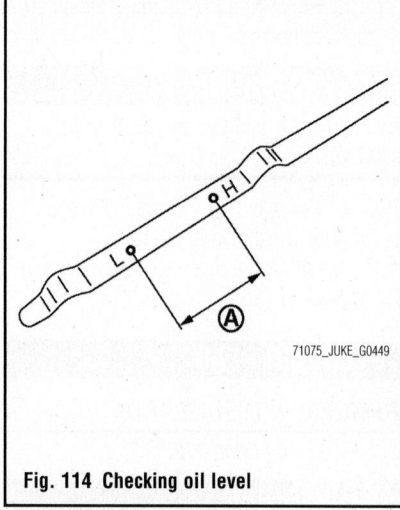

Fig. 114 Checking oil level

2. Using oil filter wrench [SST: KV10115801 (J-38956)] (A), remove oil filter.

❊❊ WARNING

Oil filter is provided with relief valve. Use genuine NISSAN oil filter or equivalent.

❊❊ CAUTION

Be careful not to get burned when engine and engine oil may be hot.

❊❊ WARNING

When removing, prepare a shop cloth to absorb any engine oil leakage or spillage.

➡Completely wipe off any engine oil that adheres to engine and vehicle.

To install:

3. Remove foreign materials adhering to the oil filter installation surface.

4. Apply new engine oil to the oil seal contact surface of new oil filter.

5. Screw oil filter manually until it touches the installation surface, then tighten it by 2/3 turn (A). Or tighten to specification (13 ft. lbs. (17.7 Nm)).

EXHAUST MANIFOLD

REMOVAL & INSTALLATION

See Figures 116 and 117.

1. Drain engine coolant.
2. Remove front tube.
3. Remove turbocharger.
4. Remove exhaust manifold cover.
5. Remove exhaust manifold.
   a. Loosen nuts in reverse order as shown.
6. Remove the gasket.

❊❊ WARNING

Cover engine openings to avoid entry of foreign materials.

To install:

7. Install gasket to cylinder head.
8. Install exhaust manifold with the following procedure.
   a. Tighten nuts in numerical order.

➡Tighten nuts No. 1 and No. 4 in two steps.

➡The numerical order No. 9 and No. 12 shows the second step.

9. Install the remaining parts in the reverse order of removal.

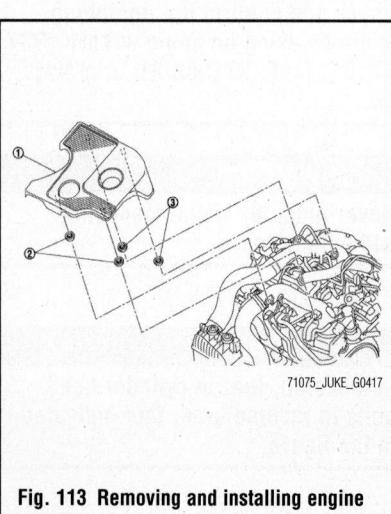

Fig. 113 Removing and installing engine cover

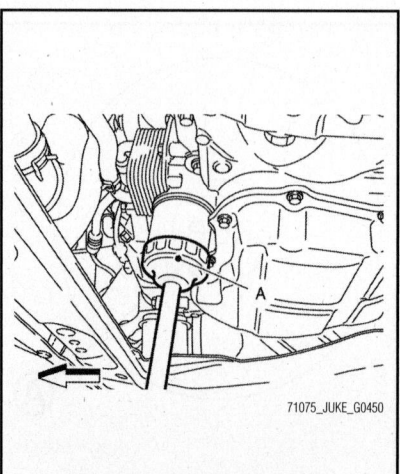

Fig. 115 Using oil filter wrench (A) to remove oil filter

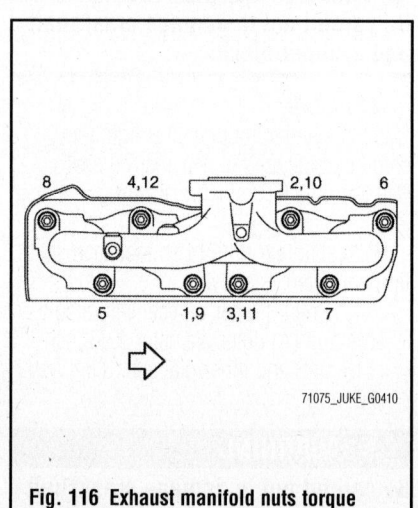

Fig. 116 Exhaust manifold nuts torque sequence

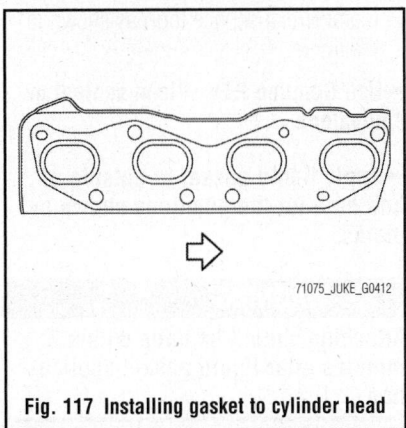

Fig. 117 Installing gasket to cylinder head

INTAKE MANIFOLD

REMOVAL & INSTALLATION

See Figures 118 through 120.

1. Remove engine cover.
2. Pull out oil level gauge.

※※ WARNING

Cover the oil level gauge guide openings to avoid entry of foreign materials.

3. Disconnect turbocharger boost sensor (with intake air temperature sensor 2) harness connector.
4. Remove air inlet tube assembly.
5. Disconnect water hoses from electric throttle control actuator as follows:
   a. Attach plug to prevent engine coolant leakage when engine coolant is not drained.

※※ WARNING

Perform this step when the engine is cold.

➡**This step is not required when removing only intake manifold.**

6. Disconnect electric throttle control actuator harness connector.
7. Remove electric throttle control actuator.

※※ WARNING

Handle carefully to avoid any shock to electric throttle control actuator. Do not disassemble the electric throttle control actuator.

8. Remove EVAP vacuum tank.
9. Disconnect EVAP canister purge volume control solenoid valve harness connector, and then remove bracket with EVAP

canister purge volume control solenoid valve.
10. Remove vacuum gallery.
11. Disconnect PCV hose (intake manifold side).
12. Remove intake manifold (1) with the following procedure.
   a. Loosen and remove intake manifold mounting bolt (A).
   b. Loosen mounting bolts in reverse order as shown in the figure.

➡**Disregard the numerical order No.6 in removal.**

※※ WARNING

Cover engine openings to avoid entry of foreign materials.

To install:

13. Note the following, and install in the reverse order of the removal procedure.
14. Make sure the gasket is properly installed in the installation groove of intake manifold.
15. Install intake manifold with the following procedure:
   a. Tighten in numerical order as shown in the figure.

➡**Tighten bolt the No.1 in two steps.**

➡**The numerical order No.6 shows the second step.**

16. Install in the reverse order of removal after this step.

※※ WARNING

Since check valve is built-in inside vacuum hose (1), install the vacuum hose fitting the arrow (A) to intake manifold side (2).

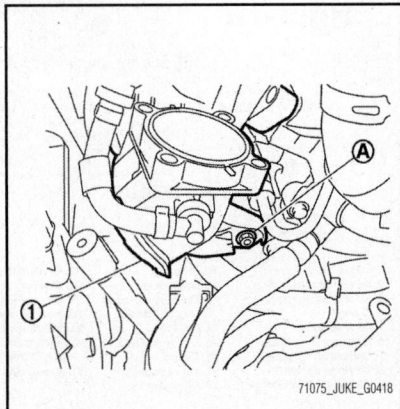

Fig. 118 Removing intake manifold (1) mounting bolt (A)

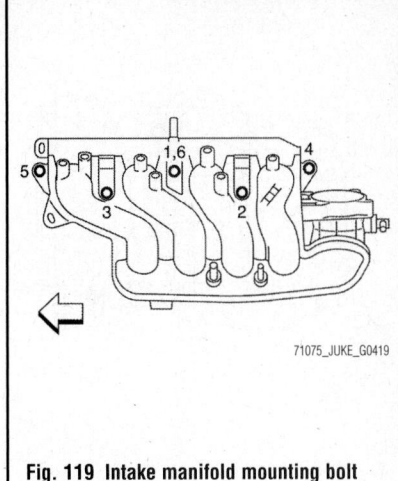

Fig. 119 Intake manifold mounting bolt torque sequence

OIL PAN

REMOVAL & INSTALLATION

Upper

See Figures 121 through 126.

1. Remove oil pan (lower).
2. Remove oil filter.
3. Remove front cover, timing chain, oil pump drive chain, and other related parts.
4. Remove oil level gauge and oil level gauge guide.
5. Remove oil pump.

➡**The oil pan (upper) can be removed and installed without removing the oil pump.**

6. Remove oil pan (upper) with the following procedure:
   a. Loosen bolts in reverse order as shown in the figure.
   b. Insert a screwdriver shown by the

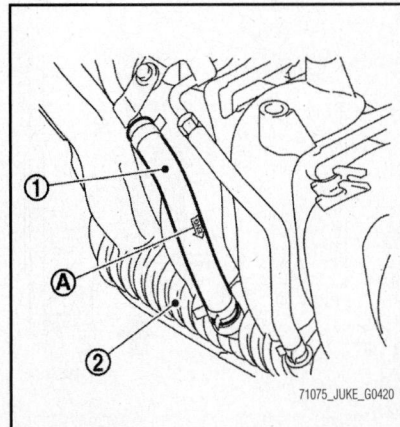

Fig. 120 Check valve is built in inside vacuum hose (1), install vacuum hose fitting the arrow (A) to intake manifold side (2)

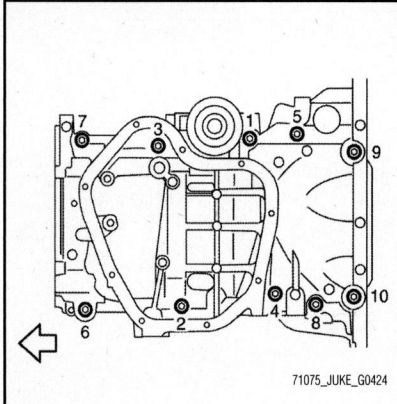

Fig. 121 Oil pan bolt installation sequence (arrow: engine front)

black arrow in the figure and open up a crack between oil pan (upper) and cylinder block.

✳✳ WARNING

A more adhesive liquid gasket is applied compared to previous types when shipped, so it should not be forced off the position not specified.

c. Insert seal cutter [SST: KV10111100 (J-37228)] between oil pan (upper) and cylinder block, and slide it by tapping on the side of the tool with a hammer.

✳✳ WARNING

Never damage the mating surface.

7. Remove O-ring between cylinder block and oil pan (upper).

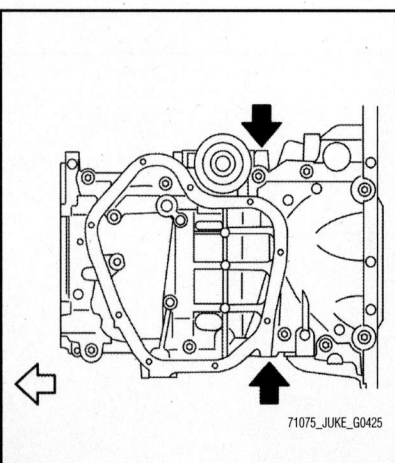

Fig. 122 Identifying screwdriver insertion points (arrow: engine front)

To install:

✳✳ WARNING

Do not reuse O-rings.

8. Install oil pan (upper) with the following procedure:

a. Use a scraper to remove old liquid gasket from mating surfaces.

➡**Remove the old liquid gasket from mating surface of cylinder block.**

➡**Remove old liquid gasket from the bolt holes and threads.**

✳✳ WARNING

Never scratch or damage the mating surfaces when cleaning off old liquid gasket.

b. Apply a continuous bead of liquid gasket (D) with a tube presser

(commercial service tool) as shown in the figure.

➡**Use Genuine RTV silicon sealant or equivalent.**

➡**Apply liquid gasket to outside of bolt hole for the positions shown by marks.**

✳✳ WARNING

Attaching should be done within 5 minutes after liquid gasket application.

c. Install new O-ring at cylinder block side.

✳✳ WARNING

Install avoiding misalignment of O-ring.

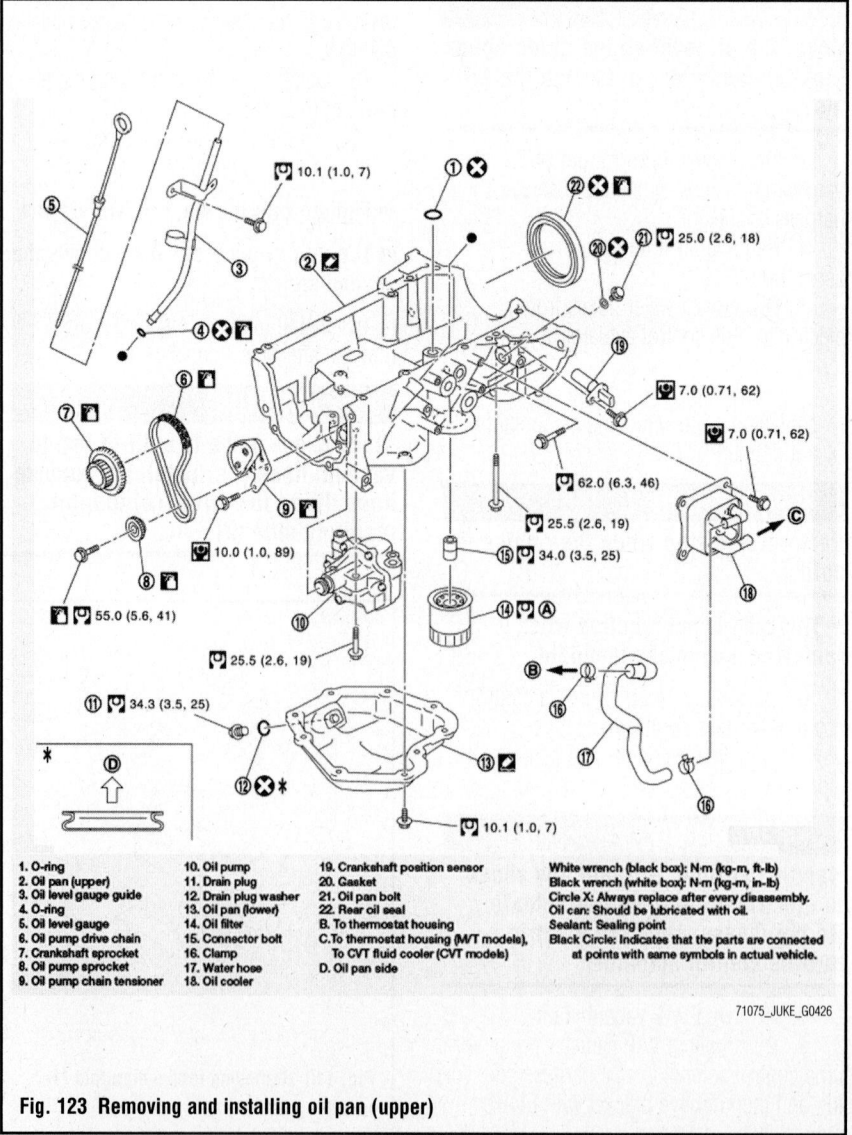

1. O-ring	10. Oil pump
2. Oil pan (upper)	11. Drain plug
3. Oil level gauge guide	12. Drain plug washer
4. O-ring	13. Oil pan (lower)
5. Oil level gauge	14. Oil filter
6. Oil pump drive chain	15. Connector bolt
7. Crankshaft sprocket	16. Clamp
8. Oil pump sprocket	17. Water hose
9. Oil pump chain tensioner	18. Oil cooler

19. Crankshaft position sensor	
20. Gasket	
21. Oil pan bolt	
22. Rear oil seal	
B. To thermostat housing	
C. To thermostat housing (M/T models), To CVT fluid cooler (CVT models)	
D. Oil pan side	

White wrench (black box): N·m (kg-m, ft-lb)
Black wrench (white box): N·m (kg-m, in-lb)
Circle X: Always replace after every disassembly.
Oil can: Should be lubricated with oil.
Sealant: Sealing point
Black Circle: Indicates that the parts are connected at points with same symbols in actual vehicle.

Fig. 123 Removing and installing oil pan (upper)

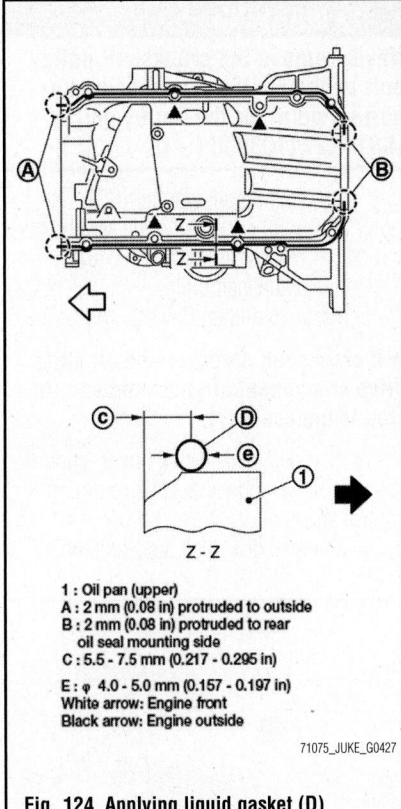

Fig. 124 Applying liquid gasket (D)

1 : Oil pan (upper)
A : 2 mm (0.08 in) protruded to outside
B : 2 mm (0.08 in) protruded to rear
    oil seal mounting side
C : 5.5 - 7.5 mm (0.217 - 0.295 in)
E : φ 4.0 - 5.0 mm (0.157 - 0.197 in)
White arrow: Engine front
Black arrow: Engine outside

71075_JUKE_G0427

d. Tighten bolts in numerical order as shown in the figure.

9. Install rear oil seal with the following procedure.

✳✳ WARNING

The installation of rear oil seal should be completed within 5 minutes after installing oil pan (upper).

➡️**Always replace rear oil seal with new one.**

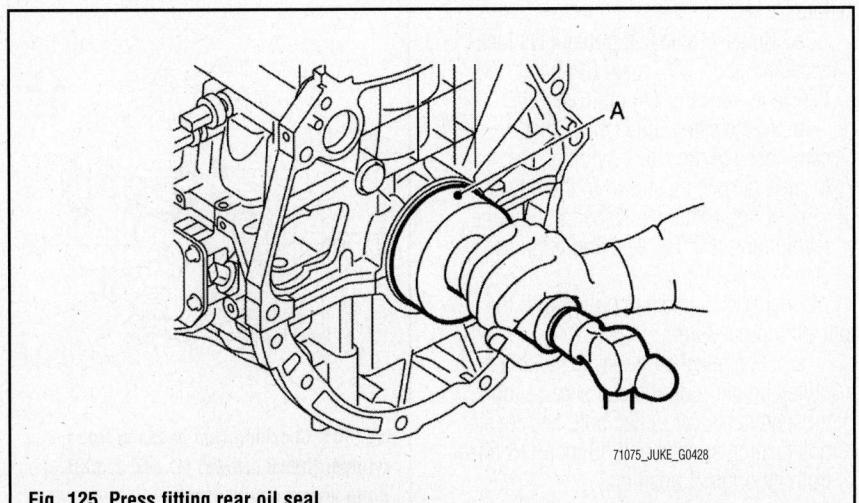

Fig. 125 Press fitting rear oil seal

71075_JUKE_G0428

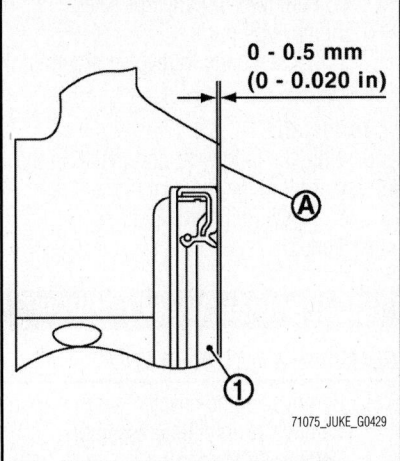

Fig. 126 Identifying oil seal press fit dimensions

71075_JUKE_G0429

✳✳ WARNING

Do not touch oil seal lip.

   a. Wipe off liquid gasket protruding to the rear oil seal mounting part of oil pan (upper) and cylinder block using a scraper.

   b. Apply engine oil to entire outside area of rear oil seal.

   c. Press-fit the rear oil seal using a suitable drift (A) with outer diameter 115 mm (4.53 in) and inner diameter 90 mm (3.54 in).

✳✳ WARNING

Press-fit to the specified dimensions as shown in the figure.

✳✳ WARNING

Never touch the grease applied to the oil seal lip.

✳✳ WARNING

Be careful not to damage the rear oil seal mounting part of oil pan (upper) and cylinder block or the crankshaft.

✳✳ WARNING

Press-fit straight, checking that rear oil seal does not curl or tilt.

➡️**The standard surface of the dimension is the rear end surface of cylinder block.**

10. Install in the reverse order of removal, for the rest of parts.

OIL PUMP

REMOVAL & INSTALLATION

See Figures 127 through 129.

1. Remove engine assembly.
2. Remove oil pan (lower).
3. Remove front cover, and other related parts.
4. Remove oil pump sprocket with the following procedure:

➡️**Add matching mark if necessary for easier installation.**

   a. Push oil pump drive chain tensioner (1) in the direction show in the figure.

   b. Insert a stopper pin (A) into the body hole (B).

   c. Remove oil pump chain tensioner.

➡️**When the holes on lever and tensioner body cannot be aligned, align**

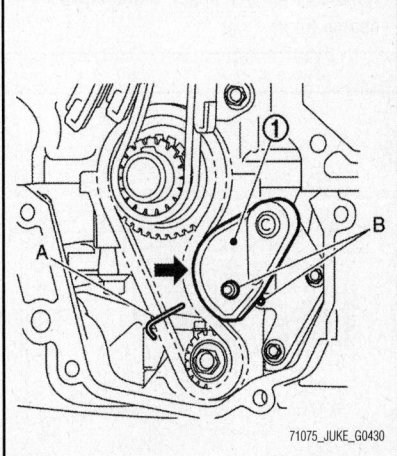

71075_JUKE_G0430

Fig. 127 Pushing oil pump drive chain tensioner (1) in direction shown and insert stopper pin (A) into body hole (B) to remove tensioner

these holes by slightly moving the oil pump chain tensioner slack guide.

d. Hold the WAF part of oil pump shaft [WAF: 10 mm (0.39 in)] (A), and then loosen the oil pump sprocket bolt and remove it.

✳✳ WARNING

Secure the oil pump shaft with the WAF part.

✳✳ WARNING

Never loosen the oil pump sprocket bolt by tightening the oil pump drive chain.

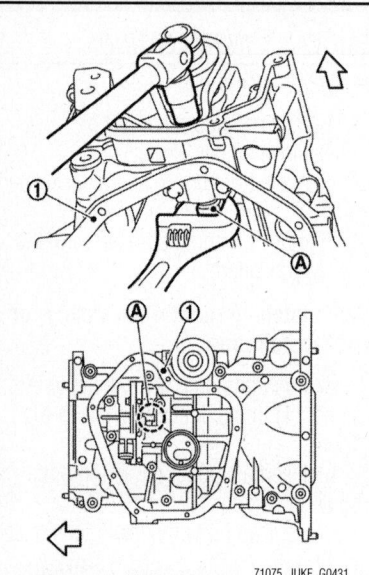

Fig. 128 Holding WAF part of oil pump shaft (A) and loosening sprocket bolt to remove (1: oil pan upper, White arrow: engine front)

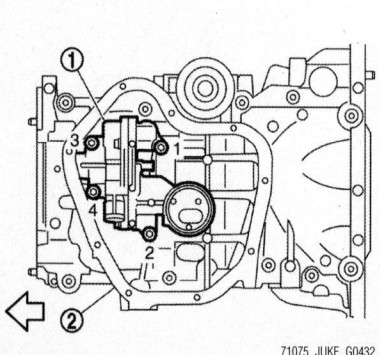

Fig. 129 Oil pump (1) bolt installation sequence (2: oil pan upper, arrow: engine front)

e. Remove oil pump sprocket.

5. Remove oil pump.

a. Loosen bolts in reverse order as shown.

To install:

6. Note the following, and install in the reverse order of the removal procedure:

a. Tighten oil pump bolts in sequence.

ROCKER ARM (VALVE) COVER

REMOVAL & INSTALLATION

1. Remove engine cover.
2. Remove air inlet tube assembly.
3. Remove PCV hose.
4. Remove rocker cover protector.

To install:

5. To install, reverse the removal procedure. Tighten cover to 15 ft. lbs. (25 Nm).

TIMING CHAIN COVER, CHAIN, TENSIONER, & SPROCKETS

REMOVAL & INSTALLATION

See Figures 130 through 144.

✳✳ WARNING

The rotating direction in the text indicates all directions seen from the engine front.

1. Drain engine oil.

✳✳ WARNING

Perform this step when engine is cold.

2. Remove the intake manifold and rocker cover.

3. Set No. 1 cylinder at TDC on its compression stroke with the following procedure:

a. Rotate crankshaft pulley (1) clockwise and align TDC mark (no paint) (B) to timing indicator (A) on front cover.

b. At the same time, check that the cam noses of the No. 1 cylinder are located (arrow) as shown in the figure.

c. If not, rotate crankshaft pulley one revolution (360 degrees) and align as shown in the figure.

4. Remove crankshaft pulley with the following procedure:

a. Fix crankshaft pulley (1) with a pulley holder (commercial service tool), loosen crankshaft pulley bolt, and locate bolt seating surface at 0.39 inch (10 mm) from its original position.

✳✳ WARNING

Never remove the crankshaft pulley bolt as they will be used as a supporting point for the pulley puller [SST: KV11103000 (-)].

b. Attach a pulley puller [SST: KV11103000 (-)] (A) in the M6 thread hole on crankshaft pulley (1), and remove crankshaft pulley.

5. Remove oil pan (lower).

➡ If crankshaft sprocket and oil pump drive component are not removed, this step is unnecessary.

6. Remove intake valve timing control solenoid valve and exhaust valve timing control valve.

7. Remove drive belt auto-tensioner.

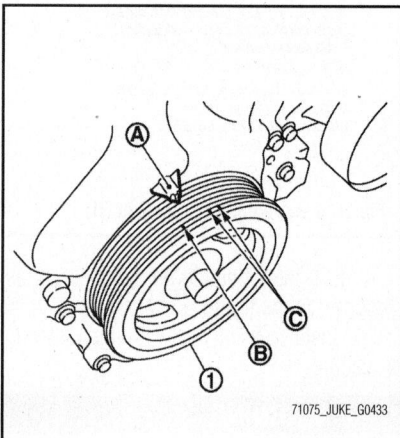

Fig. 130 Rotating crankshaft pulley (1) clockwise and aligning TDC mark (no paint) (B) to timing indicator (A) on front cover (C: white paint mark, not used for service)

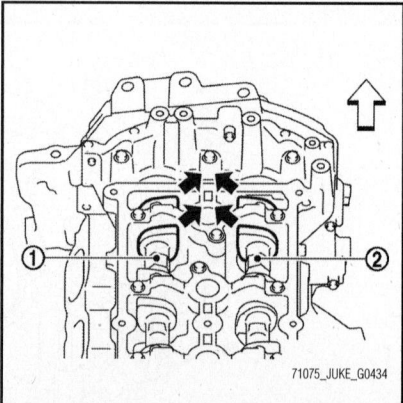

Fig. 131 Checking cam noses of No. 1 cylinder (black arrows) (1: INT, 2: EXH, white arrow: engine front)

Fig. 132 Fixing crankshaft pulley (1) with a pulley holder (A)

8. Remove front cover with the following procedure:

a. Loosen mounting bolts in reverse order as shown in the figure.

b. Cut liquid gasket by prying the position (arrows) shown in the figure, and then remove the front cover.

✳✳ WARNING

Be careful not to damage the mating surface.

✳✳ WARNING

A more adhesive liquid gasket is applied compared to previous types when shipped, so it should not be forced off the position not specified.

9. Remove front oil seal from front cover.

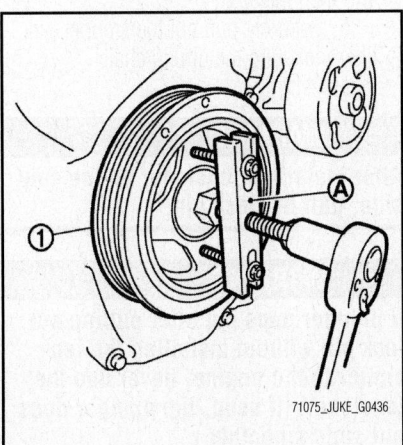

Fig. 133 Attaching pulley puller (A) in M6 thread hole on crankshaft pulley (1)

✳✳ WARNING

Be careful not to damage front cover.

a. Lift up front oil seal using a screwdriver.

10. Remove valve timing control cover, if necessary.

a. Loosen mounting bolts in reverse order as shown in the figure.

➡ **Disregard the numerical order No.1 in removal.**

11. Remove timing chain tensioner with the following procedure:

a. Insert a wire (A) (e.g: clip) into the top groove with the timing chain tensioner plunger pressed.

➡ **Timing chain tensioner plunger is securely fixed by inserting a wire (e.g. clip).**

b. Remove timing chain tensioner (1).

12. Remove slack guide (2), tension guide (3) and timing chain (1).

✳✳ WARNING

Never rotate each crankshaft and camshaft individually while timing chain is removed. It causes interference between valve and piston.

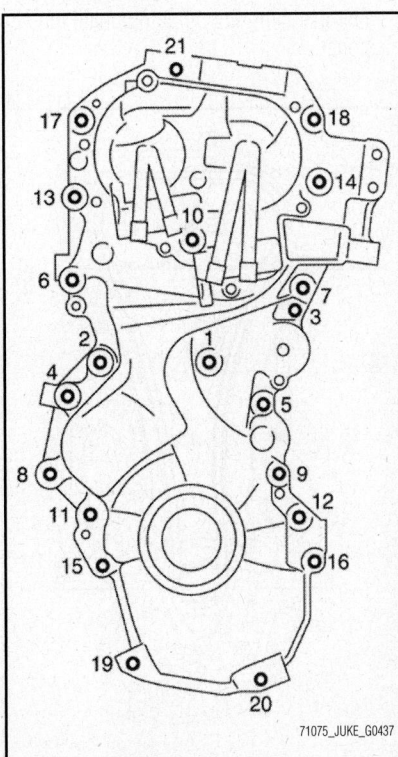

Fig. 134 Front cover mounting bolt tightening sequence

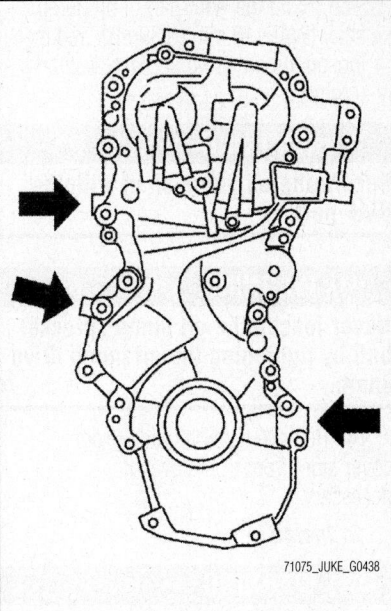

Fig. 135 Cutting liquid gasket at prying points

➡ **If timing chain is difficult to remove, remove camshaft sprocket (EXH) first to remove timing chain.**

13. Remove crankshaft sprocket and oil pump drive component with the following procedure:

a. Push oil pump drive chain tensioner (1) in the direction show in the figure.

b. Insert a stopper pin (A) into the body hole (B).

c. Remove oil pump chain tensioner.

➡ **When the holes on lever and tensioner body cannot be aligned, align these holes by slightly moving the oil pump chain tensioner slack guide.**

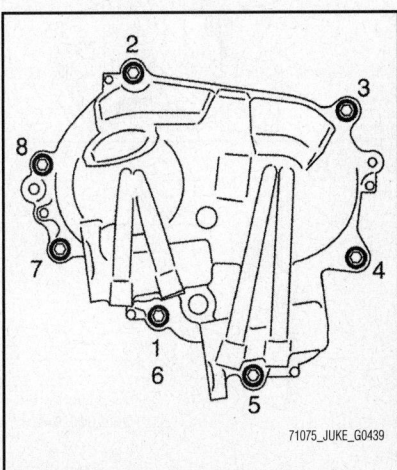

Fig. 136 Valve timing control cover bolt tightening sequence

d. Hold the WAF part of oil pump shaft [WAF: 10 mm (0.39 in)], and then loosen the oil pump sprocket bolt and remove it.

※※ WARNING
Secure the oil pump shaft with the WAF part.

※※ WARNING
Never loosen the oil pump sprocket bolt by tightening the oil pump drive chain.

14. Remove tension guide (front cover side) from front cover, if necessary.

To install:

※※ WARNING
Do not reuse O-rings.

➡The figure shows the relationship between the matching Mark on each timing chain and that on the corresponding sprocket, with the components installed.

15. Check that crankshaft key points straight up.
16. If the tension guide (front cover side) is removed, install it to the front cover.

➡**Check the joint condition by sound or feeling.**

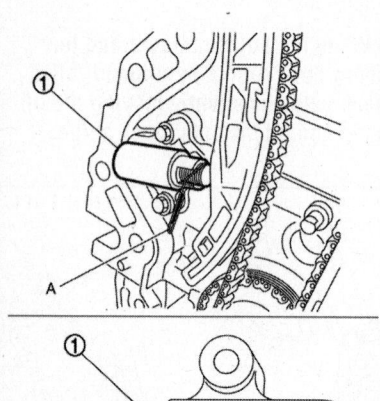

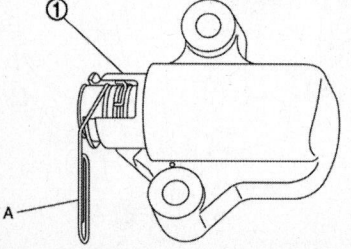

Fig. 137 Inserting wire (A) and removing timing chain tensioner (1)

17. Install crankshaft sprocket, oil pump sprocket, and oil pump drive chain.
    a. Install it by aligning matching marks on each sprockets and oil pump drive chain.
    b. If these matching marks are not aligned, rotate the oil pump shaft slightly to correct the position.

※※ WARNING
Check matching mark position of each sprockets after installing the oil pump drive chain.

18. Hold the WAF part of oil pump shaft [WAF: 10 mm (0.39 in)], and then tighten the oil pump shaft sprocket bolt.

※※ WARNING
Secure the oil pump shaft with the WAF part.

※※ WARNING
Never loosen the oil pump shaft sprocket bolt by tightening the oil pump drive chain.

19. Install oil pump chain tensioner.
    a. Fix the face oil pump tensioner at the most compressed position using a stopper pin, and then install it.
    b. Securely pull out the stopper pin after installing the oil pump chain tensioner.

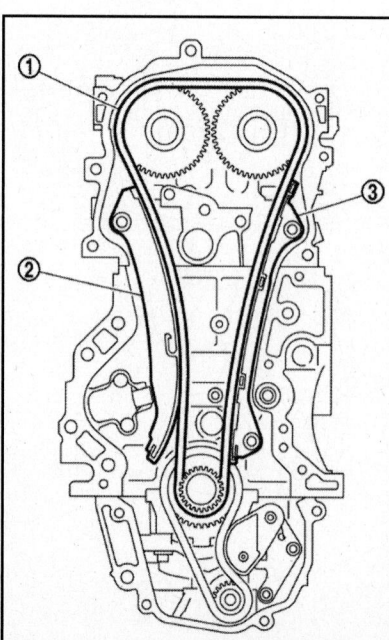

Fig. 138 Removing slack guide (2), tension guide (3) and timing chain (1)

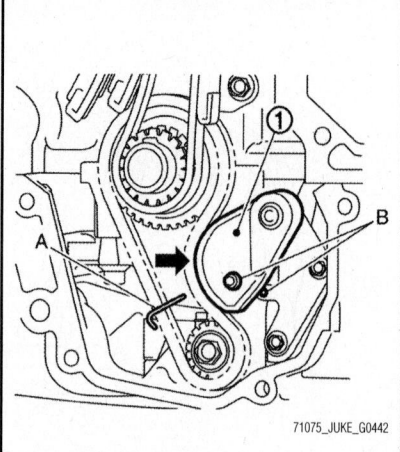

Fig. 139 Pushing oil pump drive chain tensioner (1) and inserting stopper pin (A) into body hole (B)

c. Check matching mark position of oil pump drive chain and each sprockets again.
20. Align the matching marks of each sprockets with the matching marks of timing chain.
    a. If these matching marks are not aligned, rotate the camshaft slightly by holding the hexagonal portion to correct the position.

※※ WARNING
Check matching mark position of each sprocket and timing chain again after installing the timing chain.

21. Install the slack guide and the tension guide.
22. Install timing chain tensioner.
    a. Fix the plunger at the most compressed position using a stopper pin, and then install it.
    b. Securely pull out the stopper pin after installing the timing chain tensioner.

※※ WARNING
After installing tensioner on the cam side, pull out lock pin.

※※ WARNING
If plunger pops out after pulling out lock pin without installing the tensioner to the engine, never use the tensioner. (If used, the plunger does not slide smoothly.)

23. After installation, pick up and move ratchet clip toward the tip of the plunger and

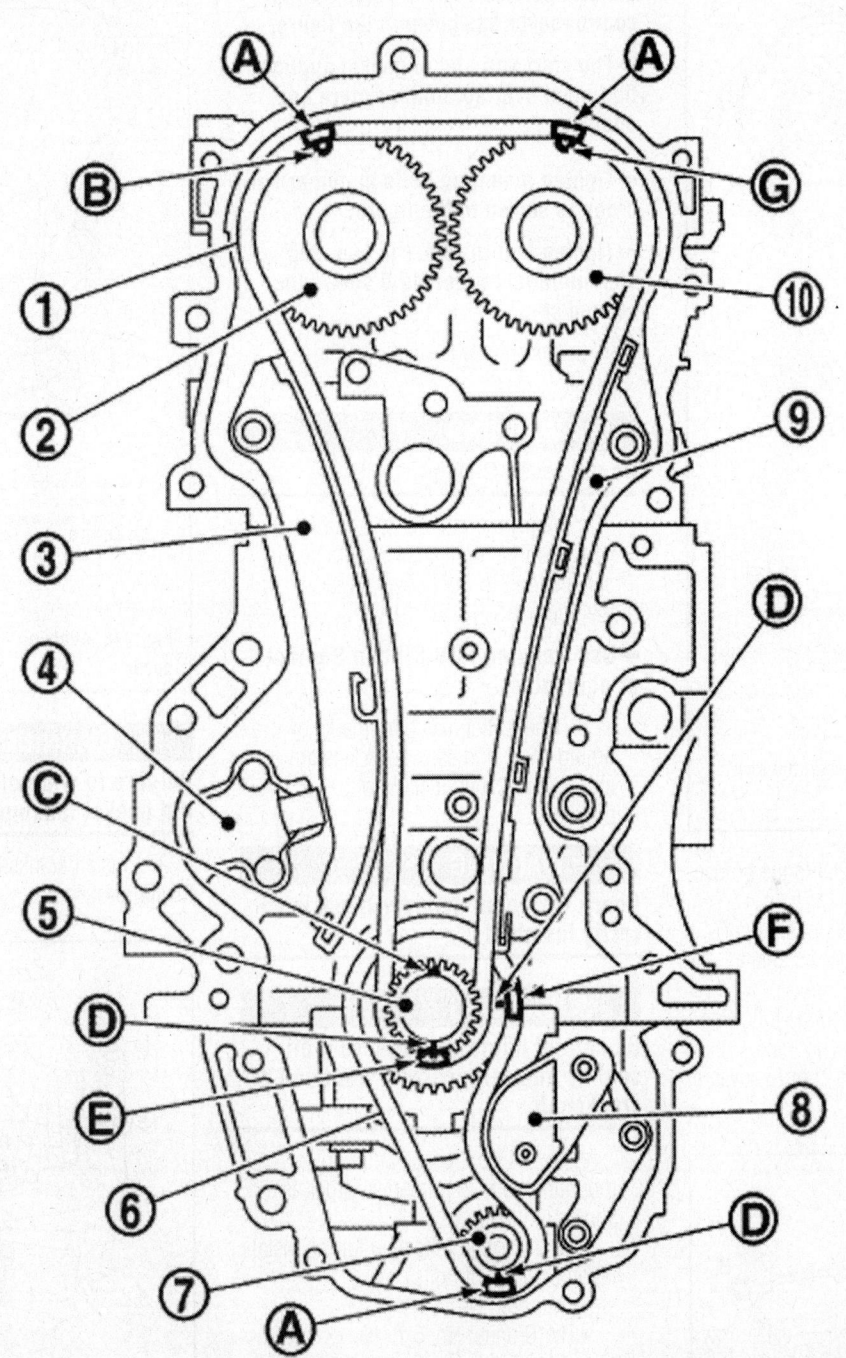

1 : Timing chain
2 : Camshaft sprocket (EXH)
3 : Slack guide
4 : Timing chain tensioner
5 : Crankshaft sprocket

6 : Oil pump drive chain
7 : Oil pump sprocket
8 : Oil pump drive chain tensioner
9 : Tension guide
10 : Camshaft sprocket (INT)

A : Matching mark (dark blue link)
B : Matching mark (outer groove)
C : Crankshaft key position (straight up)
D : Matching mark (stamping)
E : Matching mark (white link)
F : Matching mark (yellow link)
G : Matching mark (outer groove)

71075_JUKE_G0443

Fig. 140 Identifying matching mark on each timing chain and corresponding sprocket, with components installed

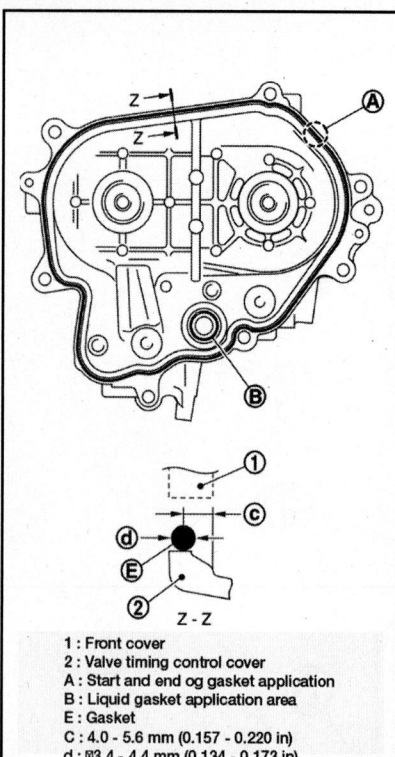

Fig. 141 Applying liquid gasket to valve timing control cover

1 : Front cover
2 : Valve timing control cover
A : Start and end og gasket application
B : Liquid gasket application area
E : Gasket
C : 4.0 - 5.6 mm (0.157 - 0.220 in)
d : ⌀3.4 - 4.4 mm (0.134 - 0.173 in)

71075_JUKE_G0444

position the tensioner parallel to the groove of the plunger.

24. Check matching mark position of timing chain and each sprockets again.

25. Install front oil seal.

26. Install front cover with the following procedure:

a. Install valve timing control cover, if removed.

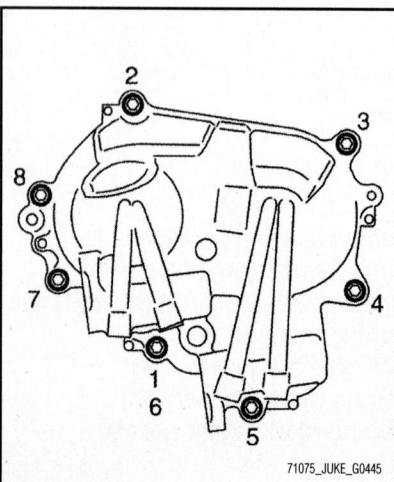

Fig. 142 Mounting bolt tightening sequence

71075_JUKE_G0445

➡ Apply a continuous bead of liquid gasket (E) with a tube presser (commercial service tool) to valve timing control cover as shown in the figure.

➡ The start and end of gasket application must overlap 5mm or more one another.

➡ Tighten mounting bolts in numerical order as shown in the figure.

➡ Tighten bolt the No.1 in two step. The numerical order No.6 shows the second step.

b. Install new O-ring to cylinder block.

✳ WARNING

Do not reuse O-rings.

c. Apply a continuous bead of liquid gasket with a tube presser (commercial service tool) to front cover as shown in the figure.

➡ Use Genuine RTV Silicon Sealant or equivalent.

d. Check that matching marks of timing chain and each sprockets are still aligned. Then install front cover.

✳ WARNING

Check O-ring on cylinder block is correctly installed.

✳ WARNING

Be careful not to damage front oil seal by interference with front end of crankshaft.

e. Install front cover, and tighten mounting bolts in numerical order as shown in the figure.

f. Refer to the following for the installation position of bolts.

- M6 bolt: No. 1
- M10 bolts: No. 6,7, 10, 13, 21
- M12 bolts: No. 2, 4, 8, 11
- M8 bolts: Except the above

✳ WARNING

Attaching should be done within 5 minutes after liquid gasket application.

g. After all bolts are tightened, retighten them to specified torque in numerical order as shown in the figure.

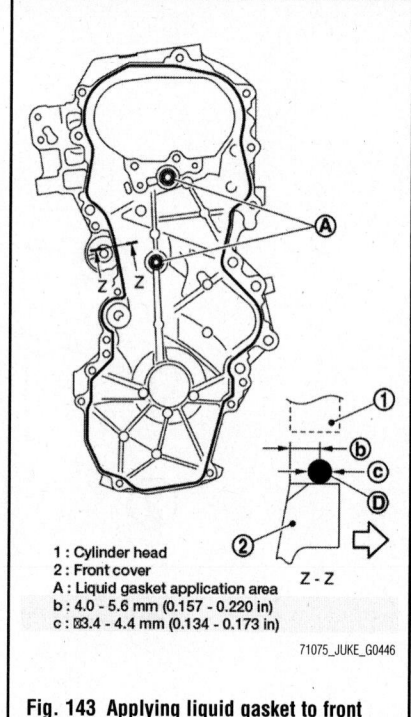

1 : Cylinder head
2 : Front cover
A : Liquid gasket application area
b : 4.0 - 5.6 mm (0.157 - 0.220 in)
c : ⌀3.4 - 4.4 mm (0.134 - 0.173 in)

71075_JUKE_G0446

Fig. 143 Applying liquid gasket to front cover

✳ WARNING

Be sure to wipe off any excessive liquid gasket leaking.

27. Install crankshaft pulley with the following procedure:

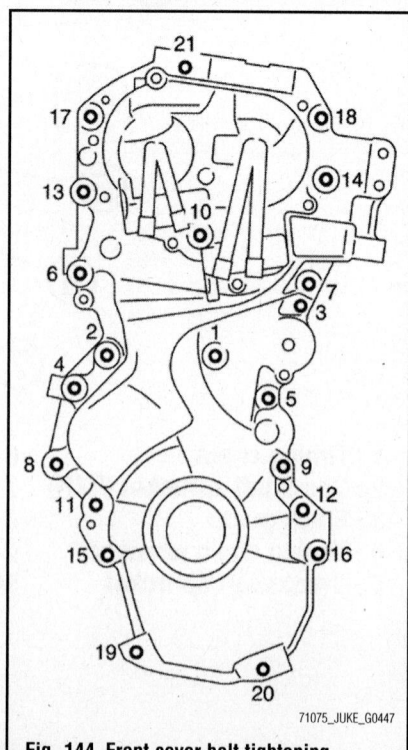

71075_JUKE_G0447

Fig. 144 Front cover bolt tightening sequence

a. When inserting crankshaft pulley with a plastic hammer, tap on its center portion (not circumference).

※※ WARNING

Never damage front oil seal lip section.

b. Secure crankshaft pulley with a pulley holder (commercial service tool).

c. Apply new engine oil to thread and seat surfaces of crankshaft pulley bolt.

d. Tighten crankshaft pulley bolt.

e. Put a paint mark on crankshaft pulley, matching with any one of six easy to recognize angle marks on crankshaft pulley bolt flange.

f. Turn another 60 degrees clockwise (angle tightening).

→Check the tightening angle with movement of one angle mark.

g. Check that crankshaft rotates clockwise smoothly.

28. Install remaining parts in the reverse order of removal.

TURBOCHARGER

REMOVAL & INSTALLATION

See Figure 145.

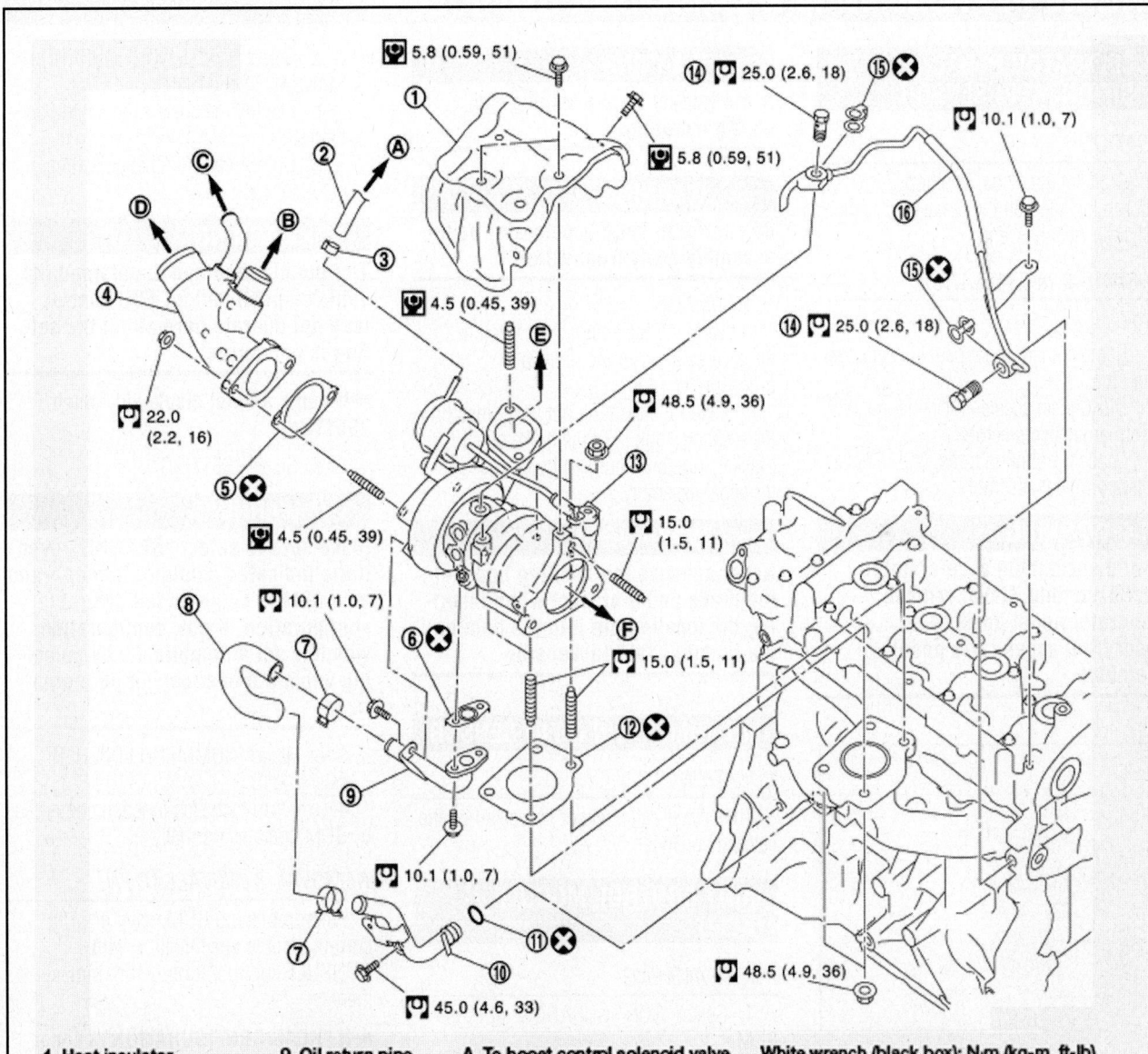

1. Heat insulator
2. Actuator hose
3. Clamp
4. Turbocharger inlet tube
5. Gasket
6. Gasket
7. Clamp
8. Oil outlet hose
9. Oil return pipe
10. Oil supply tube
11. O-ring
12. Gasket
13. Turbocharger
14. Eye bolt
15. Gasket
16. Oil supply tube

A. To boost control solenoid valve
B. To air inlet hose
C. To PCV hose
D. To air duct
E. To air inlet tube assembly
F. To catalyst converter

White wrench (black box): N·m (kg-m, ft-lb)
Black wrench (white box): N·m (kg-m, in-lb)
Circle X: Always replace after every disassembly.

71075_JUKE_G0448

Fig. 145 Removing and installing turbocharger

1. Drain engine coolant.
2. Remove engine cover.
3. Remove air cleaner cover assembly and air cleaner body assembly.
4. Remove air inlet tube 3.
5. Remove cowl top extension.
6. Disconnect heated oxygen sensor 2 harness connector.
7. Remove front tube.
8. Remove catalyst converter.

9. Remove turbocharger assembly as follows:
   a. Remove heat insulator.
   b. Disconnect water hose from turbocharger.
   c. Remove oil supply tube.
   d. Remove mounting nuts of turbocharger.

✳✳ WARNING

Never deform each turbocharger

piping when pulling out the assembly.

To install:

✳✳ WARNING

Do not reuse O-rings.

10. Installation is the reverse of the removal procedure.

ENGINE PERFORMANCE & EMISSION CONTROLS

ACCELERATOR PEDAL POSITION (APP) SENSOR

LOCATION

The Accelerator Pedal Position (APP) sensor is installed on the upper end of the accelerator pedal assembly.

REMOVAL & INSTALLATION

See Figure 146.

1. Remove instrument panel lower cover driver side.
2. Disconnect accelerator pedal position sensor harness connector.
3. Loosen mounting bolts, and remove accelerator pedal assembly.

✳✳ WARNING

Never disassemble accelerator pedal assembly. Never remove accelerator pedal position sensor from accelerator pedal assembly.

1. Accelerator pedal assembly
2. Brake pedal bracket
A. Locating hook
B. Locating pin
Black wrench (white box): N·m (kg-m, in-lb)

71075_JUKE_G0451

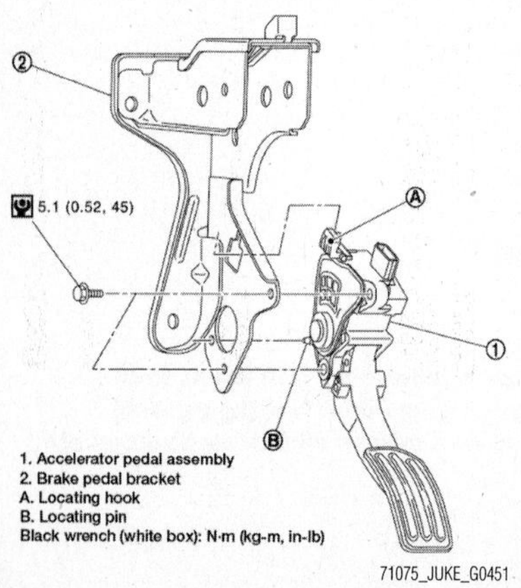

Fig. 146 Removing and installing accelerator pedal assembly

✳✳ WARNING

Avoid impact from dropping etc. during handling.

✳✳ WARNING

Be careful to keep accelerator pedal assembly away from water.

To install:

4. Note the following, and install in the reverse order of the removal procedure.
5. Insert the locating pin while inserting the locating hook in to the brake pedal bracket. Tighten mounting bolts to accelerator pedal assembly.

✳✳ WARNING

Never squeeze the locating hook into the brake pedal bracket when inserting the locating pin into the hole on the brake pedal bracket side.

AIR-FUEL RATIO (AFR) SENSOR

LOCATION

The air fuel ratio sensor is located on the catalyst converter.

BODY CONTROL MODULE (BCM)

PROGRAMMING

1. Select "CONFIGURATION" of BCM.
   a. When writing saved data, go to next step.
   b. When writing manually, go to step 3.
2. Perform "WRITE CONFIGURATION-CONFIG FILE".
   a. Work end.
3. Perform "WRITE CONFIGURATION-MANUAL SELECTION".

   a. Select "WRITE CONFIGURATION-MANUAL SELECTION".
   b. Identify the correct model and configuration.
   c. Confirm and/or change setting value for each item.

✳✳ WARNING

Thoroughly read and understand the vehicle specification. ECU control may not operate normally if the setting is not correct.

➡**If items are not displayed, touch "SETTING".**

   d. Select "SETTING".

✳✳ WARNING

Make sure to select "SETTING" even if the indicated configuration of brand new BCM is same as the desirable configuration. If not, configuration which is set automatically by selecting vehicle model cannot be memorized.

   e. When "COMMAND FINISHED", select "END".
4. Confirm that each function controlled by BCM operates normally.

REMOVAL & INSTALLATION

When replacing BCM, save or print current vehicle specification with CONSULT configuration before replacement.

➡**If "READ CONFIGURATION" cannot be used, use the "WRITE CONFIGURATION" after replacing BCM.**

1. Remove instrument lower panel.
2. Remove harness clip.
3. Remove BCM mounting screws.
4. Remove BCM and disconnect the connectors.

5. Remove relays and relay mounting bracket from BCM.

To install:

6. Installation is the reverse of the removal procedure.

❈❈ WARNING

Be sure to perform "WRITE CONFIG-URATION" when replacing BCM. Or not doing so, BCM control function does not operate normally.

➡**Be sure to perform the system initialization (NATS) when replacing BCM.**

CAMSHAFT POSITION (CMP) SENSOR

LOCATION

The Camshaft Position (CMP) sensor is located on the camshaft bracket.

CRANKSHAFT POSITION (CKP) SENSOR

LOCATION

The Crankshaft Position (CKP) sensor is located on the upper oil pan facing the gear teeth (cogs) of the signal plate.

ELECTRONIC CONTROL MODULE (ECM)

LOCATION

See Figure 147.

REMOVAL & INSTALLATION

1. Remove fusible link bracket. Keep a service area.

2. Disconnect ECM harness connector.

3. Remove ECM mounting nuts, and ECM.

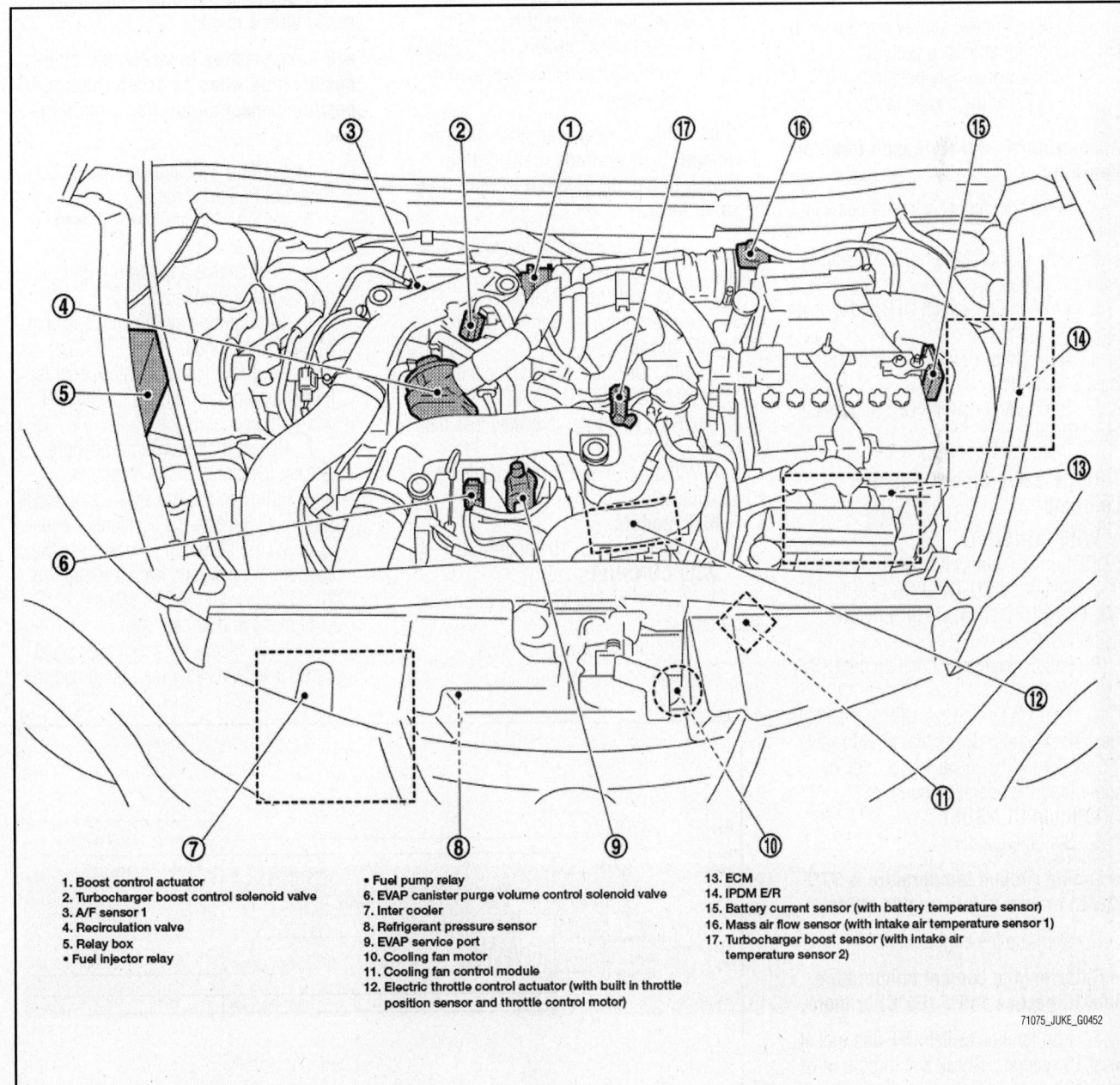

1. Boost control actuator
2. Turbocharger boost control solenoid valve
3. A/F sensor 1
4. Recirculation valve
5. Relay box
• Fuel injector relay
• Fuel pump relay
6. EVAP canister purge volume control solenoid valve
7. Inter cooler
8. Refrigerant pressure sensor
9. EVAP service port
10. Cooling fan motor
11. Cooling fan control module
12. Electric throttle control actuator (with built in throttle position sensor and throttle control motor)
13. ECM
14. IPDM E/R
15. Battery current sensor (with battery temperature sensor)
16. Mass air flow sensor (with intake air temperature sensor 1)
17. Turbocharger boost sensor (with intake air temperature sensor 2)

71075_JUKE_G0452

Fig. 147 Engine control system

To install:

4. Installation is the reverse of the removal procedure.

RESET

Initialization of NVIS (NATS) System And Registration Of All NVIS (NATS) ignition key IDs

1. Install ECM.
2. Contact backside of registered Intelligent key to push-button ignition switch, then turn power supply position to ON.

➡**To perform this step, use the key that is used before performing ECM replacement.**

3. Maintain power supply position in the ON position for at least 5 seconds.
4. Turn power supply position to OFF.
5. Check that the engine starts.

Accelerator Pedal Released Position Learning

1. Make sure that accelerator pedal is fully released.
2. Turn ignition switch ON and wait at least 2 seconds.
3. Turn ignition switch OFF and wait at least 10 seconds.
4. Turn ignition switch ON and wait at least 2 seconds.
5. Turn ignition switch OFF and wait at least 10 seconds.

Throttle Valve Closed Position Learning

With CONSULT:

1. Turn ignition switch ON.
2. Select "CLSD THL POS LEARN" in "WORK SUPPORT" mode of "ENGINE" using CONSULT.
3. Follow the instructions on the CONSULT display.
4. Turn ignition switch OFF and wait at least 10 seconds. Check that throttle valve moves during the above 10 seconds by confirming the operating sound.

Without CONSULT:

5. Start the engine.

➡**Engine coolant temperature is 77°F (25°C) or less before engine starts.**

6. Warm up the engine.

➡**Raise engine coolant temperature until it reaches 149°F (65°C) or more.**

7. Turn ignition switch OFF and wait at least 10 seconds. Check that throttle valve moves during the above 10 seconds by confirming the operating sound.

Idle Air Volume Learning

See Figure 148.

1. Perform preconditioning listed below.

❄ WARNING

Make sure that all of the following conditions are satisfied. Learning will be cancelled if any of the following conditions are missed for even a moment.

- Battery voltage: more than 12.9V (at idle)
- Engine coolant temperature: 158-212°F (70-100°C)
- Selector lever: P or N
- Electric load switch: OFF (Air conditioner, headlamp, rear window defogger)

➡**On vehicles equipped with daytime running light systems, set lighting switch to the 1st position to light only small lamps.**

- Steering wheel: Neutral (straight ahead position)
- Vehicle speed: stopped
- Transmission: warmed up

CVT models:
- With CONSULT: Drive vehicle until "ATF TEMP SEN" in "DATA MONITOR" mode of "CVT" system indicated less than 0.9V
- Without CONSULT: Drive vehicle for 10 minutes

M/T models:
- Drive vehicle for 10 minutes

With CONSULT:

2. Perform idle air volume learning procedure.

   a. Perform Accelerator Pedal Released Position Learning.

   b. Perform Throttle Valve Closed Position Learning.

   c. Start engine and warm it up to normal operating temperature.

   d. Select "IDLE AIR VOL LEARN" in "WORK SUPPORT" mode of "ENGINE".

   e. Touch "START" and wait 20 seconds.

   f. If "CMPLT" is displayed on CONSULT screen, go to step 4.

   g. If "CMPLT" is not displayed on CONSULT screen, go to step 5.

Without CONSULT:

3. Perform idle volume learning procedure.

➡**It is better to count the time accurately with a clock.**

➡**It is impossible to switch the diagnostic mode when an accelerator pedal position sensor circuit has a malfunction.**

   a. Perform Accelerator Pedal Released Position Learning.

   b. Perform Throttle Valve Closed Position Learning.

   c. Start engine and warm it up to normal operating temperature.

   d. Turn ignition switch OFF and wait at least 10 seconds.

   e. Confirm that accelerator pedal is fully released, turn ignition switch ON and wait 3 seconds.

   f. Repeat the following procedure quickly five times within 5 seconds.
- Fully depress the accelerator pedal.
- Fully release the accelerator pedal.

   g. Wait 7 seconds, fully depress the accelerator pedal and keep it for approx. 20 seconds until the MIL stops blinking and turned ON.

   h. Fully release the accelerator pedal within 3 seconds after the MIL turned ON.

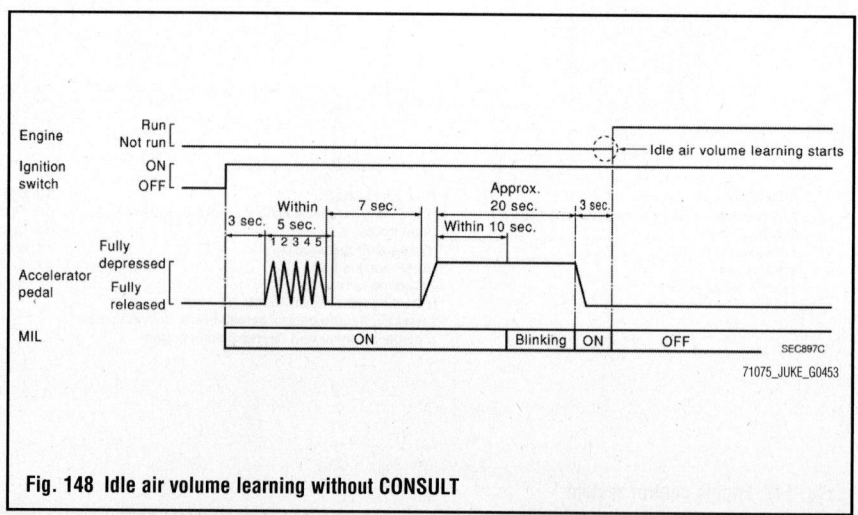

Fig. 148 Idle air volume learning without CONSULT

71075_JUKE_G0453

i.   Start engine and let it idle.

j.   Wait 20 seconds.

4.   Check idle speed and ignition timing.

a.   Rev up the engine 2 or 3 times and make sure that idle speed and ignition timing are within the specifications. Idle speed CVT with no load (in P or N position): 650 +/- 50 rpm. Idle speed M/T no load (in neutral position), 600 +-/ 50 rpm. Ignition timing CVT no load (in P or N position), 6 +/- 2°BTDC. Ignition timing M/T no load (in neutral position), 8 +/- 2°BTDC.

b.   If the inspection result is normal, inspection is complete.

c.   If the inspection result is not normal, go to next step.

5.   Detect malfunction part.

a.   Check the following:
- Throttle valve fully closed
- PCV valve operation
- Downstream of throttle valve is free from air leakage

b.   If the inspection result is normal, go to next step.

c.   If the inspection result is not normal, repair or replace malfunctioning part.

6.   Detect malfunctioning part.

a.   Engine component parts and their installation condition are questionable. Check and eliminate the cause of the incident.

b.   It is useful to perform "TROUBLE DIAGNOSIS-SPECIFICATION VALUE".

c.   If any of the following conditions occur after the engine has started, eliminate the cause of the incident and perform Idle Air Volume Learning all over again:
- Engine stalls
- Erroneous idle

G Sensor Calibration

1.   Park the vehicle on a level surface.

2.   Adjust air pressure of all tires to the specified pressure.

3.   Perform calibration.

With CONSULT:

a.   Turn ignition switch ON.

✳✳ WARNING

Never start the engine.

b.   Select "WORK SUPPORT" mode in "ENGINE".

c.   Select "G SENSOR CALBRATION".

d.   Touch "START".

✳✳ WARNING

Never swing the vehicle during "G SENSOR CALIBRATION".

e.   If "COMPLETED" is displayed, end.

f.   If "COMPLETED" is not displayed, perform steps 1 and 2 again.

ENGINE COOLANT TEMPERATURE (ECT) SENSOR

LOCATION

The Engine Coolant Temperature (ECT) sensor is located on the water outlet.

EVAP CANISTER

REMOVAL & INSTALLATION

2WD Vehicles

See Figure 149.

1.   Disconnect harness connectors (EVAP control pressure sensor and EVAP canister vent control valve) and EVAP canister hoses.

2.   Remove EVAP canister fixing bolt.

3.   Remove EVAP canister.

➡**EVAP canister vent control valve and EVAP control pressure sensor can be removed without removing the EVAP canister.**

To install:

4.   Installation is the reverse of the removal procedure. Make sure to use new O-rings.

➡**Tighten the EVAP canister fixing bolts to 83 inch lbs. (9.4 Nm).**

AWD Vehicles

See Figures 150 and 151.

1.   Remove rear stabilizer (1).

2.   Disconnect harness connectors (EVAP control pressure sensor and EVAP canister vent control valve) and EVAP canister hoses.

3.   Remove EVAP canister fixing bolt.

4.   Remove EVAP canister.

➡**EVAP canister vent control valve and EVAP control pressure sensor can be removed without removing the EVAP canister.**

To install:

5.   Installation is the reverse of the removal procedure. Make sure to use new O-rings.

➡**Tighten the EVAP canister fixing bolts to 83 inch lbs. (9.4 Nm).**

HEATED OXYGEN SENSOR (HO2S)

LOCATION

The heated oxygen sensor is located on the catalyst converter.

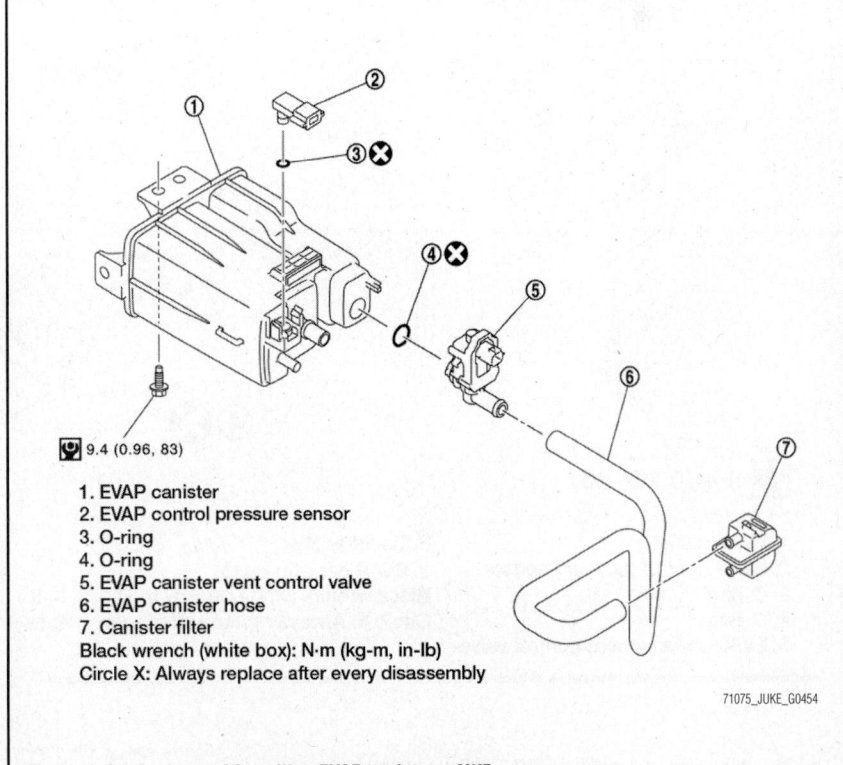

⬛ 9.4 (0.96, 83)

1. EVAP canister
2. EVAP control pressure sensor
3. O-ring
4. O-ring
5. EVAP canister vent control valve
6. EVAP canister hose
7. Canister filter
Black wrench (white box): N·m (kg-m, in-lb)
Circle X: Always replace after every disassembly

71075_JUKE_G0454

Fig. 149 Removing and installing EVAP canister—2WD

71075_JUKE_G0455

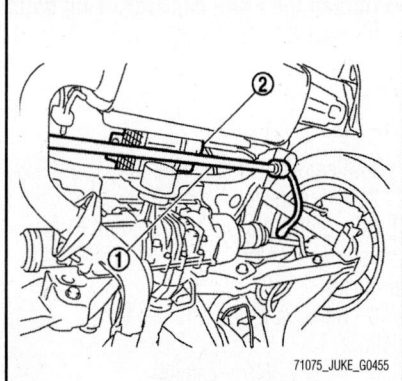

Fig. 150 Removing rear stabilizer (1) from in front the EVAP canister (2)

INTAKE AIR TEMPERATURE (IAT)/MASS AIR FLOW (MAF) SENSOR

LOCATION

See Figure 152.

The Intake Air Temperature (IAT)/Mass Air Flow (MAF) sensor is located on the air cleaner cover assembly.

KNOCK SENSOR (KS)

LOCATION

The Knock Sensor is attached to the cylinder block.

PCV VALVE

LOCATION

The PCV valve is located on the rocker cover.

THROTTLE CONTROL ACTUATOR

LOCATION

The throttle control actuator is located on the throttle body.

71075_JUKE_G0457

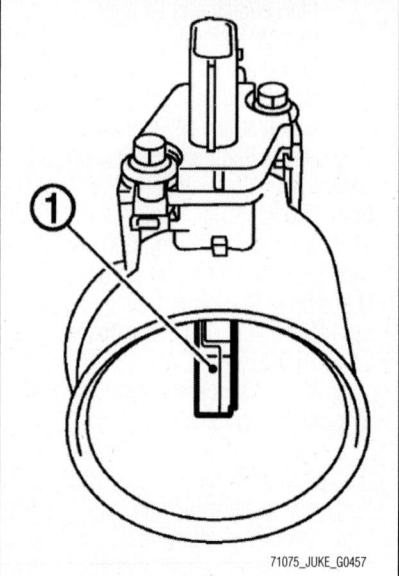

Fig. 152 Intake Air Temperature (IAT)/Mass Air Flow (MAF) sensor (1) location

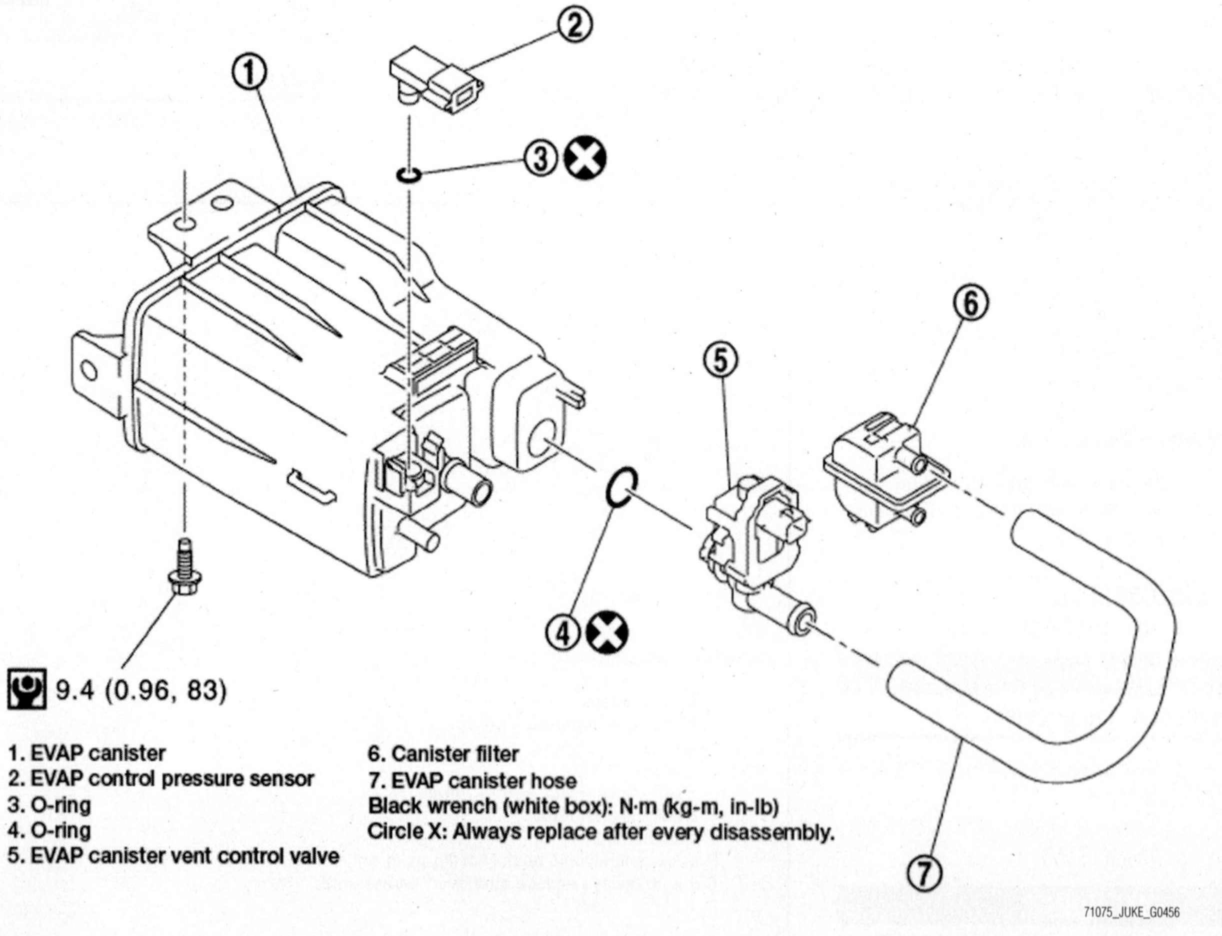

9.4 (0.96, 83)
1. EVAP canister
2. EVAP control pressure sensor
3. O-ring
4. O-ring
5. EVAP canister vent control valve
6. Canister filter
7. EVAP canister hose
Black wrench (white box): N·m (kg-m, in-lb)
Circle X: Always replace after every disassembly.
71075_JUKE_G0456

Fig. 151 Removing and installing EVAP canister—AWD

FUEL SYSTEM SERVICE PRECAUTIONS

Safety is the most important factor when performing not only fuel system maintenance but any type of maintenance. Failure to conduct maintenance and repairs in a safe manner may result in serious personal injury or death. Maintenance and testing of the vehicle's fuel system components can be accomplished safely and effectively by adhering to the following rules and guidelines.

• To avoid the possibility of fire and personal injury, always disconnect the negative battery cable unless the repair or test procedure requires that battery voltage be applied.

• Always relieve the fuel system pressure prior to disconnecting any fuel system component (injector, fuel rail, pressure regulator, etc.), fitting or fuel line connection. Exercise extreme caution whenever relieving fuel system pressure to avoid exposing skin, face and eyes to fuel spray. Please be advised that fuel under pressure may penetrate the skin or any part of the body that it contacts.

• Always place a shop towel or cloth around the fitting or connection prior to loosening to absorb any excess fuel due to spillage. Ensure that all fuel spillage (should it occur) is quickly removed from engine surfaces. Ensure that all fuel soaked cloths or towels are deposited into a suitable waste container.

• Always keep a dry chemical (Class B) fire extinguisher near the work area.

• Do not allow fuel spray or fuel vapors to come into contact with a spark or open flame.

• Always use a back-up wrench when loosening and tightening fuel line connection fittings. This will prevent unnecessary stress and torsion to fuel line piping.

• Always replace worn fuel fitting O-rings with new Do not substitute fuel hose or equivalent where fuel pipe is installed.

Before servicing the vehicle, make sure to also refer to the precautions in the beginning of this section as well.

RELIEVING FUEL SYSTEM PRESSURE

With CONSULT

1. Turn ignition switch ON.
2. Perform "FUEL PRESSURE RELEASE" in "WORK SUPPORT" mode of "ENGINE" using CONSULT.

3. Start engine.
4. After engine stalls, crank it two or three times to release all fuel pressure.
5. Turn ignition switch OFF.

Without CONSULT

1. Remove fuel pump fuse located in IPDM E/R.
2. Start engine.
3. After engine stalls, crank it two or three times to release all fuel pressure.
4. Turn ignition switch OFF.
5. Reinstall fuel pump fuse after servicing fuel system.

FUEL FILTER, FUEL LEVEL SENSOR UNIT & FUEL PUMP

REMOVAL & INSTALLATION

See Figures 153 through 156.

1. Release the fuel pressure from the fuel lines.
2. Check fuel level on a level ground. If the fuel level is 7/8 of the fuel tank (full or nearly full), draw appropriate amount of fuel from the fuel tank.
3. In the event of malfunction in fuel pump, insert a hose measuring 25 mm (0.98 in) in diameter into the filler opening to draw approximately 25 liters fuel.
   a. Open fuel filler lid.
4. Open filler cap and release the pressure inside fuel tank.
5. Remove rear seat.
6. Remove inspection hole cover.
7. Disconnect harness connector and quick connector.

8. Remove quick connector in the following procedures.
   a. Pinch quick connector square-part with your fingers, and pull out the quick connector by hand.
   b. If quick connector and tube on sender unit are stuck, push and pull several times until they move, and pull out.

✳✳ WARNING

Quick connector can be removed when the tabs are completely depressed. Never twist it more than necessary.

➡**Do not use any tools to disconnected quick connector.**

✳✳ WARNING

Keep resin tube away from heat. Be especially careful when welding near the resin tube.

✳✳ WARNING

Prevent acid liquid such as battery electrolyte, etc. from getting on resin tube.

✳✳ WARNING

Never bend or twist resin tube during installation and disconnection.

➡**To keep the connecting portion clean and to avoid damage and foreign materials, cover them completely with plastic bags or something similar.**

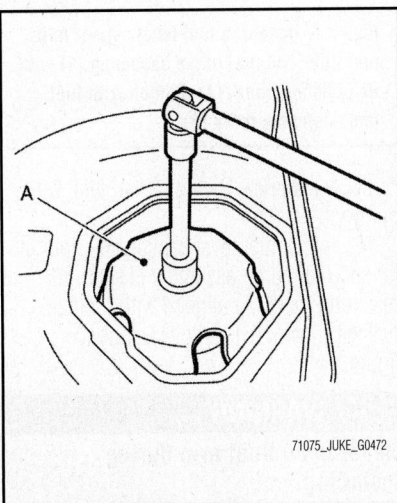

71075_JUKE_G0472

Fig. 153 Removing lock ring with lock ring wrench (A)

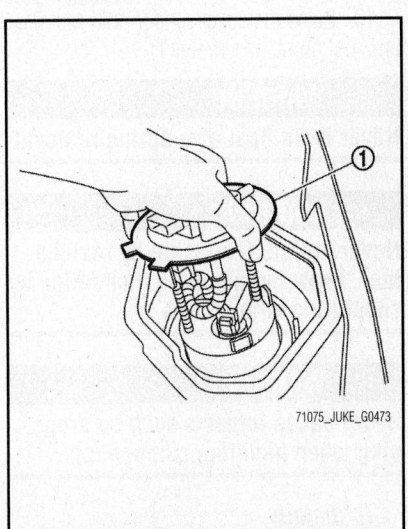

71075_JUKE_G0473

Fig. 154 Removing fuel level sensor unit, fuel filter and fuel pump assembly

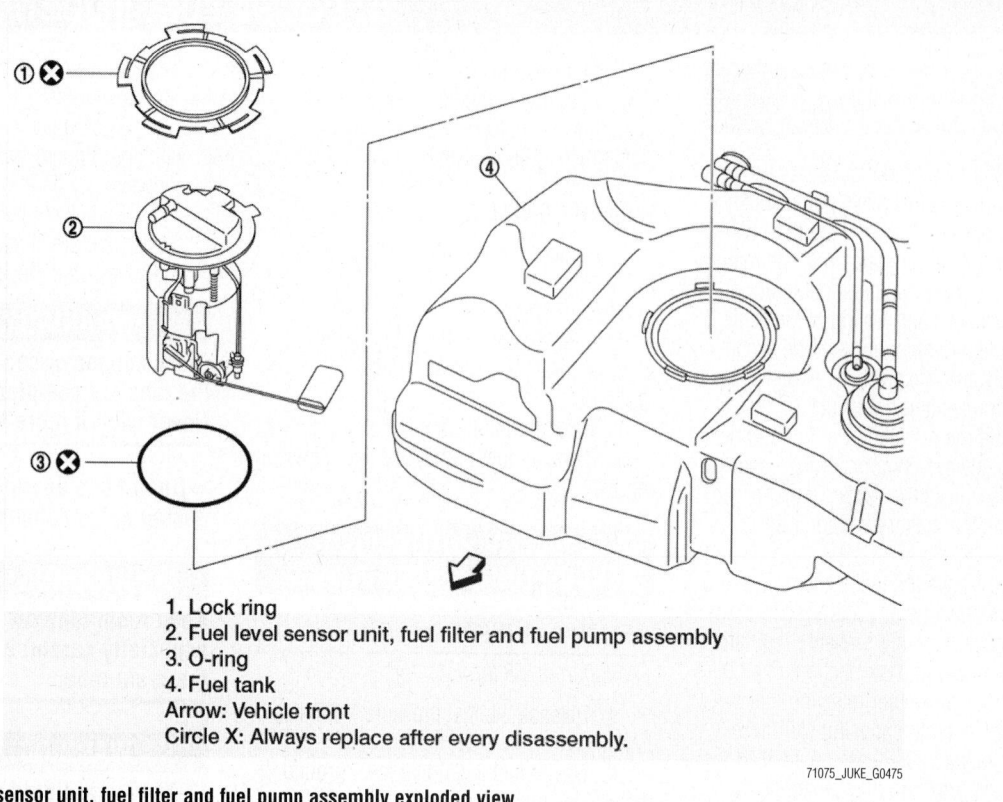

1. Lock ring
2. Fuel level sensor unit, fuel filter and fuel pump assembly
3. O-ring
4. Fuel tank
Arrow: Vehicle front
Circle X: Always replace after every disassembly.

71075_JUKE_G0475

Fig. 155 Fuel level sensor unit, fuel filter and fuel pump assembly exploded view

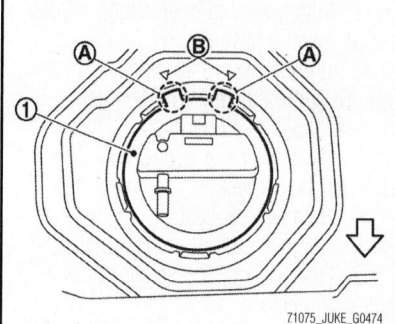

71075_JUKE_G0474

Fig. 156 Installing fuel level sensor unit, fuel filter and fuel pump assembly (1) with its matching mark (A) aligned with fuel tank matching mark (B)

✳✳ WARNING

Never insert plug, preventing damage on O-ring in quick connector.

9. Remove lock ring for fuel level sensor unit, fuel filter and fuel pump assembly with lock ring wrench (A) [SST: KV99110600 (J- 46214)] turning counterclockwise.

➡ **For reference when installing, put a matching mark on lock ring, fuel pump assembly and fuel tank.**

10. Raise fuel level sensor unit, fuel filter and fuel pump assembly (1).

✳✳ WARNING

Never bend float arm during removal.

✳✳ WARNING

Never pollute the inside by residue fuel. Draw out avoiding inclination by supporting with a cloth.

✳✳ WARNING

Never cause impacts such by dropping when handling components.

To install:
Note to the following, and install in the reverse order of removal.

11. Install new O-ring to fuel tank without twist.
12. Install fuel level sensor unit, fuel filter and fuel pump assembly (1) with its matching mark (A) aligned with fuel tank matching mark (B) as shown in the figure.

✳✳ WARNING

Never bend float arm during installing.

13. Install lock ring for fuel level sensor unit, fuel filter and fuel pump assembly with

lock ring wrench [SST: KV99110600 (J-46214)] turning clockwise.

14. Connect quick connector of fuel feed tube as per following procedures.
   a. Check the connection for damage or any foreign materials.
   b. Align the connector with the tube, then insert the connector straight into the tube until a click sound is heard.
   c. After connecting, check that the connection is secure by following method.
   • Visually confirm that the two tabs are connected to the connector.
   • Pull the tube and the connector to check they are securely connected.

FUEL RAIL & INJECTORS

REMOVAL & INSTALLATION
See Figures 157 through 165.

✳✳ CAUTION

High pressure fuel system components are between high pressure fuel pump and fuel injector.

✳✳ WARNING

Always release fuel pressure and never start the engine when performing removal and installation.

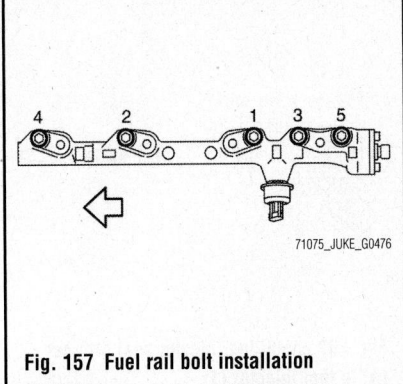

Fig. 157 Fuel rail bolt installation sequence

⁂ **CAUTION**

When removing or installing parts without releasing fuel pressure, fuel may be splashed and, if fuel contacts skin or eyes, it may cause inflammation.

⁂ **CAUTION**

Be sure to work in a well-ventilated area and furnish workshop with a CO2 fire extinguisher.

⁂ **CAUTION**

Never smoke while servicing fuel system. Keep open flames and sparks away from the work area.

⁂ **CAUTION**

To avoid the danger of being scalded, never drain engine coolant when engine is hot.

1. Release the fuel pressure.
2. Remove front bumper.
3. Remove charge air cooler.
4. Remove oil level gauge.
5. Remove intake manifold.
6. Remove alternator.
7. Remove oil level gauge guide.
8. Remove fuel rail cover, and then remove fuel rail insulator.

⁂ **WARNING**

When fuel rail insulator (separated type) and adapter cover (separated type) are removed, be sure to replace them with fuel rail insulator (united type).

9. Remove fuel tube and fuel rail connector.
10. Disconnect fuel pressure sensor harness connector.

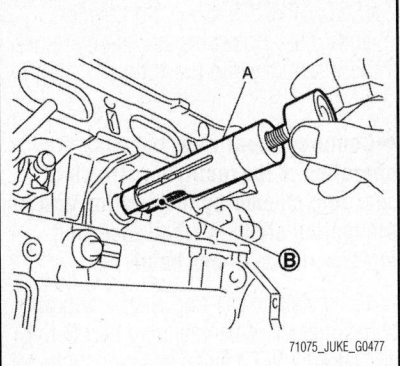

Fig. 158 Installing remover (A) to injector connector side so that cutout (B) of injector remover faces injector connector side

a. Disconnect fuel injector harness connector.
11. Remove fuel pressure sensor, if necessary.
12. Remove fuel rail.
a. Loosen mounting bolts in reverse order as shown in the figure.

⁂ **WARNING**

When removing, be careful to avoid any interference with fuel injector.

⁂ **WARNING**

Use a shop cloth to absorb any fuel leakage from fuel rail.

13. Remove fuel injector from cylinder head as per the following.

⁂ **WARNING**

Be careful with remaining fuel that may go out from fuel rail.

⁂ **WARNING**

Be careful not to damage injector nozzles during removal.

⁂ **WARNING**

Never bump or drop fuel injector. Do not disassemble the fuel injector.

a. Remove injector holder.
b. Install an remover [SST: KV10119600 (-)] (A) to the injector connector side so that cutout (B) of injector remover faces the injector connector side.

➡ **Hook pawl portion (B) of injector remover [SST: KV10119600(-)] (A) to groove portion (C) of injector.**

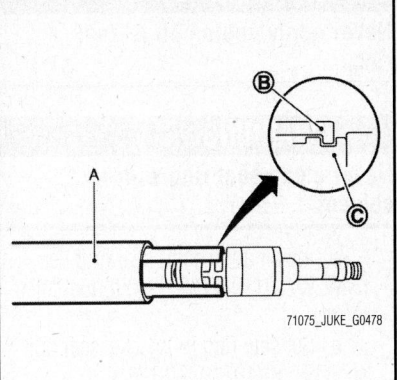

Fig. 159 Hooking pawl portion (B) of injector remover (A) to groove portion (C) of injection

c. Press down body portion of injector remover [SST: KV10119600(-)] (A) until it contacts cylinder head.
d. Tighten injector remover [SST: KV10119600 (-)] clockwise and remove injector from cylinder head.
e. Cut seal ring (1) while pinching it. Be careful not to damage injector.

To install:

⁂ **WARNING**

Do not reuse O-rings.

14. Install seal ring to fuel injector as per the following:

➡ **Handle seal ring with bare hands. Never wear gloves.**

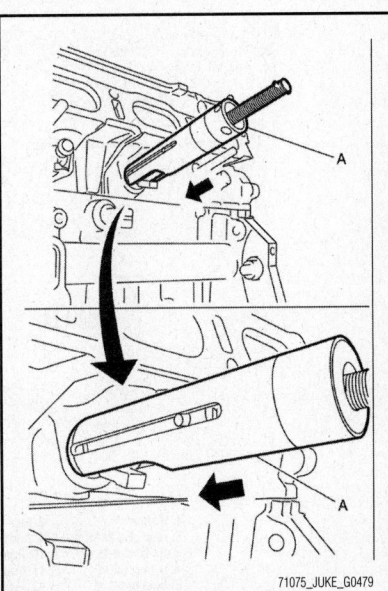

Fig. 160 Pressing down body portion of injector remover (A) until it contacts cylinder head

✸✸ WARNING

Never apply engine oil to seal ring.

✸✸ WARNING

Never clean seal ring with solvent.

   a. Install an injector seal drift set [SST: KV101197S0 (-)] (A) to fuel injector (1).

   b. Set seal ring to injector seal drift set [SST: KV101197S0 (-)].

   c. Straightly insert seal ring, which is set in step 2, to fuel injector and install.

✸✸ WARNING

Be careful that seal ring does not exceed the groove portion of fuel injector.

   d. Insert injector seal drift set [SST: KV101197S0 (-)]to injector and rotate clockwise and counterclockwise by 90°while pressing seal ring to fit it.

➡**Compress seal ring, because this operation is for rectifying stretch of seal ring caused by installation and for preventing sticking when inserting injector into cylinder head.**

   15. Install new O-ring and backup ring to fuel injector. When handing new O-ring and backup ring, paying attention to the following caution items:

✸✸ WARNING

Handle O-ring with bare hands. Never wear gloves.

✸✸ WARNING

Lubricate O-ring with new engine oil.

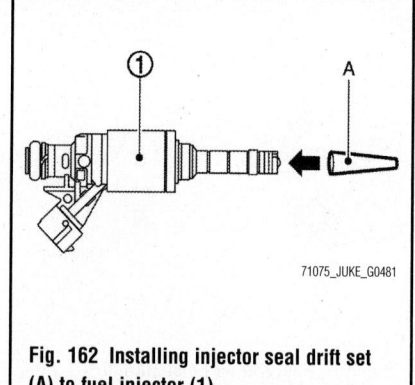

71075_JUKE_G0481

Fig. 162 Installing injector seal drift set (A) to fuel injector (1)

✸✸ WARNING

Never clean O-ring with solvent.

✸✸ WARNING

Check that O-ring and its mating part are free of foreign material.

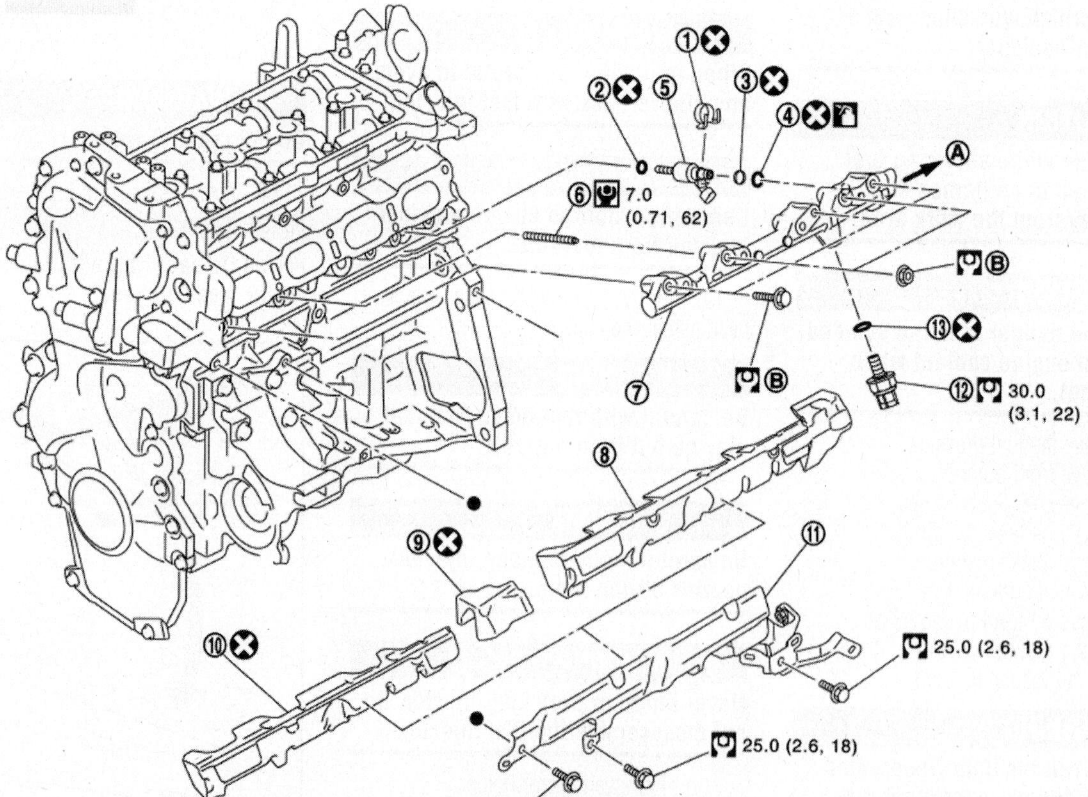

---

6. Stud bolt ☒ 7.0 (0.71, 62)

13 ☒

12 ☒ 30.0 (3.1, 22)

☒ 25.0 (2.6, 18)

☒ 25.0 (2.6, 18)

☒ 25.0 (2.6, 18)

1. Holder
2. Seal ring (white)
3. Backup ring
4. O-ring (blue)
5. Fuel injector
6. Stud bolt
7. Fuel rail
8. Fuel rail insulator (united type)
9. Adaptor cover (separated type)
10. Fuel rail insulator (separated type)
11. Fuel rail cover
12. Fuel pressure sensor
13. Gasket

A. To fuel rail connector and fuel tube.
B. Comply with the assembly procedure when tightening.
White wrench (black box): N·m (kg-m, ft-lb)
Black wrench (white box): N·m (kg-m, in-lb)
Circle X: Always replace after every disassembly.
Oil can: Should be lubricated with oil.
Black dot: Indicates that the parts is connected at points with same symbols in actual vehicle.

71075_JUKE_G0480

Fig. 161 Removing and installing fuel injector and rail

✳✳ WARNING

When installing O-ring, be careful not to scratch it with tool or fingernails. Also be careful not to twist or stretch O-ring. If O-ring was stretched while it was being attached, never insert it quickly into fuel tube.

✳✳ WARNING

Insert new O-ring straight into fuel rail. Never decenter or twist it.

✳✳ WARNING

Always install the back-up ring (1) in the right direction as instructed.

16. Install fuel injector (1) to fuel rail (2) as per the following:
   a. Install fuel injector holder (5) to fuel injector.

✳✳ WARNING

Never reuse fuel injector holder. Replace it with a new one.

✳✳ WARNING

Be careful to keep fuel injector holder from interfering with O-ring. If interference occurs, replace O-ring.

   b. Insert fuel injector into fuel rail with fuel injector holder attached.

✳✳ WARNING

Insert it while matching it to the axial center.

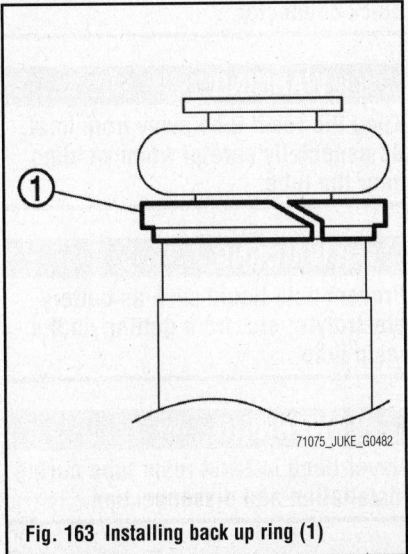

Fig. 163 Installing back up ring (1)

✳✳ WARNING

Insert so that protrusion (A) of fuel injector is aligned to cutout (B).

   c. Check that installation is complete by checking that fuel injector does not rotate or come off.

✳✳ WARNING

Check that protrusions of fuel injectors and fuel rail are aligned with cutouts of clips after installation.

17. Install fuel rail and fuel injector assembly to cylinder head.
   a. Tighten mounting bolts and nuts in two steps in numerical order as shown in the figure.
   - Step 1: Tighten to 87 inch lbs. (10 Nm)
   - Step 2: Tighten to 15 ft. lbs. (20.5 Nm)
18. Connect injector harness connector.
19. Install fuel pressure sensor, if removed.
20. Install fuel rail insulator.

✳✳ WARNING

As covering part of fuel tube connector at the back end of common rail can easily move because of its shape, do not remove it before installation.

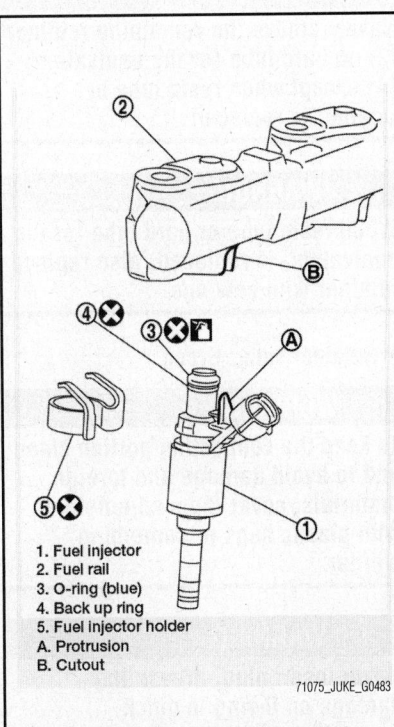

1. Fuel injector
2. Fuel rail
3. O-ring (blue)
4. Back up ring
5. Fuel injector holder
A. Protrusion
B. Cutout

Fig. 164 Installing fuel injector to fuel rail

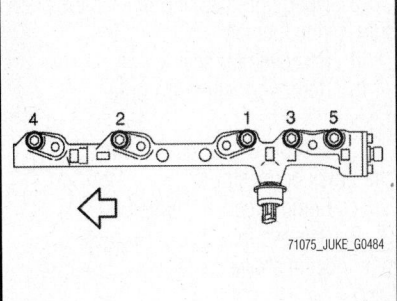

71075_JUKE_G0484

Fig. 165 Identifying mounting bolt tightening sequence (arrow: engine front)

✳✳ WARNING

Install the insulator so that it is placed under lower side of intake manifold flange.

21. Install in the reverse order of removal after this step.

FUEL TANK

DRAINING

1. Release the fuel pressure from the fuel lines.
2. Check fuel level on level ground. If fuel level is ⅞ of the fuel tank (full or nearly full), draw appropriate amount of fuel from the fuel tank.

➡**Guideline: Draw approximately 6⅝ gallons (25 liters) from a full tank condition.**

   a. In the event of malfunction in fuel pump, insert a hose measuring 0.98 inch (25 mm) in diameter into the filler opening to draw approximately 6⅝ gallons (25 liters) of fuel.

REMOVAL & INSTALLATION

2WD Vehicles

See Figures 166 through 172.

➡**Drain fuel from fuel tank if necessary.**

➡**Perform work on level place.**

1. Check fuel level on a level ground. If the fuel level is ⅞ of the fuel tank (full or nearly full), draw appropriate amount of fuel from the fuel tank.

➡**In the event of malfunction in fuel pump, insert a hose measuring 25 mm (0.98 in) in diameter into the filler opening to draw approximately 25 liters fuel.**

2. Open fuel filler lid.

3. Open filler cap and release the pressure inside fuel tank.

4. Remove rear seat.

5. Remove inspection hole cover.

6. Disconnect harness connector (1) and quick connector (2).

7. Remove center muffler.

8. Remove insulator on vehicle side located above center and main mufflers.

9. Move parking brake cable from the lower face of fuel tank. Then remove clips for parking brake cable.

10. Disconnect fuel filler hose (4) at fuel tank side.

11. Disconnect EVAP hose and vent tube at the position shown in the figure.

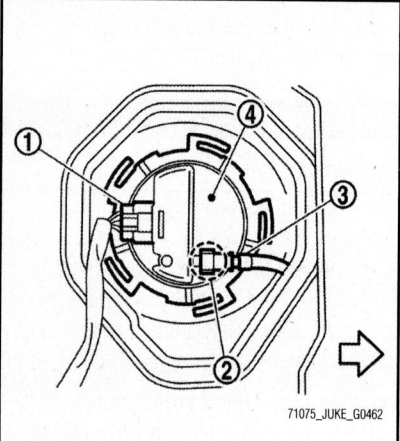

Fig. 166 Disconnecting harness connector (1) and quick connector (2)

1. Vent tube
2. Fuel tank mounting band (RH)
3. EVAP hose
A. Quick connector

71075_JUKE_G0463

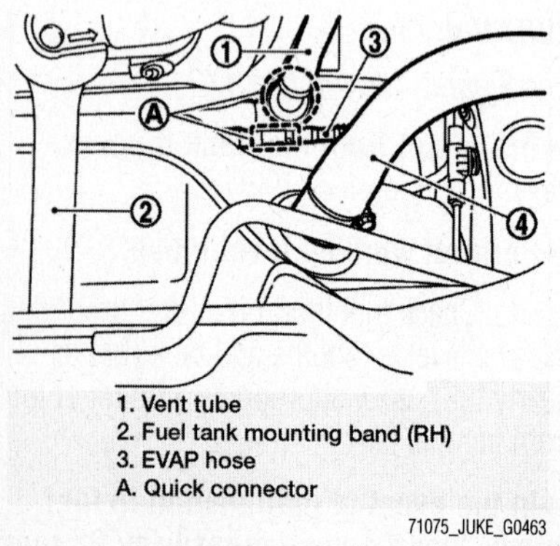

Fig. 167 Disconnecting fuel filler hose

12. Instruction for quick connector of EVAP hose and vent tube, refer to the following:

EVAP hose:

✳✳ WARNING

Quick connector (1) can be disconnected when the tabs (F) are depressed completely. Never twist it more than necessary.

✳✳ WARNING

Never use any tools to disconnected quick connector.

✳✳ WARNING

Keep resin tube (C) away from heat. Be especially careful when welding near the resin tube.

✳✳ WARNING

Prevent acid liquid such as battery electrolyte, etc. from getting on resin tube.

✳✳ WARNING

Never bend or twist resin tube during installation and disconnection.

✳✳ WARNING

Never remove the remaining retainer (2) on hard tube (or the equivalent) (A) except when resin tube or retainer is replaced.

✳✳ WARNING

When resin tube or hard tube (or the equivalent) is replaced, also replace retainer with new one

➡Retainer color: Green

✳✳ WARNING

To keep the connecting portion clean and to avoid damage and foreign materials, cover them completely with plastic bags or something similar.

✳✳ WARNING

Never insert plug, preventing damage on O-ring in quick connector.

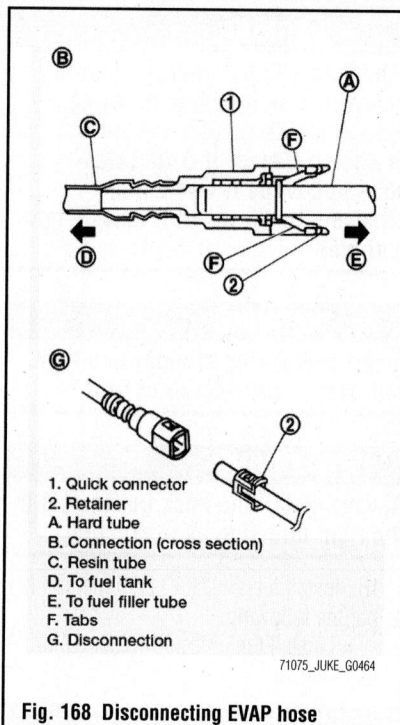

1. Quick connector
2. Retainer
A. Hard tube
B. Connection (cross section)
C. Resin tube
D. To fuel tank
E. To fuel filler tube
F. Tabs
G. Disconnection

71075_JUKE_G0464

Fig. 168 Disconnecting EVAP hose

Vent tube:

a. Pinch quick connector square-part (A) with your fingers, and pull out the quick connector (1) by hand.

b. If quick connector and tube on vehicle are stuck, push and pull several times until they move, and pull out.

✳✳ WARNING

The tube can be removed when the tabs are completely depressed. Never twist it more than necessary.

✳✳ WARNING

Never use any tools to disconnect quick connector.

✳✳ WARNING

Keep the resin tube away from heat. Be especially careful when welding near the tube.

✳✳ WARNING

Prevent acid liquid such as battery electrolyte, etc. from getting on the resin tube.

✳✳ WARNING

Never bend or twist resin tube during installation and disconnection.

⁕⁕ WARNING

To keep the connecting portion clean and to avoid damage and foreign materials, cover them completely with plastic bags or something similar.

⁕⁕ WARNING

Never insert plug, preventing damage on O-ring in quick connector.

13. Support the center part of fuel tank with transmission jack.

⁕⁕ WARNING

Securely support the fuel tank with a piece of wood.

14. Remove canister filter (1).
15. Remove fuel tank mounting bands (RH and LH).
16. Lower transmission jack carefully to remove fuel tank while holding it by hand.

⁕⁕ CAUTION

Fuel tank may be in an unstable condition because of the shape of fuel tank bottom. Never rely on jack too much. Be sure to hold tank securely.

To install:

17. Note the following, and install in the reverse order of the removal procedure:

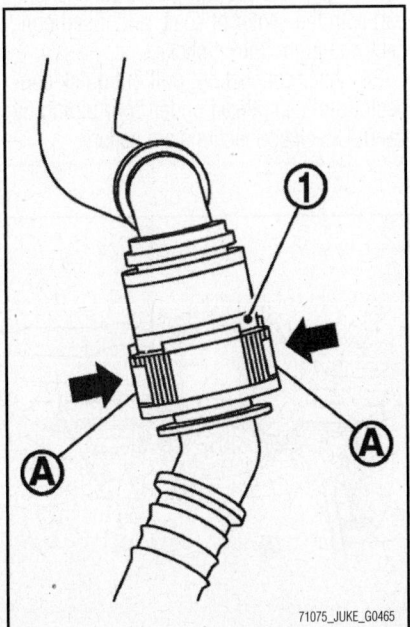

Fig. 169 Pinching quick connector square part (A) and pulling out quick connector (1)

71075_JUKE_G0465

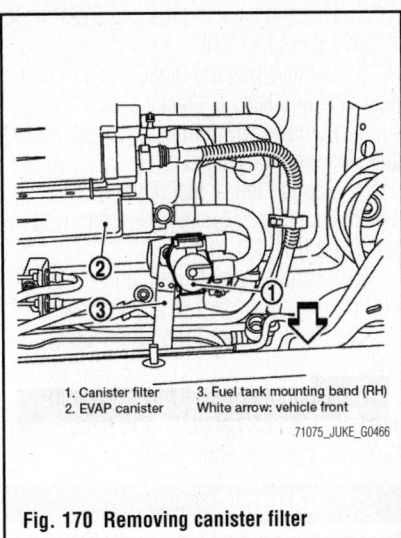

1. Canister filter	3. Fuel tank mounting band (RH)
2. EVAP canister	White arrow: vehicle front

71075_JUKE_G0466

Fig. 170 Removing canister filter

Fuel tank:

18. Temporarily tighten bolts (except No. 4) in numerical order as shown.
19. Tighten bolt No. 4 to 18 ft. lbs. (25 Nm), pressing fuel tank in the direction shown (black arrow).
20. Tighten bolts (except No. 4) to specified torque in reverse order as shown.

Fuel filler hose:

21. Surely clamp fuel hose insert fuel filler hose to the length below:
- Fuel filler hose: 1.38 inch (35 mm)
- Other hose: 0.98 inch (25 mm)

22. Be sure hose clamp is not placed on swelled area of fuel filler tube.
23. Tighten the clamp band with the top mark until the mark is on the bolt head flange.

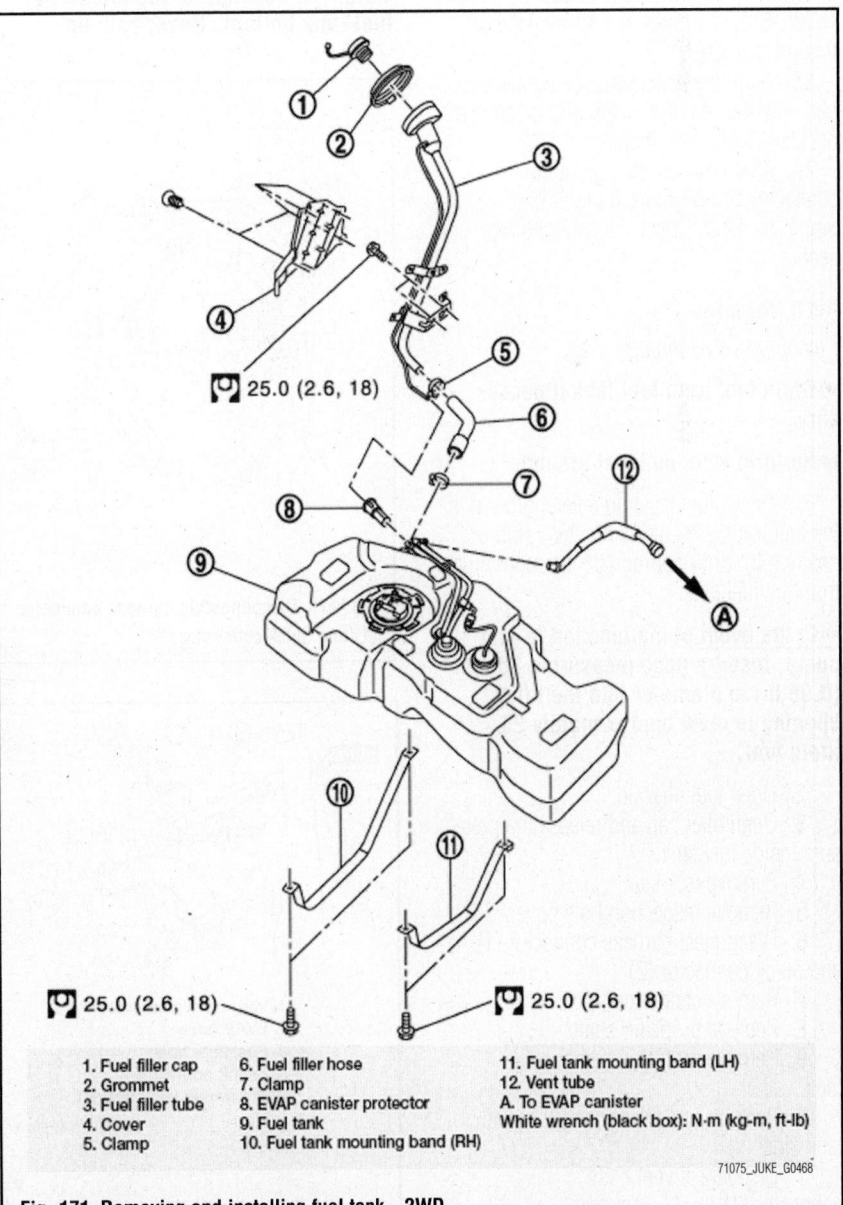

25.0 (2.6, 18)

25.0 (2.6, 18)   25.0 (2.6, 18)

1. Fuel filler cap	6. Fuel filler hose	11. Fuel tank mounting band (LH)
2. Grommet	7. Clamp	12. Vent tube
3. Fuel filler tube	8. EVAP canister protector	A. To EVAP canister
4. Cover	9. Fuel tank	White wrench (black box): N·m (kg-m, ft-lb)
5. Clamp	10. Fuel tank mounting band (RH)	

71075_JUKE_G0468

Fig. 171 Removing and installing fuel tank—2WD

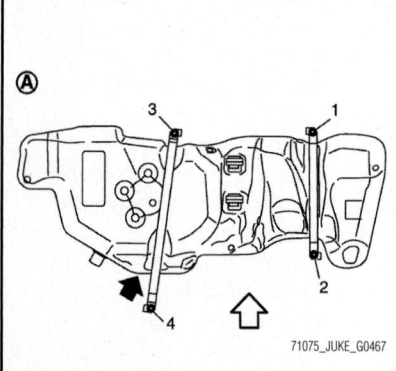

Fig. 172 Under view of fuel tank bolt tightening sequence (white arrow: vehicle front)

EVAP hose and vent tube:

24. Check connections for damage or foreign material.

25. Align the matching side connection part with the center of shaft, and insert connector straight until it clicks.

26. After connecting, pull out quick connector and centralized under floor piping by hand. Check connections are secure.

AWD Vehicles

See Figures 173 through 176.

➡ **Drain fuel from fuel tank if necessary.**

➡ **Perform work on level ground.**

1. Check fuel level on a level ground. If the fuel level is ⅞ of the fuel tank (full or nearly full), draw appropriate amount of fuel from the fuel tank.

➡ **In the event of malfunction in fuel pump, insert a hose measuring 25 mm (0.98 in) in diameter into the filler opening to draw approximately 25 liters fuel.**

2. Open fuel filler lid.

3. Open filler cap and release the pressure inside fuel tank.

4. Remove rear seat.

5. Remove inspection hole cover.

6. Disconnect harness connector (1) and quick connector (2).

7. Remove center muffler.

8. Remove propeller shaft.

9. Remove protector from fuel tank.

10. Remove fuel filler hose (3) at fuel tank side.

11. Disconnect vent hose connector (1).

12. Disconnect vent tube and EVAP hose at rear side of fuel tank.

13. Remove parking brake cable mounting bolts, and keep a service area.

14. Remove ABS harness connector from fuel tank mounting band.

15. Support left of fuel tank (1) with transmission jack (A), and them remove fuel tank band (RH) (2).

16. Support right of fuel tank (1) with transmission jack (A), and then remove fuel tank band mounting bolts (LH).

17. Supporting with hands, descend transmission jack carefully, and remove fuel tank with fuel tank band (LH).

☀ WARNING

Fuel tank may be in an unstable condition because of the shape of fuel tank bottom. Never rely on

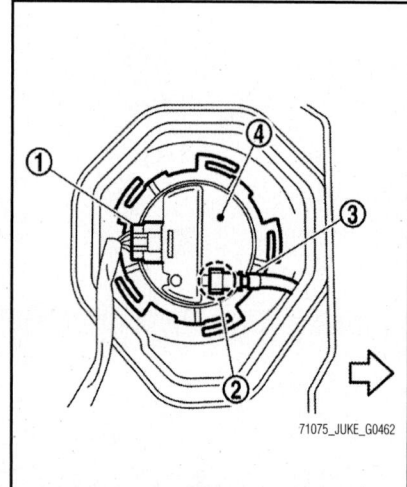

Fig. 173 Disconnecting harness connector (1) and quick connector (2)

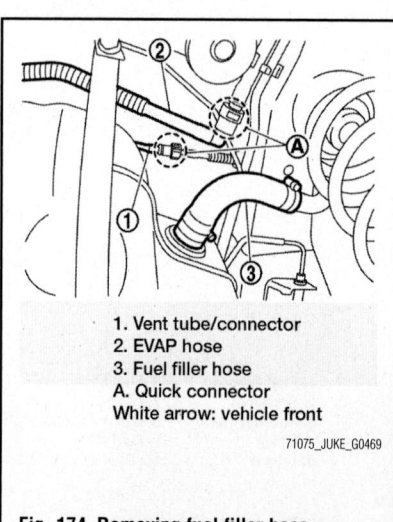

1. Vent tube/connector
2. EVAP hose
3. Fuel filler hose
A. Quick connector
White arrow: vehicle front

Fig. 174 Removing fuel filler hose

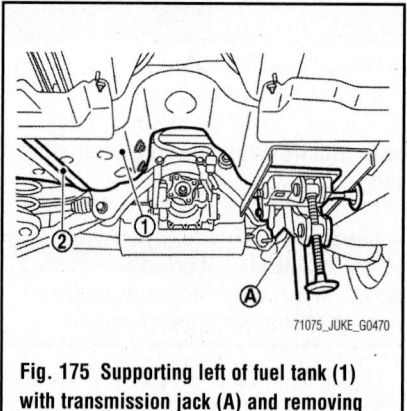

Fig. 175 Supporting left of fuel tank (1) with transmission jack (A) and removing fuel tank band (RH) (2)

jack too much. Be sure to hold tank securely.

To install:

18. Note the following, and install in the reverse order of the removal procedure:

Fuel filler hose:

19. Surely clamp fuel hose insert fuel filler hose to the length below:
- Fuel filler hose: 1.38 inch (35 mm)
- Other hose: 0.98 inch (25 mm)

20. Be sure hose clamp is not placed on swelled area of fuel filler tube.

21. Tighten the clamp band with the top mark until the mark is on the bolt head flange.

EVAP hose and vent tube:

22. Check connections for damage or foreign material.

23. Align the matching side connection part with the center of shaft, and insert connector straight until it clicks.

24. After connecting, pull out quick connector and centralized under floor piping by hand. Check connections are secure.

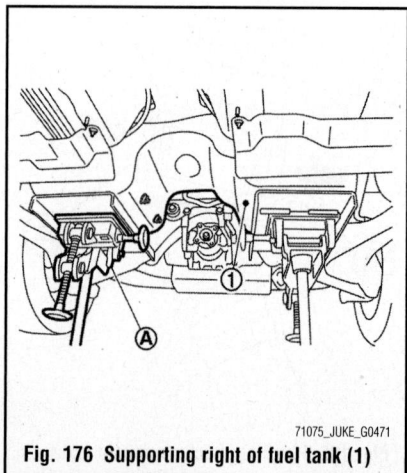

Fig. 176 Supporting right of fuel tank (1) with transmission jack (A) and removing fuel tank band (LH) (2)

HEATING & AIR CONDITIONING SYSTEM

AIR CONDITIONING UNIT

REMOVAL & INSTALLATION

See Figures 177 through 182.

☼ WARNING

Perform lubricant return operation before each refrigeration system disassembly. However, if a large amount of refrigerant or lubricant is detected, never perform lubricant return operation.

1. Use a refrigerant collecting equipment (for HFC-134a) to discharge the refrigerant.

2. Drain engine coolant from cooling system.

3. Remove cowl top and cowl top extension.

4. Remove mounting nut, and then move lower dash insulator aside.

5. Remove mounting bolt (A), and then disconnect low-pressure flexible hose (1) and high-pressure pipe (2) from expansion valve.

☼ WARNING

Cap or wrap the joint of the A/C piping and expansion valve with suitable material such as vinyl tape to prevent air from entering the system.

6. Remove the clamps, and then disconnect the heater hoses from A/C unit assembly.

7. Remove instrument panel assembly, as outlined in the Body Section.

8. Remove side ventilator duct.

9. Remove mounting bolts (A) of ground wire LH side and mounting bolt (B) of ground wire RH side.

10. Remove rear heater duct 1, if equipped.

11. Disconnect drain hose from A/C unit assembly.

12. Remove mounting nuts (A), and then remove instrument stay.

13. Remove J/B mounting screws (A), and then remove J/B.

14. Remove BCM fixing screws (A), and then remove BCM.

15. Disconnect harness connectors and clips required to remove the steering member, and then move the vehicle harness to the position without hindrance for work.

16. Move steering column assembly to a position where it does not inhibit work.

17. Remove mounting bolts (A) and (B), and then remove steering member (1) from vehicle.

18. Remove A/C unit assembly from vehicle.

To install:

19. Installation is the reverse of the removal procedure, noting the following:

   a. Replace O-rings with new ones. Then apply compressor oil to them when installing.

   b. Check for leaks when recharging refrigerant.

   c. Refill the radiator with the proper type and amount of coolant.

BLOWER MOTOR

REMOVAL & INSTALLATION

See Figure 183.

Fig. 179 Remove mounting nuts (A), and then remove instrument stay

Fig. 180 Remove J/B mounting screws (A), and then remove J/B

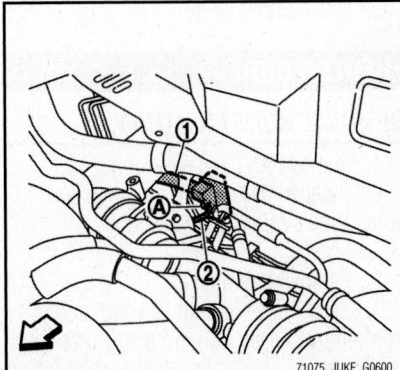

Fig. 177 Remove mounting bolt (A), and then disconnect low-pressure flexible hose (1) and high-pressure pipe (2) from expansion valve

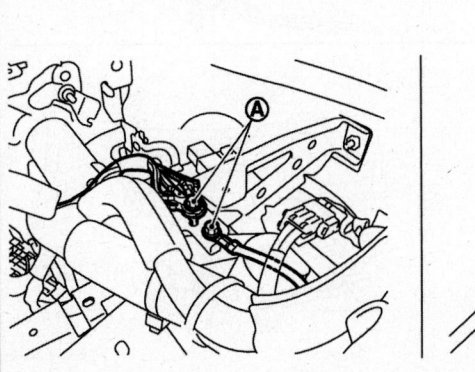

Fig. 178 Remove mounting bolts (A) of ground wire LH side and mounting bolt (B) of ground wire RH side

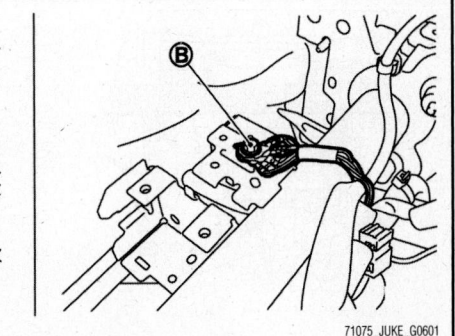

Fig. 181 Remove BCM fixing screws (A), and then remove BCM

※※ WARNING

Before servicing, turn ignition switch OFF, disconnect battery negative terminal and wait 3 minutes or more.

※※ CAUTION

Always work from the side of air bag module. Never work in front of it.

※※ WARNING

Never use the air tools or the electric tools for servicing.

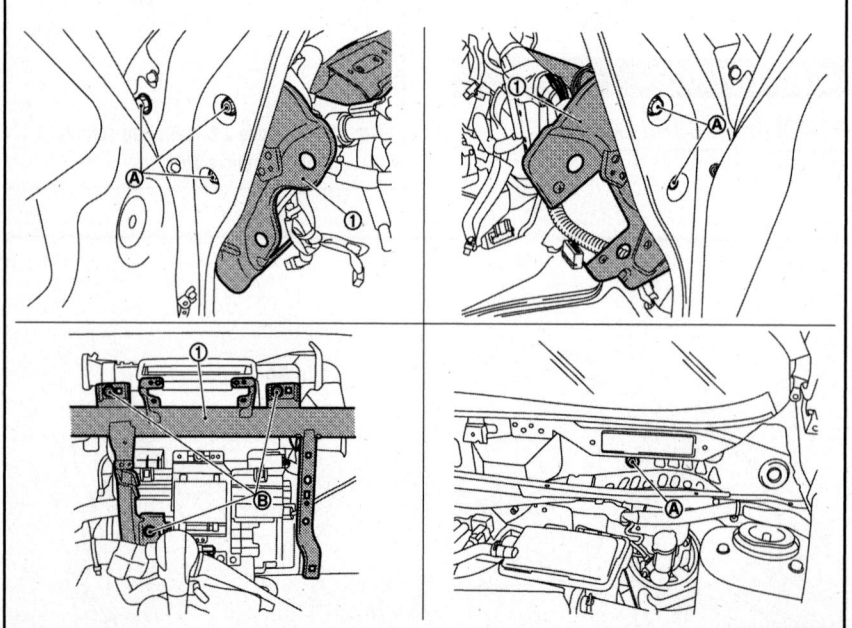

Fig. 182 Remove mounting bolts (A) and (B), and then remove steering member (1) from vehicle

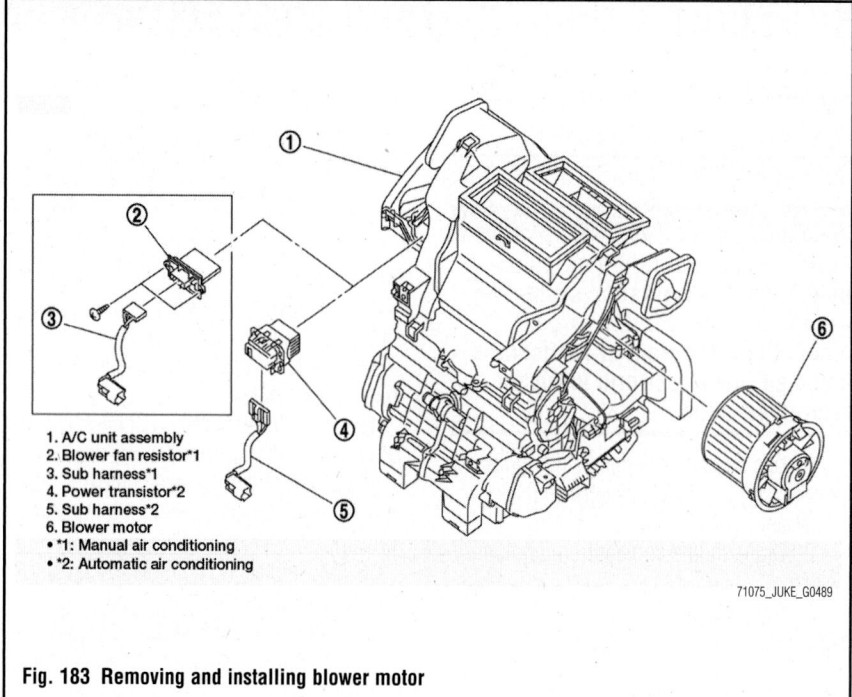

1. A/C unit assembly
2. Blower fan resistor*1
3. Sub harness*1
4. Power transistor*2
5. Sub harness*2
6. Blower motor
- *1: Manual air conditioning
- *2: Automatic air conditioning

71075_JUKE_G0489

Fig. 183 Removing and installing blower motor

1. Remove glove box assembly.
2. Disconnect blower motor harness connector and front passenger air bag module harness connector (2).
3. Remove front passenger air bag module.
4. Press flange holding hook, and then turn blower motor clockwise.

71075_JUKE_G0605

5. Pull outside, and then remove blower motor.

To install:

6. Note the following items, and then install in the reverse order of the removal procedure:

a. Never use the old mounting bolts after removal, replace with the new bolts.

b. Never damage the harness while installing.

c. If malfunction is detected by the air bag warning lamp, after repair or replacement of the malfunctioning parts, reset the memory using self-diagnosis or CONSULT-III.

d. After the work is completed, check that no system malfunction is detected by air bag warning lamp.

HEATER CORE

REMOVAL & INSTALLATION

1. Remove A/C unit assembly.
2. Remove heater pipe grommet and heater pipe support from A/C unit assembly.
3. Remove foot duct LH.
4. Slide heater core to leftward, and then remove heater core from A/C unit assembly.

To install:

5. Installation is the reverse of the removal procedure.

STEERING

STEERING GEAR & LINKAGE

REMOVAL & INSTALLATION

See Figures 184 and 185.

1. Set vehicle to the straight-ahead position.

2. Remove intermediate shaft mounting bolt (steering gear assembly side).

☀☀ WARNING

Spiral cable may be cut if steering wheel turns while separating steering column assembly and steering gear assembly. Always fix the steering wheel using string to avoid turning.

➡️Place a matching mark on both intermediate shaft and steering gear assembly before removing intermediate shaft.

☀☀ WARNING

When removing intermediate shaft, never insert a tool, such as a screwdriver, into the yoke groove to pull out the intermediate shaft. In case of the violation of the above, replace intermediate shaft with a new one.

3. Remove dash seal from vehicle.
4. Remove tires with power tool.
5. Remove steering outer socket from steering knuckle so as not to damage ball joint boot using suitable ball joint remover (commercial service tool).

☀☀ WARNING

Temporarily tighten the nut to prevent damage to threads and to prevent the ball joint remover from sudden drop turning.

6. Remove front suspension member.
7. Remove steering gear assembly.

To install:

8. Note the following, and install in the reverse order of the removal procedure:

a. Spiral cable may be cut if steering wheel turns while separating steering column assembly and steering gear assembly. Always fix the steering wheel using string to avoid turning.

b. Perform final tightening of nuts and bolts on each part under unladen conditions with tires on level ground when removing steering gear assembly. Check wheel alignment.

c. Rotate steering wheel to check for uncentered condition, binding, noise or excessive steering effort.

d. Never reuse steering outer socket mounting nut and steering gear assembly mounting nut.

e. Perform inspection after installation.

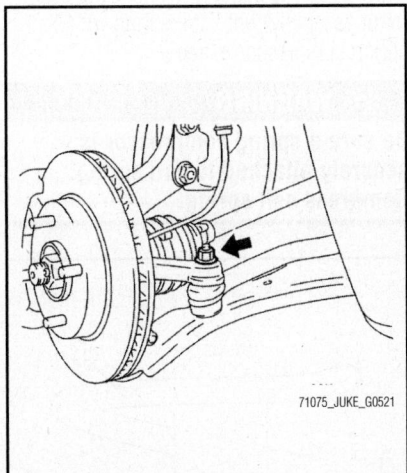

Fig. 184 Removing steering outer socket from steering knuckle

71075_JUKE_G0521

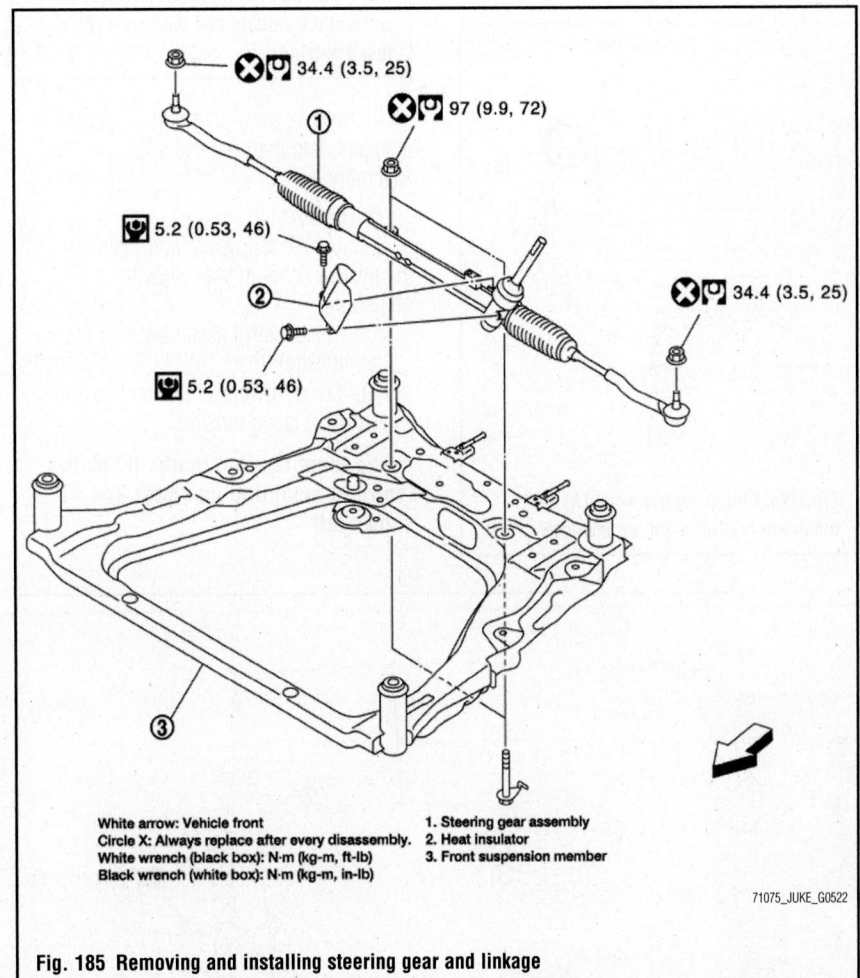

White arrow: Vehicle front
Circle X: Always replace after every disassembly.
White wrench (black box): N·m (kg-m, ft-lb)
Black wrench (white box): N·m (kg-m, in-lb)

1. Steering gear assembly
2. Heat insulator
3. Front suspension member

71075_JUKE_G0522

Fig. 185 Removing and installing steering gear and linkage

COIL SPRINGS & STRUT

REMOVAL & INSTALLATION

See Figures 186 through 188.

1. Remove tires with power tool.
2. Remove lock plate from strut assembly.
3. Remove wheel sensor.
4. Remove stabilizer connecting rod from strut assembly.
5. Remove strut mounting bolts and nuts from steering knuckle.
6. Remove grommet (A) of cowl top cover.

✳✳ WARNING

Remove mounting bolt of mounting insulator (1) from grommet hole.

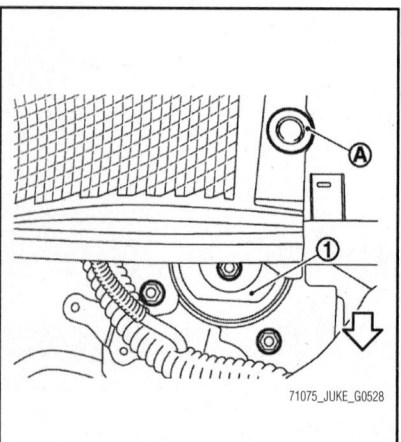

Fig. 186 Removing grommet (A) and mounting bolt of mounting insulator (1)

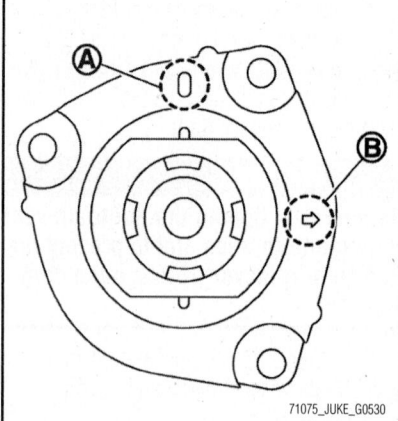

Fig. 188 Installing strut with identification mark (A) of mounting insulator faced forward of the vehicle and the arrow (B) faced outside

7. Remove mounting bolt of mounting insulator, and then remove strut assembly.

To install:

8. Note the following, and install in the reverse order of the removal procedure:

   a. Install strut assembly with the identification mark (A) of mounting insulator faced forward of the vehicle and the arrow (B) faced outside.

➡ **The identification mark "0" shows the right mounting insulator and "1" shows left.**

✳✳ WARNING

Never reuse strut mounting nut.

   b. Perform final tightening of bolts and nuts at the vehicle installation position (rubber bushing), under unladen conditions with tires on level ground.
   c. Perform inspection after installation.
   d. After replacing the strut absorber, always follow the disposal procedure to discard the strut absorber.

OVERHAUL

See Figures 189 and 193.

✳✳ WARNING

Never damage strut assembly piston rod when removing components from strut assembly.

1. Install strut attachment [SST: ST35652000 (-)] to strut assembly and secure it in a vise.

✳✳ WARNING

When installing the strut attachment to strut assembly, wrap a shop cloth around strut to protect from damage.

2. Using a spring compressor (commercial service tool), compress coil spring between spring upper seat and lower seat (strut assembly) until coil spring with a spring compressor is free.

✳✳ WARNING

Be sure a spring compressor is securely attached to coil spring. Compress coil spring.

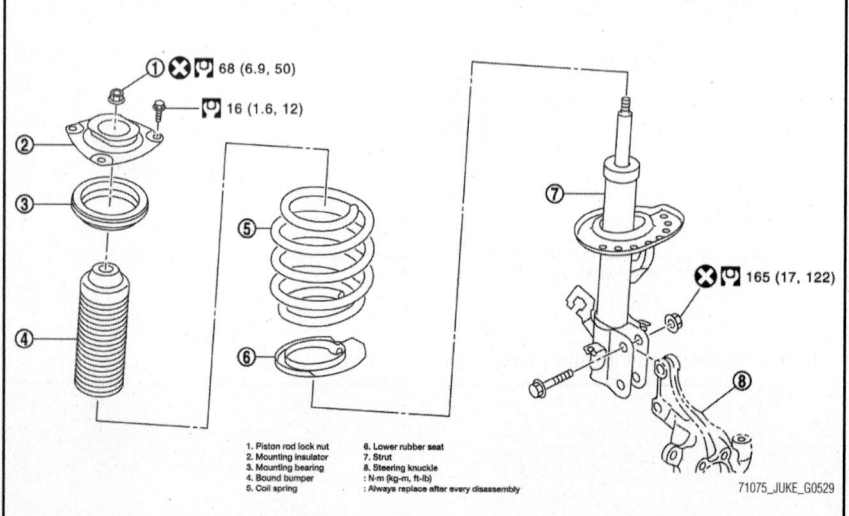

1. Piston rod lock nut
2. Mounting insulator
3. Mounting bearing
4. Bound bumper
5. Coil spring
6. Lower rubber seat
7. Strut
8. Steering knuckle
: N·m (kg-m, ft-lb)
: Always replace after every disassembly

Fig. 187 Removing and installing the front coil and strut assembly

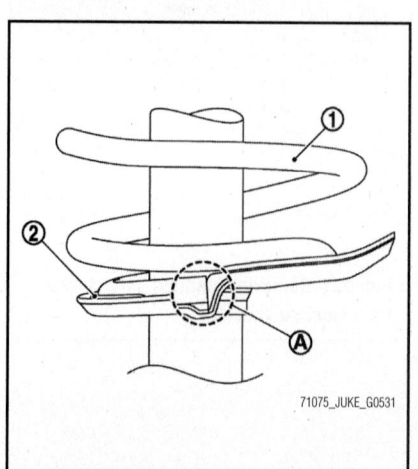

Fig. 189 Aligning lower end of coil spring (1) with (A) of lower rubber seat (2)

3. Check coil spring with a spring compressor between spring upper seat and lower seat (strut assembly) is free. And then remove piston rod lock nut while securing the piston rod tip so that piston rod does not turn.

4. Remove mounting insulator, mounting bearing, and bound bumper from strut.

5. After removing coil spring with a spring compressor (commercial service tool), then gradually release a spring compressor.

✳✳ WARNING

Loosen while making sure coil spring attachment position does not move.

6. Remove lower rubber seat.

7. Remove strut attachment [SST: ST35652000 (—)] from strut assembly.

8. Perform inspection after disassembly.

To install:

✳✳ WARNING

Never damage strut assembly piston rod when installing components from strut assembly.

9. Install strut attachment [SST: ST35652000 (—)] to strut and secure it in a vise.

✳✳ WARNING

When installing the strut attachment to strut assembly, wrap a shop cloth around strut to protect from damage.

10. Install lower rubber seat.

11. Compress coil spring using a spring compressor (commercial service tool), and install it onto strut assembly.

✳✳ WARNING

Be sure a compressor is securely attached to coil spring. Compress coil spring.

✳✳ WARNING

Be careful with the vertical direction of the coil spring.

12. Align the lower end of coil spring (1) with (A) of lower rubber seat (2) as shown in the figure.

13. Apply soapy water to bound bumper.

✳✳ WARNING

Never use machine oil.

14. Insert bound bumper into mounting insulator.

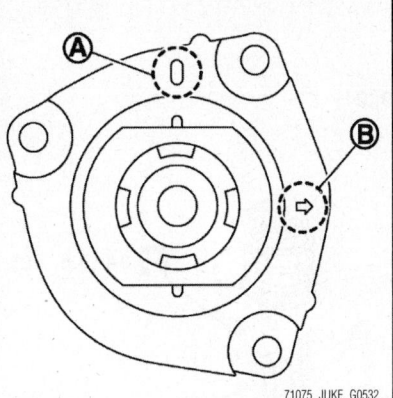

71075_JUKE_G0532

Fig. 190 Checking location of identification mark (A) of the mounting insulator and installing it with the arrow (B) facing outside of the strut

15. Install mounting bearing.

✳✳ WARNING

Never apply oils, such as grease, when installing the mounting bearing.

16. Check the location of identification mark (A) of the mounting insulator and install it with the arrow (B) faced outside of the vehicle to the strut.

➡ **The identification mark "0" shows right mounting insulator and "1" shows left.**

17. Secure piston rod tip so that piston rod does not turn, then tighten piston rod lock nut with specified torque.

✳✳ WARNING

Never reuse piston rod lock nut.

18. Gradually release a spring compressor (commercial service tool), and remove coil spring.

✳✳ WARNING

Loosen while making sure coil spring attachment position does not move.

19. Remove the strut attachment from the strut assembly.

20. Check the wheel sensor harness for proper connection.

21. Check the wheel alignment.

STABILIZER BAR

REMOVAL & INSTALLATION

See Figures 191 through 195.

1. Remove tires with power tool.

2. Remove front suspension member.

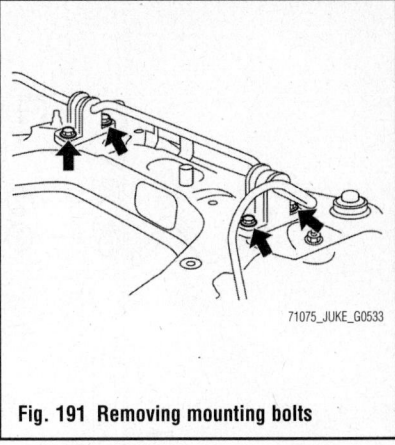

71075_JUKE_G0533

Fig. 191 Removing mounting bolts

3. Remove stabilizer connecting rod.

4. Remove mounting bolts (arrow) of stabilizer clamp, and then remove stabilizer clamp and stabilizer bushing from front suspension member.

5. Remove stabilizer bar.

6. Perform inspection after removal.

 a. Check stabilizer bar, stabilizer connecting rod, stabilizer bushing and stabilizer clamp for deformation, cracks or damage. Replace if necessary.

To install:

7. Note the following, and install in the reverse order of the removal procedure.

8. Install stabilizer clamp and stabilizer bush with notch (A) and slit (B) faced forward of the vehicle (arrow).

9. To install stabilizer clamp mounting bolt, follow the tightening
method and the numerical order shown below:

- Manual tightening: 1
- Temporary tightening: 2-3
- Final tightening 21 ft. lbs. (28 Nm): 4-5
- Arrow: vehicle front

10. To install stabilizer connecting rod (1), tighten the mounting nut with the hexagonal part (A) on the stabilizer connecting rod side fixed.

11. Perform final tightening of bolts and nuts at the vehicle installation
position (rubber bushing), under unladen conditions with tires on level ground.

12. Perform inspection after installation.

 a. Check the wheel alignment.

SUBFRAME (AXLE CARRIER)

REMOVAL & INSTALLATION

See Figures 196 through 198.

1. Separate intermediate shaft from steering gear assembly.

2. Remove tires with power tool.

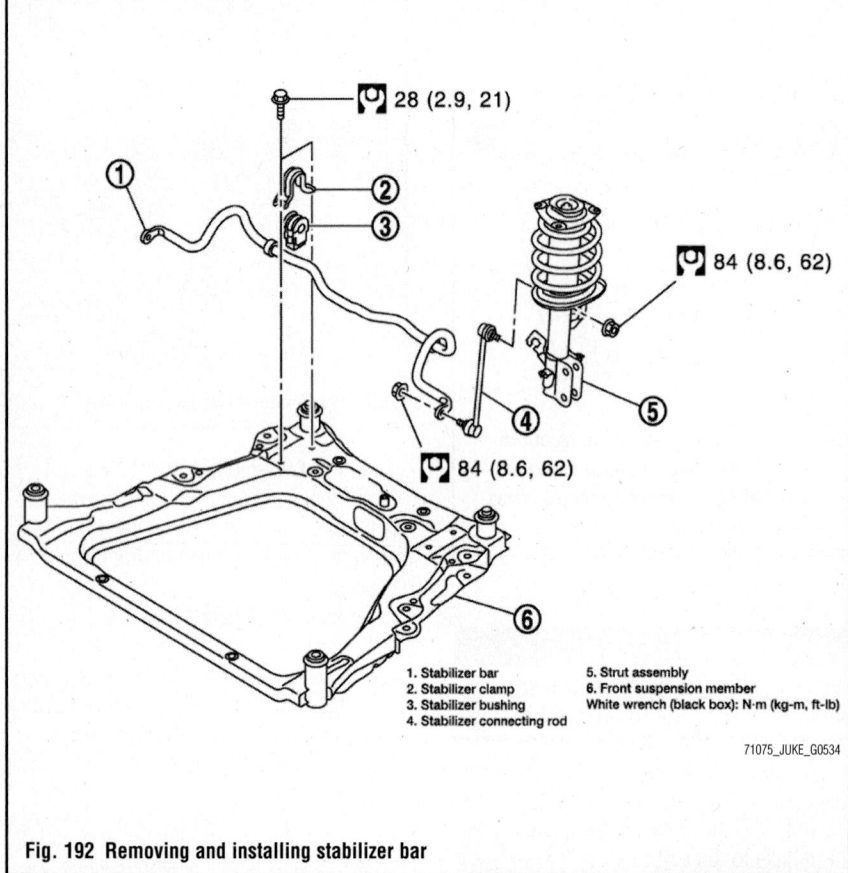

1. Stabilizer bar
2. Stabilizer clamp
3. Stabilizer bushing
4. Stabilizer connecting rod
5. Strut assembly
6. Front suspension member
White wrench (black box): N·m (kg-m, ft-lb)

71075_JUKE_G0534

Fig. 192 Removing and installing stabilizer bar

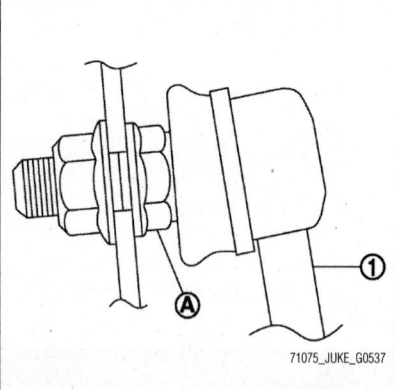

71075_JUKE_G0537

Fig. 195 Installing stabilizer connecting rod (1) and tightening mounting nut with hexagonal part (A) on the stabilizer connecting rod side fixed

front suspension member from vehicle body.

※※ WARNING

Operate while checking that jack supporting status is stable.

➡**Remove it with each component parts.**

12. Remove component parts from front suspension member.
13. Perform inspection after removal.
    a. Check front suspension member for cracks, wear or damage. Replace if necessary.

To install:

14. Note the following, and install in the reverse order of the removal procedure.
15. To install rebound stopper (1), insert it with the protrusion aligned with the hole of member stay (2).
16. To install member stay and mounting bolts of front suspension member, temporarily tighten the bolts before tightening to the specified torque, referring to the tightening method and the numerical order shown below:
    - Temporary tightening: 1-2
    - Final tightening: 3-4-5-6-7-8-9-10-11-12
    - Tightening torque: 69 ft. lbs. (94 Nm)
    a. Perform final tightening of bolts and nuts at the vehicle installation position (rubber bushing), under unladen conditions with tires on level ground.
17. Perform inspection after installation.
    a. Check wheel sensor harness for proper connector.
    b. Check wheel alignment.

3. Remove front under cover.
4. Separate stabilizer connecting rod from strut assembly.
5. Separate steering outer socket from steering knuckle.
6. Separate transverse link from steering knuckle.
7. Remove rear torque rod.
8. Set suitable jack under front suspension member.

※※ WARNING

Check the stable condition when using a jack.

9. Remove member stay and rebound stopper.
10. Remove suspension member mounting bolts, insulator, and rebound stopper rubber.
11. Gradually lower the jack to remove

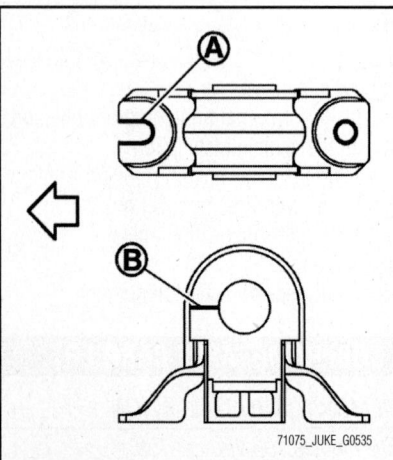

71075_JUKE_G0535

Fig. 193 Installing stabilizer clamp and stabilizer bush with notch (A) and slit (B) facing forward

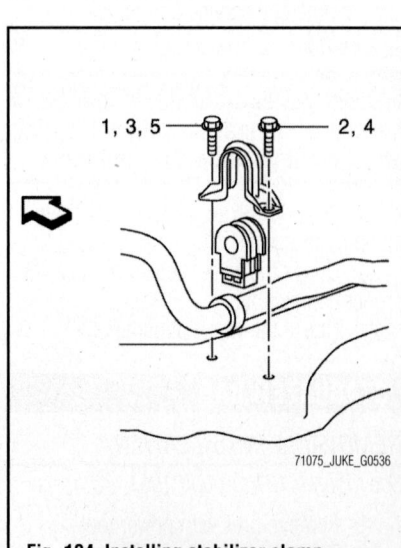

71075_JUKE_G0536

Fig. 194 Installing stabilizer clamp

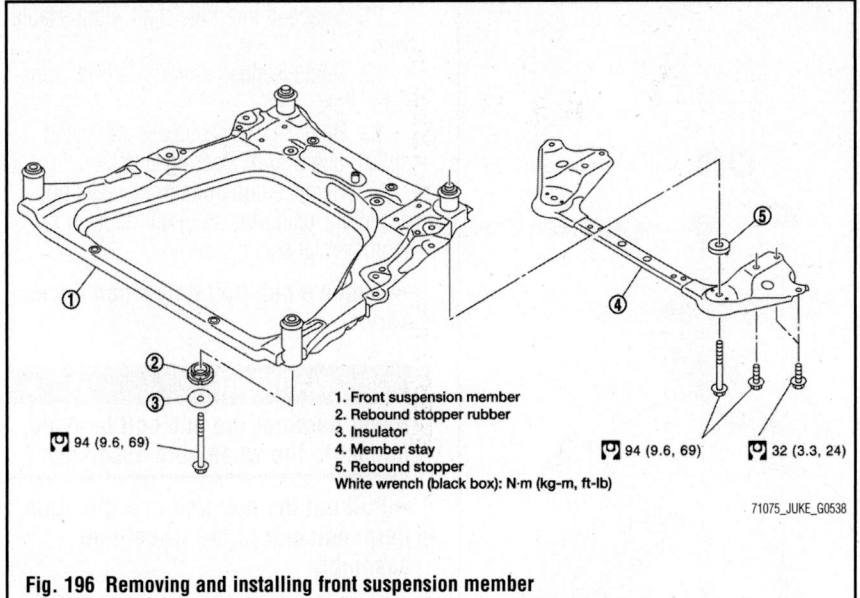

1. Front suspension member
2. Rebound stopper rubber
3. Insulator
4. Member stay
5. Rebound stopper
White wrench (black box): N·m (kg-m, ft-lb)

🔧 94 (9.6, 69)   🔧 94 (9.6, 69)   🔧 32 (3.3, 24)

71075_JUKE_G0538

Fig. 196 Removing and installing front suspension member

TRANSVERSE LINK

INSPECTION

Transverse Link

1. Check transverse link and bushing for deformation, cracks or damage.
2. Check ball joint boot for cracks or other damage, and also for grease leakage.

Swing Torque

See Figure 199.

1. Manually move ball stud to confirm it moves smoothly with no binding.
2. Move ball stud at least tem times by hand to check for smooth movement.
3. Hook a spring balance (A) at cutout on ball stud (B). Confirm spring balance measurement value is within specifications when ball stud begins moving.

- Swing torque: 5-43 inch lbs. (0.5-4.9 Nm)
- Spring balance: 3.5-33.8 lbs. (15.4-150.8 N)

4. If swing torque exceeds standard range, replace transverse link assembly.

Axial End Play

1. Move ball stud at least ten times by hand to check for smooth movement.
2. Move tip of ball stud in axial direction to check for looseness.
- Axial end play: 0 inch (0 mm)
a. If axial end play exceeds the standard value, replace transverse link assembly.

REMOVAL & INSTALLATION

See Figure 200.

1. Remove tires with power tool.

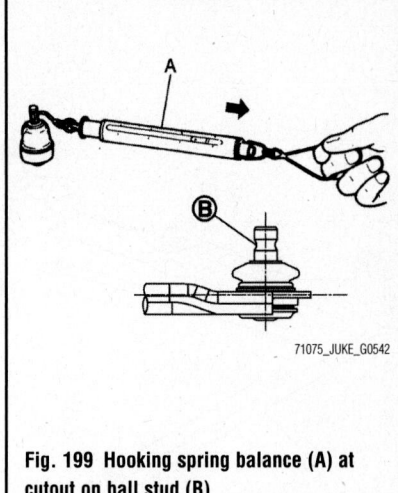

71075_JUKE_G0542

Fig. 199 Hooking spring balance (A) at cutout on ball stud (B)

2. Remove transverse link from steering knuckle.
3. Remove transverse link from suspension member.

To install:

4. Note the following, and install in the reverse order of the removal procedure.

✳✳ WARNING

Never reuse transverse link mounting nut.

5. Perform final tightening of bolts and nuts at the vehicle installation position (rubber bushing), under unladen conditions with tires on level ground.
6. Perform inspection after installation.
a. Check wheel alignment.

WHEEL HUBS & BEARINGS

REMOVAL & INSTALLATION

See Figures 201 and 202.

1. Remove tires with power tool.
2. Remove wheel sensor and sensor harness.
3. Remove lock plate from strut assembly.
4. Remove caliper assembly. Hang caliper assembly not to interfere with work.

✳✳ WARNING

Never depress brake pedal while brake caliper is removed.

5. Remove disc rotor.
a. Put matching marks on the wheel hub assembly and the disc rotor before removing the disc rotor.

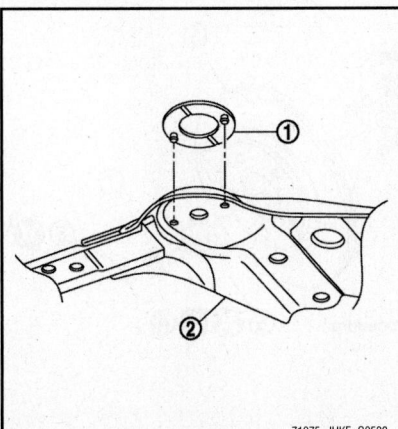

71075_JUKE_G0539

Fig. 197 Installing rebound stopper (1), inserting it with the protrusion aligned with the hole of member stay (2)

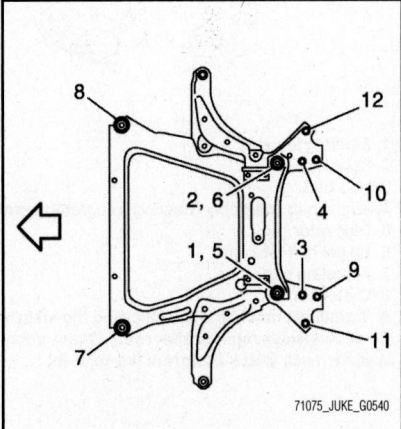

71075_JUKE_G0540

Fig. 198 Identifying member stay mounting bolt installation sequence

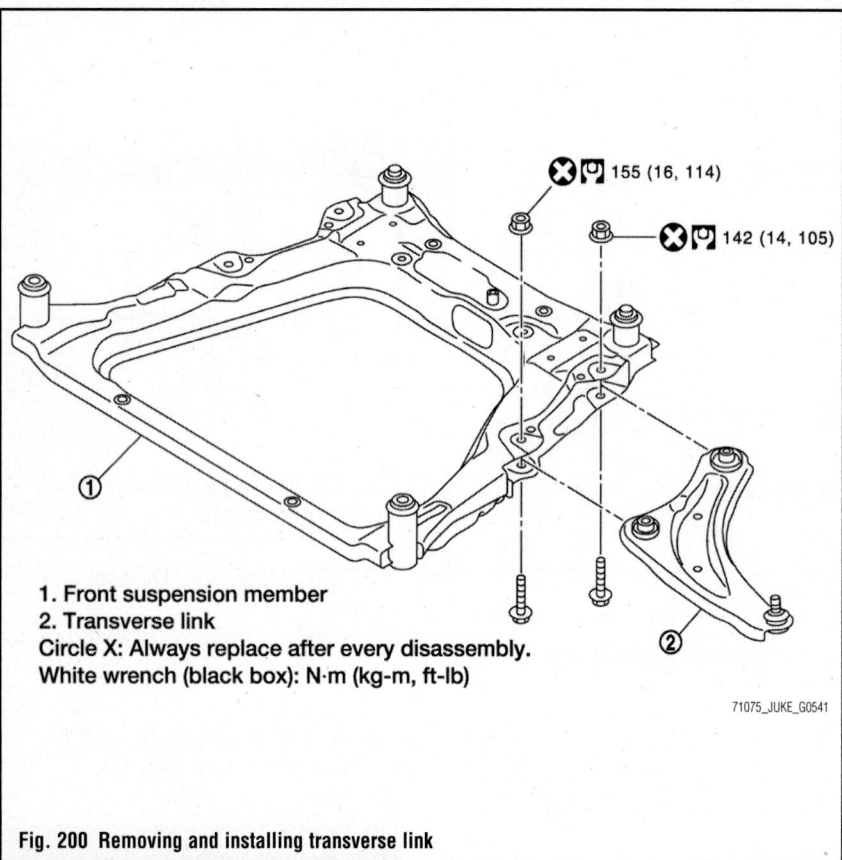

1. Front suspension member
2. Transverse link
Circle X: Always replace after every disassembly.
White wrench (black box): N·m (kg-m, ft-lb)

71075_JUKE_G0541

Fig. 200 Removing and installing transverse link

※※ WARNING

Never drop disc rotor.

6. Remove cotter pin, and adjusting cap, and then loosen wheel hub lock nut, using a hub lock nut wrench [SST: KV40104000 (—)].

7. Patch wheel hub lock nut with a piece of wood. Hammer the wood to disengage wheel hub assembly from drive shaft.

※※ WARNING

Never place drive shaft joint at an extreme angle. Also be careful not to overextend slide joint.

※※ WARNING

Never allow drive shaft to hang down without support for joint sub-assembly, shaft and the other parts.

➡Use suitable puller, if wheel hub assembly and drive shaft cannot be separated even after performing the above procedure.

8. Remove wheel hub lock nut.

9. Remove steering outer socket from steering knuckle.

10. Remove strut assembly from steering knuckle.

11. Suspend the drive shaft with suitable wire.

12. Remove steering knuckle from transverse link.

13. Remove wheel hub assembly and splash guard from steering knuckle.

14. Remove hub bolts from wheel hub assembly, using the ball joint remover (commercial service tool).

➡**Remove hub bolt only when necessary.**

※※ WARNING

Never hammer the hub bolt to avoid impact to the wheel hub assembly.

➡**Pull out the hub bolt in a direction perpendicular to the wheel hub assembly.**

15. Perform inspection after removal.

   a. Check components for deformation, cracks, and other damage.

   b. Check boots of transverse link and steering outer socket ball joint for breakage, axial play and swing torque.

To install:

16. Place a washer (A) as shown in the figure to install the hub bolts (1) by using the tightening force of the nut (B).

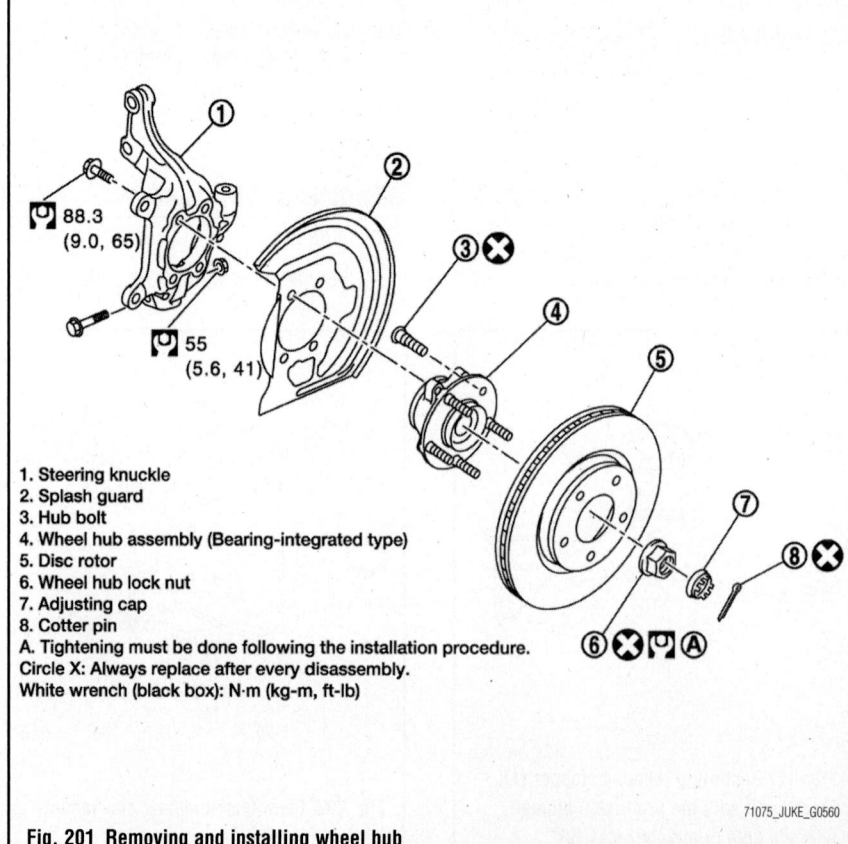

1. Steering knuckle
2. Splash guard
3. Hub bolt
4. Wheel hub assembly (Bearing-integrated type)
5. Disc rotor
6. Wheel hub lock nut
7. Adjusting cap
8. Cotter pin
A. Tightening must be done following the installation procedure.
Circle X: Always replace after every disassembly.
White wrench (black box): N·m (kg-m, ft-lb)

71075_JUKE_G0560

Fig. 201 Removing and installing wheel hub

※※ **WARNING**

Check that there is no clearance between wheel hub assembly and hub bolt.

※※ **WARNING**

Do not reuse hub bolt or steering knuckle and transverse link fixing nut.

17. Clean the matching surface of wheel hub lock nut and wheel hub assembly.

※※ **WARNING**

Never apply lubricating oil to these matching surface.

18. Clean the matching surface of drive shaft, wheel hub assembly. And then apply paste [service parts (440037S000)] to surface of joint sub-assembly of drive shaft.

※※ **WARNING**

Apply paste to cover entire flat surface of joint sub-assembly of drive shaft. (Paste amount: 0.04-0.10 oz.)

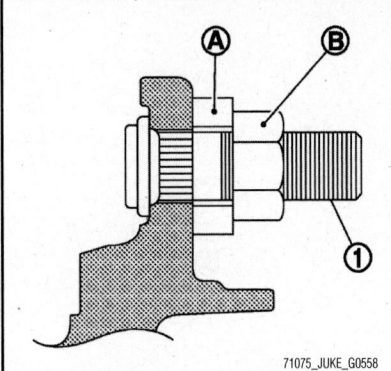

Fig. 202 Placing washer (A) as shown to install hub bolts (1) by using the tightening force of the nut (B)

※※ **WARNING**

Use the 133–136 ft. lbs. (180–185 Nm) torque for tightening the wheel hub lock nut.

※※ **WARNING**

Since the drive shaft is assembled by press-fitting, use the tightening torque range for the wheel hub lock nut.

19. Be sure to use torque wrench to tighten the wheel hub lock nut. Never use a power tool.

※※ **WARNING**

Never reuse wheel hub lock nut.

➡ Wheel hub lock nut tightening torque does not over torque for avoiding axle noise, and does not less than torque for avoiding looseness.

20. Align the matching marks that have been made during removal when reusing the disc rotor.

21. When installing a cotter pin and adjusting cap, securely bend the basal portion to prevent rattles.

※※ **WARNING**

Do not reuse cotter pin.

22. Perform the final tightening of each of parts under unladen conditions, which were removed when removing wheel hub assembly and steering knuckle.

23. Perform inspection after installation.
   a. Check wheel sensor harness for proper connection.
   b. When the wheel alignment.
   c. Adjust neutral position of steering angle sensor.

SUSPENSION

REAR SUSPENSION

COIL SPRINGS

REMOVAL & INSTALLATION

2WD Vehicles

See Figures 203 and 204.

1. Remove tires with power tool.
2. Set jack under rear suspension beam.

※※ **WARNING**

Never damage the suspension beam with a jack.

※※ **WARNING**

Check the stable condition when using a jack.

3. Remove rear shock absorber mounting bolts (lower side).
4. Slowly lower jack, then remove upper rubber seat, coil spring and lower rubber seat from rear suspension beam.
5. Perform inspection after removal.

To install:

6. Note the following, and install in the reverse order of the removal procedure:

   a. Install lower rubber seat with its protrusion (A) on the lower area aligned with the hole of rear suspension beam.
7. Securely install coil spring with the lower end of the major diameter aligned with the steps of lower rubber seat.

1. Upper rubber seat
2. Coil spring
3. Lower rubber seat
4. Rear suspension beam

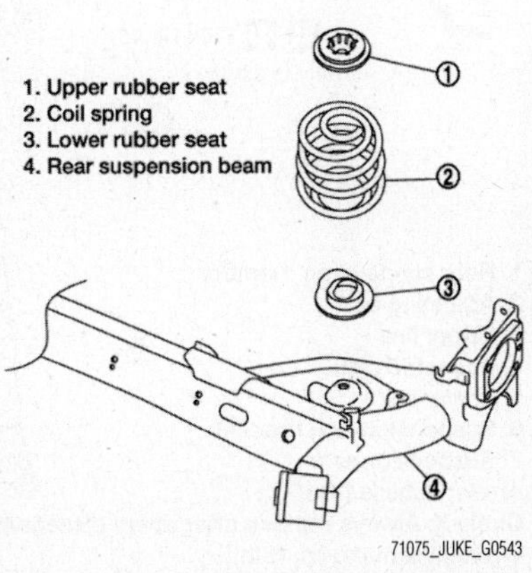

Fig. 203 Removing and installing rear coil spring—2WD

8. Perform inspection after installation.
   a. Check wheel alignment.

AWD Vehicles

See Figures 205 and 206.

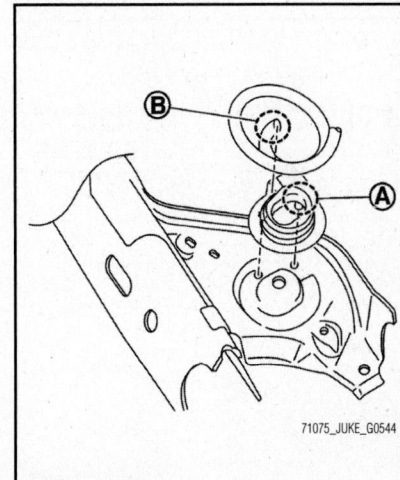

Fig. 204 Installing lower rubber seat with protrusion (A) on the lower area (B) aligned with hole of rear suspension beam

b. Remove tires with power tool.

1. Remove wheel sensor and sensor harness.

2. Set jack under suspension arm.

> ✳✳ **WARNING**
>
> **Never damage the suspension arm with a jack.**

> ✳✳ **WARNING**
>
> **Check the working condition is stable when using a jack.**

3. Separate rear shock absorber lower side form suspension arm.

4. Separate upper link from suspension arm.

5. Slowly lower jack, then remove upper rubber seat, coil spring and lower rubber seat from suspension arm.

6. Perform inspection after removal.

    a. Check rubber seat and coil spring for deformation, crack, and damage. Replace it if necessary.

To install:

7. Note the following, and install in the reverse order of the removal procedure.

8. Install the lower rubber seat a projection (A) is attached as suspension arm mounting hole (B).

9. Match up lower rubber seat indentions and suspension arm grooves and attach.

10. Perform inspection after installation.

    a. Check wheel sensor harness for proper connection.

    b. Check wheel alignment.

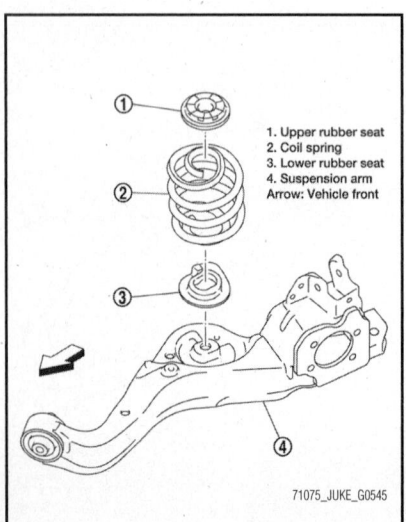

1. Upper rubber seat
2. Coil spring
3. Lower rubber seat
4. Suspension arm
Arrow: Vehicle front

71075_JUKE_G0545

Fig. 205 Removing and installing coil spring—AWD

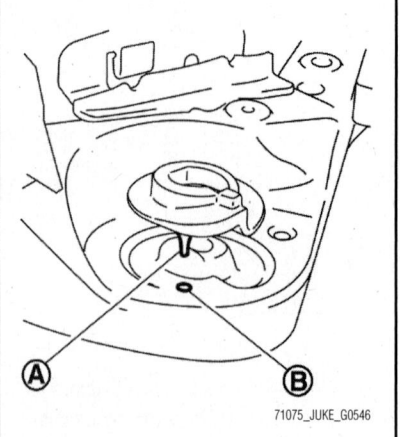

71075_JUKE_G0546

Fig. 206 Installing lower rubber seat a projection (A) at suspension arm mounting hole (B)

LOWERLINK

REMOVAL & INSTALLATION

See Figure 207.

1. Remove tires with power tool.
2. Set jack under suspension arm.

> ✳✳ **WARNING**
>
> **Never damage the suspension arm with a jack.**

> ✳✳ **WARNING**
>
> **Check the stable condition when using a jack.**

3. Remove stabilizer link.

4. Remove eccentric disc, adjusting bolt, mounting bolt, and nut, then remove lower link.

5. Perform inspection after removal.

    a. Check lower link and bushing for any deformation, cracks, or damage. Replace if necessary.

To install:

6. Note the following, and install in the reverse order of the removal procedure.

7. Perform final tightening of rear suspension member and axle installation position (rubber bushing), under unladen conditions with tires on level ground.

8. Never reuse lower link mounting nut.

9. Perform inspection after installation.

    a. Check wheel alignment.

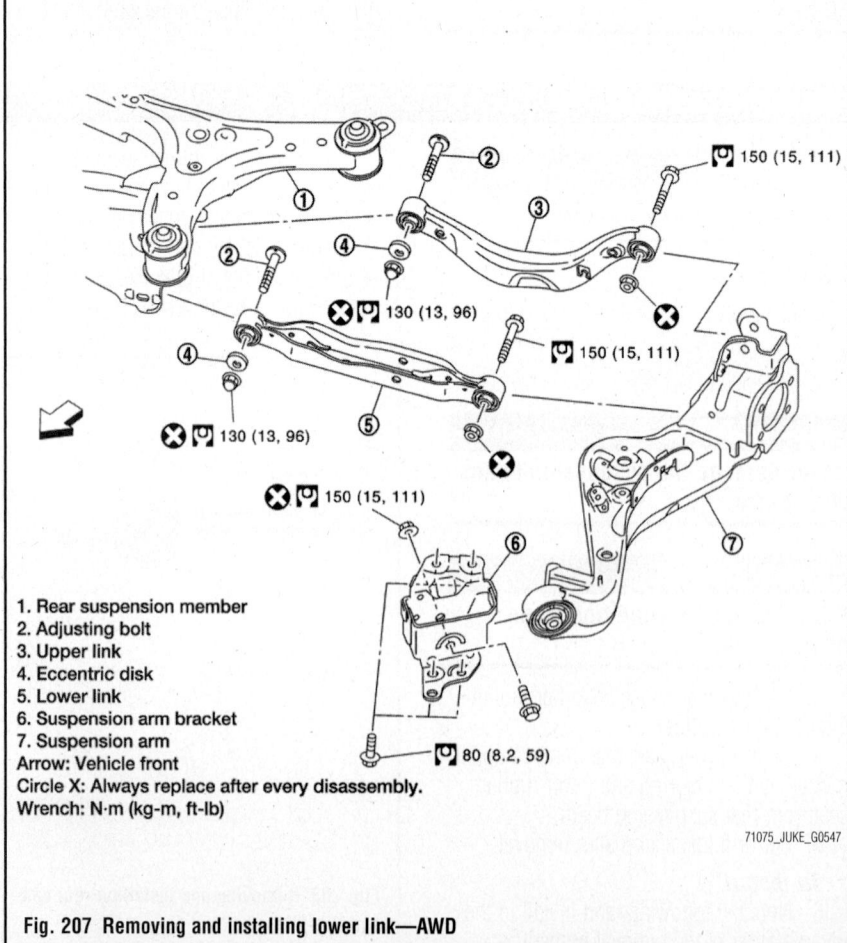

1. Rear suspension member
2. Adjusting bolt
3. Upper link
4. Eccentric disk
5. Lower link
6. Suspension arm bracket
7. Suspension arm
Arrow: Vehicle front
Circle X: Always replace after every disassembly.
Wrench: N·m (kg-m, ft-lb)

71075_JUKE_G0547

Fig. 207 Removing and installing lower link—AWD

SHOCK ABSORBERS

REMOVAL & INSTALLATION

2WD Vehicles

See Figures 208 through 214.

1. Remove tires with power tool.
2. Set suitable jack under rear suspension beam.

✳✳ WARNING

Never damage the suspension beam with a jack.

✳✳ WARNING

Check for stable condition when using a jack.

3. Remove shock absorber mounting bolt (lower side) (1).
4. Remove shock absorber mask.
5. Remove cap.
6. Remove piston rod lock nut (1), and then remove washer and bushing.

➡ **To loosen piston rod lock nut, fix the tip (A) of the piston rod.**

7. Remove shock absorber assembly.
8. Remove bushing, distance tube, washer, bound bumper cover, and bound bumper from shock absorber.
9. Perform inspection after removal.

To install:

10. Note the following, and install in the reverse order of the removal procedure.
11. To install bushings (1), securely insert protrusion (A) into the hole on the vehicle body side.
12. Install washer (1) in the direction shown in the figure.

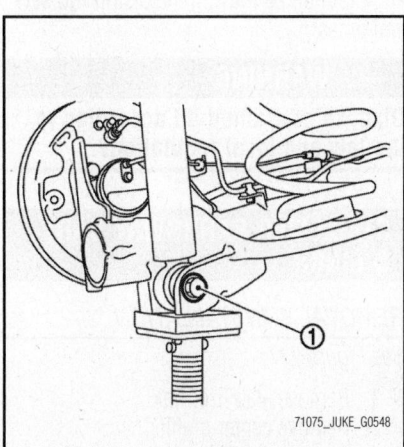

Fig. 208 Removing shock absorber mounting bolt (lower side) (1)

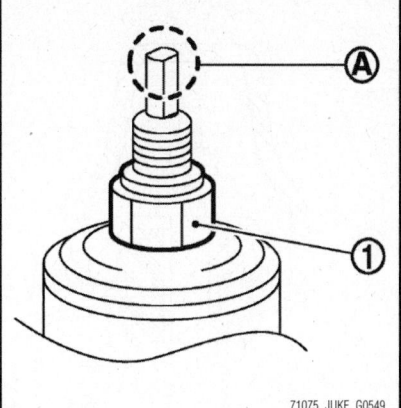

Fig. 209 Removing piston rod lock nut (1) by fixing the tip (A) of the piston rod

13. Perform final tightening of bolts and nuts at the shock absorber lower side (rubber bushing), under unladen conditions with tires on level ground.
14. Hold a head (A) of shock absorber piston rod not to have it rotate, then tighten the piston rod lock nut (1) to the specified torque.

✳✳ WARNING

Never reuse piston rod lock nut.

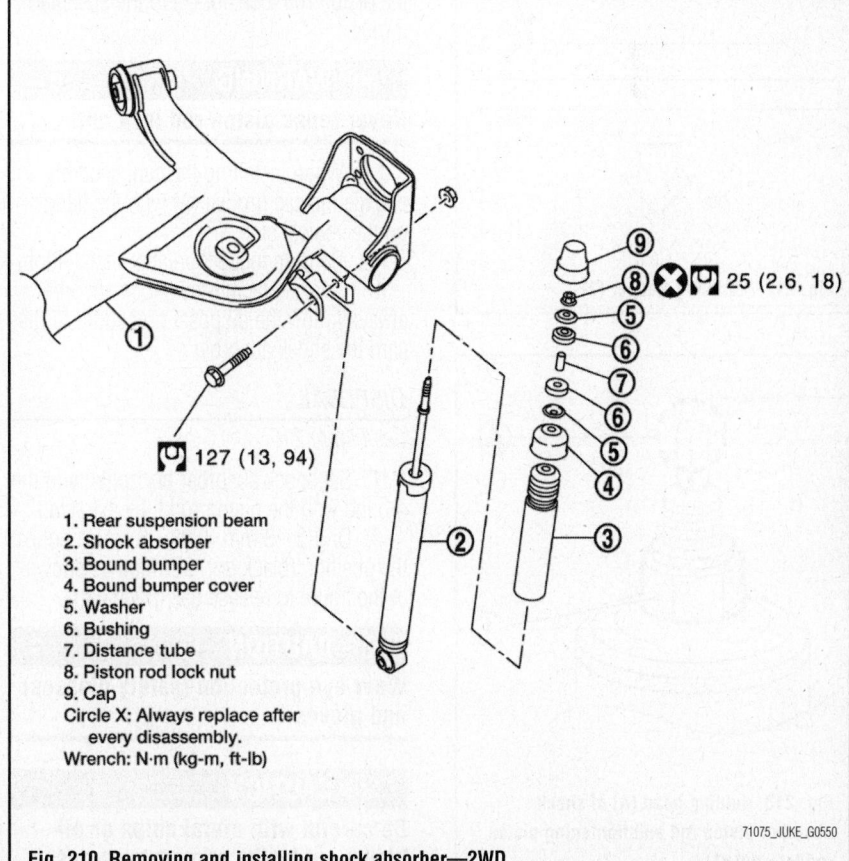

1. Rear suspension beam
2. Shock absorber
3. Bound bumper
4. Bound bumper cover
5. Washer
6. Bushing
7. Distance tube
8. Piston rod lock nut
9. Cap
Circle X: Always replace after every disassembly.
Wrench: N·m (kg-m, ft-lb)

Fig. 210 Removing and installing shock absorber—2WD

15. When installing the cap, securely engage the cap groove (A) with the flange on the vehicle side.
16. Perform inspection after installation.
17. After replacing the shock absorber, always follow the disposal procedure to discard the shock absorber.

AWD Vehicles

See Figures 211 through 215.

1. Remove tires with power tool.
2. Set suitable jack under suspension arm.

✳✳ WARNING

Never damage the suspension arm with a jack.

✳✳ WARNING

Check for stable condition when using a jack.

3. Remove shock absorber mounting bolt and nut (lower side).
4. Remove shock absorber mask.
5. Remove cap.
6. Remove piston rod lock nut (1), and then remove washer and bushing.

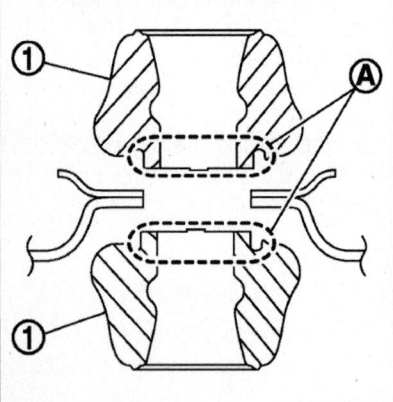

Fig. 211 Installing bushings (1), securely inserting protrusion (A) into the hole on the vehicle body side

To loosen piston rod lock nut, fix the tip (A) of the piston rod.

7. Remove shock absorber assembly.

8. Remove bushing, distance tube, bound bumper cover, and bound bumper from shock absorber.

9. Perform inspection after removal.

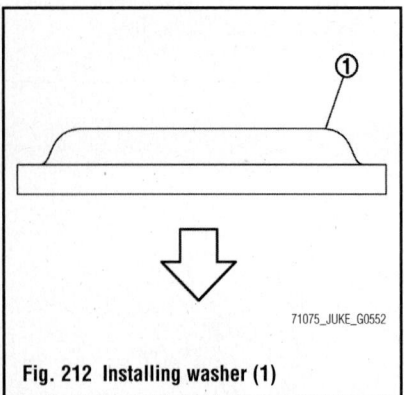

Fig. 212 Installing washer (1)

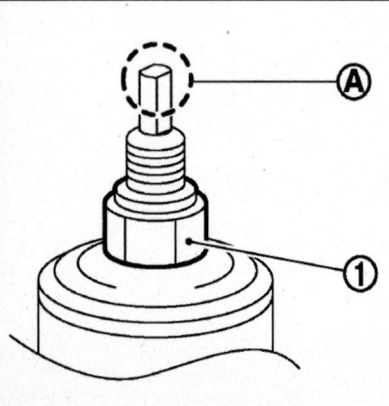

Fig. 213 Holding head (A) of shock absorber piston rod and tightening piston rod lock nut (1)

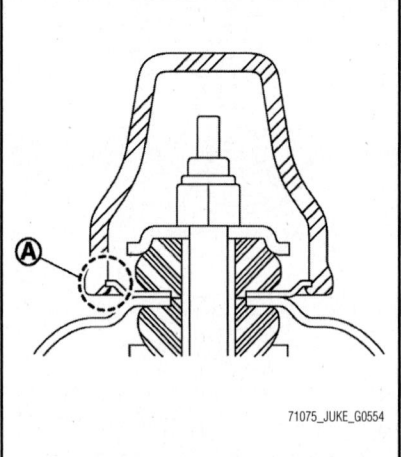

Fig. 214 Engaging cap groove (A) with the flange on the vehicle side

To install:

10. Note the following, and install in the reverse order of the removal procedure.

11. To install bushings (1), securely insert protrusion (A) into the holt on the vehicle body side.

12. Perform final tightening of bolts and nuts at the shock absorber lower side (rubber bushing), under unladen conditions with tires on level ground.

13. Hold a head (A) of shock absorber piston rod not to have it rotate, then tighten the piston rod lock nut (1) to the specified torque.

> ❊❊ **WARNING**
>
> **Never reuse piston rod lock nut.**

14. When installing the cap, securely engage the cap groove (A) with the flange on the vehicle side.

15. Perform inspection after installation.

16. After replacing the shock absorber, always follow the disposal procedure to discard the shock absorber.

DISPOSAL

See Figure 216.

1. Set shock absorber horizontally to the ground with the piston rod fully extracted.

2. Drill 2 - 3 mm (0.08 - 0.12 in) hole at the position (black dot) from top as shown in the figure to release gas gradually.

> ❊❊ **CAUTION**
>
> **Wear eye protection (safety glasses) and gloves.**

> ❊❊ **CAUTION**
>
> **Be careful with metal chips or oil blown out by the compressed gas.**

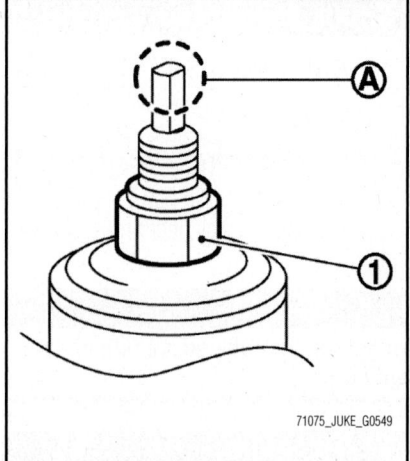

Fig. 215 Removing piston rod lock nut (1) by fixing the tip (A) of the piston rod

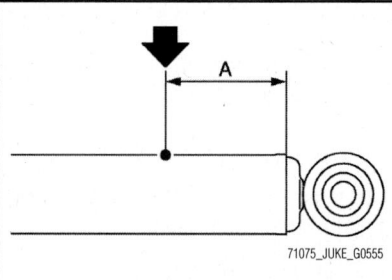

Fig. 216 Drilling 0.08–0.12 inch (2–3 mm) hole 0.79–1.18 inch (20–30 mm) (A) from top

Drill vertically in the direction shown.

Directly to the outer tube avoiding brackets.

The gas is clear, colorless, odorless, and harmless.

3. Position the drilled hole downward and drain oil by moving the piston rod several times.

> ❊❊ **WARNING**
>
> **Dispose of drained oil according to the law and local regulations.**

STABILIZER BAR (SWAY BAR) & LINKS

REMOVAL & INSTALLATION

See Figure 217.

1. Remove stabilizer link.

2. Remove center muffler.

3. Remove mounting nuts on stabilizer clamp, bushing, and stabilizer bar from suspension member.

4. Perform inspection after removal.

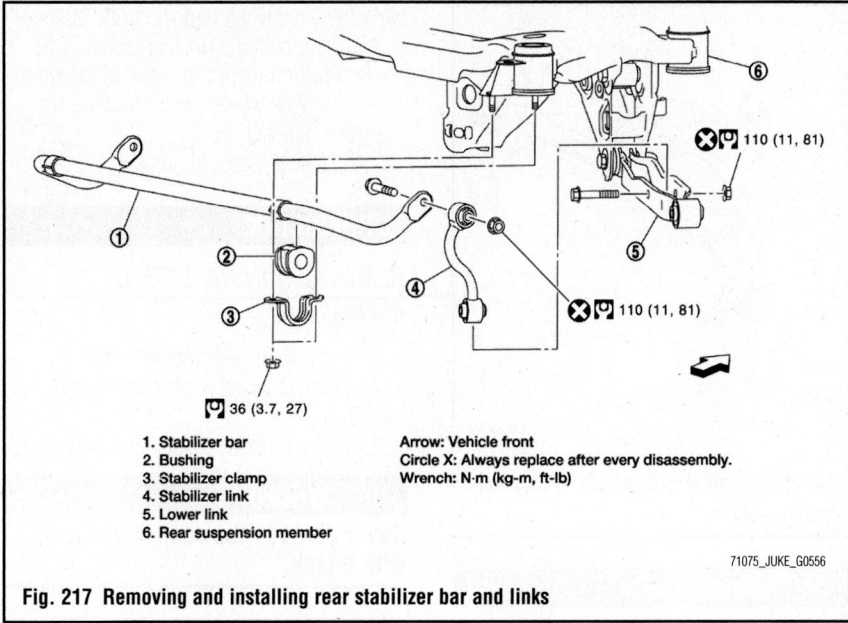

Fig. 217 Removing and installing rear stabilizer bar and links

a. Check stabilizer bar, stabilizer link, stabilizer bushing and stabilizer clamp for deformation, cracks or damage.

Replace it if necessary.

To install:

5. Note the following, and install in the reverse order of the removal procedure.

6. Perform final tightening of rear suspension member and axle installation position (rubber bushing), under unladen conditions with tires on level ground.

7. Never reuse stabilizer link mounting nut.

SUSPENSION ARM

REMOVAL & INSTALLATION

See Figure 218.

1. Remove tires with power tool.
2. Drain brake fluid.
3. Remove wheel sensor and sensor harness.
4. Remove caliper assembly. Hang caliper assembly in a place where it will not interfere with work.

❊❊ WARNING

Never depress brake pedal while brake caliper is removed.

5. Remove disc rotor.
6. Remove parking brake cable mounting bolt.
7. Separate the brake tube from the brake hose, and remove lock plate.
8. Remove wheel hub and bearing assembly.

9. Remove parking brake shoe and back plate.
10. Set jack under suspension arm.

❊❊ WARNING

Never damage the suspension arm with a jack.

❊❊ WARNING

Check for stable condition when using a jack.

11. Remove stabilizer link.
12. Remove upper link from suspension arm.
13. Remove lower link from suspension arm.
14. Remove coil spring from suspension arm.
15. Remove suspension arm bracket from vehicle.
16. Remove suspension arm from suspension arm bracket.
17. Perform inspection after removal.
   a. Check suspension arm and bushing for deformation, cracks or damage. Replace if necessary.

To install:

18. Note the following, and install in the reverse order of the removal procedure.

19. Perform final tightening of rear suspension member installation position (rubber bussing), under unladen conditions with tires on level ground.

❊❊ WARNING

Never reuse suspension arm mounting nut.

20. Perform inspection after installation.
   a. Check wheel sensor harness for proper connection.
   b. Adjust parking brake operation (stroke).
   c. Check wheel alignment.

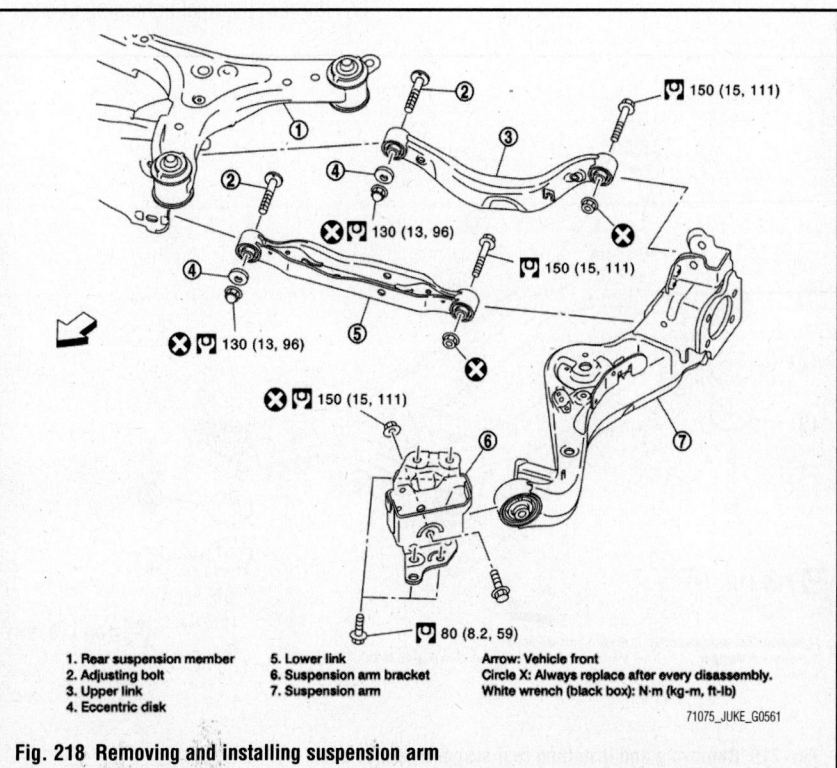

Fig. 218 Removing and installing suspension arm

SUSPENSION ASSEMBLY

REMOVAL & INSTALLATION

See Figures 219 and 220.

1. Remove tires with power tool.
2. Remove center muffler.
3. Remove propeller shaft.
4. Remove stabilizer bar.
5. Remove wheel sensor and sensor harness.
6. Remove upper link from suspension arm.
7. Remove lower link from suspension arm.
8. Remove drive shaft from rear final drive.
9. Remove rear final drive.
10. Set jack under rear suspension member.

❄❄ WARNING

Never damage the suspension member with a jack.

❄❄ WARNING

Check for stable condition when using a jack.

11. Remove rear suspension member mounting bolts, rebound stopper, and washer.
12. Slowly lower jack, then remove rear suspension member, lower link and upper link from vehicle as a unit.

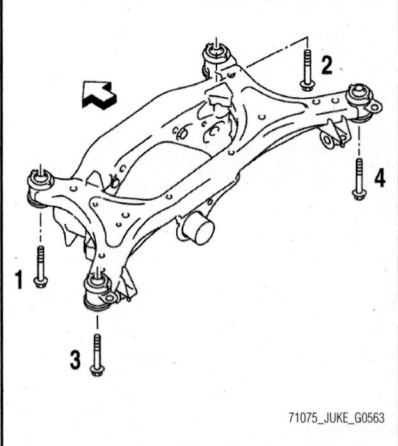

Fig. 220 Identifying mounting bolt tightening sequence

❄❄ WARNING

Operate while checking that jack supporting status is stable.

13. Remove lower link and upper link from rear suspension member.
14. Perform inspection after removal.

To install:

15. Note the following, and install in the reverse order of the removal procedure.
16. To install mounting bolts of the suspension member, temporarily tighten them in numerical order shown in the figure and tighten them to 74 ft. lbs. (100 Nm).
17. Perform the final tightening of each

parts removed when removing rear suspension assembly under unladen conditions.
18. Perform inspection after installation.
    a. Check wheel sensor harness for proper connection.
    b. Check wheel alignment.

UPPER LINK

REMOVAL & INSTALLATION

See Figure 221.

1. Remove tires with power tool.
2. Remove wheel sensor and sensor harness.
3. Set jack under suspension arm.

❄❄ WARNING

Never damage the suspension arm with a jack.

❄❄ WARNING

Check for stable condition when using a jack.

4. Remove eccentric disc, adjusting bolt, mounting bolt, and nut, then remove upper link.
5. Perform inspection after removal.

To install:

6. Note the following, and install in the reverse order of the removal procedure:
    a. Perform final tightening of rear suspension member and axle installation position (rubber bushing), under unladen conditions with tires on level ground.

❄❄ WARNING

Never reuse upper link mounting nut.

7. Perform inspection after installation:
    a. Check wheel sensor harness for proper connection.
    b. Check wheel alignment.

WHEEL HUBS & BEARINGS

REMOVAL & INSTALLATION

2WD Vehicles

See Figures 222 and 223.

1. Remove tires with power tool.
2. Remove wheel sensor.
3. Remove caliper assembly. Hang caliper assembly in a place where it will not interfere with work.

❄❄ WARNING

Never depress brake pedal while brake caliper is removed.

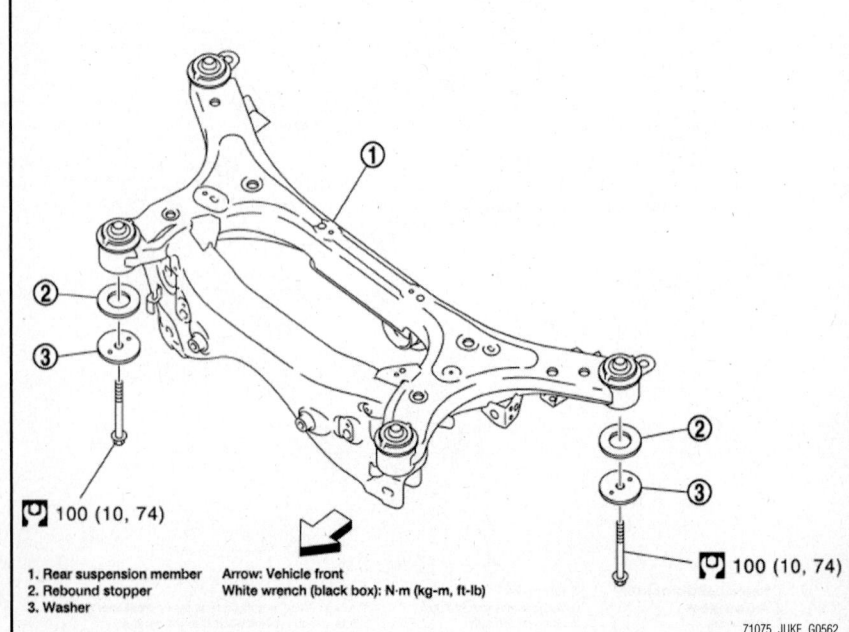

🔧 100 (10, 74)

1. Rear suspension member   Arrow: Vehicle front
2. Rebound stopper   White wrench (black box): N·m (kg-m, ft-lb)
3. Washer

🔧 100 (10, 74)

71075_JUKE_G0562

Fig. 219 Removing and installing rear suspension member

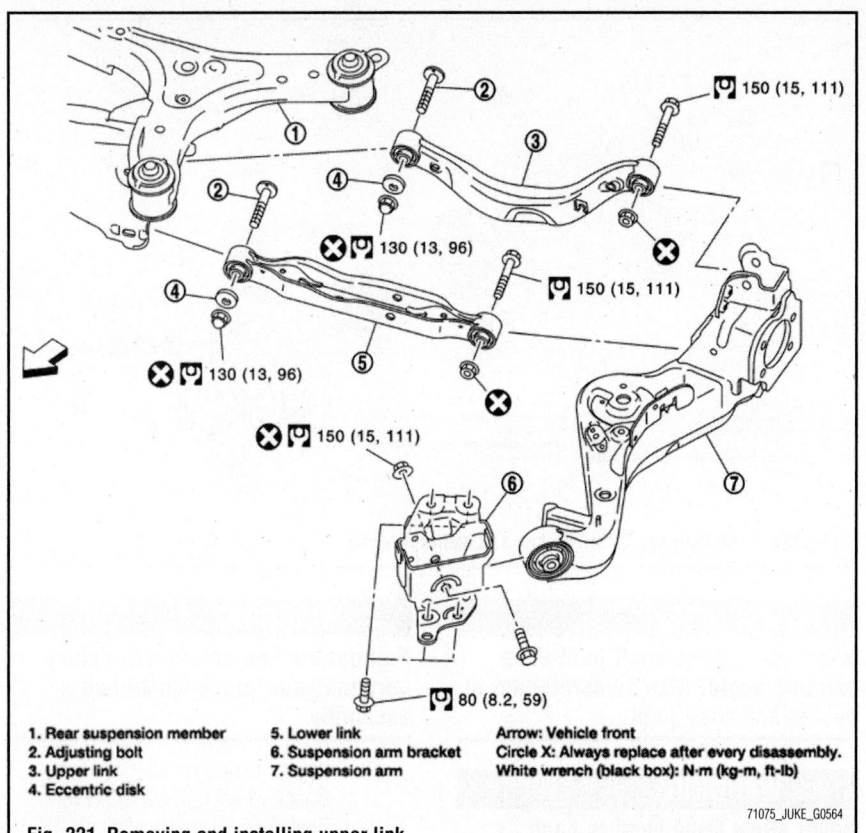

1. Rear suspension member
2. Adjusting bolt
3. Upper link
4. Eccentric disk
5. Lower link
6. Suspension arm bracket
7. Suspension arm

Arrow: Vehicle front
Circle X: Always replace after every disassembly.
White wrench (black box): N·m (kg-m, ft-lb)

71075_JUKE_G0564

Fig. 221 Removing and installing upper link

4. Remove disc rotor. If disc rotor cannot be removed, remove as follows.

※※ WARNING
Parking brake completely in the released position.

※※ WARNING
Put matching marks on the wheel hub assembly and the disc rotor before removing the disc rotor.

※※ WARNING
Never drop disc rotor.

  a. Fix the disc rotor with wheel nuts and remove the adjusting hole plug.
  b. Using suitable tool, rotate adjuster downward to
   retract and loosen brake shoe.
  c. Remove disc rotor.
5. Remove wheel hub assembly.

※※ WARNING
Never remove parking brake assembly. Protect it from falling.

6. Remove hub bolts from wheel hub assembly, using the ball joint remover (commercial service tool).

※※ WARNING
Remove hub bolt only when necessary.

※※ WARNING
Never hammer the hub bolt to avoid impact to the wheel hub assembly.

※※ WARNING
Pull out the hub bolt in a direction perpendicular to the wheel hub assembly.

7. Perform inspection after removal.
  a. Check the wheel hub assembly for wear, cracks, and damage. Replace if necessary.

To install:
8. Note the following, and install in the reverse order of the removal procedure.
9. Place a washer (A) as shown in the figure to install the hub bolts (1) by using the tightening force of the nut (B).

※※ WARNING
Check that there is no clearance between wheel hub assembly and hub bolt.

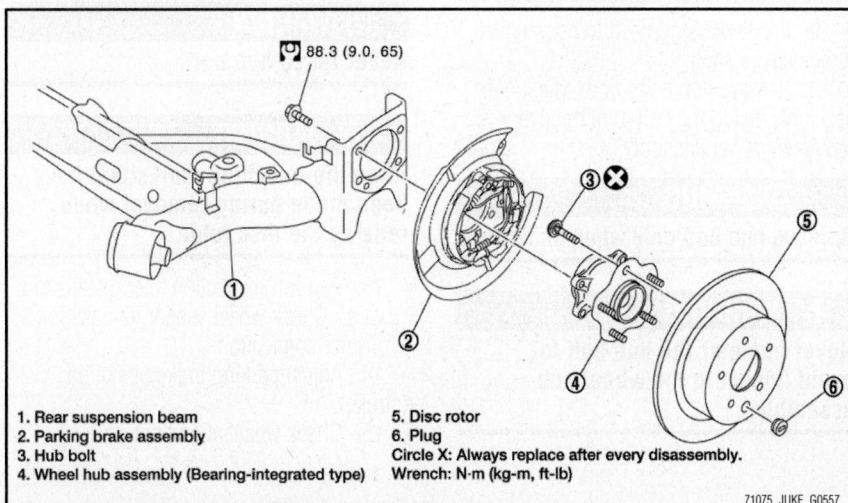

1. Rear suspension beam
2. Parking brake assembly
3. Hub bolt
4. Wheel hub assembly (Bearing-integrated type)

5. Disc rotor
6. Plug
Circle X: Always replace after every disassembly.
Wrench: N·m (kg-m, ft-lb)

71075_JUKE_G0557

Fig. 222 Removing and installing hub and bearing—2WD

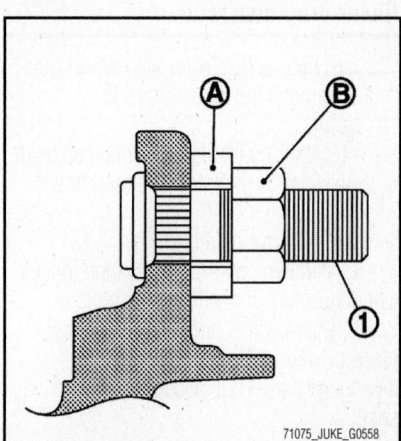

71075_JUKE_G0558

Fig. 223 Placing washer (A) as shown to install hub bolts (1) by using the tightening force of the nut (B)

> ※※ **WARNING**
>
> Never reuse hub bolt.

> ※※ **WARNING**
>
> Align the matching marks that have been made during removal when reusing the disc rotor.

10. Perform inspection after installation.
    a. Check wheel sensor harness for proper connection.
11. Adjust parking brake operation (stroke).
12. Check wheel alignment.

AWD Vehicles

See Figures 223 and 224

1. Remove tires with power tool.
2. Remove wheel sensor.
3. Remove caliper assembly. Hang caliper assembly in a place where it will not interfere with work.

> ※※ **WARNING**
>
> Never depress brake pedal while brake caliper is removed.

4. Remove disc rotor. If disc rotor cannot be removed, remove as follows.

➡ **Parking brake completely in the released position.**

> ※※ **WARNING**
>
> Put matching marks on the wheel hub assembly and the disc rotor before removing the disc rotor.

> ※※ **WARNING**
>
> Never drop disc rotor.

    a. Fix the disc rotor with wheel nuts and remove the adjusting hole plug.
    b. Using suitable tool, rotate adjuster downward to retract and loosen brake shoe.
    c. Remove disc rotor.
5. Remove cotter pin, and then loosen wheel hub lock nut with power tool.
6. Patch wheel hub lock nut with a piece of wood. Hammer the wood to disengage wheel hub assembly from drive shaft.

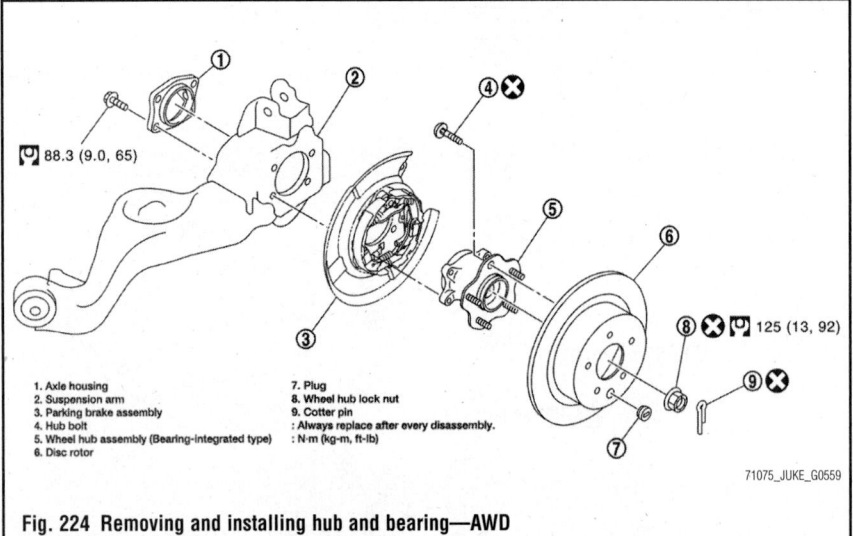

1. Axle housing
2. Suspension arm
3. Parking brake assembly
4. Hub bolt
5. Wheel hub assembly (Bearing-integrated type)
6. Disc rotor
7. Plug
8. Wheel hub lock nut
9. Cotter pin
: Always replace after every disassembly.
: N·m (kg-m, ft-lb)

88.3 (9.0, 65)
125 (13, 92)

71075_JUKE_G0559

Fig. 224 Removing and installing hub and bearing—AWD

> ※※ **WARNING**
>
> Never place drive shaft joint at an extreme angle. Also be careful not to overextend slide joint.

> ※※ **WARNING**
>
> Never allow drive shaft to hang down without support for joint sub-assembly, shaft and the other parts.

➡ **Use suitable puller, if wheel hub assembly and drive shaft cannot be separated even after performing the above procedure.**

7. Remove wheel hub lock nut.
8. Remove wheel hub assembly.

> ※※ **WARNING**
>
> Never remove parking brake assembly. Protect it from falling.

9. If axle housing need to be removed, remove drive shaft.
10. Remove hub bolts from wheel hub assembly, using the ball joint remover (commercial service tool).

> ※※ **WARNING**
>
> Remove hub bolt only when necessary.

> ※※ **WARNING**
>
> Never hammer the hub bolt to avoid impact to the wheel hub assembly.

> ※※ **WARNING**
>
> Pull out the hub bolt in a direction perpendicular to the wheel hub assembly.

11. Perform inspection after removal.
    a. Check wheel hub assembly for wear, cracks, and damage. Replace if necessary.

To install:

12. Note the following, and install in the reverse order of the removal procedure.
    a. Place a washer (A) as shown in the figure to install the hub bolts
    (1) by using the tightening force of the nut (B).

> ※※ **WARNING**
>
> Check that there is no clearance between wheel hub assembly and hub bolt.

> ※※ **WARNING**
>
> Never reuse hub bolt.

> ※※ **WARNING**
>
> Align the matching marks that have been made during removal when reusing the disc rotor.

13. Perform inspection after installation.
    a. Check wheel sensor harness for proper connection.
14. Adjust parking brake operation (stroke).
15. Check wheel alignment.

NISSAN

LEAF

12

SPECIFICATIONS AND MAINTENANCE CHARTS

ENGINE AND VEHICLE IDENTIFICATION

Mfg.
Nissan

Model Year	
Code	Year
B	2011
C	2012

71075_LEAF_C0001

CAPACITIES

Year	Model	Transmission (pts.)	Cooling System (qts.)
2011	Leaf	2.4	7.0
2012	Leaf	2.4	7.0

NOTE: All capacities are approximate. Add fluid gradually and check to be sure a proper fluid level is obtained.

71075_LEAF_C0003

FLUID SPECIFICATIONS

Year	Model	Auto. Trans.	Power Steering Fluid	Brake Master Cylinder	Cooling System
2011	Leaf	①	NISSAN PSF	DOT 3	②
2012	Leaf	①	NISSAN PSF	DOT 3	②

NA: Not Applicable

DOT: Department Of Transpotation

① Nissan Matic Fluid S

② Nissan Long Life Coolant (Blue)

71075_LEAF_C0002

WHEEL ALIGNMENT

Year	Model		Caster		Camber		Toe-in (in.)	Kingpin Inclination (Deg.)
			Range (+/-Deg.)	Preferred Setting (Deg.)	Range (+/-Deg.)	Preferred Setting (Deg.)		
2011	Leaf	F	.75	+4.83	.75	-0.42	0.08+/-0.04	11.92+/-.75
		R	—	—	.75	-0.42	0.08+/-0.04	—
2012	Leaf	F	.75	+4.83	.75	-0.42	0.08+/-0.04	11.92+/-.75
		R	—	—	.75	-0.42	0.08+/-0.04	—

Note: Specifications are taken with the following, radiator coolant. Spare tire, jack, hand tools and mats are in designated positions.

71075_LEAF_C0004

TIRE, WHEEL AND BALL JOINT SPECIFICATIONS

Year	Model	OEM Tires		Tire Pressures (psi)		Wheel Size	Ball Joint Inspection	Lug Nut Torque (ft. lbs.)
		Standard	Optional	Front	Rear			
2011	Leaf	P205/55R16 89H	NA	36	36	NS	①	80
2012	Leaf	P205/55R16 89H	NA	36	36	NS	①	80

NA: Not Applicable

NS: Not Supplied

OEM: Original Equipment Manufacturer

PSI: Pounds Per Square Inch

① 0 (0mm) inches axial end play

71075_LEAF_C0005

BRAKE SPECIFICATIONS
All measurements in inches unless noted

Year	Model		Brake Disc			Minimum Lining Thickness	Brake Caliper	
			Original Thickness	Minimum Thickness	Maximum Runout		Bracket Bolts (ft. lbs.)	Mounting Bolts (ft. lbs.)
2011	Leaf	F	1.102	1.024	0.0006	0.079	122	20
		R	0.630	0.551	0.0006	0.079	62	32
2012	Leaf	F	1.102	1.024	0.0006	0.079	122	20
		R	0.630	0.551	0.0006	0.079	62	32

F: Front

R: Rear

71075_LEAF_C0006

SCHEDULED MAINTENANCE INTERVALS (1)
Nissan—Leaf

TO BE SERVICED	TYPE OF SERVICE	7.5	15	22.5	30	37.5	45	52.5	60
Brake lines & cables	I		✓		✓		✓		✓
Brake pads, discs	I	✓	✓	✓	✓	✓	✓	✓	✓
Brake fluid	R		✓		✓		✓		✓
Charging port	I		✓		✓		✓		✓
Charging port sealing cap	I				✓				✓
Cooling system coolant ①	R								
Cabin air filter	R		✓		✓		✓		✓
Driveshaft boots	I	✓	✓	✓	✓	✓	✓	✓	✓
EV battery report	I		✓		✓		✓		✓
Heating system coolant ①	I								
Reduction gear oil	I		✓		✓		✓		✓
Steering gear, linkage, axle & suspension parts	I	✓	✓	✓	✓	✓	✓	✓	✓
Tires (rotate)	S/I	every 6,000 miles							

R: Replace   S/I: Service or Inspect

① First at 125,000, then every 75,000 miles

Follow Periodic Maintenance Schedule 1 if the driving habits frequently include one or more of the following driving conditions:

Repeated short trips of less than 5 miles (8 km).

Repeated short trips of less than 10 miles (16 km) with outside temperatures remaining below freezing

Operating in hot weather in stop-and-go "rush hour" traffic.

Extensive idling and/or low speed driving for long distances, such as police, taxi or door-to-door delivery use

Driving in dusty conditions.

Driving on rough, muddy, or salt spread roads.

Towing a trailer, using a camper or a car-top carrier.

Follow Periodic Maintenance Schedule 2 if none of driving conditions shown in Schedule 1 apply to the driving habits.

71075_LEAF_C0007

SCHEDULED MAINTENANCE INTERVALS (2)
Nissan—Leaf

TO BE SERVICED	SERVICE	7.5	15	22.5	30	37.5	45	52.5	60
Brake lines & cables	I		✓		✓		✓		✓
Brake pads, discs	I		✓		✓		✓		✓
Brake fluid	R				✓				✓
Charging port	I		✓		✓		✓		✓
Charging port sealing cap	I				✓				✓
Cooling system coolant ①	R								
Cabin air filter	R		✓		✓		✓		✓
Driveshaft boots	I		✓		✓		✓		✓
EV battery report	I		✓		✓		✓		✓
⑧③①②⑦t⑤④⓪③③ syst⑤①③⑨ ②⑨oo⑥⑧②⑦④⓪t ①	I								
Reduction gear oil	I		✓		✓		✓		✓
Steering gear, linkage, axle & suspension parts	I				✓				✓
Tires (rotate)	S/I	every 6,000 miles							

R: Replace   S/I: Service or Inspect

① First at 125,000, then every 75,000 miles

Follow Periodic Maintenance Schedule 1 if the driving habits frequently include one or more of the following driving conditions:

Repeated short trips of less than 5 miles (8 km).

Repeated short trips of less than 10 miles (16 km) with outside temperatures remaining below freezing

Operating in hot weather in stop-and-go "rush hour" traffic.

Extensive idling and/or low speed driving for long distances, such as police, taxi or door-to-door delivery use

Driving in dusty conditions.

Driving on rough, muddy, or salt spread roads.

Towing a trailer, using a camper or a car-top carrier.

Follow Periodic Maintenance Schedule 2 if none of driving conditions shown in Schedule 1 apply to the driving habits.

71075_LEAF_C0008

PRECAUTIONS

Before servicing any vehicle, please be sure to read all of the following precautions, which deal with personal safety, prevention of component damage, and important points to take into consideration when servicing a motor vehicle:

• Never open, service or drain the radiator or cooling system when the engine is hot; serious burns can occur from the steam and hot coolant.

• Observe all applicable safety precautions when working around fuel. Whenever servicing the fuel system, always work in a well-ventilated area. Do not allow fuel spray or vapors to come in contact with a spark, open flame, or excessive heat (a hot drop light, for example). Keep a dry chemical fire extinguisher near the work area. Always keep fuel in a container specifically designed for fuel storage; also, always properly seal fuel containers to avoid the possibility of fire or explosion. Refer to the additional fuel system precautions later in this section.

• Fuel injection systems often remain pressurized, even after the engine has been turned **OFF**. The fuel system pressure must be relieved before disconnecting any fuel lines. Failure to do so may result in fire and/or personal injury.

• Brake fluid often contains polyglycol ethers and polyglycols. Avoid contact with the eyes and wash your hands thoroughly after handling brake fluid. If you do get brake fluid in your eyes, flush your eyes with clean, running water for 15 minutes. If eye irritation persists, or if you have taken brake fluid internally, IMMEDIATELY seek medical assistance.

• The EPA warns that prolonged contact with used engine oil may cause a number of skin disorders, including cancer. You should make every effort to minimize your exposure to used engine oil. Protective gloves should be worn when changing oil. Wash your hands and any other exposed skin areas as soon as possible after exposure to used engine oil. Soap and water, or waterless hand cleaner should be used.

• All new vehicles are now equipped with an air bag system, often referred to as a Supplemental Restraint System (SRS) or Supplemental Inflatable Restraint (SIR) system. The system must be disabled before performing service on or around system components, steering column, instrument panel components, wiring and sensors. Failure to follow safety and disabling procedures could result in accidental air bag deployment, possible personal injury and unnecessary system repairs.

• Always wear safety goggles when working with, or around, the air bag system. When carrying a non-deployed air bag, be sure the bag and trim cover are pointed away from your body. When placing a non-deployed air bag on a work surface, always face the bag and trim cover upward, away from the surface. This will reduce the motion of the module if it is accidentally deployed. Refer to the additional air bag system precautions later in this section.

• Clean, high quality brake fluid from a sealed container is essential to the safe and proper operation of the brake system. You should always buy the correct type of brake fluid for your vehicle. If the brake fluid becomes contaminated, completely flush the system with new fluid. Never reuse any brake fluid. Any brake fluid that is removed from the system should be discarded. Also, do not allow any brake fluid to come in contact with a painted surface; it will damage the paint.

• Never operate the engine without the proper amount and type of engine oil; doing so WILL result in severe engine damage.

• Timing belt maintenance is extremely important. Many models utilize an interference-type, non-freewheeling engine. If the timing belt breaks, the valves in the cylinder head may strike the pistons, causing potentially serious (also time-consuming and expensive) engine damage. Refer to the maintenance interval charts for the recommended replacement interval for the timing belt, and to the timing belt section for belt replacement and inspection.

• Disconnecting the negative battery cable on some vehicles may interfere with the functions of the on-board computer system(s) and may require the computer to undergo a relearning process once the negative battery cable is reconnected.

• When servicing drum brakes, only disassemble and assemble one side at a time, leaving the remaining side intact for reference.

• Only an MVAC-trained, EPA-certified automotive technician should service the air conditioning system or its components.

BRAKES

GENERAL INFORMATION

PRECAUTIONS

• Certain components within the ABS system are not intended to be serviced or repaired individually.

• Do not use rubber hoses or other parts not specifically specified for and ABS system. When using repair kits, replace all parts included in the kit. Partial or incorrect repair may lead to functional problems and require the replacement of components.

• Lubricate rubber parts with clean, fresh brake fluid to ease assembly. Do not use shop air to clean parts; damage to rubber components may result.

• Use only DOT 3 brake fluid from an unopened container.

• If any hydraulic component or line is removed or replaced, it may be necessary to bleed the entire system.

• A clean repair area is essential. Always clean the reservoir and cap thoroughly before removing the cap. The slightest amount of dirt in the fluid may plug an orifice and impair the system function. Perform repairs after components have been thoroughly cleaned; use only denatured alcohol to clean components. Do not allow ABS components to come into contact with any substance containing mineral oil; this includes used shop rags.

• The Anti-Lock control unit is a microprocessor similar to other computer units in the vehicle. Ensure that the ignition switch is **OFF** before removing or installing controller harnesses. Avoid

ANTI-LOCK BRAKE SYSTEM (ABS)

static electricity discharge at or near the controller.

• If any arc welding is to be done on the vehicle, the control unit should be unplugged before welding operations begin.

SPEED SENSORS

REMOVAL & INSTALLATION

Front

See Figures 1 and 2.

1. Before servicing the vehicle, refer to the Precautions Section.
2. Remove the front wheels and tires.
3. Remove the fender protector (front).
4. Remove front wheel sensor from steering knuckle.

WARNING

To prevent damage to the parts, never rotate and never pull front wheel sensor.

5. Remove front wheel sensor harness from the vehicle.

WARNING

To prevent damage to the parts, never twist or pull front wheel sensor harness.

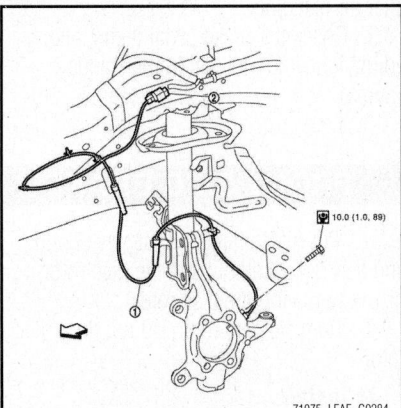

Fig. 1 Front left-hand wheel sensor (1), front left-hand wheel sensor harness connector (2)

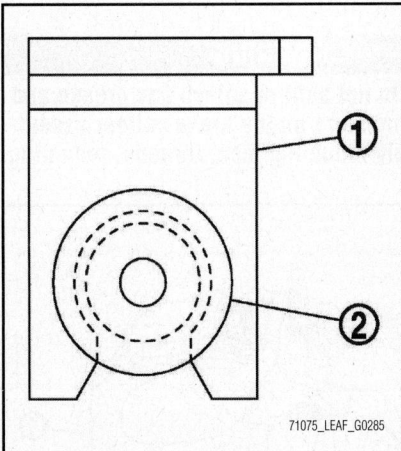

Fig. 2 Check that grommet (2) is fully inserted to bracket (1)

To install:
Installation is the reverse of removal, noting the following:

6. Check that there is no foreign material like iron powder or damage on inner surface of front wheel sensor mounting hole of steering knuckle and sensor rotor. Install after cleaning when there are foreign material like iron powder, or replace when there is a malfunction.

7. Never twist front wheel sensor harness when installing front wheel sensor. Check that grommet is fully inserted to bracket. Check that front wheel sensor harness is not twisted after installation.

8. Check that the identification line of the front wheel sensor is facing vehicle front.

Rear

See Figures 2 and 3.

1. Before servicing the vehicle, refer to the Precautions Section.

2. Remove rear wheel sensor from wheel hub and bearing assembly.

WARNING

To prevent damage to the parts, never rotate or pull rear wheel sensor.

3. Remove rear wheel sensor harness from the vehicle.

WARNING

To prevent damage to the parts, never twist and never pull rear wheel sensor harness.

To install:
Installation is the reverse of removal, noting the following:

4. Check that there is no foreign material like iron powder or damage on inner surface of rear wheel sensor mounting hole of wheel hub and bearing assembly and sensor rotor. Install after cleaning when there are foreign material like iron powder, or replace when there is a malfunction.

5. Never twist rear wheel sensor harness when installing rear wheel sensor. Check that grommet is fully inserted to bracket. Check that rear wheel sensor harness is not twisted after installation.

6. Check that the identification line of the rear wheel sensor is facing upward.

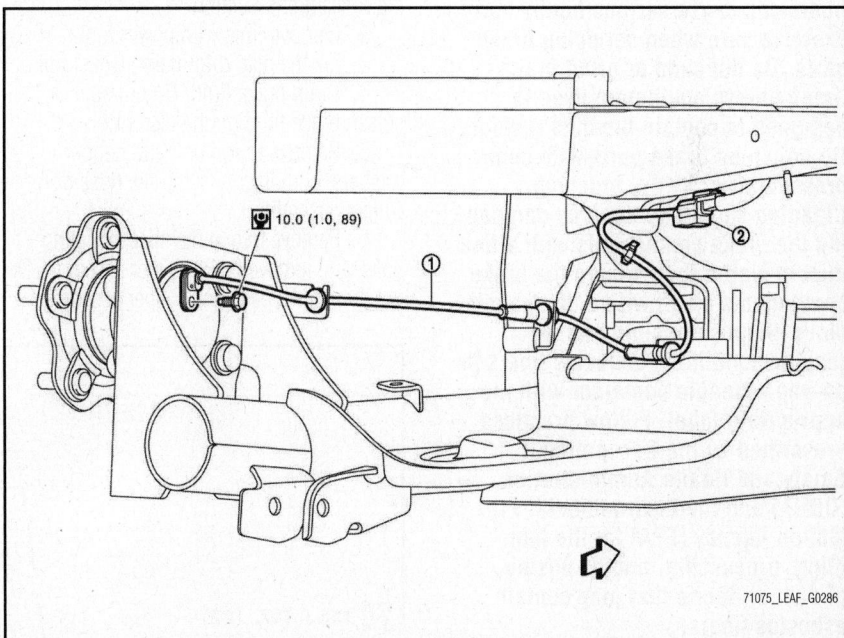

Fig. 3 Rear left-hand wheel sensor (1), rear left-hand wheel sensor harness connector (2)

BRAKES   BLEEDING THE BRAKE SYSTEM

BLEEDING PROCEDURE

BLEEDING PROCEDURE

❉❉ WARNING

Turn ON the power switch when performing the procedure. Monitor the brake fluid level in the reservoir tank while performing the air bleeding.

❉❉ WARNING

Before servicing the vehicle, refer to the Precautions Section.

❉❉ WARNING

Never reuse drained brake fluid. Never allow any oils other than the designated brake fluid to enter the system.

1. Make sure that there is no foreign material in the reservoir tank, and refill with new brake fluid.
2. Connect a vinyl tube to the rear right wheel air bleeder.
3. Fully depress the brake pedal 4 to 5 times.
4. Loosen the air bleeder and bleed air with the brake pedal depressed, then quickly tighten the bleeder valve.

5. Repeat steps 2 to 3 until all of the air is out of the brake line.
6. Tighten the air bleeder to the specified torque.
7. Perform steps 2 to 6. Occasionally fill with the brake fluid in order to keep it in the reservoir tank to at least half of the MAX line. Bleed air in the following order: rear right brake, front left brake, rear left brake, front right brake.
8. Check that the brake fluid level in the reservoir tank is within the specified range after air bleeding.
9. Check the brake pedal items, and adjust if any are not within the standard values.

BRAKES   FRONT DISC BRAKES

❉❉ CAUTION

Dust and dirt accumulating on brake parts during normal use may contain asbestos fibers from production or aftermarket brake linings. Breathing excessive concentrations of asbestos fibers can cause serious bodily harm. Exercise care when servicing brake parts. Do not sand or grind brake lining unless equipment used is designed to contain the dust residue. Do not clean brake parts with compressed air or by dry brushing. Cleaning should be done by dampening the brake components with a fine mist of water, then wiping the brake components clean with a dampened cloth. Dispose of cloth and all residue containing asbestos fibers in an impermeable container with the appropriate label. Follow practices prescribed by the Occupational Safety and Health Administration (OSHA) and the Environmental Protection Agency (EPA) for the handling, processing, and disposing of dust or debris that may contain asbestos fibers.

BRAKE CALIPER

REMOVAL & INSTALLATION
See Figure 4.

❉❉ CAUTION

Dust covering the brake must be removed with a dust collector. Never splatter the dust with an air blow gun.

❉❉ WARNING

Never depress the brake pedal. Brake fluid may splash while removing the brake hose.

1. Before servicing the vehicle, refer to the Precautions Section.
2. Remove tires with power tool.
3. Fix the disc rotor using wheel nuts.
4. Drain brake fluid. Do not spill or splash brake fluid on the disc rotor.
5. Remove union bolt and copper washers, and disconnect brake hose from caliper assembly.
6. Remove torque member mounting bolts, and remove brake caliper assembly. Do not drop brake pads and caliper assembly.

7. Put matchmarks on the wheel hub and bearing assembly and the disc rotor before removing the disc rotor.
8. Remove disc rotor. Do not drop disc rotor.

To install:
9. Align the matchmarks and install the disc rotor.
10. Install the brake caliper assembly to the vehicle and tighten the torque member mounting bolts to 122 ft. lbs. (165 Nm).

❉❉ WARNING

Do not spill or splash any grease and moisture on the brake caliper assembly mounting face, threads, mounting

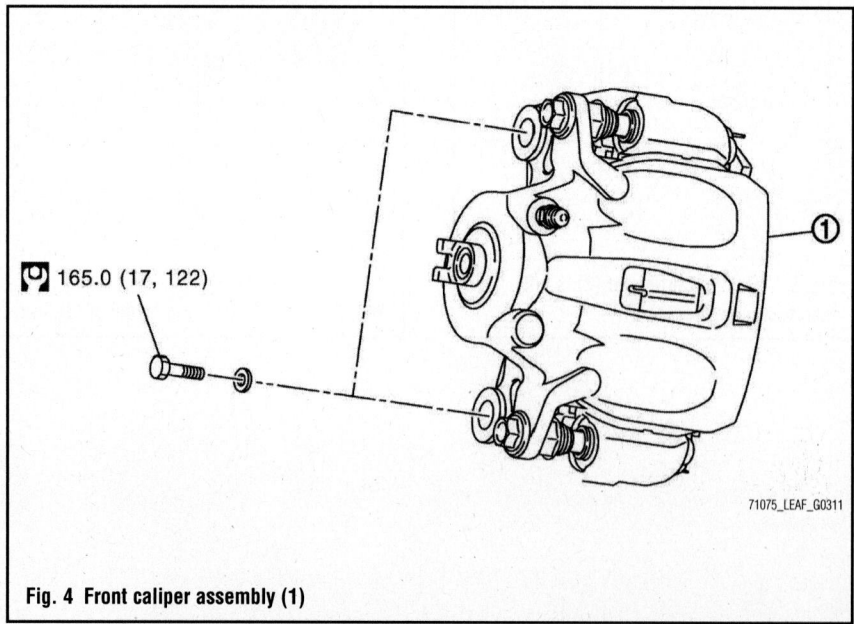

165.0 (17, 122)

71075_LEAF_G0311

Fig. 4 Front caliper assembly (1)

bolts and washers. Wipe out any
grease and moisture.

11. Install brake hose and copper washers to brake caliper assembly, and tighten
union bolts to 13 ft. lbs. (18 Nm). Do not
reuse copper washer.

12. Refill with new brake fluid and perform the air bleeding.

13. Check drag of front disc brake.

BRAKE PADS

REMOVAL & INSTALLATION

See Figure 5.

✳✳ CAUTION

**Dust covering the brake must be
removed with a dust collector. Do not
splatter the dust with an air blow
gun.**

✳✳ WARNING

**Do not depress the brake pedal while
removing the brake pads because the
piston may pop out. Do not spill or
splash brake fluid on the disc rotor.**

1. Before servicing the vehicle, refer to
the Precautions Section.

2. Remove tires with power tool.

3. Remove lower sliding pin bolt.

4. Suspend the cylinder body with suitable wire so that the brake hose will not
stretch. Remove the brake pads, shims and
shim cover from the torque member.

✳✳ WARNING

**Do not reuse the pad retainers when
removed the pad retainers from the
torque member. Do not damage the
piston boot. Do not drop the brake
pads, shims, and the shim cover.
Remember each position of the
removed brake pads.**

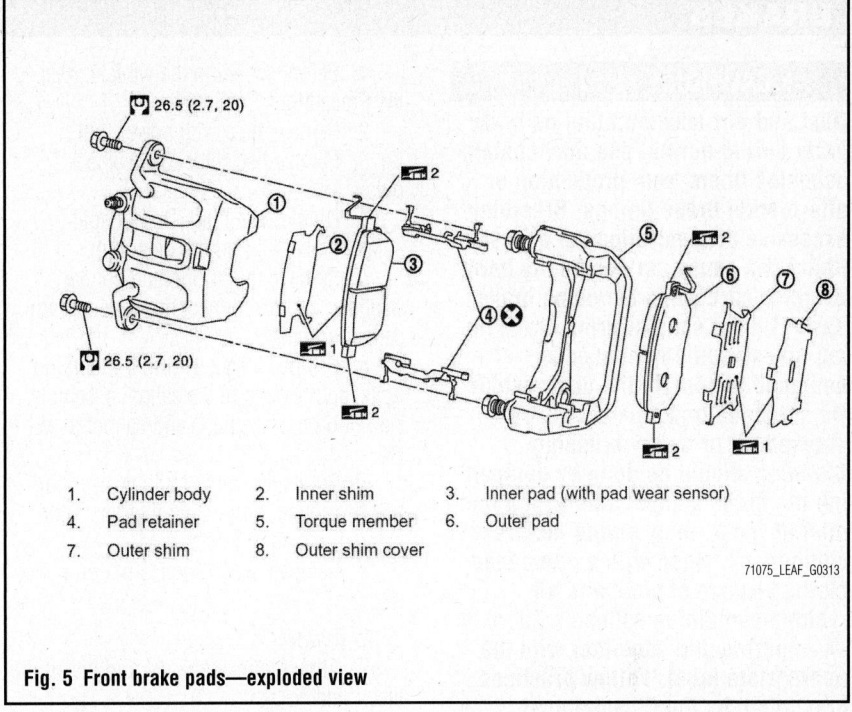

1. Cylinder body
2. Inner shim
3. Inner pad (with pad wear sensor)
4. Pad retainer
5. Torque member
6. Outer pad
7. Outer shim
8. Outer shim cover

71075_LEAF_G0313

Fig. 5 Front brake pads—exploded view

To install:

5. Install the pad retainers to the torque
member if the pad retainers have been
removed.

✳✳ WARNING

**Do not reuse the pad retainers.
Securely assemble the pad retainers
so that it will not be lifted up from
the torque member. Never deform the
pad retainers. Eliminate double-
faced adhesive tape on torque member. Remove adhesive's protective
liner on pad retainers.**

6. Apply PBC (Poly Butyl Cuprysil)
grease or silicone-based grease to the mating faces between the brake pads and pad
retainers.

7. Apply copper based brake grease to
the mating faces between the brake pads,
shims and shim covers, and install them to
the brake pad. Always replace the shims
together with the shim covers when replacing the brake pad.

8. Install the cylinder body and brake
pads to the torque member. Use a disc
brake piston tool to easily press piston.

✳✳ WARNING

**Do not damage the piston boot.
When replacing a pad with new one,
check brake fluid level in the reservoir tank because brake fluid returns
to master cylinder reservoir tank
when pressing piston in.**

9. Install the lower sliding pin bolt and
tighten it to the specified torque.

10. Depress the brake pedal several
times to check that no drag feel is present
for the front disc brake.

✳✳ CAUTION

Dust and dirt accumulating on brake parts during normal use may contain asbestos fibers from production or aftermarket brake linings. Breathing excessive concentrations of asbestos fibers can cause serious bodily harm. Exercise care when servicing brake parts. Do not sand or grind brake lining unless equipment used is designed to contain the dust residue. Do not clean brake parts with compressed air or by dry brushing. Cleaning should be done by dampening the brake components with a fine mist of water, then wiping the brake components clean with a dampened cloth. Dispose of cloth and all residue containing asbestos fibers in an impermeable container with the appropriate label. Follow practices prescribed by the Occupational Safety and Health Administration (OSHA) and the Environmental Protection Agency (EPA) for the handling, processing, and disposing of dust or debris that may contain asbestos fibers.

BRAKE CALIPERS

REMOVAL & INSTALLATION
See Figure 6.

✳✳ CAUTION

Dust covering the brake must be removed with a dust collector. Never splatter the dust with an air blow gun.

✳✳ WARNING

Never depress the brake pedal. Brake fluid may splash while removing the brake hose.

1. Before servicing the vehicle, refer to the Precautions Section.
2. Remove tires with power tool.
3. Fix the disc rotor using wheel nuts.
4. Drain brake fluid. Do not spill or splash brake fluid on the disc rotor.
5. Remove union bolt and copper washers, and disconnect brake hose from caliper assembly.
6. Remove torque member mounting bolts, and remove brake caliper assembly. Do not drop brake pads and caliper assembly.
7. Put matchmarks on the wheel hub and bearing assembly and the disc rotor before removing the disc rotor.
8. Remove disc rotor. Do not drop disc rotor.

To install:

9. Align the matchmarks and install the disc rotor.
10. Install the brake caliper assembly to the vehicle and tighten the torque member mounting bolts to 122 ft. lbs. (165 Nm).

✳✳ WARNING

Do not spill or splash any grease and moisture on the brake caliper assembly mounting face, threads, mounting bolts and washers. Wipe out any grease and moisture.

11. Install brake hose and copper washers to brake caliper assembly, and tighten union bolts to 13 ft. lbs. (18 Nm). Do not reuse copper washer.

12. Refill with new brake fluid and perform the air bleeding.
13. Check that no drag feel is present for the rear disc brake.

BRAKE PADS

REMOVAL & INSTALLATION
See Figure 7.

✳✳ CAUTION

Dust covering the brake must be removed with a dust collector. Do not splatter the dust with an air blow gun.

✳✳ WARNING

Do not depress the brake pedal while removing the brake pads or the cylinder body because the piston may pop out. Do not spill or splash brake fluid on the disc rotor.

1. Before servicing the vehicle, refer to the Precautions Section.
2. Remove tires with power tool.
3. Remove upper sliding pin bolt.
4. Suspend the cylinder body with suitable wire so that the brake hose will not stretch. Remove the brake pads, shims and shim cover from the torque member.

✳✳ WARNING

Do not deform the pad retainers if removing the pad retainers. Do not damage the piston boot. Do not drop the brake pads, shims, and the shim

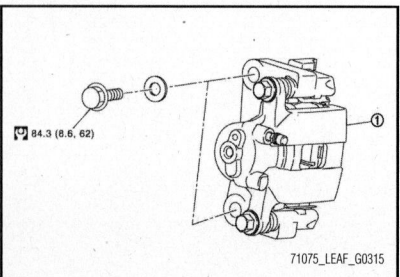

84.3 (8.6, 62)

71075_LEAF_G0315

Fig. 6 Rear caliper assembly (1)

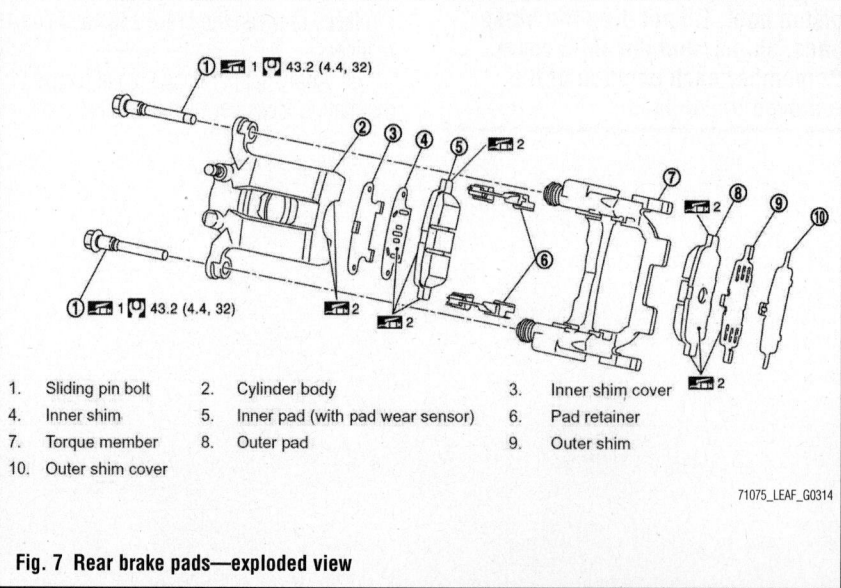

① ▭ 1 🔧 43.2 (4.4, 32)

① ▭ 1 🔧 43.2 (4.4, 32)

1.	Sliding pin bolt	2.	Cylinder body	3.	Inner shim cover
4.	Inner shim	5.	Inner pad (with pad wear sensor)	6.	Pad retainer
7.	Torque member	8.	Outer pad	9.	Outer shim
10.	Outer shim cover				

71075_LEAF_G0314

Fig. 7 Rear brake pads—exploded view

cover. Remember each position of the removed brake pads.

To install:

5. Install the pad retainers to the torque member if the pad retainers have been removed.

✳✳ WARNING

Do not reuse the pad retainers. Securely assemble the pad retainers so that it will not be lifted up from the torque member. Never deform the pad retainers. Eliminate double-

faced adhesive tape on torque member. Remove adhesive's protective liner on pad retainers.

6. Apply PBC (Poly Butyl Cuprysil) grease or silicone-based grease to the mating faces between the brake pads, the shims and pawls part of cylinder body, and install them to the brake pad. Always replace the shims together with the shim cover when replacing the brake pad.

7. Install the cylinder body and brake pads to the torque member. Use a disc brake piston tool to easily press piston.

✳✳ WARNING

Do not damage the piston boot. When replacing a pad with new one, check brake fluid level in the reservoir tank because brake fluid returns to master cylinder reservoir tank when pressing piston in.

8. Install the upper sliding pin bolt and tighten it to the specified torque.

9. Depress the brake pedal several times to check that no drag feel is present for the rear disc brake.

BRAKES | PARKING BRAKE

PARKING BRAKE SHOES

REMOVAL & INSTALLATION

See Figures 8 through 12.

✳✳ CAUTION

Dust covering the brake must be removed with a dust collector. Never splatter the dust with an air blow gun.

1. Before servicing the vehicle, refer to the Precautions Section.
2. Remove tires with power tool.
3. Remove the disc rotor.
4. Remove the upper side return spring.
5. Remove the lower side return spring.
6. Remove spring and anti-rattle pin.
7. Remove parking brake shoes, adjuster, brake strut and toggle lever.

✳✳ WARNING

The parking brake shoes for the front wheels are made of

different materials from those for the rear wheels. Do not misidentify them when removing. Do not drop the removed parts.

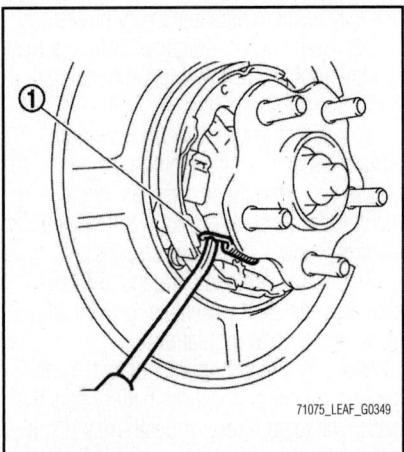

Fig. 9 Remove the lower side return spring (1)

8. Press the rear cable spring against spring tension to remove rear cable from the clamp of toggle lever. Do not bend rear cable.

9. Remove back plate, as applicable.

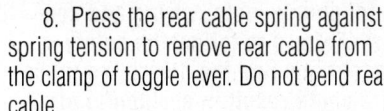
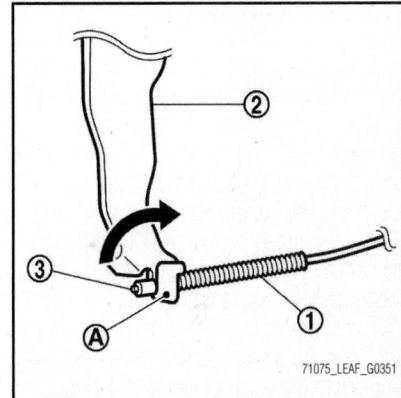

Fig. 11 Press the rear cable spring (1) against spring tension to remove rear cable (3) from the clamp (A) of toggle lever (2)

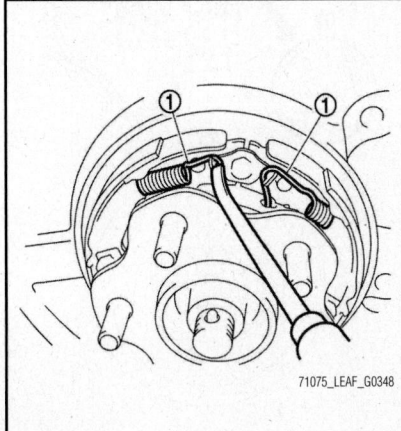

Fig. 8 Remove the upper side return spring (1)

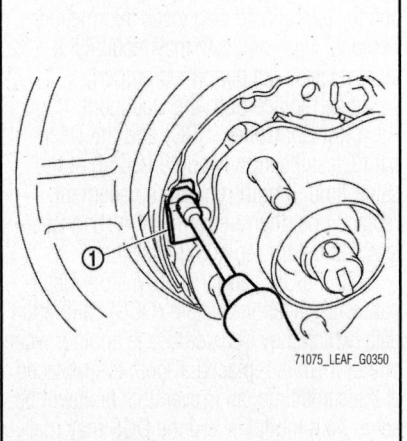

Fig. 10 Remove spring (1) and anti-rattle pin

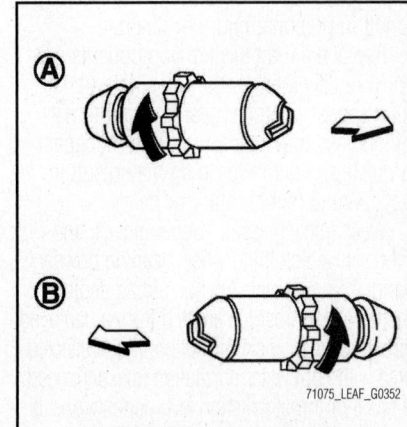

Fig. 12 A: For right side brake, B: For left side brake, White arrow: Vehicle front, Black arrow: Adjuster expands

To install:

10. Apply PBC (Poly Butyl Cuprysil) grease or silicone-based grease to the back plate and brake shoe.

11. Do not misidentify the front and rear brake shoes.

12. Assemble adjusters so that threaded part is expanded when rotating it in the direction shown.

13. Shorten adjuster by rotating it.

14. When disassembling, apply PBC (Poly Butyl Cuprysil) grease or silicone- based grease to threads.

CHASSIS ELECTRICAL

AIR BAG (SUPPLEMENTAL RESTRAINT SYSTEM)

GENERAL INFORMATION

✸✸ CAUTION

These vehicles are equipped with an air bag system. The system must be disarmed before performing service on, or around, system components, the steering column, instrument panel components, wiring and sensors. Failure to follow the safety precautions and the disarming procedure could result in accidental air bag deployment, possible injury and unnecessary system repairs.

SERVICE PRECAUTIONS

Disconnect and isolate the battery negative cable before beginning any airbag system component diagnosis, testing, removal, or installation procedures. Allow system capacitor to discharge for two minutes before beginning any component service. This will disable the airbag system. Failure to disable the airbag system may result in accidental airbag deployment, personal injury, or death.

Do not place an intact undeployed airbag face down on a solid surface. The airbag will propel into the air if accidentally deployed and may result in personal injury or death.

When carrying or handling an undeployed airbag, the trim side (face) of the airbag should be pointing towards the body to minimize possibility of injury if accidental deployment occurs. Failure to do this may result in personal injury or death.

Replace airbag system components with OEM replacement parts. Substitute parts may appear interchangeable, but internal differences may result in inferior occupant protection. Failure to do so may result in occupant personal injury or death.

Wear safety glasses, rubber gloves, and long sleeved clothing when cleaning powder residue from vehicle after an airbag deployment. Powder residue emitted from a deployed airbag can cause skin irritation. Flush affected area with cool water if irritation is experienced. If nasal or throat irritation is experienced, exit the vehicle for fresh air until the irritation ceases. If irritation continues, see a physician.

Do not use a replacement airbag that is not in the original packaging. This may result in improper deployment, personal injury, or death.

The factory installed fasteners, screws and bolts used to fasten airbag components have a special coating and are specifically designed for the airbag system. Do not use substitute fasteners. Use only original equipment fasteners listed in the parts catalog when fastener replacement is required.

During, and following, any child restraint anchor service, due to impact event or vehicle repair, carefully inspect all mounting hardware, tether straps, and anchors for proper installation, operation, or damage. If a child restraint anchor is found damaged in any way, the anchor must be replaced. Failure to do this may result in personal injury or death.

Deployed and non-deployed airbags may or may not have live pyrotechnic material within the airbag inflator.

Do not dispose of driver/passenger/curtain airbags or seat belt tensioners unless you are sure of complete deployment. Refer to the Hazardous Substance Control System for proper disposal.

Dispose of deployed airbags and tensioners consistent with state, provincial, local, and federal regulations.

After any airbag component testing or service, do not connect the battery negative cable. Personal injury or death may result if the system test is not performed first.

If the vehicle is equipped with the Occupant Classification System (OCS), do not connect the battery negative cable before performing the OCS Verification Test using the scan tool and the appropriate diagnostic information. Personal injury or death may result if the system test is not performed properly.

Never replace both the Occupant Restraint Controller (ORC) and the Occupant Classification Module (OCM) at the same time. If both require replacement, replace one, then perform the Airbag System test before replacing the other.

Both the ORC and the OCM store Occupant Classification System (OCS) calibration data, which they transfer to one another when one of them is replaced. If both are replaced at the same time, an irreversible fault will be set in both modules and the OCS may malfunction and cause personal injury or death.

If equipped with OCS, the Seat Weight Sensor is a sensitive, calibrated unit and must be handled carefully. Do not drop or handle roughly. If dropped or damaged, replace with another sensor. Failure to do so may result in occupant injury or death.

If equipped with OCS, the front passenger seat must be handled carefully as well. When removing the seat, be careful when setting on floor not to drop. If dropped, the sensor may be inoperative, could result in occupant injury, or possibly death.

If equipped with OCS, when the passenger front seat is on the floor, no one should sit in the front passenger seat. This uneven force may damage the sensing ability of the seat weight sensors. If sat on and damaged, the sensor may be inoperative, could result in occupant injury, or possibly death.

DISARMING THE SYSTEM

Before servicing, push power switch OFF, disconnect 12V battery negative terminal and wait for 5 minutes or more.

ARMING THE SYSTEM

Reconnect the 12V battery negative terminal.

CLOCKSPRING CENTERING

See Figure 13.

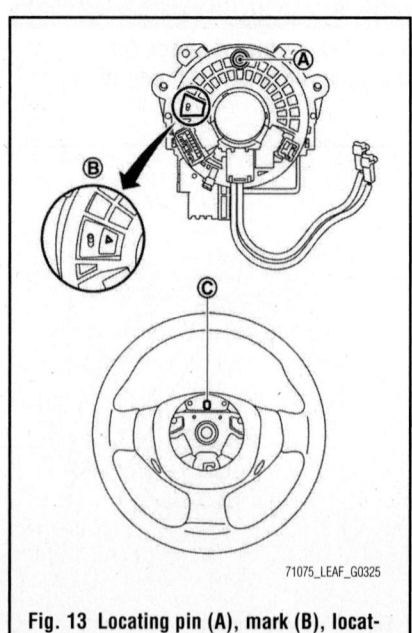

71075_LEAF_G0325

Fig. 13 Locating pin (A), mark (B), locating pin hole (C)

1. Before servicing the vehicle, refer to the Precautions Section.

2. The spiral cable may snap during steering operation if the cable is installed in an improper position. The neutral position is set as follows:

a. Carefully turn the spiral cable clockwise to the end position. Then turn it counterclockwise (about 2 and a half turns) and stop turning at the mark when the stopper insertion holes are in the same position.

b. The service part is installed in the neutral position by the stopper and can be set without adjusting after the stopper is removed.

c. Never over turn the spiral cable or go beyond the number of turns required. (This causes the cable to snap)

d. Adjust the spiral cable locating pin to the steering wheel locating pin hole.

DRIVE TRAIN

FRONT HALFSHAFT

REMOVAL & INSTALLATION

Left Side

See Figures 14 through 18.

1. Before servicing the vehicle, refer to the Precautions Section.

2. Remove tires with power tool.

3. Remove cotter pin and adjusting cap, and loosen wheel hub lock nut.

4. Patch the wheel hub lock nut with a piece of wood. Hammer the wood to disengage wheel hub assembly from half-shaft.

➡**If the wheel hub assembly and half-shaft cannot be separated even after performing the above procedure, use a suitable puller.**

5. Remove the wheel hub lock nut.

6. Remove the steering outer socket from the steering knuckle.

7. Remove the transverse link from the steering knuckle with a power tool.

8. Remove the shaft assembly from wheel hub assembly.

※※ WARNING

Do not place halfshaft joint at an extreme angle. Be careful not to overextend the slide joint. Do not allow halfshaft to hang down without support for the joint sub-assembly, shaft and the other parts.

9. Use the halfshaft attachment (SST: KV40107500) and a sliding hammer (commercial service tool) while inserting the tip of the halfshaft attachment between the shaft and reduction gear assembly, and remove the halfshaft from the reduction gear assembly. Confirm that the circular clip is attached to the halfshaft. Perform inspection after removal.

To install:

Installation is the reverse of removal, noting the following:

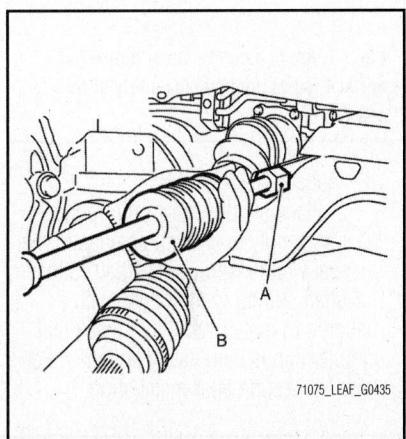

71075_LEAF_G0435

Fig. 15 Use the halfshaft attachment (A) and a sliding hammer (B) to remove halfshaft from reduction gear assembly

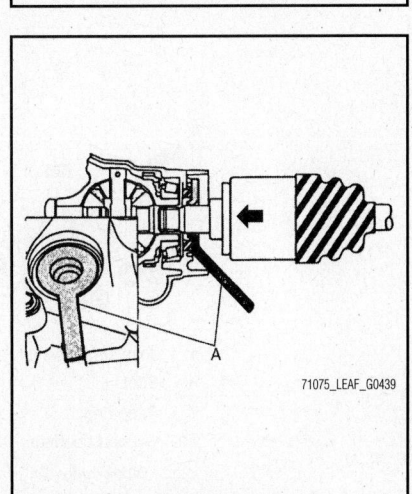

71075_LEAF_G0439

Fig. 16 Place the protector (A) on the reduction gear assembly

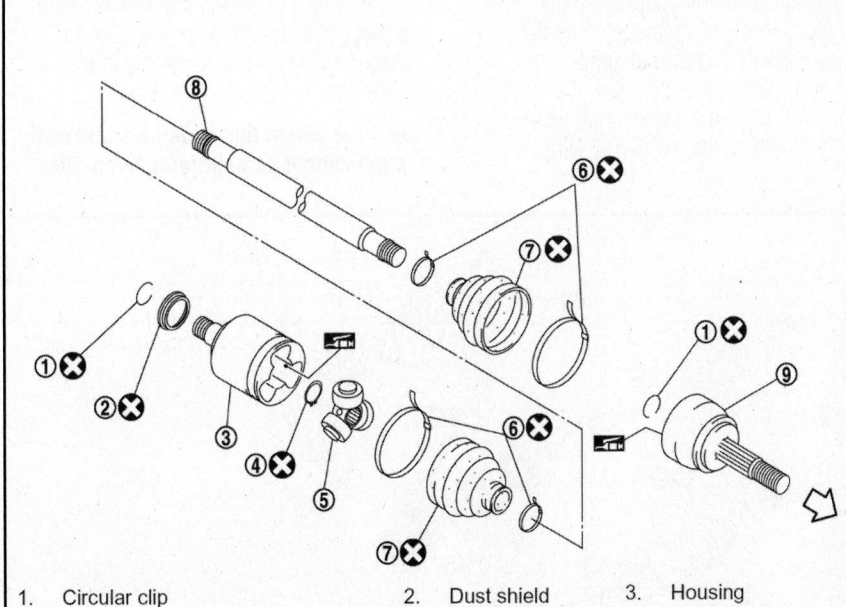

1.	Circular clip	2. Dust shield	3. Housing
4.	Snap ring	5. Spider assembly	6. Boot band
7.	Boot	8. Shaft	9. Joint sub-assembly

⟵ : Wheel side

✖ : Always replace after every disassembly.

71075_LEAF_G0441

Fig. 14 Front halfshaft components—left-side

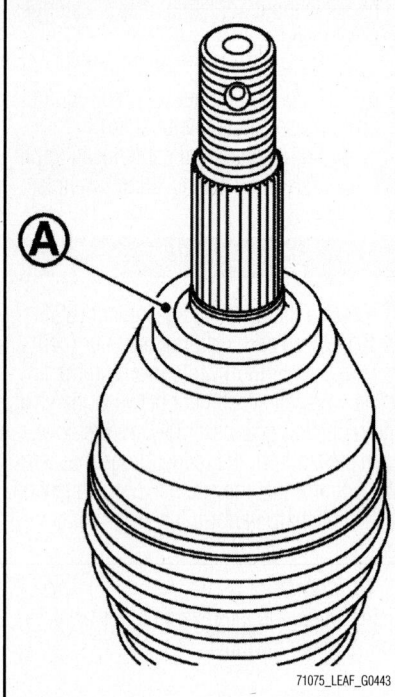

Fig. 17 Apply paste to cover entire flat surface (A) of joint subassembly of half-shaft

10. Reduction Gear Side:

a. Place the protector onto reduction gear assembly to prevent damage to the oil seal while inserting halfshaft. Slide halfshaft sliding joint and tap with a hammer to install securely. Check that circular clip is completely engaged. Perform inspection after installation.

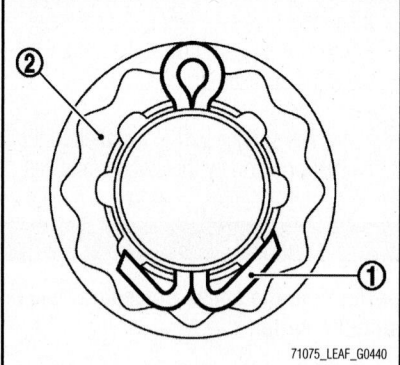

Fig. 18 Securely bend the basal portion when installing a cotter pin (1) and adjusting cap (2)

11. Wheel Hub Side:

a. Clean the matching surface of the wheel hub lock nut and wheel hub assembly.

✳✳ WARNING

Never apply lubricating oil to the matching surface.

b. Clean the matching surface of the halfshaft and wheel hub assembly, and apply paste to the surface of the halfshaft joint sub-assembly. Apply paste to cover the entire flat surface of the halfshaft joint sub-assembly.

c. Tighten the wheel hub lock nut to 133–136 ft. lbs. (180–185 Nm).

✳✳ WARNING

Since the halfshaft is assembled by press-fitting, use the tightening torque range for the wheel hub lock nut. Be sure to use a torque wrench to tighten the wheel hub lock nut. Never use a power tool. Never reuse wheel hub lock nut.

d. When reusing the disc rotor, align the matchmarks that were made during removal.

e. When installing a cotter pin and adjusting cap, securely bend the basal portion to prevent rattles. Never reuse cotter pin.

f. When tightening the parts that were removed when removing the wheel hub assembly and steering knuckle, perform final tightening under unladen conditions. Perform inspection after installation.

Right Side

See Figures 17 through 22.

1. Before servicing the vehicle, refer to the Precautions Section.

2. Remove tires with power tool.

3. Remove cotter pin and adjusting cap, and loosen wheel hub lock nut.

4. Patch the wheel hub lock nut with a piece of wood. Hammer the wood to disengage wheel hub assembly from halfshaft.

➡**If the wheel hub assembly and half-shaft cannot be separated even after**

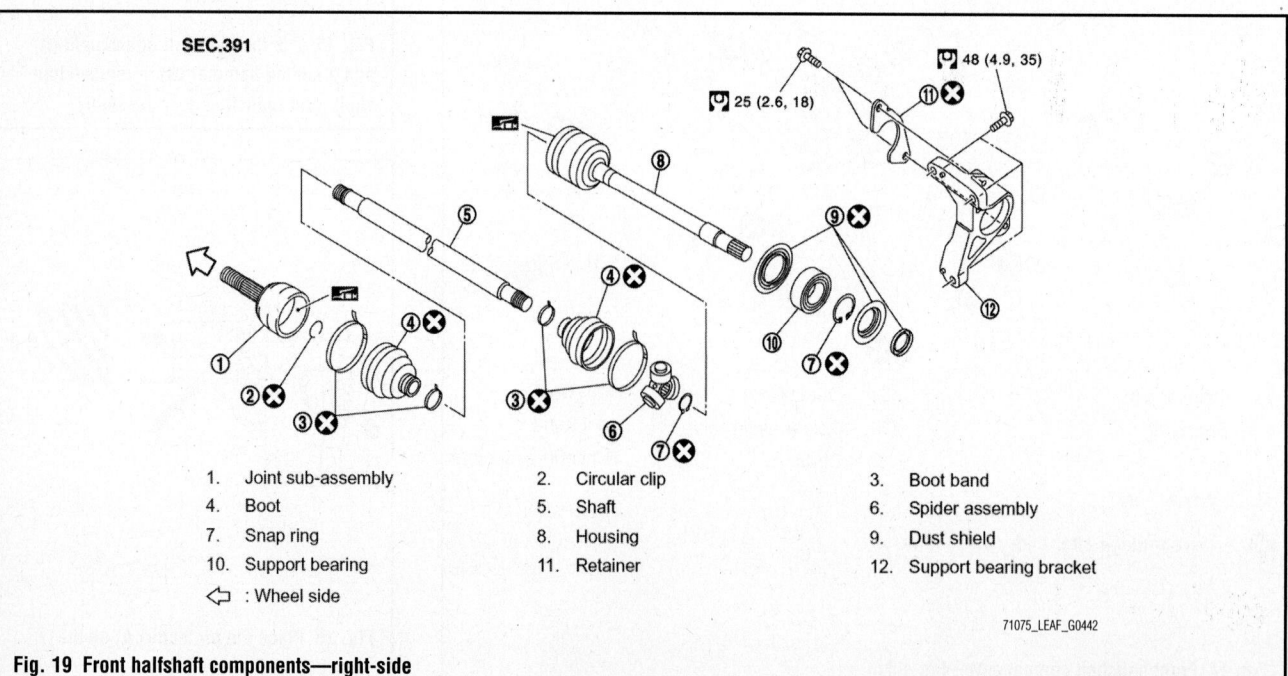

1.	Joint sub-assembly	2.	Circular clip	3.	Boot band
4.	Boot	5.	Shaft	6.	Spider assembly
7.	Snap ring	8.	Housing	9.	Dust shield
10.	Support bearing	11.	Retainer	12.	Support bearing bracket
⇦ : Wheel side					

Fig. 19 Front halfshaft components—right-side

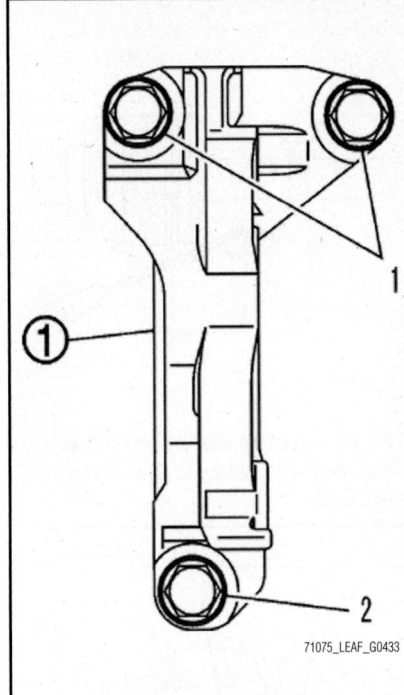

Fig. 20 Support bearing bracket (1) mounting bolt tightening order (1, 2)

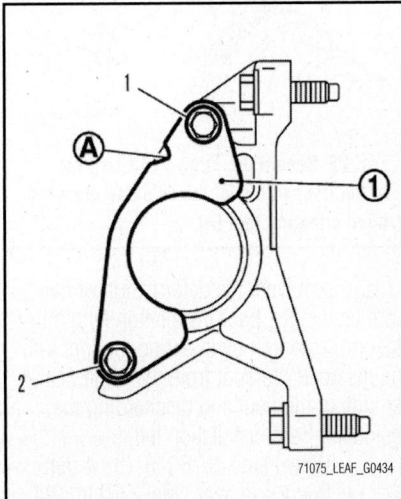

Fig. 21 Set retainer (1) so that notch (A) becomes upper side; mounting bolt tightening order (1, 2)

performing the above procedure, use a suitable puller.

5. Remove the wheel hub lock nut.

6. Remove the steering outer socket from the steering knuckle.

7. Remove the transverse link from the steering knuckle with a power tool.

8. Remove the shaft assembly from wheel hub assembly.

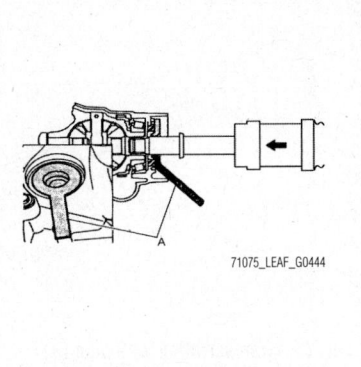

Fig. 22 Place the protector (A) on the reduction gear assembly

✳✳ WARNING
Do not place halfshaft joint at an extreme angle. Be careful not to overextend the slide joint. Do not allow halfshaft to hang down without support for the joint sub-assembly, shaft and the other parts.

9. Remove the retainer.

10. Remove the halfshaft assembly from the reduction gear assembly.

11. Remove the support bearing bracket.

12. Perform inspection after removal.

To install:

Installation is the reverse of removal, noting the following:

13. Reduction Gear Side:

a. Install the support bearing bracket as follows:

- Temporarily tighten the support bearing bracket mounting bolts in the order shown, and then tighten to 35 ft. lbs. (48 Nm).

- Set the retainer so that notch becomes upper side. Temporarily tighten mounting bolts in the order shown, and then tighten to 18 ft. lbs. (25 Nm). Never reuse the retainer.

b. Place the protector (SST: KV38107900) onto the reduction gear assembly to prevent damage to the oil seal while inserting the halfshaft.

c. Slide the halfshaft sliding joint and tap with a hammer to install securely. Perform inspection after installation.

14. Wheel Hub Side:

a. Clean the matching surface of the wheel hub lock nut and wheel hub assembly.

✳✳ WARNING
Never apply lubricating oil to the matching surface.

b. Clean the matching surface of the halfshaft and wheel hub assembly, and apply paste to the surface of the halfshaft joint sub-assembly. Apply paste to cover the entire flat surface of the halfshaft joint sub-assembly.

c. Tighten the wheel hub lock nut to 133–136 ft. lbs. (180–185 Nm).

✳✳ WARNING
Since the halfshaft is assembled by press-fitting, use the tightening torque range for the wheel hub lock nut. Be sure to use a torque wrench to tighten the wheel hub lock nut. Never use a power tool. Never reuse wheel hub lock nut.

d. When reusing the disc rotor, align the matchmarks that were made during removal.

e. When installing a cotter pin and adjusting cap, securely bend the basal portion to prevent rattles. Never reuse cotter pin.

f. When tightening the parts that were removed when removing the wheel hub assembly and steering knuckle, perform final tightening under unladen conditions. Perform inspection after installation.

CV-BOOTS

REMOVAL & INSTALLATION
See Figures 23 through 29.

1. Before servicing the vehicle, refer to the Precautions Section.

2. Fix the shaft with a vise.

✳✳ WARNING
Protect the shaft when fixing with a vise using aluminum or copper plates.

3. Remove boot bands, and then remove boot from the joint sub-assembly.

4. Screw the halfshaft puller (commercial service tool) into the joint sub-assembly screw part to a length of 1.18 in. (30 mm) or more. Support the halfshaft with one hand and pull out the joint sub-assembly from the shaft.

a. Align the halfshaft puller and halfshaft and remove them by pulling firmly and uniformly.

b. If the joint sub-assembly cannot be removed after five or more unsuccessful

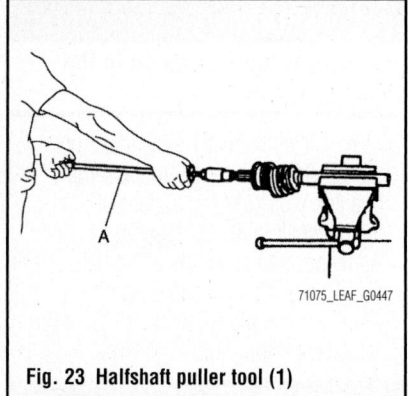

Fig. 23 Halfshaft puller tool (1)

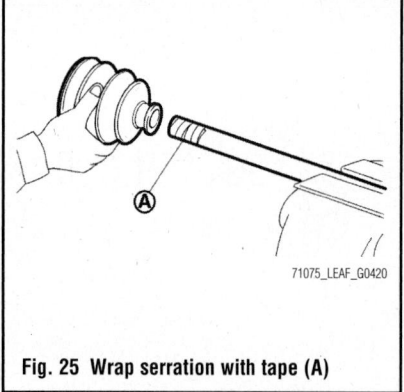

Fig. 25 Wrap serration with tape (A)

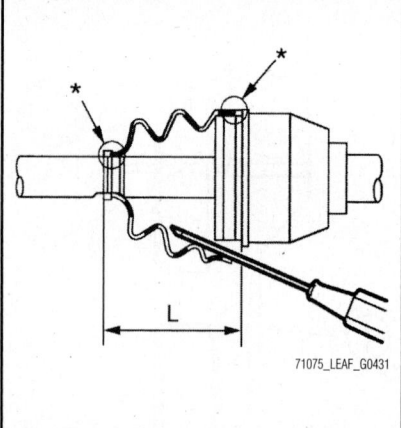

Fig. 27 Install the boot into grooves as indicated by "*" marks; L : Boot installation length

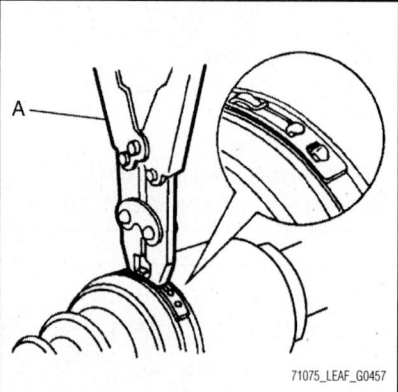

Fig. 28 Secure the large and small ends of the boot with boot bands using the boot band crimping tool (A)

attempts, replace the shaft and joint sub assembly as a set.

5. Remove the circular clip from the shaft.

6. Remove the boot from the shaft.

To assemble:

7. Clean the old grease on joint sub-assembly with paper waste.

8. Fill serration slot joint sub-assembly with NISSAN genuine grease or equivalent until the serration slot and ball groove become full to the brim. After applying grease, use a paper waste to wipe off old grease that has oozed out.

9. Install boot and boot bands to the shaft. Wrap the serration on the shaft with tape to protect the boot from damage. Never reuse boot and boot band.

10. Remove the tape wrapped around the serration.

11. Position the circular clip on the groove at the shaft edge. Do not reuse circular clip. Drive joint inserter is recommended when installing circular clip.

12. Align both center axles of the shaft edge and joint sub-assembly then assemble shaft with joint sub-assembly holding circular clip.

13. Install joint sub-assembly to shaft using a plastic hammer.

a. Check that the circular clip is properly positioned on the groove of the joint sub-assembly.

b. Confirm that the joint sub-assembly is correctly engaged while rotating halfshaft.

14. Apply the specified amount of grease into the boot inside from large diameter side of boot. Grease quantity:
- Wheel side: 3.11–3.80 oz (88–108 g)
- Reduction gear side: 5.40–5.85 oz (153–166 g)

15. Install the boot securely into the grooves.

a. If grease adheres to the boot mounting surface on the shaft or joint subassembly, boot may be removed.

b. Remove all grease from the boot mounting surface.

16. To prevent the deformation of the boot, adjust the boot installation length to the correct length by inserting the tool into the inside of the boot from the large diameter side of the boot and discharging the inside air. Boot installation length:
- Wheel side: 3.756 in. (95.4 mm)
- Reduction gear side: 3.70 in. (94 mm)

✳✳ WARNING

If the boot installation length exceeds the standard, it may cause breakage of the boot. Be careful not to touch the inside of the boot with a tip of tool.

17. Secure the large and small ends of the boot with boot bands using the boot band crimping tool (SST: KV40107300). Never reuse the boot band.

18. Secure the boot band so that dimension "A" is 0.276 in. (7 mm) or less.

19. Check that displacement does not occur when the boot is rotated with the joint

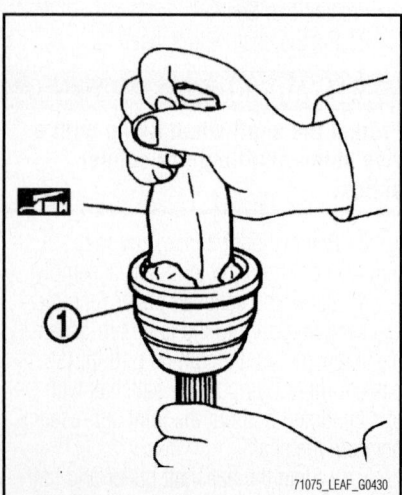

Fig. 24 Fill serration slot joint sub-assembly (1) with NISSAN genuine grease or equivalent

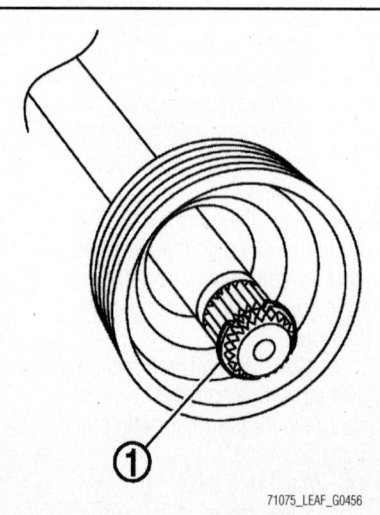

Fig. 26 Position the circular clip (1) on the groove

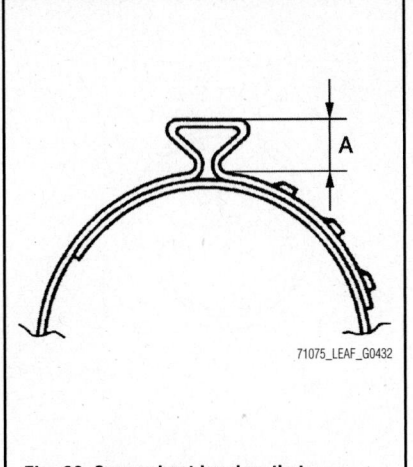

Fig. 29 Secure boot band so that dimension "A" is 0.276 in. (7mm) or less

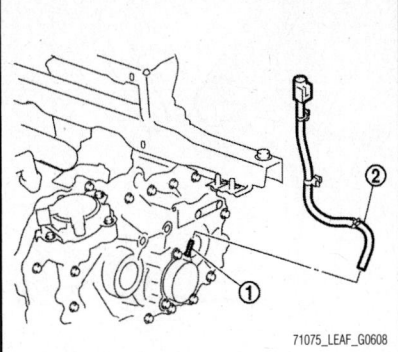

Fig. 31 Fit the breather hose (2) over the brush cover tube part (1) of the reduction gear

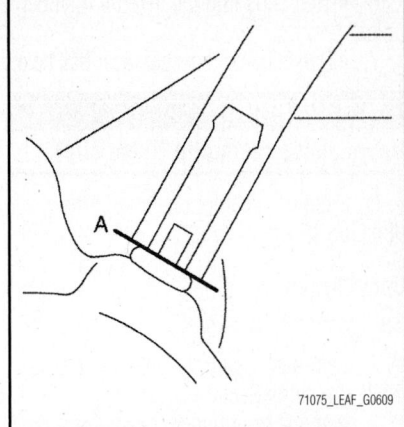

Fig. 32 Fit the breather hose onto the brush cover tube part all the way to its base (A)

sub-assembly and shaft fixed. Reinstall them using boot bands when boot installation positions become incorrect. Never reuse boot band.

REDUCTION GEAR

REMOVAL & INSTALLATION

Breather Hose

See Figures 30 through 34.

1. Before servicing the vehicle, refer to the Precautions Section.

2. Remove front under cover.

3. Using a suitable tool, remove the clip, and then pull the breather hose off the brush cover tube part of the reduction gear.

To install:

4. Position a paint mark on the breather hose toward the left side of the vehicle, and then fit the breather hose over the brush cover tube part of the reduction gear.

5. Fit the breather hose onto the brush cover tube part all the way to its base, as shown.

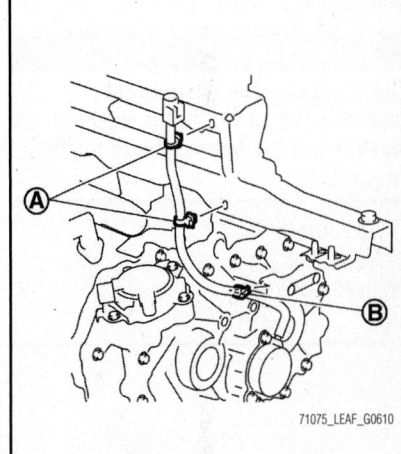

Fig. 33 Install clips (A) into the inverter member holes, and clip (B) into reduction gear bolt hole

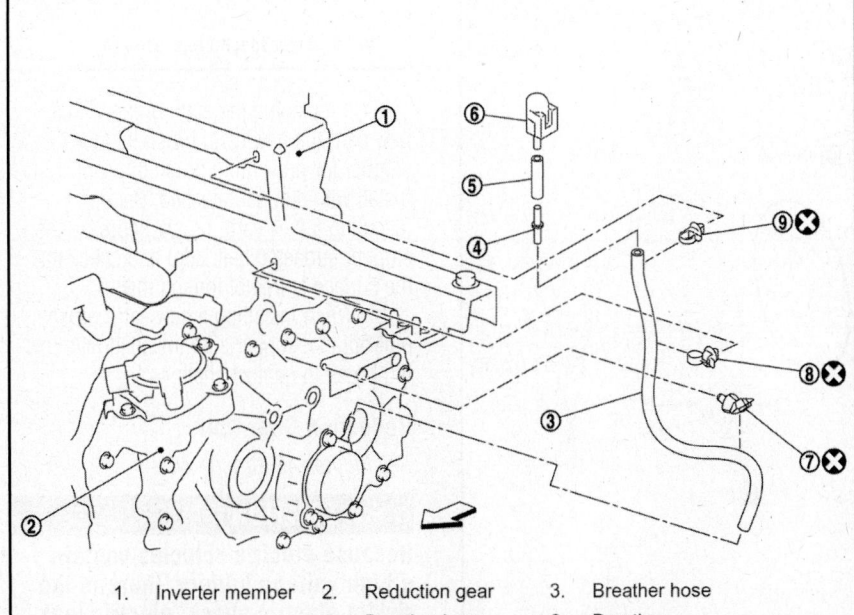

1.	Inverter member	2.	Reduction gear	3.	Breather hose
4.	Connector	5.	Breather hose	6.	Breather
7.	Clip	8.	Clip	9.	Clip

Fig. 30 Breather hose

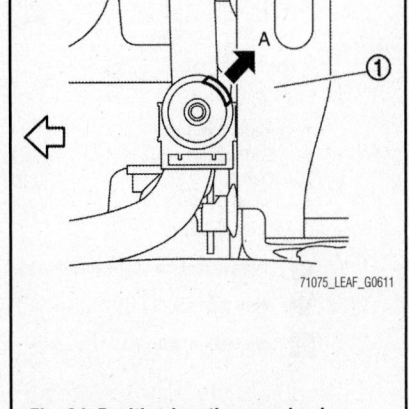

Fig. 34 Position breather opening in direction (A); inverter member (1)

6. Install clips into the inverter member holes.

7. Install clip into reduction gear bolt hole.

✳✳ WARNING

Never reuse resin clip (hose clip).

8. Position breather opening in the direction shown.

Earth Brush

See Figures 35 through 39.

1. Before servicing the vehicle, refer to the Precautions Section.

2. Remove front under cover.

3. Disconnect the breather hose from the brush cover, remove the brush cover bolts and the brush cover.

4. Remove the O-ring, brush mounting bolts, and remove the earth brush.

✳✳ WARNING

Carefully remove earth brush, because the spring in the earth brush pushes out the brush. Never touch brush area.

To install:

Installation is the reverse of removal, noting the following:

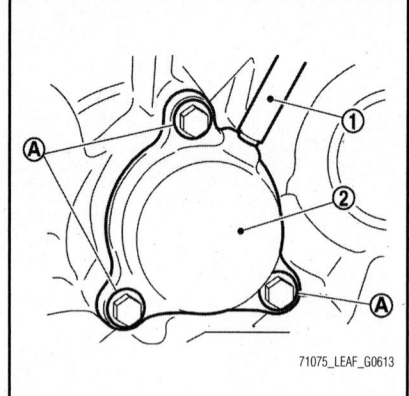

Fig. 36 Disconnect breather hose (1) from brush cover (2), and remove the brush cover bolts (A)

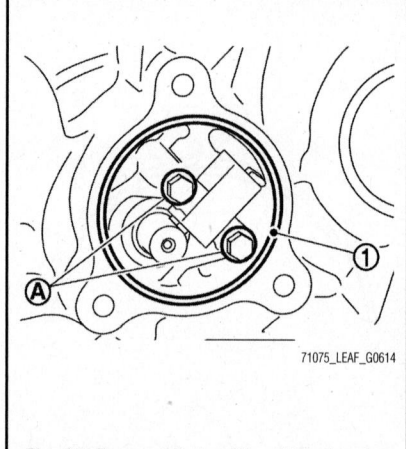

Fig. 37 Remove O-ring (1) and the brush mounting bolts (A)

5. Before installing the earth brush, degrease the shaft surface (brush contact surface) and the brush surface, and verify that there is no dust or other substance on them.

6. Do not apply oil to O-ring. Before installing the O-ring, verify that there is no oil on it. Never reuse the O-ring.

7. When assembling earth brush, without touching the brush area, press the earth brush onto the shaft and install the mounting bolt.

Fig. 38 Never touch brush area (A)

8. If replacing the earth brush with a new part, the new earth brush includes a stopper for preventing brush pop-out. Install with stopper attached. Before installing a new earth brush, degrease the stopper surface (shaft side) and check that the surface is free of foreign matter.

9. When installing a new earth brush, pull out the stopper after installation, allowing brush to contact shaft.

Reduction Gear Unit

See Figures 40 through 44.

✳✳ CAUTION

Because electric vehicles contain a high voltage battery, there is the risk of electric shock, electric leakage, or similar accidents if the high voltage component and vehicle are handled incorrectly. Be sure to follow the correct work procedures when performing inspection and maintenance.

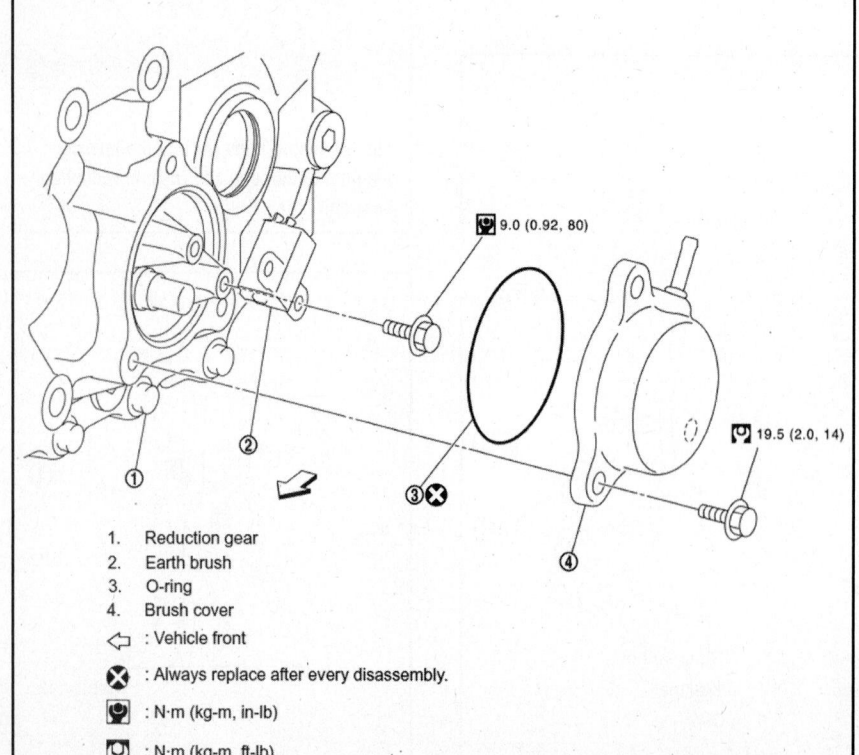

1. Reduction gear
2. Earth brush
3. O-ring
4. Brush cover
⇦ : Vehicle front
⊗ : Always replace after every disassembly.
🔧 : N·m (kg-m, in-lb)
🔧 : N·m (kg-m, ft-lb)

Fig. 35 Earth brush—exploded view

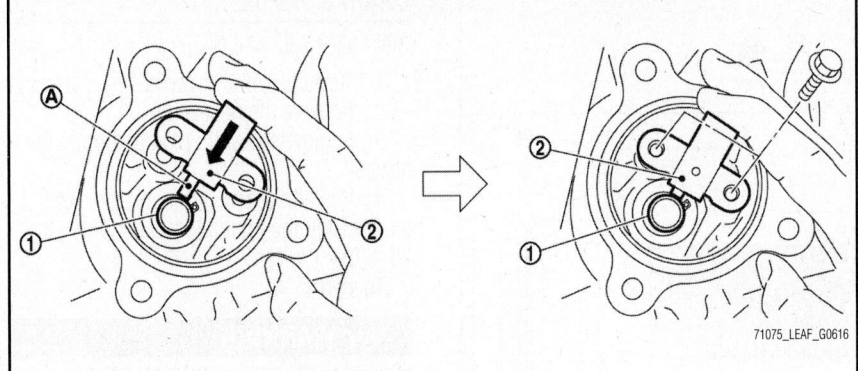

Fig. 39 When assembling earth brush, without touching the brush area (A), press earth brush (2) onto shaft (1)

Fig. 42 Motor mounting rear bracket

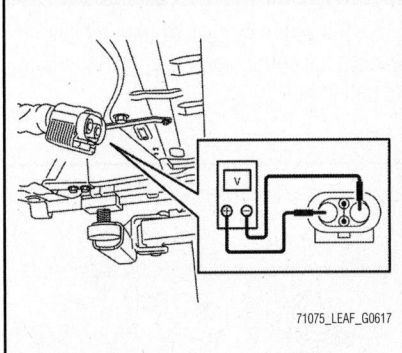

Fig. 40 Measure voltage between the high voltage harness terminals

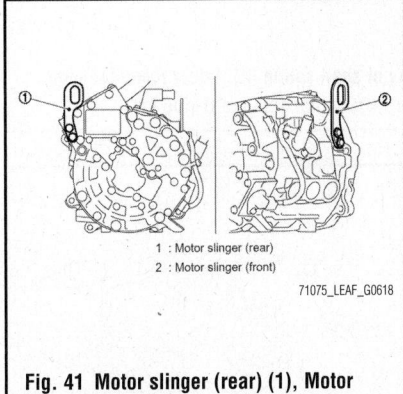

1 : Motor slinger (rear)
2 : Motor slinger (front)

Fig. 41 Motor slinger (rear) (1), Motor slinger (front) (2)

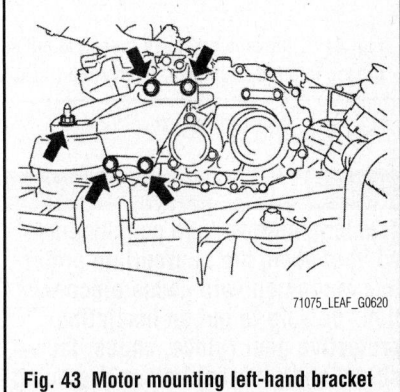

Fig. 43 Motor mounting left-hand bracket

✳✳ CAUTION

Remove the service plug in order to disconnect the high voltage circuits before performing inspection or maintenance of high voltage system harnesses and parts. To prevent the removed service plug from being connected by mistake during the procedure, always carry it in your pocket or put it in the tool box.

✳✳ CAUTION

Clearly identify the persons responsible for high voltage work and ensure that other persons do not touch the vehicle. When not working, cover high voltage parts with an insulating cover sheet or similar item to prevent other persons from contacting them.

✳✳ CAUTION

Touching high voltage components without using the appropriate protective equipment will cause electrocution. Be sure to put on insulating protective gear (glove, shoes, face

shield and glasses) before beginning work on the high voltage system.

✳✳ WARNING

There is the possibility of a malfunction occurring if the vehicle is changed to READY status while the service plug is removed. Therefore do not change the vehicle to READY status unless instructed to do so in the service instructions.

1. Before servicing the vehicle, refer to the Precautions Section.
2. Disconnect the high voltage circuit.
3. Check voltage in high voltage circuit (Check that condenser is discharged):
    a. Lift up the vehicle and remove the Li-ion battery under covers.
    b. Disconnect high voltage harness connector from the front side of the Li-ion battery.

✳✳ CAUTION

For voltage measurements, use a tester which can measure to 500 V or higher.

   c. Measure voltage between the high voltage harness connector terminals. Standard: 5 V or less
4. Drain coolant from radiator.
5. Remove traction motor inverter.
6. Drain reduction gear oil from reduction gear.
7. Remove traction motor and reduction gear from vehicle together as suspension member assembly.
8. Remove right and left front half-shafts.
9. Install motor slinger onto traction motor and tighten bolts to 21 ft. lbs. (28 Nm), then lift traction motor with hoist to hold the position of traction motor.

➡ **The traction motor does not become displaced when motor mounting and motor mounting bracket are removed.**

10. Remove motor mounting rear bracket.
11. Remove motor mounting left-hand bracket.
12. Remove the bolts securing the traction motor and reduction gear, and remove reduction gear.

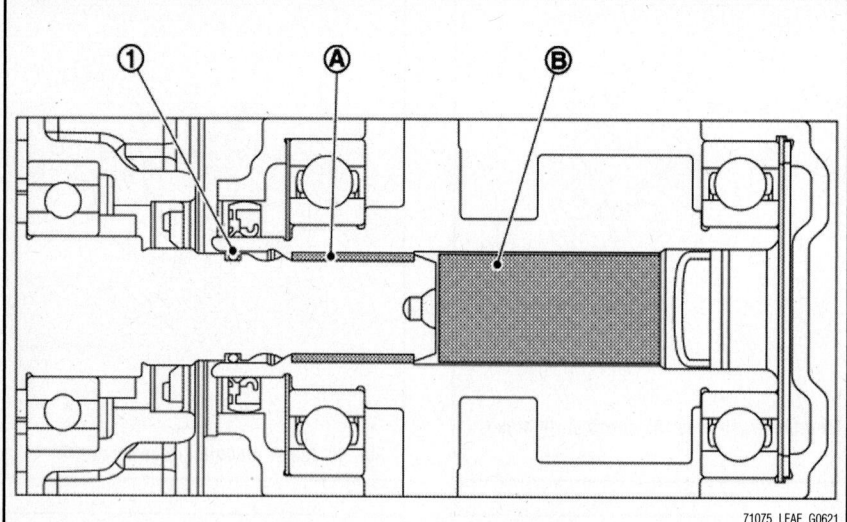

Fig. 44 Apply recommended grease to full periphery of shaft spline (A); inject recommended grease into reduction gear input shaft (inside spline) (B); do not damage O-ring (1)

To install:

❋❋ CAUTION

Touching high voltage components without using the appropriate protective equipment will cause electrocution. Be sure to put on insulating protective gear (glove, shoes, face shield and glasses) before beginning work on the high voltage system.

Installation is the reverse of removal, noting the following:

13. Be sure to reinstall high voltage harness clips in their original positions. If a clip is damaged, replace it with a new clip before installing.

14. Be sure to perform correct air bleeding after adding coolant.

➡ **Clean the grease application area to remove old grease and abrasion powder before applying grease.**

15. Before installing the reduction gear and traction motor, apply recommended grease to the full periphery of shaft spline. Inject grease [minimum 0.3 oz (8.5 g), maximum less than 0.7 oz (20 g)] into the reduction gear input shaft (inside spline). Take

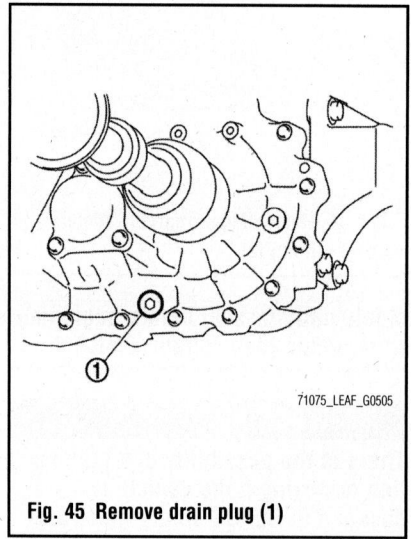

Fig. 45 Remove drain plug (1)

care to prevent damage to O-ring when installing.

16. After all parts are installed, be sure to check equipotential of traction motor, electric compressor, and traction motor inverter.

17. It is necessary to clear the P position learning value and perform the relearning of the P position after the reduction gear is removed and installed or replaced.

DRAIN & REFILL

See Figures 45 and 46.

1. Turn the power switch OFF.
2. Remove the filler plug.
3. Remove the drain plug and drain the gear oil.
4. Position a new gasket on the drain plug, install the drain plug, and tighten to 25 ft. lbs. (35 Nm).

To refill:

❋❋ WARNING

Use only Genuine NISSAN Matic Fluid S. Using reduction gear fluid other than Genuine NISSAN Matic Fluid S will damage the reduction gear.

5. Fill with new gear oil until oil level reaches the correct level near the filler plug mounting hole.

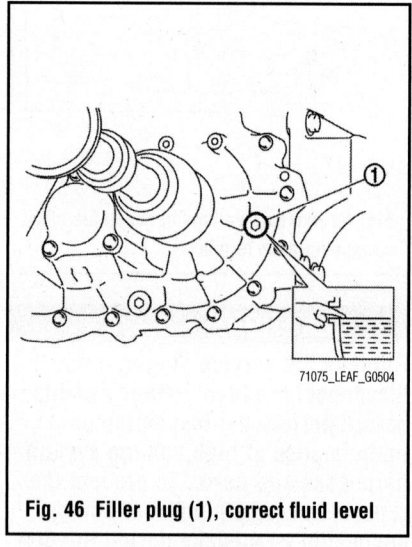

Fig. 46 Filler plug (1), correct fluid level

➡ **Do not reuse gasket. If foreign material, such as gear abrasion powder, is on the magnet of the filler plug, wipe it before installation.**

6. After refilling, check the oil level. Position a new gasket on the filler plug, install the filler plug to reduction gear, and tighten to 25 ft. lbs. (35 Nm).

ENGINE ELECTRICAL   HYBRID SYSTEM

PRECAUTIONS

See Figures 47.

ELECTRIC MEDICAL DEVICE PRECAUTIONS

❊❊ CAUTION

ANYONE USING A MEDICAL ELEC-TRIC DEVICE SUCH AS A PACEMAKER MUST NEVER PERFORM OPERA-TIONS ON THE VEHICLE. Strong magnets are used in this vehicle, and the magnetic field can disrupt the function of electric medical devices.

❊❊ CAUTION

VEHICLE CHARGING: ANYONE USING A MEDICAL ELECTRIC DEVICE SUCH AS AN IMPLANTABLE CARDIAC PACE-MAKER OR AN IMPLANTABLE CAR-DIOVERTER DEFIBRILLATOR (ICD) MUST NOT ENTER THE VEHICLE COMPARTMENT (INCLUDING LUG-GAGE ROOM) DURING NORMAL CHARGE OPERATION. Radiated electromagnetic waves generated by the on board charger at normal charge operation can disrupt the function of electric medical devices.

❊❊ CAUTION

TELEMATICS SYSTEM AND/OR INTELLIGENT KEY SYSTEM: Anyone using electric medical electric devices [implantable cardiac pace-maker or implantable cardioverter defibrillator (ICD)] OR electric med-ical electric devices other than implantable cardiac pacemaker or ICD must not approach within approximately 8.66 in. (220 mm) of interior/exterior antenna. The elec-tromagnetic wave of TCU might affect the function of electric medical devices. The electromagnetic wave of the intelligent key operations (such as at door operation, at each request switch operation, or at engine starting) might affect the function of electric medical devices. The possible effects on the devices must be checked with the device manufacturer before TCU and/or intelligent key use.

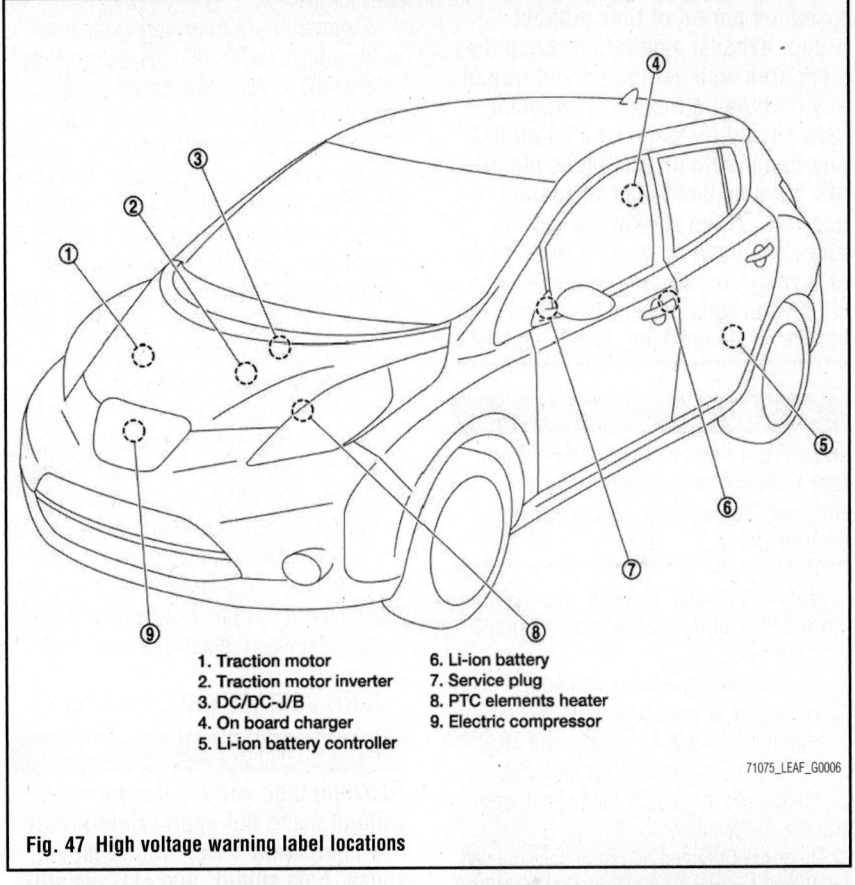

1. Traction motor
2. Traction motor inverter
3. DC/DC-J/B
4. On board charger
5. Li-ion battery controller
6. Li-ion battery
7. Service plug
8. PTC elements heater
9. Electric compressor

71075_LEAF_G0006

Fig. 47 High voltage warning label locations

GENERAL PRECAUTIONS

❊❊ CAUTION

The Supplemental Restraint System such as "AIR BAG" and "SEAT BELT PRE-TENSIONER", used along with a front seat belt, helps to reduce the risk or severity of injury to the driver and front passenger for certain types of collision. This sys-tem includes seat belt switch inputs and dual stage front air bag modules. The SRS system uses the seat belt switches to determine the front air bag deployment, and may only deploy one front air bag, depending on the severity of a collision and whether the front occupants are belted or unbelted. Before servicing components near or affected by the SRS (air bag) system, read and observe all SRS Service Precautions. Refer to Air Bags (Supplemental Restraint System), in the Chassis Electrical section. Failure to observe all precautions may result in acci-dental airbag deployment, personal injury, or death.

❊❊ CAUTION

Never use electrical test equipment on any circuit related to the SRS. SRS wiring harnesses can be identified by yellow and/or orange harnesses or harness connectors.

❊❊ CAUTION

Always observe the following items for preventing accidental activation: When working near the Air Bag Diag-nosis Sensor Unit or other Air Bag System sensors with the power switch ON, never use air or electric power tools or strike near the sen-sor(s) with a hammer. Heavy vibra-tion could activate the sensor(s) and deploy the air bag(s), possibly caus-ing serious injury.

❊❊ CAUTION

When using air or electric power tools or hammers, always switch the power switch OFF, disconnect the 12V battery, and wait at least 3 min-utes before performing any service.

✳✳ **CAUTION**

Do not operate the engine for an extended period of time without proper exhaust ventilation. Keep the work area well ventilated and free of any flammable materials. Special care should be taken when handling any flammable or poisonous materials, such as gasoline, refrigerant gas, etc. When working in a pit or other enclosed area, be sure to properly ventilate the area before working with hazardous materials. Do not smoke while working on the vehicle.

✳✳ **CAUTION**

To prevent serious burns: Avoid contact with hot metal parts. Do not remove the radiator cap when the engine is hot.

Use an approved refrigerant recovery unit any time the air conditioning system must be discharged.

When removing a heavy component such as the engine or transaxle/ transmission, be careful not to lose your balance and drop them. Also, do not allow them to strike adjacent parts, especially the brake tubes and master cylinder.

Dispose of drained oil or the solvent used for cleaning parts in an appropriate manner.

Before starting repairs which do not require battery power: Turn off power switch. Disconnect the negative battery terminal. If the battery terminals are disconnected, recorded memory of radio and each control unit is erased.

Clean all disassembled parts in the designated liquid or solvent prior to inspection or assembly.

Replace oil seals, gaskets, packings, O-rings, locking washers, cotter pins, self-locking nuts, etc. with new ones.

Replace inner and outer races of tapered roller bearings and needle bearings as a set.

Arrange the disassembled parts in accordance with their assembled locations and sequence.

Do not touch the terminals of electrical components which use microcomputers (such as ECM). Static electricity may damage internal electronic components.

After disconnecting vacuum or air hoses, attach a tag to indicate the proper connection.

Use only the fluids and lubricants specified.

Use approved bonding agent, sealants or their equivalents when required.

When repairing the fuel, oil, water, vacuum or exhaust systems, check all affected lines for leakage.

To prevent ECM from storing the diagnostic trouble codes, do not carelessly disconnect the harness connectors which are related to the engine control system and TCM (transmission control module) system.

Hose removal and installation: To prevent damage to rubber hose, do not pry off rubber hose with tapered tool or screwdriver. To reinstall the rubber hose securely, check that hose insertion length and orientation is correct. (If tube is equipped with hose stopper, insert rubber hose into tube until it butts up against hose stopper.)

Hose clamping: If old rubber hose is reused, install hose clamp in its original position (at the indentation where the old clamp was). If there is a trace of tube bulging left on the old rubber hose, align rubber hose at that position. Discard old clamps; replace with new ones. After installing plate clamps, apply force to them in the direction of the arrow, tightening rubber hose equally all around.

HIGH VOLTAGE PRECAUTIONS

✳✳ **CAUTION**

Touching high voltage components without using the appropriate protective equipment consisting of glove, shoes, face shield, and glasses will cause electrocution. Be sure to wear insulating protective equipment before beginning work on the high voltage system.

✳✳ **CAUTION**

Because hybrid vehicles and electric vehicles contain a high voltage battery, there is the risk of electric shock, electric leakage, or similar accidents if the high voltage component and vehicle are handled incorrectly. Be sure to follow the correct work procedures when performing inspection and maintenance.

✳✳ **CAUTION**

Be sure to remove the service plug in order to shut off the high voltage circuits before performing inspection or maintenance of high voltage system harnesses and parts.

✳✳ **CAUTION**

To prevent the removed service plug from being connected by mistake dur-ing the procedure, always carry it in your pocket or put it in the tool box.

✳✳ **CAUTION**

Clearly identify the persons responsible for high voltage work and ensure that other persons do not touch the vehicle. When not working, cover high voltage parts with an insulating cover sheet or similar item to prevent other persons from contacting them.

✳✳ **CAUTION**

There is the possibility of a malfunction occurring if the vehicle is changed to READY status while the service plug is removed. Therefore do not change the vehicle to READY status unless instructed to do so in the Service Manual.

✳✳ **CAUTION**

HIGH VOLTAGE HARNESS AND EQUIPMENT IDENTIFICATION: The colors of the high voltage harnesses and connectors are all orange. Orange "High Voltage" labels are applied to the Li-ion battery and other high voltage devices. Do not carelessly touch these harnesses and parts.

✳✳ **CAUTION**

HANDLING OF HIGH VOLTAGE HARNESS AND TERMINALS: Immediately insulate disconnected high voltage connectors and terminals with insulating tape.

✳✳ **CAUTION**

REGULATIONS ON WORKERS WITH MEDICAL ELECTRONICS: The vehicle contains parts that contain powerful magnets. If a person who is wearing a heart pacemaker or other medical device is close to these parts, the medical device may be affected by the magnets. Such persons must not perform work on the vehicle.

✳✳ **CAUTION**

PROHIBITED ITEMS TO CARRY DURING THE WORK: Because this vehicle uses components that contain high voltage and powerful magnetism, due not carry any metal products which may cause short circuits, or any magnetic media (cash cards, prepaid cards, etc.) which may be damaged on your person when working.

✳✳ CAUTION

HIGH VOLTAGE WARNING LABEL: At times such as when a part was replaced, or when a label had become peeled, be sure to apply the new product label in the same position and facing in the same direction.

✳✳ CAUTION

HANDLING OF INSULATION RESISTANCE TESTER: Unlike the ordinary tester, the insulation resistance tester applies 500V when measuring. If used incorrectly, there is the danger of electric shock. If used in the vehicle 12V system, there is the danger of damage to electronic devices. Read the insulation resistance tester instruction manual carefully and be sure to work safely.

INSULATED PROTECTIVE WEAR AND INSULATING TOOLS

✳✳ CAUTION

Perform a daily inspection before and after use; inspect and check for deterioration and damage, and do not use any items where abnormalities are found.

1. Inspect the insulated gloves for scratches, holes, and tears (Visual check and air leakage test):
   a. Hold and fold the glove.
   b. Fold three or four more times, preventing air from escaping from the glove.
   c. Squeeze glove to check that the glove has no holes.
2. Inspect the insulated safety boots for holes, damage, nails, metal pieces, wear or other problems on the soles. (Visual check)
3. Inspect the insulated rubber sheet for tears. (Visual inspection)
4. When performing work at locations where high voltage is applied (such as terminals), use insulated tools.

✳✳ CAUTION

The high voltage system may start automatically. If the timer air conditioner or timer charge (during EVSE connection) is set, the high voltage system starts automatically even

when the power switch is in OFF state. Check that the timer air conditioner and timer charge (during EVSE connection) are not set before starting maintenance work.

✳✳ CAUTION

Only professional technicians and/or emergency responders should handle damaged vehicles.

Indications the high voltage system is ON:
• If the READY indicator is ON, the high voltage system is active.
• If the charge indicator is ON, the high voltage system is active.
• If the air conditioning remote timer indicator (located on the HVAC controller) is ON, the high voltage system is active.
  • If the remote controlled air conditioning system is active, push the power switch to the ON position. This will turn OFF the remote controlled air conditioning system. Remote controlled air conditioning system is a feature that allows the vehicle owner to activate the air conditioning system via telematics communication (cell phone, personal computer, etc.). When this system is active, the air conditioning remote timer indicator (located on the HVAC controller) is illuminated.

✳✳ CAUTION

VEHICLE FIRE: In the case of extinguishing a fire with water, large amounts of water from a fire hydrant (if possible) must be used. DO NOT extinguish fire with a small amount of water. Small amounts of water will make toxic gas produced by a chemical reaction between the Li-ion battery electrolyte and water.

✳✳ CAUTION

In the event of a small fire, a Type ABC fire extinguisher may be used for an electrical fire caused by wiring harnesses, electrical components, etc. or oil fire.

✳✳ CAUTION

In case of vehicle fire, contact fire department immediately and extinguish the fire if possible. If you must walk away from the vehicle, notify an appropriate responder or a rescue person of the fact that the vehicle is

an electric car and contains a high voltage system and warn all others.

LI-ION BATTERY DAMAGE AND FLUID LEAKS

Li-ion Battery Electrolyte Solution Characteristics:
• Clear in color
• Sweet odor
• Similar viscosity to water
• Skin irritant
• Eye irritant - If contact with plenty of water and see a doctor immediately.
• Highly flammable
• Electrolyte liquid or fumes that have come into contact with water vapors in the air will create an oxidized substance. This substance may irritate skin and eyes. In these cases, rinse with plenty of water and see a doctor immediately.
• Since the Li-ion battery is made up of many small sealed battery modules, electrolyte solution should not leak in large quantity.

Other fluids in the vehicle (such as washer fluid, brake fluid, coolant, etc.) are the same as those in a conventional internal combustion vehicle.

✳✳ CAUTION

The Li-ion battery must be removed from the vehicle before the vehicle is scrapped. Insulate the terminals of the removed Li-ion battery with insulating tape.

PRECAUTIONS NECESSARY FOR STEERING WHEEL ROTATION AFTER BATTERY DISCONNECTION

✳✳ CAUTION

Comply with the following cautions to prevent any error and malfunction: Before removing and installing any control units, first turn the ignition switch to the LOCK position, and then disconnect both battery cables. After finishing work, confirm that all control unit connectors are connected properly, then re-connect both battery cables. Always use CONSULT to perform self-diagnosis as a part of each function inspection after finishing work. If a DTC is detected, perform trouble diagnosis according to self-diagnosis results.

For vehicle with steering lock unit, if the battery is disconnected or discharged, the steering wheel will lock and cannot be turned. If turning the steering wheel is required with the battery disconnected or discharged, follow the operation procedure below before starting the repair operation. Operation procedure:

5. Connect both battery cables.

➤Supply power using jumper cables if battery is discharged.

6. Turn the ignition switch to ACC position. (At this time, the steering lock will be released.)

7. Disconnect both battery cables. The steering lock will remain released with both battery cables disconnected and the steering wheel can be turned.

8. Perform the necessary repair operation.

9. When the repair work is completed, re-connect both battery cables. With the brake pedal released, turn the ignition switch from ACC position to ON position, then to LOCK position. (The steering wheel will lock when the ignition switch is turned to LOCK position.)

10. Perform self-diagnosis check of all control units using CONSULT.

PRECAUTIONS FOR REMOVING 12V BATTERY

> **⁕⁕ CAUTION**
>
> **Before disconnecting the 12V battery terminal, if necessary, set the parking brake, lower the windows, unlock the doors, and open the rear hatch as required. Once 12V battery is disconnected, power controls will not operate.**

11. Check that EVSE is not connected.

➤If EVSE is connected, the air conditioning system may be automatically activated by the timer A/C function.

12. Turn the power switch OFF, ON, OFF. Get out of the vehicle. Close all doors (including back door).

13. Check that the charge status indicator lamp does not blink and wait for 5 minutes or more.

➤If the battery is removed within 5 minutes after the power switch is turned OFF, plural DTCs may be detected.

14. Remove 12V battery within 1 hour after turning the power switch OFF, ON, OFF. The 12V battery automatic charge control may start automatically even when the power switch is in OFF state. Once the power switch is turned ON, OFF, the 12V battery automatic charge control does not start for approximately 1 hour.

➤After all doors (including back door) are closed, if a door (including back door) is opened before battery terminals are disconnected, start over from Step 1.

➤After turning the power switch OFF, if "Remote A/C" is activated by user operation, stop the air conditioner and start over from Step 1.

BUFFER ZONE

> **⁕⁕ CAUTION**
>
> ANYONE USING A MEDICAL ELECTRIC DEVICE SUCH AS A PACEMAKER MUST NEVER PERFORM OPERATIONS ON THE VEHICLE. Strong magnets are used in this vehicle, and the magnetic field can disrupt the function of electric medical devices.

> **⁕⁕ CAUTION**
>
> VEHICLE CHARGING: ANYONE USING A MEDICAL ELECTRIC DEVICE SUCH AS AN IMPLANTABLE CARDIAC PACEMAKER OR AN IMPLANTABLE CARDIOVERTER DEFIBRILLATOR (ICD) MUST NOT ENTER THE VEHICLE COMPARTMENT (INCLUDING LUGGAGE ROOM) DURING NORMAL CHARGE OPERATION. Radiated electromagnetic waves generated by the on board charger at normal charge operation can disrupt the function of electric medical devices.

> **⁕⁕ CAUTION**
>
> TELEMATICS SYSTEM AND/OR INTELLIGENT KEY SYSTEM: Anyone using electric medical electric devices [implantable cardiac pacemaker or implantable cardioverter defibrillator (ICD)] OR electric medical electric devices other than implantable cardiac pacemaker or ICD must not approach within approximately 8.66 in. (220 mm) of interior/exterior antenna. The electromagnetic wave of TCU might affect the function of electric medical devices. The electromagnetic wave of the intelligent key operations (such as at door operation, at each request switch operation, or at engine starting) might affect the function of electric medical devices. The possible effects on the devices must be checked with the device manufacturer before TCU and/or intelligent key use.

DISARMING THE HIGH VOLTAGE TRACTION BATTERY

DISARMING

See Figures 48 through 50.

❊❊ CAUTION

Be sure to follow the procedure below and disconnect the high voltage before performing inspection or servicing of the high voltage system.

❊❊ CAUTION

Before disconnecting the 12V battery terminal, if necessary, set the parking brake, lower the windows, unlock the doors, and open the rear hatch as required. Once 12V battery is disconnected, power controls will not operate.

Procedure for disconnecting high voltage:
1. Turn power switch OFF.

❊❊ CAUTION

The worker must keep the intelligent key on his/her person.

2. Disconnect 12V battery negative terminal.
3. Remove the service plug as follows:
    a. Put a finger to the notched part, and pull off the floor carpet.
    b. Remove inspection hole cover installation bolt and take off the inspection hole cover.
    c. Remove the service plug.
4. Wait for a minimum of approximately 10 minutes after the service plug is removed.

Fig. 48 Put a finger to the notched part (A), and pull off the floor carpet (1)

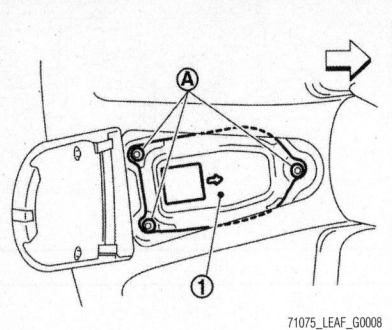

Fig. 49 Remove inspection hole cover installation bolt (A) and take off the inspection hole cover (1)

❊❊ CAUTION

Immediately insulate removed high voltage connectors and terminals with insulating tape.

❊❊ CAUTION

Be sure to put the removed service plug in your pocket and carry it with you so that another person does not accidentally connect it while work is in progress.

❊❊ CAUTION

Be sure to follow the procedure below and connect the high voltage before performing inspection or servicing of the high voltage system.

ARMING

See Figure 51.

❊❊ CAUTION

Review Precautions before proceeding.

❊❊ CAUTION

Be sure to follow the procedure below and connect the high voltage before performing inspection or servicing of the high voltage system.

1. Check that 12V battery negative terminal is disconnected.
2. Install service plug.
3. Connect 12V battery negative terminal.

DIRECT CURRENT/ ALTERNATING CURRENT INVERTER

REMOVAL & INSTALLATION

See Figures 52 through 60.

➡ **Nissan refers to this as a traction motor inverter.**

❊❊ CAUTION

Because electric vehicles contain a high voltage battery, there is the risk of electric shock, electric leakage, or similar accidents if the high voltage component and vehicle are handled incorrectly. Be sure to follow the correct work procedures when performing inspection and maintenance.

❊❊ CAUTION

Remove the service plug in order to disconnect the high voltage circuits before performing inspection or maintenance of high voltage system harnesses and parts.

❊❊ CAUTION

To prevent the removed service plug from being connected by mistake during the procedure, always carry it

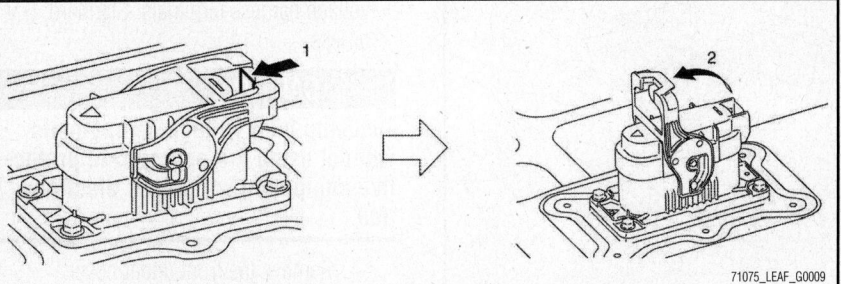

Fig. 50 Remove the service plug by pressing the locking tab (1) and rotating the handle (2) upward. Using the handle, pull the service plug completely out of its socket

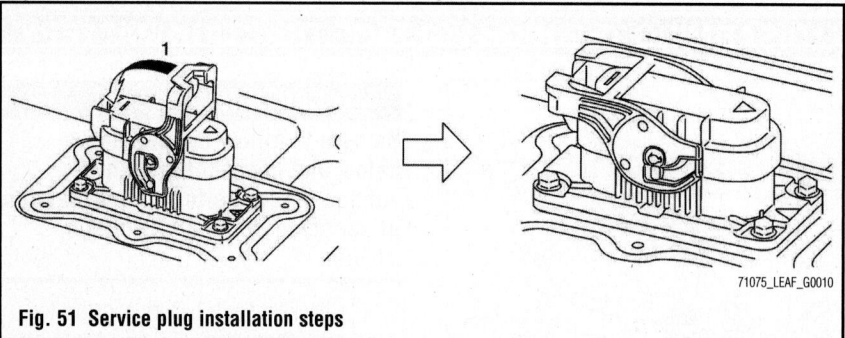

Fig. 51 Service plug installation steps

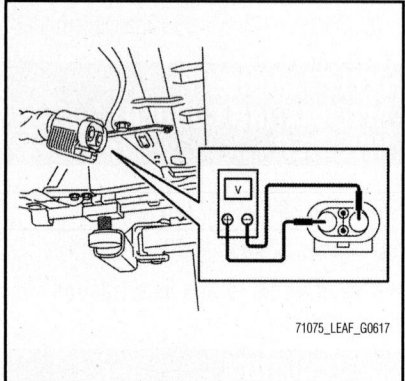

Fig. 52 Measure voltage between high voltage harness terminals

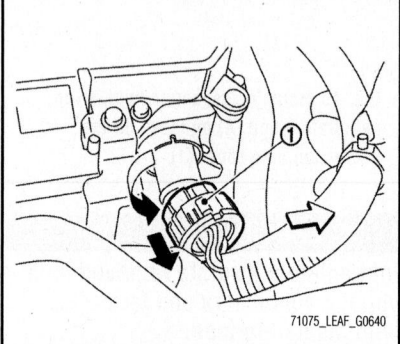

Fig. 54 Turn the traction motor inverter harness connector (1) counterclockwise to remove it

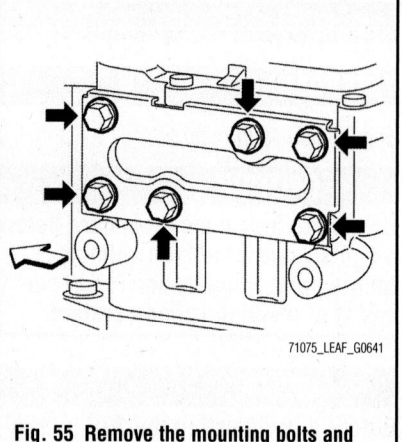

Fig. 55 Remove the mounting bolts and the 3-phase harness cover

in your pocket or put it in the tool box. Clearly identify the persons responsible for high voltage work and ensure that other persons do not touch the vehicle. When not working, cover high voltage parts with an insulating cover sheet or similar item to prevent other persons from contacting them. Wear insulating protective equipment consisting of glove, shoes, face shield and glasses.

✷✷ WARNING

There is the possibility of a malfunction occurring if the vehicle is

changed to READY status while the service plug is removed. Therefore do not change the vehicle to READY status unless instructed to do so.

1. Before servicing the vehicle, refer to the Precautions Section.
2. Disconnect the high voltage circuit.
3. Check the voltage in high voltage circuit. (Check that condenser are discharged.)
  a. Lift up the vehicle and remove the Li-ion battery under covers.
  b. Disconnect the high voltage connector from front side of Li-ion battery.

➡ For voltage measurements, use a tester which can measure to 500V or higher.

  c. Measure voltage between high voltage harness terminals. Standard: 5 V or less

✷✷ CAUTION

Touching high voltage components without using the appropriate protective equipment will cause electrocution.

4. Remove the front under cover.
5. Drain coolant from the radiator.
6. Remove the 12V battery.
7. Move the fuse box.

8. Remove the ground cable from DC/DC-J/B.
9. Remove the motor room harness clip and water hose clip which are attached to the traction motor inverter.
10. Turn the traction motor inverter harness connector of the traction motor inverter counterclockwise to remove it.
11. Remove the brake reservoir tank together with bracket, and move it in order to secure work space needed to remove traction motor inverter.
12. Remove the degas tank, and move it in order to secure work space needed to remove traction motor inverter.
13. Disconnect the water hose from the OUT side of traction motor inverter.

✷✷ WARNING

Take care that coolant does not contact the high voltage harness connectors. To prevent performance degradation, if coolant contacts a high voltage harness connector, immediately dry the high voltage connector completely with an air blow gun.

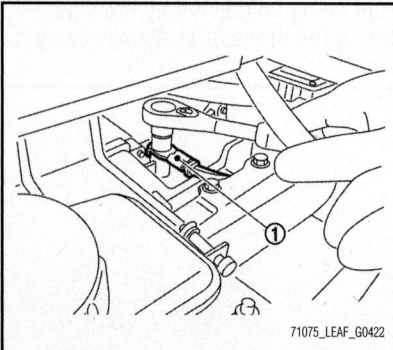

Fig. 53 Remove ground cable from DC/DC-J/B

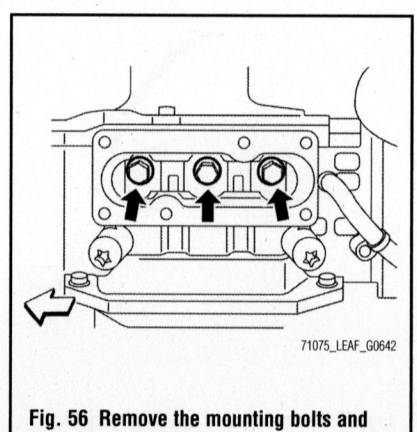

Fig. 56 Remove the mounting bolts and the 3-phase harness

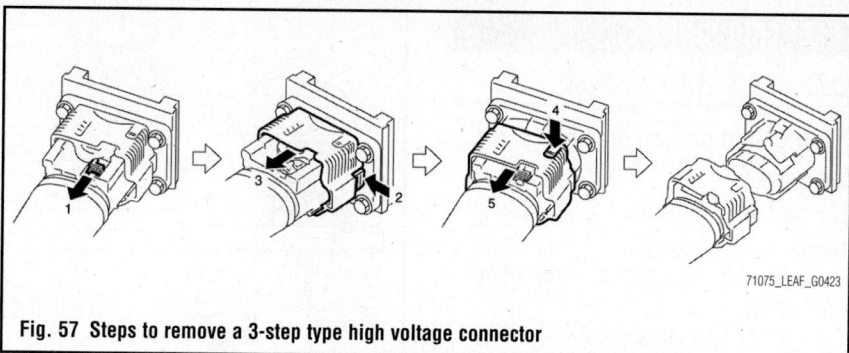

Fig. 57 Steps to remove a 3-step type high voltage connector

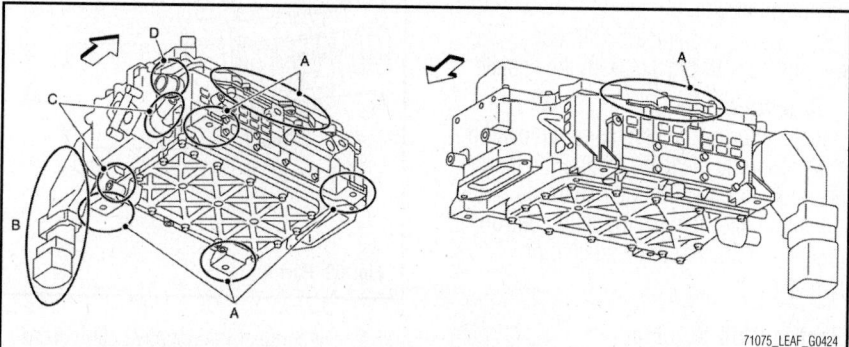

Fig. 58 Grasp part (A); do not grasp the high voltage connector (B), cooling bulge (C), or 12V system connector (D)

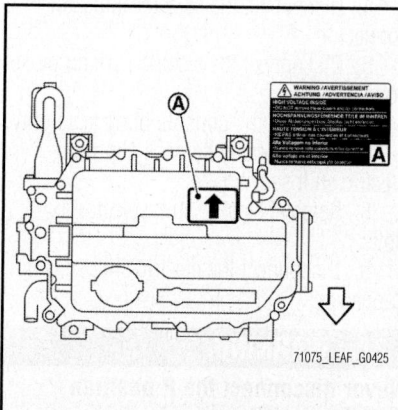

Fig. 59 Apply high voltage warning label at position (A), with top facing in the direction of arrow

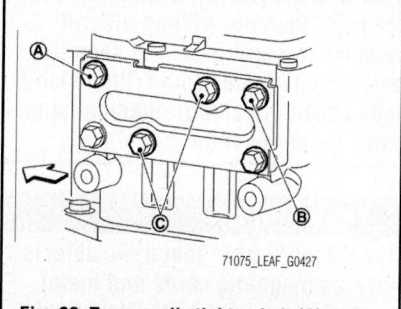

Fig. 60 Temporarily tighten bolt (A) and (B) before tightening two center bolts (C)

14. Disconnect water hose at DC/DC-J/B OUT side.

15. Remove the Torx® bolts, and then remove high voltage safety cover.

16. Remove 3-phase harness cover mounting bolts and remove 3-phase harness cover.

17. Remove 3-phase harness mounting bolts and remove 3-phase harness.

✱✱ WARNING

When removing 3-phase harness mounting bolts, take care not to drop them into traction motor inverter. Bolts cannot fall into the traction

motor inverter until 3-phase harness is pulled downwards. Therefore if bolts look likely to fall, be sure to collect them with a magnet or other means before pulling 3-phase harness out downwards. If a bolt falls into traction motor inverter, do not invert traction motor inverter. (If inverted, bolt may contact PCB inside traction motor inverter, causing damage.) Incline so that 3-phase harness bolt insertion hole faces down in order to recover the fallen bolt.

18. Remove 3-phase harness grommet mounting bolts and pull 3-phase harness out downwards.

➡ Protect the terminals of disconnected high voltage harness connector with

insulation tape so that they are not exposed.

19. Remove high voltage connector (3 step type) that is connected to DC/DC-J/B.

20. Follow the steps shown to remove a 3-step type high voltage connector.

21. Remove the traction motor inverter fastening bolts, and remove the traction motor inverter.

   a. When removing and transporting traction motor inverter, grasp part shown.

   b. Do not grasp the high voltage connector, cooling bulge, or 12V system connector.

To install:

Installation is the reverse of removal, noting the following:

22. Reinstall high voltage harness clips in their original positions. If a clip is damaged, replace it with a new clip before installing.

23. Perform correct air bleeding after adding coolant.

24. If traction motor inverter was replaced, apply high voltage warning label at the position shown, with the top facing in the direction of arrow.

   a. Before applying label, verify that there is no dust or dirt on surface of traction motor inverter.

   b. Place the ornament (NISSAN and Zero Emission) in place.

25. Follow the procedure shown and connect the 3-step type high voltage harness connector.

26. When installing 3-phase harness, take care packing does not become displaced while inserting harness into inverter.

27. To tighten 3-phase harness cover bolt, temporarily tighten bolt and shown in the figure for positioning purpose before tightening two center bolts. After this, tighten four remaining bolts.

28. To install, align the gasket tab. The gasket of the 3-phase harness cover is not reusable. Be sure to replace it with a new part.

29. After all parts are installed, check equipotential.

30. If traction motor inverter was replaced, perform Resolver Write.

RESOLVER WRITE

➡ If the traction motor inverter was replaced, then the EV system warning lamp illuminates when the power switch is turned ON, and DTC "P325C" is detected. Therefore after writing of the traction motor resolver offset is completed, verify that the EV system warning lamp has turned off and erase DTC "P325C".

1. CHECK WHICH PARTS WERE REPLACED BEFORE PERFORMING WRITING OF THE TRACTION MOTOR RESOLVER OFFSET

    a. Check the replaced parts. Which parts were replaced?
- Traction motor: GO TO STEP 2
- Traction motor inverter: GO TO STEP 3
- Traction motor and traction motor inverter: GO TO STEP 3

2. WRITING OF THE TRACTION MOTOR RESOLVER OFFSET WITH CONSULT

    a. Power switch ON.
    b. Select "Work Support" in "MOTOR CONTROL".
    c. Select "RESOLVER WRITE".
    d. Enter the traction motor resolver offset.
    e. Touch "WRITE".
    f. Is "Writing is complete" displayed?
- Yes: Power switch OFF, Power switch ON and wait 2 seconds or more, Power switch OFF to complete the work
- No: Perform STEP 2 again

3. WRITING OF THE TRACTION MOTOR RESOLVER OFFSET WITH CONSULT

    a. Power switch ON.

➡ **EV system warning lamp turns on.**

    b. Select "Work Support" in "MOTOR CONTROL".
    c. Select "RESOLVER WRITE".
    d. Enter the traction motor resolver offset.
    e. Touch "WRITE".
    f. Is "Writing is complete" displayed?
- Yes: GO TO STEP 4
- No: Perform STEP 3 again

4. STEPS AFTER WRITING OF THE TRACTION MOTOR RESOLVER OFFSET WITH CONSULT

    a. Power switch OFF.
    b. Power switch ON and wait 2 seconds or more.
    c. Verify that the EV system warning lamp is off.
    d. Select "Work Support" in "MOTOR CONTROL".
    e. Select "RESOLVER WRITE".
    f. Confirm the value is changed according to the correction value input.
    g. Perform "Self Diagnostic Results" in "MOTOR CONTROL".
    h. Erase the DTC "P325C".
    i. Power switch OFF.

ELECTRIC SHIFT

REMOVAL & INSTALLATION

Electric Shift Control Module

See Figure 61.

1. Before servicing the vehicle, refer to the Precautions Section.
2. Disconnect the negative cable from 12V battery.
3. Remove the console body assembly.
4. Disconnect the electric shift control module connector.
5. Remove the electric shift control module with bracket from the vehicle.

To install:

Installation is the reverse of removal. It is necessary to clear the P position learning value and perform the relearning of the P position after the electric shift control module is removed and installed or replaced.

Electric Shift Selector

See Figures 62 through 64.

> ※※ **CAUTION**
>
> **Part A in the figure contains a strong magnet. Persons with an electro-medical apparatus should keep it away from his/her body. Otherwise it may cause the electro-medical apparatus to malfunction.**

> ※※ **WARNING**
>
> **Keep it away from magnetic objects such as magnetic cards and metal products (e.g. watches). Never subject the electric shift selector to impact by dropping or hitting, water splash or high humidity.**

1. Before servicing the vehicle, refer to the Precautions Section.

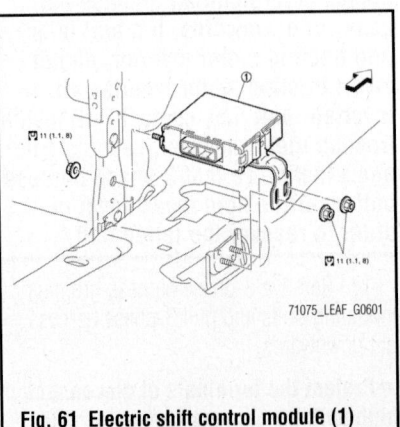

Fig. 61 Electric shift control module (1)

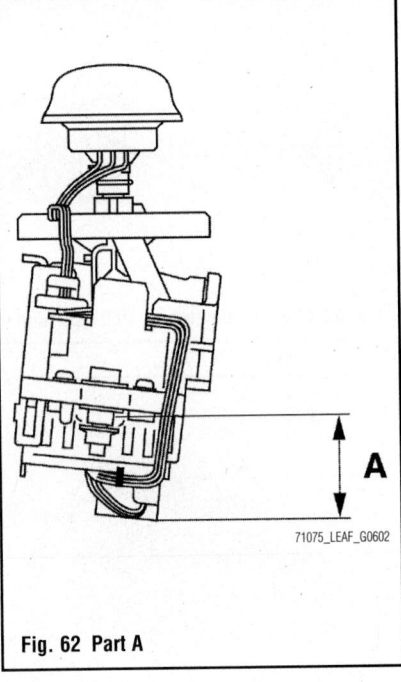

71075_LEAF_G0602

Fig. 62 Part A

2. Disconnect the negative cable from 12V battery.
3. Remove the console finisher assembly.
4. Disconnect the selector indicator connector.
5. Disconnect the electric parking brake connector.
6. Remove the console body assembly.
7. Remove body harness clip from electric shift selector.
8. Remove electric shift selector fix bolts.
9. Disconnect the electric shift sensor connector.

> ※※ **WARNING**
>
> **Never disconnect the P position switch connector.**

10. Remove the electric shift selector from the vehicle.

To install:

Installation is the reverse of removal. Check the orientation instruction on the side of the body bracket and install the part so that the direction of the arrow points toward the vehicle front.

Electric Shift Selector Indicator

See Figure 65.

1. Before servicing the vehicle, refer to the Precautions Section.
2. Disconnect the negative cable from 12V battery.
3. Remove the console finisher assembly.

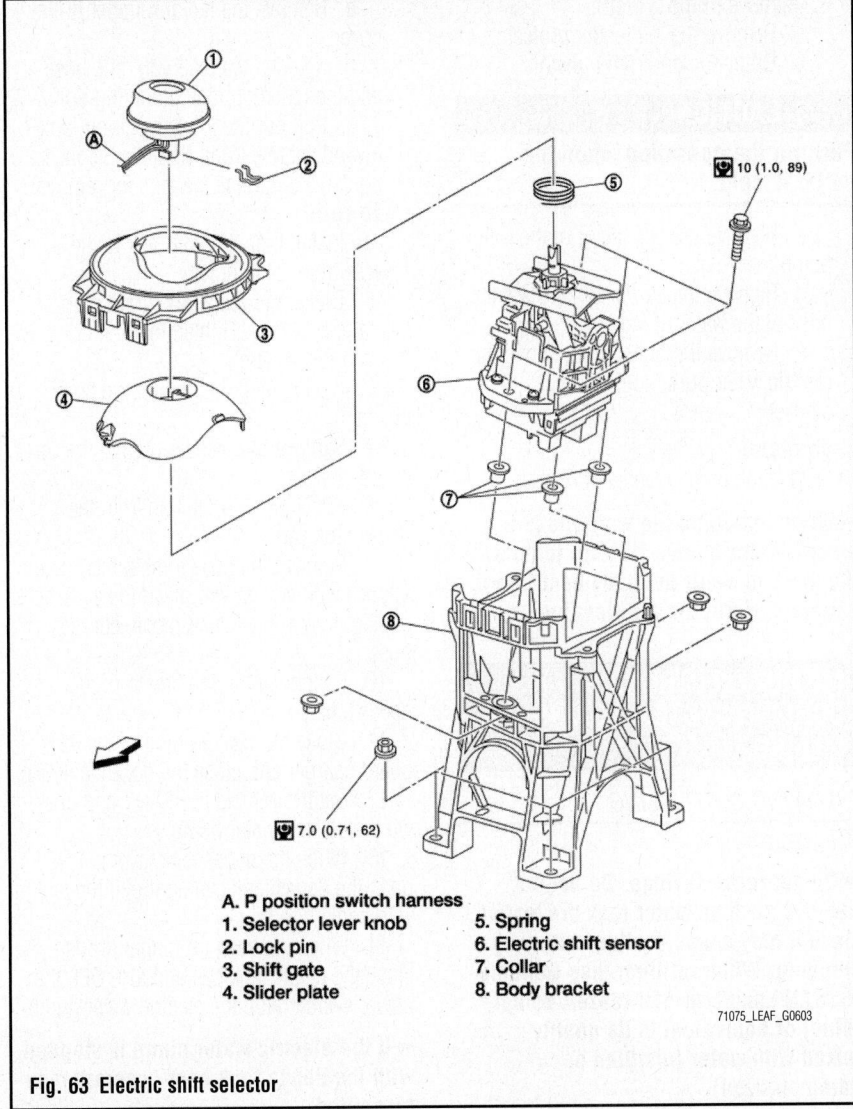

A. P position switch harness
1. Selector lever knob
2. Lock pin
3. Shift gate
4. Slider plate
5. Spring
6. Electric shift sensor
7. Collar
8. Body bracket

71075_LEAF_G0603

Fig. 63 Electric shift selector

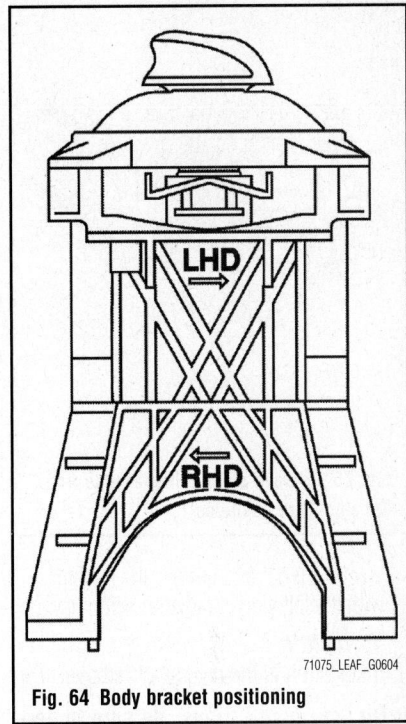

71075_LEAF_G0604

Fig. 64 Body bracket positioning

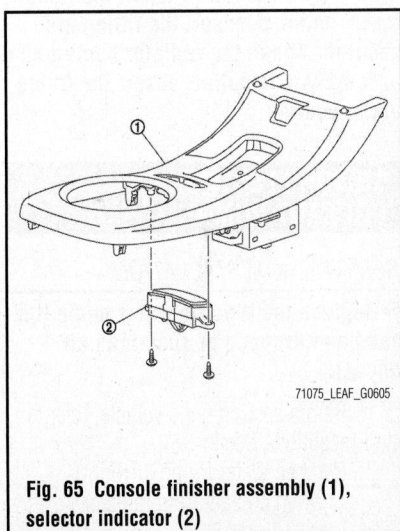

71075_LEAF_G0605

Fig. 65 Console finisher assembly (1), selector indicator (2)

4. Disconnect the selector indicator connector.

5. Disconnect the electric parking brake connector.

6. Remove the selector indicator from the console finisher assembly.

To install:

Installation is the reverse of removal. Move the selector lever and check that the light position of the selector indicator corresponds to the actual shift position.

MOTOR ELECTRONICS RADIATOR

REMOVAL & INSTALLATION

See Figure 66.

✳✳ CAUTION

Never remove the degas tank cap if a high voltage part including traction motor is hot. Hot liquid may spray out from the radiator, causing serious injury.

1. Before servicing the vehicle, refer to the Precautions Section.

2. Drain coolant from radiator drain plug.

3. Remove radiator core support upper.

4. Remove radiator hoses (upper, lower) and reservoir tank hose.

5. Remove the cooling fan shroud as follows:

a. Disconnect the harness between cooling fan and vehicle body at cooling fan control module.

b. While pressing left and right pawls of the radiator, raise the radiator cooling fan assembly in upward direction and separate fitting of the radiator and the radiator cooling fan assembly.

c. Pull out and remove the cooling fan shroud with the left side of radiator cooling fan assembly facing upward.

✳✳ WARNING

Be careful not to damage radiator core.

6. Remove the radiator assembly as follows:

a. Remove the air conditioner pipe from the clip.

b. While pressing left and right pawls of the radiator, raise the condenser upward and separate the fitting of the radiator and the condenser.

➡**Since the piping of air conditioner may be bent, never lift condenser more than necessary.**

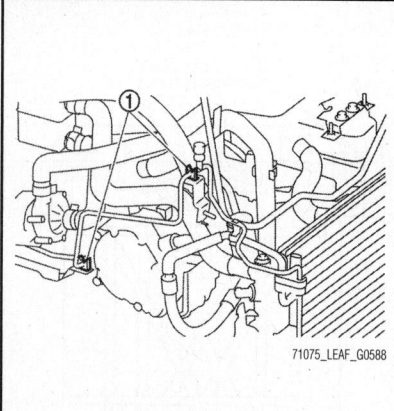

Fig. 66 Remove air conditioner pipe from the clip (1) (2 positions)

c. Pull out and remove the radiator with the left side of radiator facing upward.

To install:
Installation is the reverse of removal.

➡️Do not reuse O-rings. Be sure to perform the air bleeding. When the radiator is installed, insert the fitting area securely. When the radiator cooling fan assembly is installed, insert the fitting area securely.

MOTOR ELECTRONICS COOLANT PUMP

REMOVAL & INSTALLATION

➡️Replace the electric water pump if it has been dropped or sustained an impact.

1. Before servicing the vehicle, refer to the Precautions Section.
2. Remove Water Pump 1 (Right):
   a. Drain coolant from radiator.

❋❋ CAUTION
Perform the operation when the motor is cold.

   b. Remove the connector and water hose of the water pump.

❋❋ WARNING
Take care that coolant does not contact the high voltage harness connectors. If coolant contacts a high voltage harness connector, immediately use an air blow and fully remove the liquid.

   c. Remove the bolts and remove the electric water pump together with the bracket.

3. Remove Pump 2 (Left):
   a. Remove the left fender protector.
   b. Drain coolant from radiator.

❋❋ CAUTION
Perform the operation when the motor is cold.

   c. Remove the connector of the water pump.
   d. Remove the connector and water hose of the water pump.
   e. Remove the bolts, and remove the electric water pump together with bracket.

To install:
Installation is the reverse of removal.

➡️When installing the water hose to electric water pump, be sure to hold the electric water pump by hand. Be sure to perform the air bleeding.

MOTOR ELECTRONICS COOLING SYSTEM DRAINING & FILLING

DRAINING & REFILLING
See Figure 67.

➡️Do not reuse O-rings. Do not put additive such as water leak preventive, since it may cause cooling waterway clogging. When refilling, use Genuine NISSAN Long Life Antifreeze/Coolant (blue) or equivalent in its quality mixed with water (distilled or demineralized).

1. Before servicing the vehicle, refer to the Precautions Section.
2. Remove the Li-ion battery undercover, radiator drain plug, and degas tank cap, and drain the coolant.
3. Remove the reservoir tank and drain the coolant as follows:

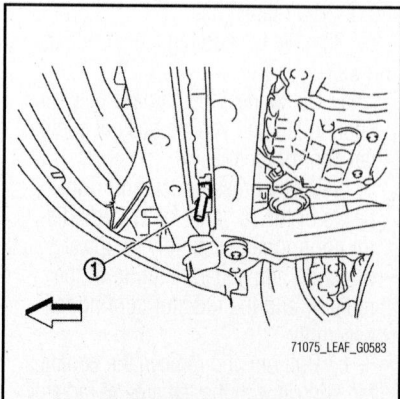

Fig. 67 Radiator drain plug (1)

   a. Remove the radiator upper grille cover.
   b. Remove the reservoir tank hose and the reservoir tank mounting bolts.
   c. Pull out the lower reservoir tank toward vehicle front. Remove the insertion area and raise the tank for removal.

To refill:
4. Install the reservoir tank in the reverse order of removal.
5. Clean and install the drain plug. Use a new O-ring. Tighten the drain plug to 11 inch lbs. (1 Nm).
6. Check the tightening of the hose clamp.
7. Remove the bleeder plug of the on board charger.
8. Fill coolant to the line from the degas tank cap.
9. Remove the hose joint at the traction motor inverter front and bleed the air. Check coolant flow visually and reconnect the hose.
10. Fill coolant to the line from the degas tank cap.
11. Close the bleeder plug of the on board charger and close the degas tank cap.
12. Set the vehicle to READY and operate the electric water pump.
13. When the degas tank level is low, open the degas tank cap and refill the tank with coolant to line.
14. When the level no longer drops, close the degas tank cap and turn OFF the power switch (stop the electric water pump).

➡️If the electric water pump is stopped with the degas tank open, coolant may be spilled.

15. Refill coolant reservoir tank to "MAX" line.

TRACTION MOTOR

INSPECTION AFTER INSTALLATION

❋❋ CAUTION
Touching high voltage components without using the appropriate protective equipment will cause electrocution. Be sure to put on insulating protective gear (glove, shoes, face shield and glasses) before beginning work on the high voltage system.

1. After installing traction motor, measure resistance (Standard: Less than 0.1 Ω):
   - Between traction motor (aluminum part) and body (ground bolt)
   - Between traction motor (aluminum part) and traction motor inverter (aluminum part)

2. If result deviates from standard values, check that no paint, oil, dirt, or other substance is adhering to bolts or conductive mounting parts. If any such substance is adhering, clean the surrounding area and remove the substance.

REMOVAL & INSTALLATION

See Figures 68 through 74.

✻✻ CAUTION

Because electric vehicles contain a high voltage battery, there is the risk of electric shock, electric leakage, or similar accidents if the high voltage component and vehicle are handled incorrectly. Be sure to follow the correct work procedures when performing inspection and maintenance.

✻✻ CAUTION

Remove the service plug in order to disconnect the high voltage circuits before performing inspection or maintenance of high voltage system

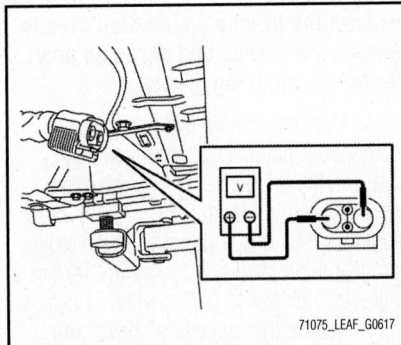

71075_LEAF_G0617

Fig. 68 Measure voltage between the high voltage harness terminals

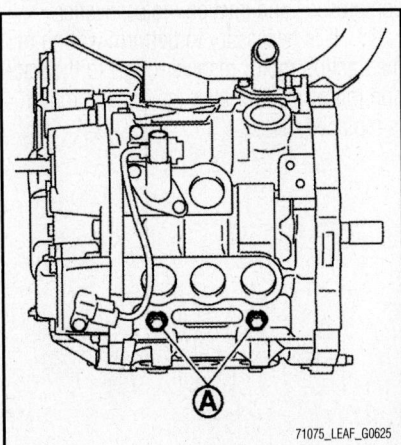

71075_LEAF_G0625

Fig. 69 Remove the drain bolt (A) of traction motor and drain coolant

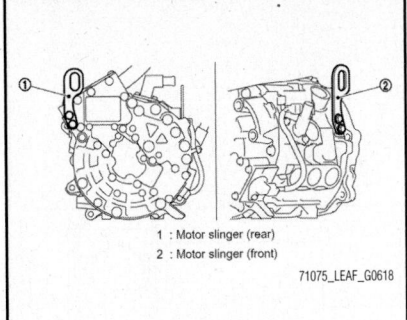

1 : Motor slinger (rear)
2 : Motor slinger (front)

71075_LEAF_G0618

Fig. 70 Motor slinger (rear) (1), Motor slinger (front) (2)

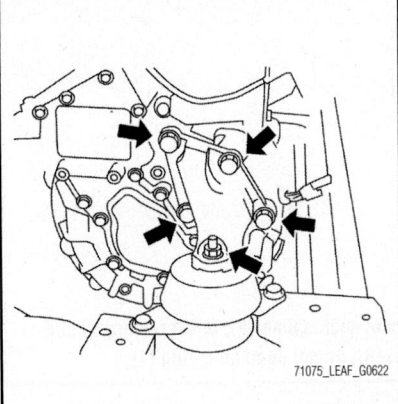

71075_LEAF_G0622

Fig. 71 Remove the right motor mounting bolts

harnesses and parts. To prevent the removed service plug from being connected by mistake during the procedure, always carry it in your pocket or put it in the tool box.

✻✻ CAUTION

Clearly identify the persons responsible for high voltage work and ensure that other persons do not touch the vehicle. When not working, cover high voltage parts with an insulating cover sheet or similar item to prevent other persons from contacting them.

✻✻ CAUTION

Touching high voltage components without using the appropriate protective equipment will cause electrocution. Be sure to put on insulating protective gear (glove, shoes, face shield and glasses) before beginning work on the high voltage system.

✻✻ WARNING

There is the possibility of a malfunction occurring if the vehicle is changed to READY status while the service plug is removed. Therefore do not change the vehicle to READY status unless instructed to do so in the service instructions.

✻✻ CAUTION

Disconnect the high voltage circuit.

---

1. Check voltage in high voltage circuit (Check that condenser is discharged):
   a. Lift up the vehicle and remove the Li-ion battery under covers.
   b. Disconnect high voltage harness connector from the front side of the Li-ion battery.

✻✻ CAUTION

For voltage measurements, use a tester which can measure to 500 V or higher.

---

   c. Measure voltage between the high voltage harness terminals. Standard: 5 V or less.
2. Drain the coolant from the cooling system.
3. Remove the drain bolt of traction motor and drain coolant.
4. Remove the traction motor inverter.
5. Drain the reduction gear oil.
6. Remove the traction motor and reduction gear from the vehicle together as suspension member assembly.
7. Remove the reduction gear from the suspension member.
8. Attach slinger to traction motor, and prepare to lift up with hoist.

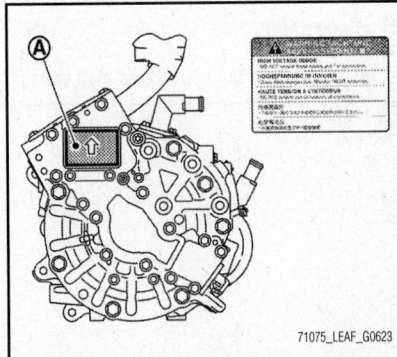

71075_LEAF_G0623

Fig. 72 If traction motor was replaced, apply high voltage warning label at position (A), with the top facing in the direction of the arrow

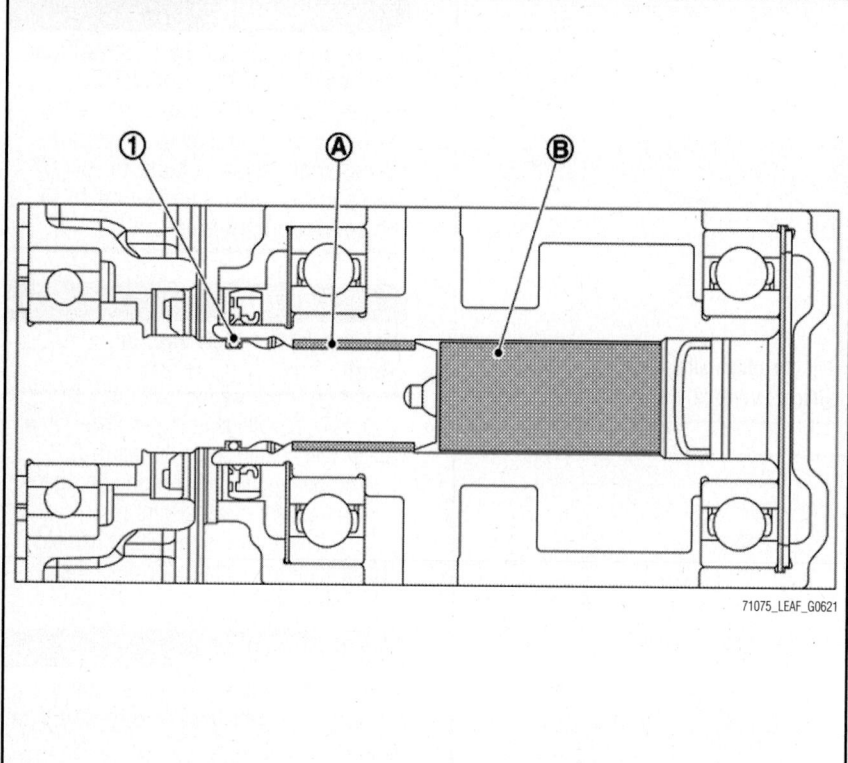

Fig. 73 Apply recommended grease to full periphery of shaft spline (A); inject recommended grease into reduction gear input shaft (inside spline) (B); do not damage O-ring (1)

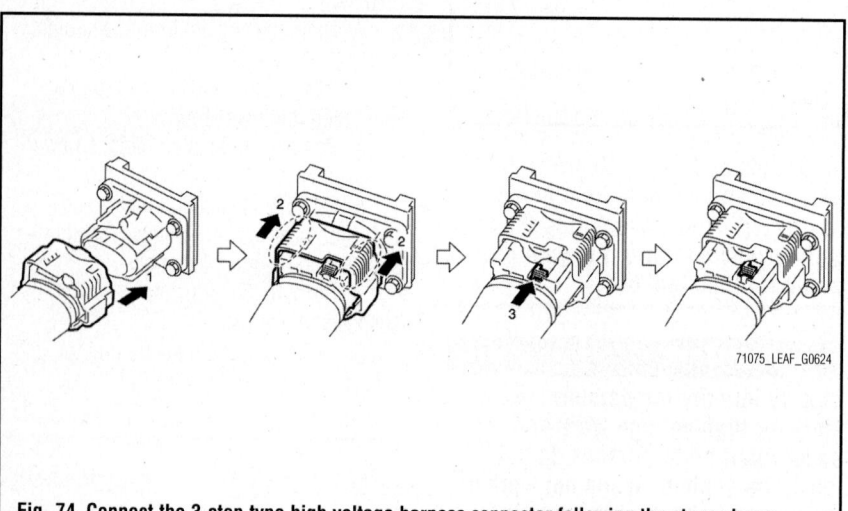

Fig. 74 Connect the 3-step type high voltage harness connector following the steps shown

9. Remove the right motor mounting bolts, lift up the traction motor with a hoist, and separate it from the suspension member.

To install:

❋❋ CAUTION

Touching high voltage components without using the appropriate protective equipment will cause electrocution. Be sure to put on insulating protective gear (glove, shoes, face shield and glasses) before beginning work on the high voltage system.

Installation is the reverse of removal, noting the following:

10. Be sure to reinstall high voltage harness clips in their original positions. If a clip is damaged, replace it with a new clip before installing.

11. Be sure to perform correct air bleeding after adding coolant.

12. If traction motor was replaced, perform resolver correction value learning.

13. If traction motor was replaced, apply high voltage warning label at position, with the top facing in the direction shown.

➡**Clean the grease application area to remove old grease and abrasion powder before applying grease.**

14. Before installing the reduction gear and traction motor, apply recommended grease to the full periphery of shaft spline. Inject grease [minimum 0.3 oz (8.5 g), maximum less than 0.7 oz (20 g)] into the reduction gear input shaft (inside spline). Take care to prevent damage to O-ring when installing.

15. Follow the procedure shown and connect the 3-step type high voltage harness connector.

16. After all parts are installed, be sure to check equipotential of traction motor, electric compressor, and traction motor inverter.

17. It is necessary to perform writing of the traction motor resolver offset to the traction motor inverter after the traction motor is replaced.

ENGINE PERFORMANCE & EMISSION CONTROLS

ACCELERATOR PEDAL POSITION (APP) SENSOR

LOCATION

See Figure 75.

REMOVAL & INSTALLATION

✳✳ CAUTION

Before servicing, push power switch OFF, disconnect 12V battery negative terminal and wait 5 minutes or more. Refer to Precautions for Removing 12V Battery. Read and observe all SRS Service Precautions. Refer to Air Bags (Supplemental Restraint System), in the Chassis Electrical section. Failure to observe all precautions may result in accidental airbag deployment, personal injury, or death.

1. Before servicing the vehicle, refer to the Precautions Section.
2. Disconnect the accelerator pedal position sensor harness connector.
3. Loosen mounting bolts, and remove accelerator pedal assembly.

✳✳ WARNING

Never disassemble accelerator pedal assembly. Never remove accelerator pedal position sensor from accelerator pedal assembly. Avoid impact from dropping etc. during handling. Be careful to keep accelerator pedal assembly away from water.

To install:

Installation is the reverse of removal. Insert the locating pin while inserting the locating hook in to the brake pedal bracket. Tighten mounting bolts to accelerator pedal assembly.

✳✳ WARNING

Never squeeze the locating hook into the brake pedal bracket when inserting the locating pin into the hole on the brake pedal bracket side.

BODY CONTROL MODULE (BCM)

ADDITIONAL SERVICE WHEN REPLACING CONTROL UNIT (BCM)

Vehicle specification needs to be written with CONSULT because it is not written after replacing BCM. Work Procedure:

1. WRITING MODE SELECTION, CONSULT Configuration:
   a. Select "CONFIGURATION" of BCM.
   b. When writing saved data GO TO 2.
   c. When writing manually GO TO 3.
2. PERFORM "WRITE CONFIGURATION - CONFIG FILE", CONSULT Configuration:
   a. Perform "WRITE CONFIGURATION - CONFIG FILE".
   b. WORK END
3. PERFORM "WRITE CONFIGURATION - MANUAL SELECTION", CONSULT Configuration:
   a. Select "WRITE CONFIGURATION - MANUAL SELECTION".
   b. Identify the correct model and configuration list.
   c. Confirm and/or change setting value for each item.

➡ **Thoroughly read and understand the vehicle specification. ECU control may not operate normally if the setting is not correct. If items are not displayed, touch "SETTING". list" for written items and setting value.**

   d. Select "SETTING".

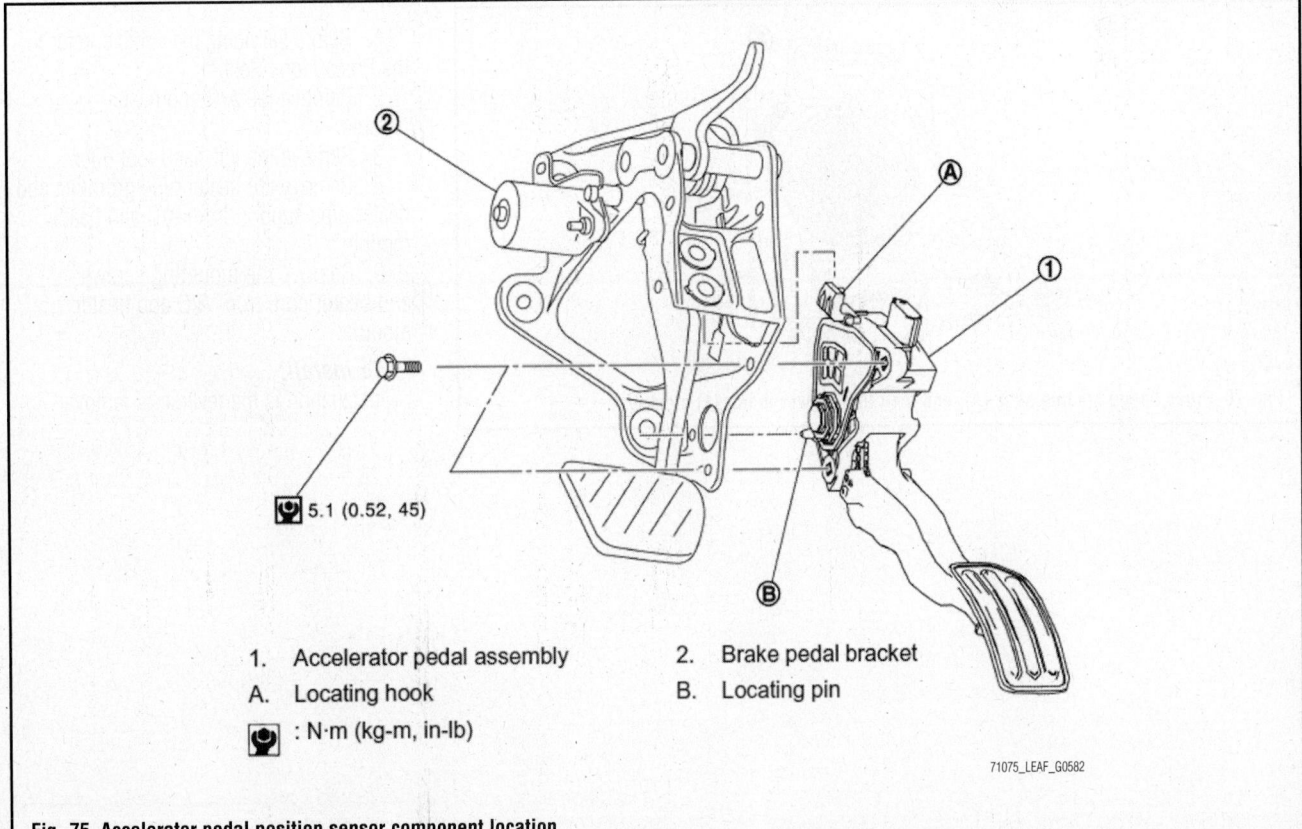

5.1 (0.52, 45)

| 1. | Accelerator pedal assembly | 2. | Brake pedal bracket |
| A. | Locating hook | B. | Locating pin |

: N·m (kg-m, in-lb)

71075_LEAF_G0582

Fig. 75 Accelerator pedal position sensor component location

→Make sure to select "SETTING" even if the indicated configuration of brand new BCM is same as the desirable configuration. If not, configuration which is set automatically by selecting vehicle model cannot be memorized.

    e. When "COMMAND FINISHED", select "END".
    f. GO TO 4.
    4. OPERATION CHECK:
    a. Confirm that each function controlled by BCM operates normally.
    b. WORK END

→Thoroughly read and understand the vehicle specification. ECU control may not operate normally if the setting is not correct. SETTING ITEMS: AUTO LIGHT WITH/WITHOUT is properly set and the THEFT ALM AREA is set to MODE2.

REMOVAL & INSTALLATION

✳✳ CAUTION

Before servicing, push power switch OFF, disconnect 12V battery negative terminal and wait 5 minutes or more. Refer to Precautions for Removing 12V Battery. Read and observe all SRS Service Precautions. Refer to Air Bags (Supplemental Restraint System), in the Chassis Electrical section. Failure to observe all precautions may result in accidental airbag deployment, personal injury, or death.

    1. Before servicing the vehicle, refer to the Precautions Section.
    2. Before replacing BCM, perform "READ CONFIGURATION" to save or print current vehicle specification. If

"READ CONFIGURATION" cannot be used, use the "WRITE CONFIGURATION - Manual selection" after replacing BCM.
    3. Remove instrument lower panel. Refer to Instrument Panel Removal & Installation in the Body Interior section.
    4. Remove fuse block (J/B).
    5. Remove harness clip.
    6. Remove BCM mounting screws.
    7. Remove BCM and disconnect the connectors.

To install:
Installation is the reverse of removal.

→Be sure to perform "WRITE CONFIGURATION" when replacing BCM. Failure to do so will cause BCM to not operate normally. Be sure to perform the system initialization (NATS) when replacing BCM.

HEATING & AIR CONDITIONING SYSTEM

BLOWER MOTOR

REMOVAL & INSTALLATION

See Figure 76.

    1. Before servicing the vehicle, refer to the Precautions Section.

    2. Remove glove box cover assembly.
    3. Disconnect blower motor harness connector.
    4. Press flange holding hook, and then turn blower motor clockwise.
    5. Remove blower motor from A/C and heater module.

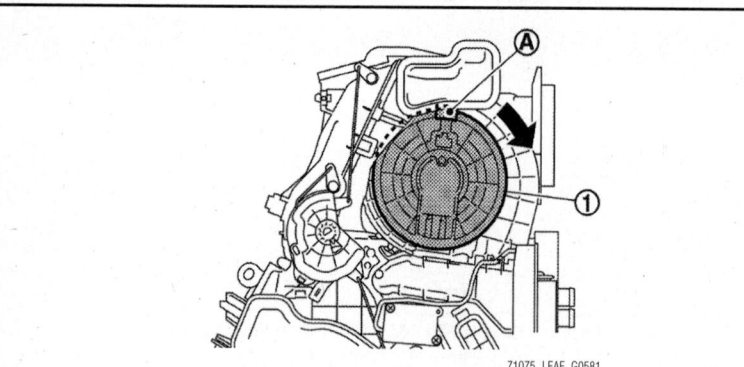

71075_LEAF_G0581

Fig. 76 Press flange holding hook (A), and then turn blower motor (1) clockwise

To install:
Installation is the reverse of removal.

HEATER CORE

REMOVAL & INSTALLATION

See Figure 77.

    1. Before servicing the vehicle, refer to the Precautions Section.
    2. Remove the A/C and heater module.
    3. Remove the left-hand foot duct.
    4. Remove the heater pipe grommet and heater pipe support from A/C and heater module.
    5. Remove the mounting screws and heater core from A/C and heater module.

To install:
Installation is the reverse of removal.

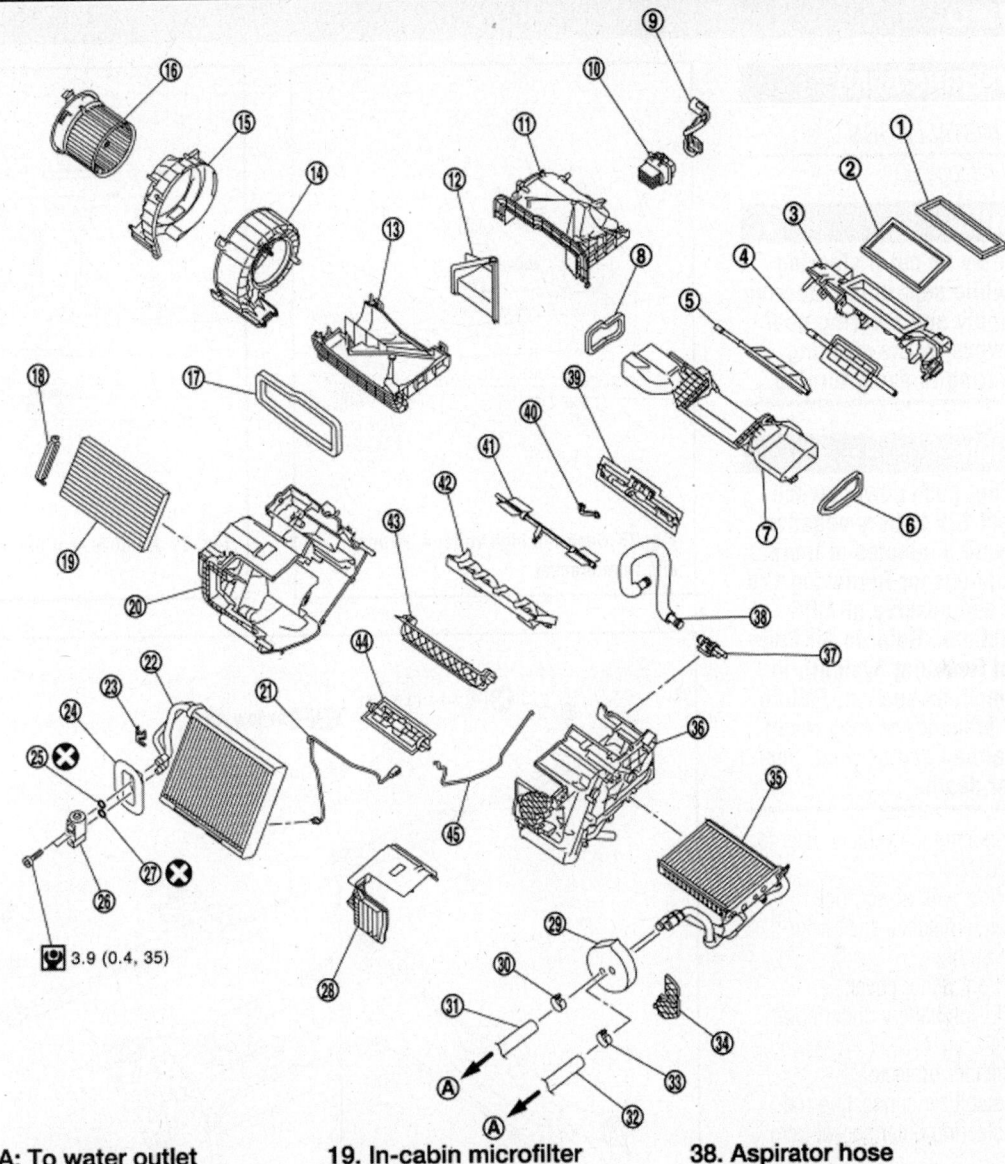

A: To water outlet
1. Ventilator seal
2. Defroster seal
3. Upper attachment case
4. Sub defroster door
5. Center ventilator and
   defroster door
6. Side ventilator seal LH
7. Lower attachment case
8. Side ventilator seal RH
9. Sub harness
10. Power transistor
11. Upper intake case
12. Intake door
13. Lower intake case
14. Blower case LH
15. Blower case RH
16. Blower motor
17. Intake seal
18. Filter cover

19. In-cabin microfilter
20. A/C unit case RH
21. Intake sensor
22. Evaporator
23. Plate
24. Expansion valve grommet
25. O-ring
26. Expansion valve
27. O-ring
28. Evaporator cover
29. Heater pipe grommet
30. Clamp
31. Heater hose
32. Heater hose
33. Clamp
34. Heater pipe support
35. Heater core
36. A/C unit case LH
37. Aspirator

38. Aspirator hose
39. Foot door
40. Foot door rod
41. Side ventilator door
42. Air mix door guide
43. Upper air mix door
44. Lower air mix door
45. Case packing

Fig. 77 Heater core (35) location in air conditioning/heater unit

STEERING

POWER STEERING GEAR

REMOVAL & INSTALLATION

See Figures 78 through 80.

> **❋❋ WARNING**
>
> Spiral cable may be cut if steering wheel turns while separating steering column assembly and steering gear assembly. Always fix the steering wheel using string to avoid turning.

> **❋❋ CAUTION**
>
> Before servicing, push power switch OFF, disconnect 12V battery negative terminal and wait 5 minutes or more. Refer to Precautions for Removing 12V Battery. Read and observe all SRS Service Precautions. Refer to Air Bags (Supplemental Restraint System), in the Chassis Electrical section. Failure to observe all precautions may result in accidental airbag deployment, personal injury, or death.

1. Before servicing the vehicle, refer to the Precautions Section.
2. Remove tires with power tool.
3. Separate intermediate shaft from steering gear assembly.
4. Remove front under cover.
5. Remove Li-ion battery under cover (front).
6. Remove fender protector.
7. Remove stabilizer connecting rod.
8. Remove steering outer socket from steering knuckle.
9. Separate transverse link from steering knuckle with power tool.
10. Set suitable jack under reduction gear and traction motor.

> **❋❋ WARNING**
>
> Never damage the reduction gear and traction motor with a jack. Check that the condition is stable when using a jack.

11. Separate high voltage harness clip from bracket.
12. Remove rear motor mounting bolt.
13. Remove (LH and RH) motor mounting bolts from front suspension member.
14. Remove member stay and rebound stopper with power tool.
15. Remove suspension member mounting bolts, washer, and rebound stopper rubber with power tool.

Fig. 78 Separate high voltage harness clip from bracket

Fig. 79 Remove rear motor mounting bolt

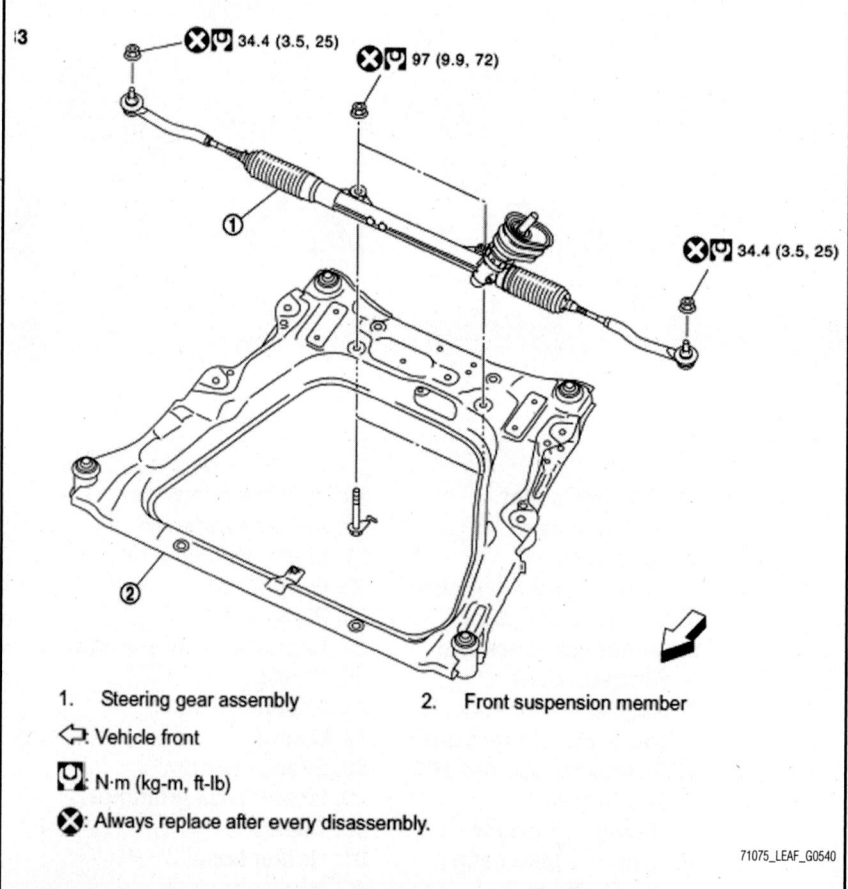

1. Steering gear assembly
2. Front suspension member

⇦ : Vehicle front

🔧 : N·m (kg-m, ft-lb)

✖ : Always replace after every disassembly.

Fig. 80 Remove steering gear assembly with power tool

16. Remove front suspension member from vehicle body.

> **❋❋ CAUTION**
>
> Check the stable condition when using a jack.

17. Remove steering gear assembly with power tool.
18. Perform inspection after removal.

To install:

Installation is the reverse of removal, noting the following:

19. Spiral cable may be cut if steering wheel turns while separating steering column assembly and steering gear assembly. Always fix the steering wheel using string to avoid turning.

20. Perform final tightening of nuts and bolts on each part under unladen conditions with tires on level ground when removing steering gear assembly. Check wheel alignment. Rotate steering wheel to check for de-

centered condition, binding, noise or excessive steering effort. Never reuse steering outer socket fixing nut and steering gear assembly mounting nut.

21. Perform inspection after installation.

SUSPENSION

KNUCKLE & SPINDLE

REMOVAL & INSTALLATION
See Figures 81 through 83.

1. Before servicing the vehicle, refer to the Precautions Section.
2. Remove tires with power tool.
3. Remove wheel sensor and sensor harness.
4. Remove lock plate from strut assembly.
5. Remove caliper assembly. Hang caliper assembly aside.

> ✳✳ **WARNING**
>
> **Never depress brake pedal while brake caliper is removed.**

6. Remove disc rotor. Put matchmarks on the wheel hub assembly and the disc rotor before removing the disc rotor.

> ✳✳ **WARNING**
>
> **Do not drop disc rotor.**

7. Remove cotter pin and adjusting cap, and then loosen wheel hub lock nut, using a hub lock nut wrench.
8. Patch wheel hub lock nut with a piece of wood. Hammer the wood to disengage wheel hub assembly from halfshaft.

> ✳✳ **WARNING**
>
> **Never place halfshaft joint at an extreme angle. Also be careful not to**

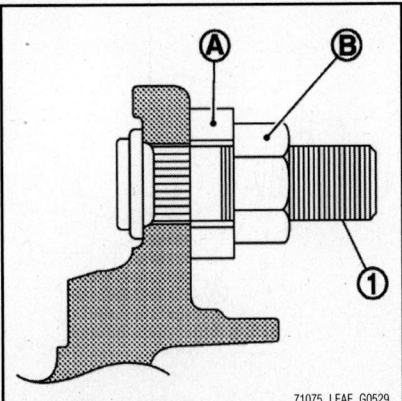

Fig. 81 Place a washer (A) as shown in the figure to install the hub bolts (1) by using the tightening force of the nut (B)

FRONT SUSPENSION

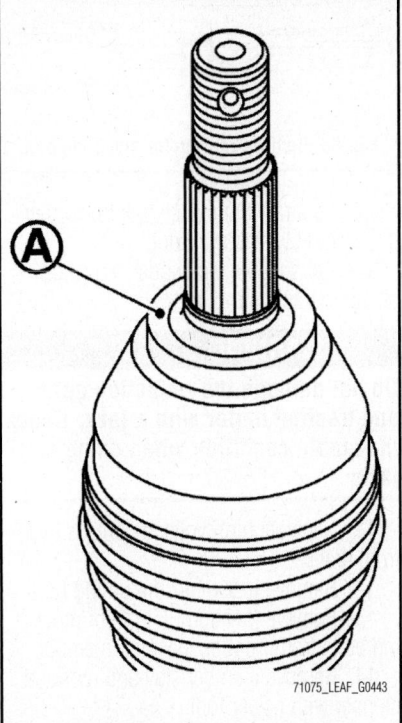

Fig. 82 Apply paste to cover entire flat surface (A) of joint subassembly of halfshaft

overextend slide joint. Never allow halfshaft to hang down without support for joint sub-assembly, shaft and the other parts. Use suitable puller, if wheel hub assembly and halfshaft cannot be separated even after performing the above procedure.

9. Remove wheel hub lock nut.
10. Separate steering knuckle from transverse link with power tool.
11. Suspend the halfshaft with suitable wire.
12. Remove wheel hub assembly and splash guard from steering knuckle with power tool.
13. Remove steering outer socket from steering knuckle.
14. Remove strut assembly from steering knuckle with power tool.
15. Remove hub bolts from wheel hub assembly, using the ball joint remover.

a. Remove hub bolt only when necessary.

➡**Never hammer the hub bolt to avoid impact to the wheel hub assembly. Pull out the hub bolt in a direction perpendicular to the wheel hub assembly.**

16. Perform inspection after removal.

To install:
Installation is the reverse of removal, noting the following:

17. Position a washer, and install the hub bolts by using the tightening force of the nut.
18. Check that there is no clearance between wheel hub assembly and hub bolt.

➡**Do not reuse hub bolt. Do not reuse steering knuckle and transverse link mounting nut.**

19. Clean the matching surface of the wheel hub lock nut and wheel hub assembly.

> ✳✳ **WARNING**
>
> **Do not apply lubricating oil to the matching surface.**

20. Clean the matching surface of the halfshaft and wheel hub assembly, and apply paste to the surface of the halfshaft joint sub-assembly. Apply paste to cover the entire flat surface of the halfshaft joint sub-assembly.
21. Tighten the wheel hub lock nut to 133–136 ft. lbs. (180–185 Nm).

➡**Be sure to use torque wrench to tighten the wheel hub lock nut. Do not use a power tool. Do not reuse wheel hub lock nut. Wheel hub lock nut tightening torque does not over torque for avoiding axle noise, and does not less than torque for avoiding looseness.**

22. When reusing the disc rotor, align the matchmarks that were made during removal.
23. When installing a cotter pin and adjusting cap, securely bend the basal portion to prevent rattles. Do not reuse cotter pin.
24. When tightening the parts that were removed when removing the wheel hub

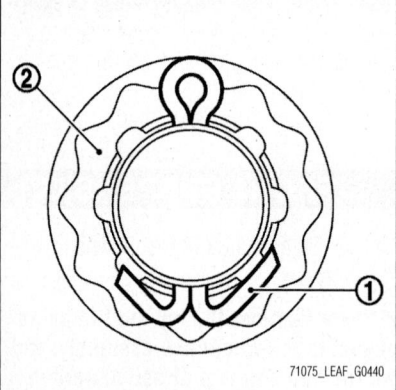

Fig. 83 Securely bend the basal portion when installing a cotter pin (1) and adjusting cap (2)

assembly and steering knuckle, perform final tightening under unladen conditions. Perform inspection after installation.

STABILIZER BAR & LINKS

REMOVAL & INSTALLATION

Stabilizer Bar

See Figures 84 through 88.

1. Before servicing the vehicle, refer to the Precautions Section.
2. Remove tires with power tool.
3. Separate intermediate shaft from steering gear assembly.
4. Remove front under cover.
5. Remove Li-ion battery under cover (front).
6. Remove fender protector.
7. Remove stabilizer connecting rod.
8. Remove steering outer socket from steering knuckle.

Fig. 84 Separate high voltage harness clip from bracket

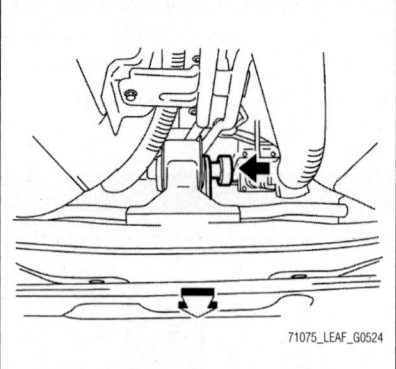

Fig. 85 Remove rear motor mounting bolt

9. Separate transverse link from steering knuckle with power tool.
10. Set suitable jack under reduction gear and traction motor.

✳✳ WARNING

Do not damage the reduction gear and traction motor with a jack. Check the stable condition when using a jack.

11. Separate high voltage harness clip from bracket.
12. Remove rear motor mounting bolt.
13. Remove (LH and RH) motor mounting bolts from front suspension member.
14. Remove member stay and rebound stopper with power tool.
15. Remove suspension member mounting bolts, washer, and rebound stopper rubber with power tool.
16. Remove front suspension member from vehicle body.

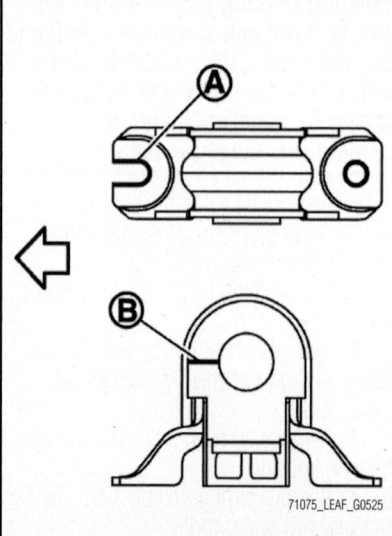

Fig. 86 Install stabilizer clamp and stabilizer bush with notch (A) and slit (B) facing to the front

✳✳ CAUTION

Check the stable condition when using a jack.

17. Remove stabilizer clamp mounting bolts, then remove stabilizer clamp and stabilizer bushing from front suspension member with power tool.
18. Remove stabilizer bar.
19. Perform inspection after removal.

To install:

Installation is the reverse of removal, noting the following:

20. Install stabilizer clamp and stabilizer bush with notch and slit facing the front of the vehicle.
21. Install the stabilizer clamp mounting bolts. Final torque: 21 ft. lbs. (28 Nm).
22. To install stabilizer connecting rod, tighten the mounting nut with the hexagonal part on the stabilizer connecting rod side fixed.

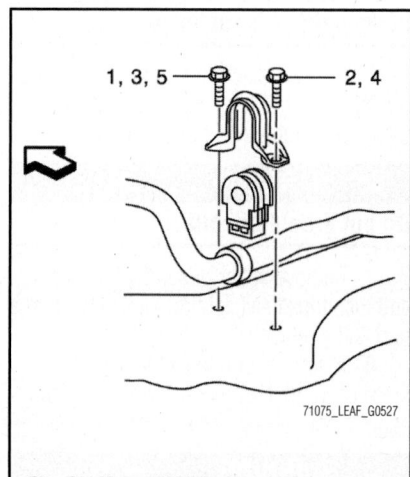

Fig. 87 Manual tightening: 1, Temporary tightening: 2, 3; Final tightening: 4, 5

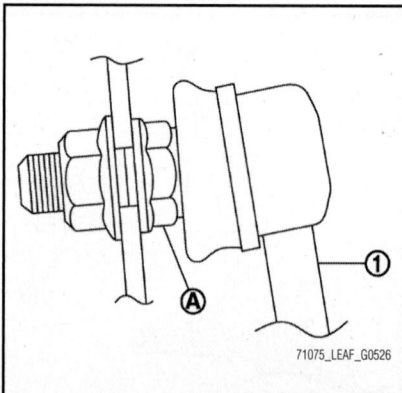

Fig. 88 Install stabilizer connecting rod (1); tighten the mounting nut with the hexagonal part (A) on the stabilizer connecting rod side fixed

➡Perform final tightening of fixing parts at the vehicle installation position (rubber bushing), under unladen conditions with tires on level ground.

23. Perform inspection after installation.

STRUTS

REMOVAL & INSTALLATION

See Figure 89.

1. Before servicing the vehicle, refer to the Precautions Section.
2. Remove tires with power tool.
3. Remove lock plate from strut assembly.
4. Remove wheel sensor.
5. Remove stabilizer connecting rod from strut assembly.
6. Remove strut mounting bolts and nuts from steering knuckle with power tool.
7. Remove cowl top cover.
8. Remove mounting bolt of mounting insulator, and then remove strut assembly.

To install:

Installation is the reverse of removal, noting the following:

9. Install strut assembly with the identification mark of mounting insulator facing the front of the vehicle and the arrow facing outside. The identification mark "0" shows the right mounting insulator and "1" shows left.

➡Do not reuse strut mounting nut. Perform final tightening of fixing parts at the vehicle installation position (rubber bushing), under unladen conditions with tires on level ground.

➡After replacing the strut, always follow the disposal procedure to discard the strut.

10. Perform inspection after installation.

OVERHAUL

See Figures 89 through 91.

✳✳ WARNING

Never damage strut assembly piston rod when removing components from strut assembly.

1. Before servicing the vehicle, refer to the Precautions Section.
2. Remove the cap.
3. Install strut attachment (SST: ST35652000) to strut assembly and secure it in a vise.

➡When installing the strut attachment to strut assembly, wrap a shop cloth around strut to protect from damage.

4. Using a spring compressor (commercial service tool), compress coil spring between spring upper seat and lower seat (strut assembly) until coil spring with a spring compressor is free.

✳✳ CAUTION

Be sure a spring compressor is securely attached to coil spring.

5. Compress coil spring.
6. Check coil spring with a spring compressor between spring upper seat and lower seat (strut assembly) is free then remove piston rod lock nut while securing the piston rod tip so that piston rod does not turn.
7. Remove mounting insulator, mounting bearing, and bound bumper from strut.
8. After removing coil spring with a spring compressor (commercial service tool), then gradually release spring compressor.

➡Loosen while making sure coil spring attachment position does not move.

9. Remove lower rubber seat.
10. Remove strut attachment (SST: ST35652000) from strut assembly.
11. Perform inspection after disassembly.

To install:

✳✳ WARNING

Never damage strut assembly piston rod when installing components from strut assembly.

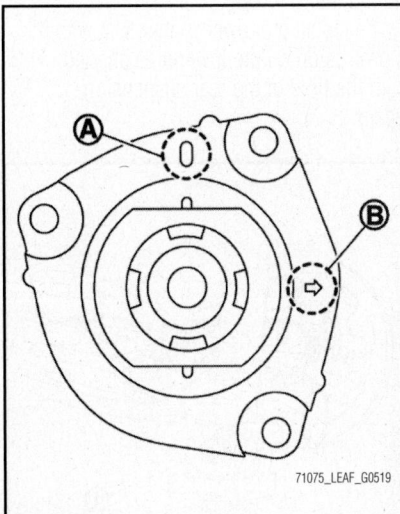

71075_LEAF_G0519

Fig. 89 Install strut assembly with the identification mark (A) of mounting insulator facing the front of the vehicle and the arrow (B) facing outside

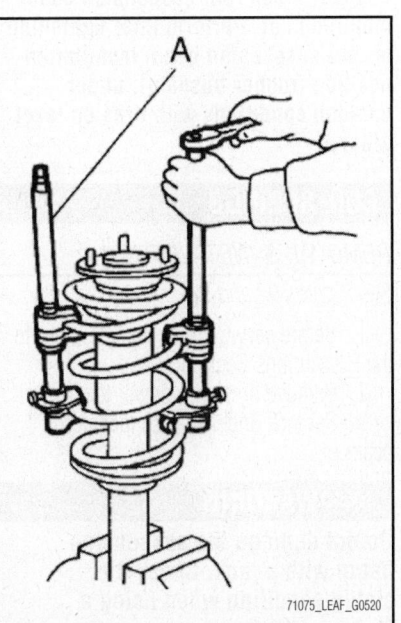

71075_LEAF_G0520

Fig. 90 Using a spring compressor (A) (commercial service tool), compress coil spring

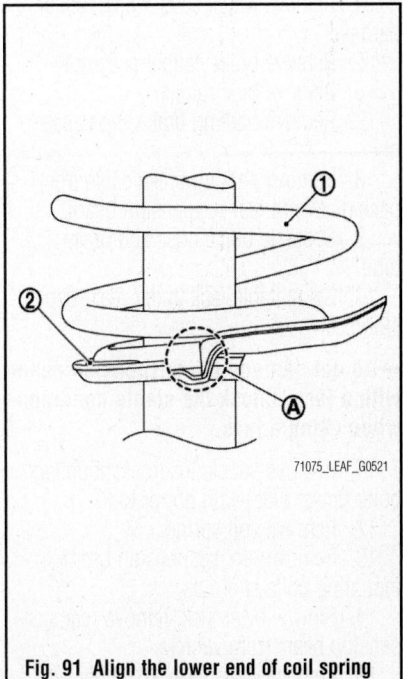

71075_LEAF_G0521

Fig. 91 Align the lower end of coil spring (1) with area "A" of lower rubber seat (2)

✳✳ WARNING

When installing the strut attachment to strut assembly, wrap a shop cloth around strut to protect from damage.

12. Install strut attachment (SST: ST35652000) to strut and secure it in a vise.

13. Install lower rubber seat.

14. Compress coil spring using a spring compressor (commercial service tool), and install it onto strut assembly.

✳✳ CAUTION

Be sure a compressor is securely attached to coil spring. Compress coil spring. Be careful with the vertical direction of the coil spring.

15. Correctly align the lower end of coil spring and the lower rubber seat.

16. Apply soapy water to bound bumper. Never use machine oil.

17. Insert bound bumper into mounting insulator.

18. Install mounting bearing. Do not apply oils, such as grease, when installing the mounting bearing.

19. Check the location of identification mark of the mounting insulator and install it with the arrow facing outside of the vehicle to the strut. The identification mark "0" shows right mounting insulator and "1" shows left.

20. Secure piston rod tip so that piston rod does not turn, then tighten piston rod lock nut with specified torque. Do not reuse piston rod lock nut.

21. Gradually release a spring compressor (commercial service tool), and remove coil spring.

22. Loosen while making sure coil spring attachment position does not move.

23. Remove the strut attachment (SST: ST35652000) from strut assembly.

24. Install the cap.

TRANSVERSE LINK

REMOVAL & INSTALLATION

1. Before servicing the vehicle, refer to the Precautions Section.

2. Remove tires with power tool.

3. Separate stabilizer connecting rod from strut assembly.

4. Separate steering outer socket from steering knuckle.

5. Remove transverse link from steering knuckle with power tool.

6. Remove transverse link from suspension member with power tool. To remove transverse link mounting nut, move stabilizer bar.

To install:

Installation is the reverse of removal. Perform inspection after installation.

➡**Do not reuse transverse link mounting nut. Perform final tightening of fixing parts at the vehicle installation position (rubber bushing), under unladen conditions with tires on level ground.**

SUSPENSION   REAR SUSPENSION

AXLE BEAM

REMOVAL & INSTALLATION

1. Before servicing the vehicle, refer to the Precautions Section.

2. Remove tires with power tool.

3. Drain brake fluid.

4. Remove wheel sensor and sensor harness.

5. Remove brake caliper assembly.

6. Remove disc rotor.

7. Remove parking brake shoe assembly.

8. Remove parking brake cable from back plate and rear suspension beam.

9. Separate brake hose and brake tube.

10. Set suitable jack under rear suspension beam.

➡**Do not damage the suspension beam with a jack. Check the stable condition when using a jack.**

11. Remove shock absorber mounting bolts (lower side) with power tool.

12. Remove coil spring.

13. Remove rear suspension beam mounting bolts and nuts.

14. Slowly lower jack, remove rear suspension beam from vehicle.

15. Remove wheel hub assembly with power tool.

16. Check rear suspension beam for deformation, cracks or damage. Replace the part if necessary.

To install:

Installation is the reverse of removal. Perform inspection after installation.

➡**Never reuse rear suspension beam mounting nut. Perform final tightening of rear suspension beam installation position (rubber bushing), under unladen conditions with tires on level ground.**

COIL SPRINGS

REMOVAL & INSTALLATION

See Figures 92 and 93.

1. Before servicing the vehicle, refer to the Precautions Section.

2. Remove tires with power tool.

3. Set jack under rear suspension beam.

✳✳ WARNING

Do not damage the suspension beam with a jack. Check the stable condition when using a jack.

4. Remove rear shock absorber mounting bolts (lower side).

5. Slowly lower jack, then remove upper rubber seat, coil spring and lower rubber seat from rear suspension beam.

6. Perform inspection after removal.

To install:

Installation is the reverse of removal, noting the following:

7. Install the lower rubber seat with its protrusion on the lower area aligned with the hole of the rear suspension beam.

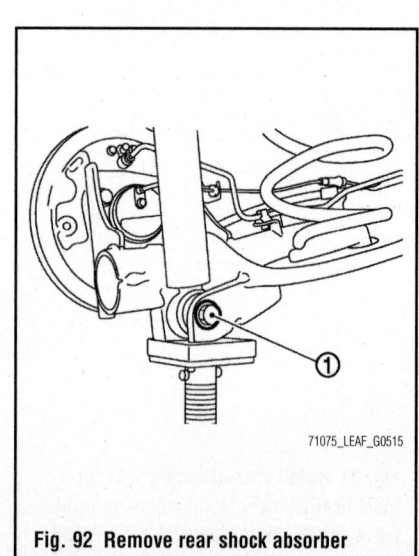

71075_LEAF_G0515

Fig. 92 Remove rear shock absorber mounting bolts (lower side) (1)

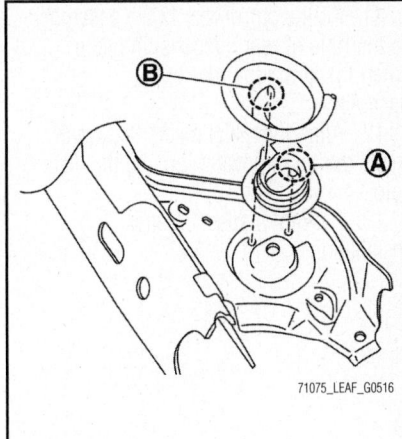

Fig. 93 Lower rubber seat protrusion (A), coil spring lower end (B)

8. Securely install coil spring with the lower end of the major diameter aligned with the steps of lower rubber seat.

9. Check rubber seat and coil spring for deformation, crack, and damage. Replace it if necessary.

SHOCK ABSORBERS

REMOVAL & INSTALLATION

See Figures 94 and 95.

1. Before servicing the vehicle, refer to the Precautions Section.

2. Remove tires with power tool.

3. Set suitable jack under rear suspension beam.

➡ **Do not damage the suspension beam with a jack. Check the stable condition when using a jack.**

4. Remove shock absorber mounting bolt (lower side) with power tool.

5. Remove shock absorber mask.

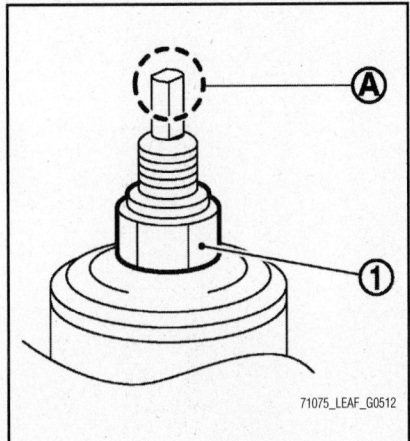

Fig. 94 Loosen piston rod lock nut (1), secure the tip (A) of the piston rod

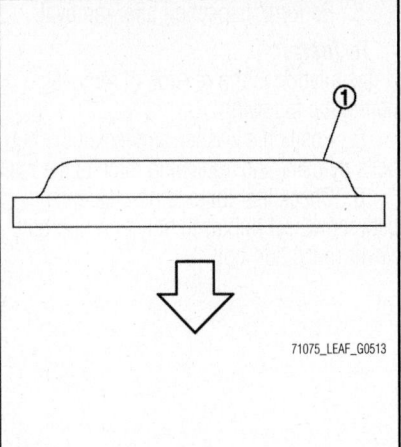

Fig. 95 Install washer (1) in the direction shown in the figure

6. Remove cap.

7. Remove piston rod lock nut, and remove the washer and bushing. To loosen piston rod lock nut, secure the tip of the piston rod.

8. Remove shock absorber assembly.

9. Remove bushing, distance tube, washer, bound bumper cover, and bound bumper from shock absorber.

10. Perform inspection after removal.

To install:

Installation is the reverse of removal, noting the following:

11. To install bushings, securely insert protrusion into the hole on the vehicle body side.

12. Install the washer toward the bushing.

➡ **Perform final tightening of bolts and nuts at the shock absorber lower side (rubber bushing), under unladen conditions with tires on level ground.**

13. Secure the tip of the shock absorber piston rod to keep it from rotating, and tighten the piston rod lock nut to the specified torque. Do not reuse piston rod lock nut.

WHEEL HUBS & BEARINGS

REMOVAL, REPACKING, & INSTALLATION

See Figure 96.

1. Before servicing the vehicle, refer to the Precautions Section.

2. Remove tires with power tool.

3. Remove the wheel sensor.

4. Remove the caliper assembly. Hang the caliper assembly in a place where it will not interfere with work.

❊❊ WARNING

Never depress brake pedal while brake caliper is removed.

5. Remove the disc rotor.

6. Remove the wheel hub assembly with power tool.

❊❊ WARNING

Never remove parking brake shoe assembly. Protect it from falling.

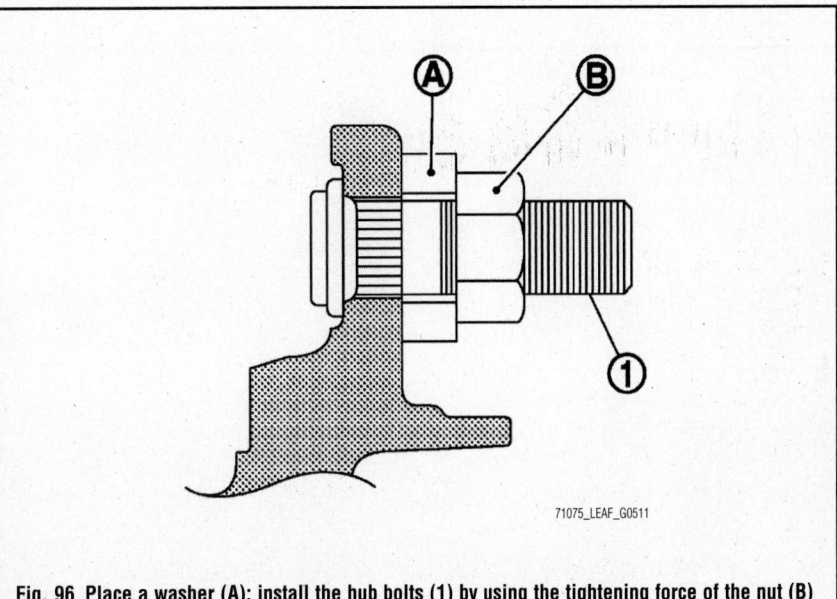

Fig. 96 Place a washer (A); install the hub bolts (1) by using the tightening force of the nut (B)

7. Remove hub bolts from wheel hub assembly, using the ball joint remover (commercial service tool).

    a. Remove hub bolt only when necessary.

    b. To avoid impact to the wheel bearing, do not hammer the hub bolt.

    c. Pull out the hub bolt in a direction perpendicular to the wheel hub assembly.

8. Perform inspection after removal.

To install:

Installation is the reverse of removal, noting the following:

9. Position a washer, and install the hub bolts by using the tightening force of the nut.

10. Check that there is no clearance between wheel hub assembly and hub bolt. Never reuse hub bolt.

11. Fit the pin of rear brake assembly to the hole of rear suspension beam when installing the rear brake assembly.

12. Align the matchmarks that were made during removal if reusing the disc rotor.

13. Perform inspection after installation.

NISSAN

Maxima

13

SPECIFICATIONS AND MAINTENANCE CHARTS

ENGINE AND VEHICLE IDENTIFICATION

Engine							Model Year	
Code ①	Liters (cc)	Cu. In.	Cyl.	Fuel Sys.	Engine Type	Eng. Mfg.	Code ②	Year
VQ35DE	3.5 (3,498)	214	6	SFI	DOHC	Nissan	B	2011
							C	2012

SFI: Sequential Fuel Injection

DOHC: Double Overhead Camshafts

① The engine code is stamped on the engine block near the starter

② 10th position of the Vehicle Identification Number (VIN)

71075_MAXI_C0001

GENERAL ENGINE SPECIFICATIONS

All measurements are given in inches.

Year	Model	Engine Displacement Liters (cc)	Engine ID/VIN	Fuel System Type	Net Horsepower @ rpm	Net Torque @ rpm (ft. lbs.)	Bore x Stroke (in.)	Compression Ratio	Oil Pressure @ rpm
2011	Maxima	3.5 (3,498)	VQ35DE	SFI	290@6,400	261@4,400	3.76 x 3.21	10.6:1	43 psi@2,000
2012	Maxima	3.5 (3,498)	VQ35DE	SFI	290@6,400	261@4,400	3.76 x 3.21	10.6:1	43 psi@2,000

SFI: Sequential Fuel Injection

71075_MAXI_C0002

ENGINE TUNE-UP SPECIFICATIONS

Year	Engine Displacement Liters	Engine ID/VIN	Spark Plug Gap (in.)	Ignition Timing (deg. BTDC) MT	AT ①	Fuel Pump (psi) ②	Idle Speed (rpm) MT	AT ①	Valve Clearance (in.) Intake ③	Exhaust ③
2011	3.5	VQ35DE	0.043	N/A	7-17	51	N/A	550-650	0.010-0.013	0.011-0.015
2012	3.5	VQ35DE	0.043	N/A	7-17	51	N/A	550-650	0.010-0.013	0.011-0.015

NOTE: The Vehicle Emission Control Information label often reflects specification changes made during production.

    The label figures must be used if they differ from those in this chart.

BTDC: Before Top Dead Center

N/A: Not Applicable

① Under no load condition (in P or N position)

② System pressure at idle

③ With engine cold

71075_MAXI_C0003

CAPACITIES

Year	Model	Engine Displacement Liters	Engine ID/VIN	Engine Oil with Filter (qts.)	Transaxle (pts.) Auto. ①	Transaxle (pts.) Manual	Drive Axle (pts.) Front	Drive Axle (pts.) Rear	Transfer Case (pts.)	Fuel Tank (gal.)	Cooling System (qts.)
2011	Maxima	3.5	VQ35DE	5.1	21.5	N/A	N/A	N/A	N/A	20.0	9.5
2012	Maxima	3.5	VQ35DE	5.1	21.5	N/A	N/A	N/A	N/A	20.0	9.5

NOTE: All capacities are approximate. Add fluid gradually and ensure a proper fluid level is obtained.

N/A: Not Applicable

① Drain and refill

71075_MAXI_C0004

FLUID SPECIFICATIONS

Year	Model	Engine Disp. Liters	Engine Oil	Manual Trans.	Auto. Trans.	Drive Axle Front	Drive Axle Rear	Transfer Case	Power Steering Fluid	Brake Master Cylinder	Cooling System
2011	Maxima	3.5	①	N/A	②	N/A	N/A	N/A	③	DOT 3	④
2012	Maxima	3.5	①	N/A	②	N/A	N/A	N/A	③	DOT 3	④

N/A: Not Applicable

DOT: Department Of Transportation

① Nissan recommends Genuine Nissan Ester Oil: 5W-30

② Use only Genuine Nissan CVT Fluid NS-2

③ Genuine Nissan PSF or equivalent (DEXRON™ VI type ATF may also be used)

④ Pre-diluted Genuine Nissan Long Life Antifreeze/Coolant (blue), or equivalent

71075_MAXI_C0005

VALVE SPECIFICATIONS

Year	Engine Displacement Liters	Engine ID/VIN	Seat Angle (deg.)	Face Angle (deg.)	Spring Test Pressure (lbs. @ in.)	Spring Free-Length (in.)	Spring Installed Height (in.)	Stem-to-Guide Clearance (in.) Intake	Stem-to-Guide Clearance (in.) Exhaust	Stem Diameter (in.) Intake	Stem Diameter (in.) Exhaust
2011	3.5	VQ35DE	①	①	84-95@ 1.0709	1.8531	1.4567	0.0008- 0.0021	0.0016- 0.0021	0.2348- 0.2354	0.2344- 0.2350
2012	3.5	VQ35DE	①	①	84-95@ 1.0709	1.8531	1.4567	0.0008- 0.0021	0.0016- 0.0021	0.2348- 0.2354	0.2344- 0.2350

① 45° 15' - 45° 45' (degrees/minutes)

71075_MAXI_C0006

CAMSHAFT SPECIFICATIONS
All measurements in inches unless noted

Year	Engine Displacement Liters	Engine Code/VIN	Journal Diameter	Brg. Oil Clearance	Shaft End-play	Runout	Journal Bore	Cam Height Intake	Cam Height Exhaust
2011	3.5	VQ35DE	①	②	0.0045-0.0074	0.0008	③	1.7904-1.7978	1.7907-1.7982
2012	3.5	VQ35DE	①	②	0.0045-0.0074	0.0008	③	1.7904-1.7978	1.7907-1.7982

NS: Not Specified

① Number 1: 1.0211 - 1.0218 inches
   Numbers 2, 3, 4: 0.9230 - 0.9238 inches

② Number 1: 0.0018 - 0.0034 inches
   Numbers 2, 3, 4: 0.0014 - 0.0030 inches

③ Camshaft bracket inner diameter
   Number 1: 1.0236 - 1.0244 inches
   Numbers 2, 3, 4: 0.9252 - 0.9260 inch

71075_MAXI_C0007

CRANKSHAFT AND CONNECTING ROD SPECIFICATIONS
All measurements are given in inches.

Year	Engine Displacement Liters	Engine ID/VIN	Crankshaft Main Brg. Journal Dia.	Crankshaft Main Brg. Oil Clearance	Crankshaft Shaft End-play	Crankshaft Thrust on No.	Connecting Rod Journal Diameter	Connecting Rod Oil Clearance	Connecting Rod Side Clearance
2011	3.5	VQ35DE	2.3603-2.3612 ①	0.0014-0.0018	0.0039-0.0098	3	2.1654-2.1659	0.0008-0.0018	0.0079-0.0138
2012	3.5	VQ35DE	2.3603-2.3612 ①	0.0014-0.0018	0.0039-0.0098	3	2.1654-2.1659	0.0008-0.0018	0.0079-0.0138

① Variance depending on Grade Number

71075_MAXI_C0008

PISTON AND RING SPECIFICATIONS
All measurements are given in inches.

Year	Engine Displacement Liters	Engine ID/VIN	Piston Clearance	Ring Gap Top Compression	Ring Gap Bottom Compression	Ring Gap Oil Control	Ring Side Clearance Top Compression	Ring Side Clearance Bottom Compression	Ring Side Clearance Oil Control
2011	3.5	VQ35DE	0.0004-0.0012	0.0091-0.0110	0.0130-0.0169	0.0079-0.0177	0.0018-0.0031	0.0012-0.0028	0.0018-0.0049
2012	3.5	VQ35DE	0.0004-0.0012	0.0091-0.0110	0.0130-0.0169	0.0079-0.0177	0.0018-0.0031	0.0012-0.0028	0.0018-0.0049

71075_MAXI_C0009

TORQUE SPECIFICATIONS
All readings in ft. lbs.

Year	Engine Disp. Liters	Engine ID/VIN	Cylinder Head Bolts	Main Bearing Bolts	Rod Bearing Bolts	Crankshaft Damper Bolts	Flywheel Bolts	Manifold Intake	Manifold Exhaust	Spark Plugs	Oil Pan Drain Plug
2011	3.5	VQ35DE	①	②	③	④	65	⑤	23	15	25
2012	3.5	VQ35DE	①	②	③	④	65	⑤	23	15	25

① Apply engine oil to bolts, refer to procedure for tightening sequence

  Step 1: Tighten to 72 ft. lbs.

  Step 2: Loosen bolts completely

  Step 3: Tighten to 29 ft. lbs.

  Step 4: Plus 103 degrees

  Step 5: Plus an additional 103 degrees

② Apply engine oil to bolts, refer to procedure for tightening sequence

  Step 1: Tighten to 24-28 ft. lbs.

  Step 2: Tighten an additional 90-95 degrees

③ Apply engine oil to bolt threads and seats of cap bolts

  Step 1: Tighten to 14-15 ft. lbs.

  Step 2: Plus 90-95 degrees

④ Apply engine oil to bolt threads

  Step 1: Tighten to 32 ft. lbs.

  Step 2: Plus 84-90 degrees

⑤ Step 1: Tighten to 66 inch lbs.

  Step 1: Tighten to 19 ft. lbs.

71075_MAXI_C0010

WHEEL ALIGNMENT

Year	Model		Caster Range (+/-Deg.)	Caster Preferred Setting (Deg.)	Camber Range (+/-Deg.)	Camber Preferred Setting (Deg.)	Toe-in (in.)
2011	Maxima	F	0.75	①	0.55	0.25	0.03 +/- 0.03
		R	N/A	N/A	0.50	②	③
2012	Maxima	F	0.75	①	0.55	0.25	0.03 +/- 0.03
		R	N/A	N/A	0.50	②	③

NOTE: Measurements given for an unladen vehicle with fuel, coolant, and engine oil full; spare tire, jack, hand tools, and mats in designated positions.

F: Front   R: Rear

N/A: Not Applicable

① With P245/45R18 tires: 4.95 degrees

  With P245/40R19 tires: 5.00 degrees

② With P245/45R18 tires: -0.42 degrees

  With P245/40R19 tires: -0.53 degrees

③ With P245/45R18 tires: 0.07 degrees +/- 0.07 degrees

  With P245/40R19 tires: 0.08 degrees +/- 0.07 degrees

71075_MAXI_C0011

TIRE, WHEEL AND BALL JOINT SPECIFICATIONS

| Year | Model | OEM Tires | | Tire Pressures (psi) | | Wheel Size | | Ball Joint Inspection | Lug Nut (ft. lbs.) |
		Standard	Optional	Front	Rear	Standard	Optional		
2011	Maxima	P245/45VR18	P245/45VR19	①	①	18 x 8J	19 x 8J	②	83
			P245/45WR19						
2012	Maxima	P245/45VR18	P245/45VR19	①	①	18 x 8J	19 x 8J	②	83
			P245/45WR19						

OEM: Original Equipment Manufacturer

PSI: Pounds Per Square Inch

① Always refer to the owner's manual and/or vehicle label: conventional tires should be inflated to 33 psi

② Measurement on spring balance (cotter pin hole position): 1.8–12.2 lbs.

Axial endplay: 0.004 inch or less

71075_MAXI_C0012

BRAKE SPECIFICATIONS

All measurements in inches unless noted

| Year | Model | | Brake Disc | | | Brake Drum Diameter | | | Pad/Lining Thickness | | Brake Caliper | |
			Original Thickness	Minimum Thickness	Max. Runout	Original Inside Diameter	Max. Wear Limit	Maximum Machine Diameter	Standard	Limit	Torque Member Bolts (ft. lbs.)	Guide Pin Bolts (ft. lbs.)
2011	Maxima	F	1.102	1.024	0.001	N/A	N/A	N/A	0.433	0.079	107	25
		R	0.630	0.551	0.002	N/A	N/A	N/A	0.335	0.039	62	32
2012	Maxima	F	1.102	1.024	0.001	N/A	N/A	N/A	0.433	0.079	107	25
		R	0.630	0.551	0.002	N/A	N/A	N/A	0.335	0.039	62	32

F: Front

R: Rear

71075_MAXI_C0013

SCHEDULED MAINTENANCE INTERVALS
Nissan—Maxima

TO BE SERVICED	TYPE OF	7.5	15	22.5	30	37.5	45	52.5	60
Engine oil & filter	R	✓	✓	✓	✓	✓	✓	✓	✓
Brake lines & cables	S/I		✓		✓		✓		✓
Brake pads, discs	I		✓		✓		✓		✓
Driveshaft boots & propeller shaft	L/I		✓		✓		✓		✓
CVT ①	I		✓		✓		✓		✓
Transfer case and differential fluid ②	I		✓		✓		✓		✓
Air cleaner filter	R				✓				✓
Drive belt (s) ③	S/I								✓
Engine coolant ④	R								✓
Spark plugs	R	Platinum plugs, every 105,000 miles							
Cabin air filter	R		✓		✓		✓		✓
Exhaust system	I				✓				✓
Evap vapor lines	I				✓				✓
Fuel lines	S/I				✓				✓
Steering gear, linkage, axle & suspension parts	I	✓	✓	✓	✓	✓	✓	✓	✓
Tires (rotate)	S/I	every 5,000-6,000 miles							
Valve clearance ⑤	S/I								

R: Replace   S/I: Service or Inspect   L: Lubricate

① If towing a trailer, using a camper or a car-top carrier, or driving on rough or muddy roads, change (not just inspect) oil at every 60,000 miles.

② If towing a trailer, using a camper or a car-top carrier, or driving on rough or muddy roads, change (not just inspect) oil at every 30,000 miles (48,000 km) or 24 months.

③ First at 60,000, then every 15,000 miles

④ After 60,000, replace every 30,000

⑤ Periodic maintenance not required, if valve noice increases, inspect valve clearance

Follow Periodic Maintenance Schedule 1 if the driving habits frequently include one or more of the following driving conditions:

Repeated short trips of less than 5 miles (8 km).

Repeated short trips of less than 10 miles (16 km) with outside temperatures remaining below freezing

Operating in hot weather in stop-and-go "rush hour" traffic.

Extensive idling and/or low speed driving for long distances, such as police, taxi or door-to-door delivery use

Driving in dusty conditions.

Driving on rough, muddy, or salt spread roads.

Towing a trailer, using a camper or a car-top carrier.

Follow Periodic Maintenance Schedule 2 if none of driving conditions shown in Schedule 1 apply to the driving habits.

PRECAUTIONS

Before servicing any vehicle, please be sure to read all of the following precautions, which deal with personal safety, prevention of component damage, and important points to take into consideration when servicing a motor vehicle:

- Never open, service or drain the radiator or cooling system when the engine is hot; serious burns can occur from the steam and hot coolant.

- Observe all applicable safety precautions when working around fuel. Whenever servicing the fuel system, always work in a well-ventilated area. Do not allow fuel spray or vapors to come in contact with a spark, open flame, or excessive heat (a hot drop light, for example). Keep a dry chemical fire extinguisher near the work area. Always keep fuel in a container specifically designed for fuel storage; also, always properly seal fuel containers to avoid the possibility of fire or explosion. Refer to the additional fuel system precautions later in this section.

- Fuel injection systems often remain pressurized, even after the engine has been turned **OFF**. The fuel system pressure must be relieved before disconnecting any fuel lines. Failure to do so may result in fire and/or personal injury.

- Brake fluid often contains polyglycol ethers and polyglycols. Avoid contact with the eyes and wash your hands thoroughly after handling brake fluid. If you do get brake fluid in your eyes, flush your eyes with clean, running water for 15 minutes. If eye irritation persists, or if you have taken

brake fluid internally, IMMEDIATELY seek medical assistance.

- The EPA warns that prolonged contact with used engine oil may cause a number of skin disorders, including cancer. You should make every effort to minimize your exposure to used engine oil. Protective gloves should be worn when changing oil. Wash your hands and any other exposed skin areas as soon as possible after exposure to used engine oil. Soap and water, or waterless hand cleaner should be used.

- All new vehicles are now equipped with an air bag system, often referred to as a Supplemental Restraint System (SRS) or Supplemental Inflatable Restraint (SIR) system. The system must be disabled before performing service on or around system components, steering column, instrument panel components, wiring and sensors. Failure to follow safety and disabling procedures could result in accidental air bag deployment, possible personal injury and unnecessary system repairs.

- Always wear safety goggles when working with, or around, the air bag system. When carrying a non-deployed air bag, be sure the bag and trim cover are pointed away from your body. When placing a non-deployed air bag on a work surface, always face the bag and trim cover upward, away from the surface. This will reduce the motion of the module if it is accidentally deployed. Refer to the additional air bag system precautions later in this section.

- Clean, high quality brake fluid from a sealed container is essential to the safe and

proper operation of the brake system. You should always buy the correct type of brake fluid for your vehicle. If the brake fluid becomes contaminated, completely flush the system with new fluid. Never reuse any brake fluid. Any brake fluid that is removed from the system should be discarded. Also, do not allow any brake fluid to come in contact with a painted surface; it will damage the paint.

- Never operate the engine without the proper amount and type of engine oil; doing so WILL result in severe engine damage.

- Timing belt maintenance is extremely important. Many models utilize an interference-type, non-freewheeling engine. If the timing belt breaks, the valves in the cylinder head may strike the pistons, causing potentially serious (also time-consuming and expensive) engine damage. Refer to the maintenance interval charts for the recommended replacement interval for the timing belt, and to the timing belt section for belt replacement and inspection.

- Disconnecting the negative battery cable on some vehicles may interfere with the functions of the on-board computer system(s) and may require the computer to undergo a relearning process once the negative battery cable is reconnected.

- When servicing drum brakes, only disassemble and assemble one side at a time, leaving the remaining side intact for reference.

- Only an MVAC-trained, EPA-certified automotive technician should service the air conditioning system or its components.

BRAKES

GENERAL INFORMATION

PRECAUTIONS

- Certain components within the ABS system are not intended to be serviced or repaired individually.

- Do not use rubber hoses or other parts not specifically specified for and ABS system. When using repair kits, replace all parts included in the kit. Partial or incorrect repair may lead to functional problems and require the replacement of components.

- Lubricate rubber parts with clean, fresh brake fluid to ease assembly. Do not use shop air to clean parts; damage to rubber components may result.

- Use only DOT 3 brake fluid from an unopened container.

- If any hydraulic component or line is removed or replaced, it may be necessary to bleed the entire system.

- A clean repair area is essential. Always clean the reservoir and cap thoroughly before removing the cap. The slightest amount of dirt in the fluid may plug an orifice and impair the system function. Perform repairs after components have been thoroughly cleaned; use only denatured alcohol to clean components. Do not allow ABS components to come into contact with any substance containing mineral oil; this includes used shop rags.

- The Anti-Lock control unit is a microprocessor similar to other computer units in the vehicle. Ensure that the ignition switch is **OFF** before removing or installing

ANTI-LOCK BRAKE SYSTEM (ABS)

controller harnesses. Avoid static electricity discharge at or near the controller.

- If any arc welding is to be done on the vehicle, the control unit should be unplugged before welding operations begin.

SPEED SENSORS

REMOVAL & INSTALLATION

Consider the following warnings:

- Be careful not to damage the wheel sensor edge and sensor rotor teeth

- When pulling out the wheel sensor, be careful to turn it as little as possible. Do not pull on the wheel sensor harness

- Check if any foreign objects, such as iron fragments, are adhered to the pick-up part of the sensor or to the inside of the

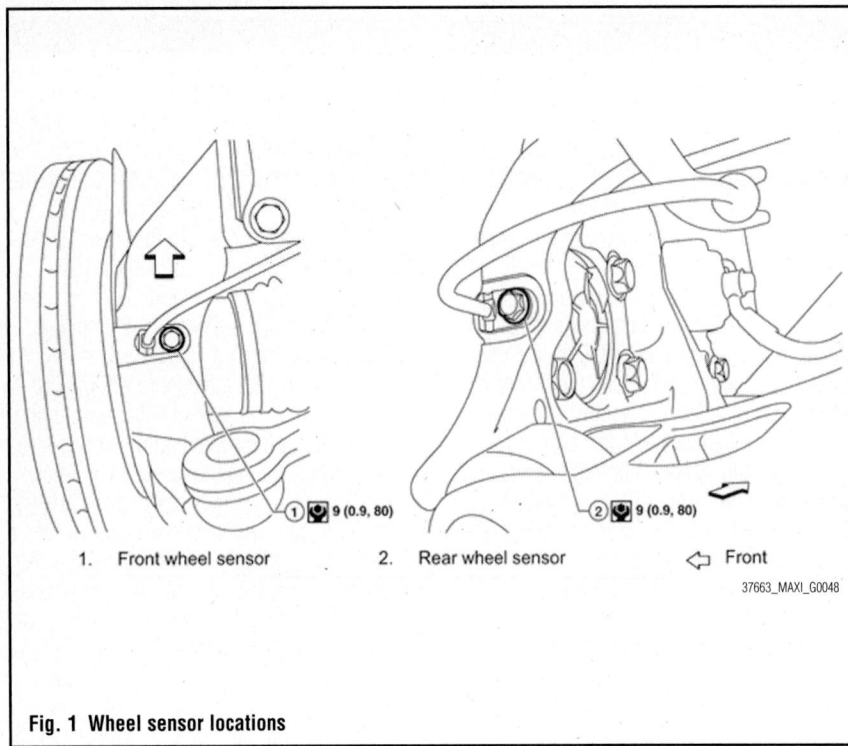

1. Front wheel sensor   2. Rear wheel sensor   ⇦ Front

37663_MAXI_G0048

Fig. 1 Wheel sensor locations

hole for the wheel sensor, or if a foreign object is caught in the surface of the mating surface for the wheel sensor. Repair as necessary and then install the wheel sensor.

Front

See Figure 1.

1. Before servicing the vehicle, refer to the Precautions.

2. Remove the front wheel and tire.

3. Partially remove the front wheel fender protector and reposition out of the way.

4. Disconnect the wheel sensor harness connector.

5. Remove the wheel sensor harness from the brackets.

6. Remove the wheel sensor bolt and wheel sensor from the front hub assembly.

7. Installation is in the reverse order of removal.

Rear

1. Before servicing the vehicle, refer to the Precautions.

2. Remove the rear wheel and tire.

3. Remove the stabilizer bar clamps and bushings using a power tool, and reposition the stabilizer bar out of the way.

4. Disconnect the wheel sensor harness connector.

5. Remove the wheel sensor harness from the brackets.

6. Remove the wheel sensor bolt and wheel sensor from the rear hub assembly.

7. Installation is in the reverse order of removal.

BRAKES

BLEEDING PROCEDURE

BLEEDING PROCEDURE

➡**Carefully monitor brake fluid level at master cylinder during bleeding operation. Fill reservoir with new brake fluid. Make sure it is full at all times while bleeding air out of system. Place a container under master cylinder to avoid spillage of brake fluid. Do not loosen the connecting portion of the actuator during air bleeding.**

1. Before servicing the vehicle, refer to the Precautions.

2. Disconnect the battery negative terminal.

3. Connect a transparent vinyl tube and container to the air bleeder valve.

4. Fully depress the brake pedal several times.

5. With the brake pedal depressed, open the air bleeder valve to release the air.

6. Close the air bleeder valve.

7. Release the brake pedal slowly.

8. Tighten the air bleeder valve to 69 inch lbs. (8 Nm).

9. Repeat steps 4 through 7 until no more air bubbles come out of the air bleeder valve.

10. Bleed the brake hydraulic system air bleeder valves in the following order:
- Right rear brake
- Left front brake
- Left rear brake
- Right front brake

MASTER CYLINDER BLEEDING

Consider the following warnings:
- Refill with new brake fluid
- Do not reuse drained brake fluid
- Do not let brake fluid splash on the painted surfaces of the body. This might damage the paint. If brake fluid is splashed on painted areas, wash it away with water immediately
- Before working, disconnect the ABS actuator and electric unit (control unit) connector or the battery negative terminal.

1. Before servicing the vehicle, refer to the Precautions.

2. Turn the ignition switch OFF and disconnect the ABS actuator and electric unit

BLEEDING THE BRAKE SYSTEM

(control unit) connector or the battery negative terminal.

3. Connect a vinyl tube to the bleed valve.

4. Depress the brake pedal, loosen the bleed valve, and gradually remove the brake fluid.

5. Make sure there is no foreign material in the reservoir tank, and refill with new brake fluid.

6. Rest a foot on the brake pedal. Loosen the bleed valve. Slowly depress the brake pedal until it stops. Tighten the bleed valve. Release the brake pedal. Repeat the process a few times, then pause to add new brake fluid to the master cylinder. Continue until the new brake fluid flows out of the bleed valve.

7. Bleed the air out of the brake hydraulic system.

BLEEDING THE ABS SYSTEM

➡**Carefully monitor the brake fluid level at the master cylinder during the bleeding operation. Fill the reservoir with new brake fluid. Make sure it is full at all times while bleeding the air** out of the system. Place a container under the master cylinder to avoid spillage of the brake fluid. Do not loosen the connecting portion of the actuator during air bleeding.

1. Before servicing the vehicle, refer to the Precautions.

2. Disconnect the battery negative terminal.

3. Connect a transparent vinyl tube and container to the air bleeder valve.

4. Fully depress the brake pedal several times.

5. With the brake pedal depressed, open the air bleeder valve to release the air.

6. Close the air bleeder valve.

7. Release the brake pedal slowly.

8. Tighten the air bleeder valve to 69 inch lbs. (8 Nm).

9. Repeat steps 4 through 7 until no more air bubbles come out of the air bleeder valve.

10. Bleed the brake hydraulic system air bleeder valves in the following order:
- Right rear brake
- Left front brake
- Left rear brake
- Right front brake

FLUID FILL PROCEDURE

❋❋ CAUTION

Brake fluid contains polyglycol ethers and polyglycols. Avoid contact with the eyes and wash your hands thoroughly after handling brake fluid. If you do get brake fluid in your eyes, flush your eyes with clean, running water for 15 minutes. If eye irritation persists, or if you have taken brake fluid internally, IMMEDIATELY seek medical assistance.

❋❋ WARNING

Clean, high quality brake fluid is essential to the safe and proper operation of the brake system. You should always buy the highest quality brake fluid that is available. If the brake fluid becomes contaminated, drain and flush the system, then refill the master cylinder with new fluid. Never reuse any brake fluid. Any brake fluid that is removed from the system should be discarded. Also, do not allow any brake fluid to come in contact with a painted surface; it will damage the paint.

❋❋ WARNING

Do not use shock absorber fluid or any other fluid which contains mineral oil. Do not use a container which has been used for mineral oil or a container which is wet from water. Mineral oil will cause swelling and distortion of rubber parts in the hydraulic brake system and water mixed into brake fluid will lower fluid the boiling point. Keep all fluid containers capped to prevent contamination.

❋❋ WARNING

Be sure to use the proper brake fluid as indicated on the reservoir cap of the vehicle or as recommended in the owner's manual of the vehicle. Use of any other fluid is strictly prohibited.

1. Before servicing the vehicle, refer to the Precautions.

2. Fill the fluid level so that it is between the MIN and MAX lines marked on the reservoir.

➡ **If the brake warning light lights sometimes during driving, replenish the fluid to the MAX level.**

3. If the fluid decreases quickly, inspect the brake system for leakage. Correct leaky points and then refill to the specified level.
- Check the master cylinder, reservoir and reservoir hose (if equipped) for cracks, damage, and brake fluid leakage. If any faulty condition exists, correct or replace needed.

4. Check that the brake fluid level is between the MAX and MIN marks on the reservoir.

BRAKES

❋❋ CAUTION

Dust and dirt accumulating on brake parts during normal use may contain asbestos fibers from production or aftermarket brake linings. Breathing excessive concentrations of asbestos fibers can cause serious bodily harm. Exercise care when servicing brake parts. Do not sand or grind brake lining unless equipment used is designed to contain the dust residue. Do not clean brake parts with compressed air or by dry brushing. Cleaning should be done by dampening the brake components with a fine mist of water, then wiping the brake components clean with a dampened cloth. Dispose of cloth and all residue containing asbestos fibers in an impermeable container with the appropriate label. Fol low practices prescribed by the Occupational Safety and Health Administration (OSHA) and the Environmental Protection Agency (EPA) for the handling, processing, and disposing of dust or debris that may contain asbestos fibers.

BRAKE CALIPER

REMOVAL & INSTALLATION
See Figure 2.

❋❋ CAUTION

Clean the dust on the caliper and brake pad with a vacuum dust collector to minimize the hazard of air borne particles or other materials.

❋❋ WARNING

When removing and installing the cylinder body, do not depress the

FRONT DISC BRAKES

brake pedal because the piston will pop out. Do not damage the piston boot. Keep the brake rotor free from grease and brake fluid. Refill the brake reservoir with new brake fluid only. Never reuse the drained brake fluid.

1. Before servicing the vehicle, refer to the Precautions.

2. Remove the front wheel and tire.

3. Secure the disc rotor using a wheel nut.

4. Drain the brake fluid.

5. Remove the union bolt and then disconnect the brake hose from the caliper assembly. Discard the copper washers.

➡ **Do not reuse the copper washers.**

6. Remove the torque member bolts, and remove the brake caliper assembly.

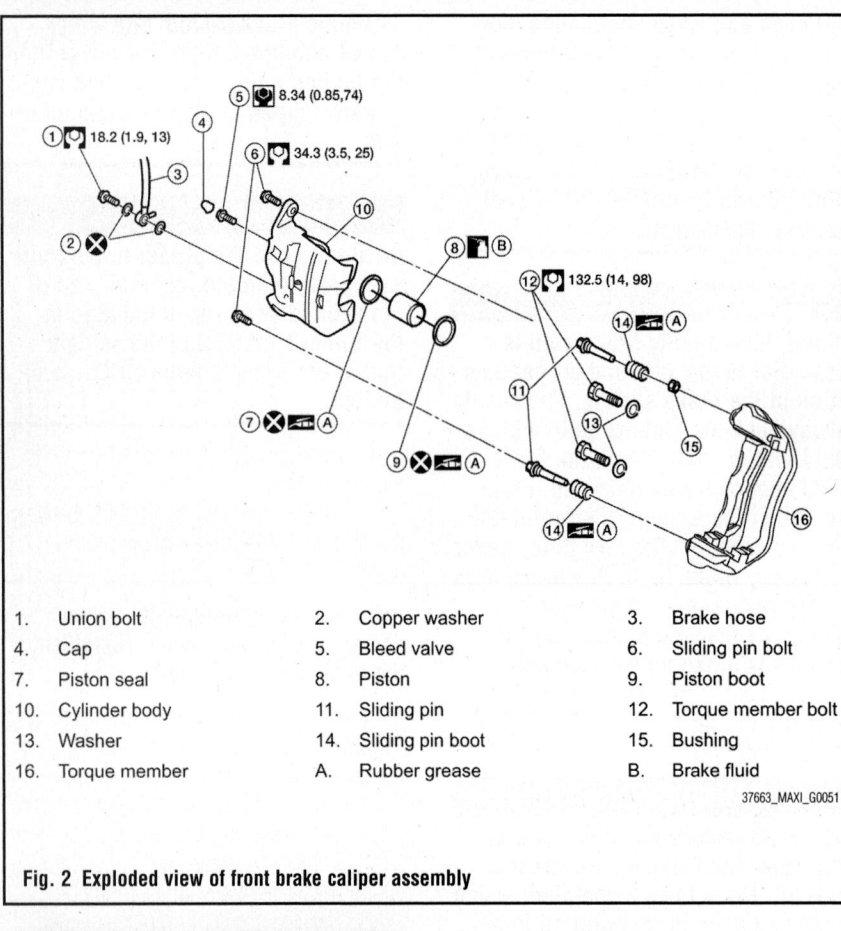

1. Union bolt
2. Copper washer
3. Brake hose
4. Cap
5. Bleed valve
6. Sliding pin bolt
7. Piston seal
8. Piston
9. Piston boot
10. Cylinder body
11. Sliding pin
12. Torque member bolt
13. Washer
14. Sliding pin boot
15. Bushing
16. Torque member
A. Rubber grease
B. Brake fluid

37663_MAXI_G0051

Fig. 2 Exploded view of front brake caliper assembly

✳✳ WARNING

Do not drop the brake pads.

7. Remove the disc rotor. If reusing the disc rotor, apply a matching mark for installation.

To install:

8. Install the disc rotor. If reusing the disc rotor, align the matching mark on the disc rotor and wheel hub assembly for installation.

➡**Align the matching marks on wheel hub assembly and disc rotor, if reusing the disc rotor.**

9. Install the brake caliper assembly, and tighten the torque member bolts to the specified torque.

➡**Do not allow oil or any moisture on the contact surfaces between the steering knuckle and caliper assembly, bolts, and washer.**

10. Install the brake hose with two new copper washers, using the L-shaped pin for alignment, and then tighten the union bolt.

➡**Do not reuse the copper washers.**

11. Refill the brake hydraulic system with new brake fluid and bleed out the air.

12. Check the front disc brakes for drag.
13. Install the front wheel and tire.

BRAKE PADS

REMOVAL & INSTALLATION

See Figure 3.

✳✳ CAUTION

Clean the dust on the caliper and brake pad with a vacuum dust collector to minimize the hazard of air borne particles or other materials.

✳✳ WARNING

While removing the brake pads, do not depress the brake pedal because the piston may pop out. It is not necessary to remove the bolts on the torque member and the brake hose except for disassembly or replacement of the caliper assembly. For brake pad removal, hang the cylinder body with a wire so as not to stretch the brake hose. Do not damage the piston boot. If any shim is subject to serious corrosion, replace it with a new one. Always replace the shim and shim cover as a set when replacing the brake pads. Keep the rotor

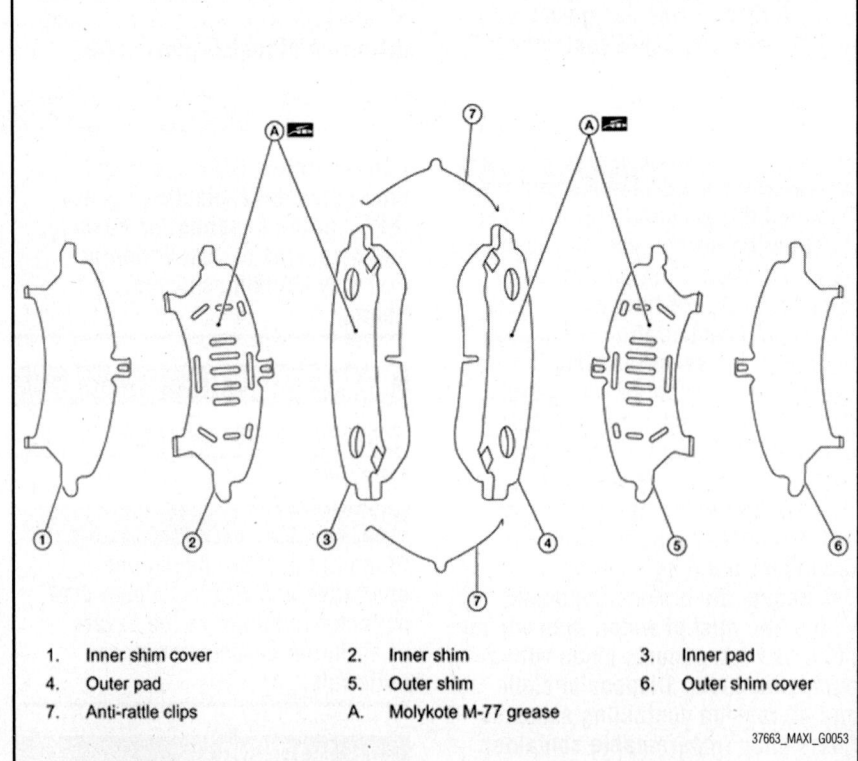

1. Inner shim cover
2. Inner shim
3. Inner pad
4. Outer pad
5. Outer shim
6. Outer shim cover
7. Anti-rattle clips
A. Molykote M-77 grease

37663_MAXI_G0053

Fig. 3 Exploded view of brake pads

free from brake fluid. Burnish the brake pads and disc rotor contacting surfaces, after refinishing or replacing rotors, after replacing pads, or if a soft pedal occurs at very low mileage.

1. Before servicing the vehicle, refer to the Precautions.
2. Remove the front wheel and tire.
3. Remove the upper and lower sliding pin bolts without disconnecting the union bolt.
4. Hang the cylinder body with a wire, and do not twist or stretch the brake hose.
5. Remove the anti-rattle clips, pads, shims, and shim covers from the torque member.

To install:

6. Apply Molykote® M-77 grease, or equivalent, between the inner and outer shims and the back of the brake pads.
7. Install the inner and outer shims and shim covers to the inner pad and outer pad.

✵✵ CAUTION

Do not get grease on the inner and outer pad friction surfaces.

8. Install the assembled inner and outer shims, shim covers, pads and anti-rattle clips to the torque member.

✵✵ CAUTION

Do not get grease on the inner and outer pad or rotor friction surfaces.

9. Press the piston into the cylinder body using a suitable tool, then install the cylinder body on the torque member.

➡**When replacing a pad with a new one, check the brake fluid level in the reservoir tank because the brake fluid returns to the master cylinder reservoir tank when pressing in the piston.**

10. Install the upper and lower sliding pin bolts and tighten it to the specified torque.
11. Check the front disc brakes for drag.
12. Install the front wheel and tire.

BRAKES

✵✵ CAUTION

Dust and dirt accumulating on brake parts during normal use may contain asbestos fibers from production or aftermarket brake linings. Breathing excessive concentrations of asbestos fibers can cause serious bodily harm. Exercise care when servicing brake parts. Do not sand or grind brake lining unless equipment used is designed to contain the dust residue. Do not clean brake parts with compressed air or by dry brushing. Cleaning should be done by dampening the brake components with a fine mist of water, then wiping the brake components clean with a dampened cloth. Dispose of cloth and all residue containing asbestos fibers in an impermeable container with the appropriate label. Follow practices prescribed by the Occupational Safety and Health Administration (OSHA) and the Environmental Protection Agency (EPA) for the handling, processing, and disposing of dust or debris that may contain asbestos fibers.

BRAKE CALIPER

REMOVAL & INSTALLATION

See Figure 4.

✵✵ CAUTION

Clean the dust on the caliper and brake pad with a vacuum dust collector to minimize the hazard of

air borne particles or other materials.

✵✵ WARNING

While removing and installing the cylinder body, do not depress the brake pedal because the piston may

REAR DISC BRAKES

pop out. Do not damage the piston boot. Keep the rotor free from grease and brake fluid. Refill the brake reservoir with new brake fluid. Never reuse drained brake fluid.

1. Before servicing the vehicle, refer to the Precautions.

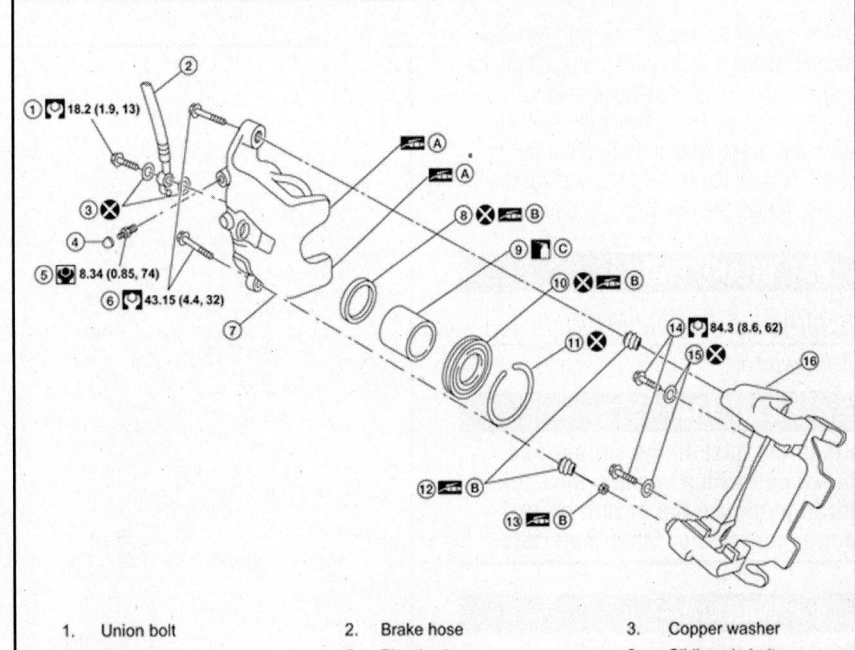

1.	Union bolt	2.	Brake hose	3.	Copper washer
4.	Cap	5.	Bleed valve	6.	Sliding pin bolt
7.	Cylinder body	8.	Piston seal	9.	Piston
10.	Piston boot	11.	Retaining ring	12.	Sliding pin boot
13.	Bushing	14.	Torque member bolt	15.	Washer
16.	Torque member	A.	Molykote M-77 grease	B.	Rubber grease
C.	Brake fluid				

37663_MAXI_G0055

Fig. 4 Exploded view of rear brake caliper assembly

2. Remove the rear wheel and tire.

3. Hold the disc rotor in place by installing a wheel nut.

4. Drain the brake fluid.

5. Remove the union bolt and copper washers, discard the copper washers.

6. Disconnect the brake hose from the cylinder body.

➡**Do not reuse the copper washers.**

7. Remove the torque member bolts, and remove the brake caliper assembly.

✳ WARNING

Do not drop brake the pads.

8. Remove the disc rotor. If reusing the disc rotor, before removing the disc rotor apply a matching mark.

To install:

9. Install the disc rotor. If reusing the disc rotor, align the matching mark to position the disc rotor on the wheel hub assembly.

10. Install the brake caliper assembly, and tighten the torque member bolts.

➡**Before installing the caliper assembly, wipe off oil and moisture on all mounting surfaces of the rear axle and caliper assembly and threads, bolts and washers.**

11. Install the brake hose with two new copper washers, using the L-shaped pin for alignment, then tighten the union bolt.

12. Refill the brake hydraulic system with new brake fluid and bleed out the air.

13. Check the rear disc brakes for drag.

14. Install the rear wheel and tire.

DISC BRAKE PADS

REMOVAL & INSTALLATION

See Figure 5.

✳ CAUTION

Clean the dust on the caliper and brake pad with a vacuum dust collector to minimize the hazard of air borne particles or other materials.

✳ WARNING

While removing and installing the cylinder body, do not depress the brake pedal because the piston may pop out. It is not necessary to remove the bolts on the torque member and the brake hose except for disassembly or replacement of the caliper assembly.

✳ WARNING

For pad removal and installation, hang the cylinder body with a wire so as not to stretch the brake hose. Do not damage the piston boot. If any shim is subject to serious corrosion, replace it with a new one. Always replace the shim and shim covers as a set when replacing the brake pads. Keep the rotor free from brake fluid.

➡**Burnish the brake pads and disc rotor mutually contacting surfaces after refinishing or replacing rotors, after replacing pads, or if a soft pedal occurs at very low mileage.**

1. Before servicing the vehicle, refer to the Precautions.

2. Remove the rear wheel and tire.

3. Remove the upper sliding pin bolt and loosen the lower sliding pin bolt to swing the cylinder body down.

4. Remove the pads, pad retainers, shims, and shim covers from the torque member.

✳ WARNING

Do not deform the pad retainers when removing them from the torque member.

To install:

5. Apply Molykote® M-77 grease, or equivalent, to between the shim covers and shims. Install the inner shim and inner shim cover to the inner pad. Install the outer shim and outer shim cover to the outer pad.

6. Apply Molykote® M-77 grease, or equivalent, between the pad retainer and pad. Install the pad retainers and pads to the torque member.

7. Press in the piston using a suitable tool, until the pads can be installed, and then install the cylinder body in the torque member.

➡**In the case of replacing a pad with new one, check the brake fluid level in the reservoir tank because brake fluid returns to the master cylinder reservoir tank when pressing in the piston.**

8. Install the upper sliding pin bolt and tighten the upper and lower sliding pin bolts.

9. Check the rear disc brakes for drag.

10. Install the rear wheel and tire.

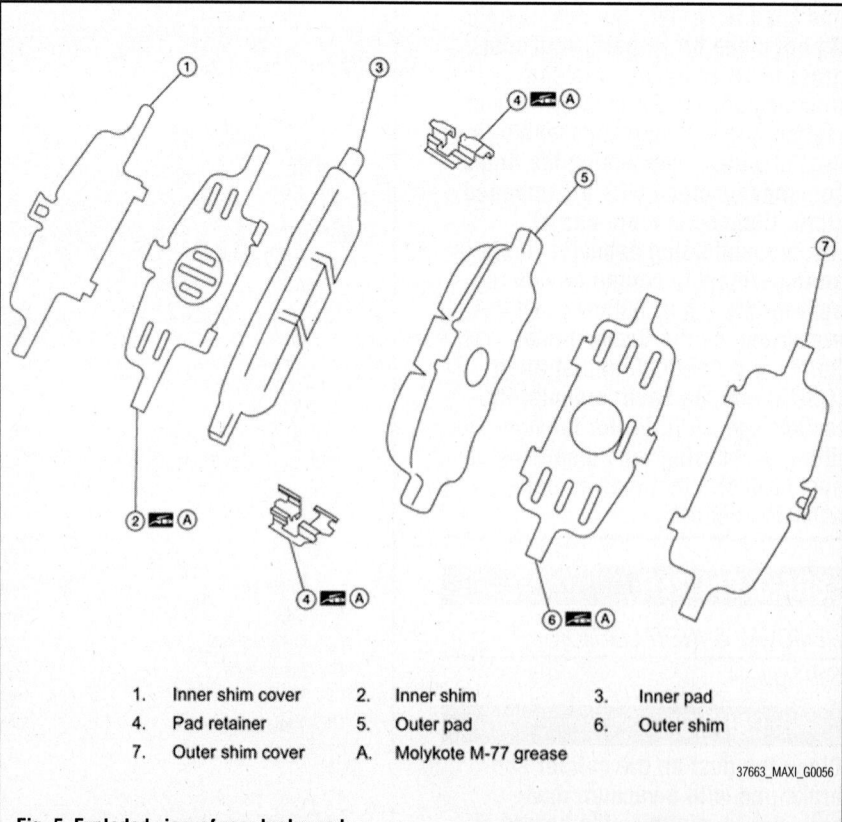

1.	Inner shim cover	2.	Inner shim	3.	Inner pad
4.	Pad retainer	5.	Outer pad	6.	Outer shim
7.	Outer shim cover	A.	Molykote M-77 grease		

37663_MAXI_G0056

Fig. 5 Exploded view of rear brake pads

BRAKES **PARKING BRAKE**

PARKING BRAKE CABLES

ADJUSTMENT

See Figures 6 and 7.

1. Before servicing the vehicle, refer to the Precautions.

2. Remove the lower instrument panel LH.

3. Partially engage the parking brake pedal to access the adjusting nut.

4. Insert a deep socket wrench to rotate adjusting the nut and loosen the cable sufficiently. Then, disengage the parking brake pedal.

5. Remove the wheel and tire using a power tool.

6. Remove the disc rotor and measure the inner diameter at the widest point using a suitable tool.

7. Transfer the measurement, less 0.24 inches (0.6mm), to the parking brake shoes and adjust accordingly.

8. Using wheel nuts, secure the disc rotor to the hub to prevent it from tilting.

9. Rotate the disc rotor to make sure there is no drag.

10. Adjust the cable as follows:

   a. Operate the pedal 10 or more times with a force of 110 lbs. (490 N).

   b. Rotate the adjusting nut with a deep socket to adjust the pedal stroke to the specification of 4–5 clicks.

   c. With the parking brake pedal completely disengaged, make sure there is no drag on the parking brake.

11. Install the disc rotor.

12. Install the wheel and tire using a power tool.

13. Install the lower instrument panel LH.

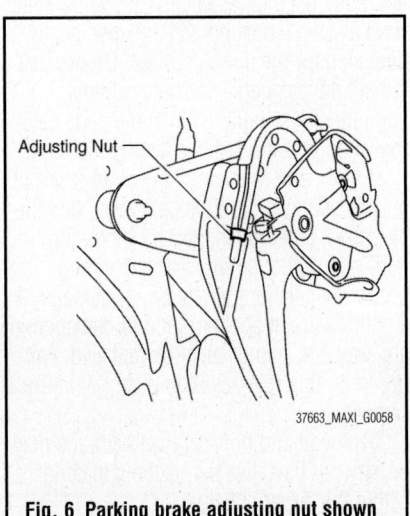

Adjusting Nut

37663_MAXI_G0058

Fig. 6 Parking brake adjusting nut shown

PARKING BRAKE SHOES

REMOVAL & INSTALLATION

See Figures 8 and 9.

✳✳ CAUTION

Clean the brakes with a vacuum dust collector to minimize the hazard of air borne particles or other materials. Clean the dust on the disc rotor and back plate using a vacuum dust collector. Do not blow with compressed air.

1. Before servicing the vehicle, refer to the Precautions.

2. Remove the rear wheel and tires using a power tool.

3. Remove the rear brake calipers.

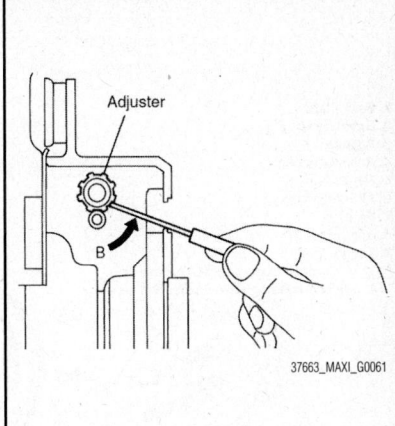

Adjuster

B

37663_MAXI_G0061

Fig. 8 Rotate the adjuster in direction (B) to retract and loosen the brake shoe, using a suitable tool

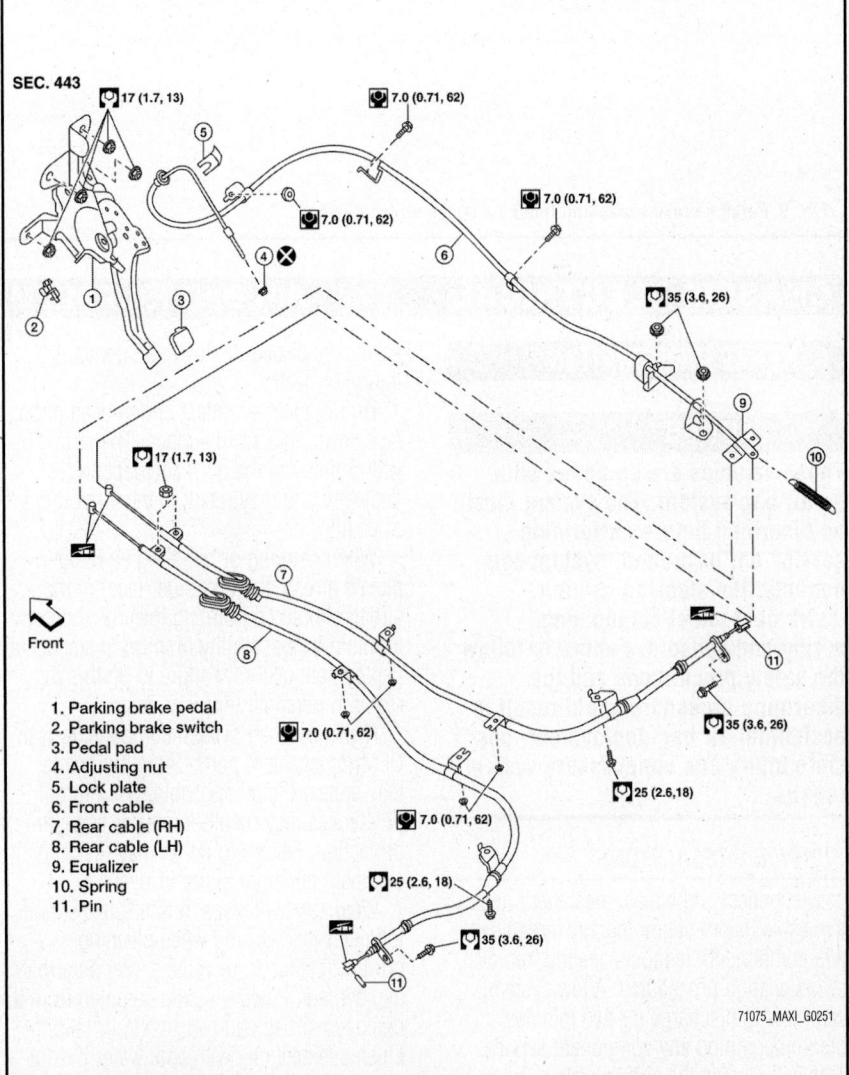

SEC. 443

17 (1.7, 13)

7.0 (0.71, 62)

7.0 (0.71, 62)

7.0 (0.71, 62)

35 (3.6, 26)

17 (1.7, 13)

Front

35 (3.6, 26)

25 (2.6, 18)

7.0 (0.71, 62)

25 (2.6, 18)

35 (3.6, 26)

1. Parking brake pedal
2. Parking brake switch
3. Pedal pad
4. Adjusting nut
5. Lock plate
6. Front cable
7. Rear cable (RH)
8. Rear cable (LH)
9. Equalizer
10. Spring
11. Pin

71075_MAXI_G0251

Fig. 7 Parking brake control component locations

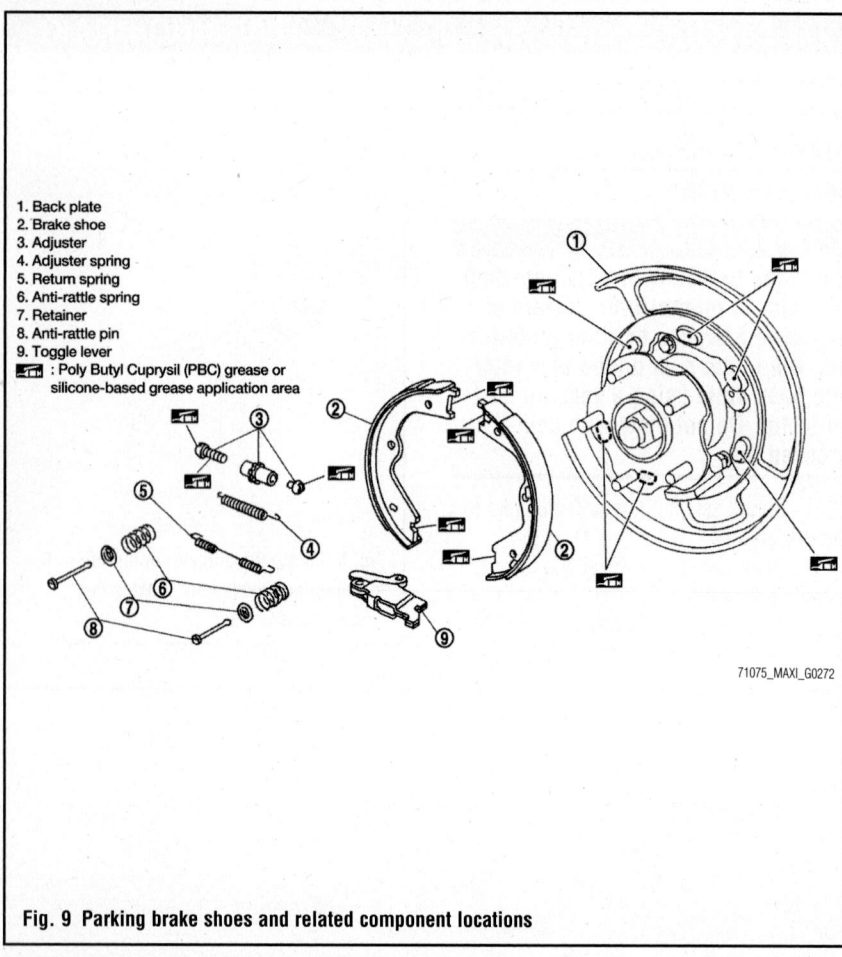

1. Back plate
2. Brake shoe
3. Adjuster
4. Adjuster spring
5. Return spring
6. Anti-rattle spring
7. Retainer
8. Anti-rattle pin
9. Toggle lever
🔧 : Poly Butyl Cuprysil (PBC) grease or silicone-based grease application area

71075_MAXI_G0272

Fig. 9 Parking brake shoes and related component locations

4. With the parking brake pedal in the fully released position, remove the disc rotor.

➡**Put matching marks on both the disc rotor and wheel hub when removing the disc rotor.**

5. If the disc rotor cannot be removed easily, remove as follows:

a. Secure the disc rotor in place with wheel nuts and remove the adjuster hole plug.

b. Rotate the adjuster in the appropriate direction to retract and loosen brake shoe. Use a suitable tool.

6. Remove the anti-rattle pins, retainers, anti-rattle springs, and return springs.

7. Remove the parking brake shoes, adjuster assembly, and toggle lever.

To install:

8. Installation is in the reverse order of removal.

9. Apply PBC (Poly Butyl Cuprysil) grease, or equivalent, to the specified points during installation.

10. Assemble the adjusters so that the threaded part is expanded when rotating it.

11. Shorten the adjuster by rotating it in the opposite direction.

12. Check the parking brake shoe sliding surface and drum inner surface for grease. Wipe off all grease adhering to the friction surfaces.

CHASSIS ELECTRICAL — AIR BAG (SUPPLEMENTAL RESTRAINT SYSTEM)

GENERAL INFORMATION

✱✱ CAUTION

These vehicles are equipped with an air bag system. The system must be disarmed before performing service on, or around, system components, the steering column, instrument panel components, wiring and sensors. Failure to follow the safety precautions and the disarming procedure could result in accidental air bag deployment, possible injury and unnecessary system repairs.

SERVICE PRECAUTIONS

Disconnect and isolate the battery negative cable before beginning any airbag system component diagnosis, testing, removal, or installation procedures. Allow system capacitor to discharge for two minutes before beginning any component service. This will disable the airbag system. Failure to disable the airbag system may result in

accidental airbag deployment, personal injury, or death.

Do not place an intact undeployed airbag face down on a solid surface. The airbag will propel into the air if accidentally deployed and may result in personal injury or death.

When carrying or handling an undeployed airbag, the trim side (face) of the airbag should be pointing towards the body to minimize possibility of injury if accidental deployment occurs. Failure to do this may result in personal injury or death.

Replace airbag system components with OEM replacement parts. Substitute parts may appear interchangeable, but internal differences may result in inferior occupant protection. Failure to do so may result in occupant personal injury or death.

Wear safety glasses, rubber gloves, and long sleeved clothing when cleaning powder residue from vehicle after an airbag deployment. Powder residue emitted from a deployed airbag can cause skin irritation. Flush affected area with cool water if irritation is experienced. If nasal or throat

irritation is experienced, exit the vehicle for fresh air until the irritation ceases. If irritation continues, see a physician.

Do not use a replacement airbag that is not in the original packaging. This may result in improper deployment, personal injury, or death.

The factory installed fasteners, screws and bolts used to fasten airbag components have a special coating and are specifically designed for the airbag system. Do not use substitute fasteners. Use only original equipment fasteners listed in the parts catalog when fastener replacement is required.

During, and following, any child restraint anchor service, due to impact event or vehicle repair, carefully inspect all mounting hardware, tether straps, and anchors for proper installation, operation, or damage. If a child restraint anchor is found damaged in any way, the anchor must be replaced. Failure to do this may result in personal injury or death.

Deployed and non-deployed airbags may or may not have live pyrotechnic material within the airbag inflator.

Do not dispose of driver/passenger/curtain airbags or seat belt tensioners unless you are sure of complete deployment. Refer to the Hazardous Substance Control System for proper disposal.

Dispose of deployed airbags and tensioners consistent with state, provincial, local, and federal regulations.

After any airbag component testing or service, do not connect the battery negative cable. Personal injury or death may result if the system test is not performed first.

If the vehicle is equipped with the Occupant Classification System (OCS), do not connect the battery negative cable before performing the OCS Verification Test using the scan tool and the appropriate diagnostic information. Personal injury or death may result if the system test is not performed properly.

Never replace both the Occupant Restraint Controller (ORC) and the Occupant Classification Module (OCM) at the same time. If both require replacement, replace one, then perform the Airbag System test before replacing the other.

Both the ORC and the OCM store Occupant Classification System (OCS) calibration data, which they transfer to one another when one of them is replaced. If both are replaced at the same time, an irreversible fault will be set in both modules and the OCS may malfunction and cause personal injury or death.

If equipped with OCS, the Seat Weight Sensor is a sensitive, calibrated unit and must be handled carefully. Do not drop or handle roughly. If dropped or damaged, replace with another sensor. Failure to do so may result in occupant injury or death.

If equipped with OCS, the front passenger seat must be handled carefully as well. When removing the seat, be careful when setting on floor not to drop. If dropped, the sensor may be inoperative, could result in occupant injury, or possibly death.

If equipped with OCS, when the passenger front seat is on the floor, no one should sit in the front passenger seat. This uneven force may damage the sensing ability of the seat weight sensors. If sat on and damaged, the sensor may be inoperative, could result in occupant injury, or possibly death.

DISARMING THE SYSTEM

✳✳ CAUTION

All SRS electrical wiring harnesses and connectors are covered with

YELLOW outer insulation. Do not use electrical test equipment on any circuit related to the SRS (air bag) sensors. When installing SRS components, always install with the arrow marks facing the front of the vehicle.

1. Before servicing the vehicle, refer to the Precautions.
2. Turn the ignition switch to the **OFF** position.
3. Disconnect both battery cables starting with the negative cable first.
4. Wait at least 3 minutes after the cables are disconnected. Be sure to insulate the battery terminal ends.

ARMING THE SYSTEM

1. Before servicing the vehicle, refer to the Precautions.
2. Make sure that the removed components are installed and/or the disconnected connectors are connected properly.
3. Turn the ignition switch to the **OFF** position.
4. Connect both battery cables starting with the positive cable first.
5. The SRS or air bag system is equipped with a self-diagnostic operation. After turning the ignition key to the ON or START position.
   a. The AIR BAG warning lamp will illuminate for 7 seconds.
   b. After 7 seconds, the AIR BAG lamp will extinguish if no malfunction is detected.
   c. If the AIR BAG lamp does not extinguish after 7 seconds, check the SRS self-diagnostic system for a malfunction.

CLOCKSPRING CENTERING

See Figure 10.

✳✳ CAUTION

Models equipped with a Supplemental Restraint System (SRS), use an inflatable air bag. Whenever working near any of the SRS components, such as the impact sensors, the air bag module, steering column, and instrument panel, disable the SRS.

✳✳ WARNING

The spiral cable may snap by the steering operation if the cable is installed in an improper position. Do not turn the spiral cable quickly or beyond the limit number of turns.

This can cause the cable to snap. The spiral cable can be turned counterclockwise about 2.5 turns from the neutral position.

✳✳ CAUTION

After the work is completed, make sure no system malfunction is detected by air bag warning lamp.

✳✳ WARNING

Do not use air tools or electric tools for servicing. Do not disassemble the spiral cable. Do not allow oil, grease, detergent or water to come in contact with the spiral cable.

✳✳ WARNING

Do not cause impact to the spiral cable by dropping it. Replace the spiral cable if it has been dropped or sustained an impact.

1. Before servicing the vehicle, refer to the Precautions.
2. Turn the ignition switch OFF, disconnect both battery terminals and wait at least 3 minutes.
3. Turn the contact coil cable assembly fully counterclockwise.

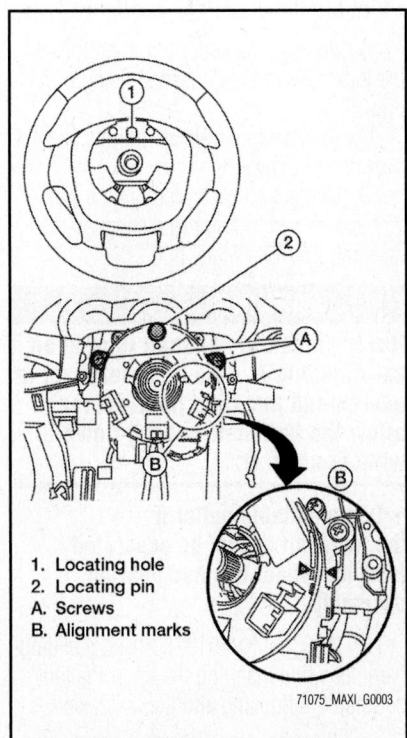

1. Locating hole
2. Locating pin
A. Screws
B. Alignment marks

71075_MAXI_G0003

Fig. 10 View of spiral cable (clockspring) alignment

4. From the fully counterclockwise position, turn the contact coil assembly clockwise about 2.5 turns.

5. Align the spiral cable correctly when installing the steering wheel. Make sure that the spiral cable is in the neutral position. The neutral position is detected by turning left 2.5 revolutions from the right end position and ending with the knob at the top.

DRIVE TRAIN

FRONT HALFSHAFT

REMOVAL & INSTALLATION

Left Side

See Figure 11.

1. Before servicing the vehicle, refer to the Precautions.

2. Remove the wheel and tire from the vehicle.

3. Remove the wheel sensor from the steering knuckle.

✳✳ WARNING

Do not pull on the wheel sensor harness.

4. Remove the brake hose lock plate from the strut assembly.

5. Remove the brake caliper torque member bolts using a power tool leaving the brake hose attached, then remove the disc rotor. Reposition the caliper aside with a wire.

➡**Avoid depressing the brake pedal while the brake caliper is removed.**

6. Remove the cotter pin, then loosen the lock nut from the halfshaft using a power tool.

7. Remove the lower strut bolts and nuts using a power tool.

8. Using a piece of wood and a hammer, tap on the lock nut to disengage the halfshaft from the wheel hub.

✳✳ WARNING

Never place the halfshaft joint at an extreme angle. Also be careful not to overextend the slide joint. Never allow the halfshaft to hang down without support.

➡**Use a suitable puller if the halfshaft cannot be separated from the wheel hub and bearing assembly.**

9. Use Tool KV40107500 and a sliding hammer while inserting the tip of the tool between the housing and the transaxle assembly.

10. Remove the halfshaft from the transaxle assembly.

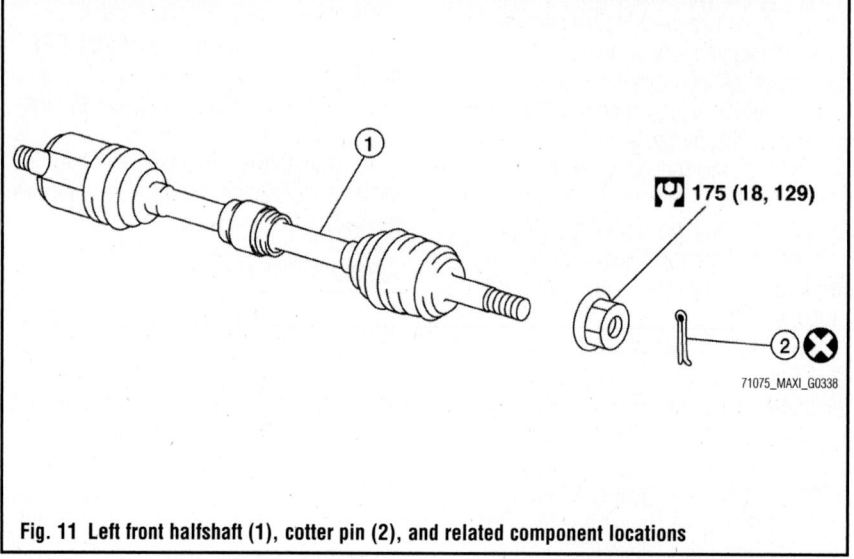

Fig. 11 Left front halfshaft (1), cotter pin (2), and related component locations

To install:

11. Installation is in the reverse order of removal.

➡**Do not reuse non-reusable parts.**

12. Install a new circlip on the halfshaft in the circular clip groove on the transaxle side. Make sure the new circlip on the halfshaft is securely fastened.

13. In order to prevent damage to the differential side oil seal, place Tool KV38107900 onto the oil seal before inserting the halfshaft. Slide the halfshaft into the slide joint and tap with a hammer to install it securely.

Right Side

See Figure 12.

1. Before servicing the vehicle, refer to the Precautions.

2. Remove the wheel and tire from the vehicle.

3. Remove the wheel sensor from the steering knuckle.

✳✳ WARNING

Do not pull on the wheel sensor harness.

4. Remove the brake hose lock plate from the strut assembly.

5. Remove the brake caliper torque member bolts using a power tool leaving the brake hose attached, then remove the disc rotor. Reposition the caliper aside with wire.

✳✳ WARNING

Avoid depressing the brake pedal while the brake caliper is removed.

6. Remove the disc rotor.

7. Remove the cotter pin, then loosen the lock nut from the halfshaft using a power tool.

8. Remove the lower strut bolts and nuts using a power tool.

9. Using a piece of wood and a hammer, tap on the lock nut to disengage the halfshaft from the wheel hub.

✳✳ WARNING

Never place the halfshaft joint at an extreme angle. Also be careful not to overextend the slide joint. Never allow the halfshaft to hang down without support.

➡**Use a suitable puller if the halfshaft cannot be separated from the wheel hub and bearing assembly.**

10. Remove the front exhaust tube.

11. Remove the bearing housing to support the bearing bracket bolts.

12. Remove the halfshaft from the transaxle assembly.

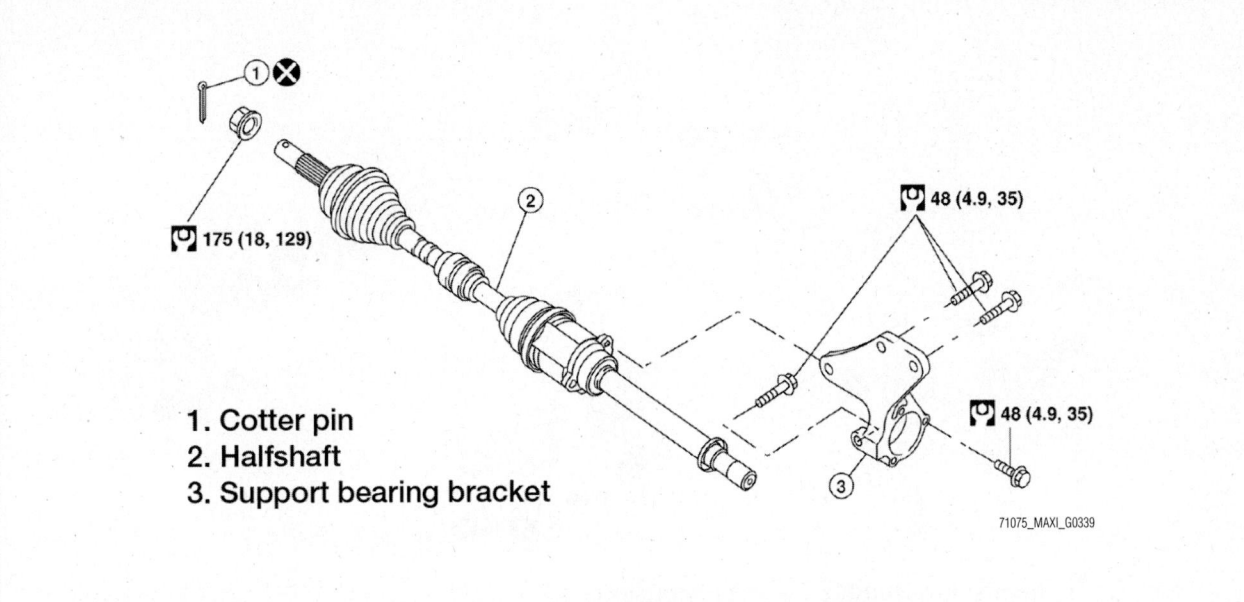

1. Cotter pin
2. Halfshaft
3. Support bearing bracket

71075_MAXI_G0339

Fig. 12 Right front halfshaft (2), cotter pin (1), and support bearing bracket (3) component locations

- Use Tool KV40107500 and sliding hammer while inserting the tip of the tool between the housing and the transaxle assembly

13. If necessary, remove the support bearing bracket.

To install:

14. Installation is in the reverse order of removal.

➡**Do not reuse non-reusable parts.**

15. Install a new circlip on the halfshaft in the circular clip groove on the transaxle side.

16. Make sure the new circlip on the halfshaft is securely fastened.

17. In order to prevent damage to the differential side oil seal, place Tool KV38107900 onto the oil seal before inserting the halfshaft. Slide the halfshaft into the slide joint and tap with a hammer to install it securely.

➡**Make sure that the circlip is completely engaged.**

18. When installing the support bearing bracket, temporarily tighten the mounting bolts, then tighten to the specified torque.

CV-BOOTS

REPLACEMENT

Left Side Halfshaft

See Figure 13.

At this time, the manufacturer does not provide specific removal and installation

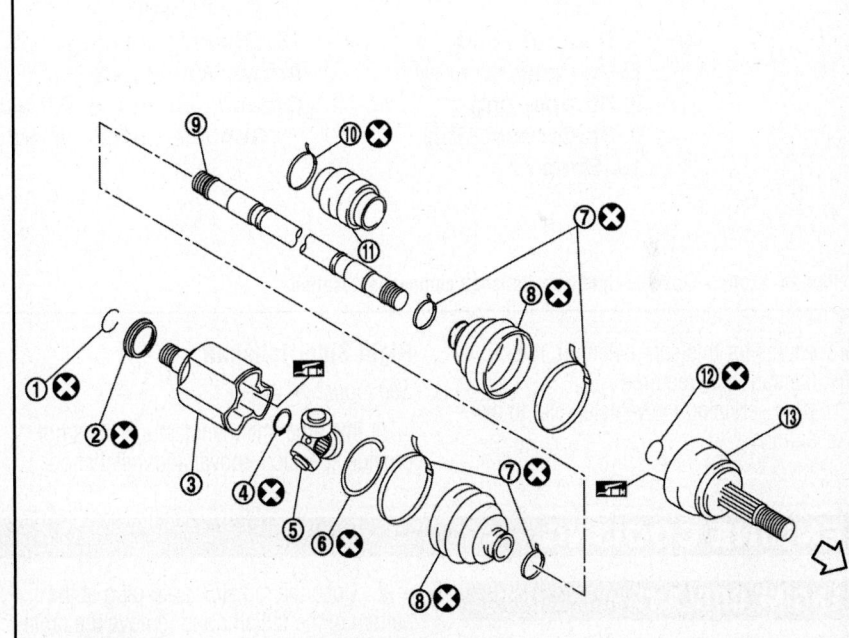

1. Circlip
2. Dust shield
3. Housing
4. Snap ring
5. Spider assembly
6. Stopper ring
7. Boot band
8. Boot
9. Shaft
10. Damper band
11. Dynamic damper
12. Circlip
13. Joint sub-assembly
Arrow: Wheel side
Grease gun image: Fill using NISSAN
     Genuine grease or equivalent

71075_MAXI_G0340

Fig. 13 Exploded view of left front halfshaft component locations

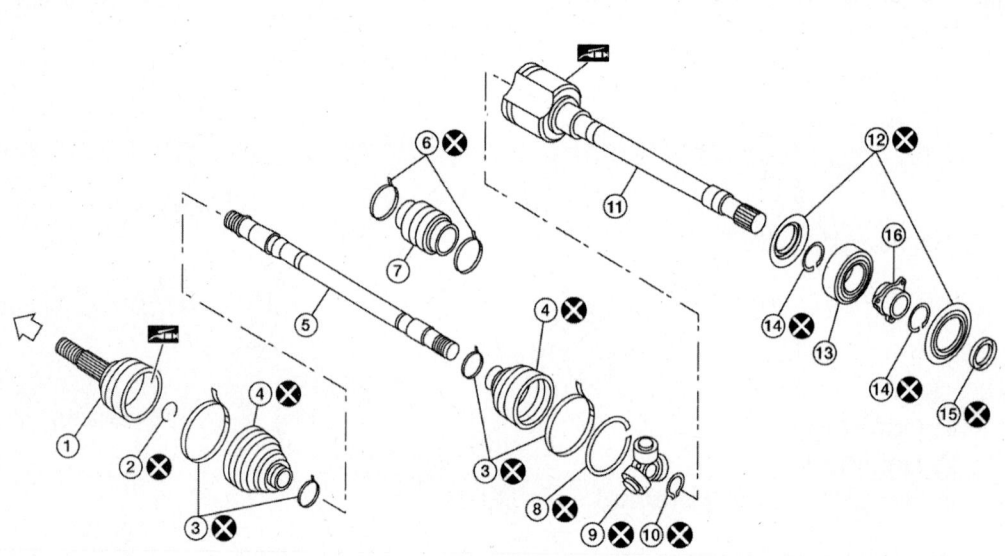

1. Joint sub-assembly
2. Circlip
3. Boot band
4. Boot
5. Shaft
6. Damper band
7. Dynamic damper
8. Stopper ring
9. Spider assembly
10. Snap ring

11. Housing
12. Dust shield
13. Support bearing
14. Snap ring
15. Dust shield
16. Bearing housing
Arrow: Wheel side
Grease gun image: Fill using NISSAN
   Genuine grease or equivalent

71075_MAXI_G0341

Fig. 14 Exploded view of right front halfshaft component locations

procedures for this component(s), refer to the illustration as required.

Before servicing the vehicle, refer to the Precautions.

Right Side Halfshaft

See Figure 14.

At this time, the manufacturer does not provide specific removal and installation

procedures for this component(s), refer to the illustration as required.

Before servicing the vehicle, refer to the Precautions.

ENGINE COOLING

ENGINE COOLANT

DRAIN & REFILL

✳✳ CAUTION

To avoid being scalded, never change the coolant when the engine is hot. Wrap a thick cloth around the cap and carefully remove the cap. First, turn the cap a quarter of a turn to release built-up pressure. Then turn the cap all the way.

Draining Engine Coolant

1. Before servicing the vehicle, refer to the Precautions.

2. Open the radiator drain plug at the bottom of the radiator and remove the radiator filler cap. This is the only step required for a partial cooling system drain.

3. If removing the heater core, remove the upper heater hose from the engine coolant outlet and apply moderate air pressure of 15 psi (1.055 kg-cm2) maximum for 30 seconds into the hose to blow out excess coolant from the core.

4. For a complete cooling system drain, remove the reservoir tank and drain the coolant, and then clean the reservoir tank before installation.

➡**Do not allow coolant to spill on the drive belts.**

5. When performing a complete cooling system drain (to remove the engine or for engine repair), remove the cylinder block front drain plug and the cylinder block RH drain plug.

6. Check the drained coolant for contaminants such as rust, corrosion, or discoloration.

7. If contaminated, flush the engine cooling system.

Refilling Engine Coolant

1. Before servicing the vehicle, refer to the Precautions.

2. Install the radiator drain plug. If the cooling system was drained completely,

install the reservoir tank and the cylinder block drain plugs.

    a. The radiator must be completely empty of coolant and water.

    b. Apply sealant to the threads of the cylinder block drain plugs. Use Genuine High Performance Thread Sealant or equivalent.

    3. If disconnected, reattach the upper radiator hose at the engine side.

    4. Set the vehicle heater controls to the full HOT and heater ON position. Turn the vehicle ignition ON with the engine OFF as necessary to activate the heater mode.

    5. Install the Tool KV991J0070 (J-45695) by installing the radiator cap adapter onto the radiator neck opening.

    6. Attach the gauge body assembly with the refill tube and the venturi assembly to the radiator cap adapter.

    7. Insert the refill hose into the coolant mixture container that is placed at floor level. Make sure the ball valve is in the closed position.

➡**Use Genuine NISSAN Engine Coolant or equivalent, mixed 50/50 with distilled water or demineralized water.**

    8. Install an air hose to the venturi assembly, the air pressure must be within specification.

➡**The compressed air supply must be equipped with an air dryer.**

    9. The vacuum gauge will begin to rise and there will be an audible hissing noise. During this process open the ball valve on the refill hose slightly. Coolant will be visible rising in the refill hose. Once the refill hose is full of coolant, close the ball valve. This will purge any air trapped in the refill hose.

    10. Continue to draw the vacuum until the gauge reaches 28 inches of vacuum. The gauge may not reach 28 inches in high altitude locations; use the vacuum specifications based on the altitude above sea level.

    11. When the vacuum gauge has reached the specified amount, disconnect the air hose and wait 20 seconds to see if the system loses any vacuum. If the vacuum level drops, perform any necessary repairs to the system and repeat steps 7–9 to bring the vacuum to the specified amount. Recheck for any leaks.

    12. Place the coolant container (with the refill hose inserted) at the same level as the top of the radiator. Then, open the ball valve on the refill hose so the coolant will be drawn up to fill the cooling system. The cooling system is full when the vacuum gauge reads zero.

➡**Do not allow the coolant container to get too low when filling in order to avoid air from being drawn into the cooling system.**

    13. Remove the Tool from the radiator neck opening.

    14. Fill the cooling system reservoir tank to the specified level and install the radiator cap. Run the engine to warm up the cooling system and top up the system as necessary.

FLUSHING

    1. Before servicing the vehicle, refer to the Precautions.

    2. Fill the radiator from the filler neck above the radiator upper hose and reservoir tank with clean water and reinstall the radiator filler cap.

    3. Run the engine and warm it up to the normal operating temperature.

    4. Rev the engine 2–3 times under a no-load condition.

    5. Stop the engine and wait until it cools down.

    6. Drain the water from the system.

    7. Repeat steps 2 through 6 until clear water begins to drain from the radiator.

ELECTRIC ENGINE FAN

REMOVAL & INSTALLATION

See Figure 15.

✳✳ CAUTION

Do not remove the radiator cap when the engine is hot. Serious burns could occur from high pressure coolant escaping from the radiator. Wrap a thick cloth around the cap. Slowly turn it a quarter turn to allow built-up pressure to escape. Carefully remove the cap by turning it all the way.

➡**When removing components such as hoses, tubes/lines, etc., cap or plug the openings to prevent fluid from spilling.**

➡**Perform this procedure when the engine is cold.**

    1. Before servicing the vehicle, refer to the Precautions.

    2. Partially drain the engine coolant from the radiator.

    3. Remove the engine room cover.

    4. Remove the Transmission Control Module (TCM).

    5. Remove the battery tray.

    6. Disconnect the radiator hose (upper).

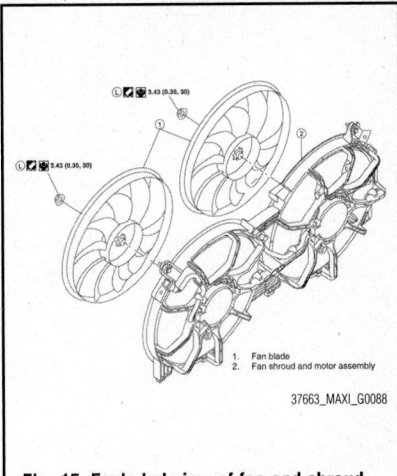

Fig. 15 Exploded view of fan and shroud assembly

    7. Disconnect the fan motor connectors.

    8. Remove the radiator cooling fan assembly.

    To install:

    9. Installation is in the reverse order of removal.

    10. Fill the cooling system with the proper type and amount of coolant.

➡**Cooling fans are controlled by the ECM.**

RADIATOR

REMOVAL & INSTALLATION

See Figure 16.

✳✳ CAUTION

Do not remove the radiator cap when the engine is hot. Serious burns could occur from high pressure coolant escaping from the radiator. Wrap a thick cloth around the cap. Slowly turn it a quarter turn to allow built-up pressure to escape. Carefully remove the cap by turning it all the way.

➡**When removing components such as hoses, tubes/lines, etc., cap or plug openings to prevent fluid from spilling.**

    1. Drain the coolant from the radiator.

    2. Remove the hoodledge covers (RH and LH).

    3. Remove the front bumper fascia.

    4. Remove the battery tray.

    5. Disconnect the coolant reservoir hose.

    6. Disconnect the radiator upper hose and lower hose.

    7. Remove the A/C condenser.

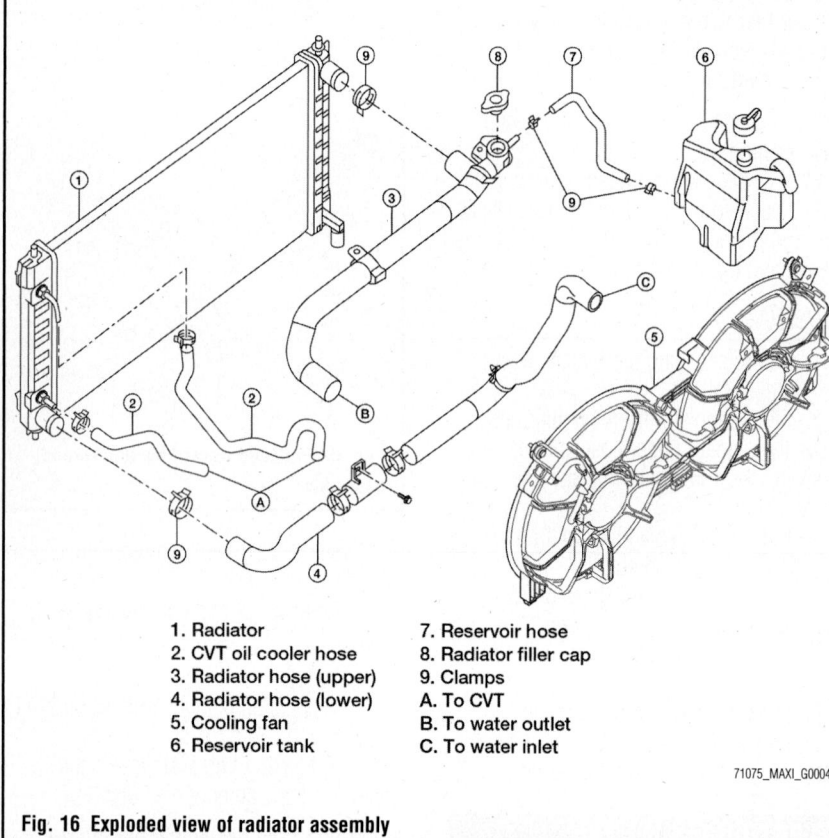

1. Radiator
2. CVT oil cooler hose
3. Radiator hose (upper)
4. Radiator hose (lower)
5. Cooling fan
6. Reservoir tank
7. Reservoir hose
8. Radiator filler cap
9. Clamps
A. To CVT
B. To water outlet
C. To water inlet

71075_MAXI_G0004

Fig. 16 Exploded view of radiator assembly

8. Disconnect the CVT oil cooler hoses.
9. Remove the radiator.

✳✳ WARNING

Do not damage or scratch the radiator core when removing.

To install:

10. Installation is in the reverse order of removal.
11. Refill the engine coolant and check for leaks.

➡**Do not spill coolant in the engine compartment. Use a shop cloth to absorb coolant.**

THERMOSTAT

REMOVAL & INSTALLATION

✳✳ CAUTION

Do not remove the radiator cap when the engine is hot. Serious burns could occur from high pressure engine coolant escaping from the radiator. Wrap a thick cloth around the cap. Slowly turn it a quarter of a turn to release built-up pressure. Carefully remove radiator cap by turning it all the way.

✳✳ CAUTION

Perform this procedure when the engine is cool.

➡**When removing components such as hoses, tubes/lines, etc., cap or plug openings to prevent fluid from spilling.**

1. Before servicing the vehicle, refer to the Precautions.
2. Drain the coolant from the radiator.
3. Disconnect the coolant reservoir hose and remove the coolant reservoir tank.
4. Disconnect the LH VTC solenoid harness connector.
5. Disconnect the lower radiator hose from the thermostat assembly.
6. Remove the thermostat assembly (water inlet).

➡**Do not disassemble the thermostat assembly (water inlet). Replace them as a unit, if necessary.**

To install:

7. Installation is in the reverse order of removal.
8. Install the thermostat with the jiggle valve facing upward.
9. Refill the engine coolant and check for leaks.

➡**Do not spill coolant in the engine compartment. Use a shop cloth to absorb coolant.**

WATER PUMP

REMOVAL & INSTALLATION

See Figures 17 through 19.

✳✳ CAUTION

Do not remove the radiator cap when the engine is hot. Serious burns could occur from high pressure coolant escaping from the radiator. Wrap a thick cloth around the cap. Slowly turn it a quarter turn to allow built-up pressure to escape. Carefully remove the cap by turning it all the way.

Please note the following:
• When removing water pump assembly, be careful not to get coolant on drive belt
• The water pump cannot be disassembled and should be replaced as a unit
• After installing the water pump, connect the hose and clamp securely, then check for leaks using a suitable tool.
• When removing components such as hoses, tubes/lines, etc., cap or plug openings to prevent fluid from spilling.

1. Before servicing the vehicle, refer to the Precautions.
2. Drain the engine coolant from the radiator.
3. Remove the RH wheel and tire.
4. Remove the fender protector side cover (RH).
5. Set No. 1 cylinder at TDC on its compression stroke.
6. Align the pointer with the TDC mark on the crankshaft pulley.
7. Remove the drive belt.
8. Remove the idler pulley and the A/C idler pulley.
9. Remove the hood ledge cover (RH).
10. Remove the water drain plug (front) on the water pump side of the cylinder block to drain the engine coolant from the engine.
11. Support the engine and remove the front engine insulator and bracket.
12. Disconnect the RH valve timing control connectors and remove the RH IVT control valve cover.
13. Remove the water pump cover.
14. Remove the timing chain tensioner assembly as follows:
   a. Pull the lever down to release the plunger stopper tab.

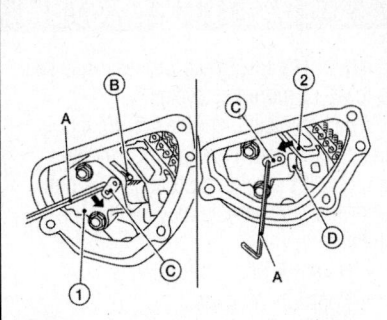

A. Stopper pin (Allen wrench)
B. Plunger stopper tab
C. Lever
D. Plunger
1. Tensioner body
2. Slack guide

37663_MAXI_G0094

Fig. 17 Removing the timing chain tensioner assembly

b. Insert the stopper pin A into the tensioner body hole to hold the lever and keep the plunger stopper tab released.

➡**An Allen wrench of 0.047 inches (1.2mm) is used for a stopper pin as an example.**

c. Compress the plunger into the tensioner body by pressing the slack guide.

d. Keep the slack guide pressed and lock the plunger in by pushing the stopper pin through the lever and into the chain tensioner body hole.

e. Remove the timing chain tensioner bolts and then remove the timing chain tensioner.

※ WARNING

Be careful not to drop the timing chain tensioner bolts inside the timing chain case.

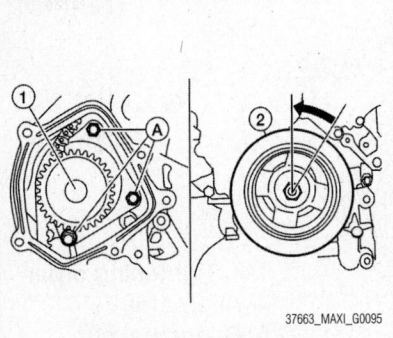

37663_MAXI_G0095

Fig. 18 Remove the 3 water pump bolts (A); make a gap between the water pump sprocket (1) and the timing chain, by carefully turning the crankshaft pulley (2) counterclockwise

15. Remove the 3 water pump bolts. Make a gap between the water pump sprocket and the timing chain, by carefully turning the crankshaft pulley counterclockwise until the timing chain loosens on the water pump sprocket.

16. Screw M8 bolts [pitch: 0.49 inch (1.25mm) length: approximately 1.97 inches (50mm)] into the water pump upper and lower bolt holes until they reach the timing chain case.

17. Remove the water pump.
Consider the following warnings:

• Place a suitable shop cloth below the water pump housing to prevent any engine coolant from dripping into the timing chain case.

• Pull the water pump straight out while preventing the vane from contacting the socket in the installation area.

• Remove the water pump without causing the sprocket to contact the timing chain.

18. Remove the M8 bolts and O-rings from the water pump.

To install:

19. Install new O-rings to the water pump.

20. Apply engine oil and coolant to the O-rings.

➡**Locate the O-ring with the white paint mark to the engine front side.**

21. Hold the timing chain to the side and install the water pump.

※ WARNING

Do not allow the cylinder block to damage the O-rings when installing the water pump.

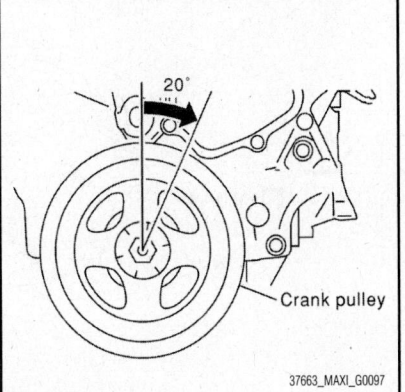

37663_MAXI_G0097

Fig. 19 Turn the crankshaft pulley approximately 20° clockwise

a. Check that the timing chain and the water pump sprocket are engaged.

b. Tighten the water pump bolts alternately and evenly.

22. Remove the dust and foreign material completely from the installation area of the timing chain tensioner and rear timing chain case.

23. Turn the crankshaft pulley approximately 20° clockwise so that the timing chain on the timing chain tensioner side is loose.

24. Apply engine oil to the oil feed hole and timing chain tensioner and install the timing chain tensioner.

25. Remove the stopper pin.

26. Install the IVT valve and cover RH (bank 1) and water pump cover.

a. Before installing, remove all traces of liquid gasket from the mating surface of the water pump cover and IVT control valve cover using a scraper.

b. Also remove traces of liquid gasket from the mating surface of the front cover.

c. Apply a continuous bead of liquid gasket to the mating surface of the IVT control valve cover and water pump cover.

27. Install the cylinder block front drain plug on the water pump side of the cylinder block.

28. Apply liquid gasket to the threads of water drain plug.

➡**Use Genuine RTV Silicone Sealant or equivalent.**

• Tighten the cylinder block front drain plug to 87 inch lbs. (10 Nm)

29. Installation of the remaining components is in the reverse order of removal.

30. After installation refill the engine coolant and check for leaks.

➡**Do not spill coolant in the engine compartment. Use a shop cloth to absorb coolant.**

31. After starting the engine, let it idle for 3 minutes, then rev the engine up to 3,000 RPM under no load to purge the air from the high-pressure chamber of the chain tensioner.

➡**The engine may produce a rattling noise. This indicates that air still remains in the chamber and is not a matter of concern.**

ENGINE ELECTRICAL   BATTERY SYSTEM

BATTERY

REMOVAL & INSTALLATION

1. Before servicing the vehicle, refer to the Precautions.
2. Loosen the battery cable assembly nuts, and disconnect both battery terminals.

✷✷ CAUTION

When disconnecting, disconnect the negative terminal first.

3. Remove the upper ECM bracket nut and bolt and the ECM upper bracket.
4. Remove the battery frame nuts and the battery frame.
5. Remove the battery.

To install:

6. Installation is in the reverse order of removal.

✷✷ CAUTION

When connecting, connect the positive terminal first.

7. Reset the electronic systems as necessary.

BATTERY RECONNECT/RELEARN PROCEDURE

Vehicles equipped with engine and transaxle computers may require a relearn procedure after the vehicle battery has been disconnected. Most vehicle computers memorize and store vehicle operational patterns. When the battery is disconnected, the information may be cleared. If the information is cleared, the computer will go into default mode in order to operate the vehicle. The vehicle computer will relearn operational patterns each time the vehicle is restarted. The relearning process may take up to 40 or more key cycles.

When a specific engine component is replaced, a relearn procedure may be required. If the relearn procedure is not performed, the vehicle may exhibit the following:

- Harsh or poor shift quality
- Poor fuel mileage
- Hesitation or stumble
- Unstable idle or stalling
- Lean or rich running conditions

If an accessory component was replaced, a relearn procedure may also be required. The following systems and components may not work properly without a relearn procedure:

- Anti-theft system
- Steering system
- Power window system
- Power sunroof system

ENGINE ELECTRICAL   CHARGING SYSTEM

ALTERNATOR

REMOVAL & INSTALLATION

See Figure 20.

1. Before servicing the vehicle, refer to the Precautions.
2. Remove the hood ledge covers (RH and LH).
3. Remove cooling fan assembly.
4. Remove the A/C compressor.
5. Remove the A/C idler pulley.
6. Disconnect the oil pressure switch.
7. Disconnect the alternator harness connectors.
8. Remove the alternator bolt and nuts, using power tools.
9. Slide the alternator out and remove.
10. Remove the alternator bracket.

To install:

11. Installation is the reverse of the removal procedure.
12. Temporarily tighten the bolts and nut, then tighten nut and bolts in the specified numerical order.

➡ **Be sure to tighten the "B" terminal nut carefully.**

13. Install the alternator and check the tension of the belt.

➡ **For this model, the power generation voltage variable control system that controls the power generation voltage of the alternator has been adopted. Therefore, the power generation volt-** age variable control system operation inspection should be performed after replacing the alternator, and then make sure that the system operates normally.

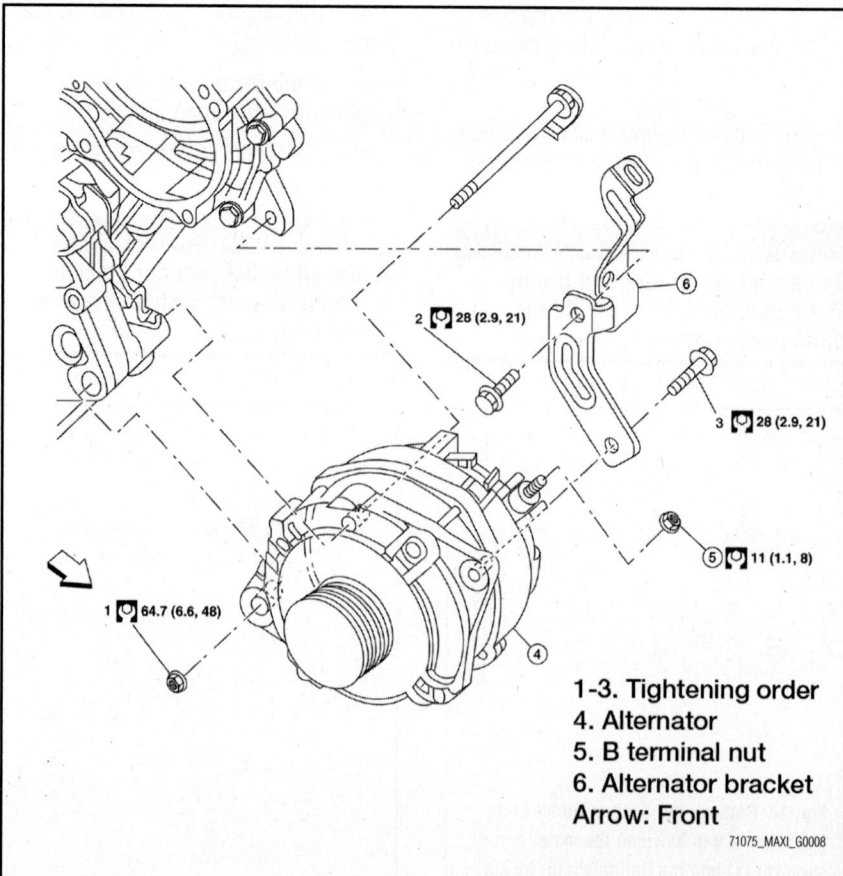

1-3. Tightening order
4. Alternator
5. B terminal nut
6. Alternator bracket
Arrow: Front

71075_MAXI_G0008

Fig. 20 Exploded view of alternator component locations

FIRING ORDER

See Figure 21.

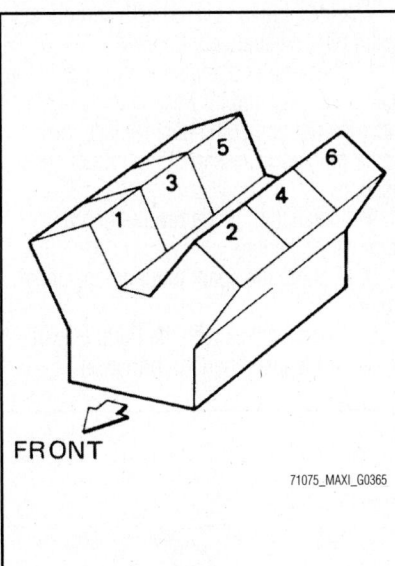

FRONT

71075_MAXI_G0365

Fig. 21 Engine Firing order: 1–2–3–4–5–6

IGNITION COILS

REMOVAL & INSTALLATION

Left Side

See Figure 22.

1. Before servicing the vehicle, refer to the Precautions.
2. Remove the engine room cover.
3. Disconnect the ignition coil connector.
4. Remove the ignition coil.

✸✸ WARNING
Never shock the ignition coil.

5. Installation is in the reverse order of removal.

Right Side

1. Before servicing the vehicle, refer to the Precautions.
2. Remove the intake manifold collector.
3. Disconnect the ignition coil connector.
4. Remove the ignition coil.

✸✸ WARNING
Never shock ignition coil.

5. Installation is in the reverse order of removal.

IGNITION TIMING

INSPECTION & ADJUSTMENT

➡**The ignition timing is not adjustable. If not within specifications, further diagnostic inspection is required. The following procedure is for viewing the ignition timing setting.**

Visually check the air cleaner, intake hoses, ducts, Exhaust Gas Recirculation (EGR) valve operation and electrical connections prior to the adjustment of the ignition timing. Correct or repair any problem as required. Be sure to inspect the throttle valve and Throttle Position (TP) sensor for proper operation.

1. Before servicing the vehicle, refer to the Precautions.
2. Locate the timing marks on the crankshaft pulley and the front of the engine.
3. Clean the timing marks.

➡**The ignition timing specification is 7–17° Before Top Dead Center (BTDC).**

4. Using chalk or white paint, color the mark on the crankshaft pulley and the mark on the scale that will indicate the correct timing when aligned with the notch on the crankshaft pulley.
5. Attach a tachometer to the engine.
6. Attach a timing light to the engine to number 1 cylinder ignition wire.
7. Turn all electrical equipment and accessories **OFF**.
8. Check to be sure all of the wires clear the fan, then, start the engine and allow it to reach normal operating temperatures.
9. Block the front wheels and set the parking brake. Shift the transmission into **NEUTRAL**. Do not stand in front of the vehicle when making adjustments.
10. Perform the following procedures:
   a. Race the engine at 2,000 RPM for about 2 minutes under a no-load condition; be sure all of the accessories are turned **OFF**.
   b. Perform on board engine diagnostics and repair any fault code.

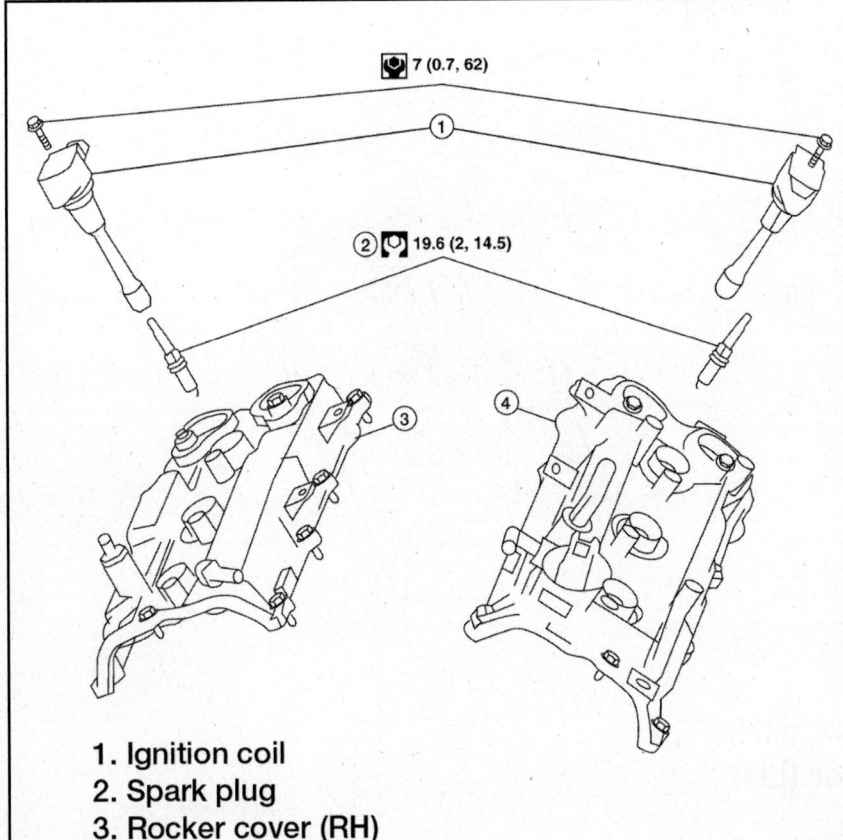

7 (0.7, 62)
2 19.6 (2, 14.5)

1. Ignition coil
2. Spark plug
3. Rocker cover (RH)
4. Rocker cover (LH)

71075_MAXI_G0009

Fig. 22 Exploded view of ignition coil spark plug component locations—left side

c. Race the engine at 2,000 RPM for about 2 minutes under a no-load condition.

d. Turn the engine **OFF** and disconnect the TP sensor.

e. Start and race the engine 2 to 3 times under no-load, then run the engine at idle speed.

11. Aim the timing light at the timing marks. If the marks on the pulley and the engine are aligned when the light flashes, the timing is correct. Turn the engine **OFF** and remove the tachometer and the timing light. If the marks are not in alignment, proceed with the following steps:

a. Turn the engine **OFF**.

b. Check the Camshaft Position (CMP) sensor (PHASE), Crankshaft Position (CKP) sensor (REF) and CKP sensor (POS). Replace if necessary.

c. If the ignition timing is still not correct, substitute a known good Electronic Control Module (ECM).

➡ **The ECM may be the cause of the problem, but this is rarely the case.**

12. Turn the engine **OFF** and remove the tachometer and the timing light.

SPARK PLUGS

REMOVAL & INSTALLATION

See Figure 23.

1. Before servicing the vehicle, refer to the Precautions.

2. Remove the ignition coil(s).

3. Remove the spark plug(s) with a suitable spark plug wrench.

To install:

Please note the following:

• Do not use a wire brush for cleaning the spark plugs

• If the plug is covered with carbon, a spark plug cleaner may be used

• Checking and adjusting the spark plug gap is not required between change intervals. Do not adjust the gap; replace the spark plug as necessary if it is out of specification.

4. Installation is the reverse of the removal procedure.

5. Tighten the spark plugs to specification.

6. Always check with the Parts Department for the latest parts information.

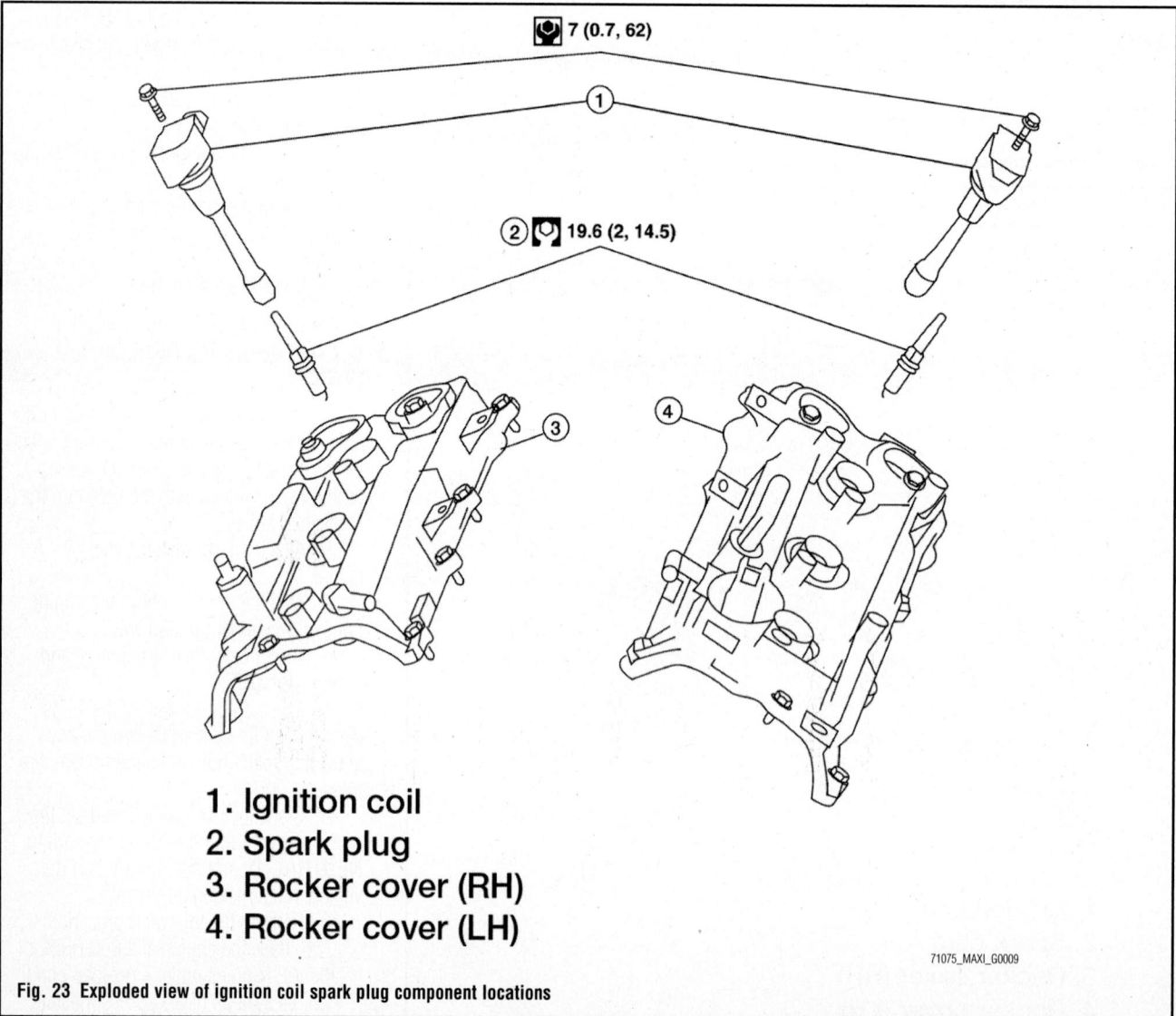

7 (0.7, 62)

①

② 19.6 (2, 14.5)

③

④

1. Ignition coil
2. Spark plug
3. Rocker cover (RH)
4. Rocker cover (LH)

71075_MAXI_G0009

Fig. 23 Exploded view of ignition coil spark plug component locations

ENGINE ELECTRICAL

STARTER

REMOVAL & INSTALLATION

See Figure 24.

1. Before servicing the vehicle, refer to the Precautions.
2. Disconnect the negative and positive battery terminals.
3. Remove the air cleaner assembly and air ducts.
4. Remove the battery tray.
5. Disconnect the battery cable and starter harness connector.
6. Remove the starter bolts and remove the starter.

To install:

7. Installation is the reverse of the removal procedure.
8. Tighten the fasteners to specification.

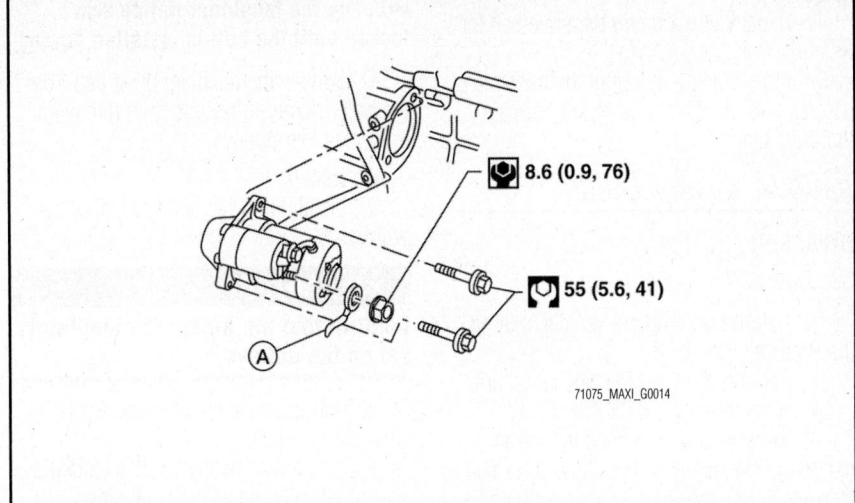

Fig. 24 Exploded view of starter and battery cable (A) component locations

71075_MAXI_G0014

ENGINE MECHANICAL

➡ **Disconnecting the negative battery cable may interfere with the functions of the on board computer systems and may require the computer to undergo a relearning process, once the negative battery cable is reconnected.**

ACCESSORY DRIVE BELT SYSTEM

ADJUSTMENT

Belt tension is not manually adjustable; it is automatically adjusted by the drive belt auto-tensioner. If the drive belt is out of adjustment, replace the drive belt as needed.

BELT ROUTINGS

See Figure 25.

INSPECTION

❄❄ CAUTION

Inspect and check the drive belts with the engine OFF.

1. Before servicing the vehicle, refer to the Precautions.
2. Check that the indicator of the drive belt auto-tensioner is within the possible use range.

➡ **Check the drive belt auto-tensioner indication when the engine is cold.**

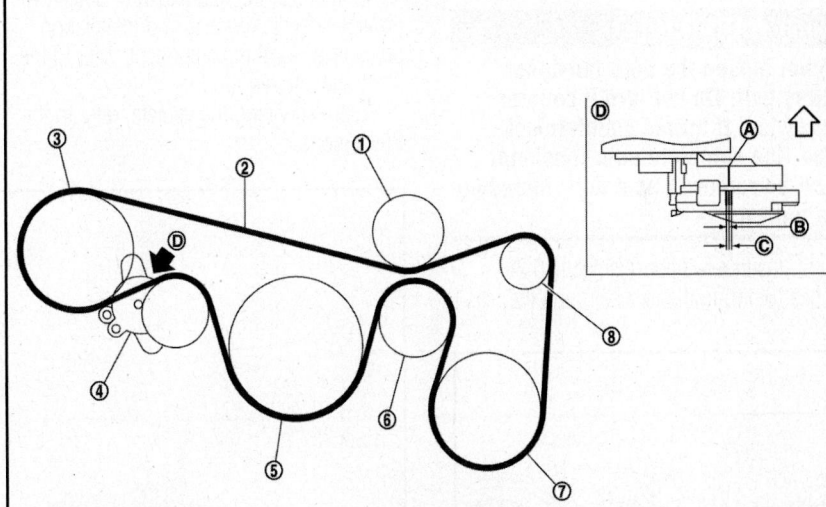

1. Idler pulley
2. Drive belt
3. Power steering oil pump
4. Drive belt auto-tensioner
5. Crankshaft pulley
6. Idler pulley
7. A/C compressor
8. Alternator

A. Indicator
B. Range when new drive belt is installed
C. Possible use range
D. View D
Arrow: Engine front

Fig. 25 Accessory drive belt routing

71075_MAXI_G0016

3. When a new drive belt is installed, the indicator should be within the new drive belt range.

4. Visually check entire the drive belt for wear, damage, or cracks.

5. If the indicator is out of the possible use range or the belt is damaged, replace the drive belt.

REMOVAL & INSTALLATION

Drive Belt

See Figure 26.

1. Before servicing the vehicle, refer to the Precautions.

2. Remove the front RH wheel and tire.

3. Remove the front RH side cover.

4. While securely holding the hexagonal part in the pulley center of the drive belt auto-tensioner, loosen the tensioner using a suitable tool.

❊❊ CAUTION

Avoid placing hand in a location where pinching may occur if the holding tool accidentally comes off.

❊❊ WARNING

Do not loosen the auto-tensioner pulley bolt. Do not turn it counter-clockwise. If turned counterclockwise, the complete auto-tensioner must be replaced as a unit, including the pulley.

5. Insert a rod approximately 0.24 inches (6mm) in diameter through the rear of the tensioner into the retaining boss to lock the tensioner pulley.

➡**Leave the tensioner pulley arm locked until the belt is installed again.**

6. Loosen the auxiliary drive belt from the water pump pulley and then remove it from the other pulleys.

To install:

7. Install the drive belt onto all of the pulleys.

❊❊ CAUTION

Confirm that the belts are completely set on the pulleys.

8. Release the tensioner, and apply tension to the belt.

9. Turn the crankshaft pulley clockwise several times to equalize the tension between each pulley.

10. Confirm that the tension of the belt at the indicator is within the allowable use range.

Drive Belt Idler Pulley

See Figure 27.

At this time, the manufacturer does not provide specific removal and installation procedures for this component, refer to the illustration as required.

Before servicing the vehicle, refer to the Precautions.

Drive Belt Tensioner

➡**The complete auto-tensioner must be replaced as a unit, including the pulley.**

1. Before servicing the vehicle, refer to the Precautions.

2. Remove the drive belt.

3. Insert a rod approximately 0.24 inch (6mm) in diameter through the rear of the tensioner into the retaining boss to lock the tensioner pulley.

4. Remove the drive belt auto-tensioner.

❊❊ WARNING

Do not loosen the auto-tensioner pulley bolt or turn it counterclockwise. If turned counterclockwise, the complete auto-tensioner must be replaced as a unit, including the pulley.

To install:

5. Installation is the reverse of the removal procedure.

6. If there is damage greater than peeled paint, replace the drive belt auto-tensioner unit.

❊❊ WARNING

Do not swap the pulley between the new and old auto-tensioner units.

Fig. 26 While securely holding the hexagonal part in the pulley center of the drive belt auto-tensioner, move in the direction of the arrow (loosening direction of tensioner) using a suitable tool

37663_MAXI_G0101

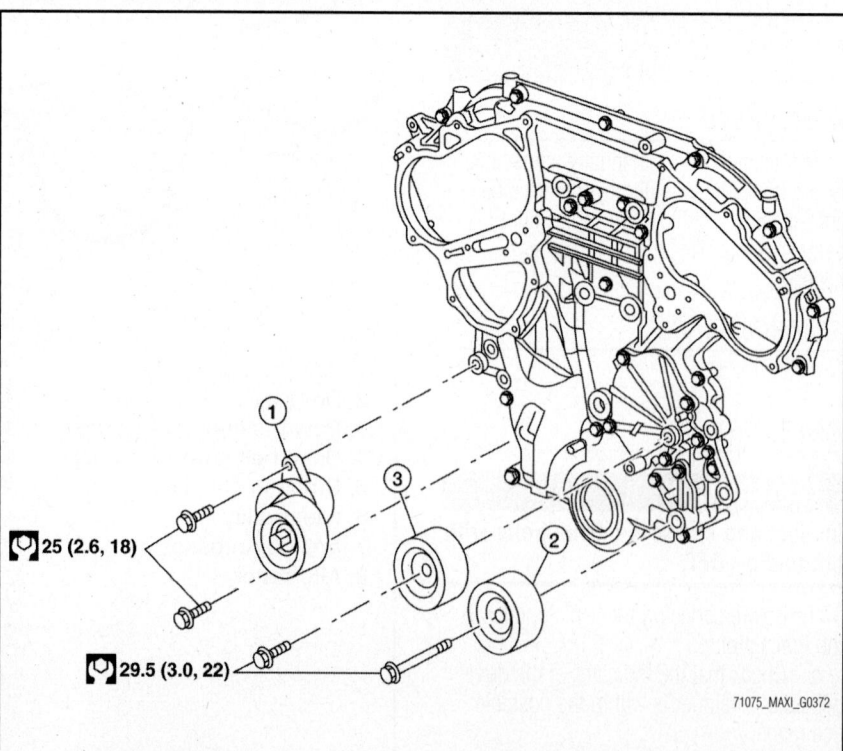

25 (2.6, 18)

29.5 (3.0, 22)

71075_MAXI_G0372

Fig. 27 View of drive belt auto-tensioner (1), idler pulley (2), and A/C idler pulley (3)

AIR CLEANER ASSEMBLY

REMOVAL & INSTALLATION

See Figure 28.

At this time, the manufacturer does not provide specific removal and installation procedures for this component, refer to the illustration as required.

Before servicing the vehicle, refer to the Precautions.

CAMSHAFT & BEARINGS

INSPECTION

Visual Check

1. Before servicing the vehicle, refer to the Precautions.
2. Check the camshaft for scratches, seizure, and wear. Replace if necessary.

Camshaft Runout

See Figure 29.

1. Before servicing the vehicle, refer to the Precautions.
2. Put a V-block on a precise flat bed and support No. 2 and No. 4 journals of the camshaft.
3. Set the dial gauges vertically to the No. 3 journal as shown.
4. Turn the camshaft in one direction slowly by hand, measure the camshaft runout on the dial gauges.
   - Runout is the largest indicator reading after one full revolution.
5. If the actual runout exceeds the limit, replace the camshaft.
   a. Camshaft Runout Standard: Less than 0.0008 inch (0.02mm)
   b. Camshaft Runout Limit: 0.0020 inch (0.05mm)

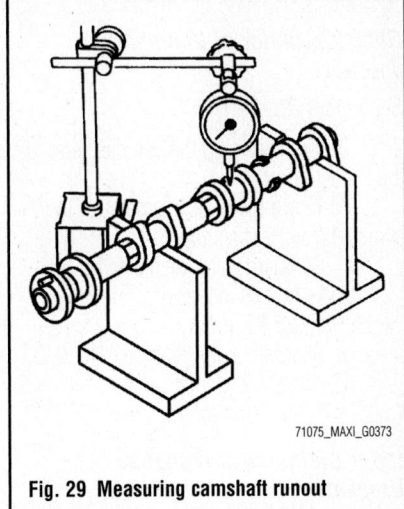

71075_MAXI_G0373

Fig. 29 Measuring camshaft runout

Camshaft Cam Lobe Height

See Figure 30.

1. Before servicing the vehicle, refer to the Precautions.
2. Measure the camshaft cam lobe height as shown.
3. If wear has reduced the lobe height below specifications, replace the camshaft.

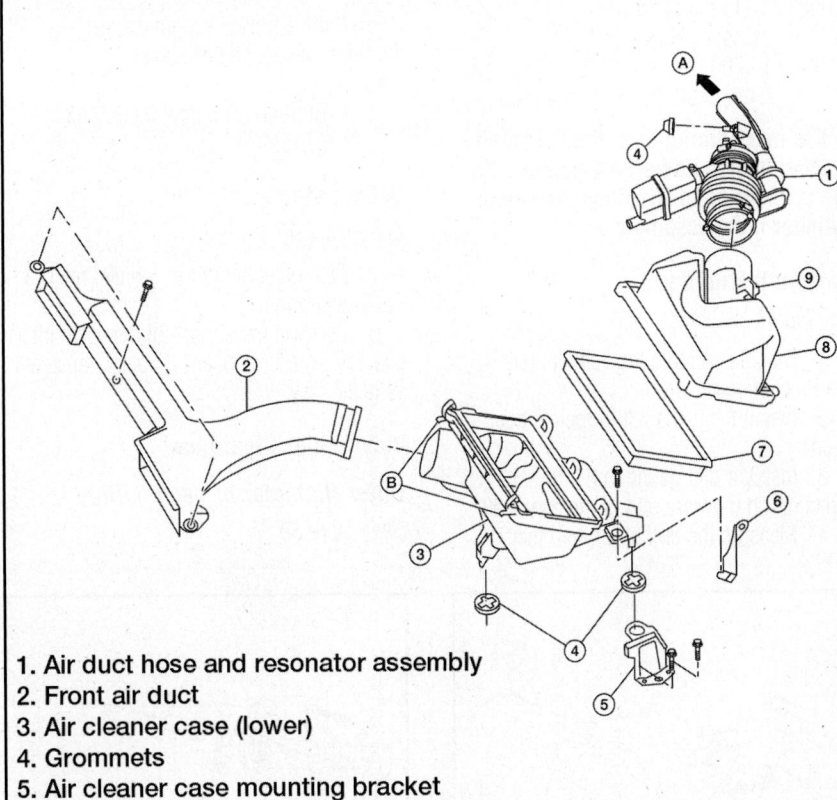

1. Air duct hose and resonator assembly
2. Front air duct
3. Air cleaner case (lower)
4. Grommets
5. Air cleaner case mounting bracket
6. Bracket
7. Air cleaner filter
8. Air cleaner case (upper)
9. Mass air flow sensor
A. To electric throttle control actuator
B. Air cleaner case side clips

71075_MAXI_G0017

Fig. 28 Exploded view of air cleaner component locations

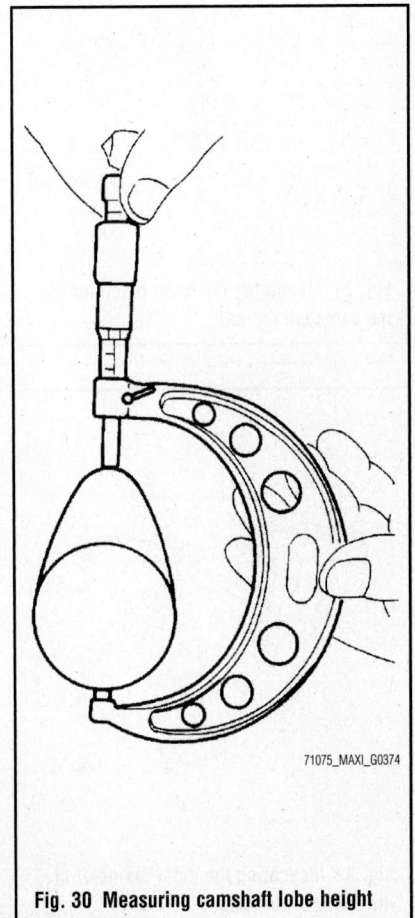

71075_MAXI_G0374

Fig. 30 Measuring camshaft lobe height

Camshaft Journal Clearance

Outer Diameter of Camshaft Journal

See Figure 31.

1. Before servicing the vehicle, refer to the Precautions.
2. Measure the outer diameter of the camshaft journal as shown.
    a. Standard outer diameter No. 1: 1.0211–1.0218 inches (25.935–25.955mm).
    b. Standard outer diameter No. 2, 3, 4: 0.9230–0.9238 inch (23.445–23.465mm).

Inner Diameter of Camshaft Bracket

See Figure 32.

1. Before servicing the vehicle, refer to the Precautions.

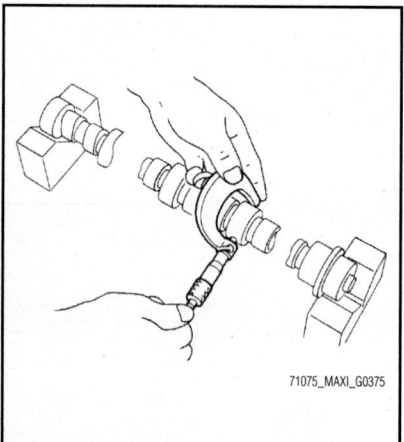

Fig. 31 Measuring the outer diameter of the camshaft journal

2. Tighten the camshaft bracket bolt to specified torque.
3. Using an inside micrometer, measure the inner diameter of the camshaft bearing.
    a. Standard inner diameter No. 1: 1.0236–1.0244 inches (26.000–26.021mm).
    b. Standard inner diameter No. 2, 3, 4: 0.9252–0.9260 inch (23.500–23.521mm).

Calculation of Camshaft Journal Clearance

1. Before servicing the vehicle, refer to the Precautions.
2. (Journal clearance) = (inner diameter of camshaft bracket) − (outer diameter of camshaft journal).
3. When out of the specified range, replace either or both the camshaft and the cylinder head.
    a. Standard No. 1: 0.0018–0.0034 inch (0.045–0.086mm).
    b. Standard No. 2, 3, 4: 0.0014–0.0030 inch (0.035–0.076mm).
    c. Limit: 0.0059 inch (0.15mm).

➡ **The inner diameter of the camshaft bracket is manufactured together with the cylinder head. Replace the whole cylinder head assembly.**

Camshaft End Play

See Figure 33.

1. Before servicing the vehicle, refer to the Precautions.
2. Install the camshaft in the cylinder head.
3. Install a dial gauge in the thrust direction on the front end of the camshaft.
4. Measure the end play when the

camshaft is moved forward/backward (in direction to the axis) as shown.
    a. If out of the specified range, replace with a new camshaft and measure again.
    b. If out of the specified range again, replace with a new cylinder head.
    • Standard: 0.0045–0.0074 inch (0.115–0.188mm)
    • Limit: 0.0094 inch (0.24mm)

Camshaft Sprocket Runout

See Figure 34.

1. Before servicing the vehicle, refer to the Precautions.
2. Put a V-block on a precise flat bed and support No. 2 and No. 4 journals of the camshaft.
3. Install the camshaft sprocket on the camshaft.
4. Measure the camshaft sprocket runout.
5. If the sprocket runout exceeds the limit, replace the camshaft sprocket.
    • Runout: Less than 0.0059 inch (0.15mm)

Valve Lifter

See Figure 35.

1. Before servicing the vehicle, refer to the Precautions.
2. Check if the surface of the valve lifter has any excessive wear or cracks, replace as necessary.

Valve Lifter Clearance

Outer Diameter of Valve Lifter

See Figure 36.

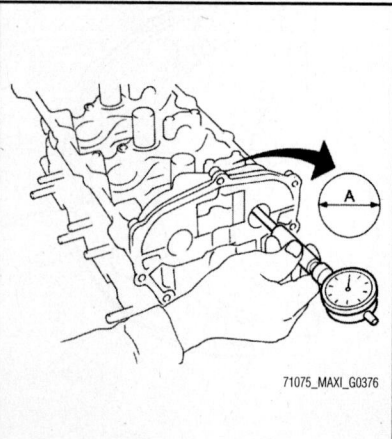

Fig. 32 Measuring the inner diameter of the camshaft bracket

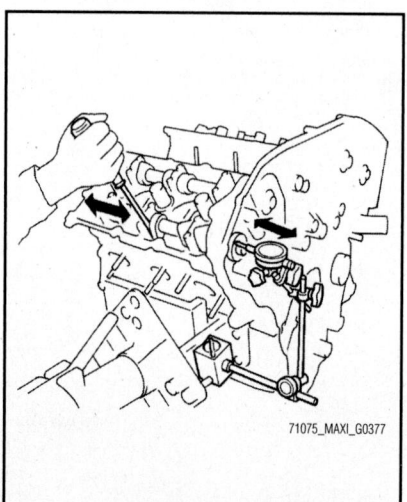

Fig. 33 Measuring camshaft end play

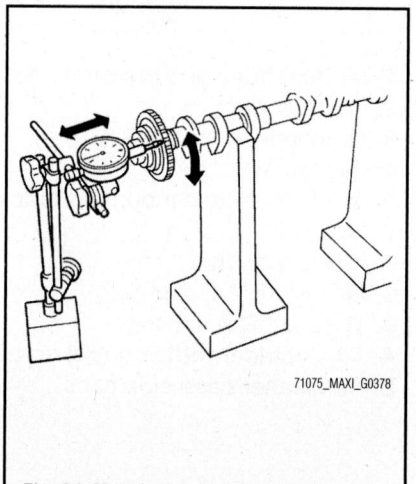

Fig. 34 Measuring camshaft sprocket runout

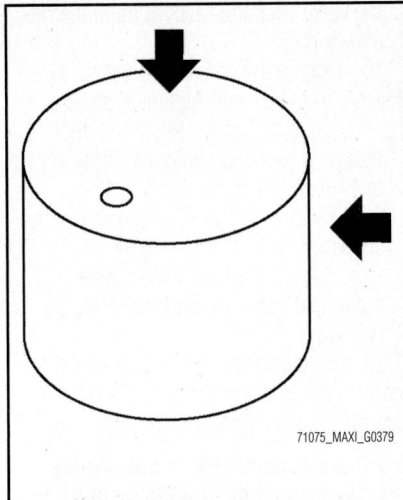

Fig. 35 Checking the valve lifter for wear

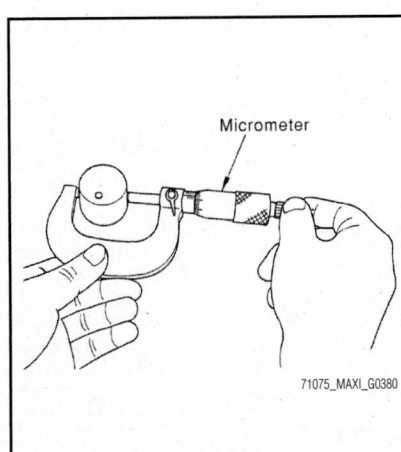

Fig. 36 Measuring the outer diameter of the valve lifter

1. Before servicing the vehicle, refer to the Precautions.

2. Measure the outer diameter of the valve lifter. If out of the specified range, replace the valve lifter.

Valve Lifter Bore Diameter

See Figure 37.

1. Before servicing the vehicle, refer to the Precautions.

2. Using an inside micrometer, measure the diameter of the valve lifter bore of the cylinder head. If out of the specified range, replace the cylinder head assembly.

REMOVAL & INSTALLATION

See Figures 38 through 43.

1. Before servicing the vehicle, refer to the Precautions.

2. Remove the timing chains.

3. Remove the camshaft position brackets.

4. Remove the intake and exhaust camshaft brackets and the camshafts.

   a. Mark the camshafts, camshaft brackets, and bolts so they are placed in the same position and direction for installation.

   b. Equally loosen the camshaft bracket bolts in several steps in the numerical order as shown.

5. Remove valve lifters, if necessary.

➡**Identify valve lifter positions to ensure proper installation.**

6. Remove the secondary timing chain tensioner from the cylinder head.

7. Remove the secondary tensioner with its stopper pin attached.

➡**The stopper pin was attached when the secondary timing chain was removed.**

   To install:

8. Before installation, remove any old Silicone RTV Sealant from component mating surfaces using a scraper.

✷✷ WARNING

Remove the old Silicone RTV Sealant from the bolt holes and threads. Do not scratch or damage the mating surfaces.

9. Before installing the front cam bracket, remove the old Silicone RTV Sealant from the mating surface using a scraper.

10. Turn the crankshaft until No. 1 piston is set at TDC on the compression stroke.

➡**The crankshaft key should line up with the right bank cylinder center line as shown.**

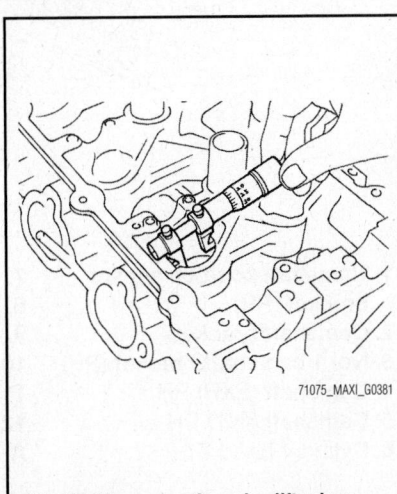

Fig. 37 Measuring the valve lifter bore diameter

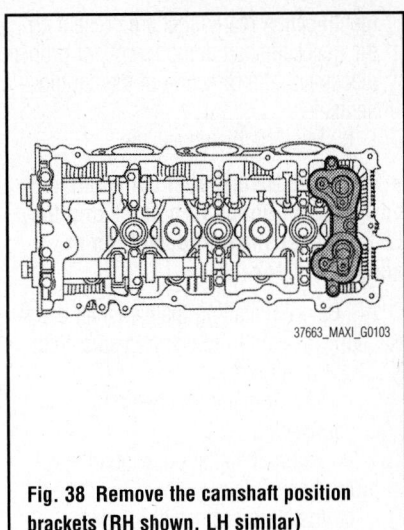

Fig. 38 Remove the camshaft position brackets (RH shown, LH similar)

11. Install the camshaft chain tensioners on both sides of the cylinder head.

12. Install the valve lifters, if removed.

➡**Install them in their original positions.**

13. Install the exhaust and intake camshafts and camshaft brackets.

   a. The intake camshaft has a drill mark on the camshaft sprocket mounting flange.

   b. Follow your identification marks made during removal, or follow the

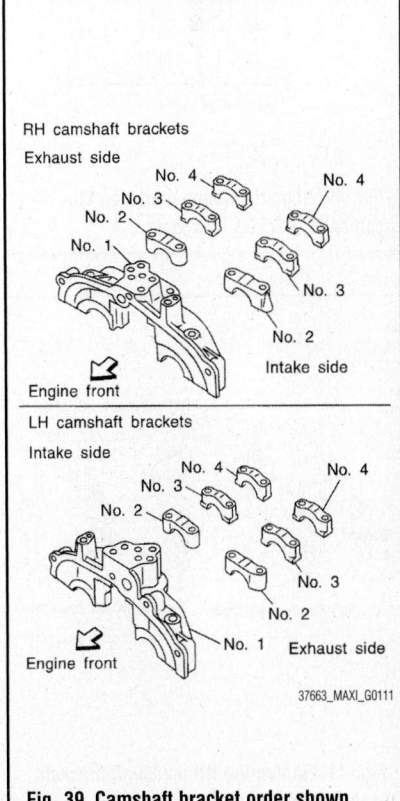

Fig. 39 Camshaft bracket order shown

identification marks that are present on the new camshaft components for proper placement and direction of the components.

   c. Position the camshafts.

➡️**Place the RH exhaust camshaft dowel pin at about 10 o'clock and the LH exhaust camshaft dowel pin at about 2 o'clock.**

14. Before installing the camshaft brackets, apply sealant to the mating surface of No. 1 camshaft bracket.

   a. Use Genuine Silicone RTV Sealant, or equivalent.

   b. Before installation, wipe off any protruding sealant.

   c. Install the camshaft brackets in their original positions and direction. Align the stamp marks as shown.

   d. If checking and adjusting any part of the valve assembly or camshaft, check

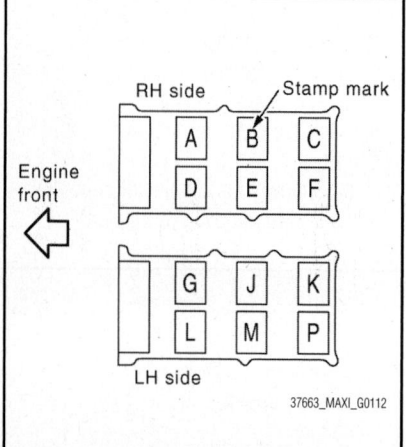

Fig. 40 Align the stamp marks on the camshaft brackets as shown

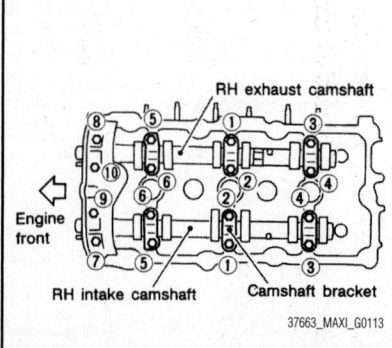

Fig. 41 Tighten the RH camshaft brackets in the numerical order shown

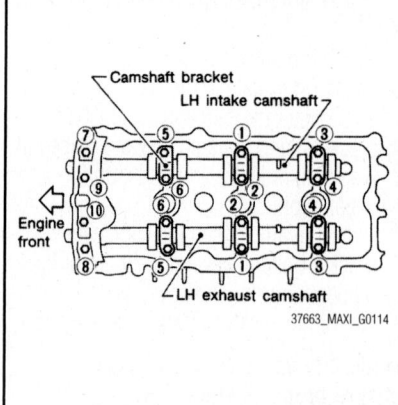

Fig. 42 Tighten the LH camshaft brackets in the numerical order shown

the valve clearance according to the reference data.

15. Tighten the camshaft brackets in 3 steps, in the proper numerical order.

   a. Step 1: Tighten No. 7–10, then tighten 1–6 in the numerical order shown to 17 inch lbs. (2 Nm).

   b. Step 2: Tighten in numerical order shown to 52 inch lbs. (6 Nm).

   c. Step 3: Tighten 1–6 in the numerical order shown to 92 inch lbs. (10 Nm).

16. Measure the difference in levels between the front end faces of the No. 1 camshaft bracket and cylinder head.

➡️**If the measurement is outside the specified range of -0.0055 inches**

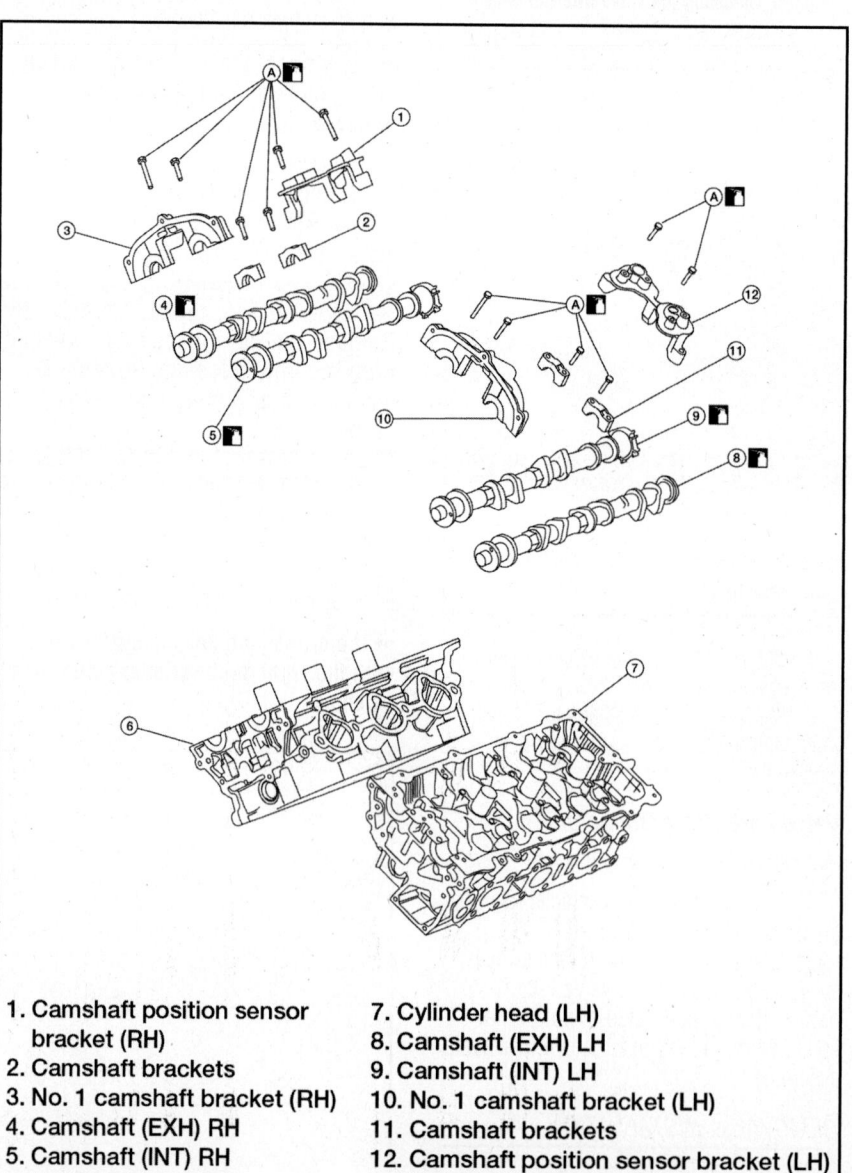

1. Camshaft position sensor bracket (RH)
2. Camshaft brackets
3. No. 1 camshaft bracket (RH)
4. Camshaft (EXH) RH
5. Camshaft (INT) RH
6. Cylinder head (RH)
7. Cylinder head (LH)
8. Camshaft (EXH) LH
9. Camshaft (INT) LH
10. No. 1 camshaft bracket (LH)
11. Camshaft brackets
12. Camshaft position sensor bracket (LH)
A. Follow installation procedure

Fig. 43 Exploded view of camshaft and related component locations

(-0.14mm), reinstall camshaft and camshaft bracket.

17. Install the camshaft position sensors (PHASE) (RH and LH bank).
18. Install the fuel rail and injectors.
19. Install the timing chains.

CRANKSHAFT DAMPER

REMOVAL & INSTALLATION

See Figure 44.

1. Before servicing the vehicle, refer to the Precautions.
2. Remove the engine under cover.
3. Remove the accessory drive belt.
4. Remove the radiator fan.
5. Lock the drive plate using Tool number: (J-50288) attached to the starter bolt hole.

> ✳✳ **WARNING**
>
> **Do not damage the ring gear teeth, or the signal plate teeth behind the ring gear, when setting the Tool.**

6. Loosen the crankshaft pulley and locate the bolt seating surface at 0.39 inch (10mm) from its original position.
7. Position a pulley puller at the recess hole of the crankshaft pulley to remove the crankshaft pulley.

> ✳✳ **WARNING**
>
> **Do not use a puller claw on the crankshaft pulley periphery.**

To install:
8. Install the crankshaft pulley and tighten the bolt in 2 steps.

➡ **Lubricate the thread and seat surface of the bolt with new engine oil.**

a. Step 1: Tighten to 33 ft. lbs. (44 Nm).
b. Step 2: Tighten an additional 84–90° using Tool number: KV10112100 (BT-8653-A), or equivalent torque wrench.

9. Remove the Tool attached to the starter bolt hole.

> ✳✳ **WARNING**
>
> **Do not damage the ring gear teeth, or the signal plate teeth behind the ring gear, when removing the Tool.**

10. Installation of the remaining components is in reverse order of removal.

CRANKSHAFT FRONT SEAL

REMOVAL & INSTALLATION

See Figure 45.

1. Before servicing the vehicle, refer to the Precautions.
2. Remove the crankshaft pulley.
3. Remove the front oil seal from the front cover using a suitable tool.

> ✳✳ **WARNING**
>
> **Be careful not to damage the front cover or the crankshaft.**

To install:
4. Apply new engine oil to the new oil seal.
5. Install the new oil seal in the correct direction.

> ✳✳ **WARNING**
>
> **Press fit straight and avoid causing burrs or tilting the oil seal.**

6. Press-fit the oil seal until it becomes flush with the timing chain case end face, using a suitable tool.
7. Make sure the garter spring in the oil seal is in position and the seal lip is not inverted.
8. Install the crankshaft pulley.

CYLINDER HEAD

REMOVAL & INSTALLATION

See Figures 46 through 50.

1. Before servicing the vehicle, refer to the Precautions.
2. Remove the engine from the vehicle.
3. Remove the rear timing chain case.
4. Remove the intake manifold.
5. Remove the intake and exhaust camshafts.
6. Remove the coolant outlet housing.
7. Remove the RH and LH cylinder head bolts, with a power tool.

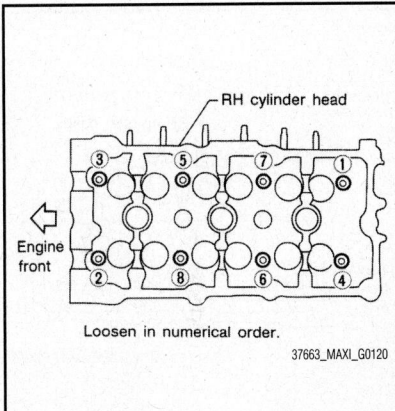

Fig. 46 Loosen the RH cylinder head bolts in the numerical order shown

Fig. 44 Crankshaft pulley shown with pulley puller installed

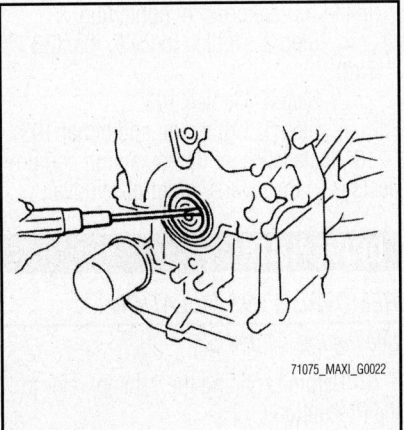

Fig. 45 Removing the front oil seal from the front cover

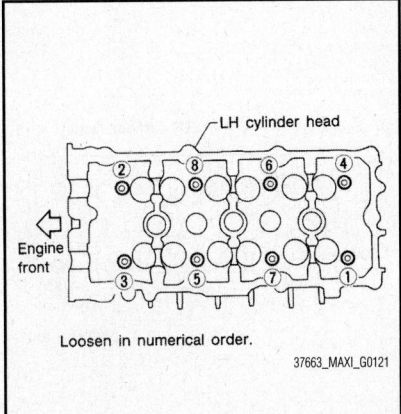

Fig. 47 Loosen the LH cylinder head bolts in the numerical order shown

a. The bolts should be loosened gradually in 3 stages.

b. Loosen the bolts in the numerical order as shown.

8. Remove the cylinder heads and gaskets.

9. Discard the cylinder head gaskets.

➡ **Use new gaskets for installation.**

To install:

10. Turn the crankshaft until No. 1 piston is set at TDC on the compression stroke.

➡ **The crankshaft key should line up with the right bank cylinder center line as shown.**

11. Install new gaskets on the cylinder heads.

✳✳ WARNING

Do not rotate crankshaft and camshaft separately or valves will strike piston heads.

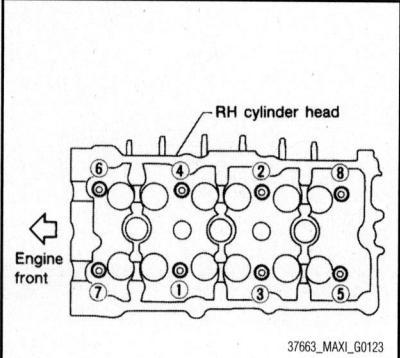

Fig. 48 RH cylinder head tightening sequence

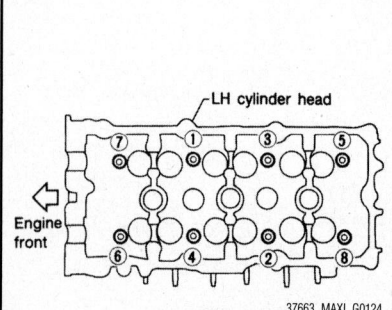

Fig. 49 LH cylinder head tightening sequence

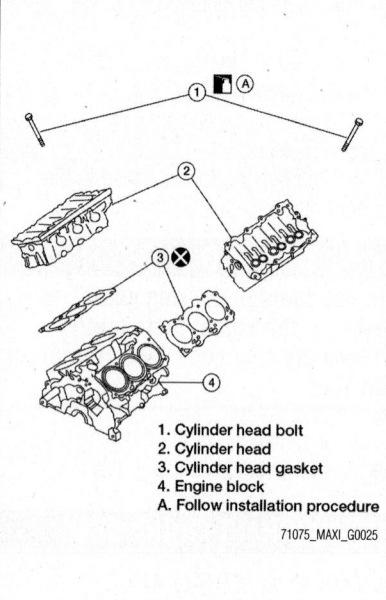

1. Cylinder head bolt
2. Cylinder head
3. Cylinder head gasket
4. Engine block
A. Follow installation procedure

71075_MAXI_G0025

Fig. 50 Cylinder head and related component locations

12. Inspect the cylinder head bolts before installing the cylinder heads.

➡ **Cylinder head bolts are tightened by the degree rotation tightening method. Whenever the size difference between d1 and d2 exceeds the limit of 0.0043 inches (0.11mm), replace the bolts with new ones.**

13. Install the cylinder heads on the cylinder block.

14. Lubricate threads and seat surfaces of the bolts with new engine oil.

15. Tighten the cylinder head bolts in 5 steps in the numerical order shown using Tool KV10112100 (BT-8653-A).

a. Step 1: Tighten to 72 ft. lbs. (98 Nm).

b. Step 2: Loosen to 0 ft. lbs. (0 Nm) in the reverse order of tightening.

c. Step 3: Tighten to 29 ft. lbs. (33 Nm).

d. Step 4: Tighten 103°.

e. Step 5: Tighten an additional 103°.

16. Installation of the remaining components is in the reverse order of removal.

DRIVEPLATE

REMOVAL & INSTALLATION

See Figures 51 and 52.

1. Before servicing the vehicle, refer to the Precautions.

2. Use a suitable tool to lock the driveplate and matchmark the driveplate before removing the bolts.

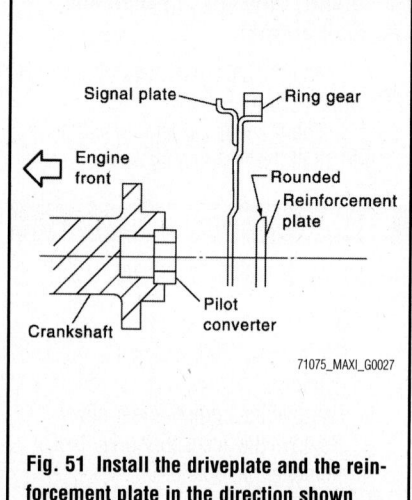

71075_MAXI_G0027

Fig. 51 Install the driveplate and the reinforcement plate in the direction shown

✳✳ WARNING

Do not damage the ring gear teeth, or the signal plate teeth behind the ring gear.

3. Loosen the bolts in diagonal order and remove the driveplate. Consider the following warnings:

• Never place the driveplate with the signal plate facing down

• When handling the signal plate, take care not to damage or scratch it

• Handle the signal plate in a manner that prevents it from becoming magnetized

To install:

4. Installation is in the reverse order of removal.

5. When installing the driveplate to the crankshaft, use the matchmark to correctly align the crankshaft side dowel pin to the driveplate side dowel pin hole.

6. Install the driveplate and the reinforcement plate in the proper direction.

7. Tighten the driveplate bolts in a diagonal pattern in 2 steps.

a. Use a suitable tool to lock the driveplate.

b. Tighten bolts to 65 ft. lbs. (88 Nm).

EXHAUST MANIFOLD

REMOVAL & INSTALLATION

See Figures 53 through 55.

✳✳ CAUTION

Perform the work when the exhaust and cooling system have completely cooled down.

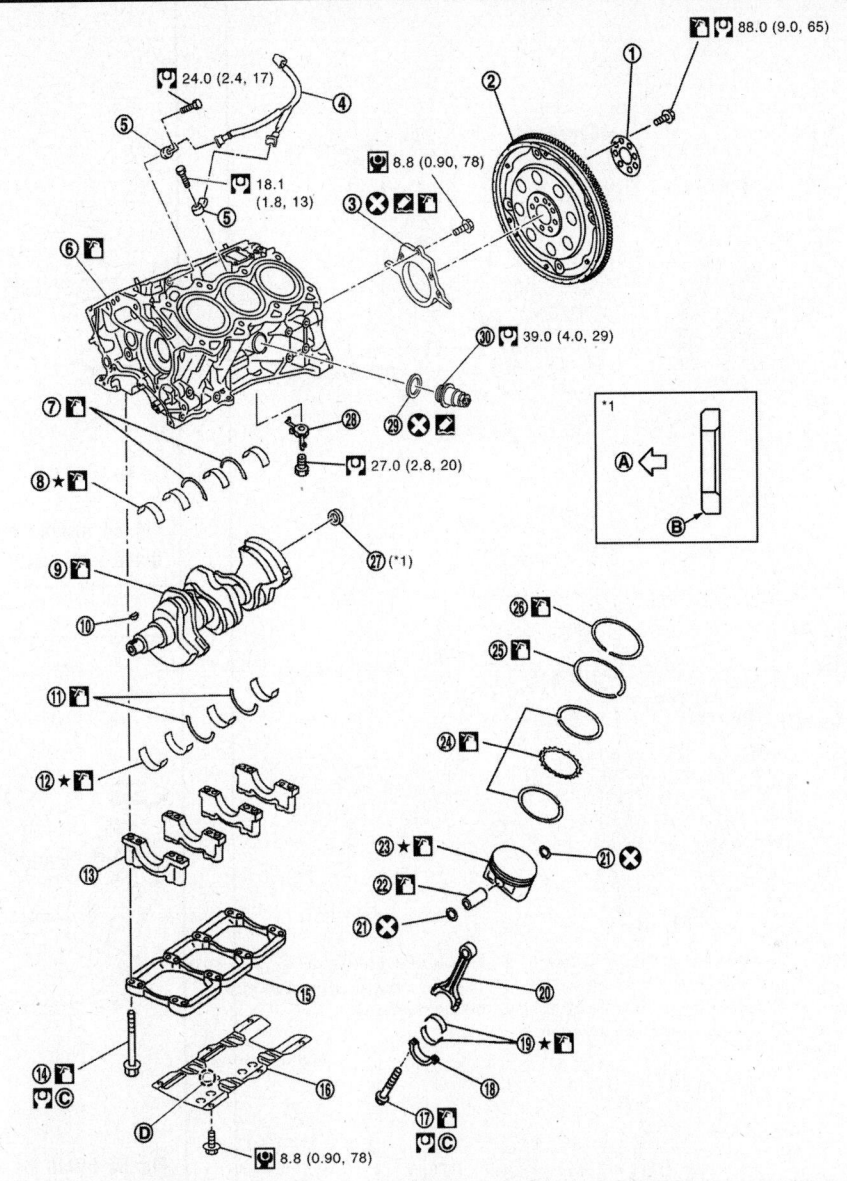

1. Reinforcement plate
2. Drive plate
3. Rear oil seal retainer
4. Sub harness
5. Knock sensor
6. Cylinder block
7. Thrust bearing (upper)
8. Main bearing (upper)
9. Crankshaft
10. Crankshaft key
11. Thrust bearing (lower)
12. Main bearing (lower)
13. Main bearing cap
14. Main bearing cap bolt
15. Main bearing beam
16. Baffle plate
17. Connecting rod bolt
18. Connecting rod bearing cap
19. Connecting rod bearing
20. Connecting rod
21. Snap ring
22. Piston pin
23. Piston
24. Oil ring
25. Second ring
26. Top ring
27. Pilot converter
28. Oil jet
29. Gasket (if equipped)
30. Cylinder block heater (if equipped)
A. Crankshaft side
B. Chamfered
C. Follow the assembly procedure
D. Front mark

71075_MAXI_G0028

Fig. 52 Exploded view of cylinder block component locations

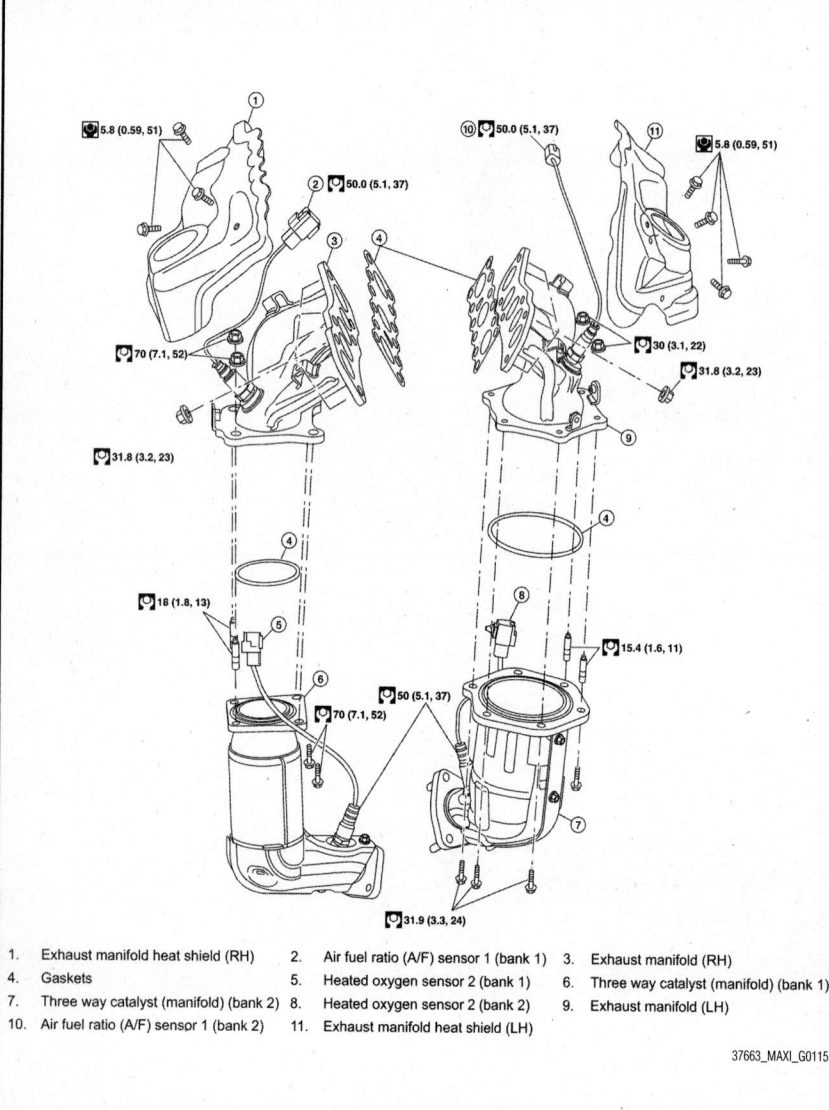

1. Exhaust manifold heat shield (RH)
2. Air fuel ratio (A/F) sensor 1 (bank 1)
3. Exhaust manifold (RH)
4. Gaskets
5. Heated oxygen sensor 2 (bank 1)
6. Three way catalyst (manifold) (bank 1)
7. Three way catalyst (manifold) (bank 2)
8. Heated oxygen sensor 2 (bank 2)
9. Exhaust manifold (LH)
10. Air fuel ratio (A/F) sensor 1 (bank 2)
11. Exhaust manifold heat shield (LH)

37663_MAXI_G0115

Fig. 53 Exploded view of exhaust manifold and three-way catalyst

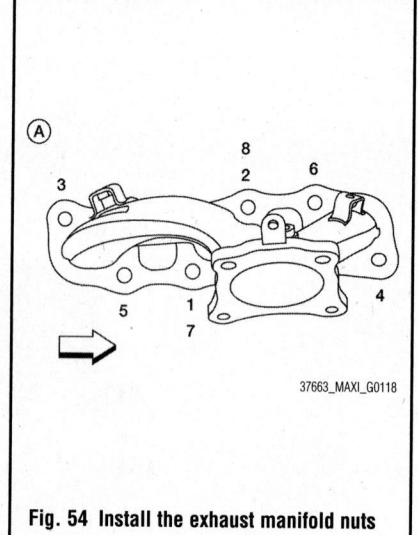

37663_MAXI_G0118

Fig. 54 Install the exhaust manifold nuts in the order as shown RH (A)

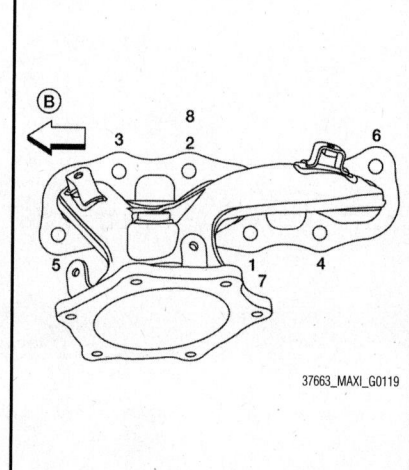

37663_MAXI_G0119

Fig. 55 Install the exhaust manifold nuts in the order as shown LH (B)

✳✳ CAUTION

When removing the front and rear engine mounting through bolts and nuts, lift the engine up slightly for safety.

1. Before servicing the vehicle, refer to the Precautions.

2. Remove the engine and CVT assembly.

3. Remove the RH and LH three-way catalyst supports.

4. Remove the heated oxygen sensor 2 (bank 1), heated oxygen sensor 2 (bank 2), Air Fuel Ratio (A/F) sensor 1 (bank 1) and A/F sensor 1 (bank 2).

   a. Remove the harness connector of each sensor, and disconnect the harness from the bracket and middle clamp.

   b. Remove both heated oxygen sensors and A/F sensors using Tool KV10114400 (J-38365) and KV991J0050 (J-44626).

✳✳ WARNING

Be careful not to damage the heated oxygen sensors or A/F sensors. Discard any heated oxygen sensor which has been dropped from a height of more than 19.7 inches (0.5m) onto a hard surface such as a concrete floor; replace with a new sensor.

5. Remove the exhaust manifold and three-way catalyst heat shields.

6. Remove the three-way catalyst (manifold) (bank 1) and three-way catalyst (manifold) (bank 2) by loosening the bolts first and then removing the nuts and through bolts.

7. Remove the exhaust manifolds RH and LH. Loosen the exhaust manifold nuts in the order as shown.

To install:

8. Installation is in the reverse order of removal.

9. Install the exhaust manifold nuts in the order as shown RH and LH and tighten to specifications.

10. Before installing a heated oxygen sensor or A/F sensor, clean the exhaust manifold threads using the oxygen sensor thread cleaner tool, and apply anti-seize lubricant.

• Oxygen sensor thread cleaner: (J-43897-18)

- Oxygen sensor thread cleaner: (J-43897-12)

※※ WARNING

Do not over-tighten the A/F sensor or heated oxygen sensors. Doing so may cause damage.

INTAKE MANIFOLD

REMOVAL & INSTALLATION

Intake Manifold Collector

See Figures 56 and 57.

※※ CAUTION

To avoid the danger of being scalded, never drain the coolant when the engine is hot.

➡The gasket for the intake manifold collector (upper) is secured with intake manifold collector (lower) bolt. When replacing the upper gasket, the lower gasket must also be replaced.

※※ WARNING

Do not remove the power valves.

➡When removing components such as hoses, tubes/lines, etc., cap or plug openings to prevent fluid from spilling.

1. Before servicing the vehicle, refer to the Precautions.
2. Remove the cowl top.
3. Remove the engine room cover.
4. Remove the front air duct and air duct hose and resonator assembly.
5. Remove the electric throttle control actuator bolts in the reverse order of the tightening sequence and remove the electric throttle control actuator and position it aside.

※※ WARNING

Handle the electric throttle control actuator carefully to avoid any shock to the actuator. Do not disassemble.

6. Disconnect the following:
   - Power brake booster vacuum hose
   - Fuel injector electrical connectors
   - PCV hose
   - Electric throttle control actuator electrical connector
   - EVAP canister purge hose

※※ WARNING

Cover any engine openings to avoid the entry of any foreign material.

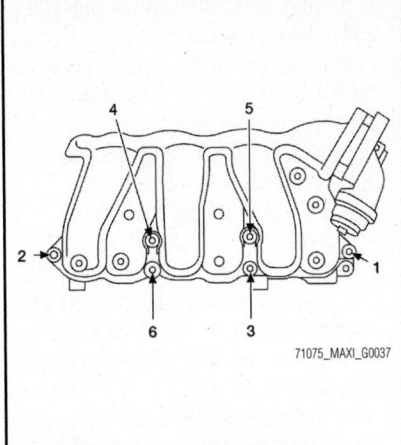

Fig. 56 Intake manifold collector bolt removal sequence

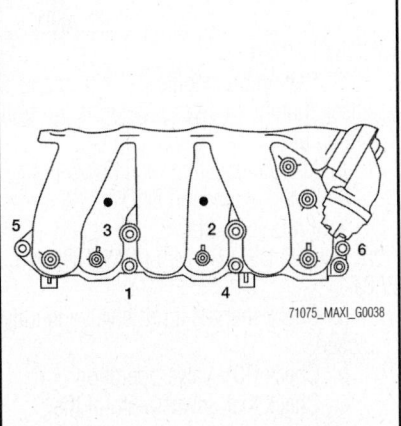

Fig. 57 Intake manifold collector bolt tightening sequence

7. Remove the EVAP canister purge volume solenoid valve bracket bolt. Position the valve aside.
8. Loosen the intake manifold collector bolts in the order shown using a power tool, and remove the intake manifold collector and gasket.
9. If necessary, remove the VIAS control solenoid valve and EVAP canister purge volume control solenoid valve.

To install:

10. Installation is in the reverse order of removal.
11. Tighten the intake manifold collector bolts in the sequence shown.
12. Tighten the electric throttle control actuator bolts in order.
13. After installation, it is necessary to re-calibrate the electric throttle control actuator as follows:
   a. Perform the "Throttle Valve Closed Position Learning" when the harness

connector of the electric throttle control actuator is disconnected.
   b. Perform the "Idle Air Volume Learning" when the electric throttle control actuator is replaced.

Throttle Valve Closed Position Learning

The Throttle Valve Closed Position Learning procedure is an operation to learn the fully closed position of the throttle valve by monitoring the throttle position sensor output signal. It must be performed each time the harness connector of electric throttle control actuator or ECM is disconnected or the electric throttle control actuator is cleaned.

With Consult Tool

1. Before servicing the vehicle, refer to the Precautions.
2. Turn the ignition switch ON.
3. Select "CLSD THL POS LEARN" in "WORK SUPPORT" mode.
4. Follow the instructions on the CONSULT display.
5. Turn the ignition switch OFF and wait at least 10 seconds.
6. Check that the throttle valve moves during the above 10 seconds by confirming the operating sound.

Without Consult Tool

7. Before servicing the vehicle, refer to the Precautions.
8. Start the engine.

➡Engine coolant temperature is 77°F (25°C) or less before engine starts.

9. Warm up the engine.

➡Raise the engine coolant temperature until it reaches 149°F (65°C) or more.

10. Turn the ignition switch OFF and wait at least 10 seconds.
11. Check that the throttle valve moves during the above 10 seconds by confirming the operating sound.

Idle Air Volume Learning

The Idle Air Volume Learning procedure is a function of the ECM to learn the idle air volume that keeps the engine idle speed within the specific range. It must be performed under the following conditions:

- Each time the electric throttle control actuator or the ECM is replaced
- The idle speed or ignition timing are out of specification

1. Before servicing the vehicle, refer to the Precautions.
2. Step 1: PRECONDITIONING

a. Before performing the Idle Air Volume Learning, check that all of the following conditions are satisfied. Learning will be cancelled if any of the following conditions are missed for even a moment.

- Battery voltage: More than 12.9 volts (at idle)
- Engine coolant temperature: 158–212°F (70–100°C)
- Selector lever: P or N
- Electric load switch: OFF (Air conditioner, head lamp, rear window defogger). On vehicles equipped with daytime light systems, if the parking brake is applied before the engine is started the head lamp will not illuminate.
- Steering wheel: Neutral (Straight-ahead position)
- Vehicle speed: Stopped
- Transmission: Warmed-up

b. With the CONSULT Tool: Drive the vehicle until "ATF TEMP SEN" in "DATA MONITOR" mode of "CVT" system indicates less than 0.9 volt.

c. Without the CONSULT Tool: Drive the vehicle for 10 minutes.

d. Will the CONSULT Tool be used?
- If Yes, go to Step 2.
- If no, got to step 3.

3. Step 2: PERFORM IDLE AIR VOLUME LEARNING With CONSULT Tool

a. Perform the Accelerator Pedal Released Position Learning procedure.

b. Perform the Throttle Valve Closed Position Learning procedure.

c. Start the engine and warm it up to normal operating temperature.

d. Select "IDLE AIR VOL LEARN" in "WORK SUPPORT" mode.

e. Touch "START" and wait 20 seconds.

f. Is "CMPLT" displayed on the CONSULT screen?
- If yes, go to Step 4.
- If no, go to Step 5.

4. Step 3: PERFORM IDLE AIR VOLUME LEARNING Without the CONSULT Tool

Please note the following:
- It is better to count the time accurately with a clock
- It is impossible to switch the diagnostic mode when an accelerator pedal position sensor circuit has a malfunction

a. Perform the Accelerator Pedal Released Position Learning procedure.

b. Perform the Throttle Valve Closed Position Learning procedure.

c. Start the engine and warm it up to normal operating temperature.

d. Turn the ignition switch OFF and wait at least 10 seconds.

e. Confirm that the accelerator pedal is fully released, turn the ignition switch ON and wait 3 seconds.

f. Repeat the following procedure quickly 5 times within 5 seconds.
- Fully depress the accelerator pedal
- Fully release the accelerator pedal

g. Wait 7 seconds, fully depress the accelerator pedal for approximately 20 seconds until the MIL stops blinking and turns ON.

h. Fully release the accelerator pedal within 3 seconds after the MIL turns ON.

i. Start the engine and let it idle.

j. Wait 20 seconds.

k. Go to Step 4.

5. Step 4: CHECK IDLE SPEED AND IGNITION TIMING

a. Rev up the engine 2–3 times and check that the idle speed and the ignition timing are within specifications.

b. Is the inspection result normal?
- If yes, inspection END
- If no, go to Step 5.

6. Step 5: DETECT MALFUNCTIONING PART-I

a. Check that the throttle valve is fully closed.

b. Check PCV valve operation.

c. Check that downstream of the throttle valve is free from air leakage.

d. Is the inspection result normal?
- If yes, got to Step 6.
- If no, repair or replace the malfunctioning part.

7. Step 6: DETECT MALFUNCTIONING PART-II

a. When the above 3 items check out OK, the engine component parts and their installation condition are questionable.

b. Check and eliminate the cause of the incident.

c. If any of the following conditions occur after the engine has started, eliminate the cause of the incident and perform Idle Air Volume Learning procedure again:
- Engine stalls
- Incorrect idle

OIL PAN

REMOVAL & INSTALLATION

Lower Oil Pan

See Figures 58 through 60.

Do not remove the oil pan until the exhaust system and cooling system have completely cooled.

1. Before servicing the vehicle, refer to the Precautions.
2. Drain the engine oil.
3. Loosen the lower oil pan bolts in the order shown.
4. Remove the lower oil pan.
   a. Insert Tool KV10111100 (J-37228) between the lower oil pan and the upper oil pan.

Be careful not to damage the mating surface. Do not insert a screwdriver, this will damage the mating surfaces.

b. Slide the Tool by tapping its side with a hammer to remove the lower oil pan from the upper oil pan.

5. If re-installing the original lower oil pan, remove the old sealant from the mating surfaces using a scraper.

6. Remove the old sealant from mating surface of the upper oil pan.

7. Remove the old sealant from the bolt holes and threads.

Do not scratch or damage the mating surfaces when cleaning off the old sealant.

8. Clean any debris from the oil strainer.

To install:

9. Apply a continuous bead of sealant to the lower oil pan.

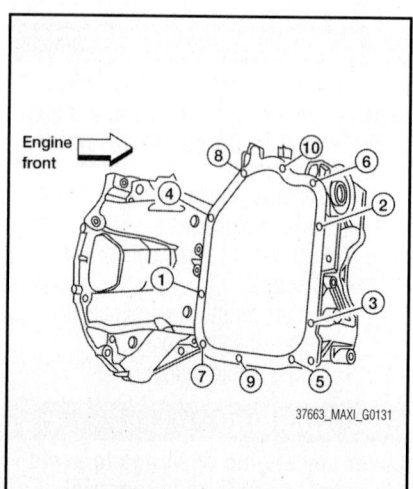

Fig. 58 Lower oil pan bolts removal order

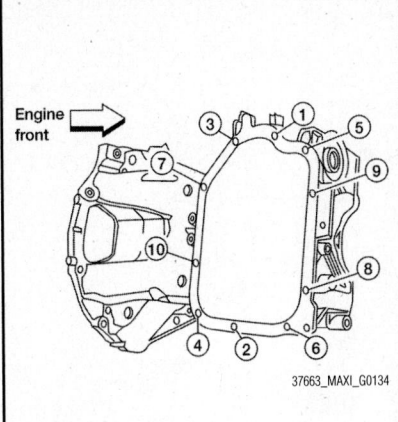

Fig. 59 Tighten the lower oil pan bolts in order

a. Use Genuine Silicone RTV Sealant, or equivalent.

b. Installation must be done within 5 minutes after applying sealant.

10. Install the lower oil pan. Tighten the lower oil pan bolts in the order shown.

11. Wait at least 30 minutes before refilling the engine with oil.

12. Start the engine and check for leaks.

13. Inspect the engine oil level.

Upper Oil Pan

See Figures 61 and 62.

> ※ **CAUTION**
>
> **Do not remove the oil pan until the exhaust system and cooling system have completely cooled off.**

> ※ **CAUTION**
>
> **When removing the front and rear engine through bolts and nuts, lift the engine up slightly for safety.**

> ※ **WARNING**
>
> **When removing the upper oil pan from the engine, first remove the Crankshaft Position Sensor (CKP). Be careful not to damage sensor edges or signal plate teeth.**

1. Before servicing the vehicle, refer to the Precautions.

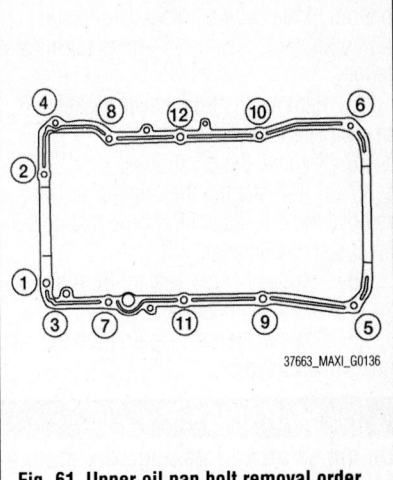

Fig. 61 Upper oil pan bolt removal order

2. Remove the engine from the vehicle.

3. Drain the engine oil.

4. Remove the oil dipstick.

5. Remove the drive belt.

6. Disconnect the A/C compressor harness connector.

7. Remove the A/C compressor bolts and remove the A/C compressor.

8. Remove the coolant pipe bolts.

9. Disconnect the coolant lines from the engine oil cooler.

10. Remove the oil filter and engine oil cooler from the upper oil pan.

11. Remove the oil pressure switch and the CKP from the upper oil pan.

12. Remove the lower oil pan.

13. Remove the upper oil pan.

a. Loosen the bolts in the order shown.

b. Insert an appropriate size tool into the notch of the upper oil pan as shown.

c. Pry off the upper oil pan by moving the tool up and down.

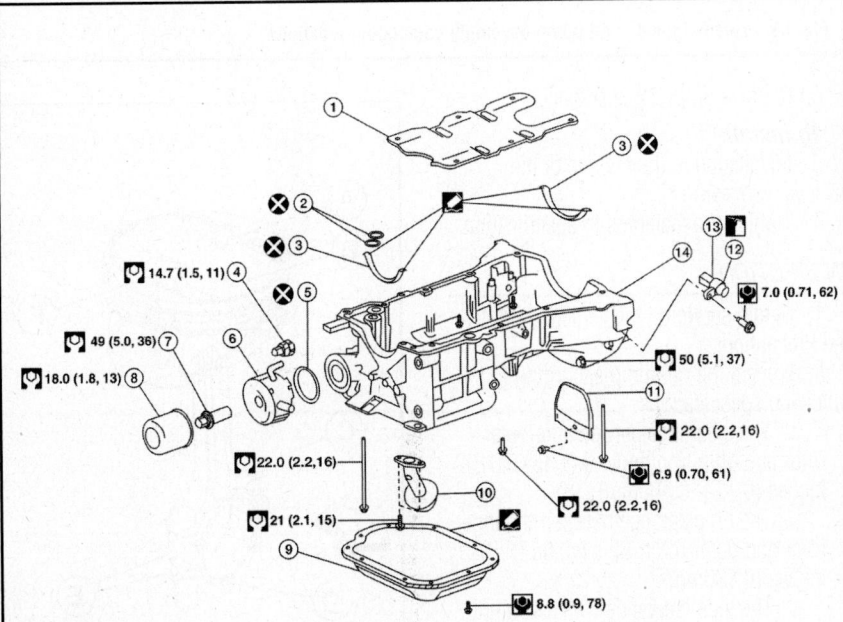

1. Oil pan baffle
2. O-ring
3. Gasket
4. Oil pressure switch
5. Oil cooler gasket
6. Oil cooler
7. Oil cooler connection
8. Oil filter
9. Lower oil pan
10. Oil strainer
11. Rear plate cover
12. Crankshaft position sensor (POS)
13. O-ring
14. Upper oil pan

Fig. 60 Exploded view of engine oil pan and related component locations

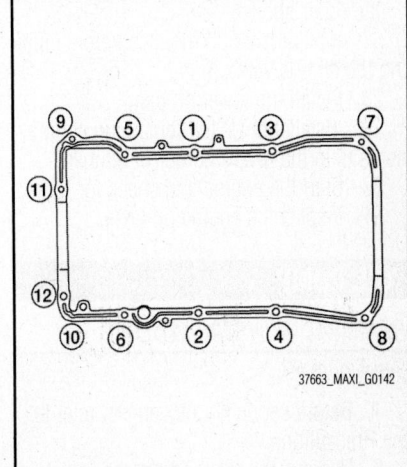

Fig. 62 Upper oil pan bolt tightening order

14. Remove the O-ring seals from the bottom of the cylinder block and the oil pump housing. Use new O-rings for installation.

15. Remove the front cover gasket and rear oil seal retainer gasket.

16. Remove the oil strainer.

17. If re-installing the original oil pan, remove the old sealant from the mating surfaces using a scraper.

18. Remove the old sealant from mating surface of the cylinder block.

19. Remove the old sealant from the bolt holes and threads.

✳✳ WARNING

Do not scratch or damage the mating surfaces when cleaning off the old sealant.

20. Clean any debris from the oil strainer.

To install:

21. Install the oil strainer and tighten the bolt to the specified torque.

22. Apply Genuine Silicone RTV Sealant or equivalent, to the front cover gasket and the rear oil seal retainer gasket as shown.

23. Install the front cover gasket and rear oil seal retainer gasket.

24. Apply a bead of sealant to the cylinder block mating surface of the upper oil pan to a limited portion as shown.

   a. Use Genuine Silicone RTV Sealant, or equivalent.

   b. Be sure the sealant is applied to the limited portion.

   c. Attaching should be done within 5 minutes after coating.

25. Install new O-rings on the cylinder block and oil pump body.

26. Install the upper oil pan.

   a. Tighten the upper oil pan bolts in the order shown.

27. Wait at least 30 minutes before refilling the engine with oil.

28. Install the lower oil pan.

29. Installation of the remaining components is in the reverse order of removal.

30. Start the engine and check for leaks.

31. Inspect the engine oil level.

OIL PUMP

REMOVAL & INSTALLATION

See Figure 63.

1. Before servicing the vehicle, refer to the Precautions.

2. Remove the engine from the vehicle.

3. Remove the upper oil pan.

4. Remove the timing chain.

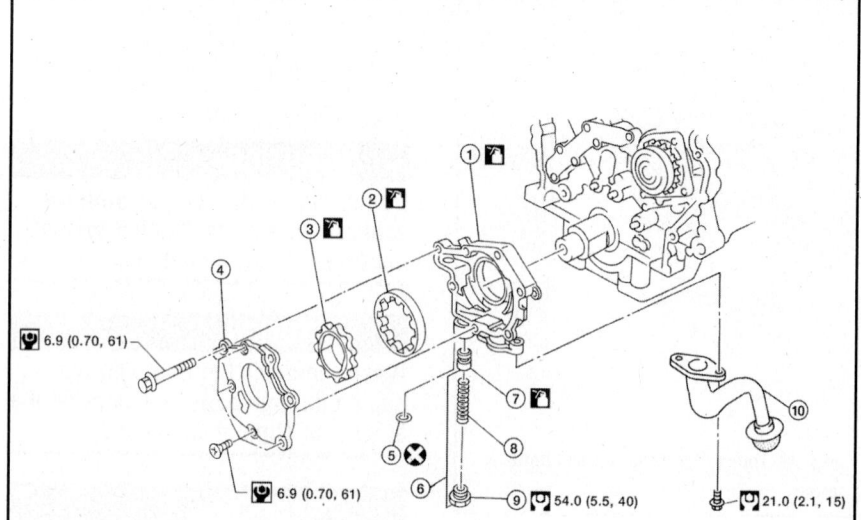

1. Oil pump housing
2. Outer rotor
3. Inner rotor
4. Oil pump cover
5. O-ring
6. Regulator valve set
7. Regulator valve
8. Spring
9. Regulator plug
10. Oil strainer

71075_MAXI_G0387

Fig. 63 Exploded view of oil pump assembly component locations

5. Remove oil pump assembly.

To install:

6. Installation is the reverse of the removal procedure.

7. Tighten the fasteners to specification.

INSPECTION

1. Before servicing the vehicle, refer to the Precautions.

2. Inspect the oil pump according to the following specifications:

   a. The clearance between the outer rotor and oil pump body: 0.0045–0.0102 inches (0.114–0.260mm).

   b. The tip clearance between the inner rotor and outer rotor: Below 0.0071 inches (0.180mm).

   c. The side clearance with a straight-edge between the inner rotor and oil pump body: 0.0012–0.0028 inches (0.030–0.070mm).

   d. The side clearance with a straight-edge between the outer rotor and oil pump body: 0.0020–0.0043 inches (0.050–0.110mm).

PISTONS & RINGS

POSITIONING

See Figures 64 through 66.

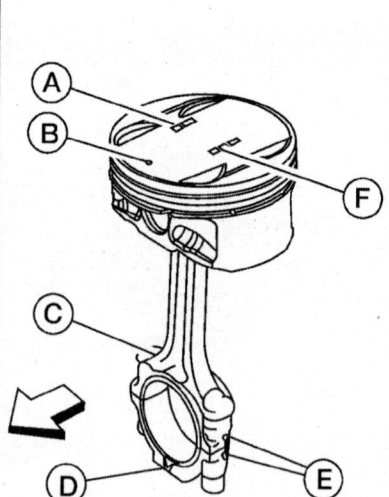

A. Piston grade number
B. Piston front mark on the crown
C. Oil hole
D. Connecting rod front mark
E. Cylinder No.
F. Pin grade number
Arrow: Front

71075_MAXI_G0040

Fig. 64 Piston and connecting rod marks shown

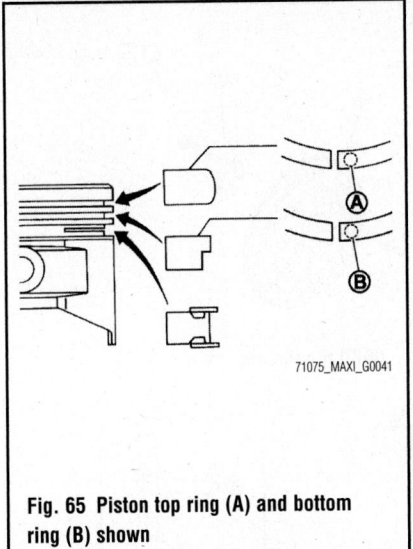

Fig. 65 Piston top ring (A) and bottom ring (B) shown

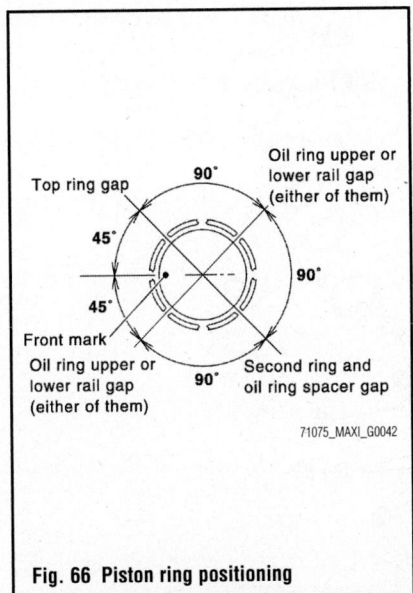

Fig. 66 Piston ring positioning

REAR MAIN SEAL

COMPONENT LOCATIONS

See Figure 67.

REMOVAL & INSTALLATION

See Figures 68 and 69.

1. Before servicing the vehicle, refer to the Precautions.

2. Remove the upper oil pan.

3. Remove the driveplate.

4. Remove the rear oil seal retainer using Tool Number: KV10111100 (J-37228), or equivalent.

Consider the following warnings:

• Be careful not to damage the mating surface

• If rear the oil retainer is removed, replace it with a new one

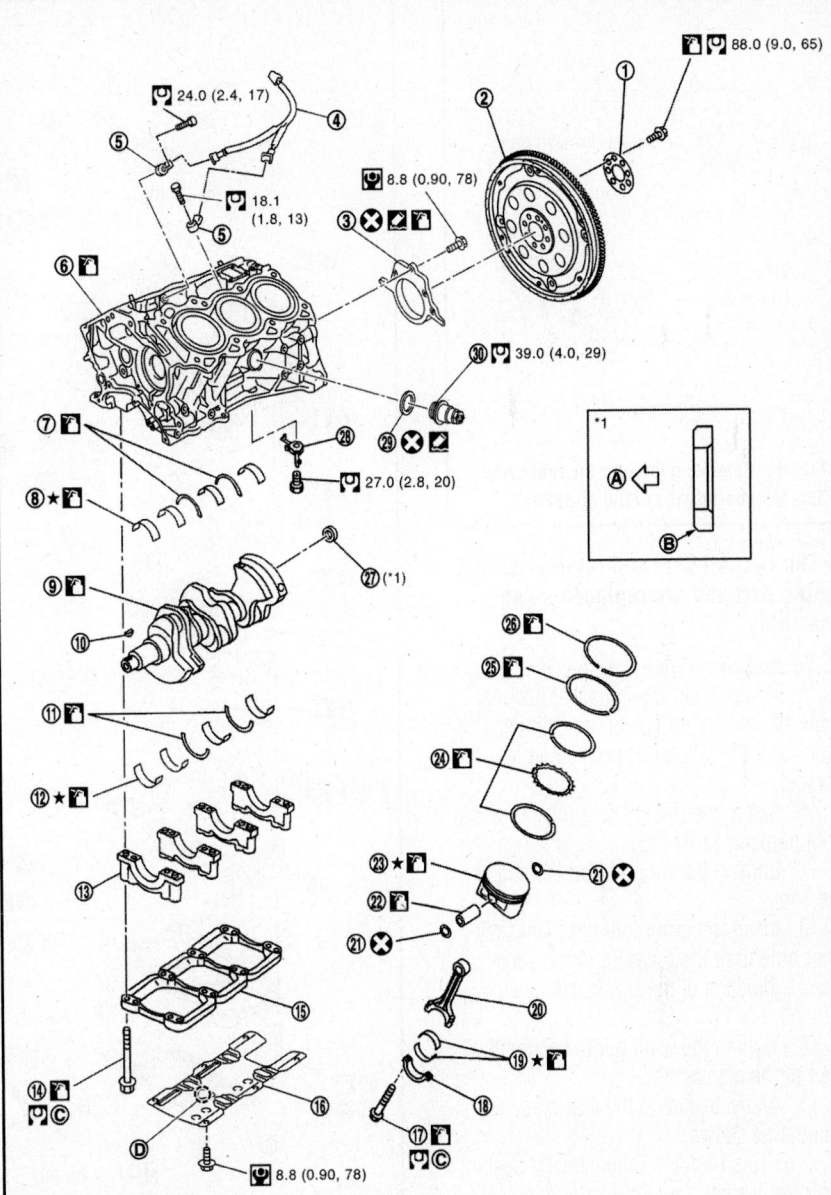

1. Reinforcement plate
2. Drive plate
3. Rear oil seal retainer
4. Sub harness
5. Knock sensor
6. Cylinder block
7. Thrust bearing (upper)
8. Main bearing (upper)
9. Crankshaft
10. Crankshaft key
11. Thrust bearing (lower)
12. Main bearing (lower)
13. Main bearing cap
14. Main bearing cap bolt
15. Main bearing beam
16. Baffle plate
17. Connecting rod bolt
18. Connecting rod bearing cap
19. Connecting rod bearing
20. Connecting rod
21. Snap ring
22. Piston pin
23. Piston
24. Oil ring
25. Second ring
26. Top ring
27. Pilot converter
28. Oil jet
29. Gasket (if equipped)
30. Cylinder block heater (if equipped)
A. Crankshaft side
B. Chamfered
C. Follow the assembly procedure
D. Front mark

Fig. 67 Exploded view of cylinder block component locations

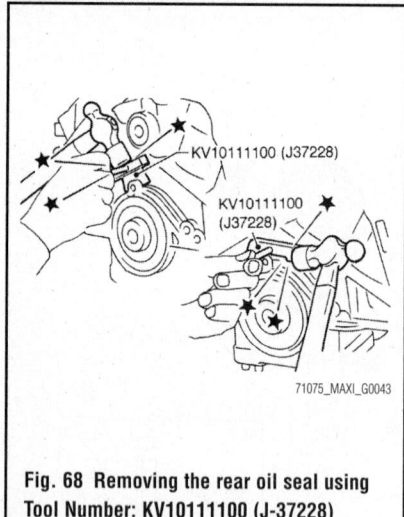

Fig. 68 Removing the rear oil seal using Tool Number: KV10111100 (J-37228)

➡The rear oil seal and retainer form a single part and are replaced as an assembly.

To install:

5. Remove old liquid gasket material from the mating surface of the cylinder block and the oil pan using a suitable scraper.

6. Install the rear oil seal retainer using Tool number: (J-47128).

7. Loosen the wing nut on the end of the Tool.

8. Insert the arbor into the crankshaft pilot hole until the outer lip of the Tool covers the edge of the crankshaft sealing surface.

9. Tighten the wing nut to secure the Tool to the crankshaft.

10. Apply sealant to the rear oil seal retainer as shown.

a. Use Genuine Silicone RTV Sealant, or equivalent.

b. Installation should be done within 5 minutes after applying liquid gasket.

c. Do not fill the engine with oil for at least 30 minutes after the components are installed to allow the sealant to cure.

11. Lubricate the sealing surface of the new rear main seal with new engine oil.

12. Slide the new rear main seal over the Tool and onto the crankshaft.

13. Loosen the wing nut and push the threaded rod into the handle to remove the Tool.

14. Tighten the rear oil seal retainer bolts to specification.

15. Installation of the remaining components is in the reverse order of removal.

Consider the following warnings:

• When replacing an engine or transmission you must make sure the dowels are installed correctly during re-assembly.

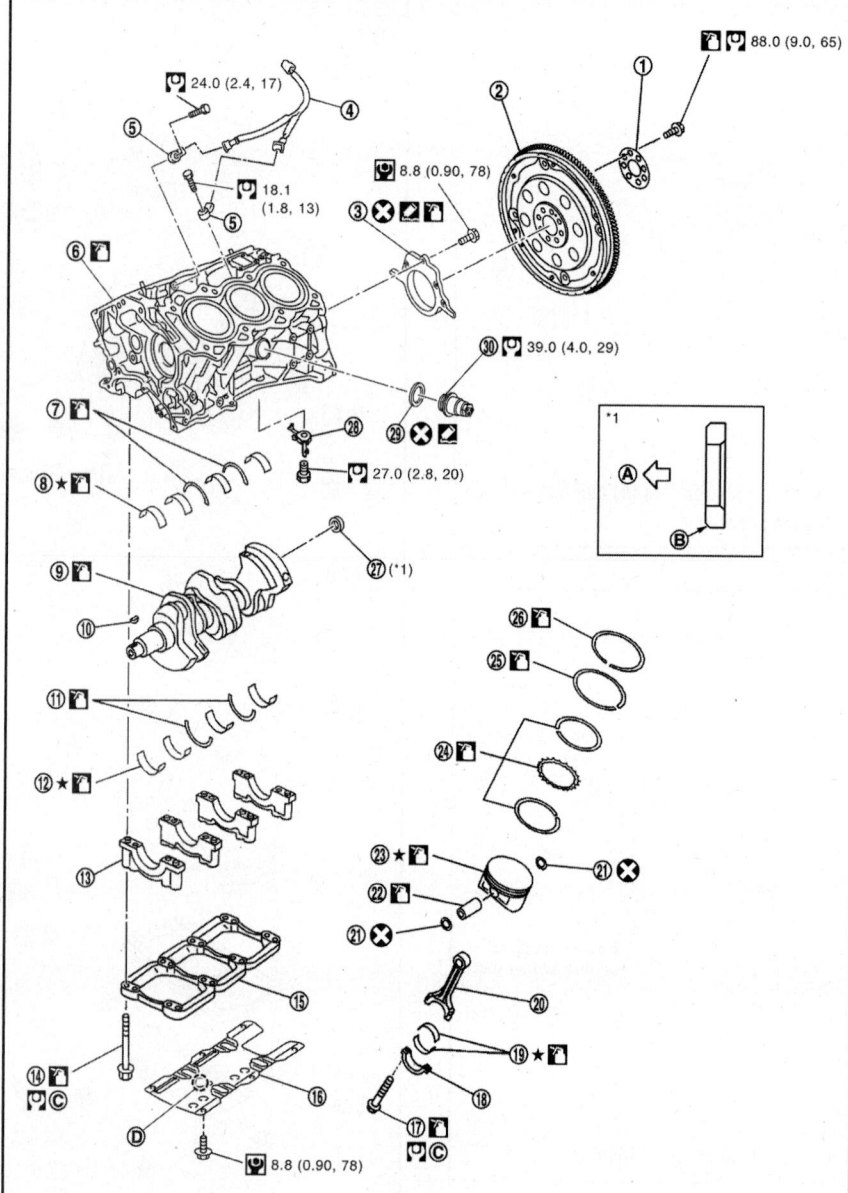

1. Reinforcement plate
2. Drive plate
3. Rear oil seal retainer
4. Sub harness
5. Knock sensor
6. Cylinder block
7. Thrust bearing (upper)
8. Main bearing (upper)
9. Crankshaft
10. Crankshaft key
11. Thrust bearing (lower)
12. Main bearing (lower)
13. Main bearing cap
14. Main bearing cap bolt
15. Main bearing beam
16. Baffle plate
17. Connecting rod bolt
18. Connecting rod bearing cap
19. Connecting rod bearing
20. Connecting rod
21. Snap ring
22. Piston pin
23. Piston
24. Oil ring
25. Second ring
26. Top ring
27. Pilot converter
28. Oil jet
29. Gasket (if equipped)
30. Cylinder block heater (if equipped)
A. Crankshaft side
B. Chamfered
C. Follow the assembly procedure
D. Front mark

Fig. 69 Exploded view of cylinder block component locations

• Improper alignment caused by missing dowels may cause vibration, oil leaks or breakage of drivetrain components.

ROCKER ARM COVER

REMOVAL & INSTALLATION

Left Side

See Figures 70 and 71.

1. Before servicing the vehicle, refer to the Precautions.
2. Remove the engine cover.
3. Remove the front air duct.
4. Remove the blow by hose from the rocker cover.
5. Remove the camshaft position sensor.

Consider the following warnings:
• Handle the camshaft position sensor carefully to avoid dropping and shocks
• Do not disassemble
• Do not allow metal powder to adhere to the magnetic part at the sensor tip
• Do not place the sensors in a location where they are exposed to magnetism
6. Disconnect the ignition coil connectors.
7. Remove the ignition coils.

※※ WARNING

Never shock the ignition coils.

8. Remove the LH rocker cover bolts from the cylinder head in sequence.

To install:

9. Installation is in the reverse order of removal.
10. Apply sealant to the areas on the front corners.

➥**Use Genuine Silicone RTV Sealant or equivalent. Use Tool number : WS39930000 or equivalent.**

a. Installation should be done within 5 minutes after applying liquid gasket.
b. Do not fill the engine with oil for at least 30 minutes after the components are installed to allow the sealant to cure.
11. Tighten the rocker cover bolts in 2 steps in order.
a. Step 1: 17 inch lbs. (2 Nm).
b. Step 2: 74 inch lbs. (8 Nm).

Right Side

1. Before servicing the vehicle, refer to the Precautions.
2. Remove the engine cover.
3. Disconnect the Mass Air Flow (MAF) sensor electrical connector and remove the air cleaner assembly and air intake tubes.
4. Remove the intake manifold collector.
5. Remove the gasket and the electric throttle control actuator.
6. Remove the ignition coils.

※※ WARNING

Never shock the ignition coils.

7. Remove the camshaft position sensor.

Consider the following warnings:
• Handle the camshaft position sensor carefully to avoid dropping and shocks
• Do not disassemble
• Do not allow metal powder to adhere to the magnetic part at sensor tip
• Do not place the sensors in a location where they are exposed to magnetism
8. Remove the RH rocker cover bolts from cylinder head in order.

To install:

9. Installation is in the reverse order of removal.
10. Apply sealant to the areas on the front corners.

➥**Use Genuine Silicone RTV Sealant or equivalent. Use Tool number : WS39930000 or equivalent.**

a. Installation should be done within 5 minutes after applying liquid gasket.
b. Do not fill the engine with oil for at least 30 minutes after the components are installed to allow the sealant to cure.
11. Tighten the rocker cover bolts in 2 steps in the tightening order.
a. Step 1: 17 inch lbs. (2 Nm).
b. Step 2: 74 inch lbs. (8 Nm).

TIMING CHAIN COVER, CHAIN, TENSIONER, & SPROCKETS

REMOVAL & INSTALLATION

See Figures 72 through 76.

Consider the following warnings:
• After removing the timing chains, do not turn the crankshaft and camshaft separately, or the valves will strike the pistons
• When installing the camshafts, chain tensioners, oil seals, or other sliding parts, lubricate the contacting surfaces with new engine oil
• Apply new engine oil to the bolt threads and seat surfaces when installing the camshaft sprockets, camshaft brackets, and crankshaft pulley
1. Before servicing the vehicle, refer to the Precautions.

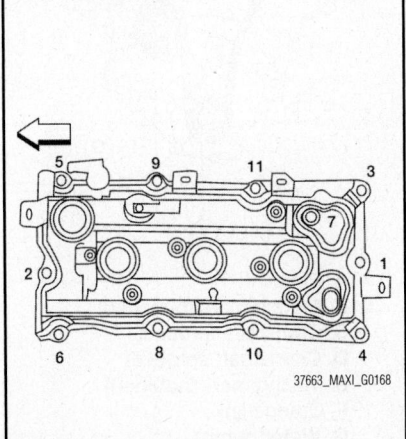

Fig. 70 LH rocker cover bolt removal sequence shown

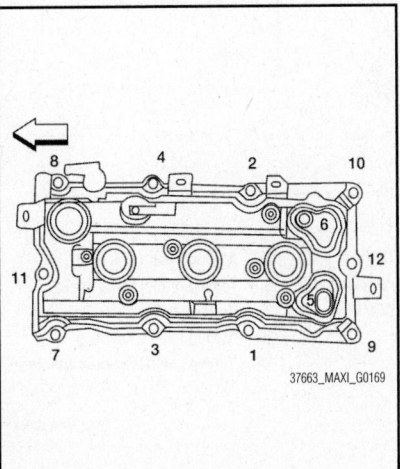

Fig. 71 LH rocker cover bolt tightening sequence shown

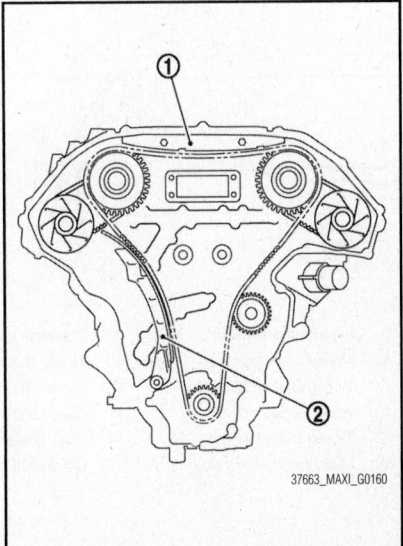

Fig. 72 Remove the internal chain guide (1) and slack guide (2)

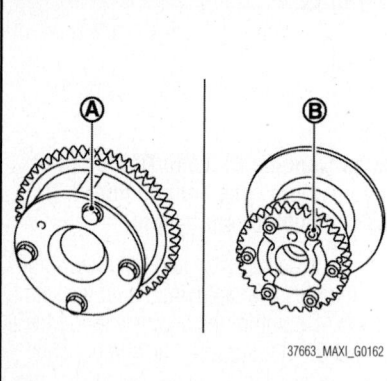

Fig. 73 Do not disassemble the camshaft sprockets; never loosen bolts (A) and (B) as shown

2. Remove the front timing chain case.

3. Remove the intake manifold collector.

4. Remove the engine oil dipstick.

5. Place paint marks on the timing chain and sprockets to indicate the correct position of the components for installation.

6. Remove the timing chain tensioner (primary).

a. Pull the lever down and release the plunger stopper tab. The plunger stopper tab can be pushed up to release the coaxial structure with the lever.

b. Insert the stopper pin into the timing chain tensioner (primary) body hole to hold the lever, and keep the tab released. An Allen wrench of 0.047 inches (1.2mm) is used for a stopper pin.

c. Insert the plunger into the tensioner body by pressing the slack guide.

d. Keep the slack guide pressed and hold it by pushing the stopper pin through the lever hole and body hole.

e. Remove the bolts and remove the timing chain tensioner (primary).

7. Remove the internal chain guide and slack guide.

8. Remove the timing chain (primary) and the crankshaft sprocket.

☀☀ WARNING

After removing the timing chains, do not turn the crankshaft and camshaft separately, or the valves will strike the pistons.

9. Attach a suitable stopper pin to the right and left timing chain tensioners (secondary).

10. Remove the timing chains (secondary) with the camshaft sprockets (INT) and (EXH).

a. Insert a metal or resin plate of 0.020 inches (0.5mm) into the guide between the timing chain (secondary) and timing chain tensioner (secondary) plunger. Remove the camshaft sprocket and the timing chain (secondary) with the timing chain removed from the guide groove.

☀☀ WARNING

The timing chain tensioner plunger can move while the stopper pin is inserted in the timing chain tensioner. The plunger can come out of the tensioner when the timing chain is removed. Use caution during removal.

➡ **Apply paint to the timing chain and camshaft sprockets for alignment during installation.**

b. Remove the camshaft sprocket (INT) and (EXH) bolts.

c. Secure the hexagonal portion of the camshaft using a wrench to loosen the bolts.

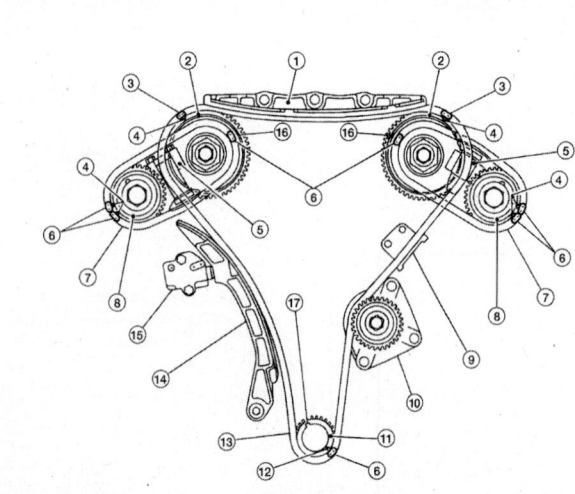

1. Internal chain guide	2. Camshaft sprocket (INT)	3. Mating mark (pink link)
4. Mating mark (punched)	5. Timing chain tensioner (secondary)	6. Mating mark (orange link)
7. Timing chain (secondary)	8. Camshaft sprocket (EXH)	9. Tension guide
10. Water pump	11. Crankshaft sprocket	12. Mating mark (notched)
13. Timing chain (primary)	14. Slack guide	15. Timing chain tensioner (primary)
16. Mating mark (back side)	17. Crankshaft key	

Fig. 74 View of timing chain assembly

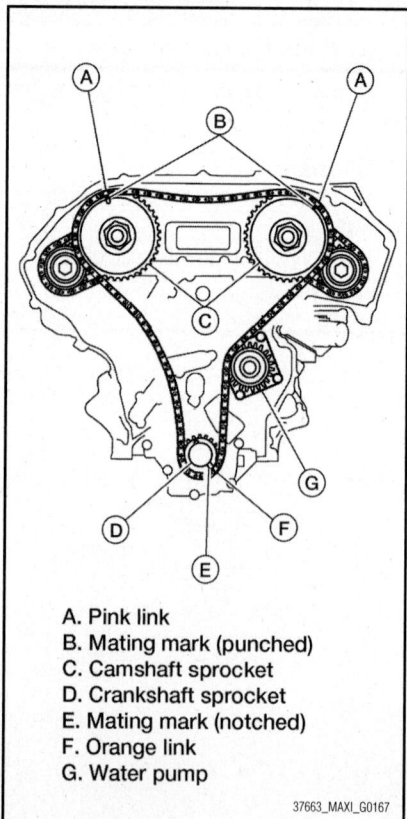

A. Pink link
B. Mating mark (punched)
C. Camshaft sprocket
D. Crankshaft sprocket
E. Mating mark (notched)
F. Orange link
G. Water pump

Fig. 75 Install the timing chain (primary)

d. Handle the sprockets as an assembly.

e. Remove the timing chains (secondary).

❋❋ WARNING

Avoid impact or dropping the camshaft sprockets. Do not disassemble the camshaft sprockets; never loosen the bolts.

11. Remove the tension guide.

12. Check for cracks and any excessive wear of the timing chain. Replace the timing chain as necessary.

To install:

➡ **The figure shows the relationship between the mating mark on each timing chain and that on the corresponding sprocket with the components installed.**

13. Install the tension guide.

14. Position the crankshaft so the No. 1 piston is set at TDC on the compression stroke.

a. Make sure that the dowel pin hole, dowel pin, and crankshaft key are located properly.

- Camshaft dowel pin hole (intake side): at cylinder head upper face side in each bank
- Camshaft dowel pin (exhaust side): at cylinder head upper face side in each bank
- Crankshaft key: at cylinder head side of RH bank

❋❋ WARNING

The hole on the small diameter side must be used for the intake camshaft sprocket dowel pin. Do not misidentify (ignore the big diameter side).

15. Install the timing chains (secondary) and camshaft sprockets.

❋❋ WARNING

The matching marks between the timing chain and sprockets slip easily. Confirm all matching mark posi-

tions repeatedly during the installation process.

a. Push the sleeve of the chain tensioner (secondary) and keep it pressed in with a stopper pin.

b. Align the matching marks on the timing chain (secondary) (orange link) with the ones on the camshaft sprockets (INT) and (EXH) (stamped), and install them.

c. The matching marks for the camshaft sprocket (INT) are on the back side of the secondary sprocket.

d. There are 2 types of matching marks—round and oval types. They should be used for the RH and LH banks, respectively.

- RH bank: use round type
- LH bank: use oval type

e. Align the dowel pin with and pin hole on the camshaft sprocket (INT) side, and dowel pin groove with the dowel pin on the camshaft sprocket (EXH) side, and install them.

- On the intake side, align the pin hole on the small diameter side of

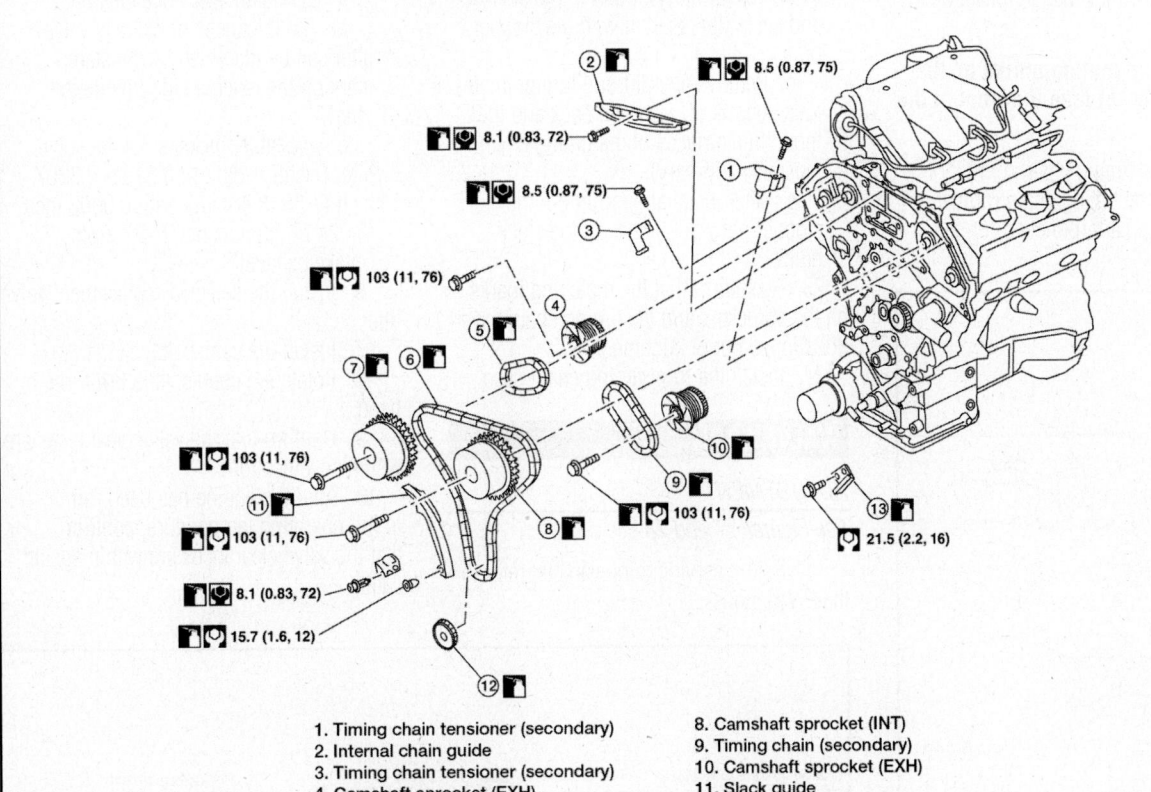

1. Timing chain tensioner (secondary)
2. Internal chain guide
3. Timing chain tensioner (secondary)
4. Camshaft sprocket (EXH)
5. Timing chain (secondary)
6. Timing chain (primary)
7. Camshaft sprocket (INT)
8. Camshaft sprocket (INT)
9. Timing chain (secondary)
10. Camshaft sprocket (EXH)
11. Slack guide
12. Crankshaft sprocket
13. Tension guide

71075_MAXI_G0055

Fig. 76 Timing chain and related component locations

the camshaft front end with the dowel pin on the back side of the camshaft sprocket, and install them.

- On the exhaust side, align the dowel pin on the camshaft front end with the dowel pin groove on the camshaft sprocket, and install them.
- The camshaft sprocket bolts must be tightened in the next step.
- Tightening the camshaft sprocket bolts by hand enough to prevent the dislocation of the dowel pins and dowel pin grooves.
- Check the mating mark (punched) on each camshaft sprocket are positioned on the mating marks (orange link) on the timing chain (secondary).

16. After confirming the mating marks are aligned, tighten the camshaft sprocket bolts.

17. Secure the camshaft using a wrench at the hexagonal portion to tighten the bolts.

18. Pull the stopper pins out from the timing chain tensioners (secondary).

19. Install the crankshaft sprocket on the crankshaft.

➡**Make sure the mating marks on the crankshaft sprocket face the front of the engine.**

20. Install the timing chain (primary).
   a. Install the timing chain (primary) so the mating mark (punched) on the

camshaft sprocket is aligned with the pink link on the timing chain, while the mating mark (notched) on the crankshaft sprocket is aligned with the orange one on the timing chain.
   b. When it is difficult to align the mating marks of the timing chain (primary) with each sprocket, gradually turn the camshaft using a wrench on the hexagonal portion to align it with the mating marks.
   c. During alignment, be careful to prevent dislocation of the mating mark alignments of the secondary timing chains.

21. Install the internal chain guide, slack guide, and timing chain (primary).

✳✳ WARNING

Do not over-tighten the slack guide bolts. It is normal for a gap to exist under the bolt seats when the bolts are tightened to specification.

22. Install the timing chain tensioner (primary) for the slack guide.
   a. When installing the timing chain tensioner (primary), push in the sleeve and keep it pressed in with the stopper pin.
   b. Remove any dirt and foreign materials completely from the back and the mounting surfaces of the timing chain tensioner (primary).
   c. After installation, pull out the stopper pin while pressing the slack guide.

23. Reconfirm that the matching marks on the sprockets and the timing chain have not slipped out of alignment.

24. Install the front timing chain case.

VALVE LASH

ADJUSTMENT

See Figures 77 and 78.

1. Before servicing the vehicle, refer to the Precautions.

➡**Adjust the valve clearance while the engine is cold. Perform the adjustment by selecting the correct head thickness of the valve lifter. Adjusting shims are not used.**

➡**The specified valve lifter thickness is the dimension at normal temperatures. Ignore dimensional differences caused by temperature. Use specifications for hot engine condition to confirm valve clearances.**

2. Remove the camshaft.

3. Remove the valve lifter that was measured as being outside the standard specifications.

4. Measure the center thickness of the removed lifter with a micrometer.

5. Use the equation below to calculate the replacement valve lifter thickness.
   a. Valve lifter thickness calculation equation: $t = t1 + (C1 - C2)$.
   - t = thickness of the replacement lifter
   - $t1$ = thickness of the removed lifter
   - $C1$ = measured valve clearance
   - $C2$ = standard valve clearance
   b. The thickness of the new valve lifter can be identified by the stamp mark on the reverse side (inside the lifter).
   c. Available thickness of the valve lifter (factory setting): 0.3102–0.3307 inch (7.88–8.40mm), with 0.0008 inch (0.02mm) increments, in 27 sizes (intake/exhaust).

6. Install the selected replacement valve lifter.

7. Install the camshaft.

8. Rotate the crankshaft a few turns by hand.

9. Confirm that the valve clearances are within specification.

10. After the engine has been run to full operating temperature, confirm that the valve clearances are within specification.

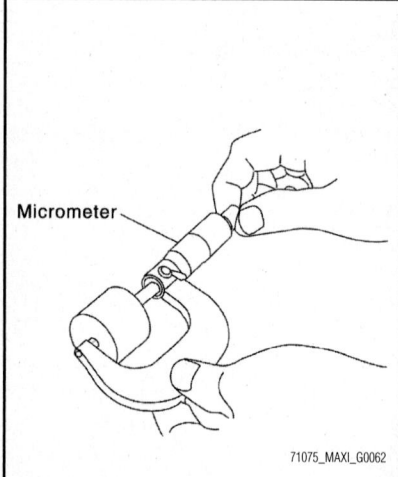

Micrometer

71075_MAXI_G0062

Fig. 77 Measure the center thickness of the removed lifter with a micrometer, as shown

Standard Valve Clearance		Cold	Hot* (reference data)
	Intake	0.26 - 0.34 mm (0.010 - 0.013 in)	0.304 - 0.416 mm (0.012 - 0.016 in)
	Exhaust	0.29 - 0.37 mm (0.011 - 0.015 in)	0.308 - 0.432 mm (0.012 - 0.017 in)

* Approximately 80°C (176°F)

71075_MAXI_G0063

Fig. 78 Standard valve clearance reference table

ENGINE PERFORMANCE & EMISSION CONTROLS

COMPONENT LOCATIONS

See Figures 79 through 84.

ACCELERATOR PEDAL POSITION (APP) SENSOR

LOCATION

See Figure 85.

The Accelerator Pedal Position (APP) sensor is an integral part of the accelerator pedal assembly.

REMOVAL & INSTALLATION

See Figure 86.

1. Before servicing the vehicle, refer to the Precautions.

2. Disconnect the battery negative terminal.

3. Disconnect the accelerator position sensor electrical connector.

4. Remove the 3 accelerator pedal nuts.

5. Remove the accelerator pedal and accelerator position sensor assembly.

Consider the following warnings:

• Do not disassemble the pedal assembly. Do not remove the accelerator pedal position sensor from the pedal assembly

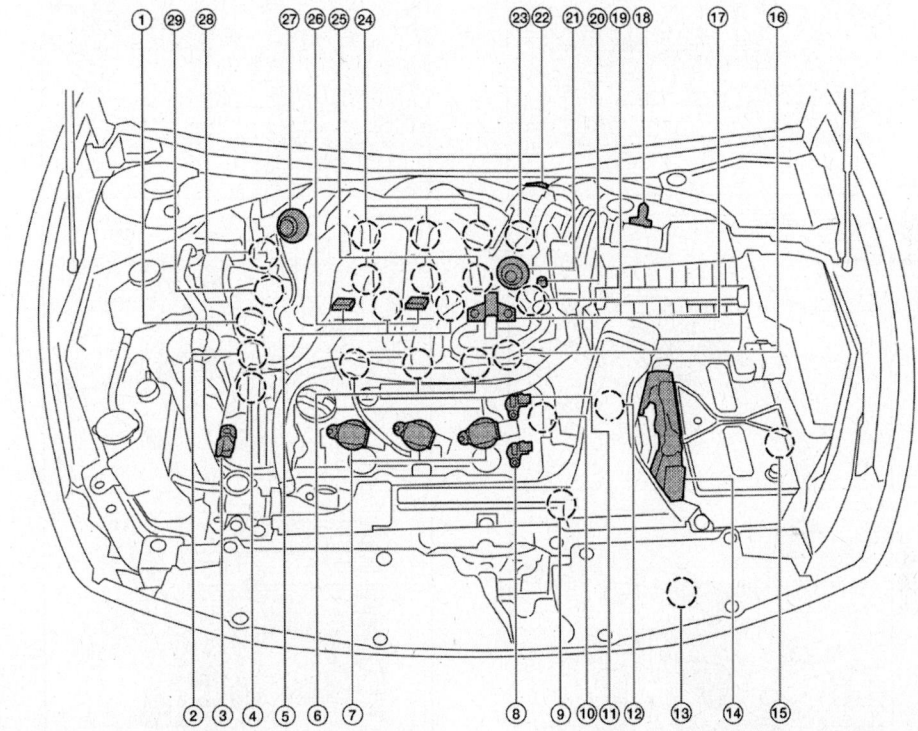

1. Intake valve timing control solenoid valve (bank 1)
2. Electronic controlled engine mount control solenoid valve
3. Exhaust valve timing control magnet retarder (bank 2)
4. Intake valve timing control solenoid valve (bank 2)
5. Knock sensor (bank 1 and 2)
6. Fuel injector (bank 2)
7. Ignition coil (with power transistor) and spark plug (bank 2)
8. Exhaust valve timing control position sensor (bank 2)
9. Crankshaft position sensor (POS)
10. Engine coolant temperature sensor
11. Camshaft position sensor (PHASE) (bank 2)
12. Transmission range switch
13. Refrigerant pressure sensor
14. ECM
15. Battery current sensor
16. Condenser-2 17. EVAP canister purge volume control solenoid valve
18. Mass air flow sensor (with intake air temperature sensor)
19. Camshaft position sensor (PHASE) (bank 1)
20. EVAP service port
21. Power valve actuator 2
22. Electric throttle control actuator
23. Exhaust valve timing control position sensor (bank 1)
24. Ignition coil (with power transistor) and spark plug (bank 1)
25. Fuel injector (bank 1)
26. VIAS control solenoid valve 1 and 2
27. Power valve actuator 1
28. Exhaust valve timing control magnet retarder (bank 1)
29. Power steering pressure sensor

71075_MAXI_G0065

Fig. 79 Engine control system component locations (view 1 of 5)

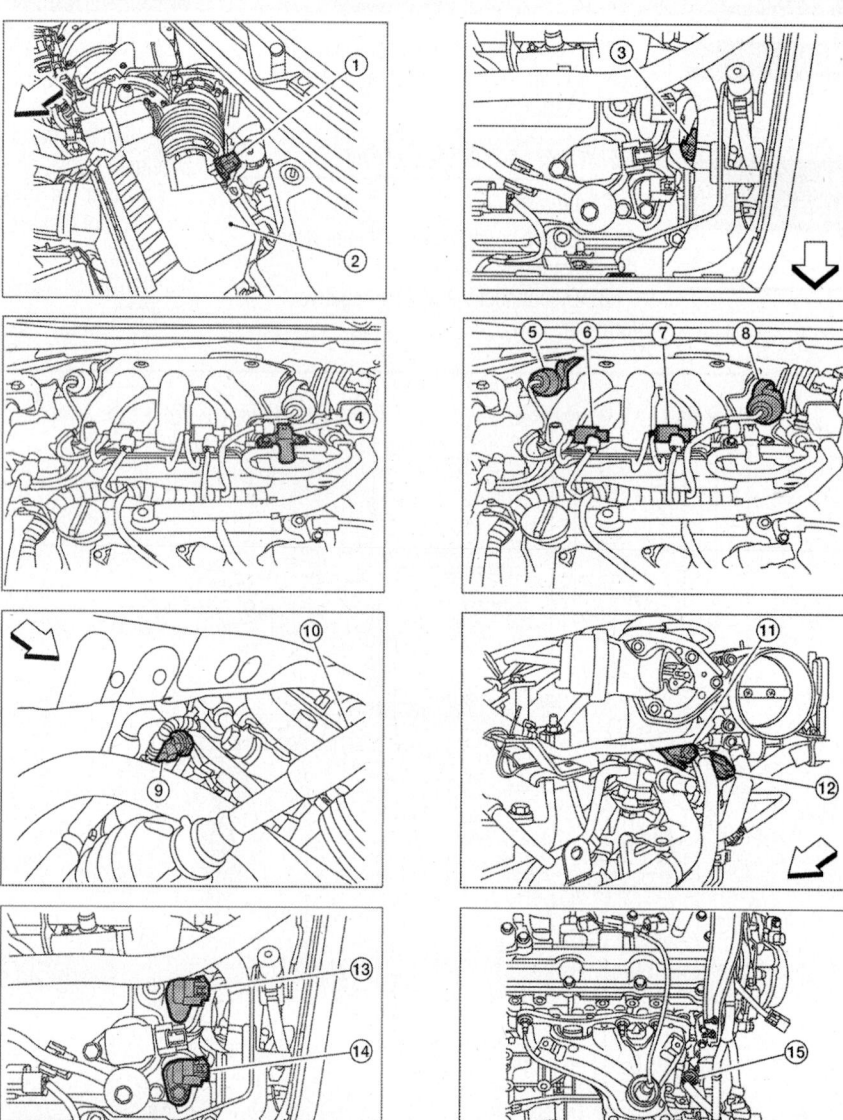

1. Mass air flow sensor (with intake air temperature sensor)
2. Air cleaner case
3. Engine coolant temperature sensor (view with engine cover removed)
4. EVAP canister purge volume control solenoid valve (view with engine cover removed)
5. Power valve actuator 1 (view with engine cover removed)
6. VIAS control solenoid valve 1
7. VIAS control solenoid valve 2
8. Power valve actuator 2
9. Power steering pressure sensor
10. Tie rod (RH) 11. Camshaft position sensor (PHASE) (bank 1) (view with air cleaner case removed)
12. Exhaust valve timing control position sensor (bank 1)
13. Camshaft position sensor (PHASE) (bank 2) (view with air cleaner case removed)
14. Exhaust valve timing control position sensor (bank 2)
15. Engine oil temperature sensor
Arrow: Vehicle front

71075_MAXI_G0066

Fig. 80 Engine control system component locations (view 2 of 5)

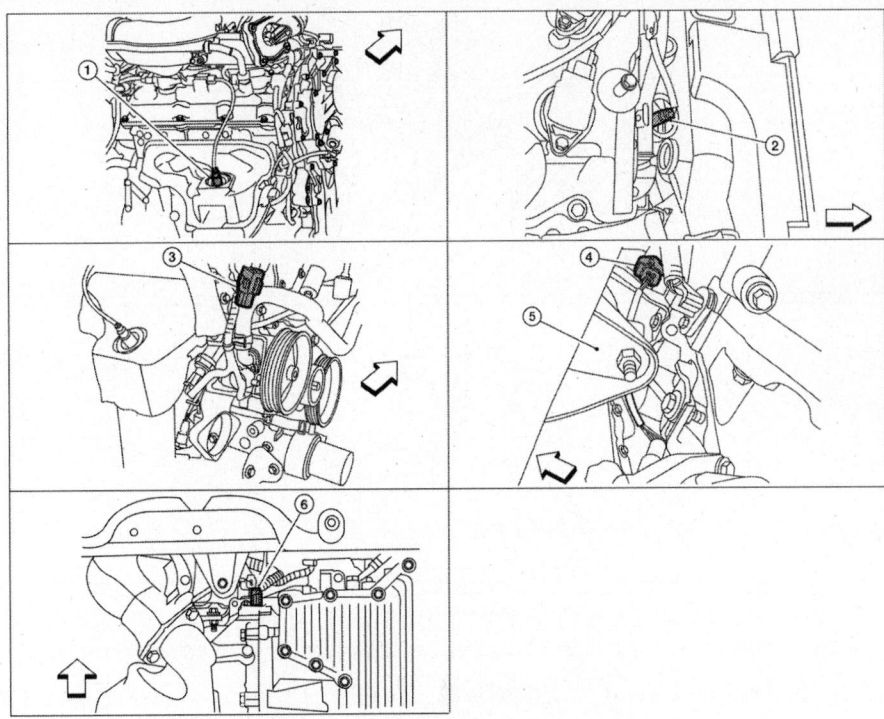

1. A/F sensor 1 (bank 1) (view with engine removed)
2. A/F sensor 1 (bank 2)
3. HO2S2 (bank 1) harness connector (view with engine removed)
4. HO2S2 (bank 2) harness connector
5. Front engine mount
6. Crankshaft position sensor (POS)
Arrow: Vehicle front

71075_MAXI_G0068

Fig. 81 Engine control system component locations (view 3 of 5)

• Avoid impact from dropping during handling
• Keep the pedal assembly away from water

To install:

6. Installation is in the reverse order of removal.

7. Align and install the accelerator pedal and accelerator position sensor assembly with the locating pins in the locating pin holes.

8. Check the accelerator pedal for smooth operation. There should be no binding or sticking when applying or releasing the accelerator pedal.

9. Check that the accelerator pedal moves through the full specified distance of pedal travel.
• Accelerator pedal total travel: 1.91–2.11 inches (48.5–53.5mm)

➥**When the harness connector of the accelerator pedal position sensor is** disconnected, perform the "Accelerator Pedal Released Position Learning."

ACCELERATOR PEDAL RELEASED POSITION LEARNING

1. Check that accelerator pedal is fully released.

2. Turn the ignition switch ON and wait at least 2 seconds.

3. Turn the ignition switch OFF and wait at least 10 seconds.

4. Turn the ignition switch ON and wait at least 2 seconds.

5. Turn the ignition switch OFF and wait at least 10 seconds.

AIR-FUEL RATIO (A/F) SENSOR

LOCATION

See Figure 87.

The Air Fuel Ratio sensors are located one each on each exhaust manifold.

REMOVAL & INSTALLATION

✳✳ CAUTION

To avoid the danger of being burned, do not touch the exhaust system when the system is hot. The Air-Fuel Ratio (AFR) sensor removal should be performed when the system is cool.

1. Before servicing the vehicle, refer to the Precautions.

2. Disconnect the AFR sensor electrical connector.

3. Remove the AFR sensor from the exhaust manifold.

To install:

4. Installation is the reverse of the removal procedure.

5. Tighten components to specification.

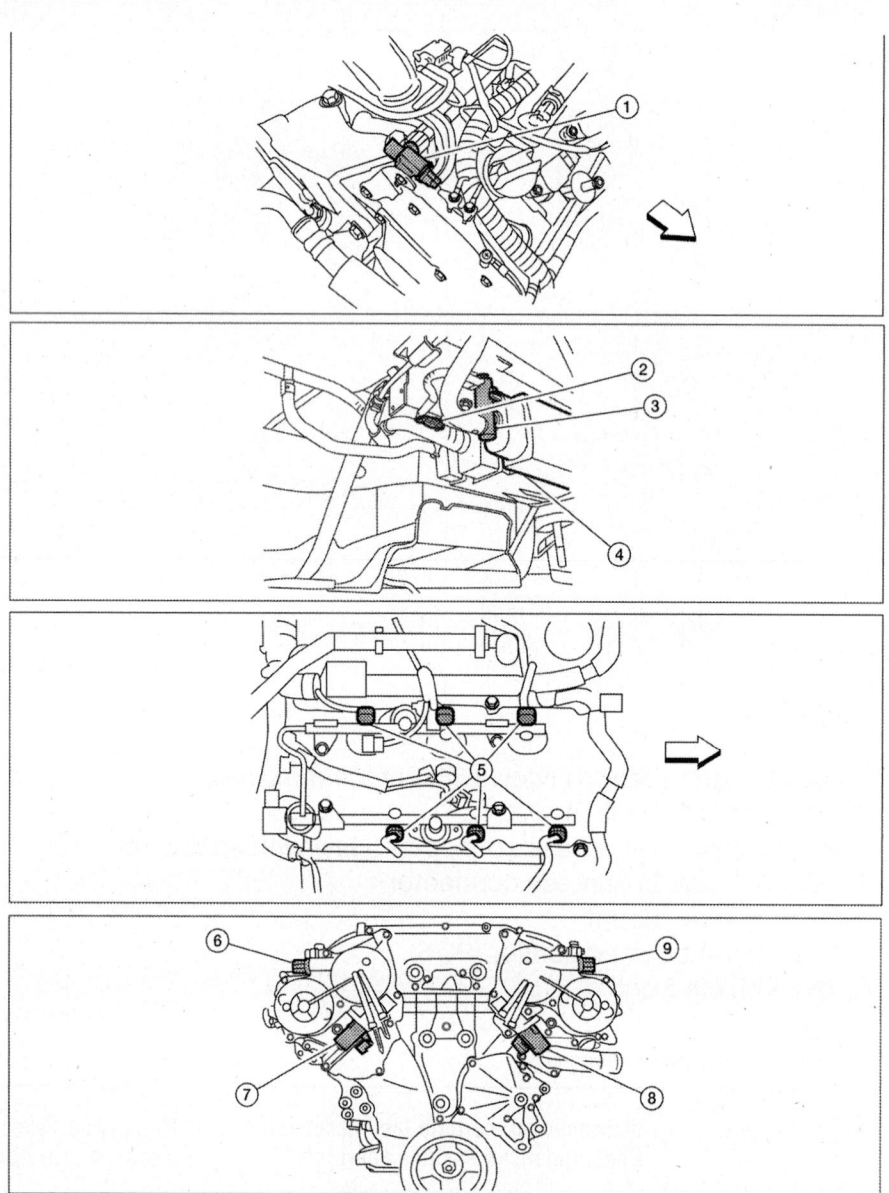

1. Electronic controlled engine mount control solenoid valve (view with engine cover removed)
2. EVAP control system pressure sensor (view with rear suspension member removed)
3. EVAP canister vent control valve
4. EVAP canister
5. Fuel injector harness connector (view with intake manifold collector removed)
6. Exhaust valve timing control magnet retarder (bank 1) (view with engine removed)
7. Intake valve timing control solenoid valve (bank 1)
8. Intake valve timing control solenoid valve (bank 2)
9. Exhaust valve timing control magnet retarder (bank 2)
Arrow: Vehicle front

71075_MAXI_G0069

Fig. 82 Engine control system component locations (view 4 of 5)

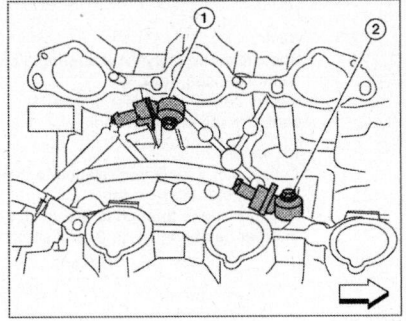

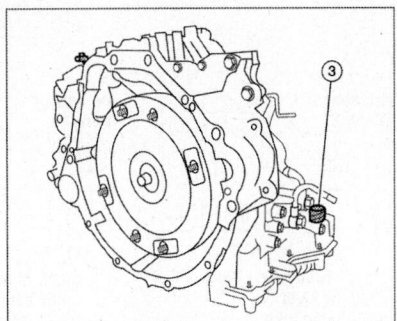

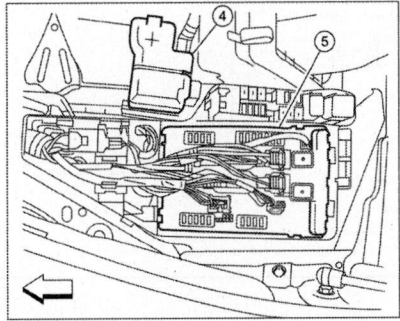

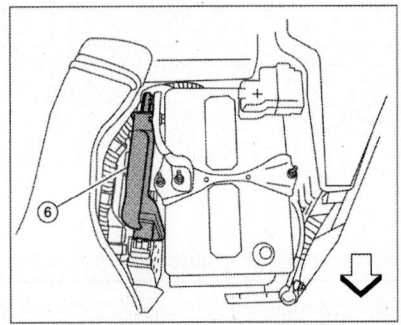

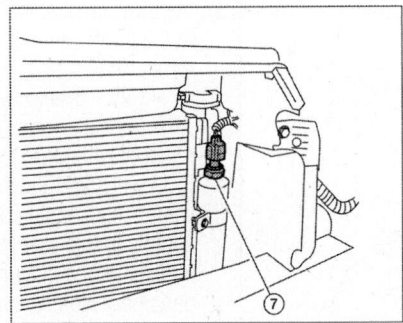

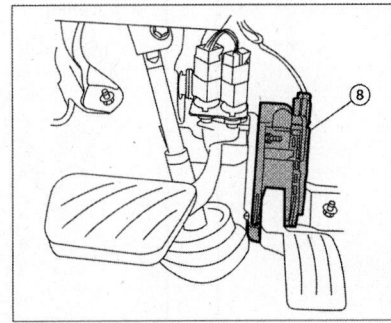

1. Knock sensor (bank 2) (view with intake manifold removed)
2. Knock sensor (bank 1)
3. Transmission range switch (view with CVT removed)
4. Battery
5. IPDM E/R
6. ECM
7. Refrigerant pressure sensor (view with front grille removed)
8. Accelerator pedal position sensor
Arrow: Vehicle front

71075_MAXI_G0070

Fig. 83 Engine control system component locations (view 5 of 5)

CAMSHAFT POSITION (CMP) SENSOR

LOCATION

The Camshaft Position (CMP) sensors are located at the rear of each cylinder head.

REMOVAL & INSTALLATION

1. Before servicing the vehicle, refer to the Precautions.
2. Loosen the fixing bolt of the sensor.
3. Disconnect Camshaft Position (CMP) sensor harness connector.
4. Remove the sensor.
5. Installation is reverse of removal.

CRANKSHAFT POSITION (CKP) SENSOR

LOCATION

See Figure 88.

REMOVAL & INSTALLATION

1. Before servicing the vehicle, refer to the Precautions.
2. Loosen the fixing bolt of the sensor.
3. Disconnect Crankshaft Position (CKP) sensor (POS) harness connector.
4. Remove the sensor.
5. Installation is the reverse of removal.

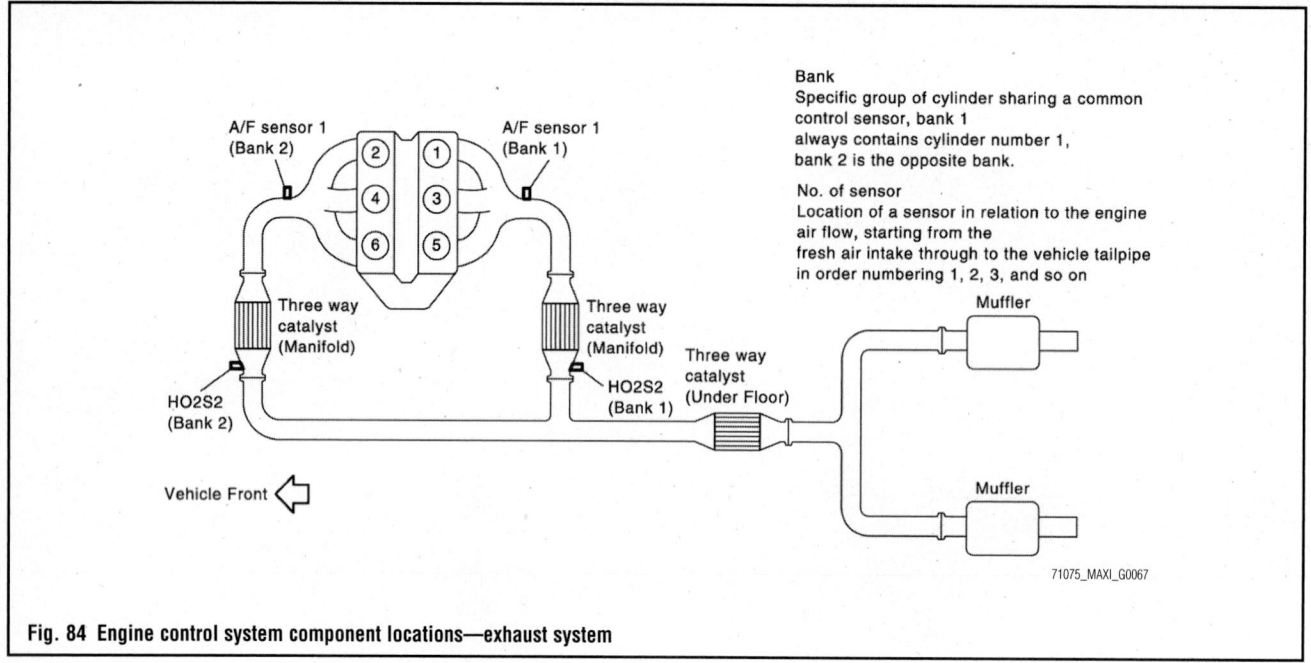

Bank
Specific group of cylinder sharing a common control sensor, bank 1 always contains cylinder number 1, bank 2 is the opposite bank.

No. of sensor
Location of a sensor in relation to the engine air flow, starting from the fresh air intake through to the vehicle tailpipe in order numbering 1, 2, 3, and so on

A/F sensor 1 (Bank 2)
A/F sensor 1 (Bank 1)
Three way catalyst (Manifold)
Three way catalyst (Manifold)
Three way catalyst (Under Floor)
HO2S2 (Bank 2)
HO2S2 (Bank 1)
Muffler
Muffler
Vehicle Front

71075_MAXI_G0067

Fig. 84 Engine control system component locations—exhaust system

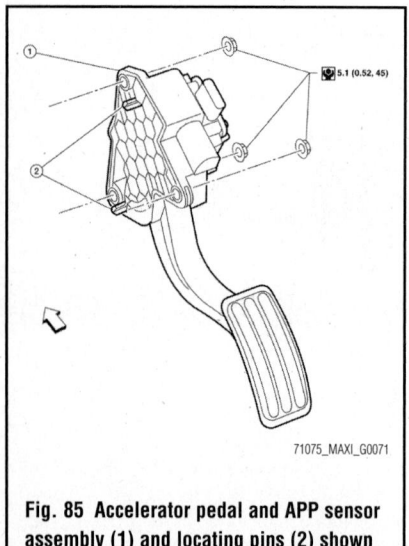

Fig. 85 Accelerator pedal and APP sensor assembly (1) and locating pins (2) shown

71075_MAXI_G0071

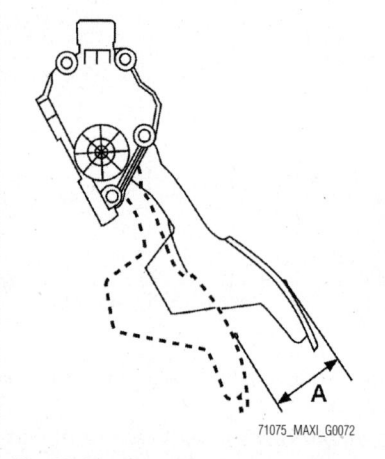

71075_MAXI_G0072

Fig. 86 Check that the accelerator pedal moves through the full specified distance of pedal travel (A)

ELECTRONIC CONTROL MODULE (ECM)

LOCATION

See Figure 89.

The ECM is located under the hood in the left front corner adjacent to the battery.

REMOVAL & INSTALLATION

Use the following precautions when servicing the Engine Control Module (ECM):

• Always turn the ignition switch OFF and disconnect the negative battery cable before any repair or inspection work. The open/short circuit of related switches, sensors, solenoid valves, etc. will cause the MIL to illuminate.

• Always connect and lock the connectors securely after work. A loose (unlocked) connector will cause the MIL to illuminate due to the open circuit. Keep the connector free from water, grease, dirt, bent terminals, etc.

• Always to route and secure the harnesses properly after work. The interference of the harness with a bracket, etc. may cause the MIL to illuminate due to the short circuit.

• Always use a 12 volt battery as the power source.

• Never attempt to disconnect the battery cables while the engine is running.

• Before connecting or disconnecting the ECM harness connector, turn the ignition switch OFF and disconnect the negative battery cable. Failure to do so may damage the ECM because battery voltage is applied to the ECM even if the ignition switch is turned OFF.

• Never disassemble the ECM.

• If a battery cable is disconnected, the memory will return to the ECM value. The ECM will start to self-control at its initial value. Thus, engine operation can vary slightly in this case. However, this is not an indication of a malfunction. Never replace parts because of a slight variation.

• If the battery is disconnected, the following emission-related diagnostic information will be lost within 24 hours: Diagnostic trouble codes—1st trip diagnostic trouble codes; Freeze frame data—1st trip freeze frame data; System Readiness Test (SRT) codes; Test values.

• When connecting the ECM harness connector, fasten it securely with a lever as far as it will go.

• When connecting or disconnecting pin connectors into or from the ECM, never damage the pin terminals. Check that there are not any bends or breaks on the ECM pin terminal when connecting the pin connectors.

• Securely connect the ECM harness connectors. A poor connection can cause an extremely high (surge) voltage to develop in the coil and condenser, thus resulting in damage to the ICs.

• Keep the engine control system harness at least 4 inches (10cm) away from adjacent harness, to prevent engine control system malfunctions due to receiving external noise, degraded operation of ICs, etc.

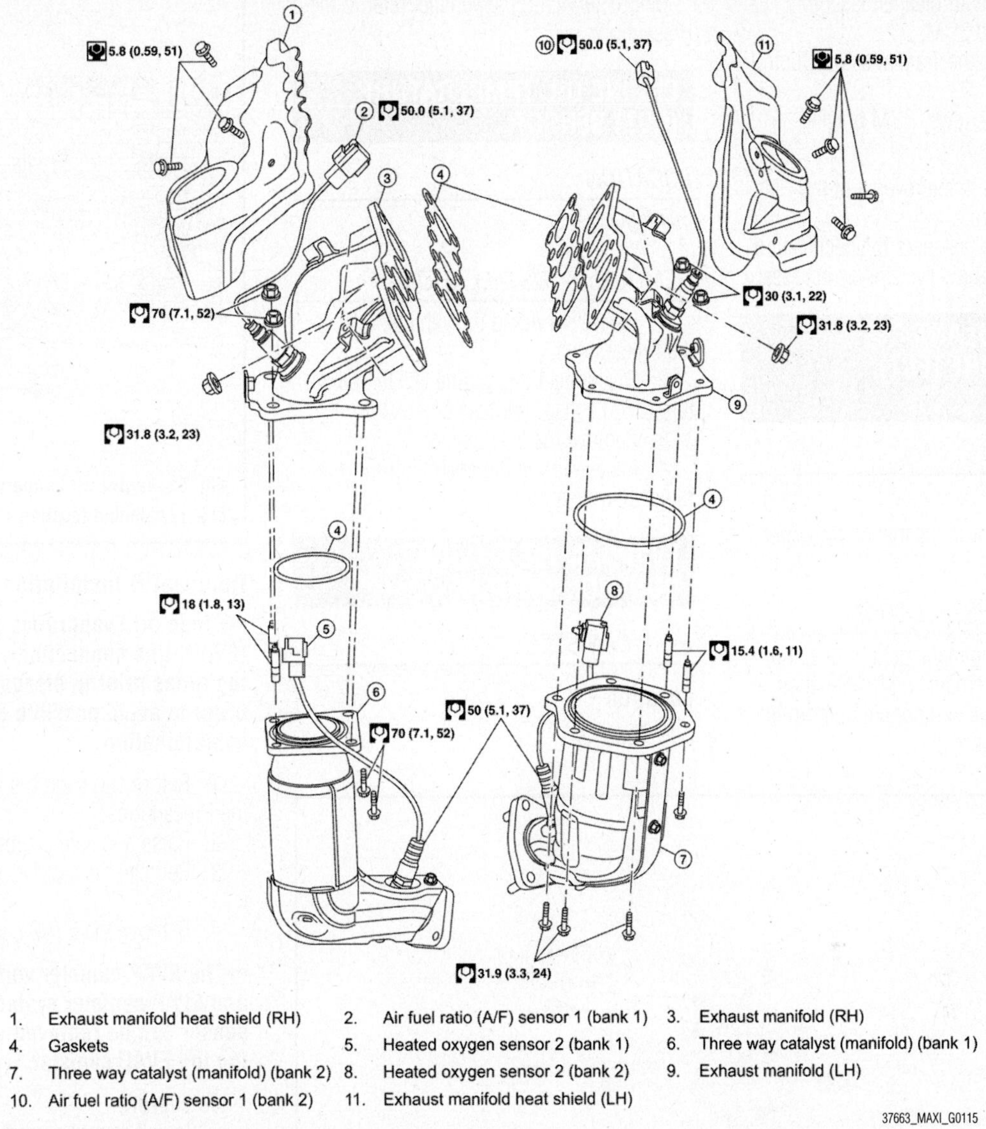

5.8 (0.59, 51) 50.0 (5.1, 37) 5.8 (0.59, 51)

50.0 (5.1, 37)

30 (3.1, 22)

31.8 (3.2, 23)

70 (7.1, 52)

31.8 (3.2, 23)

18 (1.8, 13)

15.4 (1.6, 11)

70 (7.1, 52) 50 (5.1, 37)

31.9 (3.3, 24)

1. Exhaust manifold heat shield (RH)	2. Air fuel ratio (A/F) sensor 1 (bank 1)	3. Exhaust manifold (RH)
4. Gaskets	5. Heated oxygen sensor 2 (bank 1)	6. Three way catalyst (manifold) (bank 1)
7. Three way catalyst (manifold) (bank 2)	8. Heated oxygen sensor 2 (bank 2)	9. Exhaust manifold (LH)
10. Air fuel ratio (A/F) sensor 1 (bank 2)	11. Exhaust manifold heat shield (LH)	

37663_MAXI_G0115

Fig. 87 Exploded view of exhaust manifold and three-way catalyst

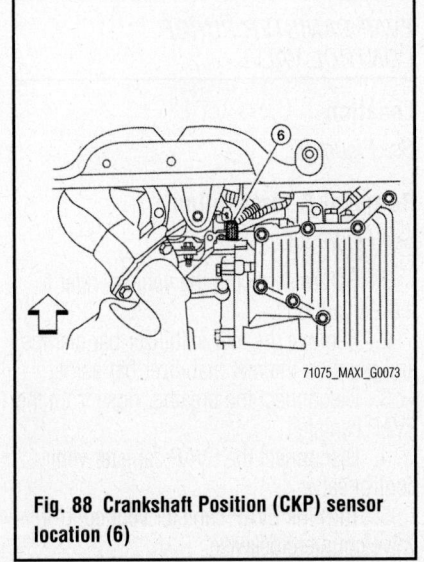

71075_MAXI_G0073

Fig. 88 Crankshaft Position (CKP) sensor location (6)

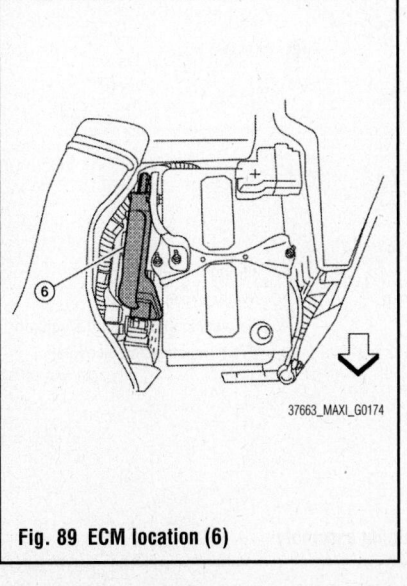

37663_MAXI_G0174

Fig. 89 ECM location (6)

• When measuring ECM signals with a circuit tester, never allow the two tester probes to contact. Accidental contact of the probes will cause a short circuit and damage the ECM power transistor.

• Never use the ECM ground terminals when measuring input/output voltage. Doing so may result in damage to the ECM's transistor. Use a ground other than the ECM terminals.

※ WARNING

As the Engine Control Module (ECM) consists of precision parts, be careful not to expose it to excessive shock.

1. Before servicing the vehicle, refer to the Precautions.

2. Remove the battery and battery tray liner.

3. Remove the air cleaner assembly.
4. Remove the ECM.
5. Disconnect the Transmission Control Module (TCM).
6. Remove the lower ECM bracket.

To install:

7. Installation is the reverse of the removal procedure.
8. Tighten the fasteners to specification.
9. Reset electronic systems as necessary.

ENGINE COOLANT TEMPERATURE (ECT) SENSOR

LOCATION

See Figure 90.

The ECT sensor is located on the water outlet.

REMOVAL & INSTALLATION

At this time, the manufacturer does not provide specific removal and installation procedures for this component, refer to the illustration as required.

Before servicing the vehicle, refer to the Precautions.

ENGINE OIL TEMPERATURE (EOT) SENSOR

LOCATION

See Figure 91.

REMOVAL & INSTALLATION

1. Before servicing the vehicle, refer to the Precautions.
2. Disconnect the Engine Oil Temperature (EOT) sensor harness connector.
3. Remove the EOT sensor.
4. Installation is the reverse of the removal procedure.

EVAPORATIVE EMISSION CONTROL SYSTEM

EVAP CANISTER

Location

See Figure 92.

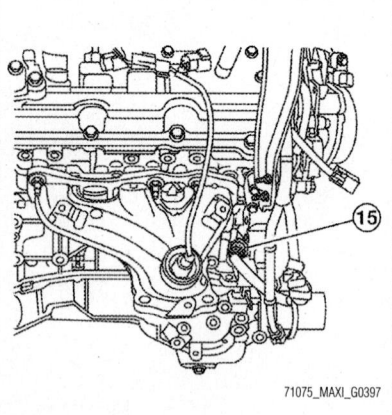

71075_MAXI_G0397

Fig. 91 Engine oil temperature sensor (15) component location

Removal & Installation

➡ **Clean all Evaporative Emission (EVAP) line connections and surrounding areas prior to disconnecting, in order to avoid possible EVAP system contamination.**

1. Before servicing the vehicle, refer to the Precautions.
2. Raise and safely support the vehicle.
3. Remove the EVAP canister fixing bolt.
4. Remove the EVAP canister.

➡ **The EVAP canister vent control valve and EVAP canister system pressure sensor can be removed without removing the EVAP canister.**

To install:

5. Install in the reverse order of removal.
6. Tighten the EVAP canister fixing bolt.

EVAP CANISTER PURGE CONTROL VALVE

Location

See Figure 92.

Removal & Installation

See Figure 93.

1. Before servicing the vehicle, refer to the Precautions.
2. Remove the rear stabilizer bar clamps and position the rear stabilizer bar aside.
3. Disconnect the breather hose from the EVAP canister.
4. Disconnect the EVAP canister vent control valve.
5. Turn the EVAP canister vent control valve counterclockwise.

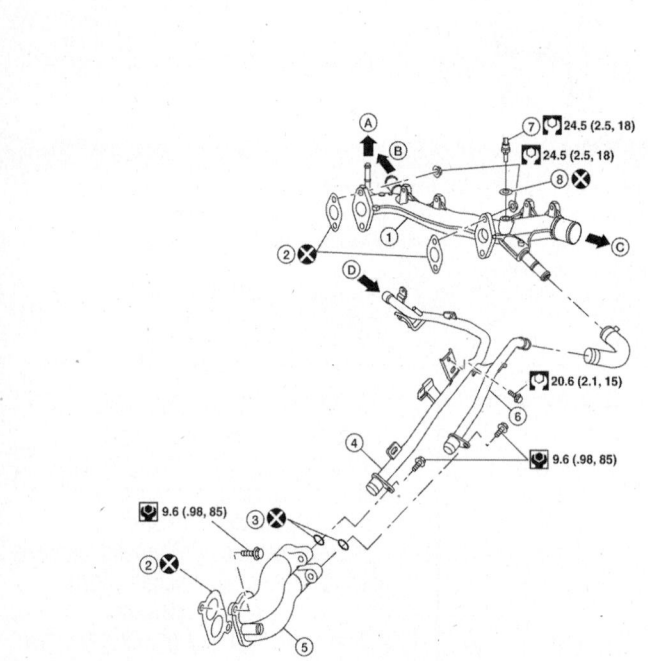

1.	Water outlet	2.	Gasket	3.	O-ring
4.	Heater pipe	5.	Water connector	6.	Water bypass pipe
7.	Engine coolant temperature sensor	8.	Washer	A.	To electric throttle control actuator
B.	To heater	C.	To radiator	D.	From transmission oil cooler

37663_MAXI_G0175

Fig. 90 Exploded view of water outlet and water piping assembly

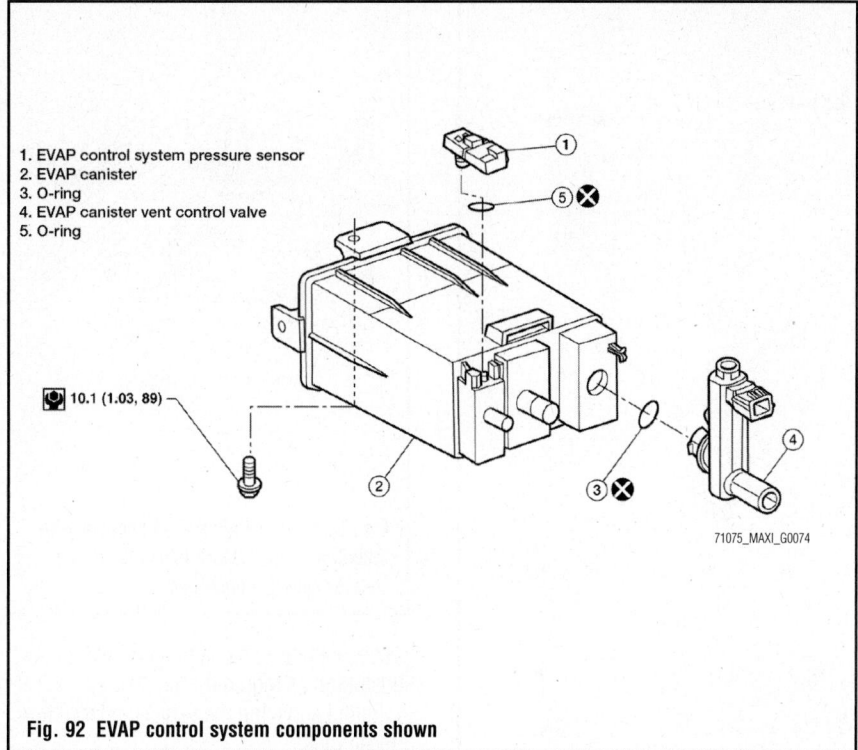

1. EVAP control system pressure sensor
2. EVAP canister
3. O-ring
4. EVAP canister vent control valve
5. O-ring

⬚ 10.1 (1.03, 89)

71075_MAXI_G0074

Fig. 92 EVAP control system components shown

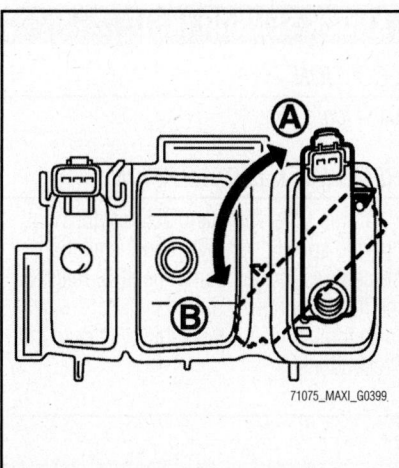

71075_MAXI_G0399

Fig. 93 Turn the EVAP canister vent control valve counterclockwise (A: lock, B: unlock)

6. Remove the EVAP canister vent control valve and O-ring.

To install:

7. Installation is the reverse of the removal procedure.

8. Do not reuse the O-ring.

FUEL TEMPERATURE SENSOR

LOCATION

See Figure 94.

REMOVAL & INSTALLATION

See Figure 95.

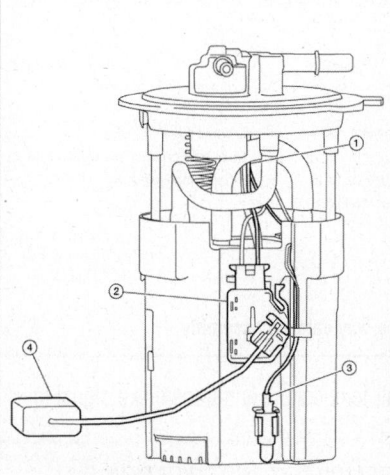

1. Harness connectors
2. Fuel level sensor unit
3. Fuel tank temperature sensor
4. Float arm assembly

71075_MAXI_G0078

Fig. 94 Fuel level sensor unit component location

➡**Before disassembly, note the proper placement of the wires to the correct terminals and correct wire routing to the terminals.**

1. Before servicing the vehicle, refer to the Precautions.

2. Disconnect the red, white, and double black wire connectors. Press the tabs on the terminals to release the locking tabs.

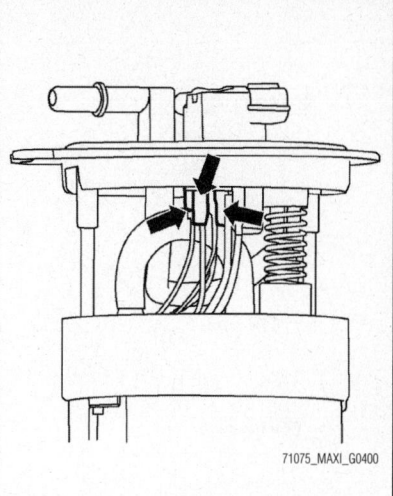

71075_MAXI_G0400

Fig. 95 Disconnect the red, white, and double black wire connectors. Press the tabs on the terminals to release the locking tabs

3. Release the 2 clips and remove the fuel tank temperature sensor from the pump assembly.

4. Release the tab and slide the fuel level sensor unit and float arm assembly up to remove.

To install:

5. Installation is the reverse of the removal procedure.

6. Ensure proper placement of the wires to the correct terminals and correct wire routing to the terminals.

7. After connecting the terminals, ensure they are securely locked and cannot be pulled out.

8. When installing the fuel level sensor unit, push down until the tab is locked into place.

HEATED OXYGEN (HO2S) SENSOR

LOCATION

See Figure 96.

The Heated Oxygen Sensors (HO2S) are located at the bottom end of the three-way catalyst for each bank.

REMOVAL & INSTALLATION

✳✳ CAUTION

To avoid the danger of being burned, do not touch the exhaust system when the system is hot. The Heated Oxygen Sensor (HO2S) removal should be performed when the system is cool.

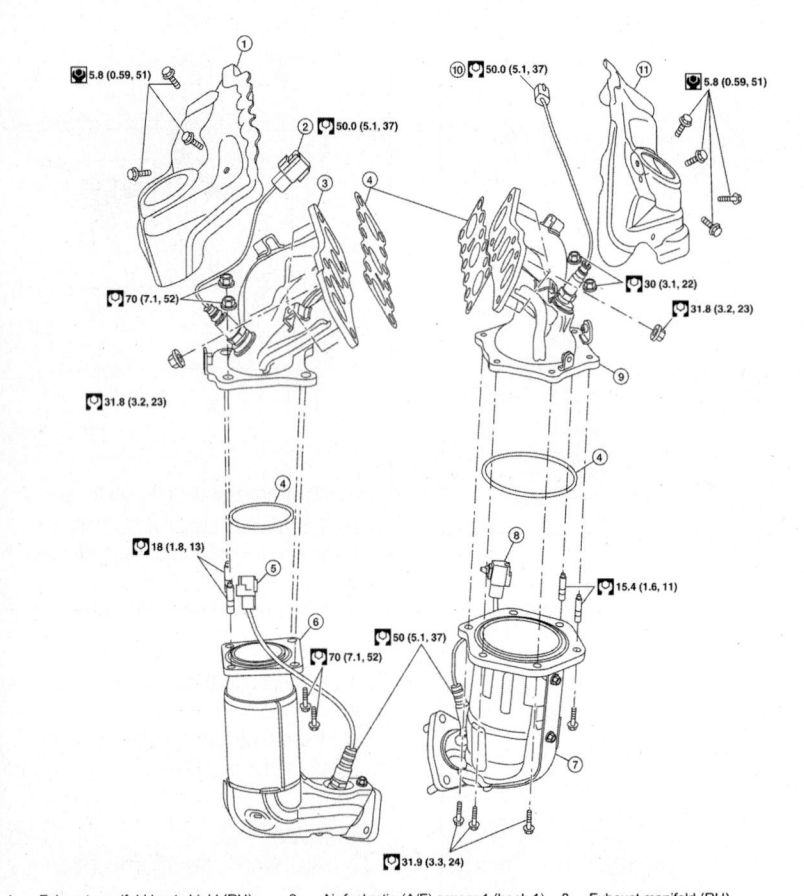

1. Exhaust manifold heat shield (RH)
2. Air fuel ratio (A/F) sensor 1 (bank 1)
3. Exhaust manifold (RH)
4. Gaskets
5. Heated oxygen sensor 2 (bank 1)
6. Three way catalyst (manifold) (bank 1)
7. Three way catalyst (manifold) (bank 2)
8. Heated oxygen sensor 2 (bank 2)
9. Exhaust manifold (LH)
10. Air fuel ratio (A/F) sensor 1 (bank 2)
11. Exhaust manifold heat shield (LH)

37663_MAXI_G0115

Fig. 96 Exploded view of exhaust manifold and three-way catalyst assembly

1. Before servicing the vehicle, refer to the Precautions.

2. Disconnect heated oxygen sensor harness connector.

3. Remove the sensor.

To install:

4. Installation is the reverse of the removal procedure.

5. Tighten the components to specification.

INTAKE AIR TEMPERATURE (IAT)/MASS AIRFLOW (MAF) SENSOR

LOCATION

See Figure 97.

The Intake Air Temperature (IAT) sensor is built into the Mass Air Flow (MAF) sensor. The sensor detects intake air temperature and transmits a signal to the ECM.

REMOVAL & INSTALLATION

Use the following precautions for this procedure.

• Do not disassemble the MAF and IAT sensor

• Do not expose the MAF and IAT sensor to any shock

• Do not clean the MAF and IAT sensor

• If the MAF and IAT sensor has been dropped, it should be replaced

• Do not blow compressed air through the MAF and IAT sensor

• Do not place a finger or any other object into the MAF and IAT sensor. Malfunction may occur

At this time, the manufacturer does not provide specific removal and installation

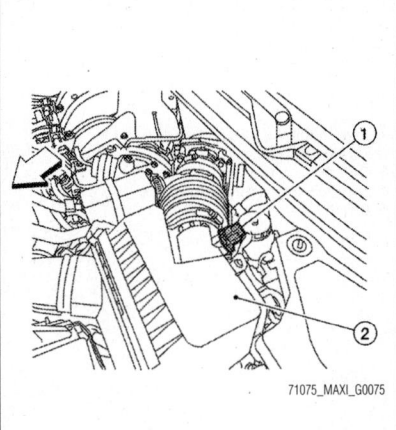

71075_MAXI_G0075

Fig. 97 Mass Air Flow (MAF) sensor with Intake Air Temperature (IAT) (1) and air cleaner case (2) locations

procedures for this component, refer to the illustration as required.

Before servicing the vehicle, refer to the Precautions.

KNOCK SENSOR (KS)

LOCATION

See Figure 98.

REMOVAL & INSTALLATION

At this time, the manufacturer does not provide specific removal and installation procedures for this component(s), refer to the illustration as required.

Before servicing the vehicle, refer to the Precautions.

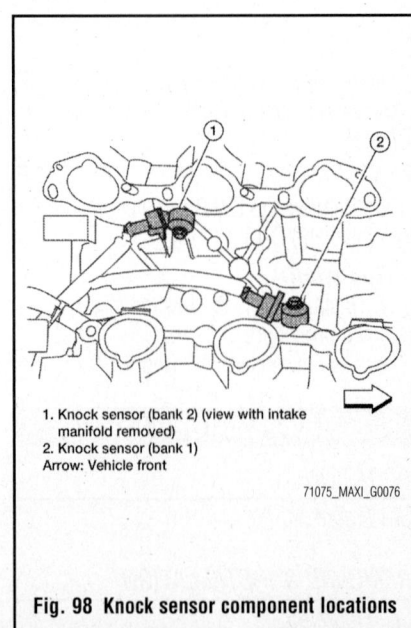

1. Knock sensor (bank 2) (view with intake manifold removed)
2. Knock sensor (bank 1)
Arrow: Vehicle front

71075_MAXI_G0076

Fig. 98 Knock sensor component locations

OUTPUT SHAFT SPEED (OSS) SENSOR

LOCATION

See Figure 99.

REMOVAL & INSTALLATION

See Figure 100.

1. Before servicing the vehicle, refer to the Precautions.
2. Disconnect the battery negative terminal.
3. Remove the hood ledge cover (LH).
4. Remove the engine room cover.
5. Remove the air duct (inlet).
6. Remove the air cleaner case.
7. Disconnect the secondary speed sensor connector.

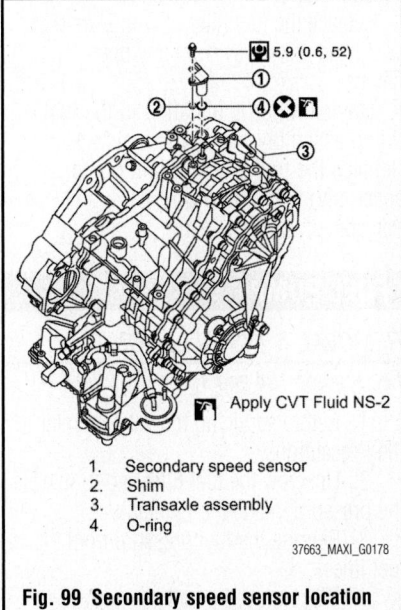

1. Secondary speed sensor
2. Shim
3. Transaxle assembly
4. O-ring

Apply CVT Fluid NS-2

37663_MAXI_G0178

Fig. 99 Secondary speed sensor location

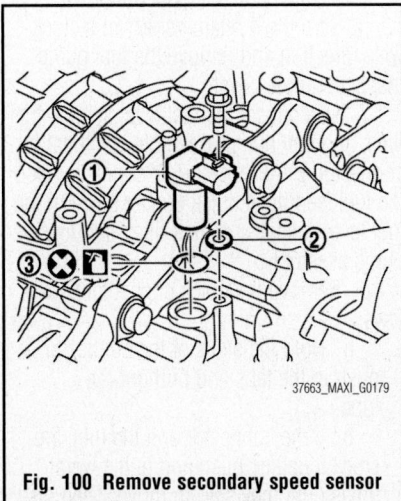

37663_MAXI_G0179

Fig. 100 Remove secondary speed sensor (1), shim (2), and O-ring (3)

8. Remove the secondary speed sensor and shim.

➡ **Do not lose the shim.**

9. Remove the O-ring from the secondary speed sensor.

To install:
10. Installation is in the reverse order of removal.
11. Do not reuse the O-ring. Apply CVT fluid to the O-ring.
12. Check for CVT fluid leakage and check the CVT fluid level.

THROTTLE POSITION SENSOR (TPS)

LOCATION

The electric throttle body includes a throttle control motor and Throttle Position Sensors (TPS). The TPS responds to the throttle valve movement. The TPS has two sensors.

REMOVAL & INSTALLATION

The Throttle Position Sensors (TPS) are integral to the electronic throttle body.

TRANSMISSION CONTROL MODULE (TCM)

LOCATION

See Figures 101 and 102.

REMOVAL & INSTALLATION

See Figure 103.

❋❋ WARNING

Never impact the TCM when removing or installing the TCM. When replacing the TCM and transaxle assembly as a set, replace the transaxle assembly first and then replace the TCM.

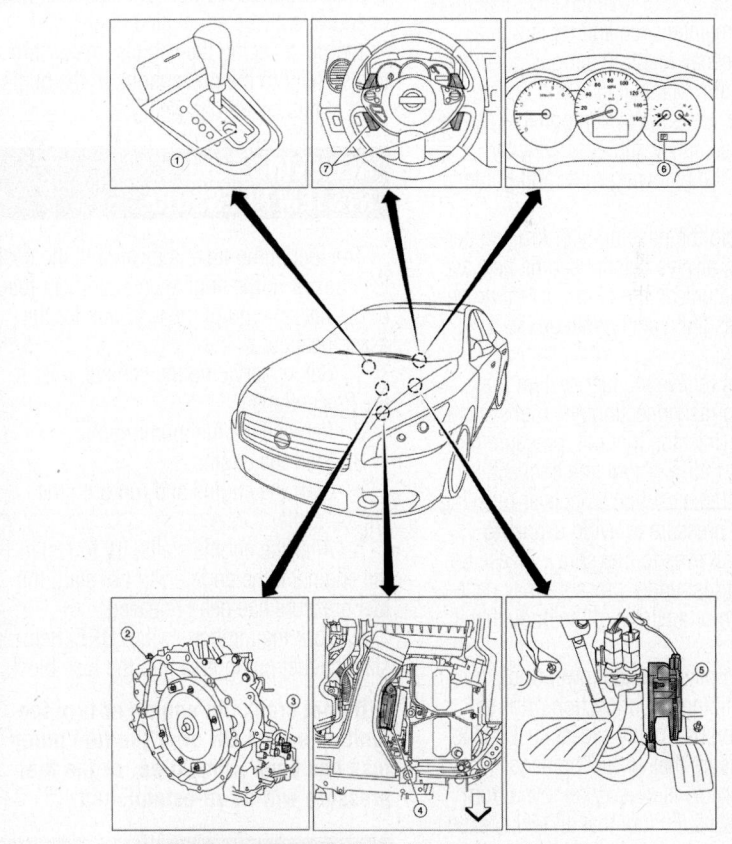

1. CVT shift selector assembly (Manual mode select switch and manual mode position select switch)
2. Secondary speed sensor
3. CVT unit harness connector
4. TCM
5. Accelerator pedal position (APP) sensor
6. Shift position indicator Manual mode indicator DS mode indicator
7. Paddle shifters

71075_MAXI_G0337

Fig. 101 CVT system component locations

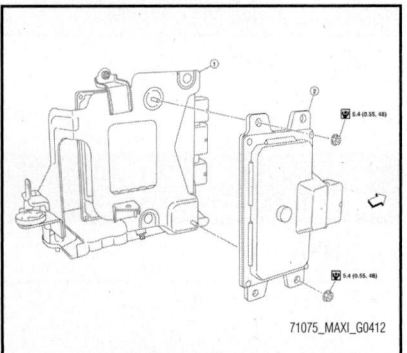

Fig. 102 Transmission Control Module (TCM) (2) and bracket (1) component locations

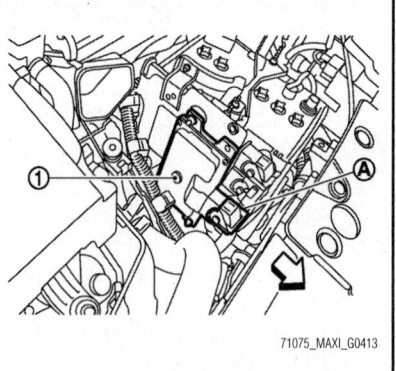

Fig. 103 View of the TCM (1) and TCM harness connector (A)

1. Before servicing the vehicle, refer to the Precautions.

2. Disconnect the battery negative terminal.

3. Remove the air duct (inlet).

4. Disconnect the TCM harness connector.

5. Remove the TCM nuts and the TCM from the bracket.

To install:

6. Installation is the reverse of the removal procedure.

7. After the TCM is replaced, check programming as needed.

FUEL   GASOLINE FUEL INJECTION SYSTEM

FUEL SYSTEM SERVICE PRECAUTIONS

Safety is the most important factor when performing not only fuel system maintenance but any type of maintenance. Failure to conduct maintenance and repairs in a safe manner may result in serious personal injury or death. Maintenance and testing of the vehicle's fuel system components can be accomplished safely and effectively by adhering to the following rules and guidelines.

• To avoid the possibility of fire and personal injury, always disconnect the negative battery cable unless the repair or test procedure requires that battery voltage be applied.

• Always relieve the fuel system pressure prior to disconnecting any fuel system component (injector, fuel rail, pressure regulator, etc.), fitting or fuel line connection. Exercise extreme caution whenever relieving fuel system pressure to avoid exposing skin, face and eyes to fuel spray. Please be advised that fuel under pressure may penetrate the skin or any part of the body that it contacts.

• Always place a shop towel or cloth around the fitting or connection prior to loosening to absorb any excess fuel due to spillage. Ensure that all fuel spillage (should it occur) is quickly removed from engine surfaces. Ensure that all fuel soaked cloths or towels are deposited into a suitable waste container.

• Always keep a dry chemical (Class B) fire extinguisher near the work area.

• Do not allow fuel spray or fuel vapors to come into contact with a spark or open flame.

• Always use a back-up wrench when loosening and tightening fuel line connection fittings. This will prevent unnecessary stress and torsion to fuel line piping.

• Always replace worn fuel fitting O-rings with new Do not substitute fuel hose or equivalent where fuel pipe is installed.

Before servicing the vehicle, make sure to also refer to the precautions in the beginning of this section as well.

RELIEVING FUEL SYSTEM PRESSURE

The fuel pump fuse is located in the dash fuse box or in the engine compartment fuse box. Check the lid of the fuse box for the exact location.

1. Before servicing the vehicle, refer to the Precautions.

2. Remove the fuel pump fuse.

3. Start the engine.

4. Start the engine and run until the engine stalls.

5. After the engine stalls, try to restart the engine. If the engine will not start, the fuel pressure has been released.

6. Turn the ignition switch **OFF**. Reinstall the fuel pump fuse into the fuse block.

➡**Do not crank the engine or turn the ignition switch ON after the fuel pump fuse has been reinstalled, or the fuel pressure will be re-established.**

FUEL FILTER

REMOVAL & INSTALLATION

✳✳ CAUTION

Replacing the fuel filter without adequate ventilation may lead to a fire.

Before servicing the vehicle, refer to the Precautions.

Replace the fuel filter only in a well-ventilated area away from any open flames.

The fuel filter is installed in the fuel pump assembly inside the fuel tank. Replace the fuel filter (or fuel pump assembly) with a new one periodically.

FUEL PUMP

REMOVAL & INSTALLATION

See Figures 104 and 105.

1. Before servicing the vehicle, refer to the Precautions.

2. Unscrew the fuel filler cap to release the pressure inside the fuel tank.

3. Release the fuel pressure from the fuel lines.

4. Disconnect the battery negative terminal.

5. Remove the rear seat bottom.

6. Turn the 4 retainers 90° in a clockwise direction and remove the fuel pump inspection hole cover.

7. Disconnect the fuel level sensor, fuel filter, and fuel pump assembly electrical connector, EVAP hose quick connector, and the fuel feed hose quick connector from the fuel level sensor unit, fuel filter, and fuel pump assembly.

8. Remove the quick connector as follows:

a. Hold the sides of the connector, push in the tabs and pull out the tube.

b. If the connector and the tube are stuck together, push and pull several times until they start to move. Then disconnect them by pulling.

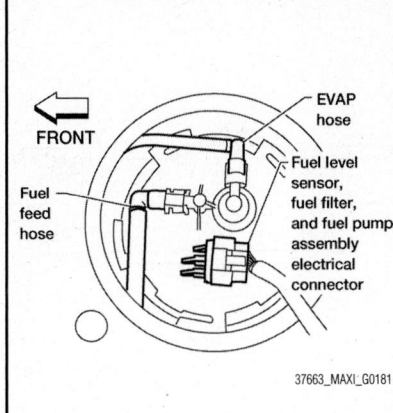

Fig. 104 Disconnect the fuel level sensor, fuel filter, and fuel pump assembly electrical connector, EVAP hose quick connector, and the fuel feed hose quick connector

Consider the following warnings:

• The tube can be removed when the tabs are completely depressed. Do not twist it more than necessary.

• Do not use any tools to remove the quick connector.

• Keep the resin tube away from heat. Be especially careful when welding near the tube.

• Prevent acid liquid such as battery electrolyte, etc. from getting on the resin tube.

• Do not bend or twist the tube during installation and removal.

• Only when the tube is replaced, remove the remaining retainer on the tube or fuel level sensor, fuel filter, and fuel pump assembly.

• When the tube or fuel level sensor, fuel filter, and fuel pump assembly is replaced,

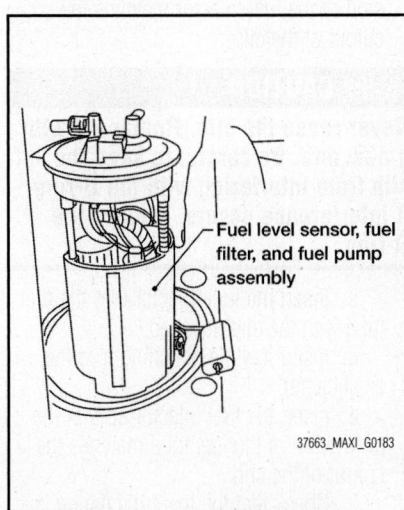

Fig. 105 Remove the fuel level sensor, fuel filter, and fuel pump assembly

also replace the retainer with a new one (green colored retainer).

• To keep the connecting portion clean and to avoid damage and foreign materials, cover them completely with plastic bags or something similar.

9. Remove the lock ring using a socket drive handle and Tool KV991J0090 (J-46214).

➡**Discard the lock ring, do not reuse the lock ring. Discard the ring seal, do not reuse the ring seal.**

10. Remove the fuel level sensor, fuel filter, and fuel pump assembly.

✳✳ WARNING

Do not bend the float arm during removal. Discard the ring seal, do not reuse the ring seal.

11. Inspect the fuel level sensor, fuel filter, and fuel pump for any defects and foreign materials. Replace as necessary.

To install:

12. Installation is in the reverse order of removal.

13. Install the fuel level sensor, fuel filter, and fuel pump assembly with the fuel feed hose facing the front of the vehicle. Use a new ring seal.

14. Connect the quick connector as follows:

a. Check the connection for damage or any foreign materials.

b. Align the connector with the tube, then insert the connector straight into the tube until a click is heard.

15. After the tube is connected, make sure the connection is secure by performing the following checks:

a. Pull the tube and the connector to make sure they are securely connected.

b. Visually confirm that the 2 retainer tabs are connected to the quick connector.

16. Use the following procedure to check for fuel leaks.

a. Turn the ignition switch to ON (without starting the engine) to apply fuel pressure to the fuel system, then check the connections for fuel leaks.

b. Start the engine and let it idle and check for fuel leaks at the fuel system connections.

FUEL RAIL & INJECTORS

REMOVAL & INSTALLATION

See Figures 106 through 109.

Consider the following cautions:

• Be sure to work in a well-ventilated area and furnish the workshop with a CO_2 fire extinguisher

• Never smoke while servicing the fuel system. Keep open flames and sparks away from the work area

• To avoid the danger of being scalded, never drain engine coolant when the engine is hot

1. Before servicing the vehicle, refer to the Precautions.

2. Remove the engine cover.

3. Release the fuel pressure.

4. Remove the front wiper arm and extension cowl top.

5. Remove the electric throttle control actuator bolts in the reverse order as shown and remove the electric throttle control actuator.

✳✳ WARNING

Handle the electric throttle control actuator carefully to avoid any shock. Do not disassemble.

6. Remove the intake manifold collector.

7. When separating the fuel feed hose and fuel tube connection, disconnect the quick connector as follows:

a. Remove the quick connector cap from the quick connector.

b. Disconnect the quick connector from the fuel tube as follows:

✳✳ WARNING

Disconnect the quick connector by using the quick connector release (commercial service tool: J- 45488), not by picking out the retainer tabs.

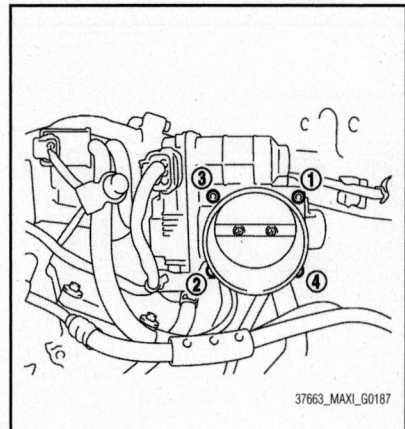

Fig. 106 Remove the electric throttle control actuator

- With the sleeve side of the quick connector release facing to the quick connector, install the quick connector release onto the fuel tube
- Insert the quick connector release into the quick connector until the sleeve contacts and goes no further. Hold the quick connector release on that position

➡**Inserting the quick connector release hard will not disconnect the quick connector. Hold the quick connector release where it contacts and go no further.**

- Draw and pull out quick connector straight from fuel tube

Consider the following warnings:

- Pull the quick connector holding the position as shown in the figure
- Never pull with lateral force applied. The O-ring inside the quick connector may be damaged
- Prepare a container and cloth beforehand as fuel may leak out
- Avoid fire and sparks
- Keep the parts away from any heat source. Especially, be careful when welding is performed
- Never expose parts to battery electrolyte or other acids
- Never bend or twist the connection between the quick connector and the fuel feed hose (with damper) during installation/removal
- To keep the connecting portion clean and to avoid damage and foreign materials entering, cover them completely with plastic bags or something similar

8. Disconnect the harness connector from the fuel injector.

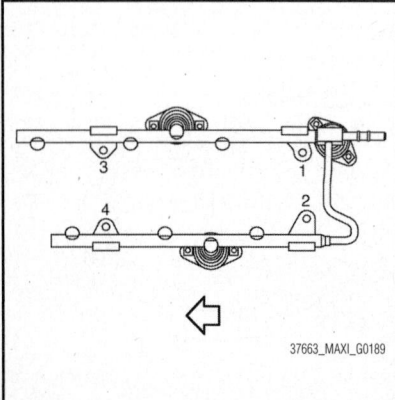

Fig. 107 Loosen the bolts in reverse order as shown, and remove the fuel tube and fuel injector assembly

9. Loosen the bolts in the reverse order as shown, and remove the fuel tube and fuel injector assembly.

➡**Do not tilt the fuel tube or the remaining fuel in the pipes may flow out from the pipes.**

10. Remove the fuel injector from the fuel tube as follows:
    a. Open and remove the clip.
    b. Remove the fuel injector from the fuel tube by pulling straight.
    Consider the following warnings:
- Be careful with the remaining fuel that may go out from the fuel tube
- Be careful not to damage the injector nozzle during removal
- Never bump or drop the fuel injector
- Never disassemble the fuel injector
11. Remove the fuel damper from the fuel tube.

To install:

12. Install the fuel damper as follows:
    a. Install a new O-ring to the fuel tube as shown.
    Consider the following warnings:
- Handle the O-ring with bare hands. Never wear gloves
- Lubricate the O-ring with new engine oil
- Never clean the O-ring with solvent
- Check that the O-ring and its mating part are free of foreign material
- When installing the O-ring, be careful not to scratch it with a tool or fingernails. Also be careful not to twist or stretch the O-ring
- Insert a new O-ring straight into the fuel tube. Never twist it
    a. Install the spacer to the fuel damper.
    b. Insert the fuel damper straight into the fuel tube.
    Consider the following warnings:
- Insert straight, checking that the axis is lined up
- Never pressure-fit with excessive force
- Insert the fuel damper until it is touching the fuel tube
    a. Tighten the bolts evenly.
    b. After tightening the bolts, check that there is no gap between the fuel damper cap and the fuel tube.
13. Install new O-rings to the fuel injectors paying attention to the following:
- The upper and lower O-rings are different. Be careful not to confuse them: the fuel tube side is BLACK; the Nozzle side is GREEN

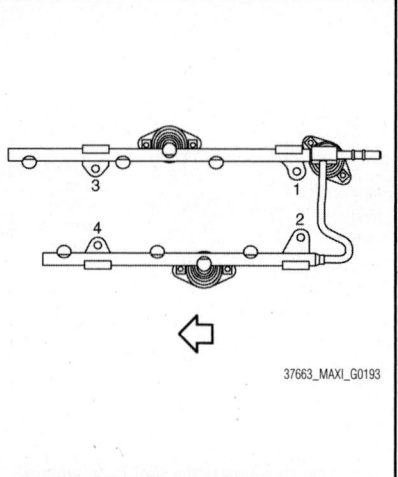

37663_MAXI_G0193

Fig. 108 Tighten the fuel injector assembly bolts in numerical order shown

- Handle the O-ring with bare hands. Never wear gloves
- Lubricate the O-ring with new engine oil
- Never clean the O-ring with solvent
- Check that the O-ring and its mating part are free of foreign material
- When installing the O-ring, be careful not to scratch it with a tool or fingernails. Also be careful not to twist or stretch the O-ring
- Insert the O-ring straight into the fuel injector. Never decenter or twist it
14. Install the fuel injector to the fuel tube as follows:
    a. Insert the clip into the clip groove on the fuel injector.
    b. Insert the clip so that the protrusion of the fuel injector matches the cutout of the clip.

✳✳ CAUTION

Never reuse the clip. Replace it with a new one. Be careful to keep the clip from interfering with the O-ring. If interference occurs, replace the O-ring.

    c. Insert the fuel injector into the fuel tube with the clip attached.
    d. Insert it while matching it to the axial center.
    e. Insert the fuel injector so that the protrusion of the fuel tube matches the cutout of the clip.
    f. Check that the fuel tube flange is securely fixed in the flange fixing groove on the clip.

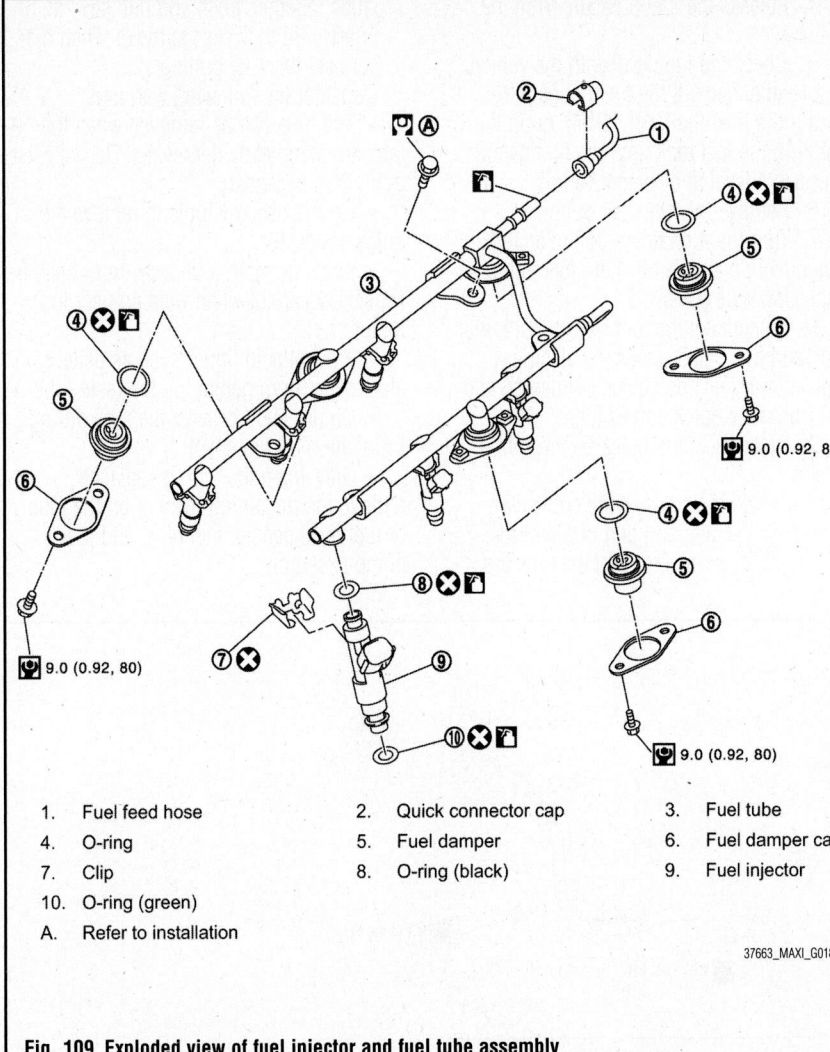

1. Fuel feed hose
4. O-ring
7. Clip
10. O-ring (green)
A. Refer to installation

2. Quick connector cap
5. Fuel damper
8. O-ring (black)

3. Fuel tube
6. Fuel damper cap
9. Fuel injector

9.0 (0.92, 80)

9.0 (0.92, 80)

9.0 (0.92, 80)

37663_MAXI_G0186

Fig. 109 Exploded view of fuel injector and fuel tube assembly

g. Check that installation is complete by checking that fuel injector does not rotate or come off.

h. Check that the protrusions of the fuel injectors and the fuel tubes are aligned with the cutouts of the clips after installation.

15. Install the fuel tube and the fuel injector assembly to the intake manifold.

※※ WARNING

Be careful not to let the tip of the injector nozzle come in contact with other parts.

16. Tighten the bolts in 2 steps in the numerical order as shown in the figure.
    a. Step 1: 7 ft. lbs. (10 Nm).
    b. Step 2: 16 ft. lbs. (22 Nm).
17. Connect the fuel injector harness.
18. Install the intake manifold collector.
19. Connect the quick connector between the fuel feed hose and the fuel tube connection with the following procedure:

a. Check that no foreign substances are deposited in and around the fuel tube and quick connector, and there is no damage.

b. Thinly apply new engine oil around the fuel tube from the tip end to the spool end.

c. Align the center to insert the quick connector straightly into the fuel tube.

d. Insert the quick connector to the fuel tube until the top spool is completely inside the quick connector, and the 2nd level spool exposes right below the quick connector.

Please note the following:
• Hold the position when inserting the fuel tube into the quick connector
• Carefully align the center to avoid the inclined insertion to prevent damage to the O-ring inside of the quick connector
• Insert until a "click" sound is heard and engagement is felt
• To avoid misidentification of the

engagement with a similar sound, be sure to perform the next step
    a. Pull the quick connector by hand holding it in position. Check it is completely engaged (connected) so that it does not come out from the fuel tube.
    b. Install the quick connector cap to the quick connector.
    c. Install the quick connector cap with the arrow on the surface facing in the direction of the quick connector (fuel feed hose side).

➡**If the quick connector cap cannot be installed smoothly, the quick connector may have not been installed correctly. Check the connection again.**

    d. Secure the fuel feed hose to the clamp of the quick connector cap.
20. Installation is in the reverse order of removal.
21. Turn the ignition switch ON (with the engine stopped). With fuel pressure applied to the fuel piping, check that there are no fuel leaks at the connection points.

➡**Use mirrors for checking points that are out of clear sight.**

22. Start the engine. With the engine speed increased, check again that there are no fuel leaks at the connection points.

※※ CAUTION

Never touch the engine immediately after stopped, as the engine becomes extremely hot.

FUEL TANK

DRAINING

1. Before servicing the vehicle, refer to the Precautions.
2. Insert fuel tubing of less than 1 inch (25mm) diameter into the fuel filler tube through the fuel filler opening to drain fuel from the fuel filler tube.
3. Disconnect the fuel filler hose from the fuel filler tube.
4. Insert fuel tubing into the fuel tank through the fuel filler hose to drain fuel from the fuel tank.

REMOVAL & INSTALLATION

See Figures 110 and 111.

1. Before servicing the vehicle, refer to the Precautions.
2. Disconnect the battery negative terminal.
3. Open the fuel filler cap to release the pressure inside the fuel tank.

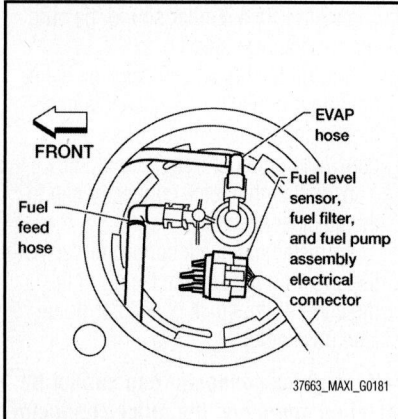

Fig. 110 Disconnect the fuel level sensor, fuel filter, and fuel pump assembly electrical connector, EVAP hose quick connector, and the fuel feed hose quick connector

4. Release the fuel pressure from the fuel line.

5. Check the fuel level with the vehicle on a level surface. If the fuel gauge indicates more than the level $7/8$ full, drain the fuel from the fuel tank until the fuel gauge indicates a level at or below $7/8$ full.

6. Remove the rear seat bottom.

7. Turn the 4 retainers 90° in a clockwise direction and remove the fuel pump inspection hole cover.

8. Disconnect the fuel level sensor, fuel filter, and fuel pump assembly electrical connector, EVAP hose quick connector, and fuel feed hose quick connector.

9. Disconnect the quick connectors as follows:

   a. Hold the sides of the connector, push in the tabs and pull out the tube.

   b. If the connector and the tube are stuck together, push and pull several times until they start to move. Then disconnect them by pulling.

Consider the following warnings:

• The tube can be removed when the tabs are completely depressed. Do not twist more than necessary

• Do not use any tools to remove the quick connector

• Keep the resin tube away from heat. Be especially careful when welding near the tube

• Prevent acid liquid such as battery electrolyte, from getting on the resin tube

• Do not bend or twist the tube during installation and removal

• Only when the tube is replaced, remove the remaining retainer on the tube or fuel level sensor, fuel filter, and fuel pump assembly

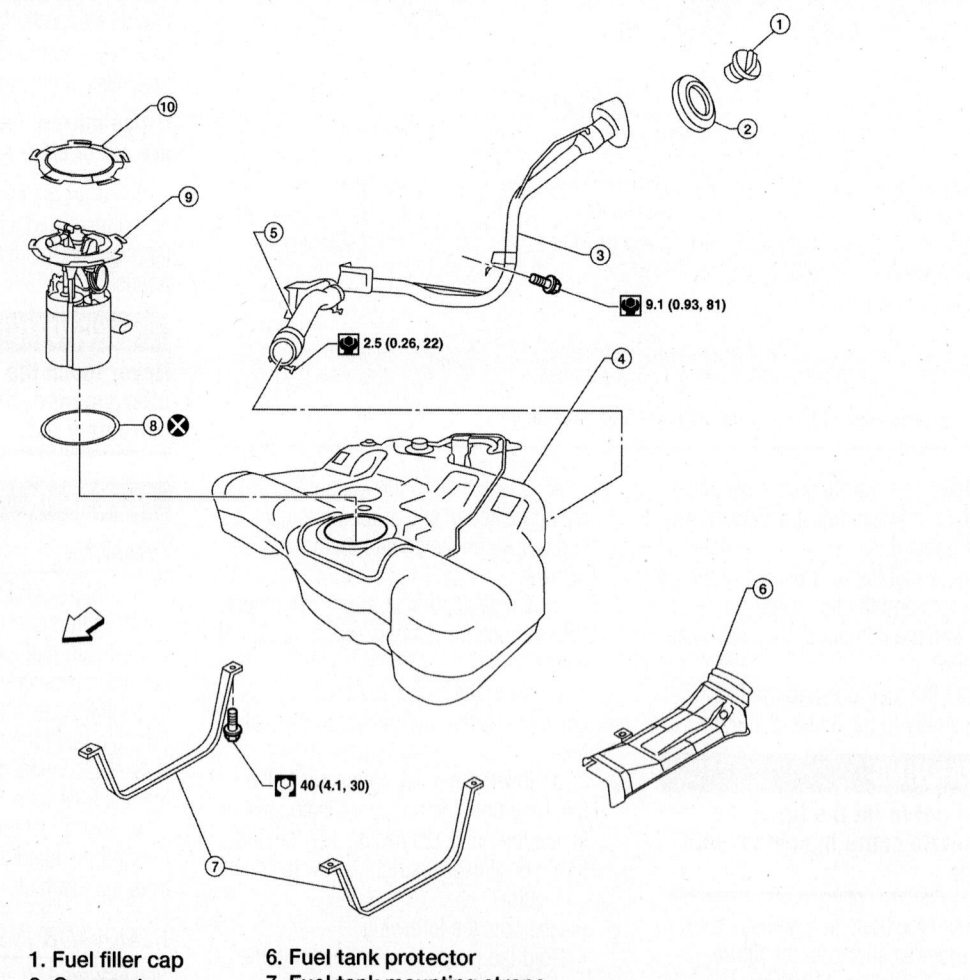

1. Fuel filler cap
2. Grommet
3. Fuel filler tube
4. Fuel tank
5. Fuel filler hose
6. Fuel tank protector
7. Fuel tank mounting straps
8. O-ring
9. Fuel level sensor, fuel filter, and fuel pump assembly
10. Lock ring
Arrow: Front

71075_MAXI_G0077

Fig. 111 Exploded view of fuel tank and related component locations

• When the tube or fuel level sensor, fuel filter, and fuel pump assembly is replaced, also replace the retainer with a new one (green colored retainer)

• To keep the connecting portion clean and to avoid damage and foreign materials, cover them completely with plastic bags or something similar

10. Remove the center exhaust tube, with the muffler.

11. Disconnect the fuel filler hose and recirculation hose at the fuel tank side.

12. Disconnect the 3 parking brake cable mounting brackets on each cable and position the cables out of the way.

13. Remove the rear stabilizer bar clamps, then allow the stabilizer bar to hang.

14. Remove the EVAP canister bolts. Then without disconnecting hoses, position the EVAP canister aside.

15. Remove the fuel tank protector.

16. Remove the fuel tank mounting strap bolts and mounting straps while supporting the fuel tank with a suitable jack.

17. Remove the fuel tank.

18. If replacing the fuel tank, remove the fuel level sensor unit, fuel filter and fuel pump assembly to transfer to the new fuel tank.

To install:

19. Install in the reverse order of removal.

20. Before tightening the fuel tank mounting straps, temporarily install the filler hose and the recirculation hose.

21. Tighten all fuel tank mounting strap bolts to specification, then tighten the hose clamps.

22. Connect the quick connector as follows:

   a. Check the connection for damage or any foreign materials.

   b. Align the connector with the tube, then insert the connector straight into the tube until a click is heard.

23. After the tube is connected, make sure the connection is secure by performing the following checks:

   a. Pull on the tube and the connector to make sure they are securely connected.

   b. Visually confirm that the 2 retainer tabs are connected to the quick connector.

24. Use the following procedure to check for fuel leaks.

   a. Turn the ignition switch ON (without starting the engine). Then check the connections for fuel leaks by applying fuel pressure to the fuel piping.

   b. Run the engine and check for fuel leaks at the fuel system tube and hose connections.

THROTTLE BODY

REMOVAL & INSTALLATION

See Figure 112.

➡**When removing components such as hoses, tubes/lines, etc., cap or plug openings to prevent fluid from spilling.**

1. Before servicing the vehicle, refer to the Precautions.

2. Remove the cowl top.

3. Remove the engine room cover.

4. Remove the front air duct and air duct hose and resonator assembly.

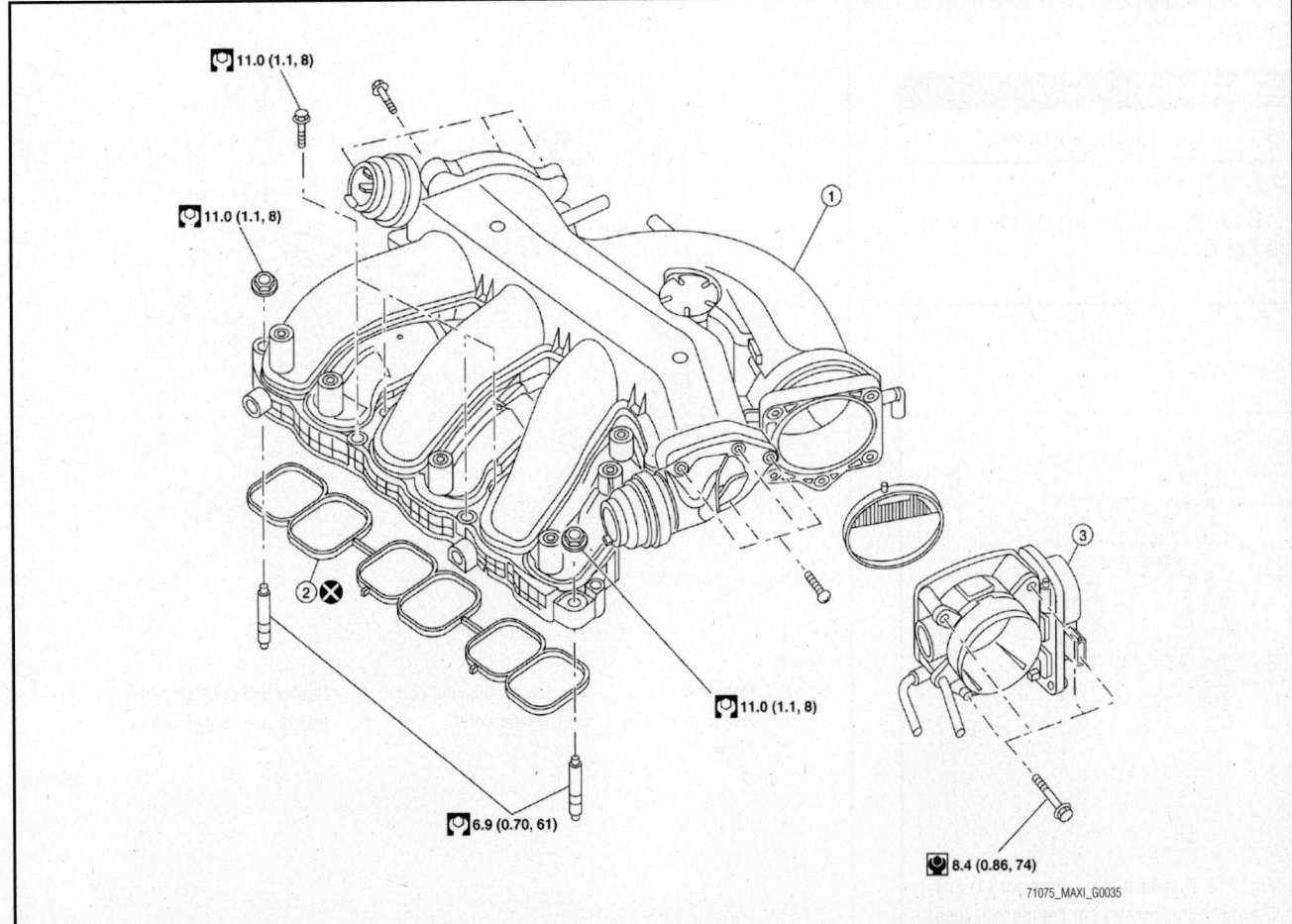

11.0 (1.1, 8)

11.0 (1.1, 8)

11.0 (1.1, 8)

6.9 (0.70, 61)

8.4 (0.86, 74)

71075_MAXI_G0035

Fig. 112 Intake manifold collector (1), gasket (2), and electric throttle control actuator (3) component locations

5. Remove the electric throttle control actuator bolts in the reverse order of the tightening sequence and remove the electric throttle control actuator.

✳✳ WARNING

Handle the electric throttle control actuator carefully to avoid any shock to the actuator. Do not disassemble.

6. Disconnect the electric throttle control actuator electrical connector.

✳✳ WARNING

Cover any engine openings to avoid the entry of any foreign material.

To install:

7. Installation is in the reverse order of removal.

8. Tighten the electric throttle control actuator bolts in order.

9. After installation, it is necessary to re-calibrate the electric throttle control actuator as follows:

   a. Perform the ?Throttle Valve Closed Position Learning? when the harness connector of the electric throttle control actuator is disconnected..

   b. Perform the ?Idle Air Volume Learning? when the electric throttle control actuator is replaced.

HEATING & AIR CONDITIONING SYSTEM

BLOWER MOTOR

REMOVAL & INSTALLATION

See Figure 113.

1. Before servicing the vehicle, refer to the Precautions.

2. Remove the glove box.

3. Disconnect the blower motor connector.

4. Remove the 3 blower motor screws and remove the blower motor from the blower unit.

5. Installation is in the reverse order of removal.

HEATER CORE

REMOVAL & INSTALLATION

See Figure 114.

1. Before servicing the vehicle, refer to the Precautions.

2. Remove the heater and cooling unit assembly.

3. Remove the heater grommet, heater pipe support, and heater pipe cover.

4. Remove the heater and cooling unit foot duct LH.

5. Remove the heater core.

To install:

6. Installation is in the reverse order of removal.

7. Make sure that the aspirator hose is securely attached to the aspirator on the heater and cooling unit foot duct LH.

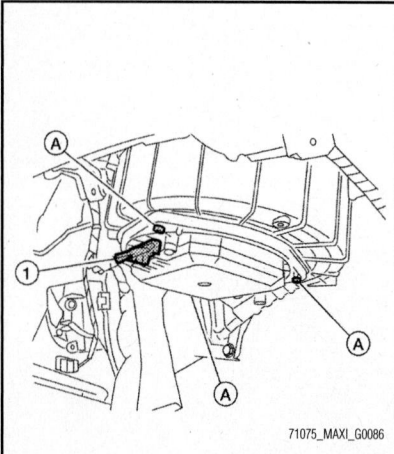

71075_MAXI_G0086

Fig. 113 Blower motor connector (1) and blower motor screws (A) component locations

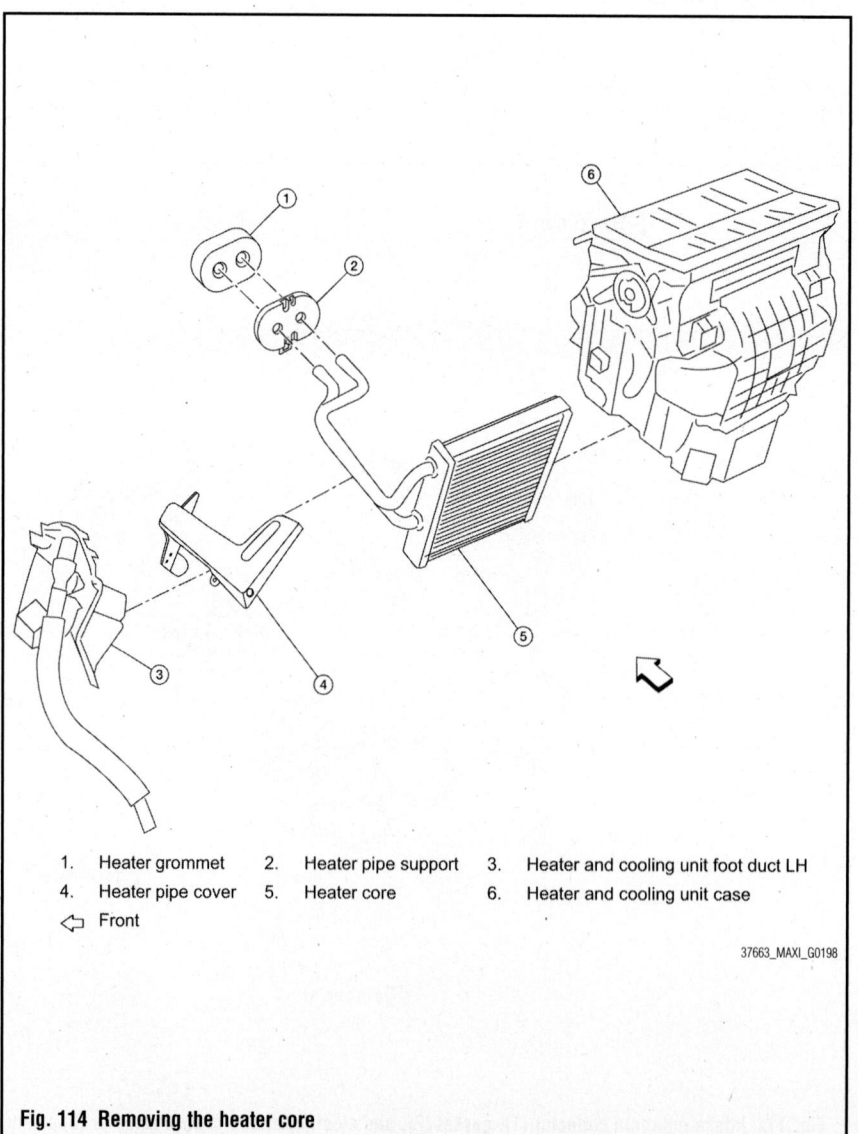

1.	Heater grommet	2.	Heater pipe support	3.	Heater and cooling unit foot duct LH
4.	Heater pipe cover	5.	Heater core	6.	Heater and cooling unit case
⟵	Front				

37663_MAXI_G0198

Fig. 114 Removing the heater core

STEERING

POWER RACK & PINION STEERING GEAR

REMOVAL & INSTALLATION

See Figures 115 and 116.

1. Before servicing the vehicle, refer to the Precautions.
2. Remove the front tires.
3. Drain the power steering fluid.
4. Disconnect the front stabilizer connecting rods from the front stabilizer and reposition the front stabilizer.
5. Remove the steering outer socket cotter pins, and then loosen the nuts.
6. Remove the steering outer sockets from the steering knuckles so as not to damage the ball joint boots using Tool HT72520000 (J-25730-A), or equivalent.

➡**Temporarily tighten the nut to prevent damage to the threads and to prevent the Tool from suddenly coming off.**

7. Remove the side bolt of the lower shaft assembly and disconnect the lower shaft assembly.
8. Remove the front exhaust tube.
9. Disconnect the SSPS valve harness connector.
10. Disconnect the high and low pressure piping from the steering gear assembly.
11. Remove the steering hydraulic piping bracket from the front suspension member.
12. Remove the bolts and nuts of the steering gear assembly, and then remove the steering gear assembly from the vehicle.

13. Check for fluid leaks or damage to the steering gear assembly. If any exist, replace the steering gear assembly.

To install:

14. Installation is in the reverse order of removal.
15. When installing the lower shaft assembly to the steering gear assembly, follow the procedure listed below.

  a. Set the rack of the steering gear in the neutral position.

➡**To get the neutral position of the rack, turn the gear sub-assembly and measure the distance of the inner socket, and then measure the intermediate position of the distance.**

  b. Align the rear cover cap projection with the marking position of the gear housing assembly.

  c. Install the slit part of the lower shaft assembly aligning with the projection of the rear cover cap. Make sure that the slit part of the lower shaft assembly is aligned with both the projection of the rear cover cap and the marking position of the gear housing assembly.

16. After installation, bleed the air from the steering hydraulic system.
17. Perform the final tightening of the nuts and bolts on each part under unladen

conditions with the tires on level ground. Check the wheel alignment.

18. Make sure that the steering wheel operates smoothly by turning it several times from full left stop to full right stop.

POWER STEERING PUMP

BLEEDING

➡**If air bleeding is not complete, the following symptoms may be observed:**

- Bubbles created in the reservoir tank
- Clicking noise heard from the oil pump
- Excessive buzzing in the oil pump

➡**Fluid noise may occur in the steering gear or oil pump. This does not affect performance or durability of the system.**

1. Before servicing the vehicle, refer to the Precautions.
2. Turn the steering wheel several times from the full left stop to the full right stop with the engine OFF.

➡**Turn the steering wheel while filling the reservoir tank with fluid so as not to lower the fluid level below the MIN line.**

1. Steering outer socket cotter pins
2. Steering outer sockets
3. Steering knuckles
4. Ball joint boots

37663_MAXI_G0209

Fig. 115 Remove the steering outer sockets from the steering knuckles

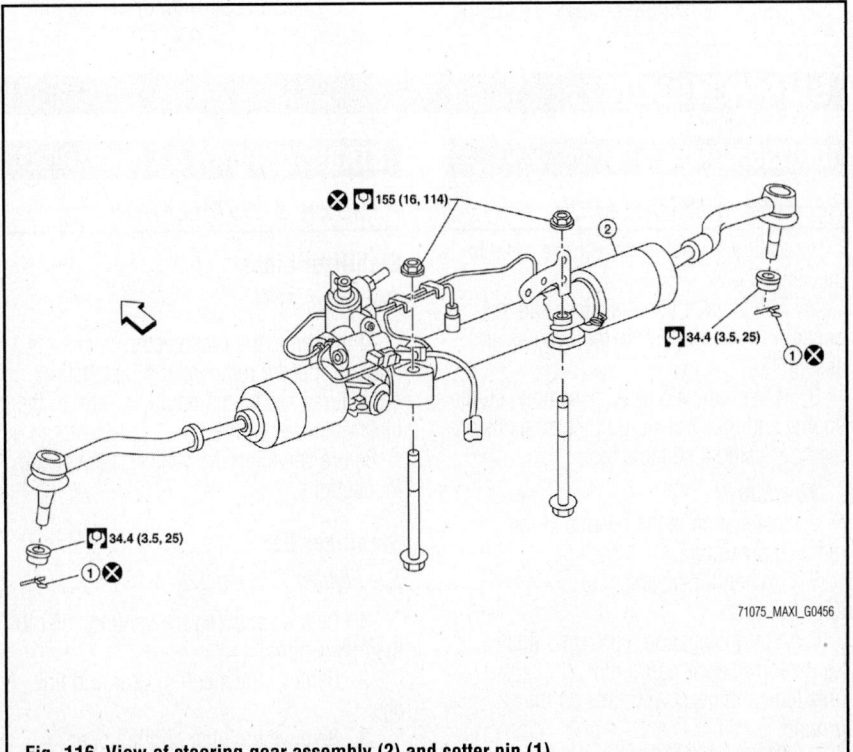

71075_MAXI_G0456

Fig. 116 View of steering gear assembly (2) and cotter pin (1)

3. Start the engine and hold the steering wheel at each lock position for 3 seconds at idle to check for fluid leakage.

4. Repeat several times at approximately 3 second intervals.

✳✳ WARNING

Do not hold the steering wheel in a locked position for more than 10 seconds. There is the possibility that the oil pump may be damaged.

5. Check the fluid for bubbles and white contamination.

6. Stop the engine if bubbles and white contamination do not drain out. Perform steps above after waiting until the bubbles and white contamination drain out.

7. Stop the engine, and then check the fluid level.

REMOVAL & INSTALLATION

See Figure 117.

1. Before servicing the vehicle, refer to the Precautions.

2. Remove the front tire (RH).

3. Remove the engine side undercover.

4. Remove the hood ledge cover (RH).

5. Drain the power steering fluid.

6. Disconnect the high pressure piping and suction hose from the power steering oil pump.

7. Loosen the drive belt.

8. Remove the drive belt from the power steering oil pump pulley.

9. Remove the power steering oil pump

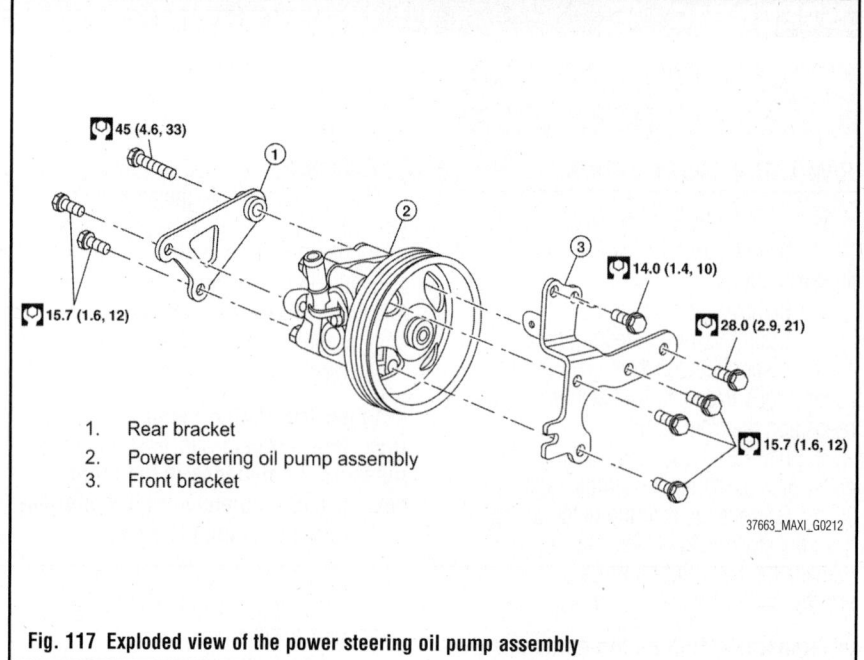

45 (4.6, 33)

15.7 (1.6, 12)

14.0 (1.4, 10)

28.0 (2.9, 21)

15.7 (1.6, 12)

1. Rear bracket
2. Power steering oil pump assembly
3. Front bracket

37663_MAXI_G0212

Fig. 117 Exploded view of the power steering oil pump assembly

bolts, and then remove the power steering oil pump.

To install:

10. Installation is in the reverse order of removal.

11. When installing the power steering oil pump, install all the bolts by hand initially, then tighten the bolts to specification.

12. Perform the following procedures after installing.

a. Check the belt tension.

b. Bleed the air from power steering system.

FLUID FILL PROCEDURE

1. Before servicing the vehicle, refer to the Precautions.

2. Clean the power steering reservoir cap and surrounding area of any debris.

3. Remove the power steering reservoir cap.

4. Fill the power steering reservoir while checking the fluid level.

5. Bleed the air from the hydraulic system.

6. Check for fluid leaks.

SUSPENSION

FRONT SUSPENSION

CROSSMEMBER

REMOVAL & INSTALLATION

1. Before servicing the vehicle, refer to the Precautions.

2. The engine, transmission, and suspension member must be removed as an assembly.

3. Once removed as an assembly, lift the engine and transmission off the suspension member using a suitable tool.

To install:

4. Installation is the reverse of the removal procedure.

5. Tighten the fasteners to specification.

6. After installation, perform a final tightening of each part under unladen conditions with tires on the ground.

7. Check wheel alignment.

STABILIZER BAR & LINKS

REMOVAL & INSTALLATION

Stabilizer Links

See Figure 118.

At this time, the manufacturer does not provide specific removal and installation procedures for this component, refer to the illustration as required.

Before servicing the vehicle, refer to the Precautions.

Stabilizer Bar

See Figures 119 and 120.

1. Before servicing the vehicle, refer to the Precautions.

2. Remove the steering gear and linkage.

3. Remove the nuts on the upper portion of the stabilizer connecting rod.

4. Remove the stabilizer clamp bolts.

5. Remove the stabilizer from the vehicle.

To install:

6. Check the stabilizer, connecting rod, bushing and clamp for deformation, cracks and damage, and replace if necessary.

7. Installation is in the reverse order of removal.

8. When installing the stabilizer, make sure that the notch in the stabilizer clips face front.

9. Make sure the slit in surface of stabilizer bushings face rear.

10. The stabilizer uses pillow ball type connecting rod. Position the ball joint with the case on the pillow ball head parallel to the stabilizer.

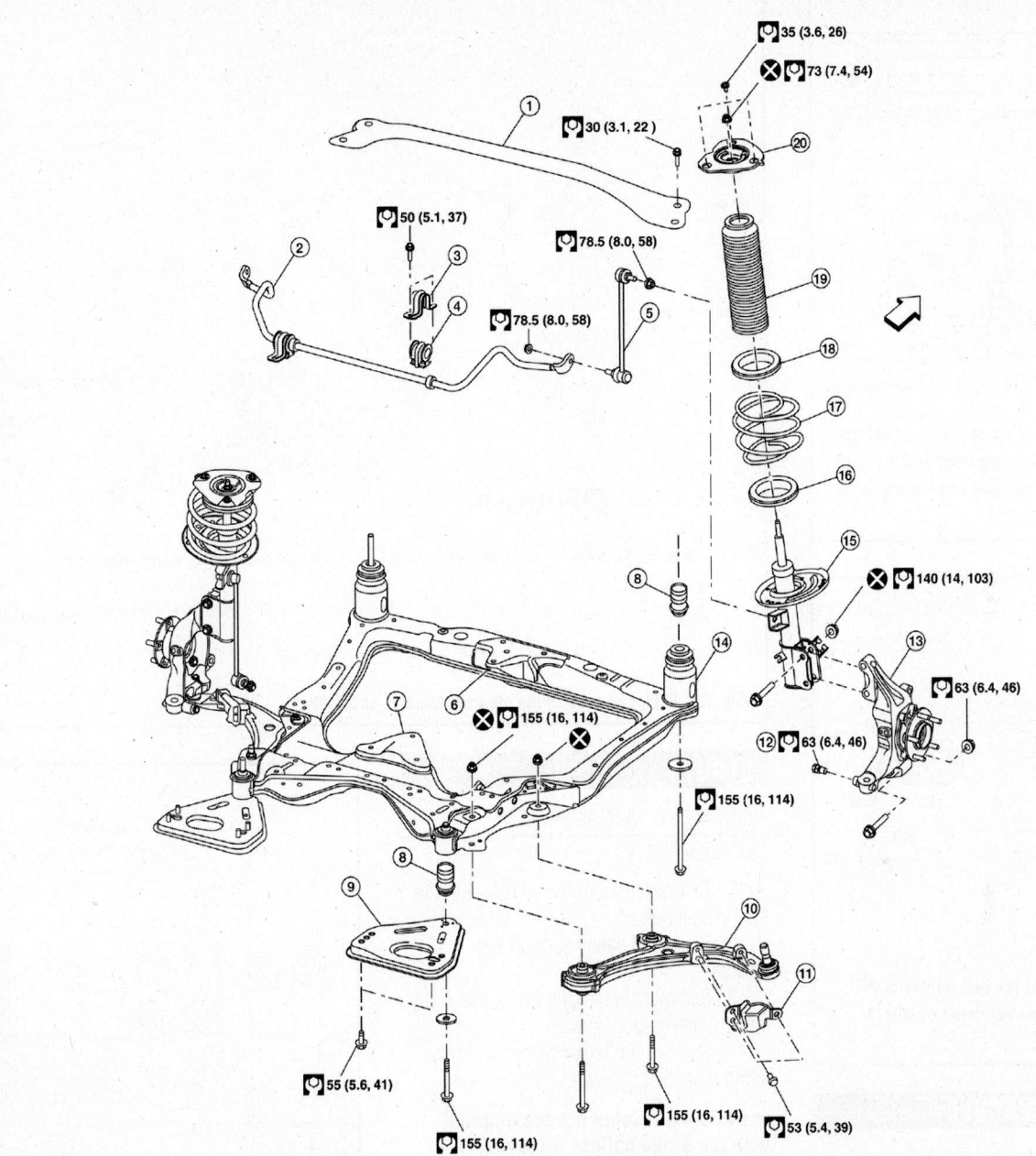

1. Strut tower bar
2. Stabilizer bar
3. Stabilizer clamp
4. Stabilizer bushing
5. Connecting rod
6. Front mount bracket
7. Rear mount bracket
8. Suspension member insulator
9. Member pin stay
10. Transverse link
11. Steering stop plate
12. Steering stop
13. Steering knuckle
14. Front suspension member
15. Strut
16. Lower rubber seat
17. Coil spring
18. Spring upper seat / strut bearing
19. Dust cover/jounce bumper
20. Strut mount insulator
Arrow: Front

71075_MAXI_G0096

Fig. 118 Exploded view of front suspension component locations

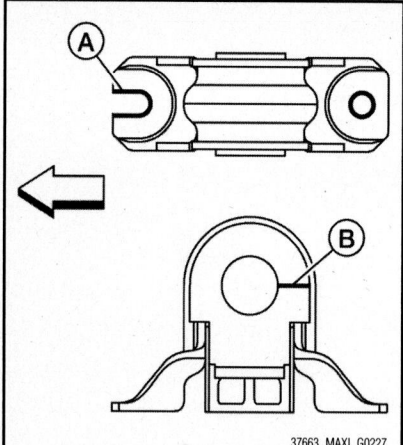

Fig. 119 Make sure that the notch (A) in the stabilizer clips face front and the slit (B) in surface of the stabilizer bushings face rear

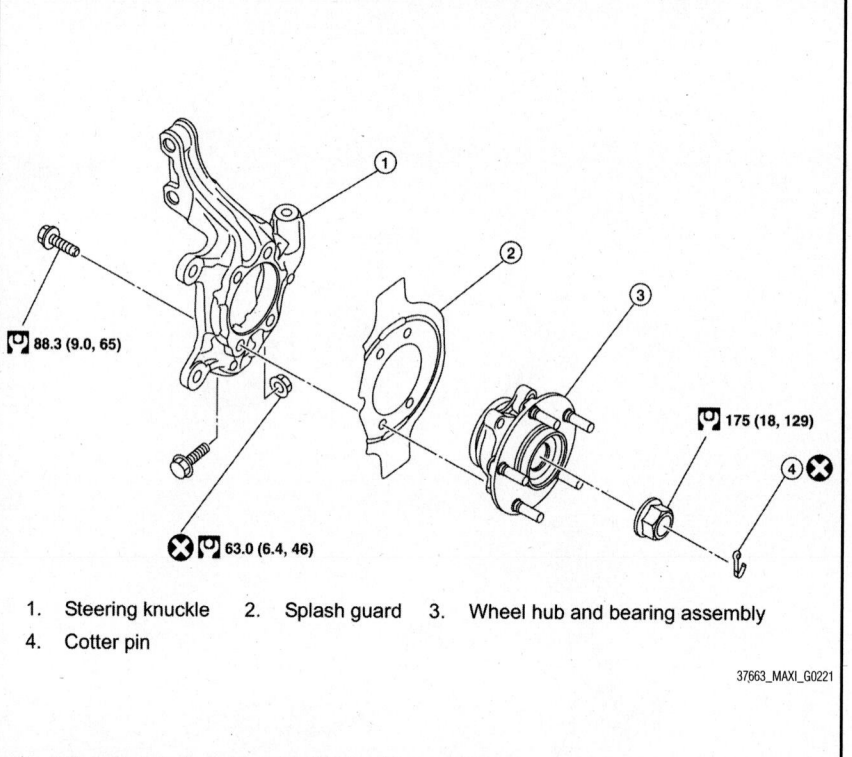

1. Steering knuckle   2. Splash guard   3. Wheel hub and bearing assembly
4. Cotter pin

Fig. 121 Exploded view of the steering knuckle assembly

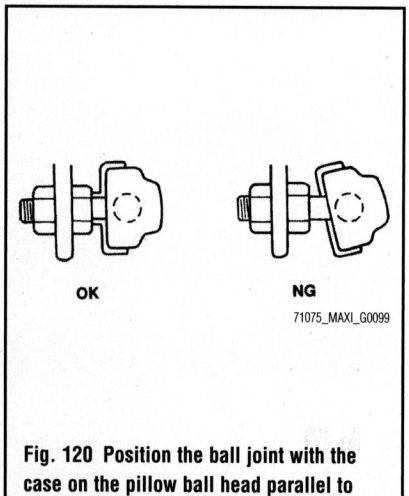

OK   NG

Fig. 120 Position the ball joint with the case on the pillow ball head parallel to the stabilizer

STEERING KNUCKLE

REMOVAL & INSTALLATION

See Figure 121.

1. Before servicing the vehicle, refer to the Precautions.
2. Remove the front wheel hub and bearing assembly.
3. Remove the steering linkage from the steering knuckle.
4. Remove the steering knuckle lower pinch bolt.
5. Remove the steering knuckle-to-strut bolts.
6. Remove the steering knuckle.

To install:
7. Installation is in the reverse order of removal.
8. Do not reuse non-reusable parts.

STRUTS

REMOVAL & INSTALLATION

See Figure 122.

1. Before servicing the vehicle, refer to the Precautions.
2. Raise and safely support the vehicle.
3. Remove the tire and wheel assembly using a power tool.
4. Remove the brake caliper and reposition it aside using wire.

➡**Avoid depressing the brake pedal with the brake caliper removed.**

5. Remove the wheel sensor electrical harness from the strut.
6. Remove the brake hose lock plate.
7. Remove the steering knuckle-to-strut bolts and nuts.
8. Remove the bolt on the strut tower bar, then the bolts on the strut tower.
9. Remove the strut from the vehicle.

To install:
10. Check the strut for any oil leakage or other damage and replace as necessary.
11. Installation is in the reverse order of removal.
12. Be sure the tab on the strut mount insulator is positioned correctly.

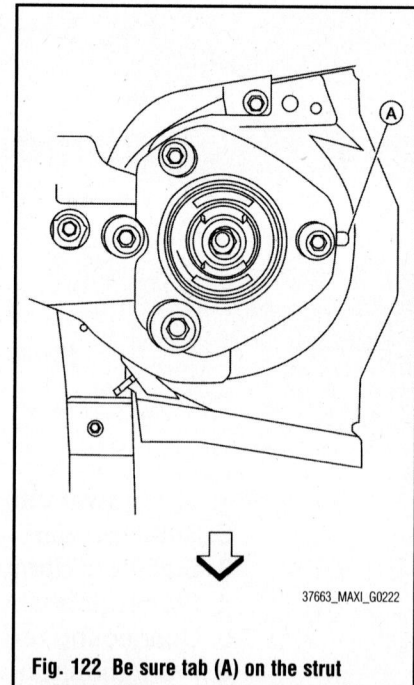

Fig. 122 Be sure tab (A) on the strut mount insulator is positioned as shown

OVERHAUL

See Figure 123.

1. Before servicing the vehicle, refer to the Precautions.
2. Install Tool ST35652000 to the strut and secure it in a vise.

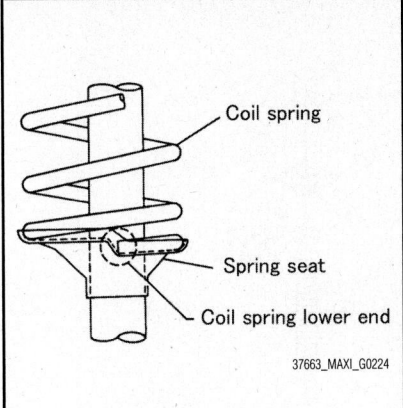

Fig. 123 Align the lower end to the spring seat as shown

✳✳ CAUTION

When installing the Tool, wrap a shop cloth around the strut to protect it from damage.

3. Slightly loosen the piston rod lock nut.

✳✳ CAUTION

Do not remove the piston rod lock nut completely. If it is removed completely, the coil spring can jump out and may cause serious damage or injury.

4. Compress the coil spring using a commercially available spring compressor.

✳✳ CAUTION

Make sure that the pawls of the 2 spring compressors are firmly hooked on the spring. The spring compressors must be tightened alternately so as not to tilt the spring.

5. Making sure the coil spring is free between the upper and lower seats, remove the piston rod lock nut.
6. Remove the small parts on the strut, strut mount insulator, spring upper seat/strut bearing. Then, remove the coil spring.
7. Remove the dust cover/jounce bumper from the strut mount insulator.
8. Gradually release the spring compressor (commercial service tool), and remove the coil spring.

To assemble:
9. Compress the coil spring using a spring compressor (commercial service tool), and install it onto the strut.

10. Face the tube side of the coil spring downward. Align the lower end to the spring seat.

✳✳ CAUTION

Be sure the spring compressor is securely attached to the coil spring. Compress the coil spring.

11. Install the dust cover/jounce bumper to the strut mount insulator.

➡**Be sure to install the dust cover/jounce bumper to the strut mount insulator securely. When installing the dust cover/jounce bumper, use soapy water. Do not use machine oil or other lubricants.**

12. Install the small parts to the strut, spring upper seat/strut bearing and strut mount insulator. Temporarily install the piston rod lock nut.

✳✳ WARNING

Do not reuse the piston rod lock nut.

13. Be sure the tab on the strut mount insulator is positioned correctly.
14. Be sure the coil spring is properly set in the spring rubber seat. Gradually release the spring compressor.

➡**Be sure upper rubber seat is properly aligned to spring upper seat and coil spring.**

15. Tighten the piston rod lock nut to the specified torque.
16. Remove the Tool from the strut.

SUSPENSION ARM

REMOVAL & INSTALLATION
See Figure 124.

1. Before servicing the vehicle, refer to the Precautions.
2. Remove the rear suspension assembly.
3. Remove the connecting rod bracket from the suspension arm.
4. Remove the 2 suspension arm bolts and nuts from the suspension member side of the suspension arm.

To install:
5. Installation is the reverse of the removal procedure.
6. Discard the cotter pin, use a new cotter pin for installation.
7. Check the rear wheel alignment and adjust if necessary.

WHEEL HUBS & BEARINGS

ADJUSTMENT

1. Before servicing the vehicle, refer to the Precautions.
2. Move the wheel hub and bearing assembly in the axial direction by hand. Make sure there is no looseness of the wheel bearing.
3. Rotate the wheel hub and make sure there are no unusual noises or other irregular conditions.
4. If there are any irregular noises or conditions, replace the wheel hub and bearing assembly. No other adjustment is possible.

REMOVAL & INSTALLATION
See Figure 125.

1. Before servicing the vehicle, refer to the Precautions.
2. Remove the wheel and tire from the vehicle.
3. Remove the wheel sensor from the steering knuckle.

✳✳ WARNING

Do not pull on the wheel sensor harness.

4. Remove the brake hose lock plate from the strut assembly.
5. Remove the brake caliper torque member bolts using a power tool leaving the brake hose attached, then remove disc rotor. Reposition caliper aside with wire.

➡**Avoid depressing the brake pedal while the brake caliper is removed.**

6. Remove the cotter pin, then loosen the lock nut from the halfshaft using a power tool.
7. Using a piece of wood and a hammer, tap on the lock nut to disengage the halfshaft from the wheel hub.

✳✳ WARNING

Never place the halfshaft joint at an extreme angle. Also be careful not to overextend the slide joint. Never allow the halfshaft to hang down without support.

➡**Use a suitable puller if the halfshaft cannot be separated from the wheel hub and bearing assembly.**

8. Remove the wheel hub and bearing assembly bolts using a power tool.

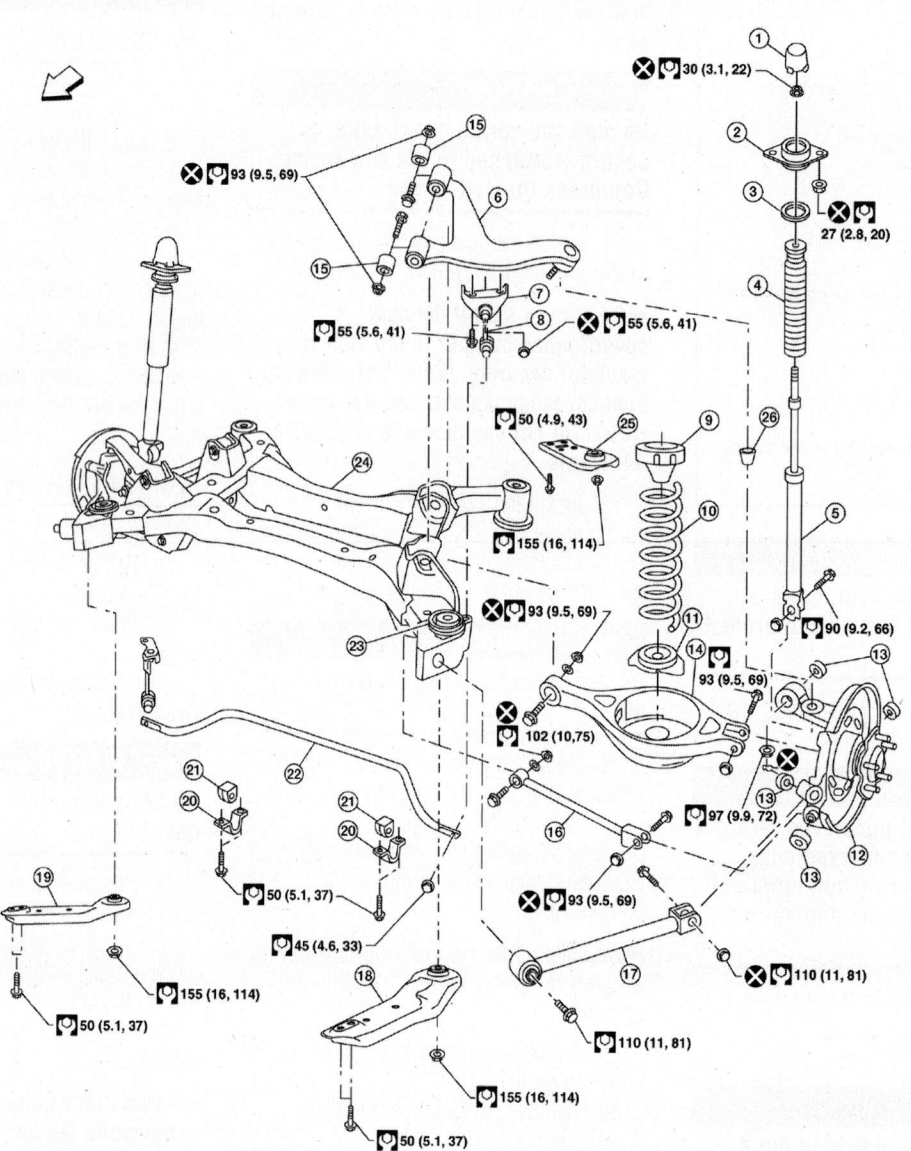

1. Cap
2. Shock absorber insulator
3. Shock absorber seal
4. Bound bumper
5. Shock absorber
6. Suspension arm
7. Connecting rod mount bracket
8. Connecting rod
9. Upper rubber seat
10. Coil spring
11. Lower rubber seat
12. Knuckle
13. Knuckle bushing
14. Rear lower link
15. Suspension arm bushing
16. Front lower link
17. Radius rod
18. Front member stay (LH)
19. Front member stay (RH)
20. Stabilizer bar clamp
21. Bushing
22. Stabilizer bar
23. Member stopper
24. Rear suspension member
25. Rear member stay (LH)
26. Ball seat
Arrow: Front

71075_MAXI_G0097

Fig. 124 Exploded view of rear suspension component locations

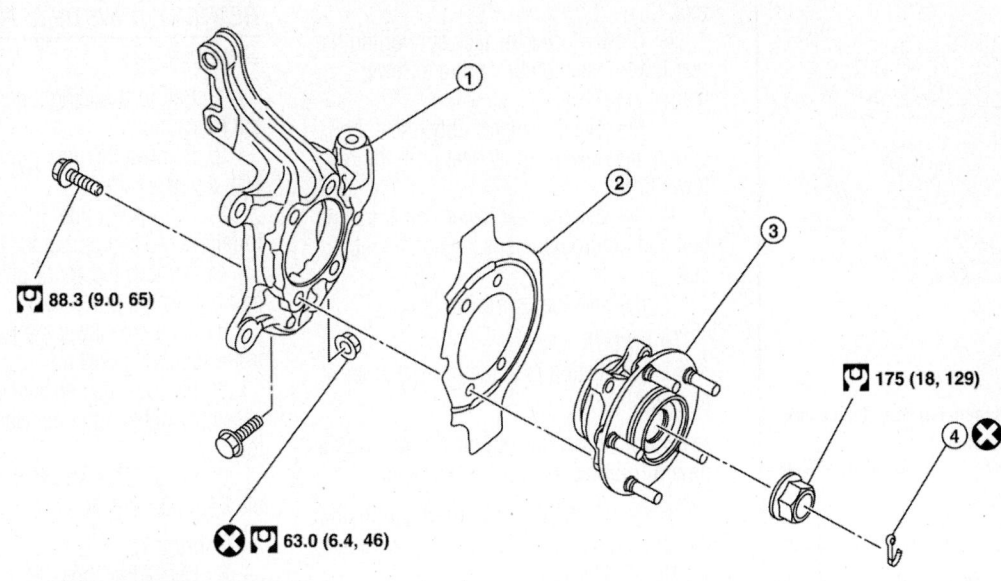

88.3 (9.0, 65)

175 (18, 129)

63.0 (6.4, 46)

1. Steering knuckle
2. Splash guard
3. Wheel hub and bearing assembly
4. Cotter pin

71075_MAXI_G0100

Fig. 125 Exploded view of steering knuckle, wheel hub and bearing assembly

9. Remove the splash guard and wheel hub and bearing assembly from the steering knuckle.

To install:

10. Installation is the reverse of the removal procedure.

11. Check for deformity, cracks, and damage on each part, replace if necessary.

⁜ **WARNING**

Do not reuse non-reusable parts.

12. When installing the wheel hub and bearing assembly to the steering knuckle, align the cutout in the toner ring cover with the wheel sensor mounting hole in the steering knuckle.

SUSPENSION

COIL SPRINGS

REMOVAL & INSTALLATION

1. Before servicing the vehicle, refer to the Precautions.
2. Remove the wheel and tire assembly.
3. Loosen the rear lower link adjusting bolt and nut on the suspension member side.
4. Support the rear lower link and knuckle by placing suitable jacks under each of them.
5. Remove the rear lower link bolt and nut from the knuckle using a power tool.
6. Slowly lower the jack supporting the rear lower link and coil spring to lower them.
7. Remove the upper rubber seat, coil

spring, and lower rubber seat from the rear lower link.

To install:

8. Installation is in the reverse order of removal.

➡**Do not reuse the adjusting nut on the rear lower link, use a new adjusting nut for installation.**

9. Check that the projecting part inside the upper rubber seat and the bracket flange are attached as shown.
10. Check that the projection part outside the upper rubber seat is directed toward the front of the vehicle.
11. Position the hollow of the lower rubber seat with the groove part of the rear lower link.

REAR SUSPENSION

12. Install the coil spring so that the side with the two paint markers is directed toward the lower side.
13. Check the rear wheel alignment and adjust if necessary.

FRONT LOWER LINK

REMOVAL & INSTALLATION

See Figure 126.

1. Before servicing the vehicle, refer to the Precautions.
2. Remove the front lower link nut and bolt from the knuckle side and the adjusting bolt and nut from the suspension member side.
3. Remove the front lower link.

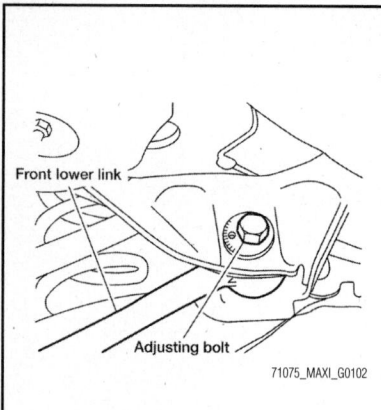

Fig. 126 Rear suspension front lower link and adjusting bolt shown

To install:

4. Installation is in the reverse order of removal.

➡**Do not reuse the adjusting nut, use a new adjusting nut for installation.**

5. Check the rear wheel alignment and adjust if necessary.

RADIUS ROD

REMOVAL & INSTALLATION

1. Before servicing the vehicle, refer to the Precautions.
2. Remove the wheel and tire assembly.
3. Remove the radius rod from the knuckle.
4. Remove the radius rod from the suspension member.

To install:

5. Installation is the reverse of the removal procedure.
6. Check the rear wheel alignment and adjust if necessary.

REAR LOWER LINK

REMOVAL & INSTALLATION

See Figure 127.

1. Before servicing the vehicle, refer to the Precautions.
2. Remove the wheel and tire assembly.
3. Loosen the rear lower link adjusting bolt and nut on the suspension member side.
4. Support the rear lower link and knuckle by placing suitable jacks under each of them.

5. Remove the rear lower link bolt and nut from the knuckle using a power tool.
6. Slowly lower the jack supporting the rear lower link and coil spring to lower them.
7. Remove the upper rubber seat, coil spring, and lower rubber seat from the rear lower link.
8. Remove the rear lower link adjusting bolt and nut from the suspension member side.
9. Remove the rear lower link.

To install:

10. Installation is in the reverse order of removal.

➡**Do not reuse the adjusting nut, use a new adjusting nut for installation.**

11. Check that the projecting part inside the upper rubber seat and the bracket flange are attached as shown.
12. Check that the projection part outside the upper rubber seat is directed toward the front of the vehicle.
13. Position the hollow of the lower rubber seat with the groove part of the rear lower link.
14. Install the coil spring so that the side with the two paint markers is directed toward the lower side.
15. Check the rear wheel alignment and adjust if necessary.

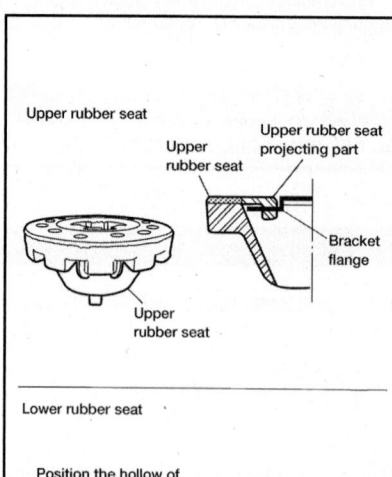
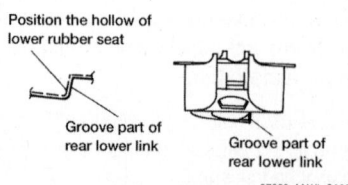

Fig. 127 Check that the projecting part inside the upper rubber seat and the bracket flange are attached as shown

SHOCK ABSORBERS

REMOVAL & INSTALLATION

See Figure 128.

1. Before servicing the vehicle, refer to the Precautions.
2. Remove the tires from the vehicle with a power tool.
3. Set a jack under the rear lower link to relieve the coil spring tension.
4. Remove the shock absorber lower end bolt with a power tool.
5. Gradually lower the jack to remove it from the rear lower link.
6. Remove the shock absorber assembly upper end nuts with a power tool.
7. Remove the shock absorber assembly from the vehicle.

To install:

8. Check the shock absorber assembly for deformation, cracks, and damage. Replace if needed.
9. Check the welded and sealed areas for oil leakage, and replace if there are leaks.
10. Installation is the reverse order of removal.

✳✳ WARNING

Do not reuse non-reusable parts.

11. Perform the final tightening of the shock absorber assembly lower side (rubber bushing) under unladen conditions with tires on level ground.
12. Check wheel alignment and adjust as needed.
13. Adjust the neutral position of the steering angle sensor after checking the wheel alignment.

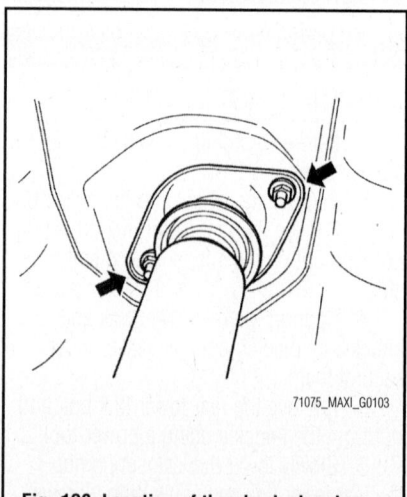

Fig. 128 Location of the shock absorber assembly upper end nuts

STABILIZER BAR & LINKS

REMOVAL & INSTALLATION

See Figure 129.

1. Before servicing the vehicle, refer to the Precautions.
2. Disconnect the stabilizer bar from the connecting rod.
3. Remove the stabilizer bar clamps and bushings.
4. Remove the stabilizer bar.
5. Installation is in the reverse order of removal.

WHEEL HUBS & BEARINGS

ADJUSTMENT

➡**The wheel hub assembly does not require maintenance. If any of the following symptoms are noted, replace the wheel hub assembly.**

- A growling noise is emitted from the wheel hub assembly while driving
- The wheel hub assembly drags or turns roughly

Wheel Hub—On-Vehicle Service

1. Before servicing the vehicle, refer to the Precautions.

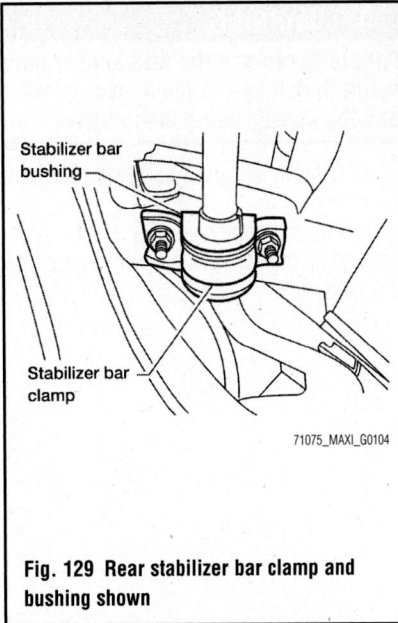

Fig. 129 Rear stabilizer bar clamp and bushing shown

2. Check the axle and suspension parts for excessive play, wear or damage.
3. Grab hold and move each rear wheel at the top and bottom to check for excessive play.

Wheel Bearing—On-Vehicle Inspection

1. Before servicing the vehicle, refer to the Precautions.

2. Check axial end play.
- Check that the wheel hub bearing axial end play is within specification: 0.004 inches (0.1mm)
3. Check that the wheel hub bearings operate smoothly.
4. Replace the wheel bearing assembly if there is axial end play or the wheel bearing does not turn smoothly.

REMOVAL & INSTALLATION

See Figure 130.

➡**The wheel hub assembly does not require maintenance. If any of the following symptoms are noted, replace the wheel hub assembly.**

- A growling noise is emitted from the wheel hub assembly while driving
- The wheel hub assembly drags or turns roughly

1. Before servicing the vehicle, refer to the Precautions.
2. Remove the rear wheel and tire.
3. Remove the brake caliper assembly and brake rotor.

➡**The brake hose does not need to be disconnected from the brake caliper.**

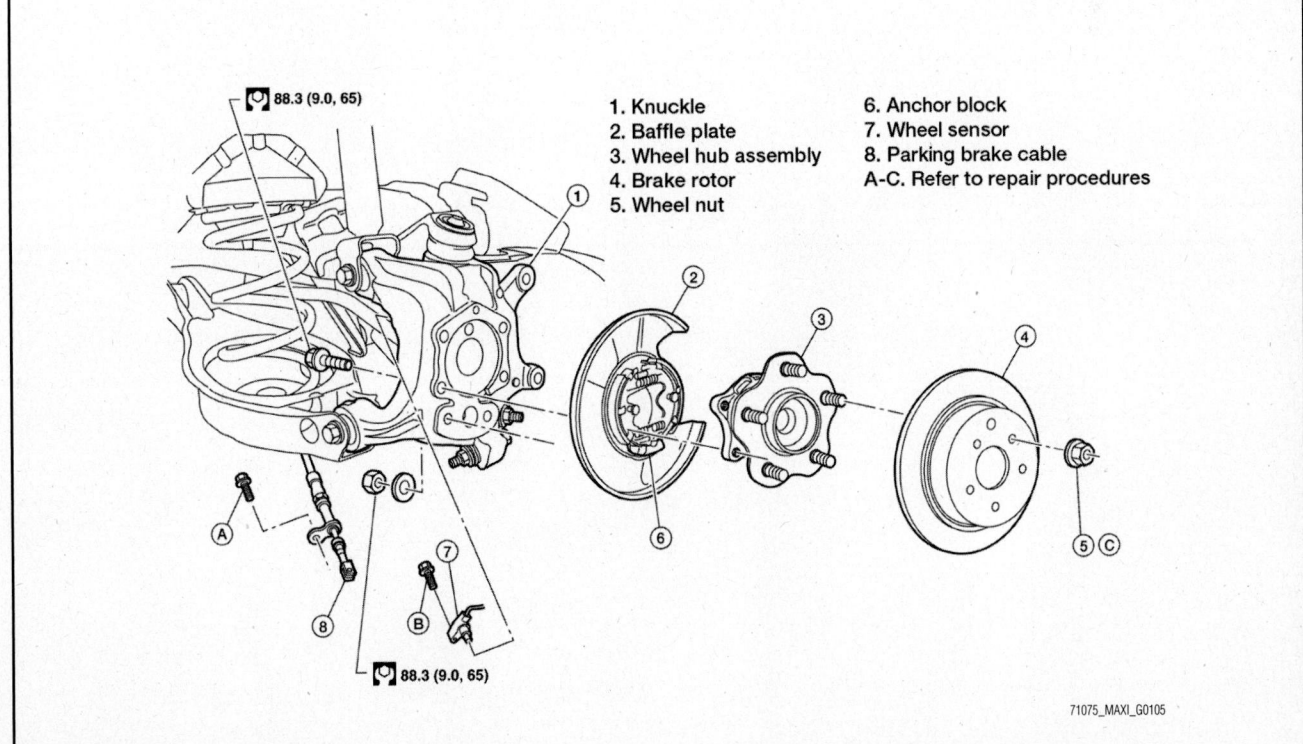

1. Knuckle
2. Baffle plate
3. Wheel hub assembly
4. Brake rotor
5. Wheel nut
6. Anchor block
7. Wheel sensor
8. Parking brake cable
A-C. Refer to repair procedures

Fig. 130 Exploded view of rear knuckle and hub assembly

❋❋ WARNING

Suspend the brake caliper assembly using wire, do not stretch the brake hose. Do not depress the brake pedal, or the caliper piston may pop out. Do not twist the brake hose.

---

4. Remove the rear ABS sensor, then move it away from the wheel hub assembly.

❋❋ WARNING

Failure to remove the ABS sensor may result in damage to the sensor wires and the sensor being inoperative.

---

5. Remove the wheel hub assembly from the knuckle.

6. Check for any deformity, cracks, or damage on the wheel hub assembly, replace if necessary.

To install:

7. Installation is in the reverse order of removal.

8. Check that the wheel bearings operate smoothly.

9. Check that the wheel hub bearing axial end play is within specification: 0.004 inches (0.1mm).

NISSAN

Murano

14

SPECIFICATIONS AND MAINTENANCE CHARTS

ENGINE AND VEHICLE IDENTIFICATION

		Engine						Model Year	
Code	Liters (cc)	Cu. In.	Cyl.	Fuel Sys.	Engine	Eng. Mfg.		Code	Year
VQ35DE	3.5 (3498)	213.45	6	MFI	DOHC	Nissan		B	2011
								C	2012

MFI: Multi-port Fuel Injection

DOHC: Double Overhead Camshaft

71075_MURA_C0001

GENERAL ENGINE SPECIFICATIONS

Year	Model	Engine Displacement Liters	Engine ID	Net Horsepower @ rpm	Net Torque @ rpm (ft. lbs.)	Bore x Stroke (in.)	Com- pression Ratio	Oil Pressure @ rpm
2011	Murano	3.5	VQ35DE	260@6000	240@4400	3.76X3.20	10.3:1	43@2000
2012	Murano	3.5	VQ35DE	260@6000	240@4400	3.76X3.20	10.3:1	43@2000

71075_MURA_C0002

ENGINE TUNE-UP SPECIFICATIONS

							Valve Clearance (in.)	
Year	Engine Displacement Liters	Engine ID	Spark Plug Gap (in.)	Ignition Timing (deg.)	Fuel Pump (psi)	Idle Speed RPM	In.	Ex.
2011	3.5	VQ35DE	0.043	7 to 17	51 ①	②	③	④
2012	3.5	VQ35DE	0.043	7 to 17	51 ①	②	③	④

NA: Not Applicable

NOTE: The Vehicle Emission Control Information label often reflects specification changes made during production. The label figures must be used

① At idle

② Idle is computer controlled and is not adjustable.

③ 0.010-0.013 cold

　0.012-0.016 hot

④ 0.011-0.015 cold

　0.012-0.017 hot

71075_MURA_C0003

CAPACITIES

Year	Model	Engine Displacement Liters	Engine ID	Engine Oil with Filter (qts.)	Transmission (pts.)	Transfer Case (pts.)	Drive Axle Front (pts.)	Drive Axle Rear (pts.)	Fuel Tank (gal.)	Cooling System (qts.)
2011	Murano	3.5	VQ35DE	4.9	21.5	0.63	NA	1.1	21.7	9.9
2012	Murano	3.5	VQ35DE	4.9	21.5	0.63	NA	1.1	21.7	9.9

NA: Not Applicable

NOTE: All capacities are approximate. Add fluid gradually and check to be sure a proper fluid level is obtained.

71075_MURA_C0005

FLUID SPECIFICATIONS

Year	Model	Engine Displ. Liters (VIN)	Engine Oil	Man. Trans.	Auto. Trans.	Drive Axle Front	Drive Axle Rear	Trans. Case	Power Steering Fluid	Brake Master Cylinder	Cooling System
2011	Murano	VQ35DE	①	NA	②	③	③	③	NISSAN PSF	DOT 3	④
2012	Murano	VQ35DE	①	NA	②	③	③	③	NISSAN PSF	DOT 3	④

NA: Not Applicable

DOT: Department Of Transpotation

① API Service SM SAE 5W-30

② Nissan CVT Fluid NS-2 2

③ GL-5 80W90

④ Nissan Long Life Antifreeze/Coolant

71075_MURA_C0004

VALVE SPECIFICATIONS

Year	Engine Displacement Liters	Engine ID	Seat Angle (deg.)	Face Angle (deg.)	Spring Test Pressure (lbs. @ in.)	Spring Installed Height (in.)	Stem-to-Guide Clearance (in.) Intake	Stem-to-Guide Clearance (in.) Exhaust	Stem Diameter (in.) Intake	Stem Diameter (in.) Exhaust
2011	3.5	VQ35DE	45.15-45.45	45	42.3@1.467	1.457	0.0008-0.0021	0.0012-0.0022	0.2348-0.2354	0.2347-0.2350
2012	3.5	VQ35DE	45.15-45.45	45	45.4@1.457	1.457	0.0008-0.0021	0.0012-0.0022	0.2348-0.2354	0.2347-0.2350

71075_MURA_C0006

CAMSHAFT SPECIFICATIONS

All measurements are given in inches.

Year	Engine Displ. Liters	Engine VIN	Journal Dia.	Brg. Oil Clearance	Shaft End-play	Runout	Lobe Height Intake	Exhaust
2011	3.5	VQ35DE	①	②	0.0045-0.0074	③	1.7900-1.7974	1.7904-1.7978
2012	3.5	VQ35DE	①	②	0.0045-0.0074	③	1.7900-1.7974	1.7904-1.7978

① No.1: 1.0211-1.0218

  No.2, No.3, No.4: 0.9230-0.9238

② No.1: 1.0018-1.0034

  No.2, No.3, No.4: 0.0014-0.0030

③ Less then 0.001 (0.02 mm)

71075_MURA_C0007

CRANKSHAFT AND CONNECTING ROD SPECIFICATIONS

All measurements are given in inches.

Year	Engine Displacement Liters	Engine ID	Crankshaft Main Brg. Journal Dia.	Main Brg. Oil Clearance	Shaft End-play	Thrust on No.	Connecting Rod Journal Diameter	Oil Clearance	Side Clearance
2011	3.5	VQ35DE	①	0.0014-0.0018	0.0039-0.0098	4	②	0.0008-0.0018	0.0079-0.0138
2012	3.5	VQ35DE	①	0.0014-0.0018	0.0039-0.0098	4	②	0.0008-0.0018	0.0079-0.0138

① There are 24 different grades, ranging from A (2.3612) to 7 (2.3603)

② Grade 0: 2.0460-2.0462

  Grade 1: 2.0457-2.0460

  Grade 2: 2.0445-2.0457

71075_MURA_C0008

PISTON AND RING SPECIFICATIONS

All measurements are given in inches.

Year	Engine Displacement Liters	Engine ID	Piston Clearance	Ring Gap Top Comp.	Bottom Comp.	Oil Control	Ring Side Clearance Top Comp.	Bottom Comp.	Oil Control
2011	3.5	VQ35DE	0.0004-0.0012	0.0091-0.0130	0.0091-0.0130	0.0079-0.0177	0.0018-0.0031	0.0012-0.0028	0.0026-0.0049
2012	3.5	VQ35DE	0.0004-0.0012	0.0091-0.0130	0.0091-0.0130	0.0079-0.0177	0.0018-0.0031	0.0012-0.0028	0.0026-0.0049

71075_MURA_C0009

TORQUE SPECIFICATIONS
All readings in ft. lbs.

Year	Engine Displacement Liters	Engine ID	Cylinder Head Bolts	Main Bearing Bolts	Rod Bearing Bolts	Crankshaft Damper Bolts	Driveplate Bolts	Manifold		Spark Plugs	Oil Pan Drain Plug
								Intake	Exhaust		
2011	3.5	VQ35DE	①	②	③	④	65	⑤	24	14-22	25
2012	3.5	VQ35DE	①	②	③	④	65	⑤	24	14-22	25

① Step 1: 72 ft. lbs.

   Step 2: Loosen all bolts completely

   Step 3: 29 ft. lbs.

   Step 4: +103 degrees

   Step 5: +103 degrees

② Step 1: 26 ft. lbs.

   Step 2: +90 degrees

③ Step 1: 14 ft. lbs.

   Step 2: +90 degrees

④ 33 ft. lbs. +90 degrees

⑤ Step 1: 5 ft. lbs

   Step 2: 19 ft. lbs.

71075_MURA_C0010

WHEEL ALIGNMENT

Year	Model		Caster		Camber		Toe-in (in.)	Kingpin Inclination (Deg.)
			Range (+/-Deg.)	Preferred Setting (Deg.)	Range (+/-Deg.)	Preferred Setting (Deg.)		
2011	Murano	F	.75	A	.75	B	0.059+/-0.4	12.75+/-.75
		R	—	—	.50	-0.72	0.106+/-.71	—
2012	Murano	F	.75	A	.75	B	0.059+/-1.0	12.75+/-.75
		R	—	—	.50	-0.72	0.106+/-.71	—

Note: Specifications are taken with the following fuel, radiator coolant and engine oil full. Spare tire, jack, hand tools and mats are in designated positions.

① Left side 4.67

   Right side 5.00

② Left side -0.25

   Right side -0.50

71075_MURA_C0011

TIRE, WHEEL AND BALL JOINT SPECIFICATIONS

Year	Model	OEM Tires		Tire Pressures (psi)		Wheel Size	Ball Joint Inspection	Lug Nut Torque (ft. lbs.)
		Standard	Optional	Front	Rear			
2011	Murano	P235/65R18	P235/55R20	33	33	7.5J	①	80
2012	Murano	P235/65R18	P235/55R20	33	33	7.5J	①	80

OEM: Original Equipment Manufacturer

PSI: Pounds Per Square Inch

① 0 (0mm) inches axial end play

71075_MURA_C0012

BRAKE SPECIFICATIONS

All measurements in inches unless noted

Year	Model		Brake Disc			Minimum Lining Thickness	Brake Caliper	
			Original Thickness	Minimum Thickness	Maximum Runout		Bracket Bolts (ft. lbs.)	Mounting Bolts (ft. lbs.)
2011	Murano	F	1.102	1.024	0.0003	0.079	122	20
		R	0.630	0.551	0.0008	0.079	62	32
2012	Murano	F	1.102	1.024	0.0003	0.079	122	20
		R	0.630	0.551	0.0008	0.079	62	32

F: Front

R: Rear

71075_MURA_C0013

SCHEDULED MAINTENANCE INTERVALS (1)
Nissan—Murano

TO BE SERVICED	TYPE OF SERVICE	7.5	15	22.5	30	37.5	45	52.5	60
Engine oil & filter	R	every 3,750 miles							
Brake lines & cables	S/I		✓		✓		✓		✓
Brake pads, discs	I	✓	✓	✓	✓	✓	✓	✓	✓
Brake fluid	R		✓		✓		✓		✓
Driveshaft boots & propeller shaft	L/I	✓	✓	✓	✓	✓	✓	✓	✓
CVT ①	I		✓		✓		✓		✓
Transfer case and differential fluid ①	I		✓		✓		✓		✓
Air cleaner filter	R				✓				✓
Drive belt (s) ②	S/I								✓
Engine coolant ③	R								✓
Spark plugs	R	Platinum plugs, every 105,000 miles							
Cabin air filter	R		✓		✓		✓		✓
Exhaust system	I	✓	✓	✓	✓	✓	✓	✓	✓
Evap vapor lines	I				✓				✓
Fuel lines	S/I				✓				✓
Steering gear, linkage, axle & suspension parts	I	✓	✓	✓	✓	✓	✓	✓	✓
Tires (rotate)	S/I	every 5,000-6,000 miles							
Valve clearance ④	S/I				✓				✓

R: Replace   S/I: Service or Inspect   L: Lubricate

① If towing a trailer, using a camper or a car-top carrier, or driving on rough or muddy roads, change (not just inspect) oil at every 60,000 miles.

② First at 60,000, then inspect every 15,000 miles

③ After 105,000, replace every 75,000

④ Periodic maintenance not required, if valve noice increases, inspect valve clearance

Follow Periodic Maintenance Schedule 1 if the driving habits frequently include one or more of the following driving conditions:

Repeated short trips of less than 5 miles (8 km).

Repeated short trips of less than 10 miles (16 km) with outside temperatures remaining below freezing

Operating in hot weather in stop-and-go "rush hour" traffic.

Extensive idling and/or low speed driving for long distances, such as police, taxi or door-to-door delivery use

Driving in dusty conditions.

Driving on rough, muddy, or salt spread roads.

Towing a trailer, using a camper or a car-top carrier.

Follow Periodic Maintenance Schedule 2 if none of driving conditions shown in Schedule 1 apply to the driving habits.

SCHEDULED MAINTENANCE INTERVALS (2)
Nissan—Murano

TO BE SERVICED	SERVICE	7.5	15	22.5	30	37.5	45	52.5	60
Engine oil & filter	R	✓	✓	✓	✓	✓	✓	✓	✓
Brake lines & cables	S/I		✓		✓		✓		✓
Brake pads, discs	I		✓		✓		✓		✓
Brake fluid	R				✓				✓
Driveshaft boots & propeller shaft	L/I		✓		✓		✓		✓
CVT ①	I		✓		✓		✓		✓
Transfer case and differential fluid ①	I		✓		✓		✓		✓
Air cleaner filter	R				✓				✓
Drive belt (s) ②	S/I								✓
Engine coolant ③	R								✓
Spark plugs	R	Platinum plugs, every 105,000 miles							
Cabin air filter	R		✓		✓		✓		✓
Exhaust system	I				✓				✓
Evap vapor lines	I				✓				✓
Fuel lines	S/I				✓				✓
Steering gear, linkage, axle & suspension parts	I				✓				✓
Tires (rotate)	S/I	every 5,000-6,000 miles							
Valve clearance ④	S/I				✓				✓

R: Replace   S/I: Service or Inspect   L: Lubricate

① If towing a trailer, using a camper or a car-top carrier, or driving on rough or muddy roads, change (not just inspect) oil at every 60,000 miles.

② First at 60,000, then inspect every 15,000 miles

③ After 105,000, replace every 75,000

④ Periodic maintenance not required, if valve noice increases, inspect valve clearance

Follow Periodic Maintenance Schedule 1 if the driving habits frequently include one or more of the following driving conditions:

Repeated short trips of less than 5 miles (8 km).

Repeated short trips of less than 10 miles (16 km) with outside temperatures remaining below freezing

Operating in hot weather in stop-and-go "rush hour" traffic.

Extensive idling and/or low speed driving for long distances, such as police, taxi or door-to-door delivery use

Driving in dusty conditions.

Driving on rough, muddy, or salt spread roads.

Towing a trailer, using a camper or a car-top carrier.

Follow Periodic Maintenance Schedule 2 if none of driving conditions shown in Schedule 1 apply to the driving habits.

71075_MURA_C0015

PRECAUTIONS

Before servicing any vehicle, please be sure to read all of the following precautions, which deal with personal safety, prevention of component damage, and important points to take into consideration when servicing a motor vehicle:

• Never open, service or drain the radiator or cooling system when the engine is hot; serious burns can occur from the steam and hot coolant.

• Observe all applicable safety precautions when working around fuel. Whenever servicing the fuel system, always work in a well-ventilated area. Do not allow fuel spray or vapors to come in contact with a spark, open flame, or excessive heat (a hot drop light, for example). Keep a dry chemical fire extinguisher near the work area. Always keep fuel in a container specifically designed for fuel storage; also, always properly seal fuel containers to avoid the possibility of fire or explosion. Refer to the additional fuel system precautions later in this section.

• Fuel injection systems often remain pressurized, even after the engine has been turned **OFF**. The fuel system pressure must be relieved before disconnecting any fuel lines. Failure to do so may result in fire and/or personal injury.

• Brake fluid often contains polyglycol ethers and polyglycols. Avoid contact with the eyes and wash your hands thoroughly after handling brake fluid. If you do get brake fluid in your eyes, flush your eyes with clean, running water for 15 minutes. If eye irritation persists, or if you have taken

brake fluid internally, IMMEDIATELY seek medical assistance.

• The EPA warns that prolonged contact with used engine oil may cause a number of skin disorders, including cancer. You should make every effort to minimize your exposure to used engine oil. Protective gloves should be worn when changing oil. Wash your hands and any other exposed skin areas as soon as possible after exposure to used engine oil. Soap and water, or waterless hand cleaner should be used.

• All new vehicles are now equipped with an air bag system, often referred to as a Supplemental Restraint System (SRS) or Supplemental Inflatable Restraint (SIR) system. The system must be disabled before performing service on or around system components, steering column, instrument panel components, wiring and sensors. Failure to follow safety and disabling procedures could result in accidental air bag deployment, possible personal injury and unnecessary system repairs.

• Always wear safety goggles when working with, or around, the air bag system. When carrying a non-deployed air bag, be sure the bag and trim cover are pointed away from your body. When placing a non-deployed air bag on a work surface, always face the bag and trim cover upward, away from the surface. This will reduce the motion of the module if it is accidentally deployed. Refer to the additional air bag system precautions later in this section.

• Clean, high quality brake fluid from a sealed container is essential to the safe and

proper operation of the brake system. You should always buy the correct type of brake fluid for your vehicle. If the brake fluid becomes contaminated, completely flush the system with new fluid. Never reuse any brake fluid. Any brake fluid that is removed from the system should be discarded. Also, do not allow any brake fluid to come in contact with a painted surface; it will damage the paint.

• Never operate the engine without the proper amount and type of engine oil; doing so WILL result in severe engine damage.

• Timing belt maintenance is extremely important. Many models utilize an interference-type, non-freewheeling engine. If the timing belt breaks, the valves in the cylinder head may strike the pistons, causing potentially serious (also time-consuming and expensive) engine damage. Refer to the maintenance interval charts for the recommended replacement interval for the timing belt, and to the timing belt section for belt replacement and inspection.

• Disconnecting the negative battery cable on some vehicles may interfere with the functions of the on-board computer system(s) and may require the computer to undergo a relearning process once the negative battery cable is reconnected.

• When servicing drum brakes, only disassemble and assemble one side at a time, leaving the remaining side intact for reference.

• Only an MVAC-trained, EPA-certified automotive technician should service the air conditioning system or its components.

BRAKES

GENERAL INFORMATION

PRECAUTIONS

• Certain components within the ABS system are not intended to be serviced or repaired individually.

• Do not use rubber hoses or other parts not specifically specified for and ABS system. When using repair kits, replace all parts included in the kit. Partial or incorrect repair may lead to functional problems and require the replacement of components.

• Lubricate rubber parts with clean, fresh brake fluid to ease assembly. Do not use shop air to clean parts; damage to rubber components may result.

• Use only DOT 3 brake fluid from an unopened container.

• If any hydraulic component or line is removed or replaced, it may be necessary to bleed the entire system.

• A clean repair area is essential. Always clean the reservoir and cap thoroughly before removing the cap. The slightest amount of dirt in the fluid may plug an orifice and impair the system function. Perform repairs after components have been thoroughly cleaned; use only denatured alcohol to clean components. Do not allow ABS components to come into contact with any substance containing mineral oil; this includes used shop rags.

• The Anti-Lock control unit is a microprocessor similar to other computer units in the vehicle. Ensure that the ignition switch is **OFF** before removing or installing con-

ANTI-LOCK BRAKE SYSTEM (ABS)

troller harnesses. Avoid static electricity discharge at or near the controller.

• If any arc welding is to be done on the vehicle, the control unit should be unplugged before welding operations begin.

SPEED SENSORS

REMOVAL & INSTALLATION

See Figures 1 and 2.

1. Before servicing the vehicle, refer to the Precautions Section.

2. Disconnect the negative battery cable.

3. Raise and support the vehicle safely.

4. Remove the tire and wheel assembly.

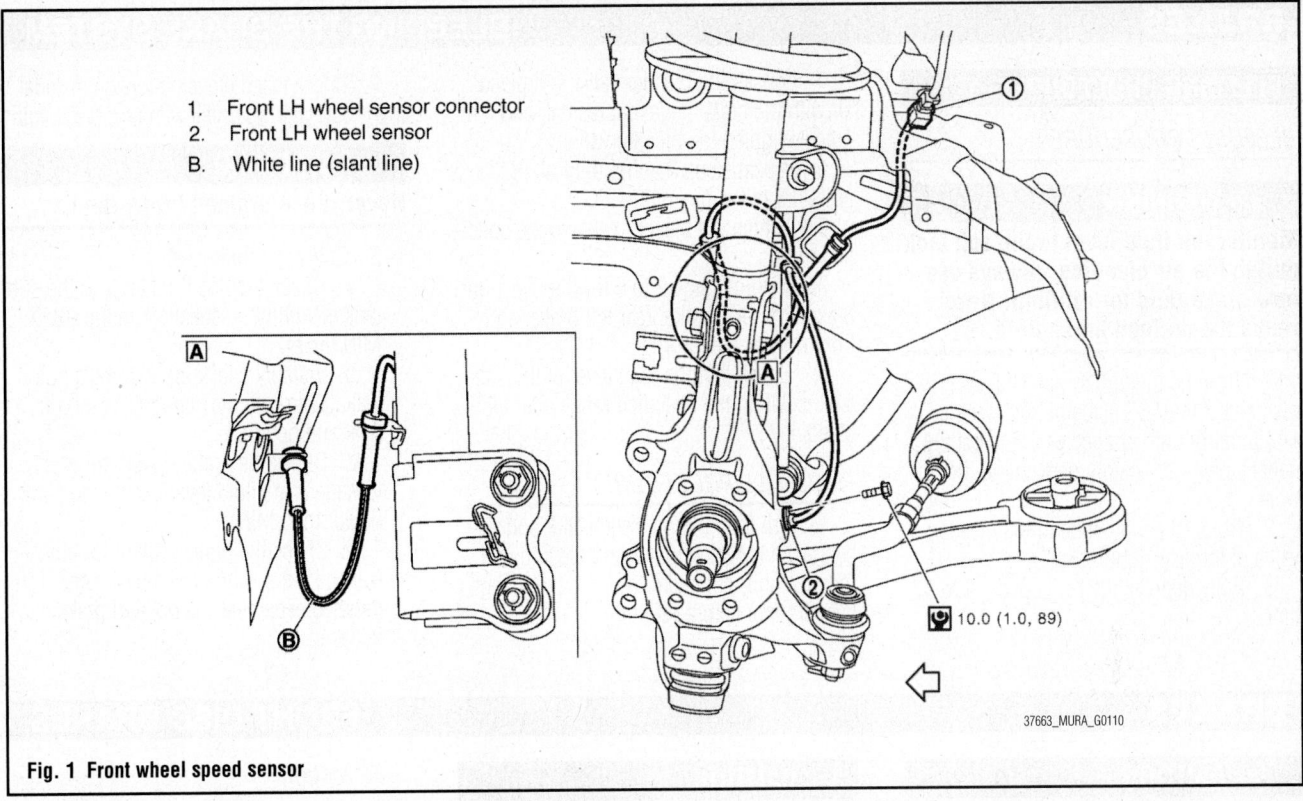

1. Front LH wheel sensor connector
2. Front LH wheel sensor
B. White line (slant line)

10.0 (1.0, 89)

37663_MURA_G0110

Fig. 1 Front wheel speed sensor

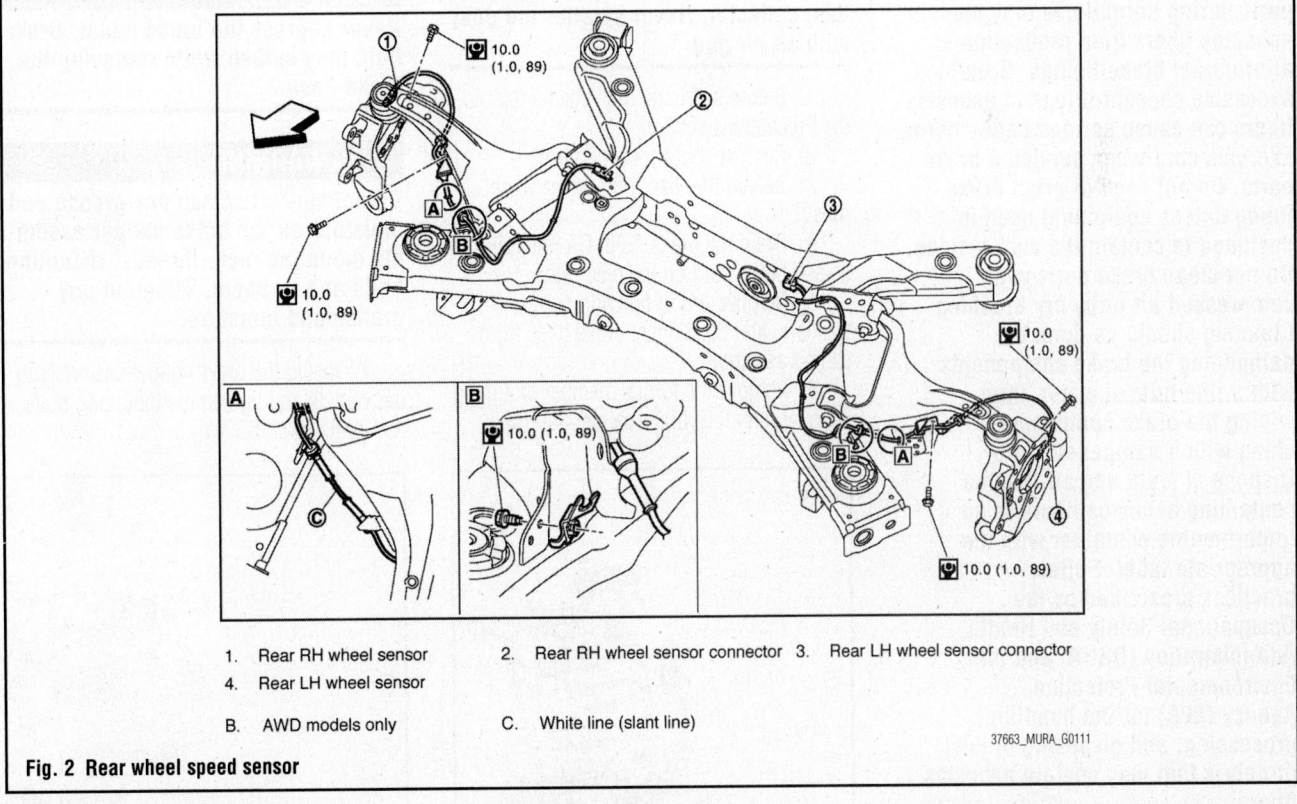

10.0 (1.0, 89)

10.0 (1.0, 89)

10.0 (1.0, 89)

10.0 (1.0, 89)

10.0 (1.0, 89)

1. Rear RH wheel sensor
2. Rear RH wheel sensor connector
3. Rear LH wheel sensor connector
4. Rear LH wheel sensor
B. AWD models only
C. White line (slant line)

37663_MURA_G0111

Fig. 2 Rear wheel speed sensor

5. Remove the wheel speed sensor mounting bolts and clip.

6. Pull the sensor out, being careful to turn it as little as possible. Do not pull on the sensor harness.

To install:

7. Before installing the sensor be sure no foreign materials such as iron fragments are adhered to the pick-up part of the sensor or to the inside of the sensor mounting hole or on the rotor mounting surface.

8. Be sure the harness is not twisted when installed.

9. Continue the installation in the reverse order of the removal procedure.

BRAKES

BLEEDING THE BRAKE SYSTEM

BLEEDING PROCEDURE

BLEEDING PROCEDURE

✳✳ WARNING

Monitor the fluid level in the sub tank during the air bleeding. Always use new brake fluid for refilling. Never reuse the drained brake fluid.

1. Turn the ignition switch OFF and disconnect the ABS actuator and electric unit (control unit) connector or the battery negative terminal before performing the work.

2. Connect a vinyl tube to the bleeder valve of the rear right brake.

3. Fully depress the brake pedal 4 to 5 times.

4. Loosen the bleeder valve and bleed air with the brake pedal depressed, and then quickly tighten the bleeder valve.

5. Repeat steps 3 and 4 until all of the air is out of the brake line.

6. Tighten the bleeder valve to the specified torque.

7. Perform steps 2 to 6 for the rear right brake, front left brake, rear left brake, and front right brake in order.

8. Check that the fluid level in the sub tank is within the specified range after air bleeding.

FLUID FILL PROCEDURE

1. Turn the ignition switch OFF and disconnect the ABS actuator and electric unit (control unit) connector or the battery negative terminal before refilling.

2. Check that there is no foreign material in the sub tank, and refill with new brake fluid.

✳✳ WARNING

Never reuse drained brake fluid.

3. Check the fluid level:

a. Check that the fluid level in the sub tank is within the specified range (MAX-MIN lines).

b. Visually check for any brake fluid leakage around the sub tank, reservoir tank and hose.

c. Check the brake system for any leakage if the fluid level is extremely low (lower than MIN).

d. Check the brake system for fluid leakage if the warning lamp remains illuminated even after the parking brake is released.

BRAKES

FRONT DISC BRAKES

✳✳ CAUTION

Dust and dirt accumulating on brake parts during normal use may contain asbestos fibers from production or aftermarket brake linings. Breathing excessive concentrations of asbestos fibers can cause serious bodily harm. Exercise care when servicing brake parts. Do not sand or grind brake lining unless equipment used is designed to contain the dust residue. Do not clean brake parts with compressed air or by dry brushing. Cleaning should be done by dampening the brake components with a fine mist of water, then wiping the brake components clean with a dampened cloth. Dispose of cloth and all residue containing asbestos fibers in an impermeable container with the appropriate label. Follow practices prescribed by the Occupational Safety and Health Administration (OSHA) and the Environmental Protection Agency (EPA) for the handling, processing, and disposing of dust or debris that may contain asbestos fibers.

BRAKE CALIPER

REMOVAL & INSTALLATION

See Figures 3 and 4.

✳✳ CAUTION

Brake dust must be removed with a dust collector. Never splatter the dust with an air gun.

1. Before servicing the vehicle, refer to the Precautions section.

2. Remove the wheels and tires.

3. Secure the disc rotor using wheel nuts.

4. Drain the brake fluid. Do not spill or splash brake fluid on the disc rotor.

5. Remove union bolt and copper washers, and disconnect brake hose from caliper assembly.

6. Remove the torque member mounting bolts, and remove brake caliper assembly.

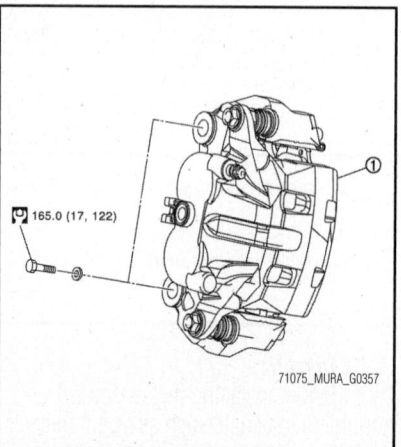

Fig. 3 Front caliper assembly (1)—CrossCabriolet models

To install:

✳✳ CAUTION

Never depress the brake pedal. Brake fluid may splash while removing the brake hose.

✳✳ WARNING

Do not spill or splash any grease and moisture on the brake caliper assembly mounting face, threads, mounting bolts and washers. Wipe out any grease and moisture.

7. Install the brake caliper assembly to the vehicle and tighten the mounting bolts to 122 ft. lbs. (165 Nm).

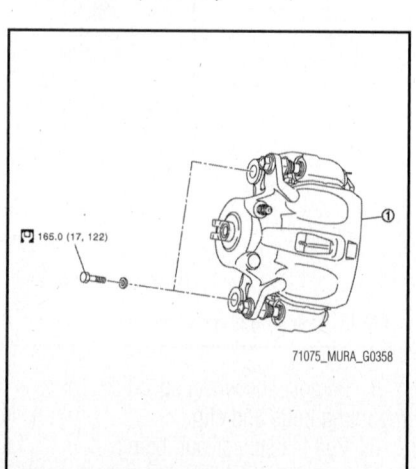

Fig. 4 Front caliper assembly (1)—Hardtop models

8. Install brake hose and copper washers to brake caliper assembly and tighten the union bolts to13 ft. lbs. (18 Nm). Do not reuse the copper washer.

9. Refill with new brake fluid and perform the air bleeding. Do not reuse drained brake fluid. Do not spill or splash brake fluid on the disc rotor.

10. Depress the brake pedal several times to check that no drag feel is present for the front disc brake. If any drag is found:

   a. Remove the brake pads.
   b. Press the pistons.

❊❊ WARNING

Do not damage the pistons boots. When replacing a pad with new one, check brake fluid level in the reservoir tank; brake fluid returns to master cylinder reservoir tank when pressing the pistons in.

   c. Install the brake pads.
   d. Depress the brake pedal several times.
   e. Check for drag again. If any drag is found, disassemble the cylinder body and replace as necessary.
   f. Burnish the contact surfaces after refinishing or replacing disc rotors, or if a soft pedal occurs at very low mileage. Refer to Front Brake Disc Adjustment/Burnishing.

11. Install the wheels and tires.

BRAKE PADS

REMOVAL & INSTALLATION

CrossCabriolet Models

1. Before servicing the vehicle, refer to the Precautions Section.

❊❊ CAUTION

Clean any dust from the brake caliper and brake pads with a vacuum dust collector. Never blow with compressed air.

❊❊ WARNING

Never depress the brake pedal while removing the brake pads because the piston may pop out. Never spill or splash brake fluid on the disc rotor.

2. Remove the wheels and tires.
3. Remove the lower sliding pin bolt.
4. Suspend the cylinder body with suitable wire so that the brake hose will not stretch.

5. Remove the brake pads, shims, shim covers, and pad retainers from the torque member.

❊❊ WARNING

Never reuse the pad retainers when the pad retainers have removed from the torque member. Do not damage the piston boots. Do not drop the brake pads, shims, and the shim covers. Remember each position of the removed brake pads.

To install:

6. Install the pad retainer to the torque member if the pad retainer has been removed.

❊❊ WARNING

Never reuse the pad retainers. Securely assemble the pad retainers so that it will not be lifted up from the torque member. Do not deform the pad retainers.

7. Apply MOLYKOTE® 7439 or equivalent to the mating faces between the brake pads and pad retainers.
8. Apply MOLYKOTE® AS880N or silicone-based grease to the mating faces between the shims and shim covers, and install them to the brake pad.

❊❊ WARNING

Always replace the shims together with the shim covers when replacing the brake pad.

9. Install the brake pads to the torque member. Both inner and outer pads have a pad return system on the pad retainer. Install pad return lever securely to pad wear sensor.
10. Install the cylinder body to the torque member.

❊❊ WARNING

Do not damage the piston boots. When replacing brake pad with a new one, check brake fluid level in the reservoir tank because brake fluid returns to master cylinder reservoir tank when pressing piston in.

➡Use a disc brake piston tool to easily press the piston.

11. Install the lower sliding pin bolt.
12. Depress the brake pedal several times to check that no drag feel is present for the front disc brake. If drag is found:

   a. Remove the brake pads.
   b. Press the pistons.

   c. Install the brake pads.
   d. Depress the brake pedal several times.
   e. Check for drag again. If any drag is found, disassemble the cylinder body and replace as necessary.
   f. Burnish the contact surfaces after refinishing or replacing disc rotors, or if a soft pedal occurs at very low mileage. Refer to Brake Disc Removal & Installation.

13. Install the wheels and tires.

Hardtop Models

1. Before servicing the vehicle, refer to the Precautions Section.

❊❊ CAUTION

Clean any dust from the brake caliper and brake pads with a vacuum dust collector. Never blow with compressed air.

❊❊ WARNING

Never depress the brake pedal while removing the brake pads because the piston may pop out. Never spill or splash brake fluid on the disc rotor.

2. Remove the wheels and tires.
3. Remove the lower sliding pin bolt.
4. Suspend the cylinder body with suitable wire so that the brake hose will not stretch.
5. Remove the brake pads, shims, shim covers, and pad retainers from the torque member.

❊❊ WARNING

Never reuse the pad retainers when the pad retainers have removed from the torque member. Do not damage the piston boots. Do not drop the brake pads, shims, and the shim covers. Remember each position of the removed brake pads.

To install:

6. Install the pad retainer to the torque member if the pad retainer has been removed.

❊❊ WARNING

Never reuse the pad retainers. Securely assemble the pad retainers so that it will not be lifted up from the torque member. Do not deform the pad retainers. Eliminate double-faced adhesive tape on torque member. Remove adhesive's protective liner on pad retainers.

7. Apply PBC (Poly Butyl Cuprysil) grease or silicone-based grease to the mating faces between the brake pads and pad retainers.

8. Apply copper based brake grease to the mating faces between the brake pads, shims and shim covers, and install them to the brake pad.

✖✖ WARNING

Always replace the shims together with the shim covers when replacing the brake pad.

9. Install the cylinder body and brake pads to the torque member.

✖✖ WARNING

Do not damage the piston boots. When replacing brake pad with a new one, check brake fluid level in the reservoir tank because brake fluid returns to master cylinder reservoir tank when pressing piston in.

➡ **Use a disc brake piston tool to easily press the piston.**

10. Install the lower sliding pin bolt.

11. Depress the brake pedal several times to check that no drag feel is present for the front disc brake. If drag is found:

a. Remove the brake pads.
b. Press the pistons.
c. Install the brake pads.
d. Depress the brake pedal several times.
e. Check for drag again. If any drag is found, disassemble the cylinder body and replace as necessary.
f. Burnish the contact surfaces after refinishing or replacing disc rotors, or if a soft pedal occurs at very low mileage. Refer to Brake Disc Removal & Installation.

12. Install the wheels and tires.

BRAKES

✖✖ CAUTION

Dust and dirt accumulating on brake parts during normal use may contain asbestos fibers from production or aftermarket brake linings. Breathing excessive concentrations of asbestos fibers can cause serious bodily harm. Exercise care when servicing brake parts. Do not sand or grind brake lining unless equipment used is designed to contain the dust residue. Do not clean brake parts with compressed air or by dry brushing. Cleaning should be done by dampening the brake components with a fine mist of water, then wiping the brake components clean with a dampened cloth. Dispose of cloth and all residue containing asbestos fibers in an impermeable container with the appropriate label. Follow practices prescribed by the Occupational Safety and Health Administration (OSHA) and the Environmental Protection Agency (EPA) for the handling, processing, and disposing of dust or debris that may contain asbestos fibers.

BRAKE CALIPER

REMOVAL & INSTALLATION

See Figure 5.

1. Before servicing the vehicle, refer to the Precautions Section.

✖✖ CAUTION

Clean any dust from the brake caliper and brake pads with a vacuum dust collector. Never blow with compressed air. Never depress the brake pedal. Brake fluid may splash while removing the brake hose.

2. Remove the tire and wheel assembly.

3. Secure the disc rotor using wheel nuts.

4. Drain brake fluid.

✖✖ WARNING

Do not spill or splash brake fluid on the disc rotor.

5. Remove union bolt and copper washers, and disconnect the brake hose from the caliper assembly.

6. Remove the torque member mounting bolts, and remove the brake caliper assembly.

To install:

7. Install the brake caliper assembly to the vehicle and tighten the mounting bolts to 62 ft. lbs. (84 Nm).

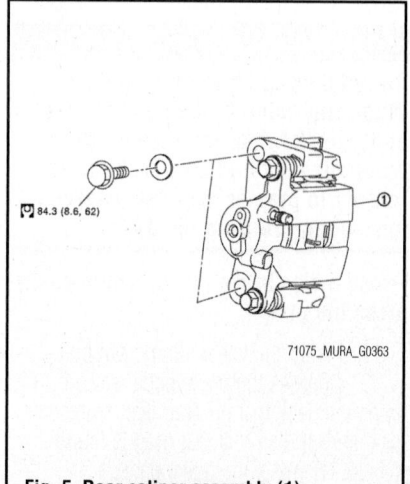

84.3 (8.6, 62)

71075_MURA_G0363

Fig. 5 Rear caliper assembly (1)

REAR DISC BRAKES

✖✖ WARNING

Never spill or splash any grease and moisture on the brake caliper assembly mounting face, threads, mounting bolts, and washers. Wipe out any grease and moisture.

8. Install the brake hose and copper washers to the brake caliper assembly and tighten the union bolt to 13 ft. lbs. (18 Nm). Never reuse copper washer.

9. Refill with new brake fluid and perform the air bleeding. Never reuse drained brake fluid.

10. Depress the brake pedal several times to check that no drag feel is present for the rear disc brake. If any drag is found:

a. Remove brake pads.
b. Press the piston. Do not damage the piston boot. When replacing a pad with new one, check brake fluid level in the reservoir tank because brake fluid returns to master cylinder reservoir tank when pressing piston in.

➡ **Use a disc brake piston tool to easily press piston.**

c. Install brake pads.
d. Depress the brake pedal several times.
e. Check for drag of rear disc brake again. If any drag is found, disassemble the cylinder body.
f. Burnish contact surfaces after refinishing or replacing disc rotors, or if a soft pedal occurs at very low mileage. Refer to Rear Brake Disc Adjustment/ Burnishing.

11. Install the tire and wheel assembly.

BRAKE PADS

REMOVAL & INSTALLATION

1. Before servicing the vehicle, refer to the Precautions Section.

> ✳✳ **CAUTION**
>
> **Clean any dust from the brake caliper and brake pads with a vacuum dust collector. Never blow with compressed air.**

> ✳✳ **WARNING**
>
> **Never depress the brake pedal while removing the brake pads because the piston may pop out. Never spill or splash brake fluid on the disc rotor.**

2. Remove the wheels and tires.
3. Remove the upper sliding pin bolt.
4. Suspend the cylinder body with suitable wire so that the brake hose will not stretch.
5. Remove the brake pads, shims and shim cover from the torque member.

> ✳✳ **WARNING**
>
> **Never reuse the pad retainers when removed the pad retainers from the torque member. Do not damage the piston boot. Do not drop the brake pads, shims, and the shim cover. Remember each position of the removed brake pads.**

To install:

6. Install the pad retainers to the torque member if the pad retainers has been removed. Securely assemble the pad retainers not to be lifted up from the torque member. Do not deform the pad retainers.

7. Apply PBC (Poly Butyl Cuprysil) grease or silicon based grease to the mating faces between the brake pads, the shims and pawls part of cylinder body, and install them to the brake pad. Always replace the shims together with the shim cover when replacing the brake pad.

8. Install cylinder body and brake pads to torque member. Do not damage the piston boot. When replacing pads with new one, check a brake fluid level in the reservoir tank because brake fluid returns to master cylinder reservoir tank when pressing piston in.

➡ **Use a disc brake piston tool to easily press piston.**

9. Install the upper sliding pin bolt and tighten it to the specified torque.

10. Depress the brake pedal several times to check that no drag feel is present for the rear disc brake. If any drag is found:

    a. Remove brake pads.
    b. Press the piston. Do not damage the piston boot. When replacing a pad with new one, check brake fluid level in the reservoir tank because brake fluid returns to master cylinder reservoir tank when pressing piston in.

➡ **Use a disc brake piston tool to easily press piston.**

    c. Install brake pads.
    d. Depress the brake pedal several times.
    e. Check for drag of rear disc brake again. If any drag is found, disassemble the cylinder body.
    f. Burnish contact surfaces after refinishing or replacing disc rotors, or if a soft pedal occurs at very low mileage. Refer to Rear Brake Disc Adjustment/Burnishing.

11. Install the tire and wheel assembly.

BRAKES

PARKING BRAKE CABLES

ADJUSTMENT

See Figure 6.

1. Operate the parking brake pedal with a force of 110 lbs. (490 N) for 10 strokes or more.

71075_QUES_G0451

Fig. 6 Adjust the parking brake pedal (1) by turning the adjusting nut (2)

2. Adjust the parking brake pedal stroke by turning the adjusting nut with a socket wrench.

> ✳✳ **WARNING**
>
> **Never reuse the adjusting nut if the nut is removed.**

3. Operate the parking brake pedal with a force of 44 ft. lbs. (196 N). Check that the pedal stroke is within 5–6 notches. (Check it by listening to the clicks of the ratchet.) When parking brake warning lamp turns ON, check that the pedal stroke is within 1 notch.

4. Rotate the disc rotor with the parking brake pedal released and check that there is no drag. If any drag is found, adjust parking brake stroke again and inspect the rear disc brake caliper.

5. With the pedal completely returned, make sure there is no drag on the rear brake. If drag is found:

    a. Remove brake pads.
    b. Press the piston. Do not damage the piston boot. When replacing a pad with new one, check brake fluid level in the reservoir tank because

PARKING BRAKE

brake fluid returns to master cylinder reservoir tank when pressing piston in.

➡ **Use a disc brake piston tool to easily press piston.**

    c. Install brake pads.
    d. Depress the brake pedal several times.
    e. Check for drag of rear disc brake again. If any drag is found, disassemble the cylinder body.
    f. Burnish contact surfaces after refinishing or replacing disc rotors, or if a soft pedal occurs at very low mileage. Refer to Rear Brake Disc Adjustment/Burnishing.

6. Install the tire and wheel assembly.

PARKING BRAKE SHOES

REMOVAL & INSTALLATION

See Figure 7.

1. Before servicing the vehicle, refer to the Precautions Section.

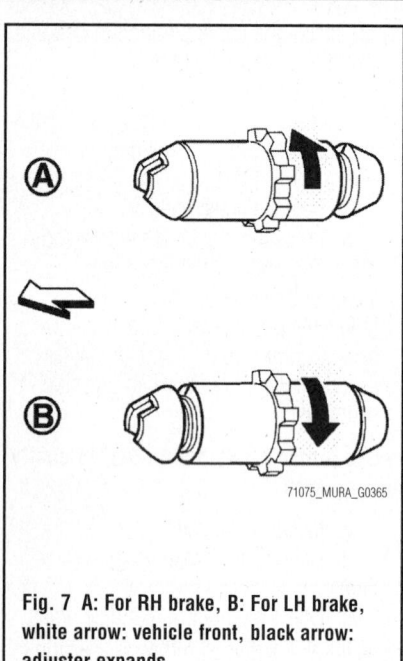

Fig. 7 A: For RH brake, B: For LH brake, white arrow: vehicle front, black arrow: adjuster expands

✳✳ CAUTION

Clean any dust from the brake caliper and brake pads with a vacuum dust collector. Never blow with compressed air.

2. Remove the rear wheels.
3. Remove the rear disc rotor. If rotor cannot be removed:
   a. Secure the rotor in place with wheel nuts and remove adjuster hole plug.
   b. Using a suitable tool, rotate the adjuster to retract and loosen the brake shoes.
4. Remove the anti-rattle pins, retainers, anti-rattle springs, adjuster spring, and return springs.
5. Remove the parking brake shoes, adjuster assembly, and toggle lever.

To install:
Install in the reverse order of removal, noting the following:

6. Apply PBC (Poly Butyl Cuprysil) grease or silicone-based grease to the back plate and brake shoe.

✳✳ WARNING

The parking brake shoes for the front side are made of different materials from those for the rear side. Never misidentify them when removing and replacing.

7. Assemble adjusters so that threaded part is expanded when rotating it in the direction shown by arrow.
8. Shorten adjuster by rotating it.
9. When disassembling, apply PBC (Poly Butyl Cuprysil) grease or silicone-based grease to threads.
10. Check parking brake shoe sliding surface and drum inner surface for grease. Wipe it off if it adheres on the surfaces.

CHASSIS ELECTRICAL

AIR BAG (SUPPLEMENTAL RESTRAINT SYSTEM)

GENERAL INFORMATION

✳✳ CAUTION

These vehicles are equipped with an air bag system. The system must be disarmed before performing service on, or around, system components, the steering column, instrument panel components, wiring and sensors. Failure to follow the safety precautions and the disarming procedure could result in accidental air bag deployment, possible injury and unnecessary system repairs.

SERVICE PRECAUTIONS

Disconnect and isolate the battery negative cable before beginning any airbag system component diagnosis, testing, removal, or installation procedures. Allow system capacitor to discharge for two minutes before beginning any component service. This will disable the airbag system. Failure to disable the airbag system may result in accidental airbag deployment, personal injury, or death.

Do not place an intact undeployed airbag face down on a solid surface. The airbag will propel into the air if accidentally deployed and may result in personal injury or death.

When carrying or handling an undeployed airbag, the trim side (face) of the airbag should be pointing towards the body to minimize possibility of injury if accidental deployment occurs. Failure to do this may result in personal injury or death.

Replace airbag system components with OEM replacement parts. Substitute parts may appear interchangeable, but internal differences may result in inferior occupant protection. Failure to do so may result in occupant personal injury or death.

Wear safety glasses, rubber gloves, and long sleeved clothing when cleaning powder residue from vehicle after an airbag deployment. Powder residue emitted from a deployed airbag can cause skin irritation. Flush affected area with cool water if irritation is experienced. If nasal or throat irritation is experienced, exit the vehicle for fresh air until the irritation ceases. If irritation continues, see a physician.

Do not use a replacement airbag that is not in the original packaging. This may result in improper deployment, personal injury, or death.

The factory installed fasteners, screws and bolts used to fasten airbag components have a special coating and are specifically designed for the airbag system. Do not use substitute fasteners. Use only original equipment fasteners listed in the parts catalog when fastener replacement is required.

During, and following, any child restraint anchor service, due to impact event or vehicle repair, carefully inspect all mounting hardware, tether straps, and anchors for proper installation, operation, or damage. If a child restraint anchor is found damaged in any way, the anchor must be replaced. Failure to do this may result in personal injury or death.

Deployed and non-deployed airbags may or may not have live pyrotechnic material within the airbag inflator.

Do not dispose of driver/passenger/curtain airbags or seat belt tensioners unless you are sure of complete deployment. Refer to the Hazardous Substance Control System for proper disposal.

Dispose of deployed airbags and tensioners consistent with state, provincial, local, and federal regulations.

After any airbag component testing or service, do not connect the battery negative cable. Personal injury or death may result if the system test is not performed first.

If the vehicle is equipped with the Occupant Classification System (OCS), do not connect the battery negative cable before performing the OCS Verification Test using the scan tool and the appropriate diagnostic information. Personal injury or death may result if the system test is not performed properly.

Never replace both the Occupant Restraint Controller (ORC) and the Occupant Classification Module (OCM) at the same time. If both require replacement, replace one, then perform the Airbag

System test before replacing the other.

Both the ORC and the OCM store Occupant Classification System (OCS) calibration data, which they transfer to one another when one of them is replaced. If both are replaced at the same time, an irreversible fault will be set in both modules and the OCS may malfunction and cause personal injury or death.

If equipped with OCS, the Seat Weight Sensor is a sensitive, calibrated unit and must be handled carefully. Do not drop or handle roughly. If dropped or damaged, replace with another sensor. Failure to do so may result in occupant injury or death.

If equipped with OCS, the front passenger seat must be handled carefully as well. When removing the seat, be careful when setting on floor not to drop. If dropped, the sensor may be inoperative, could result in occupant injury, or possibly death.

If equipped with OCS, when the passenger front seat is on the floor, no one should sit in the front passenger seat. This uneven force may damage the sensing ability of the seat weight sensors. If sat on and damaged, the sensor may be inoperative, could result in occupant injury, or possibly death.

DISARMING THE SYSTEM

Before servicing the SRS, turn the ignition switch OFF, disconnect the battery negative terminal and wait at least 3 minutes.

> **✳✳ CAUTION**
>
> **For approximately 3 minutes after the battery negative terminal is removed, it is still possible for the air bag and seat belt pre-tensioner to deploy. Therefore, do not work on any SRS connectors or wires until at least 3 minutes have elapsed.**

ARMING THE SYSTEM

Connect the battery negative terminal.

CLOCKSPRING CENTERING

> **✳✳ WARNING**
>
> **The spiral cable may snap during steering operation if the cable is installed in an improper position.**

1. Carefully turn the spiral cable clockwise to the end position.
2. Turn it counterclockwise (about 2 and a half turns) and stop turning at the mark "B" on which the stopper insertion holes are in the same position.

➡ **The service part is installed in the neutral position by the stopper and can be set without adjusting after the stopper is removed.**

> **✳✳ WARNING**
>
> **Never over-turn the spiral cable or go beyond number of turns required. (This will cause the cable to snap.)**

3. Adjust the spiral cable locating pin to the steering wheel locating pin hole.

DRIVE TRAIN

CV-BOOTS

REMOVAL & INSTALLATION

Wheel Side

See Figures 8 and 9.

1. Before servicing the vehicle, refer to the precautions section.
2. Remove the boot bands, and then remove the boot from the joint subassembly.
3. Screw the halfshaft puller (commercial service tool) into the joint subassembly screw part to a length of 1.18 in. (30 mm) or more. Support the halfshaft with one hand and pull out the joint subassembly from the housing assembly.
   a. Align the halfshaft puller and the halfshaft, and remove them by pulling firmly and uniformly.
   b. If the joint sub-assembly cannot be pulled out, try again after removing the halfshaft from vehicle.
4. Remove the circular clip from the housing assembly shaft.
5. Remove the boot from the shaft.

To install:

6. Clean the old grease on joint subassembly with paper waste.
7. Insert grease into the joint subassembly serration hole until grease begins to ooze from ball groove and serration hole. After inserting grease, use a shop cloth to wipe off old grease that has oozed out.
8. Cover the serrated part of shaft with tape. Install a new boot band and boot to shaft. Be careful not to damage the boot. Discard the old boot band and boot; replace with new.
9. Remove the protective tape wound around the serrated part of the shaft.
10. Position the circular clip on the groove at the shaft edge. Never reuse circular clip. Drive joint inserter is recommended when installing the circular clip.
11. Align both center axles of the shaft edge and the joint sub-assembly. Assemble the shaft with the joint sub-assembly holding the circular clip.
12. Install the joint sub-assembly to the shaft using a plastic hammer.
    a. Check that the circular clip is properly positioned on the groove of the joint sub-assembly.
    b. Confirm that the joint subassembly is correctly engaged while rotating the halfshaft.
13. Apply the specified amount of grease into the boot inside from large diameter side of boot. Wheel side: 6.00–6.70 oz (170–190 g)
14. Install the boot securely into the grooves as shown. If there is grease on the boot mounting surfaces of the shaft and housing, the boot may come off. Clean all grease from surfaces.
15. Make sure boot installation length "L" is the length indicated below. Boot may break if the boot installation length is less than standard value. Insert a flat-bladed screwdriver or similar tool into the large end of boot. Be careful that the screwdriver tip does not contact inside surface of boot. Bleed air from the boot to prevent boot

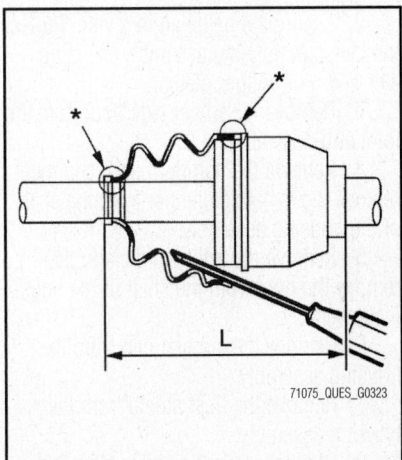

71075_QUES_G0323

Fig. 8 Install boot securely into grooves (indicated by * marks), boot installation length (L)

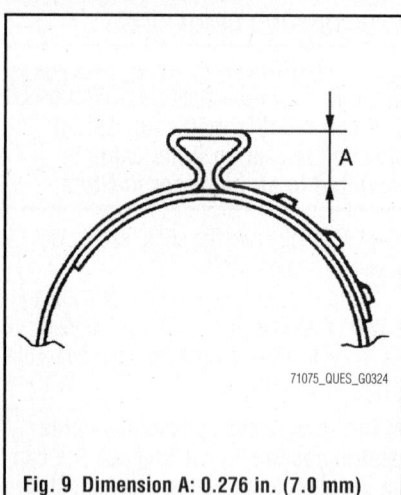

Fig. 9 Dimension A: 0.276 in. (7.0 mm) or less

deformation. Wheel side boot installation length "L": 6.24 in. (158.6 mm)

16. Secure the large and small ends of the boot with new boot bands using the boot band crimping tool (SST: KV40107300) as shown. Never reuse boot band.

   a. Secure the boot band so that dimension meets the specification as shown.

17. Check that displacement does not occur when the boot is rotated with the joint sub-assembly and the shaft of housing assembly fixed. Reinstall them using boot bands if the boot installation positions become incorrect. Never reuse boot bands.

Transaxle Side

Left Side

See Figures 10 and 11.

1. Before servicing the vehicle, refer to the Precautions Section.

2. Secure the halfshaft in a vise. Protect the halfshaft housing assembly by using aluminum or copper plates.

3. Remove the wheel side boot from the joint sub-assembly.

4. Remove the damper band, and then remove the dynamic damper from the shaft of the housing assembly.

5. Remove the boot bands, and then remove the boot from the shaft of the housing assembly.

6. Remove the circular clip from the housing assembly.

7. Remove the dust shield from the housing assembly.

8. Perform inspection after disassembly. Check the following items, and replace the parts if necessary:

   a. Dynamic Damper (Left Side): Check damper for cracks or wear.

   b. Joint Sub-Assembly. Check the following, and replace joint sub-assembly if there are any non-standard conditions of components:

- Joint sub-assembly for rough rotation and excessive axial looseness
- The inside of the joint sub-assembly for entry of foreign material
- Joint sub-assembly for compression scars, cracks, and fractures inside of joint sub-assembly

   c. Housing assembly: Replace housing assembly if there is scratching or wear of housing assembly roller contact surface. Check shaft for runout, cracks, or other damage.

   d. Support Bearing (Right Side): Check that the bearing rolls freely, and is free from noise, cracks, pitting or wear. Replace the support bearing if there are any non-standard conditions.

   e. Bearing Housing (Right Side): Check for bearing housing, cracks, or damage. Replace the support bearing bracket if there are any non-standard conditions.

To install:

9. Clean the old grease on the housing assembly with paper waste.

10. Install the dust shield to the housing assembly. Never reuse dust shield.

11. Install the circular clip to the housing. Never reuse circular clip.

12. Install the boot and boot bands to the shaft of the housing assembly. Wrap the serration on the shaft with tape to protect the boot from damage. Never reuse boot and boot band.

13. Remove protective tape wound around serrated part of shaft.

14. Apply NISSAN genuine grease to the housing assembly. Grease amount, transaxle side: 5.47–6.17 oz (155–175 g)

15. Install the boot securely into the grooves as shown. If there is grease on boot mounting surfaces of shaft and housing, boot may come off. Remove all grease from surfaces.

16. Make sure boot installation length "L" is the specified length indicated below. Boot may break if boot installation length is less than standard value. Insert a flat-bladed screwdriver or similar tool into the large end of boot. Be careful that screwdriver tip does not contact inside surface of boot. Bleed air from boot to prevent boot deformation. Transaxle side boot installation length "L": 6.44 in. (163.67 mm)

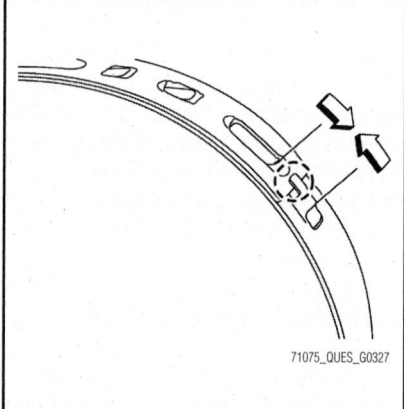

Fig. 10 Insert the tip of the band into the lower part of the pawl (marked with dotted circle)

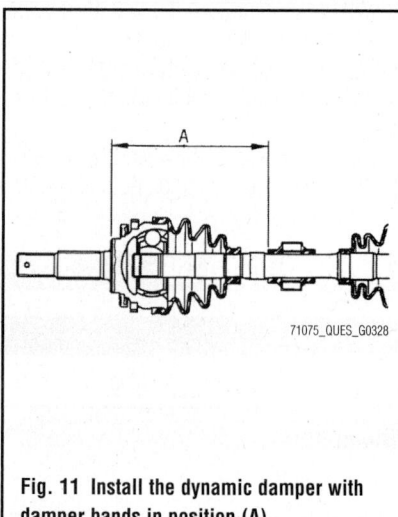

Fig. 11 Install the dynamic damper with damper bands in position (A)

✱✱ WARNING

Never reuse boot bands.

17. To install the one-touch clamp band: install the boot bands securely.

18. To install the low profile type band:

   a. Put the boot band in the groove on the halfshaft boot. Fit the pawls into the holes for temporary installation. For the large diameter side, fit the projection and guide slit at first.

   b. Pinch the projection on the band with suitable pliers to tighten the band.

   c. Insert the tip of the band into the lower part of the pawl (marked with dotted circle) as shown.

19. Check that displacement does not occur when the boot is rotated with the housing assembly fixed. If displacement occurs, reinstall the band. Never reuse boot band.

20. To install the dynamic damper:

a. Install the dynamic damper and damper bands to shaft of the housing assembly. Never reuse damper band.

b. Install the boot to the wheel side.

c. Secure the dynamic damper with damper bands in the following specified position:

- Dimension of dynamic damper 9.33–9.57 in. (237–243 mm)

d. Install the damper bands securely as shown with the one-touch clamp boot bands, above.

Right Side

1. Before servicing the vehicle, refer to the Precautions Section.

2. Secure the halfshaft in a vise. Protect the halfshaft housing assembly by using aluminum or copper plates.

3. Remove the wheel side boot from the joint sub-assembly.

4. Remove the boot bands, and then remove the boot from the housing assembly.

To install:

5. Install boot and boot bands to shaft of housing assembly. Wrap the serration on the shaft with tape (A) to protect the boot from damage. Never reuse boot and boot band.

6. Remove the protective tape wrapped around the serration on the shaft.

7. Apply NISSAN genuine grease to housing assembly. Grease amount, transaxle side: 5.47–6.17 oz (155–175 g)

8. Install the boot securely into the grooves as shown.

9. Make sure boot installation length "L" is the specified length indicated below. Boot may break if boot installation length is less than standard value. Insert a flat-bladed screwdriver or similar tool into the large end of boot. Be careful that screwdriver tip does not contact inside surface of boot. Bleed air from boot to prevent boot deformation. Transaxle side boot installation length "L": 6.44 in. (163.67 mm)

☀☀ WARNING

Never reuse boot bands.

10. Put the boot band in the groove on the halfshaft boot. Fit the pawls into the holes for temporary installation. For the large diameter side, fit the projection and guide slit at first.

11. Pinch the projection on the band with suitable pliers to tighten the band.

12. Insert the tip of the band into the lower part of the pawl (marked with dotted circle).

13. Check that displacement does not occur when the boot is rotated with the housing assembly fixed. If displacement occurs, reinstall the band. Never reuse boot band.

14. Install boot to the wheel side.

DIFFERENTIAL CARRIER

REMOVAL & INSTALLATION

See Figure 12.

1. Before servicing the vehicle, refer to the Precautions Section.

2. Remove the propeller shaft.

3. Remove the stabilizer bar.

4. Remove the EVAP canister.

5. Disconnect the rear halfshafts (final drive side).

6. Disconnect the AWD solenoid harness connector.

7. Remove the rear final drive breather hose and electric controlled coupling breather hose.

8. Support the final drive assembly with a suitable jack.

9. Remove the rear final drive mounting nut at rear suspension member.

10. Remove the final drive mounting nuts and final drive mounting bolts with a power tool. If necessary, remove the final drive mounting bracket and washer with a power tool.

☀☀ WARNING

Secure the final drive assembly to a suitable jack while removing it.

To install:

Note the following, and install in the reverse order of removal.

11. Install the breather hose to breather connector to dimension "A", as shown. Dimension A:

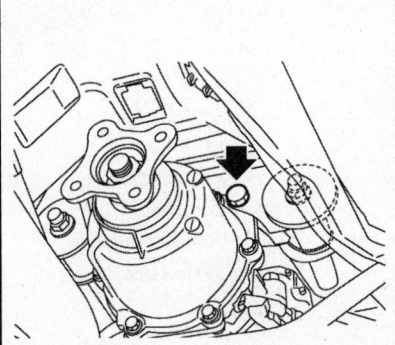

71075_MURA_G0492

Fig. 12 Remove the final drive mounting nuts and bolts; if necessary, remove the final drive mounting bracket and washer

- Final drive side: 0.79 in. (20 mm)
- Suspension member side: 0.815 in. (20.7 mm)

a. Never reuse hose clamp. Make sure there are no pinched or restricted areas on the breather hose caused by bending or winding when installing it.

b. Install the hose clamp at the final drive side with the tab facing to the vehicle front. Install the hose clamp at the suspension member side with the tab facing downward.

12. If breather connector and metal connector are removed, install breather hose, breather connector and metal connector as shown. Insert breather connector into the square hole of rear suspension. Install metal connector to rear cover, aiming painted marking to the front of vehicle.

13. Install the electric controlled coupling breather hose as shown.

a. Install electric controlled coupling breather hose at the coupling side to the metal tube of the coupling cover all the way to the point shown by the solid arrow.

b. Install electric controlled coupling breather hose at the suspension member side to dimension "L" (L: 15 mm), as shown.

c. Install the hose clip at the position 152 mm from breather hose end on the breather connector side. Make sure there are no pinched or restricted areas on the breather hose caused by bending or winding when installing it. Make sure to insert hose clip into the hole of final drive mounting bracket.

14. If the breather connector of the electric controlled coupling and metal tube are removed, install as shown.

a. Install the breather connector at the insertion side to the suspension member, facing to the vehicle front.

b. Install the metal tube to the coupling cover, facing to the vehicle front.

c. Never reuse breather connector and metal connector.

15. When oil leaks while removing final drive assembly, check oil level after the installation.

FRONT HALFSHAFT

REMOVAL & INSTALLATION

Left Side

See Figures 13 and 14.

1. Before servicing the vehicle, refer to the Precautions Section.

2. Remove the wheels and tires.

3. Remove the wheel sensor and sensor harness. Refer to Wheel Speed Sensor Removal & Installation in the ABS, Brake section.

4. Remove the lock plate from the strut assembly.

5. Remove the caliper assembly. Hang caliper assembly aside. Refer to Front Disc Brakes, Caliper Removal & Installation in the Brake section.

✳✳ WARNING

Never depress brake pedal while brake caliper is removed.

6. Remove the disc rotor. Refer to Front Disc Brakes, Rotor Removal & Installation in the Brake section.

7. Remove the cotter pin, and then loosen the wheel hub lock nut.

8. Using suitable puller, separate the halfshaft from the wheel hub and bearing assembly.

✳✳ WARNING

Never place the halfshaft joint at an extreme angle. Also be careful not to overextend the slide joint. Never allow the halfshaft to hang down without support for joint sub-assembly, housing assembly and the other parts.

9. Remove the wheel hub lock nut.

10. Remove the strut assembly from steering knuckle. Refer to Strut Removal & Installation in the Front Suspension section.

11. Remove the halfshaft from the transaxle assembly.

   a. Use the halfshaft attachment (SST: KV40107500) and a sliding hammer while inserting the tip of the halfshaft attachment between the housing assembly and the transaxle assembly.

✳✳ WARNING

Never place the halfshaft joint at an extreme angle when removing the halfshaft. Also be careful not to overextend slide joint. Confirm that the circular clip is attached to the halfshaft.

To install:

12. Note the following, and install in the reverse order of removal.

13. Always replace the differential side oil seal with a new one when installing halfshaft.

14. Place the protector (SST: KV38107900) onto the transaxle assembly to prevent damage to the oil seal while

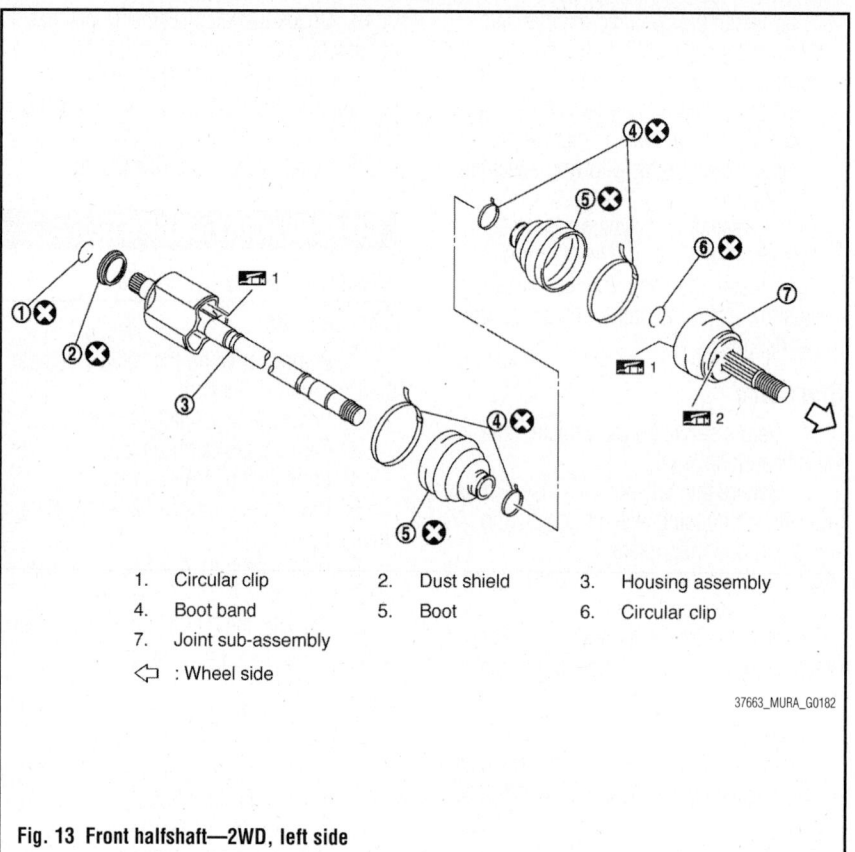

1.	Circular clip	2.	Dust shield	3.	Housing assembly
4.	Boot band	5.	Boot	6.	Circular clip
7.	Joint sub-assembly				
⟵ : Wheel side					

37663_MURA_G0182

Fig. 13 Front halfshaft—2WD, left side

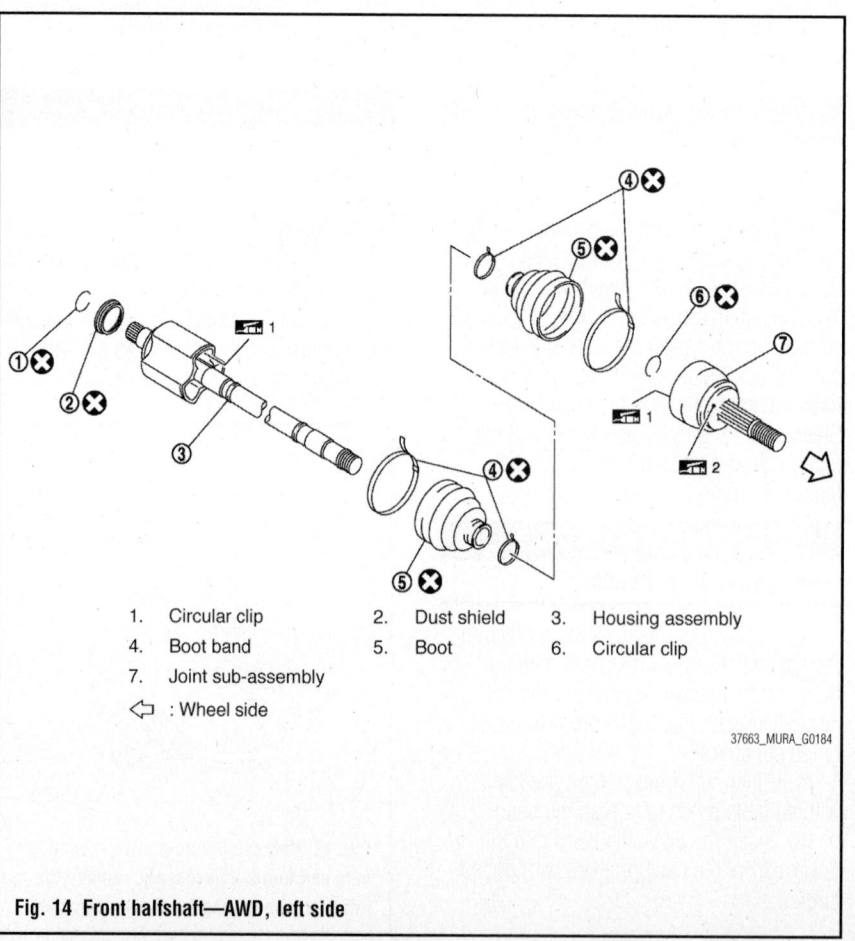

1.	Circular clip	2.	Dust shield	3.	Housing assembly
4.	Boot band	5.	Boot	6.	Circular clip
7.	Joint sub-assembly				
⟵ : Wheel side					

37663_MURA_G0184

Fig. 14 Front halfshaft—AWD, left side

inserting the halfshaft. Slide the halfshaft sliding joint and tap with a hammer to install securely.

15. Clean the matching surfaces of the wheel hub lock nut and wheel hub and bearing assembly.

16. Clean the matching surfaces of the halfshaft and wheel hub and bearing assembly.

17. Apply paste (service part: 440037S000) to cover entire flat surface of the halfshaft joint sub assembly. Amount of paste: 0.008–0.035 oz. (0.2–1.0 g)

Right Side

2WD

See Figures 15 and 16.

1. Before servicing the vehicle, refer to the Precautions Section.

2. Remove the wheels and tires.

3. Remove the wheel sensor and sensor harness. Refer to Wheel Speed Sensor Removal & Installation in the ABS, Brake section.

4. Remove the lock plate from the strut assembly.

5. Remove the caliper assembly. Hang caliper assembly aside. Refer to Front Disc Brakes, Caliper Removal & Installation in the Brake section.

6. Remove the disc rotor. Refer to Front Disc Brakes, Rotor Removal & Installation in the ABS, Brake section.

7. Remove the cotter pin, and then loosen the wheel hub lock nut.

8. Using suitable puller, separate the halfshaft from the wheel hub and bearing assembly.

9. Remove the wheel hub lock nut.

10. Remove the strut assembly from steering knuckle. Refer to Strut Removal & Installation in the Front Suspension section.

11. Remove the halfshaft from the wheel hub and bearing assembly.

12. Remove the bearing housing mounting bolts.

13. Remove the halfshaft from the transaxle assembly.

14. Remove the support bearing bracket, follow the procedure described below.

a. Remove the front exhaust tube.

b. Remove three way catalyst (Bank 1) and heated oxygen sensor harness bracket.

c. Remove support bearing bracket.

To install:

15. Note the following, and install in the reverse order of removal.

16. Always replace the differential side oil seal with a new one when installing halfshaft.

17. Place the protector (SST: KV38107900) onto the transaxle assembly to prevent damage to the oil seal while inserting the halfshaft. Slide the halfshaft sliding joint and tap with a hammer to install securely.

18. Install support bearing bracket in following procedure,

a. Temporarily tighten the mounting bolts, then tighten them to 35 ft. lbs. (48 Nm).

b. Temporarily tighten the mounting bolts, then tighten them to 35 ft. lbs. (48 Nm).

19. Clean the matching surfaces of the halfshaft and wheel hub and bearing assembly.

20. Apply paste (service part: 440037S000) to cover entire flat surface of

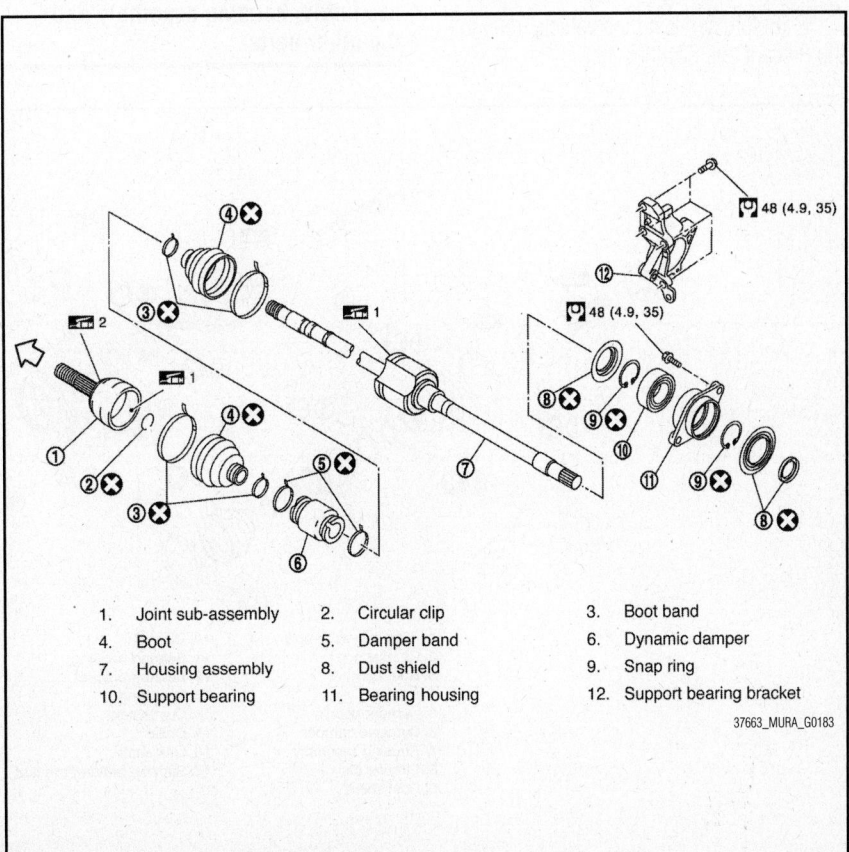

1.	Joint sub-assembly	2.	Circular clip	3.	Boot band
4.	Boot	5.	Damper band	6.	Dynamic damper
7.	Housing assembly	8.	Dust shield	9.	Snap ring
10.	Support bearing	11.	Bearing housing	12.	Support bearing bracket

37663_MURA_G0183

Fig. 15 Front halfshaft—2WD, right side

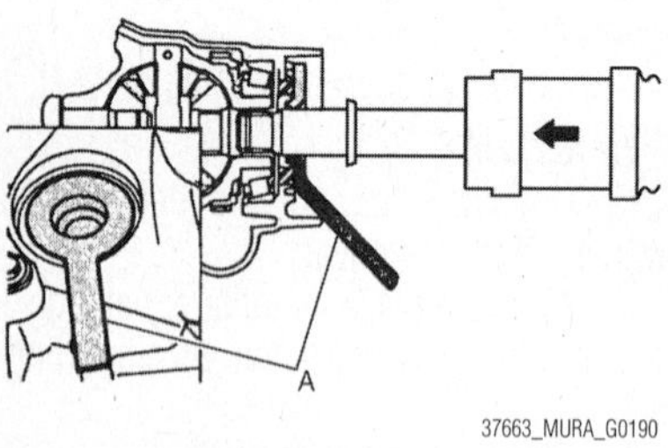

Fig. 16 Place the protector (A) (SST: KV38107900) onto the transaxle assembly

the halfshaft joint sub assembly. Amount of paste: 0.008–0.035 oz. (0.2–1.0 g)

WARNING

Never use a power tool to tighten the wheel hub lock nut. Perform the final tightening of each of parts under unladen conditions. Never reuse the cotter pin.

AWD

See Figures 16 through 20.

1. Before servicing the vehicle, refer to the Precautions Section.
2. Remove the wheels and tires.
3. Remove the wheel sensor and sensor harness. Refer to Wheel Speed Sensor Removal & Installation in the ABS, Brake section.
4. Remove the lock plate from the strut assembly.
5. Remove the caliper assembly. Hang caliper assembly aside. Refer to Front Disc Brakes, Caliper Removal & Installation in the Brake section.

WARNING

Never depress brake pedal while brake caliper is removed.

6. Remove the disc rotor. Refer to Front Disc Brakes, Rotor Removal & Installation in the ABS, Brake section.
7. Remove the cotter pin, and then loosen the wheel hub lock nut.
8. Using suitable puller, separate the halfshaft from the wheel hub and bearing assembly.

WARNING

Never place the halfshaft joint at an extreme angle. Also be careful not to overextend the slide joint. Never allow the halfshaft to hang down without support for joint sub-assembly, housing assembly and the other parts.

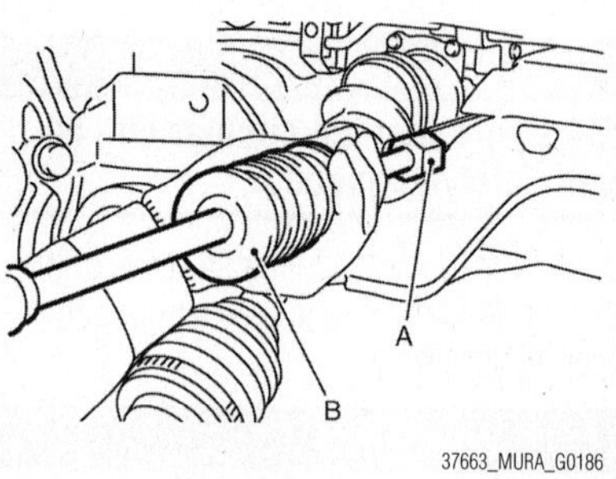

Fig. 18 Use the halfshaft attachment (A) (SST: KV40107500) and a sliding hammer (B) while inserting the tip of the halfshaft attachment between the housing assembly and the transaxle assembly

9. Remove the wheel hub lock nut.
10. Remove the strut assembly from steering knuckle. Refer to Strut Removal & Installation in the Front Suspension section.
11. Remove the halfshaft from the wheel hub and bearing assembly.
12. Remove the halfshaft from link shaft.
    a. Use the halfshaft attachment (SST: KV40107500) and a sliding hammer while inserting the tip of the halfshaft attachment between the housing assembly and link shaft assembly.

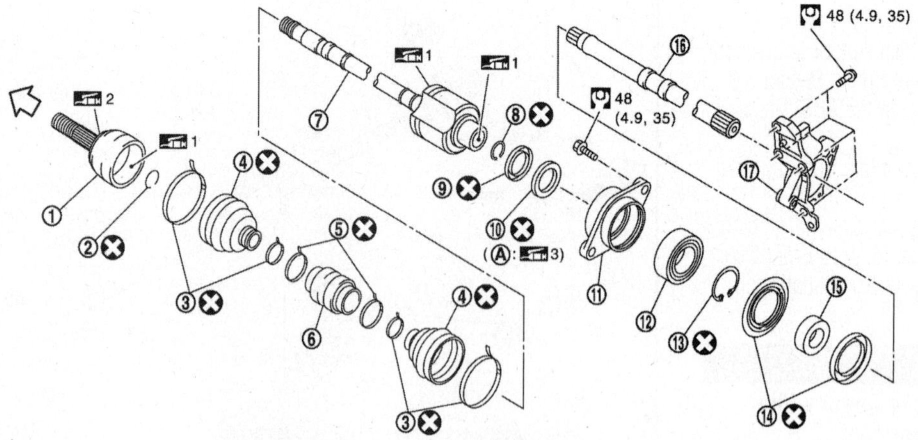

1. Joint sub-assembly
2. Circular clip
3. Boot band
4. Boot
5. Damper band
6. Dynamic damper
7. Housing assembly
8. Circular clip
9. Dust shield
10. Oil seal
11. Bearing housing
12. Bearing bracket
13. Snap ring
14. Dust shield
15. Collar
16. Link shaft
17. Support bearing bracket

37663_MURA_G0185

Fig. 17 Front halfshaft—AWD, right side

13. Remove the bearing housing mounting bolts.

14. Remove the link shaft assembly from
the support bearing bracket.

15. Remove the support bearing bracket,
follow the procedure described below.

    a. Remove the front exhaust
tube.

    b. Remove three way catalyst (Bank 1)
and heated oxygen sensor harness
bracket.

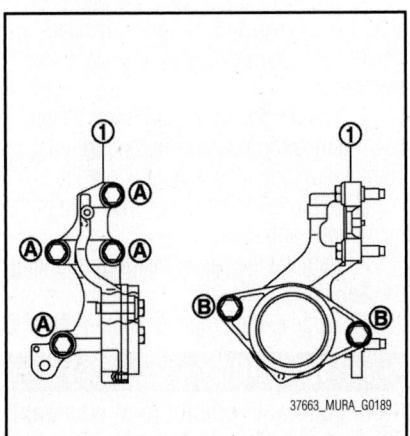

**Fig. 19 Install support bearing bracket (1)
mounting bolts (A, B)**

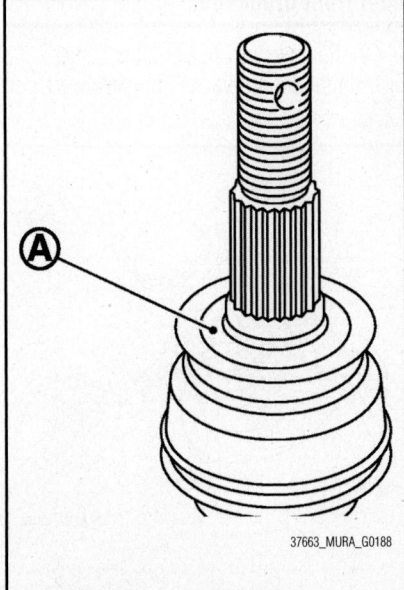

**Fig. 20 Apply paste (service parts:
440037S000) to the flat surface (A) of the
halfshaft joint sub-assembly**

    c. Remove support bearing
bracket.

To install:

16. Note the following, and install in the
reverse order of removal.

17. Always replace the differential side
oil seal with a new one when installing link
shaft.

18. Place the protector (SST:
KV38107900) onto the transaxle assembly
to prevent damage to the oil seal while
inserting the link shaft. Slide the link shaft
sliding joint and tap with a hammer to
install securely.

19. Install support bearing bracket in following procedure,

    a. Temporarily tighten the mounting
bolts, then tighten them to 35 ft. lbs.
(48 Nm).

    b. Temporarily tighten the mounting
bolts, then tighten them to 35 ft. lbs.
(48 Nm).

20. Apply NISSAN genuine grease to the
halfshaft serration (link shaft side), and
install the halfshaft onto link shaft.

21. Clean the matching surfaces of the
wheel hub lock nut and wheel hub and
bearing assembly.

➡**Never apply lubricating oil to these
matching surface.**

22. Clean the matching surfaces of the
halfshaft and wheel hub and bearing
assembly.

23. Apply paste (service part:
440037S000) to cover entire flat surface of
the halfshaft joint sub assembly. Amount of
paste: 0.008–0.035 oz. (0.2–1.0 g)

REAR HALFSHAFT

REMOVAL & INSTALLATION

See Figures 21 and 22.

1. Before servicing the vehicle, refer to
the Precautions Section.

2. Remove the wheels and tires.

3. Remove the wheel sensor and sensor
harness. Refer to Wheel Speed Sensor
Removal & Installation in the ABS, Brake
section.

4. Remove the caliper assembly. Hang
caliper assembly aside. Refer to Rear Disc
Brakes, Caliper Removal & Installation in
the Brake section.

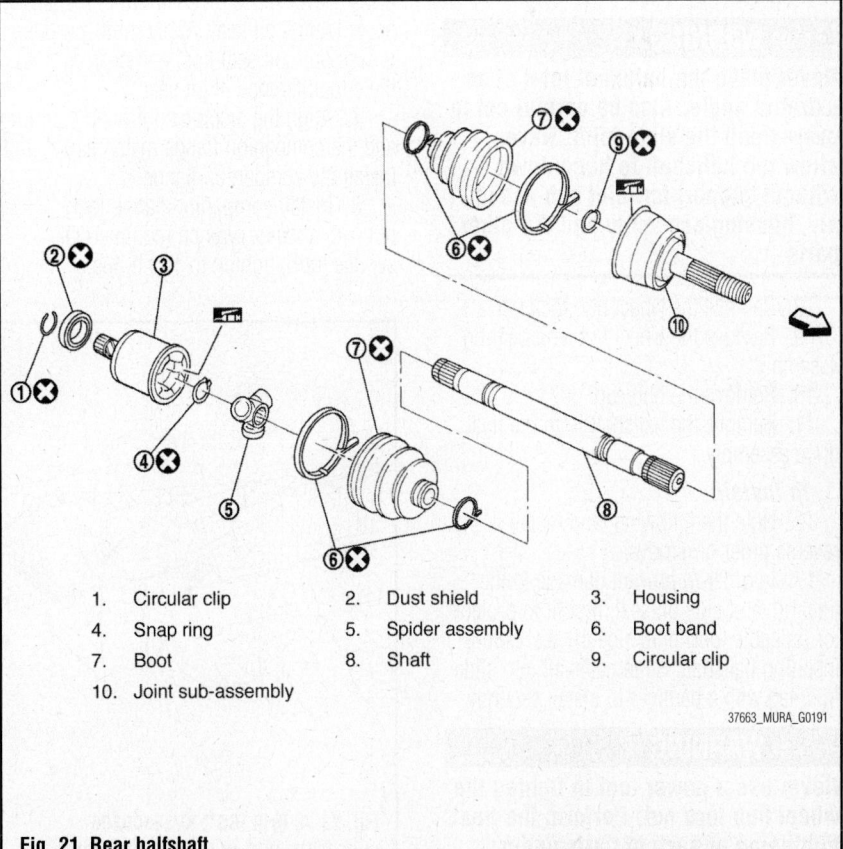

1.	Circular clip	2.	Dust shield	3.	Housing
4.	Snap ring	5.	Spider assembly	6.	Boot band
7.	Boot	8.	Shaft	9.	Circular clip
10.	Joint sub-assembly				

Fig. 21 Rear halfshaft

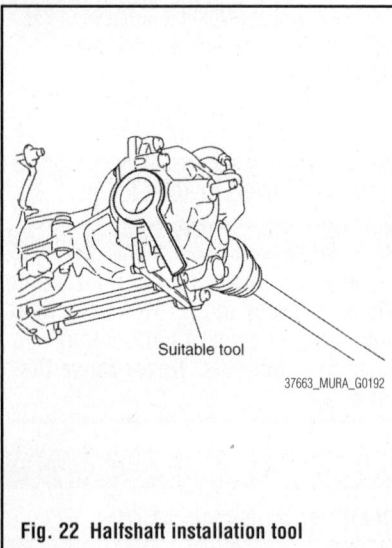

Fig. 22 Halfshaft installation tool

✳✳ WARNING

Never depress brake pedal while brake caliper is removed.

5. Remove the disc rotor. Refer to Rear Disc Brakes, Rotor Removal & Installation in the Brake section.

6. Remove the cotter pin, and then loosen the wheel hub lock nut.

7. Using suitable puller, separate the halfshaft from the wheel hub and bearing assembly.

✳✳ WARNING

Never place the halfshaft joint at an extreme angle. Also be careful not to overextend the slide joint. Never allow the halfshaft to hang down without support for joint sub-assembly, housing assembly and the other parts.

8. Remove the wheel hub lock nut.

9. Remove the wheel hub and bearing assembly.

10. Remove the hub cap.

11. Remove the halfshaft from the final drive assembly.

To install:

12. Note the following, and install in the reverse order of removal.

13. In order to prevent damage to the final drive of side oil seal, first fit to protector (suitable tool) onto side oil seal before inserting halfshaft. Slide halfshaft into slide joint tap with a hammer to install securely.

✳✳ WARNING

Never use a power tool to tighten the wheel hub lock nut. Perform the final tightening of each of parts under

unladen conditions. **Never reuse the cotter pin.**

PINION OIL SEAL

REMOVAL & INSTALLATION

See Figure 23.

1. Before servicing the vehicle, refer to the Precautions Section.

2. Remove the propeller shaft.

3. Matchmark the thread edge of electric controlled coupling and the companion flange. Use paint for matchmarks.

✳✳ WARNING

Do not damage electric controlled coupling.

4. Remove companion flange lock nut, using a flange wrench (commercial service tool). Remove companion flange.

5. Remove the front oil seal from coupling cover, using a suitable tool.

✳✳ WARNING

Be careful not to damage coupling cover.

To install:

6. Install front oil seal until it becomes flush with the coupling cover end, using the drifts. Never reuse oil seal. When installing, never incline oil seal. Apply multi-purpose grease onto oil seal lips, and gear oil onto the circumference of oil seal.

7. Align the electric controlled coupling and the companion flange matchmarks, and install the companion flange.

8. Install companion flange lock nut with a flange wrench (commercial service tool), tighten to 103 ft. lbs.

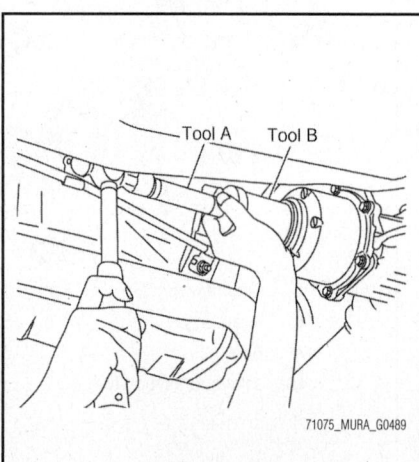

Fig. 23 A: Drift [SST: KV38100200 (J-26233)], B: Drift [SST: ST27861000]

(140 Nm). Never reuse companion flange lock nut.

9. Install the propeller shaft.

10. When oil leaks while removing, check oil level after the installation.

PROPELLER SHAFT

REMOVAL & INSTALLATION

See Figures 24 and 25.

1. Before servicing the vehicle, refer to the Precautions Section.

2. Disconnect the negative battery cable.

3. Shift the transaxle to the neutral position, and then release the parking brake.

4. Put matchmarks onto the propeller shaft flange yoke and the final drive and transfer companion flanges. Use paint for matchmarks.

5. Loosen the upper and lower center bearing mounting bracket mounting nuts. Tighten the mounting nuts temporarily.

6. Remove the propeller shaft assembly fixing bolts and nuts.

7. Remove the center bearing mounting bracket fixing nuts.

8. Remove the propeller shaft assembly.

✳✳ WARNING

If the constant velocity joint was bent during propeller shaft assembly removal, installation, or transportation, its boot may be damaged. Wrap boot interference area to metal part with shop cloth or rubber to protect boot from breakage.

9. Remove the clips and the center bearing mounting bracket (upper/lower).

Fig. 24 Put matchmarks onto the propeller shaft flange yoke and the final drive and transfer companion flanges

To install:

10. Note the following, and install in the reverse order of removal.

11. Install the upper center bearing mounting bracket with its arrow mark facing forward.

12. Adjust the position of the center bearing mounting brackets, sliding back and forth to prevent play in thrust direction of center bearing insulator. Install the upper and lower center bearing mounting brackets to the vehicle.

13. Align the matchmarks to install the propeller shaft assembly to the final drive and transfer companion flanges.

14. After assembly, perform a driving test to check propeller shaft vibration. If vibration occurs, separate the propeller

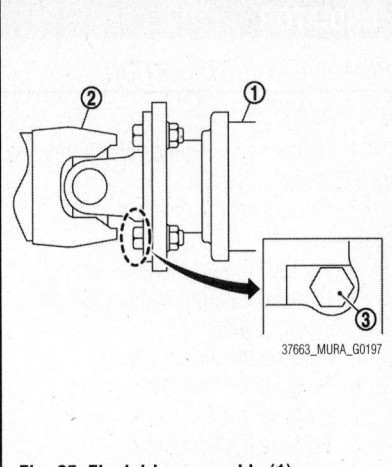

37663_MURA_G0197

Fig. 25 Final drive assembly (1), propeller shaft assembly (2), and bolts (3)

shaft from final drive. Reinstall the companion flange after rotating it by 90, 180, 270 degrees. Perform the driving test and check propeller shaft vibration again at each point.

15. After tightening the bolts and nuts to 37 ft. lbs. (50 Nm), check that the bolts on the flange side are tightened as shown.

16. If the propeller shaft assembly or final drive assembly has been replaced, connect them as follows:

    a. Install the propeller shaft while aligning its matchmark with the matchmark of the final drive on the joint as close as possible.

    b. Tighten the mounting bolts and nuts of the propeller shaft and final drive to 37 ft. lbs. (50 Nm).

ENGINE COOLING

ENGINE COOLANT

DRAIN

✳✳ CAUTION

To avoid being scalded, never change engine coolant when the engine is hot. Wrap a thick cloth around radiator cap and carefully remove radiator cap. First, turn radiator cap a quarter of a turn to release built-up pressure. Then turn radiator cap all the way.

1. Remove the engine under cover.
2. Open the radiator drain plug at the bottom of the radiator, and then remove radiator cap.

➡**When draining all of engine coolant in the system, open the water drain plugs on cylinder block.**

3. Remove the reservoir tank if necessary, and drain engine coolant and clean reservoir tank before installing.

4. Check the drained engine coolant for contaminants such as rust, corrosion or discoloration. If contaminated, flush the engine cooling system.

REFILL

1. Install the reservoir tank if removed and the radiator drain plug.

➡**Be sure to clean the drain plug and install with new O-ring. If water drain plugs on cylinder block are removed, close and tighten them.**

2. Check that each hose clamp has been firmly tightened.

3. Remove the air duct assembly and air cleaner cases (upper and lower) assembly.

4. Disconnect the heater hose.

5. Fill radiator, and reservoir tank if removed, to specified level.

✳✳ WARNING

Do not allow coolant to adhere to electronic equipments (alternator etc.).

    a. Pour engine coolant through engine coolant filler neck slowly of less than 2 ⅛ qt (2L) per minute to allow air in system to escape.

    b. When engine coolant overflows disconnect heater hose, connect heater hose, and continue filling the engine coolant.

6. Install the air duct assembly and air cleaner cases (upper and lower) assembly.

7. Install the radiator cap.

8. Warm up engine until opening thermostat. Standard for warming-up time is approximately 10 minutes at 3,000 rpm. Check thermostat opening condition by touching radiator hose (lower) to see a flow of warm water.

✳✳ WARNING

Watch water temperature gauge so as not to overheat engine.

9. Stop the engine and cool down to less than approximately 122°F (50°C).

    a. Cool down using fan to reduce the time.

    b. If necessary, refill radiator up to filler neck with engine coolant.

10. Refill reservoir tank to "MAX" level line with engine coolant.

11. Repeat steps 7 through 10 two or more times with radiator cap installed until engine coolant level no longer drops.

12. Check cooling system for leakage with engine running.

13. Warm up the engine, and check for sound of engine coolant flow while running engine from idle up to 3,000 rpm with heater temperature controller set at several position between "COOL" and "WARM".

➡**Sound may be noticeable at heater unit.**

14. Repeat step 13 three times.

15. If sound is heard, bleed air from cooling system by repeating step 5, and steps from 7 to 14 until engine coolant level no longer drops.

ELECTRIC ENGINE FAN

REMOVAL & INSTALLATION

See Figure 26.

➡**Do not remove the radiator cap when the engine is hot. Serious burns could occur from high pressure engine coolant escaping from the radiator. Be sure the engine is cold before removing the radiator cap.**

1. Before servicing the vehicle, refer to the Precautions Section.

2. Disconnect the negative battery cable.

3. Remove the engine undercover.

4. Remove the air duct (inlet).

5. Remove the oil gauge.

6. Remove the battery and the battery tray. Move the fuse and fusible link block to the side.

7. Properly drain the radiator.

➡️**Be sure the engine is cold before draining the radiator. Do not allow coolant to spill on the drive belts.**

8. Remove the radiator cap adapter and the radiator hose (upper) and radiator pipe (upper) assembly.

9. Disconnect the harness connector from the fan motors, and move the harness aside.

10. Disconnect the harness connector from crash zone sensor, and move the harness aside.

11. Remove the battery tray bracket mounting bolts, and move battery tray bracket aside.

12. Remove the cooling fan assembly.

To install:

Installation is the reverse of the removal procedure. Be sure to fill the radiator with the proper grade and type engine coolant. Start the engine, check for leaks, and correct as required.

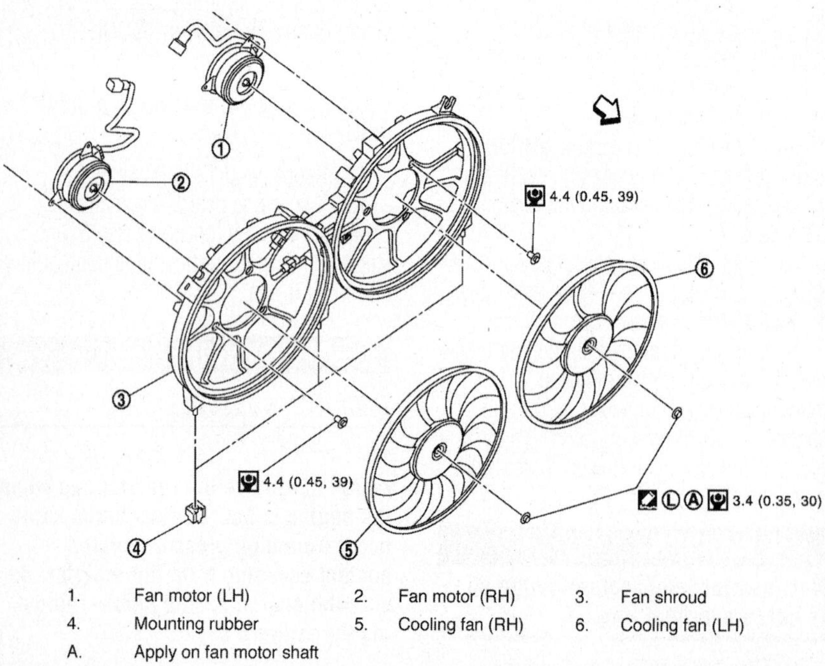

RADIATOR

REMOVAL & INSTALLATION

See Figure 27.

1. Before servicing the vehicle, refer to the Precautions Section.

2. Disconnect the negative battery cable.

3. Remove the engine undercover.

4. Remove the right and left radiator core support covers.

5. Remove the air duct (inlet).

6. Remove the front grill.

7. Remove the horn.

8. Remove the hood lock.

9. Drain engine coolant from radiator.

✳✳ CAUTION

Perform this step when the engine is cold. Do not spill engine coolant on drive belt.

10. Disconnect the reservoir tank hose from the upper radiator pipe.

11. Disconnect the CVT fluid cooler hoses from the radiator. Install blind plug to avoid leakage of CVT fluid.

12. Remove the radiator cap adapter and the upper radiator hose and the upper radiator pipe assembly.

✳✳ WARNING

Be careful not to allow engine coolant to contact drive belt.

13. Disconnect the lower radiator hose from radiator.

14. Remove the condenser. Refer to Condenser Removal & Installation in the Heating & Air Conditioning section.

✳✳ WARNING

Be careful not to damage the condenser core.

15. Remove the radiator upper clips by pulling the tabs outside to release the lock and then remove the upper mounting rubbers.

✳✳ WARNING

Never pull the tabs outside excessively.

16. Lift up and remove the radiator from the front of radiator core support.

✳✳ WARNING

Be careful not to damage or scratch the radiator core.

To install:

Note the following, and install in the reverse order of removal.

17. Install the radiator upper clips on the radiator core connection as follows:

a. Install the upper mounting rubber on the radiator mounting pin.

b. Align the radiator upper clip with the radiator core connection, then insert each the radiator upper clip straight into the radiator core connections until a click is heard.

c. After connecting the radiator upper clips, visually confirm that the radiator upper clips are connected to the radiator core connections and move the radiator upper clips and the radiator forward and backward to check they are securely connected.

THERMOSTAT

REMOVAL & INSTALLATION

See Figure 28.

➡️**Never remove the radiator cap when the engine is hot. Serious burns could occur from high-pressure engine coolant escaping from the radiator.**

1.	Fan motor (LH)	2.	Fan motor (RH)	3.	Fan shroud
4.	Mounting rubber	5.	Cooling fan (RH)	6.	Cooling fan (LH)
A.	Apply on fan motor shaft				

4.4 (0.45, 39)
4.4 (0.45, 39)
3.4 (0.35, 30)

37663_MURA_G0200

Fig. 26 Cooling fans and related components

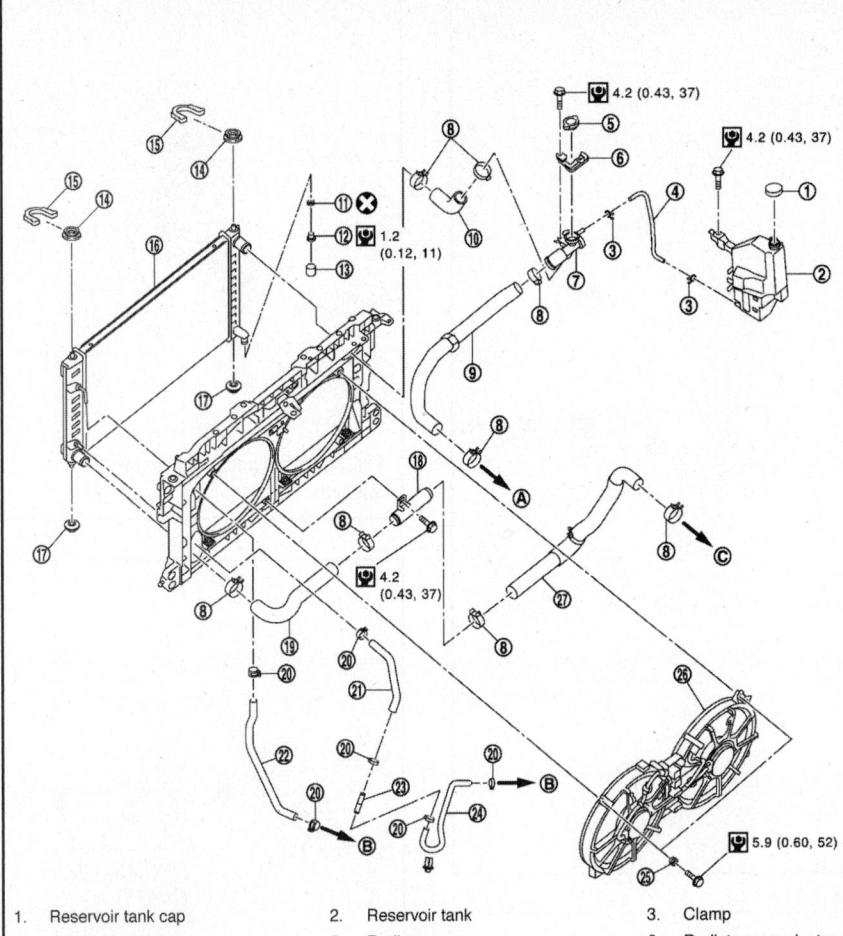

1.	Reservoir tank cap	2.	Reservoir tank	3.	Clamp
4.	Reservoir tank hose	5.	Radiator cap	6.	Radiator cap adapter
7.	Radiator pipe (upper)	8.	Clamp	9.	Radiator hose (upper)
10.	Radiator hose (upper)	11.	O-ring	12.	Drain plug
13.	Water drain hose	14.	Mounting rubber (upper)	15.	Radiator upper clip
16.	Radiator	17.	Mounting rubber (lower)	18.	Radiator pipe (lower)
19.	Radiator hose (lower)	20.	Clamp	21.	CVT fluid cooler hose
22.	CVT fluid cooler hose	23.	CVT fluid cooler pipe	24.	CVT fluid cooler hose
25.	Grommet	26.	Cooling fan assembly	27.	Radiator hose (lower)
A.	To water outlet	B.	To transaxle assembly	C.	To water inlet

37663_MURA_G0202

Fig. 27 Radiator and components

1. Before servicing the vehicle, refer to the Precautions Section.

2. Be sure the engine is cold.

3. Disconnect the negative battery cable.

4. Drain the engine coolant using the radiator drain plug and the water drain plug at the front of the cylinder block.

5. Remove the reservoir tank retaining bolts, and move it to the side.

6. Remove the intake valve timing control solenoid valve (Bank 2).

7. Disconnect the lower radiator hose from the water inlet and thermostat assembly.

8. Remove the water inlet and thermostat housing retaining bolts.

9. Remove the assembly from the engine.

➡**Do not disassemble the water inlet and thermostat assembly. Replace them as a unit, if required.**

To install:

Installation is the reverse of the removal procedure. Be sure to refill the cooling using the proper grade and type engine coolant. Start the engine and check for leaks. Start the engine and allow it to reach operation temperature. Recheck the coolant level; fill as required.

WATER PUMP

REMOVAL & INSTALLATION

See Figures 29 through 31.

✳✳ WARNING

When removing water pump assembly, be careful not to get engine coolant on drive belt. Water pump cannot be disassembled and should be replaced as a unit. After installing water pump, connect hose and clamp securely, then check for leakage.

1. Before servicing the vehicle, refer to the Precautions Section.

2. Be sure the engine is cold.

3. Disconnect the negative battery cable.

4. Remove the air duct (inlet).

5. Remove the engine cover.

6. Remove the engine undercover.

7. Remove the front wheels and tires.

8. Remove the right-hand splash guard.

9. Drain engine coolant from radiator.

✳✳ CAUTION

Perform this step when the engine is cold. Do not spill engine coolant on drive belt.

10. Remove the drive belt. Refer to Accessory Drive Belt Removal & Installation in the Engine Mechanical section.

11. Remove the idler pulleys.

12. Remove the radiator reservoir tank.

13. Remove the reservoir tank of power steering oil pump with piping connected, and move it aside.

14. Support the oil pan (lower) bottom with transmission jack.

15. Remove the right-hand engine mounting insulator, right-hand engine mounting bracket and the upper torque rod.

16. Remove the water drain plug (front) on water pump side of cylinder block to drain engine coolant from the engine.

17. Remove the valve timing control cover (Bank 1) and the water pump cover from the front timing chain case. Cut the liquid gasket for removal.

18. Remove the timing chain tensioner (primary) as follows:

   a. Remove the lower mounting bolt.

✳✳ WARNING

Be careful not to drop mounting bolt inside timing chain case.

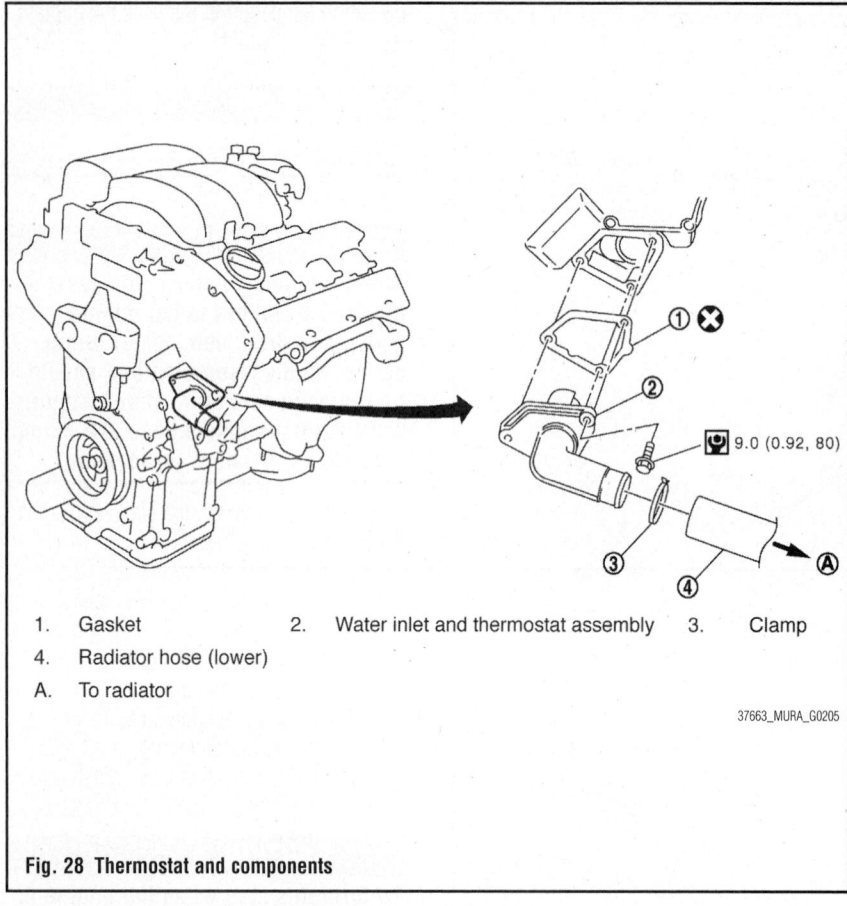

1. Gasket　　　2. Water inlet and thermostat assembly　　　3. Clamp
4. Radiator hose (lower)
A. To radiator

37663_MURA_G0205

Fig. 28 Thermostat and components

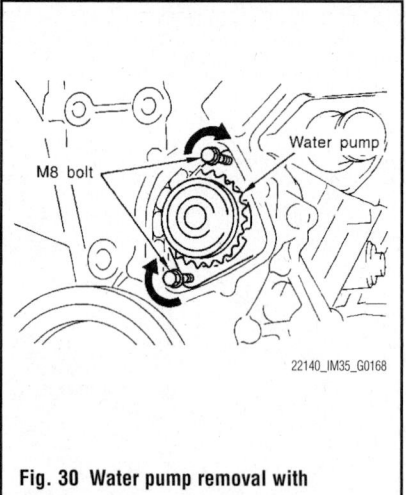

Fig. 30 Water pump removal with M8 bolts

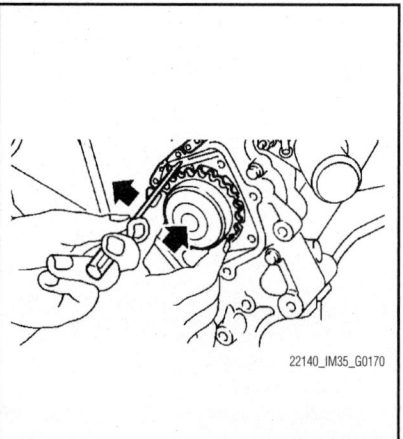

22140_IM35_G0170

Fig. 31 Installing the timing chain to the water pump

b. Loosen the upper mounting bolt slowly, and then turn chain tensioner (primary) on the mounting bolt so that plunger is fully expanded.

➡**Even if plunger is fully expanded, it is not dropped from the body of timing chain tensioner (primary).**

c. Turn the crankshaft pulley clockwise so that the timing chain on the timing chain tensioner (primary) side is loose.

d. Remove upper mounting bolt, and then remove timing chain tensioner (primary).

19. Remove water pump as follows:

a. Remove the three water pump mounting bolts. Secure a gap between the water pump gear and timing chain, by turning the crankshaft pulley counterclockwise until timing chain water pump sprocket reaches maximum looseness.

b. Screw M8 bolts: pitch: 0.0492 in (1.25 mm) length: approx. 1.97 in (50 mm); into water pumps upper and lower mounting bolt holes until they reach timing chain case. Then, alternately tighten each bolt for a half turn, and pull out water pump.

c. Pull straight out while preventing vane from contacting socket in installa-

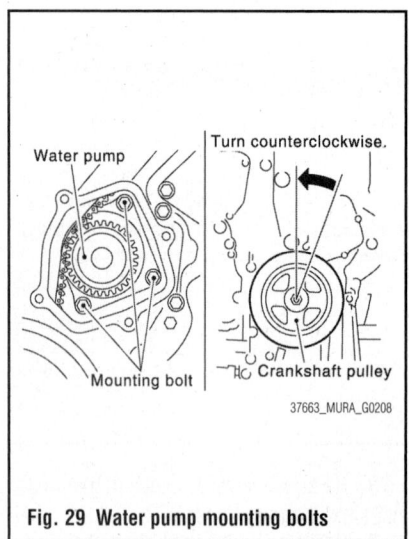

37663_MURA_G0208

Fig. 29 Water pump mounting bolts

tion area. Remove water pump without causing sprocket to contact timing chain.

d. Remove M8 bolts and O-rings from water pump.

✳✳ WARNING

Never disassemble water pump.

To install:

20. Install new O-rings to water pump.

a. Apply engine oil and engine coolant to O-rings as shown.

b. Locate O-ring with white paint mark to engine front side.

21. Install the water pump:

✳✳ WARNING

Never allow cylinder block to nip O-rings when installing water pump.

a. Check that timing chain and water pump sprocket are engaged.

b. Install the water pump by tightening mounting bolts alternately and evenly to 85 inch lbs. (10 Nm).

22. Install timing chain tensioner (primary) as follows:

a. Turn crankshaft pulley clockwise so that timing chain on the timing chain tensioner (primary) side is loose.

b. Pull plunger stopper tab up (or turn lever downward) so as to remove plunger stopper tab from the ratchet of plunger.

➡ **Plunger stopper tab and lever are synchronized.**

c. Push plunger into the inside of tensioner body.

d. Hold plunger in the fully compressed position by engaging plunger stopper tab with the tip of ratchet.

e. To secure lever, insert stopper pin through hole of lever into tensioner body hole.

f. The lever parts and the tab are synchronized. Therefore, the plunger will be secured under this condition.

➡ **The figure shows the example of 0.047 inches (1.2 mm) diameter thin screwdriver being used as the stopper pin.**

g. Install timing chain tensioner (primary). Remove dust and foreign material completely from backside of timing chain tensioner (primary) and

from installation area of rear timing chain case.

h. Remove the stopper pin.

i. Check again that timing chain and water pump sprocket are engaged.

23. Install valve timing control cover (Bank 1) and water pump cover as follows:

a. Before installing, remove all traces of old liquid gasket from mating surface of water pump cover using scraper. Also remove traces of old liquid gasket from the mating surface of front timing chain case.

b. Apply a continuous bead 0.091–0.130 in. (2–3mm) of liquid gasket with the tube presser (commercial service tool) to the mating surface of water pump cover. Use Genuine RTV Silicone Sealant or equivalent.

➡ **Attach within 5 minutes after coating.**

c. Tighten mounting bolts to 8 ft. lbs. (11 Nm).

d. Install the water drain plug.

e. Apply liquid gasket to the thread

of water drain plug (front). Use Genuine RTV Silicone Sealant or equivalent.

24. Install the right-hand engine mounting insulator, right-hand engine mounting bracket and the upper torque rod.

25. Install the reservoir tank of power steering oil pump.

26. Install the radiator reservoir tank

27. Install the idler pulleys.

28. Install the drive belt.

29. Install the right-hand splash guard.

30. Install the front wheels and tires.

31. Install the engine undercover.

32. Install the engine cover.

33. Install the air duct (inlet).

34. Refill engine coolant.

35. Connect the negative battery cable.

36. After starting engine, let idle for three minutes, then rev engine up to 3,000 rpm under no load to purge air from the high-pressure chamber of chain tensioner. Engine may produce a rattling noise. This indicates that air still remains in the chamber and is not a matter of concern.

ENGINE ELECTRICAL

BATTERY

REMOVAL & INSTALLATION

See Figure 32.

1. Before servicing the vehicle, refer to the Precautions Section.

2. Remove the air duct (inlet).

3. Remove the nut and bolt and remove the bracket from the battery fix frame.

4. Remove the cover of battery positive terminal.

5. Loosen the battery terminal nuts, and disconnect both battery cables from battery terminals.

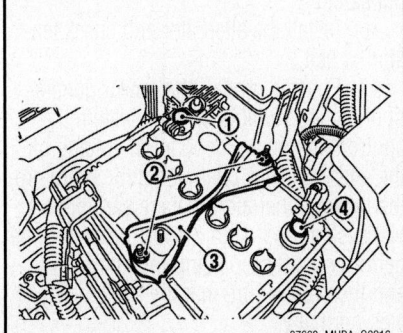

Fig. 32 Positive terminal (1), fix frame mounting nuts (2), fix frame (3), and negative terminal (4)

37663_MURA_G0216

➡ **When disconnecting, disconnect the battery cable from the negative terminal first.**

6. Remove the battery fix frame mounting nuts and the battery fix frame.

7. Remove the battery.

To install:

Install in the reverse order of removal. Tighten the fix frame mounting nuts to 35 inch lbs. (4 Nm) and the battery terminal nuts to 48 inch lbs. (5 Nm)

➡ **When connecting, connect the battery cable to the positive terminal first.**

BATTERY RECONNECT/RELEARN PROCEDURE

Perform Memory Storage

When the battery terminal is disconnected, the following functions are erased, and the indicated procedures should be performed:

• Memory (Seat, steering, mirror): Perform storing

• Intelligent Key interlock: Perform initialization, perform storing

➡ **Disconnecting the battery when DTC's are present will erase the DTC memory.**

➡ **Always perform the memory storage when the battery terminal is discon-**

BATTERY SYSTEM

nected or the driver seat control unit is replaced. The memory function will not operate normally if no memory storage is performed. Two positions for the driver seat, steering column, and outside mirror can be stored for memory operation with the following procedure. If memory is stored in the same memory switch, the previous memory will be deleted.

1. Step No. 1:
   a. Check the following conditions:
   • Ignition switch: ON
   • CVT shift selector: P position

2. Step No. 2:
   a. Adjust the driver seat and the outside mirror position manually.

   b. Push the set switch. Memory indicator for which driver seat position is already retained in memory is illuminated for 5 seconds. Memory indicator for which driver seat position is not retained in memory is illuminated for 0.5 second

   c. Push the memory switch (1 or 2) for at least 1 second within 5 seconds after pushing the set switch. To enter driver seat positions into blank memory, memory indicator will be turned on for 5 seconds. To modify driver seat positions, memory indicator will be turned OFF for 0.5 second, then turned ON for 5 seconds.

d. Confirm the operation of each part with memory operation.

Perform Intelligent Key Interlock Storage

When the battery terminal is disconnected, the following functions are erased, and the indicated procedures should be performed:

• Memory (Seat, steering, mirror): Perform storing
• Intelligent Key interlock: Perform initialization, perform storing

➡**Disconnecting the battery when DTC's are present will erase the DTC memory.**

➡**Always perform the Intelligent Key interlock function storage when the battery terminal is disconnected or the driver seat control unit is replaced. The Intelligent Key interlock function will not operate normally if no memory storage is performed. Performing the following operation associates the registered driving position with Intelligent Key. When driver door unlock operation is performed by Intelligent Key or driver door request switch, display of the registered driving position and turnout operation can be performed.**

1. Step No. 1:
   a. Check the following conditions:
   • Ignition switch: OFF
   • Initialization: done
   • Driving position: registered
2. Step No. 2:
   a. Push the set switch. Memory indicator for which driver seat position is already retained in memory is illuminated for 5 seconds.
   b. Push the Intelligent Key unlock button within 5 seconds after pushing memory switch (while the memory indicator is turned ON). From the time registration is performed, the applicable memory indicator blinks for 5 seconds.
   c. Confirm the operation of each part with memory operation and Intelligent Key interlock operation.

Steering Wheel Rotation

1. Before removing and installing any control units, first turn the push-button ignition switch to the LOCK position, and disconnect both battery cables.
2. After finishing work, confirm that all control unit connectors are connected properly, and re-connect both battery cables.
3. Always use CONSULT to perform self-diagnosis as a part of each function inspection after finishing work. If a DTC is

detected, perform trouble diagnosis according to self-diagnosis results.

4. For vehicle with steering lock unit, if the battery is disconnected or discharged, the steering wheel will lock and cannot be turned. If turning the steering wheel is required with the battery disconnected or discharged, follow the operation procedure below before starting the repair operation.
   a. Connect both battery cables. Supply power using jumper cables if battery is discharged.
   b. Turn the push-button ignition switch to ACC position. At this time, the steering lock will be released.
   c. Disconnect both battery cables. The steering lock will remain released with both battery cables disconnected and the steering wheel can be turned.
   d. Perform the necessary repair operation.
   e. When the repair work is completed, re-connect both battery cables. With the brake pedal released, turn the push-button ignition switch from ACC position to ON position, then to LOCK position. (The steering wheel will lock when the push-button ignition switch is turned to LOCK position.)
   f. Perform self-diagnosis check of all control units using CONSULT.

ENGINE ELECTRICAL

ALTERNATOR

REMOVAL & INSTALLATION
See Figure 33.

1. Before servicing the vehicle, refer to the Precautions Section.
2. Disconnect the negative battery terminal.
3. Remove the engine cover.
4. Remove the front right-hand wheel and tire.
5. Remove the right-hand splash guard.
6. Remove the air cleaner and air duct assembly. Refer to Air Cleaner Removal & Installation in the Engine Mechanical section.
7. Remove the drive belt. Refer to Accessory Drive Belt Removal & Installation in the Engine Mechanical section.
8. Remove the A/C compressor. Refer to Compressor Removal & Installation in the Heating & Air Conditioning section.

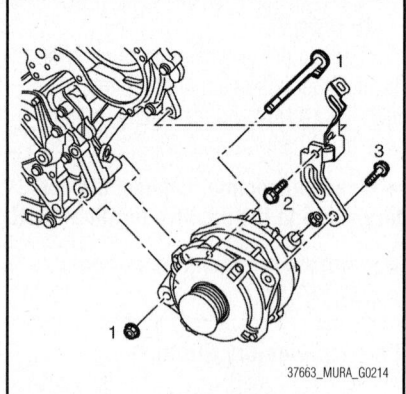

Fig. 33 Alternator bolt tightening sequence

37663_MURA_G0214

9. Remove the idler pulley.
10. Disconnect the oil pressure switch.
11. Disconnect the alternator harness connectors.

CHARGING SYSTEM

12. Remove the alternator bolt and nuts, using power tools.
13. Slide the alternator out and remove.

To install:
Installation is in the reverse order of removal, noting the following:

14. Temporarily tighten all of alternator bolt and nuts. Tighten them in numerical order shown.
15. Be sure to tighten the "B" terminal nut carefully.
16. Install the alternator and check tension of belt.
17. For this model, the power generation voltage variable control system that controls the power generation voltage of the alternator has been adopted. Therefore, the power generation voltage variable control system operation inspection should be performed after replacing the alternator, and then make sure that the system operates normally.

ENGINE ELECTRICAL

FIRING ORDER

Firing order: 1–2–3–4–5–6

IGNITION COIL(S)

REMOVAL & INSTALLATION

1. Before servicing the vehicle, refer to the Precautions Section.
2. Disconnect the negative battery cable.
3. Remove the engine cover.
4. Remove the upper and lower air cleaner cases and the air duct assembly.
5. Remove the electric throttle control actuator.
6. Remove the intake manifold collector.
7. Move aside the wiring harness, wiring harness bracket, and hoses located above ignition coil.
8. Disconnect the wiring harness connector from the ignition coil.
9. Remove the ignition coil.

✳✳ CAUTION

Do not subject the ignition coils to excessive shock or vibration.

To install:

10. Install the ignition coil on the engine.
11. Reconnect the wiring harness to the coil.

12. Reposition the wiring harness, bracket and hoses.
13. Install the air duct and the engine cover.

IGNITION TIMING

INSPECTION & ADJUSTMENT

See Figure 34.

1. Attach timing light to loop wires as shown.
2. Check ignition timing under the following conditions:
   - A/C switch: OFF
   - Electric load: OFF (Lights, heater fan & rear window defogger)
   - Steering wheel: Kept in straight-ahead position
3. Ignition timing:
   - No load (in P or N position): 12 plus or minus 5° BTDC

SPARK PLUGS

REMOVAL & INSTALLATION

1. Before servicing the vehicle, refer to the Precautions Section.
2. Disconnect the negative battery cable.
3. Remove the engine cover.
4. Remove the upper and lower air cleaner cases and the air duct assembly.
5. Remove the electric throttle control actuator.

IGNITION SYSTEM

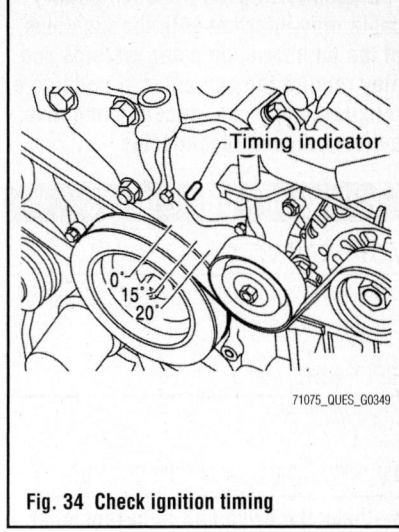

71075_QUES_G0349

Fig. 34 Check ignition timing

6. Remove the intake manifold collector.
7. Remove the ignition coil.
8. Remove the spark plug using a spark plug socket and wrench.

To install:

9. Install the ignition coil on the engine.
10. Reconnect the wiring harness to the coil.
11. Reposition the wiring harness, bracket and hoses.
12. Install the air duct and the engine cover.

ENGINE ELECTRICAL

STARTER

REMOVAL & INSTALLATION

1. Before servicing the vehicle, refer to the Precautions Section.
2. Remove the battery. Refer to Battery Removal & Installation.
3. Remove the air cleaner assembly and

air ducts. Refer to Air Cleaner Removal & Installation in the Engine Mechanical section.
4. Disconnect the following unit connectors:

   - ECM
   - TCM
   - IPDM E/R
5. Remove the battery tray.

STARTING SYSTEM

6. Disconnect the starter motor harness connectors.
7. Remove the starter motor mounting bolts, using power tools.
8. Remove the starter motor.

To install:

Installation is in the reverse order of removal. Tighten the mounting bolts to 41 ft. lbs. (55 Nm)

ENGINE MECHANICAL

➡Disconnecting the negative battery cable may interfere with the functions of the on board computer systems and may require the computer to undergo a relearning process, once the negative battery cable is reconnected.

ACCESSORY DRIVE BELTS

ADJUSTMENT

Belt tension is automatically adjusted by the auto-tensioner.

BELT ROUTINGS

See Figure 35.

INSPECTION

➡Check the drive belt auto-tensioner indication when the engine is cold.

1. Check that the indicator of drive belt auto-tensioner is within the possible use range.
2. When new drive belt is installed, the indicator should be within the range in the figure.
3. Visually check entire drive belt for wear, damage or cracks.
4. If the indicator is out of the possible use range or belt is damaged, replace drive belt.

REMOVAL & INSTALLATION

Drive Belt

See Figure 36.

1. Before servicing the vehicle, refer to the Precautions Section.
2. Disconnect the negative battery cable.
3. Remove the right-hand front wheel and tire.
4. Remove the right-hand splash guard.

✳✳ CAUTION

Avoid placing hand in a location where pinching may occur if the holding tool accidentally comes off.

✳✳ WARNING

Never loosen the hexagonal part in center of drive belt auto-tensioner pulley (Never turn it counterclockwise). If turned counterclockwise, the complete drive belt auto-tensioner must be replaced as a unit, including the pulley.

5. Hold the hexagonal part in the center of the drive belt auto-tensioner pulley securely with a box wrench. Move the

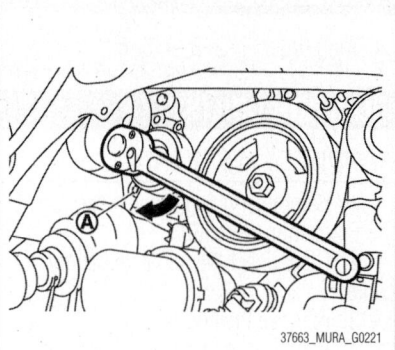

37663_MURA_G0221

Fig. 36 Insert a rod into the hole (A) of the retaining boss to fix the drive belt auto-tensioner pulley

wrench handle in the direction of arrow (loosening direction of drive belt).
6. Insert a rod approximately 0.24 in. (6 mm) in diameter such as short-length screwdriver into the hole (A) of the retaining boss to fix the drive belt auto-tensioner pulley.
7. Loosen drive belt from water pump pulley in sequence, and remove it.

To install:

✳✳ CAUTION

Avoid placing hand in a location where pinching may occur if the holding tool accidentally comes off.

✳✳ WARNING

Never loosen the hexagonal part in center of drive belt auto-tensioner pulley (Never turn it counterclockwise). If turned counterclockwise, the complete drive belt auto-tensioner must be replaced as a unit, including the pulley.

8. Hold the hexagonal part in the center of the drive belt auto-tensioner pulley securely with a box wrench. Move the wrench handle in the direction of the arrow (loosening direction of drive belt).
9. Insert a rod approximately 0.24 in. (6 mm) in diameter such as short-length screwdriver into the hole of the retaining boss to fix the drive belt auto-tensioner pulley.
10. Hook the drive belt onto all pulleys except for drive belt auto-tensioner pulley, and onto drive belt auto-tensioner pulley last.

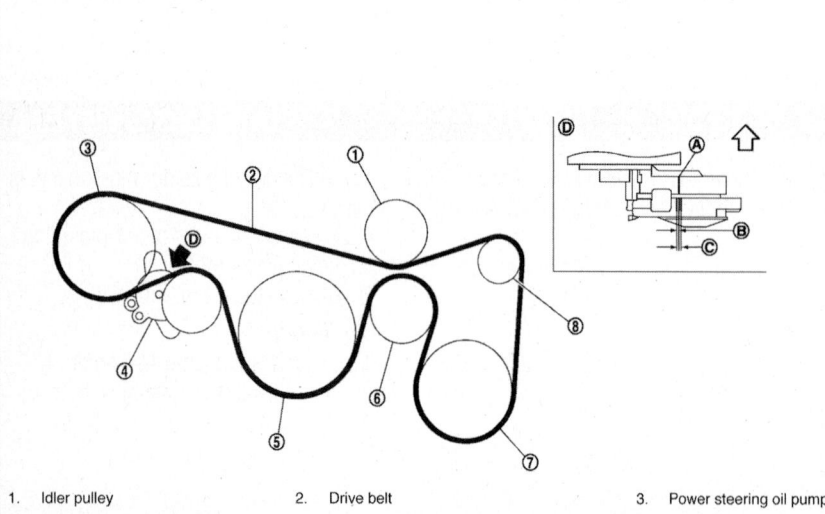

1.	Idler pulley	2.	Drive belt	3.	Power steering oil pump
4.	Drive belt auto-tensioner	5.	Crankshaft pulley	6.	Idler pulley
7.	A/C compressor	8.	Alternator		
A.	Indicator	B.	Range when new drive belt is installed	C.	Possible use range
D.	View D				
⇦ : Engine front					

37663_MURA_G0220

Fig. 35 Accessory drive belt and components

Confirm drive belt is completely set to pulleys.

Check for engine oil, working fluid and engine coolant are not adhered to drive belt and each pulley groove.

11. Release the drive belt auto-tensioner, and apply tension to drive belt.

12. Turn the crankshaft pulley clockwise several times to equalize tension between each pulley.

13. Confirm tension of drive belt at indicator is within the possible use range.

14. Install the right-hand splash guard.

15. Install the right-hand front wheel and tire.

16. Connect the negative battery cable.

Drive Belt Idler Pulley & Tensioner

1. Remove the drive belt. Keep the auto-tensioner pulley arm locked after drive belt is removed.

2. Remove the auto-tensioner and idler pulley. Keep the auto-tensioner pulley arm locked to install or remove auto-tensioner.

To install:

Install in the reverse order of removal.

AIR CLEANER

REMOVAL & INSTALLATION

See Figure 37.

1. Before servicing the vehicle, refer to the Precautions Section.

2. Disconnect the negative battery cable.

3. Remove the radiator core support covers.

4. Remove the air duct (inlet).

5. Disconnect the harness connector from mass air flow sensor.

6. Disconnect the PCV hose.

7. Remove the air cleaner cases (upper and lower) with mass air flow sensor and air duct assembly.

   a. Add matchmarks if necessary for easier installation.

   b. Remove the mass air flow sensor from air cleaner case (upper), if necessary.

To install:

8. Note the following, and install in the reverse order of removal.

9. Align marks. Attach each joint. Screw clamps firmly.

CAMSHAFT & BEARINGS

INSPECTION

See Figures 38 through 46.

1. Inspect the Camshaft Runout:

   a. Place a V-block on a precise flat table, and support No. 2 and No. 4 camshaft journals.

Never support No. 1 journal (on the side of camshaft sprocket) because it has a different diameter from the other three locations.

   b. Set a dial indicator vertically to No. 3 journal.

   c. Turn the camshaft to one direction with hands, and measure the camshaft runout on a dial indicator (Total indicator reading). If it exceeds

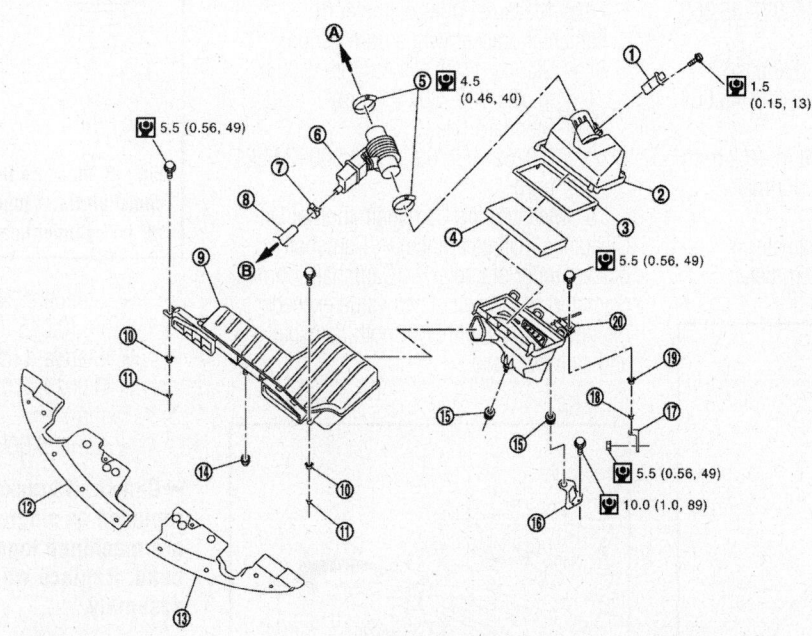

1.	Mass air flow sensor	2.	Air cleaner case (upper)	3.	Holder
4.	Air cleaner filter	5.	Clamp	6.	Air duct assembly
7.	Clamp	8.	PCV hose	9.	Air duct (inlet)
10.	Grommet	11.	Collar	12.	Radiator core support cover (RH)
13.	Radiator core support cover (LH)	14.	Grommet	15.	Grommet
16.	Bracket	17.	Bracket	18.	Collar
19.	Grommet	20.	Air cleaner case (lower)		
A.	To electric throttle control actuator	B.	To rocker cover (bank 2)		

37663_MURA_G0223

Fig. 37 Air cleaner, air duct, and related components

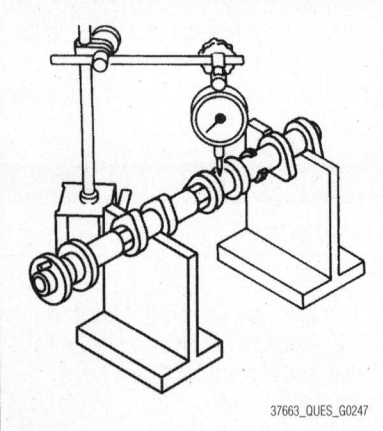

Fig. 38 Measure camshaft runout on dial gauge

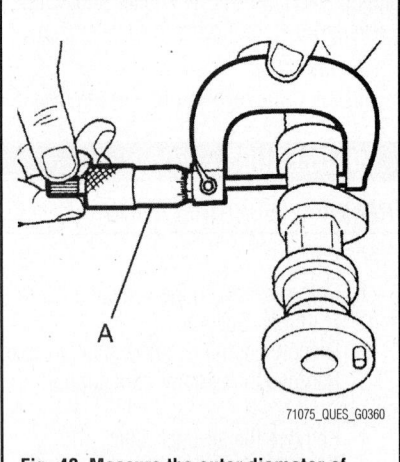

Fig. 40 Measure the outer diameter of camshaft journal with a micrometer (A)

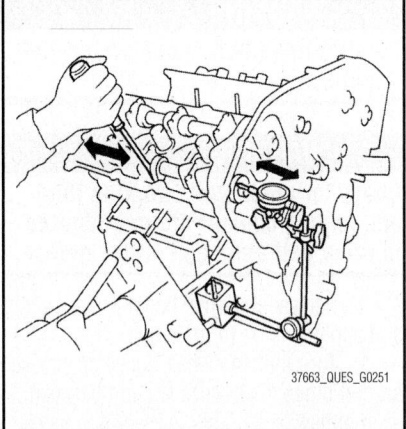

Fig. 42 Measure end play of dial gauge when camshaft is moved forward and backward

the limit, replace camshaft. Camshaft Runout:
- Standard: Less than 0.0008 in. (0.02 mm)
- Limit: 0.0020 in. (0.05 mm)

2. Measure the camshaft cam height with a micrometer. If wear exceeds the limit, replace the camshaft. Camshaft Cam Height:
- Standard cam height (Intake): 1.7900–1.7974 in. (45.465–45.655 mm)
- Standard cam height (Exhaust): 1.7904–1.7978 in. (45.475–45.665 mm)
- Cam wear limit: 0.008 in. (0.2 mm)

3. Measure the Camshaft Journal Diameter:
   a. Measure the outer diameter of camshaft journal with a micrometer:

- Standard (No. 1): 1.0211–1.0218 in. (25.935–25.955 mm)
- Standard (No. 2, 3, 4): 0.9230–0.9238 in. (23.445–23.465 mm)

4. Measure the Camshaft Bracket Inner Diameter:
   a. Tighten the camshaft bracket bolt to the specified torque. Refer to Installation.
   b. Measure inner diameter of camshaft bracket with a bore gauge.

- Standard (No. 1): 1.0236–1.0244 in. (26.000–26.021 mm)
- Standard (No. 2, 3, 4): 0.9252–0.9260 in. (23.500–23.521 mm)

5. Measure the Camshaft Journal Oil Clearance: (Oil clearance) = (Camshaft bracket inner diameter) - (Camshaft journal diameter). If the calculated value exceeds the limit, replace either or both camshaft and cylinder head.

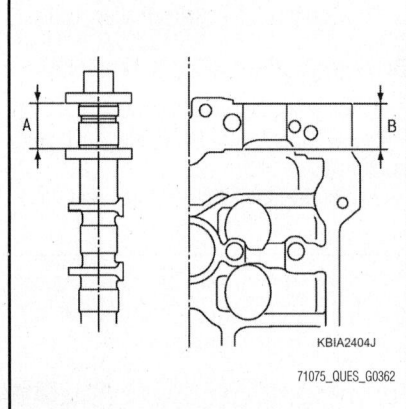

Fig. 43 Measure Dimension "A" for camshaft No. 1 journal, and Dimension "B" for cylinder head No. 1 journal bearing

- Standard (No. 1): 0.0018–0.0034 in. (0.045–0.086 mm)
- Standard (No. 2, 3, 4): 0.0014–0.0030 in. (0.035–0.076 mm)
- Limit: 0.0059 in. (0.15 mm)

➡ **Camshaft brackets cannot be replaced as single parts, because they are machined together with cylinder head. Replace whole cylinder head assembly.**

6. Measure the Camshaft End Play:
   a. Install a dial indicator in thrust direction on front end of camshaft.
   b. Measure the end play of a dial indicator when camshaft is moved forward/backward (in direction to axis).

- Standard: 0.0045–0.0074 in. (0.115–0.188 mm)
- Limit: 0.0094 in. (0.24 mm)

   c. Measure the following parts if out of the limit:

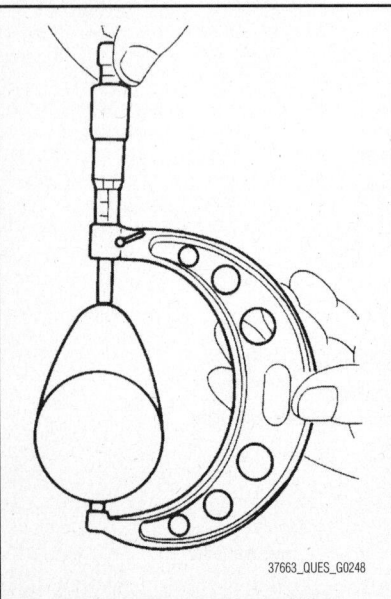

Fig. 39 Measure camshaft cam height

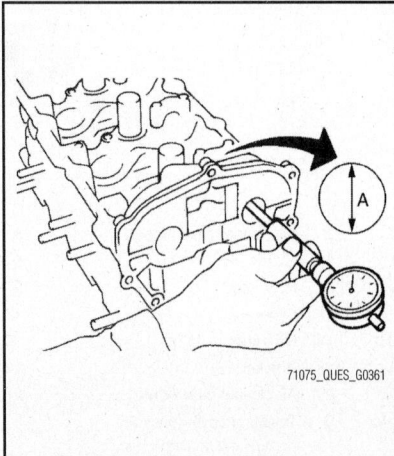

Fig. 41 Measure inner diameter (A) of camshaft bracket with a bore gauge

- Dimension "A" for camshaft No. 1 journal, Standard: 1.0827–1.0846 in. (27.500–27.548 mm)
- Dimension "B" for cylinder head No. 1 journal bearing, Standard: 1.0772–1.0781 in. (27.360–27.385 mm)

d. Refer to the standards above, and then replace the camshaft and/or cylinder head.

7. Measure the Camshaft Sprocket Runout:

a. Place a V-block on a precise flat table, and support the No. 2 and 4 journals of the camshaft.

☀☀ WARNING

Never support No. 1 journal (on the side of camshaft sprocket) because it has a different diameter from the other three locations.

b. Measure the camshaft sprocket runout with a dial indicator (Total indicator reading). If it exceeds the limit, replace camshaft sprocket. Limit: 0.0059 in. (0.15 mm)

8. Check the surface of valve lifter for wear or cracks. Replace as necessary.

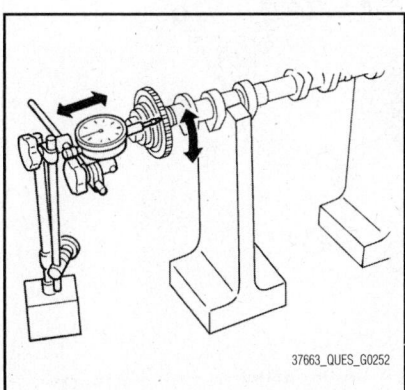

Fig. 44 Inspect the camshaft sprocket runout

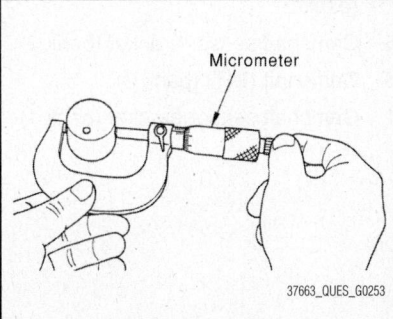

Fig. 45 Measure the valve lifter outer diameter

Fig. 46 Measure the valve lifter hole diameter

9. Measure the Valve Lifter Outer Diameter: Measure the outer diameter at ½ height of the valve lifter with a micrometer since valve lifter is in a barrel shape. Standard (Intake and exhaust): 1.3378–1.3382 in. (33.980–33.990 mm)

10. Measure the Valve Lifter Hole Diameter: Measure the inner diameter of valve lifter hole of cylinder head with an inside micrometer. Standard (Intake and exhaust): 1.3386–1.3392 in. (34.000–34.016 mm)

11. Measure the Valve Lifter Clearance: (Valve lifter clearance) = (Valve lifter hole diameter) - (Valve lifter outer diameter). Standard (Intake and exhaust): 0.0004–0.0014 in. (0.010–0.036 mm)

a. If the calculated value is out of the standard, referring to each standard of valve lifter outer diameter and valve lifter hole diameter, replace either or both valve lifter and cylinder head.

REMOVAL & INSTALLATION

See Figures 47 through 54.

1. Before servicing the vehicle, refer to the Precautions Section.

2. Disconnect the negative battery cable.

3. Drain the engine oil.

4. Drain the engine coolant from inside the engine.

5. Remove the front timing chain case, camshaft sprocket, timing chain and rear timing chain case. Refer to Timing Chain Removal & Installation.

6. Loosen camshaft sensor bracket bolts in reverse of the bolt tightening order as shown below. The order of loosening bolts in the same for bank 1 and bank 2.

7. Remove the camshaft brackets.

a. Mark the camshafts, camshaft brackets and bolts so they are placed in the same position and direction for installation.

b. Equally loosen the camshaft bracket bolts in several steps in the reverse of the installation order shown below.

8. Remove the camshafts.

9. Remove the valve lifters. Identify installation positions, and store them without mixing them up.

10. Remove the secondary timing chain tensioners from the cylinder head. Remove the secondary timing chain tensioner with its stopper pin attached.

➡**Stopper pin should be attached when the secondary timing chain is removed.**

To install:

➡**Do not reuse the washers.**

11. Install the secondary timing chain tensioners on both sides of the cylinder head.

a. Install the secondary timing chain tensioner with its stopper pin attached.

b. Install the secondary timing chain tensioner with sliding part facing downward on Bank 1 cylinder head, and with sliding part facing upward on Bank 2 cylinder head.

12. Install the valve lifters in the original position.

13. Install the camshafts.

a. Follow identification marks made during removal, or follow the identification marks that are present on new camshafts for proper placement and direction.

b. Install the camshaft so that the dowel pins on the front end face are positioned as shown. (No. 1 cylinder TDC on its compression stroke). Though the camshaft does not stop at the portion as shown in the figure, for the placement of cam nose, it is generally accepted to position the camshaft in the same direction of the figure.

14. Install the camshaft brackets.

a. Remove the foreign material completely from the camshaft bracket backside and from the cylinder head installation face.

b. Install the camshaft bracket in the original position and direction as shown.

c. Install the camshaft brackets No. 2 to 4, aligning the stamp marks as shown. There are no identification marks indicating left and right for the camshaft bracket No. 1.

15. Apply liquid gasket to the mating surface of the No. 1 camshaft bracket as

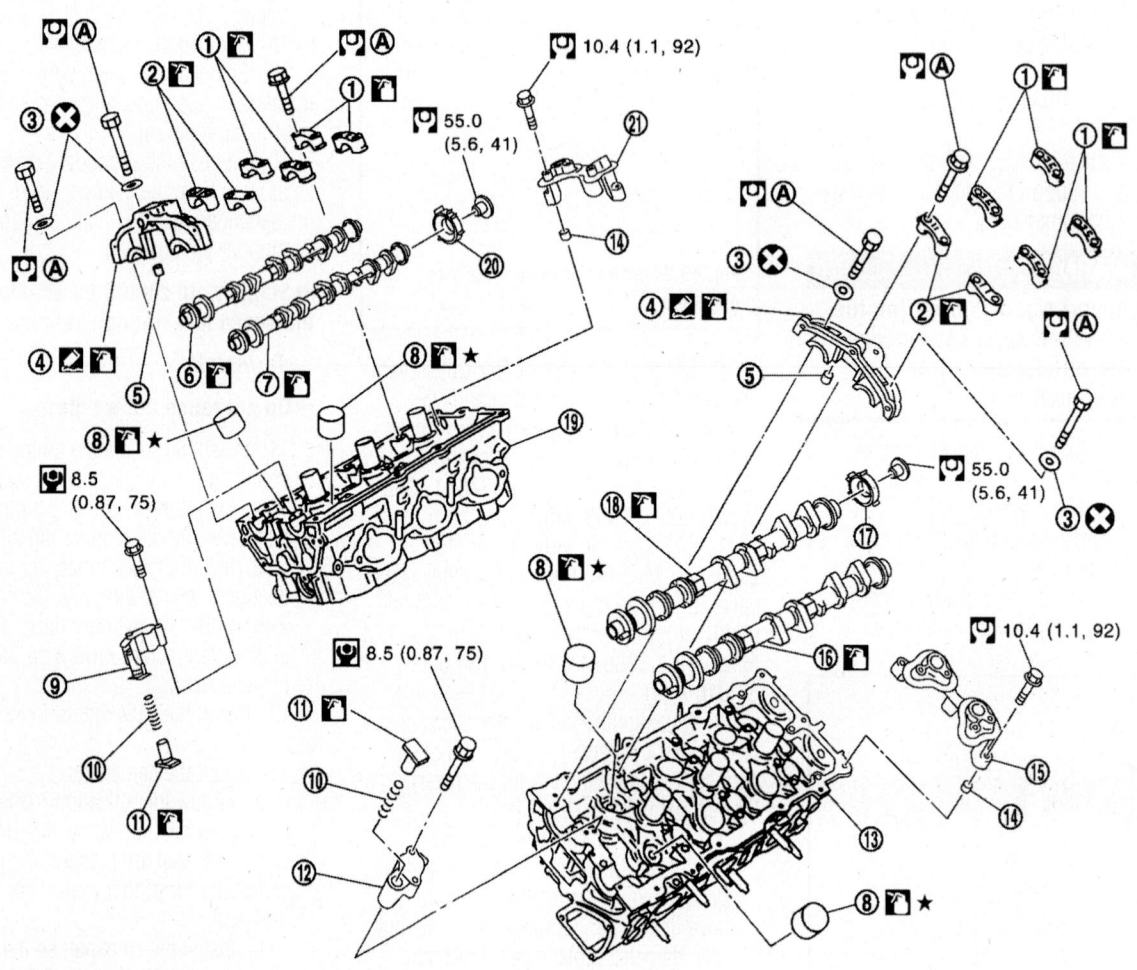

Fig. 47 Camshafts and related components

1. Camshaft bracket (No. 3, 4)
2. Camshaft bracket (No. 2)
3. Seal washer
4. Camshaft bracket (No. 1)
5. Dowel pin
6. Camshaft (EXH) (bank 1)
7. Camshaft (INT) (bank 1)
8. Valve lifter
9. Timing chain tensioner (secondary) (bank 1)
10. Spring
11. Plunger
12. Timing chain tensioner (secondary) (bank 2)
13. Cylinder head (bank 2)
14. Dowel pin
15. Camshaft sensor bracket (bank 2)
16. Camshaft (EXH) (bank 2)
17. Camshaft signal plate (bank 2)
18. Camshaft (INT) (bank 2)
19. Cylinder head (bank 1)
20. Camshaft signal plate (bank 1)
21. Camshaft sensor bracket (bank 1)

37663_MURA_G0245

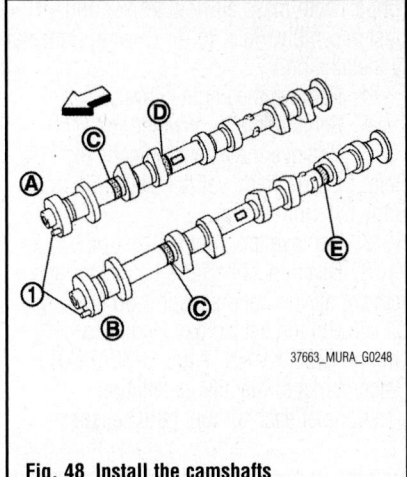

Fig. 48 Install the camshafts

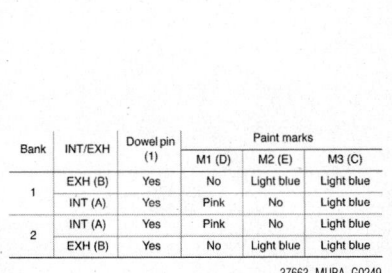

Bank	INT/EXH	Dowel pin (1)	Paint marks		
			M1 (D)	M2 (E)	M3 (C)
1	EXH (B)	Yes	No	Light blue	Light blue
	INT (A)	Yes	Pink	No	Light blue
2	INT (A)	Yes	Pink	No	Light blue
	EXH (B)	Yes	No	Light blue	Light blue

37663_MURA_G0249

Fig. 49 Camshaft identification marks

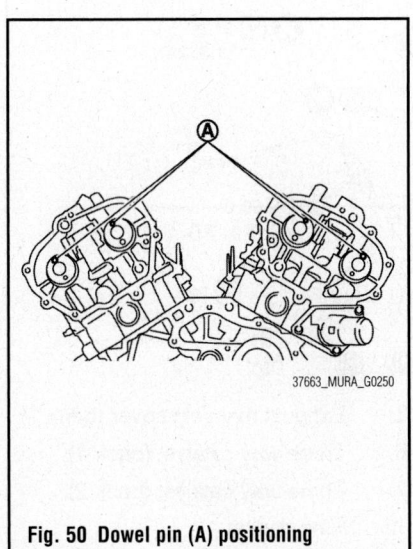

Fig. 50 Dowel pin (A) positioning

shown, on bank 1 and bank 2, as shown. Use Genuine RTV Silicone Sealant or equivalent.

16. Tighten the camshaft bracket bolts in the following steps, in numerical order as shown.

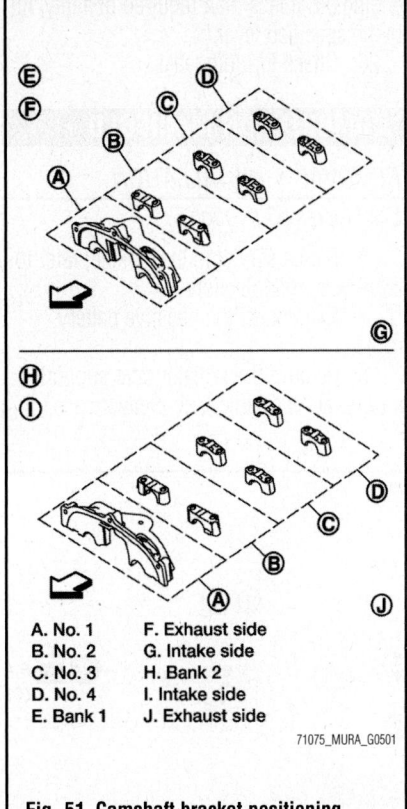

A. No. 1
B. No. 2
C. No. 3
D. No. 4
E. Bank 1
F. Exhaust side
G. Intake side
H. Bank 2
I. Intake side
J. Exhaust side

71075_MURA_G0501

Fig. 51 Camshaft bracket positioning

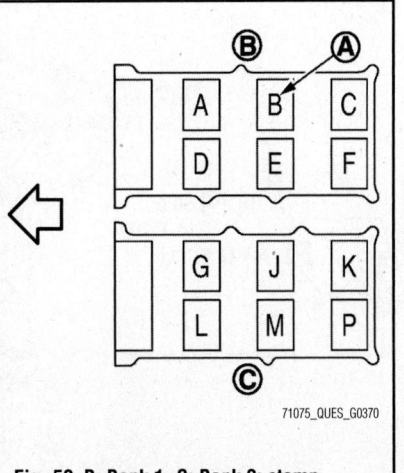

71075_QUES_G0370

Fig. 52 B: Bank 1, C: Bank 2; stamp marks (A)

a. Tighten No. 7 to 10 in numerical order as shown to 12 inch lbs. (2 Nm).

b. Tighten No. 1 to 6 in numerical order as shown to 12 inch lbs. (2 Nm).

c. Tighten No. 1 to 10 in numerical order as shown to 48 inch lbs. (6 Nm).

d. Tighten No. 1 to 10 in numerical order as shown to 8 ft. lbs. (10 Nm).

17. After tightening the mounting bolts of the No. 1 camshaft bracket, be sure to wipe off excessive liquid gasket from the parts list below.

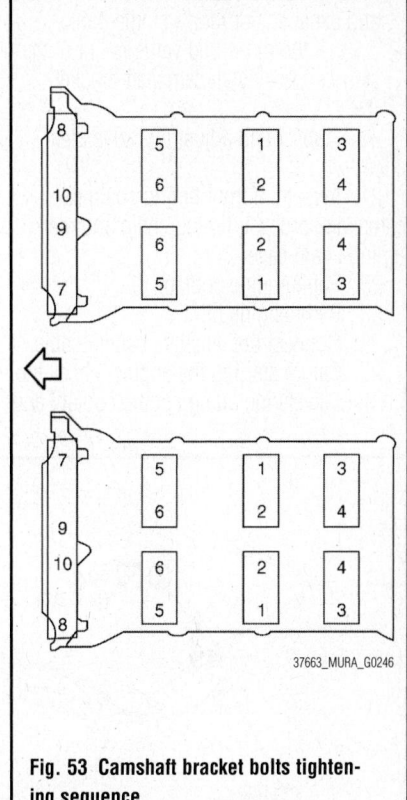

37663_MURA_G0246

Fig. 53 Camshaft bracket bolts tightening sequence

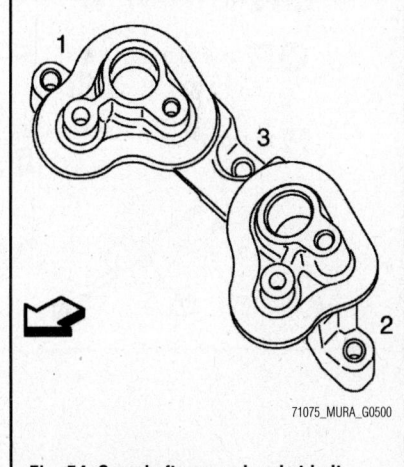

71075_MURA_G0500

Fig. 54 Camshaft sensor bracket bolt tightening sequence

• Mating surface of rocker cover
• Mating surface of rear timing chain case

18. Tighten camshaft sensor bracket bolts in numerical order as shown. The order of tightening bolts in the same for bank 1 and bank 2.

19. Measure the difference in levels between the front end faces of the camshaft bracket (No. 1) and the cylinder head. Standard: −0.0055 to 0.0055 in. (−0.14 to 0.14mm)

a. Measure two positions (both intake and exhaust side) for a single bank.

b. If the measured value is out of the standard, re-install camshaft bracket (No. 1).

20. Inspect and adjust the valve clearance.

21. Install the front timing chain case, camshaft sprocket, timing chain and rear timing chain case.

22. Refill engine coolant.

23. Refill engine oil.

24. Connect the negative battery cable.

25. Before starting the engine, check the oil/fluid levels including engine coolant and engine oil. If less than required quantity, fill to the specified level.

26. Check for fluid leaks.

COMBINATION MANIFOLD

REMOVAL & INSTALLATION

See Figures 55 through 57.

1. Before servicing the vehicle, refer to the Precautions Section.

2. Disconnect the negative battery cable.

3. Remove the radiator core support covers, air duct (inlet), air cleaner case (upper) with mass air flow sensor and air duct assembly. Refer to Air Cleaner Removal & Installation.

4. Remove the engine cover.

5. Remove the front wiper arm.

6. Remove the extension cowl top. Refer to Cowl Removal & Installation in the Body Exterior section.

7. Remove the front exhaust pipe.

8. Disconnect harness connector and remove air fuel ratio sensor 1 on both banks with the heated oxygen sensor wrench [SST: KV10117100 (J-3647-A)]. Place marks to identify installation positions of each air fuel ratio sensor

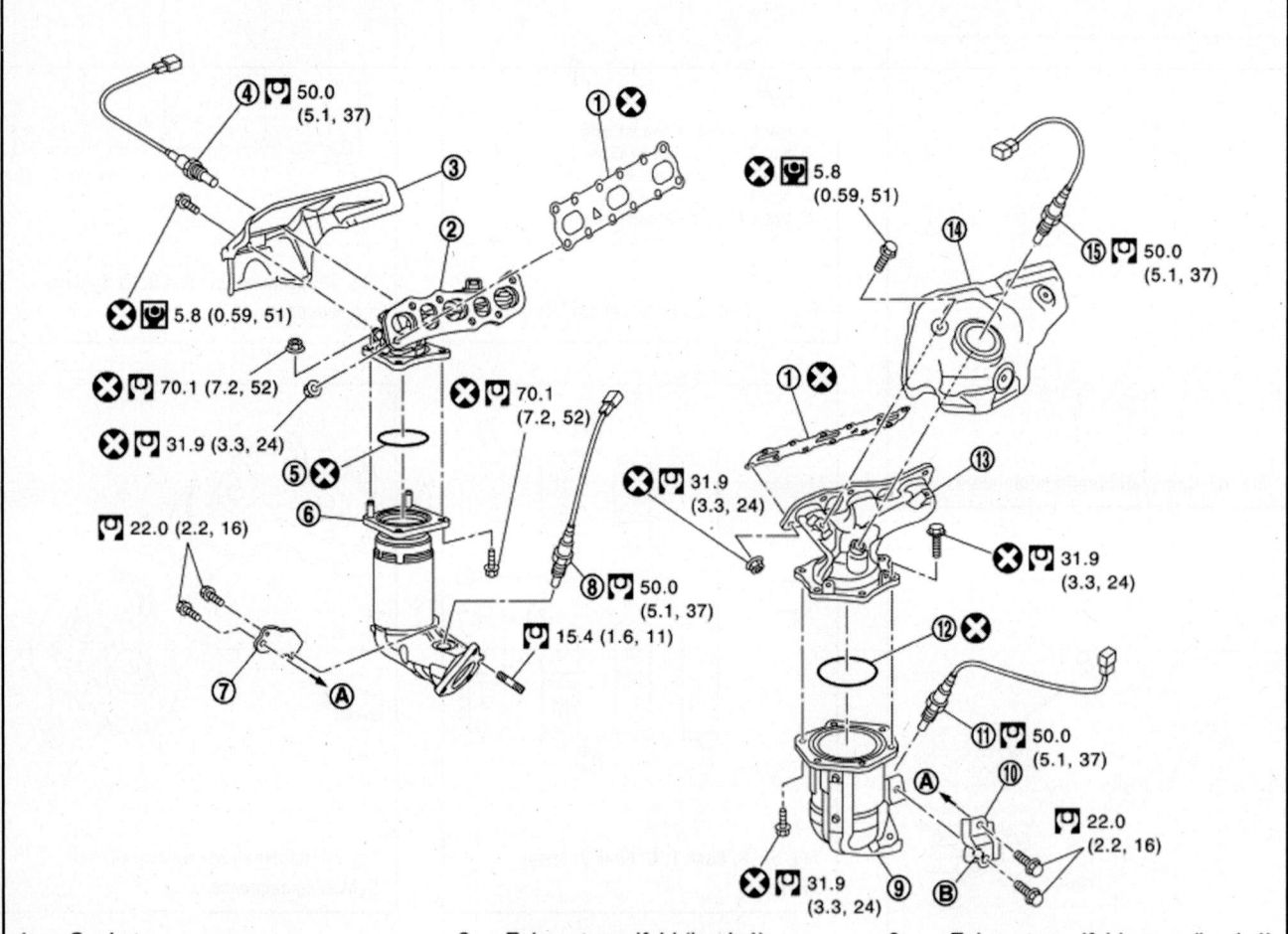

1. Gasket	2. Exhaust manifold (bank 1)	3. Exhaust manifold cover (bank 1)
4. Air fuel ratio sensor 1 (bank 1)	5. Ring gasket	6. Three way catalyst (bank 1)
7. Three way catalyst support (bank 1)	8. Heated oxygen sensor 2 (bank 1)	9. Three way catalyst (bank 2)
10. Three way catalyst support (bank 2)	11. Heated oxygen sensor 2 (bank 2)	12. Ring gasket
13. Exhaust manifold (bank 2)	14. Exhaust manifold cover (bank 2)	15. Air fuel ratio sensor 1 (bank 2)
A. To oil pan (upper)	B. Upper mark	

71075_MURA_G0504

Fig. 55 Exploded view of exhaust manifold and gasket components—CrossCabriolet models

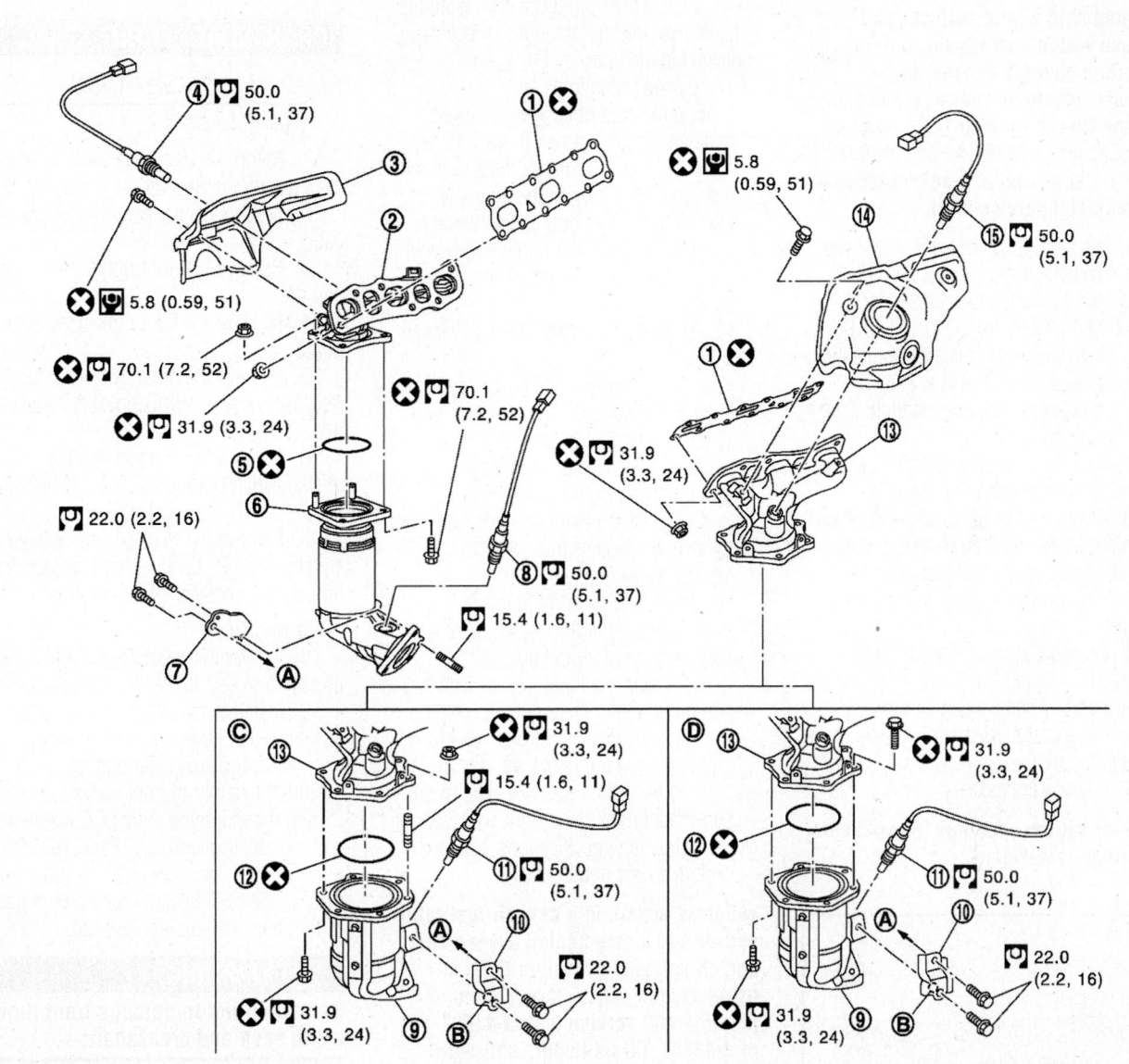

1. Gasket
2. Exhaust manifold (bank 1)
3. Exhaust manifold cover (bank 1)
4. Air fuel ratio sensor 1 (bank 1)
5. Ring gasket
6. Three way catalyst (bank 1)
7. Three way catalyst support (bank 1)
8. Heated oxygen sensor 2 (bank 1)
9. Three way catalyst (bank 2)
10. Three way catalyst support (bank 2)
11. Heated oxygen sensor 2 (bank 2)
12. Ring gasket
13. Exhaust manifold (bank 2)
14. Exhaust manifold cover (bank 2)
15. Air fuel ratio sensor 1 (bank 2)
A. To oil pan (upper)
B. Upper mark
C. Stud bolt and flange bolt type (bank 2)
D. Flange bolt type (bank 2)

37663_MURA_G0264

Fig. 56 Exploded view of exhaust manifold and gasket components—Hardtop models

1. Be careful not to damage the sensor.

➡**Discard any sensor which has been dropped onto a hard surface, and replace with a new sensor. Before installing new A/F sensor, clean exhaust system threads using Oxygen Sensor Thread Cleaner [commercial service tool (J-43897-18 or J-43897-12)] and approved anti-seize lubricant (commercial service tool).**

9. Disconnect harness connector and remove heated oxygen sensor 2 on both banks with the heated oxygen sensor wrench [SST: KV10114400 (J-38365)]. Place marks to identify installation positions of each heated oxygen sensor 2.

10. Remove the exhaust manifold covers (bank 1 and bank 2).

11. Remove three way catalyst support mounting bolts (bank 1 and bank 2).

12. Remove three way catalysts by loosening bolts first and then removing nuts. (Stud bolt and flange bolt type) Handle carefully to avoid any shock to three way catalyst.

13. Loosen the mounting nuts in the reverse of the installation order shown below. Disregard No. 7 and 8 when loosening.

14. Remove the exhaust manifolds (bank 1 and bank 2).

15. Remove the gaskets.

➡**Cover engine openings to avoid entry of foreign materials.**

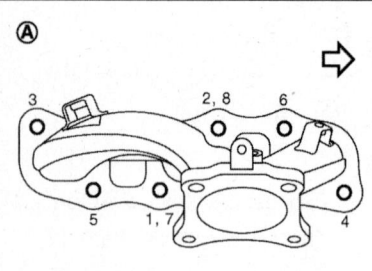

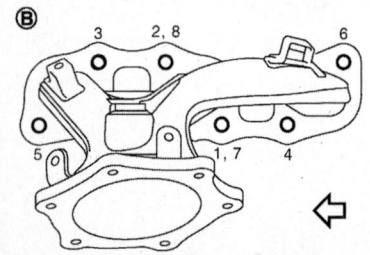

37663_MURA_G0265

Fig. 57 Exhaust manifold bolt tightening sequence; No. 7 and 8 mean double tightening of nuts No. 1 and 2—Bank 1 (A), Bank (2), engine front (arrow)

To install:

Installation is the reverse of removal, noting the following:

16. Exhaust Manifold Gasket: Install the exhaust manifold gasket in the direction indicated in the figure.

17. Exhaust Manifold:

   a. If the stud bolts were removed, install and tighten to 11 ft. lbs. (15 Nm).

   b. Install the mounting nuts in numerical order as shown in the figure. Note: No. 7 and 8 mean double tightening of nuts No. 1 and 2. For tightening specifications, refer to exploded view figure, above.

18. Three Way Catalyst (bank 2) Flange Bolt Type:

   a. Temporarily tighten the upper side bolts and lower side bolts, as shown.

   b. Tighten the upper side bolts. For tightening specifications, refer to exploded view figure.

   c. Tighten all the lower side bolts. For tightening specifications, refer to exploded view figure.

19. Three Way Catalyst Supports:

   a. Temporarily the tighten three way catalyst support mounting bolts.

   b. Tighten the three way catalyst support mounting bolts to oil pan (upper). For tightening specifications, refer to exploded view figure, above.

   c. Tighten the three way catalyst support mounting bolts to three way catalyst. For tightening specifications, refer to exploded view figure, above.

➡**Before installing a new air fuel ratio sensor and a new heated oxygen sensor, clean exhaust system threads using oxygen sensor thread cleaner (commercial service tool: J-43897-18 or J-43897-12) and apply anti-seize lubricant.**

20. Install the air fuel ratio sensor 1 and heated oxygen sensor 2 in the original positions. If the installation positions cannot be identified, check the glass tube color (air fuel ratio sensor 1: black, heated oxygen sensor: white).

☀ WARNING

Never over-torque air fuel ratio sensor and heated oxygen sensor.

21. Install the exhaust manifold covers.
22. Install the front exhaust pipe.
23. Install the extension cowl top.
24. Install the front wiper arm.
25. Install the engine cover.
26. Install the radiator core support covers, air duct (inlet), air cleaner case (upper) with mass air flow sensor and air duct assembly.

27. Connect the negative battery cable.

CRANKSHAFT FRONT SEAL

REMOVAL & INSTALLATION

See Figures 58 and 59.

1. Before servicing the vehicle, refer to the Precautions Section.

2. Disconnect the negative battery cable.

3. Remove the front right-hand wheel and tire.

4. Remove the front right-hand splash guard.

5. Remove the drive belt. Refer to Accessory Drive Belt Removal & Installation.

6. Remove the crankshaft pulley. Refer to Crankshaft Damper Removal & Installation.

7. Remove the front oil seal using a suitable tool. Be careful not to damage front timing chain case and crankshaft.

To install:

8. Apply new engine oil to both oil seal lip and dust seal lip of new front oil seal.

9. Install the front oil seal. Install so that each seal lip is oriented as shown.

   a. Using a suitable drift, press-fit until the height of front oil seal is level with the mounting surface. Suitable drift:
   • Outer diameter: 2.36 in. (60 mm)
   • Inner diameter: 1.97 in. (50 mm)

   b. Check the garter spring is in position and seal lips not inverted.

☀ WARNING

Be careful not to damage front timing chain case and crankshaft.

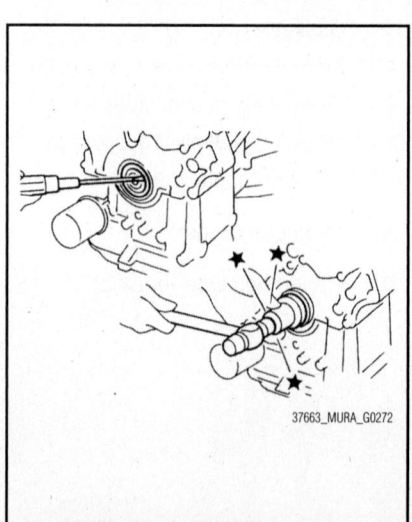

37663_MURA_G0272

Fig. 58 Remove front oil seal

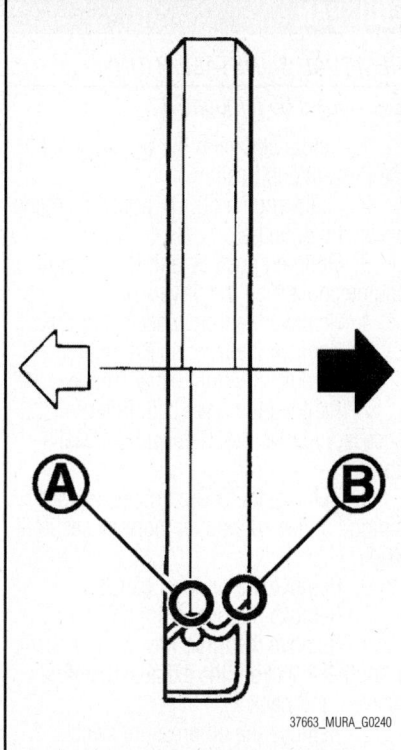

Fig. 59 Oil seal lip (A), dust seal lip (B), black arrow: engine outside, white arrow: engine inside

✳✳ WARNING

Press-fit straight and avoid causing burrs or tilting oil seal.

10. Install the crankshaft pulley.
11. Install the drive belt.
12. Install the splash guard.
13. Install the wheel and tire.
14. Refill fluids, as needed.
15. Connect the negative battery cable.
16. Check for fluid leaks.

CYLINDER HEAD

REMOVAL & INSTALLATION
See Figures 60 through 62.

1. Before servicing the vehicle, refer to the Precautions Section.
2. Disconnect the negative battery cable.
3. Remove the oil level gauge.
4. Remove the intake manifold collector.
5. Remove the rocker cover. Refer to Rocker Covers, Removal & Installation.
6. Remove the fuel tube and fuel injector assembly.
7. Remove the intake manifold. Refer to Intake Manifold Removal & Installation.
8. Remove the exhaust manifold. Refer to Combination Manifold Removal & Installation.

9. Remove the water inlet and thermostat assembly. Refer to Thermostat Removal & Installation in the Engine Cooling section.
10. Remove the water outlet, water connector, water bypass pipe, and heater pipe.
11. Remove the timing chain and rear timing chain case. Refer to Timing Chain & Sprockets, Removal & Installation.
12. Remove the camshaft. Refer to Camshaft & Valve Lifters, Removal & Installation.
13. Remove the cylinder head bolts in the reverse of the installation order shown below.
14. Remove the cylinder heads (Bank 1 and Bank 2).
15. Remove the cylinder head gaskets.

To install:

16. Install new cylinder head gaskets.
17. Turn the crankshaft until No. 1 piston is set at TDC. The crankshaft key should line up with the bank 1 cylinder center line as shown.

➡**Before installing the cylinder head, inspect cylinder head distortion.**

➡**If cylinder head bolts are being reused, check their outer diameters before installation.**

18. Install the cylinder head: follow the steps below to tighten the cylinder head bolts in numerical order as shown in the figure, using a cylinder head bolts wrench (commercial service tool).
   a. Apply new engine oil to the threads and seat surfaces of the cylinder head bolts.
   b. Tighten all cylinder head bolts to 72 ft. lbs. (98 Nm).

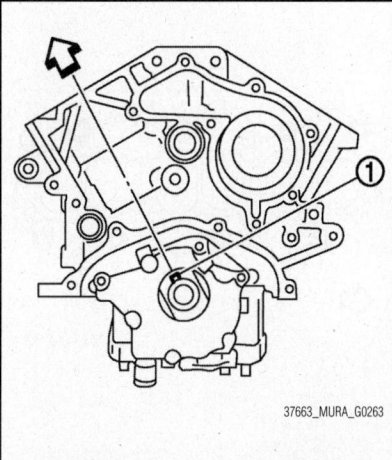

Fig. 60 The crankshaft key (1) should line up with the bank 1 cylinder center line

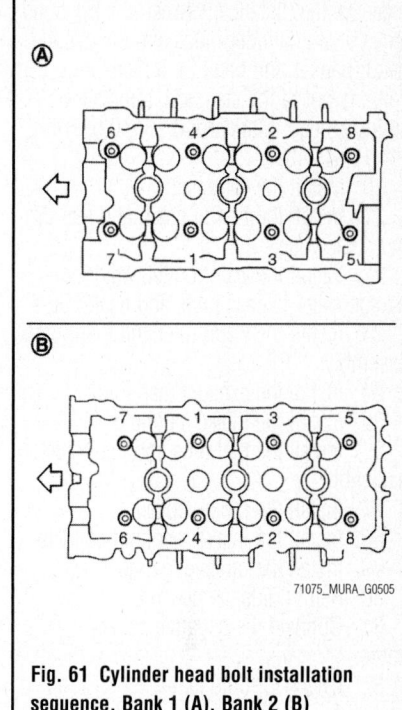

Fig. 61 Cylinder head bolt installation sequence, Bank 1 (A), Bank 2 (B)

   c. Completely loosen all cylinder head bolts, in the reverse order of that indicated in the figure.
   d. Tighten all cylinder head bolts to 29 ft. lbs. (39 Nm).
   e. Turn all cylinder head bolts 103 degrees clockwise (angle tightening). Check the tightening angle by using the angle wrench [SST: KV10112100 (BT8653-A)]. Check tightening angle indicated on the angle wrench indicator plate.
   f. Turn all cylinder head bolts 103 degrees clockwise again (angle tightening).

Fig. 62 Angle wrench (A)

19. After installing the cylinder head, measure the distance between the front end faces of the cylinder block and the cylinder head (bank 1 and bank 2). If measured value is out of the standard, reinstall the cylinder head. Standard: 0.555–0.587 in. (14.1–14.9 mm)

20. Install the camshaft.

21. Install the timing chain and rear timing chain case.

22. Install the water outlet, water connector, water bypass pipe, and heater pipe.

23. Install the water inlet and thermostat assembly.

24. Install the exhaust manifold.

25. Install the intake manifold.

26. Install the fuel tube and fuel injector assembly.

27. Install the rocker cover.

28. Install the intake manifold collector.

29. Install the oil level gauge.

30. Refill fluids, as needed.

31. Connect the negative battery cable.

32. Check for fluid leaks.

ENGINE COVER

REMOVAL & INSTALLATION

See Figure 63.

1. Before servicing the vehicle, refer to the Precautions Section.

2. Disconnect the negative battery cable.

3. Remove the radiator core support covers and air duct (inlet).

4. Remove the engine cover mounting bolts "A and B".

5. Draw and pull out the engine cover from the engine cover mounting bolts "C and D". Pull the engine cover from mounting bolt "D" holding by hand the position "E" as shown in the figure.

6. Remove the engine cover mounting bolts "C and D" if necessary.

7. Remove the engine cover.

To install:

Install in the reverse order of removal. Tighten the mounting bolts to 49 inch lbs. (6 Nm).

INTAKE MANIFOLD

REMOVAL & INSTALLATION

See Figure 64.

1. Before servicing the vehicle, refer to the Precautions Section.

2. Release the fuel pressure.

3. Disconnect the negative battery cable.

4. Remove the intake manifold collector.

5. Remove the fuel tube and fuel injector assembly.

6. Loosen the mounting nuts and bolts in the reverse of the installation order as shown below.

7. Remove the intake manifold.

8. Remove the gaskets.

✳✳ WARNING

Cover the engine openings to avoid entry of foreign materials.

To install:

9. Note the following, and install in the reverse order or removal.

a. If stud bolts were removed, install them and tighten to 8 ft. lbs. (11 Nm).

b. Tighten all mounting nuts and bolts to the specified torque in two or more steps in numerical order shown in the figure.

- 1st step: 60 inch lbs. (7 Nm)
- 2nd step and after: 19 ft. lbs. (26 Nm)

OIL PAN

REMOVAL & INSTALLATION

See Figures 65 through 72.

1. Before servicing the vehicle, refer to the Precautions Section.

2. Drain engine oil. Do not spill engine oil on the drive belt.

3. Drain engine coolant. Do not spill engine coolant on the drive belt.

4. Remove the front wheels and tires.

5. Remove the front splash guards.

6. Remove the front exhaust pipe.

7. Remove the drive belt. Refer to Accessory Drive Belt Removal & Installation.

8. Remove the A/C compressor with piping connected, and temporarily secure it aside.

9. Remove the oil level gauge.

10. Remove the right front halfshaft.

11. Remove the three way catalyst (bank 1 and bank 2) from the exhaust manifolds (bank 1 and bank 2).

12. Remove the oil pressure switch.

13. Remove the oil filter.

14. Remove the oil cooler and water pipes.

15. Support the transaxle assembly with a suitable jack. When setting the transmission jack, be careful not to allow it to collide against the drain plug.

16. Support front suspension member with a suitable jack.

17. Remove the rear engine mounting insulator.

18. Remove the left-hand engine mounting insulator mounting bolts from transaxle.

19. Remove the rear torque rod through bolts from rear torque rod bracket.

20. Remove the front suspension member stay and the mounting bolts and nuts.

21. Lower the jack for the front suspension member to the height.

22. Remove the transfer assembly.

23. Loosen the lower oil pan mounting bolts in the reverse of the installation order shown below.

24. Insert the seal cutter [SST: KV10111100 (J-37228)] between the upper and lower oil pan.

✳✳ WARNING

Be careful not to damage the mating surfaces. Never insert a screwdriver, this will damage the mating surfaces.

25. Slide the seal cutter by tapping on the side of tool with a hammer.

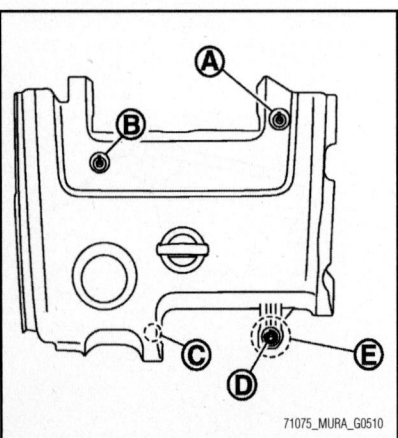

Fig. 63 Engine cover mounting bolts (A through D), holding position (E)

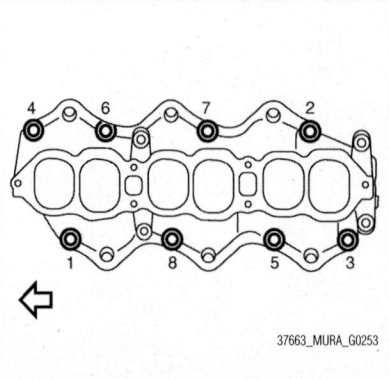

Fig. 64 Intake manifold bolt installation sequence

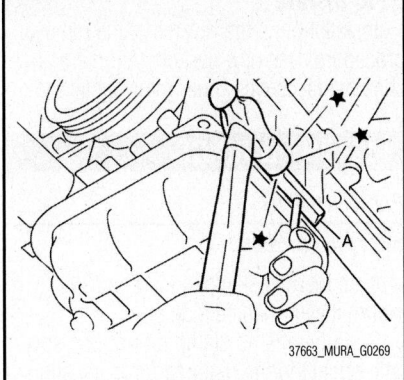

Fig. 65 Insert the seal cutter [SST: KV10111100 (J-37228)] (A) between the upper and lower oil pan

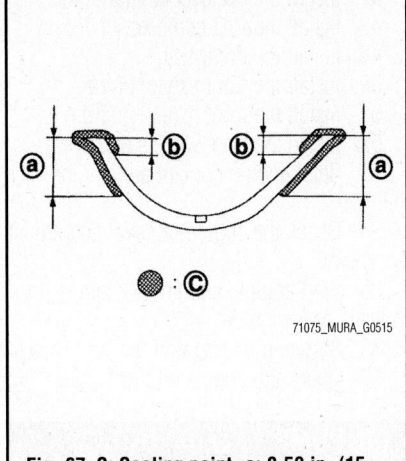

Fig. 67 C: Sealing point, a: 0.59 in. (15 mm), b: 0.20 in. (5 mm)

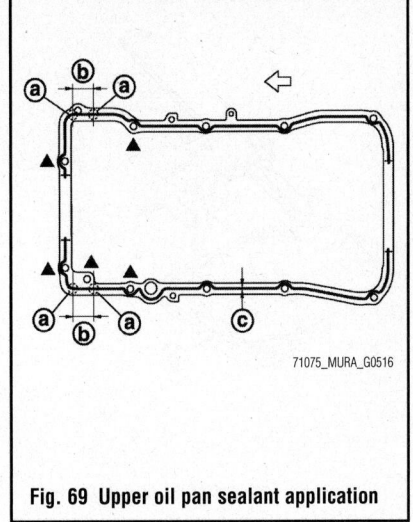

Fig. 69 Upper oil pan sealant application

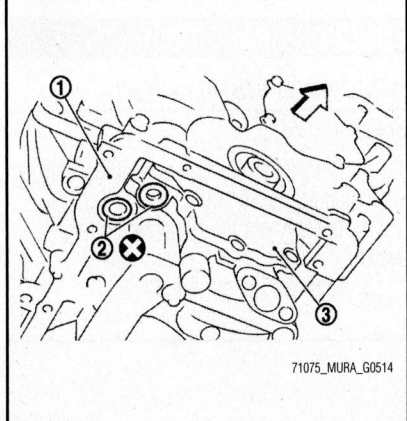

Fig. 66 Remove the O-rings (2) from bottom of cylinder block (1) and oil pump (3)

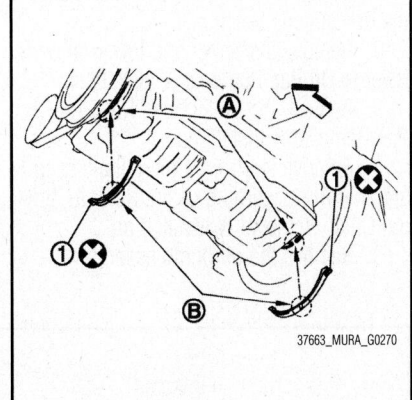

Fig. 68 Align protrusion (B) of oil pan gasket (1) with notches (A) of front timing chain case and rear oil seal retainer

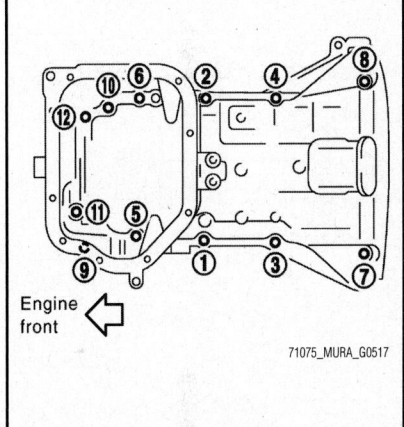

Engine front

Fig. 70 Upper oil pan bolt tightening sequence

26. Remove the lower oil pan.
27. Remove oil strainer.
28. Loosen the upper oil pan mounting bolts in the reverse of the installation order shown below.
29. Insert the seal cutter [SST: KV10111100 (J-37228)] between the upper oil pan and cylinder block. Slide seal cutter by tapping on the side of tool with a hammer.
30. Remove the upper oil pan.
31. Remove the O-rings from bottom of cylinder block and oil pump.
32. Remove the oil pan gaskets.

To install:
33. Install the upper oil pan:
 a. Use a scraper to remove old liquid gasket from upper oil pan mating surfaces. Remove old liquid gasket from mating surface of the cylinder block and the bolt holes and threads.
 b. Install new upper oil pan gaskets.

 c. Apply liquid gasket to new oil pan gaskets as shown. Use Genuine RTV Silicone Sealant or equivalent.
 d. To install, align the protrusion of the oil pan gasket with the notches of the front timing chain case and rear oil seal retainer.
 e. Install the oil pan gasket with smaller arc to the front of the timing chain case side.
 f. Install new O-rings on the bottom of the cylinder block and oil pump.
 g. Apply a continuous bead of sealant to the cylinder block mating surface of the upper oil pan to a limited portion as shown. Use RTV silicone sealant or equivalent. For bolt holes with marks (5 locations), apply liquid gasket outside the holes. Application:
 • a: 0.177–0.217 in. (4.5–5.5 mm) diameter bead
 • b: 1.38 in. (35 mm)
 • c: 0.138–0.177 in. (3.5–4.5 mm)

 h. Install the upper oil pan. Avoid misalignment of both O-rings.
 i. Tighten mounting bolts in numerical order as shown in the figure. There are three types of mounting bolts; refer to the following for locating bolts:
 • M8 × 135 mm (5.31 in.): 11
 • M8 × 92 mm (3.62 in.): 5, 7, 8
 • M8 × 25 mm (0.98 in.): Except the above
34. Install the oil strainer to the oil pump.
35. Install the lower oil pan:
 a. Use scraper to remove old liquid gasket from mating surfaces. Remove old liquid gasket from the bolt holes and thread.
 b. Apply a continuous bead of liquid gasket to the lower oil pan. Use Genuine RTV Silicone Sealant or equivalent.

➡**Attaching should be done within 5 minutes after coating.**

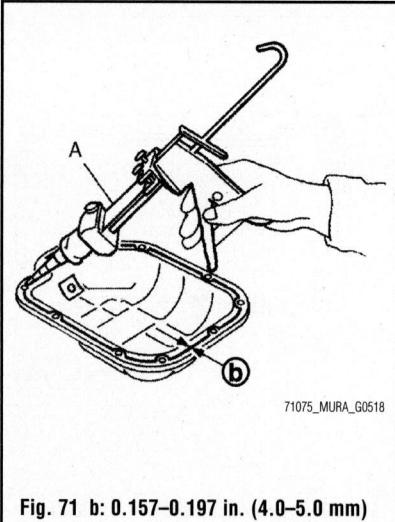

Fig. 71 b: 0.157–0.197 in. (4.0–5.0 mm)

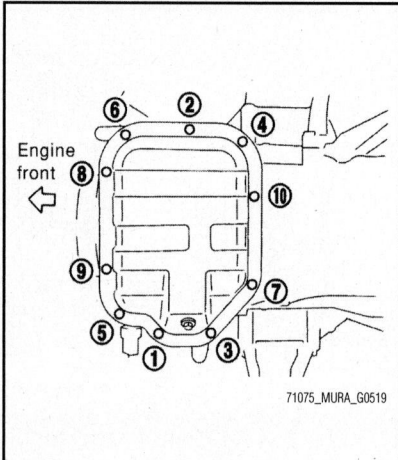

Engine front

Fig. 72 Lower oil pan bolt tightening sequence

   c. Install the lower oil pan. Install the mounting bolts in numerical order as shown in the figure and tighten to 78 inch lbs. (9 Nm).

36. Install the oil pan drain plug. Wait for at least 30 minutes after oil pan is installed to add engine oil.

37. Install the transfer assembly.

38. Install the front suspension member stay and the mounting bolts and nuts.

39. Install the rear torque rod through bolts to rear torque rod bracket.

40. Install the left-hand engine mounting insulator mounting bolts to the transaxle.

41. Install the rear engine mounting insulator.

42. Install the oil cooler and water pipes.

43. Install the oil filter.

44. Install the oil pressure switch.

45. Install the three way catalyst (bank 1 and bank 2) to the exhaust manifolds (bank 1 and bank 2).

46. Install the right front halfshaft.

47. Install the oil level gauge.

48. Install the A/C compressor.

49. Install the drive belt.

50. Install the front exhaust pipe.

51. Install the front splash guards.

52. Install the front wheels and tires.

53. Refill engine coolant and engine oil.

54. Check the engine oil level and adjust engine oil.

55. Start engine, and check there is no leakage of engine oil.

56. Stop engine and wait for 10 minutes.

57. Check the engine oil level again.

OIL PUMP

REMOVAL & INSTALLATION

See Figure 73.

1. Before servicing the vehicle, refer to the Precautions Section.

2. Remove the upper and lower oil pans. Refer to Oil Pan Removal & Installation.

3. Remove the oil strainer. Refer to Oil Pan Removal & Installation.

4. Remove the front timing chain case and the primary timing chain. Refer to Timing Chain Removal & Installation.

5. Remove the oil pump assembly.

To install:

Installation is the reverse of the removal procedure. Use new gaskets. Align crankshaft flat faces with inner rotor flat faces.

PISTONS & RINGS

POSITIONING

See Figures 74 and 75.

1. If there is stamped mark on ring, mount it with marked side up.

2. If there is no stamp on ring, no specific orientation is required for installation.

3. Stamped mark:

4. Top ring (A): —

5. Second ring (B): 2R

6. Position each ring with the gap as shown in the figure referring to the piston front mark.

ROCKER ARM COVER

REMOVAL & INSTALLATION

See Figure 76.

1. Before servicing the vehicle, refer to the Precautions Section.

2. Disconnect the negative battery cable.

3. Remove the engine cover.

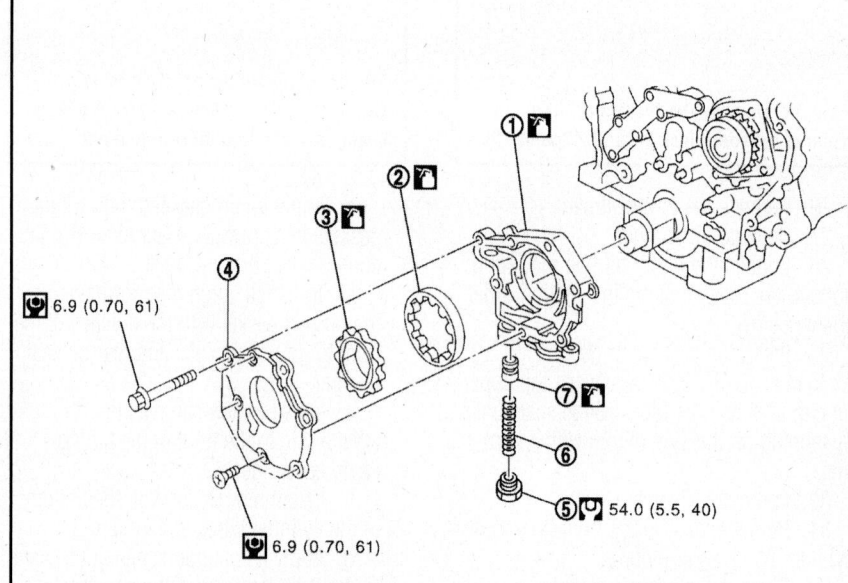

6.9 (0.70, 61)

6.9 (0.70, 61)

54.0 (5.5, 40)

1.	Oil pump body	2.	Oil pump outer rotor	3.	Oil pump inner rotor
4.	Oil pump cover	5.	Regulator valve plug	6.	Regulator valve spring
7.	Regulator valve				

Fig. 73 Oil pump

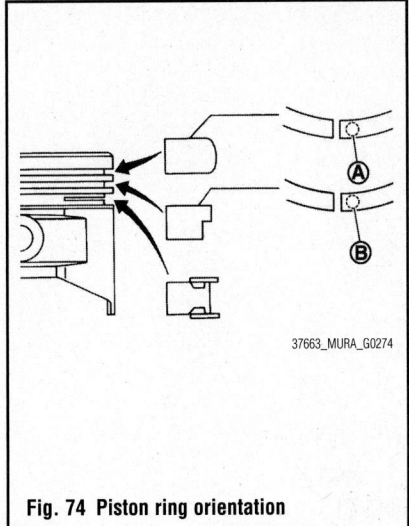

Fig. 74 Piston ring orientation

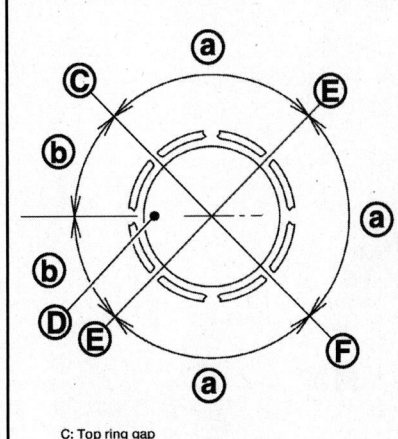

C: Top ring gap
E: Oil ring upper or lower rail gap (either of then)
F: Second ring and oil ring spacer gap
a: 90 degrees
b: 45 degrees

37663_MURA_G0275

Fig. 75 Position each ring with the gap as shown in the figure referring to the piston front mark (D)

4. Remove the upper and lower air cleaner cases and the air duct assembly. Refer to Air Cleaner Removal & Installation.

5. Remove the intake manifold collector.

6. Disconnect the PCV hose from the valve cover.

7. Remove the camshaft position sensor (PHASE) (Bank 1 and Bank 2).

➡Handle carefully to avoid dropping and shocks. Never disassemble Never allow metal powder to adhere to magnetic part at sensor tip. Never place sensors in a location where they are exposed to magnetism.

8. Remove the PCV valve and O-ring from the valve cover, if necessary.

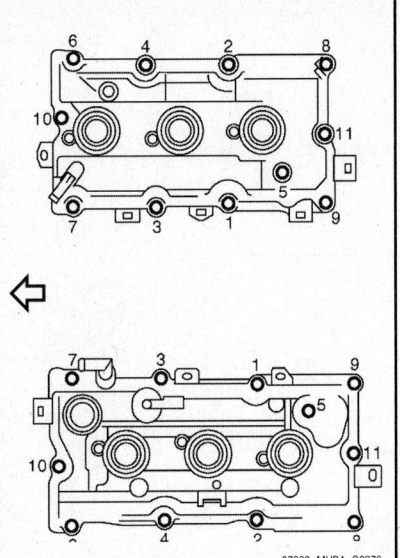

37663_MURA_G0276

Fig. 76 Valve cover bolt installation sequence

9. Remove the oil filler cap from the valve cover, if necessary.

10. Remove the ignition coil.

11. Remove the harness clips on the valve cover.

12. Loosen bolts in the reverse of the installation order.

13. Remove the valve cover gasket from valve cover.

To install:

14. Install the valve cover gasket.

15. Tighten the bolts in two steps separately in numerical order as shown in the figure.

- Step 1: 17 inch lbs. (2 Nm)
- Step 2: 74 inch lbs. (8 Nm)

16. Reconnect the harness clips.

17. Install the ignition coil.

18. Install the oil filler cap.

19. Install the PCV valve and O-ring.

20. Install the camshaft position sensor.

21. Connect the PCV hose.

22. Install the intake manifold collector.

23. Install the air cleaner cases and the air duct assembly.

24. Install the engine cover.

25. Connect the negative battery cable.

TIMING CHAIN COVER, CHAIN, TENSIONER, & SPROCKETS

REMOVAL & INSTALLATION

See Figures 77 through 85.

1. Before servicing the vehicle, refer to the Precautions Section.

2. Disconnect the negative battery cable.

3. Drain the engine oil.

4. Drain the engine coolant from inside the engine.

5. Remove the intake manifold collector.

6. Remove the rocker covers (Bank 1 and Bank 2).

7. Remove the upper and lower oil pans and oil strainer.

8. Remove the drive belt, idler pulleys and bracket.

9. Remove the power steering oil pump with piping connected, and temporarily secure it aside.

10. Separate the engine harness brackets from the front timing chain case.

11. Remove the valve timing control covers.

a. Loosen the mounting bolts in the reverse of the installation order shown below.

✳✳ WARNING

Shaft is internally jointed with intake camshaft sprocket center hole. When removing, keep it horizontal until it is completely disconnected.

12. Position the No. 1 cylinder at TDC of its compression stroke as follows:

a. Rotate the crankshaft pulley clockwise to align the timing mark (grooved line without color) with the timing indicator.

13. Check that the intake and exhaust cam noses on the No. 1 cylinder (engine front side of Bank 1) are located as shown. If not, turn crankshaft one revolution (360 degrees) and align as shown.

14. Remove the crankshaft pulley as follows:

a. Hold the crankshaft with a pulley holder.

b. Loosen the crankshaft pulley bolt and locate the bolt seating surface at 0.39 in. (10 mm) from its original position.

✳✳ WARNING

Never remove the crankshaft pulley bolt as it will be used as a supporting point for a suitable puller.

c. Place a suitable puller tab on the holes of crankshaft pulley, and pull the crankshaft pulley through.

✳✳ WARNING

Never put the suitable puller tab on the crankshaft pulley periphery, as this will damage internal damper.

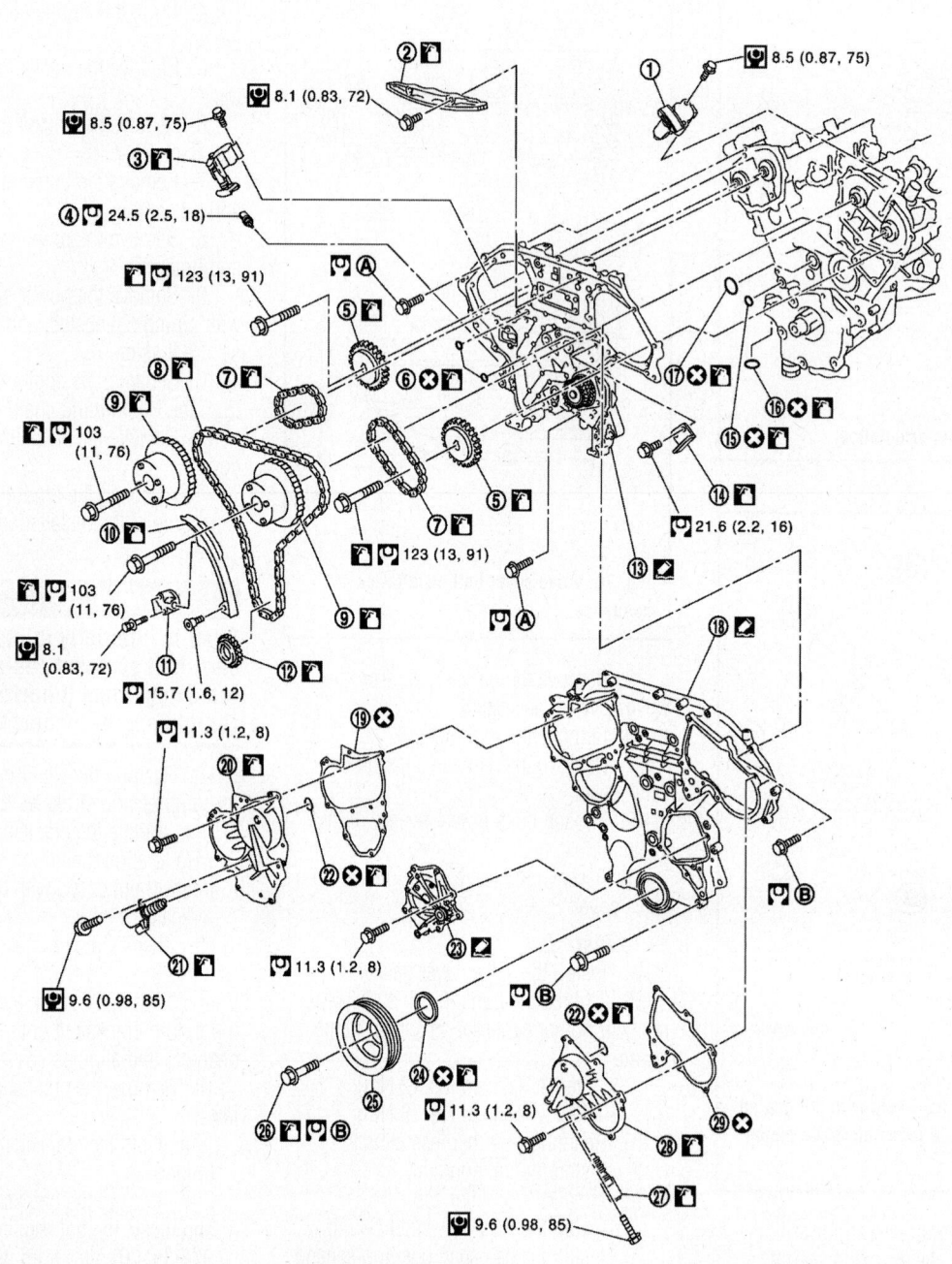

8.5 (0.87, 75)

8.1 (0.83, 72)

8.5 (0.87, 75)

24.5 (2.5, 18)

123 (13, 91)

103 (11, 76)

103 (11, 76)

8.1 (0.83, 72)

15.7 (1.6, 12)

11.3 (1.2, 8)

9.6 (0.98, 85)

123 (13, 91)

21.6 (2.2, 16)

11.3 (1.2, 8)

11.3 (1.2, 8)

9.6 (0.98, 85)

1. Secondary timing chain tensioner (Bank 2)
2. Internal chain guide
3. Secondary timing chain tensioner (Bank 1)
4. Oil temperature sensor
5. Camshaft sprocket (Exhaust)
6. O-ring
7. Secondary timing chain
8. Primary timing chain
9. Camshaft sprocket (Intake)
10. Slack guide

11. Primary timing chain tensioner
12. Crankshaft sprocket
13. Rear timing chain case
14. Tension guide
15. O-ring
16. O-ring
17. O-ring
18. Front timing chain case
19. Valve timing control cover gasket (Bank 1)
20. Valve timing control cover (Bank 1)

21. Intake valve timing control solenoid valve (Bank 1)
22. Seal ring
23. Water pump cover
24. Front oil seal
25. Crankshaft pulley
26. Crankshaft pulley bolt
27. Intake valve timing control solenoid valve (Bank 2)
28. Valve timing control cover (Bank 2)
29. Valve timing control cover gasket (Bank 2)

37663_MURA_G0224

Fig. 77 Timing chain and related components

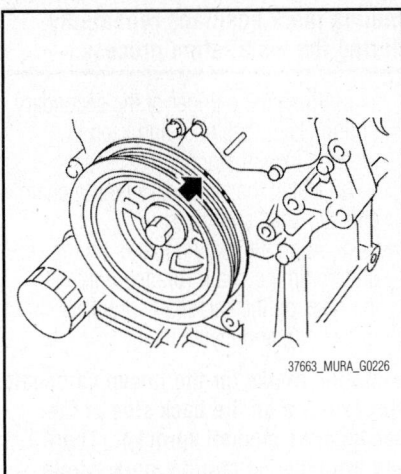

Fig. 78 Align the timing mark with timing indicator

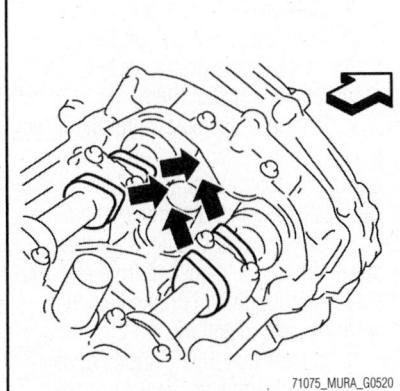

Fig. 79 Intake and exhaust cam noses on No. 1 cylinder (engine front side of bank 1)

15. Remove the front timing chain case as follows:

    a. Loosen the mounting bolts in the reverse of the installation order shown below.

    b. Insert a suitable tool into the notch at the top of the front timing chain case as shown.

    c. Pry off the case by moving the tool as shown.

    d. Use the seal cutter [SST: KV10111100 (J-37228)] to cut the liquid gasket for removal.

✲✲ WARNING

Never use screwdrivers or something similar. After removal, handle front timing chain case carefully so it does not tilt, cant, or warp under a load.

16. Remove the water pump cover from the front timing chain case.

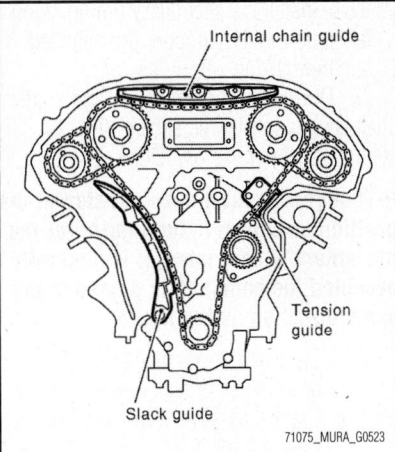

Fig. 80 Internal chain guide, tension guide and slack guide

    a. Use the seal cutter [SST: KV10111100 (J-37228)] to cut liquid gasket for removal.

17. Remove the front oil seal from the front timing chain case using a suitable tool.

18. Remove the O-ring from the rear timing chain case.

19. Remove the primary timing chain tensioner as follows:

    a. Remove the lower mounting bolt.

    b. Loosen the upper mounting bolt slowly, and turn the primary timing chain tensioner on the mounting bolt so that the plunger is fully expanded.

➥**Even if the plunger is fully expanded, it is not dropped from the body of the primary timing chain tensioner.**

    c. Remove the upper mounting bolt, and remove the primary timing chain tensioner.

20. Remove the internal chain guide, tension guide and slack guide.

➥**The tension guide can be removed after removing the primary timing chain.**

21. Remove the primary timing chain and the crankshaft sprocket.

✲✲ WARNING

After removing the primary timing chain, never turn the crankshaft and camshaft separately, or valves will strike the piston heads.

22. Remove the secondary timing chain and camshaft sprockets as follows:

    a. Attach a suitable stopper pin to the Bank 1 and Bank 2 the secondary timing chain tensioners.

➥**Use an approximately 0.02 in. (0.5 mm) diameter hard metal pin as a stopper pin.**

    b. For removal of the secondary timing chain tensioner, refer to Camshaft & Valve Lifters, Removal & Installation (removing the No. 1 camshaft bracket is required.)

    c. Remove the camshaft sprocket mounting bolts.

    d. Secure the hexagonal portion of the camshaft using a wrench to loosen the mounting bolts.

✲✲ WARNING

Never loosen the mounting bolts with securing anything other than the camshaft hexagonal portion or with tensioning the timing chain.

    e. Remove the secondary timing chain together with the camshaft sprockets.

    f. Turn the camshaft slightly to secure slackness of the timing chain on the secondary timing chain tensioner side.

    g. Insert a 0.020 in. (0.5 mm) thick metal or resin plate between the timing chain and the timing chain tensioner plunger (guide). Remove the secondary timing chain together with the camshaft sprockets with the timing chain loose from the guide groove.

✲✲ WARNING

Be careful of the plunger coming off when removing the secondary timing chain.

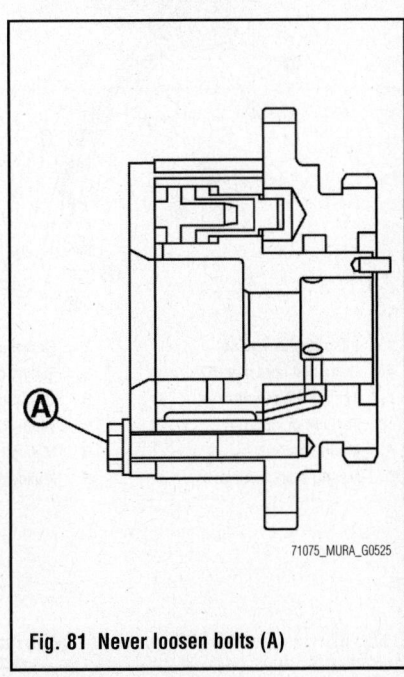

Fig. 81 Never loosen bolts (A)

→The intake camshaft sprocket is two-for-one structure of sprockets for the primary timing chain and for the secondary timing chain.

✳✳ WARNING

Handle the camshaft sprocket carefully to avoid any shock and never disassemble. Never loosen bolts as shown.

23. Remove the secondary timing chain tensioners from the cylinder head as follows, if necessary.

  a. Remove the secondary timing chain tensioners with a stopper pin attached.

24. Use a scraper to remove all traces of old liquid gasket from the front timing chain case, and opposite mating surfaces. Remove old liquid gasket from the bolt hole and thread.

25. Use a scraper to remove all traces of old liquid gasket from the water pump cover.

To install:

26. If removed, install the secondary timing chain tensioners to cylinder head as follows.

  a. Install the secondary timing chain tensioners with a stopper pin attached and new O-ring.

27. Check that the dowel pin and crankshaft key are located as shown. (No. 1 cylinder at compression TDC)

→Though camshaft does not stop at the position as shown in the figure, for the placement of cam nose, it is generally accepted the camshaft is placed in the same direction as the figure.

  • Camshaft dowel pin: At cylinder head upper face side in each bank
  • Crankshaft key: At cylinder head side of Bank 1

✳✳ WARNING

The hole on small diameter side must be used for intake side dowel pin hole. Never misidentify (ignore the big diameter side).

28. Install the secondary timing chain and camshaft sprockets as follows:

✳✳ WARNING

Mating marks between timing chain and sprockets slip easily. Confirm all mating mark positions repeatedly during the installation process.

  a. Push the plunger of the secondary timing chain tensioner and keep it pressed in with stopper pin.

  b. Install the secondary timing chain and camshaft sprockets.

  c. Align the mating marks on the secondary timing chain (orange link) with the ones on the camshaft sprockets (punched), and install them.

→Mating marks for the intake camshaft sprocket are on the back side of the secondary camshaft sprocket. There are two types of mating mark, circle and oval types. They should be used for the bank 1 and bank 2, respectively.

  • Bank 1: Use circle type
  • Bank 2: Use oval type

  d. Align the dowel pin on the camshafts with the groove or hole on the sprockets, and install them.

  e. On the intake side, align the dowel pin on the camshaft front end with the dowel pin hole on the back side of the camshaft sprocket, and install them.

  f. On the exhaust side, align the dowel pin on the camshaft front end with the dowel pin groove on the camshaft sprocket, and install them.

  g. In case that the positions of each mating mark and each dowel pin are not fit on mating parts, make fine adjustment to the position by holding the hexagonal portion on camshaft with wrench or equivalent.

  h. Mounting bolts for camshaft sprockets must be tightened in the next step. Tightening them by hand is enough to prevent the dislocation of dowel pins.

  i. It may be difficult to visually check the dislocation of mating marks during and after installation. To make the matching easier, make a mating mark on the top of sprocket teeth and its extended line in advance with paint.

  j. After confirming the mating marks are aligned, tighten the camshaft sprocket mounting bolts. Secure the camshaft using a wrench at the hexagonal portion to tighten the mounting bolts.

  k. Pull the stopper pins out from the secondary timing chain tensioners.

29. Install the tension guide.

30. Install the primary timing chain as follows:

  a. Install crankshaft sprocket.

  b. Check the mating marks on the crankshaft sprocket.

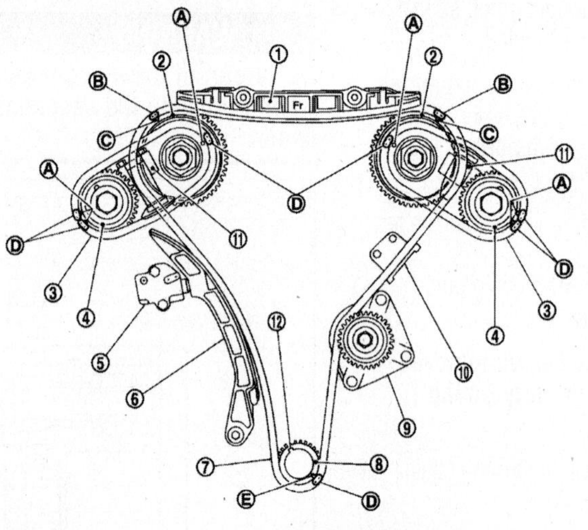

1.	Internal chain guide	2.	Camshaft sprocket (INT)	3.	Timing chain (secondary)
4.	Camshaft sprocket (EXH)	5.	Timing chain tensioner (primary)	6.	Slack guide
7.	Timing chain (primary)	8.	Crankshaft sprocket	9.	Water pump
10.	Tension guide	11.	Timing chain tensioner (secondary)	12.	Crankshaft key
A.	Mating mark	B.	Mating mark (pink link)	C.	Mating mark (punched)
D.	Mating mark (orange)	E.	Mating mark (notched)		

37663_MURA_G0232

Fig. 82 Timing chain and sprockets mating marks

c. Install the primary timing chain.

d. Install the primary timing chain so the mating mark (punched) on the intake camshaft sprocket is aligned with the pink link on the timing chain, while the mating mark (notched) on the crankshaft sprocket is aligned with the orange link on the timing chain, as shown.

e. When it is difficult to align the mating marks of the primary timing chain with each sprocket, gradually turn the camshaft using a wrench on the hexagonal portion to align it with the mating marks.

f. During alignment, be careful to prevent dislocation of the mating mark alignments of the secondary timing chains.

31. Install the internal chain guide and slack guide.

❊❊ WARNING

Never over-tighten the slack guide mounting bolt. It is normal for a gap to exist under the bolt seat when the mounting bolt is tightened to specification.

32. Install the primary timing chain tensioner with the following procedure:

a. Pull the plunger stopper tab up (or turn lever downward) so as to remove the plunger stopper tab from the ratchet of plunger. Plunger stopper tab and lever are synchronized.

b. Push the plunger into the inside of the tensioner body.

c. Hold the plunger in the fully compressed position by engaging the plunger stopper tab with the tip of the ratchet.

d. To secure the lever, insert the stopper pin through the hole of the lever into the tensioner body hole. The lever parts and the tab are synchronized. Therefore, this will secure the plunger.

e. Install the primary timing chain tensioner. Remove any dirt and foreign materials completely from the back and the mounting surfaces of the primary timing chain tensioner.

f. Pull out the stopper pin after installing, and release the plunger.

33. Check again that the mating marks on each sprocket and each timing chain have not slipped out of alignment.

34. Install new O-rings on rear timing chain case.

35. Install new front oil seal on the front timing chain case.

a. Apply new engine oil to both the oil seal lip and the dust seal lip.

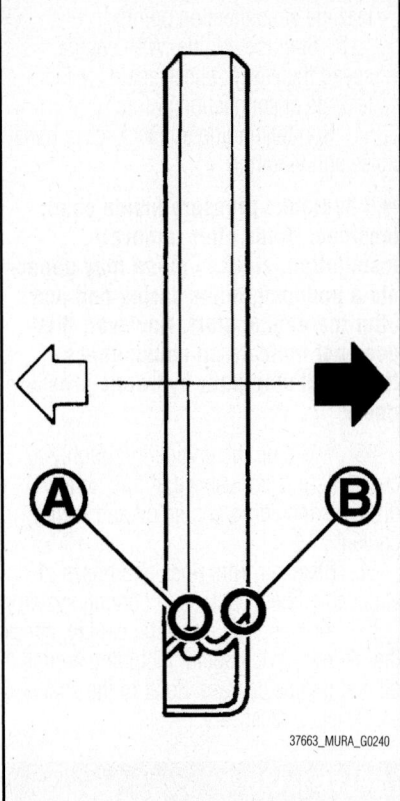

Fig. 83 Oil seal lip (A), dust seal lip (B), black arrow: engine outside, white arrow: engine inside

b. Install it so that each seal lip is oriented as shown.

c. Using a suitable drift [outer diameter: 2.36 in. (60mm)], press-fit the oil seal until it becomes flush with the front timing chain case end face.

d. Check the garter spring is in position and the seal lip is not inverted.

36. Install the water pump cover to the front timing chain case.

a. Apply a continuous bead of liquid gasket to the water pump cover as shown. Use Genuine RTV Silicone Sealant or equivalent.

37. Install the front timing chain case as follows:

a. Apply a continuous bead of liquid gasket to the front timing chain case back side as shown. Use Genuine RTV Silicone Sealant or equivalent.

38. Install the front timing chain case as to fit its dowel pin holes together with the dowel pin on the rear timing chain case.

a. Tighten the mounting bolts to the specified torque in numerical order as shown. There are two types of mounting bolt. Refer to the following for locating the bolts:

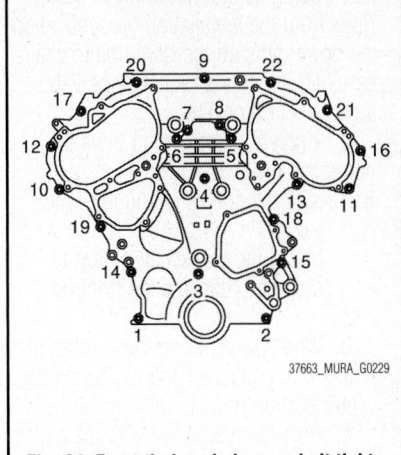

Fig. 84 Front timing chain case bolt tightening sequence

- M8 bolts (No. 1, 2): 21 ft. lbs. (28 Nm)
- M6 bolts (Except No. 1, 2): 9 ft. lbs. (13 Nm)

b. After all bolts are tightened, retighten them to the specified torque in numerical order as shown.

➥**Wipe off any excessive liquid gasket.**

c. After installing the front timing chain case, check the surface height difference between the following parts on the oil pan (upper) mounting surface: Front timing chain case to rear timing chain case: −0.006 to 0.006 in (−0.14 to 0.14 mm).

d. If not within the standard, repeat the installation procedure.

39. Install the intake valve timing control covers as follows:

a. Install new seal rings in shaft grooves.

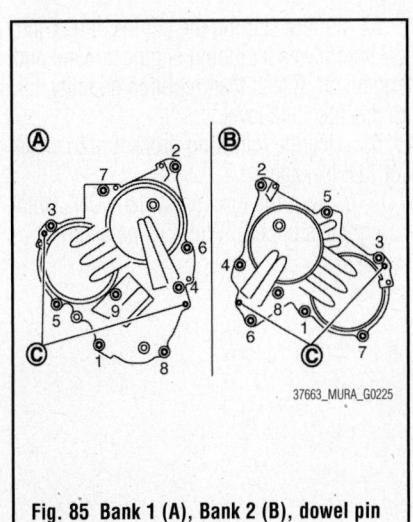

Fig. 85 Bank 1 (A), Bank 2 (B), dowel pin hole (C)

b. Being careful not to move seal rings from the installation grooves, align the dowel pins on the front timing chain case with the holes to install the valve timing control covers.

c. Tighten the mounting bolts in numerical order as shown.

40. Install the crankshaft pulley as follows:

a. Install the crankshaft pulley, taking care not to damage the front oil seal.

b. When press-fitting the crankshaft pulley with a plastic hammer, tap on its center portion (not circumference).

c. Hold the crankshaft with the pulley holder.

d. Tighten the crankshaft pulley bolt to 33 ft. lbs. (44 Nm).

e. Place a paint mark on the crankshaft pulley aligning with the angle mark on the crankshaft pulley bolt. Tighten the bolt 90 degrees (angle tightening).

41. Rotate the crankshaft pulley in normal direction (clockwise when viewed from engine front) to confirm it turns smoothly.

42. Install the valve timing control covers.

43. Install the engine harness brackets and connect the harnesses.

44. Install the power steering oil pump and piping.

45. Install the drive belt, idler pulleys and bracket.

46. Install the oil pans (lower and upper) and oil strainer.

47. Install the rocker covers.

48. Install the intake manifold collector.

49. Refill engine coolant.

50. Refill engine oil.

51. Connect the negative battery cable.

52. Before starting the engine, check the oil/fluid levels including engine coolant and engine oil. If less than required quantity, fill to the specified level.

53. Use the following procedure to check for fuel leakage:

a. Turn the ignition switch "ON" (with engine stopped). With fuel pressure applied to fuel piping, check for fuel leakage at connection points.

b. Start the engine. With engine speed increased, check again for fuel leakage at connection points.

54. Run the engine to check for unusual noise and vibration.

➡ **If hydraulic pressure inside chain tensioner drops after removal/ installation, slack in guide may generate a pounding noise during and just after the engine start. However, this does not indicate an unusualness. Noise will stop after hydraulic pressure rises.**

55. Warm up the engine thoroughly to check there is no leakage of fuel, or any oil/fluids including engine oil and engine coolant.

56. Bleed air from lines and hoses of applicable lines, such as in cooling system.

57. After cooling down the engine, check the oil/fluid levels again, including engine oil and engine coolant. Refill to the specified level, if necessary.

VALVE LASH

ADJUSTMENT

See Figure 86.

➡ **Perform adjustment depending on selected head thickness of valve lifter.**

1. Measure the valve clearance.
2. Remove the camshaft.
3. Remove the valve lifters at the locations that are out of the standard.
4. Measure the center thickness of the removed valve lifters with a micrometer.
5. Use the equation below to calculate valve lifter thickness for replacement.

- Valve lifter thickness calculation: $t = t1 + (C1 - C2)$
- t = Valve lifter thickness to be replaced
- $t1$ = Removed valve lifter thickness
- $C1$ = Measured valve clearance
- $C2$ = Standard valve clearance: Intake: 0.012 in. (0.30 mm), Exhaust: 0.013 in. (0.33 mm)

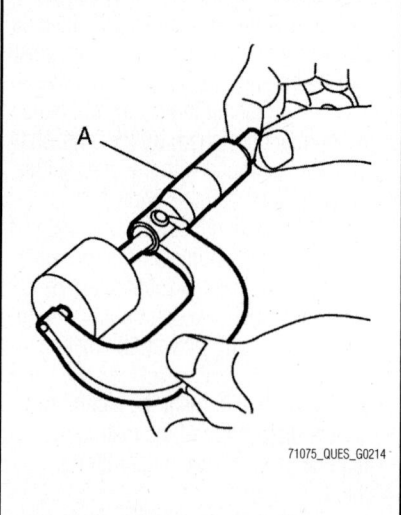

71075_QUES_G0214

Fig. 86 Measure the center thickness of the removed lifter with a micrometer (A)

➡ **Thickness of new valve lifter can be identified by stamp marks on the reverse side (inside the cylinder). Stamp mark 788P indicates 0.3102 in. (7.88 mm) in thickness.**

a. Available thickness of valve lifter: 27 sizes with range 0.3102 to 0.3307 in. (7.88 to 8.40 mm) in steps of 0.0008 in. (0.02 mm) (when manufactured at factory).

6. Install the selected valve lifter.

7. Install the camshaft.

8. Manually turn the crankshaft pulley a few turns.

9. Check that the valve clearances for cold engine are within the specifications by referring to the specified values.

a. Intake valve clearance (cold) is 0.010–0.013 in. (0.26–0.34mm) and exhaust valve clearance (cold) is 0.011–0.015 in. (0.29–0.37mm).

10. Install all removal parts in the reverse order of removal.

11. Warm up the engine, and check for unusual noise and vibration.

ENGINE PERFORMANCE & EMISSION CONTROLS

COMPONENT LOCATIONS

See Figures 87 through 90.

ACCELERATOR PEDAL POSITION (APP) SENSOR

LOCATION

See Figure 91.

REMOVAL & INSTALLATION

See Figure 92.

1. Before servicing the vehicle, refer to the Precautions Section.
2. Disconnect the negative battery cable.
3. Disconnect the accelerator pedal position sensor harness connector.
4. Remove the accelerator pedal assembly.

❋❋ WARNING

Never disengage the accelerator pedal assembly and bracket. Never disassemble the accelerator lever. Never remove the accelerator pedal position sensor from accelerator lever. Avoid impact from dropping etc. during handling. Be careful to keep the accelerator lever away from water.

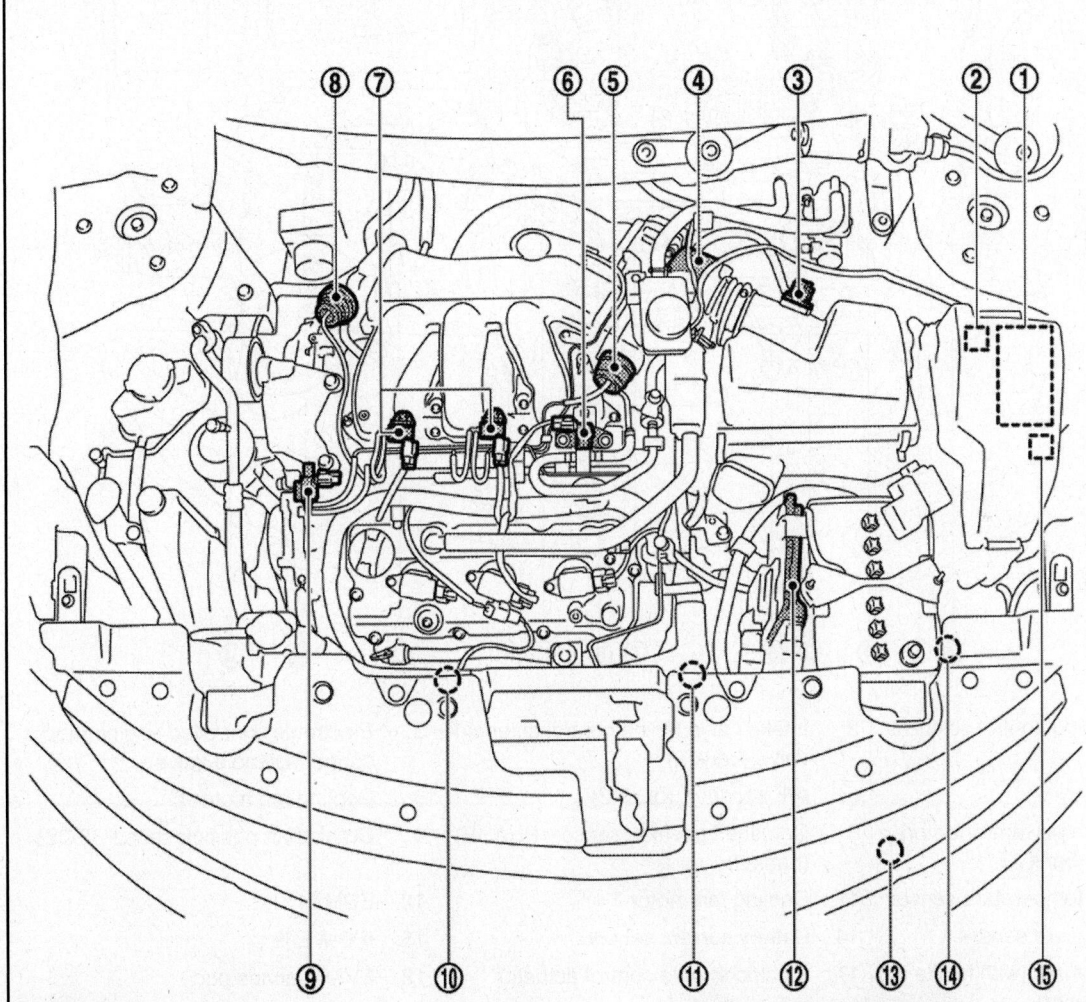

1. IPDM E/R
2. Cooling fan motor relay
3. Mass air flow sensor (with intake air temperature sensor)
4. Electric throttle control actuator
5. Power valve actuator 2
6. EVAP canister purge volume control solenoid valve
7. VIAS control solenoid valve 1 and 2
8. Power valve actuator 1
9. Electronic controlled engine mount control solenoid valve
10. Cooling fan motor-2
11. Cooling fan motor-1
12. ECM
13. Refrigerant pressure sensor
14. Battery current sensor
15. Cooling fan motor relay-2

71075_MURA_G0530

Fig. 87 Engine Performance & Emission Controls—Engine room component locations—CrossCabriolet models

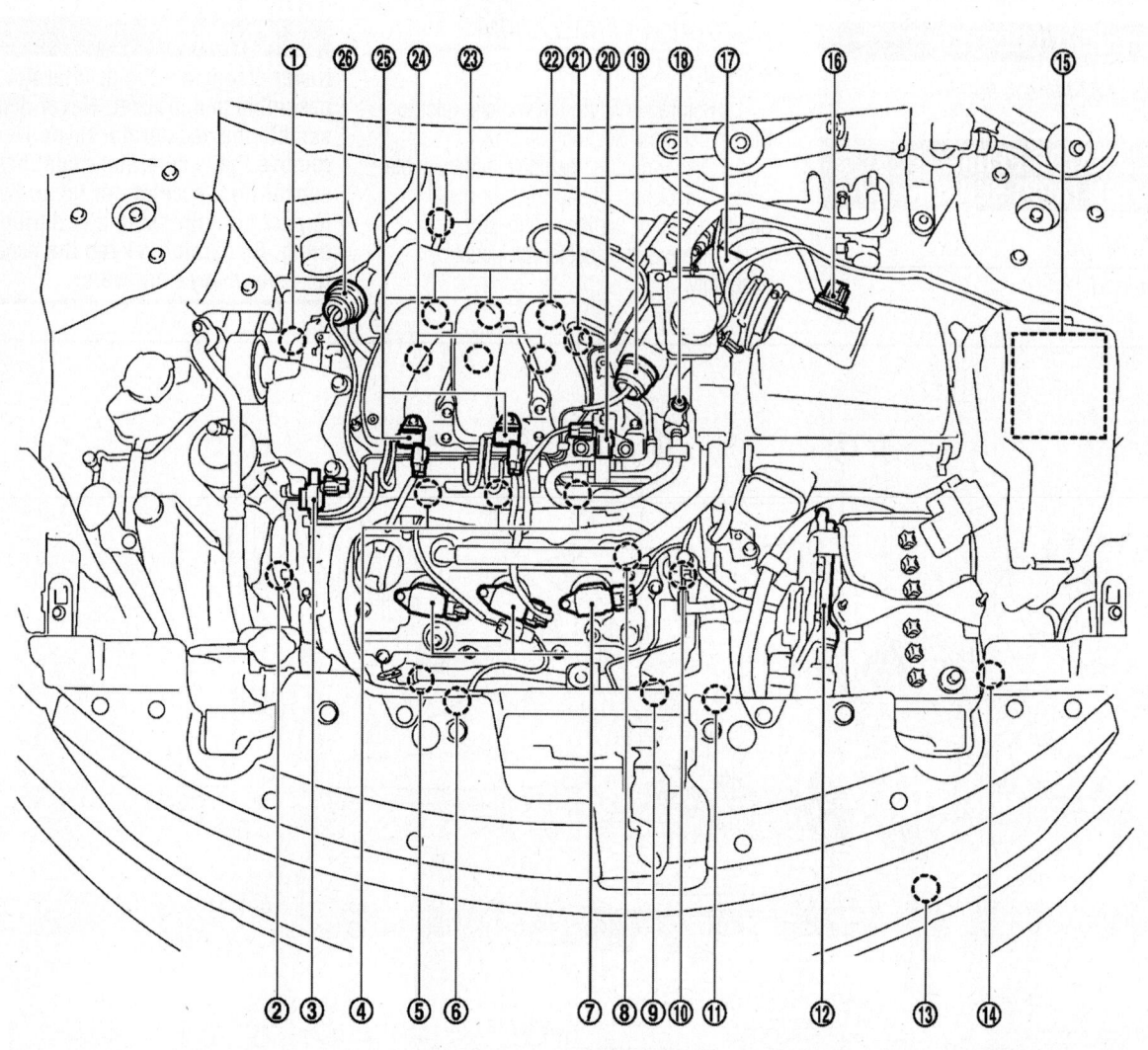

1. Intake valve timing control solenoid valve (bank 1)
2. Intake valve timing control solenoid valve (bank 2)
3. Electronic controlled engine mount control solenoid valve
4. Fuel injector (bank 2)
5. A/F sensor 1 (bank 2)
6. Cooling fan motor-2
7. Ignition coil (with power transistor) and spark plug (bank 2)
8. Camshaft position sensor (PHASE) (bank 2)
9. Crankshaft position sensor (POS)
10. Engine coolant temperature sensor
11. Cooling fan motor-1
12. ECM
13. Refrigerant pressure sensor
14. Battery current sensor
15. IPDM E/R
16. Mass air flow sensor (with intake air temperature sensor)
17. Electric throttle control actuator
18. EVAP service port
19. Power valve actuator 2
20. EVAP canister purge volume control solenoid valve
21. Camshaft position sensor (PHASE) (bank 1)
22. Ignition coil (with power transistor) and spark plug (bank 1)
23. A/F sensor 1 (bank 1)
24. Fuel injector (bank 1)
25. VIAS control solenoid valve 1 and 2
26. Power valve actuator 1

37663_MURA_G0278

Fig. 88 Engine Performance & Emission Controls—Engine room component locations—Hardtop models

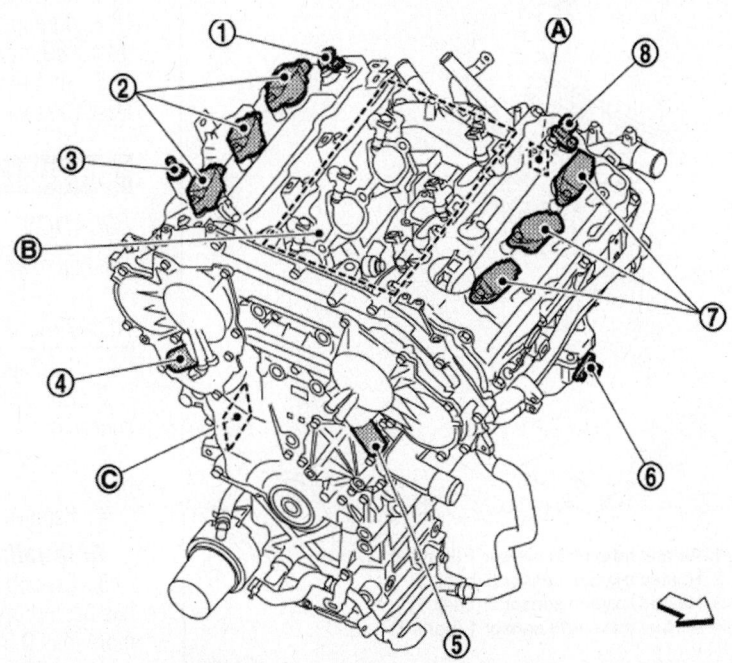

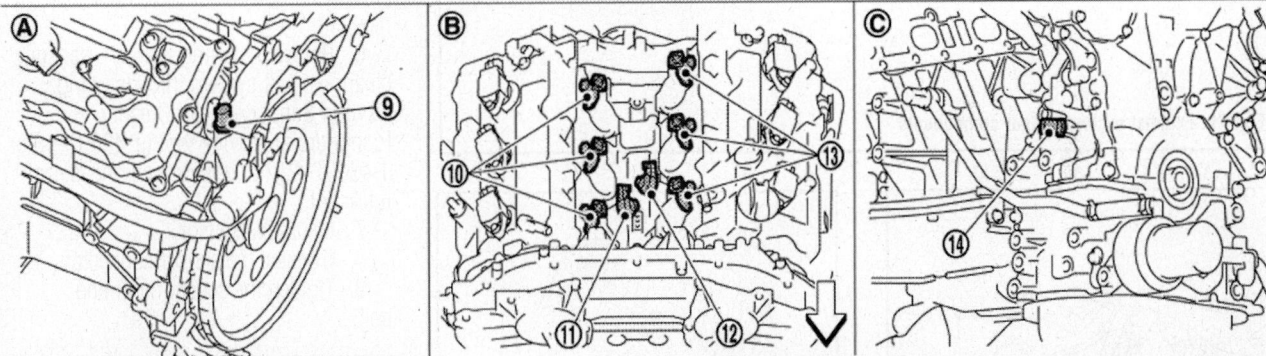

A. Engine rear upper-left
B. Engine top center
C. Engine front lower-right
1. Camshaft position sensor (PHASE) (bank 1)
2. Ignition coil (with power transistor) (bank 1)
3. PCV valve
4. Intake valve timing control solenoid valve (bank 1)
5. Intake valve timing control solenoid valve (bank 2)
6. Crankshaft position sensor
7. Ignition coil (with power transistor) (bank 2)

8. Camshaft position sensor (PHASE) (bank 2)
9. Engine coolant temperature sensor
10. Fuel injector (bank 1)
11. Knock sensor (bank 1)
12. Knock sensor (bank 2)
13. Fuel injector (bank 2)
14. Engine oil temperature sensor

71075_MURA_G0531

Fig. 89 Engine sensor components—CrossCabriolet models

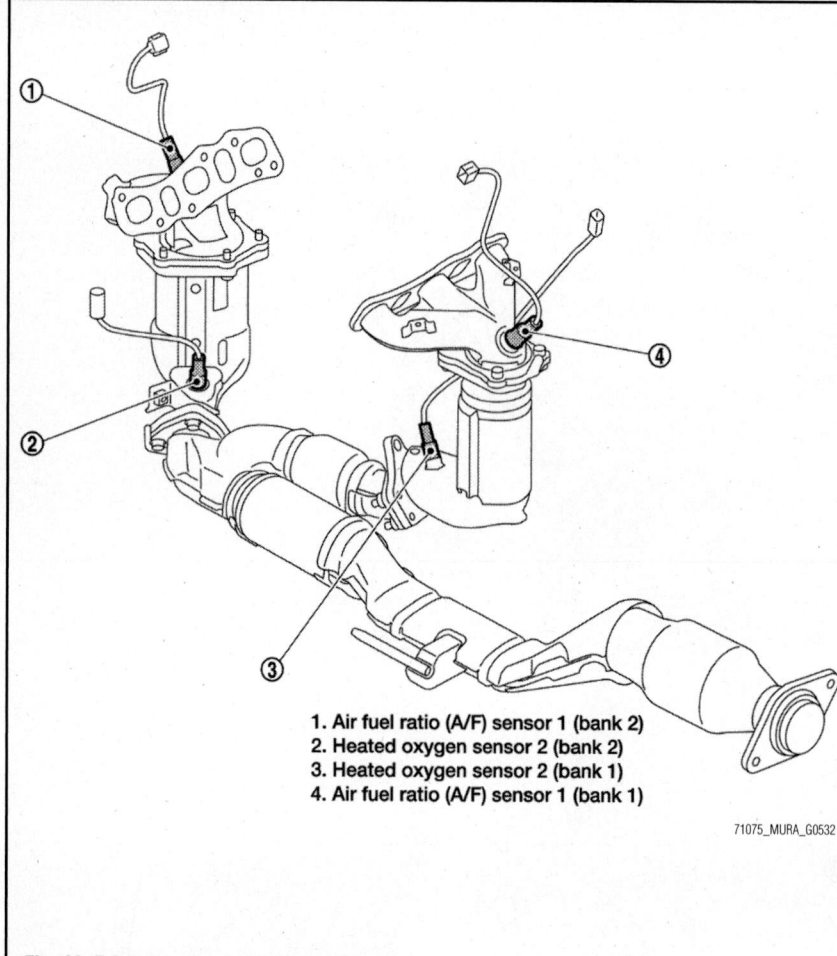

1. Air fuel ratio (A/F) sensor 1 (bank 2)
2. Heated oxygen sensor 2 (bank 2)
3. Heated oxygen sensor 2 (bank 1)
4. Air fuel ratio (A/F) sensor 1 (bank 1)

71075_MURA_G0532

Fig. 90 Exhaust system sensor components

37663_MURA_G0279

Fig. 91 Accelerator Pedal Position (APP) sensor (3)

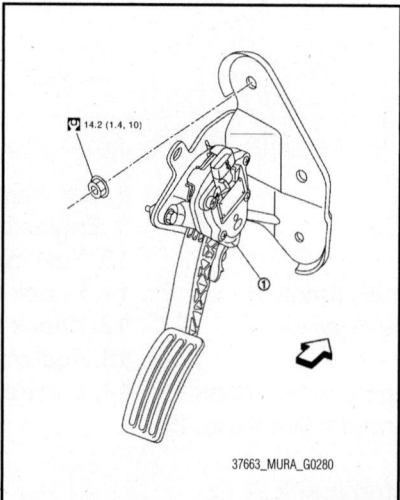

37663_MURA_G0280

Fig. 92 Accelerator pedal assembly (1)

To install:
Install in the reverse order of removal.

ACCELERATOR PEDAL RELEASED POSITION LEARING

Accelerator Pedal Released Position Learning is a function of ECM to learn the fully released position of the accelerator pedal by monitoring the accelerator pedal position sensor output signal. It must be performed each time the harness connector of the accelerator pedal position sensor or ECM is disconnected.

1. Check that accelerator pedal is fully released.
2. Turn ignition switch ON and wait at least 2 seconds.
3. Turn ignition switch OFF and wait at least 10 seconds.
4. Turn ignition switch ON and wait at least 2 seconds.
5. Turn ignition switch OFF and wait at least 10 seconds.

AIR-FUEL RATIO (A/F) SENSOR

LOCATION
See Figure 93.

REMOVAL & INSTALLATION
1. Turn ignition switch OFF.
2. Disconnect A/F sensor 1 harness connector.
3. Loosen the sensor mounting bolt.
4. Remove the sensor.

To install:
5. Discard any A/F sensor which has been dropped from a height of more than 0.5 m (19.7 in) onto a hard surface such as a concrete floor; use a new one.
6. Before installing new A/F sensor, clean exhaust system threads using Oxygen Sensor Thread Cleaner [commercial service tool (J-43897-18 or J-43897-12)] and approved anti-seize lubricant.
7. Install the sensor.
8. Connect the sensor connector.
9. Tighten the sensor mounting bolt.

CAMSHAFT POSITION (CMP) SENSOR

LOCATION
See Figure 94.

REMOVAL & INSTALLATION
1. Turn ignition switch OFF.
2. Loosen the sensor mounting bolt.
3. Disconnect the Camshaft Position (CMP) sensor (PHASE) harness connector.
4. Remove the sensor.

To install:
5. Install the CMP sensor.
6. Connect the CMP sensor connector.
7. Tighten the sensor mounting bolt.

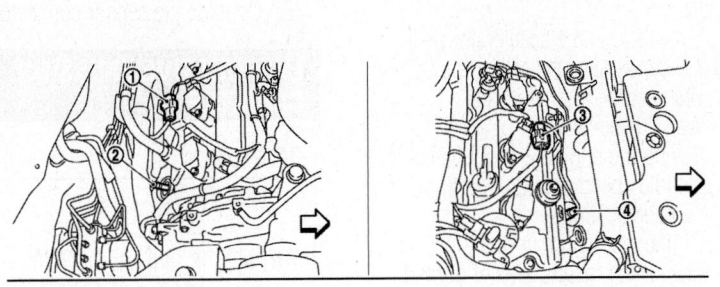

1. A/F sensor 1 (bank 1) harness connector
2. A/F sensor 1 (bank 1)
3. A/F sensor 1 (bank 2) harness connector
4. A/F sensor 1 (bank 2)
⇦ : Vehicle front

37663_MURA_G0281

Fig. 93 Air-Fuel Ratio Sensor (1) and harness connector (2), Bank 1/ Air-Fuel Ratio Sensor (4) and harness connector (3), Bank 2

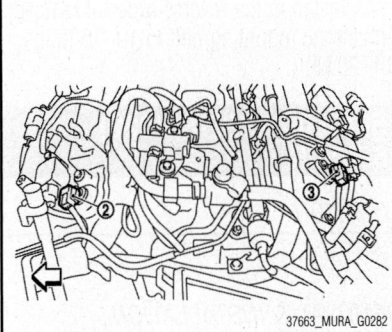

37663_MURA_G0282

Fig. 94 Camshaft Position (CMP) sensor (PHASE) Bank 1 (2), Camshaft Position (CMP) sensor (PHASE) Bank 2 (3)

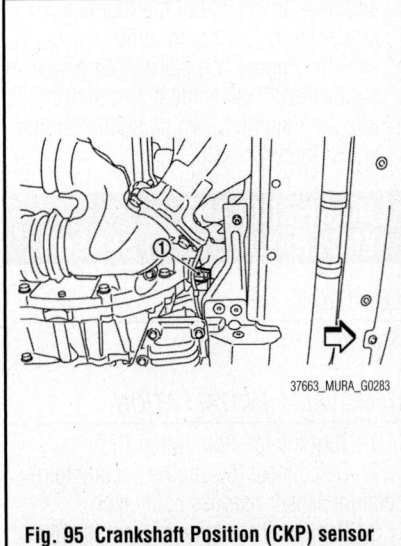

37663_MURA_G0283

Fig. 95 Crankshaft Position (CKP) sensor

CRANKSHAFT POSITION (CKP) SENSOR

LOCATION

See Figure 95.

REMOVAL & INSTALLATION

1. Turn ignition switch OFF.
2. Loosen the sensor mounting bolt.
3. Disconnect the CKP sensor harness connector.
4. Remove the sensor.

To install:

5. Install the CKP sensor.
6. Connect the CKP sensor connector.
7. Tighten the sensor mounting bolt.

ELECTRONIC CONTROL MODULE (ECM)

LOCATION

See Figure 96.

REMOVAL & INSTALLATION

See Figure 97.

1. Turn ignition switch OFF.
2. Loosen the sensor mounting bolt.
3. Disconnect the ECM harness connectors.
4. Remove the ECM.

To install:

5. Install the ECM.
6. Connect the ECM connectors.

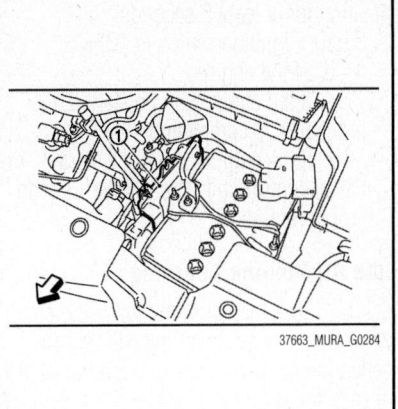

37663_MURA_G0284

Fig. 96 Electronic Control Module (ECM)

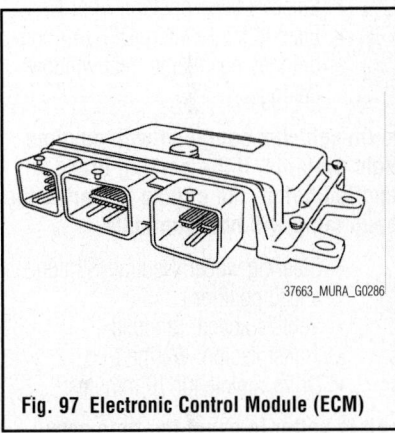

37663_MURA_G0286

Fig. 97 Electronic Control Module (ECM)

7. Tighten the sensor mounting bolt.
8. When replacing the ECM:

a. Perform the initialization of NVIS (NATS) system and registration of all NVIS (NATS) ignition key IDs. See Reset, below.

b. Perform VIN registration. The CONSULT-III scan tool is required.

c. Perform accelerator pedal released position learning. Refer to Accelerator Pedal Position (APP) sensor, Accelerator Pedal Released Position Learning.

d. Perform throttle valve closed position learning. Refer to Throttle Control Actuator, Throttle Valve Closed Position Learning.

e. Perform idle air volume learning. See Reset, below.

RESET

NVIS (NATS) System Initialization & Ignition Key ID Registration

1. Insert the registered Intelligent Key, turn ignition switch to "ON". To perform this step, use the key that has been used before performing ECM replacement.

2. Maintain ignition switch in "ON" position for at least 5 seconds.

3. Turn ignition switch to "OFF".

4. Start the engine.

   a. If the engine can be started, the procedure is complete.

   b. If the engine cannot be started, initialize control unit. The CONSULT scan tool is required.

Idle Air Volume Learning

1. Check that all of the following conditions are satisfied. Learning will be cancelled if any of the following conditions are missed for even a moment:

- Battery voltage: More than 12.9 V (At idle)
- Engine coolant temperature: 158–212°F (70–100°C)
- Selector lever position: P or N
- Electric load switch: OFF (Air conditioner, head lamp, rear window defogger)

➡On vehicles equipped with daytime light systems, if the parking brake is applied before the engine is started the head lamp will not illuminate.

- Steering wheel: Neutral (Straight-ahead position)
- Vehicle speed: Stopped
- Transmission: Warmed-up
- Drive vehicle for 10 minutes

➡It is better to count the time accurately with a clock.

➡It is impossible to switch the diagnostic mode when an accelerator pedal position sensor circuit has a malfunction.

2. Perform accelerator pedal released position learning. Refer to Accelerator Pedal Position (APP) sensor, Accelerator Pedal Released Position Learning.

3. Perform throttle valve closed position learning. Refer to Throttle Control Actuator, Throttle Valve Closed Position Learning.

4. Start engine and warm it up to normal operating temperature.

5. Turn ignition switch OFF and wait at least 10 seconds.

6. Confirm that accelerator pedal is fully released, turn ignition switch ON and wait 3 seconds.

7. Repeat the following procedure quickly 5 times within 5 seconds.

   a. Fully depress the accelerator pedal.

   b. Fully release the accelerator pedal.

8. Wait 7 seconds, fully depress the accelerator pedal for approx. 20 seconds until the MIL stops blinking and turns ON.

9. Fully release the accelerator pedal within 3 seconds after the MIL turns ON.

10. Start engine and let it idle.

11. Wait 20 seconds.

12. Rev up the engine 2 or 3 times and check that idle speed and ignition timing are within the specifications.

   a. If the results are normal, inspection in ended. If the results are not normal, check the following:

- Check that throttle valve is fully closed
- Check PCV valve operation
- Check that downstream of throttle valve is free from air leakage

   b. If the results are not normal, repair or replace malfunctioning part. If the results are normal, engine component parts and their installation condition are questionable. Check and eliminate the cause of the incident. If any of the following conditions occur after the engine has started, eliminate the cause of the incident and perform Idle Air Volume Learning again: engine stalls, Incorrect idle.

ENGINE COOLANT TEMPERATURE (ECT) SENSOR

LOCATION

See Figure 98.

REMOVAL & INSTALLATION

1. Turn the ignition switch OFF.

2. Disconnect the engine coolant temperature sensor harness connector.

3. Remove the engine coolant temperature sensor.

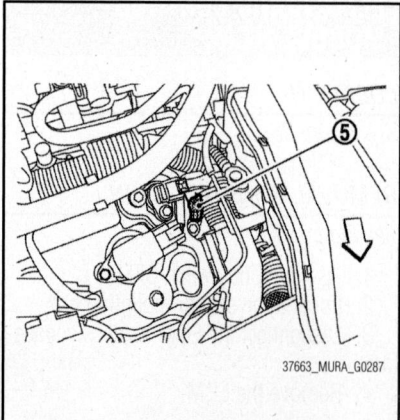

Fig. 98 Engine Coolant Temperature Sensor

To install:

4. Install the engine coolant temperature sensor.

5. Connect the sensor connector.

EVAPORATIVE EMISSION CANISTER

LOCATION

See Figure 99.

REMOVAL & INSTALLATION

See Figure 100.

1. Raise and safely support the vehicle.

2. Remove the EVAP canister mounting bolt.

3. Remove the EVAP canister.

➡The EVAP canister vent control valve and EVAP canister system pressure sensor can be removed without removing the EVAP canister.

To install:

4. Install in the reverse order of removal. Tighten the mounting bolt to 10–15 ft. lbs. (14–20 Nm).

HEATED OXYGEN (HO2S) SENSOR

LOCATION

See Figure 101.

REMOVAL & INSTALLATION

1. Turn the ignition switch OFF.

2. Disconnect the heated oxygen sensor harness connectors.

3. Remove the heated oxygen sensor with an oxygen sensor wrench [SST: KV10114400 (J-38365)]. Place marks to

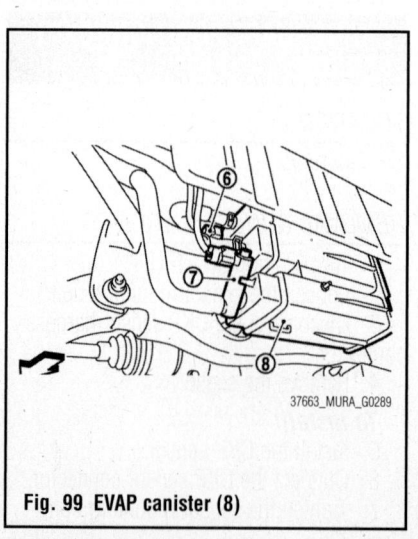

Fig. 99 EVAP canister (8)

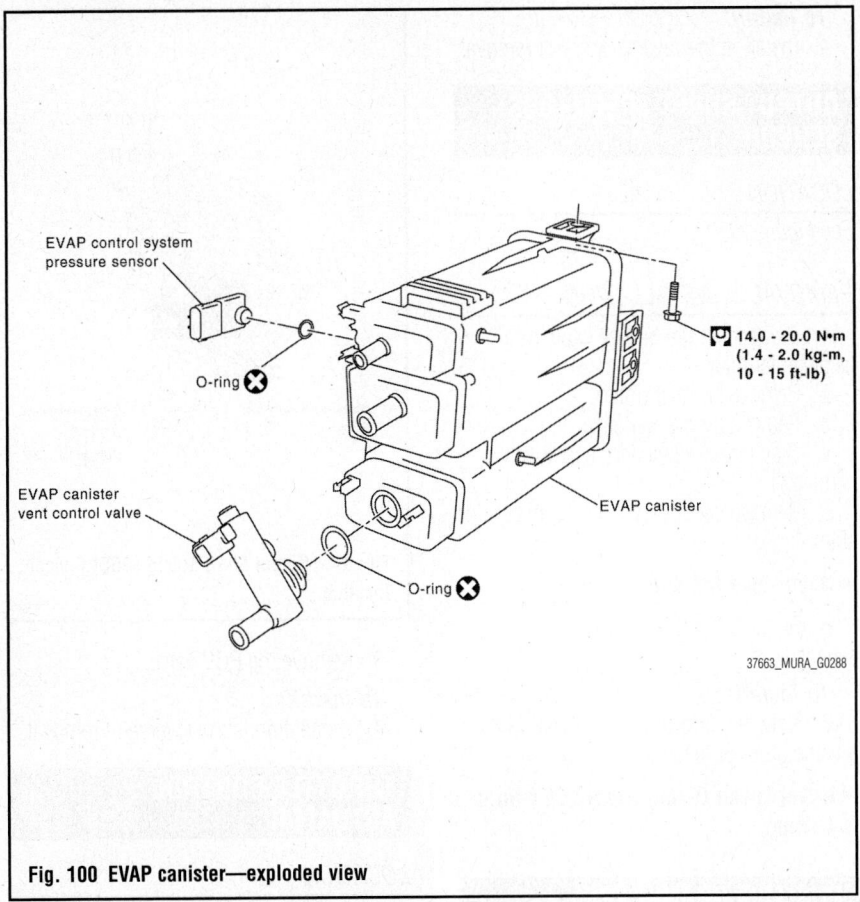

Fig. 100 EVAP canister—exploded view

EVAP control system pressure sensor

O-ring

EVAP canister vent control valve

O-ring

EVAP canister

14.0 - 20.0 N•m (1.4 - 2.0 kg-m, 10 - 15 ft-lb)

37663_MURA_G0288

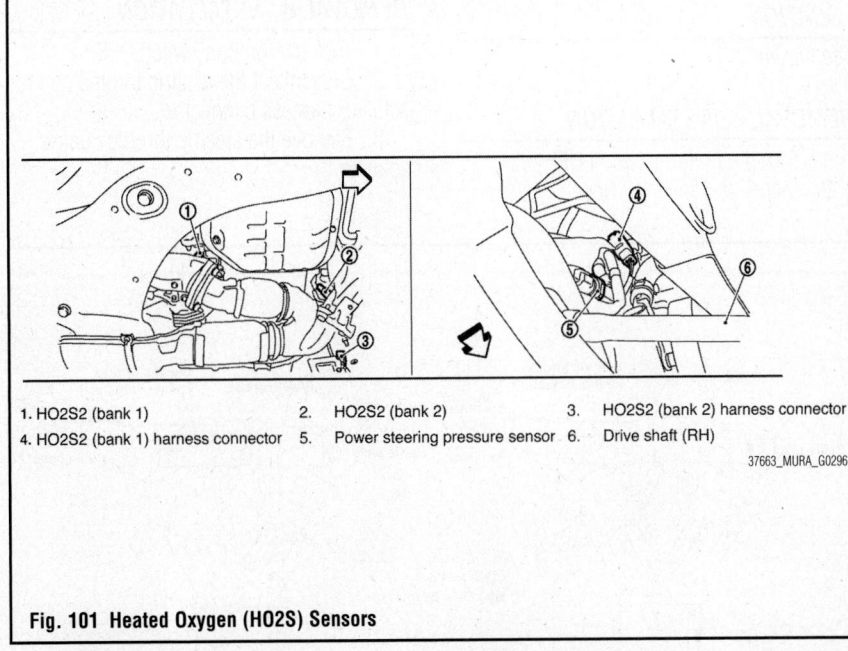

1. HO2S2 (bank 1)
2. HO2S2 (bank 2)
3. HO2S2 (bank 2) harness connector
4. HO2S2 (bank 1) harness connector
5. Power steering pressure sensor
6. Drive shaft (RH)

37663_MURA_G0296

Fig. 101 Heated Oxygen (HO2S) Sensors

identify installation positions of each heated oxygen sensor.

To install:

4. Before installing new oxygen sensor, clean exhaust system threads using oxygen sensor thread cleaner [commercial service tool (J-43897-18 or J-43897-12)] and approved anti-seize lubricant.

5. Install the heated oxygen sensors in the original positions. If the installation positions cannot be identified:

- Heated oxygen sensor 2 glass tube color: White
- Air fuel ratio sensor 1glass tube color: Black

➡Discard any heated oxygen sensor which has been dropped from a height of more than 0.5 m (19.7 in) onto a hard surface such as a concrete floor; use a new one.

> ※ **WARNING**
>
> Do not over-torque air fuel ratio sensor and heated oxygen sensor. Doing so may cause damage to air fuel ratio sensor and heated oxygen sensor, resulting in "MIL" coming on.

6. Connect the heated oxygen sensor harness connectors.

INTAKE AIR TEMPERATURE (IAT) SENSOR

LOCATION

See Figure 102.

REMOVAL & INSTALLATION

1. Turn the ignition switch OFF.
2. Disconnect the IAT sensor harness connector.
3. Remove the IAT sensor.

To install:

4. Install in the reverse order of removal.

KNOCK SENSOR (KS)

LOCATION

See Figure 103.

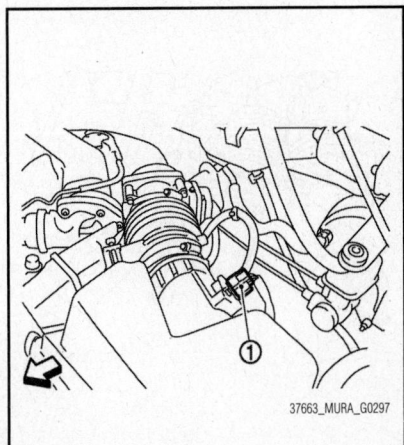

37663_MURA_G0297

Fig. 102 MAF sensor and Intake Air Temperature (IAT) sensor

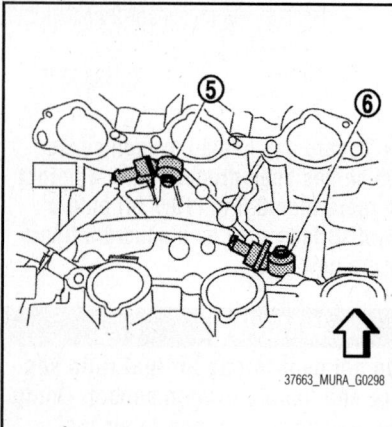

Fig. 103 Knock Sensor (KS), Bank 1 (6), Bank 2 (5)

REMOVAL & INSTALLATION

1. Turn the ignition switch OFF.
2. Disconnect the knock sensor harness connector.
3. Remove the knock sensor.

To install:

4. Install in the reverse order of removal.

MASS AIR FLOW SENSOR (HOT WIRE)

LOCATION

See Figure 104.

REMOVAL & INSTALLATION

1. Turn the ignition switch OFF.
2. Disconnect the MAF sensor harness connector.
3. Remove the MAF sensor.

To install:

4. Install in the reverse order of removal.

OUTPUT SHAFT SPEED (OSS) SENSOR

LOCATION

See Figure 105.

REMOVAL & INSTALLATION

1. Disconnect the battery cable from negative terminal.
2. Remove air duct (inlet).
3. Remove air cleaner case.
4. Disconnect secondary speed sensor connector.
5. Remove secondary speed sensor and shim.

➡ **Never lose the shim.**

6. Remove O-ring from secondary speed sensor.

To install:

7. Note the following, and install in the reverse order of removal.

➡ **Never reuse O-ring. Apply CVT fluid to O-ring.**

POSITIVE CRANKCASE VENTILATION VALVE

LOCATION

See Figure 106.

REMOVAL & INSTALLATION

1. Turn the ignition switch OFF.
2. Disconnect the PVC hose.

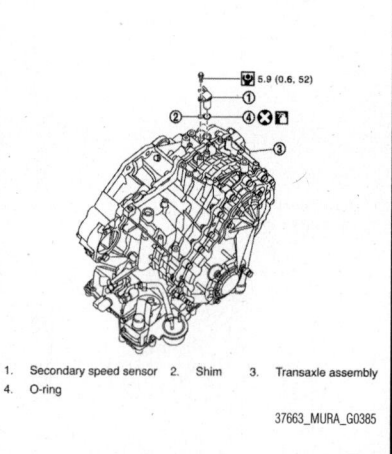

1. Secondary speed sensor   2. Shim   3. Transaxle assembly
4. O-ring

Fig. 105 Output Shaft Speed (OSS) Sensor location

3. Remove the PCV valve.

To install:

4. Installation is the reverse of removal.

THROTTLE CONTROL ACTUATOR

LOCATION

See Figure 107.

REMOVAL & INSTALLATION

1. Turn the ignition switch OFF.
2. Disconnect the electric throttle control actuator harness connector.
3. Remove the electric throttle control actuator.

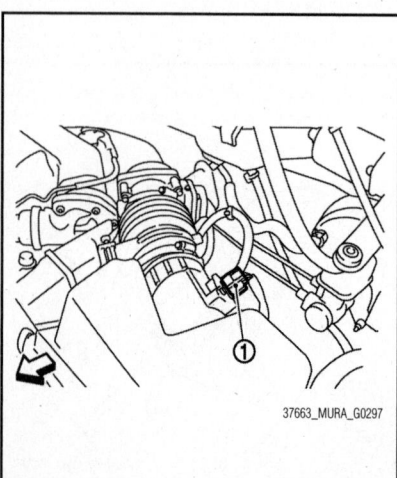

Fig. 104 Mass Air Flow (MAF) sensor and Intake Air Temperature (IAT) sensor

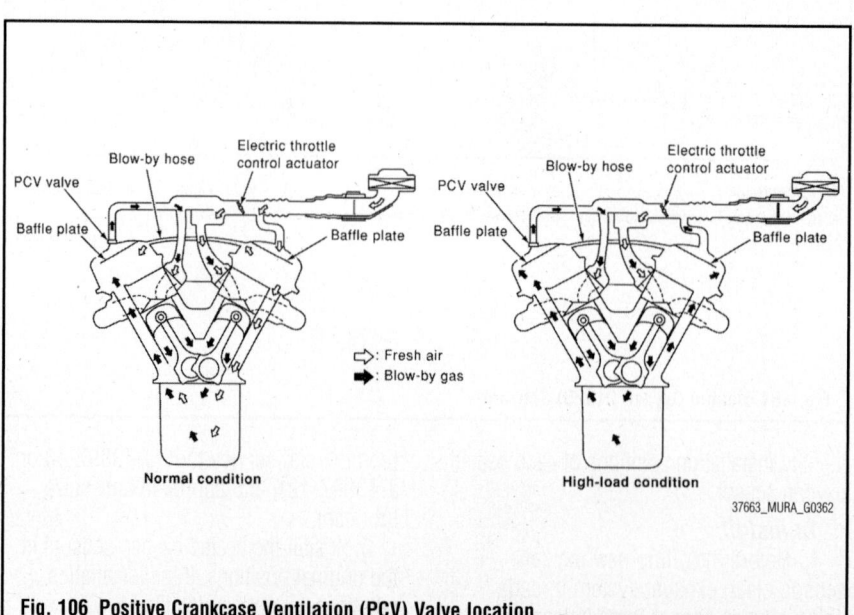

Fig. 106 Positive Crankcase Ventilation (PCV) Valve location

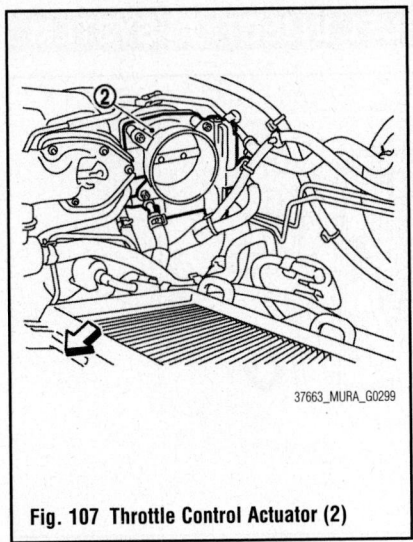

Fig. 107 Throttle Control Actuator (2)

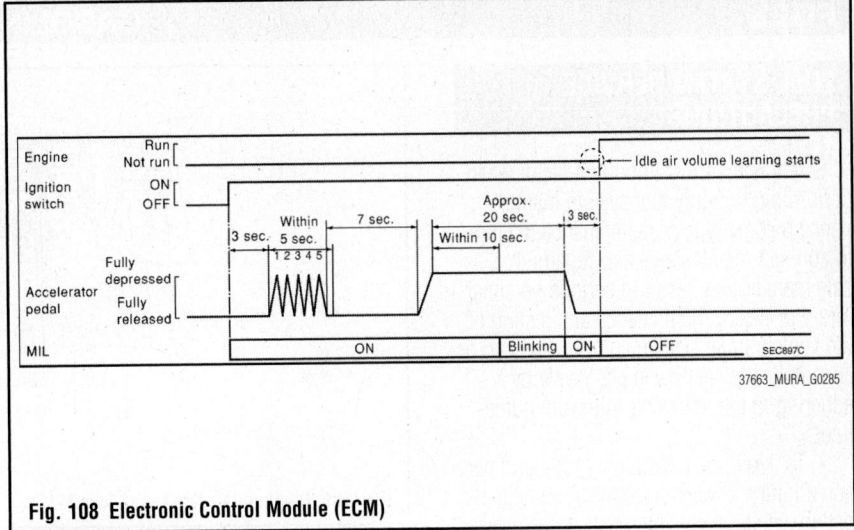

Fig. 108 Electronic Control Module (ECM)

To install:

4. Installation is the reverse of removal, noting the following:

   a. Perform throttle valve closed position learning.

   b. Perform idle air volume learning.

THROTTLE VALVE CLOSED POSITION LEARNING

1. Check that accelerator pedal is fully released.

2. Turn ignition switch ON.

3. Turn ignition switch OFF and wait at least 10 seconds.

4. Check that throttle valve moves during the above 10 seconds by confirming the operating sound.

IDLE AIR VOLUME LEARNING

See Figure 108.

1. Check that all of the following conditions are satisfied. Learning will be cancelled if any of the following conditions are missed for even a moment:

- Battery voltage: More than 12.9 V (At idle)
- Engine coolant temperature: 158–212°F (70–100°C)
- Selector lever position: P or N
- Electric load switch: OFF (Air conditioner, head lamp, rear window defogger)

➡ **On vehicles equipped with daytime light systems, if the parking brake is applied before the engine is started the head lamp will not illuminate.**

- Steering wheel: Neutral (Straight-ahead position)
- Vehicle speed: Stopped
- Transmission: Warmed-up
- Drive vehicle for 10 minutes

➡ **It is better to count the time accurately with a clock.**

➡ **It is impossible to switch the diagnostic mode when an accelerator pedal position sensor circuit has a malfunction.**

2. Perform accelerator pedal released position learning. Refer to Accelerator Pedal Position (APP) sensor, Accelerator Pedal Released Position Learning.

3. Perform throttle valve closed position learning. Refer to Throttle Control Actuator, Throttle Valve Closed Position Learning.

4. Start engine and warm it up to normal operating temperature.

5. Turn ignition switch OFF and wait at least 10 seconds.

6. Confirm that accelerator pedal is fully released, turn ignition switch ON and wait 3 seconds.

7. Repeat the following procedure quickly 5 times within 5 seconds.

   a. Fully depress the accelerator pedal.

   b. Fully release the accelerator pedal.

8. Wait 7 seconds, fully depress the accelerator pedal for approx. 20 seconds until the MIL stops blinking and turns ON.

9. Fully release the accelerator pedal within 3 seconds after the MIL turns ON.

10. Start engine and let it idle.

11. Wait 20 seconds.

12. Rev up the engine 2 or 3 times and check that idle speed and ignition timing are within the specifications.

   a. If the results are normal, inspection in ended. If the results are not normal, check the following:

- Check that throttle valve is fully closed
- Check PCV valve operation
- Check that downstream of throttle valve is free from air leakage

   b. If the results are not normal, repair or replace malfunctioning part. If the results are normal, engine component parts and their installation condition are questionable. Check and eliminate the cause of the incident. If any of the following conditions occur after the engine has started, eliminate the cause of the incident and perform Idle Air Volume Learning again: engine stalls, incorrect idle.

FUEL SYSTEM SERVICE PRECAUTIONS

Safety is the most important factor when performing not only fuel system maintenance but any type of maintenance. Failure to conduct maintenance and repairs in a safe manner may result in serious personal injury or death. Maintenance and testing of the vehicle's fuel system components can be accomplished safely and effectively by adhering to the following rules and guidelines.

• To avoid the possibility of fire and personal injury, always disconnect the negative battery cable unless the repair or test procedure requires that battery voltage be applied.

• Always relieve the fuel system pressure prior to disconnecting any fuel system component (injector, fuel rail, pressure regulator, etc.), fitting or fuel line connection. Exercise extreme caution whenever relieving fuel system pressure to avoid exposing skin, face and eyes to fuel spray. Please be advised that fuel under pressure may penetrate the skin or any part of the body that it contacts.

• Always place a shop towel or cloth around the fitting or connection prior to loosening to absorb any excess fuel due to spillage. Ensure that all fuel spillage (should it occur) is quickly removed from engine surfaces. Ensure that all fuel soaked cloths or towels are deposited into a suitable waste container.

• Always keep a dry chemical (Class B) fire extinguisher near the work area.

• Do not allow fuel spray or fuel vapors to come into contact with a spark or open flame.

• Always use a back-up wrench when loosening and tightening fuel line connection fittings. This will prevent unnecessary stress and torsion to fuel line piping.

• Always replace worn fuel fitting O-rings with new Do not substitute fuel hose or equivalent where fuel pipe is installed.

Before servicing the vehicle, make sure to also refer to the precautions in the beginning of this section as well.

RELIEVING FUEL SYSTEM PRESSURE

See Figure 109.

1. Before servicing the vehicle, refer to the Precautions Section.

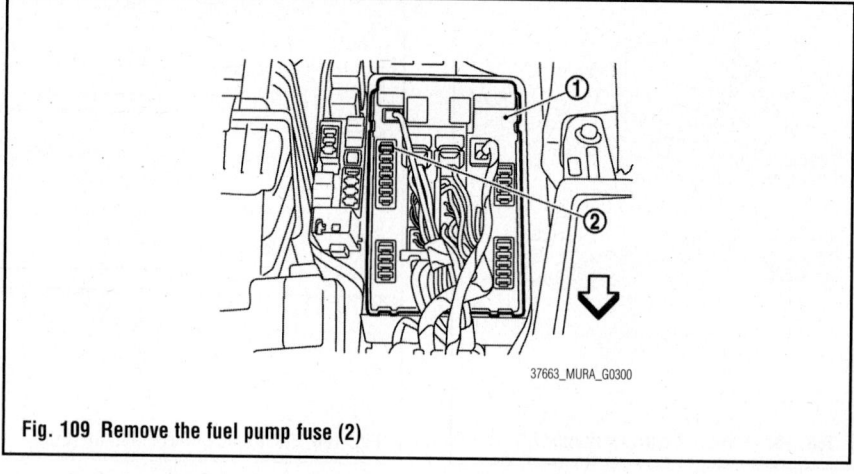

Fig. 109 Remove the fuel pump fuse (2)

2. Remove the fuel pump fuse (2) located in IPDM E/R (1).
3. Start the engine.
4. After the engine stalls, crank it 2 or 3 times to release all fuel pressure.
5. Turn the ignition switch OFF.
6. Reinstall fuel pump fuse after servicing fuel system.

FUEL PUMP

REMOVAL & INSTALLATION

See Figures 110 through 115.

1. Before servicing the vehicle, refer to the Precautions Section.
2. Check fuel level on a level ground. If the fuel level is 7/8 of the fuel tank (full or nearly full), draw appropriate amount of fuel from the fuel tank. Guideline: Draw approximately 5 1/4 gal (20 liters) from a full-tank condition. In the event of malfunction in fuel pump, insert a hose measuring 0.98 in. (25mm) in diameter into the filler opening to draw approximately 5 1/4 gal (20 liters) fuel.
3. Release the fuel pressure from the fuel lines.
4. Open the fuel filler lid.
5. Open the filler cap and release the pressure inside fuel tank.
6. Remove the rear seat cushion and reclining device assembly.
7. Peel off the floor carpet, then remove the inspection hole cover units by turning clips clockwise by 90 degrees.
8. Disconnect the harness connector, fuel feed tube and EVAP hose.
9. Disconnect the quick connector as follows:
   a. Hold the sides of connector, push in the tabs and pull out the fuel feed tube.

10. Remove the retainer for the main fuel level sensor unit, fuel filter and fuel pump assembly and sub fuel level sensor unit with the fuel tank lock ring wrench (SST) by turning counterclockwise.

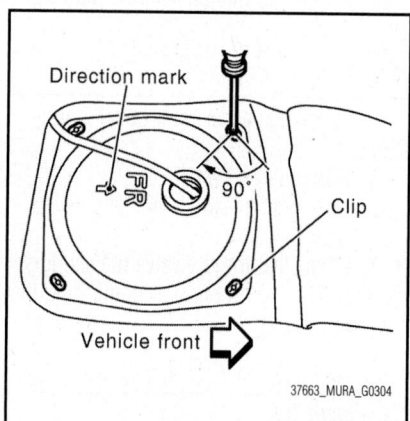

Fig. 110 Left side: Main fuel level sensor unit, fuel filter and fuel pump assembly, Right side: Sub fuel level sensor unit

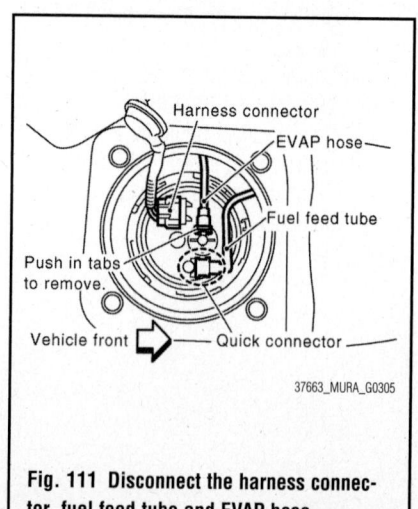

Fig. 111 Disconnect the harness connector, fuel feed tube and EVAP hose

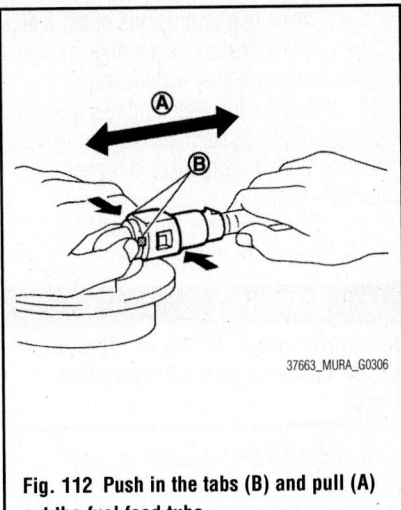

Fig. 112 Push in the tabs (B) and pull (A) out the fuel feed tube

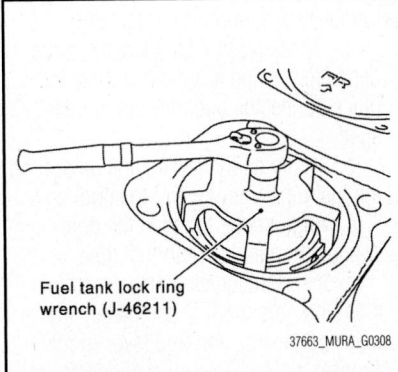

Fuel tank lock ring wrench (J-46211)

Fig. 113 Remove the retainer for the main fuel level sensor unit, fuel filter and fuel pump assembly and sub fuel level sensor unit

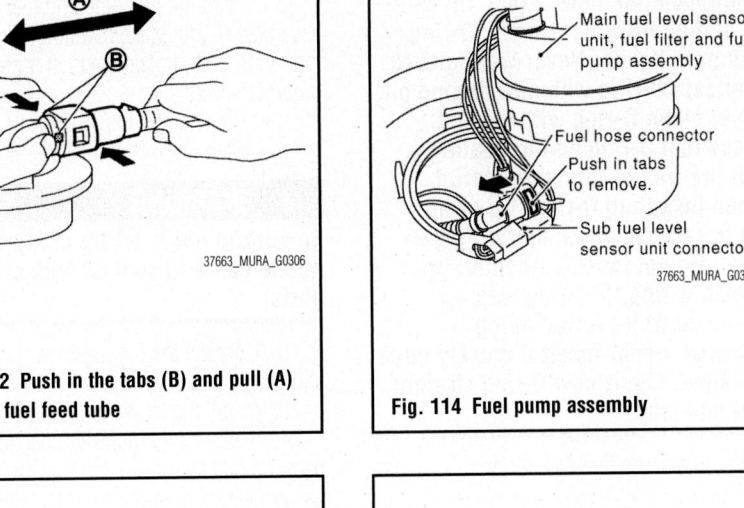

Main fuel level sensor unit, fuel filter and fuel pump assembly

Fuel hose connector
Push in tabs to remove.

Sub fuel level sensor unit connector

Fig. 114 Fuel pump assembly

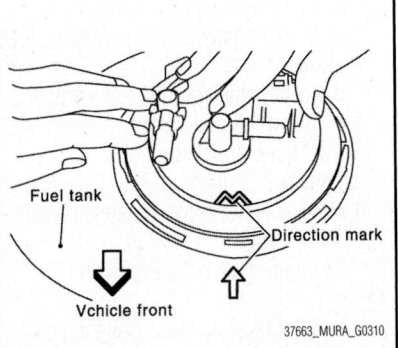

Fuel tank
Direction mark
Vehicle front

Fig. 115 Align the direction mark on main fuel level sensor unit, fuel filter and fuel pump assembly and sub fuel level sensor unit with that on fuel tank

11. Remove the main fuel level sensor unit, fuel filter and fuel pump assembly, and sub fuel level sensor unit.

   a. Raise the main fuel level sensor unit, fuel filter and fuel pump assembly, and disconnect the fuel hose connector (push in tabs and pull out) and sub fuel level sensor unit harness connector.

   b. Raise and release the sub fuel level sensor unit to remove.

To install:

12. Note to the following, and install in the reverse order of removal.

13. Install the new seal packing to the fuel tank without any twist.

14. Connect the fuel hose connector (push in until it stops) and sub fuel level sensor unit harness connector. Insert the connector until you hear a click.

15. Align the direction mark on main fuel level sensor unit, fuel filter and fuel pump assembly and sub fuel level sensor unit

with that on fuel tank as shown in the figure, and install them to fuel tank.

✳✳ WARNING

Never allow seal packing to drop. Never bend float arm during installing.

16. Install the retainer for the main fuel level sensor unit, fuel filter and fuel pump assembly and sub fuel level sensor unit with the fuel tank lock ring wrench (SST) by turning clockwise.

✳✳ WARNING

Install the retainer horizontally.

17. Connect the quick connector as follows:

   a. Check the connection for damage or any foreign materials.

   b. Align the connector with the tube, then insert the connector straight into the tube until a click sound is heard.

c. After connecting, check that the connection is secure by

FUEL RAIL & INJECTORS

REMOVAL & INSTALLATION

See Figures 116 through 118.

1. Before servicing the vehicle, refer to the Precautions Section.

2. Remove the radiator core support covers, air duct (inlet), air cleaner cases (upper and lower) with mass air flow sensor and air duct assembly. Refer to Air Cleaner Removal & Installation in the Engine Mechanical section.

3. Remove the engine cover.

4. Properly release the fuel pressure.

5. Remove the front wiper arm and extension cowl top. Refer to Cowl Top Removal & Installation in the Body Exterior section.

6. Drain the engine coolant, or when water hoses are disconnected, attach plug to prevent engine coolant leakage.

7. Remove the intake manifold collector.

8. Remove the quick connector cap from quick connector.

9. Disconnect the quick connector from the fuel tube as follows:

✳✳ CAUTION

Disconnect the quick connector by using the quick connector release (commercial service tool: J-45488), not by picking out retainer tabs.

   a. With the sleeve side of quick connector release facing toward the quick connector, install the quick connector release onto the fuel tube.

   b. Insert the quick connector release into the quick connector until the sleeve contacts and goes no further. Hold the quick connector release in that position.

✳✳ WARNING

Inserting the quick connector release hard will not disconnect the quick connector. Hold the quick connector release where it contacts and goes no further.

   c. Draw and pull out the quick connector straight from the fuel tube. Pull the quick connector holding position as shown.

✳✳ WARNING

Never pull with lateral force applied. Prepare container and cloth before as fuel will leak out. Never bend or twist connection between quick connector and fuel feed hose (with damper) during installation/removal. To keep the connecting portions clean and to avoid damage, cover them completely with plastic bags or something similar.

✳✳ CAUTION

Avoid fire and sparks. Keep parts away from heat source. Be careful if welding. Never expose parts to battery electrolyte or other acids.

10. Remove the harness connector from fuel injector.
11. Loosen the mounting bolts in reverse of the installation order shown below.
12. Remove the fuel tube and fuel injector assembly.
13. Remove the fuel injector from fuel tube with following procedure.
    a. Open and remove the clip.
    b. Remove the fuel injector from the fuel tube by pulling straight.
14. Remove the fuel damper from fuel tube.

✳✳ WARNING

Do not disassemble the fuel injector.

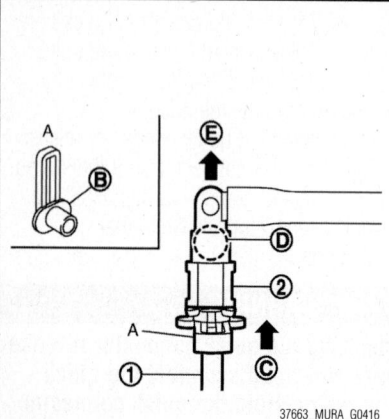

Fig. 116 Insert and retain (C) the quick connector release (A) into quick connector (2) until sleeve (B) contacts; pull the quick connector (E) from the fuel tube (1), holding position (D)

37663_MURA_G0416

To install:

✳✳ WARNING

When handling new O-ring, be careful of the following: Handle O-ring with bare hands. Never wear gloves. Lubricate O-ring with new engine oil. Never clean O-ring with solvent. Check that O-ring and its mating part are free of foreign material. When installing O-ring, be careful not to scratch it with tool or fingernails. Do not twist, de-center, or stretch O-ring. If O-ring was stretched while it was being attached, never insert it quickly into fuel tube. Insert new O-ring straight into fuel tube.

15. Install the fuel damper as follows:
    a. Install a new O-ring to the fuel tube as shown.
    b. Install spacer to the fuel damper.
    c. Insert the fuel damper straight into the fuel tube. Insert straight, checking that the axis is lined up. Never pressure-fit with excessive force. Reference value: 29 lbs. (130 N)
    d. Tighten the bolts evenly in turn.
    e. After tightening the bolts, check that there is no gap between fuel damper cap and fuel tube.
16. Install new O-rings to the fuel injector. (Fuel tube side: black, nozzle side: green)
17. Install the fuel injector to the fuel tube as follows:
    a. Insert the clip into clip mounting groove on the fuel injector. Insert clip so that protrusion of fuel injector matches cutout of clip.

✳✳ WARNING

Never reuse the clip. Replace it with new one. Be careful to keep clip from interfering with O-ring. If interference occurs, replace O-ring.

    b. Insert the fuel injector into the fuel tube with clip attached. Insert it while matching it to the axial center. Insert the fuel injector so that protrusion of fuel tube matches cutout of clip.
    c. Check that fuel tube flange is securely fixed in flange fixing groove on clip.
    d. Check that installation is complete by checking that the fuel injector does not rotate or come off.

    e. Check that protrusions of the fuel injectors and fuel tubes are aligned with cutouts of clips after installation.
18. Install the fuel tube and fuel injector assembly to the intake manifold. Tighten the mounting bolts in two steps in numerical order as shown.
- Step 1: 7 ft. lbs. (10 Nm)
- Step 2: 16 ft. lbs. (22 Nm)

✳✳ WARNING

Be careful not to let tip of injector nozzle come in contact with other parts.

19. Connect the fuel injector harness.
20. Install the intake manifold collector.
21. Connect the quick connector between fuel feed hose and fuel tube connection with the following procedure:
    a. Check no foreign substances are deposited in and around fuel tube and quick connector, and they are not damaged.
    b. Thinly apply new engine oil around the fuel tube from tip end to spool end.
    c. Align center to insert the quick connector straightly into fuel tube.
    d. Insert the quick connector to fuel tube until top spool is completely inside quick connector, and 2nd level spool exposes right below quick connector.
    e. Hold the position shown when inserting the fuel tube into quick connector.
    f. Carefully align center to avoid inclined insertion to prevent damage to O-ring inside quick connector.
    g. Insert until you hear a "click" sound and actually feel the engagement.

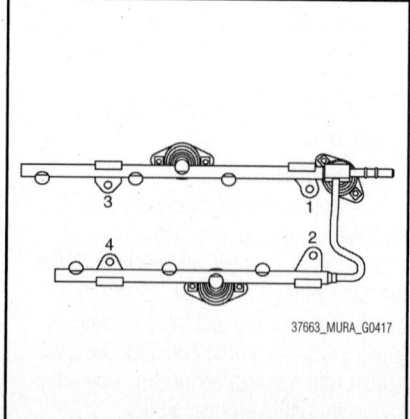

Fig. 117 Fuel injector mounting bolt installation sequence—remove in reverse order

37663_MURA_G0417

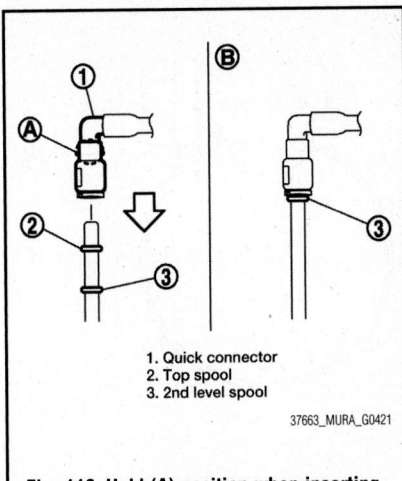

1. Quick connector
2. Top spool
3. 2nd level spool

37663_MURA_G0421

Fig. 118 Hold (A) position when inserting fuel tube; fitted (B) position

h. To ensure engagement, pull the quick connector by hand, holding position. Ensure it is completely engaged (connected) so that it does not come out from fuel tube.

i. Install the quick connector cap to quick connector.

j. Install the quick connector cap with the arrow the on surface facing in direction of quick connector (fuel feed hose side). If the quick connector cap cannot be installed smoothly, quick connector may have not been installed correctly. Check connection again.

k. Secure the fuel feed hose to the clamp of quick connector cap.

22. Install the front wiper arm and extension cowl top.

23. Install the engine cover.

24. Install the radiator core support covers, air duct (inlet), air cleaner cases (upper and lower) with mass air flow sensor and air duct assembly.

25. Add engine coolant, if necessary.

26. Turn the ignition switch ON (with engine stopped). With fuel pressure applied to fuel piping, check for fuel leakage at connection points.

27. Start the engine. With engine speed increased, check again for fuel leakage at connection points.

FUEL TANK

REMOVAL & INSTALLATION

See Figures 119 through 124.

1. Before servicing the vehicle, refer to the Precautions Section.

2. Perform the work on level place.

3. Release the fuel pressure from the fuel lines.

4. Open the fuel filler lid.

5. Open the filler cap and release the pressure inside fuel tank.

6. Remove the rear seat cushion and reclining device assembly.

7. Peel off the floor carpet, then remove the inspection hole cover units by turning clips clockwise by 90 degrees.

8. Disconnect the harness connector, fuel feed tube and EVAP hose.

9. Remove the center muffler.

10. Remove the propeller shaft. Refer to Propeller Shaft Removal & Installation in the Drive Train section.

11. Remove the rear parking brake cables.

12. Remove the rear suspension member assembly.

➡**For this service, halfshaft, final drive, and rear suspension member are required not to be separated from one another during removal.**

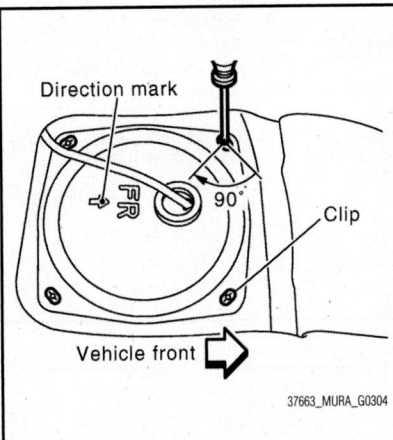

37663_MURA_G0304

Fig. 119 Left side: Main fuel level sensor unit, fuel filter and fuel pump assembly, Right side: Sub fuel level sensor unit

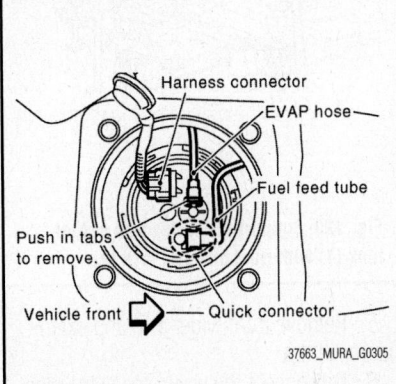

37663_MURA_G0305

Fig. 120 Left side: Main fuel level sensor unit, fuel filter and fuel pump assembly, Right side: Sub fuel level sensor unit

13. Remove the rear fuel tank protectors.

14. Disconnect the rear fuel filler hose, EVAP/Vent line hose, and EVAP (Recirculation) hose from the side other than the fuel tank side.

15. Support the lower part of fuel tank with transmission jack. Support the position that fuel tank mounting bands never engage.

16. Remove the rear fuel tank mounting bands.

17. Supporting with hands, lower the rear transmission jack carefully, and remove fuel tank.

✳✳ WARNING

Check that all connection points have been disconnected. Confirm there is no interference with vehicle.

18. Remove the rear fuel filler tube if necessary.

To install:

19. Note the following, and install in the reverse order of removal.

20. Securely clamp the fuel hoses and insert hose to the length below:
- Fuel filler hose: 1.38 in. (35 mm)
- The other hoses: 0.98 in. (25 mm)

21. Be sure the rear hose clamp is not placed on swelled area of fuel tube.

22. Tighten the clamp until the mark is on the bolt head flange.

23. Connect the quick connector as follows:

a. Check the connection for damage or any foreign materials.

b. Align the connector with the tube, then insert the connector straight into the tube until a click sound is heard.

c. After connecting, check that the connection is secure.

THROTTLE BODY

REMOVAL & INSTALLATION

See Figure 125.

1. Before servicing the vehicle, refer to the Precautions Section.

2. Remove the engine cover.

3. Remove the air cleaner cases (upper and lower) with mass air flow sensor and air duct assembly. Refer to Air Cleaner Removal & Installation.

4. Drain the engine coolant, or when water hoses are disconnected, attach plug to prevent engine coolant leakage.

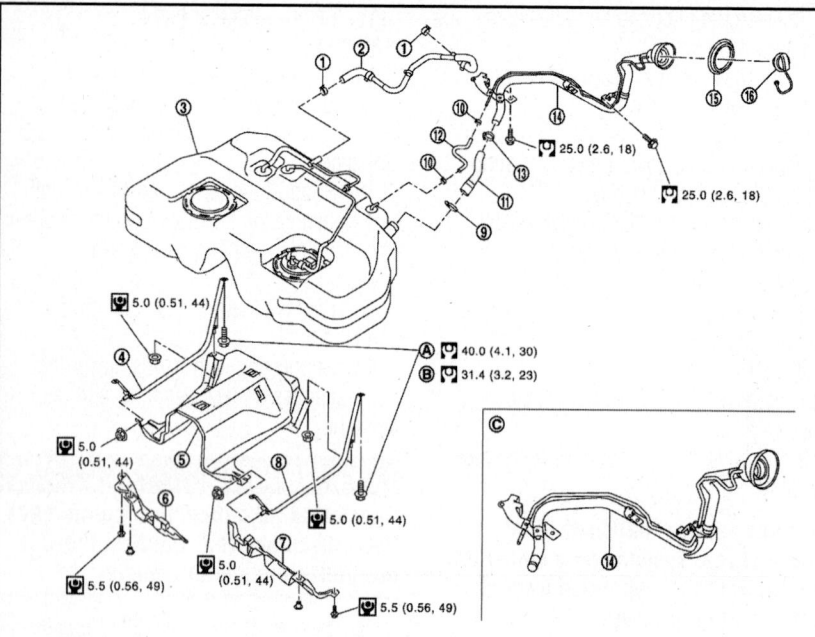

1.	Clamp	2.	EVAP/Vent line hose	3.	Fuel tank
4.	Fuel tank mounting band (RH)	5.	Fuel tank protector	6.	Fuel tank protector (RH)
7.	Fuel tank protector (LH)	8.	Fuel tank mounting band (LH)	9.	Clamp
10.	Clamp	11.	Fuel filler hose	12.	EVAP (Recirculation) hose
13.	Clamp	14.	Fuel filler tube	15.	Grommet
16.	Fuel filler cap				
A.	The measurement of hexagonal width across flat: 16 mm (0.63 in)	B.	The measurement of hexagonal width across flat: 17 mm (0.67 in)	C.	AWD 20 inch tire models (up to VIN No. JN8AZ18W79W125941)

37663_MURA_G0311

Fig. 121 Fuel tank

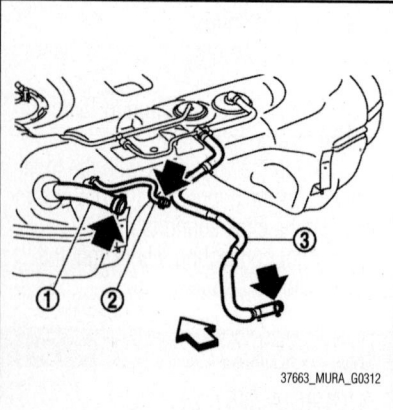

37663_MURA_G0312

Fig. 122 Disconnect fuel filler hose (1), EVAP/Vent line hose (3), and EVAP (Recirculation) hose (2)

✳✳ CAUTION

Perform this step when the engine is cold.

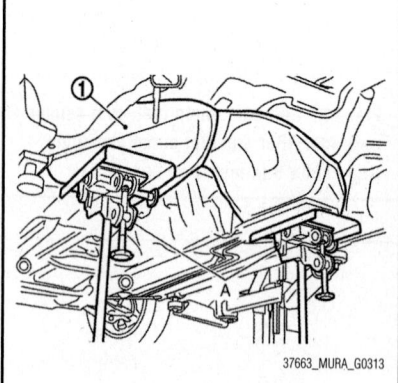

37663_MURA_G0313

Fig. 123 Support the lower part of fuel tank (1) with transmission jack (A)

5. Remove front wiper arm and extension cowl top.

6. Disconnect the water hoses the electric throttle control actuator. When engine

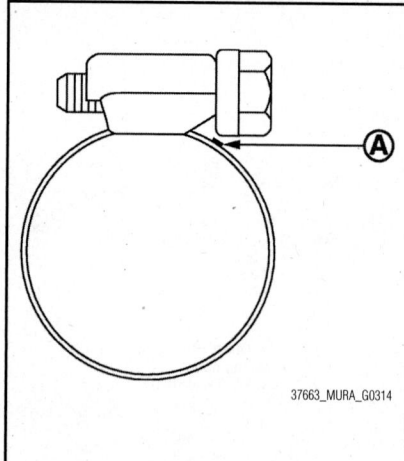

37663_MURA_G0314

Fig. 124 Tighten the clamp until the mark (A) is on the bolt head flange

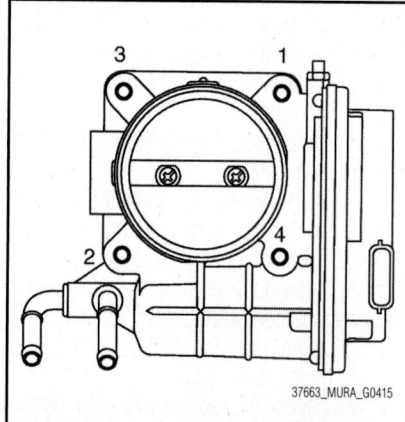

37663_MURA_G0415

Fig. 125 Throttle body mounting bolt installation sequence—remove in reverse order

coolant is not drained from the radiator, attach plug to water hoses to prevent engine coolant leakage.

7. Disconnect the electric throttle control actuator harness connector.

8. Loosen the electric throttle control actuator mounting bolts in reverse of the installation order shown.

➥**Never disassemble.**

To install:

9. Note the following, and install in the reverse order of removal.

a. Install the gasket with positioning no-protrusion surface upward or downward.

b. Tighten to 74 inch lbs. (8 Nm) in the order shown in above.

c. Perform the "Idle Air Volume Learning" and "Throttle Valve Closed Position Learning".

HEATING & AIR CONDITIONING SYSTEM

BLOWER MOTOR

REMOVAL & INSTALLATION
See Figure 126.

❋❋ **CAUTION**

Before servicing components near or affected by the SRS (air bag) system, read and observe all SRS Service Precautions. Refer to Supplemental Restraint System (SRS), in the Chassis Electrical section. Failure to observe all precautions may result in accidental airbag deployment, personal injury, or death.

1. Before servicing the vehicle, refer to the Precautions Section.
2. Disconnect the negative battery cable and wait at least 3 minutes for the SRS memory to drain.
3. Remove the right-hand instrument lower panel. Refer to Instrument Panel Removal & Installation in the Body Interior section.

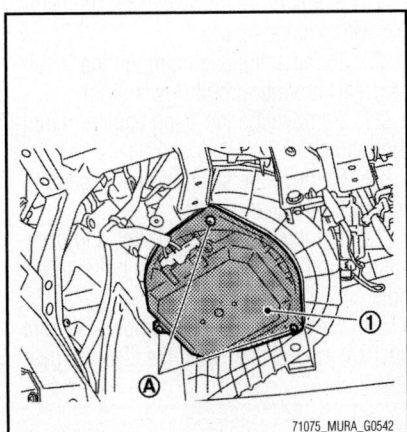

Fig. 126 Remove the mounting screws (A) and the blower motor (1)

4. Disconnect the blower motor connector.
5. Remove the mounting screws and the blower motor.

To install:
Installation is the reverse of the removal procedure.

HEATER CORE

REMOVAL & INSTALLATION
See Figures 127 and 128.

❋❋ **CAUTION**

Before servicing components near or affected by the SRS (air bag) system, read and observe all SRS Service Precautions. Refer to Air Bag (Supplemental Restraint System), in the Chassis Electrical section. Failure to observe all precautions may result in accidental airbag deployment, personal injury, or death.

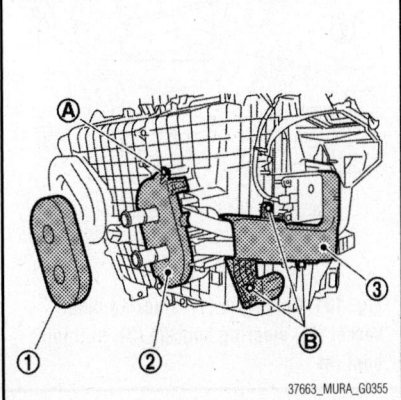

Fig. 127 Heater pipe grommet (1), heater pipe support (2) and mounting screw (A), heater pipe cover (3) and mounting screws (B)

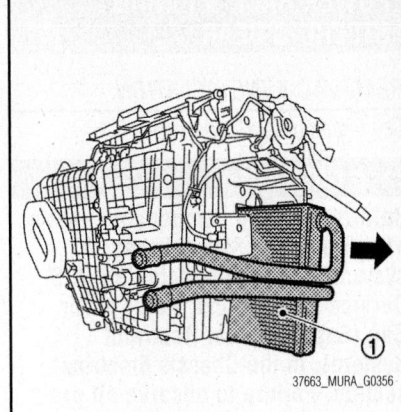

Fig. 128 Slide the heater core (1) in the direction shown by the arrow, and remove it

1. Before servicing the vehicle, refer to the Precautions Section.
2. Discharge the air conditioning system.
3. Disconnect the negative battery cable and wait at least 3 minutes for the SRS memory to drain.
4. Remove the heater and cooling unit assembly. Refer to Heater & Cooling Unit Removal & Installation.
5. Remove the left foot duct.
6. Remove the heater pipe grommet.
7. Remove the mounting screw, and remove the heater pipe support.
8. Remove the mounting screws, and remove the heater pipe cover.
9. Slide the heater core in the direction shown by the arrow, and remove it.

To install:
Installation is the reverse of the removal procedure. Replace O-rings with new ones. Apply compressor oil to the O-rings. Check for leakages when recharging refrigerant.

STEERING

POWER RACK & PINION STEERING GEAR

REMOVAL & INSTALLATION

See Figures 129 through 134.

> ※ **CAUTION**
>
> **Before servicing components near or affected by the SRS (air bag) system, read and observe all SRS Service Precautions. Refer to Air Bag (Supplemental Restraint System), in the Chassis Electrical section. Failure to observe all precautions may result in accidental airbag deployment, personal injury, or death.**

1. Before servicing the vehicle, refer to the Precautions Section.

2. Set the wheels in the straight ahead position.

3. Disconnect the negative battery cable and wait at least 3 minutes for the SRS memory to drain.

4. Remove the front road wheel and tires.

5. Remove the splash guards.

6. Remove the engine under cover.

7. Remove the front exhaust pipe.

8. For AWD vehicles, separate the rear propeller shaft (front side).

9. Remove the heat insulator from the front floor.

10. Remove the cotter pin, and loosen the nuts.

11. Using a ball joint remover, remove the steering outer socket from steering knuckle so as not to damage the ball joint boot.

12. Temporarily tighten the nut to prevent damage to the threads and to prevent the ball joint remover from suddenly coming off.

13. Remove the high pressure piping and low pressure hose of hydraulic piping, and then drain the power steering fluid.

14. Remove the steering hydraulic piping bracket from the front steering gear assembly.

15. Remove the power steering solenoid valve harness connector and harness clip.

16. Remove the lower joint mounting bolt (steering gear side).

17. Separate the lower shaft from the steering gear assembly by sliding the slide shaft.

> ※ **WARNING**
>
> **Spiral cable may be cut if steering wheel turns while separating steering column assembly and steering gear assembly. Be sure to secure steering wheel using string to avoid turning.**

18. Remove the stabilizer assembly.

19. Support the front suspension member with a suitable jack.

20. Remove the engine mounting insulator (rear) mounting bolt (lower side).

21. Remove the left-hand engine mounting insulator.

22. Remove the steering gear assembly mounting bolts and nuts.

23. Remove the member stay, and front suspension member mounting bolts and nuts.

24. Lower the jack and remove the front suspension member from the steering gear assembly.

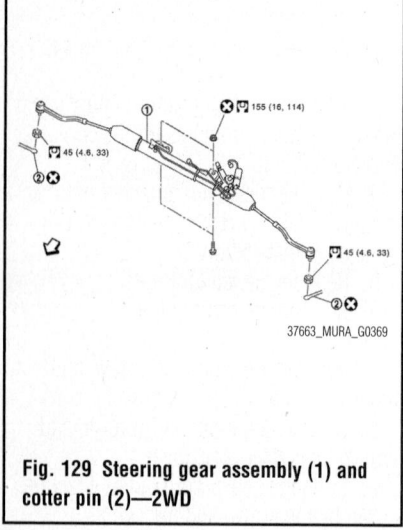

37663_MURA_G0369

Fig. 129 Steering gear assembly (1) and cotter pin (2)—2WD

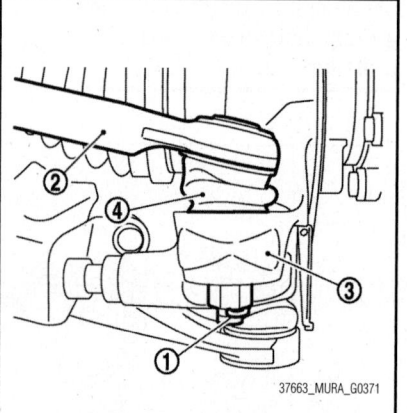

37663_MURA_G0371

Fig. 131 Cotter pin (1), steering outer socket (2), steering knuckle (3), ball joint boot (4)

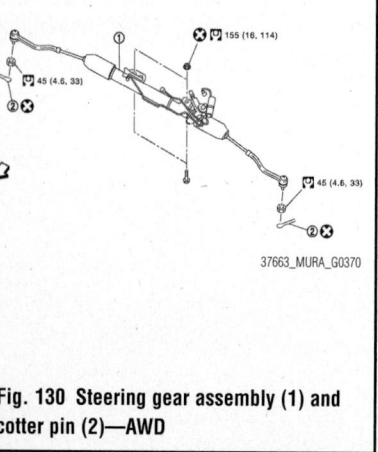

37663_MURA_G0370

Fig. 130 Steering gear assembly (1) and cotter pin (2)—AWD

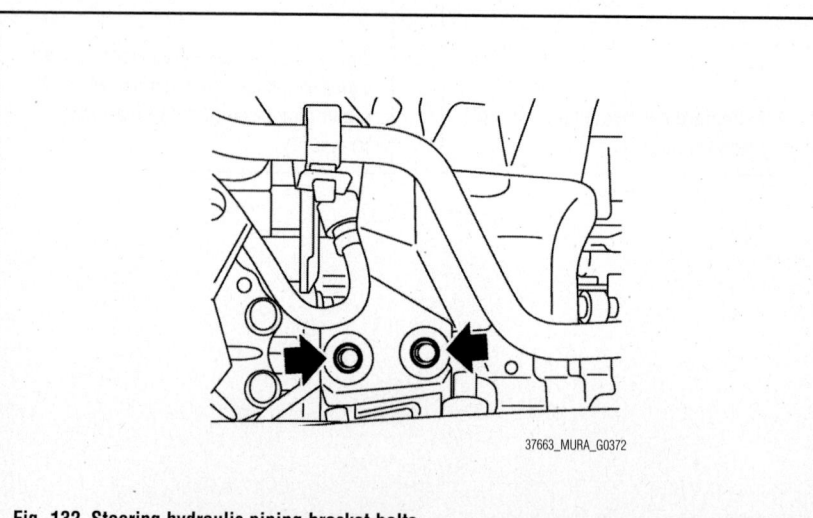

37663_MURA_G0372

Fig. 132 Steering hydraulic piping bracket bolts

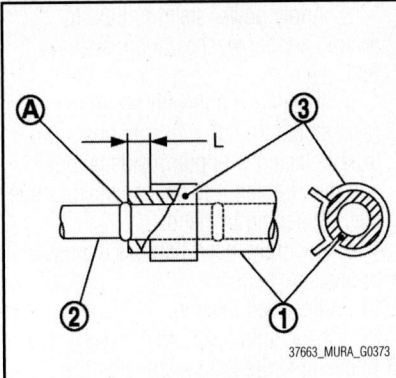

Fig. 133 Never apply fluid to the hose (1) and tube (2), insert hose until it contacts tube spool (A), and leave clearance (L) when installing clamp (3)

To install:

25. Note the following, and install in the reverse order of removal.

❊❊ WARNING

Spiral cable may be cut if steering wheel turns while separating steering column assembly and steering gear assembly. Be sure to secure steering wheel using string to avoid turning.

26. When installing low pressure hose, refer to the following:
   a. Never apply fluid to the hose and tube.
   b. Insert hose securely until it contacts spool of tube.
   c. Leave clearance when installing clamp. Standard: 0.12–0.31 in. (3–8mm)
27. When installing lower joint to steering gear assembly, follow the procedure listed below.

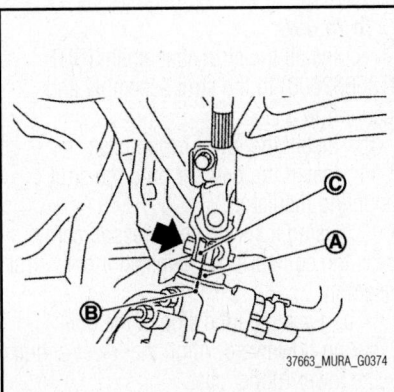

Fig. 134 Rear cover cap projection (A), gear housing assembly (B), lower joint (C)

a. Set the steering gear rack in the neutral position.

➡ **To get the neutral position of rack, turn gear-sub assembly and measure the distance of inner socket, and then measure the intermediate position of the distance.**

   b. Align rear cover cap projection with the marking position of gear housing assembly.
   c. Install the slit part of the lower joint aligning with the rear cover cap projection. Make sure that the slit part of lower joint is aligned with rear cover cap projection and the marking position of gear housing assembly.
28. After installation, bleed air from the steering hydraulic system.
29. Perform final tightening of nuts and bolts on each part under unladen conditions with tires on level ground when removing steering gear assembly. Check wheel alignment.
30. Adjust neutral position of steering angle sensor after checking wheel alignment.

POWER STEERING PUMP

BLEEDING

1. Before servicing the vehicle, refer to the Precautions Section.
2. Turn steering wheel several times from full left stop to full right stop with engine off.
3. Fill reservoir tank with a sufficient amount of fluid so that fluid level is not below the MIN line while turning steering wheel.
4. Start the engine and hold steering wheel at each lock position for 3 second at idle to check for fluid leakage.
5. Repeat step 4 above several times at approximately 3 second intervals.

❊❊ WARNING

To prevent oil pump damage, never hold the steering wheel in a locked position for more than 10 seconds.

6. Check fluid for bubbles and white contamination.
7. Stop the engine if bubbles and white contamination do not drain out. Perform step 4 and 5 above after waiting until bubbles and white contamination drain out.
8. Stop the engine, and then check fluid level.

➡ **If air bleeding is not complete, the following symptoms can be observed: bubbles in reservoir tank, clicking noise from oil pump, excessive buzzing in the oil pump.**

➡ **Fluid noise may occur in the steering gear or oil pump. This does not affect performance or durability of the system.**

REMOVAL & INSTALLATION
See Figures 135 and 136.

1. Before servicing the vehicle, refer to the Precautions section.
2. Disconnect the negative battery cable.
3. Drain power steering fluid from the reservoir tank.
4. Remove the front wheels and tires.

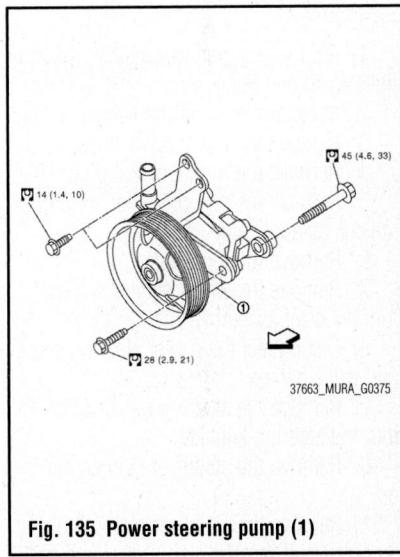

Fig. 135 Power steering pump (1)

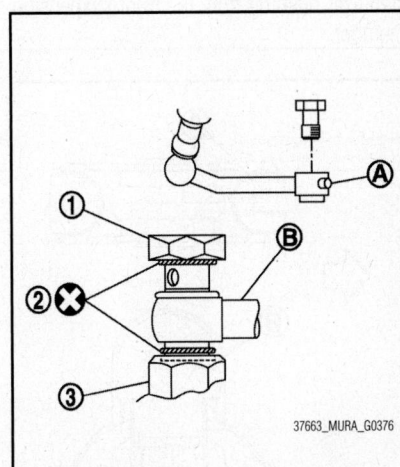

Fig. 136 When installing eye bolt (1) and copper washers (2) to oil pump (3), install the eye bolt with eye joint (B) protrusion (A) facing with pump side cutout

5. Remove the splash guard.

6. Loosen the drive belt.

7. Remove the drive belt from the oil pump pulley.

8. Remove the copper washers and eye bolt (drain fluid).

9. Remove the suction hose (drain fluid).

10. Remove the oil pump mounting bolts, and then remove oil pump. Be careful not to damage halfshaft boot.

To install:

11. Note the following, and install in the reverse order of removal.

12. When installing suction hoses, refer to the following:

a. Never apply fluid to the hose and tube.

b. Insert hose securely until it contacts spool of tube.

c. Leave clearance when installing clamp. Standard: 0.12–0.31 in. (3–8mm)

13. When installing eye bolt and copper washers to oil pump, refer to the following:

a. Do not reuse copper washer.

b. Apply power steering fluid to around copper washer, and install eye bolt.

c. Install eye bolt with eye joint (assembled to high pressure hose) protrusion facing with pump side cutout, and then tighten it to the specified torque after tightening by hand.

d. Securely insert harness connector to pressure sensor.

14. Adjust belt tension.

15. Check fluid level, fluid leakage and air bleeding hydraulic system after the installation.

SUSPENSION

STABILIZER BAR & LINKS

REMOVAL & INSTALLATION

See Figure 137.

1. Before servicing the vehicle, refer to the Precautions Section.

2. Raise and support the vehicle safely.

3. Remove the tire and wheel assembly.

4. Remove the front exhaust pipe.

5. Remove the rear propeller shaft from transfer, as applicable.

6. Remove the lock plate.

7. Remove the wheel sensor harness from the strut assembly.

8. Disconnect the power steering solenoid valve harness connector.

9. Remove the steering outer socket from the steering knuckle.

10. Remove the stabilizer connecting rod.

11. Remove the stabilizer clamp mounting bolts, and remove stabilizer clamp and stabilizer bushing from the front suspension member.

12. Remove the stabilizer bar.

To install:

Install in the reverse order of removal. Position the stabilizer clamp notch and bushing slit toward the front of the vehicle.

STRUTS

REMOVAL & INSTALLATION

See Figure 138.

1. Before servicing the vehicle, refer to the Precautions Section.

2. Raise and support the vehicle safely.

3. Remove the wheels and tires.

4. Remove the lock plate.

5. Remove the wheel sensor.

6. Remove the stabilizer connecting rod from the strut assembly.

7. Remove the strut assembly from the steering knuckle.

8. Remove the cowl top cover.

9. Remove the strut mounting insulator mounting bolts, and remove the strut assembly.

To install:

Install in the reverse order of the removal. Perform final tightening of the strut assembly lower side (rubber bushing) under unladen conditions with tires on level ground. Check and adjust the front end alignment as necessary.

➡**Be sure to replace all non reusable components with new ones.**

OVERHAUL

1. Before servicing the vehicle, refer to the Precautions Section.

✳✳ WARNING

Never damage the strut assembly piston rod when removing the components from the strut assembly.

FRONT SUSPENSION

2. Install the strut attachment (SST: ST35652000) to the strut assembly and secure it in a vise. Wrap a shop cloth around the strut to protect it from damage.

3. Using a spring compressor, compress the coil spring between the strut mounting bearing and the lower rubber seat (on strut assembly). Be sure the spring compressor is securely attached to the coil spring.

4. Make sure the coil spring between the strut mounting bearing and lower rubber seat (strut assembly) is free. Remove the piston rod lock nut while securing the piston rod tip so that piston rod does not turn.

5. Remove the strut mounting insulator, mounting bearing, and bumper from the strut.

6. After removing the coil spring with a spring compressor, gradually release the spring compressor. Make sure the coil spring attachment position does not move.

7. Remove the lower rubber seat from the strut.

8. Remove the strut attachment from the strut.

To install:

9. Install the strut attachment (SST: ST35652000) to the strut assembly and secure it in a vise.

10. Install the lower rubber seat.

11. Install the bumper onto the strut mounting insulator.

12. Using a spring compressor, compress the coil spring and install it onto strut assembly.

a. Face the tube side of the coil spring downward. Align the lower end to the lower rubber seat.

b. Be sure a compressor is securely attached to the coil spring. Compress the coil spring.

c. Set the coil spring so that its paint marks are aligned with the positions of

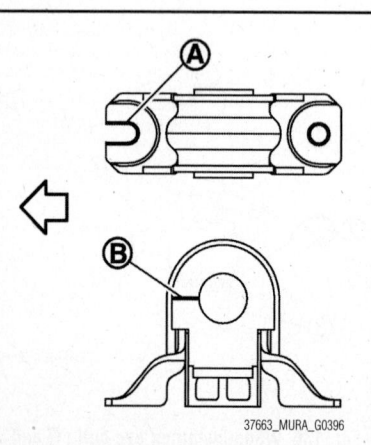

Fig. 137 Position the stabilizer clamp notch (A) and bushing slit (B) toward the front of the vehicle

37663_MURA_G0396

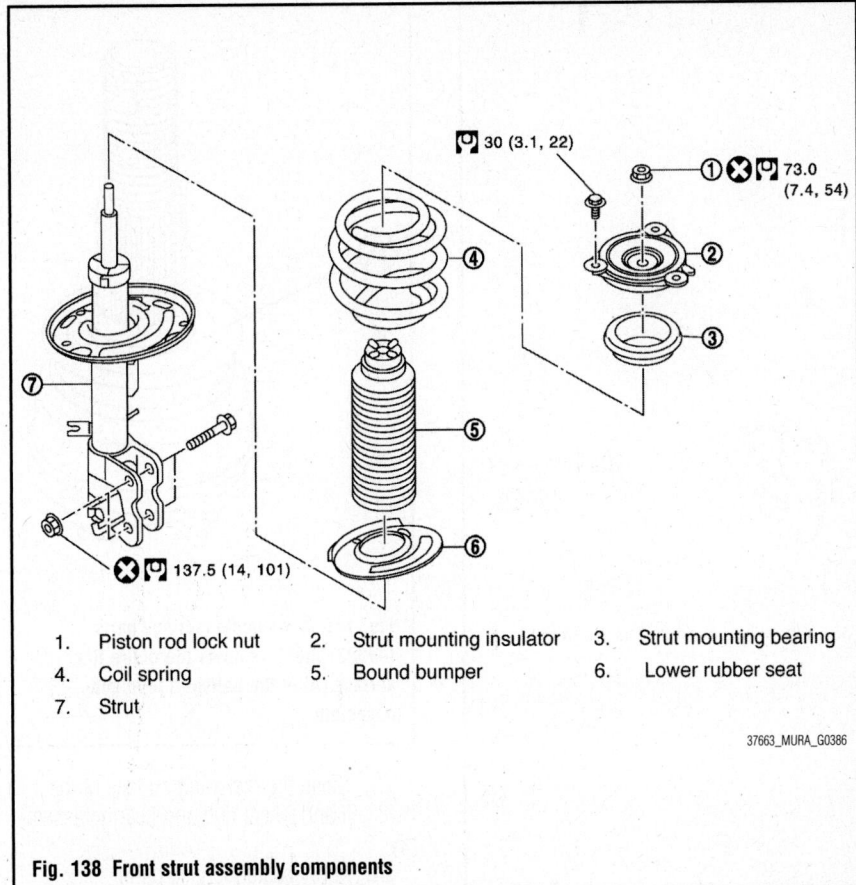

Fig. 138 Front strut assembly components

1.	Piston rod lock nut	2.	Strut mounting insulator	3.	Strut mounting bearing
4.	Coil spring	5.	Bound bumper	6.	Lower rubber seat
7.	Strut				

37663_MURA_G0386

1.25 turns and 2.25 turns from the bottom end of the coil spring.

13. Install the strut mounting bearing and the strut mounting insulator with the bumper to the strut.

14. Secure the piston rod tip so that the piston rod does not turn, then tighten the piston rod lock nut.

✻ WARNING

Never reuse the piston rod lock nut.

15. Gradually release the spring compressor and remove the coil spring. Make sure the coil spring attachment position does not move.

16. Remove the strut attachment from the strut assembly.

17. Check the wheel sensor harness.

18. Check the wheel alignment and adjust as necessary.

19. Adjust the neutral position of the steering angle sensor.

TRANSVERSE LINK

REMOVAL & INSTALLATION

See Figure 139.

1. Before servicing the vehicle, refer to the Precautions Section.

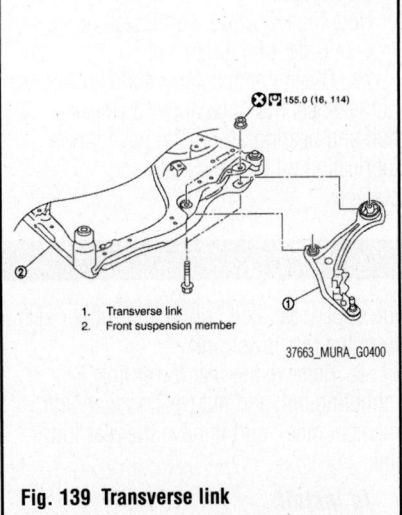

1.	Transverse link
2.	Front suspension member

37663_MURA_G0400

Fig. 139 Transverse link

2. Raise and support the vehicle safely.

3. Remove the tire and wheel assembly.

4. Remove the halfshaft.

5. Remove transverse link from steering knuckle.

6. Remove transverse link from suspension member.

To install:

Install in the reverse order of the removal. Perform final tightening of front suspension member installation position

and strut assembly lower side (rubber bushing) under unladen conditions with tires on level ground. Check the wheel alignment and adjust as necessary.

➡**Be sure to replace all non reusable components with new ones.**

WHEEL HUBS & BEARINGS

ADJUSTMENT

The front wheel bearings are part of a unitized hub and are not adjustable. Move the wheel hub in the axial direction by hand. Make sure that there is no looseness of the wheel bearing. Axial end play is 0.002 inch or less.

REMOVAL & INSTALLATION

See Figures 140 and 141.

1. Before servicing the vehicle, refer to the Precautions Section.

2. Raise and support the vehicle safely.

3. Remove the tire and wheel assembly.

4. Remove the wheel sensor and sensor harness.

5. Remove the lock plate from the strut assembly.

6. Remove the caliper assembly. Hang the caliper assembly aside.

➡**Never depress the brake pedal while the brake caliper is removed.**

7. Remove the disc rotor.

8. Remove the cotter pin, and loosen the wheel hub lock nut.

9. Patch the wheel hub lock nut with a piece of wood. Hammer the wood to disengage wheel hub and bearing assembly from halfshaft.

✻ WARNING

Never place the halfshaft joint at an extreme angle. Also be careful not to overextend the slide joint. Never allow the halfshaft to hang down without support for the joint subassembly, shaft and the other parts.

10. If the wheel hub and bearing assembly and halfshaft cannot be separated even after performing the above procedure, separate using a suitable puller.

11. Remove the wheel hub lock nut.

12. Remove the strut assembly from the steering knuckle.

13. Remove the halfshaft from the wheel hub and bearing assembly, and suspend the halfshaft with a suitable wire.

14. Temporarily tighten the strut assembly and steering knuckle.

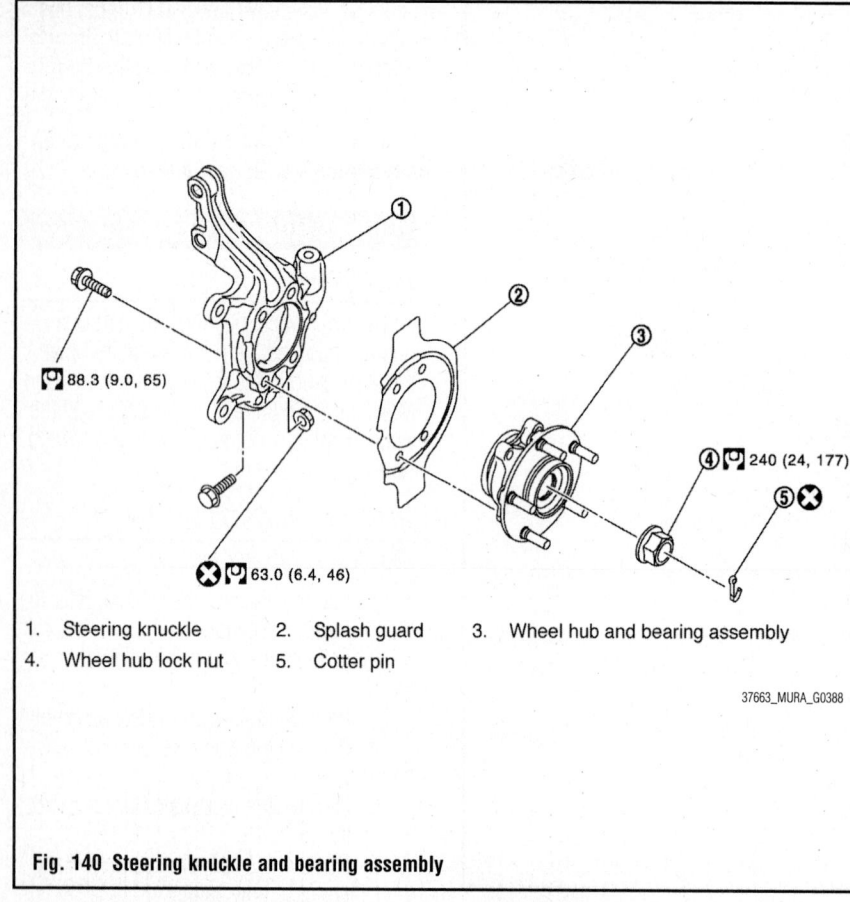

$\square$ 88.3 (9.0, 65)

$\otimes$ $\square$ 63.0 (6.4, 46)

④ $\square$ 240 (24, 177)

⑤ $\otimes$

1. Steering knuckle
2. Wheel hub lock nut
3. Wheel hub and bearing assembly
4. Splash guard
5. Cotter pin

Actual labels:

1. Steering knuckle   2. Splash guard   3. Wheel hub and bearing assembly
4. Wheel hub lock nut   5. Cotter pin

37663_MURA_G0388

Fig. 140 Steering knuckle and bearing assembly

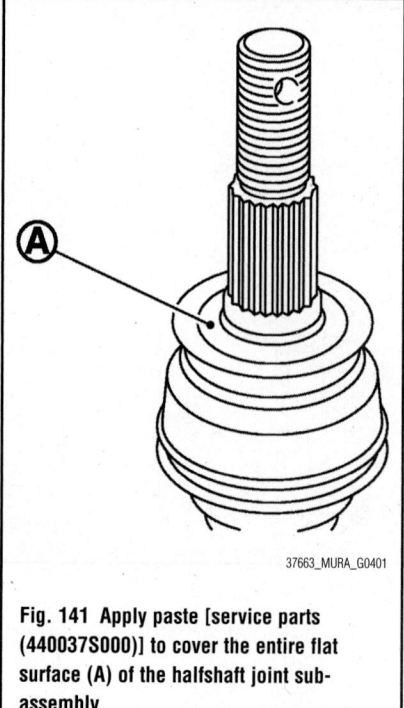

37663_MURA_G0401

Fig. 141 Apply paste [service parts (440037S000)] to cover the entire flat surface (A) of the halfshaft joint sub-assembly

15. Remove the wheel hub and bearing assembly, and remove splash guard.

16. Remove the steering outer socket from the steering knuckle.

17. Remove the steering knuckle from the transverse-link.

18. Remove the steering knuckle from the strut assembly.

To install:

Note the following, and install in the reverse order of the removal:

19. Clean the matching surface of the wheel hub lock nut and wheel hub and bearing assembly; never apply lubricating oil to these matching surfaces.

20. Clean the matching surface of the halfshaft and wheel hub and bearing assembly.

21. Apply paste [service parts (440037S000)] to cover the entire flat surface of the halfshaft joint sub-assembly. Amount: 0.04–0.10 oz. (1.0–3.0 g)

22. Never use a power tool to tighten the wheel hub lock nut.

23. Perform the final tightening of each of parts under unladen conditions.

24. Never reuse the cotter pin.

SUSPENSION

REAR SUSPENSION

COIL SPRINGS & REAR LOWER LINK

REMOVAL & INSTALLATION

See Figures 142 through 146.

1. Before servicing the vehicle, refer to the Precautions Section.

2. Raise and support the vehicle safely.

3. Remove the tire and wheel assembly.

4. Remove the stabilizer connecting rod (lower side).

5. Position a jack under the rear lower link.

6. Loosen the rear lower link mounting bolt and nut (rear suspension member side), and remove rear the lower link mounting bolt and nut (axle housing side).

7. Slowly lower the jack, and remove the upper seat, coil spring and rubber seat from the rear lower link.

8. Remove the rear lower link mounting bolt and nut (rear suspension member side), and remove the rear lower link.

To install:

Note the following, and install in the reverse order of removal:

9. Ensure that the upper seat is attached as shown.

10. For hardtop models, position the protrusion with the projection of the upper seat toward the outside of the vehicle (lateral direction). Make sure that the projection on the inside of the upper seat is securely fitted on the bracket tabs.

11. For CrossCabriolet models, when installing upper seat, align protrusion on upper seat inside to tabs of vehicle side bracket.

12. Match up rubber seat indentions and rear lower link grooves and attach.

13. Install the coil spring by aligning the lower end of the coil spring to the step between the rubber seat and the rear lower link.

14. Set the coil spring so that its paint marks are aligned with the positions of 3.5 turns (2 places) and 4.5 turns (1 place) from the bottom end of the coil spring.

15. Perform the final tightening of rear suspension member and axle installation position (rubber bushing) under unladen condition with tires on level ground.

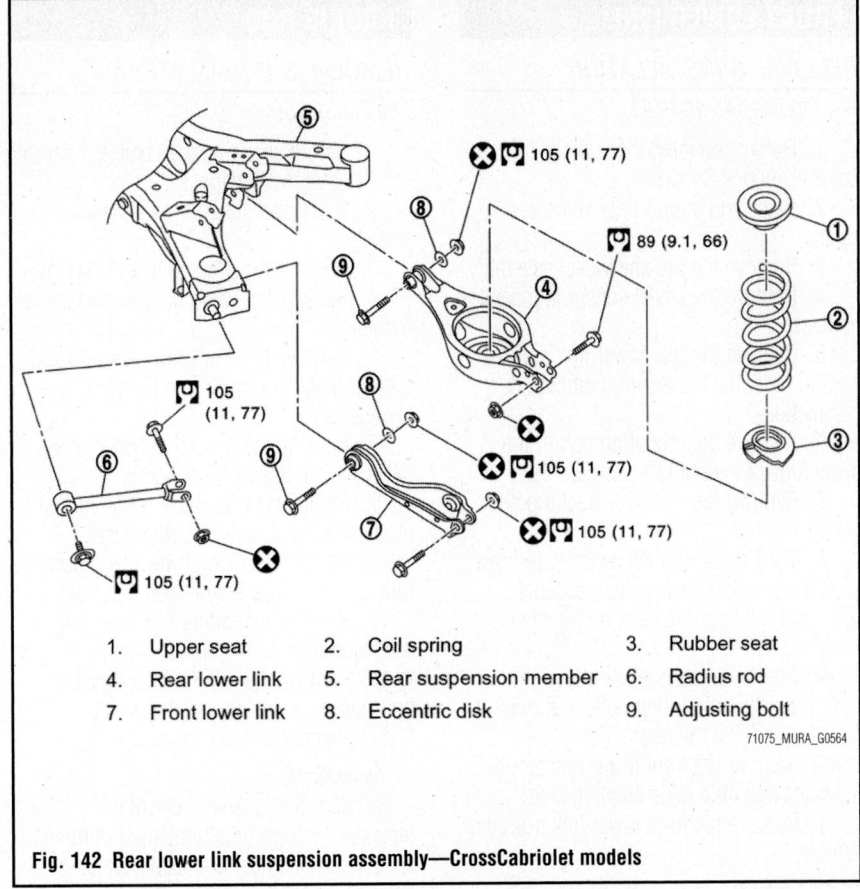

1. Upper seat
2. Coil spring
3. Rubber seat
4. Rear lower link
5. Rear suspension member
6. Radius rod
7. Front lower link
8. Eccentric disk
9. Adjusting bolt

71075_MURA_G0564

Fig. 142 Rear lower link suspension assembly—CrossCabriolet models

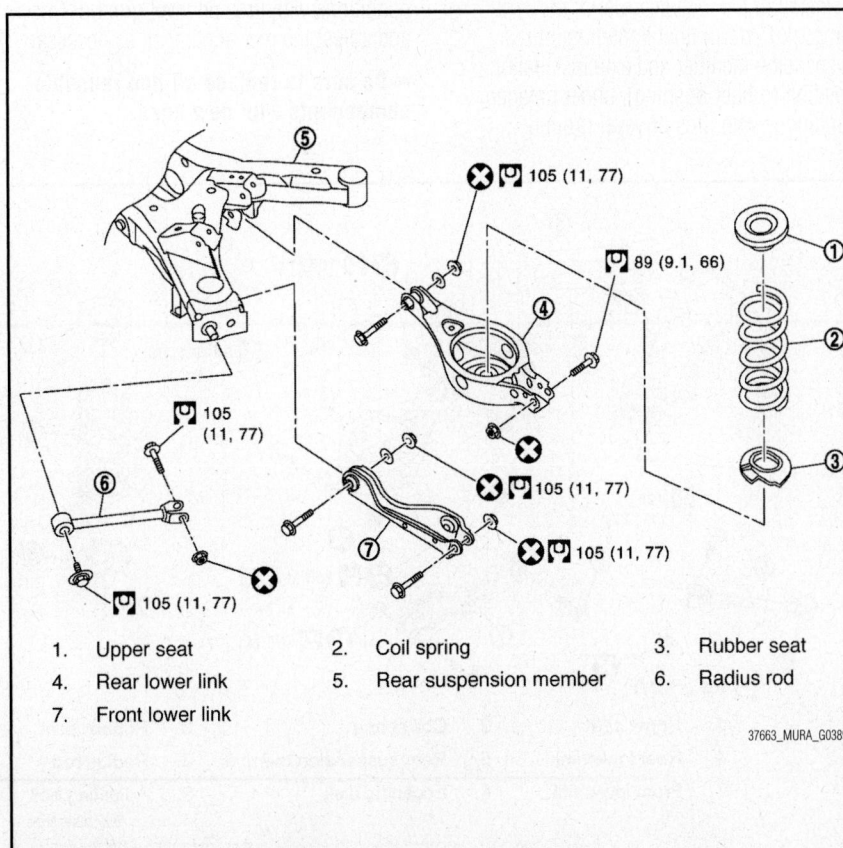

1. Upper seat
2. Coil spring
3. Rubber seat
4. Rear lower link
5. Rear suspension member
6. Radius rod
7. Front lower link

37663_MURA_G0389

Fig. 143 Rear lower link suspension assembly—Hardtop models

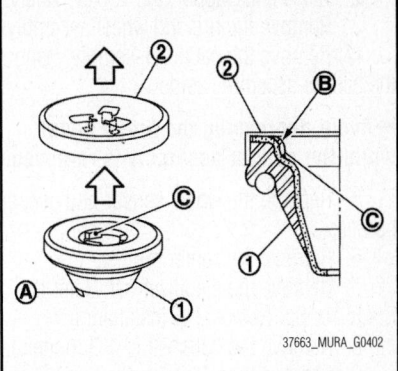

37663_MURA_G0402

Fig. 144 Upper seat (1), outside projection (A), inside projection (C), bracket (2), and tabs (B)—Hardtop models

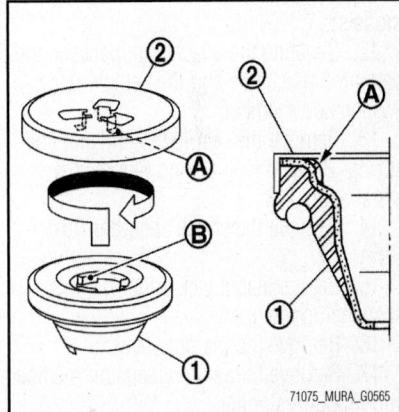

71075_MURA_G0565

Fig. 145 Upper seat (1), protrusion (B), bracket (2), and tabs (A)—CrossCabriolet models

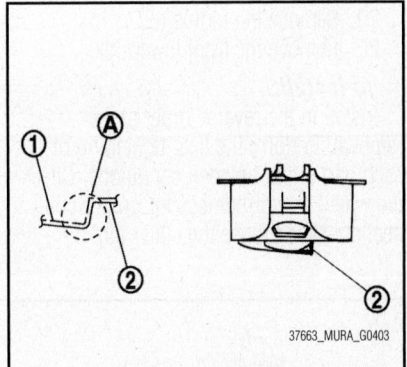

37663_MURA_G0403

Fig. 146 Align the lower end of the coil spring to the step (A) between the rubber seat (1) and the rear lower link (2)

CROSSMEMBER

REMOVAL & INSTALLATION

See Figure 147.

1. Before servicing the vehicle, refer to the Precautions Section.

2. Raise and support the vehicle safely.

3. Remove the tire and wheel assembly.

4. Remove the caliper assembly. Hang the caliper assembly aside.

➡**Avoid depressing the brake pedal while the caliper assembly is removed.**

5. Remove the wheel sensor and sensor harness.

6. Remove the center muffler.

7. Remove the stabilizer bar. Refer to Stabilizer Bar Removal & Installation.

8. Remove the halfshaft (AWD models).

9. Remove the propeller shaft (AWD models).

10. Remove the harness from the rear final drive and rear suspension member (AWD models).

11. Remove the rear final drive (AWD models).

12. Separate the attachment between the parking brake cable and the vehicle rear suspension member.

13. Remove the rear lower link and coil spring. Refer to Coil Spring Removal & Installation.

14. Remove the shock absorber (lower side).

15. Set a suitable jack under the rear suspension member.

16. Remove the member stay.

17. Remove the rear suspension member and rebound stopper.

18. Slowly lower the jack, and remove the rear suspension member, suspension arm, radius rod, front lower link and axle from the vehicle as a unit.

19. Remove the suspension arm.

20. Remove the radius rod.

21. Remove the front lower link.

To install:

Install in the reverse order of the removal. Perform the final tightening of each part under unladen conditions. Check the wheel sensor harness for proper connection. Never reuse the cotter pin.

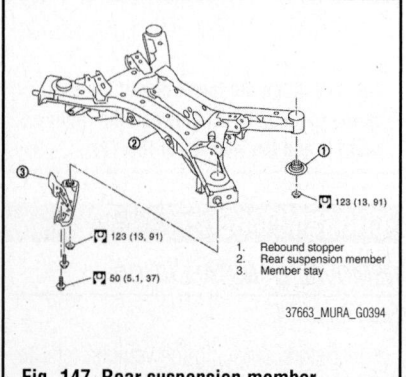

Fig. 147 Rear suspension member

FRONT LOWER LINK

REMOVAL & INSTALLATION

See Figures 148 and 149.

1. Before servicing the vehicle, refer to the Precautions Section.

2. Raise and support the vehicle safely.

3. Remove the tire and wheel assembly.

4. Remove the wheel sensor and sensor harness.

5. Remove the rear lower link and coil spring. Refer to Coil Spring Removal & Installation.

6. Remove the mounting bolt in the lower side of the shock absorber.

7. Remove the stabilizer bushing and clamp.

8. For CrossCabriolet models, remove front lower link adjusting bolt, eccentric disk, and nut (rear suspension member side)

9. For hardtop models, remove the front lower link mounting bolts and nuts (rear suspension member side).

10. Remove the front lower link mounting bolts and nuts (axle housing side).

11. Remove the front lower link from the vehicle.

To install:

Install in the reverse order of the removal. Perform final tightening of rear suspension member and axle installation position (rubber bushing), under unladen conditions with tires on level ground.

RADIUS ROD

REMOVAL & INSTALLATION

See Figure 150.

1. Before servicing the vehicle, refer to the Precautions Section.

2. Raise and support the vehicle safely.

3. Remove the tire and wheel assembly.

4. Remove the wheel sensor and sensor harness.

5. Remove the rear lower link and coil spring. Refer to Coil Spring Removal & Installation.

6. Remove the mounting bolt in the lower side of the shock absorber.

7. Remove the front lower link mounting bolt and nut (axle housing side).

8. Loosen the front lower link mounting bolt and nut (suspension member side).

9. Remove the radius rod mounting bolts and nuts (axle housing side).

10. Remove the radius rod mounting bolt (rear suspension member side).

11. Remove the radius rod.

To install:

Install in the reverse order of the removal. Perform final tightening of the rear suspension member and axle installation position (rubber bushing) under unladen conditions with tires on level ground. Check and adjust the rear alignment, as necessary.

➡**Be sure to replace all non reusable components with new ones.**

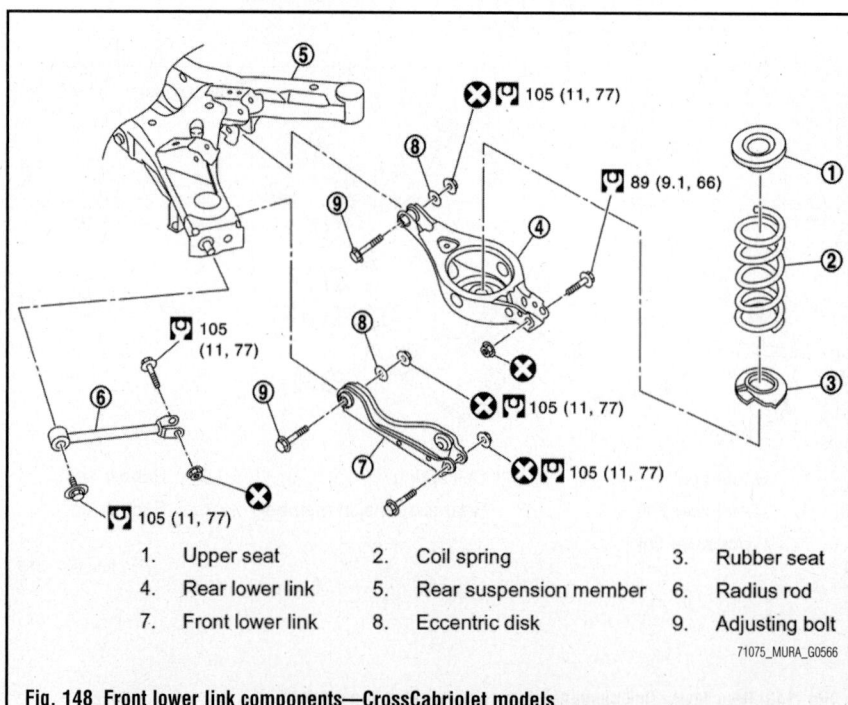

1. Upper seat	2. Coil spring	3. Rubber seat
4. Rear lower link	5. Rear suspension member	6. Radius rod
7. Front lower link	8. Eccentric disk	9. Adjusting bolt

Fig. 148 Front lower link components—CrossCabriolet models

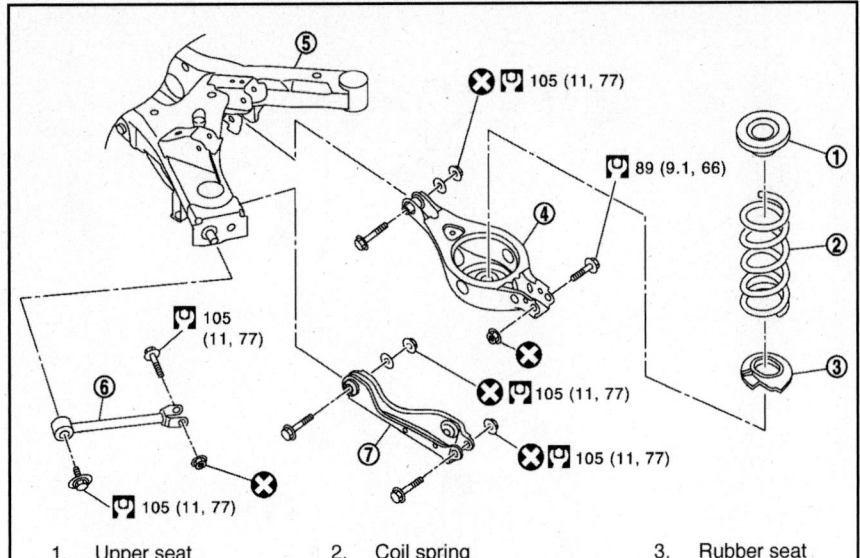

1. Upper seat
4. Rear lower link
7. Front lower link
2. Coil spring
5. Rear suspension member
3. Rubber seat
6. Radius rod

37663_MURA_G0389

Fig. 149 Front lower link components—Hardtop models

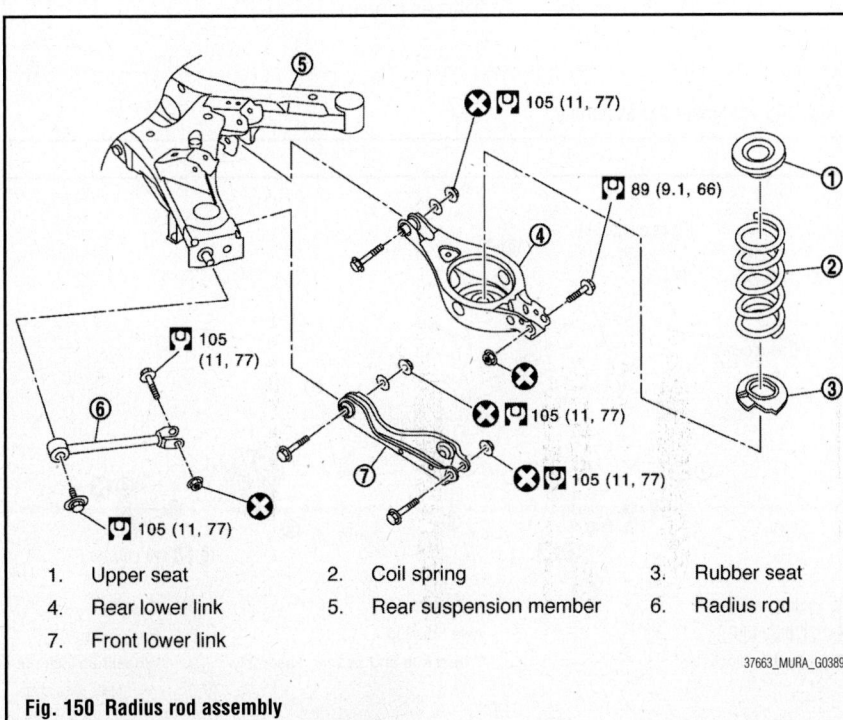

1. Upper seat
4. Rear lower link
7. Front lower link
2. Coil spring
5. Rear suspension member
3. Rubber seat
6. Radius rod

37663_MURA_G0389

Fig. 150 Radius rod assembly

REAR SUSPENSION ARM

REMOVAL & INSTALLATION

See Figure 151.

1. Before servicing the vehicle, refer to the Precautions Section.
2. Raise and support the vehicle safely.
3. Remove the tire and wheel assembly.
4. For AWD models, remove the suspension arm with the rear suspension member.
5. Remove the wheel sensor and sensor harness.
6. Remove the stabilizer connecting rod.
7. Remove the cotter pin of the suspension arm ball joint, and loosen the nut.
8. Use the ball joint remover to remove the suspension arm from the axle housing. Be careful not to damage the ball joint boot.

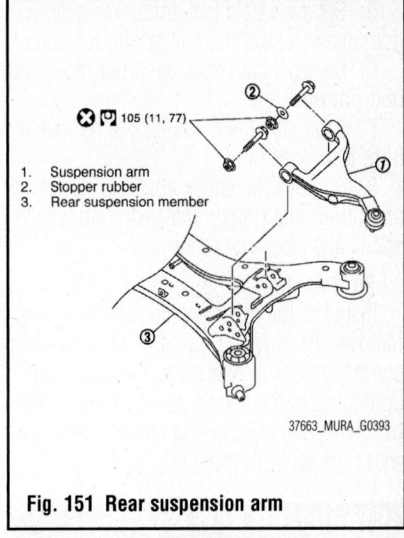

1. Suspension arm
2. Stopper rubber
3. Rear suspension member

37663_MURA_G0393

Fig. 151 Rear suspension arm

➡**Temporarily tighten the mounting nut to prevent damage to the threads and to prevent the ball joint remover from coming off.**

9. Remove the suspension arm.

To install:

Install in the reverse order of removal. Perform final tightening of rear suspension member installation position (rubber bushing), under unladen conditions with tires on level ground.

SHOCK ABSORBERS

REMOVAL & INSTALLATION

See Figure 152.

1. Before servicing the vehicle, refer to the Precautions Section.
2. Raise and support the vehicle safely.
3. Remove the tire and wheel assembly.
4. Separate the stabilizer connecting rod (lower side).

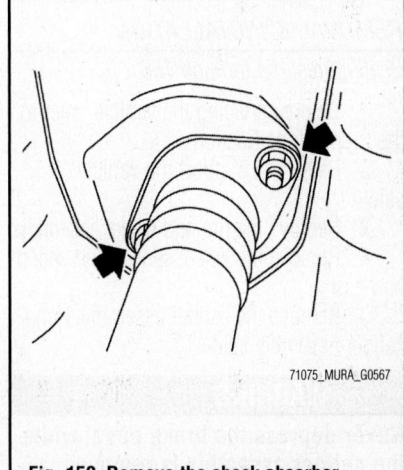

71075_MURA_G0567

Fig. 152 Remove the shock absorber assembly mounting nuts (upper side)

5. Set a suitable jack under the axle housing to relieve the coil spring tension.

6. Remove the shock absorber mounting bolt (lower side).

7. Gradually lower the jack to remove it from front lower link.

8. Remove the shock absorber assembly mounting nuts (upper side) and remove the shock absorber assembly.

To install:

Install in the reverse order of the removal. Be sure to replace all non reusable components with new ones. Perform final tightening of the shock absorber lower side (rubber bushing) under unladen conditions with tires on level ground.

STABILIZER BAR & LINKS

REMOVAL & INSTALLATION

See Figure 153.

1. Before servicing the vehicle, refer to the Precautions Section.

2. Raise and support the vehicle safely.

3. Remove the wheels and tires.

4. Remove the stabilizer connecting rod. Apply matchmarks to identify installation position.

5. Remove the mounting bolts on the stabilizer clamp and remove the stabilizer bar.

To install:

Install in the reverse order of removal. Check the matchmarks when installing. Tighten the mounting nut while holding a hexagonal part of the stabilizer connecting rod side.

WHEEL HUBS & BEARINGS

ADJUSTMENT

The rear wheel bearings are part of a unitized hub and are not adjustable.

REMOVAL & INSTALLATION

See Figures 154 through 156.

1. Before servicing the vehicle, refer to the Precautions Section.

2. Raise and support the vehicle safely.

3. Remove the tire and wheel assembly.

4. Remove the wheel sensor and sensor harness.

5. Remove the caliper assembly. Hang caliper assembly aside.

✳✳ WARNING

Never depress the brake pedal while the caliper assembly is removed.

6. Remove the disc rotor.

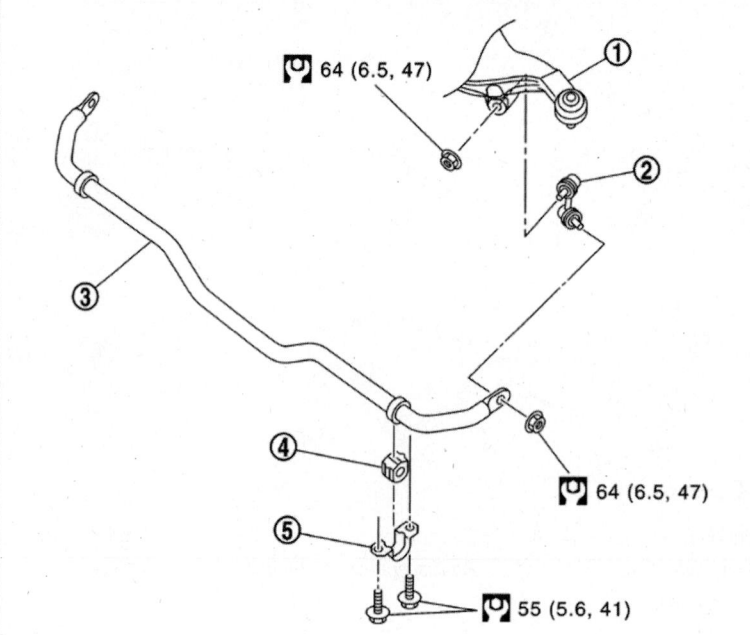

1. Suspension arm   2. Stabilizer connecting rod   3. Stabilizer bar
4. Stabilizer bushing   5. Stabilizer clamp

71075_MURA_G0568

Fig. 153 Stabilizer bar assembly

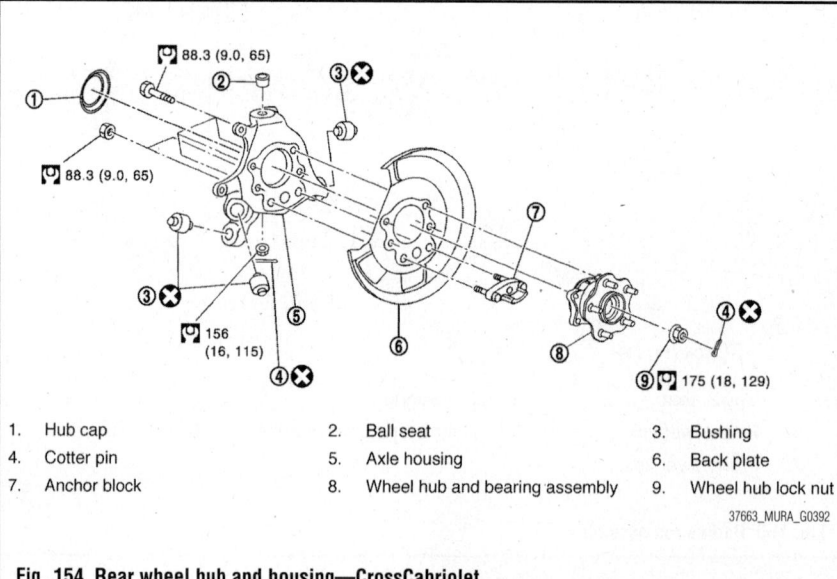

1.	Hub cap	2.	Ball seat	3.	Bushing
4.	Cotter pin	5.	Axle housing	6.	Back plate
7.	Anchor block	8.	Wheel hub and bearing assembly	9.	Wheel hub lock nut

37663_MURA_G0392

Fig. 154 Rear wheel hub and housing—CrossCabriolet

7. Remove the cotter pin, and loosen the suspension arm mounting nut of the axle housing.

8. Remove the cotter pin, and loosen the wheel hub lock nut.

9. Patch the wheel hub lock nut with a piece of wood. Hammer the wood to disengage wheel hub and bearing assembly from halfshaft. Remove the wheel hub lock nut.

✳✳ WARNING

Never place halfshaft joint at an extreme angle. Also be careful not to overextend slide joint. Never allow the halfshaft to hang down without support for the housing (or joint sub-assembly), shaft and the other parts.

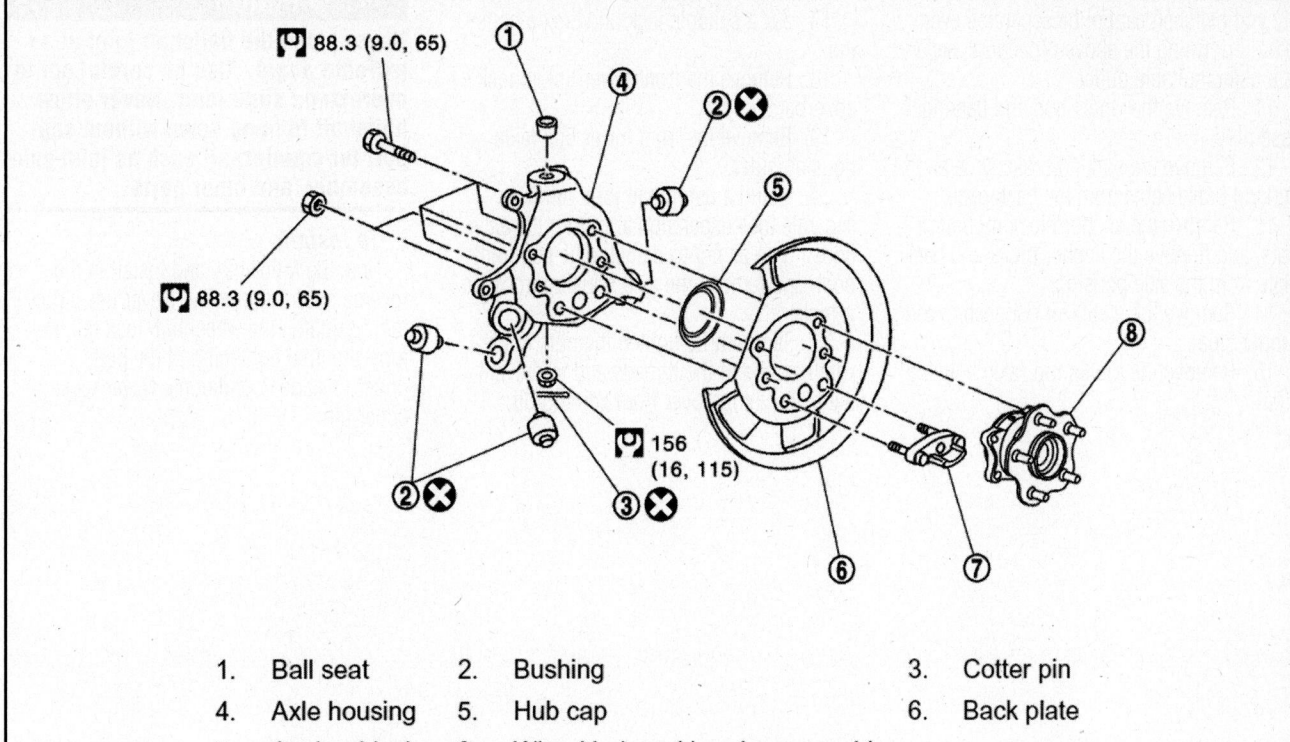

1. Ball seat
2. Bushing
3. Cotter pin
4. Axle housing
5. Hub cap
6. Back plate
7. Anchor block
8. Wheel hub and bearing assembly

71075_MURA_G0569

Fig. 155 Rear wheel hub and housing—Hardtop models, 2WD vehicles

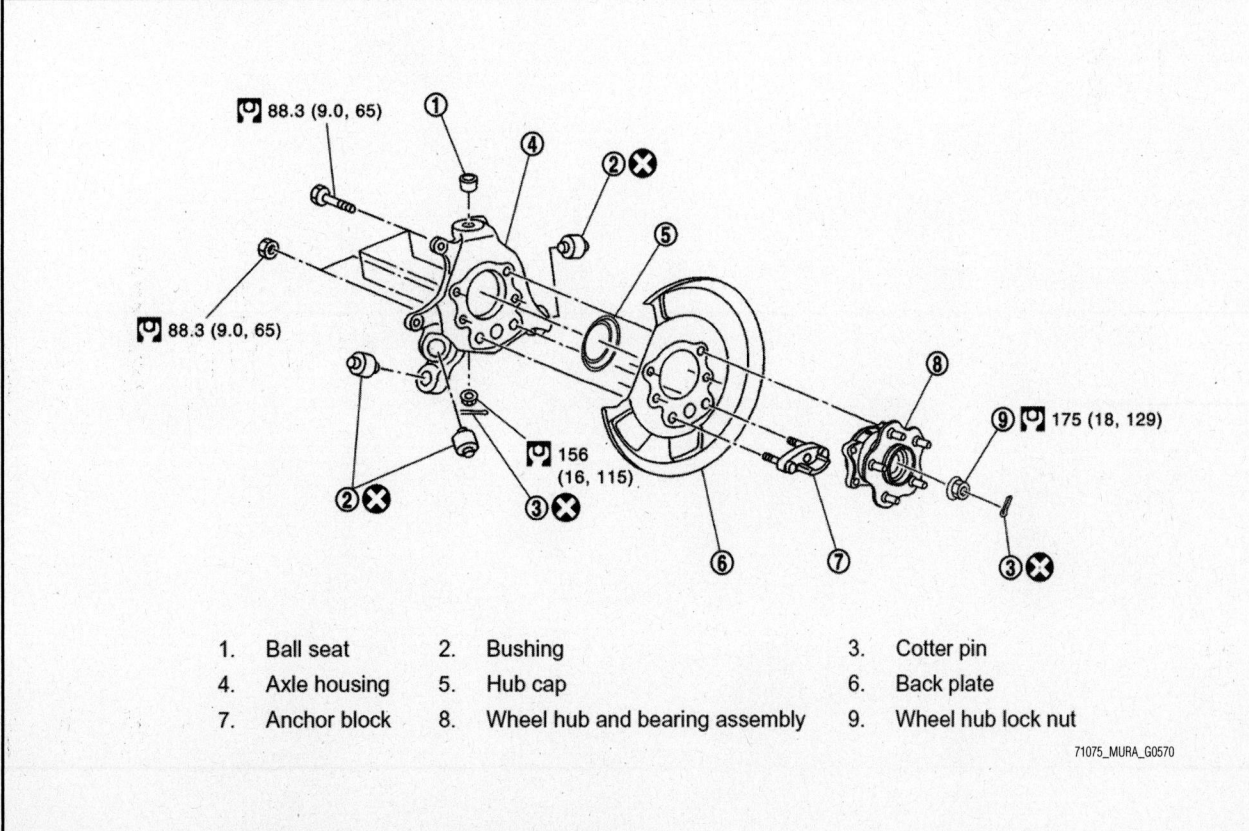

1. Ball seat
2. Bushing
3. Cotter pin
4. Axle housing
5. Hub cap
6. Back plate
7. Anchor block
8. Wheel hub and bearing assembly
9. Wheel hub lock nut

71075_MURA_G0570

Fig. 156 Rear wheel hub and housing—Hardtop models, AWD vehicles

10. If the wheel hub and bearing assembly and halfshaft cannot be separated even after performing the above procedure, separate using suitable puller.

11. Remove the wheel hub and bearing assembly.

12. Remove the parking brake shoe and parking brake cable from the back plate.

13. Remove the anchor block mounting nuts, and remove the anchor block and back plate from the axle housing.

14. Remove the stabilizer connecting rod (upper side).

15. Remove the radius rod (axle housing side).

16. Remove the coil spring.

17. Set a suitable jack under axle housing.

18. Remove the front lower link (shock absorber side).

19. Remove the front lower link (axle housing side).

20. Using a using ball joint remover, separate the suspension arm from the axle housing so as not to damage the ball joint boot, and remove the axle housing from the vehicle.

21. Temporarily tighten the nuts to prevent damage to the threads and to prevent the ball joint remover from coming off.

✵ WARNING

Never place the halfshaft joint at an extreme angle. Also be careful not to overextend slide joint. Never allow halfshaft to hang down without support for counterpart such as joint subassembly, and other parts.

To install:
Note the following, and install in the reverse order of removal. Never use a power tool to tighten the wheel hub lock nut. Perform the final tightening of the parts under un-laden conditions. Never reuse cotter pin.

NISSAN

NV

SPECIFICATIONS AND MAINTENANCE CHARTS

ENGINE AND VEHICLE IDENTIFICATION

		Engine					Model Year	
Code ①	Liters (cc)	Cu. In.	Cyl.	Fuel Sys.	Engine	Eng. Mfg.	Code ②	Year
VQ40DE	4.0 (3954)	241	6	MFI	DOHC	Nissan	C	2012
VK56DE	5.6 (5552)	339	8	MFI	DOHC	Nissan		

MFI: Multi-port Fuel Injection

DOHC: Double Overhead Camshafts

① Fourth digit of the Vehicle Identification Number (VIN) is the engine ID, A: VQ40DE engine. B: VK56DE engine.

② Tenth digit of the Vehicle Identification Number (VIN)

71075_NV15_C0001

GENERAL ENGINE SPECIFICATIONS

Year	Model	Engine Displacement Liters (cc)	Engine ID/VIN	Fuel System Type	Net Horsepower @ rpm	Net Torque @ rpm (ft. lbs.)	Bore x Stroke (in.)	Com- pression Ratio	Oil Pressure @ rpm
2012	NV 1500	4.0 (3954)	VQ40DE	MFI	261@5600	281@4000	3.76X3.62	9.7:1	43@2000
	NV 2500	4.0 (3954)	VQ40DE	MFI	261@5600	281@4000	3.76X3.62	9.7:1	43@2000
		5.6 (5552)	VQ56DE	MFI	317@5200	385@3400	3.86X3.62	9.8:1	43@2000
	Nv 3500	5.6 (5552)	VQ56DE	MFI	317@5200	385@3400	3.86X3.62	9.8:1	43@2000

MFI: Multi-port Fuel Injection

71075_NV15_C0002

ENGINE TUNE-UP SPECIFICATIONS

Year	Engine Displacement Liters (cc)	Engine ID/VIN	Spark Plug Gap (in.)	Ignition Timing (deg.)		Fuel Pump (psi)	Idle Speed (rpm)		Valve Clearance (in.)	
				MT	AT		MT	AT	In.	Ex.
2012	4.0 (3954)	VQ40DE	0.043	NA	①	③	NA	675-775	HYD	HYD
	5.6 (5552)	VQ56DE	0.043	NA	②	③	NA	550-650	HYD	HYD

NOTE: The Vehicle Emission Control Information label often reflects specification changes made during production. The label figures

must be used if they differ from those in this chart.

NA- Not Available

HYD: Hydraulic

① 20 degrees +/- 5 degrees

② 15 degrees +/- 5 degrees

③ 51 psi at idle

71075_NV15_C0003

CAPACITIES

Year	Model	Engine Displacement Liters	Engine ID/VIN	Engine Oil with Filter (qts.)	Transmission (pts.) 5-Spd	Transmission (pts.) Auto.	Transfer Case (pts.)	Drive Axle Front (pts.)	Drive Axle Rear (pts.)	Fuel Tank (gal.)	Cooling System (qts.)
2012	NV Series	4.0	VQ40DE	5.38	—	22.50	NA	NA	5.50	28	13.38
		5.6	VQ56DE	6.78	—	22.50	NA	NA	5.50	28	13.28

NOTE: All capacities are approximate. Add fluid gradually and check to be sure a proper fluid level is obtained.

NA - Not Available

71075_NV15_C0004

FLUID SPECIFICATIONS

Year	Model	Engine Displ. Liters	Engine Oil	Auto. Trans.	Drive Axle Front	Drive Axle Rear	Power Steering Fluid	Brake Master Cylinder	Cooling System
2012	NV Series	4.0	SAE 5W-30	①	NA	②	③	④	⑤
		5.6	SAE 5W-30	①	NA	②	③	④	⑤

NA - Not Available

① Genuine Nissan Matic S ATF fluid. If not available genuine Nissan Matic J ATF may be used.

② Nissan differential oil synthetic 75W-90 or API GL-5 synthetic gear oil

③ Nissan Power Steering Fluid

④ Nissan Super Heavy Duty DOT 3

⑤ Nissan Long Life antifreeze (BLUE) or equivalent

71075_NV15_C0012

VALVE SPECIFICATIONS

Year	Engine Displacement Liters	Engine ID/VIN	Seat Angle (deg.)	Face Angle (deg.)	Spring Test Pressure (lbs. @ in.)	Spring Installed Height (in.)	Stem-to-Guide Clearance (in.) Intake	Stem-to-Guide Clearance (in.) Exhaust	Stem Diameter (in.) Intake	Stem Diameter (in.) Exhaust
2012	4.0	VQ40DE	45.15-45.45	45	37-42 @1.457	1.457	0.0008-0.0021	0.0012-0.0025	0.2348-0.2354	0.2344-0.2350
	5.6	VQ56DE	45.15-45.45	45	37-42 @1.457	1.991	0.0008-0.0021	0.0012-0.0025	0.2348-0.2354	0.2344-0.2350

71075_NV15_C0006

CAMSHAFT AND BEARING SPECIFICATIONS CHART

All measurements are given in inches.

Year	Engine Displacement Liters	Engine ID/VIN	Journal Diameter	Brg. Oil Clearance	Shaft End-play	Runout	Journal Bore	Lobe Height Intake	Exhaust
2012	4.0	VQ40DE	①	②	0.0045-0.0074	0.0008	③	1.7900-1.7921	1.7746-1.7821
	5.6	VQ56DE	1.0217-1.0224	0.0012-0.0028	0.0045-0.0074	0.0008	1.0236-1.0244	1.7663-1.7738	1.7746-1.7821

① No. 1: 1.0211-1.0218
 Nos. 2-4: 0.9230-0.9238

② No. 1: 0.0018-0.0034
 Nos. 2-4: 0.0014-0.0030

③ No. 1: 1.0236-1.0244
 Nos. 2-4: 0.9252-0.9260

71075_NV15_C0014

CRANKSHAFT AND CONNECTING ROD SPECIFICATIONS

All measurements are given in inches.

Year	Engine Displacement Liters	Engine ID/VIN	Crankshaft Main Brg. Journal Dia.	Main Brg. Oil Clearance	Shaft End-play	Thrust on No.	Connecting Rod Journal Diameter	Oil Clearance	Side Clearance
2012	4.0	VQ40DE	①	0.0014-0.0018	0.0039-0.0098	4	NA	0.0013-0.0023	0.0079-0.0138
	5.6	VQ56DE	②	③	0.0039-0.0102	4	④	0.0008-0.0015	0.0079-0.0157

NA - Not Available

① There are 24 different grades, ranging from grade A (2.7549) to grade 7 (2.7540)

② There are 24 different grades, ranging from grade A (2.5182) to grade 2 (2.5174)

③ No. 1 and 5: 0.00004-0.0004
 No. 2, 3 and 4: 0.0003-0.0007

④ There are 13 different grades, ranging from 2.2441- 2.2446. Specification is for rod bearing housing.

71075_NV15_C0005

PISTON AND RING SPECIFICATIONS

All measurements are given in inches.

Year	Engine Displacement Liters	Engine ID/VIN	Piston Clearance	Ring Gap Top Comp.	Bottom Comp.	Oil Control	Ring Side Clearance Top Comp.	Bottom Comp.	Oil Control
2012	4.0	VQ40DE	0.0004-0.0012	0.0091-0.0130	0.0130-0.0189	0.0079-0.0197	0.0018-0.0031	0.0012-0.0028	0.0026-0.0053
	5.6	VQ56DE	0.0004-0.0012	0.0091-0.0130	0.098-0.0157	0.0079-0.0236	0.0014-0.0033	0.0012-0.0028	0.0006-0.0073

71075_NV15_C0007

TORQUE SPECIFICATIONS
All readings in ft. lbs.

Year	Engine Displacement Liters	Engine ID/VIN	Cylinder Head Bolts	Main Bearing Bolts	Rod Bearing Bolts	Crankshaft Damper Bolts	Flywheel Bolts	Manifold Intake	Manifold Exhaust	Spark Plugs
2012	4.0	VQ40DE	①	②	③	④	65	⑤	22	18
	5.6	VQ56DE	⑥	⑦	⑧	⑨	65	NA	25	18

NA - Not Available

① Step 1: 72 ft. lbs.

    Step 2: Loosen all bolts completely

    Step 3: 29 ft. lbs.

    Step 4: +90 degrees

    Step 5: +90 degrees

② Step 1: 26 ft. lbs.

    Step 2: +90 degrees

③ Step 1: 14 ft. lbs.

    Step 2: +90 degrees

④ 33 ft. lbs. +84 - 90 degrees

⑤ Step 1: 65 inch lbs.

    Step 2 and after: 21 ft. lbs.

    Studs 8 ft. lbs.

⑥ Step 1: 33 ft. lbs

    Step 2: +70 degrees clockwise

    Step 3: loosen in reverse order of tightening sequence

    Step 4: 33 ft. lbs.

    Step 5: +60 degrees clockwise

    Step 6: +60 degrees clockwise

⑦ Step 1: cap bolts in order 1-10: 29 ft. lbs.

    Step 2: cap sub bolts in order 11-20: 22 ft. lbs.

    Step 3: cap bolts in order 1-10: +40 degrees

    Step 4: cap sub bolts in order 11-20: +30 degrees

    Step 5: side bolts in order 21-30: 36 ft. lbs.

⑧ Step 1: 11 ft. lbs.

    Step 2: +90 degrees

⑨ 69 ft. lbs. +90 degrees

71075_NV15_C0008

WHEEL ALIGNMENT

Year	Model	Caster Range (+/-Deg.)	Caster Preferred Setting (Deg.)	Camber Range (+/-Deg.)	Camber Preferred Setting (Deg.)	Toe-in (in.)
2012	NV 1500/2500	①	①	②	②	③
	NV 3500	④	④	②	②	⑤

NA - Not Available

① Minimum: 5 degrees 35' (5.58 degrees). Nominal: 6 degrees 05' (6.08 degrees). Maximum: 6 degrees 35' (6.58 degrees).

② Minimum: -0 degrees 30' (-0.50 degrees). Nominal: 0 degrees 00' (0.00 degrees). Maximum: 0 degree 30' (0.50 degree).

③ Minimum: 5.8mm (0.23 inch). Nominal: 6.8mm (0.27 inch). Maximum: 7.8mm (0.31 inch).

④ Minimum: 5 degrees 25' (5.42 degrees). Nominal: 5 degrees 55' (5.92 degrees). Maximum: 6 degrees 25' (6.42 degrees).

⑤ Minimum: 6.0mm (0.24 inch). Nominal: 7.0mm (0.28 inch). Maximum: 8.0mm (0.31 inch).

71075_NV15_C0009

TIRE, WHEEL AND BALL JOINT SPECIFICATIONS

| Year | Model | OEM Tires | | Tire Pressures (psi) | | Wheel Size | Ball Joint Inspection | Lug Nuts (ft. lbs.) |
		Standard	Optional	Front	Rear			
2012	NV Series	LT245/70R17	none	50	80	17x7.5JJ	①	138
		LT245/75R17	none	50	80	17x7.5JJ	①	138

OEM: Original Equipment Manufacturer

PSI: Pounds Per Square Inch

① Axial play

  Upper: 0

  Lower: 0.008 in.

71075_NV15_C0010

BRAKE SPECIFICATIONS

All measurements in inches unless noted

| Year | Model | | Brake Disc | | | Minimum Lining Thickness | | Brake Caliper Bracket Bolts (ft. lbs.) | Brake Caliper Mounting Bolts (ft. lbs.) |
			Original Thickness	Minimum Thickness	Maximum Runout	Front	Rear		
2012	NV Series	F	1.496	1.437	0.0016	0.039	0.039	①	①
		R	1.181	1.122	0.0028	0.039	0.039	②	②

NA: Not Available

① Torque member mounting bolt: 118 ft. lbs.

  Lower guide pin bolt 45 ft. lbs.

② Torque member mounting bolt: 118 ft. lbs.

  Lower guide pin bolt 45 ft. lbs.

71075_NV15_C0011

SCHEDULED MAINTENANCE INTERVALS (1)
Nissan—NV Series

TO BE SERVICED	TYPE OF SERVICE	7.5	15	22.5	30	37.5	45	52.5	60
Engine oil & filter	R	every 3,750 miles							
Brake lines & cables	S/I		✓		✓		✓		✓
Brake pads, discs	I	✓	✓	✓	✓	✓	✓	✓	✓
Brake fluid	R		✓		✓		✓		✓
Driveshaft boots	I	✓	✓	✓	✓	✓	✓	✓	✓
Automatic transmission (CVT)	I						✓		✓
Air cleaner filter ①	R				✓				✓
Drive belt (s) ②	S/I								✓
Engine coolant ③	R								
Spark plugs	R	Every 105,000 miles							
Cabin air filter	R		✓		✓		✓		✓
Exhaust system	I	✓	✓	✓	✓	✓	✓	✓	✓
Evap vapor lines	I				✓				✓
Fuel lines	I				✓				✓
Steering gear, linkage, axle & suspension parts	I	✓	✓	✓	✓	✓	✓	✓	✓
Tires (rotate)	S/I	✓	✓	✓	✓	✓	✓	✓	✓
Valve clearance ④	S/I				✓				✓

R: Replace   S/I: Service or Inspect   L: Lubricate   I: Inspect

① If operating mainly in dusty conditions, more frequent maintenance may be required.

② First at 60,000, inspect every 15,000 miles and replace as necessary

③ After 105,000, replace every 75,000 thereafter

④ Periodic maintenance not required, if valve noise increases, inspect valve clearance

Follow Periodic Maintenance Schedule 1 if the driving habits frequently include one or more of the following driving conditions:

Repeated short trips of less than 5 miles (8 km).

Repeated short trips of less than 10 miles (16 km) with outside temperatures remaining below freezing

Operating in hot weather in stop-and-go "rush hour" traffic.

Extensive idling and/or low speed driving for long distances, such as police, taxi or door-to-door delivery use

Driving in dusty conditions.

Driving on rough, muddy, or salt spread roads.

Towing a trailer, using a camper or a car-top carrier.

Follow Periodic Maintenance Schedule 2 if none of driving conditions shown in Schedule 1 apply to the driving habits.

71075_NV15_C0013

SCHEDULED MAINTENANCE INTERVALS (2)
Nissan—NV Series

TO BE SERVICED	SERVICE	7.5	15	22.5	30	37.5	45	52.5	60
Engine oil & filter	R	✓	✓	✓	✓	✓	✓	✓	✓
Brake lines & cables	S/I		✓		✓		✓		✓
Brake pads, discs	I		✓		✓		✓		✓
Brake fluid	R				✓				✓
Driveshaft boots	I		✓		✓		✓		✓
Automatic transmission (CVT)	I		✓		✓		✓		✓
Air cleaner filter ①	R				✓				✓
Drive belt (s) ②	S/I								✓
Engine coolant ③	R								
Spark plugs	R				Every 105,000 miles				
Cabin air filter	R		✓		✓		✓		✓
Exhaust system	I				✓				✓
Evap vapor lines	I				✓				✓
Ful lines	I				✓				✓
Steering gear, linkage, axle & suspension parts	I				✓				✓
Tires (rotate)	S/I	✓	✓	✓	✓	✓	✓	✓	✓
Valve clearance ④	S/I				✓				✓

R: Replace   S/I: Service or Inspect   L: Lubricate   I: Inspect

① If operating mainly in dusty conditions, more frequent maintenance may be required.

② First at 60,000, inspect every 15,000 miles and replace as necessary

③ After 105,000, replace every 75,000 thereafter

④ Periodic maintenance not required, if valve noise increases, inspect valve clearance

Follow Periodic Maintenance Schedule 1 if the driving habits frequently include one or more of the following driving conditions:

Repeated short trips of less than 5 miles (8 km).

Repeated short trips of less than 10 miles (16 km) with outside temperatures remaining below freezing

Operating in hot weather in stop-and-go "rush hour" traffic.

Extensive idling and/or low speed driving for long distances, such as police, taxi or door-to-door delivery use

Driving in dusty conditions.

Driving on rough, muddy, or salt spread roads.

Towing a trailer, using a camper or a car-top carrier.

Follow Periodic Maintenance Schedule 2 if none of driving conditions shown in Schedule 1 apply to the driving habits.

71075_NV15_C0015

PRECAUTIONS

Before servicing any vehicle, please be sure to read all of the following precautions, which deal with personal safety, prevention of component damage, and important points to take into consideration when servicing a motor vehicle:

• Never open, service or drain the radiator or cooling system when the engine is hot; serious burns can occur from the steam and hot coolant.

• Observe all applicable safety precautions when working around fuel. Whenever servicing the fuel system, always work in a well-ventilated area. Do not allow fuel spray or vapors to come in contact with a spark, open flame, or excessive heat (a hot drop light, for example). Keep a dry chemical fire extinguisher near the work area. Always keep fuel in a container specifically designed for fuel storage; also, always properly seal fuel containers to avoid the possibility of fire or explosion. Refer to the additional fuel system precautions later in this section.

• Fuel injection systems often remain pressurized, even after the engine has been turned **OFF**. The fuel system pressure must be relieved before disconnecting any fuel lines. Failure to do so may result in fire and/or personal injury.

• Brake fluid often contains polyglycol ethers and polyglycols. Avoid contact with the eyes and wash your hands thoroughly after handling brake fluid. If you do get brake fluid in your eyes, flush your eyes with clean, running water for 15 minutes. If eye irritation persists, or if you have taken brake fluid internally, IMMEDIATELY seek medical assistance.

• The EPA warns that prolonged contact with used engine oil may cause a number of skin disorders, including cancer. You should make every effort to minimize your exposure to used engine oil. Protective gloves should be worn when changing oil. Wash your hands and any other exposed skin areas as soon as possible after exposure to used engine oil. Soap and water, or waterless hand cleaner should be used.

• All new vehicles are now equipped with an air bag system, often referred to as a Supplemental Restraint System (SRS) or Supplemental Inflatable Restraint (SIR) system. The system must be disabled before performing service on or around system components, steering column, instrument panel components, wiring and sensors. Failure to follow safety and disabling procedures could result in accidental air bag deployment, possible personal injury and unnecessary system repairs.

• Always wear safety goggles when working with, or around, the air bag system. When carrying a non-deployed air bag, be sure the bag and trim cover are pointed away from your body. When placing a non-deployed air bag on a work surface, always face the bag and trim cover upward, away from the surface. This will reduce the motion of the module if it is accidentally deployed. Refer to the additional air bag system precautions later in this section.

• Clean, high quality brake fluid from a sealed container is essential to the safe and proper operation of the brake system. You should always buy the correct type of brake fluid for your vehicle. If the brake fluid becomes contaminated, completely flush the system with new fluid. Never reuse any brake fluid. Any brake fluid that is removed from the system should be discarded. Also, do not allow any brake fluid to come in contact with a painted surface; it will damage the paint.

• Never operate the engine without the proper amount and type of engine oil; doing so WILL result in severe engine damage.

• Timing belt maintenance is extremely important. Many models utilize an interference-type, non-freewheeling engine. If the timing belt breaks, the valves in the cylinder head may strike the pistons, causing potentially serious (also time-consuming and expensive) engine damage. Refer to the maintenance interval charts for the recommended replacement interval for the timing belt, and to the timing belt section for belt replacement and inspection.

• Disconnecting the negative battery cable on some vehicles may interfere with the functions of the on-board computer system(s) and may require the computer to undergo a relearning process once the negative battery cable is reconnected.

• When servicing drum brakes, only disassemble and assemble one side at a time, leaving the remaining side intact for reference.

• Only an MVAC-trained, EPA-certified automotive technician should service the air conditioning system or its components.

BRAKES

GENERAL INFORMATION

PRECAUTIONS

• Certain components within the ABS system are not intended to be serviced or repaired individually.

• Do not use rubber hoses or other parts not specifically specified for and ABS system. When using repair kits, replace all parts included in the kit. Partial or incorrect repair may lead to functional problems and require the replacement of components.

• Lubricate rubber parts with clean, fresh brake fluid to ease assembly. Do not use shop air to clean parts; damage to rubber components may result.

• Use only DOT 3 brake fluid from an unopened container.

• If any hydraulic component or line is removed or replaced, it may be necessary to bleed the entire system.

• A clean repair area is essential. Always clean the reservoir and cap thoroughly before removing the cap. The slightest amount of dirt in the fluid may plug an orifice and impair the system function. Perform repairs after components have been thoroughly cleaned; use only denatured alcohol to clean components. Do not allow ABS components to come into contact with any substance containing mineral oil; this includes used shop rags.

• The Anti-Lock control unit is a microprocessor similar to other computer units in the vehicle. Ensure that the ignition switch is **OFF** before removing or installing controller harnesses. Avoid static electricity discharge at or near the controller.

• If any arc welding is to be done on the vehicle, the control unit should be unplugged before welding operations begin.

ANTI-LOCK BRAKE SYSTEM (ABS)

SPEED SENSORS

REMOVAL & INSTALLATION

Front

See Figure 1.

Before servicing the vehicle, refer to the Precautions.

➥**The front wheel speed sensor is part of the hub and bearing assembly and cannot be removed separately. To**

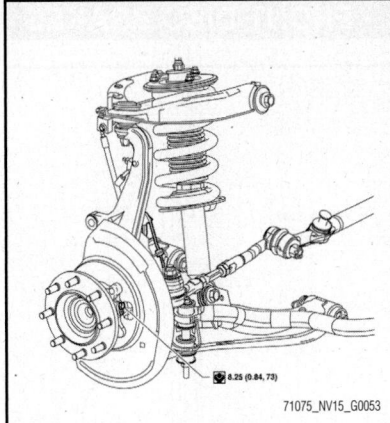

Fig. 1 ABS front wheel speed sensor and related components

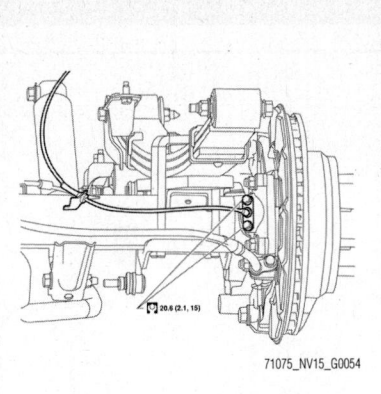

Fig. 2 ABS rear wheel speed sensor and related components

replace the wheel speed sensor the hub and bearing assembly must also be replaced.

Rear

See Figure 2.

1. Before servicing the vehicle, refer to the Precautions.

2. Disconnect the negative battery cable.
3. Raise and support the vehicle safely.
4. Remove wheel and tire.
5. Remove wheel sensor mounting bolts.
6. Pull out the wheel sensor, being careful to turn it as little as possible.

✳✳ WARNING

Be careful not to damage sensor edge and sensor rotor teeth. Do not pull on the sensor harness.

7. Disconnect wheel sensor harness electrical connector, then remove harness from mounts.

To install:

➡Be sure to use new fasteners, as required.

8. Installation is the reverse of the removal procedure.

✳✳ WARNING

Inspect wheel sensor O-ring, replace sensor assembly if damaged. Clean wheel sensor hole and mounting surface with brake cleaner and a lint-free shop rag. Be careful that dirt and debris do not enter the axle.

9. Be sure to perform the reconnect/relearn procedures.

BRAKES

BLEEDING PROCEDURE

➡Be sure that the master cylinder is full of clean fresh brake fluid before starting the bleeding process. Use only the recommended brake fluid when bleeding the system. Do not allow brake fluid to spill on painted surfaces as damage will occur.

➡Bleed air in the following order: motor/accumulator assembly, front right brake, front left brake, rear left brake and rear right brake.

➡The VDC warning lamp, ABS warning lamp and brake warning lamp turn ON and DTC code "C118E" may be detected in the self diagnosis test when the brake pedal is excessively operated, such as bleeding the system. This is not a system malfunction.

1. Before servicing the vehicle, refer to the Precautions.
2. Turn the ignition switch OFF.
3. Depress the brake pedal twenty times or more.
4. Check that there is no foreign material in the reservoir tank, refill with new brake fluid.

➡Never reuse used or drained brake fluid. Never allow oils other than brake fluid to enter the reservoir tank.

5. Turn the ignition switch ON.
6. Connect a vinyl tube to the bleed

BLEEDING THE BRAKE SYSTEM

valve. Be sure to have a catch pan handy to catch excess brake fluid.

7. Fully depress the brake pedal.
8. With the brake pedal depressed, loosen the bleed valve to let air out, then tighten it immediately.
9. Repeat the above steps until all air is removed from the system. Be sure to keep watch on the brake fluid level and replenish, as necessary.
10. Tighten the bleed valve with the pedal depressed.
11. Repeat the above until all air is out of the system.
12. Check that no drag is present for the front disc brake.
13. Be sure to perform the reconnect/relearn procedures.

✳✳ CAUTION

Dust and dirt accumulating on brake parts during normal use may contain asbestos fibers from production or aftermarket brake linings. Breathing excessive concentrations of asbestos fibers can cause serious bodily harm. Exercise care when servicing brake parts. Do not sand or grind brake lining unless equipment used is designed to contain the dust residue. Do not clean brake parts with compressed air or by dry brushing. Cleaning should be done by dampening the brake components with a fine mist of water, then wiping the brake components clean with a dampened cloth. Dispose of cloth and all residue containing asbestos fibers in an impermeable container with the appropriate label. Follow practices prescribed by the Occupational Safety and Health Administration (OSHA) and the Environmental Protection Agency (EPA) for the handling, processing, and disposing of dust or debris that may contain asbestos fibers.

BRAKE CALIPERS

REMOVAL & INSTALLATION

See Figure 3.

1. Before servicing the vehicle, refer to the Precautions.
2. Disconnect the negative battery cable.
3. Remove the fluid reservoir cap.
4. Raise and support the vehicle safely.
5. Remove the tire and wheel assembly.
6. Secure the rotor in place with the lug nuts.

➡ **Position a drain pan under the caliper assembly. Properly dispose of used fluid. Do not allow brake fluid to come in contact with painter surfaces.**

7. Separate the brake hose from the caliper assembly. Discard the copper washer.
8. Remove the torque member bolts.
9. Remove the caliper from its mounting.

To install:

➡ **Be sure to use new fasteners, as required.**

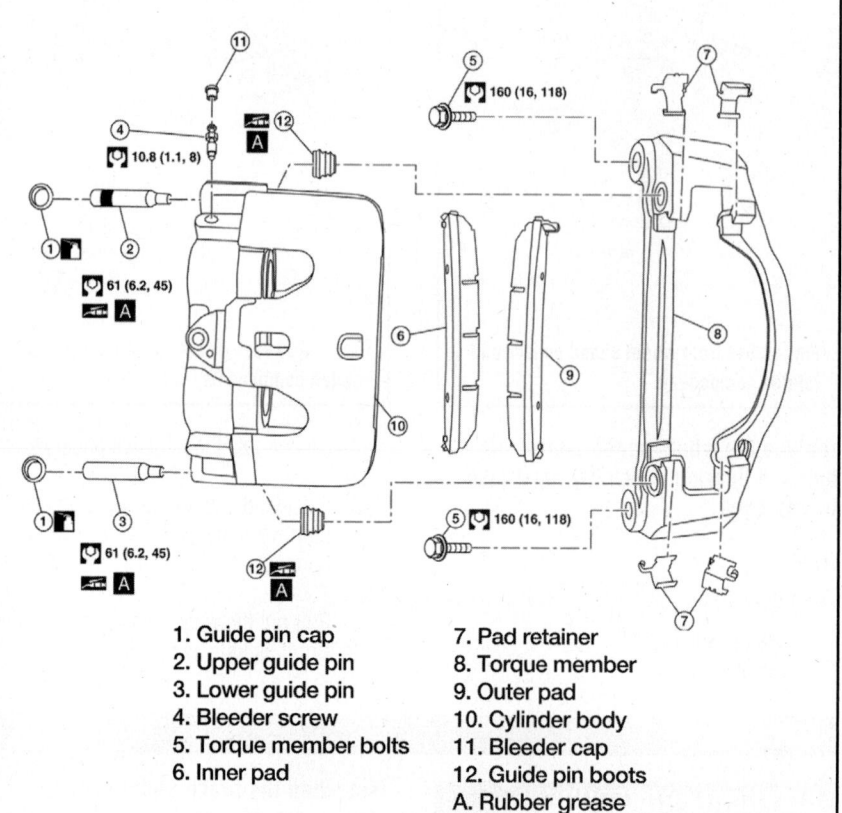

1. Guide pin cap
2. Upper guide pin
3. Lower guide pin
4. Bleeder screw
5. Torque member bolts
6. Inner pad
7. Pad retainer
8. Torque member
9. Outer pad
10. Cylinder body
11. Bleeder cap
12. Guide pin boots
A. Rubber grease

71075_NV15_G0059

Fig. 3 Front brake caliper and related components

➡ **Refill with new brake fluid. Do not reuse drained brake fluid. Do not reuse copper washers.**

10. Installation is the reverse of the removal procedure.
11. Refill with new brake fluid as necessary and bleed air.
12. Install wheel and tire.
13. Be sure to perform the reconnect/relearn procedures.

BRAKE PADS

REMOVAL & INSTALLATION

1. Before servicing the vehicle, refer to the Precautions.
2. Disconnect the negative battery cable.
3. Raise and support the vehicle safely.
4. Remove the tire and wheel assembly.
5. Remove the torque member bolts.
6. Remove the caliper from its mounting.

➡ **Do not disconnect the brake hose line. Position the caliper to the side. Do not allow it to hang by the brake hose.**

7. Remove the pads from the caliper.

To install:

➡ **Be sure to use new fasteners, as required.**

8. Installation is the reverse of the removal procedure.
9. Be sure to assemble the pad retainers so that they will not be lifted up from the torque member. Never deform the pad retainers.
10. Install wheel and tire.
11. Depress the brake pedal several times after installation to seat the pads.
12. Check and refill with new brake fluid as necessary.
13. Be sure to perform the reconnect/relearn procedures.

✳✳ CAUTION

Dust and dirt accumulating on brake parts during normal use may contain asbestos fibers from production or aftermarket brake linings. Breathing excessive concentrations of asbestos fibers can cause serious bodily harm. Exercise care when servicing brake parts. Do not sand or grind brake lining unless equipment used is designed to contain the dust residue. Do not clean brake parts with compressed air or by dry brushing. Cleaning should be done by dampening the brake components with a fine mist of water, then wiping the brake components clean with a dampened cloth. Dispose of cloth and all residue containing asbestos fibers in an impermeable container with the appropriate label. Follow practices prescribed by the Occupational Safety and Health Administration (OSHA) and the Environmental Protection Agency (EPA) for the handling, processing, and disposing of dust or debris that may contain asbestos fibers.

BRAKE CALIPER

REMOVAL & INSTALLATION

See Figure 4.

1. Before servicing the vehicle, refer to the Precautions.
2. Disconnect the negative battery cable.
3. Remove the fluid reservoir cap.
4. Raise and support the vehicle safely.
5. Remove the tire and wheel assembly.
6. Secure the rotor in place with the lug nuts.

➡Position a drain pan under the caliper assembly. Properly dispose of used fluid. Do not allow brake fluid to come in contact with painter surfaces.

7. Separate the brake hose from the caliper assembly. Discard the copper washer.
8. Remove the torque member bolts.
9. Remove the caliper from its mounting.

To install:

➡Be sure to use new fasteners, as required.

➡Refill with new brake fluid. Do not reuse drained brake fluid. Do not reuse copper washers.

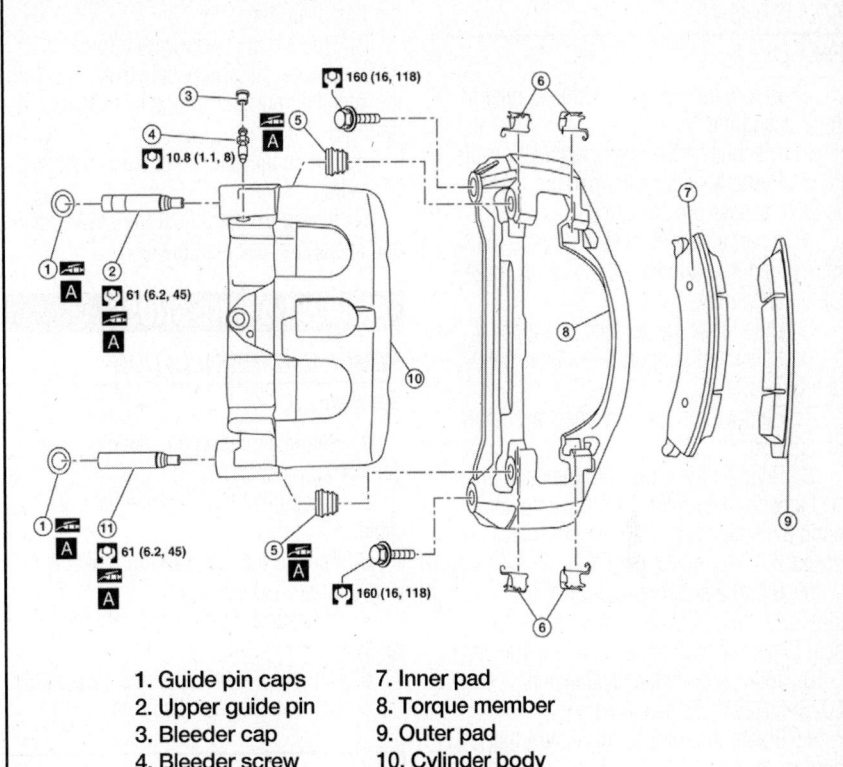

1. Guide pin caps
2. Upper guide pin
3. Bleeder cap
4. Bleeder screw
5. Guide pin boot
6. Pad retainers
7. Inner pad
8. Torque member
9. Outer pad
10. Cylinder body
11. Lower guide pin
A. Rubber grease

71075_NV15_G0060

Fig. 4 Rear brake caliper and related components

10. Installation is the reverse of removal procedure.
11. Refill with new brake fluid as necessary and bleed air.
12. Install wheel and tire.
13. Be sure to perform the reconnect/relearn procedures.

BRAKE PADS

REMOVAL & INSTALLATION

1. Before servicing the vehicle, refer to the Precautions.
2. Disconnect the negative battery cable.
3. Raise and support the vehicle safely.
4. Remove the tire and wheel assembly.
5. Remove the torque member bolts.
6. Remove the caliper from its mounting.

➡Do not disconnect the brake hose line. Position the caliper to the side. Do not allow it to hang by the brake hose.

7. Remove the pads from the caliper.

To install:

➡Be sure to use new fasteners, as required.

8. Installation is the reverse of the removal procedure.
9. Be sure to assemble the pad retainers so that they will not be lifted up from the torque member. Never deform the pad retainers.
10. Install wheel and tire.
11. Depress the brake pedal several times after installation to seat the pads.
12. Check and refill with new brake fluid as necessary.
13. Be sure to perform the reconnect/relearn procedures.

PARKING BRAKE CABLES

ADJUSTMENT

See Figure 5.

1. Before servicing the vehicle, refer to the Precautions.

2. Disconnect the negative battery cable.

3. Partially engage the parking brake pedal to access the adjusting nut.

4. Insert a deep socket and rotate the adjusting nut and loosen the parking brake cable sufficiently.

5. Disengage the parking brake pedal.

6. Raise and support the vehicle safely, as required.

7. Remove the tire and wheel assembly, as required.

8. Remove the plug from the back of the backing plate rotor. Turn the adjuster in using a suitable tool until the drum is locked.

9. Using a suitable tool rotate the adjuster in the opposite direction 14–15 teeth to set the total shoe center clearance.

10. Total center shoe clearance is 0.028–0.029 inch. See illustration.

11. Rotate the rotor to make sure that there is no drag.

12. Adjust the parking brake pedal stroke again.

13. Check the rear disc brake.

14. To adjust the cable, temporarily adjust the cable so that the parking brake pedal operating force immediately before full stroke reaches 112 lbs. or more.

15. Maintain the force (112 lbs.) for two strokes or more. Maintain parking brake pedal at 112 lbs. for forty-eight minutes or more.

16. Adjust the stroke by turning the adjusting nut. Never reuse the adjusting nut if it has been removed.

17. Operate the pedal with a force of 44 lbs. Check that the pedal stroke is within the specified number of notches. Specification is 6–7 notches.

18. Rotate the rotor to ensure that there is no drag.

19. If drag exists, adjust the stroke again and check the rear disc brake.

PARKING BRAKE SHOES

REMOVAL & INSTALLATION

See Figure 6.

1. Before servicing the vehicle, refer to the Precautions.

2. Disconnect the negative battery cable.

3. Be sure that the parking brake lever is in the released position.

4. Raise and support the vehicle safely.

5. Remove the tire and wheel assembly.

6. Remove the rear disc rotor.

7. Remove the return springs.

8. Remove the adjuster.

9. Disconnect the parking brake cable from the toggle lever.

10. Remove the retainers.

11. Remove the anti rattle pins and shoes.

To install:

➡**Be sure to use new fasteners, as required.**

12. Apply brake grease to the specified points during reassembly, see illustration for locating points.

13. Assemble the adjuster so that the threaded part expands when rotating it in the direction shown by the arrow. Shorten the adjuster by rotating it in the opposite direction shown by the arrow.

14. Continue the installation in the reverse order of the removal procedure.

15. Adjust the parking brake.

16. Perform the parking brake burnishing operation.

17. Be sure to perform the reconnect/relearn procedures.

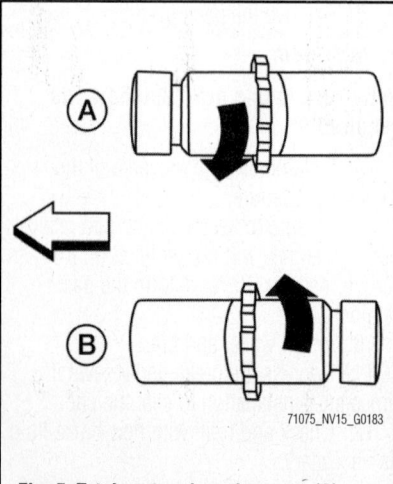

71075_NV15_G0183

Fig. 5 Total center shoe clearance (A) right, (B) left

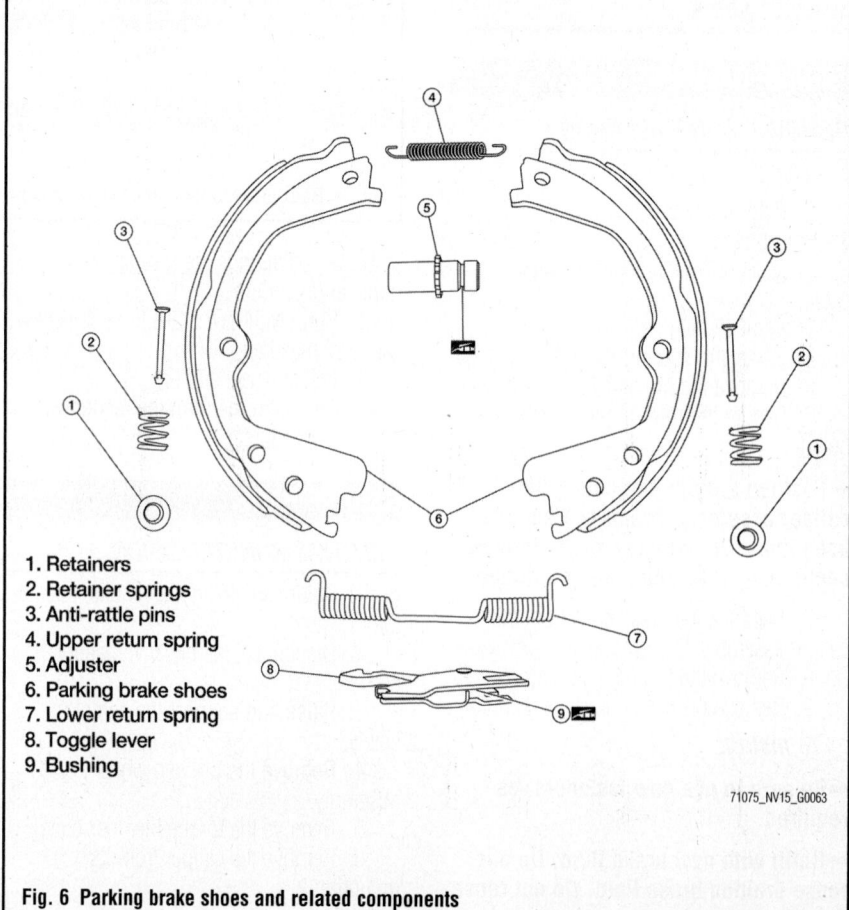

1. Retainers
2. Retainer springs
3. Anti-rattle pins
4. Upper return spring
5. Adjuster
6. Parking brake shoes
7. Lower return spring
8. Toggle lever
9. Bushing

71075_NV15_G0063

Fig. 6 Parking brake shoes and related components

CHASSIS ELECTRICAL | **AIR BAG (SUPPLEMENTAL RESTRAINT SYSTEM)**

GENERAL INFORMATION

✳✳ CAUTION

These vehicles are equipped with an air bag system. The system must be disarmed before performing service on, or around, system components, the steering column, instrument panel components, wiring and sensors. Failure to follow the safety precautions and the disarming procedure could result in accidental air bag deployment, possible injury and unnecessary system repairs.

SERVICE PRECAUTIONS

Disconnect and isolate the battery negative cable before beginning any airbag system component diagnosis, testing, removal, or installation procedures. Allow system capacitor to discharge for two minutes before beginning any component service. This will disable the airbag system. Failure to disable the airbag system may result in accidental airbag deployment, personal injury, or death.

Do not place an intact undeployed airbag face down on a solid surface. The airbag will propel into the air if accidentally deployed and may result in personal injury or death.

When carrying or handling an undeployed airbag, the trim side (face) of the airbag should be pointing towards the body to minimize possibility of injury if accidental deployment occurs. Failure to do this may result in personal injury or death.

Replace airbag system components with OEM replacement parts. Substitute parts may appear interchangeable, but internal differences may result in inferior occupant protection. Failure to do so may result in occupant personal injury or death.

Wear safety glasses, rubber gloves, and long sleeved clothing when cleaning powder residue from vehicle after an airbag deployment. Powder residue emitted from a deployed airbag can cause skin irritation. Flush affected area with cool water if irritation is experienced. If nasal or throat irritation is experienced, exit the vehicle for fresh air until the irritation ceases. If irritation continues, see a physician.

Do not use a replacement airbag that is not in the original packaging. This may result in improper deployment, personal injury, or death.

The factory installed fasteners, screws and bolts used to fasten airbag components have a special coating and are specifically designed for the airbag system. Do not use substitute fasteners. Use only original equipment fasteners listed in the parts catalog when fastener replacement is required.

During, and following, any child restraint anchor service, due to impact event or vehicle repair, carefully inspect all mounting hardware, tether straps, and anchors for proper installation, operation, or damage. If a child restraint anchor is found damaged in any way, the anchor must be replaced. Failure to do this may result in personal injury or death.

Deployed and non-deployed airbags may or may not have live pyrotechnic material within the airbag inflator.

Do not dispose of driver/passenger/curtain airbags or seat belt tensioners unless you are sure of complete deployment. Refer to the Hazardous Substance Control System for proper disposal.

Dispose of deployed airbags and tensioners consistent with state, provincial, local, and federal regulations.

After any airbag component testing or service, do not connect the battery negative cable. Personal injury or death may result if the system test is not performed first.

If the vehicle is equipped with the Occupant Classification System (OCS), do not connect the battery negative cable before performing the OCS Verification Test using the scan tool and the appropriate diagnostic information. Personal injury or death may result if the system test is not performed properly.

Never replace both the Occupant Restraint Controller (ORC) and the Occupant Classification Module (OCM) at the same time. If both require replacement, replace one, then perform the Airbag System test before replacing the other.

Both the ORC and the OCM store Occupant Classification System (OCS) calibration data, which they transfer to one another when one of them is replaced. If both are replaced at the same time, an irreversible fault will be set in both modules and the OCS may malfunction and cause personal injury or death.

If equipped with OCS, the Seat Weight Sensor is a sensitive, calibrated unit and must be handled carefully. Do not drop or handle roughly. If dropped or damaged, replace with another sensor. Failure to do so may result in occupant injury or death.

If equipped with OCS, the front passenger seat must be handled carefully as well. When removing the seat, be careful when setting on floor not to drop. If dropped, the sensor may be inoperative, could result in occupant injury, or possibly death.

If equipped with OCS, when the passenger front seat is on the floor, no one should sit in the front passenger seat. This uneven force may damage the sensing ability of the seat weight sensors. If sat on and damaged, the sensor may be inoperative, could result in occupant injury, or possibly death.

DISARMING THE SYSTEM

✳✳ CAUTION

These vehicles are equipped with an air bag system. The system must be disarmed before performing service on, or around, system components, the steering column, instrument panel components, wiring and sensors. Failure to follow the safety precautions and the disarming procedure could result in accidental air bag deployment, possible injury and unnecessary system repairs.

Before servicing the vehicle, refer to the Precautions.

1. Disconnect the negative battery cable.
2. Disconnect the positive battery cable.
3. Wait at least 3 minutes before working on the vehicle. The air bag system is designed to retain enough power to deploy the air bag for a short time after the battery has been disconnected.

ARMING THE SYSTEM

➤**Once repair work has been completed, return the ignition switch to the LOCK position, before connecting the battery cables. Always connect the positive battery cable first. At this time the steering lock mechanism will engage. Install the CONSULT-III diagnostic tool, or equivalent, and follow the directions on the screen of the tool and perform the self diagnosis check.**

➤**If the vehicle is equipped with Intelligent Key System and Nissan Anti-theft System (NATS) and battery power is interrupted steering wheel rotation will be required, follow the procedure below before starting the repair operation.**

1. Connect both battery cables.
2. Use the Intelligent Key or mechanical key to turn the ignition switch to the "ACC"

position. At this time, the steering lock will be released.

3. Disconnect both battery cables. The steering lock will remain released and the steering wheel can be rotated.

4. Perform the necessary repair operation.

5. When the repair work is completed, return the ignition switch to the "LOCK" position before connecting the battery

cables. (At this time, the steering lock mechanism will engage).

6. Perform a self-diagnosis check of all control units using CONSULT III diagnostic tool, or equivalent.

DRIVE TRAIN

DRIVESHAFT

REMOVAL & INSTALLATION

See Figures 7 and 8.

1. Before servicing the vehicle, refer to the Precautions.

2. Disconnect the negative battery cable.

3. Position the selector lever in the N position.

4. Release the parking brake.

5. Raise and support the vehicle safely.

6. Remove the engine under cover, if equipped and as required.

7. Remove the center support bracket nuts.

8. Matchmark the driveshaft and companion flange.

9. Remove the driveshaft. Discard the nuts.

To install:

➡ **Be sure to use new fasteners, as required.**

10. Installation is the reverse of the removal procedure.

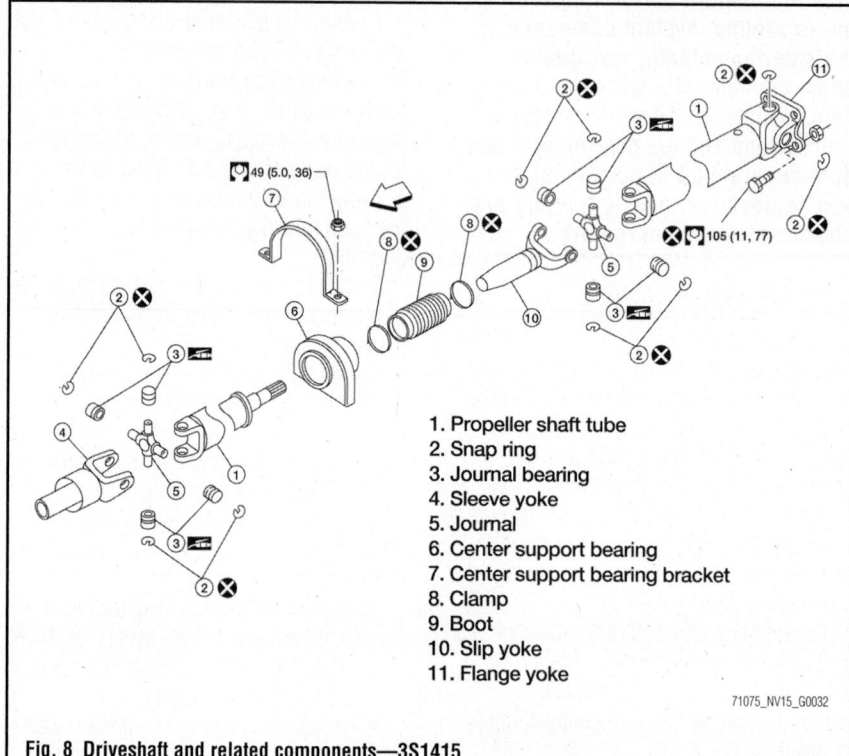

1. Propeller shaft tube
2. Snap ring
3. Journal bearing
4. Sleeve yoke
5. Journal
6. Center support bearing
7. Center support bearing bracket
8. Clamp
9. Boot
10. Slip yoke
11. Flange yoke

71075_NV15_G0032

Fig. 8 Driveshaft and related components—3S1415

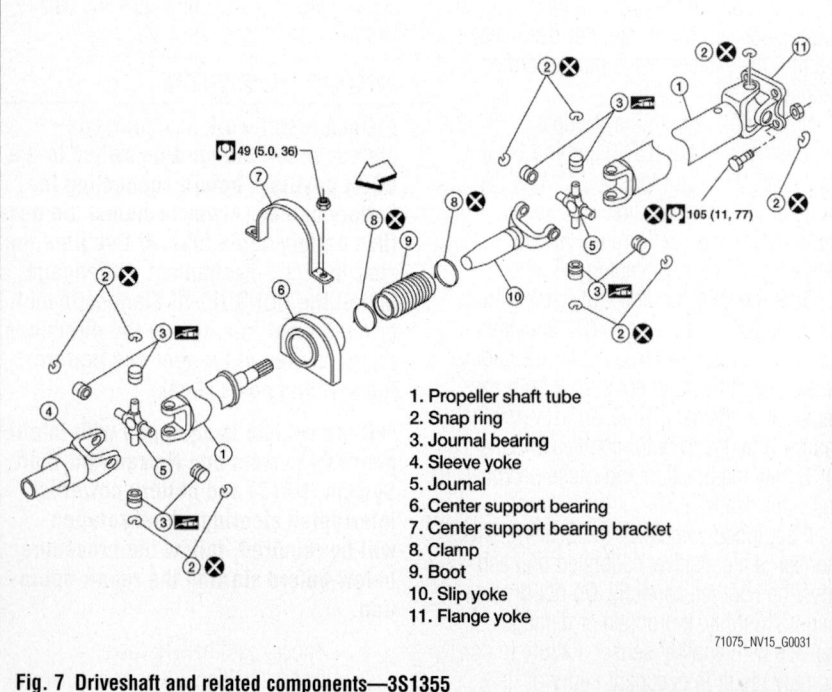

1. Propeller shaft tube
2. Snap ring
3. Journal bearing
4. Sleeve yoke
5. Journal
6. Center support bearing
7. Center support bearing bracket
8. Clamp
9. Boot
10. Slip yoke
11. Flange yoke

71075_NV15_G0031

Fig. 7 Driveshaft and related components—3S1355

11. Check for vehicle vibration, correct as required.

12. Be sure to perform the reconnect/relearn procedures.

REAR AXLE HOUSING

REMOVAL & INSTALLATION

See Figure 9.

1. Before servicing the vehicle, refer to the Precautions.

2. Disconnect the negative battery cable.

3. Raise and support the vehicle safely.

4. Drain the gear oil. Be sure to properly dispose of used oil.

5. Remove the tire and wheel assemblies.

6. Disconnect the driveshaft and position it out of the work area.

7. Disconnect the air breather hose from the assembly.

8. Disconnect the wheel sensor wire harness from the assembly.

9. Remove the tube block connectors and brake tubes from the assembly.

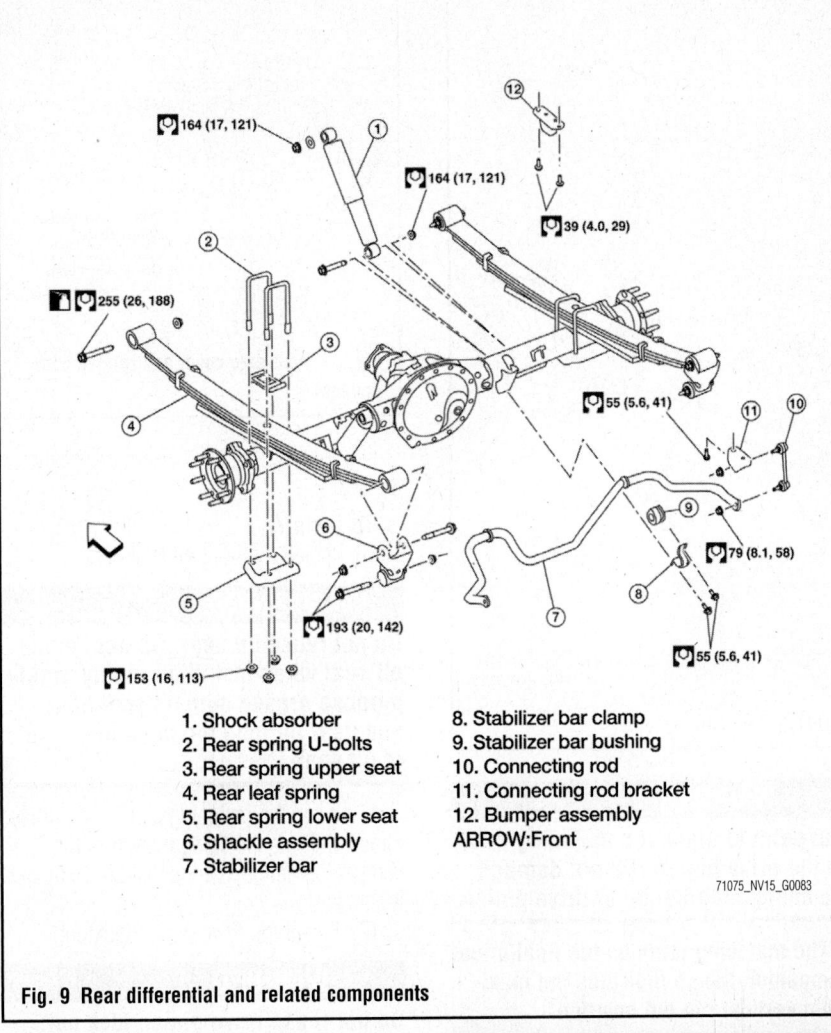

1. Shock absorber
2. Rear spring U-bolts
3. Rear spring upper seat
4. Rear leaf spring
5. Rear spring lower seat
6. Shackle assembly
7. Stabilizer bar
8. Stabilizer bar clamp
9. Stabilizer bar bushing
10. Connecting rod
11. Connecting rod bracket
12. Bumper assembly
ARROW:Front

71075_NV15_G0083

Fig. 9 Rear differential and related components

10. Remove the parking brake cable from the assembly.

11. Remove the rear stabilizer bar.

12. Remove the brake caliper with the torque member, properly support this component.

13. Position a suitable jack under the assembly.

➡**Do not position the jack under the aluminum cover.**

14. Remove the lower shock bolts.

15. Remove the leaf spring U-bolt nuts.

16. Remove the rear axle assembly retaining bolts and nuts.

17. Carefully remove the assembly from the vehicle.

To install:

➡**Be sure to use new fasteners, as required.**

18. Installation is the reverse of the removal procedure.

19. Be sure to perform the reconnect/relearn procedures.

REAR AXLE SHAFT, BEARING & SEAL

REMOVAL & INSTALLATION
See Figure 10.

1. Before servicing the vehicle, refer to the Precautions.

➡**Before removing the axle shaft, remove the wheel sensor to reposition the sensor out of the work area. Failure to do so may damage the sensor and/or cause the sensor to be inoperative.**

2. Disconnect the negative battery cable.

3. Raise and support the vehicle safely.

4. Remove the tire and wheel assembly.

5. Drain the differential, as required. Be sure to properly dispose of used gear oil.

6. Remove the wheel sensor from the housing.

7. Remove the brake rotor.

8. Disconnect the parking brake cable from the lever.

9. Remove the axle shaft nuts.

10. Remove the axle shaft, O-ring and seal.

11. Discard the O-ring and seal.

To install:

➡**Be sure to use new fasteners, as required.**

12. Installation is the reverse of the removal procedure.

13. Be sure to perform the reconnect/relearn procedures.

REAR DIFFERENTIAL HOUSING COVER

REMOVAL & INSTALLATION
See Figure 11.

1. Before servicing the vehicle, refer to the Precautions.

2. Disconnect the negative battery cable.

3. Raise and support the vehicle safely, as required.

4. Position a catch pan under the assembly.

5. Drain the fluid. Be sure to properly dispose of used fluid.

6. Remove the stabilizer bar clamps and bushings. Position the stabilizer bar out of the work area.

7. Remove the brake tube from the carrier cover and position it out of the way.

8. Remove the carrier cover retaining bolts. Remove the cover. Discard the gasket.

➡**Be sure to use new fasteners, as required.**

9. Installation is the reverse of the removal procedure.

10. Be sure to perform the reconnect/relearn procedures.

REAR PINION OIL SEAL

REMOVAL & INSTALLATION

1. Before servicing the vehicle, refer to the Precautions.

➡**The automatic back door system, if equipped, must be initialized anytime the battery has been disconnected. Close the back door. Open the back door with the automatic open feature. Do not stop the process until the back door opens completely.**

➡**On this vehicle the battery current sensor that is installed to the negative battery cable measures the charg-**

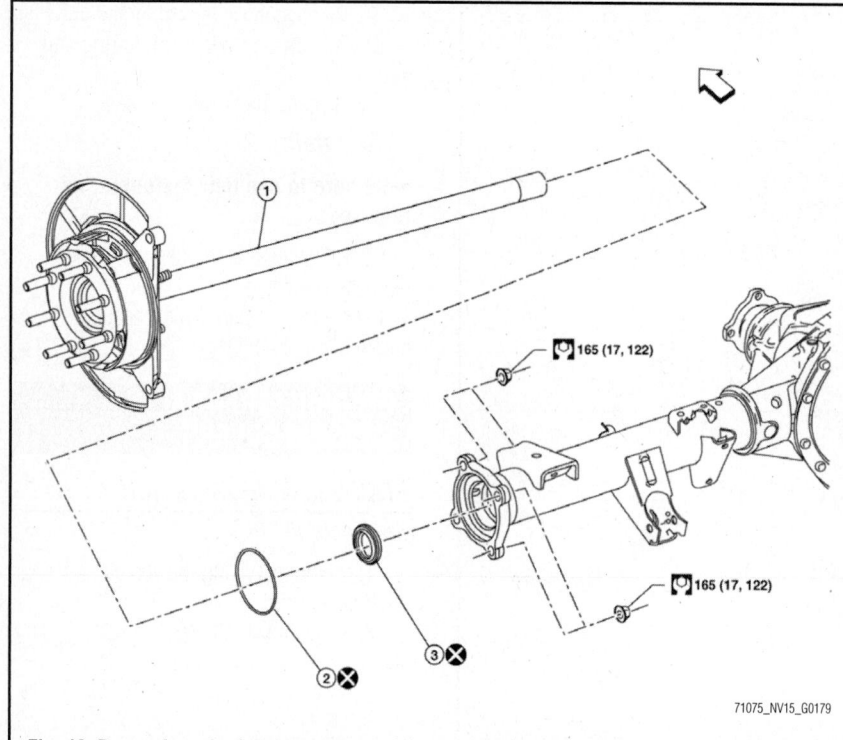

Fig. 10 Rear axle and related components (1) axle, (O-ring, (3) seal

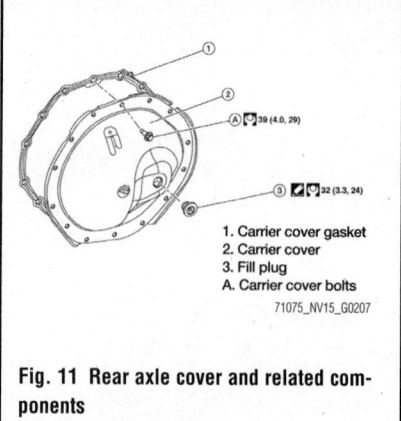

1. Carrier cover gasket
2. Carrier cover
3. Fill plug
A. Carrier cover bolts

71075_NV15_G0207

Fig. 11 Rear axle cover and related components

ing/discharging current of the battery and performs various engine controls.

2. Disconnect the negative battery cable.

3. Raise and support the vehicle safely.

4. Remove the driveshaft.

5. Remove the brake calipers and brake rotors.

6. Put a matching mark on the end of the drive pinion in line with the matching mark on the companion flange.

> ❊❊ **WARNING**
>
> **Use paint to make the matching mark on the drive pinion. Do not damage the companion flange or drive pinion.**

➥The matching mark on the final drive companion flange indicates the maximum vertical run out position.

7. Remove the drive pinion lock nut.

8. Remove the companion flange using suitable tool.

9. Remove the front oil seal. Discard the seal.

To install:

10. Install the front oil seal.

> ❊❊ **WARNING**
>
> **Do not reuse oil seal. Do not incline oil seal when installing. Apply multi-purpose grease onto oil seal lips, and gear oil onto the circumference of oil seal.**

11. Align the matching mark of the drive pinion with the matching mark B of the companion flange, then install the companion flange.

12. Install the drive pinion lock nut.

> ❊❊ **WARNING**
>
> **Do not reuse drive pinion lock nut.**

13. Install the rear propeller shaft.

14. Be sure to perform the reconnect/relearn procedures.

ENGINE COOLING

BELT-DRIVEN ENGINE FAN

REMOVAL & INSTALLATION

V6 Engine

See Figure 12.

1. Before servicing the vehicle, refer to the Precautions.

2. Disconnect the negative battery cable.

3. Remove the grille. Remove the front undercover.

4. Remove the air cleaner assembly.

5. Drain the cooling system. Be sure to properly dispose of used coolant.

6. Remove reservoir tank hose and water hose.

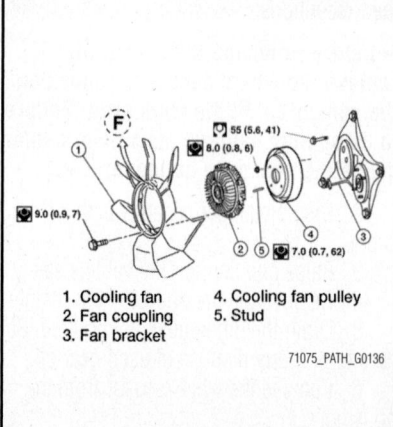

1. Cooling fan
2. Fan coupling
3. Fan bracket
4. Cooling fan pulley
5. Stud

71075_PATH_G0136

Fig. 12 Cooling fan and related components—V6 engine

7. Unclip the transmission lines from the radiator shroud.

8. Remove the reservoir tank bracket. Remove the tank.

9. Disconnect the motor driven cooling fan connector.

10. Remove the left and right lower radiator shrouds.

11. Remove the upper radiator shroud bolts. Remove the shroud.

12. Remove the upper radiator hose.

13. Remove the drive belt.

14. Remove the cooling fan retaining bolts. Remove the cooling fan from the engine. As required, remove the pulley.

To install:

➥Be sure to use new fasteners, as required.

15. Installation is the reverse of the removal procedure.

16. Start and warm up the engine. Visually make sure that there are no leaks of the engine coolant.

17. Be sure to perform the reconnect/relearn procedures.

V8 Engine

See Figure 13.

1. Before servicing the vehicle, refer to the Precautions.

2. Disconnect the negative battery cable.

3. Remove the engine cover and air cleaner assembly, as required.

4. Remove the left and right lower radiator shrouds.

5. Remove the drive belt.

6. Remove the cooling fan retaining bolts. Remove the cooling fan from the engine. As required, remove the pulley.

To install:

7. Install cooling fan with its front mark "F" facing front of engine.

8. Check for leaks of the engine coolant.

9. Start and warm up the engine. Visually make sure that there are no leaks of the engine coolant.

10. Be sure to perform the reconnect/relearn procedures.

ENGINE COOLANT

DRAIN & REFILL

See Figure 14.

➡You will need tool KV991J0070, or equivalent, and a compressed air supply to perform this procedure.

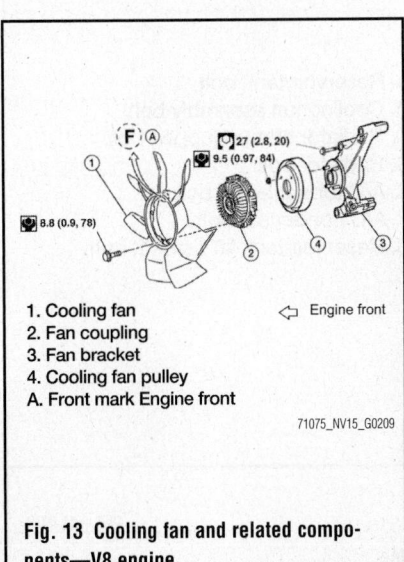

1. Cooling fan
2. Fan coupling
3. Fan bracket
4. Cooling fan pulley
A. Front mark Engine front

⇦ Engine front

71075_NV15_G0209

Fig. 13 Cooling fan and related components—V8 engine

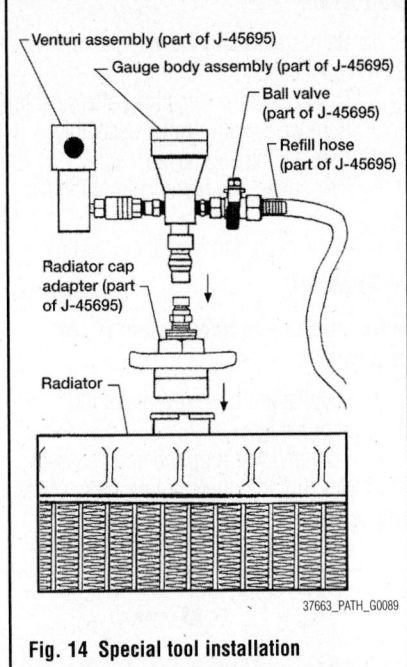

- Venturi assembly (part of J-45695)
- Gauge body assembly (part of J-45695)
- Ball valve (part of J-45695)
- Refill hose (part of J-45695)
- Radiator cap adapter (part of J-45695)
- Radiator

37663_PATH_G0089

Fig. 14 Special tool installation

1. Before servicing the vehicle, refer to the Precautions.

2. Disconnect the negative battery cable.

➡Never drain the engine coolant when the engine is hot. Wrap a thick cloth around the cap and carefully remove it. First turn the cap a quarter of a turn to release any pressure, and then turn it all the way. Do not allow coolant to come in contact with the drive belts.

3. Turn the ignition switch ON and set the temperature control lever all the way to the HOT position, or the highest temperature position. Wait ten seconds and turn the ignition switch OFF.

4. Raise and safely support the vehicle.

5. Remove the undercover, if equipped and as required.

6. Open the radiator drain plug and carefully drain the coolant into a suitable container. Be sure to properly dispose of used coolant.

7. Discard the O-ring.

8. Open the water drain plugs on the cylinder block and drain the coolant into a suitable container. Be sure to properly dispose of used coolant.

9. As necessary, remove the reservoir tank and drain the coolant. Clean the tank before reinstalling.

10. If removed install the reservoir tank.

11. Install the drain plug. Be sure to use new O-ring/gaskets, as required.

12. Apply sealant to the drain plugs. Install the engine drain plugs.

13. Check that all hose clamps are tight.

14. Set the heater controls to the full HOT position and heater ON position. Turn the ignition ON with the engine OFF as necessary to activate the heater mode.

15. Remove the vented reservoir cap and replace it with a non-vented cap before filling the system.

16. Install tool KV991J0070 or equivalent, see illustration.

17. Insert the refill hose into the coolant mixture (placed at floor level). Be sure that the ball valve is in the closed position. Install the air hose to the venture assembly. The air pressure must be within specification (80–119 psi).

➡The compressed air supply must be equipped with an air dryer.

18. The vacuum gauge will begin to rise and there will be an audible hissing noise. During this process open the ball valve on the refill hose, slightly. Coolant will be visible rising in the refill hose. Once the refill hose is full of coolant, close the ball valve. This will purge any trapped air in the refill hose.

19. Continue to draw vacuum until the gauge reaches 28 inches of vacuum.

➡The gauge may not reach specification in high altitude applications. If not see the following specifications. Altitude above sea level: 0-100m (328 ft.) 28 inches of vacuum. Altitude above sea level: 300m (984 ft.) 27 inches of vacuum. Altitude above sea level: 500m (1641 ft.) 26 inches of vacuum. Altitude above sea level: 1000m (3281 ft.) 24–25 inches of vacuum.

20. When the proper specification has been reached, disconnect the air hose and wait 20 seconds to see if the system looses vacuum. If the level drops perform any necessary repairs and repeat the procedure.

21. Place the coolant container (with the refill hose inserted) at the same level as the top of the radiator. Then open the ball valve on the refill hose so the coolant will be drawn up to fill the cooling system. The cooling system is full when the vacuum gauge reads zero.

➡Do not allow the coolant container to get too low when filling, to avoid air from being drawn into the system.

22. Remove the toll from the radiator neck opening and install the radiator cap.

23. Remove the non vented reservoir cap.

24. Fill the reservoir tank to specification with the proper coolant mixture.

25. Be sure to perform the reconnect/relearn procedures.

FLUID RECOMMENDATIONS

Be sure to use genuine Nissan (blue) coolant when filling/servicing the cooling system.

LEVEL CHECK

Before servicing the vehicle, refer to the Precautions.

Check that the coolant reservoir tank is level and within the MIN and MAX marks. Adjust coolant level, as necessary.

ELECTRIC ENGINE FAN

REMOVAL & INSTALLATION

V6 Engine

See Figure 15.

1. Before servicing the vehicle, refer to the Precautions.
2. Disconnect the negative battery cable.
3. Remove the air cleaner assembly.
4. Remove the engine fan.
5. Disconnect the harness from the fan motor.
6. Remove the fan from its mounting.

To install:

➡**Be sure to use new fasteners, as required.**

7. Installation is the reverse of the removal procedure.
8. Start and warm up the engine. Visually make sure that there are no leaks of the engine coolant.
9. Be sure to perform the reconnect/relearn procedures.

V8 Engine

1. Before servicing the vehicle, refer to the Precautions.
2. Disconnect the negative battery cable.
3. Remove the air cleaner assembly.
4. Remove the engine fan.
5. Disconnect the harness from the fan motor.
6. Remove the fan from its mounting.

To install:

➡**Be sure to use new fasteners, as required.**

7. Installation is the reverse of the removal procedure.
8. Start and warm up the engine. Visually make sure that there are no leaks of the engine coolant.

9. Be sure to perform the reconnect/relearn procedures.

RADIATOR

REMOVAL & INSTALLATION

See Figures 16 and 17.

1. Before servicing the vehicle, refer to the Precautions.
2. Disconnect the negative battery cable.
3. Remove the air cleaner assembly.
4. Remove the grille. Remove the front undercover.
5. Drain the cooling system. Be sure to properly dispose of used coolant.
6. Remove reservoir tank hose and water hose.

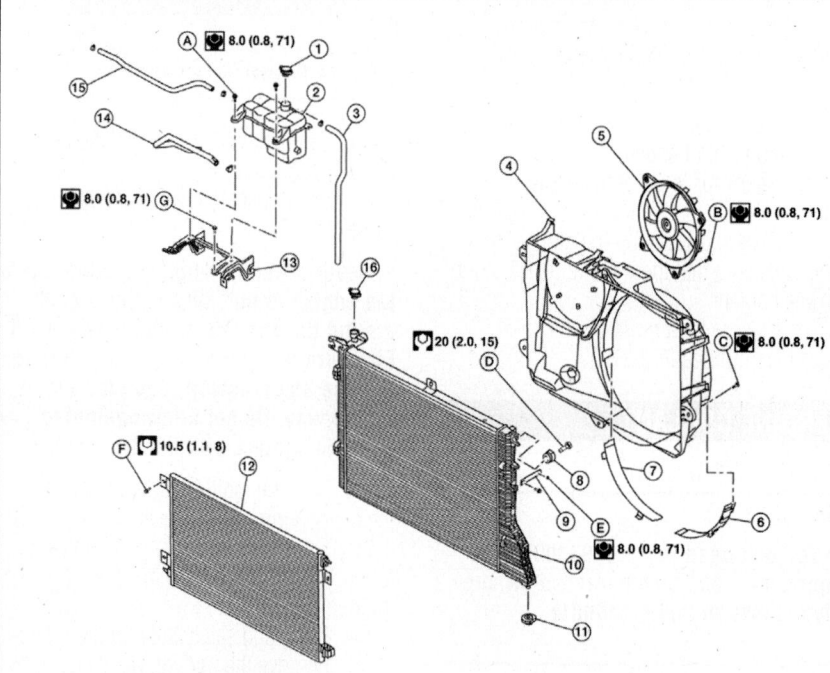

1. Reservoir tank cap
2. Reservoir tank
3. Water hose
4. Radiator shroud (upper)
5. Cooling fan assembly
6. Radiator shroud LH (lower)
7. Radiator shroud RH (lower)
8. Mounting rubber (upper)
9. A/C tube bracket
10. Radiator
11. Mounting rubber (lower)
12. A/C condenser
13. Reservoir tank lift bracket
14. Water hose
15. Reservoir tank hose
16. Radiator cap

A. Reservoir tank bolt
B. Cooling fan assembly bolt
C. Radiator shroud (upper) bolt
D. Radiator bolt
E. A/C tube bracket bolt
F. A/C condenser bolt
G. Reservoir tank lift bracket bolt

71075_NV15_G0024

Fig. 16 Radiator and related components—V6 engine

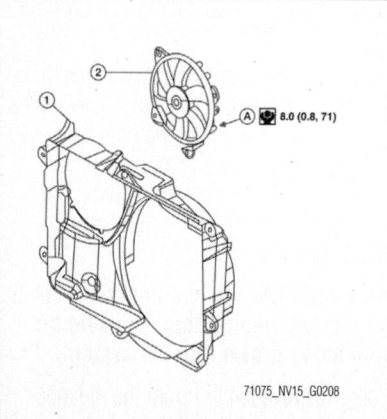

71075_NV15_G0208

Fig. 15 Electric cooling fan and related components (1) shroud, (2) fan, (A) bolts

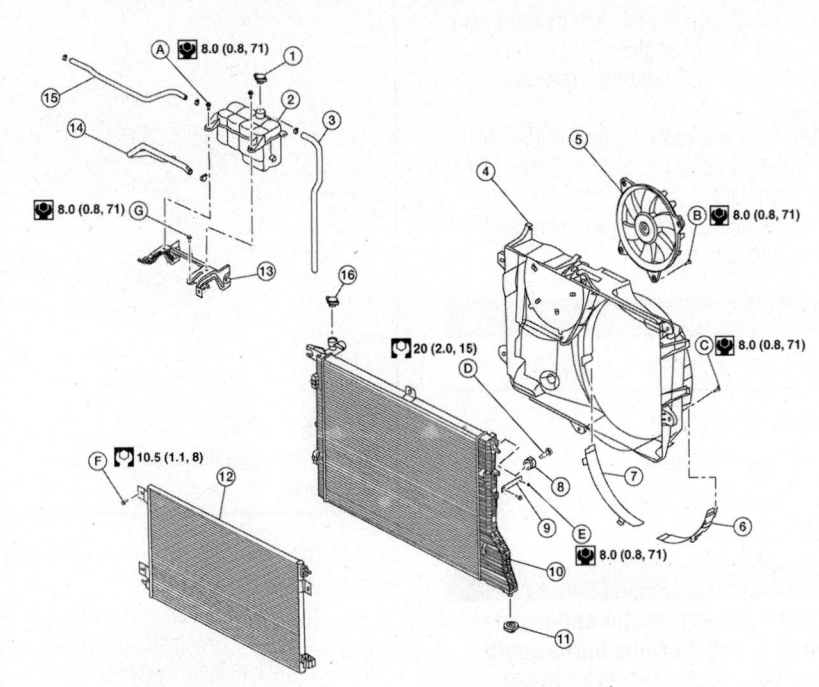

1. Coolant reservoir cap
2. Coolant reservoir tank
3. Coolant reservoir overflow hose
4. Radiator shroud (upper)
5. Engine cooling fan (motor driven)
6. Radiator shroud LH (lower)
7. Radiator shroud RH (lower)
8. Mounting rubber (upper)
9. A/C bracket
10. Radiator
11. Mounting rubber (lower)
12. A/C condenser
13. Coolant reservoir tank bracket
14. Water hose
15. Coolant reservoir tank hose
16. Radiator cap

A. Reservoir tank bolt
B. Engine cooling fan (motor driven) bolt
C. Radiator shroud (upper) bolt
D. Radiator mounting bolt
E. A/C bracket bolt
F. A/C condenser bolt
G. Coolant reservoir tank bracket bolt

71075_NV15_G0025

Fig. 17 Radiator and related components—V8 engine

7. Unclip the transmission lines from the radiator shroud.
8. Remove the reservoir tank bracket. Remove the tank.
9. Disconnect the motor driven cooling fan connector.
10. Remove the left and right lower radiator shrouds.
11. Remove the upper radiator shroud bolts. Remove the shroud.
12. Remove the upper radiator hose. Remove the lower radiator hose.
13. Remove the radiator mounting bolts.
14. Lift and pull the component rearward to disengage the lower rubber mounting from the core support center.
15. Remove the radiator from the vehicle.

To install:

➡ Be sure to use new fasteners, as required.

16. Installation is the reverse of the removal procedure.
17. Be sure to fill the cooling system with the proper grade and type engine coolant.
18. Be sure to perform the reconnect/relearn procedures.

THERMOSTAT

REMOVAL & INSTALLATION

V6 Engine

See Figure 18.

1. Before servicing the vehicle, refer to the Precautions.
2. Disconnect the negative battery cable.
3. Remove the grille. Remove the front undercover.
4. Remove the air cleaner assembly.
5. Drain the cooling system. Be sure to properly dispose of used coolant.
6. Remove reservoir tank hose and water hose.
7. Unclip the transmission lines from the radiator shroud.
8. Remove the reservoir tank bracket. Remove the tank.
9. Disconnect the motor driven cooling fan connector.
10. Remove the left and right lower radiator shrouds.
11. Remove the upper radiator shroud bolts. Remove the shroud.
12. Remove the upper radiator hose.
13. Remove the coolant reservoir hose and water hose.

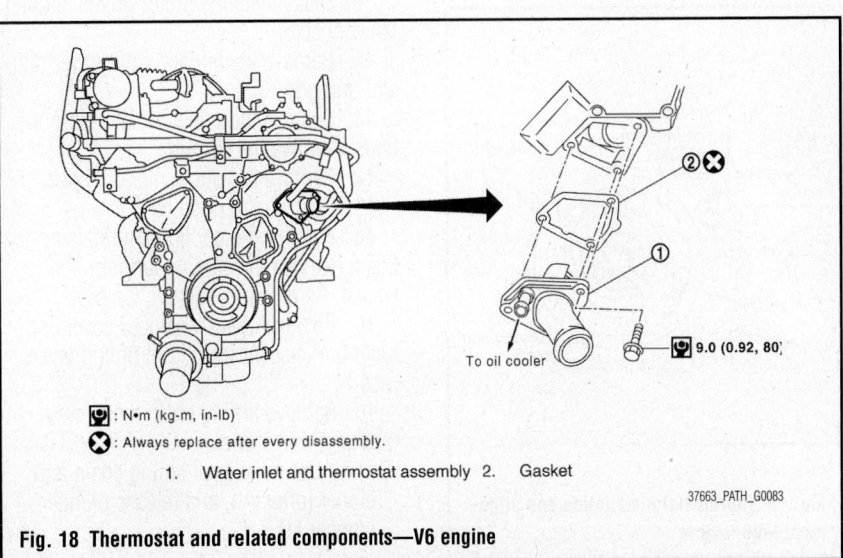

N•m (kg-m, in-lb)

: Always replace after every disassembly.

1. Water inlet and thermostat assembly   2. Gasket

37663_PATH_G0083

Fig. 18 Thermostat and related components—V6 engine

14. Disconnect the lower radiator hose and the oil cooler hose from the water inlet and thermostat assembly.

15. Remove the water inlet and the thermostat assembly. Discard the gasket.

To install:

➡**Be sure to use new fasteners, as required.**

16. Install cooling fan with its front mark "F" facing front of engine.

17. Installation is the reverse of the removal procedure.

18. Start and warm up the engine. Visually make sure that there are no leaks of the engine coolant.

19. Be sure to perform the reconnect/relearn procedures.

V8 Engine

See Figure 19.

1. Before servicing the vehicle, refer to the Precautions.

2. Disconnect the negative battery cable.

3. Remove the engine cover and air cleaner assembly, as required.

4. Drain the cooling system. Be sure to properly dispose of used engine coolant.

5. Remove reservoir tank hose and water hose.

6. Unclip the transmission lines from the radiator shroud.

7. Remove the reservoir tank bracket. Remove the tank.

8. Remove the upper radiator hose.

9. Disconnect the water suction hose from the water inlet.

10. Remove the water inlet.

11. Remove the thermostat and O-ring. Discard the O-ring.

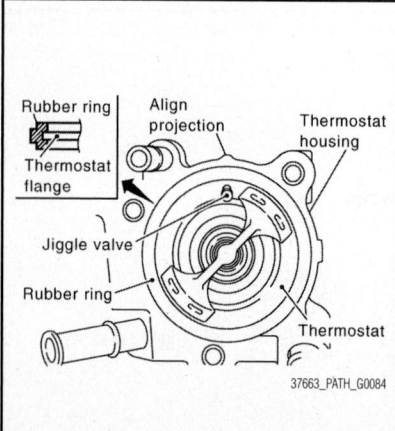

Fig. 19 Thermostat installation and alignment—V8 engine

To install:

12. Install cooling fan with its front mark "F" facing front of engine.

13. Check for leaks of the engine coolant.

14. Start and warm up the engine. Visually make sure that there are no leaks of the engine coolant.

15. Be sure to perform the reconnect/relearn procedures.

WATER PUMP

REMOVAL & INSTALLATION

V6 Engine

See Figures 20 and 21.

1. Before servicing the vehicle, refer to the Precautions.

✷✷ CAUTION

Do not remove radiator cap when engine is hot. Serious burns could occur from high-pressure engine coolant escaping from radiator. Wrap a thick cloth around the cap. Slowly turn it a quarter of a turn to release built-up pressure. Carefully remove radiator cap by turning it all the way.

2. Disconnect the negative battery cable.

3. Remove the grille. Remove the front undercover.

4. Remove the air cleaner assembly.

5. Drain the cooling system. Be sure to properly dispose of used coolant.

6. Remove reservoir tank hose and water hose.

7. Unclip the transmission lines from the radiator shroud.

8. Remove the reservoir tank bracket. Remove the tank.

9. Disconnect the motor driven cooling fan connector.

10. Remove the left and right lower radiator shrouds.

11. Remove the upper radiator shroud bolts. Remove the shroud.

12. Remove the upper radiator hose.

13. Remove the drive belt.

14. Remove the cooling fan retaining bolts. Remove the cooling fan from the engine. As required, remove the pulley.

15. Remove chain tensioner cover and water pump cover from front timing chain case.

16. Remove timing chain tensioner (primary) as follows:

　a. Loosen clip of timing chain tensioner (primary), and release plunger stopper (1).

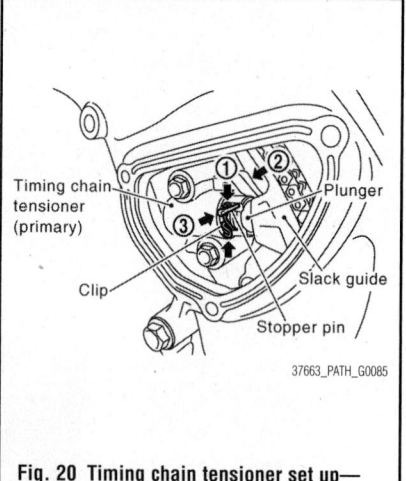

Fig. 20 Timing chain tensioner set up—V6 engine

　b. Insert plunger into tensioner body by pressing slack guide (2).

　c. Keep slack guide pressed and hold plunger in by pushing stopper pin through the tensioner body hole and plunger groove (3).

　d. Turn crankshaft pulley clockwise so that timing chain on the timing chain tensioner (primary) side is loose.

　e. Remove bolts and remove timing chain tensioner (primary).

✷✷ WARNING

Be careful not to drop bolts inside timing chain case.

17. Remove water pump as follows:

　a. Remove three water pump bolts. Secure a gap between water pump gear and timing chain, by turning crankshaft pulley counterclockwise until timing chain looseness on water pump sprocket becomes maximum.

　b. Screw M8 bolts [pitch: 1.25 mm (0.049 in.) length: approx. 50 mm (1.97 in.)] into water pumps upper and lower bolt holes until they reach timing chain case. Then, alternately tighten each bolt for a half turn, and pull out water pump.

✷✷ WARNING

Pull straight out while preventing vane from contacting socket in installation area. Remove water pump without causing sprocket to contact timing chain.

　c. Remove M8 bolts and O-rings from water pump.

✷✷ WARNING

Do not disassemble water pump.

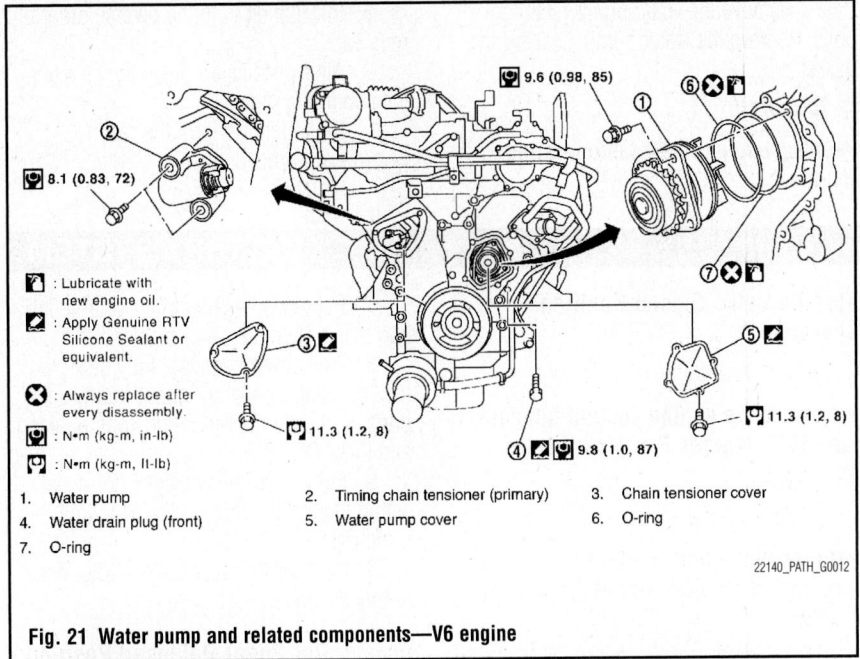

Fig. 21 Water pump and related components—V6 engine

Key (Fig. 21):

- : Lubricate with new engine oil.
- : Apply Genuine RTV Silicone Sealant or equivalent.
- : Always replace after every disassembly.
- : N•m (kg-m, in-lb)
- : N•m (kg-m, ft-lb)

1. Water pump
2. Timing chain tensioner (primary)
3. Chain tensioner cover
4. Water drain plug (front)
5. Water pump cover
6. O-ring
7. O-ring

22140_PATH_G0012

➡ Do not reuse O-rings.

To install:

➡ Be sure to use new fasteners, as required.

18. Install new O-rings to water pump.

➡ Apply engine oil to O-rings. Locate O-ring with white paint mark to engine front side.

19. Install water pump.

✳✳ WARNING

Do not allow timing chain case to pinch O-rings when installing water pump.

a. Make sure that timing chain and water pump sprocket are engaged.

b. Insert water pump by tightening bolts alternately and evenly.

20. Install timing chain tensioner (primary) as follows:

a. Remove dust and foreign material completely from backside of timing chain tensioner (primary) and from installation area of rear timing chain case.

b. Turn crankshaft pulley clockwise so that timing chain on the timing chain tensioner (primary) side is loose.

c. Install timing chain tensioner (primary) with its stopper pin attached.

✳✳ WARNING

Be careful not to drop bolts inside timing chain case.

d. Remove stopper pin.

e. Make sure again that timing chain and water pump sprocket are engaged.

21. Install chain tensioner cover and water pump cover as follows:

a. Before installing, remove all traces of old liquid gasket from mating surface of water pump cover and chain tensioner cover using scraper. Also remove traces of old liquid gasket from the mating surface of front timing chain case.

b. Apply a continuous bead of liquid gasket, to mating surface of chain tensioner and water pump cover.

✳✳ WARNING

Attaching should be done within 5 minutes after coating.

c. Tighten bolts to specified torque.

22. Refill engine coolant system.

23. Installation of the remaining components is in the reverse order of removal after this step.

a. After starting engine, let idle for three minutes, then rev engine up to 3,000 rpm under no load to purge air from the high-pressure chamber of chain tensioner. Engine may produce a rattling noise. This indicates that air still remains in the chamber and is not a matter of concern.

24. Be sure to perform the reconnect/relearn procedures.

V8 Engine

See Figure 22.

1. Before servicing the vehicle, refer to the Precautions.

✳✳ CAUTION

Do not remove radiator cap when engine is hot. Serious burns could occur from high-pressure engine coolant escaping from radiator. Wrap a thick cloth around the cap. Slowly turn it a quarter of a turn to release built-up pressure. Carefully remove radiator cap by turning it all the way.

2. Disconnect the negative battery cable.

3. Remove the engine cover and air cleaner assembly, as required.

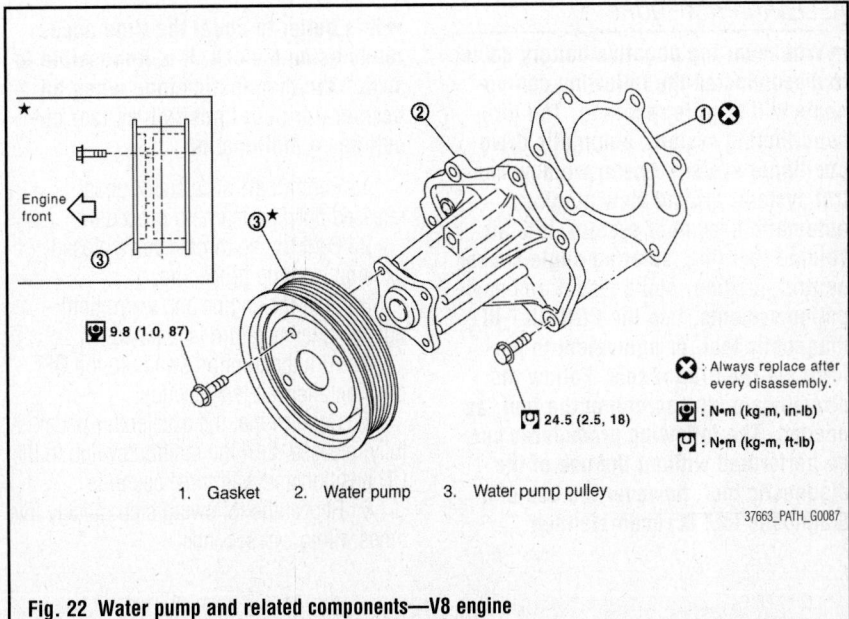

1. Gasket
2. Water pump
3. Water pump pulley

- : Always replace after every disassembly.
- : N•m (kg-m, in-lb)
- : N•m (kg-m, ft-lb)

37663_PATH_G0087

Fig. 22 Water pump and related components—V8 engine

4. Remove the left and right lower radiator shrouds.

5. Remove the drive belt.

6. Remove the cooling fan retaining bolts. Remove the cooling fan from the engine. As required, remove the pulley.

7. Remove the water pump retaining bolts. Remove the water pump. Discard the gasket.

To install:

➡**Be sure to use new fasteners, as required.**

8. Installation is in the reverse order of removal.

9. After installation bleed the air from the cooling system.

10. Be sure to perform the reconnect/relearn procedures.

ENGINE ELECTRICAL

BATTERY

REMOVAL & INSTALLATION

1. Before servicing the vehicle, refer to the Precautions.

2. Disconnect the negative battery cable.

3. Remove the battery cover, if equipped.

4. Disconnect both cables from the battery.

➡**Always disconnect the negative battery cable, first.**

5. Remove the battery clamp nuts and battery clamp.

6. Remove the battery.

To install:

7. Installation is the reverse of the removal procedure.

8. Perform reconnect/relearn procedures.

➡**Always connect the positive battery cable first.**

9. Tighten the battery terminal nut to 30 inch lbs. (3.4 Nm). Tighten the clamp nuts to 35 inch lbs. (3.92 Nm).

BATTERY RECONNECT/ RELEARN PROCEDURE

➡**Whenever the negative battery cable is disconnected the following components will require resetting. The air conditioning system, automatic drive positioner system, power window control system, around view monitor, automatic back door system, idle air volume learning, steering angle sensor neutral position, audio visual and navigation systems. Use the CONSULT-III diagnostic tool, or equivalent to perform the required resets. Follow the directions on the screen of the tool, as needed. The following procedures can be performed without the use of the diagnostic tool, however the use of a diagnostic tool is recommended.**

Throttle Valve Closed Position Learning

1. Start the engine.

➡**Be sure the engine coolant temperature is 77 degrees F or less, before starting.**

2. Warm up the engine.

➡**Raise the engine coolant temperature until it reaches 149 degrees F, or higher.**

3. Turn the ignition switch to the OFF position and wait ten seconds.

4. Check that the throttle valve moves during the above ten seconds by confirming the operating sound.

Idle Air Volume Learning

➡**Before performing this procedure be sure that the following conditions are met. Drive the vehicle for ten minutes. Battery voltage is more than 12.9 volts at idle. Engine coolant temperature is between 158–221 degrees F. Selector lever is in either P or N. All accessories are OFF. The steering wheel is in the straight head position. The vehicle is stopped. The transmission is warmed up.**

➡**It is better to count the time accurately using a clock. It is impossible to switch the diagnostic mode when an accelerator pedal position sensor circuit has a malfunction.**

1. Perform the accelerator pedal released position learning procedure.

2. Perform the throttle valve closed position learning procedure.

3. Start the engine and warm until operating temperature is reached.

4. Turn the ignition switch to the OFF position and wait ten seconds.

5. Confirm that the accelerator pedal is fully released, turn the ignition switch to the ON position and wait three seconds.

6. Repeat the following step quickly five times within five seconds.

BATTERY SYSTEM

7. Fully depress the accelerator pedal and than fully release the accelerator pedal.

8. Wait seven seconds, fully depress the accelerator pedal for about twenty seconds, wait until the MIL light stops blinking and turns ON.

9. Fully release the accelerator pedal within three seconds after the MIL light turns ON.

10. Start the engine and let it idle. Wait twenty seconds.

Accelerator Pedal Released Position Learning

1. Check that the accelerator pedal is fully released.

2. Turn the ignition switch to the ON position and wait at least two seconds.

3. Turn the ignition switch to the OFF position and wait at least ten seconds.

4. Turn the ignition switch to the ON position and wait at least two seconds.

5. Turn the ignition switch to the OFF position and wait at least ten seconds.

Steering Wheel Rotation

➡**If the vehicle is equipped with Intelligent Key System and Nissan Anti-theft System (NATS) and battery power is interrupted steering wheel rotation will be required, follow the procedure below before starting the repair operation.**

1. Connect both battery cables.

2. Use the Intelligent Key or mechanical key to turn the ignition switch to the "ACC" position. At this time, the steering lock will be released.

3. Disconnect both battery cables. The steering lock will remain released and the steering wheel can be rotated.

4. Perform the necessary repair operation.

5. When the repair work is completed, return the ignition switch to the "LOCK" position before connecting the battery cables. (At this time, the steering lock mechanism will engage).

6. Perform a self-diagnosis check of all control units using CONSULT III diagnostic tool, or equivalent.

ENGINE ELECTRICAL

ALTERNATOR

REMOVAL & INSTALLATION

See Figures 23 and 24.

1. Before servicing the vehicle, refer to the Precautions.
2. Disconnect the negative battery cable.
3. Remove the front undercover.
4. Remove the engine cover and air cleaner assembly, as required.
5. On V6 engine, drain the cooling system. Be sure to dispose of used coolant. Remove the coolant reservoir tank. Remove the upper fan shroud.
6. Remove the drive belt.
7. Remove alternator stay.
8. On V8 engine, remove the lower alternator bolt.

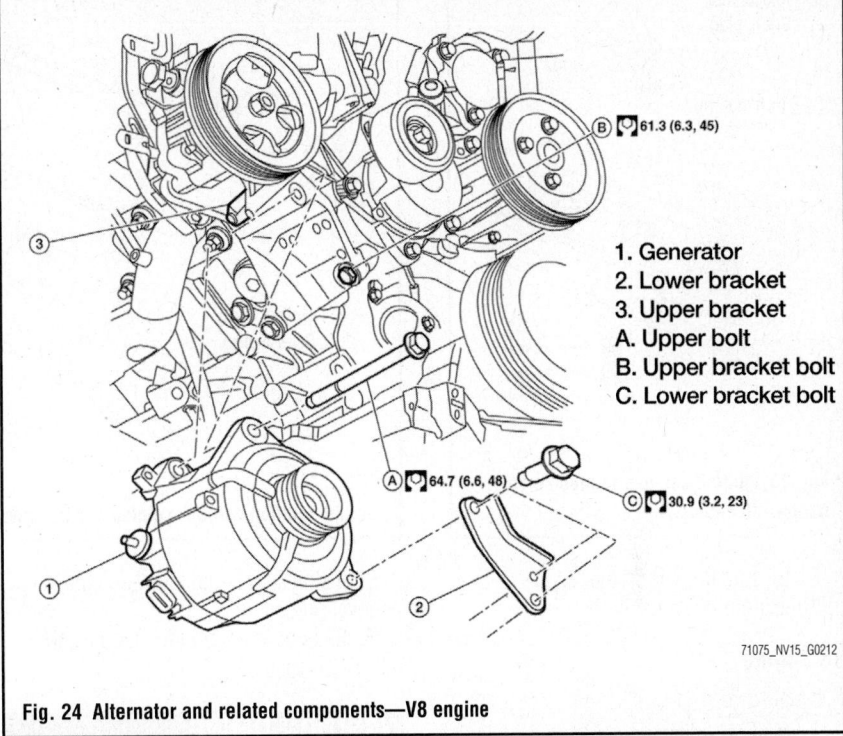

1. Generator
2. Lower bracket
3. Upper bracket
A. Upper bolt
B. Upper bracket bolt
C. Lower bracket bolt

B 61.3 (6.3, 45)
A 64.7 (6.6, 48)
C 30.9 (3.2, 23)

71075_NV15_G0212

Fig. 24 Alternator and related components—V8 engine

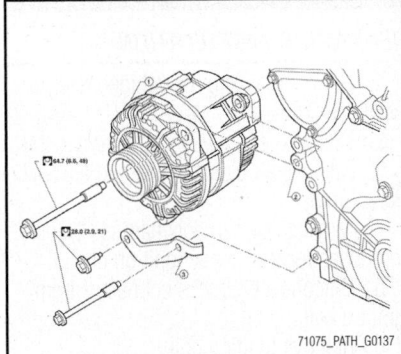

71075_PATH_G0137

Fig. 23 Alternator and related components (1) component, (2) block, (3) stay—V6 engine

9. Remove the alternator upper bolt.
10. Disconnect the alternator harness connectors.
11. Remove the alternator.

To install:
12. Installation is in the reverse order of removal.
13. Install the alternator and check tension of drive belt.

➡**The power generation variable voltage control system that controls** the power generation voltage of the alternator has been adopted. Therefore, the power generation variable voltage control system inspection should be performed after replacing the alternator in order to ensure that the system operates normally.

14. Be sure to perform the reconnect/relearn procedures.

ENGINE ELECTRICAL

IGNITION COIL MODULE

REMOVAL & INSTALLATION

V6 Engine

Left Bank

1. Before servicing the vehicle, refer to the Precautions.
2. Disconnect the negative battery cable.
3. Remove air cleaner case and air duct, as required.
4. Move aside harness, harness bracket, and hoses located above ignition coil.
5. Disconnect harness connector from ignition coil.
6. Remove ignition coil.

✳✳ WARNING
Do not shock it.

To install:
➡**Be sure to use new fasteners, as required.**

7. Installation is the reverse of the removal procedure.
8. Be sure to perform the reconnect/relearn procedures.

Right Bank

See Figure 25.

1. Before servicing the vehicle, refer to the Precautions.
2. Disconnect the negative battery cable.

3. Remove the air cleaner assembly, as required.
4. Remove intake manifold collector.
5. Move aside harness, harness bracket, and hoses located above ignition coil.
6. Disconnect harness connector from ignition coil.
7. Remove ignition coil.

✳✳ WARNING
Do not shock it.

To install:
➡**Be sure to use new fasteners, as required.**

8. Installation is the reverse of the removal procedure.

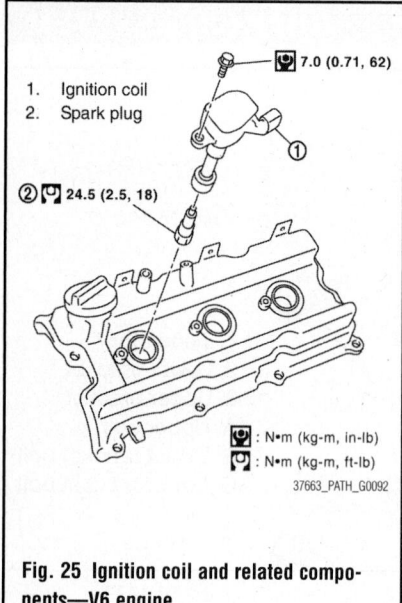

1. Ignition coil
2. Spark plug

🔧 7.0 (0.71, 62)

② 🔧 24.5 (2.5, 18)

🔧 : N•m (kg-m, in-lb)
🔧 : N•m (kg-m, ft-lb)
37663_PATH_G0092

Fig. 25 Ignition coil and related components—V6 engine

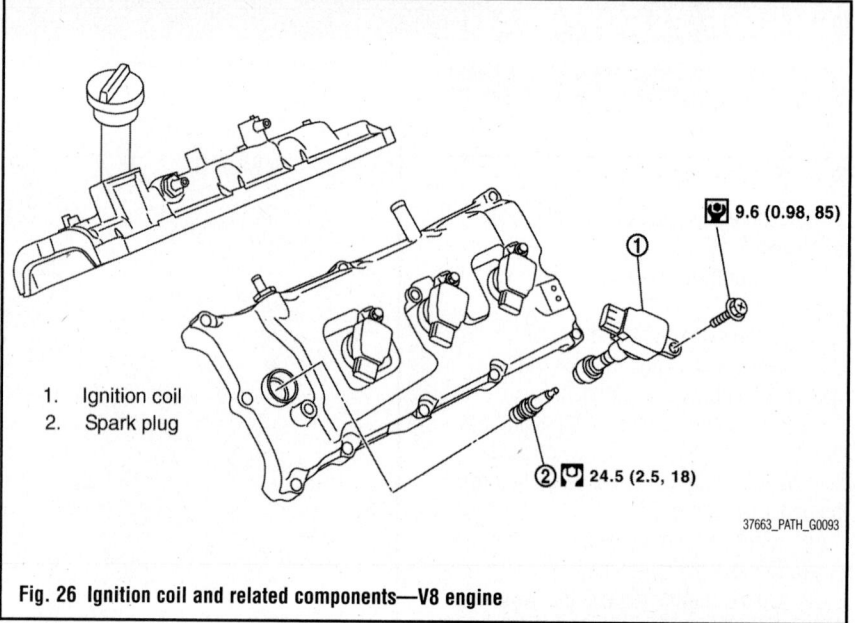

1. Ignition coil
2. Spark plug

🔧 9.6 (0.98, 85)

② 🔧 24.5 (2.5, 18)

37663_PATH_G0093

Fig. 26 Ignition coil and related components—V8 engine

9. Be sure to perform the reconnect/relearn procedures.

V8 Engine

See Figure 26.

1. Before servicing the vehicle, refer to the Precautions.
2. Disconnect the negative battery cable.
3. Remove the air duct and resonator assembly, as required.
4. Disconnect the harness connector from the ignition coil.
5. Remove the ignition coil.

❊❊ WARNING
Do not shock ignition coil.

To install:

➡**Be sure to use new fasteners, as required.**

6. Installation is the reverse of the removal procedure.
7. Be sure to perform the reconnect/relearn procedures.

IGNITION TIMING

INSPECTION & ADJUSTMENT

Ignition timing is controlled by the ECM. No adjustment is necessary or possible.

1. Before servicing the vehicle, refer to the Precautions.
2. Remove the number one ignition coil.
3. Connect the number one ignition coil and spark plug with a suitable high tension wire.
4. Attach the timing light clamp to the wire.
5. Check the ignition timing.

SPARK PLUGS

REMOVAL & INSTALLATION

1. Before servicing the vehicle, refer to the Precautions.
2. Disconnect the negative battery cable.
3. Remove air cleaner case and air duct, as required.
4. Move aside harness, harness bracket, and hoses located above ignition coil.
5. Disconnect harness connector from ignition coil.
6. Remove ignition coil.
7. Remove the spark plug.

To install:

➡**Be sure to use new fasteners, as required.**

8. Installation is the reverse of the removal procedure.
9. Be sure to perform the reconnect/relearn procedures.

ENGINE ELECTRICAL

STARTER

REMOVAL & INSTALLATION

V6 Engine

See Figure 27.

1. Before servicing the vehicle, refer to the Precautions.
2. Disconnect the negative battery cable.
3. Raise and safely support the vehicle.
4. Remove engine undercover, as necessary.

5. Remove the right tire and wheel assembly.
6. Remove the right fender protector.
7. Remove exhaust manifold cover from exhaust manifold (right bank) to gain access to starter cover bolts.
8. Remove starter cover bolts and starter cover.
9. Disconnect terminal "1" connector and terminal "2" nut.
10. Remove the two starter bolts.
11. Remove the starter.

STARTING SYSTEM

To install:

➡**Be sure to use new fasteners, as required.**

12. Installation is the reverse of the removal procedure.
13. Be sure to perform the reconnect/relearn procedures.

V8 Engine

See Figure 28.

1. Before servicing the vehicle, refer to the Precautions.

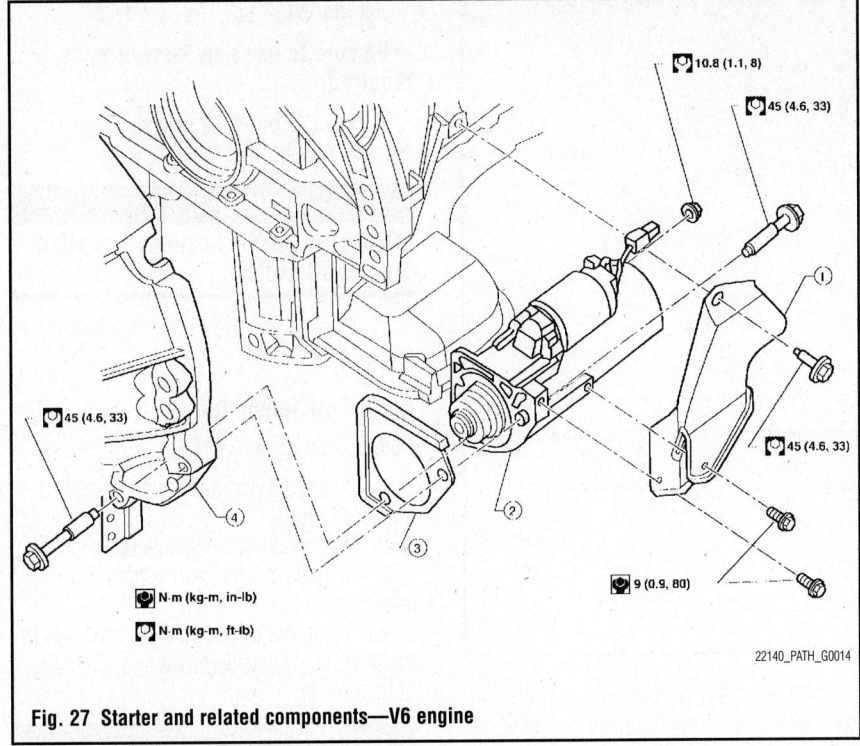

Fig. 27 Starter and related components—V6 engine

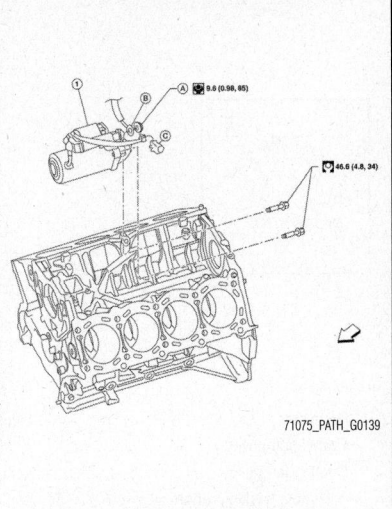

Fig. 28 Starter and related components (1) starter, (A) nut, (B) cable, (C) connector—V8 engine

2. Disconnect the negative battery cable.

3. Remove the engine cover and air cleaner assembly, as required.

4. Remove the intake manifold.

5. Disconnect terminal "1" connector and terminal "2" nut.

6. Remove the two starter bolts.

7. Remove the starter.

To install:

➡Be sure to use new fasteners, as required.

8. Installation is the reverse of the removal procedure.

9. Be sure to perform the reconnect/relearn procedures.

ENGINE MECHANICAL

➡Disconnecting the negative battery cable may interfere with the functions of the on board computer systems and may require the computer to undergo a relearning process, once the negative battery cable is reconnected.

ACCESSORY DRIVE BELTS

ADJUSTMENT

Belt tensioning is not necessary, as it is automatically adjusted by auto tensioner.

BELT ROUTINGS

See Figures 29 and 30.

INSPECTION

Inspect the drive belt for signs of glazing or cracking. A glazed belt will be perfectly smooth from slippage, while a good belt will have a slight texture of fabric visible. Cracks will usually start at the inner edge of the belt and run outward. All worn or damaged drive belts should be replaced immediately.

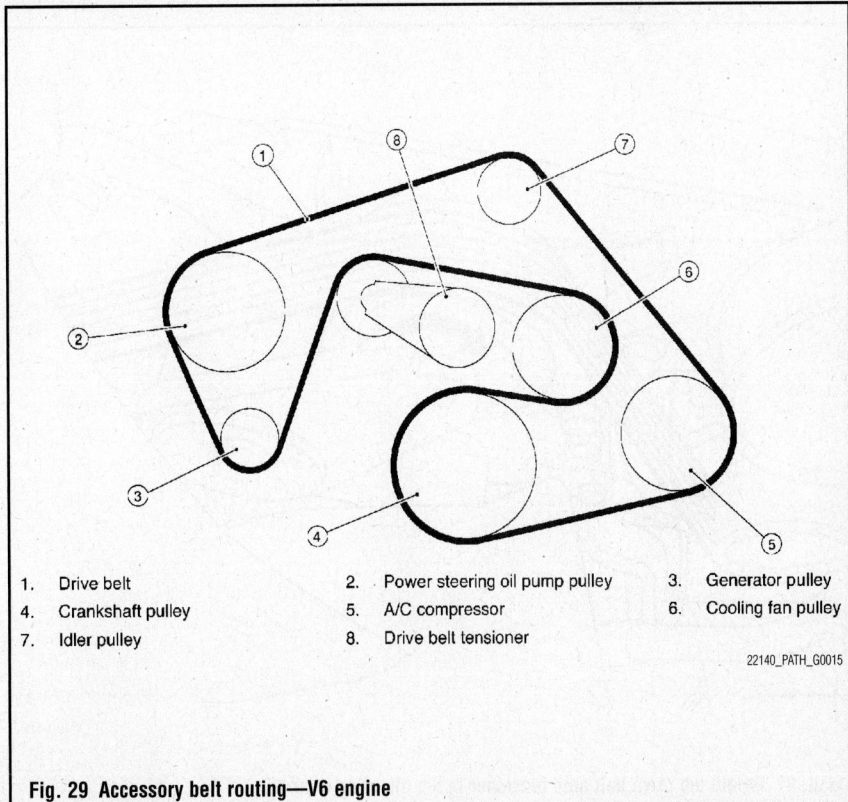

1.	Drive belt	2.	Power steering oil pump pulley	3.	Generator pulley
4.	Crankshaft pulley	5.	A/C compressor	6.	Cooling fan pulley
7.	Idler pulley	8.	Drive belt tensioner		

Fig. 29 Accessory belt routing—V6 engine

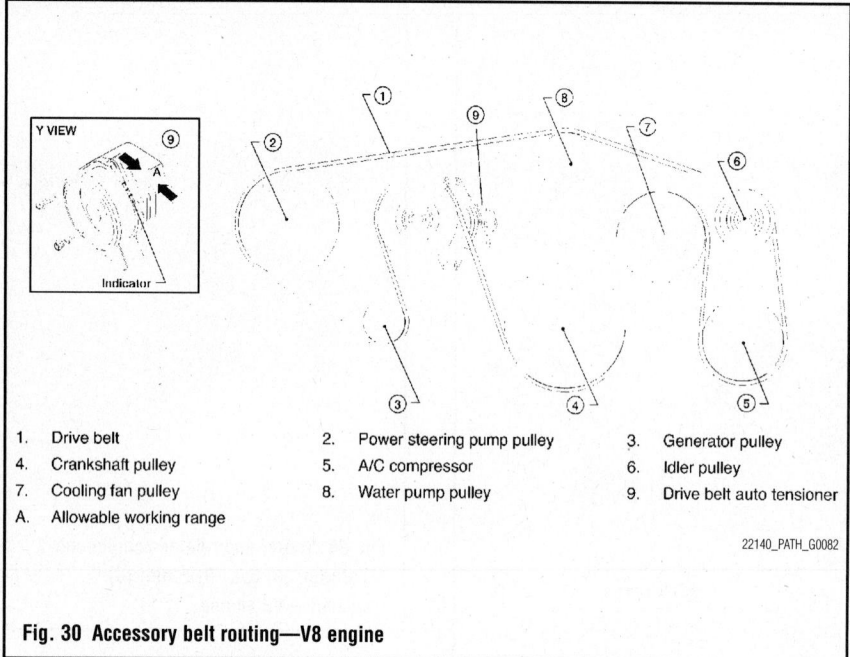

1. Drive belt
4. Crankshaft pulley
7. Cooling fan pulley
A. Allowable working range

2. Power steering pump pulley
5. A/C compressor
8. Water pump pulley

3. Generator pulley
6. Idler pulley
9. Drive belt auto tensioner

22140_PATH_G0082

Fig. 30 Accessory belt routing—V8 engine

REMOVAL & INSTALLATION

Drive Belt

See Figure 31.

1. Before servicing the vehicle, refer to the Precautions.
2. Disconnect the negative battery cable.
3. Remove air duct and resonator assembly (inlet).

4. Rotate the drive belt auto tensioner in the direction of arrow (loosening direction of tensioner).

✳✳ CAUTION

Avoid placing hand in a location where pinching may occur if the tool accidentally comes off.

5. Remove the drive belt.

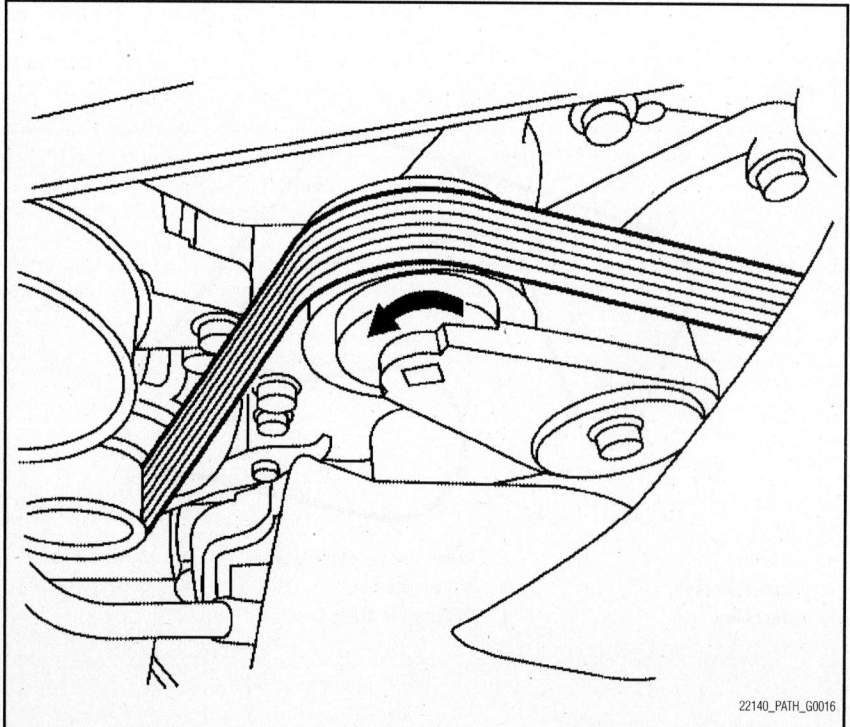

22140_PATH_G0016

Fig. 31 Rotate the drive belt auto tensioner in the direction of arrow

To install:

➡ **Be sure to use new fasteners, as required.**

6. Installation is the reverse of the removal procedure.

✳✳ WARNING

Make sure belt is securely installed around all pulleys.

7. Be sure to perform the reconnect/relearn procedures.

Drive Belt Tensioner

See Figures 32 and 33.

1. Before servicing the vehicle, refer to the Precautions.
2. Disconnect the negative battery cable.
3. Remove air duct and resonator assembly (inlet).
4. Rotate the drive belt auto tensioner in the direction of arrow (loosening direction of tensioner).

✳✳ CAUTION

Avoid placing hand in a location where pinching may occur if the tool accidentally comes off.

5. Remove the drive belt.
6. Remove the component retaining bolts. Remove the component from its mounting.

To install:

➡ **Be sure to use new fasteners, as required.**

7. Installation is the reverse of the removal procedure.

✳✳ WARNING

Make sure belt is securely installed around all pulleys.

8. Be sure to perform the reconnect/relearn procedures.

AIR CLEANER ASSEMBLY

REMOVAL & INSTALLATION

See Figures 34 and 35.

1. Before servicing the vehicle, refer to the Precautions.
2. Disconnect the negative battery cable.
3. Remove the component retaining screws/brackets/clips etc.
4. Disconnect the air duct and resonator from the air cleaner case.
5. Disconnect the electrical connections, as required.

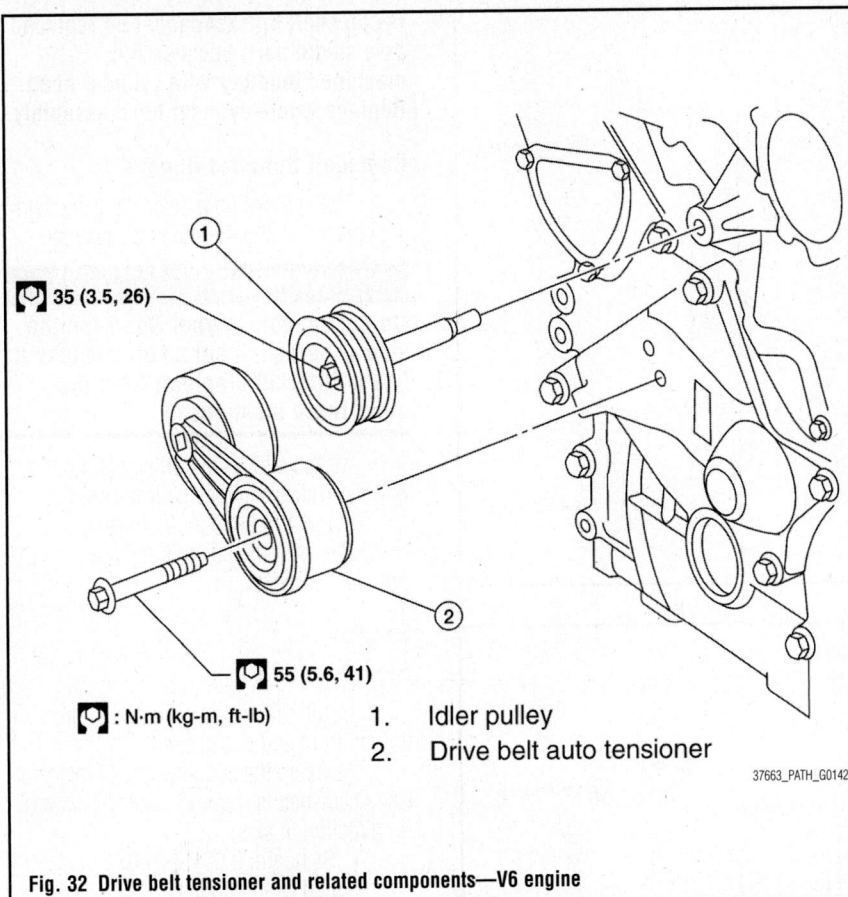

35 (3.5, 26)

55 (5.6, 41)

: N·m (kg-m, ft-lb)

1. Idler pulley
2. Drive belt auto tensioner

37663_PATH_G0142

Fig. 32 Drive belt tensioner and related components—V6 engine

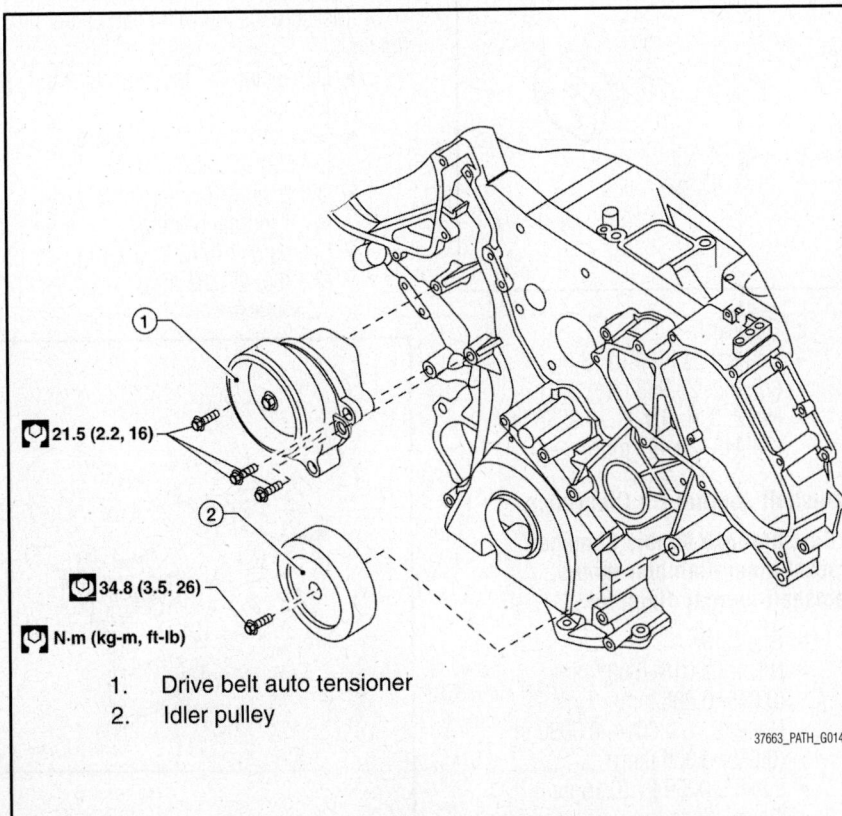

21.5 (2.2, 16)

34.8 (3.5, 26)

: N·m (kg-m, ft-lb)

1. Drive belt auto tensioner
2. Idler pulley

37663_PATH_G0143

Fig. 33 Drive belt tensioner and related components—V8 engine

6. Remove the component from the vehicle.

> *To install:*

→**Be sure to use new fasteners, as required.**

7. Installation is the reverse of the removal procedure.

8. Be sure to perform the reconnect/relearn procedures.

CAMSHAFT & BEARINGS

INSPECTION

Cam Height

1. Before servicing the vehicle, refer to the Precautions.
2. Remove the camshafts.
3. Measure the cam height with a micrometer.
4. Measure the camshaft cam height with micrometer.
   a. Standard:
   - Intake: 1.7900–1.7974 in. (45.465–45.655 mm)
   - Exhaust: 1.7746–1.7821 in. (45.075–45.265 mm)
   b. Limit:
   - Intake: 1.7821 in. 45.265 mm)
   - Exhaust: 1.7667 in. (44.875 mm)
5. If wear exceeds the limit, replace camshaft.

Camshaft Bracket Inner Diameter

See Figure 36.

1. Tighten camshaft bracket bolt with the specified torque.
2. Tighten camshaft bracket bolts in the following steps, in numerical order as shown.
   - Step 1 (bolts 7–10): 17 inch lbs. (1.96 Nm)
   - Step 2 (bolts 1–6): 17 inch lbs. (1.96 Nm)
   - Step 3: 52 inch lbs. (5.88 Nm)
   - Step 4: 92 inch lbs. (10.4 Nm)
3. Measure the inner diameter "A" of camshaft bracket with bore gauge.
   a. Standard:
   - No. 1: 1.0236–1.0244 in. (26.000–26.021 mm)
   - No. 2, 3, 4 : 0.9252–0.9260 in. (23.500–3.521 mm)

Camshaft Journal Diameter

1. Before servicing the vehicle, refer to the Precautions.
2. Remove the camshafts.
3. Measure the outer diameter of camshaft journal with micrometer and record the result.

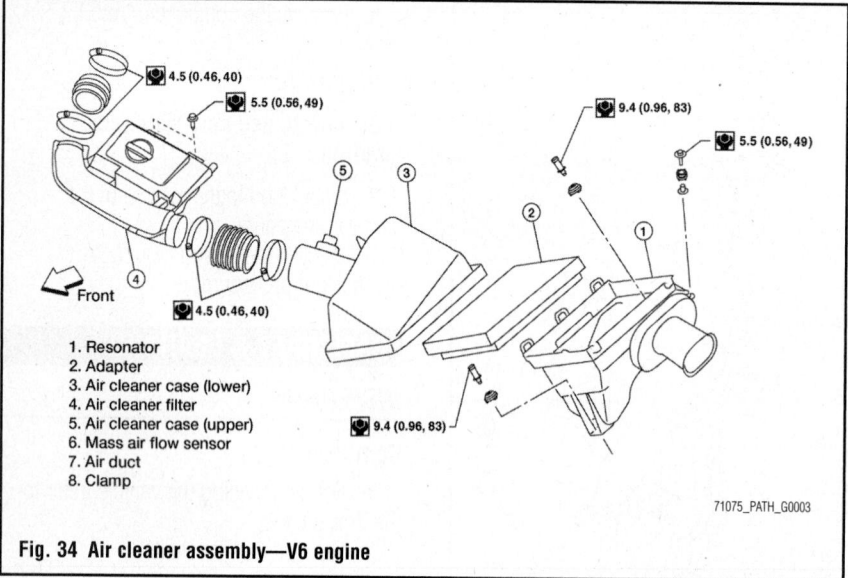

1. Resonator
2. Adapter
3. Air cleaner case (lower)
4. Air cleaner filter
5. Air cleaner case (upper)
6. Mass air flow sensor
7. Air duct
8. Clamp

71075_PATH_G0003

Fig. 34 Air cleaner assembly—V6 engine

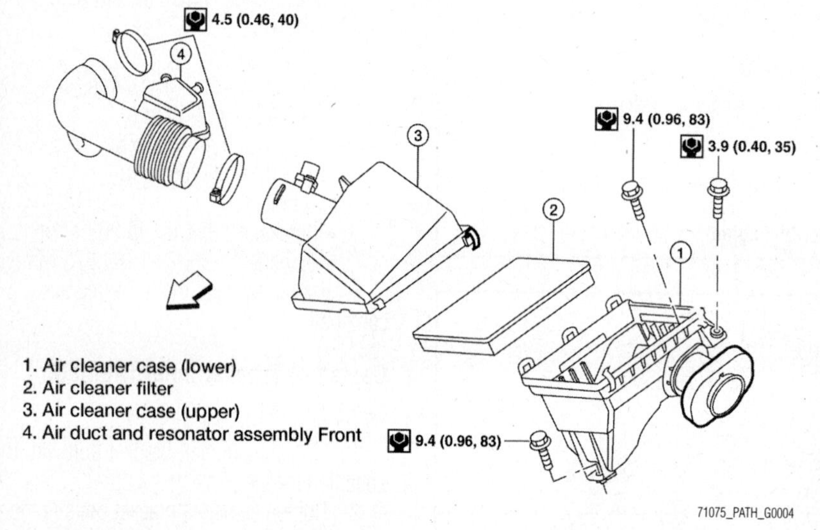

1. Air cleaner case (lower)
2. Air cleaner filter
3. Air cleaner case (upper)
4. Air duct and resonator assembly Front

71075_PATH_G0004

Fig. 35 Air cleaner assembly—V8 engine

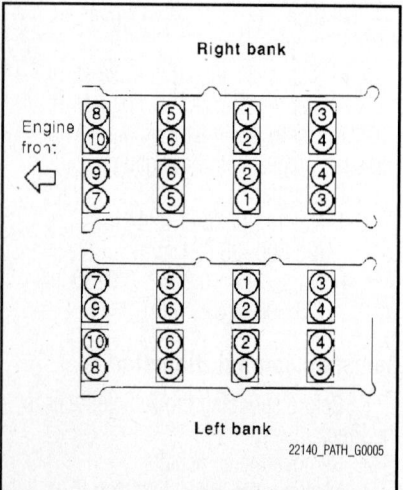

22140_PATH_G0005

Fig. 36 Camshaft bracket bolts tightening sequence

a. Standard:
- No. 1: 1.0211–1.0218 in. (25.935–25.955 mm)
- No. 2, 3, 4: 0.9230–0.9238 in. (23.445–23.465 mm)

Camshaft Journal Oil Clearance

➡Oil clearance equals Camshaft bracket inner diameter minus Camshaft journal diameter.

b. Standard:
- No. 1: 0.0018–0.0034 in. (0.045–0.086 mm)
- No. 2, 3, 4: 0.0014–0.0030 in. (0.035–0.076 mm)
- Limit: 0.0059 in. (0.15 mm)

1. If the calculated value exceeds the limit, replace either or both camshaft and cylinder head.

➡Camshaft bracket cannot be replaced as a single part, because it is machined together with cylinder head. Replace whole cylinder head assembly.

Camshaft Sprocket Runout

1. Put V-block on precise flat table, and support No. 2 and 4 journal of camshaft.

✳✳ WARNING

Do not support journal No. 1 (on the side of camshaft sprocket) because it has a different diameter from the other three locations.

2. Measure the camshaft sprocket run out with dial indicator. (Total indicator reading)
- Limit: 0.0059 in. (0.15 mm)

3. If it exceeds the limit, replace camshaft sprocket.

Endplay

See Figures 37 and 38.

1. Install dial indicator in thrust direction on front end of camshaft.
2. Measure the end play of dial indicator when camshaft is moved forward/backward (in direction to axis).
- Standard: 0.0045–0.0074 in. (0.115–0.188 mm)
- Limit: 0.0094 in. (0.24 mm)

3. Measure the following parts if out of the limit.
- Dimension "A" for camshaft No. 1 journal
- Standard: 1.0827–1.0846 in. (27.500–27.548 mm)
- Dimension "B" for cylinder head No. 1 journal bearing
- Standard: 1.0772–1.0781 in. (27.360–27.385 mm)

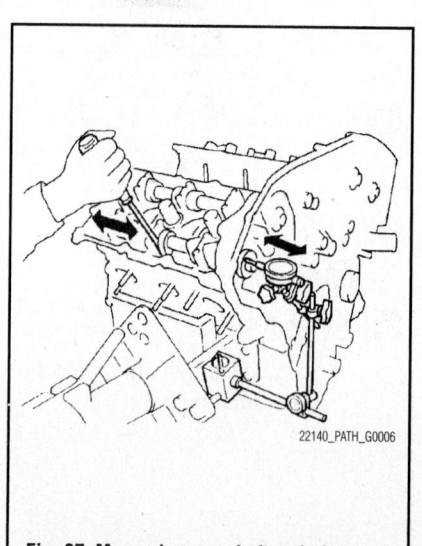

22140_PATH_G0006

Fig. 37 Measuring camshaft end play

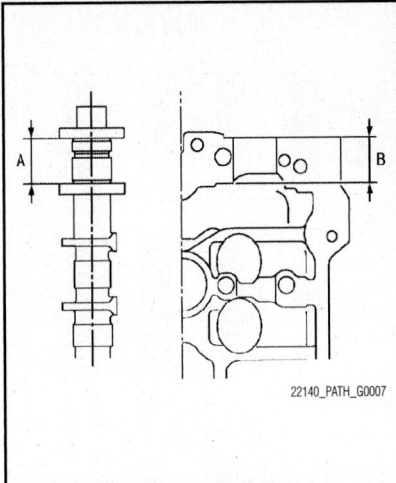

Fig. 38 Measure the following parts if out of the limit

4. Refer to the standards above, and then replace camshaft and/or cylinder head.

Runout

1. Put V-block on precise flat table, and support No. 2 and 4 journal of camshaft.

❊❊ WARNING

Do not support journal No. 1 (on the side of camshaft sprocket) because it has a different diameter from the other three locations.

2. Set dial indicator vertically to No. 3 journal.

3. Turn camshaft to one direction with hands, and measure the camshaft run out on dial indicator. (Total indicator reading)
- Standard: Less than 0.0008 in. (0.02 mm)
- Limit: 0.0020 in. (0.05 mm)

4. If it exceeds the limit, replace camshaft.

REMOVAL & INSTALLATION

V6 Engine

See Figures 39 through 43.

1. Before servicing the vehicle, refer to the Precautions.

2. Disconnect the negative battery cable.

3. Remove the front fender protectors. Remove the air cleaner assembly.

4. Remove the intake manifold collector. Remove the valve covers.

5. Remove front timing chain case, camshaft sprocket, timing chain and rear timing chain case.

6. Remove Camshaft Position (CMP) sensor (PHASE) (right and left banks) from cylinder head back side.

❊❊ WARNING

Handle carefully to avoid dropping and shocks. Do not disassemble. Do not allow metal powder to adhere to magnetic part at sensor tip. Do not place sensors in a location where they are exposed to magnetism.

7. Remove intake valve timing control solenoid valves.

➡**Discard intake valve timing control solenoid valve gaskets and use new gaskets for installation.**

8. Remove camshaft brackets.

a. Mark camshafts, camshaft brackets and bolts so they are placed in the same position and direction for installation.

b. Equally loosen camshaft bracket bolts in several steps in reverse order of the tightening sequence.

9. Remove camshafts.

10. Remove valve lifters.

➡**Identify installation positions, and store them without mixing them up.**

11. Remove timing chain tensioner (secondary) from cylinder head with its stopper pin attached.

➡**Stopper pin was attached when timing chain (secondary) was removed.**

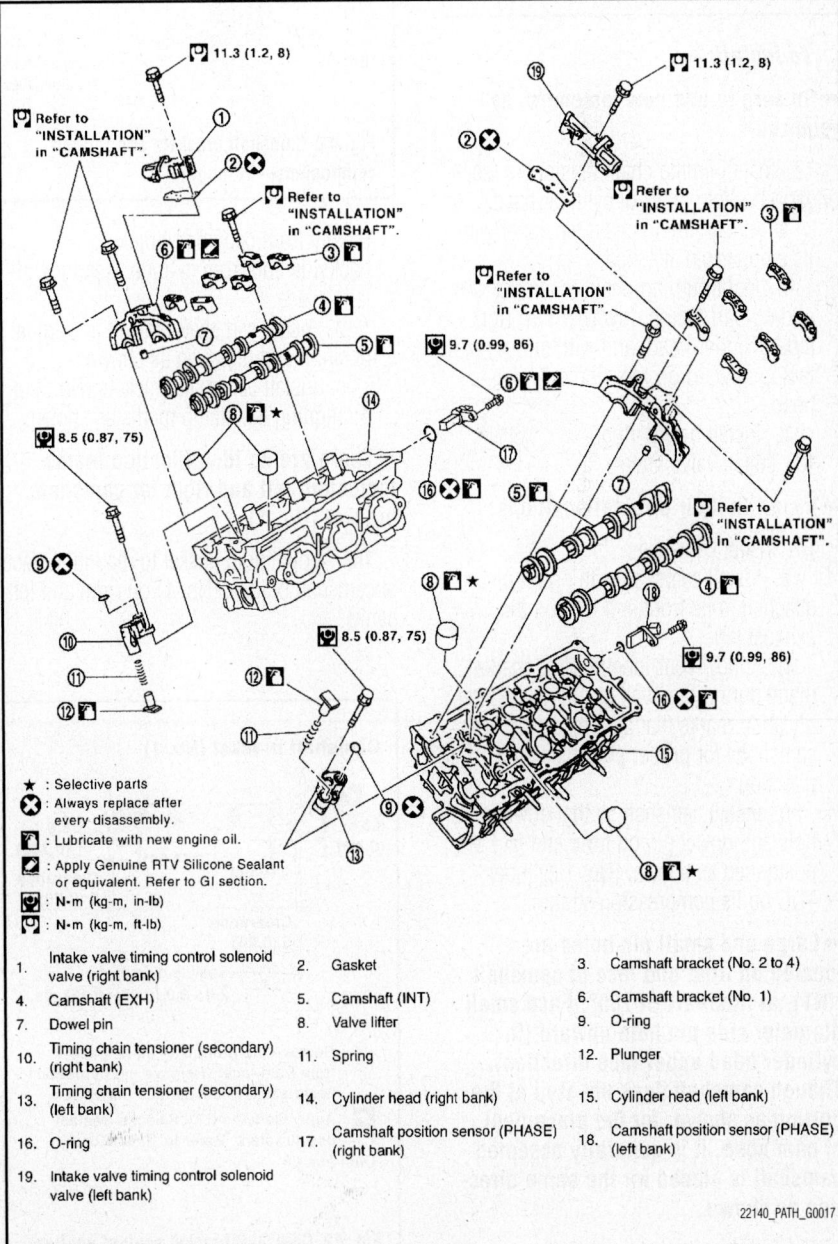

★ : Selective parts
✖ : Always replace after every disassembly.
🔧 : Lubricate with new engine oil.
🔧 : Apply Genuine RTV Silicone Sealant or equivalent. Refer to GI section.
🔧 : N·m (kg-m, in-lb)
🔧 : N·m (kg-m, ft-lb)

1.	Intake valve timing control solenoid valve (right bank)	2.	Gasket	3.	Camshaft bracket (No. 2 to 4)
4.	Camshaft (EXH)	5.	Camshaft (INT)	6.	Camshaft bracket (No. 1)
7.	Dowel pin	8.	Valve lifter	9.	O-ring
10.	Timing chain tensioner (secondary) (right bank)	11.	Spring	12.	Plunger
13.	Timing chain tensioner (secondary) (left bank)	14.	Cylinder head (right bank)	15.	Cylinder head (left bank)
16.	O-ring	17.	Camshaft position sensor (PHASE) (right bank)	18.	Camshaft position sensor (PHASE) (left bank)
19.	Intake valve timing control solenoid valve (left bank)				

Fig. 39 Camshaft and related components—V6 Engine

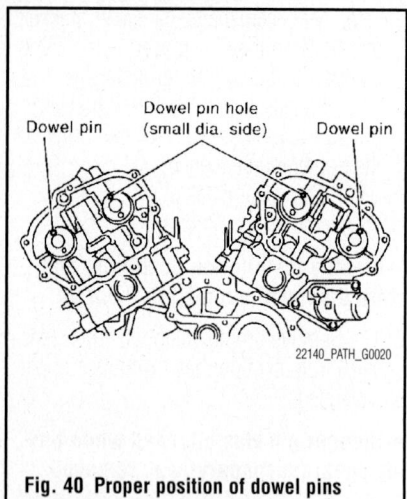

Fig. 40 Proper position of dowel pins during camshaft installation—V6 engine

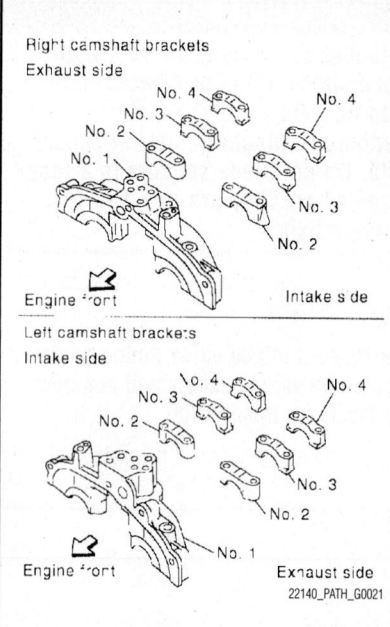

Fig. 41 Camshaft brackets and related components—V6 engine

To install:

➡ **Be sure to use new fasteners, as required.**

12. Install timing chain tensioners (secondary) on both sides of cylinder head.

   a. Install timing chain tensioner with its stopper pin attached.

   b. Install timing chain tensioner with sliding part facing downward on right-side cylinder head, and with sliding part facing upward on left-side cylinder head.

   c. Install new O-rings.

13. Install valve lifters.

➡ **Install in their original positions.**

14. Install camshafts.

   a. Install camshaft with dowel pin attached to its front end face on the exhaust side.

   b. Follow your identification marks made during removal, or follow the identification marks that are present on new camshafts for proper placement and direction.

   c. Install camshaft so that dowel pin hole and dowel pin on front end face are positioned as shown. (No. 1 cylinder TDC on its compression stroke).

➡ **Large and small pin holes are located on front end face of camshaft (INT), at intervals of 180°. Face small diameter side pin hole upward (in cylinder head upper face direction). Though camshaft does not stop at the portion as shown, for the placement of cam nose, it is generally accepted camshaft is placed for the same direction as shown.**

15. Install camshaft brackets.

   a. Remove foreign material com-

pletely from camshaft bracket backside and from cylinder head installation face.

   b. Install camshaft bracket in original position and direction as shown.

   c. Install camshaft brackets (No. 2 to 4) aligning the stamp marks as shown.

➡ **There are no identification marks indicating left and right for camshaft bracket (No. 1).**

16. Apply liquid gasket to mating surface of camshaft bracket (No. 1) on right and left banks.

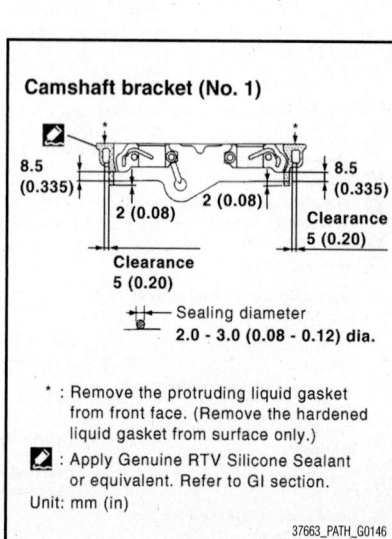

Fig. 42 Camshaft bracket sealant application points—V6 engine

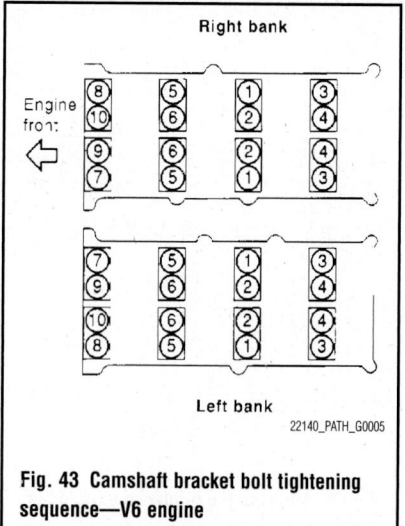

Fig. 43 Camshaft bracket bolt tightening sequence—V6 engine

➡ **Use Genuine RTV Silicone Sealant or equivalent.**

17. Tighten camshaft bracket bolts in numerical order as shown.

18. Tighten camshaft bracket bolts in the following steps:

- Step 1: Bolts 7–10: 17 inch lbs. (1.96 Nm)
- Step 2: Bolts 1–6: 17 inch lbs. (1.96 Nm)
- Step 3: All bolts: 52 inch lbs. (5.88 Nm)
- Step 4: All bolts: 92 inch lbs. (10.4 Nm)

19. Measure the difference in levels between front end faces of camshaft bracket (No. 1) and cylinder head.

   a. Standard: -0.0055 to 0.0055 in. (-0.14 to 0.14 mm)

   b. Measure two positions (both intake and exhaust side) for a single bank.

   c. If the measured value is out of the standard, re-install camshaft bracket (No. 1).

20. Check and adjust the valve clearance, as required.

21. Installation of the remaining components is in the reverse order of removal.

22. Be sure to perform the reconnect/relearn procedures.

V8 Engine

See Figures 44 through 60.

➡ **Do not remove the engine assembly to perform this procedure.**

1. Before servicing the vehicle, refer to the Precautions.

2. Disconnect the negative battery cable.

3. Remove the engine undercover. Remove the air cleaner assembly.

4. Remove the reservoir tank and water hose. Remove the PCV hoses.

5. Remove the left and right lower radiator shrouds.

6. Remove the upper radiator shroud.

7. Remove the lower radiator hose, if removing the left camshaft assembly.

8. Remove the power steering reservoir and bracket, if removing the right camshaft assembly.

9. Remove the drive belt.

10. Remove the RH bank and LH bank rocker covers.

11. Obtain compression TDC of No. 1 cylinder as follows:

a. Turn the crankshaft pulley clockwise to align the TDC identification notch (without paint mark) with the timing indicator on the front cover.

b. At this time, make sure both intake and exhaust cam lobes of No. 1 cylinder (top front on LH bank) point outside.

c. If they do not point outside, turn crankshaft pulley once more.

12. Remove the intake valve control solenoid cover RH bank (A) and intake valve control solenoid cover LH bank (B) as follows:

a. Loosen and remove the bolts, see illustration for proper removal sequence

b. Cut the liquid gasket and remove the covers.

✳✳ WARNING

Do not damage mating surfaces.

13. Paint alignment marks on the RH bank (A) timing chain links (C) and LH bank (B) timing chain links (D) and align with the camshaft sprocket alignment marks (E) and (F).

14. Remove the LH bank timing chain tensioner using the following steps.

✳✳ CAUTION

Plunger, spring, and spring seat pop out when squeezing return-proof clip without holding plunger head. It may cause serious injuries. Always hold plunger head when removing.

a. Squeeze return-proof clip ends using suitable tool and push the plunger into the tensioner body.

b. Secure plunger using stopper pin.

➡**Stopper pin is made from hard wire approximately 1 mm (0.04 in.) in diameter.**

c. Remove the bolts and the timing chain tensioner.

➡**Stop plunger in the fully extended position using return-proof clip (1) if stopper pin is removed. Push the plunger (2) into the tensioner body while squeezing the return-proof**

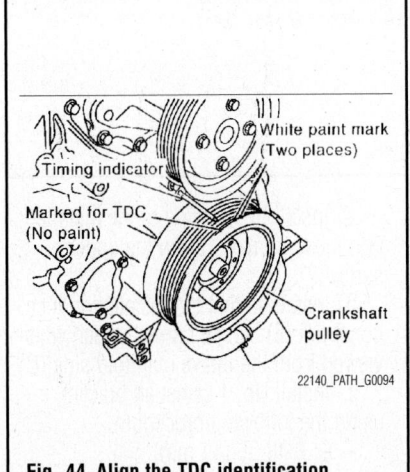
Fig. 44 Align the TDC identification notch—V8 engine

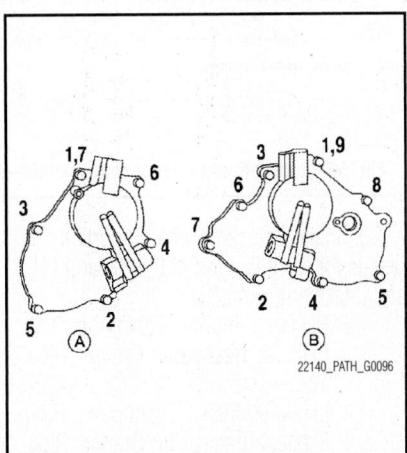
Fig. 46 Loosen and remove the bolts—V8 engine

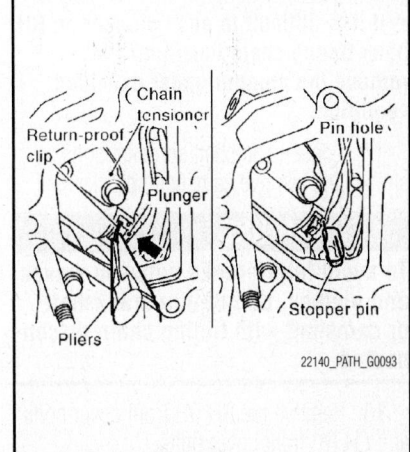
Fig. 48 Push the plunger into the tensioner body—V8 engine

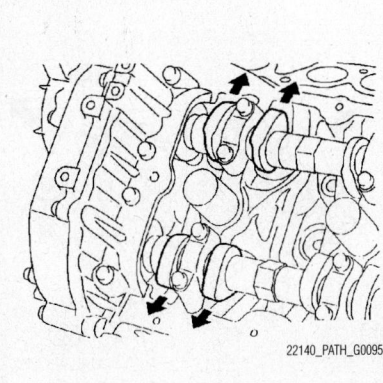
Fig. 45 Intake and exhaust cam lobes of No. 1 cylinder—V8 engine

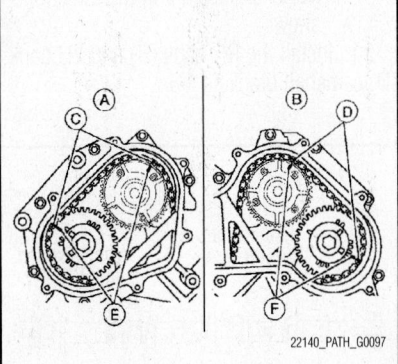
Fig. 47 Paint alignment marks on the RH bank (A) timing chain links (C) and LH bank (B) timing chain links—V8 engine

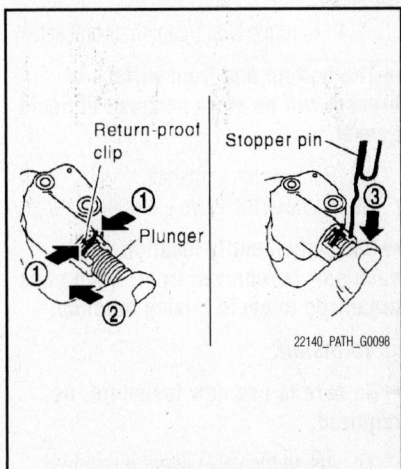
Fig. 49 Stop plunger in the fully extended position—V8 engine

clip (1). Secure it using stopper pin (3).

15. Remove the RH bank timing chain tensioner cover from the front cover.

✳✳ WARNING

Do not damage mating surfaces.

16. Remove the RH bank timing chain tensioner using the following steps.

✳✳ CAUTION

Plunger, spring, and spring seat pop out when squeezing return-proof clip without holding plunger head. It may cause serious injuries. Always hold plunger head when removing.

   a. Squeeze return-proof clip ends using suitable tool and push the plunger into the tensioner body.
   b. Secure plunger using stopper pin.
   c. Remove the bolts and the RH bank timing chain tensioner (A).

➡**If it is difficult to push plunger on RH bank timing chain tensioner (A), remove the plunger under extended condition.**

17. Loosen camshaft sprocket bolts as shown and remove camshaft sprockets.

✳✳ WARNING

To avoid interference between valves and pistons, do not turn crankshaft or camshaft with timing chain disconnected.

18. Remove the RH (A) front cover bolts and LH (B) front cover bolts.
19. Remove RH (A) camshaft bracket bolts and LH (C) camshaft bracket bolts in the reverse order of the tightening sequence.
   a. Remove No. 1 camshaft bracket.

➡**The bottom and front surface of bracket will be stuck because of liquid gasket.**

20. Remove the camshaft.
21. Remove the valve lifters if necessary.

➡**Correctly identify location where each part is removed from. Keep parts organized to avoid mixing them up.**

   To install:

➡**Be sure to use new fasteners, as required.**

22. Install the valve lifters if removed.

➡**Install removed parts in their original locations.**

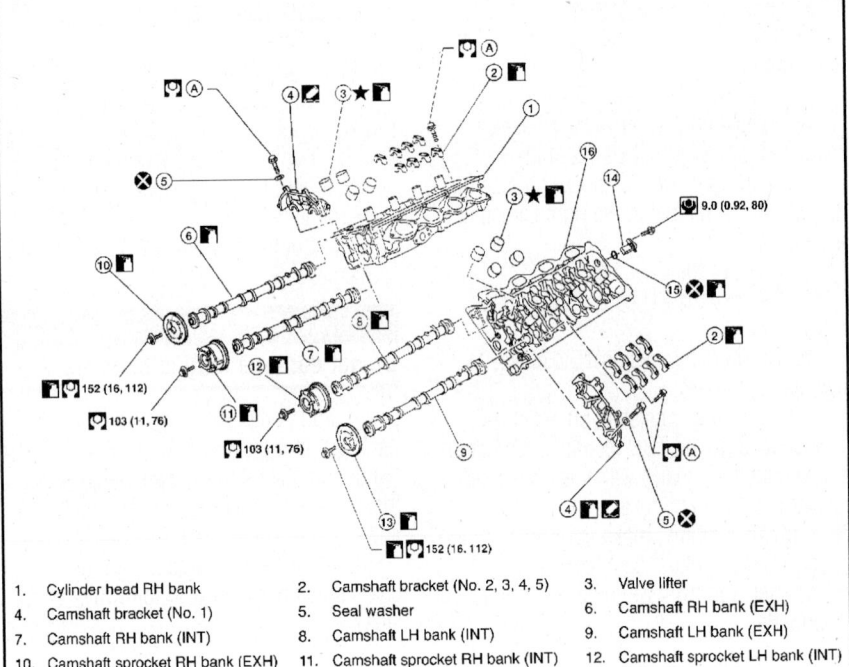

1. Cylinder head RH bank	2. Camshaft bracket (No. 2, 3, 4, 5)	3. Valve lifter
4. Camshaft bracket (No. 1)	5. Seal washer	6. Camshaft RH bank (EXH)
7. Camshaft RH bank (INT)	8. Camshaft LH bank (INT)	9. Camshaft LH bank (EXH)
10. Camshaft sprocket RH bank (EXH)	11. Camshaft sprocket RH bank (INT) (VTC)	12. Camshaft sprocket LH bank (INT) (VTC)
13. Camshaft sprocket LH bank (EXH)	14. Camshaft position sensor (PHASE)	15. O-ring
16. Cylinder head LH bank		

22140_PATH_G0083

Fig. 50 Camshaft and related components—V8 Engine

23. Install the camshafts. Important details for identification of the RH and LH, and intake and exhaust.
- RH Bank: Intake: Front paint: Pink; Exhaust: Rear paint: Orange; Rib: Yes
- LH Bank: Intake: Front paint: Pink; Exhaust: Rear paint: Orange; Rib: No
- Install so that the RH bank (B) dowel pins (A) and LH bank (C) dowel pins (A) at the front of the camshaft face are in the direction shown.
24. Install the RH bank (B) and LH bank (D) camshaft brackets (A).

   a. Install by referring to the installation location mark (E) on the upper surface.
   b. Install so that the installation location mark (E) can be correctly read when viewed from the intake manifold side (C).
   c. Install No. 1 camshaft bracket using the following procedure:
- C: 0.43 in. (11 mm)
- D: 0.079–0.118 in. (2.0–3.0 mm) diameter

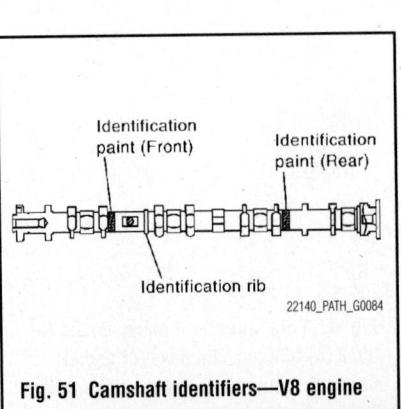

Identification paint (Front)

Identification paint (Rear)

Identification rib

22140_PATH_G0084

Fig. 51 Camshaft identifiers—V8 engine

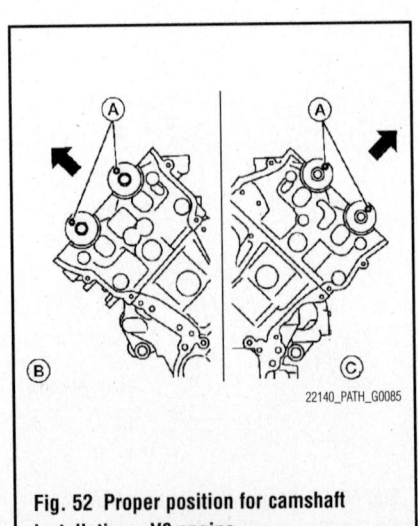

22140_PATH_G0085

Fig. 52 Proper position for camshaft installation—V8 engine

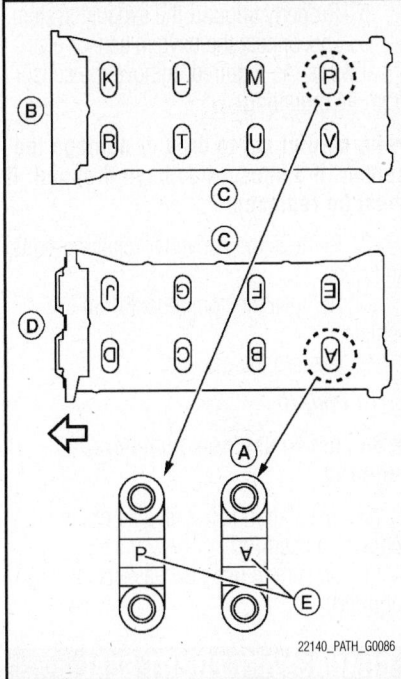

Fig. 53 Install the RH bank (B) and LH bank (D) camshaft brackets (A)—V8 engine

- Apply liquid gasket to No. 1 camshaft bracket (A) and (B) as shown.

✳✳ WARNING

After installation, be sure to wipe off any excessive liquid gasket outside of application (C) and (D) both on RH and LH sides.

- Remove completely any excess of liquid gasket inside bracket.

d. Apply liquid gasket (C) to the back side of the LH (A) bank front

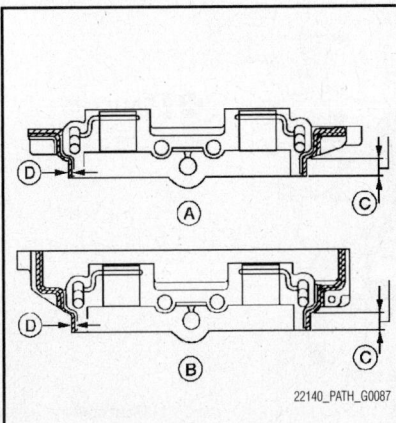

Fig. 54 Apply liquid gasket to No. 1 camshaft bracket (A) and (B)—V8 engine

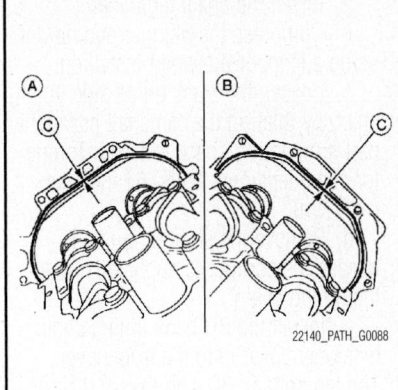

Fig. 55 Apply liquid gasket (C) to the back side of the LH (A) bank front cover and RH (B) bank front cover—V8 engine

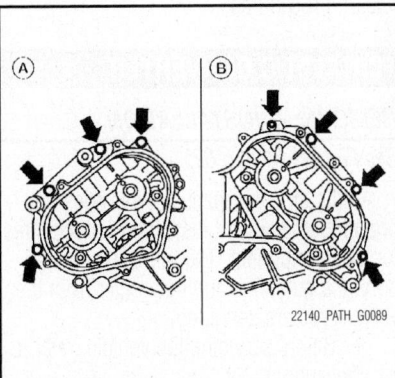

Fig. 56 Temporarily tighten the RH (A) and LH (B) front cover bolts—V8 engine

cover and RH (B) bank front cover as shown.

e. C: 0.102–0.142 in. (2.6–3.6 mm) diameter

f. Position No. 1 camshaft bracket close to the mounting position, and then install it to prevent from touching liquid gasket applied to each surface.

g. Temporarily tighten the RH (A) and LH (B) front cover bolts (4 for each bank) as shown.

25. Tighten the camshaft bracket bolts as follows:

a. Step 1 (bolts 9–12): 17 inch lbs. (2.0 Nm)

b. Step 2 (bolts 1–8): 17 inch lbs. (2.0 Nm)

c. Step 3 (all bolts): 52 inch lbs. (5.9 Nm)

d. Step 4 (all bolts): 92 inch lbs. (10.4 Nm)

✳✳ WARNING

After tightening the camshaft bracket bolts, be sure to wipe off excessive

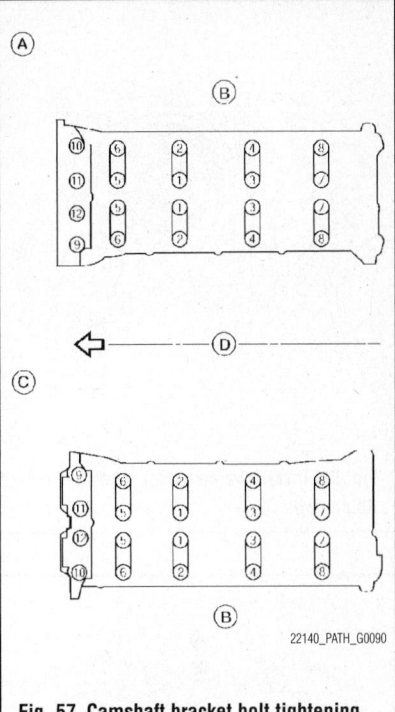

Fig. 57 Camshaft bracket bolt tightening sequence—V8 engine

liquid gasket from the parts listed below.

- Mating surface of rocker cover
- Mating surface of front cover
- A: RH bank
- B: Exhaust side
- C: LH bank
- D: Intake side

e. Tighten the RH (A) and LH (B) front cover bolts (4 for each bank) to 8 ft. lbs. (11.0 Nm)

26. Install the camshaft sprockets using the following procedure:

➡️**A: LH bank shown.**

a. Install the camshaft sprockets aligning them with the matching marks painted on the timing chain (B) and the camshaft sprockets (C) before removal. Align the camshaft sprocket key groove with the dowel pin on the camshaft front edge at the same time. Then temporarily tighten camshaft sprocket bolts.

b. Install the intake VTC (A) and exhaust (B) side camshaft sprockets by selectively using the groove of the dowel pin according to the bank for the exhaust (B) side camshaft sprockets. (Common part used for both exhaust banks.)

c. Lock the hexagonal part of the camshaft in the same way as for removal, and tighten the camshaft sprocket bolts.

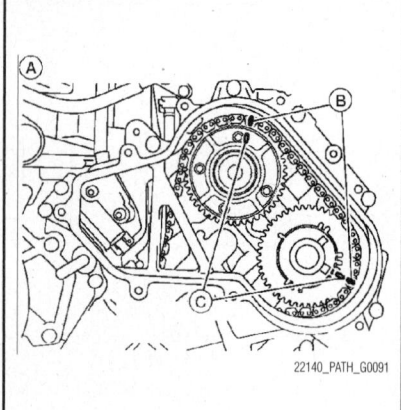

Fig. 58 Install the camshaft sprockets—V8 engine

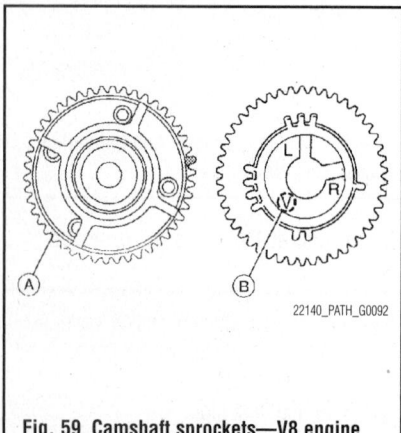

Fig. 59 Camshaft sprockets—V8 engine

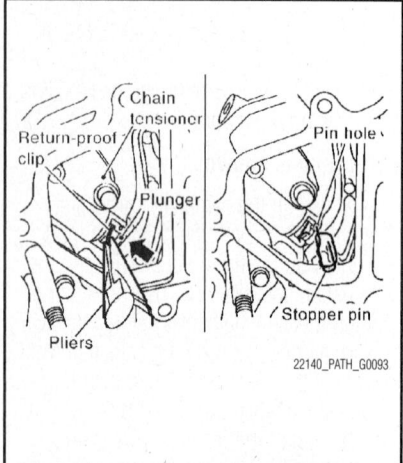

Fig. 60 Install the chain tensioner—V8 engine

d. Check again that the timing alignment mark on the timing chain and on each sprocket are aligned.

27. Install the chain tensioner using the following procedure:

→LH is shown.

a. Install the chain tensioner.

b. Compress the plunger and hold it using a stopper pin when installing.

c. Loosen the slack guide side timing chain by rotating the camshaft hexagonal part if mounting space is small. Torque for chain tensioner bolts: 61 inch lbs. (6.9 Nm).

d. Remove the stopper pin and release the plunger to apply tension to the timing chain.

e. Install the RH bank timing chain tensioner cover onto the front cover. Tighten bolts to 80 inch lbs. (9.0 Nm).

28. Check and adjust valve clearances, as required.

29. Installation of the remaining components is in the reverse order of removal.

30. Be sure to perform the reconnect/relearn procedures.

CATALYTIC CONVERTER

REMOVAL & INSTALLATION

See Figures 61 and 62.

At this time the manufacturer does not provide removal and installation procedures for this component. The following procedure is a guideline and may differ from the vehicle you are servicing.

1. Before servicing the vehicle, refer to the Precautions.

2. Disconnect the negative battery cable.

3. Remove the air cleaner assembly, as required.

4. Raise and safely support the vehicle.

5. Properly support the exhaust system.

6. Disconnect the oxygen sensor electrical wires. As required, remove the sensor from its mounting.

→Be careful not to drop or damage the sensor. If a sensor has been dropped, it must be replaced.

7. Remove the converter retaining bolts and nuts.

8. Remove the component from the vehicle.

9. Discard the gaskets.

To install:

→Be sure to use new fasteners, as required.

10. Installation is the reverse of the removal procedure.

11. Be sure to use new gaskets, as required.

CRANKSHAFT FRONT SEAL

REMOVAL & INSTALLATION

V6 Engine

1. Before servicing the vehicle, refer to the Precautions.

2. Disconnect the negative battery cable.

3. Remove the engine undercover. Remove the air cleaner assembly.

4. Remove drive belts.

5. Remove engine cooling fan assembly.

6. Remove crankshaft pulley.

7. Remove front oil seal.

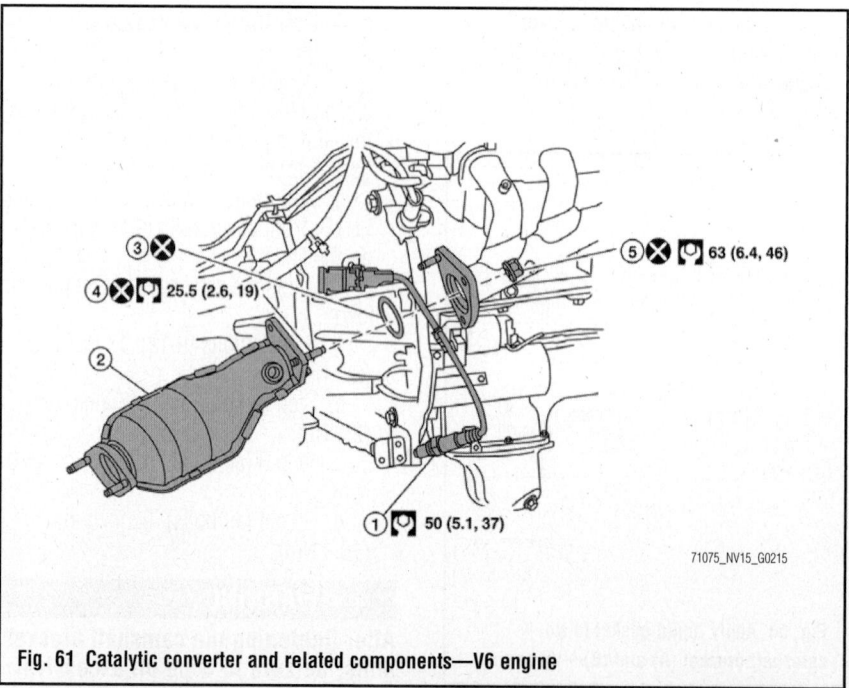

Fig. 61 Catalytic converter and related components—V6 engine

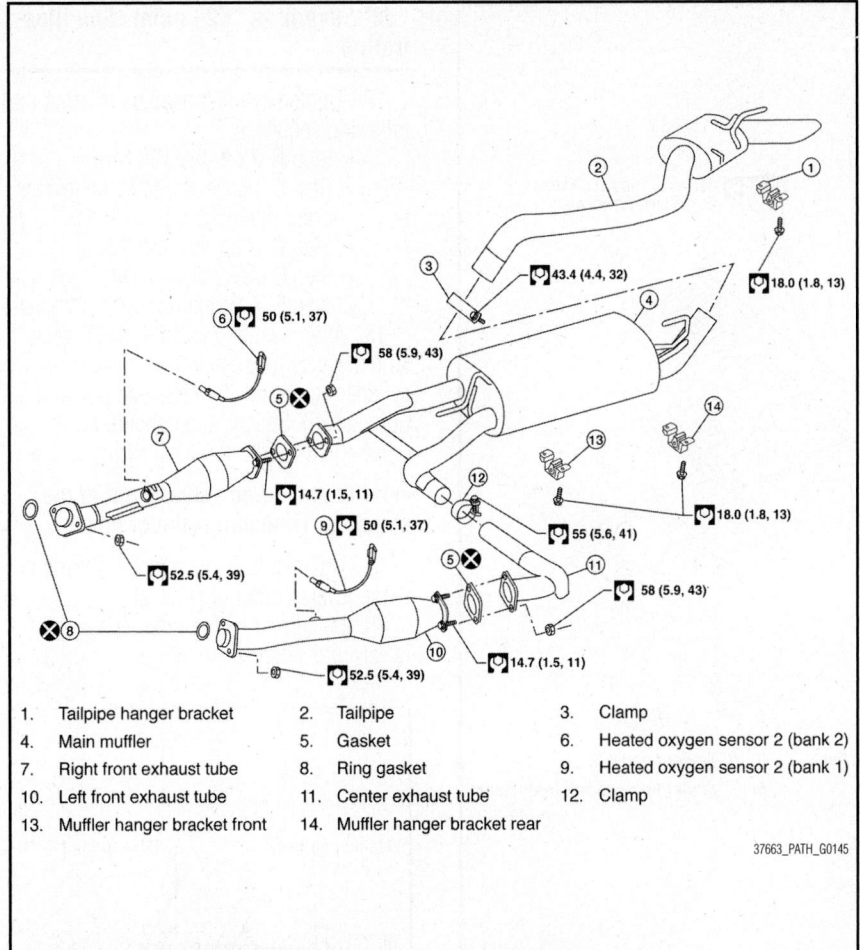

Fig. 62 Catalytic converters and related components—V8 engine

1. Tailpipe hanger bracket
2. Tailpipe
3. Clamp
4. Main muffler
5. Gasket
6. Heated oxygen sensor 2 (bank 2)
7. Right front exhaust tube
8. Ring gasket
9. Heated oxygen sensor 2 (bank 1)
10. Left front exhaust tube
11. Center exhaust tube
12. Clamp
13. Muffler hanger bracket front
14. Muffler hanger bracket rear

37663_PATH_G0145

⁑ WARNING

Be careful not to damage front timing chain case and crankshaft.

To install:

8. Apply new engine oil to both oil seal lip and dust seal lip of new front oil seal.
9. Install front oil seal.
   a. Install front oil seal so that each seal lip is oriented as shown.
   b. Press-fit until the height of front oil seal is level with the mounting surface using suitable tool.
   c. Suitable drift: outer diameter 2.36 in. (60 mm), inner diameter 1.97 in. (50 mm).

⁑ WARNING

Be careful not to damage front timing chain case and crankshaft. Press-fit straight and avoid causing burrs or tilting oil seal.

10. Installation is in the reverse order of removal after this step.
11. Be sure to perform the reconnect/relearn procedures.

V8 Engine

See Figure 63.

1. Before servicing the vehicle, refer to the Precautions.
2. Disconnect the negative battery cable.
3. Remove the engine undercover.
4. Remove the air duct and resonator assembly and the air cleaner case (upper).
5. Remove the drive belt.
6. Remove the radiator shroud assembly.
7. Remove the cooling fan (crankshaft driven type).
8. Remove the crankshaft pulley.
9. Remove the front oil seal using suitable tool.

⁑ WARNING

Do not damage front cover and oil pump drive spacer.

To install:

10. Apply new engine oil to both the oil seal lip and dust seal lip of the new front oil seal.

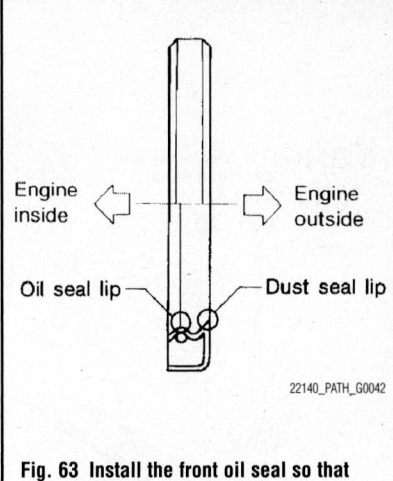

22140_PATH_G0042

Fig. 63 Install the front oil seal so that each seal lip is oriented as shown.

11. Install the front oil seal.
   a. Install the front oil seal so that each seal lip is oriented as shown.
   b. Press-fit until the front oil seal is level with the front cover using suitable tool.

⁑ WARNING

Do not damage front cover and crankshaft. Press-fit straight and avoid causing burrs or tilting oil seal.

12. Installation of the remaining components is in the reverse order of removal.
13. Be sure to perform the reconnect/relearn procedures.

CYLINDER HEAD

REMOVAL & INSTALLATION

V6 Engine

See Figures 64 through 67.

1. Before servicing the vehicle, refer to the Precautions.
2. Disconnect the negative battery cable.
3. Remove the cowl top cover. Remove the air cleaner assembly.
4. Remove camshaft.
5. Remove intake manifold.
6. Remove exhaust manifold.
7. Remove the oil pan.
8. Remove the rear timing chain case.
9. Remove the power steering pump and bracket. Remove the alternator and bracket.
10. Remove water inlet and thermostat assembly.
11. Remove water outlet, water pipe and heater pipe.
12. Remove cylinder head bolts in reverse order of the tightening sequence.

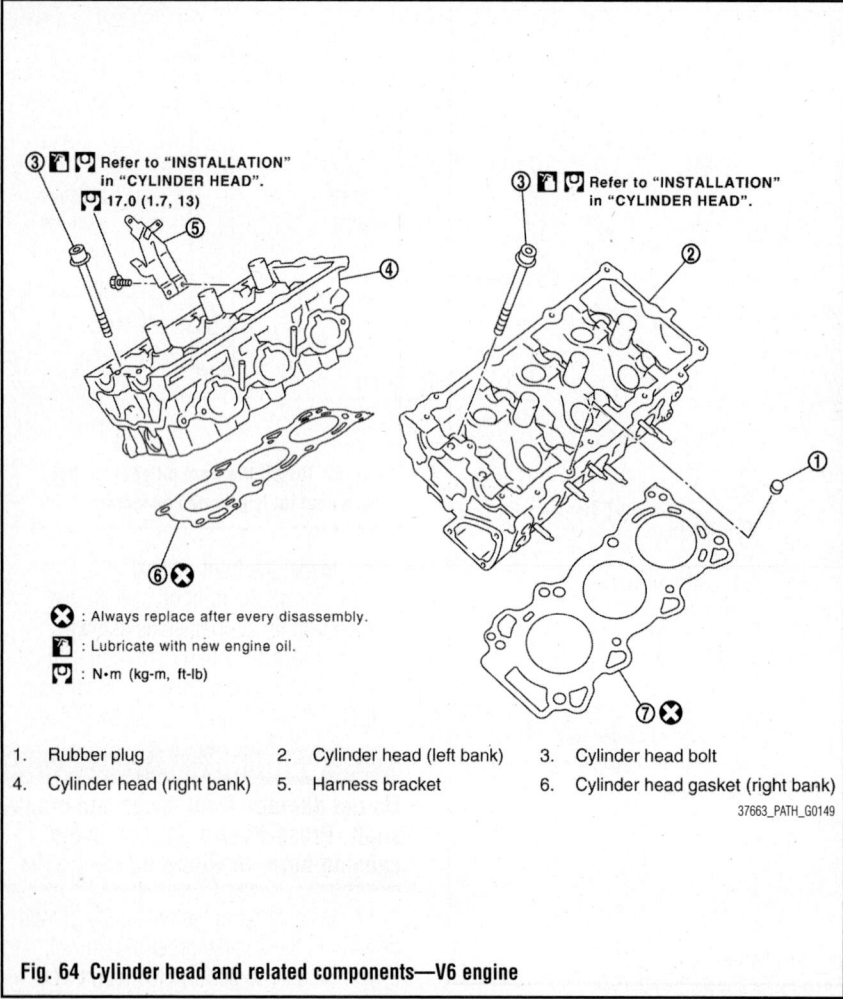

③ 🅿 🔧 Refer to "INSTALLATION" in "CYLINDER HEAD".
🔧 17.0 (1.7, 13)

③ 🅿 🔧 Refer to "INSTALLATION" in "CYLINDER HEAD".

❌ : Always replace after every disassembly.
🅿 : Lubricate with new engine oil.
🔧 : N•m (kg-m, ft-lb)

1. Rubber plug	2. Cylinder head (left bank)	3. Cylinder head bolt
4. Cylinder head (right bank)	5. Harness bracket	6. Cylinder head gasket (right bank)

37663_PATH_G0149

Fig. 64 Cylinder head and related components—V6 engine

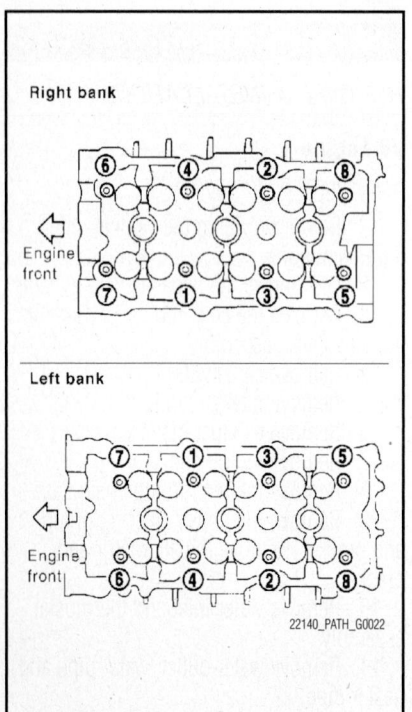

Right bank

Engine front

Left bank

Engine front

22140_PATH_G0022

Fig. 65 Cylinder head bolt tightening sequence—V6 engine

13. Remove cylinder head gaskets. Discard the gaskets.

To install:

➡ **Be sure to use new fasteners, as required.**

14. Install new cylinder head gasket.
15. Turn crankshaft until No. 1 piston is set at TDC.

➡ **Crankshaft key should line up with the right bank cylinder center line.**

16. Install cylinder head and tighten cylinder head bolts in numerical order as shown.

❋❋ WARNING

If cylinder head bolts re-used, check their outer diameters before installation. These bolts are tightened using the plastic zone tightening method. Whenever the size difference between "d1" and "d2" exceeds the limit, replace the bolt. Specification: limit ("d1"-"d2") 0.0043 inch (0.11mm). If reduction of outer diameter appears in a position other than

"d2", use it as "d2" point. See illustration.

17. Tighten cylinder head bolts using the following sequence:
  • Step A: 72 ft. lbs. (98 Nm)
  • Step B: Loosen bolts in the reverse order of tightening.
  • Step C: 29 ft. lbs. (39.2 Nm)
  • Step D: An additional 90° clockwise
  • Step E: An additional 90° clockwise

18. After installing cylinder head, measure distance between front end faces of cylinder block and cylinder head (left and right banks). Specification should be 0.555–0.587 inch.

➡ **If the measured value is out of the standard, re-install cylinder head.**

19. Installation of the remaining parts is in the reverse order of removal.
20. Be sure to perform the reconnect/relearn procedures.

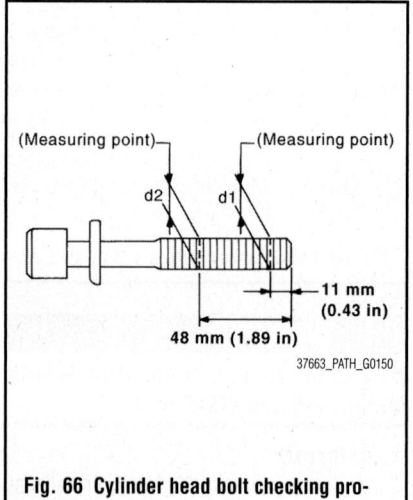

(Measuring point)   (Measuring point)

d2   d1

11 mm (0.43 in)
48 mm (1.89 in)

37663_PATH_G0150

Fig. 66 Cylinder head bolt checking procedure—V6 engine

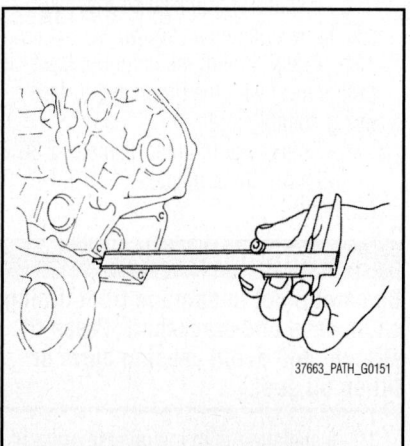

37663_PATH_G0151

Fig. 67 Cylinder head/cylinder block installation measurement point location—V6 engine

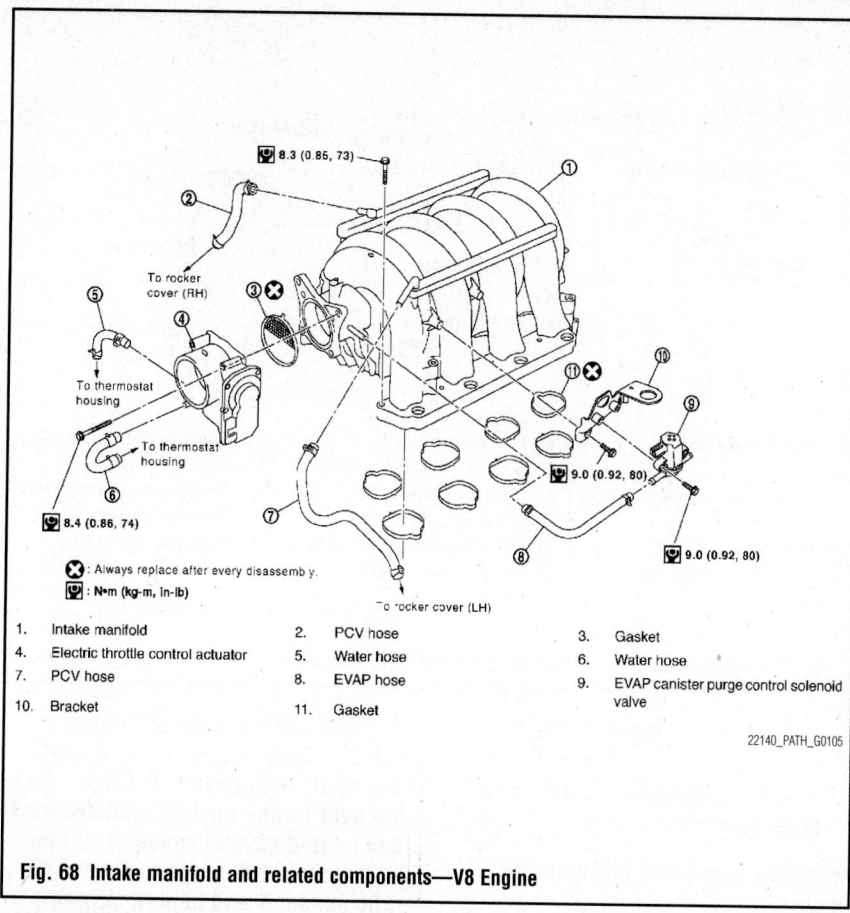

X : Always replace after every disassembly.
🔧 : N•m (kg-m, in-lb)

1.	Intake manifold	2.	PCV hose	3.	Gasket
4.	Electric throttle control actuator	5.	Water hose	6.	Water hose
7.	PCV hose	8.	EVAP hose	9.	EVAP canister purge control solenoid valve
10.	Bracket	11.	Gasket		

22140_PATH_G0105

Fig. 68 Intake manifold and related components—V8 Engine

V8 Engine

See Figures 68 through 71.

1. Before servicing the vehicle, refer to the Precautions.
2. Disconnect the negative battery cable.
3. Remove the cowl top cover and air cleaner assembly.
4. Remove the timing chains.
5. Remove the alternator. Remove the power steering pimp.
6. Remove the intake manifold.
7. Remove the water inlet and thermostat assembly.
8. Remove the oil pan, upper and oil strainer.
9. Remove the camshafts.
10. Remove the oil pump.
11. Remove the mudguards and fender protectors.
12. Remove the exhaust manifolds.
13. Remove the cylinder head retaining bolts, in the reverse order of the tightening sequence.

To install:

➡ **Be sure to use new fasteners, as required.**

14. Install the cylinder head with a new gasket.

15. Tighten the bolts in sequence to specification.

➡ **Cylinder head bolts are tightened by plastic zone tightening method. When-ever the size difference between d1 and d2 exceeds the limit, replace the bolt. Limit is 0.0091 inch (0.23mm). See illustration.**

16. Continue the installation in the reverse order of the removal procedure.
17. Start the engine and check for leaks.
18. Be sure to perform the reconnect/relearn procedures.

EXHAUST MANIFOLD

REMOVAL & INSTALLATION

V6 Engine

Left Bank

See Figures 72 and 73.

1. Before servicing the vehicle, refer to the Precautions.
2. Disconnect the negative battery cable.
3. Remove the air cleaner assembly.
4. Remove the engine undercover.
5. Drain the engine coolant. Be sure to properly dispose of used coolant.

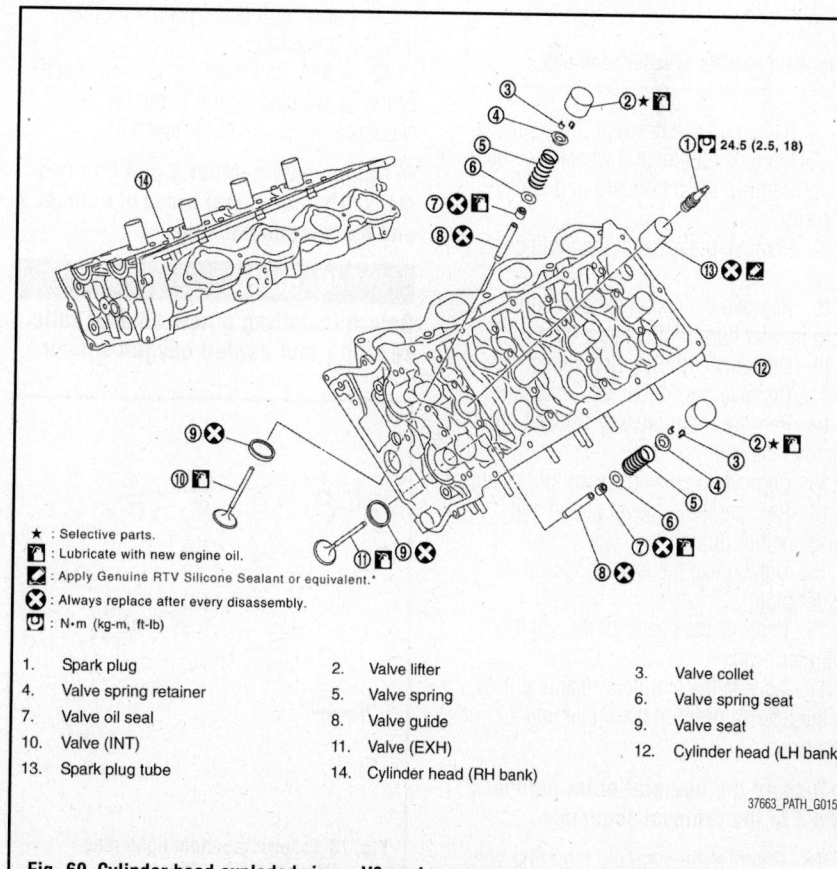

★ : Selective parts.
🛢 : Lubricate with new engine oil.
▨ : Apply Genuine RTV Silicone Sealant or equivalent.*
X : Always replace after every disassembly.
🔧 : N•m (kg-m, ft-lb)

1.	Spark plug	2.	Valve lifter	3.	Valve collet
4.	Valve spring retainer	5.	Valve spring	6.	Valve spring seat
7.	Valve oil seal	8.	Valve guide	9.	Valve seat
10.	Valve (INT)	11.	Valve (EXH)	12.	Cylinder head (LH bank)
13.	Spark plug tube	14.	Cylinder head (RH bank)		

37663_PATH_G0155

Fig. 69 Cylinder head exploded view—V8 engine

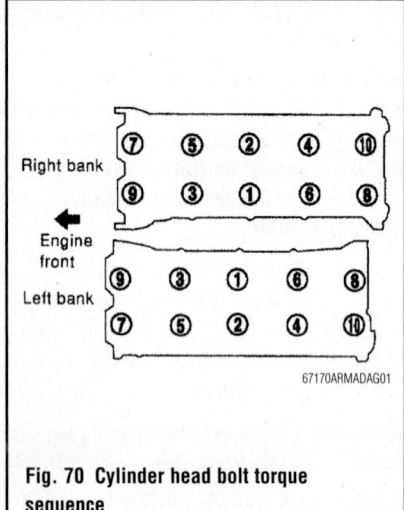

Fig. 70 Cylinder head bolt torque sequence

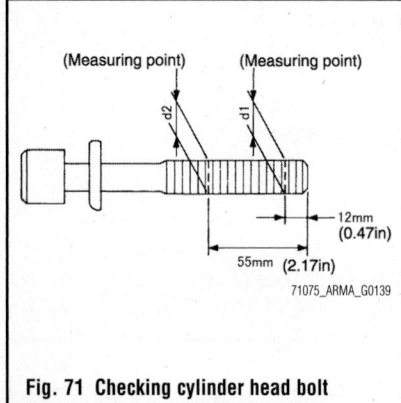

Fig. 71 Checking cylinder head bolt

6. Raise and safely support the vehicle.

7. Remove the tire and wheel assembly.

8. Remove the mudguard and fender protector.

9. Remove the exhaust manifold cover bolts.

10. Remove the center exhaust tube, main muffler and front exhaust tube.

11. Disconnect the oxygen sensor connector. Remove the sensor, as necessary.

12. Remove the three way catalyst nuts. Remove the catalyst.

13. Remove the exhaust manifold cover.

14. Remove the oil level gauge and gauge holder guide.

15. Disconnect the water hoses at the heater pipe.

16. Remove the heater pipe from the cylinder head.

17. Loosen the manifold retaining nuts in the reverse order of the tightening sequence.

➡**Discard the numeral order number 7 and 8 in the removal sequence.**

18. Remove the manifold retaining bolts and nuts.

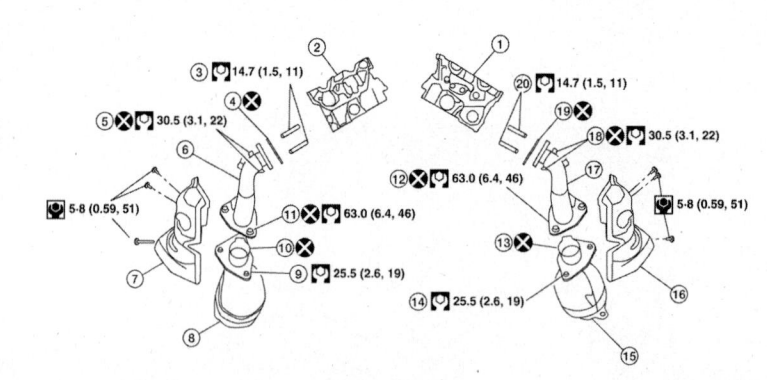

1. Cylinder head (LH)
2. Cylinder head (RH)
3. Exhaust manifold studs (RH)
4. Gasket
5. Exhaust manifold nuts (RH)
6. Exhaust manifold (RH)
7. Exhaust manifold cover (RH)
8. Three way catalyst (RH)
9. Three way catalyst studs (RH)
10. Seal ring
11. Three way catalyst nuts (RH)
12. Three way catalyst nuts (LH)
13. Seal ring
14. Three way catalyst studs (LH)
15. Three way catalyst (LH)
16. Exhaust manifold cover (LH)
17. Exhaust manifold (LH)
18. Exhaust manifold nuts (LH)

Fig. 72 Exhaust manifolds and related components—V6 engine

19. Remove the component from the vehicle.

20. Discard the gasket. Discard the nuts.

To install:

➡**Be sure to use new fasteners, as required.**

21. Installation is the reverse of the removal procedure.

22. Tighten the retaining nuts to specification in two passes and in the proper sequence. Be sure to use new nuts.

➡**Tighten nuts number 1 and 2 in two steps. The numerical order of number 7 and 8 show second step.**

❋❋ WARNING

Before installing a new air fuel ratio sensor 1 and heated oxygen sensor 2, clean exhaust system threads using oxygen sensor thread cleaner and apply anti-seize lubricant. _ Do not over torque air fuel ratio sensor 1 and heated oxygen sensor 2. Doing so may cause damage to air fuel ratio sensor 1 and heated oxygen sensor 2, resulting in the "MIL" coming on.

23. Be sure to perform the reconnect/relearn procedures.

Right Bank

See Figure 74.

1. Before servicing the vehicle, refer to the Precautions.

2. Disconnect the negative battery cable.

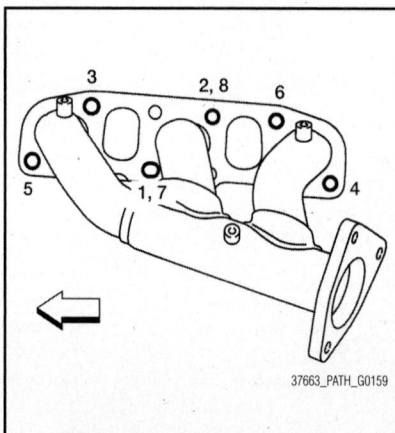

Fig. 73 Exhaust manifold tightening sequence (left side)—V6 engine

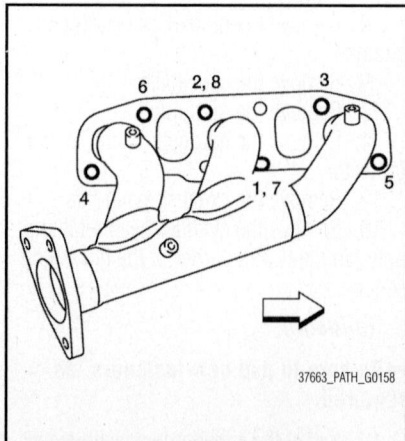

Fig. 74 Exhaust manifold tightening sequence (right side)—V6 engine

3. Remove the engine undercover and remove the air cleaner assembly, as required.

4. Raise and safely support the vehicle.

5. Remove the tire and wheel assembly.

6. Remove the mudguard and fender protector.

7. Remove the exhaust manifold cover bolts.

8. Remove the center exhaust tube, main muffler and front exhaust tube.

9. Disconnect the oxygen sensor connector. Remove the sensor, as necessary.

10. Remove the three way catalyst nuts. Remove the catalyst.

11. Remove the heat shield from the lower dash panel.

12. Remove the support bolts from the transmission filler pipe.

13. Loosen the manifold retaining nuts in the reverse order of the tightening sequence.

➡**Discard the numeral order number 7 and 8 in the removal sequence.**

14. Remove the manifold retaining bolts and nuts.

15. Remove the component from the vehicle.

16. Discard the gasket. Discard the nuts.

To install:

➡**Be sure to use new fasteners, as required.**

17. Installation is the reverse of the removal procedure.

18. Tighten the retaining nuts to specification in two passes and in the proper sequence. Be sure to use new nuts.

➡**Tighten nuts number 1 and 2 in two steps. The numerical order of number 7 and 8 show second step.**

✳✳ WARNING

Before installing a new air fuel ratio sensor 1 and heated oxygen sensor 2, clean exhaust system threads using oxygen sensor thread cleaner and apply anti-seize lubricant. Do not over torque air fuel ratio sensor 1 and heated oxygen sensor 2. Doing so may cause damage to air fuel ratio sensor 1 and heated oxygen sensor 2, resulting in the "MIL" coming on.

19. Be sure to perform the reconnect/relearn procedures.

V8 Engine

See Figures 75 and 76.

1. Before servicing the vehicle, refer to the Precautions.

✳✳ CAUTION

Perform the work when the exhaust and cooling system have cooled sufficiently.

2. Disconnect the negative battery cable.

3. Remove the air cleaner assembly, as required.

4. Raise and safely support the vehicle.

5. Remove the front tire and wheel assemblies.

6. Remove the mudguard and fender protector.

7. Remove the main muffler assembly and center exhaust tube.

8. Remove the front exhaust tubes.

9. Remove the LH and RH air fuel ratio A/F sensors.

a. Remove the harness connector of each air fuel ratio A/F sensor, and harness from bracket and middle clamp.

✳✳ WARNING

Do not damage the air fuel ratio A/F sensors. Discard any air fuel ratio A/F sensor which has been dropped

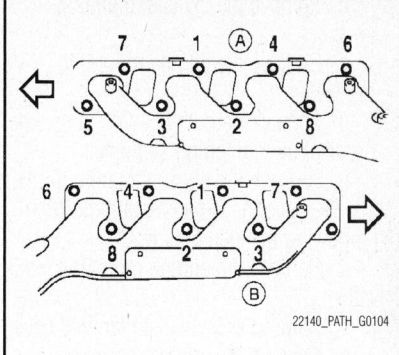

Fig. 76 Exhaust manifold retaining nut tightening sequence—V8 engine

from a height of more than 19.7 in. (0.5m) onto a hard surface such as a concrete floor. Replace it with a new one.

10. If removing the left manifold, remove the drive belt. Remove the A/C compressor and position it out of the work area. Do not disconnect the refrigerant lines. Disconnect the upper steering shaft joint.

11. Support the engine using a suitable tool, as required.

12. Remove the exhaust manifold (LH) (A) following the steps below.

a. Remove the engine mounting insulator.

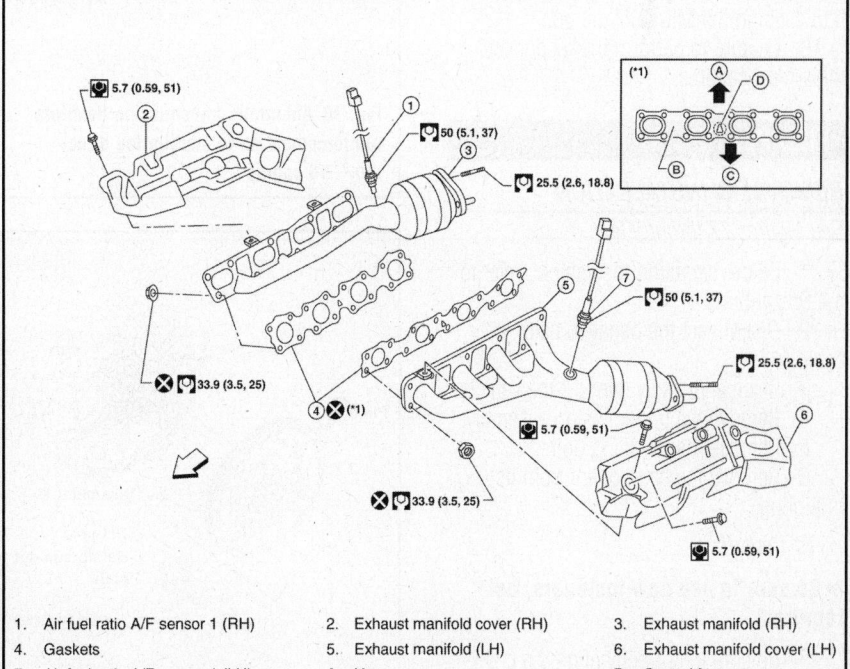

1.	Air fuel ratio A/F sensor 1 (RH)	2.	Exhaust manifold cover (RH)	3.	Exhaust manifold (RH)
4.	Gaskets	5.	Exhaust manifold (LH)	6.	Exhaust manifold cover (LH)
7.	Air fuel ratio A/F sensor 1 (LH)	A.	Up	B.	Coated face

37663_PATH_G0160

Fig. 75 Exhaust manifolds and related components—V8 engine

b. Remove the exhaust manifold cover.

c. Remove the engine mounting bracket.

d. Loosen the nuts LH side in reverse order of the tightening sequence.

e. Remove the exhaust manifold (LH).

13. Remove the exhaust manifold (RH) (B) following the steps below.

a. Remove the engine mounting insulator.

b. Remove the exhaust manifold cover.

c. Remove the engine mounting bracket.

d. Remove the oil level gauge guide.

e. Loosen the nuts RH side in reverse order of the tightening sequence.

f. Remove the exhaust manifold (RH).

14. Discard the gaskets. Discard the nuts.

To install:

➡**Be sure to use new fasteners, as required.**

15. Installation is in the reverse order of removal.

16. Install new exhaust manifold gasket with the top of the triangular up mark on it facing up and its coated face (gray side) toward the exhaust manifold side.

17. Tighten the retaining nuts to specification and in the proper sequence.

18. Do not over tighten the sensors. Doing so may interfere with sensor operation causing the MIL to come on.

19. Be sure to perform the reconnect/relearn procedures.

FLEXPLATE

REMOVAL & INSTALLATION

See Figures 77 through 80.

1. Before servicing the vehicle, refer to the Precautions.

2. Disconnect the negative battery cable.

3. Raise and safely support the vehicle.

4. Remove the transmission assembly.

5. Remove the retaining bolts.

6. Remove the component from its mounting.

To install:

➡**Be sure to use new fasteners, as required.**

7. Be sure to correctly align the crankshaft side dowel pin and the flexplate side dowel pin hole (A).

➡**If not aligned correctly the MIL light will illuminate.**

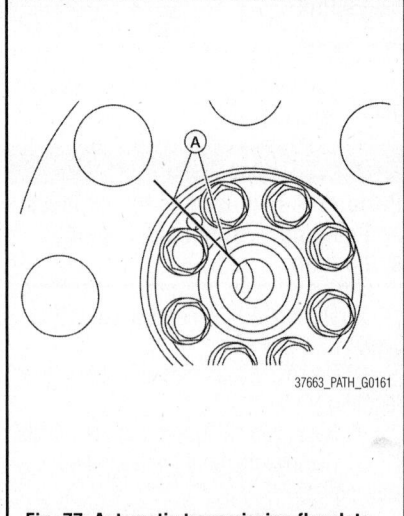

Fig. 77 Automatic transmission flexplate to transmission alignment—V6 engine

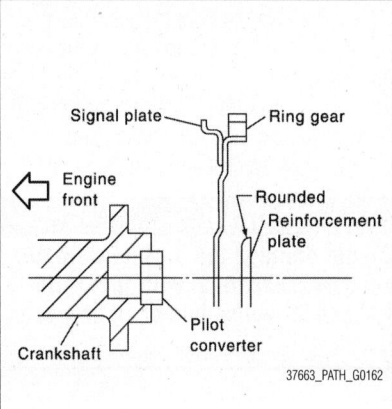

Fig. 78 Automatic transmission flexplate reinforcement plate installation direction—V6 engine

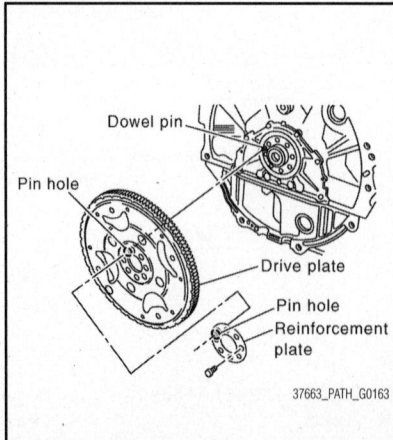

Fig. 79 Automatic transmission flexplate to transmission alignment—V8 engine

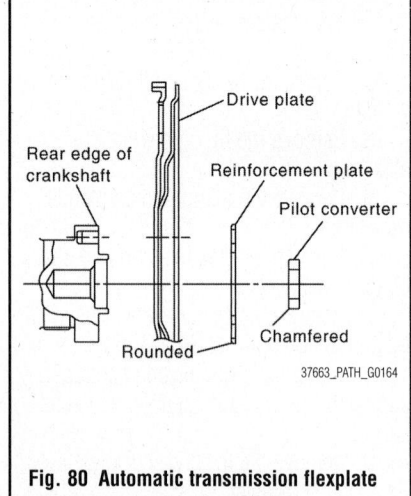

Fig. 80 Automatic transmission flexplate reinforcement plate installation direction—V8 engine

8. Install the flexplate and reinforcement plate in the direction shown. See illustration.

9. Tighten the retaining bolts to specification in a crisscross pattern.

10. Continue the installation in the reverse order of the removal procedure.

11. Be sure to perform the reconnect/relearn procedures.

FRONT COVER & SEAL

REMOVAL & INSTALLATION

V6 Engine

See Figures 81 through 87, 00.

1. Before servicing the vehicle, refer to the Precautions.

2. Properly release the fuel pressure.

3. Disconnect the negative battery cable.

4. Remove the engine the air cleaner assembly, as required.

5. Remove the engine undercover. Remove the fender protector.

6. Drain engine oil.

❊❊ CAUTION

Perform this step when engine is cold. Do not spill engine oil on drive belts.

7. Drain engine coolant from radiator.

❊❊ CAUTION

Perform this step when engine is cold. Do not spill engine coolant on drive belts.

8. Remove the upper radiator shroud.

9. Remove drive belts.

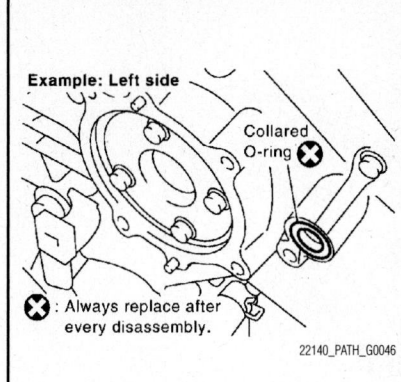

Fig. 81 Remove collared O-rings from front timing chain case (left and right side)—V6 engine

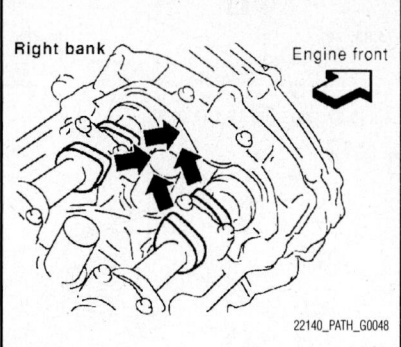

Fig. 82 Make sure that intake and exhaust cam noses on No. 1 cylinder (engine front side of right bank) are located as shown—V6 engine

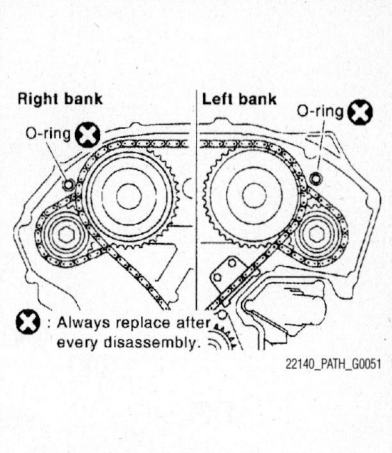

Fig. 83 Remove O-rings from rear timing chain case—V6 engine

10. Remove radiator cooling fan assembly.

11. Separate engine harnesses removing their brackets from front timing chain case.

12. Remove power steering oil pump from bracket with piping connected, and temporarily secure it aside.

13. Remove power steering oil pump bracket.

14. Remove alternator.

15. Remove water bypass hose, water hose clamp and idler pulley bracket from front timing chain case.

16. Remove the electric throttle control.

17. Remove right and left intake valve timing control covers.

 a. Loosen bolts in reverse order of tightening sequence.

 b. Cut liquid gasket for removal.

☀ WARNING

Shaft is internally jointed with camshaft sprocket (INT) center hole. When removing, keep it horizontal until it is completely disconnected.

18. Remove collared O-rings from front timing chain case (left and right side).

19. Remove rocker covers (right and left banks).

➡**When only timing chain (primary) is removed, rocker cover does not need to be removed.**

20. Obtain No. 1 cylinder at TDC of its compression stroke as follows:

➡**When timing chain is not removed/installed, this step is not required.**

 a. Rotate crankshaft pulley clockwise to align timing mark (A) with timing indicator (B).

 b. Make sure that intake and exhaust cam noses on No. 1 cylinder (engine front side of right bank) are located as shown.

 c. If not, turn crankshaft one revolution (360°) and align as shown.

➡**When only timing chain (primary) is removed, rocker cover does not need to be removed. To make sure that No. 1 cylinder is at its compression TDC, remove front timing chain case first. Then check mating marks on camshaft sprockets.**

21. Remove crankshaft pulley as follows:
 a. Remove starter motor.
 b. Loosen crankshaft pulley bolt and locate bolt seating surface as 0.39 in. (10 mm) from its original position.

☀ WARNING

Do not remove crankshaft pulley bolt. Keep loosened crankshaft pulley bolt in place protect removed crankshaft pulley from dropping.

 c. Pull crankshaft pulley with both hands to remove it.

22. Loosen two bolts in front of oil pan (upper) in the reverse order of the tightening sequence.

23. Remove front timing chain case as follows:

 a. Loosen bolts in reverse order of the tightening sequence.
 b. Insert suitable tool into the notch at the top of the front timing chain case.
 c. Pry off case by moving tool.
 d. Cut liquid gasket for removal.

☀ WARNING

Do not use screwdriver or something similar. After removal, handle front

timing chain case carefully so it does not tilt, cant, or warp under a load.

24. Remove O-rings from rear timing chain case.

To install:

➡**Be sure to use new fasteners, as required.**

25. Install front timing chain case as follows:

 a. Apply a continuous bead of liquid gasket to front timing chain case back side as shown.

 b. Install new O-rings on rear timing chain case.

 c. Assemble front timing chain case as follows:

 • Fit lower end of front timing chain case tightly onto top face of oil pan (upper). From the fitting point, make entire front timing chain case contact rear timing chain case completely.

 • Since front timing chain case is offset for difference of bolt holes, tighten bolts temporarily while holding front timing chain case from front and top.

 • Same as the previous step, insert dowel pin while holding front timing chain case from front and top completely.

 d. Tighten bolts to the specified torque in numerical order as shown.

➡**There are two type of bolts. Refer to the following for locating bolts.**

 • 1–5: 0.39 in. (10 mm)
 • 6–25: 0.24 in. (6 mm)
 • Bolt position torque specification: 1–5: 41 ft. lbs. (55.0 Nm); 6–25: 9 ft. lbs. (12.7 Nm)

 e. After all bolts tightened, retighten

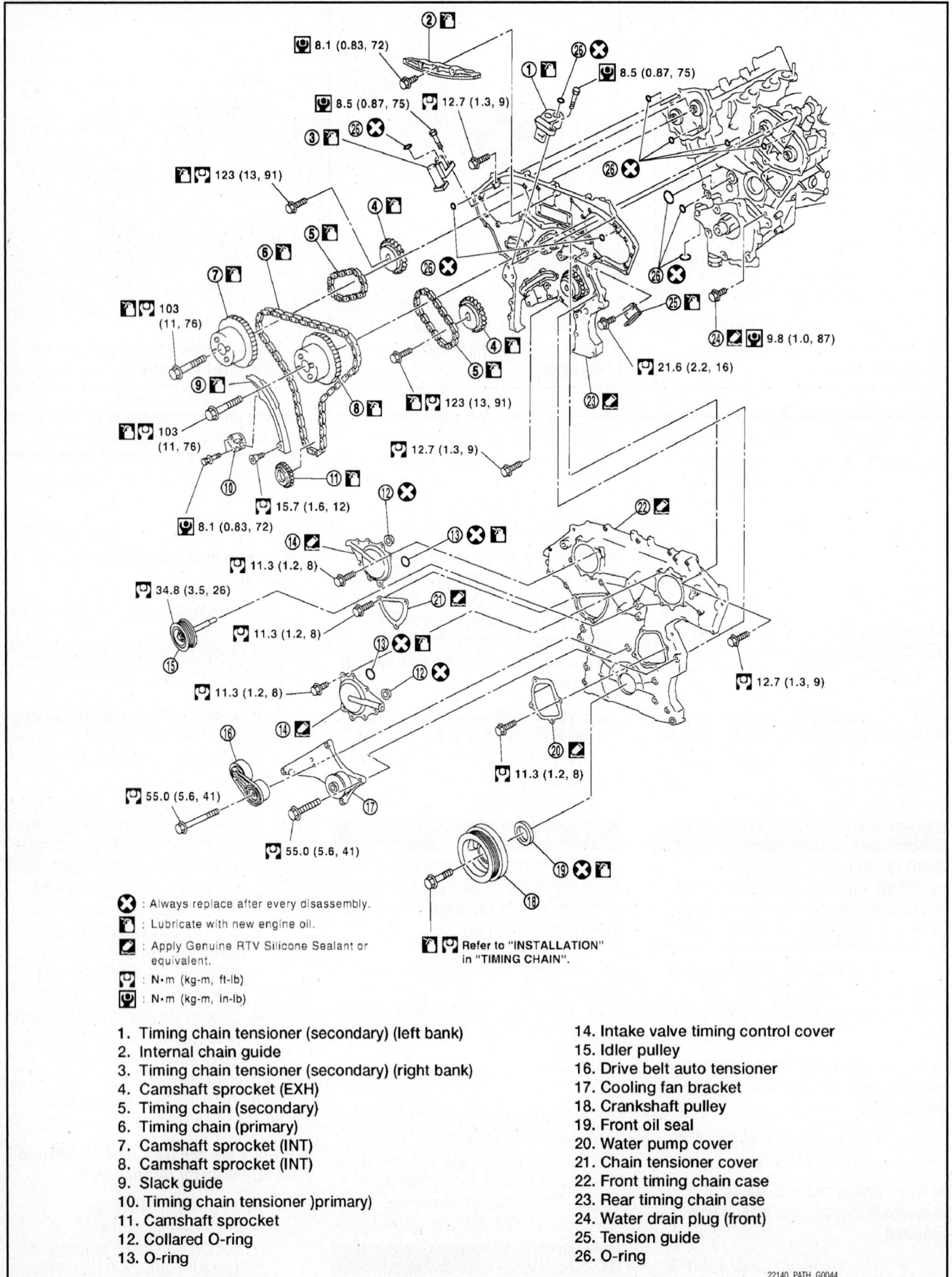

8.1 (0.83, 72)
8.5 (0.87, 75)
12.7 (1.3, 9)
8.5 (0.87, 75)
123 (13, 91)
103 (11, 76)
103 (11, 76)
8.1 (0.83, 72)
15.7 (1.6, 12)
12.7 (1.3, 9)
123 (13, 91)
21.6 (2.2, 16)
9.8 (1.0, 87)
11.3 (1.2, 8)
34.8 (3.5, 26)
11.3 (1.2, 8)
11.3 (1.2, 8)
11.3 (1.2, 8)
12.7 (1.3, 9)
55.0 (5.6, 41)
55.0 (5.6, 41)

❌ : Always replace after every disassembly.
🛢 : Lubricate with new engine oil.
✎ : Apply Genuine RTV Silicone Sealant or equivalent.
🔧 : N•m (kg-m, ft-lb)
🔧 : N•m (kg-m, in-lb)
🛢 🔧 Refer to "INSTALLATION" in "TIMING CHAIN".

1. Timing chain tensioner (secondary) (left bank)
2. Internal chain guide
3. Timing chain tensioner (secondary) (right bank)
4. Camshaft sprocket (EXH)
5. Timing chain (secondary)
6. Timing chain (primary)
7. Camshaft sprocket (INT)
8. Camshaft sprocket (INT)
9. Slack guide
10. Timing chain tensioner)primary)
11. Camshaft sprocket
12. Collared O-ring
13. O-ring

14. Intake valve timing control cover
15. Idler pulley
16. Drive belt auto tensioner
17. Cooling fan bracket
18. Crankshaft pulley
19. Front oil seal
20. Water pump cover
21. Chain tensioner cover
22. Front timing chain case
23. Rear timing chain case
24. Water drain plug (front)
25. Tension guide
26. O-ring

22140_PATH_G0044

Fig. 84 Timing chain/front cover and related components—V6 engine

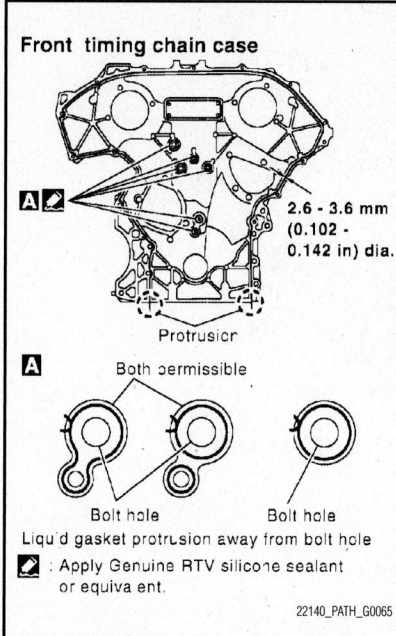

Fig. 85 Timing cover sealant application points—V6 engine

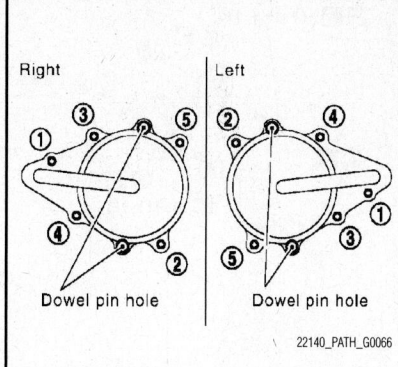

Fig. 87 Timing control covers bolt tightening sequence—V6 engine

them to the specified torque in numerical order as shown.

26. Install two bolts in front of oil pan (upper).

27. Install right and left intake valve timing control covers as follows:

a. Install new seal rings in shaft grooves.

b. Apply a continuous bead of liquid gasket to intake valve timing control covers.

c. Install new collared O-rings in front timing chain case oil hole (left and right sides).

d. Being careful not to move seal ring from the installation groove, align dowel pins on front timing chain case with the

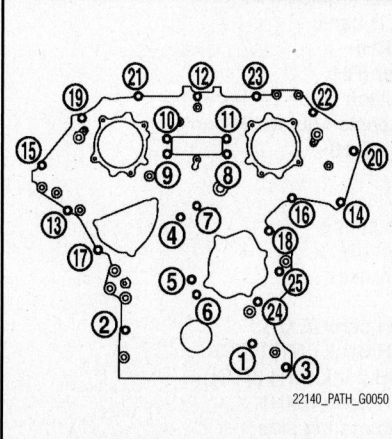

Fig. 86 Timing cover bolt tightening sequence—V6 engine

holes to install intake valve timing control covers.

e. Tighten bolts in numerical order as shown.

28. Install crankshaft pulley as follows:

a. Install crankshaft pulley, taking care not to damage front oil seal.

➡**When press-fitting crankshaft pulley with plastic hammer, tap on its center portion (not circumference).**

b. Tighten crankshaft pulley bolt. Crankshaft bolt torque: 33ft. lbs. (44.1 Nm)

c. Put a paint mark on crankshaft pulley aligning with angle mark on crankshaft pulley bolt. Then, further retighten bolt by 60°(equivalent to one graduation).

29. Rotate crankshaft pulley in normal direction (clockwise when viewed from front) to confirm it turns smoothly.

30. Install oil pans (upper and lower).

31. Install rocker covers (right and left banks).

32. Installation of the remaining components is in the reverse order of removal after this step.

33. The following are procedures for checking fluid leaks, lubricant leaks and exhaust gases leaks.

a. Before starting engine, check oil/fluid levels including engine coolant and engine oil. If less than required quantity, fill to the specified level.

34. Use procedure below to check for fuel leakage.

a. Turn ignition switch "ON" (with engine stopped). With fuel pressure applied to fuel piping, check for fuel leakage at connection points.

b. Start engine. With engine speed increased, check again for fuel leakage at connection points.

c. Run engine to check for unusual noise and vibration.

➡**If hydraulic pressure inside timing chain tensioner drops after removal/installation, slack in the guide may generate a pounding noise during and just after engine start. However, this is normal. Noise will stop after hydraulic pressure rises.**

d. Warm up engine thoroughly to make sure there is no leakage of fuel, exhaust gases, or any oil/fluids including engine oil and engine coolant.

e. Bleed air from lines and hoses of applicable lines, such as in cooling system.

f. After cooling down engine, again check oil/fluid levels including engine oil and engine coolant. Refill to the specified level, if necessary.

35. Be sure to perform the reconnect/relearn procedures.

V8 Engine

See Figures 88 through 92.

1. Before servicing the vehicle, refer to the Precautions.

➡**If removing the timing chain and associated parts, start with those on the LH bank. The procedure for removing parts on the RH bank is omitted because it is the same as that for removal on the LH bank. To install timing chain and associated parts, start with those on the RH bank. The procedure for installing parts on the LH bank is omitted because it is the same as that for installation on the RH bank.**

2. Disconnect the negative battery cable.

3. Remove the air cleaner assembly.

4. Drain the cooling system. Be sure to properly dispose of used coolant.

5. Remove the upper oil pan.

6. Remove the camshaft position sensor.

7. Remove the drive belt. Remove the tensioner assembly.

8. Remove the crankshaft pulley bracket.

9. Remove the water pump pulley.

10. Remove the power steering pump reservoir tank and bracket. Position it aside.

11. Remove the lower radiator hose and pipe assembly. Remove the suction hoses and pipe.

12. Remove the Intake valve control solenoid valve cover (RH) (A) and Intake valve control solenoid valve cover (LH) (B) as follows:

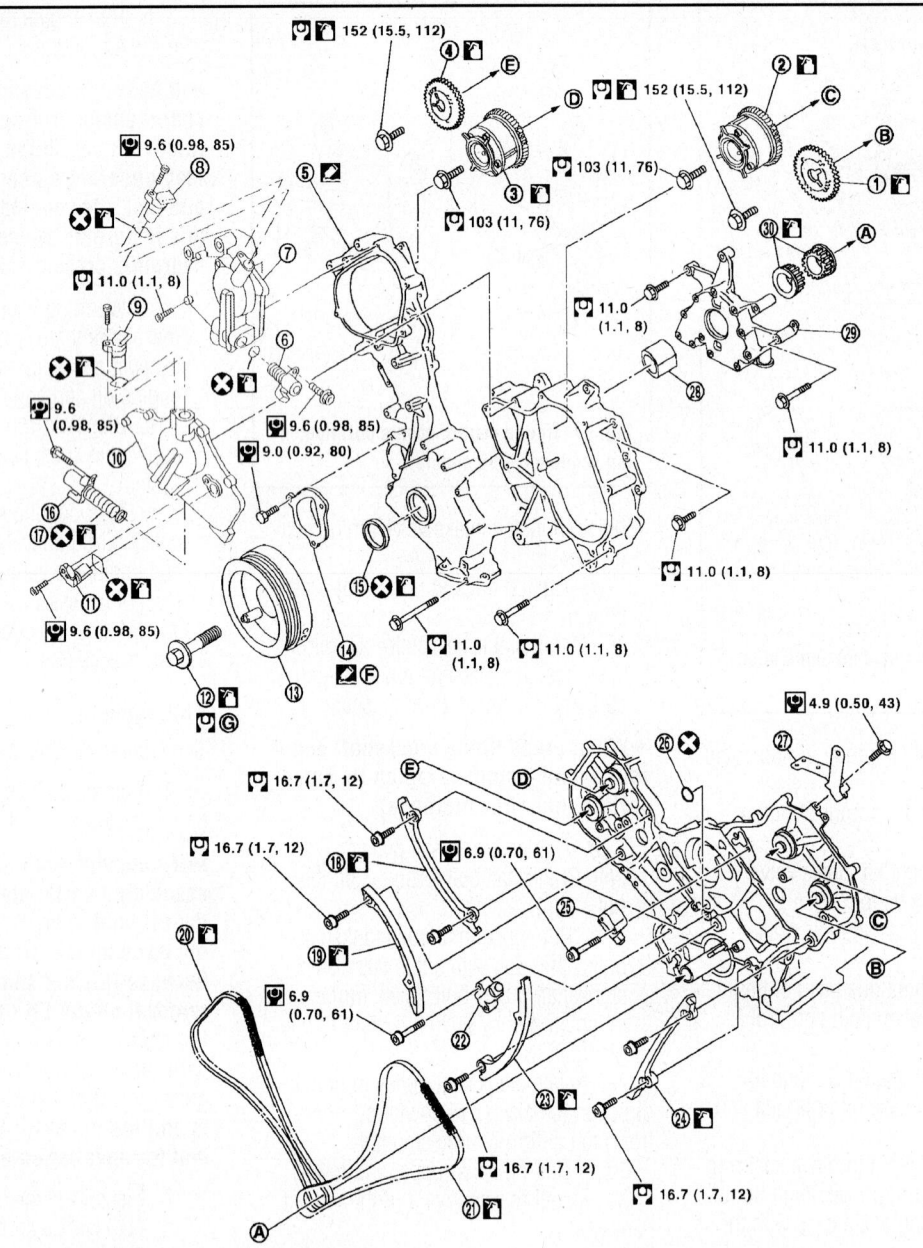

1. Camshaft sprocket LH bank (EXH)
2. Camshaft sprocket LH bank (INT) (VTC)
3. Camshaft sprocket RH bank (INT) (VTC)
4. Camshaft sprocket RH bank (EXH)
5. Front cover
6. Intake valve control solenoid valve (LH)
7. Intake valve control solenoid valve cover (LH)
8. Intake valve timing control position sensor (LH)
9. Intake valve timing control position sensor (RH)
10. Intake valve control solenoid valve cover (RH)
11. Camshaft position sensor (PHASE)
12. Crankshaft pulley bolt
13. Crankshaft pulley
14. Chain tensioner cover
15. Front oil seal
16. Intake valve control solenoid valve (RH)
17. O-ring
18. Timing chain tension guide RH bank

19. Timing chain slack guide (RH)
20. Timing chain LH bank
21. Timing chain (RH)
22. Chain tensioner (RH)
23. Timing chain slack guide LH bank
24. Timing chain tension guide LH bank
25. Chain tensioner (LH)
26. O-ring
27. Bracket
28. Oil pump drive spacer
29. Oil pump assembly
30. Crankshaft sprocket
A. To crankshaft
B. To camshaft LH bank (EXH)
C. To camshaft LH bank (INT) (VTC)
D. To camshaft RH bank (INT) (VTC)
E. To camshaft RH bank (EXH)
F. Apply sealant to mating side

22140_PATH_G0120

Fig. 88 Timing chain/front cover and related components—V8 engine

a. Loosen and remove the bolts. See illustration.

b. Cut the liquid gasket and remove the covers.

☀ WARNING

Do not damage mating surfaces.

13. Obtain compression TDC of No. 1 cylinder as follows:

a. Turn the crankshaft pulley clockwise to align the TDC identification notch (without paint mark) with the timing indicator on the front cover.

b. At this time, make sure both intake and exhaust cam lobes of No. 1 cylinder (top front on LH bank) point outside. If they do not point outside, turn crankshaft pulley once more.

14. Remove the crankshaft pulley.

a. Loosen the crankshaft pulley bolts using a hammer handle to secure the crankshaft.

b. Remove the crankshaft pulley from the crankshaft.

c. Remove the crankshaft pulley.

➡**The dimension between the centers of the two bolt holes is 61 mm (2.40 in.).**

15. Remove the front cover.

a. Loosen and remove the bolts in the reverse order of the tightening sequence.

b. Cut the liquid gasket and remove the covers.

☀ WARNING

Do not damage mating surfaces.

16. Remove the front oil seal using suitable tool.

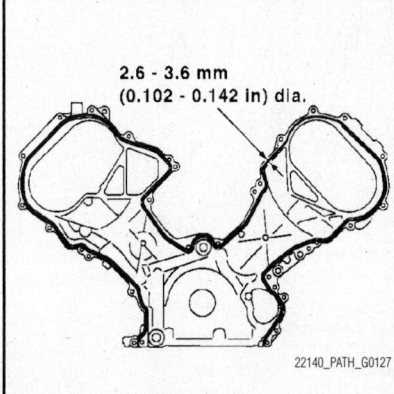

Fig. 89 Timing cover sealant application—V8 engine

☀ WARNING

Do not damage front cover.

To install:

➡**Be sure to use new fasteners, as required.**

17. Install the front oil seal using suitable tool.

☀ WARNING

Do not scratch or make burrs on the circumference of the oil seal.

18. Install the chain tensioner cover.

19. Install the front cover as follows:

a. Install a new O-ring on the cylinder block.

b. Apply liquid gasket as shown.

c. Check again that the timing alignment marks on the timing chain and on each sprocket are aligned. Then install the front cover.

d. Install the bolts in the numerical order shown.

e. After tightening, re-tighten to the specified torque.

☀ WARNING

Be sure to wipe off any excessive liquid gasket leaking onto surface mating with oil pan.

20. Install the Intake valve control solenoid valve cover (RH) (A) and Intake valve control solenoid valve cover (LH) (B) as follows:

a. Cross mark (C) that cannot be seen after assembly.

b. Apply liquid gasket (D) as shown.

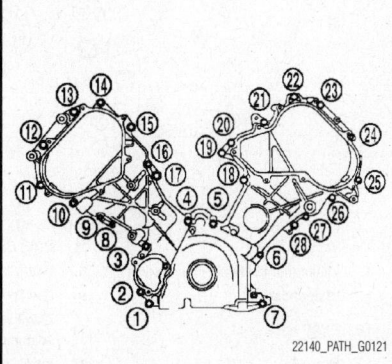

Fig. 90 Timing cover bolt tightening sequence—V8 engine

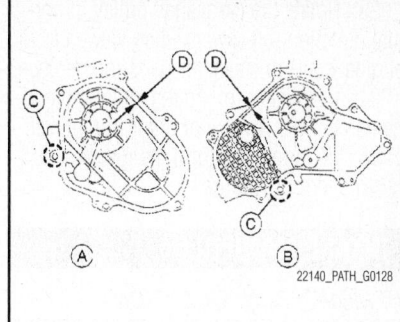

Fig. 91 Intake valve control valve sealant application (D)—V8 engine

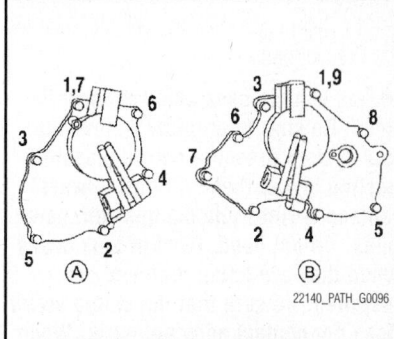

Fig. 92 Intake valve control solenoid bolt tightening sequence—V8 engine

☀ WARNING

The start and end of the application of the liquid gasket should be crossed at a position that cannot be seen after attaching the Intake valve control solenoid valve cover.

c. Install the bolts in the numerical order shown.

21. Install the crankshaft pulley.

a. Install the key of the crankshaft.

b. Insert the pulley by lightly tapping it.

☀ WARNING

Do not tap pulley on the side surface where the belt is installed (outer circumference).

22. Tighten the crankshaft pulley bolt.

a. Lock the crankshaft using suitable tool, then tighten the bolt.

b. Perform the following steps for angular tightening:

c. Apply engine oil onto the threaded parts of the bolt and seating area.

d. Select the one most visible notch of the four on the bolt flange. Corresponding to the selected notch, put a alignment mark (such as paint) on the crankshaft pulley.

23. Rotate the crankshaft pulley in normal direction (clockwise when viewed from engine front) to check for parts interference.

24. Installation of the remaining components is in the reverse of order of removal.

25. Be sure to perform the reconnect/relearn procedures.

INTAKE MANIFOLD

REMOVAL & INSTALLATION

V6 Engine

Upper Manifold

See Figures 93 through 96.

1. Before servicing the vehicle, refer to the Precautions.

➡**This vehicle uses quick connect fittings. Be sure to properly relieve the fuel system pressure before disconnecting any of these fittings. Always replace O-rings and clamps with new ones. Do not bend, twist or kink hoses when they are being removed or installed. Be sure that the clamp screw does not contact adjacent parts. When tightening the high pressure rubber hose clamp make sure the clamp end is 0.12 inch from the hose end. After connecting these fittings make sure that the connectors are secure. Check for fuel leakage at these connections turn the ignition key to the ON position (do not start the engine), correct as required. Start the engine, raise the idle, and verify that there are no fuel leaks, correct as required.**

2. Properly relieve the fuel system pressure.

3. Disconnect the negative battery cable.

4. Remove engine undercover.

5. Remove air cleaner case (upper) with mass air flow sensor and air duct assembly.

6. Remove electric throttle control actuator as follows:

　a. Drain engine coolant, or when water hoses are disconnected, attach plug to prevent engine coolant leakage.

　b. Disconnect water hoses from electric throttle control actuator.

➡**When engine coolant is not drained from radiator, attach plug to water hoses to prevent engine coolant leakage.**

　c. Disconnect harness connector.

　d. Loosen bolts in reverse order of the tightening sequence.

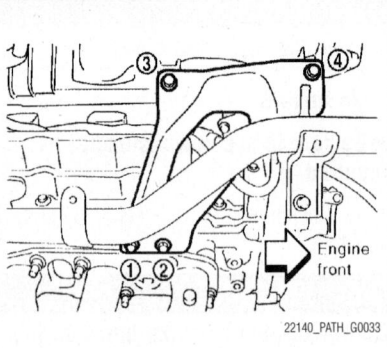

Fig. 93 Loosen bolts in reverse order as shown to remove intake manifold collector support—V6 engine

7. Remove the following parts:
- Vacuum hose (to brake booster)
- PCV hose

8. Loosen bolts in reverse order as shown to remove intake manifold collector support.

9. Disconnect EVAP hoses and harness connector from EVAP canister purge volume control solenoid valve.

10. Remove EVAP canister purge volume control solenoid valve.

11. Remove VIAS control solenoid valve and vacuum tank.

12. Loosen nuts and bolts in reverse order of the tightening sequence.

13. Loosen nuts and bolts in reverse order as shown to remove intake manifold collector (upper intake manifold).

14. Remove gaskets. Discard the gaskets.

To install:

➡**Be sure to use new fasteners, as required.**

15. Installation is in the reverse order of removal.

16. Note the following:
- If stud bolts were removed from cylinder head, install them and tighten to 8 ft. lbs. (11 Nm).
- Tighten all nuts and bolts to the specified torque in two or more steps in numerical order as shown.
- 1st step: 5 ft. lbs. (7.4 Nm)
- 2nd step and after: 21 ft. lbs. (29.0 Nm)

17. Tighten the throttle body retaining bolts to specification and in the proper sequence.

18. Using the CONSULT-III diagnostic tool or equivalent, perform the throttle valve closed position learning and the idle air volume learning procedures.

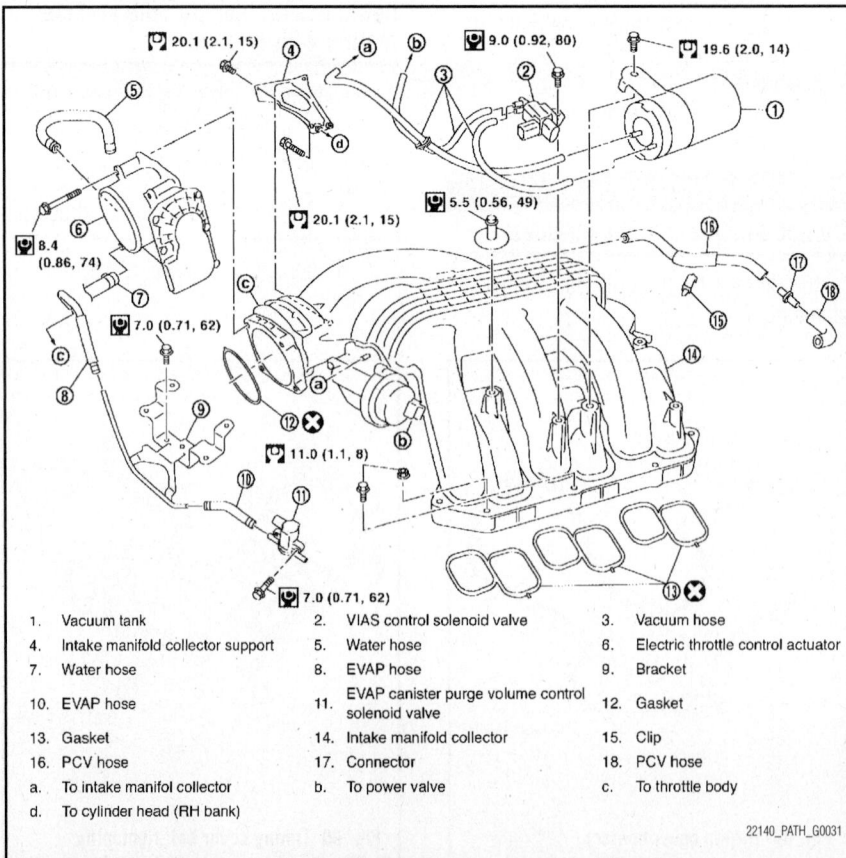

1.	Vacuum tank	2.	VIAS control solenoid valve	3.	Vacuum hose
4.	Intake manifold collector support	5.	Water hose	6.	Electric throttle control actuator
7.	Water hose	8.	EVAP hose	9.	Bracket
10.	EVAP hose	11.	EVAP canister purge volume control solenoid valve	12.	Gasket
13.	Gasket	14.	Intake manifold collector	15.	Clip
16.	PCV hose	17.	Connector	18.	PCV hose
a.	To intake manifol collector	b.	To power valve	c.	To throttle body
d.	To cylinder head (RH bank)				

Fig. 94 Upper intake manifold (collector) and related components—V6 engine

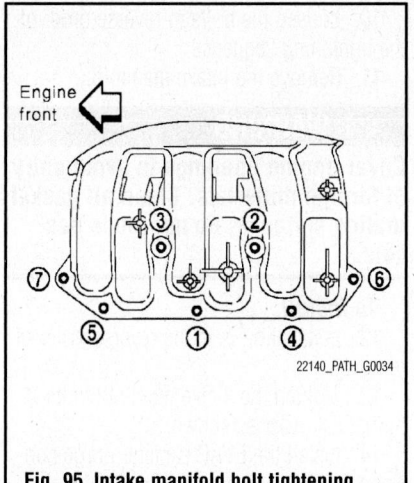

Fig. 95 Intake manifold bolt tightening sequence—V6 engine

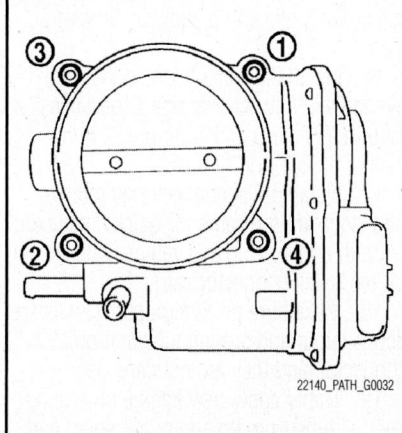

Fig. 96 Throttle body bolt tightening sequence—V6 engine

19. Be sure to perform the reconnect/relearn procedures.

Lower Manifold

See Figures 97 and 98.

1. Before servicing the vehicle, refer to the Precautions.

2. Disconnect the negative battery cable.

➡This vehicle uses quick connect fittings. Be sure to properly relieve the fuel system pressure before disconnecting any of these fittings. Always replace O-rings and clamps with new ones. Do not bend, twist or kink hoses when they are being removed or installed. Be sure that the clamp screw does not contact adjacent parts. When tightening the high pressure rubber hose clamp make sure the clamp end is 0.12 inch from the hose end. After connecting these fittings make sure

that the connectors are secure. Check for fuel leakage at these connections turn the ignition key to the ON position (do not start the engine), correct as required. Start the engine, raise the idle, and verify that there are no fuel leaks, correct as required.

3. Properly relieve the fuel system pressure.

4. Disconnect the negative battery cable.

5. Remove engine undercover.

6. Remove air cleaner case (upper) with mass air flow sensor and air duct assembly.

7. Remove the upper intake manifold (collector).

8. Remove the fuel tube and fuel injector assembly.

9. Remove the intake manifold retaining bolts, in the reverse order of the tightening sequence.

10. Remove the manifold from the engine. Discard the gaskets.

To install:

➡Be sure to use new fasteners, as required.

11. Installation is the reverse of the removal procedure.

12. Tighten the retaining bolts to the proper torque and in the proper sequence. If the stud bolts were removed from the cylinder head, tighten to 8 ft. lbs. Tighten all bolts in two or three steps as follows. Step one 65 inch lbs. Step two and after 21 ft. lbs. See illustration

13. Be sure to perform the reconnect/relearn procedures.

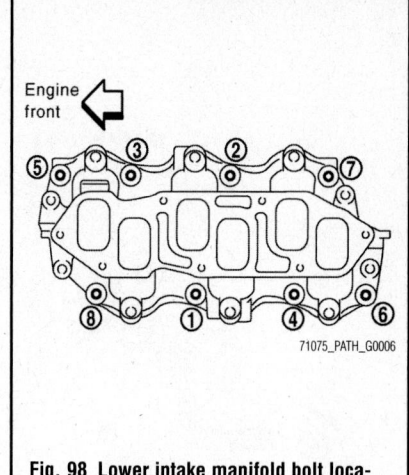

Fig. 98 Lower intake manifold bolt location and tightening sequence—V6 engine

V8 Engine

See Figures 99 through 102.

1. Before servicing the vehicle, refer to the Precautions.

➡This vehicle uses quick connect fittings. Be sure to properly relieve the fuel system pressure before disconnecting any of these fittings. Always replace O-rings and clamps with new ones. Do not bend, twist or kink hoses when they are being removed or installed. Be sure that the clamp screw does not contact adjacent parts. When tightening the high pressure rubber hose clamp make sure the clamp end is 0.12 inch from the hose end. After connecting these fittings make sure that the connectors are secure. Check for fuel leakage at these connections turn the ignition key to the ON position (do not start the engine), correct as

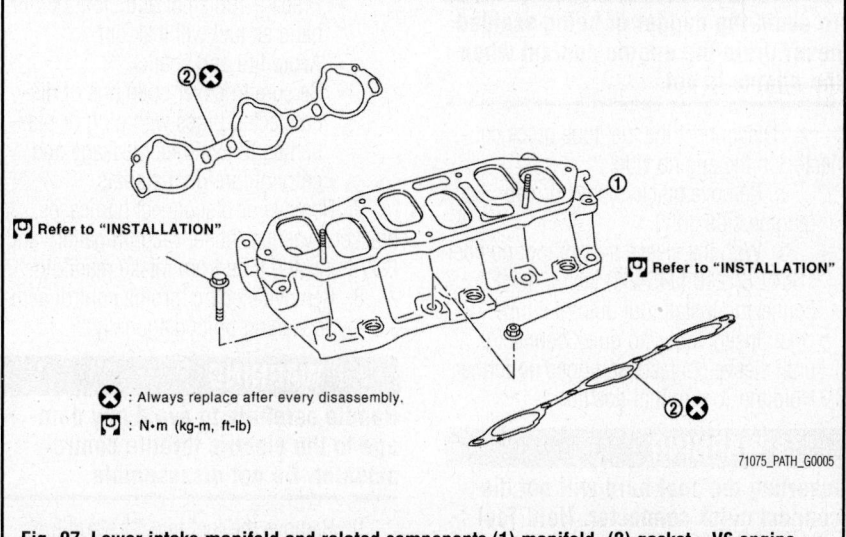

Fig. 97 Lower intake manifold and related components (1) manifold, (2) gasket—V6 engine

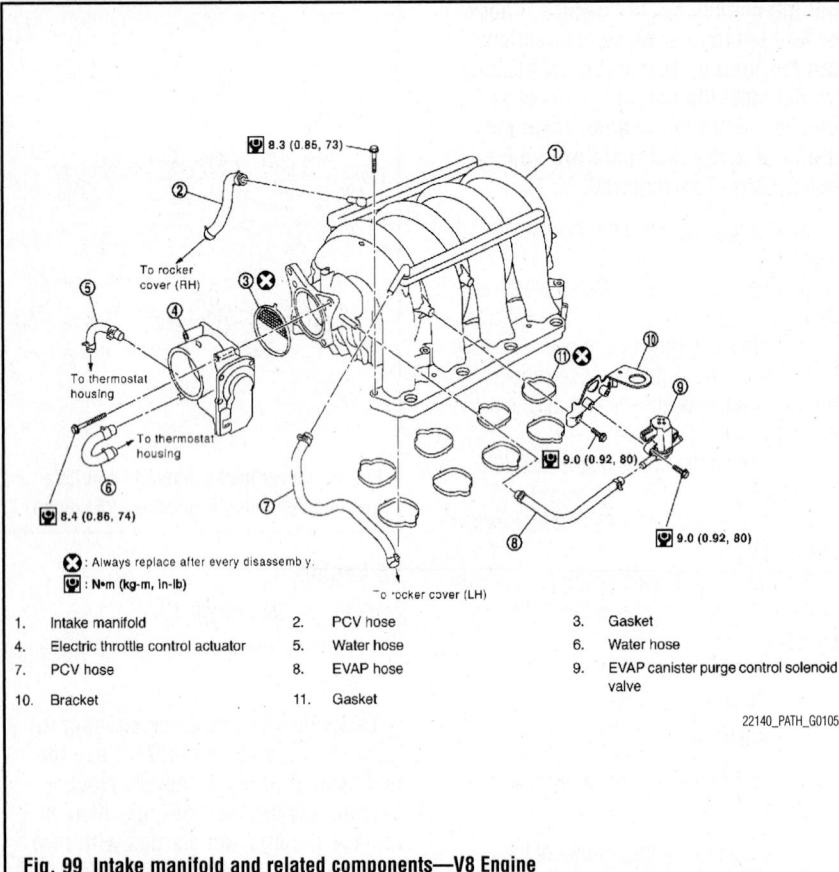

Fig. 99 Intake manifold and related components—V8 Engine

⊗ : Always replace after every disassembly.

🔧 : N•m (kg-m, In-lb)

1. Intake manifold
2. PCV hose
3. Gasket
4. Electric throttle control actuator
5. Water hose
6. Water hose
7. PCV hose
8. EVAP hose
9. EVAP canister purge control solenoid valve
10. Bracket
11. Gasket

8.3 (0.85, 73)
8.4 (0.86, 74)
9.0 (0.92, 80)
9.0 (0.92, 80)

To rocker cover (RH)
To thermostat housing
To thermostat housing
To rocker cover (LH)

22140_PATH_G0105

required. Start the engine, raise the idle, and verify that there are no fuel leaks, correct as required.

2. Properly discharge the fuel system pressure.

3. Disconnect the negative battery cable.

4. Remove the air cleaner assembly.

5. Partially drain the engine coolant.

❄❄ CAUTION

To avoid the danger of being scalded, never drain the engine coolant when the engine is hot.

6. Disconnect the fuel tube quick connector on the engine side.

a. Remove quick connector cap (engine side only).

b. With the sleeve side of tool number 16641 6N210 (J45488) facing quick connector, install tool onto fuel tube.

c. Insert Tool into quick connector until sleeve contacts and goes no further. Hold the Tool in that position.

❄❄ WARNING

Inserting the Tool hard will not disconnect quick connector. Hold Tool where it contacts and goes no further.

d. Draw and pull out quick connector straight from fuel tube.

❄❄ CAUTION

Heed the following cautions:

- Pull quick connector holding "A" position in illustration.
- Do not pull with lateral force applied. O-ring inside quick connector may be damaged.
- Prepare container and cloth beforehand as fuel will leak out.
- Avoid fire and sparks.
- Be sure to cover openings of disconnected pipes with plug or plastic bag to avoid fuel leakage and entry of foreign materials.

7. Remove or disconnect harnesses, brackets, vacuum hose, vacuum gallery and PCV hose and tube from intake manifold.

8. Remove electric throttle control actuator by loosening bolts diagonally.

❄❄ WARNING

Handle carefully to avoid any damage to the electric throttle control actuator. Do not disassemble.

9. Remove the fuel injectors and fuel tube assembly.

10. Loosen the bolts in reverse order of the tightening sequence.

11. Remove the intake manifold.

❄❄ WARNING

Cover engine openings to avoid entry of foreign materials. Clean all gasket mating surfaces, do not reuse gaskets.

To install:

12. Installation is in the reverse order of removal.

13. Tighten the intake manifold bolts in numerical order as shown.

14. Install the EVAP canister purge control solenoid valve connector with it facing front of engine.

15. Tighten the electronic throttle control actuator bolts of the electric throttle control actuator equally and diagonally in several steps.

16. Install the water hose so that its overlap width for connection is between 1.06 in. (27 mm) and 1.26 in. (32 mm) (target: 1.06 in. or 27 mm).

17. Install quick connector as follows (the steps are the same for quick connectors on both engine side and vehicle side except for the quick connector cap).

18. Make sure no foreign substances are deposited in and around tube and quick connector, and they are not damaged.

19. Thinly apply new engine oil around the fuel tube from tip end to the spool end.

20. Align center to insert quick connector straight into fuel tube.

a. Insert until the paint mark for engagement identification (white) goes completely inside quick connector so that you cannot see it from the straight side of the connected part. Use a mirror

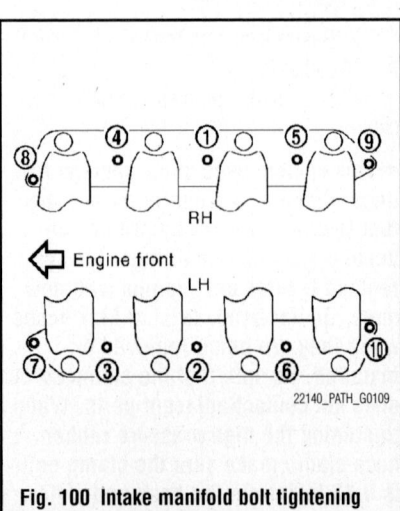

Engine front
RH
LH

22140_PATH_G0109

Fig. 100 Intake manifold bolt tightening sequence—V8 engine

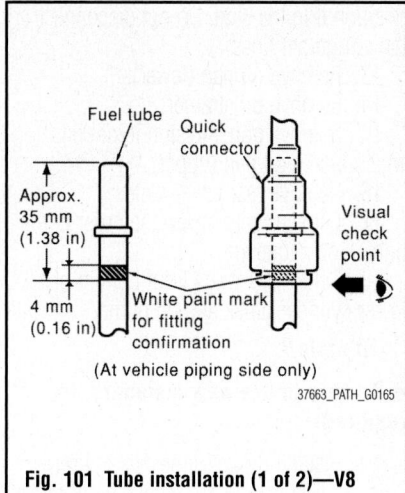

Fig. 101 Tube installation (1 of 2)—V8 engine

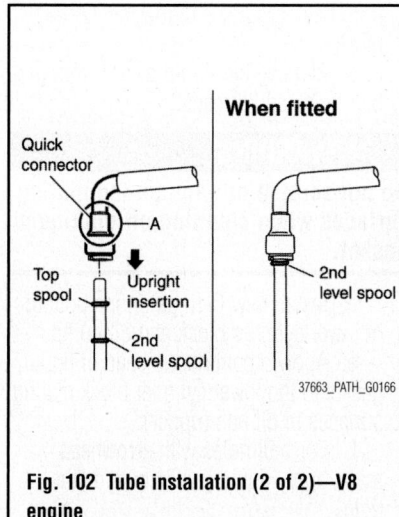

Fig. 102 Tube installation (2 of 2)—V8 engine

to check this where it is not possible to view directly from the straight side, such as quick connector on vehicle side.

   b. Insert fuel tube into quick connector until top spool is completely inside quick connector, and 2nd level spool exposes right below quick connector on engine side.

✳✳ WARNING

Hold "A" position in illustration when inserting fuel tube into quick connector. Carefully align center to avoid inclined insertion to prevent damage to O-ring inside quick connector. Insert until you hear a "click" sound and actually feel the engagement. To avoid misidentification of engagement with a similar sound, be sure to perform the next step.

   21. Pull quick connector by hand holding "A" position. Make sure it is completely

engaged (connected) so that it does not come out from fuel tube.

➡**Recommended pulling force is 50 N (5.1 kg, 11.2 lb).**

   22. Install the quick connector cap on the quick connector joint (on engine side only).

   23. Install the fuel hose and tube to hose clamps.

   24. Refill the engine coolant.

   25. Using the CONSULT-III diagnostic tool or equivalent, perform the throttle valve closed position learning and the idle air volume learning procedures.

   26. Be sure to perform the reconnect/relearn procedures.

OIL PAN

REMOVAL & INSTALLATION

V6 Engine

Lower Oil Pan

See Figures 103 through 105.

   1. Before servicing the vehicle, refer to the Precautions.

   2. Disconnect the negative battery cable.

✳✳ CAUTION

To avoid the danger of being scalded, do not drain engine oil when engine is hot.

   3. Remove the air cleaner assembly, as required.

   4. Raise and safely support the vehicle.

   5. Drain engine oil.

   6. Remove oil pan (lower) as follows:

   a. Loosen bolts in reverse order of the tightening sequence.

   b. Remove oil pan (lower).

✳✳ WARNING

Be careful not to damage the mating surfaces. Do not insert screwdriver, this will damage the mating surfaces.

➡**Slide seal cutter (1) by tapping on the side (2) of the tool with hammer.**

To install:

   7. Install oil pan (lower) as follows:

   a. Use scraper to remove old liquid gasket from mating surfaces.

   b. Also remove old liquid gasket from mating surface of oil pan (upper).

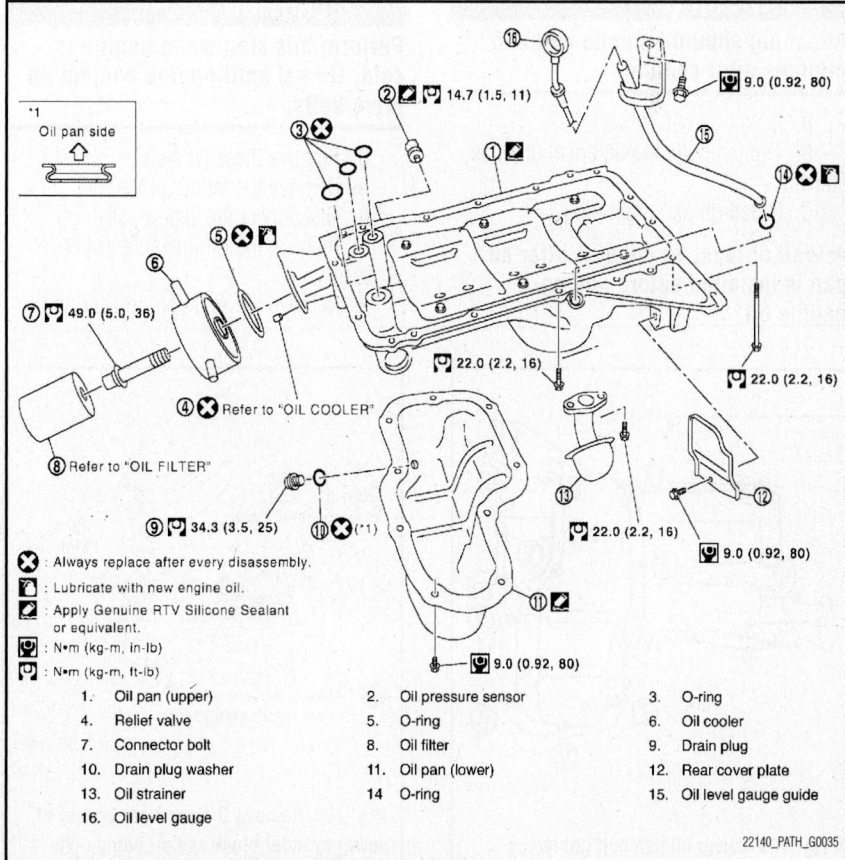

Fig. 103 Oil pan and related components—V6 engine

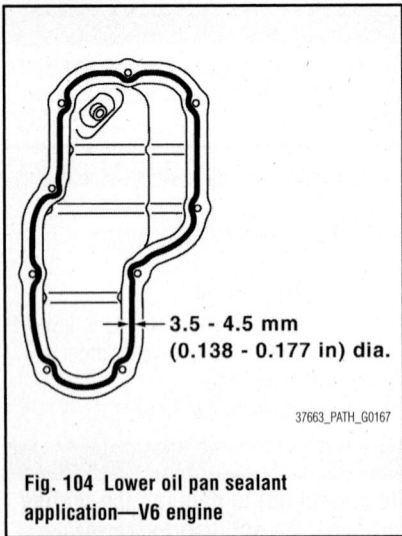

Fig. 104 Lower oil pan sealant application—V6 engine

3.5 - 4.5 mm
(0.138 - 0.177 in) dia.

37663_PATH_G0167

c. Remove old liquid gasket from the bolt holes and threads.

> ❊❊ **WARNING**
>
> **Do not scratch or damage the mating surfaces when cleaning off old liquid gasket.**

d. Apply a continuous bead of liquid gasket to the oil pan (lower).

> ❊❊ **WARNING**
>
> **Attaching should be done within 5 minutes after coating.**

8. Install oil pan (lower).
9. Tighten bolts in numerical order as shown.
10. Install oil pan drain plug.

➡ **Wait at least 30 minutes after oil pan is installed before adding engine oil.**

11. Check engine oil level and adjust engine oil.
12. Start engine, and check there is no leak of engine oil.
13. Stop engine and wait for 10 minutes.
14. Check engine oil level again.

Upper Oil Pan

See Figures 106 through 108.

1. Before servicing the vehicle, refer to the Precautions.
2. Disconnect the negative battery cable.
3. Remove the air cleaner assembly.
4. Raise and safely support the vehicle.
5. Remove engine undercover.
6. Drain the engine oil. Be sure to properly dispose of used oil.

> ❊❊ **CAUTION**
>
> **To avoid the danger of being scalded, do not drain engine oil when engine is hot. Perform this step when engine is cold. Do not spill engine oil on drive belts.**

7. Drain engine coolant. Be sure to properly dispose of used coolant.

> ❊❊ **CAUTION**
>
> **Perform this step when engine is cold. Do not spill engine coolant on drive belts.**

8. Remove the drive belt.
9. Remove the fender protector.
10. Disconnect the starter assembly.
11. Remove the oil pressure switch, if necessary.
12. Remove the A/C compressor and position it to the side. Do not disconnect the refrigerant lines.
13. Remove oil pan (lower).
14. Remove oil strainer.
15. Remove transmission joint bolts which pierce oil pan (upper).
16. Remove rear cover plate.
17. Loosen bolts in reverse order of the tightening sequence.
18. Remove O-rings from bottom of lower cylinder block and oil pump.

To install:

➡ **Be sure to use new fasteners, as required.**

19. Install oil pan (upper) as follows:
a. Use scraper to remove old liquid gasket from mating surfaces.
b. Also remove the old liquid gasket from mating surface of lower cylinder block.
c. Remove old liquid gasket from the bolt holes and threads.

> ❊❊ **WARNING**
>
> **Do not scratch or damage the mating surfaces when cleaning off old liquid gasket.**

d. Install new O-rings on the bottom of lower cylinder block and oil pump.
e. Apply a continuous bead of liquid gasket to the lower cylinder block mating surfaces of oil pan (upper).
f. For bolt holes with arrowhead mark, apply liquid gasket outside the hole.
g. Apply a bead of 4.5 to 5.5 mm (0.177 to 0.217 in) in diameter to area "A".

➡ **Attaching should be done within 5 minutes after coating.**

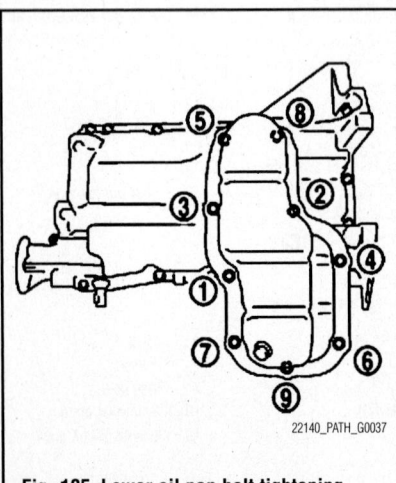

Fig. 105 Lower oil pan bolt tightening sequence—V6 engine

22140_PATH_G0037

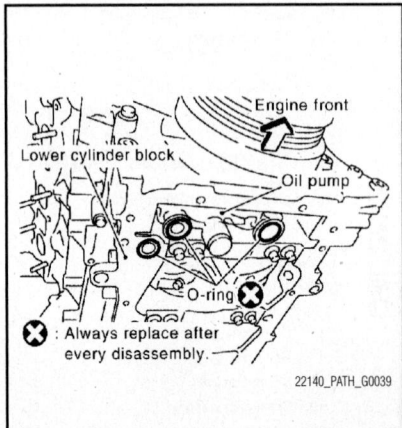

Engine front
Lower cylinder block
Oil pump
O-ring
❊ : Always replace after every disassembly.

22140_PATH_G0039

Fig. 106 Remove O-rings from bottom of lower cylinder block and oil pump—V6 engine

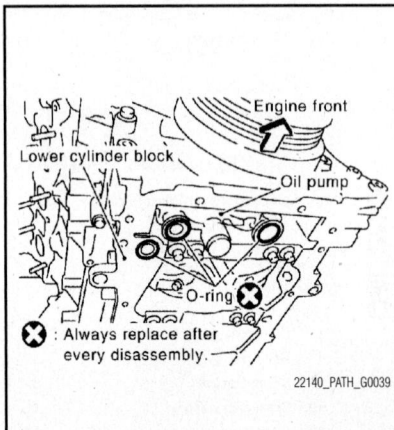

35 mm (1.38 in)
A
A
3.5 - 4.5 mm
(0.138 - 0.177 in) dia.
A
A
35 mm (1.38 in)
Engine front

22140_PATH_G0040

Fig. 107 Apply a continuous bead of liquid gasket to the lower cylinder block mating surfaces of oil pan (upper)—V6 engine

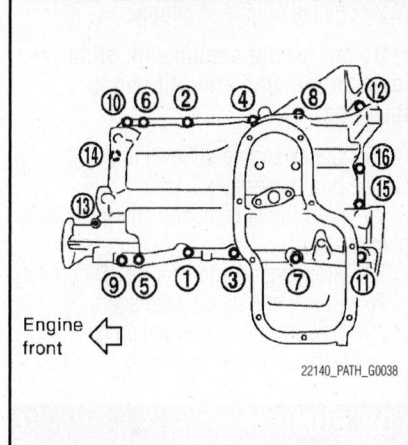

Fig. 108 Upper oil pan bolt tightening sequence—V6 engine

Install avoiding misalignment of both oil pan gaskets and O-rings.

20. Tighten bolts in numerical order as shown.

➡ **There are two types of bolts. Refer to the following for locating bolts.**

- M8x4 in. (100 mm): holes 7, 11, 12, 13
- M8x1 in. (25 mm): All other holes

21. Tighten transmission joint bolts.
22. Install oil strainer to oil pan (upper).
23. Install oil pan (upper).
24. Check engine oil level and adjust engine oil.
25. Start engine, and check there is no leak of engine oil.
26. Stop engine and wait for 10 minutes.
27. Check engine oil level again.
28. Be sure to perform the reconnect/relearn procedures.

V8 Engine

See Figures 109 through 114.

➡ **The engine must be removed from the vehicle to perform this procedure. Be sure that the engine is secured in a suitable holding fixture before performing this procedure.**

1. Before servicing the vehicle, refer to the Precautions.
2. Disconnect the negative battery cable.
3. Drain the engine oil. Be sure to properly dispose of used engine oil.

To avoid the danger of being scalded, never drain the engine oil when the engine is hot.

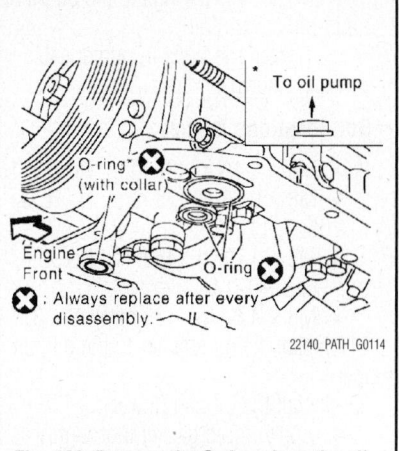

Fig. 109 Remove the O-rings from the oil pump and front cover—V8 engine

4. Remove the engine undercover.
5. Remove the air cleaner assembly.
6. Remove the fender protector and mud guard.
7. Remove the upper and lower fan shrouds.
8. Remove the oil cooler and O-ring. Discard the O-ring.
9. Remove the radiator hoses.
10. Remove the cooling fan.
11. Remove the power steering pump

reservoir tank and bracket. Position it to the side out of the work area.
12. Remove the A/C compressor. Position it to the side out of the work area. Do not disconnect the refrigerant lines.
13. Remove the alternator and bracket.
14. Remove the oil pressure switch.
15. Remove the oil pan (lower) using the following steps.
    a. Remove the oil pan (lower) bolts.
    b. Insert tool between the lower oil pan and the upper oil pan.

Be careful not to damage the mating surface.

    c. Tap seal cutter to insert it (1) and then slide it by tapping on the side (2) of the tool as shown.
16. Remove the oil cooler assembly.
17. Remove the oil strainer from the oil pan (upper).
18. Remove the oil pan (upper) using the following steps.
    a. Remove the oil pan (upper) bolts in the reverse order of the installation sequence.

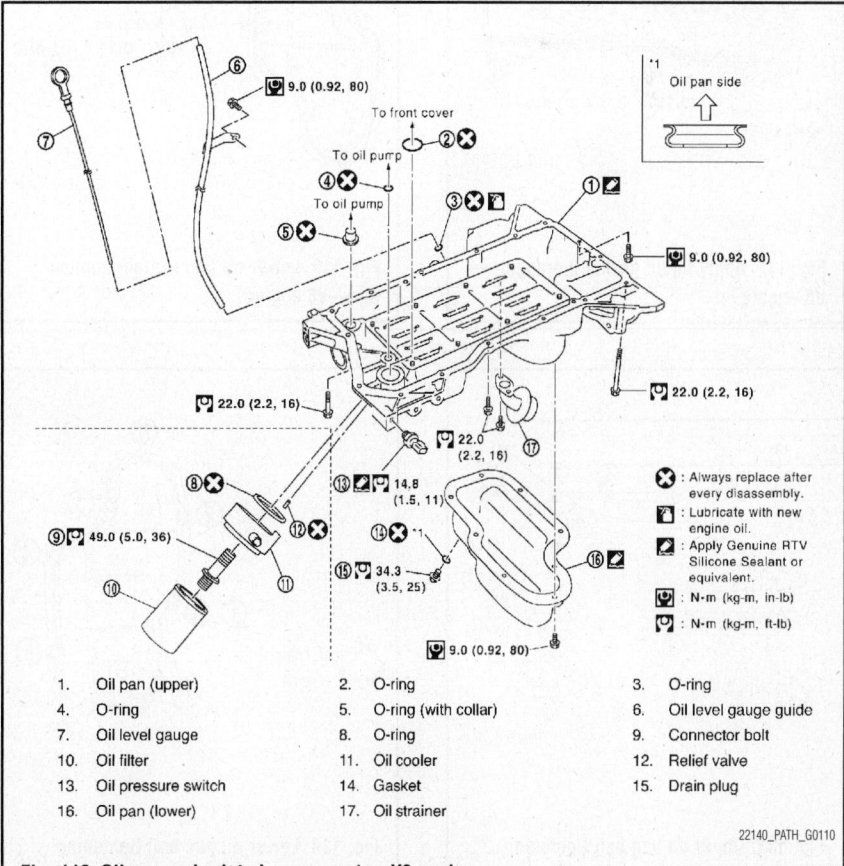

1.	Oil pan (upper)	2.	O-ring	3.	O-ring
4.	O-ring	5.	O-ring (with collar)	6.	Oil level gauge guide
7.	Oil level gauge	8.	O-ring	9.	Connector bolt
10.	Oil filter	11.	Oil cooler	12.	Relief valve
13.	Oil pressure switch	14.	Gasket	15.	Drain plug
16.	Oil pan (lower)	17.	Oil strainer		

Fig. 110 Oil pan and related components—V8 engine

b. Remove the oil pan (upper) from the cylinder block by prying.

❋❋ WARNING
Do not damage mating surface.

19. Remove the O-rings from the oil pump and front cover.

➡**Do not reuse O-rings.**

To install:

➡**Be sure to use new fasteners, as required.**

20. Install the oil pan (upper) using the following steps.
   a. Apply liquid gasket thoroughly as shown.

❋❋ WARNING
Apply liquid gasket to outside of bolt hole for the hole shown by star (*).

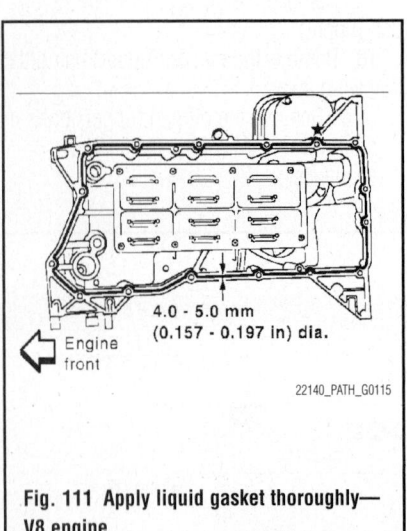

Fig. 111 Apply liquid gasket thoroughly—V8 engine

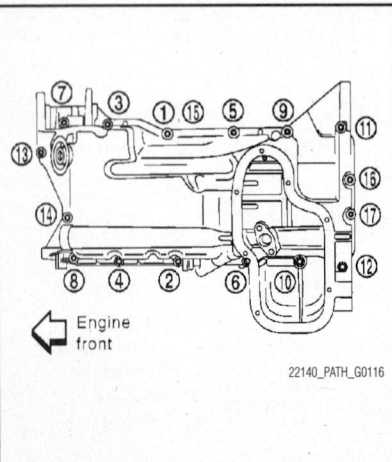

Fig. 112 Upper oil pan bolt tightening sequence—V8 engine

b. Install new O-rings to the oil pump and front cover side.
   c. Tighten the bolts in numerical order as shown.

➡**Bolt locations by size:**
   - M6 × 1.18 in. (30 mm): No. 15, 16
   - M8 × 0.98 in. (25 mm): No. 1, 3, 5, 7, 11, 13
   - M8 × 1.77 in. (45 mm): No. 2, 4, 6, 8, 10, 14
   - M8 × 4.84 in. (123 mm): No. 9, 12

21. Install the oil strainer to the oil pan (upper).
22. Install the oil pan (lower).
   a. Apply liquid gasket thoroughly as shown.

❋❋ WARNING
Attaching should be done within 5 minutes after coating.

b. Tighten the oil pan (lower) bolts in numerical order as shown.
23. Install the oil pan drain plug.

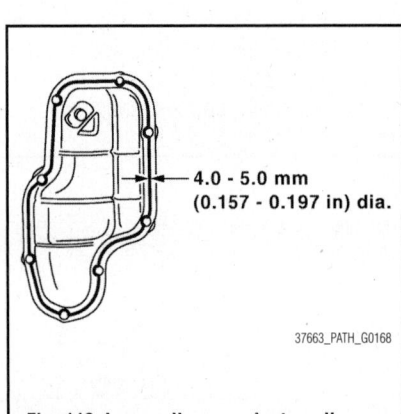

Fig. 113 Lower oil pan sealant application—V8 engine

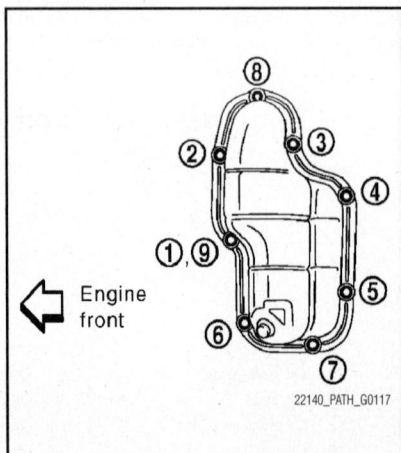

Fig. 114 Lower oil pan bolt tightening sequence—V8 engine

24. Install engine assembly.

➡**Do not fill the engine with oil for at least 30 minutes after oil pan is installed.**

25. Check engine oil level and add engine oil if necessary.
26. Start the engine, and check for leaks of engine oil.
27. Stop engine and wait for 10 minutes.
28. Check engine oil level again.
29. Be sure to perform the reconnect/relearn procedures.

OIL PUMP

REMOVAL & INSTALLATION

V6 Engine
See Figure 115.

1. Before servicing the vehicle, refer to the Precautions.
2. Disconnect the negative battery cable.
3. Remove the air cleaner assembly.
4. Raise and safely support the vehicle.
5. Drain the engine oil. Be sure to properly dispose of used oil.
6. Remove front timing chain case and timing chain (primary).
7. Remove oil pump assembly.

To install:

➡**Be sure to use new fasteners, as required.**

8. Installation is in the reverse order of removal, paying attention to the following:
9. When installing, align crankshaft flat faces with inner rotor flat faces.
10. Check the engine oil level.
11. Start engine, and check there are no leaks of engine oil.
12. Stop engine and wait for 10 minutes.
13. Check the engine oil level and add engine oil
14. Be sure to perform the reconnect/relearn procedures.

V8 Engine
See Figures 116 and 117.

1. Before servicing the vehicle, refer to the Precautions.
2. Disconnect the negative battery cable.
3. Remove the air cleaner assembly, as required.
4. Remove the front cover.
5. Remove the oil pump drive spacer.
6. Remove the oil pump assembly.

To install:

➡**Be sure to use new fasteners, as required.**

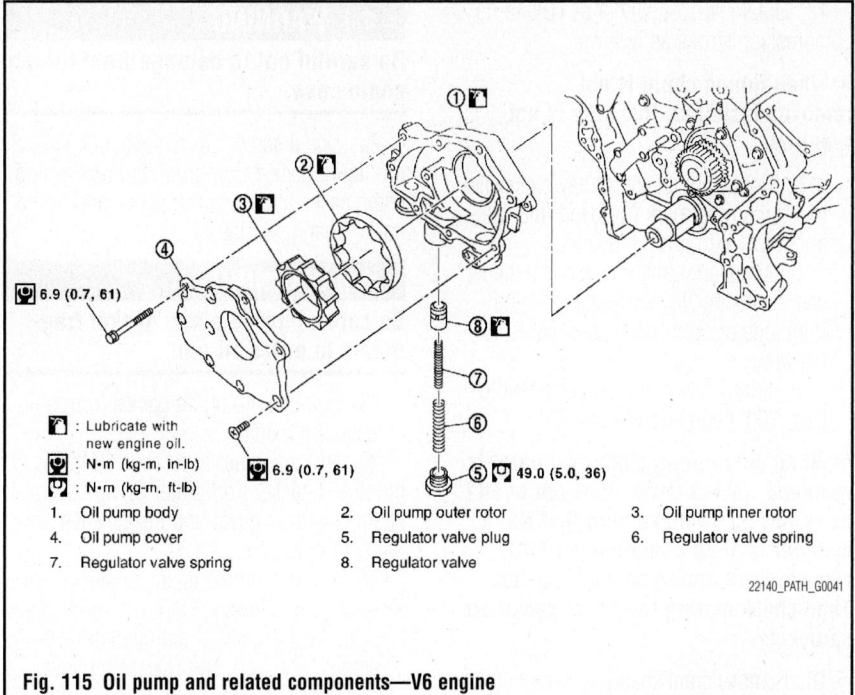

: Lubricate with new engine oil.
: N•m (kg-m, in-lb)
: N•m (kg-m, ft-lb)

6.9 (0.7, 61)

6.9 (0.7, 61)

49.0 (5.0, 36)

1. Oil pump body
2. Oil pump outer rotor
3. Oil pump inner rotor
4. Oil pump cover
5. Regulator valve plug
6. Regulator valve spring
7. Regulator valve spring
8. Regulator valve

22140_PATH_G0041

Fig. 115 Oil pump and related components—V6 engine

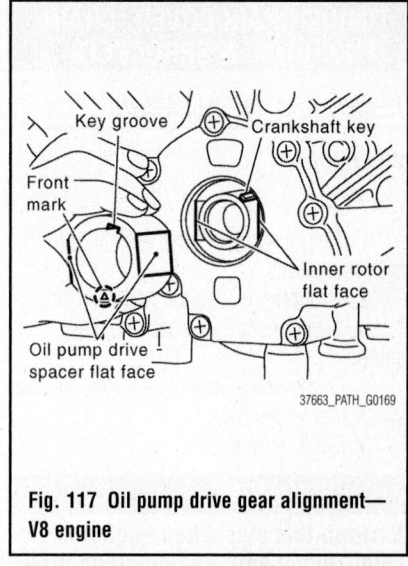

Fig. 117 Oil pump drive gear alignment—V8 engine

37663_PATH_G0169

PISTONS & RINGS

POSITIONING

See Figures 118 and 119.

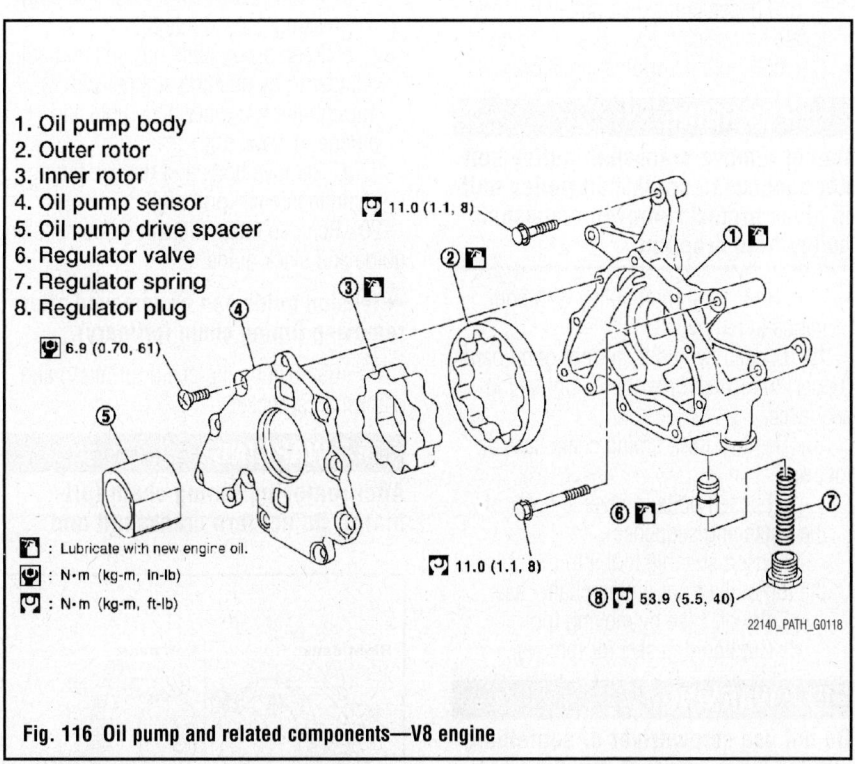

1. Oil pump body
2. Outer rotor
3. Inner rotor
4. Oil pump sensor
5. Oil pump drive spacer
6. Regulator valve
7. Regulator spring
8. Regulator plug

11.0 (1.1, 8)

6.9 (0.70, 61)

: Lubricate with new engine oil.
: N•m (kg-m, in-lb)
: N•m (kg-m, ft-lb)

11.0 (1.1, 8)

53.9 (5.5, 40)

22140_PATH_G0118

Fig. 116 Oil pump and related components—V8 engine

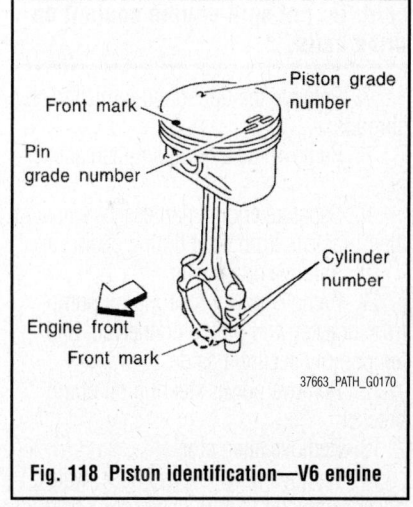

37663_PATH_G0170

Fig. 118 Piston identification—V6 engine

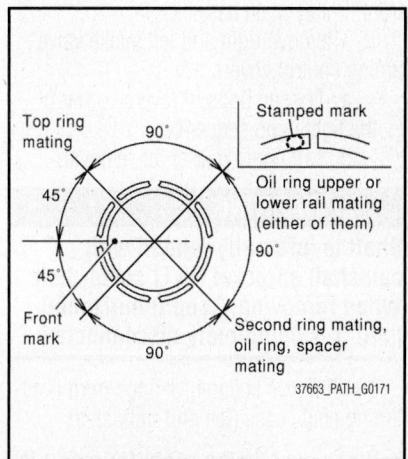

37663_PATH_G0171

Fig. 119 Piston identification and alignment—V8 engine

7. Installation is in the reverse order of removal, paying attention of the following:
• When inserting the oil pump drive spacer, align the crankshaft key and the flat face of the inner rotor.
• If they are not aligned, rotate the oil pump inner rotor by hand.
• Make sure that each part is aligned and tap lightly until it reaches the end.

8. Check the engine oil level.
9. Start the engine and check for engine oil leaks.
10. Stop the engine and wait 10 minutes.
11. Check the engine oil level and adjust the engine oil level as required.
12. Be sure to perform the reconnect/relearn procedures.

TIMING CHAIN COVER, CHAIN, TENSIONER, & SPROCKETS

REMOVAL & INSTALLATION

V6 Engine

See Figures 120 through 132.

1. Before servicing the vehicle, refer to the Precautions.
2. Properly release the fuel pressure.
3. Disconnect the negative battery cable.
4. Remove the undercover. Remove the air cleaner assembly.
5. Drain engine oil.

> ⁂ **CAUTION**
>
> **Perform this step when engine is cold. Do not spill engine oil on drive belts.**

6. Drain engine coolant from radiator.

> ⁂ **CAUTION**
>
> **Perform this step when engine is cold. Do not spill engine coolant on drive belts.**

7. Remove the upper and lower radiator shrouds.
8. Remove radiator cooling fan assembly.
9. Separate engine harnesses removing their brackets from front timing chain case.
10. Remove drive belts.
11. Remove power steering oil pump from bracket with piping connected, and temporarily secure it aside.
12. Remove power steering oil pump bracket.
13. Remove alternator.
14. Remove water bypass hose, water hose clamp and idler pulley bracket from front timing chain case.
15. Remove right and left intake valve timing control covers.
   a. Loosen bolts in reverse order of the tightening sequence.
   b. Cut liquid gasket for removal.

> ⁂ **WARNING**
>
> **Shaft is internally jointed with camshaft sprocket (INT) center hole. When removing, keep it horizontal until it is completely disconnected.**

16. Remove collared O-rings from front timing chain case (left and right side).

➡ **When only timing chain (primary) is removed, rocker cover does not need to be removed.**

17. Obtain No. 1 cylinder at TDC of its compression stroke as follows:

➡ **When timing chain is not removed/installed, this step is not required.**

   a. Rotate crankshaft pulley clockwise to align timing mark (grooved line without color) with timing indicator.
   b. Make sure that intake and exhaust cam noses on No. 1 cylinder (engine front side of right bank) are located as shown.
   c. If not, turn crankshaft one revolution (360°) and align as shown.

➡ **When only timing chain (primary) is removed, rocker cover does not need to be removed. To make sure that No. 1 cylinder is at its compression TDC, remove front timing chain case first. Then check mating marks on camshaft sprockets.**

18. Remove crankshaft pulley as follows:
   a. Remove starter motor.
   b. Loosen crankshaft pulley bolt and locate bolt seating surface as 0.39 in. (10 mm) from its original position.

> ⁂ **WARNING**
>
> **Do not remove crankshaft pulley bolt. Keep loosened crankshaft pulley bolt in place protect removed crankshaft pulley from dropping.**

   c. Pull crankshaft pulley with both hands to remove it.
19. Loosen two bolts in front of oil pan (upper) in reverse order of the tightening sequence.
20. Remove front timing chain case as follows:
   a. Loosen bolts in reverse order of the tightening sequence.
   b. Insert suitable tool into the notch at the top of the front timing chain case.
   c. Pry off case by moving tool.
   d. Cut liquid gasket for removal.

> ⁂ **WARNING**
>
> **Do not use screwdriver or something similar. After removal, handle front timing chain case carefully so it does not tilt, cant, or warp under a load.**

21. Remove O-rings from rear timing chain case.
22. Remove water pump cover and chain tensioner cover from front timing chain case, if necessary.
23. Remove front oil seal from front timing chain case using suitable tool.

> ⁂ **WARNING**
>
> **Be careful not to damage front timing chain case.**

24. Use a scraper to remove all traces of old liquid gasket from front and rear timing chain cases and oil pan (upper), and liquid gasket mating surfaces.

> ⁂ **WARNING**
>
> **Be careful not to allow gasket fragments to enter oil pan.**

25. Remove old liquid gasket from bolt holes and threads.
26. Use a scraper to remove all traces of old liquid gasket from water pump cover, chain tensioner cover and intake valve timing control covers.
27. Remove timing chain tensioner (primary) as follows:
   a. Loosen clip of timing chain tensioner (primary), and release plunger stopper (1).
   b. Insert plunger into tensioner body by pressing slack guide (2).
   c. Keep slack guide pressed and hold plunger in by pushing stopper pin through the tensioner body hole and plunger groove (3).
   d. Remove bolts and remove timing chain tensioner (primary).
28. Remove internal chain guide, tension guide and slack guide.

➡ **Tension guide can be removed after removing timing chain (primary).**

29. Remove timing chain (primary) and crankshaft sprocket.

> ⁂ **WARNING**
>
> **After removing timing chain (primary), do not turn crankshaft and**

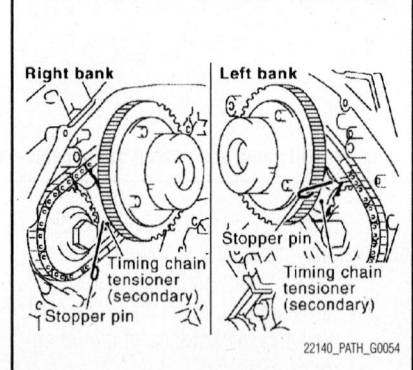

Fig. 120 Attach suitable stopper pin to the right and left timing chain tensioners (secondary)—V6 engine

camshaft separately, or valves will strike the piston heads.

30. Remove timing chain (secondary) and camshaft sprockets as follows:

a. Attach suitable stopper pin to the right and left timing chain tensioners (secondary).

➡**Use approximately 0.02 in. (0.5 mm) diameter hard metal pin as a stopper pin.**

b. Remove camshaft sprocket (INT and EXH) bolts.

➡**Secure the hexagonal portion of camshaft using wrench to loosen bolts.**

✳ WARNING

Do not loosen bolts with securing anything other than the camshaft hexagonal portion or with tensioning the timing chain.

c. Remove timing chain (secondary) together with camshaft sprockets.

- Turn camshaft slightly to secure slackness of timing chain on timing chain tensioner (secondary) side.
- Insert 0.5 mm (0.020 in) thick metal or resin plate between timing chain and timing chain tensioner plunger (guide). Remove timing chain (secondary) together with camshaft sprockets with timing chain loose from guide groove.

✳ WARNING

Be careful of plunger coming off when removing timing chain (secondary). This is because plunger of

timing chain tensioner (secondary) moves during operation, leading to coming off of fixed stopper pin.

➡**Camshaft sprocket (INT) is a one piece integrated design sprockets for timing chain (primary) and for timing chain (secondary).**

✳ WARNING

When handling camshaft sprocket (INT), be careful of the following: Handle carefully to avoid any shock to camshaft sprocket. Do not disassemble. (Do not loosen bolts "A" as shown).

31. Remove water pump.
32. Remove rear timing chain case as follows:

a. Loosen and remove bolts in reverse order of the tightening sequence.

b. Cut liquid gasket and remove rear timing chain case.

✳ WARNING

Do not remove plate metal cover of oil passage.

✳ WARNING

After removal, handle rear timing chain case carefully so it does not tilt, cant, or warp under a load.

33. Remove O-rings from cylinder head and camshaft bracket (No. 1).
34. Remove O-rings from cylinder block.
35. Remove timing chain tensioners (secondary) from cylinder head if necessary.

a. Remove camshaft brackets (No. 1).

b. Remove timing chain tensioners (secondary) with stopper pin attached.

36. Use scraper to remove all traces of

old liquid gasket from front and rear timing chain cases, and opposite mating surfaces. Remove old liquid gasket from bolt hole and thread.

37. Use scraper to remove all traces of liquid gasket from water pump cover, chain tensioner cover and intake valve timing control covers.

38. Check for cracks and any excessive wear at link plates and roller links of timing chain.

39. Replace timing chain as necessary.

To install:

➡**Be sure to use new fasteners, as required.**

➡**The figure below shows the relationship between the mating mark on each timing chain and that on the corresponding sprocket, with the components installed.**

40. Install timing chain tensioners (secondary) to cylinder head if removed.

a. Install timing chain tensioners (secondary) with stopper pin attached and new O-ring.

b. Install camshaft brackets (No. 1).

41. Install rear timing chain case as follows:

a. Install new O-rings onto cylinder block.

b. Install new O-rings to cylinder head and camshaft bracket (No. 1).

c. Apply liquid gasket to rear timing chain case back side as shown.

✳ WARNING

For "A" in the figure, completely wipe out liquid gasket extended on a portion touching at engine coolant. Apply liquid gasket on installation position of water pump and cylinder head very completely

d. Align rear timing chain case with dowel pins (right and left) on cylinder block and install rear timing chain case.

➡**Make sure O-rings stay in place during installation to cylinder block, cylinder head and camshaft bracket (No. 1).**

e. Tighten bolts in numerical order as shown.

➡**There are two type of bolts. Refer to the following for locating bolts.**

- 0.79 in. (20 mm): 1, 2, 3, 6, 7, 8, 9, 10
- 0.63 in. (16 mm): Except the above
- Rear timing case bolt torque: 9 ft lbs. (12.7 Nm)

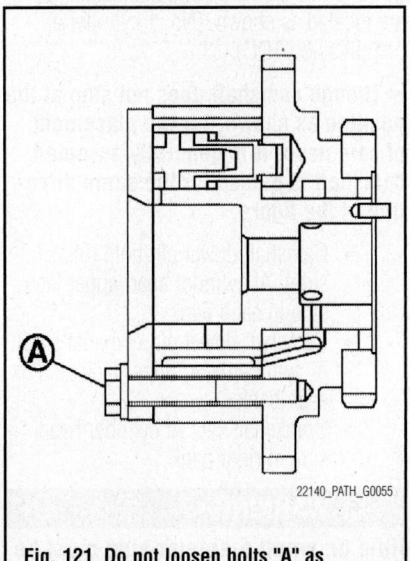

22140_PATH_G0055

Fig. 121 Do not loosen bolts "A" as shown—V6 engine

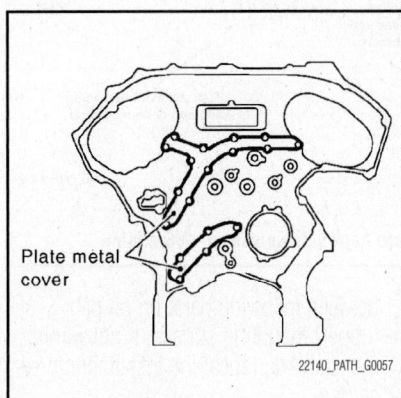

Plate metal cover

22140_PATH_G0057

Fig. 122 Do not remove plate metal cover of oil passage—V6 engine

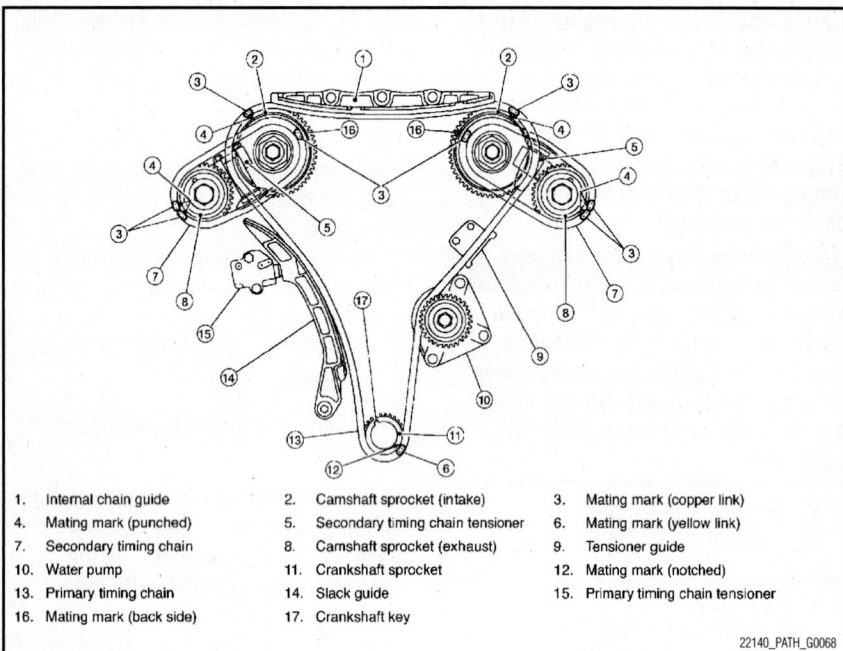

1. Internal chain guide
2. Camshaft sprocket (intake)
3. Mating mark (copper link)
4. Mating mark (punched)
5. Secondary timing chain tensioner
6. Mating mark (yellow link)
7. Secondary timing chain
8. Camshaft sprocket (exhaust)
9. Tensioner guide
10. Water pump
11. Crankshaft sprocket
12. Mating mark (notched)
13. Primary timing chain
14. Slack guide
15. Primary timing chain tensioner
16. Mating mark (back side)
17. Crankshaft key

22140_PATH_G0068

Fig. 123 Timing chain assembly with mating marks—V6 engine

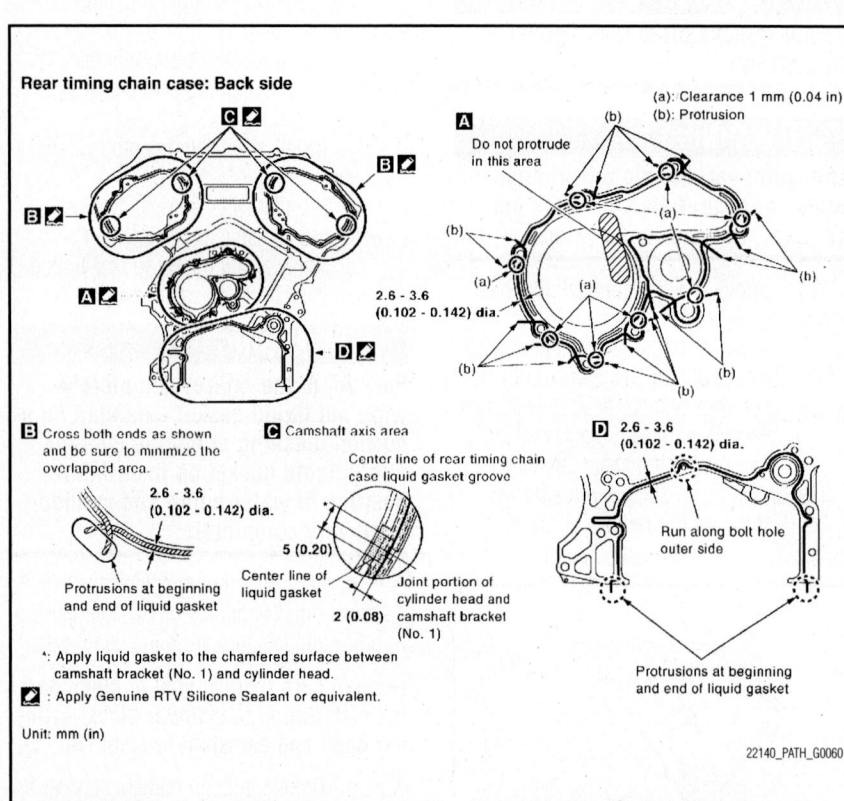

Rear timing chain case: Back side

(a): Clearance 1 mm (0.04 in)
(b): Protrusion

Do not protrude in this area

2.6 - 3.6 (0.102 - 0.142) dia.

B Cross both ends as shown and be sure to minimize the overlapped area.

2.6 - 3.6 (0.102 - 0.142) dia.

Protrusions at beginning and end of liquid gasket

C Camshaft axis area

Center line of rear timing chain case liquid gasket groove

5 (0.20)

Center line of liquid gasket

2 (0.08)

Joint portion of cylinder head and camshaft bracket (No. 1)

D 2.6 - 3.6 (0.102 - 0.142) dia.

Run along bolt hole outer side

Protrusions at beginning and end of liquid gasket

*: Apply liquid gasket to the chamfered surface between camshaft bracket (No. 1) and cylinder head.

☑ : Apply Genuine RTV Silicone Sealant or equivalent.

Unit: mm (in)

22140_PATH_G0060

Fig. 124 Apply liquid gasket to rear timing chain case back side as shown—V6 engine

f. After all bolts are tightened, retighten them to the specified torque in numerical order as shown.

g. If liquid gasket protrudes, wipe it off immediately.

h. After installing rear timing chain case, check the surface height difference between following parts on oil pan (upper) mounting surface. If not within the standard, repeat the installation procedure.

i. Standard: Rear timing chain case to lower cylinder block: -0.0094–0.0055 in. (-0.24–0.14 mm)

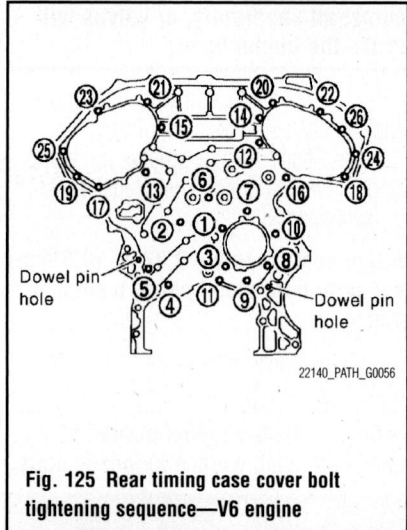

22140_PATH_G0056

Fig. 125 Rear timing case cover bolt tightening sequence—V6 engine

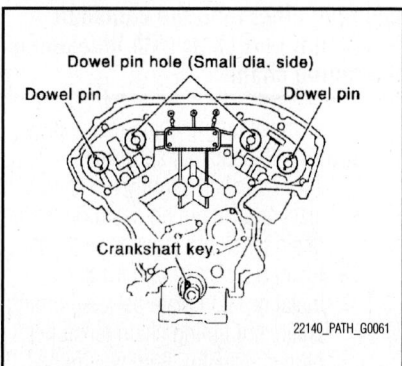

Dowel pin hole (Small dia. side)

Dowel pin

Dowel pin

Crankshaft key

22140_PATH_G0061

Fig. 126 Make sure that dowel pin hole, dowel pin of camshaft and crankshaft key are located as shown. (No. 1 cylinder at compression TDC)—V6 engine

42. Install water pump with new O-rings.

43. Make sure that dowel pin hole, dowel pin of camshaft and crankshaft key are located as shown. (No. 1 cylinder at compression TDC).

➡Though camshaft does not stop at the position as shown, for the placement of cam nose, it is generally accepted camshaft is placed for the same direction of the figure.

- Camshaft dowel pin hole (intake side): At cylinder head upper face side in each bank.
- Camshaft dowel pin (exhaust side): At cylinder head upper face side in each bank.
- Crankshaft key: At cylinder head side of right bank.

✲✲ WARNING

Hole on small diameter side must be used for intake side dowel pin hole.

Do not misidentify (ignore big diameter side).

44. Install timing chains (secondary) and camshaft sprockets as follows:

✳✳ WARNING

Mating marks between timing chain and sprockets slip easily. Confirm all mating mark positions repeatedly during the installation process.

a. Push plunger of timing chain tensioner (secondary) and keep it pressed in with stopper pin.

b. Install timing chains (secondary) and camshaft sprockets (INT and EXH).

c. Align the mating marks on timing chain (secondary) (copper color link) with the ones on camshaft sprockets (INT and EXH) (punched), and install them.

➡ **Mating marks for camshaft sprocket (INT) are on the back side of camshaft sprocket (secondary).**

➡ **There are two types of mating marks, circle and oval types.**

- Right bank: Use circle type.
- Left bank: Use oval type.
- They should be used for the right and left banks, respectively.

d. Align dowel pin and pin hole on camshafts with the groove and dowel pin on sprockets, and install them.

e. On the intake side, align pin hole on the small diameter side of the camshaft front end with dowel pin on the back side of camshaft sprocket, and install them.

f. On the exhaust side, align dowel pin on camshaft front end with pin groove on camshaft sprocket, and install them.

g. In case that positions of each mating mark and each dowel pin are not fit on mating parts, make fine adjustment to the position holding the hexagonal portion on camshaft with wrench or equivalent.

h. Bolts for camshaft sprockets must be tightened in the next step. Tightening them by hand is enough to prevent the dislocation of dowel pins.

i. It may be difficult to visually check the dislocation of mating marks during and after installation. To make the matching easier, make a mating mark on the top of sprocket teeth and its extended line in advance with paint.

j. After confirming the mating marks are aligned, tighten camshaft sprocket bolts.

➡ **Secure camshaft using wrench at the hexagonal portion to tighten bolts.**

k. Pull stopper pins out from timing chain tensioners (secondary).

45. Install tension guide.

46. Install timing chain (primary) as follows:

a. Install crankshaft sprocket.

➡ **Make sure the mating marks on crankshaft sprocket face the front of engine.**

b. Install the primary timing chain.

- Water pump (G).
- Install primary timing chain so the mating mark punched (B) on camshaft sprocket is aligned with the copper link (A) on the timing chain, while the mating mark notched (E) on the crankshaft sprocket (D) is aligned with the yellow link (F) on the timing chain, as shown.
- When it is difficult to align mating marks (A) with (B) and (E) with (F) of the primary timing chain with each sprocket, gradually turn the camshaft using a wrench on the hexagonal portion to align it with the mating marks.
- During alignment, be careful to prevent dislocation of mating mark alignments of the secondary timing chains.

47. Install internal chain guide, slack guide and timing chain tensioner (primary).

✳✳ WARNING

Do not overtighten slack guide bolts. It is normal for a gap to exist under the bolt seats when bolts are tightened to specification.

- When installing timing chain tensioner (primary), push in plunger and keep it pressed in with stopper pin.
- Remove any dirt and foreign materials completely from the back and the mounting surfaces of timing chain tensioner (primary).

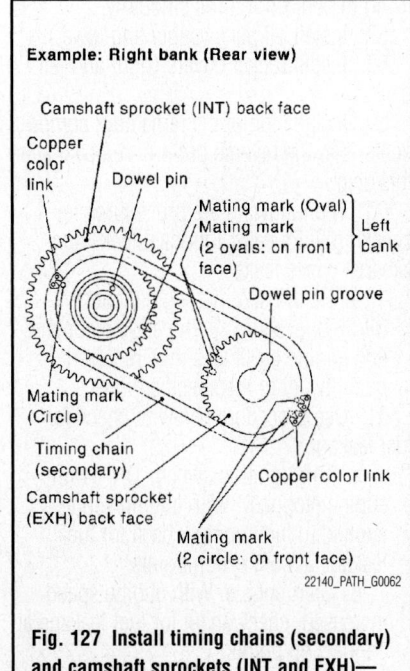

Example: Right bank (Rear view)

Camshaft sprocket (INT) back face
Copper color link
Dowel pin
Mating mark (Oval)
Mating mark (2 ovals: on front face) } Left bank
Dowel pin groove
Mating mark (Circle)
Timing chain (secondary)
Copper color link
Camshaft sprocket (EXH) back face
Mating mark (2 circle: on front face)

22140_PATH_G0062

Fig. 127 Install timing chains (secondary) and camshaft sprockets (INT and EXH)—V6 engine

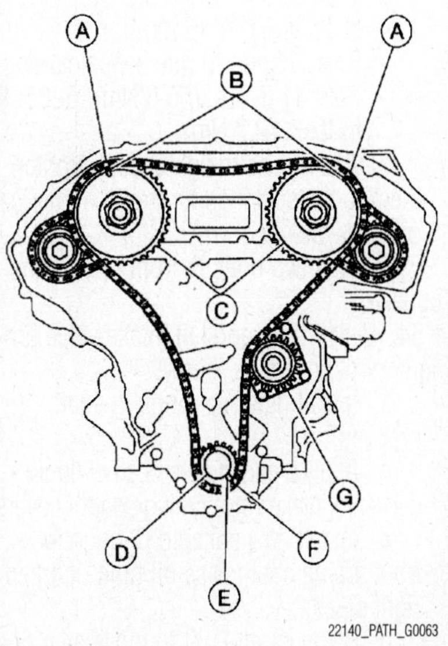

A   A
B
C
G
D   F
E

22140_PATH_G0063

Fig. 128 Install the primary timing chain—V6 engine

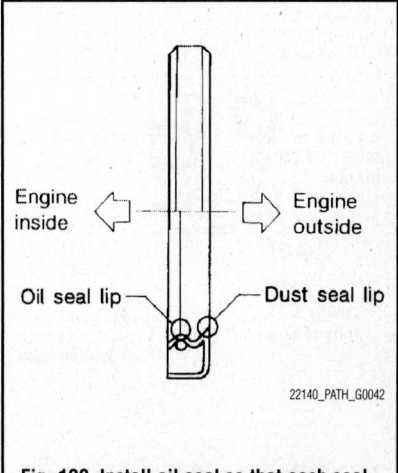

Engine inside ⇐ ⇒ Engine outside
Oil seal lip — Dust seal lip

22140_PATH_G0042

Fig. 129 Install oil seal so that each seal lip is oriented as shown—V6 engine

- After installation, pull out stopper pin by pressing slack guide.

48. Make sure again that the mating marks on camshaft sprockets and timing chain have not slipped out of alignment.

49. Install new O-rings on rear timing chain case.

50. Install new front oil seal on front timing chain case.

   a. Apply new engine oil to both oil seal lip and dust seal lip.

   b. Install it so that each seal lip is oriented as shown.

   c. Press-fit oil seal until it becomes flush with front timing chain case end face using suitable drift with outer diameter: 2.36 in. (60 mm).

   d. Make sure the garter spring is in position and seal lip is not inverted.

51. Install water pump cover and chain tensioner cover to front timing chain case.

   a. Apply a continuous bead of liquid gasket to front timing chain case as shown.

52. Install front timing chain case as follows:

   a. Apply a continuous bead of liquid gasket to front timing chain case back side as shown.

   b. Install new O-rings on rear timing chain case.

   c. Assemble front timing chain case as follows:

- Fit lower end of front timing chain case tightly onto top face of oil pan (upper). From the fitting point, make entire front timing chain case contact rear timing chain case completely.
- Since front timing chain case is offset for difference of bolt holes, tighten bolts temporarily while

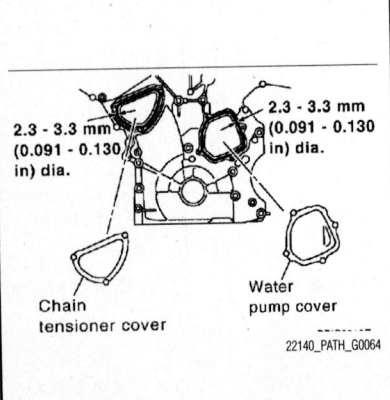

Fig. 130 Apply a continuous bead of liquid gasket to front timing chain—V6 engine

---

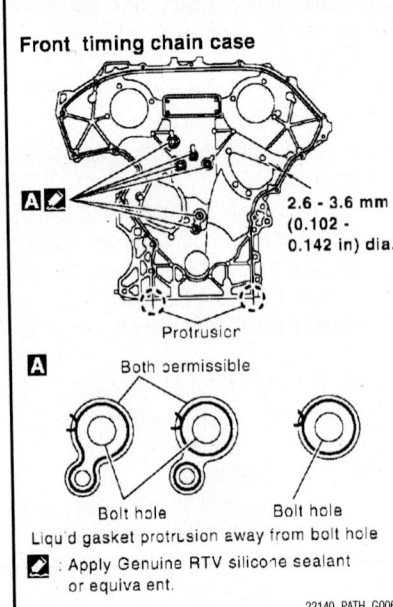

Fig. 131 Apply a continuous bead of liquid gasket to front timing chain case back side—V6 engine

holding front timing chain case from front and top.

- Same as the previous step, insert dowel pin while holding front timing chain case from front and top completely.

   d. Tighten bolts to the specified torque in numerical order as shown.

➡ **There are two type of bolts. Refer to the following for locating bolts.**

- 1–5: 0.39 in. (10 mm)
- 6–25: 0.24 in. (6 mm)
- Bolt position torque specification: 1–5: 41 ft. lbs. (55.0 Nm); 6–25: 9 ft. lbs. (12.7 Nm)

   e. After all bolts tightened, retighten them to the specified torque in numerical order as shown.

53. Install two bolts in front of oil pan (upper).

54. Install right and left intake valve timing control covers as follows:

   a. Install new seal rings in shaft grooves.

   b. Apply a continuous bead of liquid gasket to intake valve timing control covers.

   c. Install new collared O-rings in front timing chain case oil hole (left and right sides).

   d. Being careful not to move seal ring from the installation groove, align dowel pins on front timing chain case with the holes to install intake valve timing control covers.

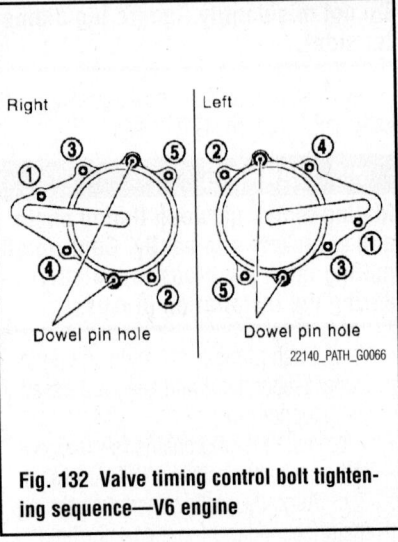

Fig. 132 Valve timing control bolt tightening sequence—V6 engine

   e. Tighten bolts in numerical order as shown.

55. Install crankshaft pulley as follows:

   a. Install crankshaft pulley, taking care not to damage front oil seal.

➡ **When press-fitting crankshaft pulley with plastic hammer, tap on its center portion (not circumference).**

   b. Tighten crankshaft pulley bolt. Crankshaft bolt torque: 33 ft. lbs. (44.1 Nm)

   c. Put a paint mark on crankshaft pulley aligning with angle mark on crankshaft pulley bolt. Then, further retighten bolt by 60° (equivalent to one graduation).

56. Rotate crankshaft pulley in normal direction (clockwise when viewed from front) to confirm it turns smoothly.

57. Install oil pans (upper and lower).

58. Install rocker covers (right and left banks).

59. Installation of the remaining components is in the reverse order of removal after this step.

60. The following are procedures for checking fluid leaks, lubricant leaks and exhaust gases leaks.

   a. Before starting engine, check oil/fluid levels including engine coolant and engine oil. If less than required quantity, fill to the specified level.

61. Use procedure below to check for fuel leakage.

   a. Turn ignition switch "ON" (with engine stopped). With fuel pressure applied to fuel piping, check for fuel leakage at connection points.

   b. Start engine. With engine speed increased, check again for fuel leakage at connection points.

   c. Run engine to check for unusual noise and vibration.

➡If hydraulic pressure inside timing chain tensioner drops after removal/ installation, slack in the guide may generate a pounding noise during and just after engine start. However, this is normal. Noise will stop after hydraulic pressure rises.

   d. Warm up engine thoroughly to make sure there is no leakage of fuel, exhaust gases, or any oil/fluids including engine oil and engine coolant.

   e. Bleed air from lines and hoses of applicable lines, such as in cooling system.

   f. After cooling down engine, again check oil/fluid levels including engine oil and engine coolant. Refill to the specified level, if necessary.

62. Be sure to perform the reconnect/ relearn procedures.

V8 Engine

See Figures 133 through 138.

1. Before servicing the vehicle, refer to the Precautions.
2. Disconnect the negative battery cable.
3. Remove the air cleaner assembly.
4. Drain the cooling system. Be sure to properly dispose of used coolant.
5. Remove the upper oil pan.
6. Remove the camshaft position sensor.
7. Remove the drive belt. Remove the tensioner assembly.
8. Remove the crankshaft pulley bracket.
9. Remove the water pump pulley.
10. Remove the power steering pump reservoir tank and bracket. Position it aside.
11. Remove the lower radiator hose and pipe assembly. Remove the suction hoses and pipe.
12. Remove the Intake valve control solenoid valve cover (RH) (A) and Intake valve control solenoid valve cover (LH) (B) as follows:
   a. Loosen and remove the bolts. See illustration.
   b. Cut the liquid gasket and remove the covers.

❊❊ WARNING

Do not damage mating surfaces.

13. Obtain compression TDC of No. 1 cylinder as follows:
   a. Turn the crankshaft pulley clockwise to align the TDC identification notch (without paint mark) with the timing indicator on the front cover.

   b. At this time, make sure both intake and exhaust cam lobes of No. 1 cylinder (top front on LH bank) point outside. If they do not point outside, turn crankshaft pulley once more.

14. Remove the crankshaft pulley.
   a. Loosen the crankshaft pulley bolts using a hammer handle to secure the crankshaft.
   b. Remove the crankshaft pulley from the crankshaft.
   c. Remove the crankshaft pulley.

➡The dimension between the centers of the two bolt holes is 61 mm (2.40 in.).

15. Remove the front cover.
   a. Loosen and remove the bolts in the reverse order of the tightening sequence.
   b. Cut the liquid gasket and remove the covers.

❊❊ WARNING

Do not damage mating surfaces.

16. Remove the front oil seal using suitable tool.

❊❊ WARNING

Do not damage front cover.

17. Remove the oil pump drive spacer.

➡Hold and remove the flat space of the oil pump drive spacer by pulling it forward.

18. Remove the oil pump.
19. Remove the chain tensioner on the LH bank using the following steps.

➡To remove the timing chain and associated parts, start with those on the LH bank. The procedure for remov-

ing parts on the RH bank is omitted because it is the same as that for the LH bank.

   a. Squeeze the return-proof clip ends using suitable tool and push the plunger into the tensioner body.
   b. Secure the plunger using stopper pin.
   c. Remove the bolts and chain tensioner.

❊❊ CAUTION

Plunger, spring, and spring seat pop out when (squeezing) return-proof clip without holding plunger head. It may cause serious injuries. Always hold plunger head when removing.

➡Stop the plunger in the fully extended position by using the return-proof clip (1) if the stopper pin is removed. Push the plunger (2) into the tensioner body while squeezing the return-proof clip (1). Secure it using stopper pin (3).

20. Remove the timing chain tension guide and timing chain slack guide.
21. Remove the timing chain and crankshaft sprocket.
22. Loosen the camshaft sprocket bolts as shown and remove the camshaft sprocket.

❊❊ WARNING

To avoid interference between valves and pistons, do not turn crankshaft or camshaft when timing chain is disconnected.

23. Repeat the same procedure to remove the RH timing chain and associated parts.

   To install:

➡Be sure to use new fasteners, as required.

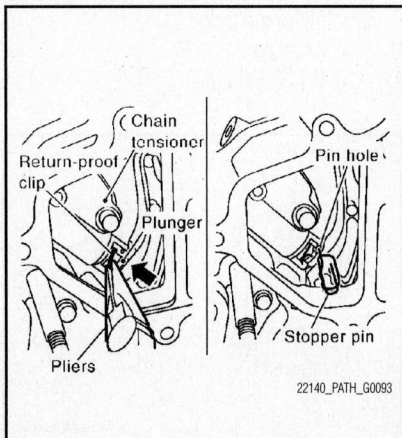

22140_PATH_G0093

Fig. 133 Secure the plunger using stopper pin—V8 engine

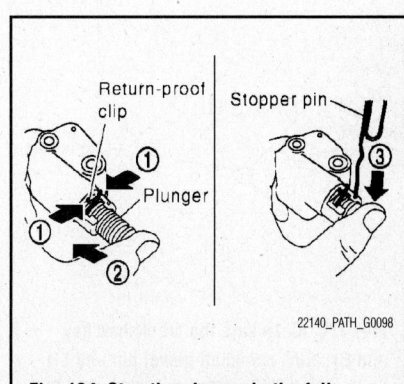

22140_PATH_G0098

Fig. 134 Stop the plunger in the fully extended position—V8 engine

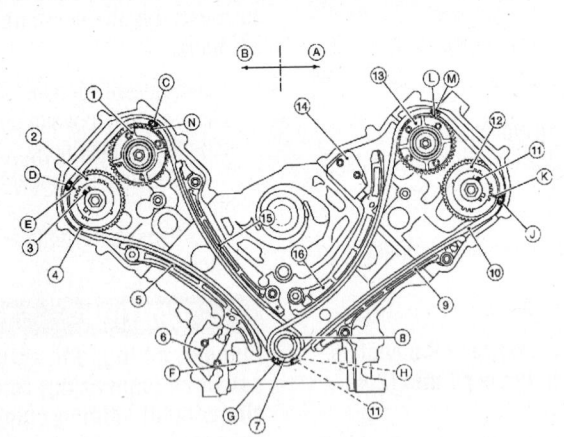

1. RH bank Camshaft sprocket (INT) (VTC)
2. RH bank Camshaft sprocket (EXH)
3. RH bank camshaft dowel pin
4. Timing chain
5. RH bank Timing chain slack guide
6. Primary timing chain tensioner
7. Crankshaft sprocket
8. Crankshaft key
9. LH Timing chain tension guide
10. Timing chain
11. LH Camshaft dowel pin
12. LH bank Camshaft sprocket (EXH)
13. LH bank Camshaft sprocket (INT) (VTC)
14. Secondary timing chain tensioner
15. RH bank timing chain tension guide
16. LH timing chain slack guide

A. LH bank
B. RH bank
C. Alignment mark (Link color: copper)
D. Alignment mark (Link color: copper)
E. Alignment mark (Identification mark)
F. Alignment mark for LH bank (Notch)
G. Alignment mark for LH bank (Link color: Yellow)
H. Alignment mark for RH bank (Link color: Yellow)
J. Alignment mark (Link color: copper)
K. Alignment mark (Identification mark)
L. Alignment mark (Identification mark)
M. Alignment mark (Link color: copper)
N. Alignment mark (Identification mark)

22140_PATH_G0123

Fig. 135 Timing chains and related components—V8 engine

➡ **The above figure shows the relationship between the mating mark on each timing chain and that of the corresponding sprocket, with the components installed.**

To install the timing chain and associated parts, start with those on the RH bank. The procedure for installing parts on the LH bank is omitted because it is the same as that for installation on the RH bank.

24. Make sure the crankshaft key and RH bank camshaft dowel pin and LH bank camshaft dowel pin are facing in the direction shown.

25. Install the camshaft sprockets.

    a. Install the intake camshaft sprocket (VTC) (A) and exhaust camshaft sprockets (B) by selectively using the groove of the dowel pin according to the bank. (Common part used for both exhaust banks.)

    b. Lock the hexagonal part of the camshaft in the same way as for removal, and tighten the bolts.

    • A: Intake
    • B = V: Exhaust

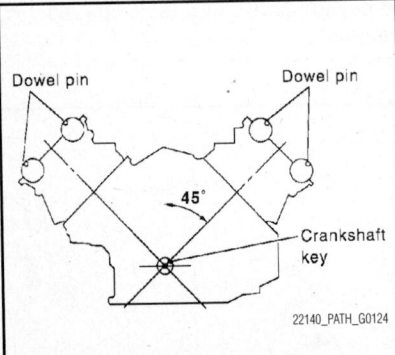

Fig. 136 Make sure the crankshaft key and RH bank camshaft dowel pin and LH bank camshaft dowel pin are facing in the direction—V8 engine

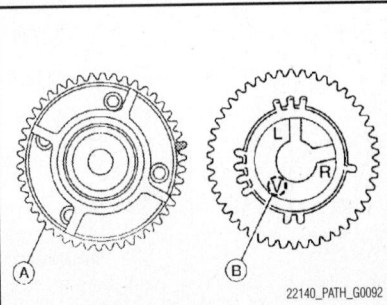

Fig. 137 Install the camshaft sprockets— V8 engine

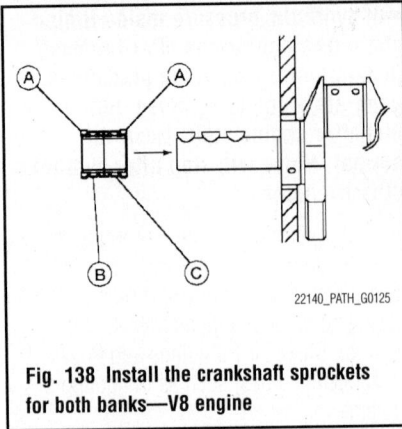

Fig. 138 Install the crankshaft sprockets for both banks—V8 engine

26. Install the crankshaft sprockets for both banks.

    a. Install LH bank crankshaft sprocket (B) and RH bank crankshaft sprocket (C) so that their flange side (A) (the larger diameter side without teeth) faces in the direction shown.

➡ **The same parts are used but facing directions are different.**

27. Install the timing chains and associated parts.

    a. Align the alignment mark on each sprocket and the timing chain for installation.

✳✳ WARNING

Before installing timing chain tensioner, it is possible to change the position of alignment mark on timing chain and each sprocket. After the alignment marks are aligned, keep them aligned by holding them by hand.

    b. Install the slack guides and tension guides onto the correct side by checking the identification mark on the surface.

    c. Install the timing chain tensioner with the plunger locked in with the stopper pin.

✳✳ WARNING

Before and after the installation of the timing chain tensioner, make sure that the alignment mark on the timing chain is not out of alignment.

    d. After installing the timing chain tensioner, remove the stopper pin to release the tensioner. Make sure the tensioner is released.

    e. To avoid chain-link skipping of the timing chain, do not move crankshaft or camshafts until the front cover is installed.

28. In the same way as for the RH bank, install the timing chain and associated parts on the LH bank.

29. Install the oil pump.

30. Install the oil pump drive spacer as follows:

    a. Install so that the front mark on the front edge of the oil pump drive spacer faces the front of the engine.

    b. Insert the oil pump drive spacer according to the directions of the crankshaft key and the two flat surfaces of the oil pump inner rotor.

    c. If the positional relationship does not allow the insertion, rotate the oil pump inner rotor to allow the oil pump drive spacer to be inserted.

31. Install the front oil seal using suitable tool.

✷✷ WARNING

Do not scratch or make burrs on the circumference of the oil seal.

32. Install the chain tensioner cover.

33. Install the front cover as follows:

    a. Install a new O-ring on the cylinder block.

    b. Apply liquid gasket as shown.

    c. Check again that the timing alignment marks on the timing chain and on each sprocket are aligned. Then install the front cover.

    d. Install the bolts in the numerical order shown.

    e. After tightening, re-tighten to the specified torque.

✷✷ WARNING

Be sure to wipe off any excessive liquid gasket leaking onto surface mating with oil pan.

34. Install the Intake valve control solenoid valve cover (RH) (A) and Intake valve control solenoid valve cover (LH) (B) as follows:

    a. Cross mark (C) that cannot be seen after assembly.

    b. Apply liquid gasket (D) as shown.

✷✷ WARNING

The start and end of the application of the liquid gasket should be crossed at a position that cannot be seen after attaching the Intake valve control solenoid valve cover.

    c. Install the bolts in the numerical order shown.

35. Install the crankshaft pulley.

    a. Install the key of the crankshaft.

    b. Insert the pulley by lightly tapping it.

✷✷ WARNING

Do not tap pulley on the side surface where the belt is installed (outer circumference).

36. Tighten the crankshaft pulley bolt.

    a. Lock the crankshaft using suitable tool, then tighten the bolt.

    b. Perform the following steps for angular tightening:

    c. Apply engine oil onto the threaded parts of the bolt and seating area.

    d. Select the one most visible notch of the four on the bolt flange. Corresponding to the selected notch, put a alignment mark (such as paint) on the crankshaft pulley.

37. Rotate the crankshaft pulley in normal direction (clockwise when viewed from engine front) to check for parts interference.

38. Installation of the remaining components is in the reverse of order of removal.

39. Be sure to perform the reconnect/relearn procedures.

VALVE CYLINDER HEAD COVERS

REMOVAL & INSTALLATION

V6 Engine

Left Side

See Figures 139 through 141.

1. Before servicing the vehicle, refer to the Precautions.

2. Disconnect the negative battery cable.

3. Remove the air cleaner assembly.

4. Separate engine harness removing their brackets from rocker covers.

5. Remove harness bracket from cylinder head, if necessary.

6. Disconnect and remove the intake valve timing control solenoid valve.

7. Remove ignition coil.

8. Remove PCV hoses from rocker cover.

9. Disconnect and remove the VTC solenoid.

10. Remove oil filler cap from rocker cover (LH), if necessary.

11. Loosen bolts in reverse order of the tightening sequence.

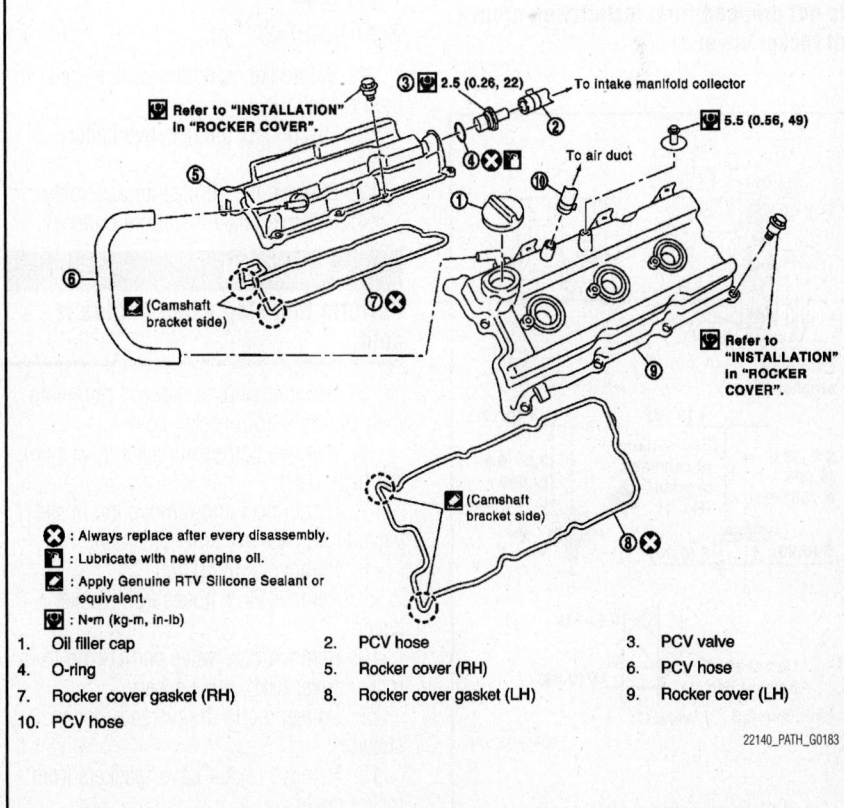

⊗ : Always replace after every disassembly.
▣ : Lubricate with new engine oil.
◨ : Apply Genuine RTV Silicone Sealant or equivalent.
▧ : N•m (kg-m, in-lb)

1. Oil filler cap	2. PCV hose	3. PCV valve
4. O-ring	5. Rocker cover (RH)	6. PCV hose
7. Rocker cover gasket (RH)	8. Rocker cover gasket (LH)	9. Rocker cover (LH)
10. PCV hose		

22140_PATH_G0183

Fig. 139 Valve covers and related components—V6 engine

12. Remove the retaining bolts. Discard the gasket.

13. Remove rocker cover gaskets from rocker covers.

14. Use scraper to remove all traces of liquid gasket from cylinder head and camshaft bracket (No. 1).

※※ WARNING

Do not scratch or damage the mating surface when cleaning off old liquid gasket.

To install:

➡**Be sure to use new fasteners, as required.**

15. Apply liquid gasket to joint part among rocker cover, cylinder head and camshaft bracket (No. 1) as follows:

    a. The figure shows an example of LH side [zoomed in shows camshaft bracket (No. 1)].

    b. Apply liquid gasket to joint part of camshaft bracket "a" (No. 1) and cylinder head.

    c. Apply liquid gasket "b" to the figure "a" squarely.

16. Install new rocker cover gasket to rocker cover.

17. Install rocker cover.

➡**Check to be sure rocker cover gasket is not dropped from installation groove of rocker cover.**

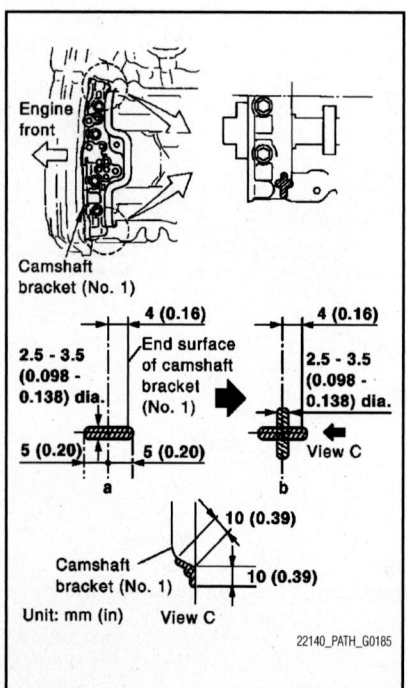

Fig. 140 Valve cover sealant application—V6 engine

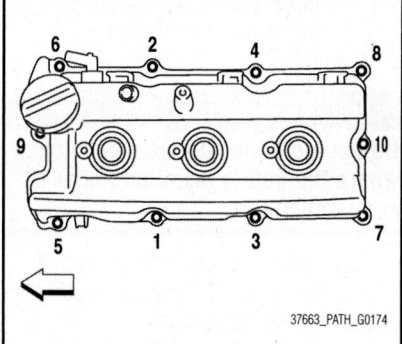

37663_PATH_G0174

Fig. 141 Valve cover bolt tightening sequence (left)—V6 engine

18. Tighten bolts in two steps in numerical order as shown.

    a. 1st step: 17 inch lbs. (1.96 Nm)

    b. 2nd step: 74 inch lbs. (8.33 Nm)

19. Install oil filer cap to rocker cover (left bank), if removed.

20. Install PCV hose.

    a. Insert PCV hose by 0.98 to 1.18 in. (25 to 30 mm) from connector end.

    b. When installing, be careful not to twist or come in contact with other parts.

21. Installation of the remaining components is in the reverse order of removal.

22. Be sure to perform the reconnect/relearn procedures.

Right Side

See Figure 142.

1. Before servicing the vehicle, refer to the Precautions.

2. Disconnect the negative battery cable.

3. Remove the air cleaner assembly.

4. Remove intake manifold collector.

※※ CAUTION

Perform this step when engine is cold.

5. Separate engine harness removing their brackets from rocker covers.

6. Remove harness bracket from cylinder head (RH).

7. Disconnect and remove the intake valve timing control solenoid valve.

8. Remove ignition coil.

9. Remove PCV hoses from rocker cover.

10. Remove PCV valve and O-ring from rocker cover (RH), if necessary.

11. Loosen bolts in reverse order as shown.

12. Remove rocker cover gaskets from rocker covers.

13. Use scraper to remove all traces of

liquid gasket from cylinder head and camshaft bracket (No. 1).

※※ WARNING

Do not scratch or damage the mating surface when cleaning off old liquid gasket.

To install:

➡**Be sure to use new fasteners, as required.**

14. Apply liquid gasket to joint part among rocker cover, cylinder head and camshaft bracket (No. 1) as follows:

    a. Apply liquid gasket to joint part of camshaft bracket "a" (No. 1) and cylinder head.

    b. Apply liquid gasket "b" to the figure "a" squarely.

15. Install new rocker cover gasket to rocker cover.

16. Install rocker cover.

➡**Check to be sure rocker cover gasket is not dropped from installation groove of rocker cover.**

17. Tighten bolts in two steps in numerical order as shown.

    a. 1st step: 17 inch lbs. (1.96 Nm)

    b. 2nd step: 74 inch lbs. (8.33 Nm)

18. Install new O-ring and PCV valve to rocker cover (RH), if removed.

19. Install PCV hose.

    a. Insert PCV hose by 0.98 to 1.18 in. (25 to 30 mm) from connector end.

    b. When installing, be careful not to twist or come in contact with other parts.

20. Installation of the remaining components is in the reverse order of removal.

21. Be sure to perform the reconnect/relearn procedures.

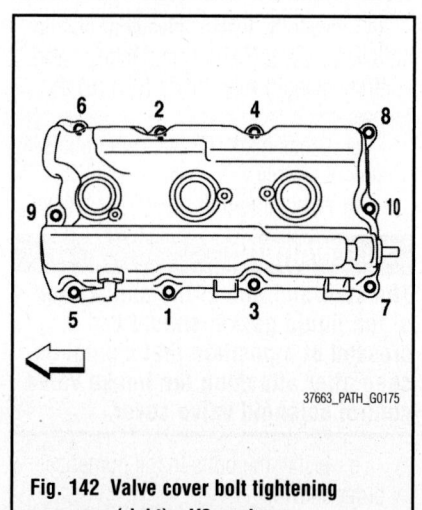

37663_PATH_G0175

Fig. 142 Valve cover bolt tightening sequence (right)—V6 engine

V8 Engine

See Figures 143 through 145.

1. Before servicing the vehicle, refer to the Precautions.

2. Disconnect the negative battery cable.

3. Remove the air duct and resonator assembly.

4. Move the harness on the upper rocker cover and its peripheral aside.

5. Remove the ignition coils.

6. Remove the PCV hose from the PCV control valves.

7. Loosen the bolts in reverse order of the tightening sequence (LH) (A) or (RH) (B).

✳✳ WARNING

Do not hold the rocker cover (RH) (B) by the oil filler neck.

To install:

➡**Be sure to use new fasteners, as required.**

8. Apply liquid gasket to the joint part of the cylinder head and camshaft bracket following the steps below.

➡**Illustration shows an example of (LH) side (zoomed in shows No. 1 camshaft bracket).**

a. Apply liquid gasket to the joint part of No. 1 camshaft bracket and cylinder head "a".

b. Apply liquid gasket 90° "b" to illustration "a".

9. Install the rocker cover (LH) (A) or (RH) (B).

a. Make sure the new rocker cover gasket is installed in the groove of the rocker cover (LH) (A) or (RH) (B).

b. Tighten the bolts in two steps in the numerical order shown.

c. 1st step: 18 inch lbs. (2.0 Nm)

d. 2nd step: 73 inch lbs. (8.3 Nm)

✳✳ WARNING

Do not hold the rocker cover (RH) (B) by the oil filler neck.

10. Install the PCV hoses.

➡**Remove foreign materials from inside the hose using compressed air.**

a. Insert PCV hose by 0.98 to 1.18 in. (25 to 30 mm) from connector end.

11. Installation of the remaining components is in the reverse order of removal.

12. Be sure to perform the reconnect/relearn procedures.

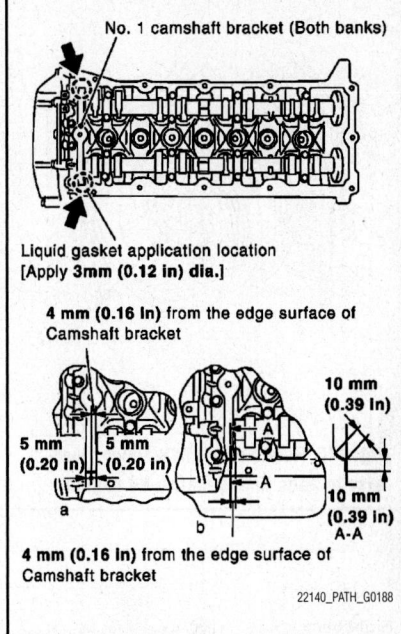

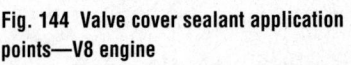

Fig. 144 Valve cover sealant application points—V8 engine

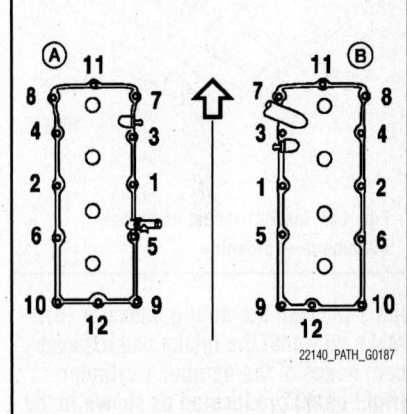

Fig. 145 Valve cover bolt tightening sequence—V8 engine

VALVE LASH

ADJUSTMENT

V6 Engine

See Figures 146 through 154.

1. Before servicing the vehicle, refer to the Precautions.

2. Disconnect the negative battery cable.

3. Remove the air cleaner assembly.

4. Remove the valve covers.

5. Position the number 1 cylinder at TDC on the compression stroke.

➡**To accomplish this, rotate the crankshaft pulley clockwise to align the timing**

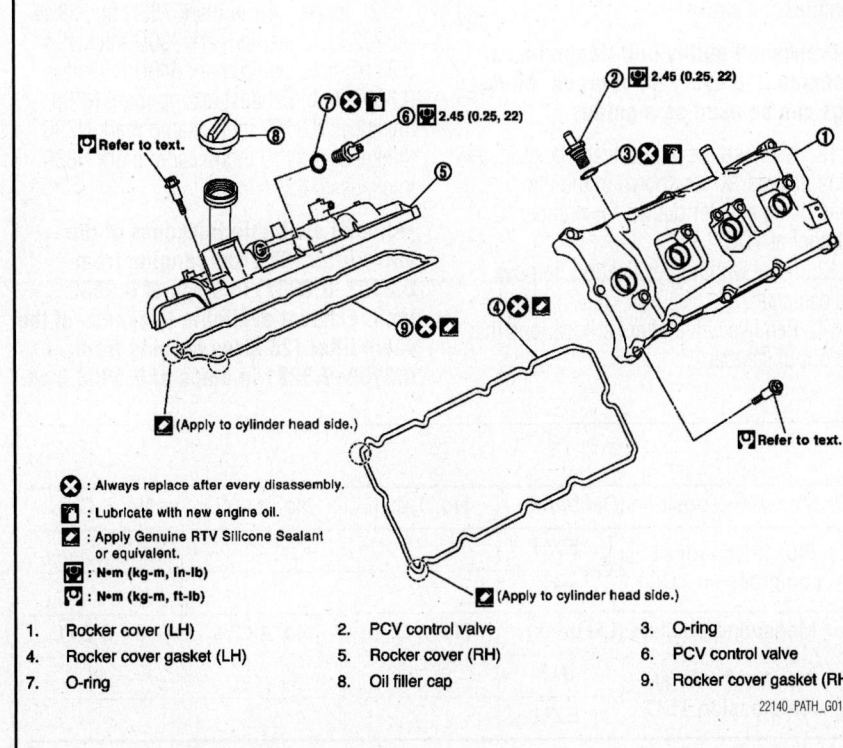

1. Rocker cover (LH)
2. PCV control valve
3. O-ring
4. Rocker cover gasket (LH)
5. Rocker cover (RH)
6. PCV control valve
7. O-ring
8. Oil filler cap
9. Rocker cover gasket (RH)

Fig. 143 Valve covers and related components—V8 engine

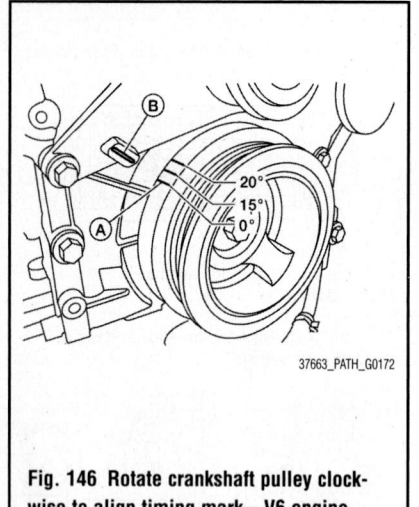

Fig. 146 Rotate crankshaft pulley clockwise to align timing mark—V6 engine

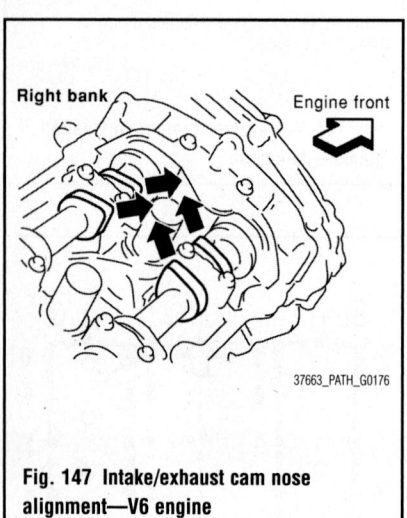

Fig. 147 Intake/exhaust cam nose alignment—V6 engine

mark (A) with the timing indicator (B). Make sure that the intake and exhaust cam noses of the number 1 cylinder (right bank) are located as shown in the illustration. If not rotate the crankshaft 360 degrees.

6. Use a feeler gauge and check the clearance between the valve lifter and camshaft.

7. Specification should be Cold: Intake 0.010–0.013, Exhaust 0.011–0.015. Hot: Intake 0.012–0.016, Exhaust 0.012–0.017.

8. Measure the valve clearance at locations marked "X" as shown in the illustrations using a feeler gauge, for number 1 cylinder at TDC.

9. Rotate the crankshaft 240 degrees clockwise (viewed from engine front) to align number 3 cylinder at TDC on the compression stroke.

➡**Crankshaft pulley bolt flange has a stamped line every 60 degrees. Markings can be used as a guide.**

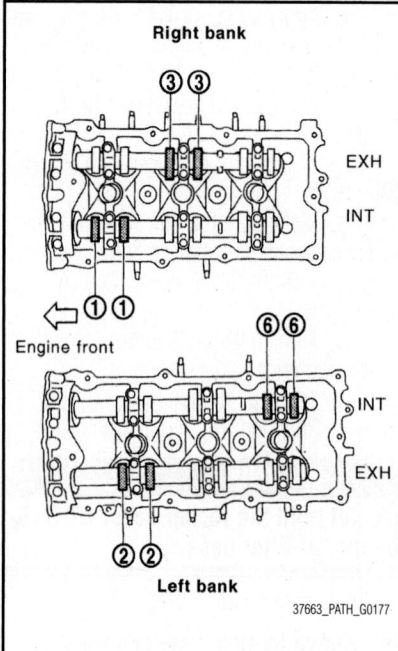

Fig. 148 Clearance measurement number 1 cylinder at TDC (1 of 2)—V6 engine

10. Measure the valve clearance at locations marked "X" as shown in the illustrations using a feeler gauge, for number 3 cylinder at TDC.

11. Rotate the crankshaft 240 degrees clockwise (viewed from engine front) to align number 5 cylinder at TDC on the compression stroke.

➡**Crankshaft pulley bolt flange has a stamped line every 60 degrees. Markings can be used as a guide.**

12. Measure the valve clearance at locations marked "X" as shown in the illustrations using a feeler gauge, for number 5 cylinder at TDC.

13. If not within specification, remove the camshaft.

14. Remove valve lifters at locations that are out of standard.

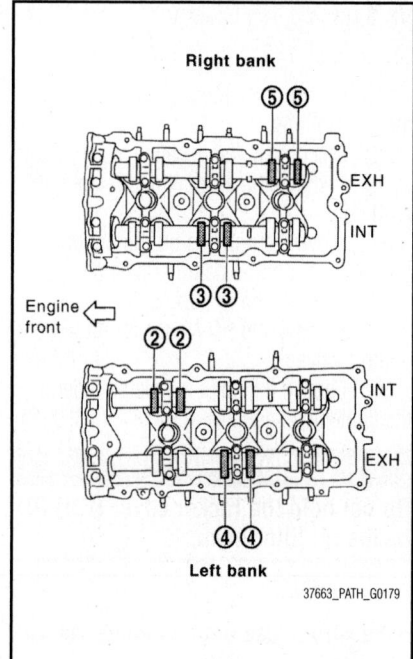

Fig. 150 Clearance measurement number 3 cylinder at TDC (1 of 2)—V6 engine

15. Measure center thickness with a micrometer. Replace lifter as required after performing replacement calculation. See illustration.

16. Thickness of the new lifter can be identified by stamp marks on the reverse side (inside the cylinder).

17. Intake: stamp mark 788U thickness 0.3102 inch, stamp mark 790U thickness 0.3110 inch, stamp mark 840U thickness 0.3307 inch. Exhaust: stamp mark N788 thickness 0.3102 inch, stamp mark N790 thickness 0.3110 inch, stamp mark N836 thickness 0.3291 inch.

➡**Intake available thickness of the valve lifter (27 sizes ranging from 0.3102–0.3307) in steps of 0.0008 inch. Exhaust available thickness of the valve lifter (25 sizes ranging from 0.3102–0.3291 in steps of 0.0008 inch.**

Measuring position (RH bank)		No. 1 CYL.	No. 3 CYL.	No. 5 CYL.
No. 1 cylinder at compression TDC	EXH		×	
	INT	×		
Measuring position (LH bank)		No. 2 CYL.	No. 4 CYL.	No. 6 CYL.
No. 1 cylinder at compression TDC	INT			×
	EXH	×		

Fig. 149 Clearance measurement number 1 cylinder at TDC (2 of 2)—V6 engine

Measuring position (RH bank)		No. 1 CYL.	No. 3 CYL.	No. 5 CYL.
No. 3 cylinder at compression TDC	EXH			×
	INT		×	
Measuring position (LH bank)		No. 2 CYL.	No. 4 CYL.	No. 6 CYL.
No. 3 cylinder at compression TDC	INT	×		
	EXH		×	

37663_PATH_G0180

Fig. 151 Clearance measurement number 3 cylinder at TDC (2 of 2)—V6 engine

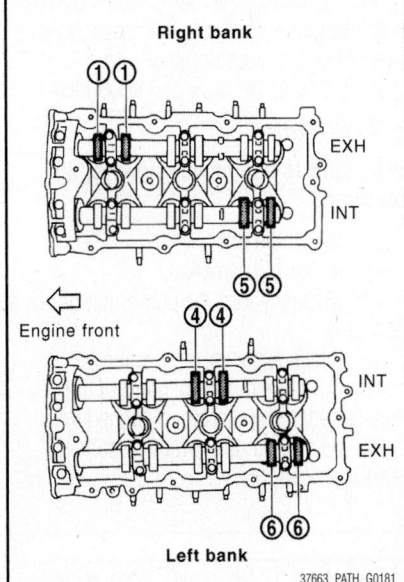

Fig. 152 Clearance measurement number 5 cylinder at TDC (1 of 2)—V6 engine

Valve lifter thickness calculation: $t = t_1 + (C_1 - C_2)$

t = Valve lifter thickness to be replaced
t_1 = Removed valve lifter thickness
C_1 = Measured valve clearance
C_2 = Standard valve clearance:

Intake : 0.26 - 0.34 mm (0.010 - 0.013 in)*
Exhaust : 0.29 - 0.37 mm (0.011 - 0.015 in)*
*: Approximately 20°C (68°F)

37663_PATH_G0183

Fig. 154 Valve lifter replacement thickness measurement

✳✳ CAUTION

Install identification letter at the end and top, "U" and "N" at each of the proper positions. Be careful of incorrect installation between intake and exhaust.

To install:

➡Be sure to use new fasteners, as required.

- Measure the valve clearances at locations marked "×" as shown in the table below (locations indicated in the illustration) with feeler gauge.
- No. 5 cylinder at compression TDC

Measuring position (RH bank)		No. 1 CYL.	No. 3 CYL.	No. 5 CYL.
No. 5 cylinder at compression TDC	EXH	×		
	INT			×
Measuring position (LH bank)		No. 2 CYL.	No. 4 CYL.	No. 6 CYL.
No. 5 cylinder at compression TDC	INT		×	
	EXH			×

37663_PATH_G0182

Fig. 153 Clearance measurement number 5 cylinder at TDC (2 of 2)—V6 engine

18. Install the lifter.
19. Install the camshaft.
20. Manually turn the crankshaft pulley a few turns.
21. Recheck the valve clearance.
22. Continue the installation in the reverse order of the removal procedure.
23. Be sure to perform the reconnect/relearn procedures.

V8 Engine

See Figures 155 through 159.

1. Before servicing the vehicle, refer to the Precautions.
2. Disconnect the negative battery cable.
3. Remove the air cleaner assembly.
4. Remove the valve covers.
5. Position the number 1 cylinder at TDC on the compression stroke.

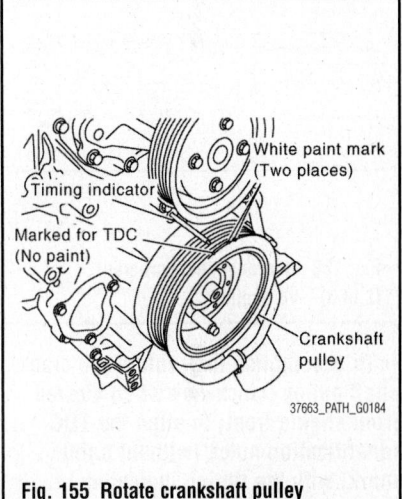

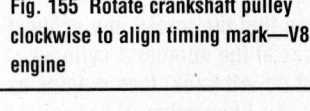

37663_PATH_G0184

Fig. 155 Rotate crankshaft pulley clockwise to align timing mark—V8 engine

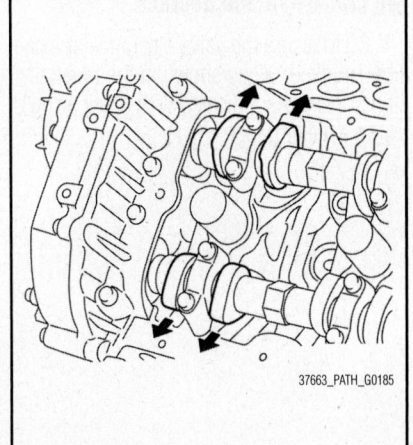

37663_PATH_G0185

Fig. 156 Intake/exhaust cam nose alignment—V8 engine

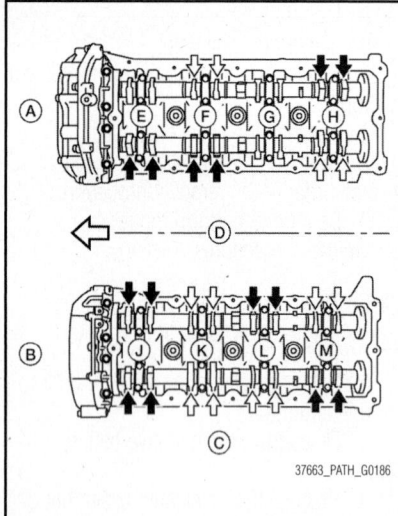

Fig. 157 Clearance measurement (1 of 3)—V8 engine

Measuring position (RH bank)		No. 2 cyl (E)	No. 4 cyl (F)	No. 6 cyl (G)	No. 8 cyl (H)
No. 1 cylinder at TDC	EXH				×
	INT	×	×		
Measuring position (LH bank)		No. 1 cyl (J)	No. 3 cyl (K)	No. 5 cyl (L)	No. 7 cyl (M)
No. 1 cylinder at TDC	INT	×		×	
	EXH	×			×

37663_PATH_G0187

Fig. 158 Clearance measurement (2 of 3)—V8 engine

➡ To accomplish this, rotate the crankshaft pulley (clockwise when viewed from engine front) to align the TDC identification notch (without paint mark) with the timing indicator.). Make sure that the intake and exhaust cam noses of the number 1 cylinder (top front on left bank) face outside as shown in the illustration. If not rotate the crankshaft 360 degrees.

6. Measure the valve clearance at locations marked "X" as shown in the illustrations (locations indicated with black arrow), using a feeler gauge.

➡ White arrow: engine front. Black arrow: measurable at number 1 cylinder at TDC compression stroke. White arrow: measurable at number 3 cylinder at TDC compression stroke. A: right side. B: left side. C: exhaust. D: intake.

7. Rotate the crankshaft 270 degrees clockwise to align number 3 cylinder at TDC on the compression stroke.

8. Measure the valve clearance at locations marked "X" as shown in the illustrations (locations indicated with black arrow), using a feeler gauge.

➡ White arrow: engine front. Black arrow: measurable at number 1 cylinder at TDC compression stroke. White arrow: measurable at number 3 cylinder at TDC compression stroke. A: right side. B: left side. C: exhaust. D: intake.

9. Rotate the crankshaft pulley 90 degrees from the position of number 3 cylinder at TDC on the compression stroke (clockwise by 360 degrees from the position of number 1 cylinder at TDC on the compression stroke) to measure the intake and exhaust valve clearances of the number 6 cylinder and the exhaust valve clearance of the number 2 cylinder.

➡ White arrow: engine front. A: right side. B: left side. C: exhaust. D: intake.

E: number 2 cylinder. F: number 4 cylinder. G: number 6 cylinder. H: number 8 cylinder. J: number 1 cylinder. K: number 3 cylinder. L: number 5 cylinder. M: number 7 cylinder.

10. If not within specification, remove the camshaft.

11. Remove valve lifters at locations that are out of standard.

12. Measure center thickness with a micrometer. Replace lifter as required after performing replacement calculation. See illustration.

13. Thickness of the new lifter can be identified by stamp marks on the reverse side (inside the cylinder).

14. Stamp mark N788 indicates 0.3102 inch thickness. Available thickness of the valve lifter (25 sizes ranging from 0.3102–0.3291 in steps of 0.0008 inch.

To install:

➡ Be sure to use new fasteners, as required.

15. Install the lifter.
16. Install the camshaft.
17. Manually turn the crankshaft pulley a few turns.
18. Recheck the valve clearance.
19. Continue the installation in the reverse order of the removal procedure.
20. Be sure to perform the reconnect/relearn procedures.

Measuring position (RH bank)		No. 2 cyl (E)	No. 4 cyl (F)	No. 6 cyl (G)	No. 8 cyl (H)
No. 3 cylinder at TDC	EXH		×		
	INT				×
Measuring position (LH bank)		No. 1 cyl (J)	No. 3 cyl (K)	No. 5 cyl (L)	No. 7 cyl (M)
No. 3 cylinder at TDC	INT		×		×
	EXH		×	×	

37663_PATH_G0189

Fig. 159 Clearance measurement (3 of 3)—V8 engine

ENGINE PERFORMANCE & EMISSION CONTROLS

COMPONENT LOCATIONS

See Figures 160 through 169.

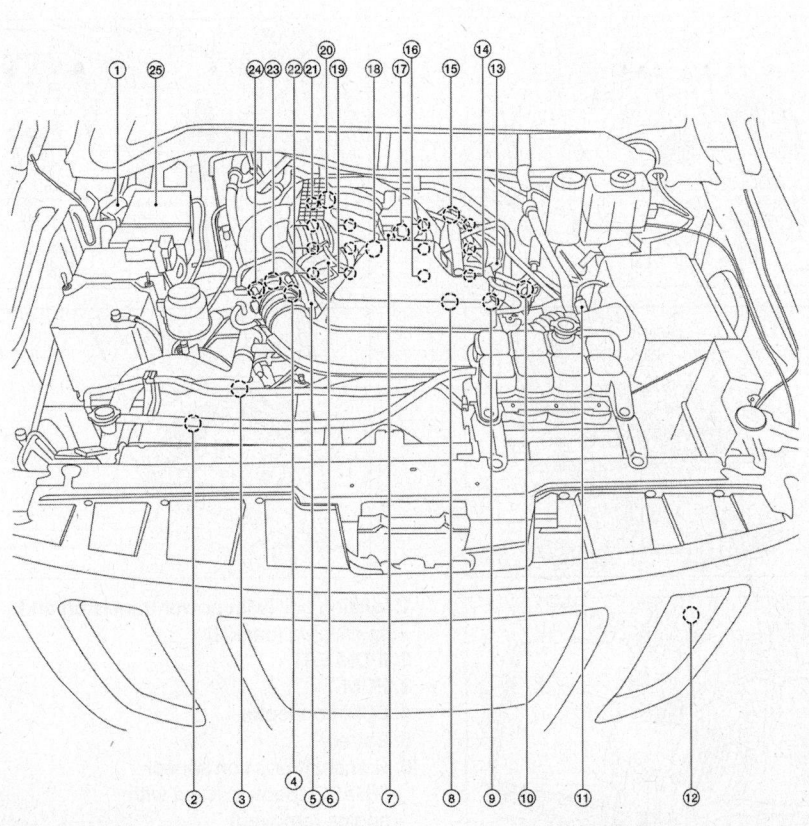

1. ECM
2. Cooling fan motor
3. Power steering pressure sensor
4. Intake valve timing control solenoid valve (bank 1)
5. Electric throttle control actuator
6. Power valve actuator
7. VIAS control solenoid valve
8. EVAP canister purge volume control solenoid valve
9. Intake valve timing control solenoid valve (bank 2)
10. Air fuel ratio (A/F) sensor 1 (bank 2)
11. Mass air flow sensor (with intake air temperature sensor)
12. Refrigerant pressure sensor
13. EVAP service port
14. Ignition coil (with power transistor) and spark plug (bank 2)
15. Camshaft position sensor (PHASE) (bank 2)
16. Fuel injector (bank 2)
17. Knock sensor (bank 2)
18. Knock sensor (bank 1)
19. Fuel injector (bank 1)
20. Engine coolant temperature sensor
21. Camshaft position sensor (PHASE) (bank 1)
22. Ignition coil (with power transistor) and spark plug (bank 1)
23. Crankshaft position sensor (POS)
24. Air fuel ratio (A/F) sensor 1 (bank 1)
25. IPDM E/R

71075_NV15_G0064

Fig. 160 Electronic engine control system component locations (1 of 5)—V6

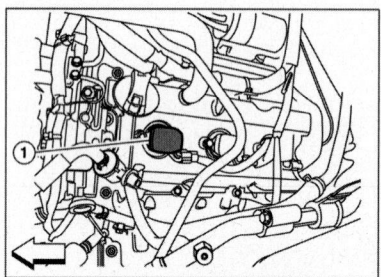
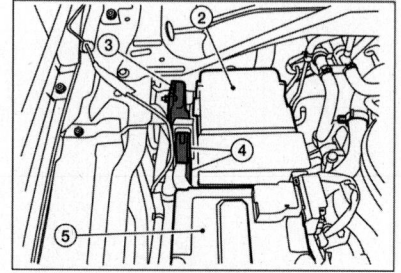
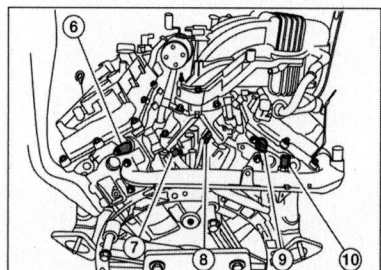
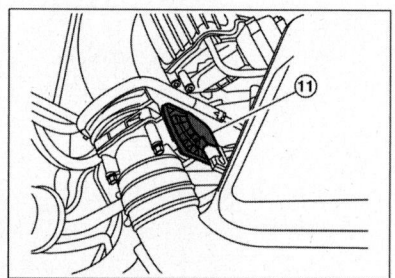
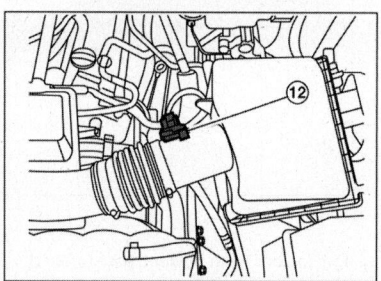

1. Ignition coil (with power transistor) and spark plug (bank 2)
2. IPDM E/R
3. ECM
4. ECM connector
5. Battery
6. Camshaft position sensor (PHASE) (bank 2) (view with engine removed)
7. Knock sensor (bank 2)
8. Knock sensor (bank 1)
9. Camshaft position sensor (PHASE) (bank 1)
10. Engine coolant temperature sensor
11. Electric throttle control actuator
12. Mass air flow sensor (with intake air temperature sensor)

71075_NV15_G0065

Fig. 161 Electronic engine control system component locations (2 of 5)—V6 engine

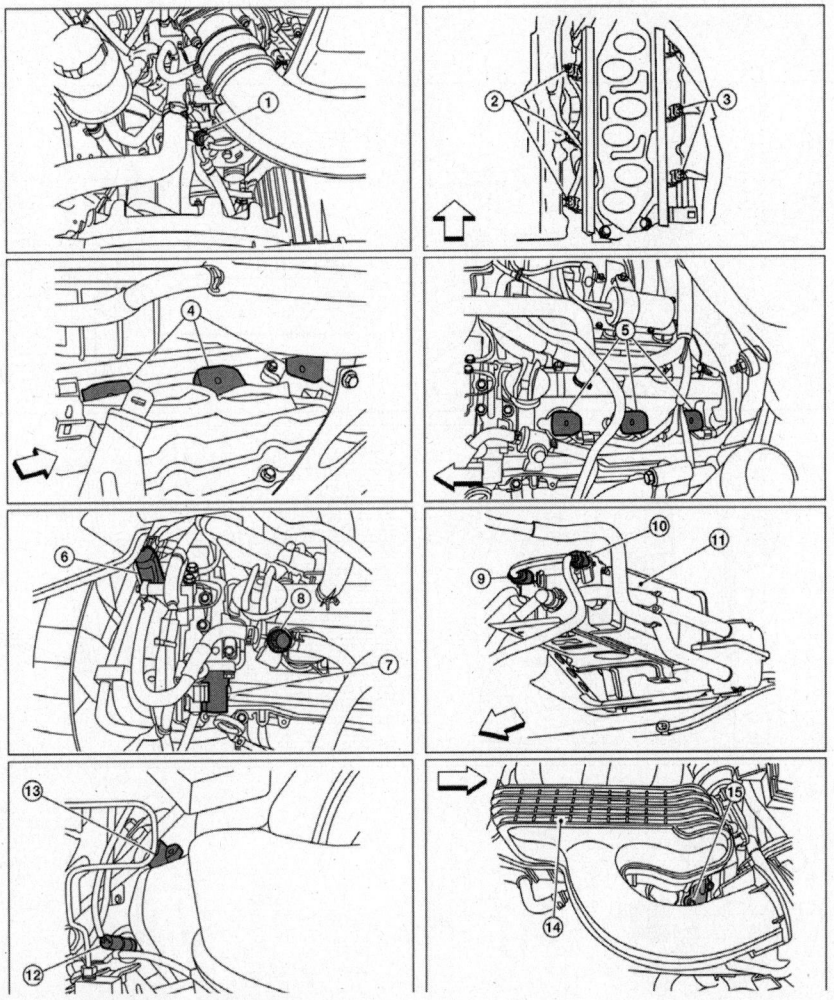

1. Power steering pressure sensor
2. Injector harness connector (bank 2) (view with intake manifold collector removed)
3. Injector harness connector (bank 1) (view with intake manifold collector removed)
4. Ignition coils (bank 1) (with power transistor)
5. Ignition coils (bank 2) (with power transistor)
6. EVAP canister purge volume control solenoid valve
7. Intake valve timing control solenoid valve (bank 2)
8. EVAP service port
9. EVAP control system pressure sensor
10. EVAP canister vent control valve
11. EVAP canister
12. Air fuel ratio (A/F) sensor (bank 1) (view through fender cover RH)
13. Crankshaft position sensor (POS)
14. Intake manifold collector
15. Intake valve timing control solenoid valve (bank 1)
ARROW: Vehicle front

71075_NV15_G0066

Fig. 162 Electronic engine control system component locations (3 of 5)—V6 engine

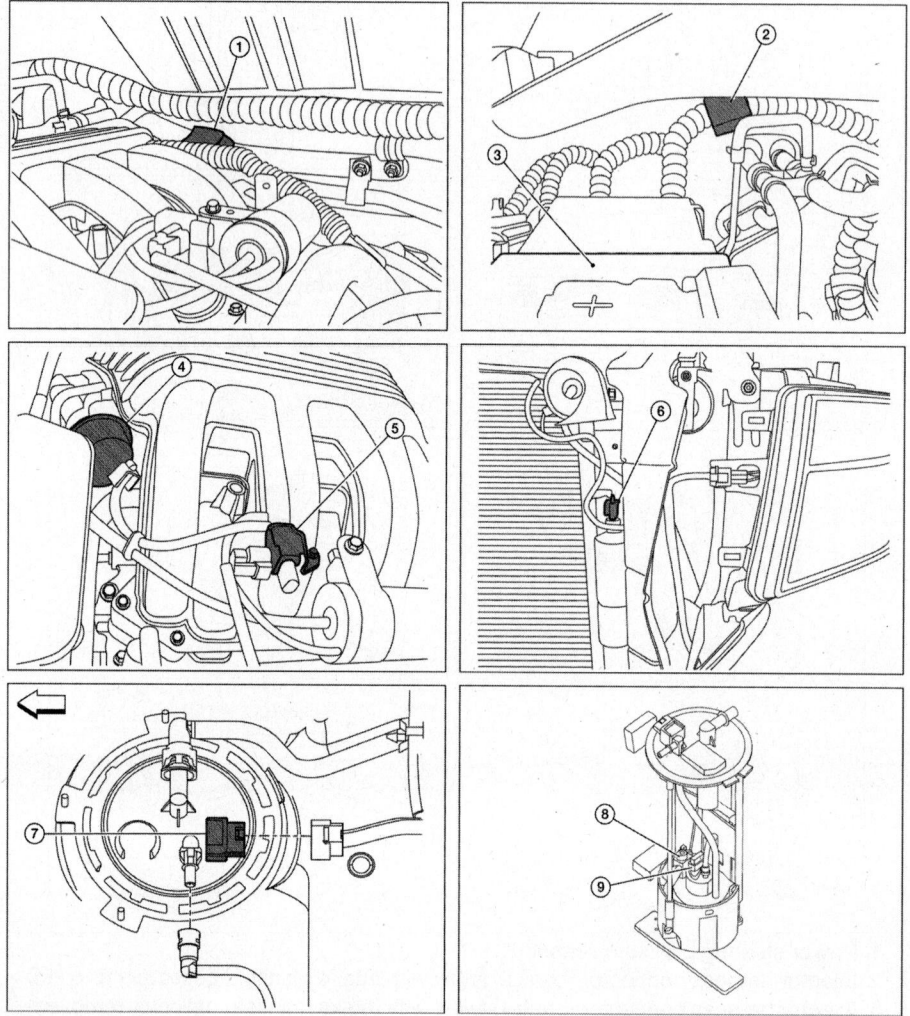

1. Condenser-1
2. Condenser-2
3. Battery
4. Power valve actuator
5. VIAS control solenoid valve
6. Refrigerant pressure sensor (view with front grille removed)
7. Fuel level sensor unit and fuel pump harness connector (view with fuel tank removed)
8. Fuel pressure regulator
9. Fuel pump, fuel level sensor unit and fuel filter
ARROW: Vehicle front

71075_NV15_G0067

Fig. 163 Electronic engine control system component locations (4 of 5)—V6 engine

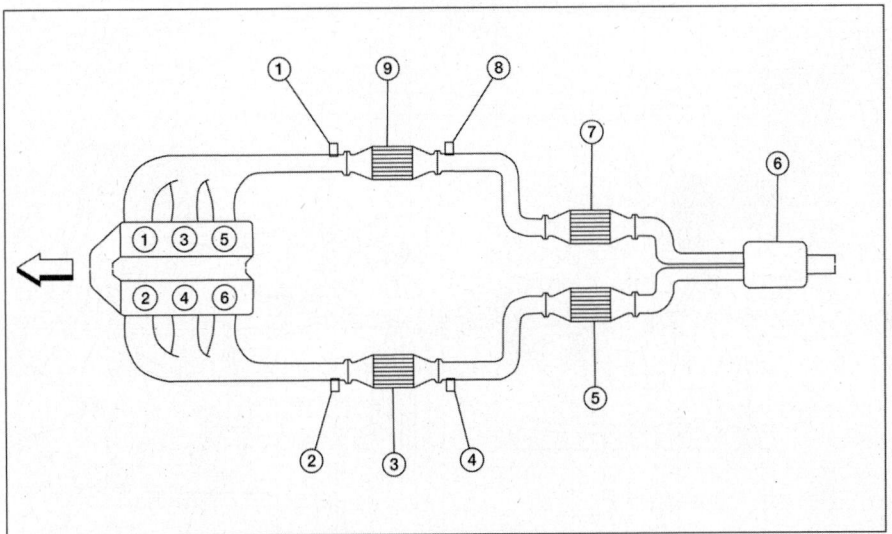

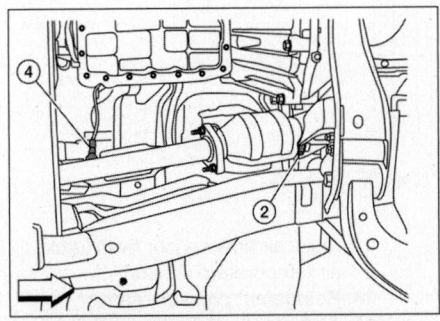

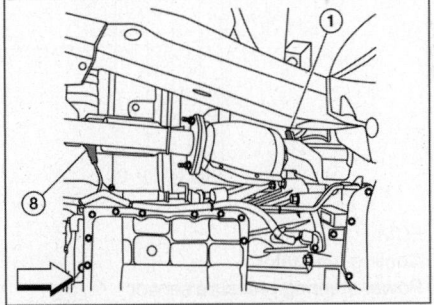

1. Air fuel ratio (A/F) sensor 1 (bank 1)
2. Air fuel ratio (A/F) sensor 1 (bank 2)
3. Three-way catalyst (manifold) (bank 2)
4. Heated oxygen sensor 2 (bank 2)
5. Three-way catalyst (under floor) (bank 2)

6. Muffler
7. Three-way catalyst (under floor) (bank 1)
8. Heated oxygen sensor 2 (bank 1)
9. Three-way catalyst (manifold) (bank 1)
ARROW: Vehicle front

71075_NV15_G0068

Fig. 164 Electronic engine control system component locations (5 of 5)—V6 engine

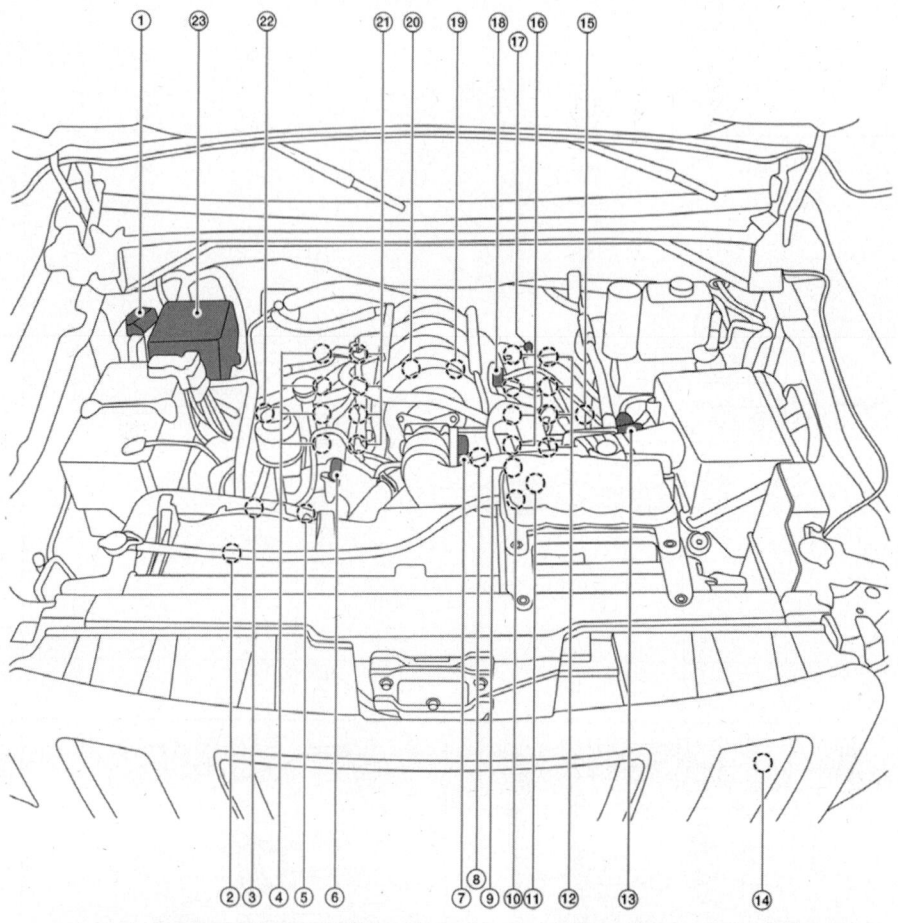

1. ECM
2. Cooling fan motor
3. Power steering pressure sensor
4. Ignition coil (with power transistor) (bank 2)
5. Intake valve timing control solenoid valve (bank 2)
6. Intake valve timing control position sensor (bank 2)
7. Electric throttle control actuator
8. Engine coolant temperature sensor
9. Intake valve timing control position sensor (bank 1)
10. Intake valve timing control solenoid valve (bank 1)
11. Camshaft position sensor (PHASE)
12. Ignition coil (with power transistor) (bank 1)

13. Mass air flow sensor (with intake air temperature sensor)
14. Refrigerant pressure sensor
15. Air fuel ratio (A/F) sensor 1 (bank 1)
16. Fuel injector (bank 1)
17. EVAP service port
18. EVAP canister purge volume control solenoid valve
19. Knock sensor (bank 1)
20. Knock sensor (bank 2)
21. Fuel injector (bank 2)
22. Air fuel ratio (A/F) sensor 1 (bank 2)
23. IPDM E/R

71075_NV15_G0070

Fig. 165 Electronic engine control system component locations (1 of 5)—V8 engine

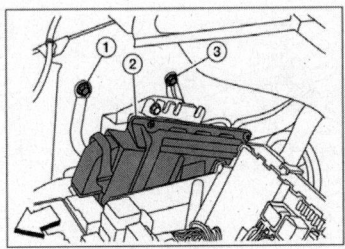
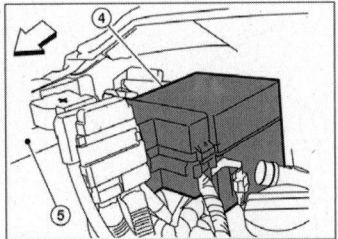
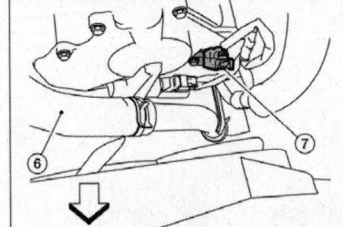
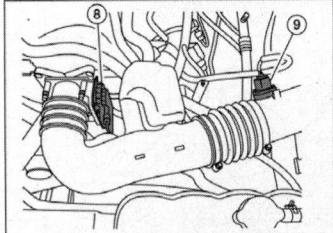
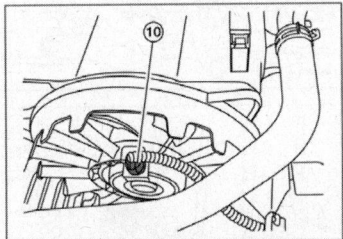
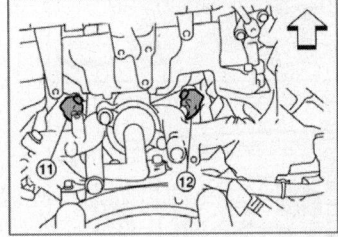
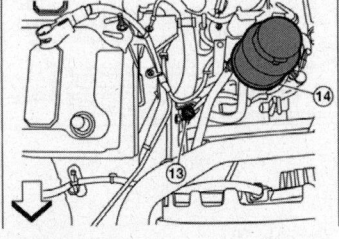

1. Body ground
2. ECM
3. Body ground
4. IPDM E/R
5. Battery
6. Radiator hose
7. Camshaft position sensor (PHASE)
8. Electric throttle control actuator
9. Mass air flow sensor (with intake air temperature sensor)
10. Cooling fan motor harness connector
11. Knock sensor (bank 1) (view with engine removed)
12. Knock sensor (bank 2) (view with engine removed)
13. Power steering pressure sensor (view with battery disconnected)
14. Power steering fluid reservoir
ARROW: Vehicle front

71075_NV15_G0071

Fig. 166 Electronic engine control system component locations (2 of 5)—V8 engine

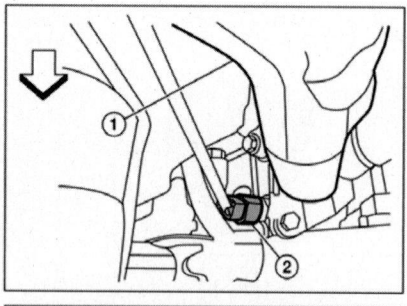
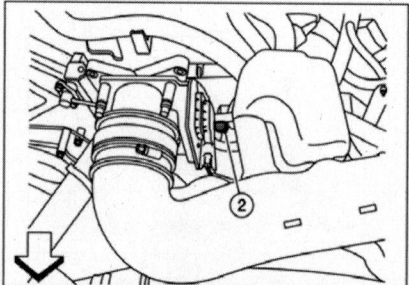
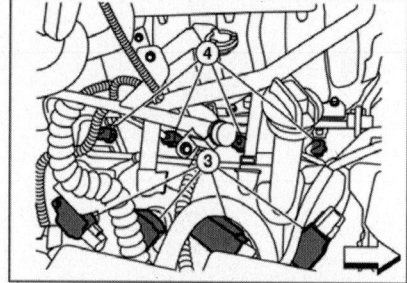
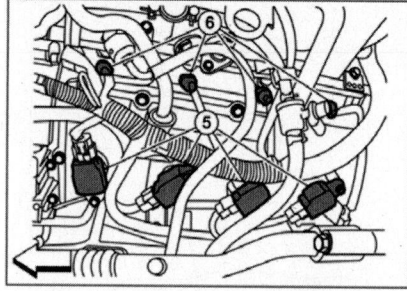
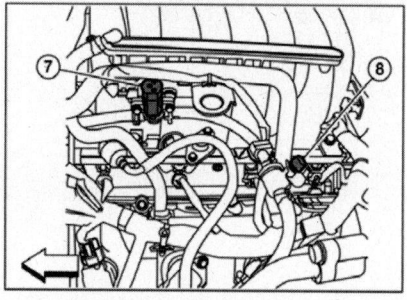
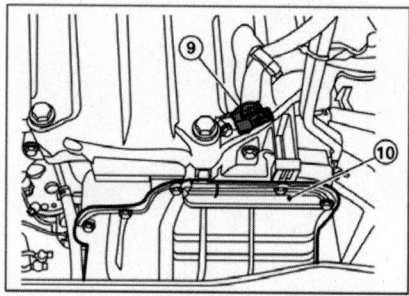

1. Intake manifold
2. Engine coolant temperature sensor
3. Ignition coils (with power transistor) (bank 2)
4. Injector harness connectors (bank 2)
5. Ignition coils (with power transistor) (bank 1)
6. Injector harness connectors (bank 1)
7. EVAP canister purge volume control solenoid valve
   (view with engine cover removed)
8. EVAP service port
9. Crankshaft position sensor (POS) (view from under the vehicle)
10. Engine oil pan (view from under the vehicle)
ARROW: Vehicle front

71075_NV15_G0072

Fig. 167 Electronic engine control system component locations (3 of 5)—V8 engine

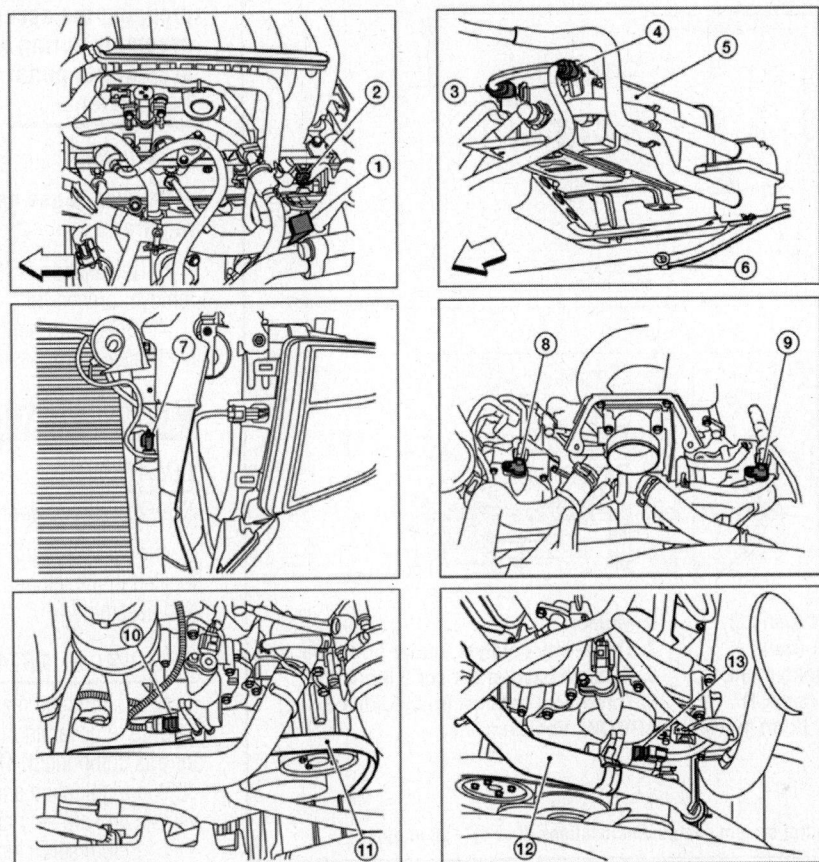

1. Condenser-1
2. Injector #7
3. EVAP control system pressure sensor
4. EVAP canister vent control valve
5. EVAP canister
6. Rear suspension member
7. Refrigerant pressure sensor (view with front grille removed)
8. Intake valve timing control position sensor (bank 2)
  (view with engine cover and intake air duct removed)
9. Intake valve timing control position sensor (bank 1)
10. Intake valve timing control solenoid valve (bank 2)
  (view with engine cover and intake air duct removed)
11. Drive belt
12. Radiator hose (view with engine cover and intake air duct removed)
13. Intake valve timing control solenoid valve (bank 1)
ARROW: Vehicle front

71075_NV15_G0073

Fig. 168 Electronic engine control system component locations (4 of 5)—V8 engine

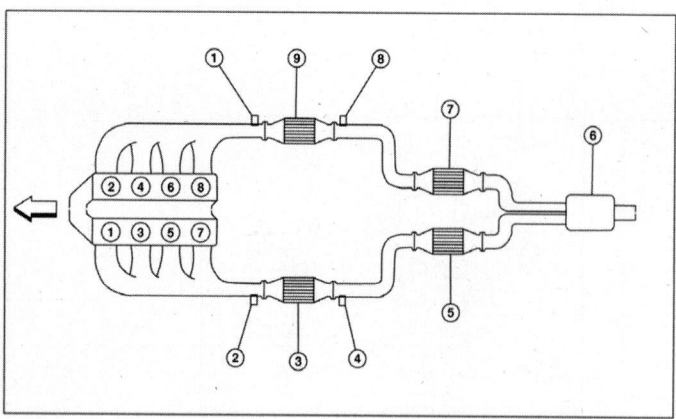

1. Air fuel ratio (A/F) sensor 1 (bank 2)
2. Air fuel ratio (A/F) sensor 1 (bank 1)
3. Three-way catalyst (manifold) (bank 1)
4. Hearted oxygen sensor 2 (bank 1)
5. Three-way catalyst (under floor) (bank 1)
6. Muffler
7. Three-way catalyst (under floor) (bank 2)
8. Hearted oxygen sensor 2 (bank 2)
9. Three-way catalyst (manifold) (bank 2)
ARROW: Vehicle front

71075_NV15_G0074

Fig. 169 Electronic engine control system component locations (5 of 5)—V8 engine

ACCELERATOR PEDAL POSITION (APP) SENSOR

LOCATION

The Accelerator Pedal Position (APP) sensor is installed on the upper end of the accelerator pedal assembly.

REMOVAL & INSTALLATION

See Figure 170.

1. Before servicing the vehicle, refer to the Precautions.
2. Disconnect the negative battery cable.
3. Disconnect the Accelerator Pedal Position (APP) sensor electrical connector.
  a. Pull the connector lock back to unlock the connector from the APP sensor.
  b. Pull up on the connector to disconnect it from the APP sensor.
4. Remove the two upper and one lower accelerator pedal nuts.
5. Remove the accelerator pedal assembly.

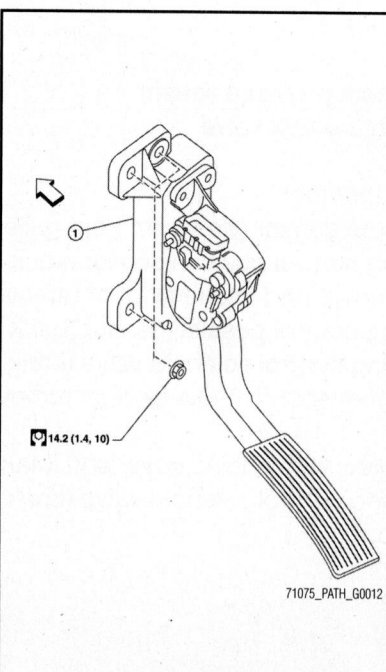

14.2 (1.4, 10)

71075_PATH_G0012

Fig. 170 Accelerator pedal and related components (1) assembly

> ✳✳ **CAUTION**
> Do not disassemble the accelerator pedal assembly. Do not remove the APP sensor from the accelerator pedal bracket. Avoid damage from dropping the accelerator pedal assembly during handling. Keep the accelerator pedal assembly away from water.

To install:

➡ Be sure to use new fasteners, as required.

6. Installation is the reverse of the removal procedure.
7. Be sure to perform the reconnect/relearn procedures.

AIR-FUEL RATIO (A/F) SENSOR

LOCATION

See Figures 171 and 172.

The Air Fuel Ratio (A/F) sensors are located in the left and right exhaust manifold assembly.

REMOVAL & INSTALLATION

At this time the manufacturer does not provide removal and installation procedures for this component. The following procedure is a guideline and may differ from the vehicle you are servicing.

1. Before servicing the vehicle, refer to the Precautions.
2. Disconnect the negative battery cable.
3. Remove the air cleaner assembly, as required.
4. Remove engine undercover, if equipped and required..
5. Raise and support the vehicle, as required.
6. Disconnect harness connector.
7. Remove the sensor from its mounting.

To install:

➡ Be sure to use new fasteners, as required.

8. Installation is the reverse of the removal procedure.
9. Be sure to perform the reconnect/relearn procedures.

CAMSHAFT POSITION (CMP) SENSOR

LOCATION

Camshaft Position (CMP) sensors are located on the right and left bank cylinder heads at the back side.

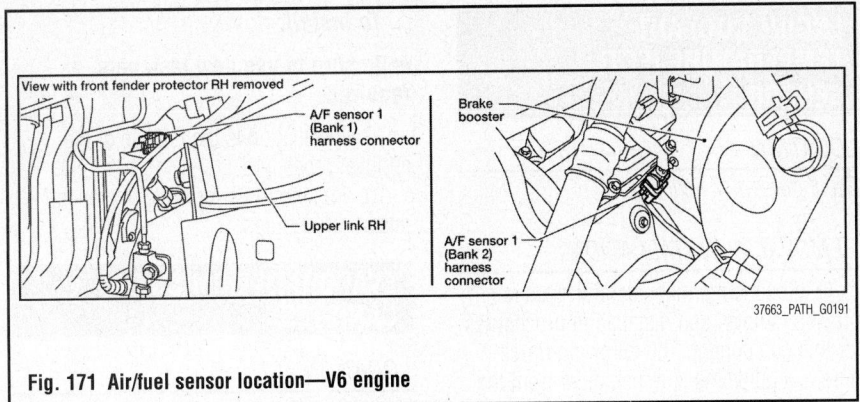

Fig. 171 Air/fuel sensor location—V6 engine

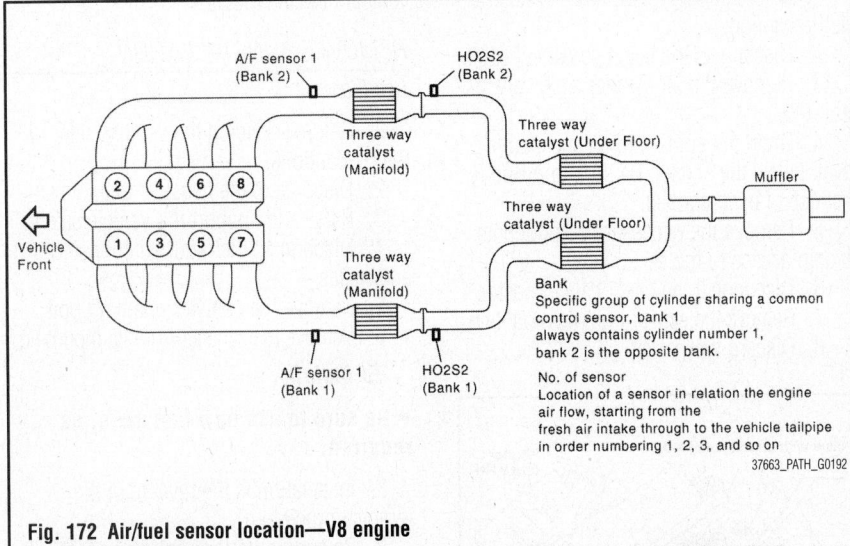

Fig. 172 Air/fuel sensor location—V8 engine

REMOVAL & INSTALLATION

1. Before servicing the vehicle, refer to the Precautions.

2. Disconnect the negative battery cable.

3. Remove air intake duct, as required.

4. Loosen the fixing bolt of the sensor.

5. Disconnect the electrical connector.

6. Remove the bolt securing the sensor.

To install:

➡ **Be sure to use new fasteners, as required.**

7. Installation is the reverse of the removal procedure.

8. Be sure to perform the reconnect/ relearn procedures.

CRANKSHAFT POSITION (CKP) SENSOR

LOCATION

The Crankshaft Position (CKP) sensor is located on the A/T assembly facing the gear teeth (cogs) of the signal plate.

REMOVAL & INSTALLATION

See Figure 173.

1. Before servicing the vehicle, refer to the Precautions.

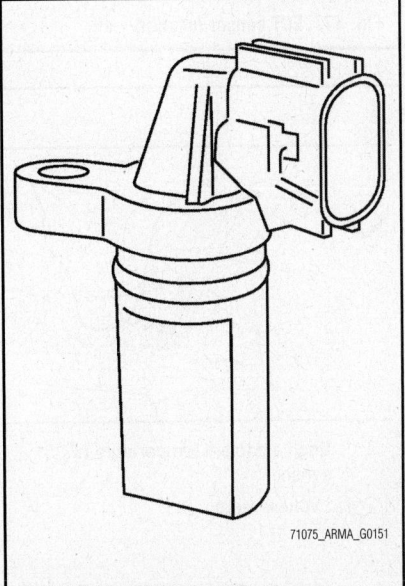

Fig. 173 Crankshaft position sensor

2. Disconnect the negative battery cable.

3. Raise and support the vehicle safely.

4. Loosen the fixing bolt of the sensor.

5. Disconnect Crankshaft Position (CKP) sensor (POS) harness connector.

6. Remove the sensor.

7. Visually check the sensor for chipping.

To install:

➡ **Be sure to use new fasteners, as required.**

8. Installation is the reverse of the removal procedure.

9. Be sure to perform the reconnect/ relearn procedures.

ELECTRONIC CONTROL MODULE (ECM)

LOCATION

The Electronic Control Module (ECM) is located in the engine room passenger side near the battery .

REMOVAL & INSTALLATION

See Figures 174 through 176.

At this time the manufacturer does not provide removal and installation procedures for this component. The following procedure is a guideline and may differ from the vehicle you are servicing.

1. Before servicing the vehicle, refer to the Precautions.

2. Disconnect the negative battery cable.

3. Remove the air cleaner assembly, if required.

4. Remove the necessary components to gain access to the ECM.

5. Disconnect the electrical connectors.

6. Remove the ECM retaining screws/bolts/clips etc.

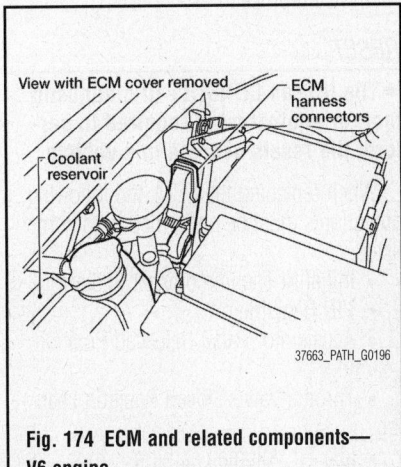

Fig. 174 ECM and related components— V6 engine

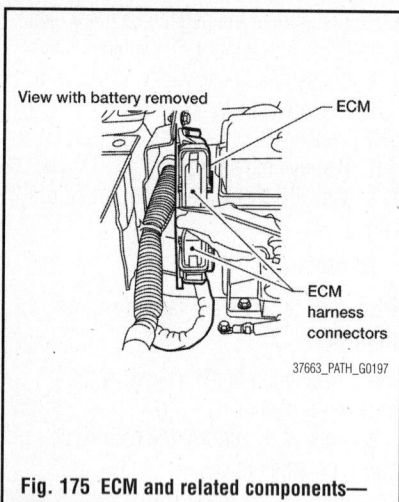

Fig. 175 ECM and related components—V8 engine

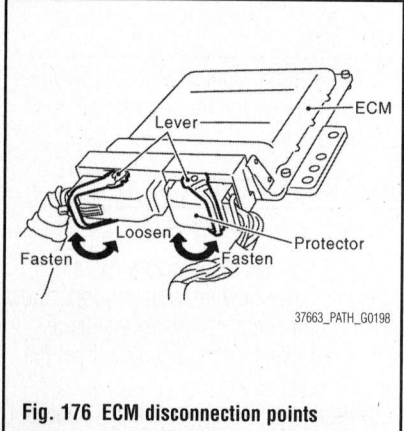

Fig. 176 ECM disconnection points

7. Remove the component from its mounting.

To install:

➡**Be sure to use new fasteners, as required.**

8. Installation is the reverse of the removal procedure.

9. Be sure to perform the reconnect/relearn procedures.

RESET

➡**The Nissan CONSULT III diagnostic tool, or equivalent is required to perform the resets used on this vehicle.**

When replacing the ECM, the following procedures must be performed in the order listed.

- Initialize The Immobilizer System
- VIN Registration
- Accelerator Pedal Released Position Learning
- Throttle Valve Closed Position Learning
- Idle Air Volume Learning

ENGINE COOLANT TEMPERATURE (ECT) SENSOR

LOCATION

See Figures 177 and 178.

REMOVAL & INSTALLATION

At this time the manufacturer does not provide removal and installation procedures for this component. The following procedure is a guideline and may differ from the vehicle you are servicing.

1. Before servicing the vehicle, refer to the Precautions.
2. Disconnect the negative battery cable.
3. Remove the air cleaner assembly, as required.
4. Drain the coolant to an acceptable level, below the sensor. Be sure to properly dispose of used coolant.
5. Remove the necessary components to gain access to the sensor.
6. Disconnect the electrical connector.
7. Remove the sensor from its mounting.
8. Discard the gasket.

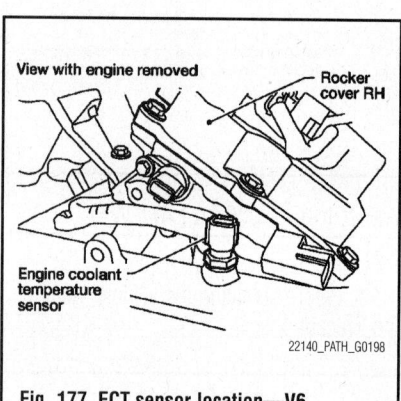

Fig. 177 ECT sensor location—V6

To install:

➡**Be sure to use new fasteners, as required.**

9. Installation is the reverse of the removal procedure.
10. Be sure to perform the reconnect/relearn procedures.

EVAPORATIVE EMISSION CANISTER

LOCATION

This component is located under the vehicle near the fuel tank.

REMOVAL & INSTALLATION

See Figure 179.

1. Before servicing the vehicle, refer to the Precautions.
2. Disconnect the negative battery cable.
3. Raise and support the vehicle safely.
4. Disconnect the required connectors and hoses.
5. Remove the canister retaining bolt.
6. Remove the canister from its mounting.

To install:

➡**Be sure to use new fasteners, as required.**

7. Installation is the reverse of the removal procedure.
8. Be sure to perform the reconnect/relearn procedures.

HEATED OXYGEN (HO2S) SENSOR

LOCATION

The Heated Oxygen (HO2S) sensors are located after the exhaust manifold converter

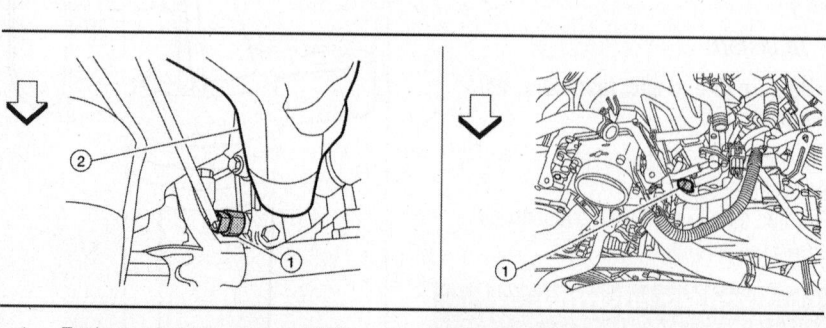

1. Engine coolant temperature (ETC) sensor
2. Intake manifold
⇦ : Vehicle front

Fig. 178 ECT sensor location—V8

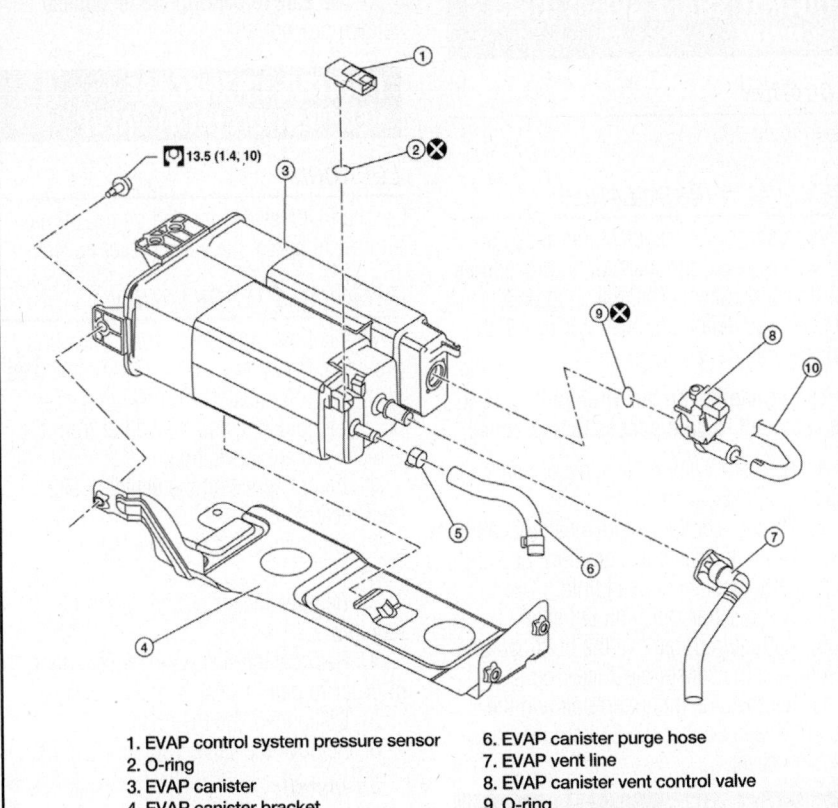

1. EVAP control system pressure sensor
2. O-ring
3. EVAP canister
4. EVAP canister bracket
5. Hose clamp
6. EVAP canister purge hose
7. EVAP vent line
8. EVAP canister vent control valve
9. O-ring
10. EVAP canister vent control valve hose

71075_NV15_G0217

Fig. 179 EVAP canister and related components

assembly, in the lower part of the exhaust system.

REMOVAL & INSTALLATION

At this time the manufacturer does not provide removal and installation procedures for this component. The following procedure is a guideline and may differ from the vehicle you are servicing.

1. Before servicing the vehicle, refer to the Precautions.
2. Disconnect the negative battery cable.
3. Remove air cleaner assembly, as required.
4. Remove engine undercover, if equipped and as required..
5. Raise and support the vehicle, as required.
6. Disconnect harness connector.
7. Remove the sensor from its mounting.

To install:

➡**Be sure to use new fasteners, as required.**

8. Installation is the reverse of the removal procedure.
9. Be sure to perform the recon-nect/relearn procedures.

INTAKE AIR TEMPERATURE (IAT)/MASS AIRFLOW (MAF) SENSOR

LOCATION

See Figure 180.

The Intake Air Temperature (IAT) sensor is built-into Mass Air Flow (MAF) sensor.

REMOVAL & INSTALLATION

1. Before servicing the vehicle, refer to the Precautions.

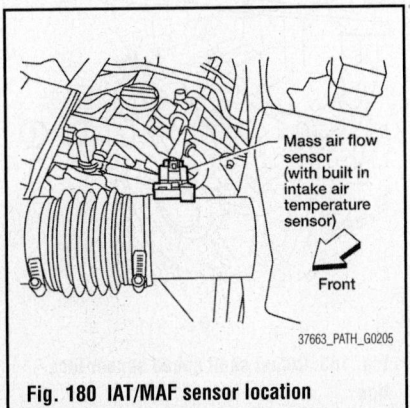

Mass air flow sensor (with built in intake air temperature sensor)

Front

37663_PATH_G0205

Fig. 180 IAT/MAF sensor location

2. Disconnect the negative battery cable.
3. Disconnect harness connector from mass air flow sensor.
4. Disconnect PCV hose.
5. Remove air cleaner case/Mass Air Flow (MAF) sensor assembly and air duct assembly disconnecting their joints. Add marks as necessary for easier installation.

✳✳ WARNING

Handle MAF sensor with care. Do not shock it. Do not disassemble it. Do not touch its sensor.

To install:

➡**Be sure to use new fasteners, as required.**

6. Installation is the reverse of the removal procedure.
7. Be sure to perform the reconnect/relearn procedures.

KNOCK SENSOR (KS)

LOCATION

See Figures 181 and 182.

The Knock (KS) sensors are mounted under the intake manifold on the cylinder block.

REMOVAL & INSTALLATION

V6 Engine

At this time the manufacturer does not provide removal and installation procedures for this component. The following procedure is a guideline and may differ from the vehicle you are servicing.

1. Before servicing the vehicle, refer to the Precautions.

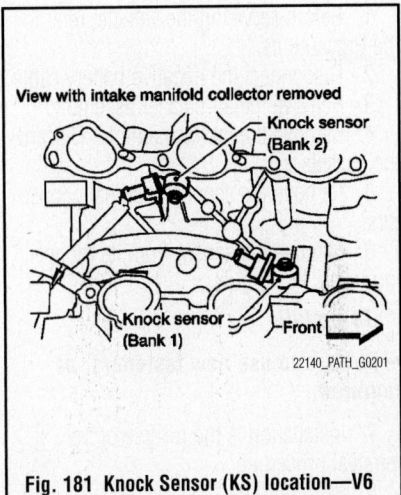

View with intake manifold collector removed

Knock sensor (Bank 2)

Knock sensor (Bank 1)

Front

22140_PATH_G0201

Fig. 181 Knock Sensor (KS) location—V6 engine

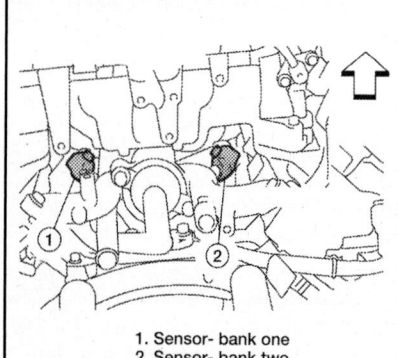

1. Sensor- bank one
2. Sensor- bank two

37663_PATH_G0206

Fig. 182 Knock Sensor (KS) location (view with engine removed from vehicle)—V8 engine

2. Disconnect the negative battery cable.
3. Remove the air cleaner assembly.
4. Remove the upper intake manifold and lower intake manifold. Discard the gaskets.
5. Remove the sensor electrical connectors.
6. Remove the sensor from its mounting.

To install:

➡**Be sure to use new fasteners, as required.**

7. Installation is the reverse of the removal procedure.
8. Be sure to perform the reconnect/relearn procedures.

V8 Engine

At this time the manufacturer does not provide removal and installation procedures for this component. The following procedure is a guideline and may differ from the vehicle you are servicing.

1. Before servicing the vehicle, refer to the Precautions.
2. Disconnect the negative battery cable.
3. Remove the air cleaner assembly.
4. Remove the intake manifold. Discard the gaskets.
5. Remove the sensor electrical connectors.
6. Remove the sensor from its mounting.

To install:

➡**Be sure to use new fasteners, as required.**

7. Installation is the reverse of the removal procedure.
8. Be sure to perform the reconnect/relearn procedures.

OUTPUT SHAFT SPEED (OSS) SENSOR

LOCATION

See Figure 183.

REMOVAL & INSTALLATION

At this time the manufacturer does not provide removal and installation procedures for this component. The following procedure is a guideline and may differ from the vehicle you are servicing.

➡**The transmission tail shaft will have to be removed to gain access to this sensor.**

1. Before servicing the vehicle, refer to the Precautions.
2. Disconnect the negative battery cable.
3. Drain the transmission fluid. Be sure to properly dispose of used fluid.
4. Remove transmission tail shaft.
5. As required, remove the fluid pan and disconnect the sensor electrical connector.
6. Remove the mounting bolt and the Output Speed Sensor (OSS) from the transmission case.

✳✳ WARNING

Do not subject the OSS to impact by dropping or hitting it. Do not disassemble the OSS. Do not allow metal filings or any foreign material to get on the sensors front edge magnetic area. Do not place in an area affected by magnetism.

To install:

➡**Be sure to use new fasteners, as required.**

7. Installation is the reverse of the removal procedure.

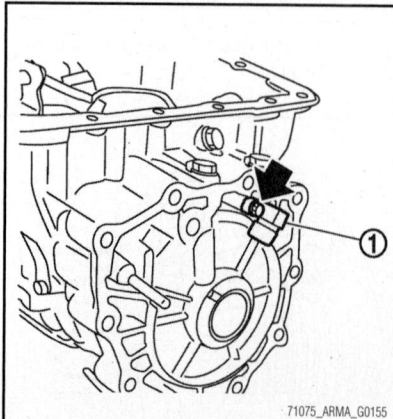

71075_ARMA_G0155

Fig. 183 Output shaft speed sensor location

8. Be sure to perform the reconnect/relearn procedures.

POSITIVE CRANKCASE VENTILATION (PCV) VALVE

LOCATION

The PCV valve is located on top of the engine, in one of the valve rocker covers.

REMOVAL & INSTALLATION

At this time the manufacturer does not provide removal and installation procedures for this component. The following procedure is a guideline and may differ from the vehicle you are servicing.

1. Before servicing the vehicle, refer to the Precautions.
2. Disconnect the negative battery cable.
3. Remove the air cleaner assembly, as required.
4. Remove the necessary components in order to gain access to the component.
5. Disconnect the PCV hose.
6. Remove the valve from its mounting.

To install:

➡**Be sure to use new fasteners, as required.**

7. Installation is the reverse of the removal procedure.
8. Be sure to perform the reconnect/relearn procedures.

THROTTLE POSITION SENSOR (TPS)

LOCATION

The Throttle Position(TPS) sensor is integral to the electric Throttle Control actuator. The Throttle Control actuator is mounted at the front of the intake manifold.

REMOVAL & INSTALLATION

See Figure 184.

At this time the manufacturer does not provide removal and installation procedures for this component. The following procedure is a guideline and may differ from the vehicle you are servicing.

1. Before servicing the vehicle, refer to the Precautions.
2. Disconnect the negative battery cable.
3. Drain the cooling system, as required. Be sure to properly dispose of used engine coolant.
4. Remove the air cleaner assembly.

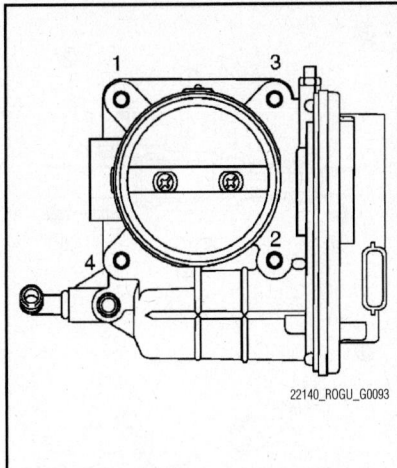

Fig. 184 Throttle body retaining bolt tightening sequence

5. Disconnect harness connector.

6. Disconnect water hoses, as required.

7. Loosen the throttle body assembly mounting bolts in reverse order of the tightening sequence.

To install:

➡ **Be sure to use new fasteners, as required.**

8. Install the throttle body assembly with a new gasket.

9. Tighten the mounting bolts in sequence to 74 inch lbs. (8.4 Nm).

10. Reconnect the water hose.

11. Reconnect the harness connector.

12. Reconnect the air intake duct.

13. Fill the cooling system with the proper grade and type engine coolant.

14. Be sure to perform the reconnect/relearn procedures.

TRANSMISSION FLUID TEMPERATURE (TFT) SENSOR

LOCATION

See Figure 185.

REMOVAL & INSTALLATION

1. Before servicing the vehicle, refer to the Precautions.

2. Disconnect the negative battery cable.

3. Remove the transmission oil pan and oil pan gasket.

➡ **Check for foreign materials in the oil pan to help determine any cause of malfunction. If the A/T fluid is very dark, smells burned, or contains foreign particles, the frictional material (clutches, band) may need replacement. A tacky film that will not wipe clean indicates varnish build up. Varnish can cause valves, servo, and clutches to stick and can inhibit pump pressure.**

4. If frictional material is detected, perform Automatic Transmission (A/T) fluid cooler cleaning.

5. Disconnect the A/T fluid temperature sensor 2 connector.

✳✳ WARNING

Do not damage the connector.

6. Disconnect the output speed sensor connector.

7. Straighten the terminal clip to free the output speed sensor harness.

8. Remove the bolts from the control valve with the TCM.

9. Remove the control valve with the TCM from the transmission case.

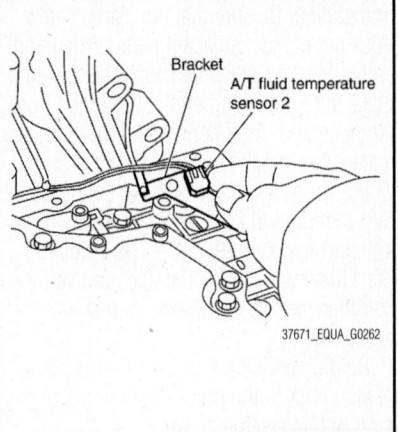

Fig. 185 Transmission Fluid Temperature (TFT) sensor location

✳✳ WARNING

When removing, be careful with the manual valve notch and manual plate height. Remove it vertically.

10. Remove the A/T fluid temperature sensor 2 with the bracket from the control valve with the TCM.

To install:

➡ **Be sure to use new fasteners, as required.**

11. Installation is the reverse of the removal procedure.

12. Replace the O-ring with a new one and lubricate it with transmission fluid.

13. Install the transmission oil pan and tighten the bolts to 70 inch lbs. (8 Nm) in the sequence shown.

FUEL   GASOLINE FUEL INJECTION SYSTEM

FUEL SYSTEM SERVICE PRECAUTIONS

Safety is the most important factor when performing not only fuel system maintenance but any type of maintenance. Failure to conduct maintenance and repairs in a safe manner may result in serious personal injury or death. Maintenance and testing of the vehicle's fuel system components can be accomplished safely and effectively by adhering to the following rules and guidelines.

• To avoid the possibility of fire and personal injury, always disconnect the negative battery cable unless the repair or test procedure requires that battery voltage be applied.

• Always relieve the fuel system pressure prior to disconnecting any fuel system component (injector, fuel rail, pressure regulator, etc.), fitting or fuel line connection. Exercise extreme caution whenever relieving fuel system pressure to avoid exposing skin, face and eyes to fuel spray. Please be advised that fuel under pressure may penetrate the skin or any part of the body that it contacts.

• Always place a shop towel or cloth around the fitting or connection prior to loosening to absorb any excess fuel due to spillage. Ensure that all fuel spillage (should it occur) is quickly removed from engine surfaces. Ensure that all fuel soaked cloths or towels are deposited into a suitable waste container.

• Always keep a dry chemical (Class B) fire extinguisher near the work area.

• Do not allow fuel spray or fuel vapors to come into contact with a spark or open flame.

• Always use a back-up wrench when loosening and tightening fuel line connection fittings. This will prevent unnecessary stress and torsion to fuel line piping.

• Always replace worn fuel fitting O-rings with new Do not substitute fuel hose or equivalent where fuel pipe is installed.

• This vehicle uses quick connect fittings. Be sure to properly relieve the fuel system pressure before disconnecting any of these fittings. Always replace O-rings and clamps with new ones. Do not bend, twist or kink hoses when they are being removed or installed. Be sure that the clamp screw does not contact adjacent parts. When tightening the high pressure rubber hose clamp make sure the clamp end is 0.12 inch from the hose end. After connecting these fittings make sure that the connectors are secure. Check for fuel leakage at these connections turn the ignition key to the ON position (do not start the engine), correct as required. Start the engine, raise the idle, and verify that there are no fuel leaks, correct as required.

Before servicing the vehicle, make sure to also refer to the precautions in the beginning of this section as well.

RELIEVING FUEL SYSTEM PRESSURE

WITH CONSULT-III®

1. Turn ignition switch **ON**.
2. Perform "FUEL PRESSURE RELEASE" in "WORK SUPPORT" mode with CONSULT-III®.
3. Start engine.
4. After engine stalls, turn over the engine two or three times to release all fuel pressure.
5. Turn ignition switch **OFF**.

WITHOUT CONSULT-III®

See Figures 186 through 188.

1. Before servicing the vehicle, refer to the Precautions.
2. Remove fuel pump fuse located in IPDM E/R.
3. Start engine.
4. After engine stalls, crank it two or three times to release all fuel pressure.
5. Turn ignition switch OFF.

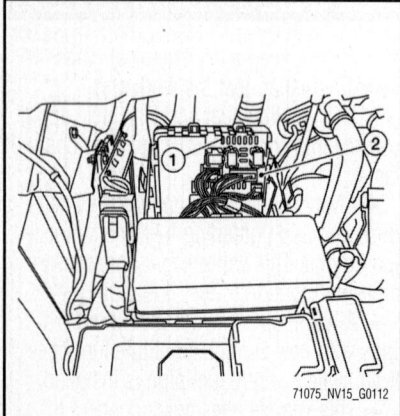

Fig. 186 Fuel system fuse location (1) fuse, (2) IPDM E/R—V6 engine

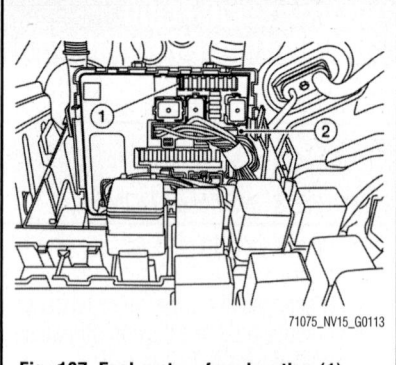

Fig. 187 Fuel system fuse location (1) fuse, (2) IPDM E/R—V8 engine

Fig. 188 IPDM E/R component location (1) IPDM E/R

6. Reinstall fuel pump fuse after servicing fuel system.
7. Disconnect the negative battery cable.

FUEL FILTER

REMOVAL & INSTALLATION

The fuel filter is part of the fuel pump module.

FUEL PUMP MODULE

TESTING

➡**The Nissan CONSULT III diagnostic tool, or equivalent is required to perform diagnostic testing of the fuel pump module that is used on this vehicle.**

REMOVAL & INSTALLATION

See Figures 189 and 190.

1. Before servicing the vehicle, refer to the Precautions.
2. Properly relieve the fuel system pressure.
3. Disconnect the negative battery cable.
4. Remove the fuel tank from the vehicle.
5. Disconnect the fuel level sensor, fuel filter, and fuel pump assembly electrical connector, and the fuel feed hose.

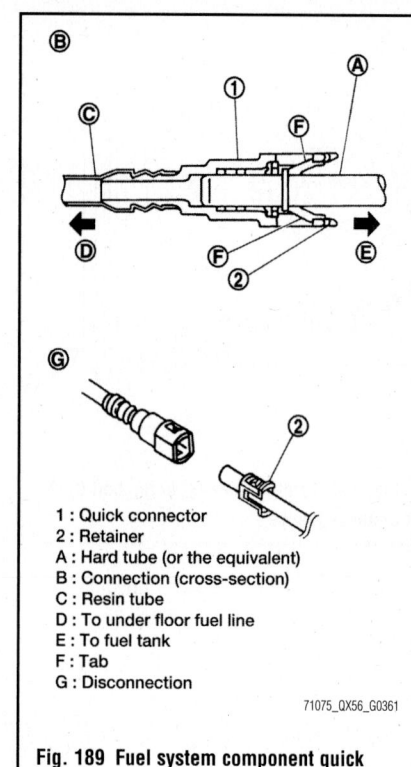

1 : Quick connector
2 : Retainer
A : Hard tube (or the equivalent)
B : Connection (cross-section)
C : Resin tube
D : To under floor fuel line
E : To fuel tank
F : Tab
G : Disconnection

Fig. 189 Fuel system component quick disconnect fitting

6. Remove the lock ring.
7. Remove the fuel level sensor, fuel filter, and fuel pump assembly.

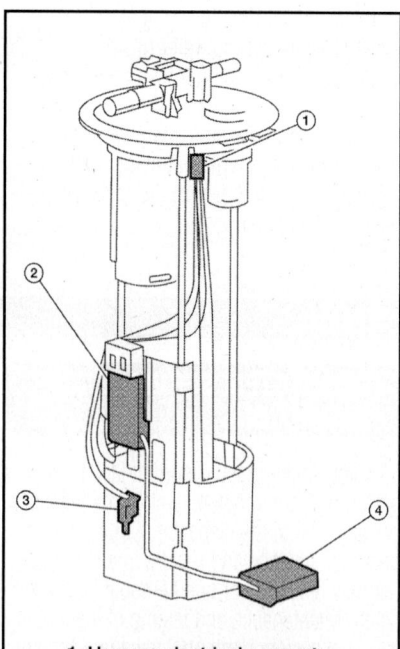

1. Harness electrical connector
2. Sending unit module
3. Fuel temperature sensor
4. Float arm assembly

Fig. 190 Fuel pump module and related components

Do not bend the float arm during removal. Avoid impacts such as dropping when handling the components.

To install:

8. Installation is in the reverse order of removal.

9. Turn the ignition switch ON but do not start engine, then check the fuel pipe and hose connections for leaks while applying fuel pressure.

10. Start the engine and rev it above idle, then check that there are no fuel leaks at any of the fuel pipe and hose connections.

11. Be sure to perform the reconnect/relearn procedures.

FUEL RAIL & INJECTORS

REMOVAL & INSTALLATION

V6 Engine

See Figures 191 through 193.

1. Before servicing the vehicle, refer to the Precautions.

➡This vehicle uses quick connect fittings. Be sure to properly relieve the fuel system pressure before disconnecting any of these fittings. Always replace O-rings and clamps with new ones. Do not bend, twist or kink hoses when they are being removed or installed. Be sure that the clamp screw does not contact adjacent parts. When tightening the high pressure rubber hose clamp make sure the clamp end is 0.12 inch from the hose end. After connecting these fittings make sure that the connectors are secure. Check for fuel leakage at these connections turn the ignition key to the ON position (do not start the engine), correct as required. Start the engine, raise the idle, and verify that there are no fuel leaks, correct as required.

2. Properly relieve the fuel system pressure.

3. Disconnect the negative battery cable.

4. Remove the air cleaner assembly.

5. Remove intake manifold collector.

Perform this step when engine is cold.

6. Disconnect the fuel quick connector on the engine side.

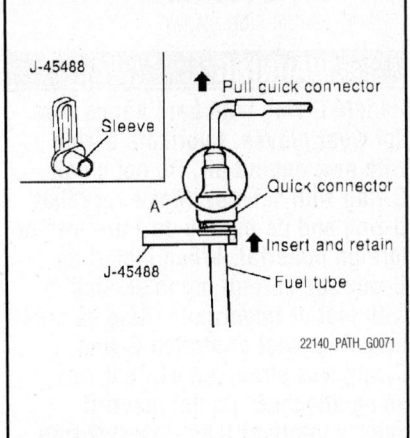

Fig. 191 Using Tool No. 45488—V6 engine

a. Using Tool No. 45488, perform the following steps to disconnect the quick connector.

b. Remove quick connector cap.

c. With the sleeve side of tool facing quick connector, install tool onto fuel tube.

d. Insert tool into quick connector until sleeve contacts and goes no further. Hold the tool on that position.

Inserting the tool hard will not disconnect quick connector. Hold tool where it contacts and goes no further.

e. Pull the quick connector straight out from the fuel tube.

Pull quick connector holding it at the A position. Do not pull with lateral force applied. O-ring inside quick connector may be damaged. Prepare container and cloth beforehand as fuel will leak out. Avoid fire and sparks. Be sure to cover openings of disconnected pipes with plug or plastic bag to avoid fuel leakage and entry of foreign materials.

7. Remove PCV hose between rocker covers (right and left banks).

8. Disconnect harness connector from fuel injector.

9. Loosen bolts in reverse order as shown, and remove fuel tube and fuel injector assembly.

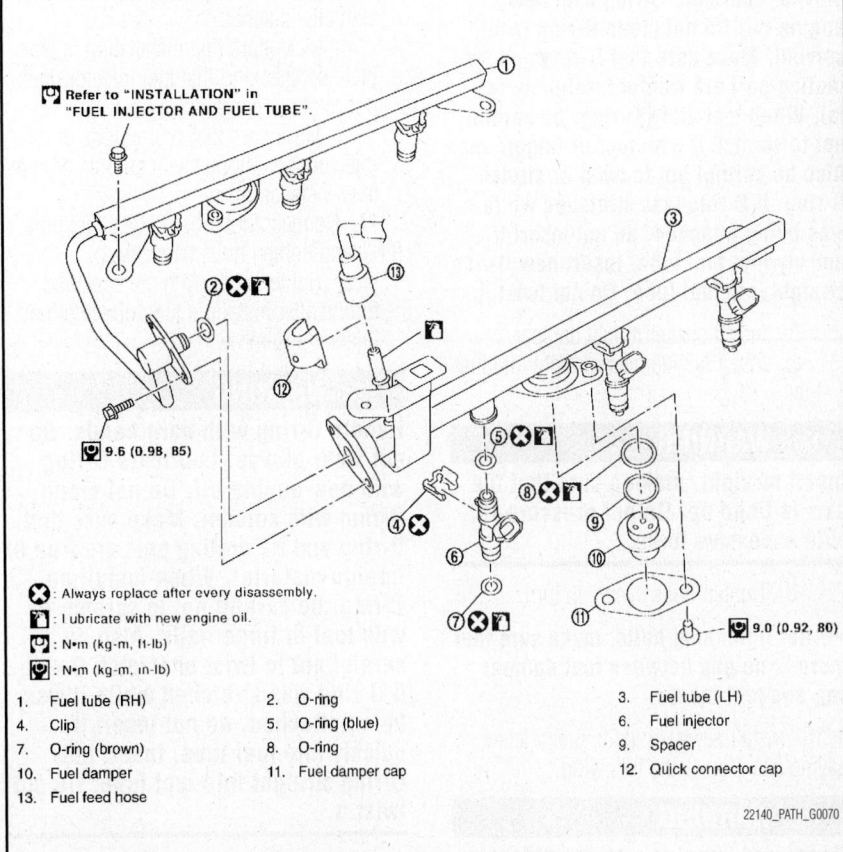

1. Fuel tube (RH)
2. O-ring
3. Fuel tube (LH)
4. Clip
5. O-ring (blue)
6. Fuel injector
7. O-ring (brown)
8. O-ring
9. Spacer
10. Fuel damper
11. Fuel damper cap
12. Quick connector cap
13. Fuel feed hose

Fig. 192 Fuel injector rail and related components—V6 Engine

❄❄ CAUTION

Do not tilt it, or remaining fuel in pipes may flow out from pipes.

10. Remove bolts which connects fuel tube (RH) and fuel tube (LH).

11. Remove fuel injector from fuel tube as follows:

    a. Carefully open and remove clip.

    b. Remove fuel injector from fuel tube by pulling straight.

❄❄ CAUTION

Be careful with remaining fuel that may go out from fuel tube. Be careful not to damage injector nozzles during removal. Do not bump or drop fuel injector. Do not disassemble fuel injector.

12. Disconnect fuel tube (RH) from fuel tube (LH).

13. Loosen bolts, to remove fuel damper cap and fuel damper, if necessary.

To install:

14. Install fuel damper as follows:

    a. Install new O-ring to fuel tube.

➡**When handling new O-rings handle O-ring with bare hands. Do not wear gloves. Lubricate O-ring with new engine oil. Do not clean O-ring with solvent. Make sure that O-ring and its mating part are free of foreign material. When installing O-ring, be careful not to scratch it with tool or fingernails. Also be careful not to twist or stretch O-ring. If O-ring was stretched while it was being attached, do not insert it quickly into fuel tube. Insert new O-ring straight into fuel tube. Do not twist it.**

    b. Install spacer to fuel damper.

    c. Insert fuel damper straight into fuel tube.

❄❄ WARNING

Insert straight, making sure that the axis is lined up. Do not pressure-fit with excessive force.

    d. Tighten bolts evenly in turn.

➡**After tightening bolts, make sure that there is no gap between fuel damper cap and fuel tube.**

15. Install new O-rings to fuel injector, paying attention to the following.

❄❄ WARNING

Upper and lower O-ring are different. Be careful not to confuse them.

- Fuel tube side: Blue
- Nozzle side: Brown

❄❄ WARNING

Handle O-ring with bare hands. Do not wear gloves. Lubricate O-ring with new engine oil. Do not clean O-ring with solvent. Make sure that O-ring and its mating part are free of foreign material. When installing O-ring, be careful not to scratch it with tool or fingernails. Also be careful not to twist or stretch O-ring. If O-ring was stretched while it was being attached, do not insert it quickly into fuel tube. Insert O-ring straight into fuel injector. Do not twist it.

16. Install fuel injector to fuel tube as follows:

    a. Insert clip into clip mounting groove on fuel injector.

❄❄ WARNING

Do not reuse clip. Replace it with a new one. Be careful to keep clip from interfering with O-ring. If interference occurs, replace O-ring.

    b. Insert fuel injector into fuel tube with clip attached.

    c. Make sure that installation is complete by checking that fuel injector does not rotate or come off.

    d. Make sure that protrusions of fuel injectors are aligned with cutouts of clips after installation.

17. Connect fuel tube (RH) to fuel tube (LH), and tighten bolts temporarily.

    a. Tighten bolts with the specified torque after installing fuel tube and fuel injector assembly.

❄❄ WARNING

Handle O-ring with bare hands. Do not wear gloves. Lubricate O-ring with new engine oil. Do not clean O-ring with solvent. Make sure that O-ring and its mating part are free of foreign material. When installing O-ring, be careful not to scratch it with tool or fingernails. Also be careful not to twist or stretch O-ring. If O-ring was stretched while it was being attached, do not insert it quickly into fuel tube. Insert new O-ring straight into fuel tube. Do not twist it.

18. Install fuel tube and fuel injector assembly to intake manifold.

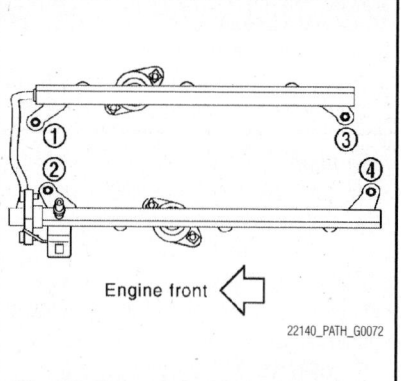

Engine front

22140_PATH_G0072

Fig. 193 Tighten bolts in two steps in numerical order as shown—V6 engine

❄❄ WARNING

Be careful not to let tip of injector nozzle come in contact with other parts.

19. Tighten bolts in two steps in numerical order as shown.

    a. Fuel injector tube assembly bolts:

- 1st step: 7 ft. lbs. (10.1 Nm)
- 2nd step: 16 ft. lbs. (22.0 Nm)

20. Tighten bolts which connects fuel tube (RH) and fuel tube (LH) with the specified torque.

21. Connect fuel injector harness connector.

22. Install intake manifold collector.

23. Installation of the remaining components is in the reverse order of removal.

24. Turn ignition switch "ON" (with engine stopped). With fuel pressure applied to fuel piping, check for fuel leakage at connection points.

➡**Use mirrors for checking at points out of clear sight.**

25. Start engine. With engine speed increased, check again for fuel leakage at connection points.

26. Be sure to perform the reconnect/relearn procedures.

V8 Engine

See Figures 194 through 198.

1. Before servicing the vehicle, refer to the Precautions.

2. Properly relieve the fuel system pressure.

3. Disconnect the negative battery cable.

4. Remove the air cleaner assembly, as required.

5. Disconnect the fuel injector harness connectors.

6. Disconnect the fuel hose assembly from the fuel tubes (RH and LH).

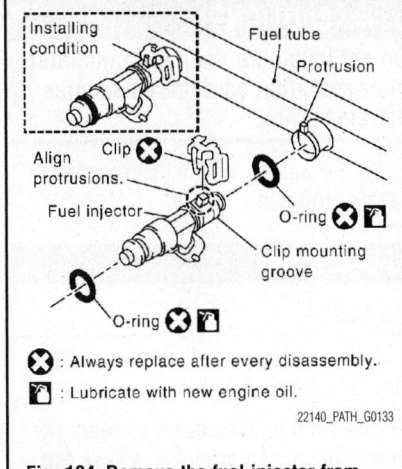

Fig. 194 Remove the fuel injector from the fuel tube—V8 engine

※※ **CAUTION**

While hoses are disconnected, plug them to prevent fuel from draining. Do not separate the fuel connector and fuel hose.

7. Remove the fuel injectors with the fuel tube assembly.

8. Remove the fuel injector from the fuel tube using the following steps:

a. Spread open and remove the clip.

b. Remove the fuel injector from the fuel tube by pulling straight out.

※※ **CAUTION**

Be careful with remaining fuel that may leak out from fuel tube. Do not damage injector nozzles during removal. Do not bump or drop fuel injectors. Do not disassemble fuel injectors.

9. Remove the fuel damper from each fuel tube.

To install:

10. Install the fuel damper to each fuel tube using the following steps:

a. Apply engine oil to the new O-ring and set it into the cup of the fuel tube.

※※ **WARNING**

Handle O-ring with bare hands. Never wear gloves. Lubricate new O-ring with new engine oil. Do not clean O-ring with solvent. Make sure that O-ring and its mating part are free of foreign material. When installing O-ring, do not scratch it with tool or fingernails. Do not twist or stretch the O-ring.

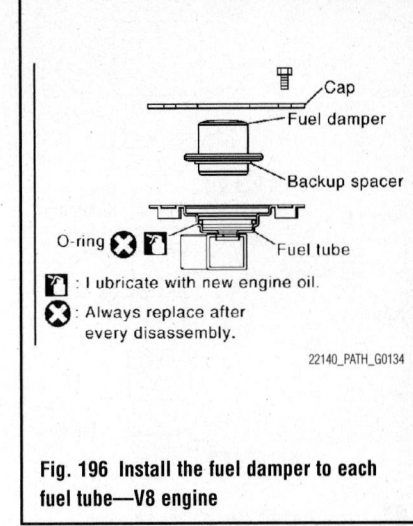

Fig. 196 Install the fuel damper to each fuel tube—V8 engine

b. Make sure that the backup spacer is in the O-ring connecting surface of the fuel damper.

➡The backup spacer is part of the fuel damper assembly.

c. Insert the fuel damper until it seats on the fuel tube.

※※ **CAUTION**

Insert straight, making sure that the axis is lined up. Do not pressure-fit with excessive force. Install the cap, and then tighten the bolts evenly. After tightening the bolts, make sure that there is no gap between the cap and fuel tube.

11. Install new O-rings to the fuel injector paying attention to the items below.

※※ **WARNING**

Upper and lower O-rings are different colors. Handle O-ring with bare hands. Never wear gloves. Lubricate new O-ring with new engine oil. Do not clean O-ring with solvent. Make sure that O-ring and its mating part are free of foreign material. When installing O-ring, be careful not to scratch it with tool or fingernails. Also be careful not to twist or stretch O-ring. If O-ring was stretched while it was being attached, do not insert it quickly into fuel tube. Insert O-ring straight into fuel tube. Do not angle or twist it.

12. Install the fuel injector to the fuel tube using the following steps.

a. Insert new clip into clip mounting groove on the fuel injector.

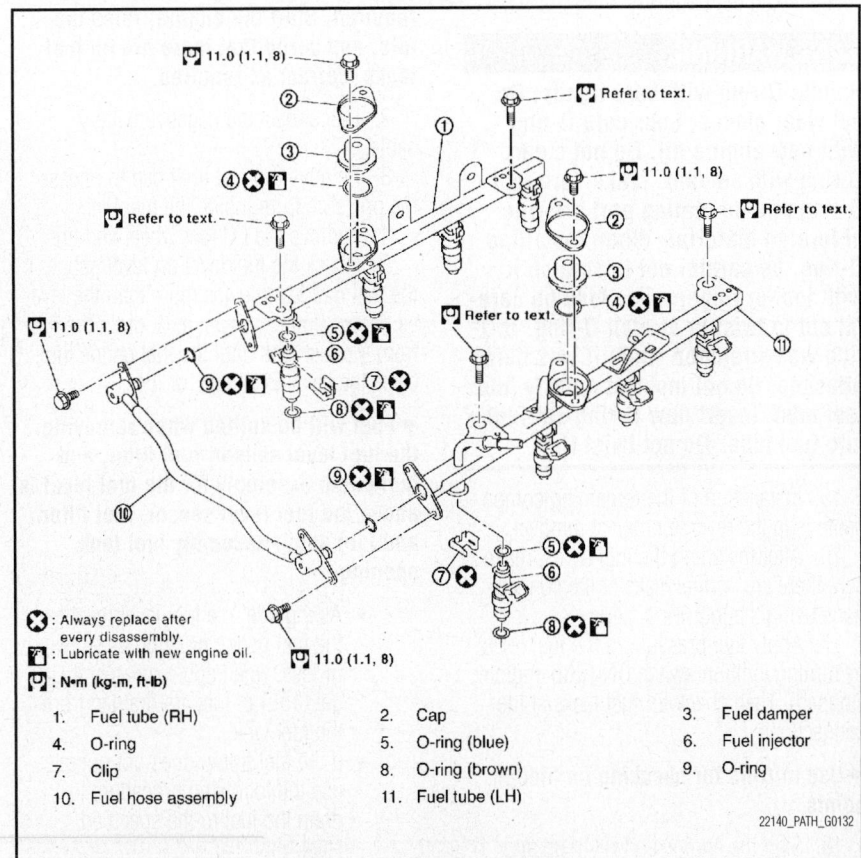

1. Fuel tube (RH)
2. Cap
3. Fuel damper
4. O-ring
5. O-ring (blue)
6. Fuel injector
7. Clip
8. O-ring (brown)
9. O-ring
10. Fuel hose assembly
11. Fuel tube (LH)

Fig. 195 Fuel injector rail and related components—V8 engine

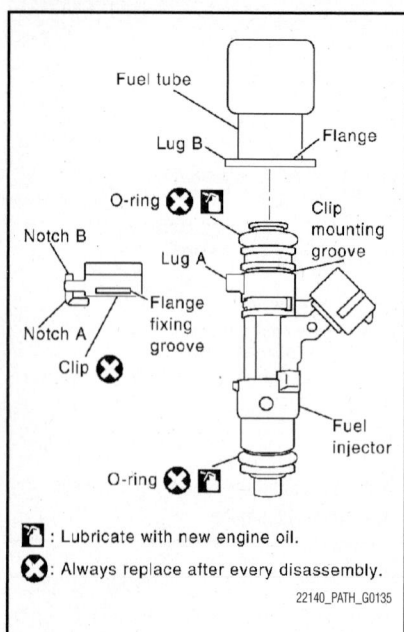

: Lubricate with new engine oil.

: Always replace after every disassembly.

22140_PATH_G0135

Fig. 197 Install the fuel injector to the fuel tube—V8 engine

- Insert clip so that lug A of fuel injector matches notch A of the clip.

✳✳ WARNING

Do not reuse clip. Replace it with a new one. Do not allow the clip to interfere with the O-ring. If interference occurs, replace the O-ring.

b. Insert the fuel injector into the fuel tube with the clip attached.
- Insert it while matching it to the axial center.
- Insert fuel injector so that lug B of fuel tube matches notch B of the clip.

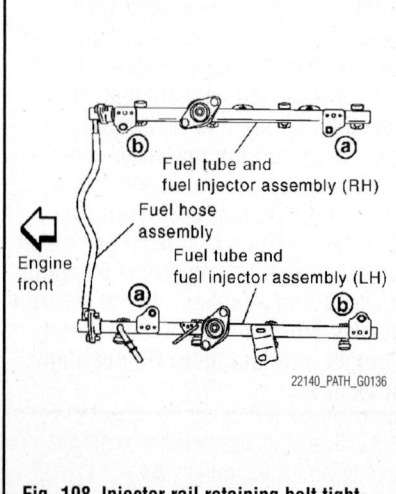

22140_PATH_G0136

Fig. 198 Injector rail retaining bolt tightening sequence—V8 engine

- Make sure that the fuel tube flange is securely seated in the flange fixing groove on the clip.

c. Make sure that installation is complete by checking that the fuel injector does not rotate or come off.

- Make sure that the protrusions of the fuel injectors are aligned with the cutouts of the clips after installation.

13. Install the fuel tube and fuel injector assembly to the intake manifold.

✳✳ WARNING

Do not let the tip of the injector nozzle come in contact with other parts.

a. Tighten fuel tube assembly bolts A to B in two steps:
- Step 1: 9 ft. lbs. (12.8 Nm)
- Step 2: 18 ft. lbs. (24.5 Nm)

14. Install the fuel hose assembly.

a. Insert connector's straight, making sure that the axis is lined up with fuel tube side to prevent O-ring from being damaged.

b. Tighten bolts evenly in several steps.

c. Make sure that there is no gap between the flange and fuel tube after tightening the bolts.

✳✳ WARNING

Handle O-ring with bare hands. Do not wear gloves. Lubricate O-ring with new engine oil. Do not clean O-ring with solvent. Make sure that O-ring and its mating part are free of foreign material. When installing O-ring, be careful not to scratch it with tool or fingernails. Also be careful not to twist or stretch O-ring. If O-ring was stretched while it was being attached, do not insert it quickly into fuel tube. Insert new O-ring straight into fuel tube. Do not twist it.

15. Installation of the remaining components is in the reverse order of removal.

16. After installing the fuel tubes, make sure there are no fuel leaks at the connections using the following steps.

17. Apply fuel pressure to the fuel lines by turning ignition switch ON (with engine stopped). Then check for fuel leaks at the connections.

➡**Use mirrors for checking on hidden points.**

18. Start the engine and rev it up and check for fuel leaks at the connections.

✳✳ CAUTION

Do not touch the engine immediately after stopping, as engine becomes extremely hot.

19. Be sure to perform the reconnect/relearn procedures.

FUEL TANK

DRAINING

1. Before servicing the vehicle, refer to the Precautions.

➡**This vehicle uses quick connect fittings. Be sure to properly relieve the fuel system pressure before disconnecting any of these fittings. Always replace O-rings and clamps with new ones. Do not bend, twist or kink hoses when they are being removed or installed. Be sure that the clamp screw does not contact adjacent parts. When tightening the high pressure rubber hose clamp make sure the clamp end is 0.12 inch from the hose end. After connecting these fittings make sure that the connectors are secure. Check for fuel leakage at these connections turn the ignition key to the ON position (do not start the engine), correct as required. Start the engine, raise the idle, and verify that there are no fuel leaks, correct as required.**

2. Disconnect the negative battery cable.

3. Remove the fuel filler cap to release the pressure from inside the fuel tank.

4. Remove the LH rear wheel and tire.

5. Check the fuel level on level gauge. If the fuel gauge indicates more than the level as shown (full or almost full), drain the fuel from the fuel tank until the fuel gauge indicates the level as shown, or less.

➡**Fuel will be spilled when removing the fuel level sensor, fuel filter, and fuel pump assembly for the fuel level is above the fuel level sensor, fuel filter, and fuel pump assembly fuel tank opening.**

- As a guide, the fuel level reaches the fuel gauge position as shown, or less, when approximately 4 US gal (15L) of fuel are drained from the fuel tank.
- If the fuel pump does not operate, use the following procedure to drain the fuel to the specified level.

a. Insert a suitable hose of less than

15 mm (0.59 in.) diameter into the fuel filler pipe through the fuel filler opening to drain the fuel from fuel filler pipe.

b. Remove the fuel filler pipe shield.

c. Disconnect the fuel filler hose from the fuel filler pipe.

d. Insert a suitable hose into the fuel tank through the fuel filler hose to drain the fuel from the fuel tank.

6. Release the fuel pressure from the fuel lines.

REMOVAL & INSTALLATION

See Figure 199.

1. Before servicing the vehicle, refer to the Precautions.

➡ **This vehicle uses quick connect fittings. Be sure to properly relieve the fuel system pressure before disconnecting any of these fittings. Always replace O-rings and clamps with new ones. Do not bend, twist or kink hoses when they are being removed or installed. Be sure that the clamp screw does not contact adjacent parts. When tightening the high pressure rubber hose clamp make sure the clamp end is 0.12 inch from the hose end. After connecting these fittings make sure that the connectors are secure. Check for fuel leakage at these connections turn the ignition key to the ON position (do not start the engine), correct as required. Start the engine, raise the idle, and verify that there are no fuel leaks, correct as required.**

2. Properly release the fuel system pressure.

3. Disconnect the negative battery cable.

4. Remove the fuel filler cap to release the pressure from inside the fuel tank.

5. Check the fuel level on level gauge. If the fuel gauge indicates more than the level as shown (full or almost full), drain the fuel from the fuel tank until the fuel gauge indicates the level as shown, or less.

➡ **Fuel will be spilled when removing the fuel level sensor, fuel filter, and fuel pump assembly for the fuel level is above the fuel level sensor, fuel filter, and fuel pump assembly fuel tank opening.**

- As a guide, the fuel level reaches the fuel gauge position as shown, or less, when approximately 4 US gal (15L) of fuel are drained from the fuel tank.
- If the fuel pump does not operate, use the following procedure to drain the fuel to the specified level.

a. Insert a suitable hose of less than 15 mm (0.59 in.) diameter into the fuel filler pipe through the fuel filler opening to drain the fuel from fuel filler pipe.

b. Remove the fuel filler pipe shield.

c. Disconnect the fuel filler hose from the fuel filler pipe.

d. Insert a suitable hose into the fuel tank through the fuel filler hose to drain the fuel from the fuel tank.

6. Release the fuel pressure from the fuel lines.

7. Disconnect the lower fuel filler hose from the fuel tank, the EVAP hose, and the vent pipe quick connector.

a. Disconnect the fuel feed hose from the molded clip in the side of the fuel tank.

- Disconnect the quick connector as follows:
- Hold the sides of the connector, push in the tabs and pull out the tube.

- If the connector and the tube are stuck together, push and pull several times until they start to move. Then disconnect them by pulling.

❈❈ **WARNING**

The quick connector can be disconnected when the tabs are completely depressed. Do not twist the quick connector more than necessary. Do not use any tools to disconnect the quick connector. Keep the resin tube away from heat. Be especially careful when welding near the tube. Do not bend or twist the resin tube during connection.

8. Remove the fuel tank shield, if equipped.

9. Remove the driveshaft.

10. Support the fuel tank using a suitable lift jack.

11. Remove the fuel tank strap bolts

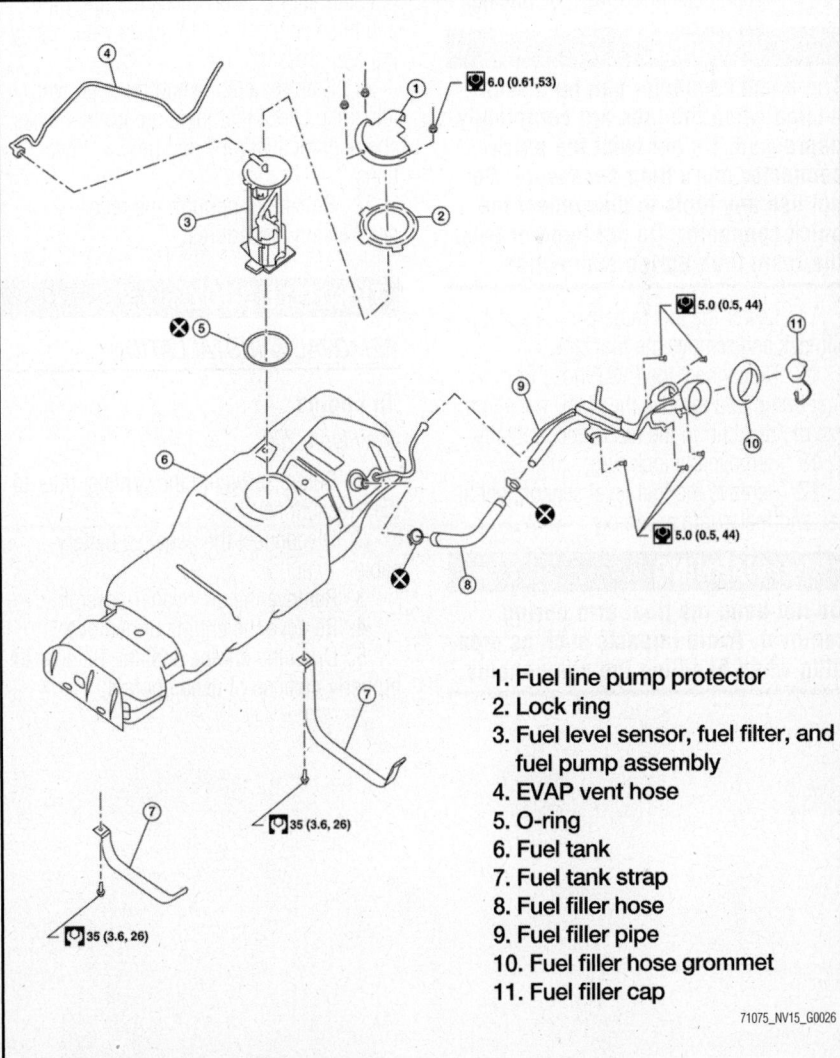

1. Fuel line pump protector
2. Lock ring
3. Fuel level sensor, fuel filter, and fuel pump assembly
4. EVAP vent hose
5. O-ring
6. Fuel tank
7. Fuel tank strap
8. Fuel filler hose
9. Fuel filler pipe
10. Fuel filler hose grommet
11. Fuel filler cap

71075_NV15_G0026

Fig. 199 Fuel tank and related components

while supporting the fuel tank with a suitable lift jack.

12. Remove the fuel tank straps and slowly lower the fuel tank to access the top of the fuel level sensor, fuel filter and fuel pump assembly.

✳✳ CAUTION

Do not lower the fuel tank too far to prevent damage to the fuel feed hose and the fuel level sensor, fuel filter and fuel pump assembly connector.

13. Disconnect the fuel level sensor, fuel filter, and fuel pump assembly electrical connector, and the fuel feed hose.

   a. Disconnect the quick connector as follows:

- Hold the sides of the connector, push in the tabs and pull out the tube.
- If the connector and the tube are stuck together, push and pull several times until they start to move. Then disconnect them by pulling.

✳✳ WARNING

The quick connector can be disconnected when the tabs are completely depressed. Do not twist the quick connector more than necessary. Do not use any tools to disconnect the quick connector. Do not bend or twist the resin tube during connection.

14. Lower the fuel tank using a suitable lift jack and remove the fuel tank.

15. Disconnect the EVAP hose from the fuel pump and remove the EVAP hose from the molded clip in the top of the fuel tank.

16. Remove the lock ring.

17. Remove the fuel level sensor, fuel filter, and fuel pump assembly.

✳✳ WARNING

Do not bend the float arm during removal. Avoid impacts such as dropping when handling the components.

To install:

18. Installation is in the reverse order of removal.

19. Connect the quick connector as follows:

- Check the connection for any damage or foreign materials.
- Align the connector with the pipe, then insert the connector straight into the pipe until a click is heard.
- Pull the tube and the connector to make sure they are securely connected.
- Visually inspect the connector to make sure the two retainer tabs are securely connected.

✳✳ WARNING

Do not bend the float arm during installation. Avoid impacts such as dropping when handling the components.

20. Turn the ignition switch ON but do not start engine, then check the fuel pipe and hose connections for leaks while applying fuel pressure.

21. Start the engine and rev it above idle, then check that there are no fuel leaks at any of the fuel pipe and hose connections.

22. Be sure to perform the reconnect/relearn procedures.

THROTTLE BODY

REMOVAL & INSTALLATION

V6 Engine

See Figure 200.

1. Before servicing the vehicle, refer to the Precautions.

2. Disconnect the negative battery cable.

3. Remove the air cleaner assembly.

4. Remove the engine undercover.

5. Drain the engine coolant. Be sure to properly dispose of used coolant.

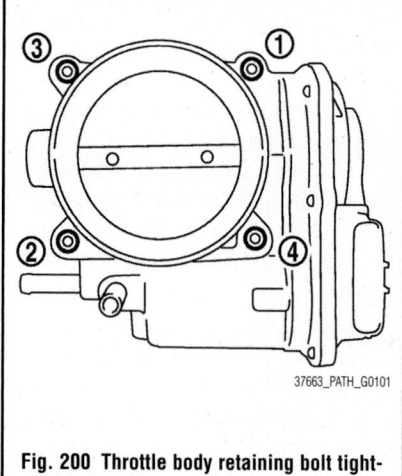

37663_PATH_G0101

Fig. 200 Throttle body retaining bolt tightening sequence

6. Disconnect the hoses from the unit.

7. Disconnect the electrical connectors.

8. Remove the retaining bolts in the reverse order of the tightening sequence.

9. Remove the component from its mounting. Discard the gasket.

To install:

➡**Be sure to use new fasteners, as required.**

10. Installation is the reverse of the removal procedure.

11. Tighten the retaining bolts to specification and in the proper sequence.

12. Using the CONSULT-III diagnostic tool or equivalent, perform the throttle valve closed position learning and the idle air volume learning procedures.

13. Be sure to perform the reconnect/relearn procedures.

V8 Engine

At this time the manufacturer does not provide removal and installation procedures for this component.

Before servicing the vehicle, refer to the Precautions.

HEATING & AIR CONDITIONING SYSTEM

BLOWER MOTOR

REMOVAL & INSTALLATION

See Figure 201.

1. Before servicing the vehicle, refer to the Precautions.
2. Disconnect the negative battery cable.
3. Disconnect the vent hose.
4. Remove the component retaining screws.
5. Remove the component from its mounting.

To install:

➡**Be sure to use new fasteners, as required.**

6. Installation is the reverse of the removal procedure.
7. Be sure to perform the reconnect/relearn procedures.

HEATER CORE

REMOVAL & INSTALLATION

See Figure 202.

1. Before servicing the vehicle, refer to the Precautions.
2. Disconnect the negative battery cable.
3. Properly discharge the refrigerant from the A/C system.
4. Remove the heating/cooling unit assembly.
5. Separate the blower motor housing from the heating/cooling unit.
6. Remove the heater core pipe cover and the heater case side cover.
7. Remove the heater core.

➡**If equipped with In-cabin filter, replace it as these filters get contaminated with coolant leaking from the heater core and must be replaced when the core is serviced.**

To install:

➡**Be sure to use new fasteners, as required.**

8. Installation is the reverse of the removal procedure.
9. Be sure to properly recharge the air conditioning system.
10. Be sure to perform the reconnect/relearn procedures.

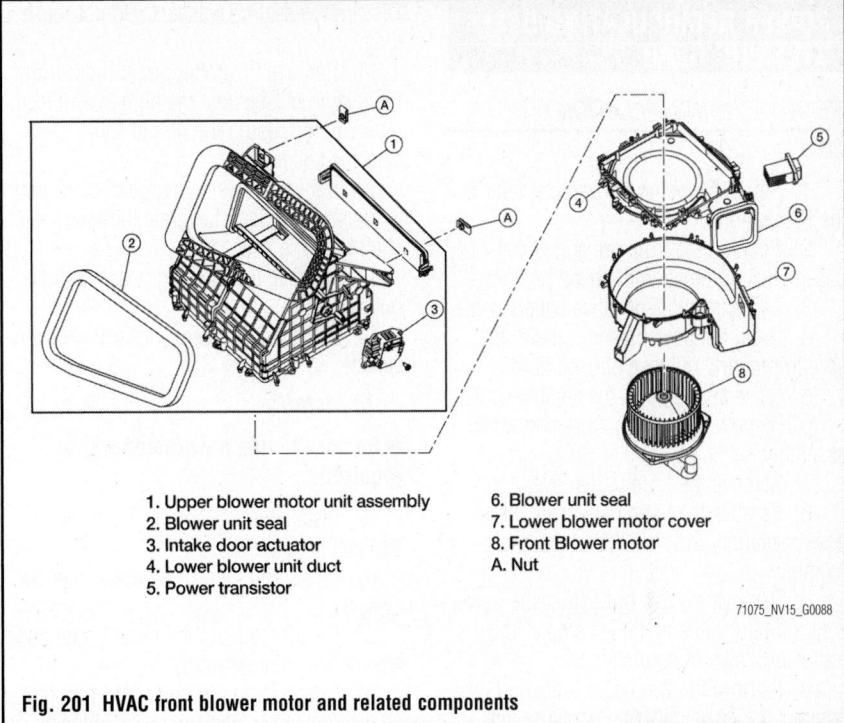

1. Upper blower motor unit assembly
2. Blower unit seal
3. Intake door actuator
4. Lower blower unit duct
5. Power transistor
6. Blower unit seal
7. Lower blower motor cover
8. Front Blower motor
A. Nut

71075_NV15_G0088

Fig. 201 HVAC front blower motor and related components

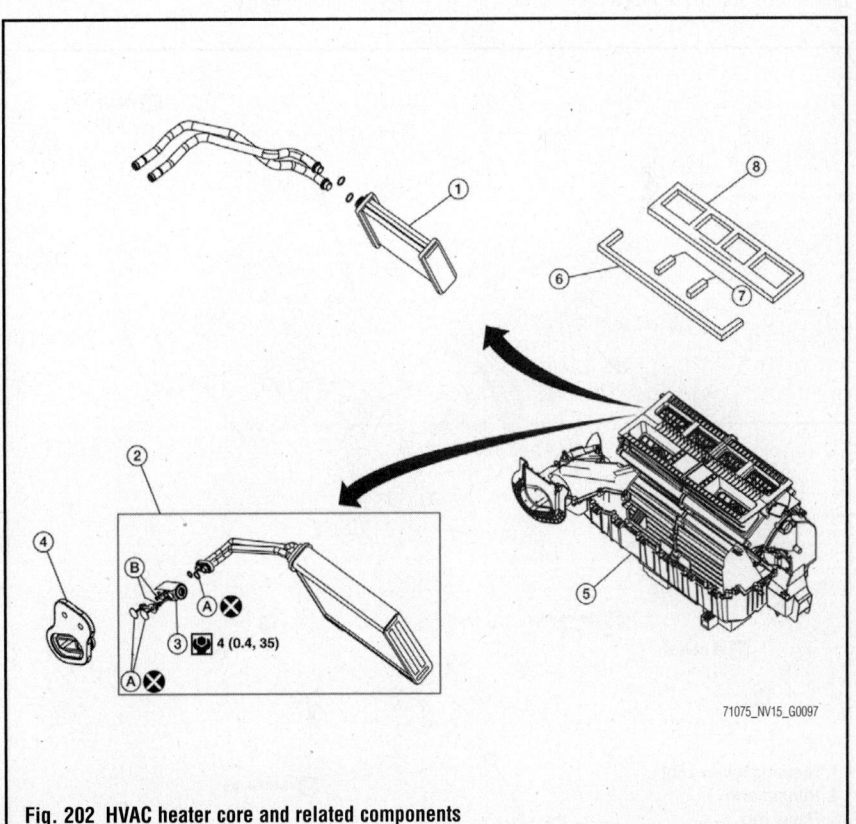

71075_NV15_G0097

Fig. 202 HVAC heater core and related components

STEERING

POWER RECIRCULATING BALL STEERING GEAR

REMOVAL & INSTALLATION

See Figure 203.

1. Before servicing the vehicle, refer to the Precautions.
2. Position the front tire and wheel assemblies in the straight ahead position.
3. Disconnect the negative battery cable.
4. Drain the power steering fluid. Be sure to properly dispose of used fluid.
5. Raise and support the vehicle safely.
6. Remove the left front tire and wheel assembly.
7. Remove the front undercover.
8. Remove the three bolts from the power steering pressure line at the crossmember.
9. Remove the two bolts from the power steering return line at the left side of the frame and radiator core support.
10. Remove the two banjo bolts and disconnect the power steering pressure line and return line from the steering gear.

Discard the sealing washers. Do not reuse them!

11. Remove the cotter pin and nut from the relay rod. Separate the pitman arm from the relay rod, using the proper tools. Discard the cotter pin.
12. Remove the steering gear lower joint at the steering gear. Separate the lower joint shaft from the gear.
13. Remove the steering gear retaining bolts.
14. Remove the steering gear from the vehicle.

To install:

→**Be sure to use new fasteners, as required.**

15. Installation is the reverse of the removal procedure.
16. Be sure to bleed the power steering system.
17. Be sure to use the proper grade and type power steering fluid.
18. Check for leaks, correct as required.
19. Be sure to perform the reconnect/relearn procedures.

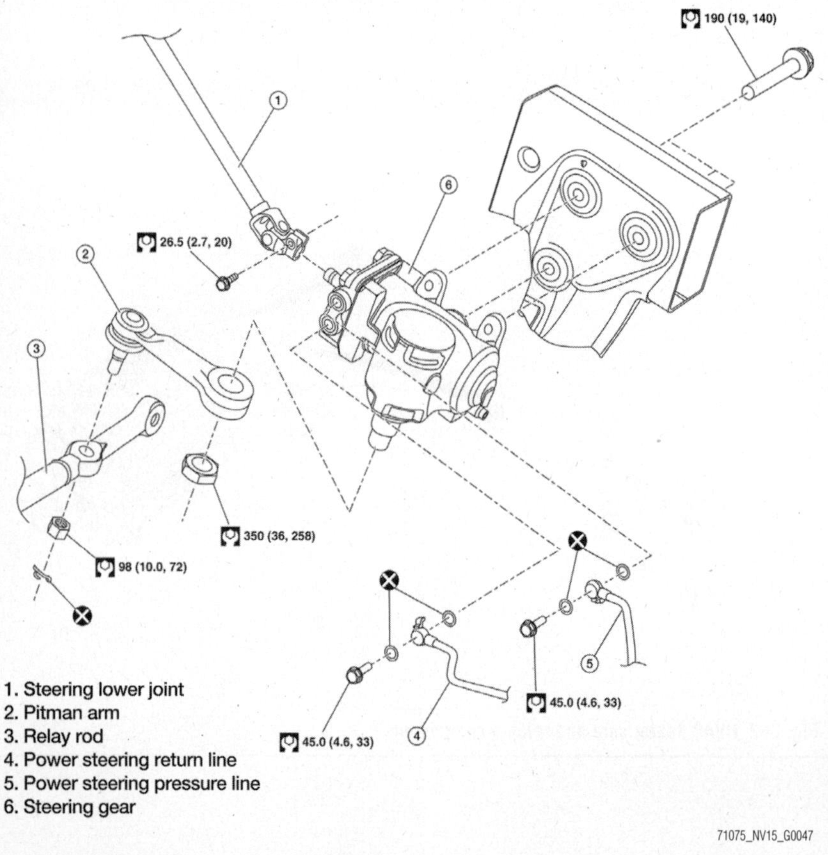

POWER STEERING PUMP

BLEEDING

1. Before servicing the vehicle, refer to the Precautions.

→**When the vehicle is stationary or while the steering wheel is being turned slowly, some noise may be heard from the oil pump or gear. This noise is normal and does not affect any system.**

2. Check for fluid leakage.
3. Start the engine and turn the steering wheel fully to the right and left several times.

❊❊ CAUTION
Do not allow steering fluid reservoir tank to go below the MIN level line. Check tank frequently and add fluid as needed.

4. Run the engine at idle speed. Hold the steering wheel at each "locked" position for three seconds.

❊❊ WARNING
Do not hold steering wheel in the locked position for more than 10 seconds. (There is the possibility that oil pump may be damaged.)

5. Repeat step 3 several times at about three second intervals.
6. Check for air bubbles, cloudy fluid and fluid leakage.
7. If air bubbles or cloudiness exists, perform steps 3 and 4 again until air bubbles and cloudiness do not exist.
8. Stop the engine and check fluid level.

REMOVAL & INSTALLATION

See Figures 204 through 208.

1. Before servicing the vehicle, refer to the Precautions.
2. Disconnect the negative battery cable.
3. Remove the air cleaner assembly, as required.
4. Drain the power steering fluid. Be sure to properly dispose of used fluid.
5. Remove the drive belt.
6. Remove the right fender protector.
7. Disconnect the pressure sensor electrical connector.
8. Remove the high pressure and low pressure piping from the power steering oil pump. Discard the sealing washers.
9. Remove the power steering oil pump bolts, then remove the power steering pump.

190 (19, 140)

26.5 (2.7, 20)

350 (36, 258)

98 (10.0, 72)

45.0 (4.6, 33)

45.0 (4.6, 33)

1. Steering lower joint
2. Pitman arm
3. Relay rod
4. Power steering return line
5. Power steering pressure line
6. Steering gear

71075_NV15_G0047

Fig. 203 Power steering gear and related components

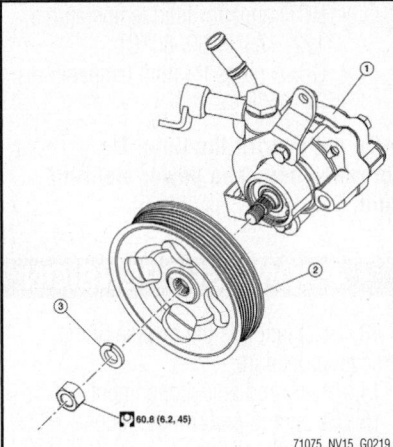

Fig. 204 Power steering pump and related components (1) pump, (2) pulley, (3) washer—V6 engine

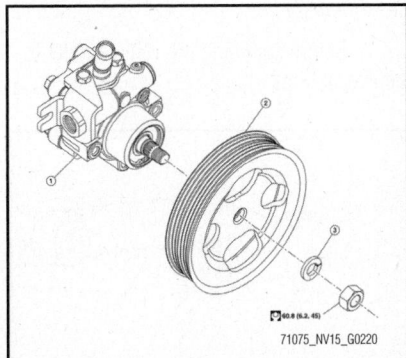

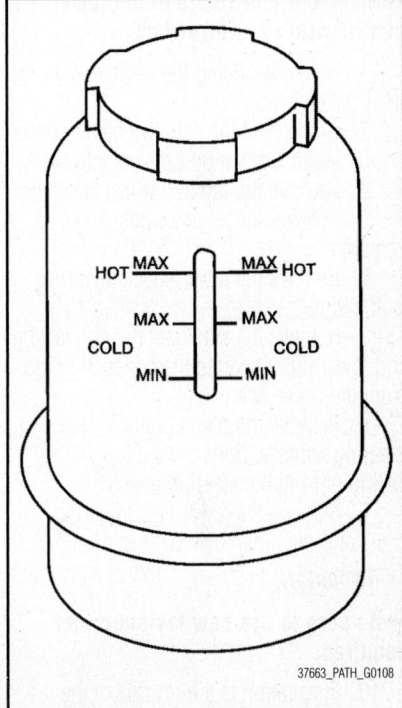

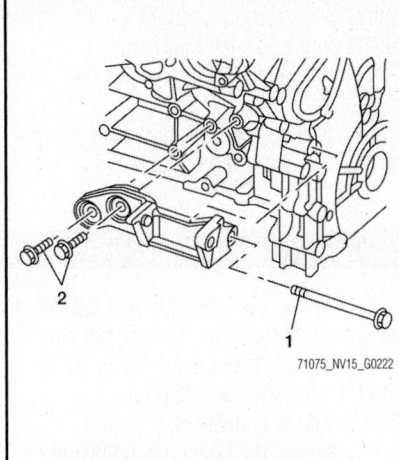

Fig. 207 Power steering pump bracket bolt tightening sequence—V6 engine

11. Tighten the pump retaining bolts in the sequence shown in the illustration and to specification. For the V6 engine, pump bolt one 48 ft. lbs., bolt two 21 ft. lbs. For the V6 engine bracket bolt one 45 ft. lbs., bolt two 45 ft. lbs. For the V8 engine, pump bolt one 48 ft. lbs., bolt two 23 ft. lbs.

12. After installation, bleed the air from the hydraulic circuit thoroughly.

13. Be sure to perform the reconnect/relearn procedures.

FLUID FILL PROCEDURE

See Figures 209 and 210.

Fig. 205 Power steering pump and related components (1) pump, (2) pulley, (3) washer—V8 engine

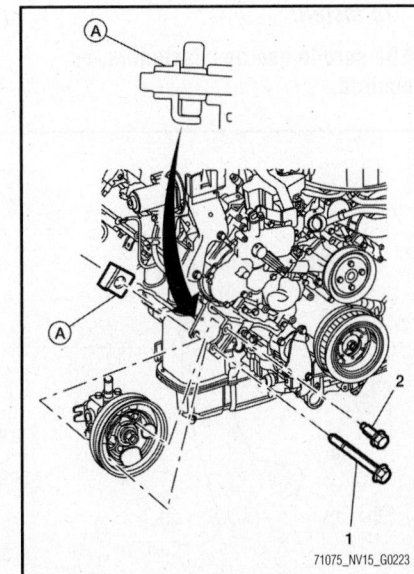

Fig. 208 Power steering pump bolt tightening sequence (A) hook—V8 engine

Fig. 209 Power steering fluid reservoir markings

Fig. 206 Power steering pump bolt tightening sequence—V6 engine

To install:

➡Be sure to use new fasteners, as required.

10. Installation is the reverse of the removal procedure.

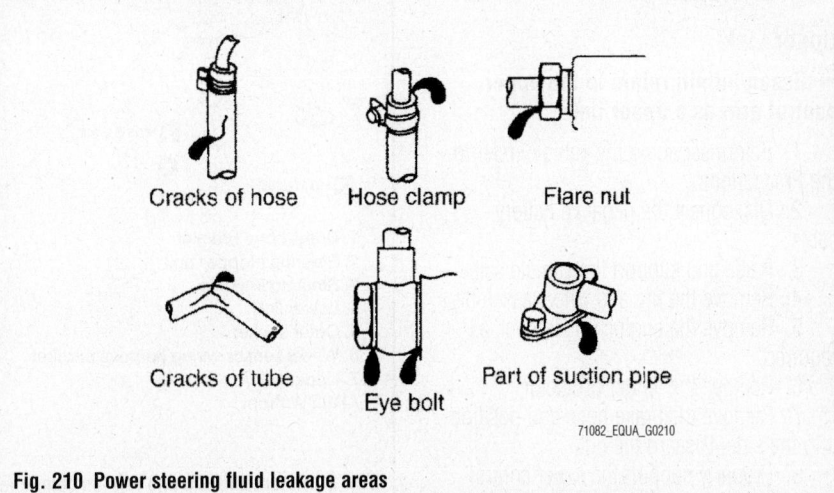

Cracks of hose   Hose clamp   Flare nut

Cracks of tube   Eye bolt   Part of suction pipe

Fig. 210 Power steering fluid leakage areas

※※ CAUTION

Used fluid is considerably more dangerous than new fluid. Avoid skin contact with used fluid.

1. Before servicing the vehicle, refer to the Precautions.

2. Inspect the power steering fluid level in the power steering reservoir. Do allow the fluid to drop below the MIN marking.

3. Remove the power steering reservoir cap.

4. Add fluid, as necessary, referring to the scale on the reservoir tank.

- HOT range for fluid temperatures: 122–176°F (50–80°C)
- COLD range for fluid temperatures: 32–86°F (0–30°C)

➡**Do not overfill the fluid. Do not reuse any used power steering fluid.**

SUSPENSION

CONTROL LINKS

REMOVAL & INSTALLATION

Lower Link

➡**Nissan/Infiniti refers to the lower control arm as a lower link.**

1. Before servicing the vehicle, refer to the Precautions.
2. Disconnect the negative battery cable.
3. Raise and support the vehicle safely.
4. Remove the tire and wheel assembly.
5. Remove the engine undercover, as required.
6. Remove the lower shock absorber bolt and nut.
7. Remove the stabilizer bar connecting rod lower nut. Separate the connecting rod from the lower link.
8. Remove the pinch bolt from the steering knuckle. Separate the lower link ball joint from the steering knuckle.
9. Remove the lower link adjusting bolts and nuts. Remove the lower link.

To install:

➡**Be sure to use new fasteners, as required.**

10. Installation is the reverse of the removal procedure.
11. Check and adjust alignment.
12. Be sure to perform the reconnect/relearn procedures.

Upper Link

➡**Nissan/Infiniti refers to the upper control arm as a upper link.**

1. Before servicing the vehicle, refer to the Precautions.
2. Disconnect the negative battery cable.
3. Raise and support the vehicle safely.
4. Remove the tire and wheel assembly.
5. Remove the engine undercover, as required.
6. Remove the fender protector.
7. Remove the brake hose and position it to the side. Discard the nut.
8. Properly support the lower control arm, as required.

9. Remove the cotter pin and nut from the upper link ball joint. Discard the pin.
10. Separate the upper link ball joint stud from the knuckle using tool ST2902001, or equivalent.
11. Remove the upper link bolts and nuts.
12. Remove the component from its mounting.

To install:

➡**Be sure to use new fasteners, as required.**

FRONT SUSPENSION

13. Installation is the reverse of the removal procedure.
14. Check and adjust alignment.
15. Be sure to perform the reconnect/relearn procedures.

KNUCKLE & SPINDLE

REMOVAL & INSTALLATION
See Figure 211.

1. Before servicing the vehicle, refer to the Precautions.

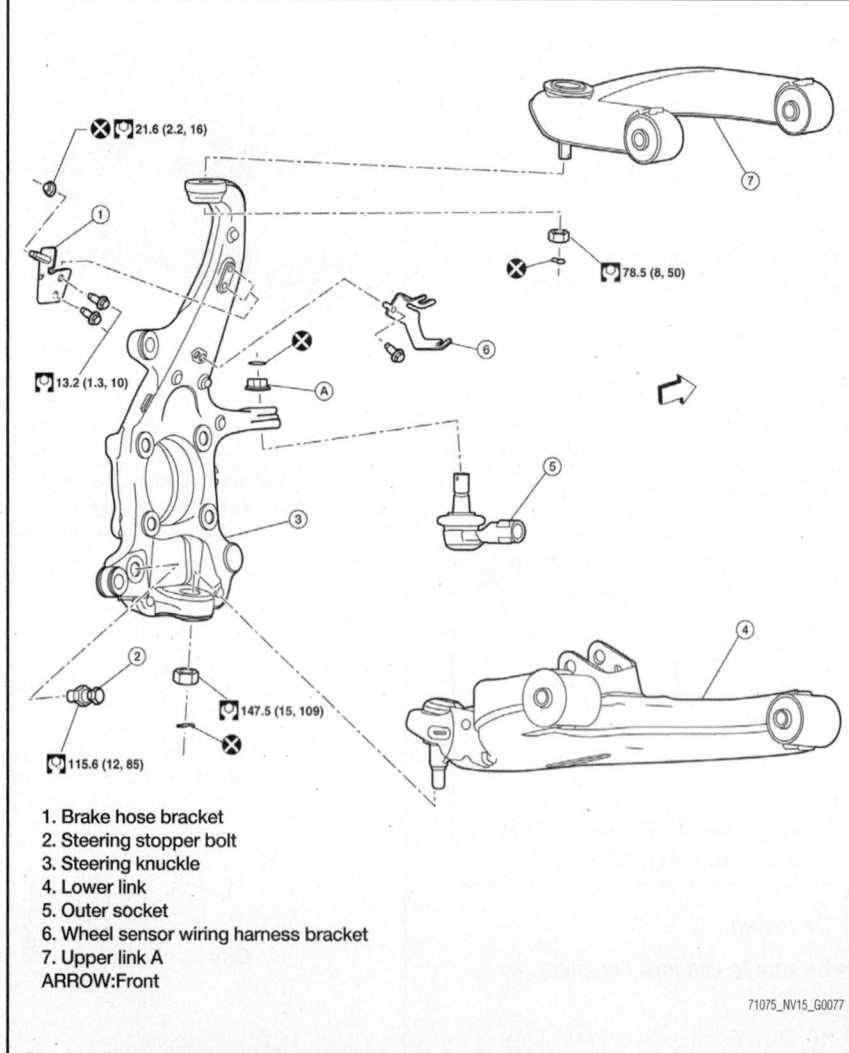

1. Brake hose bracket
2. Steering stopper bolt
3. Steering knuckle
4. Lower link
5. Outer socket
6. Wheel sensor wiring harness bracket
7. Upper link A
ARROW:Front

71075_NV15_G0077

Fig. 211 Front knuckle and related components

2. Disconnect the negative battery cable.

3. Raise and safely support the vehicle.

4. Remove wheel and tire from vehicle.

5. Remove the engine undercover, as required.

6. Disconnect the wheel speed sensor. Do not remove the wheel sensor from the hub/bearing assembly. Remove the hub/bearing and bearing assembly.

7. Remove the cotter pin and locknut from the outer socket. Discard the pin. Separate the outer socket from the knuckle assembly, using the proper tool.

8. Remove the cotter pin and locknut from the upper link ball joint. Discard the pin. Separate the upper link ball joint from the knuckle using the proper tools.

9. Remove the cotter pin and locknut from the lower link ball joint. Discard the pin. Separate the lower link ball joint from the knuckle using the proper tools.

10. Remove the steering knuckle from the vehicle.

To install:

➡**Be sure to use new fasteners, as required.**

11. Installation is the reverse of the removal procedure.

12. Perform wheel alignment.

13. Be sure to perform the reconnect/relearn procedures.

LOWER CONTROL ARMS

REMOVAL & INSTALLATION

➡**Nissan/Infiniti refers to the lower control arm as a lower link.**

1. Before servicing the vehicle, refer to the Precautions.

2. Disconnect the negative battery cable.

3. Raise and support the vehicle safely.

4. Remove the tire and wheel assembly.

5. Remove the engine undercover, as required.

6. Remove the lower shock absorber bolt and nut.

7. Remove the stabilizer bar connecting rod lower nut. Separate the connecting rod from the lower link.

8. Remove the pinch bolt from the steering knuckle. Separate the lower link ball joint from the steering knuckle.

9. Remove the lower link adjusting bolts and nuts. Remove the lower link.

To install:

➡**Be sure to use new fasteners, as required.**

10. Installation is the reverse of the removal procedure.

11. Check and adjust alignment.

12. Be sure to perform the reconnect/relearn procedures.

SHOCK ABSORBERS

REMOVAL & INSTALLATION

See Figure 212.

➡**The following procedure is for both the front shock absorber and the front coil spring. Both components are removed as a unit.**

1. Before servicing the vehicle, refer to the Precautions.

2. Disconnect the negative battery cable.

3. Raise and support the vehicle safely. Remove the wheel and tire.

4. Remove the engine undercover, as required.

5. Remove connecting rod upper joints from stabilizer bar.

6. Swing stabilizer bar down, repositioning it out of the way to access shock absorber lower mount.

7. Remove the shock absorber lower bolt and nut.

8. Remove the three shock absorber upper mounting nuts.

9. Remove the coil spring and shock absorber assembly.

To install:

➡**Be sure to use new fasteners, as required.**

10. Installation is the reverse of the removal procedure.

11. Be sure to perform the reconnect/relearn procedures.

STABILIZER BAR

REMOVAL & INSTALLATION

See Figures 213 and 214.

1. Before servicing the vehicle, refer to the Precautions.

2. Disconnect the negative battery cable.

3. Raise and safely support the vehicle.

4. Remove engine undercover.

5. Remove connecting rod nuts.

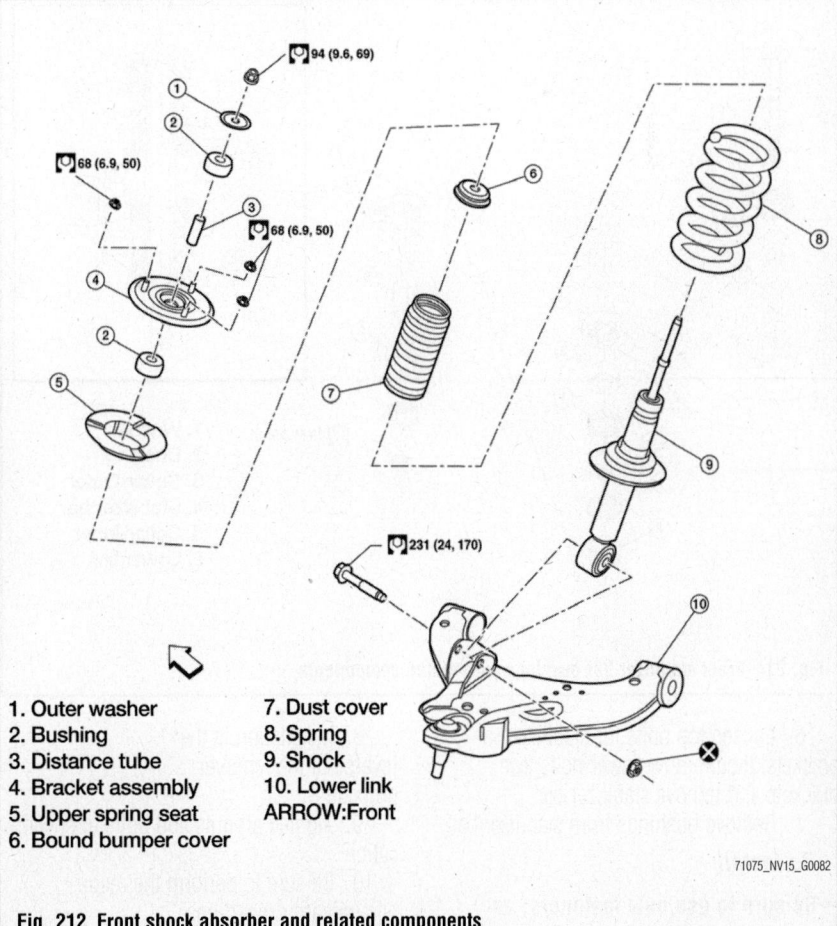

1. Outer washer
2. Bushing
3. Distance tube
4. Bracket assembly
5. Upper spring seat
6. Bound bumper cover
7. Dust cover
8. Spring
9. Shock
10. Lower link
ARROW:Front

71075_NV15_G0082

Fig. 212 Front shock absorber and related components

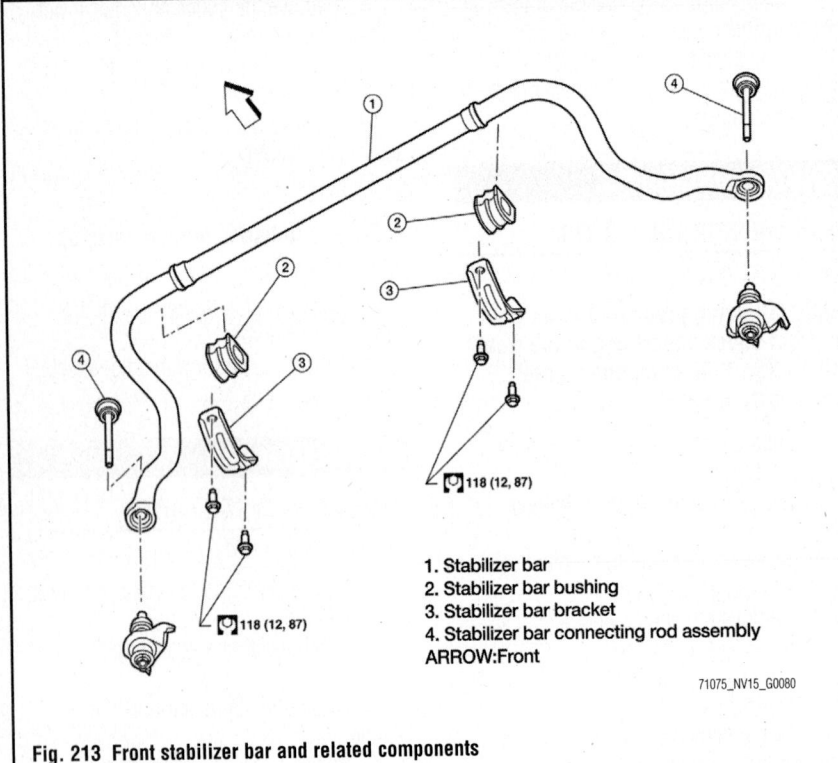

1. Stabilizer bar
2. Stabilizer bar bushing
3. Stabilizer bar bracket
4. Stabilizer bar connecting rod assembly
ARROW:Front

71075_NV15_G0080

Fig. 213 Front stabilizer bar and related components

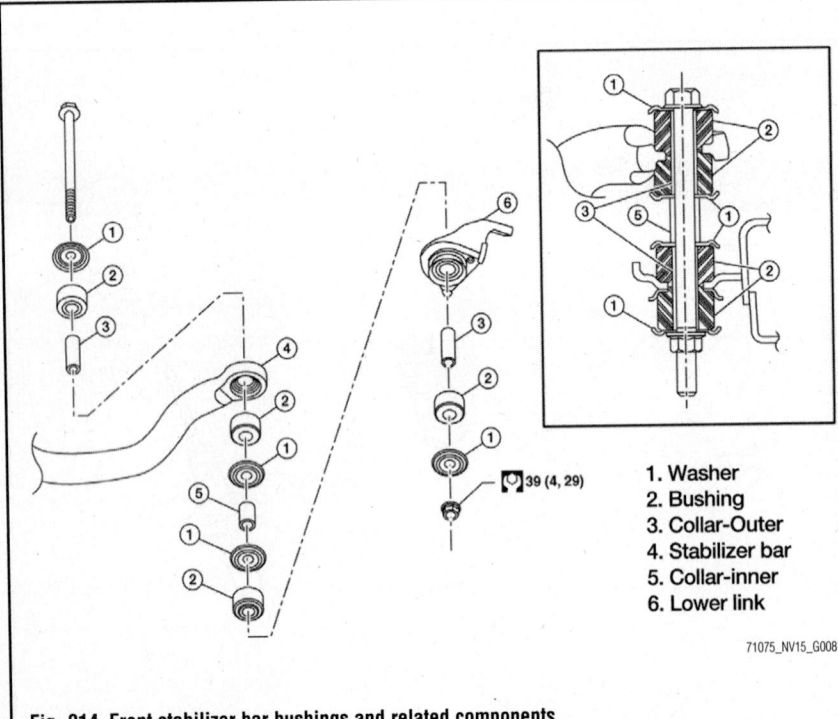

1. Washer
2. Bushing
3. Collar-Outer
4. Stabilizer bar
5. Collar-inner
6. Lower link

71075_NV15_G0081

Fig. 214 Front stabilizer bar bushings and related components

6. Loosen top bolts for stabilizer bar brackets, then remove lower bolts from brackets and remove stabilizer bar.

7. Remove bushings from stabilizer bar.

To install:

➡ **Be sure to use new fasteners, as required.**

8. Installation is the reverse of the removal procedure.

9. Tighten all nuts and bolts to specification.

10. Be sure to perform the reconnect/relearn procedures.

UPPER CONTROL ARMS

REMOVAL & INSTALLATION

➡ **Nissan/Infiniti refers to the upper control arm as a upper link.**

1. Before servicing the vehicle, refer to the Precautions.

2. Disconnect the negative battery cable.

3. Raise and support the vehicle safely.

4. Remove the tire and wheel assembly.

5. Remove the engine undercover, as required.

6. Remove the fender protector.

7. Remove the brake hose and position it to the side. Discard the nut.

8. Properly support the lower control arm, as required.

9. Remove the cotter pin and nut from the upper link ball joint. Discard the pin.

10. Separate the upper link ball joint stud from the knuckle using tool ST2902001, or equivalent.

11. Remove the upper link bolts and nuts.

12. Remove the component from its mounting.

To install:

➡ **Be sure to use new fasteners, as required.**

13. Installation is the reverse of the removal procedure.

14. Check and adjust alignment.

15. Be sure to perform the reconnect/relearn procedures.

WHEEL HUB & BEARING

REMOVAL & INSTALLATION

See Figure 215.

1. Before servicing the vehicle, refer to the Precautions.

2. Disconnect the negative battery cable.

3. Raise and safely support the vehicle.

4. Remove the engine undercover, as required.

5. Remove wheel and tire.

6. Without disassembling the hydraulic lines, remove caliper torque member bolts.

7. Reposition brake caliper aside with wire. Do not allow the caliper to hang by the brake line.

✳✳ WARNING

Do not press brake pedal while brake caliper is removed.

8. Disconnect the wheel sensor harness connector. Remove the harness from the mounting brackets.

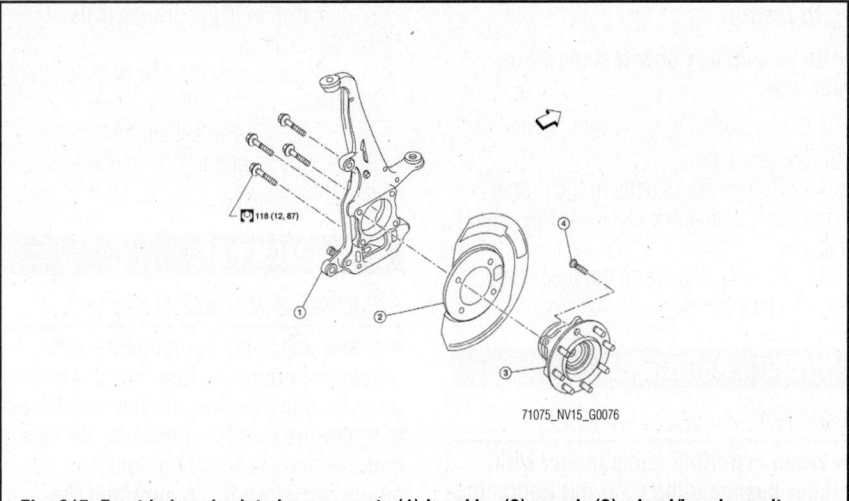

Fig. 215 Front hub and related components (1) knuckle, (2) guard, (3) wheel/bearing unit

9. Remove the wheel hub/bearing mounting bolts.

10. Remove the component from the vehicle.

To install:

➡**Be sure to use new fasteners, as required.**

11. Installation is the reverse of the removal procedure.

12. Use new bolts when installing the wheel hub and bearing assembly.

13. When installing disc rotor on wheel hub and bearing assembly, position the disc rotor according to alignment mark.

14. Check and adjust alignment.

15. Be sure to perform the reconnect/relearn procedures.

SUSPENSION   REAR SUSPENSION

LEAF SPRINGS

REMOVAL & INSTALLATION

See Figures 216 through 218.

➡**When installing components with rubber bushings the final nut tightening must be done under unladen conditions with the tires on level ground. Be sure that the fuel, radiator coolant and engine oil are full. Be sure that the spare tire, jack, factory hand tools and floor mats are in their designated positions.**

1. Before servicing the vehicle, refer to the Precautions.

2. Disconnect the negative battery cable.

3. Raise and safely support the vehicle.

4. Remove the tire and wheel assemblies.

5. Support the rear axle, using a suitable jack, to relieve spring tension from the spring. The axle weight must be supported, but there should be no compression in the spring.

6. Remove the four rear spring U-bolt nuts. Remove the rear spring lower seat.

7. Remove the U-bolts and the rear spring upper seat.

8. Remove the body plug, to accommodate the removal of the spring front bolt, left side only.

9. Remove the spring front nut and bolt.

10. Loosen the spring shackle upper nut and bolt.

11. Remove the shackle to frame lower nut and bolt.

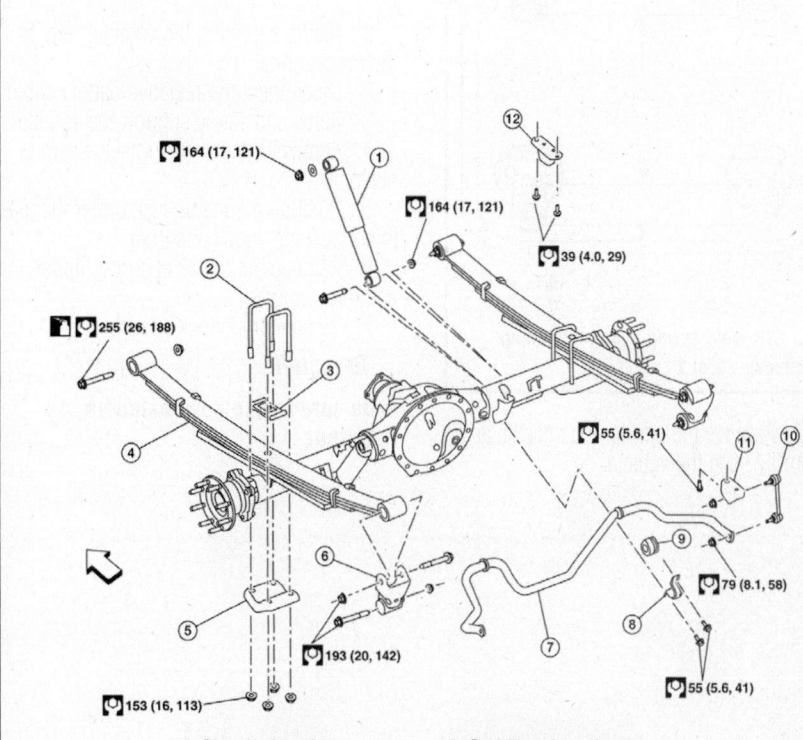

1. Shock absorber
2. Rear spring U-bolts
3. Rear spring upper seat
4. Rear leaf spring
5. Rear spring lower seat
6. Shackle assembly
7. Stabilizer bar
8. Stabilizer bar clamp
9. Stabilizer bar bushing
10. Connecting rod
11. Connecting rod bracket
12. Bumper assembly
ARROW:Front

Fig. 216 Rear leaf springs and related components

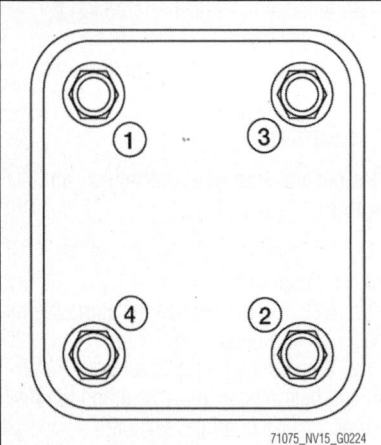

Fig. 217 Rear spring U-bolt tightening sequence—1 of 2

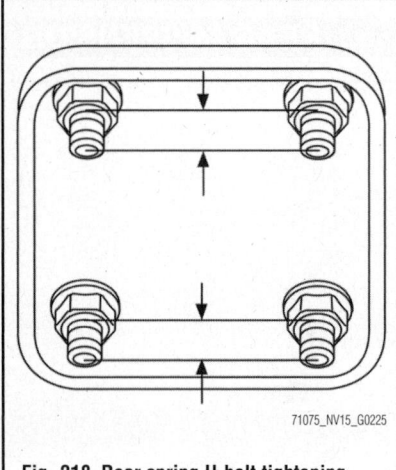

Fig. 218 Rear spring U-bolt tightening sequence—2 of 2

12. Remove the spring and shackle as an assembly from the vehicle.

To install:

➡**Be sure to use new fasteners, as required.**

13. Installation is the reverse of the removal procedure.

14. Tighten the U-bolts in the proper sequence diagonally and evenly. See illustrations.

15. Be sure to perform the reconnect/relearn procedures.

SHOCK ABSORBERS

REMOVAL & INSTALLATION

➡**When installing components with rubber bushings the final nut tightening must be done under unladen conditions with the tires on level ground. Be sure that the fuel, radiator coolant and engine oil are full. Be sure that the spare tire, jack, factory hand tools and floor mats are in their designated positions.**

1. Before servicing the vehicle, refer to the Precautions.

2. Disconnect the negative battery cable.

3. Raise and safely support the vehicle.

4. Remove the tire and wheel assemblies.

5. Position a suitable jack under the axle to support the shock absorber.

6. Remove the shock absorber upper and lower end bolts.

7. Remove the shock absorber.

To install:

➡**Be sure to use new fasteners, as required.**

8. Installation is the reverse of the removal procedure.

   a. Tighten the nuts and bolts to specification.

   b. Check the wheel alignment.

9. Be sure to perform the reconnect/relearn procedures.

STABILIZER BAR

REMOVAL & INSTALLATION

➡**When installing components with rubber bushings the final nut tightening must be done under unladen conditions with the tires on level ground. Be sure that the fuel, radiator coolant and engine oil are full. Be sure that the spare tire, jack, factory hand tools and floor mats are in their designated positions.**

1. Before servicing the vehicle, refer to the Precautions.

2. Disconnect the negative battery cable.

3. Raise and safely support the vehicle.

4. Remove the tire and wheel assemblies.

5. Disconnect the stabilizer bar ends from the connecting rods.

6. Remove the stabilizer bar clamps, and remove the stabilizer bar bushings.

7. Remove the stabilizer bar.

To install:

➡**Be sure to use new fasteners, as required.**

8. Installation is the reverse of the removal procedure.

9. Be sure to perform the reconnect/relearn procedures.

SPECIFICATIONS AND MAINTENANCE CHARTS

ENGINE AND VEHICLE IDENTIFICATION

Engine							Model Year	
Code ①	Liters (cc)	Cu. In.	Cyl.	Fuel Sys.	Engine	Eng. Mfg.	Code ②	Year
VQ40DE	4.0 (3954)	241	6	MFI	DOHC	Nissan	B	2011
VK56DE	5.6 (5552)	339	8	MFI	DOHC	Nissan	C	2012

MFI; Multi-port Fuel Injection

DOHC: Double Overhead Camshafts

① Fourth digit of the Vehicle Identification Number (VIN) is the engine ID, A: VQ40DE engine. B: VK56DE engine.

② Tenth digit of the Vehicle Identification Number (VIN)

71075_PATH_C0001

GENERAL ENGINE SPECIFICATIONS

Year	Model	Engine Displacement Liters (cc)	Engine ID/VIN	Fuel System Type	Net Horsepower @ rpm	Net Torque @ rpm (ft. lbs.)	Bore x Stroke (in.)	Compression Ratio	Oil Pressure @ rpm
2011	Pathfinder	4.0 (3954)	VQ40DE	MFI	266@5600	288@4000	3.76X3.62	9.7:1	43@2000
		5.6 (5552)	VQ56DE	MFI	310@5200	388@3400	3.86X3.62	9.8:1	43@2000
2012	Pathfinder	4.0 (3954)	VQ40DE	MFI	266@5600	288@4000	3.76X3.62	9.7:1	43@2000
		5.6 (5552)	VQ56DE	MFI	310@5200	388@3400	3.86X3.62	9.8:1	43@2000

MFI: Multi-port Fuel Injection

71075_PATH_C0002

ENGINE TUNE-UP SPECIFICATIONS

Year	Engine Displacement Liters (cc)	Engine ID/VIN	Spark Plug Gap (in.)	Ignition Timing (deg.) MT	AT	Fuel Pump (psi)	Idle Speed (rpm) MT	AT	Valve Clearance (in.) In.	Ex.
2011	4.0 (3954)	VQ40DE	0.043	NA	①	②	NA	575-675	HYD	HYD
	5.6 (5552)	VQ56DE	0.043	NA	①	②	NA	600-700	HYD	HYD
2012	4.0 (3954)	VQ40DE	0.043	NA	①	②	NA	575-675	HYD	HYD
	5.6 (5552)	VQ56DE	0.043	NA	①	②	NA	600-700	HYD	HYD

NOTE: The Vehicle Emission Control Information label often reflects specification changes made during production. The label figures must be used if they differ from those in this chart.

NA- Not Available

HYD: Hydraulic

① 15 degrees +/- 5 degrees

② 51 psi at idle

71075_PATH_C0003

CAPACITIES

Year	Model	Engine Displacement Liters	Engine ID/VIN	Engine Oil with Filter (qts.)	Transmission (pts.) 5-Spd	Transmission (pts.) Auto.	Transfer Case (pts.)	Drive Axle Front (pts.)	Drive Axle Rear (pts.)	Fuel Tank (gal.)	Cooling System (qts.)
2011	Pathfinder	4.0	VQ40DE	5.38	—	21.56	①	1.75	3.00	21.2	②
		5.6	VQ56DE	6.78	—	22.50	①	3.38	3.75	21.2	②
2012	Pathfinder	4.0	VQ40DE	5.38	—	21.56	①	1.75	3.00	21.2	②
		5.6	VQ56DE	6.78	—	22.50	①	3.38	3.75	21.2	②

NOTE: All capacities are approximate. Add fluid gradually and check to be sure a proper fluid level is obtained.

① Model ATX14B: 6.36 pts. Model TX15B: 4.36 pts.

② Without rear A/C: 10.8 qts. With rear A/C: 14.18 qts.

71075_PATH_C0004

FLUID SPECIFICATIONS

Year	Model	Engine Displ. Liters	Engine Oil	Auto. Trans.	Drive Axle Front	Drive Axle Rear	Transfer Case	Power Steering Fluid	Brake Master Cylinder	Cooling System
2011	Pathfinder	4.0	SAE 5W-30	①	②	②	③	④	⑤	⑥
		5.6	SAE 5W-30	①	②	②	③	④	⑤	⑥
2012	Pathfinder	4.0	SAE 5W-30	①	②	②	③	④	⑤	⑥
		5.6	SAE 5W-30	①	②	②	③	④	⑤	⑥

① Genuine Nissan Matic S ATF fluid. If not available genuine Nissan Matic J ATF may be used.

② Front: Nissan differential oil hypoid super GL-5 80W-90 or API GL-5 viscosity SAE 80W-90 gear oil.

   Rear: Nissan differential oil synthetic 75W-90 or API GL-5 synthetic gear oil, viscosity SAE 75W-90.

③ Nissan Matic D ATF

④ Nissan Power Steering Fluid

⑤ Nissan Super Heavy Duty DOT 3

⑥ Nissan Long Life antifreeze (BLUE) or equivalent

71075_PATH_C0012

VALVE SPECIFICATIONS

Year	Engine Displacement Liters	Engine ID/VIN	Seat Angle (deg.)	Face Angle (deg.)	Spring Test Pressure (lbs. @ in.)	Spring Installed Height (in.)	Stem-to-Guide Clearance (in.) Intake	Stem-to-Guide Clearance (in.) Exhaust	Stem Diameter (in.) Intake	Stem Diameter (in.) Exhaust
2011	4.0	VQ40DE	45.15-45.45	45	37-42 @1.457	1.457	0.0008-0.0021	0.0012-0.0025	0.2348-0.2354	0.2344-0.2350
	5.6	VQ56DE	45.15-45.45	45	37-42 @1.457	1.991	0.0008-0.0021	0.0012-0.0025	0.2348-0.2354	0.2344-0.2350
2012	4.0	VQ40DE	45.15-45.45	45	37-42 @1.457	1.457	0.0008-0.0021	0.0012-0.0025	0.2348-0.2354	0.2344-0.2350
	5.6	VQ56DE	45.15-45.45	45	37-42 @1.457	1.991	0.0008-0.0021	0.0012-0.0025	0.2348-0.2354	0.2344-0.2350

71075_PATH_C0006

CAMSHAFT AND BEARING SPECIFICATIONS CHART

All measurements are given in inches.

Year	Engine Displacement Liters	Engine ID/VIN	Journal Diameter	Brg. Oil Clearance	Shaft End-play	Runout	Journal Bore	Lobe Height	
								Intake	Exhaust
2011	4.0	VQ40DE	①	②	0.0045-0.0074	0.0008	③	1.7900-1.7921	1.7746-1.7821
	5.6	VQ56DE	1.0217-1.0224	0.0012-0.0028	0.0045-0.0074	0.0008	1.0236-1.0244	1.7663-1.7738	1.7746-1.7821
2012	4.0	VQ40DE	①	②	0.0045-0.0074	0.001	③	1.7900-1.7921	1.7746-1.7821
	5.6	VQ56DE	1.0278-1.0224	0.0012-0.0028	0.0045-0.0074	0.0008	1.0236-1.0244	1.7663-1.7738	1.7746-1.7821

① No. 1: 1.0211-1.0218
  Nos. 2-4: 0.9230-0.9238

② No. 1: 0.0018-0.0034
  Nos. 2-4: 0.0014-0.0030

③ No. 1: 1.0236-1.0244
  Nos. 2-4: 0.9252-0.9260

71075_PATH_C0014

CRANKSHAFT AND CONNECTING ROD SPECIFICATIONS

All measurements are given in inches.

Year	Engine Displacement Liters	Engine ID/VIN	Crankshaft				Connecting Rod		
			Main Brg. Journal Dia.	Main Brg. Oil Clearance	Shaft End-play	Thrust on No.	Journal Diameter	Oil Clearance	Side Clearance
2011	4.0	VQ40DE	①	0.0014-0.0018	0.0039-0.0098	4	NA	0.0013-0.0023	0.0079-0.0138
	5.6	VQ56DE	②	③	0.0039-0.0102	4	④	0.0008-0.0015	0.0079-0.0157
2012	4.0	VQ40DE	①	0.0014-0.0018	0.0039-0.0098	4	NA	0.0013-0.0023	0.0079-0.0138
	5.6	VQ56DE	②	③	0.0039-0.0102	4	④	0.0008-0.0015	0.0079-0.0157

NA - Not Available

① There are 24 different grades, ranging from grade A (2.7549) to grade 7 (2.7540)

② There are 24 different grades, ranging from grade A (2.5182) to grade 2 (2.5174)

③ No. 1 and 5: 0.00004-0.0004

  No. 2, 3 and 4: 0.0003-0.0007

④ There are 13 different grades, ranging from 2.2441- 2.2446. Specification is for rod bearing housing.

71075_PATH_C0005

PISTON AND RING SPECIFICATIONS

All measurements are given in inches.

Year	Engine Displacement Liters	Engine ID/VIN	Piston Clearance	Ring Gap			Ring Side Clearance		
				Top Comp.	Bottom Comp.	Oil Control	Top Comp.	Bottom Comp.	Oil Control
2011	4.0	VQ40DE	0.0004- 0.0012	0.0091- 0.0130	0.0130- 0.0189	0.0079- 0.0197	0.0018- 0.0031	0.0012- 0.0028	0.0026- 0.0053
	5.6	VQ56DE	0.0004- 0.0012	0.0091- 0.0130	0.098- 0.0157	0.0079- 0.0236	0.0014- 0.0033	0.0012- 0.0028	0.0006- 0.0073
2012	4.0	VQ40DE	0.0004- 0.0012	0.0091- 0.0130	0.0130- 0.0189	0.0079- 0.0197	0.0018- 0.0031	0.0012- 0.0028	0.0026- 0.0053
	5.6	VQ56DE	0.0004- 0.0012	0.0091- 0.0130	0.098- 0.0157	0.0079- 0.0236	0.0014- 0.0033	0.0012- 0.0028	0.0006- 0.0073

71075_PATH_C0007

TORQUE SPECIFICATIONS

All readings in ft. lbs.

Year	Engine Displacement Liters	Engine ID/VIN	Cylinder Head Bolts	Main Bearing Bolts	Rod Bearing Bolts	Crankshaft Damper Bolts	Flywheel Bolts	Manifold		Spark Plugs
								Intake	Exhaust	
2011	4.0	VQ40DE	①	②	③	④	65	⑤	22	18
	5.6	VQ56DE	⑥	⑦	⑧	⑨	65	NA	25	18
2012	4.0	VQ40DE	①	②	③	④	65	⑤	22	18
	5.6	VQ56DE	⑥	⑦	⑧	⑨	65	NA	25	18

NA - Not Available

① Step 1: 72 ft. lbs.

Step 2: Loosen all bolts completely

Step 3: 29 ft. lbs.

Step 4: +90 degrees

Step 5: +90 degrees

② Step 1: 26 ft. lbs.

Step 2: +90 degrees

③ Step 1: 14 ft. lbs.

Step 2: +90 degrees

④ 33 ft. lbs. +84 - 90 degrees

⑤ Step 1: 65 inch lbs.

Step 2 and after: 21 ft. lbs.

Studs 8 ft. lbs.

⑥ Step 1: 33 ft. lbs

Step 2: +70 degrees clockwise

Step 3: loosen in reverse order of tightening sequence

Step 4: 33 ft. lbs.

Step 5: +60 degrees clockwise

Step 6: +60 degrees clockwise

⑦ Step 1: cap bolts in order 1-10: 29 ft. lbs.

Step 2: cap sub bolts in order 11-20: 22 ft. lbs.

Step 3: cap bolts in order 1-10: +40 degrees

Step 4: cap sub bolts in order 11-20: +30 degrees

Step 5: side bolts in order 21-30: 36 ft. lbs.

⑧ Step 1: 11 ft. lbs.

Step 2: +90 degrees

⑨ 69 ft. lbs. +90 degrees

71075_PATH_C0008

WHEEL ALIGNMENT

Year	Model		Caster Range (+/-Deg.)	Caster Preferred Setting (Deg.)	Camber Range (+/-Deg.)	Camber Preferred Setting (Deg.)	Toe-in (in.)
2011	Pathfinder	2WD	①	①	②	②	NA
		4WD	③	③	④	②	NA
2012	Pathfinder	2WD	①	①	②	②	NA
		4WD	③	③	④	②	NA

NA - Not Available

① Minimum: 2 degrees 15' (2.25 degrees). Nominal: 3 degrees 0' (3.00 degrees). Maximum: 3 degrees 45' (3.75 degrees).

② Minimum: -0 degrees 30' (-0.50 degrees). Nominal: 0 degrees 15' (0.25 degrees). Maximum: 1 degree 00' (1.00 degree).

③ Minimum: 2 degrees 00' (2.00 degrees). Nominal: 2 degrees 45' (2.75 degrees). Maximum: 3 degrees 30' (3.50 degrees).

④ Minimum: -0 degrees 15' (-0.25 degrees). Nominal: 0 degrees 30' (0.50 degrees). Maximum: 1 degree 15' (1.25 degrees).

71075_PATH_C0009

TIRE, WHEEL AND BALL JOINT SPECIFICATIONS

Year	Model	OEM Tires Standard	OEM Tires Optional	Tire Pressures (psi) Front	Tire Pressures (psi) Rear	Wheel Size	Ball Joint Inspection	Lug Nuts (ft. lbs.)
2011	Pathfinder	P245/75R16	none	35	35	16x7JJ	①	98
		P265/65R17	none	35	35	17x7.5JJ	①	98
		P265/60R18	none	35	35	18x8JJ	①	98
2012	Pathfinder	P245/75R16	none	35	35	16x7JJ	①	98
		P265/65R17	none	35	35	17x7.5JJ	①	98
		P265/60R18	none	35	35	18x8JJ	①	98

OEM: Original Equipment Manufacturer

PSI: Pounds Per Square Inch

① Axial play

   Upper: 0

   Lower: 0.008 in.

71075_PATH_C0010

BRAKE SPECIFICATIONS

All measurements in inches unless noted

Year	Model		Brake Disc Original Thickness	Brake Disc Minimum Thickness	Brake Disc Maximum Runout	Minimum Lining Thickness Front	Minimum Lining Thickness Rear	Brake Caliper Bracket Bolts (ft. lbs.)	Brake Caliper Mounting Bolts (ft. lbs.)
2011	Pathfinder	F	1.102	1.024	0.0006	0.079	0.079	①	①
		R	0.079	0.630	0.0006	0.079	0.079	②	②
2012	Pathfinder	F	1.102	1.024	0.0006	0.079	0.079	①	①
		R	0.079	0.630	0.0006	0.079	0.079	②	②

NA: Not Available

① Torque member mounting bolt: 136 ft. lbs.

   Sliding pin bolt 20 ft. lbs.

② Torque member mounting bolt: 76 ft. lbs.

   Sliding pin bolt 20 ft. lbs.

71075_PATH_C0011

SCHEDULED MAINTENANCE INTERVALS (1)
Nissan—Pathfinder

TO BE SERVICED	TYPE OF SERVICE	7.5	15	22.5	30	37.5	45	52.5	60
Engine oil & filter	R	every 3,750 miles							
Brake lines & cables	S/I		✓		✓		✓		✓
Brake pads, discs	I	✓	✓	✓	✓	✓	✓	✓	✓
Brake fluid	R		✓		✓		✓		✓
Driveshaft boots	I	✓	✓	✓	✓	✓	✓	✓	✓
Automatic transmission (CVT)	I						✓		✓
Air cleaner filter ①	R				✓				✓
Drive belt (s) ②	S/I								✓
Engine coolant ③	R								
Spark plugs	R	Every 105,000 miles							
Cabin air filter	R		✓		✓		✓		✓
Exhaust system	I	✓	✓	✓	✓	✓	✓	✓	✓
Evap vapor lines	I				✓				✓
Fuel lines	I				✓				✓
Steering gear, linkage, axle & suspension parts	I	✓	✓	✓	✓	✓	✓	✓	✓
Tires (rotate)	S/I	✓	✓	✓	✓	✓	✓	✓	✓
Valve clearance ④	S/I				✓				✓

R: Replace   S/I: Service or Inspect   L: Lubricate   I: Inspect

① If operating mainly in dusty conditions, more frequent maintenance may be required.

② First at 60,000, inspect every 15,000 miles and replace as necessary

③ After 105,000, replace every 75,000 thereafter

④ Periodic maintenance not required, if valve noise increases, inspect valve clearance

Follow Periodic Maintenance Schedule 1 if the driving habits frequently include one or more of the following driving conditions:

Repeated short trips of less than 5 miles (8 km).

Repeated short trips of less than 10 miles (16 km) with outside temperatures remaining below freezing

Operating in hot weather in stop-and-go "rush hour" traffic.

Extensive idling and/or low speed driving for long distances, such as police, taxi or door-to-door delivery use

Driving in dusty conditions.

Driving on rough, muddy, or salt spread roads.

Towing a trailer, using a camper or a car-top carrier.

Follow Periodic Maintenance Schedule 2 if none of driving conditions shown in Schedule 1 apply to the driving habits.

SCHEDULED MAINTENANCE INTERVALS (2)
Nissan—Pathfinder

TO BE SERVICED	SERVICE	7.5	15	22.5	30	37.5	45	52.5	60
Engine oil & filter	R	✓	✓	✓	✓	✓	✓	✓	✓
Brake lines & cables	S/I		✓		✓		✓		✓
Brake pads, discs	I		✓		✓		✓		✓
Brake fluid	R				✓				✓
Driveshaft boots	I		✓		✓		✓		✓
Automatic transmission (CVT)	I		✓		✓		✓		✓
Air cleaner filter ①	R				✓				✓
Drive belt (s) ②	S/I								✓
Engine coolant ③	R								
Spark plugs	R	Every 105,000 miles							
Cabin air filter	R		✓		✓		✓		✓
Exhaust system	I				✓				✓
Evap vapor lines	I				✓				✓
Ful lines	I				✓				✓
Steering gear, linkage, axle & suspension parts	I				✓				✓
Tires (rotate)	S/I	✓	✓	✓	✓	✓	✓	✓	✓
Valve clearance ④	S/I				✓				✓

R: Replace   S/I: Service or Inspect   L: Lubricate   I: Inspect

① If operating mainly in dusty conditions, more frequent maintenance may be required.

② First at 60,000, inspect every 15,000 miles and replace as necessary

③ After 105,000, replace every 75,000 thereafter

④ Periodic maintenance not required, if valve noise increases, inspect valve clearance

Follow Periodic Maintenance Schedule 1 if the driving habits frequently include one or more of the following driving conditions:

Repeated short trips of less than 5 miles (8 km).

Repeated short trips of less than 10 miles (16 km) with outside temperatures remaining below freezing

Operating in hot weather in stop-and-go "rush hour" traffic.

Extensive idling and/or low speed driving for long distances, such as police, taxi or door-to-door delivery use

Driving in dusty conditions.

Driving on rough, muddy, or salt spread roads.

Towing a trailer, using a camper or a car-top carrier.

Follow Periodic Maintenance Schedule 2 if none of driving conditions shown in Schedule 1 apply to the driving habits.

71075_PATH_C0015

PRECAUTIONS

Before servicing any vehicle, please be sure to read all of the following precautions, which deal with personal safety, prevention of component damage, and important points to take into consideration when servicing a motor vehicle:

• Never open, service or drain the radiator or cooling system when the engine is hot; serious burns can occur from the steam and hot coolant.

• Observe all applicable safety precautions when working around fuel. Whenever servicing the fuel system, always work in a well-ventilated area. Do not allow fuel spray or vapors to come in contact with a spark, open flame, or excessive heat (a hot drop light, for example). Keep a dry chemical fire extinguisher near the work area. Always keep fuel in a container specifically designed for fuel storage; also, always properly seal fuel containers to avoid the possibility of fire or explosion. Refer to the additional fuel system precautions later in this section.

• Fuel injection systems often remain pressurized, even after the engine has been turned **OFF**. The fuel system pressure must be relieved before disconnecting any fuel lines. Failure to do so may result in fire and/or personal injury.

• Brake fluid often contains polyglycol ethers and polyglycols. Avoid contact with the eyes and wash your hands thoroughly after handling brake fluid. If you do get brake fluid in your eyes, flush your eyes with clean, running water for 15 minutes. If eye irritation persists, or if you have taken brake fluid internally, IMMEDIATELY seek medical assistance.

• The EPA warns that prolonged contact with used engine oil may cause a number of skin disorders, including cancer. You should make every effort to minimize your exposure to used engine oil. Protective gloves should be worn when changing oil. Wash your hands and any other exposed skin areas as soon as possible after exposure to used engine oil. Soap and water, or waterless hand cleaner should be used.

• All new vehicles are now equipped with an air bag system, often referred to as a Supplemental Restraint System (SRS) or Supplemental Inflatable Restraint (SIR) system. The system must be disabled before performing service on or around system components, steering column, instrument panel components, wiring and sensors. Failure to follow safety and disabling procedures could result in accidental air bag deployment, possible personal injury and unnecessary system repairs.

• Always wear safety goggles when working with, or around, the air bag system. When carrying a non-deployed air bag, be sure the bag and trim cover are pointed away from your body. When placing a non-deployed air bag on a work surface, always face the bag and trim cover upward, away from the surface. This will reduce the motion of the module if it is accidentally deployed. Refer to the additional air bag system precautions later in this section.

• Clean, high quality brake fluid from a sealed container is essential to the safe and proper operation of the brake system. You should always buy the correct type of brake fluid for your vehicle. If the brake fluid becomes contaminated, completely flush the system with new fluid. Never reuse any brake fluid. Any brake fluid that is removed from the system should be discarded. Also, do not allow any brake fluid to come in contact with a painted surface; it will damage the paint.

• Never operate the engine without the proper amount and type of engine oil; doing so WILL result in severe engine damage.

• Timing belt maintenance is extremely important. Many models utilize an interference-type, non-freewheeling engine. If the timing belt breaks, the valves in the cylinder head may strike the pistons, causing potentially serious (also time-consuming and expensive) engine damage. Refer to the maintenance interval charts for the recommended replacement interval for the timing belt, and to the timing belt section for belt replacement and inspection.

• Disconnecting the negative battery cable on some vehicles may interfere with the functions of the on-board computer system(s) and may require the computer to undergo a relearning process once the negative battery cable is reconnected.

• When servicing drum brakes, only disassemble and assemble one side at a time, leaving the remaining side intact for reference.

• Only an MVAC-trained, EPA-certified automotive technician should service the air conditioning system or its components.

BRAKES

GENERAL INFORMATION

PRECAUTIONS

• Certain components within the ABS system are not intended to be serviced or repaired individually.

• Do not use rubber hoses or other parts not specifically specified for and ABS system. When using repair kits, replace all parts included in the kit. Partial or incorrect repair may lead to functional problems and require the replacement of components.

• Lubricate rubber parts with clean, fresh brake fluid to ease assembly. Do not use shop air to clean parts; damage to rubber components may result.

• Use only DOT 3 brake fluid from an unopened container.

• If any hydraulic component or line is removed or replaced, it may be necessary to bleed the entire system.

• A clean repair area is essential. Always clean the reservoir and cap thoroughly before removing the cap. The slightest amount of dirt in the fluid may plug an orifice and impair the system function. Perform repairs after components have been thoroughly cleaned; use only denatured alcohol to clean components. Do not allow ABS components to come into contact with any substance containing mineral oil; this includes used shop rags.

• The Anti-Lock control unit is a microprocessor similar to other computer units in the vehicle. Ensure that the ignition switch

ANTI-LOCK BRAKE SYSTEM (ABS)

is **OFF** before removing or installing controller harnesses. Avoid static electricity discharge at or near the controller.

• If any arc welding is to be done on the vehicle, the control unit should be unplugged before welding operations begin.

SPEED SENSORS

REMOVAL & INSTALLATION
See Figure 1.

1. Before servicing the vehicle, refer to the Precautions.
2. Disconnect the negative battery cable.
3. Raise and support the vehicle safely.
4. Remove wheel and tire.

5. Remove wheel sensor mounting screw.

➡When removing front wheel sensor, first remove the disc rotor to gain access to the front wheel sensor mounting bolt. When removing rear wheel sensor, first remove spare tire.

6. Pull out the wheel sensor, being careful to turn it as little as possible.

✳✳ WARNING

Be careful not to damage sensor edge and sensor rotor teeth. Do not pull on the sensor harness.

7. Disconnect wheel sensor harness electrical connector, then remove harness from mounts.

To install:

➡Be sure to use new fasteners, as required.

8. Installation is the reverse of the removal procedure.

9. Be sure to perform the reconnect/relearn procedures.

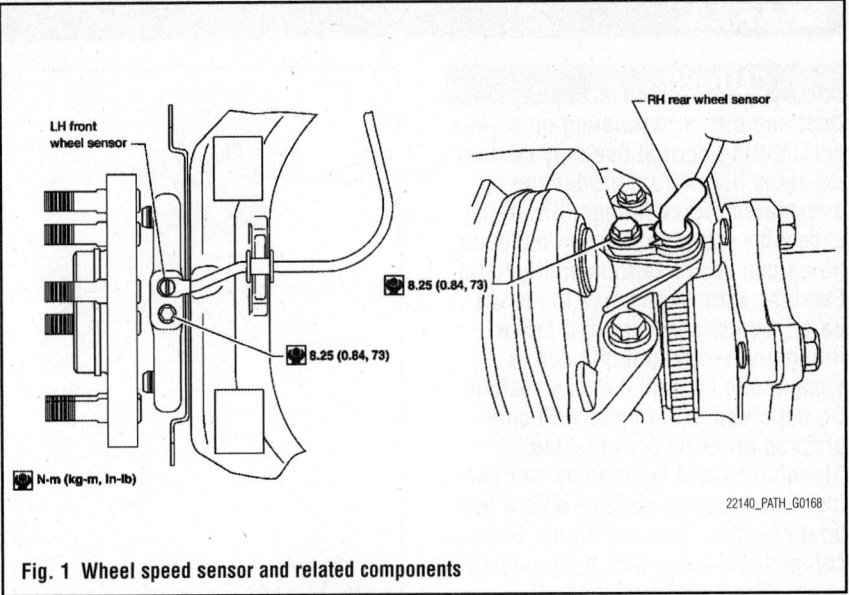

Fig. 1 Wheel speed sensor and related components

✳✳ WARNING

Inspect wheel sensor O-ring, replace sensor assembly if damaged. Clean wheel sensor hole and mounting surface with brake cleaner and a lint-free shop rag. Be careful that dirt and debris do not enter the axle.

➡Apply a coat of suitable grease to the wheel sensor O-ring and mounting hole.

10. Tighten wheel sensor bolts to 73 inch lbs. (8.25 Nm).

BRAKES

BLEEDING THE BRAKE SYSTEM

BLEEDING PROCEDURE

BLEEDING PROCEDURE

➡Be sure that the master cylinder is full of clean fresh brake fluid before starting the bleeding process. Use only the recommended brake fluid when bleeding the system. Do not allow brake fluid to spill on painted surfaces as damage will occur.

1. Before servicing the vehicle, refer to the Precautions.

✳✳ CAUTION

While bleeding, monitor the master cylinder brake fluid level.

2. Turn the ignition switch OFF. Disconnect the ABS actuator and electric control unit connector.

3. Connect a vinyl tube to the rear right bleed valve. Be sure to have a catch pan handy to catch excess brake fluid.

4. Fully depress the brake pedal four or five times.

5. With the brake pedal depressed, loosen the bleed valve to let air out, then tighten it immediately.

6. Repeat the above steps until all air is removed from the system. Be sure to keep watch on the brake fluid level and replenish, as necessary.

7. Tighten the bleed valve.

8. Repeat the above steps at each wheel, with the master cylinder reservoir tank filled at least half way.

9. Bleed the remaining components in the following order: front left, rear left and front right.

10. Be sure to perform the reconnect/relearn procedures.

BLEEDING THE ABS SYSTEM

➡Be sure that the master cylinder is full of clean fresh brake fluid before starting the bleeding process. Use only the recommended brake fluid when bleeding the system. Do not allow brake fluid to spill on painted surfaces as damage will occur.

1. Before servicing the vehicle, refer to the Precautions.

✳✳ CAUTION

While bleeding, monitor the master cylinder brake fluid level.

2. Disconnect the negative battery cable.

3. Turn the ignition switch OFF. Disconnect the ABS actuator and electric control unit connector.

4. Connect a vinyl tube to the rear right bleed valve. Be sure to have a catch pan handy to catch excess brake fluid.

5. Fully depress the brake pedal four or five times.

6. With the brake pedal depressed, loosen the bleed valve to let air out, then tighten it immediately.

7. Repeat the above steps until all air is removed from the system. Be sure to keep watch on the brake fluid level and replenish, as necessary.

8. Tighten the bleed valve.

9. Repeat the above steps at each wheel, with the master cylinder reservoir tank filled at least half way.

10. Bleed the remaining components in the following order: front left, rear left and front right.

11. Be sure to perform the reconnect/relearn procedures.

BRAKES

FRONT DISC BRAKES

✳✳ CAUTION

Dust and dirt accumulating on brake parts during normal use may contain asbestos fibers from production or aftermarket brake linings. Breathing excessive concentrations of asbestos fibers can cause serious bodily harm. Exercise care when servicing brake parts. Do not sand or grind brake lining unless equipment used is designed to contain the dust residue. Do not clean brake parts with compressed air or by dry brushing. Cleaning should be done by dampening the brake components with a fine mist of water, then wiping the brake components clean with a dampened cloth. Dispose of cloth and all residue containing asbestos fibers in an impermeable container with the appropriate label. Follow practices prescribed by the Occupational Safety and Health Administration (OSHA) and the Environmental Protection Agency (EPA) for the handling, processing, and disposing of dust or debris that may contain asbestos fibers.

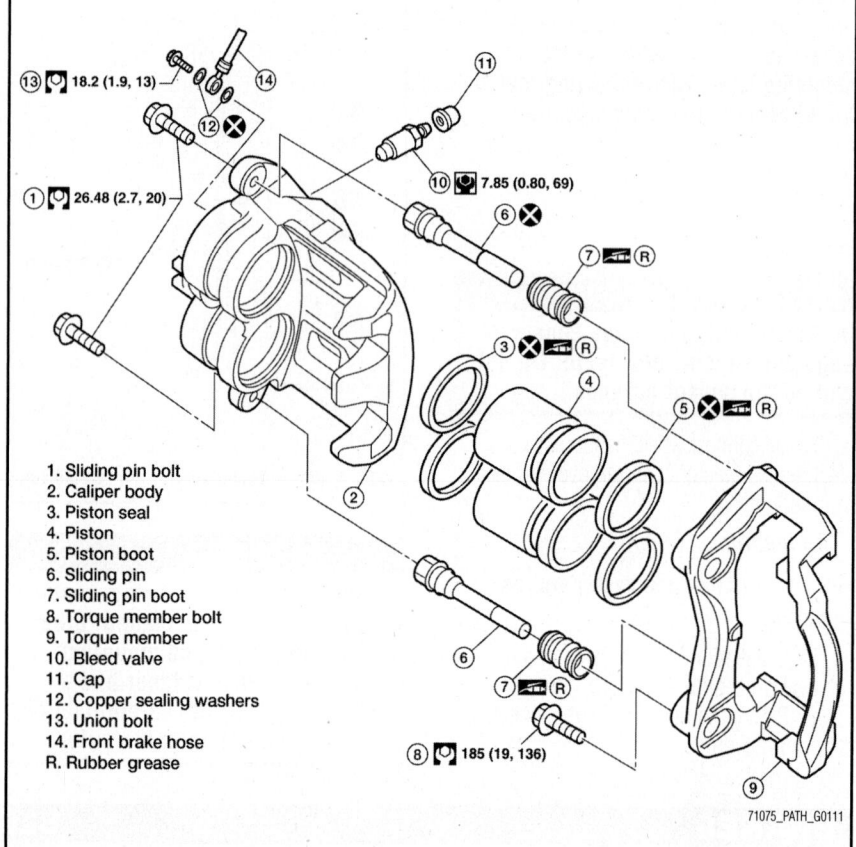

1. Sliding pin bolt
2. Caliper body
3. Piston seal
4. Piston
5. Piston boot
6. Sliding pin
7. Sliding pin boot
8. Torque member bolt
9. Torque member
10. Bleed valve
11. Cap
12. Copper sealing washers
13. Union bolt
14. Front brake hose
R. Rubber grease

71075_PATH_G0111

Fig. 2 Front brake caliper and related components

BRAKE CALIPERS

REMOVAL & INSTALLATION

See Figure 2.

1. Before servicing the vehicle, refer to the Precautions.
2. Disconnect the negative battery cable.
3. Raise and support the vehicle safely. Remove wheel and tire.
4. Drain brake fluid as necessary.

➡ Do not remove union bolt unless removing cylinder body from vehicle.

5. Remove union bolt as necessary and torque member bolts, then remove cylinder body from the vehicle.
6. Position cylinder body aside using suitable wire, as necessary.
7. When servicing brake caliper, remove sliding pin bolts and caliper from torque member.
8. Remove torque member.
9. Remove disc rotor.

To install:

➡ Be sure to use new fasteners, as required.

✳✳ CAUTION

Refill with new brake fluid. Do not reuse drained brake fluid.

10. Install disc rotor.
11. Install torque member and tighten to specification.
12. Install sliding pin bolts, if removed.
13. Install cylinder body, then tighten sliding pin bolts to specification.

✳✳ CAUTION

When attaching cylinder body to the vehicle, wipe any oil off knuckle spindle, washers and cylinder body attachment surfaces.

14. Install brake hose to cylinder body, if removed, then tighten union bolt to 24 ft. lbs. (32 Nm).

✳✳ CAUTION

Do not reuse copper washers for union bolt. Attach brake hose to cylinder body together with union bolt and washers.

15. Refill with new brake fluid as necessary and bleed air.
16. Install wheel and tire.
17. Be sure to perform the reconnect/relearn procedures.

BRAKE PADS

REMOVAL & INSTALLATION

See Figure 3.

1. Before servicing the vehicle, refer to the Precautions.
2. Disconnect the negative battery cable.
3. Raise and support the vehicle safely. Remove wheel and tire.
4. Remove master cylinder reservoir cap.
5. Remove lower sliding pin bolt.
6. Suspend cylinder body with a wire and remove pads, shim, shim covers, and retainers from torque member.

To install:

➡ Be sure to use new fasteners, as required.

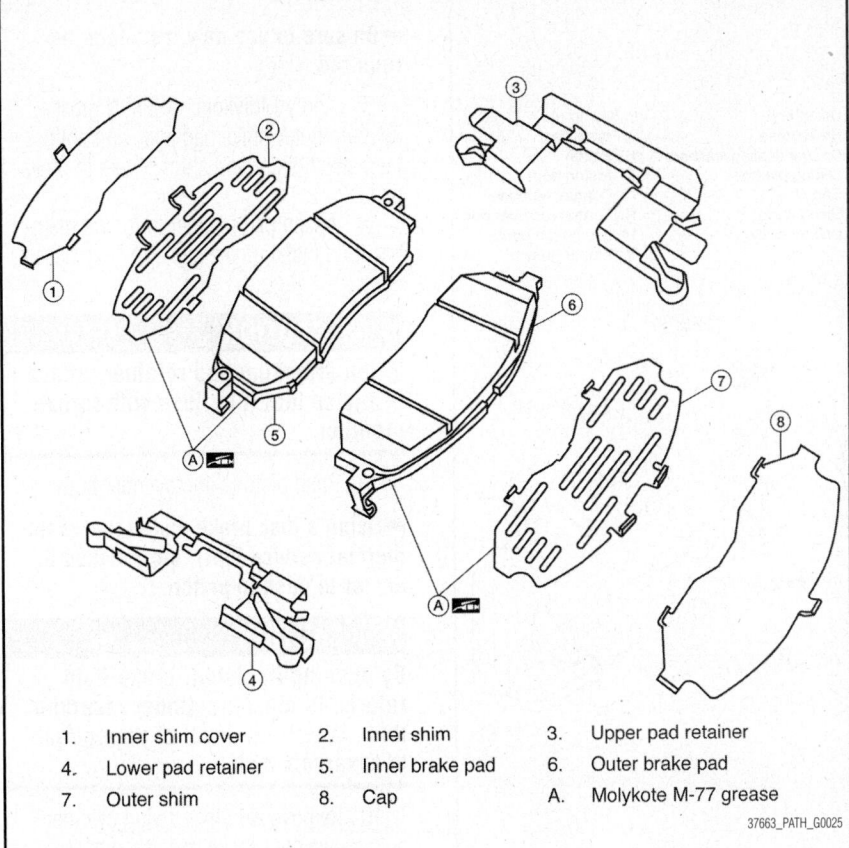

1. Inner shim cover
2. Inner shim
3. Upper pad retainer
4. Lower pad retainer
5. Inner brake pad
6. Outer brake pad
7. Outer shim
8. Cap
A. Molykote M-77 grease

37663_PATH_G0025

Fig. 3 Front brake pads and related components

7. Apply Molykote® AS880N grease between outer brake pad plate and shim, then attach shim and shim covers to brake pads.

8. Attach pad retainer to torque member, then install brake pad and shim assemblies.

✳✳ CAUTION

When attaching pad retainer, attach it firmly so that it is flush with torque member.

9. Push pistons into cylinder body.

➡**Using a disc brake piston tool (commercial service tool), etc., makes it easier to push in piston.**

✳✳ CAUTION

By pushing in piston, brake fluid returns to master cylinder reservoir tank. Watch the level of the surface of reservoir tank.

10. Remove wire then swing cylinder body down over brake pad assemblies.

11. Install lower sliding pin bolt and tighten to specification.

12. Check brake for drag.

13. Inspect fluid level, then install master cylinder reservoir cap.

14. Install wheel and tire.

15. Be sure to perform the reconnect/relearn procedures.

BRAKES

✳✳ CAUTION

Dust and dirt accumulating on brake parts during normal use may contain asbestos fibers from production or aftermarket brake linings. Breathing excessive concentrations of asbestos fibers can cause serious bodily harm. Exercise care when servicing brake parts. Do not sand or grind brake lining unless equipment used is designed to contain the dust residue. Do not clean brake parts with compressed air or by dry brushing. Cleaning should be done by dampening the brake components with a fine mist of water, then wiping the brake components clean with a dampened cloth. Dispose of cloth and all residue containing asbestos fibers in an impermeable container with the appropriate label. Follow practices prescribed by the Occupational Safety and Health Administration (OSHA) and the Environmental Pro-

tection Agency (EPA) for the handling, processing, and disposing of dust or debris that may contain asbestos fibers.

BRAKE CALIPER

REMOVAL & INSTALLATION

See Figure 4.

1. Before servicing the vehicle, refer to the Precautions.

2. Disconnect the negative battery cable.

3. Raise and support the vehicle safely. Remove wheel and tire.

4. Drain brake fluid.

5. Remove union bolt and mounting bolts, and remove cylinder body.

6. Remove torque member.

7. Remove disc rotor.

To install:

✳✳ CAUTION

Refill with new brake fluid. Do not reuse drained brake fluid.

REAR DISC BRAKES

8. Install disc rotor.

9. Install torque member and tighten to specification.

10. Install cylinder body to the vehicle, and tighten mounting bolts to specification.

✳✳ CAUTION

Before installing cylinder body to the vehicle, wipe off mounting surface of cylinder body.

11. Install brake hose to cylinder body and tighten union bolt to 19 ft. lbs. (26 Nm).

✳✳ CAUTION

Do not reuse copper washer for union bolt. Securely attach brake hose to protrusion on cylinder body.

12. Refill new brake fluid and bleed air.

13. Install wheels and tires to the vehicle.

14. Be sure to perform the reconnect/relearn procedures.

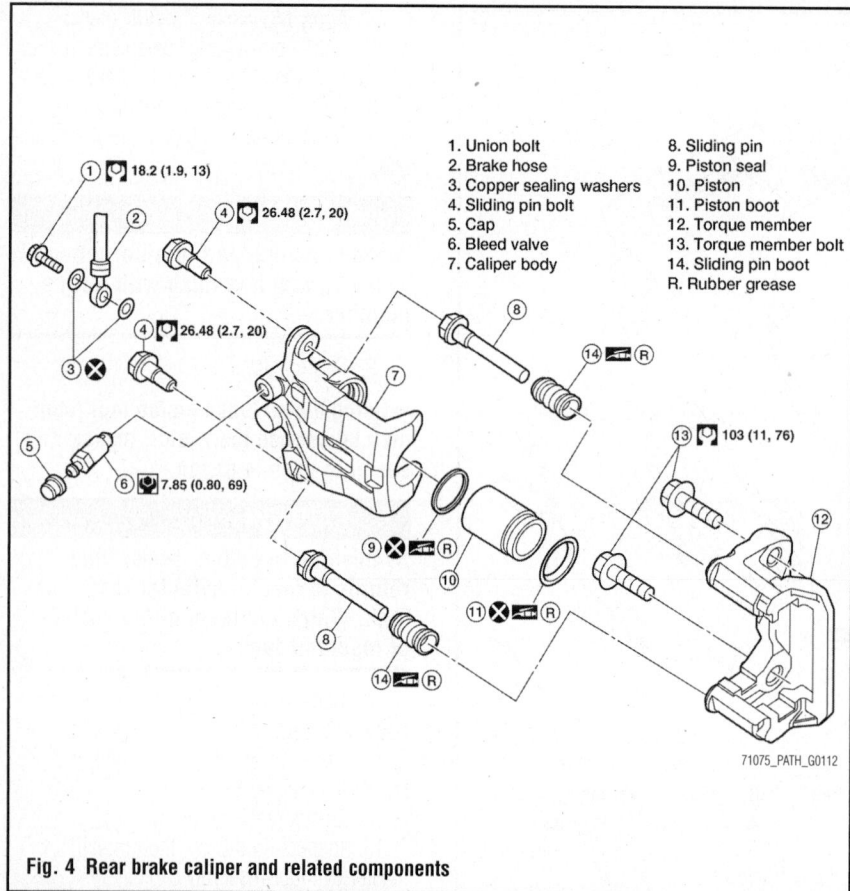

1. Union bolt
2. Brake hose
3. Copper sealing washers
4. Sliding pin bolt
5. Cap
6. Bleed valve
7. Caliper body
8. Sliding pin
9. Piston seal
10. Piston
11. Piston boot
12. Torque member
13. Torque member bolt
14. Sliding pin boot
R. Rubber grease

18.2 (1.9, 13)
26.48 (2.7, 20)
26.48 (2.7, 20)
7.85 (0.80, 69)
103 (11, 76)

71075_PATH_G0112

Fig. 4 Rear brake caliper and related components

BRAKE PADS

REMOVAL & INSTALLATION

1. Before servicing the vehicle, refer to the Precautions.
2. Disconnect the negative battery cable.
3. Raise and support the vehicle safely. Remove wheel and tire.
4. Remove master cylinder reservoir cap.
5. Remove lower sliding pin bolt.
6. Suspend cylinder body with a wire and remove pads, shim, shim covers, and retainers from torque member.

To install:

➡ **Be sure to use new fasteners, as required.**

7. Apply Molykote® AS880N grease between outer brake pad plate and shim, then attach shim and shim covers to brake pads.
8. Attach pad retainer to torque member, then install brake pad and shim assemblies.

✳✳ CAUTION

When attaching pad retainer, attach it firmly so that it is flush with torque member.

9. Push pistons into cylinder body.

➡ **Using a disc brake piston tool (commercial service tool), etc., makes it easier to push in piston.**

✳✳ CAUTION

By pushing in piston, brake fluid returns to master cylinder reservoir tank. Watch the level of the surface of reservoir tank.

10. Remove wire then swing cylinder body down over brake pad assemblies.
11. Install lower sliding pin bolt and tighten to specification.
12. Check brake for drag.
13. Inspect fluid level, then install master cylinder reservoir cap.
14. Install wheel and tire.
15. Be sure to perform the reconnect/relearn procedures.

BRAKES

PARKING BRAKE CABLES

ADJUSTMENT

1. Before servicing the vehicle, refer to the Precautions.
2. Disconnect the negative battery cable.
3. Remove front pillar lower finisher.
4. Remove lower instrument panel LH.
5. Pull rearward to release lower instrument panel LH.
6. Disconnect lower instrument panel LH harness connectors.
7. Partially engage parking brake pedal to access adjusting nut.
8. Insert a deep socket wrench to rotate adjusting nut and loosen cable until tension is sufficiently released. Then, disengage the parking brake pedal.

9. Remove the wheel and tire.
10. Remove the rotor and measure inner diameter at widest point.
11. Transfer measurement less 0.6 mm to the parking brake shoes and adjust accordingly.
12. Using wheel nuts, secure the disc to the hub to prevent it from tilting.
13. Rotate disc rotor to make sure there is no drag.
14. Adjust cable as follows:
    a. Operate pedal 10 or more times with a force of 110 lbs. (490 N).
    b. Rotate adjusting nut with deep socket to adjust pedal stroke to 4 or 5 notches under a force of 44 lbs. (196 N).
    c. With parking brake pedal completely disengaged, make sure there is no drag on the parking brake.

PARKING BRAKE

PARKING BRAKE SHOES

REMOVAL & INSTALLATION

See Figure 5.

1. Before servicing the vehicle, refer to the Precautions.
2. Disconnect the negative battery cable.
3. Remove the tire and wheel assembly.
4. Be sure that the parking brake lever is in the released position.
5. Remove the rear disc rotor.
6. Remove the return springs.
7. Remove the adjuster.
8. Disconnect the parking brake cable from the toggle lever.
9. Remove the retainers.
10. Remove the anti rattle pins and shoes.

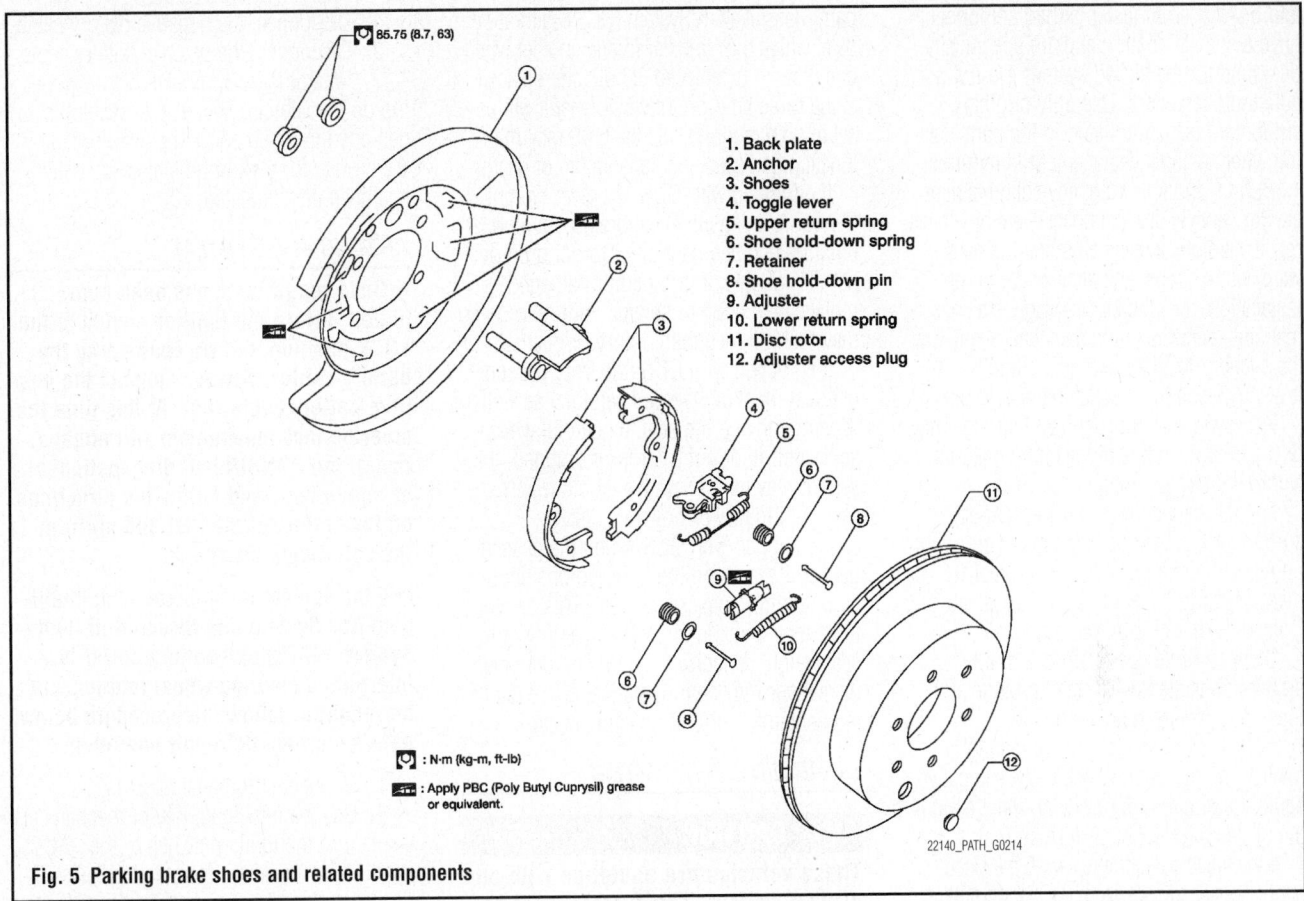

1. Back plate
2. Anchor
3. Shoes
4. Toggle lever
5. Upper return spring
6. Shoe hold-down spring
7. Retainer
8. Shoe hold-down pin
9. Adjuster
10. Lower return spring
11. Disc rotor
12. Adjuster access plug

85.75 (8.7, 63)

: N·m (kg-m, ft-lb)

: Apply PBC (Poly Butyl Cuprysil) grease or equivalent.

22140_PATH_G0214

Fig. 5 Parking brake shoes and related components

To install:

➡**Be sure to use new fasteners, as required.**

11. Apply brake grease to the specified points during reassembly.

12. Assemble the adjuster so that the threaded part expands when rotating it in the direction shown by the arrow. Shorten the adjuster by rotating it in the opposite direction shown by the arrow.

13. Continue the installation in the reverse order of the removal procedure.

14. Adjust the parking brake.

15. Perform the parking brake burnishing operation.

16. Be sure to perform the reconnect/relearn procedures.

CHASSIS ELECTRICAL

AIR BAG (SUPPLEMENTAL RESTRAINT SYSTEM)

GENERAL INFORMATION

✳✳ CAUTION

These vehicles are equipped with an air bag system. The system must be disarmed before performing service on, or around, system components, the steering column, instrument panel components, wiring and sensors. Failure to follow the safety precautions and the disarming procedure could result in accidental air bag deployment, possible injury and unnecessary system repairs.

SERVICE PRECAUTIONS

Disconnect and isolate the battery negative cable before beginning any airbag system component diagnosis, testing, removal, or installation procedures. Allow system capacitor to discharge for two minutes before beginning any component service. This will disable the airbag system. Failure to disable the airbag system may result in accidental airbag deployment, personal injury, or death.

Do not place an intact undeployed airbag face down on a solid surface. The airbag will propel into the air if accidentally deployed and may result in personal injury or death.

When carrying or handling an undeployed airbag, the trim side (face) of the airbag should be pointing towards the body to minimize possibility of injury if accidental deployment occurs. Failure to do this may result in personal injury or death.

Replace airbag system components with OEM replacement parts. Substitute parts may appear interchangeable, but internal differences may result in inferior occupant protection. Failure to do so may result in occupant personal injury or death.

Wear safety glasses, rubber gloves, and long sleeved clothing when cleaning powder residue from vehicle after an airbag deployment. Powder residue emitted from a deployed airbag can cause skin irritation. Flush affected area with cool water if irritation is experienced. If nasal or throat irritation is experienced, exit the vehicle for fresh air until the irritation ceases. If irritation continues, see a physician.

Do not use a replacement airbag that is not in the original packaging. This may result in improper deployment, personal injury, or death.

The factory installed fasteners, screws

and bolts used to fasten airbag components have a special coating and are specifically designed for the airbag system. Do not use substitute fasteners. Use only original equipment fasteners listed in the parts catalog when fastener replacement is required.

During, and following, any child restraint anchor service, due to impact event or vehicle repair, carefully inspect all mounting hardware, tether straps, and anchors for proper installation, operation, or damage. If a child restraint anchor is found damaged in any way, the anchor must be replaced. Failure to do this may result in personal injury or death.

Deployed and non-deployed airbags may or may not have live pyrotechnic material within the airbag inflator.

Do not dispose of driver/passenger/curtain airbags or seat belt tensioners unless you are sure of complete deployment. Refer to the Hazardous Substance Control System for proper disposal.

Dispose of deployed airbags and tensioners consistent with state, provincial, local, and federal regulations.

After any airbag component testing or service, do not connect the battery negative cable. Personal injury or death may result if the system test is not performed first.

If the vehicle is equipped with the Occupant Classification System (OCS), do not connect the battery negative cable before performing the OCS Verification Test using the scan tool and the appropriate diagnostic information. Personal injury or death may result if the system test is not performed properly.

Never replace both the Occupant Restraint Controller (ORC) and the Occupant Classification Module (OCM) at the same time. If both require replacement, replace one, then perform the Airbag System test before replacing the other.

Both the ORC and the OCM store Occupant Classification System (OCS) calibration data, which they transfer to one another when one of them is replaced. If both are replaced at the same time, an irreversible fault will be set in both modules and the OCS may malfunction and cause personal injury or death.

If equipped with OCS, the Seat Weight Sensor is a sensitive, calibrated unit and must be handled carefully. Do not drop or handle roughly. If dropped or damaged, replace with another sensor. Failure to do so may result in occupant injury or death.

If equipped with OCS, the front passenger seat must be handled carefully as well. When removing the seat, be careful when setting on floor not to drop. If dropped, the sensor may be inoperative, could result in occupant injury, or possibly death.

If equipped with OCS, when the passenger front seat is on the floor, no one should sit in the front passenger seat. This uneven force may damage the sensing ability of the seat weight sensors. If sat on and damaged, the sensor may be inoperative, could result in occupant injury, or possibly death.

DISARMING THE SYSTEM

> **✳✳ CAUTION**
>
> **These vehicles are equipped with an air bag system. The system must be disarmed before performing service on, or around, system components, the steering column, instrument panel components, wiring and sensors. Failure to follow the safety precautions and the disarming procedure could result in accidental air bag deployment, possible injury and unnecessary system repairs.**

Before servicing the vehicle, refer to the Precautions.

1. Disconnect the negative battery cable.
2. Disconnect the positive battery cable.
3. Wait at least 3 minutes before working on the vehicle. The air bag system is designed to retain enough power to deploy the air bag for a short time after the battery has been disconnected.

ARMING THE SYSTEM

➡ **Once repair work has been completed, return the ignition switch to the LOCK position, before connecting the battery cables. Always connect the positive battery cable first. At this time the steering lock mechanism will engage. Install the CONSULT-III diagnostic tool, or equivalent, and follow the directions on the screen of the tool and perform the self diagnosis check.**

➡ **If the vehicle is equipped with Intelligent Key System and Nissan Anti-theft System (NATS) and battery power is interrupted steering wheel rotation will be required, follow the procedure below before starting the repair operation.**

1. Connect both battery cables.
2. Use the Intelligent Key or mechanical key to turn the ignition switch to the _ACC_ position. At this time, the steering lock will be released.
3. Disconnect both battery cables. The steering lock will remain released and the steering wheel can be rotated.
4. Perform the necessary repair operation.
5. When the repair work is completed, return the ignition switch to the _LOCK_ position before connecting the battery cables. (At this time, the steering lock mechanism will engage).
6. Perform a self-diagnosis check of all control units using CONSULT III diagnostic tool, or equivalent.

DRIVE TRAIN

AXLE HOUSING

REMOVAL & INSTALLATION

See Figure 6.

1. Before servicing the vehicle, refer to the Precautions.
2. Disconnect the negative battery cable.
3. Raise and safely support the vehicle.
4. Remove the under cover, if equipped.
5. Remove the halfshafts.
6. Remove the crossmember.

7. Remove the front halfshaft.
8. Disconnect the vent hose.
9. Properly support the assembly, using the proper service jack.
10. Remove the retaining bolts and carefully remove the assembly from the vehicle.

To install:

➡ **Be sure to use new fasteners, as required.**

11. Installation is the reverse of the removal procedure.
12. Be sure to perform the reconnect/relearn procedures.

DRIVESHAFT

REMOVAL & INSTALLATION

See Figures 7 through 9.

1. Before servicing the vehicle, refer to the Precautions.
2. Disconnect the negative battery cable.
3. Position the selector lever in the N position.
4. Release the parking brake.
5. Raise and support the vehicle safely. Remove the engine under cover, if equipped and as required.

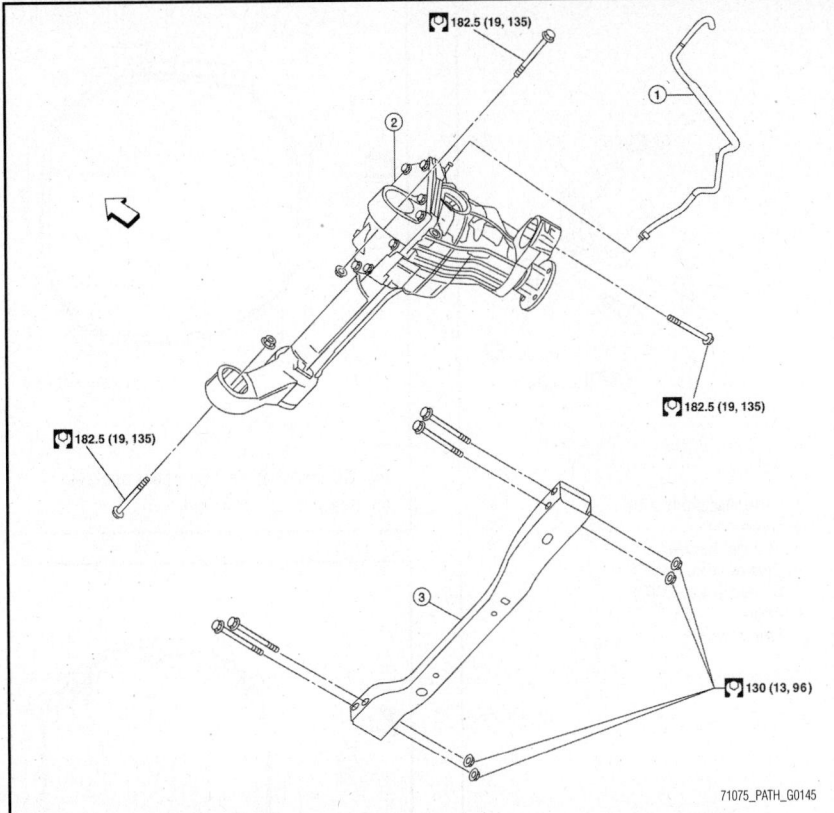

Fig. 6 Front drive axle attaching points and related components (1) hose, (2) unit, (3) crossmember—R180A unit

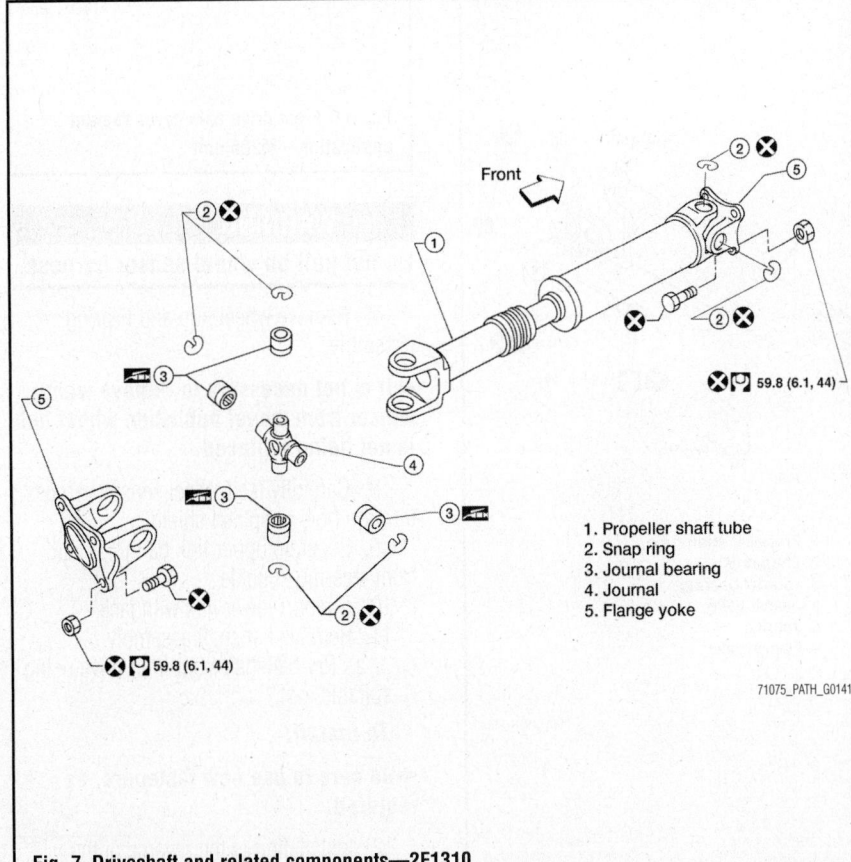

1. Propeller shaft tube
2. Snap ring
3. Journal bearing
4. Journal
5. Flange yoke

71075_PATH_G0141

Fig. 7 Driveshaft and related components—2F1310

6. Matchmark the driveshaft and companion flange.

7. Remove the driveshaft. Discard the nuts.

To install:

→ Be sure to use new fasteners, as required.

8. Installation is the reverse of the removal procedure.

9. Check for vehicle vibration, correct as required.

10. Be sure to perform the reconnect/relearn procedures.

FRONT DIFFERENTIAL HOUSING COVER

REMOVAL & INSTALLATION

See Figures 10 and 11.

1. Before servicing the vehicle, refer to the Precautions.

2. Disconnect the negative battery cable.

3. Raise and support the vehicle safely, as required.

4. Remove the front axle assembly from the vehicle. Position the unit in a suitable holding fixture.

To install:

→ Be sure to use new fasteners, as required.

5. Installation is the reverse of the removal procedure.

→ Be sure to properly remove all old RTV sealant on all mating surfaces.

6. Apply a thin bead (0.12 inch) of genuine RTV sealant, or equivalent, to the mating surface of the carrier cover.

7. Tighten the cover bolts.

8. Be sure to perform the reconnect/relearn procedures.

FRONT HALFSHAFT

REMOVAL & INSTALLATION

See Figures 12 and 13.

1. Before servicing the vehicle, refer to the Precautions.

2. Disconnect the negative battery cable.

3. Raise and support the vehicle safely.

4. Remove wheel and tire.

5. Remove rear engine undercover.

6. Remove wheel sensor harness from mount on knuckle, then disconnect wheel sensor harness connector.

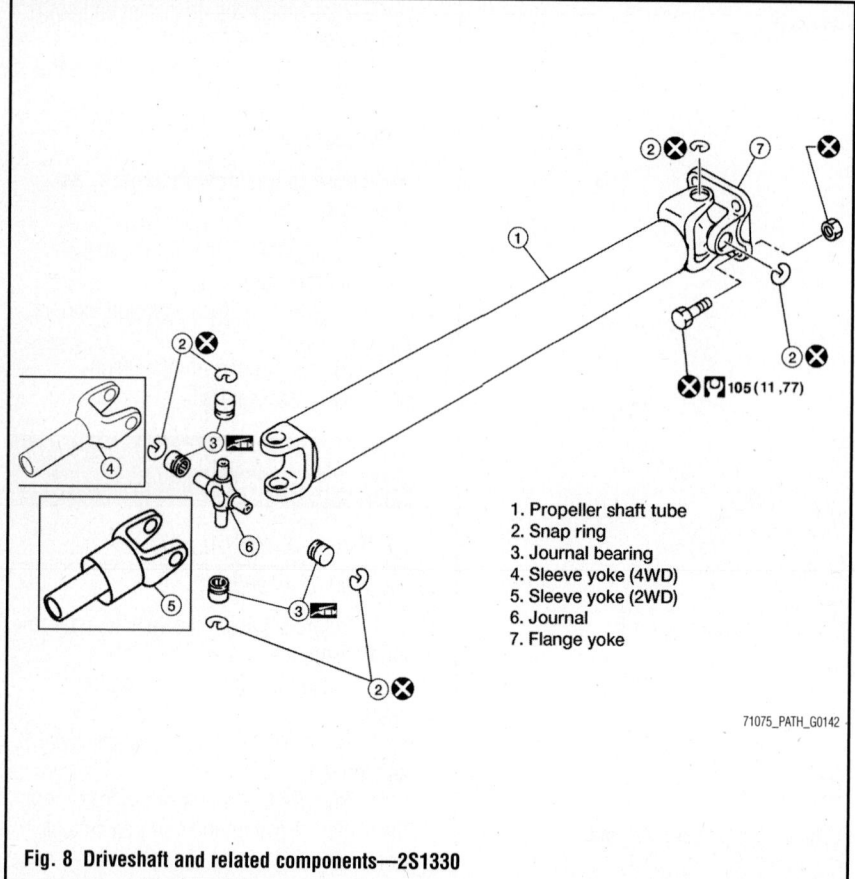

1. Propeller shaft tube
2. Snap ring
3. Journal bearing
4. Sleeve yoke (4WD)
5. Sleeve yoke (2WD)
6. Journal
7. Flange yoke

105 (11 ,77)

71075_PATH_G0142

Fig. 8 Driveshaft and related components—2S1330

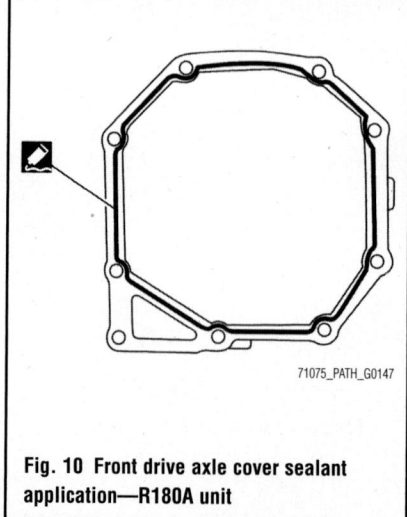

71075_PATH_G0147

Fig. 10 Front drive axle cover sealant application—R180A unit

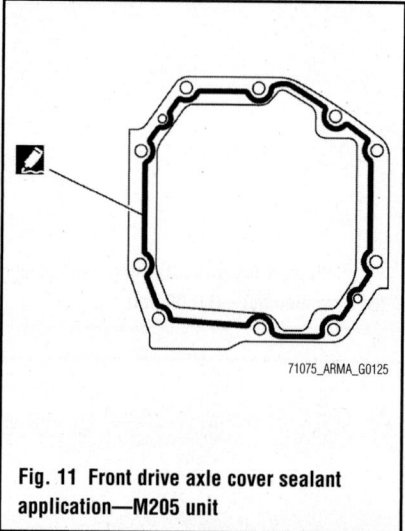

71075_ARMA_G0125

Fig. 11 Front drive axle cover sealant application—M205 unit

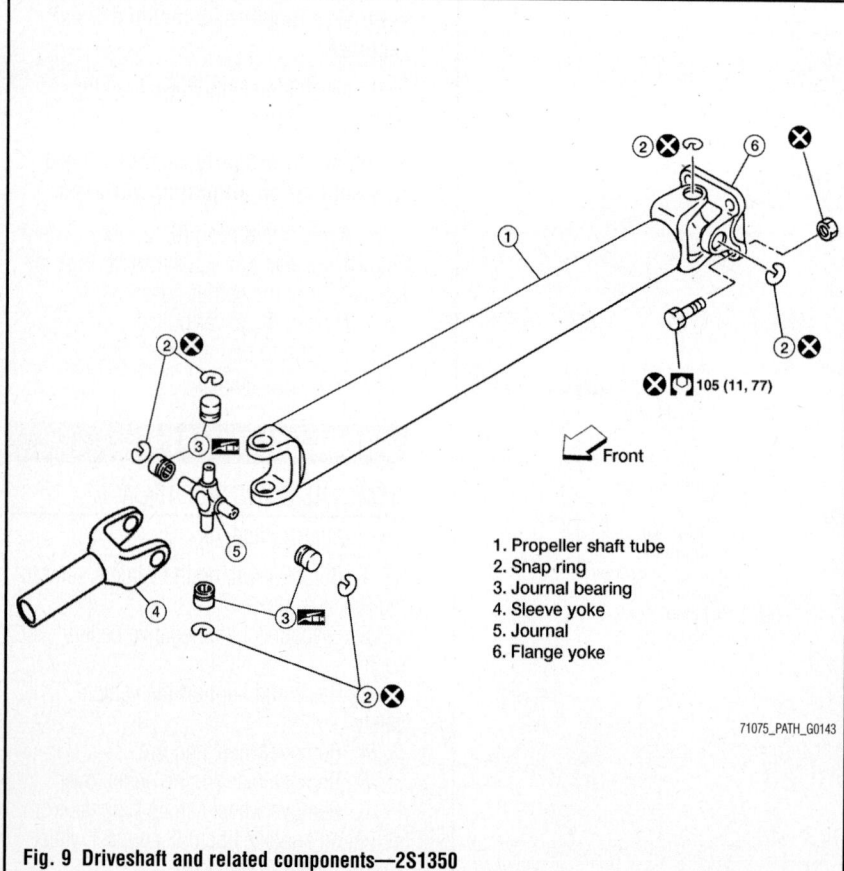

105 (11, 77)

Front

1. Propeller shaft tube
2. Snap ring
3. Journal bearing
4. Sleeve yoke
5. Journal
6. Flange yoke

71075_PATH_G0143

Fig. 9 Driveshaft and related components—2S1350

✳✳ WARNING
Do not pull on wheel sensor harness.

7. Remove wheel hub and bearing assembly.

➡**It is not necessary to remove wheel sensor from wheel hub when wheel hub is not being replaced.**

8. Carefully feed wheel sensor harness through hole in splash shield.
9. Separate upper link ball joint stud from steering knuckle.
10. Support lower link with jack.
11. Remove halfshaft assembly.
   a. Pry halfshaft front final drive using suitable tool.

To install:

➡**Be sure to use new fasteners, as required.**

12. Installation is the reverse of the removal procedure.

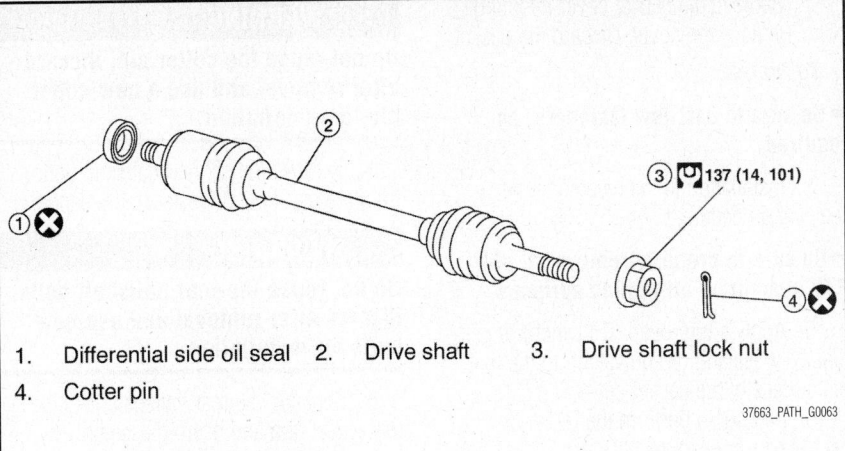

1. Differential side oil seal 2. Drive shaft 3. Drive shaft lock nut
4. Cotter pin

37663_PATH_G0063

Fig. 12 Front halfshaft and related components—V6 engine

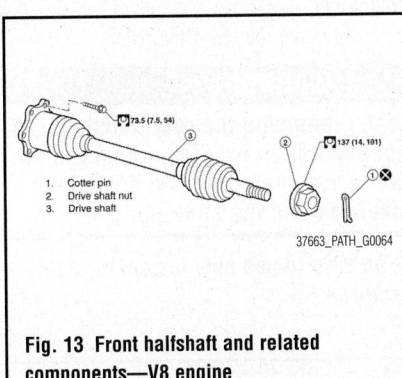

1. Cotter pin
2. Drive shaft nut
3. Drive shaft

37663_PATH_G0064

Fig. 13 Front halfshaft and related components—V8 engine

13. Be sure to perform the reconnect/relearn procedures.

FRONT PINION OIL SEAL

REMOVAL & INSTALLATION

1. Before servicing the vehicle, refer to the Precautions.
2. Disconnect the negative battery cable.
3. Raise and support the vehicle safely.
4. Remove the front wheels and tires.
5. Without disassembling the hydraulic lines, remove the caliper torque member bolts.
6. Reposition the brake caliper aside using suitable wire.

> **⁕⁕ WARNING**
>
> **Do not press the brake pedal while brake caliper is removed.**

7. Remove the ABS sensor harness from the mount on the knuckle.

> **⁕⁕ CAUTION**
>
> **Do not pull on the ABS sensor harness.**

8. Support the lower link using a suitable jack.
9. Separate the upper link ball joint stud from the steering knuckle.

> **⁕⁕ WARNING**
>
> **Support the lower link using a jack.**

10. Remove the engine undercover.
11. Remove the RH and LH drive shafts from the front final drive.

> **⁕⁕ WARNING**
>
> **Do not reuse the front final drive side oil seals.**

12. Disconnect the front driveshaft shaft from the front final drive. Then reposition the front driveshaft shaft aside using suitable wire.
13. Measure the drive pinion bearing preload with the front oil seal resistance.

➡**Record the preload measurement.**

14. Remove the drive pinion lock nut.
15. Put a matching mark on the end of the drive pinion in line with the matching mark B on the companion flange.

> **⁕⁕ WARNING**
>
> **Use paint to make the matching mark on the drive pinion. Do not damage the companion flange or drive pinion.**

16. Remove the companion flange.
17. Remove the front oil seal.

To install:

18. Apply multi-purpose grease to the front oil seal lips and gear oil onto the circumference of the new oil seal.
19. Drive the front oil seal in evenly.

> **⁕⁕ WARNING**
>
> **Do not reuse oil seal. Do not incline oil seal when installing. Apply multi-purpose grease onto oil seal lips and gear oil onto the circumference of oil seal.**

20. Align the matching mark of the drive pinion with the matching mark B of the companion flange, then install the companion flange.
21. Apply gear oil on the threads of the drive pinion and the seating surface of the new drive pinion lock nut.
22. Install the new drive pinion lock nut. Tighten to the specified torque.

> **⁕⁕ WARNING**
>
> **Do not reuse drive pinion lock nut.**

23. Measure the drive pinion bearing preload with the front oil seal resistance.

➡**Drive pinion bearing preload should equal the measurement taken during removal plus an additional 5 inch lbs. (0.56 Nm). If the drive pinion bearing preload is low, tighten the drive pinion lock nut in 5ft. lbs (6.8 Nm) increments until the drive pinion preload is met.**

> **⁕⁕ WARNING**
>
> **Never loosen the drive pinion nut to decrease drive pinion bearing preload. Do not exceed specified preload. If preload torque is exceeded a new collapsible spacer must be installed. If maximum torque is reached prior to reaching the required preload, the collapsible spacer may have been damaged. Replace the collapsible spacer.**

24. Install new side oil seals into the front final drive assembly.
25. Install the RH and LH drive shafts to the front final drive.

> **⁕⁕ WARNING**
>
> **When installing the drive shaft assembly into the front final drive assembly, do not damage the side oil seal.**

26. Install the remaining components in the reverse order of removal.

> **⁕⁕ WARNING**
>
> **Check the final drive gear oil level after installation.**

27. Tighten the upper link ball joint stud nut to specifications.

28. Tighten the wheel nuts to specification.

29. Be sure to perform the reconnect/relearn procedures.

REAR AXLE HOUSING

REMOVAL & INSTALLATION

1. Before servicing the vehicle, refer to the Precautions.

2. Disconnect the negative battery cable.

3. Remove the spare tire and wheel assembly.

4. Raise and support the vehicle safely.

5. Drain the gear oil. Be sure to properly dispose of used oil.

6. Remove the tire and wheel assemblies.

7. Remove the driveshaft.

8. Remove the rear stabilizer bar.

9. Disconnect the rear halfshaft from the rear axle assembly. Position it aside using mechanics wire or equivalent.

10. Disconnect the breather hose from the axle cover.

11. Position a suitable jack under the assembly.

➡**Do not position the jack under the aluminum cover.**

12. Remove the rear axle assembly retaining bolts and nuts.

13. Carefully remove the assembly from the vehicle.

To install:

➡**Be sure to use new fasteners, as required.**

14. Installation is the reverse of the removal procedure.

15. Be sure to perform the reconnect/relearn procedures.

REAR DIFFERENTIAL HOUSING COVER

REMOVAL & INSTALLATION

1. Before servicing the vehicle, refer to the Precautions.

2. Disconnect the negative battery cable.

3. Raise and support the vehicle safely, as required.

4. Position a catch pan under the assembly.

5. Drain the fluid. Be sure to properly dispose of used fluid.

6. Remove the rear final drive assembly from the vehicle.

7. Remove the carrier cover retaining bolts. Remove the cover. Discard the gasket.

To install:

➡**Be sure to use new fasteners, as required.**

8. Installation is the reverse of the removal procedure.

➡**Be sure to properly remove all old RTV sealant on all mating surfaces.**

9. Apply a thin bead (0.12 inch) of genuine RTV sealant, or equivalent, to the mating surface of the carrier cover.

10. Be sure to perform the reconnect/relearn procedures.

REAR HALFSHAFT

REMOVAL & INSTALLATION

See Figure 14.

1. Before servicing the vehicle, refer to the Precautions.

2. Disconnect the negative battery cable.

3. Raise and support the vehicle safely. Remove the wheel and tire assembly.

4. Remove the cotter pin and discard, then remove the lock nut from the drive shaft.

✳✳ WARNING

Do not reuse the cotter pin, discard after removal and use a new cotter pin for installation.

5. Remove the six rear halfshaft bolts from the rear final drive assembly flange.

✳✳ WARNING

Do not reuse the rear halfshaft bolts, discard after removal and use new bolts for installation.

6. Separate the rear halfshaft from the rear wheel hub and bearing assembly by lightly tapping the end of the rear drive shaft with a suitable hammer and wood block. If it is difficult to separate, use a suitable puller.

7. Remove the rear drive shaft.

✳✳ WARNING

When removing the rear halfshaft, do not bend at an excessive angle to the rear drive shaft joint. Do not excessively extend the slide joint.

➡**Be sure to use new fasteners, as required.**

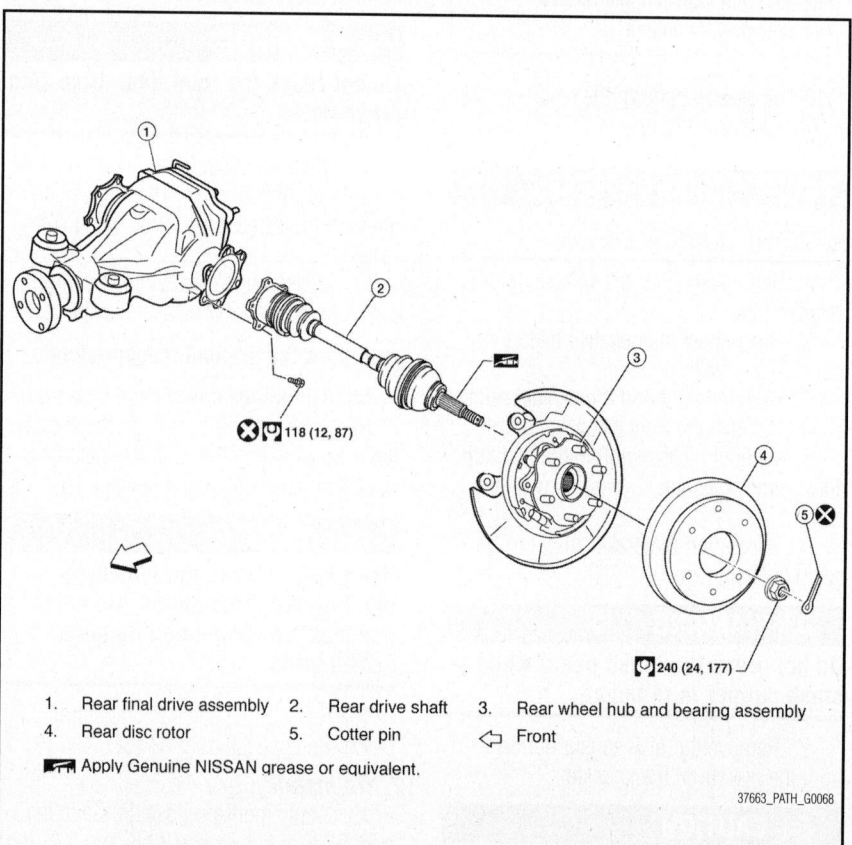

1.	Rear final drive assembly	2.	Rear drive shaft
4.	Rear disc rotor	5.	Cotter pin

3. Rear wheel hub and bearing assembly

⇦ Front

Apply Genuine NISSAN grease or equivalent.

⊗ 118 (12, 87)

240 (24, 177)

37663_PATH_G0068

Fig. 14 Rear halfshaft and related components

8. Installation is in the reverse order of removal.

> ⁑ **WARNING**
>
> **Do not reuse the drive shaft inside flange bolts and washers, discard after removal and use new bolts and washers for installation.**

> ⁑ **WARNING**
>
> **Do not reuse the cotter pin, discard after removal and use a new cotter pin for installation.**

REAR PINION OIL SEAL

REMOVAL & INSTALLATION

1. Before servicing the vehicle, refer to the Precautions.
2. Disconnect the negative battery cable.

3. Raise and support the vehicle safely.
4. Remove the rear driveshaft.
5. Put a matching mark on the end of the drive pinion in line with the matching mark B on the companion flange.

> ⁑ **WARNING**
>
> **Use paint to make the matching mark on the drive pinion. Do not damage the companion flange or drive pinion.**

➡ **The matching mark B on the final drive companion flange indicates the maximum vertical run out position.**

6. Remove the drive pinion lock nut.
7. Remove the companion flange using suitable tool.
8. Remove the front oil seal.

To install:

9. Install the front oil seal.

> ⁑ **WARNING**
>
> **Do not reuse oil seal. Do not incline oil seal when installing. Apply multi-purpose grease onto oil seal lips, and gear oil onto the circumference of oil seal.**

10. Align the matching mark of the drive pinion with the matching mark B of the companion flange, then install the companion flange.
11. Install the drive pinion lock nut.

> ⁑ **WARNING**
>
> **Do not reuse drive pinion lock nut.**

12. Install the rear propeller shaft.
13. Be sure to perform the reconnect/relearn procedures.

ENGINE COOLING

BELT-DRIVEN ENGINE FAN

REMOVAL & INSTALLATION

V6 Engine

See Figure 15.

1. Before servicing the vehicle, refer to the Precautions.
2. Disconnect the negative battery cable.
3. Remove the engine cover and air cleaner assembly, as required.
4. Remove the air dam.
5. Remove the engine undercover.
6. Drain the cooling system. Be sure to properly dispose of used coolant.
7. Remove air duct.
   a. Disconnect harness connector from Mass Air Flow (MAF) sensor.
   b. Disconnect PCV hose.
   c. Remove air cleaner case/mass air flow sensor assembly and air duct assembly disconnecting their joints.
8. Remove the reservoir tank hose from the radiator. Remove the upper radiator hose.
9. Remove the upper and lower radiator shrouds.
10. Remove drive belts.
   a. Rotate the drive belt auto tensioner in the direction of arrow (loosening direction of tensioner).

> ⁑ **CAUTION**
>
> **Avoid placing hand in a location where pinching may occur if the tool accidentally comes off.**

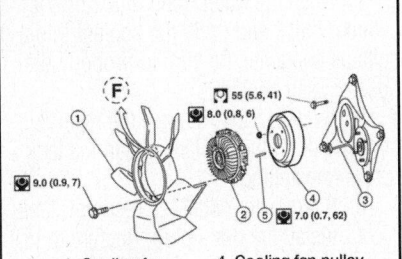

1. Cooling fan
2. Fan coupling
3. Fan bracket
4. Cooling fan pulley
5. Stud

71075_PATH_G0136

Fig. 15 Cooling fan and related components—V6 engine

   b. Remove the drive belt.
11. Remove cooling fan.

To install:

➡ **Be sure to use new fasteners, as required.**

12. Installation is the reverse of the removal procedure.
13. Start and warm up the engine. Visually make sure that there are no leaks of the engine coolant.
14. Be sure to perform the reconnect/relearn procedures.

V8 Engine

See Figure 16.

1. Before servicing the vehicle, refer to the Precautions.

2. Disconnect the negative battery cable.
3. Remove the engine cover and air cleaner assembly, as required.
4. Remove the air dam.
5. Remove the engine front undercover.
6. Partially drain engine coolant from radiator.

> ⁑ **CAUTION**
>
> **Perform this step when engine is cold. Do not spill engine coolant on drive belts.**

7. Remove the air duct and resonator assembly.
8. Remove reservoir tank hose from radiator.
9. Remove reservoir tank hose from engine.
10. Removal radiator hose (upper) from radiator.

> ⁑ **WARNING**
>
> **Do not spill engine coolant on drive belts.**

11. Remove the radiator shroud (lower) and position aside.
   a. Release the tabs, pull radiator shroud (lower) rearwards and down to remove.
12. Remove the radiator shroud (upper) bolts and remove the radiator shroud (upper).
13. Remove the drive belt.

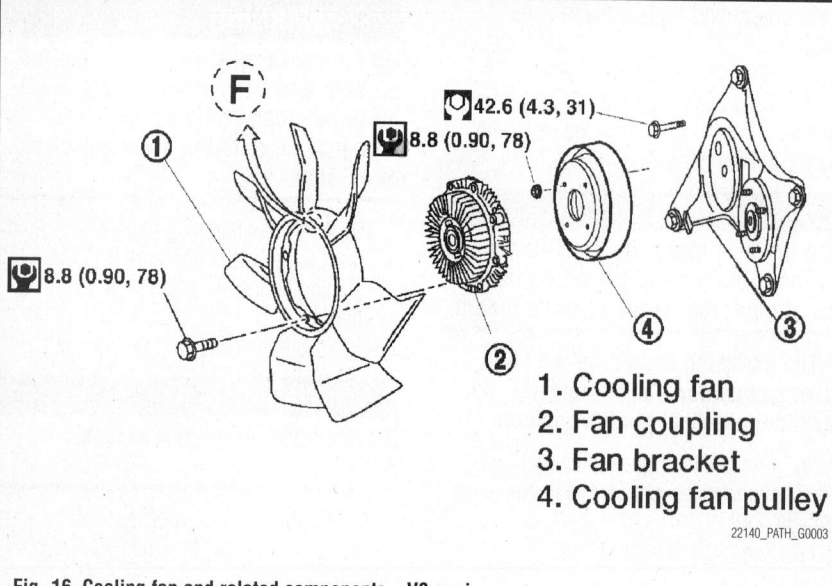

1. Cooling fan
2. Fan coupling
3. Fan bracket
4. Cooling fan pulley

22140_PATH_G0003

Fig. 16 Cooling fan and related components—V8 engine

To install:

14. Install cooling fan with its front mark "F" facing front of engine.

15. Check for leaks of the engine coolant.

16. Start and warm up the engine. Visually make sure that there are no leaks of the engine coolant.

17. Be sure to perform the reconnect/relearn procedures.

ENGINE COOLANT

DRAIN & REFILL

➡ **You will need tool KV991J0070, or equivalent, and a compressed air supply to perform this procedure.**

1. Before servicing the vehicle, refer to the Precautions.

2. Disconnect the negative battery cable.

➡ **Never drain the engine coolant when the engine is hot. Wrap a thick cloth around the cap and carefully remove it. First turn the cap a quarter of a turn to release any pressure, and then turn it all the way. Do not allow coolant to come in contact with the drive belts.**

3. Turn the ignition switch ON and set the temperature control lever all the way to the HOT position, or the highest temperature position. Wait ten seconds and turn the ignition switch OFF.

4. Raise and safely support the vehicle.

5. Remove the undercover, if equipped and as required.

6. Open the radiator drain plug and carefully drain the coolant into a suitable container. Be sure to properly dispose of used coolant.

7. Discard the O-ring.

8. Open the water drain plugs on the cylinder block and drain the coolant into a suitable container. Be sure to properly dispose of used coolant.

9. As necessary, remove the reservoir tank and drain the coolant. Clean the tank before reinstalling.

10. If removed install the reservoir tank.

11. Install the drain plug. Be sure to use new O-ring/gaskets, as required.

12. Apply sealant to the drain plugs. Install the engine drain plugs.

13. Check that all hose clamps are tight.

14. Set the heater controls to the full HOT position and heater ON position. Turn the ignition ON with the engine OFF as necessary to activate the heater mode.

15. Remove the vented reservoir cap and replace it with a non-vented cap before filling the system.

16. Install tool KV991J0070 or equivalent,

17. Insert the refill hose into the coolant mixture (placed at floor level). Be sure that the ball valve is in the closed position. Install the air hose to the venture assembly. The air pressure must be within specification (80–120 psi).

➡ **The compressed air supply must be equipped with an air dryer.**

18. The vacuum gauge will begin to rise and there will be an audible hissing noise. During this process open the ball valve on the refill hose, slightly. Coolant will be visible rising in the refill hose. Once the refill hose is full of coolant, close the ball valve. This will purge any trapped air in the refill hose.

19. Continue to draw vacuum until the gauge reaches 28 inches of vacuum.

➡ **The gauge may not reach specification in high altitude applications. If not see the following specifications. Altitude above sea level: 0-100m (328 ft.) 28 inches of vacuum. Altitude above sea level: 300m (984 ft.) 27 inches of vacuum. Altitude above sea level: 500m (1641 ft.) 26 inches of vacuum. Altitude above sea level: 1000m (3281 ft.) 24-25 inches of vacuum.**

20. When the proper specification has been reached, disconnect the air hose and wait 20 seconds to see if the system looses vacuum. If the level drops perform any necessary repairs and repeat the procedure.

21. Place the coolant container (with the refill hose inserted) at the same level as the top of the radiator. Then open the ball valve on the refill hose so the coolant will be drawn up to fill the cooling system. The cooling system is full when the vacuum gauge reads zero.

➡ **Do not allow the coolant container to get too low when filling, to avoid air from being drawn into the system.**

22. Remove the toll from the radiator neck opening and install the radiator cap.

23. Remove the non vented reservoir cap.

24. Fill the reservoir tank to specification with the proper coolant mixture.

25. Be sure to perform the reconnect/relearn procedures.

ELECTRIC ENGINE FAN

REMOVAL & INSTALLATION

V6 Engine

1. Before servicing the vehicle, refer to the Precautions.

2. Disconnect the negative battery cable.

3. Remove the engine cover and air cleaner assembly, as required.

4. Remove the air dam.

5. Remove the engine undercover.

6. Drain the cooling system. Be sure to properly dispose of used coolant.

7. Remove air duct.

   a. Disconnect harness connector from Mass Air Flow (MAF) sensor.

   b. Disconnect PCV hose.

   c. Remove air cleaner case/mass air

flow sensor assembly and air duct assembly disconnecting their joints.

8. Remove the reservoir tank hose from the radiator. Remove the upper radiator hose.

9. Remove the upper and lower radiator shrouds.

10. Disconnect harness connector from fan motor.

11. Remove the bolt and remove the fan grille and motor assembly.

To install:

➡**Be sure to use new fasteners, as required.**

12. Installation is the reverse of the removal procedure.

13. Start and warm up the engine. Visually make sure that there are no leaks of the engine coolant.

14. Be sure to perform the reconnect/relearn procedures.

V8 Engine

See Figure 17.

1. Before servicing the vehicle, refer to the Precautions.

2. Disconnect the negative battery cable.

3. Remove the engine cover and air cleaner assembly, as required.

4. Remove the air dam.

5. Remove the engine undercover.

6. Loosen the lower fan motor nuts.

7. Disconnect harness connector from fan motor.

8. Remove the upper fan motor bolts.

9. Remove the fan grille and motor assembly.

To install:

➡**Be sure to use new fasteners, as required.**

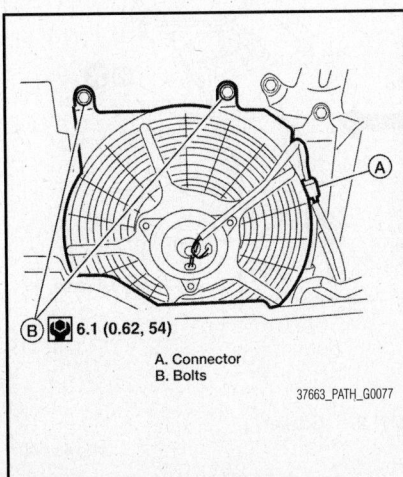

A. Connector
B. Bolts

37663_PATH_G0077

Fig. 17 Cooling fan and related components—electric fan (V8 engine)

10. Installation is the reverse of the removal procedure.

11. Start and warm up the engine. Visually make sure that there are no leaks of the engine coolant.

12. Be sure to perform the reconnect/relearn procedures.

RADIATOR

REMOVAL & INSTALLATION

See Figures 18 and 19.

1. Before servicing the vehicle, refer to the Precautions.

✳✳ CAUTION

Do not remove radiator cap when engine is hot. Serious burns could occur from high-pressure engine coolant escaping from radiator. Wrap a thick cloth around the cap. Slowly turn it a quarter of a turn to release built-up pressure. Carefully remove radiator cap by turning it all the way.

2. Disconnect the negative battery cable.

3. Remove the engine cover. Remove the air cleaner assembly.

a. Remove bolts.

b. Lift up on engine cover firmly to dislodge snap fit mounts.

4. Drain engine coolant from radiator.

✳✳ CAUTION

Perform this step when engine is cold. Do not spill engine coolant on drive belts.

a. Disconnect harness connector from mass air flow sensor.

b. Disconnect PCV hose.

c. Remove air cleaner case/mass air flow sensor assembly and air duct assembly disconnecting their joints.

5. Remove reservoir tank hose.

6. Removal radiator hoses (upper and lower) and reservoir tank hose.

✳✳ WARNING

Be careful not to allow engine coolant to contact drive belts.

7. Remove radiator cooling fan assembly.

8. Disconnect A/T fluid cooler hoses.

a. Install blind plug to avoid leakage of A/T fluid.

9. Remove the front grille.

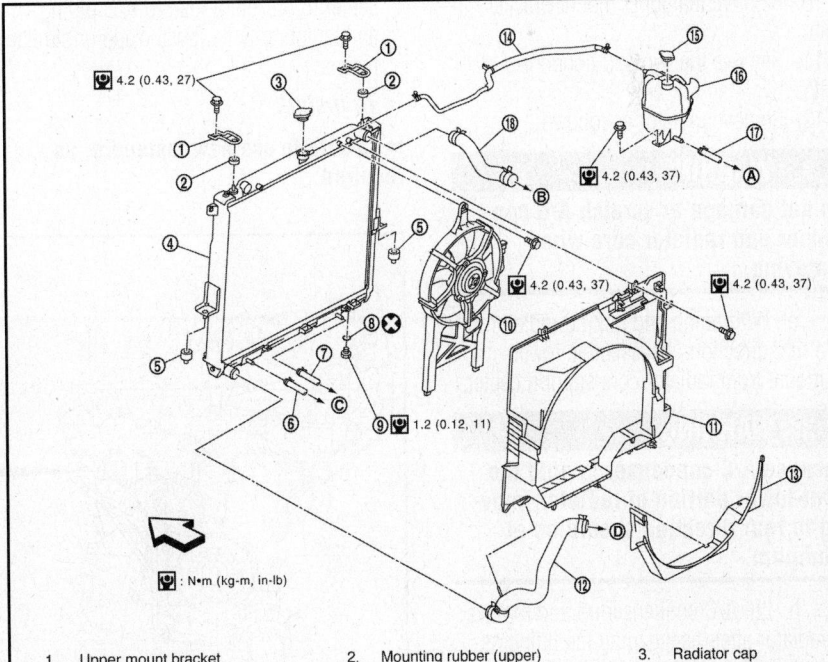

1.	Upper mount bracket	2.	Mounting rubber (upper)	3.	Radiator cap
4.	Radiator	5.	Mounting rubber (lower)	6.	A/T fluid cooler hose
7.	A/T fluid cooler hose	8.	O-ring	9.	Drain plug
10.	Cooling fan assembly	11.	Radiator shroud (upper)	12.	Radiator hose (lower)
13.	Radiator shroud (lower)	14.	Reservoir tank hose	15.	Reservoir tank cap
16.	Reservoir tank	17.	Water hose	18.	Radiator hose (upper)
A.	To heater return tube	B.	To water pipe	C.	To A/T cooler tube

37663_PATH_G0081

Fig. 18 Radiator and related components—V6 engine

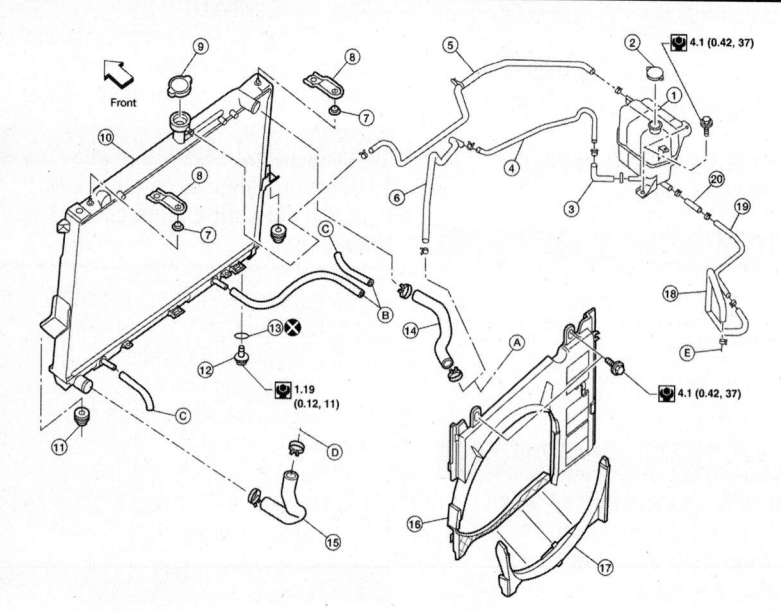

1.	Reservoir tank	2.	Reservoir tank cap	3.	By-pass hose
4.	By-pass tube	5.	Reservoir tank hose	6.	By-pass hose
7.	Mounting rubber (upper)	8.	Upper mount bracket	9.	Radiator cap
10.	Radiator	11.	Mounting rubber (lower)	12.	Radiator drain plug
13.	O-ring	14.	Radiator hose (upper)	15.	Radiator hose (lower)
16.	Radiator shroud (upper)	17.	Radiator shroud (lower)	18.	Heater by-pass hose
19.	Heater by-pass tube	20.	Heater by-pass hose	A.	To thermostat housing
B.	To A/T fluid cooler tube	C.	To transmission auxiliary cooler	D.	To water suction pipe

37663_PATH_G0082

Fig. 19 Radiator and related components—V8 engine

10. Remove the upper mount bracket bolts.

11. Remove the two A/C condenser bolts.

12. Remove radiator as follows:

✳✳ WARNING
Do not damage or scratch A/C condenser and radiator core when removing.

a. With lifting and pulling radiator in a rear direction, disassemble lower mount from radiator core support center.

✳✳ WARNING
Because A/C condenser is onto the front-lower portion of radiator, moving to rear direction should be at minimum.

b. Lift A/C condenser up and remove radiator after disengaging the fitting as front-bottom surface.

✳✳ WARNING
Lifting A/C condenser should be minimum to prevent a load to A/C piping.

c. After removing radiator, put A/C condenser on radiator core support center to prevent a load to A/C piping, and temporarily fix it with rope or similar means.

To install:

➡Be sure to use new fasteners, as required.

13. Installation is the reverse of the removal procedure.

14. Be sure to fill the cooling system with the proper grade and type engine coolant.

15. Be sure to perform the reconnect/relearn procedures.

THERMOSTAT

REMOVAL & INSTALLATION
See Figures 20 and 21.

1. Before servicing the vehicle, refer to the Precautions.

✳✳ CAUTION
Do not remove radiator cap when engine is hot. Serious burns could occur from high-pressure engine coolant escaping from radiator. Wrap a thick cloth around the cap. Slowly turn it a quarter of a turn to release built-up pressure. Carefully remove radiator cap by turning it all the way.

2. Disconnect the negative battery cable.

3. Remove the engine cover. Remove the air cleaner assembly.

4. Completely drain engine coolant.

✳✳ CAUTION
Perform this step when engine is cold. Do not spill engine coolant on drive belts.

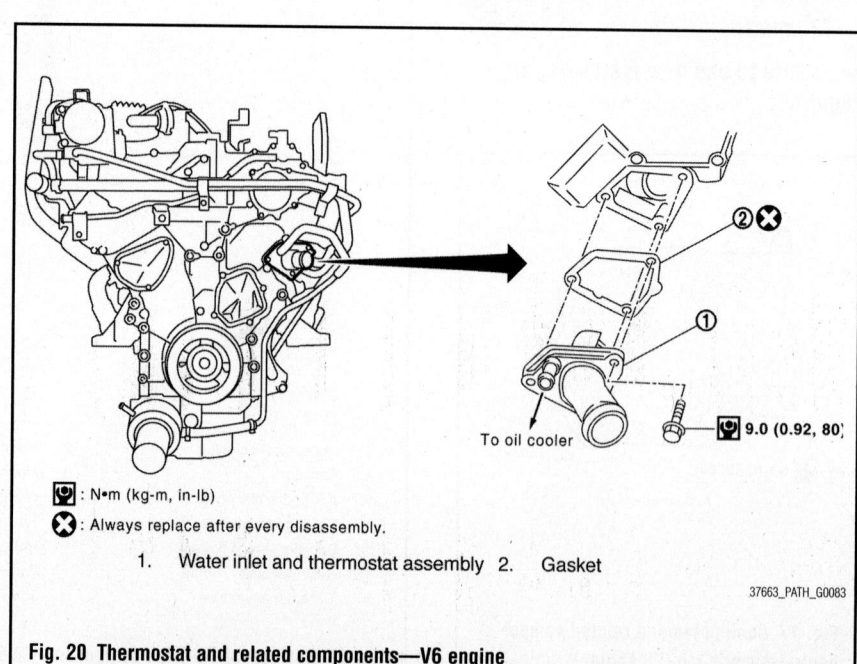

🔧 : N•m (kg-m, in-lb)

✕ : Always replace after every disassembly.

1. Water inlet and thermostat assembly   2. Gasket

37663_PATH_G0083

Fig. 20 Thermostat and related components—V6 engine

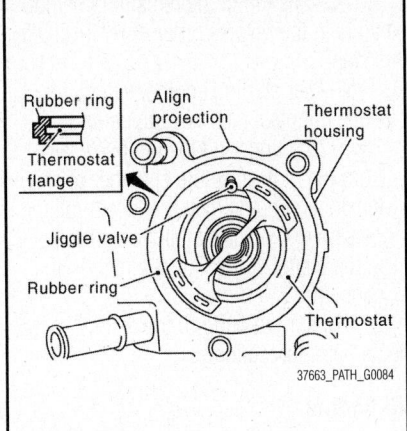

Fig. 21 Thermostat installation and alignment—V8 engine

5. Disconnect radiator hose (lower) and oil cooler hose from water inlet and thermostat assembly.

6. Remove water inlet and thermostat assembly.

✳✳ WARNING

Do not disassemble water inlet and thermostat assembly. Replace them as a unit, if necessary.

To install:

➡ **Be sure to use new fasteners, as required.**

7. Installation is the reverse of the removal procedure.

✳✳ WARNING

Be careful not to spill engine coolant over engine room. Use rag to absorb engine coolant.

8. Start and warm up engine. Visually check there are no leaks of engine coolant.

9. Be sure to fill the cooling system with the proper grade and type engine coolant.

10. Be sure to perform the reconnect/relearn procedures.

WATER PUMP

REMOVAL & INSTALLATION

V6 Engine

See Figure 22.

1. Before servicing the vehicle, refer to the Precautions.

✳✳ CAUTION

Do not remove radiator cap when engine is hot. Serious burns could occur from high-pressure engine coolant escaping from radiator. Wrap a thick cloth around the cap. Slowly turn it a quarter of a turn to release built-up pressure. Carefully remove radiator cap by turning it all the way.

2. Disconnect the negative battery cable.

3. Remove the engine cover. Remove the air cleaner assembly.

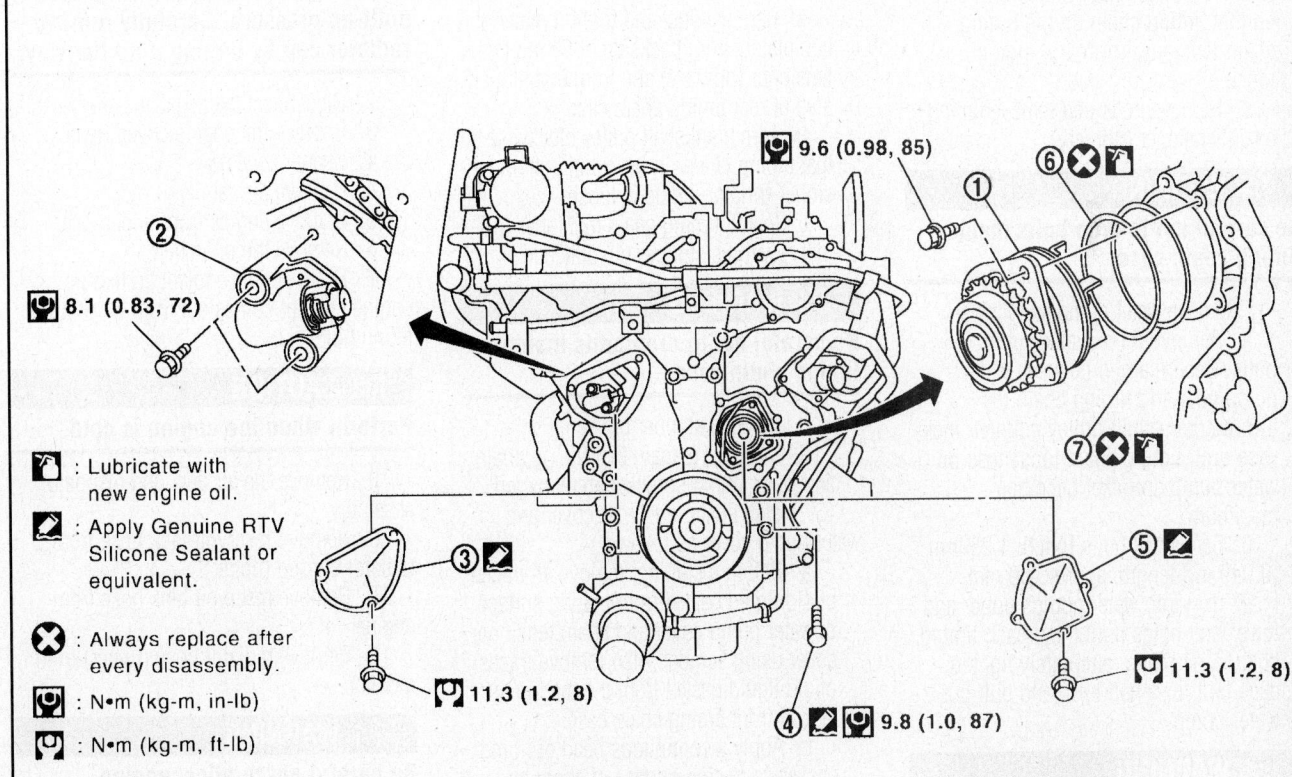

1. Water pump
2. Timing chain tensioner (primary)
3. Chain tensioner cover
4. Water drain plug (front)
5. Water pump cover
6. O-ring
7. O-ring

Fig. 22 Water pump and related components—V6 engine

4. Remove air dam.
5. Remove undercover.
6. Remove drive belts.
7. Drain engine coolant.

✳✳ CAUTION

Perform this step when engine is cold. Do not spill engine coolant on drive belts.

8. Remove radiator hoses (upper and lower) and cooling fan assembly.
9. Remove chain tensioner cover and water pump cover from front timing chain case.
10. Remove timing chain tensioner (primary) as follows:
    a. Loosen clip of timing chain tensioner (primary), and release plunger stopper (1).
    b. Insert plunger into tensioner body by pressing slack guide (2).
    c. Keep slack guide pressed and hold plunger in by pushing stopper pin through the tensioner body hole and plunger groove (3).
    d. Turn crankshaft pulley clockwise so that timing chain on the timing chain tensioner (primary) side is loose.
    e. Remove bolts and remove timing chain tensioner (primary).

✳✳ WARNING

Be careful not to drop bolts inside timing chain case.

11. Remove water pump as follows:
    a. Remove three water pump bolts. Secure a gap between water pump gear and timing chain, by turning crankshaft pulley counterclockwise until timing chain looseness on water pump sprocket becomes maximum.
    b. Screw M8 bolts [pitch: 1.25 mm (0.049 in.) length: approx. 50 mm (1.97 in.)] into water pumps upper and lower bolt holes until they reach timing chain case. Then, alternately tighten each bolt for a half turn, and pull out water pump.

✳✳ WARNING

Pull straight out while preventing vane from contacting socket in installation area. Remove water pump without causing sprocket to contact timing chain.

    c. Remove M8 bolts and O-rings from water pump.

✳✳ WARNING

Do not disassemble water pump.

➡ Do not reuse O-rings.

  To install:

➡ Be sure to use new fasteners, as required.

12. Install new O-rings to water pump.

➡ Apply engine oil to O-rings. Locate O-ring with white paint mark to engine front side.

13. Install water pump.

✳✳ WARNING

Do not allow timing chain case to pinch O-rings when installing water pump.

    a. Make sure that timing chain and water pump sprocket are engaged.
    b. Insert water pump by tightening bolts alternately and evenly.
14. Install timing chain tensioner (primary) as follows:
    a. Remove dust and foreign material completely from backside of timing chain tensioner (primary) and from installation area of rear timing chain case.
    b. Turn crankshaft pulley clockwise so that timing chain on the timing chain tensioner (primary) side is loose.
    c. Install timing chain tensioner (primary) with its stopper pin attached.

✳✳ WARNING

Be careful not to drop bolts inside timing chain case.

    d. Remove stopper pin.
    e. Make sure again that timing chain and water pump sprocket are engaged.
15. Install chain tensioner cover and water pump cover as follows:
    a. Before installing, remove all traces of old liquid gasket from mating surface of water pump cover and chain tensioner cover using scraper. Also remove traces of old liquid gasket from the mating surface of front timing chain case.
    b. Apply a continuous bead of liquid gasket, to mating surface of chain tensioner and water pump cover.

✳✳ WARNING

Attaching should be done within 5 minutes after coating.

    c. Tighten bolts to specified torque.
16. Refill engine coolant system.

17. Installation of the remaining components is in the reverse order of removal after this step.
    a. After starting engine, let idle for three minutes, then rev engine up to 3,000 rpm under no load to purge air from the high-pressure chamber of chain tensioner. Engine may produce a rattling noise. This indicates that air still remains in the chamber and is not a matter of concern.
18. Be sure to perform the reconnect/relearn procedures.

V8 Engine
See Figure 23.

1. Before servicing the vehicle, refer to the Precautions.

✳✳ CAUTION

Do not remove radiator cap when engine is hot. Serious burns could occur from high-pressure engine coolant escaping from radiator. Wrap a thick cloth around the cap. Slowly turn it a quarter of a turn to release built-up pressure. Carefully remove radiator cap by turning it all the way.

2. Disconnect the negative battery cable.
3. Remove the engine cover. Remove the air cleaner assembly.
4. Remove air dam.
5. Remove engine front undercover.
6. Remove the drive belt.
7. Drain engine coolant so that no engine coolant comes out from water pump fitting hole.

✳✳ CAUTION

Perform when the engine is cold.

8. Remove the air duct and resonator assembly.
9. Remove reservoir tank hose from radiator shroud (upper).
10. Remove reservoir tank hose from engine.
11. Remove radiator hose (upper) from radiator.

✳✳ WARNING

Be careful not to allow engine coolant to contact drive belts.

12. Remove the radiator shroud (lower) and position aside.
    a. Release the tabs, pull radiator shroud (lower) rearwards and down to remove.
13. Remove the radiator shroud (upper)

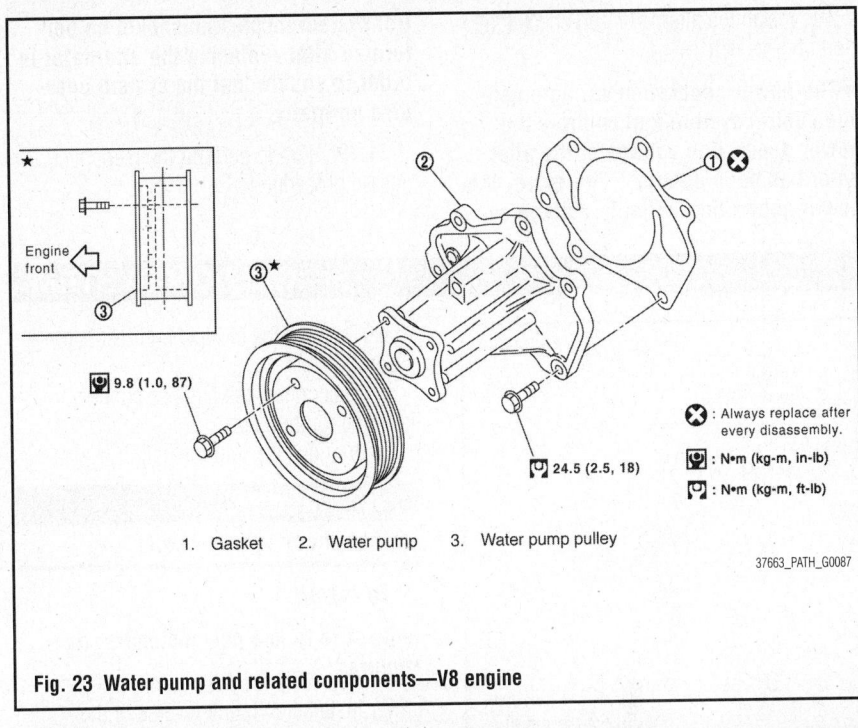

$9.8 (1.0, 87)$

$24.5 (2.5, 18)$

⊗ : Always replace after every disassembly.

🔧 : N•m (kg-m, in-lb)

🔧 : N•m (kg-m, ft-lb)

1. Gasket   2. Water pump   3. Water pump pulley

37663_PATH_G0087

Fig. 23 Water pump and related components—V8 engine

bolts and remove the radiator shroud (upper) (A).

14. Remove the engine cooling fan (crankshaft driven type).

15. Remove the water pump pulley.

16. Remove the water pump.

  a. Engine coolant will leak from the cylinder block, so have a receptacle ready below.

✳✳ WARNING

Handle water pump vane so that it does not contact any other parts.

To install:

➡**Be sure to use new fasteners, as required.**

17. Installation is in the reverse order of removal.

18. After installation bleed the air from the cooling system.

19. Be sure to perform the recon- nect/relearn procedures.

ENGINE ELECTRICAL

BATTERY

REMOVAL & INSTALLATION

1. Before servicing the vehicle, refer to the Precautions.

2. Disconnect the negative battery cable.

3. Remove the battery cover, if equipped.

4. Disconnect both cables from the battery.

➡**Always disconnect the negative bat- tery cable, first.**

5. Remove the battery clamp nuts and battery clamp.

6. Remove the battery.

To install:

7. Installation is the reverse of the removal procedure.

BATTERY SYSTEM

8. Perform reconnect/relearn proce- dures.

➡**Always connect the positive battery cable first.**

9. Tighten the battery terminal nut to 30 inch lbs. (3.4 Nm). Tighten the clamp nuts to 35 inch lbs. (3.92 Nm).

ENGINE ELECTRICAL

ALTERNATOR

REMOVAL & INSTALLATION

See Figures 24 and 25.

1. Before servicing the vehicle, refer to the Precautions.

2. Disconnect the negative battery cable.

3. Remove the engine cover and air cleaner assembly, as required.

CHARGING SYSTEM

4. Remove alternator stay.

5. On V8 engine, remove the lower alternator bolt.

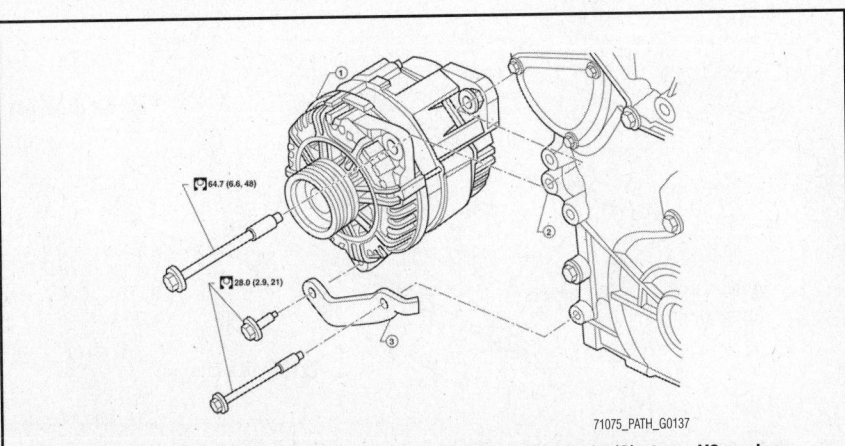

$64.7 (6.6, 48)$

$28.0 (2.9, 21)$

71075_PATH_G0137

Fig. 24 Alternator and related components (1) component, (2) block, (3) stay—V6 engine

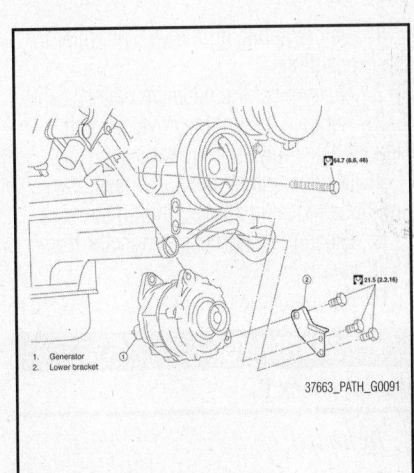

$64.7 (6.6, 48)$

$21.5 (2.2, 16)$

1. Generator
2. Lower bracket

37663_PATH_G0091

Fig. 25 Alternator and related components—V8 engine

6. Remove the alternator upper bolt.

7. Disconnect the alternator harness connectors.

8. Remove the alternator.

To install:

9. Installation is in the reverse order of removal.

10. Install the alternator and check tension of drive belt.

➡**The power generation variable voltage control system that controls the power generation voltage of the alternator has been adopted. Therefore, the power generation variable voltage con-** trol system inspection should be performed after replacing the alternator in order to ensure that the system operates normally.

11. Be sure to perform the reconnect/relearn procedures.

ENGINE ELECTRICAL

IGNITION COIL MODULE

REMOVAL & INSTALLATION

V6 Engine

Left Bank

1. Before servicing the vehicle, refer to the Precautions.

2. Disconnect the negative battery cable.

3. Remove air cleaner case and air duct, as required.

4. Move aside harness, harness bracket, and hoses located above ignition coil.

5. Disconnect harness connector from ignition coil.

6. Remove ignition coil.

✳✳ WARNING

Do not shock it.

To install:

➡**Be sure to use new fasteners, as required.**

7. Installation is the reverse of the removal procedure.

8. Be sure to perform the reconnect/relearn procedures.

Right Bank

See Figure 26.

1. Before servicing the vehicle, refer to the Precautions.

2. Disconnect the negative battery cable.

3. Remove the engine cover and air cleaner assembly, as required.

4. Move aside harness, harness bracket, and hoses located above ignition coil.

5. Disconnect harness connector from ignition coil.

6. Remove ignition coil.

✳✳ WARNING

Do not shock it.

To install:

➡**Be sure to use new fasteners, as required.**

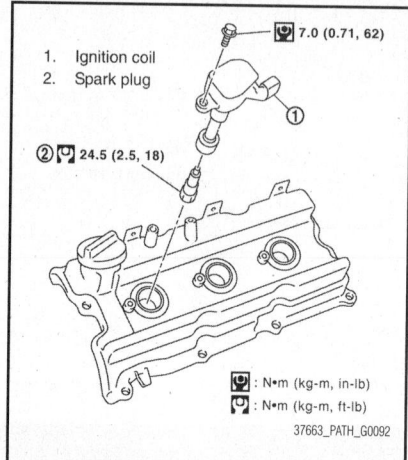

1. Ignition coil
2. Spark plug

⊡ 7.0 (0.71, 62)

② ⊡ 24.5 (2.5, 18)

⊡ : N•m (kg-m, in-lb)
⊡ : N•m (kg-m, ft-lb)

37663_PATH_G0092

Fig. 26 Ignition coil and related components—V6 engine

7. Installation is the reverse of the removal procedure.

8. Be sure to perform the reconnect/relearn procedures.

V8 Engine

See Figure 27.

1. Before servicing the vehicle, refer to the Precautions.

2. Disconnect the negative battery cable.

3. Remove the engine room cover.

IGNITION SYSTEM

4. Remove the air duct and resonator assembly.

5. Disconnect the harness connector from the ignition coil.

6. Remove the ignition coil.

✳✳ WARNING

Do not shock ignition coil.

To install:

➡**Be sure to use new fasteners, as required.**

7. Installation is the reverse of the removal procedure.

8. Be sure to perform the reconnect/relearn procedures.

IGNITION TIMING

INSPECTION & ADJUSTMENT

Ignition timing is controlled by the ECM. No adjustment is necessary or possible.

SPARK PLUGS

REMOVAL & INSTALLATION

V6 Engine

1. Before servicing the vehicle, refer to the Precautions.

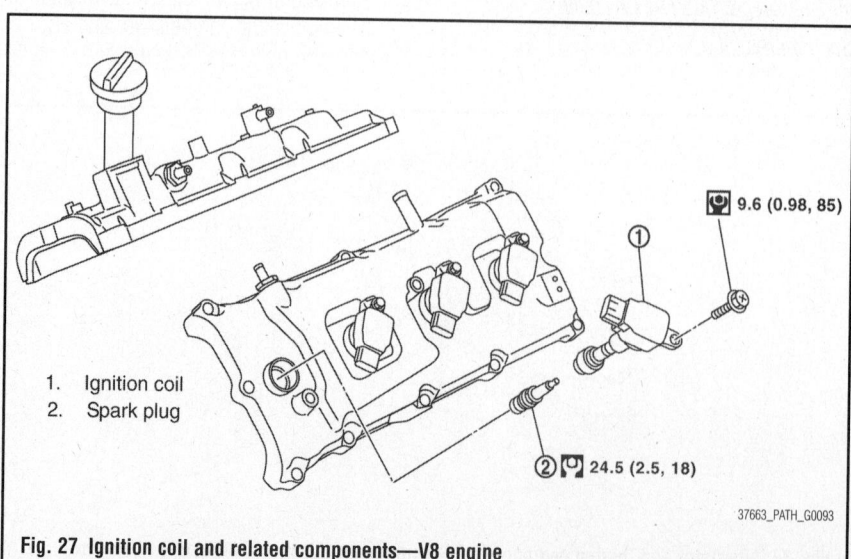

⊡ 9.6 (0.98, 85)

①

② ⊡ 24.5 (2.5, 18)

1. Ignition coil
2. Spark plug

37663_PATH_G0093

Fig. 27 Ignition coil and related components—V8 engine

2. Disconnect the negative battery cable.

3. Remove engine cover.

4. Remove air cleaner case and air duct.

5. Move aside harness, harness bracket, and hoses located above ignition coil.

6. Disconnect harness connector from ignition coil.

7. Remove ignition coil.

8. Remove the spark plug.

To install:

➡**Be sure to use new fasteners, as required.**

9. Installation is the reverse of the removal procedure.

10. Be sure to perform the reconnect/relearn procedures.

V8 Engine

1. Before servicing the vehicle, refer to the Precautions.

2. Disconnect the negative battery cable.

3. Remove the engine room cover.

4. Remove the air duct and resonator assembly.

5. Disconnect the harness connector from the ignition coil.

6. Remove the ignition coil.

✼✼ WARNING

Do not shock ignition coil.

7. Remove spark plug.

To install:

➡**Be sure to use new fasteners, as required.**

8. Installation is the reverse of the removal procedure.

9. Be sure to perform the reconnect/relearn procedures.

ENGINE ELECTRICAL

STARTER

REMOVAL & INSTALLATION

V6 Engine

See Figure 28.

1. Before servicing the vehicle, refer to the Precautions.

2. Disconnect the negative battery cable.

3. Remove the engine cover and air cleaner assembly, as required.

4. Remove engine undercover.

5. Raise and safely support the vehicle, as necessary.

6. Remove exhaust manifold cover from exhaust manifold (right bank) to gain access to starter cover bolts.

7. Remove starter cover bolts and starter cover.

8. Disconnect terminal "1" connector and terminal "2" nut.

9. Remove the two starter bolts.

10. Remove the starter.

To install:

➡**Be sure to use new fasteners, as required.**

11. Installation is the reverse of the removal procedure.

12. Be sure to perform the reconnect/relearn procedures.

STARTING SYSTEM

V8 Engine

See Figure 29.

1. Before servicing the vehicle, refer to the Precautions.

2. Disconnect the negative battery cable.

3. Remove the engine cover and air cleaner assembly, as required.

4. Remove the intake manifold.

5. Disconnect terminal "1" connector and terminal "2" nut.

6. Remove the two starter bolts.

To install:

➡**Be sure to use new fasteners, as required.**

7. Installation is the reverse of the removal procedure.

8. Be sure to perform the reconnect/relearn procedures.

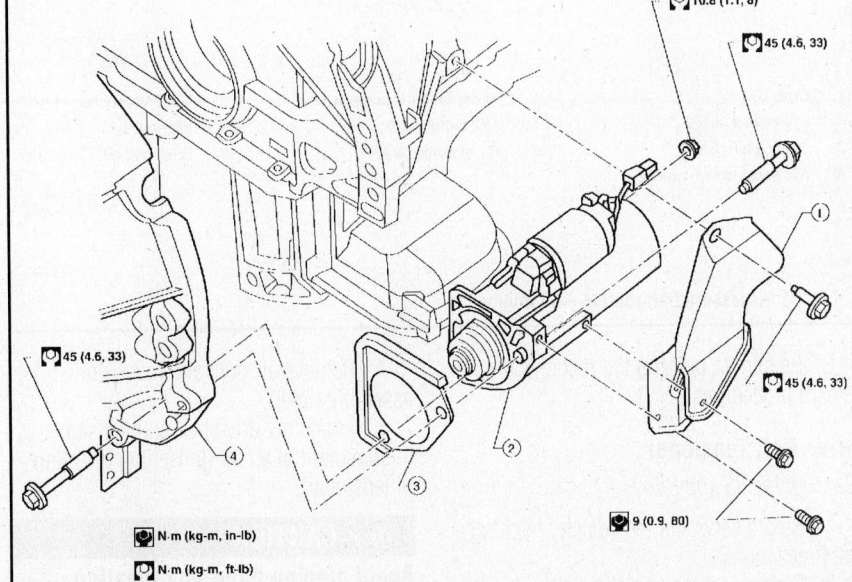

10.8 (1.1, 8)
45 (4.6, 33)
45 (4.6, 33)
45 (4.6, 33)
9 (0.9, 80)

N·m (kg-m, in-lb)
N·m (kg-m, ft-lb)

22140_PATH_G0014

Fig. 28 Starter and related components—V6 engine

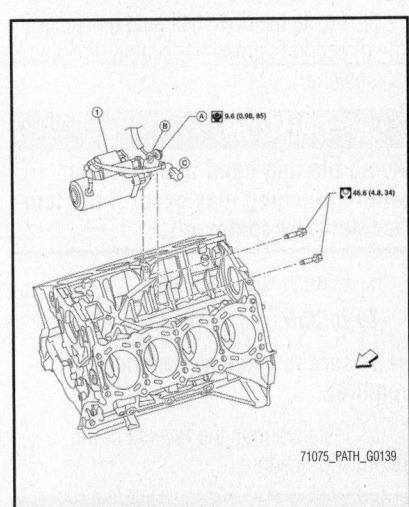

9.6 (0.98, 85)
46.6 (4.8, 34)

71075_PATH_G0139

Fig. 29 Starter and related components (1) starter, (A) nut, (B) cable, (C) connector—V8 engine

ENGINE MECHANICAL

➡Disconnecting the negative battery cable may interfere with the functions of the on board computer systems and may require the computer to undergo a relearning process, once the negative battery cable is reconnected.

ACCESSORY DRIVE BELTS

ADJUSTMENT

Belt tensioning is not necessary, as it is automatically adjusted by auto tensioner.

BELT ROUTINGS

See Figures 30 and 31.

INSPECTION

Inspect the drive belt for signs of glazing or cracking. A glazed belt will be perfectly smooth from slippage, while a good belt will have a slight texture of fabric visible. Cracks will usually start at the inner edge of the belt and run outward. All worn or damaged drive belts should be replaced immediately.

REMOVAL & INSTALLATION

Drive Belt

See Figure 32.

1. Before servicing the vehicle, refer to the Precautions.
2. Disconnect the negative battery cable.
3. Remove the engine cover, as necessary.
4. Remove air duct and resonator assembly (inlet).
5. Rotate the drive belt auto tensioner in the direction of arrow (loosening direction of tensioner).

✳✳ CAUTION

Avoid placing hand in a location where pinching may occur if the tool accidentally comes off.

6. Remove the drive belt.

To install:

➡**Be sure to use new fasteners, as required.**

7. Installation is the reverse of the removal procedure.

✳✳ WARNING

Make sure belt is securely installed around all pulleys.

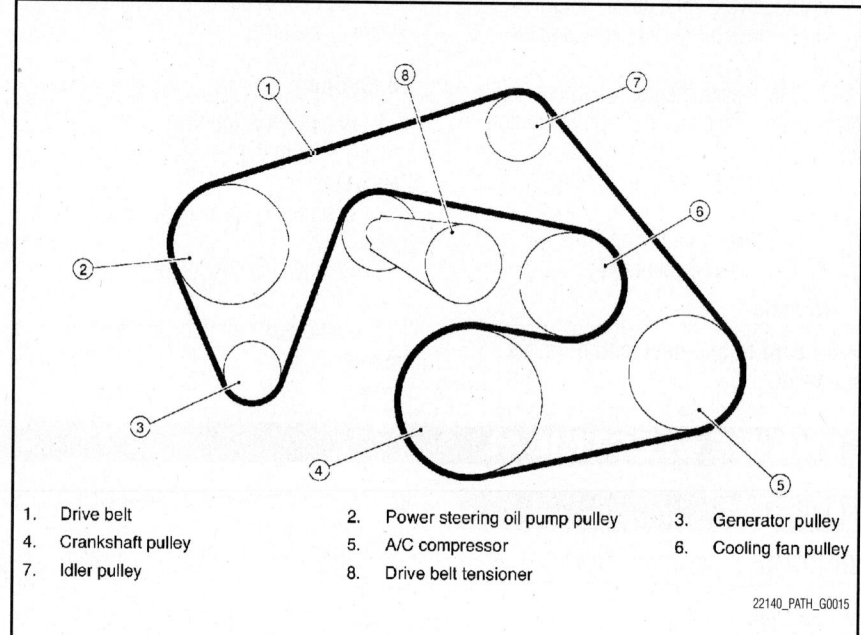

1.	Drive belt	2.	Power steering oil pump pulley	3.	Generator pulley
4.	Crankshaft pulley	5.	A/C compressor	6.	Cooling fan pulley
7.	Idler pulley	8.	Drive belt tensioner		

22140_PATH_G0015

Fig. 30 Accessory belt routing—V6 engine

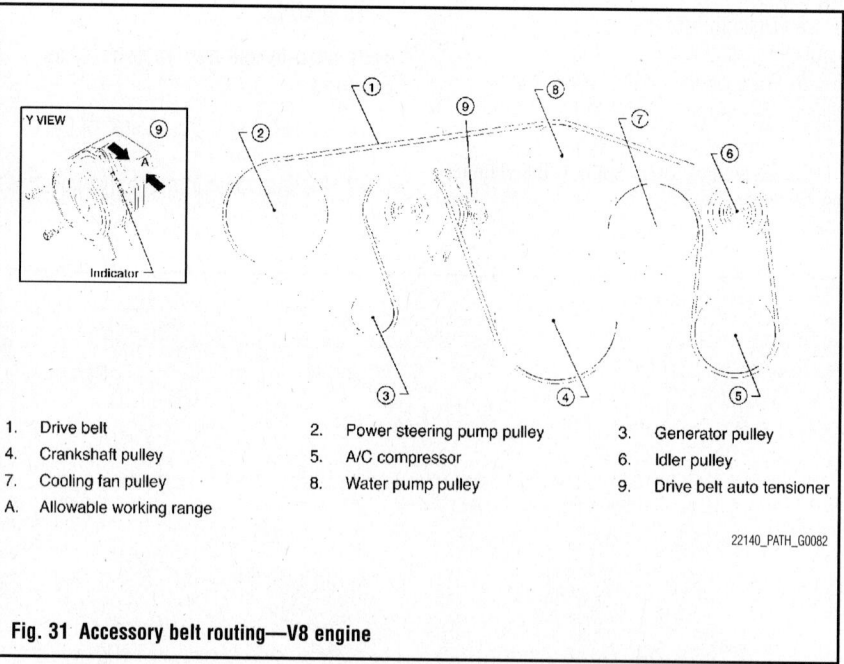

1.	Drive belt	2.	Power steering pump pulley	3.	Generator pulley
4.	Crankshaft pulley	5.	A/C compressor	6.	Idler pulley
7.	Cooling fan pulley	8.	Water pump pulley	9.	Drive belt auto tensioner
A.	Allowable working range				

22140_PATH_G0082

Fig. 31 Accessory belt routing—V8 engine

8. Be sure to perform the reconnect/relearn procedures.

Drive Belt Tensioner

See Figures 33 and 34.

1. Before servicing the vehicle, refer to the Precautions.
2. Disconnect the negative battery cable.
3. Remove the engine cover, as necessary.

4. Remove air duct and resonator assembly (inlet).
5. Rotate the drive belt auto tensioner in the direction of arrow (loosening direction of tensioner).

✳✳ CAUTION

Avoid placing hand in a location where pinching may occur if the tool accidentally comes off.

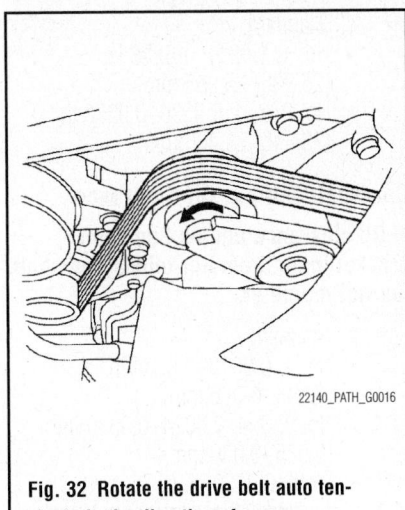

Fig. 32 Rotate the drive belt auto tensioner in the direction of arrow

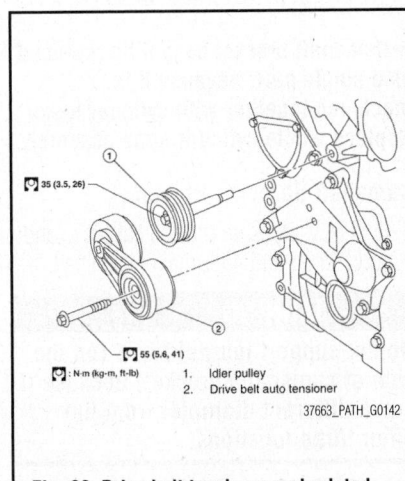

Fig. 33 Drive belt tensioner and related components—V6 engine

6. Remove the drive belt.

7. Remove the component retaining bolts. Remove the component from its mounting.

To install:

➡**Be sure to use new fasteners, as required.**

8. Installation is the reverse of the removal procedure.

✳✳ WARNING

Make sure belt is securely installed around all pulleys.

9. Be sure to perform the reconnect/relearn procedures.

AIR CLEANER

REMOVAL & INSTALLATION

See Figures 35 and 36.

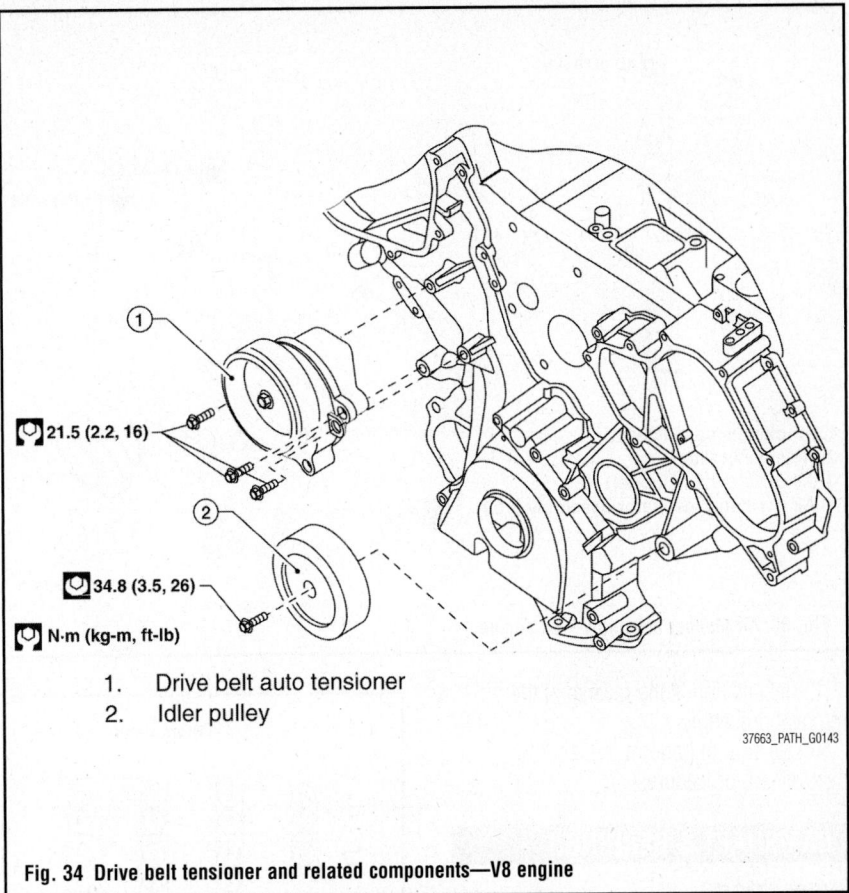

1. Drive belt auto tensioner
2. Idler pulley

Fig. 34 Drive belt tensioner and related components—V8 engine

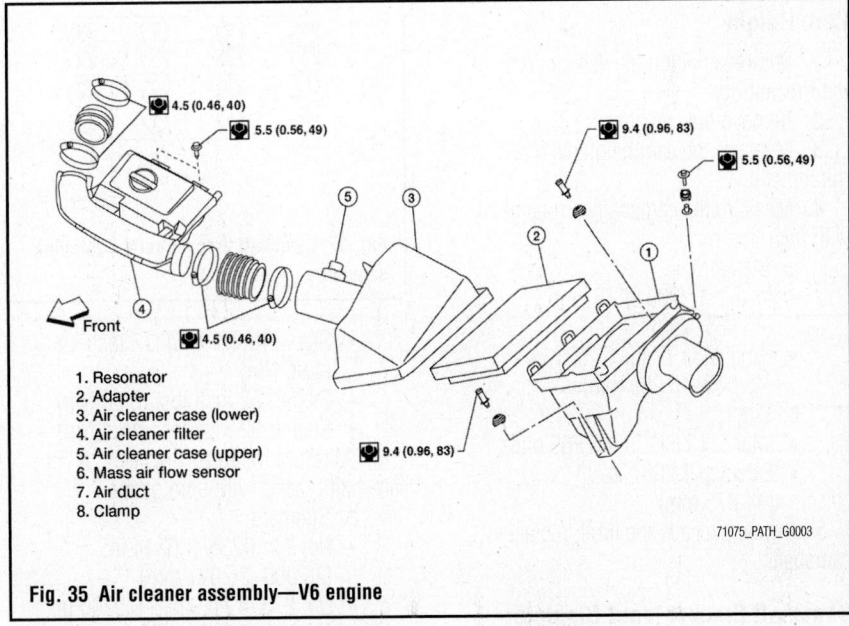

1. Resonator
2. Adapter
3. Air cleaner case (lower)
4. Air cleaner filter
5. Air cleaner case (upper)
6. Mass air flow sensor
7. Air duct
8. Clamp

Fig. 35 Air cleaner assembly—V6 engine

1. Before servicing the vehicle, refer to the Precautions.

2. Disconnect the negative battery cable.

3. Remove the engine cover, if equipped.

4. Remove the component retaining screws/brackets/clips etc.

5. Disconnect the air duct and resonator from the air cleaner case.

6. Disconnect the electrical connections, as required.

7. Remove the component from the vehicle.

To install:

➡**Be sure to use new fasteners, as required.**

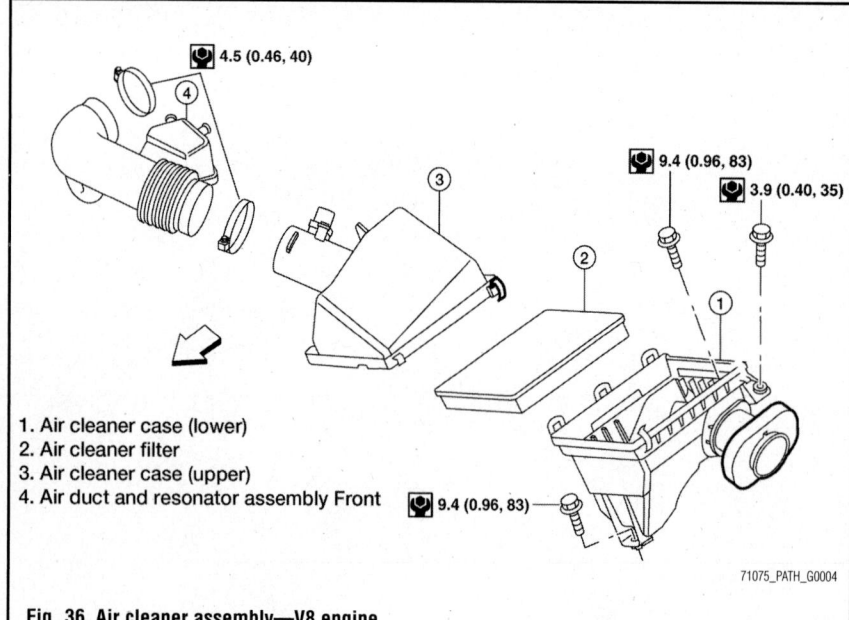

4.5 (0.46, 40)

9.4 (0.96, 83)

3.9 (0.40, 35)

9.4 (0.96, 83)

1. Air cleaner case (lower)
2. Air cleaner filter
3. Air cleaner case (upper)
4. Air duct and resonator assembly Front

71075_PATH_G0004

Fig. 36 Air cleaner assembly—V8 engine

8. Installation is the reverse of the removal procedure.

9. Be sure to perform the reconnect/relearn procedures.

CAMSHAFT & BEARINGS

INSPECTION

Cam Height

1. Before servicing the vehicle, refer to the Precautions.

2. Remove the camshafts.

3. Measure the cam height with a micrometer.

4. Measure the camshaft cam height with micrometer.

   a. Standard:
- Intake: 1.7900–1.7974 in. (45.465–45.655 mm)
- Exhaust: 1.7746–1.7821 in. (45.075–45.265 mm)

   b. Limit:
- Intake: 1.7821 in. 45.265 mm)
- Exhaust: 1.7667 in. (44.875 mm)

5. If wear exceeds the limit, replace camshaft.

Camshaft Bracket Inner Diameter

See Figure 37.

1. Tighten camshaft bracket bolt with the specified torque.

2. Tighten camshaft bracket bolts in the following steps, in numerical order as shown.
- Step 1 (bolts 7–10): 17 inch lbs. (1.96 Nm)

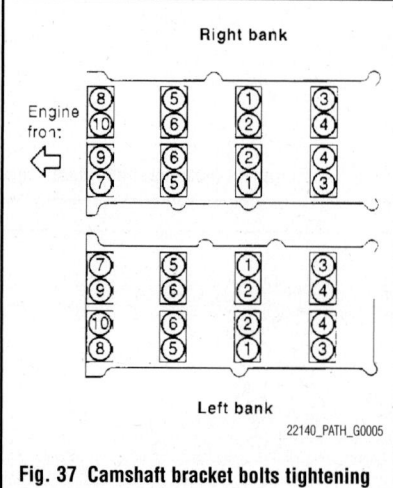

Right bank

Engine front

Left bank

22140_PATH_G0005

Fig. 37 Camshaft bracket bolts tightening sequence

- Step 2 (bolts 1–6): 17 inch lbs. (1.96 Nm)
- Step 3: 52 inch lbs. (5.88 Nm)
- Step 4: 92 inch lbs. (10.4 Nm)

3. Measure the inner diameter "A" of camshaft bracket with bore gauge.

   a. Standard:
- No. 1: 1.0236–1.0244 in. (26.000–26.021 mm)
- No. 2, 3, 4 : 0.9252–0.9260 in. (23.500–3.521 mm)

Camshaft Journal Diameter

1. Before servicing the vehicle, refer to the Precautions.

2. Remove the camshafts.

3. Measure the outer diameter of camshaft journal with micrometer and record the result.

   a. Standard:
- No. 1: 1.0211–1.0218 in. (25.935–25.955 mm)
- No. 2, 3, 4: 0.9230–0.9238 in. (23.445–23.465 mm)

Camshaft Journal Oil Clearance

➡**Oil clearance equals Camshaft bracket inner diameter minus Camshaft journal diameter.**

   b. Standard:
- No. 1: 0.0018–0.0034 in. (0.045–0.086 mm)
- No. 2, 3, 4: 0.0014–0.0030 in. (0.035–0.076 mm)
- Limit: 0.0059 in. (0.15 mm)

1. If the calculated value exceeds the limit, replace either or both camshaft and cylinder head.

➡**Camshaft bracket cannot be replaced as a single part, because it is machined together with cylinder head. Replace whole cylinder head assembly.**

Camshaft Sprocket Runout

1. Put V-block on precise flat table, and support No. 2 and 4 journal of camshaft.

☀ WARNING

Do not support journal No. 1 (on the side of camshaft sprocket) because it has a different diameter from the other three locations.

2. Measure the camshaft sprocket run out with dial indicator. (Total indicator reading)
- Limit: 0.0059 in. (0.15 mm)

3. If it exceeds the limit, replace camshaft sprocket.

Endplay

See Figures 38 and 39.

1. Install dial indicator in thrust direction on front end of camshaft.

2. Measure the end play of dial indicator when camshaft is moved forward/backward (in direction to axis).
- Standard: 0.0045–0.0074 in. (0.115–0.188 mm)
- Limit: 0.0094 in. (0.24 mm)

3. Measure the following parts if out of the limit.
- Dimension "A" for camshaft No. 1 journal
- Standard: 1.0827–1.0846 in. (27.500–27.548 mm)
- Dimension "B" for cylinder head No. 1 journal bearing
- Standard: 1.0772–1.0781 in. (27.360–27.385 mm)

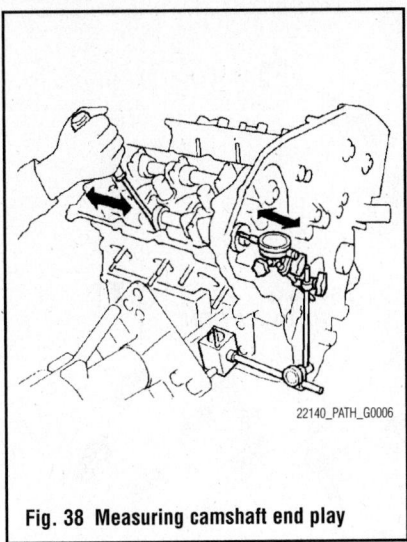

Fig. 38 Measuring camshaft end play

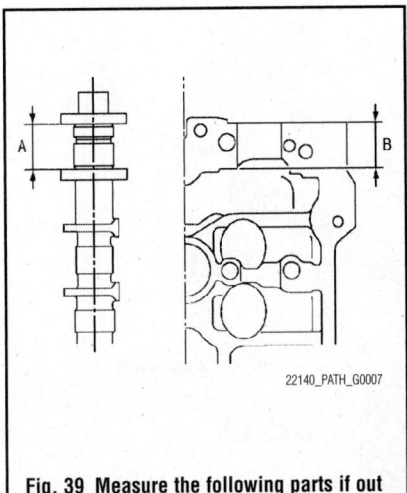

Fig. 39 Measure the following parts if out of the limit

4. Refer to the standards, and then replace camshaft and/or cylinder head.

Runout

1. Put V-block on precise flat table, and support No. 2 and 4 journal of camshaft.

❊❊ WARNING

Do not support journal No. 1 (on the side of camshaft sprocket) because it has a different diameter from the other three locations.

2. Set dial indicator vertically to No. 3 journal.
3. Turn camshaft to one direction with hands, and measure the camshaft run out on dial indicator. (Total indicator reading)
- Standard: Less than 0.0008 in. (0.02 mm)
- Limit: 0.0020 in. (0.05 mm)

4. If it exceeds the limit, replace camshaft.

REMOVAL & INSTALLATION

V6 Engine

See Figures 40 through 42.

1. Before servicing the vehicle, refer to the Precautions.
2. Disconnect the negative battery cable.
3. Remove the engine cover. Remove the air cleaner assembly.
4. Remove front timing chain case, camshaft sprocket, timing chain and rear timing chain case.
5. Remove Camshaft Position (CMP) sensor (PHASE) (right and left banks) from cylinder head back side.

❊❊ WARNING

Handle carefully to avoid dropping and shocks. Do not disassemble. Do not allow metal powder to adhere to magnetic part at sensor tip. Do not place sensors in a location where they are exposed to magnetism.

6. Remove intake valve timing control solenoid valves.

➡**Discard intake valve timing control solenoid valve gaskets and use new gaskets for installation.**

7. Remove camshaft brackets.
   a. Mark camshafts, camshaft brackets and bolts so they are placed in the same position and direction for installation.
   b. Equally loosen camshaft bracket bolts in several steps in reverse order of the tightening sequence.
8. Remove camshafts.
9. Remove valve lifters.

➡**Identify installation positions, and store them without mixing them up.**

10. Remove timing chain tensioner (secondary) from cylinder head with its stopper pin attached.

➡**Stopper pin was attached when timing chain (secondary) was removed.**

To install:

➡**Be sure to use new fasteners, as required.**

11. Install timing chain tensioners (secondary) on both sides of cylinder head.
   a. Install timing chain tensioner with its stopper pin attached.

   b. Install timing chain tensioner with sliding part facing downward on right-side cylinder head, and with sliding part facing upward on left-side cylinder head.
   c. Install new O-rings.
12. Install valve lifters.

➡**Install in their original positions.**

13. Install camshafts.
   a. Install camshaft with dowel pin attached to its front end face on the exhaust side.
   b. Follow your identification marks made during removal, or follow the identification marks that are present on new camshafts for proper placement and direction.
   c. Install camshaft so that dowel pin hole and dowel pin on front end face. (No. 1 cylinder TDC on its compression stroke).

➡**Large and small pin holes are located on front end face of camshaft (INT), at intervals of 180°. Face small diameter side pin hole upward (in cylinder head upper face direction). Though camshaft does not stop at the portion as shown, for the placement of cam nose, it is generally accepted camshaft is placed for the same direction.**

14. Install camshaft brackets.
   a. Remove foreign material completely from camshaft bracket backside and from cylinder head installation face.
   b. Install camshaft bracket in original position and direction.
   c. Install camshaft brackets (No. 2 to 4) aligning the stamp marks.

➡**There are no identification marks indicating left and right for camshaft bracket (No. 1).**

15. Apply liquid gasket to mating surface of camshaft bracket (No. 1) on right and left banks.

➡**Use Genuine RTV Silicone Sealant or equivalent.**

16. Tighten camshaft bracket bolts in numerical order as shown.
17. Tighten camshaft bracket bolts in the following steps:
- Step 1: Bolts 7–10: 17 inch lbs. (1.96 Nm)
- Step 2: Bolts 1–6: 17 inch lbs. (1.96 Nm)
- Step 3: All bolts: 52 inch lbs. (5.88 Nm)

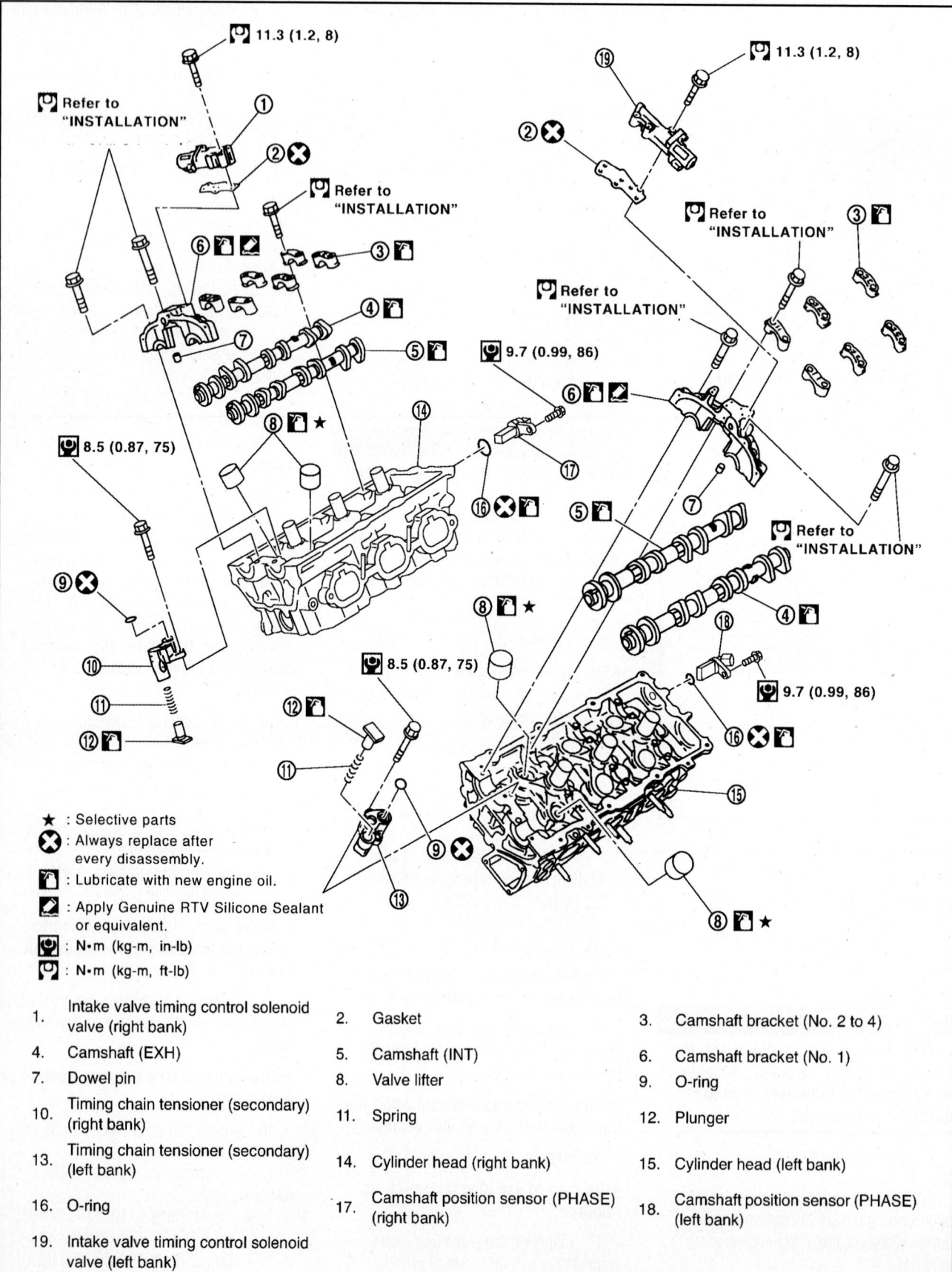

Fig. 40 Camshaft and related components—V6 Engine

★ : Selective parts

✗ : Always replace after every disassembly.

🖐 : Lubricate with new engine oil.

✍ : Apply Genuine RTV Silicone Sealant or equivalent.

🔧 : N•m (kg-m, in-lb)

🔧 : N•m (kg-m, ft-lb)

1.	Intake valve timing control solenoid valve (right bank)	2.	Gasket	3.	Camshaft bracket (No. 2 to 4)
4.	Camshaft (EXH)	5.	Camshaft (INT)	6.	Camshaft bracket (No. 1)
7.	Dowel pin	8.	Valve lifter	9.	O-ring
10.	Timing chain tensioner (secondary) (right bank)	11.	Spring	12.	Plunger
13.	Timing chain tensioner (secondary) (left bank)	14.	Cylinder head (right bank)	15.	Cylinder head (left bank)
16.	O-ring	17.	Camshaft position sensor (PHASE) (right bank)	18.	Camshaft position sensor (PHASE) (left bank)
19.	Intake valve timing control solenoid valve (left bank)				

22140_PATH_G0017

Fig. 40 Camshaft and related components—V6 Engine

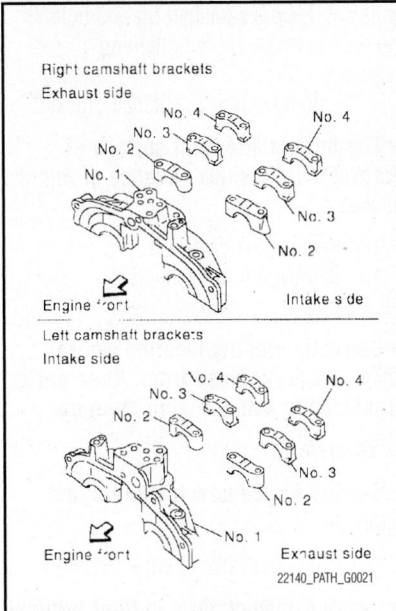

Fig. 41 Camshaft brackets and related components—V6 engine

- Step 4: All bolts: 92 inch lbs. (10.4 Nm)

18. Measure the difference in levels between front end faces of camshaft bracket (No. 1) and cylinder head.

   a. Standard: -0.0055 to 0.0055 in. (-0.14 to 0.14 mm)

   b. Measure two positions (both intake and exhaust side) for a single bank.

   c. If the measured value is out of the standard, re-install camshaft bracket (No. 1).

19. Check and adjust the valve clearance, as required.

20. Installation of the remaining components is in the reverse order of removal.

21. Be sure to perform the reconnect/relearn procedures.

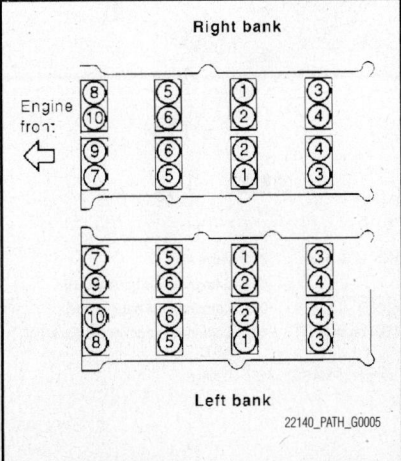

Fig. 42 Camshaft bracket bolt tightening sequence—V6 engine

V8 Engine

See Figures 43 through 59.

➡**Do not remove the engine assembly to perform this procedure.**

1. Before servicing the vehicle, refer to the Precautions.

2. Disconnect the negative battery cable.

3. Remove the engine cover. Remove the air cleaner assembly.

4. Remove the RH bank and LH bank rocker covers.

5. Obtain compression TDC of No. 1 cylinder as follows:

   a. Turn the crankshaft pulley clockwise to align the TDC identification notch (without paint mark) with the timing indicator on the front cover.

   b. At this time, make sure both intake and exhaust cam lobes of No. 1 cylinder (top front on LH bank) point outside.

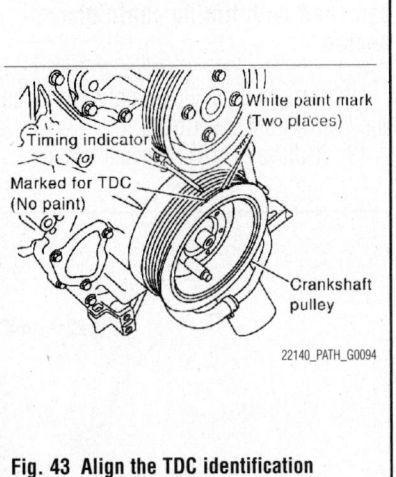

Fig. 43 Align the TDC identification notch—V8 engine

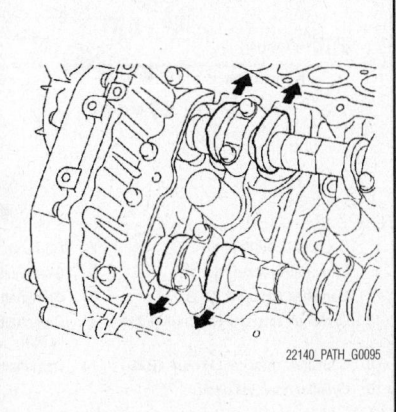

Fig. 44 Intake and exhaust cam lobes of No. 1 cylinder—V8 engine

   c. If they do not point outside, turn crankshaft pulley once more.

6. Remove the intake valve control solenoid cover RH bank (A) and intake valve control solenoid cover LH bank (B) as follows:

   a. Loosen and remove the bolts

   b. Cut the liquid gasket and remove the covers.

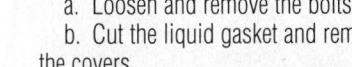
✳✳ WARNING

Do not damage mating surfaces.

7. Paint alignment marks on the RH bank (A) timing chain links (C) and LH bank (B) timing chain links (D) and align with the camshaft sprocket alignment marks (E) and (F).

8. Remove the LH bank timing chain tensioner using the following steps.

✳✳ CAUTION

Plunger, spring, and spring seat pop out when squeezing return-proof clip without holding plunger head. It may cause serious injuries. Always hold plunger head when removing.

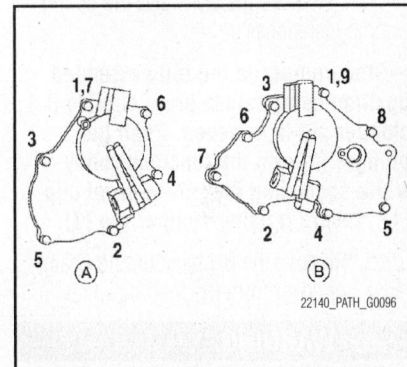

Fig. 45 Loosen and remove the bolts—V8 engine

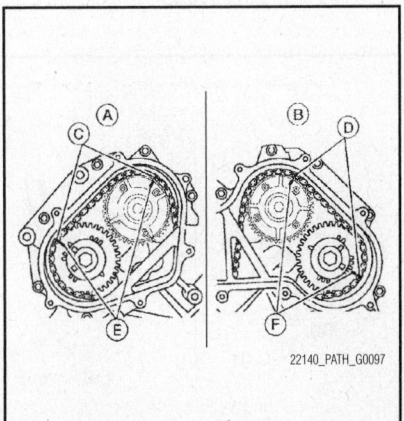

Fig. 46 Paint alignment marks on the RH bank (A) timing chain links (C) and LH bank (B) timing chain links—V8 engine

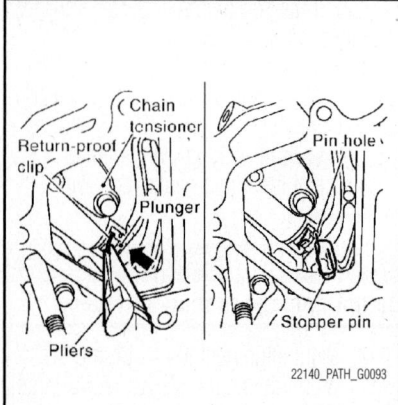

Fig. 47 Push the plunger into the tensioner body—V8 engine

a. Squeeze return-proof clip ends using suitable tool and push the plunger into the tensioner body.

b. Secure plunger using stopper pin.

➡ **Stopper pin is made from hard wire approximately 1 mm (0.04 in.) in diameter.**

c. Remove the bolts and the timing chain tensioner.

➡ **Stop plunger in the fully extended position using return-proof clip (1) if stopper pin is removed. Push the plunger (2) into the tensioner body while squeezing the return-proof clip (1). Secure it using stopper pin (3).**

9. Remove the RH bank timing chain tensioner cover from the front cover.

✷✷ WARNING

Do not damage mating surfaces.

10. Remove the RH bank timing chain tensioner using the following steps.

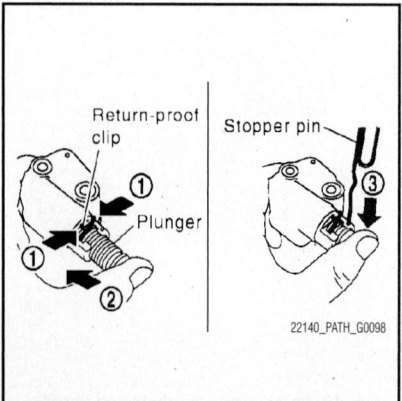

Fig. 48 Stop plunger in the fully extended position—V8 engine

✷✷ CAUTION

Plunger, spring, and spring seat pop out when squeezing return-proof clip without holding plunger head. It may cause serious injuries. Always hold plunger head when removing.

a. Squeeze return-proof clip ends using suitable tool and push the plunger into the tensioner body.

b. Secure plunger using stopper pin.

c. Remove the bolts and the RH bank timing chain tensioner (A).

➡ **If it is difficult to push plunger on RH bank timing chain tensioner (A), remove the plunger under extended condition.**

11. Loosen camshaft sprocket bolts as shown and remove camshaft sprockets.

✷✷ WARNING

To avoid interference between valves and pistons, do not turn crankshaft or camshaft with timing chain disconnected.

12. Remove the RH (A) front cover bolts and LH (B) front cover bolts.

13. Remove RH (A) camshaft bracket

bolts and LH (C) camshaft bracket bolts in the reverse order of the tightening sequence.

a. Remove No. 1 camshaft bracket.

➡ **The bottom and front surface of bracket will be stuck because of liquid gasket.**

14. Remove the camshaft.

15. Remove the valve lifters if necessary.

➡ **Correctly identify location where each part is removed from. Keep parts organized to avoid mixing them up.**

To install:

➡ **Be sure to use new fasteners, as required.**

16. Install the valve lifters if removed.

➡ **Install removed parts in their original locations.**

17. Install the camshafts. Important details for identification of the RH and LH, and intake and exhaust.

- RH Bank: Intake: Front paint: Pink; Exhaust: Rear paint: Orange; Rib: Yes
- LH Bank: Intake: Front paint: Pink; Exhaust: Rear paint: Orange; Rib: No

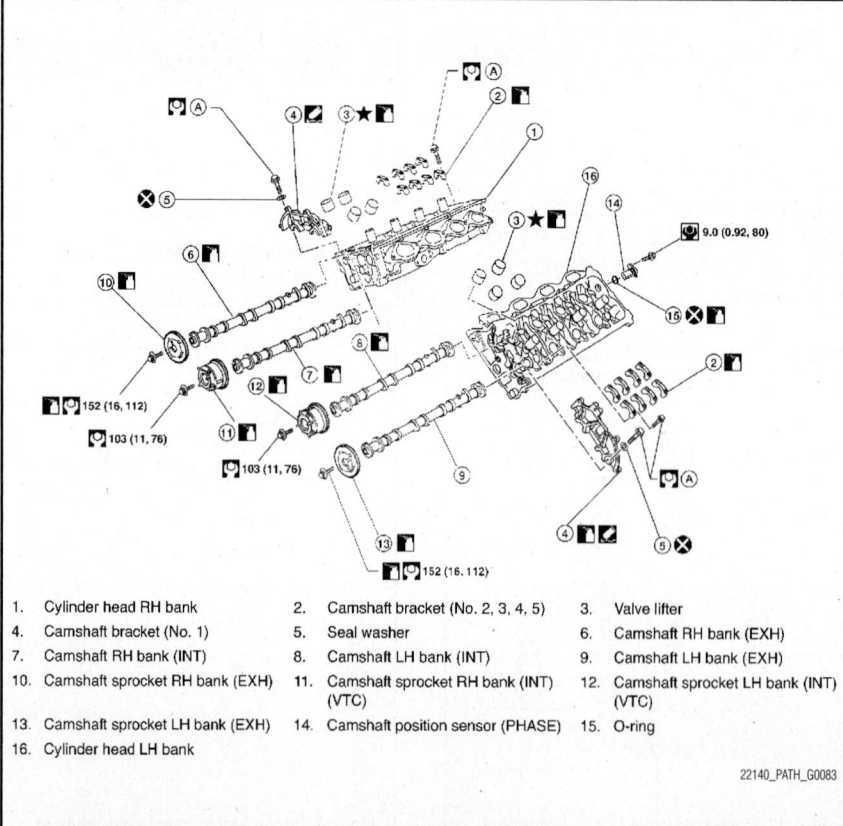

1. Cylinder head RH bank	2. Camshaft bracket (No. 2, 3, 4, 5)	3. Valve lifter
4. Camshaft bracket (No. 1)	5. Seal washer	6. Camshaft RH bank (EXH)
7. Camshaft RH bank (INT)	8. Camshaft LH bank (INT)	9. Camshaft LH bank (EXH)
10. Camshaft sprocket RH bank (EXH)	11. Camshaft sprocket RH bank (INT) (VTC)	12. Camshaft sprocket LH bank (INT) (VTC)
13. Camshaft sprocket LH bank (EXH)	14. Camshaft position sensor (PHASE)	15. O-ring
16. Cylinder head LH bank		

Fig. 49 Camshaft and related components—V8 Engine

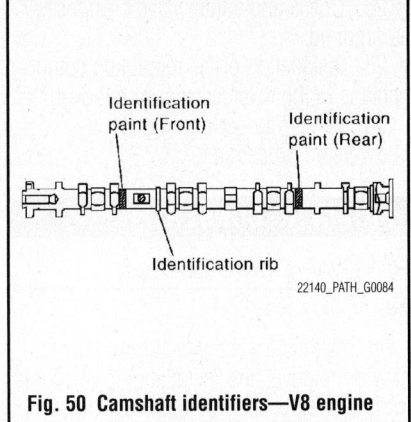

Fig. 50 Camshaft identifiers—V8 engine

- Install so that the RH bank (B) dowel pins (A) and LH bank (C) dowel pins (A) at the front of the camshaft face are in the direction shown.

18. Install the RH bank (B) and LH bank (D) camshaft brackets (A).

    a. Install by referring to the installation location mark (E) on the upper surface.

    b. Install so that the installation location mark (E) can be correctly read when viewed from the intake manifold side (C).

    c. Install No. 1 camshaft bracket using the following procedure:

- C: 0.43 in. (11 mm)
- D: 0.079–0.118 in. (2.0–3.0 mm) diameter
- Apply liquid gasket to No. 1 camshaft bracket (A) and (B) as shown.

❋❋ WARNING

After installation, be sure to wipe off any excessive liquid gasket outside

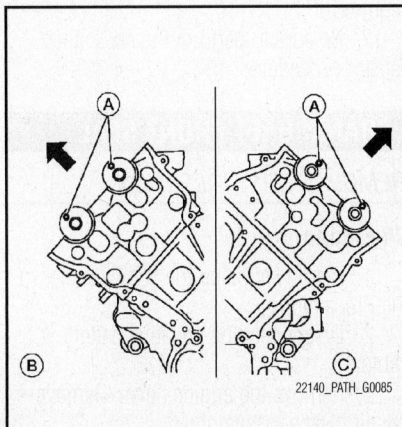

Fig. 51 Proper position for camshaft installation—V8 engine

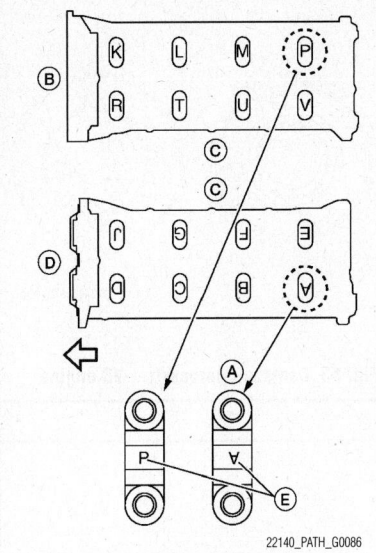

Fig. 52 Install the RH bank (B) and LH bank (D) camshaft brackets (A)—V8 engine

of application (C) and (D) both on RH and LH sides.

- Remove completely any excess of liquid gasket inside bracket.

    d. Apply liquid gasket (C) to the back side of the LH (A) bank front cover and RH (B) bank front cover as shown.

    e. C: 0.102–0.142 in. (2.6–3.6 mm) diameter

    f. Position No. 1 camshaft bracket close to the mounting position, and then install it to prevent from touching liquid gasket applied to each surface.

    g. Temporarily tighten the RH (A) and LH (B) front cover bolts (4 for each bank) as shown.

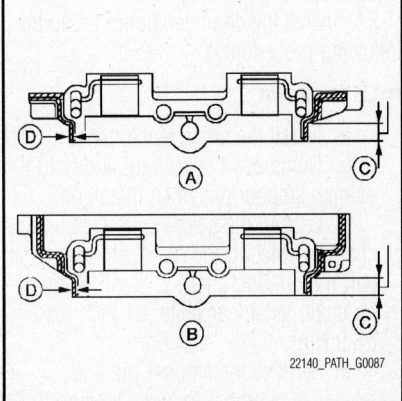

Fig. 53 Apply liquid gasket to No. 1 camshaft bracket (A) and (B)—V8 engine

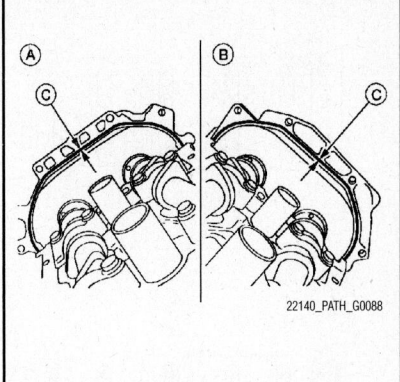

Fig. 54 Apply liquid gasket (C) to the back side of the LH (A) bank front cover and RH (B) bank front cover—V8 engine

19. Tighten the camshaft bracket bolts as follows:

    a. Step 1 (bolts 9–12): 17 inch lbs. (2.0 Nm)

    b. Step 2 (bolts 1–8): 17 inch lbs. (2.0 Nm)

    c. Step 3 (all bolts): 52 inch lbs. (5.9 Nm)

    d. Step 4 (all bolts): 92 inch lbs. (10.4 Nm)

❋❋ WARNING

After tightening the camshaft bracket bolts, be sure to wipe off excessive liquid gasket from the parts listed below.

- Mating surface of rocker cover
- Mating surface of front cover
- A: RH bank
- B: Exhaust side
- C: LH bank
- D: Intake side

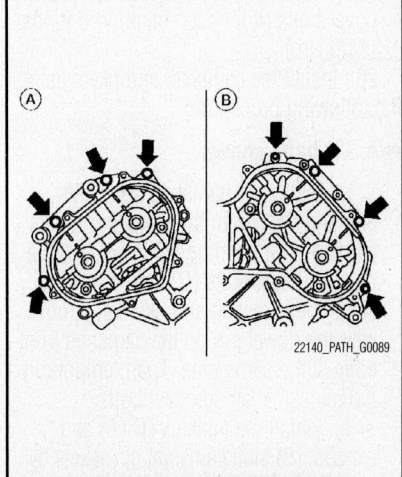

Fig. 55 Temporarily tighten the RH (A) and LH (B) front cover bolts—V8 engine

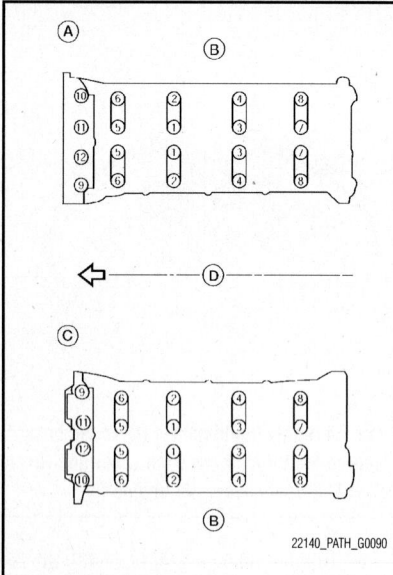

Fig. 56 Camshaft bracket bolt tightening sequence—V8 engine

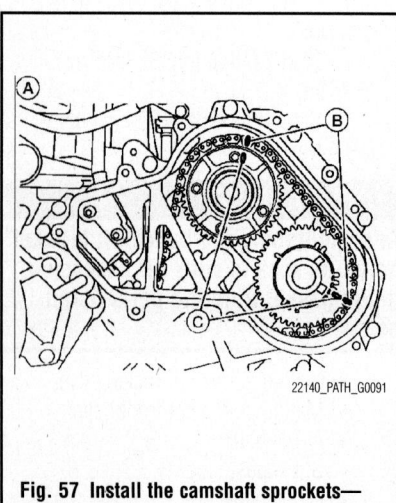

Fig. 57 Install the camshaft sprockets—V8 engine

e. Tighten the RH (A) and LH (B) front cover bolts (4 for each bank) to 8 ft. lbs. (11.0 Nm)

20. Install the camshaft sprockets using the following procedure:

➡ **A: LH bank shown.**

a. Install the camshaft sprockets aligning them with the matching marks painted on the timing chain (B) and the camshaft sprockets (C) before removal. Align the camshaft sprocket key groove with the dowel pin on the camshaft front edge at the same time. Then temporarily tighten camshaft sprocket bolts.

b. Install the intake VTC (A) and exhaust (B) side camshaft sprockets by selectively using the groove of the dowel pin according to the bank for the exhaust

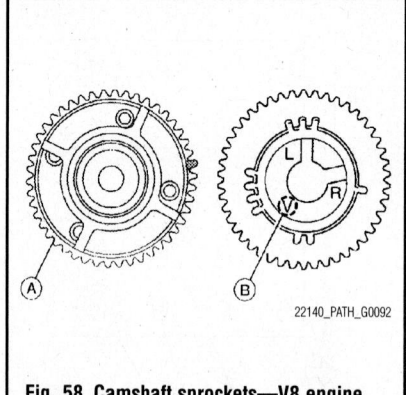

Fig. 58 Camshaft sprockets—V8 engine

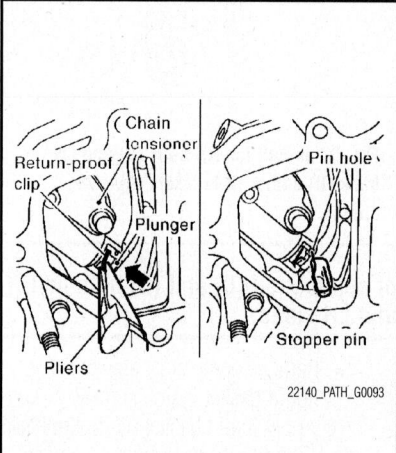

Fig. 59 Install the chain tensioner—V8 engine

(B) side camshaft sprockets. (Common part used for both exhaust banks.)

c. Lock the hexagonal part of the camshaft in the same way as for removal, and tighten the camshaft sprocket bolts.

d. Check again that the timing alignment mark on the timing chain and on each sprocket are aligned.

21. Install the chain tensioner using the following procedure:

➡ **LH is shown.**

a. Install the chain tensioner.

b. Compress the plunger and hold it using a stopper pin when installing.

c. Loosen the slack guide side timing chain by rotating the camshaft hexagonal part if mounting space is small. Torque for chain tensioner bolts: 61 inch lbs. (6.9 Nm).

d. Remove the stopper pin and release the plunger to apply tension to the timing chain.

e. Install the RH bank timing chain tensioner cover onto the front cover. Tighten bolts to 80 inch lbs. (9.0 Nm).

22. Check and adjust valve clearances, as required.

23. Installation of the remaining components is in the reverse order of removal.

24. Be sure to perform the reconnect/relearn procedures.

CATALYTIC CONVERTER

REMOVAL & INSTALLATION

See Figures 60 and 61.

At this time the manufacturer does not provide removal and installation procedures for this component. The following procedure is a guideline and may differ from the vehicle you are servicing.

1. Before servicing the vehicle, refer to the Precautions.

2. Disconnect the negative battery cable.

3. Remove the engine cover and remove the air cleaner assembly, as required.

4. Raise and safely support the vehicle.

5. Properly support the exhaust system.

6. Disconnect the oxygen sensor electrical wires. As required, remove the sensor from its mounting.

➡ **Be careful not to drop or damage the sensor. If a sensor has been dropped, it must be replaced.**

7. Remove the converter retaining bolts and nuts.

8. Remove the component from the vehicle.

9. Discard the gaskets.

To install:

➡ **Be sure to use new fasteners, as required.**

10. Installation is the reverse of the removal procedure.

11. Be sure to use new gaskets, as required.

12. Be sure to perform the reconnect/relearn procedures.

CRANKSHAFT FRONT SEAL

REMOVAL & INSTALLATION

V6 Engine

1. Before servicing the vehicle, refer to the Precautions.

2. Disconnect the negative battery cable.

3. Remove the engine cover. Remove the air cleaner assembly.

4. Remove drive belts.

5. Remove engine cooling fan assembly.

6. Remove crankshaft pulley.

7. Remove front oil seal.

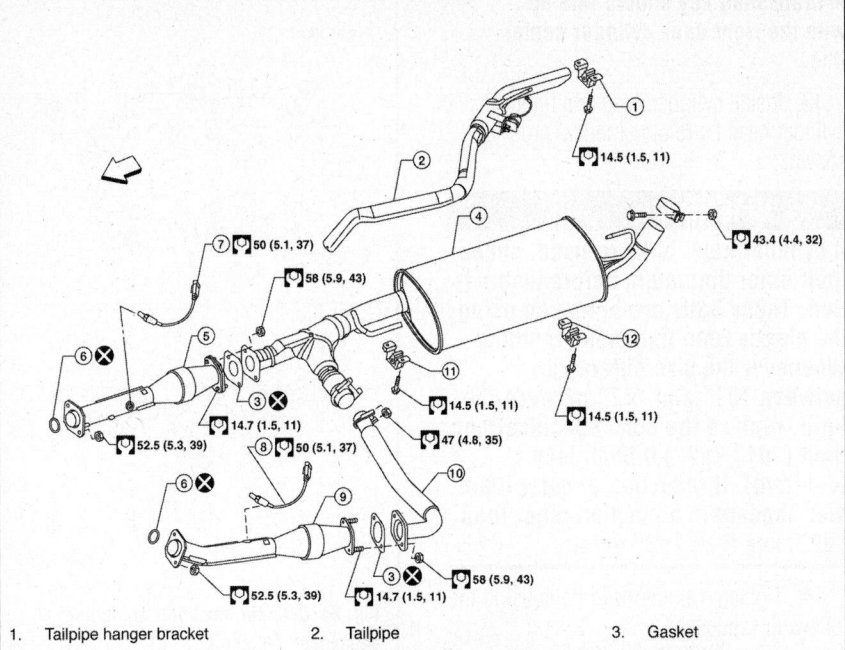

1. Tailpipe hanger bracket
2. Tailpipe
3. Gasket
4. Main muffler
5. Right front exhaust tube
6. Ring gasket
7. Heated oxygen sensor 2 (bank 1)
8. Heated oxygen sensor 2 (bank 2)
9. Left front exhaust tube
10. Center exhaust tube
11. Muffler hanger bracket front
12. Muffler hanger bracket rear
⟵ Front

37663_PATH_G0144

Fig. 60 Catalytic converters and related components—V6 engine

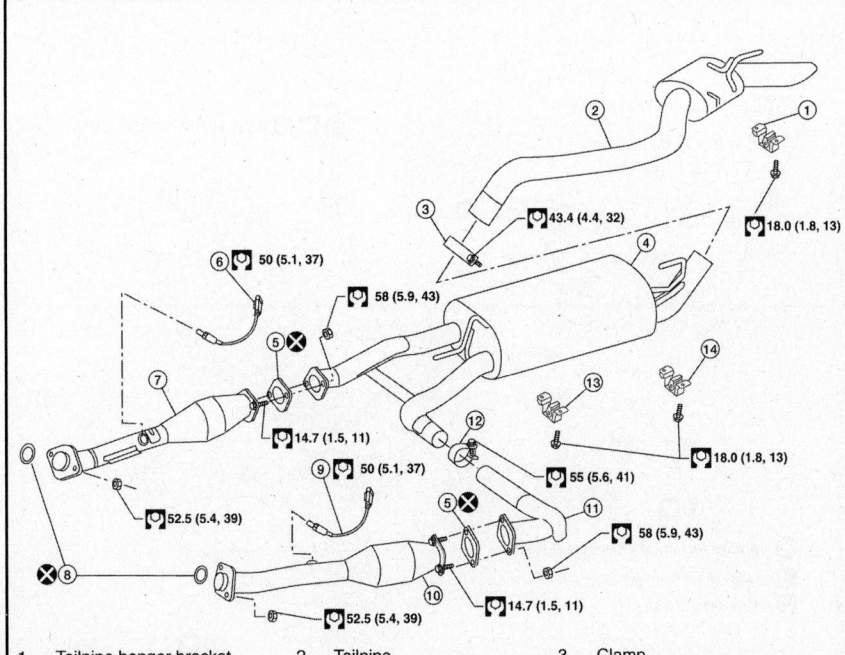

1. Tailpipe hanger bracket
2. Tailpipe
3. Clamp
4. Main muffler
5. Gasket
6. Heated oxygen sensor 2 (bank 2)
7. Right front exhaust tube
8. Ring gasket
9. Heated oxygen sensor 2 (bank 1)
10. Left front exhaust tube
11. Center exhaust tube
12. Clamp
13. Muffler hanger bracket front
14. Muffler hanger bracket rear

37663_PATH_G0145

Fig. 61 Catalytic converters and related components—V8 engine

---

❊❊ WARNING

Be careful not to damage front timing chain case and crankshaft.

To install:

8. Apply new engine oil to both oil seal lip and dust seal lip of new front oil seal.

9. Install front oil seal.

a. Install front oil seal so that each seal lip is oriented as shown.

b. Press-fit until the height of front oil seal is level with the mounting surface using suitable tool.

c. Suitable drift: outer diameter 2.36 in. (60 mm), inner diameter 1.97 in. (50 mm).

❊❊ WARNING

Be careful not to damage front timing chain case and crankshaft. Press-fit straight and avoid causing burrs or tilting oil seal.

10. Installation is in the reverse order of removal after this step.

11. Be sure to perform the reconnect/relearn procedures.

V8 Engine

See Figure 62.

1. Before servicing the vehicle, refer to the Precautions.

2. Disconnect the negative battery cable.

3. Remove the air dam.

4. Remove the engine undercover.

5. Remove the engine cover.

6. Remove the air duct and resonator assembly and the air cleaner case (upper).

7. Remove the drive belt.

8. Remove the radiator assembly.

9. Remove the cooling fan (crankshaft driven type).

10. Remove the crankshaft pulley.

11. Remove the front oil seal using suitable tool.

❊❊ WARNING

Do not damage front cover and oil pump drive spacer.

To install:

12. Apply new engine oil to both the oil seal lip and dust seal lip of the new front oil seal.

13. Install the front oil seal.

a. Install the front oil seal so that each seal lip is oriented as shown.

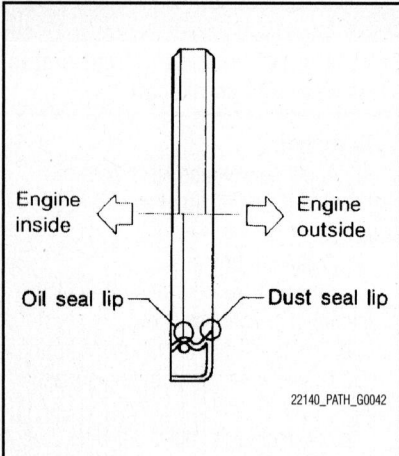

Fig. 62 Install the front oil seal so that each seal lip is oriented as shown.

b. Press-fit until the front oil seal is level with the front cover using suitable tool.

> ✳✳ WARNING
>
> **Do not damage front cover and crankshaft. Press-fit straight and avoid causing burrs or tilting oil seal.**

14. Installation of the remaining components is in the reverse order of removal.

15. Be sure to perform the reconnect/relearn procedures.

CYLINDER HEAD

REMOVAL & INSTALLATION

V6 Engine

See Figures 63 and 64.

1. Before servicing the vehicle, refer to the Precautions.

2. Disconnect the negative battery cable.

3. Remove the engine cover. Remove the air cleaner assembly.

4. Remove camshaft.

5. Remove intake manifold.

6. Remove exhaust manifold.

7. Remove water inlet and thermostat assembly.

8. Remove water outlet, water pipe and heater pipe.

9. Remove cylinder head bolts in reverse order of the tightening sequence.

10. Remove cylinder head gaskets. Discard the gaskets.

To install:

11. Install new cylinder head gasket.

12. Turn crankshaft until No. 1 piston is set at TDC.

➡**Crankshaft key should line up with the right bank cylinder center line.**

13. Install cylinder head and tighten cylinder head bolts in numerical order as shown.

> ✳✳ WARNING
>
> **If cylinder head bolts re-used, check their outer diameters before installation. These bolts are tightened using the plastic zone tightening method. Whenever the size difference between "d1" and "d2" exceeds the limit, replace the bolt. Specification: limit ("d1"-"d2") 0.0043 inch (0.11mm). If reduction of outer diameter appears in a position other than "d2", use it as "d2" point.**

14. Tighten cylinder head bolts using the following sequence:
  - Step A: 72 ft. lbs. (98 Nm)
  - Step B: Loosen bolts in the reverse order of tightening.
  - Step C: 29 ft. lbs. (39.2 Nm)
  - Step D: An additional 90° clockwise
  - Step E: An additional 90° clockwise

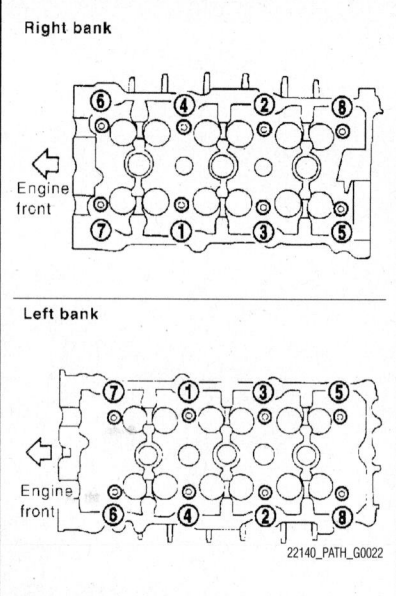

Fig. 64 Cylinder head bolt tightening sequence—V6 engine

15. After installing cylinder head, measure distance between front end faces of cylinder block and cylinder head (left and right banks). Specification should be 0.555–0.587 inch.

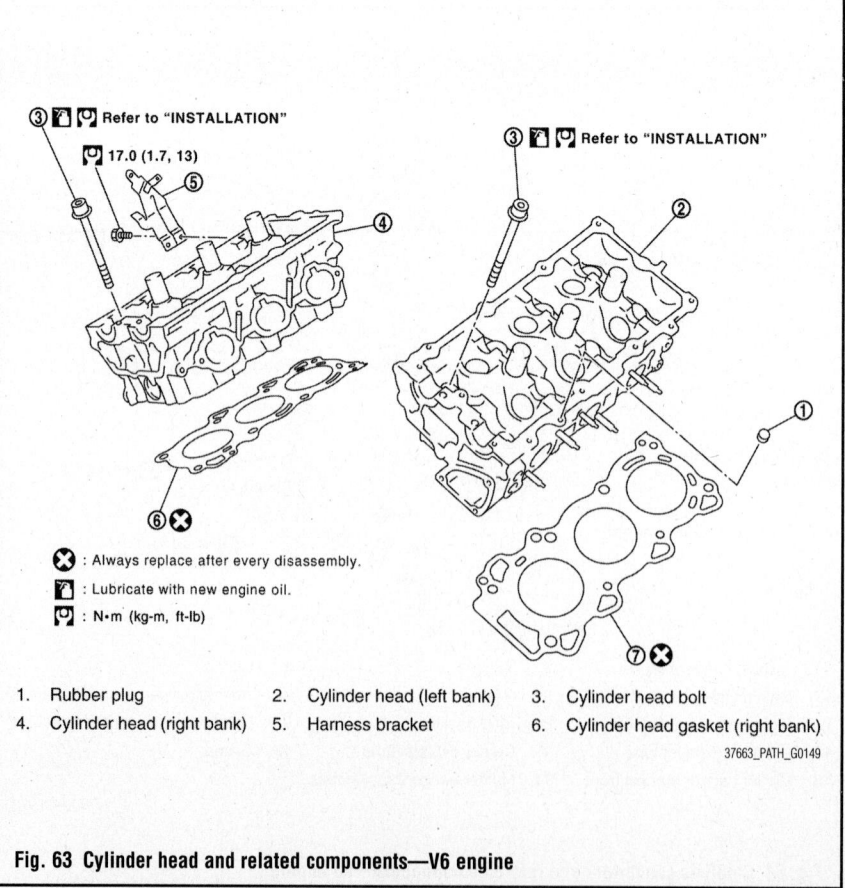

1. Rubber plug
2. Cylinder head (left bank)
3. Cylinder head bolt
4. Cylinder head (right bank)
5. Harness bracket
6. Cylinder head gasket (right bank)

Fig. 63 Cylinder head and related components—V6 engine

➡ If the measured value is out of the standard, re-install cylinder head.

16. Installation of the remaining parts is in the reverse order of removal.

17. Be sure to perform the reconnect/relearn procedures.

V8 Engine

See Figure 65.

➡ The engine must be removed from the vehicle to perform this procedure. Be sure that the engine is secured in a suitable holding fixture before performing this procedure.

1. Before servicing the vehicle, refer to the Precautions.

2. Disconnect the negative battery cable.

3. Remove the engine room cover and air cleaner assembly.

4. Remove or disconnect the following:
   • Engine assembly
   • Belt tensioner
   • Idler pulley
   • Thermostat housing and hose
   • Oil pan and strainer
   • Fuel tube and injector assembly
   • Intake manifold
   • Ignition coil
   • Rocker cover
   • Crankshaft pulley
   • Front engine cover
   • Oil pump
   • Timing chain
   • Camshaft sprockets
   • Camshafts
   • Cylinder head, removing bolts in reverse order of installation sequence

To install:

➡ Be sure to use new fasteners, as required.

5. Install the cylinder head with a new gasket.

6. Tighten the bolts in sequence to specification.

➡ Cylinder head bolts are tightened by plastic zone tightening method. Whenever the size difference between d1 and d2 exceeds the limit, replace the bolt. Limit is 0.0091 inch (0.23mm).

7. Install or connect the following:
   • Camshaft
   • Camshaft sprockets
   • Timing chain
   • Oil pump
   • Front engine cover
   • Crankshaft pulley
   • Rocker cover

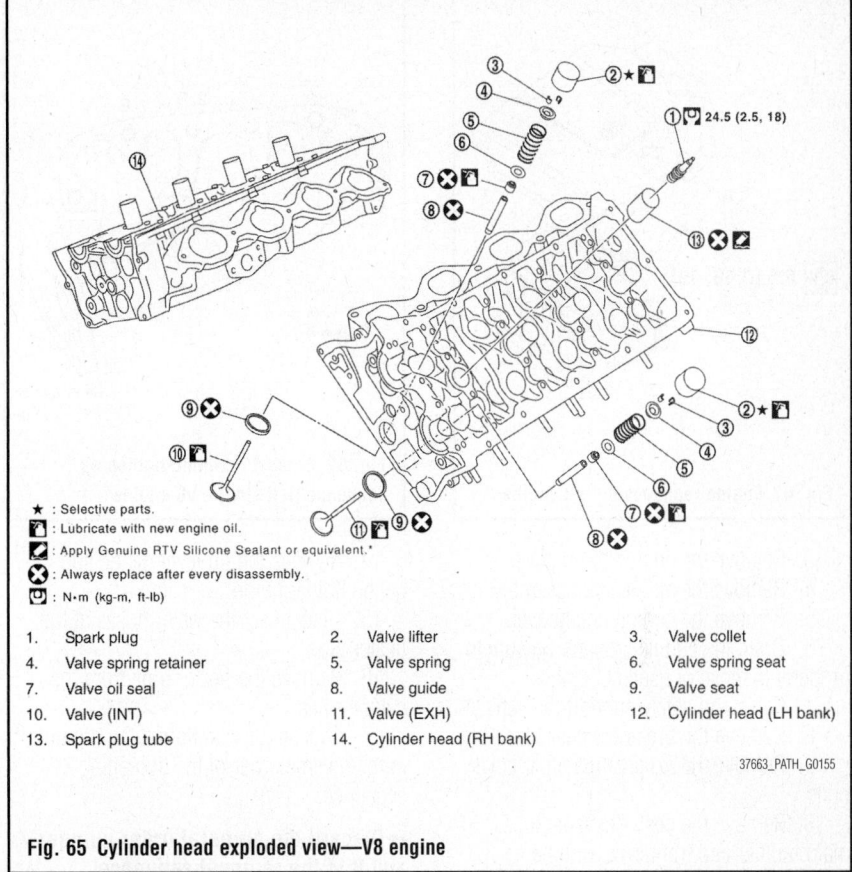

★ : Selective parts.
🛢 : Lubricate with new engine oil.
✐ : Apply Genuine RTV Silicone Sealant or equivalent.*
✕ : Always replace after every disassembly.
🔧 : N·m (kg-m, ft-lb)

1.	Spark plug	2.	Valve lifter	3.	Valve collet
4.	Valve spring retainer	5.	Valve spring	6.	Valve spring seat
7.	Valve oil seal	8.	Valve guide	9.	Valve seat
10.	Valve (INT)	11.	Valve (EXH)	12.	Cylinder head (LH bank)
13.	Spark plug tube	14.	Cylinder head (RH bank)		

37663_PATH_G0155

Fig. 65 Cylinder head exploded view—V8 engine

   • Ignition coil
   • Intake manifold
   • Fuel tube and injector assembly
   • Oil pain and strainer
   • Thermostat housing and hose
   • Idler pulley
   • Belt tensioner
   • Engine assembly

8. Start the engine and check for leaks.

9. Be sure to perform the reconnect/relearn procedures.

ENGINE COVER

REMOVAL & INSTALLATION

See Figures 66 and 67.

1. Before servicing the vehicle, refer to the Precautions.

2. Disconnect the negative battery cable.

3. Remove the cover retaining bolts.

4. Lift up on the cover to disengage the snap fit mounts.

To install:

➡ Be sure to use new fasteners, as required.

5. Installation is the reverse of the removal procedure.

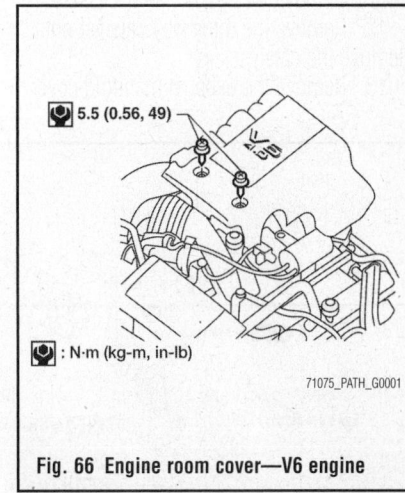

🔧 5.5 (0.56, 49)

🔧 : N·m (kg-m, in-lb)

71075_PATH_G0001

Fig. 66 Engine room cover—V6 engine

EXHAUST MANIFOLD

REMOVAL & INSTALLATION

V6 Engine

Left Bank

See Figures 68 and 69.

1. Before servicing the vehicle, refer to the Precautions.

2. Disconnect the negative battery cable.

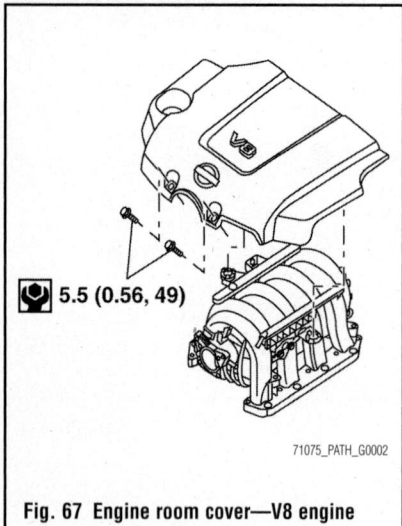

⚙ 5.5 (0.56, 49)

71075_PATH_G0002

Fig. 67 Engine room cover—V8 engine

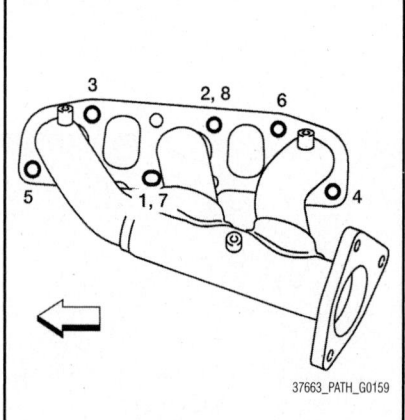

37663_PATH_G0159

Fig. 69 Exhaust manifold tightening sequence (left side)—V6 engine

3. Remove the engine room cover.
4. Remove the air cleaner assembly.
5. Remove the engine undercover.
6. Drain the engine coolant. Be sure to properly dispose of used coolant.
7. Raise and safely support the vehicle.
8. Remove the tire and wheel assembly.
9. Remove the exhaust manifold cover bolts.
10. Remove the center exhaust tube, main muffler and front exhaust tube.
11. Disconnect the oxygen sensor connector. Remove the sensor, as necessary.
12. Remove the three way catalyst nuts. Remove the catalyst.
13. Remove the exhaust manifold cover.

14. Remove the oil level gauge and gauge holder guide.
15. Disconnect the water hoses at the heater pipe.
16. Remove the heater pipe from the cylinder head.
17. Loosen the manifold retaining nuts in the reverse order of the tightening sequence.

➡ **Discard the numeral order number 7 and 8 in the removal sequence.**

18. Remove the manifold retaining bolts and nuts.
19. Remove the component from the vehicle.

20. Discard the gasket. Discard the nuts.

To install:

➡ **Be sure to use new fasteners, as required.**

21. Installation is the reverse of the removal procedure.
22. Tighten the retaining nuts to specification in two passes and in the proper sequence. Be sure to use new nuts.

➡ **Tighten nuts number 1 and 2 in two steps. The numerical order of number 7 and 8 show second step.**

✳✳ WARNING

Before installing a new air fuel ratio sensor 1 and heated oxygen sensor 2, clean exhaust system threads using oxygen sensor thread cleaner and apply anti-seize lubricant. _ Do not over torque air fuel ratio sensor 1 and heated oxygen sensor 2. Doing so may cause damage to air fuel ratio sensor 1 and heated oxygen sensor 2, resulting in the "MIL" coming on.

23. Be sure to perform the reconnect/relearn procedures.

Right Bank
See Figure 70.

1. Before servicing the vehicle, refer to the Precautions.

✳✳ CAUTION

Perform the work when the exhaust and cooling system have cooled sufficiently.

2. Disconnect the negative battery cable.
3. Remove the engine cover and remove the air cleaner assembly, as required.
4. Raise and safely support the vehicle.
5. Remove the tire and wheel assembly.
6. Remove the exhaust manifold cover bolts.
7. Remove the center exhaust tube, main muffler and front exhaust tube.
8. Disconnect the oxygen sensor connector. Remove the sensor, as necessary.
9. Remove the three way catalyst nuts. Remove the catalyst.
10. Remove the heat shield from the lower dash panel.
11. Remove the support bolts from the transmission filler pipe.
12. Loosen the manifold retaining nuts in the reverse order of the tightening sequence.

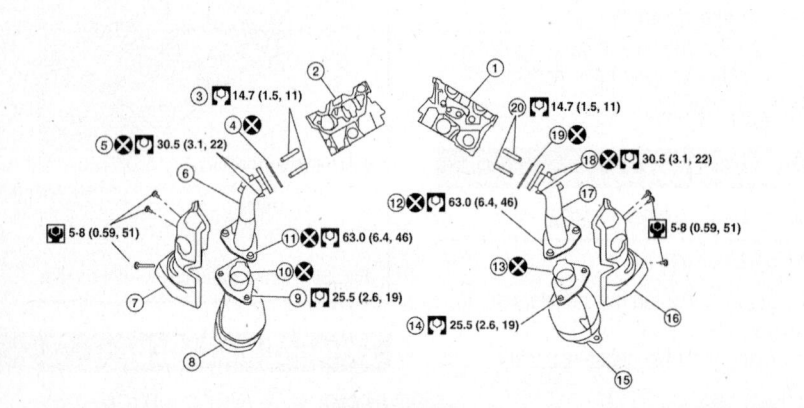

1.	Cylinder head (LH)	2.	Cylinder head (RH)	3.	Exhaust manifold studs (RH)
4.	Gasket	5.	Exhaust manifold nuts (RH)	6.	Exhaust manifold (RH)
7.	Exhaust manifold cover (RH)	8.	Three way catalyst (RH)	9.	Three way catalyst studs (RH)
10.	Seal ring	11.	Three way catalyst nuts (RH)	12.	Three way catalyst nuts (LH)
13.	Seal ring	14.	Three way catalyst studs (LH)	15.	Three way catalyst (LH)
16.	Exhaust manifold cover (LH)	17.	Exhaust manifold (LH)	18.	Exhaust manifold nuts (LH)

37663_PATH_G0157

Fig. 68 Exhaust manifolds and related components—V6 engine

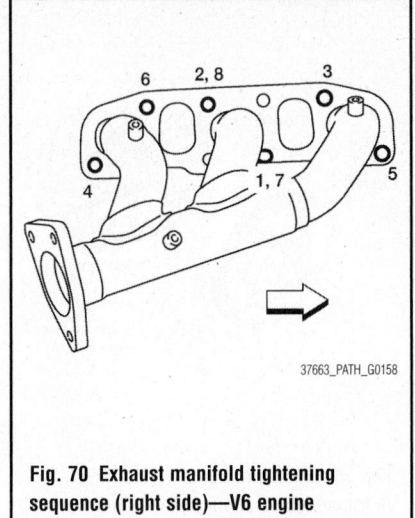

Fig. 70 Exhaust manifold tightening sequence (right side)—V6 engine

➡**Discard the numeral order number 7 and 8 in the removal sequence.**

13. Remove the manifold retaining bolts and nuts.

14. Remove the component from the vehicle.

15. Discard the gasket. Discard the nuts.

To install:

➡**Be sure to use new fasteners, as required.**

16. Installation is the reverse of the removal procedure.

17. Tighten the retaining nuts to specification in two passes and in the proper sequence. Be sure to use new nuts.

➡**Tighten nuts number 1 and 2 in two steps. The numerical order of number 7 and 8 show second step.**

❊❊ WARNING

Before installing a new air fuel ratio sensor 1 and heated oxygen sensor 2, clean exhaust system threads using oxygen sensor thread cleaner and apply anti-seize lubricant. Do not over torque air fuel ratio sensor 1 and heated oxygen sensor 2. Doing so may cause damage to air fuel ratio sensor 1 and heated oxygen sensor 2, resulting in the "MIL" coming on.

18. Be sure to perform the reconnect/relearn procedures.

V8 Engine

See Figures 71 and 72.

1. Before servicing the vehicle, refer to the Precautions.

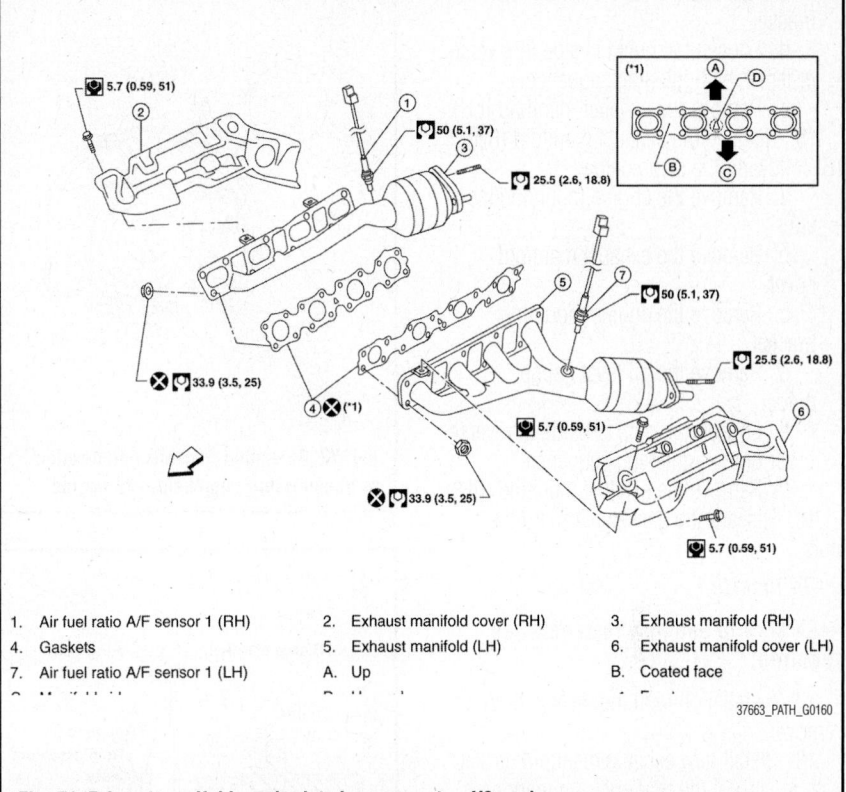

1.	Air fuel ratio A/F sensor 1 (RH)	2.	Exhaust manifold cover (RH)	3.	Exhaust manifold (RH)
4.	Gaskets	5.	Exhaust manifold (LH)	6.	Exhaust manifold cover (LH)
7.	Air fuel ratio A/F sensor 1 (LH)	A.	Up	B.	Coated face

Fig. 71 Exhaust manifolds and related components—V8 engine

❊❊ CAUTION

Perform the work when the exhaust and cooling system have cooled sufficiently.

2. Disconnect the negative battery cable.

3. Remove the engine cover and remove the air cleaner assembly, as required.

4. Raise and safely support the vehicle.

5. Remove the air dam.

6. Remove the engine undercover.

7. Remove front final drive assembly (4WD).

8. Remove the main muffler assembly and center exhaust tube.

9. Remove the front exhaust tubes.

10. Remove front tires.

11. Remove fender protector.

12. Remove the LH and RH air fuel ratio A/F sensors.

a. Remove the harness connector of each air fuel ratio A/F sensor, and harness from bracket and middle clamp.

❊❊ WARNING

Do not damage the air fuel ratio A/F sensors. Discard any air fuel ratio A/F sensor which has been dropped from a height of more than 19.7 in. (0.5m) onto a hard surface such as a concrete floor. Replace it with a new one.

13. Support the engine using a suitable tool.

14. Remove the exhaust manifold (LH) (A) following the steps below.

a. Remove the engine mounting insulator.

b. Remove the exhaust manifold cover.

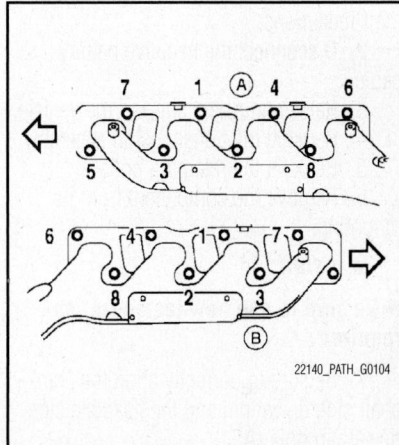

Fig. 72 Exhaust manifold retaining nut tightening sequence—V8 engine

c. Remove the engine mounting bracket.

d. Loosen the nuts LH side in reverse order of the tightening sequence.

e. Remove the exhaust manifold (LH).

15. Remove the exhaust manifold (RH) (B) following the steps below.

a. Remove the engine mounting insulator.

b. Remove the exhaust manifold cover.

c. Remove the engine mounting bracket.

d. Remove the oil level gauge guide.

e. Loosen the nuts RH side in reverse order of the tightening sequence.

f. Remove the exhaust manifold (RH).

16. Discard the gaskets. Discard the nuts.

To install:

→**Be sure to use new fasteners, as required.**

17. Installation is in the reverse order of removal.

18. Install new exhaust manifold gasket with the top of the triangular up mark on it facing up and its coated face (gray side) toward the exhaust manifold side.

19. Tighten the retaining nuts to specification and in the proper sequence.

20. Do not over tighten the sensors. Doing so may interfere with sensor operation causing the MIL to come on.

21. Be sure to perform the reconnect/relearn procedures.

DRIVEPLATE

REMOVAL & INSTALLATION

See Figures 73 through 76.

1. Before servicing the vehicle, refer to the Precautions.

2. Disconnect the negative battery cable.

3. Raise and safely support the vehicle.

4. Remove the transmission assembly.

5. Remove the retaining bolts.

6. Remove the component from its mounting.

To install:

→**Be sure to use new fasteners, as required.**

7. Be sure to correctly align the crankshaft side dowel pin and the flexplate side dowel pin hole (A).

→**If not aligned correctly the MIL light will illuminate.**

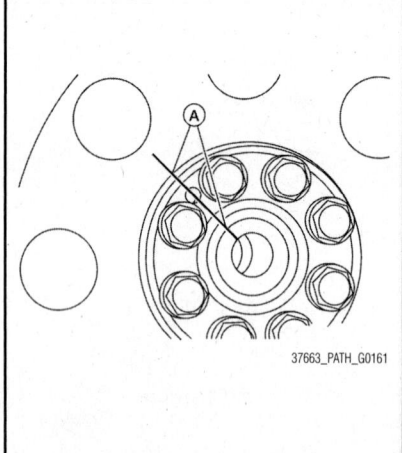

Fig. 73 Automatic transmission flexplate to transmission alignment—V6 engine

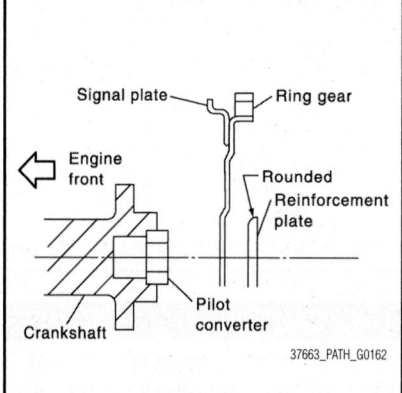

Fig. 74 Automatic transmission flexplate reinforcement plate installation direction—V6 engine

8. Install the flexplate and reinforcement plate in the direction shown.

9. Tighten the retaining bolts to specification in a crisscross pattern.

10. Continue the installation in the reverse order of the removal procedure.

11. Be sure to perform the reconnect/relearn procedures.

FRONT COVER & SEAL

REMOVAL & INSTALLATION

V6 Engine

See Figures 77 through 81.

1. Before servicing the vehicle, refer to the Precautions.

2. Properly release the fuel system pressure.

3. Disconnect the negative battery cable.

4. Remove the engine cover and

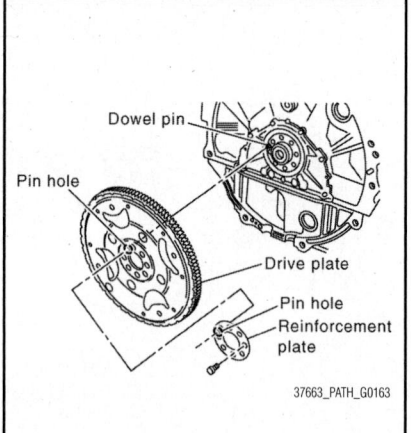

Fig. 75 Automatic transmission flexplate to transmission alignment—V8 engine

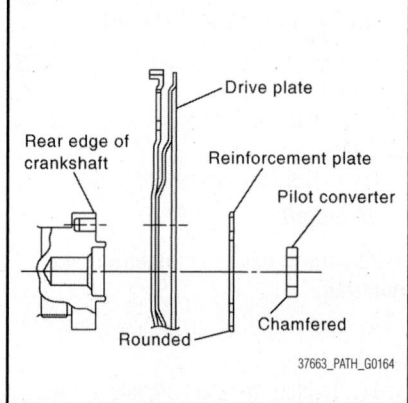

Fig. 76 Automatic transmission flexplate reinforcement plate installation direction—V8 engine

remove the air cleaner assembly, as required.

5. Remove engine cover.

6. Drain engine oil.

※ CAUTION

Perform this step when engine is cold. Do not spill engine oil on drive belts.

7. Drain engine coolant from radiator.

※ CAUTION

Perform this step when engine is cold. Do not spill engine coolant on drive belts.

8. Remove radiator cooling fan assembly.

9. Separate engine harnesses removing their brackets from front timing chain case.

10. Remove drive belts.

11. Remove power steering oil pump

from bracket with piping connected, and temporarily secure it aside.

12. Remove power steering oil pump bracket.

13. Remove alternator.

14. Remove water bypass hose, water hose clamp and idler pulley bracket from front timing chain case.

15. Remove right and left intake valve timing control covers.

    a. Loosen bolts in reverse order of tightening sequence.

    b. Cut liquid gasket for removal.

✳✳ WARNING

Shaft is internally jointed with camshaft sprocket (INT) center hole. When removing, keep it horizontal until it is completely disconnected.

16. Remove collared O-rings from front timing chain case (left and right side).

17. Remove rocker covers (right and left banks).

➡**When only timing chain (primary) is removed, rocker cover does not need to be removed.**

18. Obtain No. 1 cylinder at TDC of its compression stroke as follows:

➡**When timing chain is not removed/installed, this step is not required.**

    a. Rotate crankshaft pulley clockwise to align timing mark (A) with timing indicator (B).

    b. Make sure that intake and exhaust cam noses on No. 1 cylinder (engine front side of right bank) are located as shown.

    c. If not, turn crankshaft one revolution (360°) and align as shown.

➡**When only timing chain (primary) is removed, rocker cover does not need to be removed. To make sure that No. 1 cylinder is at its compression TDC, remove front timing chain case first. Then check mating marks on camshaft sprockets.**

19. Remove crankshaft pulley as follows:

    a. Remove starter motor.

    b. Loosen crankshaft pulley bolt and locate bolt seating surface as 0.39 in. (10 mm) from its original position.

✳✳ WARNING

Do not remove crankshaft pulley bolt. Keep loosened crankshaft pulley bolt in place protect removed crankshaft pulley from dropping.

    c. Pull crankshaft pulley with both hands to remove it.

20. Loosen two bolts in front of oil pan (upper) in the reverse order of the tightening sequence.

21. Remove front timing chain case as follows:

    a. Loosen bolts in reverse order of the tightening sequence.

    b. Insert suitable tool into the notch at the top of the front timing chain case.

    c. Pry off case by moving tool.

    d. Cut liquid gasket for removal.

✳✳ WARNING

Do not use screwdriver or something similar. After removal, handle front timing chain case carefully so it does not tilt, cant, or warp under a load.

22. Remove O-rings from rear timing chain case.

To install:

➡**Be sure to use new fasteners, as required.**

23. Install front timing chain case as follows:

    a. Apply a continuous bead of liquid gasket to front timing chain case back side as shown.

    b. Install new O-rings on rear timing chain case.

    c. Assemble front timing chain case as follows:

- Fit lower end of front timing chain case tightly onto top face of oil pan (upper). From the fitting point, make entire front timing chain case contact rear timing chain case completely.
- Since front timing chain case is offset for difference of bolt holes, tighten bolts temporarily while holding front timing chain case from front and top.
- Same as the previous step, insert dowel pin while holding front timing chain case from front and top completely.

    d. Tighten bolts to the specified torque in numerical order as shown.

➡**There are two type of bolts. Refer to the following for locating bolts.**

- 1–5: 0.39 in. (10 mm)
- 6–25: 0.24 in. (6 mm)
- Bolt position torque specification: 1–5: 41 ft. lbs. (55.0 Nm); 6–25: 9 ft. lbs. (12.7 Nm)

    e. After all bolts tightened, retighten them to the specified torque in numerical order as shown.

24. Install two bolts in front of oil pan (upper).

25. Install right and left intake valve timing control covers as follows:

    a. Install new seal rings in shaft grooves.

    b. Apply a continuous bead of liquid gasket to intake valve timing control covers.

    c. Install new collared O-rings in front timing chain case oil hole (left and right sides).

    d. Being careful not to move seal ring from the installation groove, align dowel pins on front timing chain case with the holes to install intake valve timing control covers.

    e. Tighten bolts in numerical order as shown.

26. Install crankshaft pulley as follows:

    a. Install crankshaft pulley, taking care not to damage front oil seal.

➡**When press-fitting crankshaft pulley with plastic hammer, tap on its center portion (not circumference).**

    b. Tighten crankshaft pulley bolt. Crankshaft bolt torque: 33ft. lbs. (44.1 Nm)

    c. Put a paint mark on crankshaft pulley aligning with angle mark on crankshaft pulley bolt. Then, further retighten bolt by 60°(equivalent to one graduation).

27. Rotate crankshaft pulley in normal direction (clockwise when viewed from front) to confirm it turns smoothly.

28. Install oil pans (upper and lower).

29. Install rocker covers (right and left banks).

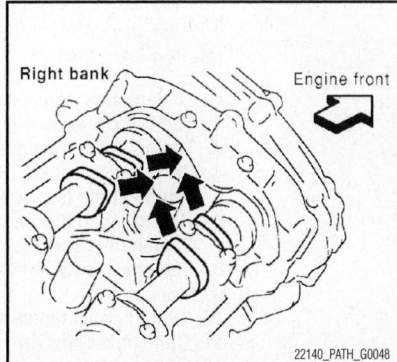

22140_PATH_G0048

Fig. 77 Make sure that intake and exhaust cam noses on No. 1 cylinder (engine front side of right bank) are located as shown— V6 engine

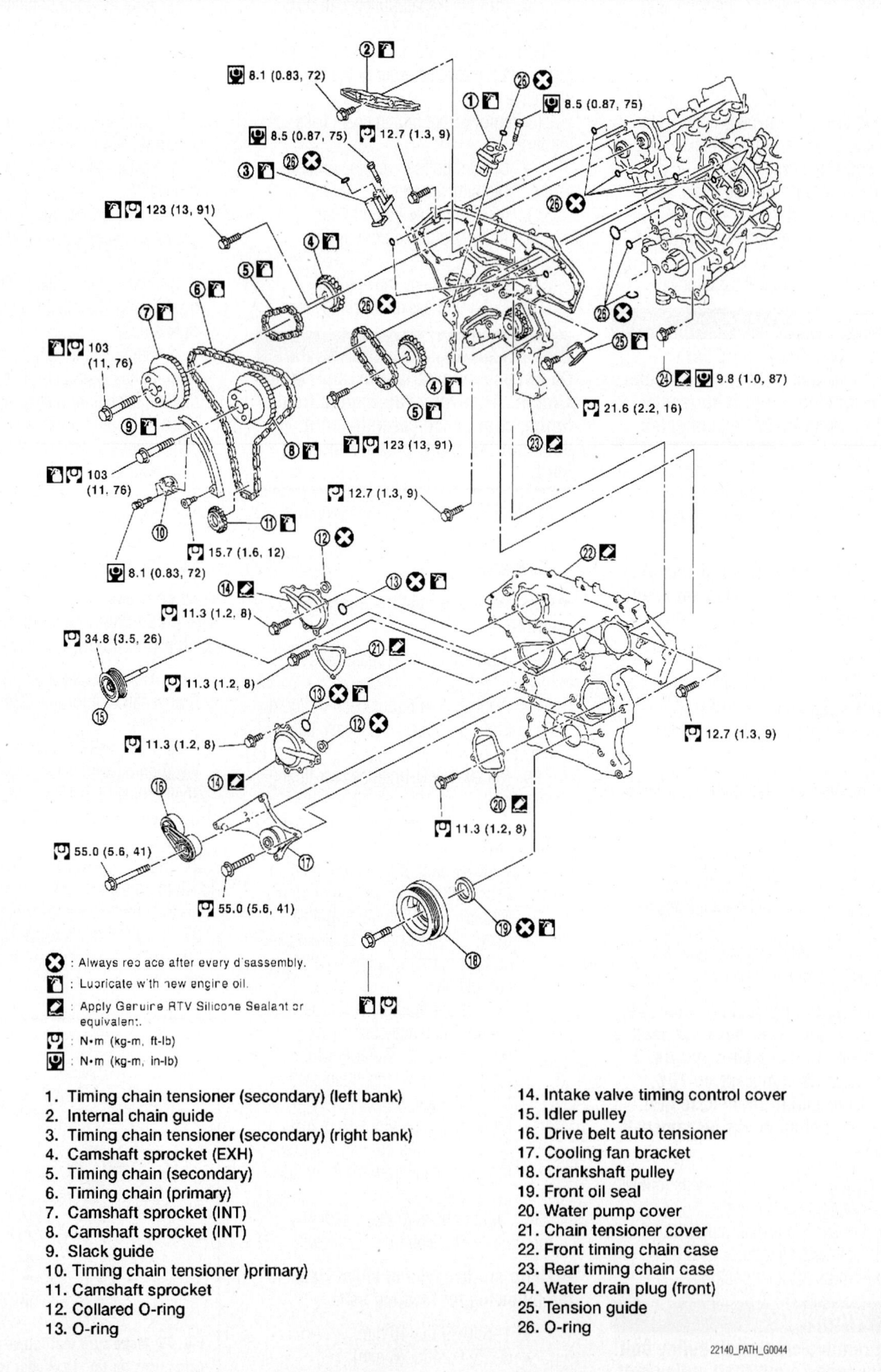

Fig. 78 Timing chain/front cover and related components—V6 engine

⊗ : Always replace after every disassembly.

[lubricate icon] : Lubricate with new engine oil.

[sealant icon] : Apply Genuine RTV Silicone Sealant or equivalent.

[torque icon] : N•m (kg-m, ft-lb)

[torque icon] : N•m (kg-m, in-lb)

1. Timing chain tensioner (secondary) (left bank)
2. Internal chain guide
3. Timing chain tensioner (secondary) (right bank)
4. Camshaft sprocket (EXH)
5. Timing chain (secondary)
6. Timing chain (primary)
7. Camshaft sprocket (INT)
8. Camshaft sprocket (INT)
9. Slack guide
10. Timing chain tensioner)primary)
11. Camshaft sprocket
12. Collared O-ring
13. O-ring

14. Intake valve timing control cover
15. Idler pulley
16. Drive belt auto tensioner
17. Cooling fan bracket
18. Crankshaft pulley
19. Front oil seal
20. Water pump cover
21. Chain tensioner cover
22. Front timing chain case
23. Rear timing chain case
24. Water drain plug (front)
25. Tension guide
26. O-ring

22140_PATH_G0044

Fig. 78 Timing chain/front cover and related components—V6 engine

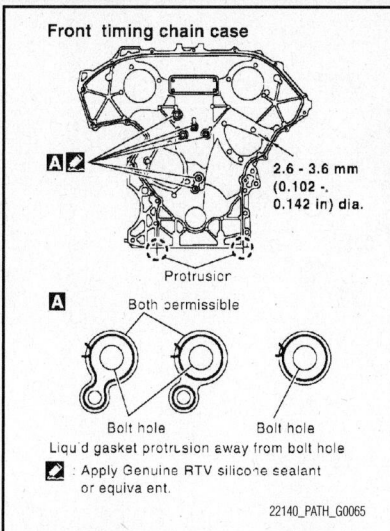

Fig. 79 Timing cover sealant application points—V6 engine

30. Installation of the remaining components is in the reverse order of removal after this step.

31. The following are procedures for checking fluid leaks, lubricant leaks and exhaust gases leaks.

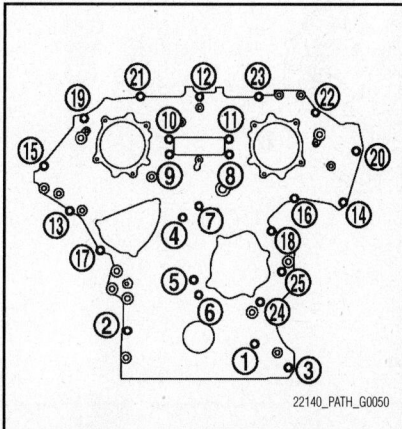

Fig. 80 Timing cover bolt tightening sequence—V6 engine

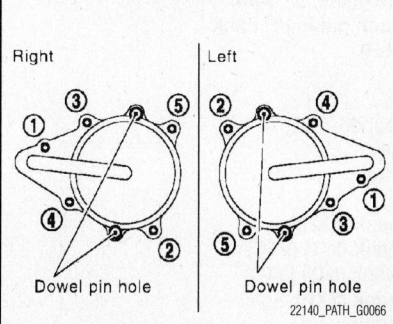

Fig. 81 Timing control covers bolt tightening sequence—V6 engine

a. Before starting engine, check oil/fluid levels including engine coolant and engine oil. If less than required quantity, fill to the specified level.

32. Use procedure below to check for fuel leakage.

a. Turn ignition switch "ON" (with engine stopped). With fuel pressure applied to fuel piping, check for fuel leakage at connection points.

b. Start engine. With engine speed increased, check again for fuel leakage at connection points.

c. Run engine to check for unusual noise and vibration.

➡**If hydraulic pressure inside timing chain tensioner drops after removal/installation, slack in the guide may generate a pounding noise during and just after engine start. However, this is normal. Noise will stop after hydraulic pressure rises.**

d. Warm up engine thoroughly to make sure there is no leakage of fuel, exhaust gases, or any oil/fluids including engine oil and engine coolant.

e. Bleed air from lines and hoses of applicable lines, such as in cooling system.

f. After cooling down engine, again check oil/fluid levels including engine oil and engine coolant. Refill to the specified level, if necessary.

33. Be sure to perform the reconnect/relearn procedures.

V8 Engine

See Figures 82 through 84.

In order to perform this repair procedure, the engine must first be removed from the vehicle. Be sure to properly position the assembly in a suitable holding fixture before performing and repair procedures.

1. Before servicing the vehicle, refer to the Precautions.

2. Disconnect the negative battery cable.

3. Remove the engine cover. Remove the air cleaner assembly.

➡**To remove timing chain and associated parts, start with those on the LH bank. The procedure for removing parts on the RH bank is omitted because it is the same as that for removal on the LH bank.**

To install timing chain and associated parts, start with those on the RH bank. The procedure for installing parts on the LH bank is omitted because it is the same as that for installation on the RH bank.

4. Remove the engine from the vehicle. Position the assembly in a suitable holding fixture.

5. Remove the following components and related parts:
- Drive belt auto tensioner and idler pulley.
- Thermostat housing and water hose.
- Power steering oil pump bracket.
- Oil pan (lower), (upper) and oil strainer.
- Ignition coil.
- Rocker cover.

6. Remove the Intake valve control solenoid valve cover (RH) (A) and Intake valve control solenoid valve cover (LH) (B) as follows:

a. Loosen and remove the bolts.

b. Cut the liquid gasket and remove the covers.

✳✳ WARNING
Do not damage mating surfaces.

7. Obtain compression TDC of No. 1 cylinder as follows:

a. Turn the crankshaft pulley clockwise to align the TDC identification notch (without paint mark) with the timing indicator on the front cover.

b. At this time, make sure both intake and exhaust cam lobes of No. 1 cylinder (top front on LH bank) point outside. If they do not point outside, turn crankshaft pulley once more.

8. Remove the crankshaft pulley.

a. Loosen the crankshaft pulley bolts using a hammer handle to secure the crankshaft.

b. Remove the crankshaft pulley from the crankshaft.

c. Remove the crankshaft pulley.

➡**The dimension between the centers of the two bolt holes is 61 mm (2.40 in.).**

9. Remove the front cover.

a. Loosen and remove the bolts in the reverse order of the tightening sequence.

b. Cut the liquid gasket and remove the covers.

✳✳ WARNING
Do not damage mating surfaces.

10. Remove the front oil seal using suitable tool.

✳✳ WARNING
Do not damage front cover.

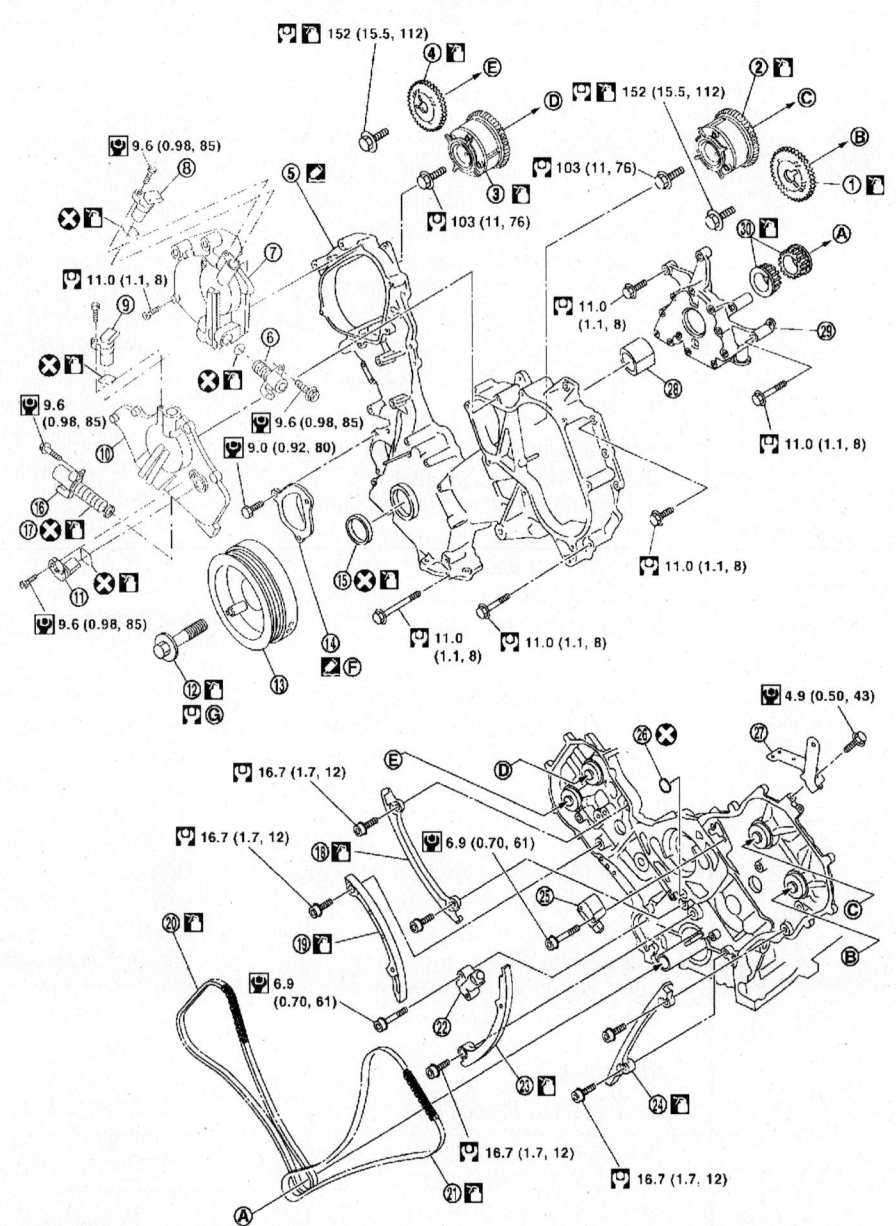

1. Camshaft sprocket LH bank (EXH)
2. Camshaft sprocket LH bank (INT) (VTC)
3. Camshaft sprocket RH bank (INT) (VTC)
4. Camshaft sprocket RH bank (EXH)
5. Front cover
6. Intake valve control solenoid valve (LH)
7. Intake valve control solenoid valve cover (LH)
8. Intake valve timing control position sensor (LH)
9. Intake valve timing control position sensor (RH)
10. Intake valve control solenoid valve cover (RH)
11. Camshaft position sensor (PHASE)
12. Crankshaft pulley bolt
13. Crankshaft pulley
14. Chain tensioner cover
15. Front oil seal
16. Intake valve control solenoid valve (RH)
17. O-ring
18. Timing chain tension guide RH bank

19. Timing chain slack guide (RH)
20. Timing chain LH bank
21. Timing chain (RH)
22. Chain tensioner (RH)
23. Timing chain slack guide LH bank
24. Timing chain tension guide LH bank
25. Chain tensioner (LH)
26. O-ring
27. Bracket
28. Oil pump drive spacer
29. Oil pump assembly
30. Crankshaft sprocket
A. To crankshaft
B. To camshaft LH bank (EXH)
C. To camshaft LH bank (INT) (VTC)
D. To camshaft RH bank (INT) (VTC)
E. To camshaft RH bank (EXH)
F. Apply sealant to mating side

22140_PATH_G0120

Fig. 82 Timing chain/front cover and related components—V8 engine

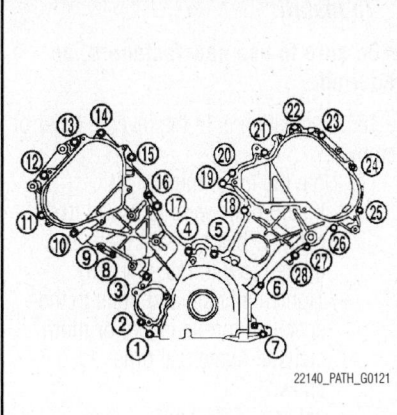

Fig. 83 Timing cover bolt tightening sequence—V8 engine

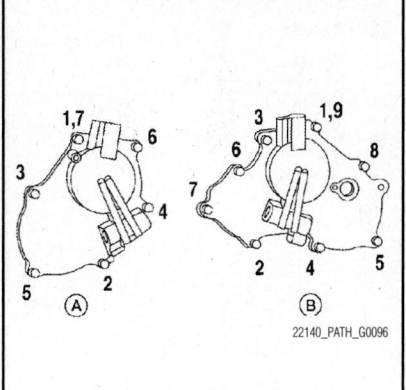

Fig. 84 Intake valve control solenoid bolt tightening sequence—V8 engine

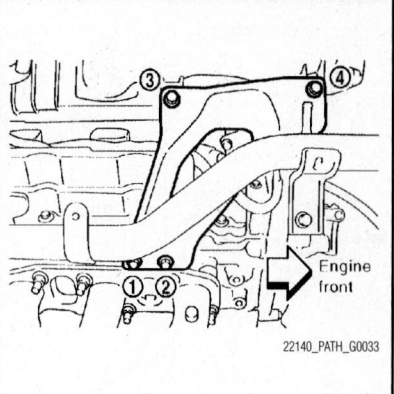

Fig. 85 Loosen bolts in reverse order as shown to remove intake manifold collector support—V6 engine

To install:

➡Be sure to use new fasteners, as required.

11. Install the front oil seal using suitable tool.

❉❉ **WARNING**

Do not scratch or make burrs on the circumference of the oil seal.

12. Install the chain tensioner cover.
13. Install the front cover as follows:
   a. Install a new O-ring on the cylinder block.
   b. Apply liquid gasket as shown.
   c. Check again that the timing alignment marks on the timing chain and on each sprocket are aligned. Then install the front cover.
   d. Install the bolts in the numerical order shown.
   e. After tightening, re-tighten to the specified torque.

❉❉ **WARNING**

Be sure to wipe off any excessive liquid gasket leaking onto surface mating with oil pan.

14. Install the Intake valve control solenoid valve cover (RH) (A) and Intake valve control solenoid valve cover (LH) (B) as follows:
   a. Cross mark (C) that cannot be seen after assembly.
   b. Apply liquid gasket (D) as shown.

❉❉ **WARNING**

The start and end of the application of the liquid gasket should be crossed at a position that cannot be seen after attaching the Intake valve control solenoid valve cover.

   c. Install the bolts in the numerical order shown.
15. Install the crankshaft pulley.
   a. Install the key of the crankshaft.
   b. Insert the pulley by lightly tapping it.

❉❉ **WARNING**

Do not tap pulley on the side surface where the belt is installed (outer circumference).

16. Tighten the crankshaft pulley bolt.
   a. Lock the crankshaft using suitable tool, then tighten the bolt.
   b. Perform the following steps for angular tightening:
   c. Apply engine oil onto the threaded parts of the bolt and seating area.
   d. Select the one most visible notch of the four on the bolt flange. Corresponding to the selected notch, put a alignment mark (such as paint) on the crankshaft pulley.
17. Rotate the crankshaft pulley in normal direction (clockwise when viewed from engine front) to check for parts interference.
18. Installation of the remaining components is in the reverse of order of removal.
19. Be sure to perform the reconnect/relearn procedures.

INTAKE MANIFOLD

REMOVAL & INSTALLATION

Upper Manifold

V6 Engine

See Figures 85 through 88.

1. Before servicing the vehicle, refer to the Precautions.

➡This vehicle uses quick connect fittings. Be sure to properly relieve the fuel system pressure before disconnecting any of these fittings. Always replace O-rings and clamps with new ones. Do not bend, twist or kink hoses when they are being removed or installed. Be sure that the clamp screw does not contact adjacent parts. When tightening the high pressure rubber hose clamp make sure the clamp end is 0.12 inch from the hose end. After connecting these fittings make sure that the connectors are secure. Check for fuel leakage at these connections turn the ignition key to the ON position (do not start the engine), correct as required. Start the engine, raise the idle, and verify that there are no fuel leaks, correct as required.

2. Properly relieve the fuel system pressure.
3. Disconnect the negative battery cable.
4. Remove engine cover.
5. Remove air cleaner case (upper) with mass air flow sensor and air duct assembly.
6. Remove electric throttle control actuator as follows:
   a. Drain engine coolant, or when water hoses are disconnected, attach plug to prevent engine coolant leakage.
   b. Disconnect water hoses from electric throttle control actuator.

➡When engine coolant is not drained from radiator, attach plug to water hoses to prevent engine coolant leakage.

   c. Disconnect harness connector.
   d. Loosen bolts in reverse order of the tightening sequence.

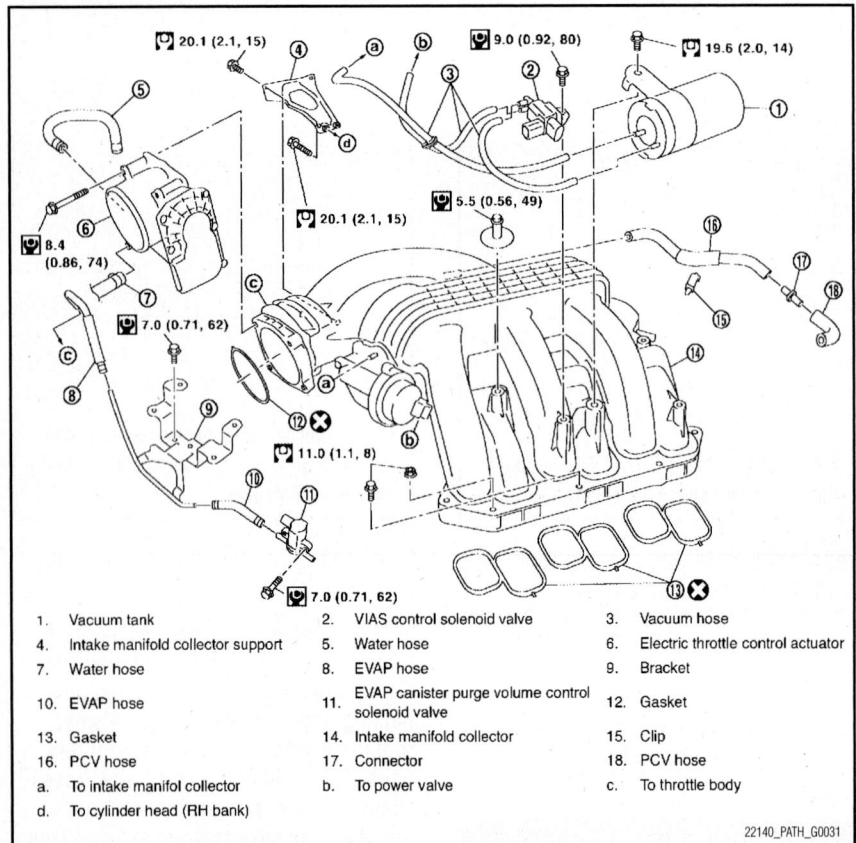

1. Vacuum tank
2. VIAS control solenoid valve
3. Vacuum hose
4. Intake manifold collector support
5. Water hose
6. Electric throttle control actuator
7. Water hose
8. EVAP hose
9. Bracket
10. EVAP hose
11. EVAP canister purge volume control solenoid valve
12. Gasket
13. Gasket
14. Intake manifold collector
15. Clip
16. PCV hose
17. Connector
18. PCV hose
a. To intake manifol collector
b. To power valve
c. To throttle body
d. To cylinder head (RH bank)

22140_PATH_G0031

Fig. 86 Upper intake manifold and related components—V6 engine

7. Remove the following parts:
- Vacuum hose (to brake booster)
- PCV hose

8. Loosen bolts in reverse order as shown to remove intake manifold collector support.

9. Disconnect EVAP hoses and harness connector from EVAP canister purge volume control solenoid valve.

10. Remove EVAP canister purge volume control solenoid valve.

11. Remove VIAS control solenoid valve and vacuum tank.

12. Loosen nuts and bolts in reverse order of the tightening sequence.

13. Remove fuel tube and fuel injector assembly.

14. Loosen nuts and bolts in reverse order as shown to remove intake manifold.

15. Remove gaskets. Discard the gaskets.

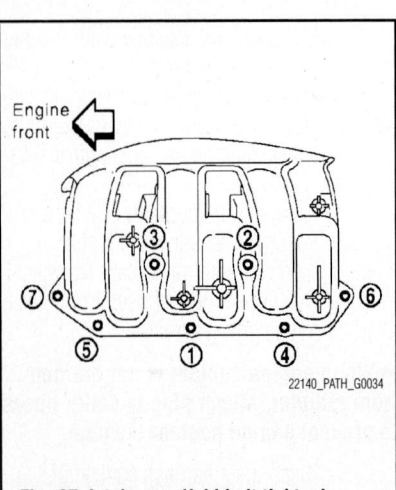

22140_PATH_G0034

Fig. 87 Intake manifold bolt tightening sequence—V6 engine

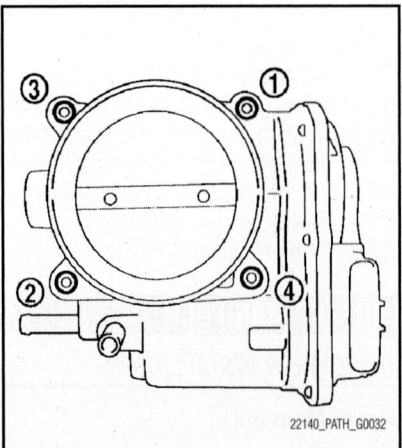

22140_PATH_G0032

Fig. 88 Throttle body bolt tightening sequence—V6 engine

To install:

➡ Be sure to use new fasteners, as required.

16. Installation is in the reverse order of removal.

17. Note the following:
- If stud bolts were removed from cylinder head, install them and tighten to 8 ft. lbs. (11 Nm).
- Tighten all nuts and bolts to the specified torque in two or more steps in numerical order as shown.
- 1st step: 5 ft. lbs. (7.4 Nm)
- 2nd step and after: 21 ft. lbs. (29.0 Nm)

18. Tighten the throttle body retaining bolts to specification and in the proper sequence.

19. Using the CONSULT-III diagnostic tool or equivalent, perform the throttle valve closed position learning and the idle air volume learning procedures.

20. Be sure to perform the reconnect/relearn procedures.

Lower Manifold

See Figures 89 and 90.

1. Before servicing the vehicle, refer to the Precautions.

➡ This vehicle uses quick connect fittings. Be sure to properly relieve the fuel system pressure before disconnecting any of these fittings. Always replace O-rings and clamps with new ones. Do not bend, twist or kink hoses when they are being removed or installed. Be sure that the clamp screw does not contact adjacent parts. When tightening the high pressure rubber hose clamp make sure the clamp end is 0.12 inch from the hose end. After connecting these fittings make sure that the connectors are secure. Check for fuel leakage at these connections turn the ignition key to the ON position (do not start the engine), correct as required. Start the engine, raise the idle, and verify that there are no fuel leaks, correct as required.

2. Properly relieve the fuel system pressure.

3. Disconnect the negative battery cable.

4. Remove engine cover.

5. Remove air cleaner case (upper) with mass air flow sensor and air duct assembly.

6. Remove the upper intake manifold.

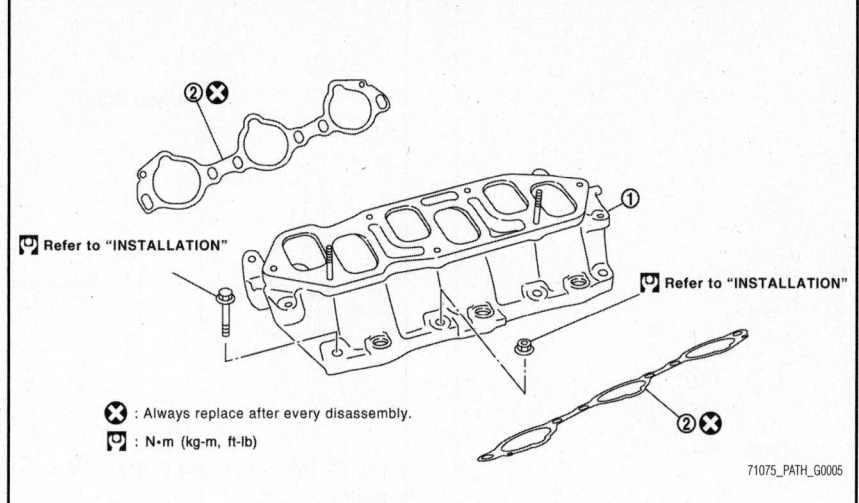

Fig. 89 Lower intake manifold and related components (1) manifold, (2) gasket—V6 engine

7. Remove the fuel tube and fuel injector assembly.

8. Remove the intake manifold retaining bolts, in the reverse order of the tightening sequence.

9. Remove the manifold from the engine. Discard the gaskets.

To install:

➡**Be sure to use new fasteners, as required.**

10. Installation is the reverse of the removal procedure.

11. Tighten the retaining bolts to the proper torque and in the proper sequence. If the stud bolts were removed from the cylinder head, tighten to 8 ft. lbs. Tighten all bolts in two or three steps as follows. Step one 65 inch lbs. Step two and after 21 ft. lbs.

12. Be sure to perform the reconnect/relearn procedures.

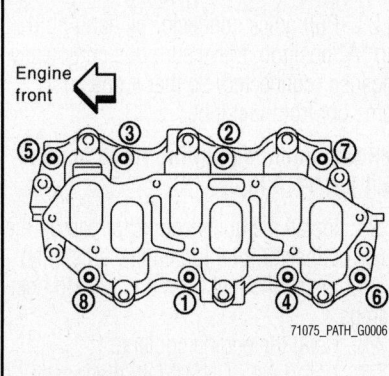

Fig. 90 Lower intake manifold bolt location and tightening sequence—V6 engine

V8 Engine

See Figures 91 through 94.

1. Before servicing the vehicle, refer to the Precautions.

➡**This vehicle uses quick connect fittings. Be sure to properly relieve the fuel system pressure before disconnecting any of these fittings. Always replace O-rings and clamps with new ones. Do not bend, twist or kink hoses when they are being removed or installed. Be sure that the clamp screw does not contact adjacent parts. When tightening the high pressure rubber hose clamp make sure the clamp end is 0.12 inch from the hose end. After connecting these fittings make sure that the connectors are secure. Check for fuel leakage at these connections turn the ignition key to the ON position (do not start the engine), correct as required. Start the engine, raise the idle, and verify that there are no fuel leaks, correct as required.**

2. Properly discharge the fuel system pressure.

3. Disconnect the negative battery cable.

4. Remove the engine cover and remove the air cleaner assembly.

5. Remove the air dam.

6. Remove the engine undercover.

7. Partially drain the engine coolant.

✳✳ CAUTION

To avoid the danger of being scalded, never drain the engine coolant when the engine is hot.

8. Disconnect the fuel tube quick connector on the engine side.

   a. Remove quick connector cap (engine side only).

   b. With the sleeve side of tool number 16641 6N210 (J45488) facing quick connector, install tool onto fuel tube.

   c. Insert Tool into quick connector until sleeve contacts and goes no further. Hold the Tool in that position.

✳✳ WARNING

Inserting the Tool hard will not disconnect quick connector. Hold Tool where it contacts and goes no further.

   d. Draw and pull out quick connector straight from fuel tube.

✳✳ CAUTION

Heed the following cautions:

- Pull quick connector holding "A" position in illustration.
- Do not pull with lateral force applied. O-ring inside quick connector may be damaged.
- Prepare container and cloth beforehand as fuel will leak out.
- Avoid fire and sparks.
- Be sure to cover openings of disconnected pipes with plug or plastic bag to avoid fuel leakage and entry of foreign materials.

9. Remove or disconnect harnesses, brackets, vacuum hose, vacuum gallery and PCV hose and tube from intake manifold.

10. Remove electric throttle control actuator by loosening bolts diagonally.

✳✳ WARNING

Handle carefully to avoid any damage to the electric throttle control actuator. Do not disassemble.

11. Remove the fuel injectors and fuel tube assembly.

12. Loosen the bolts in reverse order of the tightening sequence.

13. Remove the intake manifold.

✳✳ WARNING

Cover engine openings to avoid entry of foreign materials. Clean all gasket mating surfaces, do not reuse gaskets.

To install:

14. Installation is in the reverse order of removal.

15. Tighten the intake manifold bolts in numerical order as shown.

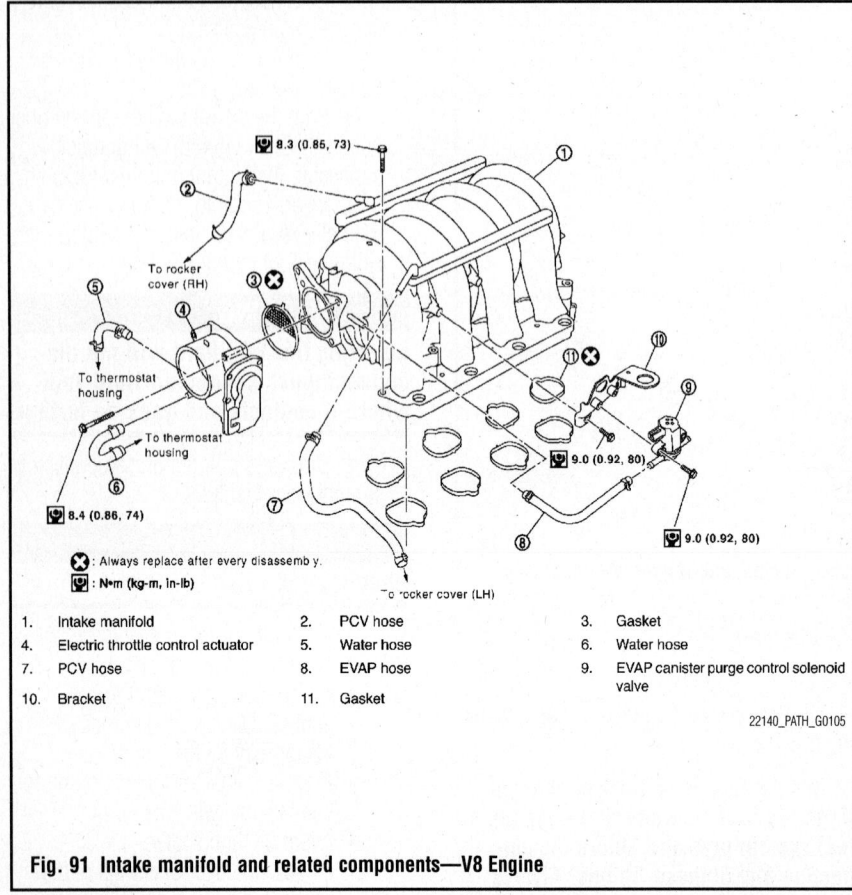

Fig. 91 Intake manifold and related components—V8 Engine

1.	Intake manifold	2.	PCV hose	3.	Gasket
4.	Electric throttle control actuator	5.	Water hose	6.	Water hose
7.	PCV hose	8.	EVAP hose	9.	EVAP canister purge control solenoid valve
10.	Bracket	11.	Gasket		

16. Install the EVAP canister purge control solenoid valve connector with it facing front of engine.

17. Tighten the electronic throttle control actuator bolts of the electric throttle control actuator equally and diagonally in several steps.

18. Install the water hose so that its overlap width for connection is between 1.06 in. (27 mm) and 1.26 in. (32 mm) (target: 1.06 in. or 27 mm).

19. Install quick connector as follows (the steps are the same for quick connectors on both engine side and vehicle side except for the quick connector cap).

20. Make sure no foreign substances are deposited in and around tube and quick connector, and they are not damaged.

21. Thinly apply new engine oil around the fuel tube from tip end to the spool end.

22. Align center to insert quick connector straight into fuel tube.

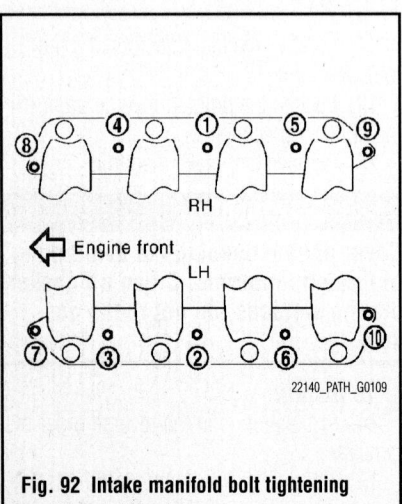

Fig. 92 Intake manifold bolt tightening sequence—V8 engine

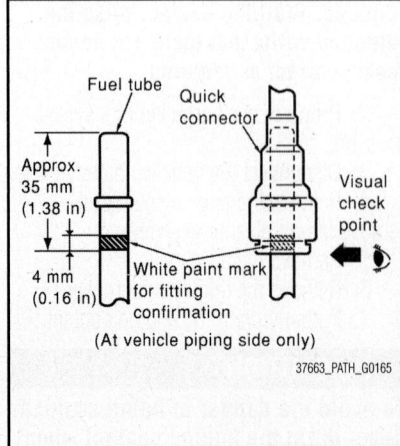

Fig. 93 Tube installation (1 of 2)—V8 engine

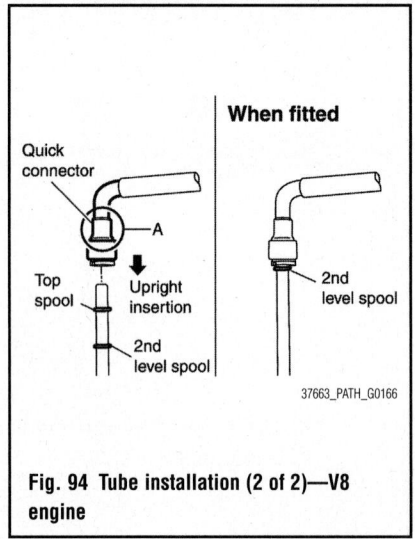

Fig. 94 Tube installation (2 of 2)—V8 engine

a. Insert until the paint mark for engagement identification (white) goes completely inside quick connector so that you cannot see it from the straight side of the connected part. Use a mirror to check this where it is not possible to view directly from the straight side, such as quick connector on vehicle side.

b. Insert fuel tube into quick connector until top spool is completely inside quick connector, and 2nd level spool exposes right below quick connector on engine side.

❊❋ WARNING

Hold "A" position in illustration when inserting fuel tube into quick connector. Carefully align center to avoid inclined insertion to prevent damage to O-ring inside quick connector. Insert until you hear a "click" sound and actually feel the engagement. To avoid misidentification of engagement with a similar sound, be sure to perform the next step.

23. Pull quick connector by hand holding "A" position. Make sure it is completely engaged (connected) so that it does not come out from fuel tube.

➡ **Recommended pulling force is 50 N (5.1 kg, 11.2 lb).**

24. Install the quick connector cap on the quick connector joint (on engine side only).

25. Install the fuel hose and tube to hose clamps.

26. Refill the engine coolant.

27. Using the CONSULT-III diagnostic tool or equivalent, perform the throttle valve closed position learning and the idle air volume learning procedures.

28. Be sure to perform the reconnect/relearn procedures.

OIL PAN

REMOVAL & INSTALLATION

V6 Engine

Lower Oil Pan

See Figure 95.

1. Before servicing the vehicle, refer to the Precautions.
2. Disconnect the negative battery cable.

3. Remove the engine cover and remove the air cleaner assembly, as required.
4. Raise and safely support the vehicle.
5. Drain engine oil.
6. Remove oil pan (lower) as follows:
   a. Loosen bolts in reverse order of the tightening sequence.
   b. Remove oil pan (lower).

➡**Slide seal cutter (1) by tapping on the side (2) of the tool with hammer.**

To install:

7. Install oil pan (lower) as follows:
   a. Use scraper to remove old liquid gasket from mating surfaces.
   b. Also remove old liquid gasket from mating surface of oil pan (upper).

*1
Oil pan side

② 🔧 14.7 (1.5, 11)

③

① 🔧

⑤ ❌ 🔧

⑥

⑦ 🔧 49.0 (5.0, 36)

④ ❌ Refer to "OIL COOLER" in LU section.

⑧ Refer to "OIL FILTER" in LU section.

⑨ 🔧 34.3 (3.5, 25)

⑩ ❌ (*1)

⑯

🔧 9.0 (0.92, 80)

⑮

⑭ ❌ 🔧

🔧 22.0 (2.2, 16)

🔧 22.0 (2.2, 16)

⑬

🔧 22.0 (2.2, 16)

⑫

🔧 9.0 (0.92, 80)

⑪ 🔧

🔧 9.0 (0.92, 80)

❌ : Always replace after every disassembly.

🔧 : Lubricate with new engine oil.

🔧 : Apply Genuine RTV Silicone Sealant or equivalent.

🔧 : N•m (kg-m, in-lb)

🔧 : N•m (kg-m, ft-lb)

1. Oil pan (upper)	2. Oil pressure sensor	3. O-ring
4. Relief valve	5. O-ring	6. Oil cooler
7. Connector bolt	8. Oil filter	9. Drain plug
10. Drain plug washer	11. Oil pan (lower)	12. Rear cover plate
13. Oil strainer	14. O-ring	15. Oil level gauge guide
16. Oil level gauge		

22140_PATH_G0035

Fig. 95 Oil pan and related components—V6 engine

c. Remove old liquid gasket from the bolt holes and threads.

> ✳✳ **WARNING**
>
> **Do not scratch or damage the mating surfaces when cleaning off old liquid gasket.**

d. Apply a continuous bead of liquid gasket to the oil pan (lower).

> ✳✳ **WARNING**
>
> **Attaching should be done within 5 minutes after coating.**

8. Install oil pan (lower).
9. Tighten bolts in numerical order as shown.
10. Install oil pan drain plug.

➡ **Wait at least 30 minutes after oil pan is installed before adding engine oil.**

11. Check engine oil level and adjust engine oil.
12. Start engine, and check there is no leak of engine oil.
13. Stop engine and wait for 10 minutes.
14. Check engine oil level again.

Upper Oil Pan

See Figure 96.

1. Before servicing the vehicle, refer to the Precautions.
2. Disconnect the negative battery cable.

> ✳✳ **CAUTION**
>
> **To avoid the danger of being scalded, do not drain engine oil when engine is hot.**

3. Remove engine cover.
4. Remove air duct.
5. Raise and safely support the vehicle. Drain engine oil.

> ✳✳ **CAUTION**
>
> **Perform this step when engine is cold. Do not spill engine oil on drive belts.**

6. Drain engine coolant.

> ✳✳ **CAUTION**
>
> **Perform this step when engine is cold. Do not spill engine coolant on drive belts.**

7. Remove front final drive (4WD).
8. Disconnect steering gear lower joint shaft bolt and steering gear mounting nuts and bolts, position out of the way.
9. Remove the starter motor.

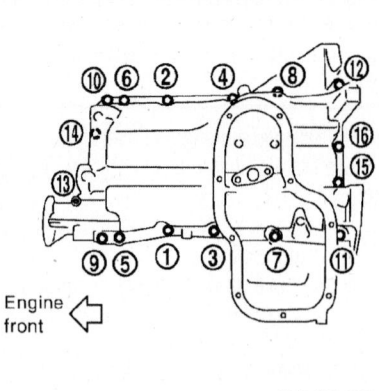

Fig. 96 Upper oil pan bolt tightening sequence—V6 engine

22140_PATH_G0038

10. Disconnect A/T fluid cooler tube brackets and position out of the way.
11. Remove oil filter, as necessary.
12. Remove oil cooler.
13. Remove oil pan (lower).
14. Remove oil strainer.
15. Remove transmission joint bolts which pierce oil pan (upper).
16. Remove rear cover plate.
17. Loosen bolts in reverse order of the tightening sequence.
18. Remove O-rings from bottom of lower cylinder block and oil pump.

To install:

➡ **Be sure to use new fasteners, as required.**

19. Install oil pan (upper) as follows:
a. Use scraper to remove old liquid gasket from mating surfaces.
b. Also remove the old liquid gasket from mating surface of lower cylinder block.
c. Remove old liquid gasket from the bolt holes and threads.

> ✳✳ **WARNING**
>
> **Do not scratch or damage the mating surfaces when cleaning off old liquid gasket.**

d. Install new O-rings on the bottom of lower cylinder block and oil pump.
e. Apply a continuous bead of liquid gasket to the lower cylinder block mating surfaces of oil pan (upper).
f. For bolt holes with arrowhead mark, apply liquid gasket outside the hole.
g. Apply a bead of 4.5 to 5.5 mm (0.177 to 0.217 in) in diameter to area "A".

➡ **Attaching should be done within 5 minutes after coating.**

> ✳✳ **WARNING**
>
> **Install avoiding misalignment of both oil pan gaskets and O-rings.**

20. Tighten bolts in numerical order as shown.

➡ **There are two types of bolts. Refer to the following for locating bolts.**

- M8x4 in. (100 mm): holes 7, 11, 12, 13
- M8x1 in. (25 mm): All other holes
21. Tighten transmission joint bolts.
22. Install oil strainer to oil pan (upper).
23. Install oil pan (upper).
24. Check engine oil level and adjust engine oil.
25. Start engine, and check there is no leak of engine oil.
26. Stop engine and wait for 10 minutes.
27. Check engine oil level again.
28. Be sure to perform the reconnect/relearn procedures.

V8 Engine

See Figures 97 through 101.

➡ **The engine must be removed from the vehicle to perform this procedure. Be sure that the engine is secured in a suitable holding fixture before performing this procedure.**

1. Before servicing the vehicle, refer to the Precautions.
2. Disconnect the negative battery cable.

> ✳✳ **CAUTION**
>
> **To avoid the danger of being scalded, never drain the engine oil when the engine is hot.**

3. Remove the engine. Position the assembly in a suitable holding fixture.
4. Remove the oil pan (lower) using the following steps.
a. Remove the oil pan (lower) bolts.
b. Insert tool between the lower oil pan and the upper oil pan.

> ✳✳ **WARNING**
>
> **Be careful not to damage the mating surface.**

c. Tap seal cutter to insert it (1) and then slide it by tapping on the side (2) of the tool as shown.
5. Remove the oil cooler assembly.

6. Remove the oil strainer from the oil pan (upper).

7. Remove the oil pan (upper) using the following steps.

   a. Remove the oil pan (upper) bolts in the reverse order of the installation sequence.

   b. Remove the oil pan (upper) from the cylinder block by prying.

❈❈ WARNING

Do not damage mating surface.

8. Remove the O-rings from the oil pump and front cover.

➡**Do not reuse O-rings.**

To install:

➡**Be sure to use new fasteners, as required.**

9. Install the oil pan (upper) using the following steps.

   a. Apply liquid gasket thoroughly as shown.

❈❈ WARNING

Apply liquid gasket to outside of bolt hole for the hole shown by star (*).

   b. Install new O-rings to the oil pump and front cover side.

   c. Tighten the bolts in numerical order as shown.

➡**Bolt locations by size:**

- M6 × 1.18 in. (30 mm): No. 15, 16
- M8 × 0.98 in. (25 mm): No. 1, 3, 5, 7, 11, 13
- M8 × 1.77 in. (45 mm): No. 2, 4, 6, 8, 10, 14

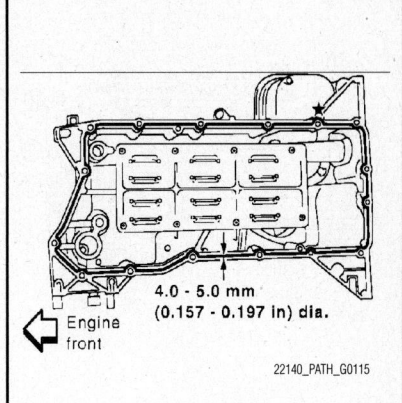

4.0 - 5.0 mm (0.157 - 0.197 in.) dia.

Engine front

22140_PATH_G0115

Fig. 98 Apply liquid gasket thoroughly—V8 engine

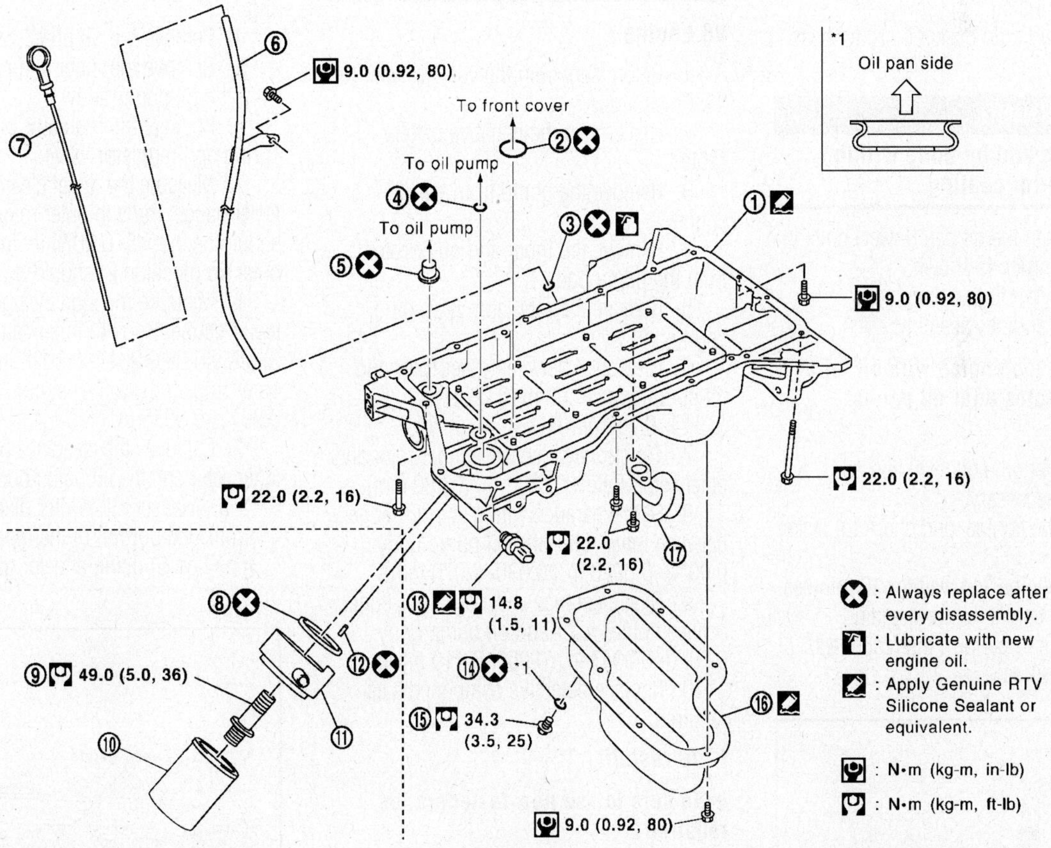

9.0 (0.92, 80)

To front cover

To oil pump

To oil pump

Oil pan side

9.0 (0.92, 80)

22.0 (2.2, 16)

22.0 (2.2, 16)

22.0 (2.2, 16)

14.8 (1.5, 11)

49.0 (5.0, 36)

34.3 (3.5, 25)

9.0 (0.92, 80)

❌ : Always replace after every disassembly.

🛢 : Lubricate with new engine oil.

🔧 : Apply Genuine RTV Silicone Sealant or equivalent.

⬒ : N·m (kg-m, in-lb)

⬓ : N·m (kg-m, ft-lb)

1.	Oil pan (upper)	2.	O-ring	3.	O-ring
4.	O-ring	5.	O-ring (with collar)	6.	Oil level gauge guide
7.	Oil level gauge	8.	O-ring	9.	Connector bolt
10.	Oil filter	11.	Oil cooler	12.	Relief valve
13.	Oil pressure switch	14.	Gasket	15.	Drain plug
16.	Oil pan (lower)	17.	Oil strainer		

22140_PATH_G0110

Fig. 97 Oil pan and related components—V8 engine

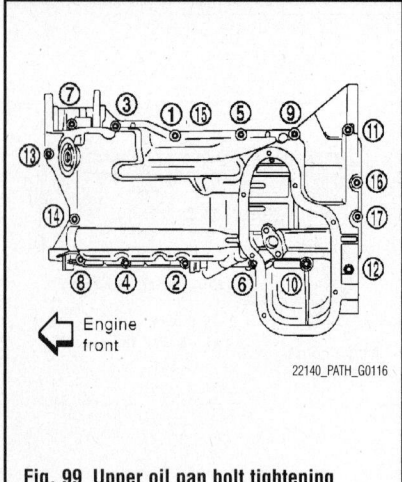

Fig. 99 Upper oil pan bolt tightening sequence—V8 engine

• M8 × 4.84 in. (123 mm): No. 9, 12

10. Install the oil strainer to the oil pan (upper).

11. Install the oil pan (lower).

a. Apply liquid gasket thoroughly as shown.

※※ WARNING

Attaching should be done within 5 minutes after coating.

b. Tighten the oil pan (lower) bolts in numerical order as shown.

12. Install the oil pan drain plug.

13. Install engine assembly.

➡ **Do not fill the engine with oil for at least 30 minutes after oil pan is installed.**

14. Check engine oil level and add engine oil if necessary.

15. Start the engine, and check for leaks of engine oil.

16. Stop engine and wait for 10 minutes.

17. Check engine oil level again.

18. Be sure to perform the reconnect/relearn procedures.

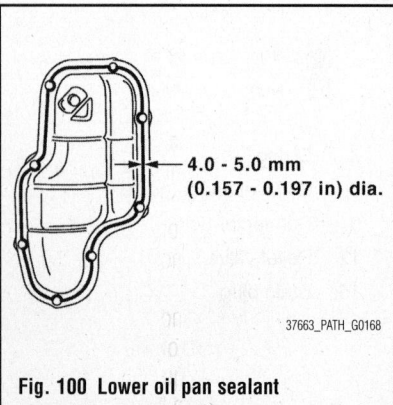

Fig. 100 Lower oil pan sealant application—V8 engine

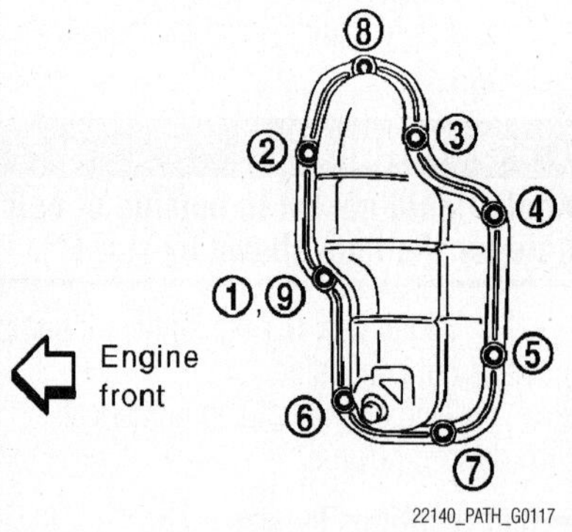

OIL PUMP

INSPECTION

V6 Engine

1. Before servicing the vehicle, refer to the Precautions.

2. Disconnect the negative battery cable.

3. Remove the pump from the engine.

4. Remove the inner and outer rotor from the pump body.

5. Remove the regulator valve plug, spring and valve.

6. Clearance between outer rotor and oil pump body: 0.0045–0.0079 in. (0.114–0.200 mm).

7. Tip clearance between inner rotor and outer rotor: Below 0.0071 in. (0.180 mm).

8. Side clearance with a straight-edge between inner rotor and oil pump body: 0.0012–0.0028 in. (0.030–0.070 mm).

9. Side clearance with a straight-edge between outer rotor and oil pump body: 0.0020–0.0043 in. (0.050–0.110 mm).

10. Replace defective components as required.

To install:

➡ **Be sure to use new fasteners, as required.**

11. Installation is the reverse of the removal procedure.

12. Be sure to perform the reconnect/relearn procedures.

V8 Engine

See Figures 102 through 107.

1. Before servicing the vehicle, refer to the Precautions.

2. Disconnect the negative battery cable.

3. Remove the oil pump from the engine.

4. Remove the oil pump cover.

5. Remove the inner and outer rotors from the oil pump body.

6. Remove the regulator plug, regulator spring and regulator valve.

7. Measure the radial clearance using a feeler gauge. Body to outer rotor (position 1) should be 0.0045–0.0079 in. Inner rotor to outer tip (position 2) should be 0.0071 in.

8. Measure the side clearance using a feeler gauge. Body to inner rotor (position 3) should be 0.0012–0.0028 in. Body to outer rotor (position 4) should be 0.0012–0.0035 in.

9. Calculate the clearance between the inner rotor and the oil pump body as follows:

a. Measure the outer diameter of the protruded portion of the inner rotor (position 5) using a feeler gauge.

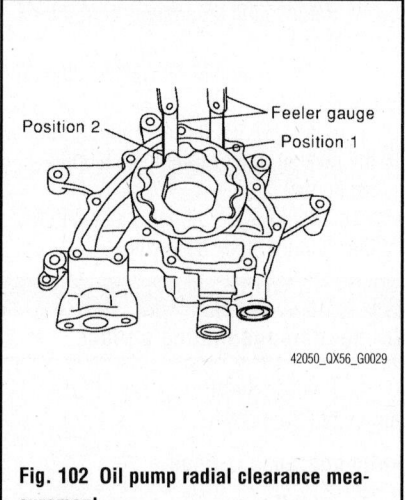

Fig. 102 Oil pump radial clearance measurement

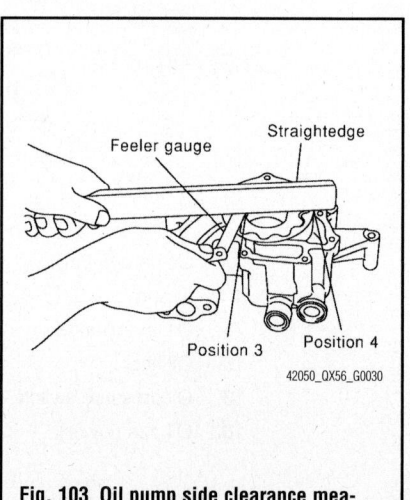

Fig. 103 Oil pump side clearance measurement

Fig. 101 Lower oil pan bolt tightening sequence—V8 engine

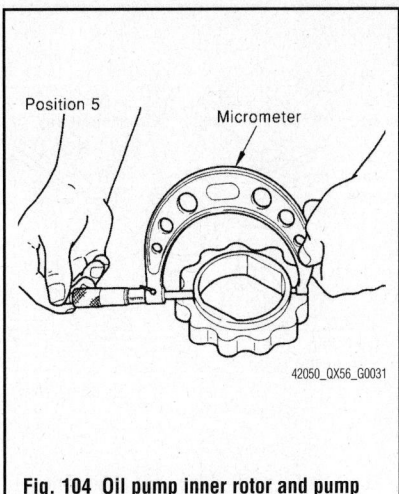

Fig. 104 Oil pump inner rotor and pump body clearance measurement

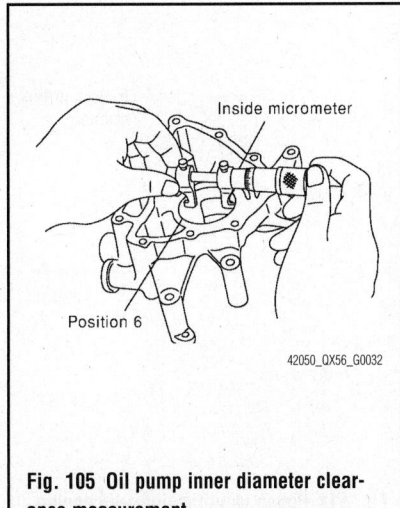

Fig. 105 Oil pump inner diameter clearance measurement

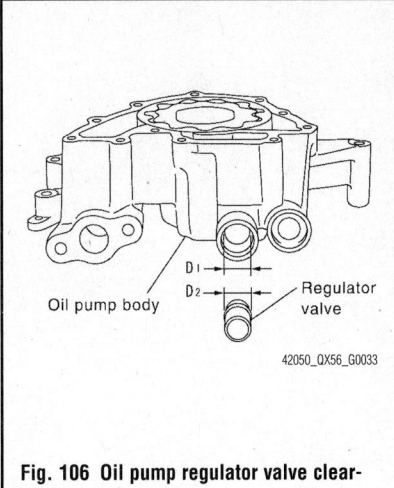

Fig. 106 Oil pump regulator valve clearance measurement

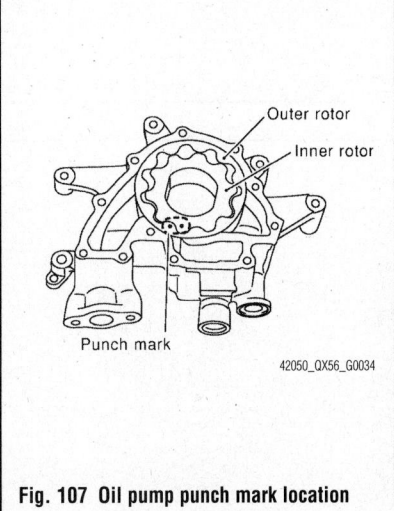

Fig. 107 Oil pump punch mark location

Measure the inner diameter of the oil pump body to the brazed portion (position 6) using a feeler gauge. Calculate the clearance using the following formula. Clearance=Inner diameter of oil pump body, minus outer diameter of inner rotor. Inner rotor to brazed portion of housing clearance specification is 0.0018–0.0036 in.

10. Check the regulator valve to oil pump cover clearance using the following formula: Clearance=D1 (valve hole diameter) minus D2 (outer diameter of valve). Regulator valve to oil pump specification should be 0.0016–0.0038 in.

➡**Coat the valve with clean engine oil. Check that it falls smoothly into the regulator valve hole by its own weight.**

11. Assemble the oil pump in the reverse order.

12. Install the inner and outer rotor with

the punched marks on the oil pump cover side.

To install:

➡**Be sure to use new fasteners, as required.**

13. Installation is the reverse of the removal procedure.

14. Be sure to perform the reconnect/relearn procedures.

REMOVAL & INSTALLATION

V6 Engine

See Figure 108.

1. Before servicing the vehicle, refer to the Precautions.

2. Disconnect the negative battery cable.

3. Remove the engine cover and remove the air cleaner assembly, as required.

4. Raise and safely support the vehicle.

5. Remove the front final drive assembly, if equipped with 4WD.

6. Drain the engine oil. Be sure to properly dispose of used oil.

7. Remove oil pans (lower and upper). Discard the gaskets.

8. Remove front timing chain case and timing chain (primary).

9. Remove oil pump assembly.

To install:

➡**Be sure to use new fasteners, as required.**

10. Installation is in the reverse order of removal, paying attention to the following:

11. When installing, align crankshaft flat faces with inner rotor flat faces.

12. Check the engine oil level.

13. Start engine, and check there are no leaks of engine oil.

14. Stop engine and wait for 10 minutes.

15. Check the engine oil level and add engine oil

16. Be sure to perform the reconnect/relearn procedures.

V8 Engine

See Figures 109 and 110.

1. Before servicing the vehicle, refer to the Precautions.

2. Disconnect the negative battery cable.

3. Remove the engine cover and remove the air cleaner assembly, as required.

4. Remove the front final drive assembly, if equipped with 4WD.

5. Remove front cover.

6. Remove the oil pump drive spacer.

7. Remove the oil pump assembly.

To install:

➡**Be sure to use new fasteners, as required.**

8. Installation is in the reverse order of removal, paying attention of the following:

- When inserting the oil pump drive spacer, align the crankshaft key and the flat face of the inner rotor.
- If they are not aligned, rotate the oil pump inner rotor by hand.
- Make sure that each part is aligned and tap lightly until it reaches the end.

9. Check the engine oil level.

10. Start the engine and check for engine oil leaks.

11. Stop the engine and wait 10 minutes.

12. Check the engine oil level and adjust the engine oil level as required.

13. Be sure to perform the reconnect/relearn procedures.

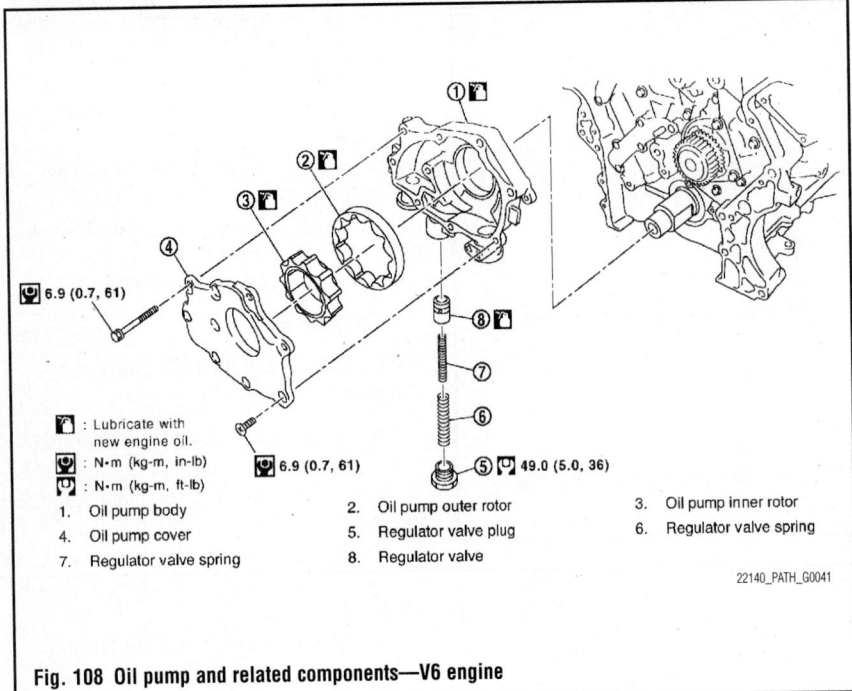

**: Lubricate with new engine oil.
**: N·m (kg-m, in-lb)
**: N·m (kg-m, ft-lb)

1. Oil pump body
4. Oil pump cover
7. Regulator valve spring

2. Oil pump outer rotor
5. Regulator valve plug
8. Regulator valve

3. Oil pump inner rotor
6. Regulator valve spring

6.9 (0.7, 61)   6.9 (0.7, 61)   49.0 (5.0, 36)

22140_PATH_G0041

Fig. 108 Oil pump and related components—V6 engine

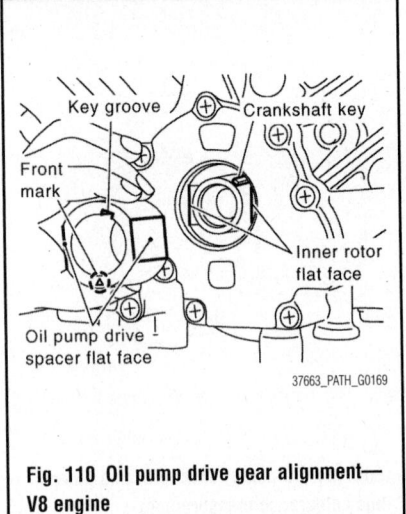

37663_PATH_G0169

Fig. 110 Oil pump drive gear alignment— V8 engine

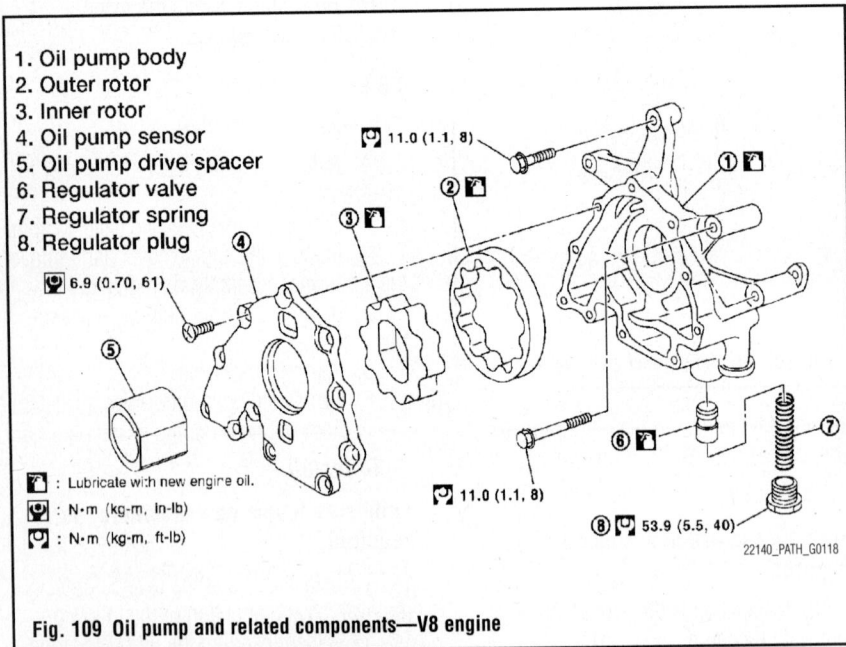

1. Oil pump body
2. Outer rotor
3. Inner rotor
4. Oil pump sensor
5. Oil pump drive spacer
6. Regulator valve
7. Regulator spring
8. Regulator plug

11.0 (1.1, 8)

6.9 (0.70, 61)

11.0 (1.1, 8)

53.9 (5.5, 40)

**: Lubricate with new engine oil.
**: N·m (kg-m, in-lb)
**: N·m (kg-m, ft-lb)

22140_PATH_G0118

Fig. 109 Oil pump and related components—V8 engine

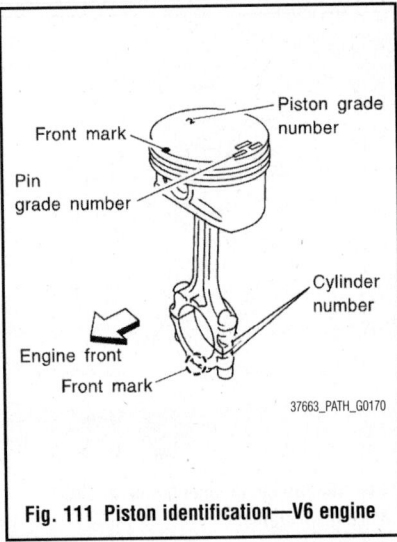

37663_PATH_G0170

Fig. 111 Piston identification—V6 engine

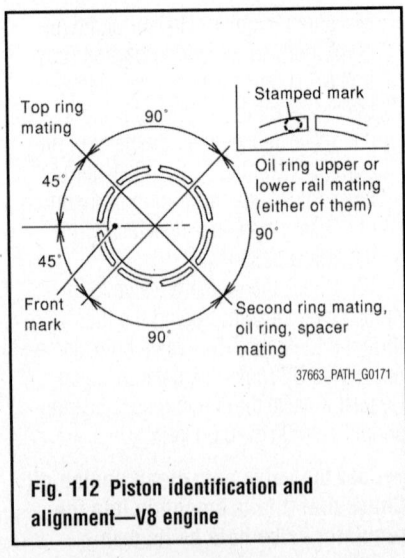

37663_PATH_G0171

Fig. 112 Piston identification and alignment—V8 engine

PISTONS & RINGS

POSITIONING

See Figures 111 and 112.

REAR MAIN SEAL

REMOVAL & INSTALLATION

V6 Engine

See Figures 113 and 114.

1. Before servicing the vehicle, refer to the Precautions.

2. Disconnect the negative battery cable.

3. Remove transmission assembly.

4. Matchmark and remove the flexplate.

5. Remove rear oil seal with a suitable tool.

✴✴ WARNING

Be careful not to damage crankshaft and cylinder block.

To install:

6. Apply new engine oil to new rear oil seal joint surface and seal lip.

7. Install rear oil seal so that each seal lip is oriented as shown.

8. Press in rear oil seal to the position as shown.

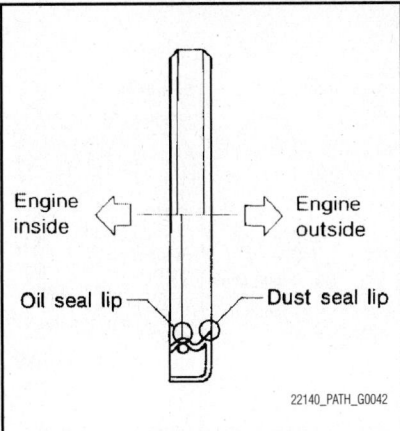

Fig. 113 Install rear oil seal so that each seal lip is oriented as shown—V6 engine

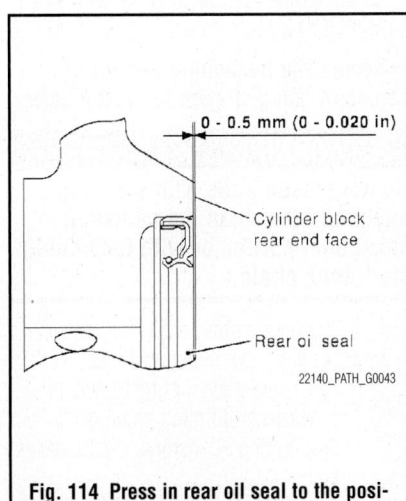

Fig. 114 Press in rear oil seal to the position as shown—V6 engine

✳✳ WARNING

Be careful not to damage crankshaft and cylinder block. Press-fit oil seal straight to avoid causing burrs or tilting. Do not touch grease applied onto oil seal lip.

9. Installation of the remaining components is in the reverse order of removal.

10. Be sure to perform the reconnect/relearn procedures.

V8 Engine

See Figure 115.

1. Before servicing the vehicle, refer to the Precautions.
2. Disconnect the negative battery cable.
3. Remove transmission assembly.
4. Matchmark and remove the flexplate.
5. Lock the drive plate using tool A.

✳✳ WARNING

Do not damage the drive plate. Especially, avoid deforming and damaging the signal plate teeth (circumference position). Keep magnetic materials away from signal plate.

6. Remove the drive plate.

✳✳ WARNING

Place the drive plate with the signal plate surface facing upward.

➡**Remove the bolts diagonally.**

7. Remove the rear oil seal using suitable tool.

✳✳ WARNING

Do not damage crankshaft or oil seal retainer surface.

To install:

8. Apply new engine oil to both the oil seal lip and dust seal lip of the new rear oil seal.
9. Install the rear oil seal.
   a. Install the rear oil seal so that each seal lip is oriented as shown.
   b. Press-fit the rear oil seal using suitable tool.

✳✳ WARNING

Do not damage the crankshaft or oil seal retainer. Press-fit the oil seal straight to avoid causing burrs or tilting. Do not touch grease applied onto the oil seal lip. Do not damage or scratch the outer circumference of the rear oil seal. Tap until flush with the front edge of the oil seal retainer.

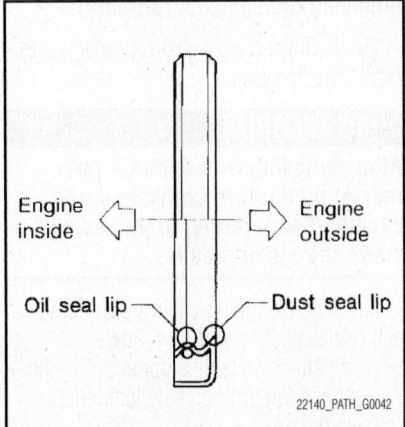

Fig. 115 Install the rear oil seal so that each seal lip is oriented as shown—V8 engine

10. Installation of the remaining components is in the reverse order of removal.

TIMING CHAIN COVER, CHAIN, TENSIONER, & SPROCKETS

REMOVAL & INSTALLATION

V6 Engine

See Figures 116 through 128.

1. Before servicing the vehicle, refer to the Precautions.
2. Properly release the fuel system pressure.
3. Disconnect the negative battery cable.
4. Remove the engine cover. Remove the air cleaner assembly.
5. Drain engine oil.

✳✳ CAUTION

Perform this step when engine is cold. Do not spill engine oil on drive belts.

6. Drain engine coolant from radiator.

✳✳ CAUTION

Perform this step when engine is cold. Do not spill engine coolant on drive belts.

7. Remove radiator cooling fan assembly.
8. Separate engine harnesses removing their brackets from front timing chain case.
9. Remove drive belts.
10. Remove power steering oil pump from bracket with piping connected, and temporarily secure it aside.
11. Remove power steering oil pump bracket.
12. Remove alternator.
13. Remove water bypass hose, water hose clamp and idler pulley bracket from front timing chain case.
14. Remove right and left intake valve timing control covers.
   a. Loosen bolts in reverse order of the tightening sequence.
   b. Cut liquid gasket for removal.

✳✳ WARNING

Shaft is internally jointed with camshaft sprocket (INT) center hole. When removing, keep it horizontal until it is completely disconnected.

15. Remove collared O-rings from front timing chain case (left and right side).

16. Remove rocker covers (right and left banks).

➡**When only timing chain (primary) is removed, rocker cover does not need to be removed.**

17. Obtain No. 1 cylinder at TDC of its compression stroke as follows:

➡**When timing chain is not removed/installed, this step is not required.**

   a. Rotate crankshaft pulley clockwise to align timing mark (grooved line without color) with timing indicator.

   b. Make sure that intake and exhaust cam noses on No. 1 cylinder (engine front side of right bank) are located as shown.

   c. If not, turn crankshaft one revolution (360°) and align as shown.

➡**When only timing chain (primary) is removed, rocker cover does not need to be removed. To make sure that No. 1 cylinder is at its compression TDC, remove front timing chain case first. Then check mating marks on camshaft sprockets.**

18. Remove crankshaft pulley as follows:
   a. Remove starter motor.
   b. Loosen crankshaft pulley bolt and locate bolt seating surface as 0.39 in. (10 mm) from its original position.

✳✳ WARNING

Do not remove crankshaft pulley bolt. Keep loosened crankshaft pulley bolt in place protect removed crankshaft pulley from dropping.

   c. Pull crankshaft pulley with both hands to remove it.

19. Loosen two bolts in front of oil pan (upper) in reverse order of the tightening sequence.

20. Remove front timing chain case as follows:
   a. Loosen bolts in reverse order of the tightening sequence.
   b. Insert suitable tool into the notch at the top of the front timing chain case.
   c. Pry off case by moving tool.
   d. Cut liquid gasket for removal.

✳✳ WARNING

Do not use screwdriver or something similar. After removal, handle front timing chain case carefully so it does not tilt, cant, or warp under a load.

21. Remove O-rings from rear timing chain case.

22. Remove water pump cover and chain tensioner cover from front timing chain case, if necessary.

23. Remove front oil seal from front timing chain case using suitable tool.

✳✳ WARNING

Be careful not to damage front timing chain case.

24. Use a scraper to remove all traces of old liquid gasket from front and rear timing chain cases and oil pan (upper), and liquid gasket mating surfaces.

✳✳ WARNING

Be careful not to allow gasket fragments to enter oil pan.

25. Remove old liquid gasket from bolt holes and threads.

26. Use a scraper to remove all traces of old liquid gasket from water pump cover, chain tensioner cover and intake valve timing control covers.

27. Remove timing chain tensioner (primary) as follows:
   a. Loosen clip of timing chain tensioner (primary), and release plunger stopper (1).
   b. Insert plunger into tensioner body by pressing slack guide (2).
   c. Keep slack guide pressed and hold plunger in by pushing stopper pin through the tensioner body hole and plunger groove (3).
   d. Remove bolts and remove timing chain tensioner (primary).

28. Remove internal chain guide, tension guide and slack guide.

➡**Tension guide can be removed after removing timing chain (primary).**

29. Remove timing chain (primary) and crankshaft sprocket.

✳✳ WARNING

After removing timing chain (primary), do not turn crankshaft and camshaft separately, or valves will strike the piston heads.

30. Remove timing chain (secondary) and camshaft sprockets as follows:
   a. Attach suitable stopper pin to the right and left timing chain tensioners (secondary).

➡**Use approximately 0.02 in. (0.5 mm) diameter hard metal pin as a stopper pin.**

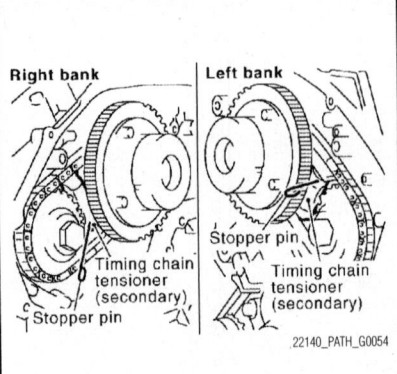

Fig. 116 Attach suitable stopper pin to the right and left timing chain tensioners (secondary)—V6 engine

   b. Remove camshaft sprocket (INT and EXH) bolts.

➡**Secure the hexagonal portion of camshaft using wrench to loosen bolts.**

✳✳ WARNING

Do not loosen bolts with securing anything other than the camshaft hexagonal portion or with tensioning the timing chain.

   c. Remove timing chain (secondary) together with camshaft sprockets.
   • Turn camshaft slightly to secure slackness of timing chain on timing chain tensioner (secondary) side.
   • Insert 0.5 mm (0.020 in) thick metal or resin plate between timing chain and timing chain tensioner plunger (guide). Remove timing chain (secondary) together with camshaft sprockets with timing chain loose from guide groove.

✳✳ WARNING

Be careful of plunger coming off when removing timing chain (secondary). This is because plunger of timing chain tensioner (secondary) moves during operation, leading to coming off of fixed stopper pin.

➡**Camshaft sprocket (INT) is a one piece integrated design sprockets for timing chain (primary) and for timing chain (secondary).**

✳✳ WARNING

When handling camshaft sprocket (INT), be careful of the following: Handle carefully to avoid any shock to camshaft sprocket. Do not disas-

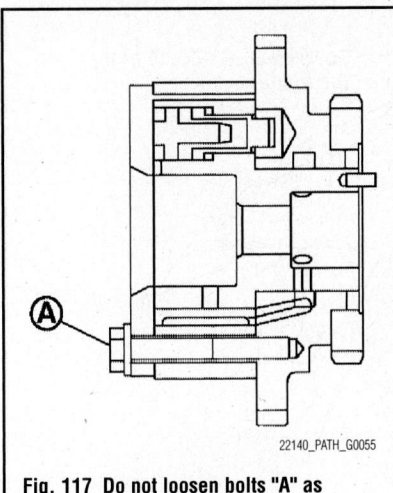

Fig. 117 Do not loosen bolts "A" as shown—V6 engine

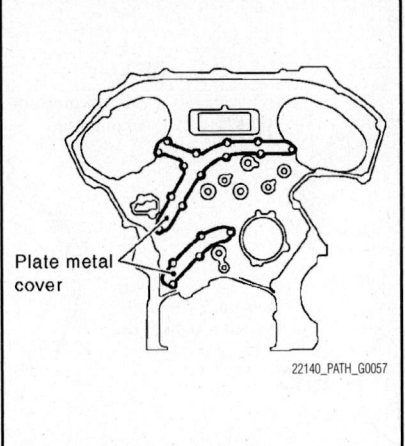

Fig. 118 Do not remove plate metal cover of oil passage—V6 engine

semble. (Do not loosen bolts "A" as shown).

31. Remove water pump.
32. Remove rear timing chain case as follows:
   a. Loosen and remove bolts in reverse order of the tightening sequence.
   b. Cut liquid gasket and remove rear timing chain case.

✳✳ WARNING

Do not remove plate metal cover of oil passage.

✳✳ WARNING

After removal, handle rear timing chain case carefully so it does not tilt, cant, or warp under a load.

33. Remove O-rings from cylinder head and camshaft bracket (No. 1).
34. Remove O-rings from cylinder block.
35. Remove timing chain tensioners (secondary) from cylinder head if necessary.
   a. Remove camshaft brackets (No. 1).
   b. Remove timing chain tensioners (secondary) with stopper pin attached.
36. Use scraper to remove all traces of old liquid gasket from front and rear timing chain cases, and opposite mating surfaces. Remove old liquid gasket from bolt hole and thread.
37. Use scraper to remove all traces of liquid gasket from water pump cover, chain tensioner cover and intake valve timing control covers.
38. Check for cracks and any excessive wear at link plates and roller links of timing chain.
39. Replace timing chain as necessary.

To install:

➡ Be sure to use new fasteners, as required.

➡ The figure below shows the relationship between the mating mark on each timing chain and that on the corresponding sprocket, with the components installed.

40. Install timing chain tensioners (secondary) to cylinder head if removed.
   a. Install timing chain tensioners (secondary) with stopper pin attached and new O-ring.
   b. Install camshaft brackets (No. 1).
41. Install rear timing chain case as follows:

a. Install new O-rings onto cylinder block.
   b. Install new O-rings to cylinder head and camshaft bracket (No. 1).
   c. Apply liquid gasket to rear timing chain case back side as shown.

✳✳ WARNING

For "A" in the figure, completely wipe out liquid gasket extended on a portion touching at engine coolant. Apply liquid gasket on installation position of water pump and cylinder head very completely

d. Align rear timing chain case with dowel pins (right and left) on cylinder block and install rear timing chain case.

➡ Make sure O-rings stay in place during installation to cylinder block, cylinder head and camshaft bracket (No. 1).

e. Tighten bolts in numerical order as shown.

➡ There are two type of bolts. Refer to the following for locating bolts.

- 0.79 in. (20 mm): 1, 2, 3, 6, 7, 8, 9, 10
- 0.63 in. (16 mm): Except the above
- Rear timing case bolt torque: 9 ft lbs. (12.7 Nm)

f. After all bolts are tightened, retighten them to the specified torque in numerical order as shown.

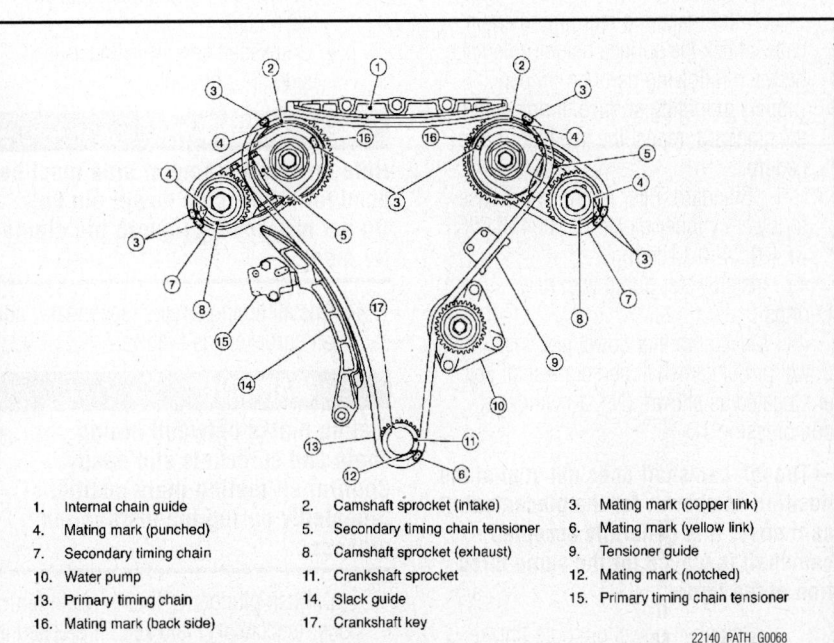

1.	Internal chain guide	2.	Camshaft sprocket (intake)	3.	Mating mark (copper link)
4.	Mating mark (punched)	5.	Secondary timing chain tensioner	6.	Mating mark (yellow link)
7.	Secondary timing chain	8.	Camshaft sprocket (exhaust)	9.	Tensioner guide
10.	Water pump	11.	Crankshaft sprocket	12.	Mating mark (notched)
13.	Primary timing chain	14.	Slack guide	15.	Primary timing chain tensioner
16.	Mating mark (back side)	17.	Crankshaft key		

Fig. 119 Timing chain assembly with mating marks—V6 engine

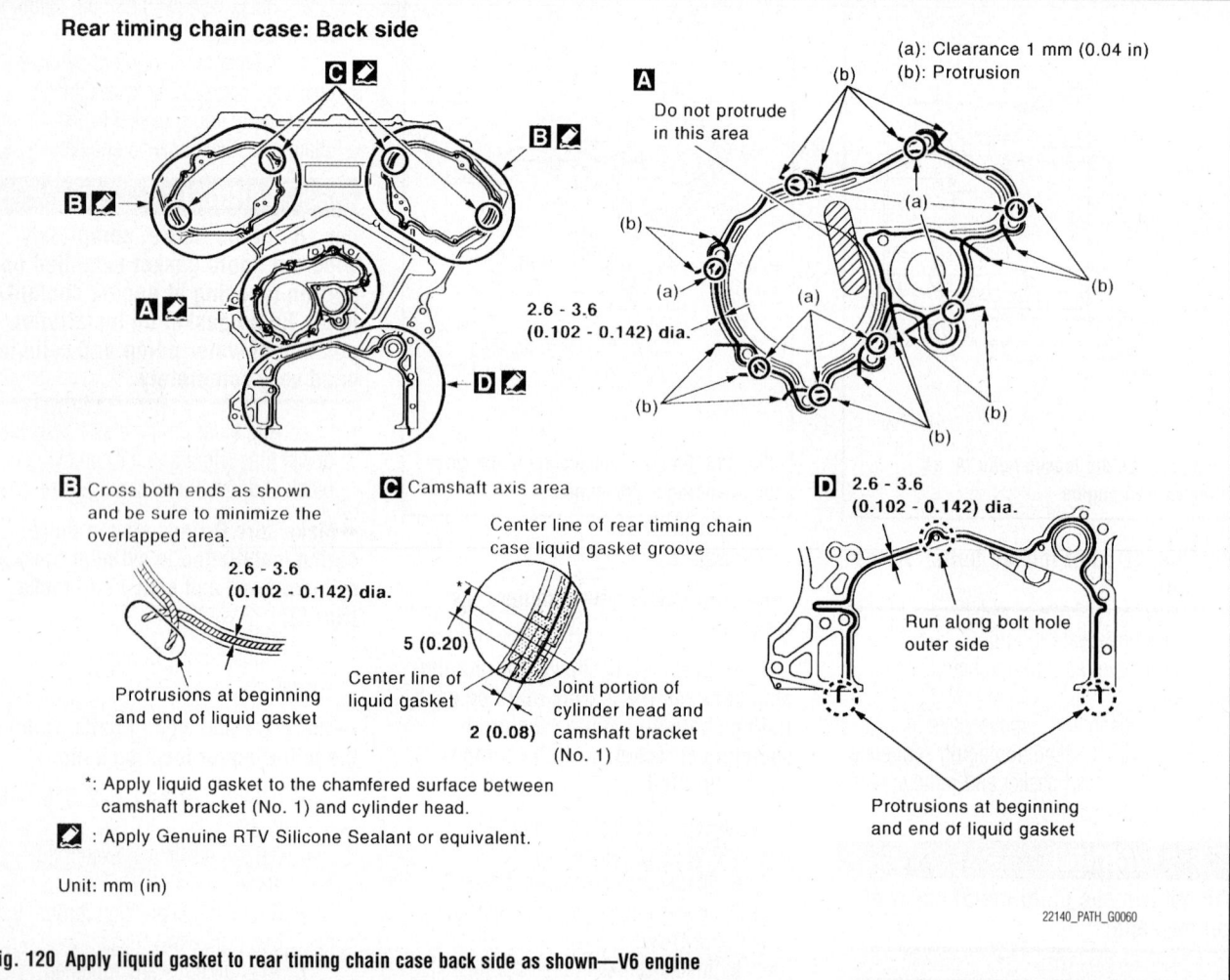

Rear timing chain case: Back side

(a): Clearance 1 mm (0.04 in)
(b): Protrusion

A Do not protrude in this area

2.6 - 3.6
(0.102 - 0.142) dia.

B Cross both ends as shown and be sure to minimize the overlapped area.

2.6 - 3.6
(0.102 - 0.142) dia.

Protrusions at beginning and end of liquid gasket

C Camshaft axis area

Center line of rear timing chain case liquid gasket groove

Center line of liquid gasket

5 (0.20)

2 (0.08)

Joint portion of cylinder head and camshaft bracket (No. 1)

D 2.6 - 3.6
(0.102 - 0.142) dia.

Run along bolt hole outer side

Protrusions at beginning and end of liquid gasket

*: Apply liquid gasket to the chamfered surface between camshaft bracket (No. 1) and cylinder head.

✎ : Apply Genuine RTV Silicone Sealant or equivalent.

Unit: mm (in)

22140_PATH_G0060

Fig. 120 Apply liquid gasket to rear timing chain case back side as shown—V6 engine

g. If liquid gasket protrudes, wipe it off immediately.

h. After installing rear timing chain case, check the surface height difference between following parts on oil pan (upper) mounting surface. If not within the standard, repeat the installation procedure.

i. Standard: Rear timing chain case to lower cylinder block: -0.0094–0.0055 in. (-0.24–0.14 mm)

42. Install water pump with new O-rings.

43. Make sure that dowel pin hole, dowel pin of camshaft and crankshaft key are located as shown. (No. 1 cylinder at compression TDC).

➡Though camshaft does not stop at the position as shown, for the placement of cam nose, it is generally accepted camshaft is placed for the same direction of the figure.

- Camshaft dowel pin hole (intake side): At cylinder head upper face side in each bank.

- Camshaft dowel pin (exhaust side): At cylinder head upper face side in each bank.
- Crankshaft key: At cylinder head side of right bank.

✳✳ WARNING

Hole on small diameter side must be used for intake side dowel pin hole. Do not misidentify (ignore big diameter side).

44. Install timing chains (secondary) and camshaft sprockets as follows:

✳✳ WARNING

Mating marks between timing chain and sprockets slip easily. Confirm all mating mark positions repeatedly during the installation process.

a. Push plunger of timing chain tensioner (secondary) and keep it pressed in with stopper pin.

b. Install timing chains (secondary) and camshaft sprockets (INT and EXH).

c. Align the mating marks on timing chain (secondary) (copper color link) with the ones on camshaft sprockets (INT and EXH) (punched), and install them.

➡Mating marks for camshaft sprocket (INT) are on the back side of camshaft sprocket (secondary).

➡There are two types of mating marks, circle and oval types.

- Right bank: Use circle type.
- Left bank: Use oval type.
- They should be used for the right and left banks, respectively.

d. Align dowel pin and pin hole on camshafts with the groove and dowel pin on sprockets, and install them.

e. On the intake side, align pin hole on the small diameter side of the camshaft front end with dowel pin on the back side of camshaft sprocket, and install them.

f. On the exhaust side, align dowel pin on camshaft front end with pin

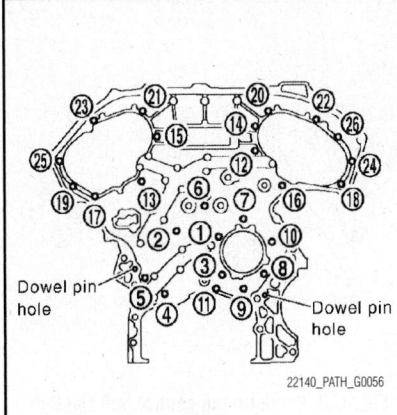

Fig. 121 Rear timing case cover bolt tightening sequence—V6 engine

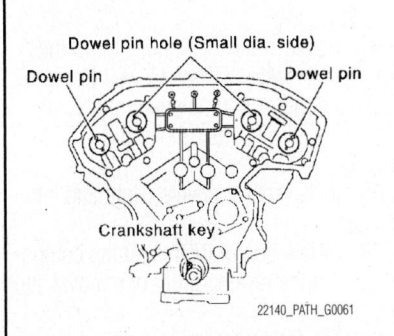

Fig. 122 Make sure that dowel pin hole, dowel pin of camshaft and crankshaft key are located as shown. (No. 1 cylinder at compression TDC)—V6 engine

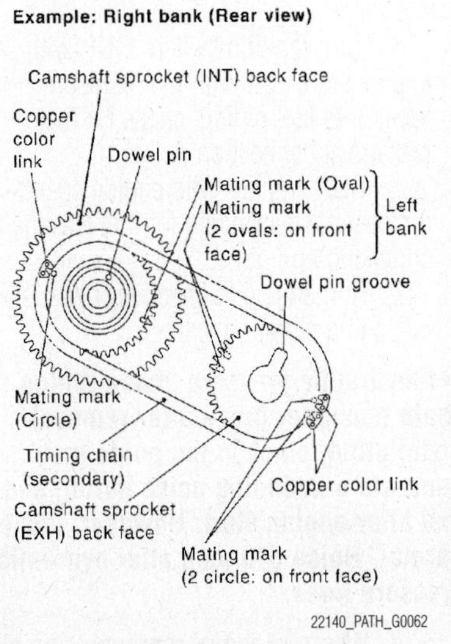

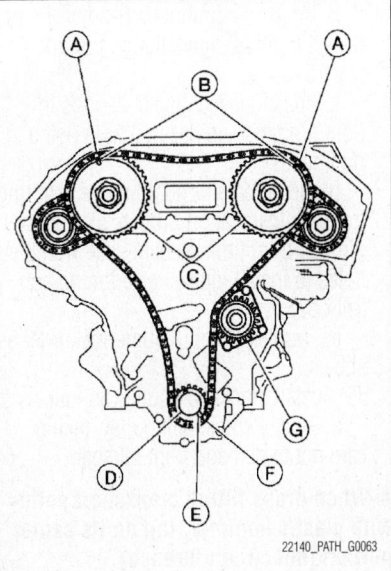

Fig. 123 Install timing chains (secondary) and camshaft sprockets (INT and EXH)—V6 engine

groove on camshaft sprocket, and install them.

g. In case that positions of each mating mark and each dowel pin are not fit on mating parts, make fine adjustment to the position holding the hexagonal portion on camshaft with wrench or equivalent.

h. Bolts for camshaft sprockets must be tightened in the next step. Tightening them by hand is enough to prevent the dislocation of dowel pins.

i. It may be difficult to visually check the dislocation of mating marks during and after installation. To make the matching easier, make a mating mark on the top of sprocket teeth and its extended line in advance with paint.

j. After confirming the mating marks are aligned, tighten camshaft sprocket bolts.

➡**Secure camshaft using wrench at the hexagonal portion to tighten bolts.**

k. Pull stopper pins out from timing chain tensioners (secondary).

45. Install tension guide.

46. Install timing chain (primary) as follows:

a. Install crankshaft sprocket.

➡**Make sure the mating marks on crankshaft sprocket face the front of engine.**

b. Install the primary timing chain.
• Water pump (G).
• Install primary timing chain so the mating mark punched (B) on

camshaft sprocket is aligned with the copper link (A) on the timing chain, while the mating mark notched (E) on the crankshaft sprocket (D) is aligned with the yellow link (F) on the timing chain, as shown.
• When it is difficult to align mating marks (A) with (B) and (E) with (F) of the primary timing chain with each sprocket, gradually turn the camshaft using a wrench on the hexagonal portion to align it with the mating marks.
• During alignment, be careful to prevent dislocation of mating mark alignments of the secondary timing chains.

47. Install internal chain guide, slack guide and timing chain tensioner (primary).

⁂ WARNING

Do not overtighten slack guide bolts. It is normal for a gap to exist under the bolt seats when bolts are tightened to specification.

• When installing timing chain tensioner (primary), push in plunger and keep it pressed in with stopper pin.
• Remove any dirt and foreign materials completely from the back and the mounting surfaces of timing chain tensioner (primary).
• After installation, pull out stopper pin by pressing slack guide.

48. Make sure again that the mating marks on camshaft sprockets and timing chain have not slipped out of alignment.

49. Install new O-rings on rear timing chain case.

50. Install new front oil seal on front timing chain case.

a. Apply new engine oil to both oil seal lip and dust seal lip.

b. Install it so that each seal lip is oriented as shown.

c. Press-fit oil seal until it becomes flush with front timing chain case end face using suitable drift with outer diameter: 2.36 in. (60 mm).

d. Make sure the garter spring is in position and seal lip is not inverted.

51. Install water pump cover and chain tensioner cover to front timing chain case.

a. Apply a continuous bead of liquid gasket to front timing chain case as shown.

52. Install front timing chain case as follows:

a. Apply a continuous bead of liquid gasket to front timing chain case back side as shown.

Fig. 124 Install the primary timing chain—V6 engine

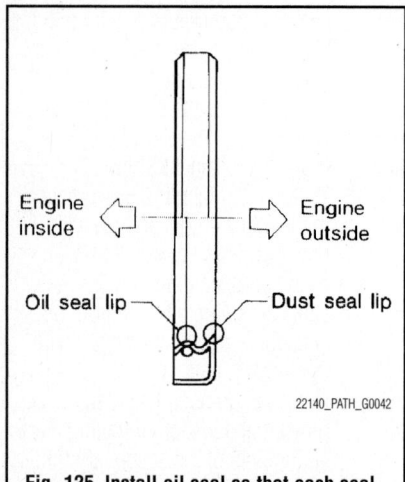

Fig. 125 Install oil seal so that each seal lip is oriented as shown—V6 engine

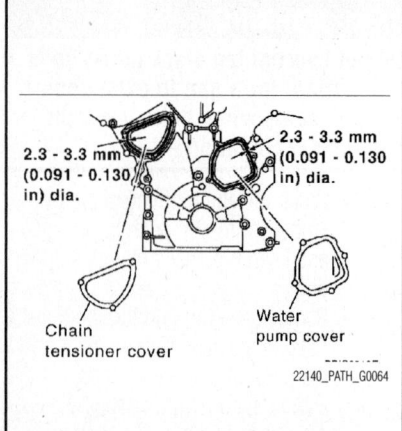

Fig. 126 Apply a continuous bead of liquid gasket to front timing chain—V6 engine

b. Install new O-rings on rear timing chain case.

c. Assemble front timing chain case as follows:

- Fit lower end of front timing chain case tightly onto top face of oil pan (upper). From the fitting point, make entire front timing chain case contact rear timing chain case completely.
- Since front timing chain case is off-set for difference of bolt holes, tighten bolts temporarily while holding front timing chain case from front and top.
- Same as the previous step, insert dowel pin while holding front timing chain case from front and top completely.

d. Tighten bolts to the specified torque in numerical order as shown.

➡ There are two type of bolts. Refer to the following for locating bolts.

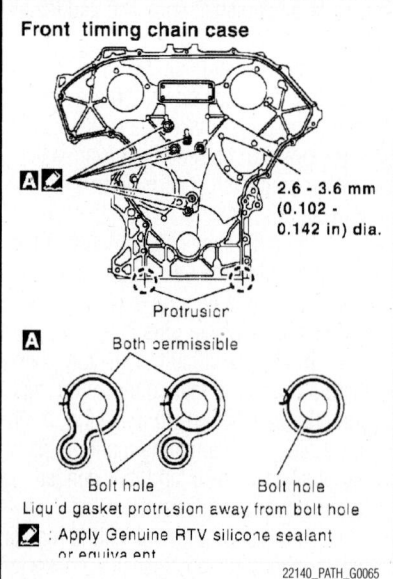

Fig. 127 Apply a continuous bead of liquid gasket to front timing chain case back side—V6 engine

- 1–5: 0.39 in. (10 mm)
- 6–25: 0.24 in. (6 mm)
- Bolt position torque specification: 1–5: 41 ft. lbs. (55.0 Nm); 6–25: 9 ft. lbs. (12.7 Nm)

e. After all bolts tightened, retighten them to the specified torque in numerical order as shown.

53. Install two bolts in front of oil pan (upper).

54. Install right and left intake valve timing control covers as follows:

a. Install new seal rings in shaft grooves.

b. Apply a continuous bead of liquid gasket to intake valve timing control covers.

c. Install new collared O-rings in front timing chain case oil hole (left and right sides).

d. Being careful not to move seal ring from the installation groove, align dowel pins on front timing chain case with the holes to install intake valve timing control covers.

e. Tighten bolts in numerical order as shown.

55. Install crankshaft pulley as follows:

a. Install crankshaft pulley, taking care not to damage front oil seal.

➡ When press-fitting crankshaft pulley with plastic hammer, tap on its center portion (not circumference).

b. Tighten crankshaft pulley bolt. Crankshaft bolt torque: 33 ft. lbs. (44.1 Nm)

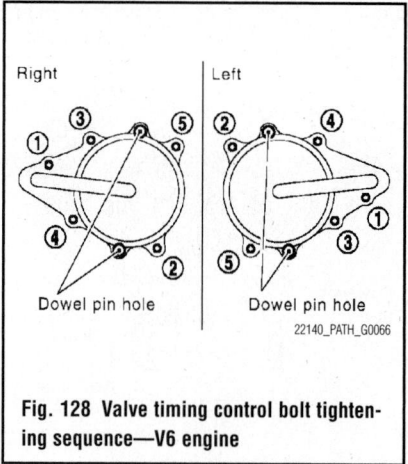

Fig. 128 Valve timing control bolt tightening sequence—V6 engine

c. Put a paint mark on crankshaft pulley aligning with angle mark on crankshaft pulley bolt. Then, further retighten bolt by 60° (equivalent to one graduation).

56. Rotate crankshaft pulley in normal direction (clockwise when viewed from front) to confirm it turns smoothly.

57. Install oil pans (upper and lower).

58. Install rocker covers (right and left banks).

59. Installation of the remaining components is in the reverse order of removal after this step.

60. The following are procedures for checking fluid leaks, lubricant leaks and exhaust gases leaks.

a. Before starting engine, check oil/fluid levels including engine coolant and engine oil. If less than required quantity, fill to the specified level.

61. Use procedure below to check for fuel leakage.

a. Turn ignition switch "ON" (with engine stopped). With fuel pressure applied to fuel piping, check for fuel leakage at connection points.

b. Start engine. With engine speed increased, check again for fuel leakage at connection points.

c. Run engine to check for unusual noise and vibration.

➡ If hydraulic pressure inside timing chain tensioner drops after removal/installation, slack in the guide may generate a pounding noise during and just after engine start. However, this is normal. Noise will stop after hydraulic pressure rises.

d. Warm up engine thoroughly to make sure there is no leakage of fuel, exhaust gases, or any oil/fluids including engine oil and engine coolant.

e. Bleed air from lines and hoses of applicable lines, such as in cooling system.

f. After cooling down engine, again check oil/fluid levels including engine oil and engine coolant. Refill to the specified level, if necessary.

62. Be sure to perform the reconnect/relearn procedures.

V8 Engine

See Figures 129 through 134.

1. Before servicing the vehicle, refer to the Precautions.

2. Properly release the fuel system pressure.

3. Disconnect the negative battery cable.

4. Remove the engine cover. Remove the air cleaner assembly.

➡**To remove timing chain and associated parts, start with those on the LH bank. The procedure for removing parts on the RH bank is omitted because it is the same as that for removal on the LH bank.**

To install timing chain and associated parts, start with those on the RH bank. The procedure for installing parts on the LH bank is omitted because it is the same as that for installation on the RH bank.

5. Remove the engine from the vehicle. Position the assembly in a suitable holding fixture.

6. Remove the following components and related parts:
   - Drive belt auto tensioner and idler pulley
   - Thermostat housing and water hose
   - Power steering oil pump bracket
   - Oil pan (lower), (upper) and oil strainer
   - Ignition coil
   - Rocker cover

7. Remove the Intake valve control solenoid valve cover (RH) (A) and Intake valve control solenoid valve cover (LH) (B) as follows:

a. Loosen and remove the bolts in the order of the tightening sequence.

b. Cut the liquid gasket and remove the covers.

❊❊ WARNING

Do not damage mating surfaces.

8. Obtain compression TDC of No. 1 cylinder as follows:

a. Turn the crankshaft pulley clockwise to align the TDC identification notch

(without paint mark) with the timing indicator on the front cover.

b. At this time, make sure both intake and exhaust cam lobes of No. 1 cylinder (top front on LH bank) point outside. If they do not point outside, turn crankshaft pulley once more.

9. Remove the crankshaft pulley.

a. Loosen the crankshaft pulley bolts using a hammer handle to secure the crankshaft.

b. Remove the crankshaft pulley from the crankshaft.

c. Remove the crankshaft pulley.

➡**The dimension between the centers of the two bolt holes is 61 mm (2.40 in.).**

10. Remove the front cover.

a. Loosen and remove the bolts in the reverse order of the tightening sequence.

b. Cut the liquid gasket and remove the covers.

❊❊ WARNING

Do not damage mating surfaces.

11. Remove the front oil seal using suitable tool.

❊❊ WARNING

Do not damage front cover.

12. Remove the oil pump drive spacer.

➡**Hold and remove the flat space of the oil pump drive spacer by pulling it forward.**

13. Remove the oil pump.

14. Remove the chain tensioner on the LH bank using the following steps.

➡**To remove the timing chain and associated parts, start with those on the LH bank. The procedure for removing parts on the RH bank is omitted because it is the same as that for the LH bank.**

a. Squeeze the return-proof clip ends using suitable tool and push the plunger into the tensioner body.

b. Secure the plunger using stopper pin.

c. Remove the bolts and chain tensioner.

❊❊ CAUTION

Plunger, spring, and spring seat pop out when (squeezing) return-proof clip without holding plunger head. It may cause serious injuries. Always hold plunger head when removing.

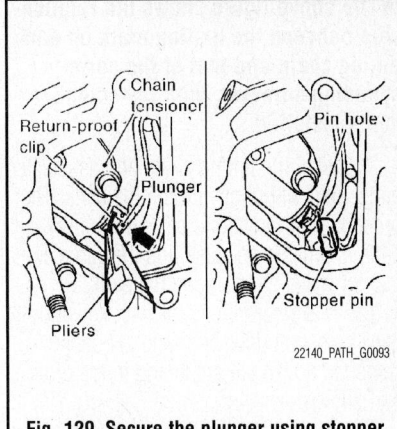

Fig. 129 Secure the plunger using stopper pin—V8 engine

➡**Stop the plunger in the fully extended position by using the return-proof clip (1) if the stopper pin is removed. Push the plunger (2) into the tensioner body while squeezing the return-proof clip (1). Secure it using stopper pin (3).**

15. Remove the timing chain tension guide and timing chain slack guide.

16. Remove the timing chain and crankshaft sprocket.

17. Loosen the camshaft sprocket bolts as shown and remove the camshaft sprocket.

❊❊ WARNING

To avoid interference between valves and pistons, do not turn crankshaft or camshaft when timing chain is disconnected.

18. Repeat the same procedure to remove the RH timing chain and associated parts.

To install:

➡**Be sure to use new fasteners, as required.**

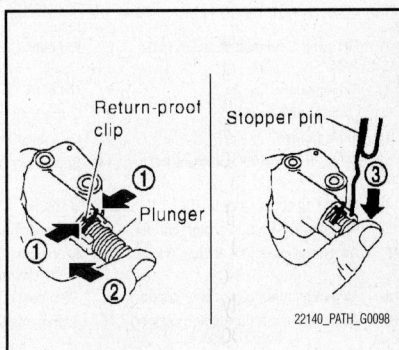

Fig. 130 Stop the plunger in the fully extended position—V8 engine

➡ The above figure shows the relationship between the mating mark on each timing chain and that of the corresponding sprocket, with the components installed.

To install the timing chain and associated parts, start with those on the RH bank. The procedure for installing parts on the LH bank is omitted because it is the same as that for installation on the RH bank.

19. Make sure the crankshaft key and RH bank camshaft dowel pin and LH bank camshaft dowel pin are facing in the direction shown.

20. Install the camshaft sprockets.

a. Install the intake camshaft sprocket (VTC) (A) and exhaust camshaft sprockets (B) by selectively using the groove of the dowel pin according to the bank. (Common part used for both exhaust banks.)

b. Lock the hexagonal part of the camshaft in the same way as for removal, and tighten the bolts.
- A: Intake
- B = V: Exhaust

21. Install the crankshaft sprockets for both banks.

a. Install LH bank crankshaft sprocket (B) and RH bank crankshaft sprocket (C)

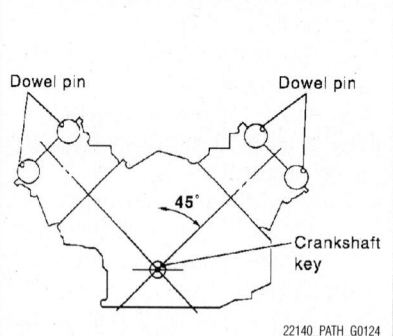

Fig. 132 Make sure the crankshaft key and RH bank camshaft dowel pin and LH bank camshaft dowel pin are facing in the direction—V8 engine

so that their flange side (A) (the larger diameter side without teeth) faces in the direction shown.

➡ **The same parts are used but facing directions are different.**

22. Install the timing chains and associated parts.

a. Align the alignment mark on each sprocket and the timing chain for installation.

⁂⁂ WARNING

Before installing timing chain tensioner, it is possible to change the position of alignment mark on timing chain and each sprocket. After the alignment marks are aligned, keep them aligned by holding them by hand.

b. Install the slack guides and tension guides onto the correct side by checking the identification mark on the surface.

c. Install the timing chain tensioner with the plunger locked in with the stopper pin.

⁂⁂ WARNING

Before and after the installation of the timing chain tensioner, make sure that the alignment mark on the timing chain is not out of alignment.

d. After installing the timing chain tensioner, remove the stopper pin to release the tensioner. Make sure the tensioner is released.

e. To avoid chain-link skipping of the timing chain, do not move crankshaft or camshafts until the front cover is installed.

23. In the same way as for the RH bank, install the timing chain and associated parts on the LH bank.

24. Install the oil pump.

25. Install the oil pump drive spacer as follows:

a. Install so that the front mark on the front edge of the oil pump drive spacer faces the front of the engine.

b. Insert the oil pump drive spacer according to the directions of the crankshaft key and the two flat surfaces of the oil pump inner rotor.

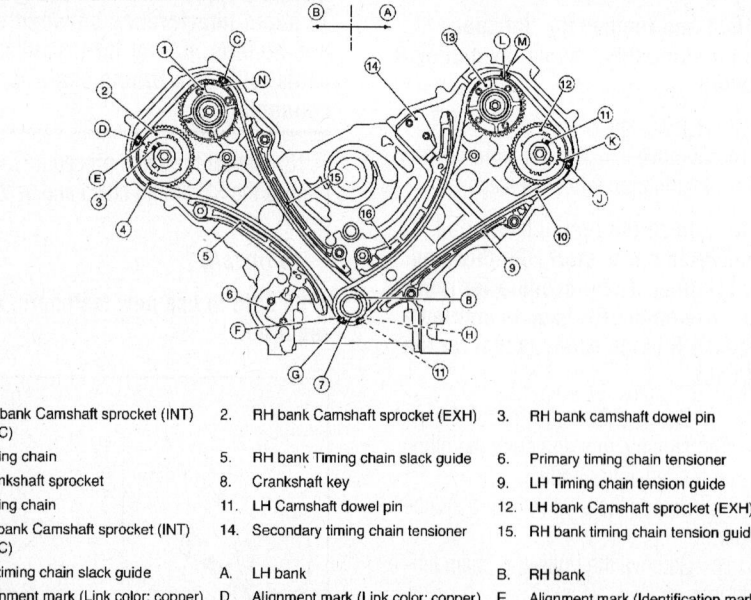

1. RH bank Camshaft sprocket (INT) (VTC)
2. RH bank Camshaft sprocket (EXH)
3. RH bank camshaft dowel pin
4. Timing chain
5. RH bank Timing chain slack guide
6. Primary timing chain tensioner
7. Crankshaft sprocket
8. Crankshaft key
9. LH Timing chain tension guide
10. Timing chain
11. LH Camshaft dowel pin
12. LH bank Camshaft sprocket (EXH)
13. LH bank Camshaft sprocket (INT) (VTC)
14. Secondary timing chain tensioner
15. RH bank timing chain tension guide
16. LH timing chain slack guide
A. LH bank
B. RH bank
C. Alignment mark (Link color: copper)
D. Alignment mark (Link color: copper)
E. Alignment mark (Identification mark)
F. Alignment mark for LH bank (Notch)
G. Alignment mark for LH bank (Link color: Yellow)
H. Alignment mark for RH bank (Link color: Yellow)
J. Alignment mark (Link color: copper)
K. Alignment mark (Identification mark)
L. Alignment mark (Identification mark)
M. Alignment mark (Link color: copper)
N. Alignment mark (Identification mark)

Fig. 131 Timing chains and related components—V8 engine

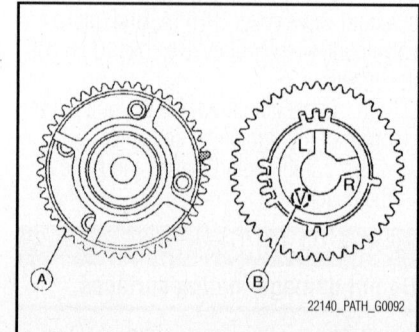

Fig. 133 Install the camshaft sprockets— V8 engine

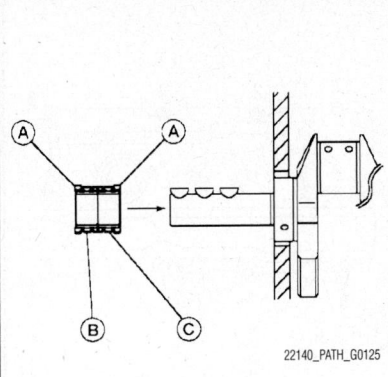

Fig. 134 Install the crankshaft sprockets for both banks—V8 engine

c. If the positional relationship does not allow the insertion, rotate the oil pump inner rotor to allow the oil pump drive spacer to be inserted.

26. Install the front oil seal using suitable tool.

☀ WARNING
Do not scratch or make burrs on the circumference of the oil seal.

27. Install the chain tensioner cover.
28. Install the front cover as follows:
    a. Install a new O-ring on the cylinder block.
    b. Apply liquid gasket as shown.
    c. Check again that the timing alignment marks on the timing chain and on each sprocket are aligned. Then install the front cover.
    d. Install the bolts in the numerical order shown.
    e. After tightening, re-tighten to the specified torque.

☀ WARNING
Be sure to wipe off any excessive liquid gasket leaking onto surface mating with oil pan.

29. Install the Intake valve control solenoid valve cover (RH) (A) and Intake valve control solenoid valve cover (LH) (B) as follows:
    a. Cross mark (C) that cannot be seen after assembly.
    b. Apply liquid gasket (D) as shown.

☀ WARNING
The start and end of the application of the liquid gasket should be crossed at a position that cannot be

seen after attaching the Intake valve control solenoid valve cover.

    c. Install the bolts in the numerical order shown.
30. Install the crankshaft pulley.
    a. Install the key of the crankshaft.
    b. Insert the pulley by lightly tapping it.

☀ WARNING
Do not tap pulley on the side surface where the belt is installed (outer circumference).

31. Tighten the crankshaft pulley bolt.
    a. Lock the crankshaft using suitable tool, then tighten the bolt.
    b. Perform the following steps for angular tightening:
    c. Apply engine oil onto the threaded parts of the bolt and seating area.
    d. Select the one most visible notch of the four on the bolt flange. Corresponding to the selected notch, put a alignment mark (such as paint) on the crankshaft pulley.
32. Rotate the crankshaft pulley in normal direction (clockwise when viewed from engine front) to check for parts interference.
33. Installation of the remaining components is in the reverse of order of removal.
34. Be sure to perform the reconnect/relearn procedures.

VALVE CYLINDER HEAD COVERS

REMOVAL & INSTALLATION

V6 Engine

Left Side
See Figures 135 through 137.

1. Before servicing the vehicle, refer to the Precautions.
2. Disconnect the negative battery cable.
3. Remove the engine cover. Remove the air cleaner assembly.
4. Remove engine cover.
5. Separate engine harness removing their brackets from rocker covers.
6. Remove harness bracket from cylinder head, if necessary.
7. Remove ignition coil.
8. Remove PCV hoses from rocker covers.
9. Remove oil filler cap from rocker cover (LH), if necessary.
10. Loosen bolts in reverse order of the tightening sequence.

11. Remove the retaining bolts. Discard the gasket.
12. Remove rocker cover gaskets from rocker covers.
13. Use scraper to remove all traces of liquid gasket from cylinder head and camshaft bracket (No. 1).

☀ WARNING
Do not scratch or damage the mating surface when cleaning off old liquid gasket.

To install:

➡ **Be sure to use new fasteners, as required.**

14. Apply liquid gasket to joint part among rocker cover, cylinder head and camshaft bracket (No. 1) as follows:
    a. The figure shows an example of LH side [zoomed in shows camshaft bracket (No. 1)].
    b. Apply liquid gasket to joint part of camshaft bracket "a" (No. 1) and cylinder head.
    c. Apply liquid gasket "b" to the figure "a" squarely.
15. Install new rocker cover gasket to rocker cover.
16. Install rocker cover.

➡ **Check to be sure rocker cover gasket is not dropped from installation groove of rocker cover.**

17. Tighten bolts in two steps in numerical order as shown.
    a. 1st step: 17 inch lbs. (1.96 Nm)
    b. 2nd step: 74 inch lbs. (8.33 Nm)
18. Install oil filer cap to rocker cover (left bank), if removed.
19. Install PCV hose.
    a. Insert PCV hose by 0.98 to 1.18 in. (25 to 30 mm) from connector end.
    b. When installing, be careful not to twist or come in contact with other parts.
20. Installation of the remaining components is in the reverse order of removal.
21. Be sure to perform the reconnect/relearn procedures.

Right Side
See Figure 138.

1. Before servicing the vehicle, refer to the Precautions.
2. Disconnect the negative battery cable.
3. Remove the engine cover. Remove the air cleaner assembly.

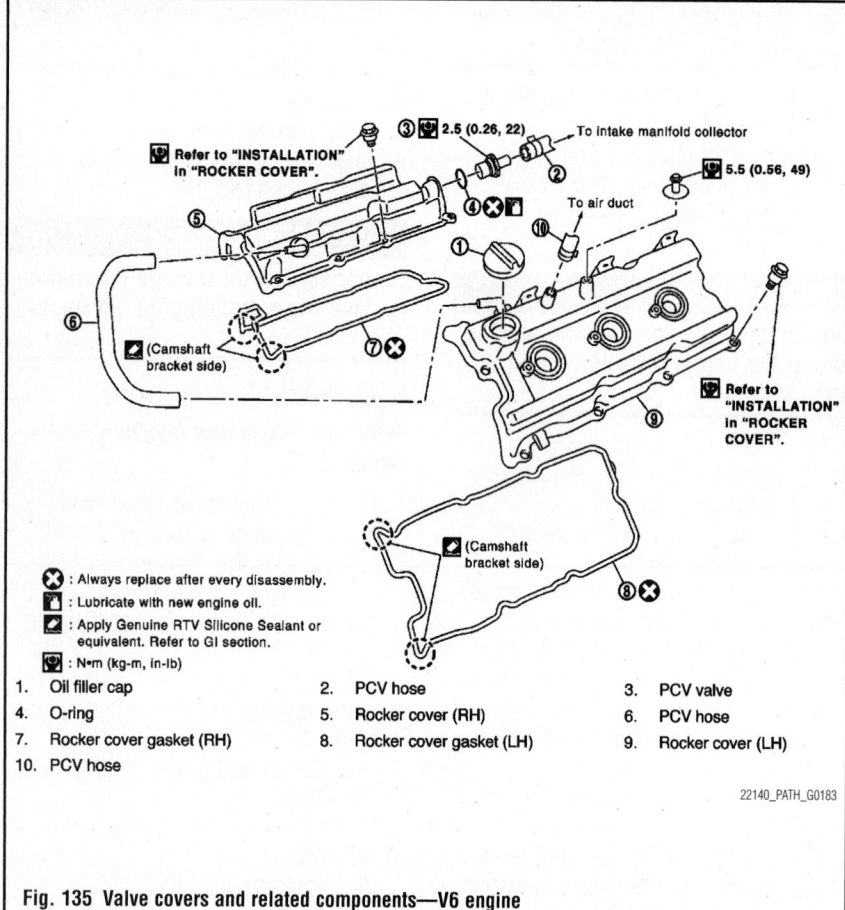

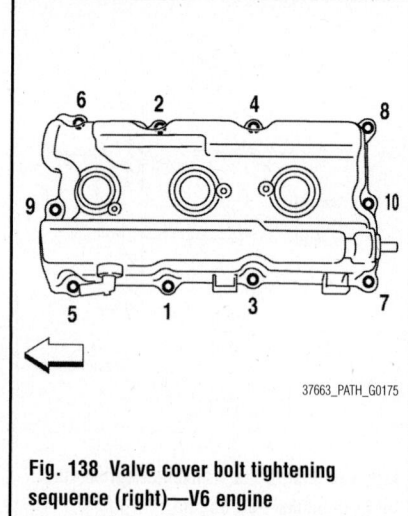

Fig. 138 Valve cover bolt tightening sequence (right)—V6 engine

9. Loosen bolts in reverse order as shown.

10. Remove rocker cover gaskets from rocker covers.

11. Use scraper to remove all traces of liquid gasket from cylinder head and camshaft bracket (No. 1).

✳✳ WARNING

Do not scratch or damage the mating surface when cleaning off old liquid gasket.

To install:

➡**Be sure to use new fasteners, as required.**

12. Apply liquid gasket to joint part among rocker cover, cylinder head and camshaft bracket (No. 1) as follows:

    a. Apply liquid gasket to joint part of camshaft bracket "a" (No. 1) and cylinder head.

    b. Apply liquid gasket "b" to the figure "a" squarely.

13. Install new rocker cover gasket to rocker cover.

14. Install rocker cover.

➡**Check to be sure rocker cover gasket is not dropped from installation groove of rocker cover.**

15. Tighten bolts in two steps in numerical order as shown.

    a. 1st step: 17 inch lbs. (1.96 Nm)

    b. 2nd step: 74 inch lbs. (8.33 Nm)

16. Install new O-ring and PCV valve to rocker cover (RH), if removed.

17. Install PCV hose.

    a. Insert PCV hose by 0.98 to 1.18 in. (25 to 30 mm) from connector end.

    b. When installing, be careful not to twist or come in contact with other parts.

✖ : Always replace after every disassembly.

⬡ : Lubricate with new engine oil.

▨ : Apply Genuine RTV Silicone Sealant or equivalent. Refer to GI section.

▧ : N•m (kg-m, in-lb)

1. Oil filler cap
2. PCV hose
3. PCV valve
4. O-ring
5. Rocker cover (RH)
6. PCV hose
7. Rocker cover gasket (RH)
8. Rocker cover gasket (LH)
9. Rocker cover (LH)
10. PCV hose

Fig. 135 Valve covers and related components—V6 engine

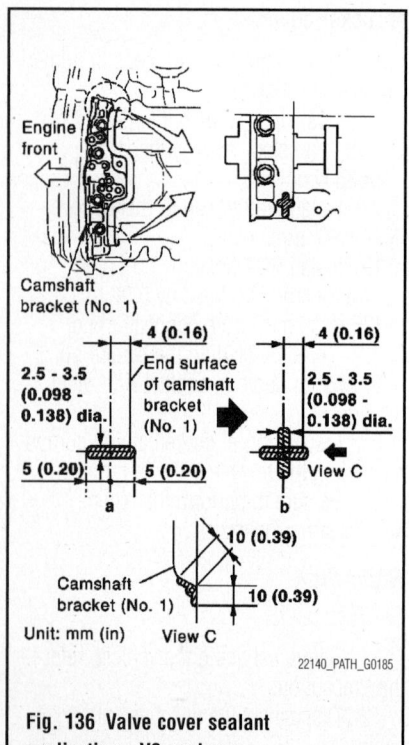

Fig. 136 Valve cover sealant application—V6 engine

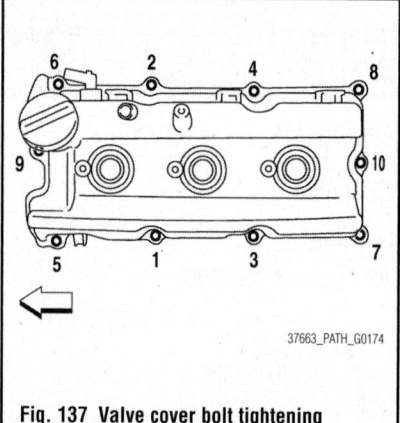

Fig. 137 Valve cover bolt tightening sequence (left)—V6 engine

✳✳ CAUTION

Perform this step when engine is cold.

4. Separate engine harness removing their brackets from rocker covers.

5. Remove harness bracket from cylinder head (RH).

6. Remove ignition coil.

7. Remove PCV hoses from rocker cover.

8. Remove PCV valve and O-ring from rocker cover (RH), if necessary.

18. Installation of the remaining components is in the reverse order of removal.

19. Be sure to perform the reconnect/relearn procedures.

V8 Engine

See Figures 139 through 141.

1. Before servicing the vehicle, refer to the Precautions.

2. Disconnect the negative battery cable.

3. Remove the engine room cover.

4. Remove the air duct and resonator assembly.

5. Move the harness on the upper rocker cover and its peripheral aside.

6. Remove the ignition coils.

7. Remove the PCV hose from the PCV control valves.

8. Loosen the bolts in reverse order of the tightening sequence (LH) (A) or (RH) (B).

✳✳ WARNING

Do not hold the rocker cover (RH) (B) by the oil filler neck.

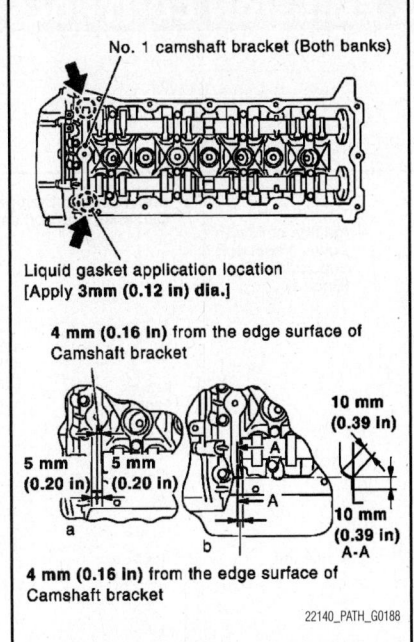

Fig. 140 Valve cover sealant application points—V8 engine

To install:

➡**Be sure to use new fasteners, as required.**

9. Apply liquid gasket to the joint part of the cylinder head and camshaft bracket following the steps below.

➡**Illustration shows an example of (LH) side (zoomed in shows No. 1 camshaft bracket).**

   a. Apply liquid gasket to the joint part of No. 1 camshaft bracket and cylinder head "a".

   b. Apply liquid gasket 90° "b" to illustration "a".

10. Install the rocker cover (LH) (A) or (RH) (B).

   a. Make sure the new rocker cover gasket is installed in the groove of the rocker cover (LH) (A) or (RH) (B).

   b. Tighten the bolts in two steps in the numerical order shown.

   c. 1st step: 18 inch lbs. (2.0 Nm)

   d. 2nd step: 73 inch lbs. (8.3 Nm)

✳✳ WARNING

Do not hold the rocker cover (RH) (B) by the oil filler neck.

11. Install the PCV hoses.

➡**Remove foreign materials from inside the hose using compressed air.**

   a. Insert PCV hose by 0.98 to 1.18 in. (25 to 30 mm) from connector end.

12. Installation of the remaining components is in the reverse order of removal.

13. Be sure to perform the reconnect/relearn procedures.

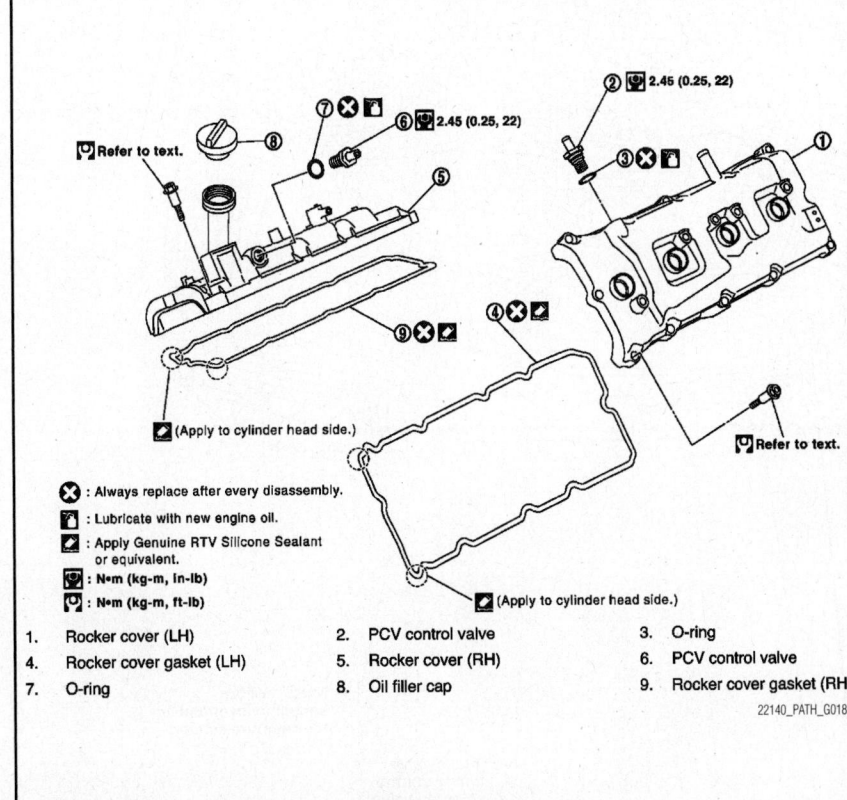

1. Rocker cover (LH)
2. PCV control valve
3. O-ring
4. Rocker cover gasket (LH)
5. Rocker cover (RH)
6. PCV control valve
7. O-ring
8. Oil filler cap
9. Rocker cover gasket (RH)

✘ : Always replace after every disassembly.

🛢 : Lubricate with new engine oil.

✐ : Apply Genuine RTV Silicone Sealant or equivalent.

⬡ : N•m (kg-m, in-lb)

⬡ : N•m (kg-m, ft-lb)

Fig. 139 Valve covers and related components—V8 engine

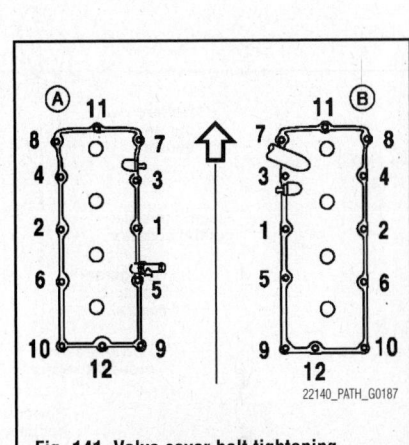

Fig. 141 Valve cover bolt tightening sequence—V8 engine

ENGINE PERFORMANCE & EMISSION CONTROLS

COMPONENT LOCATIONS

See Figures 142 through 144.

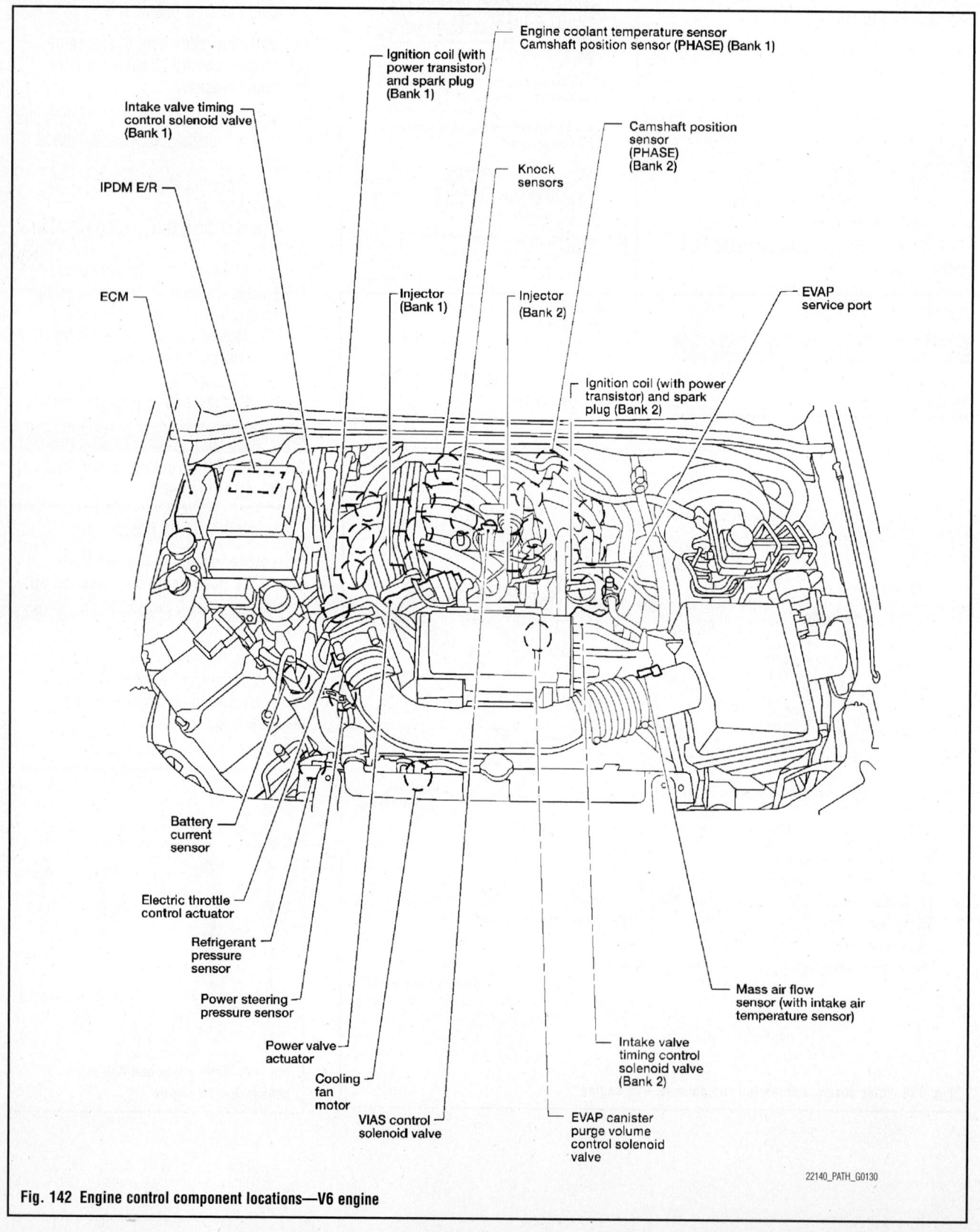

Fig. 142 Engine control component locations—V6 engine

22140_PATH_G0130

ACCELERATOR PEDAL POSITION (APP) SENSOR

LOCATION

See Figure 145.

The Accelerator Pedal Position (APP) sensor is installed on the upper end of the accelerator pedal assembly.

REMOVAL & INSTALLATION

See Figure 146.

1. Before servicing the vehicle, refer to the Precautions.

➡If equipped with adjustable pedals move the pedals to the front most position.

2. Disconnect the negative battery cable.
3. Disconnect the Accelerator Pedal Position (APP) sensor electrical connector.
   a. Pull the connector lock back to unlock the connector from the APP sensor.
   b. Pull up on the connector to disconnect it from the APP sensor.
4. Remove the two upper and one lower accelerator pedal nuts.
5. Remove the accelerator pedal assembly.

✹✹ CAUTION

Do not disassemble the accelerator pedal assembly. Do not remove the APP sensor from the accelerator pedal bracket. Avoid damage from dropping the accelerator pedal assembly during handling. Keep the accelerator pedal assembly away from water.

To install:

➡Be sure to use new fasteners, as required.

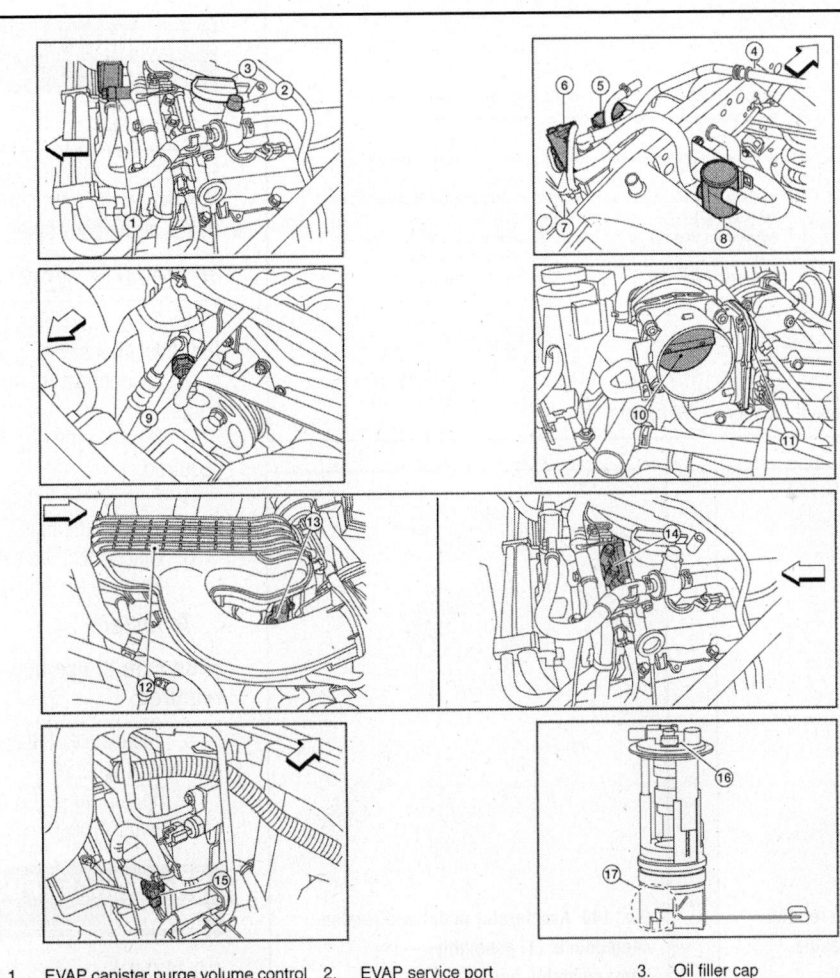

1. EVAP canister purge volume control solenoid valve (view with engine cover removed)	2. EVAP service port	3. Oil filler cap
4. Fuel filler pipe (top of frame view)	5. EVAP control system pressure sensor	6. EVAP canister vent control valve
7. EVAP canister	8. Drain filter	9. Power steering pressure sensor
10. Throttle valve (view with intake air duct removed)	11. Electric throttle control actuator	12. Intake manifold collector
13. Intake valve timing control solenoid valve (bank 1)	14. Intake valve timing control solenoid valve (bank 2) (view with engine cover and intake air duct removed)	15. Cooling fan motor harness connector (view with battery removed)
16. Fuel pump, fuel level sensor unit and fuel filter	17. Fuel pressure regulator	
⇦ : Front		

37663_PATH_G0204

Fig. 143 EVAP system component locations—V6 engine

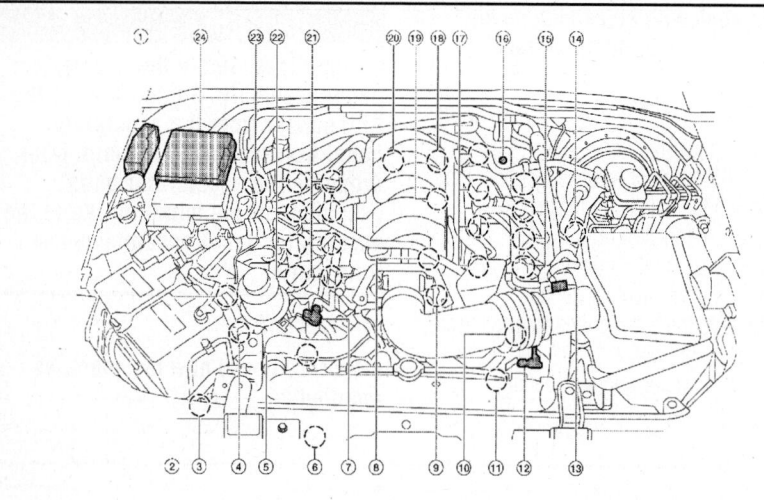

1. ECM
2. Battery current sensor
3. Refrigerant pressure sensor
4. Power steering pressure sensor
5. Intake valve timing control solenoid valve (bank 2)
6. Cooling fan motor
7. Intake valve timing control position sensor (bank 2)
8. Engine coolant temperature sensor
9. Electric throttle control actuator
10. Intake valve timing control position sensor (bank 1)
11. Intake valve timing control solenoid valve (bank 1)
12. Camshaft position sensor (PHASE)
13. Mass air flow sensor (with intake air temperature sensor)
14. A/F sensor 1 (bank 1)
15. Ignition coil (with power transistor) and spark plug (bank 1)
16. EVAP service port
17. Fuel injector (bank 1)
18. Knock sensor (bank 1)
19. EVAP canister purge volume control solenoid valve
20. Knock sensor (bank 2)
21. Fuel injector (bank 2)
22. Ignition coil (with power transistor) and spark plug (bank 2)
23. A/F sensor 1 (bank 2)
24. IPDM E/R

22140_PATH_G0131

Fig. 144 Engine control component locations—V8 engine

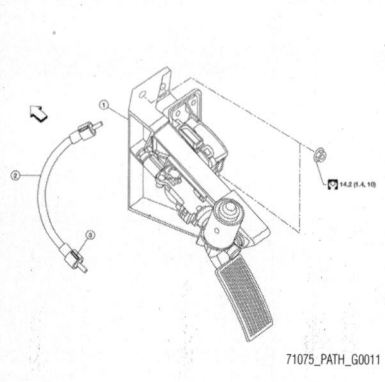

Fig. 145 Accelerator pedal and related components (1) assembly, (2) cable, (3) tab

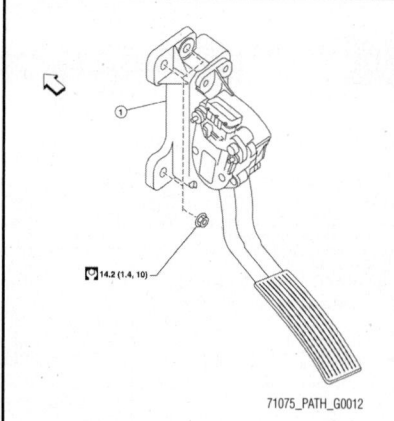

Fig. 146 Accelerator pedal and related components (1) assembly— non-adjustable pedals

6. Installation is the reverse of the removal procedure.
7. Be sure to perform the reconnect/relearn procedures.

AIR-FUEL RATIO (A/F) SENSOR

LOCATION

See Figures 147 and 148.

The Air Fuel Ratio (AF) sensors are located in the left and right exhaust manifold assembly.

REMOVAL & INSTALLATION

At this time the manufacturer does not provide removal and installation procedures for this component. The following procedure is a guideline and may differ from the vehicle you are servicing.

1. Before servicing the vehicle, refer to the Precautions.
2. Disconnect the negative battery cable.
3. Remove the engine cover and the air cleaner case, as required.

4. Remove engine undercover, if equipped and required..
5. Raise and support the vehicle, as required.
6. Disconnect harness connector.
7. Remove the sensor from its mounting.

To install:

➡ Be sure to use new fasteners, as required.

8. Installation is the reverse of the removal procedure.
9. Be sure to perform the reconnect/relearn procedures.

CAMSHAFT POSITION (CMP) SENSOR

LOCATION

Camshaft Position (CMP) sensors are located on the right and left bank cylinder heads at the back side.

REMOVAL & INSTALLATION

1. Before servicing the vehicle, refer to the Precautions.
2. Disconnect the negative battery cable.
3. Remove the engine cover as required.
4. Remove air intake duct, as required.
5. Loosen the fixing bolt of the sensor.
6. Disconnect the electrical connector.
7. Remove the bolt securing the sensor.

To install:

➡ Be sure to use new fasteners, as required.

8. Installation is the reverse of the removal procedure.
9. Be sure to perform the reconnect/relearn procedures.

CRANKSHAFT POSITION (CKP) SENSOR

LOCATION

The Crankshaft Position (CKP) sensor (POS) is located on the A/T assembly facing the gear teeth (cogs) of the signal plate.

REMOVAL & INSTALLATION

See Figure 149.

1. Before servicing the vehicle, refer to the Precautions.
2. Disconnect the negative battery cable.
3. Raise and support the vehicle safely.
4. Loosen the fixing bolt of the sensor.

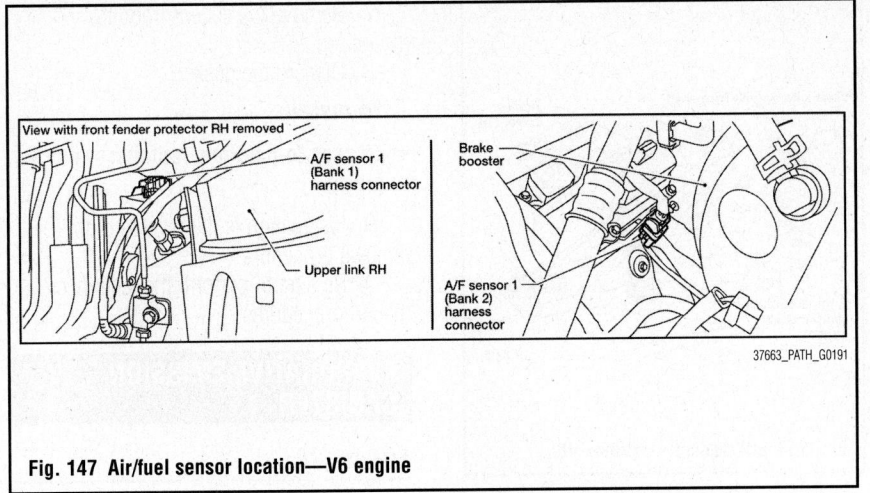

Fig. 147 Air/fuel sensor location—V6 engine

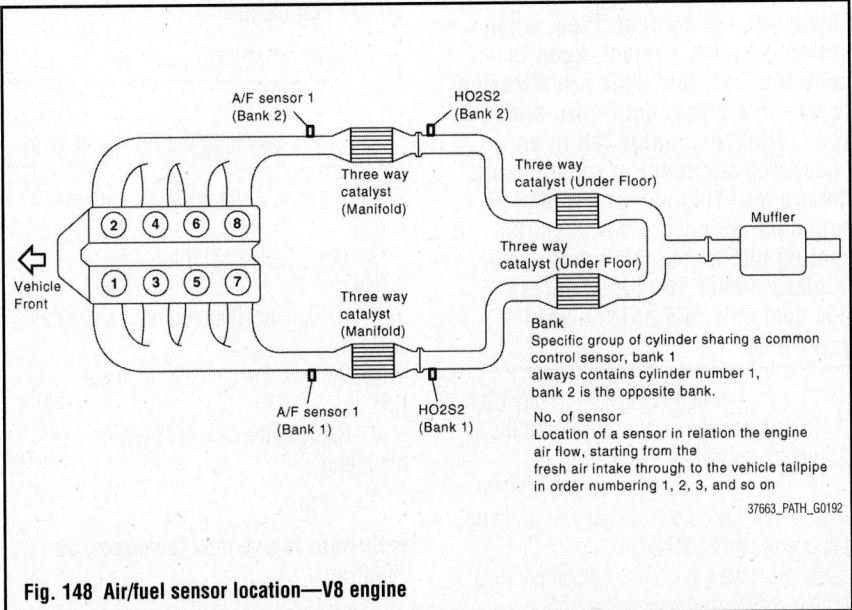

Fig. 148 Air/fuel sensor location—V8 engine

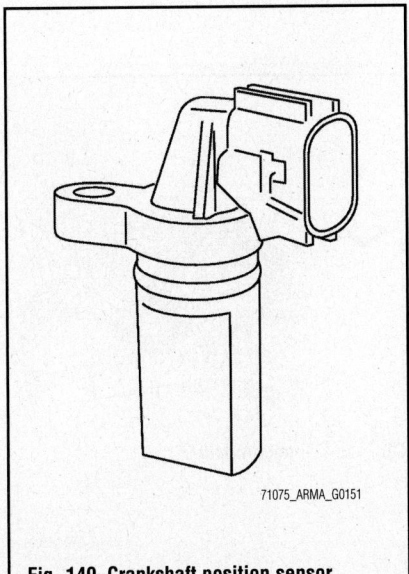

Fig. 149 Crankshaft position sensor

5. Disconnect Crankshaft Position (CKP) sensor (POS) harness connector.

6. Remove the sensor.

7. Visually check the sensor for chipping.

To install:

➡**Be sure to use new fasteners, as required.**

8. Installation is the reverse of the removal procedure.

9. Be sure to perform the reconnect/relearn procedures.

ELECTRONIC CONTROL MODULE (ECM)

LOCATION

On the V6 engine the Electronic Control Module (ECM) is located in the engine room passenger side behind reservoir tank.

On the V8 engine the Electronic Control Module (ECM) is located in the engine room passenger side behind the battery.

REMOVAL & INSTALLATION

See Figures 150 through 152.

At this time the manufacturer does not provide removal and installation procedures for this component. The following procedure is a guideline and may differ from the vehicle you are servicing.

1. Before servicing the vehicle, refer to the Precautions.

2. Disconnect the negative battery cable.

3. Remove the engine cover, if required.

4. Remove the air cleaner assembly, if required.

5. Remove the necessary components to gain access to the ECM.

6. Disconnect the electrical connectors.

7. Remove the ECM retaining screws/bolts/clips etc.

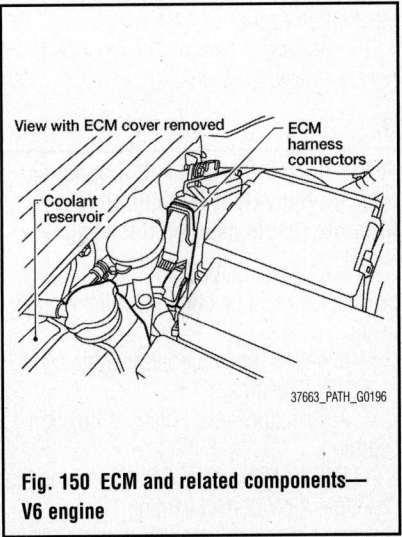

Fig. 150 ECM and related components—V6 engine

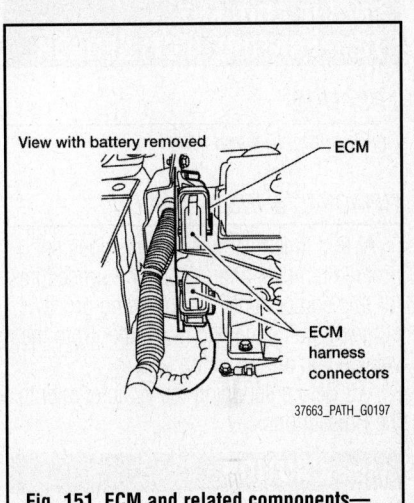

Fig. 151 ECM and related components—V8 engine

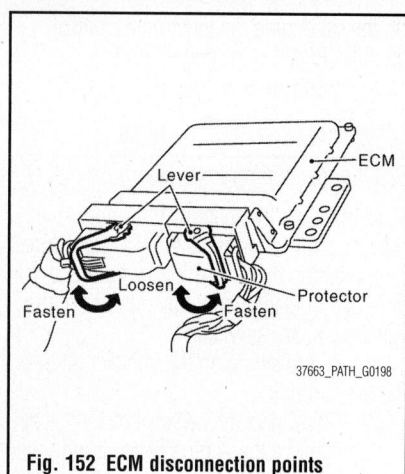

Fig. 152 ECM disconnection points

8. Remove the component from its mounting.

To install:

➡**Be sure to use new fasteners, as required.**

9. Installation is the reverse of the removal procedure.

10. Be sure to perform the reconnect/relearn procedures.

RESET

➡**The Nissan CONSULT III diagnostic tool, or equivalent is required to perform the resets used on this vehicle.**

When replacing the ECM, the following procedures must be performed in the order listed.

- Initialize The Immobilizer System
- VIN Registration
- Accelerator Pedal Released Position Learning
- Throttle Valve Closed Position Learning
- Idle Air Volume Learning

ENGINE COOLANT TEMPERATURE (ECT) SENSOR

LOCATION
See Figures 153 and 154.

REMOVAL & INSTALLATION

At this time the manufacturer does not provide removal and installation procedures for this component. The following procedure is a guideline and may differ from the vehicle you are servicing.

1. Before servicing the vehicle, refer to the Precautions.

✳✳ CAUTION

Never open, service or drain the radiator or cooling system when hot;

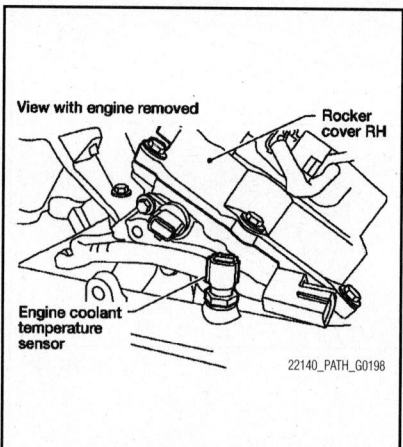

Fig. 153 ECT sensor location—V6

serious burns can occur from the steam and hot coolant. Also, when draining engine coolant, keep in mind that cats and dogs are attracted to ethylene glycol antifreeze and could drink any that is left in an uncovered container or in puddles on the ground. This will prove fatal in sufficient quantities. Always drain coolant into a sealable container. Coolant should be reused unless it is contaminated or is several years old.

2. Disconnect the negative battery cable.

3. Remove the engine cover and the air cleaner assembly, as required.

4. Drain the coolant to an acceptable level, below the sensor. Be sure to properly dispose of used coolant.

5. Remove the necessary components to gain access to the sensor.

6. Disconnect the electrical connector.

7. Remove the sensor from its mounting.

8. Discard the gasket.

To install:

➡**Be sure to use new fasteners, as required.**

9. Installation is the reverse of the removal procedure.

10. Be sure to perform the reconnect/relearn procedures.

EVAPORATIVE EMISSION CANISTER

LOCATION

This component is located under the vehicle near the fuel tank.

REMOVAL & INSTALLATION
See Figure 155.

1. Before servicing the vehicle, refer to the Precautions.

2. Disconnect the negative battery cable.

3. Raise and support the vehicle safely.

4. Disconnect the required connectors and hoses.

5. Remove the canister retaining bolt.

6. Remove the canister from its mounting.

To install:

➡**Be sure to use new fasteners, as required.**

7. Installation is the reverse of the removal procedure.

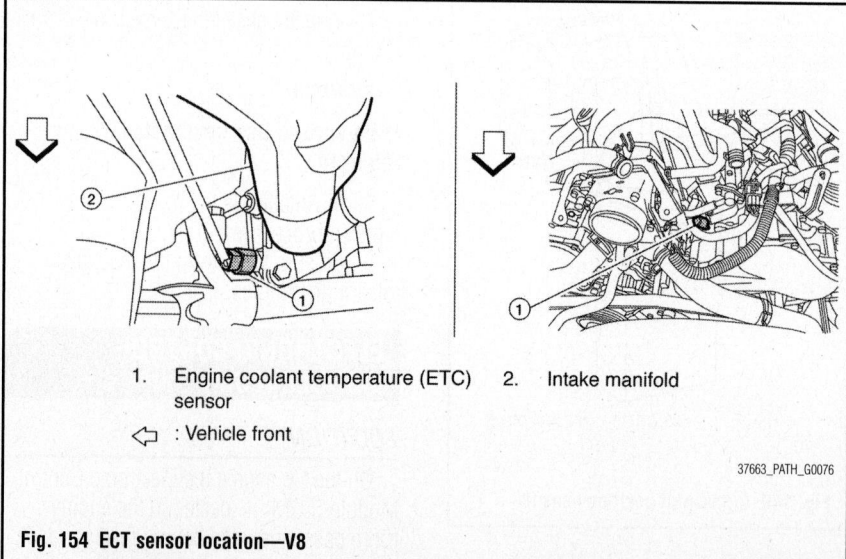

| 1. | Engine coolant temperature (ETC) sensor | 2. | Intake manifold |

⇦ : Vehicle front

Fig. 154 ECT sensor location—V8

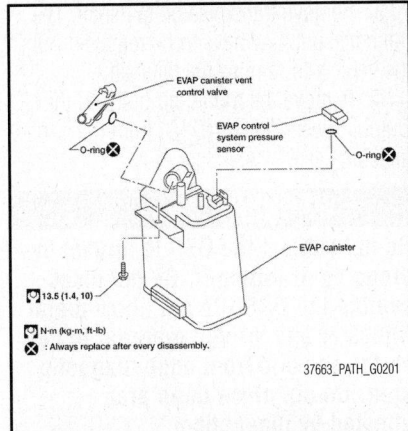

Fig. 155 EVAP canister and related components

HEATED OXYGEN (HO2S) SENSOR

LOCATION

The Heated Oxygen (HO2S) sensors are located after the exhaust manifold converter assembly, in the lower part of the exhaust system.

REMOVAL & INSTALLATION

At this time the manufacturer does not provide removal and installation procedures for this component. The following procedure is a guideline and may differ from the vehicle you are servicing.

1. Before servicing the vehicle, refer to the Precautions.
2. Disconnect the negative battery cable.
3. Remove the engine cover. Remove air cleaner case and air duct.
4. Remove engine undercover, if equipped and as required..
5. Raise and support the vehicle, as required.
6. Disconnect harness connector.
7. Remove the sensor from its mounting.

To install:

➡**Be sure to use new fasteners, as required.**

8. Installation is the reverse of the removal procedure.
9. Be sure to perform the reconnect/relearn procedures.

INTAKE AIR TEMPERATURE (IAT)/MASS AIRFLOW (MAF) SENSOR

LOCATION

See Figure 156.

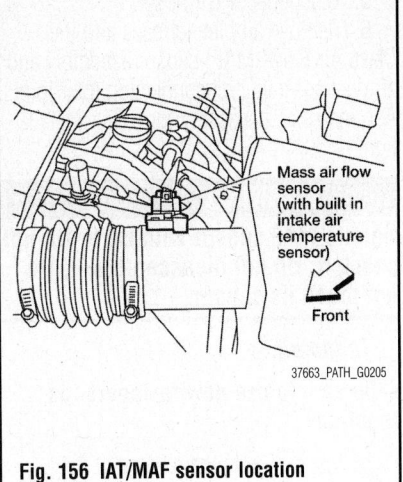

Fig. 156 IAT/MAF sensor location

The Intake Air Temperature (IAT) sensor is built-into Mass Air Flow (MAF) sensor.

REMOVAL & INSTALLATION

1. Before servicing the vehicle, refer to the Precautions.
2. Disconnect the negative battery cable.
3. Remove the engine cover.
4. Disconnect harness connector from mass air flow sensor.
5. Disconnect PCV hose.
6. Remove air cleaner case/Mass Air Flow (MAF) sensor assembly and air duct assembly disconnecting their joints. Add marks as necessary for easier installation.

❋❋ WARNING

Handle MAF sensor with care. Do not shock it. Do not disassemble it. Do not touch its sensor.

To install:

➡**Be sure to use new fasteners, as required.**

7. Installation is the reverse of the removal procedure.
8. Be sure to perform the reconnect/relearn procedures.

KNOCK SENSOR (KS)

LOCATION

See Figures 157 and 158.

The Knock (KS) sensors are mounted under the intake manifold on the cylinder block.

REMOVAL & INSTALLATION

V6 Engine

At this time the manufacturer does not provide removal and installation procedures

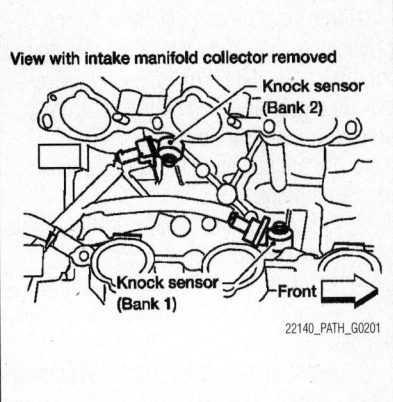

Fig. 157 Knock Sensor (KS) location—V6 engine

for this component. The following procedure is a guideline and may differ from the vehicle you are servicing.

1. Before servicing the vehicle, refer to the Precautions.
2. Disconnect the negative battery cable.
3. Remove the engine cover. Remove the air cleaner assembly.
4. Remove the upper intake manifold and lower intake manifold. Discard the gaskets.
5. Remove the sensor electrical connectors.
6. Remove the sensor from its mounting.

To install:

➡**Be sure to use new fasteners, as required.**

7. Installation is the reverse of the removal procedure.
8. Be sure to perform the reconnect/relearn procedures.

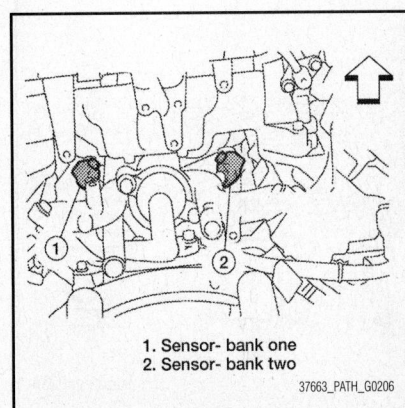

Fig. 158 Knock Sensor (KS) location (view with engine removed from vehicle)—V8 engine

V8 Engine

At this time the manufacturer does not provide removal and installation procedures for this component. The following procedure is a guideline and may differ from the vehicle you are servicing.

1. Before servicing the vehicle, refer to the Precautions.
2. Disconnect the negative battery cable.
3. Remove the engine cover.
4. Remove the air cleaner assembly.
5. Remove the intake manifold. Discard the gaskets.
6. Remove the sensor electrical connectors.
7. Remove the sensor from its mounting.

To install:

➡**Be sure to use new fasteners, as required.**

8. Installation is the reverse of the removal procedure.
9. Be sure to perform the reconnect/relearn procedures.

MASS AIR FLOW (MAF) SENSOR

LOCATION

See Figure 159.

The Mass Air Flow (MAF) sensor is positioned in the air cleaner assembly.

REMOVAL & INSTALLATION

1. Before servicing the vehicle, refer to the Precautions.
2. Disconnect the negative battery cable.
3. Remove the engine cover assembly.
4. Disconnect harness connector from mass air flow sensor.

5. Disconnect PCV hose.
6. Remove air cleaner case and the Mass Air Flow (MAF) sensor assembly and air duct assembly disconnecting their joints. Add marks as necessary for easier installation.

✲✲ WARNING

Handle MAF sensor with care. Do not shock it. Do not disassemble it. Do not touch its sensor.

To install:

➡**Be sure to use new fasteners, as required.**

7. Installation is the reverse of the removal procedure.
8. Be sure to perform the reconnect/relearn procedures.

OUTPUT SHAFT SPEED (OSS) SENSOR

LOCATION

See Figure 160.

REMOVAL & INSTALLATION

At this time the manufacturer does not provide removal and installation procedures for this component. The following procedure is a guideline and may differ from the vehicle you are servicing.

➡**The transmission tail shaft will have to be removed to gain access to this sensor. The transmission may have to be removed from the vehicle to remove the tailshaft.**

1. Before servicing the vehicle, refer to the Precautions.
2. Disconnect the negative battery cable.

3. Remove transmission tail shaft. The transmission may have to be removed from the vehicle to remove the tailshaft.
4. Remove the mounting bolt and the Output Speed Sensor (OSS) from the transmission case.

✲✲ WARNING

Do not subject the OSS to impact by dropping or hitting it. Do not disassemble the OSS. Do not allow metal filings or any foreign material to get on the sensors front edge magnetic area. Do not place in an area affected by magnetism.

To install:

➡**Be sure to use new fasteners, as required.**

5. Installation is the reverse of the removal procedure.
6. Be sure to perform the reconnect/relearn procedures.

THROTTLE POSITION SENSOR (TPS)

LOCATION

See Figure 161.

The Throttle Position (TPS) sensor is integral to the electric Throttle Control actuator. The Throttle Control actuator is mounted at the front of the intake manifold.

REMOVAL & INSTALLATION

At this time the manufacturer does not provide removal and installation procedures for this component. The following procedure is a guideline and may differ from the vehicle you are servicing.

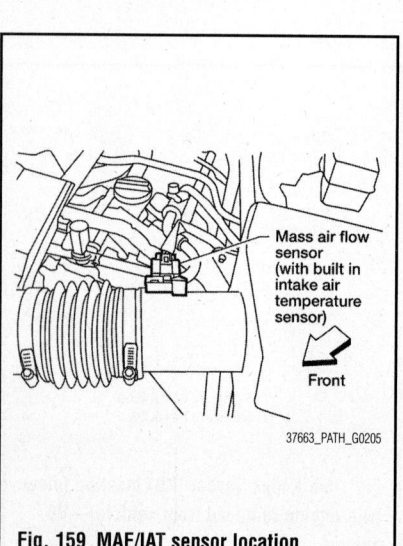

Mass air flow sensor (with built in intake air temperature sensor)

Front

37663_PATH_G0205

Fig. 159 MAF/IAT sensor location

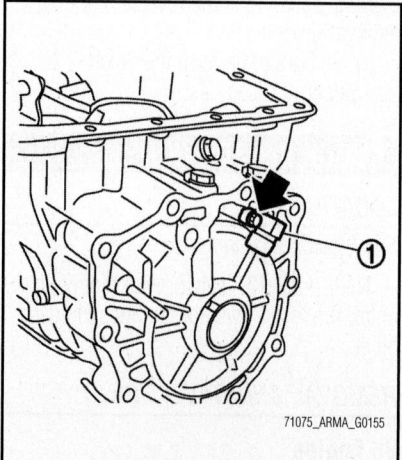

71075_ARMA_G0155

Fig. 160 Output shaft speed sensor location

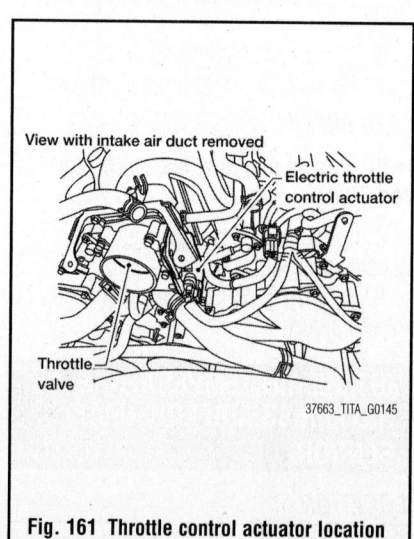

View with intake air duct removed

Electric throttle control actuator

Throttle valve

37663_TITA_G0145

Fig. 161 Throttle control actuator location

1. Before servicing the vehicle, refer to the Precautions.

2. Disconnect the negative battery cable.

3. Drain the cooling system, as required. Be sure to properly dispose of used engine coolant.

4. Remove the engine room cover. Remove the air intake duct.

5. Disconnect harness connector.

6. Disconnect water hoses, as required.

7. Loosen the throttle body assembly mounting bolts in reverse order of the tightening sequence.

To install:

➡**Be sure to use new fasteners, as required.**

8. Install the throttle body assembly with a new gasket.

9. Tighten the mounting bolts in sequence to 74 inch lbs. (8.4 Nm).

10. Reconnect the water hose.

11. Reconnect the harness connector.

12. Reconnect the air intake duct.

13. Fill the cooling system with the proper grade and type engine coolant.

14. Be sure to perform the reconnect/relearn procedures.

TRANSMISSION FLUID TEMPERATURE (TFT) SENSOR

LOCATION

See Figure 162.

REMOVAL & INSTALLATION

1. Before servicing the vehicle, refer to the Precautions.

2. Disconnect the negative battery cable.

3. Remove the transmission oil pan and oil pan gasket.

➡**Check for foreign materials in the oil pan to help determine any cause of malfunction. If the A/T fluid is very dark, smells burned, or contains foreign particles, the frictional material (clutches, band) may need replacement. A tacky film that will not wipe clean indicates varnish build up. Varnish can cause valves, servo, and clutches to stick and can inhibit pump pressure.**

4. If frictional material is detected, perform Automatic Transmission (A/T) fluid cooler cleaning.

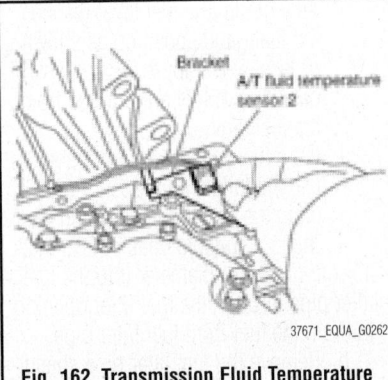

37671_EQUA_G0262

Fig. 162 Transmission Fluid Temperature (TFT) sensor location

5. Disconnect the A/T fluid temperature sensor 2 connector.

❋❋ WARNING

Do not damage the connector.

6. Disconnect the output speed sensor connector.

7. Straighten the terminal clip to free the output speed sensor harness.

8. Remove the bolts from the control valve with the TCM.

9. Remove the control valve with the TCM from the transmission case.

❋❋ WARNING

When removing, be careful with the manual valve notch and manual plate height. Remove it vertically.

10. Remove the A/T fluid temperature sensor 2 with the bracket from the control valve with the TCM.

To install:

➡**Be sure to use new fasteners, as required.**

11. Installation is the reverse of the removal procedure.

12. Replace the O-ring with a new one and lubricate it with transmission fluid.

13. Install the transmission oil pan and tighten the bolts to 70 inch lbs. (8 Nm) in the sequence shown.

FUEL   GASOLINE FUEL INJECTION SYSTEM

FUEL SYSTEM SERVICE PRECAUTIONS

Safety is the most important factor when performing not only fuel system maintenance but any type of maintenance. Failure to conduct maintenance and repairs in a safe manner may result in serious personal injury or death. Maintenance and testing of the vehicle's fuel system components can be accomplished safely and effectively by adhering to the following rules and guidelines.

• To avoid the possibility of fire and personal injury, always disconnect the negative battery cable unless the repair or test procedure requires that battery voltage be applied.

• Always relieve the fuel system pressure prior to disconnecting any fuel system component (injector, fuel rail, pressure regulator, etc.), fitting or fuel line connection. Exercise

extreme caution whenever relieving fuel system pressure to avoid exposing skin, face and eyes to fuel spray. Please be advised that fuel under pressure may penetrate the skin or any part of the body that it contacts.

• Always place a shop towel or cloth around the fitting or connection prior to loosening to absorb any excess fuel due to spillage. Ensure that all fuel spillage (should it occur) is quickly removed from engine surfaces. Ensure that all fuel soaked cloths or towels are deposited into a suitable waste container.

• Always keep a dry chemical (Class B) fire extinguisher near the work area.

• Do not allow fuel spray or fuel vapors to come into contact with a spark or open flame.

• Always use a back-up wrench when loosening and tightening fuel line connection fittings. This will prevent unnecessary stress and torsion to fuel line piping.

• Always replace worn fuel fitting O-rings with new Do not substitute fuel hose or equivalent where fuel pipe is installed.

Before servicing the vehicle, make sure to also refer to the precautions in the beginning of this section as well.

RELIEVING FUEL SYSTEM PRESSURE

WITH CONSULT-III®

1. Turn ignition switch **ON**.

2. Perform "FUEL PRESSURE RELEASE" in "WORK SUPPORT" mode with CONSULT-III®.

3. Start engine.

4. After engine stalls, turn over the engine two or three times to release all fuel pressure.

5. Turn ignition switch **OFF**.

WITHOUT CONSULT-III®

1. Before servicing the vehicle, refer to the Precautions.

➡This vehicle uses quick connect fittings. Be sure to properly relieve the fuel system pressure before disconnecting any of these fittings. Always replace O-rings and clamps with new ones. Do not bend, twist or kink hoses when they are being removed or installed. Be sure that the clamp screw does not contact adjacent parts. When tightening the high pressure rubber hose clamp make sure the clamp end is 0.12 inch from the hose end. After connecting these fittings make sure that the connectors are secure. Check for fuel leakage at these connections turn the ignition key to the ON position (do not start the engine), correct as required. Start the engine, raise the idle, and verify that there are no fuel leaks, correct as required.

2. Remove fuel pump fuse located in IPDM E/R.
3. Start engine.
4. After engine stalls, crank it two or three times to release all fuel pressure.
5. Turn ignition switch OFF.
6. Reinstall fuel pump fuse after servicing fuel system.
7. Disconnect the negative battery cable.

FUEL PUMP MODULE

REMOVAL & INSTALLATION
See Figure 163.

1. Before servicing the vehicle, refer to the Precautions.

➡This vehicle uses quick connect fittings. Be sure to properly relieve the fuel system pressure before disconnecting any of these fittings. Always replace O-rings and clamps with new ones. Do not bend, twist or kink hoses when they are being removed or installed. Be sure that the clamp screw does not contact adjacent parts. When tightening the high pressure rubber hose clamp make sure the clamp end is 0.12 inch from the hose end. After connecting these fittings make sure that the connectors are secure. Check for fuel leakage at these connections turn the ignition key to the ON position (do not start the engine), correct as required. Start the engine, raise the

idle, and verify that there are no fuel leaks, correct as required.

2. Properly relieve the fuel system pressure.
3. Disconnect the negative battery cable.
4. Remove the fuel filler cap to release the pressure from inside the fuel tank.
5. Remove the LH rear wheel and tire.
6. Check the fuel level on level gauge. If the fuel gauge indicates more than the level as shown (full or almost full), drain the fuel from the fuel tank until the fuel gauge indicates the level as shown, or less.

➡Fuel will be spilled when removing the fuel level sensor, fuel filter, and fuel pump assembly for the fuel level is above the fuel level sensor, fuel filter, and fuel pump assembly fuel tank opening.

- As a guide, the fuel level reaches the fuel gauge position as shown, or less, when approximately 4 US gal (15L) of fuel are drained from the fuel tank.
- If the fuel pump does not operate, use the following procedure to drain the fuel to the specified level.
  a. Insert a suitable hose of less than 15 mm (0.59 in.) diameter into the fuel filler pipe through the fuel filler opening to drain the fuel from fuel filler pipe.
  b. Remove the fuel filler pipe shield.
  c. Disconnect the fuel filler hose from the fuel filler pipe.
  d. Insert a suitable hose into the fuel tank through the fuel filler hose to drain the fuel from the fuel tank.

7. Release the fuel pressure from the fuel lines.
8. Disconnect the lower fuel filler hose from the fuel tank, the EVAP hose, and the vent pipe quick connector.
  a. Disconnect the fuel feed hose from the molded clip in the side of the fuel tank.
  - Disconnect the quick connector as follows:
  - Hold the sides of the connector, push in the tabs and pull out the tube.
  - If the connector and the tube are stuck together, push and pull several times until they start to move. Then disconnect them by pulling.

✳✳ WARNING
The quick connector can be disconnected when the tabs are completely depressed. Do not twist the quick

connector more than necessary. Do not use any tools to disconnect the quick connector. Keep the resin tube away from heat. Be especially careful when welding near the tube. Do not bend or twist the resin tube during connection.

9. Remove the four bolts and remove the fuel tank shield.
10. Remove the driveshaft shaft.
11. Support the fuel tank using a suitable lift jack.
12. Remove the three fuel tank strap bolts while supporting the fuel tank with a suitable lift jack.
13. Remove the fuel tank straps and slowly lower the fuel tank to access the top of the fuel level sensor, fuel filter and fuel pump assembly.

✳✳ CAUTION
Do not lower the fuel tank too far to prevent damage to the fuel feed hose and the fuel level sensor, fuel filter and fuel pump assembly connector.

14. Disconnect the fuel level sensor, fuel filter, and fuel pump assembly electrical connector, and the fuel feed hose.
  a. Disconnect the quick connector as follows:
  - Hold the sides of the connector, push in the tabs and pull out the tube.
  - If the connector and the tube are stuck together, push and pull several times until they start to move. Then disconnect them by pulling.

✳✳ WARNING
The quick connector can be disconnected when the tabs are completely depressed. Do not twist the quick connector more than necessary. Do not use any tools to disconnect the quick connector. Do not bend or twist the resin tube during connection.

15. Lower the fuel tank using a suitable lift jack and remove the fuel tank.
16. Disconnect the EVAP hose from the fuel pump and remove the EVAP hose from the molded clip in the top of the fuel tank.
17. Remove the lock ring.
18. Remove the fuel level sensor, fuel filter, and fuel pump assembly.

✳✳ WARNING
Do not bend the float arm during removal. Avoid impacts such as dropping when handling the components.

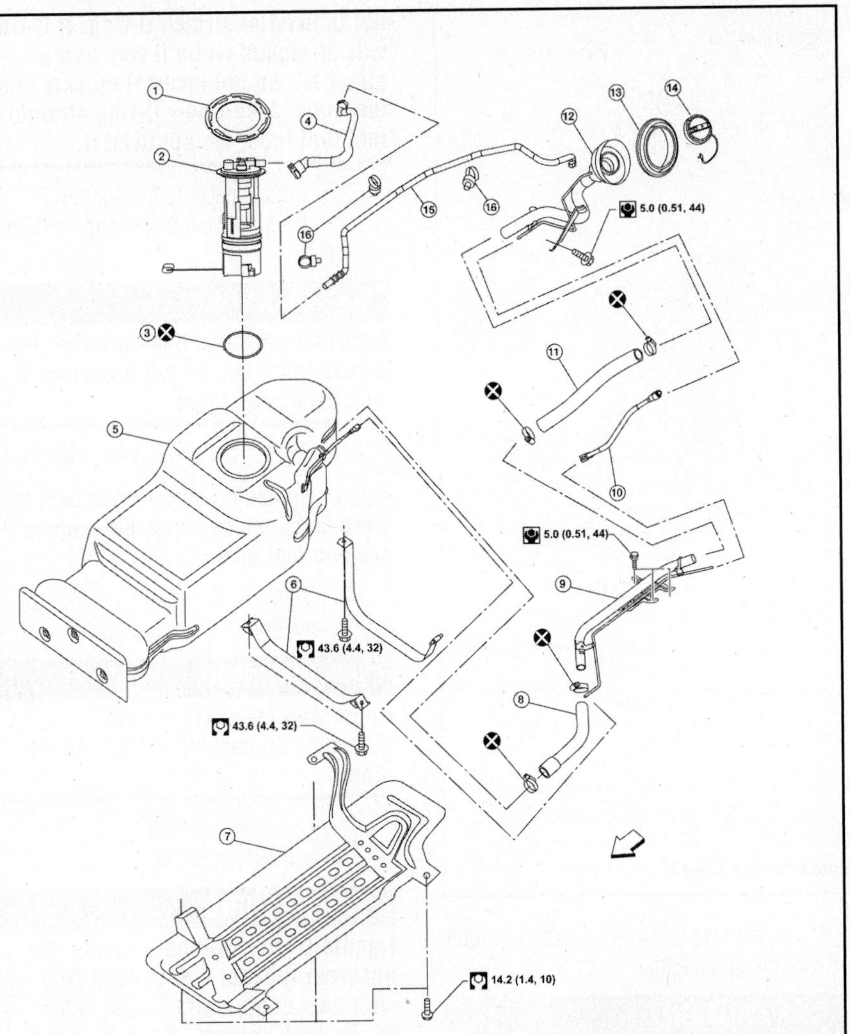

#		#		#	
1.	Lock ring	2.	Fuel level sensor, fuel filter, and fuel pump assembly	3.	Fuel level sensor, fuel filter, and fuel pump assembly O-ring
4.	EVAP hose	5.	Fuel tank	6.	Fuel tank straps
7.	Fuel tank shield (if equipped)	8.	Lower fuel filler hose	9.	Fuel filler pipe and vent pipe
10.	Vent hose	11.	Upper fuel filler hose	12.	Fuel filler pipe and cup
13.	Fuel filler hose grommet	14.	Fuel filler cap	15.	EVAP canister hose
16.	clamp	⇐	Front		

37663_PATH_G0097

Fig. 163 Fuel pump module and related components

To install:

19. Installation is in the reverse order of removal.

20. Connect the quick connector as follows:
- Check the connection for any damage or foreign materials.
- Align the connector with the pipe, then insert the connector straight into the pipe until a click is heard.
- Pull the tube and the connector to make sure they are securely connected.
- Visually inspect the connector to make sure the two retainer tabs are securely connected.

※※ WARNING

Do not bend the float arm during installation. Avoid impacts such as dropping when handling the components.

21. Turn the ignition switch ON but do not start engine, then check the fuel pipe and hose connections for leaks while applying fuel pressure.

22. Start the engine and rev it above idle, then check that there are no fuel leaks at any of the fuel pipe and hose connections.

23. Be sure to perform the reconnect/relearn procedures.

FUEL INJECTORS AND RAILS

REMOVAL & INSTALLATION

V6 Engine

See Figures 164 through 166.

1. Before servicing the vehicle, refer to the Precautions.

➡This vehicle uses quick connect fittings. Be sure to properly relieve the fuel system pressure before disconnecting any of these fittings. Always replace O-rings and clamps with new ones. Do not bend, twist or kink hoses when they are being removed or installed. Be sure that the clamp screw does not contact adjacent parts. When tightening the high pressure rubber hose clamp make sure the clamp end is 0.12 inch from the hose end. After connecting these fittings make sure that the connectors are secure. Check for fuel leakage at these connections turn the ignition key to the ON position (do not start the engine), correct as required. Start the engine, raise the idle, and verify that there are no fuel leaks, correct as required.

2. Properly relieve the fuel system pressure.

3. Disconnect the negative battery cable.

4. Remove the engine cover. Remove the air cleaner assembly.

5. Remove intake manifold collector.

※※ CAUTION

Perform this step when engine is cold.

6. Disconnect the fuel quick connector on the engine side.
   a. Using Tool No. 45488, perform the following steps to disconnect the quick connector.
   b. Remove quick connector cap.
   c. With the sleeve side of tool facing quick connector, install tool onto fuel tube.
   d. Insert tool into quick connector until sleeve contacts and goes no further. Hold the tool on that position.

※※ WARNING

Inserting the tool hard will not disconnect quick connector. Hold tool where it contacts and goes no further.

   e. Pull the quick connector straight out from the fuel tube.

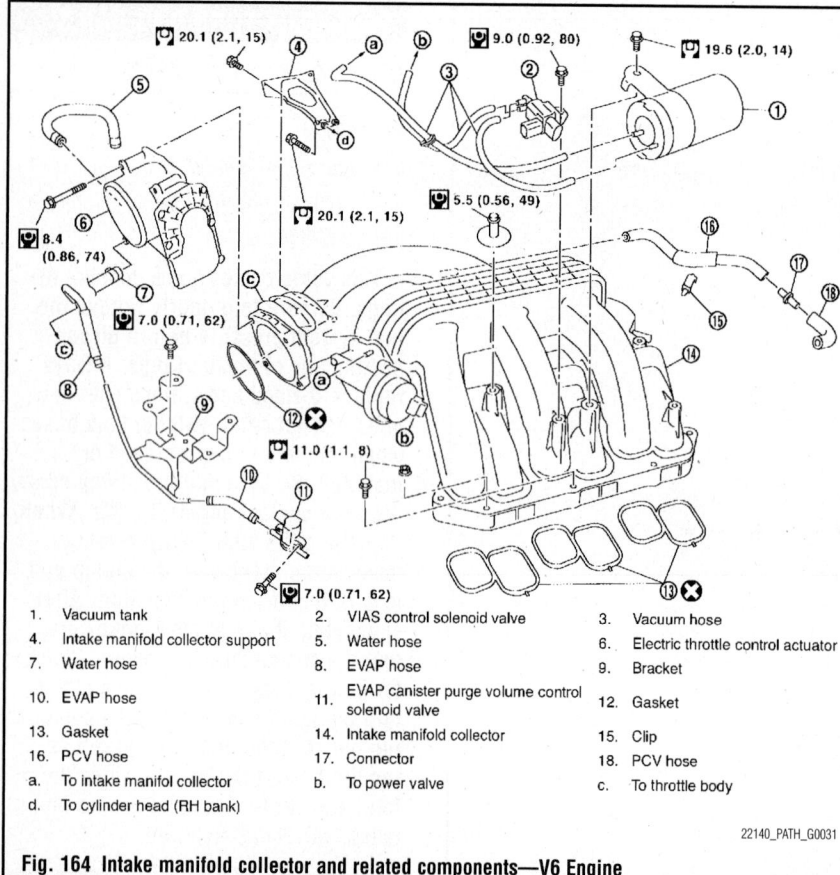

1. Vacuum tank
2. VIAS control solenoid valve
3. Vacuum hose
4. Intake manifold collector support
5. Water hose
6. Electric throttle control actuator
7. Water hose
8. EVAP hose
9. Bracket
10. EVAP hose
11. EVAP canister purge volume control solenoid valve
12. Gasket
13. Gasket
14. Intake manifold collector
15. Clip
16. PCV hose
17. Connector
18. PCV hose
a. To intake manifol collector
b. To power valve
c. To throttle body
d. To cylinder head (RH bank)

22140_PATH_G0031

Fig. 164 Intake manifold collector and related components—V6 Engine

✳✳ CAUTION

Pull quick connector holding it at the A position. Do not pull with lateral force applied. O-ring inside quick connector may be damaged. Prepare container and cloth beforehand as fuel will leak out. Avoid fire and sparks. Be sure to cover openings of disconnected pipes with plug or plastic bag to avoid fuel leakage and entry of foreign materials.

7. Remove PCV hose between rocker covers (right and left banks).
8. Disconnect harness connector from fuel injector.
9. Loosen bolts in reverse order as shown, and remove fuel tube and fuel injector assembly.

✳✳ CAUTION

Do not tilt it, or remaining fuel in pipes may flow out from pipes.

10. Remove bolts which connects fuel tube (RH) and fuel tube (LH).
11. Remove fuel injector from fuel tube as follows:
    a. Carefully open and remove clip.

b. Remove fuel injector from fuel tube by pulling straight.

✳✳ CAUTION

Be careful with remaining fuel that may go out from fuel tube. Be careful not to damage injector nozzles during removal. Do not bump or drop fuel injector. Do not disassemble fuel injector.

12. Disconnect fuel tube (RH) from fuel tube (LH).
13. Loosen bolts, to remove fuel damper cap and fuel damper, if necessary.

To install:
14. Install fuel damper as follows:
    a. Install new O-ring to fuel tube.

➡ When handling new O-rings, be careful of the following caution:

✳✳ WARNING

Handle O-ring with bare hands. Do not wear gloves. Lubricate O-ring with new engine oil. Do not clean O-ring with solvent. Make sure that O-ring and its mating part are free of foreign material. When installing O-ring, be careful not to scratch it with tool or fingernails. Also be careful

not to twist or stretch O-ring. If O-ring was stretched while it was being attached, do not insert it quickly into fuel tube. Insert new O-ring straight into fuel tube. Do not twist it.

    b. Install spacer to fuel damper.
    c. Insert fuel damper straight into fuel tube.

✳✳ WARNING

Insert straight, making sure that the axis is lined up. Do not pressure-fit with excessive force.

    d. Tighten bolts evenly in turn.

➡ After tightening bolts, make sure that there is no gap between fuel damper cap and fuel tube.

15. Install new O-rings to fuel injector, paying attention to the following.

✳✳ WARNING

Upper and lower O-ring are different. Be careful not to confuse them.

- Fuel tube side: Blue
- Nozzle side: Brown

✳✳ WARNING

Handle O-ring with bare hands. Do not wear gloves. Lubricate O-ring with new engine oil. Do not clean O-ring with solvent. Make sure that O-ring and its mating part are free of foreign material. When installing O-ring, be careful not to scratch it with tool or fingernails. Also be careful not to twist or stretch O-ring. If O-ring was stretched while it was being attached, do not insert it quickly into fuel tube. Insert O-ring straight into fuel injector. Do not twist it.

16. Install fuel injector to fuel tube as follows:
    a. Insert clip into clip mounting groove on fuel injector.

✳✳ WARNING

Do not reuse clip. Replace it with a new one. Be careful to keep clip from interfering with O-ring. If interference occurs, replace O-ring.

    b. Insert fuel injector into fuel tube with clip attached.
    c. Make sure that installation is complete by checking that fuel injector does not rotate or come off.

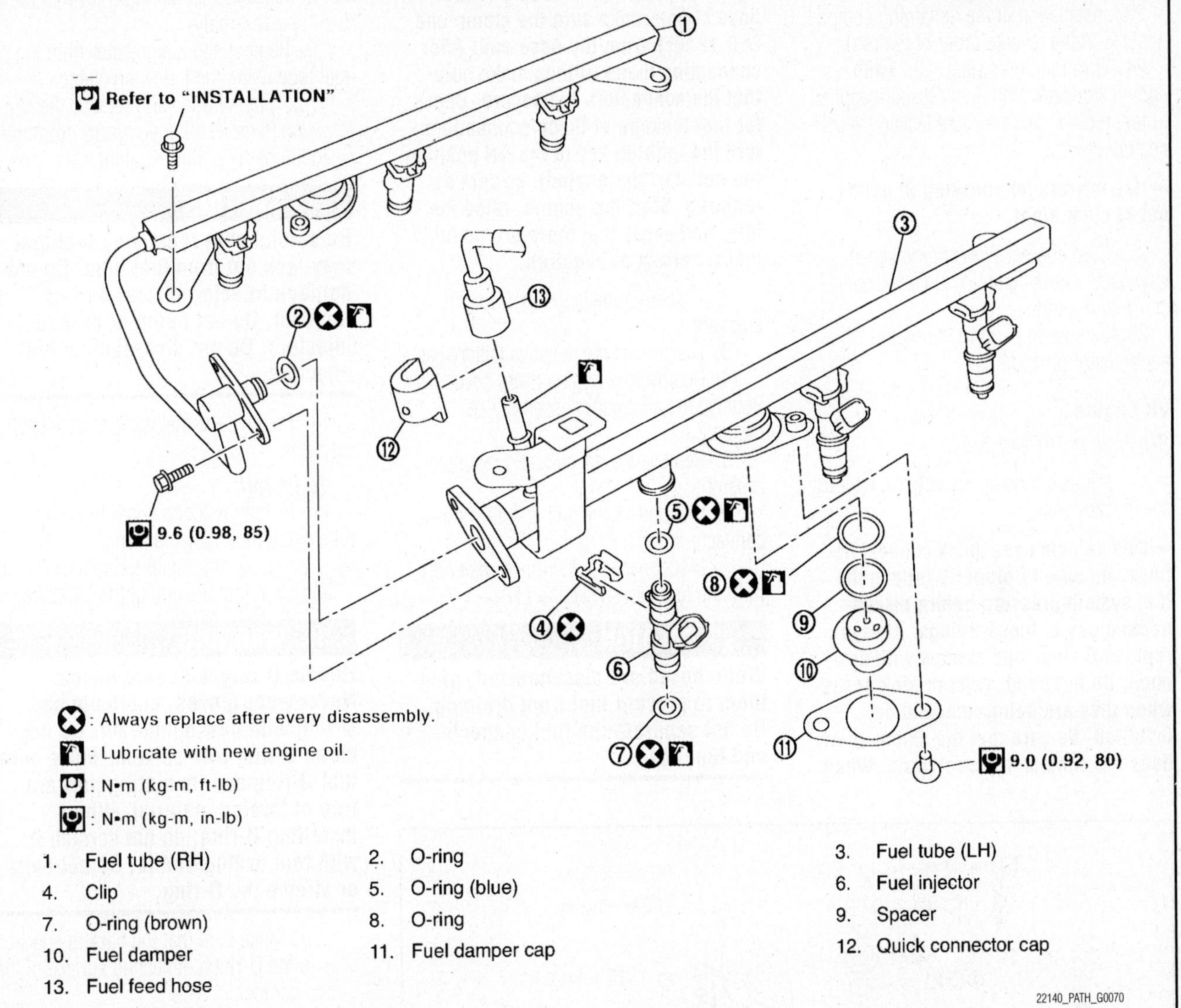

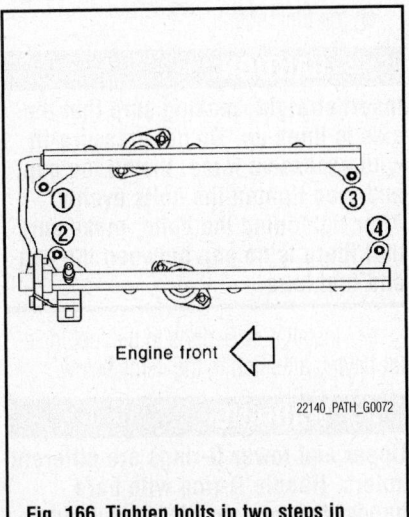

Refer to "INSTALLATION"

9.6 (0.98, 85)

9.0 (0.92, 80)

⊗ : Always replace after every disassembly.

🛢 : Lubricate with new engine oil.

🔧 : N•m (kg-m, ft-lb)

🔧 : N•m (kg-m, in-lb)

1.	Fuel tube (RH)	2.	O-ring	3.	Fuel tube (LH)
4.	Clip	5.	O-ring (blue)	6.	Fuel injector
7.	O-ring (brown)	8.	O-ring	9.	Spacer
10.	Fuel damper	11.	Fuel damper cap	12.	Quick connector cap
13.	Fuel feed hose				

22140_PATH_G0070

Fig. 165 Fuel injector rail and related components—V6 Engine

Engine front ⟵

22140_PATH_G0072

Fig. 166 Tighten bolts in two steps in numerical order as shown—V6 engine

d. Make sure that protrusions of fuel injectors are aligned with cutouts of clips after installation.

17. Connect fuel tube (RH) to fuel tube (LH), and tighten bolts temporarily.

a. Tighten bolts with the specified torque after installing fuel tube and fuel injector assembly.

✷✷ WARNING

Handle O-ring with bare hands. Do not wear gloves. Lubricate O-ring with new engine oil. Do not clean O-ring with solvent. Make sure that O-ring and its mating part are free of foreign material. When installing O-ring, be careful not to scratch it with tool or fingernails. Also be careful not to twist or stretch O-ring. If O-ring was stretched while it was

being attached, do not insert it quickly into fuel tube. Insert new O-ring straight into fuel tube. Do not twist it.

18. Install fuel tube and fuel injector assembly to intake manifold.

✷✷ WARNING

Be careful not to let tip of injector nozzle come in contact with other parts.

19. Tighten bolts in two steps in numerical order as shown.

a. Fuel injector tube assembly bolts:
- 1st step: 7 ft. lbs. (10.1 Nm)
- 2nd step: 16 ft. lbs. (22.0 Nm)

20. Tighten bolts which connects fuel tube (RH) and fuel tube (LH) with the specified torque.

21. Connect fuel injector harness connector.

22. Install intake manifold collector.

23. Installation of the remaining components is in the reverse order of removal.

24. Turn ignition switch "ON" (with engine stopped). With fuel pressure applied to fuel piping, check for fuel leakage at connection points.

➡Use mirrors for checking at points out of clear sight.

25. Start engine. With engine speed increased, check again for fuel leakage at connection points.

26. Be sure to perform the reconnect/relearn procedures.

V8 Engine

See Figures 167 and 168.

1. Before servicing the vehicle, refer to the Precautions.

➡This vehicle uses quick connect fittings. Be sure to properly relieve the fuel system pressure before disconnecting any of these fittings. Always replace O-rings and clamps with new ones. Do not bend, twist or kink hoses when they are being removed or installed. Be sure that the clamp screw does not contact adjacent parts. When

tightening the high pressure rubber hose clamp make sure the clamp end is 0.12 inch from the hose end. After connecting these fittings make sure that the connectors are secure. Check for fuel leakage at these connections turn the ignition key to the ON position (do not start the engine), correct as required. Start the engine, raise the idle, and verify that there are no fuel leaks, correct as required.

2. Properly relieve the fuel system pressure.

3. Disconnect the negative battery cable.

4. Remove the engine room cover. Remove the air cleaner assembly, as required.

5. Remove the air duct and resonator assembly.

6. Disconnect the fuel injector harness connectors.

7. Disconnect the fuel hose assembly from the fuel tubes (RH and LH).

✳✳ CAUTION

While hoses are disconnected, plug them to prevent fuel from draining. Do not separate the fuel connector and fuel hose.

8. Remove the fuel injectors with the fuel tube assembly.

9. Remove the fuel injector from the fuel tube using the following steps:

a. Spread open and remove the clip.

b. Remove the fuel injector from the fuel tube by pulling straight out.

✳✳ CAUTION

Be careful with remaining fuel that may leak out from fuel tube. Do not damage injector nozzles during removal. Do not bump or drop fuel injectors. Do not disassemble fuel injectors.

10. Remove the fuel damper from each fuel tube.

To install:

11. Install the fuel damper to each fuel tube using the following steps:

a. Apply engine oil to the new O-ring and set it into the cup of the fuel tube.

✳✳ WARNING

Handle O-ring with bare hands. Never wear gloves. Lubricate new O-ring with new engine oil. Do not clean O-ring with solvent. Make sure that O-ring and its mating part are free of foreign material. When installing O-ring, do not scratch it with tool or fingernails. Do not twist or stretch the O-ring.

b. Make sure that the backup spacer is in the O-ring connecting surface of the fuel damper.

➡The backup spacer is part of the fuel damper assembly.

c. Insert the fuel damper until it seats on the fuel tube.

✳✳ CAUTION

Insert straight, making sure that the axis is lined up. Do not pressure-fit with excessive force. Install the cap, and then tighten the bolts evenly. After tightening the bolts, make sure that there is no gap between the cap and fuel tube.

12. Install new O-rings to the fuel injector paying attention to the items below.

✳✳ WARNING

Upper and lower O-rings are different colors. Handle O-ring with bare hands. Never wear gloves. Lubricate new O-ring with new engine oil. Do not clean O-ring with solvent. Make

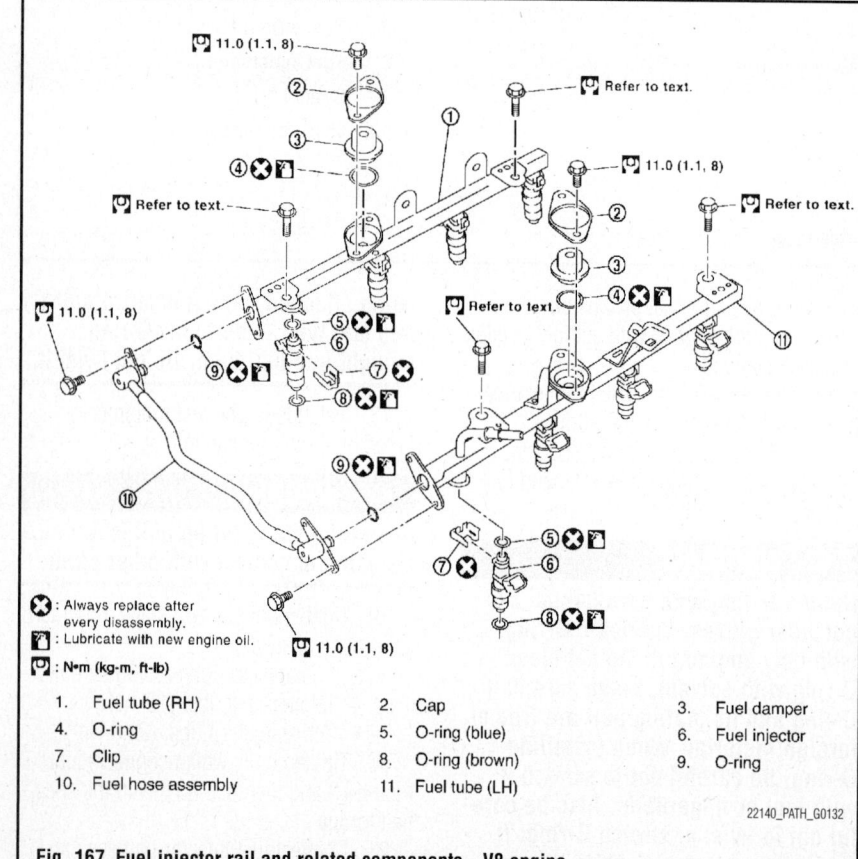

🔧 : Always replace after every disassembly.
🔩 : Lubricate with new engine oil.
🔲 : N•m (kg-m, ft-lb)

1. Fuel tube (RH)	2. Cap	3. Fuel damper
4. O-ring	5. O-ring (blue)	6. Fuel injector
7. Clip	8. O-ring (brown)	9. O-ring
10. Fuel hose assembly	11. Fuel tube (LH)	

22140_PATH_G0132

Fig. 167 Fuel injector rail and related components—V8 engine

sure that O-ring and its mating part are free of foreign material. When installing O-ring, be careful not to scratch it with tool or fingernails. Also be careful not to twist or stretch O-ring. If O-ring was stretched while it was being attached, do not insert it quickly into fuel tube. Insert O-ring straight into fuel tube. Do not angle or twist it.

13. Install the fuel injector to the fuel tube using the following steps.

    a. Insert new clip into clip mounting groove on the fuel injector.
    - Insert clip so that lug A of fuel injector matches notch A of the clip.

✳✳ WARNING
Do not reuse clip. Replace it with a new one. Do not allow the clip to interfere with the O-ring. If interference occurs, replace the O-ring.

    b. Insert the fuel injector into the fuel tube with the clip attached.
    - Insert it while matching it to the axial center.
    - Insert fuel injector so that lug B of fuel tube matches notch B of the clip.
    - Make sure that the fuel tube flange is securely seated in the flange fixing groove on the clip.
    c. Make sure that installation is complete by checking that the fuel injector does not rotate or come off.
    - Make sure that the protrusions of the fuel injectors are aligned with the cutouts of the clips after installation.

14. Install the fuel tube and fuel injector assembly to the intake manifold.

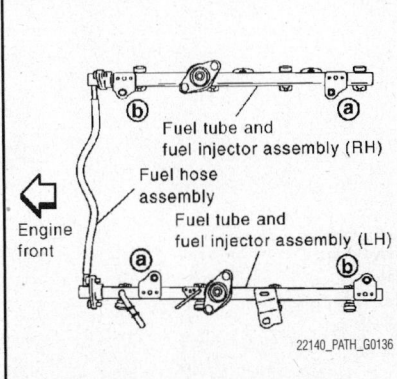

Fig. 168 Injector rail retaining bolt tightening sequence—V8 engine

✳✳ WARNING
Do not let the tip of the injector nozzle come in contact with other parts.

    a. Tighten fuel tube assembly bolts A to B in two steps:
    - Step 1: 9 ft. lbs. (12.8 Nm)
    - Step 2: 18 ft. lbs. (24.5 Nm)
15. Install the fuel hose assembly.

    a. Insert connector's straight, making sure that the axis is lined up with fuel tube side to prevent O-ring from being damaged.
    b. Tighten bolts evenly in several steps.
    c. Make sure that there is no gap between the flange and fuel tube after tightening the bolts.

✳✳ WARNING
Handle O-ring with bare hands. Do not wear gloves. Lubricate O-ring with new engine oil. Do not clean O-ring with solvent. Make sure that O-ring and its mating part are free of foreign material. When installing O-ring, be careful not to scratch it with tool or fingernails. Also be careful not to twist or stretch O-ring. If O-ring was stretched while it was being attached, do not insert it quickly into fuel tube. Insert new O-ring straight into fuel tube. Do not twist it.

16. Installation of the remaining components is in the reverse order of removal.
17. After installing the fuel tubes, make sure there are no fuel leaks at the connections using the following steps.
18. Apply fuel pressure to the fuel lines by turning ignition switch ON (with engine stopped). Then check for fuel leaks at the connections.

➡ **Use mirrors for checking on hidden points.**

19. Start the engine and rev it up and check for fuel leaks at the connections.

✳✳ CAUTION
Do not touch the engine immediately after stopping, as engine becomes extremely hot.

20. Be sure to perform the reconnect/relearn procedures.

FUEL TANK

DRAINING

1. Before servicing the vehicle, refer to the Precautions.

➡This vehicle uses quick connect fittings. Be sure to properly relieve the fuel system pressure before disconnecting any of these fittings. Always replace O-rings and clamps with new ones. Do not bend, twist or kink hoses when they are being removed or installed. Be sure that the clamp screw does not contact adjacent parts. When tightening the high pressure rubber hose clamp make sure the clamp end is 0.12 inch from the hose end. After connecting these fittings make sure that the connectors are secure. Check for fuel leakage at these connections turn the ignition key to the ON position (do not start the engine), correct as required. Start the engine, raise the idle, and verify that there are no fuel leaks, correct as required.

2. Disconnect the negative battery cable.
3. Remove the fuel filler cap to release the pressure from inside the fuel tank.
4. Remove the LH rear wheel and tire.
5. Check the fuel level on level gauge. If the fuel gauge indicates more than the level as shown (full or almost full), drain the fuel from the fuel tank until the fuel gauge indicates the level as shown, or less.

➡Fuel will be spilled when removing the fuel level sensor, fuel filter, and fuel pump assembly for the fuel level is above the fuel level sensor, fuel filter, and fuel pump assembly fuel tank opening.

- As a guide, the fuel level reaches the fuel gauge position as shown, or less, when approximately 4 US gal (15L) of fuel are drained from the fuel tank.
- If the fuel pump does not operate, use the following procedure to drain the fuel to the specified level.
    a. Insert a suitable hose of less than 15 mm (0.59 in.) diameter into the fuel filler pipe through the fuel filler opening to drain the fuel from fuel filler pipe.
    b. Remove the fuel filler pipe shield.
    c. Disconnect the fuel filler hose from the fuel filler pipe.
    d. Insert a suitable hose into the fuel tank through the fuel filler hose to drain the fuel from the fuel tank.
6. Release the fuel pressure from the fuel lines.

REMOVAL & INSTALLATION
See Figure 169.

1. Before servicing the vehicle, refer to the Precautions.

➡This vehicle uses quick connect fittings. Be sure to properly relieve the fuel system pressure before disconnecting any of these fittings. Always replace O-rings and clamps with new ones. Do not bend, twist or kink hoses when they are being removed or installed. Be sure that the clamp screw does not contact adjacent parts. When tightening the high pressure rubber hose clamp make sure the clamp end is 0.12 inch from the hose end. After connecting these fittings make sure that the connectors are secure. Check for fuel leakage at these connections turn the ignition key to the ON position (do not start the engine), correct as required. Start the engine, raise the idle, and verify that there are no fuel leaks, correct as required.

2. Disconnect the negative battery cable.
3. Remove the fuel filler cap to release the pressure from inside the fuel tank.
4. Remove the LH rear wheel and tire.
5. Check the fuel level on level gauge. If the fuel gauge indicates more than the level as shown (full or almost full), drain the fuel from the fuel tank until the fuel gauge indicates the level as shown, or less.

➡Fuel will be spilled when removing the fuel level sensor, fuel filter, and fuel pump assembly for the fuel level is above the fuel level sensor, fuel filter, and fuel pump assembly fuel tank opening.

• As a guide, the fuel level reaches the fuel gauge position as shown, or less, when approximately 4 US gal (15L) of fuel are drained from the fuel tank.
• If the fuel pump does not operate, use the following procedure to drain the fuel to the specified level.
a. Insert a suitable hose of less than 15 mm (0.59 in.) diameter into the fuel filler pipe through the fuel filler opening to drain the fuel from fuel filler pipe.
b. Remove the fuel filler pipe shield.
c. Disconnect the fuel filler hose from the fuel filler pipe.

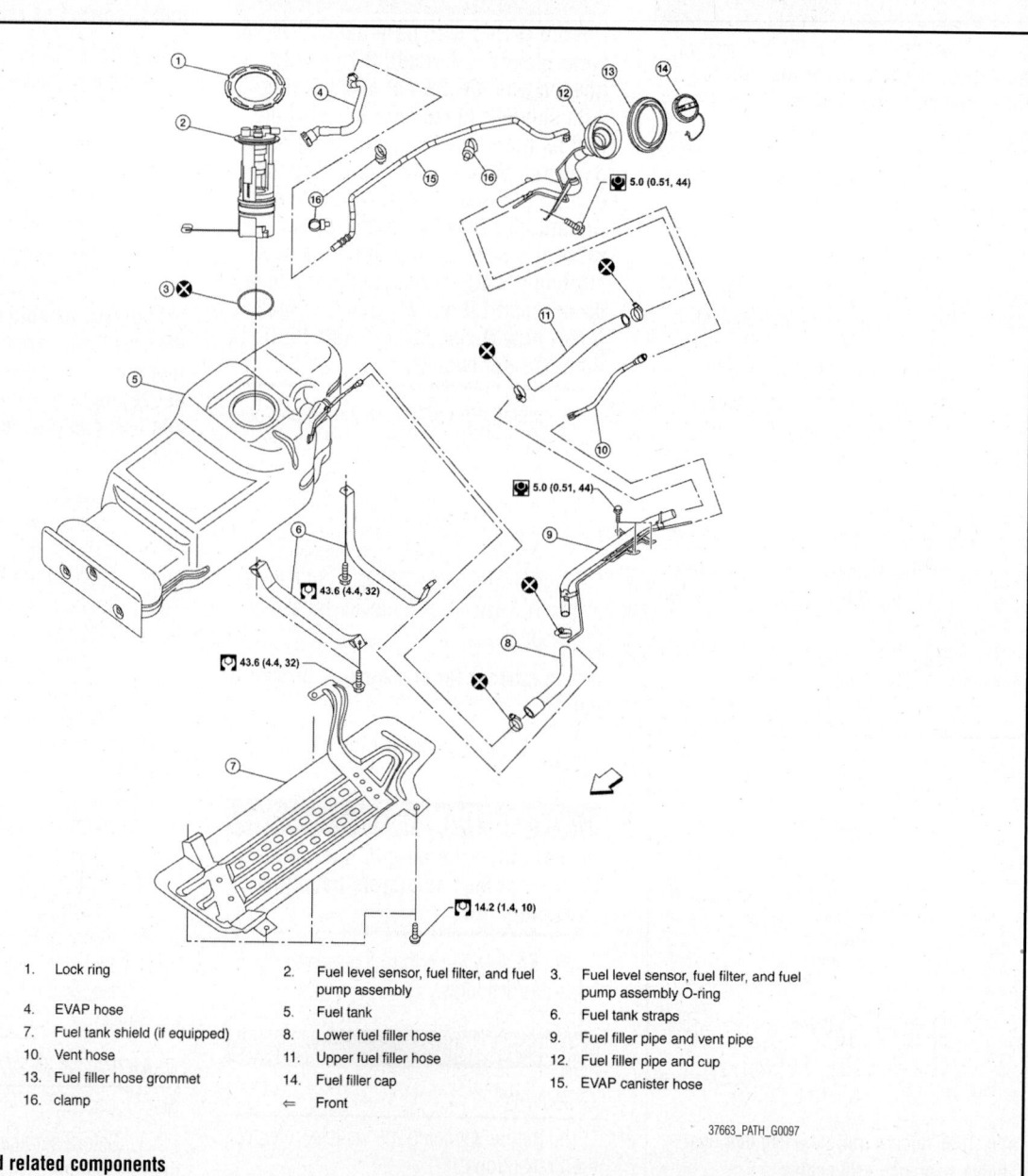

1.	Lock ring	2.	Fuel level sensor, fuel filter, and fuel pump assembly	3.	Fuel level sensor, fuel filter, and fuel pump assembly O-ring
4.	EVAP hose	5.	Fuel tank	6.	Fuel tank straps
7.	Fuel tank shield (if equipped)	8.	Lower fuel filler hose	9.	Fuel filler pipe and vent pipe
10.	Vent hose	11.	Upper fuel filler hose	12.	Fuel filler pipe and cup
13.	Fuel filler hose grommet	14.	Fuel filler cap	15.	EVAP canister hose
16.	clamp	⇐	Front		

37663_PATH_G0097

Fig. 169 Fuel tank and related components

d. Insert a suitable hose into the fuel tank through the fuel filler hose to drain the fuel from the fuel tank.

6. Release the fuel pressure from the fuel lines.

7. Disconnect the lower fuel filler hose from the fuel tank, the EVAP hose, and the vent pipe quick connector.

a. Disconnect the fuel feed hose from the molded clip in the side of the fuel tank.

- Disconnect the quick connector as follows:
- Hold the sides of the connector, push in the tabs and pull out the tube.
- If the connector and the tube are stuck together, push and pull several times until they start to move. Then disconnect them by pulling.

✳✳ WARNING

The quick connector can be disconnected when the tabs are completely depressed. Do not twist the quick connector more than necessary. Do not use any tools to disconnect the quick connector. Keep the resin tube away from heat. Be especially careful when welding near the tube. Do not bend or twist the resin tube during connection.

8. Remove the four bolts and remove the fuel tank shield.

9. Remove the driveshaft shaft.

10. Support the fuel tank using a suitable lift jack.

11. Remove the three fuel tank strap bolts while supporting the fuel tank with a suitable lift jack.

12. Remove the fuel tank straps and slowly lower the fuel tank to access the top of the fuel level sensor, fuel filter and fuel pump assembly.

✳✳ CAUTION

Do not lower the fuel tank too far to prevent damage to the fuel feed hose and the fuel level sensor, fuel filter and fuel pump assembly connector.

13. Disconnect the fuel level sensor, fuel filter, and fuel pump assembly electrical connector, and the fuel feed hose.

a. Disconnect the quick connector as follows:

- Hold the sides of the connector, push in the tabs and pull out the tube.

- If the connector and the tube are stuck together, push and pull several times until they start to move. Then disconnect them by pulling.

✳✳ WARNING

The quick connector can be disconnected when the tabs are completely depressed. Do not twist the quick connector more than necessary. Do not use any tools to disconnect the quick connector. Do not bend or twist the resin tube during connection.

14. Lower the fuel tank using a suitable lift jack and remove the fuel tank.

15. Disconnect the EVAP hose from the fuel pump and remove the EVAP hose from the molded clip in the top of the fuel tank.

16. Remove the lock ring.

17. Remove the fuel level sensor, fuel filter, and fuel pump assembly.

✳✳ WARNING

Do not bend the float arm during removal. Avoid impacts such as dropping when handling the components.

To install:

18. Installation is in the reverse order of removal.

19. Connect the quick connector as follows:

- Check the connection for any damage or foreign materials.
- Align the connector with the pipe, then insert the connector straight into the pipe until a click is heard.
- Pull the tube and the connector to make sure they are securely connected.
- Visually inspect the connector to make sure the two retainer tabs are securely connected.

✳✳ WARNING

Do not bend the float arm during installation. Avoid impacts such as dropping when handling the components.

20. Turn the ignition switch ON but do not start engine, then check the fuel pipe and hose connections for leaks while applying fuel pressure.

21. Start the engine and rev it above idle, then check that there are no fuel leaks

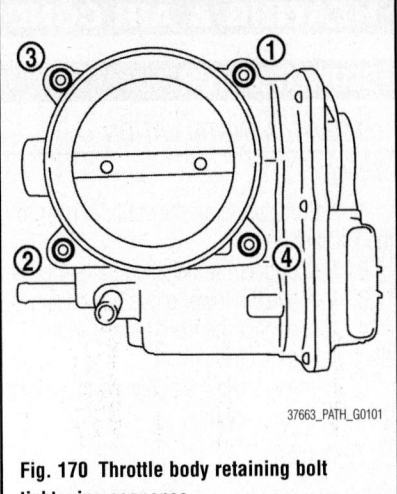

Fig. 170 Throttle body retaining bolt tightening sequence

at any of the fuel pipe and hose connections.

22. Be sure to perform the reconnect/relearn procedures.

THROTTLE BODY

REMOVAL & INSTALLATION

V6 Engine

See Figure 170.

1. Before servicing the vehicle, refer to the Precautions.

2. Disconnect the negative battery cable.

3. Remove the air dam.

4. Remove the engine room cover.

5. Remove the air cleaner assembly.

6. Remove the engine undercover.

7. Drain the engine coolant. Be sure to properly dispose of used coolant.

8. Disconnect the hoses from the unit.

9. Disconnect the electrical connectors.

10. Remove the retaining bolts in the reverse order of the tightening sequence.

11. Remove the component from its mounting. Discard the gasket.

To install:

➡**Be sure to use new fasteners, as required.**

12. Installation is the reverse of the removal procedure.

13. Tighten the retaining bolts to specification and in the proper sequence.

14. Using the CONSULT-III diagnostic tool or equivalent, perform the throttle valve closed position learning and the idle air volume learning procedures.

15. Be sure to perform the reconnect/relearn procedures.

HEATING & AIR CONDITIONING SYSTEM

BLOWER MOTOR

REMOVAL & INSTALLATION

See Figure 171.

1. Before servicing the vehicle, refer to the Precautions.
2. Disconnect the negative battery cable.
3. Remove the lower glove box assembly.
4. Disconnect the front blower motor electrical connector.
5. Remove the three screws and remove the front blower motor.

To install:

➡**Be sure to use new fasteners, as required.**

6. Installation is the reverse of the removal procedure.
7. Be sure to perform the reconnect/relearn procedures.

HEATER CORE

REMOVAL & INSTALLATION

See Figure 172.

1. Before servicing the vehicle, refer to the Precautions.
2. Disconnect the negative battery cable.
3. Properly discharge the refrigerant from the A/C system.
4. Remove the front heater and cooling unit assembly.
5. Remove the three screws and remove the front heater core cover.

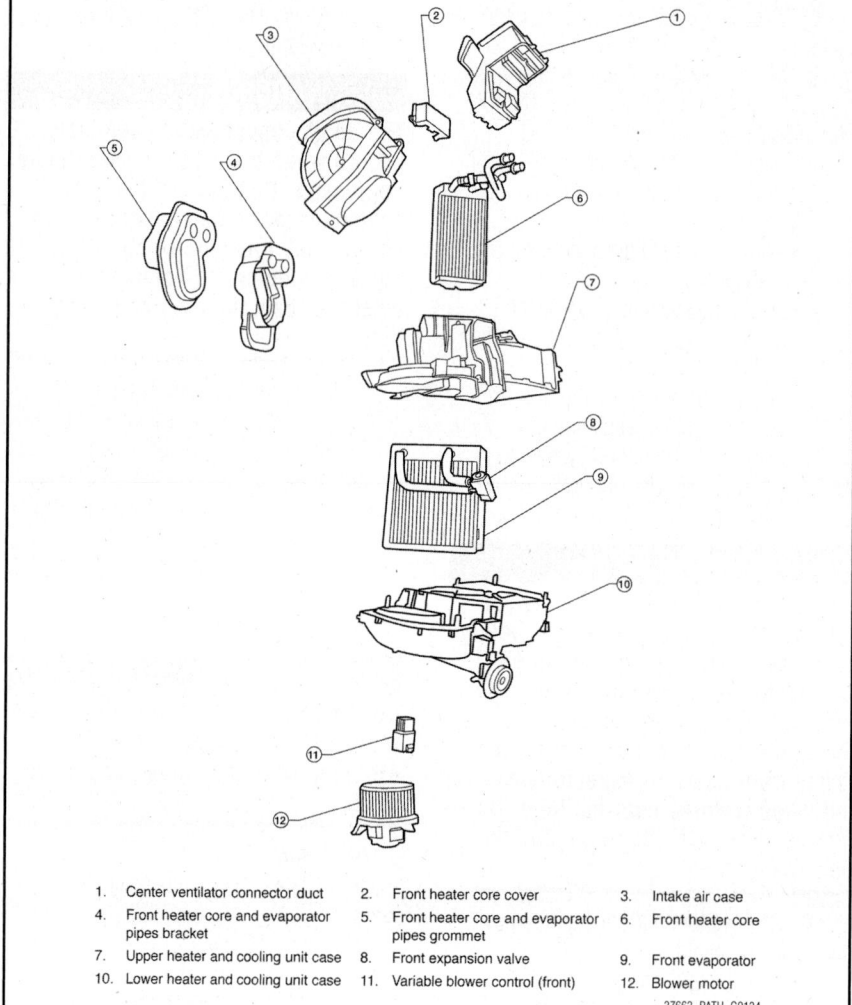

1.	Center ventilator connector duct	2.	Front heater core cover	3.	Intake air case
4.	Front heater core and evaporator pipes bracket	5.	Front heater core and evaporator pipes grommet	6.	Front heater core
7.	Upper heater and cooling unit case	8.	Front expansion valve	9.	Front evaporator
10.	Lower heater and cooling unit case	11.	Variable blower control (front)	12.	Blower motor

37663_PATH_G0134

Fig. 172 Heater core and related components

6. Remove the front heater core and evaporator pipe bracket.
7. Remove the front heater core.

➡**If the in-cabin micro filters are contaminated from coolant leaking from the front heater core, replace the in-cabin micro filters with new ones before installing the new front heater core.**

To install:

➡**Be sure to use new fasteners, as required.**

8. Installation is the reverse of the removal procedure.
9. Be sure to properly recharge the air conditioning system.
10. Be sure to perform the reconnect/relearn procedures.

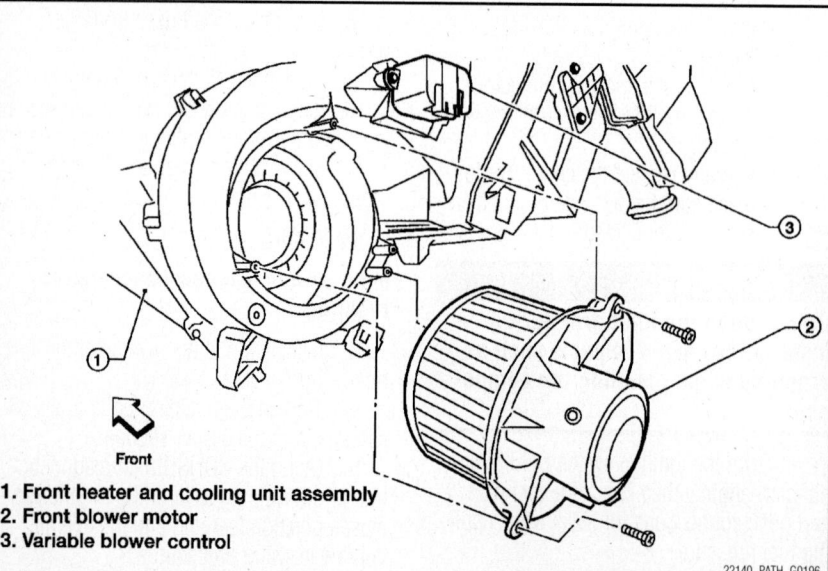

1. **Front heater and cooling unit assembly**
2. **Front blower motor**
3. **Variable blower control**

Front

22140_PATH_G0196

Fig. 171 Front blower motor and related components

STEERING

POWER RACK & PINION STEERING GEAR

REMOVAL & INSTALLATION

See Figures 173 and 174.

✳✳ CAUTION

Spiral cable may snap due to steering operation if the steering column is separated from the steering gear assembly. Therefore secure the steering wheel to avoid turning.

1. Before servicing the vehicle, refer to the Precautions.
2. Disconnect the negative battery cable.
3. Set front wheels in the straight ahead position.
4. Remove the engine cover and air cleaner assembly, as required.
5. Raise and safely support the vehicle. Remove the front tires from the vehicle.
6. Remove the undercover.
7. On 4WD, remove the front final drive, then support the drive shafts, using suitable wire.
8. Remove the stabilizer bar brackets and reposition the stabilizer bar.
9. Remove the cotter pins at the steering outer sockets.

✳✳ WARNING

Do not reuse the cotter pins.

10. Loosen the outer socket nuts.
11. Remove the steering outer sockets from the steering knuckles, then remove the nuts.

✳✳ WARNING

Do not damage the outer socket boots. Do not damage the outer socket threads. Thread the ball joint nut onto the end of the outer socket during removal.

12. Remove the high pressure and low pressure piping from the steering gear assembly, then drain the fluid from the piping.
13. Remove the bolt from the lower joint of the lower joint shaft, then separate the lower joint from the steering gear assembly.

✳✳ WARNING

Do not damage the lower joint.

14. Remove the nuts and bolts of the steering gear assembly, then remove the steering gear assembly from the vehicle.

To install:

15. Installation is in the reverse order of removal.
16. With the steering wheel in the straight ahead position, align the slit of the lower joint with the projection on the dust cover. Insert the joint until surface "A" contacts surface "B".
17. After removing/installing or replacing steering components, check wheel alignment.
18. Bleed the air from the steering hydraulic system.
19. Check that the steering wheel turns smoothly to the left and right locks.
20. Check that the number of turns are the same from the straight-forward position to the left and right locks.
21. Check that the steering wheel is in the neutral position when driving straight ahead.
22. Adjust the steering angle sensor neutral position, using the CONSULT-III diagnostic tool, or equivalent.
23. Be sure to perform the reconnect/relearn procedures.

POWER STEERING PUMP

BLEEDING

1. Before servicing the vehicle, refer to the Precautions.

➡ When the vehicle is stationary or while the steering wheel is being turned slowly, some noise may be heard from the oil pump or gear. This noise is normal and does not affect any system.

2. Check for fluid leakage.
3. Start the engine and turn the steering wheel fully to the right and left several times.

✳✳ CAUTION

Do not allow steering fluid reservoir tank to go below the MIN level line. Check tank frequently and add fluid as needed.

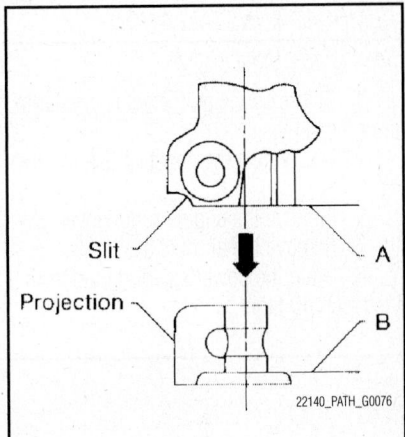

Slit
Projection
A
B

22140_PATH_G0076

Fig. 174 Insert the joint until surface "A" contacts surface "B"

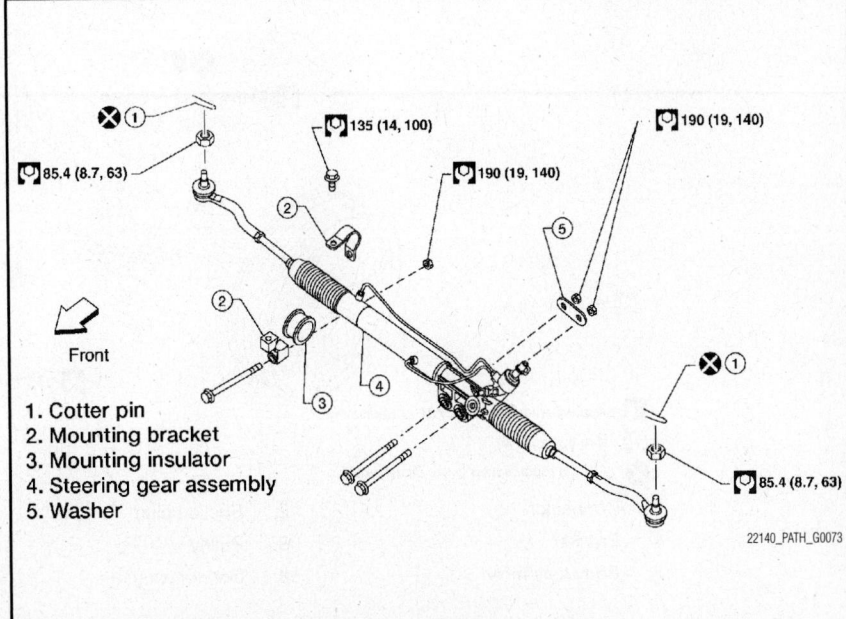

1. Cotter pin
2. Mounting bracket
3. Mounting insulator
4. Steering gear assembly
5. Washer

Front

⊗① 85.4 (8.7, 63)
135 (14, 100)
190 (19, 140)
190 (19, 140)
⊗①
85.4 (8.7, 63)

22140_PATH_G0073

Fig. 173 Power steering gear and related components

4. Run the engine at idle speed. Hold the steering wheel at each "locked" position for three seconds.

※※ WARNING

Do not hold steering wheel in the locked position for more than 10 seconds. (There is the possibility that oil pump may be damaged.)

5. Repeat step 3 several times at about three second intervals.

6. Check for air bubbles, cloudy fluid and fluid leakage.

7. If air bubbles or cloudiness exists, perform steps 3 and 4 again until air bubbles and cloudiness do not exist.

8. Stop the engine and check fluid level.

REMOVAL & INSTALLATION

See Figures 175 and 176.

1. Before servicing the vehicle, refer to the Precautions.

2. Disconnect the negative battery cable.

3. Remove the engine room cover.

4. Remove the air duct assembly.

5. Drain the power steering fluid from the reservoir tank.

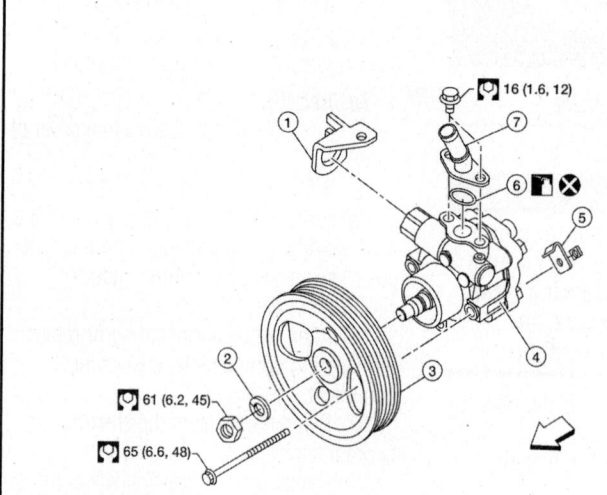

16 (1.6, 12)

61 (6.2, 45)

65 (6.6, 48)

1.	Bracket	2.	Spring washer	3.	Pulley
4.	Power steering pump	5.	High pressure hose bracket	6.	O-ring
7.	Suction pipe	⇐	Front		Apply Genuine NISSAN PSF

37663_PATH_G0107

Fig. 176 Power steering pump and related components—V8 engine

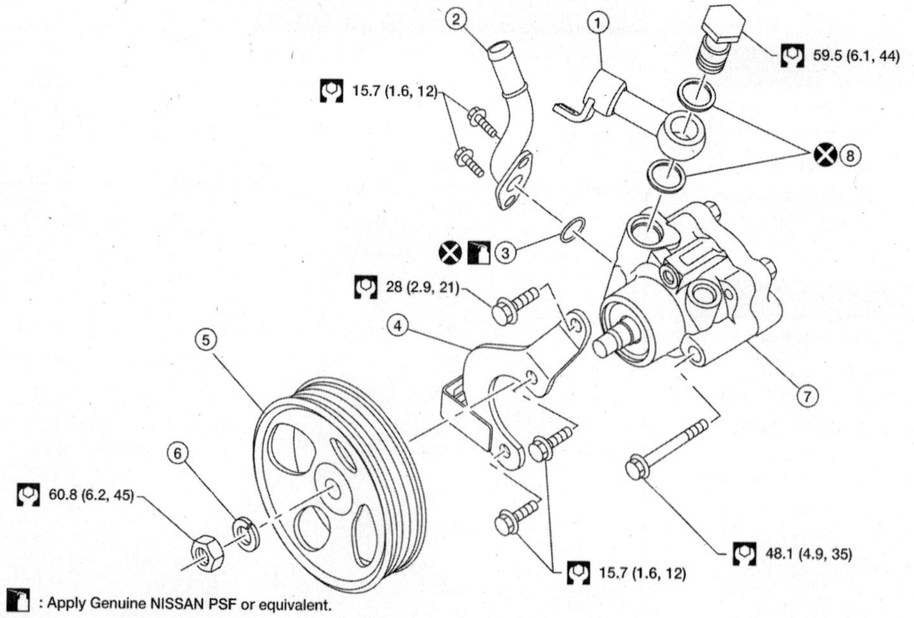

15.7 (1.6, 12)

59.5 (6.1, 44)

28 (2.9, 21)

60.8 (6.2, 45)

48.1 (4.9, 35)

15.7 (1.6, 12)

: Apply Genuine NISSAN PSF or equivalent.

: N·m (kg-m, ft-lb)

: Always replace after every disassembly.

1.	Connector	2.	Suction pipe	3.	O-ring
4.	Bracket	5.	Pulley	6.	Lock washer
7.	Body assembly	8.	Copper washers		

37663_PATH_G0106

Fig. 175 Power steering pump and related components—V6 engine

6. Remove the serpentine drive belt from the auto tensioner and power steering oil pump.

7. Disconnect the pressure sensor electrical connector.

8. Remove the high pressure and low pressure piping from the power steering oil pump.

9. Remove the power steering oil pump bolts, then remove the power steering pump.

To install:

➡**Be sure to use new fasteners, as required.**

10. Installation is the reverse of the removal procedure.

11. After installation, bleed the air from the hydraulic circuit thoroughly.

12. Be sure to perform the reconnect/relearn procedures.

FLUID FILL PROCEDURE

See Figure 177.

✳✳ CAUTION

Used fluid is considerably more dangerous than new fluid. Avoid skin contact with used fluid.

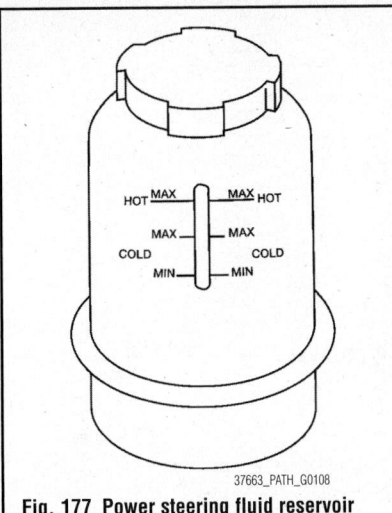

37663_PATH_G0108

Fig. 177 Power steering fluid reservoir markings

---

1. Before servicing the vehicle, refer to the Precautions.

➡**The automatic back door system must be initialized anytime the battery has been disconnected. Close the back door. Open the back door with the automatic open feature. Do not stop the process until the back door opens completely.**

2. Inspect the power steering fluid level in the power steering reservoir. Do allow the fluid to drop below the MIN marking.

3. Remove the power steering reservoir cap.

4. Add fluid, as necessary, referring to the scale on the reservoir tank.
- HOT range for fluid temperatures: 122–176°F (50–80°C)
- COLD range for fluid temperatures: 32–86°F (0–30°C)

➡**Do not overfill the fluid. Do not reuse any used power steering fluid.**

SUSPENSION

CONTROL LINKS

REMOVAL & INSTALLATION

Lower Link

See Figure 178.

1. Before servicing the vehicle, refer to the Precautions.

2. Disconnect the negative battery cable.

3. Raise and support the vehicle safely.

4. Remove the tire and wheel assembly.

5. Remove the engine undercover, as required.

6. Remove the lower shock absorber bolt and nut.

7. Remove the stabilizer bar connecting rod lower nut. Separate the connecting rod from the lower link.

8. On 4WD vehicles, remove the driveshaft.

9. Remove the pinch bolt from the steering knuckle. Separate the lower link ball joint from the steering knuckle.

10. Remove the lower link adjusting bolts and nuts. Remove the lower link.

➡**Some vehicles may be equipped with straight (non adjustable) lower link bolts and washers. In order to adjust camber and caster on these vehicles first replace the lower link bolts and washers with adjustable (cam) bolts and washers.**

To install:

➡**Be sure to use new fasteners, as required.**

11. Installation is the reverse of the removal procedure.

12. Check and adjust alignment.

13. Be sure to perform the reconnect/relearn procedures.

Upper Link

1. Before servicing the vehicle, refer to the Precautions.

2. Disconnect the negative battery cable.

3. Raise and support the vehicle safely.

4. Remove the tire and wheel assembly.

5. Remove the engine undercover, as required.

6. Support the lower link using the proper jack.

7. For left side, remove the bolt from the lower joint of the lower joint shaft. Reposition the lower joint out of the way. Do not damage the lower joint.

8. Remove the cotter pin and nut from the upper link ball joint. Discard the pin.

9. Separate the upper link ball joint stud from the knuckle using tool ST2902001, or equivalent.

FRONT SUSPENSION

10. Remove the upper link bolts and nuts.

11. Remove the component from its mounting.

To install:

➡**Be sure to use new fasteners, as required.**

12. Installation is the reverse of the removal procedure.

13. Check and adjust alignment.

14. Be sure to perform the reconnect/relearn procedures.

KNUCKLE & SPINDLE

REMOVAL & INSTALLATION

See Figure 179.

1. Before servicing the vehicle, refer to the Precautions.

2. Disconnect the negative battery cable.

3. Raise and safely support the vehicle.

4. Remove wheel and tire from vehicle.

5. Remove the engine undercover, as required.

6. Without disassembling the hydraulic lines, remove brake caliper. Reposition it aside with wire.

➡**Avoid depressing brake pedal while brake caliper is removed.**

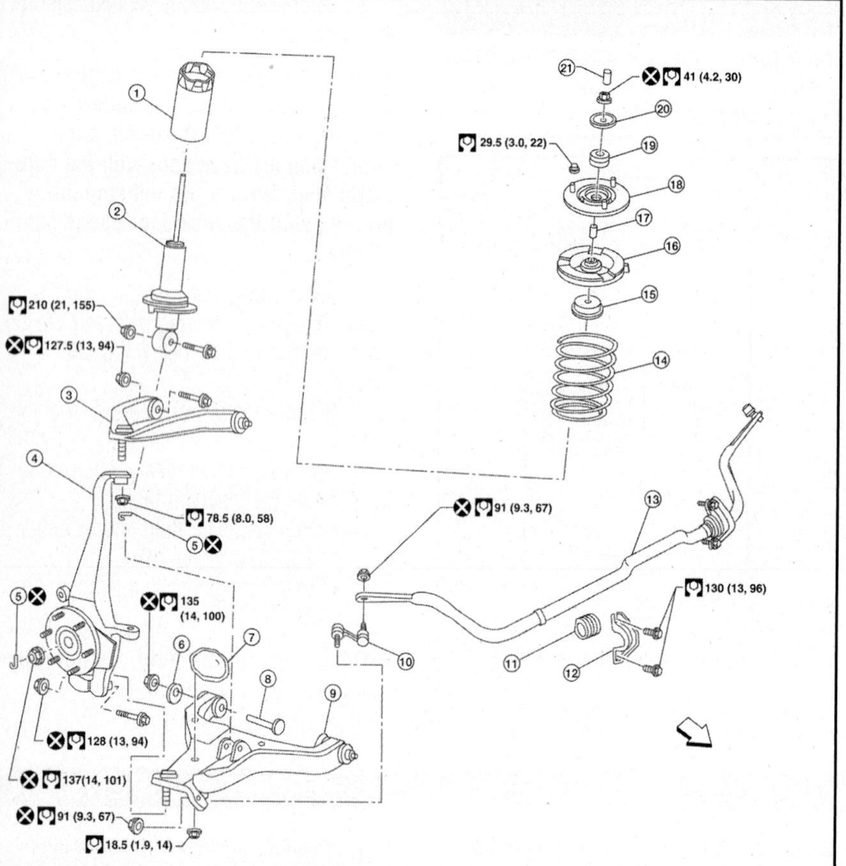

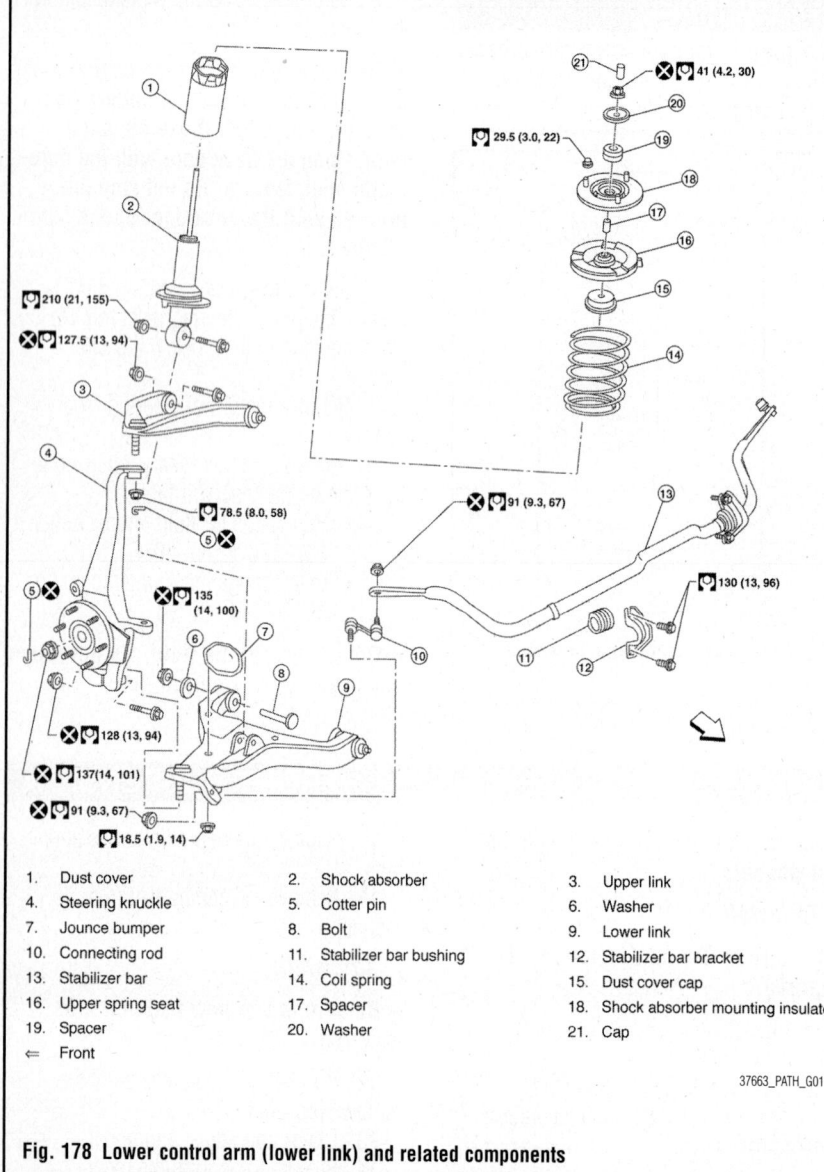

1. Dust cover	2. Shock absorber	3. Upper link
4. Steering knuckle	5. Cotter pin	6. Washer
7. Jounce bumper	8. Bolt	9. Lower link
10. Connecting rod	11. Stabilizer bar bushing	12. Stabilizer bar bracket
13. Stabilizer bar	14. Coil spring	15. Dust cover cap
16. Upper spring seat	17. Spacer	18. Shock absorber mounting insulator
19. Spacer	20. Washer	21. Cap
⇐ Front		

37663_PATH_G0110

Fig. 178 Lower control arm (lower link) and related components

7. Put alignment marks on disc rotor and wheel hub and bearing assembly, then remove disc rotor.
8. Disconnect wheel sensor and remove bracket from steering knuckle.

⁂ WARNING

Do not pull on wheel sensor harness.

9. On 4WD models, remove cotter pin, then remove lock nut from drive shaft.
10. Remove steering outer socket cotter pin at steering knuckle, then loosen nut.
11. Disconnect steering outer socket from steering knuckle. Be careful not to damage outer socket boot.

⁂ WARNING

To prevent damage to threads and to prevent Tool from coming off suddenly, temporarily tighten nut.

12. Remove wheel hub and bearing assembly bolts.
13. Remove splash guard and wheel hub and bearing assembly from steering knuckle.

⁂ WARNING

Do not pull on wheel sensor harness.

14. Remove cotter pin and nut from upper link ball joint.
15. Separate upper link ball joint from steering knuckle.
16. Remove pinch bolt from steering knuckle.
17. Separate lower link ball joint from steering knuckle.
18. Remove steering knuckle from vehicle.

To install:

➡Be sure to use new fasteners, as required.

19. Installation is the reverse of the removal procedure.

⁂ WARNING

Always replace drive shaft lock nut and cotter pin.

➡When installing disc rotor on wheel hub and bearing assembly, align the marks.

20. Perform wheel alignment.
21. Be sure to perform the reconnect/relearn procedures.

LOWER BALL JOINTS

REMOVAL & INSTALLATION

At this time the manufacturer does not provide removal and installation procedures for this component. The lower ball joint is part of the lower control arm (lower link) assembly.

➡Some vehicles may be equipped with straight (non adjustable) lower link bolts and washers. In order to adjust camber and caster on these vehicles, first replace the lower link bolts and washers with adjustable (cam) bolts and washers.

➡Nissan/Infiniti refers to the lower control arm as a lower link.

LOWER CONTROL ARMS

REMOVAL & INSTALLATION

1. Before servicing the vehicle, refer to the Precautions.
2. Disconnect the negative battery cable.
3. Raise and support the vehicle safely.
4. Remove the tire and wheel assembly.
5. Remove the engine undercover, as required.
6. Remove the lower shock absorber bolt and nut.
7. Remove the stabilizer bar connecting rod lower nut. Separate the connecting rod from the lower link.
8. On 4WD vehicles, remove the driveshaft.
9. Remove the pinch bolt from the steering knuckle. Separate the lower link ball joint from the steering knuckle.

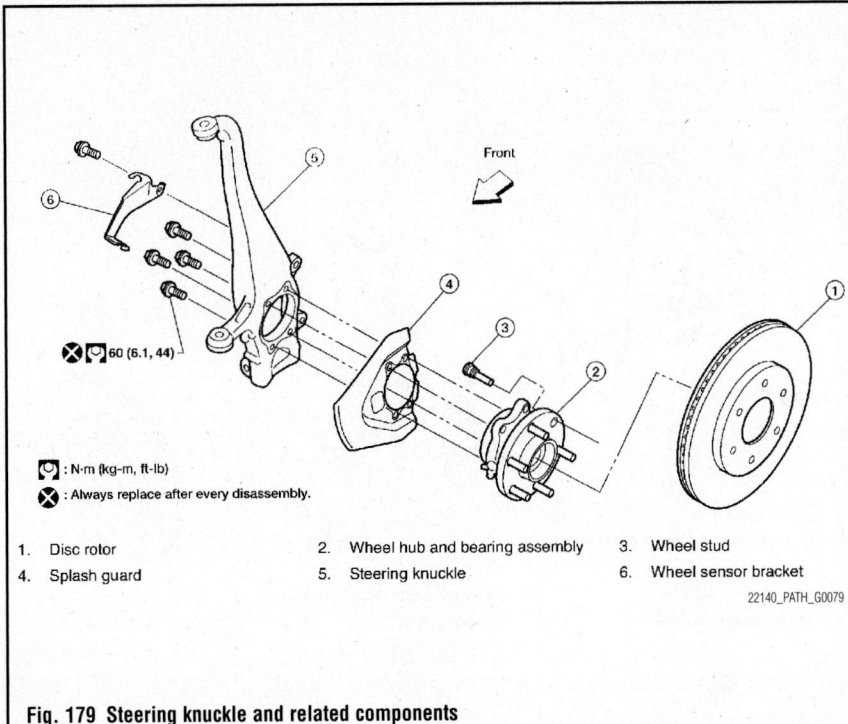

Front

60 (6.1, 44)

: N·m (kg-m, ft-lb)

: Always replace after every disassembly.

1. Disc rotor
4. Splash guard
2. Wheel hub and bearing assembly
5. Steering knuckle
3. Wheel stud
6. Wheel sensor bracket

22140_PATH_G0079

Fig. 179 Steering knuckle and related components

10. Remove the lower link adjusting bolts and nuts. Remove the lower link.

➡**Some vehicles may be equipped with straight (non adjustable) lower link bolts and washers. In order to adjust camber and caster on these vehicles first replace the lower link bolts and washers with adjustable (cam) bolts and washers.**

To install:

➡**Be sure to use new fasteners, as required.**

11. Installation is the reverse of the removal procedure.
12. Check and adjust alignment.
13. Be sure to perform the reconnect/relearn procedures.

SHOCK ABSORBERS

REMOVAL & INSTALLATION

1. Before servicing the vehicle, refer to the Precautions.
2. Disconnect the negative battery cable.
3. Raise and support the vehicle safely. Remove the wheel and tire.
4. Remove the engine undercover, as required.
5. Support the lower link using a suitable jack.
6. Remove connecting rod upper joints from stabilizer bar.

7. Swing stabilizer bar down, repositioning it out of the way to access shock absorber lower mount.
8. Remove the shock absorber lower bolt and nut.
9. Remove the three shock absorber upper mounting nuts.
10. Remove the coil spring and shock absorber assembly.
11. Turn steering knuckle out to gain enough clearance for removal.

To install:

➡**Be sure to use new fasteners, as required.**

12. Installation is the reverse of the removal procedure.
13. Be sure to perform the reconnect/relearn procedures.

STABILIZER BAR

REMOVAL & INSTALLATION

1. Before servicing the vehicle, refer to the Precautions.
2. Disconnect the negative battery cable.
3. Raise and safely support the vehicle.
4. Remove the front valance center.
5. Remove engine undercover.
6. Remove connecting rod nuts.
7. Loosen top bolts for stabilizer bar brackets, then remove lower bolts from brackets and remove stabilizer bar.

8. Remove bushings from stabilizer bar.

To install:

➡**Be sure to use new fasteners, as required.**

9. Installation is the reverse of the removal procedure.
10. Tighten all nuts and bolts to specification.
11. Be sure to perform the reconnect/relearn procedures.

UPPER BALL JOINTS

REMOVAL & INSTALLATION

At this time the manufacturer does not provide removal and installation procedures for this component. The lower ball joint is part of the lower control arm (lower link) assembly.

➡**Some vehicles may be equipped with straight (non adjustable) lower link bolts and washers. In order to adjust camber and caster on these vehicles, first replace the lower link bolts and washers with adjustable (cam) bolts and washers.**

➡**Nissan/Infiniti refers to the lower control arm as a lower link.**

UPPER CONTROL ARMS

REMOVAL & INSTALLATION

1. Before servicing the vehicle, refer to the Precautions.
2. Disconnect the negative battery cable.
3. Raise and support the vehicle safely.
4. Remove the tire and wheel assembly.
5. Remove the engine undercover, as required.
6. Support the lower link using the proper jack.
7. For left side, remove the bolt from the lower joint of the lower joint shaft. Reposition the lower joint out of the way. Do not damage the lower joint.
8. Remove the cotter pin and nut from the upper link ball joint. Discard the pin.
9. Separate the upper link ball joint stud from the knuckle using tool ST2902001, or equivalent.
10. Remove the upper link bolts and nuts.
11. Remove the component from its mounting.

To install:

➡ **Be sure to use new fasteners, as required.**

12. Installation is the reverse of the removal procedure.

13. Check and adjust alignment.

14. Be sure to perform the reconnect/relearn procedures.

WHEEL HUB & BEARING (SEALED UNIT)

REMOVAL & INSTALLATION

See Figure 180.

1. Before servicing the vehicle, refer to the Precautions.

2. Disconnect the negative battery cable.

3. Raise and safely support the vehicle.

4. Remove the engine undercover, as required.

5. Remove wheel and tire.

6. Without disassembling the hydraulic lines, remove caliper torque member bolts.

7. Reposition brake caliper aside with wire.

✳✳ WARNING

Do not press brake pedal while brake caliper is removed.

8. Put alignment mark on disc rotor and wheel hub and bearing assembly, then remove disc rotor.

9. Remove cotter pin, then remove lock nut from drive shaft.

10. Remove driveshaft from wheel hub and bearing assembly.

11. Remove wheel sensor from wheel hub and bearing assembly.

   a. Inspect the wheel sensor O-ring, replace the wheel sensor assembly if damaged.

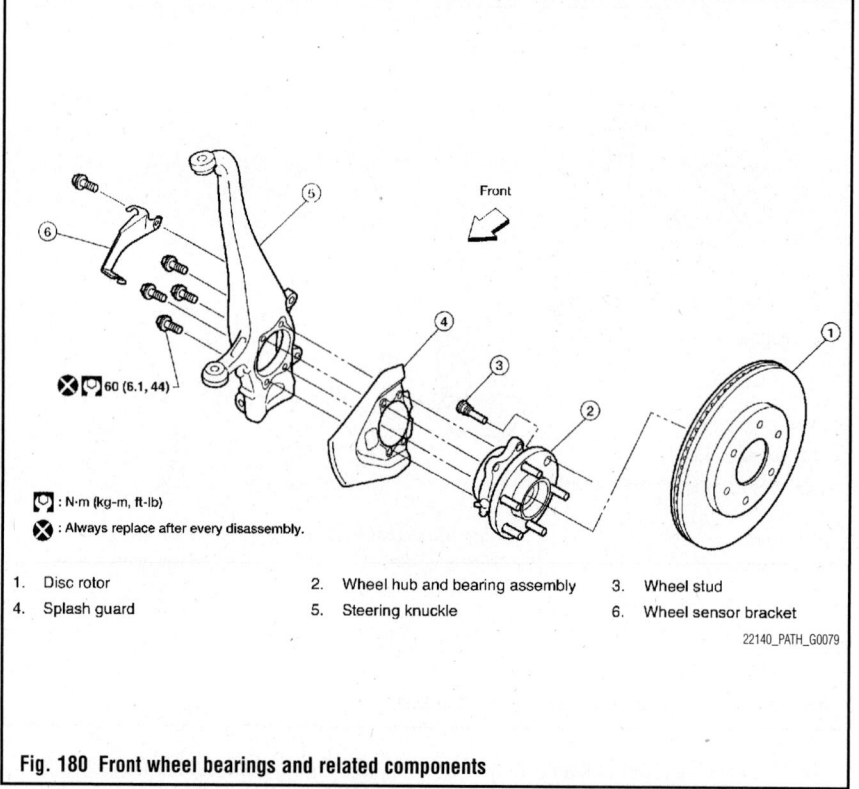

: N·m (kg-m, ft-lb)

: Always replace after every disassembly.

1.	Disc rotor	2.	Wheel hub and bearing assembly
4.	Splash guard	5.	Steering knuckle
3.	Wheel stud	6.	Wheel sensor bracket

22140_PATH_G0079

Fig. 180 Front wheel bearings and related components

   b. Clean the wheel sensor hole and mounting surface with a suitable brake cleaner and clean lint-free shop rag. Be careful that dirt and debris do not enter the axle bearing area.

   c. Apply a coat of suitable grease to the wheel sensor O-ring and mounting hole.

✳✳ WARNING

Do not pull on the wheel sensor harness.

12. Remove wheel hub and bearing assembly bolts.

13. Remove splash guard and wheel hub and bearing assembly from steering knuckle.

14. Carefully remove wheel sensor and harness through hole in splash guard.

To install:

➡ **Be sure to use new fasteners, as required.**

15. Installation is the reverse of the removal procedure.

16. Use new bolts when installing the wheel hub and bearing assembly.

17. When installing disc rotor on wheel hub and bearing assembly, position the disc rotor according to alignment mark.

18. Check and adjust alignment.

19. Be sure to perform the reconnect/relearn procedures.

COIL SPRINGS

REMOVAL & INSTALLATION

See Figure 181.

1. Before servicing the vehicle, refer to the Precautions.
2. Disconnect the negative battery cable.
3. Raise and support the vehicle safely.
4. Remove the wheel and tire assembly.
5. If removing the LH rear lower link and coil spring, remove the spare wheel and tire assembly.
6. Set a suitable jack to relieve the coil spring tension and support the rear lower link.
7. Loosen the rear lower link adjusting bolt and nut connected to the rear suspension member without removing the adjusting bolt and nut.
8. Remove the rear lower link pinch bolt and nut from the knuckle.
9. Slowly lower the rear lower link using the suitable jack to release the coil spring tension.
10. Remove the upper rubber seat, coil spring and lower rubber seat from the rear lower link.

To install:

➡ **Be sure to use new fasteners, as required.**

11. Installation is the reverse of the removal procedure.
   a. When installing the upper and lower rubber seats for the rear coil springs, the arrow embossed on the rubber seats must point out toward the wheel and tire assembly.
   b. Perform the final tightening of the rear lower link nuts and bolts (with rubber bushings) under no-load conditions with tires on level ground.

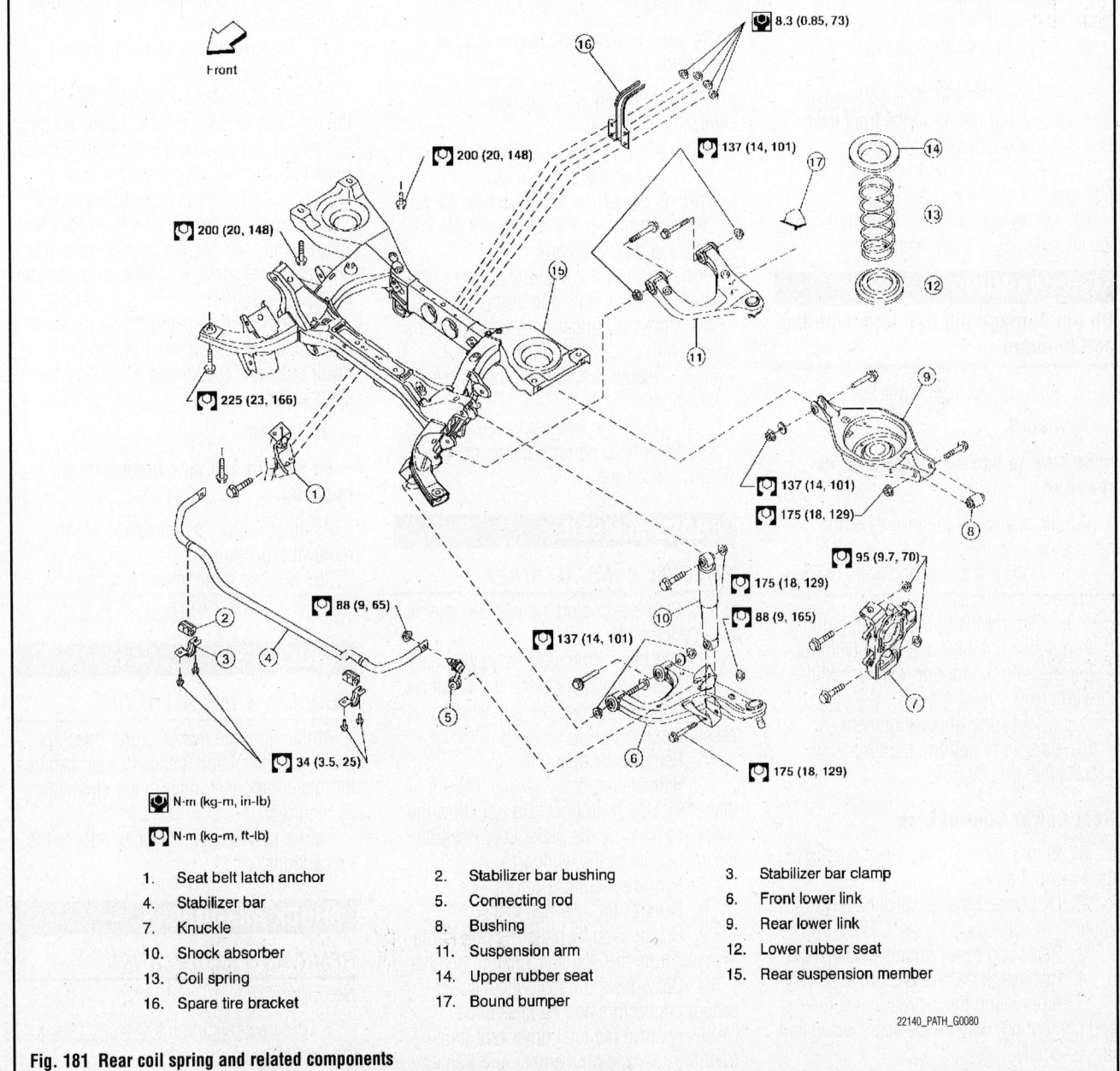

1. Seat belt latch anchor
2. Stabilizer bar bushing
3. Stabilizer bar clamp
4. Stabilizer bar
5. Connecting rod
6. Front lower link
7. Knuckle
8. Bushing
9. Rear lower link
10. Shock absorber
11. Suspension arm
12. Lower rubber seat
13. Coil spring
14. Upper rubber seat
15. Rear suspension member
16. Spare tire bracket
17. Bound bumper

Fig. 181 Rear coil spring and related components

22140_PATH_G0080

c. Tighten the nuts and bolts to specification.

d. Check the wheel alignment.

12. Be sure to perform the reconnect/relearn procedures.

CONTROL LINKS

REMOVAL & INSTALLATION

Front Lower Link

1. Before servicing the vehicle, refer to the Precautions.

2. Disconnect the negative battery cable.

3. Raise and safely support the vehicle.

4. Remove the wheel and tire assembly.

5. Remove the stabilizer bar.

6. Set a suitable jack under the rear lower link to relieve the coil spring tension.

7. Remove the shock absorber lower end bolt.

8. Remove the adjusting bolt and nut, and the bolt and nut, from the front lower link and rear suspension member.

9. Remove the front lower link pinch bolt and nut on the knuckle side.

10. Disconnect the front lower link from the knuckle using a soft hammer.

✳✳ WARNING

Do not damage the ball joint with the soft hammer.

11. Remove the front lower link.

To install:

➡**Be sure to use new fasteners, as required.**

12. Installation is the reverse of the removal procedure.

a. Tighten the nuts and bolts to specification.

b. Perform the final tightening of the front lower link nuts and bolts (with rubber bushings) under no-load conditions with tires on level ground.

c. Check the wheel alignment.

13. Be sure to perform the reconnect/relearn procedures.

Rear Lower Control Link

1. Before servicing the vehicle, refer to the Precautions.

2. Disconnect the negative battery cable.

3. Raise and safely support the vehicle.

4. Remove the wheel and tire assembly.

5. If removing the LH rear lower link and coil spring, remove the spare wheel and tire assembly.

6. Set a suitable jack to relieve the coil spring tension and support the rear lower link.

7. Loosen the rear lower link adjusting bolt and nut connected to the rear suspension member without removing the adjusting bolt and nut.

8. Remove the rear lower link pinch bolt and nut from the knuckle.

9. Slowly lower the rear lower link using the suitable jack to release the coil spring tension.

10. Remove the upper rubber seat, coil spring and lower rubber seat from the rear lower link.

11. Remove the rear lower link adjusting bolt and nut from the rear suspension member.

12. Remove the rear lower link.

To install:

➡**Be sure to use new fasteners, as required.**

13. Installation is the reverse of the removal procedure.

a. When installing the upper and lower rubber seats for the rear coil springs, the arrow embossed on the rubber seats must point out toward the wheel and tire assembly.

b. Perform the final tightening of the rear lower link nuts and bolts (with rubber bushings) under no-load conditions with tires on level ground.

c. Tighten the nuts and bolts to specification.

d. Check the wheel alignment.

14. Be sure to perform the reconnect/relearn procedures.

CROSSMEMBER

REMOVAL & INSTALLATION

1. Before servicing the vehicle, refer to the Precautions.

2. Disconnect the negative battery cable.

3. Raise and safely support the vehicle.

4. Remove the rear tire and wheel assemblies.

5. Remove the spare tire.

6. Remove the brake caliper. Do not disconnect the fluid lines. Do not allow the caliper to hang by the brake line. Position the caliper out of the work area.

7. Remove the brake rotors.

8. Remove the rear halfshafts.

9. Disconnect the parking brake cable brackets from the rear suspension member.

10. Disconnect the rear wheel speed sensor connectors and harness clips.

11. Remove the final drive vent tube from the suspension member and frame.

12. Remove the final drive assembly.

13. Position a suitable jack to support each of the rear lower links and the coil spring tension.

14. Remove both of the rear lower link outer bolts. Lower the jack to carefully remove the rear coil springs and the upper and lower rubber seats.

15. Remove the two bolts to disconnect the seat belt latch anchor from the rear suspension member.

16. Disconnect both the connecting rods from the rear stabilizer bar.

17. Position a suitable jack under each front lower link for support. Remove the front shock upper end bolts and lower end bolts.

18. Remove both shocks.

19. Position a suitable jack under the rear suspension member.

20. Remove the six rear suspension member mounting bolts.

21. Slowly lower the suitable jack and the rear suspension member to access the fuel filler tube bracket, then remove the two bolts to disconnect the fuel filler tube bracket from the rear suspension member.

22. Slowly lower the jack to remove rear suspension member, suspension arm, front and rear lower links, knuckles, and stabilizer bar as an assembly.

23. Remove the suspension arm, spare tire bracket, stabilizer bar, front and rear lower links and the knuckles from the rear suspension member.

To install:

➡**Be sure to use new fasteners, as required.**

24. Installation is the reverse of the removal procedure.

25. Be sure to perform the reconnect/relearn procedures.

KNUCKLES

REMOVAL & INSTALLATION

At this time the manufacturer does not provide removal and installation procedures for this component, refer to the illustration as required.

Before servicing the vehicle, refer to the Precautions.

SHOCK ABSORBERS

REMOVAL & INSTALLATION

See Figure 181.

1. Before servicing the vehicle, refer to the Precautions.

2. Disconnect the negative battery cable.

3. Raise and safely support the vehicle.

4. Remove the wheel and tire assembly.

5. Position a suitable jack under the front lower link to support the shock absorber.

6. Remove the shock absorber upper and lower end bolts.

7. Remove the shock absorber.

To install:

➡**Be sure to use new fasteners, as required.**

8. Installation is the reverse of the removal procedure.
   a. Tighten the nuts and bolts to specification.
   b. Check the wheel alignment.

9. Be sure to perform the reconnect/relearn procedures.

STABILIZER BAR

REMOVAL & INSTALLATION

See Figure 182.

1. Before servicing the vehicle, refer to the Precautions.

2. Disconnect the negative battery cable.

3. Raise and support the vehicle safely. Remove the tire and wheel assemblies.

4. Disconnect the stabilizer bar ends from the connecting rods.

5. Remove the stabilizer bar clamps, and remove the stabilizer bar bushings.

6. Remove the stabilizer bar.

To install:

➡**Be sure to use new fasteners, as required.**

7. Installation is the reverse of the removal procedure.

8. Install the stabilizer bar bushings and clamps so they are positioned outside of the sideslip prevention clamp on the stabilizer bar.

9. Be sure to perform the reconnect/relearn procedures.

UPPER CONTROL ARMS

REMOVAL & INSTALLATION

1. Before servicing the vehicle, refer to the Precautions.

2. Disconnect the negative battery cable.

3. Raise and safely support the vehicle.

4. Remove the rear tire and wheel assemblies.

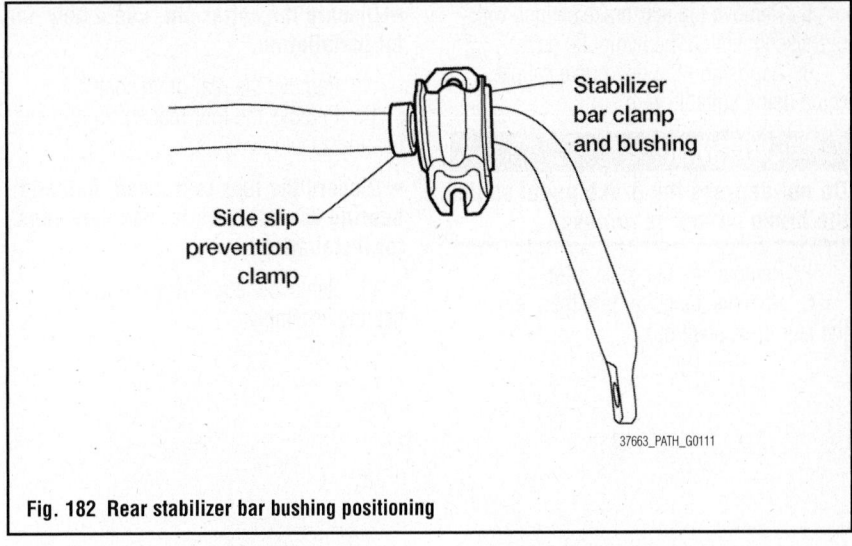

Fig. 182 Rear stabilizer bar bushing positioning

5. Remove the brake caliper. Do not disconnect the fluid lines. Do not allow the caliper to hang by the brake line. Position the caliper out of the work area.

6. Position a suitable jack under the front lower link to support the knuckle.

7. Remove the shock absorber.

8. Remove the suspension arm upper nuts and bolts on the suspension member.

9. Remove the suspension arm pinch bolt and nut on the knuckle side.

10. Disconnect the suspension arm from the knuckle using a soft hammer. Do not damage the ball joint with the hammer.

11. Remove the suspension arm.

To install:

➡**Be sure to use new fasteners, as required.**

12. Installation is the reverse of the removal procedure.

13. Be sure to perform the reconnect/relearn procedures.

WHEEL HUB & BEARING (SEALED UNIT)

REMOVAL & INSTALLATION

See Figure 183.

1. Before servicing the vehicle, refer to the Precautions.

2. Disconnect the negative battery cable.

3. Raise and support the vehicle safely.

4. Remove the wheel and tire assembly.

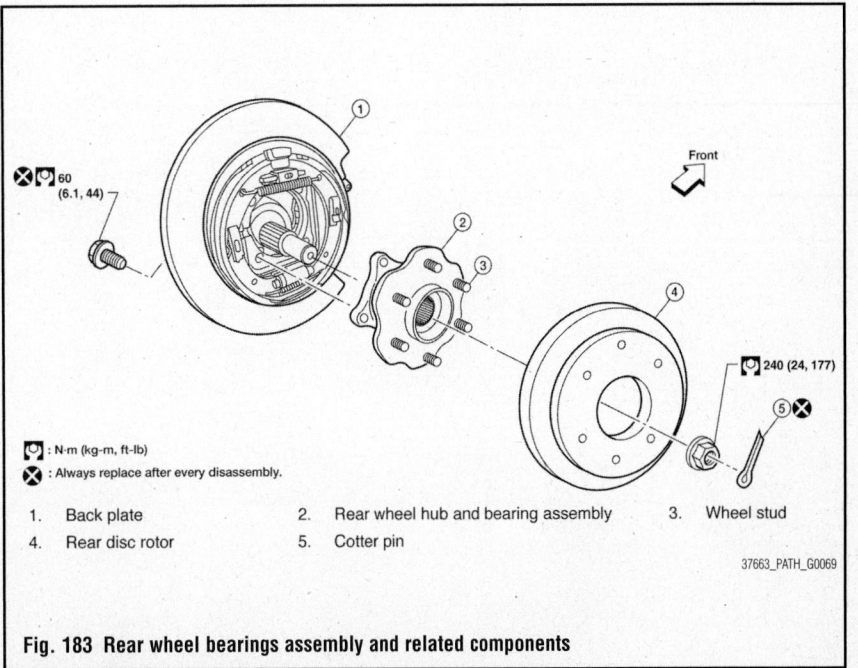

1. Back plate
2. Rear wheel hub and bearing assembly
3. Wheel stud
4. Rear disc rotor
5. Cotter pin

Fig. 183 Rear wheel bearings assembly and related components

5. Remove the rear brake caliper, without disconnecting the hydraulic hose.

6. Reposition the rear brake caliper aside using suitable wire.

✳✳ WARNING

Do not depress the brake pedal while the brake caliper is removed.

7. Remove the rear disc rotor.

8. Remove the cotter pin, then remove the rear drive shaft nut.

➡**Discard the cotter pin, use a new one for installation.**

9. Remove the rear drive shaft.

10. Remove the four rear wheel hub and bearing assembly bolts.

➡**Discard the four rear wheel hub and bearing assembly bolts, use new ones for installation.**

11. Remove the rear wheel hub and bearing assembly.

To install:

➡**Be sure to use new fasteners, as required.**

12. Installation is the reverse of the removal procedure.

   a. Use a new cotter pin for installation.

   b. Use new rear wheel hub and bearing assembly bolts for installation.

13. Be sure to perform the reconnect/relearn procedures.

NISSAN

Quest

17

SPECIFICATIONS AND MAINTENANCE CHARTS

VEHICLE AND ENGINE IDENTIFICATION CHART

		Engine Code					Model Year	
Code	Liters (cc)	Cu. In.	Cyl.	Fuel Sys.	Engine Type	Eng. Mfg.	Code ①	Year
VQ35DE	3.5 (3498)	213	6	MFI	DOHC	Nissan	B	2011
							C	2012

MFI: Multiport Fuel Injection

71075_QUES_C0001

GENERAL ENGINE SPECIFICATIONS

Year	Engine Displacement Liters (VIN)	Net Horsepower @ rpm	Net Torque @ rpm (ft. lbs.)	Bore x Stroke (in.)	Com-pression Ratio	Oil Pressure @ rpm
2011	3.5 (VQ35DE)	260@6000	240@4400	3.76X3.20	10.3	43@2000
2012	3.5 (VQ35DE)	260@6000	240@4400	3.76X3.20	10.3	43@2000

71075_QUES_C0002

ENGINE TUNE-UP SPECIFICATIONS

Year	Engine Displacement Liters (VIN)	Spark Plug Gap (in.)	Ignition Timing (deg.) MT	Ignition Timing (deg.) AT	Fuel Pump (psi)	Idle Speed (rpm) MT	Idle Speed (rpm) AT	Valve Clearance In.	Valve Clearance Ex.
2011	3.5 (VQ35DE)	0.043	NA	10-14 B	51 ②	NA	600-700	HYD	HYD
2012	3.5 (VQ35DE)	0.043	NA	10-14 B	51 ②	NA	600-700	HYD	HYD

NA: Not Applicable

NOTE: The Vehicle Emission Control Information label must be used if they differ from those in this chart.

HYD: Hydraulic

① Before top dead center

② System pressure at idle

71075_QUES_C0003

CAPACITIES

Year	Model	Engine Displacement Liters (VIN)	Engine Oil with Filter (qts.)	Transmission (pts.)			Drive Axle		Fuel Tank (gal.)	Cooling System (qts.) ①
				4-Spd	5-Spd	Auto.	Front (pts.)	Rear (pts.)		
2011	Quest	3.5 (VQ35DE)	4.9	NA	NA	22	NA	NA	20	12
2012	Quest	3.5 (VQ35DE)	4.9	NA	NA	22	NA	NA	20	12

NA: Not Applicable

NOTE: All capacities are approximate. Add fluid gradually and check to be sure a proper fluid level is obtained.

① Includes reservoir tank.

71075_QUES_C0005

FLUID SPECIFICATIONS

Year	Model	Engine Displ. Liters (VIN)	Engine Oil	Man. Trans.	Auto. Trans.	Drive Axle		Trans. Case	Power Steering Fluid	Brake Master Cylinder	Cooling System
						Front	Rear				
2011	Quest	VQ35DE	①	NA	②	NA	NA	NA	Nissan E-PSF	DOT 3	③
2012	Quest	VQ35DE	①	NA	②	NA	NA	NA	Nissan E-PSF	DOT 3	③

NA: Not Applicable

DOT: Department Of Transpotation

① API Service SM SAE 5W-30

② NISSAN CV Fluid NS-2

③ Nissan Long Life Antifreeze/Coolant

71075_QUES_C0004

VALVE SPECIFICATIONS

Year	Engine Displacement Liters (VIN)	Seat Angle (deg.)	Face Angle (deg.)	Spring Test Pressure (lbs. @ in.)	Spring Installed Height (in.)	Stem-to-Guide Clearance (in.)		Stem Diameter (in.)	
						Intake	Exhaust	Intake	Exhaust
2011	3.5 (VQ35DE)	45.15-45.45	45	37.3-42.3@ 1.457	1.457	0.0008-0.0021	0.0012-0.0022	0.2348-0.2354	0.2347-0.2350
2012	3.5 (VQ35DE)	45.15-45.45	45	37.3-42.3@ 1.457	1.457	0.0008-0.0021	0.0012-0.0022	0.2348-0.2354	0.2347-0.2350

71075_QUES_C0006

CAMSHAFT SPECIFICATIONS

All measurements are given in inches.

Year	Engine Displacement Liters	Engine VIN	Journal Dia.	Brg. Oil Clearance	Shaft End-play	Runout	Lobe Height	
							Intake	Exhaust
2011	3.5	VQ35DE	①	②	0.0045-0.0074	③	1.7900-1.7974	1.7904-1.7978
2012	3.5	VQ35DE	①	②	0.0045-0.0074	③	1.7900-1.7974	1.7904-1.7978

① No.1: 1.0211-1.0218

No.2, No.3, No.4: 0.9230-0.9238

② No.1: 1.0018-1.0034

No.2, No.3, No.4: 0.0014-0.0030

③ Less then 0.0008 (0.02 mm)

71075_QUES_C0007

CRANKSHAFT AND CONNECTING ROD SPECIFICATIONS

All measurements are given in inches.

Year	Engine Displacement Liters (VIN)	Crankshaft				Connecting Rod		
		Main Brg. Journal Dia.	Main Brg. Oil Clearance	Shaft End-play	Thrust on No.	Journal Diameter	Oil Clearance	Side Clearance
2011	3.5 (VQ35DE)	①	0.0014-0.0018	0.0039-0.0098	3	②	0.0008-0.0018	0.0079-0.0138
2012	3.5 (VQ35DE)	①	0.0014-0.0018	0.0039-0.0098	3	②	0.0008-0.0018	0.0079-0.0138

① There are 24 different grades, ranging from A (2.3612) to 7 (2.3603)

② Grade 0: 2.0460-2.0462

Grade 1: 2.0457-2.0460

Grade 2: 2.0445-2.0457

71075_QUES_C0008

PISTON AND RING SPECIFICATIONS

All measurements are given in inches.

Year	Engine Displacement Liters (VIN)	Piston Clearance	Ring Gap			Ring Side Clearance		
			Top Compression	Bottom Compression	Oil Control	Top Compression	Bottom Compression	Oil Control
2011	3.5 (VQ35DE)	0.0004-0.0012	0.0091-0.0130	0.0091-0.0130	0.0079-0.0177	0.0018-0.0031	0.0012-0.0028	0.0026-0.0049
2012	3.5 (VQ35DE)	0.0004-0.0012	0.0091-0.0130	0.0091-0.0130	0.0079-0.0177	0.0018-0.0031	0.0012-0.0028	0.0026-0.0049

71075_QUES_C0009

TORQUE SPECIFICATIONS
All readings in ft. lbs.

Year	Engine Displacement Liters (VIN)	Cylinder Head Bolts	Main Bearing Bolts	Rod Bearing Bolts	Crankshaft Damper Bolts	Flywheel Bolts	Manifold		Spark Plugs	Oil Pan Drain Plug
							Intake	Exhaust		
2011	3.5 (VQ35DE)	①	②	③	④	65	8	11	14	25
2012	3.5 (VQ35DE)	①	②	③	④	65	8	11	14	25

① Step 1: 72 ft. lbs.

   Step 2: Loosen all bolts completely

   Step 3: 29 ft. lbs.

   Step 4: +103 degrees

   Step 5: +103 degrees

② Step 1: 26 ft. lbs.

   Step 2: +90 degrees

③ Step 1: 14 ft. lbs.

   Step 2: +90 degrees

④ 33 ft. lbs. +90 degrees

71075_QUES_C0010

WHEEL ALIGNMENT

Year	Model		Caster		Camber		Toe-in (in.)
			Range (Deg.)	Preferred Setting (Deg.)	Range (Deg.)	Preferred Setting (Deg.)	
2011	Quest ①○ ②	F	③	③	④	④	⑤
		R	NA	NA	⑥	⑥	⑦
2012	Quest ①○ ②	F	③	③	④	④	⑤
		R	NA	NA	⑥	⑥	⑦

NA: Not Applicable

① Vehicle unladen

② Specifications are decimal degrees

③ Left Side Minimum: 3.92, Right Side Minimum: 4.09

   Left Side Nominal: 4.67, Right Sidfe Nominal: 4.83

   Left Side Maximum: 5.41, Right Side Maximum: 5.58

   Left and right difference -0.30 to 0.80

④ Left Minimum: - 1.00, Right Side Minimum: -1.25

   Left Side Nominal: - 0.25, Right side Nominal: -.050

   Left Side Maximum: 0.50, Right Side Maximum: 0.25

   Left and right difference -0.30 to 0.80

⑤ Minimum: 0.016

   Nominal: 0.024

   Maximum: 0.063

⑥ Minimum: -1.10

   Nominal: -0.60

   Maximum: -0.10

⑦ Minimum: 0.047

   Nominal: 0.110

   Maximum: 0.173

71075_QUES_C0012

TIRE, WHEEL AND BALL JOINT SPECIFICATIONS

Year	Model	OEM Tires		Tire Pressures (psi)		Wheel Size	Ball Joint Inspection	Lug Nut (ft. lbs.)
		Standard	Optional	Front	Rear			
2011	Quest	P225/65R16	P235/55R18	35	35	NA	①	80
2012	Quest	P225/65R16	P235/55R18	35	35	NA	①	80

NA: Not Available

OEM: Original Equipment Manufacturer

PSI: Pounds Per Square Inch

① Replace if any measurable movement is found.

71075_QUES_C0013

BRAKE SPECIFICATIONS
All measurements in inches unless noted

Year	Model		Brake Disc			Minimum Lining Thickness	Brake Caliper	
			Original Thickness	Minimum Thickness	Maximum Runout		Bracket-to-Hub Bolt (ft. lbs.)	Mounting Pin or Bolt (ft. lbs.)
2011	Quest	F	1.102	1.024	0.0016	0.079	90	54
		R	0.630	0.551	0.0020	0.079	62	32
2012	Quest	F	1.102	1.024	0.0016	0.079	90	54
		R	0.630	0.551	0.0020	0.079	62	32

F: Front

R: Rear

71075_QUES_C0011

SCHEDULED MAINTENANCE INTERVALS (1)
Nissan—Quest

TO BE SERVICED	TYPE OF SERVICE	7.5	15	22.5	30	37.5	45	52.5	60
Engine oil & filter	R	every 3,750 miles							
Brake lines & cables	S/I		✓		✓		✓		✓
Brake pads, discs	I	✓	✓	✓	✓	✓	✓	✓	✓
Brake fluid	R		✓		✓		✓		✓
Driveshaft boots	I	✓	✓	✓	✓	✓	✓	✓	✓
Automatic transmission (CVT)	I						✓		✓
Air cleaner filter ①	R				✓				✓
Drive belt (s) ②	S/I								✓
Engine coolant ③	R								
Spark plugs	R	Every 105,000 miles							
Cabin air filter	R		✓		✓		✓		✓
Exhaust system	I	✓	✓	✓	✓	✓	✓	✓	✓
Evap vapor lines	I				✓				✓
Fuel lines	I				✓				✓
Steering gear, linkage, axle & suspension parts	I	✓	✓	✓	✓	✓	✓	✓	✓
Tires (rotate)	S/I	✓	✓	✓	✓	✓	✓	✓	✓
Valve clearance ④	S/I				✓				✓

R: Replace   S/I: Service or Inspect   L: Lubricate   I: Inspect

① If operating mainly in dusty conditions, more frequent maintenance may be required.

② First at 60,000, inspect every 15,000 miles and replace as necessary

③ After 105,000, replace every 75,000 thereafter

④ Periodic maintenance not required, if valve noise increases, inspect valve clearance

Follow Periodic Maintenance Schedule 1 if the driving habits frequently include one or more of the following driving conditions:

Repeated short trips of less than 5 miles (8 km).

Repeated short trips of less than 10 miles (16 km) with outside temperatures remaining below freezing

Operating in hot weather in stop-and-go "rush hour" traffic.

Extensive idling and/or low speed driving for long distances, such as police, taxi or door-to-door delivery use

Driving in dusty conditions.

Driving on rough, muddy, or salt spread roads.

Towing a trailer, using a camper or a car-top carrier.

Follow Periodic Maintenance Schedule 2 if none of driving conditions shown in Schedule 1 apply to the driving habits.

71075_QUES_C0014

SCHEDULED MAINTENANCE INTERVALS (2)
Nissan—Quest

TO BE SERVICED	TYPE OF	7.5	15	22.5	30	37.5	45	52.5	60
Engine oil & filter	R	✓	✓	✓	✓	✓	✓	✓	✓
Brake lines & cables	S/I		✓		✓		✓		✓
Brake pads, discs	I		✓		✓		✓		✓
Brake fluid	R				✓				✓
Driveshaft boots	I		✓		✓		✓		✓
Automatic transmission (CVT)	I		✓		✓		✓		✓
Air cleaner filter ①	R				✓				✓
Drive belt (s) ②	S/I								✓
Engine coolant ③	R								
Spark plugs	R				Every 105,000 miles				
Cabin air filter	R		✓		✓		✓		✓
Exhaust system	I				✓				✓
Evap vapor lines	I				✓				✓
Ful lines	I				✓				✓
Steering gear, linkage, axle & suspension parts	I				✓				✓
Tires (rotate)	S/I	✓	✓	✓	✓	✓	✓	✓	✓
Valve clearance ④	S/I				✓				✓

R: Replace   S/I: Service or Inspect   L: Lubricate   I: Inspect

① If operating mainly in dusty conditions, more frequent maintenance may be required.

② First at 60,000, inspect every 15,000 miles and replace as necessary

③ After 105,000, replace every 75,000 thereafter

④ Periodic maintenance not required, if valve noise increases, inspect valve clearance

Follow Periodic Maintenance Schedule 1 if the driving habits frequently include one or more of the following driving conditions:

  Repeated short trips of less than 5 miles (8 km).

  Repeated short trips of less than 10 miles (16 km) with outside temperatures remaining below freezing

  Operating in hot weather in stop-and-go "rush hour" traffic.

  Extensive idling and/or low speed driving for long distances, such as police, taxi or door-to-door delivery use

  Driving in dusty conditions.

  Driving on rough, muddy, or salt spread roads.

  Towing a trailer, using a camper or a car-top carrier.

PRECAUTIONS

Before servicing any vehicle, please be sure to read all of the following precautions, which deal with personal safety, prevention of component damage, and important points to take into consideration when servicing a motor vehicle:

• Never open, service or drain the radiator or cooling system when the engine is hot; serious burns can occur from the steam and hot coolant.

• Observe all applicable safety precautions when working around fuel. Whenever servicing the fuel system, always work in a well-ventilated area. Do not allow fuel spray or vapors to come in contact with a spark, open flame, or excessive heat (a hot drop light, for example). Keep a dry chemical fire extinguisher near the work area. Always keep fuel in a container specifically designed for fuel storage; also, always properly seal fuel containers to avoid the possibility of fire or explosion. Refer to the additional fuel system precautions later in this section.

• Fuel injection systems often remain pressurized, even after the engine has been turned **OFF**. The fuel system pressure must be relieved before disconnecting any fuel lines. Failure to do so may result in fire and/or personal injury.

• Brake fluid often contains polyglycol ethers and polyglycols. Avoid contact with the eyes and wash your hands thoroughly after handling brake fluid. If you do get brake fluid in your eyes, flush your eyes with clean, running water for 15 minutes. If eye irritation persists, or if you have taken

brake fluid internally, IMMEDIATELY seek medical assistance.

• The EPA warns that prolonged contact with used engine oil may cause a number of skin disorders, including cancer. You should make every effort to minimize your exposure to used engine oil. Protective gloves should be worn when changing oil. Wash your hands and any other exposed skin areas as soon as possible after exposure to used engine oil. Soap and water, or waterless hand cleaner should be used.

• All new vehicles are now equipped with an air bag system, often referred to as a Supplemental Restraint System (SRS) or Supplemental Inflatable Restraint (SIR) system. The system must be disabled before performing service on or around system components, steering column, instrument panel components, wiring and sensors. Failure to follow safety and disabling procedures could result in accidental air bag deployment, possible personal injury and unnecessary system repairs.

• Always wear safety goggles when working with, or around, the air bag system. When carrying a non-deployed air bag, be sure the bag and trim cover are pointed away from your body. When placing a non-deployed air bag on a work surface, always face the bag and trim cover upward, away from the surface. This will reduce the motion of the module if it is accidentally deployed. Refer to the additional air bag system precautions later in this section.

• Clean, high quality brake fluid from a sealed container is essential to the safe and

proper operation of the brake system. You should always buy the correct type of brake fluid for your vehicle. If the brake fluid becomes contaminated, completely flush the system with new fluid. Never reuse any brake fluid. Any brake fluid that is removed from the system should be discarded. Also, do not allow any brake fluid to come in contact with a painted surface; it will damage the paint.

• Never operate the engine without the proper amount and type of engine oil; doing so WILL result in severe engine damage.

• Timing belt maintenance is extremely important. Many models utilize an interference-type, non-freewheeling engine. If the timing belt breaks, the valves in the cylinder head may strike the pistons, causing potentially serious (also time-consuming and expensive) engine damage. Refer to the maintenance interval charts for the recommended replacement interval for the timing belt, and to the timing belt section for belt replacement and inspection.

• Disconnecting the negative battery cable on some vehicles may interfere with the functions of the on-board computer system(s) and may require the computer to undergo a relearning process once the negative battery cable is reconnected.

• When servicing drum brakes, only disassemble and assemble one side at a time, leaving the remaining side intact for reference.

• Only an MVAC-trained, EPA-certified automotive technician should service the air conditioning system or its components.

BRAKES

GENERAL INFORMATION

PRECAUTIONS

• Certain components within the ABS system are not intended to be serviced or repaired individually.

• Do not use rubber hoses or other parts not specifically specified for and ABS system. When using repair kits, replace all parts included in the kit. Partial or incorrect repair may lead to functional problems and require the replacement of components.

• Lubricate rubber parts with clean, fresh brake fluid to ease assembly. Do not use shop air to clean parts; damage to rubber components may result.

• Use only DOT 3 brake fluid from an unopened container.

• If any hydraulic component or line is removed or replaced, it may be necessary to bleed the entire system.

• A clean repair area is essential. Always clean the reservoir and cap thoroughly before removing the cap. The slightest amount of dirt in the fluid may plug an orifice and impair the system function. Perform repairs after components have been thoroughly cleaned; use only denatured alcohol to clean components. Do not allow ABS components to come into contact with any substance containing mineral oil; this includes used shop rags.

• The Anti-Lock control unit is a microprocessor similar to other computer units in the vehicle. Ensure that the ignition switch is **OFF** before removing or installing con-

ANTI-LOCK BRAKE SYSTEM (ABS)

troller harnesses. Avoid static electricity discharge at or near the controller.

• If any arc welding is to be done on the vehicle, the control unit should be unplugged before welding operations begin.

SPEED SENSORS

REMOVAL & INSTALLATION

Front

1. Before servicing the vehicle, refer to the Precautions.
2. Remove the front wheels and tires.
3. Remove the front wheel sensor from the steering knuckle.
4. Remove the front wheel sensor harness from the vehicle.

✳ WARNING

Do not rotate, twist or pull the wheel speed sensor or harness during removal or installation.

To install:

✳ WARNING

Before installing the wheel sensor, make sure there is no damage and there are no foreign materials (such as iron fragments) on inner surface of the front wheel sensor mounting hole of steering knuckle and sensor rotor. Clean or replace as necessary.

Installation is the reverse of removal. Check that the identification line of the front wheel sensor is facing the vehicle front.

Rear

1. Before servicing the vehicle, refer to the Precautions.
2. Remove the wheel and tire.
3. Remove the rear wheel sensor from the rear axle housing.
4. Remove the rear wheel sensor harness from the vehicle.

✳ WARNING

Do not rotate, twist or pull the wheel speed sensor or harness during removal or installation.

To install:

✳ WARNING

Before installing the wheel sensor, make sure there is no damage and there are no foreign materials (such as iron fragments) on inner surface of the rear wheel sensor mounting hole of wheel hub and bearing assembly and sensor rotor. Clean or replace as necessary.

Installation is the reverse of removal. Check that grommet is fully inserted to bracket, and check that the identification line of the rear wheel sensor is facing the vehicle front.

BRAKES

BLEEDING THE BRAKE SYSTEM

BLEEDING PROCEDURE

BLEEDING PROCEDURE

✳ WARNING

Carefully monitor brake fluid level at the sub tank during bleeding operation.

✳ WARNING

Fill the sub tank with new brake fluid. Make sure it is full at all times while bleeding the air out of system.

➡Place a container under the sub tank to avoid spilling brake fluid.

➡Do not loosen the line fittings at the ABS actuator during air bleeding.

1. Turn ignition switch OFF and disconnect ABS actuator and control unit connector or negative battery terminal.

2. Connect a transparent vinyl tube and container to air bleeder valve.
3. Fully depress brake pedal several times.
4. With brake pedal depressed, open air bleeder valve to release air.
5. Close air bleeder valve.
6. Release brake pedal slowly.
7. Tighten the air bleeder valve to 71 inch lbs. (8 Nm).
8. Repeat steps 2 through 7 until no more air bubbles come out of air bleeder valve.
9. Bleed the brake hydraulic system air bleeder valves in the following
    order:
    - Right rear brake
    - Left front brake
    - Left rear brake
    - Right front brake

FLUID FILL PROCEDURE

1. Before servicing the vehicle, refer to the Precautions.
2. Turn the ignition switch OFF and disconnect the ABS actuator and electric unit (control unit) harness connector or the battery negative terminal.
3. Check that there is no foreign material in the reservoir tank and sub tank, and refill with new brake fluid. Never allow oils other than brake fluid to enter the reservoir tank.
4. Loosen the bleeder valve, slowly depress the brake pedal to the full stroke, and then release the pedal. Repeat this operation at intervals of 2 or 3 seconds until new brake fluid is discharged. Close the bleeder valve with the brake pedal depressed.
5. Repeat the same work on each wheel.
6. Perform the air bleeding.

BRAKES

FRONT DISC BRAKES

✳ CAUTION

Dust and dirt accumulating on brake parts during normal use may contain asbestos fibers from production or aftermarket brake linings. Breathing excessive concentrations of asbestos fibers can cause serious bodily harm. Exercise care when servicing brake parts. Do not sand or grind brake lining unless equipment used is designed to contain the dust residue. Do not clean brake parts with compressed air or by dry brushing. Cleaning should be done by dampen-ing the brake components with a fine mist of water, then wiping the brake components clean with a dampened cloth. Dispose of cloth and all residue containing asbestos fibers in an impermeable container with the appropriate label. Follow practices prescribed by the Occupational Safety and Health Administration (OSHA) and the Environmental Protection Agency (EPA) for the handling, processing, and disposing of dust or debris that may contain asbestos fibers.

BRAKE CALIPER

REMOVAL & INSTALLATION

See Figure 1.

1. Before servicing the vehicle, refer to the Precautions.
2. Remove the wheel and tire.
3. Secure the disc rotor using wheel nuts.
4. Drain the brake fluid.
5. Separate brake hose from caliper assembly.
6. Remove the torque member mounting bolts, and remove brake caliper assembly.

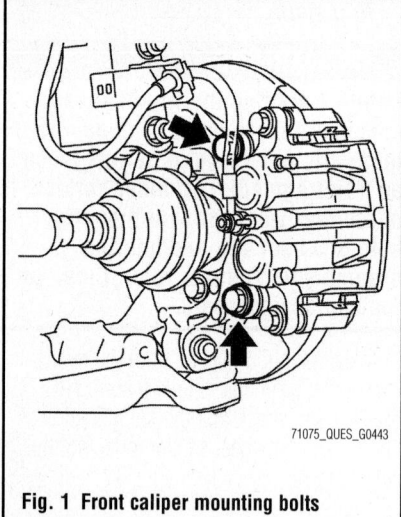

71075_QUES_G0443

Fig. 1 Front caliper mounting bolts

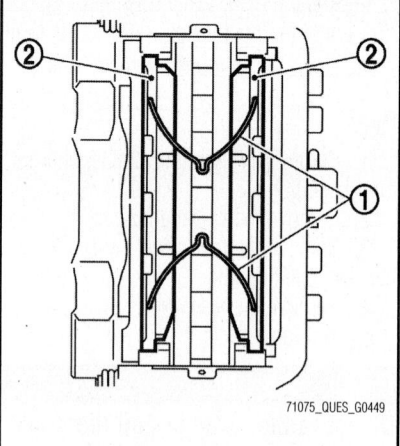

71075_QUES_G0449

Fig. 2 Install the pad return spring (1) to the brake pad (2)

To install:

➡ **Clean any dirt, grease, moisture, etc., from the brake caliper assembly mounting face, threads, mounting bolts and washers.**

7. Install the caliper assembly to the axle housing, and tighten the torque member mounting bolts to 90 ft. lbs. (121 Nm).

8. Install brake hose to the caliper assembly using a new copper washer, and tighten the bolt to 13 ft. lbs. (18 Nm).

9. Bleed the brakes.

10. Check for drag. If any drag is found:

   a. Remove brake pads.

   b. Press the pistons.

   c. Install the brake pads.

   d. Securely depress the brake pedal several times.

   e. Check for drag again. If any drag is found, disassemble the cylinder body and replace if necessary.

11. Install the wheel and tire.

BRAKE PADS

REMOVAL & INSTALLATION

See Figure 2.

❊❊ WARNING

While removing pad assemblies, do not depress brake pedal because piston will pop out.

1. Before servicing the vehicle, refer to the Precautions.

2. Remove the wheel and tire.

3. Remove sliding pin bolt (lower side) from cylinder body.

4. Suspend the cylinder body with suitable wire so that the brake hose will not stretch.

5. Remove the brake pad retainers and the pads. When removing the pad retainer from torque member, carefully lift the pad retainer upward.

6. Remove pad return spring. Be careful not to damage the pad return spring.

7. Remove the brake pads from the torque member. Remember each position of the removed brake pads. Be careful not to damage pad retainer, piston boot, pads, shims, and the shim covers.

8. Inspect the shims, shim covers, pad retainers and the torque member for rust. Clean or replace.

To install:

9. Install the pad retainers to the torque member if the pad retainers have been removed. Securely assemble the pad retainers so they will not be lifted up from the torque member.

10. Apply MOLYKOTE® AS880N or silicone-based grease to the mating faces between the brake pads and the shims, and install the shims and shim covers to the brake pads.

❊❊ WARNING

Always replace the shim and shim cover together when replacing the brake pad.

11. Apply MOLYKOTE® 7439 or equivalent to the mating faces between the brake pads and the pad retainers.

12. Install the brake pads to the torque member.

13. Install the pad return spring to the brake pad. Insert the pad return spring in to the pad return spring hole on the brake pad.

14. Install cylinder body to torque member. Do not damage the piston boot. When replacing brake pad with new one, check a brake fluid level in the reservoir tank.

➡ **Use a disc brake piston tool to press piston.**

15. Apply rubber grease to sliding pin bolt (lower side), and install the sliding pin bolt.

16. Depress the brake pedal several times to check for drag.

17. Install wheel and tire.

18. Burnish the brake contact surfaces between disc rotor and brake pads when refinishing or replacing brake rotors, after replacing pads or linings, or if a soft pedal occurs at very low mileage:

❊❊ CAUTION

Be careful of vehicle speed because the brake does not operate firmly/securely until pad and disc rotor are securely fitted. Perform this procedure under safe road and traffic conditions, and use extreme caution.

   a. Drive vehicle on straight, flat road.

   b. Depress brake pedal with the power to stop vehicle within 3 to 5 seconds until the vehicle stops.

   c. Drive without depressing brake for a few minutes to cool the brake.

   d. Repeat the above 3 steps until the pad and disc rotor are securely fitted.

BRAKES | **REAR DISC BRAKES**

Dust and dirt accumulating on brake parts during normal use may contain asbestos fibers from production or aftermarket brake linings. Breathing excessive concentrations of asbestos fibers can cause serious bodily harm. Exercise care when servicing brake parts. Do not sand or grind brake lining unless equipment used is designed to contain the dust residue. Do not clean brake parts with compressed air or by dry brushing. Cleaning should be done by dampening the brake components with a fine mist of water, then wiping the brake components clean with a dampened cloth. Dispose of cloth and all residue containing asbestos fibers in an impermeable container with the appropriate label. Follow practices prescribed by the Occupational Safety and Health Administration (OSHA) and the Environmental Protection Agency (EPA) for the handling, processing, and disposing of dust or debris that may contain asbestos fibers.

BRAKE CALIPÉR

REMOVAL & INSTALLATION

See Figure 3.

1. Before servicing the vehicle, refer to the Precautions.
2. Remove the wheel and tire.
3. Secure the disc rotor using wheel nuts.
4. Drain the brake fluid.
5. Separate brake hose from caliper assembly.
6. Remove the torque member mounting bolts, and remove brake caliper assembly.

To install:

➡ Clean any dirt, grease, moisture, etc., from the brake caliper assembly mounting face, threads, mounting bolts and washers.

7. Install the caliper assembly to the axle housing, and tighten the torque member mounting bolts to 90 ft. lbs. (121 Nm).
8. Install brake hose.
9. Bleed the brakes.
10. Check for drag. If any drag is found:
    a. Remove brake pads.
    b. Press the pistons.

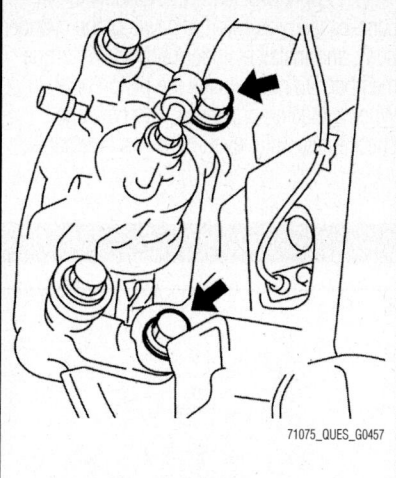

Fig. 3 Rear caliper mounting bolts

c. Install the brake pads.
d. Securely depress the brake pedal several times.
e. Check for drag again. If any drag is found, disassemble the cylinder body and replace if necessary.
11. Install the wheel and tire.

DISC BRAKE PADS

REMOVAL & INSTALLATION

See Figures 4 and 5.

✳✳ WARNING

Never depress the brake pedal while removing the brake pads because the piston may pop out.

1. Before servicing the vehicle, refer to the Precautions.
2. Remove the wheels and tires.
3. Remove sliding pin bolt (upper side).
4. Suspend the cylinder body with suitable wire so that the brake hose will not stretch.
5. Remove the brake pads from the torque member. Remember each position of the removed brake pads.

✳✳ WARNING

Do not deform the pad retainer when removing the pad retainer from the torque member. Do not damage the piston boot. Do not drop the brake pads, shims, and the shim covers.

6. Perform inspection after removal:

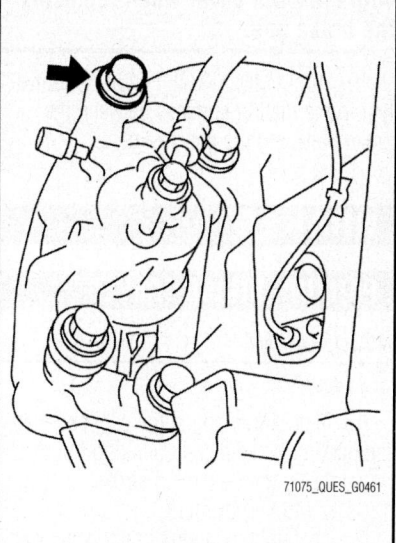

Fig. 4 Remove sliding pin bolt (upper side)

a. Replace the shims and the shim covers if rust is excessively attached.
b. Eliminate rust on the pad retainers and the torque member. Replace them if rust is excessively attached.

To install:

7. Install the pad retainers to the torque member if the pad retainers has been removed. Securely assemble the pad retainers so that it will not be lifted up from the torque member.
8. Apply MOLYKOTE® AS880N or silicone-based grease to the mating surfaces between the brake pads and the shims, and install the shims and shim covers to the brake pads.

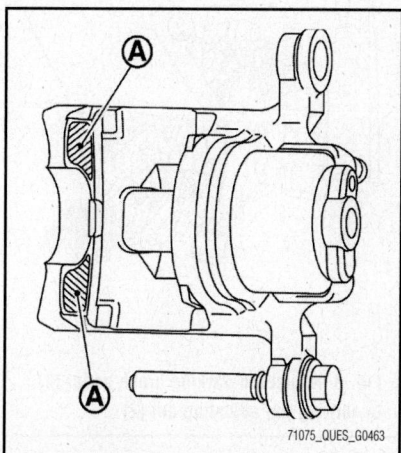

Fig. 5 Apply grease to the pawls (A) of cylinder body

> ❉❉ **WARNING**
>
> **Always replace the shim together with the shim cover when replacing the brake pad.**

9. Apply MOLYKOTE® 7439 or equivalent to the mating surfaces between the brake pads and the pad retainers.

10. Install the brake pads to the torque member.

11. Apply MOLYKOTE® AS880N or silicone-based grease to the pawls of the cylinder body, and install cylinder body to the torque member. Do not damage the piston boot. When replacing brake pad with new one, check brake fluid level in the reservoir tank.

➡ **Use a disc brake piston tool to easily press piston.**

12. Apply rubber grease to sliding pin bolt (upper side), and install the sliding pin bolt (upper side) and tighten.

13. Check brake for drag.

14. Install the wheels and tires.

BRAKES

PARKING BRAKE CABLES

ADJUSTMENT

See Figure 6.

1. Temporarily adjust the cable so that the parking brake pedal operating force immediately before the full stroke reaches 110 lbs. (490 N) or more.

2. Operate the parking brake pedal with a force of 110 lbs. (490 N) for 10 strokes or more.

3. Adjust the parking brake pedal stroke by turning the adjusting nut with a socket wrench.

> ❉❉ **WARNING**
>
> **Never reuse the adjusting nut if the nut is removed.**

4. Operate the parking brake pedal with a force of 44 ft. lbs. (196 N). Check that the pedal stroke is within 7–8 notches. (Check it by listening to the clicks of the ratchet.)

5. When parking brake warning lamp turns ON, check that the pedal stroke is

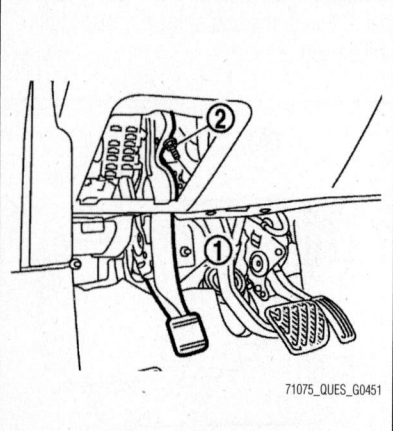

Fig. 6 Adjust the parking brake pedal (1) by turning the adjusting nut (2)

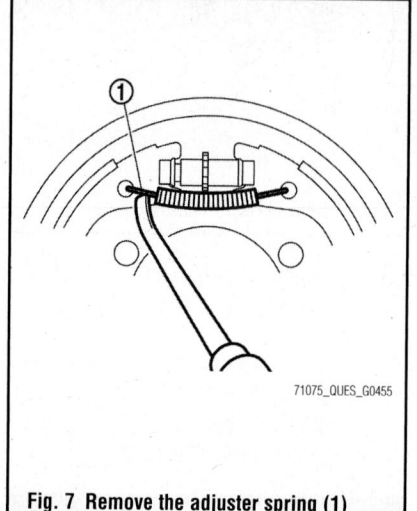

71075_QUES_G0455

Fig. 7 Remove the adjuster spring (1)

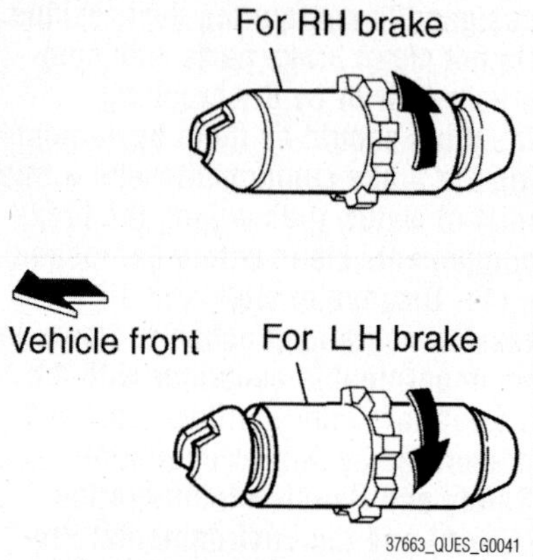

within 1 notch. (Check it by listening to the clicks of the ratchet.)

6. Rotate the disc rotor to check that there is no drag. Install the plug. If any drag is found, adjust parking brake stroke again and inspect the rear disc brake caliper.

7. With the pedal completely returned, make sure there is no drag on the rear brake.

PARKING BRAKE SHOES

REMOVAL & INSTALLATION

See Figures 7 and 8.

1. Before servicing the vehicle, refer to the Precautions.

2. Remove the rear tires.

3. Remove the disc rotor.

4. While pushing and rotating the anti-rattle pin, remove the anti-rattle pin return, retainer and spring.

> ❉❉ **WARNING**
>
> **Never drop the removed parts.**

5. Remove the adjuster spring.

6. Remove the return spring.

PARKING BRAKE

For RH brake

Vehicle front   For L H brake

37663_QUES_G0041

Fig. 8 Assemble adjusters so that threaded part is expanded when rotating it in the direction shown by arrow

7. Remove adjuster, parking brake shoes, toggle lever and anchor block.

> ❉❉ **WARNING**
>
> **The parking brake shoes for the front wheels are made of different materials from those for the rear wheels. Do not misidentify them when removing and replacing.**

To install:

Installation is the reverse of removal, noting the following:

8. Apply PBC (Poly Butyl Cuprysil) grease or silicone-based grease to the back plate and brake shoe.

9. Assemble adjusters so that threaded part is expanded when rotating it in the direction shown.

10. Shorten the adjuster by rotating it.

11. When disassembling the adjuster, apply PBC grease or silicone based grease to the threads.

12. Check that the component parts of the parking brake shoe are properly installed.

13. Check the brake shoe sliding surface and drum inner surface for grease, and wipe it off.

CHASSIS ELECTRICAL · AIR BAG (SUPPLEMENTAL RESTRAINT SYSTEM)

GENERAL INFORMATION

�֎ CAUTION

These vehicles are equipped with an air bag system. The system must be disarmed before performing service on, or around, system components, the steering column, instrument panel components, wiring and sensors. Failure to follow the safety precautions and the disarming procedure could result in accidental air bag deployment, possible injury and unnecessary system repairs.

SERVICE PRECAUTIONS

Disconnect and isolate the battery negative cable before beginning any airbag system component diagnosis, testing, removal, or installation procedures. Allow system capacitor to discharge for two minutes before beginning any component service. This will disable the airbag system. Failure to disable the airbag system may result in accidental airbag deployment, personal injury, or death.

Do not place an intact undeployed airbag face down on a solid surface. The airbag will propel into the air if accidentally deployed and may result in personal injury or death.

When carrying or handling an undeployed airbag, the trim side (face) of the airbag should be pointing towards the body to minimize possibility of injury if accidental deployment occurs. Failure to do this may result in personal injury or death.

Replace airbag system components with OEM replacement parts. Substitute parts may appear interchangeable, but internal differences may result in inferior occupant protection. Failure to do so may result in occupant personal injury or death.

Wear safety glasses, rubber gloves, and long sleeved clothing when cleaning powder residue from vehicle after an airbag deployment. Powder residue emitted from a deployed airbag can cause skin irritation. Flush affected area with cool water if irritation is experienced. If nasal or throat irritation is experienced, exit the vehicle for fresh air until the irritation ceases. If irritation continues, see a physician.

Do not use a replacement airbag that is not in the original packaging. This may result in improper deployment, personal injury, or death.

The factory installed fasteners, screws and bolts used to fasten airbag components have a special coating and are specifically designed for the airbag system. Do not use substitute fasteners. Use only original equipment fasteners listed in the parts catalog when fastener replacement is required.

During, and following, any child restraint anchor service, due to impact event or vehicle repair, carefully inspect all mounting hardware, tether straps, and anchors for proper installation, operation, or damage. If a child restraint anchor is found damaged in any way, the anchor must be replaced. Failure to do this may result in personal injury or death.

Deployed and non-deployed airbags may or may not have live pyrotechnic material within the airbag inflator.

Do not dispose of driver/passenger/curtain airbags or seat belt tensioners unless you are sure of complete deployment. Refer to the Hazardous Substance Control System for proper disposal.

Dispose of deployed airbags and tensioners consistent with state, provincial, local, and federal regulations.

After any airbag component testing or service, do not connect the battery negative cable. Personal injury or death may result if the system test is not performed first.

If the vehicle is equipped with the Occupant Classification System (OCS), do not connect the battery negative cable before performing the OCS Verification Test using the scan tool and the appropriate diagnostic information. Personal injury or death may result if the system test is not performed properly.

Never replace both the Occupant Restraint Controller (ORC) and the Occupant Classification Module (OCM) at the same time. If both require replacement, replace one, then perform the Airbag System test before replacing the other.

Both the ORC and the OCM store Occupant Classification System (OCS) calibration data, which they transfer to one another when one of them is replaced. If both are replaced at the same time, an irreversible fault will be set in both modules and the OCS may malfunction and cause personal injury or death.

If equipped with OCS, the Seat Weight Sensor is a sensitive, calibrated unit and must be handled carefully. Do not drop or handle roughly. If dropped or damaged, replace with another sensor. Failure to do so may result in occupant injury or death.

If equipped with OCS, the front passenger seat must be handled carefully as well.

When removing the seat, be careful when setting on floor not to drop. If dropped, the sensor may be inoperative, could result in occupant injury, or possibly death.

If equipped with OCS, when the passenger front seat is on the floor, no one should sit in the front passenger seat. This uneven force may damage the sensing ability of the seat weight sensors. If sat on and damaged, the sensor may be inoperative, could result in occupant injury, or possibly death.

DISARMING THE SYSTEM

To disarm the **SRS** system turn the ignition switch to the **OFF** position. Disconnect both battery cables starting with the negative cable first and wait at least 3 minutes after the cables are disconnected.

ARMING THE SYSTEM

To rearm the **SRS** system, turn the ignition switch to the **OFF** position. Connect both battery cables starting with the positive cable first.

CLOCKSPRING CENTERING

See Figure 9.

1. Carefully turn the spiral cable clockwise to the end position. Turn it counter-

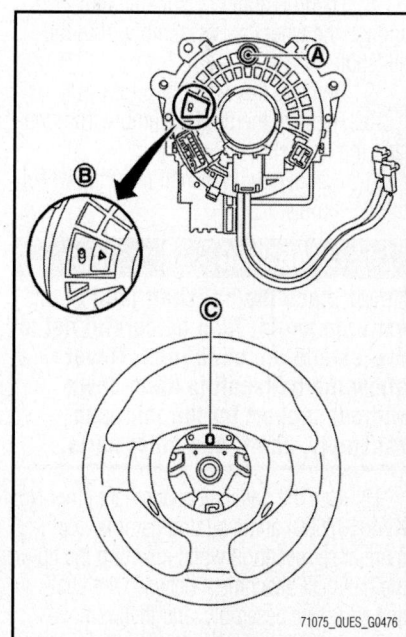

71075_QUES_G0476

Fig. 9 Turn the spiral cable counterclockwise, and stop turning at the mark (B), when the stopper insertion holes are in the same position; adjust the spiral cable locating pin (A) to the steering wheel locating pin hole (C)

clockwise (about 2 and a half turns) and stop turning at the mark when the stopper insertion holes are in the same position.

2. The service part is installed in the neutral position by the stopper and can be set without adjusting after the stopper is removed.

3. Never over-turn the spiral cable or go beyond the number of turns required. (This causes the cable to snap)

4. Adjust the spiral cable locating pin to the steering wheel locating pin hole.

5. Fix the driver air bag module harnesses to the harness fixing hook.

DRIVE TRAIN

FRONT HALFSHAFT

REMOVAL & INSTALLATION

Left

See Figure 10.

1. Before servicing the vehicle, refer to the Precautions.
2. Remove the wheel and tire.
3. Remove the wheel sensor.
4. Remove the lock plate from strut assembly.
5. Remove the caliper assembly. Hang the caliper assembly aside, so as not to interfere with work.

✳✳ WARNING

Never depress brake pedal while brake caliper is removed.

6. Remove the disc rotor.
7. Remove the cotter pin, and loosen the wheel hub lock nut.
8. Using a puller or suitable tool, disengage the wheel hub assembly from the halfshaft.
9. Remove the wheel hub lock nut.
10. Remove the strut assembly from the steering knuckle.
11. Remove the halfshaft from the wheel hub assembly.

✳✳ WARNING

Never place the halfshaft joint at an extreme angle. Also be careful not to overextend the slide joint. Never allow the halfshaft to hang down without support for the joint subassembly, shaft, and other parts.

12. Use the halfshaft attachment (Tool No. KV40107500) and a sliding hammer (commercial service tool) while inserting the tip of the halfshaft attachment between the shaft and transaxle assembly, and then remove halfshaft from the transaxle assembly.

13. Perform an inspection after removal; check the following items, and replace the part if necessary:

a. Move the joint up/down, left/right, and in the axial directions. Check for motion that is not smooth, and check for significant looseness.

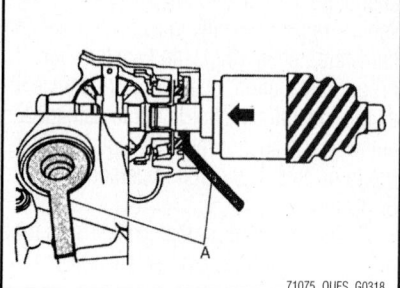

71075_QUES_G0318

Fig. 10 Place the protector tool (A) onto the transaxle assembly to prevent damaging the oil seal—left side

b. Check the boot for cracks, damage, and leakage of grease.

c. Check the support bearing bracket for cracks, deformation and other damage.

To install:

14. Installation is in the reverse order of removal, noting the following:

a. Always replace the transaxle side oil seal with new one when installing the halfshaft. Place the protector [SST: KV38107900] onto the transaxle assembly to prevent damage to the oil seal while inserting the halfshaft. Slide the halfshaft sliding joint and tap with a hammer to install securely. Check that circular clip is completely engaged.

b. Clean the matching surface of the wheel hub lock nut and the wheel hub assembly. Never apply lubricating oil to these matching surfaces.

c. Clean the matching surface of the halfshaft and wheel hub assembly. Apply paste [service part (440037S000)] to the surface of the halfshaft joint subassembly. Apply paste to cover the entire flat surface of the halfshaft joint subassembly. Paste amount: 0.04–0.10 oz (1.0–3.0 g)

d. Tighten the wheel hub lock nut to 152–155 ft. lbs. (206–211 Nm).

✳✳ WARNING

Since the halfshaft is assembled by press-fitting, use the tightening torque range for the wheel hub lock nut. Be sure to use a torque wrench to tighten the wheel hub lock nut. Never use a power tool. Never reuse wheel hub lock nut.

➡**Wheel hub lock nut tightening torque does not over torque for avoiding axle noise, and does not less than torque for avoiding looseness.**

e. When reusing the disc rotor, align the matchmarks that were made during removal.

f. When installing a cotter pin, securely bend the basal portion to prevent rattles. Never reuse cotter pin. Bend the cotter pin at the root sufficiently to prevent any looseness.

g. Perform the final tightening of each of part under unladen conditions.

h. Check the wheel sensor harness for proper connection.

i. Check the wheel alignment.

j. Adjust the neutral position of steering angle sensor.

Right

See Figure 11.

1. Before servicing the vehicle, refer to the Precautions.
2. Remove the wheel and tire.
3. Remove the wheel sensor.
4. Remove the lock plate from strut assembly.
5. Remove the caliper assembly. Hang the caliper assembly aside, so as not to interfere with work.

✳✳ WARNING

Never depress brake pedal while brake caliper is removed.

6. Remove the disc rotor.
7. Remove the cotter pin, and loosen the wheel hub lock nut.
8. Using a puller or suitable tool, disengage the wheel hub assembly from the halfshaft.
9. Remove the wheel hub lock nut.
10. Remove the strut assembly from the steering knuckle.
11. Remove the halfshaft from the wheel hub assembly.

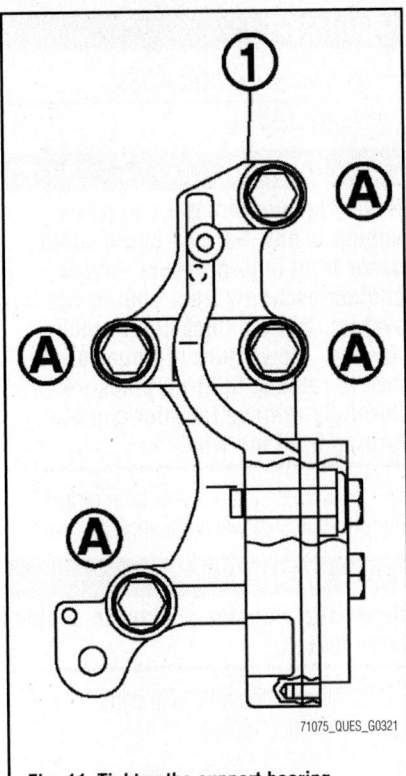

Fig. 11 Tighten the support bearing bracket (1) mounting bolts (A)

❄❄ WARNING

Never place the halfshaft joint at an extreme angle. Also be careful not to overextend the slide joint. Never allow the halfshaft to hang down without support for the joint sub-assembly, shaft, and other parts.

12. Remove the splash guard.
13. Remove the bearing housing mounting bolts.

14. Remove the halfshaft assembly from the transaxle assembly.
15. Remove the heat insulator from the front suspension member.
16. Remove the front exhaust tube.
17. Remove the three way catalyst (bank 1) and heated oxygen sensor harness bracket.
18. Remove the support bearing bracket.
19. Perform inspection after removal:
20. Perform an inspection after removal; check the following items, and replace the part if necessary:
   a. Move the joint up/down, left/right, and in the axial directions. Check for motion that is not smooth, and check for significant looseness.
   b. Check the boot for cracks, damage, and leakage of grease.
   c. Check the support bearing bracket for cracks, deformation and other damage.

To install:

21. Installation is in the reverse order of removal, noting the following:
   a. Always replace the transaxle side oil seal with new one when installing the halfshaft. Place the protector [SST: KV38107900] onto the transaxle assembly to prevent damage to the oil seal while inserting the halfshaft. Slide the halfshaft sliding joint and tap with a hammer to install securely. Check that circular clip is completely engaged.
   b. Temporarily tighten the support bearing bracket mounting bolts, and then tighten them to 35 ft. lbs. (48 Nm).
   c. Insert the halfshaft into the transaxle assembly; temporarily tighten the bearing housing mounting bolts, and then tighten them to 35 ft. lbs. (48 Nm).
   d. Clean the matching surface of the wheel hub lock nut and the wheel hub assembly. Never apply lubricating oil to these matching surfaces.
   e. Clean the matching surface of the halfshaft and wheel hub assembly. Apply paste [service part (440037S000)] to the surface of the halfshaft joint sub-assembly. Apply paste to cover the entire flat surface of the halfshaft joint sub-assembly. Paste amount: 0.04–0.10 oz (1.0–3.0 g)
   f. Tighten the wheel hub lock nut to 152–155 ft. lbs. (206–211 Nm).

❄❄ WARNING

Since the halfshaft is assembled by press-fitting, use the tightening torque range for the wheel hub lock nut. Be sure to use a torque wrench to tighten the wheel hub lock nut. Never use a power tool. Never reuse wheel hub lock nut.

➥**Wheel hub lock nut tightening torque does not over torque for avoiding axle noise, and does not less than torque for avoiding looseness.**

   g. When reusing the disc rotor, align the matchmarks that were made during removal.
   h. When installing a cotter pin, securely bend the basal portion to prevent rattles. Never reuse cotter pin. Bend the cotter pin at the root sufficiently to prevent any looseness.
   i. Perform the final tightening of each of part under unladen conditions.
   j. Check the wheel sensor harness for proper connection.
   k. Check the wheel alignment.
   l. Adjust the neutral position of steering angle sensor.

ENGINE COOLING

ENGINE COOLANT

DRAIN & REFILL

❄❄ CAUTION

To avoid being scalded, never change the coolant when the engine is hot. Wrap a thick cloth around cap and carefully remove the cap. First, turn the cap a quarter of a turn to release built-up pressure. Then push down and turn the cap all the way to remove.

1. Remove engine undercover.
2. Open the radiator drain plug at the bottom of radiator and remove the radiator cap. When draining all of engine coolant in the system, open water drain plugs on cylinder block.
3. Remove the reservoir tank if necessary. Drain engine coolant and clean the reservoir tank before installing.

❄❄ WARNING

Do not allow coolant to spill on the drive belts.

4. Check the drained engine coolant for contaminants such as rust, corrosion or discoloration. If contaminated, flush the engine cooling system.

To refill:

❄❄ WARNING

Do not reuse O-rings. Do not put additive such as water leak preventive, since it may cause cooling waterway clogging. Use Genuine NISSAN Long Life Antifreeze/Coolant (blue) or equivalent, mixed with water (distilled or demineralized).

5. Install the reservoir tank, if removed, and the radiator drain plug. Be sure to clean drain plug and install with new O-ring. Tighten to 11 inch lbs. (1 Nm). If the water drain plugs on cylinder block are removed, close and tighten them.

6. Check that each hose clamp has been firmly tightened.

7. Remove the air duct assembly and air cleaner case (upper and lower) assemblies.

8. Remove the heater pipe air bleeder plug, and the rear heater air bleeder plug, as applicable.

9. Fill the radiator, and reservoir tank, if removed, to specified level. Slowly pour engine coolant through the coolant filler neck [(less than 2 ⅛ qt (2 L) a minute] to allow air in the system to escape. When coolant comes out from the heater pipe air bleeder plug, tighten the bleeder plug.

10. Install the air duct assembly and air cleaner case (upper and lower) assemblies.

11. Install the radiator cap.

12. Start the engine.

13. Maintain the engine at 1,800 rpm for approximately 10 seconds. Stop the engine.

14. Remove the radiator cap with the engine cold (approx. 50°C or less), and check the cooling water level. If the fluid level is low, refill with cooling water to the lip of radiator.

15. Repeat steps 7 to 9 four times.

16. Cap the radiator, and start the engine.

17. Warm up the engine until the thermostat opens. Standard for warming-up time is approximately 10 minutes at 3,000 rpm. Check thermostat opening condition by touching radiator hose (lower) for a flow of warm water.

❈❈ WARNING

Watch water temperature gauge so as not to overheat engine.

18. Stop the engine and cool down to less than approximately 122°F (50°C). Cool down using fan to reduce the time. If necessary, refill the radiator up to filler neck with engine coolant.

19. Refill the reservoir tank to "MAX" level line with engine coolant.

20. Repeat steps 7 through 10 two or more times with radiator cap installed until engine coolant level no longer drops.

21. Check the cooling system for leakage with the engine running.

22. Warm up the engine, and check for sound of engine coolant flow while running engine from idle up to 3,000 rpm with heater temperature controller set at several positions between "COOL" and "WARM". Sound may be noticeable at heater unit.

23. Repeat step 13 three times.

24. If sound is heard, bleed air from cooling system by repeating step 5 and steps 7 to 14 until engine coolant level no longer drops.

FLUSHING

1. Install the reservoir tank, if removed, and the radiator drain plug. Be sure to clean drain plug and install with new O-ring. Tighten to 11 inch lbs. (1 Nm). If water drain plugs on cylinder block are removed, close and tighten them.

2. Remove the air duct assembly and air cleaner case (upper and lower) assemblies.

3. Remove the air bleeder plug.

4. Fill with cooling water until it overflows from the bleeder plug.

5. Install the air bleeder plug.

6. Fill the radiator and reservoir tank with water and reinstall the radiator cap. When water overflows the disconnected heater hose, connect the heater hose, and continue filling with water.

7. Install the air duct assembly and air cleaner case (upper and lower) assemblies.

8. Run the engine and warm it up to normal operating temperature.

9. Rev the engine two or three times under no-load.

10. Stop the engine and wait until it cools down.

11. Drain water from the system.

12. Repeat steps 1 through 11 until clear water begins to drain from radiator.

ELECTRIC ENGINE FAN

REMOVAL & INSTALLATION

1. Before servicing the vehicle, refer to the Precautions.

2. Disconnect the negative battery cable.

3. Remove the following parts:
- Engine under cover
- Air duct (inlet)
- Oil level gauge
- Battery and battery tray

4. Drain engine coolant from radiator. Perform this step engine is cold.

❈❈ WARNING

Never spill engine coolant on drive belt.

5. Remove the radiator cap adapter and the radiator hoses (upper) and radiator pipe (upper) assembly.

6. Disconnect the harness connector from the crash zone sensor, and move the harness aside.

7. Disconnect the harness connector from the right-hand and left-hand fan motors, and move the harness aside.

8. Remove the cooling fan assembly.

To install:
Installation is the reverse of removal.

RADIATOR

REMOVAL & INSTALLATION
See Figure 12.

❈❈ CAUTION

Never remove radiator cap when engine is hot. Serious burns could occur from high-pressure engine coolant escaping from engine cooling system. Wrap a thick cloth around the cap. Slowly turn it a quarter of a turn to release built-up pressure. Carefully remove radiator cap by turning it all the way.

1. Drain engine coolant from radiator. Perform this step when the engine is cold.

❈❈ WARNING

Never allow engine coolant to contact drive belt.

2. Remove the following parts:
- Air duct (inlet)
- Front grille
- Hood lock

3. Disconnect the reservoir tank hose from radiator pipe (upper).

4. Disconnect the CVT fluid cooler hoses from the radiator. Install the blind plug to avoid leakage of CVT fluid.

5. Separate the low pressure flexible hose from low pressure pipe, and move the separated hose.

6. Remove the radiator cap adapter and radiator hoses (upper and lower) and the radiator pipe (upper) assembly.

7. Remove the condenser. Be careful not to damage condenser core.

8. Remove the radiator upper clips by pulling the tabs to release the lock, and then remove the mounting rubbers (upper).

9. Lift up and remove the radiator from the front of the radiator core support.

To install:
Installation is the reverse of removal, noting the following:

10. Do not reuse O-ring.

11. When installing the radiator core support (upper), check that the radiator and air condenser upper and lower mounts are securely inserted in each mounting hole of the radiator core support (upper and lower).

12. Install each radiator upper clips on radiator core connection as follows:

a. Install the mounting rubbers (upper) on the radiator mounting pins.

b. Align the radiator upper clips with

Radiator hose	Hose end	Paint mark	Position of hose clamp*
	Rdiator side	Right side	C
Radiator hose (upper)	Radiator side (radiator pipe)	—	B
	Engine side (radiator pipe)	—	C
	Engine side	Upper	D
	Radiator side	Upper	A
Rdiator hose (lower)	Radiator side (radiator pipe)	—	A
	Engine side (radiator pipe)	—	A
	Engine side	Upper	E

*:Refer to the illustrations for the specific position each hose clamp tab.

71075_QUES_G0038

Fig. 12 Hose clamp positions (*: Refer to the illustrations for the specific position each hose clamp tab)

the radiator core connection, and insert the radiator upper clips straight into the radiator core connections until a click is heard.

c. After connecting the radiator upper clips, use the following method to ensure they are fully connected:
- Visually confirm radiator upper clips are connected to radiator core connections
- Move radiator upper clips and the radiator forward and backward to check they are securely connected

13. Insert the reservoir tank straight to the mounting position.

14. When installing the radiator hose, insert the hose all the way to the stopper, or by 1.30 in. (33 mm) (hose without a stopper).

a. For the orientation of the hose clamp pawl, refer to the chart and figure.

b. The angle created by the hose clamp pawl and the specified line must be within plus or minus 15° as shown in the figure.

c. To install the hose clamps, check that the dimension from the end of the paint mark on the radiator hose to the hose clamp is within (−0.04) − (0.04) in. [(−1) − (+1) mm].

15. After installation, check for leakage of engine coolant using the radiator cap tester adapter (commercial service tool) and the radiator cap tester (commercial service tool). Start and warm up the engine. Visually check that there is no leakage of engine coolant and CVT fluid.

THERMOSTAT

REMOVAL & INSTALLATION

See Figure 13.

1. Before servicing the vehicle, refer to the Precautions.

2. Disconnect the negative battery cable.

3. Drain the engine coolant from radiator drain plug at the bottom of radiator, and from water drain plug at the front of cylinder block. Perform this step when the engine is cold.

❋❋ WARNING

Never spill engine coolant on drive belt.

4. Remove the radiator reservoir tank, and move it aside.

5. Remove the camshaft sprocket cover (bank 2).

6. Disconnect the radiator hose (lower) from the water inlet and thermostat assembly.

7. Remove the water inlet and thermostat assembly and gasket.

❋❋ WARNING

Never disassemble the water inlet and thermostat assembly. Replace them as a unit, if necessary.

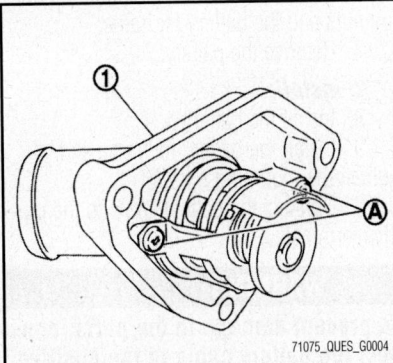

71075_QUES_G0004

Fig. 13 Remove water inlet and thermostat assembly (1) and gasket; Never loosen these screws (A)

To install:

Installation is the reverse of removal. Be careful not to spill engine coolant over engine room. Use rag to absorb engine coolant.

WATER PUMP

REMOVAL & INSTALLATION

See Figure 14.

1. Before servicing the vehicle, refer to the Precautions.

2. Disconnect the negative battery cable.

❋❋ WARNING

When removing the water pump assembly, be careful not to get engine coolant on the drive belt. The water pump cannot be disassembled and should be replaced as a unit.

3. Remove the following parts:
- Cowl top extension
- Engine under cover
- Front road wheel and tire
- Splash guard (Right-hand)

4. Drain engine coolant from the radiator. Perform this step when the engine is cold.

5. Remove the drive belt.

6. Remove the idler pulleys and harness bracket.

7. Remove the torque rod (upper).

8. Remove the water drain plug (front) on the water pump side of the cylinder block to drain engine coolant from the engine.

9. Remove the valve timing control cover (bank 1) and the water pump cover from the front timing chain case. Use the seal cutter [SST: KV10111100 (J-37228)] to cut the liquid gasket for removal.

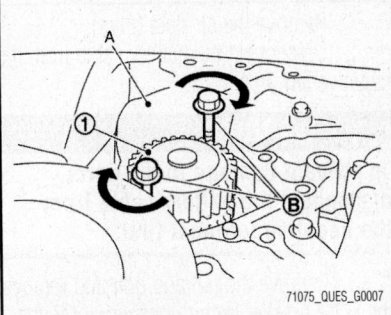

71075_QUES_G0007

Fig. 14 Screw M8 x 1.25 bolts (B) (threads length: approx. 50 mm) into the upper and lower bolt holes, and alternately tighten the bolts tighten each bolt for a half turn and pull out water pump (1)

10. Remove the primary timing chain tensioner.

11. Remove the water pump as follows:

➡ **To simplify the procedure, use a 2.36 in. (60 mm) x 5.51 in. (140 mm) thin plastic sheet.**

a. Remove three water pump mounting bolts.

b. Turn the crankshaft pulley counterclockwise to slack the water pump side timing chain for allowance. With the timing chain around water pump sprocket held with fingers, turn the crank pulley counterclockwise to the position where the timing chain cannot be pulled in the cylinder head direction.

c. Insert the plastic sheet between the timing chain and the water pump sprocket.

d. Screw M8 x 1.25 bolts (threads length: approx. 50 mm) into the upper and lower bolt holes used for mounting the water pump, and tighten the bolts alternately by half turns all the way to pull out water pump. Pull the water pump out straight to protect the mounting part from damage caused by vane. Do not allow the sprocket to interfere with the timing chain.

12. Remove the M8 bolt and two O-rings from the water pump.

To install:

※ **WARNING**

After installing the water pump, connect the hose and clamp securely,

and check for leakage using the radiator cap tester (commercial service tool) and the radiator cap tester adapter (commercial service tool).

※ **WARNING**

Do not reuse the O-rings. Each O-ring is made of different material. Always install appropriate O-ring to the correct position.

13. Install new O-rings to the water pump. Clean the mounting grooves of the O-ring to remove foreign matter. Install the O-ring with a white identification mark to the engine front side (sprocket side) and the O-ring with no identification mark to the engine rear side (vane side).

14. Install the water pump.

a. Insert the water pump by widening the timing chain with the sheet so as not to allow the O-ring to interfere with the timing chain and the corner of the cylinder block.

- Apply engine oil around the O-ring with a white identification mark
- Apply LLC around the O-ring with no identification mark
- Insert the water pump directly to the cylinder block
- Do not allow the O-ring to be displaced
- Press into the water pump by tightening the mounting bolts alternately and evenly
- Check the timing chain and the water pump sprocket for proper engagement

15. Install the primary timing chain tensioner.

16. Install the cam sprocket cover (Bank 1).

17. Install the water pump cover:

a. Remove old liquid gasket on the back of the cover, the front timing chain case mounting surface, mounting bolt threads, and the bolt hole, by using a scraper. Clean the mounting surface with white gasoline.

b. Apply liquid gasket evenly without any gap and overlap.

c. Tighten the mounting bolts.

18. Install the water drain plug (front). Apply liquid gasket to the threads and tighten the screw.

19. Install the remaining items in the reverse order of removal.

➡ **When starting the engine or immediately after an engine start after the removal/installation procedure of chain tensioner (primary), slapping sound may be heard. This is not a malfunction. The slapping sound is generated due to decrease in oil pressure of the chain tensioner (primary) and eliminated as oil pressure increases.**

20. Check for leakage of engine coolant using the radiator cap tester adapter (commercial service tool) and the radiator cap tester (commercial service tool).

21. Start and warm up the engine. Visually check that there is no leakage of engine coolant.

ENGINE ELECTRICAL

BATTERY

REMOVAL & INSTALLATION

1. Remove the air duct (inlet).
2. Disconnect the battery cable from the negative terminal.

※ **WARNING**

To prevent damage to the parts, disconnect the battery cable from the negative terminal first.

3. Remove the nut and bolt and remove the ECM bracket from the battery fix frame.
4. Remove the battery positive terminal cover.

5. Disconnect the battery cable from the positive terminal.
6. Remove the battery fix frame mounting nuts and the battery fix frame.
7. Remove the battery.

To install:

8. Install the battery.
9. Install the battery fix frame and tighten the mounting nuts.
10. Connect the battery cable to the positive terminal.

※ **WARNING**

To prevent damage to the parts, connect the battery cable to the positive terminal first.

BATTERY SYSTEM

11. Install the battery positive terminal cover.
12. Install the ECM bracket and tighten the nut and bolt.
13. Connect the negative battery cable.
14. After connecting the battery cables, to securely supply battery voltage, ensure that they are tightly clamped to battery terminals for good contact.
15. To securely supply battery voltage, check battery terminal for poor connection caused by corrosion.
16. Reset electronic systems as necessary.

ENGINE ELECTRICAL

CHARGING SYSTEM

ALTERNATOR

REMOVAL & INSTALLATION

1. Disconnect the negative battery cable.
2. Remove the air duct (inlet).
3. Remove the reservoir tank.
4. Disconnect the alternator harness connector.
5. Remove the "B" terminal harness nut, and then disconnect "B" terminal harness.
6. Remove the alternator mounting bolt (upper).
7. Remove the engine under cover.
8. Remove the right front wheel.
9. Remove the right-hand splash guard.
10. Remove the drive belt.
11. Remove the idler pulley.
12. Remove the compressor. Never disconnect low-pressure flexible hose and high-pressure flexible hose from compressor.
13. Remove the water pipe. Never disconnect water hose from water pipe.
14. Remove the return tube. Never disconnect return tube from return hose assembly.
15. Remove the alternator mounting bolt (lower) and alternator mounting nut (lower).
16. Remove the alternator from the right side of the vehicle. Be careful not to contact with and damage surrounding parts when removing alternator from the vehicle.

To install:
Installation is the reverse of removal, noting the following:
17. Temporarily tighten all of alternator bolts and nut. Next, tighten them in numerical order as shown.
18. Check the belt tension.
19. Be careful to tighten "B" terminal nut carefully.
20. Inspect the power generation voltage variable control system operation, and make sure the system operates normally.

ENGINE ELECTRICAL

IGNITION SYSTEM

FIRING ORDER

Firing order: 1–2–3–4–5–6

IGNITION COILS

REMOVAL & INSTALLATION

1. Before servicing the vehicle, refer to the Precautions.
2. Disconnect the negative battery cable.
3. Remove the engine cover.
4. Remove air cleaner cases (upper and lower) and air duct assembly.
5. Remove intake manifold collector.
6. Remove ignition coil.

To install:
Installation is the reverse of removal.

IGNITION TIMING

INSPECTION

See Figure 15.

1. Attach timing light to loop wires as shown.

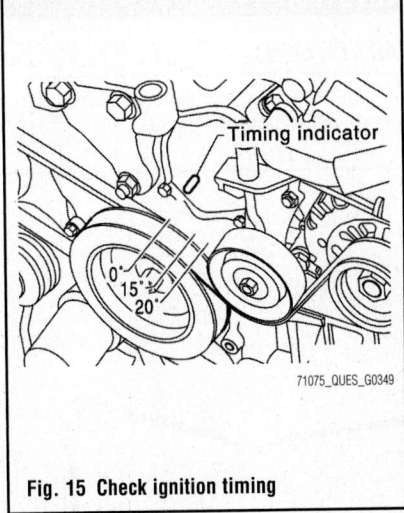

71075_QUES_G0349

Fig. 15 Check ignition timing

2. Check ignition timing under the following conditions:
- A/C switch: OFF
- Electric load: OFF (Lights, heater fan & rear window defogger)
- Steering wheel: Kept in straight-ahead position
3. Ignition timing:
- No load (in P or N position): 12 plus or minus 2° BTDC

SPARK PLUGS

REMOVAL & INSTALLATION

1. Before servicing the vehicle, refer to the Precautions.
2. Disconnect the negative battery cable.
3. Remove the engine cover.
4. Remove air cleaner cases (upper and lower) and air duct assembly.
5. Remove intake manifold collector.
6. Remove ignition coil.
7. Remove spark plug with a spark plug wrench.

To install:
Installation is the reverse of removal.

ENGINE ELECTRICAL

STARTER

REMOVAL & INSTALLATION

1. Before servicing the vehicle, refer to the Precautions.
2. Disconnect the negative battery cable.
3. Remove the battery.
4. Remove air duct (inlet) and air cleaner assembly.
5. Disconnect the TCM and ECM harness connectors.
6. Remove harness fixing clips and harness mounting bolt.

7. Remove ECM bracket mounting bolts, and then remove ECM bracket.
8. Remove battery tray mounting bolts, and then remove battery tray.
9. Disconnect "S" terminal harness connector
10. Remove "B" terminal harness nut, and then disconnect "B" terminal harness.
11. Remove the left-hand splash guard.
12. Remove starter motor mounting bolts form the left side of the vehicle and the engine room.
13. Remove starter motor form the vehicle.

To install:
Installation is the reverse of removal.

➡ Be careful to tighten "B" terminal nut to the specified torque. To prevent damage to the parts, connect the battery cable to the positive terminal first. To prevent damage to the vehicle, after connecting battery cables, ensure that they are tightly clamped to battery terminals for good contact. To prevent damage to the parts, check battery terminal for poor connection caused by corrosion.

➡ Reset electronic systems as necessary.

ENGINE MECHANICAL

➡ Disconnecting the negative battery cable may interfere with the functions of the on board computer systems and may require the computer to undergo a relearning process, once the negative battery cable is reconnected.

ACCESSORY DRIVE BELT SYSTEM

ADJUSTMENT

Belt tension is automatically adjusted by the drive belt auto-tensioner.

BELT ROUTINGS

See Figure 16.

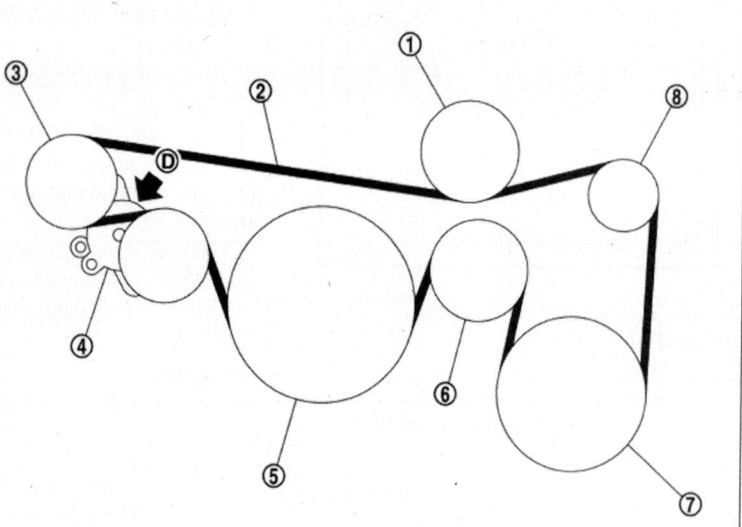

1. Idler pulley
2. Drive belt
3. Idler pulley
4. Drive belt auto-tensioner
5. Crankshaft pulley
6. Idler pulley
7. A/C compressor
8. Alternator
A. Indicator
B. Possible use range
C. Range when new drive belt is installed
D. View D
◁ : Engine front

71075_QUES_G0501

Fig. 16 Drive belt routing and tension

INSPECTION

✳✳ CAUTION

Be sure to perform this step when engine is stopped.

1. Check that the indicator of the drive belt auto-tensioner is within the possible use range.

a. Check the drive belt auto-tensioner indication when the engine is cold.

b. When a new drive belt is installed, the indicator should be within the range in the figure.

c. Visually check entire drive belt for wear, damage or cracks.

d. If the indicator is out of the possible use range or belt is damaged, replace drive belt.

REMOVAL & INSTALLATION

Drive Belt

1. Before servicing the vehicle, refer to the Precautions.

2. Disconnect the negative battery terminal.

3. Remove the front wheel and tire (RH).

4. Remove the splash guard (RH).

5. Hold the center part of the drive belt auto-tensioner pulley (the hexagonal portion) with a box wrench. Move the wrench handle in the direction shown (loosening direction of drive belt).

✳✳ CAUTION

Avoid placing hand in a location where pinching may occur if the holding tool accidentally comes off.

✳✳ WARNING

Never loosen the center part (the hexagonal portion) of the drive belt auto-tensioner pulley (Never turn it counterclockwise). If turned counterclockwise, the complete drive belt auto-tensioner must be replaced as a unit, including the pulley.

6. Insert an approximately 0.24 in. (6 mm) diameter rod, such as short-length screwdriver into the hole of the retaining boss to secure the drive belt auto-tensioner pulley.

7. Loosen drive belt from water pump pulley in sequence, and remove it.

To install:

8. Hold the center part of the drive belt auto-tensioner pulley (the hexagonal por-

tion) with a box wrench. Move the wrench handle in the direction shown above (loosening direction of drive belt).

9. Insert an approximately 0.24 in. (6 mm) diameter rod, such as short-length screwdriver into the hole of the retaining boss to secure the drive belt auto-tensioner pulley.

10. Hook the drive belt onto all the pulleys except for the drive belt auto-tensioner pulley.

11. Hook the drive belt onto the drive belt auto-tensioner pulley.

12. Confirm drive belt is completely set to pulleys. Ensure that the drive belt and each pulley groove is free of engine oil, working fluid and engine coolant.

13. Release the drive belt auto-tensioner, and apply tension to the drive belt.

14. Turn the crankshaft pulley clockwise several times to equalize the tension between each pulley.

15. Confirm the tension of the drive belt at the indicator is within the possible use range.

Drive Belt Auto-Tensioner & Idler Pulley

See Figure 17.

1. Remove drive belt.

2. Remove auto-tensioner and idler pulley.

To install:

Installation is the reverse of removal.

AIR CLEANER ASSEMBLY

REMOVAL & INSTALLATION

1. Before servicing the vehicle, refer to the Precautions.

2. Disconnect the negative battery terminal.

3. Remove the inlet air duct.

4. Disconnect the harness connector from the Mass Air Flow (MAF) sensor.

5. Remove the PCV hose at the air duct assembly side.

6. Remove the upper and lower air cleaner cases with the MAF sensor and air duct assembly. Add matchmarks if necessary for easier installation.

7. Remove the MAF sensor from the upper air cleaner case, if necessary. Handle the mass air flow sensor with care.

To install:

8. Install the MAF sensor from the upper air cleaner case, if removed.

9. Install the upper and lower air cleaner cases with the MAF sensor and air duct assembly.

a. Insert the pawls of the upper air cleaner case into the 3 notches of the lower air cleaner case and secure with clips.

b. Check that the upper air cleaner case is securely installed with no backlash.

10. Install the PCV hose at the air duct assembly side.

11. Connect the MAF sensor harness connector.

12. Install the inlet air duct.

13. Connect the negative battery cable.

CAMSHAFT & BEARINGS

INSPECTION

See Figures 18 through 25.

1. Inspect the Camshaft Runout:

a. Put V block on precise flat table, and support No. 2 and No. 4 camshaft journals.

✳✳ WARNING

Never support No. 1 journal (on the side of camshaft sprocket) because it has a different diameter from the other three locations.

b. Set a dial indicator vertically to No. 3 journal.

c. Turn the camshaft to one direction with hands, and measure the camshaft runout on a dial indicator (Total indicator reading). If it exceeds the limit, replace camshaft. Camshaft Runout:

- Standard: Less than 0.0008 in. (0.02mm)
- Limit: 0.0020 in. (0.05mm)

2. Measure the camshaft cam height with a micrometer. If wear exceeds the limit, replace the camshaft. Camshaft Cam Height:

- Standard cam height (Intake): 1.7900–1.7974 in. (45.465–45.655 mm)
- Standard cam height (Exhaust): 1.7904–1.7978 in. (45.475–45.665 mm)
- Cam wear limit: 0.008 in. (0.2 mm)

3. Measure the Camshaft Journal Diameter:

a. Measure the outer diameter of camshaft journal with a micrometer:

- Standard (No. 1): 1.0211–1.0218 in. (25.935–25.955 mm)
- Standard (No. 2, 3, 4): 0.9230–0.9238 in. (23.445–23.465 mm)

4. Measure the Camshaft Bracket Inner Diameter:

a. Tighten the camshaft bracket bolt to the specified torque.

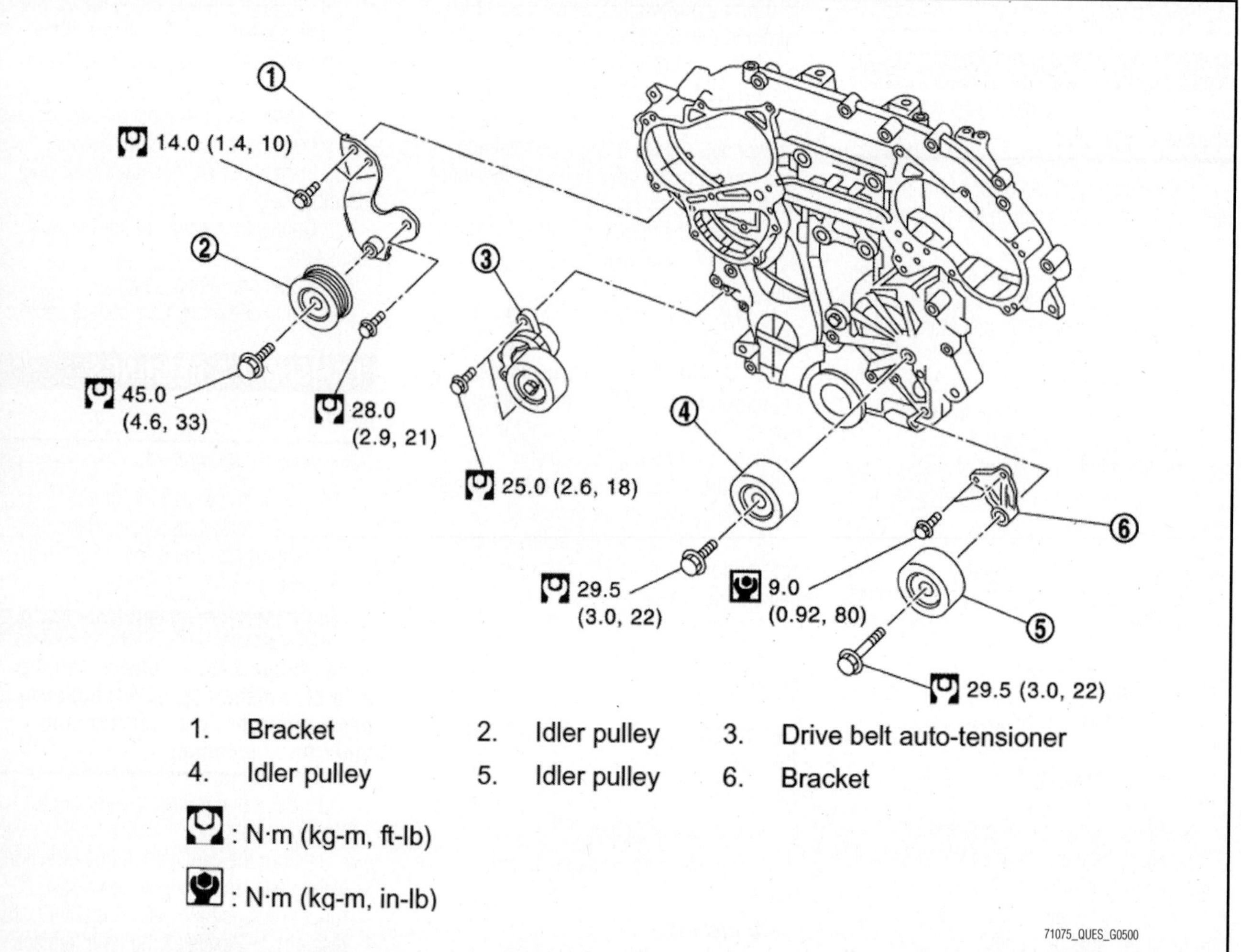

🔧 14.0 (1.4, 10)

🔧 45.0 (4.6, 33)

🔧 28.0 (2.9, 21)

🔧 25.0 (2.6, 18)

🔧 29.5 (3.0, 22)

🔧 9.0 (0.92, 80)

🔧 29.5 (3.0, 22)

1. Bracket
2. Idler pulley
3. Drive belt auto-tensioner
4. Idler pulley
5. Idler pulley
6. Bracket

🔧 : N·m (kg-m, ft-lb)

🔧 : N·m (kg-m, in-lb)

71075_QUES_G0500

Fig. 17 Drive belt auto-tensioner and idler pulley

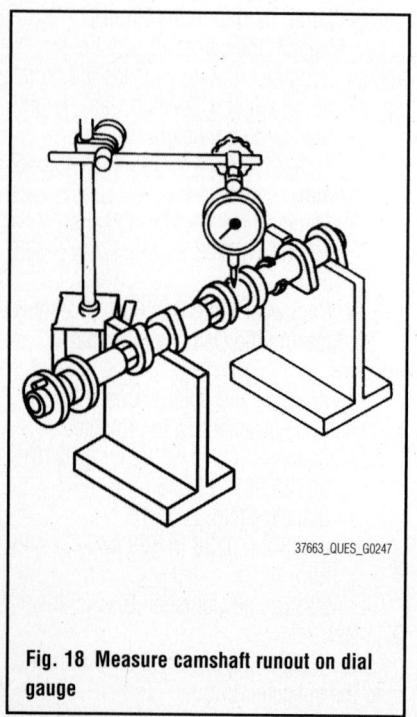

37663_QUES_G0247

Fig. 18 Measure camshaft runout on dial gauge

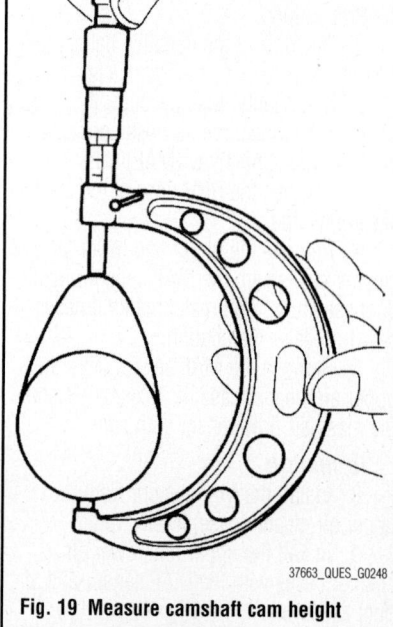

37663_QUES_G0248

Fig. 19 Measure camshaft cam height

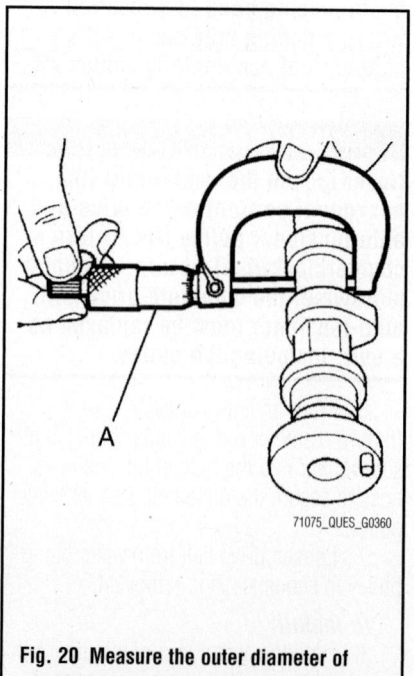

A

71075_QUES_G0360

Fig. 20 Measure the outer diameter of camshaft journal with a micrometer (A)

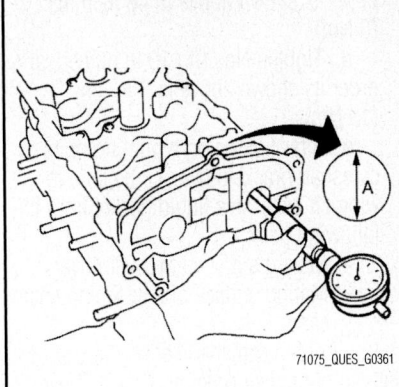

Fig. 21 Measure inner diameter (A) of camshaft bracket with a bore gauge

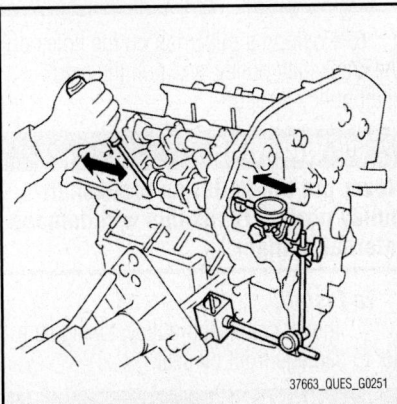

Fig. 22 Measure end play of dial gauge when camshaft is moved forward and backward

b. Measure inner diameter of camshaft bracket with a bore gauge.
- Standard (No. 1): 1.0236–1.0244 in. (26.000–26.021 mm)
- Standard (No. 2, 3, 4): 0.9252–0.9260 in. (23.500–23.521 mm)

5. Measure the Camshaft Journal Oil Clearance: (Oil clearance) = (Camshaft bracket inner diameter) - (Camshaft journal diameter). If the calculated value exceeds the limit, replace either or both camshaft and cylinder head.
- Standard (No. 1): 0.0018–0.0034 in. (0.045–0.086 mm)
- Standard (No. 2, 3, 4): 0.0014–0.0030 in. (0.035–0.076 mm)
- Limit: 0.0059 in. (0.15 mm)

➡Camshaft brackets cannot be replaced as single parts, because they are machined together with cylinder head. Replace whole cylinder head assembly.

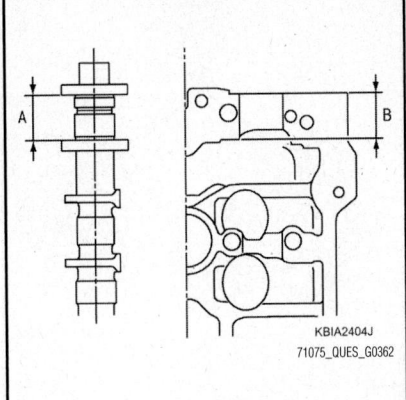

Fig. 23 Measure Dimension "A" for camshaft No. 1 journal, and Dimension "B" for cylinder head No. 1 journal bearing

6. Measure the Camshaft End Play:
a. Install a dial indicator in thrust direction on front end of camshaft.
b. Measure the end play of a dial indicator when camshaft is moved forward/backward (in direction to axis).
- Standard: 0.0045–0.0074 in. (0.115–0.188 mm)
- Limit: 0.0094 in. (0.24 mm)

c. Measure the following parts if out of the limit:
- Dimension "A" for camshaft No. 1 journal, Standard: 1.0827–1.0846 in. (27.500–27.548 mm)
- Dimension "B" for cylinder head No. 1 journal bearing, Standard: 1.0772–1.0781 in. (27.360–27.385 mm)

7. Measure the Camshaft Sprocket Runout:
a. Put V-block on precise flat table, and support the No. 2 and 4 journals of the camshaft.

✳✳ WARNING

Never support No. 1 journal (on the side of camshaft sprocket) because it has a different diameter from the other three locations.

b. Measure the camshaft sprocket runout with a dial indicator (Total indicator reading). If it exceeds the limit, replace camshaft sprocket. Limit: 0.0059 in. (0.15 mm)

8. Check the surface of valve lifter for wear or cracks. Replace as necessary.

9. Measure the Valve Lifter Outer Diameter: Measure the outer diameter at ½ height of the valve lifter with a micrometer since valve lifter is in a barrel shape. Standard (Intake and exhaust): 1.3378–1.3382 in. (33.980–33.990 mm)

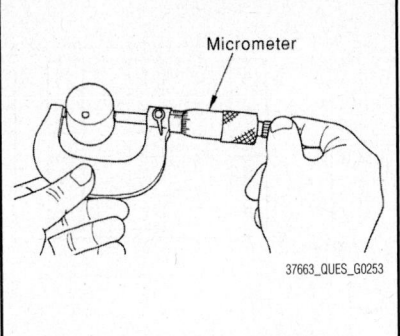

Fig. 24 Measure the valve lifter outer diameter

10. Measure the Valve Lifter Hole Diameter: Measure the inner diameter of valve lifter hole of cylinder head with an inside micrometer. Standard (Intake and exhaust): 1.3386–1.3392 in. (34.000–34.016 mm)

11. Measure the Valve Lifter Clearance: (Valve lifter clearance) = (Valve lifter hole diameter) – (Valve lifter outer diameter). Standard (Intake and exhaust): 0.0004–0.0014 in. (0.010–0.036 mm)

a. If the calculated value is out of the standard, referring to each standard of valve lifter outer diameter and valve lifter hole diameter, replace either or both valve lifter and cylinder head.

REMOVAL & INSTALLATION

See Figures 26 and 27.

1. Before servicing the vehicle, refer to the Precautions.
2. Disconnect the negative battery terminal.
3. Remove the front timing chain case, camshaft sprocket, timing chain and the rear timing chain case.

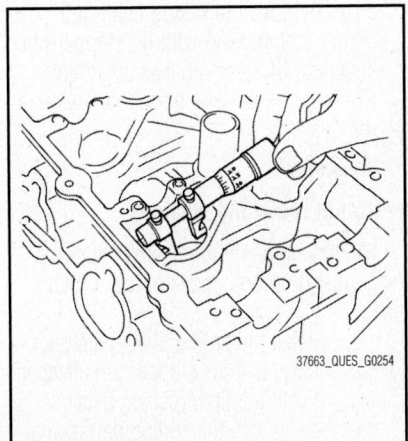

Fig. 25 Measure the valve lifter hole diameter

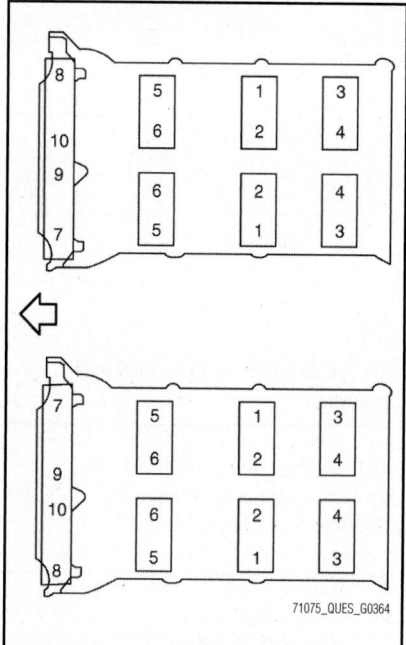

Fig. 26 Camshaft bolt bracket tightening sequence—reverse for removal

4. Loosen camshaft sensor bracket bolts in the reverse order as shown. The order of loosening bolts in the same for bank 1 and bank 2.

5. Remove the intake and exhaust camshaft brackets. Mark the camshafts, camshaft brackets, and bolts so they are placed in the same position and direction for installation.

6. Evenly loosen the camshaft bracket bolts in several steps in the reverse of the order shown.

7. Remove the camshafts.

8. Remove the valve lifters. Identify installation positions, and store them without mixing them up.

9. Remove the timing chain tensioners (secondary) from cylinder head.

   a. Remove the timing chain tensioner (secondary) with its stopper pin attached. Stopper pin was attached when the timing chain (secondary) was removed.

To install:

➡**Do not reuse the washers.**

10. Install the timing chain tensioners (secondary) on both sides of the cylinder head.

   a. Install the timing chain tensioner (secondary) with its stopper pin attached.

   b. Install the timing chain tensioner (secondary) with the sliding part facing downward on the cylinder head (bank 1), and with the sliding part facing upward on the cylinder head (bank 2).

Fig. 27 Install the camshaft so the dowel pins (A) on front end face are positioned as shown

11. Install the valve lifters in the original position.

12. Install the camshafts.

   a. Follow the identification marks made during removal, or follow the identification marks that are present on new camshafts for proper placement and direction.

13. Install the camshaft so the dowel pins on front end face are positioned as shown (No. 1 cylinder TDC on its compression stroke). Though camshaft does not stop at the portion as shown in the figure, for the placement of cam nose, it is generally accepted that the camshaft is placed for the same direction of the figure.

14. Install the camshaft brackets.

   a. Remove foreign material completely from the camshaft bracket backside and from the cylinder head installation face.

   b. Install the camshaft brackets in the original position and direction as shown.

   c. Install the camshaft brackets (No. 2 to 4), aligning the stamp marks as shown. There are no identification marks indicating left and right for the camshaft bracket (No. 1).

   d. Apply liquid gasket to the mating surface of the camshaft bracket (No. 1) on bank 1 and bank 2, as shown. Use Genuine RTV Silicone Sealant or equivalent.

15. Tighten the camshaft bracket bolts in the following steps, in numerical order as shown above.

   a. Tighten No. 7 to 10 in order as shown above to 12 inch lbs. (2 Nm).

   b. Tighten No. 1 to 6 in order as shown above to 12 inch lbs. (2 Nm).

   c. Tighten No. 1 to 10 in numerical

order as shown above to 48 inch lbs. (6 Nm).

   d. Tighten No. 1 to 10 in numerical order as shown above to 8 ft. lbs. (10 Nm).

   e. After tightening mounting bolts of camshaft brackets (No. 1), be sure to wipe off excessive liquid gasket from the following parts:
   • Mating surface of rocker cover
   • Mating surface of rear timing chain case

16. Tighten camshaft sensor bracket bolts in numerical order as shown above. The order of tightening bolts in the same for bank 1 and bank 2.

17. Inspect and adjust the valve clearance.

18. Install the front timing chain case, camshaft sprocket, timing chain and the rear timing chain case.

19. Connect the negative battery terminal.

CRANKSHAFT DAMPER

REMOVAL & INSTALLATION
See Figure 28.

1. Before servicing the vehicle, refer to the Precautions.

2. Disconnect the negative battery terminal.

3. Remove the components necessary to access the crankshaft pulley.

4. Secure the crankshaft with the pulley holder.

5. Loosen the crankshaft pulley bolt and locate the bolt seating surface at 0.39 in. (10 mm) from its original position.

✳✳ WARNING

Never remove crankshaft pulley bolt as it will be used as a supporting point for suitable puller.

6. Position a puller tab on the holes of the crankshaft pulley, and pull the crankshaft pulley through.

✳✳ WARNING

Never put the puller on crankshaft pulley periphery, as this will damage internal damper.

To install:

7. Install crankshaft pulley, taking care not to damage front oil seal.

✳✳ WARNING

When press-fitting crankshaft pulley with a plastic hammer, tap on its center portion (not circumference).

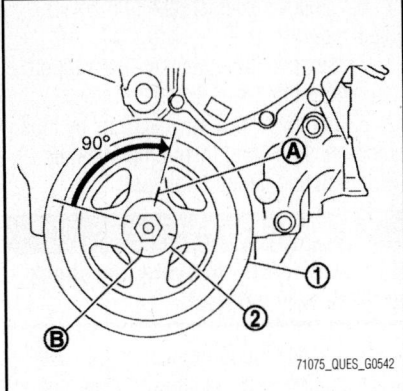

Fig. 28 Place a paint mark (A) on crankshaft pulley (1) aligning with the angle mark (B) on crankshaft pulley bolt (2)

8. Position the crankshaft with the pulley holder and tighten the crankshaft pulley bolt to 33 ft. lbs. (44 Nm).

9. Place a paint mark on crankshaft pulley aligning with the angle mark on crankshaft pulley bolt. Tighten the bolt 90 degrees (angle tightening).

10. Rotate the crankshaft pulley in normal direction (clockwise when viewed from front) to confirm it turns smoothly.

11. Install the remaining components.

CRANKSHAFT FRONT SEAL

REMOVAL & INSTALLATION

See Figure 29.

1. Remove the following parts:
   - Right-hand wheel and tire
   - Right-hand splash guard
   - Drive belt
   - Crankshaft pulley
2. Remove the front oil seal using a suitable tool.

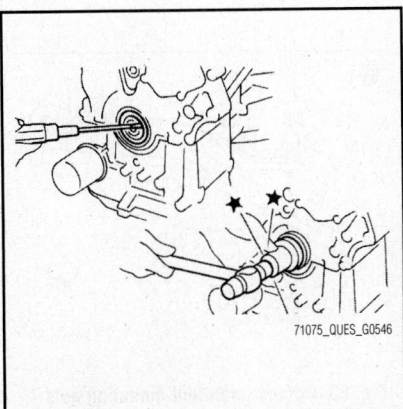

Fig. 29 Front oil seal removal/installation

✳✳ WARNING

Be careful not to damage front timing chain case and crankshaft.

To install:

3. Apply new engine oil to both oil seal lip and dust seal lip of new front oil seal.

4. Install front oil seal so that each seal lip is oriented as shown.

5. Using a suitable drift, press-fit until the height of front oil seal is level with the mounting surface. Suitable drift: outer diameter 2.36 in. (60 mm), inner diameter 1.97 in. (50 mm).

6. Check the garter spring is in position and seal lips not inverted.

✳✳ WARNING

Be careful not to damage front timing chain case and crankshaft. Press-fit straight and avoid causing burrs or tilting oil seal.

7. Install the following parts:
   - Crankshaft pulley
   - Drive belt
   - Right-hand splash guard
   - Right-hand wheel and tire

CRANKSHAFT REAR COVER & SEAL

REMOVAL & INSTALLATION

See Figures 30 and 31.

1. Remove the transaxle assembly.
2. Remove the drive plate.
3. Remove the oil pan (upper).
4. Use a seal cutter (SST) to cut away liquid gasket and remove the rear oil seal retainer. Be careful not to damage mating surfaces. Remove the rear oil seal and retainer as an assembly.

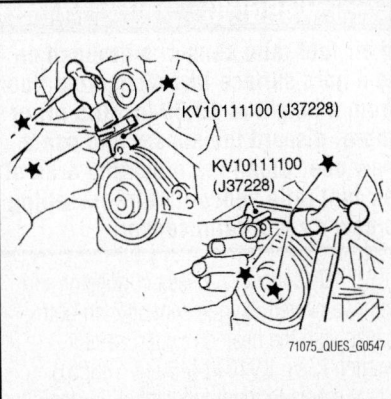

Fig. 30 Rear oil seal removal/installation

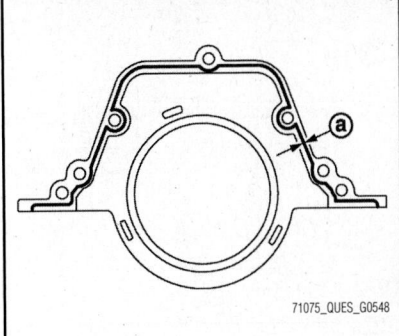

Fig. 31 Sealant application (a): 0.08–0.12 in. (2.0–3.0 mm)

To install:

5. Remove old liquid gasket on mating surfaces of the cylinder block and oil pan (upper) using a scraper.

6. Apply new engine oil to both the oil seal lip and dust seal lip of the new oil seal retainer.

7. Apply a continuous bead of liquid gasket with the tube presser to rear the oil seal retainer as shown. Use Genuine RTV Silicone Sealant or equivalent. Attach within 5 minutes after coating.

8. Install the rear oil seal retainer to the cylinder block. Ensure the garter spring is in position and seal lips not inverted.

9. Install the oil pan (upper).
10. Install the drive plate.
11. Install the transaxle assembly.

CYLINDER HEAD

REMOVAL & INSTALLATION

See Figure 32.

1. Remove the following parts:
   - Oil level gauge
   - Intake manifold cover (collector)
   - Rocker cover
   - Fuel tube and fuel injector assembly
   - Intake manifold
   - Exhaust manifold
   - Water inlet and thermostat assembly
   - Water outlet, water connector, water bypass pipe, and heater pipe
   - Front timing chain case, timing chain and rear timing chain case
   - Camshaft

2. Remove the cylinder head bolts in the reverse of the tightening sequence shown below, using a cylinder head bolt wrench and a power tool to remove cylinder heads (bank 1 and bank 2).

3. Remove the cylinder head gaskets.

To install:

4. Turn the crankshaft until the No. 1 piston is set at TDC. The crankshaft key

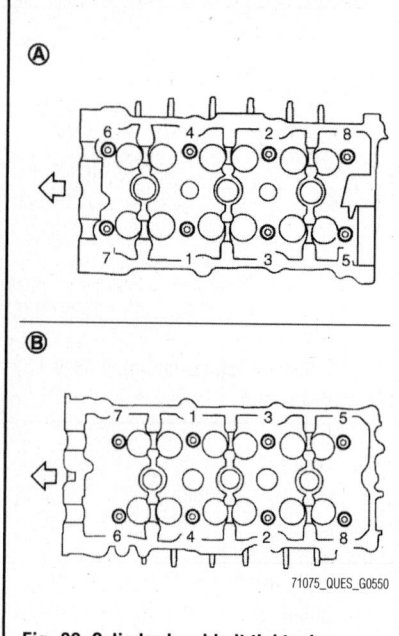

Fig. 32 Cylinder head bolt tightening sequence; A: Bank 1, B: Bank 2, arrow: engine front

should line up with the bank 1 cylinder center line as shown.

5. Install new cylinder head gaskets.

❋❋ WARNING

If cylinder head bolts are reused, check their outer diameters before installation. Before installing cylinder head, inspect cylinder head distortion.

6. Install the cylinder head. Follow the steps below to tighten the cylinder head bolts in numerical order as shown.

   a. Apply new engine oil to the threads and seat surfaces of cylinder head bolts.

   b. Tighten all cylinder head bolts to 72 ft. lbs. (98 Nm).

   c. Completely loosen all cylinder head bolts in reverse of the tightening order.

   d. Tighten all cylinder head bolts to 29 ft. lbs. (39 Nm).

   e. Turn all cylinder head bolts 103 degrees clockwise (angle tightening). Check the tightening angle by using the angle wrench [SST: KV10112100 (BT8653-A)].

   f. Check tightening angle indicated on the angle wrench indicator plate.

   g. Turn all cylinder head bolts 103 degrees clockwise again (angle tightening).

7. After installing the cylinder head, measure distance between front end faces of cylinder block and cylinder head (bank 1 and bank 2). Standard: 0.555–0.587 in.

(14.1–14.9 mm). If measured value is out of the standard, reinstall cylinder head.

8. Remove the following parts:
- Camshaft
- Front timing chain case, timing chain and rear timing chain case
- Water outlet, water connector, water bypass pipe, and heater pipe
- Water inlet and thermostat assembly
- Exhaust manifold
- Intake manifold
- Fuel tube and fuel injector assembly
- Rocker cover
- Intake manifold cover (collector)
- Oil level gauge

EXHAUST MANIFOLD

REMOVAL & INSTALLATION
See Figures 33 and 34.

❋❋ CAUTION

Perform the work when the exhaust and cooling system have completely cooled down.

1. Remove the following Bank 1 side parts:
- Air cleaner case (upper) with mass air flow sensor and air duct assembly
- Engine cover
- Front wiper arm
- Extension cowl top
- Heat insulator

2. Remove the Bank 2 side engine under cover.

3. Remove exhaust front tube.

4. Disconnect harness connector and remove air fuel ratio sensor 1 on both banks with the heated oxygen sensor wrench [SST: KV10117100 (J-3647-A)]. Place marks to identify installation positions of each air fuel ratio sensor 1.

❋❋ WARNING

If air fuel ratio sensor is dropped on to a hard surface like a concrete floor from a height of 19.69 in. (0.5 m) or more, discard the sensor and use a new one. Clean the mounting area of air fuel ratio sensor before installing a new air fuel ratio sensor.

5. Disconnect harness connector and remove heated oxygen sensor 2 on both banks with the heated oxygen sensor wrench [SST: KV10114400 (J-38365)]. Place marks to identify installation positions of each heated oxygen sensor 2.

6. Remove exhaust manifold covers (bank 1 and bank 2).

7. Remove three way catalyst support mounting bolts (bank 1 and bank 2).

8. Remove three way catalysts by loosening bolts first and then removing nuts (Stud bolt and flange bolt type).

❋❋ WARNING

Handle carefully to avoid any shock to three way catalyst.

9. Loosen mounting nuts in the reverse of installation order as shown below, and remove exhaust manifolds (bank 1 and bank 2). Disregard No. 7 and 8 when loosening.

10. Remove gaskets.

❋❋ WARNING

Cover engine openings to avoid entry of foreign materials.

To install:

Installation is the reverse of removal, noting the following:

11. Install the exhaust manifold gasket in the direction indicated in the figure.

12. If the exhaust manifold stud bolts were removed, install them and tighten to 11 ft. lbs. (15 Nm).

13. Tighten the mounting nuts in numerical order as shown. Note: No. 7 and 8 mean double tightening of nuts No. 1 and 2.

14. Install the three way catalyst (bank 2) as follows, referring to the com-

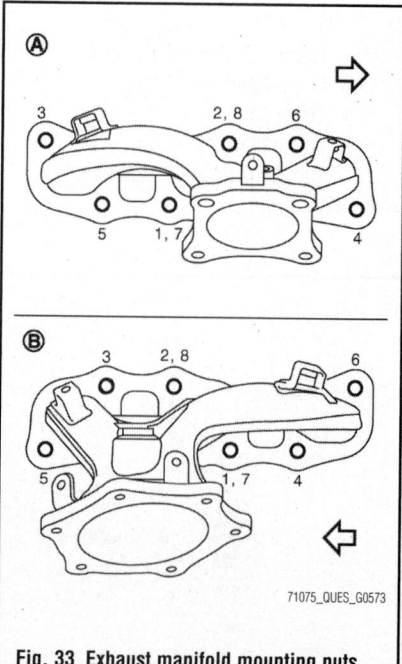

Fig. 33 Exhaust manifold mounting nuts tightening sequence

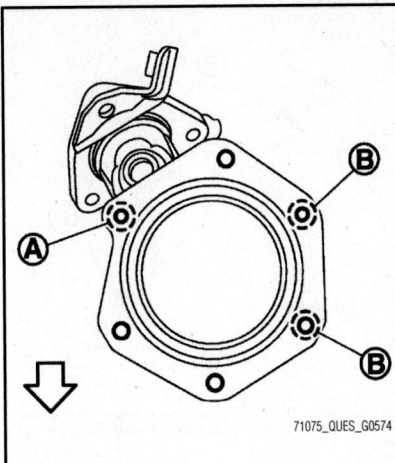

Fig. 34 Three way catalyst (bank 2) bolts

ponent location figure for tightening specifications:

    a. Temporarily tighten the upper side bolts "B" and lower side bolt "A" shown.

    b. Tighten the upper side bolts "B" shown.

    c. Tighten all the lower side bolts (the other four except "B").

15. Install the three way catalyst supports as follows:

    a. Temporarily tighten three way catalyst support mounting bolts.

    b. Tighten three way catalyst support mounting bolts to oil pan (upper).

    c. Tighten three way catalyst support mounting bolts to three way catalyst.

➡**Before installing a new air fuel ratio sensor and a new heated oxygen sensor, clean exhaust system threads using oxygen sensor thread cleaner (commercial service tool: J-43897-18 or J-43897-12) and apply anti-seize lubricant. Never over torque air fuel ratio sensor and heated oxygen sensor. Doing so may cause damage to air fuel ratio sensor and heated oxygen sensor. Prevent rust preventives from adhering to the sensor body.**

16. Install air fuel ratio sensor 1 and heated oxygen sensor 2 in the original position. If the installation positions cannot be identified, note the following glass tube colors:

- Air fuel ratio sensor 1: Black
- Heated oxygen sensor 2: White

INTAKE MANIFOLD

REMOVAL & INSTALLATION

See Figure 35.

❊❊ CAUTION

To avoid the danger of being scalded, never drain engine coolant when the engine is hot.

1. Release the fuel pressure.
2. Disconnect the negative battery cable.
3. Remove the intake manifold cover (collector).
4. Remove the fuel tube and fuel injector assembly.
5. Remove the intake manifold mounting nuts and bolts in reverse of the tightening order, shown below.

    a. Matchmark the intake manifold and cylinder head with paint to assist with installing in the right direction.

    b. Cover the engine openings with tape to prevent entry of foreign materials.

6. Remove the gaskets.

To install:

7. Install new gaskets.
8. Align the matchmarks on the intake manifold and the cylinder head.
9. If stud bolts were removed, install them and tighten to 8 ft. lbs. (11 Nm).
10. Tighten all intake manifold mounting nuts and bolts to the specified torque in two or more steps in the order shown.

- 1st step: 5 ft. lbs. (7 Nm)
- 2nd step and after: 19 ft. lbs. (26 Nm)

11. Install the fuel tube and fuel injector assembly.
12. Install the intake manifold cover (collector) and gasket.
13. Connect the negative battery cable.

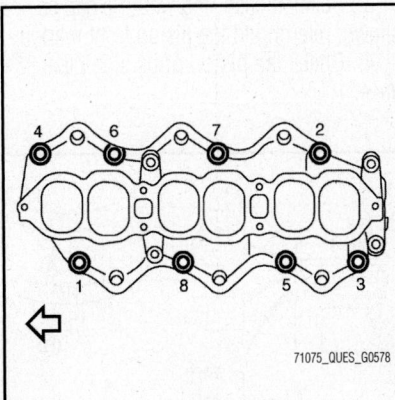

Fig. 35 Intake manifold bolt tightening sequence; arrow: engine front

OIL COOLER

REMOVAL & INSTALLATION

See Figure 36.

1. Before servicing the vehicle, refer to the Precautions.
2. Remove the right front wheel.
3. Remove the right-hand splash guard. If removing oil cooler only, it is not necessary to remove the splash guard.
4. If removing the water pipes, drain the engine coolant from radiator and cylinder block.
5. Remove the oil filter.

❊❊ WARNING

Never spill engine oil on drive belt.

6. Disconnect water hoses from the oil cooler. Perform this step when the engine is cold. If removing oil cooler only, pinch water hoses near oil cooler to prevent engine coolant from spilling out. Position a container to catch any coolant that comes out.

❊❊ WARNING

Never spill coolant on drive belt.

7. Remove the connector bolt and the oil cooler.
8. Remove the water hoses if necessary.

To install:

❊❊ WARNING

Do not reuse O-rings.

9. Install the water hoses, if removed.
10. Check that no foreign objects are adhering to the installation surfaces of the oil cooler and oil pan (upper).

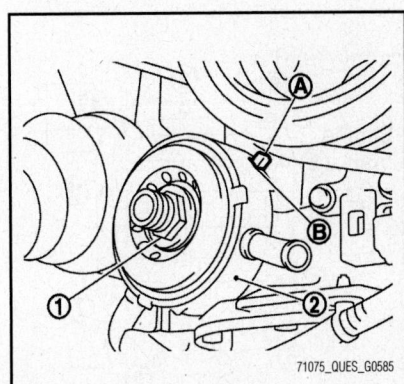

Fig. 36 Align the cutout (B) on the oil cooler (2) with the protrusion (A) on the oil pan (upper) side, and tighten the connector bolt (1)

11. Align the cutout on the oil cooler with the protrusion on the oil pan (upper) side, and tighten the connector bolt to 29 ft. lbs. (39 Nm).

12. Connect the water hoses, if disconnected.

13. Install the oil filter.

14. Refill the engine coolant.

15. Install the splash guard, if removed.

16. Install the right front wheel.

OIL PAN

REMOVAL & INSTALLATION

See Figures 37.

1. Before servicing the vehicle, refer to the Precautions.

2. Disconnect the negative battery cable.

3. Drain engine oil.

4. Remove the mounting bolts. Remove the bolts in the reverse of tightening sequence, shown below.

5. Insert the seal cutter [SST: KV10111100 (J-37228)] between oil pan (upper) and oil pan (lower). Be careful not to damage the mating surfaces. Never insert a screwdriver, this will damage the mating surfaces.

6. Slide the seal cutter by tapping on the side of tool with a hammer.

7. Remove oil pan (lower).

To install:

8. Use scraper to remove old liquid gasket from mating surfaces. Remove old liquid gasket from the bolt holes and thread. Never scratch or damage the mating surfaces when cleaning off old liquid gasket.

9. Apply a continuous bead of liquid gasket with the tube presser to the oil pan (lower). Use Genuine RTV Silicone Sealant

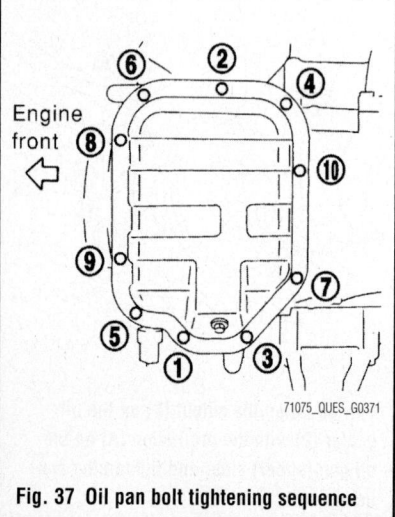

Fig. 37 Oil pan bolt tightening sequence

or equivalent. Attaching should be done within 5 minutes after coating.

10. Install oil pan (lower). Tighten mounting bolts in numerical order as shown.

11. Install oil pan drain plug with new gasket. Tighten to 25 ft. lbs. (34 Nm).

12. Wait at least 30 minutes after oil pan is installed to refill engine oil.

13. Connect the negative battery cable.

OIL PUMP

REMOVAL & INSTALLATION

1. Remove oil pan (lower and upper) and oil strainer.

2. Remove front timing chain case and timing chain (primary).

3. Remove oil pump assembly.

To install:

Installation is the reverse of removal. When installing, align crankshaft flat faces with oil pump inner rotor flat faces.

PISTONS & RINGS

POSITIONING

See Figures 38 and 39.

1. Using a piston ring expander, install the piston rings.

✳✳ WARNING

Be careful not to damage piston. Be careful not to damage piston rings by expending them excessively.

2. If there is stamped mark on ring, mount it with marked side up. If there is no stamp on ring, no specific orientation is required for installation. Stamped mark:

- Top ring: —
- Second ring : 2 R

3. Position each ring with the gap as shown, referring to the piston front mark.

4. Check the piston rings side clearance.

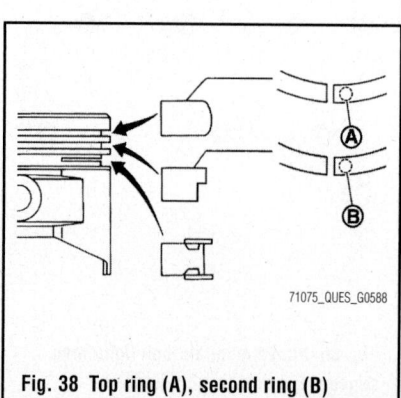

Fig. 38 Top ring (A), second ring (B)

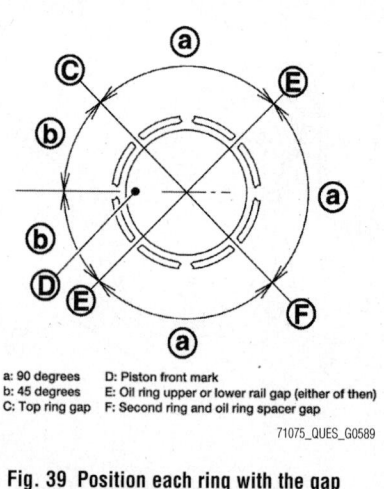

a: 90 degrees   D: Piston front mark
b: 45 degrees   E: Oil ring upper or lower rail gap (either of then)
C: Top ring gap   F: Second ring and oil ring spacer gap

71075_QUES_G0589

Fig. 39 Position each ring with the gap as shown, referring to the piston front mark (D)

ROCKER ARM COVER

REMOVAL & INSTALLATION

See Figures 40 and 41.

1. Before servicing the vehicle, refer to the Precautions.

2. Disconnect the negative battery cable.

3. Remove the engine cover.

4. Remove the air cleaner cases (upper and lower) and air duct assembly.

5. Remove the intake manifold collector.

6. Disconnect the PCV hose from rocker cover.

7. Remove camshaft position sensor (PHASE) (bank 1 and bank 2).

8. Remove PCV valve and O-ring from rocker cover, if necessary.

9. Remove oil filler cap from rocker cover, if necessary.

10. Remove ignition coil.

11. Remove harness clips on the rocker cover.

12. Loosen the rocker cover bolts in reverse of the tightening order, shown below.

13. Remove rocker cover gasket from rocker cover.

14. Use scraper to remove all traces of liquid gasket from cylinder head and camshaft bracket (No. 1).

To install:

➡Do not reuse O-rings.

15. Apply liquid gasket to the position shown. Use Genuine RTV silicone sealant or equivalent.

16. Install rocker cover gasket to rocker cover.

17. Install rocker cover. Be sure rocker

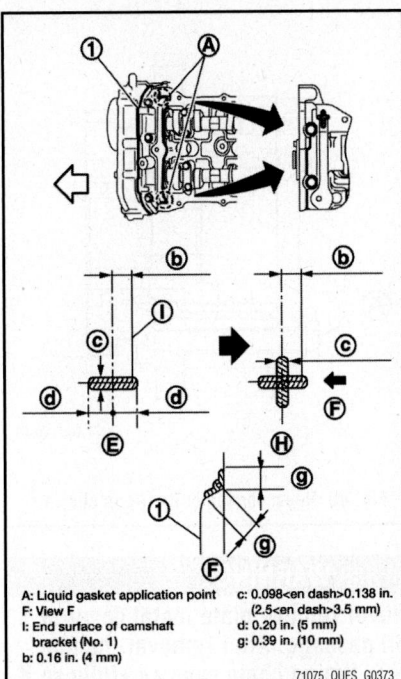

A: Liquid gasket application point
F: View F
I: End surface of camshaft bracket (No. 1)
b: 0.16 in. (4 mm)
c: 0.098<en dash>0.138 in. (2.5<en dash>3.5 mm)
d: 0.20 in. (5 mm)
g: 0.39 in. (10 mm)

71075_QUES_G0373

Fig. 40 Refer to figure (E) to apply liquid gasket to joint part of camshaft bracket (No. 1) (1) and cylinder head; refer to figure (H) to apply liquid gasket in 90 degrees to figure

cover gasket is not dropped from the installation groove of rocker cover.

18. Tighten bolts in two steps separately in numerical order as shown.
- First pass: 17 inch lbs. (2 Nm)
- Second pass: 74 inch lbs. (8 Nm)

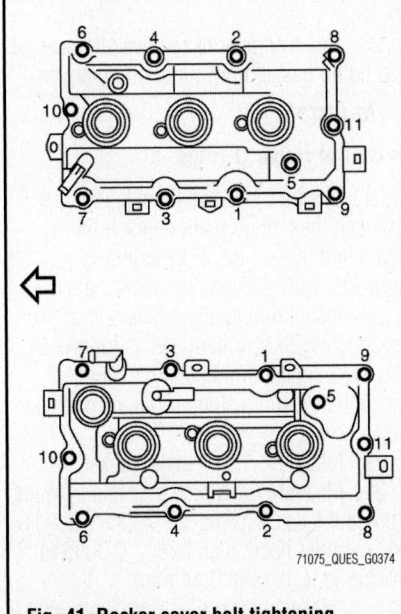

71075_QUES_G0374

Fig. 41 Rocker cover bolt tightening sequence

TIMING CHAIN COVER, CHAIN, TENSIONER, & SPROCKETS

REMOVAL & INSTALLATION
See Figures 42 through 56.

1. Before servicing the vehicle, refer to the Precautions.
2. Disconnect the negative battery cable.
3. Remove intake manifold collector.
4. Remove rocker covers (bank 1 and bank 2).
5. Remove oil pans (lower and upper) and oil strainer.
6. Remove drive belt, idler pulleys and bracket.
7. Separate engine harness removing their brackets from front timing chain case.
8. Remove camshaft sprocket covers. Loosen mounting bolts in reverse of the tightening order, shown below.

➡**Shaft is internally jointed with camshaft sprocket (INT) center hole. When removing, keep it horizontal until it is completely disconnected.**

9. Position No. 1 cylinder at TDC of its compression stroke as follows:
   a. Rotate crankshaft pulley clockwise to align timing mark (grooved line without color) with timing indicator.
   b. Check that intake and exhaust cam noses on No. 1 cylinder (engine front side of bank 1) are located as shown. If not, turn crankshaft one revolution (360 degrees) and align as shown.
10. Remove crankshaft pulley.
11. Remove front timing chain case as follows:
    a. Loosen mounting bolts in reverse of the tightening order, shown below.

71075_QUES_G0376

Fig. 42 Align timing mark (grooved line without color) with timing indicator

b. Insert a suitable tool into the notch at the top of front timing chain case as shown.
c. Pry off case by moving the tool as shown. Use the seal cutter [SST: KV10111100 (J-37228)] to cut liquid gasket for removal.

❋❋ WARNING

Never use screwdrivers or something similar. After removal, handle front timing chain case carefully so it does not tilt, cant, or warp under a load.

12. Remove water pump cover from front timing chain case. Use the seal cutter [SST: KV10111100 (J-37228)] to cut liquid gasket for removal.
13. Remove front oil seal from front timing chain case using a suitable tool. Use a screwdriver for removal. Be careful not to damage front timing chain case.
14. Remove O-ring from rear timing chain case.
15. Remove timing chain tensioner (primary) as follows:
    a. Remove lower mounting bolt.
    b. Loosen upper mounting bolt slowly, and then turn timing chain tensioner (primary) on the mounting bolt so that plunger is fully expanded. Even if plunger is fully expanded, it is not dropped from the body of timing chain tensioner (primary).
    c. Remove upper mounting bolt, and then remove timing chain tensioner (primary).
16. Remove internal chain guide, tension guide and slack guide.

➡**Tension guide can be removed after removing timing chain (primary).**

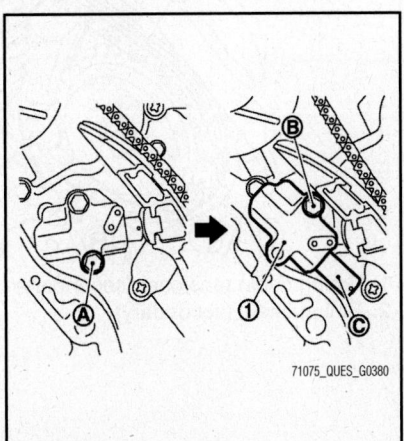

71075_QUES_G0380

Fig. 43 Lower mounting bolt (A), upper mounting bolt (B), timing chain tensioner (primary) (1), plunger (C)

> ※※ **WARNING**
>
> After removing timing chain (primary), never turn crankshaft and camshaft separately, or valves will strike the piston heads.

17. Remove timing chain (primary) and crankshaft sprocket.

18. Remove timing chain (secondary) and camshaft sprockets as follows:

   a. Attach a suitable stopper pin to the bank 1 and bank 2 timing chain tensioners (secondary).

➡Use approximately 0.020 in. (0.5 mm) dia. hard metal pin as a stopper pin.

   b. Remove camshaft sprockets (INT and EXH) mounting bolts. Secure the hexagonal portion of camshaft using a wrench to loosen mounting bolts.

> ※※ **WARNING**
>
> Never loosen mounting bolts with securing anything other than the camshaft hexagonal portion or with tensioning the timing chain.

   c. Remove timing chain (secondary) together with camshaft sprockets. Turn camshaft slightly to secure slackness of timing chain on timing chain tensioner (secondary) side.

   d. Insert 0.020 in. (0.5 mm) thick metal or resin plate between timing chain

and timing chain tensioner plunger (guide). Remove timing chain (secondary) together with camshaft sprockets with timing chain loose from guide groove.

> ※※ **WARNING**
>
> Be careful of plunger coming-off when removing timing chain (secondary). This is because plunger of timing chain tensioner (secondary) moves during operation, leading to coming-off of fixed stopper pin.

➡Camshaft sprocket (INT) is two-for-one structure of sprockets for timing chain (primary) and for timing chain (secondary). Figure is shown as an example of bank 1. When handling camshaft sprocket (INT), be careful of the following caution: Handle carefully to avoid any shock to camshaft sprocket. Never disassemble.

19. Remove water pump.
20. Remove oil pump.
21. Remove rear timing chain case as follows:

   a. Loosen and remove mounting bolts in reverse of the tightening order, shown below.

   b. Cut liquid gasket using the seal cutter [SST: KV10111100 (J- 37228)] and remove rear timing chain case.

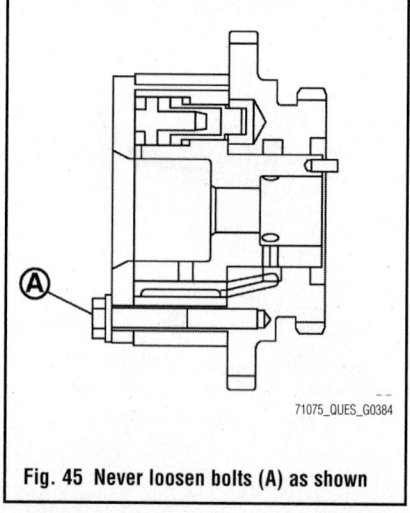

71075_QUES_G0384

Fig. 45 Never loosen bolts (A) as shown

> ※※ **WARNING**
>
> Never remove plate metal cover of oil passage. After removal, handle rear timing chain case carefully so it does not tilt, cant, or warp under a load.

22. Remove oil temperature sensor from timing chain case (rear) if necessary.

23. Remove O-rings from cylinder block.

24. Remove timing chain tensioners (secondary) from cylinder head as follows, if necessary.

   a. Remove camshaft brackets (No. 1).

   b. Remove timing chain tensioners (secondary) with a stopper pin attached.

25. Use a scraper to remove all traces of old liquid gasket from front timing chain case, and opposite mating surfaces. Remove old liquid gasket from bolt hole and thread.

26. Use a scraper to remove all traces of old liquid gasket from water pump cover.

To install:

➡Do not reuse O-rings.

27. The figure shows the relationship between the mating mark on each timing chain and that on the corresponding sprocket, with the components installed.

28. Install timing chain tensioners (secondary) to cylinder head as follows if removed. Install timing chain tensioners (secondary) with a stopper pin attached and new O-ring.

29. Install No.1 camshaft bracket.

30. Measure difference in levels between front end faces of camshaft bracket (No. 1) and cylinder head. Standard : _0.0055 to 0.0055 in. (_0.14 to 0.14 mm)

   a. Measure two positions (both intake and exhaust side) for a single bank.

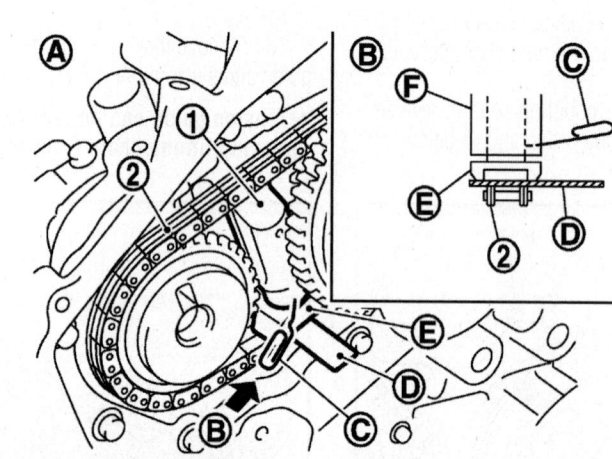

1. Timing chain tensioner (secondary)
2. Timing chain (secondary)

A. Bank 1
B. View B
C. Stopper pin
D. Plate
E. Timing chain tensioner plunger (guide)
F. Timing chain tensioner (body)

71075_QUES_G0383

Fig. 44 Remove timing chain (secondary) together with camshaft sprockets

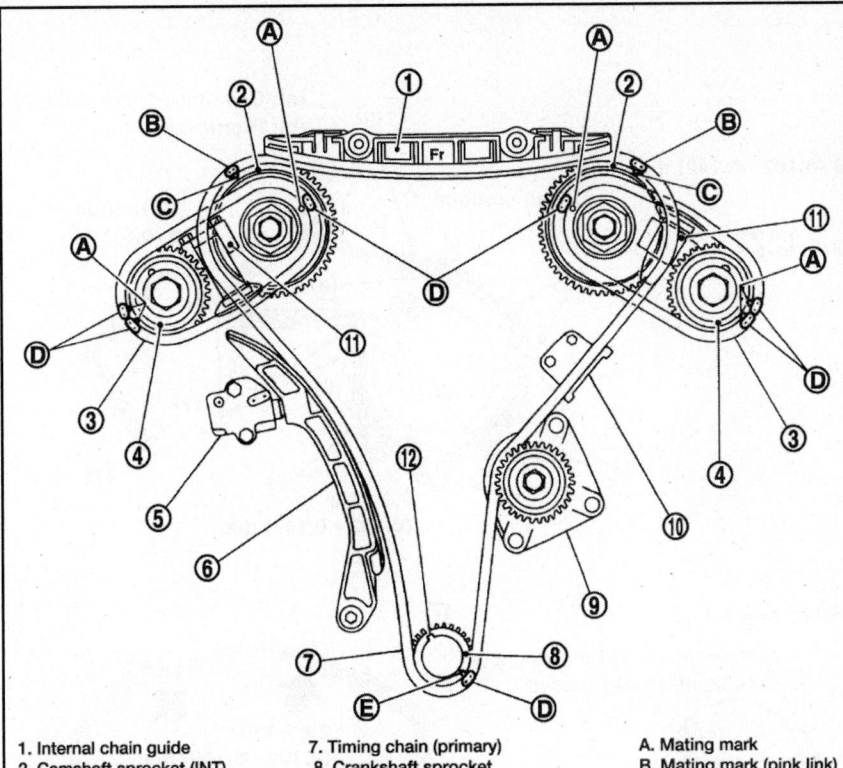

1. Internal chain guide
2. Camshaft sprocket (INT)
3. Timing chain (secondary)
4. Camshaft sprocket (EXH)
5. Timing chain tensioner (primary)
6. Slack guide
7. Timing chain (primary)
8. Crankshaft sprocket
9. Water pump
10. Tension guide
11. Timing chain tensioner (secondary)
12. Crankshaft key

A. Mating mark
B. Mating mark (pink link)
C. Mating mark (punched)
D. Mating mark (orange)
E. Mating mark (notched)

71075_QUES_G0387

Fig. 46 The figure shows the relationship between the mating mark on each timing chain and that on the corresponding sprocket, with the components installed

b. If the measured value is out of the standard, reinstall camshaft bracket (No. 1).

31. Install rear timing chain case as follows:

a. Install new O-rings onto cylinder block.

b. Apply liquid gasket with the tube presser (commercial service tool) to rear timing chain case back side as shown in the figure. Use Genuine RTV Silicone Sealant or equivalent.

c. For "A" in the figure, completely wipe out liquid gasket extended on a portion touching at engine coolant.

d. Apply liquid gasket on installation position of water pump and cylinder head very completely.

e. Align rear timing chain case with dowel pins (bank 1 and bank 2) on cylinder block and install rear timing chain case. Check O-rings stay in place during installation to cylinder block and cylinder head.

f. Tighten mounting bolts in numerical order as shown to 9 ft. lbs. (13 Nm). There are two types of mounting bolts. Refer to the following for locating bolts:

• Bolt length 0.79 in.(20 mm): Bolt position 1, 2, 3, 6, 7, 8, 9, 10
• Bolt length 0.63 in.(16 mm): Except the above

g. After all bolts are tightened, retighten them to the specified in numerical order as shown in the figure. If liquid gasket protrudes, wipe it off immediately.

h. After installing rear timing chain case, check the surface height difference between the following parts on the oil pan (upper) mounting surface. Rear timing chain case to cylinder block:

• Standard: −0.009 to 0.006 in. (−0.24 to 0.14 mm)

i. If not within the standard, repeat the installation procedure.

32. Install water pump.

33. Install oil pump.

34. Check that dowel pin (A) and crankshaft key (1) are located as shown. (No. 1 cylinder at compression TDC)

➡**Though camshaft does not stop at the position as shown in the figure, for the placement of cam nose, it is generally accepted camshaft is placed for the same direction of the figure.**

Hole on small dia. side must be used for intake side dowel pin hole. Never misidentify (ignore big dia. side).

35. Install timing chain (secondary) and camshaft sprockets (INT and EXH) as follows:

✳✳ WARNING

Mating marks between timing chain and sprockets slip easily. Confirm all mating mark positions repeatedly during the installation process.

a. Push plunger of timing chain tensioner (secondary) and keep it pressed in with stopper pin.

b. Install timing chain (secondary) and camshaft sprockets (INT and EXH).

c. Align the mating marks on timing chain (secondary) (orange link) with the ones on camshaft sprockets (INT and EXH) (punched), and install them. Mating marks for camshaft sprocket (INT) are on the back side of camshaft sprocket (secondary). There are two types of mating mark, circle and oval types. They should be used for the bank 1 and bank 2, respectively.

• Bank 1: Use circle type
• Bank 2: Use oval type

d. Align dowel pin on camshafts with the groove or hole on sprockets, and install them.

e. On the intake side, align dowel pin on the camshaft front end with dowel pin hole on the back side of camshaft sprocket, and install them.

f. On the exhaust side, align dowel pin on camshaft front end with dowel pin groove on camshaft sprocket, and install them.

g. In case that positions of each mating mark and each dowel pin are not fit on mating parts, make fine adjustment to the position holding the hexagonal portion on camshaft with wrench or equivalent.

h. Mounting bolts for camshaft sprockets must be tightened in the next step. Tightening them by hand is enough to prevent the dislocation of dowel pins.

i. It may be difficult to visually check the dislocation of mating marks during and after installation. To make the matching easier, make a mating mark on the top of sprocket teeth and its extended line in advance with paint.

j. After confirming the mating marks are aligned, tighten camshaft sprocket mounting bolts. Secure camshaft using

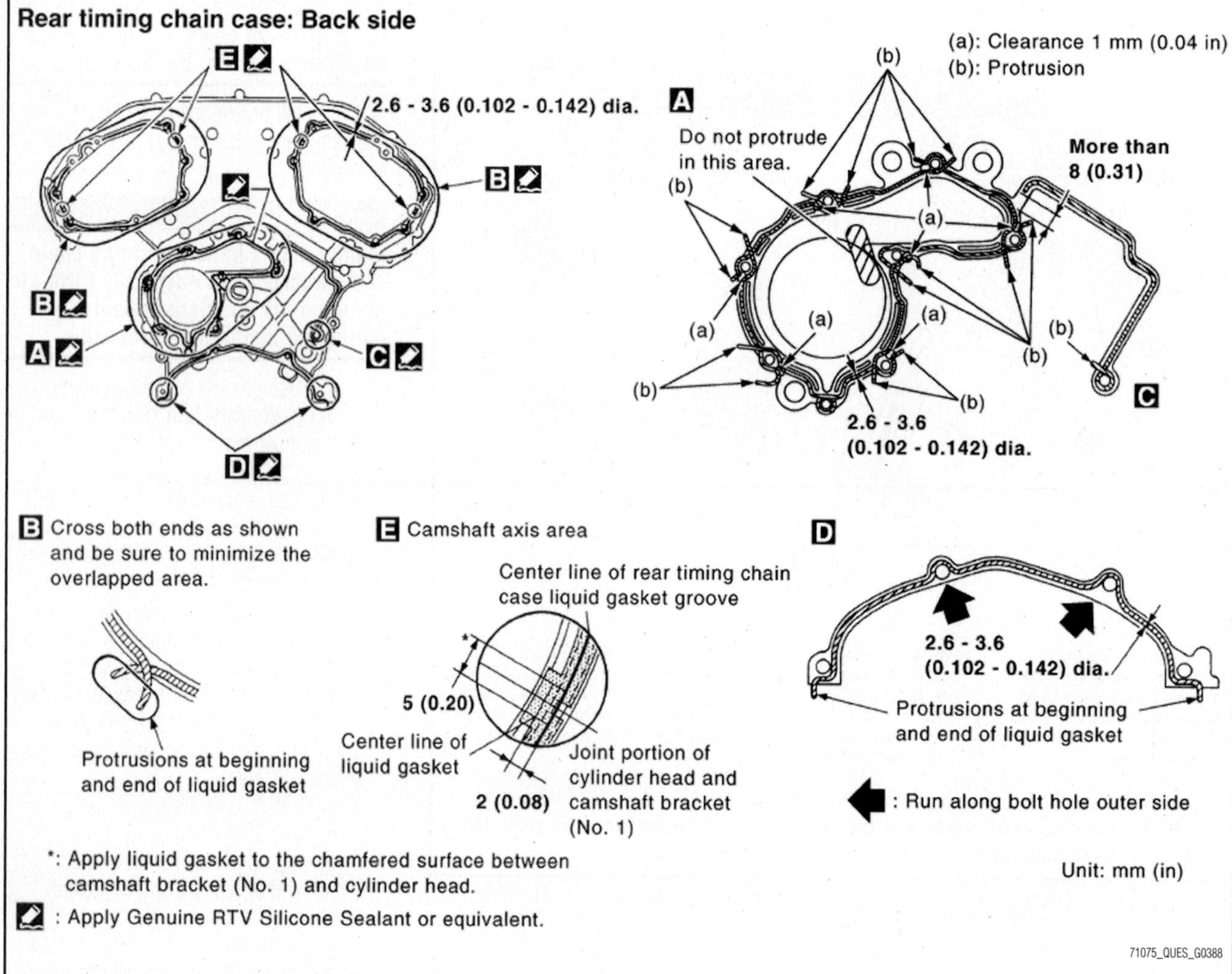

Rear timing chain case: Back side

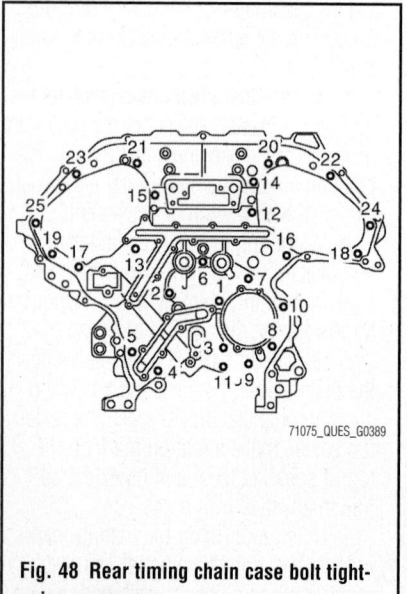

E 2.6 - 3.6 (0.102 - 0.142) dia.

B

B

A

C

D

(a): Clearance 1 mm (0.04 in)
(b): Protrusion

A

Do not protrude
in this area.

(b)

(b)

(a)

(a)

(a)

(b)

(a)

(a)

(b)

(b)

(b)

(b)

**More than
8 (0.31)**

C

2.6 - 3.6
(0.102 - 0.142) dia.

B Cross both ends as shown
and be sure to minimize the
overlapped area.

Protrusions at beginning
and end of liquid gasket

E Camshaft axis area

Center line of rear timing chain
case liquid gasket groove

5 (0.20)

Center line of
liquid gasket

2 (0.08)

Joint portion of
cylinder head and
camshaft bracket
(No. 1)

D

2.6 - 3.6
(0.102 - 0.142) dia.

Protrusions at beginning
and end of liquid gasket

◄ : Run along bolt hole outer side

*: Apply liquid gasket to the chamfered surface between
camshaft bracket (No. 1) and cylinder head.

◢ : Apply Genuine RTV Silicone Sealant or equivalent.

Unit: mm (in)

71075_QUES_G0388

Fig. 47 Rear timing chain case sealant application

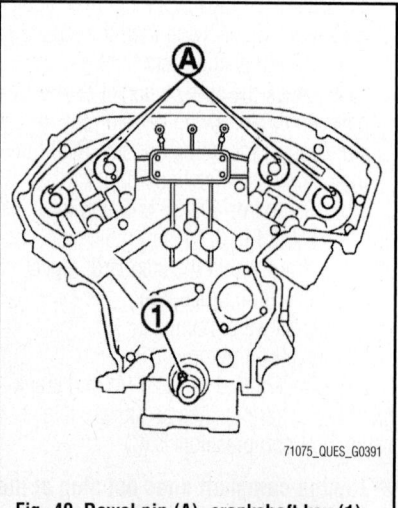

71075_QUES_G0389

**Fig. 48 Rear timing chain case bolt tight-
ening sequence**

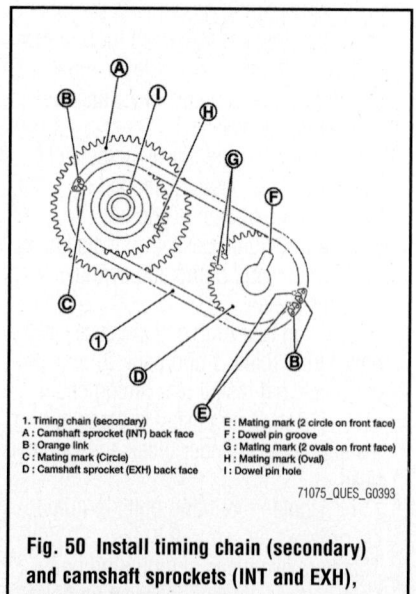

A

1

71075_QUES_G0391

**Fig. 49 Dowel pin (A), crankshaft key (1);
Camshaft dowel pin: At cylinder head
upper face side in each bank; Crankshaft
key: At cylinder head side of bank 1**

1. Timing chain (secondary)
A : Camshaft sprocket (INT) back face
B : Orange link
C : Mating mark (Circle)
D : Camshaft sprocket (EXH) back face
E : Mating mark (2 circle on front face)
F : Dowel pin groove
G : Mating mark (2 ovals on front face)
H : Mating mark (Oval)
I : Dowel pin hole

71075_QUES_G0393

**Fig. 50 Install timing chain (secondary)
and camshaft sprockets (INT and EXH),
Bank 1 (rear view)**

wrench at the hexagonal portion to tighten mounting bolts.

    k. Pull stopper pins out from timing chain tensioners (secondary).

36. Install tension guide.

37. Install timing chain (primary) as follows:

    a. Install crankshaft sprocket. Check the mating marks on crankshaft sprocket face the front of the engine.

    b. Install timing chain (primary). Install timing chain (primary) so the mating mark (punched) on camshaft sprocket (INT) is aligned with the pink link on timing chain, while the mating mark (notched) on crankshaft sprocket is aligned with the orange link on timing chain, as shown.

    c. When it is difficult to align mating marks of timing chain (primary) with each sprocket, gradually turn camshaft using wrench on the hexagonal portion to align it with the mating marks.

    d. During alignment, be careful to prevent dislocation of mating mark alignments of timing chains (secondary).

38. Install internal chain guide and slack guide.

✳✳ WARNING

Never over tighten slack guide mounting bolt. It is normal for a gap to exist under the bolt seat when mounting bolt is tightened to specification.

39. Install the timing chain tensioner (primary) with the following procedure:

    a. Pull plunger stopper tab up (or turn lever downward) so as to remove plunger stopper tab from the ratchet of plunger.

➡**Plunger stopper tab and lever are synchronized.**

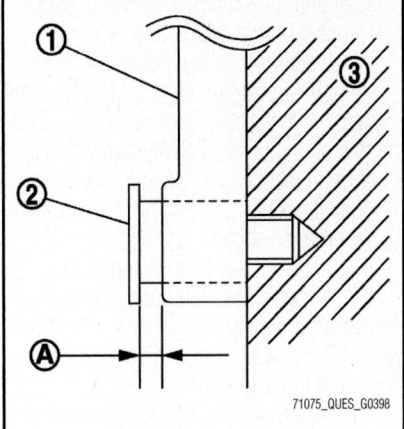

Fig. 52 Slack guide (1), mounting bolt (2), gap (A), cylinder block (3)

    b. Push plunger into the inside of tensioner body.

    c. Hold plunger in the fully compressed position by engaging plunger stopper tab with the tip of ratchet.

    d. To secure lever, insert stopper pin through hole of lever into tensioner body hole . The lever parts and the tab are synchronized. Therefore, the plunger will be secured under this condition. Figure shows the example of 0.047 in. (1.2 mm) diameter thin screwdriver being used as the stopper pin.

    e. Install timing chain tensioner (primary). Remove any dirt and foreign materials completely from the back and the mounting surfaces of timing chain tensioner (primary).

    f. Pull out stopper pin after installing, and then release plunger.

40. Check again that the mating marks on each sprocket and each timing chain have not slipped out of alignment.

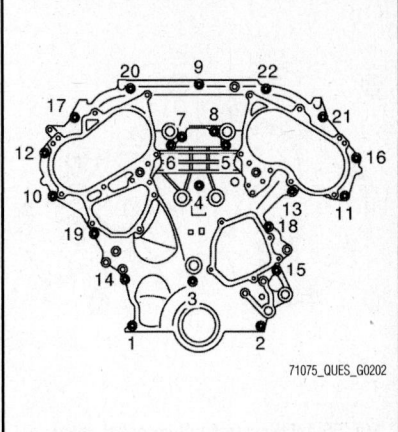

Fig. 54 Front timing chain case bolt tightening sequence

41. Install new O-rings on rear timing chain case. Do not reuse O-rings.

42. Install new front oil seal on front timing chain case.

43. Install front timing chain case as follows:

    a. Apply a continuous bead of liquid gasket with the tube presser (commercial service tool) to front timing chain case back side as shown.

    b. Install front timing chain case as to fit its dowel pin hole together dowel pin on rear timing chain case.

    c. Tighten mounting bolts to the specified torque in numerical order as shown. There are two types of mounting bolt. Refer to the following for locating bolts:

- M8 bolts (1, 2): 21 ft. lbs. (28 Nm)
- M68 bolts (Except the above): 9 ft. lbs. (13 Nm)

    d. After all bolts are tightened, retighten them to the specified torque in numerical order as shown. Be sure to wipe

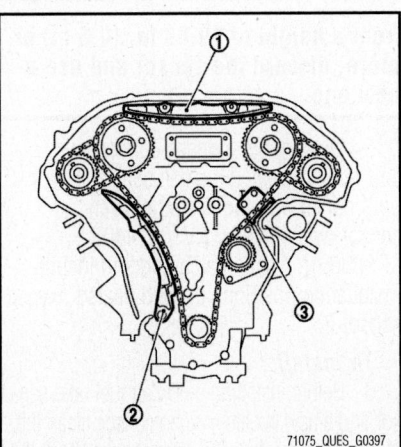

Fig. 51 Chain guide (1), slack guide (2), tension guide (3)

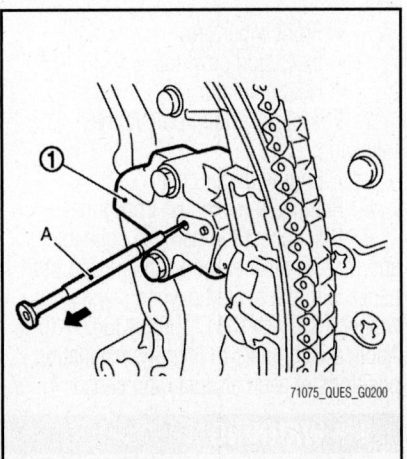

Fig. 53 Timing chain tensioner (primary) (1), stopper pin (A)

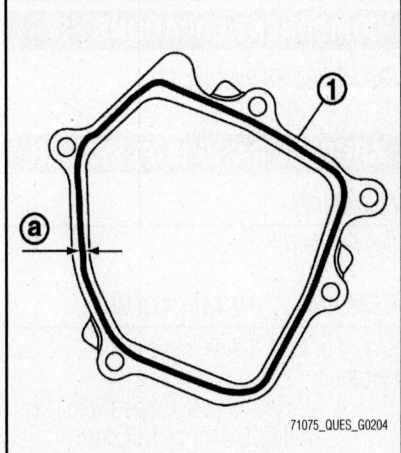

Fig. 55 Water pump cover (1) sealant application

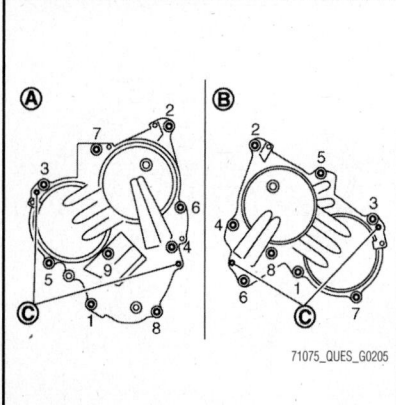

Fig. 56 Intake camshaft sprocket cover bolt tightening sequence; A: Bank 1, B: Bank 2, C: Dowel pin hole

off any excessive liquid gasket leaking on surface mating with oil pan (upper).

e. After installing front timing chain case, check the surface height difference between the following parts on the oil pan (upper) mounting surface. Standard:
- Front timing chain case to rear timing chain case: −0.006 to 0.006 in. (−0.14 to 0.14 mm)

f. If not within the standard, repeat the installation procedure.

g. Install water pump cover to front timing chain case. Apply a continuous bead of liquid gasket with the tube presser to water pump cover as shown. Use Genuine RTV Silicone Sealant or equivalent.

44. Install intake camshaft sprocket covers as follows:

a. Install new seal rings in shaft grooves.

b. Install camshaft sprocket cover to timing chain case (front). To insert, align

shaft and the shaft hole center of cam sprocket on the intake side. Securely install seal ring to each shaft groove.

c. Being careful not to move seal rings from the installation grooves, align dowel pins on front timing chain case with the holes to install camshaft sprocket covers.

d. Tighten mounting bolts in numerical order as shown.

45. Install oil pan (upper and lower).

46. Install crankshaft pulley. Rotate crankshaft pulley in normal direction (clockwise when viewed from engine front) to confirm it turns smoothly.

47. The remainder of installation is the reverse of removal.

48. Before starting engine, check oil/fluid levels including engine coolant and engine oil.

49. Check for fluid leakage.

VALVE LASH

ADJUSTMENT

See Figure 57.

Perform adjustment depending on selected head thickness of valve lifter.

1. Measure the valve clearance.
2. Remove camshaft.
3. Remove valve lifters at the locations that are out of the standard.
4. Measure the center thickness of the removed valve lifters with a micrometer.
5. Use the equation below to calculate valve lifter thickness for replacement.
- Valve lifter thickness calculation: $t = t1 + (C1 − C2)$
- t = Valve lifter thickness to be replaced
- $t1$ = Removed valve lifter thickness
- $C1$ = Measured valve clearance

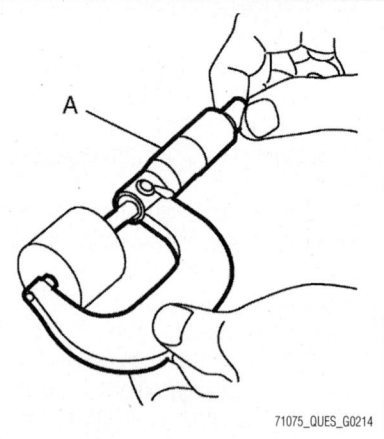

Fig. 57 Measure the center thickness of the removed lifter with a micrometer (A)

- $C2$ = Standard valve clearance: Intake: 0.012 in. (0.30 mm), Exhaust: 0.013 in. (0.33 mm)

Thickness of a new valve lifter can be identified by stamp marks on the reverse side (inside the cylinder). Stamp mark 788P indicates 0.3102 in. (7.88 mm) in thickness. Available thickness of valve lifter: 27 sizes with range 0.3102 to 0.3307 in. (7.88 to 8.40 mm) in steps of 0.0008 in. (0.02 mm) (when manufactured at factory).

6. Install selected valve lifter.
7. Install camshaft.
8. Manually turn crankshaft pulley a few turns.
9. Check that the valve clearances for cold engine are within the specifications by referring to the specified values.
10. Install all removal parts in the reverse order of removal.
11. Warm up the engine, and check for unusual noise and vibration.

ENGINE PERFORMANCE & EMISSION CONTROLS

COMPONENT LOCATIONS

See Figures 58 through 62.

AIR-FUEL RATIO (A/F) SENSOR

LOCATION

See Figure 63.

REMOVAL & INSTALLATION

1. For bank 1 side, remove the following:
- Air cleaner case (upper) with mass air flow sensor and air duct assembly
- Engine cover

- Front wiper arm
- Extension cowl top
- Heat insulator

2. For bank 2 side, remove the following:
- Engine under cover

3. Remove the exhaust front tube.

4. Disconnect the harness connector and remove air fuel ratio sensor 1 on both banks with the heated oxygen sensor wrench [SST: KV10117100 (J-3647-A)]. Apply matchmarks to identify installation positions of each air fuel ratio sensor 1.

❊❊ WARNING

If air fuel ratio sensor is dropped on a hard surface like a concrete floor

from a height of 19.69 in. (0.5 m) or more, discard the sensor and use a new one.

5. Disconnect harness connector and remove heated oxygen sensor 2 on both banks with the heated oxygen sensor wrench [SST: KV10114400 (J-38365)]. Apply matchmarks to identify installation positions of each heated oxygen sensor 2.

To install:

6. Before installing a new air fuel ratio sensor and a new heated oxygen sensor, clean the mounting area of air fuel ratio sensor. Clean the exhaust system threads using oxygen sensor thread cleaner (commercial service tool: J-

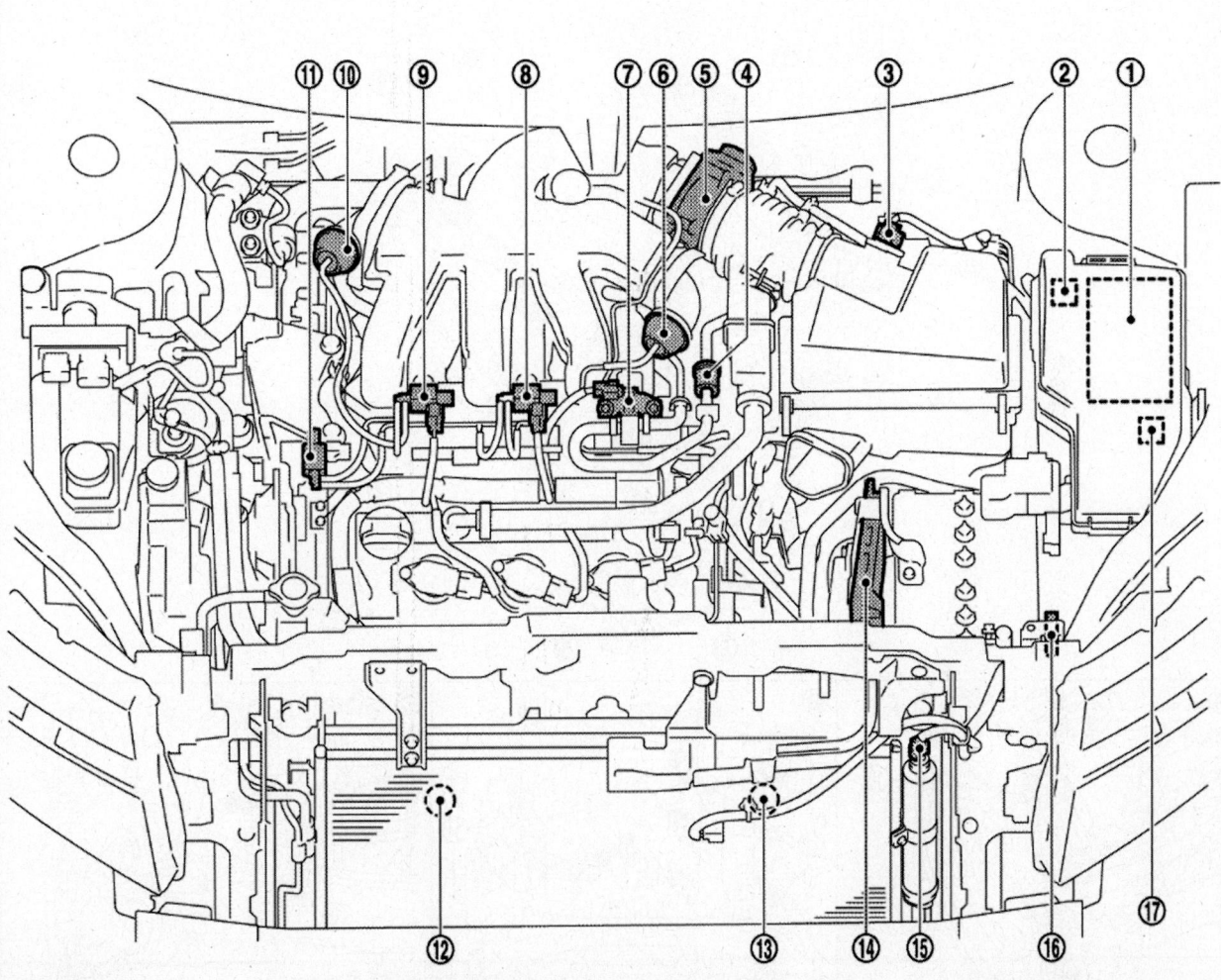

1. IPDM E/R
2. Cooling fan motor relay-3
3. Mass air flow sensor (with intake air temperature sensor)
4. EVAP service port
5. Electric throttle control actuator
6. Power valve actuator 2
7. EVAP canister purge volume control solenoid valve
8. VIAS control solenoid valve 2
9. VIAS control solenoid valve 1
10. Power valve actuator 1
11. Electronic controlled engine mount control solenoid valve
12. Cooling fan motor-2
13. Cooling fan motor-1
14. ECM
15. Refrigerant pressure sensor
16. Battery current sensor (With Battery Temperature Sensor)
17. Cooling fan motor relay-2

71075_QUES_G0219

Fig. 58 Engine Performance & Emission Controls—Engine room component locations

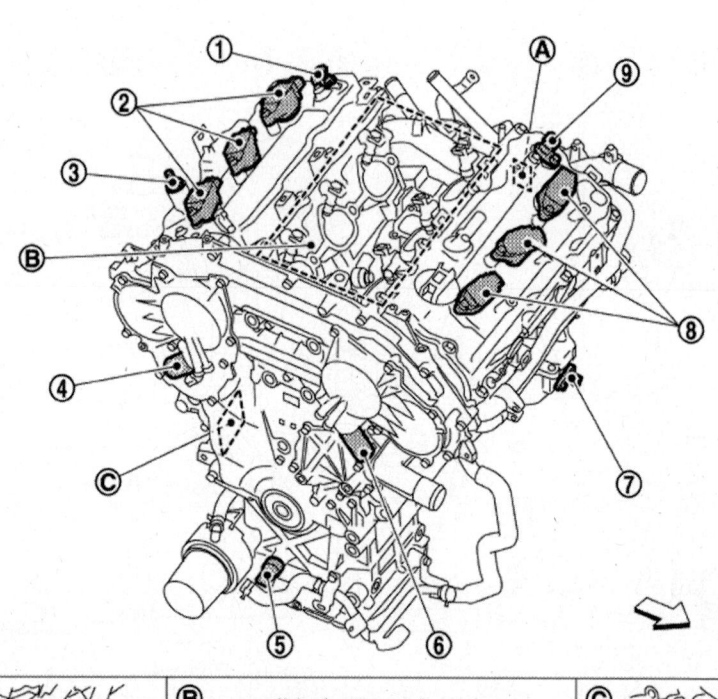

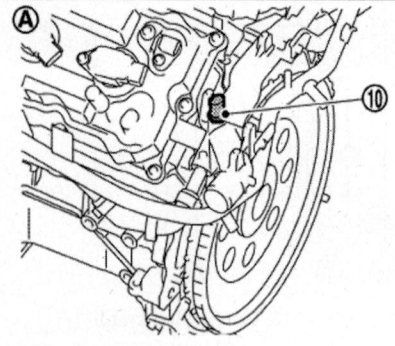

A. Engine rear upper-left
B. Engine top center
C. Engine front lower-right
1. Camshaft position sensor (PHASE) (bank 1)
2. Ignition coil (with power transistor) (bank 1)
3. PCV valve
4. Intake valve timing control solenoid valve (bank 1)
5. Engine oil pressure sensor
6. Intake valve timing control solenoid valve (bank 2)
7. Crankshaft Position Sensor (POS)
8. Ignition coil (with power transistor) (bank 2)
9. Camshaft position sensor (PHASE) (bank 2)
10. Engine coolant temperature sensor
11. Fuel injector (bank 1)
12. Knock sensor (bank 1)
13. Knock sensor (bank 2)
14. Fuel injector (bank 2)
15. Engine oil temperature sensor

71075_QUES_G0220

Fig. 59 Engine Performance & Emission Controls—Engine component locations

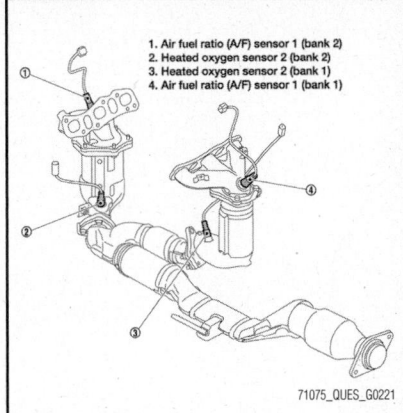

1. Air fuel ratio (A/F) sensor 1 (bank 2)
2. Heated oxygen sensor 2 (bank 2)
3. Heated oxygen sensor 2 (bank 1)
4. Air fuel ratio (A/F) sensor 1 (bank 1)

71075_QUES_G0221

Fig. 60 Engine Performance & Emission Controls—Exhaust system component locations

43897-18 or J-43897-12) and apply anti-seize lubricant (commercial service tool).

7. Install air fuel ratio sensor 1 in the original position. If the installation position cannot be identified, note the following glass tube colors:

- Air fuel ratio sensor 1: Black

8. Do not over-torque air fuel ratio sensor. Doing so may cause damage to air fuel ratio sensor, resulting in "MIL" coming on. Do not allow rust preventives to adhere to the sensor body.

ELECTRONIC CONTROL MODULE (ECM)

REMOVAL & INSTALLATION
See Figure 64.

✳✳ CAUTION

Before servicing components near or affected by the SRS (air bag) system, read and observe all SRS Service Precautions. Failure to observe all precautions may result in accidental airbag deployment, personal injury, or death.

1. Before servicing, turn ignition switch OFF, disconnect battery negative terminal and wait 3 minutes or more.
2. Remove the battery.
3. Disconnect the ECM harness connectors.
4. Remove the ECM mounting nuts, and then remove ECM.

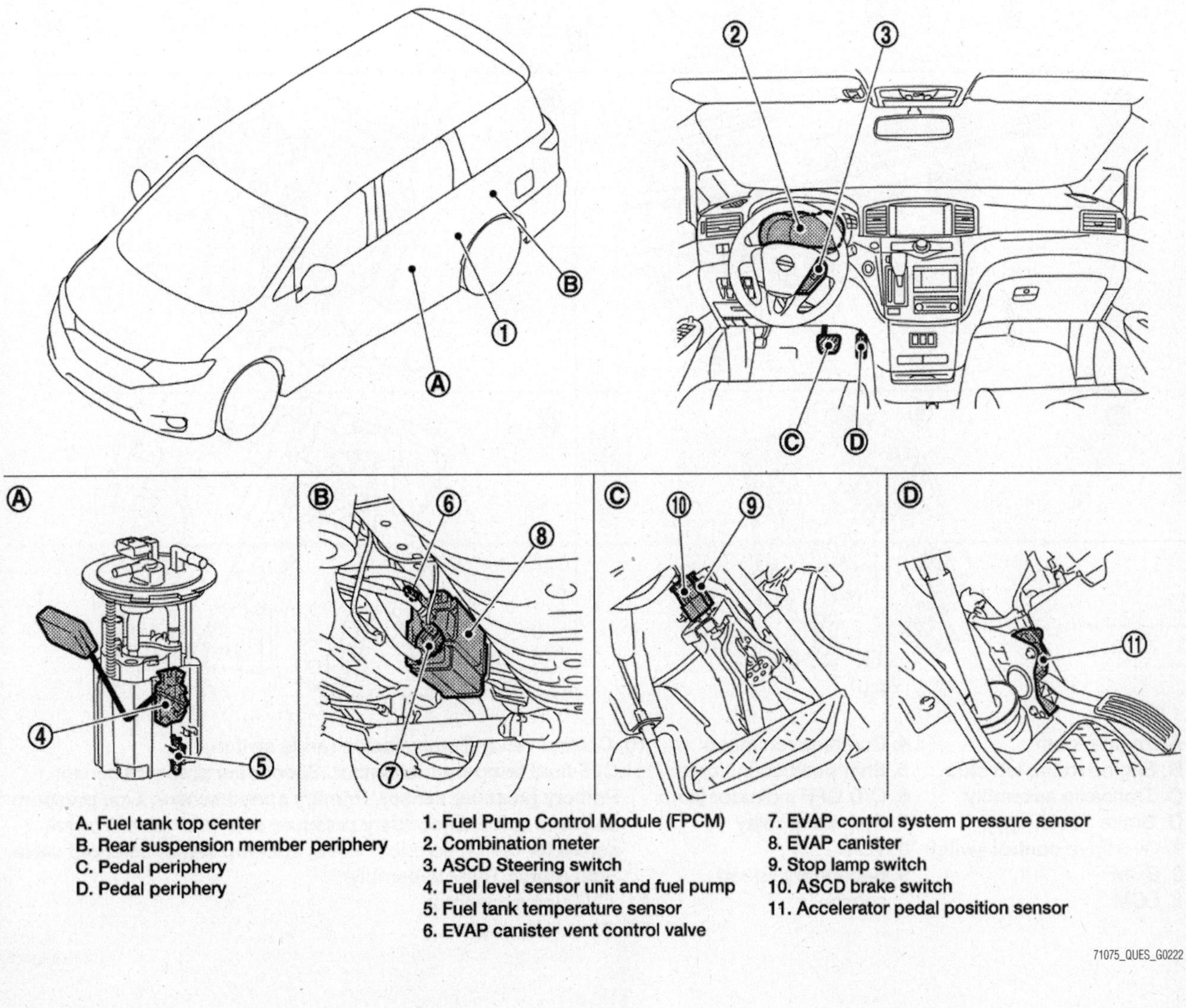

A. Fuel tank top center
B. Rear suspension member periphery
C. Pedal periphery
D. Pedal periphery

1. Fuel Pump Control Module (FPCM)
2. Combination meter
3. ASCD Steering switch
4. Fuel level sensor unit and fuel pump
5. Fuel tank temperature sensor
6. EVAP canister vent control valve

7. EVAP control system pressure sensor
8. EVAP canister
9. Stop lamp switch
10. ASCD brake switch
11. Accelerator pedal position sensor

71075_QUES_G0222

Fig. 61 Engine Performance & Emission Controls—Body component locations

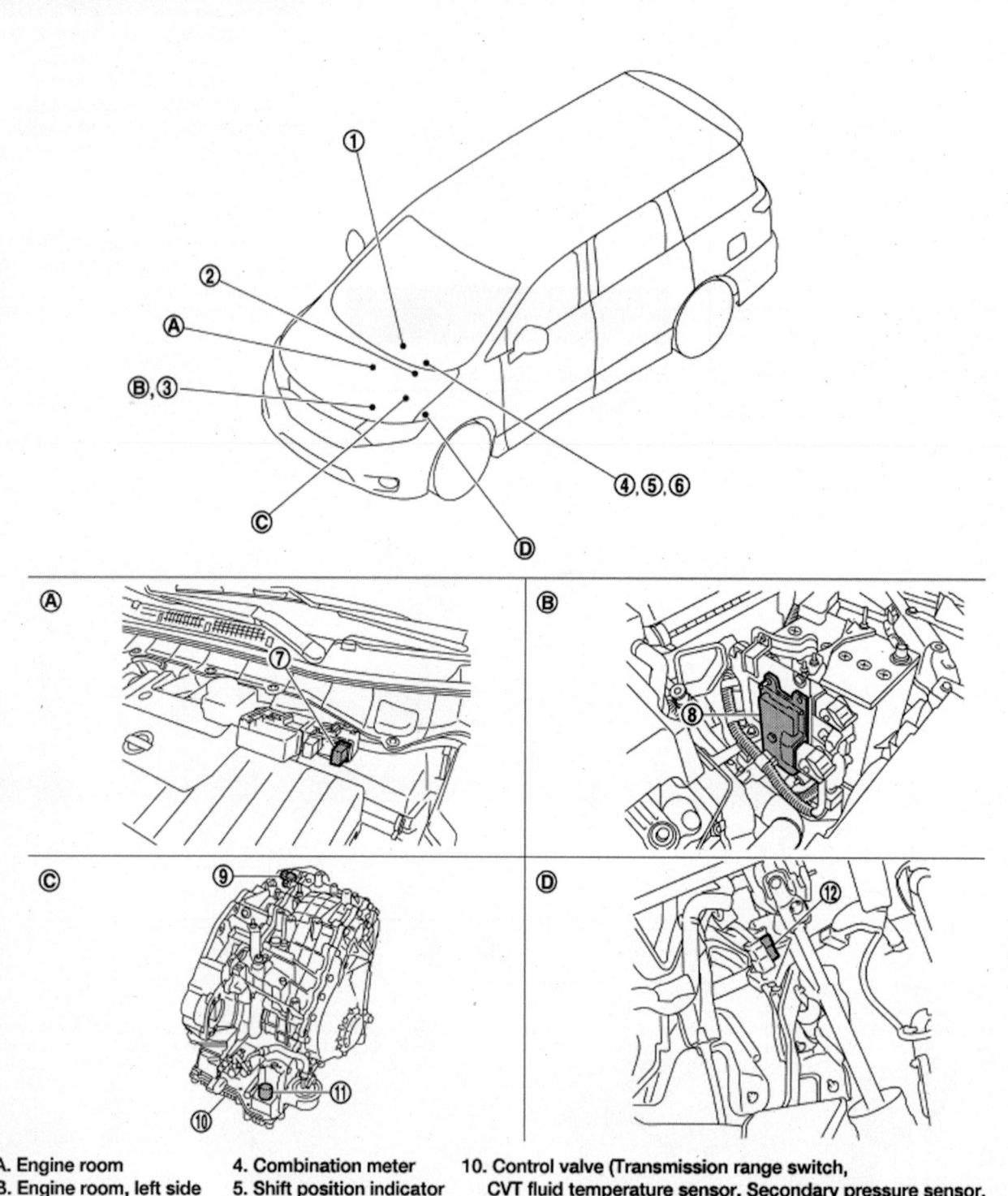

A. Engine room
B. Engine room, left side
C. Transaxle assembly
D. Brake pedal, upper
1. Overdrive control switch
2. BCM
3. ECM
4. Combination meter
5. Shift position indicator
6. O/D OFF indicator lamp
7. Stop lamp relay
8. TCM
9. Secondary speed sensor
10. Control valve (Transmission range switch, CVT fluid temperature sensor, Secondary pressure sensor, Primary pressure sensor, Primary speed sensor, Line pressure solenoid valve, Secondary pressure solenoid valve, Torque converter clutch solenoid valve, Lock-up select solenoid valve, Step motor, ROM assembly)
11. CVT unit connector
12. Stop lamp switch

71075_QUES_G0245

Fig. 62 CVT Control System: component locations

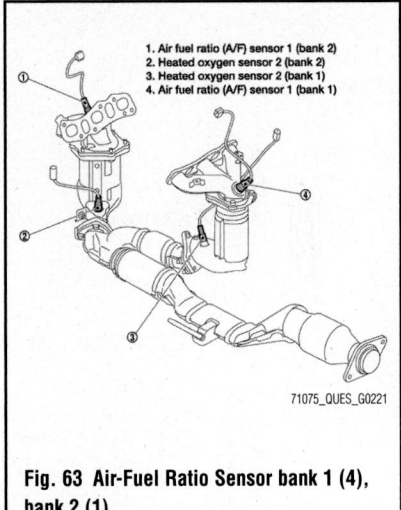

1. Air fuel ratio (A/F) sensor 1 (bank 2)
2. Heated oxygen sensor 2 (bank 2)
3. Heated oxygen sensor 2 (bank 1)
4. Air fuel ratio (A/F) sensor 1 (bank 1)

71075_QUES_G0221

Fig. 63 Air-Fuel Ratio Sensor bank 1 (4), bank 2 (1)

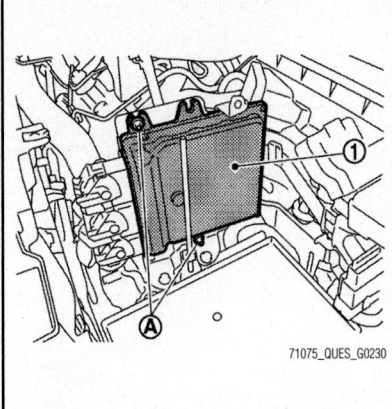

71075_QUES_G0230

Fig. 64 Remove the ECM mounting nuts (1), and then remove ECM (A)

To install:

Installation is the reverse of removal.

RESET

When replacing ECM, the following procedure must be performed:

1. Perform initialization of NVIS (NATS) system and registration of all NVIS (NATS) ignition Key IDs. Perform ECM Recommunicating Function:

   a. Install ECM.

   b. Contact backside of registered Intelligent key to push-button ignition switch, then turn power supply position

   to ON. To perform this step, use the key that is used before performing ECM replacement.

   c. Maintain power supply position in the ON position for at least 5 seconds.

   d. Turn power supply position to OFF.

   e. Check that the engine starts.

2. Perform VIN Registration:

   a. Turn ignition switch ON with engine stopped.

   b. Select "VIN REGISTRATION" in "WORK SUPPORT" mode.

   c. Follow the instructions on the CONSULT display.

3. Perform Accelerator Pedal Released Position Learning:

   a. Check that accelerator pedal is fully released.

   b. Turn ignition switch ON and wait at least 2 seconds.

   c. Turn ignition switch OFF and wait at least 10 seconds.

   d. Turn ignition switch ON and wait at least 2 seconds.

   e. Turn ignition switch OFF and wait at least 10 seconds.

4. Perform Throttle Valve Closed Position Learning:

   a. With Consult:
   - Turn ignition switch ON
   - Select "CLSD THL POS LEARN" in "WORK SUPPORT" mode
   - Follow the instructions on the CONSULT display
   - Turn ignition switch OFF and wait at least 10 seconds. Check that throttle valve moves during the above 10 seconds by confirming the operating sound

   b. Without Consult:
   - Start the engine. Note: Coolant temperature is less than 77°F (25°C) before engine starts

   c. Warm up the engine. Note: Warm up the engine until "COOLAN TEMP/S" on "DATA MONITOR" of CONSULT reaches more than 149°F (65°C).

   d. Turn ignition switch OFF and wait at least 10 seconds. Note: Check that throttle valve moves during the above 10 seconds by confirming the operating sound.

5. Perform Idle Air Volume Learning.

   a. Check that all of the following conditions are satisfied. Learning will be cancelled if any of the following conditions are missed for even a moment.
   - Battery voltage: More than 12.9 V (At idle)
   - Engine coolant temperature: 158–212°F (70–100°C)
   - Selector lever position: P or N
   - Electric load switch: OFF (Air conditioner, head lamp, rear window defogger); On vehicles equipped

with daytime light systems, if the parking brake is applied before the engine is started the head lamp will not illuminate.
   - Steering wheel: Neutral (Straight-ahead position)
   - Vehicle speed: Stopped
   - Transmission: Warmed-up
   - With CONSULT: Drive vehicle until "ATF TEMP SEN" in "DATA MONITOR" mode of "CVT" system indicates less than 0.9 V.
   - Without CONSULT: Drive vehicle for 10 minutes.

6. To Perform Idle Air Volume Learning With CONSULT:

   a. Perform Accelerator Pedal Released Position Learning.

   b. Perform Throttle Valve Closed Position Learning.

   c. Start engine and warm it up to normal operating temperature.

   d. Select "IDLE AIR VOL LEARN" in "WORK SUPPORT" mode.

   e. Touch "START" and wait 20 seconds.

   f. If "CMPLT" is displayed on CONSULT screen, Rev up the engine 2 or 3 times and check that idle speed and ignition timing are within the specifications.

   g. If "CMPLT" is not displayed on CONSULT screen, Check that throttle valve is fully closed, Check PCV valve operation, Check that downstream of throttle valve is free from air leakage.

7. To Perform Idle Air Volume Learning Without CONSULT:

➡**It is better to count the time accurately with a clock. It is impossible to switch the diagnostic mode when an accelerator pedal position sensor circuit has a malfunction.**

   a. Perform Accelerator Pedal Released Position Learning.

   b. Perform Throttle Valve Closed Position Learning.

   c. Start engine and warm it up to normal operating temperature.

   d. Turn ignition switch OFF and wait at least 10 seconds.

   e. Confirm that accelerator pedal is fully released, turn ignition switch ON and wait 3 seconds.

   f. Repeat the following procedure quickly 5 times within 5 seconds.
   - Fully depress the accelerator pedal.

- Fully release the accelerator pedal.
  g. Wait 7 seconds, fully depress the accelerator pedal for approx. 20 seconds until the MIL stops blinking and turns ON.
  h. Fully release the accelerator pedal within 3 seconds after the MIL turns ON.
  i. Start engine and let it idle.
  j. Wait 20 seconds.

EVAPORATIVE EMISSION CONTROL SYSTEM

EVAP CANISTER

Location

See Figures 65 and 66.

Removal & Installation

1. Disconnect the harness connectors (EVAP control pressure sensor and EVAP canister vent control valve) and the EVAP canister hoses.
2. Remove the EVAP canister mounting bolt.
3. Remove the EVAP canister.

➡**EVAP canister vent control valve and EVAP control pressure sensor can be removed without removing the EVAP canister.**

To install:

Installation is the reverse of removal. Tighten the EVAP canister mounting bolt to 13 ft. lbs. (17 Nm).

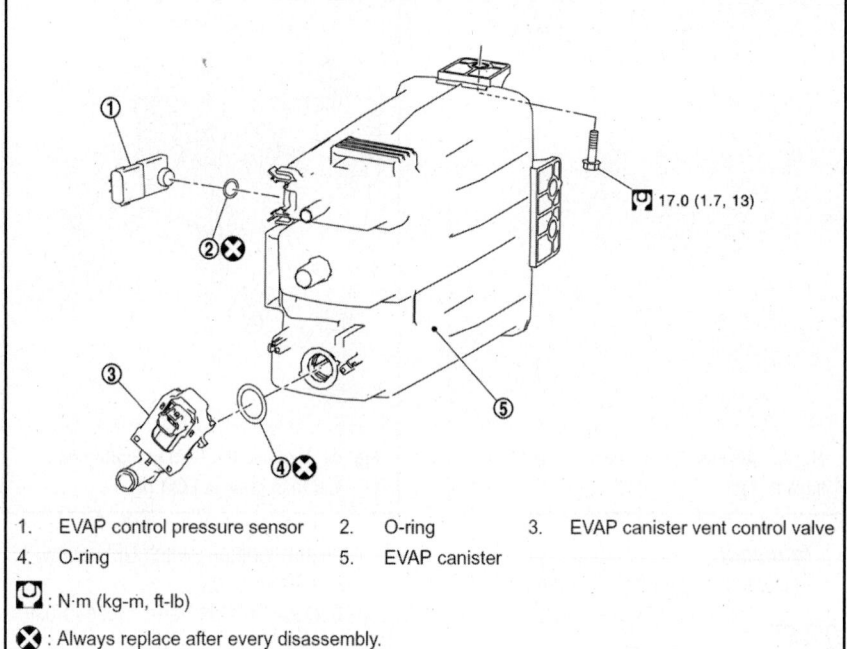

1. EVAP control pressure sensor
2. O-ring
3. EVAP canister vent control valve
4. O-ring
5. EVAP canister

🔧 : N·m (kg-m, ft-lb)

❌ : Always replace after every disassembly.

71075_QUES_G0234

Fig. 66 EVAP canister—exploded view

EVAP CANISTER PURGE VOLUME CONTROL SOLENOID VALVE.

Location

See Figures 00 and 67

FUEL TANK TEMPERATURE SENSOR

LOCATION

See Figure 68.

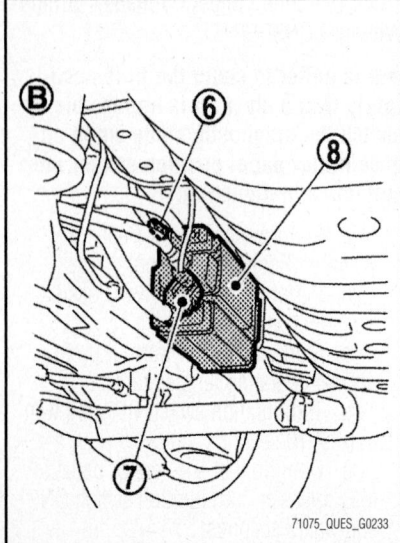

71075_QUES_G0233

Fig. 65 EVAP canister (8), EVAP canister vent control valve (6),EVAP control system pressure sensor (7)

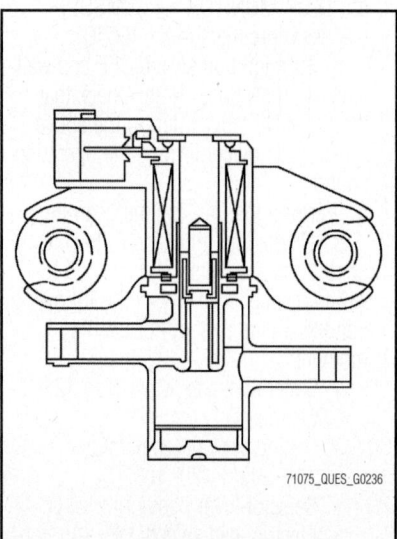

71075_QUES_G0236

Fig. 67 EVAP canister purge volume control solenoid valve

HEATED OXYGEN (HO2S) SENSOR

LOCATION

See Figure 69.

REMOVAL & INSTALLATION

1. For bank 1 side, remove the following:
   - Air cleaner case (upper) with mass air flow sensor and air duct assembly.
   - Engine cover.
   - Front wiper arm.
   - Extension cowl top.
   - Heat insulator.
2. For bank 2 side, remove the following:
   - Engine under cover.

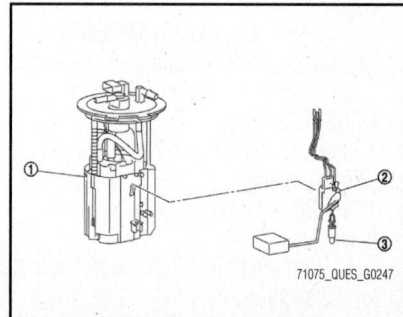

71075_QUES_G0247

Fig. 68 Fuel filter and fuel pump assembly (1), fuel level sensor unit (2), fuel tank temperature sensor (3)

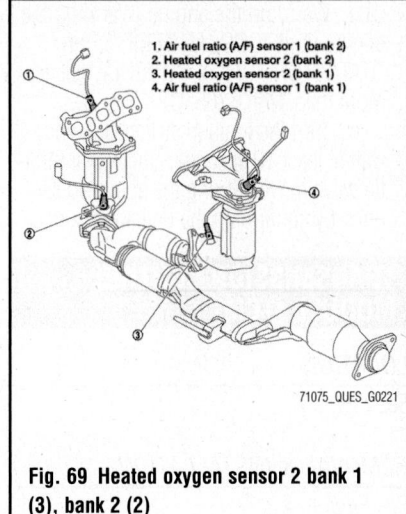

1. Air fuel ratio (A/F) sensor 1 (bank 2)
2. Heated oxygen sensor 2 (bank 2)
3. Heated oxygen sensor 2 (bank 1)
4. Air fuel ratio (A/F) sensor 1 (bank 1)

71075_QUES_G0221

Fig. 69 Heated oxygen sensor 2 bank 1 (3), bank 2 (2)

3. Remove exhaust front tube.

4. Disconnect harness connector and remove heated oxygen sensor 2 on both banks with the heated oxygen sensor wrench [SST: KV10114400 (J-38365)]. Apply matchmarks to identify installation positions of each heated oxygen sensor 2.

To install:

5. Before installing a new heated oxygen sensor, clean exhaust system threads using oxygen sensor thread cleaner (commercial service tool: J-43897-18 or J-43897-12) and apply anti-seize lubricant (commercial service tool).

6. Install the heated oxygen sensor 2 in the original position. If the installation positions cannot be identified, note the following glass tube colors:

   • Heated oxygen sensor 2: White

7. Do not over-torque the heated oxygen sensor. Doing so may cause damage to the heated oxygen sensor, resulting in "MIL" coming on. Do not allow rust preventives to adhere to the sensor body.

INTAKE AIR TEMPERATURE (IAT)/MASS AIRFLOW (MAF) SENSOR

LOCATION

See Figure 70.

THROTTLE CONTROL ACTUATOR

REMOVAL & INSTALLATION

See Figure 71.

1. Remove the engine cover. Be careful not to damage or scratch engine cover.

2. Remove air cleaner cases (upper and lower).

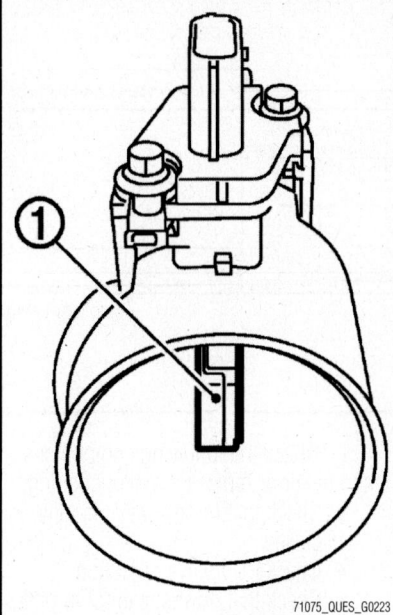

71075_QUES_G0223

Fig. 70 Intake Air Temperature/Mass Airflow Sensor (1)

3. Remove the electric throttle control actuator. Loosen the mounting bolts in reverse of the tightening order, shown below. Handle carefully. Never disassemble.

To install:

4. Tighten the electric throttle control actuator mounting bolts in the order shown.

5. Install the air cleaner cases.

6. Install the engine cover.

7. Perform the "Throttle Valve Closed Position Learning" when the harness connector of electric throttle control actuator is disconnected.

8. Perform the "Idle Air Volume Learning" and "Throttle Valve Closed Position Learning" when the electric throttle control actuator is replaced.

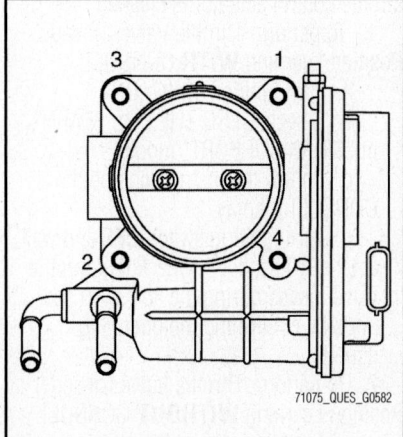

71075_QUES_G0582

Fig. 71 Electric throttle control actuator bolt tightening sequence

IDLE AIR VOLUME LEARNING

See Figure 72.

Idle Air Volume Learning is a function of ECM to learn the idle air volume that keeps engine idle speed within the specific range. It must be performed under the following conditions:

• Each time the electric throttle control actuator or ECM is replaced

• Idle speed or ignition timing is out of the specification

1. Check that all of the following conditions are satisfied. Learning will be cancelled if any of the following conditions are missed for even a moment.

• Battery voltage: More than 12.9 V (At idle)

• Engine coolant temperature: 158–212°F (70–100°C)

• Selector lever position: P or N

• Electric load switch: OFF (Air conditioner, head lamp, rear window defogger); on vehicles equipped with daytime light systems, if the parking brake is applied before the engine is started the head lamp will not illuminate

• Steering wheel: Neutral (Straight-ahead position)

• Vehicle speed: Stopped

• Transmission: Warmed-up. **WITH** CONSULT: Drive vehicle until "ATF TEMP SEN" in "DATA MONITOR" mode of "CVT" system indicates less than 0.9 V. **WITHOUT** CONSULT: Drive vehicle for 10 minutes.

2. To perform Idle Air Volume Learning **WITH** CONSULT:

   a. Perform accelerator pedal released position learning.

   b. Perform throttle valve closed position learning.

   c. Start engine and warm it up to normal operating temperature.

   d. Select "IDLE AIR VOL LEARN" in "WORK SUPPORT" mode.

   e. Touch "START" and wait 20 seconds.

   f. If "CMPLT" **IS** displayed on CONSULT screen, rev up the engine 2 or 3 times and check that idle speed and ignition timing are within the specifications. If they are not within specifications, go to the next step.

   g. If "CMPLT" **IS NOT** displayed on CONSULT screen, check the following components, and repair or replace if malfunctioning:

• Check that throttle valve is fully closed

• Check PCV valve operation

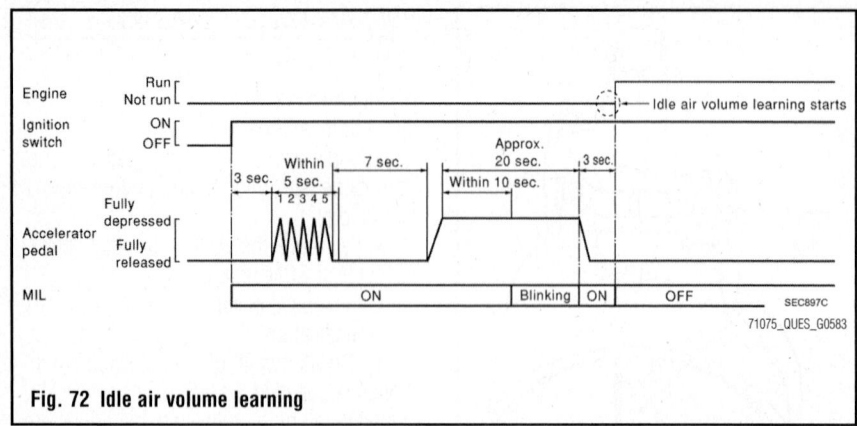

Fig. 72 Idle air volume learning

- Check that downstream of throttle valve is free from air leakage

h. If the previous step did not detect any malfunctioning parts, the engine component parts and their installation condition are questionable. Diagnose and eliminate the cause of the incident. If any of the following conditions occur after the engine has started, eliminate the cause of the incident and perform Idle Air Volume Learning again:
- Engine stalls
- Incorrect idle

3. To perform Idle Air Volume Learning **WITHOUT** CONSULT:

➡It is better to count the time accurately with a clock. It is impossible to switch the diagnostic mode when an accelerator pedal position sensor circuit has a malfunction.

a. Perform accelerator pedal released position learning.

b. Perform throttle valve closed position learning.

c. Start engine and warm it up to normal operating temperature.

d. Turn ignition switch OFF and wait at least 10 seconds.

e. Confirm that accelerator pedal is fully released, turn ignition switch ON and wait 3 seconds.

f. Repeat the following procedure quickly 5 times within 5 seconds:
- Fully depress the accelerator pedal
- Fully release the accelerator pedal

g. Wait 7 seconds, fully depress the accelerator pedal for approx. 20 seconds until the MIL stops blinking and turns ON.

h. Fully release the accelerator pedal within 3 seconds after the MIL turns ON.

i. Start engine and let it idle.

j. Wait 20 seconds.

k. Rev up the engine 2 or 3 times and check that idle speed and ignition timing are within the specifications. If they are not within specifications, go to the next step.

l. Check the following components, and repair or replace if malfunctioning:
- Check that throttle valve is fully closed
- Check PCV valve operation
- Check that downstream of throttle valve is free from air leakage

m. If the previous step did not detect any malfunctioning parts, the engine component parts and their installation condition are questionable. Diagnose and eliminate the cause of the incident. If any of the following conditions occur after the engine has started, eliminate the cause of the incident and perform Idle Air Volume Learning again:
- Engine stalls
- Incorrect idle

THROTTLE VALVE CLOSED POSITION LEARNING

Throttle Valve Closed Position Learning is a function of ECM to learn the fully closed position of the throttle valve by monitoring the throttle position sensor output signal. It must be performed each time the harness connector of electric throttle control actuator or ECM is disconnected or electric throttle control actuator is cleaned.

1. To perform Throttle Valve Closed Position Learning **WITH** CONSULT:

a. Turn ignition switch ON.

b. Select "CLSD THL POS LEARN" in "WORK SUPPORT" mode.

c. Follow the instructions on the CONSULT display.

d. Turn ignition switch OFF and wait at least 10 seconds. Check that throttle valve moves during the above 10 seconds by confirming the operating sound.

2. To perform Throttle Valve Closed Position Learning **WITHOUT** CONSULT:

a. Start the engine. Coolant temperature is less than 77°F (25°C) before engine starts.

b. Warm up the engine. Warm up the engine until "COOLAN TEMP/S" on "DATA MONITOR" of CONSULT reaches more than 149°F (65°C).

c. Turn the ignition switch OFF and wait at least 10 seconds. Check that throttle valve moves during the above 10 seconds by confirming the operating sound.

TRANSMISSION CONTROL MODULE (TCM)

LOCATION
See Figure 73.

REMOVAL & INSTALLATION
See Figure 74.

➡When replacing TCM and transaxle assembly as a set, replace transaxle assembly first and then replace TCM.

1. Disconnect the battery cable from the negative terminal.

2. Remove air duct (inlet).

3. Move battery harness to a place to keep the harness clear of working area.

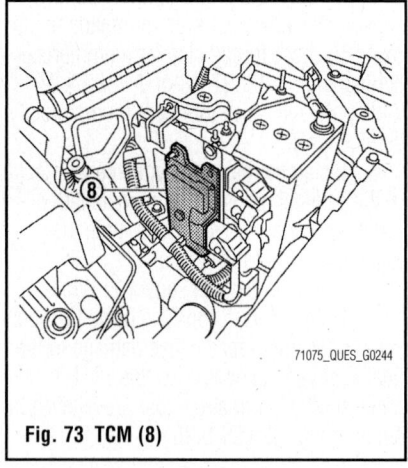

Fig. 73 TCM (8)

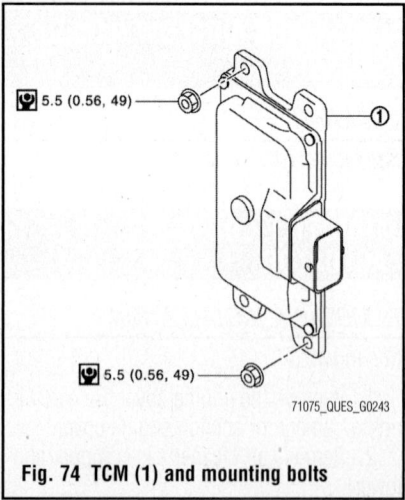

Fig. 74 TCM (1) and mounting bolts

4. Disconnect TCM connector.
5. Remove TCM.

To install:

Installation is the reverse of removal. After installation, perform ADDITIONAL SERVICE WHEN REPLACING TCM.

ADDITIONAL SERVICE WHEN REPLACING TCM

1. Load Calibration Data:
   a. Shift the selector lever to the "P" position.
   b. Turn ignition switch ON.
   c. Check that "P" is displayed on shift position indicator on combination meter.

➡**Displayed approximately 1–2 seconds after the selector lever is moved to the "P" position.**

   d. If the shift position indicator displays "P", store calibration data. If the shift position indicator does not display "P", detect the malfunctioning item.
2. Detect Malfunctioning Item:
   a. Check the following items:
   • Harness between the TCM and the ROM assembly inside the transaxle assembly is open or shorted
   • Disconnected, loose, bent, collapsed, or otherwise abnormal connector housing terminals

   b. If the inspection result is normal, load calibration data. If the inspection result is not normal, repair or replace the malfunctioning parts.
3. Store Calibration Data:
   a. Turn ignition switch OFF and wait for 5 seconds.
   b. Turn ignition switch ON.
   c. If the shift position indicator displays "P" at the same time when turning ON the ignition switch, the work is complete. If the shift position indicator does not display "P" at the same time when turning ON the ignition switch, check the harness between battery and TCM harness connector terminal.

FUEL   GASOLINE FUEL INJECTION SYSTEM

FUEL SYSTEM SERVICE PRECAUTIONS

Safety is the most important factor when performing not only fuel system maintenance but any type of maintenance. Failure to conduct maintenance and repairs in a safe manner may result in serious personal injury or death. Maintenance and testing of the vehicle's fuel system components can be accomplished safely and effectively by adhering to the following rules and guidelines.

• To avoid the possibility of fire and personal injury, always disconnect the negative battery cable unless the repair or test procedure requires that battery voltage be applied.

• Always relieve the fuel system pressure prior to disconnecting any fuel system component (injector, fuel rail, pressure regulator, etc.), fitting or fuel line connection. Exercise extreme caution whenever relieving fuel system pressure to avoid exposing skin, face and eyes to fuel spray. Please be advised that fuel under pressure may penetrate the skin or any part of the body that it contacts.

• Always place a shop towel or cloth around the fitting or connection prior to loosening to absorb any excess fuel due to spillage. Ensure that all fuel spillage (should it occur) is quickly removed from engine surfaces. Ensure that all fuel soaked cloths or towels are deposited into a suitable waste container.

• Always keep a dry chemical (Class B) fire extinguisher near the work area.

• Do not allow fuel spray or fuel vapors to come into contact with a spark or open flame.

• Always use a back-up wrench when

loosening and tightening fuel line connection fittings. This will prevent unnecessary stress and torsion to fuel line piping.

• Always replace worn fuel fitting O-rings with new Do not substitute fuel hose or equivalent where fuel pipe is installed.

Before servicing the vehicle, refer to the Precautions.

RELIEVING FUEL SYSTEM PRESSURE

See Figure 75.

1. With CONSULT:
   a. Turn the ignition switch ON.
   b. Perform "FUEL PRESSURE RELEASE" in "WORK SUPPORT" mode with CONSULT.
   c. Start the engine.
   d. After the engine stalls, crank it 2 or 3 times to release all fuel pressure.
   e. Turn the ignition switch OFF.

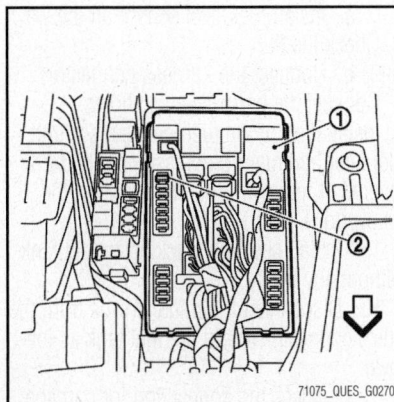

Fig. 75 Remove the fuel pump fuse (2) located in IPDM E/R (1)

71075_QUES_G0270

2. Without CONSULT:
   a. Remove the fuel pump fuse located in IPDM E/R.
   b. Start the engine.
   c. After the engine stalls, crank it 2 or 3 times to release all fuel pressure.
   d. Turn the ignition switch OFF.
   e. Reinstall the fuel pump fuse after servicing fuel system.

FUEL LEVEL SENSOR UNIT, FUEL FILTER & FUEL PUMP ASSEMBLY

REMOVAL & INSTALLATION
See Figures 76 through 81.

1. Before servicing the vehicle, refer to the Precautions.
2. Release the fuel pressure from the fuel lines.
3. Check fuel level on a level ground. If the fuel level is ⅞ of the fuel tank (full or nearly full), draw appropriate amount of fuel from the fuel tank. Guideline: Draw approximately 20 liters from a full-tank condition.
   a. In the event of malfunction in fuel pump, insert a hose measuring 0.79 in. (20 mm) in diameter into the filler opening to draw approximately 20 liters fuel.
4. Open fuel filler lid.
5. Open filler cap and release the pressure inside fuel tank.
6. Slide the left second seat in the forward direction.
7. Peel off floor carpet, then remove inspection hole cover units by turning clips clockwise by 90 degrees.
8. Disconnect harness connector, fuel feed tube and EVAP hose.
   a. To keep the connecting portion clean and to avoid damage and foreign

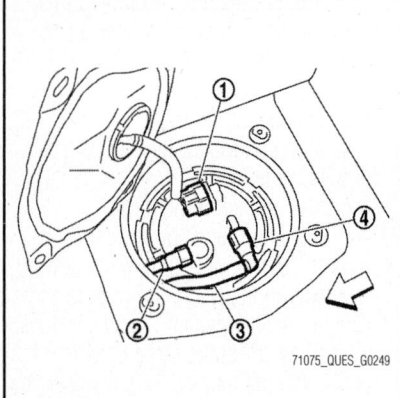

Fig. 76 Harness connector (1), fuel feed tube (3), EVAP hose (2), quick connector (4)

materials, cover them completely with plastic bags or something similar.

9. Remove retainer for fuel level sensor unit, fuel filter and fuel pump assembly with fuel tank lock ring wrench (SST: KV10119900) by turning counterclockwise.

   a. To reduce impact caused by removal operation, use long spinner handle [handle length: 23.62 in. (60 cm) or more] and slowly turn it counterclockwise. To prevent lock ring wrench from being detached, securely hold down spinner handle by hand.

10. Pull up the fuel level sensor unit, fuel filter and fuel pump assembly to the position shown in the figure.

✳✳ WARNING

To prevent the transfer tube from becoming disconnected, never pull the assembly more than necessary.

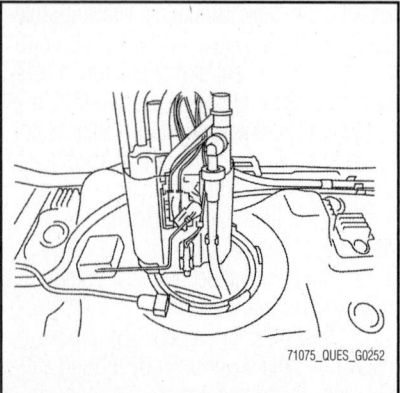

Fig. 77 Pull up the fuel level sensor unit, fuel filter and fuel pump assembly to the position shown

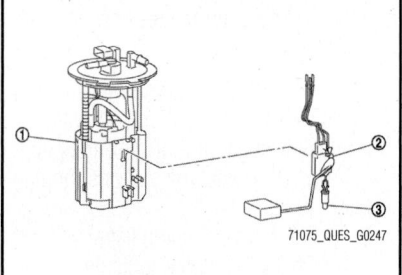

Fig. 78 Fuel filter and fuel pump assembly (1), fuel level sensor unit (2), fuel tank temperature sensor (3)

11. Disconnect transfer tube from the fuel level sensor unit, fuel filter and fuel pump assembly by holding and pulling quick connector in the direction shown.

✳✳ WARNING

Never pull transfer tube because it is inseparable from the fuel tank assembly.

12. Remove fuel level sensor unit, fuel filter and fuel pump assembly.

13. To disassemble the fuel level sensor unit, fuel filter and fuel pump assembly, and remove the fuel level sensor unit from fuel filter and fuel pump assembly:

   a. Disconnect harness connector. Press into the pawl of the mounting area to pull out connector.

   b. Press into pawl of fuel level sensor mounting area.

   c. With the fuel level sensor unit unlocked, slide it in the direction shown by the arrow.

To install:

14. To reassemble the fuel level sensor unit, fuel filter and fuel pump assembly, and install the fuel level sensor unit:

   a. Install fuel level sensor all the way. Check the fit.

   b. Connect the harness connector. Connect the harnesses as shown.

Installation of the fuel level sensor unit, fuel filter and fuel pump assembly is the reverse of removal procedure, noting the following:

15. Install new seal packing to fuel tank without any twist.

16. Disconnect the transfer tube from the quick connector within the fuel tank as follows:

   a. Check the connection for damage or any foreign materials.

   b. Eliminate distortion from the transfer tube in the fuel tank, referring to the figure.

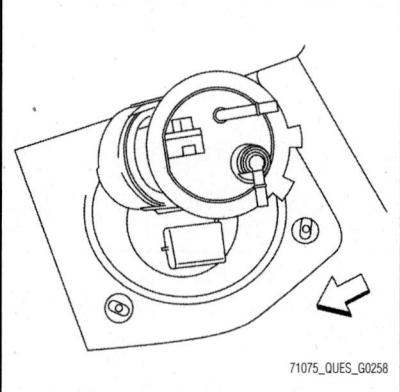

Fig. 79 Float, fuel level sensor unit, fuel filter and fuel pump positioning; arrow: vehicle front

   c. Align the connector with the tube, then insert the connector straight into the tube until a click sound is heard.

   d. Pull the tube and the connector to check they are securely connected.

17. With the float of fuel level sensor unit, fuel filter and fuel pump assembly positioned as shown in the figure, insert the fuel pump assembly into the fuel tank. Never bend float arm during installing.

18. After the insertion, turn the fuel level sensor unit, fuel filter and fuel pump assembly by 90 degrees in the direction shown.

19. Through the opening of the fuel tank, check that the transfer hose is positioned as shown in step one. Check that the transfer tube does not interfere with the float and the float arm.

20. Align protrusion of fuel level sensor unit, fuel filter and fuel pump assembly with the matching mark of fuel tank to install the fuel gauge fuel filter fuel pump assembly.

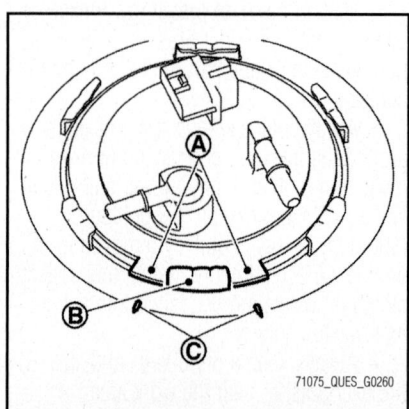

Fig. 80 Align protrusion (A) of fuel level sensor unit, fuel filter and fuel pump assembly with the matching mark (C) of fuel tank; Retainer mounting pawl (B)

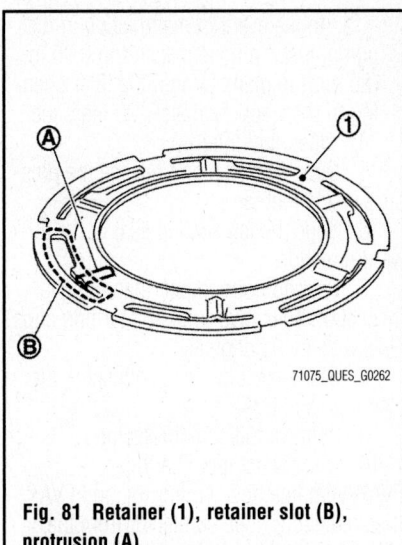

Fig. 81 Retainer (1), retainer slot (B), protrusion (A)

✳✳ WARNING

Do not allow seal packing to drop. Do not bend float arm during installing.

21. Tighten the retainer with a lock ring wrench (SST: KV10119900). Horizontally press the lock ring wrench to protect the retainer from tilting, and tighten until the retainer protrusion overlaps the retainer mounting pawl. If it's hard to tighten retainer, inspect the retainer.

22. To inspect the retainer:

a. Visually check that slot of retainer and its surrounding parts (6 points) are free from protrusions (e.g. burr).

b. If the slot and its surrounding parts have protrusions, replace retainer with a new one.

➡The inside of the slot includes locating protrusion (one for each slot) used for the fit between retainer and retainer mounting pawl.

23. To inspect for fuel leakage:

a. Turn the ignition switch "ON" (with engine stopped), then check connections for leakage by applying fuel pressure to the fuel piping.

b. Start the engine and let it idle, and check there is no fuel leakage at the fuel system connections.

FUEL RAIL & INJECTORS

REMOVAL & INSTALLATION

See Figures 82 and 83.

1. Before servicing the vehicle, refer to the Precautions.

2. Disconnect the negative battery cable.

3. Remove the air duct (inlet); air cleaner cases (upper and lower) with mass air flow sensor and air duct assembly.

4. Remove the engine cover.

5. Release the fuel pressure.

6. Drain the engine coolant, or when water hoses are disconnected, attach plug to prevent engine coolant leakage. Perform this step when the engine is cold.

7. Remove the intake manifold collector.

8. Prepare container and cloth beforehand as fuel will leak out.

9. Disconnect the quick connector by using the quick connector release (service tool: J- 45488), not by picking out retainer tabs. Disconnect the quick connector from the fuel tube as follows:

a. Remove the quick connector cap.

b. With the sleeve side of tool facing quick connector, install the tool on to fuel tube.

c. Insert the tool into quick connector until sleeve contacts and goes no further. Hold the tool in that position. Inserting the tool hard will not disconnect quick connector.

d. Pull the quick connector straight out from the fuel tube. Pull the quick connector holding it at the position shown.

✳✳ CAUTION

Avoid fire and sparks. Keep parts away from heat source. Especially, be careful when welding is performed nearby. Never expose parts to battery electrolyte or other acids.

✳✳ WARNING

Never pull with lateral force applied. O-ring inside quick connector may be damaged. Never bend or twist connection between quick connector and fuel feed hose (with damper) during installation/removal. To keep the connecting portions clean and to avoid damage, cover them completely with plastic bags, etc. or something similar.

10. Disconnect the harness connector from the fuel injector.

11. Loosen the mounting bolts in the reverse of tightening order, shown below, and remove the fuel tube and fuel injector assembly. Never tilt the fuel tube, or remaining fuel in pipes may flow out from pipes.

12. Remove the fuel injector from the fuel tube as follows:

a. Open and remove the clip.

b. Pull the injector from fuel tube by pulling straight out. Be careful with remaining fuel that may go out from fuel tube.

✳✳ WARNING

Never disassemble fuel injector. Be careful not to damage injector nozzle, and do not bump or drop fuel injector.

13. If applicable, remove the fuel damper and O-ring from fuel tube.

To install:

✳✳ WARNING

When handling new O-ring, note the following: Do not reuse O-rings. Handle O-ring with bare hands; do not wear gloves. Do not clean O-ring with solvent. Lubricate O-ring with new engine oil. Be careful not to damage the O-rings and O-ring sealing surfaces. Do not scratch, twist or stretch O-rings.

14. Install the fuel damper as follows:

a. Install new O-ring to the fuel tube as shown. Insert new O-ring straight into fuel tube. Do not twist.

b. Install the spacer to the fuel damper.

c. Insert the fuel damper straight into the fuel tube. Insert straight, checking that the axis is lined up. Do not pressure-fit with excessive force. Reference value: 29.2 lbs. (130 N)

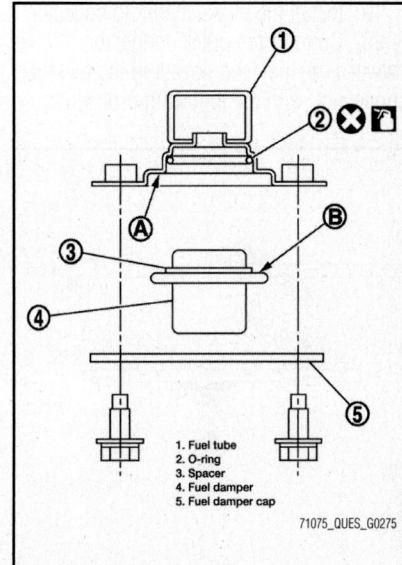

1. Fuel tube
2. O-ring
3. Spacer
4. Fuel damper
5. Fuel damper cap

Fig. 82 Insert fuel damper until (B) is touching (A) of fuel tube

d. Tighten the bolts evenly in turn. After tightening bolts, check that there is no gap between fuel damper cap and fuel tube.

15. Install new O-rings to the fuel injector. Upper and lower O-ring are different; be careful not to confuse them.
- Fuel tube side: Black
- Nozzle side: Green

16. Install the fuel injector to the fuel tube as follows:

a. Insert the clip into the clip mounting groove on the fuel injector. Insert the clip so the protrusion of the fuel injector matches the cutout of the clip. Use a new clip. Be careful to keep the clip from interfering with the O-ring. If interference occurs, replace O-ring.

b. Insert the fuel injector into the fuel tube with the clip attached. Insert it while matching it to the axial center. Insert the fuel injector so that the protrusion of the fuel tube matches the cutout of the clip. Check that the fuel tube flange is securely fixed in the flange groove on the clip.

c. Check that installation is complete by checking that the fuel injector does not rotate or come off. Check that the protrusions of the fuel injectors and fuel tubes are aligned with the cutouts of the clips.

17. Install the fuel tube and fuel injector assembly to the intake manifold. Be careful not to let tip of injector nozzle come in contact with other parts. Tighten mounting bolts in two steps in numerical order as shown.
- 1st step: 7 ft. lbs. (10 Nm)
- 2nd step: 16 ft. lbs. (22 Nm)

18. Connect the fuel injector harness.

19. Install the intake manifold collector.

20. Connect the quick connector between the fuel feed hose and the fuel tube connection with the following procedure:

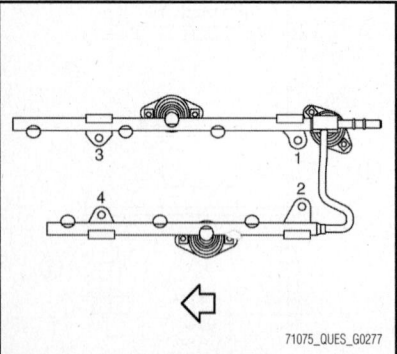

Fig. 83 Fuel tube and fuel injector assembly bolt tightening sequence

71075_QUES_G0277

a. Make sure there is no damage, and no dirt or foreign objects are around the tube and quick connector.

b. Thinly apply new engine oil around fuel tube from tip end to spool end.

c. Align the center to insert the quick connector straight onto the fuel tub until top spool is completely inside quick connector, and 2nd level spool exposes right below the quick connector. Hold the position shown while inserting the fuel tube into the quick connector.

d. Insert until you hear a "click" sound and actually feel the engagement.

e. To ensure engagement, pull quick connector by hand, holding the position.

f. Install the quick connector cap with the arrow on the surface facing in the direction of the quick connector (fuel feed hose side).

✳✳ WARNING

If quick connector cap cannot be installed smoothly, quick connector may have not been installed correctly. Check connection again.

g. Secure the fuel feed hose to the clamp of the quick connector cap.

21. Install the engine cover.

22. Install the air duct (inlet), air cleaner cases (upper and lower) with mass air flow sensor and air duct assembly.

23. Connect the negative battery cable.

24. Refill engine coolant.

25. Make sure there is no fuel leakage at connections as follows:

a. Apply fuel pressure to fuel lines by turning ignition switch **ON** with the engine **OFF** and check for fuel leaks at connections. Use mirrors for checking on connections out of the direct line of sight.

b. Start the engine and rev it up and check for fuel leaks at connections.

FUEL TANK

REMOVAL & INSTALLATION

See Figures 84 through 86.

1. Before servicing the vehicle, refer to the Precautions.

2. Release the fuel pressure from the fuel lines.

3. Check fuel level on a level ground. If the fuel level is 7/8 of the fuel tank (full or nearly full), draw appropriate amount of fuel from the fuel tank. Guideline: Draw approximately 20 liters from a full-tank condition.

a. In the event of malfunction in fuel pump, insert a hose measuring 0.79 in. (20 mm) in diameter into the filler opening to draw approximately 20 liters fuel.

4. Open fuel filler lid.

5. Open filler cap and release the pressure inside fuel tank.

6. Slide the left second seat in the forward direction.

7. Peel off floor carpet, then remove inspection hole cover units by turning clips clockwise by 90 degrees.

8. Disconnect harness connector, fuel feed tube and EVAP hose.

9. Remove fuel tank protector 1.

10. Disconnect fuel filler hose, EVAP/Vent line hose connector, and EVAP (Recirculation) hose connector from the side other than the fuel tank side.

11. Support the lower part of fuel tank with transmission jacks.

✳✳ WARNING

Securely support the flat bottom face. Horizontally support the fuel tank by using two mission jacks. Support the positions that fuel tank mounting bands do not support.

12. Remove the fuel tank mounting bands.

13. Supporting with hands, lower the transmission jack carefully, and remove the fuel tank. Check that all connection points have been disconnected. Confirm there is no interference with vehicle.

14. Remove the fuel tank protectors 2, 3 and 4. Fuel tank protector 2 and 3 are not reusable. Never remove them unless it is necessary.

15. When removing fuel filler tube, slide rear suspension member assembly downward, as follows:

a. Secure the bottom face of the rear final drive assembly with a mission jack.

b. Remove rear suspension member mounting nut, and carefully lower the jack to the position close to the height that rear suspension member becomes almost detached from the stud.

To install:

Installation is the reverse of removal, noting the following:

16. Clamp the fuel hoses and insert the hose to 1.38 in. (35 mm). Be sure hose clamp is not placed on swelled area of fuel tube.

17. Tighten the clamp hand with the top mark until the mark is on the bolt head flange.

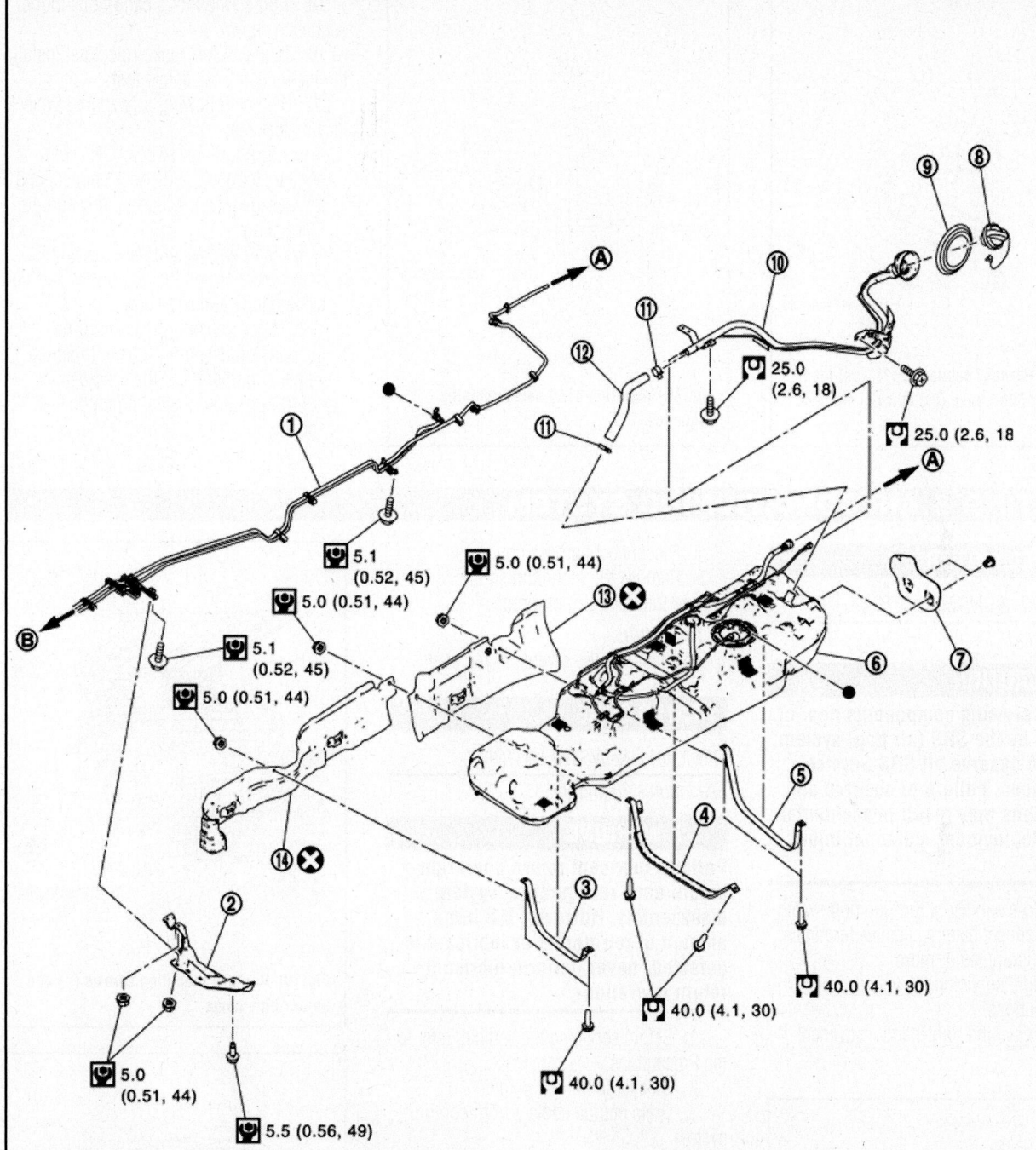

5.1 (0.52, 45)

5.0 (0.51, 44)

5.0 (0.51, 44)

5.1 (0.52, 45)

5.0 (0.51, 44)

25.0 (2.6, 18)

25.0 (2.6, 18)

40.0 (4.1, 30)

40.0 (4.1, 30)

40.0 (4.1, 30)

5.0 (0.51, 44)

5.5 (0.56, 49)

1. EVAP/Vent line tube
2. Fuel tank protector 1
3. Fuel tank mounting band 1
4. Fuel tank mounting band 2
5. Fuel tank mounting band 3
6. Fuel tank

7. Fuel tank protector 4
8. Fuel filler cap
9. Grommet
10. Fuel filler tube
11. Clamp
12. Fuel filler hose
13. Fuel tank protector 3

A. To EVAP canister
B. To engine
Tightening specifications:
White wrench on black background: Nm (kg m, ft. lbs.)
Black wrench on white background: Nm (kg m, inch lbs.)

71075_QUES_G0265

Fig. 84 Fuel tank components

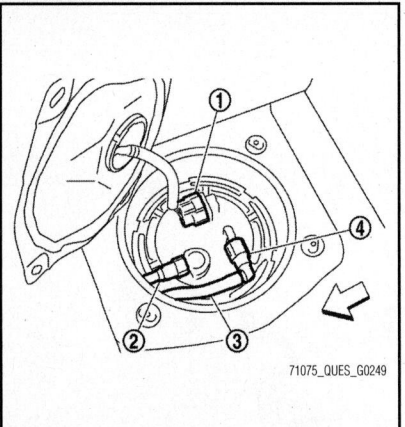

Fig. 85 Harness connector (1), fuel feed tube (3), EVAP hose (2), quick connector (4)

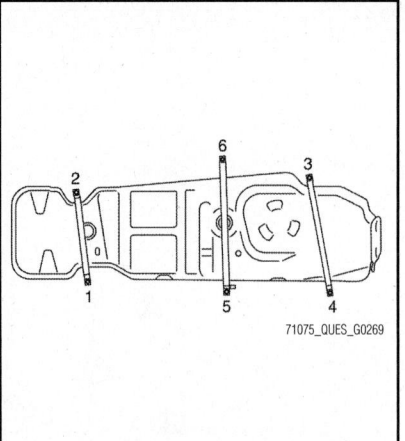

Fig. 86 Mounting band bolt tightening sequence

18. Tighten mounting band bolts in the sequence shown.

19. To install fuel tank protectors, install the fuel tank to the vehicle first.

20. Use the following procedure to check for fuel leakage.

a. Turn ignition switch "ON" (with engine stopped), and check connections for leakage by applying fuel pressure to fuel piping.

b. Start engine and rev it up and check there are no fuel leakage at the fuel system tube and hose connections.

c. After removing/installing rear suspension assembly, check to adjust wheel alignment and then, adjust neutral position of steering angle sensor.

HEATING & AIR CONDITIONING SYSTEM

BLOWER MOTOR

REMOVAL & INSTALLATION
See Figure 87.

❋❋ CAUTION

Before servicing components near or affected by the SRS (air bag) system, read and observe all SRS Service Precautions. Failure to observe all precautions may result in accidental airbag deployment, personal injury, or death.

1. Before servicing, turn ignition switch OFF, disconnect battery negative terminal and wait 3 minutes or more.

2. Before servicing the vehicle, refer to the Precautions.

3. Remove the right-hand instrument panel.

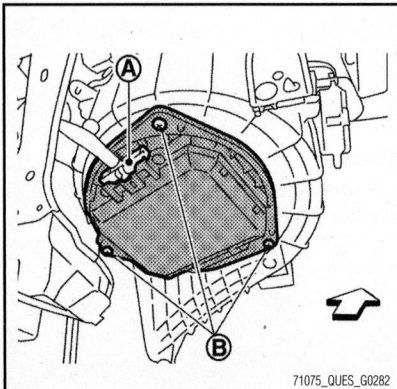

Fig. 87 Disconnect harness connector (A), remove the mounting screws (B), and remove front blower motor

4. Disconnect harness connector.

5. Remove the mounting screws, and remove the front blower motor.

To install:
Installation is the reverse of removal.

HEATER CORE

REMOVAL & INSTALLATION
See Figures 88 and 89.

❋❋ WARNING

Perform lubricant return operation before each refrigeration system disassembly. However, if a large amount of refrigerant or lubricant is detected, never perform lubricant return operation.

1. Before servicing the vehicle, refer to the Precautions.

2. Discharge the refrigerant.

3. Drain engine coolant from cooling system.

4. Remove front A/C unit assembly.

5. Separate blower unit assembly and heater and cooling unit assembly.

6. Remove heater pipe grommet.

7. Remove the mounting screw and mounting pawls.

8. Remove heater pipe support from heater and cooling unit assembly.

9. Remove mounting screws, and remove heater pipe cover.

10. Pull out heater core toward heater and cooling unit assembly to remove.

To install:
Installation is the reverse of removal.

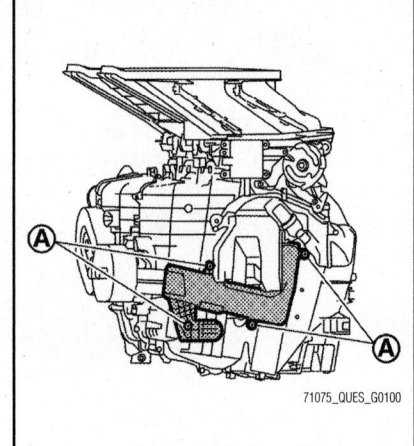

Fig. 88 Remove mounting screws (A) and heater pipe cover

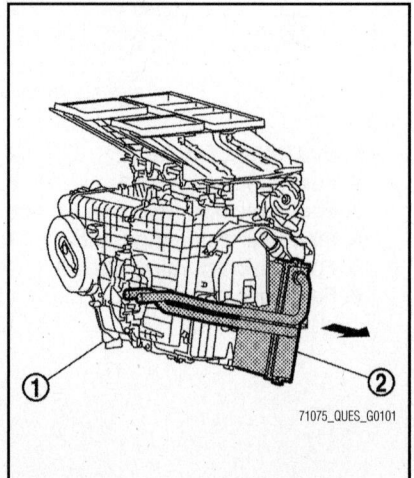

Fig. 89 Pull out heater core (2) toward heater and cooling unit assembly (1)

STEERING

POWER RACK & PINION STEERING GEAR

REMOVAL & INSTALLATION

See Figure 90.

1. Position the vehicle to the straight-ahead position.
2. Remove tires with power tool.
3. Remove engine under cover.
4. Remove exhaust front tube.
5. Remove heat insulator from front suspension member.
6. Remove cotter pin, and then loosen the nut.
7. Remove steering outer socket from steering knuckle so as not to damage ball joint boot using a ball joint remover. Temporarily tighten the nut to prevent damage to threads and to prevent the ball joint remover from suddenly coming off.
8. Remove high pressure piping and return hose of hydraulic piping, and then drain power steering fluid.
9. Remove intermediate shaft mounting bolt (steering gear side), and separate intermediate shaft from steering gear assembly.

   a. Place a matchmark on both intermediate shaft and steering gear assembly before removing intermediate shaft.

   b. When removing intermediate shaft, never insert a tool, such as a screwdriver, into the yoke groove to pull out the intermediate shaft. In case of the violation of the above, replace intermediate shaft with a new one.

   c. Spiral cable may be cut if steering wheel turns while separating the intermediate shaft and steering gear assembly. Be sure to secure steering wheel to avoid turning.

10. Remove the stabilizer connecting rod.
11. Position a jack under front suspension member. At this step, the jack must be set only for supporting the removal procedure. Do not damage the front suspension member with a jack. Check the stable condition when using a jack.
12. Remove the engine mounting insulator (rear) mounting bolts.
13. Remove the right-hand engine mounting insulator.
14. Remove steering gear assembly mounting bolts, and nuts.
15. Remove suspension member stay, and remove suspension member mounting bolts, and nuts.
16. Lower the jack.
17. Remove steering gear assembly.

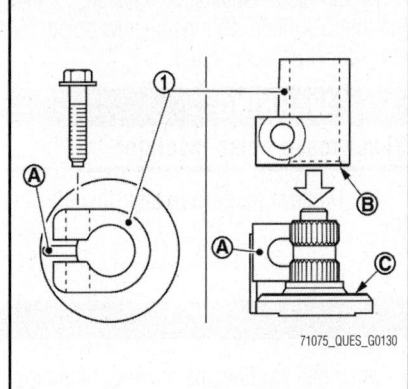

Fig. 90 Intermediate shaft positioning

To install:

18. Installation is the reverse of removal, noting the following:

> ※※ **WARNING**
>
> **Spiral cable may be cut if steering wheel turns while separating intermediate shaft and steering gear assembly. Be sure to secure the steering wheel to avoid turning.**

19. Never apply fluid to the hose and tube. Insert hose securely until it contacts spool of tube. Leave clearance "L" when installing clamp. L: 0.12–0.31 in. (3–8 mm)
20. When installing intermediate shaft to steering gear assembly, follow the procedure listed below.

   a. Set rack of steering gear in the neutral position. To get the neutral position of rack, turn gear-sub assembly and measure the distance of inner socket, and then measure the intermediate position of the distance.

   b. Align slit of intermediate shaft with rear cover cap projection, insert intermediate shaft end face until contacts steering gear assembly end face.

   c. When tightening the mounting bolt of the intermediate shaft (steering gear assembly side), manually tighten the bolt and check that there is no hook and scratch. Check that the bolt is properly placed in the groove of the steering gear assembly before tightening the bolt to the specified torque.

21. Perform inspection after installation.

POWER STEERING PUMP

BLEEDING

➡ **Fluid noise may occur in the steering gear or oil pump. This does not affect performance or durability of the system.**

1. Turn steering wheel several times from full left stop to full right stop with engine stopped.

> ※※ **WARNING**
>
> **Fill reservoir tank with a sufficient amount of fluid so that fluid level is not below the MIN line while turning steering wheel.**

2. Start the engine and hold steering wheel at each lock position for 3 seconds at idle to check for fluid leakage.
3. Repeat step 2 above several times at approximately 3 seconds intervals.

> ※※ **WARNING**
>
> **Never hold the steering wheel in a locked position for more than 10 seconds. (There is the possibility that oil pump may be damaged.)**

4. Check fluid for bubbles and white contamination. Stop the engine if bubbles and white contamination do not drain out. Perform step 2 and 3 above after waiting until bubbles and white contamination drain out.
5. Stop the engine, and then check fluid level.

REMOVAL & INSTALLATION

1. Disconnect the battery negative terminal.

> ※※ **WARNING**
>
> **Noise insulator is a functional part, never remove (Iron cover type).**

2. Remove reservoir tank of radiator.
3. Remove washer tank inlet.
4. Disconnect each connector and grand cable from power steering control module.
5. Remove high-pressure flexible hose.
6. Remove return hose, and drain power steering fluid.

> ※※ **WARNING**
>
> **Never reuse drained power steering fluid. Always use the specified fluid.**

7. Remove high-pressure piping and O-ring.
8. Remove hood side cover.
9. Remove oil pump assembly mounting bolts and nuts, and then remove oil pump assembly.

10. Remove brackets.

11. Perform inspection after removal.

To install:

12. Installation is the reverse of removal, noting the following:

13. When installing high-pressure piping, securely install a new O-ring to high-pressure piping.

14. Never apply fluid to the hose and tube. Insert hose securely until it contacts spool of tube. Leave clearance "L" when installing clamp. L: 0.12–0.31 in. (3–8 mm)

15. If noise insulator is removed, clean mounting surface, and install new noise insulator (Iron cover type).

✸✸ WARNING

Never reuse noise insulator.

16. Perform inspection after installation.

FLUID FILL PROCEDURE

If the fluid is below the MIN line, remove the cap and fill through the opening.

✸✸ WARNING

Use of a power steering fluid other than Genuine NISSAN E-PSF will prevent the power steering system from operating properly.

SUSPENSION

CROSSMEMBER

REMOVAL & INSTALLATION

See Figure 91.

1. Remove wheels and tires.

2. Remove the engine and the transaxle assembly from the vehicle together with the front suspension member.

3. Remove the following parts from front suspension member:
- Engine assembly
- Transaxle assembly
- Steering gear assembly
- Steering hydraulic line
- Stabilizer bar
- Transverse link

4. Check the transverse link and bushing for deformation, cracks or damage.

5. Check the ball joint boot for cracks or other damage, and also for grease leakage.

6. Inspect the swing torque:

a. Manually move ball stud to confirm it moves smoothly with no binding.

b. Move ball stud at least ten times by hand to check for smooth movement.

c. Hook a spring balance at cutout on ball stud. Confirm spring balance measurement value is within specifications when ball stud begins moving. Standard range:
- Swing torque: 5–43 inch lbs. (0.5–4.9 Nm)
- Measurement on spring balance: 3–24 lbs. (11.1–108.9 N)

d. If swing torque exceeds standard range, replace transverse link assembly.

7. Inspect the Axial End Play:

a. Move ball stud at least ten times by hand to check for smooth movement.

b. Move tip of ball stud in axial direction to check for looseness. Standard value:
- Axial end play: 0 in. (0 mm)

c. If axial end play exceeds the standard value, replace transverse link assembly.

To install:

8. Installation is the reverse of removal, noting the following:

a. Perform final tightening of bolts and nuts at the vehicle installation position (rubber bushing), under unladen conditions with tires on level ground.

b. Check wheel sensor harness for proper connector.

9. Check wheel alignment.

LOWER TRANSVERSE LINK

REMOVAL & INSTALLATION

1. Remove the wheels and tires.

2. Remove the drive shaft from the wheel hub assembly.

3. Separate the transverse link from the steering knuckle.

4. Remove the transverse link from the suspension member.

5. Check the transverse link and bushing for deformation, cracks or damage.

6. Check the ball joint boot for cracks or other damage, and also for grease leakage.

7. Inspect the swing torque:

a. Manually move ball stud to confirm it moves smoothly with no binding.

FRONT SUSPENSION

b. Move ball stud at least ten times by hand to check for smooth movement.

c. Hook a spring balance at cutout on ball stud. Confirm spring balance measurement value is within specifications when ball stud begins moving. Standard range:
- Swing torque: 5–43 inch lbs. (0.5–4.9 Nm)
- Measurement on spring balance: 3–24 lbs. (11.1–108.9 N)

d. If swing torque exceeds standard range, replace transverse link assembly.

8. Inspect the Axial End Play:

a. Move ball stud at least ten times by hand to check for smooth movement.

b. Move tip of ball stud in axial direction to check for looseness. Standard value:
- Axial end play: 0 in. (0 mm)

c. If axial end play exceeds the standard value, replace transverse link assembly.

To install:

9. Installation is the reverse of removal, noting the following:

a. Never reuse transverse link mounting nut.

b. Perform final tightening of bolts and nuts at the vehicle installation position (rubber bushing), under unladen conditions with tires on level ground.

c. Check wheel sensor harness for proper connector.

10. Check wheel alignment.

STABILIZER BAR

REMOVAL & INSTALLATION

See Figure 92.

1. Remove wheels and tires.

2. Remove exhaust front tube.

3. Remove wheel sensor harness from strut assembly.

4. Separate steering outer socket from steering knuckle.

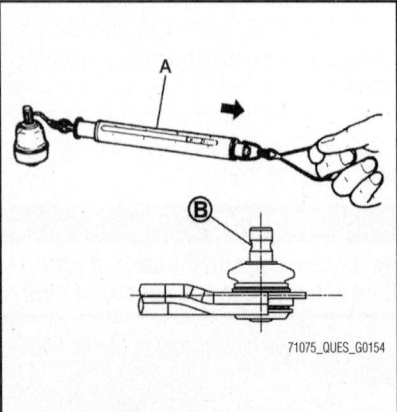

71075_QUES_G0154

Fig. 91 Hook a spring balance (A) at cutout on ball stud (B)

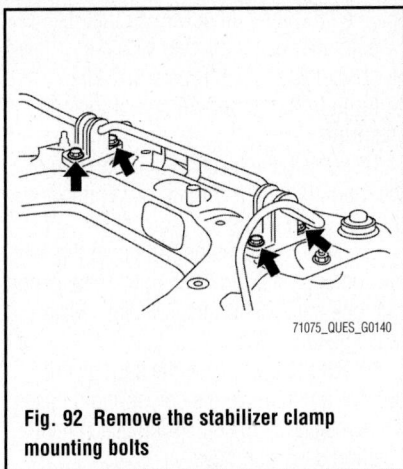

Fig. 92 Remove the stabilizer clamp mounting bolts

5. Remove stabilizer connecting rod.
6. Remove the halfshaft (right side).
7. Remove the stabilizer clamp mounting bolts, and remove stabilizer clamp and stabilizer bushing from the front suspension member.
8. Remove stabilizer bar.
9. Perform inspection after removal:
   a. Check stabilizer bar, stabilizer connecting rod, stabilizer bushing and stabilizer clamp for deformation, cracks or damage. Replace it if necessary.

To install:
10. Install the stabilizer bar.
11. Install stabilizer clamp and stabilizer bush with notch and slit faced forward of the vehicle.
12. Install the halfshaft.
13. To install stabilizer connecting rod, tighten the mounting nut with the hexagonal part on the stabilizer connecting rod side fixed.
14. Install the steering outer socket to steering knuckle.
15. Install the wheel sensor harness to strut assembly.
16. Install the exhaust front tube.
17. Install the wheels and tires.
18. Perform final tightening of bolts and nuts at the vehicle installation position (rubber bushing), under unladen conditions with tires on level ground.
19. Check wheel sensor harness for proper connector.
20. Check wheel alignment.

STRUTS

REMOVAL & INSTALLATION
See Figure 93.

1. Remove wheels and tires.
2. Remove lock plate from strut assembly.
3. Remove wheel sensor.

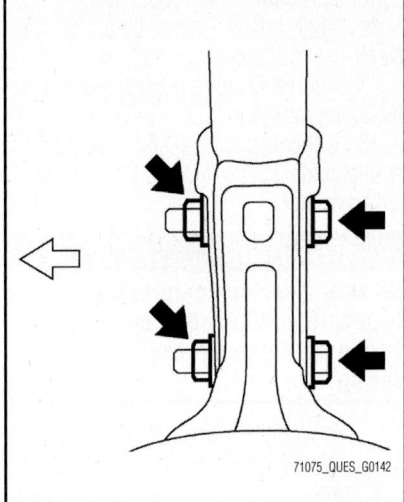

Fig. 93 Remove strut mounting bolts and nuts

4. Separate stabilizer connecting rod from strut assembly.
5. Remove strut mounting bolts and nuts from steering knuckle.
6. Remove cowl top extension.
7. Remove the mounting insulator mounting bolt, and then remove strut assembly.

To install:
8. Install strut assembly with the protrusion of mounting insulator facing the outside of the vehicle. Do not reuse strut mounting nut.
9. Install the cowl top extension.
10. Install the strut mounting bolts and nuts to steering knuckle.
11. Install the stabilizer connecting rod to strut assembly.
12. Install the wheel sensor.
13. Install the lock plate to strut assembly.
14. Install the wheels and tires.
15. Perform final tightening of fixing parts at the vehicle installation position (rubber bushing), under unladen conditions with tires on level ground.
   a. Check wheel sensor harness for proper connector.
16. Check wheel alignment.
17. After replacing the strut, always follow the disposal procedure to discard the strut.

OVERHAUL
See Figure 94.

1. Install strut attachment (SST: ST35652000) to the strut assembly and secure it in a vise. When installing the strut attachment to strut assembly, wrap a shop cloth around strut to protect from damage.

2. Using a spring compressor, compress coil spring between spring upper seat and lower seat (strut assembly) until coil spring with a spring compressor is free.

✳✳ **CAUTION**

Be sure a spring compressor is securely attached to coil spring. Compress coil spring.

3. Check that the coil spring with a spring compressor between spring upper seat and lower seat (strut assembly) is free. Remove piston rod lock nut while securing the piston rod tip so that piston rod does not turn.
4. Remove the mounting insulator, mounting bearing, and bound bumper from the strut.
5. After removing the coil spring with a spring compressor, gradually release the spring compressor.

✳✳ **CAUTION**

Loosen while making sure coil spring attachment position does not move.

6. Remove the lower rubber seat.
7. Remove the strut attachment from the strut.
8. Perform inspection after disassembly. Check the following items, and replace the parts if necessary:
   a. Strut:
   • Strut for deformation, cracks or damage

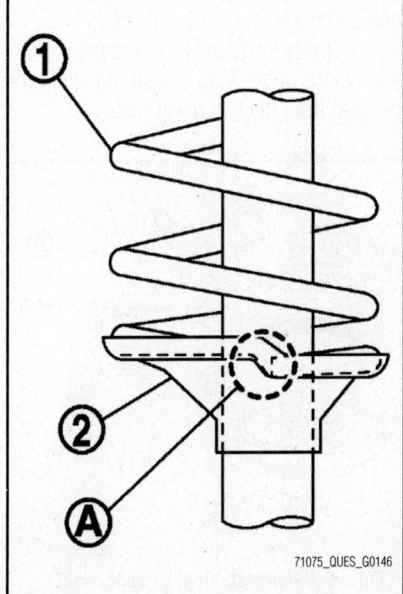

Fig. 94 Align the lower end of coil spring (1) with "A" of lower rubber seat (2)

- Piston rod for damage, uneven wear or distortion
- Oil leakage

b. Check strut mounting insulator and bound bumper for cracks, wear or damage.

c. Check coil spring for cracks, wear or damage.

To assemble:

※ WARNING

Never damage strut assembly piston rod when installing components from strut assembly.

9. Install the strut attachment to strut and secure it in a vise. Wrap a shop cloth around strut to protect from damage.

10. Install the lower rubber seat.
11. Apply soapy water to bound bumper. Never use machine oil.
12. Insert the bound bumper into the mounting insulator.
13. Compress the coil spring using a spring compressor and install it onto the strut assembly.

※ CAUTION

Be sure a spring compressor is securely attached to coil spring before compressing the coil spring.

a. Align the lower end of the coil spring with the lower rubber seat as shown.

b. Set the coil spring so that its paint marks are aligned with the positions of 1.25 turns and 2.25 turns from the bottom end of the coil spring.

14. Check the location of protrusion of the mounting insulator and install it facing outside of the vehicle.

15. Secure the piston rod tip so that piston rod does not turn, and tighten the piston rod lock nut to 54 ft. lbs. (73 Nm). Never reuse piston rod lock nut.

16. Gradually release the spring compressor and remove the coil spring. Loosen while making sure coil spring attachment position does not move.

17. Remove the strut attachment from the strut assembly.

SUSPENSION

COIL SPRINGS

REMOVAL & INSTALLATION

See Figures 95 and 96.

1. Remove wheels and tires.
2. Position a suitable jack under rear lower link.

➥At this step, the jack must be set only for supporting the removal procedure. Do not damage the rear lower link with a jack. Check the stable condition when using a jack.

3. Loosen the rear lower link adjusting bolt and nut from the rear suspension member, and remove rear lower link mounting bolt and nut from the axle housing.

4. Slowly lower the jack, and then remove the upper seat, coil spring, and rubber seat from the rear lower link.

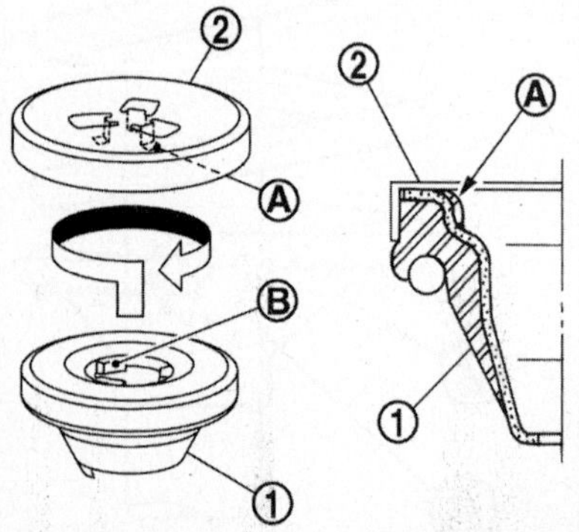

5. Remove the rear lower link.
6. Perform inspection after removal:

a. Check rear lower link, rubber seat, upper seat, and coil spring for deformation, crack, and damage. Replace it if necessary.

To install:

7. Installation is the reverse of removal, noting the following:

a. Install adjusting bolt so that the graduation marks on adjusting bolt are positioned downward.

b. When installing the upper seat, align the protrusion on the upper seat inside to the vehicle side bracket tabs.

c. Fit the rubber seat to the step of the rear lower link.

d. Fit the bottom end of coil spring to the step of rubber seat. Be careful with the vertical direction of the coil spring.

e. Perform final tightening of mounting parts at the vehicle installation position (rubber bushing), under

REAR SUSPENSION

unladen conditions with tires on level ground.

8. Check wheel alignment.

CROSSMEMBER

REMOVAL & INSTALLATION

1. Remove wheels and tires.
2. Remove the brake caliper assembly. Hang caliper assembly aside so as not to interfere with work. Never depress brake pedal while brake caliper is removed.

3. Remove the wheel sensor and sensor harness.

4. Remove the center muffler.
5. Separate the parking brake cable from the rear suspension member.

6. Remove the rear lower link and coil spring.

7. Separate the shock absorber from the front lower link.

8. Position a jack under the rear suspension member.

➥At this step, the jack must be set only for supporting the removal procedure. Do not damage the rear suspension member with a jack. Check the stable condition when using a jack.

9. Remove the rear suspension member stay.

10. Remove the rear suspension member mounting nuts and rebound stopper.

11. Slowly lower the jack, and then remove the rear suspension member, front lower link, radius rod, suspension arm, and axle housing from vehicle as a unit.

12. Remove the following parts from rear suspension member:

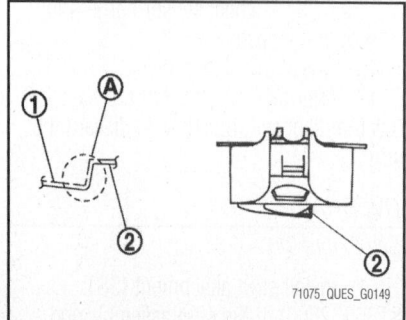

71075_QUES_G0148

Fig. 95 When installing the upper seat (1), align the protrusion (B) on the upper seat inside to the vehicle side bracket (2) tabs (A)

71075_QUES_G0149

Fig. 96 Fit the rubber seat (1) to the step (A) of the rear lower link (2)

- Axle housing
- Suspension arm
- Radius rod
- Front lower link

13. Perform inspection after removal:

a. Check rear suspension member for deformation, cracks, or any other damage. Replace it if necessary.

To install:

14. Installation is the reverse of removal, noting the following:

a. Perform final tightening of the mounting parts at the vehicle installation position (rubber bushing), under unladen conditions with tires on level ground.

b. Check the wheel sensor harness for proper connection.

c. Adjust the parking brake operation (stroke).

15. Check the wheel alignment.

FRONT LOWER LINK

REMOVAL & INSTALLATION

1. Remove wheels and tires.
2. Remove wheel sensor and sensor harness.
3. Remove the coil spring and rear lower link.
4. Separate shock absorber from front lower link.
5. Remove eccentric disc, adjusting bolt, mounting bolt, and nut, then remove front lower link.
6. Perform inspection after removal:

a. Check front lower link and bushing for any deformation, cracks, or damage. Replace it if necessary.

To install:

7. Installation is the reverse of removal, noting the following:

a. Install the adjusting bolt so that the graduation marks on adjusting bolt are positioned downward.

b. Perform final tightening of fixing parts at the vehicle installation position (rubber bushing), under unladen conditions with tires on level ground.

c. Never reuse front lower link mounting nut.

d. Check wheel sensor harness for proper connection.

8. Check wheel alignment.

RADIUS ROD

REMOVAL & INSTALLATION

1. Remove the wheels and tires.
2. Remove the wheel sensor and sensor harness.

3. Remove the rear lower link and coil spring.
4. Separate the shock absorber from the front lower link.
5. Separate the front lower link from the axle housing.
6. Loosen the front lower link mounting bolt and nut from the suspension member.
7. Remove the radius rod mounting bolts and nuts from the axle housing.
8. Remove the radius rod mounting bolt from the rear suspension member, and remove the radius rod.
9. Perform inspection after removal:

a. Check radius rod and bushing for any deformation, cracks, or damage. Replace if necessary.

To install:

10. Installation is the reverse of removal, noting the following:

a. Perform final tightening of fixing parts at the vehicle installation position (rubber bushing), under unladen conditions with tires on level ground.

b. Never reuse radius rod mounting nut.

c. Check wheel sensor harness for proper connection.

11. Check wheel alignment.

SHOCK ABSORBERS

REMOVAL & INSTALLATION

See Figure 97.

1. Remove wheels and tires.
2. Position a jack under axle housing.
3. Separate the shock absorber mounting bolt from the front lower link.
4. Slowly lower the jack, and remove the shock absorber from the front lower link.
5. Remove the mounting bracket mount-

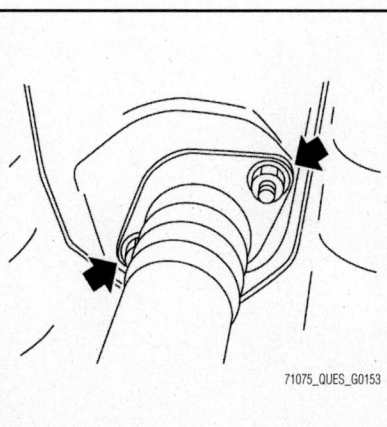

71075_QUES_G0153

Fig. 97 Remove the mounting bracket mounting nuts

ing nuts, and remove the shock absorber assembly.

To install:

6. Installation is the reverse of removal, noting the following:

a. Perform final tightening of fixing parts at the vehicle installation position (rubber bushing), under unladen conditions with tires on level ground.

7. Check wheel alignment.

8. After replacing the shock absorber, always follow the disposal procedure to discard the shock absorber.

SUSPENSION ARM

REMOVAL & INSTALLATION

1. Remove tires with power tool.
2. Remove wheel sensor and sensor harness.
3. Separate suspension arm from front lower link.
4. Remove mounting bolt and nut, and then remove stopper rubber and suspension arm.
5. Perform inspection after removal. Check the following items, and replace the parts if necessary:

a. Suspension arm:
- Suspension arm and bushing for deformation, cracks or damage
- Ball joint boot for cracks or other damage, and also for grease leakage

b. Swing torque:
- Manually move ball stud to confirm it moves smoothly with no binding
- Move ball stud at least ten times by hand to check for smooth movement
- Hook spring balance at cotter pin mounting hole; confirm spring balance measurement value is within specifications when ball stud begins moving
- Swing torque standard range: 5–30 inch lbs. (0.5–3.4 Nm)

c. If swing torque exceeds standard range, replace suspension arm assembly.

d. Axial end play:
- Manually move ball stud to confirm it moves smoothly with no binding
- Move ball stud at least ten times by hand to check for smooth movement
- Move tip of ball stud in axial direction to check for looseness
- Axial end play standard value: 0 in. (0 mm)

e. If axial end play exceeds the standard value, replace suspension arm assembly.

To install:

6. Note the following, and install in the reverse order of removal.

a. Install stopper rubber to rear suspension member together with suspension arm.

b. Perform final tightening of fixing parts at the vehicle installation position (rubber bushing), under unladen conditions with tires on level ground.

c. Never reuse suspension arm mounting nut.

7. Perform inspection after installation:

a. Check wheel sensor harness for proper connection.

b. Check wheel alignment.

SPECIFICATIONS AND MAINTENANCE CHARTS

ENGINE AND VEHICLE IDENTIFICATION

Engine							Model Year	
Code ①	Liters (cc)	Cu. In.	Cyl.	Fuel Sys.	Engine Type	Eng. Mfg.	Code ②	Year
A	2.5 (2488)	151.82	4	MPI	DOHC	Nissan	B	2011
							C	2012

① 4th position of VIN

② 10th position of VIN

71075_ROGU_C0001

GENERAL ENGINE SPECIFICATIONS

All measurements are given in inches.

Year	Model	Engine Displacement Liters (cc)	Engine ID	Fuel System Type	Net Horsepower @ rpm	Net Torque @ rpm (ft. lbs.)	Bore x Stroke (in.)	Com- pression Ratio	Oil Pressure @ rpm
2011	Rogue	2.5	A	MPI	170@6000	175@4400	3.50x3.94	9.6:1	43@2000
2012	Rogue	2.5	A	MPI	170@6000	175@4400	3.50x3.94	9.6:1	43@2000

71075_ROGU_C0002

ENGINE TUNE-UP SPECIFICATIONS

Year	Engine Displacement Liters	Engine ID	Spark Plug Gap (in.)	Ignition Timing (deg.) MT	Ignition Timing (deg.) AT	Fuel Pump (psi)	Idle Speed (rpm) MT	Idle Speed (rpm) AT	Valve Clearance Intake	Valve Clearance Exhaust
2011	2.5	A	0.043	N/A	10 BTDC ①	51	N/A	700+/-50 ①	0.009-0.013	0.010-0.013
2012	2.5	A	0.043	N/A	10 BTDC ①	51	N/A	700+/-50 ①	0.009-0.013	0.010-0.013

NOTE: The Vehicle Emission Control Information label often reflects specification changes made during production.

The label figures must be used if they differ from those in this chart.

① Component is computer controlled and is not adjustible.

N/A Not applicable

71075_ROGU_C0003

CAPACITIES

Year	Model	Engine Displacement Liters	Engine ID/VIN	Engine Oil with Filter	Transmission/axle (pts.) Auto.	Manual	Drive Axle (pts.) Front	Rear ①	Transfer Case (pts.)	Fuel Tank (gal.)	Cooling System (qts.)
2011	Rogue	2.5	A	4.9	②	N/A	N/A	0.6	1.2	15.8	7.9 ③
2012	Rogue	2.5	A	4.9	②	N/A	N/A	0.6	1.2	15.8	7.9 ③

NOTE: All capacities are approximate. Add fluid gradually and ensure a proper fluid level is obtained.

N/A Not applicable

① Synthetic GL-5 (75W-90) or equivalent

② 2WD (8qts) and AWD (9qts)

③ The use of genuine Nissan engine coolant is recommended or similar ethylene glycol based non-silicate, non-amine, non- nitrite, and non- borat coolant

71075_ROGU_C0004

FLUID SPECIFICATIONS

Year	Model	Engine Disp. Liters	Engine Oil	Manual Trans.	Auto. Trans. ①	Drive Axle Front	Rear ②	Transfer Case	Power Steering Fluid	Brake Master Cylinder	Cooling System ③
2011	Rogue	2.5	5W-30	N/A	Nissan	N/A	80W-90	N/A	NA	DOT 3	Nissan
2012	Rogue	2.5	5W-30	N/A	Nissan	N/A	80W-90	N/A	NA	DOT 3	Nissan

NA Not available

DOT: Department Of Transpotation

N/A Not applicable

① Using transmission fluid other than genuine Nissan CVT fluid NS-2 will damage the CVT

② GL-5 (80W-90) or equivalent

③ The use of ggenuine Nissan engine coolant is recommended

71075_ROGU_C0005

VALVE SPECIFICATIONS

Year	Engine Displacement Liters	Engine ID/VIN	Seat Angle (deg.)	Face Angle (deg.)	Spring Test Pressure (lbs. @ in.)	Spring Free- Length (in.)	Spring Installed Height (in.)	Stem-to-Guide Clearance (in.) Intake	Exhaust	Stem Diameter (in.) Intake	Exhaust
2011	2.5	A	45	NS	NA	①	1.390	0.0008- 0.0021	0.0012- 0.0025	0.234- 0.235	0.234- 0.235
2012	2.5	A	45	NS	NA	①	1.390	0.0008- 0.0021	0.0012- 0.0025	0.234- 0.235	0.234- 0.235

NA: Not Available

① Intake (1.7213-1.7291) and Exhuast (1.7831-1.7909)

71075_ROGU_C0006

CAMSHAFT SPECIFICATIONS

All measurements in inches unless noted

Year	Engine Displacement Liters	Engine Code	Journal Diameter	Brg. Oil Clearance	Shaft End-play	Runout	Journal Bore	Lobe Height Intake	Lobe Height Exhaust
2011	2.5	A	①	0.0018-0.0034	0.0045-0.0074	0.0008	NA	②	1.7313-1.7388
2012	2.5	A	①	0.0018-0.0034	0.0045-0.0074	0.0008	NA	②	1.7313-1.7388

NA: Not Available

① Journal 1: 1.0998-1.1006
All Others: 0.9226-0.9234

② Non-California: 1.7644-1.7718
California: 1.7722-1.7797

71075_ROGU_C0007

CRANKSHAFT AND CONNECTING ROD SPECIFICATIONS

All measurements are given in inches.

Year	Engine Displacement Liters	Engine ID/VIN	Crankshaft Main Brg. Journal Dia.	Crankshaft Main Brg. Oil Clearance	Crankshaft Shaft End-play	Crankshaft Thrust on No.	Connecting Rod Journal Diameter	Connecting Rod Oil Clearance	Connecting Rod Side Clearance
2011	2.5	A	2.1636-2.1645	①	0.0039-0.0102	3	1.7699-1.7706	0.0014-0.0018	0.0079-0.0138
2012	2.5	A	2.1636-2.1645	①	0.0039-0.0102	3	1.7699-1.7706	0.0014-0.0018	0.0079-0.0138

① Nos. 1, 3 and 5: 0.0005-0.0009 in.
Nos. 2 and 4: 0.0007-0.0011

71075_ROGU_C0008

PISTON AND RING SPECIFICATIONS

All measurements are given in inches.

Year	Engine Displacement Liters	Engine ID/VIN	Piston Clearance	Ring Gap Top Compression	Ring Gap Bottom Compression	Ring Gap Oil Control	Ring Side Clearance Top Compression	Ring Side Clearance Bottom Compression	Ring Side Clearance Oil Control
2011	2.5	A	0.0004-0.0012	0.0091-0.0130	0.0130-0.0189	0.0079-0.0177	0.0016-0.0031	0.0012-0.0028	0.0018-0.0049
2012	2.5	A	0.0004-0.0012	0.0091-0.0130	0.0130-0.0189	0.0079-0.0177	0.0016-0.0031	0.0012-0.0028	0.0018-0.0049

71075_ROGU_C0009

TORQUE SPECIFICATIONS
All readings in ft. lbs.

Year	Engine Disp. Liters	Engine ID	Cylinder Head Bolts	Main Bearing Bolts	Rod Bearing Bolts	Crankshaft Damper Bolts	Flywheel Bolts	Manifold Intake	Exhaust	Spark Plugs	Oil Pan Drain Plug
2011	2.5	A	①	②	③	④	76-83	13-15	29-32	14	25
2012	2.5	A	①	②	③	④	76-83	13-15	29-32	14	25

① Step 1: 37 ft. lbs.

Step 2: 60-66 degrees

Step 3: Loosen bolts completely, then retorque to 26-32 ft. lbs.

Step 4: Turn each bolt, in sequence, an additional 75-80 degrees

Step 5: Repeat, turn each bolt, in sequence, an additional 75-80 degrees

② Bolt Nos. 1-10:

Step 1: 27-31 ft. lbs.

Step 2: Torque an additional 60-65 degrees

Bolt Nos. 11-14: Torque last, to 17-20 ft. lbs.

③ Step 1: 20 ft. lbs.

Step 2: Loosen bolts completely

Step 3: 14 ft. lbs.

Step 4: 85-95 degrees

④ Step 1: 29-36 ft. lbs.

Step 2: 60-66 degrees

71075_ROGU_C0010

WHEEL ALIGNMENT

Year	Model		Caster Range (+/-Deg.)	Preferred Setting (Deg.)	Camber Range (+/-Deg.)	Preferred Setting (Deg.)	Toe-in (in.)
2011	Rogue	F	0.75	+4.75	0.75	①	0.08+/-0.04
		R	NA	NA	0.50	-0.92	0.08+/-0.08
2012	Rogue	F	0.75	+4.75	0.75	①	0.08+/-0.04
		R	NA	NA	0.50	-0.92	0.08+/-0.08

NA: Not Applicable

F: Front

R: Rear

① Left side front camber -0.25

Right side front front camber -0.50

71075_ROGU_C0011

TIRE, WHEEL AND BALL JOINT SPECIFICATIONS

Year	Model	OEM Tires		Tire Pressures (psi)		Wheel Size	Ball Joint Inspection	Lug Nut (ft. lbs.)
		Standard	Optional	Front	Rear			
2011	Rogue	P215/70R16	225/60R17	33	33	6.5-JJ and 7-JJ	①	80
2012	Rogue	P215/70R16	225/60R17	33	33	6.5-JJ and 7-JJ	①	80

OEM: Original Equipment Manufacturer

PSI: Pounds Per Square Inch

STD: Standard

OPT: Optional

① Replace if any measurable movement is found.

71075_ROGU_C0012

BRAKE SPECIFICATIONS

All measurements in inches unless noted

Year	Model		Brake Disc			Minimum Pad/Lining Thickness	Brake Caliper	
			Original Thickness	Minimum Thickness	Max. Runout		Bracket Bolts (ft. lbs.)	Mounting Bolts (ft. lbs.)
2011	Rogue	F	1.024	0.945	0.0014	0.079	122	25
		R	0.630	0.551	0.0028	0.059	62	32
2012	Rogue	F	1.024	0.945	0.0014	0.079	122	25
		R	0.630	0.551	0.0028	0.059	62	32

F: Front

R: Rear

71075_ROGU_C0013

SCHEDULED MAINTENANCE INTERVALS (1)
Nissan—Rogue

TO BE SERVICED	TYPE OF	7.5	15	22.5	30	37.5	45	52.5	60
Engine oil & filter	R	every 3,750 miles							
Brake lines & cables	S/I		✓		✓		✓		✓
Brake pads, discs	I		✓		✓		✓		✓
Brake fluid	R				✓				✓
CVT	I		✓		✓		✓		✓
Driveshaft boots	I		✓		✓		✓		✓
Automatic transmission (CVT)	I		✓		✓		✓		✓
Air cleaner filter ①	R				✓				✓
Drive belt (s) ②	S/I								✓
Engine coolant ③	R								
Spark plugs	R	Platinum plugs, every 105,000 miles							
Cabin air filter	R		✓		✓		✓		✓
Exhaust system	I				✓				✓
Evap vapor lines	I				✓				✓
Ful lines	I				✓				✓
Steering gear, linkage, axle & suspension parts	I				✓				✓
Tires (rotate)	S/I	every 5,000-6,000 miles							
Valve clearance ④	S/I				✓				✓

R: Replace   S/I: Service or Inspect   L: Lubricate   I: Inspect

① If operating mainly in dusty conditions, more frequent maintenance may be required.

② First at 60,000, inspect every 15,000 miles and replace as necessary

③ After 105,000, replace every 75,000 thereafter

④ Periodic maintenance not required, if valve noise increases, inspect valve clearance

Follow Periodic Maintenance Schedule 1 if the driving habits frequently include one or more of the following driving conditions:

Repeated short trips of less than 5 miles (8 km).

Repeated short trips of less than 10 miles (16 km) with outside temperatures remaining below freezing

Operating in hot weather in stop-and-go "rush hour" traffic.

Extensive idling and/or low speed driving for long distances, such as police, taxi or door-to-door delivery use

Driving in dusty conditions.

Driving on rough, muddy, or salt spread roads.

Towing a trailer, using a camper or a car-top carrier.

Follow Periodic Maintenance Schedule 2 if none of driving conditions shown in Schedule 1 apply to the driving habits.

71075_ROGU_C0014

SCHEDULED MAINTENANCE INTERVALS (2)
Nissan—Rogue

TO BE SERVICED	TYPE OF	7.5	15	22.5	30	37.5	45	52.5	60
Engine oil & filter	R	✓	✓	✓	✓	✓	✓	✓	✓
Brake lines & cables	S/I		✓		✓		✓		✓
Brake pads, discs	I		✓		✓		✓		✓
Brake fluid	R				✓				✓
Driveshaft boots	I		✓		✓		✓		✓
Automatic transmission (CVT)	I		✓		✓		✓		✓
Air cleaner filter ①	R				✓				✓
Drive belt (s) ②	S/I								✓
Engine coolant ③	R								
Spark plugs	R	Platinum plugs, every 105,000 miles							
Cabin air filter	R		✓		✓		✓		✓
Exhaust system	I				✓				✓
Evap vapor lines	I				✓				✓
Ful lines	I				✓				✓
Steering gear, linkage, axle & suspension parts	I				✓				✓
Tires (rotate)	S/I	every 5,000-6,000 miles							
Valve clearance ④	S/I				✓				✓

R: Replace   S/I: Service or Inspect   L: Lubricate   I: Inspect

① If operating mainly in dusty conditions, more frequent maintenance may be required.

② First at 60,000, inspect every 15,000 miles and replace as necessary

③ After 105,000, replace every 75,000 thereafter

④ Periodic maintenance not required, if valve noise increases, inspect valve clearance

Follow Periodic Maintenance Schedule 1 if the driving habits frequently include one or more of the following driving conditions:

Repeated short trips of less than 5 miles (8 km).

Repeated short trips of less than 10 miles (16 km) with outside temperatures remaining below freezing

Operating in hot weather in stop-and-go "rush hour" traffic.

Extensive idling and/or low speed driving for long distances, such as police, taxi or door-to-door delivery use

Driving in dusty conditions.

Driving on rough, muddy, or salt spread roads.

Towing a trailer, using a camper or a car-top carrier.

Follow Periodic Maintenance Schedule 2 if none of driving conditions shown in Schedule 1 apply to the driving habits.

71075_ROGU_C0015

PRECAUTIONS

Before servicing any vehicle, please be sure to read all of the following precautions, which deal with personal safety, prevention of component damage, and important points to take into consideration when servicing a motor vehicle:

• Never open, service or drain the radiator or cooling system when the engine is hot; serious burns can occur from the steam and hot coolant.

• Observe all applicable safety precautions when working around fuel. Whenever servicing the fuel system, always work in a well-ventilated area. Do not allow fuel spray or vapors to come in contact with a spark, open flame, or excessive heat (a hot drop light, for example). Keep a dry chemical fire extinguisher near the work area. Always keep fuel in a container specifically designed for fuel storage; also, always properly seal fuel containers to avoid the possibility of fire or explosion. Refer to the additional fuel system precautions later in this section.

• Fuel injection systems often remain pressurized, even after the engine has been turned **OFF**. The fuel system pressure must be relieved before disconnecting any fuel lines. Failure to do so may result in fire and/or personal injury.

• Brake fluid often contains polyglycol ethers and polyglycols. Avoid contact with the eyes and wash your hands thoroughly after handling brake fluid. If you do get brake fluid in your eyes, flush your eyes with clean, running water for 15 minutes. If eye irritation persists, or if you have taken brake fluid internally, IMMEDIATELY seek medical assistance.

• The EPA warns that prolonged contact with used engine oil may cause a number of skin disorders, including cancer. You should make every effort to minimize your exposure to used engine oil. Protective gloves should be worn when changing oil. Wash your hands and any other exposed skin areas as soon as possible after exposure to used engine oil. Soap and water, or waterless hand cleaner should be used.

• All new vehicles are now equipped with an air bag system, often referred to as a Supplemental Restraint System (SRS) or Supplemental Inflatable Restraint (SIR) system. The system must be disabled before performing service on or around system components, steering column, instrument panel components, wiring and sensors. Failure to follow safety and disabling procedures could result in accidental air bag deployment, possible personal injury and unnecessary system repairs.

• Always wear safety goggles when working with, or around, the air bag system. When carrying a non-deployed air bag, be sure the bag and trim cover are pointed away from your body. When placing a non-deployed air bag on a work surface, always face the bag and trim cover upward, away from the surface. This will reduce the motion of the module if it is accidentally deployed. Refer to the additional air bag system precautions later in this section.

• Clean, high quality brake fluid from a sealed container is essential to the safe and proper operation of the brake system. You should always buy the correct type of brake fluid for your vehicle. If the brake fluid becomes contaminated, completely flush the system with new fluid. Never reuse any brake fluid. Any brake fluid that is removed from the system should be discarded. Also, do not allow any brake fluid to come in contact with a painted surface; it will damage the paint.

• Never operate the engine without the proper amount and type of engine oil; doing so WILL result in severe engine damage.

• Timing belt maintenance is extremely important. Many models utilize an interference-type, non-freewheeling engine. If the timing belt breaks, the valves in the cylinder head may strike the pistons, causing potentially serious (also time-consuming and expensive) engine damage. Refer to the maintenance interval charts for the recommended replacement interval for the timing belt, and to the timing belt section for belt replacement and inspection.

• Disconnecting the negative battery cable on some vehicles may interfere with the functions of the on-board computer system(s) and may require the computer to undergo a relearning process once the negative battery cable is reconnected.

• When servicing drum brakes, only disassemble and assemble one side at a time, leaving the remaining side intact for reference.

• Only an MVAC-trained, EPA-certified automotive technician should service the air conditioning system or its components.

BRAKES

GENERAL INFORMATION

PRECAUTIONS

• Certain components within the ABS system are not intended to be serviced or repaired individually.

• Do not use rubber hoses or other parts not specifically specified for and ABS system. When using repair kits, replace all parts included in the kit. Partial or incorrect repair may lead to functional problems and require the replacement of components.

• Lubricate rubber parts with clean, fresh brake fluid to ease assembly. Do not use shop air to clean parts; damage to rubber components may result.

• Use only DOT 3 brake fluid from an unopened container.

• If any hydraulic component or line is removed or replaced, it may be necessary to bleed the entire system.

• A clean repair area is essential. Always clean the reservoir and cap thoroughly before removing the cap. The slightest amount of dirt in the fluid may plug an orifice and impair the system function. Perform repairs after components have been thoroughly cleaned; use only denatured alcohol to clean components. Do not allow ABS components to come into contact with any substance containing mineral oil; this includes used shop rags.

• The Anti-Lock control unit is a microprocessor similar to other computer units in the vehicle. Ensure that the ignition switch is **OFF** before removing or installing controller harnesses. Avoid static electricity discharge at or near the controller.

ANTI-LOCK BRAKE SYSTEM (ABS)

• If any arc welding is to be done on the vehicle, the control unit should be unplugged before welding operations begin.

SPEED SENSORS

REMOVAL & INSTALLATION

Front
See Figure 1.

1. Before servicing the vehicle, refer to the Precautions.
2. Disconnect the negative battery cable.

✳✳ WARNING

Avoiding twisting the sensor harness as much as possible when removing it. Pull sensors out without pulling front or rear wheel hub. This is to

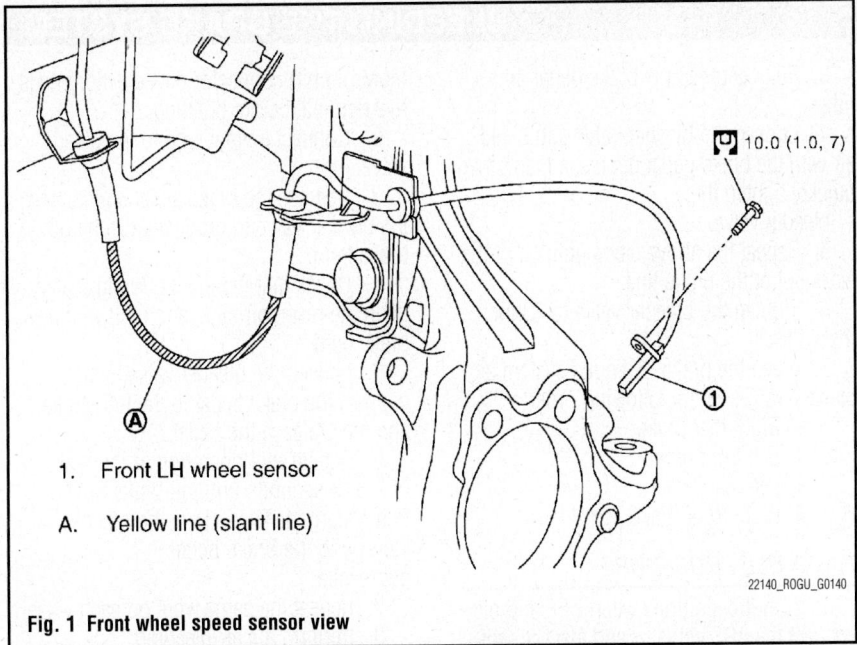

1. Front LH wheel sensor

A. Yellow line (slant line)

22140_ROGU_G0140

Fig. 1 Front wheel speed sensor view

avoid damage to sensor wiring and loss of sensor function. When you see the harness of the wheel sensor from the front side of the vehicle ensure that the yellow lines are not twisted. Avoiding twisting the sensor harness as much as possible when removing it. Pull sensors out without pulling on the sensor harness.

3. Disconnect the wheel speed sensor harness.
4. Remove the mounting bolt.
5. Remove the front speed sensor.

To install:
6. Install the front speed sensor.
7. Install the mounting bolt and tighten to 7 ft. lbs. (10 Nm).
8. Reconnect the wheel speed sensor harness.
9. Connect the negative battery cable.

Rear

See Figure 2.

1. Before servicing the vehicle, refer to the Precautions.
2. Disconnect the negative battery cable.

※※ **WARNING**

Avoiding twisting the sensor harness as much as possible when

removing it. Pull sensors out without pulling front or rear wheel hub. This is to avoid damage to sensor wiring and loss of sensor function. When you see the harness of the wheel sensor from the front side of the vehicle ensure that the yellow lines are not twisted. Avoiding twisting the sensor harness as much as possible, when removing it. Pull sensors out without pulling on the sensor harness.

3. Disconnect the wheel speed sensor harness.
4. Remove the mounting bolt.
5. Remove the front speed sensor.

To install:
6. Install the front speed sensor.
7. Install the mounting bolt and tighten to 7 ft. lbs. (10 Nm).
8. Reconnect the wheel speed sensor harness.
9. Connect the negative battery cable.

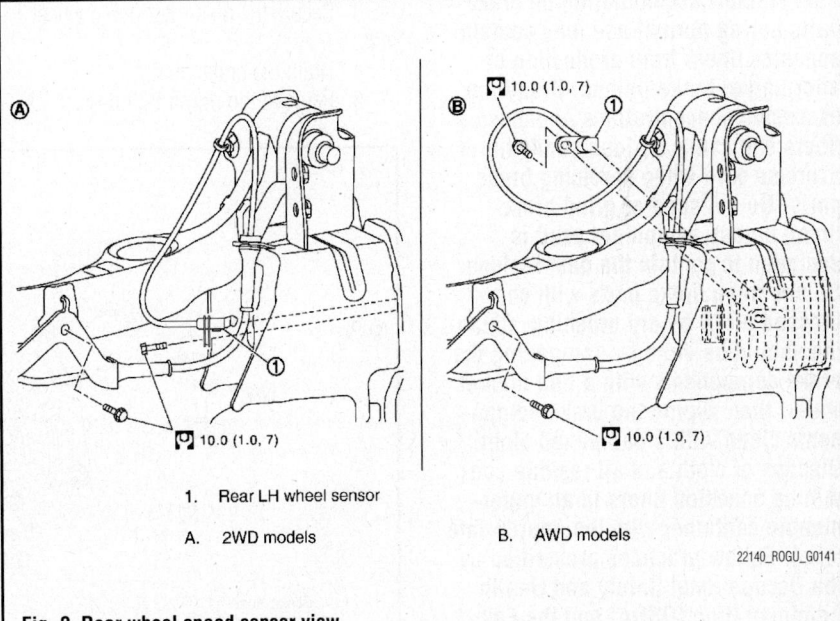

1. Rear LH wheel sensor

A. 2WD models

B. AWD models

22140_ROGU_G0141

Fig. 2 Rear wheel speed sensor view

BRAKES   BLEEDING THE BRAKE SYSTEM

BLEEDING PROCEDURE

MASTER CYLINDER BLEEDING

> **※※ WARNING**
>
> **Carefully monitor brake fluid level in the reservoir tank during bleeding operation.**

> **※※ WARNING**
>
> **Fill the sub tank with new brake fluid. Make sure it is full at all times while bleeding the air out of system.**

1. Turn ignition switch OFF and disconnect ABS actuator and control unit connector or negative battery terminal.
2. Connect a vinyl tube to the bleeder valve.

3. Fully depress the brake pedal 4 to 5 times.
4. Loosen the bleeder valve and bleed air with the brake pedal depressed, and then quickly tighten the bleeder valve.
5. Repeat the above steps until all of the air is out of the brake line.
6. Tighten the bleeder valve to 73 inch lbs. (8 Nm).
7. Bleed the brake hydraulic system air bleeder valves in the following order:
   - Right rear brake
   - Left front brake
   - Left rear brake
   - Right front brake

FLUID FILL PROCEDURE

1. Turn the ignition switch OFF and disconnect the ABS actuator and electric unit

(control unit) connector or the battery negative terminal before draining.
2. Connect a vinyl tube to the bleed valve.
3. Depress the brake pedal and loosen the bleeder valve to gradually discharge brake fluid.
4. Check that there is no foreign material in the reservoir tank, and refill with new brake fluid.
5. Loosen the bleeder valve, slowly depress the brake pedal to the full stroke, and then release the pedal.
6. Repeat this operation at intervals of 2 or 3 seconds until all brake fluid is discharged. Then close the bleeder valve with the brake pedal depressed.
7. Repeat the same work on each wheel.
8. Perform the air bleeding.

BRAKES   FRONT DISC BRAKES

> **※※ CAUTION**
>
> **Dust and dirt accumulating on brake parts during normal use may contain asbestos fibers from production or aftermarket brake linings. Breathing excessive concentrations of asbestos fibers can cause serious bodily harm. Exercise care when servicing brake parts. Do not sand or grind brake lining unless equipment used is designed to contain the dust residue. Do not clean brake parts with compressed air or by dry brushing. Cleaning should be done by dampening the brake components with a fine mist of water, then wiping the brake components clean with a dampened cloth. Dispose of cloth and all residue containing asbestos fibers in an impermeable container with the appropriate label. Follow practices prescribed by the Occupational Safety and Health Administration (OSHA) and the Environmental Protection Agency (EPA) for the handling, processing, and disposing of dust or debris that may contain asbestos fibers.**

BRAKE CALIPERS

REMOVAL & INSTALLATION

See Figure 3.

1. Before servicing the vehicle, refer to the Precautions.

2. Remove the front wheels and tires.
3. Secure the disc rotor using wheel nuts.
4. Drain the brake fluid.
5. Remove the union bolt and

copper washers, and disconnect the brake hose from the caliper assembly.
6. Remove the torque member mounting bolts, and remove the brake caliper assembly.

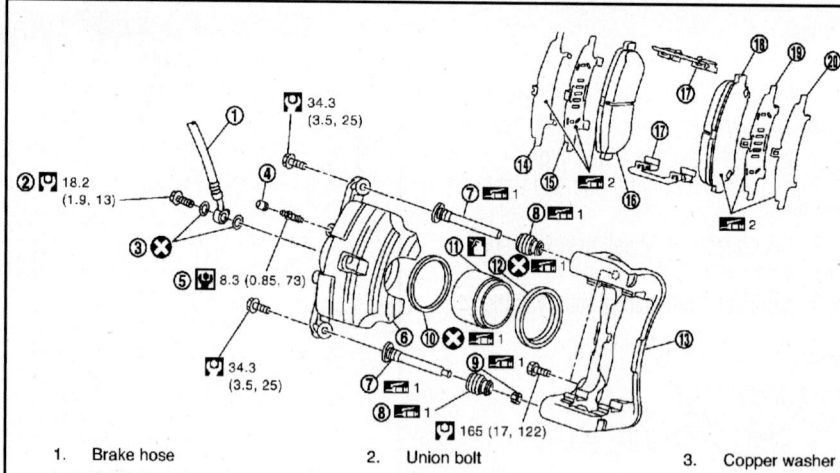

1.	Brake hose	2.	Union bolt	3.	Copper washer
4.	Cap	5.	Bleeder valve	6.	Cylinder body
7.	Sliding pin	8.	Sliding pin boot	9.	Bushing
10.	Piston seal	11.	Piston	12.	Piston boot
13.	Torque member	14.	Inner shim cover	15.	Inner shim
16.	Inner pad (only RH side with pad wear sensor)	17.	Pad retainer	18.	Outer pad
19.	Outer shim	20.	Outer shim cover		

⌷1: Apply rubber grease.

⌷2: Apply copper based brake grease.

⌷: Apply brake fluid.

Fig. 3 Front disc brake components

22140_ROGU_G0020

To install:

7. Install the brake caliper assembly to the vehicle and tighten the torque member mounting bolts to 122 ft. lbs. (165 Nm).

8. Install the brake hose and copper washers to the brake caliper assembly, and tighten union bolt to 13 ft. lbs. (18 Nm).

9. Refill with new brake fluid and perform the air bleeding.

10. Check that no drag is present for the front disc brake.

11. Install the wheels and tires.

BRAKE PADS

REMOVAL & INSTALLATION

See Figure 4.

1. Before servicing the vehicle, refer to the Precautions.

2. Remove the front wheels and tires.

3. Remove the lower sliding pin bolt.

4. Suspend the cylinder body with suitable wire so that the brake hose will not stretch.

5. Remove the brake pad from the torque member.

To install:

6. Install the pad retainer to the torque member if the pad retainers have been removed.

7. Apply copper based brake grease and install them to the brake pad.

8. Install the cylinder body and brake pads to the torque member.

9. Use a disc brake piston tool to easily press the piston back into the caliper.

10. Install the lower sliding pin bolt and tighten.

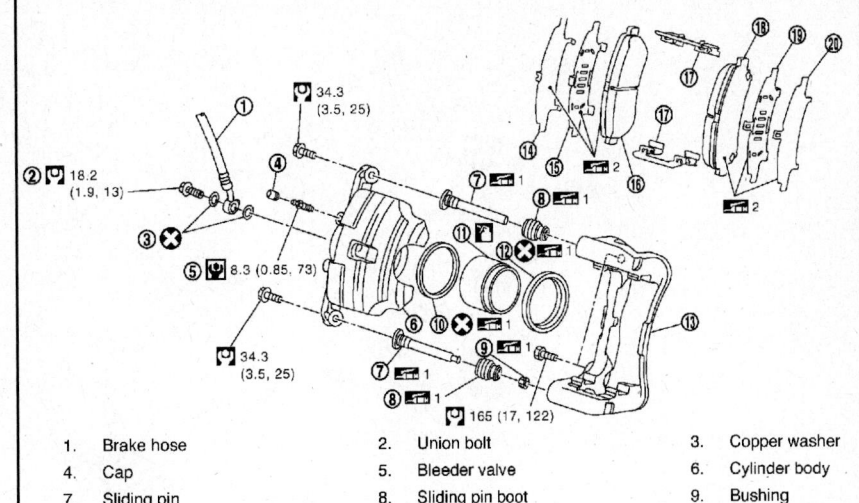

1.	Brake hose	2.	Union bolt	3.	Copper washer
4.	Cap	5.	Bleeder valve	6.	Cylinder body
7.	Sliding pin	8.	Sliding pin boot	9.	Bushing
10.	Piston seal	11.	Piston	12.	Piston boot
13.	Torque member	14.	Inner shim cover	15.	Inner shim
16.	Inner pad (only RH side with pad wear sensor)	17.	Pad retainer	18.	Outer pad
19.	Outer shim	20.	Outer shim cover		

1: Apply rubber grease.

2: Apply copper based brake grease.

: Apply brake fluid.

22140_ROGU_G0020

Fig. 4 Exploded view of the front brake pads and related components

11. Depress the brake pedal several times to check that no drag feel is present for the front disc brake.

12. Install the wheels and tires.

13. Brake burnishing procedure is as follows:

   a. Drive the vehicle on a straight, flat road.

   b. Depress the brake pedal with the power to stop vehicle within 3 to 5 seconds until the vehicle stops.

   c. Drive without depressing the brakes for a few minutes to cool down the brake system.

   d. Repeat steps A–C until pad and disc rotor are securely fitted.

BRAKES

REAR DISC BRAKES

✳✳ CAUTION

Dust and dirt accumulating on brake parts during normal use may contain asbestos fibers from production or aftermarket brake linings. Breathing excessive concentrations of asbestos fibers can cause serious bodily harm. Exercise care when servicing brake parts. Do not sand or grind brake lining unless equipment used is designed to contain the dust residue. Do not clean brake parts with compressed air or by dry brushing. Cleaning should be done by dampening the brake components with a fine mist of water, then wiping the brake components clean with a dampened cloth. Dis-pose of cloth and all residue containing asbestos fibers in an impermeable container with the appropriate label. Follow practices prescribed by the Occupational Safety and Health Administration (OSHA) and the Environmental Protection Agency (EPA) for the handling, processing, and disposing of dust or debris that may contain asbestos fibers.

BRAKE CALIPER

REMOVAL & INSTALLATION

See Figure 5.

1. Before servicing the vehicle, refer to the Precautions.

2. Clean any dust from the brake caliper and brake pads with a vacuum dust collector. Never blow with compressed air.

3. Remove the rear wheels and tires.

4. Secure the disc rotor using wheel nuts.

5. Drain brake fluid.

6. Remove the union bolt and copper washers and disconnect the brake hose from the caliper assembly.

7. Remove the torque member mounting bolts, and remove the brake caliper assembly.

To install:

8. Install the brake caliper assembly to the vehicle and tighten the torque member mounting bolts to 62 ft. lbs. (84 Nm).

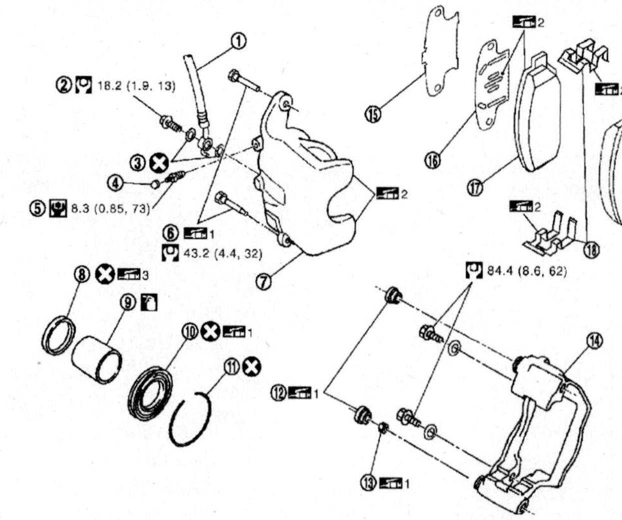

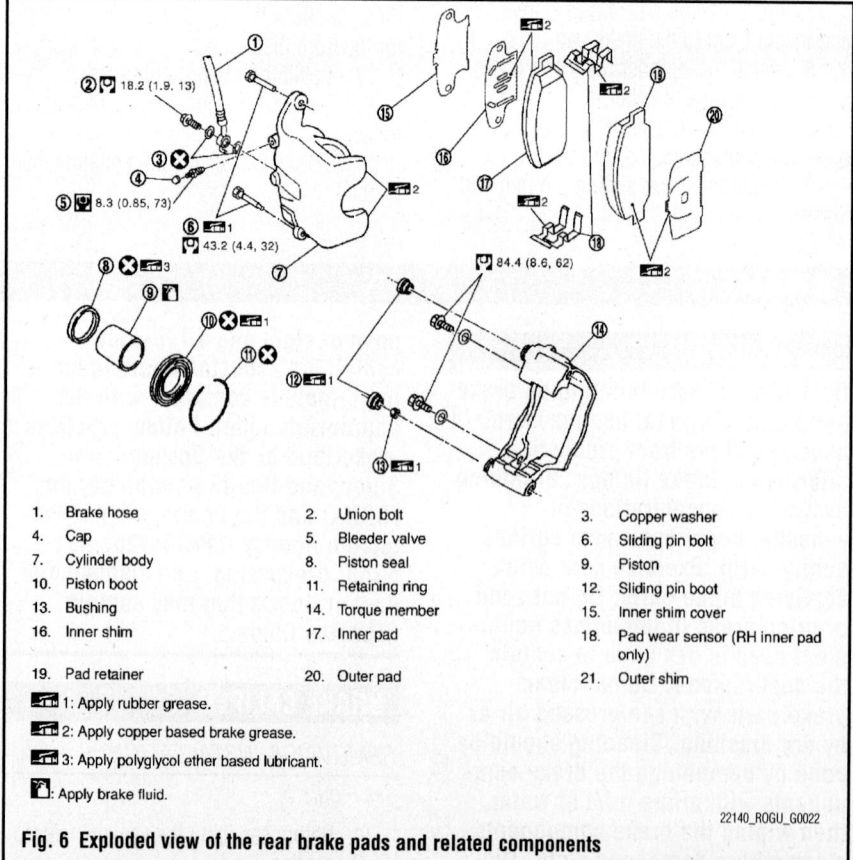

1. Brake hose
4. Cap
7. Cylinder body
10. Piston boot
13. Bushing
16. Inner shim

19. Pad retainer

2. Union bolt
5. Bleeder valve
8. Piston seal
11. Retaining ring
14. Torque member
17. Inner pad

20. Outer pad

3. Copper washer
6. Sliding pin bolt
9. Piston
12. Sliding pin boot
15. Inner shim cover
18. Pad wear sensor (RH inner pad only)

21. Outer shim

1: Apply rubber grease.

2: Apply copper based brake grease.

3: Apply polyglycol ether based lubricant.

: Apply brake fluid.

22140_ROGU_G0022

Fig. 5 Rear disc brake components

9. Install brake hose and copper washers to brake caliper assembly, and tighten union bolt to 13 ft. lbs. (18 Nm).

10. Refill with new brake fluid and perform the air bleeding.

11. Check that no drag is present for the rear disc brake.

12. Install the wheels and tires.

DISC BRAKE PADS

REMOVAL & INSTALLATION

See Figure 6.

1. Before servicing the vehicle, refer to the Precautions.

2. Remove the rear wheels and tires.

3. Remove lower sliding pin bolt.

4. Suspend the cylinder body with suitable wire so that the brake hose will not stretch.

5. Remove the brake pad from the torque member.

To install:

6. Install the pad retainer to the torque member if the pad retainers have been removed.

7. Apply copper based brake grease and install them to the brake pad.

1. Brake hose
4. Cap
7. Cylinder body
10. Piston boot
13. Bushing
16. Inner shim

19. Pad retainer

2. Union bolt
5. Bleeder valve
8. Piston seal
11. Retaining ring
14. Torque member
17. Inner pad

20. Outer pad

3. Copper washer
6. Sliding pin bolt
9. Piston
12. Sliding pin boot
15. Inner shim cover
18. Pad wear sensor (RH inner pad only)

21. Outer shim

1: Apply rubber grease.

2: Apply copper based brake grease.

3: Apply polyglycol ether based lubricant.

: Apply brake fluid.

22140_ROGU_G0022

Fig. 6 Exploded view of the rear brake pads and related components

8. Install the cylinder body and brake pads to the torque member.

9. Use a disc brake piston tool to easily press the piston back into the caliper.

10. Install the lower sliding pin bolt and tighten it to 32 ft. lbs. (43 Nm).

11. Depress the brake pedal several times to check that no drag feel is present for the front disc brake.

12. Install the wheels and tires.

13. Brake burnishing procedure is as follows:

a. Drive the vehicle on a straight, flat road.

b. Depress the brake pedal with the power to stop vehicle within 3 to 5 seconds until the vehicle stops.

c. Drive without depressing the brakes for a few minutes to cool down the brake system.

d. Repeat steps A—C until pad and disc rotor are securely fitted.

BRAKES

PARKING BRAKE

PARKING BRAKE CABLES

ADJUSTMENT

See Figure 7.

1. Adjust the cable with the following procedure:

a. Operate the parking brake pedal with a force of 110 lbs. (490 N) for 10 strokes or more.

b. Adjust the parking brake pedal stroke by turning the adjusting nut with a deep socket wrench.

c. Operate the parking brake pedal with a force of 44 lbs. (196 N). Check that the pedal stroke is within the specified number of notches. (Check it by listening to the clicks of the ratchet.)

✳✳ WARNING

Never reuse the adjusting nut if the nut is removed.

2. Install the rear tires with power tool.

PARKING BRAKE SHOES

REMOVAL & INSTALLATION

See Figure 8.

1. Before servicing the vehicle, refer to the Precautions.

2. Remove the rear wheels and tires.

3. Remove the disc rotor with parking brake completely in the released position.

4. Put matchmarks on the disc rotor and the wheel hub and bearing assembly when reusing the disc rotor.

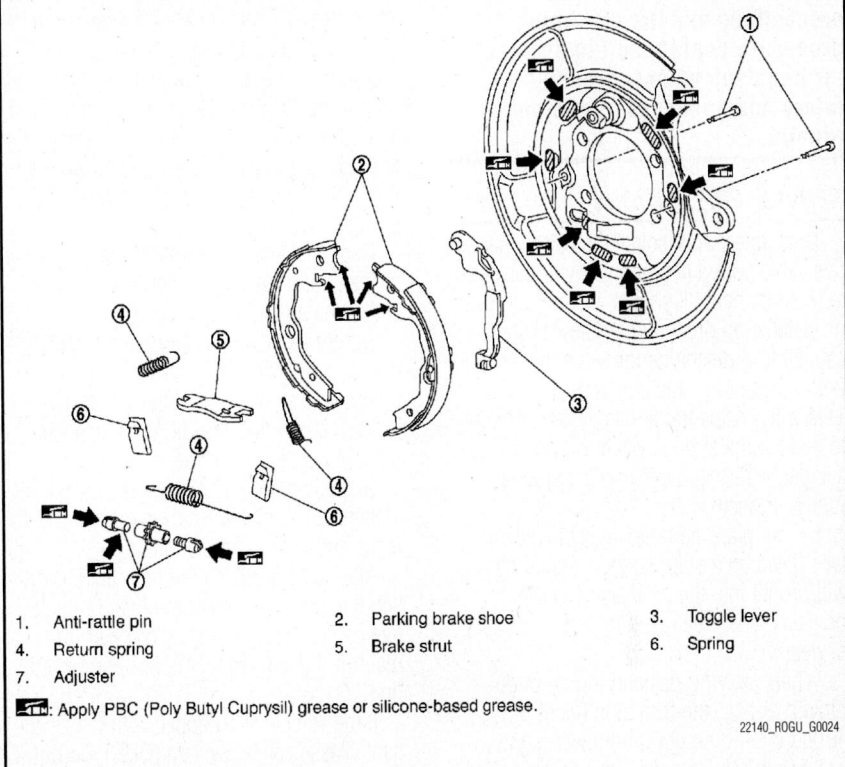

1. Anti-rattle pin	2. Parking brake shoe	3. Toggle lever
4. Return spring	5. Brake strut	6. Spring
7. Adjuster		

🔩: Apply PBC (Poly Butyl Cuprysil) grease or silicone-based grease.

22140_ROGU_G0024

Fig. 8 Exploded view of the parking brake shoes

5. If disc rotor cannot be removed, remove as follows:

a. Secure the disc rotor with wheel nuts and remove the adjusting hole plug.

b. Using suitable tool, rotate the adjuster to retract and loosen parking brake shoe.

6. Remove anti-rattle pins, springs, and return springs.

7. Remove brake strut, adjuster, parking brake shoes and toggle lever.

To install:

8. Install in the reverse order of removal, noting the following:

a. Apply PBC (Poly Butyl Cuprysil) grease or silicone-based grease to the back plate and brake shoe.

b. Assemble adjusters so that threaded part is expanded when rotated.

c. Shorten the adjuster by rotating it.

d. When disassembling, apply PBC (Poly Butyl Cuprysil) grease or silicone-based grease to threads.

Item	Standard
Number of notches [under force of 196 N (20 kg, 44 lb)]	6 – 7 notches
Number of notches when brake warning lamp turns ON	1 notch

37663_ROGU_G0049

Fig. 7 Parking brake adjustment

CHASSIS ELECTRICAL   AIR BAG (SUPPLEMENTAL RESTRAINT SYSTEM)

GENERAL INFORMATION

✳✳ CAUTION

These vehicles are equipped with an air bag system. The system must be disarmed before performing service on, or around, system components, the steering column, instrument panel components, wiring and sensors. Failure to follow the safety precautions and the disarming procedure could result in accidental air bag deployment, possible injury and unnecessary system repairs.

SERVICE PRECAUTIONS

Disconnect and isolate the battery negative cable before beginning any airbag system component diagnosis, testing, removal, or installation procedures. Allow system capacitor to discharge for two minutes before beginning any component service. This will disable the airbag system. Failure to disable the airbag system may result in accidental airbag deployment, personal injury, or death.

Do not place an intact undeployed airbag face down on a solid surface. The airbag will propel into the air if accidentally deployed and may result in personal injury or death.

When carrying or handling an undeployed airbag, the trim side (face) of the airbag should be pointing towards the body to minimize possibility of injury if accidental deployment occurs. Failure to do this may result in personal injury or death.

Replace airbag system components with OEM replacement parts. Substitute parts may appear interchangeable, but internal differences may result in inferior occupant protection. Failure to do so may result in occupant personal injury or death.

Wear safety glasses, rubber gloves, and long sleeved clothing when cleaning powder residue from vehicle after an airbag deployment. Powder residue emitted from a deployed airbag can cause skin irritation. Flush affected area with cool water if irritation is experienced. If nasal or throat irritation is experienced, exit the vehicle for fresh air until the irritation ceases. If irritation continues, see a physician.

Do not use a replacement airbag that is not in the original packaging. This may

result in improper deployment, personal injury, or death.

The factory installed fasteners, screws and bolts used to fasten airbag components have a special coating and are specifically designed for the airbag system. Do not use substitute fasteners. Use only original equipment fasteners listed in the parts catalog when fastener replacement is required.

During, and following, any child restraint anchor service, due to impact event or vehicle repair, carefully inspect all mounting hardware, tether straps, and anchors for proper installation, operation, or damage. If a child restraint anchor is found damaged in any way, the anchor must be replaced. Failure to do this may result in personal injury or death.

Deployed and non-deployed airbags may or may not have live pyrotechnic material within the airbag inflator.

Do not dispose of driver/passenger/curtain airbags or seat belt tensioners unless you are sure of complete deployment. Refer to the Hazardous Substance Control System for proper disposal.

Dispose of deployed airbags and tensioners consistent with state, provincial, local, and federal regulations.

After any airbag component testing or service, do not connect the battery negative cable. Personal injury or death may result if the system test is not performed first.

If the vehicle is equipped with the Occupant Classification System (OCS), do not connect the battery negative cable before performing the OCS Verification Test using the scan tool and the appropriate diagnostic information. Personal injury or death may result if the system test is not performed properly.

Never replace both the Occupant Restraint Controller (ORC) and the Occupant Classification Module (OCM) at the same time. If both require replacement, replace one, then perform the Airbag System test before replacing the other.

Both the ORC and the OCM store Occupant Classification System (OCS) calibration data, which they transfer to one another when one of them is replaced. If both are replaced at the same time, an irreversible fault will be set in both modules and the OCS may malfunction and cause personal injury or death.

If equipped with OCS, the Seat Weight Sensor is a sensitive, calibrated unit and must be handled carefully. Do not drop or

handle roughly. If dropped or damaged, replace with another sensor. Failure to do so may result in occupant injury or death.

If equipped with OCS, the front passenger seat must be handled carefully as well. When removing the seat, be careful when setting on floor not to drop. If dropped, the sensor may be inoperative, could result in occupant injury, or possibly death.

If equipped with OCS, when the passenger front seat is on the floor, no one should sit in the front passenger seat. This uneven force may damage the sensing ability of the seat weight sensors. If sat on and damaged, the sensor may be inoperative, could result in occupant injury, or possibly death.

DISARMING THE SYSTEM

Before servicing the SRS, turn ignition switch OFF, disconnect both battery cables, and wait at least 3 minutes.

✳✳ CAUTION

For approximately 3 minutes after the cables are removed, it is still possible for the air bag and seat belt pretensioner to deploy. Therefore, do not work on any SRS connectors or wires until at least 3 minutes have elapsed.

ARMING THE SYSTEM

Reconnect the battery cables to rearm the SRS system.

CLOCKSPRING CENTERING

1. Before servicing the vehicle, refer to the Precautions.

2. The spiral cable may snap during steering operation if the cable is installed in an improper position.

3. The neutral position is set as follows:
   a. Turn carefully the spiral cable clockwise to the end position.
   b. Then turn it counterclockwise (about 2 and a half turns) and stop turning at the point (B) on which the stopper insertion holes are in the same position.
   c. The service part is installed in the neutral position by the stopper and can be set without adjusting after the stopper is removed.

✳✳ WARNING

Do not over turn the spiral cable or go beyond number of turns required.

(These will cause the cable to snap.)

    d. Adjust the spiral cable locating pin (A) to the steering wheel locating pin hole (C).

    In the case that a malfunction is detected by the air bag warning lamp, reset by the self-diagnosis function and delete the memory by CONSULT-III®.

    4. In the case that a malfunction is still detected after the above operation, perform self-diagnosis to repair malfunctions.

    5. After the work is completed, check that no system malfunction is detected by air bag warning lamp.

DRIVE TRAIN

FRONT HALFSHAFT

REMOVAL & INSTALLATION

2WD Models

Left Side Axle

See Figure 9.

    1. Before servicing the vehicle, refer to the Precautions.

    2. Remove the wheels and tires.

    3. Remove the wheel sensor from the steering knuckle.

> ❋❋ **WARNING**
>
> **Never pull on the wheel sensor harness.**

    4. Remove the lock plate from the strut assembly.

    5. Remove the torque member mounting bolts. Hang torque member aside.

    6. Remove the brake disc rotor. Put matchmarks on the wheel hub and bearing assembly and the disc rotor before removing the disc rotor. Never drop the brake disc rotor.

    7. Remove the cotter pin and loosen the hub lock nut.

    8. Patch the hub lock nut with a piece of wood. Hammer the wood to disengage the wheel hub and bearing assembly from the drive shaft.

> ❋❋ **WARNING**
>
> **Never place drive shaft joint at an extreme angle. Also, be careful not to overextend slide joint. Never allow drive shaft to hang down without support for housing (or joint sub-assembly), shaft and the other parts.**

    9. If wheel hub and drive shaft cannot be separated after performing the above procedure, use a suitable puller.

    10. Remove the hub lock nut.

    11. Remove if wheel hub and drive shaft cannot be separated even after performing the above procedure transverse link from steering knuckle.

    12. Loosen the steering outer socket nut.

    13. Remove the steering outer socket from the steering knuckle using the ball joint remover so as to not damage ball joint boot.

    14. Remove drive shaft from transaxle assembly.

    15. Use the drive shaft attachment SST: KV40107500 and a sliding hammer while inserting the tip of the drive shaft attachment between the housing and transaxle assembly.

➡ **Never place drive shaft joint at an extreme angle when removing drive shaft. Also, be careful not to overextend slide joint. Confirm that the circular clip is attached to the drive shaft.**

    To install:

    16. Install in the reverse order of removal, noting the following:

    a. Place the axle protector SST: KV38107900 onto transaxle assembly to prevent damage to the oil seal while inserting drive shaft. Slide drive shaft sliding joint and tap with a hammer to install securely.

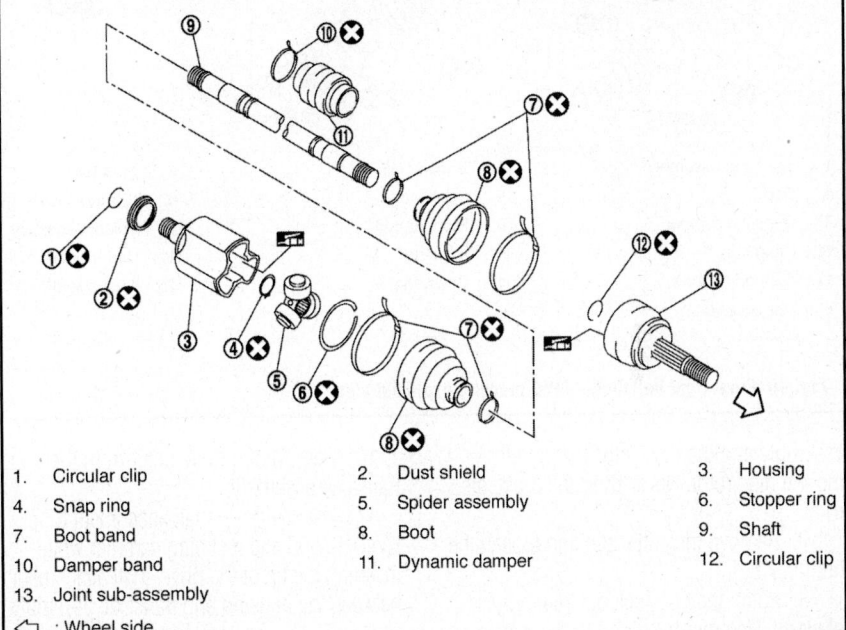

1.	Circular clip	2.	Dust shield	3.	Housing
4.	Snap ring	5.	Spider assembly	6.	Stopper ring
7.	Boot band	8.	Boot	9.	Shaft
10.	Damper band	11.	Dynamic damper	12.	Circular clip
13.	Joint sub-assembly				
⬅	: Wheel side				

37663_ROGU_G0075

Fig. 9 Front left halfshaft—exploded view

> ❋❋ **WARNING**
>
> **Make sure that circular clip is completely engaged.**

    b. Tighten the axle nut to 92 ft. lbs. (125 Nm). Install the cotter pin.

Right Side Axle

See Figure 10.

    1. Before servicing the vehicle, refer to the Precautions.

    2. Remove the wheels and tires.

    3. Remove the wheel sensor from the steering knuckle.

> ❋❋ **WARNING**
>
> **Never pull on the wheel sensor harness.**

    4. Remove the lock plate from the strut assembly.

    5. Remove the torque member mounting bolts. Hang torque member aside.

    6. Remove the brake disc rotor. Put matchmarks on the wheel hub and bearing

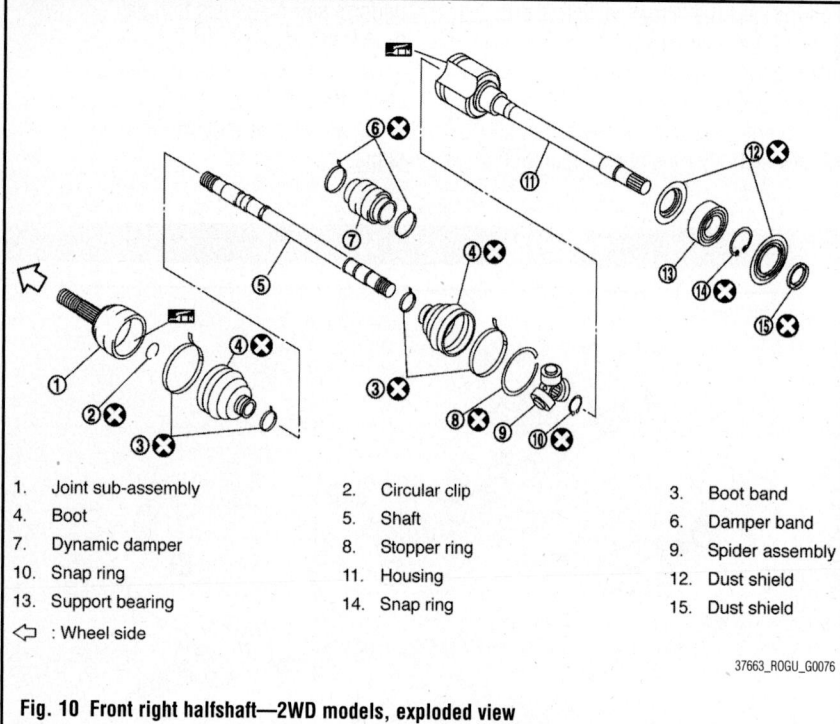

1.	Joint sub-assembly	2.	Circular clip	3.	Boot band
4.	Boot	5.	Shaft	6.	Damper band
7.	Dynamic damper	8.	Stopper ring	9.	Spider assembly
10.	Snap ring	11.	Housing	12.	Dust shield
13.	Support bearing	14.	Snap ring	15.	Dust shield
⇦	: Wheel side				

37663_ROGU_G0076

Fig. 10 Front right halfshaft—2WD models, exploded view

assembly and the disc rotor before removing the disc rotor. Never drop the brake disc rotor.

7. Remove the cotter pin and loosen the hub lock nut.

8. Patch the hub lock nut with a piece of wood. Hammer the wood to disengage the wheel hub and bearing assembly from the drive shaft.

❋❋ WARNING

Never place drive shaft joint at an extreme angle. Also, be careful not to overextend slide joint. Never allow drive shaft to hang down without support for housing (or joint sub-assembly), shaft and the other parts.

9. If wheel hub and drive shaft cannot be separated after performing the above procedure, use a suitable puller.

10. Remove the hub lock nut.

11. Remove if wheel hub and drive shaft cannot be separated even after performing the above procedure transverse link from steering knuckle.

12. Loosen the steering outer socket nut.

13. Remove the steering outer socket from the steering knuckle using the ball joint remover so as to not damage ball joint boot.

14. Remove the plate mounting bolts and plate.

15. If necessary, remove the support bearing bracket mounting bolts and the support bearing bracket.

16. Remove the drive shaft from the transaxle assembly.

17. Use the drive shaft attachment SST: KV40107500 and a sliding hammer while inserting the tip of the drive shaft attachment between the housing and transaxle assembly.

❋❋ WARNING

Never place drive shaft joint at an extreme angle when removing drive shaft. Also, be careful not to overextend slide joint.

To install:

18. Install in the reverse order of removal, noting the following:

a. Place the axle protector onto the transaxle assembly to prevent damage to the oil seal while inserting drive shaft. Slide drive shaft sliding joint and tap with a hammer to install securely.

b. Tighten the axle nut to 92 ft. lbs. (125 Nm). Install the cotter pin.

c. Make sure that circular clip is completely engaged.

d. When installing support bearing bracket:
- (Temporarily) tighten mounting bolts.
- Set plate so that notch becomes upper side. (Temporarily) tighten mounting bolts.

❋❋ WARNING

Never reuse plate.

AWD Models

Left Side Axle

See Figure 11.

1. Before servicing the vehicle, refer to the Precautions.

2. Remove the wheel and tires.

3. Remove the wheel sensor from the steering knuckle.

❋❋ WARNING

Never pull on the wheel speed sensor harness.

4. Remove the lock plate from the strut assembly.

5. Remove the torque member mounting bolts. Hang the torque member aside.

6. Remove the brake disc rotor. Put matchmarks on the wheel hub and bearing assembly and the disc rotor before removing the disc rotor. Never drop brake disc rotor.

7. Remove cotter pin, and loosen hub lock nut.

8. Patch hub lock nut with a piece of wood. Hammer the wood to disengage wheel hub and bearing assembly from drive shaft.

❋❋ WARNING

Never place drive shaft joint at an extreme angle. Also, be careful not to overextend slide joint. Never allow drive shaft to hang down without support for housing (or joint sub-assembly), shaft and the other parts.

9. If the wheel hub and drive shaft cannot be separated after performing the above procedure, use a suitable puller.

10. Remove the hub lock nut.

11. Remove the transverse link from steering knuckle.

12. Loosen the steering outer socket steering outer socket nut.

13. Remove the steering outer socket from the steering knuckle using the ball joint remover so as not to damage ball joint boot.

14. Temporarily tighten the nut to prevent damage to threads and to prevent the ball joint remover from suddenly coming off.

15. Remove the drive shaft from the transaxle assembly.

16. Use the drive shaft attachment (A) SST: KV40107500 and a sliding hammer (B) while inserting tip of the drive shaft attachment between housing and transaxle assembly.

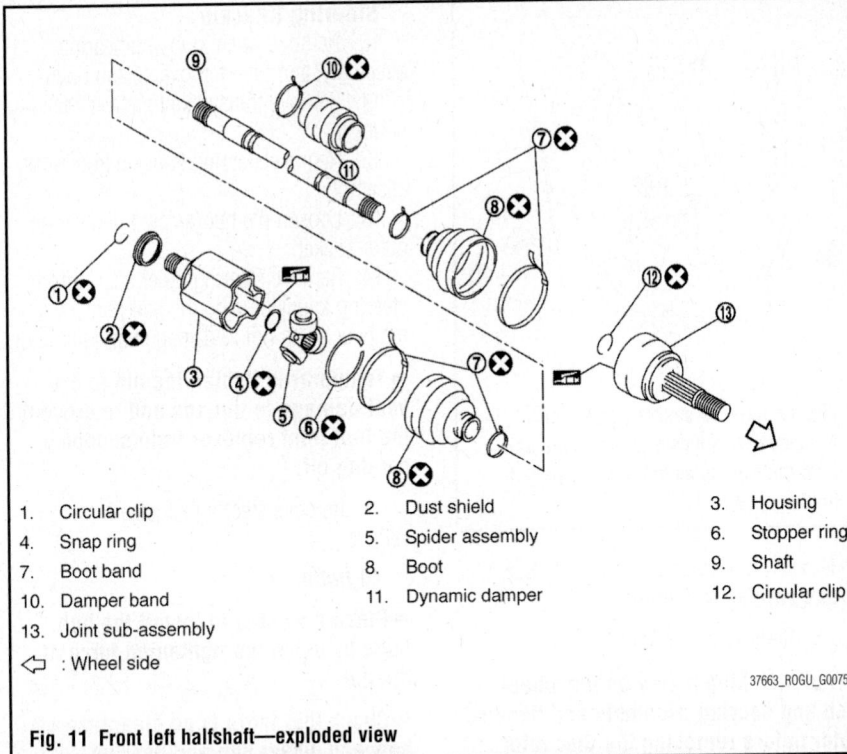

1. Circular clip
2. Dust shield
3. Housing
4. Snap ring
5. Spider assembly
6. Stopper ring
7. Boot band
8. Boot
9. Shaft
10. Damper band
11. Dynamic damper
12. Circular clip
13. Joint sub-assembly
◁ : Wheel side

37663_ROGU_G0075

Fig. 11 Front left halfshaft—exploded view

※※ **WARNING**

Never place drive shaft joint at an extreme angle when removing drive shaft. Also be careful not to overextend slide joint. Confirm that the circular clip is attached to the drive shaft.

To install:

17. Install in the reverse order of removal, noting the following:

a. Always replace the differential side oil seal with new one when installing drive shaft.

b. Place the protector (SST: KV38107900) onto transaxle assembly to prevent damage to the oil seal while inserting drive shaft. Slide drive shaft sliding joint and tap with a hammer to install securely.

※※ **WARNING**

Make sure that circular clip is completely engaged.

c. Tighten the axle nut to 92 ft. lbs. (125 Nm). Install the cotter pin.

Right Side Axle

See Figure 12.

1. Before servicing the vehicle, refer to the Precautions.
2. Remove the wheel and tires.
3. Remove the wheel sensor from the steering knuckle.

※※ **WARNING**

Never pull on the wheel speed sensor harness.

4. Remove the lock plate from the strut assembly.

5. Remove the torque member mounting bolts. Hang the torque member aside.

6. Remove the brake disc rotor. Put matchmarks on the wheel hub and bearing assembly and the disc rotor before removing the disc rotor. Never drop brake disc rotor.

7. Remove cotter pin, and loosen hub lock nut.

8. Patch hub lock nut with a piece of wood. Hammer the wood to disengage wheel hub and bearing assembly from drive shaft.

※※ **WARNING**

Never place drive shaft joint at an extreme angle. Also, be careful not to overextend slide joint. Never allow drive shaft to hang down without support for housing (or joint sub-assembly), shaft and the other parts.

9. If the wheel hub and drive shaft cannot be separated after performing the above procedure, use a suitable puller.

10. Remove the hub lock nut.

11. Remove the transverse link from the steering knuckle.

12. Loosen the steering outer socket steering outer socket nut.

13. Remove the steering outer socket from the steering knuckle using the ball joint remover so as not to damage ball joint boot.

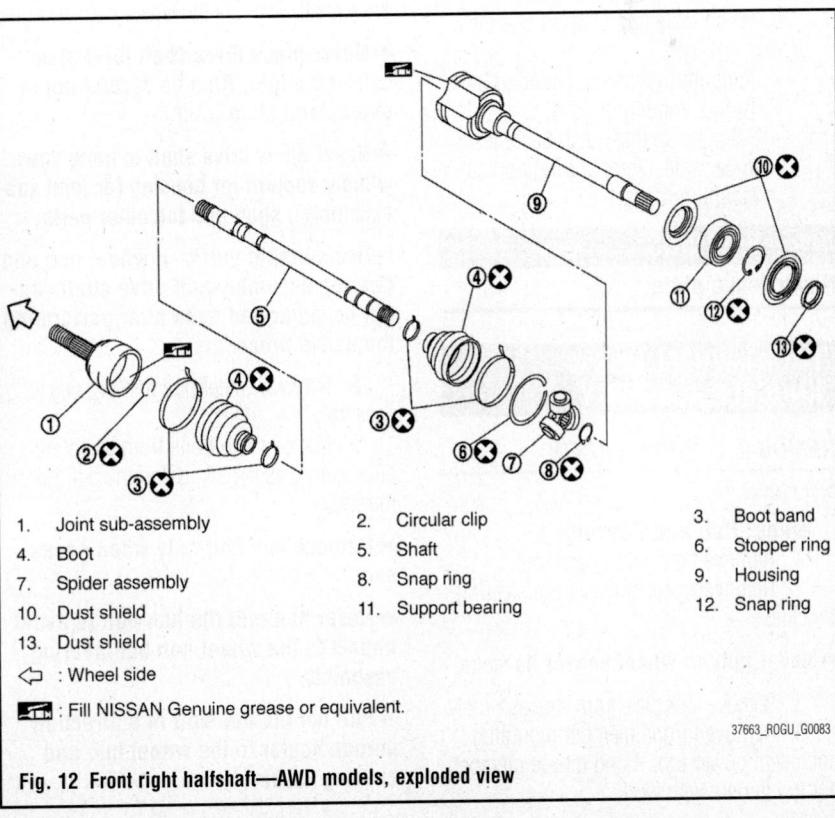

1. Joint sub-assembly
2. Circular clip
3. Boot band
4. Boot
5. Shaft
6. Stopper ring
7. Spider assembly
8. Snap ring
9. Housing
10. Dust shield
11. Support bearing
12. Snap ring
13. Dust shield
◁ : Wheel side

⬛ : Fill NISSAN Genuine grease or equivalent.

37663_ROGU_G0083

Fig. 12 Front right halfshaft—AWD models, exploded view

14. Temporarily tighten the nut to prevent damage to threads and to prevent the ball joint remover from suddenly coming off.

15. Remove plate mounting bolts and plate.

16. If necessary, remove the support bearing bracket mounting bolts and the support bearing bracket.

17. Remove drive shaft from transaxle assembly.

18. Use the drive shaft attachment (SST: KV40107500) and a sliding hammer while inserting tip of the drive shaft attachment between housing and transaxle assembly.

✼✼ WARNING

Never place drive shaft joint at an extreme angle when removing drive shaft. Also, be careful not to over-extend slide joint.

To install:

19. Install in the reverse order of removal, noting the following:

   a. Always replace differential side oil seal with new one when installing drive shaft.

   b. Place the protector (SST: KV38107900) onto the transaxle assembly to prevent damage to the oil seal while inserting drive shaft. Slide drive shaft sliding joint and tap with a hammer to install securely.

   c. Tighten the axle nut to 92 ft. lbs. (125 Nm). Install the cotter pin.

   d. When installing support bearing bracket.

   • Temporarily tighten mounting bolts.
   • Tighten mounting bolts.
   • Set plate so that notch becomes upper side. (Temporarily) tighten mounting bolts.

✼✼ WARNING

Never reuse plate.

FRONT WHEEL HUB & KNUCKLE ASSEMBLY

REMOVAL & INSTALLATION

See Figure 13.

Wheel Hub and Bearing

1. Remove tires with power tool.
2. Remove wheel sensor from steering knuckle.

➡**Never pull on wheel sensor harness.**

3. Remove lock plate from strut assembly.
4. Remove torque member mounting bolts with power tool. Hang torque member not to interfere with work.

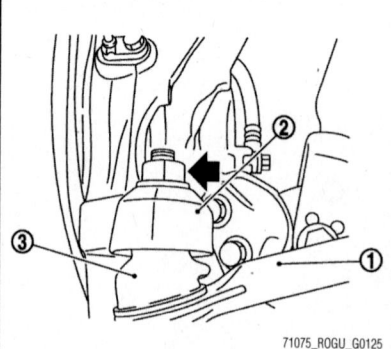

Fig. 13 Remove steering outer socket (1) from steering knuckle (2) using the ball joint remover so as not to damage ball joint boot (3).

71075_ROGU_G0125

➡**Never depress brake pedal while brake caliper is removed.**

5. Remove disc rotor.

➡**Put matching marks on the wheel hub and bearing assembly and the disc rotor before removing the disc rotor.**

➡**Never drop disc rotor.**

6. Remove cotter pin and adjusting cap (if equipped), then loosen wheel hub lock nut with a power tool.
7. Patch wheel hub lock nut with a piece of wood. Hammer the wood to disengage wheel hub and bearing assembly from drive shaft. Remove the wheel hub lock nut.

➡**Never place drive shaft joint at an extreme angle. Also be careful not to overextend slide joint.**

➡**Never allow drive shaft to hang down without support for housing (or joint sub-assembly), shaft and the other parts.**

➡**Use suitable puller, if wheel hub and bearing assembly and drive shaft cannot be separated even after performing the above procedure.**

8. Remove wheel hub and bearing assembly.
9. Remove hub bolts from wheel hub and bearing assembly, using the ball joint remover.

➡**Remove hub bolt only when necessary.**

➡**Never hammer the hub bolt to avoid impact to the wheel hub and bearing assembly.**

➡**Pull out the hub bolt in a direction perpendicular to the wheel hub and bearing assembly.**

Steering Knuckle

10. Remove wheel hub and bearing assembly, and then remove splash guard.
11. Remove transverse link from steering knuckle.
12. Remove steering knuckle from strut assembly.
13. Loosen the nut (arrow) of steering outer socket.
14. Remove steering outer socket from steering knuckle using the ball joint remover so as not to damage ball joint boot.

➡**Temporarily tighten the nut to prevent damage to threads and to prevent the ball joint remover from suddenly coming off.**

15. Remove steering knuckle from vehicle.

To install:

➡**Place a washer to install the hub bolts by using the tightening force of the nut.**

➡**Check that there is no clearance between wheel hub and bearing assembly, and hub bolt.**

➡**Never reuse hub bolt.**

➡**Clean the matching surface of wheel hub lock nut and wheel hub and bearing assembly. (With adjusting cap for wheel hub lock nut)**

➡**Never apply lubricating oil to these matching surface.**

➡**Clean the matching surface of drive shaft, wheel hub and bearing assembly. And then apply paste [service parts (440037S000)] to surface of joint sub-assembly of drive shaft. (With adjusting cap for wheel hub lock nut).**

➡**Apply paste to cover entire flat surface of joint sub-assembly of drive shaft.**

➡**Use the following torque range for tightening the wheel hub lock nut. (With adjusting cap for wheel hub lock nut)**

➡**Since the drive shaft is assembled by press-fitting, use the tightening torque range for the wheel hub lock nut.**

➡**Be sure to use torque wrench to tighten the wheel hub lock nut. Never use a power tool.**

➡**Never reuse wheel hub lock nut.**

➡**Wheel hub lock nut tightening torque does not over torque for avoiding axle noise, and does not less than torque for avoiding looseness.**

➡ **Align the matching marks made during removal when reusing the disc rotor.**

➡ **When installing a cotter pin and adjusting cap, securely bend the basal portion to prevent rattles. (With adjusting cap for wheel hub lock nut).**

➡ **Never reuse cotter pin, steering knuckle, and transverse link fixing nut.**

➡ **Perform the final tightening of each of parts under unladen conditions, which were removed when removing wheel hub and bearing assembly and steering knuckle.**

Steering Knuckle

16. Install steering knuckle on vehicle.

➡ **Temporarily tighten the nut to prevent damage to threads and to prevent the ball joint remover from suddenly coming off.**

17. Install steering outer socket from steering knuckle using the ball joint remover so as not to damage ball joint boot.

18. Tighten the nut (arrow) of steering outer socket.

19. Install steering knuckle to strut assembly.

20. Install transverse link to steering knuckle.

21. Install wheel hub and bearing assembly, and then install splash guard.

Wheel Hub and Bearing

22. Install hub bolts from wheel hub and bearing assembly, using the ball joint remover.

23. Install wheel hub and bearing assembly.

24. Patch wheel hub lock nut with a piece of wood. Hammer the wood to engage wheel hub and bearing assembly to drive shaft. Install the wheel hub lock nut.

25. Install cotter pin and adjusting cap (if equipped), then tighten wheel hub lock nut with a power tool.

26. Install disc rotor.

27. Install torque member mounting bolts with power tool. Hang torque member not to interfere with work.

28. Install lock plate from strut assembly.

29. Install wheel sensor from steering knuckle.

30. Install tires.

REAR HALFSHAFT

REMOVAL & INSTALLATION

See Figure 14.

1. Before servicing the vehicle, refer to the Precautions.

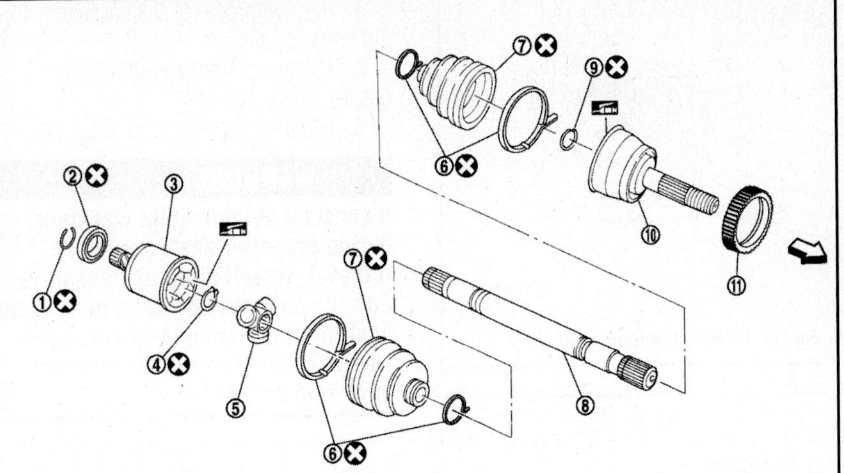

1. Circular clip
2. Dust shield
3. Housing
4. Snap ring
5. Spider assembly
6. Boot band
7. Boot
8. Shaft
9. Circular clip
10. Joint sub-assembly
11. Sensor rotor
⬅ : Wheel side

37663_ROGU_G0085

Fig. 14 Rear halfshaft—exploded view

2. Remove the wheels and tires.
3. Remove the torque member mounting bolts. Hang torque member aside.
4. Remove the disc rotor.
5. Remove the cotter pin and loosen the hub lock nut.
6. Patch the hub lock nut with a piece of wood. Hammer the wood to disengage the wheel hub and bearing assembly from the drive shaft.

✳✳ WARNING

Never place drive shaft joint at an extreme angle. Also be careful not to overextend slide joint. Never allow drive shaft to hang down without support for housing (or joint sub-assembly), shaft and the other parts.

7. If the wheel hub and bearing assembly and drive shaft cannot be separated after performing the above procedure, use a suitable puller.
8. Remove the hub lock nut.
9. Remove the wheel sensor from the axle housing.
10. Remove the stabilizer link.
11. Set a suitable jack under the suspension arm.
12. Remove the shock absorber from the suspension arm.
13. Remove the upper link from the suspension arm.
14. Remove the lower link from suspension arm.
15. Remove the drive shaft from the final drive assembly.

To install:

16. Install in the reverse order of removal, noting the following:

   a. Align the matchmarks made during removal, if reusing the disc rotor.

   b. Tighten the axle nut to 92 ft. lbs. (125 Nm).

   c. Perform final tightening of bolts and nuts at suspension arm (rubber bushing), under relaxed conditions with tires on level ground.

REAR PINION OIL SEAL

REMOVAL & INSTALLATION

See Figures 15 and 16.

1. Before servicing the vehicle, refer to the Precautions.
2. Remove rear propeller shaft.
3. Put matching mark on the thread edge of electric controlled coupling. The matching mark should be in line with the matching mark on companion flange.
4. Remove the companion flange lock nut, using a flange wrench (commercial service tool).
5. Remove the companion flange.
6. Remove the front oil seal from coupling cover, using a flat-bladed screwdriver.

To install:

7. Apply multi-purpose grease onto oil seal lips, and gear oil onto the circumference of oil seal.
8. Install front oil seal until it becomes flush with the coupling cover. Never reuse the oil seal. Never incline the oil the seal.

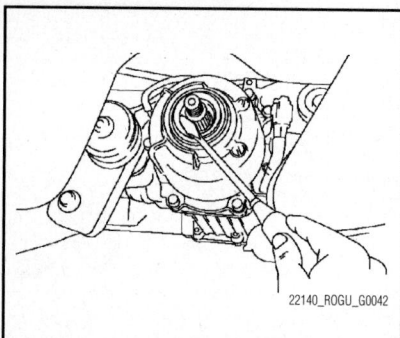

Fig. 15 Remove the front oil seal

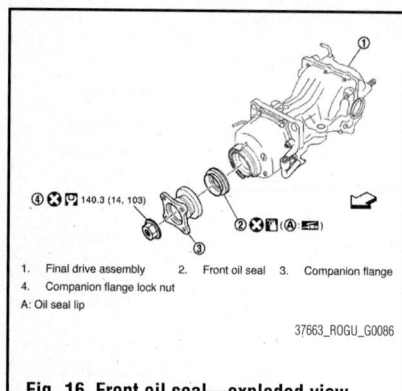

1. Final drive assembly 2. Front oil seal 3. Companion flange
4. Companion flange lock nut
A: Oil seal lip

Fig. 16 Front oil seal—exploded view

9. Align the matchmark of the electric controlled coupling with the matchmark of companion flange, and install the companion flange.

10. Install the companion flange lock nut with a flange wrench (commercial service tool) and tighten to 103 ft. lbs. (140 Nm).

✴✴ WARNING

Never reuse companion flange lock nut.

11. Install the rear propeller shaft.
12. If oil leaks while removing, check oil level after the installation.

REAR PROPELLER SHAFT

REMOVAL & INSTALLATION

See Figures 17 and 18.

1. Shift the transaxle to the neutral position, and then release the parking brake.
2. Remove the following:
   - Muffler assembly
   - Exhaust center pipe
3. Put matchmarks onto propeller shaft flange yoke and final drive and transfer companion flanges. For matching marks, use paint. Never damage propeller shaft flange yoke and transfer companion flange.
4. Loosen mounting nuts of center bearing mounting brackets (upper/lower). Tighten mounting nuts temporarily.

5. Remove propeller shaft assembly fixing bolts and nuts.
6. Remove center bearing mounting bracket fixing nuts.
7. Remove propeller shaft assembly.

✴✴ WARNING

If constant velocity joint was bent during propeller shaft assembly removal, installation, or transportation, its boot may be damaged. Wrap boot interference area to metal part with shop cloth or rubber to protect boot from breakage.

8. Remove clips and center bearing mounting bracket (upper/lower).

To install:

9. Note the following, and install in the reverse order of removal.
   a. Install center bearing mounting bracket (upper) with its arrow mark facing forward.
   b. Adjust position of center bearing mounting bracket (upper), center bearing mounting bracket (lower) sliding back and forth to prevent play in thrust direction of center

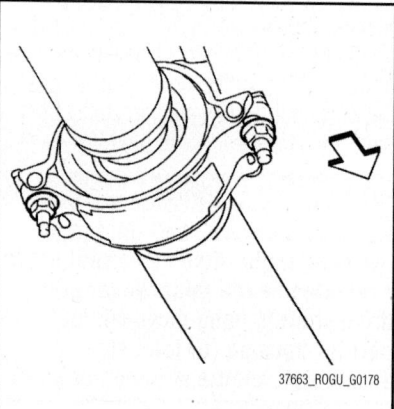

Fig. 17 Loosen mounting nuts of center bearing mounting brackets (upper/lower)

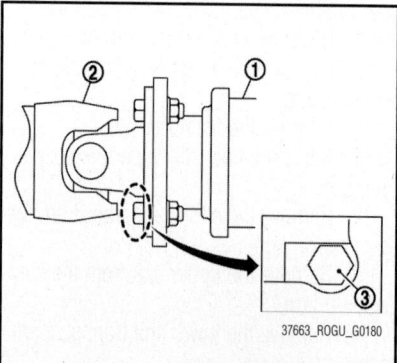

Fig. 18 Make sure that the bolts (3) on the flange side is tightened as shown

bearing insulator. Install center bearing mounting bracket (upper/lower) to vehicle.
   c. Align matching marks to install propeller shaft assembly to final drive and transfer companion flanges.
   d. After assembly, perform a driving test to check propeller shaft vibration. If vibration occurred, separate propeller shaft from final drive. Reinstall companion flange after rotating it by 90, 180, and 270 degrees. Then perform driving test and check propeller shaft vibration again at each point.
   e. After tightening the bolts and nuts to the specified torque, make sure that the bolts on the flange side is tightened.
   f. If propeller shaft assembly or final drive assembly has been replaced, connect them as follows:
   - Face the companion flange mark of the final drive upward. With the mark faced upward, couple the propeller shaft and the final drive so that the matching mark of propeller shaft can be positioned as closest as possible with the matching mark of the final drive companion flange.
   - Tighten mounting bolts and nuts of propeller shaft and final drive to the specified torque.

REAR WHEEL HUB AND BEARING

REMOVAL & INSTALLATION

AWD

See Figures 19 and 20.

1. Remove tires with power tool.
2. Remove wheel sensor from axle housing.

➥Never pull on wheel sensor harness.

3. Remove torque member mounting bolts. Hang torque member not to interfere with work.

➥Never depress brake pedal while brake caliper is removed.

4. Remove disc rotor.

➥Put matching marks on the wheel hub and bearing assembly and the disc rotor before removing the disc rotor.

➥Never drop disc rotor.

5. Remove cotter pin, and then loosen wheel hub lock nut with power tool.

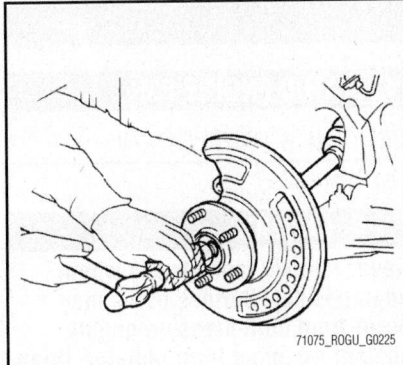

Fig. 19 wheel hub lock nut with a piece of wood. Hammer the wood to disengage wheel hub and bearing assembly from drive shaft. Remove the wheel hub lock nut

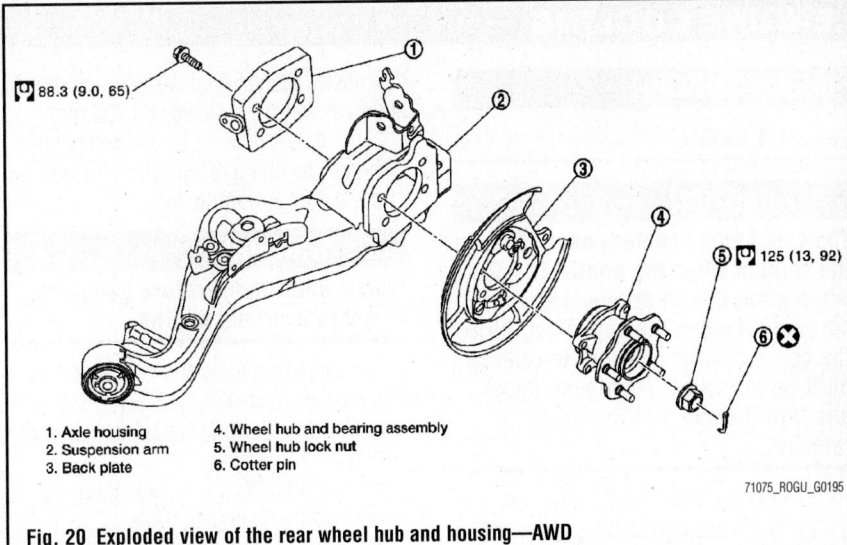

1. Axle housing
2. Suspension arm
3. Back plate
4. Wheel hub and bearing assembly
5. Wheel hub lock nut
6. Cotter pin

71075_ROGU_G0195

Fig. 20 Exploded view of the rear wheel hub and housing—AWD

6. Patch wheel hub lock nut with a piece of wood. Hammer the wood to disengage wheel hub and bearing assembly from drive shaft. Remove the wheel hub lock nut.

✴✴ WARNING
Never place drive shaft joint at an extreme angle. Also be careful not to overextend slide joint.

➡️**ever allow drive shaft to hang down without support for housing (or joint sub-assembly), shaft and the other parts.**

➡️**Use suitable puller, if wheel hub and bearing assembly and drive shaft cannot be separated even after performing the above procedure.**

7. Remove wheel hub and bearing assembly.

To install:

➡️**Align the matching marks made during removal when reusing the disc rotor.**

8. Install wheel hub and bearing assembly.

➡️**Use suitable puller, if wheel hub and bearing assembly and drive shaft cannot be separated even after performing the above procedure.**

➡️**Never allow drive shaft to hang down without support for housing (or joint sub-assembly), shaft and the other parts.**

✴✴ WARNING
Never place drive shaft joint at an extreme angle. Also be careful not to overextend slide joint.

9. Patch wheel hub lock nut with a piece of wood. Hammer the wood to engage wheel hub and bearing assembly from drive shaft. Install the wheel hub lock nut.

10. Tighten wheel hub lock nut with power tool, and then install cotter pin

11. Install disc rotor.

➡️**Align matching marks made during removal on the wheel hub and bearing assembly and the disc rotor before installing the disc rotor.**

➡️**Never drop disc rotor.**

12. Install torque member mounting bolts. Hang torque member not to interfere with work.

➡️**Never depress brake pedal while brake caliper is removed.**

13. Install wheel sensor to axle housing.

➡️**Never pull on wheel sensor harness.**

14. Install tires with power tool.

2WD
See Figure 21.

1. Remove tires with power tool.
2. Remove wheel sensor from hub and bearing assembly.

➡️**Never pull on wheel sensor harness.**

3. Remove torque member mounting bolts with power tool. Hang torque member not to interfere with work.

➡️**Never depress brake pedal while brake caliper is removed.**

4. Remove disc rotor.

➡️**Put matching marks on the wheel hub and bearing assembly and the disc rotor before removing the disc rotor.**

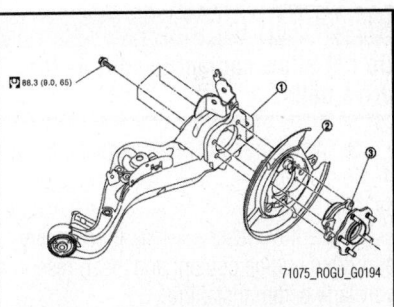

71075_ROGU_G0194

Fig. 21 Exploded view of the rear wheel hub–2WD Suspension arm (1), Back plate (2), Wheel hub and bearing assembly (3)

➡️**Never drop disc rotor.**

5. Remove wheel hub and bearing assembly.

To install:

➡️**Align the matching marks made during removal when reusing the disc rotor.**

6. Install wheel hub and bearing assembly.

7. Install disc rotor.

➡️**Never drop disc rotor.**

➡️**Align matching marks on the wheel hub and bearing assembly and the disc rotor before removing the disc rotor.**

8. Install torque member mounting bolts with power tool. Hang torque member not to interfere with work.

➡️**Never depress brake pedal while brake caliper is removed.**

9. Install wheel sensor to hub and bearing assembly.

➡️**Never pull on wheel sensor harness.**

10. Install tires with power tool.

ENGINE COOLING

ENGINE COOLANT

DRAIN & REFILL

✳✳ CAUTION

To avoid being scalded, never change the coolant when the engine is hot. Wrap a thick cloth around cap and carefully remove the cap. First, turn the cap a quarter of a turn to release built-up pressure. Then push down and turn the cap all the way to remove.

1. Remove engine undercover.
2. Open radiator drain plug at the bottom of radiator and remove the radiator filler cap.

✳✳ WARNING

Do not allow coolant to spill on the drive belts.

3. When draining all of engine coolant in the system, open water drain plugs on cylinder block.
4. Remove reservoir tank if necessary, and drain engine coolant and clean reservoir tank before installing.
5. Check the drained coolant for contaminants such as rust, corrosion or discoloration. If contaminated, flush the engine cooling system.

To refill:

6. Install the radiator drain plug. If the cooling system was drained completely, install the reservoir tank and the cylinder block drain plugs and tighten to specification.
7. Check that each hose clamp has been firmly tightened.
8. Remove air duct assembly, and move electric throttle control actuator aside.
9. Disconnect heater hose.
10. Lift up the heater hose end approximately 100 mm (3.94 in) higher than the height at installation.
11. Fill radiator to specified level.
    a. Pour engine coolant through engine coolant filler neck slowly to allow air in system to escape.
    b. When engine coolant overflows disconnected heater hose, connect heater hose, and continue filling the engine coolant.
12. Refill reservoir tank to "MAX" level line with engine coolant.
13. Install radiator cap.
14. Install air duct assembly and electric throttle control actuator.
15. Warm up engine until opening

thermostat. Standard for warming-up time is approximately 10 minutes at 3,000 rpm.
    a. Check thermostat opening condition by touching radiator hose (lower) to see a flow of warm water.

✳✳ WARNING

Watch water temperature gauge so as not to overheat engine.

16. Stop the engine and cool down to less than approximately 122°F (50°C).
    a. Cool down using fan to reduce the time.
    b. If necessary, refill radiator up to filler neck with engine coolant.
17. Refill reservoir tank to "MAX" level line with engine coolant.
18. Repeat steps 5 through 10 two or more times with radiator cap installed until engine coolant level no longer drops.
19. Check cooling system for leakage with engine running.
20. Warm up the engine, and check for sound of engine coolant flow while running engine from idle up to 3,000 rpm with heater temperature controller set at several position between "COOL" and "WARM". Sound may be noticeable at heater unit.
    a. Repeat step 14 three times.
    b. If sound is heard, bleed air from cooling system by repeating step 5 through 10 until engine coolant level.

ELECTRIC ENGINE FAN

REMOVAL & INSTALLATION
See Figure 22.

✳✳ CAUTION

Never remove radiator cap when engine is hot. Serious burns may occur from high-pressure engine coolant escaping from radiator. Wrap a thick cloth around the radiator cap. Slowly turn it a quarter of a turn to release built-up pressure.

1. Before servicing the vehicle, refer to the Precautions.
2. Remove engine under cover.
3. Drain engine coolant from radiator.
4. Remove air duct (inlet).
5. Remove radiator hose (upper) and reservoir tank hose.
6. Disconnect harness connector from fan motor, and move harness to aside.
7. Remove cooling fan assembly.

✳✳ WARNING

Be careful not to damage or scratch on radiator core when removing.

8. Remove cooling fan mounting nuts, and then remove the left-hand and right-hand cooling fans.

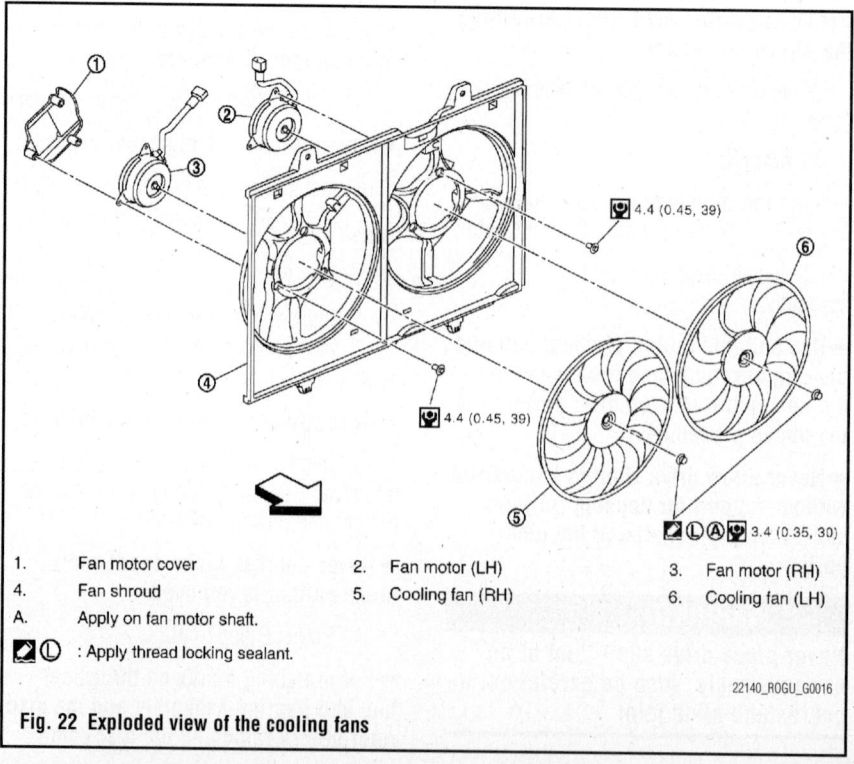

1.	Fan motor cover	2.	Fan motor (LH)	3. Fan motor (RH)
4.	Fan shroud	5.	Cooling fan (RH)	6. Cooling fan (LH)
A.	Apply on fan motor shaft.			

🔧Ⓛ : Apply thread locking sealant.

22140_ROGU_G0016

Fig. 22 Exploded view of the cooling fans

9. Remove fan motor cover and left-hand and right-hand fan motors.

To install:

10. Install in the reverse order of removal and note the following:

a. Add engine coolant and bleed the cooling system.

b. Pressure test engine and check for leaks.

c. Check the fan operation.

RADIATOR

REMOVAL & INSTALLATION

See Figure 23.

✳✳ CAUTION

Never remove radiator cap when engine is hot. Serious burns may occur from high-pressure engine coolant escaping from radiator. Wrap a thick cloth around the radiator cap. Slowly turn it a quarter of a turn to release built-up pressure.

1. Before servicing the vehicle, refer to the Precautions.

2. Remove engine under cover.

3. Drain engine coolant from radiator.

4. Remove air duct (inlet).

5. Remove radiator hose (upper) and reservoir tank hose.

6. Disconnect harness connector from fan motor, and move harness to aside.

7. Remove cooling fan assembly.

8. Remove the lower radiator hose.

9. Remove radiator upper clips by pulling the tabs outside to release the lock.

10. Remove the radiator.

✳✳ WARNING

Be careful not to damage or scratch on radiator core when removing.

To install:

11. Install in the reverse order of removal and note the following:

a. Add engine coolant and bleed the cooling system.

b. Pressure test engine and check for leaks.

THERMOSTAT AND WATER CONTROL VALVE

REMOVAL & INSTALLATION

See Figure 24.

✳✳ CAUTION

Never remove radiator cap when engine is hot. Serious burns may

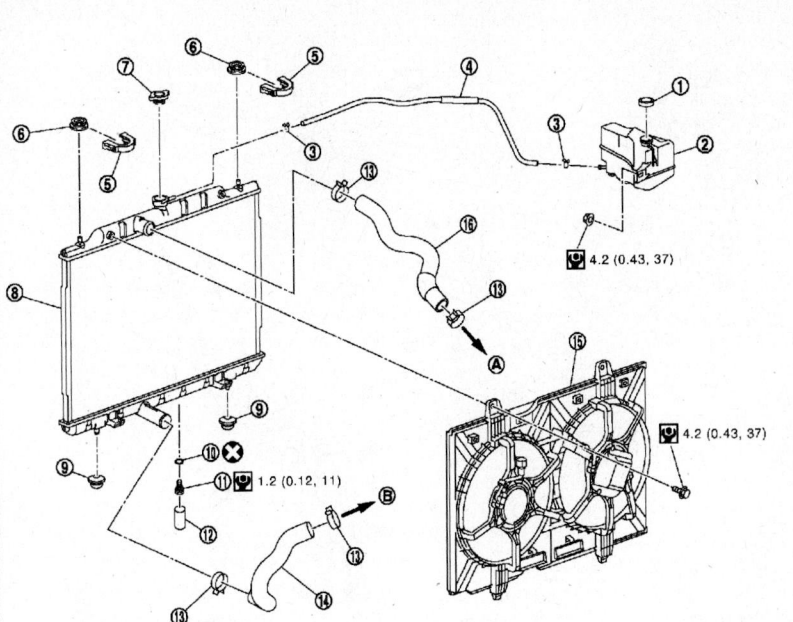

1. Reservoir tank cap
2. Reservoir tank
3. Clamp
4. Reservoir tank hose
5. Radiator upper clip
6. Mounting rubber (upper)
7. Radiator cap
8. Radiator
9. Mounting rubber (lower)
10. O-ring
11. Drain plug
12. Water drain hose
13. Clamp
14. Radiator hose (lower)
15. Cooling fan assembly
16. Radiator hose (upper)
A. To water outlet
B. To water inlet

4.2 (0.43, 37)

4.2 (0.43, 37)

1.2 (0.12, 11)

22140_ROGU_G0159

Fig. 23 Exploded view of the radiator fan and related components

occur from high-pressure engine coolant escaping from radiator. Wrap a thick cloth around the radiator cap. Slowly turn it a quarter of a turn to release built-up pressure.

1. Before servicing the vehicle, refer to the Precautions.

2. Remove the battery.

3. Disconnect engine room harness connectors at unit sides TCM and ECM, and then move it aside.

4. Remove the battery tray.

5. Remove the air duct and air cleaner case assembly.

6. Drain the engine coolant.

7. Disconnect the lower radiator hose at water inlet side.

8. Disconnect water hose at water inlet side (Type 1).

9. Remove water inlet and thermostat.

10. Remove the water control valve with the following procedure:

a. Disconnect radiator hose (upper) at water control valve housing (water outlet) side.

b. Disconnect harness connector from engine coolant temperature sensor.

c. Remove the CVT fluid level gauge and CVT fluid charging pipe.

d. Disconnect water hoses.

e. Disconnect the air fuel ratio sensor 1 and heated oxygen sensor 2 harness connectors, and remove harness clips from heater pipe.

f. Remove the heater pipe and heater hose.

g. After removing the water control valve housing (water outlet), remove water control valve.

To install:

11. Install in the reverse order of removal, noting the following:

a. Refer to the exploded view for tightening specifications.

b. Install thermostat and water control valve with making rubber ring groove fit to thermostat flange and water control valve flange.

➡**The same procedure is applied for installation of thermostat.**

c. Install the thermostat with jiggle valve facing upwards. The position deviation may be within the range of 20 degrees.

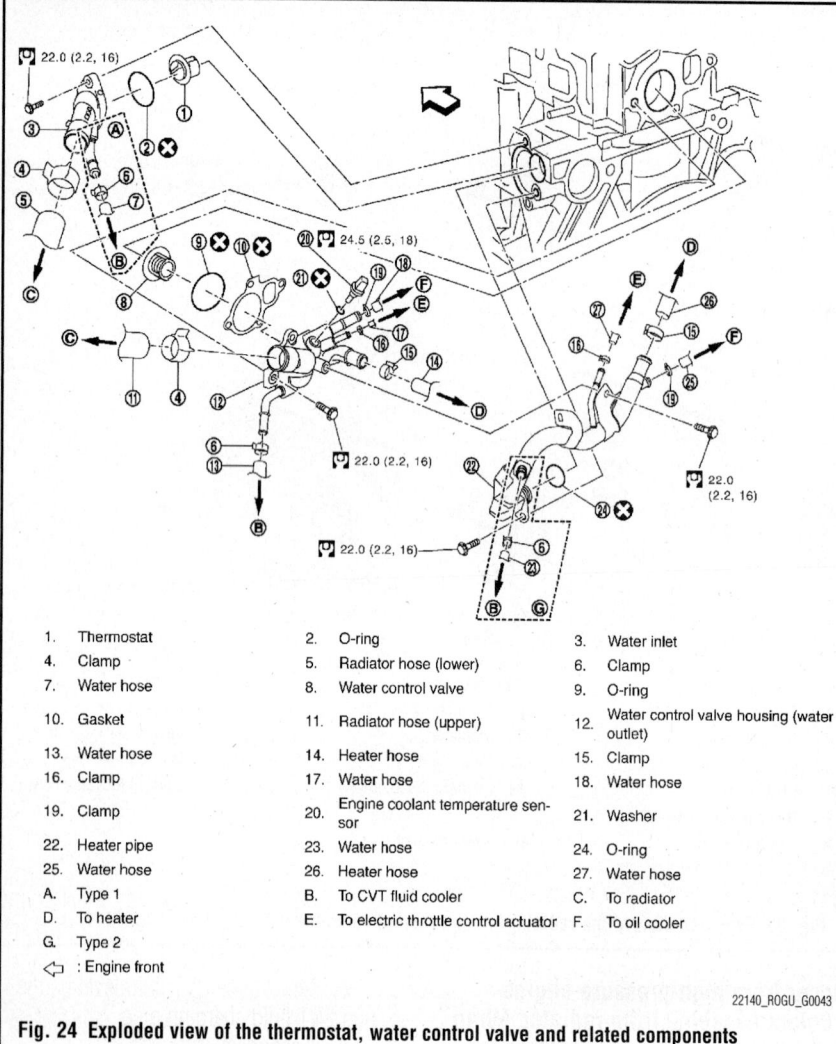

1.	Thermostat	2.	O-ring	3.	Water inlet
4.	Clamp	5.	Radiator hose (lower)	6.	Clamp
7.	Water hose	8.	Water control valve	9.	O-ring
10.	Gasket	11.	Radiator hose (upper)	12.	Water control valve housing (water outlet)
13.	Water hose	14.	Heater hose	15.	Clamp
16.	Clamp	17.	Water hose	18.	Water hose
19.	Clamp	20.	Engine coolant temperature sensor	21.	Washer
22.	Heater pipe	23.	Water hose	24.	O-ring
25.	Water hose	26.	Heater hose	27.	Water hose
A.	Type 1	B.	To CVT fluid cooler	C.	To radiator
D.	To heater	E.	To electric throttle control actuator	F.	To oil cooler
G.	Type 2				
⟨⊐	: Engine front				

22140_ROGU_G0043

Fig. 24 Exploded view of the thermostat, water control valve and related components

1. Before servicing the vehicle, refer to the Precautions.

2. Drain the engine coolant.

3. Remove the following parts:
   • Drive belt
   • Drive belt auto-tensioner
   • Alternator
   • Water pump

4. Engine coolant will leak from the cylinder block, so have a receptacle ready below.

5. Remove water pump housing with the following procedure:

   a. Remove exhaust manifold cover.

   b. Remove oil level gauge and oil level gauge guide.

   c. Remove the mounting bolts for water pipe.

   d. Remove water pump housing.

   e. Remove exhaust manifold and three way catalyst assembly.

   f. Remove the water pipe.

To install:

6. Install in the reverse order of removal, noting the following:

   a. When inserting water pipe end into cylinder block, apply a neutral detergent to O-ring, and then insert it immediately.

   b. Tighten water pump mounting bolts to 18 ft. lbs. (25 Nm).

   c. Refer to the exploded water pump view for additional tightening specification.

d. Install water control valve with the arrow facing up and the frame center part facing upwards. The position deviation may be within the range of 20 degrees.

WATER PUMP

REMOVAL & INSTALLATION

See Figure 25.

❊❊ CAUTION

Never remove radiator cap when engine is hot. Serious burns may occur from high-pressure engine coolant escaping from radiator. Wrap a thick cloth around the radiator cap. Slowly turn it a quarter of a turn to release built-up pressure.

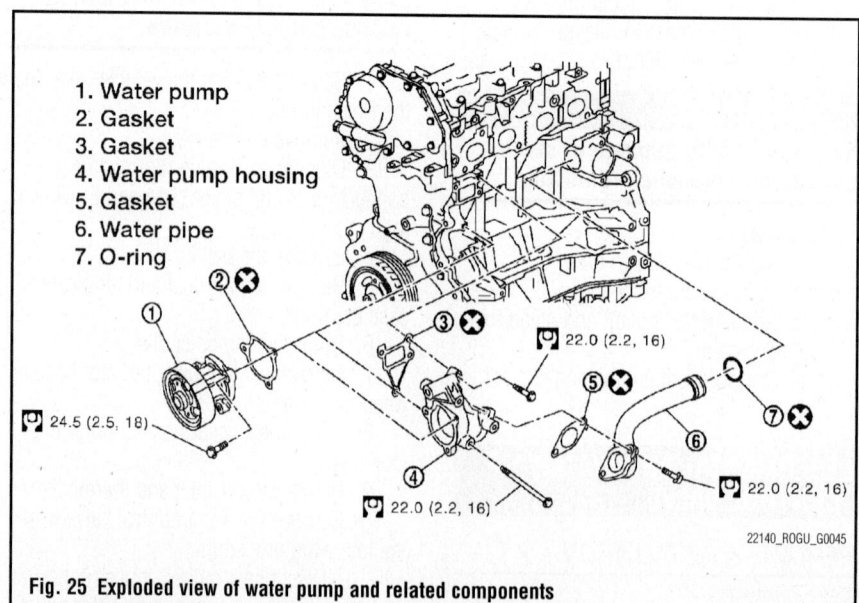

1. Water pump
2. Gasket
3. Gasket
4. Water pump housing
5. Gasket
6. Water pipe
7. O-ring

22140_ROGU_G0045

Fig. 25 Exploded view of water pump and related components

ENGINE ELECTRICAL

BATTERY

REMOVAL & INSTALLATION

1. Remove cover of battery positive terminal.
2. Loosen battery terminal nuts, and disconnect both battery cables from battery terminals.

> ❊❊ **CAUTION**
>
> **When disconnecting, disconnect the battery cable from the negative terminal first.**

3. Remove battery fix frame mounting nuts and battery fix frame.
4. Remove battery.

To install:

5. Install battery.
6. Install battery fix frame mounting nuts and battery fix frame. Tighten fix frame mounting nuts to 48 inch lb. (5.4 Nm).

> ❊❊ **CAUTION**
>
> **When connecting, connect the battery cable to the positive terminal first.**

7. Tighten battery terminal nuts, and connect both battery cables to battery terminals. Tighten battery terminal nuts to 48 inch lb. (5.4 Nm).
8. Install cover of battery positive terminal.
9. Reset electronic systems as necessary.

Battery Terminal With Fusible Link

See Figures 26 and 27.

1. Disconnect the battery cable from the negative terminal.
2. Remove cover of battery positive terminal.
3. Remove harness mounting nut to disconnect harness connectors.
4. Remove fusible link holder mounting nut to remove battery terminal with fusible link.

To install:

5. Install battery terminal with fusible link, and then install fusible link holder mounting nut. Tighten the fusible link holder mounting nut to 120 inch lb. (13.2 Nm).
6. Install harness mounting nut to connect harness connectors. Tighten harness mounting nut to 120 inch lb. (13.2 Nm).
7. Install cover of battery positive terminal.
8. Connect the battery cable to the negative terminal.

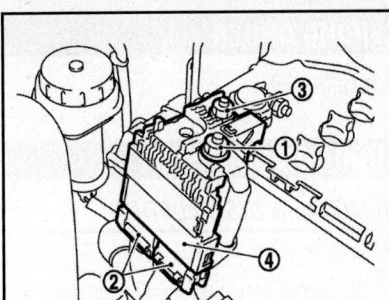

1. Harness mounting nut
2. Harness connectors
3. Fusible link holder mounting nut
4. Battery terminal with fusible link

71075_ROGU_G0226

Fig. 26 Remove harness mounting nut to disconnect harness connectors. Remove fusible link holder mounting nut to remove battery terminal with fusible link.

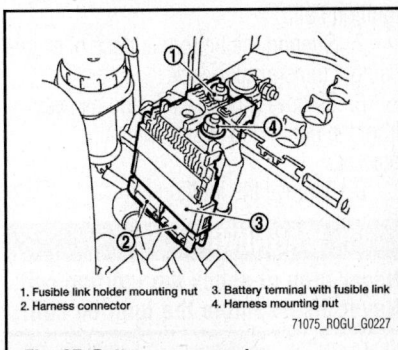

1. Fusible link holder mounting nut
2. Harness connector
3. Battery terminal with fusible link
4. Harness mounting nut

71075_ROGU_G0227

Fig. 27 Battery components

ENGINE ELECTRICAL

ALTERNATOR

REMOVAL & INSTALLATION

See Figure 28.

1. Before servicing the vehicle, refer to the Precautions.
2. Disconnect the battery cable from the negative terminal.
3. Remove drive belt.
4. Disconnect the alternator connector.
5. Remove the "B" terminal nut and "B" terminal harness.
6. Remove harness bracket.
7. Remove the upper alternator mounting bolt.
8. Remove lower alternator mounting bolt.
9. Remove the alternator upward from the vehicle.

To install:

10. Install in the reverse order of removal, noting the following:
   a. Install the alternator and tighten the mounting bolts to 48 ft. lbs. (65 Nm).

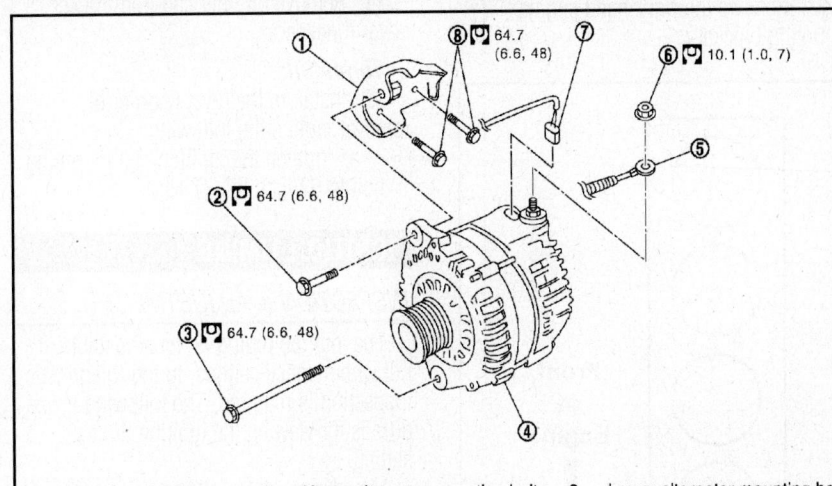

1. Alternator bracket	2. Upper alternator mounting bolt	3. Lower alternator mounting bolt
4. Alternator	5. B terminal harness	6. B terminal nut
7. Alternator connector	8. Alternator bracket mounting bolts	

22140_ROGU_G0046

Fig. 28 Exploded view of the alternator and related mounting components

   b. Install the terminal nut "B" and tighten to 7 ft. lbs. (10 Nm).

   c. Connect the negative battery cable.
   d. Check tension of the drive belt.

ENGINE ELECTRICAL

IGNITION SYSTEM

FIRING ORDER

See Figure 29.

IGNITION COIL PACK

REMOVAL & INSTALLATION

See Figure 30.

1. Before servicing the vehicle, refer to the Precautions.
2. Remove air duct and resonator assembly.
3. Remove the electric throttle control actuator without disconnecting water hose.
4. Loosen the intake manifold mounting bolts and nuts.
5. Remove the intake manifold.
6. Disconnect harness connector from ignition coil.
7. Support the bottom surface of engine using a transmission jack.
8. Remove ground cable and harness from the right-hand engine mounting bracket.
9. Remove the ignition coil.

✴✴ WARNING

Never drop or shock the ignition coil. Never disassemble the ignition coil.

10. Disconnect PCV hose from rocker cover.
11. Remove the right-hand engine mounting insulator.
12. Remove the right-hand engine mounting bracket.

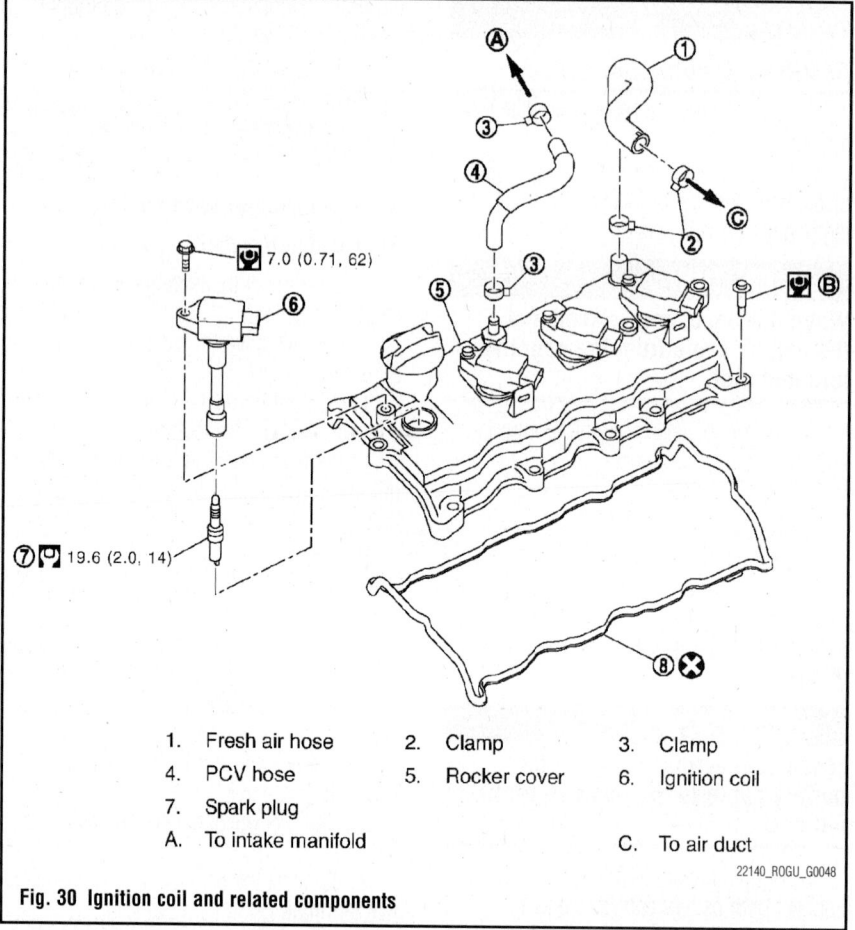

1.	Fresh air hose	2.	Clamp	3.	Clamp
4.	PCV hose	5.	Rocker cover	6.	Ignition coil
7.	Spark plug				
A.	To intake manifold			C.	To air duct

22140_ROGU_G0048

Fig. 30 Ignition coil and related components

13. Remove PCV valve and O-ring from rocker cover, if necessary.
14. Remove oil filler cap from rocker cover if needed.

To install:

15. Install in the reverse order of removal, noting the following:

   a. Tighten the ignition coil mounting bolt to 62 inch lbs. (7 Nm).

IGNITION TIMING

INSPECTION & ADJUSTMENT

The ignition timing is not adjustable. If not within specifications, further diagnostic inspection is required. The following procedure is for viewing the ignition timing setting.

Visually check the air cleaner, intake hoses, ducts, Exhaust Gas Recirculation (EGR) valve operation and electrical connections prior to the adjustment of the ignition timing. Correct or repair any problem as required. Be sure to inspect the throttle valve and Throttle Position (TP) sensor for proper operation.

1. Before servicing the vehicle, refer to the Precautions.
2. Locate the timing marks on the crankshaft pulley and the front of the engine.
3. Clean the timing marks.
4. The ignition timing specifications are as follows:

   • 10–20 degrees Before Top Dead Center (BTDC)

5. Using chalk or white paint, color the mark on the crankshaft pulley and the mark on the scale, which will indicate the correct timing when aligned with the notch on the crankshaft pulley.
6. Attach a tachometer to the engine.
7. Attach a timing light to the engine to number 1 cylinder ignition coil wire.
8. Turn **OFF** all the electrical equipment and accessories.
9. Check to be sure all of the wires clear the fan, then, start the engine and allow it to reach normal operating temperatures.
10. Block the front wheels and set the parking brake. Shift the transmission into **NEUTRAL.**

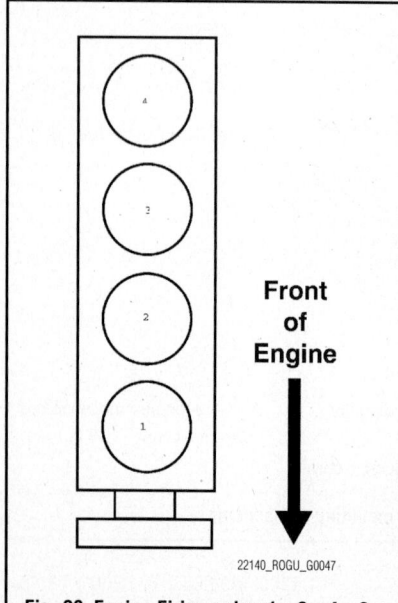

Front of Engine

22140_ROGU_G0047

Fig. 29 Engine Firing order: 1—3—4—2

11. Perform the following procedures:

a. Race the engine at 2000 rpm for about 2 minutes under a no-load condition; be sure all of the accessories are turned **OFF**.

b. Perform on board engine diagnostics and repair any fault code.

c. Race the engine at 2000 rpm for about 2 minutes under a no-load condition.

d. Turn the engine **OFF** and disconnect the TP sensor.

e. Start and race the engine 2–3 times under no-load, then run the engine at idle speed.

12. Aim the timing light at the timing marks. If the marks on the pulley and the engine are aligned when the light flashes, the timing is correct. Turn the engine **OFF** and remove the tachometer and the timing light. If the marks are not in alignment, proceed with the following steps:

a. Turn the engine **OFF**.

b. Check the Camshaft Position (CMP) sensor (PHASE), Crankshaft Position (CKP) sensor (REF) and CKP sensor (POS). Replace if necessary.

c. Check that all the timing chain and gears are correctly aligned.

d. If the ignition timing is still not correct, substitute a known good Electronic Control Module (ECM).

➡ **The ECM may be the cause of the problem, but this is rarely the case.**

e. Turn the engine **OFF** and remove the tachometer and the timing light.

SPARK PLUGS

REMOVAL & INSTALLATION

1. Before servicing the vehicle, refer to the Precautions.

2. Remove air duct and resonator assembly.

3. Remove the electric throttle control actuator without disconnecting water hose.

4. Loosen the intake manifold mounting bolts and nuts.

5. Remove the intake manifold.

6. Disconnect harness connector from ignition coil.

7. Support the bottom surface of engine using a transmission jack.

8. Remove ground cable and harness from the right-hand engine mounting bracket.

9. Remove the ignition coils.

10. Remove the spark plugs.

To install:

11. Install in the reverse order of removal, noting the following:

a. Tighten the spark plugs to 14 ft. lbs. (20 Nm).

b. Tighten the ignition coil mounting bolts to 62 inch lbs. (7 Nm).

ENGINE ELECTRICAL

STARTING SYSTEM

STARTER

REMOVAL & INSTALLATION

2WD Models

See Figure 31.

1. Before servicing the vehicle, refer to the Precautions.

2. Disconnect the battery cable from the negative terminal.

3. Remove the terminal nut and terminal harness.

4. Disconnect "S" connector.

5. Remove the starter motor mounting bolts.

6. Remove the starter motor downward from the vehicle.

To install:

7. Install in the reverse order of removal, noting the following:

a. Tighten the starter mounting bolts to 37 ft. lbs. (50 Nm).

b. Tighten terminal nut "B" carefully to 76 inch lbs. (8.6 Nm).

8. Connect the negative battery cable.

AWD Models

See Figures 31 and 32.

1. Before servicing the vehicle, refer to the Precautions.

2. Disconnect the battery cable from the negative terminal.

3. Remove the front wheel and tire.

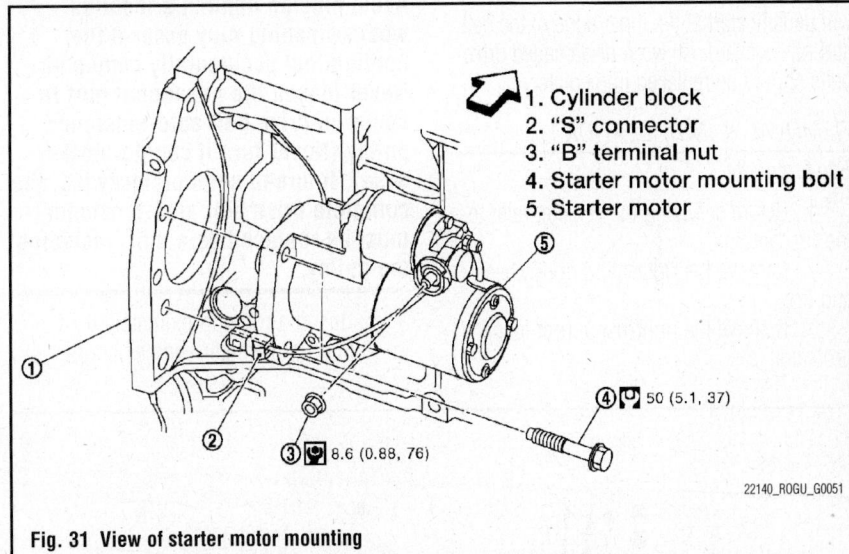

1. Cylinder block
2. "S" connector
3. "B" terminal nut
4. Starter motor mounting bolt
5. Starter motor

④ 50 (5.1, 37)

③ 8.6 (0.88, 76)

22140_ROGU_G0051

Fig. 31 View of starter motor mounting

4. Remove the "B" terminal nut and "B" terminal harness.

5. Disconnect "S" connector.

6. Remove the starter motor mounting bolts.

7. Slide the alternator out and remove.

To install:

8. Install in the reverse order of removal, noting the following:

a. Tighten the starter mounting bolts to 37 ft. lbs. (50 Nm).

b. Tighten the terminal nut "B" carefully to 76 inch lbs. (9 Nm).

c. Connect the negative battery cable.

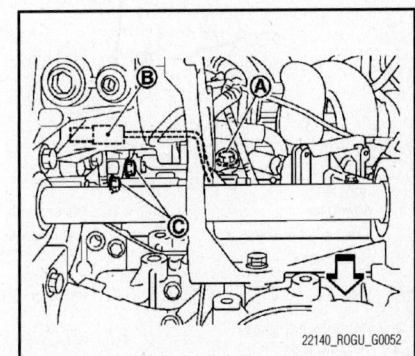

22140_ROGU_G0052

Fig. 32 Remove the terminal nut (A), connector (B) and mounting bolts (C)

ENGINE MECHANICAL

➡Disconnecting the negative battery cable may interfere with the functions of the on board computer systems and may require the computer to undergo a relearning process, once the negative battery cable is reconnected.

ACCESSORY DRIVE BELT SYSTEM

ADJUSTMENT

Belt tension is not manually adjustable, it is automatically adjusted by the drive belt auto-tensioner.

BELT ROUTINGS

See Figure 33.

INSPECTION

Inspect the drive belt for signs of glazing or cracking. A glazed belt will be perfectly smooth from slippage, while a good belt will have a slight texture of fabric visible. Cracks will usually start at the inner edge of the belt and run outward. All worn or damaged drive belts should be replaced immediately.

REMOVAL & INSTALLATION

See Figure 34.

1. Before servicing the vehicle, refer to the Precautions.
2. Remove the right-hand front wheel and tire.
3. Remove the right-hand front fender protector.

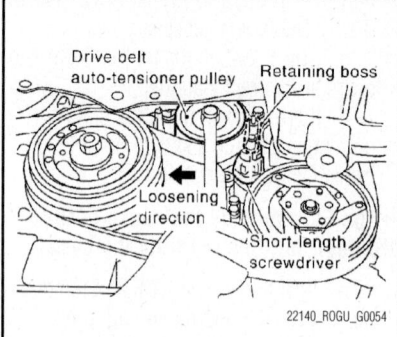

Fig. 34 Drive belt removal direction shown

4. Hold the hexagonal part in center of drive belt auto-tensioner pulley with a box wrench securely. Then move the wrench handle in the direction of arrow (loosening direction of tensioner).

⁕⁕ WARNING

Avoid placing hand in a location where pinching may occur if the holding tool accidentally comes off. Never loosen the hexagonal part in center of drive belt auto-tensioner pulley (Never turn it counterclockwise). If turned counterclockwise, the complete drive belt auto-tensioner must be replaced as a unit, including the pulley.

5. Insert a rod approximately 0.24 in. (6 mm) in diameter such as short-length screwdriver into the hole of the retaining boss to fix drive belt auto-tensioner pulley.

6. Loosen drive belt from water pump pulley in sequence, and remove it.

To install:

7. Hold the hexagonal part in center of drive belt auto-tensioner pulley with a box wrench securely. Then move the wrench handle in the direction of arrow (loosening direction of tensioner).

8. Insert a rod approximately 0.24 in. (6 mm) in diameter such as short-length screwdriver into the hole of retaining boss to fix drive belt auto-tensioner pulley.

9. Hook drive belt onto all pulleys except for water pump, and then onto water pump pulley finally.

10. Confirm drive belt is completely set to pulleys.

11. Release the drive belt auto-tensioner, and apply tension to drive belt.

12. Turn the crankshaft pulley clockwise several times to equalize tension between each pulley.

13. Confirm tension of drive belt at indicator (notch on fixed side) is within the possible use range.

AIR CLEANER ASSEMBLY

REMOVAL & INSTALLATION

See Figure 35.

1. Before servicing the vehicle, refer to the Precautions.
2. Remove the air duct (inlet).
3. Remove the battery.
4. Disconnect the harness connectors and remove the bracket.
5. Disconnect the mass air flow sensor harness connector.
6. Disconnect the fresh air hose.
7. Remove the air cleaner case and mass air flow sensor assembly, air duct and resonator assembly and air duct disconnecting their joints. Add matchmarks if necessary for easier installation.
8. Remove the mass air flow sensor from air cleaner case, if necessary.

To install:

9. Install in the reverse order of removal. Align marks. Attach each joint. Screw clamps firmly.

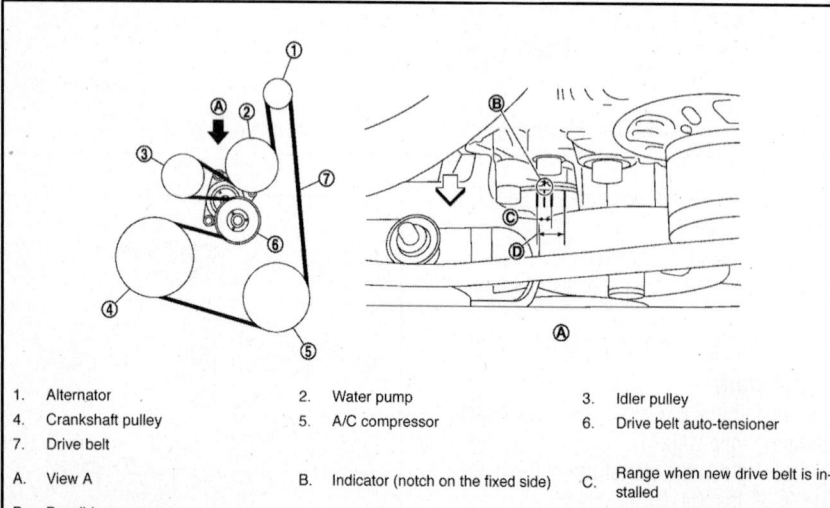

1.	Alternator	2.	Water pump	3.	Idler pulley
4.	Crankshaft pulley	5.	A/C compressor	6.	Drive belt auto-tensioner
7.	Drive belt				
A.	View A	B.	Indicator (notch on the fixed side)	C.	Range when new drive belt is installed
D.	Possible use range				
⟵	: Engine front				

Fig. 33 Accessory drive belt

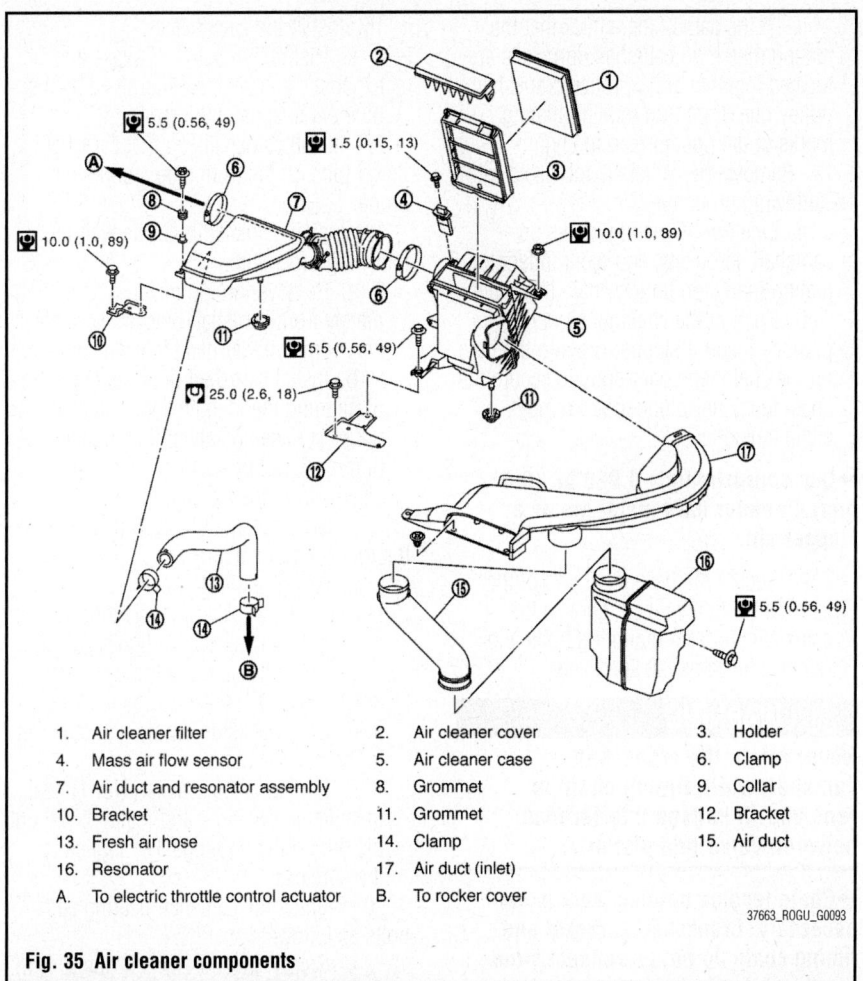

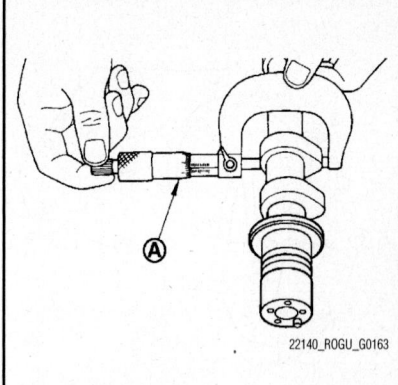

Fig. 35 Air cleaner components

1. Air cleaner filter
2. Air cleaner cover
3. Holder
4. Mass air flow sensor
5. Air cleaner case
6. Clamp
7. Air duct and resonator assembly
8. Grommet
9. Collar
10. Bracket
11. Grommet
12. Bracket
13. Fresh air hose
14. Clamp
15. Air duct
16. Resonator
17. Air duct (inlet)
A. To electric throttle control actuator
B. To rocker cover

37663_ROGU_G0093

Fig. 37 Measure the camshaft cam height with a micrometer (A)

22140_ROGU_G0163

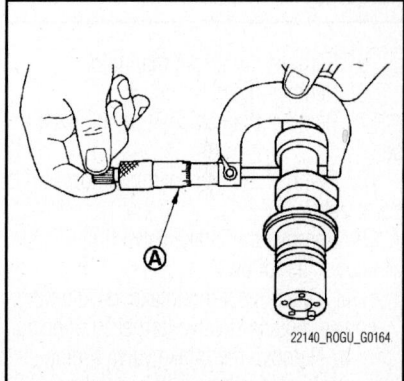

Fig. 38 Measure the camshaft Journal Diameter with a micrometer (A)

22140_ROGU_G0164

CAMSHAFT AND VALVE LIFTERS

INSPECTION

Camshaft Runout

See Figure 36.

1. Put V-block on a precise flat table, and support No. 2 and 5 journal of camshaft.

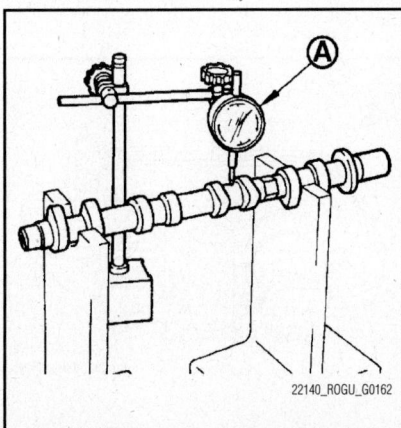

22140_ROGU_G0162

Fig. 36 Set dial indicator (A) vertically to No. 3 journal

2. Set dial indicator vertically to No. 3 journal.

3. Turn camshaft to one direction with hands, and measure the camshaft runout on dial indicator (Total indicator reading).

4. If out of standard, replace camshaft.

Camshaft Cam Height

See Figure 37.

1. Measure the camshaft cam height with a micrometer.

2. If out of standard, replace camshaft.

Camshaft Journal Diameter

See Figure 38.

1. If out of standard, replace camshaft.

Camshaft Bracket Inner Diameter

See Figure 39.

1. Tighten camshaft bracket bolts with specified torque.

2. Measure the inner diameter of camshaft bracket with an inside micrometer.

Valve Lifter Outer Diameter

1. Measure the outer diameter of valve lifter with a micrometer.

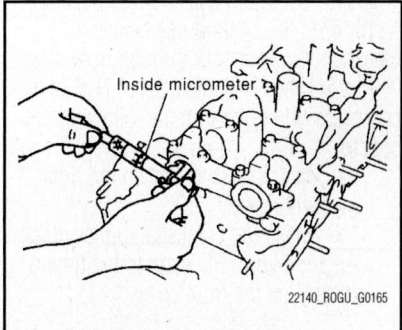

Fig. 39 Measure the inner diameter of camshaft bracket with an inside micrometer

22140_ROGU_G0165

2. Specifications for outer diameter are 1.3378–1.3382 inches (33.98–33.99 mm).

REMOVAL & INSTALLATION

See Figures 40 through 43.

1. Before servicing the vehicle, refer to the Precautions.

2. Disconnect the negative battery cable.

3. Relieve the fuel system pressure.

4. Drain the coolant from the engine and radiator.

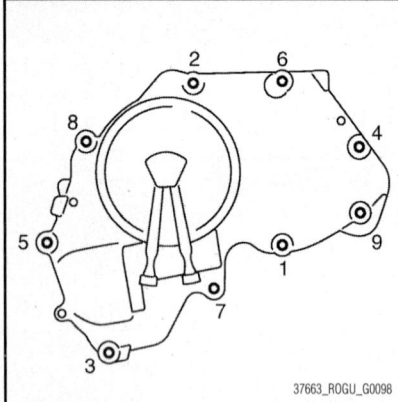

Fig. 40 Intake valve timing control cover bolt tightening sequence—reverse the sequence for removal

5. Remove the intake manifold.

6. Remove the valve cover.

7. Remove the camshaft position sensor.

8. Remove the camshaft position sensor bracket.

9. Remove the intake valve timing control cover, as follows:

   a. Disconnect the intake valve timing control solenoid valve harness connector.

   b. Remove the intake valve timing control solenoid valve, if necessary.

   c. Loosen the bolts.

   d. Use a seal cutter [SST: KV10111100 (J-37228)] or equivalent tool to cut liquid gasket for removal.

10. Pull the chain guide between camshaft sprockets out through front cover.

11. Set the No. 1 cylinder at TDC on its compression stroke with the following procedure:

   a. Open splash guard on the right-hand undercover.

   b. Rotate the crankshaft pulley clockwise and align TDC mark to the timing indicator on the front cover.

   c. At the same time, check that the mating marks on camshaft sprockets are located together. If not, rotate crankshaft pulley one more turn to align mating marks to the positions in the figure.

12. Remove the camshaft sprockets with the following procedure:

   a. Line up the mating marks on camshaft sprockets, and paint indelible mating marks on timing chain link plate.

   b. Push in the chain tensioner plunger. Insert a stopper pin into the hole on the chain tensioner body to secure the chain tensioner plunger and remove chain tensioner.

➡**Use approximately 0.020 in. (0.5 mm) diameter hard metal pin as a stopper pin.**

   c. Secure the hexagonal part of the camshaft with a wrench. Loosen the camshaft sprocket mounting bolts and remove the camshaft sprockets.

✳✳ WARNING

Never rotate the crankshaft or camshaft while timing chain is removed. It causes interference between valve and piston.

➡**Chain tension holding work is not necessary. Crankshaft sprocket and timing chain do not disconnect structurally while front cover is attached.**

13. Loosen the mounting bolts, and remove camshaft brackets and camshafts.

   a. Remove camshaft bracket (No. 1) by slightly tapping it with a plastic hammer.

14. Remove the valve lifters.

   a. Identify installation positions, and store them without mixing them up.

To install:

15. Install the valve lifters in the original positions.

16. Install the camshafts.

   a. Distinction between intake and exhaust camshafts is determined by the different shapes of the rear end.

17. Install camshafts so that camshaft dowel pins on the front side are in their proper position.

18. Install camshaft brackets with the following procedure:

   a. Remove foreign material completely from camshaft bracket backside and from cylinder head installation face.

   b. Install camshaft brackets (No. 2 to 5) aligning the identification marks on upper surface. Install so that identification mark can be correctly read when viewed from the exhaust side.

19. Install camshaft bracket (No. 1) with the following procedure:

   a. Apply liquid gasket to camshaft bracket (No. 1). After installation, be sure to wipe off any excessive liquid gasket.

   b. Apply liquid gasket to camshaft bracket (No. 1) contact surface on the front cover backside. Apply liquid gasket to the outside of bolt hole on front cover.

   c. For camshaft bracket (No. 1) near installation position, and install it without disturbing the liquid gasket applied to the surfaces.

20. Tighten the camshaft bearing caps bolts as follows:

   a. Step 1, bolts 9–11: 1 ft. lbs (2 Nm).

   b. Step 2, bolts 1–8: 1 ft. lbs (2 Nm).

   c. Step 3, bolts 1–11: 4 ft. lbs. (6 Nm).

   d. Step 4, bolts 1–11: 8 ft. lbs. (10 Nm).

   e. After tightening mounting bolts of camshaft brackets, be sure to wipe off excessive liquid gasket from the following:

- Mating surface of rocker cover
- Mating surface of front cover (When installed without front cover)

21. Install the camshaft sprockets.

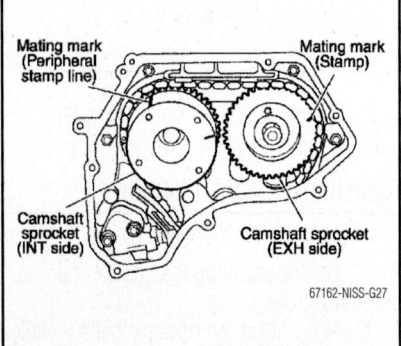

Fig. 41 Camshaft sprocket alignment marks

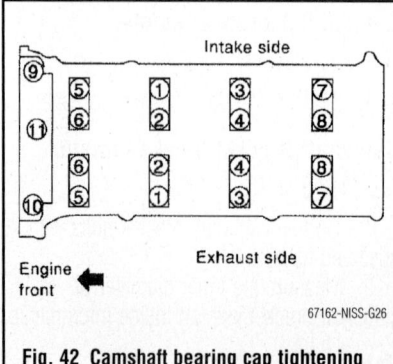

Fig. 42 Camshaft bearing cap tightening sequence—reverse the sequence for removal

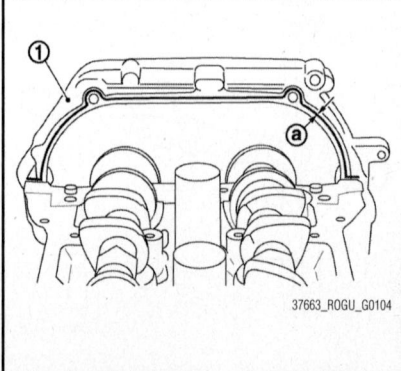

Fig. 43 1: Front Cover, a: 0.134–0.173 in. (3.4–4.4mm)

a. Align the mating marks on each camshaft sprocket with the ones painted on timing chain link plate during removal.

b. Aligned mating marks could slip. Therefore, after matching them, hold the timing chain in place by hand. Before and after installing chain tensioner, check again that mating marks have not slipped.

Before installation of chain tensioner, it is possible to re-match the marks on timing chain with the ones on each sprocket.

22. Install the chain tensioner.

a. After installation, pull the stopper pin off completely, and check that chain tensioner plunger is released.

23. Install the chain guide.

24. Install the intake valve timing control cover with the following procedure:

a. Install the intake valve timing control solenoid valve to intake valve timing control cover if removed.

b. Install the oil rings to the camshaft sprocket (INT) insertion points on back-side of intake valve timing control cover.

c. Install new O-ring to front cover.

d. Apply liquid gasket with a tube presser to intake valve timing control cover.

➡ **Attaching should be done within 5 minutes after liquid gasket application.**

e. Tighten the mounting bolts in numerical order.

25. Install the camshaft position sensor bracket.

a. Apply liquid gasket with a tube presser the camshaft position sensor bracket. After installation, be sure to wipe off any excessive liquid gasket leaking from part "b".

➡ **Attaching should be done within 5 minutes after liquid gasket application.**

b. Tighten mounting bolts in numerical order.

26. Install camshaft position sensor.

27. Inspect and adjust valve clearance.

28. Install the valve cover.

29. Install the intake manifold.

30. Connect the negative battery cable.

31. Fill the cooling system.

32. Start the vehicle, check for leaks and repair if necessary.

CRANKSHAFT DAMPER

REMOVAL & INSTALLATION

See Figures 44 and 45.

1. Before servicing the vehicle, refer to the Precautions.

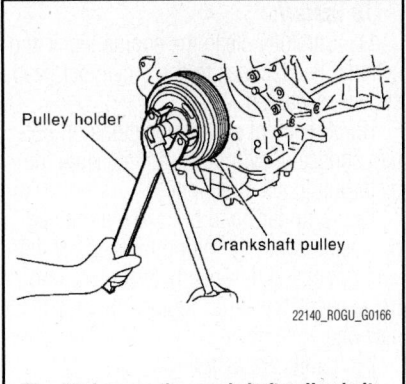

Fig. 44 Loosen the crankshaft pulley bolt

2. Remove engine under cover.

3. Remove front tire.

4. Remove front fender protector.

5. Fix crankshaft pulley with a pulley holder (commercial service tool), loosen crankshaft pulley bolt, and locate bolt seating surface at 0.39 inch (10 mm) from its original position.

6. Attach a pulley puller (commercial service tool) in the M 6 thread hole on crankshaft pulley, and then remove crankshaft pulley.

To install:

7. Secure crankshaft pulley with a pulley holder (commercial service tool), and tighten crankshaft pulley.

8. Install front oil seal so that each seal lip is oriented properly.

9. Press-fit front oil seal until it is flush with front end surface of front cover using a suitable drift with outer diameter 2.20 inches (56 mm) and inner diameter 1.89 inches (48 mm).

✳✳ WARNING

Be careful not to damage front cover and crankshaft. Press-fit oil seal straight to avoid causing burrs or tilting.

10. Insert crankshaft pulley by aligning with crankshaft key.

11. When inserting crankshaft pulley with a plastic hammer, tap on its center portion (not circumference).

12. Secure crankshaft pulley with a pulley holder (commercial service tool).

13. Perform angle tightening with the following procedure:

a. Apply new engine oil to thread and seat surfaces of crankshaft pulley bolt.

b. Tighten crankshaft pulley bolt to 31 ft. lbs. (42 Nm).

c. Put a paint mark on crankshaft pulley, mating with any one of six easy to recognize angle marks on bolt flange.

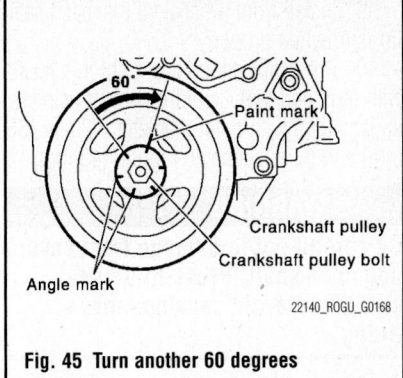

Fig. 45 Turn another 60 degrees clockwise

d. Turn another 60 degrees clockwise (angle tightening).

14. Check the tightening angle with movement of one angle mark.

15. Install all removed parts in the reverse order of removal.

CRANKSHAFT FRONT SEAL

REMOVAL & INSTALLATION

See Figure 46.

1. Before servicing the vehicle, refer to the Precautions.

2. Remove the engine under cover.

3. Remove front tire.

4. Remove front fender protector.

5. Remove the drive belt.

6. Remove the crankshaft pulley.

7. Remove the front oil seal with suitable tool.

To install:

8. Apply new engine oil to new front oil seal joint surface and seal lip.

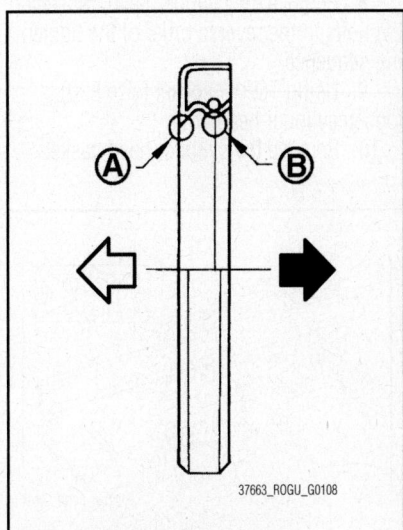

Fig. 46 A: Dust seal lip, B: Oil seal lip, White arrow: Engine outside, Black arrow: Engine inside.

9. Install front oil seal so that each seal lip is oriented properly.

10. Press-fit front oil seal until it is flush with front end surface of front cover using a suitable drift with outer diameter 2.20 in. (56 mm) and inner diameter 1.89 in. (48 mm).

✳✳ WARNING

Be careful not to damage front cover and crankshaft. Press-fit oil seal straight to avoid causing burrs or tilting.

11. Install all removed parts in the reverse order of removal.

CYLINDER HEAD

REMOVAL & INSTALLATION

See Figures 47 through 50.

1. Before servicing the vehicle, refer to the Precautions.
2. Release fuel pressure.
3. Drain the engine coolant and engine oil.
4. Remove the following components and related parts:
   - Exhaust manifold and three way catalyst assembly
   - Intake manifold and fuel tube assembly
   - Water control valve and water control valve housing (water outlet)
5. Remove the front cover and timing chain.
6. Remove the camshafts.
7. Securely support bottom of cylinder block with a jack or equivalent tool, and release the hoist that was supporting it.
8. Remove the cylinder head, loosening the bolts in the reverse order of the tightening sequence.
9. Using TORX® socket (size E20), loosen cylinder head bolts.
10. Remove the cylinder head gasket.

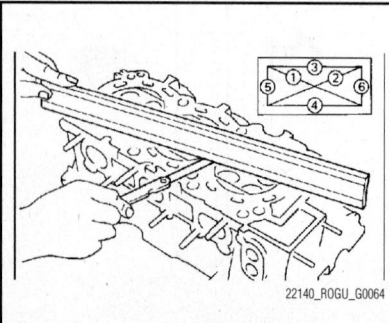

Fig. 47 Cylinder head distortion measurement locations

To install:

11. Carefully clean the engine block and cylinder head, check to see if cylinder head surface is warped.

12. At each of several locations on bottom surface of cylinder head, measure the distortion in six directions.

13. Cylinder head bolts are tightened by plastic zone tightening method. Whenever the size difference between (A) and (B) exceeds the limit, replace them with new one.

14. Limits are as follows:
   - Limit ("B" minus "A") : 0.0091 in. (0.23 mm)
   - c: 2.165 in. (55 mm)
   - d: 0.472 in. (12 mm)
15. Install the cylinder head gasket.
16. Tighten cylinder head bolts in numerical order with the following procedure, and install cylinder head.

✳✳ WARNING

If cylinder head bolts are reused, check their outer diameters before installation.

17. Apply new engine oil to threads and seating surface of mounting bolts.
18. Tighten all bolts to 37 ft. lbs. (50 Nm).

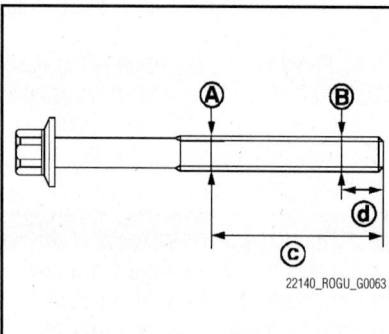

Fig. 48 Cylinder bolt measurement location view

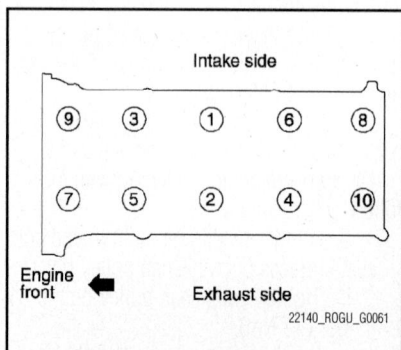

Fig. 49 Tighten cylinder head bolts in numerical order as shown

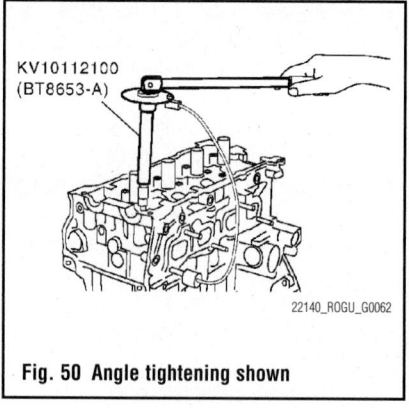

Fig. 50 Angle tightening shown

19. Turn all bolts 60 degrees clockwise (angle tightening).
20. Completely loosen.
21. In this step, loosen bolts in reverse order of that indicated in the figure.
22. Tighten all bolts as follows to 29 ft. lbs. (39 Nm):
   a. Turn all bolts 75 degrees clockwise (angle tightening).
   b. Turn all bolts 75 degrees clockwise again (angle tightening).

✳✳ WARNING

Check and confirm the tightening angle by using an angle wrench (SST) or protractor. Avoid judgment by visual inspection without the tool.

23. Installation is in the reverse order of removal procedure after this step.

EXHAUST MANIFOLD

REMOVAL & INSTALLATION

See Figures 51 and 52.

1. Remove air fuel ratio sensor 1 and heated oxygen sensor 2 with the following procedure:
   a. Disconnect harness connector of air fuel ratio sensor 1 and heated oxygen sensor 2 and harness from bracket and middle clamp.
   b. Using heated oxygen sensor wrench [SST: KV10117100 (J-3647-A)], remove air fuel ratio sensor 1 and heated oxygen sensor 2.

✳✳ WARNING

Be careful not to damage air fuel ratio sensor 1 and heated oxygen sensor 2.

➡Discard any air fuel ratio sensor 1 and heated oxygen sensor 2 which has been dropped onto a hard surface such as a concrete floor. Replace with a new one.

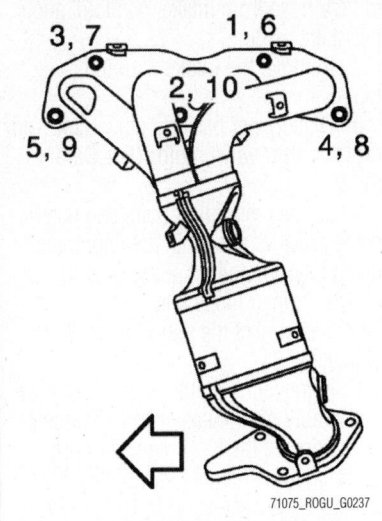

Fig. 51 Loosen nuts in reverse order as shown to remove exhaust manifold and three way catalyst assembly. Disregard No. 6 to 10 when loosening. Arrow=Engine front

➡The accompanying illustration shows air fuel ratio sensor 1 as an example.

2. Remove exhaust front tube.
3. Remove alternator.
4. Remove exhaust manifold cover (upper).
5. Loosen nuts in reverse order to remove exhaust manifold and three way catalyst assembly.

➡Disregard No. 6 to 10 when loosening.

6. Remove gasket.

➡**Cover engine openings to avoid entry of foreign materials.**

7. Remove exhaust manifold cover (lower) and three way catalyst cover from exhaust manifold and three way catalyst assembly.

To install:
Exhaust Manifold
8. If stud bolts were removed, install them and tighten to 132 inch lb. (14.7 Nm).
9. Tighten nuts in numerical order.

➡**No. 6 to 10 mean double tightening of bolts No. 1 and 5.**

Air Fuel Ratio Sensor 1, Heated Oxygen Sensor 2

➡**Before installing a new air fuel ratio sensor 1 and heated oxygen sensor 2, clean exhaust system threads using heated oxygen sensor thread cleaner and apply anti-seize lubricant (commercial service tool: J-43897-18 or J-43897-12).**

⁂ **WARNING**

Never over torque the air fuel ratio sensor 1 and heated oxygen sensor 2. Doing so may cause damage to the air fuel ratio sensor 1 and heated oxygen sensor 2, resulting in the "MIL" coming on.

➡**Prevent rust preventives from adhering to the sensor body.**

10. Install exhaust manifold cover (lower) and three way catalyst cover from exhaust manifold and three way catalyst assembly.
11. Install gasket.

➡**Cover engine openings to avoid entry of foreign materials.**

12. Tighten nuts to install exhaust manifold and three way catalyst assembly.

➡**Disregard No. 6 to 10 when tightening.**

13. Install exhaust manifold cover (upper).
14. Install alternator.
15. Install exhaust front tube.

⁂ **WARNING**

Be careful not to damage air fuel ratio sensor 1 and heated oxygen sensor 2.

➡**Discard any air fuel ratio sensor 1 and heated oxygen sensor 2 which has been dropped onto a hard surface such as a concrete floor. Replace with a new one.**

➡**The accompanying illustration shows air fuel ratio sensor 1 as an example**

16. Install air fuel ratio sensor 1 and heated oxygen sensor 2 with the following procedure:
    a. Using heated oxygen sensor wrench [SST: KV10117100 (J-3647-A)], install air fuel ratio sensor 1 and heated oxygen sensor 2.
    b. Connect harness connector of air fuel ratio sensor 1 and heated oxygen sensor 2 and harness to bracket and middle clamp.

FLYWHEEL

REMOVAL & INSTALLATION
See Figure 53.

1. Before servicing the vehicle, refer to the Precautions.
2. Disconnect the negative battery cable.
3. Remove the transaxle assembly.
4. Secure the flywheel and remove the mounting bolts.
5. Remove the flywheel.

To install:
6. Install the flywheel.
7. Secure the flywheel and tighten in sequence the mounting bolts to 80 ft. lbs. (108 Nm).

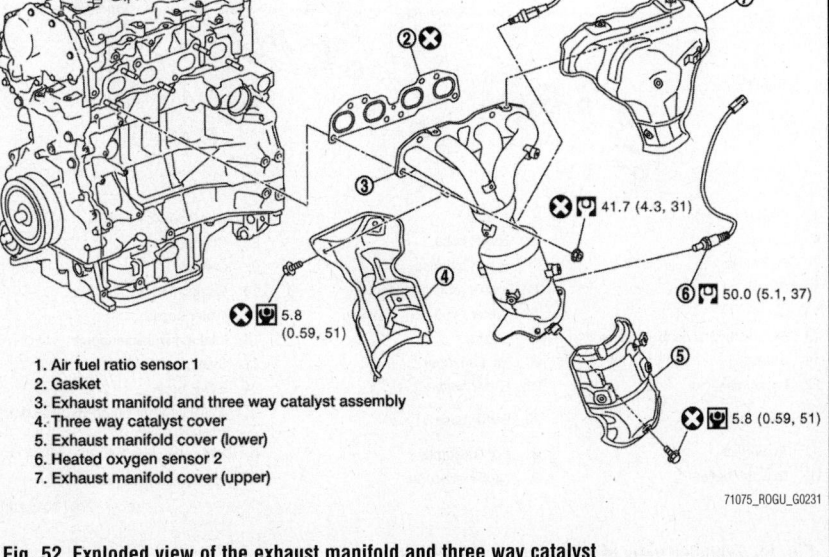

1. Air fuel ratio sensor 1
2. Gasket
3. Exhaust manifold and three way catalyst assembly
4. Three way catalyst cover
5. Exhaust manifold cover (lower)
6. Heated oxygen sensor 2
7. Exhaust manifold cover (upper)

Fig. 52 Exploded view of the exhaust manifold and three way catalyst

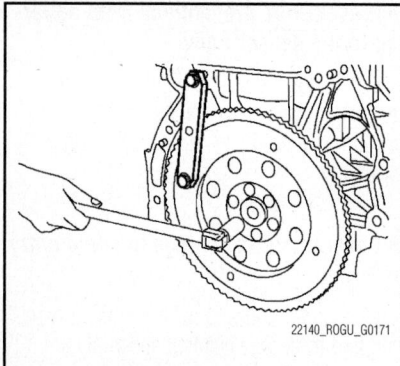

Fig. 53 Secure the flywheel and tighten mounting bolts

8. Install the transaxle assembly.
9. Connect the negative battery cable.

INTAKE MANIFOLD

REMOVAL & INSTALLATION

See Figures 54 through 56.

1. Before servicing the vehicle, refer to the Precautions.
2. Release the fuel pressure.
3. Remove the cowl top cover.
4. Remove the air cleaner case and mass air flow sensor assembly and air duct and resonator assembly.
5. Remove the electric throttle control actuator with the following procedure:
    a. Disconnect the harness connector.
    b. Loosen the mounting bolts and remove electric throttle control actuator and gasket.

❋❋ WARNING

Handle carefully to avoid any shock to electric throttle control actuator. Never disassemble.

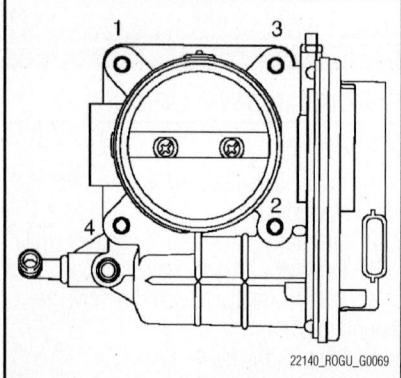

Fig. 54 Loosen mounting bolts in reverse order as shown in the figure, and remove electric throttle control actuator and gasket

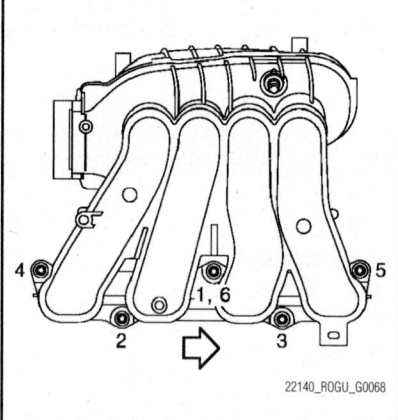

Fig. 55 Loosen mounting bolts and nuts in reverse order as shown in the figure

➡**When removing only the intake manifold, move electric throttle control actuator without disconnecting the water hose.**

6. Disconnect harness, vacuum hose and PCV hose from intake manifold, and move them aside.
7. Remove the intake manifold support.
8. Disconnect harness connector from tumble control valve motor (For California).
9. Loosen mounting bolts and nuts in reverse order as shown in the figure, and remove intake manifold and gasket. Disregard No. 6 when loosening.
10. Disconnect the sub-harness from fuel injector.
11. Remove the fuel tube and fuel injector assembly from intake manifold adaptor.
12. Remove the EVAP canister purge volume control solenoid valve from intake manifold, if necessary.

To install:

13. Install in the reverse order of removal, noting the following:
    a. If the stud bolts were removed, install them and tighten to 83 inch lbs. (9 Nm).
    b. No. 6 refers to double tightening of bolt No. 1.

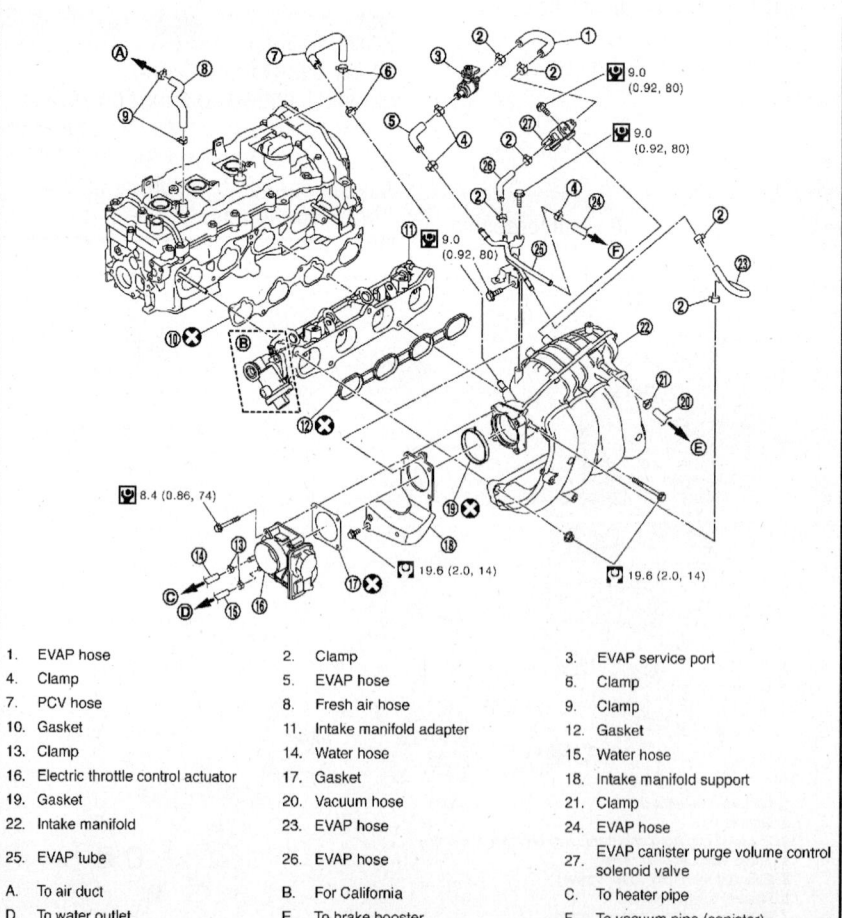

1. EVAP hose	2. Clamp	3. EVAP service port
4. Clamp	5. EVAP hose	6. Clamp
7. PCV hose	8. Fresh air hose	9. Clamp
10. Gasket	11. Intake manifold adapter	12. Gasket
13. Clamp	14. Water hose	15. Water hose
16. Electric throttle control actuator	17. Gasket	18. Intake manifold support
19. Gasket	20. Vacuum hose	21. Clamp
22. Intake manifold	23. EVAP hose	24. EVAP hose
25. EVAP tube	26. EVAP hose	27. EVAP canister purge volume control solenoid valve
A. To air duct	B. For California	C. To heater pipe
D. To water outlet	E. To brake booster	F. To vacuum pipe (canister)

Fig. 56 Exploded view of intake manifold—Rogue

c. Tighten the mounting bolts equally to 14 ft. lbs. (20 Nm), and diagonally in several steps and in numerical order in the reverse order of removal.

d. Tighten the electric throttle body to 74 inch lbs. (8 Nm).

e. Perform the "Throttle Valve Closed Position Learning" when harness connector of electric throttle control actuator is disconnected.

f. Perform the "Idle Air Volume Learning" and "Throttle Valve Closed Position Learning" when electric throttle control actuator is replaced.

THROTTLE VALVE CLOSED POSITION LEARNING

The Throttle Valve Closed Position Learning is a function of ECM to learn the fully closed position of the throttle valve by monitoring the throttle position sensor output signal. It must be performed each time harness connector of electric throttle control actuator or ECM is disconnected.

1. Check that accelerator pedal is fully released.
2. Turn ignition switch ON.
3. Turn ignition switch OFF and wait at least 10 seconds.
4. Check that throttle valve moves during the above 10 seconds by confirming the operating sound.

OIL PAN

REMOVAL & INSTALLATION

Lower Oil Pan

See Figures 57 and 58.

1. Before servicing the vehicle, refer to the Precautions.
2. Drain the engine oil.
3. Loosen the bolts in reverse of tightening sequence,.
4. Insert seal cutter (SST) between oil pan (upper) and oil pan (lower). Be careful not to damage the mating surface.

To install:

5. Use a scraper to remove old liquid gasket from mating surfaces.
6. Remove old liquid gasket from mating surface of oil pan (upper).
7. Remove old liquid gasket from the bolt holes and threads.

➡ **Never scratch or damage the mating surface when cleaning off liquid gasket.**

8. Apply a continuous bead of liquid gasket with a tube presser (commercial service tool).

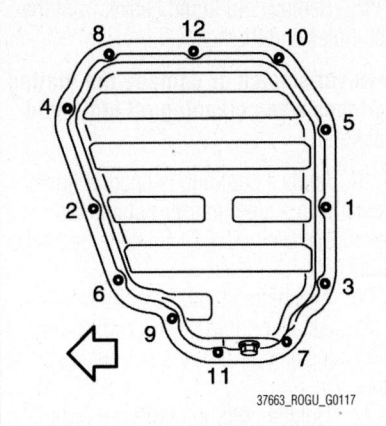

Fig. 57 Power oil pan bolt tightening sequence—reverse the sequence for bolt removal

37663_ROGU_G0117

9. Attaching should be done within 5 minutes after liquid gasket application.
10. Tighten bolts in numerical order.
11. Install oil pan drain plug.
12. Pour engine oil at least 30 minutes after oil pan is installed.
13. Check engine oil level and adjust engine oil.
14. Start engine, and check there is no leaks of engine oil.

15. Stop engine and wait for 10 minutes.
16. Check engine oil level again.

Upper Oil Pan

See Figures 59 and 60.

1. Before servicing the vehicle, refer to the Precautions.
2. Remove the undercover.
3. Drain the engine oil.
4. Remove the oil pan (lower).
5. Remove the oil level gauge and guide.
6. Disconnect the steering lower joint at steering gear assembly side, and release the steering lower shaft.
7. Disconnect the steering outer sockets from steering knuckle.
8. Remove the rear torque rod.
9. Remove the stabilizer connecting rod.
10. Remove the front suspension member.
11. Remove the A/C compressor without disconnecting A/C piping, and temporarily fasten it on vehicle with a rope.
12. Remove the oil strainer.
13. Loosen the bolts in reverse of tightening sequence.

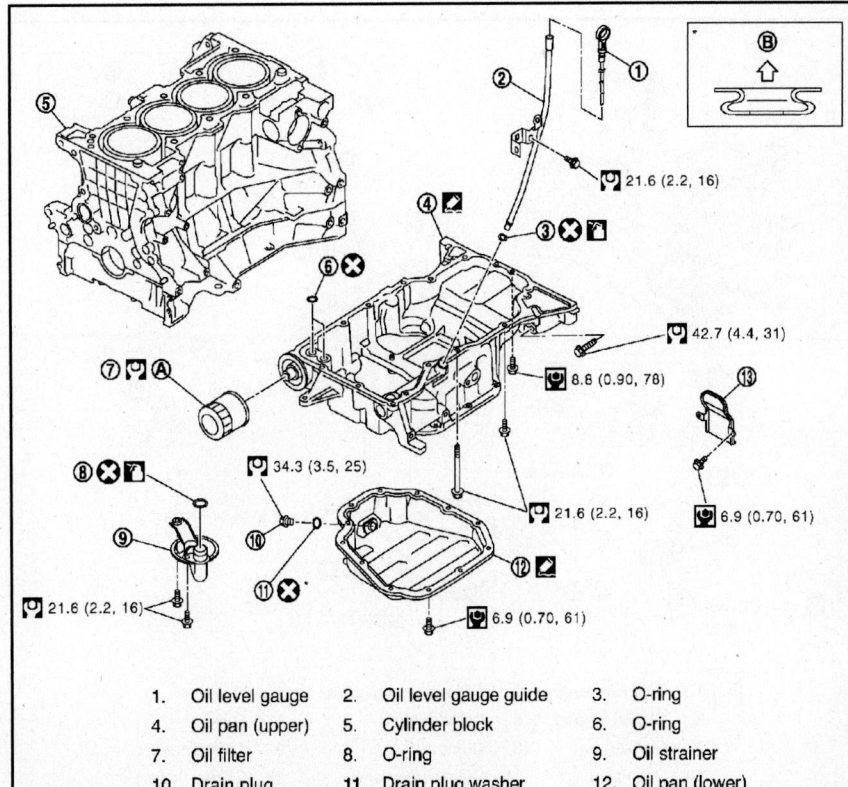

1.	Oil level gauge	2.	Oil level gauge guide	3.	O-ring
4.	Oil pan (upper)	5.	Cylinder block	6.	O-ring
7.	Oil filter	8.	O-ring	9.	Oil strainer
10.	Drain plug	11.	Drain plug washer	12.	Oil pan (lower)
13.	Rear plate cover	B.	Oil pan side		

22140_ROGU_G0073

Fig. 58 Exploded view of upper and lower oil pan and mounting bolt tightening specifications

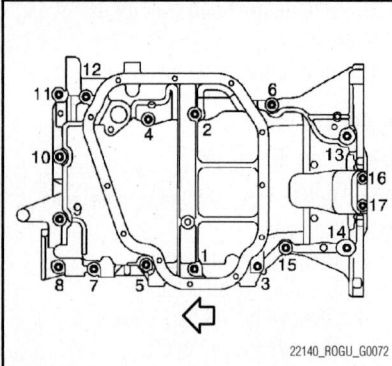

Fig. 59 Bolt tightening sequence, loosen bolts in reverse order

14. Insert the seal cutter (SST) between oil pan (upper) and lower cylinder block, and slide it by tapping on the side of the tool with a hammer.

➡**Be careful not to damage the mating surface.**

15. Remove the O-rings at front cover side.

To install:

16. Use a scraper to remove old liquid gasket from mating surfaces.

17. Remove the old liquid gasket from mating surface of cylinder block.

18. Remove old liquid gasket from the bolt holes and threads.

➡**Never scratch or damage the mating surfaces when cleaning off old liquid gasket.**

19. Apply a continuous bead of liquid gasket with a tube presser outside the holes. Use Genuine RTV Silicone Sealant or equivalent.

20. Attaching should be done within 5 minutes after liquid gasket application.

21. Install new O-rings at front cover side.

22. Tighten bolts in numerical order .

23. Refer to the following for locating bolts:

- M6 × 20 mm (0.79 inch): No. 16, 17
- M8 × 25 mm (0.98 inch): No. 4, 6, 11, 13, 14, 15
- M8 × 60 mm (2.36 inch): No. 7, 8, 9, 10
- M8 × 100 mm (3.94 inch): No. 1, 2, 3, 5, 12

24. Install the oil strainer.
25. Install the front suspension member.
26. Install the oil pan (lower).
27. Install the oil pan drain plug.
28. Install in the reverse order of removal procedure after this step.

29. Pour engine oil at least 30 minutes after oil pan is installed.
30. Check engine oil level and adjust engine oil.
31. Start engine, and check there is no leaks of engine oil.
32. Stop engine and wait for 10 minutes.
33. Check engine oil level again.

OIL PUMP

REMOVAL & INSTALLATION

See Figure 61.

1. Before servicing the vehicle, refer to the Precautions.
2. Disconnect the negative battery cable.
3. Drain the engine oil.
4. Remove the engine front cover.
5. Remove the oil pump. Oil pump is built into front cover.

To install:

6. Install the oil pump.
7. Connect the negative battery cable.
8. Fill the engine with clean oil.
9. Start the vehicle; check for leaks and repair if necessary.

PISTONS & RINGS

POSITIONING

See Figure 62.

REAR MAIN SEAL

REMOVAL & INSTALLATION

See Figures 63 and 64.

1. Before servicing the vehicle, refer to the Precautions.
2. Remove the transaxle assembly.
3. Remove the drive plate.
4. Remove rear oil seal with a suitable tool.

To install:

5. Apply new engine oil to new rear oil seal joint surface and seal lip.
6. Install rear oil seal.
7. Press in rear oil seal.
8. Press-fit rear oil seal with a suitable drift [outer diameter 4.02 in. (102 mm), inner diameter 3.39 in. (86 mm)].

❊❊ WARNING

Be careful not to damage crankshaft and cylinder block. Press-fit oil seal straight to avoid causing burrs or tilting. Never touch grease applied onto oil seal lip.

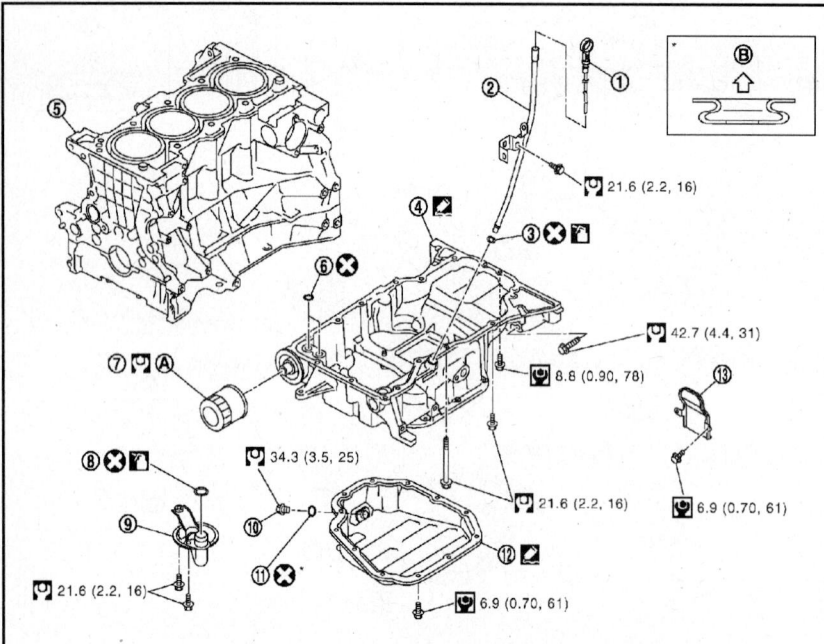

1.	Oil level gauge	2.	Oil level gauge guide	3.	O-ring
4.	Oil pan (upper)	5.	Cylinder block	6.	O-ring
7.	Oil filter	8.	O-ring	9.	Oil strainer
10.	Drain plug	11.	Drain plug washer	12.	Oil pan (lower)
13.	Rear plate cover	B.	Oil pan side		

Fig. 60 Exploded view of upper and lower oil pan and mounting bolt tightening specifications

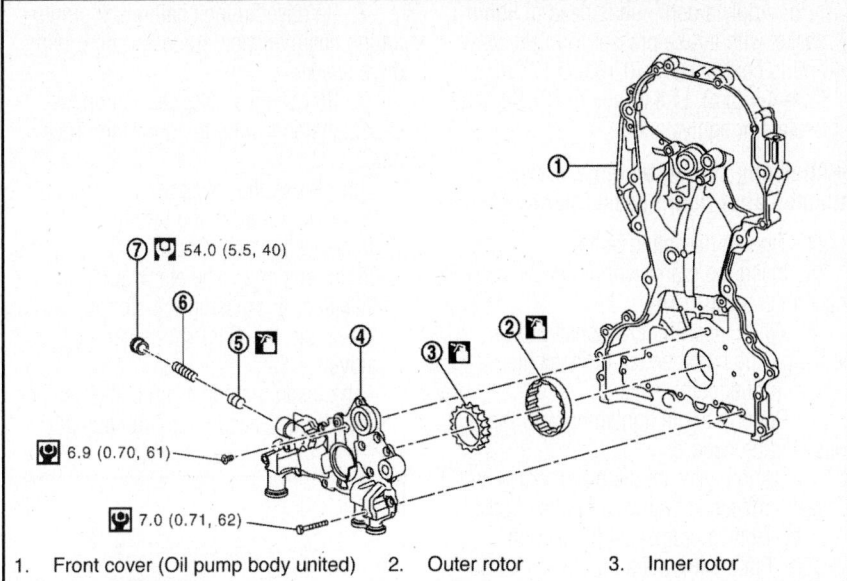

1. Front cover (Oil pump body united) 2. Outer rotor 3. Inner rotor
4. Oil pump cover 5. Regulator valve 6. Regulator valve spring
7. Regulator valve plug

37663_ROGU_G0120

Fig. 61 Oil pump components

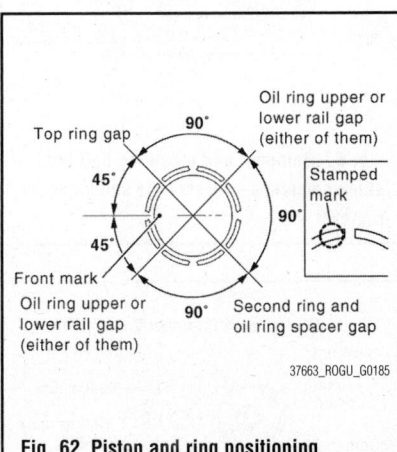

37663_ROGU_G0185

Fig. 62 Piston and ring positioning

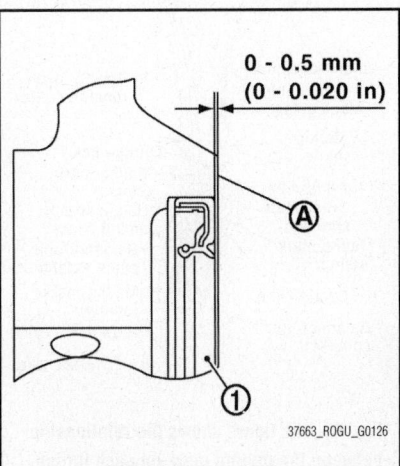

37663_ROGU_G0126

Fig. 63 Rear oil seal (1), rear end cylinder block surface (A)

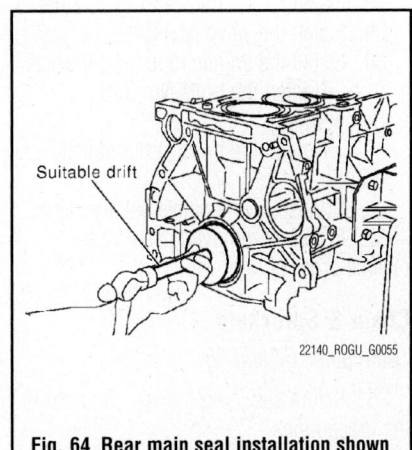

22140_ROGU_G0055

Fig. 64 Rear main seal installation shown

9. Install the drive plate.
10. Install the transaxle assembly.

TIMING CHAIN FRONT COVER, CHAIN, TENSIONER, & SPROCKETS

REMOVAL & INSTALLATION

Front Cover

See Figures 65 and 66.

1. Before servicing the vehicle, refer to the Precautions.
2. Disconnect the negative battery cable.
3. Remove PCV hose.
4. Remove intake manifold.
5. Remove ignition coil.
6. Remove drive belt.

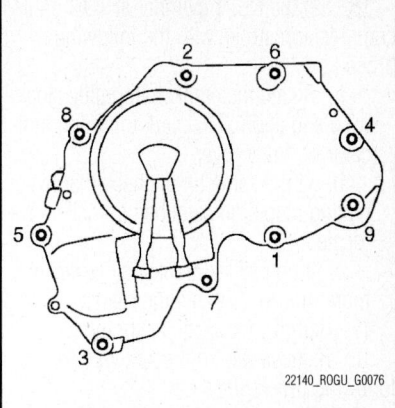

22140_ROGU_G0076

Fig. 65 Intake valve timing control cover bolt tightening sequence—reverse the sequence for removal

7. Remove drive belt auto-tensioner.
8. Remove engine mounting bracket.
9. Remove rocker cover.
10. Remove oil pan (lower).
11. Remove the oil pan (upper), and oil strainer.
12. Remove intake valve timing control cover.
13. Loosen bolts in reverse order.
14. Use a seal cutter tool to cut liquid gasket for removal.

❊❊ WARNING

Be careful not to damage mounting surface.

15. Pull chain guide between camshaft sprockets out through front cover.

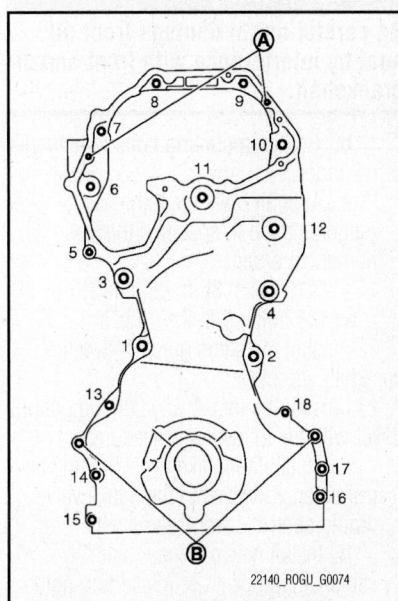

22140_ROGU_G0074

Fig. 66 Front cover tightening sequence with dowel alignment shown (A) and (B)—reverse the sequence for removal

16. Set the No. 1 cylinder at TDC on its compression stroke with the following procedure:

a. Rotate the crankshaft pulley clockwise and align TDC mark to timing indicator on front cover.

b. At the same time, check that the mating marks on camshaft sprockets are located.

c. If not, rotate crankshaft pulley one more turn to align mating marks.

17. Remove the crankshaft pulley.

18. Remove the front cover with the following procedure:

a. Remove the mounting bolts in reverse order.

b. Use a seal cutter to cut liquid gasket for removal.

19. If front oil seal needs to be replaced, lift it with a suitable tool, and remove it.

To install:

20. Install the front oil seal to front cover.

21. Install the front cover with the following procedure:

a. Install O-rings to cylinder head and cylinder block.

b. Apply a continuous bead of liquid gasket with a tube presser to the front cover. For bolt holes with triangle marks (5 locations), apply liquid gasket outside the holes. Attaching should be done within 5 minutes after coating.

c. Check that mating marks of timing chain and each sprocket are still aligned, and install the front cover.

✳✳ WARNING

Be careful not to damage front oil seal by interference with front end of crankshaft.

d. Tighten mounting bolts in numerical order .

e. After all bolts are tightened, retighten them to specified torque in numerical order .
- M10 bolts: 36 ft. lbs. (49 Nm)
- M6 bolts: 9 ft. lbs. (13 Nm)

22. Install the chain guide between camshaft sprockets.

23. Install the intake valve timing control cover with the following procedure:

a. Install the intake valve timing control solenoid valves to the intake valve timing control cover if removed.

b. Install new oil rings to the camshaft sprocket (INT) insertion points on backside of intake valve timing control cover.

c. Install new O-ring to front cover.

d. Apply a continuous bead of liquid gasket with a tube presser to intake valve timing control cover, 0.134–0.173 in. (3.4–4.4 mm). Use Genuine RTV Silicone Sealant or equivalent.

➡**Attaching should be done within 5 minutes after liquid gasket application.**

24. Tighten mounting bolts.

25. Insert the crankshaft pulley by aligning with crankshaft key.

26. When inserting crankshaft pulley with a plastic hammer, tap on its center portion (not circumference).

27. Perform angle tightening with the following procedure:

a. Apply new engine oil to thread and seat surfaces of crankshaft pulley bolt.

b. Tighten crankshaft pulley bolt.

c. Put a paint mark on crankshaft pulley, mating with any one of six easy to recognize angle marks on bolt flange.

d. Turn another 60 degrees clockwise (angle tightening).

e. Check the tightening angle with movement of one angle mark.

28. Install the oil pans and oil strainer.

29. Install the valve covers.

30. Install the engine mounting bracket.

a. Tighten the bolts No. 3, 5 (temporarily).

b. Tighten the bolts in numerical order.

31. Connect the negative battery cable.

32. Install all removed parts in the reverse order of removal.

Chain & Sprockets

See Figures 67 and 68.

1. Before servicing the vehicle, refer to the Precautions.

2. Disconnect the negative battery cable.

3. Remove the timing chain front cover.

4. Remove timing chain and camshaft sprockets with the following procedure:

5. Push in the chain tensioner plunger. Insert a stopper pin into hole on chain tensioner body to secure chain tensioner plunger and remove chain tensioner. Use approximately 0.02 in. (0.5 mm) diameter hard metal pin as a stopper pin.

6. Secure the hexagonal part of the camshaft with a wrench. Loosen the camshaft sprocket mounting bolts and remove timing chain and camshaft sprockets.

✳✳ WARNING

Never rotate the crankshaft or camshaft while timing chain is removed. It causes interference between valve and piston.

7. Remove timing chain slack guide, timing chain tension guide and oil pump drive spacer.

8. Remove the balancer unit timing chain tensioner with the following procedure:

a. Press the stopper tab in the direction that will push the timing chain slack guide toward balancer unit timing chain tensioner. The slack guide is released by pressing the stopper tab. As the result, the slack guide can be moved.

b. Insert a stopper pin into tensioner body hole to secure the timing chain

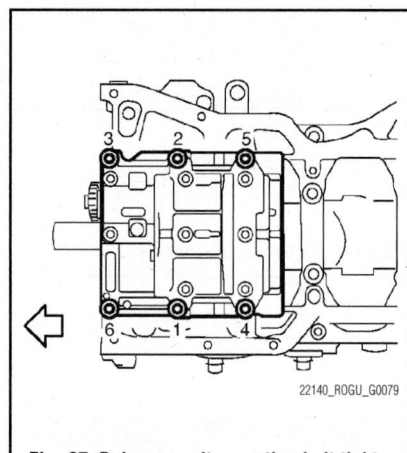

22140_ROGU_G0079

Fig. 67 Balancer unit mounting bolt tightening sequence—reverse the sequence for removal

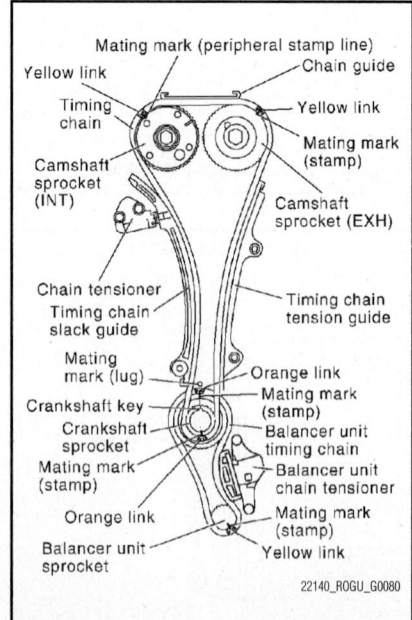

22140_ROGU_G0080

Fig. 68 The figure shows the relationship between the mating mark on each timing chain and that on the corresponding sprocket

slack guide. Use a hard metal pin with the diameter of approximately 0.047 in. (1.2 mm) as a stopper pin.

c. Remove the balancer unit timing chain tensioner.

d. When the holes on the lever and tensioner body cannot be aligned, align these holes by slightly moving the slack guide.

9. Remove the balancer unit timing chain and crankshaft sprocket.

10. Remove the mounting bolts in reverse order, and remove balancer unit.

✳✳ WARNING

Never disassemble balancer unit.

11. The figure shows the relationship between the mating mark on each timing chain and that on the corresponding sprocket, with the components installed.

To install:

12. Check that crankshaft key points straight up.

13. Tighten mounting bolts in numerical order , and install the balancer unit:

a. Apply new engine oil to threads and seat surfaces of mounting bolts.

b. Tighten No. 1–5 bolts to 31 ft. lbs. (42 Nm).

c. Tighten the No. 6 bolt to 27 ft. lbs. (36 Nm).

d. Turn No. 1–5 bolts 120 degrees clockwise (angle tightening).

➡**Use the angle wrench [SST: KV10112100 (BT8653-A)] to check tightening angle. Never make judgment by visual inspection.**

e. Turn No. 6 bolt 90 degrees clockwise (angle tightening).

f. Completely loosen all bolts.

g. In this step, loosen bolts in reverse order.

h. Repeat the steps b to e.

14. Install the crankshaft sprocket and balancer unit timing chain.

a. Check that crankshaft sprocket is positioned with mating marks on cylinder block and crankshaft sprocket meeting at the top.

b. Install it by aligning mating marks on each sprocket and balancer unit timing chain.

15. Install the balancer unit timing chain tensioner. Be careful not to let mating marks of each sprocket and timing chain slip.

16. After installation, check the mating marks have not slipped, then remove stopper pin and release tensioner sleeve.

17. Install the timing chain and related parts.

a. Install by aligning mating marks on each sprocket and timing chain .

b. Before and after installing chain tensioner, check again to ensure that mating marks have not slipped.

18. After installing chain tensioner, remove stopper pin, and check that tensioner moves freely.

➡**After the mating marks are aligned, keep them aligned by holding them by hand. To avoid skipped teeth, never rotate crankshaft and camshaft until front cover is installed.**

19. Install the timing chain front cover.

VALVE COVERS

REMOVAL & INSTALLATION

See Figures 69 and 70.

1. Before servicing the vehicle, refer to the Precautions.

2. Disconnect the negative battery cable.

3. Remove air duct and resonator assembly.

4. Remove electric throttle control actuator without disconnecting water hose.

5. Loosen the intake manifold mounting bolts and nuts.

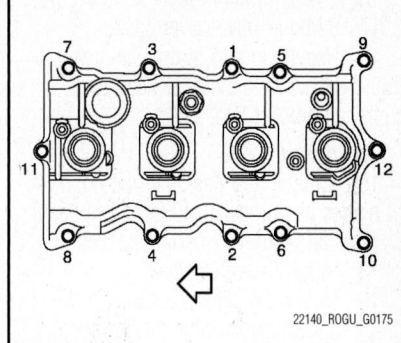

Fig. 69 Valve cover bolt tightening sequence—reverse order for removal

6. Remove the intake manifold.

7. Disconnect harness connector from ignition coil.

➡**Support the bottom surface of engine using a transmission jack.**

8. Remove the ground cable and harness from engine mounting bracket.

9. Remove the ignition coil.

10. Disconnect PCV hose from rocker cover.

11. Remove engine mounting insulator (RH).

12. Remove engine mounting bracket.

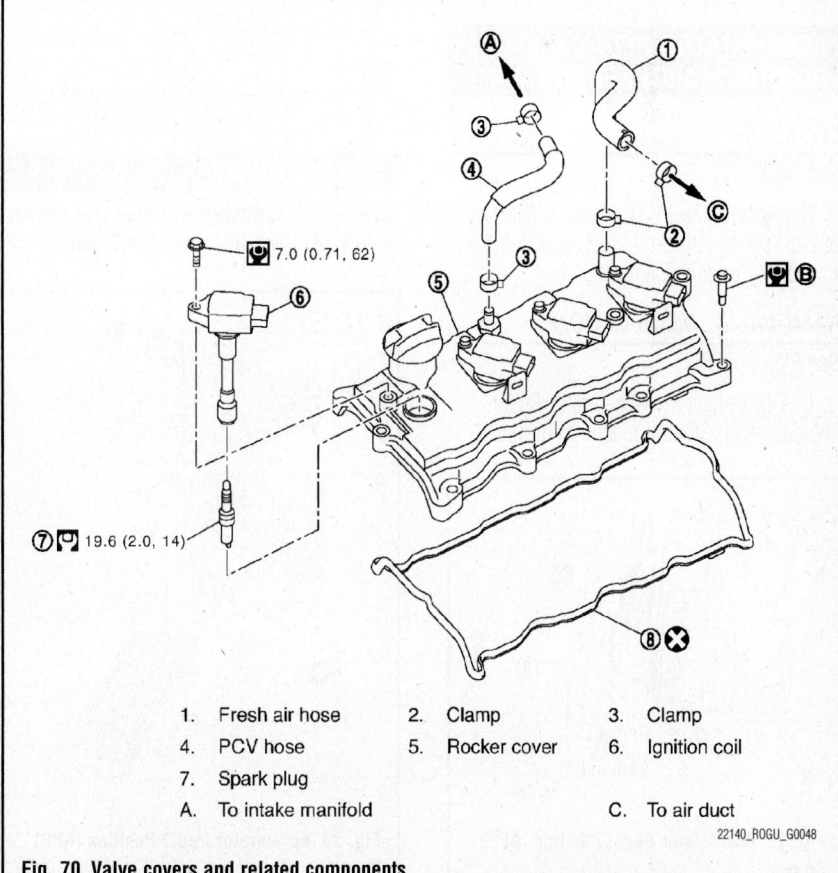

1.	Fresh air hose	2.	Clamp	3.	Clamp
4.	PCV hose	5.	Rocker cover	6.	Ignition coil
7.	Spark plug				
A.	To intake manifold			C.	To air duct

Fig. 70 Valve covers and related components

13. Loosen bolts in reverse of the order.

14. Remove valve cover gasket.

15. Use scraper to remove all traces of liquid gasket from cylinder head and camshaft bracket (No. 1).

To install:

16. Apply liquid gasket to the position. (4 places of cylinder head front and back).

17. Install the valve cover gasket.

18. Install the valve cover.

19. Check if rocker cover gasket has dropped from the installation groove of rocker cover.

20. Tighten bolts in two steps separately in numerical order.
- Step 1: 17 inch lbs. (2 Nm)
- Step 2: 74 inch lbs. (8 Nm)

21. Install in the reverse order of removal after this step.

VALVE LASH

ADJUSTMENT

See Figure 71.

1. Perform adjustments depending on selected head thickness of valve lifter.

2. Remove the camshaft.

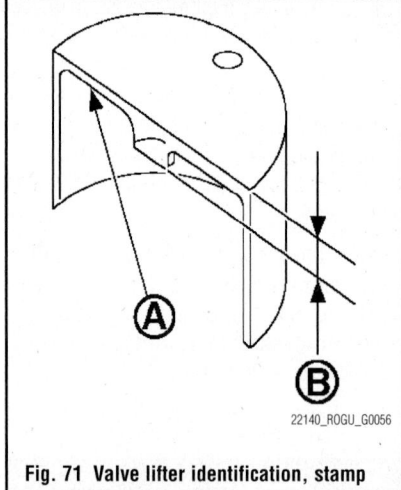

Fig. 71 Valve lifter identification, stamp mark (A), thickness (B)

3. Remove valve lifters at the locations that are out of the standard.

4. Measure the center thickness of the removed valve lifters with a micrometer.

5. Use the equation below to calculate valve lifter thickness for replacement:
- Valve lifter thickness calculation: $t = t1 + (C1 - C2)$

- t = Valve lifter thickness to be replaced
- t1 = Removed valve lifter thickness
- C1 = Measured valve clearance
- C2 = Standard valve clearance:
  Intake: 0.011 in. (0.28 mm),
  Exhaust: 0.012 in. (0.30 mm)

6. Thickness of new valve lifter can be identified by stamp mark on the reverse side (inside the cylinder).

7. Stamp mark "788" indicates 0.3102 inch (7.88 mm) in thickness.

➡**Available thickness of valve lifter: 26 sizes range 0.3102 to 0.3299 inch (7.88 to 8.38 mm) in steps of 0.0008 inch (0.02 mm) (when manufactured at factory).**

8. Install the selected valve lifter.

9. Install the camshaft.

10. Manually rotate the crankshaft pulley a few rotations.

11. Check that the valve clearances for cold engine are within specifications by referring to the specified values.

12. Install all removed parts in the reverse order of removal.

13. Warm up the engine, and check for unusual noise and vibration.

ENGINE PERFORMANCE & EMISSION CONTROLS

ACCELERATOR PEDAL POSITION (APP) SENSOR

LOCATION

See Figure 72.

The Accelerator Pedal Position (APP) sensor is installed on the upper end of the accelerator pedal assembly.

REMOVAL & INSTALLATION

See Figure 73.

1. Disconnect the Accelerator Pedal Position (APP) sensor electrical harness.

2. Remove the accelerator pedal mounting nuts.

3. Loosen the brake pedal mounting nuts only. Do not remove assembly.

✵ WARNING

Never disassemble accelerator lever. Never remove accelerator lever.

Avoid impact from dropping etc., during handling. Be careful to keep accelerator lever away from water.

To install:

4. Install in the reverse order of removal.

AIR/FUEL RATIO (A/F) SENSOR

LOCATION

See Figure 74.

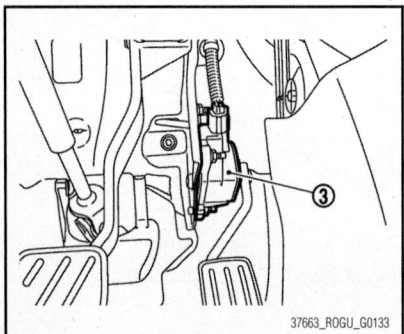

Fig. 72 Accelerator Pedal Position (APP) Sensor

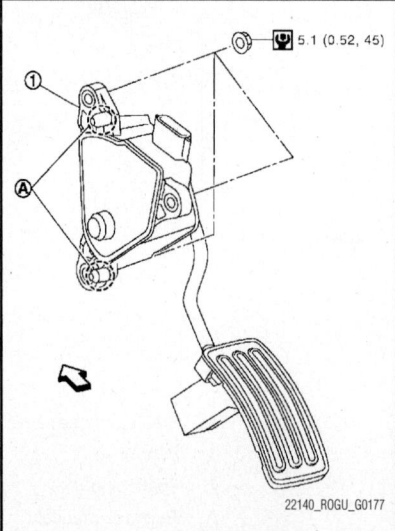

Fig. 73 Accelerator Pedal Position (APP) Sensor

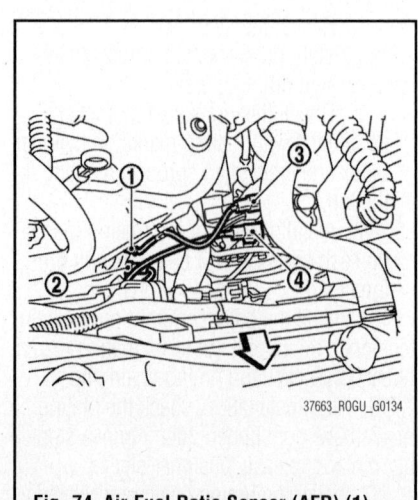

Fig. 74 Air-Fuel Ratio Sensor (AFR) (1)

CAMSHAFT POSITION (CMP) SENSOR

LOCATION

See Figure 75.

REMOVAL & INSTALLATION

1. Before servicing the vehicle, refer to the Precautions.
2. Disconnect the negative battery cable.
3. Remove the air cleaner duct and resonator.
4. Remove the Camshaft Position (CMP) sensor mounting bolt.
5. Remove the CMP sensor harness connector.
6. Remove the CMP sensor.

To install:
7. Installation is the reverse of removal.

CRANKSHAFT POSITION (CKP) SENSOR

LOCATION

See Figure 76.

REMOVAL & INSTALLATION

1. Before servicing the vehicle, refer to the Precautions.

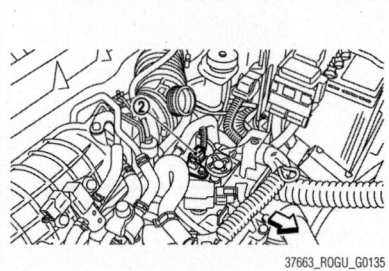

Fig. 75 Camshaft Position (CMP) Sensor (2)

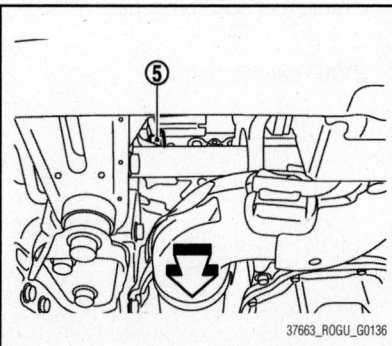

Fig. 76 Crankshaft Position (CMP) Sensor (5)

2. Disconnect the negative battery cable.
3. Disconnect the Crankshaft Position (CKP) sensor harness connector.
4. Remove the CKP sensor mounting bolt.
5. Remove the CKP sensor.

To install:
6. Installation is the reverse of removal.

ELECTRONIC CONTROL MODULE (ECM)

LOCATION

See Figure 77.

ACCELERATOR PEDAL RELEASED POSITION LEARNING

1. Check that accelerator pedal is fully released.
2. Turn ignition switch ON and wait at least 2 seconds.
3. Turn ignition switch OFF and wait at least 10 seconds.
4. Turn ignition switch ON and wait at least 2 seconds.
5. Turn ignition switch OFF and wait at least 10 seconds.

IDLE AIR VOLUME LEARNING

See Figures 78 and 79.

The Idle Air Volume Learning is a function of ECM to learn the idle air volume that keeps each engine idle speed within the specific range. It must be performed under the following conditions:

- Each time the electric throttle control actuator or ECM is replaced
- Idle speed or ignition timing is out of the specification
1. Check that all of the following conditions are satisfied. Learning will be cancelled if any of the following conditions are missed for even a moment:
  - Battery voltage: More than 12.9 V (At idle)
  - Engine coolant temperature: 158–212°F (70–100°C)
  - Selector lever: P or N
  - Electric load switch: OFF (Air conditioner, headlamp, and rear window defogger)
  - On vehicles equipped with daytime light systems, if the parking brake is applied before the engine is started, the headlamp will not illuminate.

- Steering wheel: Neutral (Straight-ahead position)
- Vehicle speed: Stopped
- Transmission: Warmed-up
- With CONSULT-III: Drive vehicle until "ATF TENP SEN" in "DATA MONITOR" mode of "CVT" system indicates less than 0.9V.
- Without CONSULT-III: Drive vehicle for 10 minutes.
2. If a CONSULT-III scan tool is not available, perform Accelerator Pedal Released Position Learning.
3. Perform Throttle Valve Closed Position Learning.
4. Start engine and warm it up to normal operating temperature.
5. Turn ignition switch OFF and wait at least 10 seconds.
6. Confirm that accelerator pedal is fully released, turn ignition switch ON and wait 3 seconds.
7. Repeat the following procedure quickly five times within 5 seconds.
  a. Fully depress the accelerator pedal.
  b. Fully release the accelerator pedal.
8. Wait 7 seconds, fully depress the accelerator pedal for approx. 20 seconds until the MIL stops blinking and turns ON.
9. Fully release the accelerator pedal within 3 seconds after the MIL turns ON.
10. Start engine and let it idle.
11. Wait 20 seconds.
12. Rev up the engine two or three times and check that idle speed and ignition timing are within the specifications. If the inspection result is not normal, find the malfunctioning part.

ENGINE COOLANT TEMPERATURE (ECT) SENSOR

LOCATION

See Figure 80.

REMOVAL & INSTALLATION

1. Before servicing the vehicle, refer to the Precautions.
2. Disconnect the negative battery cable.
3. Disconnect the Engine Coolant Temperature (ECT) sensor harness connector.
4. Remove the ECT sensor.

To install:
5. Installation is the reverse of removal.

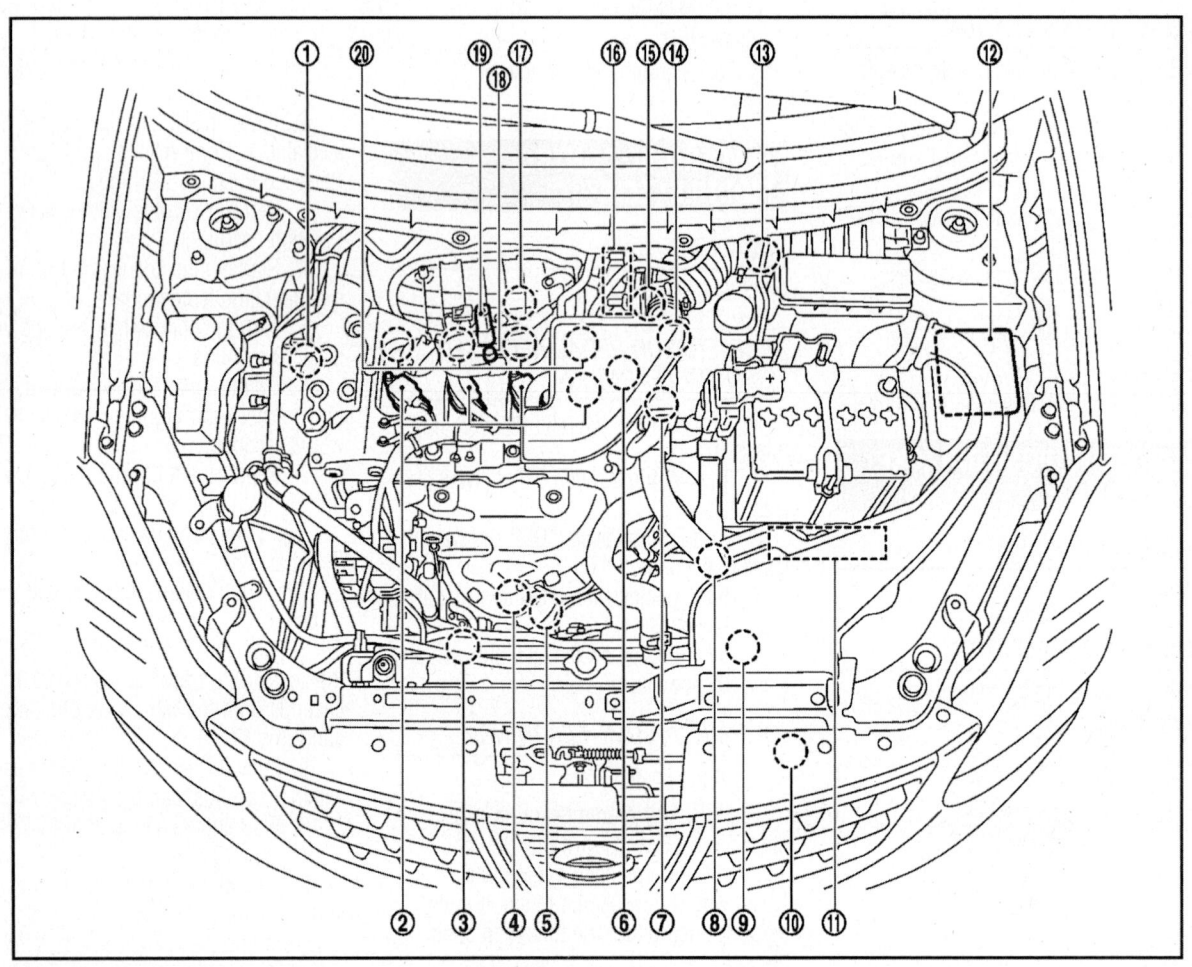

1. Intake valve timing control solenoid valve
2. Ignition coil (with power transistor) and spark plug
3. Cooling fan motor-2
4. Air fuel ratio (A/F) sensor 1
5. Heated oxygen sensor 2
6. Camshaft position sensor (PHASE)
7. Engine coolant temperature sensor
8. Transmission range switch
9. Cooling fan motor-1
10. Refrigerant pressure sensor
11. ECM
12. IPDM E/R
13. Mass air flow sensor (with intake air temperature sensor)
14. Tumble control valve actuator
15. Crankshaft position sensor (POS)
16. Electric throttle control actuator (with built in throttle position sensor and throttle control motor)
17. Knock sensor
18. EVAP service port
19. EVAP canister purge volume control solenoid valve
20. Fuel injector

37663_ROGU_G0132

Fig. 77 Electronic Control Module (ECM) (11)

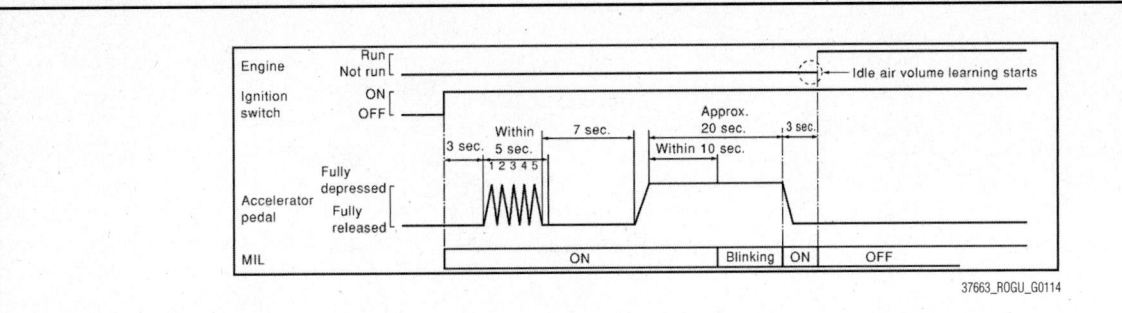

Fig. 78 Idle Air Volume Learning

Idle Speed

Transmission	Condition	Specification
CVT	No load* (in P or N position)	700 ± 50 rpm

*: Under the following conditions
- A/C switch: OFF
- Electric load: OFF (Lights, heater fan & rear window defogger)
- Steering wheel: Kept in straight-ahead position

Ignition Timing

Transmission	Condition	Specification
CVT	No load* (in P or N position)	10 ± 5° BTDC

*: Under the following conditions
- A/C switch: OFF
- Electric load: OFF (Lights, heater fan & rear window defogger)
- Steering wheel: Kept in straight-ahead position

Fig. 79 Idle Speed and Ignition Timing

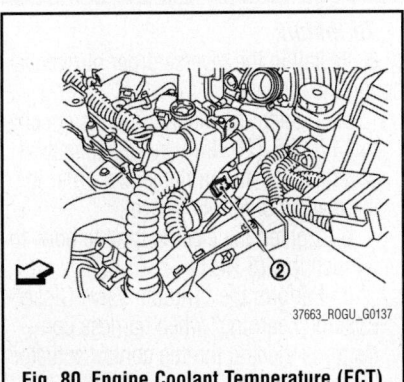

Fig. 80 Engine Coolant Temperature (ECT) Sensor (2)

EVAPORATIVE EMISSION CANISTER

LOCATION
See Figure 81.

HEATED OXYGEN (HO2S) SENSOR

LOCATION
See Figure 82.

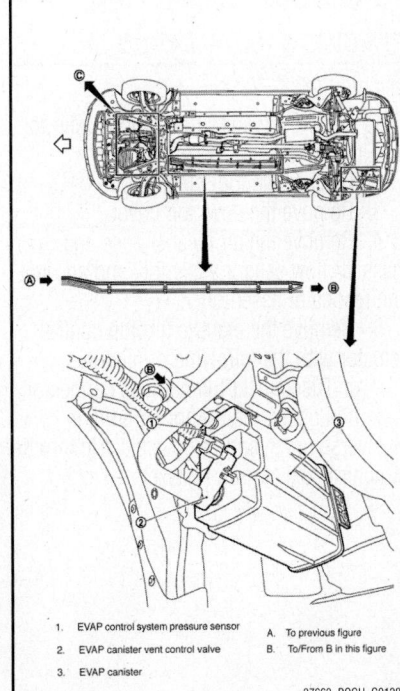

1. EVAP control system pressure sensor
2. EVAP canister vent control valve
3. EVAP canister

A. To previous figure
B. To/From B in this figure

Fig. 81 EVAP Canister (3)

REMOVAL & INSTALLATION
1. Raise the vehicle.
2. Remove the front engine under cover.
3. Remove the Heated Oxygen (HO2S) sensor harness connector.
4. Remove the HO2S sensor.

To install:
5. Installation is the reverse of removal.

INTAKE AIR TEMPERATURE (IAT)/MASS AIRFLOW (MAF) SENSOR

LOCATION
See Figure 83.

REMOVAL & INSTALLATION
1. Remove the Mass Air Flow (MAF) sensor (with Intake Air Temperature sensor) harness connector.
2. Remove the retaining screws and remove the sensor.

To install:
3. Install the sensor and tighten the retaining screws.

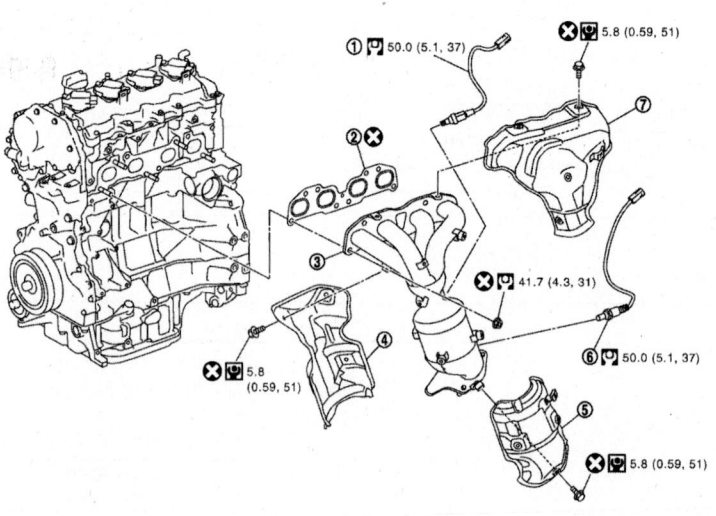

1. Air fuel ratio sensor 1
2. Gasket
3. Exhaust manifold and three way catalyst assembly
4. Three way catalyst cover
5. Exhaust manifold cover (lower)
6. Heated oxygen sensor 2
7. Exhaust manifold cover (upper)

37663_ROGU_G0112

Fig. 82 Exhaust manifold and three way catalyst

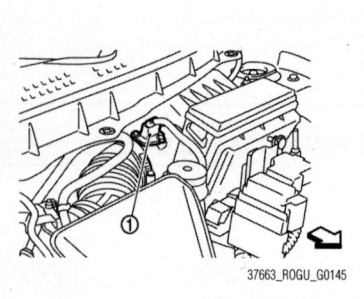

37663_ROGU_G0145

Fig. 83 Mass Air Flow (MAF) sensor and Intake Air Temperature (IAT) sensor (1)

4. Install the MAF sensor (with IAT sensor) harness connector.

KNOCK SENSOR (KS)

LOCATION

See Figure 84.

THROTTLE CONTROL ACTUATOR (TAC)

LOCATION

See Figure 85.

REMOVAL & INSTALLATION

See Figures 86 and 87.

1. Before servicing the vehicle, refer to the Precautions.
2. Release the fuel pressure.
3. Remove the cowl top cover.
4. Remove the air cleaner case and mass air flow sensor assembly and air duct and resonator assembly.
5. Remove the electric throttle control actuator with the following procedure:
   a. Disconnect the harness connector.
   b. Loosen the mounting bolts in reverse order, and remove electric throttle control actuator and gasket.

☀ WARNING

Handle carefully to avoid any shock to electric throttle control actuator. Never disassemble.

To install:

6. Install in the reverse order of removal, noting the following:
   a. Tighten the mounting bolts equally to 14 ft. lbs. (20 Nm), and diagonally in several steps and in numerical order in the reverse order of removal.
   b. Tighten the electric throttle body to 74 inch lbs. (8 Nm).
   c. Perform the "Throttle Valve Closed Position Learning" when harness connector of electric throttle control actuator is disconnected.
   d. Perform the "Idle Air Volume Learning" and "Throttle Valve Closed Position Learning" when electric throttle control actuator is replaced.

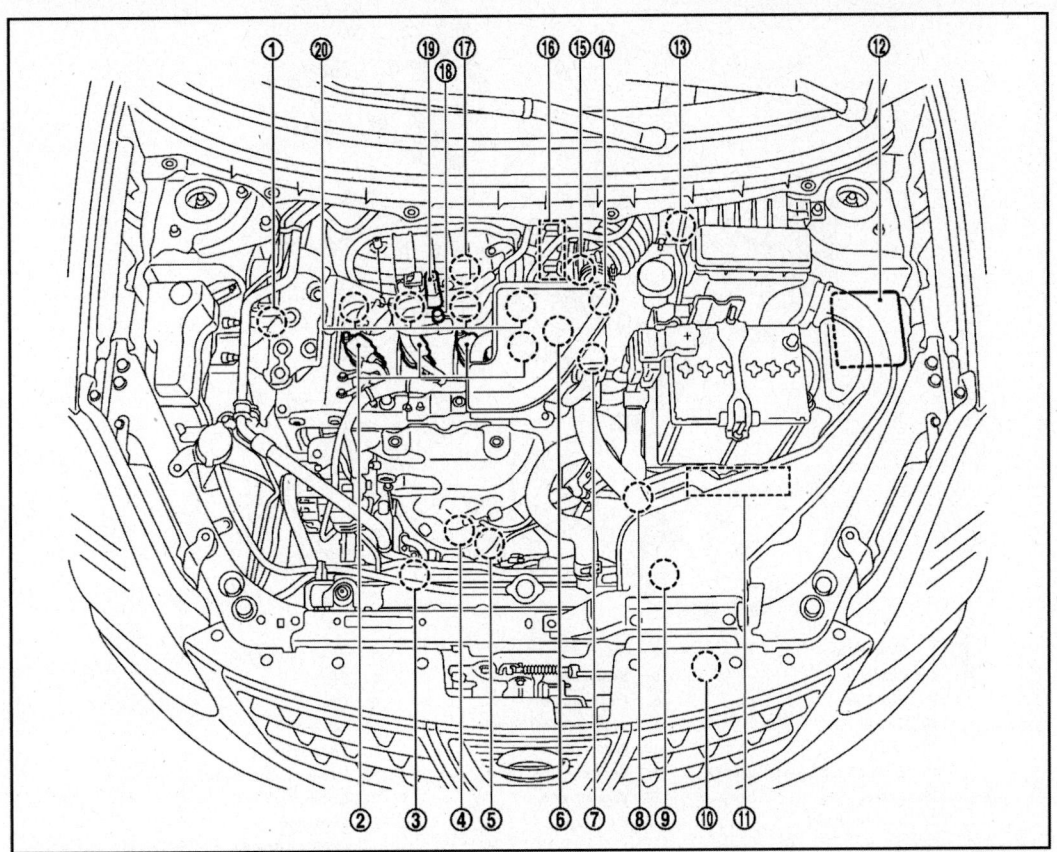

1. Intake valve timing control solenoid valve
2. Ignition coil (with power transistor) and spark plug
3. Cooling fan motor-2
4. Air fuel ratio (A/F) sensor 1
5. Heated oxygen sensor 2
6. Camshaft position sensor (PHASE)
7. Engine coolant temperature sensor
8. Transmission range switch
9. Cooling fan motor-1
10. Refrigerant pressure sensor
11. ECM
12. IPDM E/R
13. Mass air flow sensor (with intake air temperature sensor)
14. Tumble control valve actuator
15. Crankshaft position sensor (POS)
16. Electric throttle control actuator (with built in throttle position sensor and throttle control motor)
17. Knock sensor
18. EVAP service port
19. EVAP canister purge volume control solenoid valve
20. Fuel injector

37663_ROGU_G0132

Fig. 84 Knock Sensor (KS) (17)

37663_ROGU_G0144

Fig. 85 Throttle Control Actuator (TAC) (1)

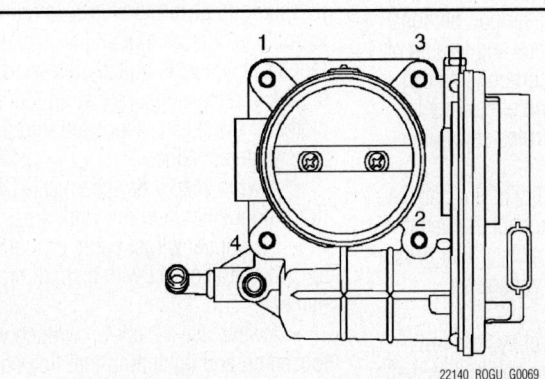

22140_ROGU_G0069

Fig. 86 Loosen mounting bolts in reverse order as shown in the figure, and remove electric throttle control actuator and gasket

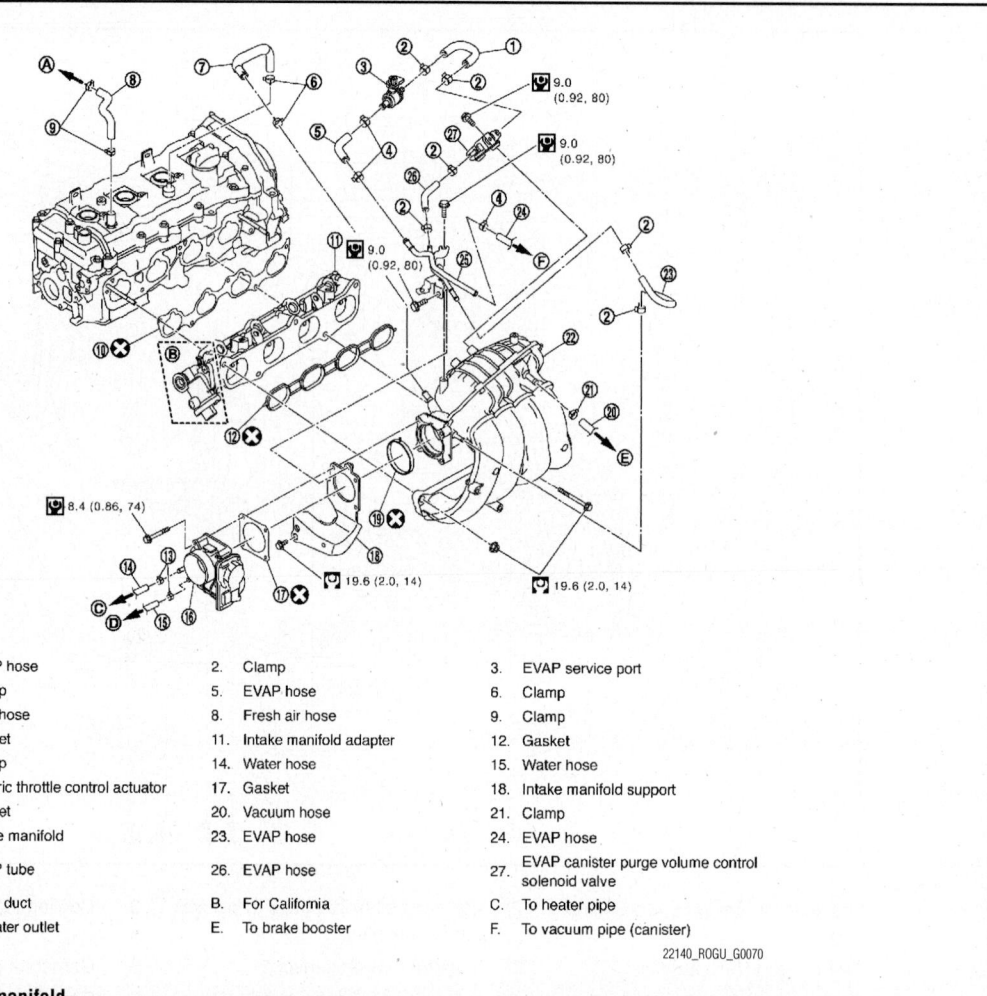

1.	EVAP hose	2.	Clamp	3.	EVAP service port
4.	Clamp	5.	EVAP hose	6.	Clamp
7.	PCV hose	8.	Fresh air hose	9.	Clamp
10.	Gasket	11.	Intake manifold adapter	12.	Gasket
13.	Clamp	14.	Water hose	15.	Water hose
16.	Electric throttle control actuator	17.	Gasket	18.	Intake manifold support
19.	Gasket	20.	Vacuum hose	21.	Clamp
22.	Intake manifold	23.	EVAP hose	24.	EVAP hose
25.	EVAP tube	26.	EVAP hose	27.	EVAP canister purge volume control solenoid valve
A.	To air duct	B.	For California	C.	To heater pipe
D.	To water outlet	E.	To brake booster	F.	To vacuum pipe (canister)

22140_ROGU_G0070

Fig. 87 Exploded view of intake manifold

FUEL   GASOLINE FUEL INJECTION SYSTEM

FUEL SYSTEM SERVICE PRECAUTIONS

Safety is the most important factor when performing not only fuel system maintenance but any type of maintenance. Failure to conduct maintenance and repairs in a safe manner may result in serious personal injury or death. Maintenance and testing of the vehicle's fuel system components can be accomplished safely and effectively by adhering to the following rules and guidelines.

• To avoid the possibility of fire and personal injury, always disconnect the negative battery cable unless the repair or test procedure requires that battery voltage be applied.

• Always relieve the fuel system pressure prior to disconnecting any fuel system component (injector, fuel rail, pressure regulator, etc.), fitting or fuel line connection. Exercise extreme caution whenever relieving fuel system pressure to avoid exposing skin, face and eyes to fuel spray. Please be advised that fuel under pressure may penetrate the skin or any part of the body that it contacts.

• Always place a shop towel or cloth around the fitting or connection prior to loosening to absorb any excess fuel due to spillage. Ensure that all fuel spillage (should it occur) is quickly removed from engine surfaces. Ensure that all fuel soaked cloths or towels are deposited into a suitable waste container.

• Always keep a dry chemical (Class B) fire extinguisher near the work area.

• Do not allow fuel spray or fuel vapors to come into contact with a spark or open flame.

• Always use a back-up wrench when loosening and tightening fuel line connection fittings. This will prevent unnecessary stress and torsion to fuel line piping.

• Always replace worn fuel fitting O-rings with new Do not substitute fuel hose or equivalent where fuel pipe is installed.

Before servicing the vehicle, make sure to also refer to the precautions.

RELIEVING FUEL SYSTEM PRESSURE

1. Remove fuel pump fuse located in IPDM E/R. or unplug the fuel pump connector.
2. Start the engine.
3. After engine stalls, crank it two or three times to release all fuel pressure.
4. Turn ignition switch OFF.
5. Reinstall fuel pump fuse after servicing fuel system.

FUEL PUMP MODULE

REMOVAL & INSTALLATION

See Figures 88 through 91.

1. Before servicing the vehicle, refer to the Precautions.

2. Disconnect the negative battery cable.

3. Check fuel level on fuel gauge. If fuel gauge indicates more than ¾ of a tank, drain fuel tank to around a half a tank.

4. In case fuel pump does not operate, perform the following procedure:

a. Insert hose of less than 0.79 in. (20 mm) diameter into fuel filler tube through fuel filler opening to draw fuel from fuel filler tube.

b. Disconnect fuel filler hose from fuel filler tube.

c. Insert hose into fuel tank through fuel filler hose to draw fuel from fuel tank.

5. Release the fuel pressure from the fuel lines.

6. Open the fuel filler lid.

7. Open fuel filler cap and release the pressure inside the fuel tank.

8. Remove rear seat cushion.

9. Remove inspection hole cover.

10. Using a screwdriver, remove it by turning clips clockwise by 90 degrees.

11. Disconnect the harness connector and fuel line quick connector.

12. Using lock ring wrench [SST: KV991J0090 (J-46214)], remove lock ring.

➡**For reference when installing, put a matching mark on lock ring, fuel pump assembly and fuel tank.**

13. Raise fuel level sensor unit, fuel filter and fuel pump assembly, and disconnect fuel tube and harness connector.

❊❊ WARNING

Never bend float arm during removal. Never pollute the inside by residue fuel. Draw out avoiding inclination by supporting with a cloth. Never cause impact such as dropping when handling components.

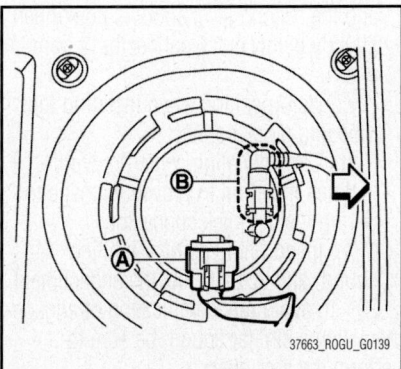

Fig. 88 Disconnect the harness connector (A) and the quick connector (B)

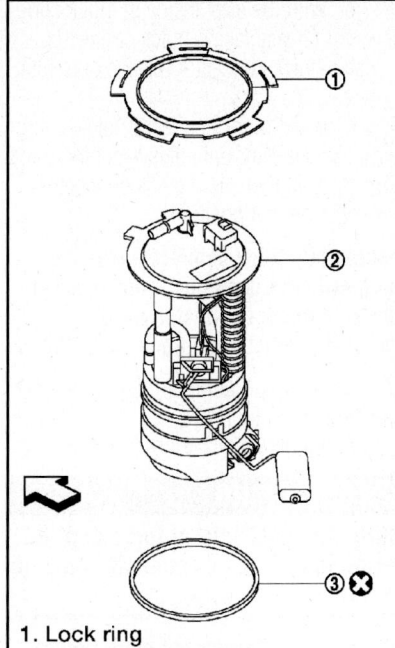

1. Lock ring
2. Fuel level sensor unit, fuel filter and fuel pump assembly
3. O-ring

22140_ROGU_G0102

Fig. 89 2WD models fuel level sensor assembly

To install:

14. Install O-ring to fuel tank without any twist.

15. Align with vehicle front. Install fuel level sensor unit (1) to fuel tank.

16. Connect quick connector of fuel feed tube using the following procedures:

a. Check the connection for damage or any foreign materials.

b. Align the connector with the tube, then insert the connector straight into the tube until a "click" sound is heard.

17. After connecting, check that the connection is secured with following procedures:

a. Visually confirm that the two tabs are connected to the connector.

b. Pull the tube and the connector to check that they are securely connected.

18. Connect the negative battery cable.

19. Before installing inspection hole cover, check that the connecting part has no fuel leakage.

20. Install inspection hole covers with the front mark (arrow) facing front of vehicle.

21. Lock clips by turning counterclockwise.

22. Install the rear seat cushion.

23. Install the fuel cap.

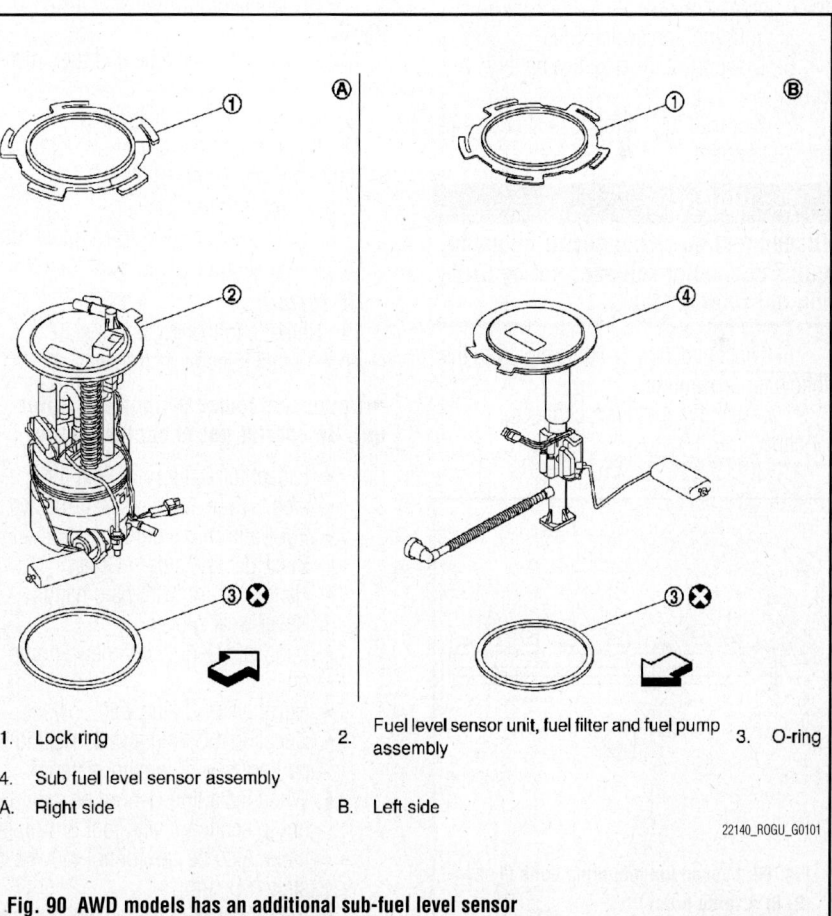

1. Lock ring
2. Fuel level sensor unit, fuel filter and fuel pump assembly
3. O-ring
4. Sub fuel level sensor assembly
A. Right side
B. Left side

22140_ROGU_G0101

Fig. 90 AWD models has an additional sub-fuel level sensor

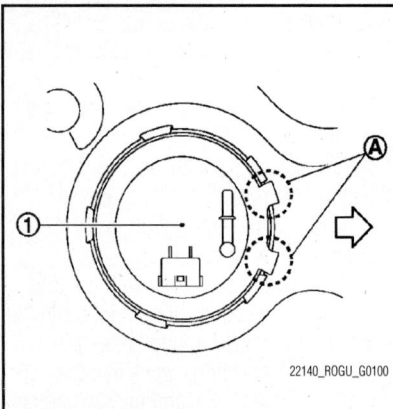

22140_ROGU_G0100

Fig. 91 Align (A) with vehicle front as shown in the figure and install fuel level sensor unit (1) to fuel tank

FUEL RAIL & INJECTORS

REMOVAL & INSTALLATION

See Figure 92.

1. Before servicing the vehicle, refer to the Precautions.
2. Remove fuel pump fuse located in IPDM E/R.
3. Start the engine.
4. After engine stalls, crank it two or three times to release all fuel pressure.
5. Turn ignition switch OFF.
6. Disconnect the negative battery cable.
7. Reinstall fuel pump fuse after servicing fuel system.

✳✳ WARNING

Disconnect quick connector by using quick connector release, not by picking out retainer tabs.

8. Disconnect quick connector with the following procedure:
   a. Remove the quick connector cap.

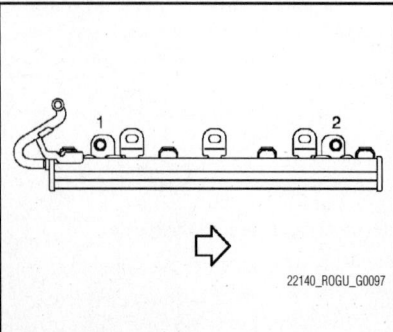

22140_ROGU_G0097

Fig. 92 Loosen the mounting bolts (1) and (2) in reverse order

b. With the sleeve side of quick connector release facing quick connector, install quick connector release onto fuel tube.
c. Insert quick connector release into quick connector until sleeve contacts and goes no further. Hold quick connector release on that position.

➡**Inserting quick connector release hard will not disconnect quick connector. Hold quick connector release where it contacts and goes no further.**

d. Draw and pull out quick connector straight from fuel tube.
e. Pull the quick connector.

✳✳ WARNING

Never pull with lateral force applied. O-ring inside quick connector may be damaged.

➡**Prepare container and cloth beforehand as fuel will leak out.**

➡**Keep clean the connecting portion and to avoid damage and foreign materials, cover them completely with plastic bags or something similar.**

9. Remove the intake manifold.
10. Disconnect sub-harness for fuel injector.
11. Remove the fuel tube and fuel injector assembly.
12. Loosen the mounting bolts.
13. Remove fuel injector from fuel tube with the following procedure:
   a. Open and remove clip.
   b. Remove fuel injector from fuel tube by pulling straight.

To install:

14. Note the following, and install O-rings to fuel injector as follows:

➡**Upper and lower O-rings are different. Be careful not to confuse them.**

- Except for California: Fuel tube side—Blue, Nozzle side—Brown
- For California: Fuel tube side—Black, Nozzle side—Green
- Handle O-ring with bare hands. Never wear gloves.
- Lubricate O-ring with new engine oil.
- Never clean O-ring with solvent.
- Check that O-ring and its mating part are free of foreign material.
- When installing O-ring, be careful not to scratch it with tool or fingernails. Also be careful not to twist or stretch O-ring.
- If O-ring was stretched while it was

being attached, never insert it quickly into fuel tube.
- Insert O-ring straight into fuel tube. Never decenter or twist it.

15. Install fuel injector to fuel tube with the following procedure:
   a. Insert clip into clip mounting groove on fuel injector.
   b. Insert clip so that protrusion of fuel injector matches cutout of clip.

✳✳ WARNING

Never reuse clip. Replace it with a new one. Be careful to keep clip from interfering with O-ring. If interference occurs, replace O-ring.

c. Insert fuel injector into fuel tube with clip attached.
d. Insert it while matching it to the axial center.
e. Insert fuel injector so that protrusion of fuel tube matches cutout of clip.
f. Check that fuel tube flange is securely fixed in flange fixing groove on clip.
g. Check that installation is complete by making sure that fuel injector does not rotate or come off.

16. Install fuel tube and fuel injector assembly with the following procedure:
   a. Insert the tip of each fuel injector into intake manifold adapter.
   b. Tighten mounting bolts in numerical order

17. Connect sub-harness for fuel injector.
18. Install the intake manifold.
19. Note the following, and connect quick connector to install fuel feed hose:
   a. Check the connection for foreign material and damage.
   b. Align center to insert quick connector straight into fuel tube. Insert fuel tube into quick connector until the top spool on fuel tube is inserted completely and the second level spool is positioned slightly below quick connector bottom end.
   c. Hold position when inserting fuel tube into quick connector.
   d. Carefully align center to avoid inclined insertion to prevent damage to O-ring inside quick connector.
   e. Insert until you hear a (click sound) and actually feel the engagement.

20. To avoid misidentification of engagement with a similar sound, be sure to perform the next step:
   a. Before clamping fuel feed hose with hose clamps, pull quick connector hard by hand holding position.

b. Check it is completely engaged (connected) so that it does not come out from fuel feed tube.

21. Install the quick connector cap to quick connector connection.

22. Install so that the arrow mark on the side faces up.

23. Install fuel feed hose to hose clamp.

24. Install in the reverse order of removal procedure after this step.

FUEL TANK

REMOVAL & INSTALLATION

2WD Models

See Figures 93 and 94.

1. Before servicing the vehicle, refer to the Precautions.

2. Disconnect the negative battery cable.

3. Check the fuel level on the fuel gauge. If fuel gauge indicates more than ¾ of a tank, drain fuel tank to around a half a tank.

4. In case fuel pump does not operate, perform the following procedure:

a. Insert hose of less than 0.79 inch (20 mm) diameter into fuel filler tube through fuel filler opening to draw the fuel from fuel filler tube.

b. Disconnect fuel filler hose from fuel filler tube.

c. Insert hose into fuel tank through fuel filler hose to draw fuel from fuel tank.

5. Release the fuel pressure from the fuel lines.

6. Open the fuel filler lid.

7. Open fuel filler cap and release the pressure inside the fuel tank.

8. Remove rear seat cushion.

9. Remove inspection hole cover.

10. Disconnect fuel lines and harness connectors.

11. Remove the muffler assembly.

12. Remove the protector from the fuel tank.

13. Remove the vent hose at rear side of fuel tank.

14. Disconnect the EVAP tube at rear side of fuel tank.

15. Remove fuel filler hose at fuel filler tube side.

16. Remove the suspension bar.

17. Remove parking brake cable mounting bolts and separate the parking brake cable from suspension arm.

18. Support center of fuel tank with transmission jack.

19. Securely support the fuel tank with a piece of wood.

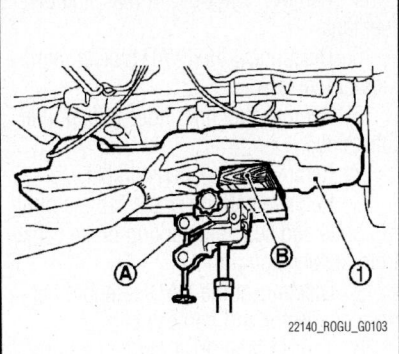

Fig. 93 Fuel tank (1) supported with transmission jack (A) and block of wood (B)

20. Remove fuel tank band (RH and LH).

21. Lower the transmission jack carefully to remove fuel tank while holding it by hand.

To install:

22. Reverse the removal at this point and note the following:

a. Tighten the fuel tank bands to 23 ft. lbs. (31 Nm).

b. Tighten the fuel tank protector to 44 inch lbs. (5 Nm).

23. Connect the negative battery cable.

AWD Models

See Figures 95 and 96.

1. Before servicing the vehicle, refer to the Precautions.

2. Disconnect the negative battery cable.

3. Check the fuel level on the fuel gauge. If fuel gauge indicates more than ¾ of a tank, drain fuel tank to around a half a tank.

4. In case fuel pump does not operate, perform the following procedure:

a. Insert hose of less than 0.79 in. (20 mm) diameter into fuel filler tube through fuel filler opening to draw the fuel from fuel filler tube.

b. Disconnect fuel filler hose from fuel filler tube.

c. Insert hose into fuel tank through fuel filler hose to draw fuel from fuel tank.

5. Release the fuel pressure from the fuel lines.

6. Open the fuel filler lid.

7. Open fuel filler cap and release the pressure inside the fuel tank.

8. Remove rear seat cushion.

9. Remove inspection hole cover.

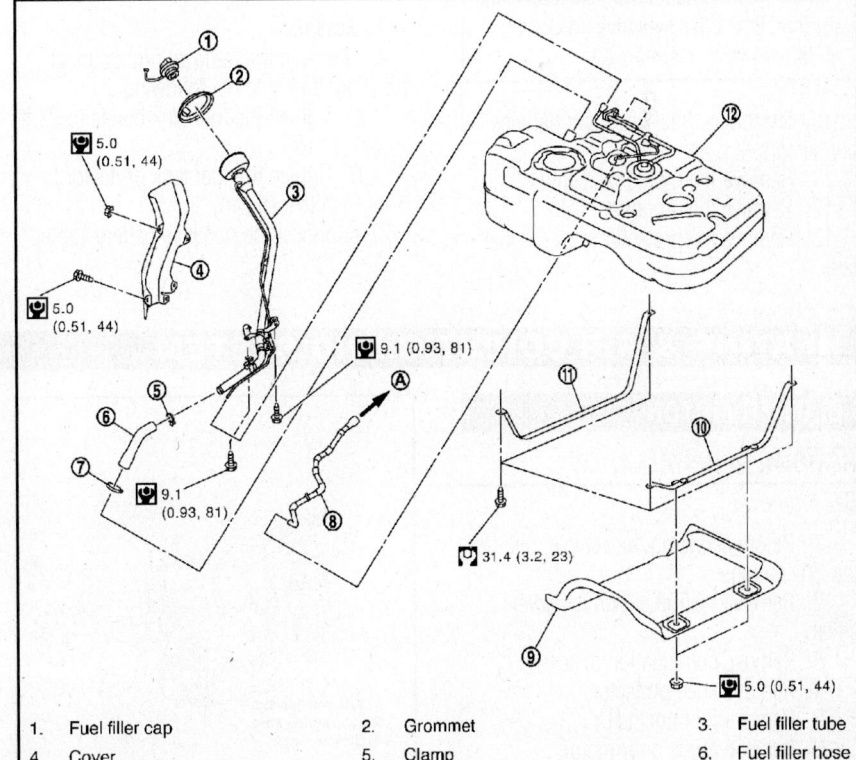

1.	Fuel filler cap	2.	Grommet	3.	Fuel filler tube
4.	Cover	5.	Clamp	6.	Fuel filler hose
7.	Clamp	8.	Vent hose	9.	Protector
10.	Fuel tank band (LH)	11.	Fuel tank band (RH)	12.	Fuel tank
A.	To EVAP canister				

Fig. 94 Exploded view of the 2WD fuel tank

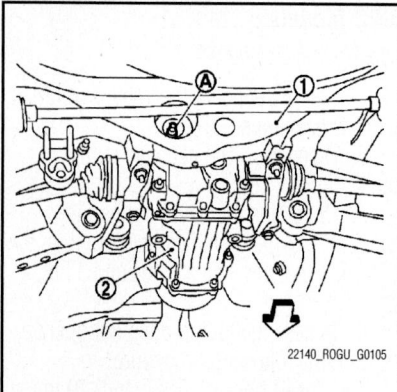

Fig. 95 Loosen final drive mounting nut (A) at rear suspension member (1)

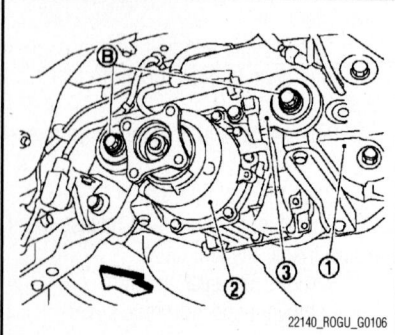

Fig. 96 Remove final drive mounting bolts (B) from final drive mounting bracket (3) to tilt final drive assembly (2)

10. Disconnect fuel lines and harness connectors.
11. Remove the muffler assembly.
12. Remove the propeller shaft.
13. Remove the protector from the fuel tank.

14. Remove vent hose at rear side of fuel tank.
15. Disconnect the EVAP tube at rear side of fuel tank.
16. Remove fuel filler hose at fuel filler tube side.
17. Remove the suspension bar.
18. Remove parking brake cable mounting bolts and separate parking brake cable from suspension arm.
19. Disconnect the AWD solenoid harness connector and harness clip.
20. Loosen final drive mounting nut at rear suspension member.

➡ **Never remove final drive mounting nut.**

21. Remove final drive mounting bolts from final drive mounting bracket (3) to tilt final drive assembly.

➡ **Final drive assembly does not have to be removed from the vehicle.**

22. Support center of fuel tank with transmission jack.
23. Securely support the fuel tank with a piece of wood.
24. Remove fuel tank band (RH and LH).
25. Lower the transmission jack carefully to remove fuel tank while holding it by hand.

To install:
26. Reverse the removal procedure at this point and note the following:
   a. Tighten the fuel tank bands to 23 ft. lbs. (31 Nm).
   b. Tighten the fuel tank protector to 44 inch lbs. (5 Nm).
27. Connect the negative battery cable.

THROTTLE BODY

REMOVAL & INSTALLATION
See Figure 97.

1. Before servicing the vehicle, refer to the Precautions.
2. Remove the air intake duct.
3. Disconnect harness connector.
4. Disconnect water hose.
5. Loosen the throttle body assembly mounting bolts in reverse order as shown in the figure.

To install:
6. Install the throttle body assembly with a new gasket.
7. Tighten the mounting bolts in sequence to 74 inch lbs. (8.4 Nm)
8. Reconnect the water hose.
9. Reconnect the harness connector.
10. Reconnect the air intake duct.

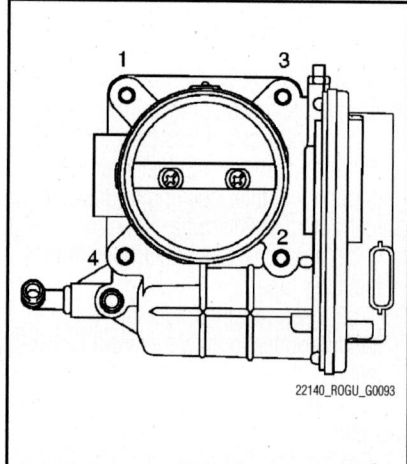

Fig. 97 Loosen mounting bolts in reverse order as shown

HEATING & AIR CONDITIONING SYSTEM

BLOWER MOTOR

REMOVAL & INSTALLATION

See Figure 98.

1. Before servicing the vehicle, refer to the Precautions.
2. Remove instrument driver lower cover.
3. Remove combination meter.
4. Remove knee protector.
5. Remove foot duct LH.
6. Remove mode door motor.
7. Remove brake pedal assembly and accelerator pedal.
8. Disconnect blower motor connector.
9. Press flange holding hook. Turn blower motor counterclockwise.

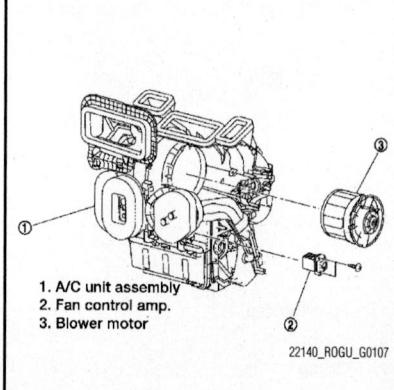

1. A/C unit assembly
2. Fan control amp.
3. Blower motor

Fig. 98 Blower motor removal

10. Pull outside and remove blower motor.

✳✳ WARNING

The balance is adjusted when blower fan and blower motor are assembled, so do not replace the individual parts.

To install:
11. Installation is the reverse of removal.

HEATER CORE

REMOVAL & INSTALLATION
See Figures 99 and 100.

1. Before servicing the vehicle, refer to the Precautions.

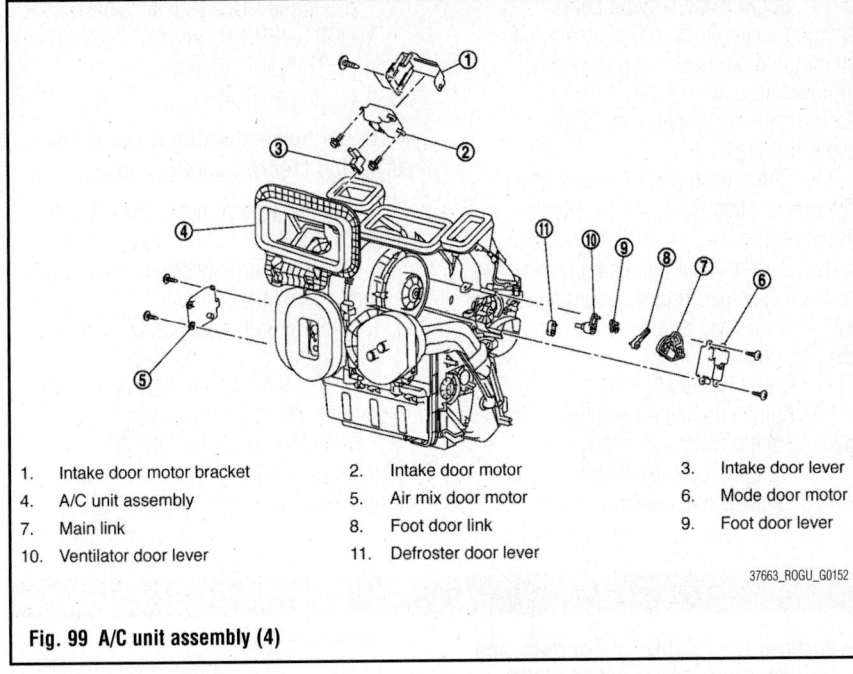

1. Intake door motor bracket
2. Intake door motor
3. Intake door lever
4. A/C unit assembly
5. Air mix door motor
6. Mode door motor
7. Main link
8. Foot door link
9. Foot door lever
10. Ventilator door lever
11. Defroster door lever

37663_ROGU_G0152

Fig. 99 A/C unit assembly (4)

2. Remove the A/C unit assembly.
3. Remove heater packing.
4. Remove heater pipe flange.

5. Remove mounting screw, and then remove heater pipe clamp.
6. Slide heater core to leftward (as shown in the figure).

37663_ROGU_G0161

Fig. 100 Remove heater packing (1), heater pipe flange (2), mounting screw (A), heater pipe clamp (3), and heater core (4)

To install:

7. Installation is the reverse order of removal.

STEERING

POWER STEERING GEAR

REMOVAL & INSTALLATION

See Figures 101 and 102.

1. Before servicing the vehicle, refer to the Precautions.
2. Set vehicle to the straight-ahead position.
3. Disconnect the negative battery cable.
4. Remove the upper cover.
5. Remove dash seal.
6. Remove hole cover.
7. Remove the bolt of intermediate shaft (lower side), and then remove intermediate shaft from steering gear pinion shaft.
8. Remove tires with a power tool.
9. Remove steering outer socket from steering knuckle so as not to damage ball joint boot using suitable ball joint remover.
10. Temporarily tighten the nut to prevent damage to threads and to prevent the ball joint remover from suddenly coming off.
11. Remove front suspension member.
12. Remove the steering gear assembly.

To install:

13. Note the following, and install in the reverse order of removal:

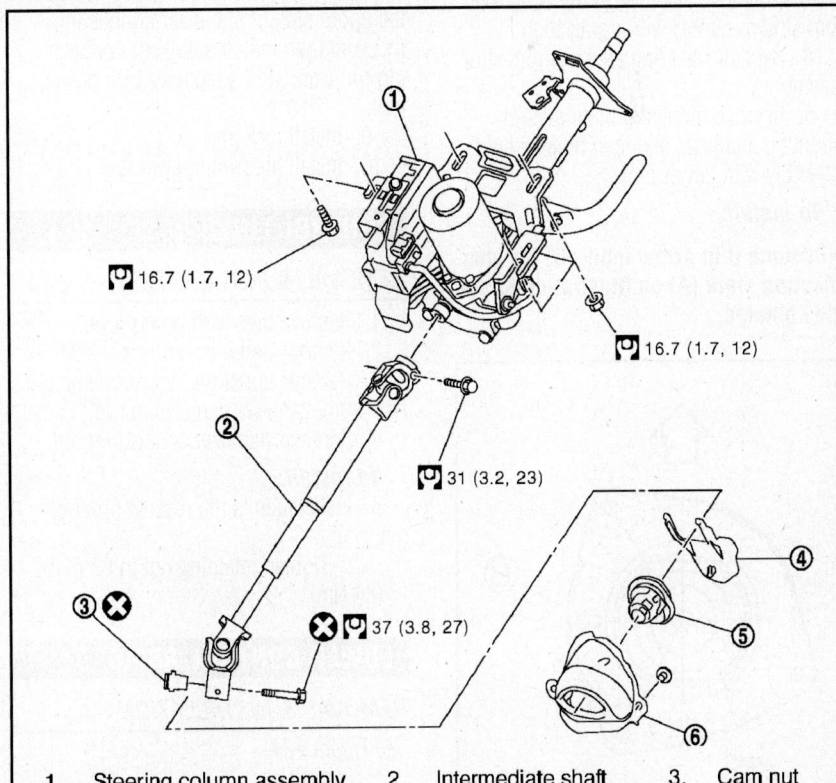

1. Steering column assembly
2. Intermediate shaft
3. Cam nut
4. Upper cover
5. Dash seal
6. Hole cover

22140_ROGU_G0111

Fig. 101 Remove the bolt of intermediate shaft (lower side)

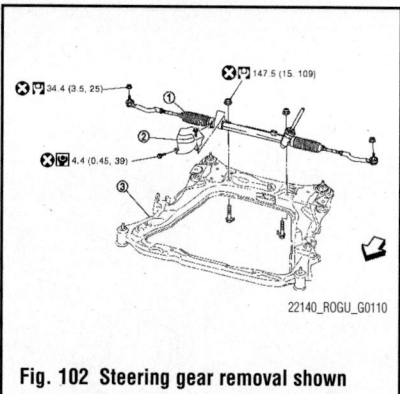

Fig. 102 Steering gear removal shown

a. Tighten the steering gear mounting nuts to 109 ft. lbs. (147.5 Nm).

b. Tighten the tie rod end mounting nuts to 25 ft. lbs. (34.4 Nm).

14. Spiral cable may be cut if steering wheel turns while separating steering column assembly and steering gear assembly. Be sure to secure steering wheel using string to avoid turning.

15. Check each part of dash seal for damage or other malfunctions. Replace if necessary.

16. Perform final tightening of nuts and bolts on each part under relaxed conditions with tires on level ground when removing steering gear assembly.

17. Check wheel alignment.

18. Adjust the neutral position of the steering angle sensor as follows:

a. Stop the vehicle with front wheels in straight-ahead position.

b. On the CONSULT-III® screen, touch "WORK SUPPORT" and "ST ANG SEN ADJUSTMENT" in order.

c. Touch "START".

➡**Do not touch steering wheel while adjusting steering angle sensor.**

d. After approximately 10 seconds, touch "END".

e. After approximately 60 seconds, it ends automatically.

f. Turn ignition switch OFF, then turn it ON again.

g. Run vehicle with front wheels in straight-ahead position, then stop.

h. Select "DATA MONITOR". Then make sure "STR ANGLE SIG" is within 0 plus or minus 2.5°.

SUSPENSION

COIL SPRINGS

REMOVAL & INSTALLATION

See Figure 103.

1. Remove tires with power tool.
2. Remove lock plat.
3. Remove cap and mounting nut on the upper side of stabilizer connecting rod, and then remove stabilizer connecting rod from strut assembly with power tool.
4. Separate steering knuckle from strut assembly.
5. Remove mounting bolts of strut mounting insulator, and then remove strut assembly with power tool.

To install:

➡**Become it in arrow mark (B) for identification mark (A) an illustration to the body outside.**

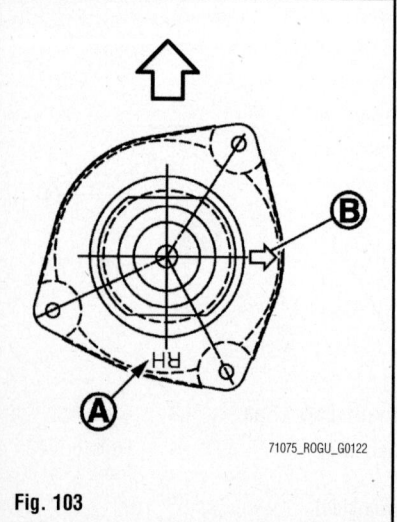

Fig. 103

➡**Perform final tightening of bolts and nuts, under unladen conditions with tires on level ground.**

6. Install mounting bolts of strut mounting insulator, and then install strut assembly with power tool.
7. Install steering knuckle to strut assembly.
8. Install cap and mounting nut on the upper side of stabilizer connecting rod, and then install stabilizer connecting rod from strut assembly with power tool.
9. Install lock plat.
10. Install tires with power tool.

CONTROL LINKS

REMOVAL & INSTALLATION

1. Remove tires with power tool.
2. Remove under cover from vehicle.
3. Remove upper and lower retaining nuts from stabilizer connecting rod.
4. Remove stabilizer connecting rod.

To install:

5. Installation is the reverse order of removal.

a. Tighten retaining nut to 62 ft. lbs. (84 Nm)

LOWER CONTROL ARMS

REMOVAL & INSTALLATION

See Figure 104.

1. Before servicing the vehicle, refer to the Precautions.
2. Raise the vehicle.
3. Remove the tire.

FRONT SUSPENSION

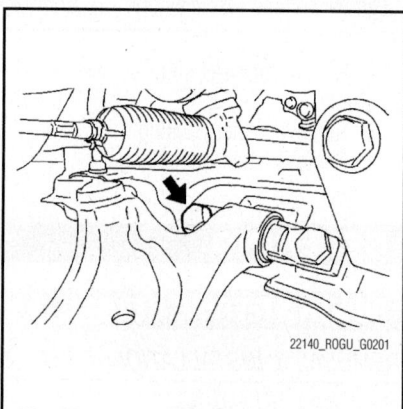

Fig. 104 Lower control arm to suspension member mounting bolts

4. Remove the lower control arm from the steering knuckle.

5. Remove the lower control arm from the suspension member.

➡**The transverse link cannot be pulled out because the mounting bolt of transverse link at the rear of the mounting area located on the front side of vehicle hits against the stabilizer bar. Therefore, get stabilizer bar out of the way to remove the transverse link.**

To install:

6. Note the following, and install in the reverse order of removal:

a. Perform final tightening of bolts and nuts at the front suspension member, under relaxed conditions with tires on level ground.

b. Tighten the front control arm mounting bolts to 126 ft. lbs. (171 Nm).

c. Tighten the rear control arm bolt and nut to 104 ft. lbs. (142 Nm).

STABILIZER BAR

REMOVAL & INSTALLATION

See Figures 105 and 106.

1. Remove tires power tool.
2. Remove under cover from vehicle.
3. Remove steering outer socket from steering knuckle.
4. Remove stabilizer connecting rod.
5. Remove rear torque rod.
6. Separate intermediate shaft from steering gear.
7. Set suitable jack under front suspension member.
8. Remove front suspension member stay from vehicle.
9. Gradually lower jack front suspension member in order to remove stabilizer mounting bolts.
10. Remove mounting bolts of stabilizer clamp, and then remove stabilizer clamp and stabilizer bushing from front suspension member.
11. Remove stabilizer bar.

To install:

12. Note the following, and install in the reverse order of removal.

   a. Install stabilizer clamp that notch is toward vehicle front side.

   b. Install stabilizer bushing that slit is toward vehicle front side.

STEERING KNUCKLE

REMOVAL & INSTALLATION

See Figure 107.

1. Before servicing the vehicle, refer to the Precautions.
2. Remove tires with power tool.
3. Remove wheel sensor from steering knuckle.

❊❊ WARNING

Never pull on the wheel sensor harness.

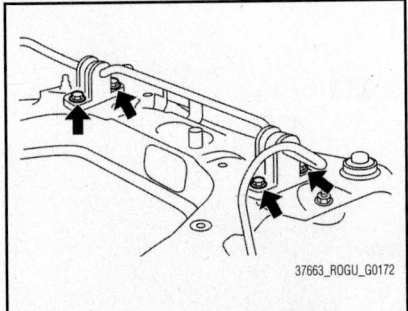

Fig. 105 Remove the stabilizer clamp mounting

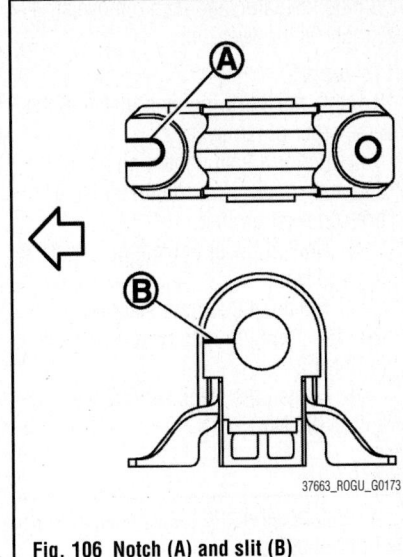

Fig. 106 Notch (A) and slit (B)

4. Remove lock plate from strut assembly.
5. Remove torque member mounting bolts with power tool. Hang torque member so as not to interfere with work.
6. Remove disc rotor. Put matching marks on the wheel hub and bearing assembly and the disc rotor before removing the disc rotor.
7. Remove cotter pin, and then loosen hub lock nut with power tool and disengage wheel hub and bearing assembly from drive shaft.
8. Remove the hub lock nut.

➡**Use suitable puller, if wheel hub and bearing assembly and driveshaft cannot be separated even after performing the above procedure.**

9. Remove wheel hub and bearing assembly, and then remove splash guard.
10. Remove transverse link from steering knuckle.
11. Remove the steering knuckle from strut assembly.
12. Loosen the nut of steering outer socket.
13. Remove steering outer socket from steering knuckle using the ball joint remover so as not to damage ball joint boot.
14. Temporarily tighten the nut to prevent damage to threads and to prevent the ball joint remover from suddenly coming off.
15. Remove the steering knuckle from vehicle.

To install:

16. Note the following, and install in the reverse order of the removal:

   a. Align the matching marks made during removal when reusing the disc rotor.

   b. Install removed wheel hub and bearing assembly and steering knuckle and perform the final tightening of each part under relaxed conditions on the level surface.

   c. Tighten the stabilizer link mounting nuts to 62 ft. lbs. (84 Nm).

   d. Tighten the lower strut knuckle bolt to 104 ft. lbs. (141 Nm).

   e. Tighten the tie rod end mounting nuts to 25 ft. lbs. (34 Nm).

   f. Tighten the axle nut to 92 ft. lbs. (125 Nm).

   g. Install the front wheels, lower the vehicle and perform a front end alignment.

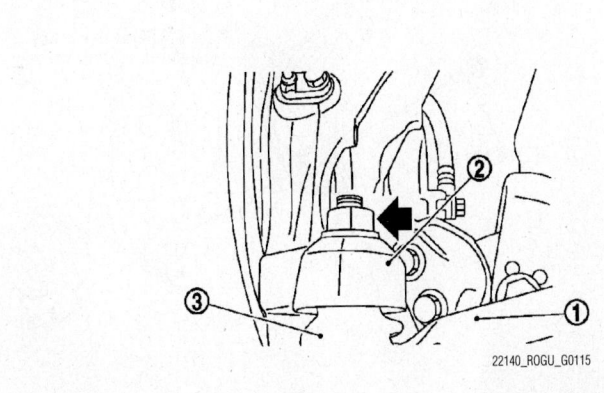

Fig. 107 Remove steering outer socket (1) from steering knuckle (2) using the ball joint remover so as not to damage ball joint boot (3)

STRUTS

REMOVAL & INSTALLATION

See Figure 108.

1. Before servicing the vehicle, refer to the Precautions.
2. Raise the vehicle.
3. Remove the tire.
4. Remove the brake hose lock plate clip.
5. Remove cap and mounting nut on the upper side of stabilizer connecting rod, and then remove stabilizer connecting rod from strut assembly with power tool.
6. Separate steering knuckle from strut assembly.
7. Remove mounting bolts of strut

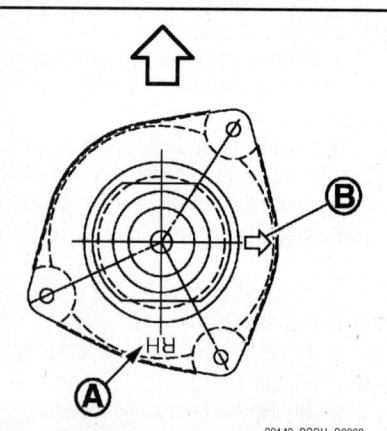

Fig. 108 Identification mark (A), Arrow mark (B) and Large arrow front of the vehicle

mounting insulator, and then remove strut assembly with power tool.

To install:

8. Note the following, and install in the reverse order of removal:

   a. Perform final tightening of bolts and nuts, under relaxed conditions with tires on level ground.

   b. Tighten the strut plate nuts to 14 ft. lbs. (19 Nm).

   c. Tighten the strut to steering knuckle mounting nuts to 104 ft. lbs. (142 Nm).

   d. Tighten the stabilizer link nut to 62 ft. lbs. (84 Nm).

OVERHAUL

1. Before servicing the vehicle, refer to the Precautions.
2. Raise the vehicle.
3. Remove the tire.
4. Remove the strut assembly.
5. Install the strut assembly into a spring compressor service tool.
6. Compress the coil spring. Make sure coil spring with a spring compressor between strut mounting bearing and lower rubber seat (strut assembly) is free.
7. Remove piston rod lock nut while securing the piston rod tip so that piston rod does not turn.
8. Remove strut mounting insulator and strut mounting bearing, and bound bumper from strut.
9. After remove coil spring with a spring compressor, and then gradually release a spring compressor.
10. Remove the strut.

To install:

11. Install the strut.
12. Install lower rubber seat.
13. Install bound bumper onto strut mounting insulator.
14. Compress coil spring using a spring compressor (commercial service tool), and install it onto strut assembly.
15. Install strut mounting bearing and strut mounting insulator with bound bumper to strut.
16. Secure piston rod tip so that piston rod does not turn, then tighten to 58 ft. lbs. (79 Nm).
17. Gradually remove the spring pressure.
18. Remove the strut assembly from service tool.
19. Install the strut assembly to the vehicle.
20. Install the tire and lower the vehicle.

WHEEL HUBS & BEARINGS (SEALED UNIT)

ADJUSTMENT

No adjustment is necessary or possible.

REMOVAL & INSTALLATION

See Figure 109.

1. Remove tires with power tool.
2. Remove wheel sensor from steering knuckle.
3. Remove lock plate from strut assembly.
4. Remove torque member mounting

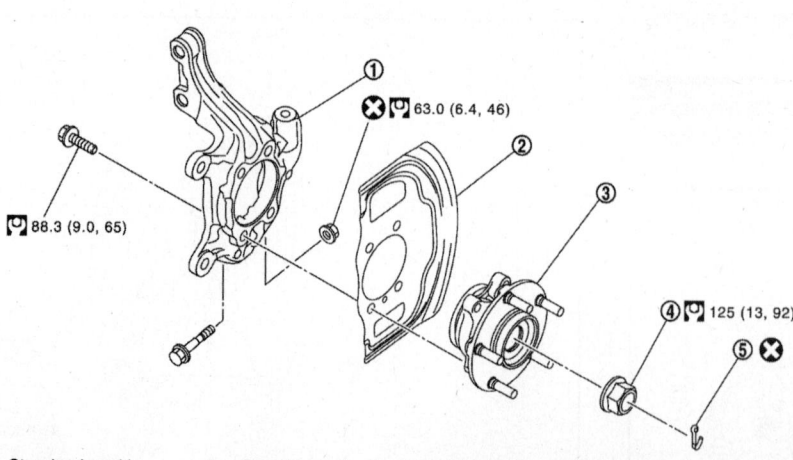

1. Steering knuckle
2. Splash guard
3. Wheel hub and bearing assembly
4. Wheel hub lock nut
5. Cotter pin

Fig. 109 Wheel hub and bearing assembly

bolts with power tool. Hang torque member not to interfere with work.

❋❋ WARNING

Never depress brake pedal while brake caliper is removed.

5. Remove disc rotor.
6. Remove cotter pin, and then loosen wheel hub lock nut with power tool.
7. Patch wheel hub lock nut with a piece of wood.

8. Hammer the wood to disengage wheel hub and bearing assembly from drive shaft. Remove the wheel hub lock nut.

❋❋ WARNING

Never place drive shaft joint at an extreme angle. Also be careful not to overextend slide joint. Never allow drive shaft to hang down without support for housing (or joint sub-assembly), shaft and the other parts.

➡**Use suitable puller, if wheel hub and bearing assembly and drive shaft cannot be separated even after performing the above procedure.**

9. Remove wheel hub and bearing assembly.

To install:

10. Installation is the reverse of removal.

SUSPENSION

CONTROL ARMS/LINKS

REMOVAL & INSTALLATION

1. Before servicing the vehicle, refer to the Precautions.
2. Remove tires with power tool.
3. Set suitable jack under suspension arm to relieve the coil spring tension.
4. Remove the shock absorber.
5. Remove the coil spring.

To install:

➡**Perform final tightening of the bolts and nuts at the shock absorber lower side (rubber bushing), under relaxed conditions with tires on level ground.**

6. Install the coil spring.
7. Raise the suspension arm to line up shock mounting holes.
8. Install the shock absorber.
9. Tighten the top mounting nut and bolt to 89 ft. lbs. (120 Nm).
10. Tighten the bottom mounting nut and bolt to 89 ft. lbs. (120 Nm).

SHOCK ABSORBERS

REMOVAL & INSTALLATION

1. Before servicing the vehicle, refer to the Precautions.
2. Remove tires with power tool.
3. Set suitable jack under suspension arm to relieve the coil spring tension.
4. Remove the shock absorber.

To install:

➡**Perform final tightening of the bolts and nuts at the shock absorber lower side (rubber bushing), under relaxed conditions with tires on level ground.**

5. Raise the suspension arm to line up shock mounting holes.
6. Install the shock absorber.
7. Tighten the top mounting nut and bolt to 89 ft. lbs. (120 Nm).
8. Tighten the bottom mounting nut and bolt to 89 ft. lbs. (120 Nm).

STABILIZER BAR

REMOVAL & INSTALLATION

1. Before servicing the vehicle, refer to the Precautions.
2. Remove the stabilizer links.
3. Remove the main muffler.
4. Remove the mounting nuts on stabilizer clamp and stabilizer bar from suspension member.

To install:

5. Install the mounting nuts on stabilizer clamp and stabilizer bar to suspension member.
6. Tighten the stabilizer clamp nuts to 26 ft. lbs. (36 Nm).
7. Install the main muffler.
8. Tighten the stabilizer link mounting bolts to 81 ft. lbs. (110 Nm).

REAR SUSPENSION

WHEEL HUBS & BEARINGS (SEALED UNIT)

REMOVAL & INSTALLATION

See Figure 110.

1. Before servicing the vehicle, refer to the Precautions.
2. Remove tires with power tool.
3. Remove wheel sensor from axle housing.

❋❋ WARNING

Never pull on the wheel sensor harness.

4. Remove torque member mounting bolts. Hang torque member not to interfere with work.
5. Remove disc rotor. Put matching marks on the wheel hub and bearing assembly and the disc rotor before removing the disc rotor.
6. Remove cotter pin, and then loosen hub lock nut with power tool (AWD).
7. Patch hub lock nut with a piece of wood. Hammer the wood to disengage

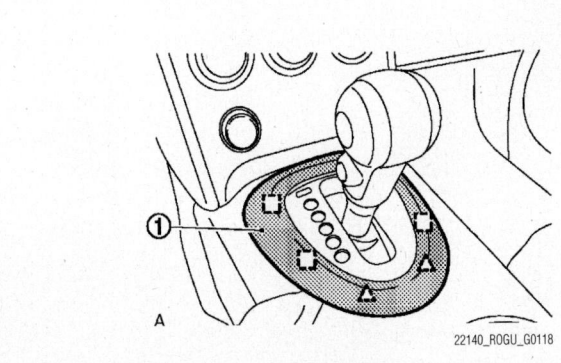

22140_ROGU_G0118

Fig. 110 Remove rear wheel hub and bearing assembly

wheel hub and bearing assembly from drive shaft.

8. Remove the hub lock nut (AWD).

✳✳ WARNING

Never place drive shaft joint at an extreme angle. Also be careful not to overextend slide joint. Never allow drive shaft to hang down without

support for housing (or joint sub-assembly), shaft and the other parts (AWD).

9. Use suitable puller if wheel hub and bearing assembly and drive shaft cannot be separated even after performing the above procedure.

10. Remove rear wheel hub and bearing assembly.

To install:

11. Reverse the removal procedure and note the following:

a. Tighten the hub and bearing assembly mounting bolts to 65 ft. lbs. (88.3 Nm).

b. Tighten the axle locking nut to 92 ft. lbs. (125 Nm) (AWD).

c. Install new cotter pin (AWD).

NISSAN

Sentra

SPECIFICATIONS AND MAINTENANCE CHARTS

ENGINE AND VEHICLE IDENTIFICATION

		Engine						Model Year	
Code ①	Liters (cc)	Cu. In.	Cyl.	Fuel Sys.	Engine Type	Eng. Mfg.		Code ②	Year
MR20DE	2.0 (1997)	122	4	MFI	DOHC	Nissan		B	2011
QR25DE	2.5 (2488)	152	4	MFI	DOHC	Nissan		C	2012

① 4th position of VIN

② 10th position of VIN

71075_SENT_C0001

GENERAL ENGINE SPECIFICATIONS

All measurements are given in inches.

Year	Model	Engine Disp. Liters	Engine ID	Fuel System Type	Net Horsepower @ rpm	Net Torque @ rpm (ft. lbs.)	Bore x Stroke (in.)	Compression Ratio	Oil Pressure @ rpm
2011	Sentra	2.0	MR20DE	SMPI	140@5100	147@4800	3.31x3.55	9.7:1	29@2000
	Sentra SE-R	2.5	QR25DE	SMPI	177@6000	172@2800	3.5x3.94	9.5:1	43@2000
	Sentra SE-R Spec V	2.5	QR25DE	SMPI	200@6600	180@5200	3.5x3.94	10.5:1	43@2000
2012	Sentra	2.0	MR20DE	SMPI	140@5100	147@4800	3.31x3.55	9.7:1	29@2000
	Sentra SE-R	2.5	QR25DE	SMPI	177@6000	172@2800	3.5x3.94	9.5:1	43@2000
	Sentra SE-R Spec V	2.5	QR25DE	SMPI	200@6600	180@5200	3.5x3.94	10.5:1	43@2000

71075_SENT_C0002

ENGINE TUNE-UP SPECIFICATIONS

Year	Engine Disp. Liters	Engine ID	Spark Plug Gap (in.)	Ignition Timing (deg.) ①		Fuel Pump (psi) ②	Idle Speed (rpm) ③		Valve Clearance ④	
				MT	AT		MT	AT	Intake	Exhaust
2011	2.0	MR20DE	0.043	6 +/- 5	6 +/- 5	51	675 +/- 50	700 +/- 50	0.010-0.013	0.011-0.015
	2.5	QR25DE	0.043	10 +/- 5	10 +/- 5	51	800 +/- 50	650 +/- 50	0.009-0.013	0.010-0.013
2012	2.0	MR20DE	0.043	6 +/- 5	6 +/- 5	51	675 +/- 50	700 +/- 50	0.010-0.013	0.011-0.015
	2.5	QR25DE	0.043	10 +/- 5	10 +/- 5	51	800 +/- 50	650 +/- 50	0.009-0.013	0.010-0.013

B: Before top dead center

① A/T - in P or N position; M/T - in Neutral position

② At idle

③ A/T - In P or N position; M/T - in Neutral position

④ Engine cold

71075_SENT_C0003

CAPACITIES

Year	Model	Engine Displacement Liters	Engine ID	Engine Oil with Filter (qts.)	Transaxle (pts.) Auto.	Transaxle (pts.) Manual	Fuel Tank (gal.)	Cooling System (qts.) ①
2011	Sentra	2.0	MR20DE	4.0	15.5	8.5	14.5	7.4
	Sentra SE-R/ SE-R Spec V	2.5	QR25DE	4.5	15.75	7.25	14.5	②
2012	Sentra	2.0	MR20DE	4.0	15.5	8.5	14.5	7.4
	Sentra SE-R/ SE-R Spec V	2.5	QR25DE	4.5	15.75	7.25	14.5	②

NOTE: All capacities are approximate. Add fluid gradually and ensure a proper fluid level is obtained.

① With reservoir tank at MAX level.

② M/T 7.25, A/T 7.5

71075_SENT_C0004

FLUID SPECIFICATIONS

Year	Model	Engine Displacement Liters	Engine ID	Engine Oil	Manual Trans.	Auto. Trans.	Brake Master Cylinder	Cooling System
2011	Sentra	2.0	MR20DE	5W-30	75W-80	Nissan NS-2	DOT 3	N-LL
	Sentra SE-R SE-R Spec V	2.5	QR25DE	5W-30	75W-85	Nissan NS-2	DOT 3	N-LL
2012	Sentra	2.0	MR20DE	5W-30	75W-80	Nissan NS-2	DOT 3	N-LL
	Sentra SE-R SE-R Spec V	2.5	QR25DE	5W-30	75W-85	Nissan NS-2	DOT 3	N-LL

DOT: Department Of Transpotation

N-LL: Nissan Long Life Anitfreeze/Coolant (blue) or equivalent

NA: Not available

71075_SENT_C0005

VALVE SPECIFICATIONS

Year	Engine Displacement Liters	Engine ID	Seat Angle (deg.)	Spring Test Pressure (lbs. @ in.)	Spring Free-Length (in.)	Spring Installed Height (in.)	Stem-to-Guide Clearance (in.) Intake	Stem-to-Guide Clearance (in.) Exhaust	Stem Diameter (in.) Intake	Stem Diameter (in.) Exhaust
2011	2.0	MR20DE	45.15-45.45	34-39@ 1.39	①	1.390	0.0008-0.0021	0.0012-0.0025	0.2152 0.2157	0.2148 0.2154
	2.5	QR25DE	45.15-45.45	34-39@ 1.39	②	1.390	0.0008-0.0021	0.0012-0.0025	0.2348-0.2354	0.2344-0.2350
2012	2.0	MR20DE	45.15-45.45	34-39@ 1.39	①	1.390	0.0008-0.0021	0.0012-0.0025	0.2152 0.2157	0.2148 0.2154
	2.5	QR25DE	45.15-45.45	34-39@ 1.39	②	1.390	0.0008-0.0021	0.0012-0.0025	0.2348-0.2354	0.2344-0.2350

① Intake (1.7677-1.7755) and Exhuast (1.8007-1.8086)

② Intake (1.7213-1.7291) and Exhaust (1.7831-1.7909)

71075_SENT_C0006

CAMSHAFT SPECIFICATIONS
All measurements in inches unless noted

Year	Engine Displacement Liters	Engine ID	Journal Diameter	Brg. Oil Clearance	Shaft End-play	Runout	Journal Bore	Lobe Height Intake	Lobe Height Exhaust
2011	2.0	MR20DE	①	0.0018-0.0034	0.0030-0.006	0.0008	NA	1.7560-1.7635	1.6997-1.7072
	2.5	QR25DE	②	0.0018-0.0034	0.0045-0.0074	0.0016	NA	1.7644-1.7718	1.7313-1.7388
2012	2.0	MR20DE	①	0.0018-0.0034	0.003-0.006	0.0008	NA	1.7560-1.7635	1.6997-1.7072
	2.5	QR25DE	②	0.0018-0.0034	0.0045-0.0074	0.0016	NA	1.7644-1.7718	1.7313-1.7388

NA: Not available

① Journal 1: 1.0998 - 1.1006

  All Others: 0.9823 - 0.9381

② Journal 1: 1.0998 - 1.1006

  All Others: 0.9226 - 0.9234

71075_SENT_C0007

CRANKSHAFT AND CONNECTING ROD SPECIFICATIONS
All measurements are given in inches.

Year	Engine Disp. Liters	Engine ID	Crankshaft Main Brg. Journal Dia.	Crankshaft Main Brg. Oil Clearance	Crankshaft Shaft End-play	Crankshaft Thrust on No.	Connecting Rod Journal Diameter	Connecting Rod Oil Clearance	Connecting Rod Side Clearance
2011	2.0	MR20DE	2.0456-2.0464	①	0.0039-0.0102	3	1.7305-1.7311	0.0015-0.0019	0.0079-0.0138
	2.5	QR25DE	2.1636-2.1645	②	0.0039-0.0102	3	1.7699-1.7706	0.0014-0.0018	0.0079-0.0138
2012	2.0	MR20DE	2.0456-2.0464	①	0.0039-0.0102	3	1.7305-1.7311	0.0015-0.0019	0.0079-0.0138
	2.5	QR25DE	2.1636-2.1645	②	0.0039-0.0102	3	1.7699-1.7706	0.0014-0.0018	0.0079-0.0138

① No. 1, 4, 5 : 0.0009-0.0013

  No. 2, 3 : 0.0005-0.0009

② No. 1, 3, and 5: 0.0005 - 0.0009

  No. 2 and 4: 0.0007 - 0.0011

71075_SENT_C0008

PISTON AND RING SPECIFICATIONS

All measurements are given in inches.

Year	Engine Disp. Liters	Engine ID	Piston Clearance	Ring Gap			Ring Side Clearance		
				Top Compression	Bottom Compression	Oil Control	Top Compression	Bottom Compression	Oil Control
2011	2.0	MR20DE	0.0008-0.0016	0.008-0.0120	0.020-0.0260	0.006-0.0180	0.002-0.0030	0.001-0.0030	0.001-0.0070
	2.5	QR25DE	0.0004-0.0012	0.0083-0.0122	0.0146-0.0205	0.0079-0.0177	0.0018-0.0031	0.0012-0.0028	0.0018-0.0049
2012	2.0	MR20DE	0.0008-0.0016	0.008-0.0120	0.020-0.0260	0.006-0.0180	0.002-0.0030	0.001-0.0030	0.001-0.0070
	2.5	QR25DE	0.0004-0.0012	0.0083-0.0122	0.0146-0.0205	0.0079-0.0177	0.0018-0.0031	0.0012-0.0028	0.0018-0.0049

71075_SENT_C0009

TORQUE SPECIFICATIONS

All readings in ft. lbs.

Year	Engine Disp. Liters	Engine ID	Cylinder Head Bolts	Main Bearing Bolts	Rod Bearing Bolts	Crankshaft Damper Bolts	Flywheel Bolts	Manifold		Spark Plugs	Oil Pan Drain Plug
								Intake	Exhaust		
2011	2.0	MR20DE	⑤	⑥	⑦	④	80	20	25	14	25
	2.5	QR25DE	①	②	③	④	76-83	13-15	29-32	14	25
2012	2.0	MR20DE	⑤	⑥	⑦	④	80	20	25	14	25
	2.5	QR25DE	①	②	③	④	76-83	13-15	29-32	14	25

① Step 1: 37 ft. lbs.

Step 2: 60-66 degrees

Step 3: Loosen bolts completely, then retorque to 26-32 ft. lbs.

Step 4: Turn each bolt, in sequence, an additional 75-80 degrees

Step 5: Repeat, turn each bolt, in sequence, an additional 75-80 degrees

② Bolt Nos. 1-10:

Step 1: 27-31 ft. lbs.

Step 2: Torque an additional 60-65 degrees

Bolt Nos. 11-14: Torque last, to 17-20 ft. lbs.

③ Step 1: 20 ft. lbs.

Step 2: Loosen bolts completely

Step 3: 14 ft. lbs.

Step 4: 85-95 degrees

④ Step 1: 29-36 ft. lbs.

Step 2: 60-66 degrees

⑤ Step 1: Tighten all in sequence to 30 ft. lbs.

Step 2: Tighten an additional 100 degrees

Step 3: Loosen all bolts in reverse order

Step 4: Repeat steps 1 and 2

Step 5: Tighten an additional 100 degrees

⑥ Step 1: 25 ft. lbs.

Step 2: Tighten an additional 60 degrees

⑦ Step 1: 20 ft. lbs.

Step 2: loosen bolts

Step 3: 14 ft. lbs.

71075_SENT_C0010

WHEEL ALIGNMENT

Year	Model		Caster Range (+/-Deg.)	Caster Preferred Setting (Deg.)	Camber Range (+/-Deg.)	Camber Preferred Setting (Deg.)	Toe-in (in.)
2011	Sentra	F	0.75	+4.75	0.75	①	0.08+/-0.04
		R	NA	NA	0.50	-0.92	0.08+/-0.08
2012	Sentra	F	0.75	+4.75	0.75	①	0.08+/-0.04
		R	NA	NA	0.50	-0.92	0.08+/-0.08

71075_SENT_C0011

TIRE, WHEEL AND BALL JOINT SPECIFICATIONS

Year	Model	OEM Tires Standard	OEM Tires Optional	Tire Pressures (psi) Front	Tire Pressures (psi) Rear	Wheel Size	Ball Joint Inspection	Lug Nut (ft. lbs.)
2011	Sentra	②	N/A	33	33	③	①	83
	Sentra SE-R	②	N/A	35	35	17 x 7.0JJ	①	83
	Sentra SE-R Spec V	②	N/A	35	35	17 x 7.0JJ	①	83
2012	Sentra	②	N/A	33	33	③	①	83
	Sentra SE-R	②	N/A	35	35	17 x 7.0JJ	①	83
	Sentra SE-R Spec V	②	N/A	35	35	17 x 7.0JJ	①	83

OEM: Original Equipment Manufacturer

PSI: Pounds Per Square Inch

NA: Information not available

N/A: Not applicable

① Replace if any measurable movement is found.

② Sentra: P205/60HR15

   Sentra S, SR, SL: P205/55HR16

   Sentra SE-R: P225/45VR17

   Sentra SE-R Spec V: P225/45WR17

③ Sentra: 15 x 6.5JJ; Sentra S, SR, SL - 16 x 6.5JJ

71075_SENT_C0012

BRAKE SPECIFICATIONS

All measurements in inches unless noted

Year	Model		Brake Disc Original Thickness	Brake Disc Minimum Thickness	Brake Disc Max. Runout	Brake Drum Diameter Original Inside Diameter	Brake Drum Diameter Max. Wear Limit	Minimum Pad/Lining Thickness Front	Minimum Pad/Lining Thickness Rear	Brake Caliper Bracket Bolts (ft. lbs.)	Brake Caliper Mounting Bolts (ft. lbs.)
2011	Sentra	F	0.945	0.866	0.0014	N/A	N/A	0.079	0.114	122	25
		R	N/A	N/A	N/A	9.000	9.055	N/A	0.059	62	32
	Sentra SE-R	F	1.024	0.945	0.0014	N/A	N/A	0.079	—	122	25
		R	0.354	0.315	0.0028	N/A	N/A	—	0.079	62	32
	Sentra SE-R Spec V	F	1.102	1.024	0.0014	N/A	N/A	0.079	—	122	25
		R	0.354	0.315	0.0028	N/A	N/A	—	0.079	62	32
2012	Sentra	F	0.945	0.866	0.0014	N/A	N/A	0.079	0.114	122	25
		R	N/A	N/A	N/A	9.000	9.055	N/A	0.059	62	32
	Sentra SE-R	F	1.024	0.945	0.0014	N/A	N/A	0.079	—	122	25
		R	0.354	0.315	0.0028	N/A	N/A	—	0.079	62	32
	Sentra SE-R Spec V	F	1.102	1.024	0.0014	N/A	N/A	0.079	—	122	25
		R	0.354	0.315	0.0028	N/A	N/A	—	0.079	62	32

F: Front

R: Rear

NS: Information not specified

N/A: Not applicable

71075_SENT_C0013

SCHEDULED MAINTENANCE INTERVALS
Nissan—Sentra

TO BE SERVICED	TYPE OF SERVICE	VEHICLE MILEAGE INTERVAL (x1000)												
		7.5	15	22.5	30	37.5	45	52.5	60	67.5	75	82.5	90	97.5
Engine oil & filter	R	✓	✓	✓	✓	✓	✓	✓	✓	✓	✓	✓	✓	✓
Brake lines & cables	S/I		✓		✓		✓		✓		✓		✓	
Brake pads, discs, drums & linings	S/I		✓		✓		✓		✓		✓		✓	
Driveshaft boots	S/I		✓		✓		✓		✓		✓		✓	
Exhaust system	S/I				✓				✓				✓	
Transaxle fluid	S/I		✓		✓		✓		✓		✓		✓	
Air cleaner filter	R				✓				✓				✓	
Spark plugs (except platinum)	R				✓				✓				✓	
Spark plugs (iridium and platinum)	R	Replace every 105,000 miles												
Steering gear & linkage, axle & suspension parts	S/I				✓				✓				✓	
Engine coolant	R	Replace every 60,000 miles, then every 30,000 miles												
Inverter coolant	R	Replace every 60,000 miles, then every 30,000 miles												
Drive belts	S/I								✓					
Fuel lines	S/I								✓					
Vapor lines	S/I								✓					
Cabin microfilter	R		✓		✓		✓		✓		✓		✓	
Valve adjustment	S/I	As needed												

R: Replace   S/I: Service or Inspect

FREQUENT OPERATION MAINTENANCE (SEVERE SERVICE)

If a vehicle is operated under any of the following conditions it is considered severe service:

- Extremely dusty areas.

- 50% or more of the vehicle operation is in 32°C (90°F) or higher temperatures, or constant operation in temperatures below 0°C (32°F).

- Prolonged idling (vehicle operation in stop and go traffic).

- Frequent short running periods (engine does not warm to normal operating temperatures).

- Police, taxi, delivery usage or trailer towing usage.

Oil & oil filter: change every 3750 miles.

Brake pads & discs: service or inspect every 7500 miles.

Driveshaft boots: service or inspect every 7500 miles.

Exhaust system: service or inspect every 7500 miles.

Steering gear & linkage, axle & suspension parts: service or inspect every 7500 miles.

Steering linkage ball joints & front suspension ball joints: service or inspect every 7500 miles.

Air cleaner filter: service or inspect every 15,000 miles.

71075_SENT_C0014

PRECAUTIONS

Before servicing any vehicle, please be sure to read all of the following precautions, which deal with personal safety, prevention of component damage, and important points to take into consideration when servicing a motor vehicle:

• Never open, service or drain the radiator or cooling system when the engine is hot; serious burns can occur from the steam and hot coolant.

• Observe all applicable safety precautions when working around fuel. Whenever servicing the fuel system, always work in a well-ventilated area. Do not allow fuel spray or vapors to come in contact with a spark, open flame, or excessive heat (a hot drop light, for example). Keep a dry chemical fire extinguisher near the work area. Always keep fuel in a container specifically designed for fuel storage; also, always properly seal fuel containers to avoid the possibility of fire or explosion. Refer to the additional fuel system precautions later in this section.

• Fuel injection systems often remain pressurized, even after the engine has been turned **OFF**. The fuel system pressure must be relieved before disconnecting any fuel lines. Failure to do so may result in fire and/or personal injury.

• Brake fluid often contains polyglycol ethers and polyglycols. Avoid contact with the eyes and wash your hands thoroughly after handling brake fluid. If you do get brake fluid in your eyes, flush your eyes with clean, running water for 15 minutes. If eye irritation persists, or if you have taken brake fluid internally, IMMEDIATELY seek medical assistance.

• The EPA warns that prolonged contact with used engine oil may cause a number of skin disorders, including cancer. You should make every effort to minimize your exposure to used engine oil. Protective gloves should be worn when changing oil. Wash your hands and any other exposed skin areas as soon as possible after exposure to used engine oil. Soap and water, or waterless hand cleaner should be used.

• All new vehicles are now equipped with an air bag system, often referred to as a Supplemental Restraint System (SRS) or Supplemental Inflatable Restraint (SIR) system. The system must be disabled before performing service on or around system components, steering column, instrument panel components, wiring and sensors. Failure to follow safety and disabling procedures could result in accidental air bag deployment, possible personal injury and unnecessary system repairs.

• Always wear safety goggles when working with, or around, the air bag system. When carrying a non-deployed air bag, be sure the bag and trim cover are pointed away from your body. When placing a non-deployed air bag on a work surface, always face the bag and trim cover upward, away from the surface. This will reduce the motion of the module if it is accidentally deployed. Refer to the additional air bag system precautions later in this section.

• Clean, high quality brake fluid from a sealed container is essential to the safe and proper operation of the brake system. You should always buy the correct type of brake fluid for your vehicle. If the brake fluid becomes contaminated, completely flush the system with new fluid. Never reuse any brake fluid. Any brake fluid that is removed from the system should be discarded. Also, do not allow any brake fluid to come in contact with a painted surface; it will damage the paint.

• Never operate the engine without the proper amount and type of engine oil; doing so WILL result in severe engine damage.

• Timing belt maintenance is extremely important. Many models utilize an interference-type, non-freewheeling engine. If the timing belt breaks, the valves in the cylinder head may strike the pistons, causing potentially serious (also time-consuming and expensive) engine damage. Refer to the maintenance interval charts for the recommended replacement interval for the timing belt, and to the timing belt section for belt replacement and inspection.

• Disconnecting the negative battery cable on some vehicles may interfere with the functions of the on-board computer system(s) and may require the computer to undergo a relearning process once the negative battery cable is reconnected.

• When servicing drum brakes, only disassemble and assemble one side at a time, leaving the remaining side intact for reference.

• Only an MVAC-trained, EPA-certified automotive technician should service the air conditioning system or its components.

BRAKES

ANTI-LOCK BRAKE SYSTEM (ABS)

GENERAL INFORMATION

PRECAUTIONS

• Certain components within the ABS system are not intended to be serviced or repaired individually.

• Do not use rubber hoses or other parts not specifically specified for and ABS system. When using repair kits, replace all parts included in the kit. Partial or incorrect repair may lead to functional problems and require the replacement of components.

• Lubricate rubber parts with clean, fresh brake fluid to ease assembly. Do not use shop air to clean parts; damage to rubber components may result.

• Use only DOT 3 brake fluid from an unopened container.

• If any hydraulic component or line is removed or replaced, it may be necessary to bleed the entire system.

• A clean repair area is essential. Always clean the reservoir and cap thoroughly before removing the cap. The slightest amount of dirt in the fluid may plug an orifice and impair the system function. Perform repairs after components have been thoroughly cleaned; use only denatured alcohol to clean components. Do not allow ABS components to come into contact with any substance containing mineral oil; this includes used shop rags.

• The Anti-Lock control unit is a microprocessor similar to other computer units in the vehicle. Ensure that the ignition switch is **OFF** before removing or installing controller harnesses. Avoid static electricity discharge at or near the controller.

• If any arc welding is to be done on the vehicle, the control unit should be unplugged before welding operations begin.

WHEEL SPEED SENSOR

REMOVAL & INSTALLATION

See Figures 1 and 2.

1. Pay attention to the following when removing wheel sensor.

    a. As much as possible, avoid rotating wheel sensor when removing it. Pull wheel sensors out without pulling on sensor harness.

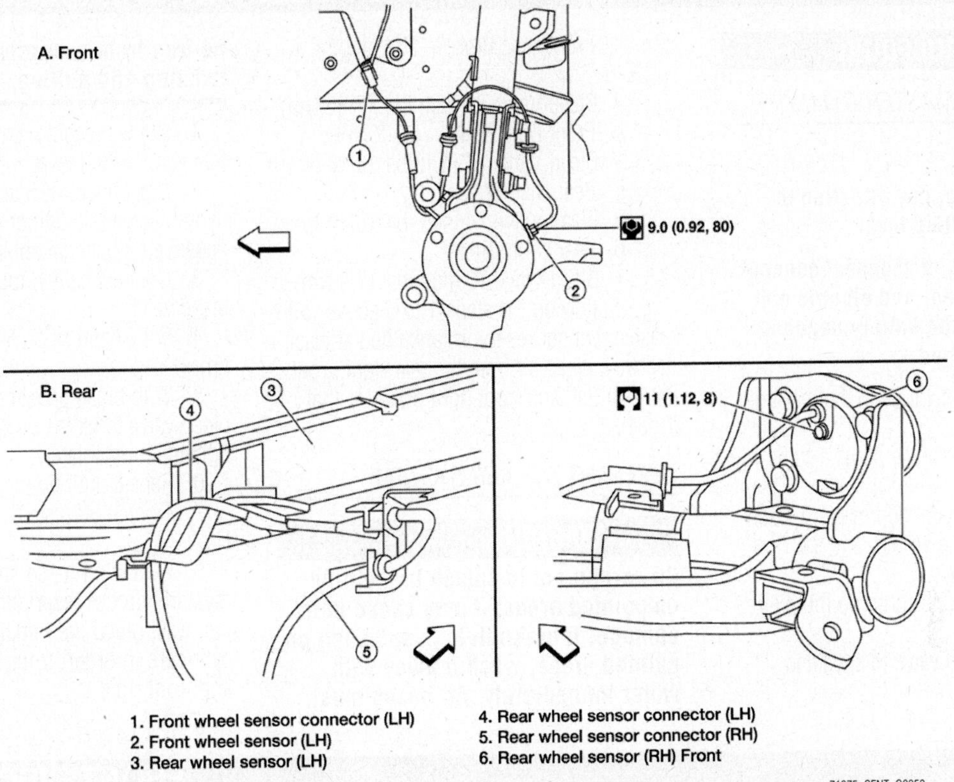

Fig. 1 Exploded view of the wheel sensor components

1. Front wheel sensor connector (LH)
2. Front wheel sensor (LH)
3. Rear wheel sensor (LH)
4. Rear wheel sensor connector (LH)
5. Rear wheel sensor connector (RH)
6. Rear wheel sensor (RH) Front

71075_SENT_G0353

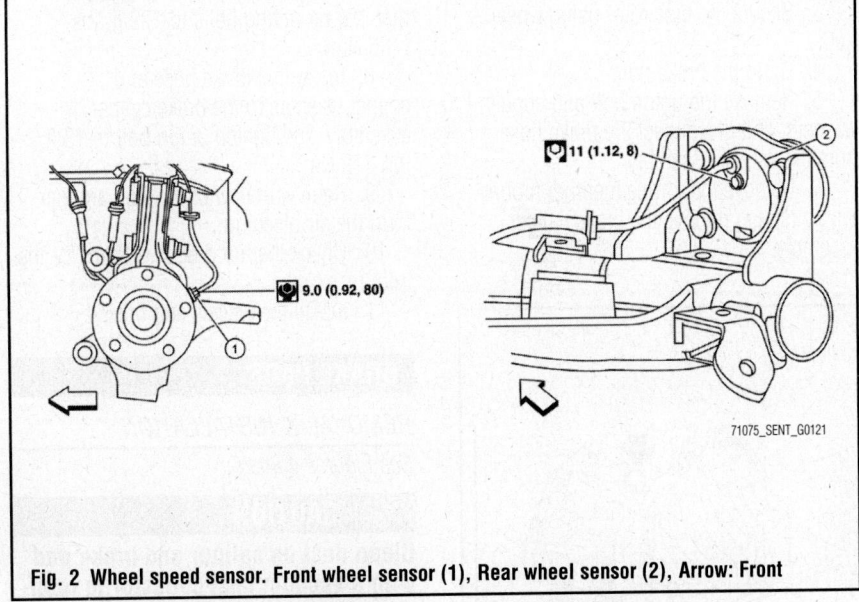

Fig. 2 Wheel speed sensor. Front wheel sensor (1), Rear wheel sensor (2), Arrow: Front

71075_SENT_G0121

✳✳ WARNING

Take care to avoid damaging wheel sensor edges or rotor teeth. Remove wheel sensor first before removing front or rear wheel hub. This is to avoid damage to wheel sensor wiring and loss of sensor function.

To install:

2. Pay attention to the following when installing wheel sensor. Tighten wheel sensor bolts to the specified torques.

3. When installing, make sure there is no foreign material such as iron chips on and in the mounting hole of the wheel sensor. Make sure no foreign material has been caught in the sensor rotor. Remove any foreign material and clean the mount.

4. When installing front wheel sensor, press rubber grommets of strut bracket and body all the way in until they get locked, and be careful not to apply a twist to harness. Harness should not be twisted after installation. (Install it with harness paint mark on body side grommet facing front of vehicle, and the strut side grommet facing outside of vehicle.)

5. When installing rear wheel sensor, press rubber grommets of suspension arm bracket and harness of side member all the way in until they get locked, and be careful not to apply a twist to harness. Harness should not be twisted after installation. (Aim the paint mark upward of vehicle.)

BRAKES

BLEEDING THE BRAKE SYSTEM

BLEEDING PROCEDURE

BLEEDING THE MASTER CYLINDER

In-Vehicle

➡ While bleeding, pay attention to master cylinder fluid level.

➡ Before working, disconnect connectors of ABS actuator and electric unit (control unit) or the battery negative terminal if equipped.

1. Connect a vinyl tube to the rear right bleed valve.
2. Fully depress brake pedal 4 to 5 times.
3. With brake pedal depressed, loosen bleed valve to let the air out, and then tighten it immediately.
4. Repeat steps 2, 3 until no more air comes out.
5. Tighten bleed valve to specified torque.

a. Front Disc Brake—2.0L engine & SE-R
- Bleed Valve 69 inch lbs. (7.85 Nm)
b. Front Disc Brake—SE-R SPEC-V
- Bleed Valve 74 inch lbs. (8.34 Nm)
c. Rear Disc Brake
- Bleed Valve 74 inch lbs. (8.34 Nm)
d. Rear Drum Brake
- Bleed Valve 70 inch lbs. (7.9 Nm)

6. Following the steps 1 to 5 above, with master cylinder reservoir tank filled at least half way, bleed air from the rear right, front left, rear left, and front right brake, in that order.

BLEEDING THE ABS SYSTEM

✳✳ WARNING

Be careful not to splash brake fluid on painted areas; it may cause paint damage. If brake fluid is splashed on painted areas, wash it away with water immediately. All hoses must

be free from excessive bending, twisting and pulling.

1. While bleeding, pay attention to master cylinder fluid level.
2. Disconnect ABS actuator and the hydraulic electric unit or disconnect the negative (-) battery cable.
3. Connect a vinyl tube to the rear right bleed valve.
4. Fully depress brake pedal 4 to 5 times.
5. With brake pedal depressed, loosen bleed valve to let the air out, and then tighten it immediately.
6. Repeat until no more air comes out.
7. Tighten bleed to 61–78 inch lbs (7–9 Nm).
8. Following these steps, with master cylinder reservoir filled at least half way, bleed the remaining brake cylinders in order: front left, rear left, and front right.

BRAKES

✳✳ CAUTION

Dust and dirt accumulating on brake parts during normal use may contain asbestos fibers from production or aftermarket brake linings. Breathing excessive concentrations of asbestos fibers can cause serious bodily harm. Exercise care when servicing brake parts. Do not sand or grind brake lining unless equipment used is designed to contain the dust residue. Do not clean brake parts with compressed air or by dry brushing. Cleaning should be done by dampening the brake components with a fine mist of water, then wiping the brake components clean with a dampened cloth. Dispose of cloth and all residue containing asbestos fibers in an impermeable container with the appropriate label. Follow practices prescribed by the Occupational Safety and Health Administration (OSHA) and the Environmental Protection Agency (EPA) for the handling, processing, and disposing of dust or debris that may contain asbestos fibers.

BRAKE CALIPERS

REMOVAL & INSTALLATION

See Figure 3.

1. Before servicing the vehicle, refer to the Precautions Section.
2. Remove the front wheels and tires.
3. Secure the disc rotor using wheel nuts.
4. Drain the brake fluid.
5. Remove the union bolt and copper washers, and disconnect the brake hose from the caliper assembly.
6. Remove the torque member mounting bolts, and remove the brake caliper assembly.

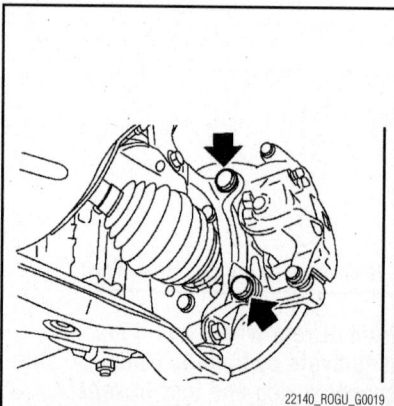

22140_ROGU_G0019

Fig. 3 Remove the torque member mounting bolts

FRONT DISC BRAKES

To install:

7. Install the brake caliper assembly to the vehicle and tighten the torque member mounting bolts to 122 ft. lbs. (165 Nm).
8. Install the brake hose and copper washers to the brake caliper assembly, and tighten union bolt to 13 ft. lbs. (18 Nm).
9. Refill with new brake fluid and perform the air bleeding.
10. Check that no drag is present for the front disc brake.
11. Install the wheels and tires.

BRAKE PADS

REMOVAL & INSTALLATION

See Figures 4 and 5.

✳✳ CAUTION

Clean dust on caliper and brake pad with a vacuum dust collector to minimize the hazard of air borne particles or other materials.

➡ While removing caliper, do not depress brake pedal because piston will pop out.

➡ It is not necessary to remove bolts on torque member and brake hose except for disassembly or replacement of caliper assembly. In this case, hang caliper with a wire so as not to stretch brake hose.

☼ WARNING

Do not damage piston boot.

☼ WARNING

If any shim is subject to serious corrosion, replace it with a new one.

➡Always replace shim and shim cover as a set when replacing brake pads.

➡Keep rotor and pads free from brake fluid and grease.

➡Burnish the brake pads and disc rotor mutually contacting surfaces, after refinishing or replacing rotors, after replacing pads, or if a soft pedal occurs at very low mileage.

   1. Remove front wheels and tires using power tools.
   2. Remove lower sliding pin bolt.
   3. Swing cylinder body up and support cylinder body with a suitable wire. Remove pads, shims and pad retainers from torque member.

➡When removing pad retainer from torque member, lift pad retainer in the direction shown by, so as not to deform it.

To install:

   4. Apply Molykote AS-880N grease or equivalent to the shims. Install shims to pads.

➡Securely install shims according to mounting direction of pads.

   5. Apply Molykote M-7439 grease or equivalent to pad contact surfaces on pad

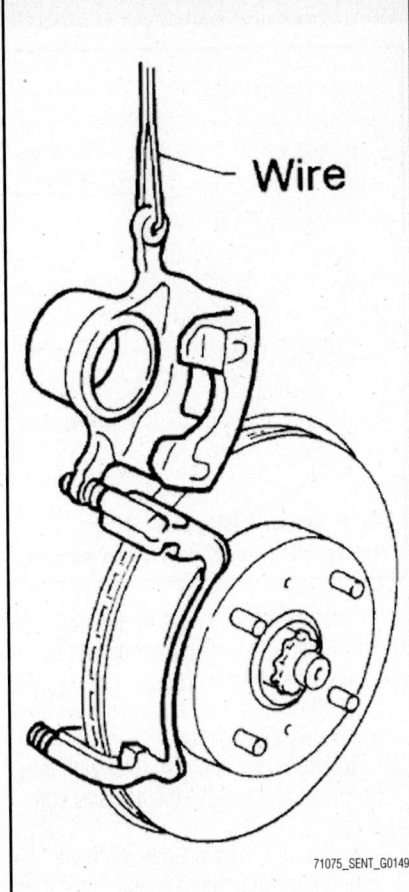

Fig. 4 Swing cylinder body up and support cylinder body with a suitable wire as shown. Remove pads, shims and pad retainers from torque member

retainers. Install pad retainers and pads to the torque member.

➡When installing pad retainer, attach it firmly so that it is not lifted up from torque member, as shown.

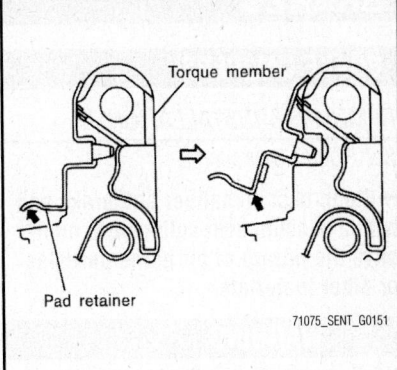

Fig. 5 When removing pad retainer from torque member, lift pad retainer in the direction shown by arrow, so as not to deform it

➡If equipped, both inner and outer pads have a pad return system on the pad retainer. Install pad return lever securely to pad wear sensor.

   6. On SE-R models, apply Molykote AS-880N grease to inside of cylinder fingers.
   7. Install the cylinder body to torque member.
     a. Press the piston into the cylinder body using a suitable tool.

➡Check the brake fluid level in the reservoir tank because brake fluid returns to master cylinder reservoir tank when pressing piston in.

   8. Install lower sliding pin bolt, and tighten to the following specified torque:
     a. 2.0L & SE-R
     • Sliding pin bolt–20 ft. lbs. (26.48 Nm)
     b. SE-R SPEC-V
     • Sliding pin bolt–25 ft. lbs. (34.3 Nm)
   9. Check front brakes for drag.
  10. Install front wheels and tires.

BRAKES | **REAR DISC BRAKES**

BRAKE CALIPERS

REMOVAL & INSTALLATION

See Figures 6 through 8.

➡ Clean dust on caliper and brake pad with a vacuum dust collector to minimize the hazard of air borne particles or other materials.

✳ WARNING

While removing caliper, do not depress brake pedal because piston will pop out.

➡ It is not necessary to remove bolts on torque member and brake hose except for disassembly or replacement of caliper assembly. In this case, hang caliper with a wire so as not to stretch brake hose.

✳ WARNING

Do not damage piston boot.

✳ WARNING

If any shim is subject to serious corrosion, replace it with a new one.

➡ Always replace shim and shim cover as a set when replacing brake pads.

➡ Keep rotor and pads free from brake fluid and grease.

➡ Burnish the brake pads and disc rotor mutually contacting surfaces, after refinishing or replacing rotors, after replacing pads, or if a soft pedal occurs at very low mileage.

➡ When removing components such as hoses, tubes/lines, etc., cap or plug openings to prevent fluid from spilling.

1. Partially drain the brake fluid.
2. Remove rear wheels and tires using power tool.
3. Secure disc rotor using wheel nuts.
4. Remove the union bolt to disconnect the rear brake hose. Discard the copper sealing washers.

✳ WARNING

Do not reuse the copper sealing washers.

5. Remove sliding pins and remove cylinder body from torque member.

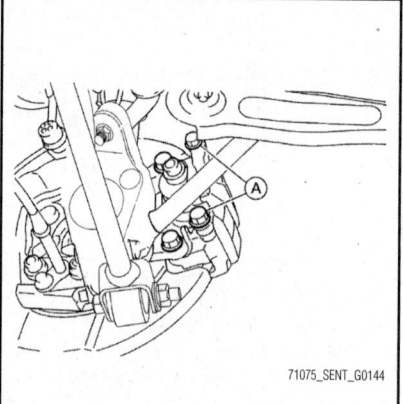

Fig. 6 Remove sliding pins (A) and remove cylinder body from torque member

6. If necessary, remove torque member and disc rotor. If the disc rotor cannot be removed, remove as follows:
   a. Make sure parking brake lever is completely disengaged.
   b. Hold down the disc rotor with the wheel nut and remove the adjuster hole plug.
   c. Insert a flat-bladed screwdriver through the plug opening and rotate the star wheel on the adjuster assembly in the direction to loosen and retract the brake shoes.
   d. Prior to removing disc rotor, make alignment mark using a marker between the hub and disc rotor.
   e. Remove wheel nut and rotor.

To install:

7. If necessary, install disc rotor and torque member.

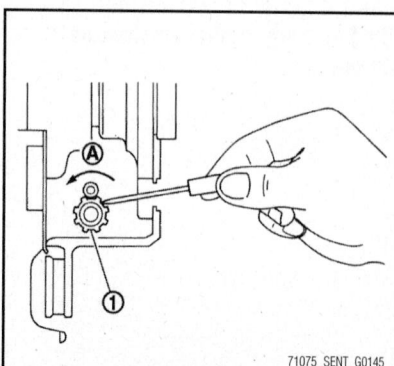

Fig. 7 Insert a flat-bladed screwdriver through the plug opening and rotate the star wheel (1) on the adjuster assembly in the direction (A) shown to loosen and retract the brake shoes

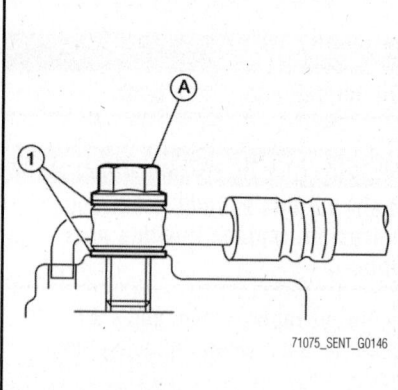

Fig. 8 Install brake hose with new copper sealing washers (1) to cylinder body and tighten union bolt (A)

   a. Align marks made during removal on the hub and disc rotor.
8. Install cylinder body to the torque member and tighten sliding pins to the following specified torque.
   • Sliding pins–32 ft. lbs. (43.15 Nm)

➡ Before installing cylinder body, wipe off oil and grease on mounting surfaces of cylinder body.

9. Install brake hose with new copper sealing washers to cylinder body and tighten union bolt to the following specified torque.
   a. Tighten the union bolt to 14 ft. lbs. (18.5 Nm)

➡ Do not reuse the copper sealing washers.

➡ Align brake hose protrusion to groove on cylinder body.

10. Adjust the parking brake if necessary. Service".
11. Bleed the brake system.
12. Install rear wheels and tires.

BRAKE PADS

REMOVAL & INSTALLATION

See Figures 9 through 12.

✳ CAUTION

Clean dust on caliper and brake pad with a vacuum dust collector to minimize the hazard of air borne particles or other materials.

➡ While removing caliper, do not depress brake pedal because piston will pop out.

➡It is not necessary to remove bolts on torque member and brake hose except for disassembly or replacement of caliper assembly. In this case, hang caliper with a wire so as not to stretch brake hose.

✷✷ WARNING

Do not damage piston boot.

✷✷ WARNING

If any shims have serious corrosion, replace with a new one.

➡Always replace shim and shim cover as a set when replacing brake pads.

➡Keep rotor and pads free from brake fluid and grease.

➡Burnish the brake pads and disc rotor mutually contacting surfaces, after refinishing or replacing rotors, after replacing pads, or if a soft pedal occurs at very low mileage.

   1. Remove rear wheels and tires using power tool.
   2. Remove lower sliding pin.
   3. Swing cylinder body up and support cylinder body with a suitable wire as shown. Remove pads, shim, covers and pad retainers from torque member.

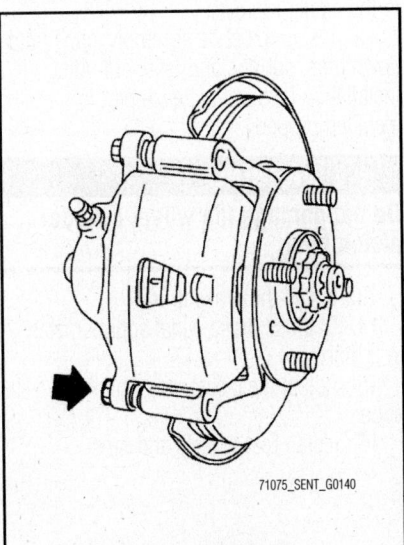

71075_SENT_G0140

Fig. 9 Remove lower sliding pin

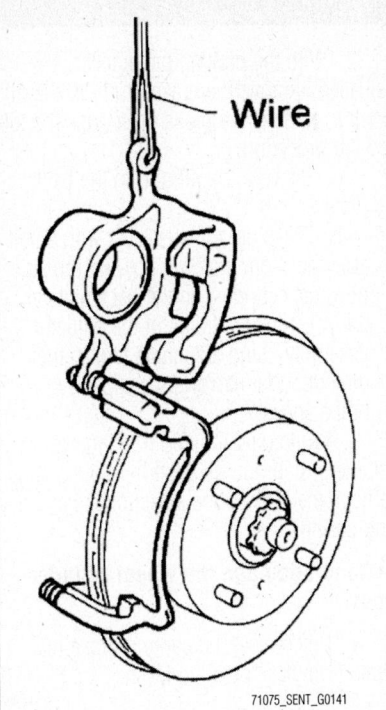

71075_SENT_G0141

Fig. 10 Swing cylinder body up and support cylinder body with a suitable wire as shown. Remove pads, shim, covers and pad retainers from torque member

✷✷ WARNING

When removing pad retainer from torque member, lift pad retainer in the direction shown by arrow, so as not to deform it.

   To install:
   4. Apply Molykote AS-880N grease to cylinder body fingers, and between the inner shim and inner shim cover. Apply Molykote AS-880N grease inside the outer cover.
   5. Attach the inner shim and inner shim cover to the inner pad, and the outer cover to the outer pad.
   6. Install the pad retainers and assembled pads on the torque member.

➡When installing pad retainer, attach it firmly so that it is not lifted up from torque member.

   7. Install the cylinder body to the torque member.

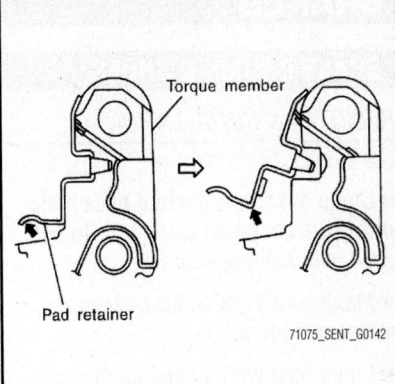

71075_SENT_G0142

Fig. 11 When removing pad retainer from torque member, lift pad retainer in the direction shown by arrow, so as not to deform it

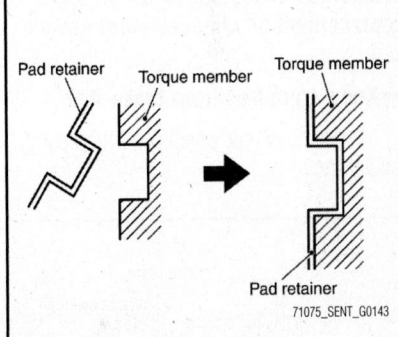

71075_SENT_G0143

Fig. 12 Install the pad retainers and assembled pads on the torque member. When installing pad retainer, attach it firmly so that it is not lifted up from torque member, as shown.

   a. Press the piston into the cylinder body using a suitable tool.

➡Check the brake fluid level in the reservoir tank because brake fluid returns to master cylinder reservoir tank when pressing piston in.

   8. Install the lower sliding pin and tighten to the following specified torque.
   a. Tighten the lower sliding pin to 32 ft. lbs. (43.15 Nm).
   9. Check rear brakes for drag.
   10. Install rear wheels and tires.

BRAKE DRUMS & SHOES

REMOVAL & INSTALLATION

See Figures 13 and 14.

➡ Clean dust on drum and back plate with a vacuum dust collector. Do not blow with compressed air.

➡ Make sure parking brake lever is released completely.

➡ Clean dust with a vacuum dust collector to minimize the hazard of air borne particles or other materials.

➡ While removing brake shoes, do not depress brake pedal because wheel cylinder pistons will pop out.

➡ It is not necessary to disconnect brake tube except for disassembly or replacement of wheel cylinder assembly.

➡ Keep drum free from brake fluid.

1. Remove rear wheels and tires using power tools.

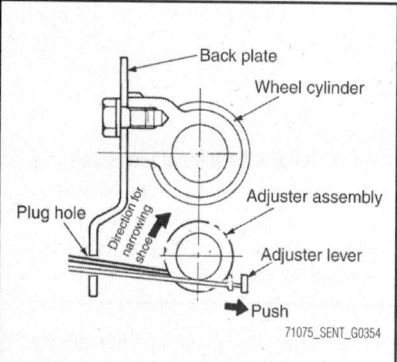

Fig. 13 Press up adjuster lever with a suitable tool from the plug hole (at the side of wheel cylinder) on the back plate as shown. Turn the frame of the adjuster assembly using a suitable tool in the direction that narrows the expanded brake shoes.

2. With the parking brake lever released, remove the brake drum. If it is difficult to remove the brake drum, use the following procedure.

   a. Remove the plug from the back plate.

   b. Press up adjuster lever with a suitable tool from the plug hole (at the side of wheel cylinder) on the back plate as shown. Turn the frame of the adjuster assembly using a suitable tool in the direction that narrows the expanded brake shoes.

3. While pushing and rotating the retainers, pull out the shoe hold pins, and remove the brake shoe assembly, retainers and springs.

➡ Do not damage the wheel cylinder boot.

4. Disconnect the parking brake rear cable from the operating lever.

➡ Do not bend the parking brake cable.

5. Disassemble the brake shoe assembly

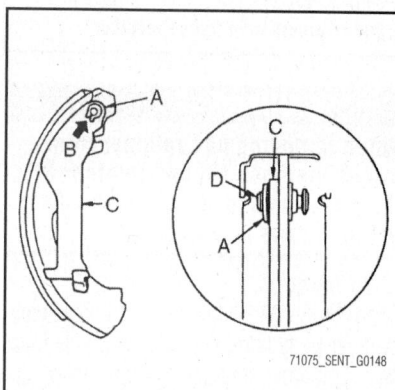

Fig. 14 Install the operating lever (C) using the following procedure. Position operating lever (C) on brake shoe. Install retainer ring (A) on operating lever (C), and crimp them until their contact points (B) are met. Pin (D)

(brake shoes, springs, adjuster and adjuster lever).

6. Remove the retainer ring with a suitable tool to separate the operating lever from brake shoe.
   - Contact point
   - Pin

To install:

7. Install the operating lever using the following procedure.

   a. Position operating lever on brake shoe.

   b. Install retainer ring on operating lever, and crimp them until their contact points are met.

8. Apply NISSAN brake grease (KRF00 00005) to brake shoes sliding surfaces (the shaded areas) and other parts on the back plate as indicated by arrows.

9. Apply NISSAN brake grease (KRF00 00005) to screw and confirm the difference between right and left wheel for assembling when disassembled.

   a. Right rear wheel thread cutting direction: right-hand screw.

   b. Left rear wheel thread cutting direction: left-hand screw.

10. Assemble the shoe, adjuster, adjuster lever and springs on the shoe assembly.

11. Connect the parking brake rear cable to the operating lever.

12. Install the shoe assembly, shoe hold down pins, springs and retainers. After installation be sure that each part is installed properly.

✳✳ WARNING

Do not damage the wheel cylinder piston boot.

13. Install the brake drum.

14. Depress brake pedal approximately 2 to 3 times.

15. Adjust the clearance of brake shoe.

16. Install rear wheels and tires.

BRAKES   **PARKING BRAKE**

ADJUSTMENTS

CONTROL ASSEMBLY

See Figure 15.

1. Engage parking brake lever to access adjusting nut hole below grip.
2. Insert a deep socket wrench onto adjusting nut. Rotate adjusting nut to fully loosen cable, and then release parking brake lever.
3. Depress the foot brake about 10 times and adjust the rear shoe clearance (rear drum brake type only).

➡ **Be sure to securely depress the foot brake.**

4. Adjust the parking brake shoe clearance (rear drum in disc brake type only).
5. Rotate brake drum or disc rotor to make sure that there is no drag.

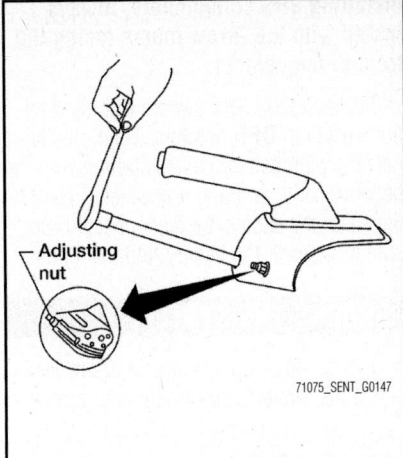

Adjusting nut

71075_SENT_G0147

Fig. 15 Insert a deep socket wrench onto adjusting nut. Rotate adjusting nut to fully loosen cable, and then release parking brake lever.

6. Adjust parking brake cable with the following procedure.
   a. Operate parking brake lever with the specified force about 10 times.
   - Rear drum brake type 44 ft. lbs. (196 Nm).
   - Rear drum in disc brake type 66 ft. lbs. (294 Nm).
   b. Engage parking brake lever to access adjusting nut hole below grip.
   c. Rotate adjusting nut to adjust parking brake lever stroke using a deep socket wrench.
   d. Operate parking brake lever with the specified force, make sure the parking brake lever stroke is within the specified number of notches. Check it by listening and counting ratchet clicks.
   e. Make sure that there is no drag on rear brake or parking brake shoe with parking brake lever completely released.

CHASSIS ELECTRICAL   AIR BAGS (SUPPLEMENTAL RESTRAINT SYSTEM)

PRECAUTIONS

Disconnect and isolate the battery negative cable before beginning any airbag system component diagnosis, testing, removal, or installation procedures. Allow system capacitor to discharge for three minutes before beginning any component service. This will disable the airbag system. Failure to disable the airbag system may result in accidental airbag deployment, personal injury, or death.

Do not place an intact undeployed airbag face down on a solid surface. The airbag will propel into the air if accidentally deployed and may result in personal injury or death.

When carrying or handling an undeployed airbag, the trim side (face) of the airbag should be pointing towards the body to minimize possibility of injury if accidental deployment occurs. Failure to do this may result in personal injury or death.

Replace airbag system components with OEM replacement parts. Substitute parts may appear interchangeable, but internal differences may result in inferior occupant protection. Failure to do so may result in occupant personal injury or death.

Wear safety glasses, rubber gloves, and long sleeved clothing when cleaning powder residue from vehicle after an airbag deployment. Powder residue emitted from a deployed airbag can cause skin irritation. Flush affected area with cool water if irrita-

tion is experienced. If nasal or throat irritation is experienced, exit the vehicle for fresh air until the irritation ceases. If irritation continues, see a physician.

Do not use a replacement airbag that is not in the original packaging. This may result in improper deployment, personal injury, or death.

The factory installed fasteners, screws and bolts used to fasten airbag components have a special coating and are specifically designed for the airbag system. Do not use substitute fasteners. Use only original equipment fasteners listed in the parts catalog when fastener replacement is required.

During, and following, any child restraint anchor service, due to impact event or vehicle repair, carefully inspect all mounting hardware, tether straps, and anchors for proper installation, operation, or damage. If a child restraint anchor is found damaged in any way, the anchor must be replaced. Failure to do this may result in personal injury or death.

Deployed and non-deployed airbags may or may not have live pyrotechnic material within the airbag inflator.

Do not dispose of driver/passenger/curtain airbags or seat belt tensioners unless you are sure of complete deployment. Refer to the Hazardous Substance Control System for proper disposal.

Dispose of deployed airbags and tensioners consistent with state, provincial, local, and federal regulations.

After any airbag component testing or service, do not connect the battery negative cable. Personal injury or death may result if the system test is not performed first.

If the vehicle is equipped with the Occupant Classification System (OCS), do not connect the battery negative cable before performing the OCS Verification Test using the scan tool and the appropriate diagnostic information. Personal injury or death may result if the system test is not performed properly.

Never replace both the Occupant Restraint Controller (ORC) and the Occupant Classification Module (OCM) at the same time. If both require replacement, replace one, then perform the Airbag System test before replacing the other.

Both the ORC and the OCM store Occupant Classification System (OCS) calibration data, which they transfer to one another when one of them is replaced. If both are replaced at the same time, an irreversible fault will be set in both modules and the OCS may malfunction and cause personal injury or death.

If equipped with OCS, the Seat Weight Sensor is a sensitive, calibrated unit and must be handled carefully. Do not drop or handle roughly. If dropped or damaged, replace with another sensor. Failure to do so may result in occupant injury or death.

If equipped with OCS, the front passenger seat must be handled carefully as well. When removing the seat, be careful when setting on floor not to drop. If dropped, the sensor may be inoperative, could result in occupant injury, or possibly death.

If equipped with OCS, when the passenger front seat is on the floor, no one should sit in the front passenger seat. This uneven force may damage the sensing ability of the seat weight sensors. If sat on and damaged, the sensor may be inoperative, could result in occupant injury, or possibly death.

DISARMING THE SYSTEM

➡**All Supplemental Restraint System (SRS) electrical wiring harnesses and connectors are covered with YELLOW outer insulation. Do not use electrical test equipment on any circuit related to the SRS (air bag) sensors. When installing SRS components, always install with the arrow marks facing the front of the vehicle.**

To disarm the SRS system turn the ignition switch to **OFF** position. Then, disconnect the both battery cables starting with the negative cable first and wait at least 10 minutes after the cables are disconnected. Be sure to insulate the battery terminal ends.

ARMING THE SYSTEM

To arm the Supplemental Restraint System (SRS) system turn the ignition switch to **OFF** position. Connect the both battery cables starting with the positive cable first.

➡**The SRS or air bag system is equipped with a self-diagnostic operation. After turning the ignition key to the ON or START position, the AIR BAG warning lamp will illuminate for 7 seconds. After 7 seconds, the AIR BAG lamp will extinguish if no malfunction is detected. If the AIR BAG lamp does not extinguish after 7 seconds, check the SRS self-diagnostic system for a malfunction.**

DRIVETRAIN

CONTINUALLY VARIABLE TRANSMISSION (CVT)

DRAIN & REFILL

1. Remove the drain plug from the oil pan and drain the CVT fluid.
2. Remove the drain plug gasket from the drain plug.
3. Install the drain plug gasket to the drain plug. Never reuse the drain plug gasket.
4. Install the drain plug to the oil pan.
5. Fill the CVT fluid from the CVT fluid charging pipe to the specified level.

✳✳ WARNING

Use only Genuine NISSAN CVT Fluid NS-2. Never mix with other fluid. Using CVT fluid other than Genuine NISSAN CVT Fluid NS-2 will deteriorate in driveability and CVT durability, and may damage the CVT, which is not covered by the warranty.

➡**When filling the CVT fluid, take care not to scatter heat generating parts such as exhaust. Sufficiently shake the container of CVT fluid before using. Delete CVT fluid deterioration date with CONSULT-III after changing CVT fluid.**

6. With the engine warmed up, drive the vehicle in an urban area.

➡**When ambient temperature is 68°F (20°C), it takes about 10 minutes for the CVT fluid to warm up to 122 to 176°F (50 to 80°C).**

7. Check the CVT fluid level and condition.
8. Repeat steps 1 to 5 if CVT fluid has been contaminated.

CLUTCH HYDRAULIC SYSTEM BLEEDING

BLEEDING PROCEDURE
See Figure 16.

✳✳ WARNING

Two people are required to bleed the clutch cylinder. Do not use vacuum assist or any type of power bleeder. It will not purge all the air from this system.

1. Fill master cylinder reservoir with new brake fluid.
2. Connect a clear tube to the bleeding connector on the Concentric Slave Cylinder (CSC).
3. Push and release the clutch pedal slowly and fully 15 times, waiting 3 seconds between each cycle.

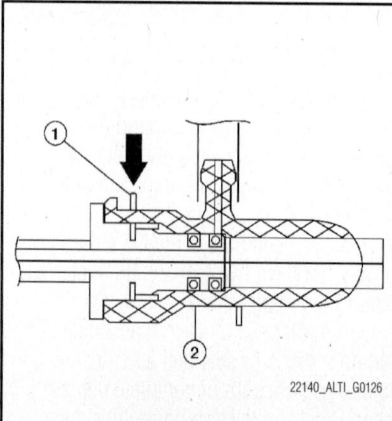

22140_ALTI_G0126

Fig. 16 Push and hold the locking pin (1) to let the bleeding connector (2) slide open

4. Push in the lock pin on the bleeding connector and hold it in.

✳✳ WARNING

Hold the lock pin in to prevent the bleeding connector from separating when fluid pressure is applied.

5. Slide the bleeding connector out ³⁄₁₆ (5mm), then press the clutch pedal and hold it down.
6. Push the bleeding connector back in and release the clutch pedal.
7. Repeat until no bubbles are observed in the fluid flow.

CLUTCH MASTER CYLINDER

REMOVAL & INSTALLATION
See Figures 17 through 19.

➡**When removing components such as hoses, tubes/lines, etc., cap or plug openings to prevent fluid from spilling.**

1. Remove the battery and battery tray.
2. Remove the engine room cover.
3. Remove the air cleaner and air duct.
4. Remove engine undercover.
5. Drain clutch fluid from reservoir tank and remove hose.

➡**Do not spill clutch fluid onto painted surfaces. If it spills, wipe up immediately and wash the affected area with water.**

6. Remove master cylinder rod end from clutch pedal assembly.
7. Remove lock pin from connector of master cylinder and separate clutch tube.
8. Rotate master cylinder clockwise by 45° and remove from the vehicle.

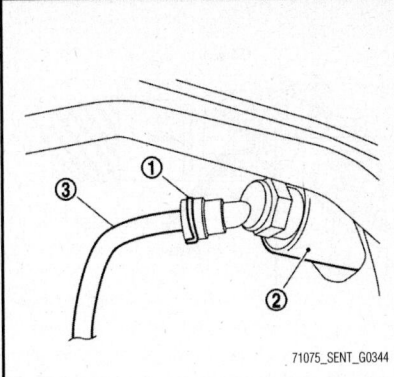

Fig. 17 Remove lock pin (1) from connector of master cylinder (2) and separate clutch tube (3).

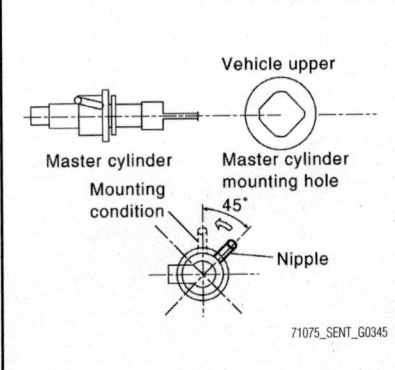

Fig. 18 Tilt master cylinder clockwise by 45°and insert it in the mounting hole. Rotate counterclockwise to secure it. At this time, nipple is in the up position

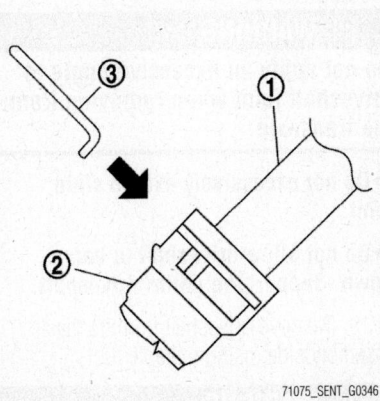

Fig. 19 Install clutch tube (1) fully into connector of master cylinder (2). 4. Install lock pin (3) fully into connector of master cylinder (2)

To install:

9. Tilt master cylinder clockwise by 45° and insert it in the mounting hole. Rotate counterclockwise to secure it. At this time, nipple is in the up position.

10. Install master cylinder rod end to clutch pedal.

11. Install clutch tube fully into connector of master cylinder.

12. Install lock pin fully into connector of master cylinder.

13. Fill with new clutch fluid and bleed air from the system.

14. After completing this procedure, inspect clutch pedal operation.

15. Install the air cleaner and air duct.

16. Install the battery and battery tray.

17. Install the engine room cover.

18. Install the engine undercover.

FRONT HALFSHAFT

REMOVAL & INSTALLATION

Left Side

See Figures 20 through 25.

➡**The manufacturer refers to this component as a "driveshaft".**

On-Vehicle Inspection and Service

1. Check driveshaft mounting point and joint for looseness and other damage.

2. Check boot for cracks and other damage.

3. Replace or repair components as necessary.

➡**When removing components such as hoses, tubes/lines, etc., cap or plug openings to prevent fluid from spilling.**

4. Remove wheel and tire using power tool.

5. Remove wheel sensor from steering knuckle.

> ✳✳ **WARNING**
>
> **Do not pull on wheel sensor harness.**

6. Remove cotter pin, then loosen lock nut using power tool.

> ✳✳ **WARNING**
>
> **Temporarily leave the lock nut installed to prevent damage to threads.**

7. Remove transverse link ball joint nut and bolt. Then, separate transverse link from steering knuckle.

8. Separate the driveshaft from the wheel hub and bearing assembly by lightly tapping the end of the driveshaft using a

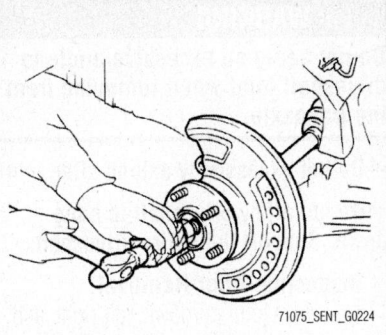

Fig. 20 Separate the driveshaft from the wheel hub and bearing assembly by lightly tapping the end of the driveshaft using a hammer or suitable tool and wood block, and then remove lock nut

hammer or suitable tool and wood block, and then remove lock nut.

➡**Use a suitable puller if wheel hub and bearing assembly and driveshaft cannot be separated after performing the above procedure.**

9. Remove the driveshaft from the wheel hub and bearing assembly.

> ✳✳ **WARNING**
>
> **Do not apply an excessive angle to driveshaft joint when removing from the wheel hub and bearing assembly,**

➡**Do not excessively extend slide joint.**

➡**Do not allow driveshaft to hang down. Support the entire driveshaft.**

10. Remove driveshaft from transaxle assembly side, using suitable tool.

➡**Make sure that circlip is attached on the edge.**

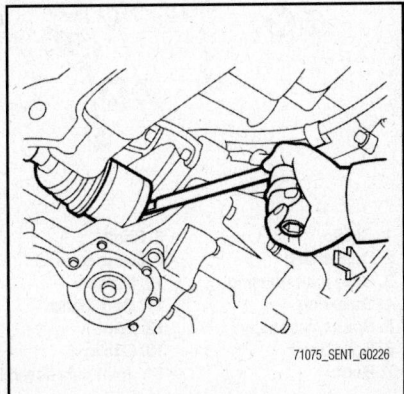

Fig. 21 Remove driveshaft from transaxle assembly side, using suitable tool as shown.

⁂ WARNING

Do not apply an excessive angle to driveshaft joint when removing from the transaxle.

➡ Do not excessively extend slide joint.

➡ Do not allow driveshaft to hang down. Support the entire driveshaft.

Inspection After Removal

11. Move joint up/down, left/right, and in axial direction. Check for any rough movement or significant looseness.

12. Check boot for cracks or other damage, and for grease leakage.

13. If damaged, disassemble driveshaft to verify damage, and repair or replace as necessary.

To install:

➡ When removing components such as hoses, tubes/lines, etc., cap or plug openings to prevent fluid from spilling.

14. Install new circlip on driveshaft in the circlip groove on transaxle side.

➡ Make sure the new circlip on the driveshaft is securely fastened.

⁂ WARNING

In order to prevent damage to differential side oil seal, place Tool onto oil seal before inserting driveshaft as shown. Slide driveshaft into slide joint and tap with a hammer to install securely.

➡ After inserting driveshaft, try to pull the driveshaft out of the transaxle by hand. If it pulls out, the circlip is not properly meshed with the transaxle side gear.

➡ Apply grease Molykote M77 to contact surface between wheel husband driveshaft. Use sufficient grease to completely coat contact area.

15. Tighten lock nut.

➡ Do not reuse non-reusable parts.

16. Tighten ball joint to the specification as shown in the accompanying illustration.

17. Tighten the wheel nuts to 83 ft. lbs. (113 Nm).

18. Check transaxle fluid level.

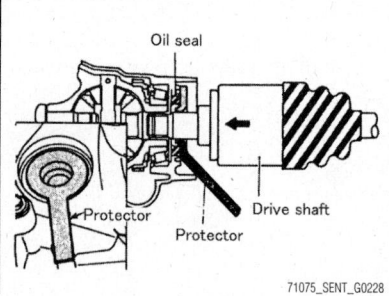

71075_SENT_G0228

Fig. 23 In order to prevent damage to differential side oil seal, place Tool [KV38105500 (J-33904)] onto oil seal before inserting driveshaft as shown. Slide driveshaft into slide joint and tap with a hammer to install securely

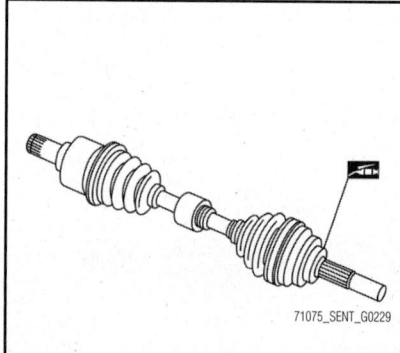

71075_SENT_G0229

Fig. 24 Apply grease Molykote M77 to contact surface between wheel hub and driveshaft. Use sufficient grease to completely coat contact area

➡ Make sure that circlip is attached on the edge.

⁂ WARNING

Do not apply an excessive angle to driveshaft joint when removing from the transaxle.

➡ Do not excessively extend slide joint.

➡ Do not allow driveshaft to hang down. Support the entire driveshaft.

19. Remove driveshaft from transaxle assembly side, using suitable tool.

⁂ WARNING

Do not apply an excessive angle to driveshaft joint when removing from the wheel hub and bearing assembly,

➡ Do not excessively extend slide joint.

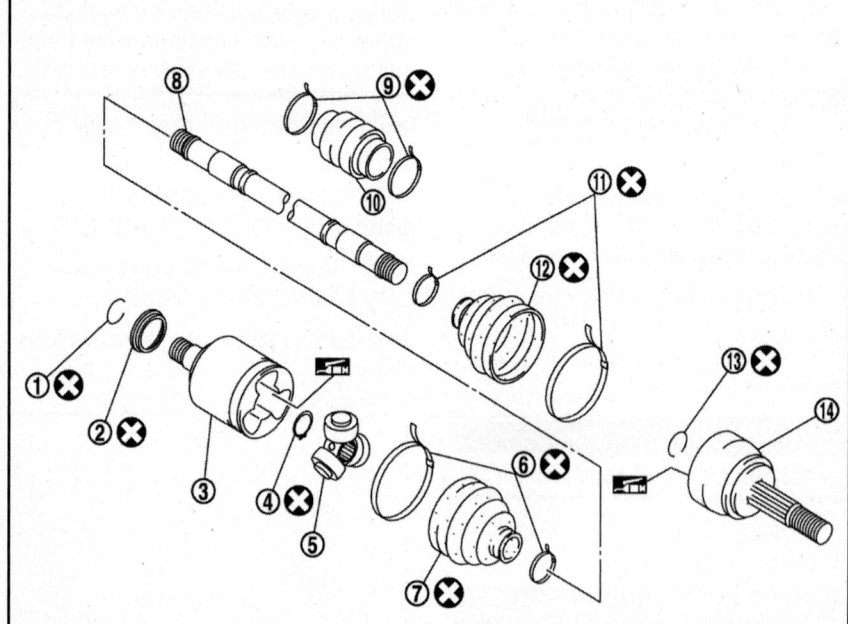

1. Circlip
2. Dust shield
3. Slide joint housing
4. Snap ring
5. Spider assembly
6. Boot band
7. Boot
8. Shaft
9. Damper band
10. Damper
11. Boot band
12. Boot
13. Circlip
14. Joint sub-assembly

71075_SENT_G0230

Fig. 22 Exploded view of the front driveshaft—left side disassembly and assembly

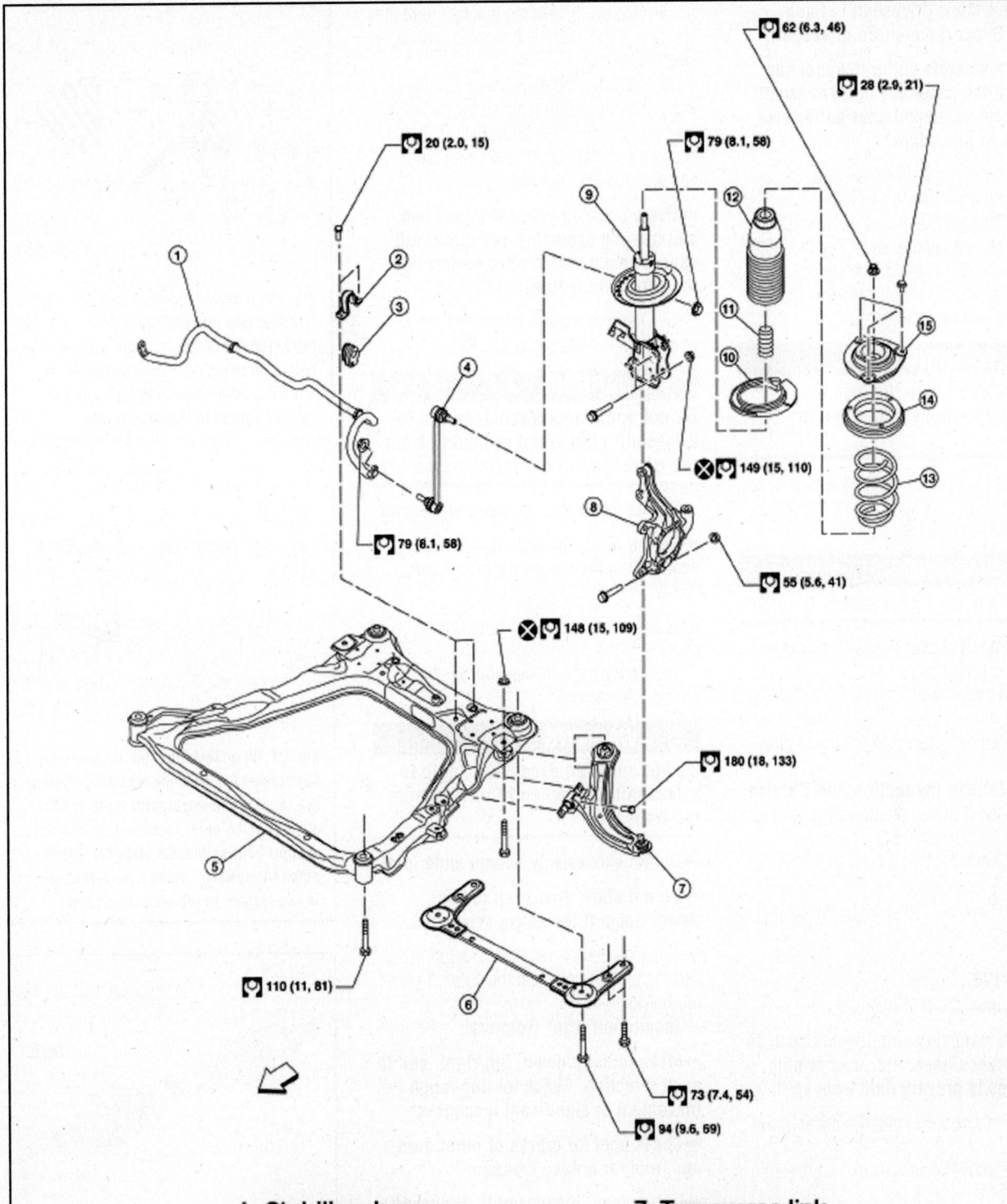

1. Stabilizer bar
2. Stabilizer clamp
3. Stabilizer bushing
4. Stabilizer connecting rod
5. Front suspension member
6. Member stay
7. Transverse link
8. Steering knuckle
9. Strut
10. Coil spring insulator
11. Bound bumper
12. Dust cover

71075_SENT_G0225

Fig. 25 Exploded view of the front suspension assembly

➡Do not allow driveshaft to hang down. Support the entire driveshaft.

➡Use a suitable puller if wheel hub and bearing assembly and driveshaft cannot be separated after performing the above procedure.

20. Remove the driveshaft from the wheel hub and bearing assembly.

21. Separate the driveshaft from the wheel hub and bearing assembly by lightly tapping the end of the driveshaft using a hammer or suitable tool and wood block, and then remove lock nut.

❋❋ WARNING

Temporarily leave the lock nut installed to prevent damage to threads.

22. Remove transverse link ball joint nut and bolt. Then, separate transverse link from steering knuckle.

❋❋ WARNING

Do not pull on wheel sensor harness.

23. Remove cotter pin, then loosen lock nut using power tool.

24. Remove wheel sensor from steering knuckle.

25. Remove wheel and tire using power tool.

On-Vehicle Inspection and Service

26. Check driveshaft mounting point and joint for looseness and other damage.

27. Check boot for cracks and other damage.

28. Replace or repair components as necessary.

Right Side

See Figures 20, 26 through 29.

➡When removing components such as hoses, tubes/lines, etc., cap or plug openings to prevent fluid from spilling.

1. Remove wheel and tire using power tool.

2. Remove wheel sensor from steering knuckle.

❋❋ WARNING

Do not pull on wheel sensor harness.

3. Remove cotter pin, then loosen lock nut using power tool.

❋❋ WARNING

Temporarily leave the lock nut installed to prevent damage to threads.

4. Remove transverse link ball joint nut and bolt. Then, remove transverse link from steering knuckle.

5. Separate the driveshaft from the wheel hub and bearing assembly by lightly tapping the end of the driveshaft using a hammer or suitable tool and wood block, and then remove lock nut.

➡Use a suitable puller if wheel hub and bearing assembly and driveshaft cannot be separated after performing the above procedure.

6. Remove the driveshaft from the wheel hub and bearing assembly.

❋❋ WARNING

Do not apply an excessive angle to driveshaft joint when removing from the wheel hub and bearing assembly,

➡Do not excessively extend slide joint.

➡Do not allow driveshaft to hang down. Support the entire driveshaft.

7. Remove the support bearing bracket plate bolts and support bearing bracket plate.

8. Remove the driveshaft from the transaxle assembly.

❋❋ WARNING

Do not apply an excessive angle to driveshaft joint when removing from the transaxle.

➡Do not excessively extend slide joint.

➡Do not allow driveshaft to hang down. Support the entire driveshaft.

9. If necessary, remove the support bearing bracket bolts and the support bearing bracket.

Inspection After Removal

➡Move joint up/down, left/right, and in axial direction. Check for any rough movement or significant looseness.

➡Check boot for cracks or other damage, and for grease leakage.

➡If damaged, disassemble driveshaft to verify damage, and repair or replace as necessary.

➡Check for cracks or other damage to the support bearing bracket, replace as necessary.

To install:

❋❋ WARNING

In order to prevent damage to differential side oil seal, place Tool onto

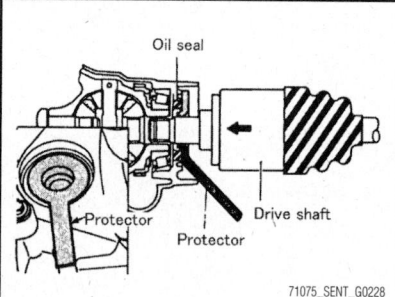

Fig. 26 In order to prevent damage to differential side oil seal, place Tool [KV38105500 (J-33904)] onto oil seal before inserting driveshaft as shown. Slide driveshaft into slide joint and tap with a hammer to install securely

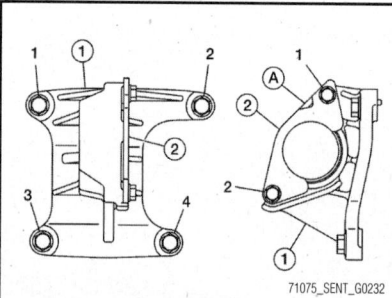

Fig. 27 When installing the support bearing bracket (1), note the following: Tighten the support bearing bracket bolts in two stages, in the order as shown. Install the support bearing bracket plate (2) with the notch (A) upward. Tighten the plate bolts in two stages, in the order as shown.

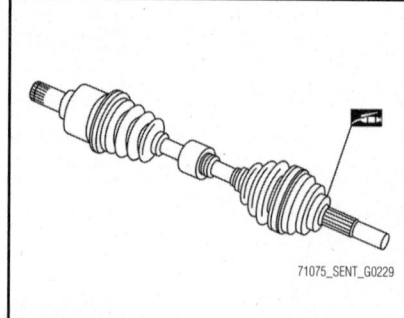

Fig. 28 Apply grease Molykote M77 to contact surface between wheel hub and driveshaft. Use sufficient grease to completely coat contact area

oil seal before inserting driveshaft as shown. Slide driveshaft into slide joint and tap with a hammer to install securely.

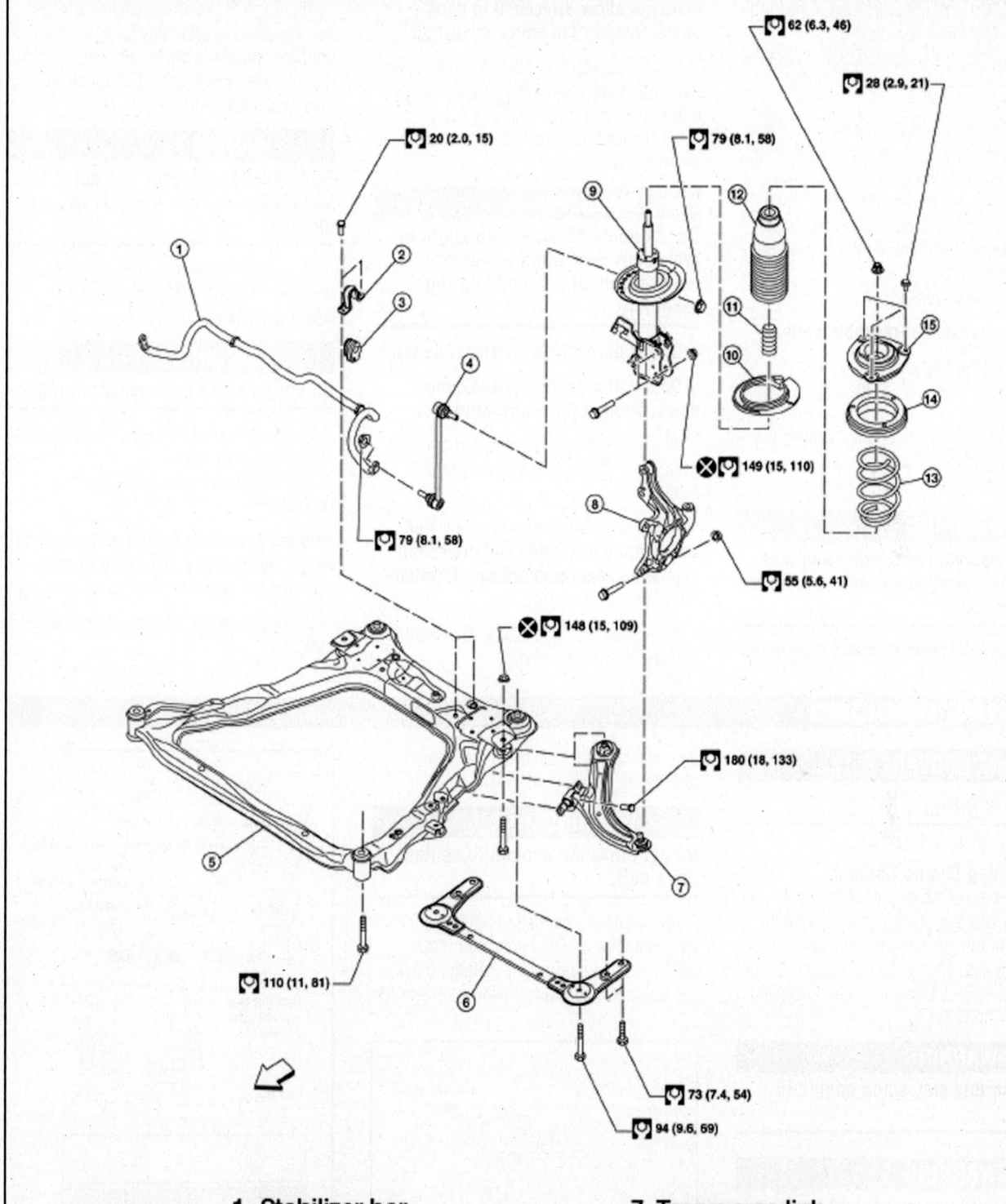

1. Stabilizer bar
2. Stabilizer clamp
3. Stabilizer bushing
4. Stabilizer connecting rod
5. Front suspension member
6. Member stay
7. Transverse link
8. Steering knuckle
9. Strut
10. Coil spring insulator
11. Bound bumper
12. Dust cover

71075_SENT_G0225

Fig. 29 Exploded view of the front suspension assembly

10. When installing the support bearing bracket, note the following:

    a. Tighten the support bearing bracket bolts in two stages, in the order as shown.

    b. Install the support bearing bracket plate with the notch upward. Tighten the plate bolts in two stages.

11. Apply grease Molykote M77 to contact surface between wheel hub and driveshaft. Use sufficient grease to completely coat contact area.

    a. Tighten lock nut.

➡**Do not reuse non-reusable parts.**

12. Tighten ball joint to the specification as shown in the accompanying illustration.

13. Tighten the wheel nuts. Tighten the wheel nuts to 83 ft. lbs. (113 Nm).

14. Check transaxle fluid level.

> ✳✳ **WARNING**
>
> **Do not apply an excessive angle to driveshaft joint when removing from the transaxle.**

➡**Do not excessively extend slide joint.**

➡**Do not allow driveshaft to hang down. Support the entire driveshaft.**

15. If necessary, install the support bearing bracket bolts and the support bearing bracket.

16. Install the driveshaft to the transaxle assembly.

> ✳✳ **WARNING**
>
> **Do not apply an excessive angle to driveshaft joint when removing from the wheel hub and bearing assembly,**

➡**Do not excessively extend slide joint.**

➡**Do not allow driveshaft to hang down. Support the entire driveshaft.**

17. Install the support bearing bracket plate bolts and support bearing bracket plate.

➡**Use a suitable puller if wheel hub and bearing assembly and driveshaft cannot be separated after performing the above procedure.**

18. Install the driveshaft to the wheel hub and bearing assembly.

19. Replace the driveshaft to the wheel hub and bearing assembly by lightly tapping the end of the driveshaft using a hammer or suitable tool and wood block, and then install lock nut.

> ✳✳ **WARNING**
>
> **Temporarily leave the lock nut installed to prevent damage to threads.**

20. Install transverse link ball joint nut and bolt. Then, install transverse link to steering knuckle.

> ✳✳ **WARNING**
>
> **Do not pull on wheel sensor harness.**

21. Install cotter pin, then tighten lock nut using power tool.

22. Install wheel sensor to steering knuckle.

➡**When installing components such as hoses, tubes/lines, etc., cap or plug openings to prevent fluid from spilling.**

23. Install wheel and tire using power tool.

ENGINE COOLING

ENGINE COOLANT

DRAIN & REFILL

See Figures 30 through 32.

Draining Engine Coolant

1. Remove the engine undercover.

2. Open the radiator drain plug at the bottom of the radiator, and remove the radiator filler cap. This is the only step required when partially draining the cooling system (radiator only).

> ✳✳ **WARNING**
>
> **Perform this step when engine is cold.**

> ✳✳ **WARNING**
>
> **Do not spill engine coolant on drive belt.**

3. Follow this step for heater core removal/replacement only. Disconnect the upper heater hose at the engine side and apply moderate air pressure 15 psi maximum air pressure into the hose for 30 seconds to blow the excess coolant out of the heater core.

4. When draining all of the coolant in the system, remove the reservoir tank and drain the coolant, then clean the reservoir tank before installation.

> ✳✳ **WARNING**
>
> **Do not allow the coolant to contact drive belt.**

5. When draining all of the coolant in the system for engine removal or repair, open the drain plug on the cylinder block.

6. Check the drained engine coolant for

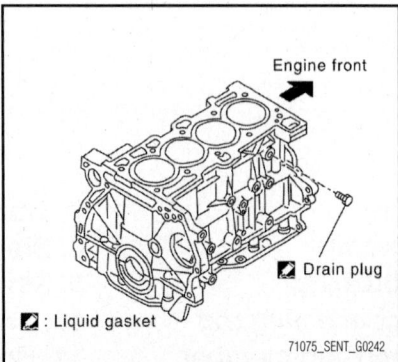

Fig. 30 When draining all of the coolant in the system for engine removal or repair, open the drain plug on the cylinder block

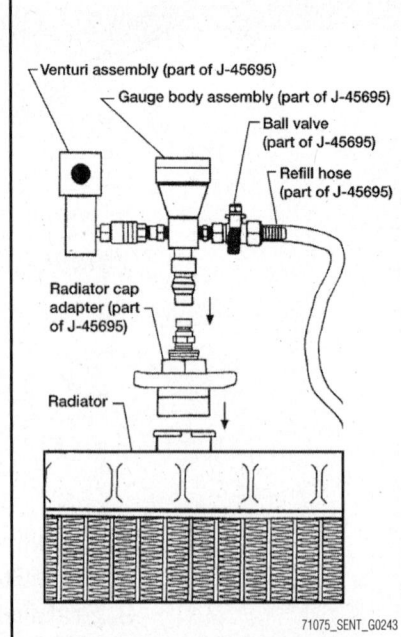

Fig. 31 Install the Tool [KV991J0070 (J-45695)] by installing the radiator cap adapter onto the radiator neck opening. Then attach the gauge body assembly with the refill tube and the venturi assembly to the radiator cap adapter.

contaminants such as rust, corrosion or discoloration. Flush the engine cooling system if the coolant is contaminated. Follow the "FLUSHING COOLING SYSTEM" procedure.

Refilling Engine Coolant

7. Install the radiator drain plug. Install the reservoir tank and cylinder block drain plug as necessary.

a. The radiator must be completely empty of coolant and water.

b. Apply sealant to the threads of the cylinder block drain plugs. Use Genuine High Performance
Thread Sealant or equivalent.

8. If disconnected, reattach the upper radiator hose at the engine side.

9. Set the vehicle heater controls to the full HOT and heater ON position. Turn the vehicle ignition ON with the engine OFF as necessary to activate the heater mode.

10. Install the Tool by installing the radiator cap adapter onto the radiator neck opening. Then attach the gauge body assembly with the refill tube and the venturi assembly to the radiator cap adapter.

11. Insert the refill hose into the coolant mixture container that is placed at floor level. Make sure the ball valve is in the closed position.

a. Use the recommended coolant or equivalent. Pre-diluted Genuine NISSAN Long Life Anti-Freeze/Coolant (blue) concentrate is recommended by manufacturer. If an equivalent coolant other than Pre-diluted Genuine NISSAN Long Life Anti-Freeze/Coolant (blue) is used, follow the coolant manufacturer's instructions to maintain minimum freeze protection to 34°F (-37°C). The use of other types of coolant solutions other than Pre-diluted Genuine NISSAN Long Life Anti-Freeze/Coolant (blue) or equivalent may damage the engine cooling system.

12. Install an air hose to the venturi assembly, the air pressure must be within specification. Compressed air supply pressure of 80-119 psi.

➡**The compressed air supply must be equipped with an air dryer.**

13. The vacuum gauge will begin to rise and there will be an audible hissing noise. During this process open the ball valve on the refill hose slightly. Coolant will be visible rising in the refill hose. Once the refill hose is full of coolant, close the ball valve. This will purge any air trapped in the refill hose.

Altitude above sea level	Vacuum gauge reading
0 - 100 m (328 ft)	: 28 inches of vacuum
300 m (984 ft)	: 27 inches of vacuum
500 m (1,641 ft)	: 26 inches of vacuum
1,000 m (3,281 ft)	: 24 - 25 inches of vacuum

71075_SENT_G0245

Fig. 32 Vacuum specifications

14. Continue to draw the vacuum until the gauge reaches 28 inches of vacuum. The gauge may not reach 28 inches in high altitude locations, use the vacuum specifications based on the altitude above sea level.

15. When the vacuum gauge has reached the specified amount, disconnect the air hose and wait 20 seconds to see if the system loses any vacuum. If the vacuum level drops, perform any necessary repairs to the system and repeat steps 6 - 8 to bring the vacuum to the specified amount. Recheck for any leaks.

16. Place the coolant container (with the refill hose inserted) at the same level as the top of the radiator. Then open the ball valve on the refill hose so the coolant will be drawn up to fill the cooling system. The cooling system is full when the vacuum gauge reads zero.

➡**Do not allow the coolant container to get too low when filling, to avoid air from being drawn into the cooling system.**

17. Remove the Tool from the radiator neck opening.

18. Fill the cooling system reservoir tank to the specified level and install the radiator cap. Run the engine to warm up the cooling system and top up the system as necessary.

19. Install the engine undercover.

Flushing Cooling System

20. Fill radiator and reservoir tank with water and install radiator cap.

21. Run engine until it reaches normal operating temperature.

22. Rev the engine two or three times under no-load.

23. Stop the engine and wait until it cools down.

24. Drain water from the cooling system.

25. Repeat steps 1 through 5 until clear water begins to drain from the radiator.

FLUID RECOMMENDATIONS

For the Sentra, Nissan Long Life (N-LL) Anti-freeze/Coolant (blue) is recommended.

LEVEL CHECK

1. Check if the reservoir tank engine coolant level is within MIN to MAX when the engine is cool.

2. Adjust the engine coolant level as necessary.

ELECTRIC ENGINE FAN

REMOVAL & INSTALLATION

See Figures 33 and 34.

✳✳ WARNING

Never remove the radiator cap when the engine is hot. Serious burns could occur from high pressure engine coolant escaping from the radiator. Wrap a thick cloth around the cap. Slowly turn it a quarter of a turn to release built-up pressure. Carefully remove radiator cap by turning it all the way.

➡**When removing components such as hoses, or tubes/lines, etc., cap or plug openings to prevent fluid from spilling.**

1. Partially drain engine coolant from radiator.

➡**Perform this step when engine is cold.**

➡**Do not spill engine coolant on drive belt.**

2. Remove air duct (inlet).

3. Disconnect radiator hose (upper) at radiator side.

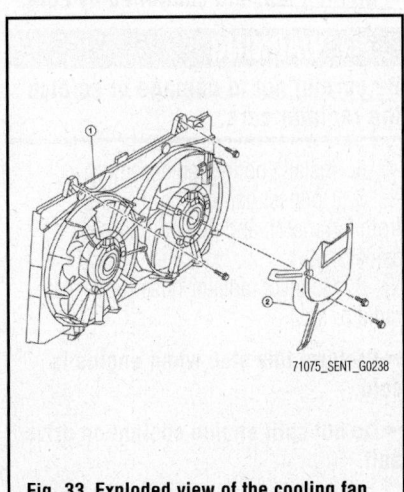

71075_SENT_G0238

Fig. 33 Exploded view of the cooling fan assembly (1), and radiator shroud (2); if equipped

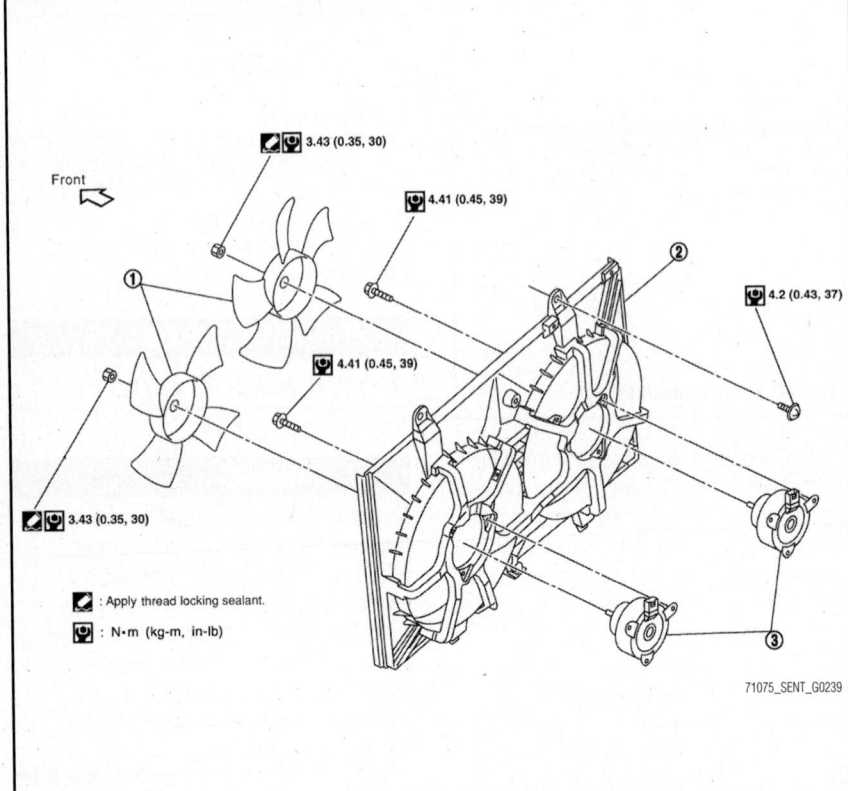

Fig. 34 Exploded view of the cooling fan components. Fan blade (1), fan shroud (2), fan motor (3)

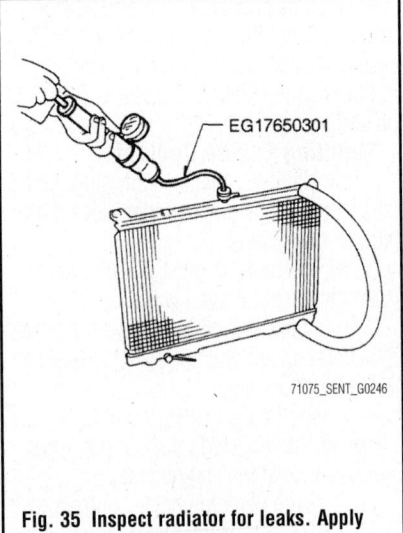

Fig. 35 Inspect radiator for leaks. Apply pressure using suitable tool (Specified pressure value [23 psi], and Tool EG17650301 [J-33984-A]).

4. Disconnect harness connectors from fan motor, and position harness aside.
5. Remove cooling fan assembly.

✳✳ WARNING
Be careful not to damage or scratch the radiator core.

To install:

➡ **Cooling fans are controlled by ECM.**

✳✳ WARNING
Be careful not to damage or scratch the radiator core.

6. Install cooling fan assembly.
7. Connect harness connectors from fan motor, and position harness aside.
8. Connect radiator hose (upper) at radiator side.

➡ **Perform this step when engine is cold.**

➡ **Do not spill engine coolant on drive belt.**

9. Install air duct (inlet).
10. Fill engine coolant from radiator.

RADIATOR

REMOVAL & INSTALLATION
See Figure 35.

1. Drain engine coolant from radiator.
   a. Perform this step when engine is cold.
   b. Do not spill engine coolant on drive belt
2. Remove front air duct.
3. Disconnect radiator upper and lower hoses.
4. Disconnect reservoir tank hose.
5. Disconnect harness connector from fan motors and move harness to aside.
6. Remove the cooling fan assembly to radiator bolts and remove cooling fan assembly.
7. Remove radiator upper mounts.
8. Move radiator assembly to the rearward direction of vehicle, and then lift it upward to remove.

✳✳ WARNING
Do not damage or scratch A/C condenser and radiator core when removing.

Inspection After Removal
9. Inspect radiator for leaks as follows:
   a. Apply pressure using suitable tool and Tool.

✳✳ WARNING
To prevent the risk of the hose coming undone while under pressure, securely fasten it down with a hose clamp.

   b. Check for leakage.

To install:

✳✳ WARNING
Do not damage or scratch A/C condenser and radiator core when installing.

10. Move radiator assembly to the forward direction of vehicle, and then lift it downward to install.
11. Install radiator upper mounts.
12. Install the cooling fan assembly to radiator bolts and install cooling fan assembly.
13. Connect harness connector to fan motors and move harness into position.
14. Connect reservoir tank hose.
15. Connect radiator upper and lower hoses.
16. Install front air duct.
17. Fill engine coolant to radiator.
   a. Do not spill engine coolant on drive belt
   b. Perform this step when engine is cold.

THERMOSTAT

REMOVAL & INSTALLATION

See Figures 36 and 37.

➡️There are two thermostats on these engines. The lower one is referred to as a "thermostat," while the upper one is referred to as a "water control valve."

1. Before servicing the vehicle, refer to the Precautions Section.
2. Remove front air duct and engine undercover as necessary.
3. Drain the cooling system.
4. Remove radiator lower hose.
5. Remove engine coolant inlet and thermostat.

To install:

6. Fit a new rubber ring on the thermostat, making sure the flange seats properly inside the ring.
7. Install the thermostat with the jiggle valve facing upwards. The position deviation may be within the range of +/- 10°.
8. To complete installation, reverse remaining removal procedure
9. Fill the cooling system.

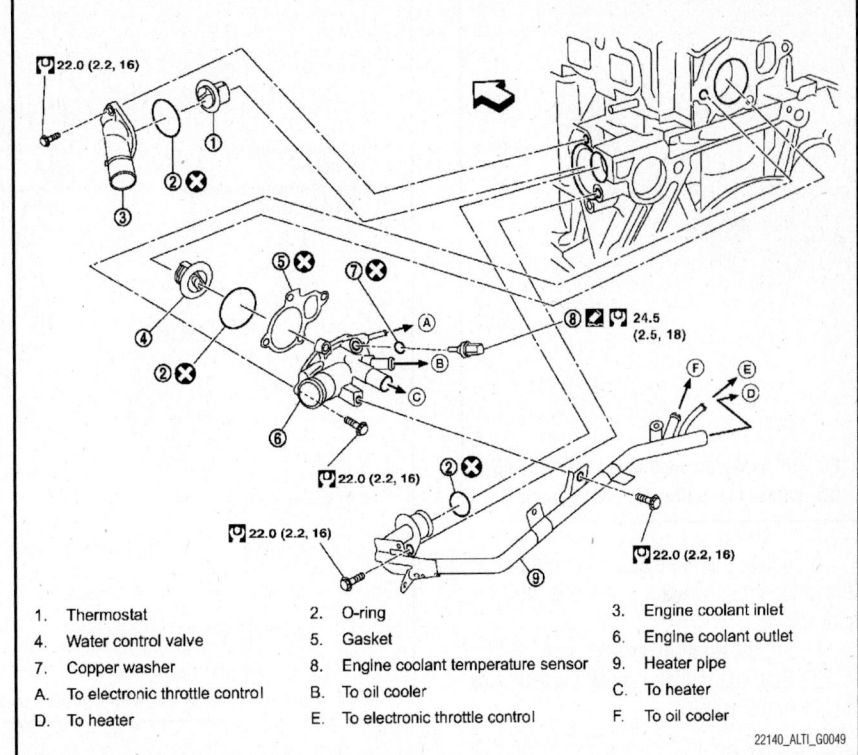

1.	Thermostat	2.	O-ring	3.	Engine coolant inlet
4.	Water control valve	5.	Gasket	6.	Engine coolant outlet
7.	Copper washer	8.	Engine coolant temperature sensor	9.	Heater pipe
A.	To electronic throttle control	B.	To oil cooler	C.	To heater
D.	To heater	E.	To electronic throttle control	F.	To oil cooler

22140_ALTI_G0049

Fig. 37 Thermostat (1) and water valve (4) housings—2.5L engine

10. After installation, run engine for a few minutes and check for leaks.

WATER PUMP

REMOVAL & INSTALLATION

2.0L Engine

See Figure 38.

1. Before servicing the vehicle, refer to the precautions in the beginning of this section.
2. Drain the cooling system.
3. To remove the drive belt tensioner:
   - Working underneath the vehicle, place a wrench on the tensioner idler pulley nut.
   - Push the wrench clockwise and insert a short screwdriver in the hole that appears to the right of the tensioner to hold the tensioner in place.
   - Remove the drive belt
4. Loosen the water pump bolts and remove the pump.
5. Remove all traces of gasket material from sealing surfaces.

To install:

6. Install a new gasket:
7. Install the water pump and torque the bolts to 18 ft. lbs. (25 Nm).

Engine front

1. Thermostat housing
2. Water hose (models with oil cooler)
3. Radiator hose (lower)
4. Water inlet
5. Thermostat
6. Rubber ring
7. Gasket
A. To oil cooler
B. To radiator

🔧 : N•m (kg-m, ft-lb)

22140_ALTI_G0008

Fig. 36 Thermostat and components—2.0L engine with manual transmission shown

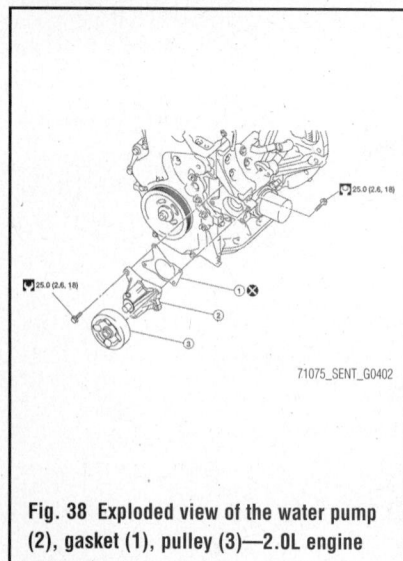

Fig. 38 Exploded view of the water pump (2), gasket (1), pulley (3)—2.0L engine

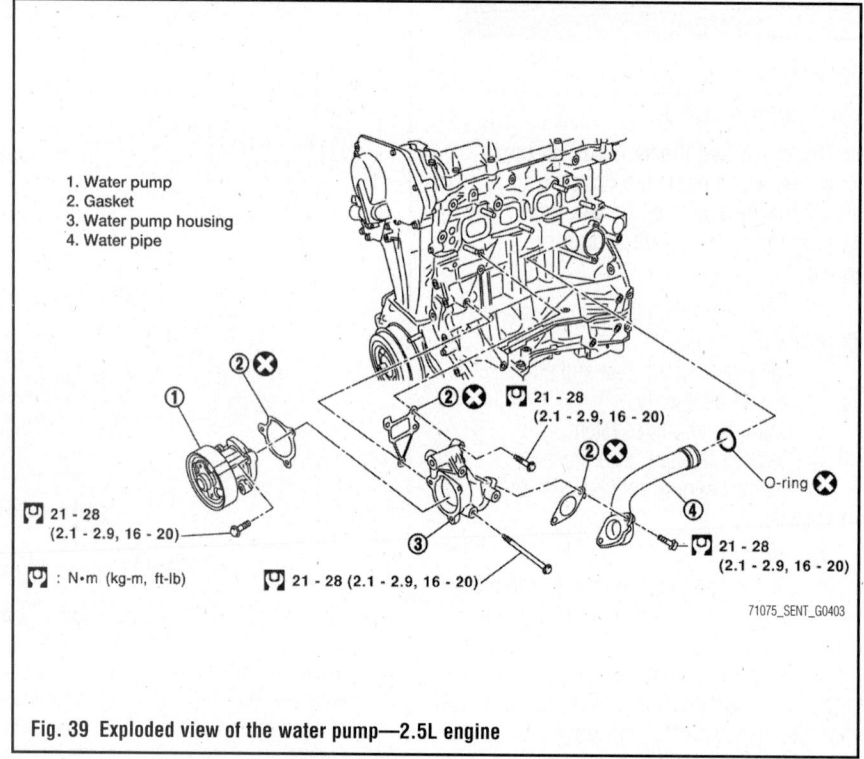

1. Water pump
2. Gasket
3. Water pump housing
4. Water pipe

Fig. 39 Exploded view of the water pump—2.5L engine

8. Install the drive belt and hold the tensioner with a wrench to remove the screwdriver.

9. Fill the cooling system.

10. Start the vehicle, check for leaks and repair if necessary.

2.5L Engine

See Figure 39.

1. Before servicing the vehicle, refer to the precautions in the beginning of this section.

2. Drain the cooling system.

3. Remove the engine undercover.

4. A special tool is needed to remove the drive belt tensioner. To remove the tensioner:

- Working underneath the vehicle, place tool J-46535 on the tensioner idler pulley and push clockwise.

✳✳ WARNING

Do not loosen the belt tensioner pulley bolt or turn the tensioner counterclockwise. If the bolt is loosened, the tensioner assembly must be replaced.

- Insert a short screwdriver into the tensioner retaining boss to lock the tensioner in place
- Remove the drive belt

5. Remove the coolant reservoir.

6. Remove the Intelligent Power Distribution Module (IPDM) by unlocking the pawls and unplugging the connector.

7. Raise and safely support the vehicle and remove the right front wheel.

8. Remove the inner fender.

9. Remove the ground strap.

➡The alternator and exhaust system may interfere with removal of the water pipe.

10. Loosen the water pump bolts and remove the pump.

11. Remove all traces of gasket material from sealing surfaces.

To install:

12. Install a new gasket:

13. Install the water pump and torque the bolts to 20 ft. lbs. (28 Nm).

14. Install the drive belt and hold the tensioner with the special tool to remove the screwdriver.

15. Reinstall the remaining components.

16. Fill the cooling system.

17. Start the engine, check for leaks and repair if necessary.

ENGINE ELECTRICAL   **BATTERY SYSTEM**

BATTERY

REMOVAL & INSTALLATION

2.0L Engine

See Figure 40.

1. Loosen battery terminal nuts, and disconnect both negative and positive battery terminals.

> ❊❊ **WARNING**
>
> **Disconnect the battery negative terminal first.**

2. Remove battery frame nuts and battery frame.
3. Remove battery.

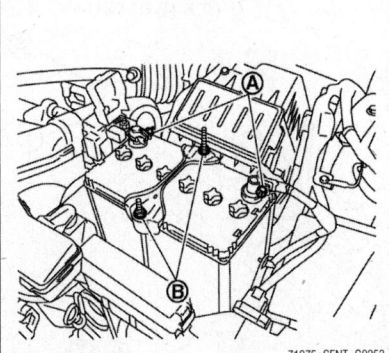

71075_SENT_G0253

Fig. 40 Loosen battery terminal nuts (A), and disconnect both negative and positive battery terminals. Remove battery frame nuts (B) and battery frame

To install:

4. Install battery.
5. Install battery frame nuts and battery frame.

> ❊❊ **WARNING**
>
> **Connect the battery positive terminal first.**

6. Tighten battery terminal nuts, and connect both negative and positive battery terminals.

   a. Tighten the battery frame nuts to 48 inch lbs. (5.4 Nm), and the battery terminal nuts to 48 inch lbs. (5.4 Nm).

2.5L Engine

See Figure 41.

1. Loosen battery terminal nuts, and disconnect both negative and positive battery terminals.

> ❊❊ **WARNING**
>
> **Disconnect the battery negative terminal first.**

2. Remove battery hold-down wedge bolt and battery wedge bracket.
3. Remove battery.
4. Remove battery tray liner, if necessary.

To install:

5. Install battery tray liner, if necessary.
6. Install battery.

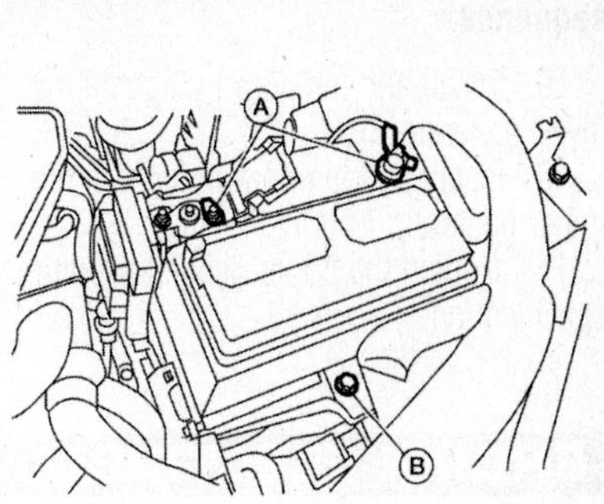

71075_SENT_G0254

Fig. 41 Loosen battery terminal nuts (A), and disconnect both negative and positive battery terminals. Remove battery hold-down wedge bolt (B) and battery wedge bracket.

7. Install battery hold-down wedge bolt and battery wedge bracket.

> ❊❊ **WARNING**
>
> **Connect the battery positive terminal first.**

8. Tighten battery terminal nuts, and connect both negative and positive battery terminals.

   a. Tighten the battery edge bracket bolt to 22 ft. lbs. (30 Nm), and the battery terminal nuts to 48 inch lbs. (5.4 Nm).

9. Reset electronic systems as necessary.

ENGINE ELECTRICAL   **CHARGING SYSTEM**

GENERATOR

REMOVAL & INSTALLATION

2.0L Engine

1. Disconnect the battery negative terminal.
2. Remove RH front wheel and tire.
3. Remove splash shield RH.
4. Remove drive belt.
5. Disconnect generator connector.
6. Remove "B" terminal nut.
7. Remove generator bolts.
8. Remove generator assembly from the vehicle.

To install:

➡**Generator bolts must be tightened in sequence.**

9. Install and temporarily tighten the lower generator bolt.
10. Install and temporarily tighten the upper generator bolt.
11. Tighten the upper generator bolt to specification.
12. Tighten the lower generator bolt to specification.

➡**Slide bushing must contact engine bracket after generator is installed.**

13. Install generator assembly to the vehicle.
14. Install generator bolts.
15. Install "B" terminal nut.

➡**Be sure to tighten "B" terminal nut carefully.**

16. Connect generator connector.
17. Install drive belt.
18. Install splash shield RH.
19. Install RH front wheel and tire.
20. Connect the battery negative terminal.

2.5L Engine

1. Disconnect the battery negative terminal.
2. Remove splash shield RH.
3. Remove drive belt.
4. Disconnect generator connector.
5. Remove "B" terminal nut.
6. Remove harness bracket and position aside.
7. Remove generator bolts.
8. Remove generator assembly from the vehicle.

To install:

➡**Generator bolts must be tightened in sequence.**

9. Install and temporarily tighten the lower generator bolt.
10. Install and temporarily tighten the upper generator bolt.
11. Tighten the upper generator bolt to specification.

12. Tighten the lower generator bolt to specification.

➡**Slide bushing must contact engine bracket after generator is installed.**

13. Install generator assembly from the vehicle.
14. Install generator bolts.
15. Install harness bracket and position aside.

16. Install "B" terminal nut.

➡**Be sure to tighten "b" terminal nut carefully.**

17. Connect generator connector.
18. Install drive belt.
19. Install splash shield RH.
20. Connect the battery negative terminal.

ENGINE ELECTRICAL   IGNITION SYSTEM

FIRING ORDERS

Firing order: 1–3–4–2. Distributorless ignition system (one coil on each cylinder)

IGNITION COIL

REMOVAL & INSTALLATION

2.0L Engine

See Figure 42.

1. Remove the intake manifold.
2. Remove the four ignition coils.

➡**Handle the ignition coil carefully and avoid impacts.**

➡**Never disassemble the ignition coil.**

To install:
3. Install the four ignition coils.
4. Install the intake manifold.

2.5L Engine

See Figure 43.

1. Remove the engine room cover.
2. Remove resonator.
3. Disconnect the harness connector from the ignition coil.
4. Remove the ignition coil.

➡**Do not drop or shock the ignition coil it.**

To install:
5. Install the ignition coil.
6. Connect the harness connector to the ignition coil.
7. Install resonator.
8. Install the engine room cover.

IGNITION TIMING

INSPECTION & ADJUSTMENT

2.0L Engine

See Figures 44 and 45.

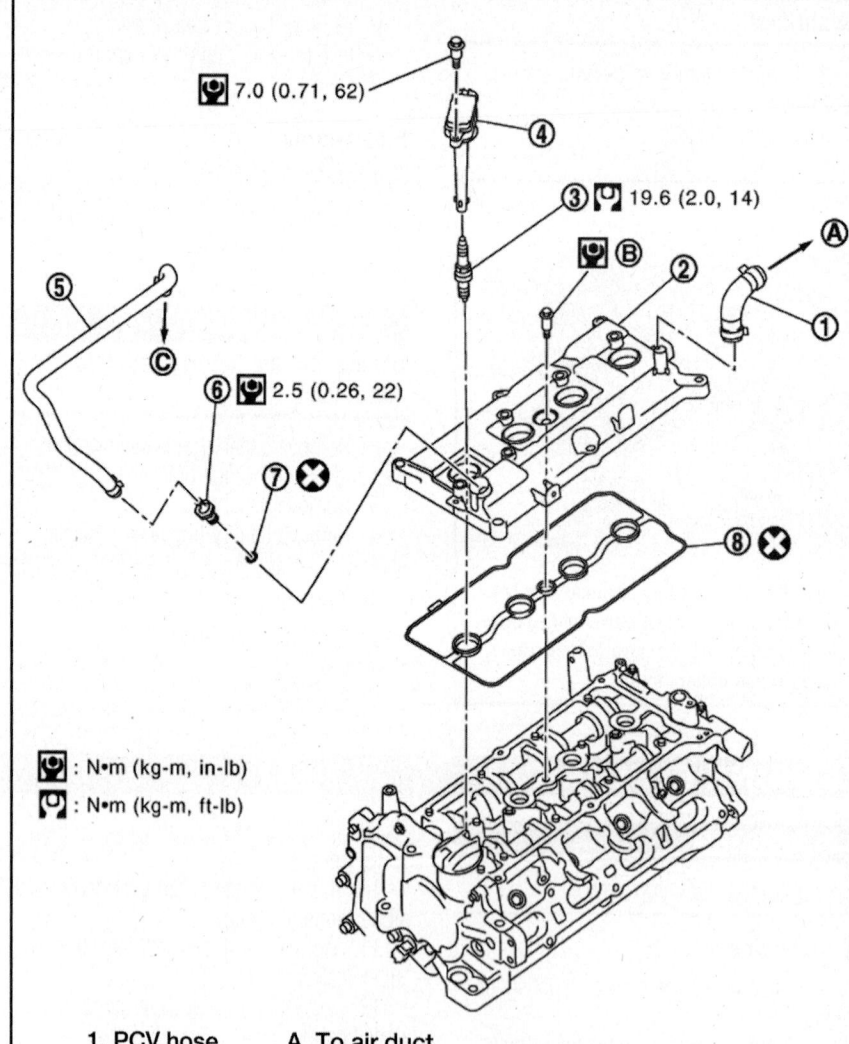

1. PCV hose
2. Rocker cover
3. Spark plug
4. Ignition coil
5. PCV hose
6. PCV valve
7. O-ring
8. Gasket

A. To air duct
B. Rocker cover bolts. Tighten in two steps:
    1st step : 17 inch lb. (1.96 Nm)
    2nd step : 73 inch lb. (8.33 Nm)
C. To intake manifold

🔧 : N•m (kg-m, in-lb)
🔧 : N•m (kg-m, ft-lb)

71075_SENT_G0388

Fig. 42 Exploded view of the ignition coil, spark plug, and rocker cover components

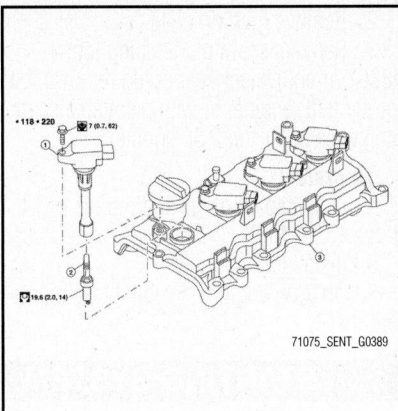

Fig. 43 Exploded view of the ignition coil (1), Spark plug (2), Rocker cover (3)— 2.5L engine

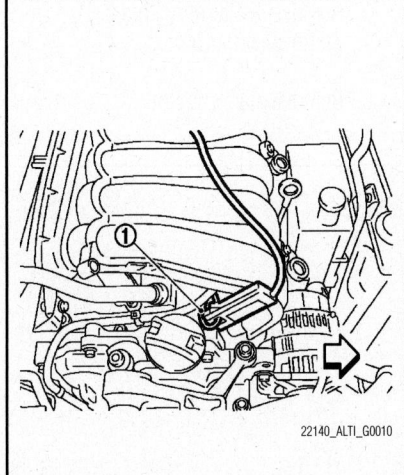

Fig. 44 Some vehicles have a timing loop for connecting a standard timing light

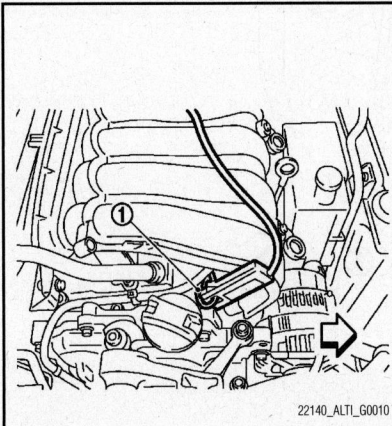

Fig. 45 Add an extension wire between the No. 4 ignition coil and spark plug to connect timing light

1. There are two different ways to connect a standard timing light. If equipped with a loop wire as shown, simply attach the pick-up clamp and check ignition timing at the crankshaft pulley.

2. If no loop wire is installed:
- Remove No. 4 ignition coil
- Make up a high-tension extension wire using a suitable sparkplug wire.
- Connect No. 4 ignition coil to No. 4 spark plug with high-tension extension wire and attach timing light clamp to this wire.

3. Check ignition timing at the crankshaft pulley. Timing should be 6°± 5°BTDC (in Neutral), but there is no adjustment.

2.5L Engine

See Figures 46 and 47.

The ignition timing is not adjustable. If not within specifications, further diagnostic inspection is required. The following procedure is for viewing the ignition timing setting.

Visually check the air cleaner, intake hoses, ducts, Exhaust Gas Recirculation (EGR) valve operation and electrical connections prior to the adjustment of the ignition timing. Correct or repair any problem as required. Be sure to inspect the throttle valve and Throttle Position (TP) sensor for proper operation.

1. Before servicing the vehicle, refer to the Precautions Section.

2. Locate the timing marks on the crankshaft pulley and the front of the engine.

3. Clean the timing marks.

4. The ignition timing specifications are as follows:

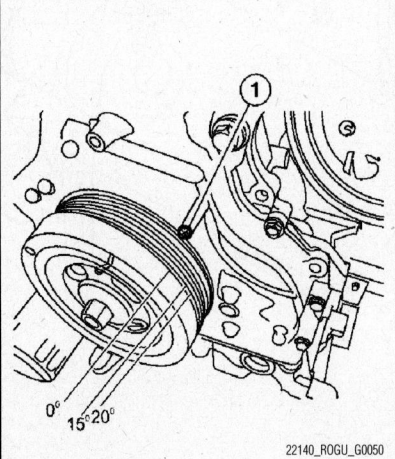

Fig. 46 Locate the timing marks on the crankshaft pulley

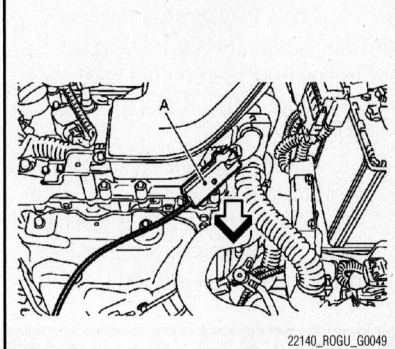

Fig. 47 Attach a timing light to the engine to number 1 cylinder ignition coil wire

- 10–20 degrees Before Top Dead Center (BTDC)

5. Using chalk or white paint, color the mark on the crankshaft pulley and the mark on the scale, which will indicate the correct timing when aligned with the notch on the crankshaft pulley.

6. Attach a tachometer to the engine.

7. Attach a timing light to the engine to number 1 cylinder ignition coil wire.

8. Turn **OFF** all the electrical equipment and accessories.

9. Check to be sure all of the wires clear the fan, then, start the engine and allow it to reach normal operating temperatures.

10. Block the front wheels and set the parking brake. Shift the transmission into **NEUTRAL.**

11. Perform the following procedures:

a. Race the engine at 2000 rpm for about 2 minutes under a no-load condition; be sure all of the accessories are turned **OFF**.

b. Perform on board engine diagnostics and repair any fault code.

c. Race the engine at 2000 rpm for about 2 minutes under a no-load condition.

d. Turn the engine **OFF** and disconnect the TP sensor.

e. Start and race the engine 2–3 times under no-load, then run the engine at idle speed.

12. Aim the timing light at the timing marks. If the marks on the pulley and the engine are aligned when the light flashes, the timing is correct. Turn the engine **OFF** and remove the tachometer and the timing light. If the marks are not in alignment, proceed with the following steps:

a. Turn the engine **OFF**.

b. Check the Camshaft Position (CMP) sensor (PHASE), Crankshaft Position (CKP) sensor (REF) and CKP sensor (POS). Replace if necessary.

c. Check that all the timing chain and gears are correctly aligned.

d. If the ignition timing is still not correct, substitute a known good Electronic Control Module (ECM).

➡ **The ECM may be the cause of the problem, but this is rarely the case.**

e. Turn the engine **OFF** and remove the tachometer and the timing light.

SPARK PLUGS

REMOVAL & INSTALLATION

1. On 2.0L engines, remove intake manifold

2. Remove ignition coils.

3. Remove spark plugs using tool J-48891 or equivalent spark plug removal tool for the 2.0L engine, or with a suitable spark plug wrench for the 2.5L engine.

To install:

4. Install spark plugs and tighten to 14 ft. lbs. (19.6 Nm).

5. Install ignition coils

6. On 2.0L engines, install intake manifold

ENGINE ELECTRICAL

STARTER

REMOVAL & INSTALLATION

2.0L Engine

See Figure 48.

1. Disconnect the battery negative terminal.

2. Remove "S" terminal nut.

3. Remove "B" terminal nut.

4. Remove starter motor bolts.

5. Remove starter motor.

To install:

6. Install starter motor.

7. Install starter motor bolts.

8. Install "B" terminal nut.

9. Install "S" terminal nut.

10. Connect the battery negative terminal.

STARTING SYSTEM

2.5L Engine

See Figures 49 and 50.

1. Disconnect the battery negative terminal.

2. Raise vehicle.

3. Remove "B" terminal nut.

4. Remove starter motor bolts.

5. Remove starter motor.

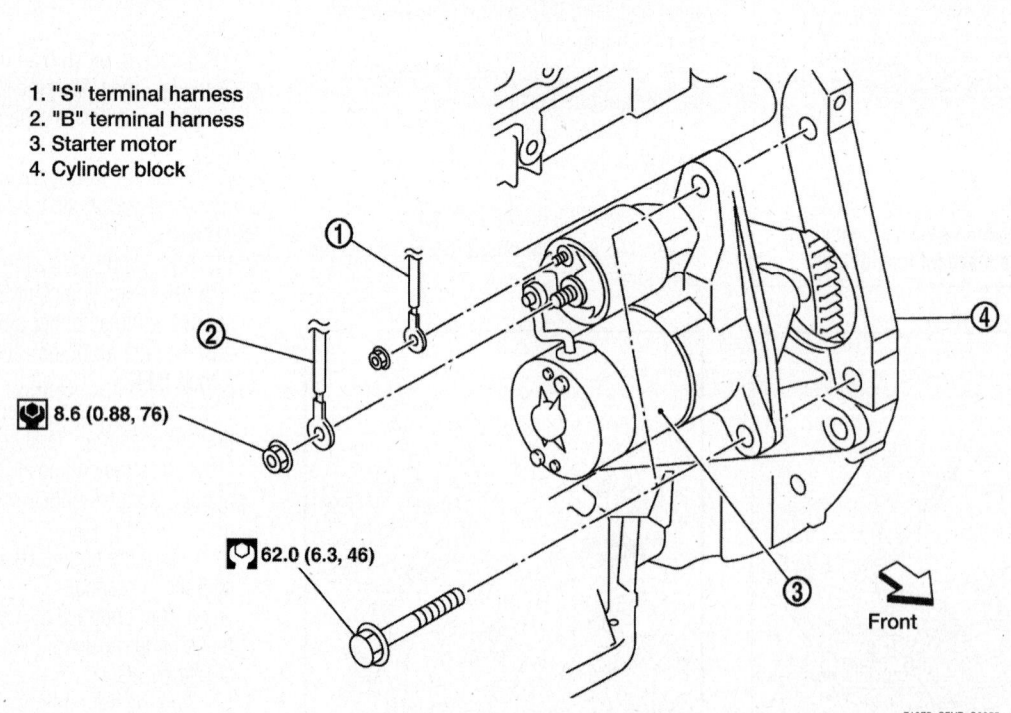

1. "S" terminal harness
2. "B" terminal harness
3. Starter motor
4. Cylinder block

8.6 (0.88, 76)

62.0 (6.3, 46)

Front

71075_SENT_G0355

Fig. 48 Exploded view of the starter motor—2.0L engine

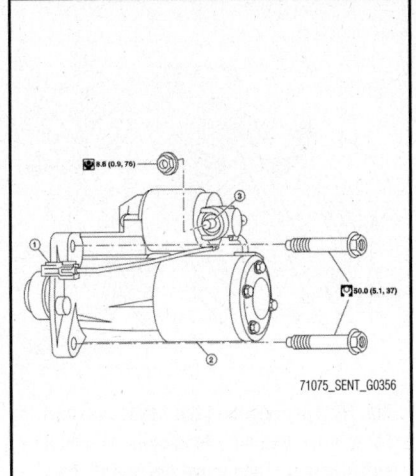

Fig. 49 Exploded view of the starter motor
(2), "S" terminal harness (1), "B" terminal
(3)—CVT with 2.5L engine

To install:

6. Install starter motor.
7. Install starter motor bolts.
8. Install "B" terminal nut.
9. Lower vehicle.
10. Connect the battery negative terminal.

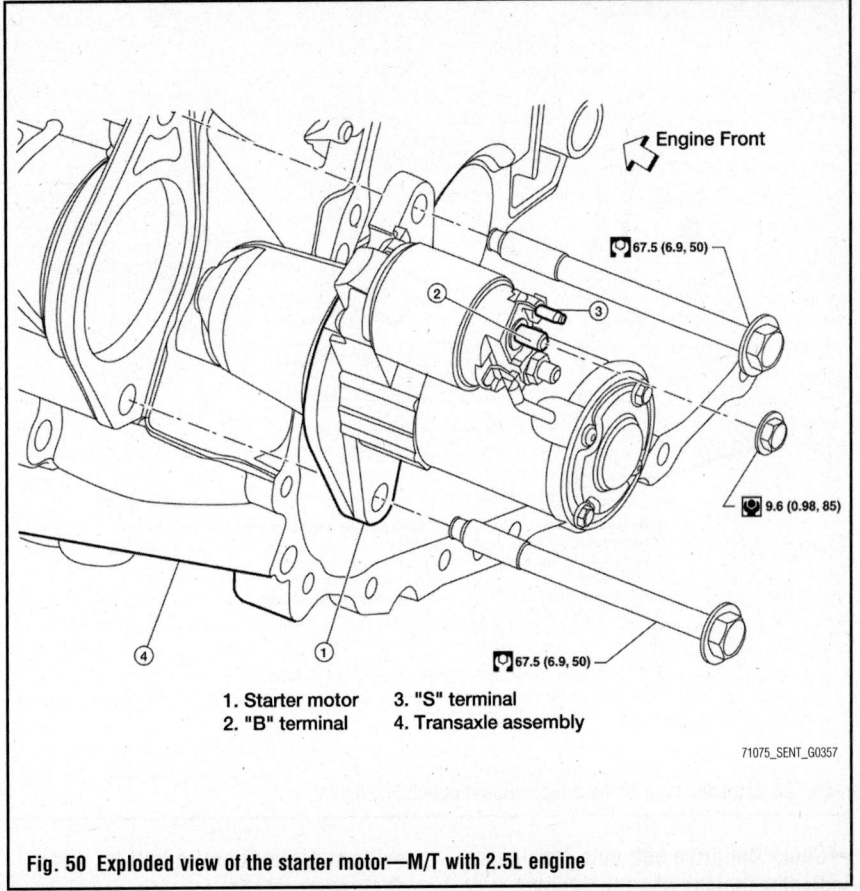

1. Starter motor
2. "B" terminal
3. "S" terminal
4. Transaxle assembly

Fig. 50 Exploded view of the starter motor—M/T with 2.5L engine

ENGINE MECHANICAL

➡Disconnecting the negative battery cable may interfere with the functions of the on board computer systems and may require the computer to undergo a relearning process, once the negative battery cable is reconnected.

➡Make sure that the indicator (notch on fixed side) of drive belt auto-tensioner is within the possible use range.

➡On vehicles not equipped with A/C, there is an idler pulley in the position for the drive belt routing.

ACCESSORY DRIVE BELT SYSTEM

ADJUSTMENT

Belt tension is not manually adjustable, it is automatically adjusted by the drive belt auto-tensioner.

BELT ROUTING

See Figures 51 and 52.

INSPECTION

2.0L Engine

See Figure 51.

> ✳✳ **CAUTION**
>
> Inspect the drive belt only when the engine is stopped.

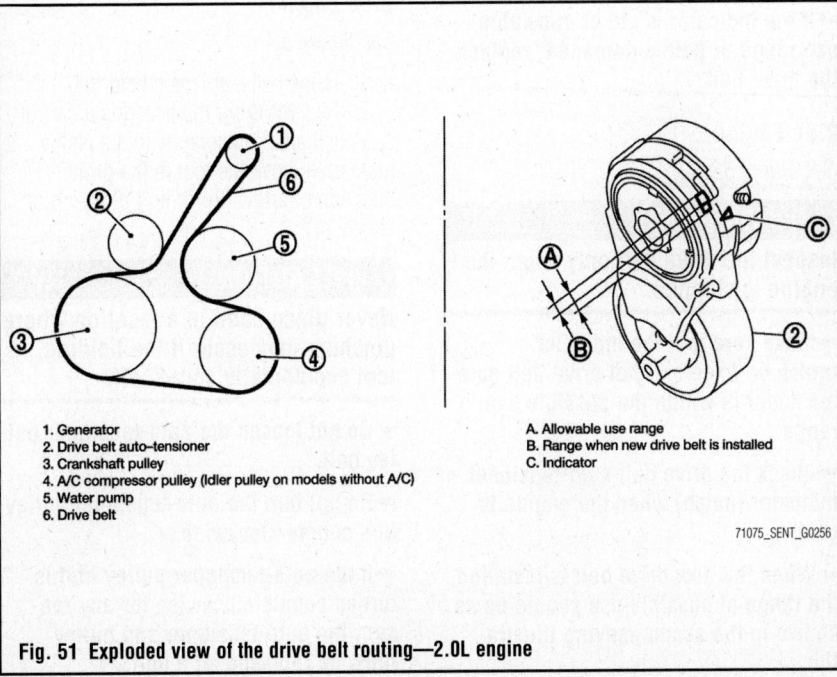

1. Generator
2. Drive belt auto-tensioner
3. Crankshaft pulley
4. A/C compressor pulley (Idler pulley on models without A/C)
5. Water pump
6. Drive belt

A. Allowable use range
B. Range when new drive belt is installed
C. Indicator

Fig. 51 Exploded view of the drive belt routing—2.0L engine

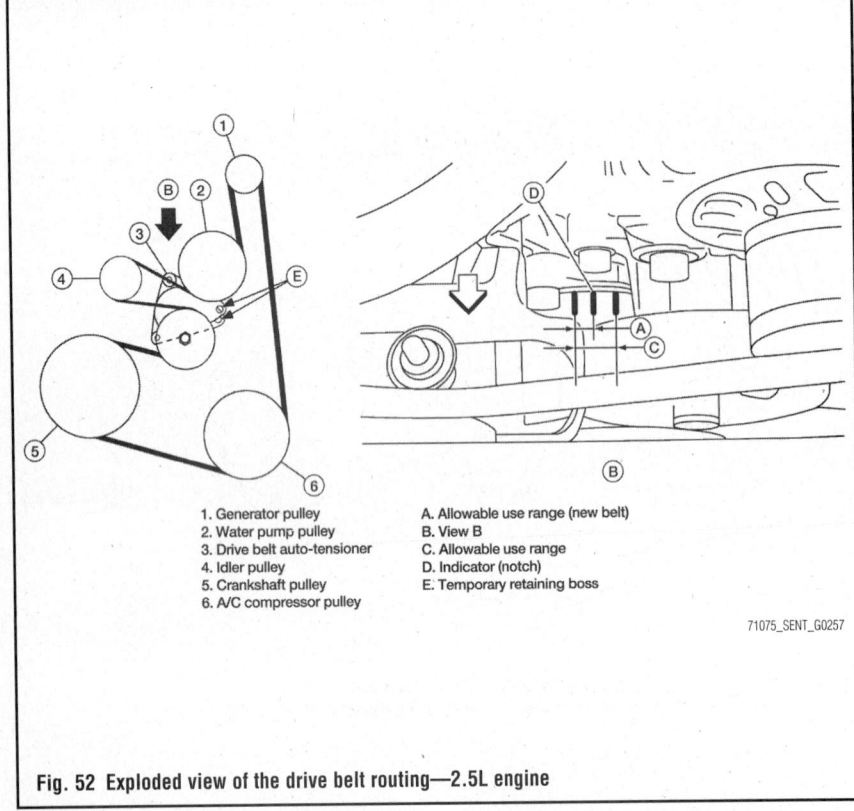

1. Generator pulley
2. Water pump pulley
3. Drive belt auto-tensioner
4. Idler pulley
5. Crankshaft pulley
6. A/C compressor pulley

A. Allowable use range (new belt)
B. View B
C. Allowable use range
D. Indicator (notch)
E. Temporary retaining boss

71075_SENT_G0257

Fig. 52 Exploded view of the drive belt routing—2.5L engine

➡ Check the drive belt auto-tensioner indicator (notch) when the engine is cold.

➡ When the new drive belt is installed, the range of possible use should be as shown in the accompanying illustration.

➡ Visually check entire belt for wear, damage or cracks.

➡ If the indicator is out of allowable use range or belt is damaged, replace the drive belt.

2.5L Engine

See Figure 52.

✳✳ WARNING

Inspect the drive belt only when the engine is stopped.

➡ Make sure that the indicator (notch on fixed side) of drive belt auto-tensioner is within the possible use range.

➡ Check the drive belt auto-tensioner indicator (notch) when the engine is cold.

➡ When the new drive belt is installed, the range of possible use should be as shown in the accompanying illustration.

➡ Visually check entire belt for wear, damage or cracks.

➡ If the indicator is out of allowable use range or belt is damaged, replace the drive belt.

REMOVAL & INSTALLATION

Drive Belt

2.0L Engine

See Figure 53.

1. Remove the splash shield RH.
2. Securely hold the hexagonal part of drive belt auto-tensioner with a suitable tool. Then move the tool in the proper direction of arrow (loosening direction of tensioner).

✳✳ CAUTION

Never place hand in a location where pinching may occur if the holding tool accidentally comes off.

➡ Do not loosen the auto-tensioner pulley bolt.

➡ Do not turn the auto-tensioner pulley bolt counterclockwise.

➡ If the auto-tensioner pulley bolt is turned counterclockwise for any reason, the auto-tensioner and pulley must be replaced as a unit.

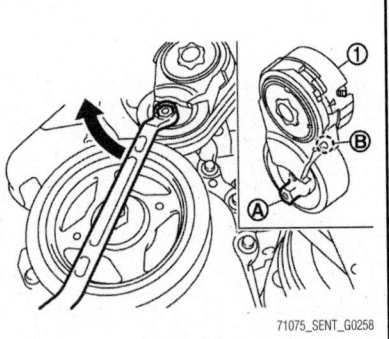

71075_SENT_G0258

Fig. 53 Securely hold the hexagonal part (A) of drive belt auto-tensioner (1) with a suitable tool. Then move the tool in the direction of arrow (loosening direction of tensioner). Insert a rod approximately 6 mm (0.24 in) in diameter into the hole (B) of the retaining boss to lock drive belt auto-tensioner

3. Insert a rod approximately 0.24 inch (6 mm) in diameter into the hole of the retaining boss to lock drive belt auto-tensioner.
   a. Leave tensioner pulley arm locked until belt is installed again.
4. Remove drive belt.

To install:

5. Install drive belt.

➡ Confirm drive belt is set completely and correctly on all of the pulleys.

6. Release drive belt auto-tensioner, and apply tension to drive belt.

✳✳ CAUTION

Never place hand in a location where pinching may occur if the holding tool accidentally comes off.

➡ Ensure the drive belt and each groove on the pulley are free from all engine oil, coolant and other contaminants.

7. Turn crankshaft pulley clockwise several times to equalize tension between each pulley.
8. Confirm tension of drive belt at indicator is within the allowable use range.
9. Install the splash shield RH.

2.5L Engine

See Figure 54.

1. Remove the splash shield (RH).
2. Securely hold the hexagonal part in pulley center of drive belt auto-tensioner,

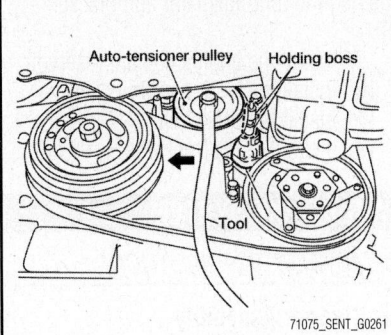

Fig. 54 Securely hold the hexagonal part in pulley center of drive belt auto-tensioner, move in the direction of arrow (loosening direction of tensioner) using Tool.

move in the proper direction using Tool (J-46535).

✳✳ CAUTION

Never place hand in a location where pinching may occur if the holding tool accidentally comes off.

➡Do not loosen the auto-tensioner pulley bolt.

➡Do not turn the auto-tensioner pulley bolt counterclockwise.

➡If the auto-tensioner pulley bolt is turned counterclockwise for any reason, the auto-tensioner and pulley must be replaced as a unit.

3. Insert a rod approximately 0.24 inch (6 mm) in diameter through the rear of tensioner into retaining boss to lock tensioner pulley.

   a. Leave tensioner pulley arm locked until belt is installed again.

4. Remove drive belt.

To install:

5. Install the drive belt onto all of the pulleys except for the water pump pulley. Install the drive belt onto water pump pulley last.

➡Confirm drive belt is completely set on the pulleys.

6. Release drive belt auto-tensioner, and apply tension to drive belt.

✳✳ CAUTION

Never place hand in a location where pinching may occur if the holding tool accidentally comes off.

➡Ensure the drive belt and each groove on the pulley are free from all engine oil, coolant and other contaminants.

7. Turn crankshaft pulley clockwise several times to equalize tension between each pulley.

8. Confirm tension of drive belt at indicator is within the allowable use range.

9. Install the splash shield (RH).

Drive Belt Tensioner

2.0L Engine

See Figure 55.

➡The complete auto-tensioner must be replaced as a unit, including the pulley.

1. Remove front air duct.
2. Disconnect battery negative terminal.
3. Remove drive belt.
4. Support the engine and remove the torque rod (RH), engine mounting insulator (RH) and engine mounting bracket (RH).
5. Release the fixed drive belt auto-tensioner pulley.
6. Loosen bolt and remove drive belt auto-tensioner.

➡Use TORX socket (size T50).

7. Remove idler pulley and bracket (models without A/C).

➡Do not loosen the auto-tensioner pulley bolt.

➡Do not turn the auto-tensioner pulley bolt counterclockwise.

➡If the auto-tensioner pulley bolt is turned counterclockwise for any reason, the auto-tensioner and pulley must be replaced as a unit.

To install:

➡The complete auto-tensioner must be replaced as a unit, including the pulley.

✳✳ WARNING

If there is damage greater than peeled paint, replace drive belt auto-tensioner units

✳✳ WARNING

Install the drive belt auto-tensioner carefully so as to not damage the water pump pulley.

➡Do not swap the pulley between the new and old auto-tensioner units.

➡The auto-tensioner and pulley must be replaced as a unit.

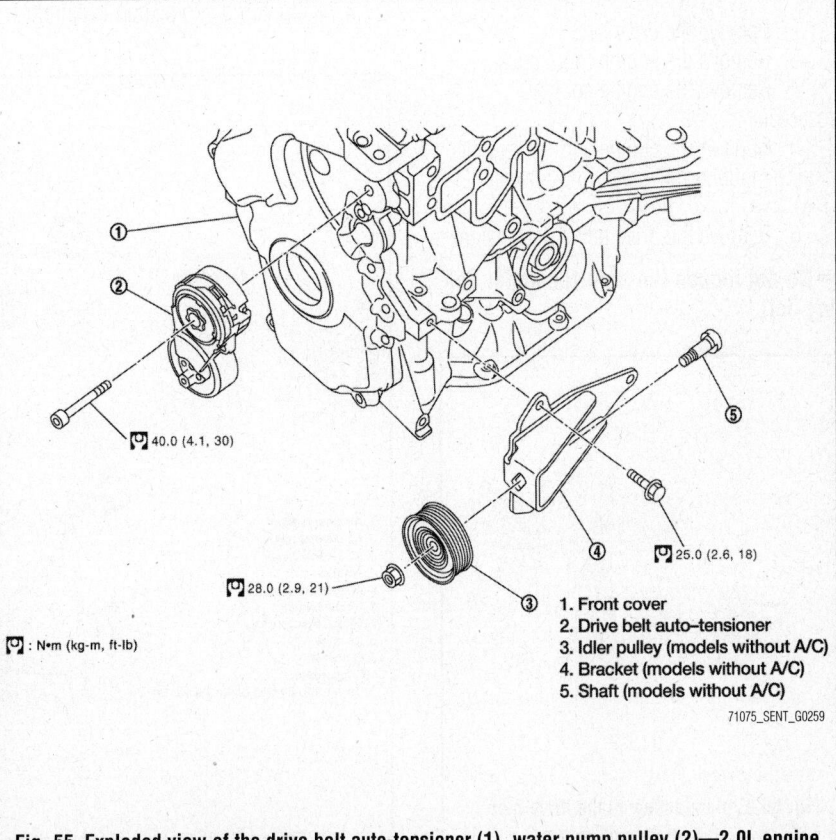

40.0 (4.1, 30)

28.0 (2.9, 21)

25.0 (2.6, 18)

⬡ : N•m (kg-m, ft-lb)

1. Front cover
2. Drive belt auto-tensioner
3. Idler pulley (models without A/C)
4. Bracket (models without A/C)
5. Shaft (models without A/C)

Fig. 55 Exploded view of the drive belt auto-tensioner (1), water pump pulley (2)—2.0L engine

➡️If the auto-tensioner pulley bolt is turned counterclockwise for any reason, the auto-tensioner and pulley must be replaced as a unit.

➡️Do not turn the auto-tensioner pulley bolt counterclockwise.

➡️Do not loosen the auto-tensioner pulley bolt.

➡️Use TORX® socket (size T50).

8. Install idler pulley and bracket (models without A/C).

9. Tighten bolt and install drive belt auto-tensioner.

10. Engage the fixed drive belt auto-tensioner pulley.

11. Support the engine and install the torque rod (RH), engine mounting insulator (RH) and engine mounting bracket (RH).

12. Install drive belt.

13. Connect battery negative terminal.

14. Install front air duct.

2.5L Engine

See Figure 56.

➡️The complete auto-tensioner must be replaced as a unit, including the pulley.

1. Remove the RH front wheel and tire.

2. Remove the drive belt.

3. Remove the engine room cover.

4. Remove the engine coolant reservoir.

5. Support the engine and remove the engine mounting insulator and bracket (RH).

6. Remove the drive belt auto-tensioner.

➡️Do not loosen the auto-tensioner pulley bolt.

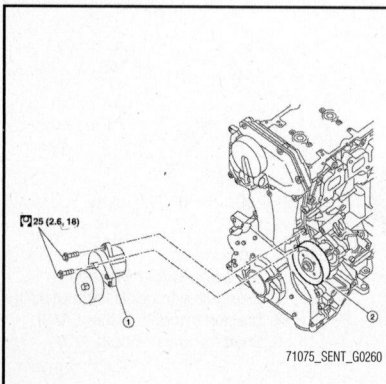

Fig. 56 Exploded view of the drive belt auto-tensioner—2.5L engine

➡️Do not turn the auto-tensioner pulley bolt counterclockwise.

➡️If the auto-tensioner pulley bolt is turned counterclockwise for any reason, the auto-tensioner and pulley must be replaced as a unit.

To install:

➡️The complete auto-tensioner must be replaced as a unit, including the pulley.

➡️If there is damage greater than peeled paint, replace drive belt auto-tensioner units

➡️Install the drive belt auto-tensioner carefully so not to damage the water pump pulley.

➡️Do not swap the pulley between the new and old auto-tensioner units

➡️If the auto-tensioner pulley bolt is turned counterclockwise for any reason, the auto-tensioner and pulley must be replaced as a unit.

➡️Do not turn the auto-tensioner pulley bolt counterclockwise.

➡️Do not loosen the auto-tensioner pulley bolt.

7. Install the drive belt auto-tensioner.

8. Support the engine and install the

engine mounting insulator and bracket (RH).

9. Install the engine coolant reservoir.

10. Install the engine room cover.

11. Install the drive belt.

12. Install the RH front wheel and tire.

AIR INTAKE SYSTEM

REMOVAL & INSTALLATION

Air Cleaner Assembly

2.0L Engine

See Figure 57.

1. Remove the engine room cover.

2. Remove the air duct (inlet).

3. Remove the air cleaner filter from the air cleaner case.

4. Remove the air duct [between air duct (inlet) and air cleaner case] from the air cleaner case.

5. Remove the PCV hose.

6. Remove the air duct (between air cleaner case and electric throttle control actuator).

a. Add marks as necessary for easier installation.

7. Remove air cleaner case with the following procedure.

a. Remove battery.

b. Disconnect the brake fluid level sensor.

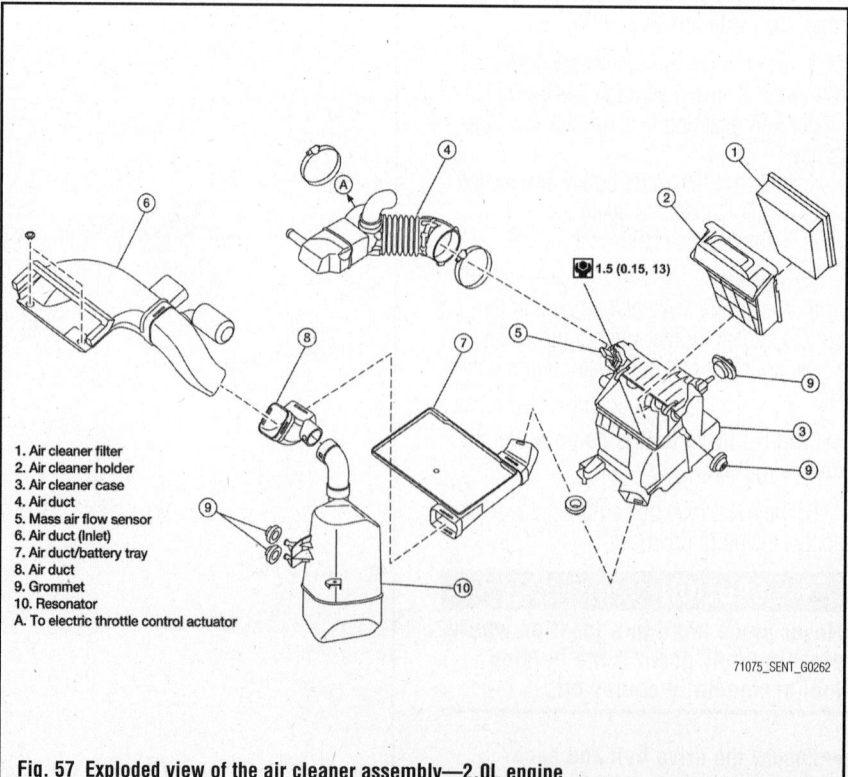

1. Air cleaner filter
2. Air cleaner holder
3. Air cleaner case
4. Air duct
5. Mass air flow sensor
6. Air duct (Inlet)
7. Air duct/battery tray
8. Air duct
9. Grommet
10. Resonator
A. To electric throttle control actuator

1.5 (0.15, 13)

71075_SENT_G0262

Fig. 57 Exploded view of the air cleaner assembly—2.0L engine

c. Disconnect and remove the ECM.

d. Disconnect harness connector from Mass Air Flow (MAF) sensor.

e. Remove the air cleaner case.

8. Remove the Mass Air Flow (MAF) sensor from the air cleaner case, as necessary.

➡**Handle the Mass Air Flow (MAF) sensor with care:**

➡**Do not shock it.**

➡**Do not disassemble it.**

➡**Do not touch the internal sensor.**

To install:

➡**Align marks.**

➡**Attach each joint securely.**

➡**Screw clamps firmly.**

9. Install the Mass Air Flow (MAF) sensor to the air cleaner case, as necessary.

10. Install air cleaner case with the following procedure.

a. Install the air cleaner case.

b. Connect harness connector from Mass Air Flow (MAF) sensor.

c. Connect and install the ECM.

d. Connect the brake fluid level sensor.

e. Install battery.

11. Install the air duct (between air cleaner case and electric throttle control actuator).

a. Add marks as necessary for easier installation.

12. Install the PCV hose.

13. Install the air duct [between air duct (inlet) and air cleaner case] to the air cleaner case.

14. Install the air cleaner filter to the air cleaner case.

15. Install the air duct (inlet).

16. Install the engine room cover.

2.5L Engine

See Figure 58.

1. Remove the battery tray.

2. Remove the engine room cover.

3. Disconnect the Mass Air Flow (MAF) sensor electrical connector.

4. Disconnect the tube clamp at the electric throttle control actuator and the fresh air intake tube.

5. Remove air cleaner to electric throttle control actuator tube, air cleaner case, with Mass Air Flow (MAF) sensor attached.

a. Add marks as necessary for easier installation.

6. Remove Mass Air Flow (MAF) sensor from air cleaner case, as necessary.

➡**Handle the Mass Air Flow (MAF) sensor with care:**

➡**Do not shock it.**

➡**Do not disassemble it.**

➡**Do not touch the internal sensor.**

7. Remove the resonator in the fender, if necessary.

a. Remove the left front wheel and tire.

b. Partially remove the left front fender protector.

c. Remove the left hand side splash shield.

d. Remove the engine undercover.

To install:

➡**Align marks.**

➡**Attach each joint securely.**

➡**Screw clamps firmly.**

8. Install the resonator in the fender, if necessary.

a. Install the engine undercover.

b. Install the left hand side splash shield.

c. Partially install the left front fender protector.

d. Install the left front wheel and tire.

➡**Handle the Mass Air Flow (MAF) sensor with care:**

➡**Do not shock it.**

➡**Do not disassemble it.**

➡**Do not touch the internal sensor.**

9. Install Mass Air Flow (MAF) sensor to air cleaner case, as necessary.

10. Install air cleaner to electric throttle control actuator tube, air cleaner case, with Mass Air Flow (MAF) sensor attached.

a. Use marks as necessary for easier installation.

11. Connect the tube clamp at the electric throttle control actuator and the fresh air intake tube.

12. Connect the Mass Air Flow (MAF) sensor electrical connector.

13. Install the engine room cover.

14. Install the battery tray.

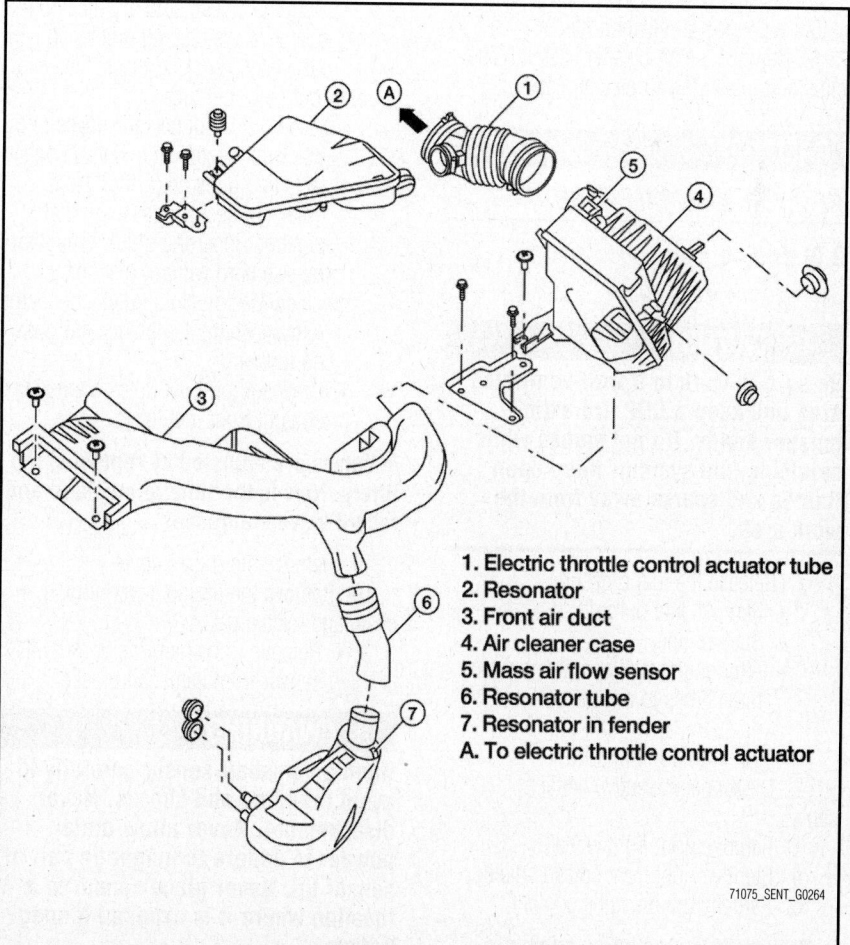

1. Electric throttle control actuator tube
2. Resonator
3. Front air duct
4. Air cleaner case
5. Mass air flow sensor
6. Resonator tube
7. Resonator in fender
A. To electric throttle control actuator

71075_SENT_G0264

Fig. 58 Exploded view of the air cleaner assembly—2.5L engine

Air Filter Element

2.0L Engine

1. Unclip the tabs of both ends of the air cleaner cover.
2. Remove the air cleaner filter and holder assembly from the air cleaner case.
3. Remove the air cleaner filter from the holder.

Inspection After Removal

It is necessary to replace it at the recommended intervals, more often under dusty driving conditions.

To install:

4. Install the air cleaner filter to the holder.
5. Install the air cleaner filter and holder assembly to the air cleaner case.
6. Clip the tabs of both ends of the air cleaner cover.

2.5L Engine

1. Depress the air cleaner case lid side clips and remove the air cleaner case lid.
2. Remove the air cleaner filter.
3. Install a new air cleaner filter.

To install:

4. Install a new air cleaner filter.
5. Install the air cleaner filter.
6. Depress the air cleaner case lid side clips and install the air cleaner case lid.

CAMSHAFT & VALVE LIFTERS

REMOVAL & INSTALLATION

2.0L Engine

See Figures 59 and 60.

❊❊ CAUTION

Be sure to work in a well-ventilated area and keep a CO2 fire extinguisher handy. Do not smoke while servicing fuel system. Keep open flames and sparks away from the work area.

1. Release the fuel pressure.
   - Remove fuel pump fuse.
   - Start engine.
   - After engine stalls, crank it two or three times to release all fuel pressure.
   - Turn ignition switch OFF.
2. Disconnect negative battery cable.
3. Remove right front wheel.
4. Remove right front splash shield.
5. Partially drain engine coolant.

➡Perform this step when engine is cold.

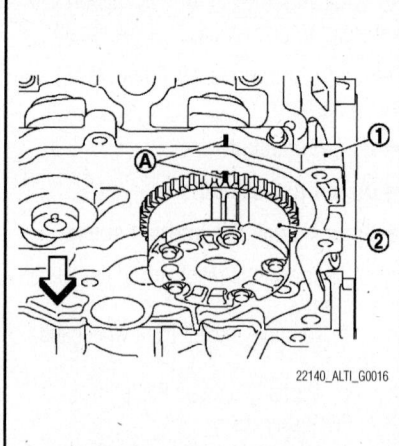

Fig. 59 Align match marks on intake camshaft sprocket and camshaft bracket.

6. Remove the intake manifold
7. Remove ignition coils, spark plugs and rocker cover.
8. Remove fuel tube and fuel injector assembly
   - Remove quick connector cap from quick connector connection.
   - Disconnect fuel feed hose from hose clamp.
   - With the sleeve side of quick connector release facing quick connector, install quick connector release tool onto fuel tube.
   - Insert quick connector release into quick connector until sleeve contacts and goes no further. Hold quick connector release on that position. Inserting quick connector release hard will not disconnect quick connector. Hold quick connector release where it contacts and goes no further.
   - Draw and pull out quick connector straight from fuel tube.

➡Valves are adjusted by replacing the lifters. Now is the time to measure and record valve clearances.

9. Remove the rocker cover.
10. Remove the timing cover, timing chain and related parts.
    - Remove Camshaft Position (CMP) sensor from camshaft bracket.

❊❊ WARNING

Handle camshaft sensor carefully to avoid dropping and shocks. Never disassemble. Never allow metal powder to adhere to magnetic part at sensor tip. Never place sensor in a location where it is exposed to magnetism.

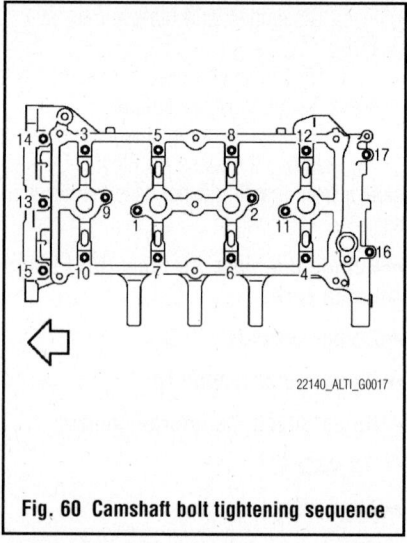

Fig. 60 Camshaft bolt tightening sequence

11. Align the match marks on the intake camshaft sprocket and the camshaft bracket. This prevents the knock pin of the camshaft from engaging with the incorrect pin hole when installing the camshaft sprocket.
12. Hold the camshaft with a wrench and loosen camshaft sprocket bolts to remove camshaft sprocket.

❊❊ WARNING

Never rotate crankshaft or camshaft while timing chain is removed. It causes interference between valve and piston. Never loosen the sprocket bolts without holding the camshaft securely with a wrench.

13. Loosen bolts in reverse of tightening order.
14. Cut liquid gasket by prying at the right front and left rear corners of the camshaft bracket, then remove the camshaft bracket.

❊❊ WARNING

Be careful not to damage the mating surface. A more adhesive liquid gasket is applied compared to previous types when shipped, so it should not be forced off the position not specified.

15. Carefully lift out the camshafts.

To install:

16. Install valve lifters in their original positions.
17. Install camshafts. Position the dowel pins at the sprocket end at the 12 o'clock position.
18. Carefully clean the camshaft bracket and apply a bead of RTV sealer around the outer edges and spark plug holes.

19. Fit the camshaft bracket into position and install the bolts. The long bolts go into holes 13, 14 and 15.

20. Tighten all bolts in numerical order in three steps. Do not over tighten the bolts.
- Step 1: 17 inch lbs. (2 Nm)
- Step 2: 52 inch lbs. (6 Nm)
- Step 3: 84 inch lbs. (9.5 Nm)

21. Install the intake camshaft sprocket, making sure to align the match marks. Hold the camshaft with a wrench and tighten the bolt to 26 ft. lbs. (35 Nm) plus 67°.

※※ WARNING

Use an angle-measuring wrench.

22. Install the exhaust camshaft sprocket, hold the camshaft with a wrench and torque the bolt to 65 ft. lbs. (88 Nm).

23. Install timing chain and related parts.

24. Inspect and adjust valve clearance.

25. Installation of the remaining components is in the reverse order of removal.

2.5L Engine

See Figures 61 through 65.

※※ CAUTION

Be sure to work in a well-ventilated area and keep a CO2 fire extinguisher handy. Do not smoke while servicing fuel system. Keep open flames and sparks away from the work area.

1. Disconnect the negative battery cable.

2. Support the engine using a suitable hoist or jack.

3. Remove the right engine mount and brackets.

4. Remove the rocker cover.

5. Remove the power steering reservoir.

6. Remove the coolant overflow reservoir.

7. Disconnect variable timing control solenoid.

8. Loosen the camshaft timing control cover bolts in reverse order of the tightening sequence.

9. Set the No.1 cylinder at TDC on its compression stroke with the following procedure:
- Open the splash cover under the engine.
- Rotate crankshaft pulley clockwise to align timing mark for TDC with timing indicator on front cover.
- Make sure the mating marks on camshaft sprockets are lined up with the yellow links in the timing

chain as shown. If not, rotate crankshaft pulley one more turn to line up the mating marks to the yellow links.

10. Pull the timing chain guide out between the camshaft sprockets through front cover.

11. Remove camshaft sprockets with the following procedure.

※※ WARNING

Do not rotate the crankshaft or camshaft while the timing chain is removed. This is an interference engine.

➡**Chain tension holding work is not necessary. Crankshaft sprocket and timing chain do not disconnect while front cover is attached.**

Note the following:
- Make sure marks on camshaft sprockets are aligned with the yellow links in the timing chain
- Paint an indelible mating mark on the sprocket and timing chain link plate.
- Push in the chain tensioner plunger and insert a 0.020 in. (0.5 mm) pin into the hole on tensioner body to hold the chain tensioner.
- Remove the timing chain tensioner.

12. Hold the hexagonal part of camshaft with a wrench and loosen the camshaft sprocket mounting bolts to remove the camshaft sprockets.

13. Loosen the camshaft bearing cap bolts in reverse of the order and remove the caps. A rubber mallet may be needed.

14. Remove the camshafts.

15. When removing the valve lifters, keep them in order so they can be installed in the same locations.

To install:

16. Install the valve lifters.

17. Lubricate and carefully fit the camshafts into place. The back of the intake camshaft has the signal plate for the Camshaft Position (CMP) sensor. Make sure the dowel pins on the front side are positioned as shown.

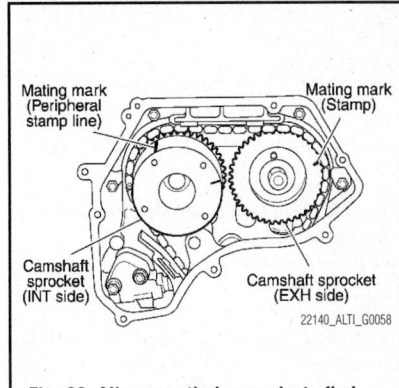

Fig. 62 Align cam timing marks to find TDC

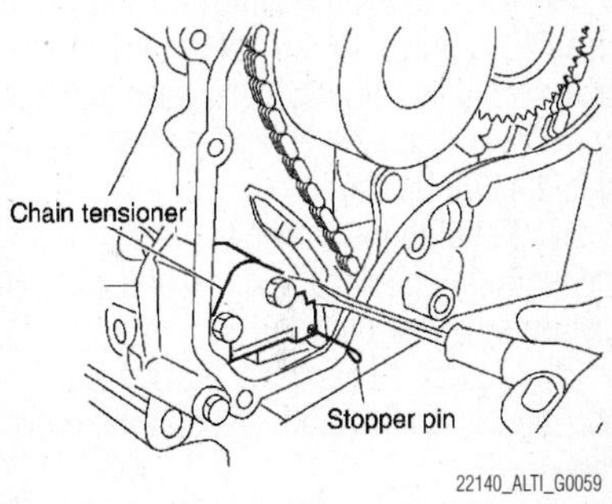

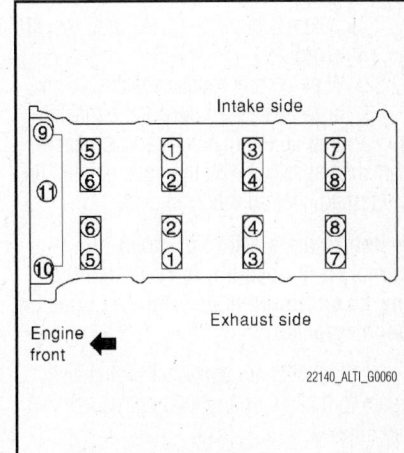

Fig. 63 Push the chain tensioner plunger in and insert a pin to keep it retracted

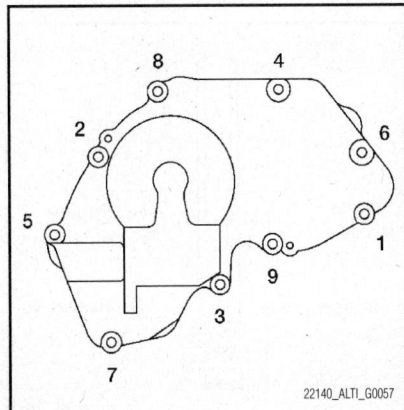

Fig. 61 Valve timing control cover bolt tightening sequence—2.5L engine

Fig. 64 Camshaft bearing cap bolt tightening order—2.5L engine

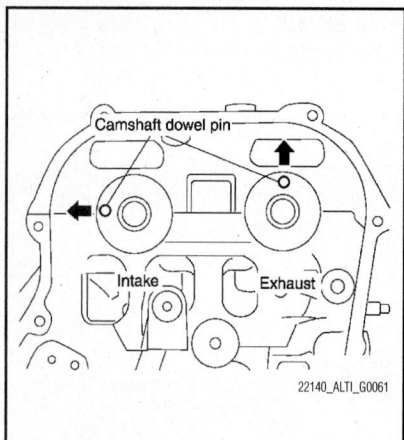

Fig. 65 Make sure camshaft dowel pins are positioned as shown.

18. Install the camshaft bearing caps so the numbers or letters can be read from the exhaust side of the cylinder head. Caps 2, 3, 4 and 5 go on the intake camshaft, front to rear of engine. Caps A, B, C and D are front-to-rear exhaust camshaft caps. Make the bolts only finger tight at this time.

19. Make sure all sealing surfaces are clean and apply RTV sealant to the bracket where it contacts the cylinder head and the front cover. The sealant bead should be outside the bolt holes.

20. Position the camshaft bracket near the mounting position and install it without disturbing the sealant.

21. Tighten all camshaft bracket and bearing cap bolts in four steps in the order shown.
- Step 1, bolts 9–11: 17 inch lbs. (2 Nm)
- Step 2, bolts 1–8: 17 inch lbs. (2 Nm)
- Step 3, bolts 1–11 52 inch lbs. (6 Nm)
- Step 4, bolts 1–11: 92 inch lbs. (10 Nm)

22. Wipe off any excess sealant.

23. Install camshaft sprockets, making sure to line up the mating marks on each camshaft sprocket with the ones painted on the timing chain during removal.

➡ **Before installation of chain tensioner, it is possible to re-match the marks on timing chain with the ones on each sprocket.**

24. Install chain tensioner and check again to make sure that mating marks have not slipped.

25. Remove the stopper pin from the tensioner and check the timing marks again.

26. Install chain guide.

27. Install camshaft timing control cover with the following procedure.
- Install control solenoid valve to intake valve timing control cover.
- Install new O-ring to front cover side.

28. Apply RTV sealant to the cover. The bead should be inside the bolt holes.
- Install the control cover and tighten the bolts in numerical order to 9 ft. lbs. (12 Nm).

29. Check and adjust valve clearances.

30. Installation of the remaining components is in the reverse order.

CRANKSHAFT FRONT SEAL

REMOVAL & INSTALLATION

2.0L Engine

See Figure 66.

1. Remove front timing cover.
2. Carefully pry out old seal.
3. Using a suitable tool, press-fit the new seal until it is flush with the front of the cover. Make sure the seal is straight and not curled.

2.5L Engine

See Figure 66.

1. Raise and safely support the vehicle and remove the right front wheel.
2. Remove the engine under cover.
3. Remove the drive belts.
4. Remove the crankshaft pulley.
5. Carefully pry the front oil seal from the front cover. Be careful not to scratch front cover or crankshaft.

To install:

6. Apply new engine oil to new oil seal and fit it into place with the longer lip towards the outside of the engine.

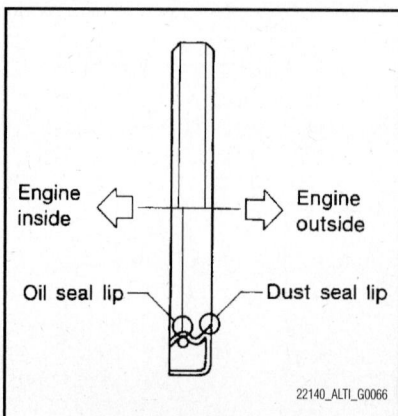

Fig. 66 Front and rear main seals are installed with the longer lip towards the outside of the engine

7. Press the seal straight in using an appropriately sized drift. Make sure the garter spring in the oil seal is in position and seal lip is not inverted.

8. Install crankshaft pulley and toque the bolts to 31 ft. lbs. plus an additional 60°.

9. Installation of the remaining components is in reverse order of removal.

CYLINDER HEAD

REMOVAL & INSTALLATION

2.0 Engine

See Figures 67 through 70.

✳✳ CAUTION

Be sure to work in a well-ventilated area and furnish workshop with a CO_2 fire extinguisher.

✳✳ CAUTION

Do not smoke while servicing fuel system. Keep open flames and sparks away from the work area.

1. Drain engine coolant and engine oil.
2. Remove the exhaust manifold.
3. Remove the intake manifold.
4. Remove the fuel tube and fuel injector assembly.
5. Remove the water inlet and thermostat.
6. Remove the water outlet.
7. Remove the timing chain front cover.
8. Remove the camshaft.
9. Remove cylinder head:
   a. Loosen bolts in reverse order as shown.
   b. Using TORX® socket (size E18), loosen cylinder head bolts.
10. Remove cylinder head gasket.

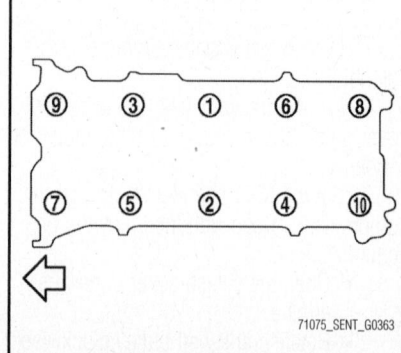

Fig. 67 Remove cylinder head. Loosen bolts in reverse order as shown. Using TORX® socket (size E18), loosen cylinder head bolts.

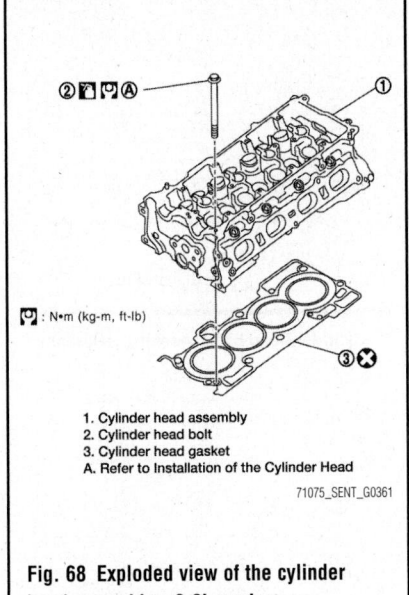

1. Cylinder head assembly
2. Cylinder head bolt
3. Cylinder head gasket
A. Refer to Installation of the Cylinder Head

71075_SENT_G0361

Fig. 68 Exploded view of the cylinder head assembly—2.0L engine

To install:

11. Install cylinder head gasket.
12. Apply new engine oil to threads and seating surface of bolts.

➡**If cylinder head bolts are re-used, check their outer diameters before installation. Follow the "Cylinder Head Bolts Outer Diameter" procedure as shown in Inspection After Removal in the Cylinder Head Section.**

13. Install cylinder head, follow the steps below to tighten cylinder head bolts in numerical order as shown in the accompanying illustration.
    a. Step a : 30 ft. lbs. (40 Nm)
    b. Step b : 100° clockwise
    c. Step c : Loosen to 0 Nm in the reverse order of tightening.
    d. Step d : 30 ft. lbs. (40 Nm)

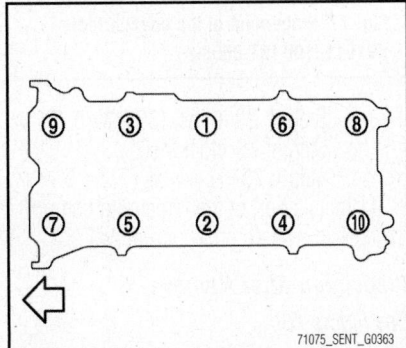

71075_SENT_G0363

Fig. 69 Install cylinder head, follow the steps to tighten cylinder head bolts in numerical order as shown in the accompanying illustration [Arrow=Engine front]

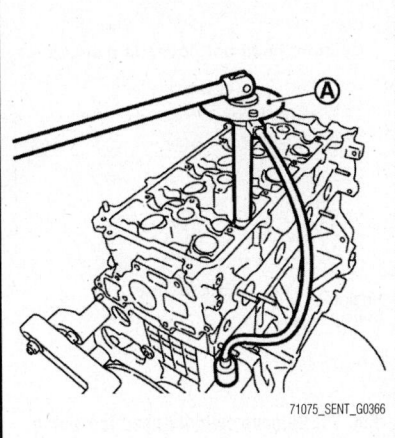

71075_SENT_G0366

Fig. 70 Check and confirm the tightening angle by using Tool: KV10112100 (BT-8653-A) (A) or protractor. Never judge by visual inspection without the tool

    e. Step e : 100° clockwise
    f. Step f : 100° clockwise

➡**Check and confirm the tightening angle by using Tool: KV10112100 (BT-8653-A) or protractor. Never judge by visual inspection without the tool.**

14. Installation of the remaining components is in the reverse order of removal.

Inspection After Removal
See Figures 71 and 72.

1. Cylinder head bolts are tightened by plastic zone tightening method. Whenever the size difference between "d1" and "d2" exceeds the limit, replace them with a new one.
2. If reduction of outer diameter appears in a position other than "d2", use it as "d2" point.

Cylinder Head Distortion

➡**When performing this inspection, cylinder block distortion should be also checked.**

3. Wipe off engine oil and remove water scale (like deposit), gasket, sealant, carbon, etc. with a scraper.

➡**Use utmost care not to allow gasket debris to enter passages for engine oil or water.**

4. At each of several locations on bottom surface of cylinder head, measure the distortion in six directions using straightedge and feeler gauge.
    a. If it exceeds the limit, replace cylinder head.

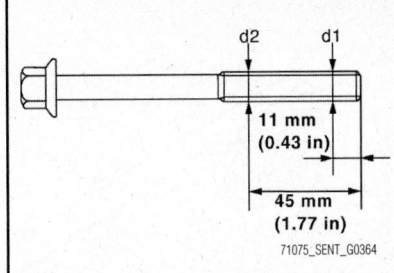

71075_SENT_G0364

Fig. 71 Cylinder head bolts are tightened by plastic zone tightening method. Whenever the size difference between "d1" and "d2" exceeds the limit, replace them with a new one. If reduction of outer diameter appears in a position other than "d2", use it as "d2" point.

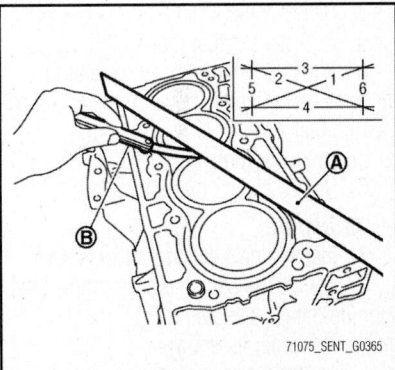

71075_SENT_G0365

Fig. 72 At each of several locations on bottom surface of cylinder head, measure the distortion in six directions using straightedge (A) and feeler gauge (B). Limit: 0.004 in. (0.1 mm)

Cylinder Block Top Surface Distortion Check
See Figure 73.

1. Using a scraper, remove gasket on the cylinder block surface, and also remove engine oil, scale, carbon, or other contamination.

➡**Be careful not to allow gasket flakes to enter engine oil or engine coolant passages.**

2. Measure the distortion on the cylinder block upper face at some different points in six directions with a straight edge and feeler gauge.
3. If it exceeds the limit, replace cylinder block.

2.5L Engine
See Figures 74 through 77.

1. Remove the timing chain.
2. Remove the camshafts.

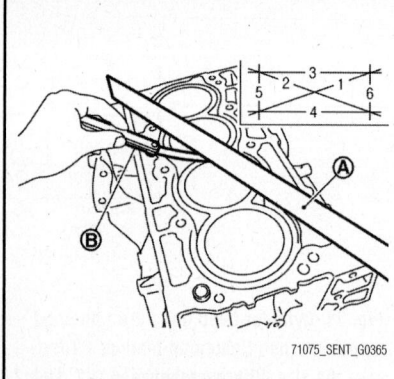

Fig. 73 Measure the distortion on the cylinder block upper face at some different points in six directions with a straight edge (A) and feeler gauge (B).

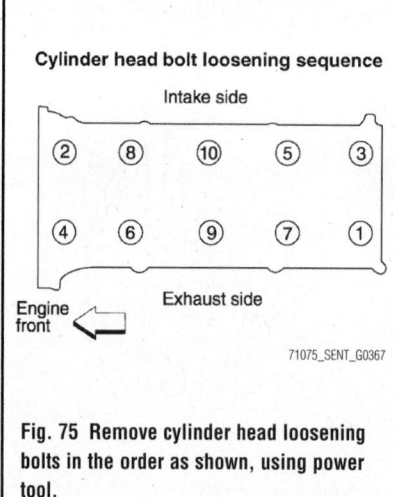

Fig. 75 Remove cylinder head loosening bolts in the order as shown, using power tool.

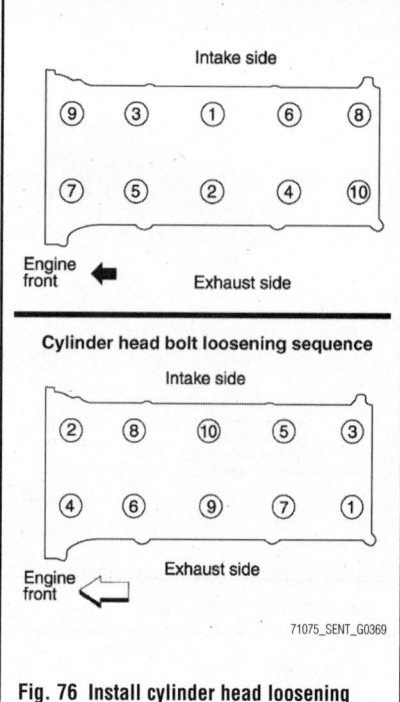

Fig. 76 Install cylinder head loosening bolts in the order as shown, using tool KV10112100 (BT-8653-A).

3. Remove spark plugs.
4. Remove exhaust manifold.
5. Remove cylinder head loosening bolts in the order as shown in the accompanying illustration
using power tool.
6. Remove water outlet if necessary.

To install:

7. Install a new cylinder head gasket.
8. Follow the steps below to tighten the cylinder head bolts using Tool, in the numerical order as shown:

  a. Apply new engine oil to the threads and the seating surfaces of bolts.

➡ If cylinder head bolts are re-used, check their outer diameters before installation. Follow the "Outer Diameter of Cylinder Head Bolts" procedure in Inspection After Removal in the Cylinder Head Section.

➡ Check and confirm the tightening angle by using angle wrench or protractor. Avoid judgment by visual inspection without the tool KV10112100 (BT-8653-A).

  b. Step a: 37 ft. lbs. (98.1 Nm)
  c. Step c: Loosen to 0 Nm in the reverse order of tightening.

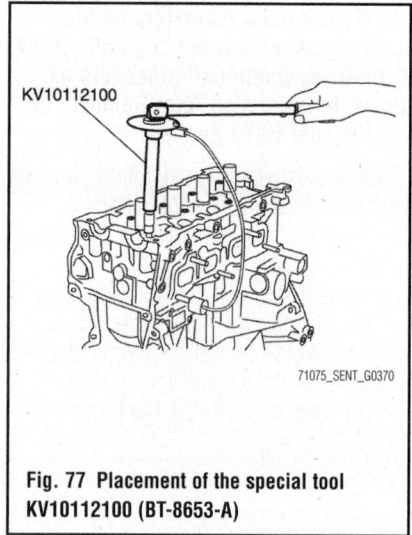

Fig. 77 Placement of the special tool KV10112100 (BT-8653-A)

  d. Step d: 29 ft. lbs. (39.2 Nm)
  e. Step e: 75°clockwise
  f. Step f: 75°clockwise

9. Installation of the remaining components is in reverse order of removal.

Inspection After Removal

See Figure 78.

1. Cylinder head bolts are tightened by plastic zone tightening method. Whenever the size difference between $d1$ and $d2$ exceeds the limit, replace the bolts with new ones. Limit ($d1 - d2$): 0.0091 in. (0.23 mm) or less.

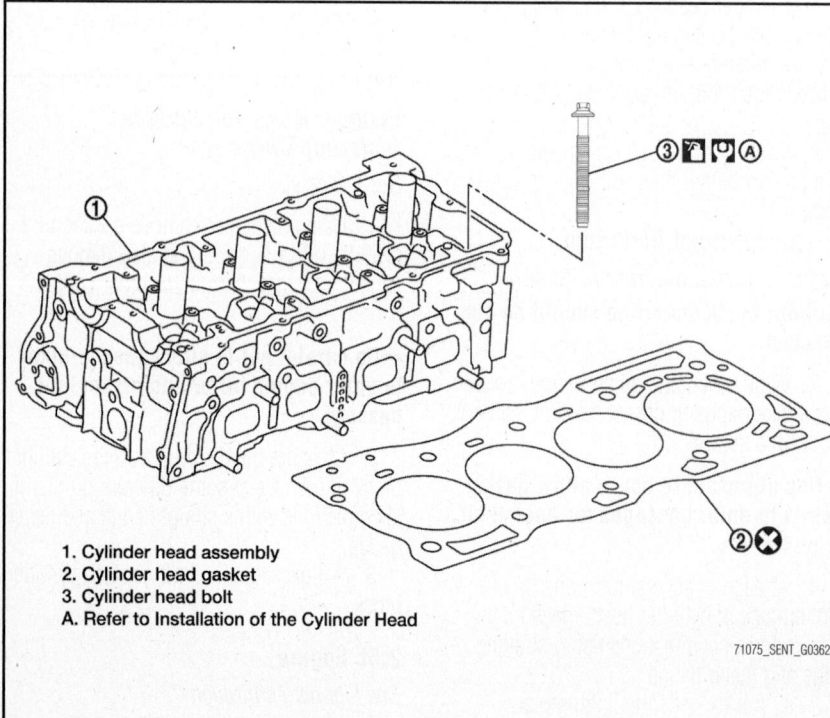

1. Cylinder head assembly
2. Cylinder head gasket
3. Cylinder head bolt
A. Refer to Installation of the Cylinder Head

Fig. 74 Exploded view of the cylinder head assembly—2.5L engine

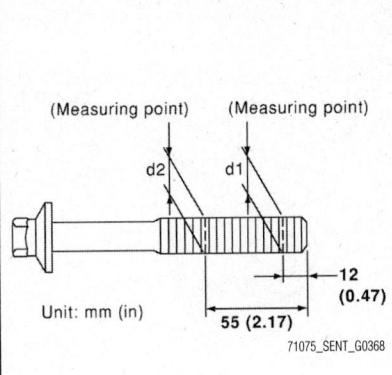

Fig. 78 Cylinder head bolts are tightened by plastic zone tightening method. Whenever the size difference between d1 and d2 exceeds the limit, replace the bolts with new ones. Limit (d1 - d2): 0.0091 in. (0.23 mm) or less

2. If reduction of outer diameter appears in a position other than d2, use it as d2 point.

ENGINE OIL & FILTER

OIL LEVEL CHECK

See Figure 79.

➡Before starting engine, park vehicle on a level surface and check the engine oil level. If engine is already started, stop it and allow 10 minutes before checking.

1. Pull out oil level gauge and wipe it clean.
2. Insert oil level gauge and make sure the engine oil level is within the range.
3. If it is out of range, add oil as necessary.

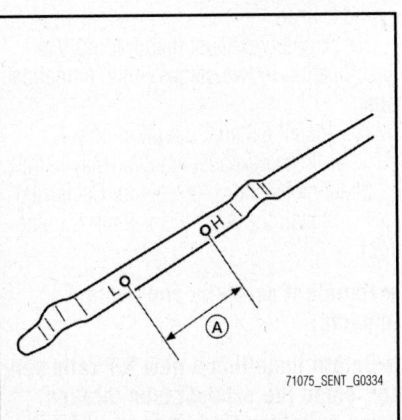

Fig. 79 Insert oil level gauge and make sure the engine oil level is within the range (A) as shown

OIL & FILTER CHANGE

2.0L Engine
See Figures 80 through 82.

Engine Oil

※※ CAUTION

Be careful not to burn yourself, as engine oil may be hot.

※※ CAUTION

Prolonged and repeated contact with used engine oil may cause skin cancer; try to avoid direct skin contact with used engine oil. If skin contact is made, wash thoroughly with soap or hand cleaner as soon as possible.

1. Park vehicle on a level surface and check for engine oil leakage from engine components.
2. Warm up the engine.
3. Stop engine and wait for 10 minutes.
4. Remove oil filler cap and then remove oil pan drain plug.
5. Drain the engine oil.
6. Install the oil pan drain plug with a new copper sealing washer.

➡Do not reuse copper sealing washers.

➡Be sure to clean drain plug and install with a new copper sealing washer.

7. Refill with new engine oil.

➡The refill capacity depends on the engine oil temperature and drain time. Use these specifications for reference only.

Fig. 80 Remove oil filter (A) using Tool KV10115801 (J-38956) [Arrow: front]

➡Always use oil level gauge to determine the proper amount of engine oil in the engine.

8. Warm up engine and check area around drain plug and oil filter for engine oil leakage.
9. Stop engine and wait for 10 minutes.
10. Check the engine oil level. Adjust as necessary.

➡Do not overfill the engine with oil.

Oil Filter
11. Remove engine undercover.
12. Drain engine oil.
13. Remove oil filter using Tool KV10115801 (J-38956).

※※ CAUTION

Be careful not to get burned, engine and engine oil may be hot.

➡When removing, prepare a shop cloth to absorb any engine oil leakage or spillage.

➡Do not allow engine oil to adhere to drive belt.

➡Completely wipe off any engine oil that adheres to the engine and the vehicle.

➡The oil filter has a built in pressure relief valve. Use Genuine NISSAN oil filter or equivalent.

To install:
14. Remove foreign materials adhering to the oil filter installation surface.
15. Apply new engine oil to the oil seal contact surface of the new oil filter.
16. Screw the new oil filter manually until it touches the installation surface, then

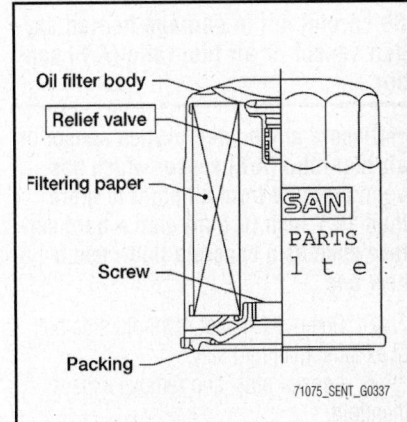

Fig. 81 The oil filter has a built in pressure relief valve. Use Genuine NISSAN oil filter or equivalent

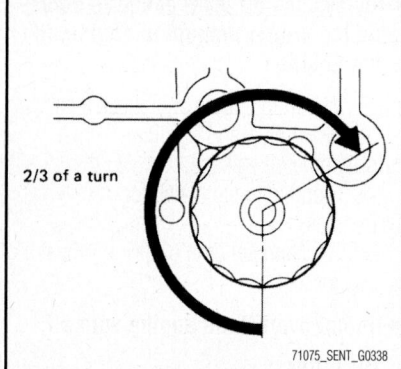

Fig. 82 Screw the new oil filter manually until it touches the installation surface, then tighten it by 2/3 turn. Or tighten to specification.

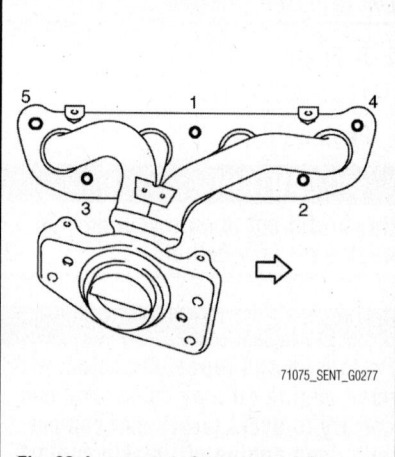

Fig. 83 Loosen nuts in reverse of the order shown and remove exhaust manifold

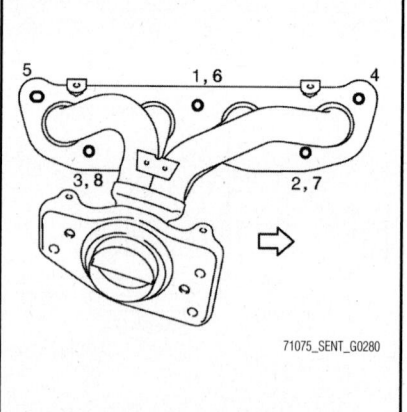

Fig. 86 Tighten exhaust manifold nuts to specification in two stages in the numerical order as shown (Arrow: Engine front).

tighten it by 2/3 turn. Or tighten to specification.

17. Refill engine with new engine oil.
18. Install engine undercover.

Inspection After Installation

19. Check the engine oil level.
20. Start engine, and make sure there are no leaks of engine oil.
21. Stop engine and wait for 10 minutes.
22. Check the engine oil level and adjust as necessary.

EXHAUST MANIFOLD

REMOVAL & INSTALLATION

2.0L Engine

See Figures 83 through 87.

1. Remove cowl top.
2. Remove exhaust front tube.
3. Remove exhaust manifold cover.
4. Remove the A/F ratio sensor 1.

❊❊ WARNING

Be careful not to damage heated oxygen sensor or air fuel ratio (A/F) sensor.

➡Discard any heated oxygen sensor or air fuel ratio (A/F) sensor which has been dropped from a height of more than 19.7 inch (0.5 m) onto a hard surface such as a concrete floor; use a new one.

5. Remove exhaust manifold side bolt of exhaust manifold stay.
6. Loosen nuts, and remove exhaust manifold.

➡Cover engine openings to avoid entry of foreign materials. Inspection After Removal

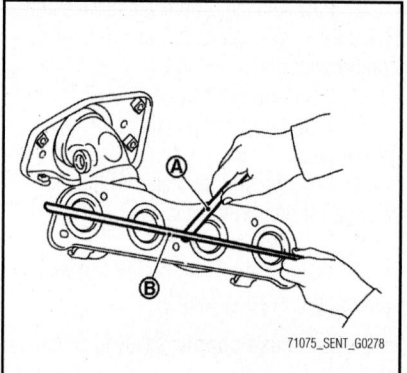

Fig. 84 Using straightedge (B) and feeler gauge (A), check the surface distortion of exhaust manifold mating surface in each exhaust port and entire part.

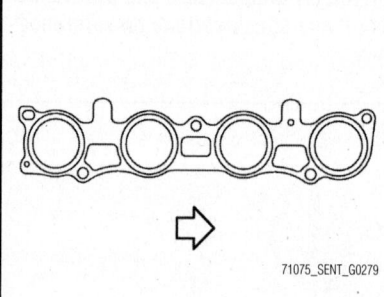

Fig. 85 Install exhaust manifold gasket to cylinder head as shown (Arrow: Engine front).

7. Surface Distortion
   a. Using straightedge and feeler gauge, check the surface distortion of exhaust manifold mating surface in each exhaust port and entire part.

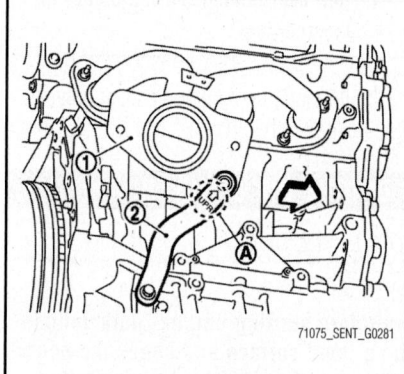

Fig. 87 Install exhaust manifold (1) stay. Install exhaust manifold stay (2) in the direction (Arrow: Engine front), (A: Upper mark) as shown (except for California).

   b. If it exceeds the limit, replace exhaust manifold.

To install:

8. Install exhaust manifold gasket to cylinder head.
9. Tighten exhaust manifold nuts to specification in two stages in the numerical order.
10. Install exhaust manifold stay.
    a. Install exhaust manifold stay in the direction as shown (except for California).
11. Install the A/F ratio sensor 1, using Tool.

➡Handle it carefully and avoid impacts.

➡Before installing a new A/F ratio sensor, clean the exhaust tube threads using suitable tool and approved anti-seize lubricant.

➡Do not over-tighten the A/F ratio sensor. Doing so may damage the A/F ratio sensor, resulting in the MIL coming on.

12. Installation of the remaining parts is in the reverse order of removal.

2.5L Engine

See Figures 88 through 91.

1. Remove the engine undercover.
2. Remove the splash shield (RH).
3. Remove the front air duct.
4. Remove generator and generator bracket.
5. Remove the exhaust front tube.
6. Remove oil level indicator tube.
7. Disconnect the electrical connector of heated oxygen sensor 1 or air fuel ratio (A/F) sensor 1, and unhook the harness from the bracket and middle clamp on the cover.
8. Remove the heated oxygen sensor 1 or air fuel ratio (A/F) sensor 1.

✳✳ WARNING

Be careful not to damage heated oxygen sensor or air fuel ratio (A/F) sensor.

➡ Discard any heated oxygen sensor or air fuel ratio (A/F) sensor which has been dropped from a height of more than 0.5 m (19.7 in) onto a hard surface such as a concrete floor; use a new one.

9. Remove the lower exhaust manifold covers.
10. Remove the upper exhaust manifold cover.
11. Loosen the nuts in reverse order as shown, on the exhaust manifold and three way catalyst.
12. Remove the exhaust manifold and three way catalyst assembly and gasket. Discard the gasket.

Inspection After Removal
13. Surface Distortion
    a. Use a reliable straightedge and feeler gauge to check the flatness of exhaust manifold fitting surface.

To install:
14. Install a new gasket and then the exhaust manifold. Tighten the nuts in the proper sequence and to the specifications shown in the accompanying illustration.
15. Clean the A/F sensor and heated oxygen sensor threads with the Tools (Oxygen sensor thread cleaner : J-43897-18, Oxygen sensor thread cleaner : J-43897-12), then apply the anti-seize lubricant to the threads before installing the A/F sensor and heated oxygen sensors.

✳✳ WARNING

Do not over-tighten the A/F sensors and heated oxygen sensors. Doing so may cause damage to the A/F sensors and heated oxygen sensors, resulting in a malfunction and the MIL coming on.

16. Install the exhaust manifold and three way catalyst assembly and gasket.
17. Install the upper exhaust manifold cover.
18. Install the lower exhaust manifold covers.
19. Install the heated oxygen sensor 1 or air fuel ratio (A/F) sensor 1.

✳✳ WARNING

Be careful not to damage heated oxygen sensor or air fuel ratio (A/F) sensor.

➡ Discard any heated oxygen sensor or air fuel ratio (A/F) sensor which has been dropped from a height of more than 0.5 m (19.7 in) onto a hard surface such as a concrete floor; use a new one.

20. Connect the electrical connector of heated oxygen sensor 1 or air fuel ratio (A/F) sensor 1, and unhook the harness to the bracket and middle clamp on the cover.
21. Install oil level indicator tube.
22. Install the exhaust front tube.
23. Install generator and generator bracket.
24. Install the front air duct.
25. Install the splash shield (RH).
26. Install the engine undercover.

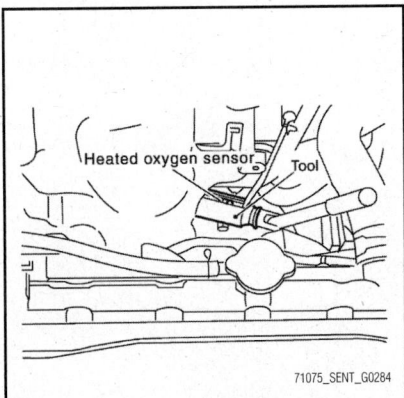

Fig. 88 Remove the heated oxygen sensor 1 or air fuel ratio (A/F) sensor 1 using Tools: Tool numbers : KV991J0050 (J-44626), KV10117100 (J-36471-A)

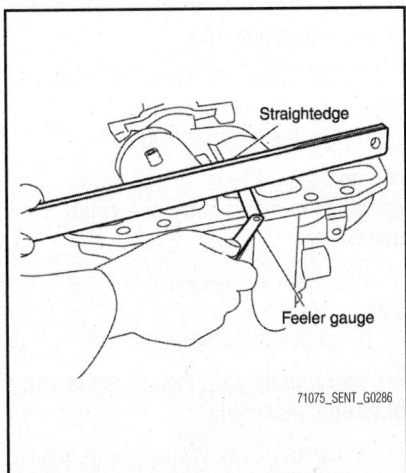

Fig. 90 Use a reliable straightedge and feeler gauge to check the flatness of exhaust manifold fitting surface

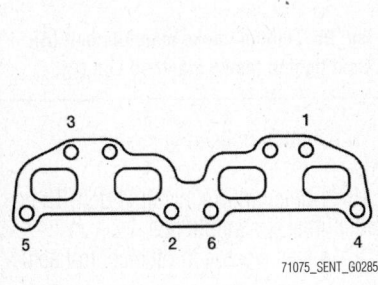

Fig. 89 Loosen the nuts in reverse order as shown, on the exhaust manifold and three way catalyst

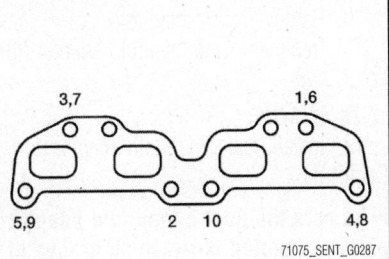

Fig. 91 Tightening Exhaust Manifold Nuts. Tighten the nuts in the numerical order shown, to specification.

INTAKE MANIFOLD

REMOVAL & INSTALLATION

2.0L Engine

See Figures 92 through 97.

> ❋❋ **CAUTION**
>
> **To avoid the danger of being scalded, never drain the coolant when the engine is hot.**

1. Remove engine room cover.
2. Remove the air duct (inlet) and air ducts.
3. Disconnect the EVAP canister purge volume control solenoid valve.
4. Partially drain engine coolant from the radiator.

> ❋❋ **CAUTION**
>
> **Perform this step when engine is cold.**

➡This step is unnecessary when putting plugs to water hoses (to electric throttle control actuator)

  a. Disconnect water hoses from electric throttle control actuator.
  b. Remove electric throttle control actuator.

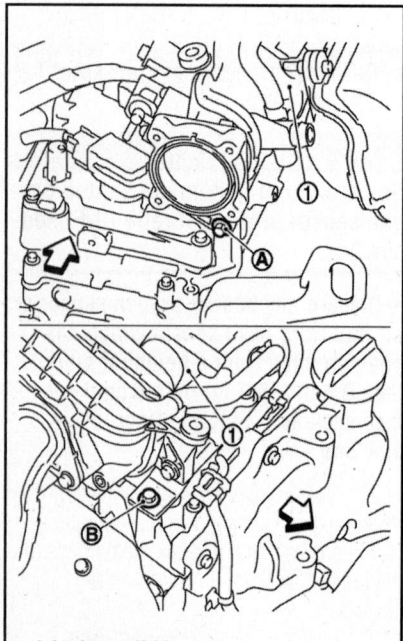

1. Intake manifold
A. Bolt 14 ft. lb. (19.6 Nm)
B. Bolt 14 ft. lb. (19.6 Nm)
Arrow: Engine front

71075_SENT_G0295

Fig. 92 Loosen and remove intake manifold (1) bolts (A) (B). (Arrow: Engine front)

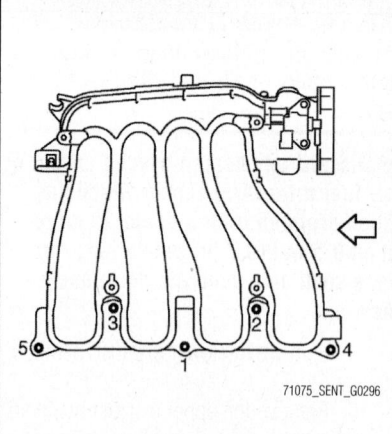

Fig. 93 Loosen bolts in reverse of the order shown (Arrow: Engine front)

> ❋❋ **CAUTION**
>
> **Handle carefully to avoid any shock to electric throttle control actuator.**

➡Never disassemble.

5. Remove the PCV hose and the vacuum hose.
6. Remove oil level gauge.

➡Cover the oil level gauge guide openings to avoid entry of foreign materials.

7. Loosen and remove intake manifold bolts.
8. Loosen bolts in reverse order.

➡Cover engine openings to avoid entry of foreign materials.

9. Remove EVAP canister purge volume control solenoid valve from intake manifold, if necessary.

> ❋❋ **WARNING**
>
> **Handle it carefully and avoid impacts.**

10. Remove intake manifold.
11. Remove intake manifold adapter (for California).

  To install:
12. Install intake manifold adapter (for California).

➡Be sure the intake manifold adapter gasket is seated correctly in groove of intake manifold adapter (for California).

13. Install intake manifold.

➡Be sure the intake manifold gasket is seated correctly in groove of intake manifold (except for California).

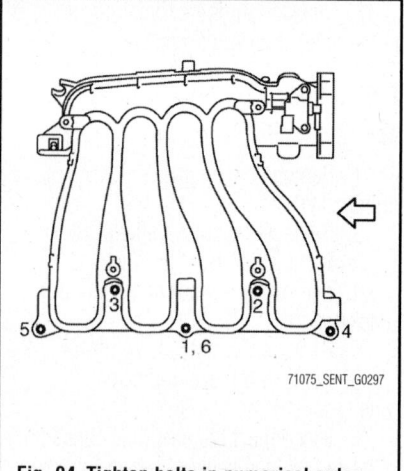

Fig. 94 Tighten bolts in numerical order as shown (Arrow: Engine front)

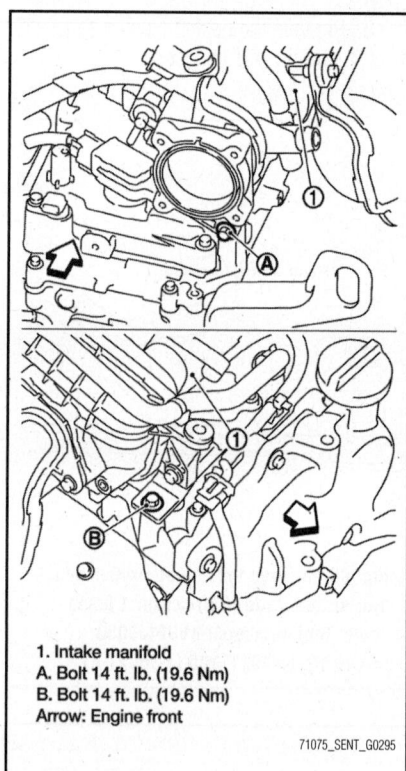

1. Intake manifold
A. Bolt 14 ft. lb. (19.6 Nm)
B. Bolt 14 ft. lb. (19.6 Nm)
Arrow: Engine front

71075_SENT_G0295

Fig. 95 Tighten intake manifold bolt (A). Then tighten intake manifold bolt (B).

14. Tighten bolts in numerical order.
15. Tighten intake manifold bolt. Then tighten intake manifold bolt.
16. Install electric throttle control actuator.
17. Install water hoses, to electric throttle control actuator (M/T models).

➡The clamp shall not interfere with the bulged section.

18. Install water hoses, to electric throttle control actuator (CVT models).

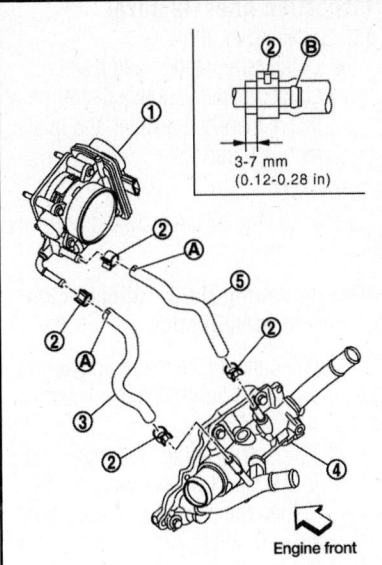

1. Electric throttle control actuator
2. Clamp
3. Water hose
4. Water outlet
5. Water hose
A. Paint Mark
B. Bulged section

71075_SENT_G0298

Fig. 96 Install water hoses (3), (5) to electric throttle control actuator as shown (M/T models). The clamp (2) shall not interfere with the bulged section (B).

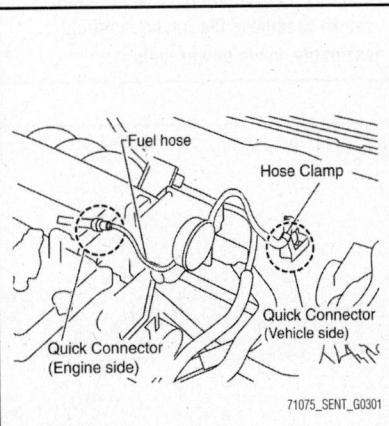

1. Electric throttle control actuator
2. Clamp
3. Water hose
4. Water outlet
5. Water hose
A. Paint Mark

71075_SENT_G0299

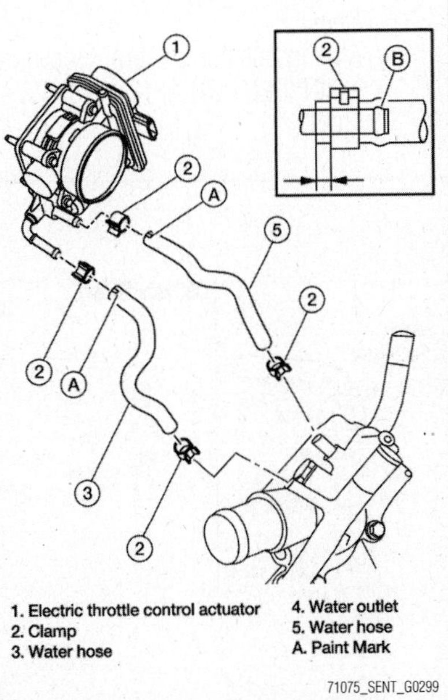

Fig. 97 Install water hoses (3), (5) to electric throttle control actuator as shown (CVT models). The clamp (2) shall not interfere with the bulged section (B).

➡ The clamp shall not interfere with the bulged section.

19. Installation of the remaining components is in the reverse order of removal.

Inspection After Installation

20. Check for engine coolant leaks.
21. Start and warm up the engine. Visually check for engine coolant leaks.

2.5L Engine

See Figures 98 through 105.

❄ WARNING

To avoid the danger of being scalded, never drain the coolant when the engine is hot.

1. Remove the engine room cover.
2. Release the fuel pressure.
3. Drain engine coolant from the radiator.

❄ CAUTION

Perform this step when engine is cold.

➡ This step necessary only when removing electric throttle control actuator from the vehicle.

4. Remove the battery tray.
5. Disconnect the MAF sensor electrical connector.
6. Remove air cleaner case and air duct assembly.
7. Remove cowl top.
8. Disconnect the following components at the intake side:
   - PCV hose

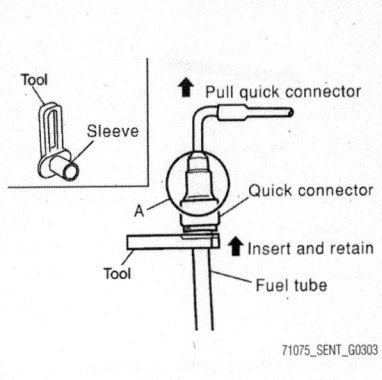

71075_SENT_G0301

Fig. 98 Disconnect the fuel quick connector on the engine side. Using Tool (16441 6N210 [J-45488]) perform the following steps to disconnect the quick connector

71075_SENT_G0303

Fig. 99 With the sleeve side of Tool facing quick connector, install Tool onto fuel tube. Insert Tool into quick connector until sleeve contacts and goes no further. Hold the Tool on that position. Pull the quick connector straight out from the fuel tube.

- EVAP canister purge volume control solenoid
- Electric throttle control actuator
- Brake booster vacuum hose

9. Disconnect the fuel quick connector on the engine side.
   - Using Tool perform the following steps to disconnect the quick connector.
   a. Remove quick connector cap.
   b. With the sleeve side of Tool facing quick connector, install Tool onto fuel tube.
   c. Insert Tool into quick connector until sleeve contacts and goes no further. Hold the Tool on that position.

➡ Inserting the Tool hard will not disconnect quick connector. Hold Tool where it contacts and goes no further.

   d. Pull the quick connector straight out from the fuel tube.

➡ Pull quick connector holding it at the "A" position, as shown.

❄ WARNING

Do not pull with lateral force applied. O-ring inside quick connector may be damaged.

➡ Do not reuse O-ring.

➡ Prepare container and cloth beforehand as fuel will leak out.

❄ WARNING

Avoid fire and sparks.

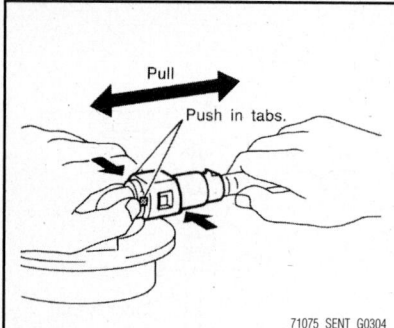

Fig. 100 Remove quick connector cap. Hold the sides of the connector, push in tabs and pull out the tube. (The figure is shown for reference only.) If the connector and the tube are stuck together, push and pull several times until they start to move. Then disconnect them by pulling.

➡Be sure to cover openings of disconnected pipes with plug or plastic bag to avoid fuel leakage and entry of foreign materials.

10. When removing fuel hose quick connector at vehicle piping side, perform as follows.

    a. Remove quick connector cap.

    b. Hold the sides of the connector, push in tabs and pull out the tube. (The figure is shown for reference only.)

    • If the connector and the tube are stuck together, push and pull several times until they start to move. Then disconnect them by pulling.

➡The tube can be removed when the tabs are completely depressed. Do not twist it more than necessary.

➡Do not use any tools to remove the quick connector.

➡Keep the resin tube away from heat. Be especially careful when welding near the tube.

➡Prevent acid liquid such as battery electrolyte etc. from getting on the resin tube.

➡Do not bend or twist the tube during installation and removal.

➡Do not remove the remaining retainer on tube.

➡When the tube is replaced, also replace the retainer with a new one. Retainer color: Green.

✳✳ WARNING

To keep clean the connecting portion and to avoid damage and foreign materials, cover them completely with plastic bags or something similar.

11. Remove EVAP canister purge volume control solenoid valve.

12. Loosen mounting bolts diagonally, reposition the electric throttle control actuator and position aside without disconnecting the coolant hoses.

✳✳ CAUTION

Handle carefully to avoid any shock to electric throttle control actuator.

➡Never disassemble.

13. Disconnect intake manifold collector harness, and vacuum hose.

➡Cover engine openings to avoid entry of foreign materials.

14. Loosen the bolts in the order shown to remove the intake manifold assembly, using power tools.

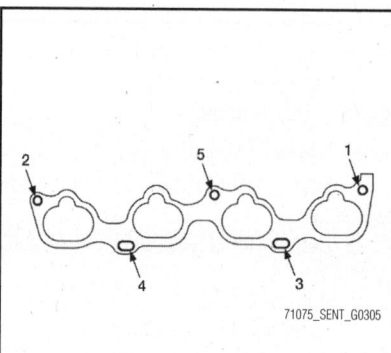

Fig. 101 Loosen the bolts in the order shown to remove the intake manifold assembly, using power tools

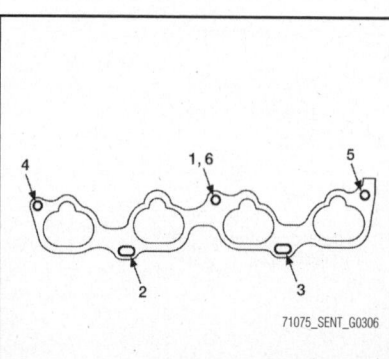

Fig. 102 Tighten the intake manifold bolts and nuts in the numerical order as shown

Inspection After Removal

15. Surface Distortion

    • Using straightedge and feeler gauge, inspect surface distortion of intake manifold adapter and intake manifold surface.

To install:

16. Tighten the intake manifold bolts and nuts.

➡After tightening No.5, retighten the No.1 bolt to specification.

17. Tighten the intake manifold support bolts following the tightening sequence below:

    • Temporarily tighten the intake manifold support (rear) bolts.

    • Tighten the electric throttle control actuator bolts.

    • Tighten the intake manifold support (rear) bolts.

    • Temporarily tighten the intake manifold support (front) bolts (intake manifold side).

    • Tighten the intake manifold support (front) bolts (engine side).

    • Tighten the intake manifold support (front) bolts (intake manifold side).

18. Installation of Electric Throttle Control Actuator:

    • Tighten the mounting bolts of electric throttle control actuator equally and diagonally in several steps.

    • After installation perform procedure in "INSPECTION AFTER INSTALLATION".

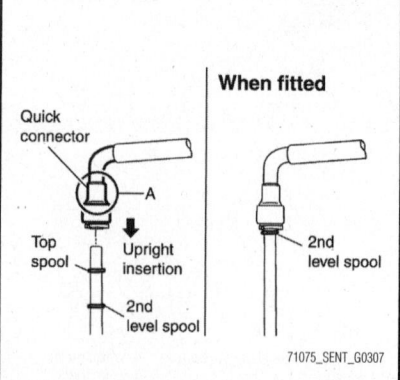

Fig. 103 Align center to insert quick connector straight into fuel tube. Insert fuel tube into quick connector until the top spool on fuel tubes is inserted completely and the second level spool is positioned slightly below the quick connector bottom end

Connecting Quick Connector On The Fuel Hose (Engine Side)

19. Make sure no foreign substances are deposited in and around the fuel tube and quick connector, and there is no damage to them.

20. Thinly apply new engine oil around the fuel tube tip end.

21. Align center to insert quick connector straight into fuel tube.

- Insert fuel tube into quick connector until the top spool on fuel tubes is inserted completely and the second level spool is positioned slightly below the quick connector bottom end.

➡**Hold at position "A" as shown, when inserting the fuel tube into the quick connector.**

❊❊ WARNING

Carefully align to center to avoid inclined insertion to prevent damage to the O-ring inside the quick connector.

➡**Do not reuse O-ring.**

➡**Insert the fuel tube until you hear a "click" sound and actually feel the engagement.**

➡**To avoid misidentification of engagement with a similar sound, be sure to perform the next step.**

22. Before clamping the fuel hose with the hose clamp, pull the quick connector hard by hand, holding at the "A" position, as shown. Make sure it is completely engaged (connected) so that it does not come off of the fuel tube.

➡**Recommended pulling force is 134.4 inch lbs. (50 N).**

23. Install quick connector cap on quick connector joint.

- Direct arrow mark on quick connector cap to upper side (fuel hose side).

24. Install fuel hose to hose clamp.

Connecting Quick Connector On The Fuel Hose (Vehicle Piping Side)

25. Make sure no foreign substances are deposited in and around the fuel tube and quick connector, and there is no damage to them.

26. Align center to insert quick connector straight into fuel tube.

- Insert fuel tube until a click is heard.
- Install quick connector cap on quick connector joint. Direct arrow mark on quick connector cap upper side.
- Install fuel hose to hose clamp.

Inspection After Installation

Make sure there is no fuel leakage at connections as follows:

27. Apply fuel pressure to fuel lines by turning ignition switch ON (with engine stopped). Then check for fuel leaks at connections.

28. Start the engine and rev it up and check for fuel leaks at connections.

❊❊ CAUTION

Do not touch engine immediately after stopping as engine is extremely hot.

➡**Use mirrors for checking on connections out of the direct line of sight.**

- Perform procedures for "Throttle Valve Closed Position Learning" after finishing repairs.
- If electric throttle control actuator is replaced, perform procedures for "Idle Air Volume Learning" after finishing repairs.

OIL PAN

REMOVAL & INSTALLATION

2.0L Engine

See Figures 106 and 107.

1. Drain engine oil.
2. Remove engine and transaxle assembly.
3. Remove oil filter.
4. Remove lower oil pan bolts in reverse order.
5. After removing the pan, clean off the sealant being careful not to damage the mating surfaces.
6. Remove the flywheel.
7. Remove the front cover, timing chain, and the oil pump drive chain
8. Remove oil pan (lower) bolts in reverse order as shown in the accompanying illustration.

To install:

❊❊ WARNING

The rear oil seal should be installed within 5 minutes after installing upper oil pan. Always replace rear oil seal with new one. Never touch oil seal lip.

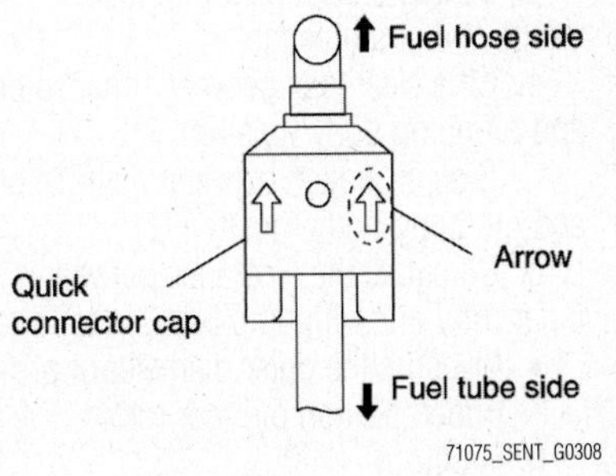

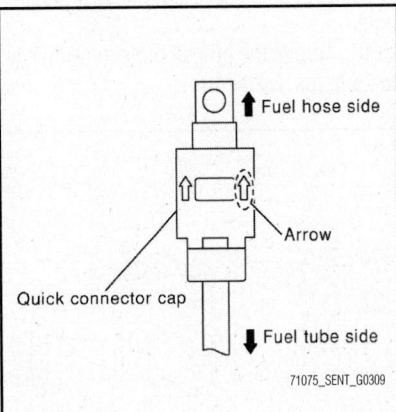

Fig. 104 Install quick connector cap on quick connector joint. Direct arrow mark on quick connector cap to upper side (fuel hose side).

Fig. 105 Align center to insert quick connector straight into fuel tube. Insert fuel tube until a click is heard. Install quick connector cap on quick connector joint. Direct arrow mark on quick connector cap upper side.

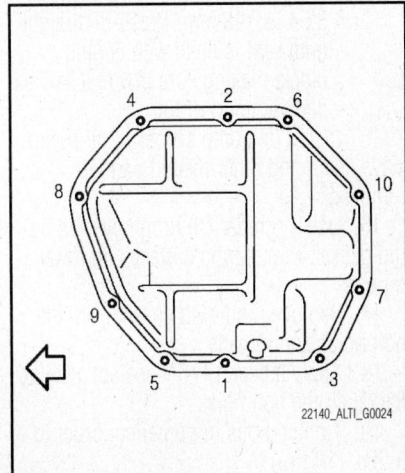

Fig. 106 Lower oil pan bolt torque sequence

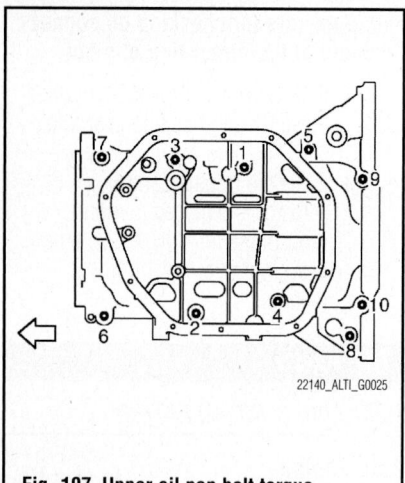

Fig. 107 Upper oil pan bolt torque sequence

9. Carefully scrape old liquid gasket from mating surfaces on oil pan and cylinder block.

10. Remove old liquid gasket from the bolt holes and threads.

11. Apply RTV liquid gasket to the oil pan. The bead should be outside the four center bolt holes and inside the four corner bolt holes.

12. Install new oil filter passage O-ring on the cylinder block. Make sure it's properly aligned.

13. Fit the pan into place on the block and tighten bolts in numerical order to 19 ft. lbs. (25 Nm).

14. Install rear oil seal with the following procedure.

- Wipe off liquid gasket protruding from the block/pan assembly.
- Apply engine oil to entire outside area of rear oil seal.
- Press-fit the rear oil seal using a press tool with outer diameter 4.53 in. (115mm) and inner diameter 3.54 in. (90 mm) Press-fit straight until seal is flush with engine block, making sure that rear oil seal does not curl or tilt.

15. Install oil pump sprocket, oil pump drive chain and other related parts if removed.

16. Use a scraper (A) to remove old liquid gasket from mating surfaces on lower oil pan.

17. Remove old liquid gasket from the bolt holes and threads.

18. Apply a bead of RTV sealant, staying inside all the bolt holes.

19. Tighten bolts in numerical order to 7 ft. lbs. (10 Nm).

20. Install oil filter, hand tight only.

21. Installation of the remaining components is in the reverse order of removal.

2.5L Engine

See Figure 108.

1. Raise and safely support the vehicle and drain the oil.

2. Remove the front exhaust pipe.

3. Remove power steering cooler hose bracket from suspension member.

4. Remove the front subframe for clearance to remove the oil pan.

- Remove nut on lower portion of stabilizer bar connecting rod from lower suspension arm.
- Remove suspension arm from subframe and swing it outward.
- Remove front exhaust pipe.
- Support engine or transmission with a jack.
- Remove steering gear bolts. Remove steering gear and power steering tube bracket from suspension member. Hang steering gear.
- Set a jack under subframe and remove the nuts.

5. Remove the lower oil pan bolts and remove the pan.

6. Remove the oil strainer.

7. Remove rear plate cover and four engine-to transaxle bolts.

8. Loosen the upper oil pan bolts and remove the upper oil pan. Note the different bolt lengths.

To install:

9. Carefully clean all sealing surfaces and apply a bead of RTV sealant to the upper oil pan. Install new O-rings and fit the pan into place. Start the bolts finger tight.

10. Torque the oil pan bolts in sequence to 16 ft. lbs. (22 Nm).

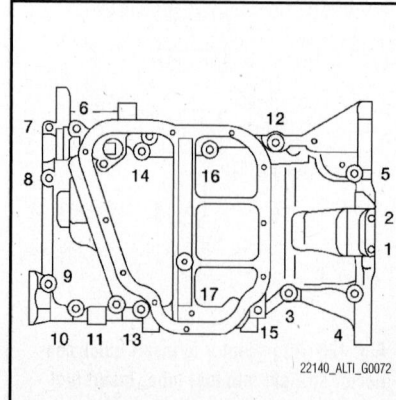

Fig. 108 Upper oil pan bolt torque sequence—2.5L engine

11. Torque the pan-to-transaxle bolts to 31 ft. lbs. (46 Nm).

12. Install the strainer and torque the bolts

13. Apply RTV to the lower oil pan to 16 ft. lbs. (22 Nm).

14. Install the subframe:

- Raise the subframe into place and start all fasteners.
- Torque the large subframe bolts to 107 ft. lbs. (145 Nm).
- Torque the suspension arm bolts to 114 ft. lbs. (155 Nm).
- Reconnect steering gear and stabilizer bar.

15. Install the remaining components and refill the engine with oil.

16. Run the engine to check for leaks.

OIL PUMP

REMOVAL & INSTALLATION

2.5L Engine

See Figures 109 through 114.

1. Remove the oil pump cover.

2. Remove inner rotor and outer rotor from front cover.

3. After removing regulator plug, remove regulator spring and regulator valve.

INSPECTION AFTER DISASSEMBLY

4. Measure the clearance of the oil pump parts.

a. Measure clearance with feeler gauge.

b. Clearance between outer rotor and oil pump body (position 1). Standard: 0.0045-0.0070 in. (0.114-0.179 mm).

c. Tip clearance between inner rotor and outer rotor (position 2). Standard: 0.0067-0.0087 in. (0.170-0.220 mm).

d. Measure clearance with feeler gauge and straightedge.

e. Side clearance between inner rotor and oil pump body (position 3).

f. Side clearance between outer rotor and oil pump body (position 4).

g. Calculate the clearance between inner rotor and oil pump body as follows:

- Measure the outer diameter of protruded portion of inner rotor (Position 5).
- Measure the inner diameter of oil pump body with inside micrometer (Position 6). (Clearance) = (Inner diameter of oil pump body) - (Outer diameter of inner rotor).
- Regulator valve clearance: (Clearance) = D1(Valve hole diameter) - D2 (Outer diameter of valve)

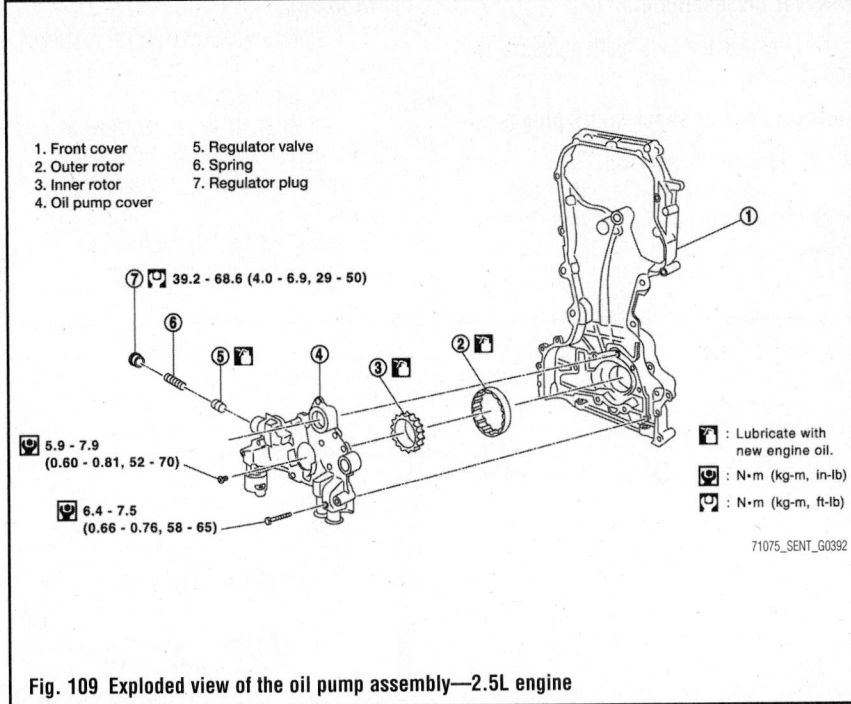

1. Front cover
2. Outer rotor
3. Inner rotor
4. Oil pump cover
5. Regulator valve
6. Spring
7. Regulator plug

⑦ 39.2 - 68.6 (4.0 - 6.9, 29 - 50)

5.9 - 7.9
(0.60 - 0.81, 52 - 70)

6.4 - 7.5
(0.66 - 0.76, 58 - 65)

: Lubricate with new engine oil.

: N·m (kg-m, in-lb)

: N·m (kg-m, ft-lb)

71075_SENT_G0392

Fig. 109 Exploded view of the oil pump assembly—2.5L engine

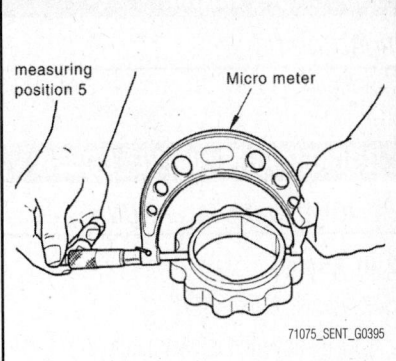

Fig. 112 Calculate the clearance between inner rotor and oil pump body. Measure the outer diameter of protruded portion of inner rotor (Position 5).

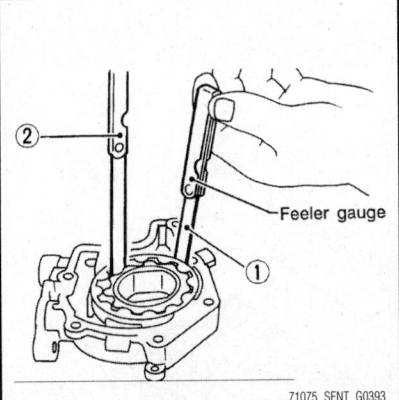

Feeler gauge

71075_SENT_G0393

Fig. 110 Measure the clearance of the oil pump parts. Measure clearance with feeler gauge. Clearance between outer rotor and oil pump body (position 1); Standard: 0.0045-0.0070 in. (0.114-0.179 mm). Tip clearance between inner rotor and outer rotor (position 2); Standard: 0.0067-0.0087 in. (0.170-0.220 mm). —2.5L engine

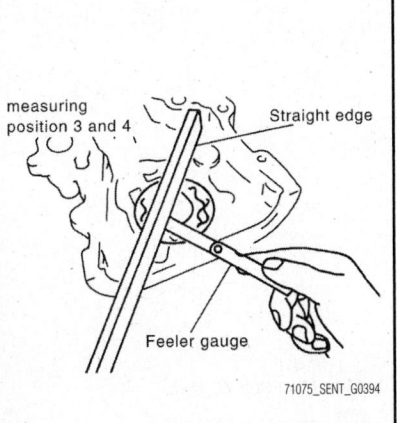

measuring position 3 and 4

Straight edge

Feeler gauge

71075_SENT_G0394

Fig. 111 Measure clearance with feeler gauge and straightedge. Side clearance between inner rotor and oil pump body (position 3); Standard: 0.0012-0.0028 in. (0.030-0.070 mm). Side clearance between outer rotor and oil pump body (position 4). Standard: 0.0024-0.0043 in (0.060-0.110 mm)—2.5L engine

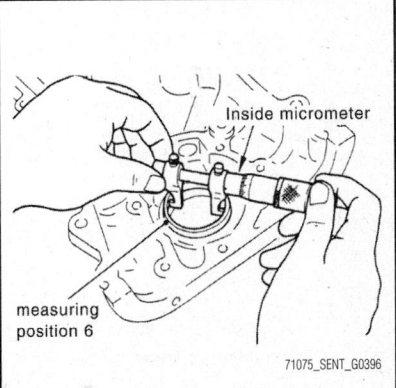

measuring position 6

measuring position 6

Inside micrometer

71075_SENT_G0396

Fig. 113 Measure the inner diameter of oil pump body with inside micrometer (Position 6). (Clearance) = (Inner diameter of oil pump body) - (Outer diameter of inner rotor). Standard: 0.0014-0.0028 in. (0.035-0.070 mm)

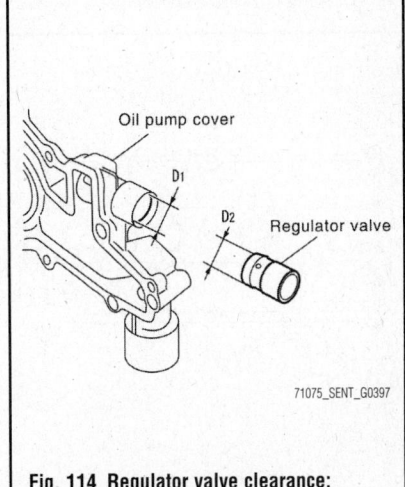

Oil pump cover

D1

D2

Regulator valve

71075_SENT_G0397

Fig. 114 Regulator valve clearance: (Clearance) = D1(Valve hole diameter) - D2 (Outer diameter of valve)

➡Coat regulator valve with engine oil.

Check that it falls smoothly into the valve hole by its ownweight.

To install:

➡**Before installation, apply new engine oil to the parts as shown in the accompanying illustration.**

➡**Install the inner rotor and outer rotor with the punched marks on the oil pump cover side.**

5. Before installing regulator plug, install regulator spring and regulator valve.

6. Install inner rotor and outer rotor from front cover.

7. Install the oil pump cover.

PISTONS & RINGS

POSITIONING

See Figure 115.

ROCKER ARM (VALVE) COVER

REMOVAL & INSTALLATION

2.0L Engine

See Figures 116 and 117.

1. Remove the intake manifold.
2. Remove the four ignition coils.

➡ **Handle intake manifold carefully and avoid impacts.**

➡ **Never disassemble.**

3. Remove the four spark plugs using Tool J-48891

➡ **Never drop or shock spark plugs.**

4. Remove rocker cover.
   a. Loosen bolts in reverse order as shown in the accompanying illustration.

To install:

5. Install rocker cover gasket to rocker cover.
6. Install rocker cover.
7. Tighten bolts in proper order as shown in the accompanying illustration.
8. Install the four spark plugs using Tool J-48891
9. Install the four ignition coils.
10. Install the intake manifold.

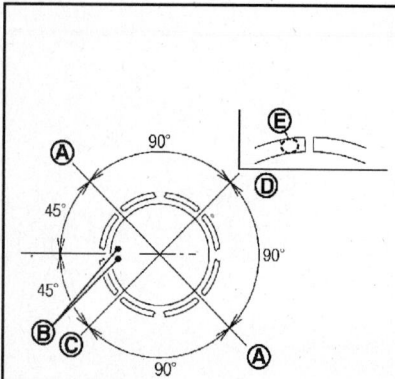

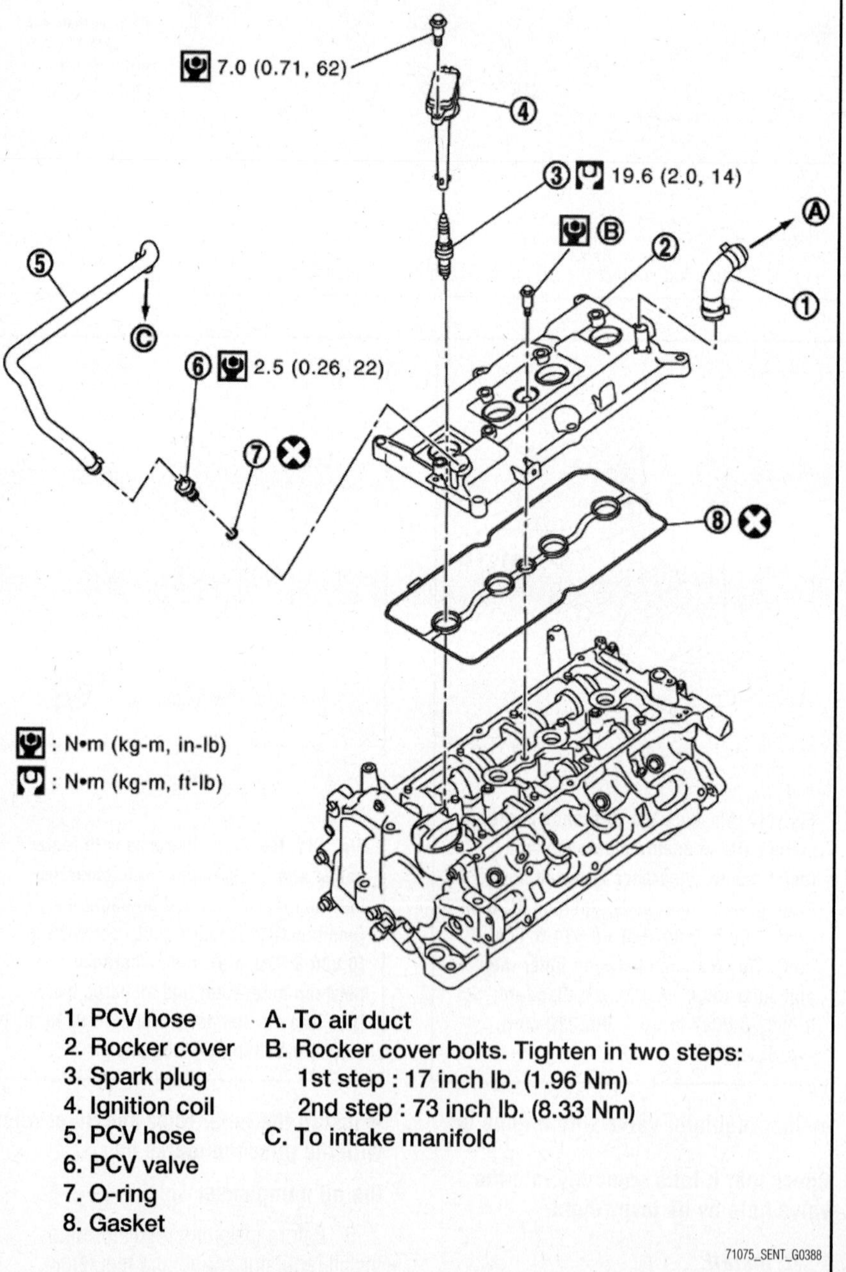

A : Oil ring upper or lower rail gap
B : Front mark
C : Second ring and oil ring spacer gap
D : Top ring gap
E : Stamped mark

22140_ALTI_G0026

Fig. 115 Piston ring positioning—all engines

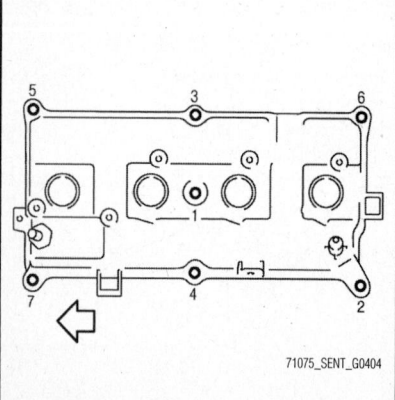

71075_SENT_G0404

Fig. 116 Rocker arm cover tightening sequence—loosen in the reverse order

7.0 (0.71, 62) ④

③ 19.6 (2.0, 14)

Ⓑ ② Ⓐ ①

⑥ 2.5 (0.26, 22)

⑦ ✕

⑧ ✕

◨ : N•m (kg-m, in-lb)

◧ : N•m (kg-m, ft-lb)

1. PCV hose
2. Rocker cover
3. Spark plug
4. Ignition coil
5. PCV hose
6. PCV valve
7. O-ring
8. Gasket

A. To air duct
B. Rocker cover bolts. Tighten in two steps:
   1st step : 17 inch lb. (1.96 Nm)
   2nd step : 73 inch lb. (8.33 Nm)
C. To intake manifold

71075_SENT_G0388

Fig. 117 Exploded view of the ignition coil, spark plug, and rocker cover components

2.5L Engine

See Figures 118 through 121.

1. Disconnect the battery negative terminal.

2. Remove the engine room cover.

3. Remove engine coolant reservoir.

4. Remove the ignition coil.

5. Install a suitable jack under engine.

6. Remove engine mounting insulator and bracket (RH).

7. Remove PCV hose.

8. Loosen the bolts in the numerical order as shown using power tool.

9. Remove the rocker cover. Remove the oil filler cap and PCV valve if necessary, to transfer to the new rocker cover.

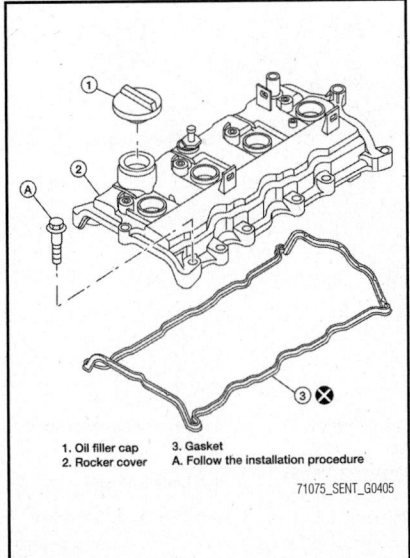

1. Oil filler cap
2. Rocker cover
3. Gasket
A. Follow the installation procedure

71075_SENT_G0405

Fig. 118 Exploded view of the rocker cover—2.5L engine

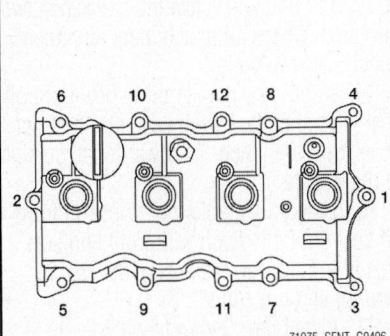

71075_SENT_G0406

Fig. 119 Remove the rocker cover. Remove the oil filler cap and PCV valve if necessary, to transfer to the new rocker cover.

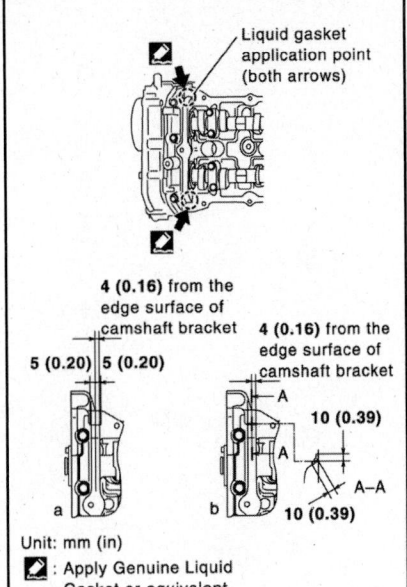

Liquid gasket application point (both arrows)

4 (0.16) from the edge surface of camshaft bracket

5 (0.20) 5 (0.20)

4 (0.16) from the edge surface of camshaft bracket

A

10 (0.39)

A

A–A

a   b   10 (0.39)

Unit: mm (in)

: Apply Genuine Liquid Gasket or equivalent.

71075_SENT_G0407

Fig. 120 Apply RTV Silicone Sealant to the joint part of the cylinder head and camshaft bracket using the following the steps: 1. Follow illustration "a" to apply sealant to joint part of No.1 camshaft bracket and cylinder head. 2.Follow illustration "b" to apply sealant in a 90°angle to the illustration "a".

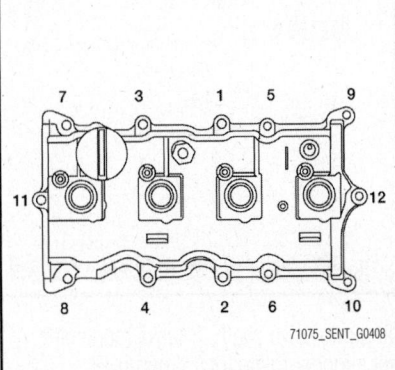

71075_SENT_G0408

Fig. 121 Tighten the rocker cover bolts in two steps, in the numerical order as shown in the accompanying illustration

To install:

10. Apply RTV Silicone Sealant or equivalent to the joint part of the cylinder head and camshaft bracket using the following the steps:

a. Follow illustration "a" to apply sealant to joint part of No.1 camshaft bracket and cylinder head.

b. Follow illustration "b" to apply sealant in a 90° angle to the illustration "a".

11. Install the rocker cover.

a. The rocker cover gasket must be securely installed in the groove in the rocker cover.

12. Tighten the rocker cover bolts in two steps, in the numerical order as shown in the accompanying illustration.

a. Step 1 : 17 inch lbs. (1.96 Nm)

b. Step 2 : 74 inch lbs. (8.33 Nm)

13. Connect the PCV hose and breather hose to the rocker cover. If necessary, install the oil filler cap and PCV valve and lubricate the PCV valve O-ring with new engine oil.

14. Install the ignition coils.

15. Install engine mounting insulator and bracket (RH).

16. Remove jack from under engine.

17. Install the ignition coil.

18. Install engine coolant reservoir.

19. Install the engine room cover.

20. Connect the battery negative terminal.

TIMING CHAIN, FRONT COVER AND SEAL, SPROCKETS

REMOVAL & INSTALLATION

2.0L Engine

See Figures 122 through 125.

1. Remove right front wheel and inner fender

2. Drain engine oil.

3. Remove the intake manifold:

4. Remove ignition coils, spark plugs and rocker cover.

5. Remove drive belt.

6. Remove water pump pulley.

7. Disconnect ground cable between engine bracket and radiator support.

8. Support the bottom surface of engine using a transmission jack, and then remove the engine bracket and insulator. (See Engine Removal and Installation)

9. Set No.1 cylinder at TDC on its compression stroke with the following procedure:

• Rotate crankshaft pulley clockwise and align TDC mark (no paint) to timing indicator on front cover.

• At the same time, make sure that the cam noses of the No.1 cylinder both pointing up.

10. Hold crankshaft pulley using suitable tool and loosen crankshaft pulley bolt. Do not remove the bolt yet.

11. Attach a pulley puller in the threaded holes on crankshaft pulley and remove the pulley.

12. If removing the crankshaft sprocket, oil pump sprocket or other related parts, remove the oil lower pan. See Oil Pan Removal and Installation for bolt sequence.

13. Remove intake cam timing control solenoid valve.

14. Remove drive belt auto-tensioner.

15. Loosen front cover bolts in reverse order.

16. Remove the front cover. Be careful not to damage the mating surfaces.

17. Remove front oil seal from front cover.

18. While facing the front of the engine, the timing chain tensioner on your left must be retracted and locked. Push in timing chain tensioner plunger and insert a 0.060 in (1.5mm) stopper pin into the body hole to retain the plunger in collapsed position.

19. Remove timing chain tensioner.

20. Remove timing chain slack guide, timing chain tension guide and timing chain.

✳✳ WARNING

Never rotate crankshaft or camshafts while timing chain is removed. This is an interference engine.

21. While facing the engine, the chain tensioner on your right must be retracted and locked.

- Fully lift up lever A and push the slack guide B into the chain tensioner (1).
- Matching the hole on lever with the hole on tensioner body, insert a 0.040 in. (1.0mm) stopper pin (C) to secure slack guide.

22. Remove chain tensioner.

23. Remove crankshaft sprocket, oil pump sprocket and oil pump drive chain as a set.

24. Remove timing chain tension guide (front cover side) from front cover if necessary.

To install:

➡ **The figure shows the relationship between the match mark on each timing chain and on the corresponding sprocket, with the components installed. Make sure the crankshaft key points straight up. There are two outer grooves on the intake camshaft sprocket. The wider one is a match mark.**

25. If the timing chain tension guide (front cover side) is removed, install it to the front cover.

26. Align the match marks and install crankshaft sprocket, oil pump sprocket and

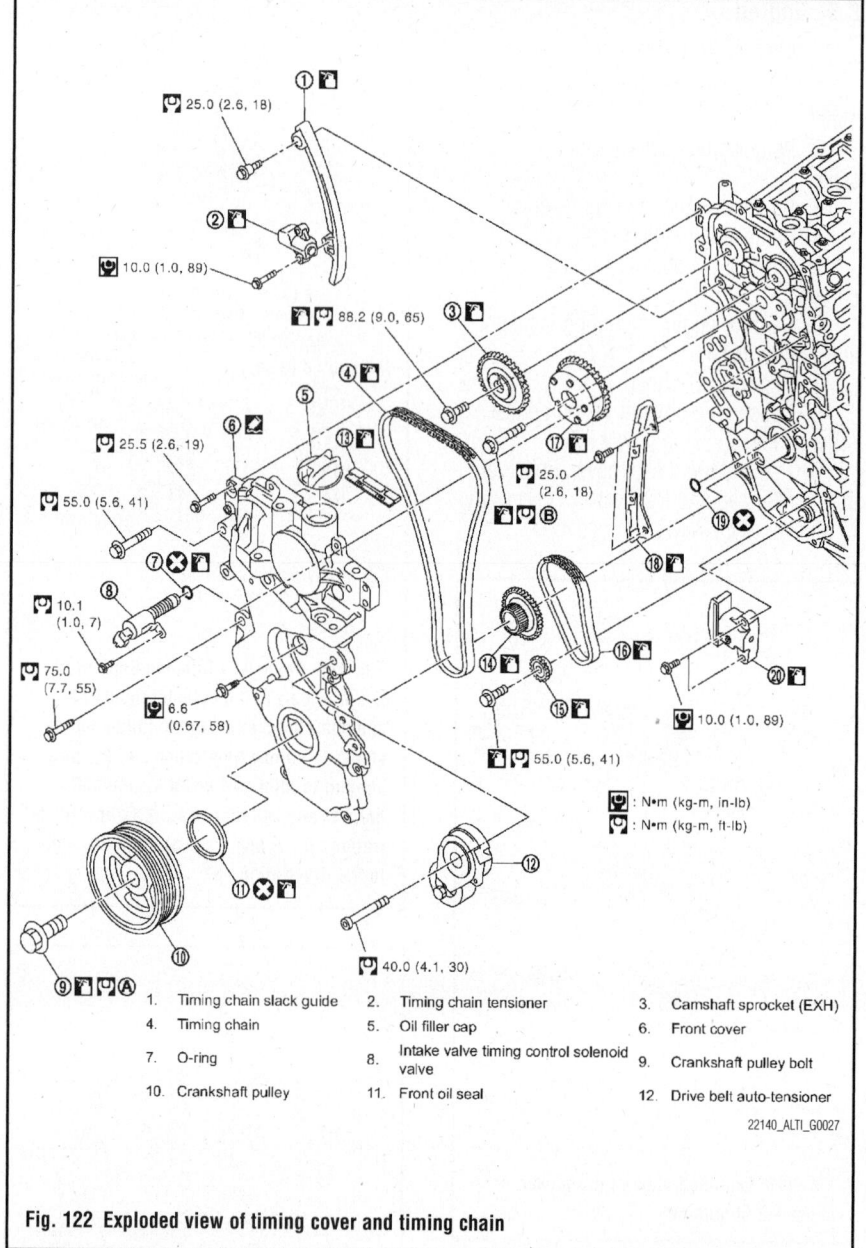

Fig. 122 Exploded view of timing cover and timing chain

1. Timing chain slack guide	2. Timing chain tensioner	3. Camshaft sprocket (EXH)
4. Timing chain	5. Oil filler cap	6. Front cover
7. O-ring	8. Intake valve timing control solenoid valve	9. Crankshaft pulley bolt
10. Crankshaft pulley	11. Front oil seal	12. Drive belt auto-tensioner

oil pump drive chain. If these marks are not aligned, rotate the oil pump as needed.

27. Make sure the oil pump chain tensioner plunger is compressed and locked a stopper pin, then install it.

28. Pull out the tensioner stopper pin and check the match mark alignment again.

29. Install the timing chain with all match marks aligned.

30. Install the timing chain tension guide and the timing chain slack guide.

31. With the plunger retracted and pinned, install timing chain tensioner.

32. Pull out the stopper pin and check all match mark alignments again.

33. Temporarily install the crankshaft

pulley and bolt and rotate the crankshaft two full turns. Check all match mark alignments again. Remove pulley.

34. Apply new engine oil to new front oil seal joint surface. Fit the seal into the front cover with the longer lip towards the outside of the engine.

35. Using a suitable tool, press-fit front oil seal until it is flush with front end surface of front cover. Make sure the seal is straight and not curled.

36. Install new O-ring to cylinder block.

37. Apply RTV sealant to the front cover. Don't forget the opening in the upper center (bolts 10 and 11).

38. Make sure O-ring on cylinder block is correctly installed.

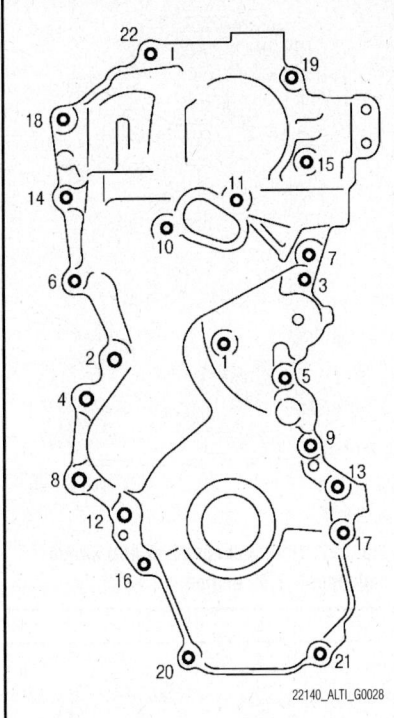

Fig. 123 Timing chain cover bolt tightening sequence

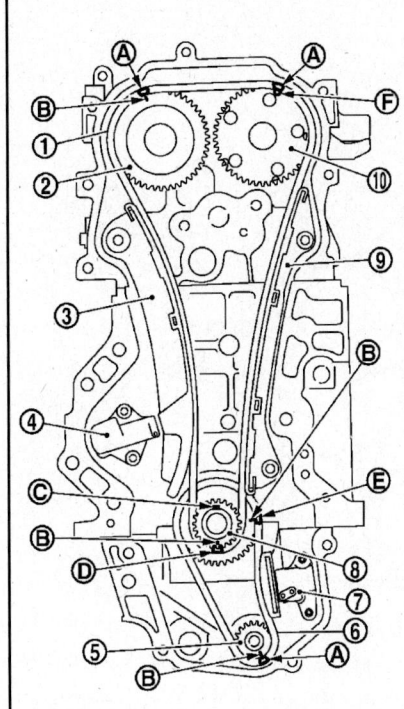

1. Timing chain
2. Exhaust camshaft sprocket
3. Timing chain slack guide
4. Timing chain tensioner
5. Oil pump sprocket
6. Oil pump drive chain
7. Chain tensioner (for oil pump)
8. Crankshaft sprocket
9. Timing chain tension guide
10. Intake camshaft sprocket
A. Match mark (dark blue link)
B. Match mark (stamping)
C. Crankshaft key position (straight up)
D. Match mark (gold link)
E. Match mark (orange link)
F. Match mark (outer groove)

Fig. 124 Timing chain assembly match marks.

39. Install front cover and bolts. Be careful not to damage front oil seal on front end of crankshaft.

40. Tighten bolts in numerical order in two steps to specified torque.

41. Wipe off any excess liquid gasket.

42. Install crankshaft pulley and apply new oil to the pulley bolt.

43. Hold crankshaft pulley using the proper tool.

44. Tighten crankshaft pulley bolt in four steps.

- 51 ft. lbs. (68.6 Nm)
- 0 ft. lbs. (0 Nm)
- 22 ft. lbs. (29.4 Nm)
- 60°

45. Make sure crankshaft rotates clockwise smoothly.

46. Installation of the remaining components is in the reverse order of removal.

2.5L Engine

See Figures 126 through 130.

1. Raise and safely support the vehicle.

2. Remove the upper and lower oil pan and strainer.

3. Remove the alternator.

4. Disconnect variable timing control solenoid harness connector.

5. Remove the coolant overflow reservoir.

6. Position the engine compartment fuse and relay box aside.

7. Remove the right engine mount and bracket.

8. Remove the camshaft timing control (IVT) cover using.

9. Pull the chain guide (between camshaft sprockets) out through front cover.

10. Set the No.1 cylinder at TDC on the compression stroke. Make sure the timing marks on the camshaft sprockets are aligned. See Camshaft Removal and Installation.

11. Hold the crankshaft pulley and loosen but do not remove the crankshaft pulley bolt.

12. Remove the crankshaft pulley with a bolt-on puller. Do not use a claw type puller.

13. Remove the timing cover bolts in reverse of the tightening sequence.

14. Push in the chain tensioner plunger and secure it with a stopper pin in the hole on the tensioner body. Remove chain tensioner.

15. Remove the timing chain.

16. If necessary, hold the hexagonal part of the camshaft with a wrench and loosen the camshaft sprocket bolt and remove the camshaft sprocket.

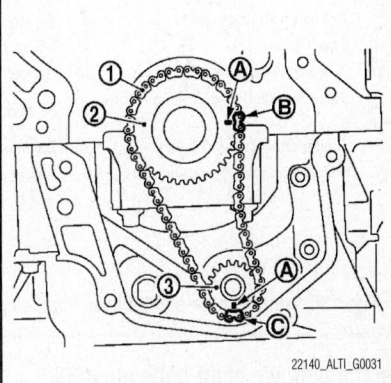

Fig. 125 The oil pump must be indexed to the crankshaft by aligning the marks on the chain and sprockets.

❊❊ WARNING

Do not rotate the crankshaft or camshafts while the timing chain is removed. This is an interference engine.

17. To remove the balance shaft chain, compress the chain guide tensioner and remove the tensioner, guide, timing chain, and oil pump drive spacer.

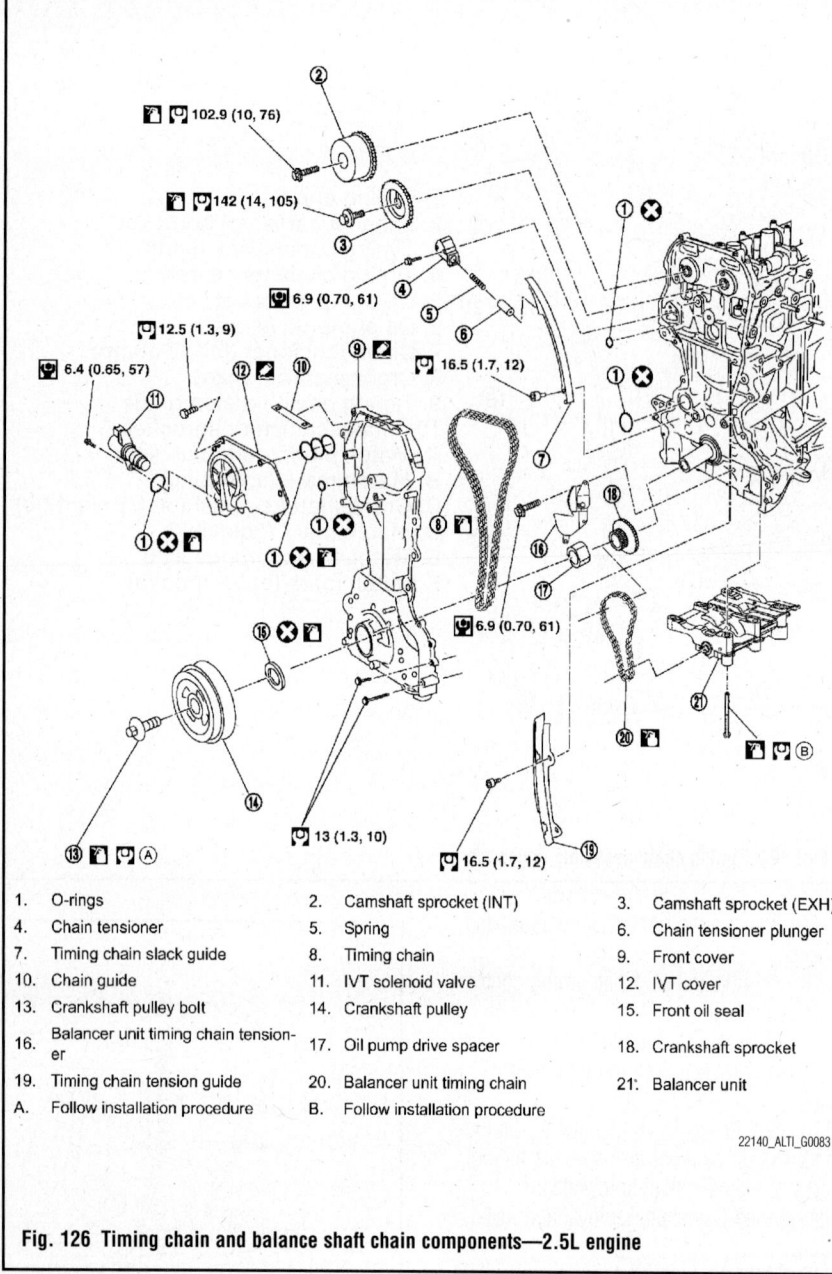

No.	Part	No.	Part	No.	Part
1.	O-rings	2.	Camshaft sprocket (INT)	3.	Camshaft sprocket (EXH)
4.	Chain tensioner	5.	Spring	6.	Chain tensioner plunger
7.	Timing chain slack guide	8.	Timing chain	9.	Front cover
10.	Chain guide	11.	IVT solenoid valve	12.	IVT cover
13.	Crankshaft pulley bolt	14.	Crankshaft pulley	15.	Front oil seal
16.	Balancer unit timing chain tensioner	17.	Oil pump drive spacer	18.	Crankshaft sprocket
19.	Timing chain tension guide	20.	Balancer unit timing chain	21.	Balancer unit
A.	Follow installation procedure	B.	Follow installation procedure		

22140_ALTI_G0083

Fig. 126 Timing chain and balance shaft chain components—2.5L engine

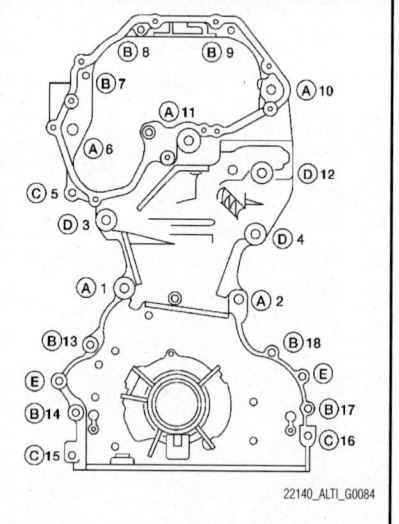

22140_ALTI_G0084

Fig. 127 Timing chain cover bolt torque sequence—2.5L engine

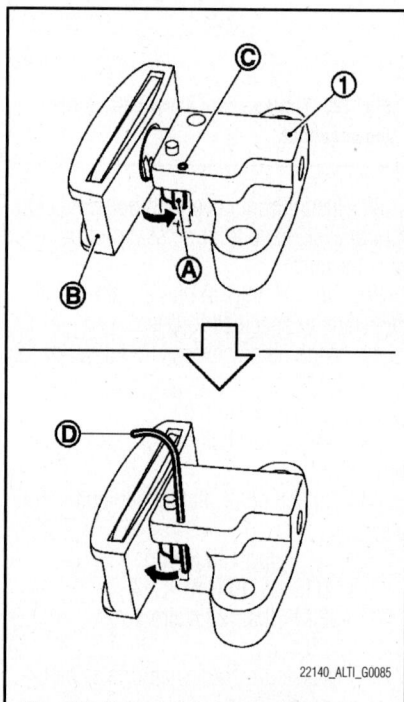

22140_ALTI_G0085

Fig. 128 Release tab A to compress chain guide tensioner B and insert pin D in hole C to lock it in place—2.5L engine

➡The balance shaft bolts must be replaced once removed.

18. Secure the left balancer shaft with a wrench and loosen the sprocket bolt.

19. Remove balancer unit timing chain, balancer unit sprocket and crankshaft sprocket.

To install:

➡There may be two color variations of the link marks (link colors) on the timing chain. There are 26 links between the gold/yellow marks on the timing chain; and 64 links between the camshaft sprocket gold/yellow link and the crankshaft sprocket orange/blue

link on the timing chain side without the tensioner.

20. Make sure the crankshaft key points straight up.

21. Use new bolts for the balance shaft assembly and oil the threads before installation. Install the balancer unit and tighten the bolts in numerical in the following steps:
- Step 1: Bolts 1-5, 31 ft. lbs. (42 Nm): Bolt 6, 27 ft. lbs. (36 Nm)
- Step 2: Bolts 1-5, 120°: Bolt 6, 90°
- Step 3: Loosen in reverse order or tightening sequence
- Step 4: Bolts 1-5, 31 ft. lbs. (42 Nm): Bolt 6, 27 ft. lbs. (36 Nm)
- Step 5: Bolts 1-5, 120°: Bolt 6, 90°

22. Install the crankshaft sprocket and timing chain for the balancer unit. Make sure the crankshaft sprocket mark is at the top to align with the mark on the block. The orange or blue link on the chain aligns with this same mark, while the gold or yellow link aligns with the mark on the balance shaft sprocket.

23. Install balancer timing chain tensioner and remove the stopper pin. Make sure the timing marks are still aligned.

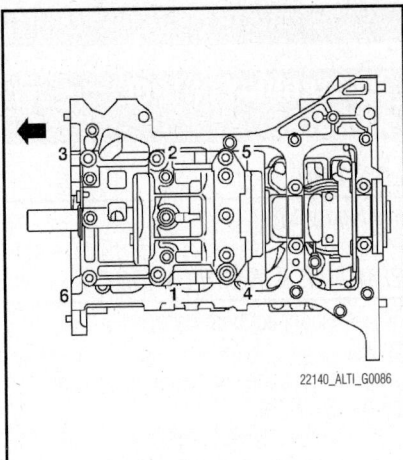

Fig. 129 Balance shaft assembly bolt torque sequence—2.5L engine

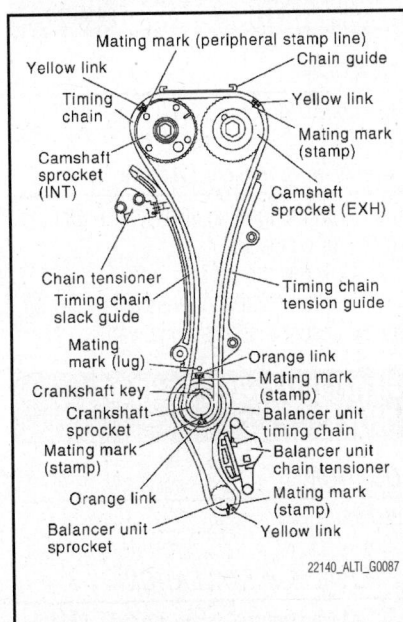

Fig. 130 Timing mark alignment—2.5L engine

24. Install the camshaft timing chain and related parts.

25. After installing timing chain tensioner, remove the stopper pin and make sure that the tensioner moves freely.

26. Rotate the crankshaft two full turns and make sure all timing marks still line up.

27. Install a new front oil seal

28. Carefully clean all sealing surfaces and apply RTV sealant to the timing cover. Install the cover and torque bolts A and D to 36 ft. lbs. (49 Nm) and bolts B and C to 9 ft. lbs. (13 Nm).

29. Carefully clean all sealing surfaces and apply RTV sealant to the rocker cover. Install the cover

30. Install IVT solenoid valve to IVT cover with a new O-ring.

31. Apply RTV sealant to the IVT cover and install it. Torque the bolts in order to 9 ft. lbs. (13 Nm).

32. Install crankshaft pulley and torque bolt to 31 ft. lbs. (42 Nm) plus 60°.

33. Install remaining components.

VALVE LASH (CLEARANCE) ADJUSTMENT

ADJUSTMENT

See Figures 131 through 133.

➡**Valves are adjusted by changing lifters. There are 26 different thicknesses of Nissan valve lifter available in sizes ranging from 3.00 to 3.50 mm (0.1181 to 0.1378 in) in steps of 0.02 mm (0.0008 in.).**

1. Remove the rocker cover.
2. Rotate crankshaft to TDC on No.1 cylinder. Align TDC mark (no paint) to timing indicator on front cover.
3. Measure and record valve clearance at the cam lobes indicated.
4. Rotate the crankshaft one full turn to TDC No. 4.
5. Measure and record valve clearance at the remaining cam lobes.
6. If adjustment is required, remove the camshafts and measure the valve lifter thickness with a micrometer. Use the equation below to calculate the replacement valve lifter thickness.

- Valve lifter thickness calculation:
  $t = t1 + (C1 - C2)$
- t = Valve lifter thickness to be replaced
- $t1$ = Removed valve lifter thickness

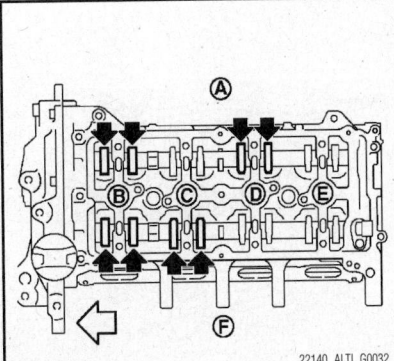

Fig. 131 With engine at TDC No. 1, check valve clearance at these cam lobes—2.0L engine

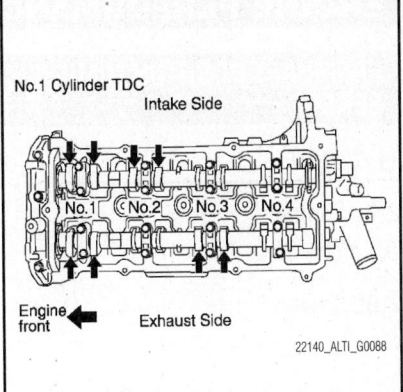

Fig. 132 With engine at TDC No. 1, check valve clearance at these cam lobes—2.5L engine

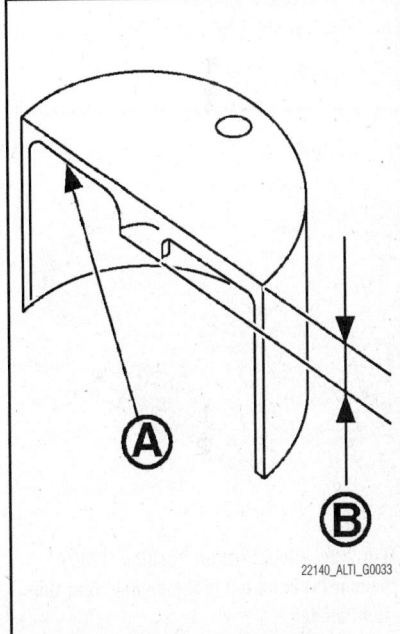

Fig. 133 Valve lifter thickness (B) is marked inside the body (A). Stamp mark "302" indicates a thickness of 3.02 mm (0.1189 in.).

- $C1$ = Measured valve clearance
- $C2$ = Specified valve clearance

7. Thickness of new valve lifters can be identified by a stamp mark inside the lifter body.
8. Install the selected valve lifters.
9. Install the camshafts and check valve clearance again.
10. Install timing chain and related parts.
11. Manually rotate crankshaft pulley a few rotations and check valve clearance again.
12. Installation of the remaining components is the reverse of removal.

ENGINE PERFORMANCE & EMISSION CONTROLS

CAMSHAFT POSITION (CMP) SENSOR

LOCATION

2.0L Engine

See Figure 134.

2.5L Engine

See Figure 135.

REMOVAL & INSTALLATION

1. Unplug the connector.
2. Remove the bolt to remove the sensor.
3. During installation, use a new O-ring and make sure the sensor is clean and free of metal filings on the magnet end.

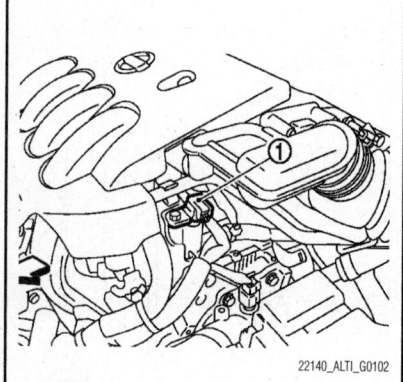

Fig. 134 The Camshaft Position (CMP) sensor (1) is on top of the engine near the ignition coil.

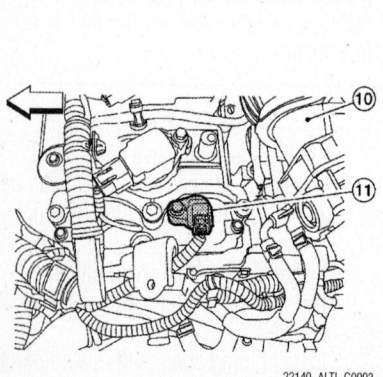

Fig. 135 The Camshaft Position (CMP) sensor (11) is located on top of the engine at the flywheel end.

CRANKSHAFT POSITION (CKP) SENSOR

LOCATION

2.0L Engine

See Figure 136.

2.5L Engine

See Figure 137.

REMOVAL & INSTALLATION

1. Unplug the connector.
2. Remove the bolt to remove the sensor.
3. During installation, use a new O-ring and make sure the sensor is clean and free of metal filings on the magnet end.

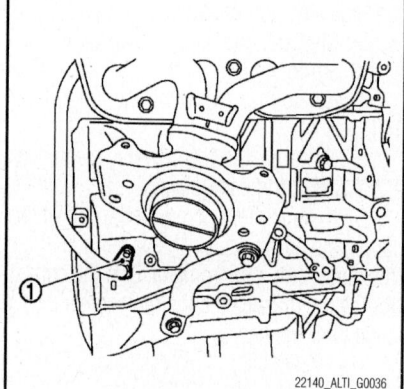

Fig. 136 Crankshaft Position (CKP) sensor (1) is on the rear of the cylinder block.

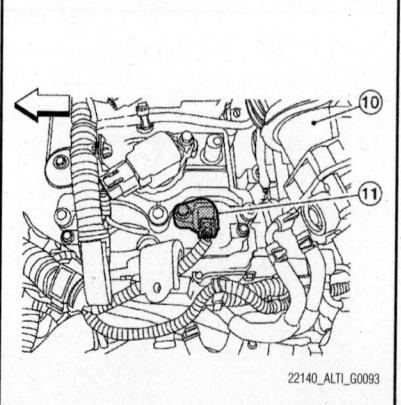

Fig. 137 The Crankshaft Position (CKP) sensor (11) is located on the oil pan.

ELECTRONIC CONTROL MODULE

LOCATION

See Figure 138.

REMOVAL & INSTALLATION

1. Disconnect both battery cables.
2. Remove the connector cover and remove the ECM.
3. Unplug the connector.
4. When installing, take care not to bend the pins in the connector.

ENGINE COOLANT TEMPERATURE SENSOR

LOCATION

See Figure 139.

REMOVAL & INSTALLATION

1. With the ignition switch **OFF**, unplug the sensor connector.
2. Unscrew the sensor.
3. During installation, use a new O-ring and make sure the sensor is clean.

HEATED OXYGEN SENSOR (HO2S)

LOCATION

See Figure 140.

REMOVAL & INSTALLATION

1. Make sure the exhaust manifold is cool.

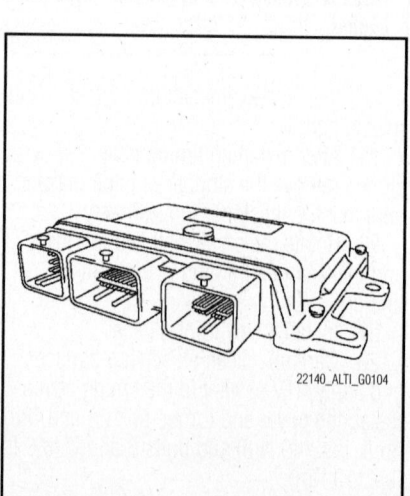

Fig. 138 ECM is mounted in the engine compartment.

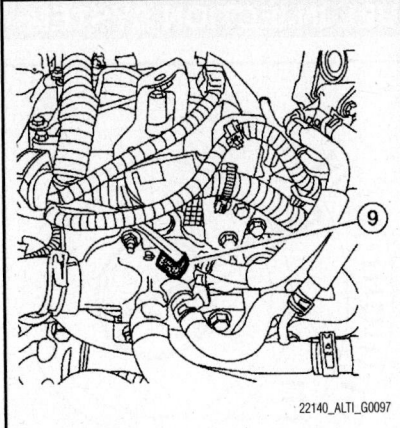

Fig. 139 The ECT sensor (9) is in the water outlet assembly near the upper radiator hose connection.

2. Unplug the connector and remove the sensor with an oxygen sensor wrench or socket.

3. When installing, lightly coat the threads with anti-seize and torque to 23 ft. lbs. (31 Nm). Do not over tighten or the Malfunction Indicator Light (MIL) may turn on.

KNOCK SENSOR (KS)

LOCATION

2.0L Engine

See Figure 141.

2.5L Engine

See Figure 142.

REMOVAL & INSTALLATION

1. Remove the bolt to remove the sensor from the engine block.
2. During installation, do not over tighten the bolt or the sensor will not function properly.

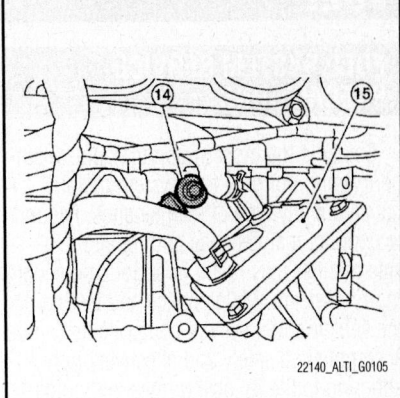

Fig. 142 The Knock Sensor (KS) (14) is on the front of the engine block near the oil cooler.

MASS AIR FLOW (MAF) SENSOR

LOCATION

See Figure 143.

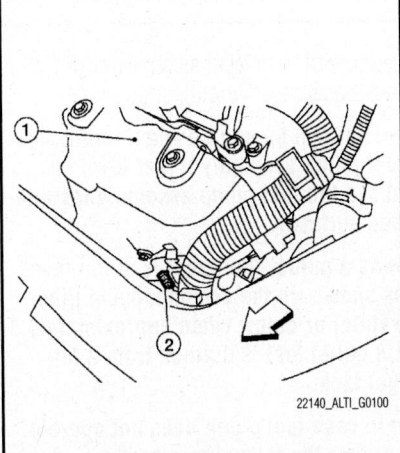

Fig. 140 The primary oxygen sensor (air/fuel ratio sensor) (2) is in the intake manifold below the heat shield. 2.5L engine shown.

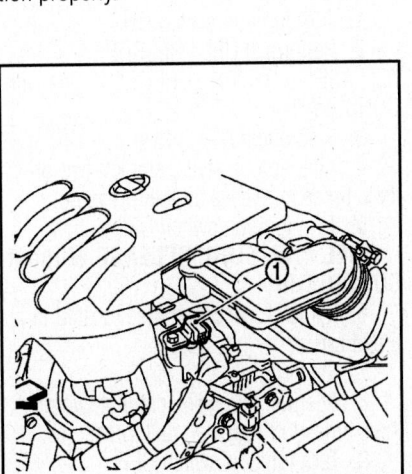

Fig. 141 The Knock Sensor (KS) (1) is mounted on the engine block below the intake manifold.

Fig. 143 The Mass Air Flow (MAF) sensor (7) is in the air filter housing—2.5L engine shown

FUEL SYSTEM SERVICE PRECAUTIONS

Safety is the most important factor when performing not only fuel system maintenance but any type of maintenance. Failure to conduct maintenance and repairs in a safe manner may result in serious personal injury or death. Maintenance and testing of the vehicle's fuel system components can be accomplished safely and effectively by adhering to the following rules and guidelines.

• To avoid the possibility of fire and personal injury, always disconnect the negative battery cable unless the repair or test procedure requires that battery voltage be applied.

• Always relieve the fuel system pressure prior to disconnecting any fuel system component (injector, fuel rail, pressure regulator, etc.), fitting or fuel line connection. Exercise extreme caution whenever relieving fuel system pressure to avoid exposing skin, face and eyes to fuel spray. Please be advised that fuel under pressure may penetrate the skin or any part of the body that it contacts.

• Always place a shop towel or cloth around the fitting or connection prior to loosening to absorb any excess fuel due to spillage. Ensure that all fuel spillage (should it occur) is quickly removed from engine surfaces. Ensure that all fuel soaked cloths or towels are deposited into a suitable waste container.

• Always keep a dry chemical (Class B) fire extinguisher near the work area.

• Do not allow fuel spray or fuel vapors to come into contact with a spark or open flame.

• Always use a back-up wrench when loosening and tightening fuel line connection fittings. This will prevent unnecessary stress and torsion to fuel line piping.

• Always replace worn fuel fitting O-rings with new Do not substitute fuel hose or equivalent where fuel pipe is installed.

Before servicing the vehicle, make sure to also refer to the precautions in the beginning of this section as well.

RELIEVING FUEL SYSTEM PRESSURE

See Figure 144.

FUEL PRESSURE RELEASE With CONSULT-III

Fig. 144 Remove fuel pump fuse (1) located in IPDM E/R (2) (Arrow=Vehicle front)

1. Turn ignition switch ON.
2. Perform "FUEL PRESSURE RELEASE" in "WORK SUPPORT" mode with CONSULT-III.
3. Start engine.
4. After engine stalls, crank it two or three times to release all fuel pressure.
5. Turn ignition switch OFF.

FUEL PRESSURE RELEASE Without CONSULT-III

6. Remove fuel pump fuse located in IPDM E/R.
7. Start engine.
8. After engine stalls, crank it two or three times to release all fuel. pressure.
9. Turn ignition switch OFF.
10. Reinstall fuel pump fuse after servicing fuel system.

FUEL FILTER

REMOVAL & INSTALLATION

➡The fuel filter is part of the fuel pump assembly and not serviced separately. The fuel pump is inside the fuel tank and can be accessed through an open under the rear seat.

FUEL LEVEL SENDING UNIT

REMOVAL & INSTALLATION

See Figures 145 through 152.

1. Refer to the Precautions before working on the fuel system.
2. Check fuel level with the vehicle on a level surface. If the fuel gauge indicates more than the level as shown (7/8 full), drain fuel from the fuel tank until the fuel

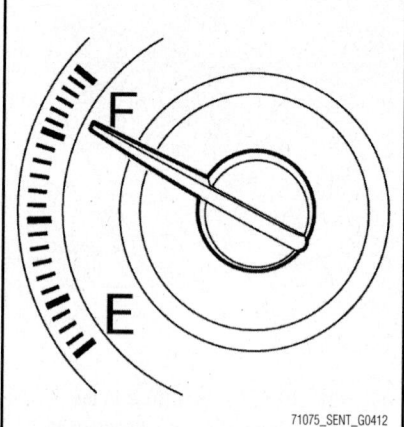

Fig. 145 Check fuel level with the vehicle on a level surface. If the fuel gauge indicates more than the level as shown (7/8 full), drain fuel from the fuel tank until the fuel gauge indicates level as shown (7/8 full) in the accompanying illustration

gauge indicates level as shown (7/8 full) in the accompanying illustration.

➡Fuel will be spilled when removing fuel pump assembly if fuel level is above the fuel pump assembly installation surface.

➡As a guide, fuel level is at the level as shown (in the accompanying illustration) or below when approximately 3 1/8 gal of fuel is drained from a full fuel tank.

➡In case fuel pump does not operate, perform the following procedure.

    a. Insert fuel tubing of less than 0.98 in. (25 mm) in diameter into fuel filler tube through fuel filler opening to drain fuel from fuel filler tube.

    b. Disconnect fuel filler hose from fuel filler tube.

    c. Insert fuel tubing into fuel tank through fuel filler hose to drain fuel from fuel tank.

3. Open fuel door and unscrew the fuel filler cap to release the pressure inside the fuel tank.
4. Release the fuel pressure from the fuel lines.
5. Disconnect battery negative terminal.
6. Remove rear seat bottom.
7. Turn the three retainers (A) 90°in a clockwise direction and remove the fuel pump inspection hole cover.

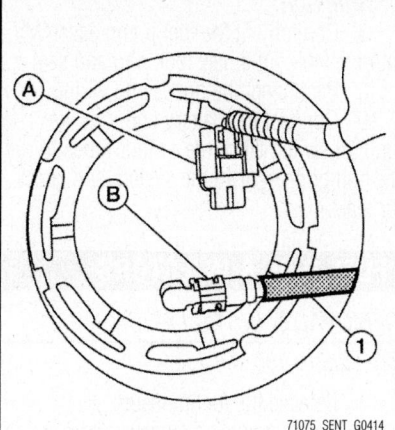

Fig. 146 Disconnect electrical connector
(A) and fuel feed hose quick connector (B)
from the fuel feed hose (1)

8. Disconnect electrical connector and
fuel feed hose quick connector.
  a. Fuel feed hose.
  b. Disconnect the quick connector
using the following procedure:
  • Hold the sides of the connector,
push in tabs and pull out the
tube.
  • If quick connector and tube on fuel
pump assembly are stuck, push
and pull several times until they
move. Disconnect them by
pulling.

➡The tube can be removed when the
tabs are completely depressed. Do not
twist it more than necessary.

➡Do not use any tools to remove the
quick connector.

➡Keep resin tube away from heat. Be
especially careful when welding near
the resin tube.

➡Prevent acid liquid such as battery
electrolyte, from getting on resin
tube.

➡Do not bend or twist resin tube during installation and removal.

❋❋ WARNING

**To keep the connecting portion clean,
free of foreign materials and to avoid
damage, cover them completely with
plastic bags or something similar.**

❋❋ WARNING

**Do not insert plug to prevent damage
to O-ring in quick connector.**

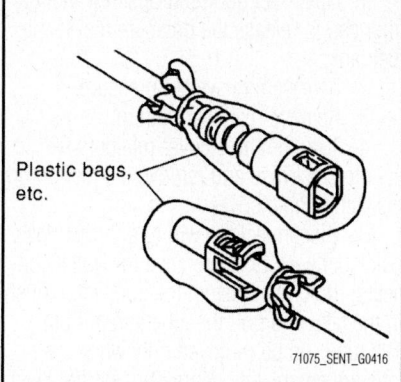

Fig. 147 Cover the quick connectors with
plastic bags to prevent foreign contaminants from entering the fuel system

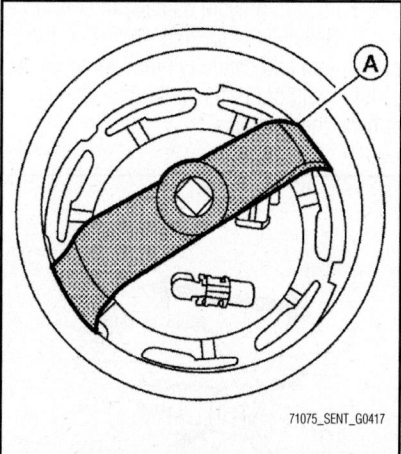

Fig. 148 Remove the lock ring using the
special tool shown (A)

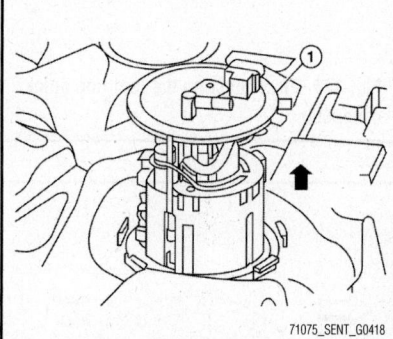

Fig. 149 Remove the fuel level sensor
unit, fuel filter and fuel pump assembly
(1)

9. Remove the lock ring using Tool J-
45722 as shown in the accompanying illustration.
10. Remove fuel level sensor unit, fuel
filter and fuel pump assembly.

➡Do not bend float arm during
removal.

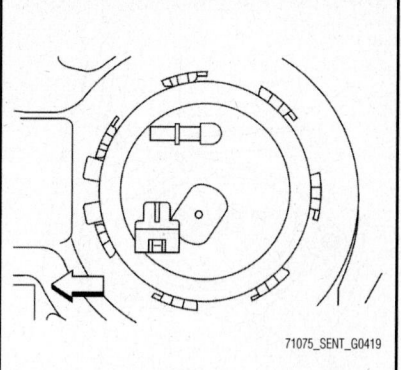

Fig. 150 Install the fuel level sensor unit,
fuel filter and fuel assembly (arrow points
to vehicle front)

➡Do not allow foreign materials to fall
into fuel tank. Use a lint free cloth
when handling components.

➡Avoid impacts such as dropping
when handling components.

INSPECTION AFTER REMOVAL
Make sure that the fuel level sensor unit,
fuel filter and fuel pump is free from defects
and foreign materials.

To install:
Fuel Level Sensor Unit, Fuel Filter
and Fuel Pump Assembly
11. Install O-ring to fuel tank without
twisting.
12. Install fuel level sensor unit, fuel filter and fuel pump assembly as shown in the
accompanying illustration.
  a. Turn the lock ring using Tool J-
45722 until the lock ring is fully rotated
into the fuel tank lock tabs.
Quick Connector
13. Check the connection for damage or
any foreign materials.

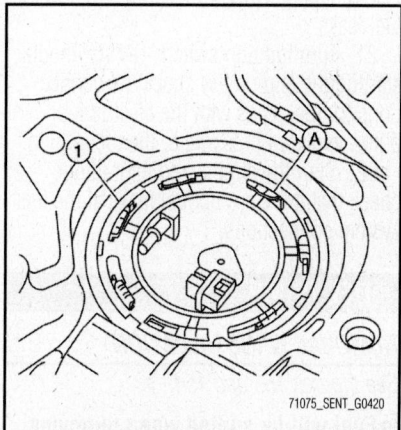

Fig. 151 Turn the lock ring (1) using Tool
until the lock ring is fully rotated into the
fuel tank lock tabs (A) as shown

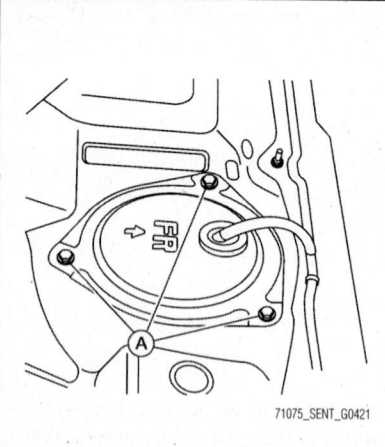

Fig. 152 Install inspection hole cover with the front mark (arrow) facing front of vehicle. Lock clips (A) by turning counterclockwise

14. Align the connector with the tube, then insert the connector straight into the tube until a "click" sound is heard.

15. After connecting, make sure that the connection is secure using the following procedure.

a. Visually confirm that the two retainer tabs are secured to the connector.

b. Pull the tube and the connector to make sure they are securely connected.

Inspection Hole Cover

16. Before installing inspection hole cover, confirm that there are no fuel leaks.

17. Install inspection hole cover with the front mark facing front of vehicle.

18. Lock clips by turning counterclockwise.

19. Install rear seat bottom.

20. Connect battery negative terminal.

INSPECTION AFTER INSTALLATION

Use the following procedure to check for fuel leaks.

21. Turn ignition switch "ON" (without starting the engine), to check the connections for fuel leaks with the electric fuel pump applying pressure to the fuel piping.

22. Start the engine and let it idle to check that there are no fuel leaks at the fuel system connections.

FUEL PUMP

REMOVAL & INSTALLATION

See Figures 153 and 154.

➡**Fuel will be spilled when removing fuel pump assembly if the tank is full. If the fuel gauge indicates more than (⅞ full), drain at least 3 ⅛ gallons (12L) from the fuel tank.**

1. Open fuel door and unscrew the fuel filler cap to release the pressure inside the fuel tank.

2. Release fuel system pressure.

3. Remove rear seat bottom.

4. Turn the three cover retainers 90° in counterclockwise and remove the fuel pump inspection hole cover.

5. Disconnect the wiring and fuel hose quick connectors. To remove the quick connector, hold the sides of the connector, push in the tabs and pull the tube straight out. The tube can be removed only when the tabs are completely depressed. Do not twist or use any tools.

6. To keep the connectors clean and to avoid damage, cover them completely with plastic bags or something similar. Do not insert plugs to prevent damage to O-ring.

7. Remove the locking ring using the correct tool and carefully lift the pump/filter/sending unit out of the tank. Take care not to bend the float arm.

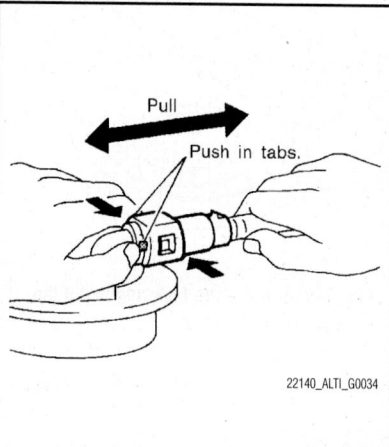

Fig. 153 Disconnecting the fuel line quick connector

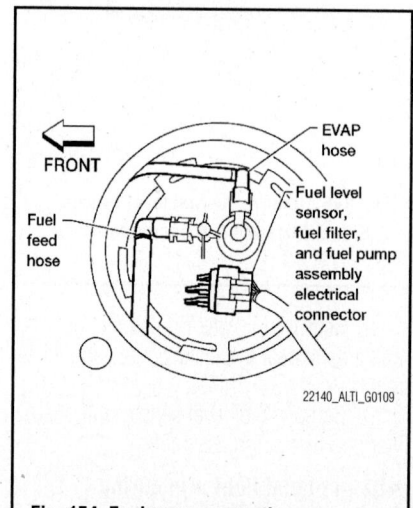

Fig. 154 Fuel pump connections

To install:

8. Carefully fit the fuel pump assembly into the tank. Fit a new lock ring and seal.

9. Reconnect the hoses and wiring.

10. Connect the battery and turn the ignition switch ON three or four times to run the pump, pressurize the system and check for leaks.

FUEL RAIL & INJECTORS

REMOVAL & INSTALLATION

See Figures 155 and 156.

1. Release the fuel pressure.

2. Remove quick connector cap from quick connector connection.

3. Disconnect fuel feed hose from hose clamp.

- With the sleeve side of quick connector release facing quick connector, install quick connector release onto fuel tube.

- Insert quick connector release into quick connector until sleeve contacts and goes no further. Hold quick connector release on that position.

✷✷ WARNING

Inserting quick connector release hard will not disconnect quick connector. Hold quick connector release where it contacts and goes no further.

- Draw and pull out quick connector straight from fuel tube.

✷✷ WARNING

Pull quick connector holding "A" position. Do not pull with lateral force applied. O-ring inside quick connector may be damaged. Do not bend or twist connection between quick connector and fuel feed hose.

- To keep clean the connecting portion and to avoid damage and foreign materials, cover them completely with plastic bags or something similar.

4. Remove intake manifold.

5. Remove fuel tube bolts.

6. Remove the fuel tube and fuel injector assembly.

✷✷ WARNING

When removing, be careful to avoid any interference with fuel injector.

7. Remove fuel injector from fuel tube with the following procedure:

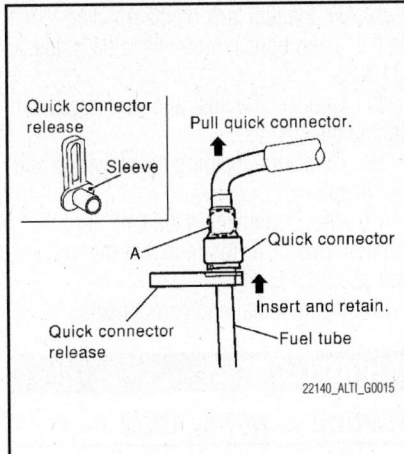

Fig. 155 Use the special tool to release the fuel line quick connector

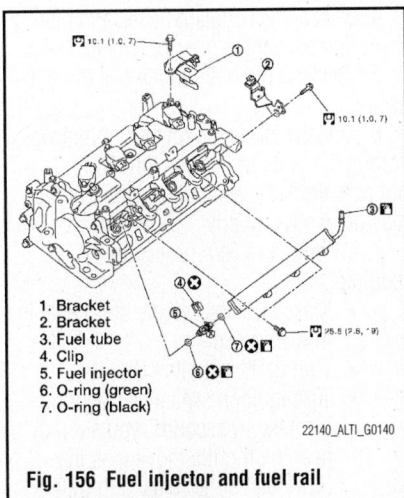

1. Bracket
2. Bracket
3. Fuel tube
4. Clip
5. Fuel injector
6. O-ring (green)
7. O-ring (black)

Fig. 156 Fuel injector and fuel rail assembly—2.0L engine

- Open and remove clip.
- Remove fuel injector from fuel tube by pulling straight. Be careful not to damage fuel injector nozzle during removal.

To install:

➡**Upper and lower injector O-rings are different. Be careful not to confuse them. Handle O-ring with bare hands. Never wear gloves.**

8. Lubricate O-ring with new engine oil and fit them onto the injector. The engine-side O-ring is green, the fuel rail O-ring is black. Be careful not to twist or stretch O-rings.

9. Slowly insert injector straight into fuel tube. Never twist it.

10. Fit fuel injectors into the cylinder head and torque the bolts to 19 ft. lbs. (25 Nm).

11. Connect the fuel hose quick connector and pressurize the fuel system to check

for leaks. The pump will run for two seconds each time the ignition switch is turned **ON**.

12. Install the intake manifold.

FUEL TANK

DRAINING

See Figures 157.

1. Refer to the Precautions before working on the fuel system.

2. Check fuel level with the vehicle on a level surface. If the fuel gauge indicates more than the level as shown (7⁄8 full), drain fuel from the fuel tank until the fuel gauge indicates level as shown (7⁄8 full) in the accompanying illustration.

➡**Fuel will be spilled when removing fuel pump assembly if fuel level is above the fuel pump assembly installation surface.**

➡**As a guide, fuel level is at the level as shown (in the accompanying illustration) or below when approximately 3 1/8 gal of fuel is drained from a full fuel tank.**

➡**In case fuel pump does not operate, perform the following procedure.**

   a. Insert fuel tubing of less than 0.98 in. (25 mm) in diameter into fuel filler tube through fuel filler opening to drain fuel from fuel filler tube.

   b. Disconnect fuel filler hose from fuel filler tube.

   c. Insert fuel tubing into fuel tank through fuel filler hose to drain fuel from fuel tank.

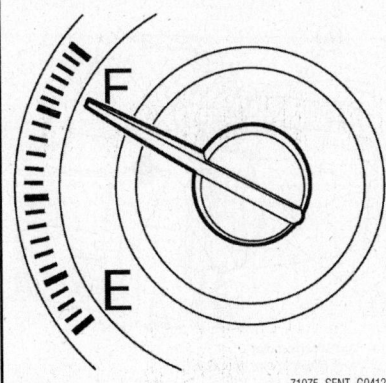

Fig. 157 Check fuel level with the vehicle on a level surface. If the fuel gauge indicates more than the level as shown (7⁄8 full), drain fuel from the fuel tank until the fuel gauge indicates level as shown (7⁄8 full) in the accompanying illustration

3. Open fuel door and unscrew the fuel filler cap to release the pressure inside the fuel tank.

4. Release the fuel pressure from the fuel lines.

REMOVAL & INSTALLATION

See Figure 158.

➡**Fuel will be spilled when removing fuel pump assembly if the tank is full. If the fuel gauge indicates more than (7⁄8 full), drain at least 3⅛ gallons (12L) from the fuel tank.**

1. Check fuel level with the vehicle on a level surface. If the fuel gauge indicates more than (7⁄8 full), drain at least 3⅛ gallons (12L) from the fuel tank .

2. Siphon fuel from fuel tank if necessary.

3. Open fuel door and unscrew the fuel filler cap to release the pressure inside the fuel tank.

4. Release the fuel pressure.

5. Remove rear seat bottom.

6. Turn the three retainers 90° in a counterclockwise direction and remove the fuel pump inspection hole cover.

7. Disconnect wiring and fuel feed hoses. To keep the connectors clean and to avoid damage, cover them completely with plastic bags or something similar. Do not insert plugs to prevent damage to O-ring.

8. Remove center exhaust pipe.

9. Remove fuel tank protector.

10. Loosen fuel filler hose clamp and remove fuel filler hose.

❄❄ WARNING

Do not remove fuel filler hose from fuel filler tube. Mark components for alignment.

11. Remove vent hose and EVAP hose at rear of fuel tank.

12. Support center of fuel tank with transmission jack.

13. Remove fuel tank bands. If they are not marked "R" and "L," mark them now.

14. Lower transmission jack carefully to remove fuel tank while supporting it by hand.

❄❄ WARNING

Fuel tank may be in an unstable position because of the shape of fuel tank bottom. Be sure to support tank securely.

To install:

15. Secure tank on a transmission jack and carefully raise it into position

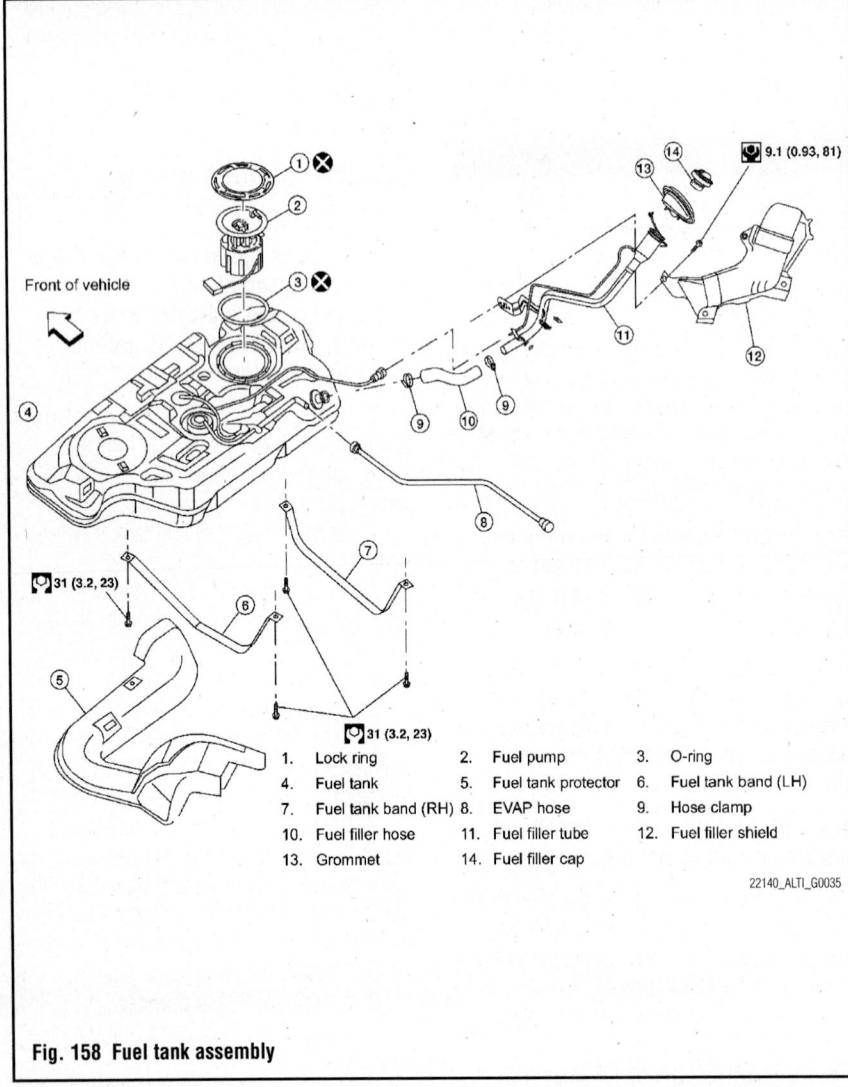

Front of vehicle

9.1 (0.93, 81)

31 (3.2, 23)

31 (3.2, 23)

1. Lock ring	2. Fuel pump	3. O-ring
4. Fuel tank	5. Fuel tank protector	6. Fuel tank band (LH)
7. Fuel tank band (RH)	8. EVAP hose	9. Hose clamp
10. Fuel filler hose	11. Fuel filler tube	12. Fuel filler shield
13. Grommet	14. Fuel filler cap	

22140_ALTI_G0035

Fig. 158 Fuel tank assembly

16. Fit the fuel tank bands (marked "R" and "L") and tighten the bolts to 23 ft. lbs. (31 Nm).

17. Connect the filler and vent hoses and tighten the clamps.

18. Connect remaining tubes, wiring and fuel lines.

19. With some fuel in the tank, turn the ignition switch ON to pressurize the system and check for leaks.

20. Install remaining components.

THROTTLE BODY

REMOVAL & INSTALLATION

1. Remove engine cover.

2. Remove air intake ducts and air filter housing as required.

3. Disconnect throttle wiring.

4. Disconnect coolant hoses and plug them immediately to prevent coolant leaks.

5. Remove the bolts to remove throttle assembly.

6. Installation is the reverse of removal. Torque the bolts to 7 ft. lbs. (10 Nm). Do not over tighten these bolts or the throttle may not work properly.

7. A throttle relearn procedure is required.

- Make sure that accelerator pedal is fully released.
- Turn ignition switch ON.
- Turn ignition switch OFF. During the next 10 seconds, you should hear the throttle moving as the ECM learns the "closed throttle" position.

HEATING & AIR CONDITIONING

BLOWER MOTOR

REMOVAL & INSTALLATION

See Figure 159.

1. Remove the instrument panel.

2. Disconnect the wiring and remove the front blower motor.

3. Remove the one screw from the front blower motor

4. Turn the front blower motor counterclockwise and remove it.

To install:

5. Turn the front blower motor clockwise and install it.

6. Install the one screw to the front blower motor

7. Connect the wiring and install the front blower motor.

8. Install the instrument panel.

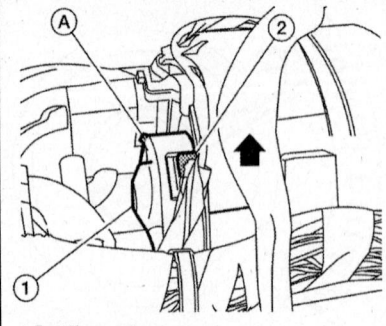

1. Front blower motor
2. Front blower motor connector
A. Screw
Arrow: Front

71075_SENT_G0311

Fig. 159 Disconnect the wiring and remove the front blower motor. Remove the one screw from the front blower motor. Turn the front blower motor counterclockwise and remove it.

HEATER CORE

REMOVAL & INSTALLATION

See Figure 160.

➡**Heater core removal requires removing the dashboard.**

1. Before servicing the vehicle, refer to the Precautions Section.

2. Position the steering wheel in the straight-ahead position.

3. Turn the ignition switch OFF.

4. Disconnect the negative (-) battery cable; then, the positive (+) battery cable.

➡**Wait at least 10 minutes after disconnecting the battery cables for the charge in the air bag circuit to dissipate before working on the air bag module(s).**

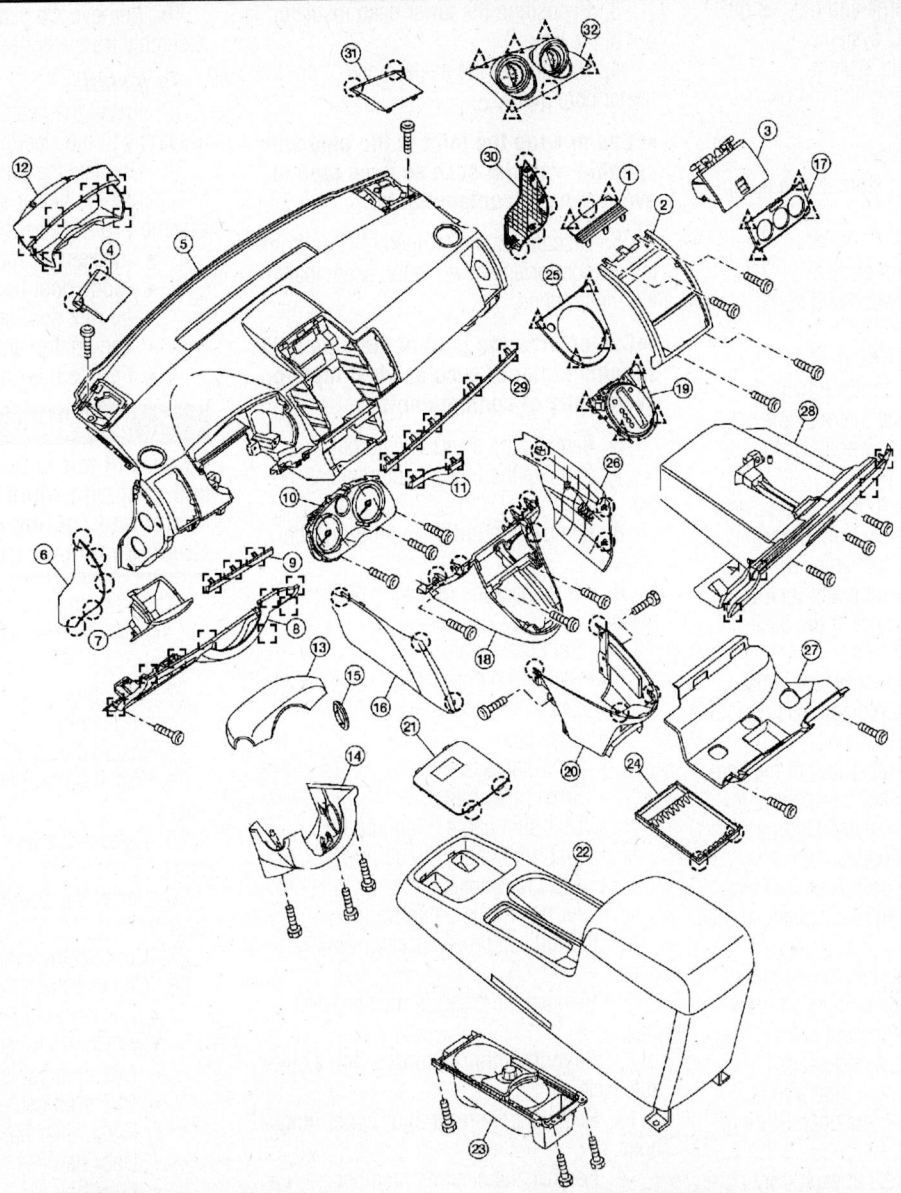

1. Cluster lid C upper mask
2. Cluster lid C
3. Cluster lid C storage bin
4. Speaker grille (LH)
5. Instrument panel
6. Instrument side mask (LH)
7. Fuse block lid storage bin
8. Instrument lower finisher
9. Instrument panel trim (LH)
10. Combination meter
11. Instrument panel trim center
12. Cluster lid A
13. Steering column cover upper
14. Steering column cover lower
15. Steering lock escutcheon
16. Instrument lower cover (LH)

17. Controller finisher
18. Instrument upper cover (center)
19. CVT finisher
20. Instrument lower cover (center)
21. Center console mat
22. Center console
23. Center console cup holder
24. Center console tray
25. MT finisher
26. Instrument lower cover (RH)
27. Glove box lower finisher
28. Glove box assembly
29. Instrument panel trim (RH)
30. Instrument side mask (RH)
31. Speaker grille (RH)

22140_ALTI_G0038

Fig. 160 Exploded view of dashboard

5. Properly discharge and recover the refrigerant from the A/C system.

6. Drain the cooling system.

7. Install the steering column by removing the following:

- lower dash cover
- airbag module: store it face up out of the way
- steering wheel
- steering column covers
- combination switch and spiral cable
- disconnect all wiring
- install the bolt to disconnect the upper end of the intermediate shaft from the power steering motor.
- install the nuts to lower the power steering assembly from the vehicle.

8. Install the center console. There are two screws under the front panel (below the brake handle) and two more at the back (open the lid).

9. Install the lower instrument panel covers and disengage the diagnostic connector and hood release handle.

10. Install the uppermost part of the center cluster trim, then install four screws to install the whole center cluster. Unplug wiring connectors as needed.

11. Install the glove box. There are two screws below the door and four inside at the top.

12. Disengage three clips to install the instrument cluster cover, then install three screws to install the instrument cluster. Unplug the connectors as necessary.

13. Install the passenger side airbag module from the steering member. Store it face up out of the way.

14. Install the speakers at each front corner of the dashboard and install the bolts securing the dashboard.

15. Install the screws at each upper and lower corner of the dashboard and in the instrument cluster opening.

16. Disconnect the antenna and install the dashboard.

17. Disconnect the refrigerant lines from the evaporator.

18. Disconnect the hoses from the heater core.

HEATER & COOLING UNIT

REMOVAL & INSTALLATION

See Figures 161 through 163.

1. Discharge the refrigerant from the A/C system.

2. Drain the engine coolant from the cooling system.

3. Reposition the lower dash insulator out of the way.

4. Disconnect the heater hoses from the heater core pipes.

➡ **Cap or wrap the joint of the pipe with suitable material such as vinyl tape to avoid entry of contaminants.**

5. Disconnect the refrigerant lines from the evaporator as shown in the accompanying illustrations.

➡ **Cap or wrap the joint of the pipe with suitable material such as vinyl tape to avoid entry of contaminants.**

6. Remove the steering column.

7. Remove the instrument panel assembly.

8. Remove the lower steering member stays.

9. Disconnect the following components:

- Shift lock cable
- Steering harness clips
- SMJ
- Fuse box
- Door harness
- Front pillar harness
- CVT shift cable (if equipped)
- M/T shift cable (if equipped)
- A/C drain hose
- Air bag module control unit

10. Disconnect the evaporator drain hose.

11. Remove the steering member bolt caps.

12. Support both front doors with a suitable jack.

13. Remove both front door upper hinge bolts.

14. Remove the steering member bolts.

15. Temporarily install the front door upper hinge bolts.

16. Remove the heater and cooling unit and the steering member from the vehicle as one unit.

❄❄ WARNING

Be careful not to damage the interior and seat trim when removing the heater and cooling unit assembly and steering member from the vehicle.

17. Disconnect the following components to separate the heater and cooling unit assembly from the steering member

- Front blower motor
- Fan control amplifier
- air mix door motor
- Mode door motor
- Intake door motor

18. Remove the heater and cooling unit assembly from the steering member.

To install:

19. Install the heater and cooling unit assembly to the steering member.

20. Connect the following components to separate the heater and cooling unit assembly to the steering member

- Intake door motor
- Mode door motor
- air mix door motor
- Fan control amplifier
- Front blower motor

❄❄ WARNING

Be careful not to damage the interior and seat trim when removing the heater and cooling unit assembly and steering member from the vehicle.

21. Install the heater and cooling unit and the steering member to the vehicle as one unit.

22. Temporarily install the front door upper hinge bolts.

23. Install the steering member bolts.

24. Install both front door upper hinge bolts.

25. Release the jack from the front doors.

26. Install the steering member bolt caps.

27. Connect the evaporator drain hose.

28. Connect the following components:

- Air bag module control unit
- A/C drain hose
- M/T shift cable (if equipped)
- CVT shift cable (if equipped)
- Front pillar harness
- Door harness
- Fuse box
- SMJ
- Steering harness clips
- Shift lock cable

29. Install the lower steering member stays.

30. Install the instrument panel assembly.

31. Install the steering column.

➡ **Cap or wrap the joint of the pipe with suitable material such as vinyl tape to avoid entry of contaminants.**

32. Connect the refrigerant lines to the evaporator as shown in the accompanying illustrations.

➡ **Cap or wrap the joint of the pipe with suitable material such as vinyl tape to avoid entry of contaminants.**

33. Connect the heater hoses to the heater core pipes.

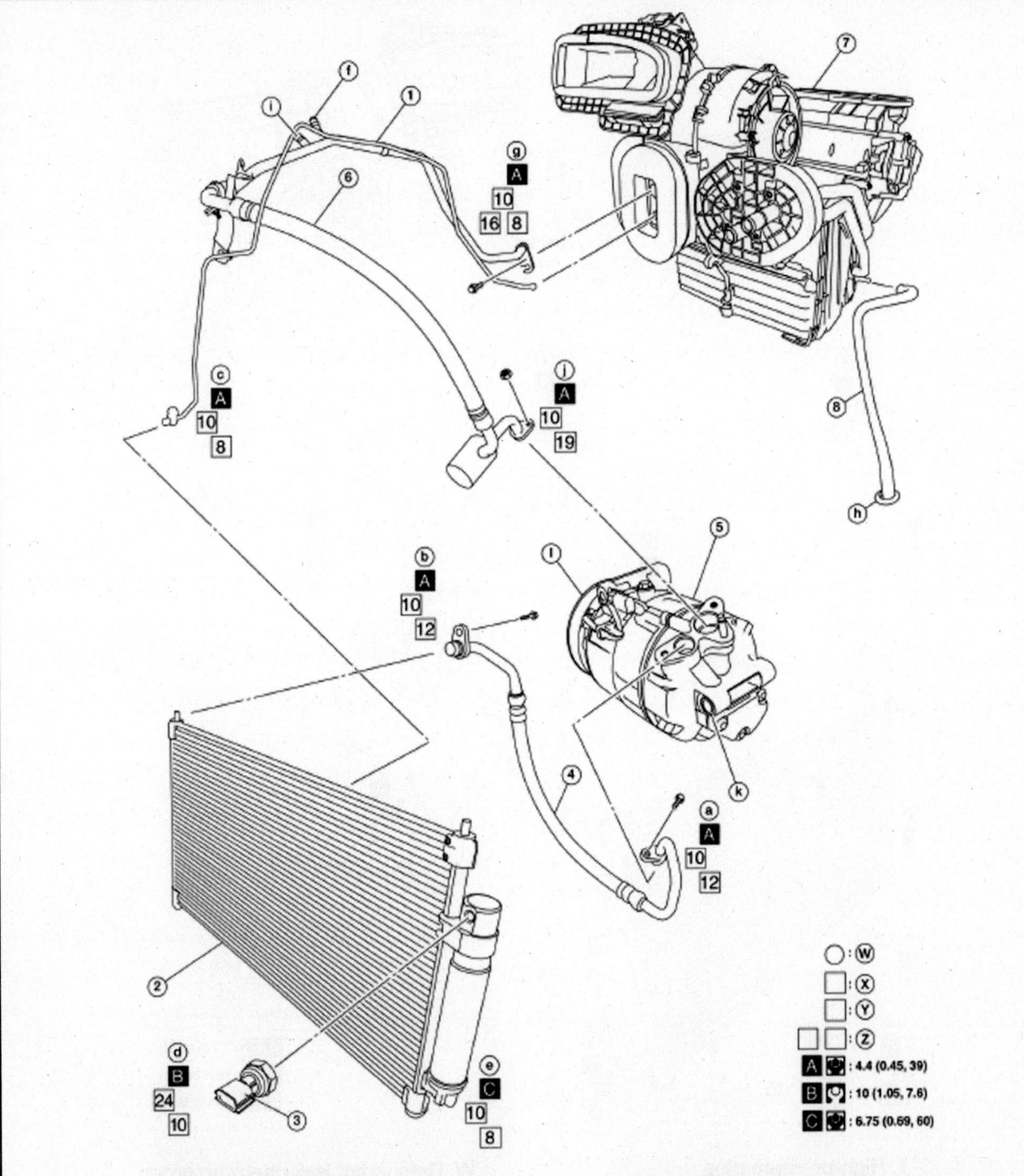

1. High-pressure pipe
2. Condenser and liquid tank assembly
3. Refrigerant pressure sensor
4. High-pressure flexible hose
5. Compressor
6. Low-pressure flexible hose
7. Heater and cooling unit assembly
8. Drain hose

W. Refrigerant leak checking order
X. Tightening torque (A-C)
Y. Wrench size
Z. O-ring size

71075_SENT_G0316

Fig. 161 Exploded view of the refrigerant lines—2.0L engine

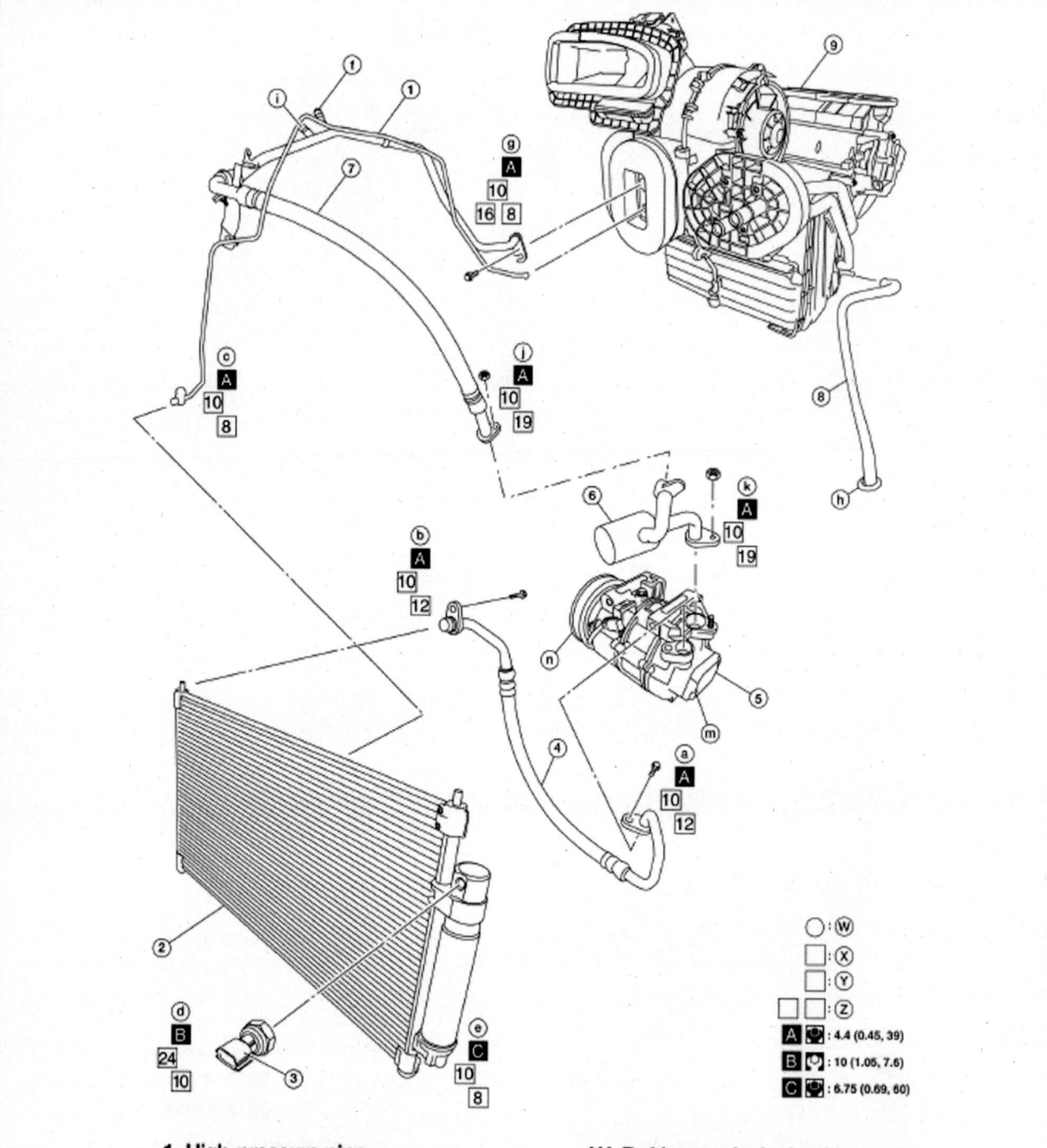

1. High-pressure pipe
2. Condenser and liquid tank assembly
3. Refrigerant pressure sensor
4. High-pressure flexible hose
5. Compressor
6. Muffler pipe
7. Low-pressure flexible hose
8. Drain hose
9. Heater and cooling unit assembly

W. Refrigerant leak checking order
X. Tightening torque (A-C)
Y. Wrench size
Z. O-ring size

71075_SENT_G0317

Fig. 162 Exploded view of the refrigerant lines—2.5L engine

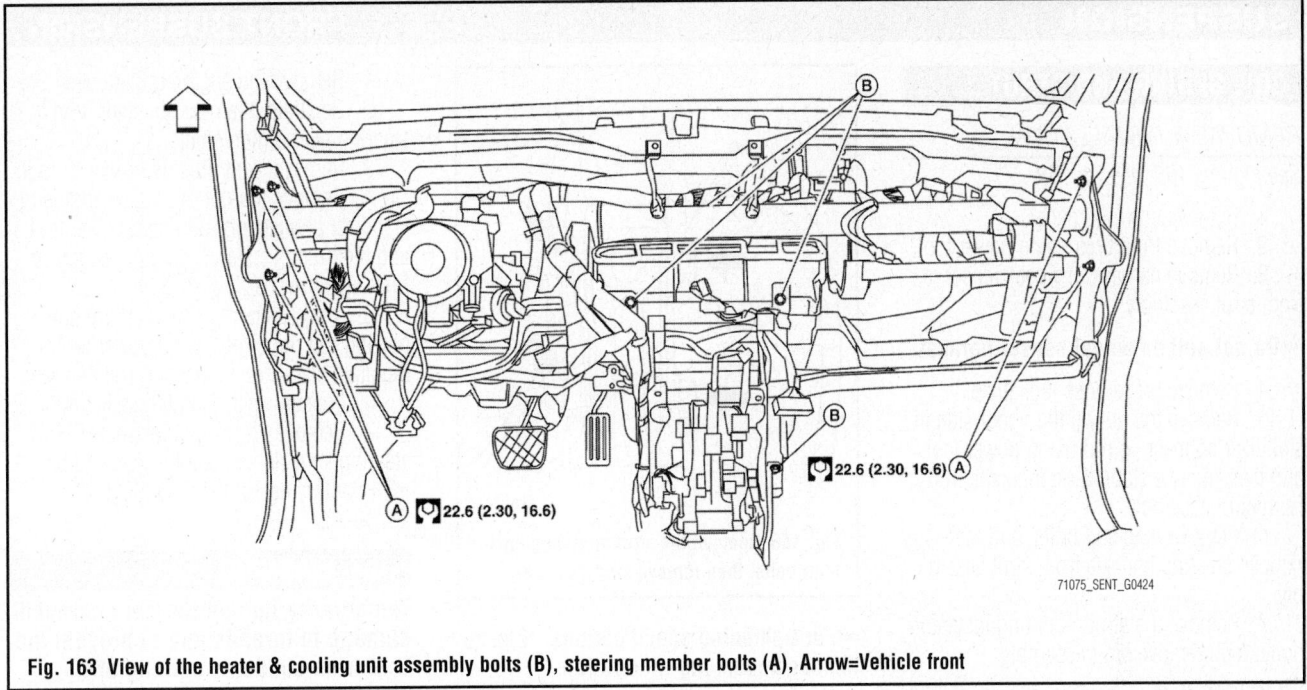

Fig. 163 View of the heater & cooling unit assembly bolts (B), steering member bolts (A), Arrow=Vehicle front

34. Reposition the lower dash insulator.

35. Fill the engine coolant to the cooling system.

36. Charge the refrigerant to the A/C system.

➥ Fill the radiator with the specified water and coolant mixture.

➥ Recharge the A/C system.

➥ Check and adjust the front door alignment as necessary.

STEERING

POWER RACK & PINION STEERING GEAR

REMOVAL & INSTALLATION

See Figure 164.

✳✳ WARNING

Spiral cable may be cut if steering wheel turns while separating steering column assembly and steering gear assembly. Be sure to secure steering wheel using string to avoid turning.

1. Set vehicle to the straight-ahead position.
2. Remove front suspension member.
3. Remove the steering gear assembly nuts and bolts.
4. Remove the steering gear assembly from the suspension member.

To install:

➥ Clean mating surface on the body side of dash panel seal when installing steering gear assembly.

➥ Perform final tightening of nuts and bolts on each part under unladen conditions with tires on level ground when

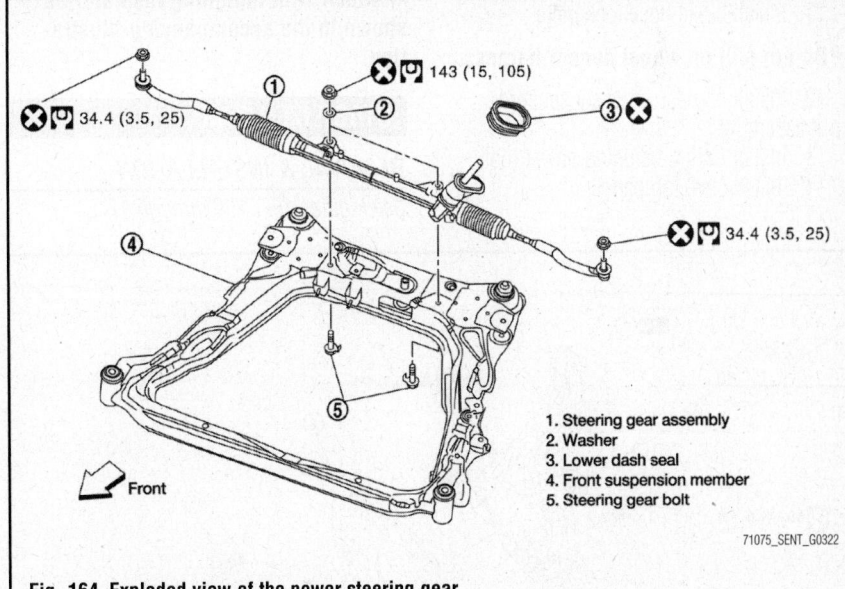

Fig. 164 Exploded view of the power steering gear

1. Steering gear assembly
2. Washer
3. Lower dash seal
4. Front suspension member
5. Steering gear bolt

removing steering gear assembly. Check wheel alignment.

5. Install the steering gear assembly from the suspension member.
6. Install the steering gear assembly nuts and bolts.
7. Install front suspension member.

8. Set vehicle to the straight-ahead position.

INSPECTION AFTER INSTALLATION

➥ Rotate steering wheel to check for decentered condition, binding, noise or excessive steering effort.

➥ Check and adjust toe as necessary.

SUSPENSION

FRONT SUSPENSION

COIL SPRING AND STRUT

REMOVAL & INSTALLATION

See Figures 165 through 168.

1. Remove cowl top panel.
2. Remove front tires using power tool.
3. Remove harness of wheel sensor from strut assembly.

➥**Do not pull on wheel sensor harness.**

4. Remove brake hose lock plate.
5. Remove the nut on the upper side of stabilizer connecting rod using power tool, and then remove stabilizer connecting rod from strut assembly.
6. Remove nuts and bolts, and then remove steering knuckle from strut assembly.
7. Remove the strut mounting insulator bolts, then remove strut assembly.

To install:

8. Install the strut mounting insulator bolts, then install strut assembly.
9. Install nuts and bolts, and then install steering knuckle to strut assembly.
10. Install the nut on the upper side of stabilizer connecting rod using power tool, and then install stabilizer connecting rod to strut assembly.
11. Install brake hose lock plate.

➥**Do not pull on wheel sensor harness.**

12. Install harness of wheel sensor to strut assembly.
13. Install front tires using power tool.
14. Install cowl top panel.

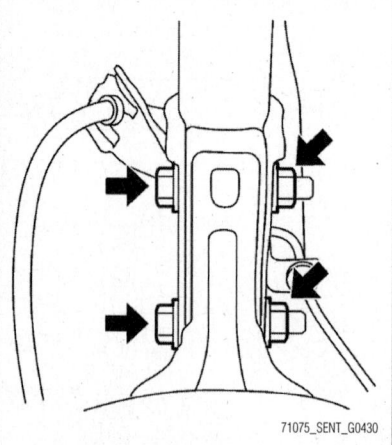

Fig. 166 Remove the strut mounting insulator bolts, then remove strut assembly.

➥**For tightening specifications, refer to the accompanying illustration.**

➥**Perform final tightening of bolts and nuts at the strut assembly lower side (rubber bushing) under unladen conditions with tires on level ground. Check wheel alignment.**

➥**Check wheel sensor harness for proper connection.**

➥**Attach strut mounting insulator as shown in the accompanying illustration.**

STABILIZER BAR

REMOVAL & INSTALLATION

See Figures 167, 169 through 171.

1. Remove the engine under cover.
2. Separate intermediate shaft from steering gear pinion shaft.
3. Remove front tires using power tool.
4. Remove the nut on the lower side of stabilizer connecting rod using power tool, and then remove stabilizer connecting rod from stabilizer bar.
5. If necessary remove stabilize connecting rod upper nut using power tool. Separate stabilizer connecting rod and strut.
6. Loosen steering outer socket nut.
7. Remove steering outer socket from steering knuckle so as not to damage ball joint boot using Tool HT72520000 (J-25730-A).

✳✳ WARNING

Temporarily tighten the nut to prevent damage to threads and to prevent the Tool from suddenly coming off.

8. Remove the heated oxygen sensor bracket from the suspension member.
9. Remove rear torque rod.
10. Set Tool under the front suspension member, then remove the bolts from the front suspension member using power tool.
11. Remove the bolts of member stay using power tool, and then remove member stay.
12. Gradually lower front suspension member in order to remove stabilizer bolts.
13. Remove the stabilizer clamp bolts, then remove stabilizer clamps and stabilizer bushing.
14. Remove stabilizer bar.
15. Inspection after removal:
    a. Check stabilizer bar, stabilizer connecting rod, stabilizer bushing, and stabilizer clamp for deformation, cracks, and damage. Replace it if necessary.

To install:

16. Install stabilizer bar.
17. Install the stabilizer clamp bolts, then install stabilizer clamps and stabilizer bushing.
18. Gradually lower front suspension member in order to install stabilizer bolts.
19. Install the bolts of member stay using power tool, and then install member stay.
20. Set Tool under the front suspension member, then install the bolts to the front suspension member using power tool.
21. Install rear torque rod.
22. Install the heated oxygen sensor bracket to the suspension member.

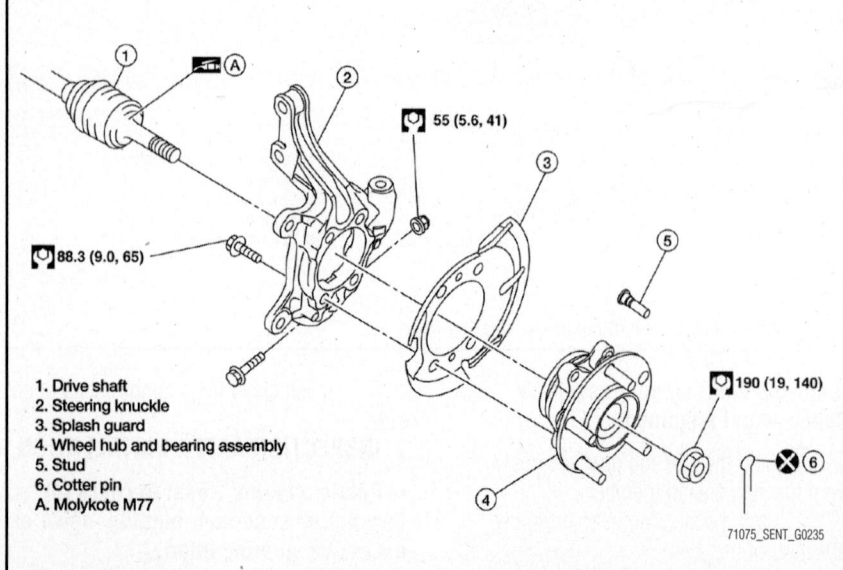

1. Drive shaft
2. Steering knuckle
3. Splash guard
4. Wheel hub and bearing assembly
5. Stud
6. Cotter pin
A. Molykote M77

55 (5.6, 41)
88.3 (9.0, 65)
190 (19, 140)

Fig. 165 Exploded view of the front wheel hub and knuckle assembly

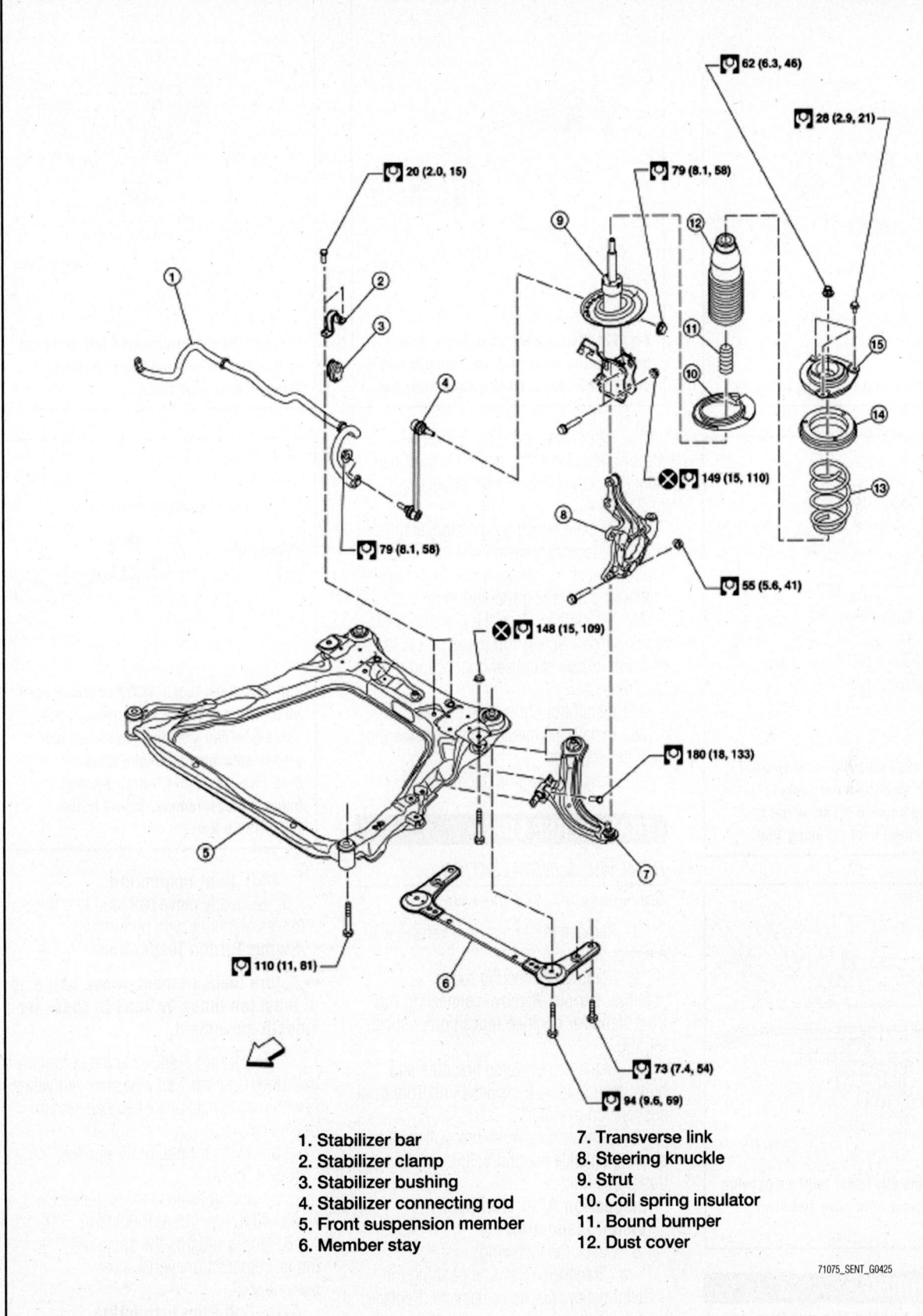

20 (2.0, 15)
62 (6.3, 46)
28 (2.9, 21)
79 (8.1, 58)
149 (15, 110)
55 (5.6, 41)
148 (15, 109)
180 (18, 133)
79 (8.1, 58)
110 (11, 81)
73 (7.4, 54)
94 (9.6, 59)

1. Stabilizer bar
2. Stabilizer clamp
3. Stabilizer bushing
4. Stabilizer connecting rod
5. Front suspension member
6. Member stay
7. Transverse link
8. Steering knuckle
9. Strut
10. Coil spring insulator
11. Bound bumper
12. Dust cover

71075_SENT_G0425

Fig. 167 Exploded view of the front suspension assembly

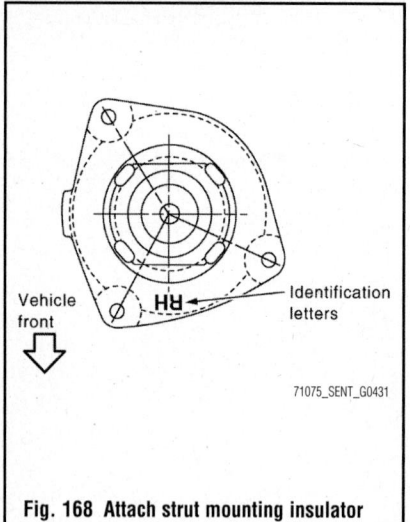

Fig. 168 Attach strut mounting insulator

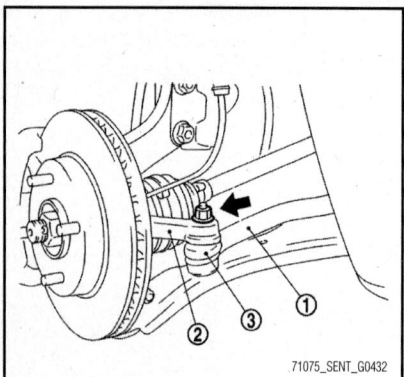

Fig. 169 Loosen steering outer socket (1) nut. Remove steering outer socket (1) from steering knuckle (2) so as not to damage ball joint boot (3) using Tool.

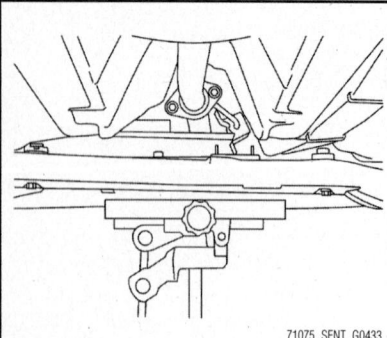

Fig. 170 Gradually lower front suspension member in order to remove stabilizer bolts.

✳✳ WARNING

Temporarily tighten the nut to prevent damage to threads and to prevent the Tool from suddenly coming off.

Fig. 171 Remove the stabilizer clamp bolts, then remove stabilizer clamps and stabilizer bushing. Remove stabilizer bar

23. Install steering outer socket to steering knuckle so as not to damage ball joint boot using Tool HT72520000 (J-25730-A).

24. Tighten steering outer socket nut.

25. If necessary install stabilize connecting rod upper nut using power tool. Engage stabilizer connecting rod and strut.

26. Install the nut on the lower side of stabilizer connecting rod using power tool, and then install stabilizer connecting rod to stabilizer bar.

27. Install front tires using power tool.

28. Install intermediate shaft to steering gear pinion shaft.

29. Install the engine under cover.

TRANSVERSE LINK

REMOVAL & INSTALLATION

See Figures 167, 172 and 173.

1. Remove front tires from vehicle with a power tool.

2. Remove connecting rod to stabilizer bar nut. Remove connecting rod from stabilizer bar then reposition connecting rod.

3. Remove transverse link nuts and bolts, then remove transverse link from front suspension member.

4. Remove transverse link ball joint to steering knuckle nut and bolt, then remove transverse link.

Inspection After Removal
Visual Inspection

5. Check the following:

   a. Transverse link and bushing for deformation, cracks or damage. Replace it if necessary.

   b. Ball joint boot for cracks or other damage, and also for grease leakage. Replace it if necessary.

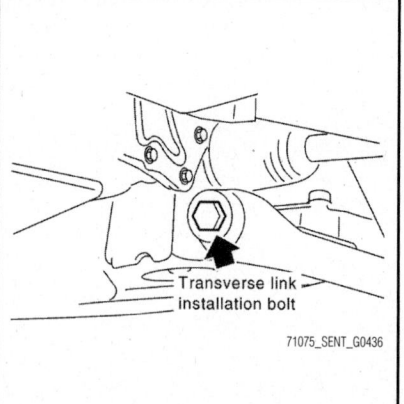

Fig. 172 Remove transverse link ball joint to steering knuckle nut and bolt, then remove transverse link.

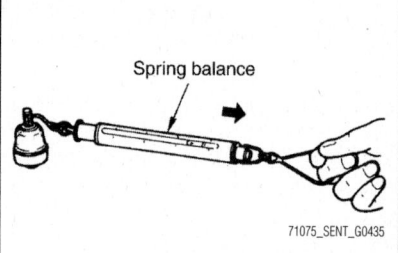

Fig. 173 Hook Tool J-44372 at the cutout on ball stud. Confirm Tool measurement value is within specifications when ball stud begins moving. Swing torque: 5-43 inch lbs. (0.5-4.9 Nm). Spring balance measurement: 3.5-40 ft. lbs. (15.4-150.8 Nm).

Ball Joint Inspection

6. Manually move ball stud to confirm it moves smoothly with no binding.

Swing Torque Inspection

➡**Before measurement, move ball stud at least ten times by hand to check for smooth movement.**

7. Hook Tool J-44372 at the cutout on ball stud. Confirm Tool measurement value is within specifications when ball stud begins moving.

   a. Swing torque: 5-43 inch lbs. (0.5-4.9 Nm).

   b. Spring balance measurement: 3.5-40 ft. lbs. (15.4-150.8 Nm).

8. If it is outside the specified range, replace transverse link assembly.

Axial End Play Inspection

9. Move tip of ball stud in axial direction to check for looseness.

10. If it is outside the specified range,

replace transverse link assembly. Axial end play 0 in (0 mm).

To install:

11. Install transverse link ball joint to steering knuckle nut and bolt, then install transverse link.

12. Install transverse link nuts and bolts, then install transverse link to front suspension member.

13. Install connecting rod to stabilizer bar nut. Install connecting rod to stabilizer bar then reposition connecting rod.

14. Install front tires to vehicle with a power tool.

➡ **Perform final tightening of bolts and at the front suspension member installation position (rubber bushing) under unladen conditions with tires on level ground. Check wheel alignment.**

SUSPENSION

COIL SPRINGS

REMOVAL & INSTALLATION

See Figures 174 through 176.

1. Set jack under rear suspension beam.

2. Remove both lower shock absorber bolts.

3. Carefully lower the suspension and remove the spring.

Inspection After Removal

4. Check coil spring and spring rubber seat for deformation, cracks, and damage, and replace it if a malfunction is detected.

To install:

5. When installing rear spring rubber seat, be sure that the flat areas are aligned.

6. When installing rear spring, be sure that the gap is less than 0.20 in (5 mm).

7. Carefully lower the suspension and install the spring.

8. Install both lower shock absorber bolts.

9. Remove jack under rear suspension beam.

10. Torque upper shock absorber bolts to 50 ft. lbs. (88 Nm) and lower bolts to 92 ft. lbs. (125 Nm).

REAR SUSPENSION BEAM

REMOVAL & INSTALLATION

See Figures 174, 177 through 179.

1. Remove rear tires from vehicle using power tool.

2. On non-SE-R vehicles, remove rear drum and brake assembly.

3. On SE-R vehicles, remove the rear disc and brake caliper assembly. Then hang brake caliper using wire.

4. Separate parking brake rear cable from rear brake and rear suspension beam.

WHEEL HUB & BEARING

REMOVAL & INSTALLATION

1. Remove the steering knuckle.

2. Remove the bolts to remove the hub and bearing assembly from the steering knuckle.

3. Installation is the reverse of removal.
   • Torque the hub assembly bolts to 65 ft. lbs. (88 Nm).

REAR SUSPENSION

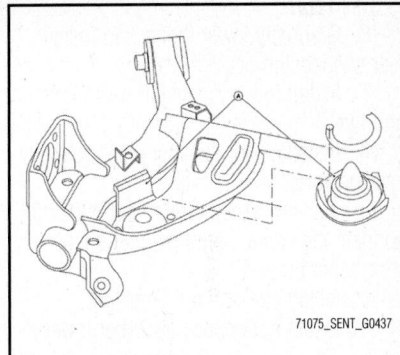
71075_SENT_G0437

Fig. 175 When installing rear spring rubber seat, be sure that the flat areas (A) are aligned.

5. Remove wheel sensor and wheel sensor harness from wheel hub and bearing assembly and rear suspension beam, if equipped.

6. On non-SE-R vehicles, remove rear brake tube from the wheel cylinder.

7. Remove wheel hub and bearing assembly and back plate.

8. Set jack under rear suspension beam.

9. Remove both shock absorber lower bolts.

10. Remove coil springs (left/right).

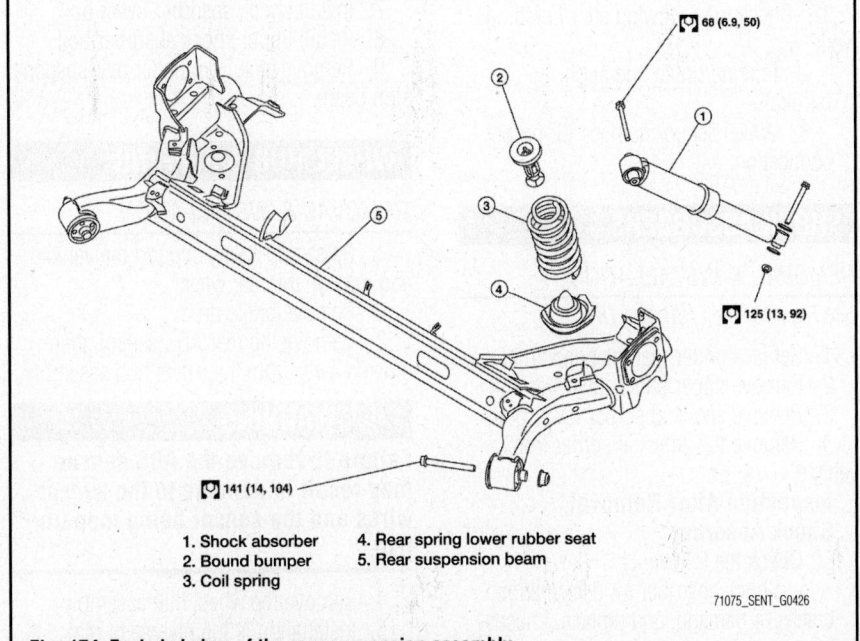

68 (6.9, 50)

125 (13, 92)

141 (14, 104)

1. Shock absorber
2. Bound bumper
3. Coil spring
4. Rear spring lower rubber seat
5. Rear suspension beam

71075_SENT_G0426

Fig. 174 Exploded view of the rear suspension assembly

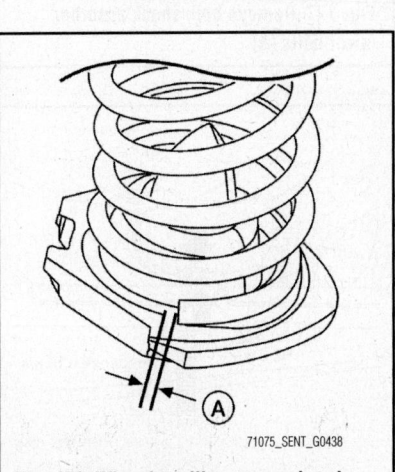
71075_SENT_G0438

Fig. 176 When installing rear spring, be sure that the gap (A) is less than 5mm (0.20 in)

11. Remove brake line retaining clip and disconnect the brake line from the rear suspension beam bracket.

12. Remove center exhaust pipe assembly and insulator.

13. Remove rear suspension beam bolt and nut.

14. Gradually lower the jack to remove rear suspension beam from vehicle.

Inspection After Removal

15. Check components for deformation, cracks, and other damage, and replace if necessary.

To install:

16. Gradually lower the jack to install rear suspension beam to vehicle.

17. Install rear suspension beam bolt and nut.

18. Install center exhaust pipe assembly and insulator.

19. Install brake line retaining clip and connect the brake line to the rear suspension beam bracket.

20. Install coil springs (left/right).

21. Install both shock absorber lower bolts.

22. Set jack under rear suspension beam.

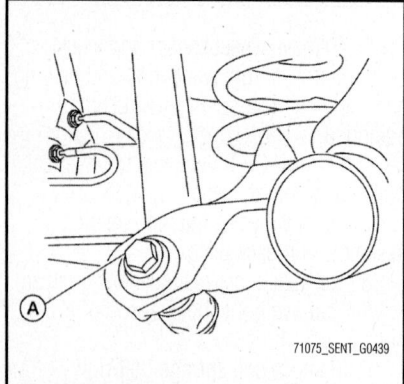

Fig. 177 Remove both shock absorber lower bolts (A)

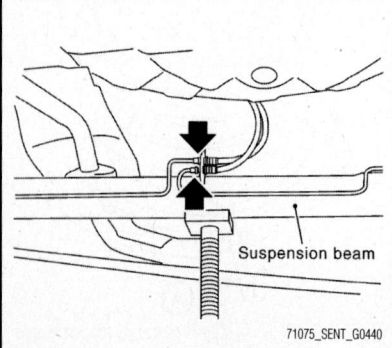

Fig. 178 Gradually lower the jack to remove rear suspension beam from vehicle

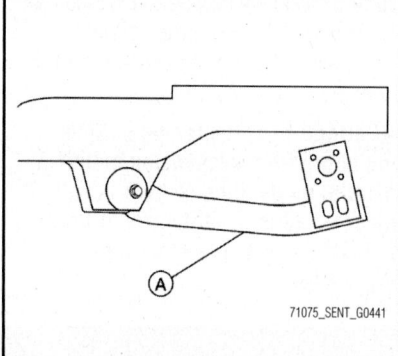

Fig. 179 Perform final tightening of rear suspension beam (A) under unladen conditions with tires on level ground.

23. Install wheel hub and bearing assembly and back plate.

24. On non-SE-R vehicles, install rear brake tube to the wheel cylinder.

25. Install wheel sensor and wheel sensor harness to wheel hub and bearing assembly and rear suspension beam, if equipped.

26. Install parking brake rear cable to rear brake and rear suspension beam.

27. On SE-R vehicles, install the rear disc and brake caliper assembly. Then release brake caliper from wire.

28. On non-SE-R vehicles, install rear drum and brake assembly.

29. Install rear tires to vehicle using power tool.

30. Perform final tightening of rear suspension beam (A) under unladen conditions with tires on level ground.

31. Refill with new brake fluid and bleed air.

32. Check the following after finishing work:

   a. Parking brake operation (stroke):

   b. Wheel sensor harness for proper connection:

SHOCK ABSORBERS

REMOVAL & INSTALLATION

See Figures 174, 180 and 181.

1. Set jack under rear suspension beam.

2. Remove upper shock absorber bolt.

3. Remove shock absorber lower bolt.

4. Remove the shock absorber from vehicle.

Inspection After Removal
Shock Absorber

5. Check the following:

   a. Shock absorber for deformation, cracks or damage, and replace if necessary.

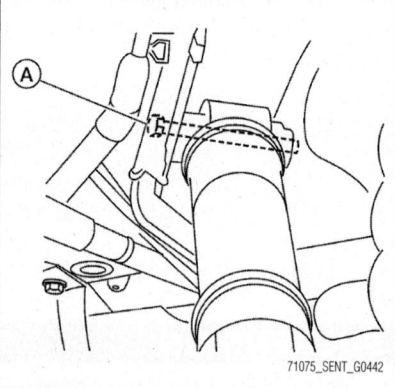

Fig. 180 Remove upper shock absorber bolt (A).

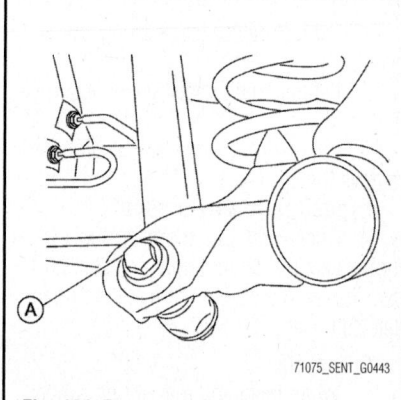

Fig. 181 Remove shock absorber lower bolt (A)

   b. Piston rod for damage, uneven wear or distortion, and replace if necessary.

To install:

6. Install the shock absorber to vehicle.

7. Install shock absorber lower bolt.

8. Install upper shock absorber bolt.

9. Remove jack from under rear suspension beam.

WHEEL HUB & BEARING

REMOVAL & INSTALLATION

1. Raise and safely support the vehicle and remove the rear wheel.

2. Remove brake drum.

3. Remove the rear ABS sensor, then move it away from the wheel hub assembly.

❊❊ WARNING

Failure to remove the ABS sensor may result in damage to the sensor wires and the sensor being inoperative.

4. Remove the wheel hub assembly.

5. Installation is the reverse of removal. Torque the hub bolts to 65 ft. lbs. (88 Nm).

NISSAN

Titan

20

SPECIFICATIONS AND MAINTENANCE CHARTS

ENGINE AND VEHICLE IDENTIFICATION

Engine								Model Year	
Code ①	Liters (cc)	Cu. In.	Cyl.	Fuel Sys.	Engine	Eng. Mfg.		Code ②	Year
VK56DE	5.6 (5552)	338.8	8	MFI	DOHC	Nissan		B	2011
								C	2012

MFI: Multi-port Fuel Injection

DOHC: Double Overhead Camshafts

① Engine VIN: A. Engine VIN: B (FFV= flex fuel vehicle).

② 10th digit of the Vehicle Identification Number (VIN)

71075_TITA_C0001

GENERAL ENGINE SPECIFICATIONS

Year	Model	Engine Displacement Liters	Engine ID	Net Horsepower @ rpm	Net Torque @ rpm (ft. lbs.)	Bore x Stroke (in.)	Com-pression Ratio	Oil Pressure @ rpm
2011	Titan	5.6	VK56DE	317@5200	385@3400	3.86X3.62	9.8:1	43@2000
2012	Titan	5.6	VK56DE	317@5200	385@3400	3.86X3.62	9.8:1	43@2000

71075_TITA_C0002

ENGINE TUNE-UP SPECIFICATIONS

Year	Engine Displacement Liters	Engine ID	Spark Plug Gap (in.)	Ignition Timing	Fuel Pump (psi) ①	Idle Speed	Valve Clearance (in.) In.	Valve Clearance (in.) Ex.
2011	5.6	VK56DE	0.043	②	51	600-700	0.010-0.013	0.011-0.015
2012	5.6	VK56DE	0.043	②	51	600-700	0.010-0.013	0.011-0.015

NOTE: The Vehicle Emission Control Information label often reflects specification changes made during production. The label figures

must be used if they differ from those in this chart.

① Approximate at idle

② 15 degrees +/- 5 degrees BTDC

71075_TITA_C0003

CAPACITIES

Year	Model	Engine Displacement Liters	Engine ID	Engine Oil with Filter (qts.)	Transmission (pts.)	Transfer Case (pts.)	Drive Axle Front (pts.)	Rear (pts.)	Fuel Tank (gal.)	Cooling System (qts.)
2011	Titan	5.6	VK56DE	6.75	22.50	4.36	3.380	4.250	28.0	12.75
2012	Titan	5.6	VK56DE	6.75	22.50	4.36	3.380	4.250	28.0	12.75

NOTE: All capacities are approximate. Add fluid gradually and check to be sure a proper fluid level is obtained.

71075_TITA_C0004

FLUID SPECIFICATIONS

Year	Model	Engine Displ. Liters	Engine Oil	Auto. Trans.	Drive Axle Front	Rear	Transfer Case	Power Steering Fluid	Brake Master Cylinder	Cooling System
2011	Titan	5.6	SAE 5W-30	①	②	②	③	④	⑤	⑥
2012	Titan	5.6	SAE 5W-30	①	②	②	③	④	⑤	⑥

① Genuine Nissan Matic S ATF fluid. If not available genuine Nissan Matic J ATF may be used.

② Front: Nissan differential oil hypoid super GL-5 80W-90 or API GL-5 viscosity SAE 80W-90 gear oil.

   Rear: Nissan differential oil API GL-5 synthetic 75W-90 gear oil.

③ Nissan Matic D ATF

④ Nissan Power Steering Fluid

⑤ Nissan Super Heavy Duty DOT 3

⑥ Nissan Long Life antifreeze (BLUE) or equivalent

71075_TITA_C0013

VALVE SPECIFICATIONS

Year	Engine Displacement Liters	Engine ID	Seat Angle (deg.)	Face Angle (deg.)	Spring Test Pressure (lbs. @ in.)	Spring Installed Height (in.)	Stem-to-Guide Clearance (in.) Intake	Exhaust	Stem Diameter (in.) Intake	Exhaust
2011	5.6	VK56DE	45.15-45.45	45	37.0@1.457	1.9913	0.0008-0.0021	0.0012-0.0025	0.2348-0.2354	0.2344-0.2350
2012	5.6	VK56DE	45.15-45.45	45	37.0@1.457	1.9913	0.0008-0.0021	0.0012-0.0025	0.2348-0.2354	0.2344-0.2350

71075_TITA_C0006

CAMSHAFT SPECIFICATIONS

All measurements are given in inches.

Year	Engine Displ. Liters	Engine ID/VIN	Journal Dia.	Brg. Oil Clearance	Shaft End-play	Runout	Journal Bore	Lobe Height Intake	Exhaust
2011	5.6	VK56DE	1.0217- 1.0224	0.0012- 0.0028	0.0045- 0.0074	0.0008	1.0236- 1.0244	1.7663- 1.7738	1.7746- 1.7821
2012	5.6	VK56DE	1.0278- 1.0224	0.0012- 0.0028	0.0045- 0.0074	0.0008	1.0236- 1.0244	1.7663- 1.7738	1.7746- 1.7821

71075_TITA_C0007

CRANKSHAFT AND CONNECTING ROD SPECIFICATIONS

All measurements are given in inches.

Year	Engine Displ. Liters	Engine ID	Crankshaft Main Brg. Journal Dia.	Main Brg. Oil Clearance	Shaft End-play	Thrust on No.	Connecting Rod Journal Diameter	Oil Clearance	Side Clearance
2011	5.6	VK56DE	①	②	0.0039- 0.0102	3	③	0.0008- 0.0015	0.0079- 0.0157
2012	5.6	VK56DE	①	②	0.0039- 0.0102	3	③	0.0008- 0.0015	0.0079- 0.0157

① There are 24 different grades, ranging from 2.5182- 2.5174

② No. 1 and 5: 0.00004- 0.0004

   No. 2, 3 and 4: 0.0003- 0.0007

③ There are 13 different grades, ranging from 2.2441- 2.2446. Specification is for rod bearing housing.

71075_TITA_C0005

PISTON AND RING SPECIFICATIONS

All measurements are given in inches.

Year	Engine Displacement Liters	Engine ID	Piston Clearance	Ring Gap Top Comp.	Bottom Comp.	Oil Control	Ring Side Clearance Top Comp.	Bottom Comp.	Oil Control
2011	5.6	VK56DE	0.0004- 0.0012	0.0091- 0.0130	0.0098- 0.0157	0.0079- 0.0236	0.0014- 0.0033	0.0012- 0.0028	0.0006- 0.0073
2012	5.6	VK56DE	0.0004- 0.0012	0.0091- 0.0130	0.0098- 0.0157	0.0079- 0.0236	0.0014- 0.0033	0.0012- 0.0028	0.0006- 0.0073

71075_TITA_C0008

TORQUE SPECIFICATIONS
All readings in ft. lbs.

Year	Engine Displacement Liters	Engine ID	Cylinder Head Bolts	Main Bearing Bolts	Rod Bearing Bolts	Crankshaft Damper Bolts	Flywheel Bolts	Manifold		Spark Plugs	Oil Pan Drain Plug
								Intake	Exhaust		
2011	5.6	VK56DE	①	②	③	④	65	NA	25	18	25
2012	5.6	VK56DE	①	②	③	④	65	NA	25	18	25

NA: not available

① Step 1: 33 ft. lbs
Step 2: +70 degrees clockwise
Step 3: loosen in reverse order of tightening sequence
Step 4: 33 ft. lbs.
Step 5: +60 degrees clockwise
Step 6: +60 degrees clockwise

② Step 1: cap bolts in order 1-10: 29 ft. lbs.
Step 2: cap sub bolts in order 11-20: 22 ft. lbs.
Step 3: cap bolts in order 1-10: +40 degrees
Step 4: cap sub bolts in order 11-20: +30 degrees
Step 5: side bolts in order 21-30: 36 ft. lbs.

③ Step 1: 11 ft. lbs.
Step 2: +90 degrees clockwise

④ Step 1: 69 ft. lbs.
Step 2: +90 degrees

71075_TITA_C0009

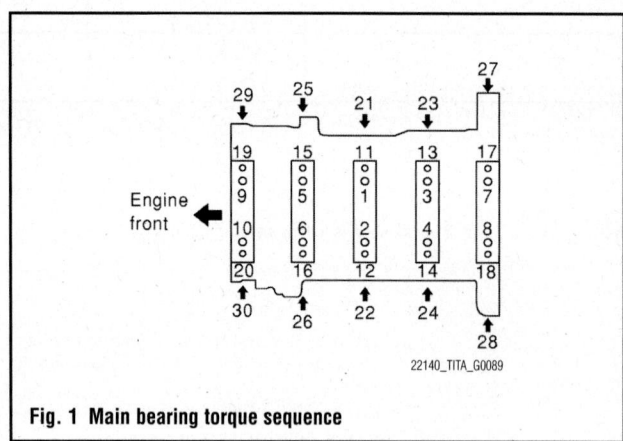

22140_TITA_G0089

Fig. 1 Main bearing torque sequence

WHEEL ALIGNMENT

			Caster		Camber		
Year	Model		Range (+/-Deg.)	Preferred Setting (Deg.)	Range (+/-Deg.)	Preferred Setting (Deg.)	Toe-in (in.)
2011	Titan	2WD	①	①	②	②	NA
		4WD	③	③	④	②	NA
2012	Titan	2WD	①	①	②	②	NA
		4WD	③	③	④	②	NA

NA - Not Available

① Minimum: 2 degrees 15' (2.25 degrees). Nominal: 3 degrees 0' (3.00 degrees). Maximum: 3 degrees 45' (3.75 degrees).

② Minimum: -0 degrees 57' (-0.95 degrees). Nominal: 0 degrees 12' (0.20 degrees). Maximum: 0 degree 33' (0.55 degree).

③ Minimum: 1 degrees 27' (1.45 degrees). Nominal: 2 degrees 12' (2.20 degrees). Maximum: 2 degrees 57' (2.95 degrees).

④ Minimum: -0 degrees 27' (-0.45 degrees). Nominal: 0 degrees 18' (0.30 degrees). Maximum: 1 degree 03' (1.05 degrees).

71075_TITA_C0010

TIRE, WHEEL AND BALL JOINT SPECIFICATIONS

		OEM Tires		Tire Pressures (psi)		Wheel Size	Ball Joint Inspection	Lug Nut Torque (ft. lbs.)
Year	Model	Standard	Optional	Front	Rear			
2011	Titan	①	None	35	35	②	③	98
2012	Titan	①	None	35	35	②	③	98

OEM: Original Equipment Manufacturer

PSI: Pounds Per Square Inch

① P265/70R18, P275/70R18 or P275/60R20

② 18x8JJ or 20x8JJ

③ Axial play

  Upper: 0

71075_TITA_C0011

BRAKE SPECIFICATIONS

All measurements in inches unless noted

			Brake Disc			Minimum Pad Thickness	Brake Caliper	
			Original Thickness	Minimum Thickness	Maximum Runout		Bracket Bolts (ft. lbs.)	Mounting Bolts (ft. lbs.)
Year	Model							
2011	Titan	F	1.181	1.102	①	0.039	②	②
		R	0.551	0.472	①	0.039	③	③
2012	Titan	F	1.181	1.102	①	0.039	②	②
		R	0.551	0.472	①	0.039	③	③

① Maximum uneven wear measured at 8 positions: 0.0006. Runout limit, attached to vehicle: Front: 0.001, Rear: 0.002.

② Torque member mounting bolt: 155

  Sliding pin bolt 53

③ Sliding pin bolt: 24

71075_TITA_C0012

SCHEDULED MAINTENANCE INTERVALS
Nissan Titan

TO BE SERVICED	TYPE OF SERVICE	7.5	15	22.5	30	37.5	45	52.5	60
Engine oil & filter	R	✓	✓	✓	✓	✓	✓	✓	✓
Brake lines & cables	S/I		✓		✓		✓		✓
Brake pads and rotors	I	✓	✓	✓	✓	✓	✓	✓	✓
Driveshaft boots & propeller shaft (4x4)	L/I		✓		✓		✓		✓
Transmission, transfer & differential gear oil	I		✓		✓		✓		✓
Air cleaner filter	R								✓
Engine coolant	R	Replace after 60,000miles. Inspect every 15,000 miles							
Spark plugs (Platinum)	R	Replace every 105,000 miles							
Drive belt(s)	S/I	Replace after 60,000miles. Inspect every 15,000 miles							
Cabin air filter	R		✓		✓		✓		✓
Exhaust system	I	✓	✓	✓	✓	✓	✓	✓	✓
Fuel lines	S/I				✓				✓
Fuel filter		Maintenance free item							
Steering gear (box) & linkage, axle & suspension parts	I				✓				✓
Tire Rotation	S	✓	✓	✓	✓	✓	✓	✓	✓
Vapor lines	S/I				✓				✓

R: Replace   S/I: Service or Inspect   L: Lubricate

FREQUENT OPERATION MAINTENANCE (SEVERE SERVICE)

If a vehicle is operated under any of the following conditions it is considered severe service:

- Extremely dusty areas.
- Rough, muddy, or salt spread roads.
- 50% or more of the vehicle constant operation is in 32°C (90°F) or higher temperatures, or temperatures below 0°C (32°F).
- Prolonged idling (vehicle operation in stop and go traffic).
- Frequent short running periods (engine does not warm to normal operating temperatures).
- Police, taxi, delivery usage or trailer towing usage.

Oil & oil filter: replace every 3750 miles.

Brake pads, discs, drums & linings: service or inspect every 7500 miles.

Driveshaft boots & propeller shaft: service or inspect every 7500 miles.

Exhaust system: service or inspect every 7500 miles.

Steering gear (box) & linkage, (steering damper-4x4), axle & suspension parts: service or inspect every 7500 miles.

Steering linkage ball joints & front suspension ball joints: service or inspect every 7500 miles.

Transfer case fluid, transmission fluid and differential gear oil: Change every 30000 miles if towing a trailer.

71075_TITA_C0014

PRECAUTIONS

Before servicing any vehicle, please be sure to read all of the following precautions, which deal with personal safety, prevention of component damage, and important points to take into consideration when servicing a motor vehicle:

• Never open, service or drain the radiator or cooling system when the engine is hot; serious burns can occur from the steam and hot coolant.

• Observe all applicable safety precautions when working around fuel. Whenever servicing the fuel system, always work in a well-ventilated area. Do not allow fuel spray or vapors to come in contact with a spark, open flame, or excessive heat (a hot drop light, for example). Keep a dry chemical fire extinguisher near the work area. Always keep fuel in a container specifically designed for fuel storage; also, always properly seal fuel containers to avoid the possibility of fire or explosion. Refer to the additional fuel system precautions later in this section.

• Fuel injection systems often remain pressurized, even after the engine has been turned **OFF**. The fuel system pressure must be relieved before disconnecting any fuel lines. Failure to do so may result in fire and/or personal injury.

• Brake fluid often contains polyglycol ethers and polyglycols. Avoid contact with the eyes and wash your hands thoroughly after handling brake fluid. If you do get brake fluid in your eyes, flush your eyes with clean, running water for 15 minutes. If eye irritation persists, or if you have taken brake fluid internally, IMMEDIATELY seek medical assistance.

• The EPA warns that prolonged contact with used engine oil may cause a number of skin disorders, including cancer. You should make every effort to minimize your exposure to used engine oil. Protective gloves should be worn when changing oil. Wash your hands and any other exposed skin areas as soon as possible after exposure to used engine oil. Soap and water, or waterless hand cleaner should be used.

• All new vehicles are now equipped with an air bag system, often referred to as a Supplemental Restraint System (SRS) or Supplemental Inflatable Restraint (SIR) system. The system must be disabled before performing service on or around system components, steering column, instrument panel components, wiring and sensors. Failure to follow safety and disabling procedures could result in accidental air bag deployment, possible personal injury and unnecessary system repairs.

• Always wear safety goggles when working with, or around, the air bag system. When carrying a non-deployed air bag, be sure the bag and trim cover are pointed away from your body. When placing a non-deployed air bag on a work surface, always face the bag and trim cover upward, away from the surface. This will reduce the motion of the module if it is accidentally deployed. Refer to the additional air bag system precautions later in this section.

• Clean, high quality brake fluid from a sealed container is essential to the safe and proper operation of the brake system. You should always buy the correct type of brake fluid for your vehicle. If the brake fluid becomes contaminated, completely flush the system with new fluid. Never reuse any brake fluid. Any brake fluid that is removed from the system should be discarded. Also, do not allow any brake fluid to come in contact with a painted surface; it will damage the paint.

• Never operate the engine without the proper amount and type of engine oil; doing so WILL result in severe engine damage.

• Timing belt maintenance is extremely important. Many models utilize an interference-type, non-freewheeling engine. If the timing belt breaks, the valves in the cylinder head may strike the pistons, causing potentially serious (also time-consuming and expensive) engine damage. Refer to the maintenance interval charts for the recommended replacement interval for the timing belt, and to the timing belt section for belt replacement and inspection.

• Disconnecting the negative battery cable on some vehicles may interfere with the functions of the on-board computer system(s) and may require the computer to undergo a relearning process once the negative battery cable is reconnected.

• When servicing drum brakes, only disassemble and assemble one side at a time, leaving the remaining side intact for reference.

• Only an MVAC-trained, EPA-certified automotive technician should service the air conditioning system or its components.

BRAKES

GENERAL INFORMATION

PRECAUTIONS

• Recommended fluid is Genuine NISSAN Super Heavy Duty Brake Fluid or equivalent.

• Do not reuse drained brake fluid.

• Be careful not to splash brake fluid on painted areas.

• To clean or wash all parts of master cylinder, disc brake caliper and wheel cylinder, use clean brake fluid.

• Do not use mineral oils such as gasoline or kerosene. They will ruin rubber parts of the hydraulic system.

• Use flare nut wrench when removing and installing brake tube.

• Always check tightening torque when installing brake lines.

ANTI-LOCK BRAKE SYSTEM (ABS)

• Before working, turn ignition switch to OFF and disconnect connectors for ABS actuator and electric unit (control unit) or battery terminals.

• Burnish the brake contact surfaces after refinishing or replacing drums or rotors, after replacing pads or linings, or if a soft pedal occurs at very low mileage.

• Clean brake pads and shoes with a waste cloth, then wipe with a dust collector.

• During ABS operation, the brake pedal may vibrate lightly and a mechanical noise may be heard. This is normal.

• Just after starting vehicle, the brake pedal may vibrate or a motor operating noise may be heard from engine compartment. This is a normal status of operation check.

• Stopping distance may be longer than that of vehicles without ABS when vehicle drives on rough, gravel, or snow-covered (fresh, deep snow) roads.

• When an error is indicated by ABS or another warning lamp, collect all necessary information from customer (what symptoms are present under what conditions) and check for simple causes before starting diagnosis. Besides electrical system inspection, check brake booster operation, brake fluid level, and fluid leaks.

• If incorrect tire sizes or types are installed on the vehicle or brake pads are not Genuine NISSAN parts, stopping distance or steering stability may deteriorate.

• If there is a radio, antenna or related wiring near control module, ABS function may have a malfunction or error.

• If aftermarket parts (car stereo, CD player, etc.) have been installed, check for incidents such as harness pinches, open circuits or improper wiring.

• If the following components are replaced with non-genuine components or modified, the VDC OFF indicator lamp and SLIP indicator lamp may turn on or the VDC system may not operate properly. Components related to suspension (shock absorbers, struts, springs, bushings, etc.), tires, wheels (exclude specified size), components related to brake system (pads, rotors, calipers, etc.), components related to engine (muffler, ECM, etc.), components related to body reinforcement (roll bar, tower bar, etc.).

• Driving with broken or excessively worn suspension components, tires or brake system components may cause the VDC OFF indicator lamp and the SLIP indicator lamp to turn on, and the VDC system may not operate properly.

• When the TCS or VDC is activated by sudden acceleration or sudden turn, some noise may occur. The noise is a result of the normal operation of the TCS and VDC.

• When driving on roads which have extreme slopes (such as mountainous roads) or high banks (such as sharp curves on a freeway), the VDC may not operate normally, or the VDC warning lamp and the SLIP indicator lamp may turn on. This is not a problem if normal operation can be resumed after restarting the engine.

• Sudden turns (such as spin turns, acceleration turns), drifting, etc. with VDC turned off may cause the yaw rate/side G sensor to indicate a problem. This is not a problem if normal operation can be resumed after restarting the engine.

• If battery is removed or steering angle sensor is disconnected, power to steering angle sensor is lost and the screen goes into steering angle sensor safe mode.

• When screen goes into steering angle sensor safe mode, perform "Adjustment of Steering Angle Sensor Neutral Position" with CONSULT-II®and check that VDC OFF indicator turns off. Additionally, perform self-diagnosis, check that only "Steering Angle Sensor Safe Mode" is shown for self-diagnostic result, and then delete the memory. (If the self-diagnostic result shows an indication other than "Steering Angle Sensor Safe Mode", repair the relevant part and restart self-diagnosis.) The steering angle sensor is released and returns to normal condition by performing the above operation.

• When checking, if only "Steering Angle Sensor Safe Mode" is shown in the self-diagnostic result and VDC OFF indicator is off, delete history of malfunction. This happens when battery power supply is lost and the screen goes into Steering Angle Sensor Safe Mode, and then screen returns to normal mode automatically by driving the vehicle in a straight forward direction [for approximately 30 seconds at 12 MPH (20 km/h) or more] after power is supplied again.

WHEEL SPEED SENSORS

REMOVAL & INSTALLATION
See Figure 2.

☀ WARNING

Be careful not to damage sensor edge and sensor rotor teeth. Do not pull on the sensor harness.

➡**When removing the front wheel sensor, first remove the disc rotor to gain access to the front wheel sensor bolt.**

1. Before servicing the vehicle, refer to the Precautions Section.

➡**If working near and/or around the SRS system and components, be sure to disable the SRS system. After disabling the system wait three minutes or more before servicing the vehicle.**

➡**Whenever the negative battery cable is disconnected the following components will require resetting. The Idle Air Volume Learning, Steering Angle Sensor Neutral Position, Sunroof Memory Reset/Initialization, Automatic Drive Positioner System, Audio presets and Navigation. Use the CONSULT-III diagnostic tool, or equivalent to perform the required resets.**

2. Disconnect the negative battery cable.

3. Remove wheel sensor bolt.

4. Pull out the sensor, being careful to turn it as little as possible.

5. Disconnect wheel sensor harness electrical connector, then remove harness from mounts.

6. Inspect wheel sensor O-ring, replace sensor assembly if damaged.

7. Clean wheel sensor hole and mounting surface with brake cleaner and a lint-free shop rag. Be careful that dirt and debris do not enter the axle.

8. Apply a coat of suitable grease to the wheel sensor O-ring and mounting hole.

To install:

➡**Be sure to use new fasteners, as required.**

9. Installation is the reverse of the removal procedure.

10. Be sure to perform the reconnect/relearn procedures.

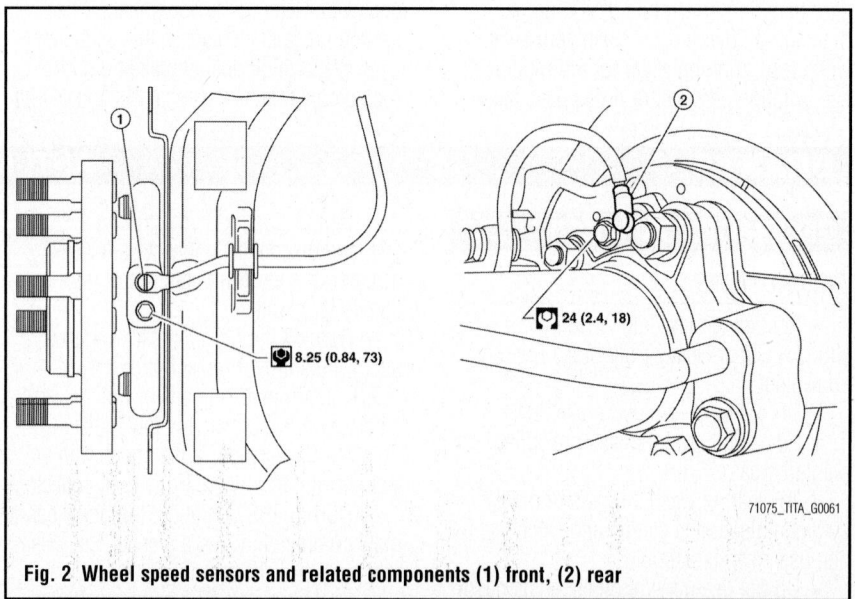

Fig. 2 Wheel speed sensors and related components (1) front, (2) rear

71075_TITA_G0061

BRAKES

BLEEDING PROCEDURE

MANUAL

See Figures 3 and 4.

➡Be sure that the master cylinder is full of clean fresh brake fluid before starting the bleeding process. Use only the recommended brake fluid when bleeding the system. Do not allow brake fluid to spill on painted surfaces as damage will occur.

   1. Before servicing the vehicle, refer to the Precautions Section.

➡If working near and/or around the SRS system and components, be sure to disable the SRS system. After disabling the system wait three minutes or more before servicing the vehicle.

➡Whenever the negative battery cable is disconnected the following components will require resetting. The Idle Air Volume Learning, Steering Angle Sensor Neutral Position, Sunroof Memory Reset/Initialization, Automatic Drive Positioner System, Audio presets and Navigation. Use the CONSULT-III diagnostic tool, or equivalent to perform the required resets.

➡The automatic back door system must be initialized anytime the battery has been disconnected. Close the back door. Open the back door with the automatic open feature. Do not stop the process until the back door opens completely.

   2. Disconnect the negative battery cable.
   3. Turn the ignition switch OFF. Disconnect the ABS actuator and electric control unit connector.
   4. Connect a vinyl tube to the rear right bleed valve. Be sure to have a catch pan handy to catch excess brake fluid.
   5. Fully depress the brake pedal four or five times.
   6. With the brake pedal depressed, loosen the bleed valve to let air out, then tighten it immediately.
   7. Repeat the above steps until all air is removed from the system. Be sure to keep watch on the brake fluid level and replenish, as necessary.
   8. Tighten the bleed valve.
   9. Repeat the above steps at each wheel, with the master cylinder reservoir tank filled at least half way.

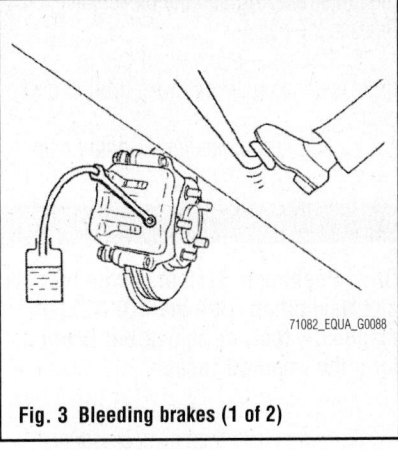

Fig. 3 Bleeding brakes (1 of 2)

   10. Bleed the brake hydraulic system air bleeder valves in the following order:
- Right rear brake
- Left front brake
- Left rear brake
- Right front brake

   11. Be sure to perform the reconnect/relearn procedures.

BLEEDING THE ABS SYSTEM

✳✳ CAUTION

Carefully monitor brake fluid level at the sub tank during bleeding operation.

✳✳ CAUTION

Fill the sub tank with new brake fluid. Make sure it is full at all times while bleeding the air out of system.

➡Place a container under the sub tank to avoid spilling brake fluid.

➡Do not loosen the line fittings at the ABS actuator during air bleeding.

   1. Before servicing the vehicle, refer to the Precautions Section.
   2. Turn ignition switch OFF and disconnect ABS actuator and control unit connector or negative battery terminal.
   3. Connect a transparent vinyl tube and container to air bleeder valve.
   4. Fully depress brake pedal several times.
   5. With brake pedal depressed, open air bleeder valve to release air.
   6. Close air bleeder valve.
   7. Release brake pedal slowly.
   8. Tighten air bleeder valve to 71 inch lbs. (8 Nm).
   9. Repeat steps 2 through 7 until no

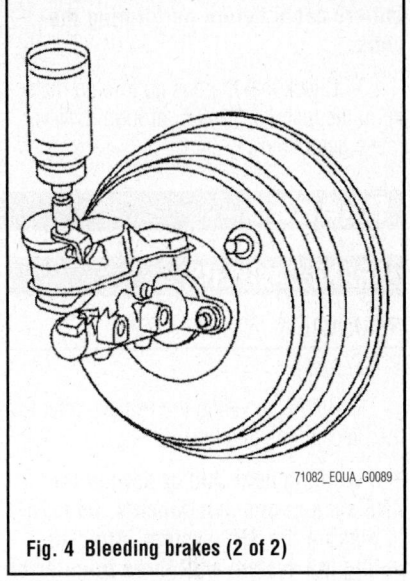

Fig. 4 Bleeding brakes (2 of 2)

more air bubbles come out of air bleeder valve.

   10. Bleed the brake hydraulic system air bleeder valves in the following order:
- Right rear brake
- Left front brake
- Left rear brake
- Right front brake

BRAKE FLUID

FLUID RECOMMENDATIONS

   When adding/changing fluid to the brake system be sure to use DOT 3 (US FMVSS No. 116, or equivalent).

LEVEL CHECK

   1. Before servicing the vehicle, refer to the Precautions Section.

➡If working near and/or around the SRS system and components, be sure to disable the SRS system. After disabling the system wait three minutes or more before servicing the vehicle.

➡Whenever the negative battery cable is disconnected the following components will require resetting. The Idle Air Volume Learning, Steering Angle Sensor Neutral Position, Sunroof Memory Reset/Initialization, Automatic Drive Positioner System, Audio presets and Navigation. Use the CONSULT-III diagnostic tool, or equivalent to perform the required resets.

   2. Disconnect the negative battery cable.

→Turn the ignition switch off and disconnect the ABS actuator and electric control unit connector, or the negative battery cable before performing the work.

3. Check that there is no foreign material in the reservoir tank or around it. Never reuse used fluid.

4. Loosen the bleeder valve.
5. Slowly depress the brake pedal to the full stroke. Tighten the bleed valve. Release the pedal.
6. Repeat at intervals of two or three seconds until all brake fluid is discharged.
7. Add new brake fluid. Continue the process until new fluid flows out of the bleed valve.
8. Bleed the air from the brake system.
9. Fill the master cylinder to specification, with the correct grade and type brake fluid.
10. Be sure to perform the reconnect/relearn procedures.

BRAKES

BRAKE CALIPERS

REMOVAL & INSTALLATION

See Figure 5.

1. Before servicing the vehicle, refer to the Precautions Section.

→If working near and/or around the SRS system and components, be sure to disable the SRS system. After disabling the system wait three minutes or more before servicing the vehicle.

→Whenever the negative battery cable is disconnected the following components will require resetting. The Idle Air Volume Learning, Steering Angle Sensor Neutral Position, Sunroof Memory Reset/Initialization, Automatic Drive Positioner System, Audio presets and Navigation. Use the CONSULT-III diagnostic tool, or equivalent to perform the required resets.

2. Disconnect the negative battery cable.
3. Drain brake fluid as necessary.
4. Raise and safely support the vehicle.
5. Remove or disconnect the following:
   • Wheel and tire assembly
   • Union bolt, discard copper washers
   • Caliper-to-torque member slide pins, or remove the caliper and torque member as an assembly
   • Brake caliper

To install:

→Be sure to use new fasteners, as required.

FRONT DISC BRAKES

6. Install the brake caliper, tighten torque member bolts to specification, and caliper slide pins to specification.
7. Install the union bolt and tighten to 13 ft. lbs. (18 Nm).
8. Fill the master cylinder and bleed the brake system.
9. Install the wheels.
10. Be sure to perform the reconnect/relearn procedures.

BRAKE PADS

REMOVAL & INSTALLATION

See Figures 6 and 7.

1. Before servicing the vehicle, refer to the Precautions Section.

→If working near and/or around the SRS system and components, be sure to disable the SRS system. After disabling the system wait three minutes or more before servicing the vehicle.

→Whenever the negative battery cable is disconnected the following components will require resetting. The Idle Air Volume Learning, Steering Angle Sensor Neutral Position, Sunroof Memory Reset/Initialization, Automatic Drive Positioner System, Audio presets and Navigation. Use the CONSULT-III diagnostic tool, or equivalent to perform the required resets.

2. Disconnect the negative battery cable.
3. Raise and support the vehicle safely.
4. Partially drain the brake system, as necessary.
5. Remove the tire and wheel assembly.
6. Remove lower sliding pin bolt.
7. Suspend brake caliper with a remove and remove brake pads and shim from torque member. Do not allow the caliper to hang by the brake line.

→When removing the pad retainer from the torque member, lift it in the direction indicated by the arrow, as shown in the illustration so that it does not deform.

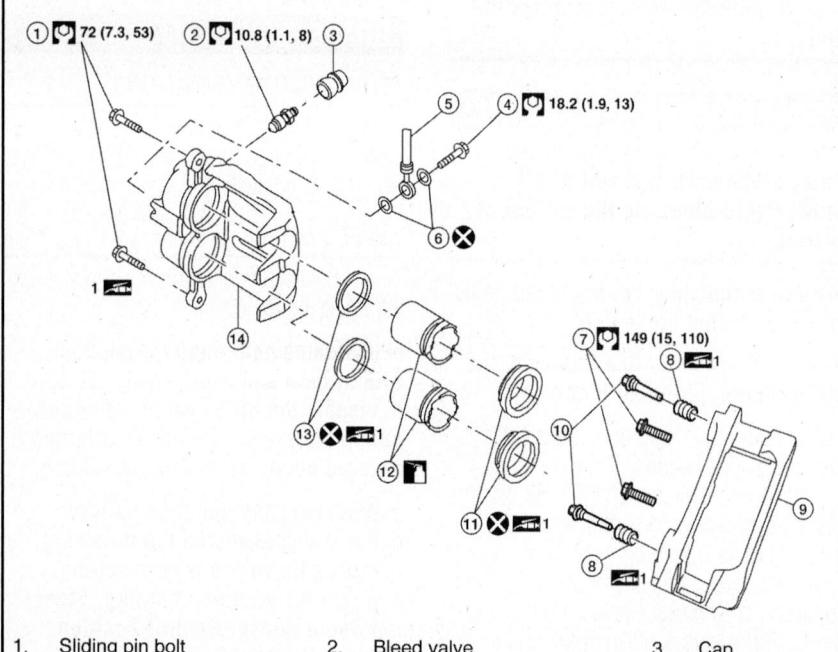

1.	Sliding pin bolt	2.	Bleed valve	3.	Cap
4.	Union bolt	5.	Brake hose	6.	Copper washer
7.	Torque member bolt	8.	Sliding pin boot	9.	Torque member
10.	Sliding pin	11.	Piston boot	12.	Piston
13.	Piston seal	14.	Cylinder body		Brake fluid

37663_TITA_G0019

Fig. 5 Front caliper and related components

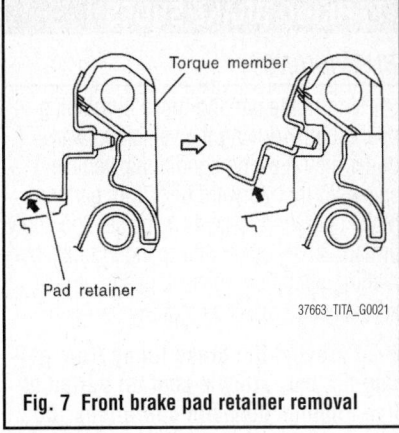

Fig. 7 Front brake pad retainer removal

To install:

➡ **Be sure to use new fasteners, as required.**

By pushing in the pistons, brake fluid returns to master cylinder reservoir tank. Watch the level of the surface of reservoir tank.

　8. Push pistons in so that the pad is firmly installed, using a suitable tool.

　9. Mount the brake caliper to torque member.

　10. Attach pad retainer to torque member.

　11. Lubricate lower sliding pin bolt with a thin layer of silicone grease and install. Torque to specification

　12. Install the tires from vehicle with power tool.

　13. Road test the vehicle.

　14. Be sure to perform the reconnect/ relearn procedures.

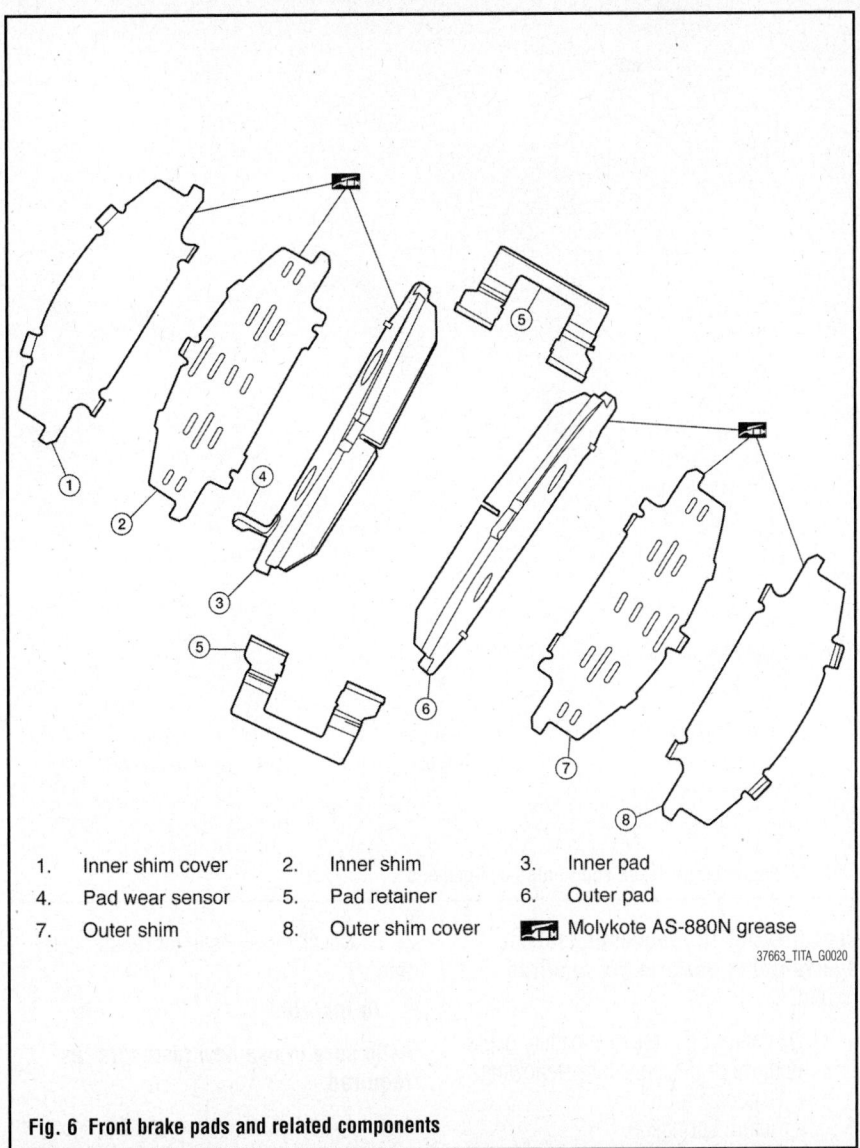

1.	Inner shim cover	2.	Inner shim	3.	Inner pad
4.	Pad wear sensor	5.	Pad retainer	6.	Outer pad
7.	Outer shim	8.	Outer shim cover		Molykote AS-880N grease

Fig. 6 Front brake pads and related components

BRAKES

ADJUSTMENTS

CABLES

　1. Before servicing the vehicle, refer to the Precautions Section.

➡ **If working near and/or around the SRS system and components, be sure to disable the SRS system. After disabling the system wait three minutes or more before servicing the vehicle.**

➡ **Whenever the negative battery cable is disconnected the following components will require resetting. The Idle Air Volume Learning, Steering Angle Sensor Neutral Position, Sunroof Memory Reset/Initialization, Automatic Drive Positioner System, Audio presets** and Navigation. Use the CONSULT-III diagnostic tool, or equivalent to perform the required resets.

　2. Disconnect the negative battery cable.

　3. Remove the left-hand lower instrument panel.

　4. Partially engage parking brake pedal to access adjusting nut.

　5. Insert a deep socket wrench to rotate adjusting nut and loosen cable sufficiently. Then, disengage the parking brake pedal.

　6. Remove the wheel and tire.

　7. Remove the disc rotor and measure inner diameter at widest point using tool J-21177-A.

　8. Transfer measurement less 0.6 mm to the parking brake shoes and adjust accordingly.

PARKING BRAKE

　9. Using wheel nuts, secure the disc rotor to the hub to prevent it from tilting.

　10. Rotate the disc rotor to make sure there is no drag.

　11. Adjust cable as follows:

　　a. Operate pedal 10 or more times with a force of 110 ft. lbs. (490 Nm).

　　b. Rotate adjusting nut with deep socket to adjust pedal stroke to specification: 3 to 4 notches; under force of 44.1 lbs. (196 N).

　　c. With parking brake pedal completely disengaged, make sure there is no drag on the parking brake.

　12. Be sure to perform the reconnect/ relearn procedures.

PARKING BRAKE SHOES

BURNISHING

Perform the parking brake burnishing operation by driving the vehicle forward under the following conditions: vehicle speed 25 mph forward direction, parking brake operating force 44.1 lbs set and apply time of 30 seconds. After parking brake burnishing operation, recheck parking brake adjustment, correct as required.

➡To prevent the brake lining from getting too hot, allow a cool off period of five minutes between operations. Do not perform excessive break-in operations, because it may cause uneven or early wear of the lining.

REMOVAL & INSTALLATION

See Figure 8.

✳✳ CAUTION

Clean the brakes with a vacuum dust collector to minimize the hazard of airborne particles or other materials.

➡Remove the disc rotor only with the parking brake pedal completely in the released position.

1. Before servicing the vehicle, refer to the Precautions Section.

➡If working near and/or around the SRS system and components, be sure to disable the SRS system. After disabling the system wait three minutes or more before servicing the vehicle.

➡Whenever the negative battery cable is disconnected the following components will require resetting. The Idle Air Volume Learning, Steering Angle Sensor Neutral Position, Sunroof Memory Reset/Initialization, Automatic Drive Positioner System, Audio presets and Navigation. Use

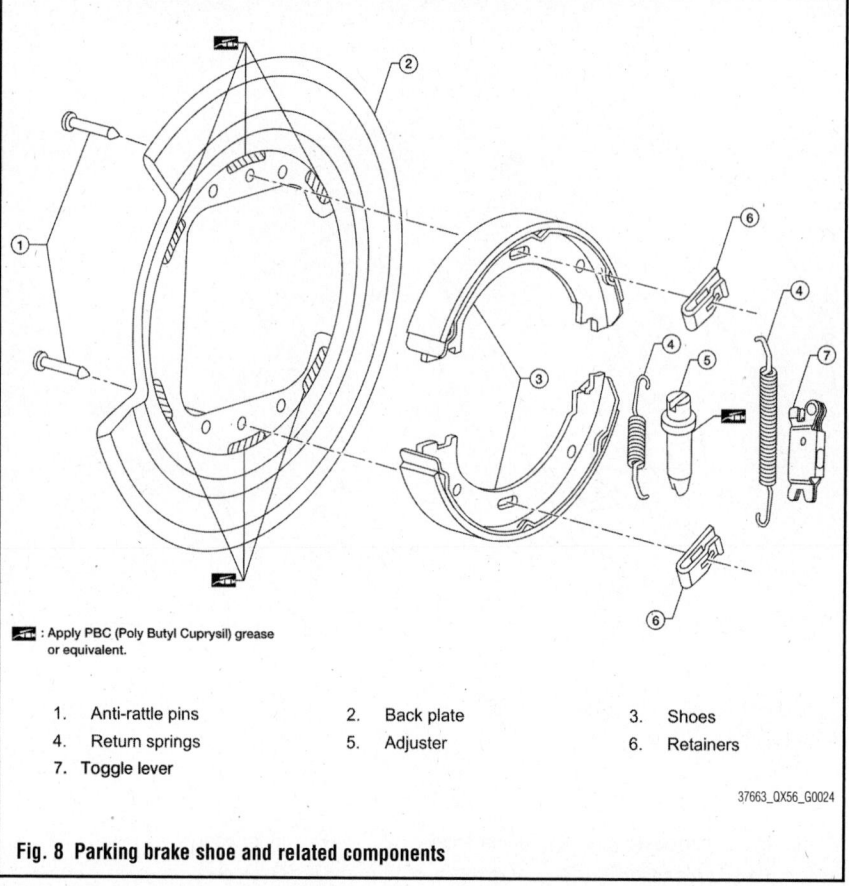

: Apply PBC (Poly Butyl Cuprysil) grease or equivalent.

1.	Anti-rattle pins	2.	Back plate	3.	Shoes
4.	Return springs	5.	Adjuster	6.	Retainers
7.	Toggle lever				

37663_QX56_G0024

Fig. 8 Parking brake shoe and related components

the CONSULT-III diagnostic tool, or equivalent to perform the required resets.

2. Disconnect the negative battery cable.
3. Remove or disconnect the following:
   - Rear disc rotor
   - Return springs
   - Adjuster
   - Rear cable from the toggle lever, if necessary
   - Retainers, anti-rattle pins and shoes
4. Check shoe sliding surface on back plate for excessive wear and damage.
5. Check anti-rattle pins for excessive wear and corrosion.
6. Check the return springs for sagging.

7. Check the adjuster for rough operation.

To install:

➡Be sure to use new fasteners, as required.

8. To install, reverse removal procedure.

➡Apply brake grease to the specified points during assembly.

➡There is a difference between the adjuster's orientation from left and right. Assemble the adjuster so the threaded part expands when rotating it.

9. Be sure to perform the reconnect/relearn procedures.

BRAKES

BRAKE CALIPERS

REMOVAL & INSTALLATION

See Figure 9.

1. Before servicing the vehicle, refer to the Precautions Section.

➡ If working near and/or around the SRS system and components, be sure to disable the SRS system. After disabling the system wait three minutes or more before servicing the vehicle.

➡ Whenever the negative battery cable is disconnected the following components will require resetting. The Idle Air Volume Learning, Steering Angle Sensor Neutral Position, Sunroof Memory Reset/Initialization, Automatic Drive Positioner System, Audio presets and Navigation. Use the CONSULT-III diagnostic tool, or equivalent to perform the required resets.

2. Disconnect the negative battery cable.

3. Raise and safely support the vehicle.

4. Remove the tire and wheel assembly.

5. Drain the brake fluid, as necessary.

6. Remove the brake hose mounting bolt and brake hose.

7. Remove the brake caliper assembly. Discard the copper washer. Do not reuse.

To install:

➡ Be sure to use new fasteners, as required.

8. Install the brake caliper assembly and tighten mounting bolts to specification.

9. Install brake hose and tighten to 13 ft. lbs. (18 Nm).

10. Bleed the air from the brake caliper.

11. Install the tires from vehicle with power tool.

12. Road test the vehicle.

13. Be sure to perform the reconnect/relearn procedures.

BRAKE PADS

REMOVAL & INSTALLATION

See Figure 10.

1. Before servicing the vehicle, refer to the Precautions Section.

➡ If working near and/or around the SRS system and components, be sure to disable the SRS system. After dis-

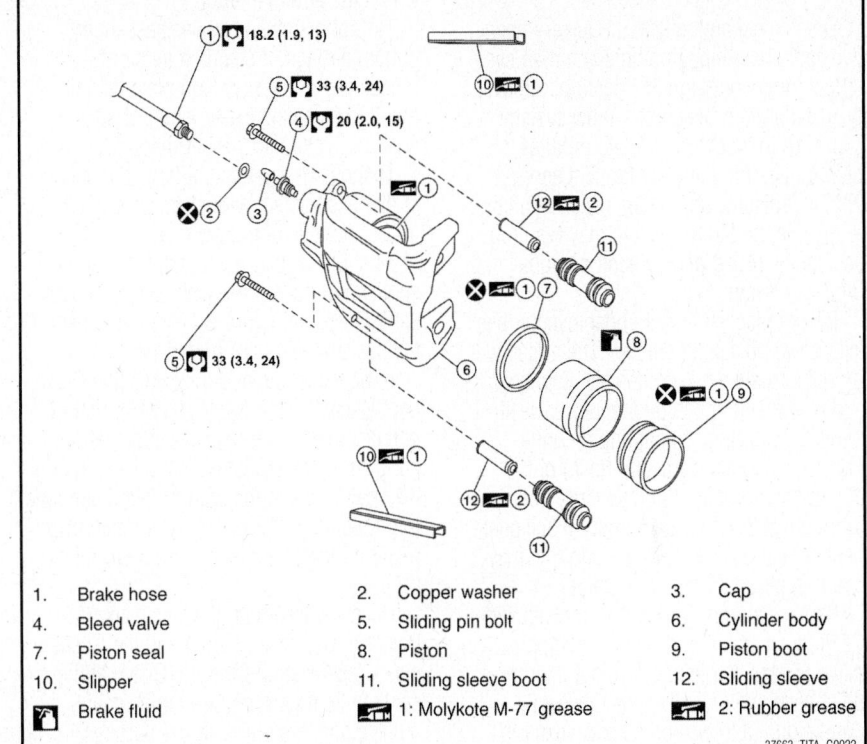

1.	Brake hose	2.	Copper washer	3.	Cap
4.	Bleed valve	5.	Sliding pin bolt	6.	Cylinder body
7.	Piston seal	8.	Piston	9.	Piston boot
10.	Slipper	11.	Sliding sleeve boot	12.	Sliding sleeve
	Brake fluid		1: Molykote M-77 grease		2: Rubber grease

37663_TITA_G0022

Fig. 9 Rear caliper and related components

abling the system wait three minutes or more before servicing the vehicle.

➡ Whenever the negative battery cable is disconnected the following components will require resetting. The Idle Air Volume Learning, Steering Angle

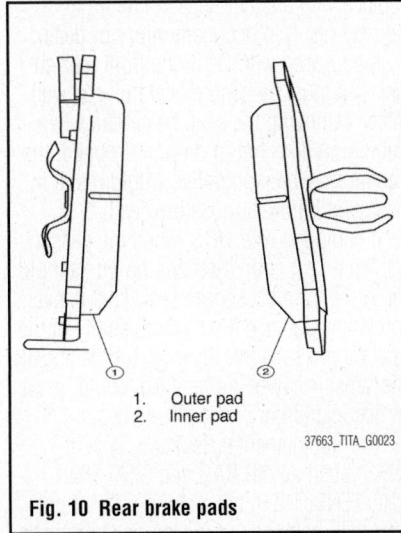

1. Outer pad
2. Inner pad

37663_TITA_G0023

Fig. 10 Rear brake pads

Sensor Neutral Position, Sunroof Memory Reset/Initialization, Automatic Drive Positioner System, Audio presets and Navigation. Use the CONSULT-III diagnostic tool, or equivalent to perform the required resets.

2. Disconnect the negative battery cable.

3. Raise and safely support the vehicle.

4. Remove the tire and wheel assembly.

5. Drain the brake fluid, as necessary.

6. Remove the sliding pin.

7. Remove the cylinder body and remove the pads.

To install:

➡ Be sure to use new fasteners, as required.

8. Apply Molykote® (M-77) grease to the knuckle slide where the brake pads contact.

9. Install pads to cylinder body.

10. Install top mounting bolt and tighten to specification.

11. Check brake for drag.

12. Install tires to the vehicle.

13. Be sure to perform the reconnect/relearn procedures.

PRECAUTIONS

Disconnect and isolate the battery negative cable before beginning any airbag system component diagnosis, testing, removal, or installation procedures. Allow system capacitor to discharge for two minutes before beginning any component service. This will disable the airbag system. Failure to disable the airbag system may result in accidental airbag deployment, personal injury, or death.

Do not place an intact undeployed airbag face down on a solid surface. The airbag will propel into the air if accidentally deployed and may result in personal injury or death.

When carrying or handling an undeployed airbag, the trim side (face) of the airbag should be pointing towards the body to minimize possibility of injury if accidental deployment occurs. Failure to do this may result in personal injury or death.

Replace airbag system components with OEM replacement parts. Substitute parts may appear interchangeable, but internal differences may result in inferior occupant protection. Failure to do so may result in occupant personal injury or death.

Wear safety glasses, rubber gloves, and long sleeved clothing when cleaning powder residue from vehicle after an airbag deployment. Powder residue emitted from a deployed airbag can cause skin irritation. Flush affected area with cool water if irritation is experienced. If nasal or throat irritation is experienced, exit the vehicle for fresh air until the irritation ceases. If irritation continues, see a physician.

Do not use a replacement airbag that is not in the original packaging. This may result in improper deployment, personal injury, or death.

The factory installed fasteners, screws and bolts used to fasten airbag components have a special coating and are specifically designed for the airbag system. Do not use substitute fasteners. Use only original equipment fasteners listed in the parts catalog when fastener replacement is required.

During, and following, any child restraint anchor service, due to impact event or vehicle repair, carefully inspect all mounting hardware, tether straps, and anchors for proper installation, operation, or damage. If a child restraint anchor is found damaged in any way, the anchor must be replaced. Failure to do this may result in personal injury or death.

Deployed and non-deployed airbags may or may not have live pyrotechnic material within the airbag inflator.

Do not dispose of driver/passenger/curtain airbags or seat belt tensioners unless you are sure of complete deployment. Refer to the Hazardous Substance Control System for proper disposal.

Dispose of deployed airbags and tensioners consistent with state, provincial, local, and federal regulations.

After any airbag component testing or service, do not connect the battery negative cable. Personal injury or death may result if the system test is not performed first.

If the vehicle is equipped with the Occupant Classification System (OCS), do not connect the battery negative cable before performing the OCS Verification Test using the scan tool and the appropriate diagnostic information. Personal injury or death may result if the system test is not performed properly.

Never replace both the Occupant Restraint Controller (ORC) and the Occupant Classification Module (OCM) at the same time. If both require replacement, replace one, then perform the Airbag System test before replacing the other.

Both the ORC and the OCM store Occupant Classification System (OCS) calibration data, which they transfer to one another when one of them is replaced. If both are replaced at the same time, an irreversible fault will be set in both modules and the OCS may malfunction and cause personal injury or death.

If equipped with OCS, the Seat Weight Sensor is a sensitive, calibrated unit and must be handled carefully. Do not drop or handle roughly. If dropped or damaged, replace with another sensor. Failure to do so may result in occupant injury or death.

If equipped with OCS, the front passenger seat must be handled carefully as well. When removing the seat, be careful when setting on floor not to drop. If dropped, the sensor may be inoperative, could result in occupant injury, or possibly death.

If equipped with OCS, when the passenger front seat is on the floor, no one should sit in the front passenger seat. This uneven force may damage the sensing ability of the seat weight sensors. If sat on and damaged, the sensor may be inoperative, could result in occupant injury, or possibly death.

The Supplemental Restraint System (SRS) such as AIR BAG and SEAT BELT PRE-TENSIONER, used along with a front seat belt, helps to reduce the risk or severity of injury to the driver and front passenger for certain types of collision. This system includes seat belt switch inputs and dual stage front air bag modules. The SRS system uses the seat belt switches to determine the front air bag deployment, and may only deploy one front air bag, depending on the severity of a collision and whether the front occupants are belted or unbelted.

✳✳ CAUTION

Improper maintenance, including incorrect removal and installation of the SRS, can lead to personal injury caused by unintentional activation of the system. Do not use electrical test equipment on any circuit related to the SRS unless instructed. SRS wiring harnesses can be identified by yellow and/or orange harnesses or harness connectors.

✳✳ CAUTION

When working near the Airbag Diagnosis Sensor Unit or other Airbag System sensors with the Ignition ON or engine running, DO NOT use air or electric power tools or strike near the sensor(s) with a hammer. Heavy vibration could activate the sensor(s) and deploy the air bag(s), possibly causing serious injury. When using air or electric power tools or hammers, always switch the Ignition OFF, disconnect the battery, and wait at least 3 minutes before performing any service.

✳✳ CAUTION

Before removing the seat belt pretensioner assembly, turn the ignition switch OFF, disconnect both battery terminals and wait at least three minutes. For approximately three minutes after the battery terminals have been removed, it is still possible for the air bag and seat belt pretensioner to deploy. Therefore, do not attempt work on any SRS connectors or wires until at least three minutes have passed. After replacing or reinstalling seat belt pre-tensioner assembly, or reconnecting seat belt pre-tensioner assembly connector, make sure entire SRS operates properly.

➡**Do not disassemble buckle or seat belt assembly. Replace anchor bolts if they are deformed or worn out. Never** oil tongue and buckle. If any component of seat belt assembly is questionable, do not repair. Replace the whole seat belt assembly. If webbing is cut, frayed, or damaged, replace seat belt assembly. When replacing seat belt assembly, use a genuine Nissan seat belt assembly.

➡After a collision inspect all seat belt assemblies including retractors and attaching hardware after any collision. Nissan recommends that all seat belt assemblies in use during a collision be replaced unless the collision was minor and the belts show no damage and continue to operate properly. Failure to do so could result in serious personal injury in an accident. Seat belt assemblies not in use during a collision should also be replaced if either damage or improper operation is noted. Seat belt pre-tensioner should be replaced even if the seat belts are not in use during a frontal collision in which the air bags are deployed. Replace any seat belt assembly (including anchor bolts) if the seat belt was in use at the time of a collision (except for minor collisions and the belts, retractors and buckles show no damage and continue to operate properly). The seat belt was damaged in an accident. (i.e., torn webbing, bent retractor or guide, etc.). The seat belt attaching point was damaged in an accident. Inspect the seat belt attaching area for damage or distortion and repair as necessary before installing a new seat belt assembly. Anchor bolts are deformed or worn out. The seat belt pre-tensioner should be replaced even if the seat belts are not in use during the collision in which the air bags are deployed.

➡Replace occupant classification system control unit and passenger front seat cushion as an assembly.

➡If the steering wheel was rotated after battery disconnect, perform the following: This Procedure is applied only to models with Intelligent Key system and NATS (NISSAN ANTI-THEFT SYSTEM). Remove and install all control units after disconnecting both battery cables with the ignition knob in the "LOCK" position. Always use CONSULT to perform self-diagnosis as a part of each function inspection after finishing work. If DTC is detected, perform trouble diagnosis according to self-diagnostic results. For models equipped with the Intelligent Key system and NATS, an electrically controlled steering lock mechanism is adopted on the key cylinder. For this reason, if the battery is disconnected or if the battery is discharged, the steering wheel will lock and steering wheel rotation will become impossible. If steering wheel rotation is required when battery power is interrupted, follow the procedure below before starting the repair operation.

RESET

1. Connect both battery cables.
2. Use the Intelligent Key or mechanical key to turn the ignition switch to the "ACC" position. At this time, the steering lock will be released.
3. Disconnect both battery cables. The steering lock will remain released and the steering wheel can be rotated.
4. Perform the necessary repair operation.
5. When the repair work is completed, return the ignition switch to the "LOCK" position before connecting the battery cables. (At this time, the steering lock mechanism will engage).
6. Perform a self-diagnosis check of all control units using CONSULT III diagnostic tool, or equivalent.

DISARMING THE SYSTEM

➡If working near and/or around the SRS system and components, be sure to disable the SRS system. After disabling the system wait three minutes or more before servicing the vehicle.

➡Whenever the negative battery cable is disconnected the following components will require resetting. The Idle Air Volume Learning, Steering Angle Sensor Neutral Position, Sunroof Memory Reset/Initialization, Automatic Drive Positioner System, Audio presets and Navigation. Use the CONSULT-III diagnostic tool, or equivalent to perform the required resets.

1. Before servicing the vehicle, refer to the Precautions Section.
2. Disconnect the negative battery cable.
3. Disconnect the positive battery cable.
4. Wait at least 3 minutes before working on the vehicle.

➡The air bag system is designed to retain enough power to deploy the air bag for a short time after the battery has been disconnected.

ARMING THE SYSTEM

1. Before servicing the vehicle, refer to the Precautions Section.
2. After repairs are complete, connect the negative battery cable.
3. Turn the ignition switch to the **ON** position and check the air bag warning light blinks for proper operation.

➡Once repair work has been completed, return the ignition switch to the LOCK position, before connecting the battery cables. At this time the steering lock mechanism will engage. Install the CONSULT-III diagnostic tool, or equivalent, and follow the directions on the screen of the tool and perform the self-diagnosis check.

DRIVELINE

AUTOMATIC TRANSMISSION

DRAIN & REFILL

See Figures 11 and 12.

1. Before servicing the vehicle, refer to the Precautions Section.

➡ If working near and/or around the SRS system and components, be sure to disable the SRS system. After disabling the system wait three minutes or more before servicing the vehicle.

➡ Whenever the negative battery cable is disconnected the following components will require resetting. The Idle Air Volume Learning, Steering Angle Sensor Neutral Position, Sunroof Memory Reset/Initialization, Automatic Drive Positioner System, Audio presets and Navigation. Use the CONSULT-III diagnostic tool, or equivalent to perform the required resets.

2. Run the engine until operating temperature is reached.
3. Stop the engine.
4. Disconnect the negative battery cable.
5. Loosen the fluid level gauge bolt.
6. Raise and support the vehicle safely.
7. Loosen the drain plug. Discard the gasket.

➡ When replacing the drain plug use a new gasket and tighten to 25 ft. lbs. (34 Nm).

8. Drain the fluid from the drain plug opening and refill with new fluid.

➡ Always refill same volume with drained fluid.

9. To replace the fluid, pour in new fluid at the transmission charging pipe with the engine idling and at the same time drain

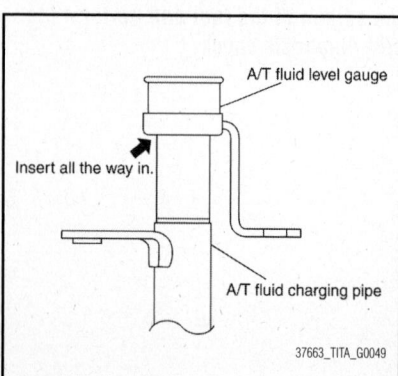

Fig. 11 Automatic transmission fluid level gauge location

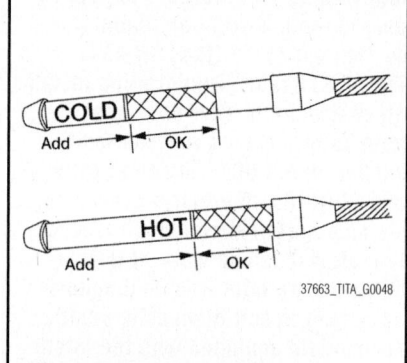

Fig. 12 Automatic transmission fluid dipstick markings

the old fluid from the radiator cooler hose return side.

10. When the color of the fluid coming out is about the same as the color of the new fluid, replacement is complete.

➡ Be sure to use the proper grade and type replacement fluid. Do not mix with other fluids. Using other than the required fluid will cause deterioration in drivability and transmission damage.

11. Run the engine at idle speed for about five minutes.
12. Check and correct the fluid level.
13. If the fluid is still dirty, repeat the process.
14. Install the removed fluid level gauge in the fluid charging pipe.
15. Install the fluid level gauge bolt. Tighten to 45 inch lbs. (5.1 Nm).
16. Be sure to perform the reconnect/relearn procedures.

FLUID LEVEL CHECK

See Figures 11 and 12.

1. Before servicing the vehicle, refer to the Precautions Section.

➡ If working near and/or around the SRS system and components, be sure to disable the SRS system. After disabling the system wait three minutes or more before servicing the vehicle.

➡ Whenever the negative battery cable is disconnected the following components will require resetting. The Idle Air Volume Learning, Steering Angle Sensor Neutral Position, Sunroof Memory Reset/Initialization, Automatic Drive Positioner System, Audio presets and Navigation. Use the CONSULT-III diagnostic tool, or equivalent to perform the required resets.

2. Run the engine until operating temperature is reached.
3. Check for external fluid leakage.
4. Loosen the level gauge bolt.

➡ Before driving, fluid level can be checked at fluid temperatures of 86—122 degrees F using the COLD range on the dipstick.

5. To COLD check the transmission, park the vehicle on a level surface and set the parking brake.
6. Start the engine and move the selector lever through each gear range. Leave the selector lever in the PARK (P) range.
7. Check the fluid level with the engine idling. Always use lint free paper, not a cloth to check the fluid level.
8. Re-insert the gauge. Remove the gauge and note the reading. As required add fluid. Do not overfill.
9. Drive the vehicle for about five minutes in urban areas.
10. Using the CONSULT-III tool, or equivalent, make the fluid temperature about 149 degrees F.
11. Recheck the fluid level, correct as required.

FLUID RECOMMENDATIONS

The automatic transmission uses Nissan approved Matic S ATF transmission fluid. If this fluid is not available Nissan Matic J ATF may be used. If information differs from the owner's manual, use the data in the owner's manual.

DRIVESHAFT

REMOVAL & INSTALLATION

Rear

See Figures 13 and 14.

1. Before servicing the vehicle, refer to the Precautions Section.

➡ If working near and/or around the SRS system and components, be sure to disable the SRS system. After disabling the system wait three minutes or more before servicing the vehicle.

➡ Whenever the negative battery cable is disconnected the following components will require resetting. The Idle Air Volume Learning, Steering Angle Sensor Neutral Position, Sunroof Memory Reset/Initialization, Automatic Drive Positioner System, Audio presets and Navigation. Use the CONSULT-III

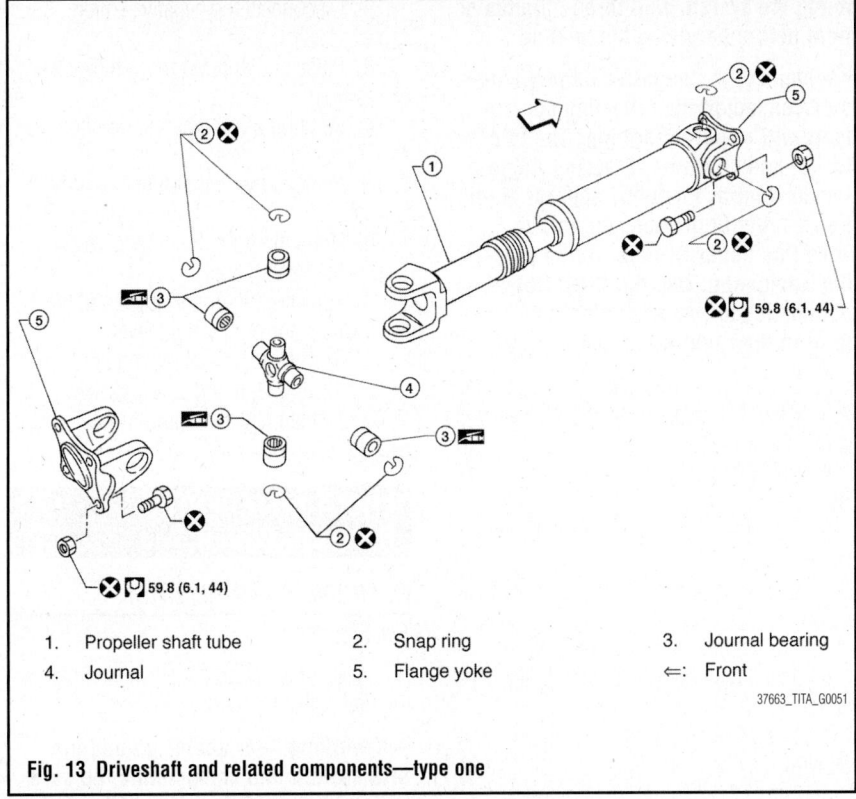

1. Propeller shaft tube
2. Snap ring
3. Journal bearing
4. Journal
5. Flange yoke
⇐: Front

37663_TITA_G0051

Fig. 13 Driveshaft and related components—type one

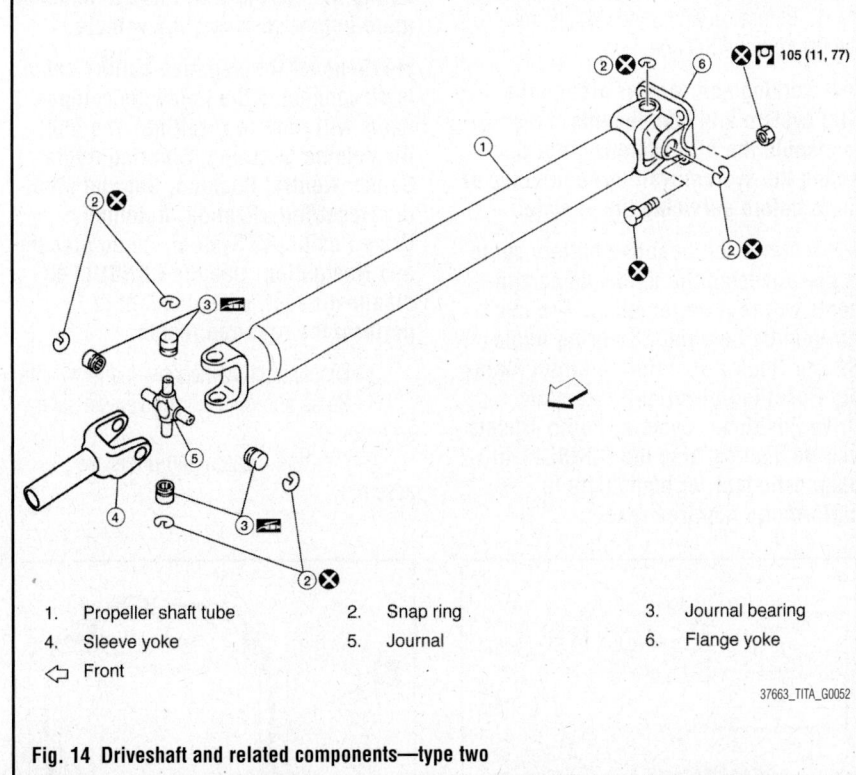

1. Propeller shaft tube
2. Snap ring
3. Journal bearing
4. Sleeve yoke
5. Journal
6. Flange yoke
◁ Front

37663_TITA_G0052

Fig. 14 Driveshaft and related components—type two

diagnostic tool, or equivalent to perform the required resets.

2. Disconnect the negative battery cable.
3. Position the selector lever in the N position.
4. Release the parking brake.
5. Raise and support the vehicle safely.
6. Remove the engine undercover.
7. Matchmark the driveshaft and companion flange.

8. Remove the center support bearing bracket nuts, if equipped.
9. Remove the driveshaft. Discard the nuts.

To install:

➡ **Be sure to use new fasteners, as required.**

10. Installation is the reverse of the removal procedure.
11. Check for vehicle vibration, correct as required.
12. Be sure to perform the reconnect/relearn procedures.

DRIVESHAFT ANGLE MEASUREMENT

See Figures 15 through 17.

1. Before servicing the vehicle, refer to the Precautions Section.

➡ **If working near and/or around the SRS system and components, be sure to disable the SRS system. After disabling the system wait three minutes or more before servicing the vehicle.**

➡ **Whenever the negative battery cable is disconnected the following components will require resetting. The Idle Air Volume Learning, Steering Angle Sensor Neutral Position, Sunroof Memory Reset/Initialization, Automatic Drive Positioner System, Audio presets and Navigation. Use the CONSULT-III diagnostic tool, or equivalent to perform the required resets.**

2. Disconnect the negative battery cable.
3. Raise and support the vehicle safely.
4. Matchmark the driveshaft and companion flange.
5. Remove the driveshaft. Discard the nuts.
6. Inspect the driveshaft runout, replace the driveshaft as necessary. See illustration.

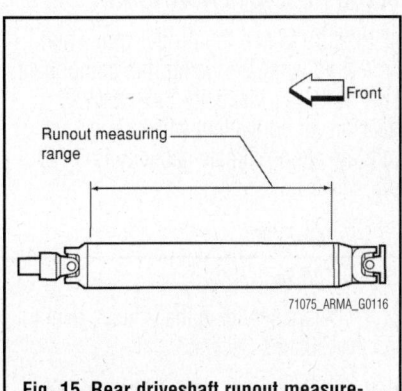

71075_ARMA_G0116

Fig. 15 Rear driveshaft runout measurement location points—one piece driveshaft

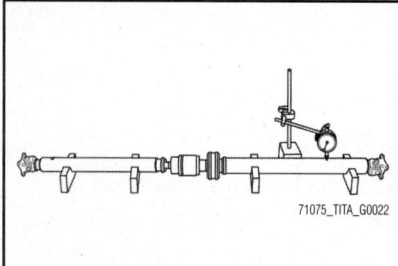

Fig. 16 Rear driveshaft runout measurement location point—two piece driveshaft

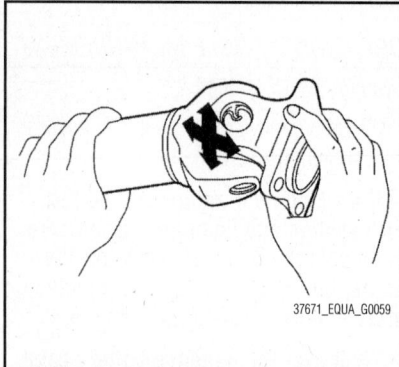

Fig. 17 Check the axial play of the joint

Runout specification should be 0.024 inch (0.60mm).

7. While holding the flange yoke on one side, check the axial play of the joint. See illustration. If the journal axial play exceeds 0.0008 inch (0.02mm), repair or replace the journal parts.

8. Check the driveshaft for dents or cracks. If damage is detected, replace the driveshaft assembly.

To install:

➡Be sure to use new fasteners, as required.

FRONT DRIVE AXLE

FLUID RECOMMENDATIONS

Be sure to use the proper grade and type fluid when servicing this component. Use API GL-5 viscosity SAE 80W-90 gear oil, or equivalent when servicing/refilling. Approximate capacity is 3 3/8 pints (1.6L).

LEVEL CHECK

See Figure 18.

1. Before servicing the vehicle, refer to the Precautions Section.

➡If working near and/or around the SRS system and components, be sure to disable the SRS system. After disabling the system wait three minutes or more before servicing the vehicle.

➡Whenever the negative battery cable is disconnected the following components will require resetting. The Idle Air Volume Learning, Steering Angle Sensor Neutral Position, Sunroof Memory Reset/Initialization, Automatic Drive Positioner System, Audio presets and Navigation. Use the CONSULT-III diagnostic tool, or equivalent to perform the required resets.

2. Disconnect the negative battery cable.
3. Raise and support the vehicle safely, as required.
4. Position a catch pan under the assembly.
5. Remove the filler plug. Discard the gasket.
6. Check for proper fluid level, see illustration.
7. Correct fluid level, as required.
8. Install the filler plug, using a new gasket.
9. Tighten filler plug to 27 ft. lbs. (36 Nm).

DRAIN & REFILL

1. Before servicing the vehicle, refer to the Precautions Section.

➡If working near and/or around the SRS system and components, be sure to disable the SRS system. After disabling the system wait three minutes or more before servicing the vehicle.

➡Whenever the negative battery cable is disconnected the following components will require resetting. The Idle Air Volume Learning, Steering Angle Sensor Neutral Position, Sunroof Memory Reset/Initialization, Automatic Drive Positioner System, Audio presets and Navigation. Use the CONSULT-III diagnostic tool, or equivalent to perform the required resets.

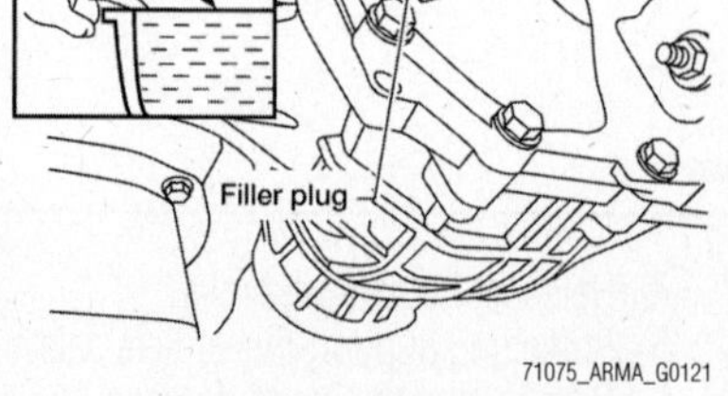

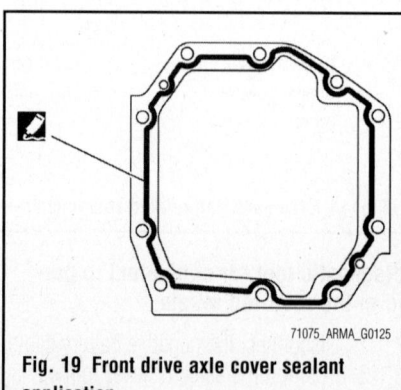

Fig. 18 Front drive axle plug location—checking

2. Disconnect the negative battery cable.
3. Raise and support the vehicle safely, as required.
4. Position a catch pan under the assembly.
5. Remove the drain plug. Discard the gasket.
6. Drain the fluid. Be sure to properly dispose of used fluid.
7. Fill with the proper grade and type fluid. Check for proper fluid level.
8. Correct fluid level, as required.
9. Install the, using a new gasket.
10. Tighten checking plug to 27 ft. lbs. (36 Nm).

FRONT DIFFERENTIAL HOUSING COVER

REMOVAL & INSTALLATION

See Figure 19.

1. Before servicing the vehicle, refer to the Precautions Section.

➡If working near and/or around the SRS system and components, be sure to disable the SRS system. After disabling the system wait three minutes or more before servicing the vehicle.

➡Whenever the negative battery cable is disconnected the following components will require resetting. The Idle Air Volume Learning, Steering Angle Sensor Neutral Position, Sunroof Memory Reset/Initialization, Automatic Drive Positioner System, Audio presets and Navigation. Use the CONSULT-III diagnostic tool, or equivalent to perform the required resets.

2. Disconnect the negative battery cable.
3. Raise and support the vehicle safely, as required.
4. Position a catch pan under the assembly.

Fig. 19 Front drive axle cover sealant application

5. Drain the fluid. Be sure to properly dispose of used fluid.

6. Remove the front axle assembly from the vehicle. Position the unit in a suitable holding fixture.

7. Remove the carrier cover retaining bolts. Remove the cover. Discard the gasket.

To install:

➡**Be sure to use new fasteners, as required.**

8. Installation is the reverse of the removal procedure.

➡**Be sure to properly remove all old RTV sealant on all mating surfaces.**

9. Apply a thin bead (0.12 inch) of genuine RTV sealant, or equivalent, to the mating surface of the carrier cover.

10. Tighten the cover bolts.

11. Be sure to perform the reconnect/relearn procedures.

FRONT HALFSHAFT

REMOVAL & INSTALLATION

See Figure 20.

1. Before servicing the vehicle, refer to the Precautions Section.

➡**If working near and/or around the SRS system and components, be sure to disable the SRS system. After disabling the system wait three minutes or more before servicing the vehicle.**

➡**Whenever the negative battery cable is disconnected the following components will require resetting. The Idle Air Volume Learning, Steering Angle Sensor Neutral Position, Sunroof Memory Reset/Initialization, Automatic Drive Positioner System, Audio presets and Navigation. Use the CONSULT-III diagnostic tool, or equivalent to perform the required resets.**

2. Disconnect the negative battery cable.

3. Raise and support the vehicle safely.

4. Remove or disconnect the following:
- Wheel and tire assembly
- Engine undercover
- ABS sensor harness on knuckle
- Brake caliper and suspend it aside
- Coil spring and shock absorber assembly

5. Separate upper ball joint stud from steering knuckle using special tool J-24319-01.

6. Remove or disconnect the following:
- Cotter pin and half shaft nut
- Halfshaft from front differential
- Halfshaft from hub and bearing assembly

To install:

➡**Be sure to use new fasteners, as required.**

7. Install or connect the following:
- Half shaft into hub
- Halfshaft into front differential
- Halfshaft nut and tighten to 101 ft. lbs. (137 Nm) and replace cotter pin
- Upper ball joint to steering knuckle
- Coil spring and shock absorber assembly
- Brake caliper
- ABS sensor
- Engine splash guard
- Wheel

8. Be sure to perform the reconnect/relearn procedures.

REAR DRIVE AXLE

FLUID RECOMMENDATIONS

Be sure to use the proper grade and type fluid when servicing this component. Use API GL-5 synthetic gear oil, Viscosity SAE 75W-90 gear oil, or equivalent when servicing/refilling the axle. Approximate capacity is 4.25 pints.

LEVEL CHECK

See Figure 21.

1. Before servicing the vehicle, refer to the Precautions Section.

➡**If working near and/or around the SRS system and components, be sure to disable the SRS system. After disabling the system wait three minutes or more before servicing the vehicle.**

➡**Whenever the negative battery cable is disconnected the following components will require resetting. The Idle Air Volume Learning, Steering Angle Sensor Neutral Position, Sunroof Memory Reset/Initialization, Automatic Drive Positioner System, Audio presets and Navigation. Use the CONSULT-III diagnostic tool, or equivalent to perform the required resets.**

2. Disconnect the negative battery cable.

3. Raise and safely support the vehicle.

4. Position a catch pan under the assembly.

5. Remove the drain plug.

6. Discard the gasket.

7. Check the oil level, correct as required.

DRAIN & REFILL

See Figure 22.

1. Before servicing the vehicle, refer to the Precautions Section.

➡**If working near and/or around the SRS system and components, be sure to disable the SRS system. After disabling the system wait three minutes or more before servicing the vehicle.**

➡**Whenever the negative battery cable is disconnected the following**

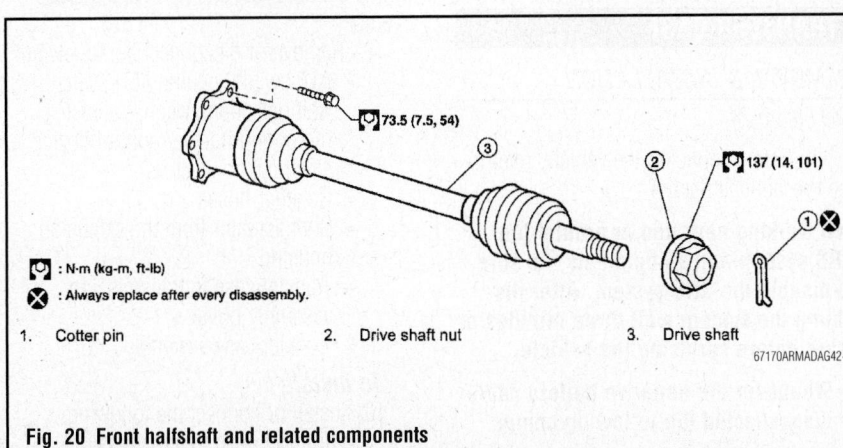

Fig. 20 Front halfshaft and related components

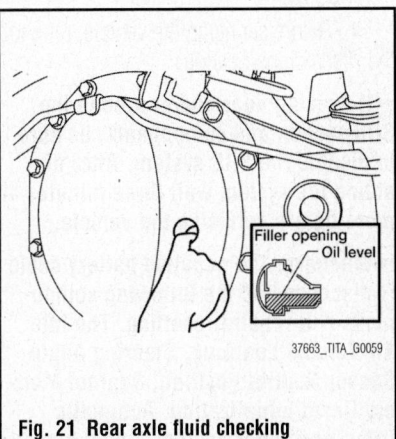

Fig. 21 Rear axle fluid checking

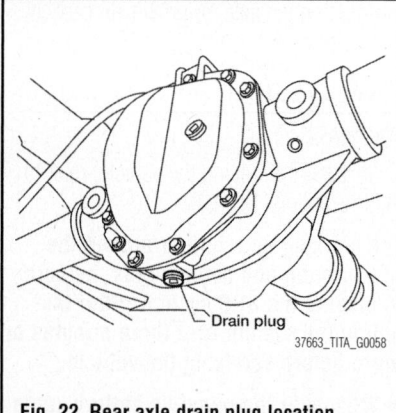

Fig. 22 Rear axle drain plug location

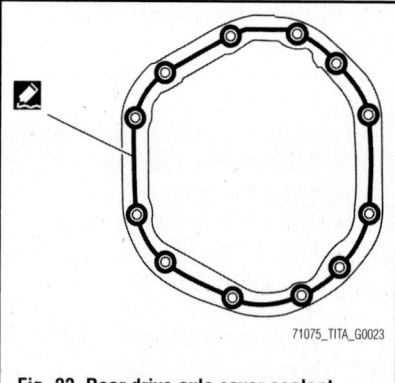

Fig. 23 Rear drive axle cover sealant application

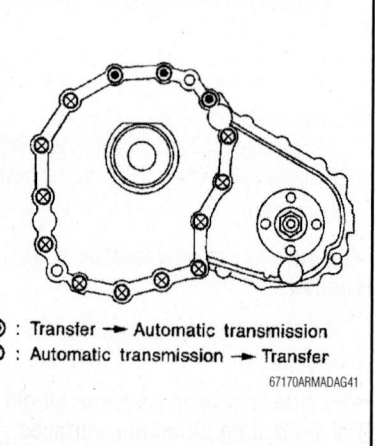

○ : Transfer → Automatic transmission
⊗ : Automatic transmission → Transfer

Fig. 24 Transfer case mounting bolt locations

components will require resetting. The Idle Air Volume Learning, Steering Angle Sensor Neutral Position, Sunroof Memory Reset/Initialization, Automatic Drive Positioner System, Audio presets and Navigation. Use the CONSULT-III diagnostic tool, or equivalent to perform the required resets.

2. Disconnect the negative battery cable.
3. Raise and safely support the vehicle.
4. Remove the drain plug and drain the fluid into a suitable container.
5. Discard the gasket.
6. Fill the unit with new gear oil until the oil level reaches the specified level near the filler plug mounting hole.

➡Be sure to use the proper grade and type gear oil.

7. Check the oil level after refilling.
8. Using a new gasket install the drain plug. Tighten to specification.
9. Be sure to perform the reconnect/relearn procedures.

REAR DIFFERENTIAL HOUSING COVER

REMOVAL & INSTALLATION
See Figure 23.

1. Before servicing the vehicle, refer to the Precautions Section.

➡If working near and/or around the SRS system and components, be sure to disable the SRS system. After disabling the system wait three minutes or more before servicing the vehicle.

➡Whenever the negative battery cable is disconnected the following components will require resetting. The Idle Air Volume Learning, Steering Angle Sensor Neutral Position, Sunroof Memory Reset/Initialization, Automatic Drive Positioner System, Audio presets

and Navigation. Use the CONSULT-III diagnostic tool, or equivalent to perform the required resets.

2. Disconnect the negative battery cable.
3. Raise and support the vehicle safely, as required.
4. Position a catch pan under the assembly.
5. Drain the fluid. Be sure to properly dispose of used fluid.
6. Disconnect the parking brake cable and brake tube.
7. Remove the carrier cover retaining bolts. Remove the cover. Discard the gasket.

To install:

➡Be sure to use new fasteners, as required.

8. Installation is the reverse of the removal procedure.

➡Be sure to properly remove all old RTV sealant on all mating surfaces.

9. Apply a thin bead (0.12 inch) of genuine RTV sealant, or equivalent, to the mating surface of the carrier cover.
10. Be sure to perform the reconnect/relearn procedures.

TRANSFER CASE

REMOVAL & INSTALLATION
See Figure 24.

1. Before servicing the vehicle, refer to the Precautions Section.

➡If working near and/or around the SRS system and components, be sure to disable the SRS system. After disabling the system wait three minutes or more before servicing the vehicle.

➡Whenever the negative battery cable is disconnected the following compo-

nents will require resetting. The Idle Air Volume Learning, Steering Angle Sensor Neutral Position, Sunroof Memory Reset/Initialization, Automatic Drive Positioner System, Audio presets and Navigation. Use the CONSULT-III diagnostic tool, or equivalent to perform the required resets.

2. Disconnect the negative battery cable.
3. Raise and safely support the vehicle.
4. Remove the engine undercovers.
5. Drain the transfer case.
6. Ensure the transfer case is set to 2WD.
7. Remove or disconnect the following:
   • Transmission splash guard
   • Center exhaust pipe and muffler
   • Front and rear driveshafts

➡Plug rear oil seal after removing rear driveshaft.

   • Transmission assembly mounting bolts
8. Support the transmission assembly with a suitable jack and remove the crossmember.
9. Remove or disconnect the following:
   • ATP switch, neutral 4LO switch, wait detection switch, transfer motor and transfer control device electrical connectors
   • Breather hoses
   • Shift actuator from the extension housing
   • Transfer case to transmission assembly bolts
   • Transfer case assembly

To install:
10. Install or connect the following:
   • Transfer case to transmission

assembly bolts, tightening to 27 ft. lbs. (36 Nm)
- Shift actuator
- Breather hoses
- ATP switch, neutral 4LO switch,

wait detection switch, transfer motor and transfer control device electrical connectors
- Support crossmember
- Transmission mounting bolts

- Driveshafts
- Muffler and center exhaust pipe
- Transmission splash guard

11. Be sure to perform the reconnect/ relearn procedures.

ENGINE COOLING

BELT-DRIVEN ENGINE FAN

REMOVAL & INSTALLATION

See Figure 25.

1. Before servicing the vehicle, refer to the Precautions Section.

➡️**If working near and/or around the SRS system and components, be sure to disable the SRS system. After disabling the system wait three minutes or more before servicing the vehicle.**

➡️**Whenever the negative battery cable is disconnected the following components will require resetting. The Idle Air Volume Learning, Steering Angle Sensor Neutral Position, Sunroof Memory Reset/Initialization, Automatic Drive Positioner System, Audio presets and Navigation. Use the CONSULT-III diagnostic tool, or equivalent to perform the required resets.**

2. Disconnect the negative battery cable.
3. Remove or disconnect the following:
- Engine cover
- Air duct and resonator assembly
- Lower radiator shroud
- Drive belt
- Cooling fan

To install:

➡️**Be sure to use new fasteners, as required.**

4. To install, reverse removal procedure.

5. Install cooling fan with its front mark "F" facing front of engine.
6. Be sure to perform the reconnect/ relearn procedures.

ENGINE COOLANT

DRAIN & REFILL

See Figures 26 through 28.

➡️**You will need tool KV991J0070, or equivalent, and a compressed air supply to perform this procedure.**

1. Before servicing the vehicle, refer to the Precautions Section.

➡️**If working near and/or around the SRS system and components, be sure to disable the SRS system. After disabling the system wait three minutes or more before servicing the vehicle.**

➡️**Whenever the negative battery cable is disconnected the following components will require resetting. The Idle Air Volume Learning, Steering Angle Sensor Neutral Position, Sunroof Memory Reset/Initialization, Automatic Drive Positioner System, Audio presets and Navigation. Use the CONSULT-III diagnostic tool, or equivalent to perform the required resets.**

2. Disconnect the negative battery cable.

➡️**Never drain the engine coolant when the engine is hot. Wrap a thick cloth around the cap and carefully remove it.**

First turn the cap a quarter of a turn to release any pressure, and then turn it all the way. Do not allow coolant to come in contact with the drive belts.

3. Turn the ignition switch ON and set the temperature control lever all the way to the HOT position, or the highest temperature position. Wait ten seconds and turn the ignition switch OFF.
4. Raise and safely support the vehicle.
5. Remove the undercover.
6. Open the radiator drain plug and carefully drain the coolant into a suitable container. Be sure to properly dispose of used coolant.
7. Discard the O-ring.
8. When draining the system for engine repair or removal, open the water drain plug on the right cylinder bank and the oil cooler

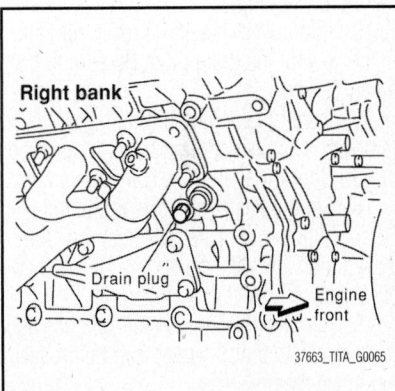

Fig. 26 Engine block drain plug location— right bank

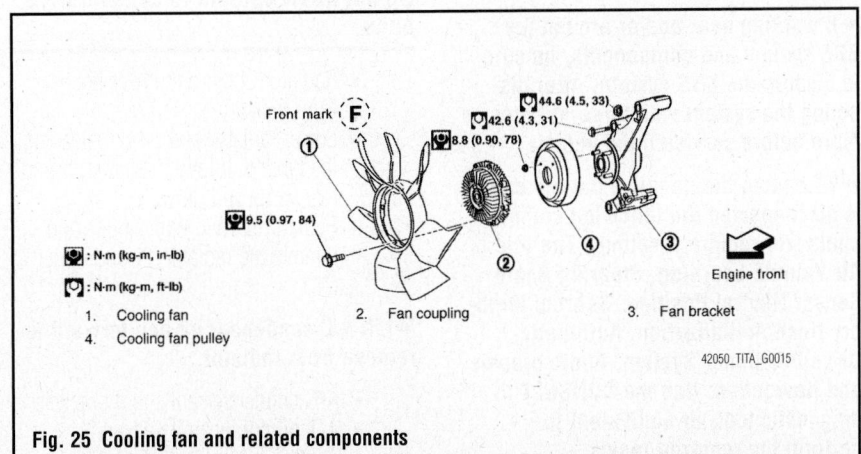

: N·m (kg-m, in-lb)

: N·m (kg-m, ft-lb)

1. Cooling fan
2. Fan coupling
3. Fan bracket
4. Cooling fan pulley

44.6 (4.5, 33)
42.6 (4.3, 31)
8.8 (0.90, 78)
9.5 (0.97, 84)

Front mark (F)

Engine front

42050_TITA_G0015

Fig. 25 Cooling fan and related components

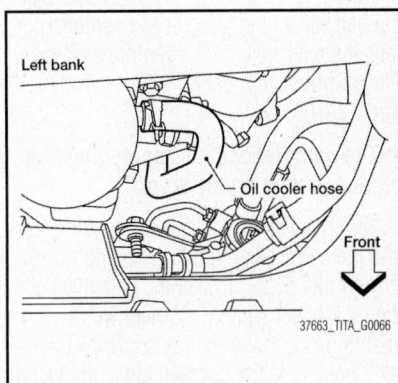

37663_TITA_G0066

Fig. 27 Engine cooler drain hose location—left bank

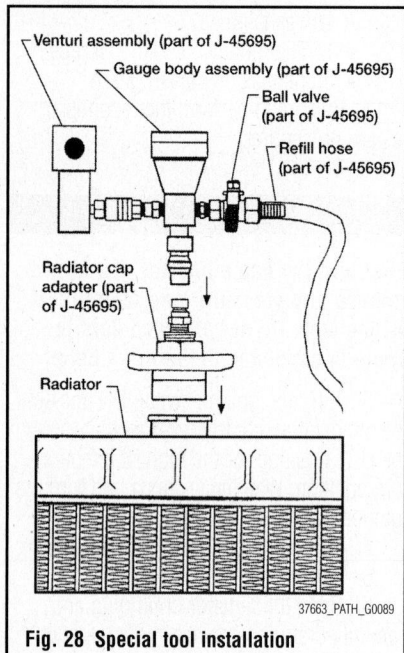

Fig. 28 Special tool installation

hose on the left bank and drain the coolant into a suitable container. Be sure to properly dispose of used coolant.

9. As necessary, remove the reservoir tank and drain the coolant. Clean the tank before reinstalling.

10. If removed install the reservoir tank.

11. Install the drain plug. Be sure to use a new O-ring.

12. Install the engine drain plug, if removed. Connect cooling hose, if removed.

13. Check that all hose clamps are tight.

14. Set the heater controls to the full HOT position and heater ON position. Turn the ignition ON with the engine OFF as necessary to activate the heater mode.

15. Remove the vented reservoir cap and replace it with a non-vented cap before filling the system.

16. Install tool KV991J0070 or equivalent, see illustration.

17. Insert the refill hose into the coolant mixture (placed at floor level). Be sure that the ball valve is in the closed position. Install the air hose to the venture assembly. The air pressure must be within specification (80–120 psi).

➡**The compressed air supply must be equipped with an air dryer.**

18. The vacuum gauge will begin to rise and there will be an audible hissing noise. During this process open the ball valve on the refill hose, slightly. Coolant will be visible rising in the refill hose. Once the refill hose is full of coolant, close the ball valve. This will purge any trapped air in the refill hose.

19. Continue to draw vacuum until the gauge reaches 28 inches of vacuum.

➡**The gauge may not reach specification in high altitude applications. If not see the following specifications. Altitude above sea level: 0-100m (328 ft.) 28 inches of vacuum. Altitude above sea level: 300m (984 ft.) 27 inches of vacuum. Altitude above sea level: 500m (1641 ft.) 26 inches of vacuum. Altitude above sea level: 1000m (3281 ft.) 24-25 inches of vacuum.**

20. When the proper specification has been reached, disconnect the air hose and wait 20 seconds to see if the system loses vacuum. If the level drops perform any necessary repairs and repeat the procedure.

21. Place the coolant container (with the refill hose inserted) at the same level as the top of the radiator. Then open the ball valve on the refill hose so the coolant will be drawn up to fill the cooling system. The cooling system is full when the vacuum gauge reads zero.

➡**Do not allow the coolant container to get too low when filling, to avoid air from being drawn into the system.**

22. Remove the toll from the radiator neck opening and install the radiator cap.

23. Remove the non-vented reservoir cap.

24. Fill the reservoir tank to specification with the proper coolant mixture.

25. Be sure to perform the reconnect/relearn procedures.

FLUID RECOMMENDATIONS

Be sure to use genuine Nissan (blue) coolant when filling/servicing the cooling system.

LEVEL CHECK

1. Before servicing the vehicle, refer to the Precautions Section.

➡**If working near and/or around the SRS system and components, be sure to disable the SRS system. After disabling the system wait three minutes or more before servicing the vehicle.**

➡**Whenever the negative battery cable is disconnected the following components will require resetting. The Idle Air Volume Learning, Steering Angle Sensor Neutral Position, Sunroof Memory Reset/Initialization, Automatic Drive Positioner System, Audio presets and Navigation. Use the CONSULT-III diagnostic tool, or equivalent to perform the required resets.**

Check that the coolant reservoir tank is level and within the MIN and MAX marks. Adjust coolant level, as necessary.

RADIATOR

REMOVAL & INSTALLATION
See Figure 29.

✱✱ CAUTION

Never remove the radiator cap when the engine is hot. Serious burns could occur from high-pressure engine coolant escaping from the radiator. Perform this procedure when engine is cold.

1. Before servicing the vehicle, refer to the Precautions Section.

➡**If working near and/or around the SRS system and components, be sure to disable the SRS system. After disabling the system wait three minutes or more before servicing the vehicle.**

➡**Whenever the negative battery cable is disconnected the following components will require resetting. The Idle Air Volume Learning, Steering Angle Sensor Neutral Position, Sunroof Memory Reset/Initialization, Automatic Drive Positioner System, Audio presets and Navigation. Use the CONSULT-III diagnostic tool, or equivalent to perform the required resets.**

2. Disconnect the negative battery cable.
3. Drain the cooling system.
4. Remove or disconnect the following:
   - Engine splash guard
   - Air intake assembly
   - A/T fluid cooler hoses, install blind plug to avoid leakage of A/T fluid

✱✱ WARNING

Do not allow coolant to contact drive belts.

   - Radiator upper and lower hoses from radiator
   - Lower radiator shroud by releasing the tabs, pull lower radiator shroud rearwards and down
   - Radiator shroud upper bolts and remove the radiator shroud upper (A)

➡**Lift A/C condenser up and forward to remove from radiator.**

   - A/C condenser bolts and brackets
   - A/T oil cooler bolts and oil

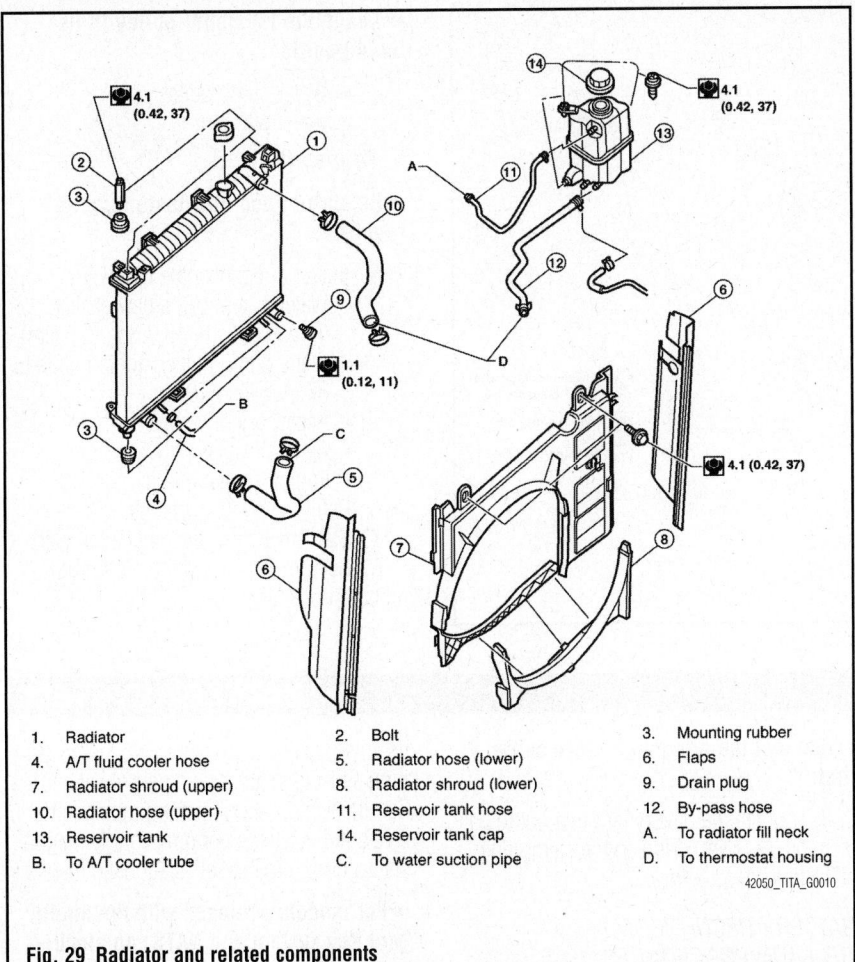

1. Radiator
4. A/T fluid cooler hose
7. Radiator shroud (upper)
10. Radiator hose (upper)
13. Reservoir tank
B. To A/T cooler tube

2. Bolt
5. Radiator hose (lower)
8. Radiator shroud (lower)
11. Reservoir tank hose
14. Reservoir tank cap
C. To water suction pipe

3. Mounting rubber
6. Flaps
9. Drain plug
12. By-pass hose
A. To radiator fill neck
D. To thermostat housing

42050_TITA_G0010

Fig. 29 Radiator and related components

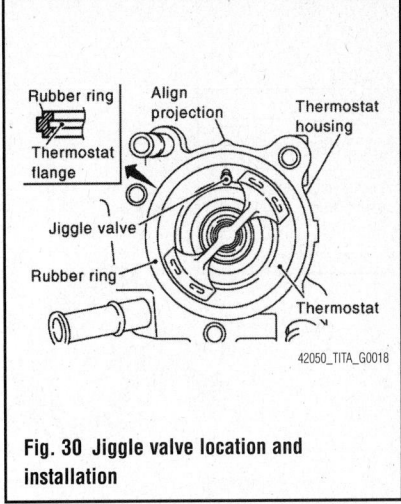

Fig. 30 Jiggle valve location and installation

cooler from radiator and position aside

✳✳ WARNING

Do not damage or scratch air conditioner condenser and radiator core when removing.

• Radiator

To install:

➡**Be sure to use new fasteners, as required.**

5. Installation is the reverse of the removal procedure.
6. Start and warm up the engine. Visually check for leaks of the engine coolant and A/T fluid.
7. Be sure to perform the reconnect/relearn procedures.

THERMOSTAT

REMOVAL & INSTALLATION

See Figure 30.

1. Before servicing the vehicle, refer to the Precautions Section.

➡**If working near and/or around the SRS system and components, be sure to disable the SRS system. After disabling the system wait three minutes or more before servicing the vehicle.**

➡**Whenever the negative battery cable is disconnected the following components will require resetting. The Idle Air Volume Learning, Steering Angle Sensor Neutral Position, Sunroof Memory Reset/Initialization, Automatic Drive Positioner System, Audio presets and Navigation. Use the CONSULT-III diagnostic tool, or equivalent to perform the required resets.**

2. Disconnect the negative battery cable.
3. Drain the cooling system.
4. Remove or disconnect the following:
• Engine cover
• Air duct and resonator assembly
• Water suction hose from the water inlet
• Water inlet and thermostat

To install:

➡**Be sure to use new fasteners, as required.**

5. To install, reverse removal procedure.
6. Install the thermostat with the whole circumference of each flange part fit securely inside the rubber ring as shown.
7. Install the thermostat with the jiggle valve facing upwards.
8. Tighten the thermostat mounting bolts to 15 ft. lbs. (20.6 Nm).
9. Install engine coolant and bleed the system.
10. Start and warm up the engine. Visually check for leaks of the engine coolant.
11. Be sure to perform the reconnect/relearn procedures.

WATER PUMP

REMOVAL & INSTALLATION

See Figure 31.

1. Before servicing the vehicle, refer to the Precautions Section.

➡**If working near and/or around the SRS system and components, be sure to disable the SRS system. After disabling the system wait three minutes or more before servicing the vehicle.**

➡**Whenever the negative battery cable is disconnected the following components will require resetting. The Idle Air Volume Learning, Steering Angle Sensor Neutral Position, Sunroof Memory Reset/Initialization, Automatic Drive Positioner System, Audio presets and Navigation. Use the CONSULT-III diagnostic tool, or equivalent to perform the required resets.**

2. Disconnect the negative battery cable.
3. Drain the cooling system.
4. Remove or disconnect the following:
• Engine splash guard
• Air intake assembly
• Accessory drive belt

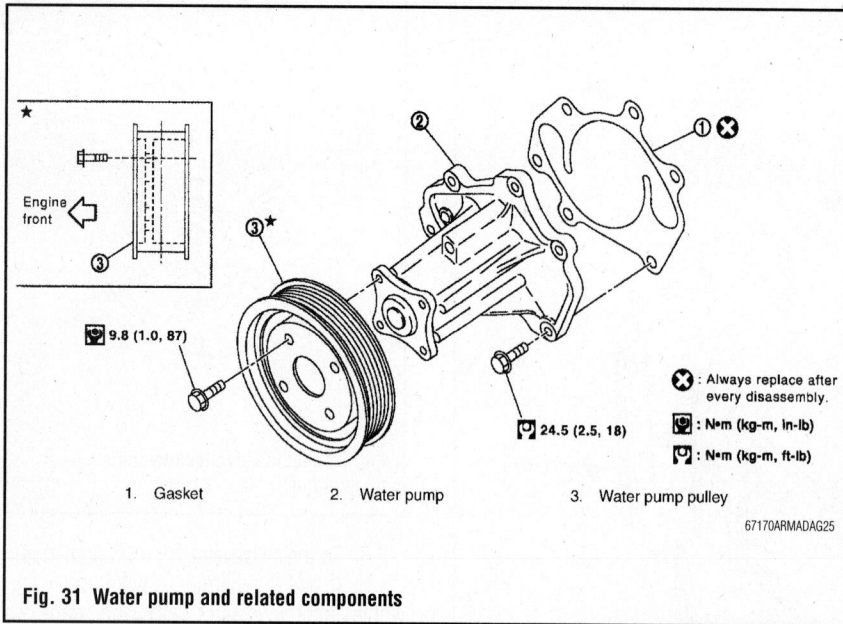

9.8 (1.0, 87)

24.5 (2.5, 18)

✕ : Always replace after every disassembly.

⚙ : N•m (kg-m, in-lb)

⚙ : N•m (kg-m, ft-lb)

1. Gasket
2. Water pump
3. Water pump pulley

67170ARMADAG25

Fig. 31 Water pump and related components

➡️Leave the tensioner pulley in its fixed position.

- Water pump pulley
- Water pump

To install:

➡️**Be sure to use new fasteners, as required.**

5. Install or connect the following:
- Water pump with a new gasket, tighten bolts to 18 ft. lbs. (25 Nm)
- Water pump pulley, tighten bolts to 87 inch lbs. (10 Nm)
- Accessory drive belt
- Air intake assembly
- Engine splash guard
6. Refill the cooling system.
7. Start the engine and check for leaks.
8. Be sure to perform the reconnect/relearn procedures.

ENGINE ELECTRICAL

BATTERY

REMOVAL & INSTALLATION

➡️**Whenever the negative battery cable is disconnected the following components will require resetting. The Idle Air Volume Learning, Steering Angle Sensor Neutral Position, Sunroof Memory Reset/Initialization, Automatic Drive Positioner System, Audio presets and Navigation. Use the CONSULT-III diagnostic tool, or equivalent to perform the required resets.**

1. Before servicing the vehicle, refer to the Precautions Section.

➡️**If working near and/or around the SRS system and components, be sure to disable the SRS system. After disabling the system wait three minutes or more before servicing the vehicle.**

2. Disconnect both cables from the battery.

➡️**Disconnect the negative battery cable, first.**

3. Remove the battery cover.
4. Remove the battery clamp nuts and battery clamp.
5. Remove the battery.

To install:

6. Installation is the reverse of the removal procedure.
7. Perform reconnect/relearn procedures.

➡️**Connect the positive battery cable first.**

8. Tighten the battery terminal nut to 31 inch lbs. (3.5 Nm). Tighten the clamp nuts to 11 ft. lbs.3 (3.5 Nm).

BATTERY RECONNECT/ RELEARN PROCEDURE

➡️**Whenever the negative battery cable is disconnected the following components will require resetting. The Idle Air Volume Learning, Steering Angle Sensor Neutral Position, Sunroof Memory Reset/Initialization, Automatic Drive Positioner System, Audio presets and Navigation. Use the CONSULT-III diagnostic tool, or equivalent to perform the required resets.**

➡️**The following systems, if equipped, require attention once the negative battery cable is disconnected. These systems are Automatic temperature control system, Automatic drive positioner, Power window control, Sunshade system and Rear view monitor. You will need the CONSULT-III diagnostic tool, or equivalent. Follow the directions on the screen of the tool, as needed.**

➡️**The following applies to vehicles equipped with Intelligent Key system and Nissan Anti-Theft System (NATS).**

Remove and install all control units after disconnecting both battery cables with the ignition knob in the LOCK position. Always

BATTERY SYSTEM

use the CONSULT-III diagnostic tool to perform self-diagnostics as a part of each function inspection after finishing repair work. If DTC's are detected, perform trouble diagnosis according to the self-diagnostic results.

➡️**For models equipped with the Intelligent Key system and NATS, an electrically controlled steering lock mechanism is adopted on the key cylinder. For this reason, if the battery is disconnected or discharged the steering wheel will lock and steering wheel rotation will become impossible. If wheel rotation is required when battery power is interrupted, follow the procedure below before starting the repair operation.**

1. Connect both battery cables.

➡️**Supply power using jumper cables if the battery is discharged.**

2. Using the Intelligent Key or mechanical key turn the ignition switch to the ACC position. At this time the steering lock will be released.
3. Disconnect both battery cables. The steering lock will remain released and the steering wheel can be rotated.
4. Perform the necessary repair operation.
5. When repair is complete, return the ignition switch to the LOCK position before connecting the battery cables. At this time the steering lock mechanism will engage.
6. Using the CONSULT-III diagnostic tool, perform self-diagnostics.

ENGINE ELECTRICAL
CHARGING SYSTEM

ALTERNATOR

REMOVAL & INSTALLATION

See Figure 32.

1. Before servicing the vehicle, refer to the Precautions Section.

➡️**If working near and/or around the SRS system and components, be sure to disable the SRS system. After disabling the system wait three minutes or more before servicing the vehicle.**

➡️**Whenever the negative battery cable is disconnected the following components will require resetting. The Idle Air Volume Learning, Steering Angle Sensor Neutral Position, Sunroof Memory Reset/Initialization, Automatic Drive Positioner System, Audio presets and Navigation. Use the CONSULT-III diagnostic tool, or equivalent to perform the required resets.**

2. Disconnect the negative battery cable.
3. Remove or disconnect the following:
   - Drive belt
   - Fan shroud
   - Lower alternator bracket
   - Alternator upper bolt

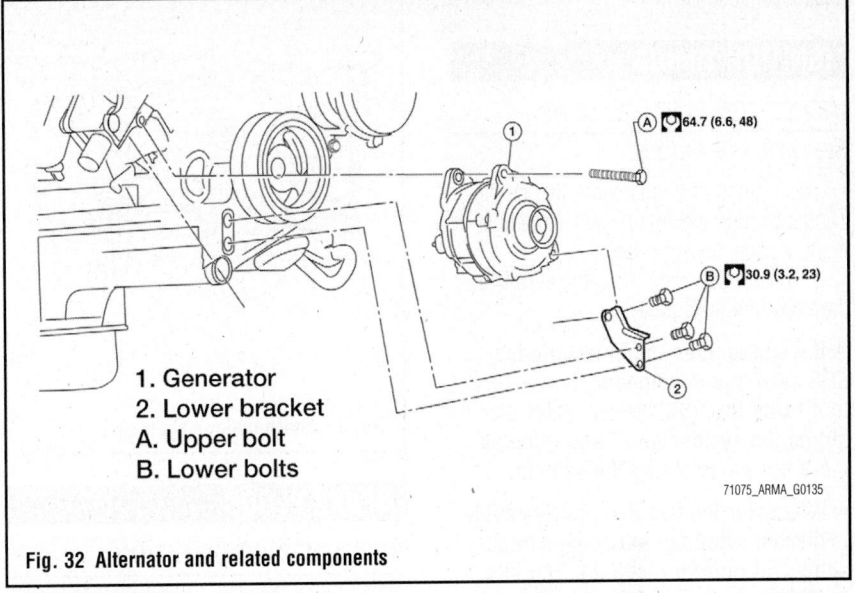

1. Generator
2. Lower bracket
A. Upper bolt
B. Lower bolts

71075_ARMA_G0135

Fig. 32 Alternator and related components

- Alternator harness connectors
- Alternator

To install:

➡️**Be sure to use new fasteners, as required.**

4. Install or connect the following:
   - Alternator
   - Alternator harness connectors

- Upper bolt, tighten to 48 ft. lbs. (65 Nm)
- Lower bracket, tighten to 16 ft. lbs (22 Nm)
- Drive belt
- Fan shroud
- Negative battery cable
5. Be sure to perform the reconnect/relearn procedures.

ENGINE ELECTRICAL
IGNITION SYSTEM

FIRING ORDERS

See Figure 33.

IGNITION COIL

REMOVAL & INSTALLATION

See Figure 34.

1. Before servicing the vehicle, refer to the Precautions Section.

➡️**If working near and/or around the SRS system and components, be sure to disable the SRS system. After disabling the system wait three minutes or more before servicing the vehicle.**

➡️**Whenever the negative battery cable is disconnected the following components will require resetting. The Idle Air Volume Learning, Steering Angle Sensor Neutral Position, Sunroof Memory Reset/Initialization, Automatic Drive Positioner System, Audio presets and Navigation. Use the CONSULT-III**

diagnostic tool, or equivalent to perform the required resets.

2. Disconnect the negative battery cable.
3. Remove the engine room cover using power tool.

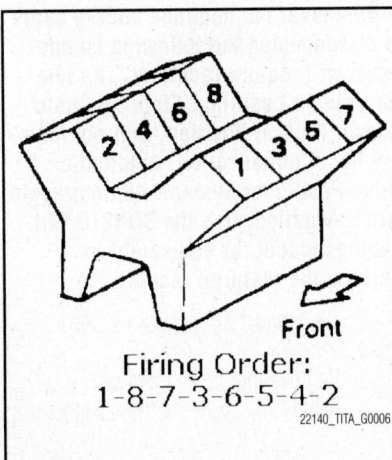

Firing Order:
1-8-7-3-6-5-4-2

22140_TITA_G0006

Fig. 33 Engine firing order—5.6L (VK56DE) engine

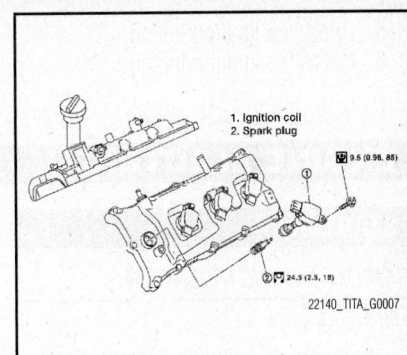

1. Ignition coil
2. Spark plug

22140_TITA_G0007

Fig. 34 Ignition coils and related components

4. Disconnect the harness connector from the ignition coil.
5. Remove the ignition coil.

To install:

➡️**Be sure to use new fasteners, as required.**

6. Install and tighten the ignition coil to 85 inch lbs. (10 Nm).

7. Install the engine room cover.

8. Be sure to perform the reconnect/relearn procedures.

IGNITION TIMING

INSPECTION & ADJUSTMENT

See Figures 35 and 36.

The ignition timing is controlled by the Engine Control Module (ECM). No adjustment is necessary or possible.

1. Before servicing the vehicle, refer to the Precautions Section.

➡**If working near and/or around the SRS system and components, be sure to disable the SRS system. After disabling the system wait three minutes or more before servicing the vehicle.**

➡**Whenever the negative battery cable is disconnected the following components will require resetting. The Idle Air Volume Learning, Steering Angle Sensor Neutral Position, Sunroof Memory Reset/Initialization, Automatic Drive Positioner System, Audio presets and Navigation. Use the CONSULT-III diagnostic tool, or equivalent to perform the required resets.**

2. Remove the number one ignition coil.

3. Connect the number one ignition coil and spark plug with a suitable high tension wire.

4. Attach the timing light clamp to the wire.

5. Check the ignition timing.

6. Check the ignition timing.

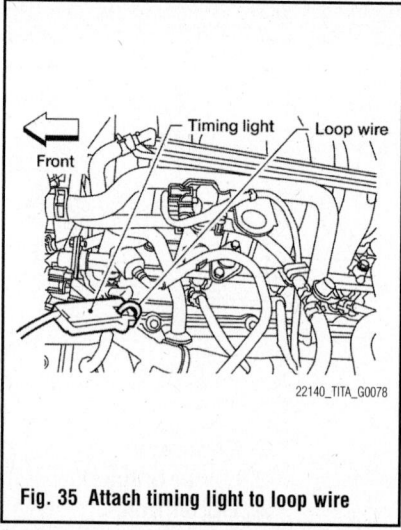

Fig. 35 Attach timing light to loop wire

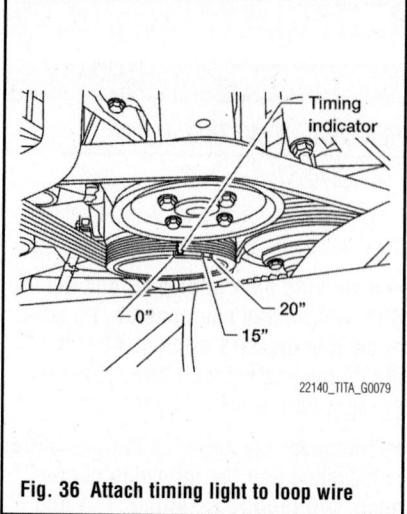

Fig. 36 Attach timing light to loop wire

SPARK PLUGS

REMOVAL & INSTALLATION

1. Before servicing the vehicle, refer to the Precautions Section.

➡**If working near and/or around the SRS system and components, be sure to disable the SRS system. After disabling the system wait three minutes or more before servicing the vehicle.**

➡**Whenever the negative battery cable is disconnected the following components will require resetting. The Idle Air Volume Learning, Steering Angle Sensor Neutral Position, Sunroof Memory Reset/Initialization, Automatic Drive Positioner System, Audio presets and Navigation. Use the CONSULT-III**

diagnostic tool, or equivalent to perform the required resets.

2. Disconnect the negative battery cable.

3. Remove the engine room cover using power tool.

4. Disconnect the harness connector from the ignition coil.

5. Remove the ignition coil.

6. Remove the spark plug.

To install:

➡**Be sure to use new fasteners, as required.**

7. Install the spark plug and tighten to 18 ft. lbs. (25 Nm).

8. Install and tighten the ignition coil to 85 inch lbs. (10 Nm).

9. Install the engine room cover.

10. Be sure to perform the reconnect/relearn procedures.

ENGINE ELECTRICAL

STARTER

REMOVAL & INSTALLATION

See Figure 37.

1. Before servicing the vehicle, refer to the Precautions Section.

➡**If working near and/or around the SRS system and components, be sure to disable the SRS system. After disabling the system wait three minutes or more before servicing the vehicle.**

➡**Whenever the negative battery cable is disconnected the following components will require resetting. The Idle Air Volume Learning, Steering Angle Sensor Neutral Position, Sunroof Memory Reset/Initialization, Automatic Drive Positioner System, Audio presets and Navigation. Use the CONSULT-III diagnostic tool, or equivalent to perform the required resets.**

2. Disconnect the negative battery cable.

3. Remove the intake manifold.

4. Remove the starter harness connectors.

STARTING SYSTEM

5. Remove the starter retaining bolts.

6. Remove the starter from its mounting.

To install:

➡**Be sure to use new fasteners, as required.**

7. Installation is the reverse of the removal procedure.

8. Tighten the retaining bolts to 34 ft. lbs. (46 Nm).

9. Tighten the terminal nut to 8 ft. lbs. (10.8 Nm).

10. Be sure to perform the reconnect/relearn procedures.

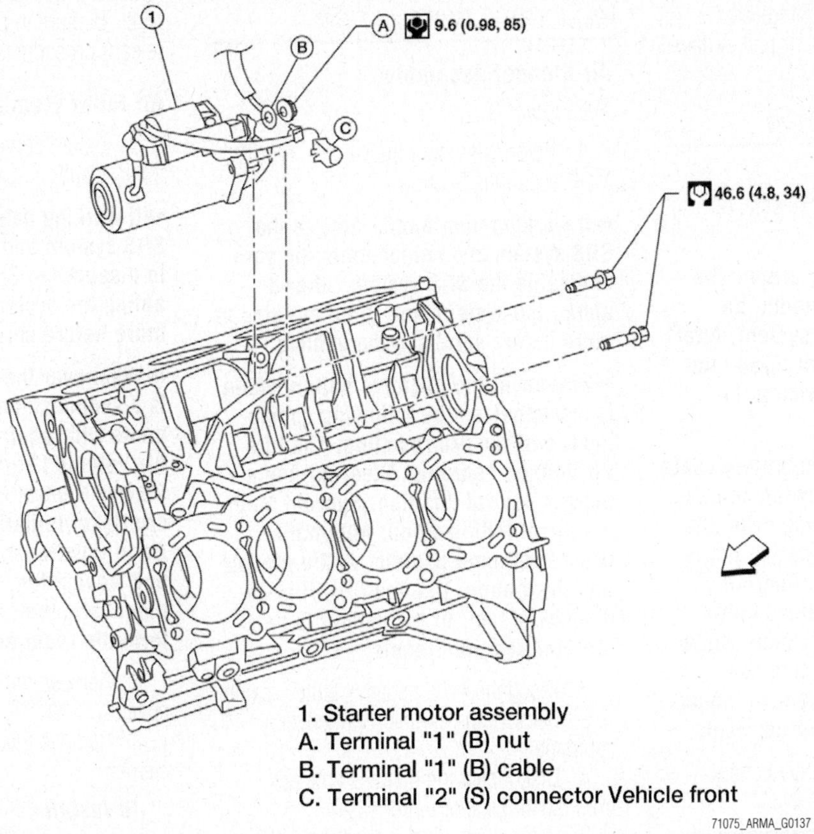

9.6 (0.98, 85)

46.6 (4.8, 34)

1. Starter motor assembly
A. Terminal "1" (B) nut
B. Terminal "1" (B) cable
C. Terminal "2" (S) connector Vehicle front

71075_ARMA_G0137

Fig. 37 Starter and related components

ENGINE MECHANICAL

ACCESSORY DRIVE BELT

ADJUSTMENT

There is no manual drive belt tension adjustment. The drive belt tension is automatically adjusted by the drive belt auto tensioner.

BELT ROUTING

See Figure 38.

INSPECTION

See Figure 38.

Remove air duct and resonator assembly when inspecting drive belt. Make sure that indicator (single line notch) of each auto tensioner is within the allowable working range "A"(between three line notches). The indicator notch is located on the moving side of the drive belt auto tensioner. Inspect the drive belt for signs of glazing or cracking. A glazed belt will be perfectly smooth from slippage, while a good belt will have a slight texture of fabric visible. Cracks will usually start at the

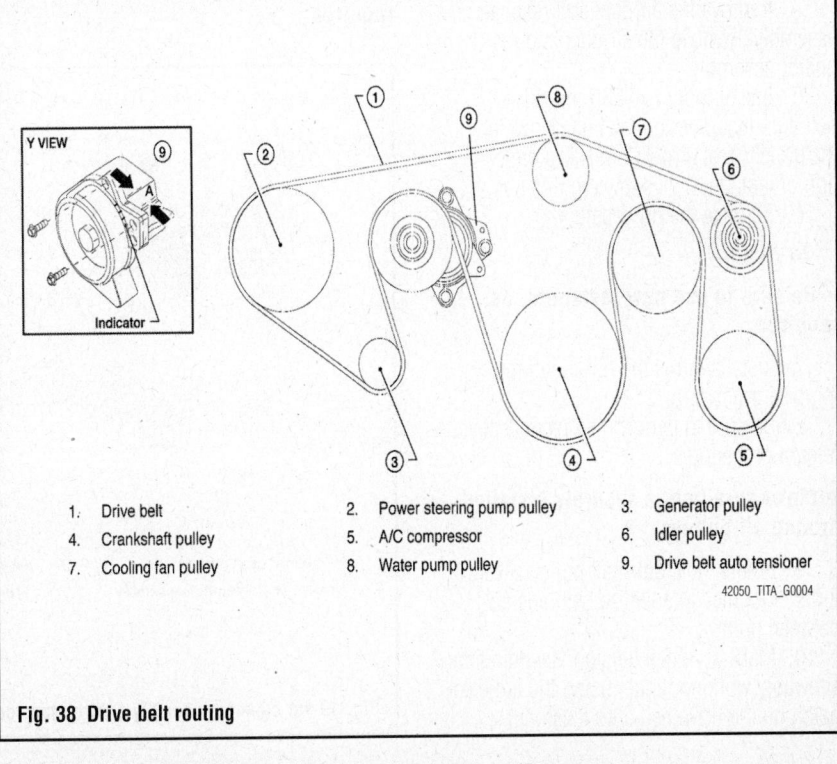

Y VIEW

Indicator

1.	Drive belt	2.	Power steering pump pulley	3.	Generator pulley
4.	Crankshaft pulley	5.	A/C compressor	6.	Idler pulley
7.	Cooling fan pulley	8.	Water pump pulley	9.	Drive belt auto tensioner

42050_TITA_G0004

Fig. 38 Drive belt routing

inner edge of the belt and run outward. All worn or damaged drive belts should be replaced immediately. If the indicator is out of allowable working range or belt is damaged, replace the belt

REMOVAL & INSTALLATION

See Figure 38.

1. Before servicing the vehicle, refer to the Precautions Section.

➡ If working near and/or around the SRS system and components, be sure to disable the SRS system. After disabling the system wait three minutes or more before servicing the vehicle.

➡ Whenever the negative battery cable is disconnected the following components will require resetting. The Idle Air Volume Learning, Steering Angle Sensor Neutral Position, Sunroof Memory Reset/Initialization, Automatic Drive Positioner System, Audio presets and Navigation. Use the CONSULT-III diagnostic tool, or equivalent to perform the required resets.

2. Disconnect the negative battery cable.

✳✳ CAUTION

Avoid placing hand in a location where pinching may occur if the holding tool accidentally comes off.

3. Remove the engine room cover.
4. Remove the air duct and resonator assembly. Remove the air duct and resonator assembly.
5. Install tool (J-46535) on drive belt auto tensioner pulley bolt, move in the direction of arrow (loosening direction of tensioner) as shown in illustration.
6. Remove the drive belt.

To install:

➡ Be sure to use new fasteners, as required.

7. Installation is the reverse of the removal procedure.
8. Be sure to perform the reconnect/relearn procedures.

➡ Make sure belt is securely installed around all pulleys.

9. Rotate the crankshaft pulley several turns clockwise to equalize belt tension between pulleys.
10. Make sure belt tension is within the allowable working range, using the indicator notch on the drive belt auto tensioner.

AIR INTAKE SYSTEM

REMOVAL & INSTALLATION

Air Cleaner Assembly

See Figure 39.

1. Before servicing the vehicle, refer to the Precautions Section.

➡ If working near and/or around the SRS system and components, be sure to disable the SRS system. After disabling the system wait three minutes or more before servicing the vehicle.

➡ Whenever the negative battery cable is disconnected the following components will require resetting. The Idle Air Volume Learning, Steering Angle Sensor Neutral Position, Sunroof Memory Reset/Initialization, Automatic Drive Positioner System, Audio presets and Navigation. Use the CONSULT-III diagnostic tool, or equivalent to perform the required resets.

2. Disconnect the negative battery cable.
3. Remove the engine room cover, if equipped.
4. Disconnect the harness connector from the air cleaner upper case.
5. Remove the air duct and resonator assembly and the air cleaner case.
6. Remove the air cleaner filter and lower air cleaner case.

To install:

➡ Be sure to use new fasteners, as required.

7. Installation is the reverse of the removal procedure.
8. Be sure to perform the reconnect/relearn procedures.

Air Filter Element

1. Before servicing the vehicle, refer to the Precautions Section.

➡ If working near and/or around the SRS system and components, be sure to disable the SRS system. After disabling the system wait three minutes or more before servicing the vehicle.

➡ Whenever the negative battery cable is disconnected the following components will require resetting. The Idle Air Volume Learning, Steering Angle Sensor Neutral Position, Sunroof Memory Reset/Initialization, Automatic Drive Positioner System, Audio presets and Navigation. Use the CONSULT-III diagnostic tool, or equivalent to perform the required resets.

2. Remove the air cleaner case upper cover.
3. Remove the air cleaner filter element.

To install:

➡ Be sure to use new fasteners, as required.

4. Installation is the reverse of the removal procedure.
5. Be sure to perform the reconnect/relearn procedures.

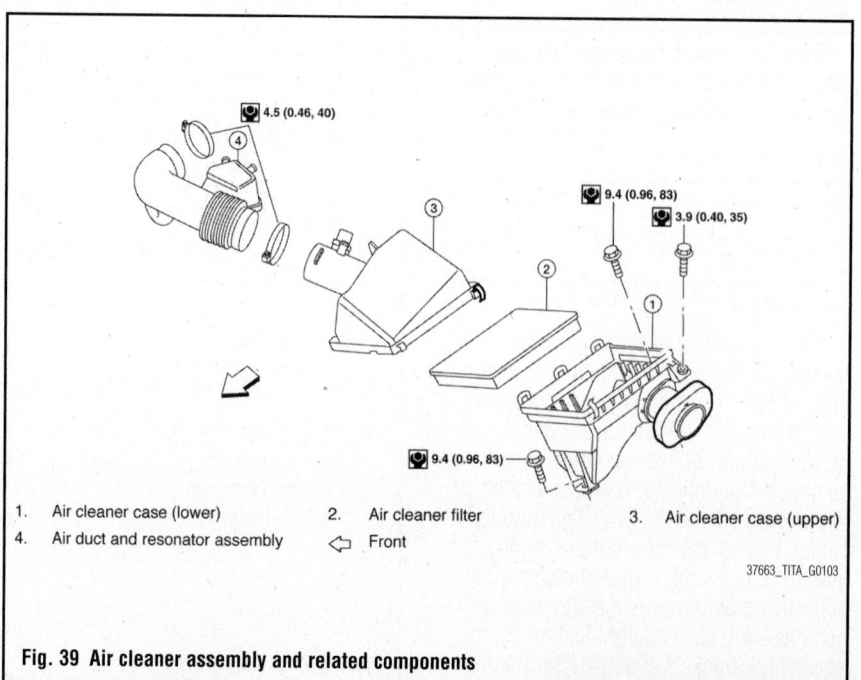

4.5 (0.46, 40)
9.4 (0.96, 83)
3.9 (0.40, 35)
9.4 (0.96, 83)

1. Air cleaner case (lower)
2. Air cleaner filter
3. Air cleaner case (upper)
4. Air duct and resonator assembly
⬅ Front

37663_TITA_G0103

Fig. 39 Air cleaner assembly and related components

CAMSHAFT & BEARINGS

REMOVAL & INSTALLATION

See Figures 40 through 56.

1. Before servicing the vehicle, refer to the Precautions Section.

➡️**If working near and/or around the SRS system and components, be sure to disable the SRS system. After disabling the system wait three minutes or more before servicing the vehicle.**

➡️**Whenever the negative battery cable is disconnected the following components will require resetting. The Idle Air Volume Learning, Steering Angle Sensor Neutral Position, Sunroof Memory Reset/Initialization, Automatic Drive Positioner System, Audio presets and Navigation. Use the CONSULT-III diagnostic tool, or equivalent to perform the required resets.**

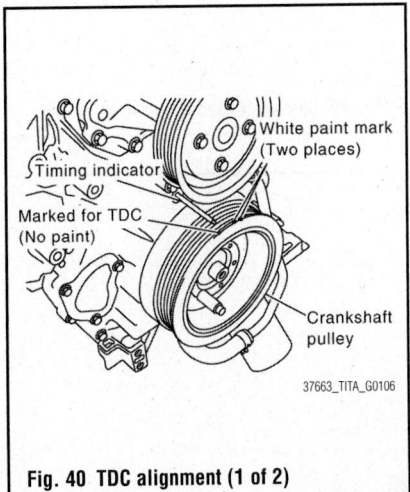

Fig. 40 TDC alignment (1 of 2)

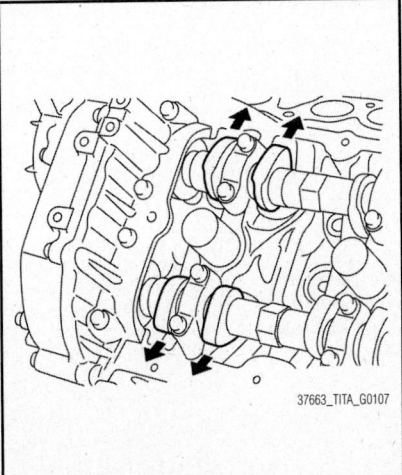

Fig. 41 TDC alignment (2 of 2)

2. Disconnect the negative battery cable.

3. Remove the power steering reservoir tank bolts. Position the unit to the side.

4. Remove the valve covers.

5. Remove the spark plugs.

6. Remove the drive belt.

7. Be sure that the number one cylinder is at TDC on the compression stroke.

➡️**Turn the crankshaft pulley clockwise to align the TDC identification notch (without paint mark) with the timing indicator on the front cover. At this time make sure that the intake and exhaust cam lobes of the number one cylinder (top front on left bank) point outside. If not turn the crankshaft pulley once more. See illustration**

8. Remove the CMP sensor.

9. Remove the intake valve timing control position sensors (right and left).

10. Remove the intake valve timing control solenoid valves (right and left).

11. Loosen and remove the intake valve timing control valve cover (right and left) bolts in the reverse order of the tightening sequence.

12. Paint alignment marks on the right bank (A) timing chain links (C) and left bank (B) timing chain links (D) and align with the camshaft sprocket alignment marks (E) and (F). See illustration.

13. To remove the left tensioner, squeeze the return proof clip ends using a suitable tool and push the plunger into the tensioner body. Secure the plunger using a stopper pin (hard wire 0.04 inch in diameter). Remove the bolts and the tensioner.

➡️**The plunger, spring and spring seat pop out when squeezing the return proof clip without holding the plunger head. It may cause serious injuries. Always hold the plunger head when removing.**

➡️**Stop the plunger in the fully extended position using the return proof clip (1) if the stopper pin is removed. Push the plunger (2) into the chain tensioner body while squeezing the return proof clip (1). Secure it using a stopper pin (3). See illustration.**

14. Remove the chain tensioner cover from the front cover, using tool

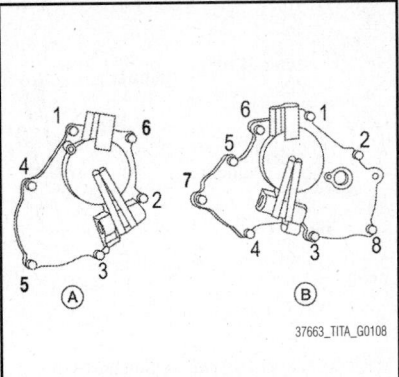

Fig. 42 Intake valve timing control solenoid cover tightening sequence

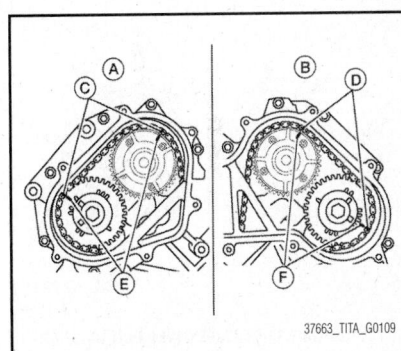

Fig. 43 Camshaft sprocket/chain link alignment

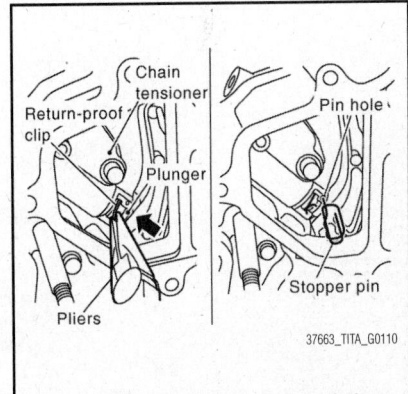

Fig. 44 Camshaft stopper pin installation

KV10111100, or equivalent. Do not damage the mating surfaces.

15. To remove the left tensioner, squeeze the return proof clip ends using a suitable tool and push the plunger into the tensioner body. Secure the plunger using a stopper pin (hard wire 0.04 inch in diameter). Remove the bolts and the tensioner.

➡️**The plunger, spring and spring seat pop out when squeezing the return proof clip without holding the plunger head. It may cause serious injuries.**

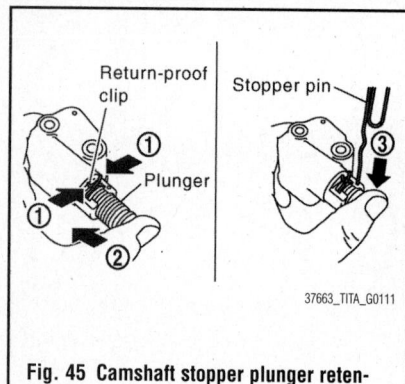

Fig. 45 Camshaft stopper plunger retention

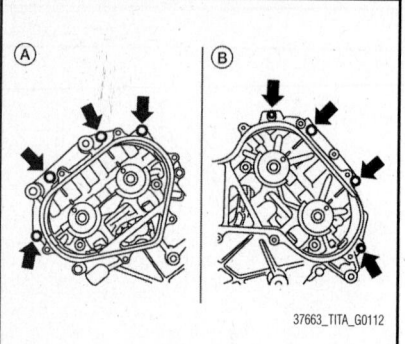

Fig. 46 Camshaft front cover bolt location (arrow)

Always hold the plunger head when removing.

➡ If it is difficult to push the plunger on the tensioner, remove the plunger under the extended condition.

16. Loosen the camshaft sprocket bolts and remove the sprockets.

➡ To avoid interference between the valves and pistons, do not turn the crankshaft or camshaft with the timing chain disconnected.

17. Remove the front cover bolts. See illustration for location (arrow).

18. Remove the camshaft bracket bolts in the reverse order of the tightening sequence. Remove the number one camshaft bracket. The bottom of the front surface of the bracket will be stuck because of liquid gasket.

19. Remove the camshaft. Remove the lifters, as necessary.

To install:

➡ Be sure to use new fasteners, as required.

20. Install the camshafts. Be sure that the camshafts are properly identified. See illustrations.

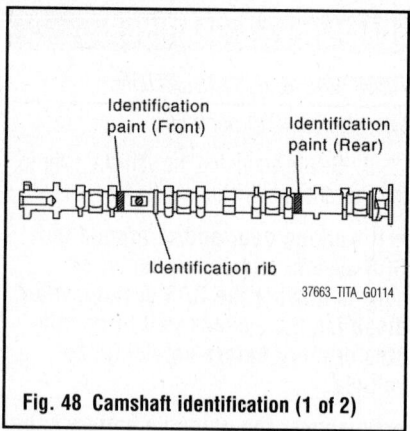

Fig. 48 Camshaft identification (1 of 2)

Bank	INT EXH	Identification paint (front)	Identification paint (rear)	Identification rib
RH	INT	Pink	—	Yes
	EXH	—	Orange	Yes
LH	INT	Pink	—	No
	EXH	—	Orange	No

37663_TITA_G0115

Fig. 49 Camshaft identification (2 of 2)

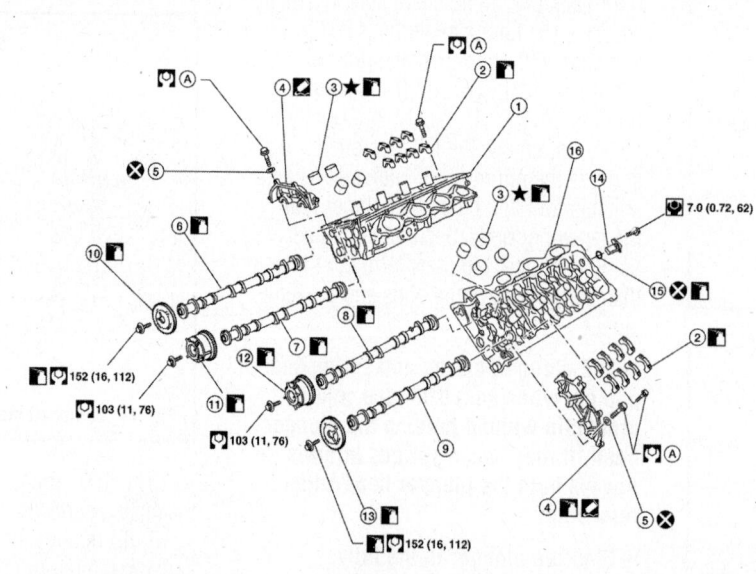

37663_TITA_G0105

1. Cylinder head RH bank
2. Camshaft bracket (No. 2, 3, 4, 5)
3. Valve lifter
4. Camshaft bracket (No. 1)
5. Seal washer
6. Camshaft RH bank EXH
7. Camshaft RH bank INT
8. Camshaft LH bank INT
9. Camshaft LH bank EXH
10. Camshaft sprocket RH bank EXH
11. Camshaft sprocket RH bank INT (VTC)
12. Camshaft sprocket LH bank INT (VTC)
13. Camshaft sprocket LH bank EXH
14. Camshaft position sensor (PHASE)
15. O-ring
16. Cylinder head LH bank

Fig. 47 Camshafts and related components

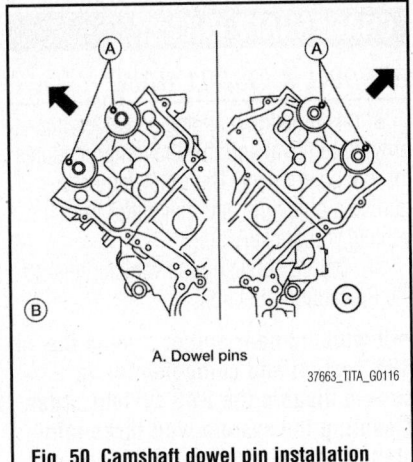

A. Dowel pins

37663_TITA_G0116

Fig. 50 Camshaft dowel pin installation

21. Install the dowel pins at the front of the camshaft. See illustration for proper direction.

22. Install the camshaft brackets.

➡**Install by referring to the illustration location mark on the upper surface. Install so that the installation mark can be correctly read when viewed from the intake manifold side.**

23. To install the number one

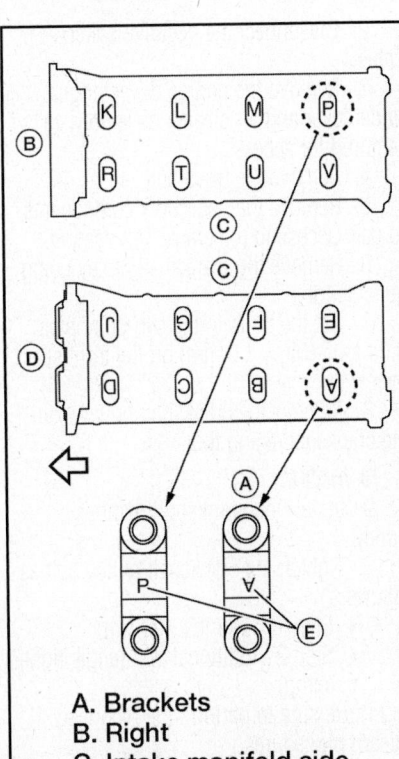

A. Brackets
B. Right
C. Intake manifold side
D. Left
E. Location mark

37663_TITA_G0117

Fig. 51 Camshaft bracket installation and identification

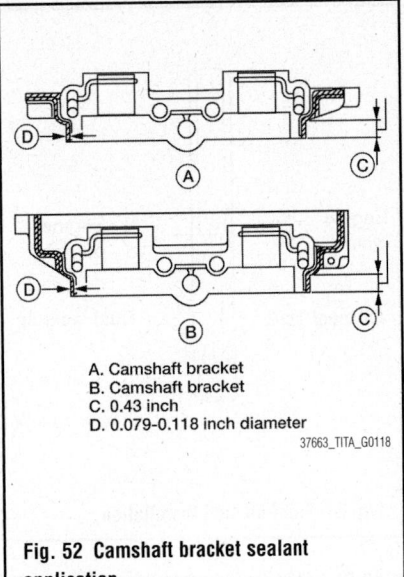

A. Camshaft bracket
B. Camshaft bracket
C. 0.43 inch
D. 0.079–0.118 inch diameter

37663_TITA_G0118

Fig. 52 Camshaft bracket sealant application

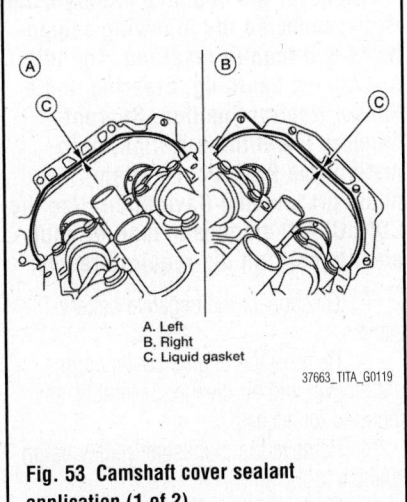

A. Left
B. Right
C. Liquid gasket

37663_TITA_G0119

Fig. 53 Camshaft cover sealant application (1 of 2)

camshaft bracket, apply liquid gasket as shown in the illustration. Be sure to wipe off any excessive gasket after installation.

24. Apply liquid gasket to the back side of the left front cover and the right front cover. Bead diameter should be 0.102–0.142 inch. Position the number one camshaft bracket close to the mounting position and then install it to prevent from touching gasket applied to each surface.

25. Temporarily tighten the right and left front cover bolts.

26. Tighten the camshaft bracket bolts to specification and in the proper sequence.

27. Tighten the right and left front cover bolts to 8 ft. lbs.

28. Install the camshaft sprockets

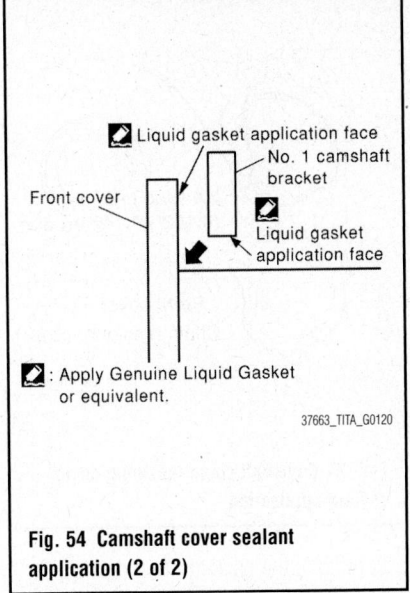

Liquid gasket application face
No. 1 camshaft bracket
Front cover
Liquid gasket application face

✐: Apply Genuine Liquid Gasket or equivalent.

37663_TITA_G0120

Fig. 54 Camshaft cover sealant application (2 of 2)

aligning them with the matching marks painted on the timing chain and the camshaft sprockets, before removal. Align the sprocket key groove with the dowel pin on the camshaft front edge at the same time. Temporarily tighten the sprocket bolts.

29. Install the intake VTC and the exhaust side camshaft sprockets by selectively using the groove of the dowel pin

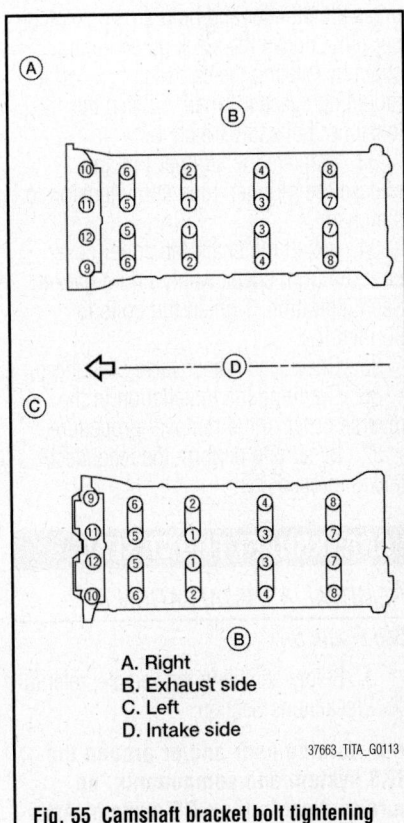

A. Right
B. Exhaust side
C. Left
D. Intake side

37663_TITA_G0113

Fig. 55 Camshaft bracket bolt tightening sequence

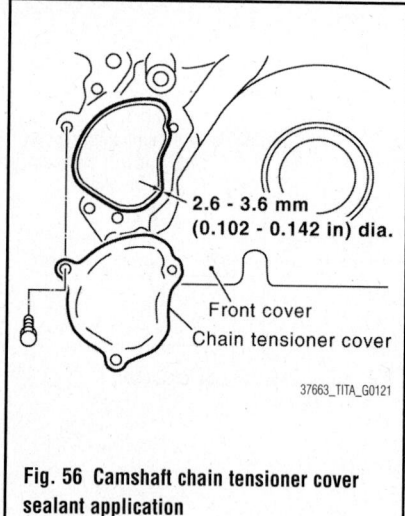

Fig. 56 Camshaft chain tensioner cover sealant application

according to the bank for the exhaust side camshaft sprockets, (common part used for both exhaust banks).

➡**Use the groove marked "R" for right bank and "L" for left bank.**

30. Lock the hex part of the camshaft in the same way as for removal. Tighten the sprocket bolts.

31. Check that the timing marks are properly aligned.

32. To install the chain tensioner, compress the plunger and hold it using a stopper pin. Loosen the slack guide timing chain by rotating the camshaft hex part if mounting space is small. Tighten the tensioner bolts to 61 inch lbs.

33. Remove the stopper pin and release the plunger, then apply tension to the chain.

34. Install the chain tensioner cover onto the front cover. Apply liquid gasket. See illustration. Tighten the bolts to 80 inch lbs.

35. Check and adjust valve clearances.

36. Continue the installation in the reverse order of the removal procedure.

37. Be sure to perform the reconnect/relearn procedures.

CRANKSHAFT FRONT SEAL

REMOVAL & INSTALLATION

See Figure 57.

1. Before servicing the vehicle, refer to the Precautions Section.

➡**If working near and/or around the SRS system and components, be sure to disable the SRS system. After disabling the system wait three min-**

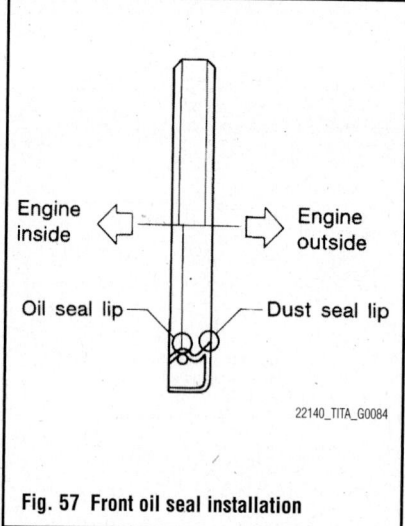

Fig. 57 Front oil seal installation

utes or more before servicing the vehicle.

➡**Whenever the negative battery cable is disconnected the following components will require resetting. The Idle Air Volume Learning, Steering Angle Sensor Neutral Position, Sunroof Memory Reset/Initialization, Automatic Drive Positioner System, Audio presets and Navigation. Use the CONSULT-III diagnostic tool, or equivalent to perform the required resets.**

2. Disconnect the negative battery cable.

3. Remove the engine cover, engine undercover and air cleaner assembly, as required for access.

4. Remove the crankshaft pulley using suitable tool.

5. Set the bolts in the two bolt holes 0.04 inch (M6 x 1.0 mm) on the front surface.

6. Remove the crankshaft pulley from the crankshaft using tool.

7. Remove the oil seal using a suitable tool.

To install:

8. Apply new engine oil to both the oil seal lip and dust seal lip of the new front oil seal.

9. Install the front oil seal so that each seal lip is oriented as shown.

10. Install the crankshaft damper pulley.

11. Tighten the crankshaft pulley bolt as follows:
   - Step 1: 69 ft. lbs. (93 Nm)
   - Step 2: Additional 90° (angle tightening)

12. Be sure to perform the reconnect/relearn procedures.

CRANKSHAFT PULLEY

REMOVAL & INSTALLATION

At this time the manufacturer does not provide removal and installation procedures for this component. The following procedure is a guideline and may differ from the vehicle you are servicing.

1. Before servicing the vehicle, refer to the Precautions Section.

➡**If working near and/or around the SRS system and components, be sure to disable the SRS system. After disabling the system wait three minutes or more before servicing the vehicle.**

➡**Whenever the negative battery cable is disconnected the following components will require resetting. The Idle Air Volume Learning, Steering Angle Sensor Neutral Position, Sunroof Memory Reset/Initialization, Automatic Drive Positioner System, Audio presets and Navigation. Use the CONSULT-III diagnostic tool, or equivalent to perform the required resets.**

2. Disconnect the negative battery cable.

3. Remove the engine cover, engine undercover and air cleaner assembly, as required for access.

4. Remove the drive belt.

5. Remove the necessary components to gain access to the crankshaft damper.

6. Remove the crankshaft pulley using suitable tool.

7. Set the bolts in the two bolt holes 0.04 inch (M6 x 1.0 mm) on the front surface.

8. Remove the crankshaft pulley from the crankshaft using tool.

To install:

9. Install the crankshaft damper pulley.

10. Tighten the crankshaft pulley bolt as follows:
   - Step 1: 69 ft. lbs. (93 Nm)
   - Step 2: Additional 90° (angle tightening)

11. Be sure to perform the reconnect/relearn procedures.

ENGINE COVER

REMOVAL & INSTALLATION

See Figure 58.

1. Before servicing the vehicle, refer to the Precautions Section.

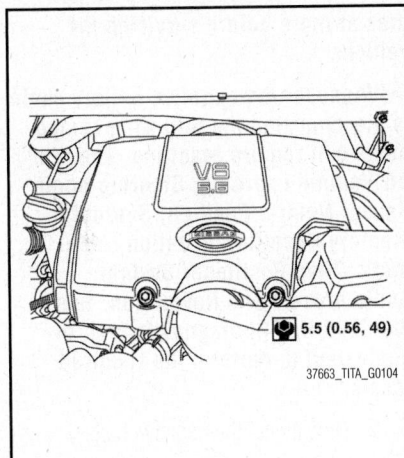

Fig. 58 Engine room cover and related components

➡If working near and/or around the SRS system and components, be sure to disable the SRS system. After disabling the system wait three minutes or more before servicing the vehicle.

➡Whenever the negative battery cable is disconnected the following components will require resetting. The Idle Air Volume Learning, Steering Angle Sensor Neutral Position, Sunroof Memory Reset/Initialization, Automatic Drive Positioner System, Audio presets and Navigation. Use the CONSULT-III diagnostic tool, or equivalent to perform the required resets.

2. Disconnect the negative battery cable.
3. Remove the cover retaining bolts.
4. Lift up on the cover to disengage the snap fit mounts.

To install:

➡Be sure to use new fasteners, as required.

5. Installation is the reverse of the removal procedure.

ENGINE OIL & FILTER

OIL LEVEL CHECK

See Figure 59.

✳ CAUTION

Prolonged and repeated contact with used engine oil may cause skin cancer. Try to avoid direct skin contact with used oil. If skin contact is made, wash thoroughly with soap or hand cleaner as soon as possible.

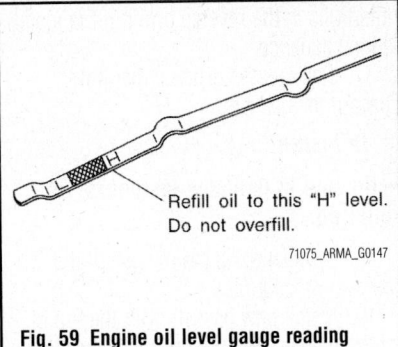

Refill oil to this "H" level. Do not overfill.

71075_ARMA_G0147

Fig. 59 Engine oil level gauge reading

Wear protective clothing, including impervious gloves where practicable. Do not use gasoline, kerosene, diesel fuel, gas oil, thinners, or solvents for cleaning skin. Where there is a risk of eye contact, eye protection should be worn, for example, chemical goggles or face shields; in addition an eye wash facility should be provided.

➡Be sure to check the engine oil with the vehicle engine OFF and parked on a level surface. If the engine was running prior to checking the oil, turn it off and wait 10 minutes.

1. Before servicing the vehicle, refer to the Precautions Section.
2. Remove the oil level gauge and wipe it clean with a shop towel. Be sure to properly dispose of the used shop towel.
3. Insert the gauge into its mounting.
4. Remove the oil level gauge and check the reading. See illustration.
5. Correct, oil level as required. Be sure to use the proper grade and type engine oil.

OIL & FILTER CHANGE

See Figure 60.

1. Before servicing the vehicle, refer to the Precautions Section.

➡If working near and/or around the SRS system and components, be sure to disable the SRS system. After disabling the system wait three minutes or more before servicing the vehicle.

➡Whenever the negative battery cable is disconnected the following components will require resetting. The Idle Air Volume Learning, Steering Angle Sensor Neutral Position, Sunroof Memory Reset/Initialization, Automatic Drive Positioner System, Audio presets and Navigation. Use the

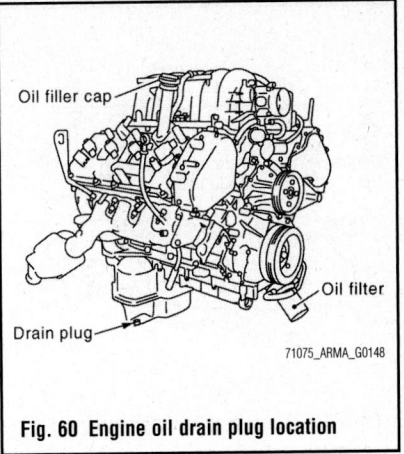

Fig. 60 Engine oil drain plug location

CONSULT-III diagnostic tool, or equivalent to perform the required resets.

2. Disconnect the negative battery cable.

➡Be sure that the engine is cold and the engine oil is cold.

3. Raise and safely support the vehicle.
4. Remove the undercover, if equipped.
5. Remove the oil drain plug. Discard the washer.
6. Drain the engine oil into a suitable container. Properly dispose of used engine oil.
7. Using oil filter wrench remove the oil filter. Discard the filter.

To install:

➡Be sure to use new fasteners, as required

8. Install the drain plug. Use a new washer.
9. Tighten to specification, 25 ft. lbs. (34 Nm).
10. Coat the oil filter seal with clean engine oil prior to installation.
11. Do not over tighten the filter to its mounting.
12. Tightening specification should be 13 ft. lbs. (17.7 Nm).
13. Fill the engine with the proper grade and type engine oil.
14. Start the engine and check for leaks, correct as required.

EXHAUST MANIFOLD

REMOVAL & INSTALLATION

See Figure 61.

1. Before servicing the vehicle, refer to the Precautions Section.

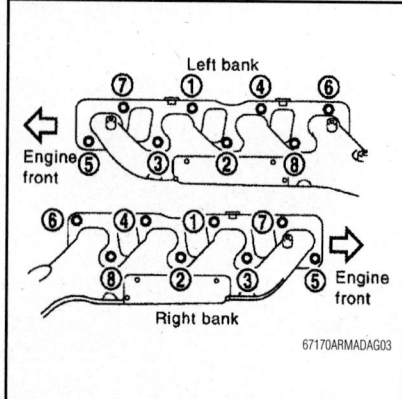

Fig. 61 Exhaust manifold bolt torque sequence

➡If working near and/or around the SRS system and components, be sure to disable the SRS system. After disabling the system wait three minutes or more before servicing the vehicle.

➡Whenever the negative battery cable is disconnected the following components will require resetting. The Idle Air Volume Learning, Steering Angle Sensor Neutral Position, Sunroof Memory Reset/Initialization, Automatic Drive Positioner System, Audio presets and Navigation. Use the CONSULT-III diagnostic tool, or equivalent to perform the required resets.

2. Disconnect the negative battery cable.
3. Raise and support the vehicle safely.
4. Remove the engine under cover, if equipped.
5. Remove the front final drive assembly, if equipped.
6. Remove the main muffler and center exhaust tube.
7. Remove the front exhaust tubes.
8. Remove the tire and wheel assemblies.
9. Remove the fender protectors.
10. Remove the A/F sensors. Do not drop the sensors. If the sensor is dropped it must be replaced.
11. Properly support the engine.
12. Remove the engine mounting insulator.
13. Remove the exhaust manifold cover.
14. Remove the engine mounting bracket.
15. On right side remove the oil level dipstick.
16. Remove the left exhaust manifold

nuts/bolts in the reverse order of the installation sequence.
17. Remove the exhaust manifold. Discard the gaskets.

To install:

➡Be sure to use new fasteners, as required.

18. Installation is the reverse of the removal procedure.
19. Install new gaskets with the top of the triangular UP mark on it facing up and its coated face (gray side) toward the exhaust manifold side.
20. Tighten the retaining nuts/bolts to specification and in the proper sequence.
21. Be sure to perform the reconnect/relearn procedures.

INTAKE MANIFOLD

REMOVAL & INSTALLATION

See Figures 62 and 63.

1. Before servicing the vehicle, refer to the Precautions Section.

➡If working near and/or around the SRS system and components, be sure to disable the SRS system. After disabling the system wait three minutes or more before servicing the vehicle.

➡Whenever the negative battery cable is disconnected the following components will require resetting. The Idle Air Volume Learning, Steering Angle Sensor Neutral Position, Sunroof Memory Reset/Initialization, Automatic Drive Positioner System, Audio presets and Navigation. Use the CONSULT-III diagnostic tool, or equivalent to perform the required resets.

2. Disconnect the negative battery cable.
3. Drain the cooling system.
4. Relieve the fuel system pressure.
5. Remove or disconnect the following:
- Engine cover
- Air intake assembly
- Fuel tube quick connector using special tool J-45488
- Wiring harnesses and brackets from manifold
- Vacuum hoses
- PCV hose and tube
- Electric throttle control actuator, loosening bolts diagonally
- Fuel injectors

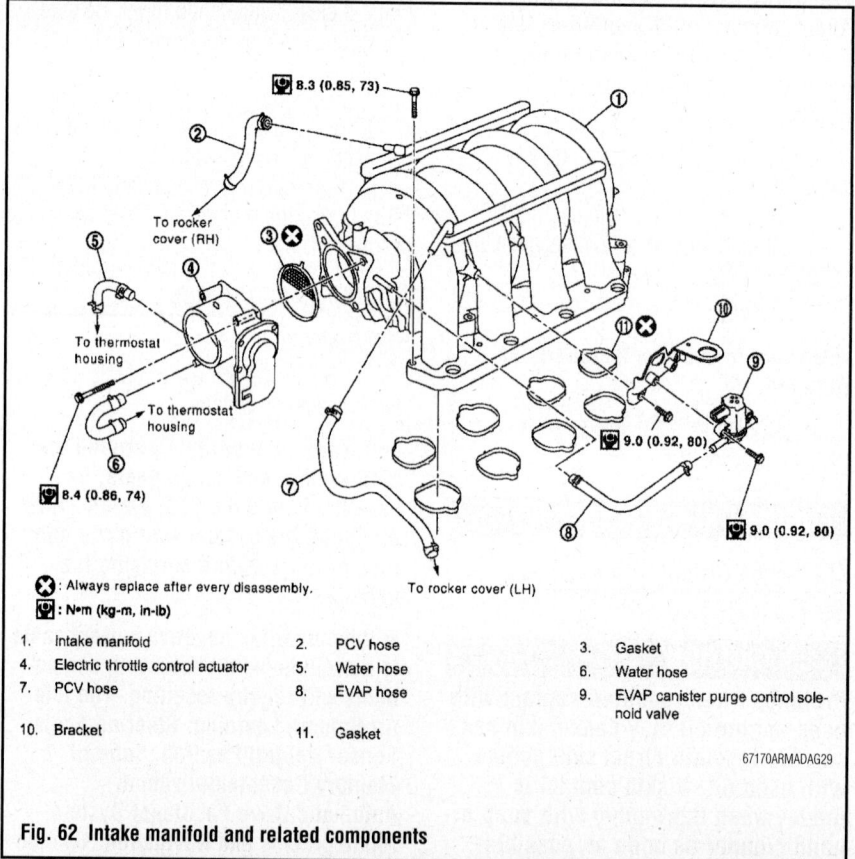

Fig. 62 Intake manifold and related components

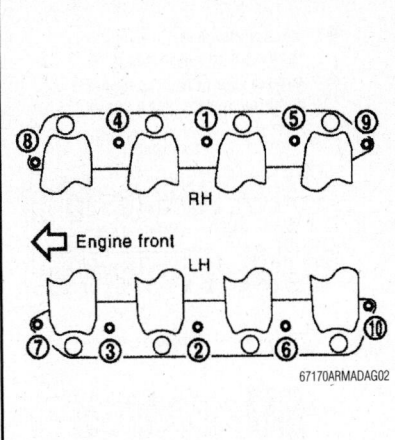

Fig. 63 Intake manifold bolt torque sequence

- Fuel tube assembly
- Intake manifold, removing bolts in reverse order of installation

To install:

➡**Be sure to use new fasteners, as required.**

6. Install the intake manifold with new gaskets. Tighten the bolts in order as shown.

7. Install or connect the following:
- Fuel tube assembly
- Fuel injectors
- Electronic throttle control actuator, tightening the bolts in several steps
- PCV hose
- Vacuum hoses
- Wiring harnesses

8. Connect the fuel tube as follows:

a. Apply a thin layer of engine oil on the tube from tip end to spool end.

b. Insert tube into quick connector past the white identification mark.

c. Insert tube into quick connector until top spool is completely inside the connector and 2nd level spool is exposed right below the connector.

d. Pull slightly on the quick connector to ensure it is fully engaged.

e. Install quick connector cap on quick connector joint.

9. Install or connect the following:
- Air intake assembly
- Engine cover

10. Refill the cooling system.

11. Start engine and check for leaks.

12. Be sure to perform the reconnect/relearn procedures.

PISTONS & RINGS

POSITIONING

See Figures 64 and 65.

ROCKER COVER

REMOVAL & INSTALLATION

See Figures 66 through 68.

1. Before servicing the vehicle, refer to the Precautions Section.

➡**If working near and/or around the SRS system and components, be sure to disable the SRS system. After disabling the system wait three minutes or more before servicing the vehicle.**

➡**Whenever the negative battery cable is disconnected the following components will require resetting. The Idle Air Volume Learning, Steering Angle Sensor Neutral Position, Sunroof Memory Reset/Initialization, Automatic Drive Positioner System, Audio presets and Navigation. Use the CONSULT-III diagnostic tool, or equivalent to perform the required resets.**

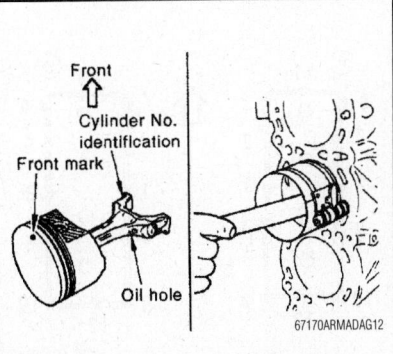

Fig. 64 Piston and rod positioning and identification

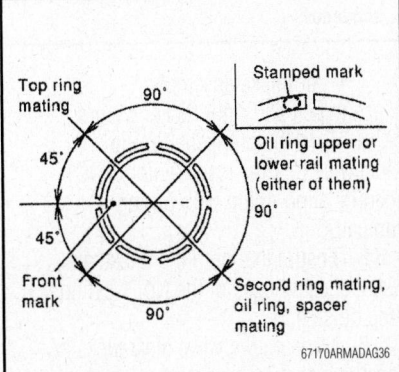

Fig. 65 Piston ring installation

---

2. Disconnect the negative battery cable.

3. Remove or disconnect the following:
- Engine room cover
- Air duct and resonator assembly, as required
- Harness on the upper rocker cover and its peripheral and aside
- Electric throttle control actuator, loosening the bolts diagonally, as required
- Ignition coils
- PCV hose from the PCV control valves

❊❊ CAUTION

Do not handle valve cover (B) by oil filler neck.

- Bolts in reverse order shown in tightening sequence
- Valve cover
- discard the gaskets

To install:

➡**Be sure to use new fasteners, as required.**

4. Use Genuine RTV Silicone Sealant or equivalent. Refer to illustration "a" to apply liquid gasket to the joint part of No.1 camshaft bracket and cylinder head.

5. Refer to illustration "b" to apply liquid gasket 90° to illustration "a".

6. Install valve cover.

7. Make sure the new rocker cover gasket is installed in the groove of the rocker cover.

8. Tighten the bolts in sequence and to specification by performing the following:
- First pass: 18 inch lbs. (2 Nm)
- Second pass: 73 inch lbs. (8 Nm)

➡**Remove foreign materials from inside the hose using compressed air. The inserted length is within 0.98–1.18 inches (25–30 mm).**

9. Install the PCV hoses

10. To complete installation, reverse remaining removal procedures.

11. Be sure to perform the reconnect/relearn procedures.

VALVE LASH (CLEARANCE) ADJUSTMENT

ADJUSTMENT

See Figures 69 and 70.

1. Before servicing the vehicle, refer to the Precautions Section.

➡**Perform the following inspection after removal, installation or replacement of camshaft or valve-related parts, or if there are unusual engine**

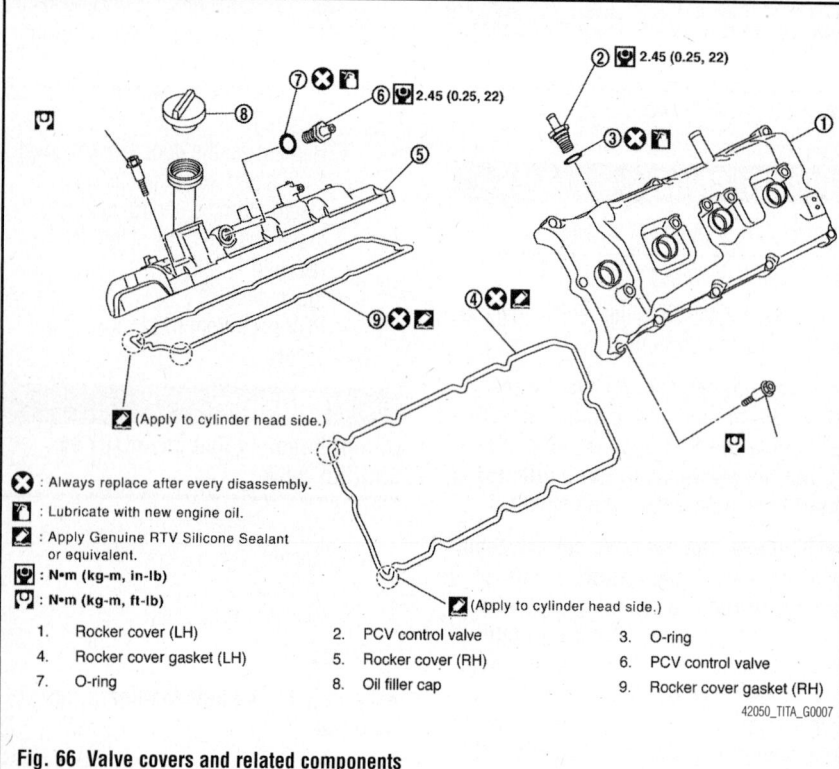

- ✕ : Always replace after every disassembly.
- ▨ : Lubricate with new engine oil.
- ▨ : Apply Genuine RTV Silicone Sealant or equivalent.
- ▨ : N•m (kg-m, in-lb)
- ▨ : N•m (kg-m, ft-lb)

1. Rocker cover (LH)
2. PCV control valve
3. O-ring
4. Rocker cover gasket (LH)
5. Rocker cover (RH)
6. PCV control valve
7. O-ring
8. Oil filler cap
9. Rocker cover gasket (RH)

42050_TITA_G0007

Fig. 66 Valve covers and related components

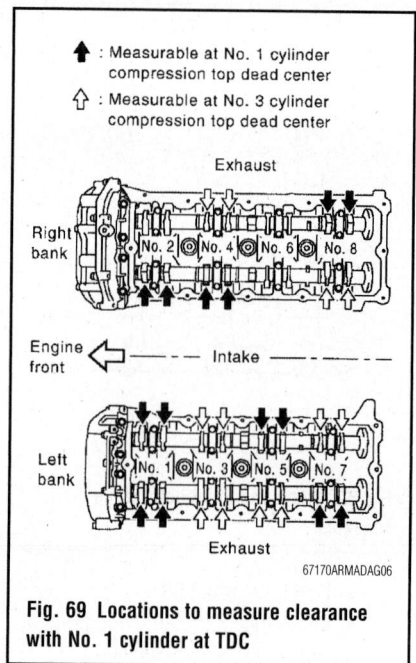

Fig. 69 Locations to measure clearance with No. 1 cylinder at TDC

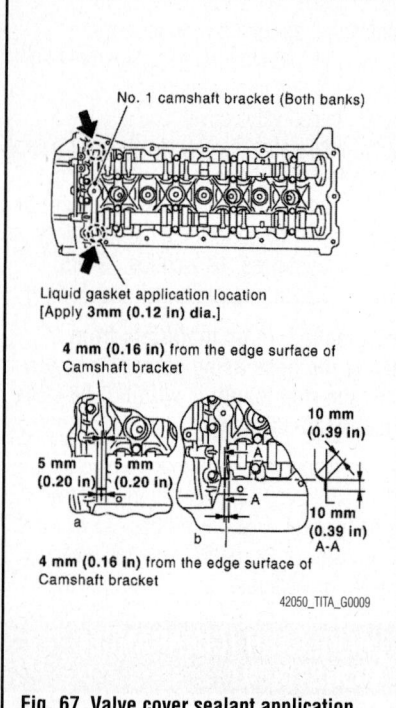

42050_TITA_G0009

Fig. 67 Valve cover sealant application

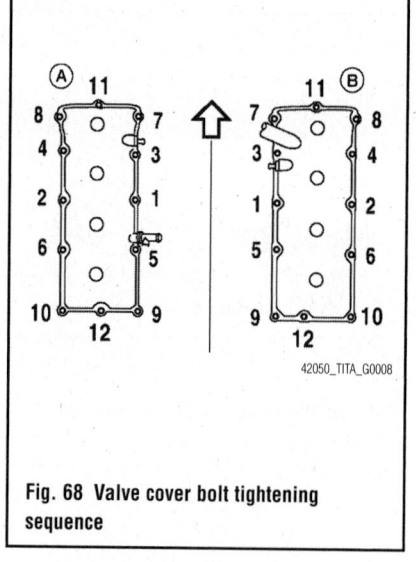

42050_TITA_G0008

Fig. 68 Valve cover bolt tightening sequence

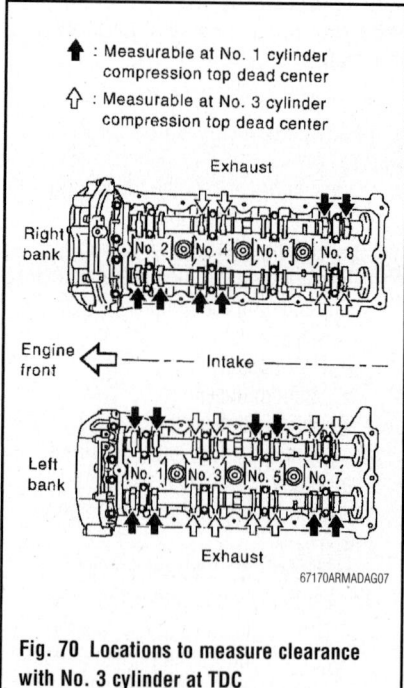

Fig. 70 Locations to measure clearance with No. 3 cylinder at TDC

conditions due to changes in valve clearance over time (starting, idling, and/or noise).

2. Run engine to operating temperature.
3. Remove or disconnect the following:
   - Battery cover, if equipped
   - Engine room cover
   - Air intake assembly
   - Left and right rocker covers
4. Turn the crankshaft pulley clockwise to Top Dead Center (TDC) identification notch with timing indicator.
5. Ensure that both the intake and exhaust cam noses of the No. 1 cylinder face outside.
6. Measure the valve clearances at locations shown in figure.
7. Turn the crankshaft pulley clockwise 270 degrees from the position of No. 1

cylinder compression to obtain No. 3 cylinder compression TDC.
8. Measure the valve clearances at locations shown in the figure.
9. Turn crankshaft pulley clockwise 90 degrees and measure the intake and exhaust valve clearance of No. 6 cylinder and exhaust valve clearance of No. 2 cylinder.
10. To adjust the valves, remove camshaft and valve lifter(s) out of specification.
11. Install replacement valve lifter(s).

12. Install the camshaft.

13. Manually turn the crankshaft pulley several turns.

14. Recheck valve clearances with engine at operating temperature.

CAMSHAFT POSITION SENSOR

LOCATION

The Camshaft Position (CMP) sensors are located on the timing cover, facing the engine.

REMOVAL & INSTALLATION

See Figure 71.

1. Before servicing the vehicle, refer to the Precautions Section.

➡ **If working near and/or around the SRS system and components, be sure to disable the SRS system. After disabling the system wait three minutes or more before servicing the vehicle.**

➡ **Whenever the negative battery cable is disconnected the following components will require resetting. The Idle Air Volume Learning, Steering Angle Sensor Neutral Position, Sunroof Memory Reset/Initialization, Automatic Drive Positioner System, Audio presets and Navigation. Use the CONSULT-III diagnostic tool, or equivalent to perform the required resets.**

2. Disconnect the negative battery cable.

3. Remove the engine cover.

4. Remove air intake duct.

5. Disconnect the camshaft position sensor.

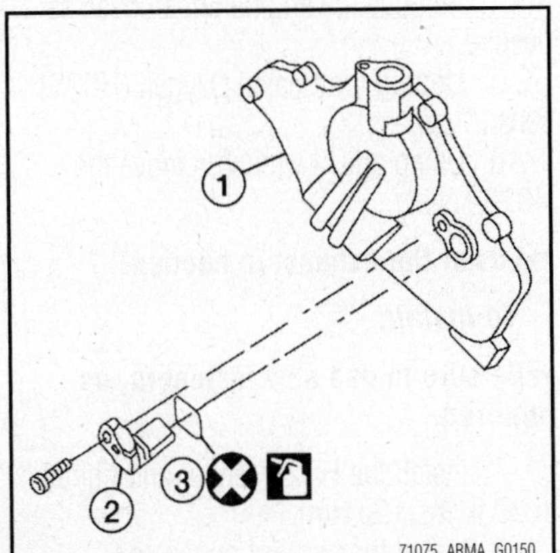

Fig. 71 Camshaft position sensor and related components (1) VTC valve cover, (2) CPS, (3) O-ring

INSPECTION

1. Before servicing the vehicle, refer to the Precautions Section.

2. Remove camshaft and valve lifter(s) out of specification.

6. Remove the bolt and the Camshaft Position (CMP) sensor.

To install:

7. Install the CMP sensor and tighten the bolt.

8. Reconnect the camshaft electrical sensor.

9. Install the air intake duct.

10. Install the engine cover.

11. Be sure to perform the reconnect/relearn procedures.

CRANKSHAFT POSITION SENSOR

LOCATION

The Crankshaft Position (CKP) sensor is located on the transmission assembly facing the gear teeth (cogs) of the signal plate.

REMOVAL & INSTALLATION

See Figure 72.

1. Before servicing the vehicle, refer to the Precautions Section.

➡ **If working near and/or around the SRS system and components, be sure to disable the SRS system. After disabling the system wait three minutes or more before servicing the vehicle.**

➡ **Whenever the negative battery cable is disconnected the following components will require resetting. The Idle Air Volume Learning, Steering Angle Sensor Neutral Position, Sunroof Memory Reset/Initialization, Automatic Drive Positioner System, Audio presets and Navigation. Use the CONSULT-III diagnostic tool, or equivalent to perform the required resets.**

2. Disconnect the negative battery cable.

3. Raise and support the vehicle safely.

4. Disconnect the Crankshaft Position (CKP) sensor connector.

5. Remove the mounting bolt and CKP sensor.

To install:

6. Install the CKP sensor and tighten the mounting bolt.

7. Reconnect the CKP sensor connector.

8. Lower the vehicle.

3. Install replacement valve lifter(s).

4. Install the camshaft.

5. Manually turn the crankshaft pulley several turns.

6. Recheck valve clearances with engine at operating temperature.

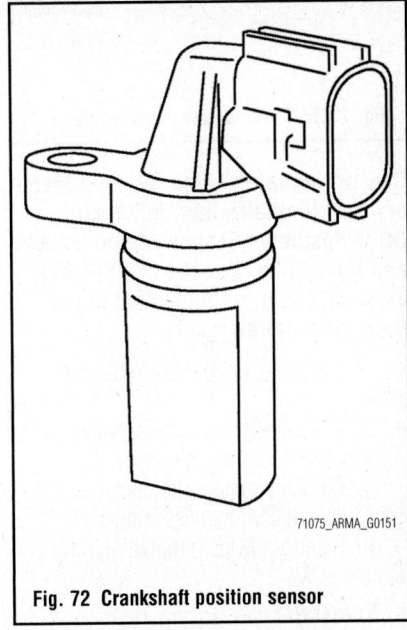

71075_ARMA_G0151

Fig. 72 Crankshaft position sensor

9. Be sure to perform the reconnect/relearn procedures.

ELECTRONIC CONTROL MODULE

LOCATION

The Electronic Control Module (ECM) is located in the engine room passenger side behind battery.

REMOVAL & INSTALLATION

See Figure 73.

At this time the manufacturer does not provide removal and installation procedures for this component. The following procedure is a guideline and may differ from the vehicle you are servicing.

1. Before servicing the vehicle, refer to the Precautions Section.

➡ **If working near and/or around the SRS system and components, be sure to disable the SRS system. After disabling the system wait three minutes or more before servicing the vehicle.**

➡ **Whenever the negative battery cable is disconnected the following components will require resetting. The Idle Air Volume Learning, Steering Angle**

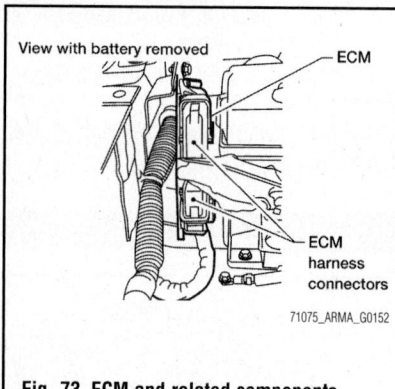

Fig. 73 ECM and related components

Sensor Neutral Position, Sunroof Memory Reset/Initialization, Automatic Drive Positioner System, Audio presets and Navigation. Use the CONSULT-III diagnostic tool, or equivalent to perform the required resets.

2. Disconnect the negative battery cable.

3. Disconnect the positive battery cable and remove the battery, as required.

4. Carefully remove the Electronic Control Module (ECM) harness connectors.

5. Remove the ECM mounting bolts and the ECM.

To install:

6. Install the ECM and mounting bolts and tighten to 62 inch lbs. (7 Nm).

7. Carefully install the ECM harness connectors.

8. Install the battery.

9. Reconnect the battery cables.

10. Be sure to perform the reconnect/relearn procedures.

RESET

➡The Nissan CONSULT III diagnostic tool, or equivalent is required to perform the resets used on this vehicle.

When replacing the ECM, the following procedures must be performed in the order listed.

• Initialize The Immobilizer System
• VIN Registration
• Accelerator Pedal Released Position Learning
• Throttle Valve Closed Position Learning
• Idle Air Volume Learning

ENGINE COOLANT TEMPERATURE SENSOR

LOCATION

The Engine Coolant Temperature (ECT) sensor is mounted in the front of the intake manifold. It is just to the right of the throttle body.

REMOVAL & INSTALLATION

See Figure 74.

✳✳ CAUTION

Never open, service or drain the radiator or cooling system when hot; serious burns can occur from the steam and hot coolant. Also, when draining engine coolant, keep in mind that cats and dogs are attracted to ethylene glycol antifreeze and could drink any that is left in an uncovered container or in puddles on the ground. This will prove fatal in sufficient quantities. Always drain coolant into a sealable container. Coolant should be reused unless it is contaminated or is several years old.

1. Before servicing the vehicle, refer to the Precautions Section.

➡If working near and/or around the SRS system and components, be sure to disable the SRS system. After disabling the system wait three minutes or more before servicing the vehicle.

➡Whenever the negative battery cable is disconnected the following components will require resetting. The Idle Air Volume Learning, Steering Angle Sensor Neutral Position, Sunroof Memory Reset/Initialization, Automatic Drive Positioner System, Audio presets and Navigation. Use the CONSULT-III diagnostic tool, or equivalent to perform the required resets.

2. Disconnect the negative battery cable.

3. Remove the engine cover.

4. Remove the intake air duct.

5. Partially drain the cooling system.

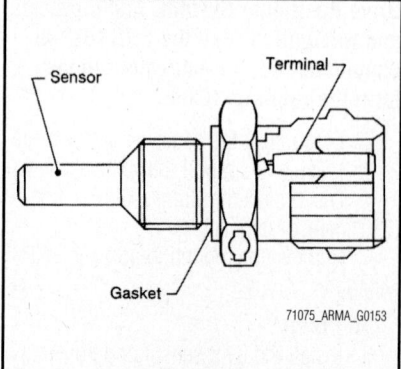

Fig. 74 Engine coolant temperature sensor

6. Disconnect the harness connector.

7. Remove the Engine Coolant Temperature (ECT) sensor.

To install:

➡Be sure to use new fasteners, as required.

8. Install the ECT sensor and carefully tighten.

9. Reconnect the harness connector.

10. Install the intake air duct.

11. Install the engine cover.

12. Refill the engine coolant.

13. Be sure to perform the reconnect/relearn procedures.

HEATED OXYGEN SENSOR (HO2S)

REMOVAL & INSTALLATION

1. Before servicing the vehicle, refer to the Precautions Section.

➡If working near and/or around the SRS system and components, be sure to disable the SRS system. After disabling the system wait three minutes or more before servicing the vehicle.

➡Whenever the negative battery cable is disconnected the following components will require resetting. The Idle Air Volume Learning, Steering Angle Sensor Neutral Position, Sunroof Memory Reset/Initialization, Automatic Drive Positioner System, Audio presets and Navigation. Use the CONSULT-III diagnostic tool, or equivalent to perform the required resets.

2. Disconnect the negative battery cable.

3. Raise and safely support the vehicle.

4. Remove the engine undercover, as needed.

5. Unplug the Heated Oxygen (HO2S) sensor harness.

6. Using an O2 wrench remove the HO2S sensor.

➡Lower the exhaust in needed.

To install:

➡Be sure to use new fasteners, as required.

7. Install the HO2S sensor and tighten to 37 ft. lbs. (50 Nm).

8. Install the harness connector.

9. Keep the harness connector and wiring away from exhaust system.

10. Be sure to perform the reconnect/relearn procedures.

INTAKE AIR TEMPERATURE/MASS AIR FLOW SENSOR

LOCATION

The Intake Air Temperature (IAT) sensor is integral to the Mass Air Flow (MAF) sensor, and is mounted on the air filter housing lid.

REMOVAL & INSTALLATION

1. Before servicing the vehicle, refer to the Precautions Section.

➡**If working near and/or around the SRS system and components, be sure to disable the SRS system. After disabling the system wait three minutes or more before servicing the vehicle.**

➡**Whenever the negative battery cable is disconnected the following components will require resetting. The Idle Air Volume Learning, Steering Angle Sensor Neutral Position, Sunroof Memory Reset/Initialization, Automatic Drive Positioner System, Audio presets and Navigation. Use the CONSULT-III diagnostic tool, or equivalent to perform the required resets.**

2. Disconnect the negative battery cable.
3. Remove the engine room cover.
4. Remove the Intake Air Temperature (IAT/MAF) sensor harness.
5. Remove the mounting screws and the IAT/MAF sensor.

To install:

6. Install the IAT/MAF sensor.
7. Install the harness connector.
8. Install the engine room cover.
9. Be sure to perform the reconnect/relearn procedures.

KNOCK SENSOR (KS)

LOCATION

The Knock (KS) sensors are mounted under the intake manifold on the cylinder block.

REMOVAL & INSTALLATION

At this time the manufacturer does not provide removal and installation procedures for this component. The intake manifold will have to be removed to service this component.

➡**If working near and/or around the SRS system and components, be sure to disable the SRS system. After disabling the system wait three minutes or more before servicing the vehicle.**

➡**Whenever the negative battery cable is disconnected the following components will require resetting. The Idle Air Volume Learning, Steering Angle Sensor Neutral Position, Sunroof Memory Reset/Initialization, Automatic Drive Positioner System, Audio presets and Navigation. Use the CONSULT-III diagnostic tool, or equivalent to perform the required resets.**

MALFUNCTION INDICATOR LIGHT

RESET PROCEDURE

Clearing diagnostic trouble codes resets the MIL.

Proper operation of the Malfunction Indicator Light (MIL):
- The MIL will illumine with the ignition switch ON and the engine OFF
- The MIL will turn OFF when the engine is started
- The MIL will remain ON if the self-diagnostic system has detected a malfunction
- The MIL may turn OFF if the malfunction is no longer present
- If the MIL is illuminated and then the engine stalls, the MIL will remain illuminated as long as the ignition switch is ON
- If the MIL is not illuminated and the engine stalls, the MIL will not illuminate until the ignition switch is cycled OFF, then ON

1. Before servicing the vehicle, refer to the Precautions Section.
2. Resetting the MIL:
- The control module turns OFF the MIL after 3 consecutive ignition cycles that the diagnostic system runs and does not fail
- The control module turns OFF the MIL after a current Diagnostic Trouble Code (DTC) clears when the diagnostic cycle runs and passes
- There may still be a history of DTC's stored in the system. These will clear after 40 consecutive warm-up cycles, if no failures are reported by any other related diagnostic system
- Manual resetting of the MIL and any DTC stored in the system, requires the use of an OBD2 scan tool connected to the Data Link Connector (DLC) for communication with the vehicle. Follow the instructions of the scan tool for both retrieval and resetting of DTC's. The scan tool can be used to command the MIL off.

➡**If the error symptoms causing the MIL to illuminate have been corrected, the MIL will return to normal operation.**

3. If a DTC is present, record the code and troubleshoot the fault.

THROTTLE POSITION SENSOR

LOCATION

See Figure 75.

The Throttle Position(TPS) sensor is integral to the electric Throttle Control actuator. The Throttle Control actuator is mounted at the front of the intake manifold.

REMOVAL & INSTALLATION

See Figure 76.

At this time the manufacturer does not provide removal and installation procedures for this component. The following procedure is a guideline and may differ from the vehicle you are servicing.

1. Before servicing the vehicle, refer to the Precautions Section.

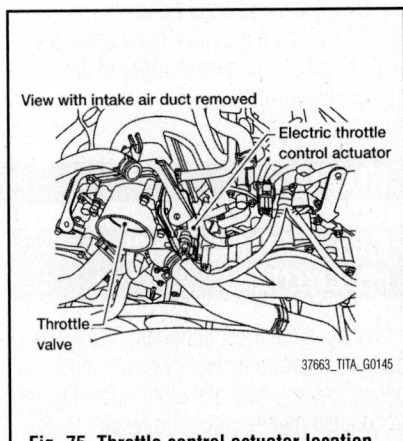

Fig. 75 Throttle control actuator location

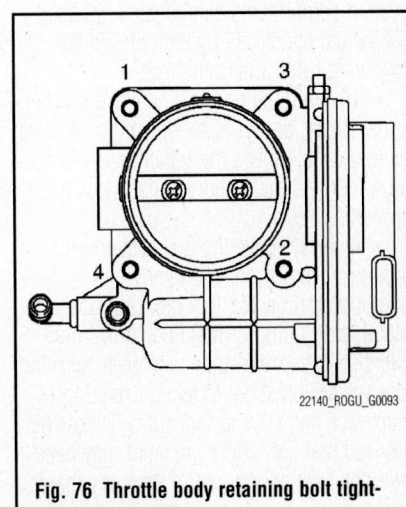

Fig. 76 Throttle body retaining bolt tightening sequence

➡If working near and/or around the SRS system and components, be sure to disable the SRS system. After disabling the system wait three minutes or more before servicing the vehicle.

➡Whenever the negative battery cable is disconnected the following components will require resetting. The Idle Air Volume Learning, Steering Angle Sensor Neutral Position, Sunroof Memory Reset/Initialization, Automatic Drive Positioner System, Audio presets and Navigation. Use the CONSULT-III diagnostic tool, or equivalent to perform the required resets.

2. Disconnect the negative battery cable.
3. Drain the cooling system, as required. Be sure to properly dispose of used engine coolant.
4. Remove the engine room cover. Remove the air intake duct.
5. Disconnect harness connector.
6. Disconnect water hoses.
7. Loosen the throttle body assembly mounting bolts in reverse order of the tightening sequence.

To install:

➡Be sure to use new fasteners, as required.

8. Install the throttle body assembly with a new gasket.
9. Tighten the mounting bolts in sequence to 74 inch lbs. (8.4 Nm).
10. Reconnect the water hose.
11. Reconnect the harness connector.
12. Reconnect the air intake duct.
13. Fill the cooling system with the proper grade and type engine coolant.
14. Be sure to perform the reconnect/relearn procedures.

VEHICLE SPEED SENSOR

LOCATION

The VSS sensor is located at the rear of the transmission case, under the tail shaft. On 4WD vehicles this component is located under the transfer case.

REMOVAL & INSTALLATION

1. Before servicing the vehicle, refer to the Precautions Section.

➡If working near and/or around the SRS system and components, be sure

to disable the SRS system. After disabling the system wait three minutes or more before servicing the vehicle.

➡Whenever the negative battery cable is disconnected the following components will require resetting. The Idle Air Volume Learning, Steering Angle Sensor Neutral Position, Sunroof Memory Reset/Initialization, Automatic Drive Positioner System, Audio presets and Navigation. Use the CONSULT-III diagnostic tool, or equivalent to perform the required resets.

2. Disconnect the negative battery cable.
3. Raise and safely support the vehicle.
4. Disconnect the sensor harness.
5. Remove the mounting bolt and the sensor.

To install:

➡Be sure to use new fasteners, as required.

6. Apply a small amount of transmission fluid to the sensor O-ring.
7. Install the Speed sensor and tighten the mounting bolt to 51 inch lbs. (5.8 Nm).
8. Be sure to perform the reconnect/relearn procedures.

FUEL   GASOLINE FUEL INJECTION SYSTEM

FUEL SYSTEM SERVICE PRECAUTIONS

Safety is the most important factor when performing not only fuel system maintenance, but any type of maintenance. Failure to conduct maintenance and repairs in a safe manner may result in serious personal injury or death. Work on a vehicle's fuel system components can be accomplished safely and effectively by adhering to the following rules and guidelines.

• To avoid the possibility of fire and personal injury, always disconnect the negative battery cable unless the repair or test procedure requires that battery voltage be applied.

• Always relieve the fuel system pressure prior to disconnecting any fuel system component (injector, fuel rail, pressure regulator, etc.) fitting or fuel line connection. Exercise extreme caution whenever relieving fuel system pressure to avoid exposing skin, face and eyes to fuel spray. Please be advised that fuel under pressure may penetrate the skin or any part of the body that it contacts.

• Always place a shop towel or cloth

around the fitting or connection prior to loosening to absorb any excess fuel due to spillage. Ensure that all fuel spillage is quickly removed from engine surfaces. Ensure that all fuel-soaked cloths or towels are deposited into a flame-proof waste container with a lid.

• Always keep a dry chemical (Class B) fire extinguisher near the work area.

• Do not allow fuel spray or fuel vapors to come into contact with a spark or open flame.

• Always use a second wrench when loosening or tightening fuel line connection fittings. This will prevent unnecessary stress and torsion on fuel piping. Always follow the proper torque specifications.

• Always replace worn fuel fitting O-rings with new ones. Do not substitute fuel hose where rigid pipe is installed.

RELIEVING FUEL SYSTEM PRESSURE

With CONSULT-II®

1. Turn ignition switch **ON**.
2. Perform "FUEL PRESSURE RELEASE"

in "WORK SUPPORT" mode with CONSULT-II®.
3. Start engine.
4. After engine stalls, turn over the engine two or three times to release all fuel pressure.
5. Turn ignition switch **OFF**.

Without CONSULT-II®

See Figure 77.

1. Before servicing the vehicle, refer to the Precautions Section.

➡If working near and/or around the SRS system and components, be sure to disable the SRS system. After disabling the system wait three minutes or more before servicing the vehicle.

➡Whenever the negative battery cable is disconnected the following components will require resetting. The Idle Air Volume Learning, Steering Angle Sensor Neutral Position, Sunroof Memory Reset/Initialization, Automatic Drive Positioner System, Audio presets and Navigation. Use the CONSULT-III diagnostic tool, or equivalent to perform the required resets.

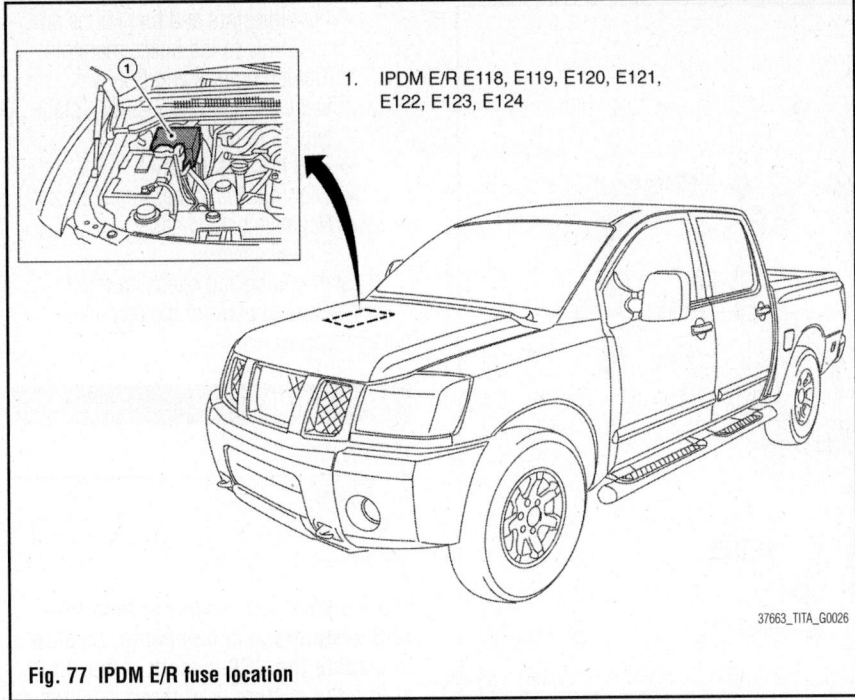

1. IPDM E/R E118, E119, E120, E121, E122, E123, E124

37663_TITA_G0026

Fig. 77 IPDM E/R fuse location

2. Remove fuel pump fuse located in IPDM E/R.

3. Start engine.

4. After engine stalls, turn over engine two or three times to release all fuel pressure.

5. Turn ignition switch **OFF**.

6. Disconnect the negative battery cable.

7. Reinstall fuel pump fuse after servicing fuel system.

8. Be sure to perform the reconnect/relearn procedures.

FUEL LEVEL SENDING UNIT

REMOVAL & INSTALLATION

The fuel level sending unit is part of the fuel pump module.

FUEL PUMP MODULE

REMOVAL & INSTALLATION

See Figure 00.

1. Before servicing the vehicle, refer to the Precautions Section.

➡**If working near and/or around the SRS system and components, be sure to disable the SRS system. After disabling the system wait three minutes or more before servicing the vehicle.**

➡**Whenever the negative battery cable is disconnected the following components will require resetting. The Idle Air Volume Learning, Steering Angle Sensor Neutral Position, Sunroof Mem-** ory Reset/Initialization, Automatic Drive Positioner System, Audio presets and Navigation. Use the CONSULT-III diagnostic tool, or equivalent to perform the required resets.

➡**Be sure to check the fuel gauge indicator. Make sure that it reads not more than half a tank. If not, properly drain fuel until the gauge reads half a tank or less.**

➡**This vehicle uses quick connect fittings. Be sure to properly relieve the fuel system pressure before disconnecting any of these fittings. Always replace O-rings and clamps with new ones. Do not bend, twist or kink hoses when they are being removed or installed. Be sure that the clamp screw does not contact adjacent parts. When tightening the high pressure rubber hose clamp make sure the clamp end is 0.12 inch from the hose end. After connecting these fittings make sure that the connectors are secure. Check for fuel leakage at these connections turn the ignition key to the ON position (do not start the engine), correct as required. Start the engine, raise the idle, and verify that there are no fuel leaks, correct as required.**

2. Disconnect the negative battery cable.

3. Relieve the fuel system pressure.

4. Remove fuel filler cap to release pressure from inside tank.

5. Disconnect fuel filler hose from fuel filler pipe.

6. Drain fuel tank through the fuel filler opening using a suitable hose.

7. Disconnect the following:
- Fuel pump line protector
- EVAP hose
- Fuel level sensor
- Fuel filter
- Fuel pump wiring harness
- Fuel supply hose

8. Using a suitable jack to support the fuel tank, remove the strap bolts and remove the fuel tank from the vehicle.

9. Remove the lock ring using special tool J-46536, or equivalent.

10. Remove the following:
- Fuel level sensor
- Fuel filter
- Fuel pump assembly

To install:

➡**Be sure to use new fasteners, as required.**

11. Install or connect the following:
- Fuel pump assembly, using new O-ring
- Fuel filter, using new filter
- Fuel level sensor, using new sensor
- Fuel pump assembly lock ring
- Fuel tank
- Fuel supply hose
- Fuel pump wiring harness
- EVAP hose
- Fuel pump line protector
- Fuel filler pipe

12. Start engine and check for leaks.

13. Be sure to perform the reconnect/relearn procedures.

FUEL RAIL & INJECTORS

REMOVAL & INSTALLATION

See Figures 78 and 79.

1. Before servicing the vehicle, refer to the Precautions Section.

➡**If working near and/or around the SRS system and components, be sure to disable the SRS system. After disabling the system wait three minutes or more before servicing the vehicle.**

➡**Whenever the negative battery cable is disconnected the following components will require resetting. The Idle Air Volume Learning, Steering Angle Sensor Neutral Position, Sunroof Memory Reset/Initialization, Automatic Drive Positioner System, Audio presets and Navigation. Use the CONSULT-III**

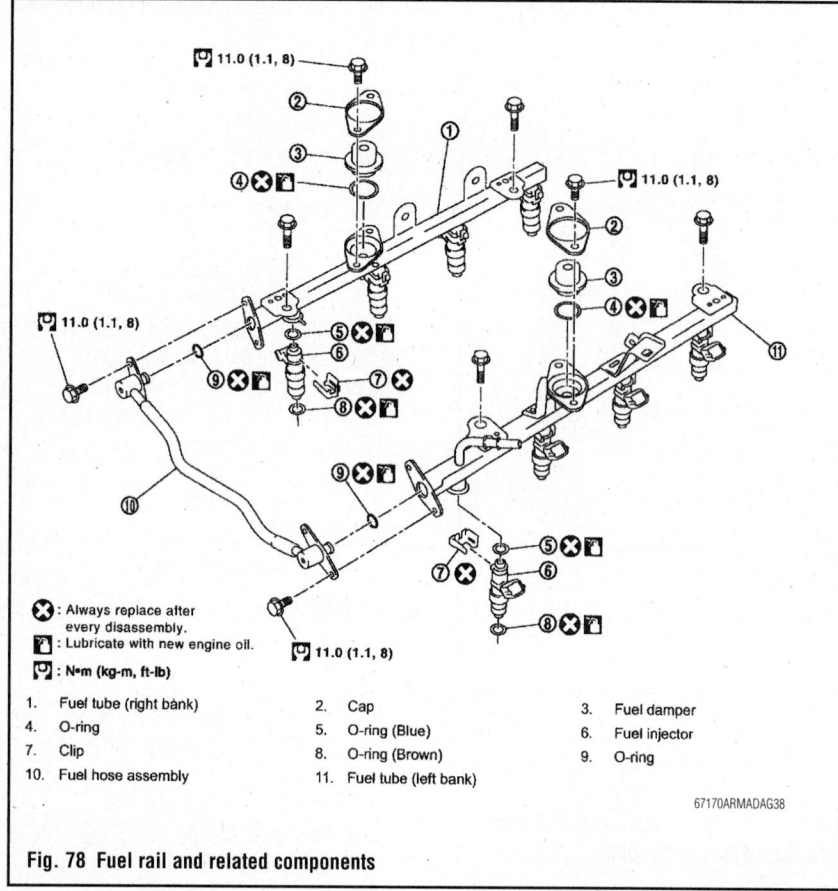

X: Always replace after every disassembly.
▣: Lubricate with new engine oil.
☐: N•m (kg-m, ft-lb)

1. Fuel tube (right bank)
2. Cap
3. Fuel damper
4. O-ring
5. O-ring (Blue)
6. Fuel injector
7. Clip
8. O-ring (Brown)
9. O-ring
10. Fuel hose assembly
11. Fuel tube (left bank)

67170ARMADAG38

Fig. 78 Fuel rail and related components

diagnostic tool, or equivalent to perform the required resets.

➡This vehicle uses quick connect fittings. Be sure to properly relieve the fuel system pressure before disconnecting any of these fittings. Always replace O-rings and clamps with new ones. Do not bend, twist or kink hoses when they are being removed or installed. Be sure that the clamp screw

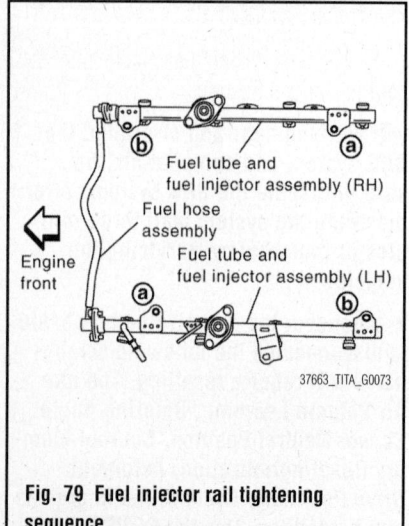

Fuel tube and fuel injector assembly (RH)
Fuel hose assembly
Engine front
Fuel tube and fuel injector assembly (LH)

37663_TITA_G0073

Fig. 79 Fuel injector rail tightening sequence

does not contact adjacent parts. When tightening the high pressure rubber hose clamp make sure the clamp end is 0.12 inch from the hose end. After connecting these fittings make sure that the connectors are secure. Check for fuel leakage at these connections turn the ignition key to the ON position (do not start the engine), correct as required. Start the engine, raise the idle, and verify that there are no fuel leaks, correct as required.

2. Disconnect the negative battery cable.
3. Remove engine cover. Remove the air cleaner assembly.
4. Relieve fuel system pressure.
5. Remove or disconnect the following:
- Fuel injector harness connectors
- Fuel hose assembly from right and left fuel rails
- Fuel injectors with fuel rail as an assembly
- Fuel injector from fuel rail

To install:

➡Be sure to use new fasteners, as required.

6. Install or connect the following:
- New clip onto the fuel injector
- Fuel injector to fuel rail

- Fuel injectors and fuel rail as an assembly to the intake manifold. Tighten bolts A and B in two stages. Stage one 9 ft. lbs., stage two 18 ft. lbs.
- Fuel hose assembly
- Fuel injector harness connectors
- Negative battery cable
- Engine cover

7. Start engine and check for leaks.
8. Be sure to perform the reconnect/relearn procedures.

FUEL TANK

DRAINING

See Figure 80.

1. Before servicing the vehicle, refer to the Precautions Section.

➡**If working near and/or around the SRS system and components, be sure to disable the SRS system. After disabling the system wait three minutes or more before servicing the vehicle.**

➡**Whenever the negative battery cable is disconnected the following components will require resetting. The Idle Air Volume Learning, Steering Angle Sensor Neutral Position, Sunroof Memory Reset/Initialization, Automatic Drive Positioner System, Audio presets and Navigation. Use the CONSULT-III diagnostic tool, or equivalent to perform the required resets.**

➡**This vehicle uses quick connect fittings. Be sure to properly relieve the fuel system pressure before disconnecting any of these fittings. Always replace O-rings and clamps with new ones. Do not bend, twist or kink hoses**

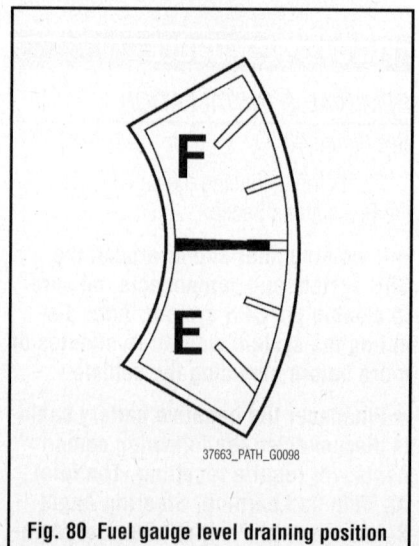

37663_PATH_G0098

Fig. 80 Fuel gauge level draining position

when they are being removed or installed. Be sure that the clamp screw does not contact adjacent parts. When tightening the high pressure rubber hose clamp make sure the clamp end is 0.12 inch from the hose end. After connecting these fittings make sure that the connectors are secure. Check for fuel leakage at these connections turn the ignition key to the ON position (do not start the engine), correct as required. Start the engine, raise the idle, and verify that there are no fuel leaks, correct as required.

2. Disconnect the negative battery cable.
3. Remove the fuel filler cap to release the pressure from inside the fuel tank.
4. Remove the LH rear wheel and tire.
5. Check the fuel level on level gauge. If the fuel gauge indicates more than the level as shown (full or almost full), drain the fuel from the fuel tank until the fuel gauge indicates the level as shown, or less.

➡Fuel will be spilled when removing the fuel level sensor, fuel filter, and fuel pump assembly for the fuel level is above the fuel level sensor, fuel filter, and fuel pump assembly fuel tank opening.

• As a guide, the fuel level reaches the fuel gauge position as shown, or less, when approximately 4 US gal (15L) of fuel are drained from the fuel tank.
• If the fuel pump does not operate, use the following procedure to drain the fuel to the specified level.
a. Insert a suitable hose of less than 15 mm (0.59 in.) diameter into the fuel filler pipe through the fuel filler opening to drain the fuel from fuel filler pipe.
b. Remove the fuel filler pipe shield.
c. Disconnect the fuel filler hose from the fuel filler pipe.
d. Insert a suitable hose into the fuel tank through the fuel filler hose to drain the fuel from the fuel tank.
6. Release the fuel pressure from the fuel lines.
7. Be sure to perform the reconnect/relearn procedures.

REMOVAL & INSTALLATION
See Figure 81.

※※ CAUTION

When replacing fuel line parts, be sure to work in a well-ventilated area and furnish workshop with a CO2 fire extinguisher. Do not smoke while

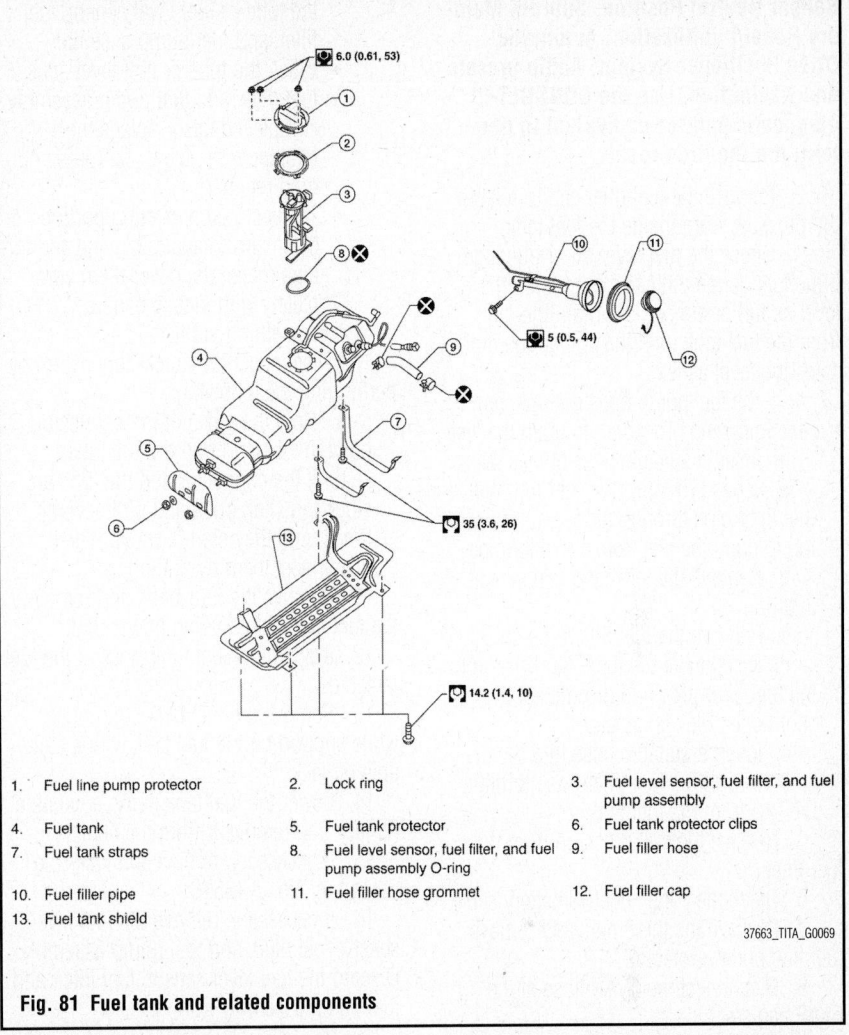

1. Fuel line pump protector
2. Lock ring
3. Fuel level sensor, fuel filter, and fuel pump assembly
4. Fuel tank
5. Fuel tank protector
6. Fuel tank protector clips
7. Fuel tank straps
8. Fuel level sensor, fuel filter, and fuel pump assembly O-ring
9. Fuel filler hose
10. Fuel filler pipe
11. Fuel filler hose grommet
12. Fuel filler cap
13. Fuel tank shield

37663_TITA_G0069

Fig. 81 Fuel tank and related components

servicing fuel system. Keep open flames and sparks away from the work area.

※※ CAUTION

Always replace O-rings and clamps with new ones. Do not kink or twist hoses when they are being installed. Do not tighten hose clamps excessively to avoid damaging hoses. Tighten high-pressure rubber hose clamp so that clamp end is 0.12 inches (3 mm) from hose end. Tightening torque specifications are the same for all rubber hose clamps. Ensure that screw does not contact adjacent parts.

➡This vehicle uses quick connect fittings. Be sure to properly relieve the fuel system pressure before disconnecting any of these fittings. Always replace O-rings and clamps with new ones. Do not bend, twist or kink hoses when they are being removed or

installed. Be sure that the clamp screw does not contact adjacent parts. When tightening the high pressure rubber hose clamp make sure the clamp end is 0.12 inch from the hose end. After connecting these fittings make sure that the connectors are secure. Check for fuel leakage at these connections turn the ignition key to the ON position (do not start the engine), correct as required. Start the engine, raise the idle, and verify that there are no fuel leaks, correct as required.

1. Before servicing the vehicle, refer to the Precautions Section.

➡If working near and/or around the SRS system and components, be sure to disable the SRS system. After disabling the system wait three minutes or more before servicing the vehicle.

➡Whenever the negative battery cable is disconnected the following components will require resetting. The Idle Air Volume Learning, Steering Angle

Sensor Neutral Position, Sunroof Memory Reset/Initialization, Automatic Drive Positioner System, Audio presets and Navigation. Use the CONSULT-III diagnostic tool, or equivalent to perform the required resets.

2. Remove the fuel filler cap to release the pressure from inside the fuel tank.

3. Check the fuel level on level gauge. If the fuel gauge indicates more than the level as full or almost full, drain the fuel from the fuel tank until the fuel gauge indicates the level as less.

4. If the fuel pump does not operate, use the following procedure to drain the fuel:

a. Insert a suitable hose of less than 0.59 inches (15 mm) diameter into the fuel filler pipe through the fuel filler opening to drain the fuel from fuel filler pipe.

b. Remove the left-hand rear wheel and tire.

c. Remove the fuel filler pipe shield.

d. Disconnect the fuel filler hose from the fuel filler pipe and disconnect the vent hose quick connector.

e. Insert a suitable hose into the fuel tank through the fuel filler hose to drain the fuel from the fuel tank.

5. Release the fuel pressure from the fuel lines.

6. Disconnect the negative battery cable.

7. Remove the three nuts and remove fuel line pump protector.

8. Disconnect the EVAP hose at the EVAP canister.

9. Disconnect the fuel level sensor, fuel filter, and fuel pump assembly electrical connector, and the fuel feed hose.

❋❋ CAUTION

Observe the following when disconnecting the quick-connectors:

- The tube can be removed when the tabs are completely depressed. Do not twist it more than necessary.
- Do not use any tools to remove the quick connector.
- Keep the resin tube away from heat. Be especially careful when welding near the tube.
- Prevent liquid acids, such as battery electrolyte, from getting on the resin tube.
- Do not bend or twist the tube during installation and removal.
- Only when the tube is replaced, remove the remaining retainer on

the tube or fuel level sensor, fuel filter, and fuel pump assembly.

- When the tube or fuel level sensor, fuel filter, and fuel pump assembly is replaced, also replace the retainer with a new one (green colored retainer).
- To keep the connecting portion clean and to avoid damage and foreign materials, cover them completely with plastic bags or something similar.

10. Disconnect the quick-connectors by performing the following:

a. Hold the sides of the connector, push in tabs and pull out the tube.

b. If the connector and the tube are stuck together, push and pull several times until they start to move. Then disconnect them by pulling.

11. Remove the four bolts and remove the fuel tank shield using power tool.

12. Disconnect fuel filler hose at the fuel tank side.

13. Remove the fuel tank strap bolts while supporting the fuel tank with a suitable lift jack.

14. Lower the fuel tank using a suitable lift jack and remove it from the vehicle.

15. If necessary, remove the lock ring using tool No. J-46536.

16. If necessary, remove the fuel level sensor, fuel filter, and fuel pump assembly. Discard the fuel level sensor, fuel filter, and fuel pump assembly O-ring.

To install:

➡ **Be sure to use new fasteners, as required.**

17. To install, reverse removal procedure.

18. Connect the quick-connectors by performing the following:

a. Check the connection for damage or any foreign materials.

b. Align the connector with the tube, then insert the connector straight into the tube until a click is heard.

c. After the tube is connected, make sure the connection is secure by pulling on the tube and the connector to make sure they are securely connected.

19. Turn the ignition switch ON but do not start engine, then check the fuel pipe and hose connections for leaks while applying fuel pressure to the system.

20. Start the engine and rev it above idle speed, then check that there are no fuel

leaks at any of the fuel pipe and hose connections.

21. Be sure to perform the reconnect/relearn procedures.

THROTTLE BODY

REMOVAL & INSTALLATION

1. Before servicing the vehicle, refer to the Precautions Section.

➡ **If working near and/or around the SRS system and components, be sure to disable the SRS system. After disabling the system wait three minutes or more before servicing the vehicle.**

➡ **Whenever the negative battery cable is disconnected the following components will require resetting. The Idle Air Volume Learning, Steering Angle Sensor Neutral Position, Sunroof Memory Reset/Initialization, Automatic Drive Positioner System, Audio presets and Navigation. Use the CONSULT-III diagnostic tool, or equivalent to perform the required resets.**

2. Disconnect the negative battery cable.

3. Partially drain the engine coolant.

4. Remove the engine room cover.

5. Remove the air duct and resonator assembly.

6. Drain the engine coolant. Be sure to properly dispose of used coolant.

7. Disconnect the hoses from the unit.

8. Remove the 4 mounting bolts.

9. Remove electric throttle control actuator by loosening bolts diagonally.

10. Remove the old gasket and discard it.

To install:

➡ **Be sure to use new fasteners, as required.**

11. Install a new gasket and the throttle body.

12. Install the 4 mounting bolts in alternate sequence and tighten to 74 inch lbs. (8.4 Nm).

13. Reconnect the hoses to the throttle body.

14. Reconnect the air duct and resonator assembly.

15. As required, fill the cooling system.

16. Install the engine cover.

17. Be sure to perform the reconnect/relearn procedures.

HEATING & AIR CONDITIONING SYSTEM

BLOWER MOTOR

REMOVAL & INSTALLATION

See Figure 82.

1. Before servicing the vehicle, refer to the Precautions Section.

➡ **If working near and/or around the SRS system and components, be sure to disable the SRS system. After disabling the system wait three minutes or more before servicing the vehicle.**

➡ **Whenever the negative battery cable is disconnected the following components will require resetting. The Idle Air Volume Learning, Steering Angle Sensor Neutral Position, Sunroof Memory Reset/Initialization, Automatic Drive Positioner System, Audio presets and Navigation. Use the CONSULT-III diagnostic tool, or equivalent to perform the required resets.**

2. Disconnect the negative battery cable.
3. Disconnect or remove the following:
   - Glove box assembly
   - Blower motor electrical connector
   - Three screws and blower motor

To install:

➡ **Be sure to use new fasteners, as required.**

4. Installation is the reverse of the removal procedure.
5. Be sure to perform the reconnect/relearn procedures.

HEATER CORE

REMOVAL & INSTALLATION

See Figure 83.

1. Before servicing the vehicle, refer to the Precautions Section.

➡ **If working near and/or around the SRS system and components, be sure to disable the SRS system. After disabling the system wait three minutes or more before servicing the vehicle.**

➡ **Whenever the negative battery cable is disconnected the following components will require resetting. The Idle Air Volume Learning, Steering Angle Sensor Neutral Position, Sunroof Memory Reset/Initialization, Automatic Drive Positioner System, Audio presets and Navigation. Use the CONSULT-III diagnostic tool, or equivalent to perform the required resets.**

2. Position the front seats in the rearmost position.
3. Disconnect the negative battery cable.
4. Properly discharge the A/C system.
5. Drain the cooling system. Be sure to properly dispose of used coolant.
6. Disconnect the heater hoses at the heater core.
7. Disconnect and plug the refrigerant lines at the evaporator core.
8. Remove the instrument panel.
9. Remove the heater/cooling unit.
10. Remove the four screws and remove the upper bracket.
11. Remove the four screws and remove the heater core cover.
12. Remove the core pipe bracket.
13. Remove the heater core from its mounting.

➡ **Be sure to replace the in cabin micro-filter.**

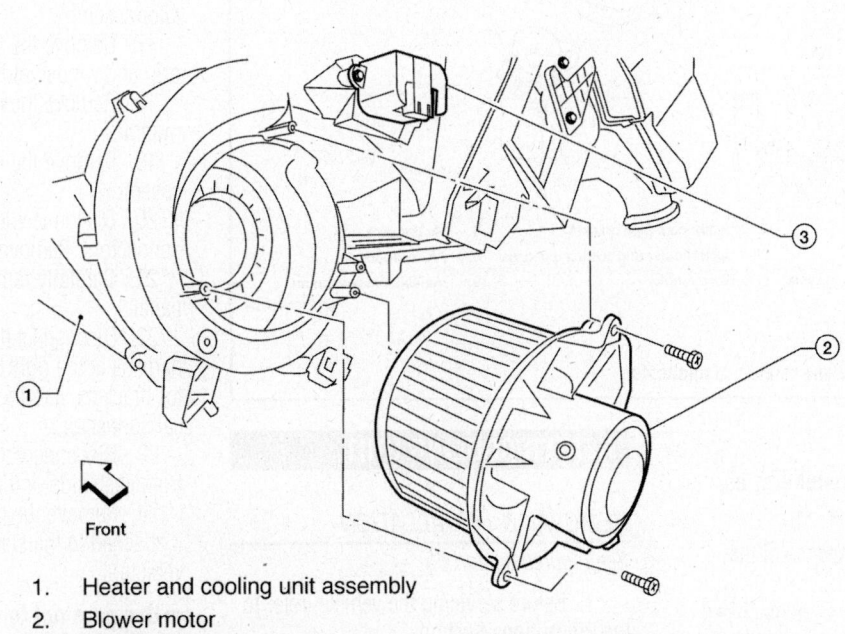

1. Heater and cooling unit assembly
2. Blower motor
3. Variable blower control or front blower motor resistor if equipped

37663_TITA_G0083

Fig. 82 Blower motor and related components

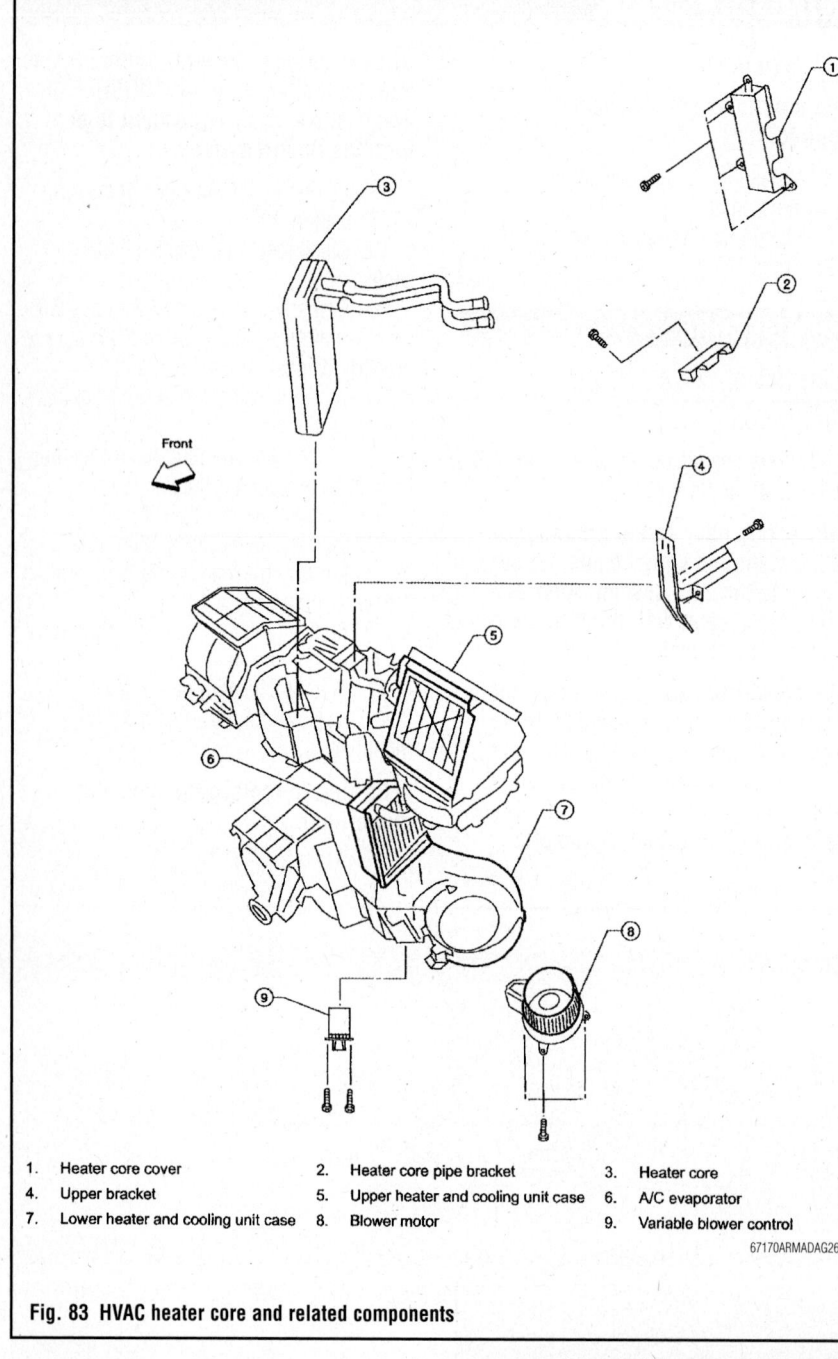

1.	Heater core cover	2.	Heater core pipe bracket	3.	Heater core
4.	Upper bracket	5.	Upper heater and cooling unit case	6.	A/C evaporator
7.	Lower heater and cooling unit case	8.	Blower motor	9.	Variable blower control

67170ARMADAG26

Fig. 83 HVAC heater core and related components

To install:

➡**Be sure to use new fasteners, as required.**

14. Installation is the reverse of the removal procedure.

15. Be sure to use new O-rings coated with clean refrigerant oil, as required.

16. Fill the cooling system with the proper grade and the coolant.

17. Properly recharge the A/C system.

18. Start the engine and check for leaks, correct as required.

19. Be sure to perform the reconnect/relearn procedures.

HEATER/COOLING UNIT

REMOVAL & INSTALLATION
See Figures 84 and 85.

1. Before servicing the vehicle, refer to the Precautions Section.

➡**If working near and/or around the SRS system and components, be sure to disable the SRS system. After disabling the system wait three minutes or more before servicing the vehicle.**

➡**Whenever the negative battery cable is disconnected the following components will require resetting. The Idle Air Volume Learning, Steering Angle Sensor Neutral Position, Sunroof Memory Reset/Initialization, Automatic Drive Positioner System, Audio presets and Navigation. Use the CONSULT-III diagnostic tool, or equivalent to perform the required resets.**

2. Position the front seats in the rear-most position.

3. Disconnect the negative battery cable.

4. Properly discharge the A/C system.

5. Drain the cooling system. Be sure to properly dispose of used coolant.

6. Disconnect the heater hoses at the heater core.

7. Disconnect and plug the refrigerant lines at the evaporator core.

8. Remove the instrument panel lower cover.

9. Remove the center console.

10. Remove the steering column.

11. Remove the combination meter.

12. Remove the audio unit.

13. Remove the display unit, if equipped.

14. Remove the lower knee protector.

15. Remove the defroster grille. Disconnect the optical sensor electrical connector.

16. Remove the left and right side ventilator assembly.

17. Remove the left and right side assist grip and windshield garnish.

18. Remove the passenger's side air bag module.

19. Remove the instrument panel pad assembly.

20. Disconnect the remaining electrical connectors. Remove the retaining bolts.

21. Carefully remove the instrument panel.

22. Disconnect the instrument panel wire harness at the right and left in-line connector brackets, and the fuse block (JB) electrical connectors.

23. Disconnect the steering member from each side of the vehicle body.

24. Remove the heater/cooling unit with it attached to the steering member from the vehicle.

➡**Use care not to damage the seats or interior trim panels.**

25. Remove the heater/cooling unit from the steering member.

To install:

➡**Be sure to use new fasteners, as required.**

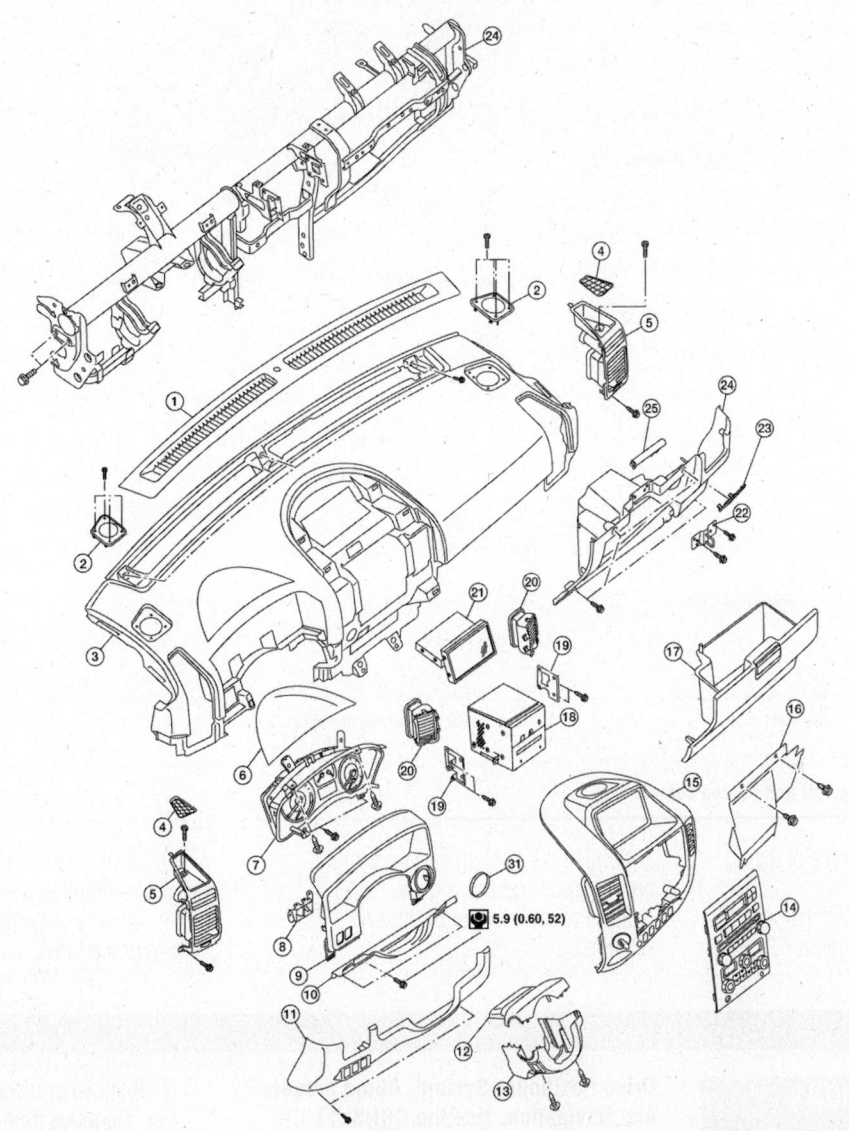

5.9 (0.60, 52)

1.	Defroster grille	2.	Speaker grille RH/LH	3.	Instrument panel and pad assembly
4.	Deck pocket mat RH/LH	5.	Side ventilator assembly RH/LH	6.	Combination meter cover
7.	Combination meter	8.	Switch assembly	9.	Cluster lid A
10.	Lower knee protector	11.	Lower instrument panel LH	12.	Steering column cover upper
13.	Steering column cover lower	14.	Clluster lid D	15.	Cluster lid C
16.	Instrument lower cover RH	17.	Glove box	18.	Audio unit
19.	Audio bracket RH/LH	20.	Center ventilator assembly RH/LH	21.	Display assembly (if equipped)
22.	Glove box lid striker	23.	Fuse block cover	24.	Lower instrument panel RH
25.	Glove box damper (if equipped)	26.	Steering member		

37663_TITA_G0005

Fig. 84 Instrument panel and related components

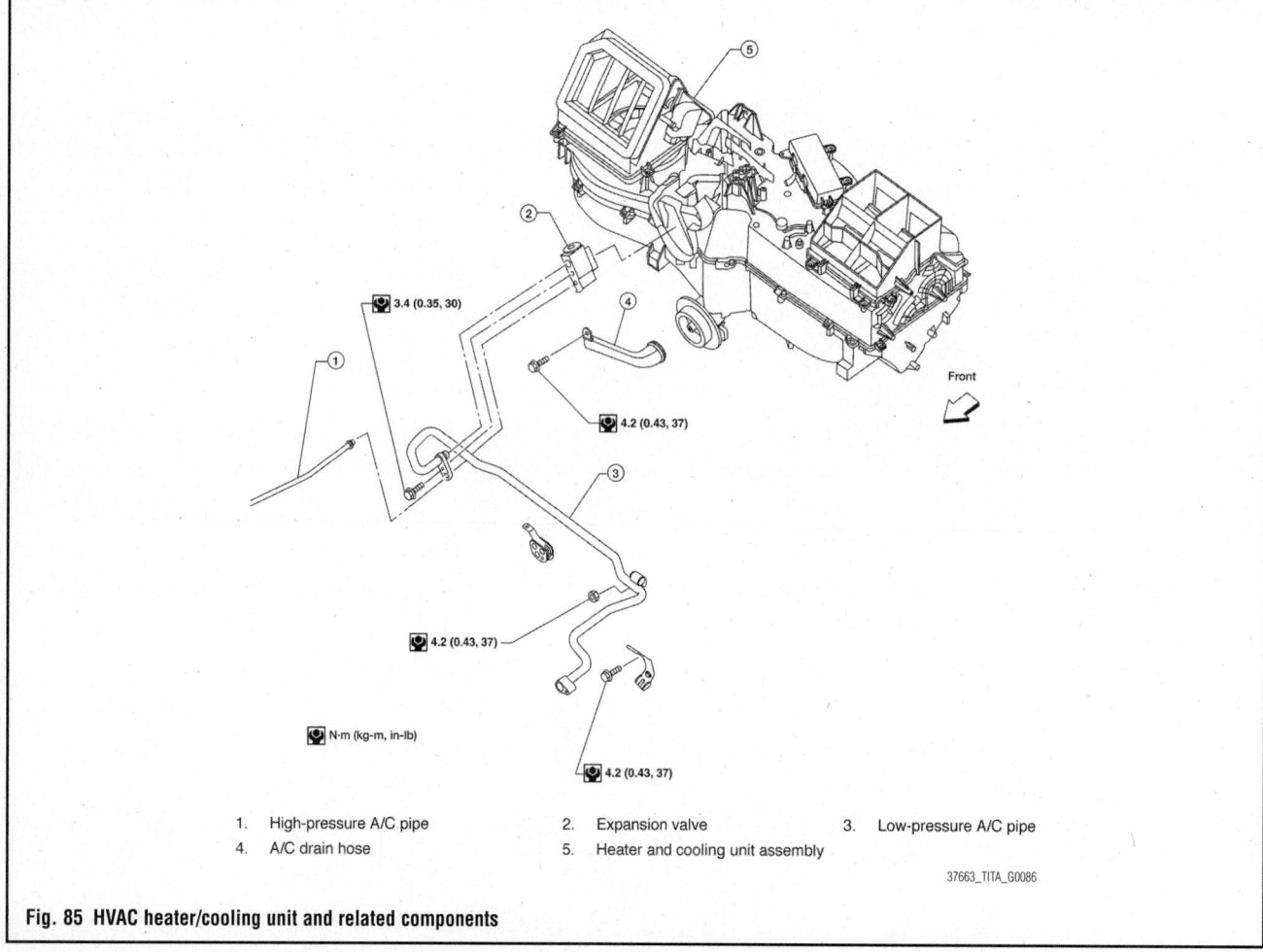

1. High-pressure A/C pipe
2. Expansion valve
3. Low-pressure A/C pipe
4. A/C drain hose
5. Heater and cooling unit assembly

37663_TITA_G0086

Fig. 85 HVAC heater/cooling unit and related components

26. Installation is the reverse of the removal procedure.
27. Be sure to use new O-rings coated with clean refrigerant oil, as required.

28. Fill the cooling system with the proper grade and the coolant.
29. Properly recharge the A/C system.

30. Start the engine and check for leaks, correct as required.
31. Be sure to perform the reconnect/relearn procedures.

STEERING

POWER RACK & PINION STEERING GEAR

REMOVAL & INSTALLATION

See Figures 86 and 87.

1. Before servicing the vehicle, refer to the Precautions Section.

➡ **If working near and/or around the SRS system and components, be sure to disable the SRS system. After disabling the system wait three minutes or more before servicing the vehicle.**

➡ **Whenever the negative battery cable is disconnected the following components will require resetting. The Idle Air Volume Learning, Steering Angle Sensor Neutral Position, Sunroof Memory Reset/Initialization, Automatic**

Drive Positioner System, Audio presets and Navigation. Use the CONSULT-III diagnostic tool, or equivalent to perform the required resets.

2. Disconnect the negative battery cable.
3. Ensure the wheels are in the straight-ahead position.
4. Remove or disconnect the following:
   • Wheels
   • Engine splash guard
5. On 4WD, remove front final drive and support the driveshafts.
6. Remove cotter pin at steering outer socket and loosen mounting nut.
7. Remove steering outer socket from steering knuckle using special tool J-25730-A.
8. On 2WD, remove stabilizer bar mounting bolts and secure the stabilizer bar.

9. Remove or disconnect the following:
   • Oil pipes from steering gear assembly
   • Lower joint mounting bolt from lower shaft
   • Mounting bolts and nuts from steering gear assembly
   • Steering gear assembly

To install:
10. Install or connect the following:
   • Steering gear assembly, and tighten nuts to specification
   • Lower joint mounting bolt
   • Oil pipes to steering gear assembly
   • Stabilizer bar, 2WD
   • Steering outer socket to steering knuckle, and tighten nut to 63 ft. lbs. (86 Nm)

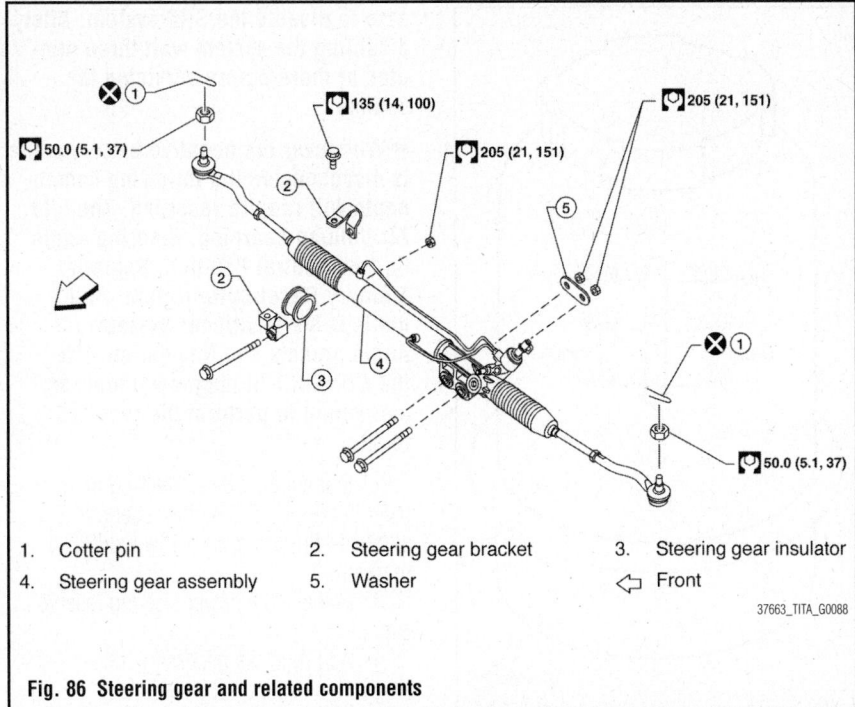

1. Cotter pin
2. Steering gear bracket
3. Steering gear insulator
4. Steering gear assembly
5. Washer
⬅ Front

37663_TITA_G0088

Fig. 86 Steering gear and related components

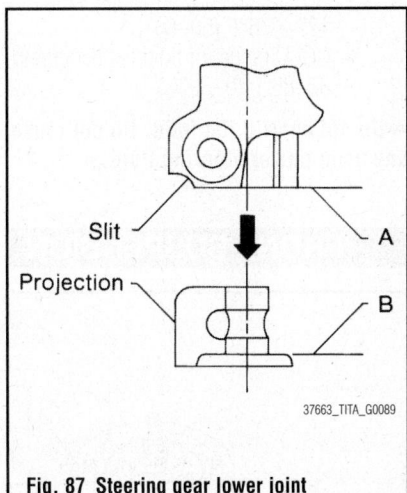

37663_TITA_G0089

Fig. 87 Steering gear lower joint alignment

- Front final drive, 4WD
- Engine splash guard
- Wheels

11. With the steering wheel in the straight ahead position, make sure that the slit of the lower joint (A) fits with the projection on the rear cover cap (B), while checking that the mark on the steering gear assembly aligns with the mark on the rear cover cap. See illustration.

12. Check the wheel alignment and adjust as necessary.

13. Adjust the steering angle sensor neutral position, using the CONSULT-III diagnostic tool, or equivalent.

14. Be sure to perform the reconnect/relearn procedures.

POWER STEERING PUMP

BLEEDING

1. Before servicing the vehicle, refer to the Precautions Section.

➡**If working near and/or around the SRS system and components, be sure to disable the SRS system. After disabling the system wait three minutes or more before servicing the vehicle.**

➡**Whenever the negative battery cable is disconnected the following components will require resetting. The Idle Air Volume Learning, Steering Angle Sensor Neutral Position, Sunroof Memory Reset/Initialization, Automatic Drive Positioner System, Audio presets and Navigation. Use the CONSULT-III diagnostic tool, or equivalent to perform the required resets.**

2. Stop the engine.
3. Turn the steering wheel fully to the right and left several times.

➡**Do not allow the fluid level in the reservoir tank to go below the MIN level line. Check and add fluid as needed.**

4. Run the engine at idle speed. Turn the steering wheel fully to the right and then fully to the left. Hold for about three seconds. Check for fluid leakage.

5. Repeat the above step several times at three second intervals.

➡**Do not hold the steering wheel in the locked position for more than ten seconds.**

6. Check for air bubbles or cloudy fluid. If found, repeat the bleeding procedure.

7. Stop the engine and check the fluid level. Correct as required.

REMOVAL & INSTALLATION

See Figure 88.

1. Before servicing the vehicle, refer to the Precautions Section.

➡**If working near and/or around the SRS system and components, be sure to disable the SRS system. After disabling the system wait three minutes or more before servicing the vehicle.**

➡**Whenever the negative battery cable is disconnected the following components will require resetting. The Idle Air Volume Learning, Steering Angle Sensor Neutral Position, Sunroof Memory Reset/Initialization, Automatic Drive Positioner System, Audio presets and Navigation. Use the CONSULT-III diagnostic tool, or equivalent to perform the required resets.**

2. Disconnect the negative battery cable.

3. Drain power steering fluid from reservoir tank.

4. Remove air duct assembly.

5. Remove power steering reservoir tank.

6. Remove serpentine drive belt from auto tensioner and power steering pump.

7. Disconnect pressure sensor electrical connector.

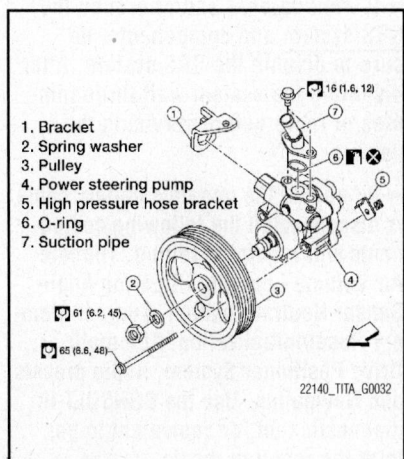

1. Bracket
2. Spring washer
3. Pulley
4. Power steering pump
5. High pressure hose bracket
6. O-ring
7. Suction pipe

22140_TITA_G0032

Fig. 88 Power steering pump and related components

8. Remove the high pressure and low pressure piping from power steering oil pump.

9. Remove bolts, then remove power steering pump.

To install:

➡Be sure to use new fasteners, as required.

10. Installation is the reverse of the removal procedure.

11. Bleed air from power steering system.

12. Be sure to perform the reconnect/relearn procedures.

FLUID RECOMMENDATIONS

Use genuine Nissan power steering fluid (PSF), or equivalent when servicing the power steering system. The system capacity is 2 1/8 pints.

FLUID FILL PROCEDURE

See Figure 89.

✳✳ CAUTION

Used fluid is considerably more dangerous than new fluid. Avoid skin contact with used fluid.

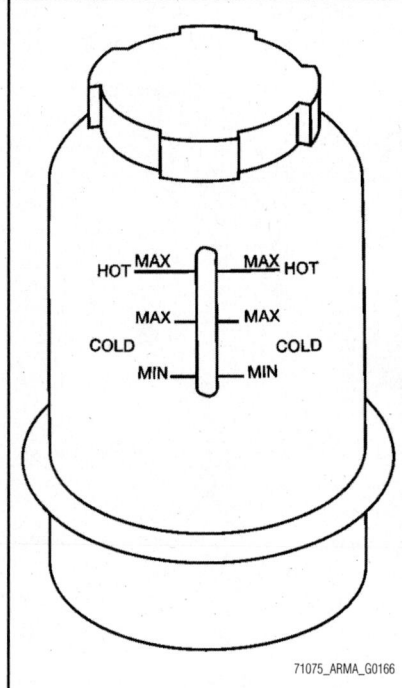

Fig. 89 Power steering reservoir tank fill markings

1. Before servicing the vehicle, refer to the Precautions Section.

➡If working near and/or around the SRS system and components, be

sure to disable the SRS system. After disabling the system wait three minutes or more before servicing the vehicle.

➡Whenever the negative battery cable is disconnected the following components will require resetting. The Idle Air Volume Learning, Steering Angle Sensor Neutral Position, Sunroof Memory Reset/Initialization, Automatic Drive Positioner System, Audio presets and Navigation. Use the CONSULT-III diagnostic tool, or equivalent to perform the required resets.

2. Inspect the power steering fluid level in the power steering reservoir. Do allow the fluid to drop below the MIN marking.

3. Remove the power steering reservoir cap.

4. Add fluid, as necessary, referring to the scale on the reservoir tank.

- HOT range for fluid temperatures: 122–176°F (50–80°C)
- COLD range for fluid temperatures: 32–86°F (0–30°C)

➡Do not overfill the fluid. Do not reuse any used power steering fluid.

SUSPENSION

FRONT SUSPENSION

COIL SPRINGS

REMOVAL & INSTALLATION

See Figures 90 and 91.

1. Before servicing the vehicle, refer to the Precautions Section.

➡If working near and/or around the SRS system and components, be sure to disable the SRS system. After disabling the system wait three minutes or more before servicing the vehicle.

➡Whenever the negative battery cable is disconnected the following components will require resetting. The Idle Air Volume Learning, Steering Angle Sensor Neutral Position, Sunroof Memory Reset/Initialization, Automatic Drive Positioner System, Audio presets and Navigation. Use the CONSULT-III diagnostic tool, or equivalent to perform the required resets.

2. Disconnect the negative battery cable.

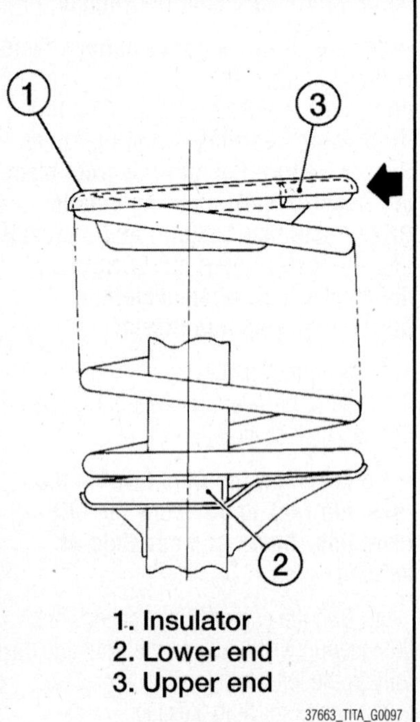

Fig. 90 Front shock positioning (1 of 2)

1. Insulator
2. Lower end
3. Upper end

37663_TITA_G0097

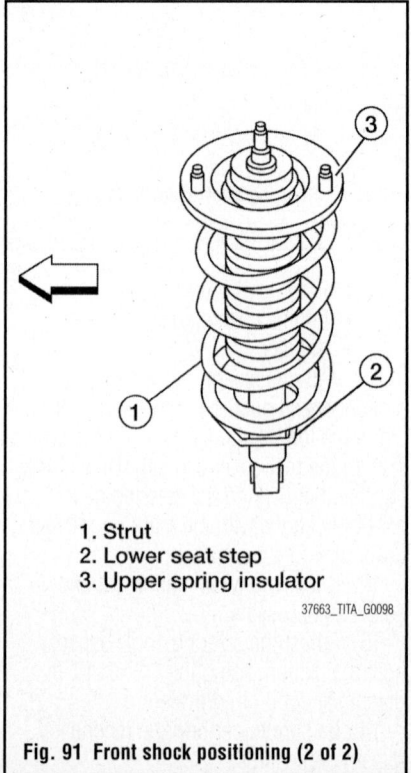

1. Strut
2. Lower seat step
3. Upper spring insulator

37663_TITA_G0098

Fig. 91 Front shock positioning (2 of 2)

3. Raise and safely support the vehicle.

4. Remove the shock.

5. Secure the shock absorber in a vice and loosen (without removing) the piston rod lock nut.

6. Install a spring compressor and tighten until the shock absorber mounting insulator can be turned by hand.

7. Remove piston rod lock nut and remove shock absorber from the coil spring.

To install:

➡**Be sure to use new fasteners, as required.**

8. Install upper mounting insulator in line with the lower shock absorber mount and step in shock absorber lower seat as shown in figure.

9. Tighten the new piston rod lock nut to 40 ft. lbs. (54 Nm).

10. Install or connect the following:
- Coil spring and shock absorber assembly
- Upper shock absorber bolts and tighten to 22 ft. lbs (30 Nm)
- Lower the shock absorber bolt and tighten to 99 ft. lbs. (134 Nm)
- Wheel and tire assembly

11. Check wheel alignment and adjust as necessary.

LOWER BALL JOINTS

REMOVAL & INSTALLATION

At this time the manufacturer does not provide removal and installation procedures for this component. The lower ball joint is part of the lower control arm (lower link) assembly.

➡**Some vehicles may be equipped with straight (nonadjustable) lower link bolts and washers. In order to adjust camber and caster on these vehicles, first replace the lower link bolts and washers with adjustable (cam) bolts and washers.**

➡**Nissan/Infiniti refers to the lower control arm as a lower link.**

LOWER CONTROL ARMS

REMOVAL & INSTALLATION

See Figures 92 and 93.

1. Before servicing the vehicle, refer to the Precautions Section.

➡**If working near and/or around the SRS system and components, be sure to disable the SRS system. After disabling the system wait three minutes or more before servicing the vehicle.**

➡**Whenever the negative battery cable is disconnected the following components will require resetting. The Idle Air Volume Learning, Steering Angle Sensor Neutral Position, Sunroof Memory Reset/Initialization, Automatic Drive Positioner System, Audio presets and Navigation. Use the CONSULT-III diagnostic tool, or equivalent to perform the required resets.**

2. Disconnect the negative battery cable.

3. Raise and support the vehicle safely.

4. Remove the wheel and tire using power tool.

5. Remove lower shock absorber bolt.

6. Remove stabilizer bar connecting rod lower nut using power tool, then separate connecting rod from lower link.

7. Remove driveshaft, if equipped.

8. Remove pinch bolt from steering knuckle using power tool, then separate lower link ball joint from steering knuckle.

9. Remove lower link cam bolts (1) and nuts, then the lower link (2).

To install:

10. Install all removed parts in the reverse order of removal procedure and note the following:

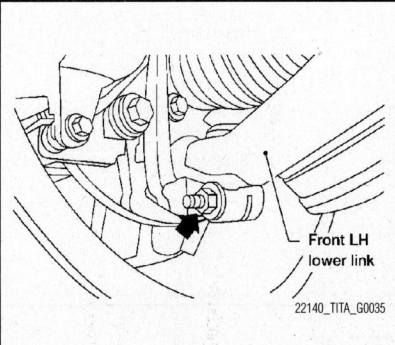

Fig. 92 Remove pinch bolt from steering knuckle

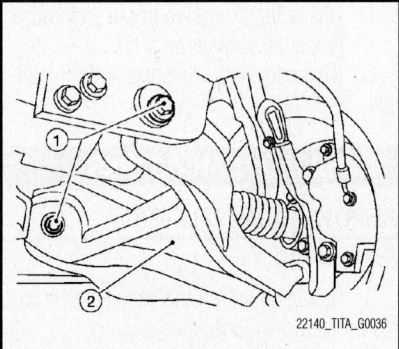

Fig. 93 Remove lower link cam bolts

a. Tighten the cam nuts only in a relaxed position to 98 ft. lbs. (133 Nm). Cam bolts are for adjustment.

b. Tighten the steering knuckle pinch bolt to 70 ft. lbs. (95 Nm).

c. After installation, check that the front wheel alignment is within specification.

➡**Some vehicles may be equipped with straight (nonadjustable) lower link bolts and washers. In order to adjust camber and caster on these vehicles, first replace the lower link bolts and washers with adjustable ones.**

11. Be sure to perform the reconnect/relearn procedures.

SHOCK ABSORBERS

REMOVAL & INSTALLATION

See Figure 94.

1. Before servicing the vehicle, refer to the Precautions Section.

➡**If working near and/or around the SRS system and components, be sure to disable the SRS system. After disabling the system wait three minutes or more before servicing the vehicle.**

➡**Whenever the negative battery cable is disconnected the following components will require resetting. The Idle Air Volume Learning, Steering Angle Sensor Neutral Position, Sunroof Memory Reset/Initialization, Automatic Drive Positioner System, Audio presets and Navigation. Use the CONSULT-III diagnostic tool, or equivalent to perform the required resets.**

2. Disconnect the negative battery cable.

3. Raise and safely support the vehicle.

4. Remove or disconnect the following:
- Wheel and tire assembly
- Lower shock absorber bolt
- Upper shock absorber bolts
- Coil spring and shock absorber assembly

5. Secure the shock absorber in a vice and loosen (without removing) the piston rod lock nut.

6. Install a spring compressor and tighten until the shock absorber mounting insulator can be turned by hand.

7. Remove piston rod lock nut and remove shock absorber from the coil spring.

To install:

➡**Be sure to use new fasteners, as required.**

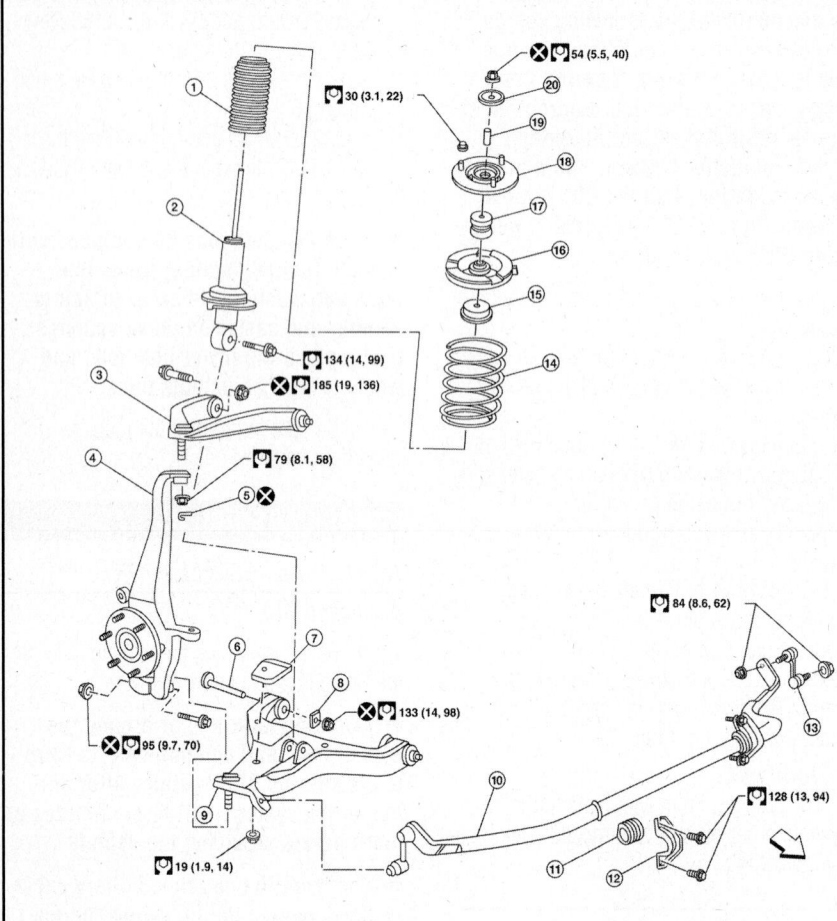

Fig. 94 Front shock and related components

1.	Dust cover	2.	Shock absorber	3.	Upper link
4.	Steering knuckle	5.	Cotter pin	6.	Bolt
7.	Jounce bumper	8.	Washer	9.	Lower link
10.	Stabilizer bar	11.	Stabilizer bar bushing	12.	Stabilizer bar mounting bracket
13.	Connecting rod	14.	Coil spring	15.	Upper seat
16.	Upper spring seat	17.	Shock absorber bushing	18.	Shock absorber mounting insulator
19.	Spacer	20.	Washer	⇦	Front

37663_TITA_G0094

8. Install upper mounting insulator in line with the lower shock absorber mount and step in shock absorber lower seat.

9. Tighten the new piston rod lock nut to 40 ft. lbs. (54 Nm).

10. Install or connect the following:
- Coil spring and shock absorber assembly
- Upper shock absorber bolts and tighten to 22 ft. lbs (30 Nm)
- Lower the shock absorber bolt and tighten to 99 ft. lbs. (134 Nm)
- Wheel and tire assembly

11. Check wheel alignment and adjust as necessary.

12. Be sure to perform the reconnect/relearn procedures.

TESTING

1. Before servicing the vehicle, refer to the Precautions Section.

2. Road test the vehicle.

3. Check for excessive bounce or roll.

4. Raise the vehicle on a lift.

5. Check for bad bushings and oil leakage.

STABILIZER BAR & LINKS

REMOVAL & INSTALLATION

See Figure 95.

1. Before servicing the vehicle, refer to the Precautions Section.

➡ **If working near and/or around the SRS system and components, be sure**
to disable the SRS system. After disabling the system wait three minutes or more before servicing the vehicle.

➡ **Whenever the negative battery cable is disconnected the following components will require resetting. The Idle Air Volume Learning, Steering Angle Sensor Neutral Position, Sunroof Memory Reset/Initialization, Automatic Drive Positioner System, Audio presets and Navigation. Use the CONSULT-III diagnostic tool, or equivalent to perform the required resets.**

2. Disconnect the negative battery cable.

3. Raise and safely support the vehicle.

4. Remove or disconnect the following:
- Engine undercover
- Stabilizer bar mounting bracket bolts and connecting rod nuts using power tool
- Bushings from stabilizer bar

5. Check stabilizer bar for twist and deformation. Replace if necessary. Check rubber bushing for cracks, wear and deterioration. Replace if necessary.

To install:

6. To install, reverse removal procedure and note the following:

a. Tighten stabilizer bar bracket mounting bolts to 94 ft. lbs. (128 Nm).

b. Tighten the connecting rod nuts to 62 ft. lbs. (84 Nm).

7. Be sure to perform the reconnect/relearn procedures.

STEERING KNUCKLE

REMOVAL & INSTALLATION

1. Before servicing the vehicle, refer to the Precautions Section.

➡ **If working near and/or around the SRS system and components, be sure to disable the SRS system. After**

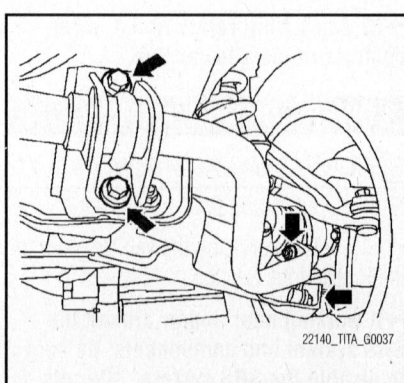

Fig. 95 Stabilizer bar mounting bracket bolts and connecting rod nuts

22140_TITA_G0037

disabling the system wait three minutes or more before servicing the vehicle.

➡Whenever the negative battery cable is disconnected the following components will require resetting. The Idle Air Volume Learning, Steering Angle Sensor Neutral Position, Sunroof Memory Reset/Initialization, Automatic Drive Positioner System, Audio presets and Navigation. Use the CONSULT-III diagnostic tool, or equivalent to perform the required resets.

2. Disconnect the negative battery cable.
3. Raise and safely support the vehicle.
4. Remove or disconnect the following:
   - Wheel and tire assembly
   - Engine splash guard

➡Disconnect wheel sensor harness connector. Do not remove wheel sensor from wheel hub and bearing assembly for this procedure.

   - Wheel and hub assembly (Remove cotter pin, then remove driveshaft nut—4WD)

⁂ WARNING

Be careful not to damage ball joint boot. Temporarily tighten nut to prevent damage to threads and to prevent tool from coming off.

   - Steering outer socket from steering knuckle using tool HT72520000 (J-25730-A)
   - Coil spring and shock absorber assembly

➡Support the lower link using a suitable jack.

   - Cotter pin and nut from upper link ball joint and discard the cotter pin

5. Separate the upper link ball joint from steering knuckle using tool ST29020001 (J-24319-01).
6. Remove pinch bolt from steering knuckle using power tool, then separate lower link ball joint from steering knuckle.
7. Remove the steering knuckle from vehicle.
8. Check for deformity, cracks and damage on each part, and replace if necessary.

To install:

➡Be sure to use new fasteners, as required.

9. To install, reverse removal procedure and note the following:
   a. Tighten the steering knuckle pinch bolt to 70 ft. lbs. (95 (Nm).
   b. Tighten the upper control arm ball joint nut to 58 ft. lbs. (79 Nm).
   c. Use new cotter pins for installation of lock nuts.
   d. For 4WD, tighten the axle nut to 101 ft. lbs. (137 Nm).
10. Be sure to perform the reconnect/relearn procedures.

UPPER BALL JOINTS

REMOVAL & INSTALLATION

At this time the manufacturer does not provide removal and installation procedures for this component. The upper ball joint is part of the upper control arm (upper link) assembly.

➡Nissan/Infiniti refers to the lower control arm as a lower link.

UPPER CONTROL ARMS

REMOVAL & INSTALLATION

1. Before servicing the vehicle, refer to the Precautions Section.

➡If working near and/or around the SRS system and components, be sure to disable the SRS system. After disabling the system wait three minutes or more before servicing the vehicle.

➡Whenever the negative battery cable is disconnected the following components will require resetting. The Idle Air Volume Learning, Steering Angle Sensor Neutral Position, Sunroof Memory Reset/Initialization, Automatic Drive Positioner System, Audio presets and Navigation. Use the CONSULT-III diagnostic tool, or equivalent to perform the required resets.

2. Disconnect the negative battery cable.
3. Remove or disconnect the following:
   - Wheel and tire assembly
   - Fender protector to access the upper link
   - Cotter pin and nut from upper ball joint
4. Separate upper ball joint stud from steering knuckle using special tool J-24319-01.
5. Remove the upper control arm mounting bolts.
6. Remove the upper arm.

To install:

7. Install or connect the following:
   - Upper control arm and tighten bolts to 107 ft. lbs. (145 Nm)
   - Upper ball joint with new cotter pin and tighten nut to 58 ft. lbs. (79 Nm)
   - Fender protector
   - Wheel and tire assembly
8. Check front end alignment, adjust as required.
9. Be sure to perform the reconnect/relearn procedures.

WHEEL HUBS & BEARINGS

REMOVAL & INSTALLATION

1. Before servicing the vehicle, refer to the Precautions Section.

➡If working near and/or around the SRS system and components, be sure to disable the SRS system. After disabling the system wait three minutes or more before servicing the vehicle.

➡Whenever the negative battery cable is disconnected the following components will require resetting. The Idle Air Volume Learning, Steering Angle Sensor Neutral Position, Sunroof Memory Reset/Initialization, Automatic Drive Positioner System, Audio presets and Navigation. Use the CONSULT-III diagnostic tool, or equivalent to perform the required resets.

2. Disconnect the negative battery cable.
3. Remove or disconnect the following:
   - Wheel and tire assembly
   - Engine undercover
   - Brake caliper without disconnecting the hydraulic lines, and reposition aside with wire
4. Install a matchmark on the brake rotor and to the wheel hub, and remove the brake rotor.
5. Remove or disconnect the following:
   - Cotter pin and lock nut from Driveshaft (4WD)
   - Driveshaft from wheel hub and bearing assembly (4WD)
   - ABS sensor
   - Wheel hub and bearing assembly bolts
   - Wheel hub and bearing assembly

To install:

➡Be sure to use new fasteners, as required.

6. Install or connect the following:
- Wheel hub and bearing assembly, using new bolts and tightening to 155 ft. lbs. (210 Nm)
- ABS sensor

- Driveshaft to wheel hub and bearing assembly
- Cotter pin and lock nut and tighten to 101 ft. lbs. (137 Nm)
- Brake rotor

- Brake caliper
- Engine splash guard
- Wheel and tire assembly

7. Be sure to perform the reconnect/relearn procedures.

SUSPENSION

REAR SUSPENSION

LEAF SPRINGS

REMOVAL & INSTALLATION

See Figure 96.

1. Before servicing the vehicle, refer to the Precautions Section.

➡**If working near and/or around the SRS system and components, be sure to disable the SRS system. After dis-** abling the system wait three minutes or more before servicing the vehicle.

➡**Whenever the negative battery cable is disconnected the following components will require resetting. The Idle Air Volume Learning, Steering Angle Sensor Neutral Position, Sunroof Memory Reset/Initialization, Automatic Drive Positioner System, Audio presets and Navigation. Use the CONSULT-III** diagnostic tool, or equivalent to perform the required resets.

2. Disconnect the negative battery cable.

3. Raise and support the vehicle safely.

4. Remove the tire and wheel assembly.

5. Support the rear differential with a suitable jack to relieve the tension from the leaf spring.

6. Remove or disconnect the following:
- Shock absorber lower mounting bolt
- Spring clip U-bolt nuts
- Spring pad
- Storage box, if equipped
- Rear shackle lower bolt
- Leaf spring front mounting bolt
- Leaf spring

To install:

➡**Be sure to use new fasteners, as required.**

7. Install or connect the following:
- Front mounting bolt and shackle lower bolt and finger tighten the nuts
- U-bolts, rear spring pad and nuts or the U-bolts

8. Tighten the U-bolt nuts diagonally and evenly to specification.

9. Install the shock absorber and finger tighten the nuts.

10. Remove the jack supporting the rear differential and bounce the rear of the vehicle to stabilize the suspension.

11. Tighten the front mount bolt to specification.

12. Tighten the rear shackle lower bolt to specification.

13. Tighten the shock absorber lower mounting bolt to specification.

14. Be sure to perform the reconnect/relearn procedures.

SHOCK ABSORBERS

REMOVAL & INSTALLATION

1. Before servicing the vehicle, refer to the Precautions Section.

➡**If working near and/or around the SRS system and components, be sure to disable the SRS system. After dis-**

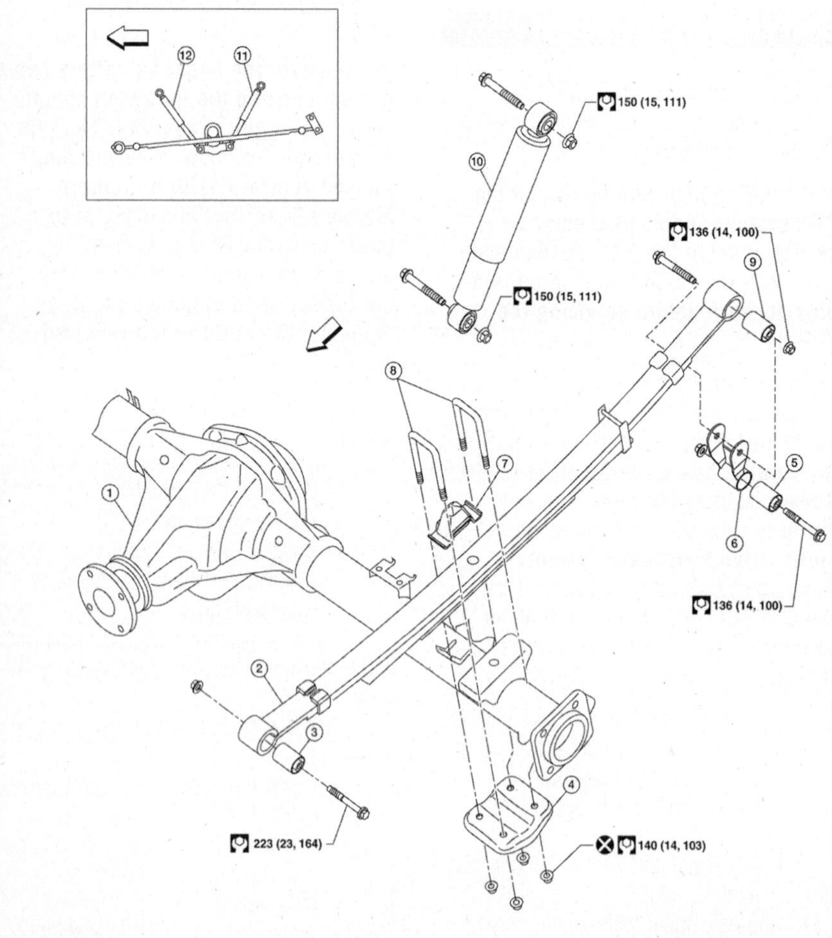

1. Rear final drive	2. Rear leaf spring	3. Rear spring bushing (front)
4. Rear spring pad	5. Rear spring shackle bushing	6. Rear spring shackle
7. Bumper	8. Rear spring clip U-bolts	9. Rear spring bushing (rear)
10. Shock absorber	11. Shock absorber (left side)	12. Shock absorber (right side)
⇦ Front		

37663_TITA_G0095

Fig. 96 Rear spring and related components

abling the system wait three minutes or more before servicing the vehicle.

➡Whenever the negative battery cable is disconnected the following components will require resetting. The Idle Air Volume Learning, Steering Angle Sensor Neutral Position, Sunroof Memory Reset/Initialization, Automatic Drive Positioner System, Audio presets and Navigation. Use the CONSULT-III diagnostic tool, or equivalent to perform the required resets.

2. Disconnect the negative battery cable.
3. Raise and support the vehicle safely.
4. Remove the tire and wheel assembly.

5. Support the rear differential with a suitable jack.
6. Remove the upper and lower shock absorber mounting bolts.
7. Remove the shock absorber.

To install:

➡Be sure to use new fasteners, as required.

8. Install the shock absorber and tighten the upper and lower mounting bolts to specification.
9. Be sure to perform the reconnect/relearn procedures.

TESTING

1. Before servicing the vehicle, refer to the Precautions Section.

2. Road test the vehicle.
3. Check for excessive bounce or roll.
4. Raise the vehicle on a lift.
5. Check for bad bushings and oil leakage.

WHEEL HUBS & BEARINGS (SEALED UNIT)

REMOVAL & INSTALLATION

See Figure 97.

1. Before servicing the vehicle, refer to the Precautions Section.

➡If working near and/or around the SRS system and components, be sure to disable the SRS system. After disabling the system wait three

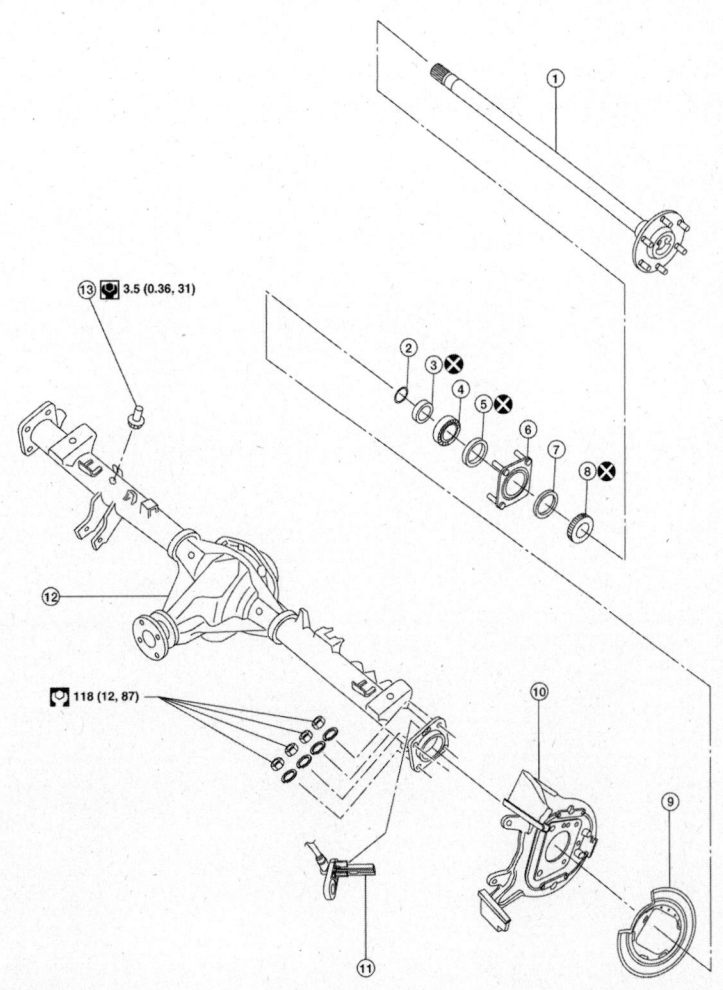

1. Axle shaft	2. Snap ring	3. Bearing ring retainer
4. Axle shaft bearing and cup	5. Axle oil seal	6. Axle shaft bearing cage
7. Seal plate	8. ABS sensor rotor	9. Back plate
10. Torque member	11. ABS sensor	12. Rear final drive
13. Breather		

37663_TITA_G0099

Fig. 97 Rear axle shaft and related components

minutes or more before servicing the vehicle.

➡Whenever the negative battery cable is disconnected the following components will require resetting. The Idle Air Volume Learning, Steering Angle Sensor Neutral Position, Sunroof Memory Reset/Initialization, Automatic Drive Positioner System,

Audio presets and Navigation. Use the CONSULT-III diagnostic tool, or equivalent to perform the required resets.

2. Disconnect the negative battery cable.

3. Remove the axle shaft.

4. Remove the component from its mounting.

To install:

➡Be sure to use new fasteners, as required.

5. Installation is the reverse of the removal procedure.

6. Be sure to perform the reconnect/relearn procedures.

SPECIFICATIONS AND MAINTENANCE CHARTS

ENGINE AND VEHICLE IDENTIFICATION

Engine							Model Year	
Code ①	Liters (cc)	Cu. In.	Cyl.	Fuel Sys.	Engine Type	Eng. Mfg.	Code ②	Year
HR16DE	1.6 (1,598)	97.51	4	MFI	DOHC	Nissan	B	2011
MR18DE	1.8 (1,797)	109.65	4	MFI	DOHC	Nissan	C	2012

MFI: Multiport Fuel Injection

DOHC: Double Overhead Camshafts

① The engine code is stamped on the engine block near the starter

② 10th position of the Vehicle Identification Number (VIN)

71075_VERS_C0001

GENERAL ENGINE SPECIFICATIONS

All measurements are given in inches.

Year	Model	Engine Displacement Liters (cc)	Engine ID/VIN	Fuel System Type	Net Horsepower @ rpm	Net Torque @ rpm (ft. lbs.)	Bore x Stroke (in.)	Com-pression Ratio	Oil Pressure @ rpm
2011	Versa	1.6 (1,598)	HR16DE	MFI	107@6,000	111@4,600	3.07 x 3.29	10.7:1	39 psi@2,000
		1.8 (1,797)	MR18DE	MFI	122@5,200	127@4,800	3.31 x 3.19	9.9:1	29 psi@2,000
2012	Versa	1.6 (1,598)	HR16DE	MFI	109@6,000	107@4,400	3.07 x 3.29	9.8:1	43 psi@2,000
		1.8 (1,797)	MR18DE	MFI	122@5,200	127@4,800	3.31 x 3.19	9.9:1	29 psi@2,000

MFI: Multiport Fuel Injection

71075_VERS_C0002

ENGINE TUNE-UP SPECIFICATIONS

Year	Engine Displacement Liters	Engine ID/VIN	Spark Plug Gap (in.)	Ignition Timing (deg. BTDC) MT ①	AT ①	Fuel Pump (psi) ②	Idle Speed (rpm) MT ①	AT ①	Valve Clearance (in.) Intake ③	Exhaust ③
2011	1.6	HR16DE	0.043	1-11	1-11	51	600-700	650-750	0.010-0.013	0.011-0.015
	1.8	MR18DE	0.043	8-18	8-18	51	650-750	650-750	0.010-0.013	0.011-0.015
2012	1.6	HR16DE	0.043	5-15	-1-9	51	600-700	600-700	0.010-0.013	0.011-0.014
	1.8	MR18DE	0.043	8-18	8-18	51	650-750	650-750	0.010-0.013	0.011-0.015

NOTE: The Vehicle Emission Control Information label often reflects specification changes made during production.

The label figures must be used if they differ from those in this chart.

BTDC: Before Top Dead Center

① Under no load condition (in P or N position)

② System pressure at idle

③ With engine cold

71075_VERS_C0003

CAPACITIES

Year	Model	Engine Displacement Liters	Engine ID/VIN	Engine Oil with Filter (qts.)	Transaxle (pts.) Auto. ①	Manual	Drive Axle (pts.) Front	Rear	Transfer Case (pts.)	Fuel Tank (gal.)	Cooling System (qts.) ②
2011	Versa	1.6	HR16DE	3.2	16.3	5.5	N/A	N/A	N/A	13.2	6.7
		1.8	MR18DE	4.3	③	4.2	N/A	N/A	N/A	13.2	7.2
2012	Versa	1.6	HR16DE	3.2	14.6	5.6	N/A	N/A	N/A	11.3	④
		1.8	MR18DE	4.3	③	4.2	N/A	N/A	N/A	11.3	7.2

NOTE: All capacities are approximate. Add fluid gradually and ensure a proper fluid level is obtained.

N/A: Not Applicable

① Drain and refill

② With reservoir tank at "MAX" level

③ Automatic transaxle (4-speed): 16.7 pints
   Continuously Variable Transaxle (CVT): 14.6 pints

④ Manual transaxle: 6.7 quarts
   Continuously Variable Transaxle (CVT): 7.6 quarts

71075_VERS_C0004

FLUID SPECIFICATIONS

Year	Model	Engine Disp. Liters	Engine Oil	Manual Trans.	Auto. Trans.	Drive Axle Front	Rear	Transfer Case	Power Steering Fluid	Brake Master Cylinder	Cooling System
2011	Versa	1.6	5W-30	①	②	N/A	N/A	N/A	N/A	③	④
		1.8	5W-30	⑤	⑥	N/A	N/A	N/A	N/A	③	④
2012	Versa	1.6	5W-30	①	⑦	N/A	N/A	N/A	N/A	③	④
		1.8	5W-30	⑤	⑧	N/A	N/A	N/A	N/A	③	④

N/A: Not Applicable

DOT: Department Of Transportation

① Genuine NISSAN Manual Transmission Fluid (MTF) HQ Multi 75W-85 or equivalent

② Genuine NISSAN Matic S ATF

③ Genuine NISSAN Super Heavy Duty Brake Fluid or equivalent DOT 3

④ Pre-diluted Genuine NISSAN Long Life Anti-freeze/Coolant (blue) or equivalent

⑤ Genuine NISSAN gear oil (Chevron Texaco ETL8997B) 75W-80, or equivalent

⑥ Automatic transaxle (4-speed): Genuine NISSAN Matic S ATF
   Continuously Variable Transaxle (CVT): Genuine NISSAN CVT Fluid NS-2

⑦ Genuine NISSAN CVT Fluid NS-2

⑧ Automatic transaxle (4-speed): Genuine NISSAN Matic D ATF
   Continuously Variable Transaxle (CVT): Genuine NISSAN CVT Fluid NS-2

71075_VERS_C0005

VALVE SPECIFICATIONS

Year	Engine Displacement Liters	Engine ID/VIN	Seat Angle (deg.)	Face Angle (deg.)	Spring Test Pressure (lbs. @ in.)	Spring Free-Length (in.)	Spring Installed Height (in.)	Stem-to-Guide Clearance (in.) Intake	Exhaust	Stem Diameter (in.) Intake	Exhaust
2011	1.6	HR16DE	①	NS	59-67@ 0.9433	1.6638	1.2756	0.0008- 0.0021	0.0012- 0.0025	0.1955- 0.1961	0.1951- 0.1957
	1.8	MR18DE	①	NS	②	③	1.3900	0.0008- 0.0021	0.0012- 0.0025	0.2152- 0.2157	0.2148- 0.2154
2012	1.6	HR16DE	①	NS	54-61@ 0.9433	1.8398	1.2756	0.0008- 0.0021	0.0012- 0.0025	0.1955- 0.1961	0.1951- 0.1957
	1.8	MR18DE	①	NS	②	③	1.3900	0.0008- 0.0021	0.0012- 0.0025	0.2152- 0.2157	0.2148- 0.2154

NS: Not Specified

① 45° 15' - 45° 45' (degrees/minutes)

② Intake: 75-85 lbs. @ 1.0377 inches
   Exhaust: 60-67 lbs. @ 1.0944 inches

③ Intake: 1.7677-1.7755 inches
   Exhaust: 1.8007-1.8086 inches

71075_VERS_C0006

CAMSHAFT SPECIFICATIONS
All measurements in inches unless noted

Year	Engine Displacement Liters	Engine Code/VIN	Journal Diameter	Brg. Oil Clearance	Shaft End-play	Runout	Journal Bore	Cam Height Intake	Cam Height Exhaust
2011	1.6	HR16DE	①	②	0.0030-0.0060	0.0008	③	1.6419-1.6494	1.5817-1.5892
	1.8	MR18DE	①	②	0.0030-0.0060	0.0008	③	1.7560-1.7635	1.6997-1.7072
2012	1.6	HR16DE	①	②	0.0030-0.0060	0.0008	③	1.6419-1.6494	1.6108-1.6183
	1.8	MR18DE	①	②	0.0030-0.0060	0.0008	③	1.7560-1.7635	1.6997-1.7072

① No. 1: 1.0998-1.1006 inches
  No. 2, 3, 4, 5: 0.9823-0.9831 inch

② No. 1: 0.0018-0.0034 inch
  No. 2, 3, 4, 5: 0.0012-0.0028 inch

③ Camshaft bracket inner diameter
  No. 1: 1.1024-1.1032 inches
  No. 2, 3, 4, 5: 0.9843-0.9851 inch

71075_VERS_C0007

CRANKSHAFT AND CONNECTING ROD SPECIFICATIONS
All measurements are given in inches.

Year	Engine Displacement Liters	Engine ID/VIN	Crankshaft Main Brg. Journal Dia.	Crankshaft Main Brg. Oil Clearance	Crankshaft Shaft End-play	Crankshaft Thrust on No.	Connecting Rod Journal Diameter	Connecting Rod Oil Clearance	Connecting Rod Side Clearance
2011	1.6	HR16DE	1.8881-1.8889 ①	0.0009-0.0013	0.0039-0.0102	3	1.5729-1.5737 ①	0.0011-0.0015	0.0079-0.0139
	1.8	MR18DE	2.0456-2.0464 ①	②	0.0039-0.0102	3	1.7304-1.7311 ①	0.0015-0.0019	0.0079-0.0138
2012	1.6	HR16DE	1.8881-1.8889 ①	0.0009-0.0013	0.0039-0.0102	3	1.5729-1.5737 ①	0.0008-0.0012	0.0079-0.0139
	1.8	MR18DE	2.0456-2.0464 ①	②	0.0039-0.0102	3	1.7304-1.7311 ①	0.0015-0.0019	0.0079-0.0138

① Variance depending on diameter Grade

② Journal No. 1, 4, 5: 0.0009-0.0013 inch
  No. 2, 3: 0.0005-0.0009 inch

71075_VERS_C0008

PISTON AND RING SPECIFICATIONS
All measurements are given in inches.

| Year | Engine Displacement Liters | Engine ID/VIN | Piston Clearance | Ring Gap | | | Ring Side Clearance | | |
				Top Compression	Bottom Compression	Oil Control	Top Compression	Bottom Compression	Oil Control
2011	1.6	HR16DE	0.0008-0.0020	0.0079-0.0118	0.0138-0.0197	0.0079-0.0236	0.0016-0.0031	0.0012-0.0028	0.0018-0.0049
	1.8	MR18DE	0.0008-0.0016	0.0079-0.0118	0.0197-0.0256	0.0059-0.0177	0.0016-0.0031	0.0012-0.0028	0.0006-0.0073
2012	1.6	HR16DE	0.0008-0.0020	0.0079-0.0118	0.0138-0.0197	0.0079-0.0177	0.0016-0.0031	0.0012-0.0028	0.0018-0.0049
	1.8	MR18DE	0.0008-0.0016	0.0079-0.0098	0.0197-0.0256	0.0059-0.0177	0.0016-0.0031	0.0012-0.0028	0.0006-0.0073

71075_VERS_C0009

TORQUE SPECIFICATIONS
All readings in ft. lbs.

| Year | Engine Disp. Liters | Engine ID/VIN | Cylinder Head Bolts | Main Bearing Bolts | Rod Bearing Bolts | Crankshaft Damper Bolts | Flywheel Bolts | Manifold | | Spark Plugs | Oil Pan Drain Plug |
								Intake	Exhaust		
2011	1.6	HR16DE	①	②	③	④	80	20	25	15	25
	1.8	MR18DE	⑤	②	③	⑥	80	20	25	15	25
2012	1.6	HR16DE	①	②	⑦	④	80	20	25	15	25
	1.8	MR18DE	⑤	②	③	⑥	80	20	25	15	25

① Apply engine oil to bolts, refer to procedure for tightening sequence
  Step 1: Tighten to 30 ft. lbs.
  Step 2: Plus 60 degrees
  Step 3: Loosen bolts completely
  Step 4: Tighten to 30 ft. lbs.
  Step 5: Plus 75 degrees
  Step 6: Plus an additional 75 degrees
② Apply engine oil to bolts, refer to procedure for tightening sequence
  Step 1: Tighten to 25 ft. lbs.
  Step 2: Plus 60 degrees
③ Apply engine oil to bolt threads and seats of cap bolts
  Step 1: Tighten to 14 ft. lbs.
  Step 2: Plus 60 degrees
④ Apply engine oil to bolt threads
  Step 1: Tighten to 26 ft. lbs.
  Step 2: Plus 60 degrees

⑤ Apply engine oil to bolts, refer to procedure for tightening sequence
  Step 1: Tighten to 30 ft. lbs.
  Step 2: Plus 100 degrees
  Step 3: Loosen bolts completely
  Step 4: Tighten to 30 ft. lbs.
  Step 5: Plus 100 degrees
  Step 6: Plus an additional 100 degrees
⑥ Apply engine oil to bolt threads
  Step 1: Tighten to 22 ft. lbs.
  Step 2: Plus 60 degrees
⑦ Apply engine oil to bolt threads and seats of cap bolts
  Step 1: Tighten to 20 ft. lbs. in several steps
  Step 2: Loosen bolts completely
  Step 3: Tighten to 14 ft. lbs. in several steps
  Step 4: Plus 60 degrees

71075_VERS_C0010

WHEEL ALIGNMENT

Year	Model		Caster Range (+/-Deg.)	Caster Preferred Setting (Deg.)	Camber Range (+/-Deg.)	Camber Preferred Setting (Deg.)	Toe-in (in.)
2011	Versa	F	0.75	①	0.55	0.0	0.05 +/- 0.05
		R	N/A	N/A	0.50	②	0.08 +/- 0.18
2012	Versa Sedan	F	0.75	3.67	0.74	-0.08	0.08 +/- 0.08
		R	N/A	N/A	0.50	-1.42	③
	Versa Hatchback	F	0.75	④	0.55	0.0	0.05 +/- 0.05
		R	N/A	N/A	0.50	-1.51	0.14 +/- 0.18

NOTE: Measurements given for an unladen vehicle with fuel, coolant, and engine oil full; spare tire, jack, hand tools, and mats in designated positions.

F: Front   R: Rear   N/A: Not Applicable

① With P185/65R14 tires: 4.50 degrees (RH); 4.33 degrees (LH)

  With P185/65R15 and P195/55R16 tires: 4.83 degrees (RH); 4.67 degrees (LH)

② With P185/65R14 tires: -1.52 degrees

  With P185/65R15 and P195/55R16 tires: -1.51 degrees

③ Nominal: In 0.17 degrees; Maximum: In 0.58 degrees; Minimum: Out 0.25 degrees

④ With P185/65R15 and P195/55R16 tires: 4.83 degrees (RH); 4.67 degrees (LH)

71075_VERS_C0011

TIRE, WHEEL AND BALL JOINT SPECIFICATIONS

Year	Model	OEM Tires Standard	OEM Tires Optional	Tire Pressures (psi) Front	Tire Pressures (psi) Rear	Wheel Size Standard	Wheel Size Optional	Ball Joint Inspection	Lug Nut (ft. lbs.)
2011	Versa	P185/65R14	NS	①	①	14 x 5.5	NS	②	83
		P185/65R15	NS	③	③	15 x 5.5	NS	②	83
		P195/55R16	NS	③	③	16 x 5.5	NS	②	83
2012	Versa Sedan	P185/65R15	NS	①	①	15 x 5.5	NS	④	83
	Versa Hatchback	P185/65R15	NS	③	③	15 x 5.5	NS	②	83
		P195/55R16	NS	③	③	16 x 5.5	NS	②	83

OEM: Original Equipment Manufacturer   PSI: Pounds Per Square Inch   NS: Not Specified

① Always refer to the owner's manual and/or vehicle label: conventional tires should be inflated to 33 psi

② Measurement on spring balance: 3.5-40.0 lbs.

  Swing torque: 5-43 inch lbs.; Axial endplay: 0.0 inch

③ Always refer to the owner's manual and/or vehicle label: conventional tires should be inflated to 35 psi

④ Measurement on spring balance: 3.5-23.5 lbs.

  Swing torque: 4.4-30 inch lbs.; Axial endplay: 0.0 inch

71075_VERS_C0012

BRAKE SPECIFICATIONS

All measurements in inches unless noted

Year	Model		Brake Disc Original Thickness	Brake Disc Minimum Thickness	Brake Disc Max. Runout	Brake Drum Diameter Original Inside Diameter	Max. Wear Limit	Maximum Machine Diameter	Pad/Lining Thickness Standard	Pad/Lining Thickness Limit	Brake Caliper Torque Member Bolts (ft. lbs.)	Brake Caliper Guide Pin Bolts (ft. lbs.)
2011	Versa Sedan	F	0.866	0.787	0.0028	N/A	N/A	N/A	0.354	0.079	62	32
		R	N/A	N/A	N/A	7.992	8.051	NS	0.157	0.059	N/A	N/A
	Versa Hatchback	F	0.945	0.866	0.0028	N/A	N/A	N/A	0.374	0.079	62	20
		R	N/A	N/A	N/A	9.000	9.055	NS	0.157	0.059	N/A	N/A
2012	Versa Sedan	F	0.870	0.787	0.0020	N/A	N/A	N/A	0.354	0.079	62	20
		R	N/A	N/A	N/A	8.000	8.039	NS	0.189	0.039	N/A	N/A
	Versa Hatchback	F	0.945	0.866	0.0028	N/A	N/A	N/A	0.374	0.079	62	20
		R	N/A	N/A	N/A	9.000	9.055	NS	0.157	0.059	N/A	N/A

F: Front   N/A: Not Applicable

R: Rear   NS: Not Specified

71075_VERS_C0013

SCHEDULED MAINTENANCE INTERVALS
Nissan—Versa

TO BE SERVICED	TYPE OF	7.5	15	22.5	30	37.5	45	52.5	60
Engine oil & filter	R	✓	✓	✓	✓	✓	✓	✓	✓
Brake lines & cables	S/I		✓		✓		✓		✓
Brake pads, discs	I		✓		✓		✓		✓
Driveshaft boots & propeller shaft	L/I		✓		✓		✓		✓
CVT ①	I		✓		✓		✓		✓
Transfer case and differential fluid ②	I		✓		✓		✓		✓
Air cleaner filter	R				✓				✓
Drive belt (s) ③	S/I								✓
Engine coolant ④	R								✓
Spark plugs	R	Platinum plugs, every 105,000 miles							
Cabin air filter	R		✓		✓		✓		✓
Exhaust system	I				✓				✓
Evap vapor lines	I				✓				✓
Fuel lines	S/I				✓				✓
Steering gear, linkage, axle & suspension parts	I	✓	✓	✓	✓	✓	✓	✓	✓
Tires (rotate)	S/I	every 5,000-6,000 miles							
Valve clearance ⑤	S/I								

R: Replace   S/I: Service or Inspect   L: Lubricate

① If towing a trailer, using a camper or a car-top carrier, or driving on rough or muddy roads, change (not just inspect) oil at every 60,000 miles.

② If towing a trailer, using a camper or a car-top carrier, or driving on rough or muddy roads, change (not just inspect) oil at every 30,000 miles (48,000 km) or 24 months.

③ First at 60,000, then every 15,000 miles

④ After 60,000, replace every 30,000

⑤ Periodic maintenance not required, if valve noice increases, inspect valve clearance

Follow Periodic Maintenance Schedule 1 if the driving habits frequently include one or more of the following driving conditions:

Repeated short trips of less than 5 miles (8 km).

Repeated short trips of less than 10 miles (16 km) with outside temperatures remaining below freezing

Operating in hot weather in stop-and-go "rush hour" traffic.

Extensive idling and/or low speed driving for long distances, such as police, taxi or door-to-door delivery use

Driving in dusty conditions.

Driving on rough, muddy, or salt spread roads.

Towing a trailer, using a camper or a car-top carrier.

Follow Periodic Maintenance Schedule 2 if none of driving conditions shown in Schedule 1 apply to the driving habits.

71075_VERS_C0014

PRECAUTIONS

Before servicing any vehicle, please be sure to read all of the following precautions, which deal with personal safety, prevention of component damage, and important points to take into consideration when servicing a motor vehicle:

• Never open, service or drain the radiator or cooling system when the engine is hot; serious burns can occur from the steam and hot coolant.

• Observe all applicable safety precautions when working around fuel. Whenever servicing the fuel system, always work in a well-ventilated area. Do not allow fuel spray or vapors to come in contact with a spark, open flame, or excessive heat (a hot drop light, for example). Keep a dry chemical fire extinguisher near the work area. Always keep fuel in a container specifically designed for fuel storage; also, always properly seal fuel containers to avoid the possibility of fire or explosion. Refer to the additional fuel system precautions later in this section.

• Fuel injection systems often remain pressurized, even after the engine has been turned **OFF**. The fuel system pressure must be relieved before disconnecting any fuel lines. Failure to do so may result in fire and/or personal injury.

• Brake fluid often contains polyglycol ethers and polyglycols. Avoid contact with the eyes and wash your hands thoroughly after handling brake fluid. If you do get brake fluid in your eyes, flush your eyes with clean, running water for 15 minutes. If eye irritation persists, or if you have taken brake fluid internally, IMMEDIATELY seek medical assistance.

• The EPA warns that prolonged contact with used engine oil may cause a number of skin disorders, including cancer. You should make every effort to minimize your exposure to used engine oil. Protective gloves should be worn when changing oil. Wash your hands and any other exposed skin areas as soon as possible after exposure to used engine oil. Soap and water, or waterless hand cleaner should be used.

• All new vehicles are now equipped with an air bag system, often referred to as a Supplemental Restraint System (SRS) or Supplemental Inflatable Restraint (SIR) system. The system must be disabled before performing service on or around system components, steering column, instrument panel components, wiring and sensors. Failure to follow safety and disabling procedures could result in accidental air bag deployment, possible personal injury and unnecessary system repairs.

• Always wear safety goggles when working with, or around, the air bag system. When carrying a non-deployed air bag, be sure the bag and trim cover are pointed away from your body. When placing a non-deployed air bag on a work surface, always face the bag and trim cover upward, away from the surface. This will reduce the motion of the module if it is accidentally deployed. Refer to the additional air bag system precautions later in this section.

• Clean, high quality brake fluid from a sealed container is essential to the safe and proper operation of the brake system. You should always buy the correct type of brake fluid for your vehicle. If the brake fluid becomes contaminated, completely flush the system with new fluid. Never reuse any brake fluid. Any brake fluid that is removed from the system should be discarded. Also, do not allow any brake fluid to come in contact with a painted surface; it will damage the paint.

• Never operate the engine without the proper amount and type of engine oil; doing so WILL result in severe engine damage.

• Timing belt maintenance is extremely important. Many models utilize an interference-type, non-freewheeling engine. If the timing belt breaks, the valves in the cylinder head may strike the pistons, causing potentially serious (also time-consuming and expensive) engine damage. Refer to the maintenance interval charts for the recommended replacement interval for the timing belt, and to the timing belt section for belt replacement and inspection.

• Disconnecting the negative battery cable on some vehicles may interfere with the functions of the on-board computer system(s) and may require the computer to undergo a relearning process once the negative battery cable is reconnected.

• When servicing drum brakes, only disassemble and assemble one side at a time, leaving the remaining side intact for reference.

• Only an MVAC-trained, EPA-certified automotive technician should service the air conditioning system or its components.

BRAKES

ANTI-LOCK BRAKE SYSTEM (ABS)

GENERAL INFORMATION

PRECAUTIONS

• Certain components within the ABS system are not intended to be serviced or repaired individually.

• Do not use rubber hoses or other parts not specifically specified for and ABS system. When using repair kits, replace all parts included in the kit. Partial or incorrect repair may lead to functional problems and require the replacement of components.

• Lubricate rubber parts with clean, fresh brake fluid to ease assembly. Do not use shop air to clean parts; damage to rubber components may result.

• Use only DOT 3 brake fluid from an unopened container.

• If any hydraulic component or line is removed or replaced, it may be necessary to bleed the entire system.

• A clean repair area is essential. Always clean the reservoir and cap thoroughly before removing the cap. The slightest amount of dirt in the fluid may plug an orifice and impair the system function. Perform repairs after components have been thoroughly cleaned; use only denatured alcohol to clean components. Do not allow ABS components to come into contact with any substance containing mineral oil; this includes used shop rags.

• The Anti-Lock control unit is a microprocessor similar to other computer units in the vehicle. Ensure that the ignition switch is **OFF** before removing or installing controller harnesses. Avoid static electricity discharge at or near the controller.

• If any arc welding is to be done on the vehicle, the control unit should be unplugged before welding operations begin.

SPEED SENSORS

REMOVAL & INSTALLATION

Front

Consider the following warnings:

• As much as possible, avoid rotating wheel sensor when removing it. Pull wheel sensors out without pulling on sensor harness.

• Take care to avoid damaging wheel sensor edges or rotor teeth. Remove wheel sensor first before removing front or rear wheel hub. This is to avoid damage to wheel sensor wiring and loss of sensor function.

1. Before servicing the vehicle, refer to the Precautions.
2. Remove the front wheel and tire.
3. Remove the front fender protector.
4. Disconnect the front wheel sensor connector.
5. Remove the front wheel sensor.

To install:

6. Installation is the reverse of the removal procedure.
7. Tighten the fasteners to specification.

✳✳ WARNING

When installing, make sure there is no foreign material such as iron chips on and in the mounting hole of the wheel sensor. Make sure no foreign material has been caught in the sensor rotor. Remove any foreign material and clean the mount.

8. When installing front wheel sensor, press rubber grommets of strut bracket and body all the way in until they get locked, and be careful not to apply a twist to harness.

➡**Harness should not be twisted after installation. (Install it with harness paint mark on body side grommet facing front of vehicle, and the strut side grommet facing outside of vehicle).**

Rear

Consider the following warnings:
• As much as possible, avoid rotating wheel sensor when removing it. Pull wheel sensors out without pulling on sensor harness
• Take care to avoid damaging wheel sensor edges or rotor teeth. Remove wheel sensor first before removing front or rear wheel hub. This is to avoid damage to wheel sensor wiring and loss of sensor function

1. Before servicing the vehicle, refer to the Precautions.

2. Disconnect the rear wheel sensor connector.
3. Remove the rear wheel sensor.

To install:

4. Installation is the reverse of the removal procedure.
5. Tighten the fasteners to specification.

✳✳ WARNING

When installing, make sure there is no foreign material such as iron chips on and in the mounting hole of the wheel sensor. Make sure no foreign material has been caught in the sensor rotor. Remove any foreign material and clean the mount.

6. When installing rear wheel sensor, press rubber grommets of suspension arm bracket and harness of side member all the way in until they get locked, and be careful not to apply a twist to harness.
7. Harness should not be twisted after installation. (Aim the paint mark upward of vehicle).

BRAKES

BLEEDING THE BRAKE SYSTEM

BLEEDING PROCEDURE

BLEEDING PROCEDURE

➡**Carefully monitor the brake fluid level at the master cylinder during the bleeding operation. Fill the reservoir with new brake fluid. Make sure it is full at all times while bleeding the air out of the system. Place a container under the master cylinder to avoid spillage of the brake fluid. Do not loosen the connecting portion of the actuator during air bleeding.**

1. Before servicing the vehicle, refer to the Precautions.
2. Disconnect the battery negative terminal.
3. Connect a transparent vinyl tube and container to the air bleeder valve.
4. Fully depress the brake pedal several times.
5. With the brake pedal depressed, open the air bleeder valve to release the air.
6. Close the air bleeder valve.
7. Release the brake pedal slowly.
8. Tighten the air bleeder valve to 74 inch lbs. (8 Nm) for front disc brakes and 70 inch lbs. (8 Nm) for rear drum brakes.
9. Repeat steps 4 through 7 until no more air bubbles come out of the air bleeder valve.

10. Bleed the brake hydraulic system air bleeder valves in the following order:
• Right rear brake
• Left front brake
• Left rear brake
• Right front brake

MASTER CYLINDER BLEEDING

Consider the following warnings:
• Refill with new brake fluid
• Do not reuse drained brake fluid
• Do not let brake fluid splash on the painted surfaces of the body. This might damage the paint. If brake fluid is splashed on painted areas, wash it away with water immediately
• Before working, disconnect the ABS actuator and electric unit (control unit) connector or the battery negative terminal.

1. Before servicing the vehicle, refer to the Precautions.
2. Turn the ignition switch OFF and disconnect the ABS actuator and electric unit (control unit) connector or the battery negative terminal.
3. Connect a vinyl tube to the bleed valve.
4. Depress the brake pedal, loosen the bleed valve, and gradually remove the brake fluid.
5. Make sure there is no foreign material in the reservoir tank, and refill with new brake fluid.

6. Rest a foot on the brake pedal. Loosen the bleed valve. Slowly depress the brake pedal until it stops. Tighten the bleed valve. Release the brake pedal. Repeat the process a few times, then pause to add new brake fluid to the master cylinder. Continue until the new brake fluid flows out of the bleed valve.
7. Bleed the air out of the brake hydraulic system.

FLUID FILL PROCEDURE

✳✳ CAUTION

Brake fluid contains polyglycol ethers and polyglycols. Avoid contact with the eyes and wash your hands thoroughly after handling brake fluid. If you do get brake fluid in your eyes, flush your eyes with clean, running water for 15 minutes. If eye irritation persists, or if you have taken brake fluid internally, IMMEDIATELY seek medical assistance.

✳✳ WARNING

Clean, high quality brake fluid is essential to the safe and proper operation of the brake system. You should always buy the highest quality brake fluid that is available. If the brake fluid becomes contaminated, drain

and flush the system, then refill the master cylinder with new fluid. Never reuse any brake fluid. Any brake fluid that is removed from the system should be discarded. Also, do not allow any brake fluid to come in contact with a painted surface; it will damage the paint.

❋❋ WARNING

Do not use shock absorber fluid or any other fluid which contains mineral oil. Do not use a container which has been used for mineral oil or a container which is wet from water. Mineral oil will cause swelling and distortion of rubber parts in the

hydraulic brake system and water mixed into brake fluid will lower fluid the boiling point. Keep all fluid containers capped to prevent contamination.

❋❋ WARNING

Be sure to use the proper brake fluid as indicated on the reservoir cap of the vehicle or as recommended in the owner's manual of the vehicle. Use of any other fluid is strictly prohibited.

1. Before servicing the vehicle, refer to the Precautions.
2. Fill the fluid level so that it is between

the MIN and MAX lines marked on the reservoir.

➡**If the brake warning light lights sometimes during driving, replenish the fluid to the MAX level.**

3. If the fluid decreases quickly, inspect the brake system for leakage. Correct leaky points and then refill to the specified level.
   • Check the master cylinder, reservoir and reservoir hose (if equipped) for cracks, damage, and brake fluid leakage. If any faulty condition exists, correct or replace needed.
4. Check that the brake fluid level is between the MAX and MIN marks on the reservoir.

BRAKES

❋❋ CAUTION

Dust and dirt accumulating on brake parts during normal use may contain asbestos fibers from production or aftermarket brake linings. Breathing excessive concentrations of asbestos fibers can cause serious bodily harm. Exercise care when servicing brake parts. Do not sand or grind brake lining unless equipment used is designed to contain the dust residue. Do not clean brake parts with compressed air or by dry brushing. Cleaning should be done by dampening the brake components with a fine mist of water, then wiping the brake components clean with a dampened cloth. Dispose of cloth and all residue containing asbestos fibers in an impermeable container with the appropriate label. Follow practices prescribed by the Occupational Safety and Health Administration (OSHA) and the Environmental Protection Agency (EPA) for the handling, processing, and disposing of dust or debris that may contain asbestos fibers.

BRAKE CALIPERS

REMOVAL & INSTALLATION
See Figures 1.

❋❋ CAUTION

Clean the dust on the caliper and brake pad with a vacuum dust collector to minimize the hazard of air borne particles or other materials.

❋❋ WARNING

When removing and installing the cylinder body, do not depress the brake pedal because the piston will pop out. Do not damage the piston boot. Keep the brake rotor free from grease and brake fluid. Refill the brake reservoir with new brake fluid only. Never reuse the drained brake fluid.

➡**When removing components such as hoses, tubes/lines, etc., cap or plug openings to prevent fluid from spilling.**

1. Before servicing the vehicle, refer to the Precautions.
2. Remove wheel and tire.
3. Secure disc rotor using wheel nuts. Put matching marks on wheel hub assembly and disc rotor, if it is necessary to remove disc rotor.
4. Drain brake fluid.
5. Remove the union bolt and discard the copper washers.
6. Remove the brake hose.

➡**Do not reuse the copper washers.**

7. Remove torque member mounting bolts from torque member, and remove caliper assembly from vehicle.

   To install:

➡**Before installing torque member to vehicle, wipe oil and grease on mounting surface of steering knuckle and torque member.**

8. Install caliper assembly to vehicle, and tighten mounting bolts to 62 ft. lbs. (84 Nm).

FRONT DISC BRAKES

9. Install brake hose to caliper assembly.
10. Refill with new brake fluid and bleed air.
11. Check front disc brake for drag.
12. Install front wheel and tire assemblies to the vehicle.

BRAKE PADS

REMOVAL & INSTALLATION
See Figure 2.

Use the following precautions when servicing the brake pads:
   • Clean dust on caliper and brake pad with a vacuum dust collector. Do not blow with compressed air
   • While removing brake pad or cylinder body, do not depress brake pedal because piston will pop out
   • It is not necessary to remove torque member mounting bolts and brake hose except for disassembly or replacement of caliper assembly. In this case, hang cylinder body with a wire so that brake hose is not under tension
   • Do not damage piston boot
   • If any shim is subject to serious corrosion, replace it with a new one
   • Keep rotor free from brake fluid
   • When replacing brake pad, replace shim with a new one
1. Before servicing the vehicle, refer to the Precautions.
2. Partially drain brake fluid reservoir.
3. Remove wheel and tire.
4. Remove sliding pin bolt (lower side).
5. Hang cylinder body with a wire, and remove pad return spring (Sedan), pads, shims and pad retainers from torque member.

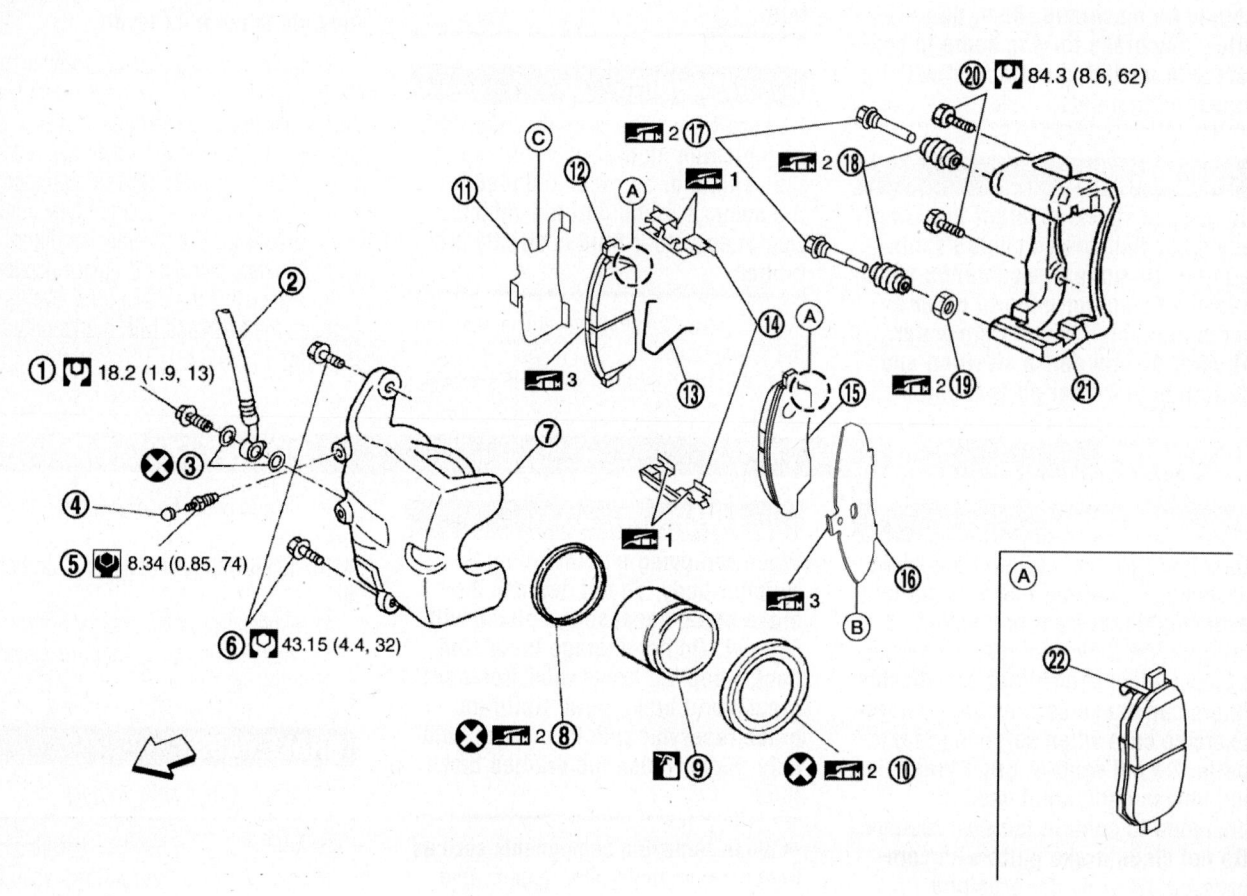

1. Union bolt
2. Brake hose
3. Copper washer
4. Cap
5. Bleed valve
6. Sliding pin bolt
7. Cylinder body
8. Piston seal
9. Piston
10. Piston boot
11. Inner shim
12. Inner pad
13. Pad return spring
14. Pad retainer
15. Outer pad
16. Outer shim
17. Sliding pin
18. Sliding pin boot
19. Bushing
20. Torque member bolt
21. Torque member
22. Pad wear sensor

⇦ : Front

: Brake fluid

1: Molykote 7439 grease   2: Rubber grease   3: Molykote AS-880N grease

71075_VERS_G0011

Fig. 1 Exploded view of front disc brake component locations—2011 Sedan

⚠ WARNING

When removing pad retainer from torque member, lift pad retainer, so as not to deform it.

To install:

6. Apply Molykote® AS-880N grease or equivalent to the shims. Install shims to pads. Securely install shims according to mounting direction of pads.

7. Apply Molykote® M7439 grease or equivalent to pad contact surface on pad retainers.

8. Install pad retainers, pad return spring (Sedan), shims and pads to the torque member.

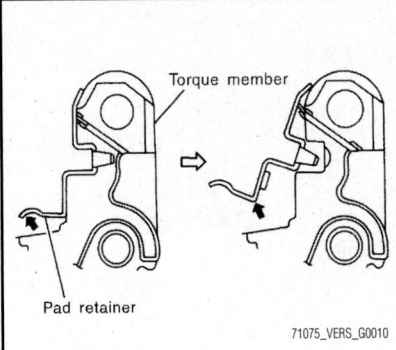

Fig. 2 When removing the pad retainer from the torque member, lift the pad retainer in the direction shown by the arrow, so as not to deform it

71075_VERS_G0010

➡**When installing pad retainer, attach it firmly so that it is not lifted up from torque member.**

9. Install cylinder body to torque member.

10. Install lower sliding pin bolt (lower side), and tighten it to specification illustrated.

➡**Use a disc brake piston tool to press the piston into the cylinder body.**

➡**Check the brake fluid level in the reservoir tank for fluid level because brake fluid returns to master cylinder reservoir tank when pressing piston in.**

11. Check brake for drag.

12. Install the wheels and tires.

   a. When installing the wheels, tighten the lug nuts in a star pattern in 2–3 incremental steps in order to prevent the wheels from developing any distortion. Tighten the wheel lug nuts to 83 ft. lbs. (113 Nm).

✳✳ WARNING

Be careful not to tighten the wheel nut at a torque exceeding the criteria for preventing strain of disc rotor.

➡**Use NISSAN genuine wheel nuts for aluminum wheels.**

13. Check brake fluid level.

BRAKES

✳✳ CAUTION

Dust and dirt accumulating on brake parts during normal use may contain asbestos fibers from production or aftermarket brake linings. Breathing excessive concentrations of asbestos fibers can cause serious bodily harm. Exercise care when servicing brake parts. Do not sand or grind brake lining unless equipment used is designed to contain the dust residue. Do not clean brake parts with compressed air or by dry brushing. Cleaning should be done by dampening the brake components with a fine mist of water, then wiping the brake components clean with a dampened cloth. Dispose of cloth and all residue containing asbestos fibers in an impermeable container with the appropriate label. Follow practices prescribed by the Occupational Safety and Health Administration (OSHA) and the Environmental Protection Agency (EPA) for the handling, processing, and disposing of dust or debris that may contain asbestos fibers.

BRAKE DRUM

REMOVAL & INSTALLATION

Except 2012 Sedan
See Figure 3.

1. Before servicing the vehicle, refer to the Precautions.
2. Remove wheel and tire.

3. With the parking brake lever released, remove the brake drum. If it is difficult to remove brake drum, remove as follows:

   a. Press up the adjuster lever with a wire, or equivalent, from the plug hole (at the side of wheel cylinder) on the back plate.

   b. Turn the frame of the adjuster assembly with a flat-bladed screw driver in the direction that narrows the frame to the narrow enlarged brake shoe.

To install:
4. Installation is the reverse of removal.
5. Inspect and adjust the brakes.

2012 Sedan
See Figure 4.

1. Before servicing the vehicle, refer to the Precautions.
2. Remove the wheel and tire assembly using power tool.
3. Remove hub cap from brake drum, using a suitable tool.
4. Remove wheel hub lock nut and brake drum.

✳✳ WARNING

Do not apply force to the brake drum to avoid damage to the wheel bearing. If the wheel bearing inner race is separated due to force, replace the wheel bearing with a new one.

To install:
5. Installation is the reverse of the removal procedure.
6. Check each mating surface for water and foreign matter. If there is any water or

foreign matter, clean the mating surface.

7. Insert brake drum to the spindle with the spindle axis arranged in a straight line. (Brake drum may be stuck at the axis if not installed in a straight line).

➡**If the brake drum becomes stuck and must be pulled out, do not use tools. Replace the wheel bearing with a new one if the brake drum cannot be pulled out without use of tools.**

➡**If the brake drum becomes stuck and the wheel bearing inner race is damaged, replace the wheel bearing with a new one.**

8. Using Tool number: ST30720000 (—), install the hub cap on the brake drum. Do not reuse the hub cap.

9. Tighten the wheel lock nut to the specified torque. Do not reuse the wheel lock nut.

BRAKE SHOES

REMOVAL & INSTALLATION

Except 2012 Sedan

1. Before servicing the vehicle, refer to the Precautions.
2. Remove brake drum.
3. While pushing and rotating the retainer, pull out shoe hold pin, and then remove shoe assembly.

✳✳ WARNING

Do not damage the wheel cylinder boot.

REAR DRUM BRAKES

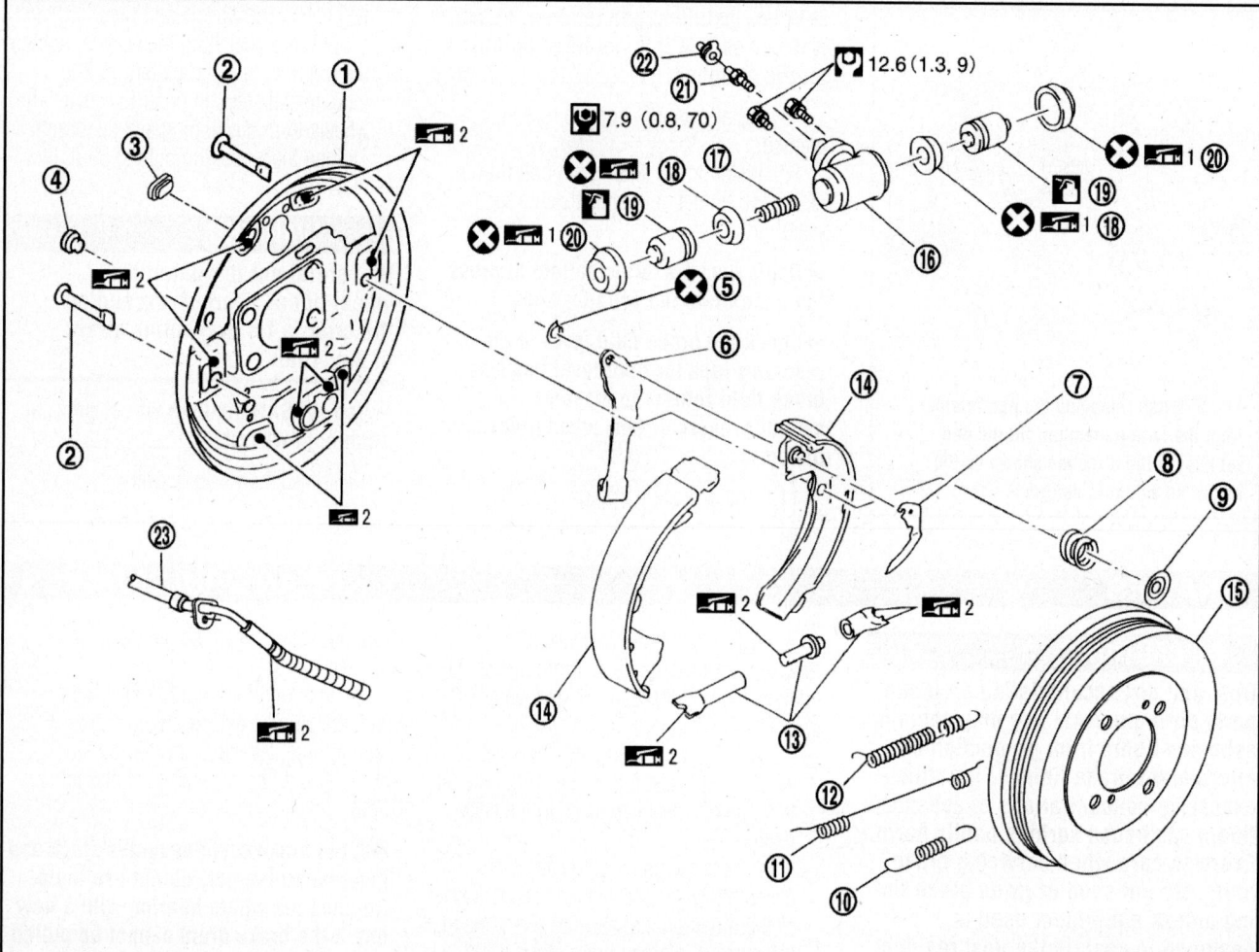

1. Back plate
2. Shoe hold pin
3. Plug
4. Plug
5. Retainer ring
6. Operating lever
7. Adjuster lever
8. Spring
9. Retainer
10. Return spring (lower side)
11. Return spring (upper side)
12. Adjuster spring
13. Adjuster
14. Brake shoe
15. Brake drum
16. Wheel cylinder
17. Spring
18. Piston seal
19. Piston
20. Boot
21. Bleed valve
22. Cap
23. Parking brake rear cable

1: Rubber grease

2: PBC (Poly Butyl Cuprysil) grease or silicone-based grease

: Brake fluid

71075_VERS_G0016

Fig. 3 Rear brake drum assembly component locations—2011 Sedan

4. Remove the parking brake rear cable from the operating lever.

❋❋ WARNING

Do not bend the parking brake cable.

5. Disassemble the shoe assembly (shoe, springs, adjuster, adjuster lever).

6. Remove retainer ring with a tool to separate operating lever from brake shoe.

To install:

7. Install operating lever to brake shoe, if necessary.

8. Install retainer ring to operating lever, and crimp them until their contact points are met.

9. Apply Poly Butyl Cuprysil (PBC) grease or silicone based grease to brake shoes sliding surfaces (the shaded areas) and other parts on the back plate as indicated by arrows.

10. Apply PBC grease or silicone based grease to screw and confirm the difference between right and left wheel for assembling when disassembled.

➡**The brake drum adjusting screw right rear wheel thread cutting direction: Right-hand screw; Left rear wheel**

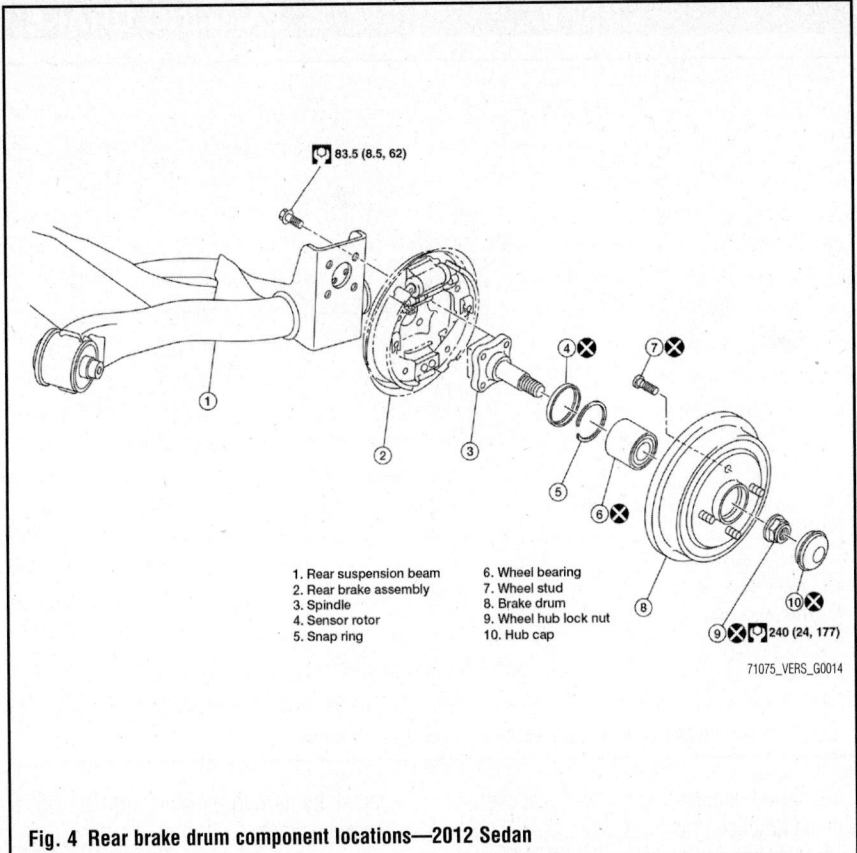

1. Rear suspension beam
2. Rear brake assembly
3. Spindle
4. Sensor rotor
5. Snap ring
6. Wheel bearing
7. Wheel stud
8. Brake drum
9. Wheel hub lock nut
10. Hub cap

83.5 (8.5, 62)

240 (24, 177)

71075_VERS_G0014

Fig. 4 Rear brake drum component locations—2012 Sedan

thread cutting direction: Left-hand screw threads.

11. Assemble the shoe, adjuster, adjuster lever, and springs to the shoe assembly.

12. Connect the parking brake rear cable to the operating lever.

13. Install the shoe assembly. After assembly, be sure that each part is installed properly. Do not damage the wheel cylinder piston boot.

14. Install the brake drum.

15. Depress the brake pedal 2–3 times.

16. Adjust the clearance of the brake shoe.

17. Install the rear wheel and tire.

2012 Sedan

✳✳ CAUTION

Clean dust from brake drum and shoe assembly with a vacuum dust collec- tor to minimize the hazard of air borne particles or other materials.

✳✳ WARNING

Never depress the brake pedal while removing the brake drum because the pistons may pop out. Never drop the removed parts. Never spill or splash brake fluid on the brake drum.

➡When removing components such as hoses, tubes/lines, etc., cap or plug openings to prevent fluid from spilling.

1. Before servicing the vehicle, refer to the Precautions.

2. Remove the wheel and tire assemblies.

3. Remove the brake drum.

➡Make sure the parking brake lever is fully released prior to removal of the brake drum. The rear wheel hub is housed inside the brake drum.

4. Remove the springs by pushing them inward toward the vehicle and rotating, this will release the shoe hold pins, and the brake shoe assembly (brake shoes, each spring, and adjuster).

✳✳ WARNING

Use care to never damage the boot of the wheel cylinder.

5. Disconnect the parking brake cable from operating lever.

✳✳ WARNING

Use care to never bend the parking brake lever.

6. Disassemble the brake shoe assembly (brake shoe, each spring, and adjuster).

To install:

7. Installation is the reverse of the removal procedure.

8. Check the difference between left and right wheel of adjuster.

9. Shorten the length of the adjuster by rotating it.

• The adjuster on the right side has right-handed screw threads and the adjuster on the left side has left-handed screw threads.

10. Apply Poly Butyl Cuprysil (PBC) silicone-based grease to the mating surfaces between the adjusters and the brake shoes, between the back plates and the brake shoes, between the wheel cylinders and brake shoes, between the brake shoe anchor areas and brake shoes.

11. Check the component parts of drum brake assembly are installed properly.

12. Check the brake shoe sliding surface and brake drum inner surface for grease. Make sure that grease does not contact the lining material.

13. Adjust the brake shoe clearance (parking brake lever stroke) after install and air bleeding.

BRAKES

PARKING BRAKE

PARKING BRAKE CABLES

ADJUSTMENT

Except 2012 Sedan

See Figure 5.

1. Before servicing the vehicle, refer to the Precautions.
2. Remove console mask cover.
3. Engage parking brake lever, then lift up the end of the trim on the lever to access the adjusting nut.
4. When replace parking brake cable, operate parking brake lever with a force of 110 ft. lbs. (490 Nm) about 10 times.
5. Engage parking brake lever, then lift up the end of the trim on the lever to access the adjusting nut.
6. Rotate adjusting nut to adjust parking brake lever stroke using a deep socket wrench.
7. Operate parking brake lever with a force of 44 ft. lbs. (196 Nm), make sure the parking brake lever stroke is within 8 to 9 notches. (Check it by listening and counting ratchet clicks).
8. Make sure that there is no drag on rear brake with parking brake lever completely released.
9. Install console mask.

2012 Sedan

1. Before servicing the vehicle, refer to the Precautions.
2. Pull the parking brake lever until the access hole is visible.
3. Insert a socket wrench onto the adjusting nut.

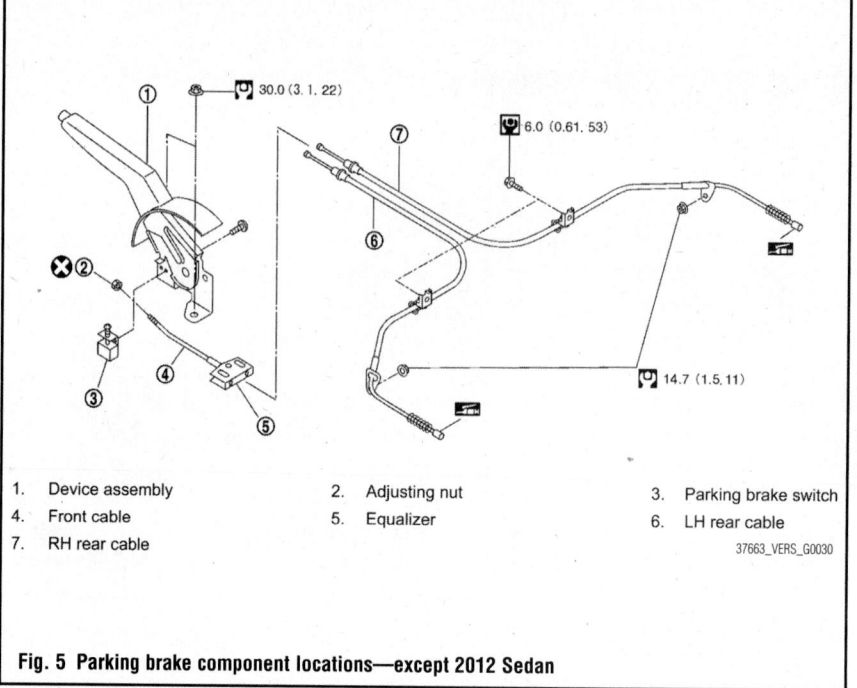

1. Device assembly
2. Adjusting nut
3. Parking brake switch
4. Front cable
5. Equalizer
6. LH rear cable
7. RH rear cable

37663_VERS_G0030

Fig. 5 Parking brake component locations—except 2012 Sedan

4. Turn the adjusting nut with a socket wrench to loosen the front cable.
5. Depress the brake pedal with a force of 44 ft. lbs. (196 N) about 10 times and adjust the brake shoe clearance.

➡**Make sure to securely operate the brake pedal.**

6. Check drag on the rear drum brake.
7. Adjust the cable with the following procedure:
   a. Operate the parking brake lever with a force of 90 ft. lbs. (400 N) for 25 minutes or longer.
   b. Adjust the parking brake lever

stroke by turning the adjusting nut with a socket wrench.

➡**Do not reuse the adjusting nut if the nut is removed.**

   c. Operate the parking brake lever with a force of 44 ft. lbs. (196 N). Check that the lever stroke is within the specified number of notches. (Check it by listening to the clicks of the ratchet).
   • Standard specification: 11–12 notches
   d. Rotate the brake drum with the parking brake lever released and verify that there is no drag present.

CHASSIS ELECTRICAL

AIR BAG (SUPPLEMENTAL RESTRAINT SYSTEM)

GENERAL INFORMATION

❋❋ CAUTION

These vehicles are equipped with an air bag system. The system must be disarmed before performing service on, or around, system components, the steering column, instrument panel components, wiring and sensors. Failure to follow the safety precautions and the disarming procedure could result in accidental air bag deployment, possible injury and unnecessary system repairs.

SERVICE PRECAUTIONS

Disconnect and isolate the battery negative cable before beginning any airbag system component diagnosis, testing, removal, or installation procedures. Allow system capacitor to discharge for two minutes before beginning any component service. This will disable the airbag system. Failure to disable the airbag system may result in accidental airbag deployment, personal injury, or death.

Do not place an intact undeployed airbag face down on a solid surface. The airbag will propel into the air if accidentally deployed and may result in personal injury or death.

When carrying or handling an undeployed airbag, the trim side (face) of the airbag should be pointing towards the body to minimize possibility of injury if accidental deployment occurs. Failure to do this may result in personal injury or death.

Replace airbag system components with OEM replacement parts. Substitute parts may appear interchangeable, but internal differences may result in inferior occupant protection. Failure to do so may result in occupant personal injury or death.

Wear safety glasses, rubber gloves, and long sleeved clothing when cleaning powder residue from vehicle after an airbag deployment. Powder residue emitted from a deployed airbag can cause skin irritation. Flush affected area with cool water if irritation is experienced. If nasal or throat irritation is experienced, exit the vehicle for fresh air until the irritation ceases. If irritation continues, see a physician.

Do not use a replacement airbag that is not in the original packaging. This may result in improper deployment, personal injury, or death.

The factory installed fasteners, screws and bolts used to fasten airbag components have a special coating and are specifically designed for the airbag system. Do not use substitute fasteners. Use only original equipment fasteners listed in the parts catalog when fastener replacement is required.

During, and following, any child restraint anchor service, due to impact event or vehicle repair, carefully inspect all mounting hardware, tether straps, and anchors for proper installation, operation, or damage. If a child restraint anchor is found damaged in any way, the anchor must be replaced. Failure to do this may result in personal injury or death.

Deployed and non-deployed airbags may or may not have live pyrotechnic material within the airbag inflator.

Do not dispose of driver/passenger/curtain airbags or seat belt tensioners unless you are sure of complete deployment. Refer to the Hazardous Substance Control System for proper disposal.

Dispose of deployed airbags and tensioners consistent with state, provincial, local, and federal regulations.

After any airbag component testing or service, do not connect the battery negative cable. Personal injury or death may result if the system test is not performed first.

If the vehicle is equipped with the Occupant Classification System (OCS), do not connect the battery negative cable before performing the OCS Verification Test using the scan tool and the appropriate diagnostic information. Personal injury or death may result if the system test is not performed properly.

Never replace both the Occupant Restraint Controller (ORC) and the Occupant Classification Module (OCM) at the same time. If both require replacement, replace one, then perform the Airbag System test before replacing the other.

Both the ORC and the OCM store Occupant Classification System (OCS) calibration data, which they transfer to one another when one of them is replaced. If both are replaced at the same time, an irreversible fault will be set in both modules and the OCS may malfunction and cause personal injury or death.

If equipped with OCS, the Seat Weight Sensor is a sensitive, calibrated unit and must be handled carefully. Do not drop or handle roughly. If dropped or damaged, replace with another sensor. Failure to do so may result in occupant injury or death.

If equipped with OCS, the front passenger seat must be handled carefully as well. When removing the seat, be careful when setting on floor not to drop. If dropped, the sensor may be inoperative, could result in occupant injury, or possibly death.

If equipped with OCS, when the passenger front seat is on the floor, no one should sit in the front passenger seat. This uneven force may damage the sensing ability of the seat weight sensors. If sat on and damaged, the sensor may be inoperative, could result in occupant injury, or possibly death.

DISARMING THE SYSTEM

> **✳✳ CAUTION**
>
> **All SRS electrical wiring harnesses and connectors are covered with yellow and/or orange outer insulation. Do not use electrical test equipment on any circuit related to the SRS (air bag) sensors. When installing SRS components, always install with the arrow marks facing the front of the vehicle.**

1. Before servicing the vehicle, refer to the Precautions.
2. Turn the ignition switch to the **OFF** position.
3. Disconnect both battery cables starting with the negative cable first.
4. Wait at least 3 minutes after the cables are disconnected. Be sure to insulate the battery terminal ends.

ARMING THE SYSTEM

1. Before servicing the vehicle, refer to the Precautions.
2. Make sure that the removed components are installed and/or the disconnected connectors are connected properly.
3. Turn the ignition switch to the **OFF** position.

4. Connect both battery cables starting with the positive cable first.
5. The SRS or air bag system is equipped with a self-diagnostic operation. After turning the ignition key to the ON or START position.
   a. The AIR BAG warning lamp will illuminate for 7 seconds.
   b. After 7 seconds, the AIR BAG lamp will extinguish if no malfunction is detected.
   c. If the AIR BAG lamp does not extinguish after 7 seconds, check the SRS self-diagnostic system for a malfunction.

CLOCKSPRING CENTERING

See Figures 6 and 7.

> **✳✳ CAUTION**
>
> **Models equipped with a Supplemental Restraint System (SRS), use an inflatable air bag. Whenever working near any of the SRS components, such as the impact sensors, the air bag module, steering column, and instrument panel, disable the SRS.**

> **✳✳ WARNING**
>
> **The spiral cable may snap by the steering operation if the cable is installed in an improper position. Do not turn the spiral cable quickly or beyond the limit number of turns. This can cause the cable to snap. The spiral cable can be turned counterclockwise about 2.5 turns from the neutral position.**

> **✳✳ CAUTION**
>
> **After the work is completed, make sure no system malfunction is detected by air bag warning lamp.**

> **✳✳ WARNING**
>
> **Do not use air tools or electric tools for servicing. Do not disassemble the spiral cable. Do not allow oil, grease, detergent or water to come in contact with the spiral cable.**

> **✳✳ WARNING**
>
> **Do not cause impact to the spiral cable by dropping it. Replace the spiral cable if it has been dropped or sustained an impact.**

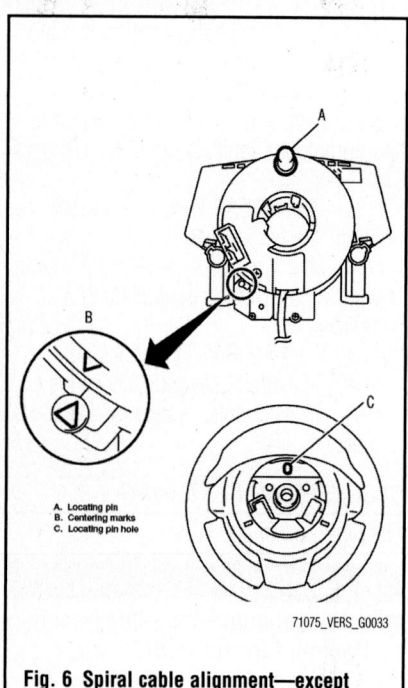

A. Locating pin
B. Centering marks
C. Locating pin hole

71075_VERS_G0033

Fig. 6 Spiral cable alignment—except 2012 Sedan

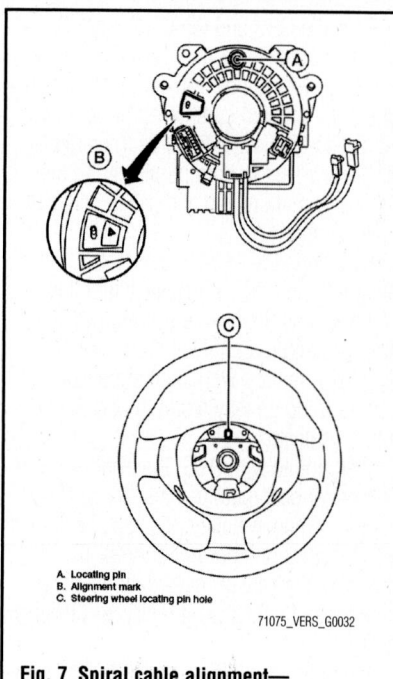

A. Locating pin
B. Alignment mark
C. Steering wheel locating pin hole

71075_VERS_G0032

Fig. 7 Spiral cable alignment—2012 Sedan

1. Before servicing the vehicle, refer to the Precautions.

2. Turn the ignition switch OFF, disconnect both battery terminals and wait at least 3 minutes.

3. Turn the contact coil cable assembly fully counterclockwise.

❊❊ WARNING

Turning with excessive force will break the coil. Turn the rotation part with a light force.

4. From the fully counterclockwise position, turn the contact coil assembly clockwise about 2.5 turns.

5. Align the spiral cable correctly when installing the steering wheel. Make sure that the spiral cable is in the neutral position. The neutral position is detected by turning left 2.5 revolutions from the right end position and ending with the knob at the top.

DRIVE TRAIN

FRONT HALFSHAFT

REMOVAL & INSTALLATION

Except 2012 Sedan

Left Side

See Figure 8.

1. Before servicing the vehicle, refer to the Precautions.

2. Remove wheel and tire.

3. Remove wheel sensor from steering knuckle.

❊❊ WARNING

Do not pull on wheel sensor harness.

4. Remove transverse link ball joint nut and bolt. Then, remove transverse link from steering knuckle.

5. Remove cotter pin, and loosen hub lock nut. Temporarily leave the hub lock nut installed to prevent damage to threads.

6. Separate the halfshaft from the wheel hub and bearing assembly by lightly tapping the end of the halfshaft using a hammer or suitable tool and wood block, and then remove hub lock nut.

7. If wheel hub and bearing assembly and halfshaft cannot be separated after performing the above procedure, use a suitable puller.

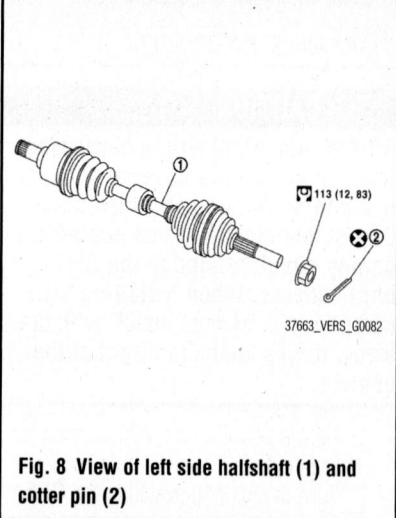

113 (12, 83)

37663_VERS_G0082

Fig. 8 View of left side halfshaft (1) and cotter pin (2)

8. Remove the halfshaft from the wheel hub and bearing assembly.

❊❊ WARNING

Do not apply an excessive angle to halfshaft joint when removing from the wheel hub and bearing assembly. Do not excessively extend slide joint. Do not allow halfshaft to hang down. Support the entire halfshaft.

9. Pry off the halfshaft from the transaxle assembly side. Make sure that the circlip is attached on the edge.

To install:

10. Installation is the reverse of the removal procedure.

11. Tighten the halfshaft nut to 83 ft. lbs. (113 Nm).

➡**Do not reuse non-reusable parts.**

12. In order to prevent damage to differential side oil seal, place tool No. KV38105500 (J-33904) onto oil seal before inserting halfshaft. Slide halfshaft into slide joint and tap with a hammer to install securely.

13. Install new circlip on halfshaft in the circlip groove on transaxle side. Make sure the new circlip on the halfshaft is securely fastened.

14. After its insertion, try to pull the flange out of the slide joint by hand. If it pulls out, the circlip is not properly meshed with the transaxle side gear.

15. Check transaxle fluid level.

Right Side

See Figure 9.

1. Before servicing the vehicle, refer to the Precautions.

2. Remove wheel and tire.

3. Remove wheel sensor from steering knuckle.

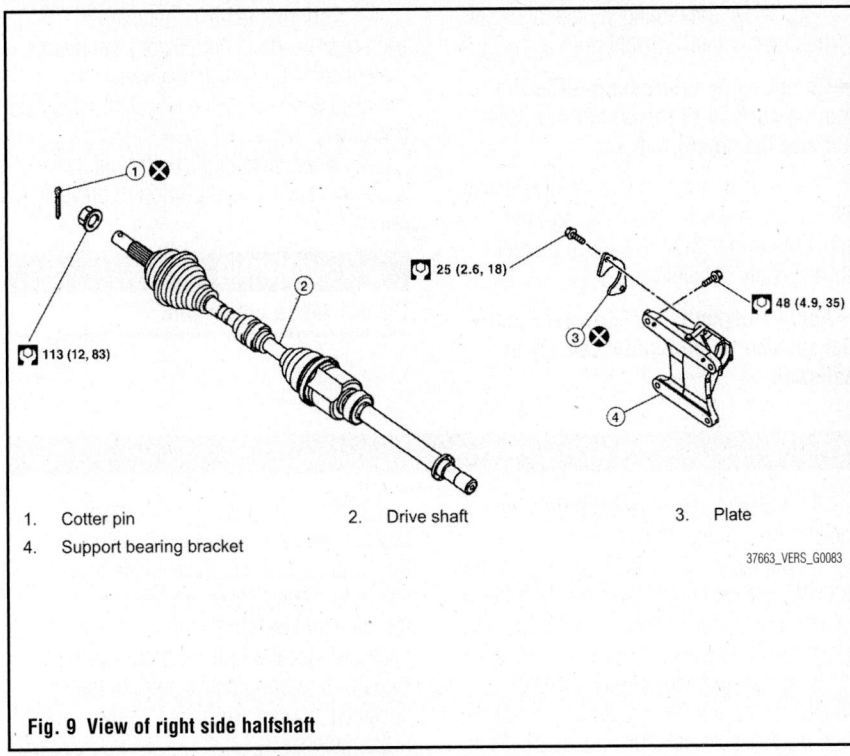

1. Cotter pin
2. Drive shaft
3. Plate
4. Support bearing bracket

37663_VERS_G0083

Fig. 9 View of right side halfshaft

✳✳ WARNING

Do not pull on wheel sensor harness.

4. Remove transverse link ball joint nut and bolt. Remove transverse link from steering knuckle.

5. Remove cotter pin and loosen hub lock nut. Temporarily leave the hub lock nut installed to prevent damage to threads.

6. Separate the halfshaft from the wheel hub and bearing assembly by lightly tapping the end of the halfshaft using a hammer or suitable tool and wood block, and remove hub lock nut.

7. If wheel hub and bearing assembly and halfshaft cannot be separated after performing the above procedure, use a suitable puller.

8. Remove the halfshaft from the wheel hub and bearing assembly.

✳✳ WARNING

Do not apply an excessive angle to halfshaft joint. Do not excessively extend slide joint. Do not allow halfshaft to hang down. Support the entire halfshaft.

9. Remove the plate bolts and plate.
10. Remove the halfshaft from the transaxle assembly.

To install:
11. Installation is the reverse of the removal procedure.

12. Tighten the halfshaft nut to 83 ft. lbs. (113 Nm).
13. Tighten the plate bolts to 18 ft. lbs. (25 Nm).

➡**Do not reuse non-reusable parts.**

14. In order to prevent damage to differential side oil seal, place tool No. KV38106700 (J-34296) onto oil seal before inserting halfshaft. Slide halfshaft into slide joint and tap with a hammer to install securely.

15. When installing the support bearing bracket, note the following:
a. Tighten the support bearing bracket bolts in 2 stages.
b. Install the plate with the notch upward. Tighten the plate bolts in 2 stages.
16. Check transaxle fluid level.

2012 Sedan

1. Before servicing the vehicle, refer to the Precautions.
2. Remove the wheel and tire assembly.
3. Remove the wheel sensor and sensor harness.
4. Remove the brake lock plate from the strut assembly.
5. Remove the brake caliper, leaving the brake caliper hydraulic lines connected. Reposition the brake caliper aside with wire.

➡**Avoid depressing the brake pedal while the brake caliper is removed.**

6. Put matching marks on the disc rotor and the wheel hub and bearing assembly, then remove the disc rotor.

➡**Put matching marks on the wheel hub and the disc rotor before removing the disc rotor. Do not drop disc rotor.**

7. Remove and discard the cotter pin, and then loosen the wheel hub lock nut, using Tool KV40104000.
8. Separate the halfshaft from the wheel hub and bearing assembly by lightly tapping the end of the halfshaft using a suitable tool and a wood block.

✳✳ WARNING

Do not place halfshaft joint at an extreme angle. Also be careful not to overextend slide joint. Do not allow halfshaft to hang down without support for joint sub-assembly, shaft and the other parts.

➡**Use suitable puller, if wheel hub and halfshaft cannot be separated even after performing the above procedure.**

9. Remove wheel hub lock nut.
10. Remove the nuts and bolts, then separate the steering knuckle from strut assembly.
11. Remove halfshaft from wheel hub.
12. Use the Tool KV40107500 and a suitable tool while inserting tip of the halfshaft attachment between shaft and transaxle assembly, and then remove halfshaft from transaxle assembly.

✳✳ WARNING

Do not place halfshaft joint at an extreme angle when removing halfshaft. Also be careful not to overextend slide joint. Confirm that the circular clip is attached to the halfshaft.

To install:
13. Installation is the reverse of the removal procedure.
14. On the transaxle side:
a. Always replace differential side oil seal with new one when installing halfshaft.
b. Place the protector KV38107900 onto transaxle assembly to prevent damage to the oil seal while inserting halfshaft. Slide halfshaft sliding joint and tap with a hammer to install securely.

➡**Check that circular clip is completely engaged.**

15. On the wheel hub side:

❋❋ WARNING

During the installation, never damage the wheel bearing seal. If damage has occurred, replace wheel bearing with a new one.

a. Do not allow paint to adhere to the wheel bearing seal.

b. Check each mating surface for water and foreign matter. If there is any water or foreign matter, clean the mating surface.

c. Clean the mating surface of wheel hub lock nut and wheel hub.

➡**Do not apply lubricating oil to the mating surface of the wheel hub lock nut and the wheel hub.**

16. Clean the mating surface of halfshaft, wheel hub, and wheel bearing. And then apply Molykote® M77 to surface of joint subassembly of halfshaft.

➡**Apply Molykote® M77 to cover entire flat surface of joint subassembly of halfshaft.**

17. Perform the final tightening of each of the parts under unladen conditions, the conditions in which they were removed when removing wheel hub and axle housing.

18. When installing a cotter pin, securely bend the end portion to prevent rattles.

❋❋ WARNING

Do not reuse cotter pin.

ENGINE COOLING

ENGINE COOLANT

DRAIN & REFILL

Except 2012 Sedan

❋❋ CAUTION

To avoid being scalded, do not change engine coolant when engine is hot. Wrap a thick cloth around radiator cap and carefully remove the cap. First, turn the cap ¼ of a turn to release built-up pressure. Then turn the cap all the way.

❋❋ WARNING

Do not spill engine coolant on drive belt.

1. Before servicing the vehicle, refer to the Precautions.

2. Open radiator drain plug at the bottom of radiator, and then remove radiator cap. When draining all of engine coolant in the system, open the water drain plug on cylinder block.

3. Remove reservoir tank as necessary, and drain engine coolant and clean reservoir tank before installing.

4. Check drained engine coolant for contaminants such as rust, corrosion, or discoloration. If contaminated, flush the engine cooling system.

To refill:

5. Install the radiator drain plug. If the cooling system was drained completely, install the reservoir tank and the cylinder block drain plugs.

a. The radiator must be completely empty of coolant and water.

b. Apply sealant to the threads of the cylinder block drain plugs. Use Genuine High Performance Thread Sealant or equivalent.

6. If disconnected, reattach the upper radiator hose at the engine side.

7. Set the vehicle heater controls to the full HOT and heater ON position. Turn the vehicle ignition ON with the engine OFF as necessary to activate the heater mode.

8. Install Tool No. KV991J0070 (J-45695) by installing the radiator cap adapter onto the radiator neck opening. Then attach the gauge body assembly with the refill tube and the venturi assembly to the radiator cap adapter.

9. Insert the refill hose into the coolant mixture container that is placed at floor level. Make sure the ball valve is in the closed position.

10. Install an air hose to the venturi assembly; the air pressure must be within specification. Compressed air supply pressure: 80–119 psi (549–824 kPa).

➡**The compressed air supply must be equipped with an air dryer.**

11. The vacuum gauge will begin to rise and there will be an audible hissing noise. During this process open the ball valve on the refill hose slightly. Coolant will be visible rising in the refill hose. Once the refill hose is full of coolant, close the ball valve. This will purge any air trapped in the refill hose.

12. Continue to draw the vacuum until the gauge reaches 28 inches of vacuum. The gauge may not reach 28 inches in high altitude locations; use the vacuum gauge reading specifications below, based on the listed altitude above sea level.

- 0–328 ft. (0–100 m): 28 inches of vacuum
- 984 ft. (300 m): 27 inches of vacuum
- 1,641 ft. (500 m): 26 inches of vacuum
- 3,281 ft. (1,000 m): 24–25 inches of vacuum

13. When the vacuum gauge has reached the specified amount, disconnect the air hose and wait 20 seconds to see if the system loses any vacuum. If the vacuum level drops, perform any necessary repairs to the system and repeat steps 6–8 to bring the vacuum to the specified amount. Recheck for any leaks.

14. Place the coolant container (with the refill hose inserted) at the same level as the top of the radiator. Then open the ball valve on the refill hose so the coolant will be drawn up to fill the cooling system. The cooling system is full when the vacuum gauge reads zero.

➡**Do not allow the coolant container to get too low when filling, to avoid air from being drawn into the cooling system.**

15. Remove the tool from the radiator neck opening.

16. Fill the cooling system reservoir tank to the specified level, and install the radiator cap. Run the engine to warm up the cooling system and top up the system as necessary.

2012 Sedan

Draining Engine Coolant

❋❋ CAUTION

Never remove the radiator cap when the engine is hot. Serious burns could occur from high-pressure engine coolant escaping from the radiator. Wrap a thick cloth around the radiator cap. Slowly turn it ¼ of a turn to release built-up pressure. Then turn it all the way.

1. Before servicing the vehicle, refer to the Precautions.

2. Remove the engine undercover.

3. Connect a drain hose to the radiator drain plug. The drain hose should have the following dimensions:
- Inside diameter: 0.31 inch (0.8mm)
- Length: 11.8 inches (300mm)

4. Open radiator drain plug at the bottom of radiator, and then remove radiator cap.

❋❋ CAUTION

Perform this step when engine is cold.

❋❋ WARNING

Do not spill engine coolant on the drive belt.

5. It is necessary to drain the cylinder block when draining all of engine coolant in the system. To drain the cylinder block, open the water drain plugs on cylinder block.

6. Remove reservoir tank if necessary, and drain engine coolant and clean reservoir tank before installing.

7. Check drained engine coolant for contaminants such as rust, corrosion or discoloration. If contaminated, flush the engine cooling system.

Refilling Engine Coolant

See Figure 10.

1. Before servicing the vehicle, refer to the Precautions.

2. Install the radiator drain plug. Install the reservoir tank and cylinder block drain plug, if removed.

➡**Be sure to clean drain plug and install with a new O-ring. Apply sealant to the threads of the cylinder block drain plug. Use Genuine High Performance Thread Sealant or equivalent.**

3. Set the vehicle heater controls to the full HOT and heater ON position. Turn the vehicle ignition ON with the engine OFF as necessary to activate the heater mode.

4. Install Tool KV991J0070 (J-45695) by installing the radiator cap adapter onto the radiator neck opening. Then attach the gauge body assembly with the refill tube and the venturi assembly to the radiator cap adapter.

5. Insert the refill hose into the coolant mixture container that is placed at floor level. Make sure the ball valve is in the closed position.

- Use recommended coolant: Pre-diluted Genuine NISSAN Long Life Anti-freeze/Coolant (blue) or equivalent

6. Install an air hose to the venturi assembly. The air pressure must be within specification.
- Compressed air supply pressure: 80–119 psi (549–824 kPa)

➡**The compressed air supply must be equipped with an air dryer.**

7. The vacuum gauge will begin to rise and there will be an audible hissing noise. During this process open the ball valve on the refill hose slightly. Coolant will be visible rising in the refill hose. Once the refill hose is full of coolant, close the ball valve. This will purge any air trapped in the refill hose.

8. Continue to draw the vacuum until the gauge reaches 28 inches of vacuum. The gauge may not reach 28 inches in high altitude locations; use the vacuum specifications based on the altitude above sea level.
- 0–328 ft. (0–100 m): 28 inches of vacuum
- 984 ft. (300 m): 27 inches of vacuum

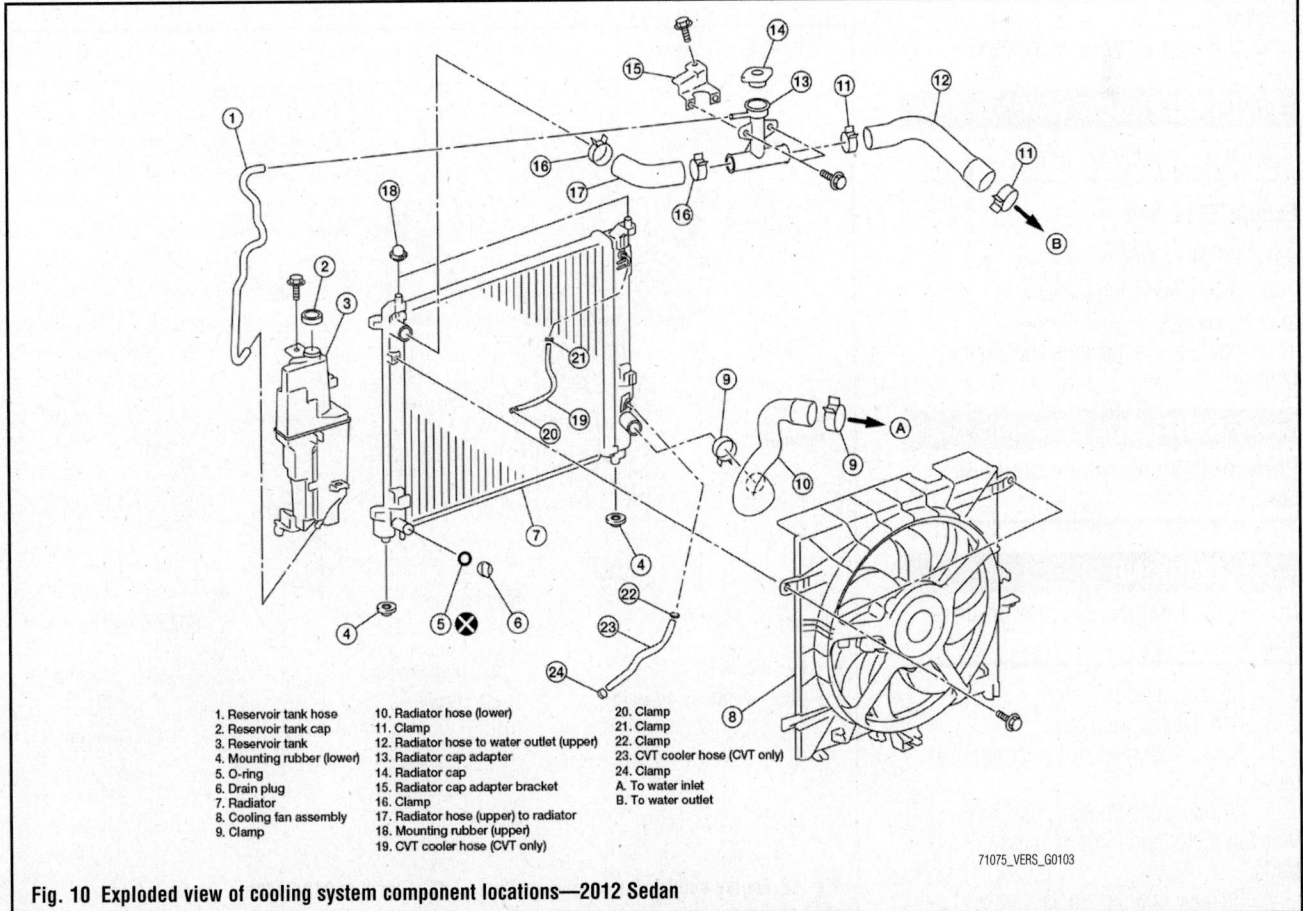

1. Reservoir tank hose	10. Radiator hose (lower)
2. Reservoir tank cap	11. Clamp
3. Reservoir tank	12. Radiator hose to water outlet (upper)
4. Mounting rubber (lower)	13. Radiator cap adapter
5. O-ring	14. Radiator cap
6. Drain plug	15. Radiator cap adapter bracket
7. Radiator	16. Clamp
8. Cooling fan assembly	17. Radiator hose (upper) to radiator
9. Clamp	18. Mounting rubber (upper)
	19. CVT cooler hose (CVT only)

20. Clamp
21. Clamp
22. Clamp
23. CVT cooler hose (CVT only)
24. Clamp
A. To water inlet
B. To water outlet

71075_VERS_G0103

Fig. 10 Exploded view of cooling system component locations—2012 Sedan

- 1,641 ft. (500 m): 26 inches of vacuum
- 3,281 ft. (1,000 m): 24–25 inches of vacuum

9. When the vacuum gauge has reached the specified amount, disconnect the air hose and wait 20 seconds to see if the system loses any vacuum. If the vacuum level drops, perform any necessary repairs to the system and repeat steps 7–9 to bring the vacuum to the specified amount. Recheck for any leaks.

10. Place the coolant container (with the refill hose inserted) at the same level as the top of the radiator. Then open the ball valve on the refill hose so the coolant will be drawn up to fill the cooling system. The cooling system is full when the vacuum gauge reads zero.

➡**Do not allow the coolant container to get too low when filling, to avoid air from being drawn into the cooling system.**

11. Remove the Tool from the radiator neck opening.

12. Fill the cooling system reservoir tank to the specified level and install the radiator cap. Run the engine to warm up the cooling system and top up the system as necessary.

13. Install the engine undercover.

ELECTRIC ENGINE FAN

REMOVAL & INSTALLATION

Except 2012 Sedan

See Figures 11 and 12.

1. Before servicing the vehicle, refer to the Precautions.

2. Partially drain engine coolant from radiator.

✶✶ CAUTION

Perform this step when engine is cold.

✶✶ WARNING

Do not spill engine coolant on drive belt.

3. Remove air duct (inlet).

4. Remove reservoir tank.

5. Disconnect radiator hose (upper) at radiator side.

6. Disconnect harness connectors from fan motor and move harness aside.

7. Remove cooling fan assembly.

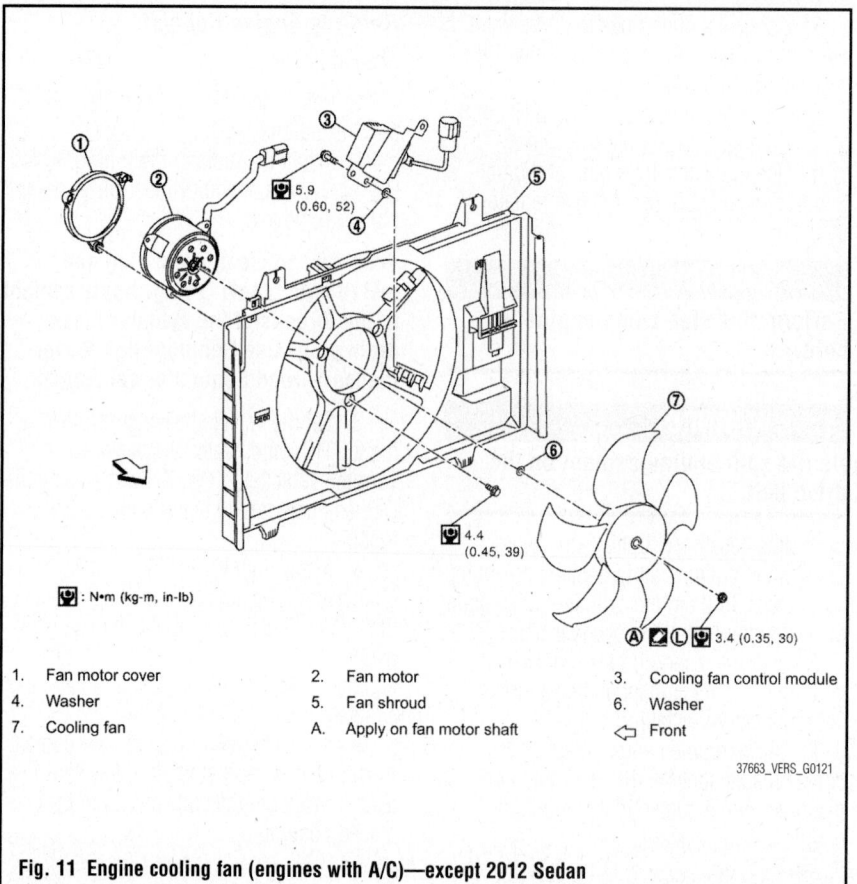

: N•m (kg-m, in-lb)

1.	Fan motor cover	2.	Fan motor	3.	Cooling fan control module
4.	Washer	5.	Fan shroud	6.	Washer
7.	Cooling fan	A.	Apply on fan motor shaft	⇦	Front

37663_VERS_G0121

Fig. 11 Engine cooling fan (engines with A/C)—except 2012 Sedan

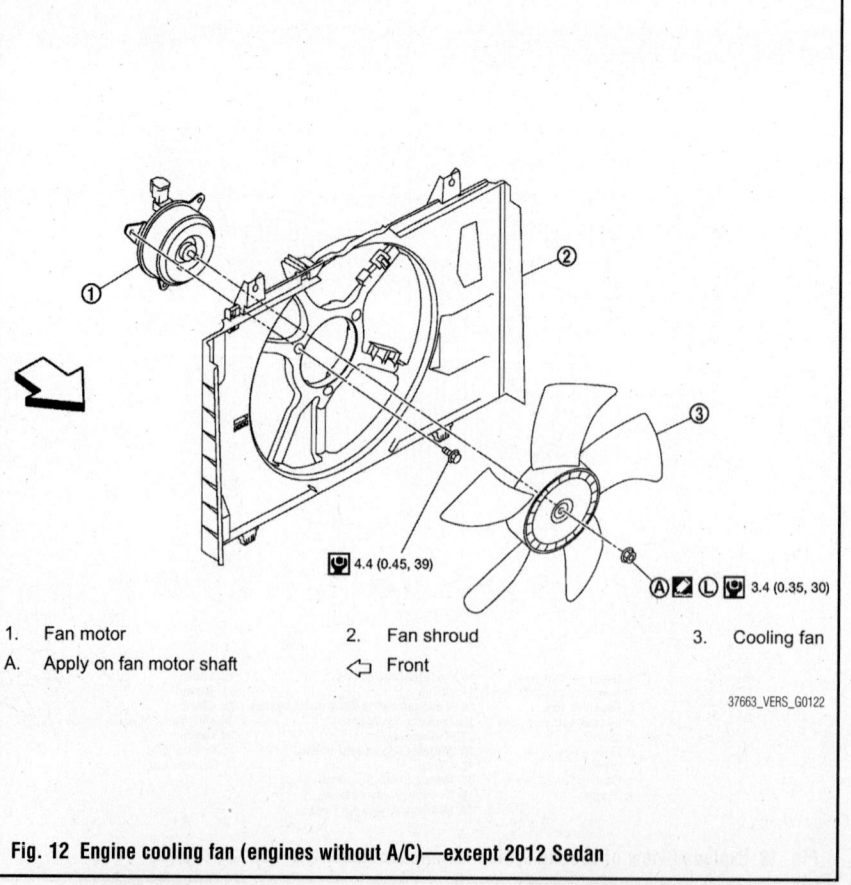

1.	Fan motor	2.	Fan shroud	3.	Cooling fan
A.	Apply on fan motor shaft	⇦	Front		

37663_VERS_G0122

Fig. 12 Engine cooling fan (engines without A/C)—except 2012 Sedan

WARNING

Be careful not to damage or scratch the radiator core.

To install:

8. Installation is the reverse of the removal procedure.

9. Tighten the fasteners to specification.

2012 Sedan

See Figure 13.

CAUTION

Never remove the radiator cap when the engine is hot. Serious burns could occur from high-pressure engine coolant escaping from the radiator. Wrap a thick cloth around the radiator cap. Slowly turn it a quarter of a turn to release built-up pressure. Carefully remove radiator cap by turning it all the way.

➡ **When removing components such as hoses, tubes/lines, etc., cap or plug openings to prevent fluid from spilling.**

1. Before servicing the vehicle, refer to the Precautions.

2. Partially the drain engine coolant from the radiator.

WARNING

Never spill engine coolant on the drive belt.

3. Remove the air duct (inlet).

4. Disconnect the battery negative terminal.

5. Loosen the following from the radiator core support (upper):
- Air bag harness
- Hood lock cable
- Radiator cap adapter
- Horn harness

6. Remove the radiator hose (upper) from the water outlet.

7. Disconnect the harness connector from the fan motor, and move the harness aside.

8. Disconnect the reservoir tank hose, and remove the reservoir tank.

9. Remove the radiator core support (upper).

10. Remove the radiator hose (upper) from the radiator.

11. Remove the cooling fan assembly.

WARNING

Be careful not to damage or scratch the radiator.

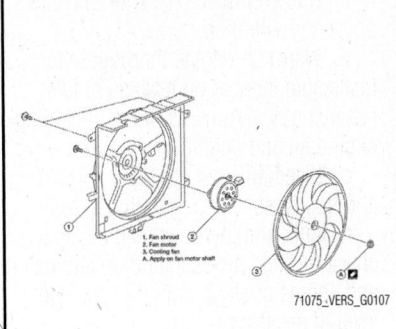

Fig. 13 Exploded view of engine fan component locations—2012 Sedan

To install:

12. Installation is the reverse of the removal procedure.

13. Tighten the fasteners to specification.

WARNING

Only use genuine NISSAN parts for the fan shroud mounting bolt and make sure to follow the specified torque to prevent radiator damage.

➡ **The cooling fan is controlled by ECM.**

WARNING

If the fan is removed from the shaft, apply genuine high strength thread locking sealant on the fan motor shaft.

RADIATOR

REMOVAL & INSTALLATION

Except 2012 Sedan
See Figures 14 and 15.

CAUTION

Do not remove radiator cap when the engine is hot. Serious burns could occur from high-pressure engine coolant escaping from radiator. Wrap a thick cloth around the cap. Slowly turn it a quarter of a turn to release built-up pressure. Carefully remove radiator cap by turning it all the way.

1. Before servicing the vehicle, refer to the Precautions.

2. Remove the engine under cover.

3. Drain engine coolant from radiator.

WARNING

Do not spill engine coolant on the drive belt.

4. Remove air duct (inlet).

5. Remove reservoir tank as follows:
   a. Disconnect reservoir tank hose.
   b. Release the tab.
   c. Lift up while removing the reservoir tank hose, and remove it.

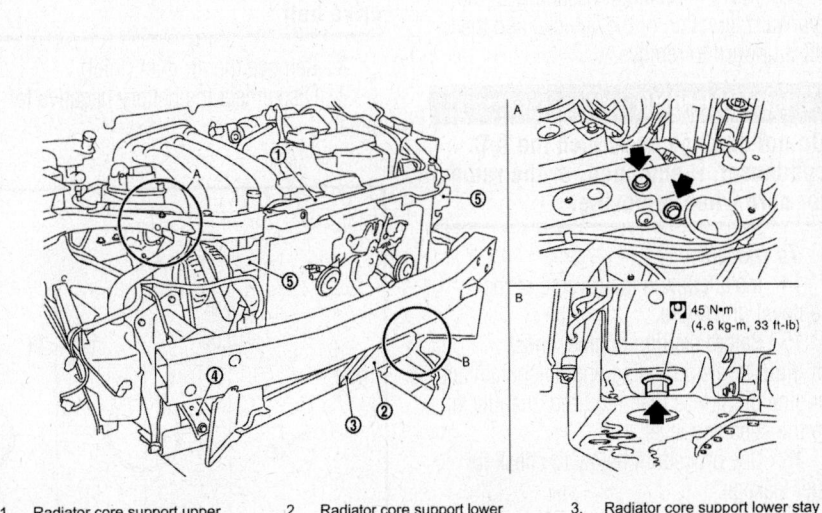

1. Radiator core support upper
2. Radiator core support lower
3. Radiator core support lower stay
4. Radiator core support side stay
5. Air guide

Fig. 14 Radiator core support components shown

Fig. 15 View of radiator assembly (1) and radiator (upper) mount part of the radiator core support (upper) (2)

6. Disconnect the harness connector from the fan motor and move the harness aside.

7. Disconnect CVT or A/T fluid cooler hoses, if equipped. Install plug to avoid leakage of CVT or A/T fluid, if necessary.

8. Remove the radiator hoses (upper and lower).

9. Remove the radiator core support cover.

10. Remove the cooling fan assembly.

11. Remove the radiator core support (upper) bolts, bolts of stationary part on the radiator core support side and clip.

12. Lift the radiator assembly from the radiator (upper) mount part of the radiator core support (upper).

13. Move the radiator assembly to the rearward direction of the vehicle, and then lift it upward to remove.

❋❋ WARNING

Do not damage or scratch the A/C condenser, if equipped, or the radiator core when removing.

To install:

14. Installation is the reverse of the removal procedure.

15. Before starting engine, check oil/fluid levels including engine coolant and engine oil. If less than required quantity, fill to the specified level.

16. Use procedure below to check for fuel leakage.

a. Turn ignition switch ON (with engine stopped). With fuel pressure applied to fuel piping, check for fuel leakage at connection points.

b. Start engine. With engine speed increased, check again for fuel leakage at connection points.

c. Run engine to check for unusual noise and vibration.

d. Warm up engine thoroughly to make sure there is no leakage of fuel, exhaust gas, or any oils/fluids including engine oil and engine coolant.

e. Bleed air from passages in lines and hoses, such as in cooling system.

f. After cooling down engine, again check oil/fluid levels including engine oil and engine coolant. Refill to specified level, if necessary.

2012 Sedan

See Figures 16 and 17.

❋❋ CAUTION

Never remove the radiator cap when the engine is hot. Serious burns could occur from high-pressure engine coolant escaping from the radiator. Wrap a thick cloth around the radiator cap. Slowly turn it a quarter of a turn to release built-up pressure. Carefully remove radiator cap by turning it all the way.

➡**When removing components such as hoses, tubes/lines, etc., cap or plug openings to prevent fluid from spilling.**

1. Before servicing the vehicle, refer to the Precautions.

2. Drain the engine coolant.

❋❋ WARNING

Never spill engine coolant on the drive belt.

3. Remove the air duct (inlet).

4. Disconnect the battery negative terminal.

5. Loosen the following from the radiator core support (upper):
- Air bag harness
- Hood lock cable
- Radiator cap adapter
- Horn harness

6. Remove the radiator hose (lower).

7. Remove the radiator hose (upper) from the water outlet.

8. Disconnect the harness connector from the fan motor, and move the harness aside.

9. Disconnect the reservoir tank hose, and remove the reservoir tank.

10. Remove the radiator core support (upper).

11. Remove the radiator hose (upper) from the radiator.

12. Remove the cooling fan assembly.

❋❋ WARNING

Be careful not to damage or scratch the radiator.

13. Disconnect the CVT cooler lines (if equipped).

14. Remove the radiator from the bottom of the vehicle.

❋❋ WARNING

When removing, do not damage or scratch the radiator core or A/C condenser (if equipped).

To install:

15. Installation is the reverse of the removal procedure.

16. Insert the radiator hose all the way to the stopper or by 1.30 inches (33mm) (hose without a stopper).

17. The angle created by the hose clamp pawl and the specified line must be within plus or minus 30°.

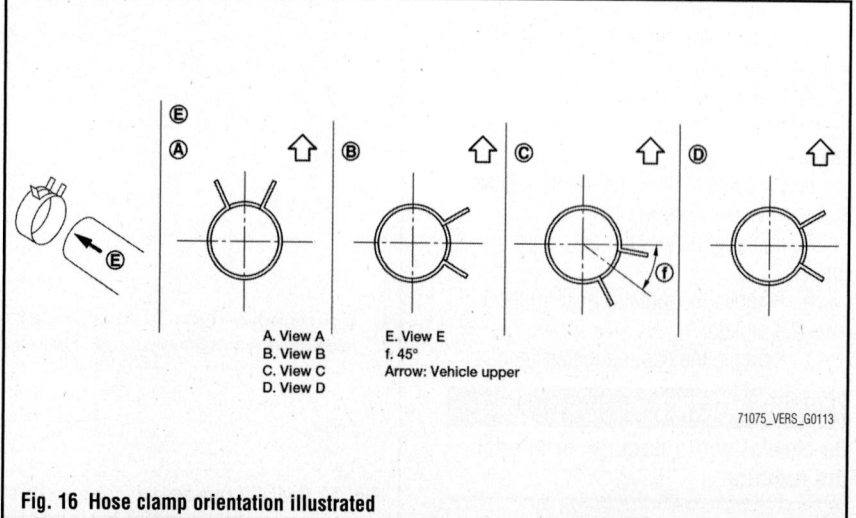

A. View A
B. View B
C. View C
D. View D

E. View E
f. 45°
Arrow: Vehicle upper

Fig. 16 Hose clamp orientation illustrated

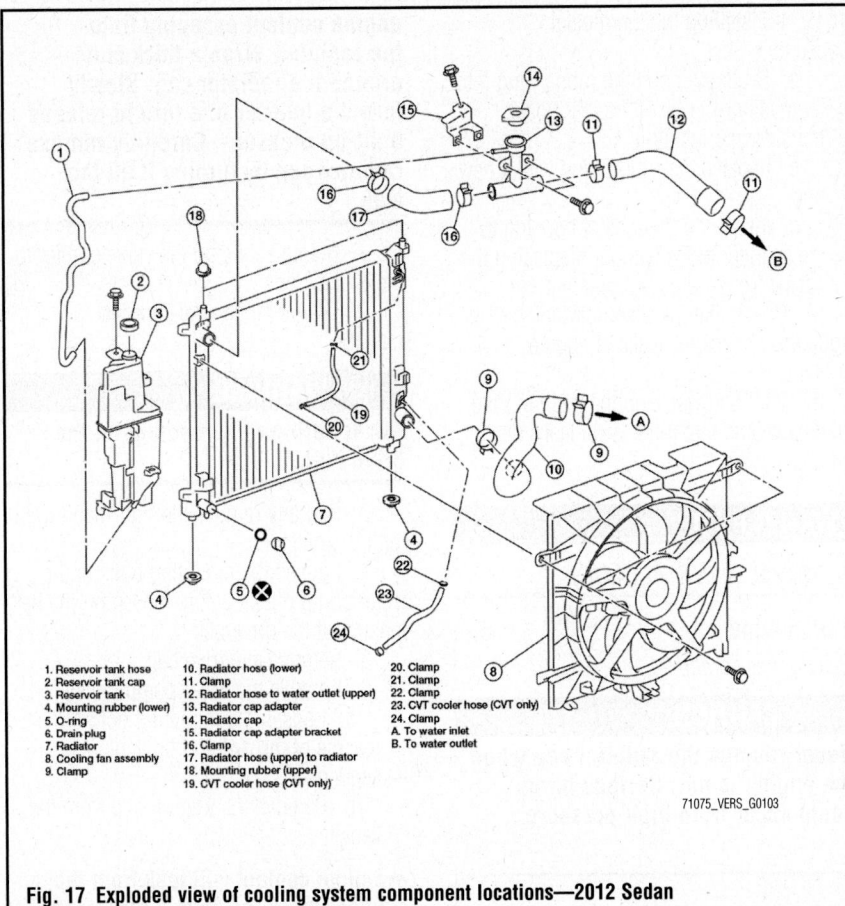

1. Reservoir tank hose
2. Reservoir tank cap
3. Reservoir tank
4. Mounting rubber (lower)
5. O-ring
6. Drain plug
7. Radiator
8. Cooling fan assembly
9. Clamp
10. Radiator hose (lower)
11. Clamp
12. Radiator hose to water outlet (upper)
13. Radiator cap adapter
14. Radiator cap
15. Radiator cap adapter bracket
16. Clamp
17. Radiator hose (upper) to radiator
18. Mounting rubber (upper)
19. CVT cooler hose (CVT only)
20. Clamp
21. Clamp
22. Clamp
23. CVT cooler hose (CVT only)
24. Clamp
A. To water inlet
B. To water outlet

71075_VERS_G0103

Fig. 17 Exploded view of cooling system component locations—2012 Sedan

1. Before servicing the vehicle, refer to the Precautions.
2. Drain engine coolant from radiator.
3. Remove air duct (inlet).
4. Remove the reservoir tank.
5. Add paint mark, then disconnect radiator hose (lower) from water inlet.
6. Remove water inlet and thermostat. Engine coolant will leak from cylinder block, so have a receptacle ready below.

To install:

7. Installation is the reverse of the removal procedure.
8. Tighten the fasteners to specification.
9. Replace the rubber ring with a new one.
10. Install the thermostat making sure the rubber ring groove fits securely to the thermostat flange.
11. Install the thermostat to the cylinder block with jiggle valve facing upwards.
12. After installation, secure the water inlet clip on the oil level gauge guide.
13. Check that the reservoir tank cap is tightened. Check for leaks of engine coolant.
14. Start and warm up the engine. Visually make sure that there is no leaks of engine coolant.

18. To install the hose clamps, check that the dimension from the end of the hose clamp on the radiator hose to the hose clamp is within 0.12–0.28 inch (3–7mm).

✷✷ WARNING

When installing, do not damage or scratch the radiator core or A/C condenser (if equipped).

➡Replace the water hose clamp if it is removed.

✷✷ WARNING

Use only genuine NISSAN mounting bolts for the cooling fan assembly and strictly follow the tightening torque. Over-tightening may damage the radiator.

19. Start and warm up the engine. Visually inspect for coolant leaks. Repair as necessary.

THERMOSTAT

REMOVAL & INSTALLATION

1.6L Engine

See Figure 18.

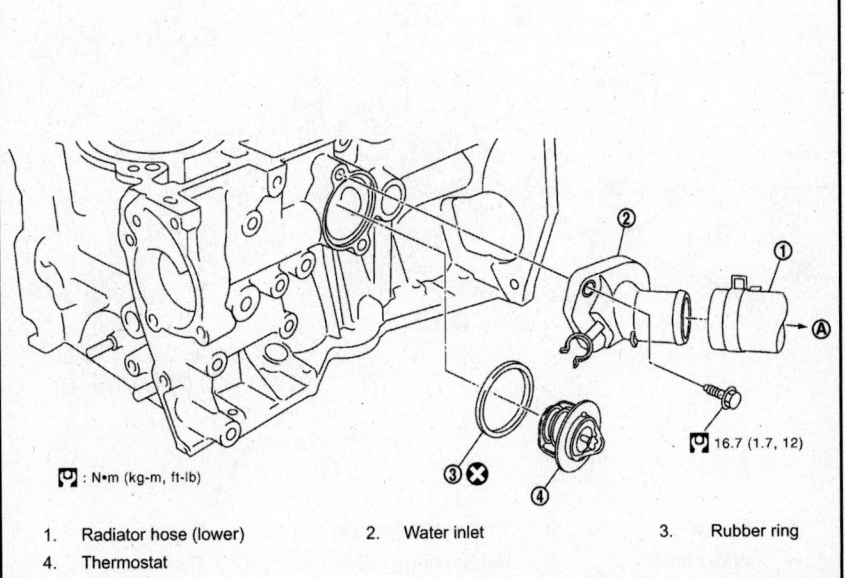

⊡ : N•m (kg-m, ft-lb)

16.7 (1.7, 12)

1. Radiator hose (lower)
2. Water inlet
3. Rubber ring
4. Thermostat
A. To radiator

37663_VERS_G0113

Fig. 18 Thermostat component locations—1.6L engine

1.8L Engine

See Figure 19.

1. Before servicing the vehicle, refer to the Precautions.
2. Drain engine coolant from radiator.
3. Remove air duct (inlet).
4. Remove the radiator hose (lower) from the engine.
5. Remove water inlet.
6. Remove thermostat.
7. Remove water pump, if necessary.
8. Remove thermostat housing, if necessary.

To install:

9. Installation is in the reverse order of removal, noting the following:
10. Replace the rubber ring with a new one.
11. Install the thermostat making sure the rubber ring groove fits securely to the thermostat flange.

➡ **Replace the rubber ring with a new one.**

12. Install the thermostat into the thermostat housing with the jiggle valve facing upwards.

13. If installing the thermostat housing:

a. Securely insert the rubber ring into the mating groove of the thermostat housing and install it.

b. Replace the rubber ring with a new one.

c. Install the thermostat housing to the cylinder block without displacing the gasket from the gasket position.

14. Check that the reservoir tank cap is tightened. Check for leaks of engine coolant.

15. Start and warm up the engine. Visually make sure that there is no leaks of engine coolant.

WATER PUMP

REMOVAL & INSTALLATION

1.6L Engine

See Figures 20 and 21.

❋❋ **CAUTION**

Never remove the radiator cap when the engine is hot. Serious burns could occur from high-pressure

engine coolant escaping from the radiator. Wrap a thick cloth around the radiator cap. Slowly turn it a quarter of a turn to release built-up pressure. Carefully remove radiator cap by turning it all the way.

1. Before servicing the vehicle, refer to the Precautions.
2. Drain engine coolant from radiator.

❋❋ **WARNING**

Never spill engine coolant on the drive belt.

3. Partially remove the right-hand front fender protector.
4. Loosen the mounting bolts of the water pump pulley before loosening the belt tension of the drive belt.
5. Remove the drive belt.
6. Remove the water pump pulley.

a. Loosen the mounting bolts in reverse of the order of tightening sequence.

b. Remove the water pump from the vehicle.

➡ **Engine coolant will leak from the cylinder block, so have a receptacle ready below.**

❋❋ **WARNING**

Do not allow the water pump vane to contact any other parts. The water pump cannot be disassembled and must be replaced as an assembly.

To install:

7. Installation is the reverse of the removal procedure.
8. Do not reuse the gasket. The sealing surface must be clean and free of dents or flaws.
9. Tighten the water pump bolts in sequence to specification.
10. When installing the water pump pulley, never install the mounting bolts to the oblong holes.
11. Bleed air from passages in lines and hoses, as needed.
12. After cooling down the engine, again check fluid levels. Refill to specified level, if necessary.
13. Check that the reservoir tank cap is tightened. Check for leaks of engine coolant.
14. Start and warm up the engine. Visually make sure that there are no leaks of engine coolant.

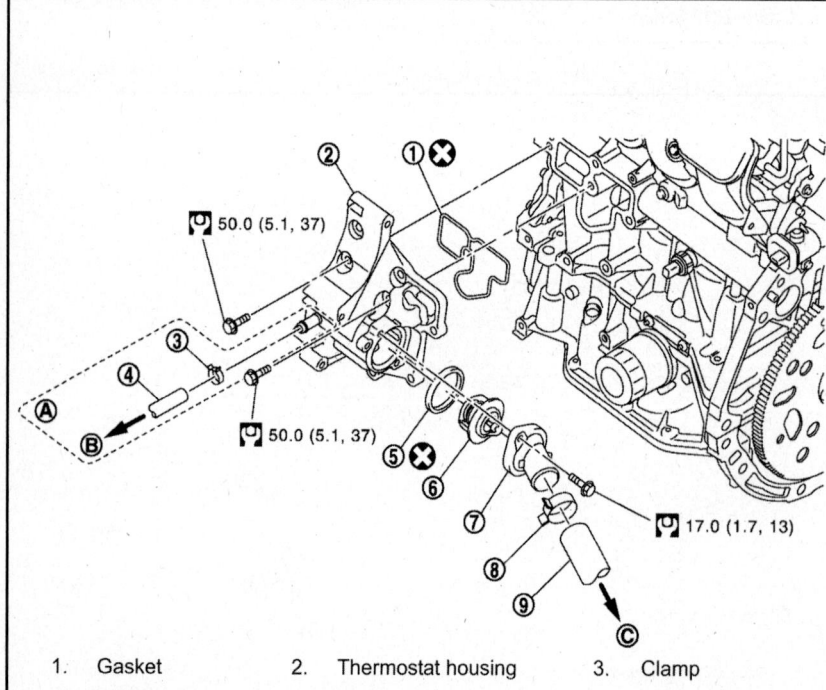

1.	Gasket	2.	Thermostat housing	3.	Clamp
4.	Water hose	5.	Rubber ring	6.	Thermostat
7.	Water inlet	8.	Clamp	9.	Radiator hose (lower)
A.	CVT models	B.	To CVT fluid cooler	C.	To radiator

37663_VERS_G0110

Fig. 19 Thermostat component locations—1.8L engine

1. Before servicing the vehicle, refer to the Precautions.

2. Disconnect the negative battery cable.

3. Drain engine coolant from radiator.

4. Remove the right-hand front fender protector.

5. Remove the drive belt.

6. Remove the alternator.

7. Remove the radiator hose (lower).

8. Remove the water pump from the vehicle.

⁕⁕ WARNING

Handle the water pump vane so that it does not contact any other parts. The water pump cannot be disassembled and should be replaced as a unit.

To install:

9. Installation is the reverse of the removal procedure.

10. Check that the reservoir tank cap is tightened. Check for leaks of engine coolant.

11. Start and warm up the engine. Visually make sure that there is no leaks of engine coolant.

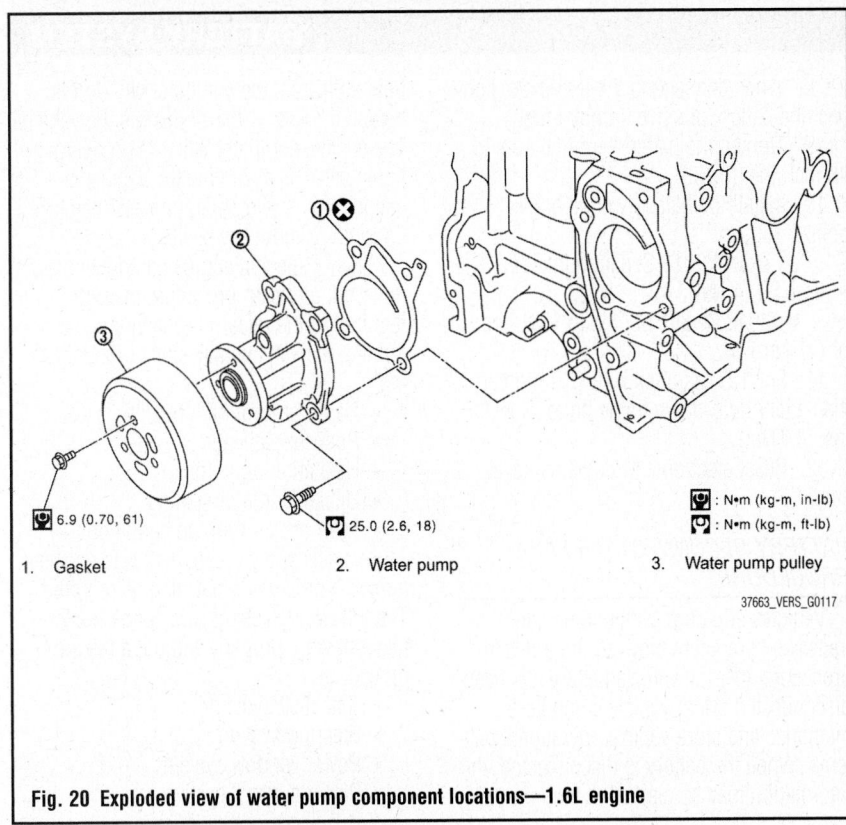

⬛ : N•m (kg-m, in-lb)
⬛ : N•m (kg-m, ft-lb)

⬛ 6.9 (0.70, 61)   ⬛ 25.0 (2.6, 18)

1. Gasket
2. Water pump
3. Water pump pulley

37663_VERS_G0117

Fig. 20 Exploded view of water pump component locations—1.6L engine

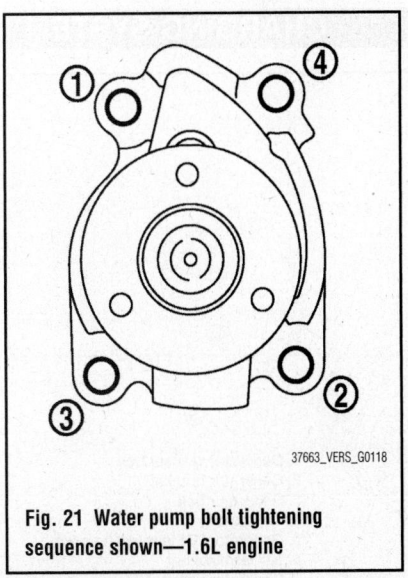

37663_VERS_G0118

Fig. 21 Water pump bolt tightening sequence shown—1.6L engine

1.8L Engine

See Figure 22.

⁕⁕ CAUTION

Never remove the radiator cap when the engine is hot. Serious burns could occur from high pressure engine coolant escaping from the radiator.

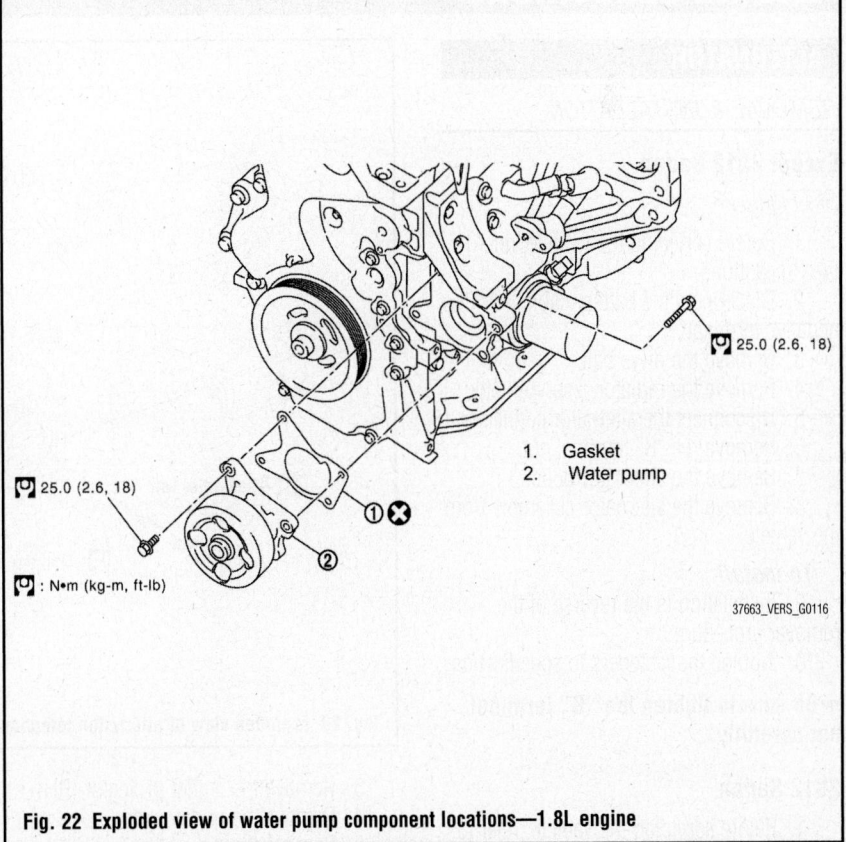

⬛ 25.0 (2.6, 18)

⬛ 25.0 (2.6, 18)

⬛ : N•m (kg-m, ft-lb)

1. Gasket
2. Water pump

37663_VERS_G0116

Fig. 22 Exploded view of water pump component locations—1.8L engine

ENGINE ELECTRICAL   **BATTERY SYSTEM**

BATTERY

REMOVAL & INSTALLATION

1. Before servicing the vehicle, refer to the Precautions.

2. Disconnect the battery negative and positive terminals.

❈❈ CAUTION

When disconnecting, disconnect the battery negative terminal first.

3. For all models EXCEPT 1.8L engines with CVT, remove the battery hold-down wedge bolt and remove battery hold-down wedge bracket.

4. For 1.8L engines with CVT, remove the battery hold-down frame bolts and remove the battery hold-down frame.

5. Remove the battery cover.

6. Remove the battery from the vehicle.

To install:

7. Installation is the reverse of the removal procedure.

8. When connecting the terminals, connect the battery positive terminal first.

9. Tighten the battery terminal nuts to 48 inch lbs. (5 Nm).

10. Install the battery wedge bracket bolts:

    a. Except 2012 Sedan: tighten to 10 ft. lbs. (14 Nm).

    b. 2012 Sedan: tighten to 13 ft. lbs. (17 Nm).

11. For 1.8L engines with CVT, tighten the battery hold-down frame bolts to 10 ft. lbs. (14 Nm).

12. Reset electronic systems as necessary.

BATTERY RECONNECT/RELEARN PROCEDURE

Vehicles equipped with engine and transaxle computers may require a relearn procedure after the vehicle battery has been disconnected. Most vehicle computers memorize and store vehicle operational patterns. When the battery is disconnected, the information may be cleared. If the information is cleared, the computer will go into default mode in order to operate the vehicle. The vehicle computer will relearn operational patterns each time the vehicle is restarted. The relearning process may take up to 40 or more key cycles.

When a specific engine component is replaced, a relearn procedure may be required. If the relearn procedure is not performed, the vehicle may exhibit the following:

- Harsh or poor shift quality
- Poor fuel mileage
- Hesitation or stumble
- Unstable idle or stalling
- Lean or rich running conditions

If an accessory component was replaced, a relearn procedure may also be required. The following systems and components may not work properly without a relearn procedure:

- Anti-theft system
- Steering system
- Power window system
- Power sunroof system

ENGINE ELECTRICAL   **CHARGING SYSTEM**

ALTERNATOR

REMOVAL & INSTALLATION

Except 2012 Sedan

See Figure 23.

1. Before servicing the vehicle, refer to the Precautions.

2. Disconnect the battery cable from the negative terminal.

3. Remove the drive belt.

4. Remove the radiator reservoir tank.

5. Disconnect the alternator connector.

6. Remove the "B" terminal nut.

7. Remove the alternator bolts.

8. Remove the alternator assembly from the vehicle.

To install:

9. Installation is the reverse of the removal procedure.

10. Tighten the fasteners to specification.

➡**Be sure to tighten the "B" terminal nut carefully.**

2012 Sedan

1. Before servicing the vehicle, refer to the Precautions.

2. Disconnect the battery cable from the negative terminal.

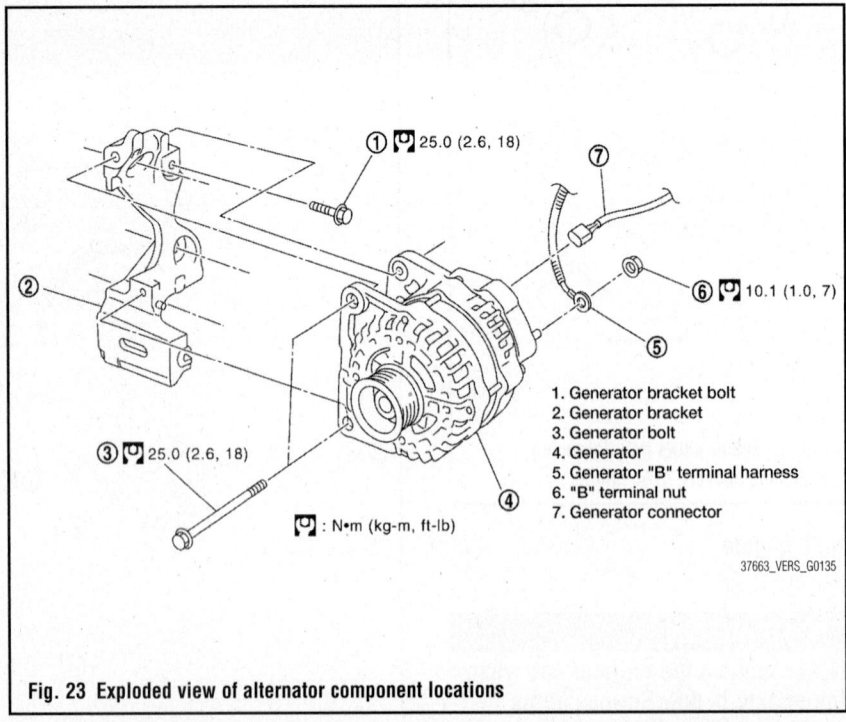

1. Generator bracket bolt
2. Generator bracket
3. Generator bolt
4. Generator
5. Generator "B" terminal harness
6. "B" terminal nut
7. Generator connector

🔧 : N•m (kg-m, ft-lb)

37663_VERS_G0135

Fig. 23 Exploded view of alternator component locations

3. Remove the fender protector (RH).

4. Remove the undercover.

5. Remove the drive belt.

6. Remove the horn bracket.

7. Disconnect the alternator connector.

8. Remove the "B" terminal nut and disconnect the "B" terminal harness.

9. Remove the alternator bolts.

10. Remove the alternator from the vehicle.

Be careful not to damage the surrounding parts when removing the alternator from the vehicle.

11. Remove the alternator bracket if necessary.

To install:

12. Install the alternator bracket, if removed, using the following procedure.

a. Install and temporarily tighten the first 2 bolts.

b. Install and tighten all bolts to specification.

13. Install the alternator.

a. Temporarily tighten the alternator bolts in order from the lower to the upper.

b. Tighten the alternator bolts to specification starting with the top bolt.

c. Install the "B" terminal harness and the "B" terminal nut.

Be sure to tighten the "B" terminal nut carefully.

14. Install and check the tension of the drive belt.

15. Installation of the remaining components is in the reverse order of removal.

ENGINE ELECTRICAL

FIRING ORDER

Firing Order for the 1.6L (HR16DE) and 1.8L (MR18DE) engines: 1–3–4–2.

IGNITION COIL

REMOVAL & INSTALLATION

Except 2012 Sedan

See Figures 24 and 25.

1. Before servicing the vehicle, refer to the Precautions.

2. Disconnect the negative battery cable.

3. Remove the intake manifold.

4. Remove the ignition coil.

Handle the ignition coil carefully and avoid impacts. Never disassemble.

IGNITION SYSTEM

To install:

5. Install the ignition coil.

6. Install the intake manifold.

7. Connect the negative battery cable.

2012 Sedan

See Figure 26.

1. Before servicing the vehicle, refer to the Precautions.

2. Disconnect the battery negative terminal.

3. Remove intake manifold.

4. Remove the ignition coil(s).

Never drop or shock an ignition coil. Never disassemble an ignition coil.

To install:

5. Installation is the reverse of the removal procedure.

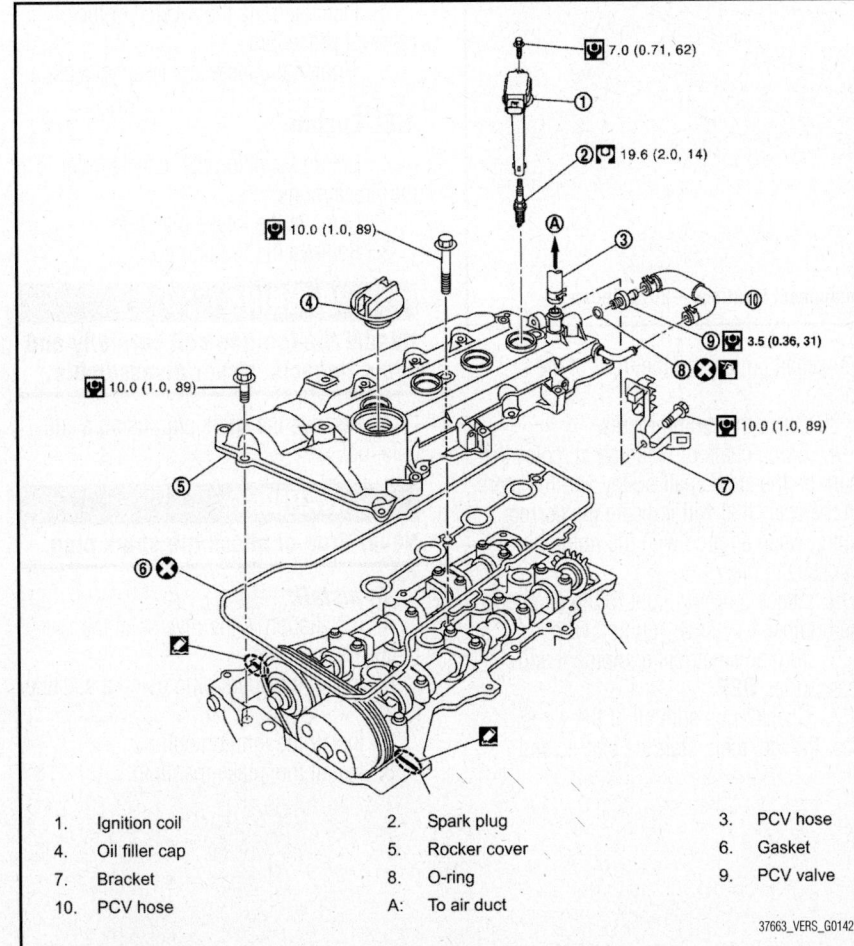

1.	Ignition coil	2.	Spark plug	3.	PCV hose
4.	Oil filler cap	5.	Rocker cover	6.	Gasket
7.	Bracket	8.	O-ring	9.	PCV valve
10.	PCV hose	A:	To air duct		

37663_VERS_G0142

Fig. 24 Ignition coil, spark plug, and related component locations—1.6L engine (except 2012 Sedan)

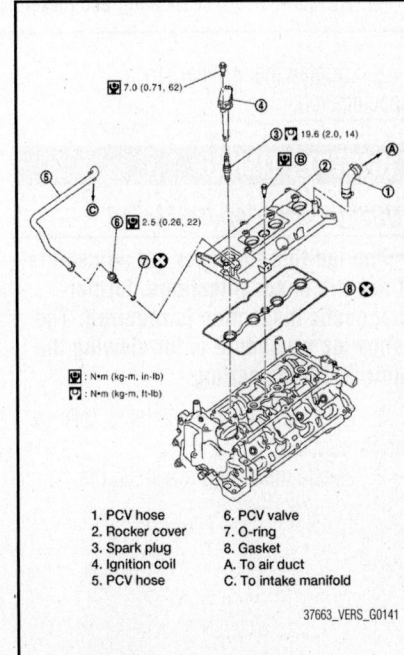

1.	PCV hose	6.	PCV valve
2.	Rocker cover	7.	O-ring
3.	Spark plug	8.	Gasket
4.	Ignition coil	A.	To air duct
5.	PCV hose	C.	To intake manifold

37663_VERS_G0141

Fig. 25 Ignition coil, spark plug, and related component locations—1.8L engine

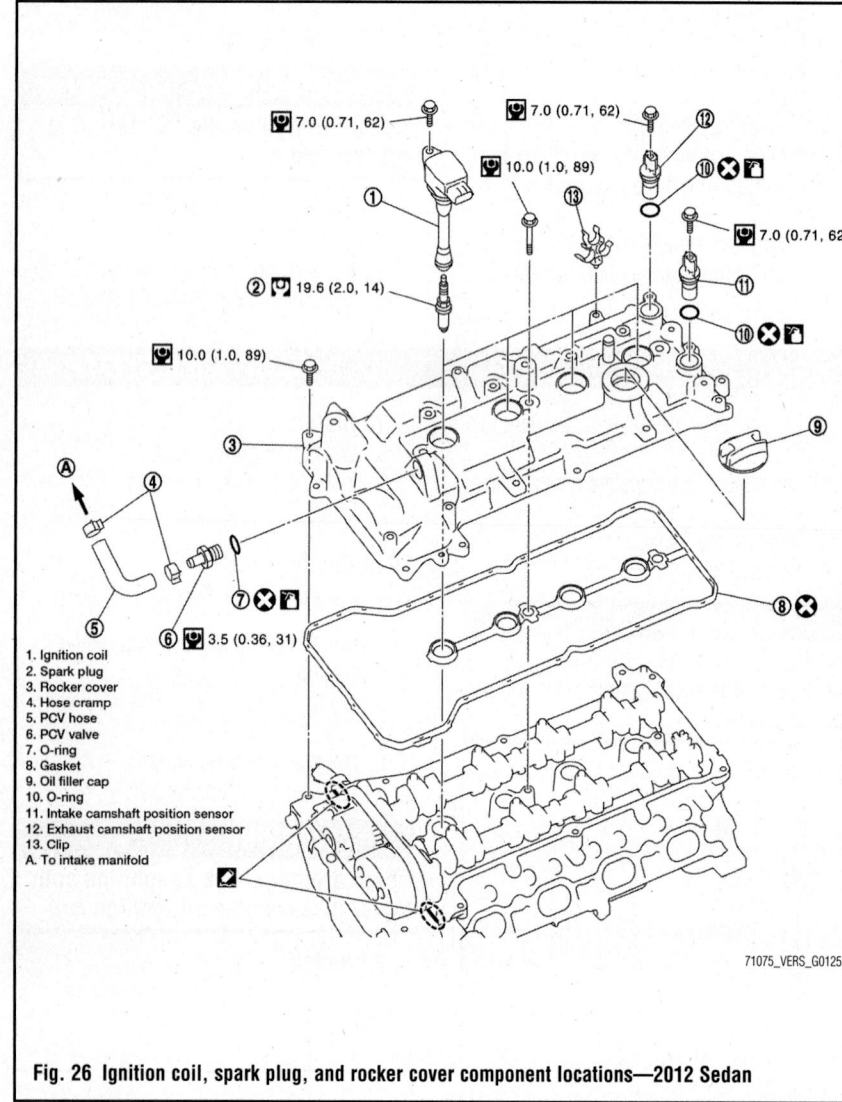

1. Ignition coil
2. Spark plug
3. Rocker cover
4. Hose cramp
5. PCV hose
6. PCV valve
7. O-ring
8. Gasket
9. Oil filler cap
10. O-ring
11. Intake camshaft position sensor
12. Exhaust camshaft position sensor
13. Clip
A. To intake manifold

71075_VERS_G0125

Fig. 26 Ignition coil, spark plug, and rocker cover component locations—2012 Sedan

6. Tighten the fasteners to specification.

IGNITION TIMING

INSPECTION & ADJUSTMENT

➡The ignition timing is not adjustable. If not within specifications, further diagnostic inspection is required. The following procedure is for viewing the ignition timing setting.

1. Before servicing the vehicle, refer to the Precautions.
2. Locate the timing marks on the crankshaft pulley and the front of the engine.
3. Clean the timing marks.
4. Using chalk or white paint, color the mark on the crankshaft pulley and the mark on the scale that will indicate the correct timing when aligned with the notch on the crankshaft pulley.
5. Attach a timing light to the engine to the number 1 cylinder ignition wire.
6. Turn all electrical equipment and accessories **OFF**.
7. Check to be sure all of the wires clear the fan, then, start the engine and

allow it to reach normal operating temperatures.
8. Check the ignition timing.
   a. 1.6L engine (2011)
   • A/T: 6 plus or minus 5° Before Top Dead Center (BTDC)
   • M/T: 6 plus or minus 5° BTDC
   b. 1.6L engine (2012)
   • CVT: 4 plus or minus 5° BTDC
   • M/T: 10 plus or minus 5° BTDC
   c. 1.8L engine
   • A/T: 13 plus or minus 5° BTDC
   • CVT: 13 plus or minus 5° BTDC
   • M/T: 13 plus or minus 5° BTDC

SPARK PLUGS

REMOVAL & INSTALLATION

1.6L Engine

1. Before servicing the vehicle, refer to the Precautions.
2. Remove the ignition coil.
3. Remove the spark plug with a suitable tool.

To install:

4. Installation is the reverse of the removal procedure.
5. Tighten the fasteners to specification.

1.8L Engine

1. Before servicing the vehicle, refer to the Precautions.
2. Remove the intake manifold.
3. Remove the ignition coil.

✳✳ WARNING

Handle the ignition coil carefully and avoid impacts. Never disassemble.

4. Remove the spark plug using a suitable tool.

✳✳ WARNING

Never drop or shock the spark plug.

To install:

5. Installation is the reverse of the removal procedure.
6. Install the spark plug using a suitable tool.
7. Install the ignition coil.
8. Install the intake manifold.

ENGINE ELECTRICAL   **STARTING SYSTEM**

STARTER

REMOVAL & INSTALLATION

See Figures 27 and 28.

1. Before servicing the vehicle, refer to the Precautions.
2. Disconnect the battery negative terminal.
3. Remove the air duct (inlet).
4. Remove the reservoir tank.
5. Remove the "S" terminal nut.
6. Remove the "B" terminal nut.
7. On the 2012 Sedan, remove the oil filter.
8. Remove the starter motor bolts.
9. Remove the starter motor.

✳✳ WARNING

Be careful to not damage surrounding parts when removing the starter motor from the vehicle.

To install:

10. Installation is the reverse of the removal procedure.

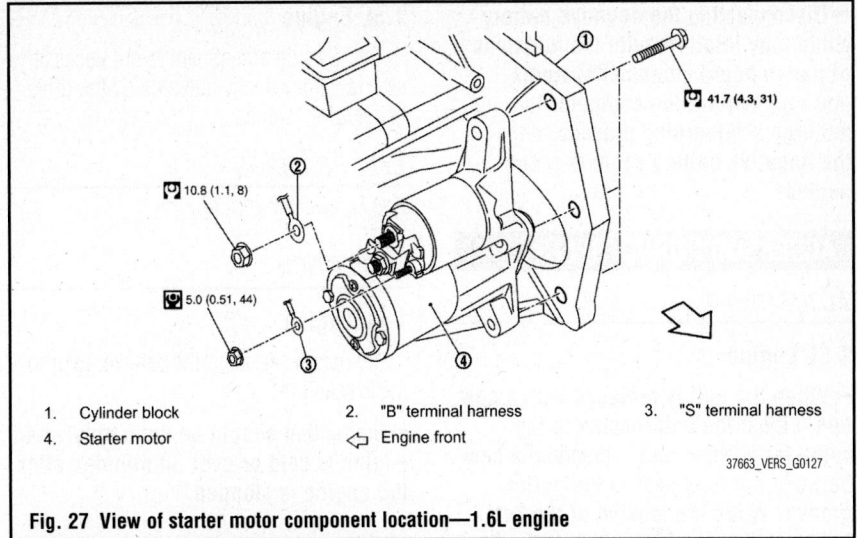

1. Cylinder block
4. Starter motor
2. "B" terminal harness
⇦ Engine front
3. "S" terminal harness

37663_VERS_G0127

Fig. 27 View of starter motor component location—1.6L engine

✳✳ WARNING

Be careful to tighten starter bolts, "S" terminal nut and "B" terminal nut to the specified torque.

11. Clean the battery terminals and battery cables to remove corrosion prior to connecting.

12. Ensure the battery cables are tightened to the specified torque.

➡Reset electronic systems as necessary.

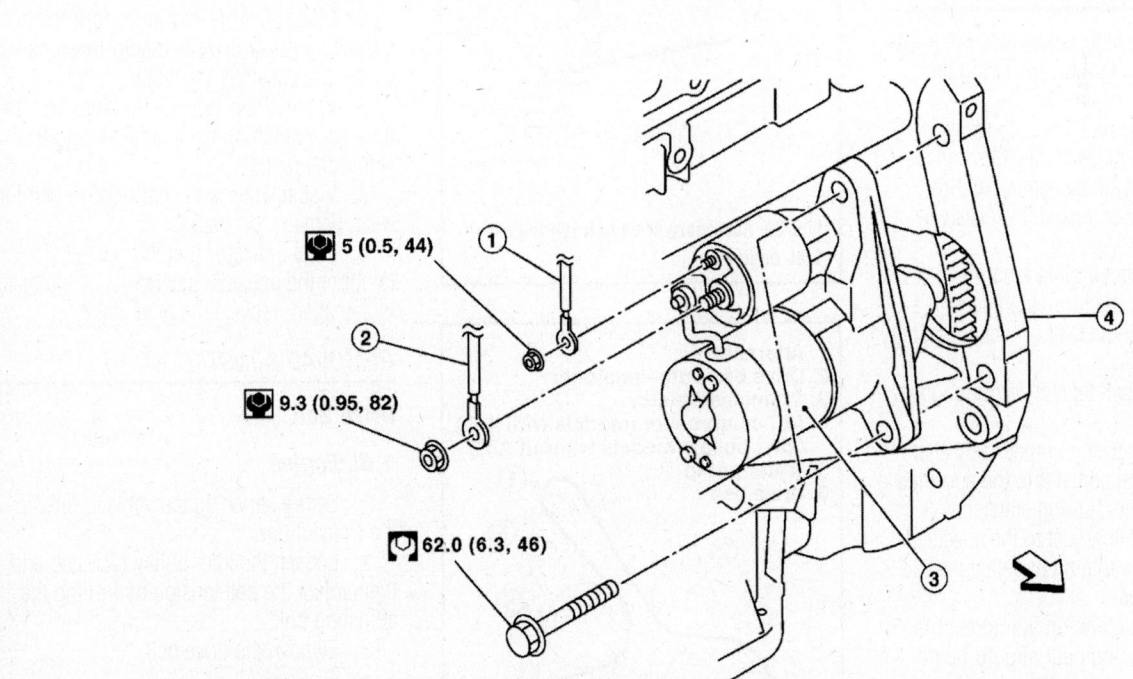

1. "S" terminal harness
4. Cylinder block
2. "B" terminal harness
⇦ Front
3. Starter motor

37663_VERS_G0128

Fig. 28 View of starter motor component location—1.8L engine

ENGINE MECHANICAL

➡Disconnecting the negative battery cable may interfere with the functions of the on board computer systems and may require the computer to undergo a relearning process, once the negative battery cable is reconnected.

ACCESSORY DRIVE BELTS

ADJUSTMENT

1.6L Engine

➡When the belt is replaced with a new one, adjust the belt tension to the value for a "New belt," because a new belt will not fully seat in the pulley groove. When the tension of the belt being used exceeds the "Limit," adjust it to the value for "After adjusted." When installing a belt, make sure it is correctly engaged with the pulley groove.

✳✳ WARNING

Never allow oil or engine coolant to get on the belt. Never twist or bend the belt strongly.

1. Loosen the idler pulley lock nut from the tightening position by 45° using the specified torque.

   a. When the lock nut is loosened excessively, the idler pulley tilts and the correct tension adjustment cannot be performed. Never loosen it excessively (more than 45°).

   b. Put a matching mark on the lock nut, and check the turning angle with a protractor. Never visually check the tightening angle only.

2. Adjust the belt tension by turning the adjusting bolt.

   a. When checking immediately after installation, first adjust it to the specified value. Then, after turning crankshaft 2 turns or more, re-adjust to the specified value to avoid variation in deflection between pulleys.

   b. When the tension adjustment is performed, the lock nut should be no more than 45° loosened. If the tension adjustment is performed when the lock nut is loosened more than the standard, the idler pulley tilts and the correct tension adjustment cannot be performed.

3. Tighten the idler pulley lock nut to 26 ft. lbs. (35 Nm).

1.8L Engine

Belt tension adjustment is not necessary, as it is automatically adjusted by the drive belt auto-tensioner.

BELT ROUTINGS

See Figures 29 and 30.

INSPECTION

1.6L Engine

1. Before servicing the vehicle, refer to the Precautions.

➡Inspection should be done only when engine is cold or over 30 minutes after the engine is stopped.

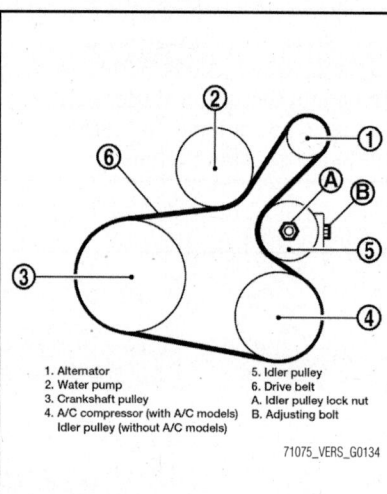

1. Alternator
2. Water pump
3. Crankshaft pulley
4. A/C compressor (with A/C models) Idler pulley (without A/C models)
5. Idler pulley
6. Drive belt
A. Idler pulley lock nut
B. Adjusting bolt

71075_VERS_G0134

Fig. 29 Accessory drive belt routing—1.6L engine

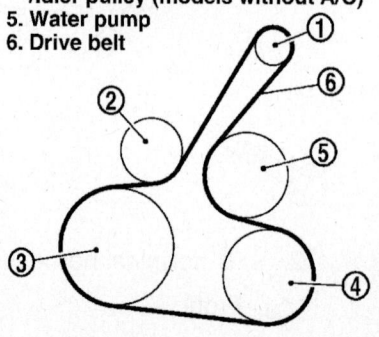

1. Alternator
2. Drive belt auto–tensioner
3. Crankshaft pulley
4. A/C compressor (models with A/C) /Idler pulley (models without A/C)
5. Water pump
6. Drive belt

22140_VERS_G0015

Fig. 30 Accessory drive belt routing—1.8L engine

2. Visually check belts for wear, damage, and cracks on inside and edges.

3. Turn the crankshaft pulley 2 times clockwise, and make sure the tension on all of the pulleys is equal before doing the test.

4. When measuring deflection, apply 22 lbs. (98 N) at the marked point.

5. Measure the belt tension and frequency with an acoustic tension gauge at the marked point.

➡When the tension and frequency are measured, the acoustic tension gauge should be used.

6. When checking immediately after installation, first adjust it to the specified value. Then, after turning the crankshaft 2 turns or more, re-adjust to the specified value to avoid variation in deflection between pulleys.

1.8L Engine

1. Before servicing the vehicle, refer to the Precautions.

➡Inspection should be done only when engine is cold or over 30 minutes after the engine is stopped.

2. Make sure that the indicator (notch on fixed side) of drive belt auto-tensioner is within the possible use range.

3. When a new drive belt is installed, the indicator (notch on fixed side) should be within the range.

4. Visually check the entire drive belt for wear, damage, or cracks.

5. If the indicator (notch on fixed side) is out of the possible use range, or the belt is damaged, replace the drive belt.

REMOVAL & INSTALLATION

Drive Belt

1.6L Engine

1. Before servicing the vehicle, refer to the Precautions.

2. Loosen the idler pulley lock nut, and then adjust the belt tension by turning the adjusting bolt.

3. Remove the drive belt.

To install:

4. Pull the idler pulley in the loosening direction, and then temporarily tighten the idler pulley lock nut to 39 inch lbs. (4 Nm).

➡Do not move the lock nut from the tightened position.

5. Install the drive belt to each pulley.

Make sure that there is no oil, grease, or coolant, etc. in the pulley grooves. Make sure that the belt is securely inside the groove on each pulley.

6. Adjust the drive belt tension by turning the adjusting bolt.

   a. Perform the belt tension adjustment with the lock nut temporarily tightened so as not to tilt the idler pulley.

   b. When checking immediately after installation, first adjust it to the specified value. Then, after turning the crankshaft 2 turns or more, re-adjust to the specified value to avoid variation in deflection between the pulleys.

7. Tighten the idler pulley lock nut to 26 ft. lbs. (35 Nm).

8. Make sure that belt tension of each belt is within the standard.

1.8L Engine

1. Before servicing the vehicle, refer to the Precautions.

2. Remove the right-hand fender protector.

3. Hold the hexagonal part of drive belt auto-tensioner with a wrench securely. Then move the wrench handle in the direction of arrow (loosening direction of tensioner).

⁂ CAUTION

Never place hand in a location where pinching may occur if the holding tool accidentally comes off.

⁂ WARNING

Do not loosen the auto-tensioner pulley bolt. (Do not turn it counterclockwise.) If turned counterclockwise, the complete auto-tensioner must be replaced as a unit, including pulley.

4. Insert a rod such as short-length screwdriver approximately 0.24 inch. (6mm) in diameter into the hole of the retaining boss to hold the drive belt auto-tensioner. Leave the tensioner pulley arm locked until the belt is installed again.

5. Remove drive belt.

To install:

6. Install the drive belt. Confirm the drive belt is completely set to pulleys.

7. Release the drive belt auto-tensioner, and apply tension to the drive belt.

8. Make sure no engine oil, working fluid, or and engine coolant has adhered to the drive belt and pulley grooves.

9. Turn the crankshaft pulley clockwise several times to equalize the tension between each pulley.

10. Confirm the tension of the drive belt

at the indicator (notch on the fixed side) is within the possible use range.

11. Install the right-hand fender protector.

Drive Belt Idler Pulley

1.6L Engine

See Figure 31.

1. Before servicing the vehicle, refer to the Precautions.

2. Remove the drive belt.

3. Remove the lock nut, and then remove the plate, idler pulley, and washer.

4. Remove the center shaft together with the spacer with the adjusting bolt.

To install:

5. Insert the center shaft into the slide groove of the spacer.

6. Fully screw in the adjusting bolt in the belt loosening direction.

7. Place the flange of the adjusting bolt and the seat of the center shaft on the spacer.

8. Place each surface of the spacer on the alternator bracket.

9. Install the washer, idler pulley, and plate, and then temporarily tighten the lock nut to 39 inch lbs. (4 Nm).

10. Installation continues in the reverse order of removal.

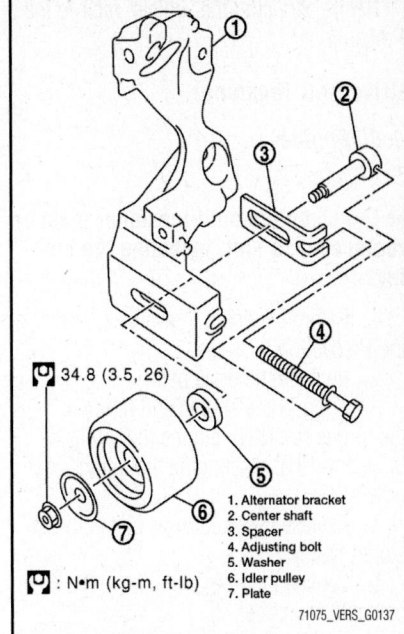

34.8 (3.5, 26)

: N•m (kg-m, ft-lb)

71075_VERS_G0137

Fig. 31 Exploded view of drive belt idler pulley—1.6L engine

Legend:
1. Alternator bracket
2. Center shaft
3. Spacer
4. Adjusting bolt
5. Washer
6. Idler pulley
7. Plate

1.8L Engine

See Figure 32.

At this time, the manufacturer does not provide specific removal and installation procedures for this component, refer to the illustration as required.

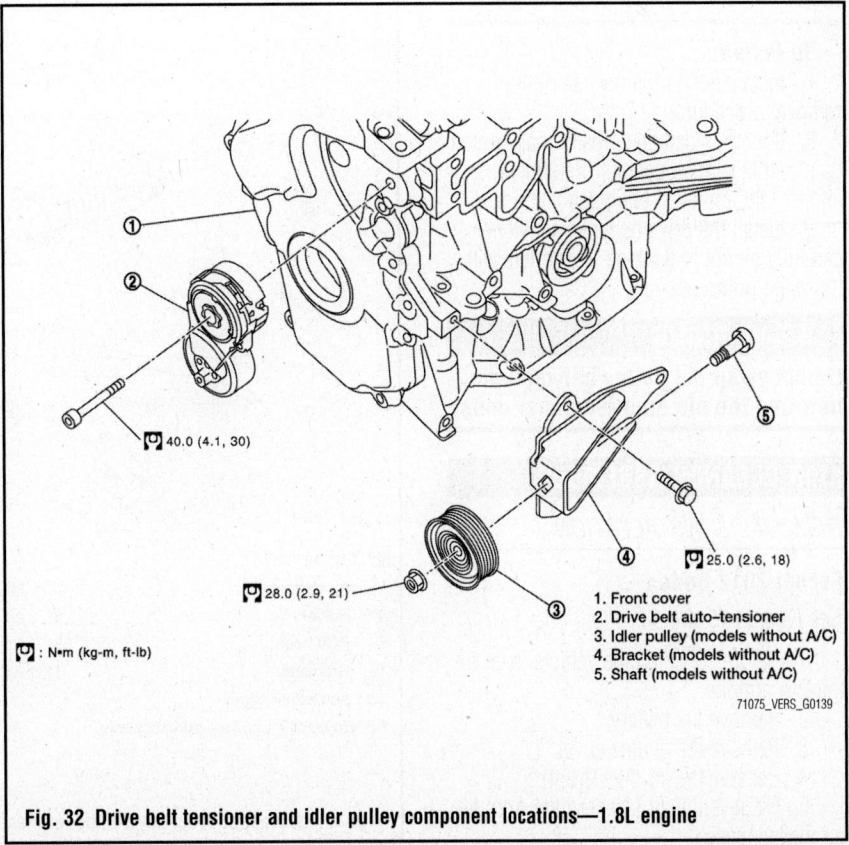

40.0 (4.1, 30)

28.0 (2.9, 21)

25.0 (2.6, 18)

: N•m (kg-m, ft-lb)

Legend:
1. Front cover
2. Drive belt auto-tensioner
3. Idler pulley (models without A/C)
4. Bracket (models without A/C)
5. Shaft (models without A/C)

71075_VERS_G0139

Fig. 32 Drive belt tensioner and idler pulley component locations—1.8L engine

Before servicing the vehicle, refer to the Precautions.

Drive Belt Tensioner

1.8L Engine

See Figure 32.

➡ **The complete auto-tensioner must be replaced as a unit, including the pulley.**

1. Before servicing the vehicle, refer to the Precautions.
2. Remove the drive belt.
3. Support the engine and remove the torque rod (RH), engine mounting insulator (RH), and engine mounting bracket (RH).
4. Release the fixed drive belt auto-tensioner pulley.
5. Loosen the bolt and remove the drive belt auto-tensioner.

➡ **Use a TORX® socket (size T50).**

6. Remove the idler pulley and bracket (models without A/C).

※ WARNING

Do not loosen the auto-tensioner pulley bolt. (Do not turn it counterclockwise). If turned counterclockwise, the complete auto-tensioner must be replaced as a unit, including the pulley.

To install:

7. Installation is the reverse of the removal procedure.
8. If there is damage greater than peeled paint, replace drive belt auto-tensioner and/or idler pulley, if equipped.
9. Install the drive belt auto-tensioner carefully so not to damage or interfere with the water pump pulley.

※ WARNING

Do not swap the pulley between the new and the old auto-tensioner units

AIR CLEANER ASSEMBLY

REMOVAL & INSTALLATION

Except 2012 Sedan

See Figures 33 and 34.

1. Before servicing the vehicle, refer to the Precautions.
2. Remove the battery.
3. Remove the engine cover.
4. Remove the air duct (front).
5. Remove the air cleaner filter from the air cleaner case.

6. Remove the air duct (inlet) from the air cleaner case.
7. Remove the PCV hose.
8. Remove the air duct. Add marks as necessary for easier installation.
9. Disconnect harness connector from mass air flow sensor.
10. Remove air cleaner case.
11. Remove the mass air flow sensor from the air cleaner case, if necessary.

※ WARNING

Handle the mass air flow sensor carefully and avoid impacts. Never touch the sensor part.

To install:

12. Installation is the reverse of the removal procedure.
13. Align marks. Attach each joint. Screw clamps firmly.

2012 Sedan

See Figure 35.

➡ **The mass air flow sensor is removable as an assembly with the air cleaner cover.**

1. Before servicing the vehicle, refer to the Precautions.

2. Remove the air duct (inlet) from the air cleaner body.
3. Disconnect the PCV hose from the air duct.
4. Remove the air duct (between the air cleaner case and electric throttle control actuator).

➡ **Add matching marks if necessary for easier installation.**

5. Remove the air cleaner assembly with the following steps:
   a. Disconnect the mass air flow sensor harness connector.
   b. Remove the 2 air cleaner body bolts.
   c. Pull up on the air cleaner assembly to disengage it from the grommet and remove the air cleaner assembly.

※ WARNING

Never shock the mass air flow sensor. Never disassemble the mass air flow sensor. Never touch the sensor of the mass air flow sensor.

To install:

6. Installation is the reverse of the removal procedure.

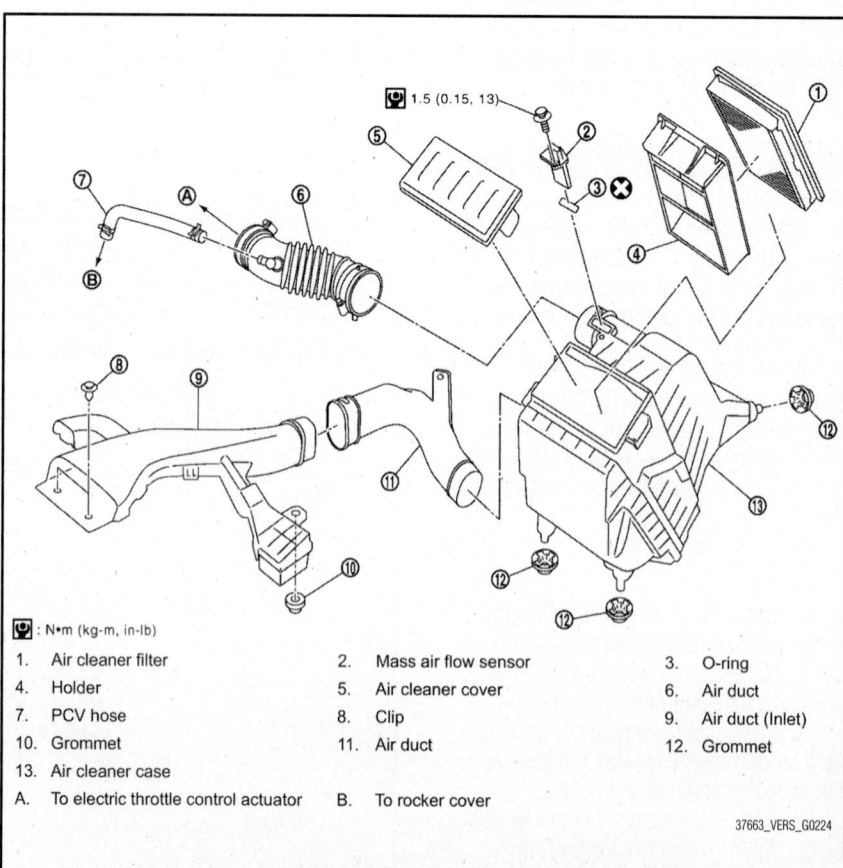

: N•m (kg-m, in-lb)

1.	Air cleaner filter	2.	Mass air flow sensor	3.	O-ring
4.	Holder	5.	Air cleaner cover	6.	Air duct
7.	PCV hose	8.	Clip	9.	Air duct (Inlet)
10.	Grommet	11.	Air duct	12.	Grommet
13.	Air cleaner case				
A.	To electric throttle control actuator	B.	To rocker cover		

37663_VERS_G0224

Fig. 33 Air cleaner assembly component locations—1.6L engine (2011 model)

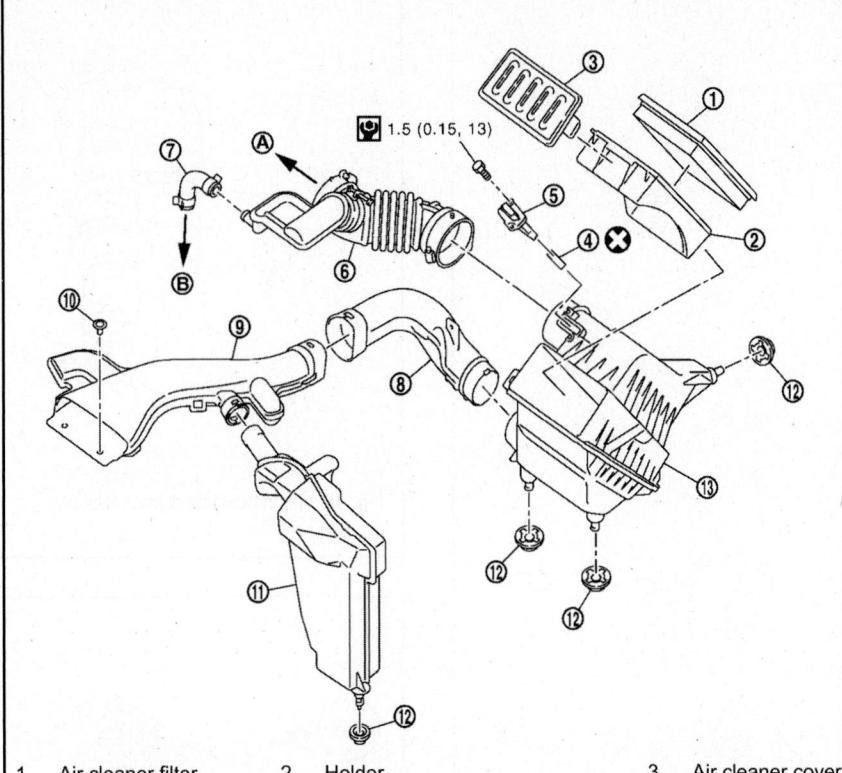

1.5 (0.15, 13)

1. Air cleaner filter	2. Holder	3. Air cleaner cover
4. Seal	5. Mass air flow sensor	6. Air duct
7. PCV hose	8. Air duct (Inlet)	9. Air duct (Front)
10. Clip	11. Resonator	12. Grommet
13. Air cleaner case	A. To electric throttle control actuator	B. To rocker cover

37663_VERS_G0225

Fig. 34 Air cleaner assembly component locations—1.8L engine

7. Inspect the air duct (inlet) and air duct for cracks, tears, or breaks. Replace the air duct (inlet) and air duct if any problems are found.

8. Align the marks, attach each joint and screw clamp firmly.

CAMSHAFT & BEARINGS

INSPECTION

See Figures 36 through 44.

1. Before servicing the vehicle, refer to the Precautions.

2. Inspect the Oil Filter:

a. Make sure that there is no foreign material on the oil filter and check it for clogging.

b. Check the oil filter for damage.

c. If there is some damage, replace the oil filter, the plug, and the washer as a set.

➡**Do not reuse the washer.**

3. Inspect camshaft runout:

a. Put V-block on a precise flat table, and support No. 2 and 5 journal of camshaft.

➡**Never support No. 1 journal (on the side of camshaft sprocket) because it has a different diameter from the other four locations.**

b. Set dial indicator vertically to No. 3 journal.

c. Turn camshaft to one direction with hands, and measure the camshaft runout on dial indicator (Total indicator reading). If it exceeds the limit, replace camshaft. Runout specifications:
- Standard: Less than 0.0008 inch (0.02mm)
- 1.6L engine Limit: 0.0040 inch (0.1mm)
- 1.8L engine Limit: 0.0020 inch (0.05mm)

4. Inspect Camshaft Cam Height:

a. Measure the camshaft cam height with a micrometer. Minimum height should be not less than the specification.

If it exceeds the limit, replace camshaft. Cam height specifications:
- 1.6L engine Standard Intake: 1.6419–1.6494 inches (41.705–41.895mm)
- 1.6L engine (2011) Standard Exhaust: 1.5817–1.5892 inches (40.175—40.365mm)
- 1.6L engine (2012) Standard Exhaust: 1.6108–1.6183 inches (40.914—41.105mm)
- 1.8L engine Standard Intake: 1.7560–1.7635 inches (44.605–44.795mm)
- 1.8L engine Standard Exhaust: 1.6997–1.7072 inches (43.175—43.365mm)
- 1.8L engine Limit Intake: 1.7482 inches (44.405mm)
- 1.8L engine Limit Exhaust: 1.6919 inches (42.975mm)

5. Inspect Camshaft Journal Outer Diameter:

a. Measure the outer diameter of camshaft journal with a micrometer. If it exceeds the limit, replace camshaft. Standard outer diameter:
- No. 1: 1.0998–1.1006 inches (27.935–27.955mm)
- No. 2, 3, 4, 5: 0.9823–0.9831 inches (24.950–24.970mm)

6. Inspect Camshaft Brackets Inner Diameter:

a. Tighten camshaft bracket bolts to the specified torque.

b. Measure inner diameter of camshaft bracket with a bore gauge.

c. Using inside micrometer, measure inner diameter of the camshaft bracket. Standard inner diameter:
- No. 1: 1.1024–1.1032 inches (28.000–28.021mm)
- No. 2, 3, 4, 5: 0.9843–0.9851 inch (25.000–25.021mm)

7. Calculate the Camshaft Journal Oil Clearance. Oil clearance = (camshaft bracket inner diameter) minus (camshaft journal diameter). Oil clearance specifications:
- Standard No. 1: 0.0018–0.0034 inch (0.045–0.086mm)
- Standard No. 2, 3, 4, 5: 0.0012–0.0028 inch (0.030–0.071mm)
- Limit: 0.0059 inch (0.15mm)

a. When outside the limit, replace either or both camshaft and cylinder head. Camshaft bracket cannot be replaced as a single part, because it is machined together with cylinder head. Replace the whole cylinder head assembly.

8. Inspect the Camshaft End Play:

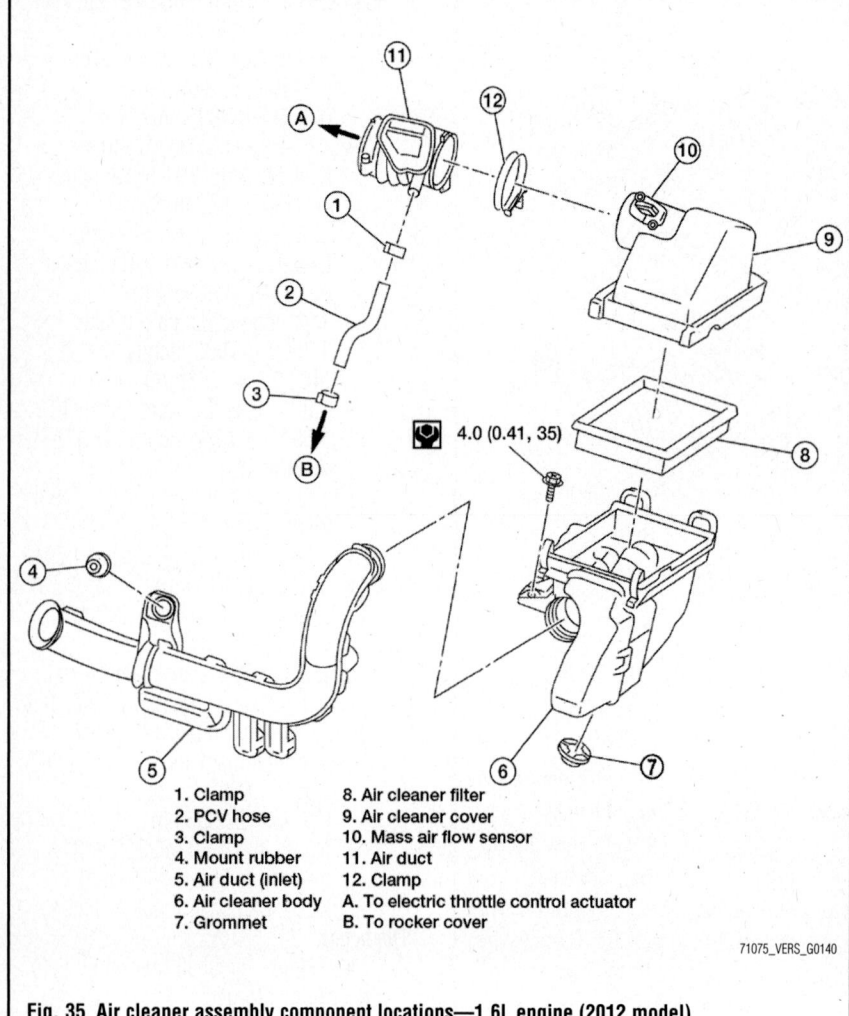

1. Clamp
2. PCV hose
3. Clamp
4. Mount rubber
5. Air duct (inlet)
6. Air cleaner body
7. Grommet
8. Air cleaner filter
9. Air cleaner cover
10. Mass air flow sensor
11. Air duct
12. Clamp
A. To electric throttle control actuator
B. To rocker cover

4.0 (0.41, 35)

71075_VERS_G0140

Fig. 35 Air cleaner assembly component locations—1.6L engine (2012 model)

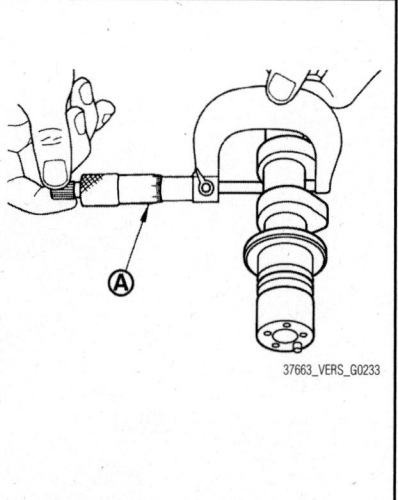

37663_VERS_G0233

Fig. 38 Inspect camshaft journal outer diameter

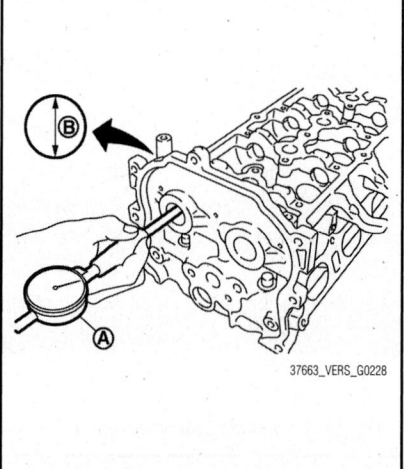

37663_VERS_G0228

Fig. 39 Measure inner diameter (B) of camshaft bracket with a bore gauge (A)

c. Measure the following parts if out of the standard:
- Dimension for groove of cylinder head No. 1 journal. Standard: 0.1575–0.1587 inch (4.000–4.030mm)
- Dimension for camshaft flange. Standard: 0.1526–0.1545 inch (3.877–3.925mm)

d. Apply the standards above, and then replace camshaft and/or cylinder head, if necessary.

9. Inspect the Camshaft Sprocket Runout:

a. Put V-block on precise flat bed and support No. 2 and No. 5 journal of camshaft.

❊❊ WARNING

Never support No. 1 journal (on the side of camshaft sprocket) because it

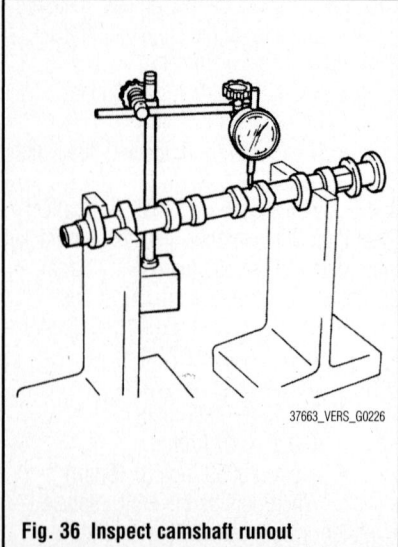

37663_VERS_G0226

Fig. 36 Inspect camshaft runout

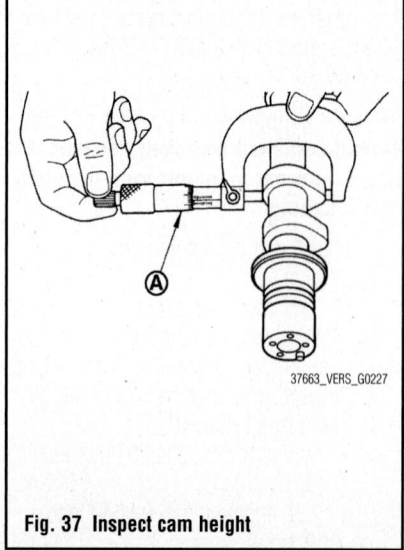

37663_VERS_G0227

Fig. 37 Inspect cam height

a. Install the camshaft in the cylinder head.

b. Install dial gauge in thrust direction on front end of camshaft. Measure end play of dial gauge when camshaft is moved forward and backward. End play specifications:
- Standard: 0.0030–0.0060 inch (0.075–0.153mm)
- Limit: 0.0094 inch (0.240mm)

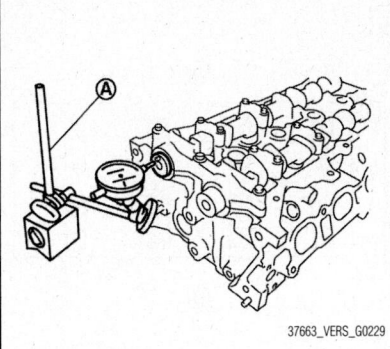

Fig. 40 Measure end play of dial gauge (A) when camshaft is moved forward and backward

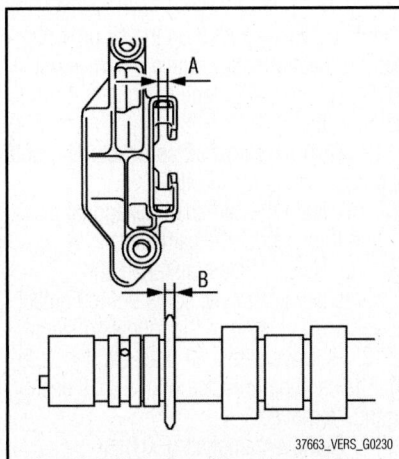

Fig. 41 Cylinder head No. 1 journal bearing: Dimension A, Camshaft flange: Dimension B

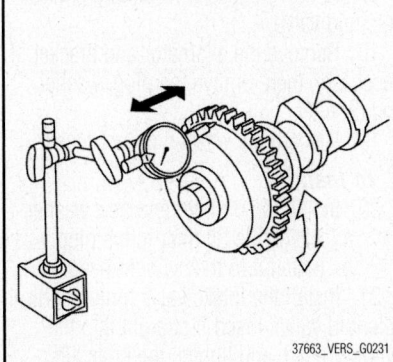

Fig. 42 Inspect the camshaft sprocket runout

has a different diameter from the other four locations.

   b. Measure camshaft sprocket runout. Limit: 0.0059 inch (0.15mm)

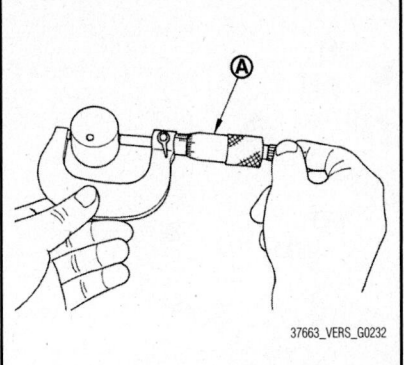

Fig. 43 Valve lifter outer diameter inspection

   c. If sprocket runout exceeds the limit, replace camshaft sprocket.
   10. Check the valve lifter surface for wear or cracks, and replace if necessary.
   11. Inspect the valve lifter outer diameter:
- 1.6L engine: 1.1802–1.1806 inches (29.977–29.987mm)
- 1.8L engine Intake: 1.3377–1.3381 inches (33.977–33.987mm)
- 1.8L engine Exhaust: 1.1802–1.1806 inches (29.977–29.987mm)

   12. Inspect the valve lifter inner diameter:
   a. Measure the inner diameter of valve lifter hole of cylinder head with an inside micrometer.
- 1.6L engine: 1.1811–1.1819 inches (30.000–30.021mm)
- 1.8L engine Intake: 1.3386–1.3394 inches (34.000–34.021mm)
- 1.8L engine Exhaust: 1.1811–1.1819 inches (30.000–30.021mm)

   13. Inspect the valve lifter clearance:
   a. Determine the valve lifter clearance by subtracting the valve lifter outer diameter from the valve lifter hole diameter. Valve lifter clearance:
- 1.6L engine Intake: 0.010–0.013 inch (0.26–0.34mm)
- 1.6L engine Exhaust: 0.011–0.015 inch (0.29–0.37mm)
- 1.8L engine: 0.0005–0.0017 inch (0.013–0.044mm)

   b. If the calculated value is out of the standard, referring to each standard of valve lifter outer diameter and valve lifter hole diameter, replace either or both valve lifter and cylinder head.

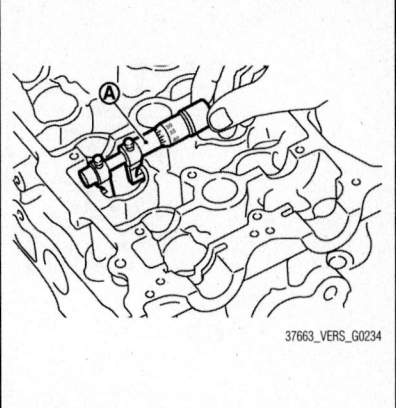

Fig. 44 Measure of the inner diameter of the valve lifter hole of the cylinder head

REMOVAL & INSTALLATION

Except 2012 Sedan

1.6L Engine

See Figure 45 and 46.

   1. Before servicing the vehicle, refer to the Precautions.
   2. Remove the timing chain.
   3. Remove the camshaft position sensor (PHASE) from the rear end of the cylinder head. Handle it carefully and avoid impacts.
   4. Remove the exhaust camshaft sprocket (EXH) bolt and camshaft sprocket (EXH).
   a. Hold the camshaft hexagonal part, and secure the camshaft.
   b. Never rotate crankshaft and camshaft separately, so as not to contact valve with piston in the following steps.
   5. Turn the camshaft sprocket (INT) to the most advanced position. Installation and removal of the camshaft sprocket (INT) must be done in the most advanced position for the following reasons:
   a. The sprocket and vane (camshaft coupling) are designed to spin and move within the range of a certain angle.
   b. With the engine stopped and the vane in the most retarded angle, it will not spin because it is locked to the sprocket side by the internal lock pin.
   c. If the camshaft sprocket bolts are turned in the situation described above, the lock pin will become damaged and cause malfunctions because of the increased horizontal load (cutting force) on the lock pin.
   6. Remove camshaft bracket (No. 1) by loosening the bolts in several steps, and removing them.
   7. Apply 44 psi (300kPa) or more of air pressure to the No. 1 journal oil hole of camshaft (INT) using an air gun. The air

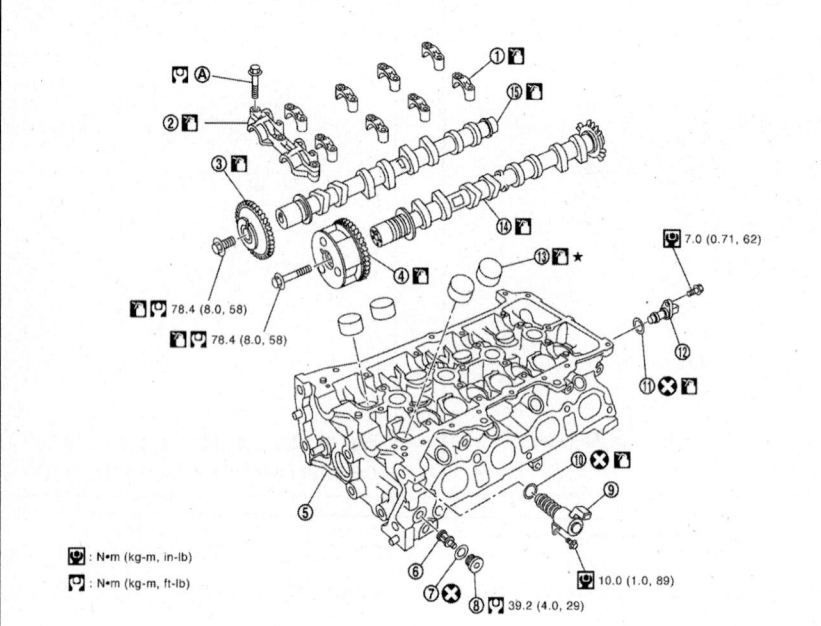

Fig. 45 Exploded view of camshafts and related component locations—1.6L engine (2011)

1. Camshaft bracket (No. 2 to 5)
2. Camshaft bracket (No.1)
3. Camshaft sprocket (EXH)
4. Camshaft sprocket (INT)
5. Cylinder head
6. Oil filter (for intake valve timing control)
7. Washer
8. Plug
9. Intake valve timing control solenoid valve
10. O-ring
11. O-ring
12. Camshaft position sensor (PHASE)
13. Valve lifter
14. Camshaft (INT)
15. Camshaft (EXH)
A. Refer to Installation procedure

37663_VERS_G0250

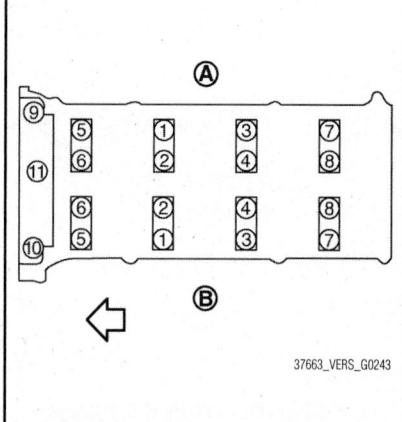

Fig. 46 Camshaft bracket bolt installation sequence, EXH side (A), INT side (B)— reverse for removal

37663_VERS_G0243

pressure is used to move the lock pin into the disengage position.

a. Apply and maintain the air pressure into the oil hole on the second groove from the front of camshaft thrust.

b. Attach the rubber nozzle narrowed to the top of the air gun to prevent air leakage from the oil hole. Securely apply the air pressure to the oil hole.

※※ WARNING

There are other oil holes in the side grooves. Never use the incorrect oil holes. Be sure not to damage the oil path with the tip of the air gun. Wipe all the oil off the air gun to prevent oil from being blown all over along with the air, and the area around the air gun should be wiped with a rag when applying air pressure.

※※ CAUTION

Eye protection should be worn as needed.

8. Hold the camshaft sprocket (INT) with hands, and then apply the power counterclockwise/clockwise alternatively.

a. Rotate the sprocket of the camshaft sprocket (INT) counterclockwise.

b. Perform the work while applying the air pressure to the oil hole.

c. If the lock pin is not released by hand, tap the camshaft sprocket (INT) lightly with a plastic hammer.

d. If the camshaft sprocket (INT) (1) is not rotated counterclockwise even if the above procedures are performed, check the air pressure and the oil hole position.

9. While doing the above, once you hear a click (the sound of the internal lock pin disengaging) from inside the camshaft sprocket (INT), start turning the camshaft sprocket (INT) in the counterclockwise direction in the most advanced angle position.

a. Keep the air pressure on.

b. If there is no click, as soon as the vane-side (camshaft side) starts moving independently of the sprocket, the lock pin has become disengaged.

c. Make sure that it is in the most advanced angle position by seeing if the stopper pin groove and the stopper pin hole are matched up.

10. Stop applying air pressure and holding the camshaft (INT).

11. Insert the stopper pin into the stopper pin holes in the camshaft sprocket (INT) and lock in the most advanced angle posi-

tion. No load is exerted on the stopper pin (spring reaction, etc.). Since it comes out easily, secure it with tape to prevent it from coming out.

12. Remove camshaft sprocket (INT) bolt and camshaft sprocket (INT).

a. Hold the camshaft hexagonal part, and then secure the camshaft.

b. Never rotate crankshaft and camshaft separately, so as not to contact valve with piston in the following steps.

13. Remove camshaft brackets (No. 2 to 5). The camshaft bracket (No. 1) has been already removed.

14. Remove camshaft (EXH).

15. Remove camshaft (INT).

16. Remove valve lifter. Identify installation positions, and store them without mixing them up.

17. Remove intake valve timing control solenoid valve.

18. Remove the alternator and bracket, if necessary, then remove the plug, washer and oil filter.

19. Discard the washer, do not reuse.

To install:

20. Install the oil filter and new washer.

a. Attach the oil filter to the plug.

b. Install it to the cylinder head.

21. Install the intake valve timing control solenoid valve. Insert it straight into the cylinder head, and tighten the bolts after positioning it securely.

22. Install the valve lifter. If it is reused, install in its original positions.

23. Put a matchmark for positioning the camshaft (INT) and the camshaft sprocket (INT). It will prevent the knock pin from engaging with the incorrect pin hole after installing the camshaft (INT) and the camshaft sprocket (INT).

a. Put the matchmarks on a line

extending from the knock pin position of camshaft (INT) front surface. Put the marks on the visible position with the camshaft sprocket installed.

b. Put the matching marks on a line extending from the knock pin hole position of camshaft sprocket (INT). Put the marks on the visible position with it installed to the camshaft.

24. Install the camshafts.

a. Distinction between the camshafts (INT and EXH) is performed by identifying the different shapes of the rear end.

b. Install the camshafts to the cylinder head so that the knock pins on the front end are in proper position.

➡ **For the placement of the cam nose, it is generally accepted that the camshaft is placed in the same direction.**

25. Install camshaft brackets (No. 2 to 5), aligning the identification marks on upper surface. Install so that identification mark can be correctly read when viewed from the INT side.

26. Tighten camshaft bracket bolts in the following steps, in numerical order.

a. Tighten No. 9 to 11 in numerical order to 17 inch lbs. (2 Nm).

b. Tighten No. 1 to 8 in numerical order to 17 inch lbs. (2 Nm).

c. Tighten all bolts in numerical order to 52 inch lbs. (6 Nm).

d. Tighten all bolts in numerical order to 92 inch lbs. (10 Nm).

27. Install the camshaft sprocket (INT) to the camshaft (INT):

a. Align the matchmark. Securely align the knock pin and the pin hole, and then install them.

b. Temporarily tighten the camshaft sprocket (INT) bolt on the front side of camshaft sprocket (INT).

c. Hold the camshaft hexagonal part to secure the camshaft and tighten the bolt.

28. Install the camshaft sprocket (EXH) to the camshaft (EXH) while aligning the matching mark and the matchmark of camshaft sprocket (EXH).

a. If the positions of knock pin and pin groove are not aligned, move the camshaft (EXH) slightly to correct these positions.

b. Hold the camshaft hexagonal part to secure the camshaft and tighten the bolt.

c. Make sure that the matchmark and each camshaft sprocket matchmark are in the correct location.

29. Install the timing chain.

30. Install the camshaft position sensor (PHASE) to the rear end of cylinder head. Tighten bolts with it seated completely.

31. Check and adjust valve clearance.

32. Installation of the remaining components is in the reverse order of removal.

1.8L Engine

See Figure 47.

1. Before servicing the vehicle, refer to the Precautions.

2. Release the fuel pressure.

3. Disconnect negative battery cable.

4. Remove the right front wheel.

5. Remove the right front fender protector.

6. Drain engine coolant.

7. Remove the intake manifold.

8. Remove the rocker cover.

9. Remove the fuel tube and fuel injector assembly.

10. Remove the front cover, timing chain, and related parts.

11. Remove the alternator.

12. Remove the camshaft position sensor (PHASE) from camshaft bracket.

13. Put matchmarks on the intake camshaft sprocket and the camshaft bracket.

14. Remove camshaft intake and exhaust sprockets.

15. Secure hexagonal part of camshaft with a wrench. Loosen camshaft sprocket bolts and remove the camshaft sprocket.

➡ **Never rotate crankshaft or camshaft while timing chain is removed. It causes interference between valve and piston.**

➡ **Never loosen the bolts with securing anything other than the camshaft hexagonal part or with tensioning the timing chain.**

16. Loosen the bolts in the reverse sequence of the tightening order.

17. Cut liquid gasket, and then remove the camshaft bracket.

➡ **Be careful not to damage the mating surface. A more adhesive liquid gasket is applied compared to previous types when shipped, so it should not be forced off the position not specified.**

18. Remove the camshafts.

19. Remove the valve lifters, if necessary. Identify installed positions, and store them without mixing them up.

To install:

20. Install the valve lifters in their original positions.

21. Install the camshafts.

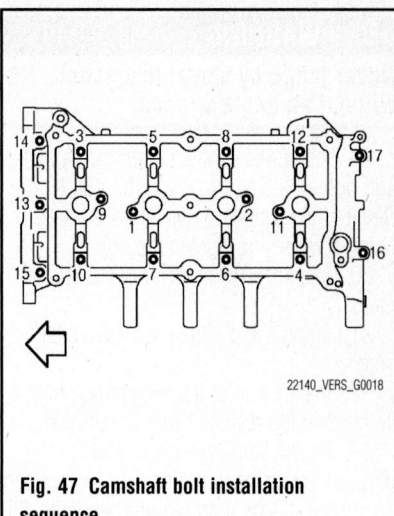

22140_VERS_G0018

Fig. 47 Camshaft bolt installation sequence

a. Clean the camshaft journal to remove any foreign material.

➡ **Distinguish between the intake and the exhaust by looking at the different shapes of the front and rear ends of the camshaft or using the identification colors.**

22. Install camshafts so that camshaft dowel pins on the front side are positioned.

23. Remove foreign material completely from the camshaft bracket backside and from the cylinder head installation face.

24. Apply liquid gasket to the camshaft bracket.

➡ **Use Genuine Silicone RTV Sealant (Tool No. WS39930000), or equivalent.**

25. Install camshaft bracket bolts.

a. Note the 2 types of M6 bolts: bolts No. 13, 14, and 15 in the figure have a thread length of 2.26 inches (57.5mm), and the remaining bolts have a thread length of 1.38 inches (35mm).

b. Tighten all the bolts in 3 steps:
• Step 1: 17 inch lbs. (2 Nm)
• Step 2: 52 inch lbs. (6 Nm)
• Step 3: 84 inch lbs. (10 Nm)

26. Install the intake camshaft sprocket to the intake camshaft.

➡ **When installing the intake camshaft sprocket, refer to the matchmark. Securely align the knock pin and the pin hole, and then install.**

27. Tighten the intake camshaft sprocket bolt to 26 ft. lbs. (35 Nm). Secure the hexagonal part of the intake camshaft using a wrench to tighten the bolt.

28. Turn 67° clockwise (angle tightening) using Tool No. KV10112100 (BT-8653-A), or equivalent.

29. Install the exhaust camshaft sprocket and tighten the bolt to 65 ft. lbs. (88 Nm). Secure the hexagonal part of the camshaft using a wrench to tighten the bolt.

30. Install the timing chain and related parts.

31. Inspect and adjust the valve clearance.

32. Installation of the remaining components is in the reverse order of removal.

33. Before starting engine, check oil/fluid levels including engine coolant and engine oil. If less than required quantity, fill to the specified level.

34. Turn the ignition switch ON (with engine stopped). With fuel pressure applied to fuel piping, check for fuel leakage at connection points.

35. Start the engine. With the engine speed increased, check again for fuel leakage at the connection points.

36. Run the engine to check for unusual noise and vibration.

37. Warm up engine thoroughly to make sure there is no leakage of fuel, exhaust gas, or any oils/fluids including engine oil and engine coolant.

38. Bleed air from passages in lines and hoses, such as in cooling system, as needed.

39. After cooling down the engine, check oil/fluid levels including engine oil and engine coolant. Refill to specified level, if necessary.

2012 Sedan

See Figures 48 through 50.

1. Before servicing the vehicle, refer to the Precautions.

➡**The rotation direction indicated in the procedure is as viewed from the engine front.**

2. Hold the bottom surface of the engine with a jack to remove the right engine mount assembly and the insulator.

3. Remove the rocker cover.

4. Place cylinder No. 1 at Top Dead Center (TDC) of its compression stroke:

   a. Rotate the crankshaft pulley clockwise and align the TDC mark (without paint mark) to the timing indicator on the front cover.

   b. Check that the matching marks on each of the camshaft sprockets are in the position. If not, rotate the crankshaft pulley one more turn to align the matching marks.

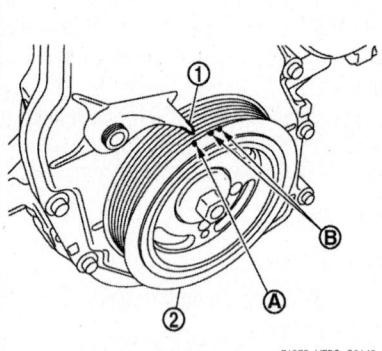

Fig. 48 View of crankshaft pulley (2) TDC mark (without paint mark) (A), timing indicator (1), and white paint mark (not used for service) (B)—1.6L engine (2012)

71075_VERS_G0145

   c. Paint matching marks on the timing chain links.

5. Remove the crankshaft pulley.

6. Remove the front cover.

7. Secure the plunger of the chain tensioner in the fully compressed position. Then, loosen the timing chain tension.

   a. Fully push down the lever of the chain tensioner from the plug hole, and then insert the stopper pin into the body side hole and secure the lever at the lowest position. The tab is released by fully pushing the lever down. As a result, the plunger can be moved.

➡**A 0.098 inch (2.5mm) hexagonal wrench may be used for a stopper pin.**

8. Turn the crankshaft pulley counterclockwise with the camshaft (EXH) held. Apply the tension to the timing chain, and then push the plunger into the inside of chain tensioner.

➡**Hold the camshaft hexagonal part, and then secure the camshaft.**

   a. Pull out the stopper pin of chain tensioner side from plug hole. Lift the lever up to align its hole position with the hole of the body.

- When the lever hole is aligned with the body hole position, the plunger is secured
- When the protrusion parts of the plunger ratchet and the tab face each other, both hole positions are not aligned. At that time, correctly engage them and align these hole positions by slightly moving the plunger.

   b. Insert the stopper pin into the body hole through the lever hole, and then secure the lever at the upper position.

9. Remove the timing chain.

10. Remove the camshaft sprocket (EXH).

   a. Hold the camshaft hexagonal part, and then secure the camshaft.

   b. Never rotate the crankshaft and camshaft separately, so as not to contact valve with piston.

➡**The timing chain, with the front cover installed, is not disengaged from the crankshaft sprocket and it is not dropped into the front cover. Therefore, the timing chain tension holding device is not necessary.**

11. Turn the camshaft sprocket (INT) to the most advanced position.

Consider the following warnings:

- Installation and removal of the camshaft sprocket (INT) must be done in the most advanced position. Make sure to follow the procedure exactly.
- The sprocket and vane (camshaft coupling) are designed to spin and move within the range of a certain angle.
- With the engine stopped and the vane in the most retarded angle, it will not spin because it is locked to the sprocket side by the internal lock pin.
- If the camshaft sprocket mounting bolts are turned in the situation described above, the lock pin will become damaged and cause malfunctions because of the increased horizontal load (cutting force) on the lock pin.

   a. Remove the camshaft bracket (No. 1). Loosen the bolts in several steps, and then remove them.

   b. Apply air pressure of 44 psi (300 kPa) or more to the No. 1 journal oil hole of camshaft (INT) using an air gun.

   c. Apply air pressure into the oil hole on the second groove from the front of camshaft thrust.

➡**Supply air pressure until the procedure says to remove it.**

   d. Attach the narrowed rubber nozzle to the top of the air gun to prevent air leakage from the oil hole. Securely apply the air pressure to the oil hole.

Consider the following warnings:

- There are other oil holes in the side grooves. Never use the incorrect oil holes
- Be sure not to damage the oil path with the tip of the air gun
- Wipe all the oil off the air gun to prevent oil from being blown all over along with the air, and the area around the air gun should be wiped with a rag when applying air pressure

➡**The air pressure is used to move the lock pin into the disengage position.**

a. Hold the camshaft sprocket (INT) with hands, and then apply power counterclockwise/clockwise alternately.

- Finally rotate the sprocket of the camshaft sprocket (INT) counterclockwise
- Perform the work while continuously applying the air pressure to the oil hole
- If the lock pin is not released, tap the camshaft sprocket (INT) lightly with a plastic hammer
- If the camshaft sprocket (INT) is not rotated counterclockwise even when the above procedures are performed, check the air pressure and the oil hole position

b. Once you hear a click (the sound of the internal lock pin disengaging) from inside the camshaft sprocket (INT), start turning the camshaft sprocket (INT) in the counterclockwise direction in the most advanced angle position.

- Keep the air pressure on
- If there is no click, as soon as the vane side (camshaft side) starts moving independently of the sprocket, the lock pin has become disengaged
- Check that it is in the most advanced angle position by seeing if the stopper pin groove and the stopper pin hole are matched up

c. Stop applying air pressure and release the camshaft (INT).

d. Insert the stopper pin into the stopper pin holes in the camshaft sprocket (INT) and lock in the most advanced angle position.

- No load is exerted on the stopper pin (spring reaction, etc.). Since it comes out easily, secure it with tape to prevent it from falling out
- The stopper pin may be a 0.098 inch (2.5mm) hexagonal wrench with the length of insertion of approximately 0.59 inch (15mm).

12. Remove the camshaft sprocket (INT):

a. Keeping the camshaft hexagonal part in place with a wrench, loosen the mounting bolts for the camshaft sprocket (INT).

Consider the following warnings:

- Never drop the stopper pin
- Use tape on the stopper pin so it does not come out
- Take care not to drop the sprocket
- Never disassemble the sprocket or loosen the 3 mounting bolts.

a. While removing the camshaft sprocket (INT), if you have taken out the stopper pin and the lock pin has been rejoined in the most retarded angle, do the following to restore it.

- Install the camshaft (INT) and tighten the mounting bolts enough to prevent air from leaking out

✳✳ WARNING

The internal lock pin may be damaged, so keep the torque on the mounting bolts to the minimum required to prevent air from escaping.

- Apply air pressure, disengage the lock pin, and turn the vane to the most advanced angle position
- Insert the stopper pin
- Remove the camshaft sprocket (INT) from the camshaft

13. Remove camshaft brackets (No. 2 to 5).

a. Loosen bolts in several steps in the reverse order of the tightening sequence.

➡**The camshaft bracket (No. 1) has been already removed.**

14. Remove the camshaft (EXH).
15. Remove the camshaft (INT).
16. Remove the valve lifters. Identify

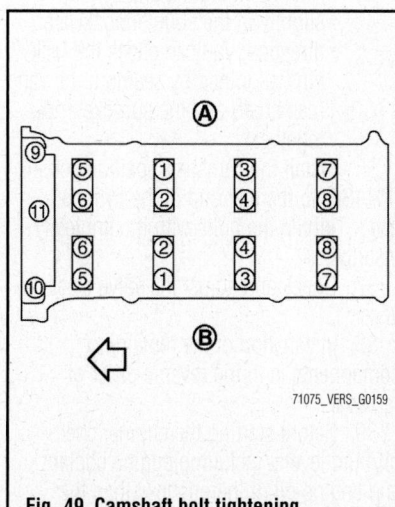

Fig. 49 Camshaft bolt tightening sequence—1.6L engine (2012)

installation positions, and store lifters without mixing them up.

17. Remove the intake valve timing control solenoid valve.

18. Remove the alternator and bracket, remove the plug, and then remove the oil filter.

19. Remove the exhaust valve timing control solenoid valve.

20. Remove the plug on the exhaust valve timing control solenoid valve and the oil filter.

To install:

➡**Do not reuse O-rings or washers.**

21. Install the oil filter for the intake and exhaust valve timing control solenoid valves.

- The oil filter and washer are assembled to the plug, and then installed into the cylinder head.

22. Install the intake and exhaust valve timing control solenoid valves.

a. Insert it straight into the cylinder head.

b. Tighten the bolts.

23. Install the valve lifters. If they are reused, install them in the original positions.

24. Put a matching mark for positioning the camshaft (INT) and the camshaft sprocket (INT).

➡**The matching mark helps prevent the knock pin from engaging with the incorrect pin hole after installing the camshaft (INT) and the camshaft sprocket (INT).**

a. Put the matching marks on a line extending from the knock pin position of camshaft (INT) front surface. Put the marks on the visible position with the camshaft sprocket installed.

b. Put the matching marks on a line extending from the knock pin hole position of camshaft sprocket (INT). Put the marks on the visible position with it installed to the camshaft.

25. Put a matching mark for positioning the camshaft (EXH) and the camshaft sprocket (EXH).

➡**The matching mark helps prevent the knock pin from engaging with the incorrect pin hole after installing the camshaft (INT) and the camshaft sprocket (EXH).**

a. Put the matching marks on a line extending from the knock pin position of camshaft (EXH) front surface. Put the marks on the visible position with the camshaft sprocket installed.

b. Put the matching marks on a line extending from the knock pin hole position of camshaft sprocket (EXH). Put the marks on the visible position with it installed to the camshaft.

26. Install the camshafts.

a. Note that the camshafts (INT and EXH) have different shapes at the rear.

b. Install the camshafts to the cylinder head so that knock pins on the front end are positioned.

27. Install the camshaft brackets (No. 2 to 5) aligning the identification marks on the upper surface. Install so that identification mark can be correctly read when viewed from the intake side.

28. Tighten the mounting bolts of the camshaft brackets in the following steps, in the proper tightening sequence.

a. Tighten No. 9 to 11 to 17 inch lbs. (2 Nm).

b. Tighten No. 1 to 8 to 17 inch lbs. (2 Nm).

c. Tighten all bolts to 52 inch lbs. (6 Nm).

d. Tighten all bolts 92 inch lbs. (10 Nm).

29. Install the camshaft sprocket (INT and EXH) to the camshaft (INT and EXH).

a. Refer to the matching mark added earlier.

b. Securely align the knock pin and the pin hole, and then install.

30. Tighten the camshaft sprocket mounting bolt (INT and EXH). Hold the camshaft hexagonal part, using a suitable tool to secure the camshaft.

31. Install the timing chain to the camshaft sprocket (INT and EXH) while aligning the matching mark (marked when timing chain is removed) and the pink link of camshaft sprocket (INT and EXH).

➡If the positions of the knock pin and pin groove are not aligned, move the camshaft (EXH) slightly to correct these positions.

32. Pull out the stopper pin, and then apply the tension to the timing chain by rotating the crankshaft pulley clockwise slightly.

33. Pull out the stopper pin of the chain tensioner.

34. Install the front cover.

35. Return the camshaft sprocket (INT) to the most retarded position.

a. Remove the stopper pin from the camshaft sprocket (INT).

b. Turn the crankshaft pulley slowly clockwise and return the camshaft sprocket (INT) to the most retarded angle position.

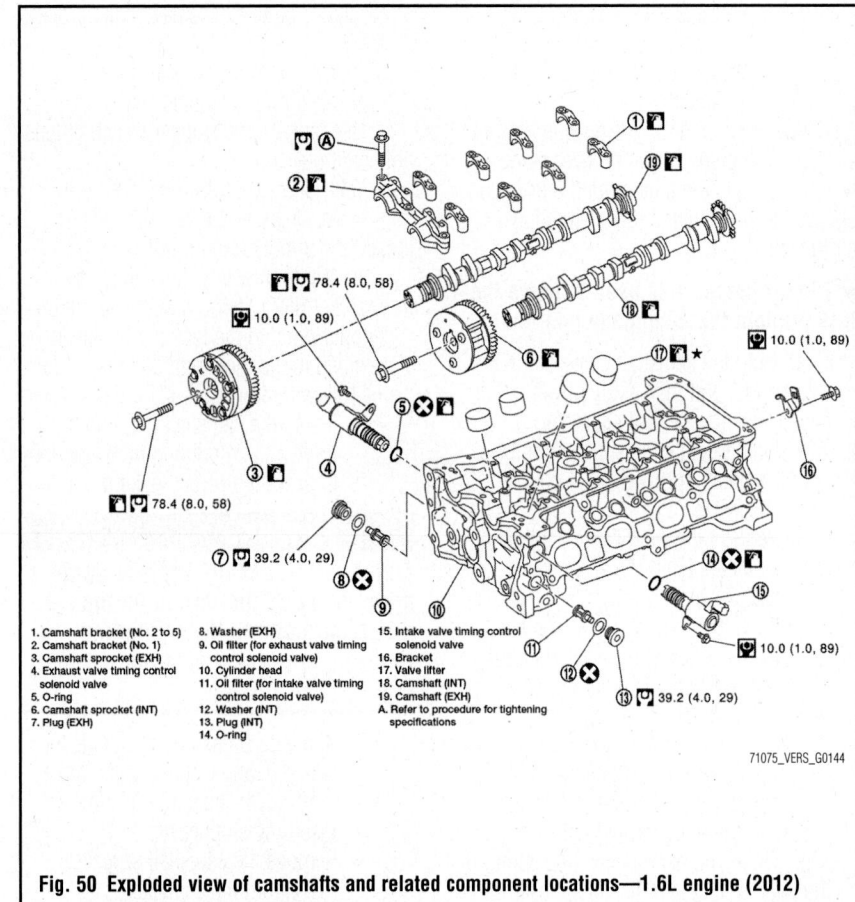

1. Camshaft bracket (No. 2 to 5)
2. Camshaft bracket (No. 1)
3. Camshaft sprocket (EXH)
4. Exhaust valve timing control solenoid valve
5. O-ring
6. Camshaft sprocket (INT)
7. Plug (EXH)
8. Washer (EXH)
9. Oil filter (for exhaust valve timing control solenoid valve)
10. Cylinder head
11. Oil filter (for intake valve timing control solenoid valve)
12. Washer (INT)
13. Plug (INT)
14. O-ring
15. Intake valve timing control solenoid valve
16. Bracket
17. Valve lifter
18. Camshaft (INT)
19. Camshaft (EXH)
A. Refer to procedure for tightening specifications

71075_VERS_G0144

Fig. 50 Exploded view of camshafts and related component locations—1.6L engine (2012)

- When first turning the crankshaft, the camshaft sprocket (INT) will turn. Once it is turned more, and the vane (camshaft) also turns, then it has reached the most retarded angle position
- The most retarded angle position can be checked by seeing if the stopper pin groove is shifted clockwise
- After spinning the crankshaft slightly in the counterclockwise direction, you can check the lock pin has joined by seeing if the vane (camshaft) and the sprocket move together

36. Install the camshaft position sensor (PHASE) to the rear end of the cylinder head. Tighten the bolts with it completely inserted.

37. Check and adjust the valve clearance.

38. Installation of the remaining components is in the reverse order of removal.

39. Before starting the engine, check oil/fluid levels, including engine coolant and engine oil. If there is less than the required quantity, fill to the specified level.

40. Turn the ignition switch ON (with engine stopped). With fuel pressure applied to fuel piping, check for fuel leakage at connection points.

41. Start the engine. With the engine speed increased, check again for fuel leakage at the connection points.

42. Run the engine to check for unusual noise and vibration.

➡If the hydraulic pressure inside the timing chain tensioner drops after removal and installation, slack in the guide may generate a pounding noise during and just after the engine start. However, this is normal. The noise should stop after the hydraulic pressure rises.

43. Warm up the engine thoroughly to make sure there is no leakage of fuel, exhaust gas, or any oils/fluids including the engine oil and engine coolant.

44. Bleed the air from the passages in the lines and hoses, such as in the cooling system.

45. After cooling down the engine, again check the oil/fluid levels including the engine oil and engine coolant. Refill to specified level, if necessary.

CRANKSHAFT FRONT SEAL

REMOVAL & INSTALLATION

See Figure 51.

1. Before servicing the vehicle, refer to the Precautions.
2. Remove the right front fender protector.
3. Remove the drive belt.
4. Remove the crankshaft pulley.
5. Remove the front oil seal using a suitable tool.

✳✳ WARNING

Be careful not to damage the front cover or crankshaft.

To install:

6. Apply new engine oil to the new front oil seal joint surface and seal lip.
7. Install the front oil seal.
8. Using a suitable drift, press-fit until the height of the front oil seal is level with the mounting surface:
   - 1.8L engine outer diameter: 57 mm (2.24 inch)
   - 1.8L engine inner diameter: 45 mm (1.77 inch)
   - 1.6L engine outer diameter: 50 mm (1.97 inch)
   - 1.6L engine inner diameter: 44 mm (1.73 inch)

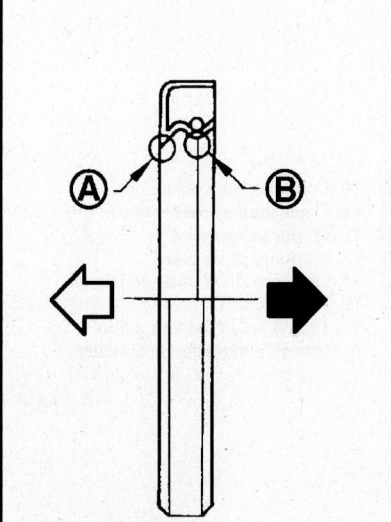

A. Dust seal lip
B. Oil seal lip
White arrow: Engine outside
Black arrow: Engine inside

22140_VERS_G0026

Fig. 51 Front oil seal lip installation positioning

✳✳ WARNING

Press-fit the oil seal straight to avoid causing burrs or tilting.

9. Installation of the remaining components is in the reverse order of removal.

CRANKSHAFT PULLEY

REMOVAL & INSTALLATION

1.6L Engine

See Figures 52 and 53.

1. Before servicing the vehicle, refer to the Precautions.
2. Disconnect the negative battery cable.
3. Remove the right front fender protector.
4. Remove the drive belt.
5. Secure crankshaft pulley using a suitable tool.

6. Loosen and pull out crankshaft pulley bolts.

✳✳ WARNING

Never remove the bolts as they are used as a supporting point for the pulley puller.

7. For 2011 models, attach Tool No. KV11103000, or equivalent, in the M6 thread holes on the crankshaft pulley, and remove the crankshaft pulley.
8. For 2012 models, attach Tool No. KV11123000, or equivalent, in the M6 thread holes on the crankshaft pulley, and remove the crankshaft pulley.

To install:

9. Insert the crankshaft pulley by aligning with the crankshaft key.
   Consider the following warnings:
   - When inserting the crankshaft pulley with a plastic hammer, tap on its center portion (not the circumference).

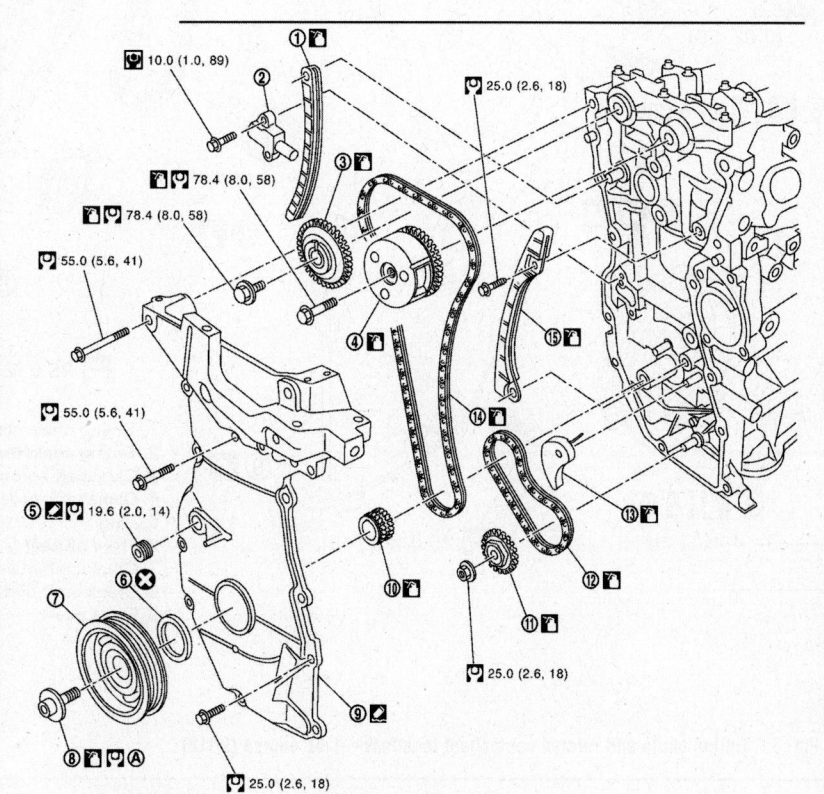

1.	Timing chain slack guide	2.	Chain tensioner (for timing chain)	3.	Camshaft sprocket (EXH)
4.	Camshaft sprocket (INT)	5.	Plug	6.	Front oil seal
7.	Crankshaft pulley	8.	Crankshaft pulley bolt	9.	Front cover
10.	Crankshaft sprocket	11.	Oil pump sprocket	12.	Oil pump drive chain
13.	Chain tensioner (for oil pump drive chain)	14.	Timing chain	15.	Timing chain tension guide
A.	Refer to Installation procedure				

37663_VERS_G0285

Fig. 52 Timing chain and related component locations—1.6L engine (2011)

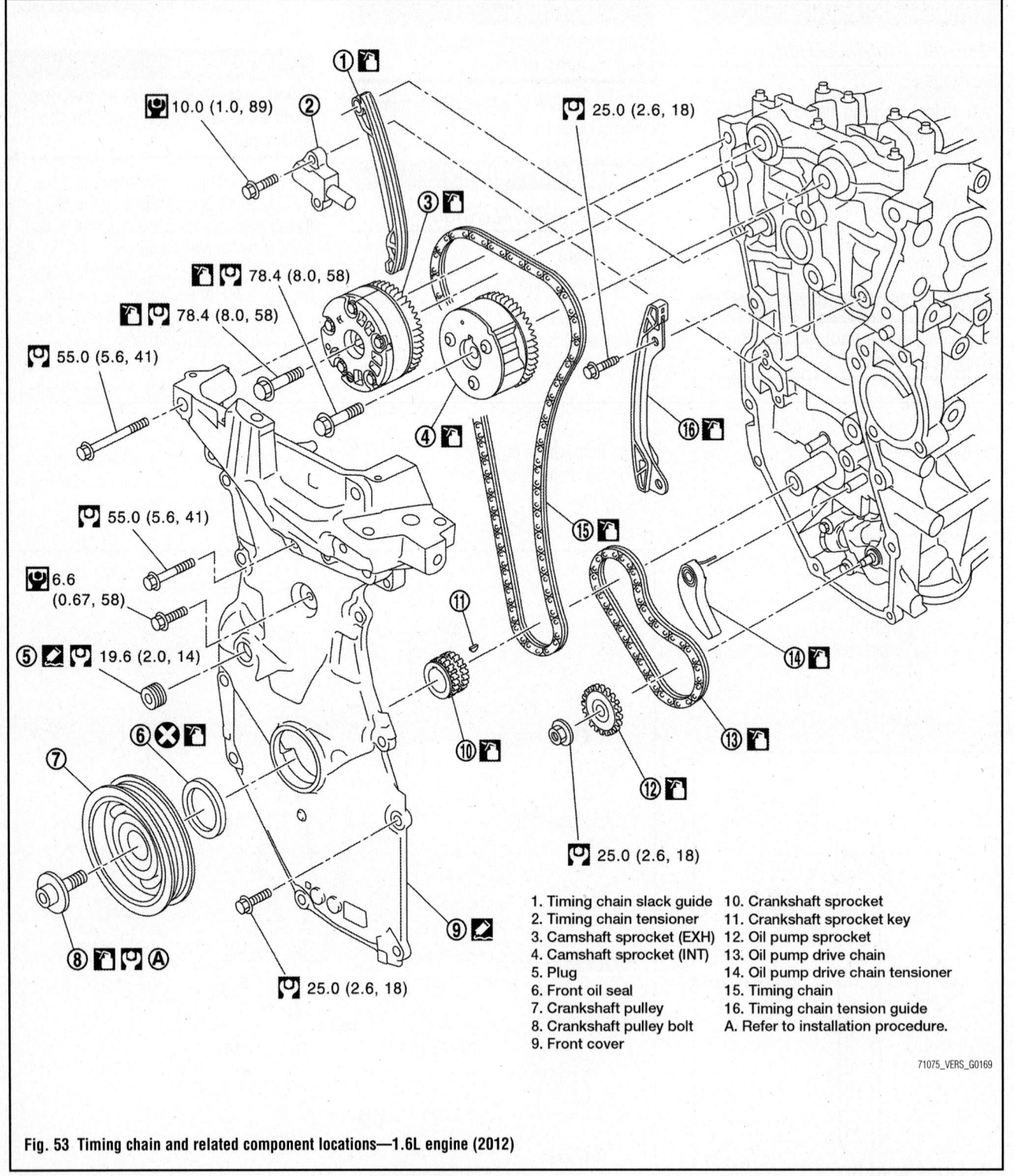

1. Timing chain slack guide
2. Timing chain tensioner
3. Camshaft sprocket (EXH)
4. Camshaft sprocket (INT)
5. Plug
6. Front oil seal
7. Crankshaft pulley
8. Crankshaft pulley bolt
9. Front cover
10. Crankshaft sprocket
11. Crankshaft sprocket key
12. Oil pump sprocket
13. Oil pump drive chain
14. Oil pump drive chain tensioner
15. Timing chain
16. Timing chain tension guide
A. Refer to installation procedure.

71075_VERS_G0169

Fig. 53 Timing chain and related component locations—1.6L engine (2012)

- Never damage the front oil seal lip section.

10. Secure the crankshaft pulley with a suitable tool.

11. Apply new engine oil to thread and seat surfaces of crankshaft pulley bolt.

12. Tighten the crankshaft pulley bolt to 26 ft. lbs. (35 Nm).

13. Put a paint mark on the crankshaft pulley, mating with any one of six easy to recognize angle marks on the crankshaft bolt flange.

14. Turn another 60° clockwise (angle tightening). Check the tightening angle with movement of one angle mark.

15. Make sure that crankshaft turns smoothly by rotating by hand clockwise.

16. Install the drive belt.

17. Install the front fender protector (RH).

18. Connect the negative battery cable.

1.8L Engine

See Figure 54.

1. Before servicing the vehicle, refer to the Precautions.

2. Disconnect the negative battery cable.

3. Remove the right front fender protector.

4. Remove the drive belt.

5. Hold the crankshaft pulley using Tool No. KV10109300, or equivalent, and loosen the crankshaft pulley bolt. Locate the bolt seating surface at 0.39 inch (10mm) from its original position. Do not remove the crankshaft pulley bolt as it will be used as a supporting point for the pulley puller.

6. Attach a pulley puller in the M6 thread holes on crankshaft pulley, and remove crankshaft pulley.

To install:

❋❋ WARNING

Never damage the front oil seal lip section. If necessary, use a plastic hammer to tap on the center portion of the crankshaft pulley to seat the crankshaft pulley. Do not tap on the circumference.

7. Apply new engine oil to the thread and seat surfaces of the crankshaft pulley bolt.

8. Secure the crankshaft pulley using Tool No. KV10109300, or equivalent.

9. Tighten the crankshaft pulley bolt in 2 steps:

   a. Step 1: Tighten to 22 ft. lbs. (29 Nm).

   b. Step 2: Tighten an additional 60° clockwise.

➥**For angle tightening, put a paint mark on crankshaft pulley matching with any one of six easy to recognize angle marks on crankshaft pulley bolt flange. Check the tightening angle with movement of one angle mark.**

10. Make sure that crankshaft rotates clockwise smoothly.

11. Installation of the remaining components is in the reverse order of removal.

CYLINDER HEAD

REMOVAL & INSTALLATION

1.6L Engine

2011 Model

See Figures 55 and 56.

1. Before servicing the vehicle, refer to the Precautions.

2. Release the fuel pressure.

3. Disconnect the negative battery cable.

4. Drain engine coolant and engine oil.

❋❋ WARNING

Do not spill engine coolant or engine oil on the drive belt.

5. Remove the right front fender protector.

6. Remove the alternator.

7. Remove the front exhaust pipe.

8. Remove the exhaust manifold.

9. Remove the intake manifold.

10. Remove the fuel tube and fuel injector assembly.

11. Remove the water outlet.

12. Remove the drive belt.

13. Remove the timing chain front cover.

14. Remove the camshaft.

15. Remove the air cleaner.

16. Remove the cylinder head bolts in reverse order of tightening sequence.

17. Remove the cylinder head gasket.

To install:

18. Install a new cylinder head gasket.

19. Tighten the cylinder head bolts using the following procedure.

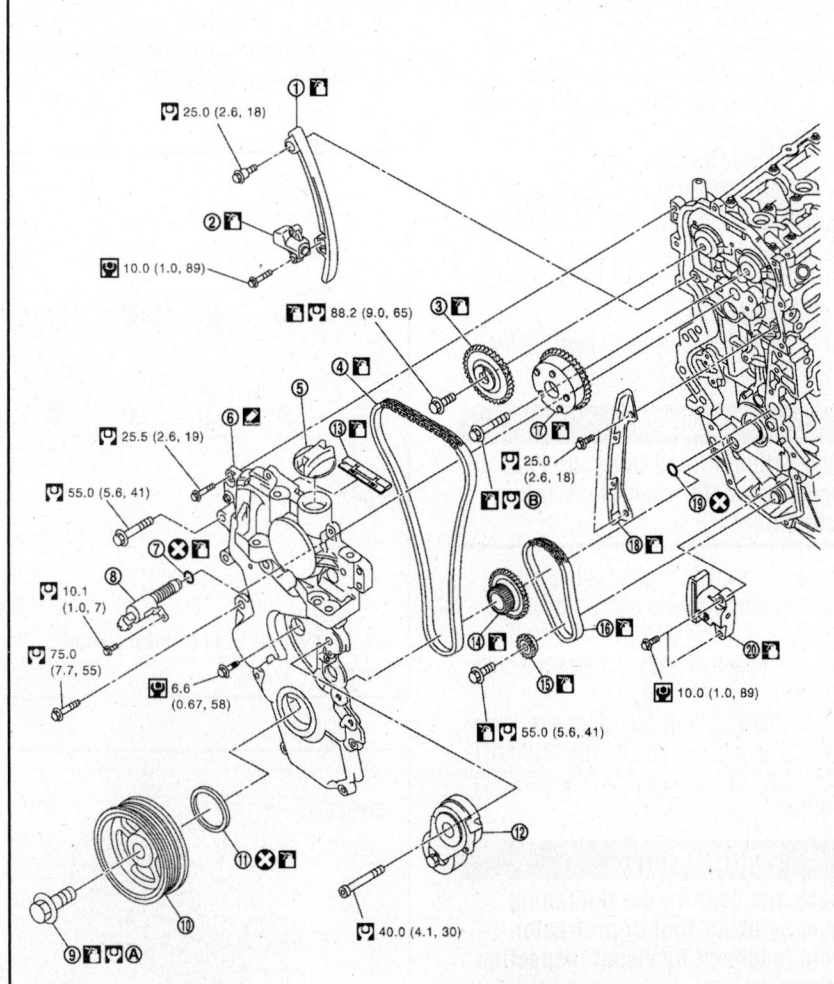

25.0 (2.6, 18)

10.0 (1.0, 89)

88.2 (9.0, 65)

25.5 (2.6, 19)

25.0 (2.6, 18)

55.0 (5.6, 41)

10.1 (1.0, 7)

75.0 (7.7, 55)

6.6 (0.67, 58)

55.0 (5.6, 41)

10.0 (1.0, 89)

40.0 (4.1, 30)

1. Timing chain slack guide
2. Timing chain tensioner
3. Camshaft sprocket (EXH)
4. Timing chain
5. Oil filler cap
6. Front cover
7. O-ring
8. Intake valve timing control solenoid valve
9. Crankshaft pulley bolt
10. Crankshaft pulley

11. Front oil seal
12. Drive belt auto-tensioner
13. Timing chain tension guide (front cover side)
14. Crankshaft sprocket
15. Oil pump sprocket
16. Oil pump drive chain
17. Camshaft sprocket (INT)
18. Timing chain tension guide
19. O-ring
20. Chain tensioner (for oil pump)

37663_VERS_G0284

Fig. 54 Timing chain and related component locations—1.8L engine

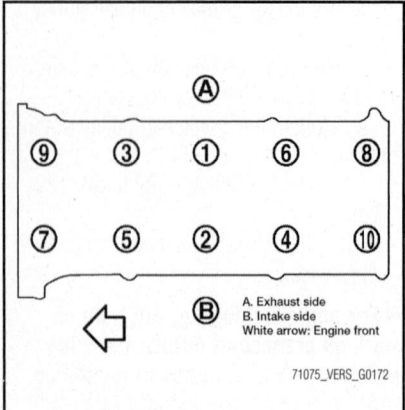

Fig. 55 Cylinder head bolt installation sequence—1.6L engine

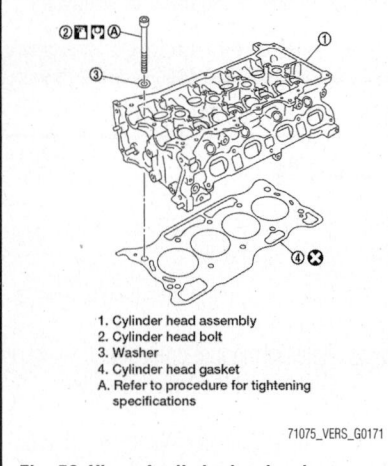

1. Cylinder head assembly
2. Cylinder head bolt
3. Washer
4. Cylinder head gasket
A. Refer to procedure for tightening specifications

71075_VERS_G0171

Fig. 56 View of cylinder head and gasket—1.6L engine

a. Apply new engine oil to threads and seating surface of bolts. If cylinder head bolts re-used, check their outer diameters before installation.

b. Tighten all bolts in the specified order to 30 ft. lbs. (40 Nm).

c. Turn all bolts an additional 60° clockwise (angle tightening) using tool No. KV10112100 (BT-8653-A), or equivalent specified order.

d. Check and confirm the tightening angle by using tightening tool or protractor. Avoid judgment by visual inspection without the tool.

e. Completely loosen all bolts in reverse order of tightening sequence.

f. Retighten all bolts in the specified order to 30 ft. lbs. (40 Nm).

g. Turn all bolts an additional 75° clockwise (angle tightening) using tool No. KV10112100 (BT-8653-A) in the specified order. Confirm the tightening angle.

h. Turn all bolts 75° clockwise again, as above.

20. Installation of the remaining components is in the reverse order of removal.

2012 Model

➡**When removing components such as hoses, tubes/lines, etc., cap or plug openings to prevent fluid from spilling.**

1. Before servicing the vehicle, refer to the Precautions.

2. Release the fuel pressure.

3. Remove air duct.

4. Remove the fuel tube and fuel injector assembly..

5. Remove the water outlet.

6. Remove the exhaust manifold..

7. Remove the timing chain front cover and timing chain.

8. Remove the camshaft.

9. Remove the cylinder head, loosening the bolts in the reverse order of the tightening sequence.

10. Remove the cylinder head gasket.

To install:

11. Install a new cylinder head gasket.

12. Tighten the cylinder head bolts.

❄❄ WARNING

If the cylinder head bolts are reused, check their outer diameters before installation.

a. Apply new engine oil to threads and seating surfaces of the mounting bolts.

b. Tighten all the cylinder head bolts to 30 ft. lbs. (40 Nm).

c. Turn all bolts 60° clockwise (angle tightening) using Tool KV10112100 (BT-8653-A), or equivalent, in the specified order.

❄❄ WARNING

Check and confirm the tightening angle by using Tool or protractor. Avoid judgment by visual inspection without the tool.

d. Completely loosen all of the bolts in the reverse order of tightening.

e. Tighten all bolts to 30 ft. lbs. (40 Nm) in tightening sequence.

f. Turn all bolts 75° clockwise (angle tightening) in tightening sequence.

g. Turn all bolts 75° clockwise again (angle tightening) in tightening sequence.

13. Installation of the remaining components is in the reverse order of removal.

1.8L Engine

See Figures 57 and 58.

1. Before servicing the vehicle, refer to the Precautions.

2. Release the fuel pressure.

3. Drain engine coolant and engine oil.

4. Remove the right front fender protector.

5. Remove the drive belt.

6. Remove the exhaust manifold.

7. Remove the intake manifold.

8. Remove the water outlet.

9. Remove the fuel tube and fuel injector assembly.

10. Remove the rocker cover.

11. Remove the timing chain front cover.

12. Remove the camshaft.

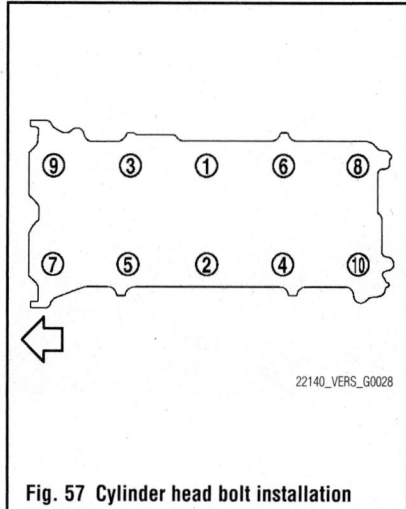

Fig. 57 Cylinder head bolt installation sequence

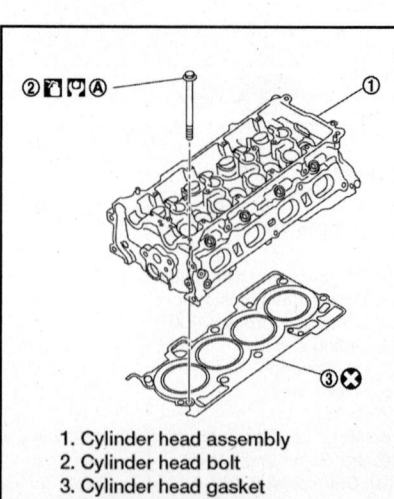

1. Cylinder head assembly
2. Cylinder head bolt
3. Cylinder head gasket
A. Refer to procedure for tightening specifications

71075_VERS_G0173

Fig. 58 View of cylinder head and gasket—1.8L engine

13. Using a TORX® socket (size E18), remove the cylinder head, loosening the bolts in reverse order of the tightening sequence.

14. Remove the cylinder head gasket.

To install:

15. Install a new cylinder head gasket.

16. Apply new engine oil to threads and seating surface of bolts. If cylinder head bolts re-used, check their outer diameters before installation.

17. Install the cylinder head and tighten the cylinder head bolts according to the tightening sequence illustrated:

   a. Step 1: Tighten to 30 ft. lbs. (40 Nm).

   b. Step 2: Tighten 100° clockwise.

   c. Step 3: Loosen all bolts to 0 ft. lbs. (0 Nm) in the reverse order of tightening.

   d. Step 4: Tighten to 30 ft. lbs. (40 Nm).

   e. Step 5: Tighten 100° clockwise.

   f. Step 6: Tighten an additional 100° clockwise.

❊❊ WARNING

Check and confirm the tightening angle by using Tool No. KV10112100 (BT-8653-A) or protractor. Never judge by visual inspection without the tool.

18. Installation of the remaining components is in the reverse order of removal.

EXHAUST MANIFOLD

REMOVAL & INSTALLATION

1.6L Engine

2011 Model

See Figures 59 and 60.

1. Before servicing the vehicle, refer to the Precautions.

2. Disconnect the negative battery cable.

3. Remove the cowl top.

4. Remove the heat insulator.

5. Remove the front exhaust pipe.

6. Remove the harness bracket of air fuel ratio sensor 1 from the cylinder head.

7. Remove the exhaust manifold covers.

8. Remove the air fuel ratio sensor 1. Handle it carefully and avoid impacts. Use Tool (Tool No. KV10117100) to remove air fuel ratio sensor 1.

➡**The exhaust manifold can be removed and installed without removing the air fuel ratio sensor 1 (disassembly of harness connector is necessary).**

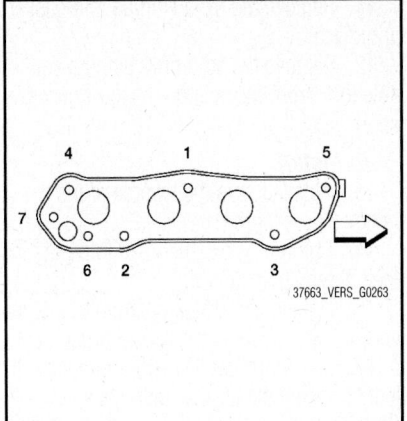

Fig. 59 Exhaust manifold bolt installation sequence—reverse for removal

9. Remove exhaust manifold side bolt of exhaust manifold stay.

10. Remove exhaust manifold.

11. Remove the gasket.

12. Cover the engine openings to avoid entry of foreign materials.

13. Remove the exhaust manifold cover from the back of the exhaust manifold.

To install:

14. Installation is in the reverse order of removal.

15. Tighten the exhaust manifold nuts to specification in 2 stages in numerical order.

16. Before installing a new air fuel ratio sensor 1, clean the exhaust tube threads using suitable tool and approved anti-seize lubricant. Use Tool (Tool No. KV10117100) to install the air fuel ratio sensor 1.

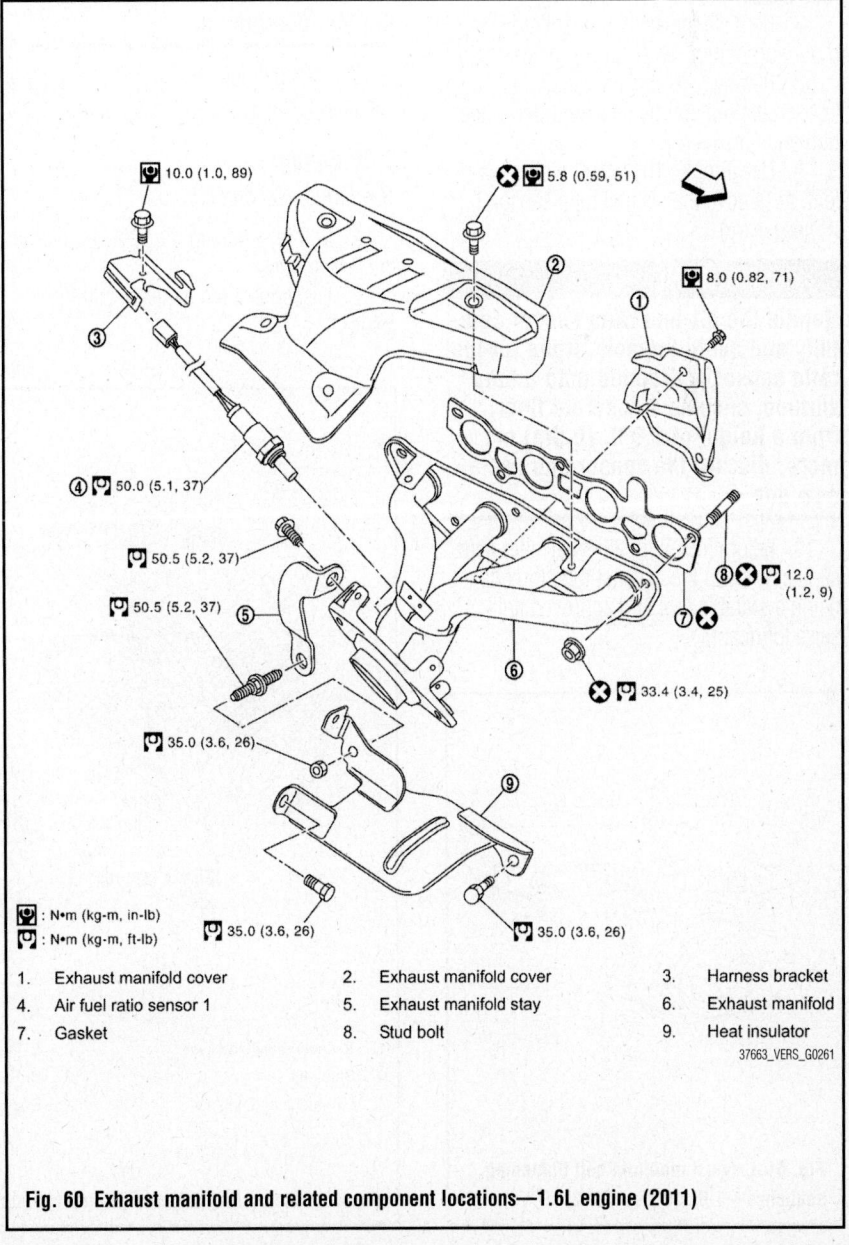

1. Exhaust manifold cover	2. Exhaust manifold cover	3. Harness bracket
4. Air fuel ratio sensor 1	5. Exhaust manifold stay	6. Exhaust manifold
7. Gasket	8. Stud bolt	9. Heat insulator

Fig. 60 Exhaust manifold and related component locations—1.6L engine (2011)

2012 Model

See Figure 61.

1. Before servicing the vehicle, refer to the Precautions.
2. Remove the air duct (inlet), air duct, and air cleaner assembly.
3. Remove the exhaust center tube and front tube.
4. Remove the air-fuel ratio sensor harness bracket from the cylinder head on the right rear side.
5. Remove the exhaust manifold cover.
6. Disconnect the harness from air-fuel ratio sensor 1.
7. Remove the exhaust manifold side mounting bolt of the exhaust manifold stay.
8. Remove the exhaust manifold. Loosen the nuts in the reverse order of the tightening sequence.
9. Use Tool KV10117100, or equivalent, to remove the air-fuel ratio sensor 1 (if necessary).

10. Before installing a new air-fuel ratio sensor 1, clean the exhaust tube threads using a suitable tool and approved anti-seize lubricant.

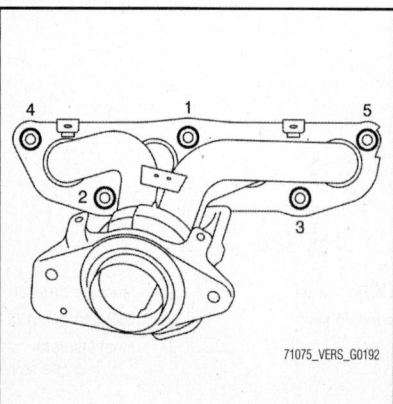

Fig. 61 Exhaust manifold bolt tightening sequence—1.6L engine (2012)

11. Remove the exhaust manifold gasket and discard.
12. Remove the stud bolt, using a suitable tool, from the cylinder head (if necessary).

To install:

13. Installation is the reverse of the removal procedure.
14. Tighten the exhaust manifold nuts in the tightening sequence.
15. Tighten the exhaust manifold nuts to specification again in numerical order.
16. Use Tool KV10117100, or equivalent, to install the air-fuel ratio sensor 1 (if removed).

17. Start the engine and raise the engine speed to check for any exhaust leaks.

1.8L Engine

See Figures 62 and 63.

1. Before servicing the vehicle, refer to the Precautions.
2. Disconnect the negative battery cable.

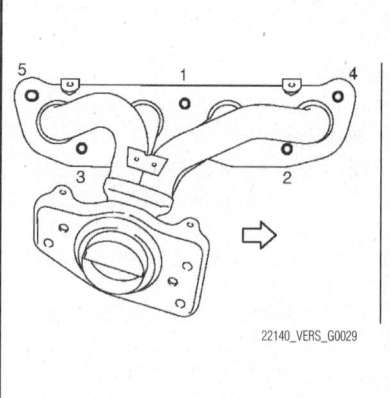

Fig. 62 Exhaust manifold bolt installation sequence—reverse for removal

3. Remove the cowl top.
4. Remove the front exhaust pipe.
5. Remove the exhaust manifold cover.
6. Remove the A/F sensor 1, using Tool No. KV991J0050 (J-44626), or equivalent. Handle it carefully and avoid impacts.
7. Remove the exhaust manifold side bolt of the exhaust manifold stay.
8. Loosen the nuts and remove the exhaust manifold.
9. Remove the gasket.
10. Cover the engine openings to avoid entry of foreign materials.

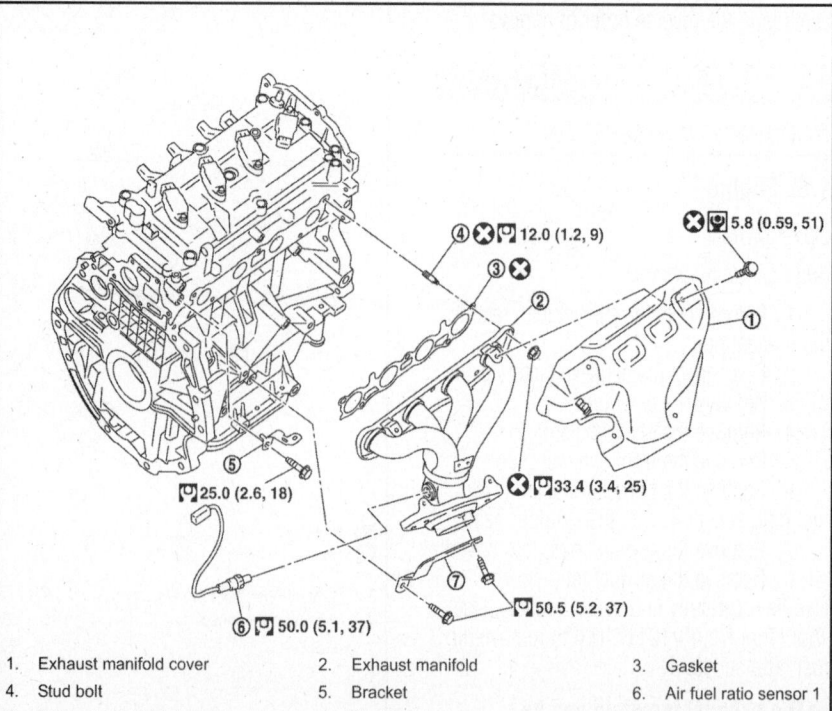

1.	Exhaust manifold cover	2.	Exhaust manifold
4.	Stud bolt	5.	Bracket
7.	Exhaust manifold stay	⟵	Engine front

3. Gasket
6. Air fuel ratio sensor 1

Fig. 63 Exhaust manifold and related component locations—1.8L engine

To install:

11. Install the exhaust manifold gasket.

12. Tighten the exhaust manifold nuts to specification in 2 stages.

13. Install the exhaust manifold stay.

14. Before installing a new A/F ratio sensor, clean the exhaust tube threads using suitable tool and approved anti-seize lubricant.

15. Install the A/F ratio sensor 1, using tool No. KV991J0050 (J-44626), or equivalent.

> ### ※※ WARNING
>
> **Do not over-tighten the air-fuel ratio sensor 1. Doing so may damage the air-fuel ratio sensor 1, resulting in the MIL coming on.**

16. Installation of the remaining parts is in the reverse order of removal.

INTAKE MANIFOLD

REMOVAL & INSTALLATION

1.6L Engine

2011 Model

See Figures 64 through 66.

1. Before servicing the vehicle, refer to the Precautions.

2. Disconnect the negative battery cable.

3. Remove the air duct (inlet) and air duct (between air cleaner case and electric throttle control actuator).

4. Remove the reservoir tank.

5. Remove the oil level gauge. Cover the oil level gauge guide openings to avoid entry of foreign materials.

6. Remove electric throttle control actuator and position aside. Handle carefully and avoid impacts. Never disassemble or adjust electric throttle control actuator.

7. Disconnect the harness connector and EVAP hose from the EVAP canister purge volume control solenoid valve.

8. Disconnect vacuum hose for brake booster from intake manifold.

9. Remove the front intake manifold support bracket and the bolt from the rear. Rear bracket is not removed. Remove bolt from intake manifold.

10. Remove the intake manifold.

11. Remove EVAP canister purge volume control solenoid valve from intake manifold, if necessary. Handle carefully and avoid impacts.

12. Remove intake manifold support

(center) from cylinder head, if necessary. The intake manifold support (center) functions as the guide when the intake manifold is installed.

To install:

13. Installation is the reverse of the removal procedure.

14. Install the gasket to the intake manifold. Align the protrusion of gasket to the groove of intake manifold.

15. Place the intake manifold into the installation position.

16. Make sure that the oil level gauge guide is not disconnected from the fixing clip of water inlet due to interference with intake manifold.

17. Tighten bolts in numerical order.

18. Install intake manifold support (front and rear).

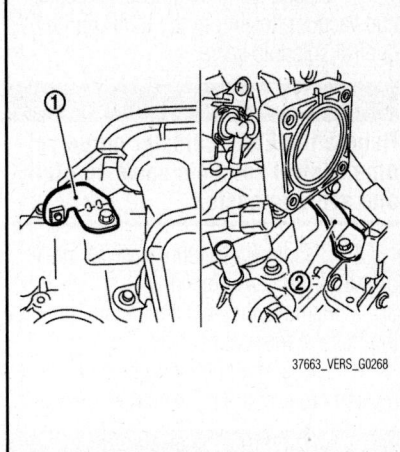

Fig. 64 Remove front intake manifold support bracket (1) and bolt from rear (2)

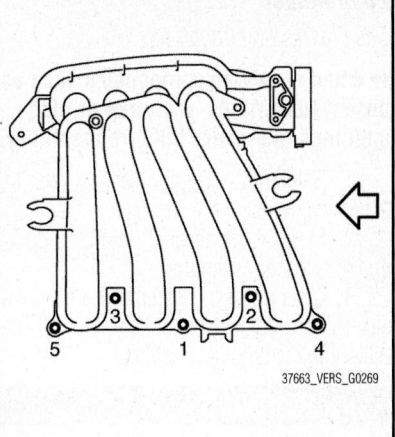

Fig. 65 Intake manifold bolt installation sequence—reverse for removal

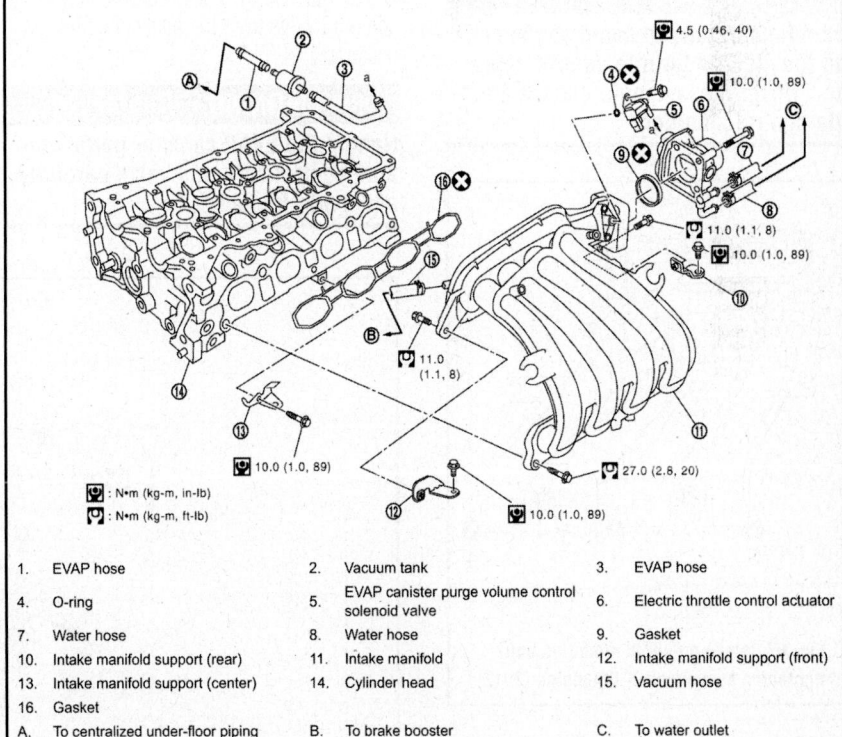

1. EVAP hose	2. Vacuum tank	3. EVAP hose
4. O-ring	5. EVAP canister purge volume control solenoid valve	6. Electric throttle control actuator
7. Water hose	8. Water hose	9. Gasket
10. Intake manifold support (rear)	11. Intake manifold	12. Intake manifold support (front)
13. Intake manifold support (center)	14. Cylinder head	15. Vacuum hose
16. Gasket		
A. To centralized under-floor piping	B. To brake booster	C. To water outlet

Fig. 66 Intake manifold and related component locations—1.6L engine (2011)

2012 Model

See Figures 67 through 69.

➡When removing components such as hoses, tubes/lines, etc., cap or plug openings to prevent fluid from spilling.

1. Before servicing the vehicle, refer to the Precautions.

2. Remove the air duct (inlet), air duct, and air cleaner assembly.

3. Disconnect the water hoses from the electric throttle control actuator. Ensure that the engine is cold for this step.

✳✳ WARNING
Never spill engine coolant on the drive belt.

4. Remove the electric throttle control actuator.

✳✳ WARNING
Handle carefully to avoid any shock to the electric throttle control actuator. Never disassemble electric throttle control actuator.

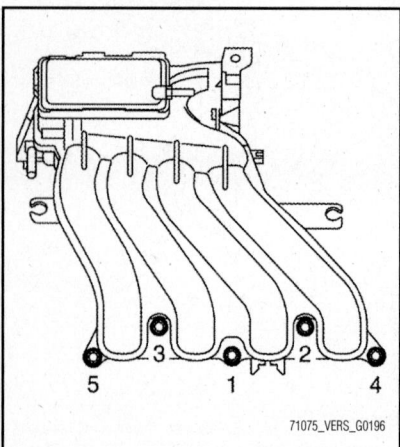

Fig. 67 Intake manifold mounting bolt tightening sequence—1.6L engine (2012)

71075_VERS_G0196

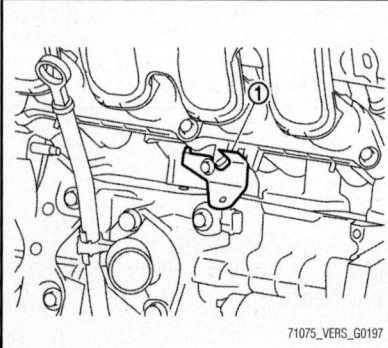

Fig. 68 Location of the intake manifold support (1)—1.6L engine (2012)

71075_VERS_G0197

5. Disconnect the harness connector and vacuum hose from the EVAP purge control solenoid valve.

✳✳ WARNING
Handle the EVAP canister purge volume control solenoid valve carefully and avoid impacts.

6. Disconnect vacuum hose for brake booster from intake manifold.

7. Remove the intake manifold mounting bolts at the rocker cover.

8. Loosen the mounting bolts in reverse order of the tightening sequence.

✳✳ WARNING
Cover the engine openings to avoid entry of foreign materials.

9. Remove the intake manifold.

10. Remove the EVAP purge control solenoid valve from the intake manifold, if necessary.

✳✳ WARNING
Handle the EVAP canister purge volume control solenoid valve carefully and avoid impacts.

11. Remove the intake manifold support, if necessary.

➡The intake manifold support functions as a guide for installing the intake manifold.

To install:

12. Installation is the reverse of the removal procedure.

13. Install the new gasket to the intake manifold. Align the protrusions used for checking gasket installation condition with the clearance grooves of the intake manifold mounting groove.

➡Do not reuse the gasket. A new gasket for the electronically-controlled throttle can be installed when the electronically-controlled throttle is installed.

14. Place the intake manifold into position.

✳✳ WARNING
Check that the oil level gauge guide is not detached from the securing clip of the water inlet due to interference of intake manifold.

15. Tighten the bolts to specification.

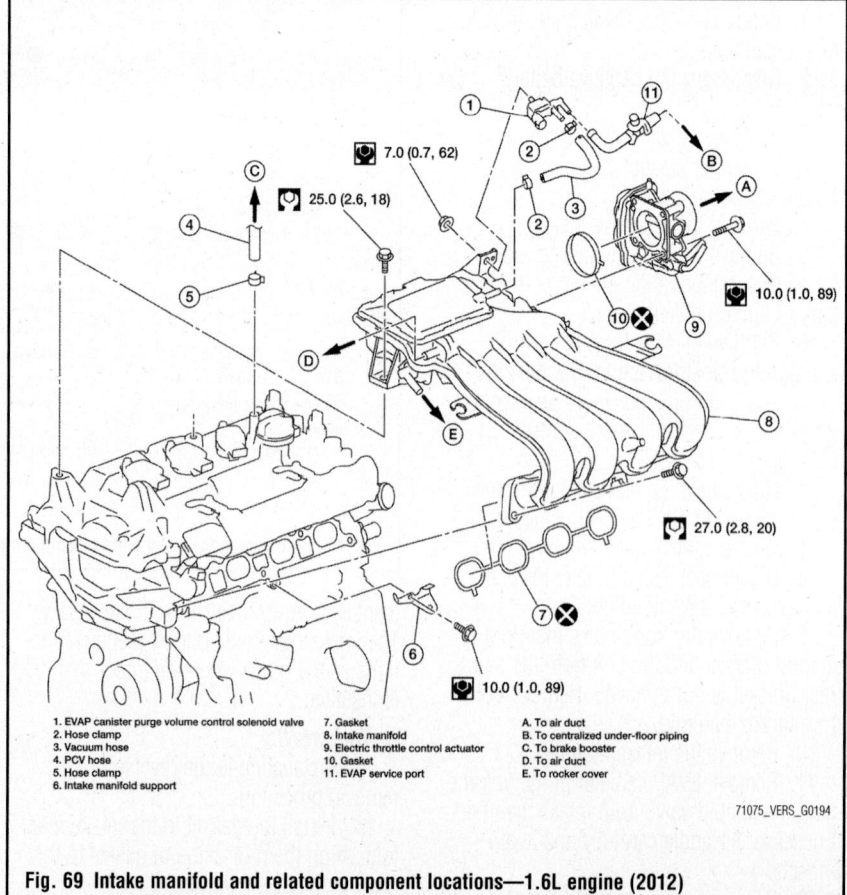

1. EVAP canister purge volume control solenoid valve
2. Hose clamp
3. Vacuum hose
4. PCV hose
5. Hose clamp
6. Intake manifold support
7. Gasket
8. Intake manifold
9. Electric throttle control actuator
10. Gasket
11. EVAP service port

A. To air duct
B. To centralized under-floor piping
C. To brake booster
D. To air duct
E. To rocker cover

Fig. 69 Intake manifold and related component locations—1.6L engine (2012)

71075_VERS_G0194

16. Tighten the intake manifold mounting bolts to specification at the rocker cover.

17. Tighten the bolts of the electric throttle control actuator equally and diagonally in several steps.

18. Perform "Throttle Valve Closed Position Learning" after repair when removing the harness connector of the electric throttle control actuator.

1.8L Engine

See Figures 70 and 71.

1. Before servicing the vehicle, refer to the Precautions.

2. Disconnect the negative battery cable.

3. Remove the engine cover.

4. Remove the air duct, air duct (inlet) and air duct (front).

5. Disconnect the EVAP canister purge volume control solenoid valve.

6. Partially drain the engine coolant. Perform this step when engine is cold. This step is unnecessary when putting plugs to water hoses (to electric throttle control actuator).

7. Disconnect the water hoses from the electronic throttle control actuator and remove the electronic throttle control actuator. Do not disassemble.

8. Remove the PCV hose and the vacuum hose.

9. Remove the oil level gauge. Cover

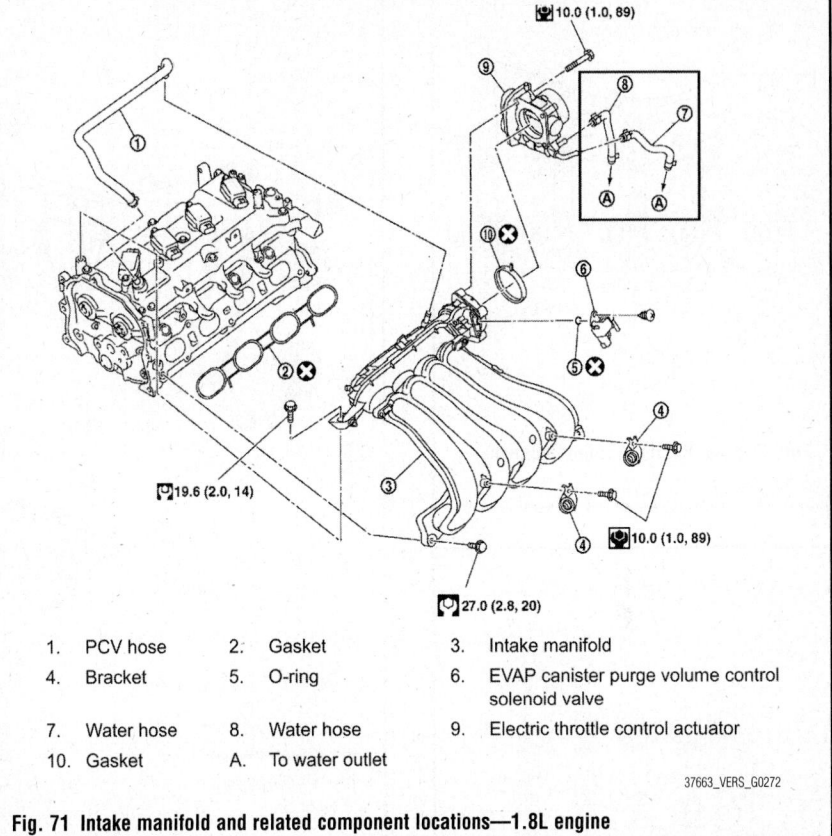

1.	PCV hose	2.	Gasket	3.	Intake manifold
4.	Bracket	5.	O-ring	6.	EVAP canister purge volume control solenoid valve
7.	Water hose	8.	Water hose	9.	Electric throttle control actuator
10.	Gasket	A.	To water outlet		

37663_VERS_G0272

Fig. 71 Intake manifold and related component locations—1.8L engine

the oil level gauge guide openings to avoid entry of foreign materials.

10. Loosen and remove the intake manifold bolts.

11. Loosen the bolts in the reverse order of the installation sequence. Cover the engine openings to avoid entry of foreign materials.

12. Remove the intake manifold.

To install:

13. Install the intake manifold. Be sure the intake manifold gasket is seated correctly in the groove of the intake manifold.

14. Tighten the intake manifold bolts in numerical order to 14 ft. lbs. (20 Nm).

15. Install the electronic throttle control actuator.

16. Install the water hoses to the electronic throttle control actuator.

17. Installation of the remaining components is in the reverse order of removal.

18. Before starting engine, check oil/fluid levels including engine coolant and engine oil. If less than required quantity, fill to the specified level.

19. Turn ignition switch ON (with engine stopped). With fuel pressure applied to fuel piping, check for fuel leakage at connection points.

20. Start engine. With engine speed

increased, check again for fuel leakage at connection points.

21. Run engine to check for unusual noise and vibration.

22. Warm up engine thoroughly to make sure there is no leakage of fuel, exhaust gas, or any oils/fluids including engine oil and engine coolant.

23. Bleed air from passages in lines and hoses, such as in cooling system.

24. After cooling down engine, again check oil/fluid levels including engine oil and engine coolant. Refill to specified level, if necessary.

MAIN BEARING TORQUE SEQUENCE

See Figures 72 and 73.

OIL PAN

REMOVAL & INSTALLATION

1.6L Engine

2011 Model

See Figure 74.

1. Before servicing the vehicle, refer to the Precautions.

2. Drain the engine oil.

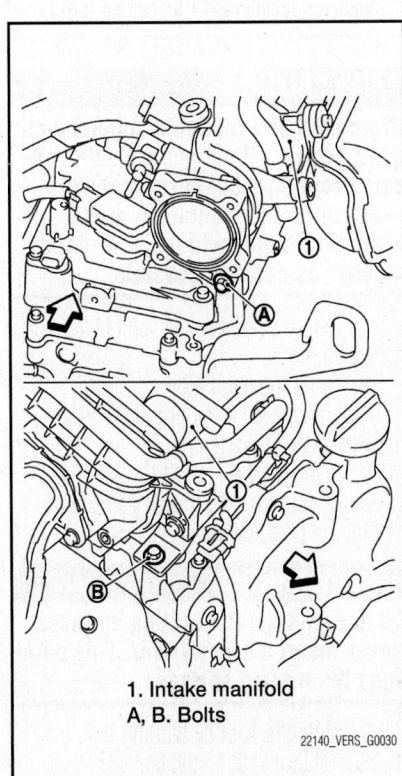

1. Intake manifold
A, B. Bolts

22140_VERS_G0030

Fig. 70 Removing the intake manifold bolts—1.8L engine

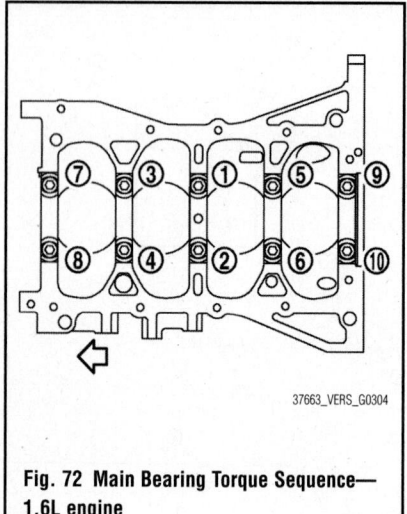

Fig. 72 Main Bearing Torque Sequence—1.6L engine

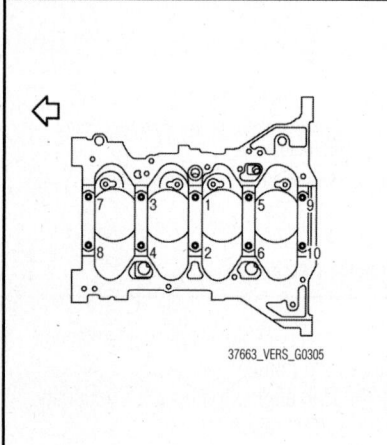

Fig. 73 Main Bearing Torque Sequence—1.8L engine

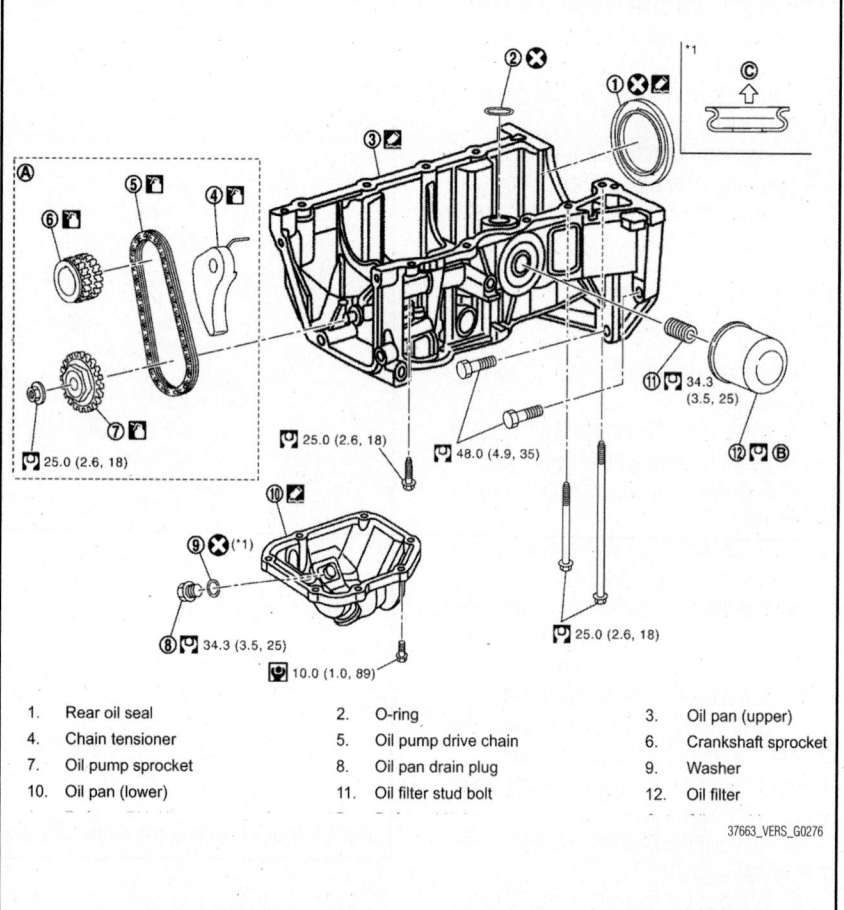

1.	Rear oil seal	2.	O-ring	3.	Oil pan (upper)
4.	Chain tensioner	5.	Oil pump drive chain	6.	Crankshaft sprocket
7.	Oil pump sprocket	8.	Oil pan drain plug	9.	Washer
10.	Oil pan (lower)	11.	Oil filter stud bolt	12.	Oil filter

Fig. 74 Exploded view of lower oil pan and related component locations—1.6L engine (2011)

3. Loosen the bolts in the reverse of the tightening order.

4. Insert the tool No. KV10111100 (J-37228) between oil pan (upper) and oil pan (lower).

❋❋ WARNING

Be careful not to damage the mating surface. A more adhesive liquid gasket is applied compared to previous types when shipped, so it should not be forced off using a flat-bladed screwdriver, etc.

5. Remove the lower oil pan.

To install:

6. Use scraper to carefully remove old liquid gasket from mating surfaces. Also remove the old liquid gasket from mating surface of oil pan (upper). Remove old liquid gasket from the bolt holes and threads.

7. Apply a continuous bead of liquid gasket with tool No. WS39930000. Bead width:

- A: 0.295–0.374 inch (7.5–9.5mm)
- B: 0.157–0.196 inch (4.0–5.0mm)

➡**Attaching should be done within 5 minutes after sealant application.**

8. Tighten bolts in numerical order.

❋❋ WARNING

Do not pour engine oil until at least 30 minutes after oil pan (lower) is installed.

9. Refill the engine oil level, check and adjust as necessary.

10. Start engine, and check there are no leaks of engine oil.

11. Stop engine and wait for 10 minutes.

12. Check the engine oil level again

2012 Model

See Figures 75 and 76.

❋❋ CAUTION

Be careful not to get burned, engine coolant and engine oil may be hot.

❋❋ CAUTION

Prolonged and repeated contact with used engine oil may cause skin cancer; avoid direct skin contact with used oil. If skin contact is made, wash thoroughly with soap or hand cleaner as soon as possible.

1. Before servicing the vehicle, refer to the Precautions.

2. Drain the engine oil.

3. Loosen the mounting bolts in the reverse order of tightening.

4. Insert Tool KV10111100 (J-37228), or equivalent, between the oil pan (upper) and the oil pan (lower).

❋❋ WARNING

Do not damage the mating surfaces. Never insert a screwdriver. This damages the mating surfaces.

5. Slide the Tool by tapping on the side of tool with a suitable tool to loosen the oil pan (lower).

6. Remove the oil pan (lower).

5
2
7
4
3
6
1
71075_VERS_G0199

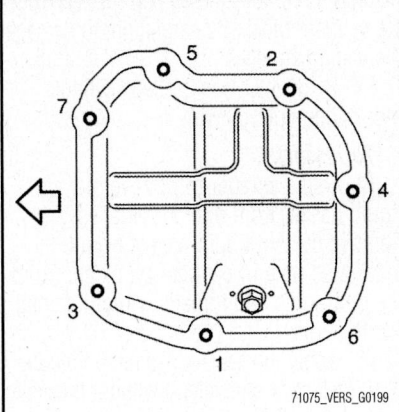

Fig. 75 Lower oil pan mounting bolt tightening sequence—1.6L engine (2012)

To install:

7. Installation is the reverse of the removal procedure.

8. Clean debris from oil pan (lower) and from the strainer.

➡Do not reuse O-rings or washers.

9. Use a scraper to remove old liquid gasket from mating surfaces.

10. Remove old liquid gasket from mating surface of oil pan (upper).

11. Remove old liquid gasket from the bolt holes and threads.

WARNING

Never scratch or damage the mating surface when cleaning off old liquid gasket.

12. Apply a continuous bead of liquid gasket. Use Genuine Silicone RTV Sealant, or equivalent.

➡The components must be installed within 5 minutes of the liquid gasket application. Do not confirm torque after the 5 minutes have elapsed. Then allow 30 minutes for the liquid gasket to set before adding oil to the engine.

13. Tighten bolts in numerical order.

14. Install oil pan drain plug.

15. Add the specified oil after waiting for at least 30 minutes.

16. Before starting engine, check oil/fluid levels, including engine coolant and engine oil. If less than required quantity, fill to the specified level.

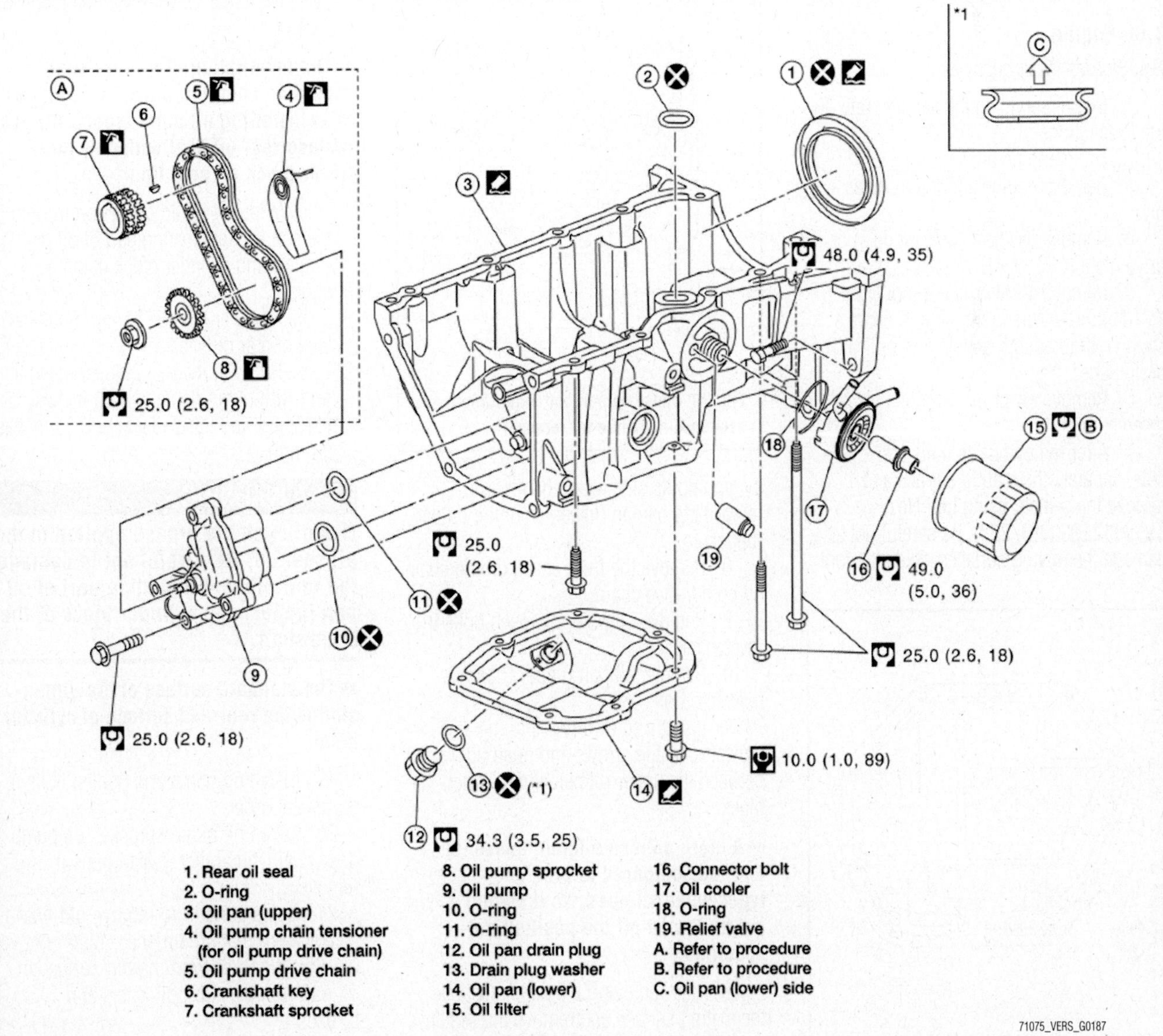

*1
C
A
5
4
6
7
2
1
3
48.0 (4.9, 35)
8
25.0 (2.6, 18)
18
15
B
17
25.0
(2.6, 18)
19
16
49.0
(5.0, 36)
11
10
25.0 (2.6, 18)
9
25.0 (2.6, 18)
10.0 (1.0, 89)
13 (*1)
14
12
34.3 (3.5, 25)
1. Rear oil seal
2. O-ring
3. Oil pan (upper)
4. Oil pump chain tensioner (for oil pump drive chain)
5. Oil pump drive chain
6. Crankshaft key
7. Crankshaft sprocket
8. Oil pump sprocket
9. Oil pump
10. O-ring
11. O-ring
12. Oil pan drain plug
13. Drain plug washer
14. Oil pan (lower)
15. Oil filter
16. Connector bolt
17. Oil cooler
18. O-ring
19. Relief valve
A. Refer to procedure
B. Refer to procedure
C. Oil pan (lower) side
71075_VERS_G0187

Fig. 76 Exploded view of engine oil pan and related component locations—1.6L engine (2012)

17. Turn ignition switch ON (with engine stopped). With fuel pressure applied to fuel piping, check for fuel leakage at connection points.

18. Start engine. With engine speed increased, check again for fuel leakage at connection points.

19. Run engine to check for unusual noise and vibration.

➡**If hydraulic pressure inside timing chain tensioner drops after removal and installation, slack in the guide may generate a pounding noise during and just after engine start. However, this is normal. Noise will stop after hydraulic pressure rises.**

20. Warm up engine thoroughly to make sure there is no leakage of fuel, exhaust gas, or any oils/fluids including engine oil and engine coolant.

1.8L Engine

See Figures 77 through 79.

1. Before servicing the vehicle, refer to the Precautions.

2. Drain engine oil.

3. Remove engine and transaxle assembly.

4. Remove flywheel (M/T models) or drive plate (CVT or A/T models).

5. Remove oil filter using tool No. KV10115801. When removing, prepare a shop cloth to absorb any engine oil leakage or spillage.

6. Remove lower oil pan bolts in reverse order.

7. After removing the bolts and nuts, separate the mating surface and remove the sealant using tool No. KV10111100 (J-37228). Be careful not to damage the mating surfaces. Slide the tool

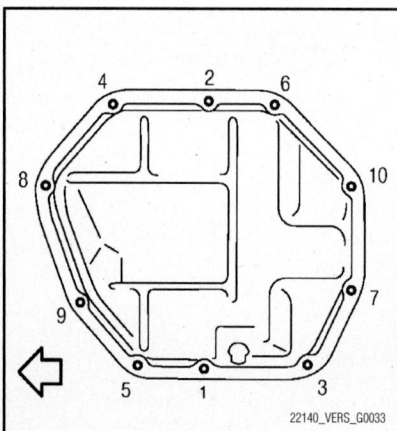

Fig. 77 Oil pan bolt installation sequence—reverse for removal

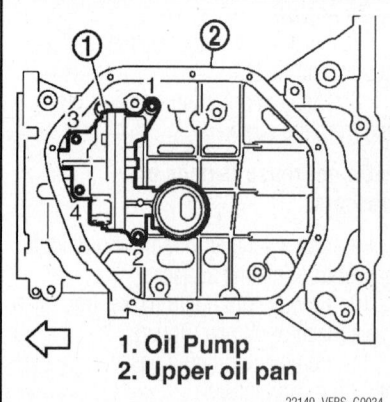

1. Oil Pump
2. Upper oil pan

Fig. 78 Oil pump bolt installation sequence, oil pump (1) and oil pan (upper) (2)—reverse for removal

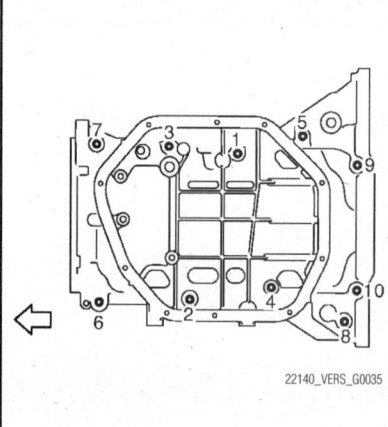

Fig. 79 Upper oil pan bolt installation sequence—reverse for removal

by tapping its side with a hammer to remove the oil pan (lower) from the oil pan (upper).

8. Remove the front cover, timing chain, oil pump drive chain.

9. Remove oil pump. Loosen bolts in reverse order.

10. Remove oil pan (upper) bolts in reverse order.

11. Insert a screwdriver into the area indicated by the arrows and open up a crack between the upper oil pan and cylinder block.

➡**A more adhesive liquid gasket is applied compared to previous types when shipped, so it should not be forced off the position not specified.**

12. After removing the bolts, separate the mating surface and remove the sealant using tool No. KV10111100 (J-37228). Slide the tool by tapping its side with a

hammer to remove the oil pan (upper) from the cylinder block. Be careful not to damage the mating surfaces.

13. Remove O-ring between cylinder block and upper oil pan.

To install:

14. Use a scraper to remove old liquid gasket from mating surfaces. Remove the old liquid gasket from the mating surface of cylinder block and from the bolt holes and threads without damaging the area.

15. Apply the sealant (Genuine Silicone RTV Sealant or equivalent) without breaks to the specified location using Tool No. WS39930000. Apply liquid gasket to outside of bolt hole.

16. Install new O-ring at cylinder block side.

17. Tighten bolts in numerical order.

18. Install rear oil seal with the following procedure:

➡**The installation of rear oil seal should be completed within 5 minutes after installing oil pan (upper). Always replace rear oil seal with new one. Never touch oil seal lip.**

a. Wipe off liquid gasket protruding to the rear oil seal mating part of oil pan (upper) and cylinder block using a scraper.

b. Apply engine oil to entire outside area of rear oil seal.

c. Press-fit the rear oil seal using a drift with outer diameter 4.53 inches (115mm) and inner diameter 3.54 inches (90mm).

⁂ **WARNING**

Never touch the grease applied to the oil seal lip. Be careful not to damage the rear oil seal mounting part of oil pan (upper) and cylinder block or the crankshaft.

➡**The standard surface of the dimension is the rear end surface of cylinder block.**

19. Install oil pump and tighten bolts in numerical order.

20. Install oil pump sprocket, oil pump drive chain, and other related parts if removed.

21. Use a scraper to remove old liquid gasket from mating surfaces. Also remove old liquid gasket from mating surface of oil pan (upper) and bolt holes and threads.

22. Apply the sealant (Genuine Silicone RTV Sealant or equivalent) without breaks to

the specified location using Tool No. WS39930000.

23. Tighten bolts in numerical order.

24. Install oil filter with the following procedure:

　a. Remove foreign materials adhering to the oil filter installation surface.

　b. Apply new engine oil to the oil seal contact surface of new oil filter.

　c. Screw oil filter manually until it touches the installation surface, and then tighten it by ⅔ turn, or tighten to 13 ft. lbs. (18 Nm).

25. The remainder of installation is the reverse of removal.

OIL PUMP

REMOVAL & INSTALLATION

1.6L Engine

2011 Sedan

The oil pump is not serviced separately, it is serviced as part of the oil pan.

2012 Sedan

See Figure 80.

1. Before servicing the vehicle, refer to the Precautions.

2. Drain engine oil.

3. Remove timing chain and oil pump drive chain.

4. Loosen bolts.

5. Remove oil pump and O-rings.

❊❊ WARNING

Do not reuse O-rings. Never disassemble oil pump.

To install:

6. Installation is the reverse of the removal procedure.

7. Install new O-rings on the oil pan (upper) before installing the oil pump.

8. Install the oil pump.

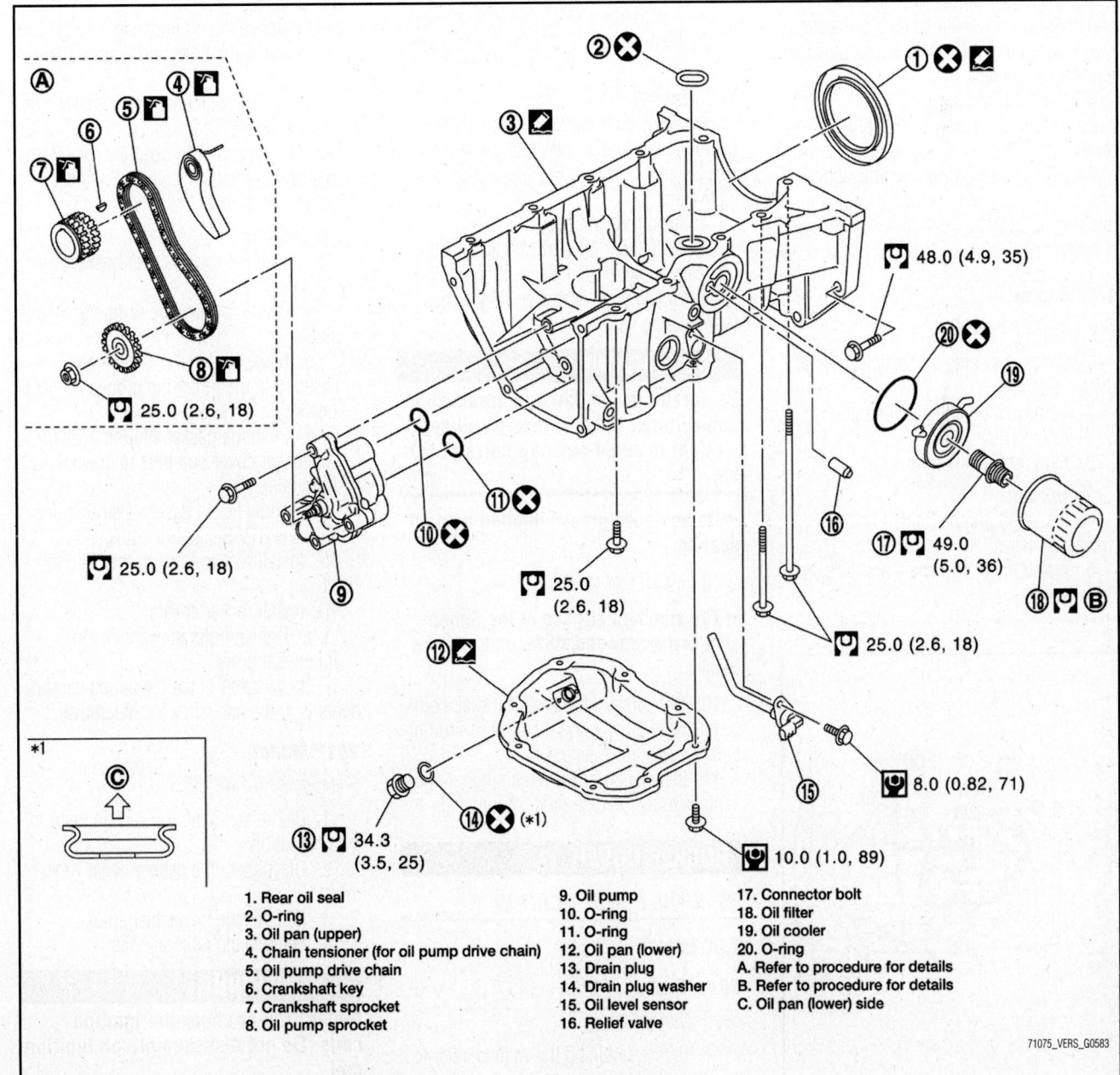

1. Rear oil seal
2. O-ring
3. Oil pan (upper)
4. Chain tensioner (for oil pump drive chain)
5. Oil pump drive chain
6. Crankshaft key
7. Crankshaft sprocket
8. Oil pump sprocket
9. Oil pump
10. O-ring
11. O-ring
12. Oil pan (lower)
13. Drain plug
14. Drain plug washer
15. Oil level sensor
16. Relief valve
17. Connector bolt
18. Oil filter
19. Oil cooler
20. O-ring
A. Refer to procedure for details
B. Refer to procedure for details
C. Oil pan (lower) side

Fig. 80 Oil pump and related component locations—1.6L engine (2012)

9. Tighten the bolts to specification.

10. Install timing chain and oil pump drive chain.

11. Before starting engine, check oil/fluid levels, including engine coolant and engine oil. If less than required quantity, fill to the specified level.

➡ **If hydraulic pressure inside timing chain tensioner drops after removal and installation, slack in the guide may generate a pounding noise during and just after engine start. However, this is normal. Noise will stop after hydraulic pressure rises.**

12. Warm up engine thoroughly to make sure there is no leakage of fuel, exhaust gas, or any oils/fluids including engine oil and engine coolant.

13. Bleed air from passages in lines and hoses, such as in cooling system.

14. After cooling down engine, again check oil/fluid levels, including engine oil and engine coolant. Refill to specified level, if necessary.

1.8L Engine

See Figures 81 and 82.

1. Before servicing the vehicle, refer to the Precautions.

2. Remove the timing chain and oil pump drive chain.

3. Remove the oil pump.

To install:

4. Install oil pump. Tighten bolts in numerical order.

5. Install the timing chain and oil pump drive chain.

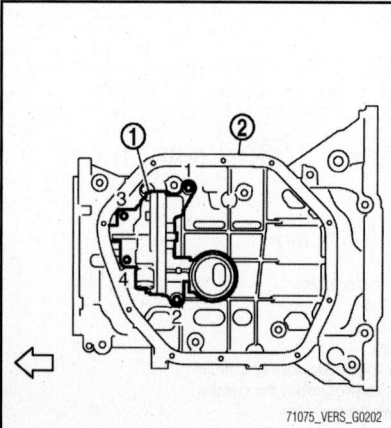

Fig. 81 Oil pump (1) bolt tightening sequence shown on the upper oil pan (2)

PISTONS & RINGS

POSITIONING

See Figures 83 and 84.

REAR MAIN SEAL

REMOVAL & INSTALLATION

See Figure 85.

1. Before servicing the vehicle, refer to the Precautions.

2. Remove transaxle assembly.

3. Remove clutch cover and clutch disk (M/T models).

4. Remove drive plate (A/T or CVT models) or flywheel (M/T models).

5. Remove rear oil seal with a suitable tool.

To install:

6. Apply the liquid gasket lightly to entire outside area of new rear oil seal. Use Genuine Silicone RTV Sealant or equivalent.

7. Install rear oil seal.

8. Install rear oil seal with a suitable tool with an outer diameter 4.53 inches (115mm) and inner diameter 3.54 inches (90mm).

❊❊ WARNING

Be careful not to damage crankshaft and cylinder block. Press-fit oil seal straight to avoid causing burrs or tilting.

➡ **Do not touch grease applied onto oil seal lip.**

9. Install rear oil seal.

➡ **The standard surface of the dimension is the rear end surface of cylinder block.**

10. After press-fitting rear oil seal, completely wipe off any liquid gasket protruding to rear end surface side.

11. Installation of the remaining components is in the reverse order of removal.

ROCKER COVER

REMOVAL & INSTALLATION

1.6L Engine

2011 Model

See Figure 86.

1. Before servicing the vehicle, refer to the Precautions.

2. Disconnect the negative battery cable.

3. Remove the intake manifold.

4. Remove the ignition coils.

❊❊ WARNING

Handle the ignition coil carefully and avoid impacts. Never disassemble.

5. Remove the spark plug using a suitable tool. Never drop or shock it.

6. Remove the right-hand ground cable.

7. Support the bottom surface of the engine using a transmission jack, and then remove the right-hand engine mounting bracket and insulator.

8. Remove the fuel tube protector.

9. Disconnect the PCV valve hose from the PCV valve.

10. Remove the oil filler cap.

11. Loosen the bolts and remove the rocker cover.

12. Remove the rocker cover gasket from the rocker cover.

13. Use a scraper to carefully remove all traces of liquid gasket from the cylinder head and front cover.

To install:

14. Install the rocker cover gasket to the rocker cover:

   a. Check for damage or foreign material.

   b. Make sure that it is securely inserted in the mounting groove of rocker cover.

   c. Push the gasket into the boss for the rocker cover bolt hole to prevent it from falling.

   d. Apply liquid gasket. Use Genuine Silicone RTV Sealant or equivalent.

15. Install rocker cover to the cylinder head.

16. Install rocker cover:

   a. Tighten bolts in two steps in numerical order.

17. Installation of the remaining components is in the reverse order of removal.

2012 Model

See Figures 87 and 88.

1. Before servicing the vehicle, refer to the Precautions.

2. Disconnect the battery negative terminal.

3. Remove the intake manifold.

4. Remove the ignition coils.

❊❊ WARNING

Never drop or shock the ignition coils. Do not disassemble an ignition coil.

5. Remove the fuel tube protector.

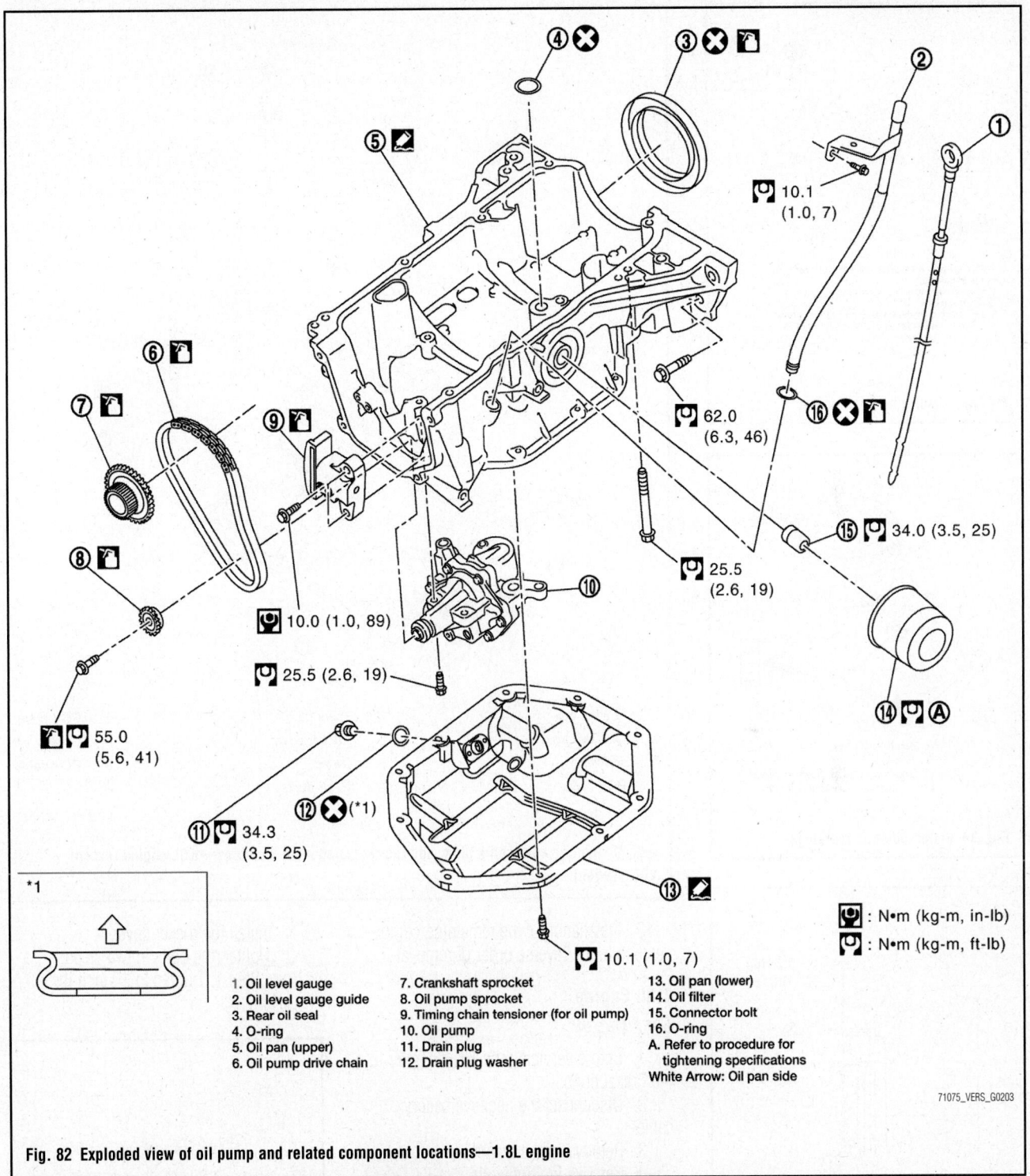

1. Oil level gauge
2. Oil level gauge guide
3. Rear oil seal
4. O-ring
5. Oil pan (upper)
6. Oil pump drive chain
7. Crankshaft sprocket
8. Oil pump sprocket
9. Timing chain tensioner (for oil pump)
10. Oil pump
11. Drain plug
12. Drain plug washer
13. Oil pan (lower)
14. Oil filter
15. Connector bolt
16. O-ring
A. Refer to procedure for tightening specifications
White Arrow: Oil pan side

: N•m (kg-m, in-lb)
: N•m (kg-m, ft-lb)

71075_VERS_G0203

Fig. 82 Exploded view of oil pump and related component locations—1.8L engine

6. Remove the PCV hose from the rocker cover.

7. Remove the PCV valve, if necessary.

8. Remove the rocker cover.

9. Remove the rocker cover gasket from the rocker cover.

10. Use a scraper to remove all traces of liquid gasket from the cylinder head and front cover.

✳✳ WARNING

Do not scratch or damage the mating surface when cleaning off old liquid gasket.

To install:

➡**Do not reuse O-ring.**

11. Install the rocker cover:

a. Press the gasket onto the bosses

for the rocker cover bolt holes to prevent the gasket from dropping off.

b. Apply liquid gasket to the cylinder head and front cover. Use Genuine RTV Silicone Sealant or equivalent.

➡**The components must be installed within 5 minutes of the liquid gasket application. Then allow 30 minutes for the liquid gasket to set before putting oil in the engine.**

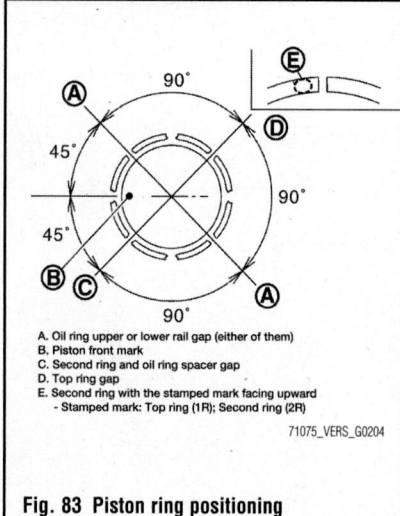

A. Oil ring upper or lower rail gap (either of them)
B. Piston front mark
C. Second ring and oil ring spacer gap
D. Top ring gap
E. Second ring with the stamped mark facing upward
   - Stamped mark: Top ring (1R); Second ring (2R)

71075_VERS_G0204

Fig. 83 Piston ring positioning

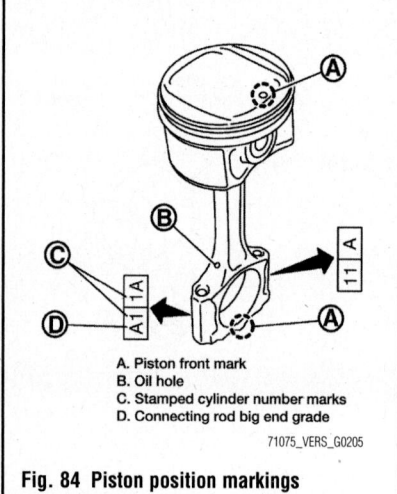

A. Piston front mark
B. Oil hole
C. Stamped cylinder number marks
D. Connecting rod big end grade

71075_VERS_G0205

Fig. 84 Piston position markings

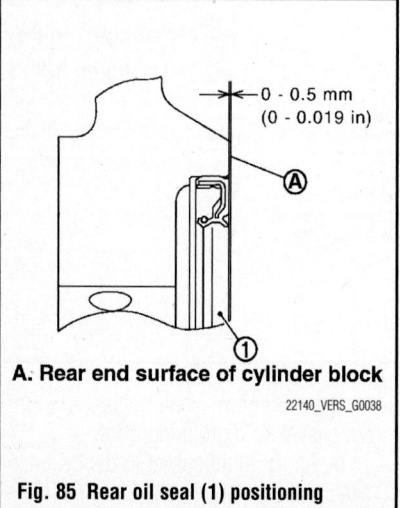

A. Rear end surface of cylinder block

22140_VERS_G0038

Fig. 85 Rear oil seal (1) positioning

c. Install the rocker cover to the cylinder head. Check that the gasket has not slipped out of position.

d. Tighten the bolts in numerical order in two separate stages.

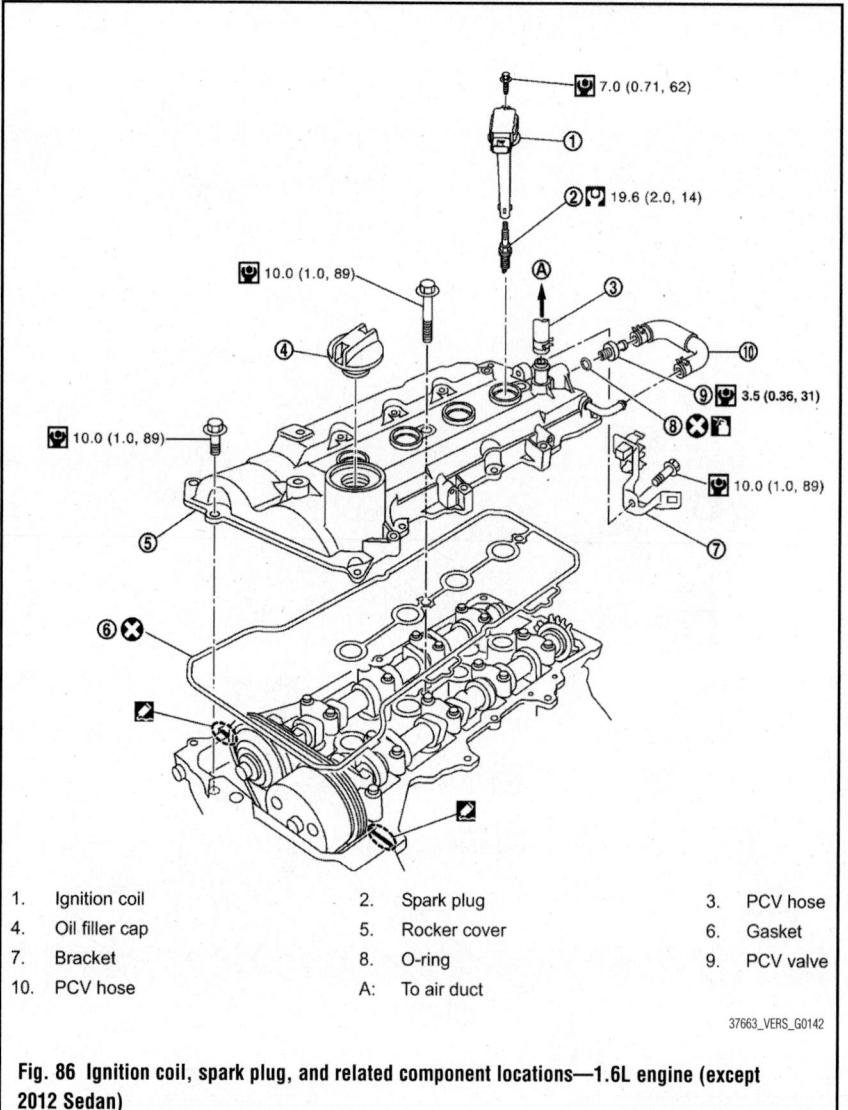

1.	Ignition coil	2.	Spark plug	3.	PCV hose
4.	Oil filler cap	5.	Rocker cover	6.	Gasket
7.	Bracket	8.	O-ring	9.	PCV valve
10.	PCV hose	A:	To air duct		

37663_VERS_G0142

Fig. 86 Ignition coil, spark plug, and related component locations—1.6L engine (except 2012 Sedan)

12. Installation of the remaining components is in the reverse order of removal.

1.8L Engine

See Figure 89.

1. Before servicing the vehicle, refer to the Precautions.
2. Disconnect the negative battery cable.
3. Remove intake manifold.
4. Remove ignition coils.

※ WARNING

Handle ignition coil carefully and avoid impacts. Never disassemble.

5. Remove the spark plugs using a suitable tool. Never drop or shock it.
6. Remove the rocker cover.

To install:

7. Install the rocker cover gasket to the rocker cover.

8. Install the rocker cover.
9. Tighten the bolts in two steps:
   a. Step 1: Tighten to 17 inch lbs. (2 Nm).

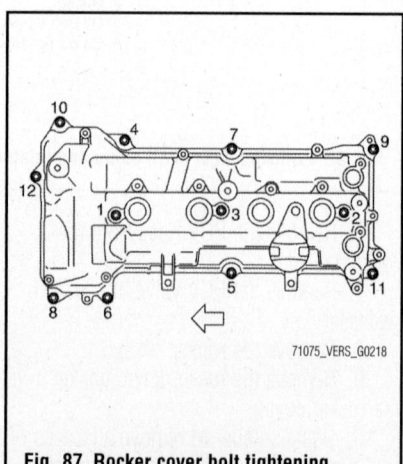

71075_VERS_G0218

Fig. 87 Rocker cover bolt tightening sequence—1.6L engine (2012)

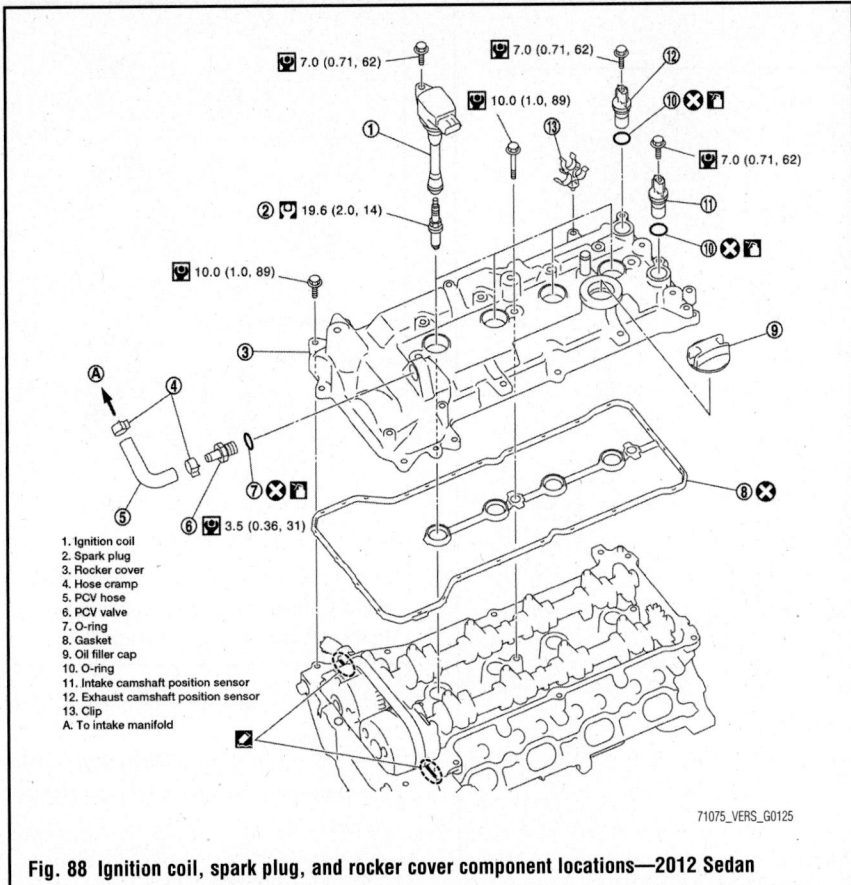

Fig. 88 Ignition coil, spark plug, and rocker cover component locations—2012 Sedan

1. Ignition coil
2. Spark plug
3. Rocker cover
4. Hose cramp
5. PCV hose
6. PCV valve
7. O-ring
8. Gasket
9. Oil filler cap
10. O-ring
11. Intake camshaft position sensor
12. Exhaust camshaft position sensor
13. Clip
A. To intake manifold

71075_VERS_G0125

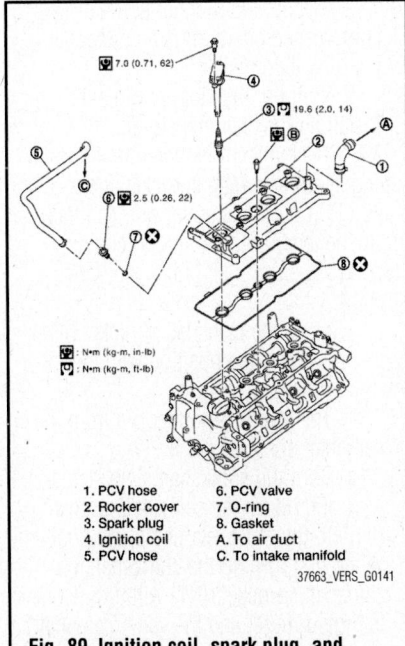

Fig. 89 Ignition coil, spark plug, and related component locations—1.8L engine

1. PCV hose
2. Rocker cover
3. Spark plug
4. Ignition coil
5. PCV hose
6. PCV valve
7. O-ring
8. Gasket
A. To air duct
C. To intake manifold

37663_VERS_G0141

b. Step 2: Tighten to 73 inch lbs. (8 Nm).

10. Install the spark plugs using a suitable tool.

11. Install the ignition coils.

12. Install the intake manifold.

13. Connect the negative battery cable.

TIMING CHAIN COVER, CHAIN, TENSIONER, & SPROCKETS

REMOVAL & INSTALLATION

1.6L Engine

2011 Model

See Figures 90 through 93.

1. Before servicing the vehicle, refer to the Precautions.

2. Disconnect the negative battery cable.

3. Drain engine oil.

4. Remove the right front wheel.

5. Remove the right front fender protector.

6. Remove the intake manifold.

7. Remove the drive belt.

8. Remove the water pump pulley.

9. Remove the ground cable (RH).

10. Support the bottom surface of engine using a transmission jack, and remove the right-hand engine bracket and insulator.

11. Remove the rocker cover.

12. Set No. 1 cylinder at TDC on its compression stroke with the following procedure:

a. Rotate crankshaft pulley clockwise and align TDC mark (no paint) to timing indicator on front cover.

b. Make sure the matching marks on each camshaft sprocket are in lined up in position. If not, rotate crankshaft pulley one more turn to align matching marks to the positions.

13. Remove crankshaft pulley.

14. Remove front cover by loosening the bolts and cutting the gasket.

15. Remove front oil seal from front cover. Remove by lifting it up using a suitable tool. Be careful not to damage the front cover.

16. Remove chain tensioner with the following procedure:

a. Fully push down the chain tensioner lever, and then push the plunger into the inside of tensioner.

b. The tab is released by fully pushing the lever down. As a result, the plunger can be moved.

c. Pull up the lever to align its hole position with the body hole position.

d. When the lever hole is aligned with the body hole position, the plunger is fixed.

e. When the protrusion parts of the plunger ratchet and the tab face each other, both hole positions are not aligned. At that time, correctly engage them and align these hole positions by slightly moving the plunger.

f. Insert the stopper pin into the body hole through the lever hole, and then fix the lever at the upper position.

g. Remove chain tensioner.

17. Remove the timing chain tension guide and the timing chain slack guide.

18. Remove the timing chain. Pull the looseness of timing chain toward the camshaft sprocket (EXH), and then remove the timing chain and start the removal from camshaft sprocket (EXH) side.

✳✳ WARNING

Never rotate crankshaft or camshaft while timing chain is removed. It causes interference between valve and piston.

19. Remove the crankshaft sprocket and the oil pump drive related parts with the following procedure:

a. Remove chain tensioner. Pull out from the shaft and spring fixing holes.

b. Hold the top of the oil pump shaft using the TORX® socket (size: E8), and

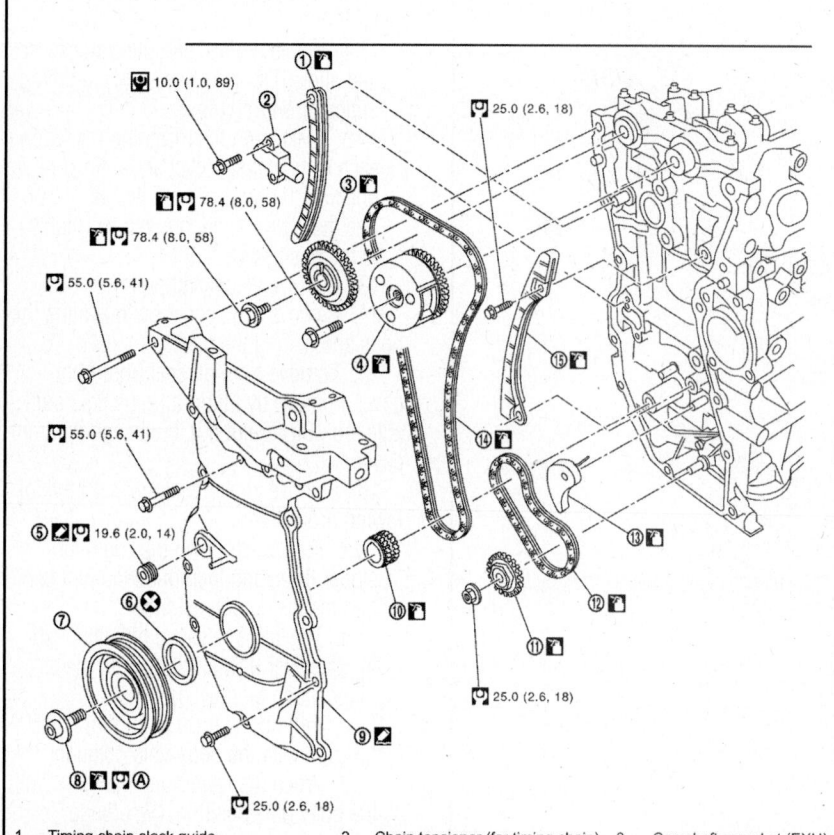

Fig. 90 Timing chain and related component locations—1.6L engine (2011)

1. Timing chain slack guide	2. Chain tensioner (for timing chain)	3. Camshaft sprocket (EXH)
4. Camshaft sprocket (INT)	5. Plug	6. Front oil seal
7. Crankshaft pulley	8. Crankshaft pulley bolt	9. Front cover
10. Crankshaft sprocket	11. Oil pump sprocket	12. Oil pump drive chain
13. Chain tensioner (for oil pump drive chain)	14. Timing chain	15. Timing chain tension guide
A. Refer to Installation procedure		

37663_VERS_G0285

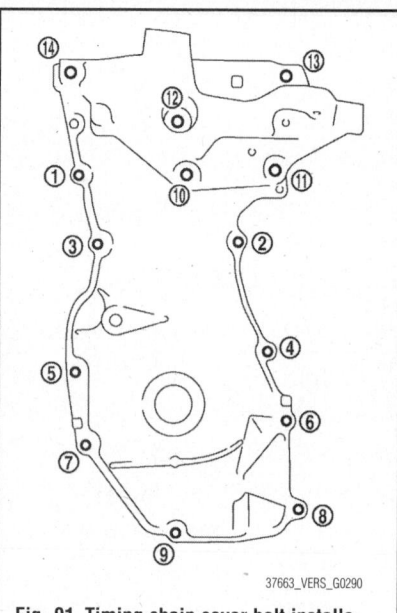

37663_VERS_G0290

Fig. 91 Timing chain cover bolt installation sequence—reverse for removal

then loosen the oil pump sprocket nut and remove it.

c. Remove the crankshaft sprocket, the oil pump drive chain, and the oil pump sprocket at the same time.

To install:

20. The illustration shows the relationship between the matching mark on each timing chain and that on the corresponding sprocket, with the components installed.

21. Install the crankshaft sprocket and the oil pump drive related parts with the following procedure:

a. Install the crankshaft sprocket, the oil pump drive chain, and the oil pump sprocket at the same time.

b. Install the crankshaft sprocket so that its invalid gear area is towards the back of the engine.

c. Install the oil pump sprocket so that its hexagonal surface faces the front of engine.

➡**There is no matching mark in the oil pump drive related parts.**

d. Hold the top of the oil pump shaft using the TORX® socket (size: E8), and then tighten the oil pump sprocket nuts.

e. Install chain tensioner. Insert the body into the shaft while inserting the spring into the fixing hole of cylinder block front surface.

f. Make sure that the tension is applied to the oil pump drive chain after installing.

22. Install timing chain with the following procedure:

a. Install by aligning matching marks on each sprocket and timing chain. If these matching marks are not aligned, rotate the camshaft slightly to correct the position.

b. Check matching mark position of each sprocket and timing chain again after installing the timing chain, keep

matching marks aligned by holding them with a hand.

c. To avoid skipped teeth, never rotate crankshaft and camshaft until front cover is installed.

23. Install timing chain tension guide and timing chain slack guide.

24. Install chain tensioner:

a. Secure the plunger at the most compressed position using a stopper pin, and then install it.

b. Pull out the stopper pin after installing the chain tensioner.

25. Check matching mark position of timing chain and each sprocket again.

26. Pull out the stopper pin, and then apply the tension to the timing chain by rotating the crankshaft pulley clockwise slightly.

27. Return the camshaft sprocket (INT) in the most retarded position with the following procedure:

a. Remove the stopper pin from the camshaft sprocket (INT).

b. Turn the crankshaft slowly clockwise and return the camshaft sprocket (INT) to the most retarded angle position. When first turning the crankshaft the camshaft sprocket (INT) will turn. Once it is turned more, and the vane (camshaft) also turns, then it has reached the most retarded angle position.

c. The most retarded angle position can be checked by seeing if the stopper pin groove is shifted clockwise.

d. After spinning the crankshaft slightly in the counterclockwise direction, you can make sure the lock pin has

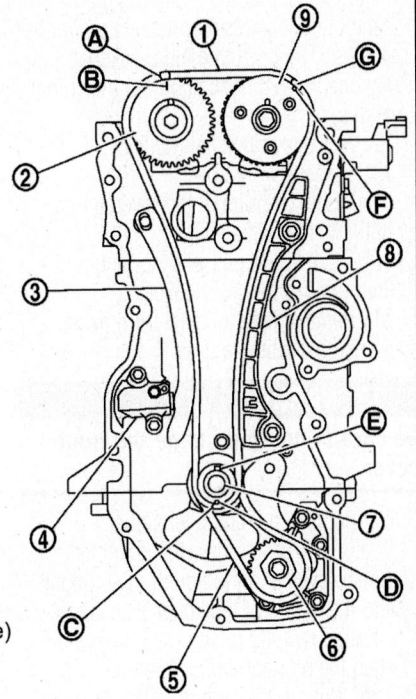

1 : Timing chain
2 : Camshaft sprocket (EXH)
3 : Timing chain slack guide
4 : Chain tensioner
5 : Oil pump drive chain
6 : Oil pump sprocket
7 : Crankshaft sprocket
8 : Timing chain tension guide
9 : Camshaft sprocket (INT)
A : Dark blue link
B : Matching mark (stamp)
C : Orange link
D : Matching mark (stamp)
E : Crankshaft key (point straight up)
F : Matching mark (peripheral stamp line)
G : Dark blue link

37663_VERS_G0297

Fig. 92 Timing chain positioning illustrated

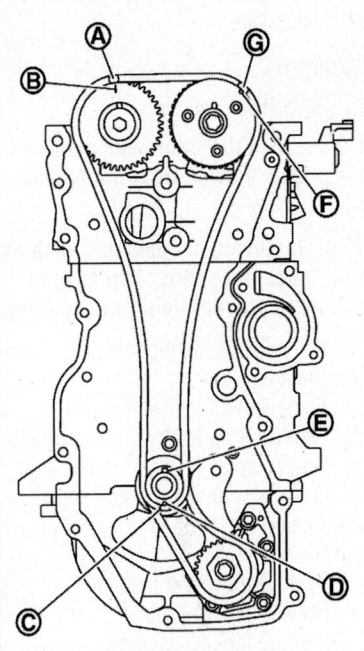

A. Dark blue link
B. Matching mark (stamp)
C. Orange link
D. Matching mark (stamp)
E. Crankshaft key (point straight up)
F. Matching mark (peripheral stamp line)
G. Dark blue link

37663_VERS_G0299

Fig. 93 Installing the timing chain

joined by seeing if the vane (camshaft) and the sprocket move together.

28. Install the front oil seal to the front cover.

29. Install front cover with the following procedure:

   a. Apply a continuous bead of liquid gasket 0.12–0.16 inch (3.0–4.0mm) wide to front of engine.

   b. Apply a continuous bead of liquid gasket 0.12–0.16 inch (3.0–4.0mm) wide to front cover. Use Genuine Silicone RTV Sealant or equivalent.

   c. Tighten bolts in numerical order.

   d. After all bolts are tightened, retighten them to specified torque in numerical order. Be sure to wipe off any excessive liquid gasket leaking to surface.

30. Insert crankshaft pulley by aligning with crankshaft key.

   a. When inserting crankshaft pulley with a plastic hammer, tap on its center portion (not circumference).

❊❊ WARNING

Never damage front oil seal lip section.

31. Tighten crankshaft pulley bolt with the following procedure:

   a. Secure crankshaft pulley with a suitable tool, and tighten crankshaft pulley bolt.

   b. Apply new engine oil to thread and seat surfaces of crankshaft pulley bolt.

   c. Tighten crankshaft pulley bolt to 26 ft. lbs. (35 Nm).

   d. Put a paint mark on crankshaft pulley, mating with any one of six easy to recognize angle marks on crankshaft bolt flange.

   e. Turn another 60 degrees clockwise (angle tightening). Check the tightening angle with movement of one angle mark.

32. Make sure that crankshaft turns smoothly by rotating by hand clockwise.

33. Install the rocker cover.

34. Install the engine mounting bracket and insulator (RH).

35. Install the ground cable.

36. Install the water pump pulley.

37. Install the drive belt.

38. Install the intake manifold.

39. Install the front fender protector (RH).

40. Install the front wheel (RH).

41. Connect the negative battery cable.

42. Before starting engine, check oil/fluid levels including engine coolant and engine oil. If less than required quantity, fill to the specified level.

43. Use procedure below to check for fuel leakage:

   a. Turn ignition switch "ON" (with engine stopped). With fuel pressure applied to fuel piping, check for fuel leakage at connection points.

   b. Start engine. With engine speed increased, check again for fuel leakage at connection points.

44. Run engine to check for unusual noise and vibration.

➡**If hydraulic pressure inside chain tensioner drops after removal/installation, slack in guide may generate a pounding noise during and just after the engine start. However, this does not mean something is wrong. Noise should stop after hydraulic pressure rises.**

45. Warm up engine thoroughly to make sure there is no leakage of fuel, or any oil/fluids including engine oil and engine coolant.

46. Bleed air from lines and hoses of applicable lines, such as in cooling system.

47. After cooling down engine, again check oil/fluid levels including engine oil and engine coolant. Refill to the specified level, if necessary.

2012 Model

See Figures 94 through 97.

➥The rotation direction indicated in the text is as viewed from the engine front.

➥When removing components such as hoses, tubes/lines, etc., cap or plug openings to prevent fluid from spilling.

1. Before servicing the vehicle, refer to the Precautions.
2. Remove front road wheel (RH).
3. Remove front fender protector (RH).
4. Drain engine oil. Perform this step when engine is cold. Do not spill engine oil on drive belt.
5. Drain coolant.
6. Remove the intake manifold.
7. Remove the drive belt.
8. Remove the rocker cover.
9. Remove the water pump pulley.
10. Support the bottom surface of engine using a transmission jack, and then remove the engine mounting bracket and insulator (RH).
11. Set No. 1 cylinder at TDC of its compression stroke:
   a. Rotate crankshaft pulley clockwise

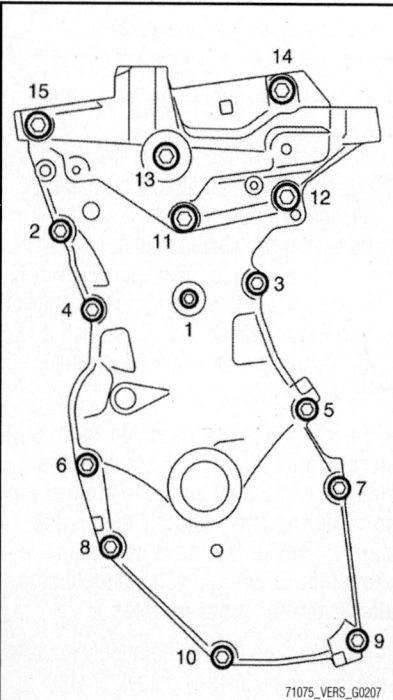

Fig. 94 Front cover bolt tightening sequence—1.6L engine (2012)

and align TDC mark (without paint mark) to timing indicator on front cover.
   b. Check that the matching marks on each of the camshaft sprockets are in the position. If not, rotate the crankshaft pulley one more turn to align the matching marks.
12. Remove crankshaft pulley.
13. Remove front cover:
   a. Loosen bolts in the reverse order of tightening sequence.
   b. Cut liquid gasket by prying and then remove the front cover.
14. Remove front oil seal from front cover using a suitable tool.

✳✳ WARNING

Be careful not to damage the front cover.

15. Remove the chain tensioner:
   a. Fully push down the chain tensioner lever, and then push the plunger into the inside of tensioner. The tab is released by fully pushing the lever down. Then the plunger can be moved.
   b. Pull up the lever to align its hole position with the body hole position.
   • When the lever hole is aligned with the body hole position, the plunger is secured.
   • When the protrusion parts of the plunger ratchet and the tab face each other, both hole positions are not aligned. At that time, correctly engage them and align these hole positions by slightly moving the plunger.
   c. Insert the stopper pin into the body hole through the lever hole, and then secure the lever at the upper position.

➥A hexagonal wrench of 0.098 inch (2.5mm) is used as a stopper pin.

   d. Remove the chain tensioner.
16. Remove the timing chain tension guide and the timing chain slack guide.
17. Remove the timing chain.
   a. Pull the timing chain slack toward the camshaft sprocket (EXH), and then remove the timing chain and start the removal from camshaft sprocket (EXH) side.
   b. Never rotate the crankshaft or camshaft while the timing chain is removed. It causes interference between valve and piston.
18. Remove the crankshaft sprocket and the oil pump drive related parts:
   a. Remove oil pump drive chain tensioner.

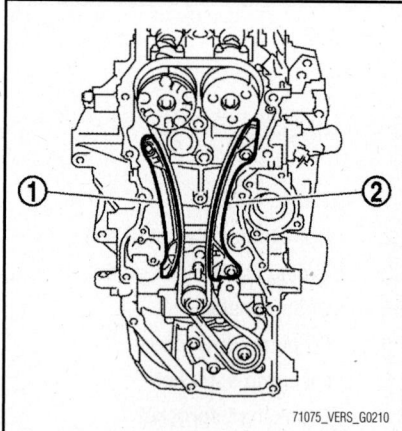

Fig. 95 Remove the timing chain tension guide (2) and the timing chain slack guide (1)—1.6L engine (2012)

   b. Pull out from the shaft and spring attaching holes.
   c. Hold the top of the oil pump shaft using the TORX® socket (size: E8), and then loosen the oil pump sprocket nut and remove it.
   d. Remove the crankshaft sprocket, the oil pump drive chain, and the oil pump sprocket at the same time.

To install:

➥For installation, follow the relationship between the matching mark on each timing chain and that of the corresponding sprocket, with the components installed.

19. Install the crankshaft sprocket and the oil pump drive related parts:
   a. Install the crankshaft sprocket, the oil pump drive chain, and the oil pump sprocket at the same time.
   • Install the crankshaft sprocket so that its invalid gear area is toward the back of the engine.

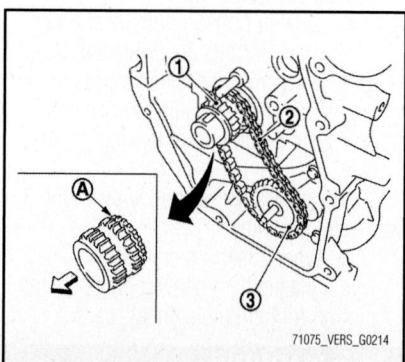

Fig. 96 View of crankshaft sprocket (1), oil pump drive chain (2), oil pump sprocket (3) and invalid gear area (A)—1.6L engine (2012)

- Install the oil pump sprocket so that its protrusion faces the front of engine.

→ **There is no matching mark in the oil pump drive related parts.**

b. Hold the top of the oil pump shaft using the TORX® socket (size: E8), and then tighten the oil pump sprocket nut.

c. Install oil pump drive chain tensioner.
- Insert the body into the shaft while inserting the spring into the attaching hole of cylinder block front surface.
- Check that the tension is applied to the oil pump drive chain after installing.

20. Install the timing chain:
  a. Install by aligning matching marks on each sprocket and timing chain.
  b. If these matching marks are not aligned, rotate the camshaft slightly to correct the position.

→ **After the matching marks are aligned, keep them aligned by holding them.**

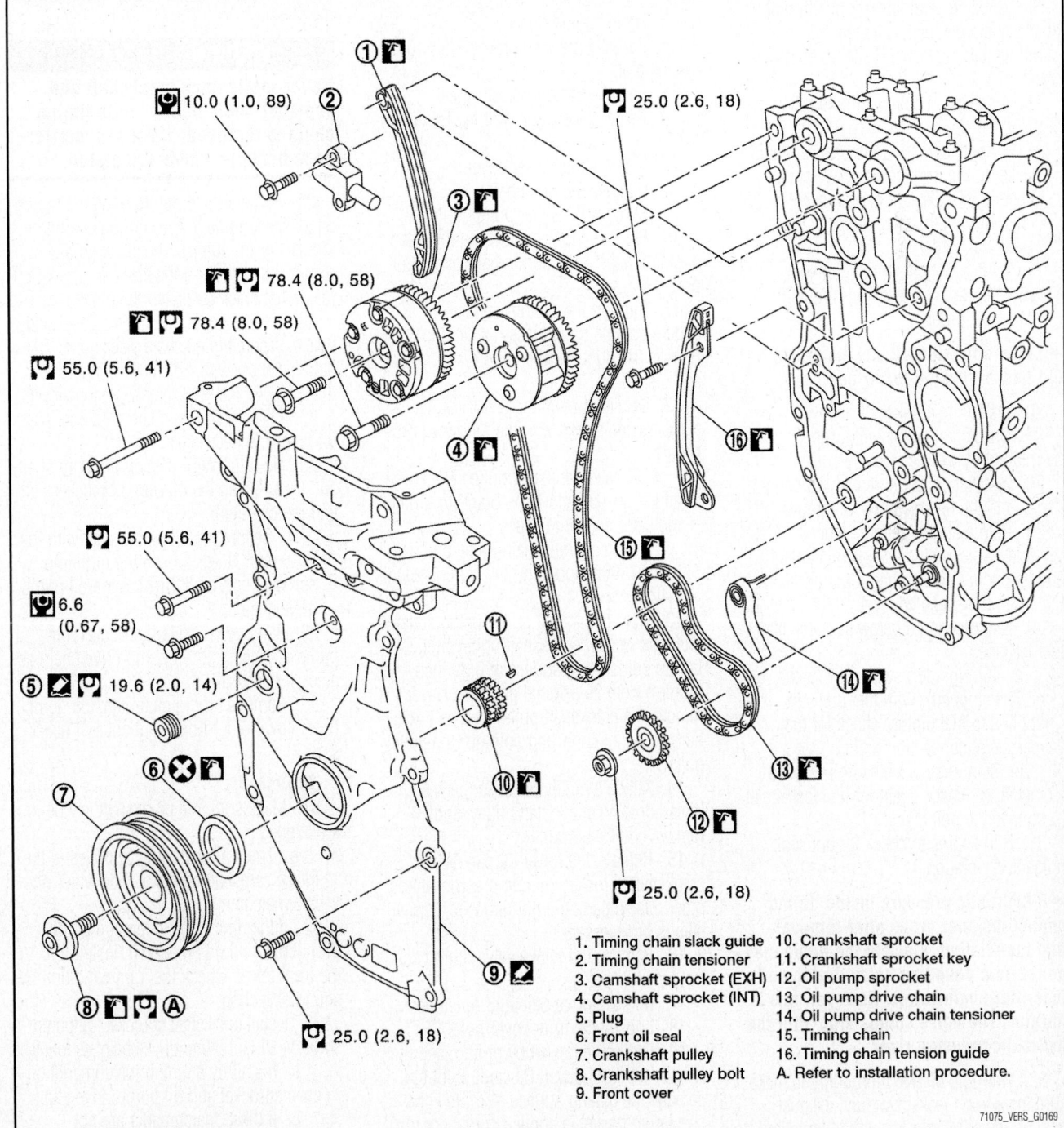

1. Timing chain slack guide
2. Timing chain tensioner
3. Camshaft sprocket (EXH)
4. Camshaft sprocket (INT)
5. Plug
6. Front oil seal
7. Crankshaft pulley
8. Crankshaft pulley bolt
9. Front cover
10. Crankshaft sprocket
11. Crankshaft sprocket key
12. Oil pump sprocket
13. Oil pump drive chain
14. Oil pump drive chain tensioner
15. Timing chain
16. Timing chain tension guide
A. Refer to installation procedure.

71075_VERS_G0169

Fig. 97 Timing chain and related component locations—1.6L engine (2012)

c. To avoid skipped teeth, never rotate crankshaft and camshaft until front cover is installed.

21. Install timing chain tension guide and timing chain slack guide.

22. Install the chain tensioner.

a. Secure the plunger at the most compressed position using a stopper pin, and then install it.

b. Pull out the stopper pin after installing the chain tensioner.

23. Check matching mark position of timing chain and each sprocket again.

24. Install the front oil seal to the front cover.

25. Install front cover with the following procedure:

a. Apply a continuous bead of liquid gasket to the front cover. Use Genuine Silicone RTV Sealant or equivalent.

b. Tighten the cover bolts in the numerical order.

c. After all bolts are tightened, retighten them to specified torque in numerical order.

➡ **Be sure to wipe off any excessive liquid gasket leaking to surface.**

26. Install the crankshaft pulley.

27. Check that crankshaft turns smoothly by rotating by hand clockwise.

28. Installation of the remaining components is in the reverse order of removal.

29. Before starting engine, check oil/fluid levels, including engine coolant and engine oil. If less than required quantity, fill to the specified level.

30. Use procedure below to check for fuel leakage.

a. Turn ignition switch ON (with engine stopped). With fuel pressure applied to fuel piping, check for fuel leakage at connection points.

b. Start engine. With engine speed increased, check again for fuel leakage at connection points.

31. Run engine to check for unusual noise and vibration.

➡ **If hydraulic pressure inside timing chain tensioner drops after removal and installation, slack in the guide may generate a pounding noise during and just after engine start. However, this is normal. The noise should stop after the hydraulic pressure rises.**

32. Warm up engine thoroughly to make sure there is no leakage of fuel, exhaust gas, or any oils/fluids including engine oil and engine coolant.

33. Bleed air from passages in lines and hoses, such as in cooling system.

34. After cooling down engine, again check oil/fluid levels including engine oil and engine coolant. Refill to specified level, if necessary.

1.8L Engine

See Figures 98 through 102.

1. Before servicing the vehicle, refer to the Precautions.

2. Disconnect the negative battery cable.

3. Drain engine oil.

4. Partially drain engine coolant from the radiator.

5. Remove the right front wheel.

6. Remove the right front fender protector.

7. Remove the rocker cover.

8. Remove the drive belt.

9. Remove the water pump pulley.

10. Remove the ground cable (between engine bracket (RH) and radiator core support).

11. Support the bottom surface of engine using a transmission jack, and remove the right-hand engine bracket and insulator.

12. Set No. 1 cylinder at TDC on its compression stroke with the following procedure:

a. Rotate crankshaft pulley clockwise and align TDC mark (no paint) to timing indicator on front cover.

b. Rrotate crankshaft pulley one revolution (360°) and align the cam noses of cylinder one.

13. Hold crankshaft pulley using suitable tool, and loosen crankshaft pulley bolt. Locate bolt seating surface at 0.39 inch (10mm) from its original position. Never remove the crankshaft pulley bolt as it will be used as a supporting point for the pulley puller.

14. Attach a pulley puller in the M6 thread hole on crankshaft pulley, and remove crankshaft pulley.

15. Remove the lower oil pan. When crankshaft sprocket, oil pump sprocket and other related parts are not removed, this step is unnecessary.

16. Remove the intake valve timing control solenoid valve.

17. Remove drive belt auto-tensioner.

18. Loosen the front cover bolts.

19. Cut liquid gasket by prying, and then remove the front cover. Be careful not to damage the mating surface. A more adhesive liquid gasket is applied compared to previous types when shipped, so it should not be forced off the position not specified.

20. Remove front oil seal from front cover. Lift up front oil seal using a suitable tool. Be careful not to damage front cover.

21. Push in timing chain tensioner plunger.

22. Insert a stopper pin into the body hole to retain the plunger in collapsed position. Use approximately 0.059 inch (1.5mm) diameter hard metal pin as a stopper pin.

23. Remove the timing chain tensioner.

24. Remove the timing chain slack guide, timing chain tension guide and timing chain.

✳✳ WARNING

Never rotate each crankshaft and camshaft individually while timing chain is removed. It causes interference between valve and piston.

25. Press stopper tab to push the timing chain slack guide toward timing chain tensioner (for oil pump). The slack guide is released by pressing the stopper tab. As a result, the slack guide can be moved.

26. Insert stopper pin into tensioner body hole to secure timing chain slack guide. Use a hard metal pin with a diameter of approximately 0.047 inch (1.2mm) as a stopper pin.

27. Remove timing chain tensioner (for oil pump), if necessary.

28. Hold the WAF part of oil pump shaft, and then loosen the oil pump sprocket bolt and remove them.

a. Secure the oil pump shaft with the WAF part. Never loosen the oil pump sprocket bolt by tightening the oil pump drive chain.

29. Remove crankshaft sprocket, oil pump sprocket and oil pump drive chain as a set, if necessary.

30. Remove timing chain tension guide (front cover side) from front cover if necessary.

To install:

31. Make sure that crankshaft key points are aligned.

a. There are two outer grooves in the intake camshaft sprocket. The wider one is a matchmark.

32. If the timing chain tension guide (front cover side) is removed, install it to the front cover. Check the joint condition by sound or feeling.

33. Install crankshaft sprocket, oil pump sprocket and oil pump drive chain, as follows:

a. Install by aligning matchmarks on each sprocket and oil pump drive chain.

b. If these matchmarks are not aligned, rotate the oil pump shaft slightly to correct the position.

c. Check matchmark position of each

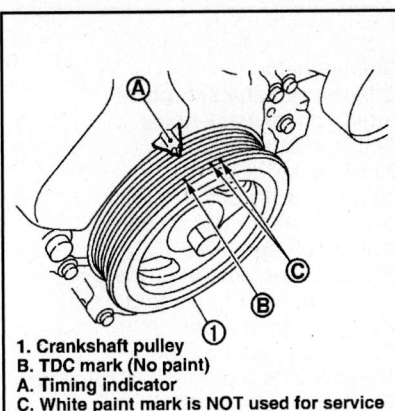

1. Crankshaft pulley
B. TDC mark (No paint)
A. Timing indicator
C. White paint mark is NOT used for service

22140_VERS_G0047

Fig. 98 Rotate crankshaft pulley and align TDC mark to timing indicator

sprocket after installing the oil pump drive chain.

34. Hold the WAF part of oil pump shaft, and then tighten the oil pump sprocket bolt.

35. Secure the oil pump shaft with the WAF part. Never loosen the oil pump sprocket bolt by tightening the oil pump drive chain.

36. Install chain tensioner (for oil pump), as follows:

   a. Secure the plunger at the most compressed position using a stopper pin, and install it.

   b. Securely pull out the stopper pin after installing the chain tensioner (for oil pump).

   c. Check matchmark position of oil pump drive chain and each sprocket again.

37. Align the matchmarks of each sprocket with the matchmarks of timing chain. There are two outer grooves in the intake camshaft sprocket. The wider one is a matchmark. If these matchmarks are not aligned, rotate the camshaft slightly by holding the hexagonal portion to correct the position.

38. Check matchmark position of each sprocket and timing chain again after installing the timing chain.

39. Install the timing chain tension guide and the timing chain slack guide.

40. Install timing chain tensioner, as follows:

   a. Secure the plunger at the most compressed position using a stopper pin, and install it.

   b. Securely pull out the stopper pin after installing the timing chain tensioner.

41. Check matchmark position of timing chain and each sprocket again.

42. Apply new engine oil to new front oil seal joint surface.

43. Using a suitable tool install front oil

seal so that each seal lip is oriented properly. Press-fit front oil seal until it is flush with front end surface of front cover below with a suitable tool. Within 0.012 inch (0.3mm) toward engine front, within 0.020 inch (0.5mm) toward engine rear. Press-fit oil seal straight to avoid causing burrs or tilting.

❊❊ WARNING

Be careful not to damage front cover and crankshaft. Never touch grease applied onto oil seal lip.

44. Install new O-ring to cylinder block.

45. Apply the sealant without breaks to the specified location using Tool No.

WS39930000. Use Genuine Silicone RTV Sealant, or equivalent.

46. Make sure that matching marks of timing chain and each sprocket are still aligned.

47. Install front cover, and tighten bolts in numerical order.

➡ Attaching should be done within 5 minutes after liquid gasket application.

48. Bolt installation positions:
   - M6 bolts: No. 1
   - M10 bolts: No. 6, 7, 10, 11, 14
   - M12 bolts: No. 2, 4, 8, 12
   - M8 bolts: Except the above

49. Tighten all bolts are in two stages to specified torque in numerical order. Be sure

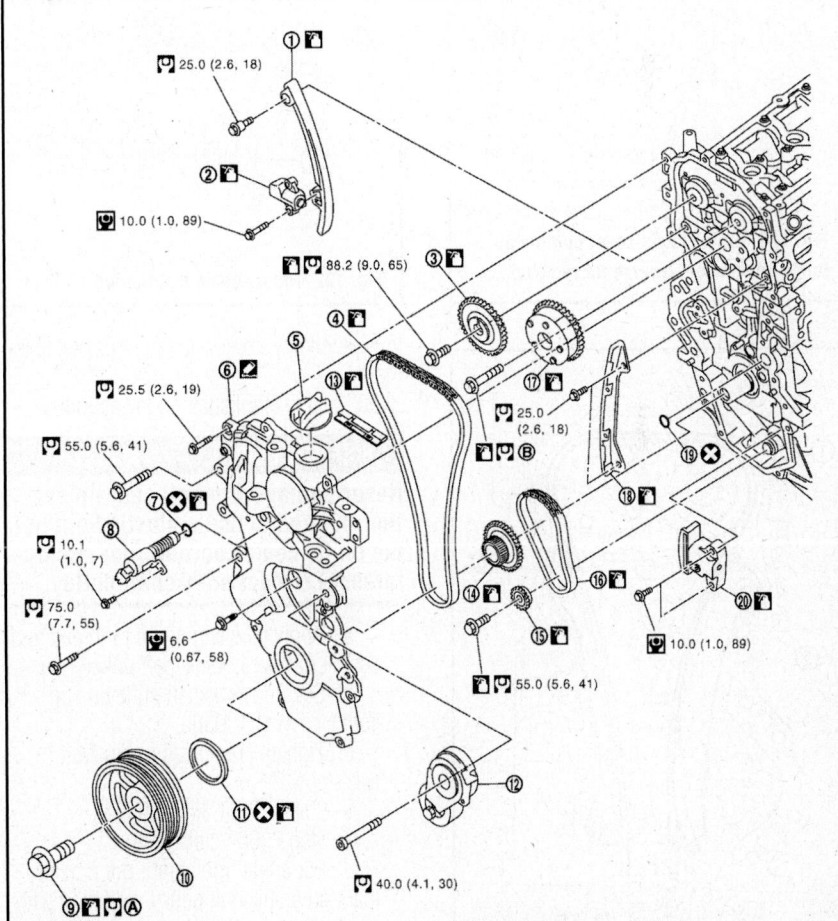

1. Timing chain slack guide
2. Timing chain tensioner
3. Camshaft sprocket (EXH)
4. Timing chain
5. Oil filler cap
6. Front cover
7. O-ring
8. Intake valve timing control solenoid valve
9. Crankshaft pulley bolt
10. Crankshaft pulley
11. Front oil seal
12. Drive belt auto-tensioner
13. Timing chain tension guide (front cover side)
14. Crankshaft sprocket
15. Oil pump sprocket
16. Oil pump drive chain
17. Camshaft sprocket (INT)
18. Timing chain tension guide
19. O-ring
20. Chain tensioner (for oil pump)

37663_VERS_G0284

Fig. 99 Timing chain and related component locations—1.8L engine

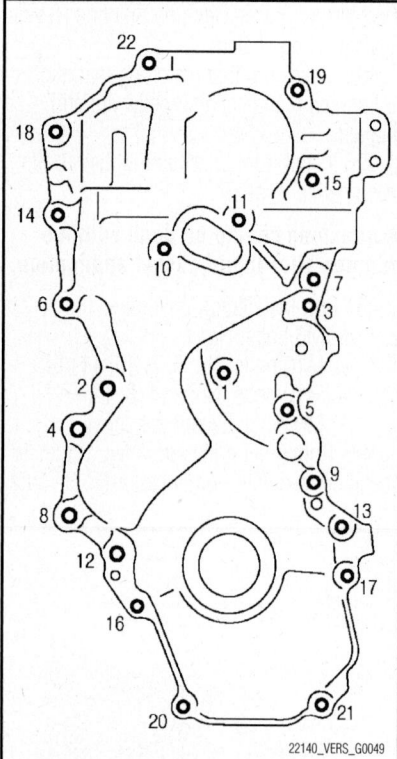

Fig. 100 Timing chain cover bolt installation sequence—reverse for removal

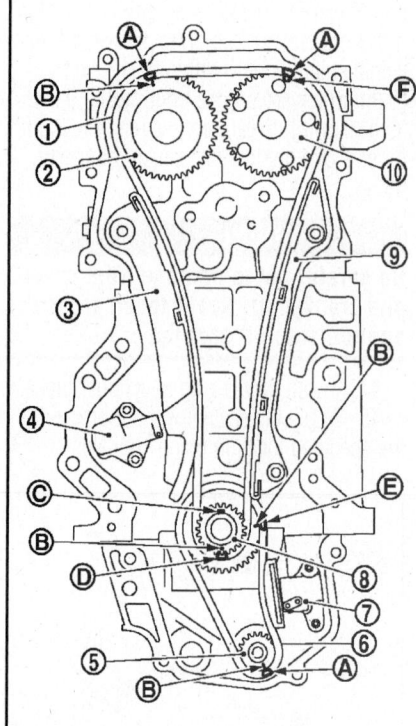

1. Timing chain
2. Exhaust camshaft sprocket
3. Timing chain slack guide
4. Timing chain tensioner
5. Oil pump sprocket
6. Oil pump drive chain
7. Chain tensioner (for oil pump)
8. Crankshaft sprocket
9. Timing chain tension guide
10. Intake camshaft sprocket
A. Matchmark (dark blue link)
B. Matchmark (stamping)
C. Crankshaft key position (straight up)
D. Matchmark (gold link)
E. Matchmark (orange link)
F. Matchmark (outer groove)

Fig. 102 Timing chain positioning

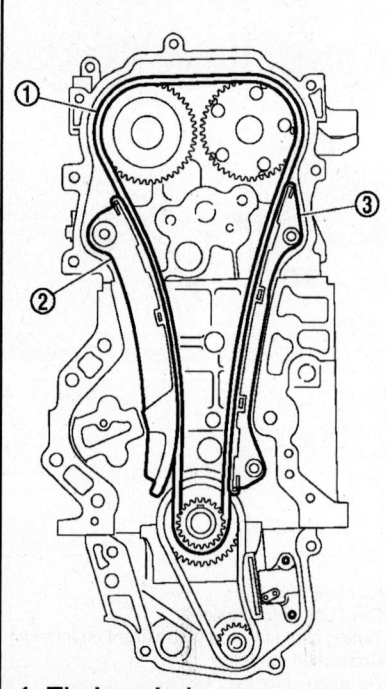

1. Timing chain
2. Timing chain slack guide
3. Timing chain tension guide

Fig. 101 Remove the timing chain guides

to wipe off any excessive liquid gasket leaking.

50. Install crankshaft pulley as follows:

✳✳ WARNING

Never damage front oil seal lip section. If needed use a plastic hammer, tap on its center portion (not circumference) to seat crankshaft pulley.

a. Apply new engine oil to thread and seat surfaces of crankshaft pulley bolt.

b. Secure crankshaft pulley using Tool No. KV10109300.

c. Tighten crankshaft pulley bolt in two steps:
  • Step 1: 22 ft. lbs. (29 Nm).
  • Step 2: 60° clockwise.

d. For angle tightening, put a paint mark on crankshaft pulley matching with any one of six easy to recognize angle marks on crankshaft pulley bolt flange.

e. Turn 60° clockwise (angle tightening).

f. Check the tightening angle with movement of one angle mark.

g. Make sure that crankshaft rotates clockwise smoothly.

51. Install the oil pan (lower).

52. Install the engine mounting bracket and insulator (RH).

53. Install the ground cable.

54. Install the water pump pulley.

55. Install the drive belt.

56. Install the rocker cover.

57. Install the front fender protector (RH).

58. Install the front wheel (RH).

59. Connect the negative battery cable.

60. Before starting engine, check oil/fluid levels including engine coolant and engine oil. If less than required quantity, fill to the specified level.

61. Use procedure below to check for fuel leakage:

a. Turn ignition switch "ON" (with engine stopped). With fuel pressure applied to fuel piping, check for fuel leakage at connection points.

b. Start engine. With engine speed increased, check again for fuel leakage at connection points.

62. Run engine to check for unusual noise and vibration.

➡If hydraulic pressure inside chain tensioner drops after removal/installation, slack in guide may generate a pounding noise during and just after the engine start. However, this does not indicate an unusualness. Noise will stop after hydraulic pressure rises.

63. Warm up engine thoroughly to make sure there is no leakage of fuel, or any oil/fluids including engine oil and engine coolant.

64. Bleed air from lines and hoses of applicable lines, such as in cooling system.

65. After cooling down engine, again check oil/fluid levels including engine oil and engine coolant. Refill to the specified level, if necessary.

VALVE LASH

ADJUSTMENT

See Figures 103 through 106.

1. Before servicing the vehicle, refer to the Precautions.

2. Measure the valve clearance with the following procedure:

3. Set No. 1 cylinder at TDC of its compression stroke by rotating the crankshaft pulley clockwise and align TDC mark (no paint) to timing indicator on front cover. At the same time, make sure that both intake and exhaust cam noses of No. 1 cylinder face inside. If they do not, rotate crankshaft pulley once more (360 degrees) and align.

4. Use a feeler gauge, measure the clearance between valve lifter and camshaft.

5. Intake valve clearance:
   - Cold: 0.010–0.013 inches (0.26–0.34 mm)
   - Hot: 0.012–0.016 inches (0.304–0.416 mm)

6. Exhaust valve clearance:
   - Cold: 0.011–0.015 inches (0.29–0.37 mm)
   - Hot: 0.012–0.017 inches (0.308–0.432 mm)

7. Set No. 4 cylinder at TDC of its compression stroke by rotating crankshaft pulley one revolution (360 degrees) and align TDC mark (no paint) to timing indicator on front cover.

8. Use a feeler gauge, measure the clearance between valve lifter and camshaft.

9. If out of standard, perform adjustment. Perform adjustment depending on selected head thickness of valve lifter.

10. Remove camshaft.

11. Remove valve lifters at the locations that are out of the standard.

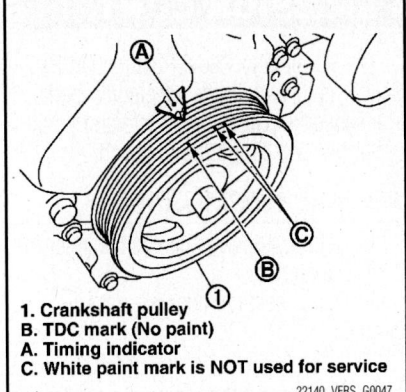

1. Crankshaft pulley
B. TDC mark (No paint)
A. Timing indicator
C. White paint mark is NOT used for service

22140_VERS_G0047

Fig. 103 Set No. 1 cylinder at TDC of its compression stroke

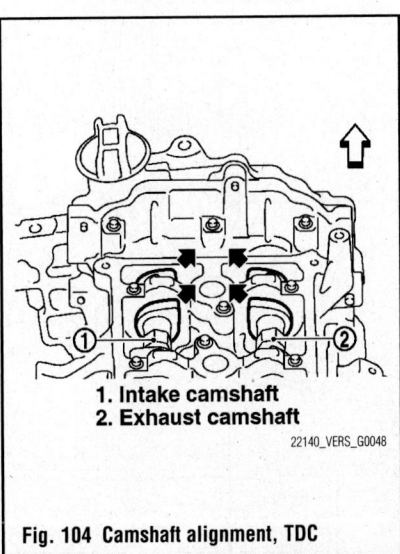

1. Intake camshaft
2. Exhaust camshaft

22140_VERS_G0048

Fig. 104 Camshaft alignment, TDC

12. Measure the center thickness of the removed valve lifters with a micrometer.

13. Use the equation below to calculate valve lifter thickness for replacement:
   - Valve lifter thickness calculation: $t = t1 + (C1 - C2)$
   - t=Valve lifter thickness to be replaced
   - t1=Removed valve lifter thickness
   - C1=Measured valve clearance
   - C2=Standard valve clearance: Intake: 0.012 inch (0.30mm), Exhaust: 0.013 inch (0.33mm).
   - Available thickness of valve lifter: 26 sizes range 0.1181–0.1378 inch

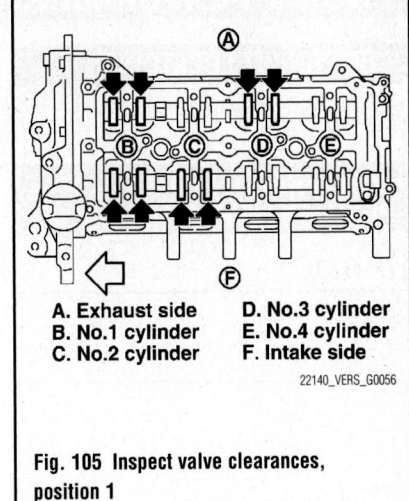

A. Exhaust side
B. No.1 cylinder
C. No.2 cylinder
D. No.3 cylinder
E. No.4 cylinder
F. Intake side

22140_VERS_G0056

Fig. 105 Inspect valve clearances, position 1

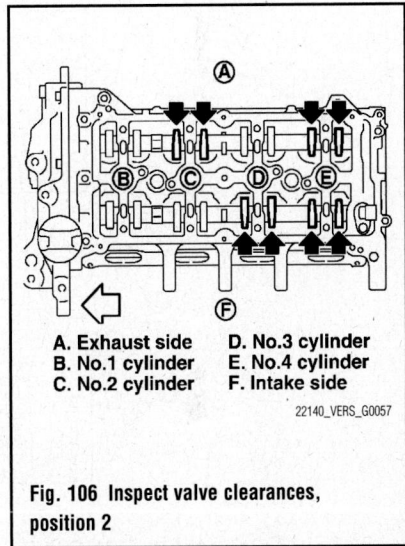

A. Exhaust side
B. No.1 cylinder
C. No.2 cylinder
D. No.3 cylinder
E. No.4 cylinder
F. Intake side

22140_VERS_G0057

Fig. 106 Inspect valve clearances, position 2

(3.00–3.50mm) in steps of 0.0008 inch (0.02mm) (when manufactured at factory)

➡ **A stamp mark of "302" indicates 0.1189 inch (3.02mm) in thickness.**

14. Install the selected valve lifter.

15. Install camshaft.

16. Install timing chain and related parts.

17. Manually rotate crankshaft pulley a few rotations.

18. Make sure that the valve clearances is within the standard.

ENGINE PERFORMANCE & EMISSION CONTROLS

COMPONENT LOCATIONS

See Figures 107 through 111.

ACCELERATOR PEDAL POSITION (APP) SENSOR

LOCATION

See Figure 112.

The Accelerator Pedal Position (APP) Sensor is located in the accelerator pedal. The sensor is part of the accelerator pedal assembly, and cannot be removed.

REMOVAL & INSTALLATION

1. Before servicing the vehicle, refer to the Precautions.
2. Disconnect the negative battery terminal.

3. Disconnect the Accelerator Pedal Position (APP) sensor harness connector.
4. Loosen mounting bolts, and remove accelerator pedal assembly.

❋❋ WARNING

Do not disassemble accelerator pedal assembly. Do not remove accelerator pedal position sensor from accelerator pedal assembly.

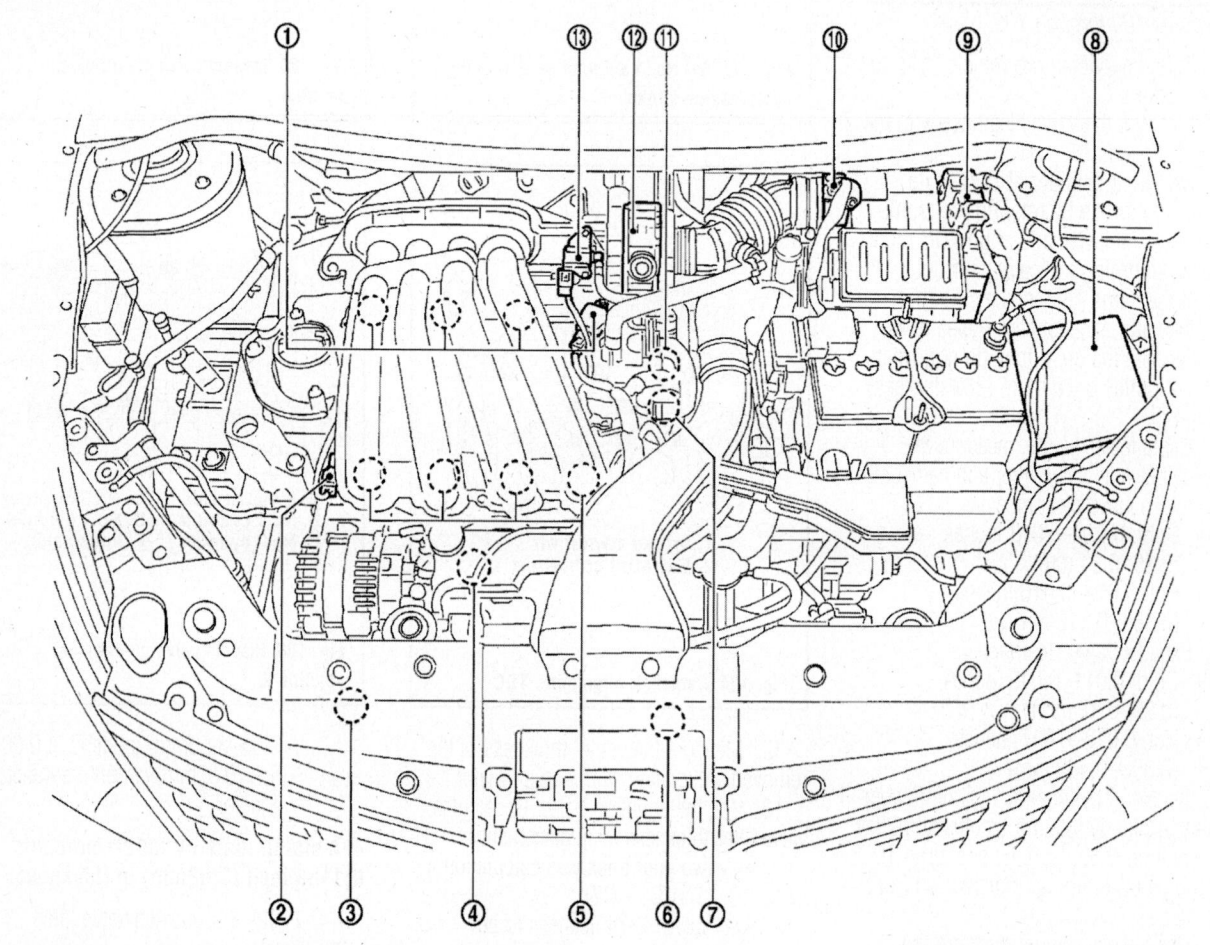

1. Ignition coil (with power transistor) and spark plug
2. Intake valve timing control solenoid valve
3. Refrigerant pressure sensor
4. Knock sensor
5. Fuel injector
6. Cooling fan motor
7. Camshaft position sensor (PHASE)
8. IPDM E/R
9. ECM
10. Mass air flow sensor (with intake air temperature sensor)
11. Engine coolant temperature sensor
12. Electric throttle control actuator (with built in throttle position sensor and throttle control motor)
13. EVAP canister purge volume control solenoid valve

37663_VERS_G0332

Fig. 107 Engine control component locations—1.6L engine (2011)

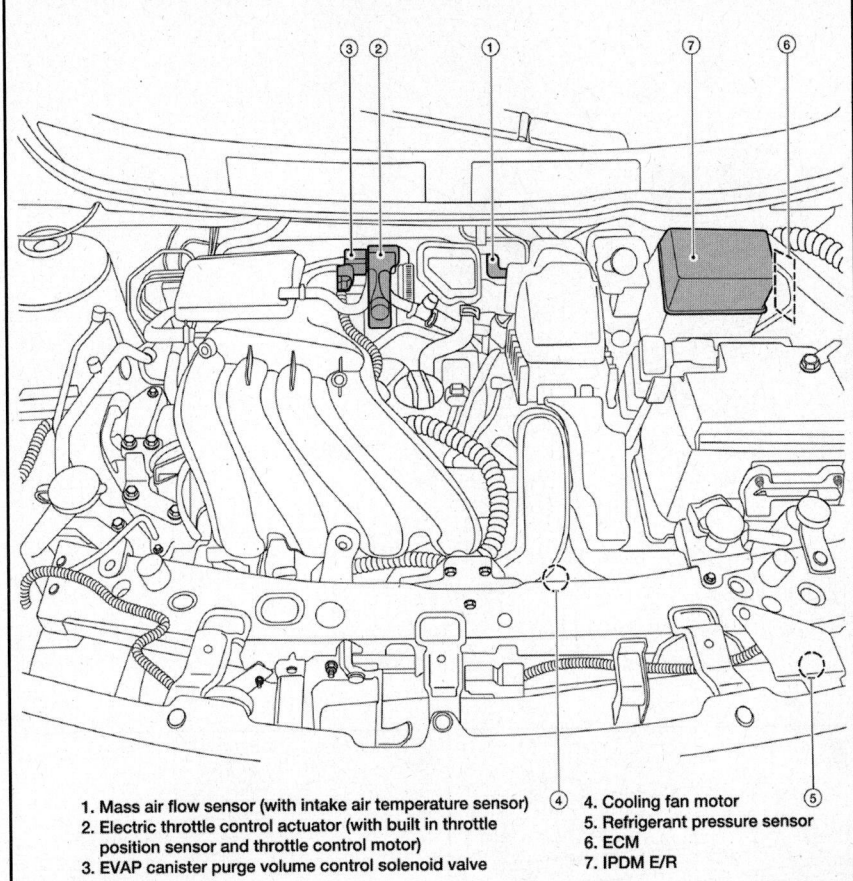

1. Mass air flow sensor (with intake air temperature sensor)
2. Electric throttle control actuator (with built in throttle position sensor and throttle control motor)
3. EVAP canister purge volume control solenoid valve
4. Cooling fan motor
5. Refrigerant pressure sensor
6. ECM
7. IPDM E/R

71075_VERS_G0226

Fig. 108 Engine control component locations (view 1 of 2)—1.6L engine (2012)

To install:

5. Installation is the reverse of the removal procedure.

6. Insert locating pin into vehicle side to position accelerator pedal assembly. Install mounting bolt to accelerator pedal assembly.

7. Align the stud bolt on the floor with the thread hole to insert accelerator pedal stopper until it contacts the face.

8. Perform accelerator pedal released position learning.

ACCELERATOR PEDAL RELEASED POSITION LEARNING

1. Before servicing the vehicle, refer to the Precautions.

2. Check that accelerator pedal is fully released.

3. Turn ignition switch ON and wait at least 2 seconds.

4. Turn ignition switch OFF and wait at least 10 seconds.

5. Turn ignition switch ON and wait at least 2 seconds.

6. Turn ignition switch OFF and wait at least 10 seconds.

AIR-FUEL RATIO (A/F) SENSOR

LOCATION

See Figures 113 and 114.

REMOVAL & INSTALLATION

1.6L Engine

✳✳ CAUTION

To avoid the danger of being burned, do not touch the exhaust system when the system is hot. The Air-Fuel Ratio (AFR) sensor removal should be performed when the system is cool.

1. Before servicing the vehicle, refer to the Precautions.

2. Remove the harness bracket of air fuel ratio sensor 1 from the cylinder head.

3. Remove exhaust manifold covers.

4. Remove the air fuel ratio sensor 1.

5. Use Tool No. KV10117100 to remove air fuel ratio sensor 1.

✳✳ CAUTION

Handle air fuel ratio sensor 1 carefully and avoid impacts.

6. Installation is the reverse of the removal procedure.

1.8L Engine

✳✳ CAUTION

To avoid the danger of being burned, do not touch the exhaust system when the system is hot. The Air-Fuel Ratio (AFR) sensor removal should be performed when the system is cool.

1. Before servicing the vehicle, refer to the Precautions.

2. Remove exhaust manifold cover.

3. Remove the A/F sensor 1, using Tool No. KV991J0050 (J-44626). Handle it carefully and avoid impacts.

4. Installation is the reverse of the removal procedure.

CAMSHAFT POSITION (CMP) SENSOR

LOCATION

See Figures 115 through 118.

REMOVAL & INSTALLATION

1. Before servicing the vehicle, refer to the Precautions.

2. Loosen the fixing bolt of the sensor.

3. Disconnect the Camshaft Position (CMP) sensor harness connector.

4. Remove the CMP sensor.

5. Installation is the reverse of the removal procedure.

CRANKSHAFT POSITION (CKP) SENSOR

LOCATION

See Figures 119 and 120.

REMOVAL & INSTALLATION

1. Before servicing the vehicle, refer to the Precautions.

2. Loosen the fixing bolt of the sensor.

3. Disconnect the Crankshaft Position (CKP) sensor harness connector.

4. Remove the sensor.

To install:

5. Installation is the reverse of the removal procedure.

6. Tighten the fixing bolt to 62 inch lbs. (7 Nm).

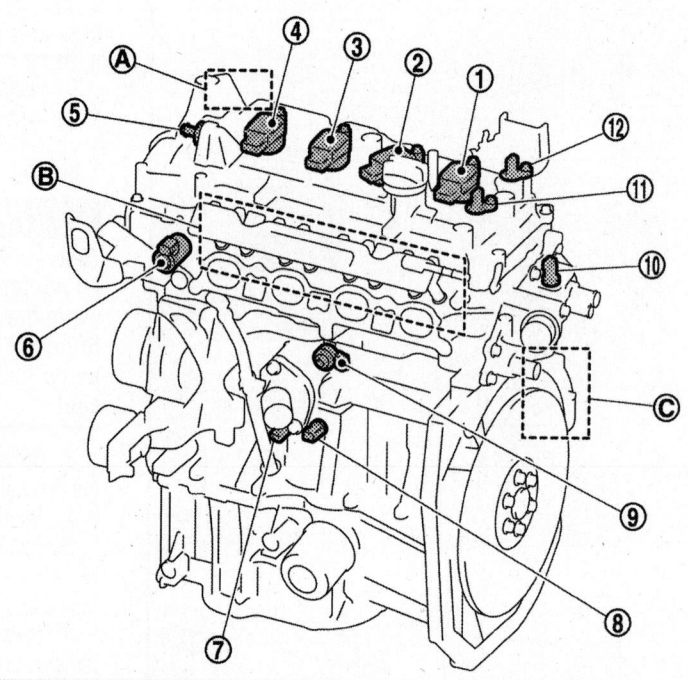

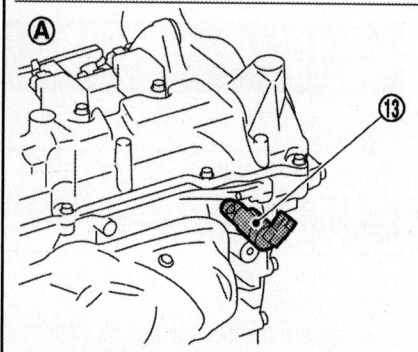

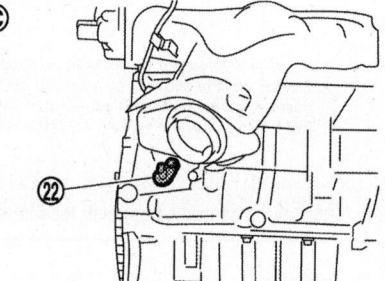

1. Ignition coil No. 4 (with power transistor)
2. Ignition coil No. 3 (with power transistor)
3. Ignition coil No. 2 (with power transistor)
4. Ignition coil No. 1 (with power transistor)
5. PCV valve
6. Intake valve timing control solenoid valve
7. Engine oil pressure sensor
8. Engine oil temperature sensor

9. Knock sensor
10. Engine coolant temperature sensor
11. Intake camshaft position sensor
12. Exhaust camshaft position sensor
13. Exhaust valve timing control solenoid valve
14. Fuel injector No. 1 (Front)
15. Fuel injector No. 1 (Rear)
16. Fuel injector No. 2 (Front)
17. Fuel injector No. 2 (Rear)

18. Fuel injector No. 3 (Front)
19. Fuel injector No. 3 (Rear)
20. Fuel injector No. 4 (Front)
21. Fuel injector No. 4 (Rear)
22. Crankshaft position sensor
A. Engine front right side
B. Left view of the engine
C. Engine rear right side

71075_VERS_G0225

Fig. 109 Engine control component locations (view 2 of 2)—1.6L engine (2012)

ELECTRONIC CONTROL MODULE (ECM)

LOCATION

See Figures 121 and 122.

The Electronic Control Module (ECM) is located in the engine room at the left side near the battery.

REMOVAL & INSTALLATION

Use the following precautions when servicing the Engine Control Module (ECM):

• Always turn the ignition switch OFF and disconnect the negative battery cable before any repair or inspection work. The open/short circuit of related switches, sensors, solenoid valves, etc. will cause the MIL to illuminate.

• Always connect and lock the connectors securely after work. A loose (unlocked) connector will cause the MIL to illuminate due to the open circuit. Keep the connector free from water, grease, dirt, bent terminals, etc.

• Always to route and secure the harnesses properly after work. The interference of the harness with a bracket, etc. may

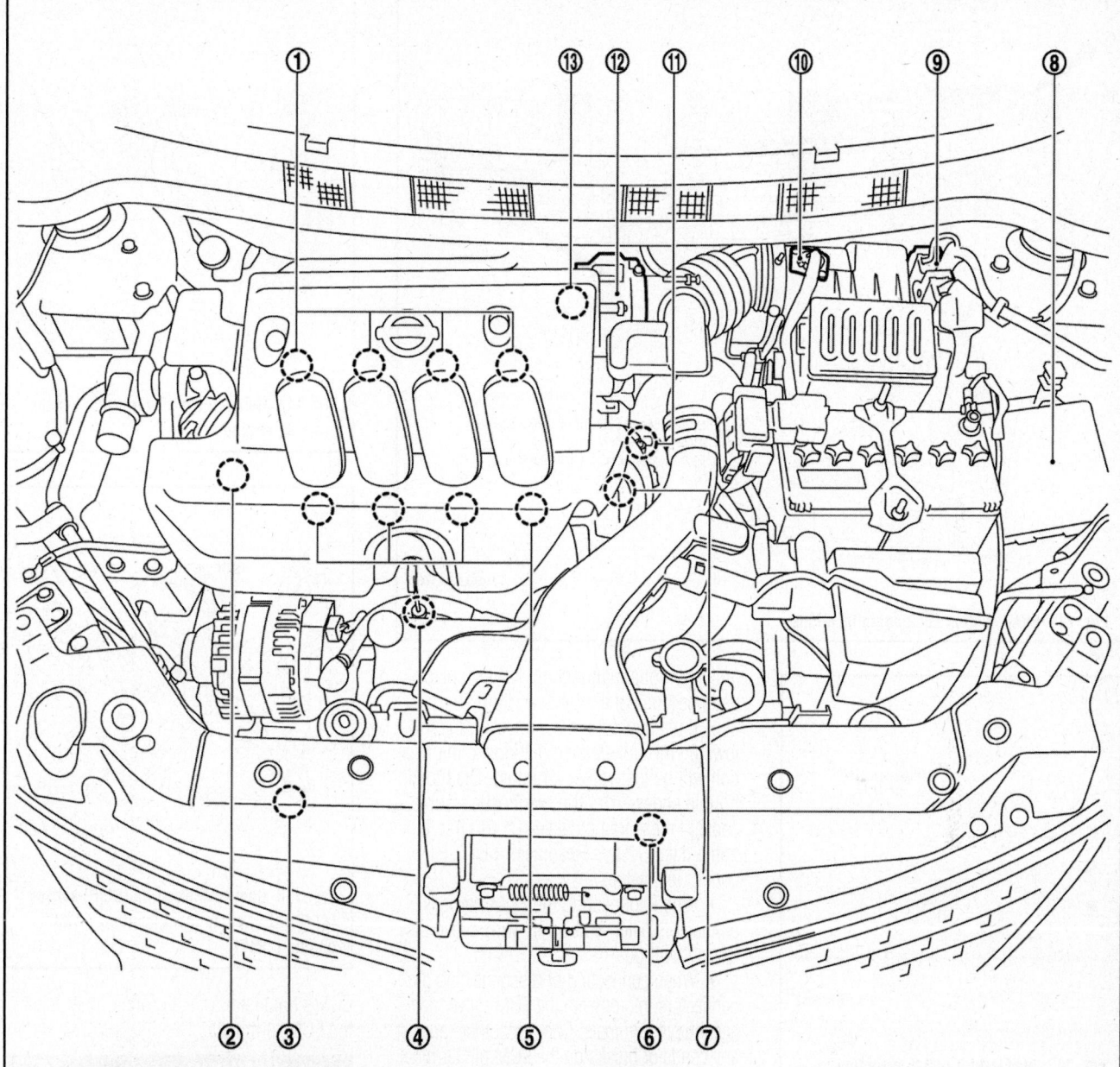

1. Ignition coil (with power transistor) and spark plug
2. Intake valve timing control solenoid valve
3. Refrigerant pressure sensor
4. Knock sensor
5. Fuel injector
6. Cooling fan motor
7. Camshaft position sensor (PHASE)
8. IPDM E/R
9. ECM
10. Mass air flow sensor (with intake air temperature sensor)
11. Engine coolant temperature sensor
12. Electric throttle control actuator (with built-in throttle position sensor, throttle control motor)
13. EVAP canister purge volume control solenoid valve

71075_VERS_G0221

Fig. 110 Engine control component locations—1.8L engine

cause the MIL to illuminate due to the short circuit.
• Always use a 12 volt battery as the power source.
• Never attempt to disconnect the battery cables while the engine is running.

• Before connecting or disconnecting the ECM harness connector, turn the ignition switch OFF and disconnect the negative battery cable. Failure to do so may damage the ECM because battery voltage is applied to the ECM even if the ignition switch is turned OFF.

• Never disassemble the ECM.
• If a battery cable is disconnected, the memory will return to the ECM value. The ECM will start to self-control at its initial value. Thus, engine operation can vary slightly in this case. However, this is not an

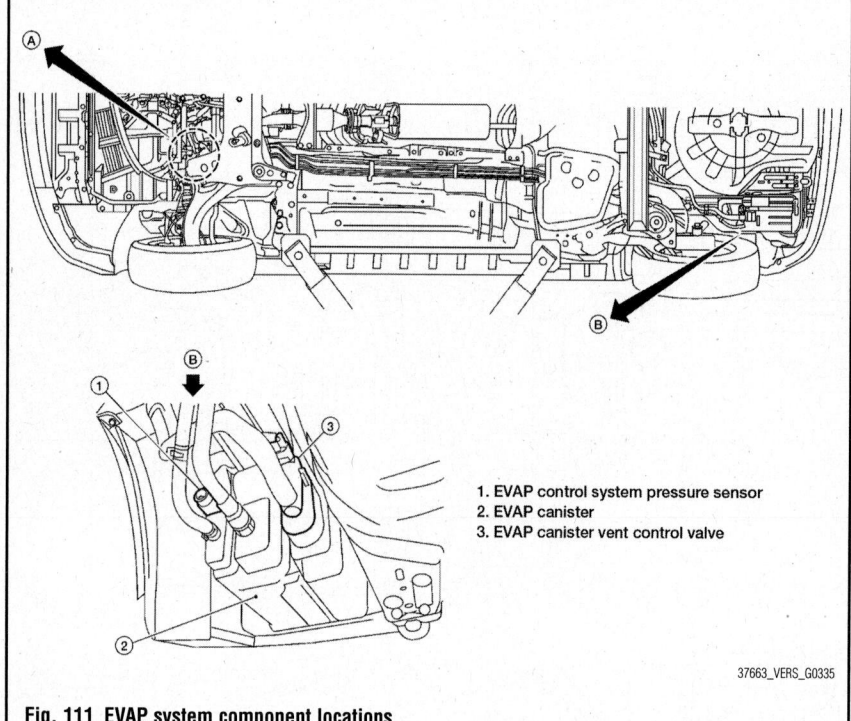

1. EVAP control system pressure sensor
2. EVAP canister
3. EVAP canister vent control valve

37663_VERS_G0335

Fig. 111 EVAP system component locations

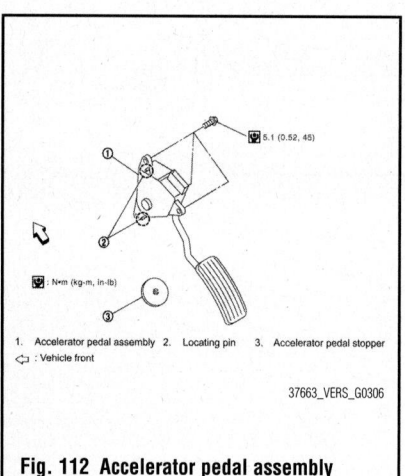

5.1 (0.52, 45)

N•m (kg-m, in-lb)

1. Accelerator pedal assembly 2. Locating pin 3. Accelerator pedal stopper
⬋ : Vehicle front

37663_VERS_G0306

Fig. 112 Accelerator pedal assembly

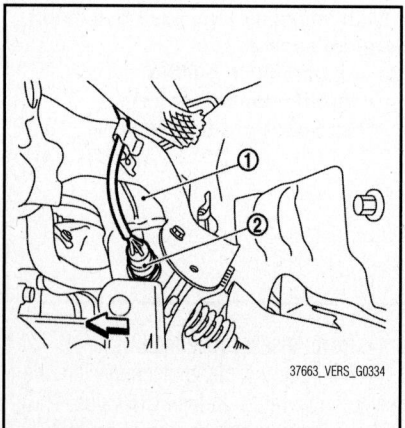

37663_VERS_G0334

Fig. 113 Air fuel ratio sensor 1 (2), exhaust manifold (1)—1.6L engine

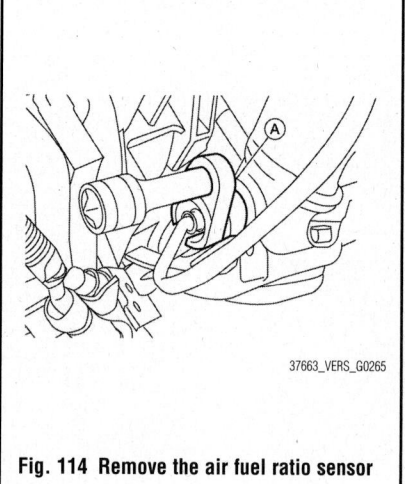

37663_VERS_G0265

Fig. 114 Remove the air fuel ratio sensor 1—1.8L engine

71075_VERS_G0222

Fig. 115 Camshaft Position (CMP) sensor (4) (PHASE) component location—1.6L engine (2011)

indication of a malfunction. Never replace parts because of a slight variation.

• If the battery is disconnected, the following emission-related diagnostic information will be lost within 24 hours: Diagnostic trouble codes—1st trip diagnostic trouble codes; Freeze frame data—1st trip freeze frame data; System Readiness Test (SRT) codes; Test values.

• When connecting the ECM harness connector, fasten it securely with a lever as far as it will go.

• When connecting or disconnecting pin connectors into or from the ECM, never damage the pin terminals. Check that there are not any bends or breaks on the ECM pin terminal when connecting the pin connectors.

• Securely connect the ECM harness connectors. A poor connection can cause an extremely high (surge) voltage to develop in the coil and condenser, thus resulting in damage to the ICs.

• Keep the engine control system harness at least 4 inches (10cm) away from adjacent harness, to prevent engine control system malfunctions due to receiving external noise, degraded operation of ICs, etc.

• When measuring ECM signals with a circuit tester, never allow the two tester probes to contact. Accidental contact of the probes will cause a short circuit and damage the ECM power transistor.

• Never use the ECM ground terminals when measuring input/output voltage. Doing so may result in damage to the ECM's transistor. Use a ground other than the ECM terminals.

❋❋ WARNING

As the Engine Control Module (ECM) consists of precision parts, be careful not to expose it to excessive shock.

1. Before servicing the vehicle, refer to the Precautions.
2. Remove the battery.
3. Remove the IPDM E/R.
4. Remove the IPDM E/R cover.
5. Disconnect the ECM harness connectors.
6. Remove the ECM mounting nuts.
7. Remove the ECM from the ECM bracket.

To install:

8. Installation is the reverse of the removal procedure.
9. Reset electronic systems as necessary.

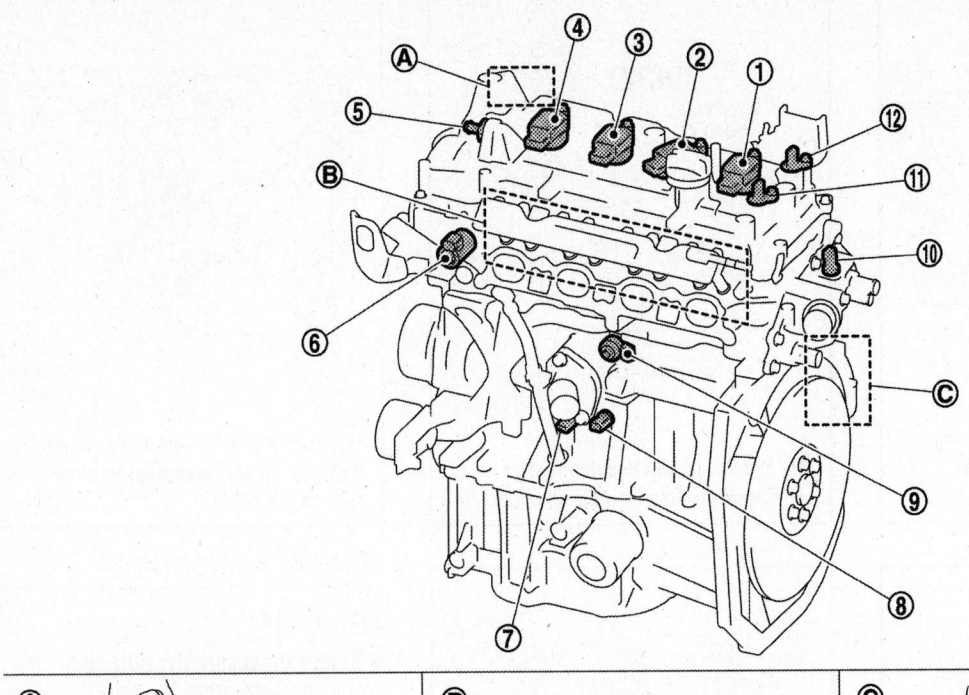

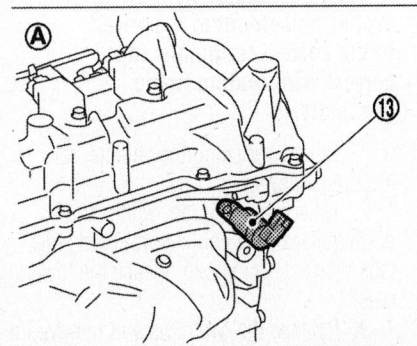

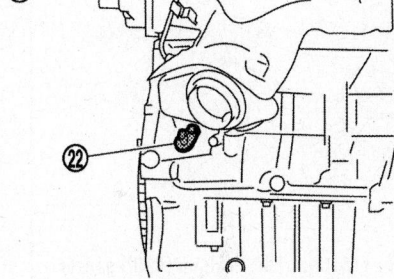

1. Ignition coil No. 4 (with power transistor)
2. Ignition coil No. 3 (with power transistor)
3. Ignition coil No. 2 (with power transistor)
4. Ignition coil No. 1 (with power transistor)
5. PCV valve
6. Intake valve timing control solenoid valve
7. Engine oil pressure sensor
8. Engine oil temperature sensor

9. Knock sensor
10. Engine coolant temperature sensor
11. Intake camshaft position sensor
12. Exhaust camshaft position sensor
13. Exhaust valve timing control solenoid valve
14. Fuel injector No. 1 (Front)
15. Fuel injector No. 1 (Rear)
16. Fuel injector No. 2 (Front)
17. Fuel injector No. 2 (Rear)

18. Fuel injector No. 3 (Front)
19. Fuel injector No. 3 (Rear)
20. Fuel injector No. 4 (Front)
21. Fuel injector No. 4 (Rear)
22. Crankshaft position sensor
A. Engine front right side
B. Left view of the engine
C. Engine rear right side

71075_VERS_G0225

Fig. 116 Engine control component locations (view 2 of 2)—1.6L engine (2012)

ENGINE COOLANT TEMPERATURE (ECT) SENSOR

LOCATION

See Figures 123 through 126.

EVAPORATIVE EMISSION CANISTER

LOCATION

Location
See Figures 127 through 129.

REMOVAL & INSTALLATION

➡**Clean all Evaporative Emission (EVAP) line connections and surrounding areas prior to disconnecting, in order to avoid possible EVAP system contamination.**

➡**The EVAP canister vent control valve and EVAP canister system pressure sensor can be removed without removing the EVAP canister.**

1. Before servicing the vehicle, refer to the Precautions.

2. Raise and safely support the vehicle.

3. Remove the EVAP canister protector cover.

4. Disconnect the EVAP control system pressure sensor harness connector and the EVAP canister vent control valve harness connector.

5. Disconnect the EVAP canister purge hose, the EVAP vent line, and the EVAP canister vent control valve hose.

6. Remove the EVAP canister bolt.

7. Remove the EVAP canister from the vehicle.

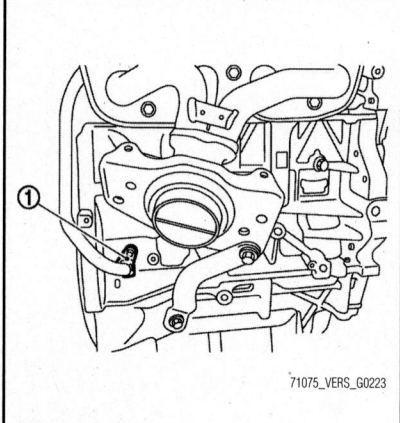

Fig. 117 Camshaft Position (CMP) sensor (1) (POS) component location—1.8L engine

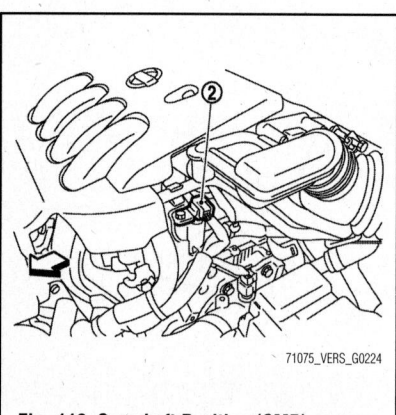

Fig. 118 Camshaft Position (CMP) sensor (2) (PHASE) component location—1.8L engine

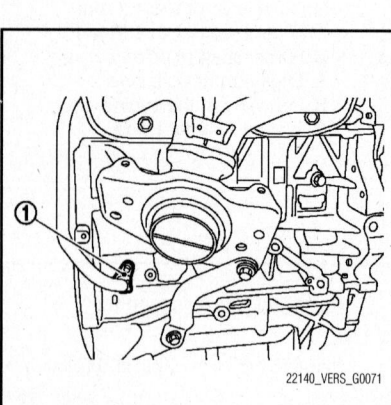

Fig. 119 Crankshaft Position (CKP) sensor (1) component location—1.6L (2011) and 1.8L engines

To install:

8. Installation is the reverse of the removal procedure.

9. During service requiring removal of O-rings, replace each O-ring with a new one during installation.

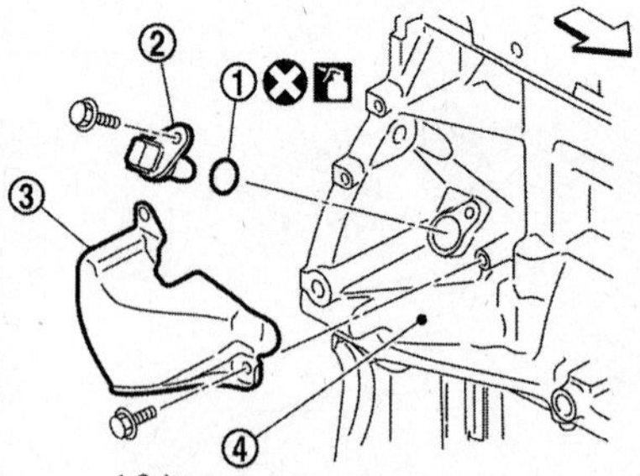

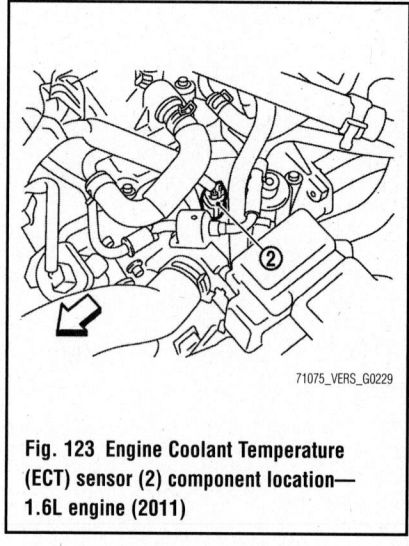

1. O-ring
2. Crankshaft position sensor (POS)
3. Crankshaft position sensor cover
4. Cylinder block
White Arrow: Engine front

Fig. 120 Crankshaft Position (CKP) sensor component location—1.6L engine (2012)

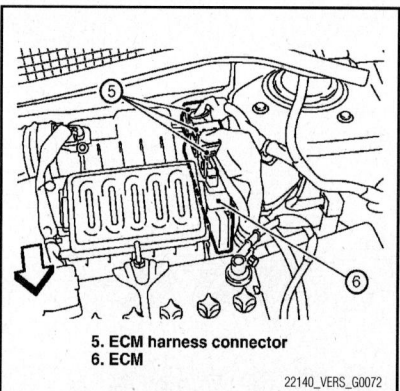

5. ECM harness connector
6. ECM

Fig. 121 Electronic Control Module (ECM) and harness location—1.6L engine (2011) and 1.8L engines

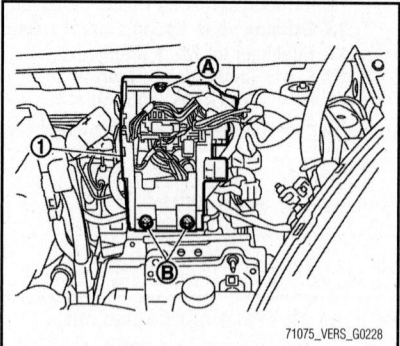

Fig. 122 Electronic Control Module (ECM) mounting nuts (A), bolts (B), and IPDM E/R cover (1) component locations—1.6L engine (2012)

FUEL TEMPERATURE SENSOR

LOCATION

2012 Sedan
See Figure 130.

Fig. 123 Engine Coolant Temperature (ECT) sensor (2) component location—1.6L engine (2011)

REMOVAL & INSTALLATION

2012 Sedan

➡Before disassembly, note the proper placement of the wires to the correct terminals and correct wire routing to the terminals.

1. Before servicing the vehicle, refer to the Precautions.

2. Disconnect the red, white, and double black wire connectors. Press the tabs on the terminals to release the locking tabs.

3. Release the two clips and remove the fuel temperature sensor from the pump assembly.

4. Release the tab and slide the level sending unit module and float arm assembly up to remove.

To install:

5. Installation is the reverse of the removal procedure.

➡Ensure proper placement of the wires to the correct terminals and correct wire routing to the terminals.

6. After connecting terminals, ensure they are securely locked and cannot be pulled out.

7. When installing the level sending unit, push down until the tab is locked into place.

HEATED OXYGEN (HO2S) SENSOR

LOCATION

See Figures 131 through 133.

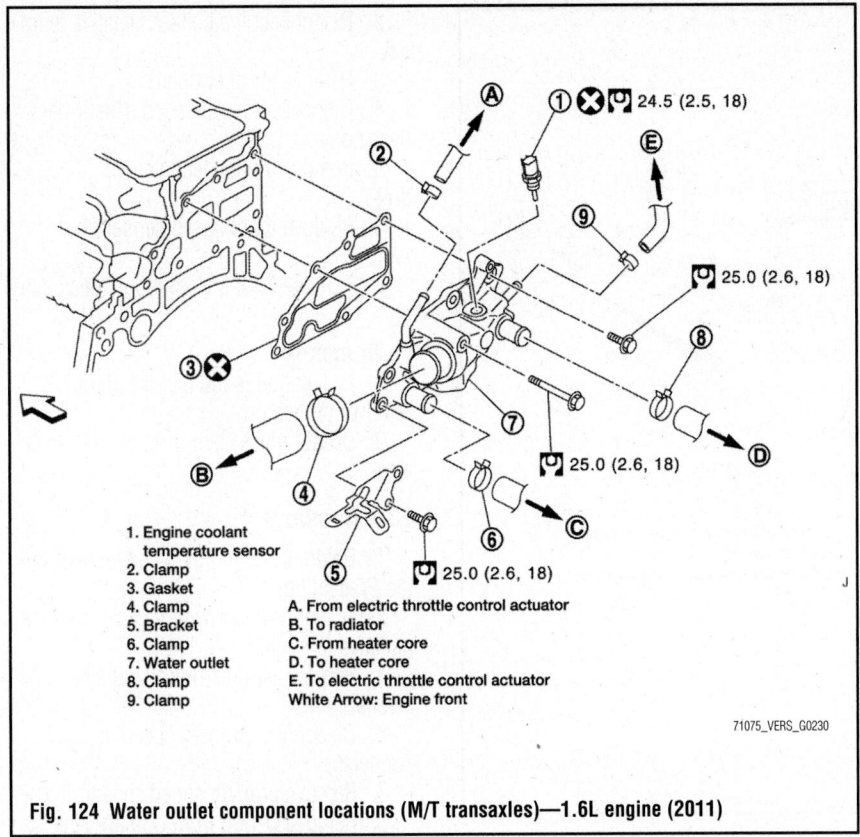

1. Engine coolant
   temperature sensor
2. Clamp
3. Gasket
4. Clamp
5. Bracket
6. Clamp
7. Water outlet
8. Clamp
9. Clamp

A. From electric throttle control actuator
B. To radiator
C. From heater core
D. To heater core
E. To electric throttle control actuator
White Arrow: Engine front

71075_VERS_G0230

Fig. 124 Water outlet component locations (M/T transaxles)—1.6L engine (2011)

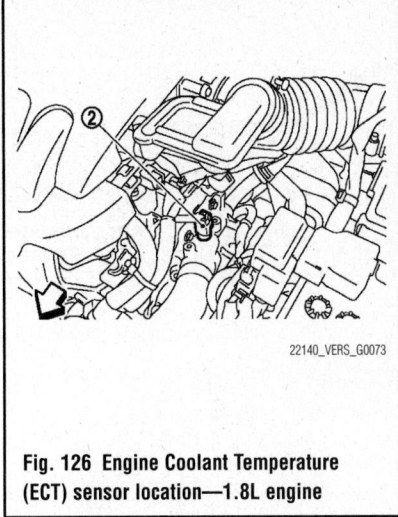

Fig. 126 Engine Coolant Temperature (ECT) sensor location—1.8L engine

(0.5m) onto a hard surface such as a concrete floor; install a new one.

✳✳ WARNING

Do not over-tighten the heated oxygen sensor. Doing so may damage the heated oxygen sensor, resulting in the MIL coming on.

REMOVAL & INSTALLATION

✳✳ CAUTION

To avoid the danger of being burned, do not touch the exhaust system when the system is hot. The Heated Oxygen Sensor (HO2S) removal should be performed when the system is cool.

1. Before servicing the vehicle, refer to the Precautions.
2. Disconnect heated oxygen sensor harness connector.
3. Remove the heated oxygen sensor using Tool number: KV10114400 (J-38365), or equivalent.

✳✳ WARNING

Be careful not to damage the heated oxygen sensor.

To install:

4. Before installing a new heated oxygen sensor, clean and apply anti-seize lubricant to exhaust system threads using suitable tool.

✳✳ WARNING

Discard any heated oxygen sensor which has been dropped from a height of more than 19.7 inches

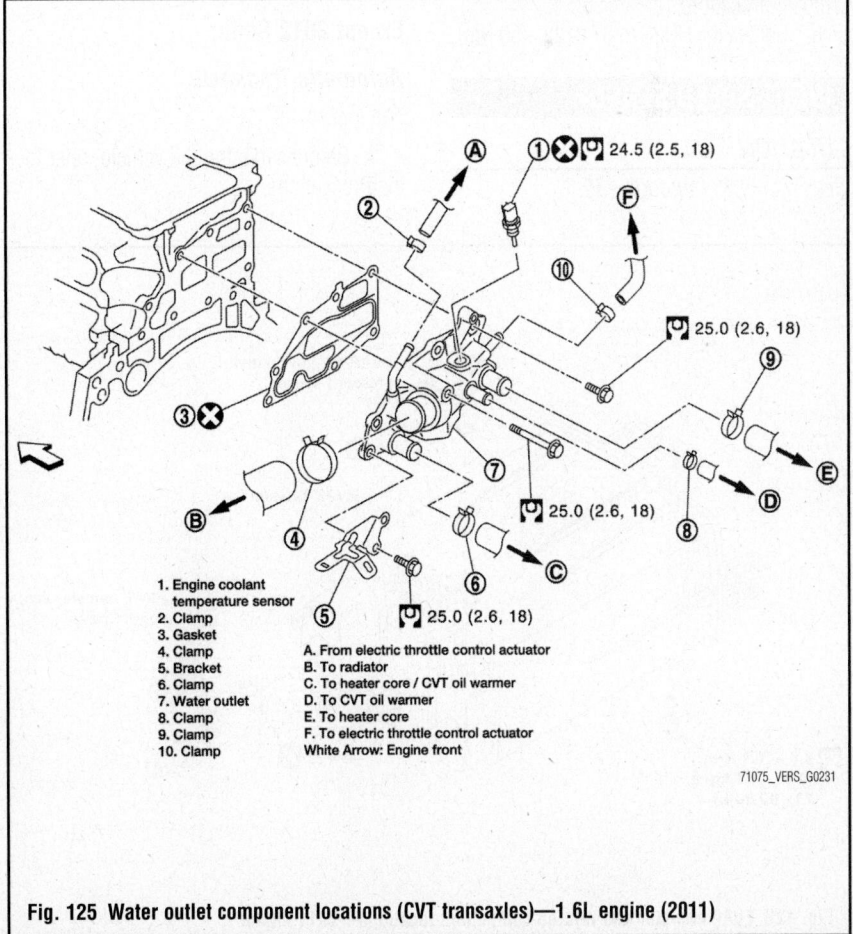

1. Engine coolant
   temperature sensor
2. Clamp
3. Gasket
4. Clamp
5. Bracket
6. Clamp
7. Water outlet
8. Clamp
9. Clamp
10. Clamp

A. From electric throttle control actuator
B. To radiator
C. To heater core / CVT oil warmer
D. To CVT oil warmer
E. To heater core
F. To electric throttle control actuator
White Arrow: Engine front

71075_VERS_G0231

Fig. 125 Water outlet component locations (CVT transaxles)—1.6L engine (2011)

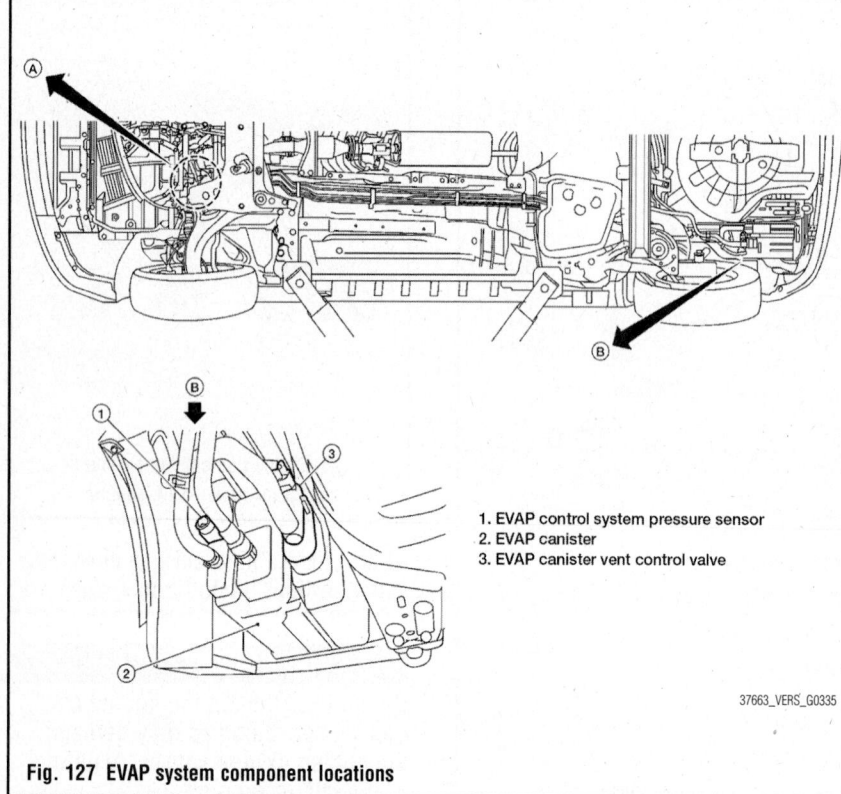

1. EVAP control system pressure sensor
2. EVAP canister
3. EVAP canister vent control valve

37663_VERS_G0335

Fig. 127 EVAP system component locations

5. Installation is the reverse of the removal procedure.
6. Tighten the HO2S to 37 ft. lbs. (50 Nm).

INPUT SPEED SENSOR

LOCATION

See Figures 134 through 136.

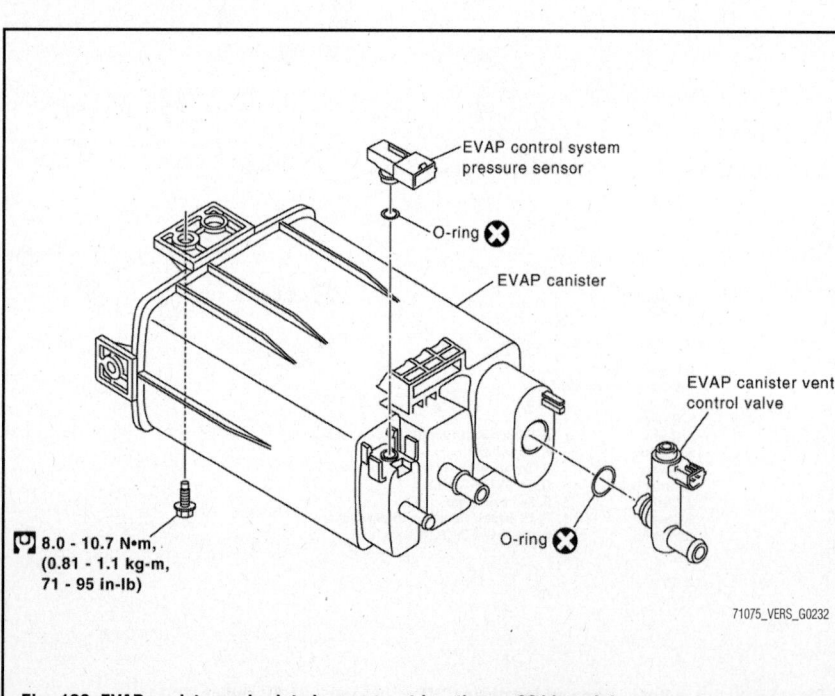

EVAP control system
pressure sensor

O-ring ✖

EVAP canister

EVAP canister vent
control valve

8.0 - 10.7 N•m,
(0.81 - 1.1 kg-m,
71 - 95 in-lb)

O-ring ✖

71075_VERS_G0232

Fig. 128 EVAP canister and related component locations—2011 model

REMOVAL & INSTALLATION

Except 2012 Sedan

Automatic Transaxle

See Figure 137.

1. Before servicing the vehicle, refer to the Precautions.

2. Disconnect the battery negative terminal.
3. Remove air duct (inlet).
4. Disconnect input speed sensor harness connector.
5. Remove input speed sensor bolt.
6. Remove input speed sensor from A/T.
7. Remove O-ring from input speed sensor.

To install:

8. Installation is the reverse of the removal procedure.
9. Do not reuse O-ring. Apply ATF to O-ring.

CVT Transaxle

1. Before servicing the vehicle, refer to the Precautions.
2. Disconnect the battery cable from the negative terminal.
3. Remove air duct (inlet) and air cleaner case.
4. Disconnect primary speed sensor connector.
5. Remove primary speed sensor.
6. Remove O-ring from primary speed sensor.

To install:

7. Installation is the reverse of the removal procedure.
8. Never reuse O-ring. Apply CVT fluid to O-ring.
9. Check for CVT fluid leakage and check CVT fluid level.

2012 Sedan

CVT Transaxle

1. Before servicing the vehicle, refer to the Precautions.
2. Remove the front LH wheel and tire.
3. Remove the LH fender protector.
4. Disconnect the primary speed sensor harness connector.
5. Remove the primary speed sensor.
6. Remove the O-ring from the primary speed sensor.

To install:

7. Installation is the reverse of the removal procedure.
8. Do not reuse O-ring. Apply Genuine NISSAN CVT Fluid NS-2 to the O-ring.
9. Check for CVT fluid leakage.
10. Adjust the CVT fluid level.

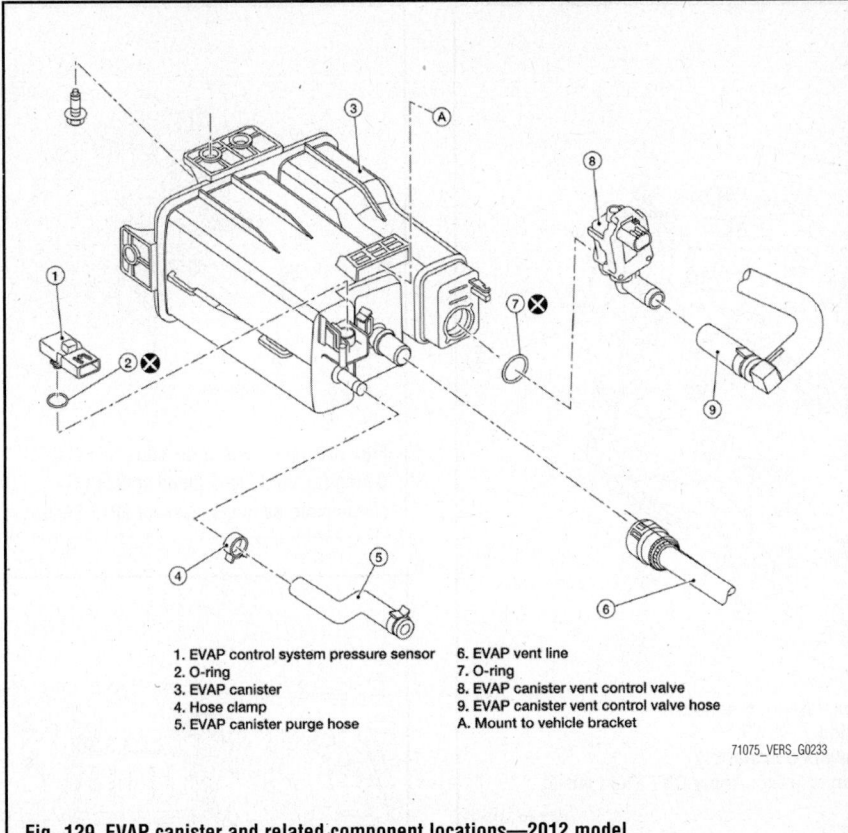

1. EVAP control system pressure sensor
2. O-ring
3. EVAP canister
4. Hose clamp
5. EVAP canister purge hose
6. EVAP vent line
7. O-ring
8. EVAP canister vent control valve
9. EVAP canister vent control valve hose
A. Mount to vehicle bracket

71075_VERS_G0233

Fig. 129 EVAP canister and related component locations—2012 model

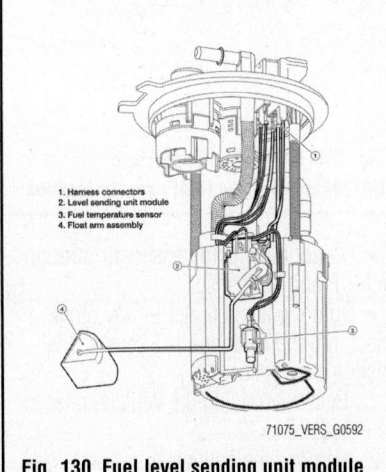

1. Harness connectors
2. Level sending unit module
3. Fuel temperature sensor
4. Float arm assembly

71075_VERS_G0592

Fig. 130 Fuel level sending unit module and related component locations—2012 Sedan

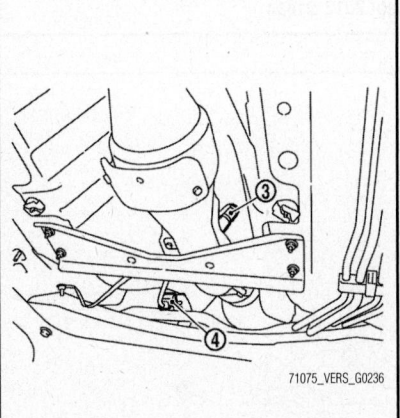

71075_VERS_G0236

Fig. 131 Heated Oxygen (HO2S) sensor 2 (3) and harness connector (4) component locations—1.6L engine (2011)

INTAKE AIR TEMPERATURE (IAT)/MASS AIRFLOW (MAF) SENSOR

LOCATION

See Figure 138.

The Intake Air Temperature (IAT) sensor is built into the Mass Air Flow (MAF) sensor. The sensor detects intake air temperature and transmits a signal to the ECM.

REMOVAL & INSTALLATION

Use the following precautions for this procedure.

• Do not disassemble the MAF and IAT sensor

• Do not expose the MAF and IAT sensor to any shock

• Do not clean the MAF and IAT sensor

• If the MAF and IAT sensor has been dropped, it should be replaced

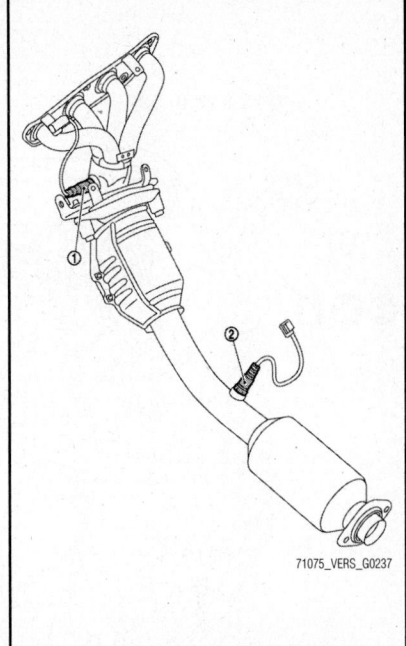

71075_VERS_G0237

Fig. 132 Heated Oxygen (HO2S) sensor 2 (2) and Air Fuel (A/F) sensor (1) component locations—1.6L engine (2012)

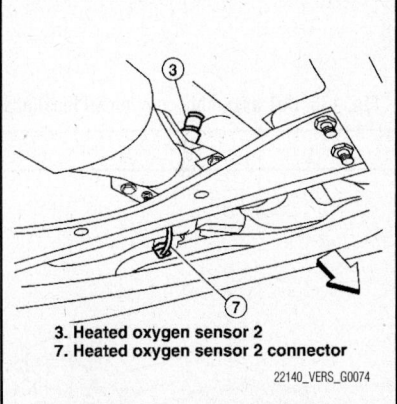

3. Heated oxygen sensor 2
7. Heated oxygen sensor 2 connector

22140_VERS_G0074

Fig. 133 Heated Oxygen (HO2S) sensor 2 and harness connector—1.8L engine

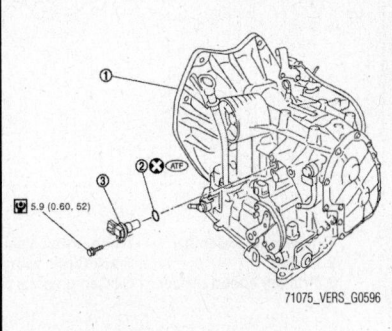

5.9 (0.60, 52)

71075_VERS_G0596

Fig. 134 Automatic Transaxle (A/T) (1), O-ring (2), and input speed sensor (3) component locations—except 2012 Sedan

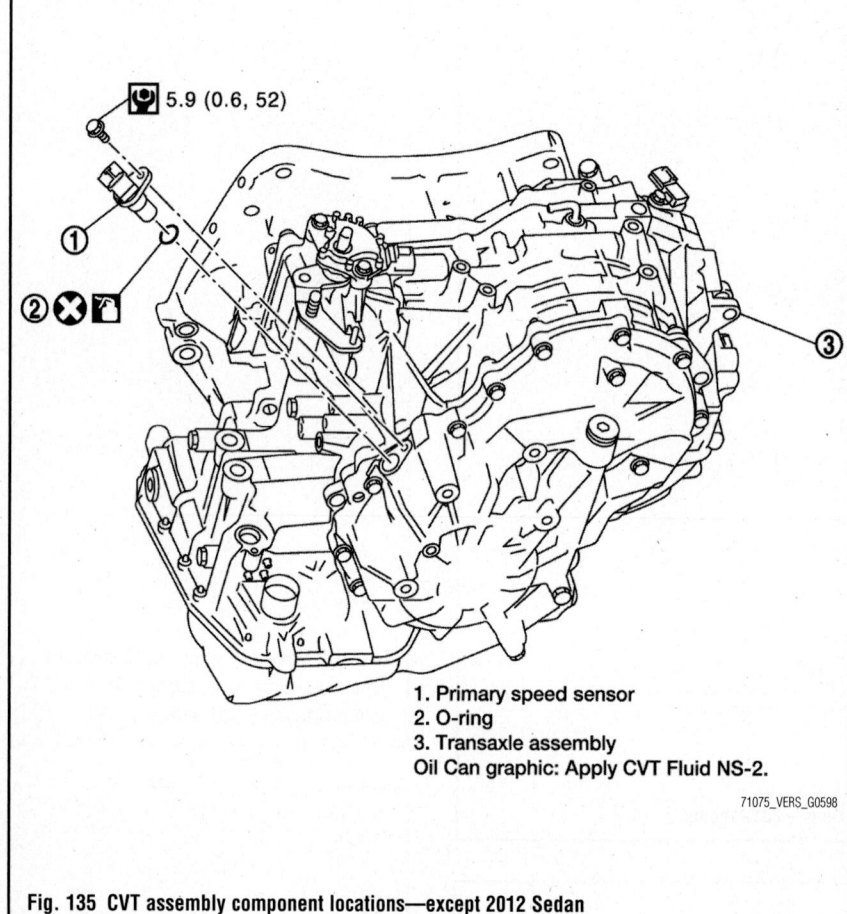

5.9 (0.6, 52)

1. Primary speed sensor
2. O-ring
3. Transaxle assembly
Oil Can graphic: Apply CVT Fluid NS-2.

71075_VERS_G0598

Fig. 135 CVT assembly component locations—except 2012 Sedan

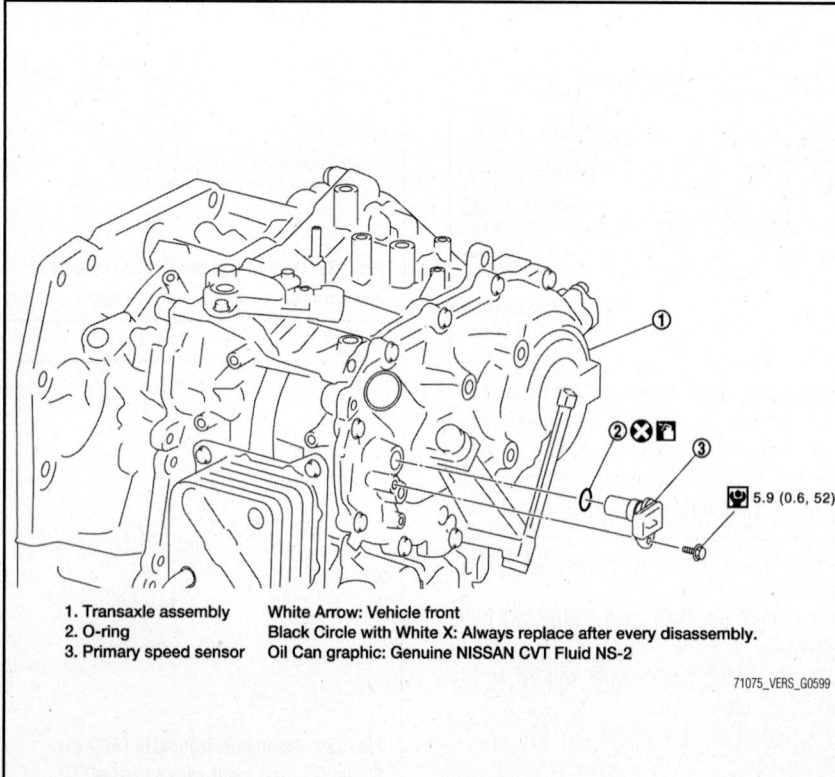

1. Transaxle assembly
2. O-ring
3. Primary speed sensor

White Arrow: Vehicle front
Black Circle with White X: Always replace after every disassembly.
Oil Can graphic: Genuine NISSAN CVT Fluid NS-2

5.9 (0.6, 52)

71075_VERS_G0599

Fig. 136 CVT assembly component locations—2012 Sedan

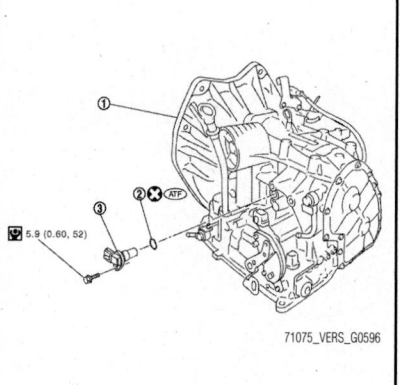

5.9 (0.60, 52)

71075_VERS_G0596

Fig. 137 Automatic Transaxle (A/T) (1), O-ring (2), and input speed sensor (3) component locations—except 2012 Sedan

22140_VERS_G0075

Fig. 138 Intake Air Temperature (IAT)/Mass Air Flow (MAF) sensor location

• Do not blow compressed air through the MAF and IAT sensor

• Do not place a finger or any other object into the MAF and IAT sensor. Malfunction may occur

1. Before servicing the vehicle, refer to the Precautions.

2. Remove the battery, as needed.

3. Remove the engine cover, if applicable.

4. Disconnect the harness connector from the Mass Air Flow (MAF) sensor.

5. Remove the air cleaner.

6. Remove the MAF sensor from the air cleaner case.

7. Installation is the reverse of removal.

KNOCK SENSOR (KS)

LOCATION

See Figure 139.

Fig. 139 Knock Sensor (KS) (2) component location

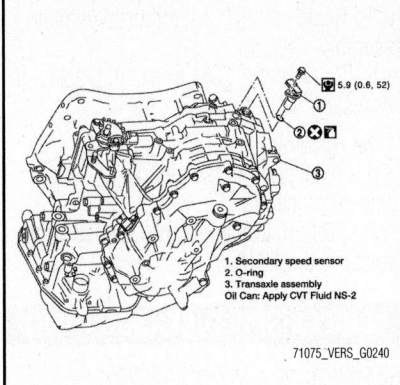

1. Secondary speed sensor
2. O-ring
3. Transaxle assembly
   Oil Can: Apply CVT Fluid NS-2

Fig. 141 Secondary speed sensor component location—CVT transaxle

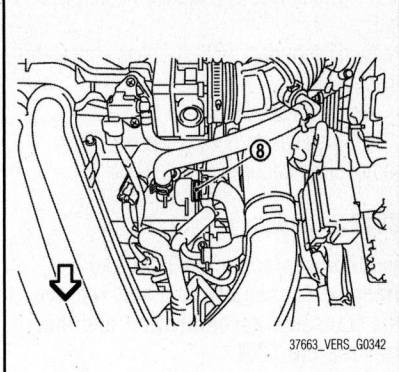

Fig. 142 Positive Crankcase Ventilation (PCV) valve (8) component location—1.6L engine

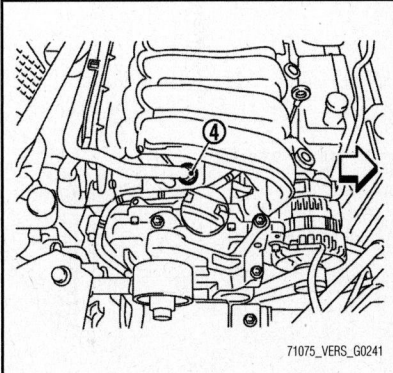

Fig. 143 Positive Crankcase Ventilation (PCV) valve (4) component location—1.8L engine

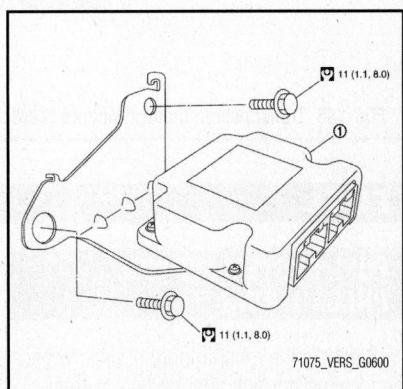

Fig. 144 Transmission Control Module (TCM) (1) component shown—except 2012 Sedan

OUTPUT SHAFT SPEED (OSS) SENSOR

LOCATION

See Figures 140 and 141.

REMOVAL & INSTALLATION

A/T Transaxle

1. Before servicing the vehicle, refer to the Precautions.
2. Disconnect the battery cable from the negative terminal.
3. Remove the air duct (inlet), air duct, and air cleaner case.
4. Remove the gusset.
5. Disconnect the output speed sensor harness connector.
6. Remove the clip.
7. Remove the output speed sensor from A/T.
8. Remove the O-ring from the output speed sensor.

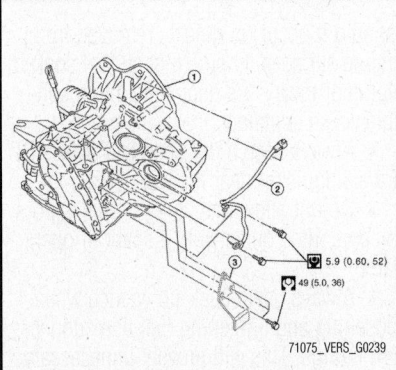

Fig. 140 Output shaft speed sensor (2), A/T transaxle case (1), and gusset (3) component locations—A/T transaxle

To install:

9. Installation is the reverse of the removal procedure.
10. Do not reuse O-ring.
11. Apply ATF to O-ring.
12. Ensure the output speed sensor harness is firmly secured with the bolt.

CVT Transaxle

1. Before servicing the vehicle, refer to the Precautions.
2. Disconnect the battery cable from the negative terminal.
3. Remove the air duct (inlet) and the air cleaner case.
4. Disconnect the secondary speed sensor connector.
5. Remove the secondary speed sensor.
6. Remove the O-ring from the secondary speed sensor.

To install:

7. Installation is the reverse of the removal procedure.
8. Never reuse the O-ring.
9. Apply CVT fluid to the O-ring.
10. Check for CVT fluid leakage and check the CVT fluid level.

POSITIVE CRANKCASE VENTILATION (PCV) VALVE

LOCATION

See Figures 142 and 143.

TRANSMISSION CONTROL MODULE (TCM)

LOCATION

See Figures 144 and 145.

REMOVAL & INSTALLATION

Except 2012 Sedan

See Figures 146 and 147.

1. Before servicing the vehicle, refer to the Precautions.

2. Disconnect the battery negative terminal.

3. Disconnect TCM harness connectors from TCM.

4. Remove TCM.

5. Installation is the reverse of the removal procedure.

2012 Sedan

➥**When replacing the TCM and transaxle assembly as a set, replace the transaxle assembly first and then replace the TCM.**

1. Before servicing the vehicle, refer to the Precautions.

2. Remove the battery.

3. Disconnect the TCM connector.

4. Remove the TCM and bracket as an assembly.

5. Remove the TCM from the bracket, if necessary.

To install:

6. Installation is the reverse of the removal procedure.

7. Perform any additional service when replacing the TCM.

VEHICLE SPEED SENSOR (VSS)

REMOVAL & INSTALLATION

1. Before servicing the vehicle, refer to the Precautions.

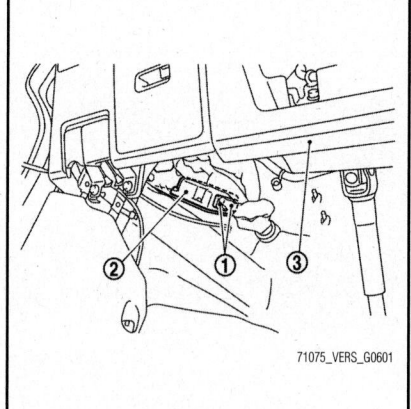

Fig. 146 View of TCM (2), harness connectors (1), and instrument lower finisher (3)—except 2012 Sedan

2. Disconnect vehicle speed sensor.

3. Remove vehicle speed sensor.

4. Installation is the reverse of the removal procedure.

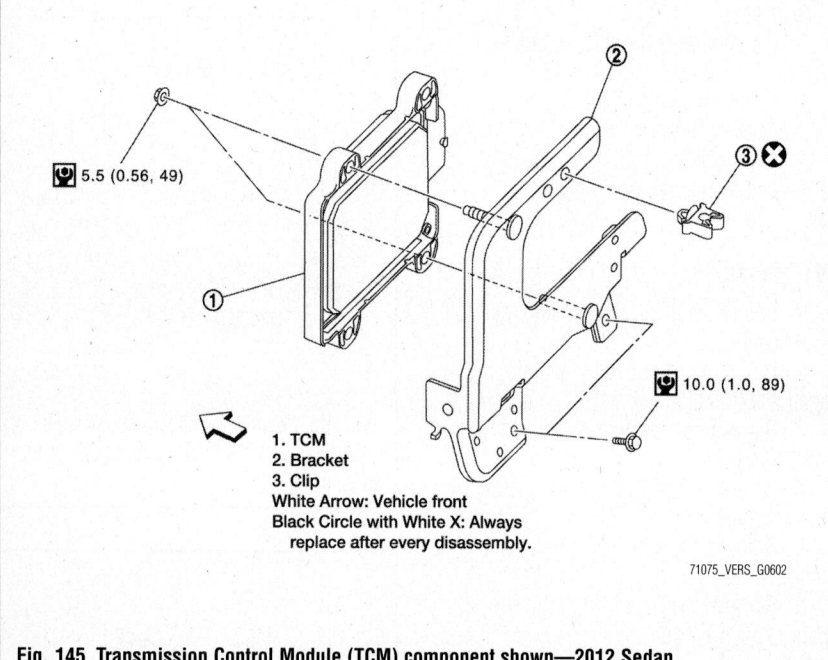

1. TCM
2. Bracket
3. Clip
White Arrow: Vehicle front
Black Circle with White X: Always replace after every disassembly.

71075_VERS_G0602

Fig. 145 Transmission Control Module (TCM) component shown—2012 Sedan

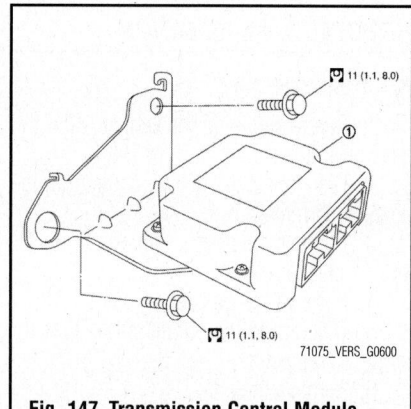

Fig. 147 Transmission Control Module (TCM) (1) component shown—except 2012 Sedan

FUEL   GASOLINE FUEL INJECTION SYSTEM

FUEL SYSTEM SERVICE PRECAUTIONS

Safety is the most important factor when performing not only fuel system maintenance but any type of maintenance. Failure to conduct maintenance and repairs in a safe manner may result in serious personal injury or death. Maintenance and testing of the vehicle's fuel system components can be accomplished safely and effectively by adhering to the following rules and guidelines.

• To avoid the possibility of fire and personal injury, always disconnect the negative battery cable unless the repair or test procedure requires that battery voltage be applied.

• Always relieve the fuel system pressure prior to disconnecting any fuel system component (injector, fuel rail, pressure regulator, etc.), fitting or fuel line connection. Exercise extreme caution whenever relieving fuel system pressure to avoid exposing skin, face and eyes to fuel spray. Please be advised that fuel under pressure may penetrate the skin or any part of the body that it contacts.

• Always place a shop towel or cloth around the fitting or connection prior to loosening to absorb any excess fuel due to spillage. Ensure that all fuel spillage (should it occur) is quickly removed from engine surfaces. Ensure that all fuel soaked cloths or towels are deposited into a suitable waste container.

• Always keep a dry chemical (Class B) fire extinguisher near the work area.

• Do not allow fuel spray or fuel vapors to come into contact with a spark or open flame.

• Always use a back-up wrench when loosening and tightening fuel line connection fittings. This will prevent unnecessary stress and torsion to fuel line piping.

• Always replace worn fuel fitting O-rings with new Do not substitute fuel hose or equivalent where fuel pipe is installed.

Before servicing the vehicle, make sure to also refer to the precautions in the beginning of this section as well.

RELIEVING FUEL SYSTEM PRESSURE

The fuel pump fuse is located in the dash fuse box or in the engine compartment fuse box. Check the lid of the fuse box for the exact location.

1. Before servicing the vehicle, refer to the Precautions.
2. Remove the fuel pump fuse.
3. Start the engine.
4. Start the engine and run until the engine stalls.
5. After the engine stalls, try to restart the engine. If the engine will not start, the fuel pressure has been released.
6. Turn the ignition switch **OFF**. Reinstall the fuel pump fuse into the fuse block.

➡Do not crank the engine or turn the ignition switch ON after the fuel pump fuse has been reinstalled, or the fuel pressure will be re-established.

FUEL FILTER

REMOVAL & INSTALLATION

✳✳ CAUTION

Replacing the fuel filter without adequate ventilation may lead to a fire.

Before servicing the vehicle, refer to the Precautions.

Replace the fuel filter only in a well-ventilated area away from any open flames.

The fuel filter is installed in the fuel pump assembly inside the fuel tank. Replace the fuel filter (or fuel pump assembly) with a new one periodically.

FUEL PUMP ASSEMBLY

REMOVAL & INSTALLATION

See Figure 148.

1. Before servicing the vehicle, refer to the Precautions.
2. Check fuel level with the vehicle on a level surface. If the fuel gauge indicates more than ⅞ full, drain fuel from the fuel tank until the fuel gauge indicates the correct level.
3. Open fuel door and unscrew the fuel filler cap to release the pressure inside the fuel tank.
4. Release the fuel pressure from the fuel lines.

5. Remove rear seat bottom.
6. Turn the four retainers 90° in a clockwise direction and remove the fuel pump inspection hole cover.
7. Disconnect electrical connector and fuel feed hose quick connector.
8. Disconnect the quick connector using the following procedure:
    a. Hold the sides of the connector, push in tabs and pull out the tube.
    b. If quick connector and tube on fuel pump assembly are stuck, push and pull several times until they move. Disconnect them by pulling.

➡The tube can be removed when the tabs are completely depressed. Do not twist it more than necessary. Do not use any tools to remove the quick connector. Keep resin tube away from heat. Be especially careful when welding near the resin tube. Prevent acid liquid such as battery electrolyte, from getting on resin tube. Do not bend or twist resin tube during installation and removal. To keep the connecting portion clean, free of foreign materials and to avoid damage, cover them completely with plastic bags or something similar. Do not insert plug to prevent damage to O-ring in quick connector.

9. Remove the lock ring using tool No. KV991J0090 (J-46214).
10. Remove fuel level sensor unit, fuel filter and fuel pump assembly.

✳✳ WARNING

Do not bend float arm during removal. Do not allow foreign materials to fall into fuel tank. Use a lint free cloth when handling components. Avoid impacts such as dropping when handling components.

To install:

11. Installation is in the reverse order of removal, noting the following:
    a. Install O-ring to fuel tank without twisting.
    b. Install fuel level sensor unit with the matchmarks aligned on the fuel tank and fuel level sensor unit.
    c. Turn the lock ring until the lock ring is fully rotated into the fuel tank lock tabs.
    d. Connect fuel feed tube quick connector by aligning the connector with the tube, then inserting the connector straight into the tube until it clicks.
    e. After connecting, visually confirm that the two retainer tabs are secured to the connector and pull the tube and the connector to make sure they are securely connected.
12. Connect electrical harness connector.
13. Before installing inspection hole cover, confirm that there are no fuel leaks, install inspection hole cover with the front mark (arrow) facing front of vehicle,

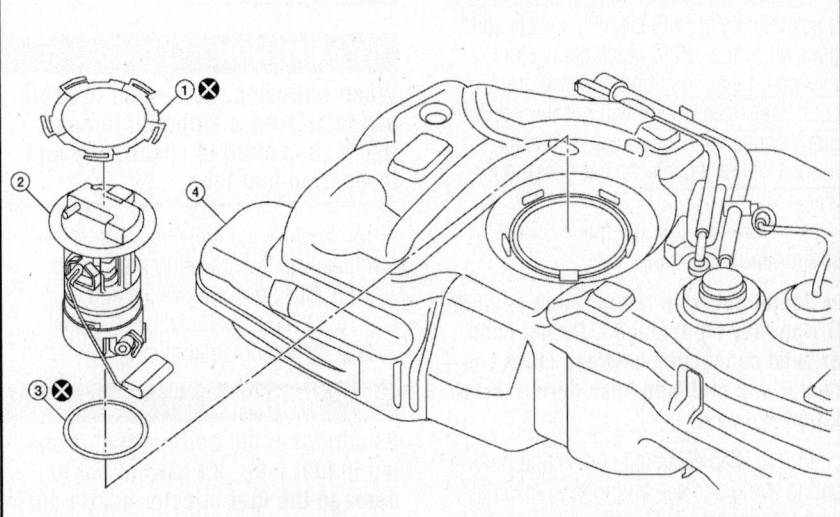

| 1. | Lock ring | 2. | Fuel level sensor unit, fuel filter and fuel pump assembly | 3. | O-ring |
| 4. | Fuel tank | | | | |

37663_VERS_G0201

Fig. 148 Fuel pump module and related components

and lock the clips by turning counterclockwise.

14. Check for fuel leaks:

a. Turn ignition switch "ON" (without starting the engine), to check the connections for fuel leaks with the electric fuel pump applying pressure to the fuel piping.

b. Start the engine and let it idle to check that there are no fuel leaks at the fuel system connections.

FUEL RAIL & INJECTORS

REMOVAL & INSTALLATION

1.6L Engine

2011 Model

See Figures 149 through 151.

1. Before servicing the vehicle, refer to the Precautions.

2. Release the fuel pressure.

3. Disconnect the negative battery cable.

4. Remove intake manifold.

5. Disconnect fuel feed hose from fuel tube. There is no fuel return path.

6. Remove quick connector cap from quick connector connection.

7. Disconnect fuel feed hose from hose clamp. There is no fuel return path.

8. With the sleeve side of quick connector release facing quick connector, install quick connector release onto fuel tube. Prepare container and cloth beforehand as fuel will leak out.

9. Insert quick connector release into quick connector until sleeve contacts and goes no further. Hold quick connector release on that position. Inserting quick connector release hard will not disconnect quick connector. Hold quick connector release where it contacts and goes no further.

10. Draw and pull out quick connector straight from fuel tube.

➡ Do not pull with lateral force applied; O-ring may be damaged. Do not bend or twist connection between quick connector and fuel feed hose during installation/removal.

11. To keep clean the connecting portion and to avoid damage and foreign materials, cover them completely with plastic bags or something similar.

12. Disconnect harness connector from fuel injector.

13. Remove fuel tube protector.

14. Remove the fuel injector and fuel tube assembly.

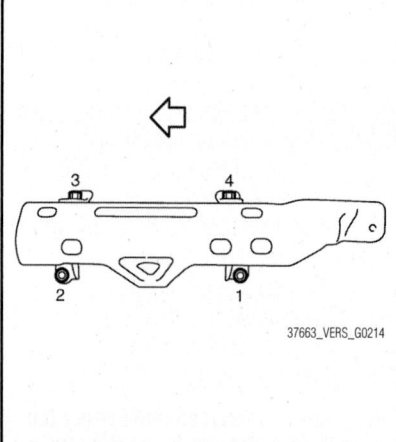

37663_VERS_G0214

Fig. 149 Fuel tube protector bolt installation sequence—reverse for removal

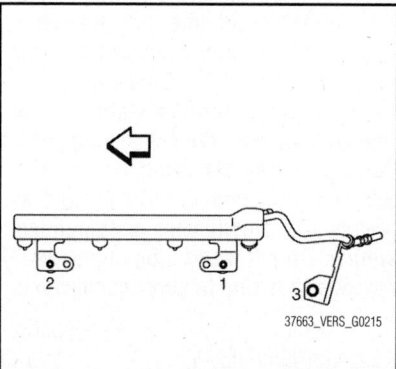

37663_VERS_G0215

Fig. 150 Fuel injector and fuel tube assembly bolt installation sequence—reverse for removal

❋❋ CAUTION

When removing, be careful to avoid any interference with fuel injector. Use a shop cloth to absorb any fuel leaks from fuel tube.

15. Remove the fuel injector from the fuel tube with the following procedure:

a. Open and remove the clip.

b. Remove fuel injector from the fuel tube by pulling straight out.

❋❋ WARNING

Be careful about spilling fuel remaining in fuel tube. Be careful not to damage the fuel injector nozzle during removal. Never bump or drop fuel injector. Never disassemble fuel injector.

To install:

16. Install O-rings to fuel injector, noting the following:

a. The upper and lower O-rings are different. Be careful not to confuse them.

• Fuel tube side: Black

• Nozzle side: Green

b. Handle O-ring with bare hands. Never wear gloves. Lubricate O-ring with new engine oil. Never clean O-ring with solvent. Make sure that O-ring and its mating part are free of foreign material. When installing O-ring, be careful not to scratch it with tool or fingernails. Also be careful not to twist or stretch O-ring. If O-ring was stretched while it was being attached, never insert it quickly into fuel tube. Insert O-ring straight into fuel tube. Never twist it.

17. Install the fuel injector onto the fuel tube with the following procedure:

a. Insert the clips into the clip mounting grooves on the fuel injector.

b. Insert clip cut-out into fuel injector protrusion. Always replace clip with new one.

c. Make sure that the clip does not interfere with the O-ring. If interference occurs, replace the O-ring.

d. Insert the fuel injector into the fuel tube with clip attached. Make sure that the axis is lined up when inserting.

e. Insert clip cut-out into fuel tube protrusion.

f. Make sure that the flange on the fuel tube fits securely in the clip flange fixing groove.

g. Make sure that installation is complete by checking that fuel injector does not rotate or come off.

18. Install the fuel tube and injector assembly onto the cylinder head. Tighten the bolts in numerical order.

❋❋ WARNING

Be careful not to let tip of injector nozzle interfere with other parts.

19. Install fuel tube protector. Tighten bolts in numerical order.

20. Connect harness connector to fuel injector.

21. Connect fuel feed hose with the following procedure:

a. Check for damage or foreign material on the fuel tube and quick connector.

b. Apply new engine oil lightly to area around the top of fuel tube.

c. Align center to insert quick connector straightly into fuel tube.

d. Insert quick connector to fuel tube until the top spool on fuel tube is inserted completely and the 2nd level spool is positioned slightly below quick connector bottom end.

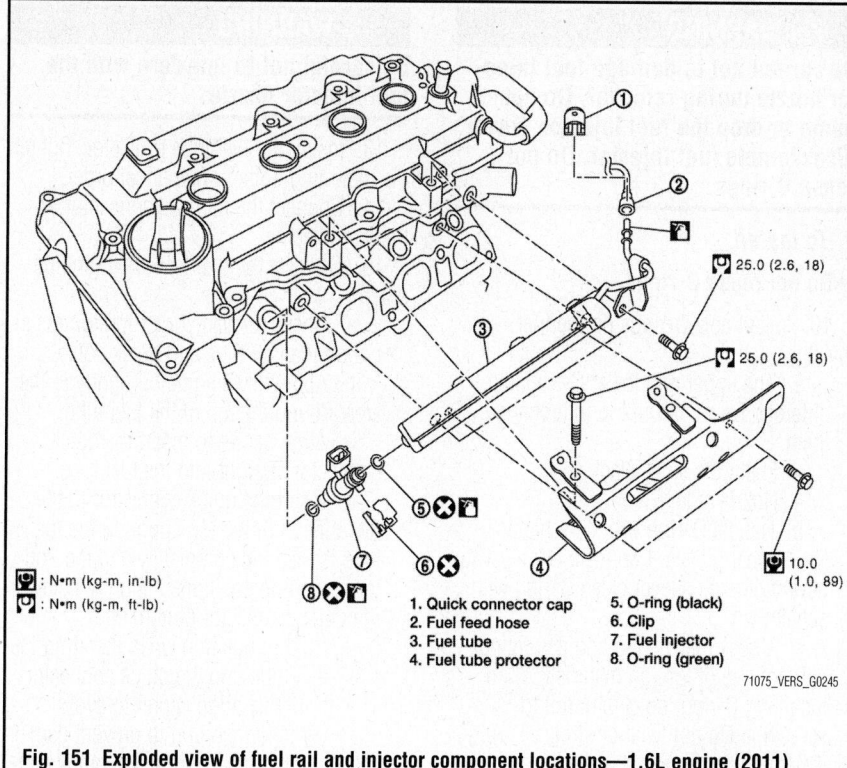

1. Quick connector cap
2. Fuel feed hose
3. Fuel tube
4. Fuel tube protector
5. O-ring (black)
6. Clip
7. Fuel injector
8. O-ring (green)

[¤] : N•m (kg-m, in-lb)
[¤] : N•m (kg-m, ft-lb)

71075_VERS_G0245

Fig. 151 Exploded view of fuel rail and injector component locations—1.6L engine (2011)

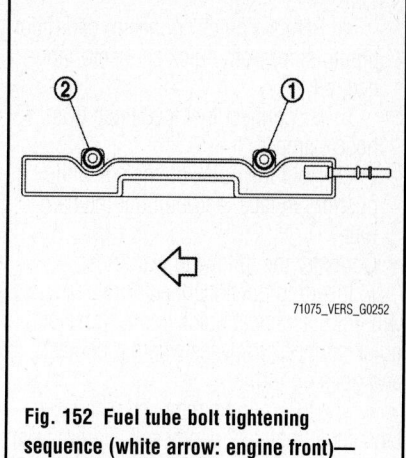

71075_VERS_G0252

Fig. 152 Fuel tube bolt tightening sequence (white arrow: engine front)— 1.6L engine (2012)

e. Carefully align center to avoid inclined insertion to prevent damage to O-ring inside quick connector.

f. Insert until you hear a "click" sound and actually feel the engagement.

g. To avoid misidentification of engagement with a similar sound, be sure to perform the next step.

h. Before clamping fuel feed hose with hose clamp, pull quick connector hard by hand holding "A" position. Make sure it is completely engaged (connected) so that it does not come out from fuel tube.

i. Install quick connector cap to quick connector connection. Install quick connector cap with the side arrow facing quick connector side (fuel feed hose side).

j. Make sure that the quick connector and fuel tube are securely engaged with the quick connector cap mounting groove.

k. If quick connector cap cannot be installed easily, the quick connector may not be connected correctly. Remove the quick connector cap, and then check the connection of quick connector again.

l. Install fuel feed hose to hose clamp.

22. Installation of the remaining components is in the reverse order of removal.

23. After installation is complete, check for fuel leaks:

a. Turn ignition switch "ON" (without starting the engine), to check the connections for fuel leaks with the electric fuel pump applying pressure to the fuel piping.

b. Start the engine and let it idle to check that there are no fuel leaks at the fuel system connections.

2012 Model
See Figure 152.

❈❈ WARNING

Do no remove or disassemble parts unless instructed.

❈❈ CAUTION

Put a "CAUTION: FLAMMABLE" sign in the workshop. Be sure to work in a well-ventilated area and furnish workshop with a CO2 fire extinguisher. Never smoke while servicing fuel system. Keep open flames and sparks away from the work area.

1. Before servicing the vehicle, refer to the Precautions.
2. Release the fuel pressure.
3. Remove the intake manifold.
4. Disconnect the fuel feed hose from the fuel tube. Disconnect quick connector:

➡**There is no fuel return path.**

a. Remove quick connector cap (engine side) from quick connector connection.

b. Disconnect fuel feed hose from hose clamp.

c. With the sleeve side of quick connector release facing quick connector, install quick connector release onto fuel tube.

d. Insert quick connector release into quick connector until sleeve contacts and goes no further. Hold quick connector release in that position.

➡**Inserting the quick connector release hard will not disconnect the quick connector. Hold the quick connector release where it contacts and goes no further.**

e. Draw and pull out quick connector straight up from fuel tube. Pull quick connector up from its position. Consider the following warnings:

• Do not pull on the quick connector with lateral force applied. The O-ring inside quick connector may be damaged. Prepare container and cloth beforehand as fuel will leak out. Discard O-ring, do not reuse

• Avoid fire and sparks. Keep parts away from heat source. Be especially careful when welding is performed. Do not expose parts to battery electrolyte or other acids

• Never bend or twist connection between quick connector and fuel feed tube during installation or removal.

• Be sure to cover openings of disconnected fuel feed hose and fuel tube with plug or plastic bag to avoid fuel leaks and entry of foreign material

5. Disconnect fuel feed hose from fuel pipe as follows:

→**There is no fuel return path.**

a. Remove quick connector cap (floor piping side) from quick connector connection.

b. Disconnect fuel feed hose from hose clamp.

c. Hold the quick connector while pushing in tabs, and pull out the hard tube.

Consider the following warnings:

• Inserting quick connector release hard will not disconnect quick connector. Hold quick connector release where it contacts and goes no further

• The tube can be removed when the tabs are completely depressed. Do not twist it more than necessary

• Do not use any tools to remove the quick connector

• Keep the resin tube away from heat. Be especially careful when welding near the tube

• Prevent acid such as battery electrolyte etc. from getting on the resin tube

• Do not bend or twist the tube during installation and removal

• Remove the remaining retainer only when the tube is replaced

• When the tube is replaced, also replace the retainer

• Be sure to cover openings of disconnected pipes with plug or plastic bag to avoid fuel leaks and entry of foreign material

6. Disconnect the harness connector from the fuel injector.

7. Remove the fuel tube protector. Loosen the mounting bolts in reverse order of tighten sequence.

8. Remove the fuel injector and the fuel tube assembly:

a. Loosen the 2 nuts in reverse order of the tightening sequence.

b. Pull the fuel tube straight out until the injector lower O-rings are clear.

c. Remove the nuts and the fuel tube.

❊❊ **WARNING**

When removing, be careful to avoid interference with fuel injectors. Use a shop cloth to absorb any fuel leaks from the fuel tube.

9. Remove the fuel injectors from the fuel tube:

a. Open and remove clip.

b. Remove fuel injectors and from fuel tube by pulling straight out.

❊❊ **CAUTION**

Be careful with remaining fuel that may leak from fuel tube.

❊❊ **WARNING**

Be careful not to damage fuel injector nozzle during removal. Do not bump or drop the fuel injector. Never disassemble fuel injector. Do not reuse O-rings.

To install:

→**Do not reuse O-rings.**

10. Install new O-rings on the fuel injector.

a. The upper and lower O-rings are different. Be careful not to interchange them.

• Fuel tube side: Black
• Nozzle side: Green

b. Handle O-ring with bare hands. Never wear gloves. Lubricate O-ring with new engine oil. Never clean O-ring with solvent.

c. Check that O-ring and its mating part are free of foreign material. When installing O-ring, be careful not to scratch it. Do not twist or stretch O-ring. If O-ring was stretched while it was being attached, allow it to retract before inserting it into the fuel tube.

d. Insert the O-ring straight into the fuel tube. Never angle or twist it.

11. Install the fuel injector to the fuel tube:

a. Insert a new clip into the clip mounting groove on the fuel injector.

• Insert the new clip so that the protrusion of the fuel injector matches the cut-out of the clip
• Never reuse the clip. Replace it with a new one
• Be careful to keep the clip from interfering with the O-ring. If interference occurs, replace the O-ring

b. Insert the fuel injector into the fuel tube with the clip attached.

• Insert the fuel injector while matching it to the axial center
• Insert the fuel injector so that the protrusion of the fuel tube matches the cut-out of the clip
• Check that the fuel tube flange is securely located in the flange groove on the clip

c. Check that installation is complete by checking that the fuel injector does not rotate or come off.

12. Set the fuel tube and the fuel injector assembly in position for installation on the cylinder head. Tighten mounting bolts in numerical order.

❊❊ **WARNING**

Be careful not to interfere with the fuel injector nozzle.

13. Install the fuel tube protector. Tighten the mounting bolts in numerical order.

14. Connect the harness connector to the fuel injector.

15. Connect the fuel feed tube (engine side):

a. Check for damage or foreign material on the fuel tube and quick connector.

b. Apply new engine oil lightly to the area around the top of the fuel tube.

c. Align center to insert the quick connector straight into the fuel tube.

d. Insert the quick connector to the fuel tube until the top spool on the fuel tube is inserted completely and the 2nd level spool is positioned slightly below the quick connector bottom end.

• Hold in position when inserting the fuel tube into the quick connector
• Carefully align center to avoid inclined insertion to prevent damage to the O-ring inside quick connector
• Insert until you hear a "click" sound and actually feel the engagement

e. To avoid misidentification of engagement with a similar sound, pull the quick connector hard by hand. Check it is completely engaged (connected) so that it does not come out from the fuel tube.

f. Install the quick connector cap (engine side) with the side arrow facing the quick connector side (fuel feed tube side).

→**Check that the quick connector and the fuel tube are securely engaged with the quick connector cap (engine side) mounting groove. If the quick connector cap (engine side) cannot be installed easily, the quick connector may not be connected correctly. Remove and reconnect.**

g. Install the fuel feed hose to the hose clamp.

16. Connect the fuel feed tube (floor piping side):

a. Check the connection for damage or any foreign materials.

b. Align the quick connector with the tube, then insert the connector straight into the centralized under floor piping until a click is heard.

c. After connecting, check that the connection is secure:

- Visually confirm that the two retainer tabs are connected to the connector
- With the fuel feed hose not fixed to the clamp, pull the quick connector hard by hand to check that the quick connector is not disconnected from the centralized under-floor piping and that the quick connector is securely connected

17. Install remaining parts in the reverse order of removal.

18. Check for fuel leaks:

a. Turn the ignition switch ON with the engine stopped. Ensure there are no fuel leaks at the fuel pipe connection points.

➡**Use mirrors for checking points out of clear sight.**

b. Start the engine and increase engine speed. Check again that there are no fuel leaks.

✳✳ CAUTION

Never touch the engine immediately after stopped, as the engine becomes extremely hot.

1.8L Engine

See Figures 153 through 155.

1. Before servicing the vehicle, refer to the Precautions.

2. Release the fuel pressure.

3. Disconnect the negative battery cable.

4. Disconnect fuel feed hose from hose clamp. There is no fuel return path.

5. Remove quick connector cap from quick connector connection.

6. With the sleeve side of quick connector release facing quick connector, install quick connector release onto fuel tube. Prepare container and cloth beforehand as fuel will leak out.

7. Insert quick connector release into quick connector until sleeve contacts and goes no further. Hold quick connector release on that position. Inserting quick connector release hard will not disconnect quick connector. Hold quick connector release where it contacts and goes no further.

8. Draw and pull out quick connector straight from fuel tube.

✳✳ WARNING

Do not pull with lateral force applied; O-ring may be damaged. Do not bend or twist connection between quick

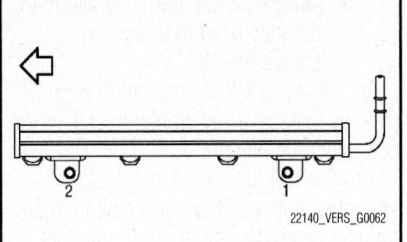

Fig. 153 Fuel tube bolt installation sequence—reverse for removal

connector and fuel feed hose during installation/removal.

9. To keep clean the connecting portion and to avoid damage and foreign materials, cover them completely with plastic bags or something similar.

10. Remove intake manifold.

11. Remove fuel tube.

12. Remove the fuel tube and fuel injector assembly.

✳✳ WARNING

When removing, be careful to avoid any interference with fuel injector. Use a shop cloth to absorb any fuel leaks from fuel tube.

13. Remove fuel injector from fuel tube with the following procedure:

a. Open and remove clip.

b. Remove fuel injector from fuel tube by pulling straight.

✳✳ CAUTION

Be careful with remaining fuel that may go out from fuel tube.

✳✳ WARNING

Be careful not to damage fuel injector nozzle during removal. Never bump or drop fuel injector. Never disassemble fuel injector.

To install:

14. Install O-rings to fuel injector, noting the following:

a. The upper and lower O-rings are different. Be careful not to confuse them.

- Fuel tube side: Black
- Nozzle side: Green

b. Handle O-ring with bare hands. Never wear gloves. Lubricate O-ring with new engine oil. Never clean O-ring with solvent. Make sure that O-ring and its mating part are free of foreign material.

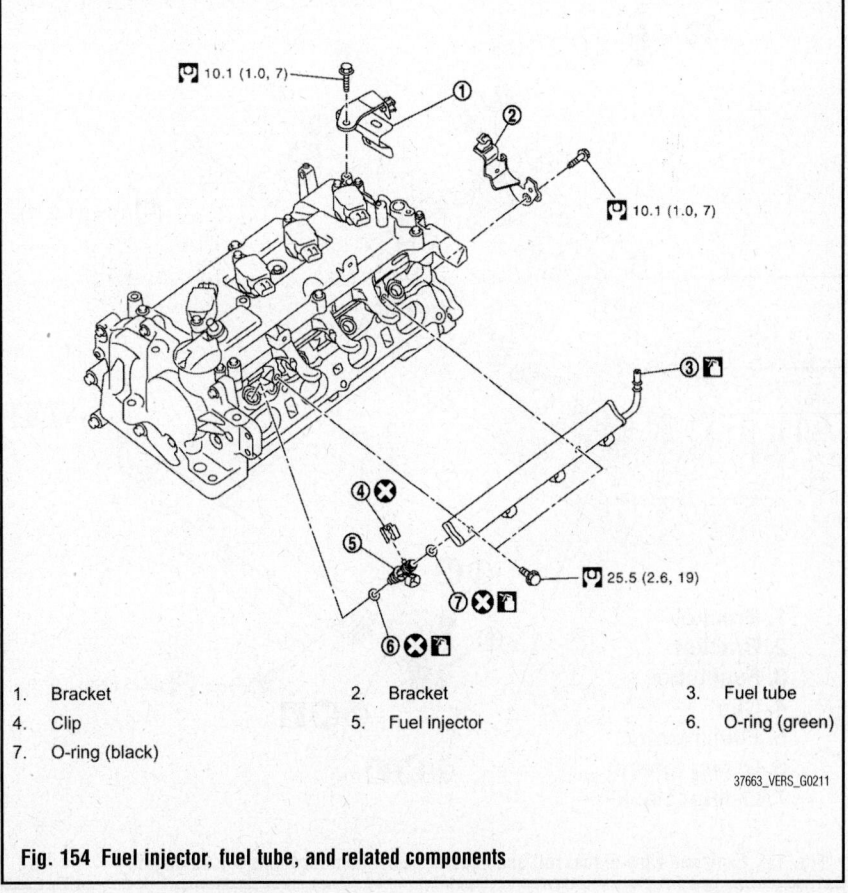

1.	Bracket	2.	Bracket	3.	Fuel tube
4.	Clip	5.	Fuel injector	6.	O-ring (green)
7.	O-ring (black)				

Fig. 154 Fuel injector, fuel tube, and related components

⁑ WARNING

When installing O-ring, be careful not to scratch it with tool or fingernails. Also be careful not to twist or stretch O-ring. If O-ring was stretched while it was being attached, never insert it quickly into fuel tube. Insert O-ring straight into fuel tube. Never twist it.

15. Install fuel injector to fuel tube with the following procedure:

   a. Insert clips into clip groove on fuel injector. Insert clip so that protrusion of fuel injector matches cutout of clip.

⁑ WARNING

Never reuse clip. Replace it with a new one. Be careful to keep clip from interfering with O-ring. If interference occurs, replace O-ring.

   b. Insert fuel injector into fuel tube with clip attached:

   • Insert it while matching it to the axial center
   • Insert fuel injector so that protrusion of fuel tube matches cut-out of clip

   • Make sure that fuel tube flange is securely fixed in flange fixing groove on clip

16. Make sure that installation is complete by making sure that fuel injector does not rotate or come off.

17. Set fuel tube and fuel injector assembly at its position for installation on cylinder head. Be careful not to interfere with fuel injector nozzle.

18. Tighten bolts in numerical order.

19. Installation of the remaining components is in the reverse order of removal.

20. After installation is complete, check for fuel leaks:

   a. Turn ignition switch "ON" (without starting the engine), to check the connections for fuel leaks with the electric fuel pump applying pressure to the fuel piping.

   b. Start the engine and let it idle to check that there are no fuel leaks at the fuel system connections.

FUEL TANK

DRAINING

1. Before servicing the vehicle, refer to the Precautions.

2. Insert fuel tubing of less than 1 inch

(25mm) diameter into the fuel filler tube through the fuel filler opening to drain fuel from the fuel filler tube.

3. Disconnect the fuel filler hose from the fuel filler tube.

4. Insert fuel tubing into the fuel tank through the fuel filler hose to drain fuel from the fuel tank.

REMOVAL & INSTALLATION

See Figure 156.

1. Before servicing the vehicle, refer to the Precautions.

2. Drain fuel from fuel tank if necessary.

⁑ CAUTION

Because fuel tank becomes unstable when installing/removing, fuel should be drained if the level exceeds specification.

3. Place vehicle on a flat and solid surface.

4. Open fuel door and unscrew the fuel filler cap to release the pressure inside the fuel tank.

5. Release the fuel pressure from the fuel lines.

6. Remove rear seat bottom.

7. Turn the four retainers 90° in a clockwise direction and remove the fuel pump inspection hole cover.

8. Disconnect electrical connector and fuel feed hose quick connector. For quick connector disconnecting instructions, refer to Fuel Pump Assembly, removal & installation.

9. Remove center exhaust tube.

10. Remove exhaust heat shields.

11. Disconnect parking brake cables from the lower surface of fuel tank and axle and position the parking brake cables out of the way.

12. Remove brake tube protector from rear axle.

13. Loosen fuel filler hose clamp and remove fuel filler hose from fuel tank. When removing fuel filler hose at the fuel filler tube, mark components for alignment.

14. Remove vent hose and EVAP hose at rear of fuel tank.

15. Disconnect vent hose and EVAP hose quick connectors using the following procedures:

   a. Pinch retaining tabs of vent hose quick connector and remove vent hose.

   b. Slide sleeve of EVAP hose quick connector and remove EVAP hose.

   c. If hoses are stuck, push and pull several times until they move freely, and disconnect.

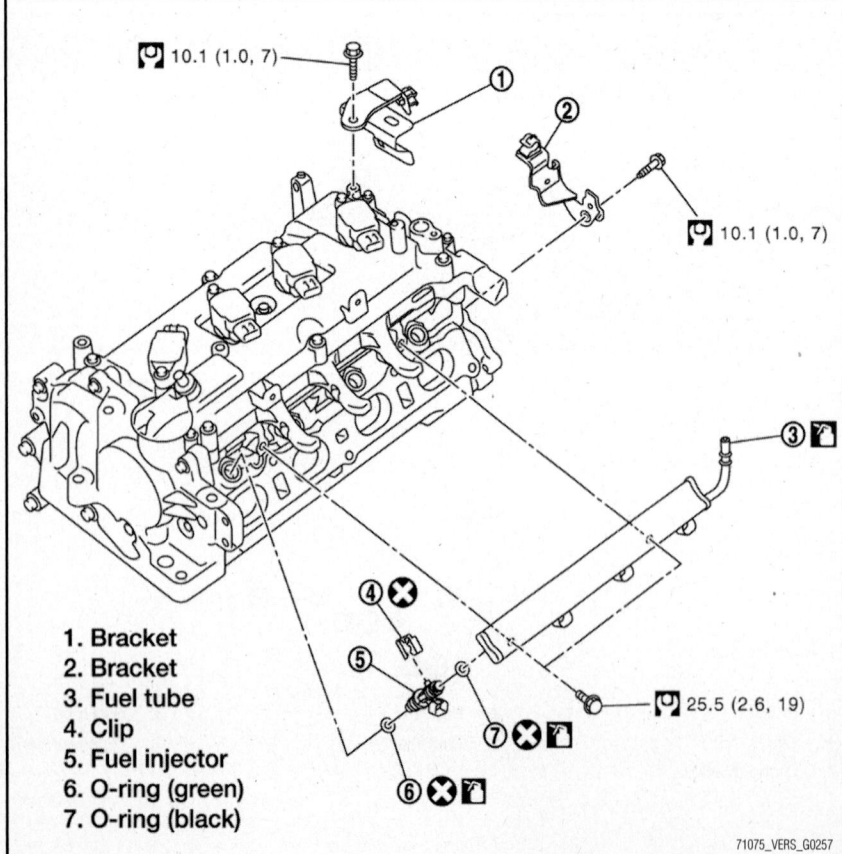

10.1 (1.0, 7)
10.1 (1.0, 7)
25.5 (2.6, 19)

1. Bracket
2. Bracket
3. Fuel tube
4. Clip
5. Fuel injector
6. O-ring (green)
7. O-ring (black)

71075_VERS_G0257

Fig. 155 Exploded view of fuel rail and injector component locations—1.8L engine

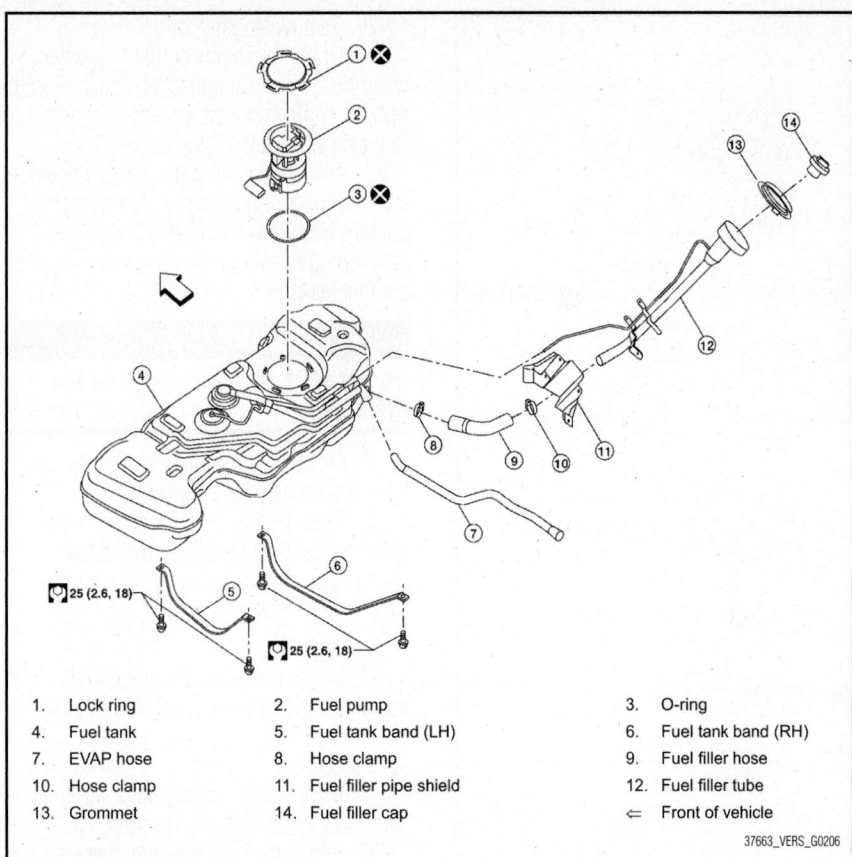

1. Lock ring
2. Fuel pump
3. O-ring
4. Fuel tank
5. Fuel tank band (LH)
6. Fuel tank band (RH)
7. EVAP hose
8. Hose clamp
9. Fuel filler hose
10. Hose clamp
11. Fuel filler pipe shield
12. Fuel filler tube
13. Grommet
14. Fuel filler cap
⇐ Front of vehicle

37663_VERS_G0206

Fig. 156 Fuel tank and related components

16. Support the center of the fuel tank with a transmission jack. Securely support the fuel tank with a suitable tool.

17. Remove fuel tank bands.

18. Lower transmission jack carefully to remove fuel tank while supporting it by hand. Fuel tank may be in an unstable position because of the shape of fuel tank bottom. Be sure to support tank securely.

To install:

19. Installation is in the reverse order of removal, noting the following:

20. Check the EVAP canister connection for damage or any foreign materials, and align the connector with the tube, then insert the connector straight into the tube until it clicks. After connecting, make sure that the connection is secure by pulling the tube and the connector.

21. Install the fuel tank bands in the proper position by referring to the identification stamp mark "R" and "L" on the end.

a. While supporting the fuel tank, install bolts 1, 3 and 4 to support the tank, but do not fully tighten.

b. Install bolt 2 while positioning the fuel tank toward the front of the vehicle. Tighten bolt 2 to specified torque.

c. Tighten bolts 1, 3 and 4 to specified torque.

✳✳ CAUTION

Do not allow fuel filler tube to contact the suspension during installation.

22. Insert fuel filler hose to 1.38 inch (35mm). Be sure hose clamp is not placed on swelled area of fuel filler tube.

a. Check the EVAP hose connections for damage or foreign material, align the matching quick connector with the center of EVAP hose, and insert quick connector straight until it clicks. Make sure connections are secure by pulling on quick connector and EVAP hose by hand.

23. Check for fuel leaks:

a. Turn ignition switch "ON" (without starting the engine), to check the connections for fuel leaks with the electric fuel pump applying pressure to the fuel piping.

b. Start the engine and let it idle to check that there are no fuel leaks at the fuel system connections.

THROTTLE BODY

REMOVAL & INSTALLATION

See Figures 157 through 159.

1. Before servicing the vehicle, refer to the Precautions.

2. Disconnect the negative battery cable.

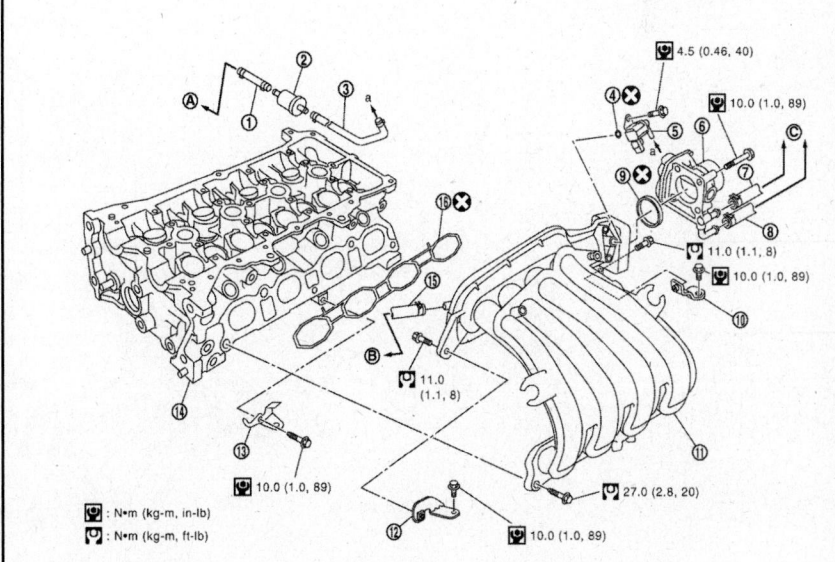

1. EVAP hose
2. Vacuum tank
3. EVAP hose
4. O-ring
5. EVAP canister purge volume control solenoid valve
6. Electric throttle control actuator
7. Water hose
8. Water hose
9. Gasket
10. Intake manifold support (rear)
11. Intake manifold
12. Intake manifold support (front)
13. Intake manifold support (center)
14. Cylinder head
15. Vacuum hose
16. Gasket
A. To centralized under-floor piping
B. To brake booster
C. To water outlet

🔧 : N•m (kg-m, in-lb)
🔧 : N•m (kg-m, ft-lb)

37663_VERS_G0220

Fig. 157 Intake manifold, throttle, and related components—1.6L engine (2011)

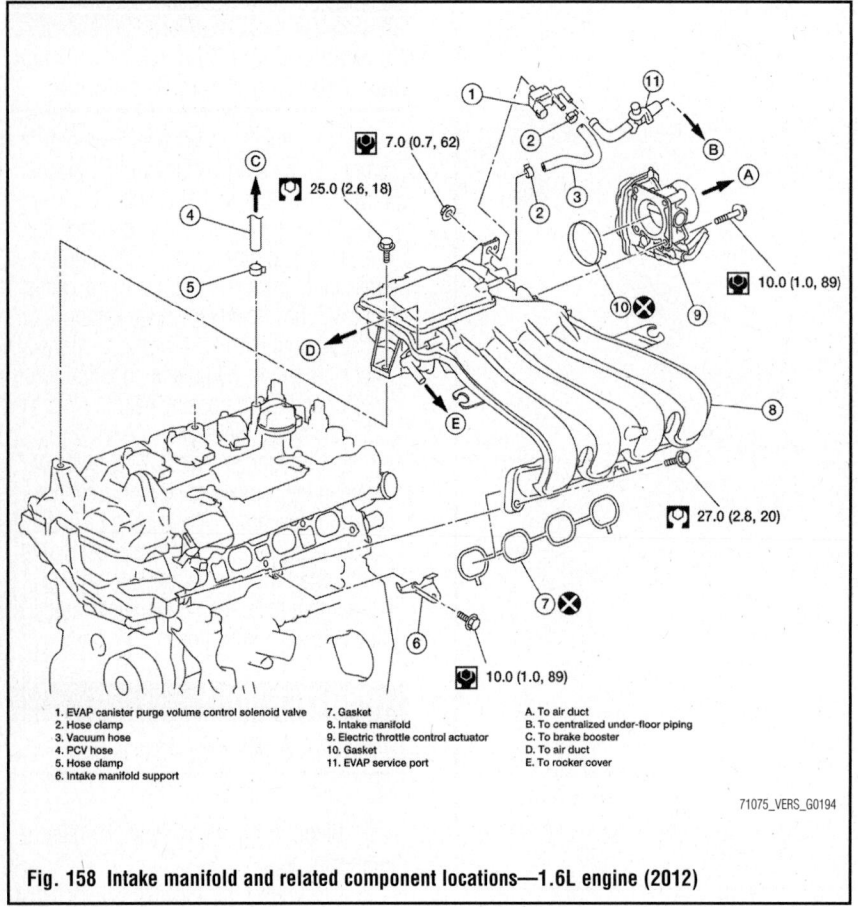

7.0 (0.7, 62)
25.0 (2.6, 18)
10.0 (1.0, 89)
27.0 (2.8, 20)
10.0 (1.0, 89)
1. EVAP canister purge volume control solenoid valve
2. Hose clamp
3. Vacuum hose
4. PCV hose
5. Hose clamp
6. Intake manifold support
7. Gasket
8. Intake manifold
9. Electric throttle control actuator
10. Gasket
11. EVAP service port
A. To air duct
B. To centralized under-floor piping
C. To brake booster
D. To air duct
E. To rocker cover
71075_VERS_G0194

Fig. 158 Intake manifold and related component locations—1.6L engine (2012)

3. Remove engine cover.
4. Drain the engine coolant, or when water hoses are disconnected, attach plug to prevent engine coolant leakage. Perform this step when the engine is cold.
5. Disconnect the water hoses the electric throttle control actuator. When engine coolant is not drained from the radiator, attach plug to water hoses to prevent engine coolant leakage.

⁂ WARNING

Do not spill engine coolant on the drive belt.

6. Disconnect the electric throttle control actuator harness connector.
7. Remove electronic throttle control actuator. Handle carefully to avoid any shock to electric throttle control actuator. Never disassemble.

To install:

8. Install electronic throttle control actuator. Tighten bolts of electric throttle control actuator equally and diagonally in several steps.
9. If applicable, install water hoses to electronic throttle control actuator.
10. Add engine coolant and check for leaks.
11. Connect the negative battery cable.

10.0 (1.0, 89)
19.6 (2.0, 14)
10.0 (1.0, 89)
27.0 (2.8, 20)

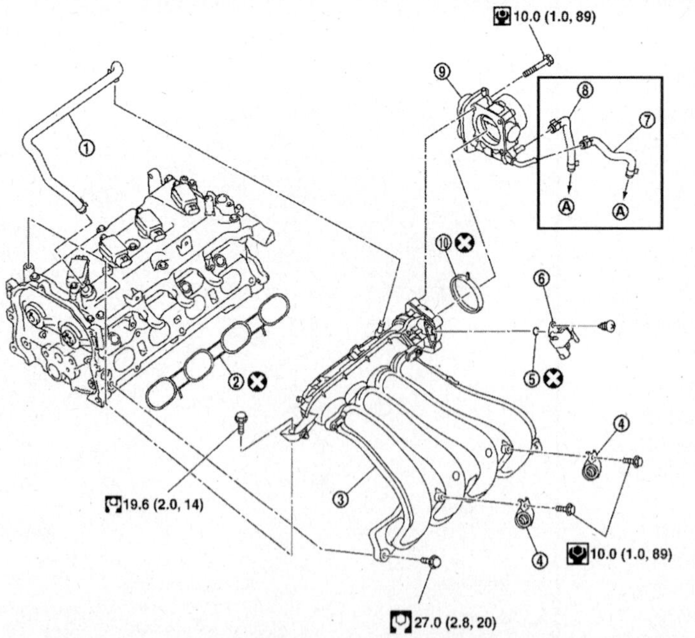

1.	PCV hose	2.	Gasket	3.	Intake manifold
4.	Bracket	5.	O-ring	6.	EVAP canister purge volume control solenoid valve
7.	Water hose	8.	Water hose	9.	Electric throttle control actuator
10.	Gasket	A.	To water outlet		

37663_VERS_G0219

Fig. 159 Intake manifold, throttle, and related components—1.8L engine

HEATING & AIR CONDITIONING SYSTEM

BLOWER MOTOR

REMOVAL & INSTALLATION

Except 2012 Sedan

See Figure 160.

> ❊❊ **CAUTION**
>
> **Before servicing components near or affected by the SRS (air bag) system, read and observe all SRS Service Precautions. Failure to observe all precautions may result in accidental airbag deployment, personal injury, or death.**

1. Before servicing the vehicle, refer to the Precautions.

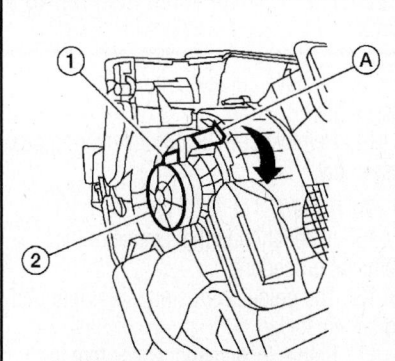

1. Blower motor connector
2. Blower motor
A. Flange holding hook

22140_VERS_G0043

Fig. 160 Blower motor shown—except 2012 Sedan

2. Disconnect the negative battery cable and wait at least 3 minutes for the SRS memory to drain.

3. Remove the instrument panel and pad.

4. Remove the right side ventilator duct.

5. Disconnect the blower motor connector.

6. Push the flange holding hook toward the blower motor, then rotate the blower motor clockwise and remove it from the A/C unit assembly.

> ❊❊ **WARNING**
>
> **When the blower fan and blower motor are assembled, the balance is adjusted, do not disassemble to replace the individual parts.**

To install:

7. Installation is the reverse of removal.

8. Rotate the blower motor until the blower motor flange holding hook locks securely into the A/C unit assembly.

2012 Sedan

See Figure 161.

> ❊❊ **CAUTION**
>
> **Before servicing components near or affected by the SRS (air bag) system, read and observe all SRS Service Precautions. Failure to observe all precautions may result in accidental airbag deployment, personal injury, or death.**

1. Before servicing the vehicle, refer to the Precautions.

2. Disconnect the negative battery cable and wait at least 3 minutes for the SRS memory to drain.

3. Remove the lower knee protector LH.

4. Remove the accelerator pedal assembly.

5. Remove the brake pedal assembly.

6. Disconnect the blower motor connectors and thermo control amp harness.

7. Remove the blower motor screw.

8. Press the flange holding hook, then turn the blower motor counterclockwise and remove the blower motor.

9. Installation is the reverse of the removal procedure.

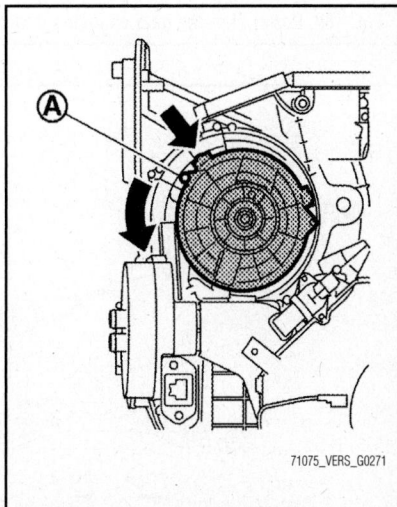

71075_VERS_G0271

Fig. 161 Blower motor and screw (A) component location—2012 Sedan

STEERING

POWER RACK & PINION STEERING GEAR

REMOVAL & INSTALLATION

Except 2012 Sedan

See Figure 162.

1. Before servicing the vehicle, refer to the Precautions.

> ❊❊ **WARNING**
>
> **Spiral cable may be cut if steering wheel turns while separating steering column assembly and steering gear assembly. Be sure to secure steering wheel using string to avoid turning.**

2. Remove the front suspension member.

3. Remove mounting bolts and nuts of steering gear assembly.

To install:

4. Installation is in the reverse order of removal.

5. Clean mounting surface on the body side of fire wall seal when installing steering gear assembly.

6. Check wheel alignment under unladen conditions with tires on level ground.

2012 Sedan

See Figure 163.

1. Before servicing the vehicle, refer to the Precautions.

2. Set vehicle to the straight-ahead position.

3. Place the steering column tilt to middle selection.

4. Loosen lower joint upper bolt.

5. Remove lower joint lower bolt.

> ❊❊ **WARNING**
>
> **Spiral cable may be cut if steering wheel turns while separating steering column assembly and steering gear assembly. Always secure the steering wheel using string to avoid turning.**

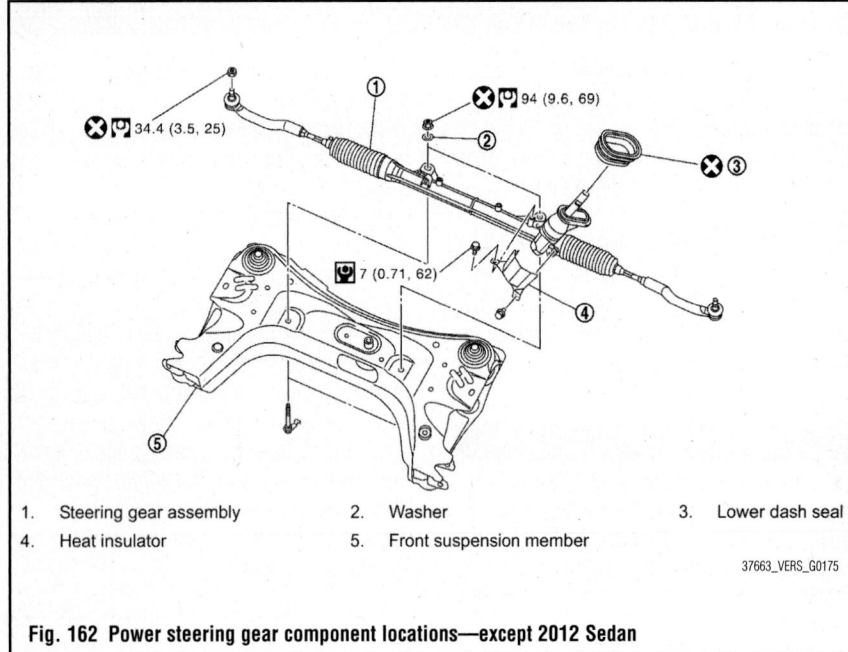

1. Steering gear assembly
2. Washer
3. Lower dash seal
4. Heat insulator
5. Front suspension member

37663_VERS_G0175

Fig. 162 Power steering gear component locations—except 2012 Sedan

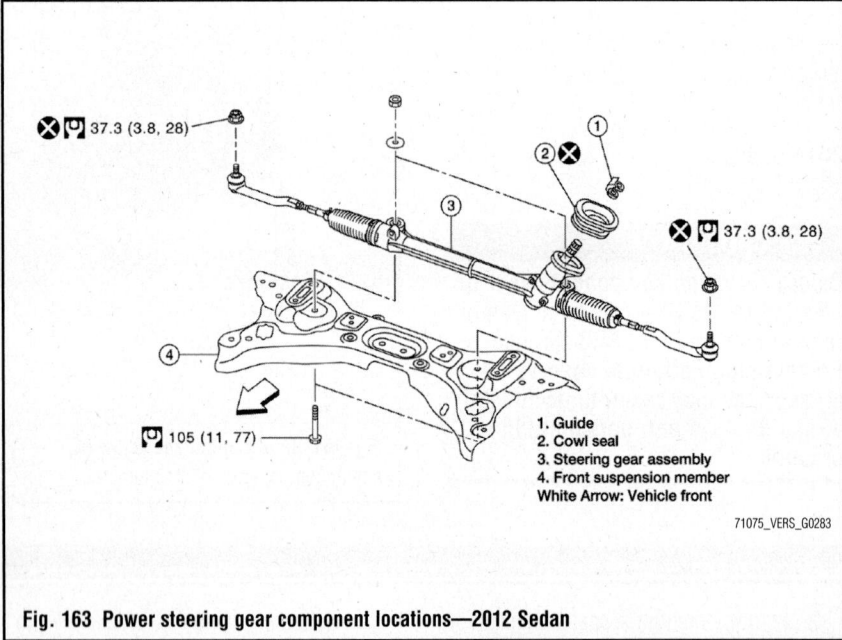

1. Guide
2. Cowl seal
3. Steering gear assembly
4. Front suspension member
White Arrow: Vehicle front

71075_VERS_G0283

Fig. 163 Power steering gear component locations—2012 Sedan

➡ Place a matching mark on both lower joint and steering gear assembly before removing lower joint.

❇❇ **WARNING**

When removing lower joint, never place any tool into the yoke groove to pull out the lower joint. Replace the lower joint with a new one if damaged.

6. Remove the wheel and tire assemblies using power tool.

7. Remove steering outer socket from steering knuckle using a suitable tool.

8. Support front suspension member with a suitable jack.

9. Remove front suspension member bolts.

10. Remove the bolts and nuts of steering gear assembly.

11. Separate cowl seal from vehicle.

12. Lower the suitable jack to the position where the steering gear assembly can be removed.

❇❇ **WARNING**

Secure front suspension member to a jack.

13. Remove guide from pinion assembly.

14. Remove cowl seal from steering gear assembly.

To install:

15. Installation is the reverse of the removal procedure.

16. The guide protrusion is not required to be reinstalled.

17. Before installation, check that the steering column tilt position is at the middle level.

18. Clean mating surface on the body side of cowl seal when installing steering gear assembly.

19. Perform final tightening of nuts and bolts on each part under unladen conditions with tires on level ground.

20. Check wheel alignment.

21. Rotate steering wheel to check that the spiral cable is centered, binding, noise or excessive steering effort.

22. Do not reuse steering outer socket nut and cowl seal.

23. Perform inspection after installation.

KNUCKLE & SPINDLE

REMOVAL & INSTALLATION

Except 2012 Sedan

See Figure 164.

1. Before servicing the vehicle, refer to the Precautions.

2. Remove wheel and tire.

3. Without disassembling the hydraulic lines, remove the torque member bolts. Then reposition the torque member and brake caliper assembly aside with wire.

➡**Do not depress brake pedal while brake caliper is removed.**

4. Put alignment marks on disc rotor and wheel hub and bearing assembly, and remove disc rotor.

5. Remove wheel sensor from steering knuckle. Do not pull on wheel sensor harness.

6. Loosen steering outer socket nut.

7. Remove steering outer socket from steering knuckle so as not to damage ball joint boot using ball joint remover or suitable tool. Temporarily leave the outer socket nut installed to prevent damage to threads and to prevent the ball joint remover or suitable tool from suddenly coming off.

8. Remove transverse link ball joint nut and bolt, and remove transverse link from steering knuckle.

9. Remove cotter pin, and loosen hub lock nut. Temporarily leave the hub lock nut installed to prevent damage to threads.

10. Separate the halfshaft from the wheel hub and bearing assembly by lightly tapping the end of the halfshaft using a hammer or suitable tool, and remove hub lock nut.

11. If wheel hub and bearing assembly and halfshaft cannot be separated after performing the above procedure, use a suitable puller.

12. Remove the halfshaft from the wheel hub and bearing assembly and support the halfshaft.

※※ WARNING

Do not apply an excessive angle to halfshaft joint when removing from the wheel hub and bearing assembly. Do not excessively extend slide joint. Do not allow halfshaft to hang down. Support the entire halfshaft.

13. Remove wheel hub and bearing assembly bolts, and then remove splash

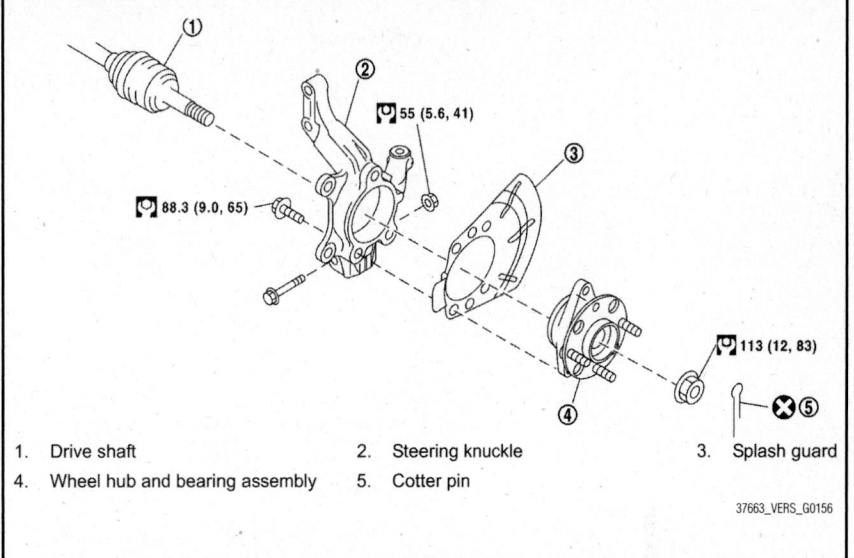

1. Drive shaft
2. Steering knuckle
3. Splash guard
4. Wheel hub and bearing assembly
5. Cotter pin

37663_VERS_G0156

Fig. 164 Front wheel hub and bearing component locations—except 2012 Sedan

guard and wheel hub and bearing assembly from steering knuckle.

14. Remove nuts and bolts, and remove steering knuckle from strut assembly.

To install:

15. Installation is the reverse order of removal.

16. Tighten the hub lock nut to 83 ft. lbs. (113 Nm).

➡**Perform the final tightening of each of parts under unladen conditions, which were removed when removing wheel hub and bearing assembly and steering knuckle.**

17. When installing disc rotor on wheel hub and bearing assembly, align the marks.

18. Check the wheel alignment.

2012 Sedan

See Figure 165.

1. Before servicing the vehicle, refer to the Precautions.

2. Remove the wheel and tire assembly.

3. Remove wheel sensor and sensor harness.

4. Remove brake caliper, leaving brake caliper hydraulic lines connected. Reposition brake caliper aside with wire.

➡**Avoid depressing brake pedal while brake caliper is removed.**

5. Put matching marks on disc rotor and wheel hub and bearing assembly, then remove disc rotor.

a. Use paint for matching parts. Do not damage the disc rotor or wheel hub assembly.

b. Do not drop disc rotor.

6. Remove and discard cotter pin, and then loosen wheel hub lock nut, using Tool number: KV40104000, or equivalent.

7. Separate the halfshaft from the wheel hub and bearing assembly by lightly tapping the end of the halfshaft using a suitable tool and a wood block.

➡**Use a suitable puller, if the wheel hub and halfshaft cannot be separated after performing the above procedure.**

8. Remove the wheel hub lock nut.

9. Remove the nut and separate the steering outer socket from steering knuckle.

10. Remove the nuts and bolts, then separate the steering knuckle from strut assembly.

11. Suspend the halfshaft with suitable wire.

※※ WARNING

Do not place halfshaft joint at an extreme angle. Also be careful not to overextend slide joint. Do not allow halfshaft to hang down without support for joint sub-assembly, shaft, and the other parts.

12. Remove steering knuckle from transverse link.

To install:

13. Installation is the reverse of the removal procedure.

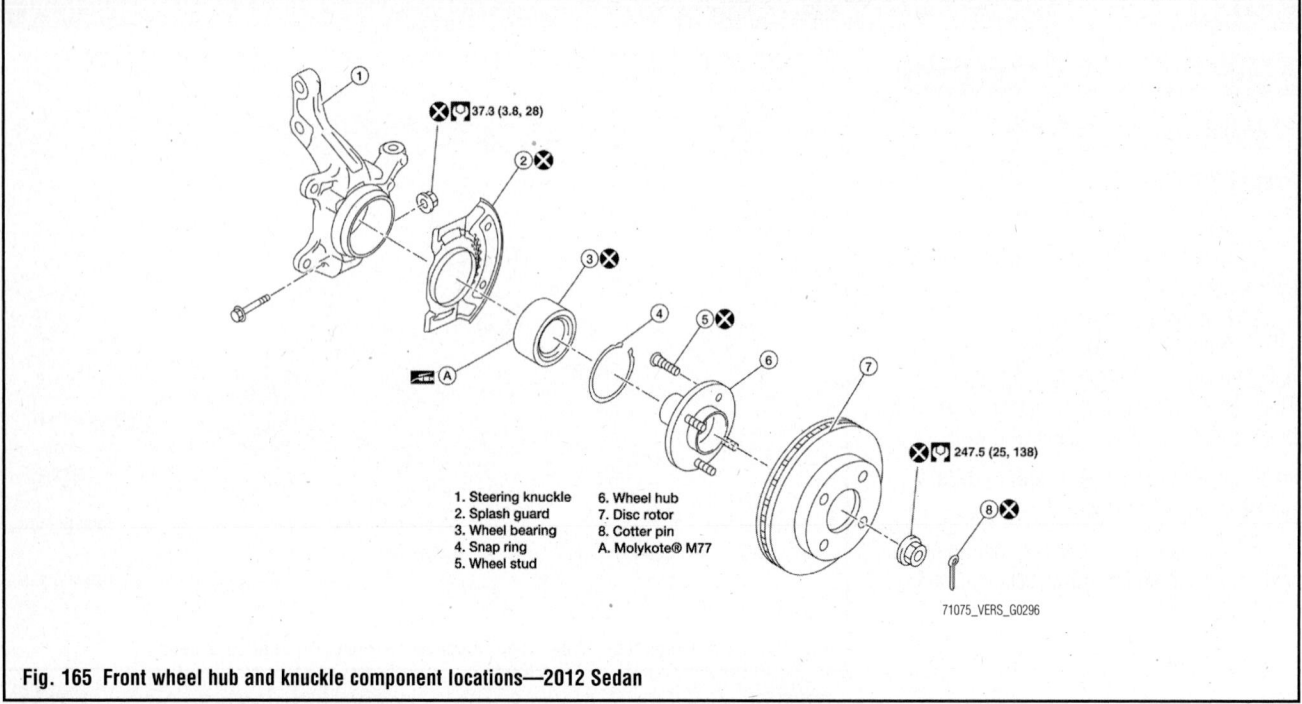

1. Steering knuckle
2. Splash guard
3. Wheel bearing
4. Snap ring
5. Wheel stud
6. Wheel hub
7. Disc rotor
8. Cotter pin
A. Molykote® M77

71075_VERS_G0296

Fig. 165 Front wheel hub and knuckle component locations—2012 Sedan

---

❄❄ WARNING

During installation, do not damage the wheel bearing seal. If damage has occurred, replace wheel bearing with a new one. Do not allow paint to adhere to the wheel bearing seal.

14. Check each mating surface for water and foreign matter. If there is any water or foreign matter, clean the mounting surface.

➡**Do not reuse steering knuckle and transverse link nut.**

15. Clean the mating surface of wheel hub lock nut and wheel hub. Do not apply lubricating oil to the mating surface of the wheel hub lock nut and the wheel hub.

16. Clean the mating surface of halfshaft, wheel hub, and wheel bearing. And then apply Molykote® M77 to surface of joint subassembly of halfshaft. Apply Molykote® M77 to cover entire flat surface of joint sub-assembly of halfshaft.

17. When reusing disc rotor, align the matching marks during removal.

18. Perform the final tightening of each part under unladen conditions, which were removed when removing wheel hub and axle housing.

19. When installing a cotter pin, securely bend the basal portion to prevent rattles.

❄❄ WARNING

Do not reuse cotter pin.

STABILIZER BAR & LINKS

REMOVAL & INSTALLATION

Stabilizer Links

Except 2012 Sedan

See Figure 166.

1. Before servicing the vehicle, refer to the Precautions.

2. Remove the front wheel and tire assemblies from the vehicle.

3. Remove the nut on the upper side and lower side of the stabilizer connecting rod.

4. Remove the stabilizer connecting rod from the strut assembly.

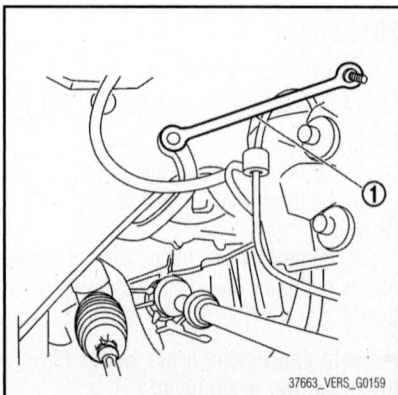

37663_VERS_G0159

Fig. 166 Remove the nut on the upper side of stabilizer connecting rod (1), and remove stabilizer connecting rod

---

To install:

5. Installation is in the reverse order of removal.

6. Tighten the stabilizer connecting rod nut to 27 ft. lbs. (37 Nm).

2012 Sedan

See Figure 167.

1. Before servicing the vehicle, refer to the Precautions.

2. Remove the front wheel and tire assemblies from the vehicle.

3. Remove the nut on the upper side and lower side of the stabilizer connecting rod.

4. Remove the stabilizer connecting rod (control link) from the strut assembly.

To install:

5. Installation is in the reverse order of removal.

6. Tighten the stabilizer connecting rod nut to 29 ft. lbs. (39 Nm).

Stabilizer Bar

Except 2012 Sedan

See Figure 168.

1. Before servicing the vehicle, refer to the Precautions.

2. Separate intermediate shaft from steering gear pinion shaft.

3. Remove wheels and tires.

4. Remove the nut on the lower side of the stabilizer connecting rod, and remove the stabilizer connecting rod from the stabilizer bar.

5. If necessary remove the stabilizer

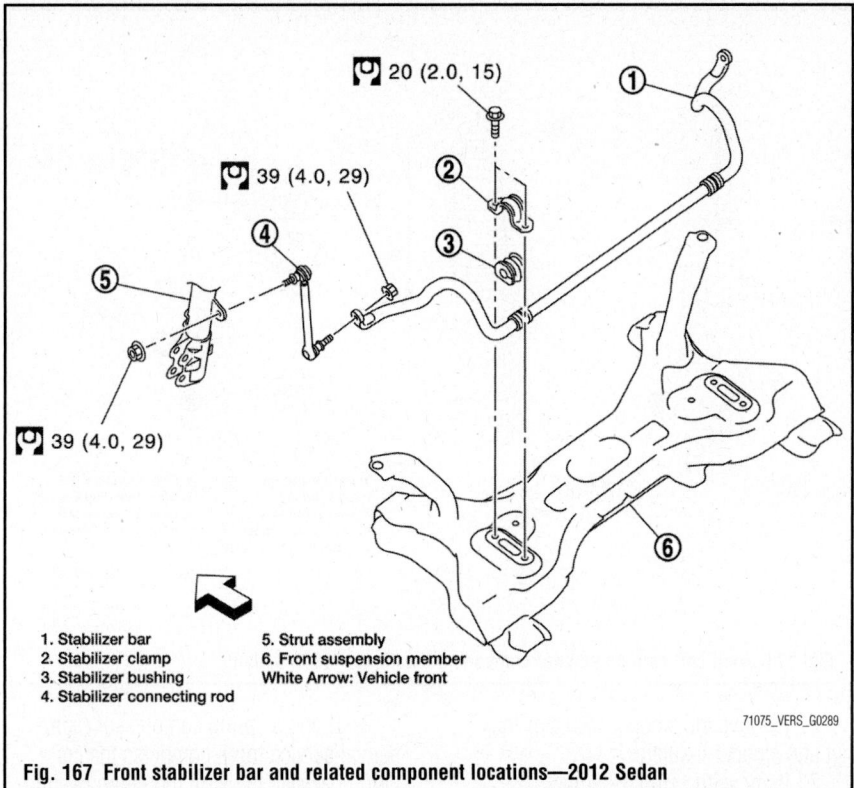

1. Stabilizer bar
2. Stabilizer clamp
3. Stabilizer bushing
4. Stabilizer connecting rod
5. Strut assembly
6. Front suspension member
White Arrow: Vehicle front

71075_VERS_G0289

Fig. 167 Front stabilizer bar and related component locations—2012 Sedan

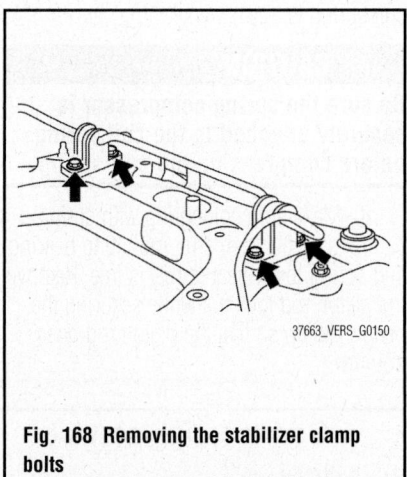

37663_VERS_G0150

Fig. 168 Removing the stabilizer clamp bolts

connecting rod upper nut. Separate the stabilizer connecting rod and the strut.

6. Loosen the steering outer socket nut.

7. Using the ball joint remover (tool No. HT72520000 (J-25730-A) or suitable tool, remove the steering outer socket from the steering knuckle so as not to damage the ball joint boot.

✳✳ WARNING

Temporarily tighten the nut to prevent damage to threads and to prevent the ball joint remover (suitable tool) from suddenly coming off.

8. Remove the rear torque rod.

9. Set a jack under the front suspension member.

10. Remove the bolts of the member stay, and remove member stay from vehicle.

11. Gradually lower the front suspension member in order to remove the stabilizer bolts. Be careful not to lower it too far. (Do not over load the links).

12. Remove the bolts of the stabilizer clamp, and remove the stabilizer clamp and stabilizer bushing from vehicle.

13. Remove the stabilizer bar from the vehicle.

To install:

14. Installation is the reverse of removal.

15. Install the stabilizer connecting rod and strut and tighten the nut to 27 ft. lbs. (37 Nm).

16. Install the stabilizer clamp and tighten the bolts to 21 ft. lbs. (28 Nm).

2012 Sedan

See Figures 168 and 169.

1. Before servicing the vehicle, refer to the Precautions.

2. Remove wheel and tire assemblies.

3. Remove stabilizer connecting rod (control links).

4. Remove the pinch bolt and separate the intermediate shaft from the lower joint.

5. Remove the engine rear torque rod.

6. Position a suitable jack under front suspension member.

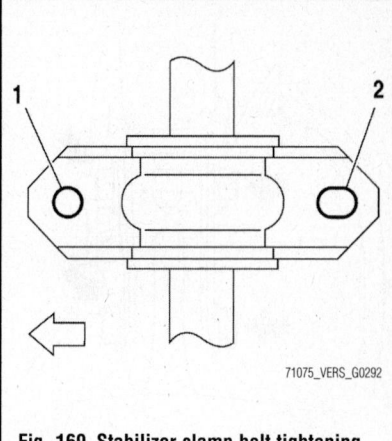

71075_VERS_G0292

Fig. 169 Stabilizer clamp bolt tightening sequence shown (white arrow: vehicle front)

✳✳ WARNING

Do not damage the front suspension member with jack.

7. Remove front suspension member bolts.

8. Gradually lower jack front suspension member in order to remove stabilizer mounting bolts.

9. Remove the stabilizer clamp bolts, stabilizer clamps and stabilizer bushings from front suspension member.

10. Remove stabilizer bar from the vehicle.

11. Check stabilizer bar, stabilizer connecting rod, stabilizer bushing, and stabilizer clamp for deformation, cracks, and damage. Replace it if necessary.

To install:

12. Installation is the reverse of the removal procedure.

13. Tighten the fasteners to specification.

14. Install the stabilizer bushing with the slit facing the rear of the vehicle.

15. To install stabilizer clamp bolt, temporarily tighten in numerical order and then tighten to 15 ft. lbs. (20 Nm).

16. Install the stabilizer connecting rod by tightening the nut with the hexagonal part on the stabilizer connecting rod side.

➡**Perform final tightening of bolts and nuts with the vehicle under unladen conditions with tires on level ground.**

17. Perform inspection after installation.

STRUTS

REMOVAL & INSTALLATION

Except 2012 Sedan

See Figure 170.

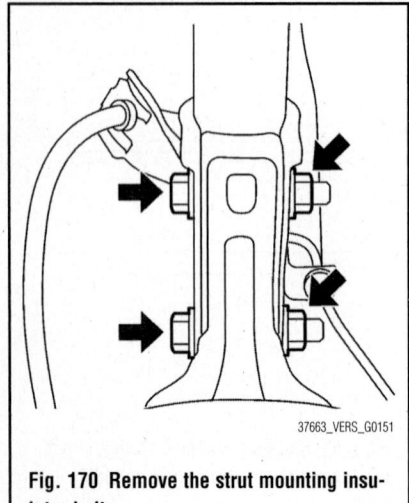

Fig. 170 Remove the strut mounting insulator bolts

1. Before servicing the vehicle, refer to the Precautions.

2. Remove the cowl top panel.

3. Remove the wheels and tires.

4. Remove the wheel sensor harness from the strut assembly. Do not pull on the wheel sensor harness.

5. Remove the brake hose lock plate.

6. Remove the nut on the upper side of the stabilizer connecting rod, and remove the stabilizer connecting rod from the strut assembly.

7. Remove the nuts and bolts, and then remove steering knuckle from the strut assembly.

8. Remove the strut mounting insulator bolts, and remove the strut assembly from vehicle.

To install:

9. Installation is in the reverse order of removal.

10. Perform final tightening of bolts and nuts at the strut assembly lower side (rubber bushing) under unladen conditions with tires on level ground. Check wheel alignment.

11. Check wheel sensor harness for proper connection.

12. Attach strut mounting insulator.

2012 Sedan

See Figure 171.

1. Before servicing the vehicle, refer to the Precautions.

2. Remove the wheel and tire assembly.

3. Remove the brake hose lock plate from the strut assembly.

4. Remove the stabilizer connecting rod from the strut assembly.

5. Remove the strut mounting bolts and nuts from the steering knuckle.

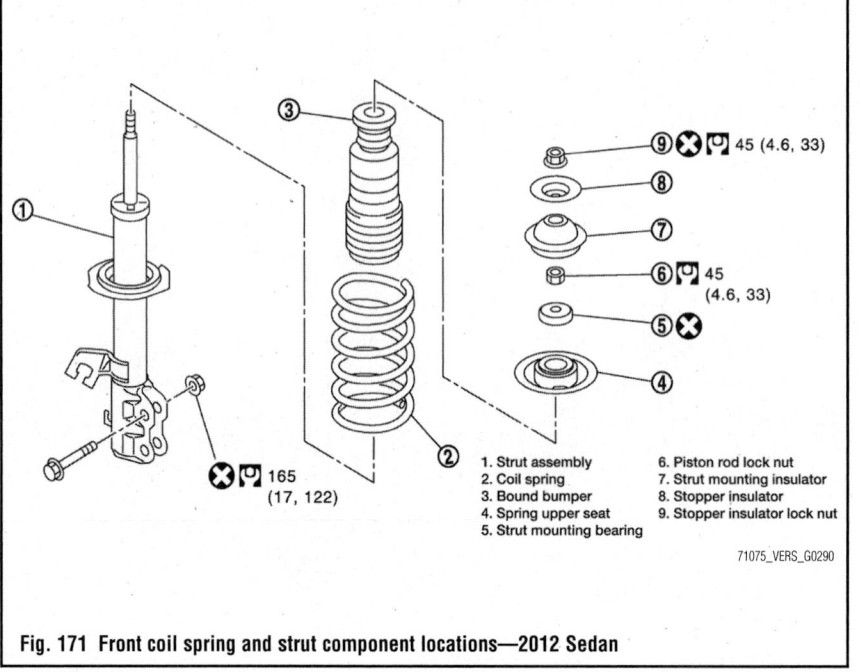

1. Strut assembly
2. Coil spring
3. Bound bumper
4. Spring upper seat
5. Strut mounting bearing
6. Piston rod lock nut
7. Strut mounting insulator
8. Stopper insulator
9. Stopper insulator lock nut

Fig. 171 Front coil spring and strut component locations—2012 Sedan

6. Remove the stopper insulator lock nut and stopper insulator.

7. Remove the strut assembly.

To install:

8. Installation is the reverse of the removal procedure.

9. Secure the head of the strut piston rod to keep it from rotating, then tighten the stopper insulator lock nut to the specified torque.

✺✺ WARNING

Do not reuse stopper insulator lock nut. Do not reuse the nuts that secure the strut to the steering knuckle and stopper insulator lock nut.

10. Perform inspection after installation.

OVERHAUL

See Figure 172.

1. Before servicing the vehicle, refer to the Precautions.

✺✺ WARNING

Do not damage the strut piston rod when removing/installing components from/to the strut assembly.

2. Install the tool (Tool number: ST35652000), or equivalent, to the strut and secure it in a vise.

✺✺ WARNING

When installing the strut attachment to strut, wrap a shop cloth around the strut to protect it from damage.

3. Using a spring compressor (commercial service tool), compress the coil spring between the strut mounting bearing and the spring lower seat (on strut) until the coil spring is free.

✺✺ CAUTION

Be sure the spring compressor is securely attached to the coil spring before compressing the coil spring.

4. Make sure coil spring with spring compressor between strut mounting bearing and spring lower seat (strut) is free. Remove the piston rod lock nut while securing the piston rod tip so that the piston rod does not turn.

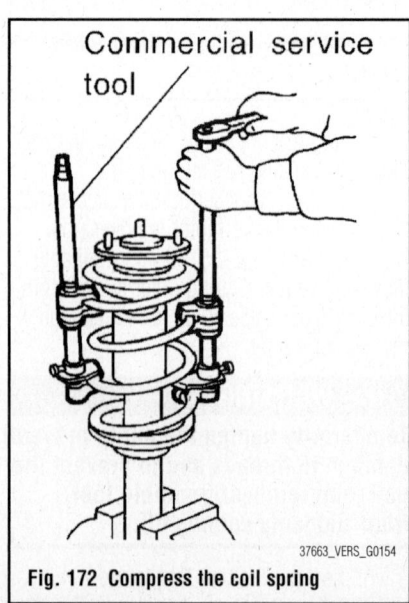

Commercial service tool

Fig. 172 Compress the coil spring

5. Remove the strut mounting insulator, strut mounting bearing, and bound bumper from strut.

6. Remove the coil spring with spring compressor, and gradually release the spring compressor.

✳✳ CAUTION

Loosen while making sure the coil spring attachment position does not move.

7. Remove the strut attachment from strut.

8. Check the following:
- Strut for deformation, cracks or damage, and replace if necessary
- Piston rod for damage, uneven wear or distortion, and replace if necessary
- For oil leakage, and replace if necessary
- Strut mounting insulator for cracks and rubber parts for wear, and replace if necessary
- Coil spring for cracks, wear or damage, and replace if necessary

To assemble:

9. Install the tool (Tool number: ST35652000) to the strut and secure it in a vise.

10. Compress the coil spring using a spring compressor (commercial service tool), and install it onto strut.

a. Face the tube side of the coil spring downward. Align the lower end to the spring lower seat.

b. Be sure the spring compressor is securely attached to the coil spring. Compress the coil spring.

11. Apply soapy water to the bound bumper. Insert the bound bumper into strut mounting insulator.

✳✳ WARNING

Do not use machine oil for a lubricant on the bound bumper.

12. Attach the strut mounting bearing and strut mounting insulator. Attach the strut mounting insulator.

13. Secure the piston rod tip so that piston rod does not turn, and tighten the piston rod lock nut to specified torque.

14. Gradually release the spring compressor, and remove the coil spring. Loosen while making sure the coil spring attachment position does not move.

15. Remove the strut attachment from the strut.

SUSPENSION ASSEMBLY

REMOVAL & INSTALLATION

Except 2012 Sedan

See Figure 173.

1. Before servicing the vehicle, refer to the Precautions.

2. Separate intermediate shaft from steering gear pinion shaft.

3. Remove wheels and tires.

4. Remove the wheel sensor from the steering knuckle. Do not pull on wheel sensor harness.

5. Remove the nut on the upper side of stabilizer connecting rod, and remove stabilizer connecting rod from strut assembly.

6. Loosen the steering outer socket nut.

7. Using the ball joint remover (tool No. HT72520000 (J-25730-A) or suitable

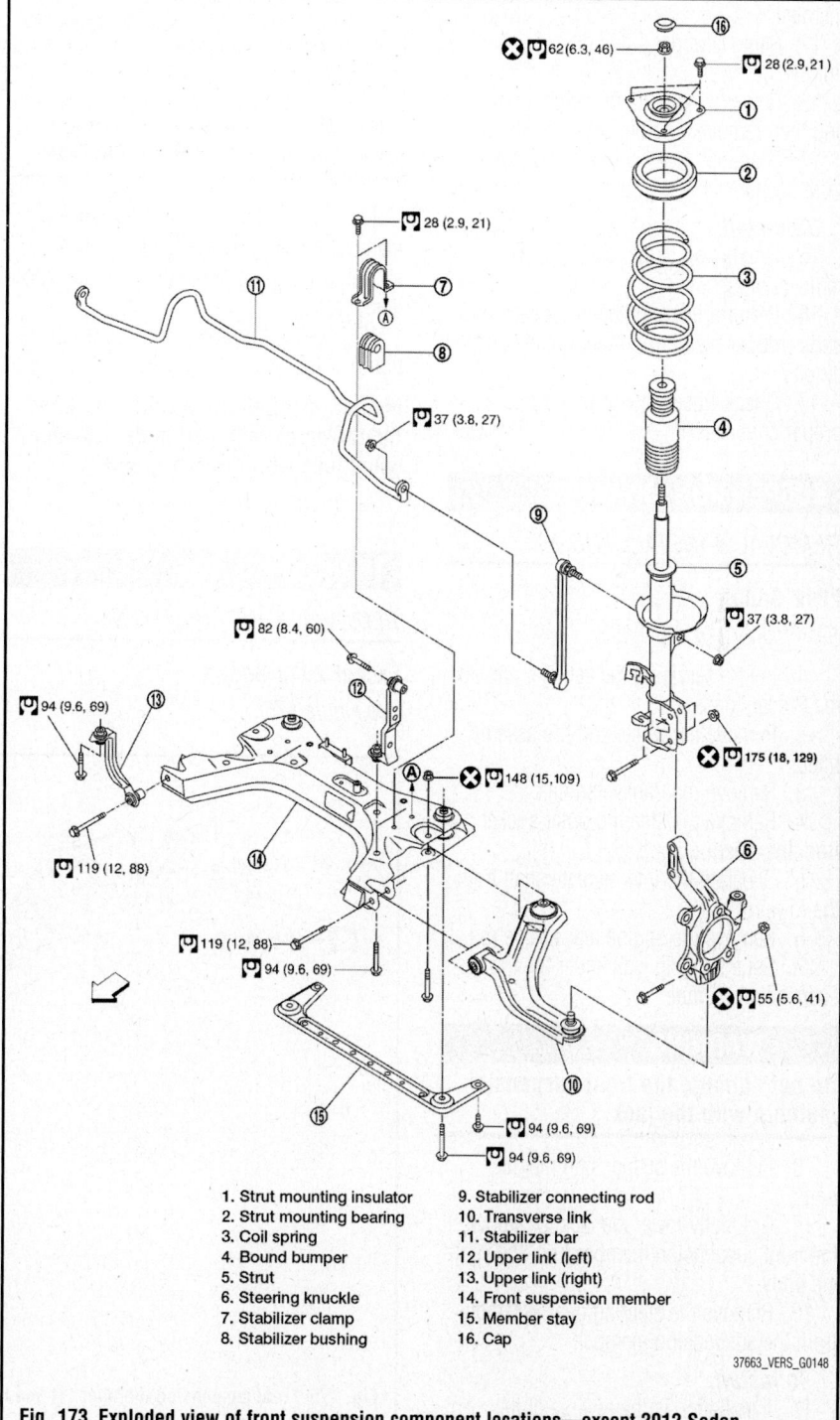

1. Strut mounting insulator
2. Strut mounting bearing
3. Coil spring
4. Bound bumper
5. Strut
6. Steering knuckle
7. Stabilizer clamp
8. Stabilizer bushing
9. Stabilizer connecting rod
10. Transverse link
11. Stabilizer bar
12. Upper link (left)
13. Upper link (right)
14. Front suspension member
15. Member stay
16. Cap

Fig. 173 Exploded view of front suspension component locations—except 2012 Sedan

37663_VERS_G0148

tool, remove the steering outer socket from the steering knuckle so as not to damage the ball joint boot.

8. Temporarily tighten the nut to prevent damage to threads and to prevent the ball joint remover (suitable tool) from suddenly coming off.

9. Remove the rear torque rod.

10. Remove transverse link ball joint nut and bolt, and remove transverse link from steering knuckle.

11. Set a jack under the front suspension member.

12. Remove upper side bolts of upper link.

13. Remove the bolts of member stay, and then remove member stay from vehicle.

14. Gradually lower a jack to remove front suspension assembly.

To install:

15. Installation is the reverse of the removal procedure.

16. Perform final tightening of each of parts (rubber bushing), under unladen conditions.

17. Check wheel sensor harness for proper connection.

SUSPENSION MEMBER

REMOVAL & INSTALLATION

2012 Sedan

See Figures 174 and 175.

1. Before servicing the vehicle, refer to the Precautions.

2. Remove the wheel and tire assemblies.

3. Remove the transverse link.

4. Remove the steering outer socket from the steering knuckle.

5. Separate the intermediate shaft from the lower joint.

6. Remove the engine rear torque rod.

7. Set a suitable jack under the front suspension member.

❄❄ WARNING

Do not damage the front suspension member with the jack.

8. Remove the suspension member bolts.

9. Gradually lower the jack to remove the front suspension member from the vehicle body.

10. Remove the steering gear assembly from the suspension member.

To install:

11. Installation is the reverse of the removal procedure.

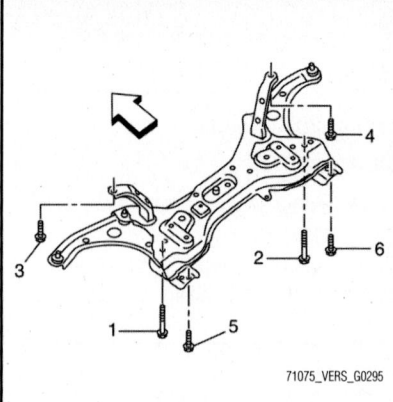

Fig. 174 Front suspension member bolt tightening sequence shown—2012 Sedan

71075_VERS_G0295

12. For installation of the suspension member, temporarily tighten the bolts in sequence and then tighten them to the specified torque.

13. Tighten the wheel nuts to specification.

➡**After installation, perform the final tightening of each part under unladen conditions with tires on ground.**

14. Check wheel alignment.

TRANSVERSE LINK

REMOVAL & INSTALLATION

Except 2012 Sedan

See Figure 176.

1. Before servicing the vehicle, refer to the Precautions.

2. Remove wheels and tires.

3. Remove transverse link ball joint nut and bolt, and remove transverse link from steering knuckle.

4. Remove transverse link nuts and bolts, and remove transverse link from front suspension member.

➡**When removing the left-hand transverse link it may be necessary to lower the suspension member in order to remove bolts to avoid contact with the transaxle.**

a. Set jack under front suspension member.

b. Loosen right-hand upper link bolts, left-hand upper link bolt (front suspension member side), front suspension member bolts (left/right). Lower the front suspension member in order to remove transverse link bolts.

5. Remove transverse link from vehicle.

To install:

6. Installation is the reverse of removal.

7. Perform the final tightening of each of parts under unladen conditions, which were removed when removing wheel hub and bearing assembly and steering knuckle. Check the wheel alignment.

8. Perform final tightening of bolts and at the front suspension member installation position (rubber bushing) under unladen conditions with tires on level ground.

9. Check wheel alignment.

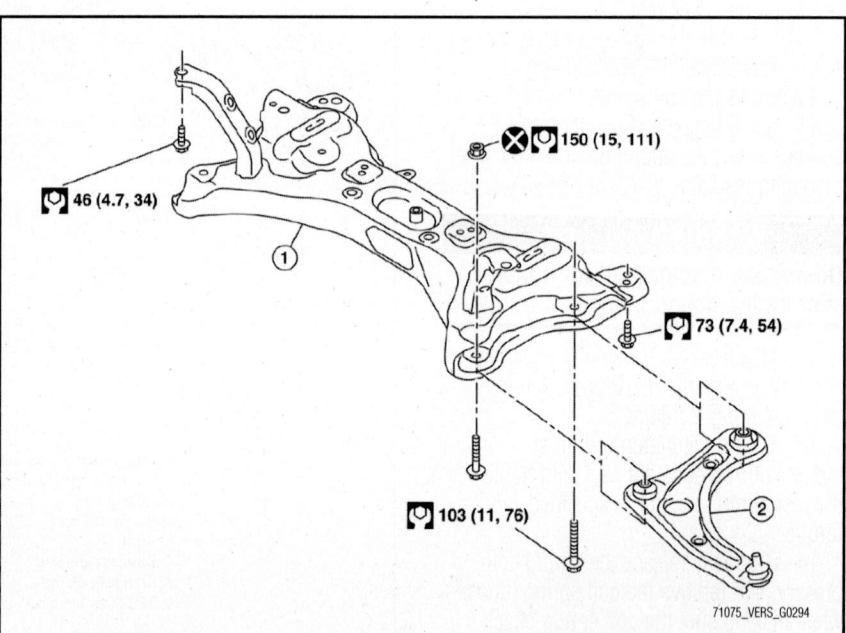

Fig. 175 Front suspension member (1) and transverse link (2) component locations—2012 Sedan

71075_VERS_G0294

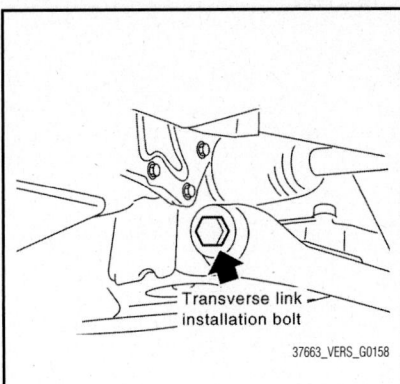

Fig. 176 Remove transverse link nuts and bolts, and remove transverse link from front suspension member

2012 Sedan

See Figure 175.

1. Before servicing the vehicle, refer to the Precautions.
2. Remove the wheel and tire assembly.
3. Remove transverse link from steering knuckle.
4. Remove transverse link from suspension member.

To install:

5. Installation is the reverse of the removal procedure.
6. Perform final tightening of bolts and at the front suspension member installation position (rubber bushing) under unladen conditions with tires on level ground.
7. Check wheel alignment.

WHEEL HUBS & BEARINGS

ADJUSTMENT

1. Before servicing the vehicle, refer to the Precautions.
2. Move the wheel hub and bearing assembly in the axial direction by hand. Make sure there is no looseness of the wheel bearing.
3. Rotate the wheel hub and make sure there are no unusual noises or other irregular conditions.
4. If there are any irregular noises or conditions, replace the wheel hub and bearing assembly. No other adjustment is possible.

REMOVAL & INSTALLATION

Except 2012 Sedan

See Figure 177.

1. Before servicing the vehicle, refer to the Precautions.

2. Remove wheel and tire.
3. Without disassembling the hydraulic lines, remove the torque member bolts. Then reposition the torque member and brake caliper assembly aside with wire.

➡**Do not depress brake pedal while brake caliper is removed.**

4. Put alignment marks on disc rotor and wheel hub and bearing assembly, and remove disc rotor.
5. Remove wheel sensor from steering knuckle. Do not pull on wheel sensor harness.
6. Loosen steering outer socket nut.
7. Remove steering outer socket from steering knuckle so as not to damage ball joint boot using ball joint remover or suitable tool. Temporarily leave the outer socket nut installed to prevent damage to threads and to prevent the ball joint remover or suitable tool from suddenly coming off.
8. Remove transverse link ball joint nut and bolt, and remove transverse link from steering knuckle.
9. Remove cotter pin, and loosen hub lock nut. Temporarily leave the hub lock nut installed to prevent damage to threads.
10. Separate the halfshaft from the wheel hub and bearing assembly by lightly tapping the end of the halfshaft using a hammer or suitable tool, and remove hub lock nut.
11. If wheel hub and bearing assembly and halfshaft cannot be separated after performing the above procedure, use a suitable puller.
12. Remove the halfshaft from the wheel hub and bearing assembly and support the halfshaft.

❋❋ WARNING

Do not apply an excessive angle to halfshaft joint when removing from the wheel hub and bearing assembly. Do not excessively extend slide joint. Do not allow halfshaft to hang down. Support the entire halfshaft.

13. Remove wheel hub and bearing assembly bolts, and then remove splash guard and wheel hub and bearing assembly from steering knuckle.
14. Remove nuts and bolts, and remove steering knuckle from strut assembly.

To install:

15. Installation is the reverse order of removal.
16. Tighten the hub lock nut to 83 ft. lbs. (113 Nm).

➡**Perform the final tightening of each of parts under unladen conditions, which were removed when removing wheel hub and bearing assembly and steering knuckle.**

17. When installing disc rotor on wheel hub and bearing assembly, align the marks.
18. Check the wheel alignment.

2012 Sedan

See Figure 178.

1. Before servicing the vehicle, refer to the Precautions.
2. Remove the wheel and tire assembly.
3. Remove wheel sensor and sensor harness.
4. Remove brake caliper, leaving brake caliper hydraulic lines connected. Reposition brake caliper aside with wire.

➡**Avoid depressing brake pedal while brake caliper is removed.**

5. Put matching marks on disc rotor and wheel hub and bearing assembly, then remove disc rotor.
   a. Use paint for matching parts. Do not damage the disc rotor or wheel hub assembly.
   b. Do not drop disc rotor.
6. Remove and discard cotter pin, and then loosen wheel hub lock nut, using Tool number: KV40104000, or equivalent.
7. Separate the halfshaft from the wheel hub and bearing assembly by lightly tapping the end of the halfshaft using a suitable tool and a wood block.

➡**Use a suitable puller, if the wheel hub and halfshaft cannot be separated after performing the above procedure.**

8. Remove the wheel hub lock nut.
9. Remove the nut and separate the steering outer socket from steering knuckle.
10. Remove the nuts and bolts, then separate the steering knuckle from strut assembly.
11. Suspend the halfshaft with suitable wire.

❋❋ WARNING

Do not place halfshaft joint at an extreme angle. Also be careful not to overextend slide joint. Do not allow halfshaft to hang down without support for joint sub-assembly, shaft, and the other parts.

12. Remove steering knuckle from transverse link.

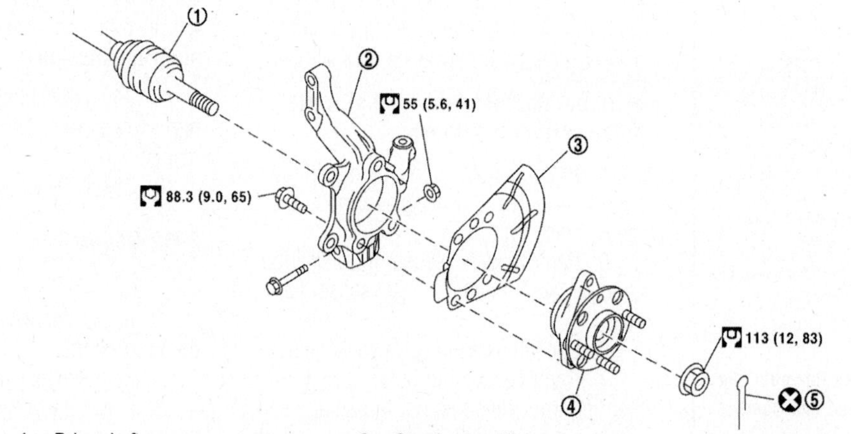
55 (5.6, 41)
88.3 (9.0, 65)
113 (12, 83)
1. Drive shaft
2. Steering knuckle
3. Splash guard
4. Wheel hub and bearing assembly
5. Cotter pin
37663_VERS_G0156

Fig. 177 Front wheel hub and bearing component locations—except 2012 Sedan

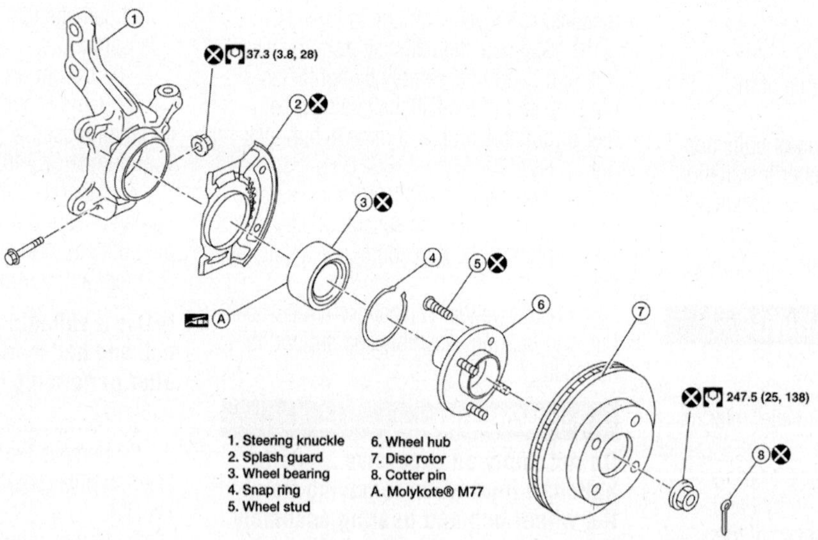
37.3 (3.8, 28)
247.5 (25, 138)
1. Steering knuckle
2. Splash guard
3. Wheel bearing
4. Snap ring
5. Wheel stud
6. Wheel hub
7. Disc rotor
8. Cotter pin
A. Molykote® M77
71075_VERS_G0296

Fig. 178 Front wheel hub and knuckle component locations—2012 Sedan

13. If necessary, remove the wheel studs from the wheel hub, using a suitable tool.

WARNING

Remove stud only when necessary. Do not hammer the wheel stud or damage to the wheel bearing may occur. Pull the wheel stud straight out to avoid damaging the stud.

To install:

14. Installation is the reverse of the removal procedure.

WARNING

During installation, do not damage the wheel bearing seal. If damage has occurred, replace wheel bearing with a new one. Do not allow paint to adhere to the wheel bearing seal.

15. Check each mating surface for water and foreign matter. If there is any water or foreign matter, clean the mounting surface.

16. Position the stud to the wheel hub flange. Place a washer and nut on the opposite end of the stud, tighten to press the stud into the wheel hub flange.

WARNING

Check that no clearance exists between wheel hub, and wheel stud

after installation. **Do not reuse wheel stud.**

➡**Do not reuse steering knuckle and transverse link nut.**

17. Clean the mating surface of wheel hub lock nut and wheel hub. Do not apply lubricating oil to the mating surface of the wheel hub lock nut and the wheel hub.

18. Clean the mating surface of halfshaft, wheel hub, and wheel bearing. And then apply Molykote® M77 to surface of joint subassembly of halfshaft. Apply Molykote® M77 to cover entire flat surface of joint subassembly of half-shaft.

19. When reusing disc rotor, align the matching marks during removal.

20. Perform the final tightening of each part under unladen conditions, which were removed when removing wheel hub and axle housing.

21. When installing a cotter pin, securely bend the basal portion to prevent rattles.

❋❋ **WARNING**

Do not reuse cotter pin.

SUSPENSION

COIL SPRINGS

REMOVAL & INSTALLATION

Except 2012 Sedan

See Figure 179.

1. Before servicing the vehicle, refer to the Precautions.
2. Remove rear wheels and tires.
3. Remove wheel sensor from wheel hub and bearing assembly. Do not pull on wheel sensor harness.
4. Set a jack under rear suspension beam.
5. Remove shock absorber lower side bolt.
6. Gradually lower the jack, and then remove coil spring and rear spring rubber seat (upper and lower).

To install:

7. Installation is the reverse of removal.
8. Tighten the lower shock absorber side bolt to 91 ft. lbs. (124 Nm).

➡**When installing spring, be sure to securely install the spring end position aligned to flush of rear spring rubber seat (lower).**

2012 Sedan

See Figure 180.

1. Before servicing the vehicle, refer to the Precautions.
2. Remove the wheel and tire assemblies.
3. Position a suitable jack under rear suspension beam.

❋❋ **WARNING**

Place the jack in the center of the suspension beam. Do not damage the suspension beam with jack.

4. Remove the lower shock absorber bolts.
5. Slowly lower jack, then remove upper rubber seat, coil spring, and lower rubber seat from rear suspension beam.

REAR SUSPENSION

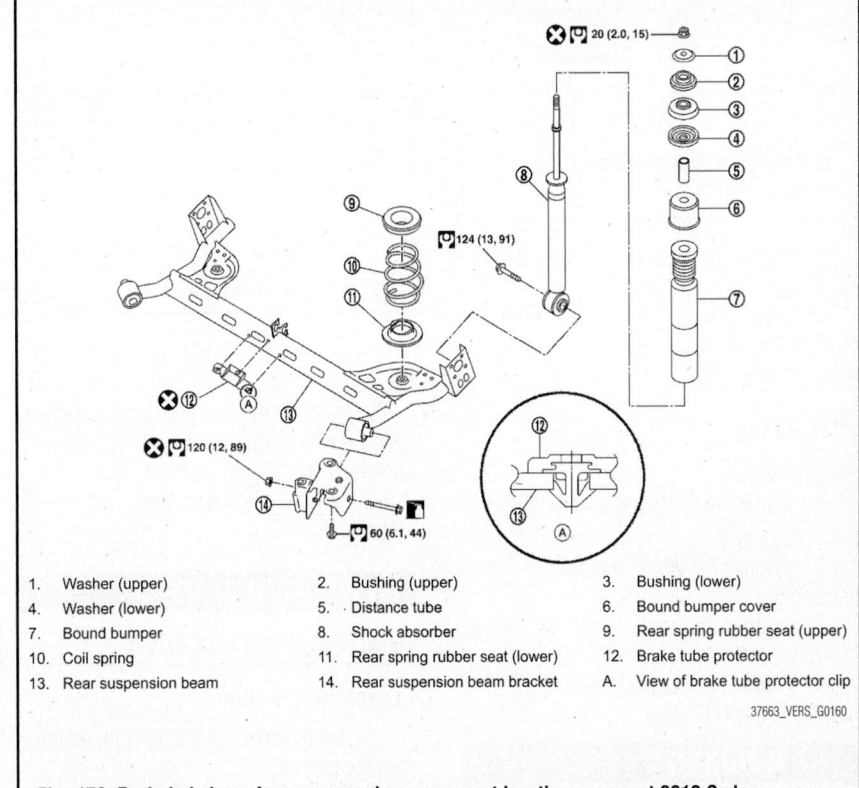

1.	Washer (upper)	2.	Bushing (upper)	3.	Bushing (lower)
4.	Washer (lower)	5.	Distance tube	6.	Bound bumper cover
7.	Bound bumper	8.	Shock absorber	9.	Rear spring rubber seat (upper)
10.	Coil spring	11.	Rear spring rubber seat (lower)	12.	Brake tube protector
13.	Rear suspension beam	14.	Rear suspension beam bracket	A.	View of brake tube protector clip

37663_VERS_G0160

Fig. 179 Exploded view of rear suspension component locations—except 2012 Sedan

To install:

6. Installation is the reverse of the removal procedure.
7. Install the lower rubber seat to the rear suspension beam mounting hole.
8. Match up lower rubber seat indentions and rear suspension beam grooves and attach.

SHOCK ABSORBERS

REMOVAL & INSTALLATION

Except 2012 Sedan

1. Before servicing the vehicle, refer to the Precautions.
2. Remove rear wheels and tires.
3. Remove wheel sensor from wheel hub and bearing assembly and rear suspen-

sion beam. Do not pull on wheel sensor harness.

4. Remove shock absorber mask from trunk side finisher using a flat-bladed screwdriver with its tip taped.
5. Set jack under rear suspension beam.
6. Remove upper nut of the shock absorber, and then remove washer (upper), bushing (upper) from shock absorber.
7. Remove shock absorber lower side bolt.
8. Gradually lower the jack, and remove the bushing (lower), washer (lower), distance tube, bound bumper cover, bound bumper and shock absorber from vehicle.

To install:

9. Installation is the reverse of removal.
10. Tighten the lower shock absorber side bolt to 91 ft. lbs. (124 Nm).

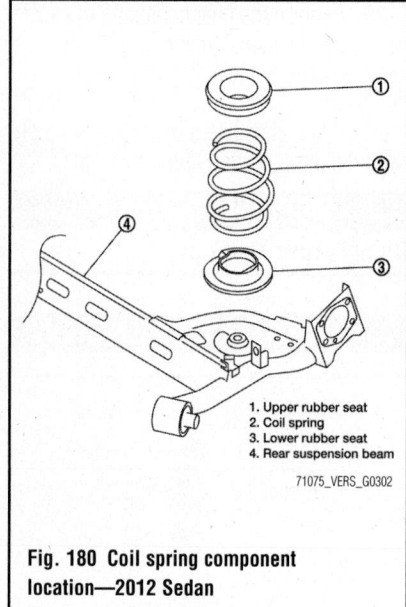

1. Upper rubber seat
2. Coil spring
3. Lower rubber seat
4. Rear suspension beam

71075_VERS_G0302

Fig. 180 Coil spring component location—2012 Sedan

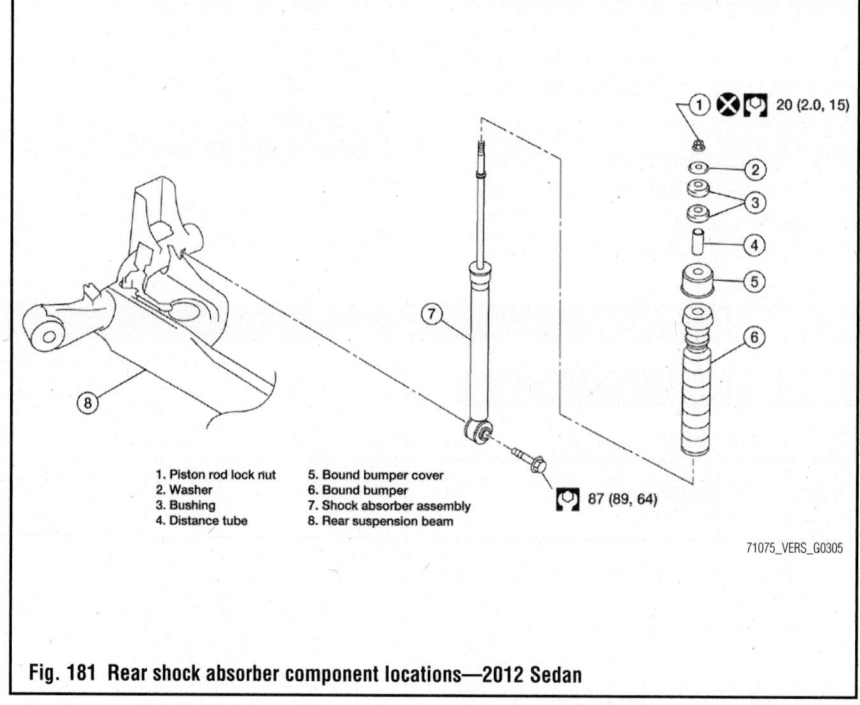

20 (2.0, 15)

87 (89, 64)

1. Piston rod lock nut
2. Washer
3. Bushing
4. Distance tube
5. Bound bumper cover
6. Bound bumper
7. Shock absorber assembly
8. Rear suspension beam

71075_VERS_G0305

Fig. 181 Rear shock absorber component locations—2012 Sedan

11. Tighten the upper shock absorber nut to 15 ft. lbs. (20 Nm).

12. When installing body side bushing (upper), install the projection to the body side hole securely.

2012 Sedan

See Figure 181.

1. Before servicing the vehicle, refer to the Precautions.

2. Remove the wheel and tire assembly.

3. Remove the wheel sensor and sensor harness.

4. Position a suitable jack under the rear suspension beam.

※※ WARNING

Place the jack in the center of the suspension beam. Do not damage the suspension beam with the jack.

5. Remove the lower shock absorber bolt.

6. Remove the trunk side finisher.

7. Remove the upper shock absorber nut, washer, and bushing.

8. Remove the shock absorber assembly.

9. Remove the bushing, distance tube, bound bumper cover, and bound bumper from the shock absorber assembly.

To install:

10. Installation is the reverse of the removal procedure.

11. Perform the final tightening of the bolts and nuts at the shock absorber lower side (rubber bushing), under unladen conditions with tires on level ground.

12. Install the bushing into the hole on the vehicle side.

13. Hold the head of the shock absorber piston rod to keep it from rotating, and tighten the piston rod lock nut.

14. Perform inspection after installation.

SUSPENSION BEAM

REMOVAL & INSTALLATION

Except 2012 Sedan

1. Before servicing the vehicle, refer to the Precautions.

2. Remove rear wheels and tires.

3. Separate parking brake rear cable from rear drum brake and rear suspension beam.

4. Remove wheel sensor and wheel sensor harness from wheel hub and bearing assembly and rear suspension beam.

5. Remove lock plate and separate brake tube from brake hose.

6. Remove wheel hub and bearing assembly and back plate.

7. Set jack under rear suspension beam.

8. Remove coil spring (left/right).

9. Remove bolts between body and rear suspension beam bracket.

10. Gradually lower the jack, and then remove rear suspension beam from vehicle.

11. Remove the rear suspension beam bracket bolt and nut, and then remove rear

suspension beam bracket from rear suspension beam.

12. Remove brake tube protector from rear suspension beam.

To install:

13. Installation is in the reverse order of removal.

14. Check components for deformation, cracks, and other damage, and replace if necessary.

15. Refill with new brake fluid and bleed air.

16. Tighten the wheel nuts to specification.

17. Check the parking brake operation (stroke).

18. Check the wheel sensor harness for proper connection.

19. Perform final tightening of rear suspension beam and rear suspension beam bracket (rubber bushing) under unladen conditions with tires on level ground.

2012 Sedan

1. Before servicing the vehicle, refer to the Precautions.

2. Remove the wheel and tire assemblies.

3. Remove the wheel sensor and sensor harness.

4. Remove the brake hose and brake pipe from the rear suspension beam.

5. Remove the rear drum brake assemblies.

6. Remove the parking brake cable from the rear suspension beam.

7. Position a suitable jack under the rear suspension beam.

✳✳ WARNING

Place the jack in the center of the suspension beam. Do not damage the suspension beam with the jack.

8. Remove the shock absorber bolts (lower side).

9. Remove the coil spring.

10. Remove the rear suspension beam bolts and nuts.

11. Slowly lower the jack, remove the rear suspension beam from the vehicle body.

✳✳ WARNING

While lowering the rear suspension beam with the jack, be sure to maintain the stability of the jack.

To install:

12. Installation is the reverse of the removal procedure.

13. Perform the final tightening of the rear suspension beam installation position (rubber bushing), under unladen conditions with the tires on level ground.

14. Check the wheel sensor harness for proper connection.

15. Adjust the parking brake.

16. Check the wheel alignment.

WHEEL HUBS & BEARINGS

REMOVAL & INSTALLATION

Except 2012 Sedan

See Figure 182.

The wheel hub assembly does not require maintenance. If any of the following symptoms are noted, replace the wheel hub assembly.

• Growling noise is emitted from the wheel hub bearing during operation

• Wheel hub bearing drags or turns roughly

1. Before servicing the vehicle, refer to the Precautions.

2. Remove the tires.

3. Remove the wheel sensor from the wheel hub and bearing assembly. Do not pull on the wheel sensor harness.

4. Remove the drum brake assembly.

5. Remove the wheel hub and bearing assembly bolts, and then remove the wheel hub and bearing assembly from the vehicle.

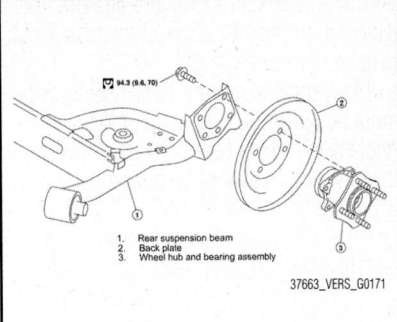

1. Rear suspension beam
2. Back plate
3. Wheel hub and bearing assembly

37663_VERS_G0171

Fig. 182 Rear wheel hub and bearing component locations—except 2012 Sedan

To install:

6. Check for any deformity, cracks, or other damage on the wheel hub assembly, replace if necessary.

7. Installation is the reverse of removal.

2012 Sedan

See Figure 183.

1. Before servicing the vehicle, refer to the Precautions.

2. Remove the wheel and tire assembly.

3. Remove the hub cap from the brake drum, using a suitable tool.

4. Remove the wheel hub lock nut and brake drum.

✳✳ WARNING

Do not apply force to the brake drum to avoid damage to the wheel bearing. If the wheel bearing inner

race is separated due to force, replace the wheel bearing with a new one.

5. Remove the wheel sensor.

6. Remove the brake shoe assembly.

7. Remove the spindle bolts. Separate the back plate and spindle from the rear suspension beam.

8. Remove the wheel studs from the brake drum using a suitable press.

✳✳ WARNING

Remove the studs only when necessary. Do not hammer the stud and avoid impact to the brake drum. Pull the stud straight out to avoid damage to the stud.

To install:

9. Installation is the reverse of the removal procedure.

✳✳ WARNING

During the installation, do not damage the wheel bearing seal. If damaged, replace the wheel bearing with a new one. Do not allow paint to adhere to the wheel bearing seal.

10. Check each mating surface for water and foreign matter. If there is any water or foreign matter, clean the mating surface.

11. Position the stud to the brake drum. Place a washer on the opposite end of the

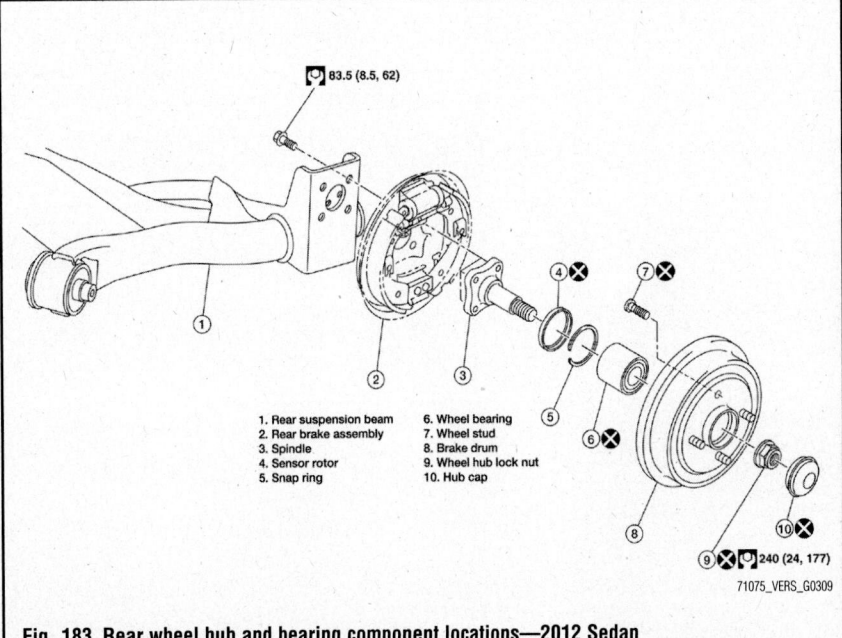

83.5 (8.5, 62)

1. Rear suspension beam
2. Rear brake assembly
3. Spindle
4. Sensor rotor
5. Snap ring
6. Wheel bearing
7. Wheel stud
8. Brake drum
9. Wheel hub lock nut
10. Hub cap

240 (24, 177)

71075_VERS_G0309

Fig. 183 Rear wheel hub and bearing component locations—2012 Sedan

stud and by use of a nut, tighten to press the stud into the brake drum.

12. Check that no clearance exists between brake drum, and stud after installation.

※※ WARNING

Do not reuse stud.

13. Insert brake drum to the spindle with the spindle axis arranged in a straight line. (Brake drum may be stuck at the axis if not installed in a straight line).

14. If the brake drum becomes stuck and must be pulled out, do not use tools. Replace the wheel bearing with a new one if the brake drum cannot be pulled out without use of tools.

15. If the brake drum becomes stuck and the wheel bearing inner race is damaged, replace the wheel bearing with a new one.

16. Using Tool number: ST30720000, or equivalent, install the hub cap on the brake drum.

➡**Do not reuse the hub cap. Do not reuse wheel lock nut.**

17. Tighten the wheel lock nut to the specified torque.

NISSAN

Xterra

22

SPECIFICATIONS AND MAINTENANCE CHARTS

ENGINE AND VEHICLE IDENTIFICATION

		Engine						Model Year	
ID	Liters (cc)	Cu. In.	Cyl.	Fuel Sys.	Engine Type	Eng. Mfg.		Code ①	Year
VQ40DE	4.0 (3954)	241	6	MFI	DOHC	Nissan		B	2011
								C	2012

MFI: Multi-port Fuel Injection

SOHC: Single Overhead Camshaft

DOHC: Double Overhead Camshafts

① 10th digit of the Vehicle Identification Number (VIN)

71075_XTER_C0001

GENERAL ENGINE SPECIFICATIONS

Year	Model	Engine Displacement Liters	Engine ID	Net Horsepower @ rpm	Net Torque @ rpm (ft. lbs.)	Bore x Stroke (in.)	Com-pression Ratio	Oil Pressure @ rpm
2011	Xterra	4.0	VQ40DE	265@5600	284@4000	3.76X3.62	9.7:1	43@2000
2012	Xterra	4.0	VQ40DE	265@5600	284@4000	3.76X3.62	9.7:1	43@2000

71075_XTER_C0002

ENGINE TUNE-UP SPECIFICATIONS

Year	Engine Displ. Liters	Engine ID	Spark Plug Gap (in.)	Ignition Timing (deg.) MT	Ignition Timing (deg.) AT	Fuel Pump (psi)	Idle Speed (rpm) MT	Idle Speed (rpm) AT ①	Valve Clearance (in.) In.	Valve Clearance (in.) Ex.
2011	4.0	VQ40DE	0.043	10-20B	10-20B	51 ②	575-675	650-750	③	④
2012	4.0	VQ40DE	0.043	10-20B	10-20B	51 ②	575-675	650-750	③	④

NOTE: The Vehicle Emission Control Information label often reflects specification changes made during production. The label figures must be used if they differ from those in this chart.

B: Before top dead center

HYD: Hydraulic

① Automatic transmission in Neutral

② At idle

③ 0.010-0.013 cold

   0.012-0.016 hot

④ 0.011-0.015 cold

   0.012-0.017 hot

71075_XTER_C0003

CAPACITIES

Year	Model	Engine Displacement Liters	Engine ID	Engine Oil with Filter (qts.)	Transmission (pts.) Manual	Transmission (pts.) Auto.	Transfer Case (pts.)	Drive Axle Front (pts.)	Drive Axle Rear (pts.)	Fuel Tank (gal.)	Cooling System (qts.)
2011	Xterra	4.0	VQ40DE	5.10	①	21.50	2.1	1.75	②	21.2	11.0
2012	Xterra	4.0	VQ40DE	5.50	①	21.50	2.1	1.75	②	21.2	11.0

NOTE: All capacities are approximate. Add fluid gradually and check to be sure a proper fluid level is obtained.

① 2WD: 8.3; 4WD: 8.9

② C200: 3.3; M226: 4.25

71075_XTER_C0004

VALVE SPECIFICATIONS

Year	Engine Displacement Liters	Engine ID	Seat Angle (deg.)	Face Angle (deg.)	Spring Test Pressure (lbs. @ in.)	Spring Installed Height (in.)	Stem-to-Guide Clearance (in.) Intake	Stem-to-Guide Clearance (in.) Exhaust	Stem Diameter (in.) Intake	Stem Diameter (in.) Exhaust
2011	4.0	VQ40DE	①	②	③	1.456	0.0008-0.0021	0.0012-0.0025	0.2348-0.2354	0.2344-0.2350
2012	4.0	VQ40DE	①	②	③	1.456	0.0008-0.0021	0.0012-0.0025	0.2348-0.2354	0.2344-0.2350

① 44 degrees 22 minutes to 45 degrees 8 minutes

② 45 degrees 15 minutes to 45 degrees 45 minutes

③ Installation: 37-42@1.457

   Valve open: 84-95@1.071

71075_XTER_C0005

CAMSHAFT SPECIFICATIONS CHART
All measurements are given in inches.

Year	Engine Displ. Liters	Engine VIN	Journal Dia.	Brg. Oil Clearance	Shaft End-play	Runout	Lobe Height Intake	Lobe Height Exhaust
2011	4.0	VQ40DE	①	②	0.0045-0.0074	0.0010	1.7900-1.7921	1.7746-1.7821
2012	4.0	VQ40DE	①	②	0.0045-0.0074	0.0010	1.7900-1.7921	1.7746-1.7821

① No.1: 1.0211-1.0218. No's. 2, 3, 4: 0.9230-0.9238

② No.1: 1.0018-0.0034. No's. 2, 3, 4: 0.0014-0.0030

71075_XTER_C0006

FLUID SPECIFICATIONS

Year	Model	Engine Displacement Liters (ID)	Engine Oil	Manual Trans.	Auto. Trans.	Transfer Case	Drive Axle Front	Drive Axle Rear	Power Steering Fluid	Brake Master Cylinder	Cooling System
2011	Xterra	4.0 (VQ40DE)	5W-30	①	②	③	④	⑤	⑥	DOT 3	⑦
2012	Xterra	4.0 (VQ40DE)	5W-30	①	②	③	④	⑤	⑥	DOT 3	⑦

NA: Not Applicable

NS: Not specified by manufacturer at date of publication

DOT: Department Of Transportation

① Nissan Manual Transmission Fluid (MTF) HQ Multi 75W-85 ir API GL-4, Viscosity SAE 75W-85

② Nissan approved Matic S ATF. If Matic S ATF is not available, NISSAN approved Matic J ATF may also be used.

③ Nissan approved Matic D AT

④ API GL-5 Viscosity SAE 80W-90. For hot climates, viscosity SAE 90 is suitable for ambient temperatures above 32 degrees F (0 degrees C).

⑤ C200 axle: API GL-5 synthetic gear oil, Viscosity SAE 75W-90

   M226 axle: API GL-5 synthetic gear oil, Viscosity SAE 75W-90

⑥ Nissan Power Steering Fluid (PSF) or equivalent. Dexron® VI type ATF may also be used

⑦ Nissan genuine coolant (blue), or equivalent

CRANKSHAFT AND CONNECTING ROD SPECIFICATIONS

All measurements are given in inches.

Year	Engine Displacement Liters	Engine ID	Crankshaft Main Brg. Journal Dia.	Crankshaft Main Brg. Oil Clearance	Crankshaft Shaft End-play	Crankshaft Thrust on No.	Connecting Rod Journal Diameter	Connecting Rod Oil Clearance	Connecting Rod Side Clearance
2011	4.0	VQ40DE	②	0.0014-0.0018	0.0039-0.0098	NA	2.2441-2.2446	0.0013-0.0023	0.0079-0.0138
2012	4.0	VQ40DE	②	0.0014-0.0018	0.0039-0.0098	NA	2.2441-2.2446	0.0013-0.0023	0.0079-0.0138

NA: Not Available

① There are 24 different grades, ranging from A (2.1645) to 7 (2.1636)

② There are 24 different grades, ranging from A (2.7549) to 7 (2.7540)

PISTON AND RING SPECIFICATIONS

All measurements are given in inches.

Year	Engine Displacement Liters	Engine ID	Piston Clearance	Ring Gap Top Comp.	Ring Gap Bottom Comp.	Ring Gap Oil Control	Ring Side Clearance Top Comp.	Ring Side Clearance Bottom Comp.	Ring Side Clearance Oil Control
2011	4.0	VQ40DE	0.0004-0.0012	0.0091-0.0130	0.0130-0.0189	0.0079-0.0197	0.0018-0.0031	0.0012-0.0028	0.0026-0.0053
2012	4.0	VQ40DE	0.0004-0.0012	0.0091-0.0130	0.0130-0.0189	0.0079-0.0197	0.0018-0.0031	0.0012-0.0028	0.0026-0.0053

TORQUE SPECIFICATIONS
All readings in ft. lbs.

Year	Engine Displacement Liters	Engine ID	Cylinder Head Bolts	Main Bearing Bolts	Rod Bearing Bolts	Crankshaft Damper Bolts	Flywheel Bolts	Manifold Intake	Manifold Exhaust	Spark Plugs	Oil Pan Drain Plug
2011	4.0	VQ40DE	①	②	14	③	65	④	⑤	14-22	25
2012	4.0	VQ40DE	①	②	14	③	65	④	⑤	14-22	25

① Step 1: 72 ft. lbs.

Step 2: loosen completely to 0 ft. lbs.

Step 3: 29 ft. lbs.

Step 4: Plus 90 degrees clockwise

Step 5: Plus 90 degrees clockwise

② Bolts: 17-24 (M8) 16 ft. lbs.

Install rear main seal

Bolts: 1-16 (M10) 26 ft. lbs.

Bolts: 1-16 (M10) Plus 90 degrees clockwise

③ Step 1: 33 ft. lbs.

Step 2: Plus 84-90 degrees clockwise

④ Intake manifold collector:

Bolts and nuts: 8 ft. lbs.

Stud bolts: 61 inch lbs.

⑤ Intake manifold:

Bolts and nuts: 5 ft. lbs. and

than to 21 ft. lbs.

Studs: 8 ft. lbs.

71075_XTER_C0009

WHEEL ALIGNMENT

Year	Model		Caster Range (+/-Deg.)	Caster Preferred Setting (Deg.)	Camber Range (+/-Deg.)	Camber Preferred Setting (Deg.)	Toe-in (in.)
2011	Xterra	2WD	①	①	②	②	③
		4WD	④	④	⑤	⑤	③
2012	Xterra	2WD	①	①	②	②	③
		4WD	④	④	⑤	⑤	③

NOTE: On 2009-2010 vehicles, fuel, coolant and engine oil must be full. Spare tire, jack, hand tools and mats must be in place.

Some 2009-2010 vehicles may be equipped with non adjustable lower link bolts and washers. In order to adjust caster and camber on these vehicles,

first replace these bolts with adjustable cam bolts and washers.

① Minimum: 2 degrees 15' (2.25 degrees)

Nominal: 3 degrees 0' (3.00 degrees)

Maximum: 3 degrees 45' (3.75 degrees)

② Minimum: 2 degrees 0' (2.00 degrees)

Nominal: 0 degrees 15' (0.25 degrees)

Maximum: 1 degrees 0' (1.00 degrees)

③ Minimum: 0.12 inches

Nominal 0.16 inches

Maximum 0.20 inches

④ Minimum: 0 degrees 15' (-0.25 degrees)

Nominal: 2 degrees 45' (2.75 degrees)

Maximum: 3 degrees 30' (3.50 degrees)

⑤ Minimum: 0 degrees 15' (-0.25 degrees)

Nominal: 0 degrees 30' (0.50 degrees)

Maximum: 1 degrees 15' (1.25 degrees)

TIRE, WHEEL AND BALL JOINT SPECIFICATIONS

Year	Model	OEM Tires Standard	OEM Tires Optional	Tire Pressures (psi) Front	Tire Pressures (psi) Rear	Wheel Size	Ball Joint Inspection	Lug Nut Torque (ft. lbs.)
2011	Xterra S	P265/70R16	None	①	①	7J	②	98
	Xterra S-O/R	P265/75R16	None	①	①	7J	②	98
	Xterra SE	P255/65R17	None	①	①	7.5J	②	98
2012	Xterra X	P265/70R16	None	①	①	7J	②	98
	Xterra S	P265/70R16	None	①	①	7J	②	98
	Xterra S-O/R	P265/75R16	None	①	①	7J	②	98
	Xterra SE	P255/65R17	None	①	①	7.5J	②	98

OEM: Original Equipment Manufacturer

PSI: Pounds Per Square Inch

① See placard on vehicle

② Replace if any measurable movement is found.

71075_XTER_C0011

BRAKE SPECIFICATIONS

All measurements in inches unless noted

Year	Model	Brake Disc Original Thickness	Brake Disc Minimum Thickness	Brake Disc Maximum Runout	Brake Drum Diameter Original Inside Diameter	Brake Drum Diameter Max. Wear Limit	Minimum Lining Thickness Front	Minimum Lining Thickness Rear	Brake Caliper Bracket Bolts (ft. lbs.)	Brake Caliper Mounting Bolts (ft. lbs.)
2011	Xterra	1.102	1.024	0.002	④	①	0.079	0.079	②	③
2012	Xterra	1.102	1.024	0.002	④	①	0.079	0.079	②	③

① Rear disc brakes: 0.630

② Front: 136 ft. lbs.

　Rear: 76 ft. lbs.

③ Front: 32 ft. lbs.

　Rear: 19 ft. lbs.

④ Rear disc brakes: 0.709

71075_XTER_C0012

SCHEDULED MAINTENANCE INTERVALS
Nissan Xterra

TO BE SERVICED	TYPE OF SERVICE	VEHICLE MILEAGE INTERVAL (x1000)												
		7.5	15	22.5	30	37.5	45	52.5	60	67.5	75	82.5	90	97.5
Engine oil & filter	R	✓	✓	✓	✓	✓	✓	✓	✓	✓	✓	✓	✓	✓
Brake lines & cables	S/I		✓		✓		✓		✓		✓		✓	
Brake pads, discs, drums & linings	S/I		✓		✓		✓		✓		✓		✓	
Driveshaft boots & propeller shaft	S/I				✓				✓				✓	
Front wheel bearings (4X2)	S/I				✓				✓				✓	
Front wheel bearings (4X4)	S/I				✓				✓				✓	
Automatic & manual transmission, transfer & differential gear oil ①	S/I		✓		✓		✓		✓		✓		✓	
Air cleaner filter	R					✓			✓				✓	
Engine coolant	R								✓				✓	
Spark plugs (platinum)	R	replace every 105,000 miles												
Drive belt(s)	S/I					✓			✓				✓	
Exhaust system	S/I					✓			✓				✓	
Fuel lines	S/I					✓			✓				✓	
Steering gear (box) & linkage, axle & suspension parts	S/I					✓			✓				✓	
Vapor lines	S/I					✓			✓				✓	
Tires (rotate)	S/I	✓	✓	✓	✓	✓	✓	✓	✓	✓	✓	✓	✓	✓
Timing belt ②	R													

R: Replace   S/I: Service or Inspect

① Differential (w/limited-slip differential) oil: replace oil every 30,000 miles, 2007-2008 vehicles.

② Timing belt: replace at 105,000 miles.

FREQUENT OPERATION MAINTENANCE (SEVERE SERVICE)

If a vehicle is operated under any of the following conditions it is considered severe service:

- Extremely dusty areas.
- 50% or more of the vehicle operation is in 32°C (90°F) or higher temperatures, or constant operation in temperatures below 0°C (32°F).
- Prolonged idling (vehicle operation in stop and go traffic).
- Frequent short running periods (engine does not warm to normal operating temperatures).
- Police, taxi, delivery usage or trailer towing usage.

Oil & oil filter: replace every 3750 miles.

Brake pads, discs, drums & linings: service or inspect every 7500 miles.

Driveshaft boots & propeller shaft: service or inspect every 7500 miles.

Exhaust system: service or inspect every 7500 miles.

Steering gear (box) & linkage, (steering damper-4X4), axle & suspension parts: service or inspect every 7500 miles.

Steering linkage ball joints & front suspension ball joints: service or inspect every 7500 miles.

71075_XTER_C0013

BRAKES   INFORMATION AND PRECAUTIONS

ANTI-LOCK SYSTEMS

- Certain components within the ABS system are not intended to be serviced or repaired individually.
- Do not use rubber hoses or other parts not specifically specified for and ABS system. When using repair kits, replace all parts included in the kit. Partial or incorrect repair may lead to functional problems and require the replacement of components.
- Lubricate rubber parts with clean, fresh brake fluid to ease assembly. Do not use shop air to clean parts; damage to rubber components may result.
- Use only DOT 3 brake fluid from an unopened container.
- If any hydraulic component or line is removed or replaced, it may be necessary to bleed the entire system.
- A clean repair area is essential. Always clean the reservoir and cap thoroughly before removing the cap. The slightest amount of dirt in the fluid may plug an orifice and impair the system function. Perform repairs after components have been thoroughly cleaned; use only denatured alcohol to clean components. Do not allow ABS components to come into contact with any substance containing mineral oil; this includes used shop rags.
- The Anti-Lock control unit is a microprocessor similar to other computer units in the vehicle. Ensure that the ignition switch is **OFF** before removing or installing controller harnesses. Avoid static electricity discharge at or near the controller.
- If any arc welding is to be done on the vehicle, the control unit should be unplugged before welding operations begin.

DISC AND DRUM SYSTEMS

> ❋❋ **CAUTION**
>
> Dust and dirt accumulating on brake parts during normal use may contain asbestos fibers from production or aftermarket brake linings. Breathing excessive concentrations of asbestos fibers can cause serious bodily harm. Exercise care when servicing brake parts. Do not sand or grind brake lining unless equipment used is designed to contain the dust residue. Do not clean brake parts with compressed air or by dry brushing. Cleaning should be done by dampening the brake components with a fine mist of water, then wiping the brake components clean with a dampened cloth. Dispose of cloth and all residue containing asbestos fibers in an impermeable container with the appropriate label. Follow practices prescribed by the Occupational Safety and Health Administration (OSHA) and the Environmental Protection Agency (EPA) for the handling, processing, and disposing of dust or debris that may contain asbestos fibers.

BRAKES   BLEEDING THE BRAKE SYSTEM

BLEEDING PROCEDURE

BLEEDING PROCEDURE

1. Before servicing the vehicle, refer to the Precautions Section.

> ❋❋ **CAUTION**
>
> While bleeding the brake system, pay attention to the master cylinder fluid level.

2. Disconnect the negative battery cable.
3. Raise and safely support the vehicle.
4. Attach a vinyl tube to the right, rear bleeder valve.
5. Depress the brake pedal fully 4 or 5 times.
6. With the brake pedal depressed, loosen the bleeder valve to let the air out, then tighten it immediately.
7. Repeat steps 3 and 4 until no more air comes out.
8. Tighten the bleeder valve.
9. Fill the master cylinder reservoir.
10. Repeat the above steps for the left front, left rear, and the right front calipers, in that order.

BRAKES   ANTI-LOCK BRAKE SYSTEM (ABS)

WHEEL SPEED SENSORS

REMOVAL & INSTALLATION

See Figure 1.

1. Before servicing the vehicle, refer to the Precautions Section.
2. Disconnect the negative battery cable.
3. Raise and support the vehicle safely.
4. Remove the tire and wheel assembly.
5. Remove the rotor.
6. Remove the wheel speed sensor mounting bolts and clip.
7. Pull the sensor out, being careful to turn it as little as possible. Do not pull on the sensor harness.

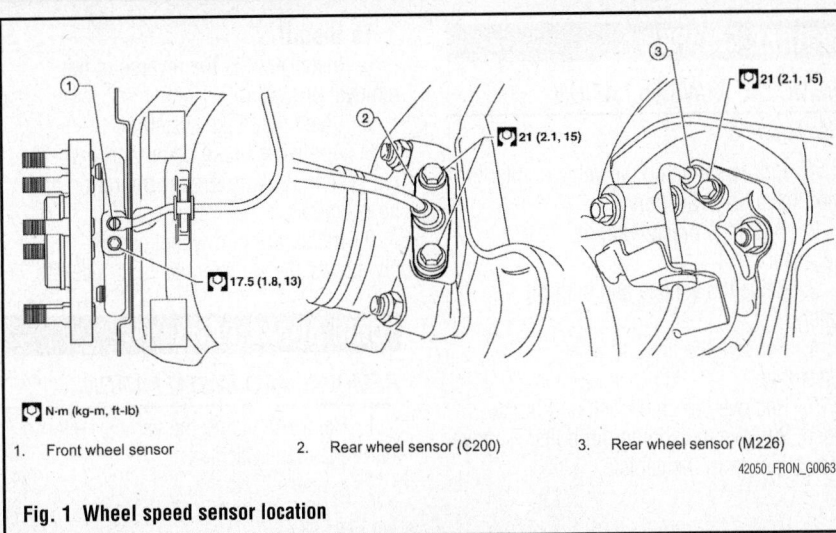

Fig. 1 Wheel speed sensor location

1. Front wheel sensor
2. Rear wheel sensor (C200)
3. Rear wheel sensor (M226)

42050_FRON_G0063

To install:

8. Before installing the sensor be sure no foreign materials such as iron fragments are adhered to the pick-up part of the sensor or to the inside of the sensor mounting hole or on the rotor mounting surface.

9. Be sure the harness is not twisted when installed.

10. Continue the installation in the reverse order of the removal procedure.

WHEEL SPEED SENSOR RINGS (TOOTHED RINGS)

REMOVAL & INSTALLATION

Front Wheel

The front sensor rotor cannot be disassembled. It is integrated with the hub and bearing assembly. To replace the sensor rotor, replace the front hub and knuckle assembly. The rear sensor rotor cannot be disassembled. It is integrated with the hub and bearing assembly. To replace the sensor rotor, replace the rear hub and bearing assembly.

Rear Wheel

C200 Axle

1. Disconnect the negative battery cable.
2. Raise and support the vehicle safely.
3. Remove the axle shaft.
4. Remove the sensor rotor using tool J25852-B or equivalent.

To install:

5. Installation is the reverse of the removal procedure.
6. Pay attention to the direction of the sensor rotor.

M226 Axle

1. Disconnect the negative battery cable.
2. Raise and support the vehicle safely.
3. Remove the axle shaft.

➡ **It is necessary to disassemble the rear axle to replace the sensor rotor.**

4. Pull the sensor off of the axle shaft using tool ST30031000, or equivalent, and a press.

To install:

5. Installation is the reverse of the removal procedure.
6. Make sure that the sensor rotor is fully seated.

➡ **Never reuse the old sensor rotor. Never reuse the old oil seal.**

BRAKES   FRONT DISC BRAKES

BRAKE CALIPER

REMOVAL & INSTALLATION

See Figure 2.

1. Before servicing the vehicle, refer to the Precautions Section.
2. Drain the brake fluid, as necessary.
3. Raise the vehicle and support safely.
4. Remove the tire and wheel assembly.
5. Remove the bolt attaching the brake hose to the caliper. Plug the brake hose to prevent brake fluid loss.
6. Remove the caliper support mounting bolts and lift the caliper assembly from the knuckle.

To install

7. Position the caliper assembly onto the knuckle and install the bolts. Make sure the rotor fits between the brake pads. Torque the bolts to specification.
8. Using new copper washers, connect the brake hose to the caliper. Torque the brake hose attaching bolt to specification.
9. Bleed the brake system.
10. Apply the brake pedal and inspect the system. Ensure proper operation and no leakage.
11. Install tire and wheel assembly. Lower the vehicle and road test.

DISC BRAKE PADS

REMOVAL AND INSTALLATION

1. Before servicing the vehicle, refer to the Precautions Section.
2. Drain the brake fluid, as necessary.
3. Raise the vehicle and support safely.
4. Remove the bottom pin from the caliper and swing the caliper cylinder body upward; support the caliper with a wire.
5. Remove the brake pad retainers, shims and the pads.

To install:

6. Compress the piston of the disc brake caliper.
7. Install the brake pads and caliper assembly.

BRAKES   REAR DISC BRAKES

BRAKE CALIPER

REMOVAL & INSTALLATION

See Figure 3.

1. Before servicing the vehicle, refer to the Precautions Section.
2. Drain the brake fluid, as necessary.
3. Raise the vehicle and support safely.
4. Remove the tire and wheel assembly.
5. Remove the union bolt and brake hose. Remove the sliding pin bolts. Remove the caliper from the vehicle.

To install

6. Installation is the reverse of the removal procedure.
7. Bleed the brake system.
8. Apply the brake pedal and inspect the system. Ensure proper operation and no leakage.
9. Install tire and wheel assembly. Lower the vehicle and road test.

DISC BRAKE PADS

REMOVAL AND INSTALLATION

1. Before servicing the vehicle, refer to the Precautions Section.
2. Drain the brake fluid, as necessary.
3. Raise the vehicle and support safely.
4. Remove the tire and wheel assembly.
5. Remove the top bolt from the caliper.
6. Swing the caliper open and remove the pads.

To install:

7. Compress the piston of the disc brake caliper.
8. Install the brake pads and caliper assembly. Tighten caliper bolts to 24 ft. lbs. (32 Nm).

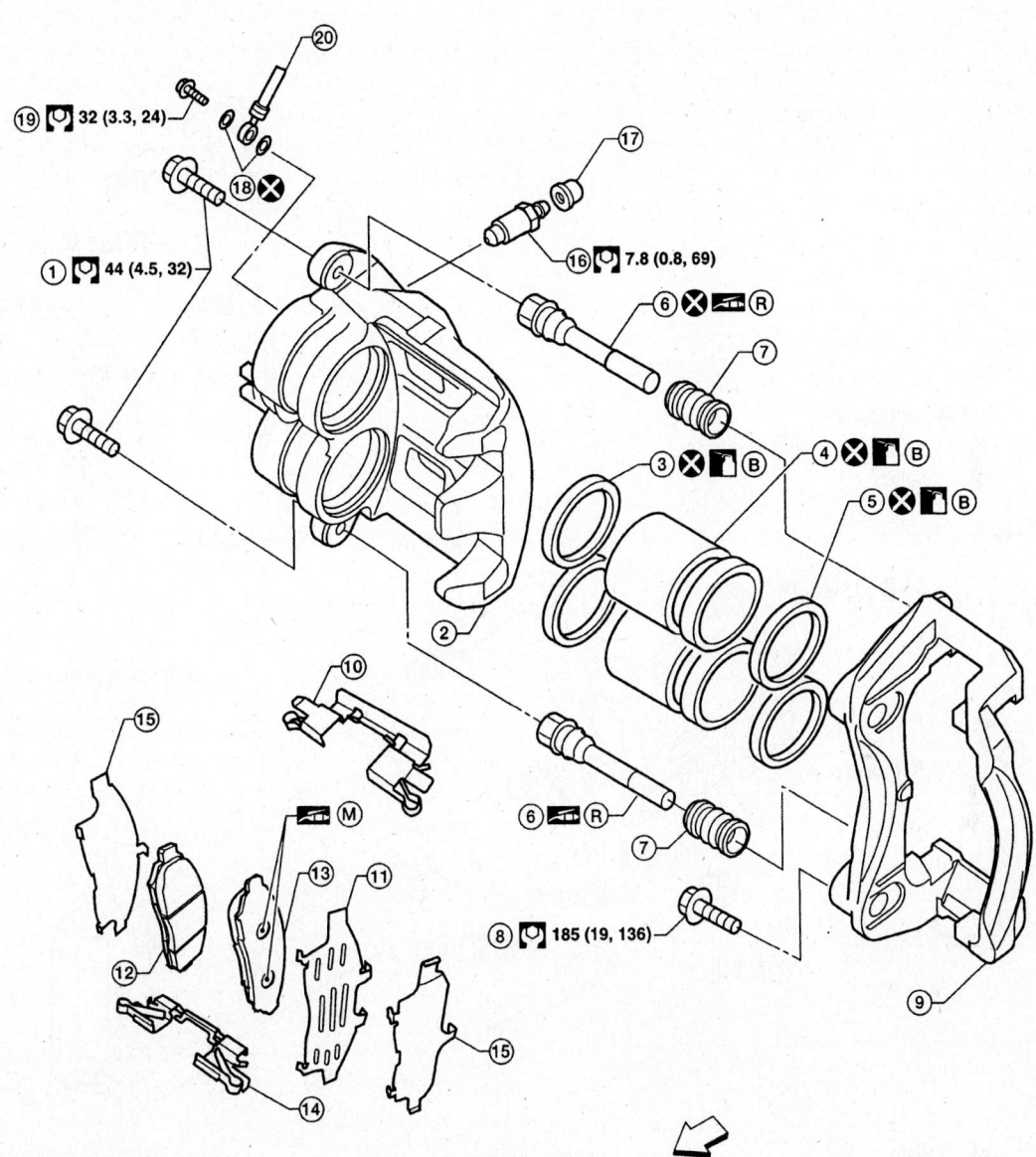

1.	Sliding pin bolt	2.	Cylinder body	3.	Piston seal
4.	Piston	5.	Piston boot	6.	Sliding pin
7.	Sliding pin boot	8.	Torque member bolt	9.	Torque member
10.	Pad retainer	11.	Inner shim	12.	Inner brake pad
13.	Outer brake pad	14.	Pad retainer	15.	Outer shim
16.	Bleed valve	17.	Cap	18.	Copper washers
19.	Union bolt	20.	Brake hose		

09482_FRON_G0121

Fig. 2 Front disc brake and related components

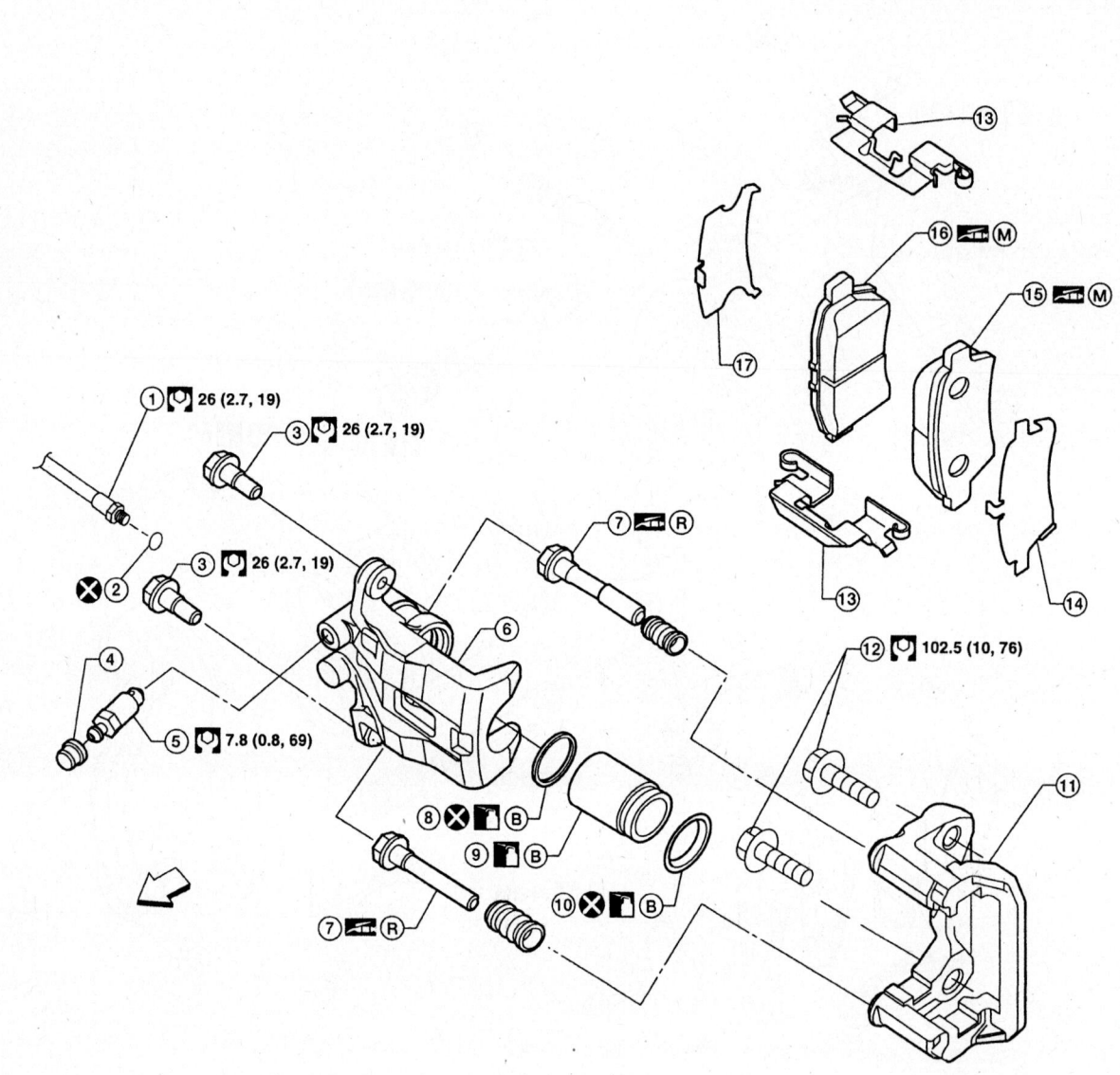

1. Brake hose
4. Cap
7. Sliding pin
10. Piston boot
13. Pad retainer
16. Inner brake pad

2. Copper washer
5. Bleed valve
8. Piston seal
11. Torque member
14. Outer shim
17. Inner shim

3. Sliding pin bolt
6. Cylinder body
9. Piston
12. Torque member bolt
15. Outer brake pad
⇐: Front

09482_FRON_G0123

Fig. 3 Rear disc brake and related components—Xterra

PARKING BRAKE CABLES

ADJUSTMENT

1. Remove rear half of the center console.

2. Rotate adjusting nut and loosen cable until tension is sufficiently released.

3. Remove the wheel and tire using power tool.

4. Remove the rotor and measure inner diameter at widest point.

5. Transfer measurement less 0.6 mm to the parking brake shoes and adjust accordingly.

6. Using wheel nuts, secure the disc to the hub to prevent it from tilting.

7. Rotate disc rotor to make sure there is no drag.

8. Operate parking brake lever 10 or more times with a force of 110 lbs. (490 N).

9. Rotate adjusting nut to adjust lever stroke to 6–8 notches with 44 lbs. (196 N) force.

10. With parking brake lever completely disengaged, make sure there is no drag on the parking brake.

PARKING BRAKE SHOES

REMOVAL & INSTALLATION

See Figures 4 and 5.

1. Before servicing the vehicle, refer to the Precautions Section.

➡**Clean the brakes with a vacuum dust collector to minimize the hazard of airborne particles or other materials.**

➡**Remove the disc rotor with the parking brake completely disengaged.**

2. Raise and safely support the vehicle.
3. Release the parking brake.
4. Remove the rear wheels.
5. Remove the rotor.
6. Remove the return springs.
7. Remove the adjuster.
8. Remove the, retainers, anti-rattle pins and shoes.
9. Remove the pin retainer. Disconnect the parking brake cable from the toggle lever.
10. Remove the back plate.

To install:

11. Installation is the reverse of the removal procedure.

12. Assemble the adjuster so that the threaded part expands when rotating it in

the direction shown by the arrow. Shorten the adjuster by rotating it in the opposite direction shown by the arrow.

13. Perform the parking brake break-in operation as follows: Safely, drive forward at approximately 25 mph (40 km/h) with the parking brake set with a force of approx. 45 lbs. (200 N) for about 30 seconds.

14. After the break-in operation, check the pedal stroke of parking brake. Readjust if necessary.

➡**To prevent lining from getting too hot, allow a cool off period of approximately 5 minutes after every break-in operation.**

15. Check and adjust the parking brake pedal stroke. Correct as required.

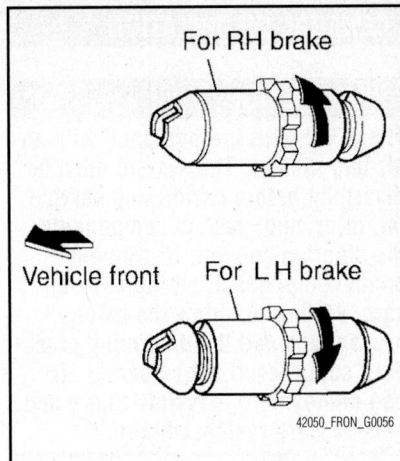

Fig. 5 Parking brake shoe adjuster identification

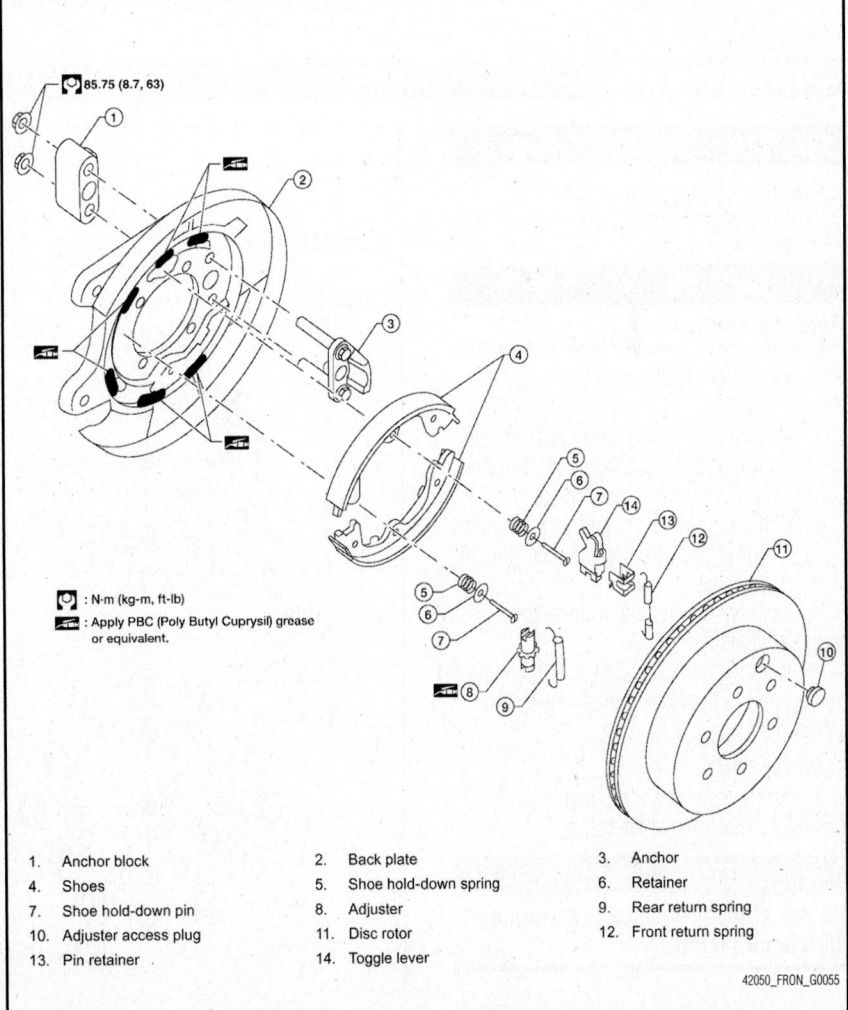

1.	Anchor block	2.	Back plate	3.	Anchor
4.	Shoes	5.	Shoe hold-down spring	6.	Retainer
7.	Shoe hold-down pin	8.	Adjuster	9.	Rear return spring
10.	Adjuster access plug	11.	Disc rotor	12.	Front return spring
13.	Pin retainer	14.	Toggle lever		

Fig. 4 Parking brake and related components

CHASSIS ELECTRICAL AIR BAG (SUPPLEMENTAL RESTRAINT SYSTEM)

GENERAL INFORMATION

✳✳ CAUTION

These vehicles are equipped with an air bag system. The system must be disarmed before performing service on, or around, system components, the steering column, instrument panel components, wiring and sensors. Failure to follow the safety precautions and the disarming procedure could result in accidental air bag deployment, possible injury and unnecessary system repairs.

PRECAUTIONS

Disconnect and isolate the battery negative cable before beginning any airbag system component diagnosis, testing, removal, or installation procedures. Allow system capacitor to discharge for two minutes before beginning any component service. This will disable the airbag system. Failure to disable the airbag system may result in accidental airbag deployment, personal injury, or death.

DISARMING THE SYSTEM

To disarm the SRS system turn the ignition switch to the off position. Then, disconnect both battery cables starting with the negative cable first and wait at least 3 minutes after the cables are disconnected.

ARMING THE SYSTEM

To rearm the SRS system, turn the ignition switch to the off position. Connect both battery cables starting with the positive cable first.

CLOCKSPRING CENTERING

1. Be sure to align the spiral cable correctly when installing the steering wheel. Make sure that the spiral cable is in the neutral position.

➡**The neutral position is detected by turning to the left 2.6 revolutions from the right end position and ending with the knob at the top. The spiral cable may snap due to steering operation if the cable is installed incorrectly. Also, with the steering linkage disconnected the cable may snap by turning the steering wheel beyond the limited number of turns (2.6 from the neutral position to both the left and right).**

DRIVE TRAIN

CLUTCH

REMOVAL & INSTALLATION
See Figure 6.

✳✳ CAUTION
Note the following:

- Do not clean the clutch disc with solvent.
- When installing, do not get grease from the main drive shaft onto the clutch disc friction surface.
- If the flywheel is removed, align the dowel pin with the smallest hole of flywheel.

1. Remove the manual transmission from the vehicle.
2. Remove the clutch cover bolts using power tool. Remove the clutch cover and clutch disc.

To install:
3. Apply recommended grease to clutch disc and main drive shaft spline.

✳✳ CAUTION
Do not allow grease to contaminate the clutch facing.

4. Install clutch disc and clutch cover. Pre-tighten the bolts and install Tool. Then tighten the clutch cover bolts evenly in two steps to the specified torque in the order shown in 6.
5. Install the manual transmission.

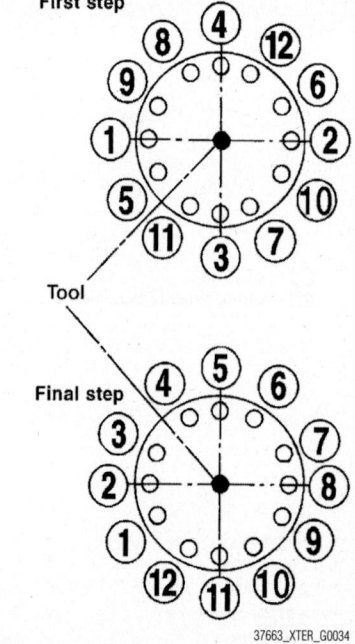

VQ40DE models

First step

Tool

Final step

37663_XTER_G0034

Fig. 6 Tightening order of the clutch disc and cover—Tool number ST20630000 (J-26366)

HYDRAULIC SYSTEM BLEEDING

BLEEDING PROCEDURE

1. Before servicing the vehicle, refer to the Precautions Section.

➡**Do not use a vacuum assist or any other type of power bleeder on this system. Use of a vacuum assist or power bleeder will not purge all the air from the system.**

2. Fill the system with the proper grade and type fluid.
3. Have an assistant pump the clutch pedal slowly several times and hold it depressed.
4. Open the slave cylinder bleeder screw and allow air to escape.
5. Close the bleeder screw before releasing the clutch pedal.
6. Repeat until all air is purged from the clutch hydraulic system.
7. Refill the reservoir to the full mark.

✳✳ CAUTION

Do not spill brake fluid on painted surfaces. If it spills, wipe up immediately and wash the affected area with water.

➡**Do not use a vacuum assist or any other type of power bleeder on this system. Use of a vacuum assist or power bleeder will not purge all the air from** the system. Monitor the fluid level in the reservoir tank to make sure it does not empty.

8. Top off reservoir with new brake fluid.

9. Connect a transparent vinyl tube and container to the air bleeder valve on the clutch operating cylinder.

10. Fully depress the clutch pedal several times.

11. With the clutch pedal depressed, open the bleeder valve to release the air.

12. Close the bleeder valve.

13. Repeat steps 3 to 5 until clear brake fluid comes out of the air bleeder valve.

14. Tighten the air bleeder to 70 inch lbs. (7.9 Nm).

FRONT AXLE SHAFT, BEARING & SEAL

REMOVAL & INSTALLATION

1. Before servicing the vehicle, refer to the Precautions Section.

2. Raise and support the vehicle safely. Remove the tire and wheel assembly.

3. Remove the rear engine cover.

4. Remove the wheel sensor harness from the mount on the knuckle. Disconnect the harness connector.

➡**Do not pull on the wheel sensor harness.**

5. Remove the wheel hub and bearing assembly.

➡**It is not necessary to remove the wheel speed sensor from the wheel hub when the wheel hub is not being replaced. Carefully feed the sensor harness through the hole in the splash shield.**

6. Separate the upper link ball joint stud from the steering knuckle using tool ST29020001 (J-24319-01) or equivalent.

7. Remove the halfshaft assembly from the vehicle by prying the halfshaft from the front final drive using the proper tool.

To install:

8. Installation is the reverse of the removal procedure.

9. Be sure to use a new differential side oil seal.

FRONT CV-JOINT

OVERHAUL

Inner CV-Joint

1. Before servicing the vehicle, refer to the Precautions Section.

2. Remove the axle halfshaft from the vehicle.

3. Remove the CV-joint boot clamps and push the boot away from the joint.

4. Match mark the housing and the shaft before separation.

5. Remove the stopper ring and pull housing off.

6. Remove the snap ring using a suitable tool, then remove the ball cage, steel ball, and inner race assembly from the shaft.

7. Remove the boot from the shaft. Remove circlip and dust cover from the housing. Clean the old grease off of the housing using paper towels.

To install:

➡**Use new circlips and boot clamps for assembly.**

8. Wrap the serrated part of the shaft with tape. Install the boot band and boot to shaft.

9. Remove the tape wound around the serrated part of the shaft.

10. Install the ball cage, steel ball, and inner race assembly on the shaft, and secure them using the snap ring.

➡**Use new snap ring.**

11. Pack the joint with 4.23–4.94 oz. grease.

➡**Ensure boot mounting surfaces do not have any grease.**

12. Install the stopper ring onto the housing.

13. Install the boot securely into the grooves.

14. Check that the overall boot installation length is 6.45–6.47 inches. Insert a flat-tip screwdriver or similar tool into the large end of the boot. Bleed air from boot to prevent boot deformation.

➡**Do not to touch the tip of the screwdriver to the inside of the boot.**

15. Install the boot clamps.

16. Install the axle halfshaft to the vehicle.

Outer CV-Joint

1. Before servicing the vehicle, refer to the Precautions Section.

2. Remove the axle halfshaft from the vehicle.

3. Remove the boot bands and slide the boot back.

4. Screw a sliding hammer or suitable tool 30 mm (1.18 in) or more into threaded part of joint sub-assembly. Pull joint sub-assembly off of shaft.

5. Remove boot from the shaft. Remove circlip from the shaft.

To install:

➡**Use new snap rings and boots for assembly.**

6. Insert the Genuine NISSAN Grease or equivalent, into the joint sub-assembly serration hole until the grease begins to ooze from the ball groove and serration hole.

7. Wrap the serrated part of the shaft with tape. Install the boot band and boot onto the shaft. Do not damage the boot.

8. Remove the tape wound around the serrated part of the shaft.

9. Attach the circlip to the shaft. The circlip must fit securely into the shaft groove. Attach the nut to the joint sub-assembly. Use a soft hammer to press-fit the circlip.

➡**Use new circlip.**

10. Pack the joint with 4.01–4.76 oz. grease.

➡**Ensure boot mounting surfaces do not have any grease.**

11. Install the boot securely into the grooves.

12. Check that the overall boot installation length is 5.32 inches. Insert a flat-tip screwdriver or similar tool into the large end of the boot. Bleed air from boot to prevent boot deformation.

➡**Do not to touch the tip of the screwdriver to the inside of the boot.**

13. Install the boot clamps.

14. Install the axle halfshaft to the vehicle.

FRONT DRIVESHAFT

REMOVAL & INSTALLATION

See Figure 7.

1. Remove wheel and tire using power tool.

2. Remove rear engine under cover using power tool.

3. Remove wheel sensor harness from mount on knuckle, then disconnect wheel sensor harness connector.

✳✳ CAUTION

Do not pull on wheel sensor harness.

4. Remove wheel hub and bearing assembly.

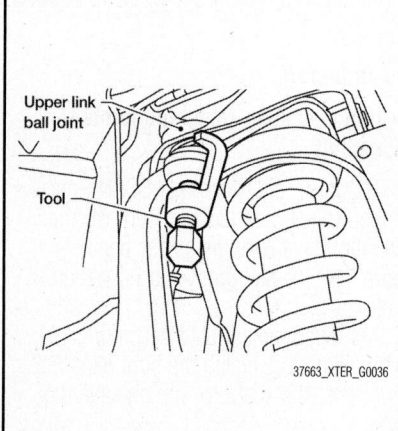

Fig. 7 Separate upper link ball joint stud from steering knuckle using tool— ST29020001 (J-24319-01)

- It is not necessary to remove wheel sensor from wheel hub when wheel hub is not being replaced.
- Carefully feed wheel sensor harness through hole in splash shield.

5. Separate upper link ball joint stud from steering knuckle using Tool.
- Support lower link with jack.

6. Remove drive shaft assembly.
- Pry drive shaft front final drive using suitable tool.
- Remove differential side oil seal.

To install:

7. Install drive shaft assembly.
- Install differential side oil seal.
- Pry drive shaft front final drive using suitable tool.

8. Separate upper link ball joint stud from steering knuckle using Tool.
- Support lower link with jack.

9. Install wheel hub and bearing assembly.
- Carefully feed wheel sensor harness through hole in splash shield.
- It is not necessary to remove wheel sensor from wheel hub when wheel hub is not being replaced.

✱✱ CAUTION

Do not pull on wheel sensor harness.

10. Install wheel sensor harness from mount on knuckle, then connect wheel sensor harness connector.

11. Install rear engine under cover using power tool.

12. Install wheel and tire using power tool.

REAR AXLE SHAFT, BEARING & SEAL

REMOVAL & INSTALLATION

See Figures 8 through 10.

1. Before servicing the vehicle, refer to the Precautions Section.

2. Remove or disconnect the following:
- Rear wheel and tire assembly
- Wheel speed sensor
- Brake rotor
- Brake caliper assembly
- Parking brake cable
- Brake fluid line
- Bearing cage and backing plate bolts
- Axle shaft assembly
- Axle seal
- Wheel speed sensor rotor
- Snap ring and shim washer
- Bearing ring retainer
- back plate and torque member
- Axle bearing studs
- Wheel bearing
- Grease catcher

To install:

➡ **Use new seals, bearings, circlips and snap rings for assembly.**

3. Install grease catcher

4. Install the wheel studs through the grease catcher into the axle shaft using a suitable press.

➡ **All six wheel studs must be pressed on at the same time and are flush with the grease catcher when installed.**

5. Position the axle bearing on the back plate and torque member.

6. Install the axle bearing studs using a suitable press to attach the axle bearing to the back plate and torque member.

➡ **Always replace the axle bearing with a new one.**

7. Install the back plate and torque member, new axle bearing and new bearing ring retainer on the axle shaft using a suitable press. Do not exceed 11 tons force.

8. Press the new bearing ring retainer on the axle shaft with the taper side positioned toward press.

➡ **Always replace the bearing ring retainer with a new one.**

9. Select the correct size shim washer. Select the size of shim washer so that the installed snap ring to shim washer clearance is 0.008 inch or less.

10. Install a new snap ring on the axle shaft.

11. Do not over spread the snap ring when installing, measure the outer diameter of the snap ring after installation and replace if the snap ring outer diameter exceeds 1.87 inch maximum.

12. Check the snap ring to shim washer clearance. Repeat previous steps as necessary.

13. Perform break-in rotation of the wheel bearing.
- Rotate the wheel bearing in the forward direction for a minimum of 10 revolutions at 50–70 RPM.
- Rotate the wheel bearing in the reverse direction for a minimum of 10 revolutions at 50–70 RPM.

14. Measure the rotational torque of the wheel bearing. Rotational torque should be 16 inch lbs. (1.8 Nm) at 8–12 RPM.

15. Inspect that the wheel bearing is free from axial play relative to the axle shaft.

16. Install a new ABS sensor rotor on the axle shaft with notch side away from press using a suitable press.

➡ **Always replace the ABS sensor rotor with a new one.**

17. Install new axle seal in housing.

18. Apply multi-purpose grease to the recess of axle case end as shown in illustration.

19. Insert tool J-34296 into the new axle oil seal as a guide. Ensure tool ends do not overlap.

20. Insert the axle shaft assembly. Tighten the axle shaft nuts evenly in a criss-cross pattern to specification. Remove the tool when the axle shaft assembly is approximately 90 percent inserted to protect the new axle oil seal.

21. Install parking brake assembly, rear caliper assembly and ABS wheel sensor.

TRANSFER CASE ASSEMBLY

REMOVAL & INSTALLATION

TX15B

See Figure 11.

1. Before servicing the vehicle, refer to the Precautions Section.

2. Disconnect the negative battery cable.

3. Switch the 4WD switch to 2WD. Set the transfer case to 2WD.

4. Raise and support the vehicle safely.

5. Remove the undercovers. Drain the transfer case fluid.

6. Remove the center exhaust tube and main muffler.

7. Remove the front and rear driveshafts. Install plug in rear oil seal.

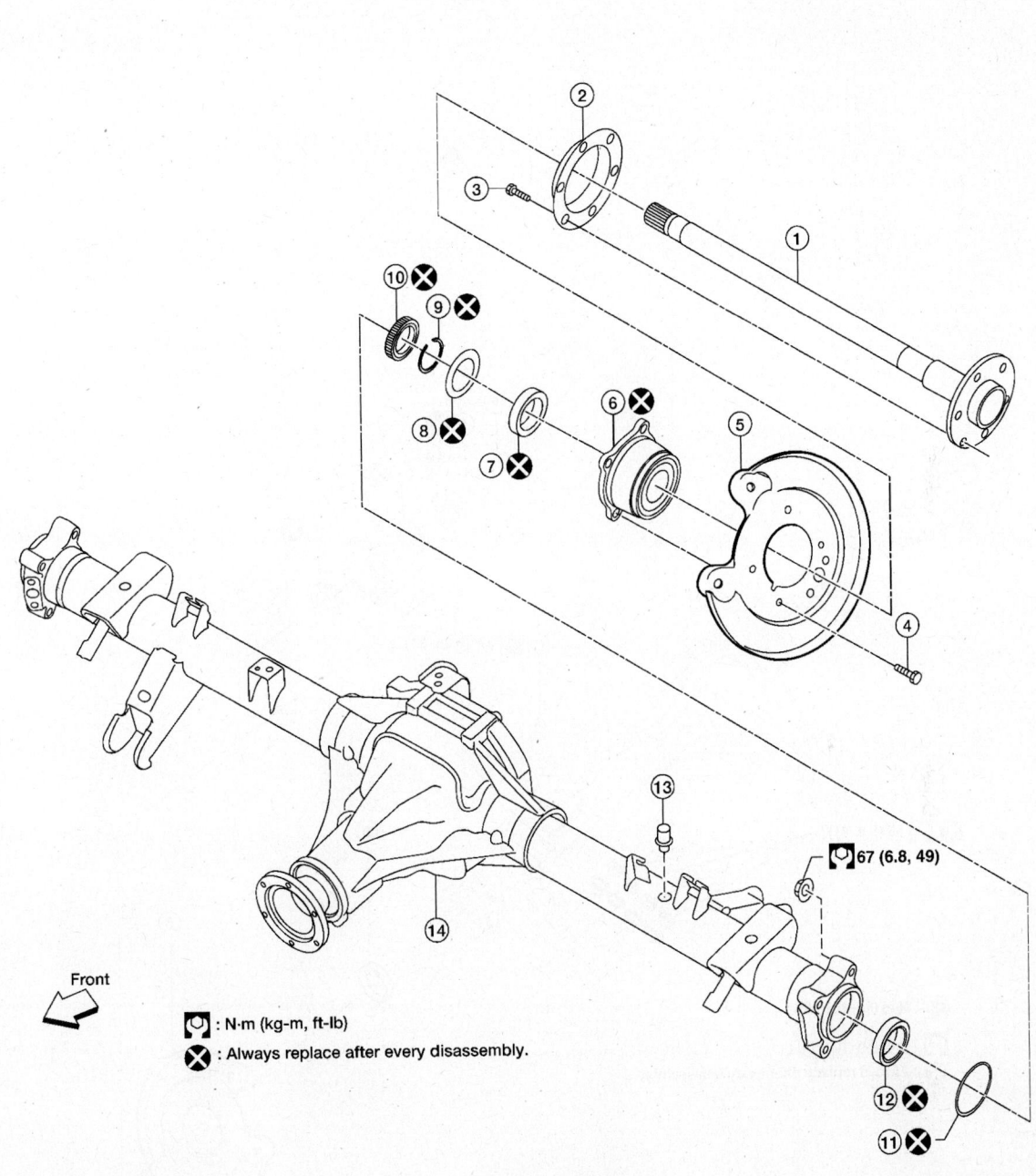

67 (6.8, 49)

Front

⬤ : N·m (kg-m, ft-lb)

✖ : Always replace after every disassembly.

1. Axle shaft	2. Grease catcher	3. Wheel stud
4. Axle bearing stud	5. Back plate and torque member	6. Axle bearing
7. Bearing ring retainer	8. Shim washer	9. Snap ring
10. ABS sensor rotor	11. O-ring	12. Axle oil seal
13. Breather	14. Rear final drive	

22140_FRON_G0001

Fig. 8 Rear axle shaft, bearing and seal—C200

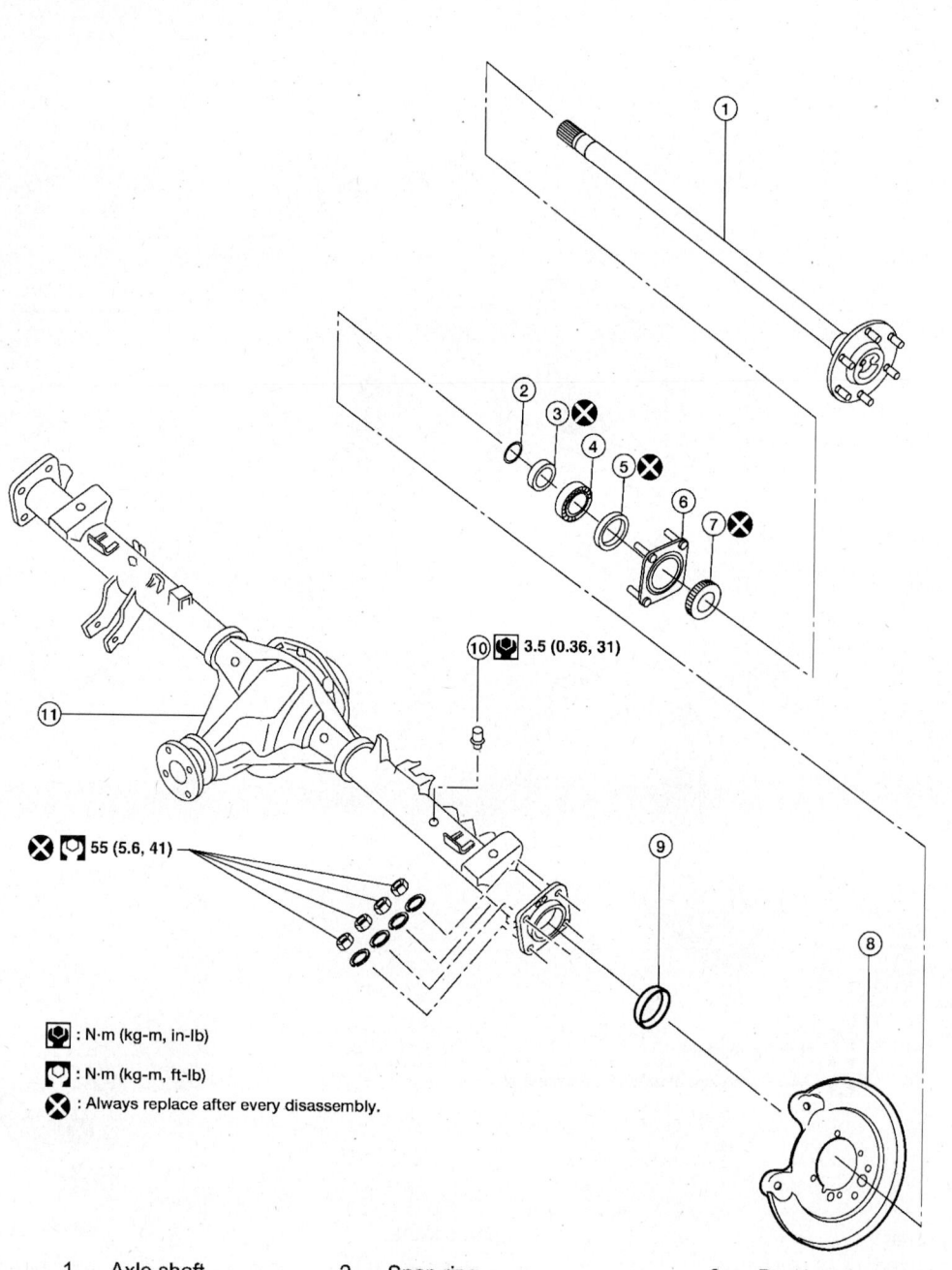

10 ⊠ 3.5 (0.36, 31)

⊠ ⊡ 55 (5.6, 41)

⊡ : N·m (kg-m, in-lb)

⊡ : N·m (kg-m, ft-lb)

⊠ : Always replace after every disassembly.

1. Axle shaft	2. Snap ring	3. Bearing ring retainer
4. Axle shaft bearing	5. Axle oil seal	6. Axle shaft bearing cage
7. ABS sensor rotor	8. Back plate and torque member	9. Axle shaft bearing cup
10. Breather	11. Rear final drive	

22140_FRON_G0002

Fig. 9 Rear axle shaft, bearing and seal—M226

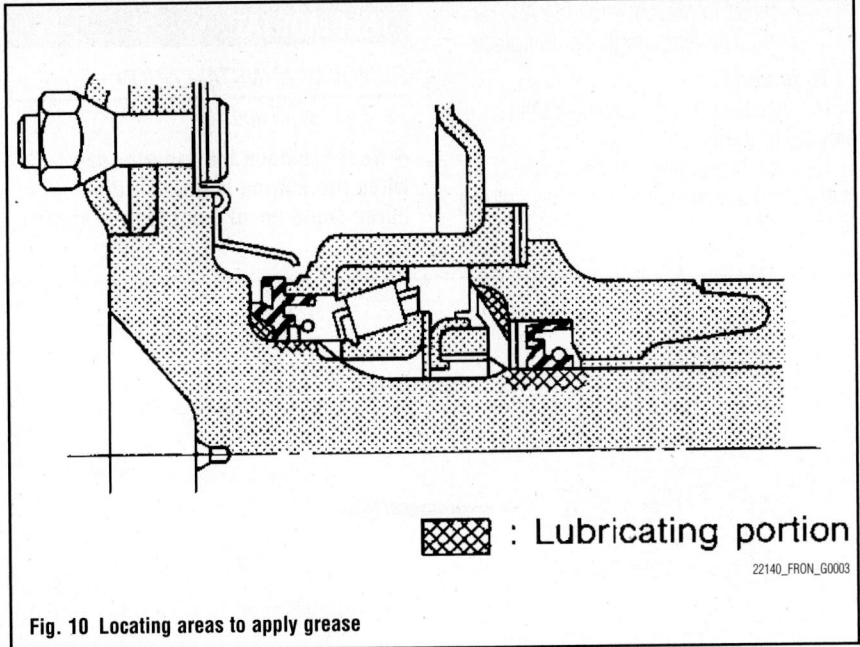

: Lubricating portion

22140_FRON_G0003

Fig. 10 Locating areas to apply grease

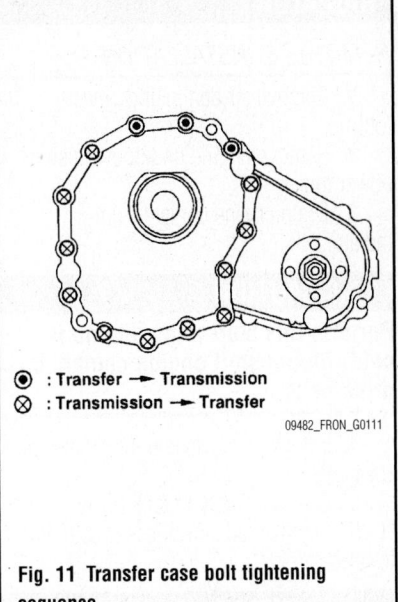

⊙ : Transfer → Transmission
⊗ : Transmission → Transfer

09482_FRON_G0111

**Fig. 11 Transfer case bolt tightening
sequence**

8. Remove the transmission-to-cross-member bolts. Properly support the transmission and transfer case assembly, using a suitable jack.

9. Remove the transmission crossmember.

➡**Support the transmission and transfer case using two suitable jacks while removing the transmission crossmember.**

10. Disconnect the ATP electrical connector, the 4LO switch connector, the wait detection switch and the transfer control device.

11. Disconnect each air breather hose from the transfer control device and the breather tube.

12. Remove the transfer case to transmission retaining bolts.

➡**Support the transmission and transfer case, using a suitable jack.**

13. Remove the transfer case from the vehicle.

To install:

14. Installation is the reverse of the removal procedure.

15. Tighten the transfer case to transmission retaining bolts to specification and in the proper sequence. Specification is 27 ft. lbs. (36.6 Nm).

16. Start the engine and check for leaks, correct as required.

ENGINE COOLING

ENGINE FAN

REMOVAL & INSTALLATION

Electric

1. Remove air dam using power tool.

2. Remove engine front undercover using power tool.

3. Partially drain engine coolant from radiator.

4. Release the radiator shroud (lower) from the radiator shroud (upper) and position aside. Release the tabs, pull radiator shroud (lower) rearwards and down.

5. Remove air duct.

6. Remove reservoir tank hose from shroud.

7. Remove the radiator shroud (upper) bolts and remove the radiator shroud (upper).

8. Disconnect fan motor harness connector.

9. Remove bolt and fan grille and motor assembly.

To install:

10. Installation is the reverse of the removal procedure.

11. Start the engine and check for leaks, correct as required.

Mechanical (Belt Driven)

1. Remove air dam using power tool.

2. Remove engine front undercover using power tool.

3. Partially drain engine coolant from radiator.

✳✳ CAUTION

Perform this step when engine is cold. Do not spill engine coolant on drive belts.

4. Remove air duct.

5. Remove reservoir tank hose from shroud.

6. Removal radiator hose (upper) from radiator.

✳✳ CAUTION

Be careful not to allow engine coolant to contact drive belts.

7. Release the radiator shroud (lower) from the radiator shroud (upper) and position aside. Release the tabs, pull radiator shroud (lower) rearwards and down.

8. Remove the radiator shroud (upper) bolts and remove the radiator shroud (upper).

9. Remove the drive belt.

10. Remove the engine cooling fan.

To install:

11. Installation is the reverse of the removal procedure.

12. Start the engine and check for leaks, correct as required.

RADIATOR

REMOVAL & INSTALLATION

1. Remove air dam using power tool.
2. Remove engine undercover using power tool.
3. Drain engine coolant from radiator.

✳✳ CAUTION

Perform this step when engine is cold. Do not spill engine coolant on drive belts.

4. Remove air duct and air cleaner case assembly.
5. Remove reservoir tank hose.
6. Remove radiator hoses (upper and lower).

✳✳ CAUTION

Be careful not to allow engine coolant to contact drive belts.

7. Disconnect A/T fluid cooler hoses and plug.
8. Remove radiator shroud (lower).
9. Remove radiator shroud (upper).
10. Remove radiator cooling fan assembly.
11. Remove the radiator mounting bracket bolts.
12. Remove the two A/C condenser bolts.
13. Remove radiator as follows:

✳✳ CAUTION

Do not damage or scratch A/C condenser and radiator core when removing.

- With lifting and pulling radiator in a rear direction, disassemble mounting rubber (lower) from radiator core support center
- Because A/C condenser is attached to the front-lower portion of radiator, moving it in the rear direction should be at a minimum
- Lift A/C condenser up and remove radiator after disengaging the fitting as front-bottom surface

✳✳ CAUTION

Lifting A/C condenser should be minimum to prevent a load to A/C piping.

- After removing radiator, put A/C condenser on radiator core support

center to prevent a load to A/C piping, and temporarily tie it in place

To install:

14. Installation is the reverse of the removal procedure.
15. Start the engine and check for coolant and transmission fluid leaks, correct as required.

THERMOSTAT

REMOVAL & INSTALLATION

See Figures 12 and 13.

➡**Never remove the radiator cap when the engine is hot. Serious burns could occur from high-pressure**

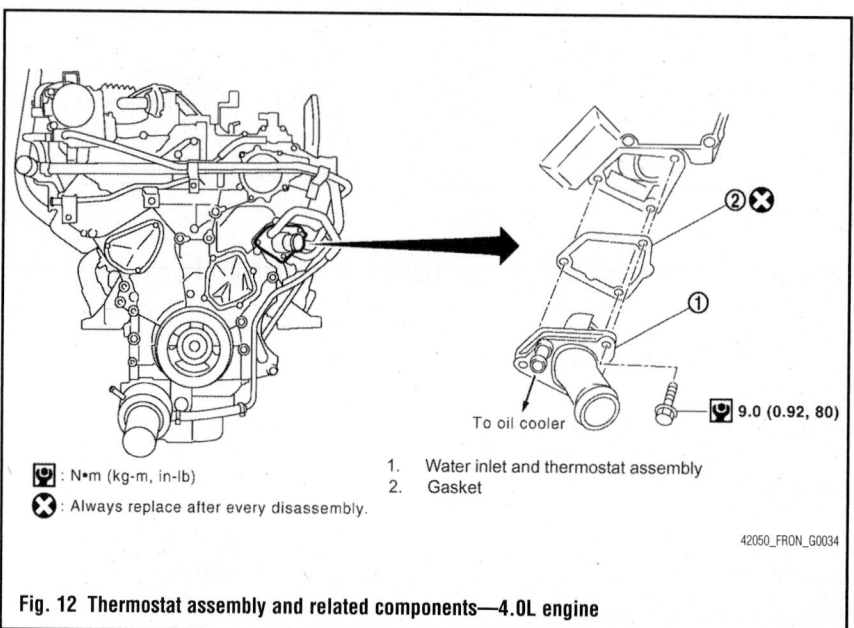

⊕ : N•m (kg-m, in-lb)

⊗ : Always replace after every disassembly.

1. Water inlet and thermostat assembly
2. Gasket

42050_FRON_G0034

Fig. 12 Thermostat assembly and related components—4.0L engine

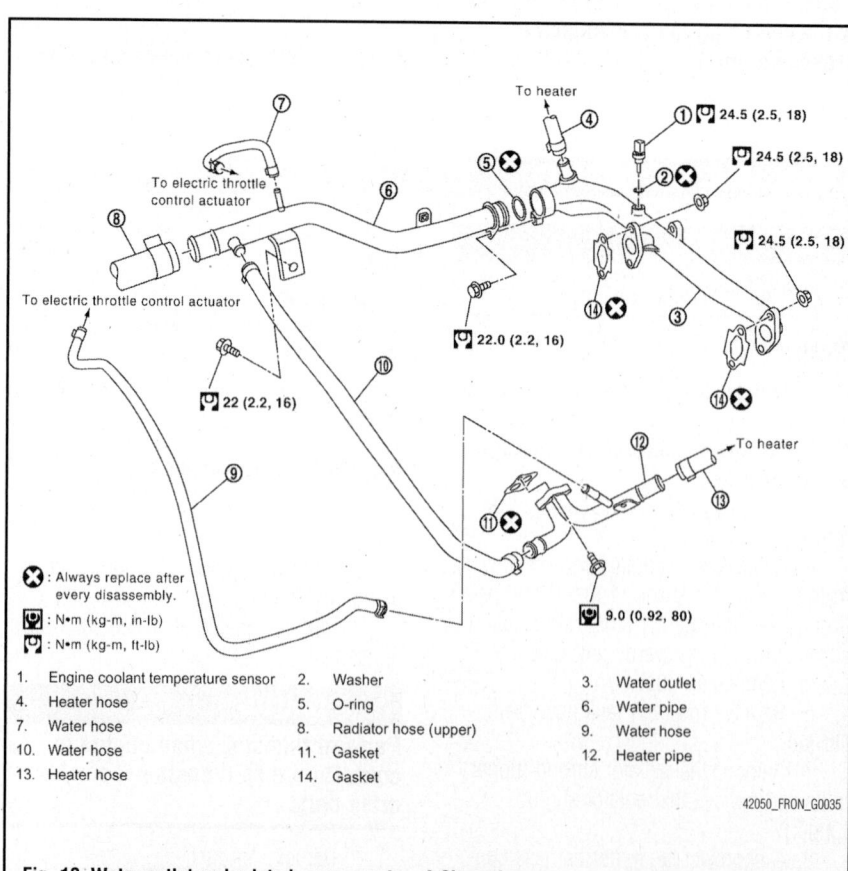

⊗ : Always replace after every disassembly.

⊕ : N•m (kg-m, in-lb)

⊟ : N•m (kg-m, ft-lb)

1. Engine coolant temperature sensor	2. Washer	3. Water outlet
4. Heater hose	5. O-ring	6. Water pipe
7. Water hose	8. Radiator hose (upper)	9. Water hose
10. Water hose	11. Gasket	12. Heater pipe
13. Heater hose	14. Gasket	

42050_FRON_G0035

Fig. 13 Water outlet and related components—4.0L engine

engine coolant escaping from the radiator.

1. Be sure the engine is cold.
2. Disconnect the negative battery cable.
3. Drain the coolant. Properly disposed of used coolant.
4. Remove the air duct and air cleaner case.
5. Disconnect radiator hose (lower) and oil cooler hose from water inlet and thermostat assembly.
6. Remove the water inlet retaining bolts.
7. Remove the water inlet and the thermostat.

➡**Do not disassemble the water inlet and thermostat assembly. Replace them as a unit, if required.**

To install:

8. Installation is the reverse of the removal procedure.
9. Be sure to refill the cooling using the proper grade and type engine coolant.
10. Start the engine and check for leaks.
11. Start the engine and allow it to reach operation temperature. Recheck the coolant level, fill as required.

WATER PUMP

REMOVAL & INSTALLATION

See Figures 14 and 15.

1. Before servicing the vehicle, refer to the Precautions Section.
2. Disconnect the negative battery cable. Drain the cooling system.
3. Remove the air dam and undercover.
4. Remove air duct and resonator.
5. Remove the drive belts.
6. Remove the radiator upper hose. Remove the cooling fan.
7. Remove the chain tensioner cover and water pump cover from the front timing case, using tool KV10111100 (J-37228) or equivalent.
8. To remove the timing chain tensioner (primary), loosen the clip of the timing chain tensioner (primary) and release the plunger stopper. Insert the plunger into the tensioner body by pressing the slack guide. Keep the slack guide pressed and hold the plunger in by pushing the stopper pin through the tensioner body hole and plunger groove. Turn the crankshaft pulley clockwise so that the timing chain on the timing chain tensioner (primary) side is loose. Remove the bolts

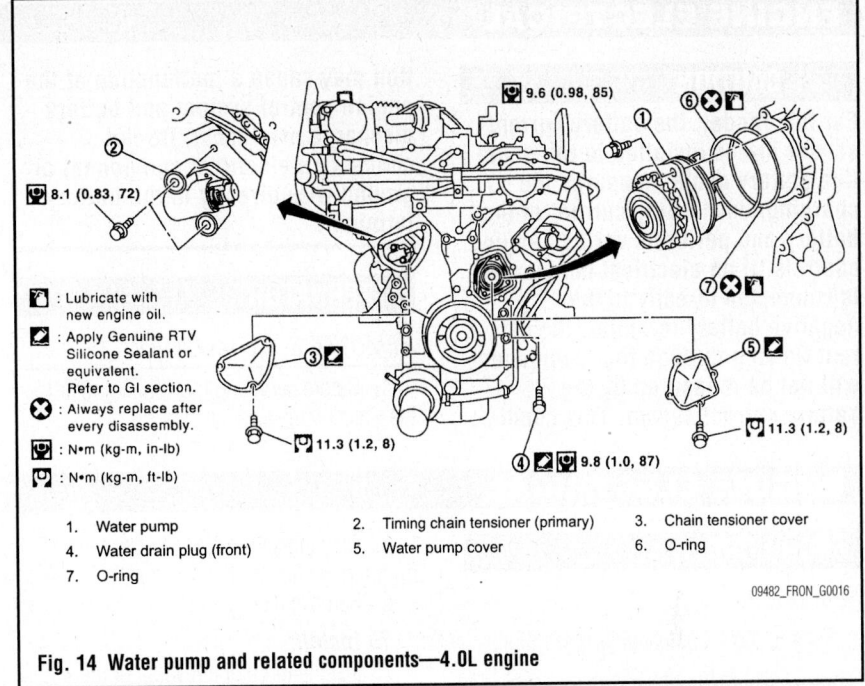

Fig. 14 Water pump and related components—4.0L engine

Legend for Fig. 14:

⬛ : Lubricate with new engine oil.

✅ : Apply Genuine RTV Silicone Sealant or equivalent. Refer to GI section.

❌ : Always replace after every disassembly.

🔧 : N•m (kg-m, in-lb)

🔧 : N•m (kg-m, ft-lb)

9.6 (0.98, 85)
8.1 (0.83, 72)
11.3 (1.2, 8)
9.8 (1.0, 87)
11.3 (1.2, 8)

1. Water pump
2. Timing chain tensioner (primary)
3. Chain tensioner cover
4. Water drain plug (front)
5. Water pump cover
6. O-ring
7. O-ring

09482_FRON_G0016

and remove the timing chain tensioner (primary).

➡**Be careful not to drop the bolts inside the timing chain case.**

9. Remove the three water pump retaining bolts. Secure a gap between the water pump gear and the timing chain, by turning the crankshaft pulley counterclockwise until timing chain looseness on the water pump sprocket is at its maximum point.
10. Screw M8 bolts approximately 1.97 inch in length into the water pumps upper and lower bolt holes until they reach the timing chain case.
11. Alternately tighten each bolt for a half turn and pull out the water pump.

➡**Pull the pump straight out while preventing the vane from contacting the**

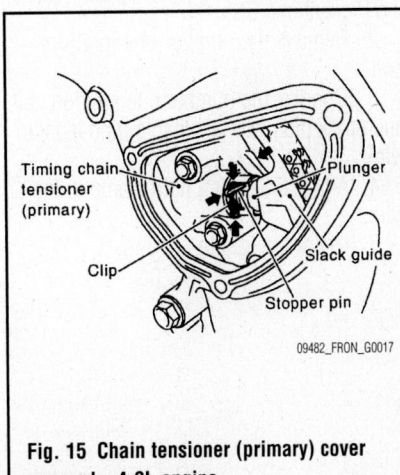

Timing chain tensioner (primary)
Plunger
Clip
Slack guide
Stopper pin

09482_FRON_G0017

Fig. 15 Chain tensioner (primary) cover removal—4.0L engine

socket in the installation area. Remove the pump without causing the sprocket to contact the timing chain.

12. Remove the M8 bolts. Remove and discard the O-rings.

To install:

13. Installation is the reverse of the removal procedure.
14. Be sure to use new gaskets and O-rings, as required. Apply engine oil to new O-rings before installation. Locate O-ring with white paint mark in forward groove.
15. When installing the water pump make sure that the timing chain and water pump sprocket are engaged. Tighten the bolts alternately and evenly to specification.
16. Before installing the chain tensioner cover and the water pump cover be sure to apply a continuous bead of sealant to the mating surfaces of the covers.

➡**Do not allow the sealant to set for more than five minutes before installing the covers.**

17. Be sure to fill the cooling system with the proper grade and type engine coolant.
18. Start the engine and check for leaks.
19. Let the engine idle for about three minutes than rev it up to 3,000 rpm's under a no load condition to purge air from the high pressure chamber of the chain tensioner. The engine may produce a rattling noise. This indicates that air still remains in the chamber and is not a matter of concern.

ENGINE ELECTRICAL

✳✳ CAUTION

For this model, the battery current sensor that is installed to the negative battery cable measures the charging/discharging current of the battery and performs various engine controls. If an electrical component is connected directly to the negative battery terminal, the current flowing through that component will not be measured by the battery current sensor. This condi-tion may cause a malfunction of the engine control system and battery discharge may occur. Do not connect an electrical component or ground wire directly to the battery terminal.

ALTERNATOR

REMOVAL & INSTALLATION

1. Before servicing the vehicle, refer to the Precautions Section.

CHARGING SYSTEM

2. Disconnect the negative battery cable.
3. Remove the drive belt.
4. Remove alternator stay.
5. Remove the upper alternator mount-ing bolt.
6. Disconnect the alternator harness electrical connectors.
7. Remove the alternator from the vehi-cle.

To install:

8. Installation is the reverse of the removal procedure.

ENGINE ELECTRICAL

FIRING ORDER

See Figure 16.

Refer to the accompanying illustration.

IGNITION COIL

REMOVAL & INSTALLATION

Left Bank

1. Disconnect the negative battery cable.
2. Remove the engine cover.
3. Remove the air cleaner case and air duct.
4. Move aside the harness, harness bracket and hoses which are located above the ignition coil.

5. Disconnect the harness connector from the ignition coil.
6. Remove the ignition coil.

To install:

7. Installation is the reverse of the removal procedure.

Right Bank

1. Disconnect the negative battery cable.
2. Remove the intake manifold collector.
3. Move aside the harness, harness bracket and hoses which are located above the ignition coil.
4. Disconnect the harness connector from the ignition coil.
5. Remove the ignition coil.

To install:

6. Installation is the reverse of the removal procedure.

IGNITION TIMING

INSPECTION

1. Before servicing the vehicle, refer to the Precautions Section.
2. Remove the number one ignition coil.
3. Connect the number one ignition coil and spark plug with a suitable high tension wire.
4. Attach the timing light clamp to the wire.
5. Check the ignition timing.

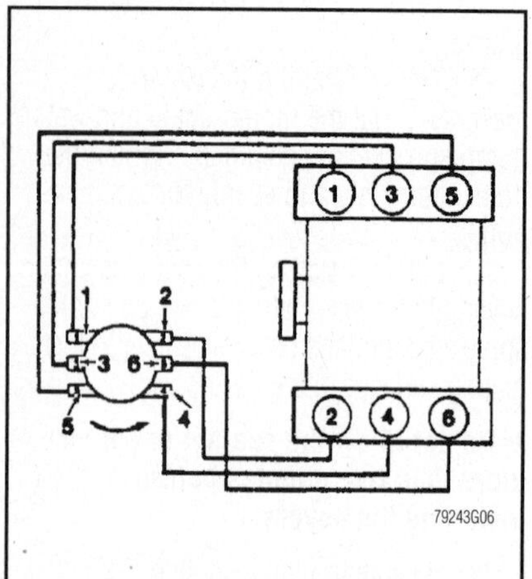

Fig. 16 Firing order: 1–2–3–4–5–6

79243G06

IGNITION SYSTEM

SPARK PLUGS

REMOVAL & INSTALLATION

See Figure 17.

1. Disconnect the negative battery cable.
2. Remove the intake manifold collector.
3. Remove the ignition coil.
4. Remove the spark plug using a spark plug socket and wrench.

To install:

5. Installation is the reverse of the removal procedure.

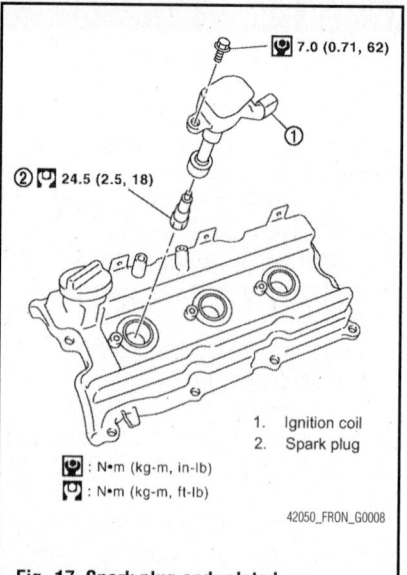

1. Ignition coil
2. Spark plug

⚙ : N•m (kg-m, in-lb)
🔧 : N•m (kg-m, ft-lb)

42050_FRON_G0008

Fig. 17 Spark plug and related compo-nents—4.0L engine

ENGINE ELECTRICAL

STARTING SYSTEM

STARTER

REMOVAL & INSTALLATION

1. Before servicing the vehicle, refer to the Precautions Section.
2. Remove or disconnect the following:
   - Negative battery cable
   - Engine under cover
   - On vehicles with 4.0L engine, remove the exhaust manifold cover to gain access to the starter retaining bolts
   - Starter harness connectors
   - Starter bolts
   - Starter motor

To install:

3. Install or connect the following:
   - Starter motor
   - air cleaner cover and the air cleaner to intake manifold collector duct, if equipped
   - Exhaust manifold cover, if equipped
   - Starter harness connectors
   - Engine under cover
   - Negative battery cable

ENGINE MECHANICAL

ACCESSORY DRIVE BELTS

ACCESSORY BELT ROUTING

See Figure 18.

Refer to the accompanying illustration.

INSPECTION

Inspect the drive belt for signs of glazing or cracking. A glazed belt will be perfectly smooth from slippage, while a good belt will have a slight texture of fabric visible. Cracks will usually start at the inner edge of the belt and run outward. All worn or damaged drive belts should be replaced immediately.

ADJUSTMENT

Belt tensioning is not necessary, as it is automatically adjusted by the drive belt auto tensioner.

REMOVAL & INSTALLATION

See Figures 19 and 20.

1. Before servicing the vehicle, refer to the Precautions section.
2. Disconnect the negative battery cable.
3. Remove the air duct and resonator assembly (inlet).
4. Rotate the drive belt auto tensioner in the direction shown in the illustration.

❄ CAUTION

Do not place your hand in a location where pinching may occur if the holding tool accidentally comes off.

5. Remove the drive belt.

To install:

6. Installation is the reverse of the removal procedure.

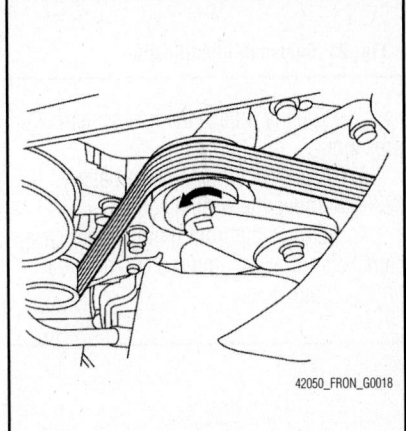

Fig. 19 Drive belt tension tool installation and removal direction—4.0L engine

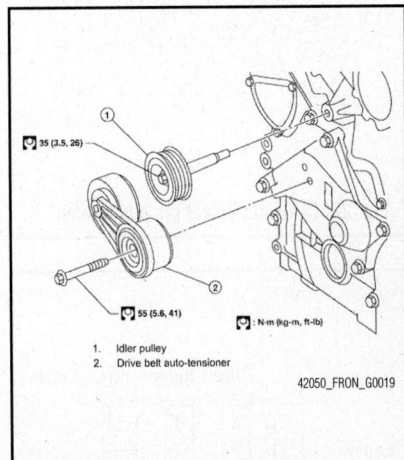

Fig. 20 Drive belt auto tensioner and related components—4.0L engine

CAMSHAFT, BEARINGS & LIFTERS

REMOVAL & INSTALLATION

See Figures 21 through 29.

1. Before servicing the vehicle, refer to the Precautions Section.

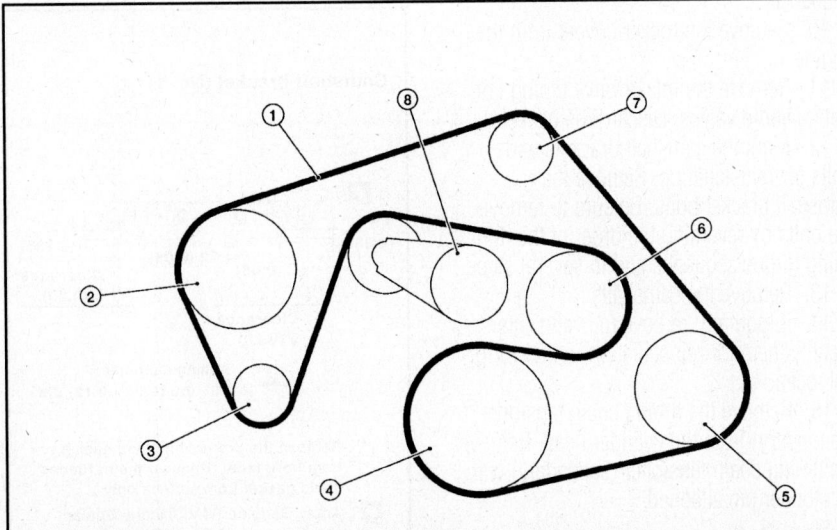

1.	Drive belt	2.	Power steering pump pulley	3.	Generator pulley
4.	Crankshaft pulley	5.	A/C compressor	6.	Cooling fan pulley
7.	Idler pulley	8.	Drive belt auto-tensioner		

Fig. 18 Accessory drive belt routing—4.0L engine

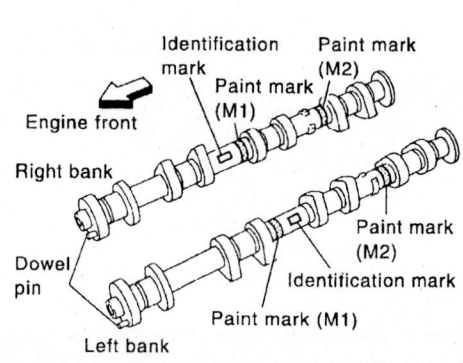

Bank	INT/EXH	Dowel pin	Paint marks		Identification mark
			M1	M2	
RH	INT	No	Green	No	RE
	EXH	Yes	No	White	RE
LH	INT	No	Green	No	LH
	EXH	Yes	No	White	LH

Fig. 21 Camshaft identification

2. Properly relieve the fuel system pressure.

3. Disconnect the negative battery cable. Remove the engine cover.

4. Remove the front timing chain case, camshaft sprocket, timing chain and rear timing chain case.

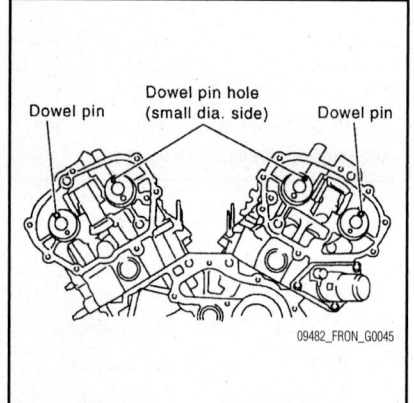

Fig. 22 Camshaft dowel pin positioning

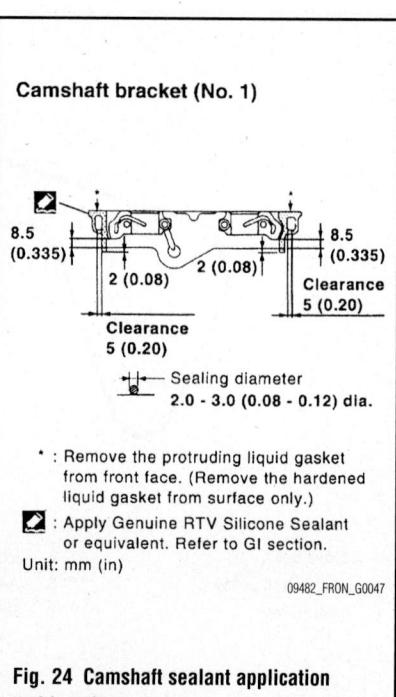

Fig. 23 Camshaft bearing cap identification

5. Remove the camshaft position sensor (PHASE) from the cylinder head back side.

➡**Handle carefully to avoid dropping and shocks. Do not disassemble. Do not place in a location where the sensor can be exposed to magnetism.**

6. Remove the intake manifold collector.

7. Separate the engine harness and remove their brackets from the rocker covers. Remove the harness bracket from the cylinder head, if necessary.

8. Remove the ignition coil. Remove the PCV hoses. Remove the oil filler cap, if necessary.

9. Loosen the rocker cover retaining bolts, in the reverse order of the tightening sequence.

10. Remove the rocker covers from the engine.

11. Remove the intake valve timing control solenoid valves. Discard the gaskets.

12. Mark the camshaft brackets and bolts for reinstallation. Remove the camshaft bracket bolts. Be sure to remove the bolts by reversing the order of the tightening torque sequence and in several steps.

13. Remove the camshafts.

14. If required, remove the valve lifters. Identify them for reinstallation in their original locations.

15. Remove the timing chain tensioner (secondary) from the cylinder head. Remove the timing chain tensioner (secondary) with its stopper pin attached.

To install:

16. Inspect the camshafts, replace as required.

17. Install the timing chain tensioners (secondary) on both sides of the cylinder head. Be sure to use new O-rings.

➡**Install the tensioner with its stopper pin attached. Install the tensioner with the sliding part facing downward on the right cylinder head and with the sliding part facing upward on the left cylinder head.**

18. Install the valve lifters, in their original bores.

19. Install the camshafts, with the dowel pin attached to its front end face on the exhaust side.

➡**Follow the identification marks for proper placement and direction.**

20. Install the camshaft so that the dowel pin hole and dowel pin on the front end face are positioned as shown in the illustration

Camshaft bracket (No. 1)

8.5 (0.335)   8.5 (0.335)
2 (0.08)   2 (0.08)
Clearance 5 (0.20)
Clearance 5 (0.20)
Sealing diameter
2.0 - 3.0 (0.08 - 0.12) dia.

* : Remove the protruding liquid gasket from front face. (Remove the hardened liquid gasket from surface only.)

▨ : Apply Genuine RTV Silicone Sealant or equivalent. Refer to GI section.

Unit: mm (in)

09482_FRON_G0047

Fig. 24 Camshaft sealant application and location

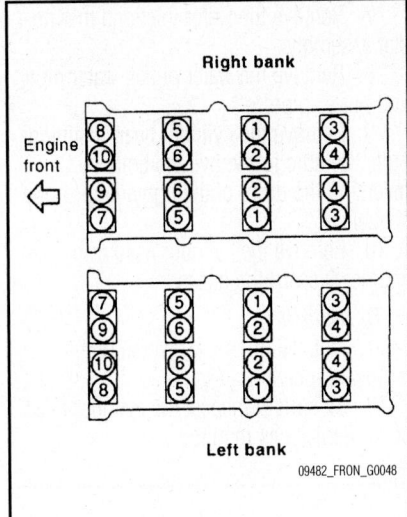

Fig. 25 Camshaft bearing bracket bolt torque sequence

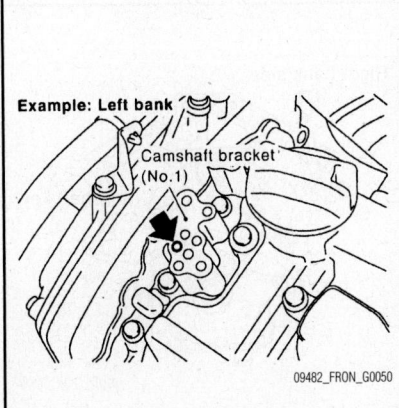

Fig. 26 Camshaft bracket and cylinder head measurement

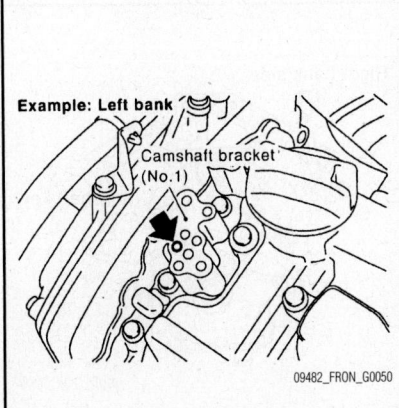

Fig. 27 Camshaft bracket (No. 1) oil hole location

(No. 1 piston at TDC on its compression stroke).

➡**Large and small pin holes are located on the front end face of the camshaft (INT), at intervals of 180 degrees. Face small diameter side pin hole upward (in cylinder head upper face direction).**

➡**Though the camshaft does not stop at the portion as shown, for placement of the cam nose, it is generally accepted that the camshaft is placed for the same direction as shown.**

21. Install the camshaft brackets in the same position that they were removed. Install brackets No. 2–No. 4 aligning the stamp marks as indicated in the illustration.

➡**There are no identification marks indicating left or right for camshaft bracket No. 1.**

22. Apply liquid gasket to the mating surfaces of camshaft bracket No. 1 as shown in the illustration on both the left and right cylinder heads. Be sure to use genuine RTV sealant or equivalent.

23. Tighten the camshaft bracket bolts in the proper sequence and to specification:
    a. Bolts 7–10 to 17 inch lbs. (1.13 to 1.92 Nm)
    b. Bolts 1–6 to 17 inch lbs. (0.67 to 1.92 Nm)
    c. All bolts to 52 inch lbs. (5.88 Nm)
    d. All bolts to 92 inch lbs. (10.39 Nm)

24. Measure the difference in levels between the front end faces of the camshaft bracket No. 1 and the cylinder head. Specification should be -0.0055–0.0055 inch. If not within specification, reinstall camshaft bracket No. 1.

➡**Measure two positions (both intake and exhaust side) for a single bank.**

25. Check and adjust valve clearance, as required.

26. Apply liquid gasket, be sure to use genuine RTV silicone sealant or equivalent, to the positions shown in the illustration. Refer to figure "a" to apply liquid gasket to joint part of camshaft bracket No. 1 and cylinder head. Refer to figure "b" to apply liquid gasket to the figure "a" squarely.

27. Install the rocker cover. Torque the retaining bolts to 17 inch lbs. and then to 74 inch lbs., in the proper sequence.

28. Continue the installation in the reverse order of the removal procedure.

29. Inspect the camshaft sprocket (INT) oil groove.

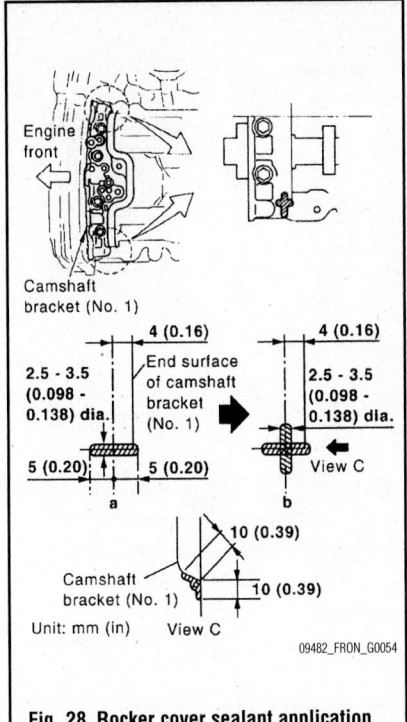

Fig. 28 Rocker cover sealant application locating points

➡**Perform this inspection only when DTC P0011 or DTC P0021 are detected in self diagnostic results of CONSULT-II.**

30. Be sure the engine is cold. Check and adjust oil level, as required.

31. Properly release the fuel system pressure. Disconnect the ignition coil and injector harness connectors.

➡**This is being done to prevent the engine from unintentionally being started while checking.**

32. Remove the intake valve timing control solenoid valve.

33. Crank the engine, and then make sure that engine oil comes out from camshaft bracket (No. 1) oil hole.

✲✲ WARNING

Be careful not to touch rotating parts, (drive belt, idler pulley, crankshaft pulley, etc) as injury could result.

➡**Oil may squirt from the intake valve timing control solenoid valve installation hole during engine cranking. Use a shop towel to prevent oil from squirting on engine components.**

34. Clean the oil groove between the oil strainer and the intake timing control solenoid valve if engine oil does not come out from camshaft bracket (No. 1) oil hole.

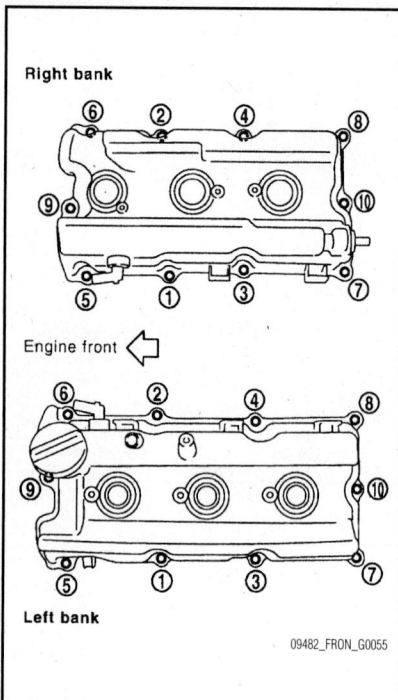

Fig. 29 Rocker cover bolt torque sequence

35. Remove the components between the intake valve timing control solenoid valve and the camshaft sprocket (INT). Check each oil groove for clogging.

36. After inspection install any removed components.

CRANKSHAFT DAMPER (BALANCER)

REMOVAL & INSTALLATION

1. Before servicing the vehicle, refer to the Precautions Section.

2. Disconnect the negative battery cable.

3. Remove the engine undercover.

4. Remove the drive belts.

5. Remove the cooling fan.

6. Loosen the crankshaft pulley retaining bolt and locate the bolt seating surface, which is about 0.39 inch from its original position.

➡**Do not remove the crankshaft pulley bolt. Keep the loosened pulley bolt in place to protect the removed crankshaft pulley from dropping.**

7. Pull the pulley with both hands and remove it from its mounting. Remove the bolt and pulley from the engine.

To install:

8. Installation is the reverse order of the removal procedure.

CRANKSHAFT FRONT SEAL

REMOVAL & INSTALLATION

1. Before servicing the vehicle, refer to the Precautions Section.

2. Disconnect the negative battery cable.

3. Remove the engine undercover.

4. Remove the drive belts.

5. Remove the cooling fan.

6. Loosen the crankshaft pulley retaining bolt and locate the bolt seating surface, which is about 0.39 inch from its original position.

➡**Do not remove the crankshaft pulley bolt. Keep the loosened pulley bolt in place to protect the removed crankshaft pulley from dropping.**

7. Pull the pulley with both hands and remove it from its mounting. Remove the bolt and pulley from the engine.

8. Using a seal removal tool, remove the oil seal from its mounting.

➡**Be careful not to damage the front cover and/or the crankshaft.**

To install:

9. Installation is the reverse order of the removal procedure.

10. Press fit until the height of the front oil seal is level with the mounting surface, using the proper tools.

CYLINDER HEAD

REMOVAL & INSTALLATION

See Figures 30 through 33.

1. Before servicing the vehicle, refer to the Precautions Section.

2. Properly relieve the fuel system pressure.

3. Disconnect the negative battery cable. Drain the cooling system.

4. Remove the camshaft.

5. Remove the intake manifold.

6. Remove the exhaust manifold.

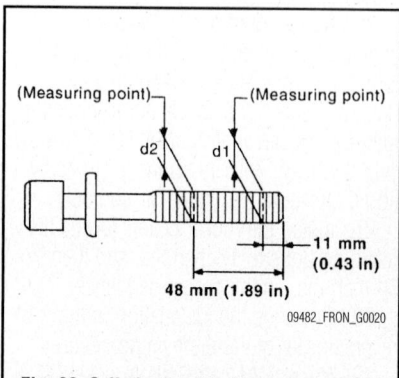

Fig. 30 Cylinder head bolt measurement

7. Remove the water inlet and thermostat assembly.

8. Remove the water outlet, water pipe and heater pipe.

9. Remove the cylinder head retaining bolts. Be sure to remove the bolts by reversing the order of the tightening torque sequence.

10. Remove the cylinder head from the engine. Discard the gasket.

To install:

11. Installation is the reverse of the removal procedure.

12. Be sure to inspect the cylinder head bolts. Replace as required.

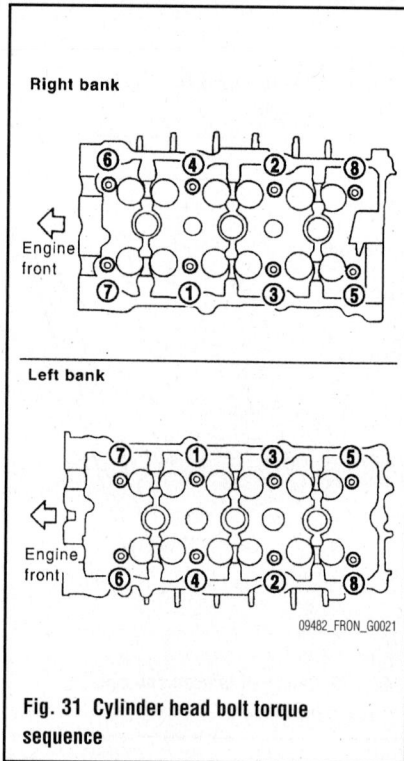

Fig. 31 Cylinder head bolt torque sequence

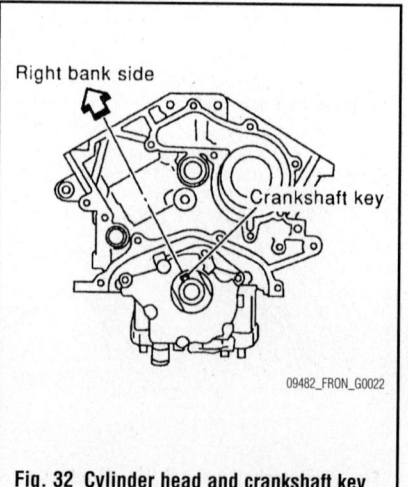

Fig. 32 Cylinder head and crankshaft key alignment

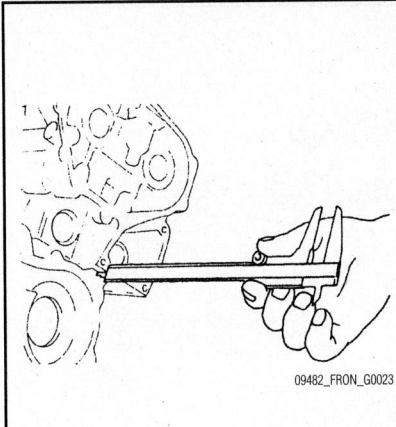

Fig. 33 Cylinder head to cylinder block installation measurement

➡Head bolts are tightened by plastic zone tightening method. Whenever the size difference between "d1" and "d2" exceeds the limit, replace the bolt. "d1"-"d2" limit is 0.0043. If reduction of the outer diameter appears in a position other than "d2", use it the "d2" point.

13. Install the new cylinder head gasket. Turn the crankshaft until the number one piston is at TDC.

➡The crankshaft key should line up with the right bank center line, see illustration.

14. Torque the cylinder head bolts to specification and in the proper sequence.

15. Measure the distance between the front end faces of the cylinder block and the cylinder head on both the left and right banks. If the measured value is not within specification reinstall the cylinder head. Specification is 0.555–0.587 inch.

16. Be sure to fill the cooling system with the proper grade and type engine coolant.

17. Start the engine and check for leaks.

EXHAUST MANIFOLD

REMOVAL & INSTALLATION

Left Side

See Figure 34.

1. Before servicing the vehicle, refer to the Precautions Section.
2. Disconnect the negative battery cable.
3. Remove air duct, PCV hose (between air duct and rocker cover) and electric throttle control actuator.
4. Disconnect the harness connector and remove the heated oxygen sensor.

➡Be careful not to damage the air fuel ratio sensor. Discard the sensor if it has been dropped from a height of more than 19.7 inches on to a hard surface.

5. Remove the front exhaust tube.
6. Remove the exhaust manifold cover.
7. Remove bracket between exhaust manifold–three way catalyst assembly and transmission assembly.
8. Remove the exhaust manifold retaining bolts. Be sure to remove the bolts by reversing the order of the tightening torque sequence.
9. Remove the exhaust manifold from the engine. Discard the gasket.

To install:

10. Installation is the reverse of the removal procedure.
11. Be sure to use new gaskets.
12. Be sure to tighten the exhaust manifold retaining bolts to specification and in the proper sequence.

➡Before installing a new air fuel sensor and heated oxygen sensor, clean threads and apply anti seize lubricant to the threads. Do not over torque the sensor, doing so may cause damage to the sensor resulting in the MIL light coming on.

Right Side

1. Before servicing the vehicle, refer to the Precautions Section.

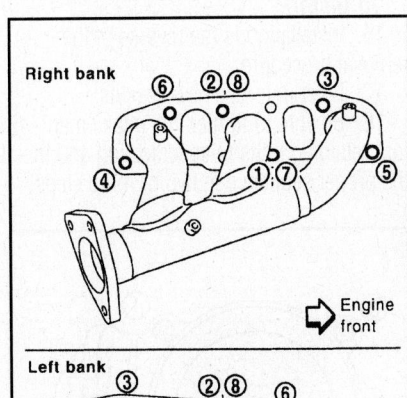

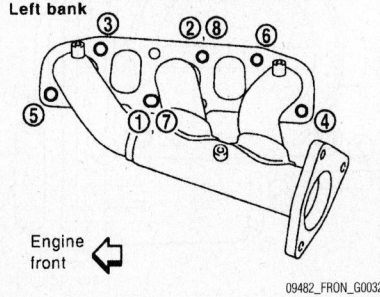

Fig. 34 Exhaust manifold bolt torque sequence

2. Disconnect the negative battery cable.
3. Remove the engine from the vehicle. Position the assembly in a suitable holding fixture.
4. Remove the exhaust manifold retaining bolts. Be sure to remove the bolts by reversing the order of the tightening torque sequence.

➡Disregard the numerical order of No.7 and No.8 in the removal process.

5. Discard the gaskets.

To install:

6. Installation is the reverse of the removal procedure.
7. Be sure to use new gaskets.
8. Be sure to tighten the exhaust manifold retaining bolts to specification and in the proper sequence.

➡Before installing a new air fuel sensor and heated oxygen sensor apply anti seize lubricant to the threads. Do not over torque the sensor, doing so may cause damage to the sensor resulting in the MIL light coming on.

INTAKE MANIFOLD

REMOVAL & INSTALLATION

See Figures 35 through 38.

➡Upper intake manifold is also referred to as intake manifold collector.

1. Before servicing the vehicle, refer to the Precautions Section.
2. Properly relieve the fuel system pressure.
3. Disconnect the negative battery cable. Drain the cooling system.
4. Remove the engine cover. Remove the air cleaner case (upper) with the mass air flow sensor and air duct assembly.
5. Disconnect the water hoses from the electric throttle control actuator. Disconnect the harness connector.
6. Remove the electric throttle control actuator retaining bolts. Be sure to remove the bolts by reversing the order of the tightening torque sequence.
7. Remove the electric throttle control actuator.
8. Remove the brake booster vacuum hose and the PCV hose. Remove the intake manifold collector support.
9. Disconnect the EVAP hoses and harness connector from the EVAP canister purge volume control solenoid valve. Remove the EVAP canister purge volume control solenoid valve.
10. Remove the VIAS control solenoid valve and vacuum tank.

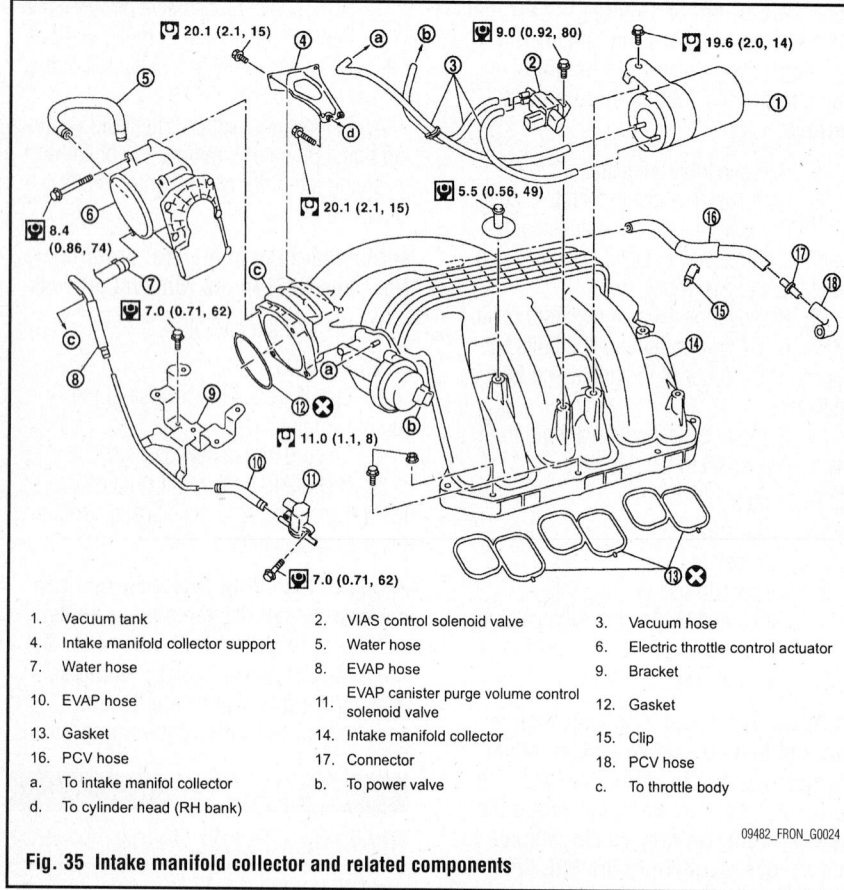

1. Vacuum tank
2. VIAS control solenoid valve
3. Vacuum hose
4. Intake manifold collector support
5. Water hose
6. Electric throttle control actuator
7. Water hose
8. EVAP hose
9. Bracket
10. EVAP hose
11. EVAP canister purge volume control solenoid valve
12. Gasket
13. Gasket
14. Intake manifold collector
15. Clip
16. PCV hose
17. Connector
18. PCV hose
a. To intake manifol collector
b. To power valve
c. To throttle body
d. To cylinder head (RH bank)

09482_FRON_G0024

Fig. 35 Intake manifold collector and related components

11. Remove the intake manifold collector retaining bolts. Be sure to remove the bolts by reversing the order of the tightening torque sequence.

12. Remove the intake manifold collector from the engine.

13. Remove the fuel tube and fuel injector assembly.

14. Remove the intake manifold retaining bolts. Be sure to remove the bolts by reversing the order of the tightening torque sequence.

15. Remove the intake manifold from the engine.

To install:

16. Installation is the reverse of the removal procedure.

17. Be sure to use new gaskets.

18. Be sure to tighten the intake manifold retaining bolts to specification and in the proper sequence in two or more steps.

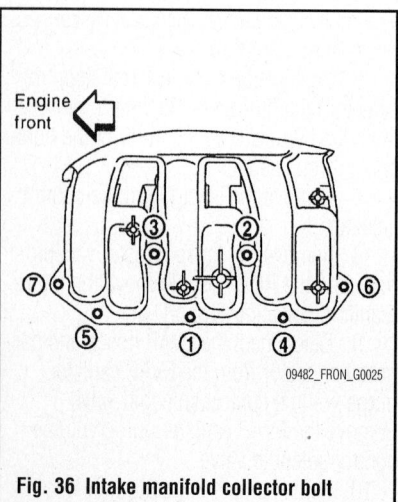

09482_FRON_G0025

Fig. 36 Intake manifold collector bolt torque sequence

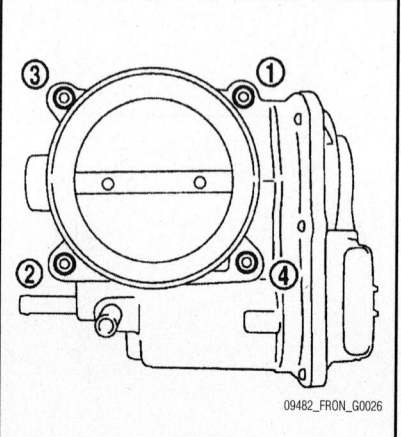

09482_FRON_G0026

Fig. 37 Electric throttle control actuator bolt torque sequence

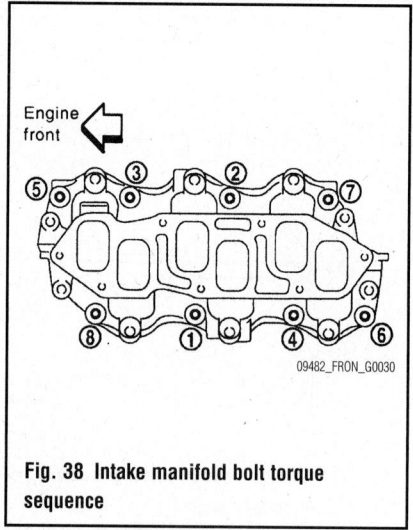

09482_FRON_G0030

Fig. 38 Intake manifold bolt torque sequence

19. Be sure to tighten the intake manifold collector retaining bolts to specification and in the proper sequence.

20. Be sure to tighten the electric throttle control actuator retaining bolts to specification and in the proper sequence.

➡See throttle valve closed position learning and idle air volume learning procedures, for relearning information.

OIL PAN

REMOVAL & INSTALLATION

Lower

See Figure 39.

1. Before servicing the vehicle, refer to the Precautions Section.

2. Disconnect the negative battery cable.

3. Remove the engine under cover. Drain the engine oil.

4. Loosen the oil pan retaining bolts, in the reverse order of the installation sequence.

5. Insert a seal cutter tool between the oil pan and the cylinder block, and slide it by tapping on the side of the tool with a hammer.

6. Remove the oil pan from the engine.

To install:

7. Be sure to clean all the oil gasket material from both the oil pan and the cylinder block surfaces, using the proper tools.

8. Apply a continuous bead of sealant 0.138–0.177 inches (3.5–4.5 mm) to the oil pan mating surface. Be sure to use genuine RTV sealant or equivalent.

9. Install the oil pan to the cylinder block. This must be done within 5 minutes after applying the liquid gasket.

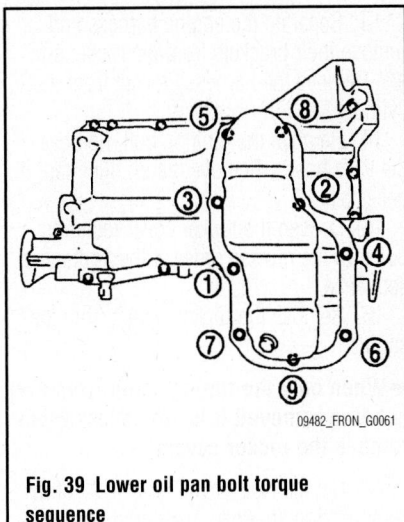

Fig. 39 Lower oil pan bolt torque sequence

10. Torque the bolts to specification and in the proper sequence.

11. Continue the installation in the reverse order of the removal procedure.

➡**Wait 30 minutes after installation of the oil pan to allow the sealant to cure before adding oil.**

12. Fill the crankcase to the correct level.

13. Start the engine and check for leaks.

Upper

See Figures 40 and 41.

1. Before servicing the vehicle, refer to the Precautions Section.

2. Disconnect the negative battery cable.

3. Remove the air duct. Remove the engine under cover.

4. Drain the engine oil. Drain the engine coolant.

5. Remove the final drive, if equipped with 4WD.

6. Disconnect the steering gear lower shaft joint bolt and steering gear nuts and bolts, position the assembly out of the way.

7. Remove the starter.

8. Disconnect the automatic transmission fluid cooler brackets, if equipped and position them out of the way.

9. Remove the oil filter, as necessary. Remove the oil cooler.

10. Remove the lower oil pan. Remove the oil strainer.

11. Remove the transmission joint bolts which pierce the oil pan.

12. Remove the rear cover plate.

13. Loosen the upper oil pan retaining bolts, in the reverse order of the installation sequence.

14. Insert a seal cutter tool between the oil pan and the cylinder block, and slide it by tapping on the side of the tool with a hammer.

15. Remove the oil pan from the engine. Remove the O-rings from the bottom lower cylinder block and oil pump.

To install:

16. Be sure to clean all the oil gasket material from both the oil pan and the cylinder block surfaces, using the proper tools.

17. Install new O-rings on the bottom lower cylinder block and oil pump.

18. Apply a continuous bead of sealant 0.138–0.177 inches (3.5–4.5 mm) to the lower cylinder block mating surfaces of the upper oil pan. Be sure to use genuine RTV sealant or equivalent.

➡**For bolt holes marked with a solid black triangle, apply liquid gasket outside the hole. Apply a bead of sealant (0.177–0.217 inch diameter) to area "A".**

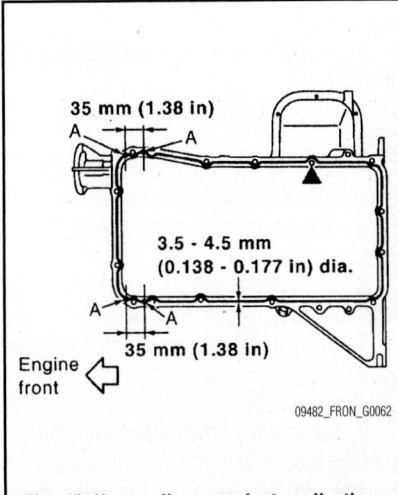

Fig. 40 Upper oil pan sealant application

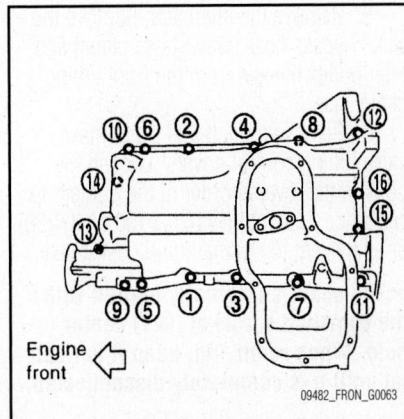

Fig. 41 Upper oil pan bolt torque sequence

19. Install the upper oil pan. This must be done within 5 minutes after applying the liquid gasket.

20. Torque the bolts to specification and in the proper sequence. There are two types of bolts M8X100 mm (3.97 inch) bolts 7, 11, 12, 13 and M8X25 mm (0.98 inch) except 7, 11, 12 and 13.

21. Tighten the transmission joint bolts.

22. Install the oil strainer to the upper oil pan.

23. Continue the installation in the reverse order of the removal procedure.

➡**Wait 30 minutes after installation of the oil pan to allow the sealant to cure before adding oil.**

24. Fill the crankcase to the correct level.

25. Start the engine and check for leaks.

OIL PUMP

REMOVAL & INSTALLATION

See Figure 42.

1. Before servicing the vehicle, refer to the Precautions Section.

2. Disconnect the negative battery cable. Drain the engine oil. Drain the engine coolant.

3. Remove the lower oil pan.

4. Remove the upper oil pan.

5. Remove the front timing chain case and timing chain (primary).

6. Remove the oil pump from the engine.

To install:

7. Installation is the reverse of removal procedure.

➡**Wait 30 minutes after installation of the oil pan to allow the sealant to cure before adding oil.**

8. Fill the crankcase to the correct level.

9. Start the engine and check for leaks.

PISTON AND RING

POSITIONING

See Figure 43.

REAR MAIN SEAL

REMOVAL & INSTALLATION

1. Before servicing the vehicle, refer to the Precautions Section.

2. Remove or disconnect the following:
   - Transmission
   - Flywheel
   - Clutch, if equipped
   - Rear main seal

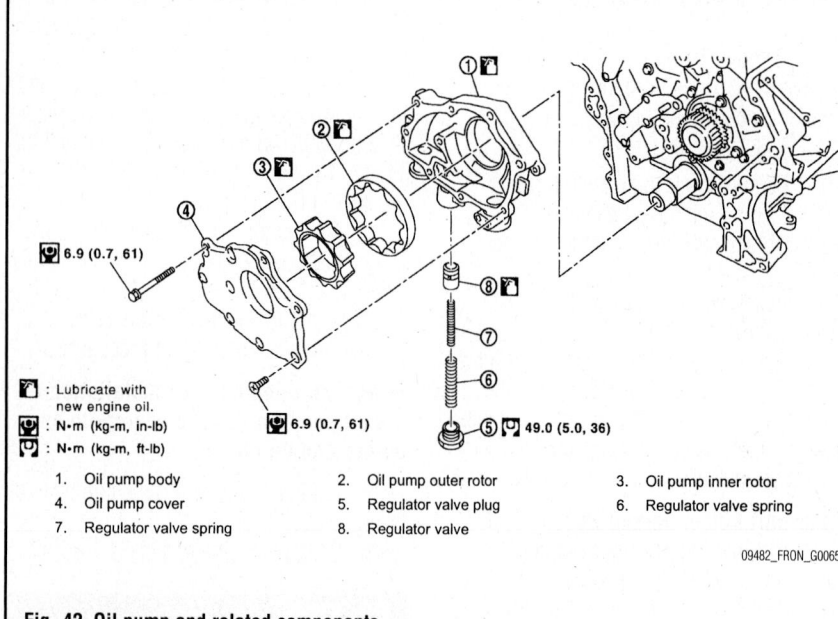

6.9 (0.7, 61)

: Lubricate with new engine oil.

N·m (kg-m, in-lb)

N·m (kg-m, ft-lb)

6.9 (0.7, 61)

49.0 (5.0, 36)

1. Oil pump body
2. Oil pump outer rotor
3. Oil pump inner rotor
4. Oil pump cover
5. Regulator valve plug
6. Regulator valve spring
7. Regulator valve spring
8. Regulator valve

09482_FRON_G0065

Fig. 42 Oil pump and related components

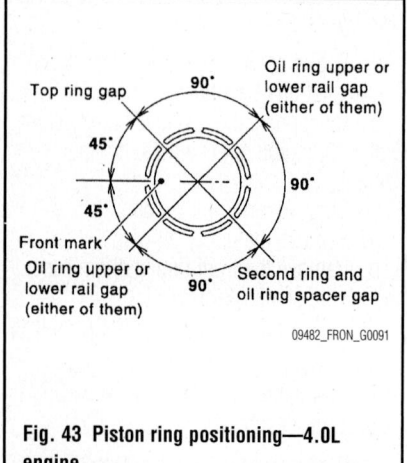

09482_FRON_G0091

Fig. 43 Piston ring positioning—4.0L engine

To install:

3. Install the seal so that it is flush with the retainer housing.

4. Install or connect the following:
- Flywheel.
- Transmission

TIMING CHAIN & SPROCKETS

REMOVAL & INSTALLATION

See Figures 44 through 59.

➡**The procedure below describes the removal and installation of the front timing case and timing chain related parts and rear timing chain case, when the upper oil pan needs to be removed or installed. When only the timing chain (primary) is being removed it is not necessary to remove the rocker covers.**

1. Before servicing the vehicle, refer to the Precautions Section.

2. Properly relieve the fuel system pressure.

3. Disconnect the negative battery cable.

4. Remove the engine cover. Drain the engine oil. Drain the engine coolant.

5. Remove the upper and lower oil pans.

6. Remove the radiator cooling fan assembly. Remove the drive belts.

7. Separate the engine wiring harnesses by removing their brackets from the front timing chain case.

8. Remove the power steering pump from the bracket with the fluid hoses attached. Position the assembly to the side. Do not disconnect the hoses. Remove the bracket.

9. Remove the alternator. Remove the water bypass hose, water hose clamp and idler pulley bracket from the front timing chain case.

10. Remove the left and right intake valve timing control covers. Loosen the bolts in the reverse order of the tightening sequence. Use tool KV10111100 (J-37228) or equivalent to cut the liquid gasket seal.

➡**The shaft is internally jointed with the camshaft sprocket (INT) center hole. When removing, keep it horizontal until it is completely disconnected.**

11. Remove the collared O-rings from the front timing chain case on both the left and right side.

12. Remove the intake manifold collector.

13. Separate the engine harness and remove their brackets from the rocker covers. Remove the harness bracket from the cylinder head, if necessary.

14. Remove the ignition coil. Remove the PCV hoses. Remove the oil filler cap, if necessary.

15. Loosen the rocker cover retaining bolts, in the reverse order of the tightening sequence.

16. Remove the rocker covers from the engine.

➡**When only the timing chain (primary) is being removed it is not necessary to remove the rocker covers.**

17. Set the No. 1 cylinder at TDC of its compression stroke by rotating the crankshaft pulley clockwise to align the timing mark (grooved line without color) with the timing indicator. Make sure that the intake and exhaust cam noses on No. 1 cylinder (engine front side on right bank) are in alignment as shown in the illustration. If not, rotate the crankshaft in the clockwise direction 360 degrees.

➡**When only the timing chain (primary) is removed, the rocker cover does not need to be removed. To be sure that the No. 1 cylinder is set at TDC on the compression stroke, remove the front timing chain case cover first, then check the mating marks on the camshaft sprockets.**

18. Remove the starter. Position tool KV10117700 (J-44716) or equivalent.

19. Loosen the crankshaft pulley retaining bolt and locate the bolt seating surface, which is about 0.39 inch from its original position.

➡**Do not remove the crankshaft pulley bolt. Keep the loosened pulley bolt in place to protect the removed crankshaft pulley from dropping.**

20. Pull the pulley with both hands and remove it from its mounting. Remove the bolt and pulley from the engine.

21. Loosen and remove the two bolts of the upper oil pan.

22. Loosen the front timing chain cover retaining bolts in the reverse order of the tightening sequence.

23. Insert a suitable tool in the notch at the top of the front timing chain case and pry off the case by moving the tool as shown in the illustration. Use tool KV10111100 (J-37228) or equivalent to cut the liquid gasket seal.

➡**Do not use a screwdriver or something similar. After removal handle the**

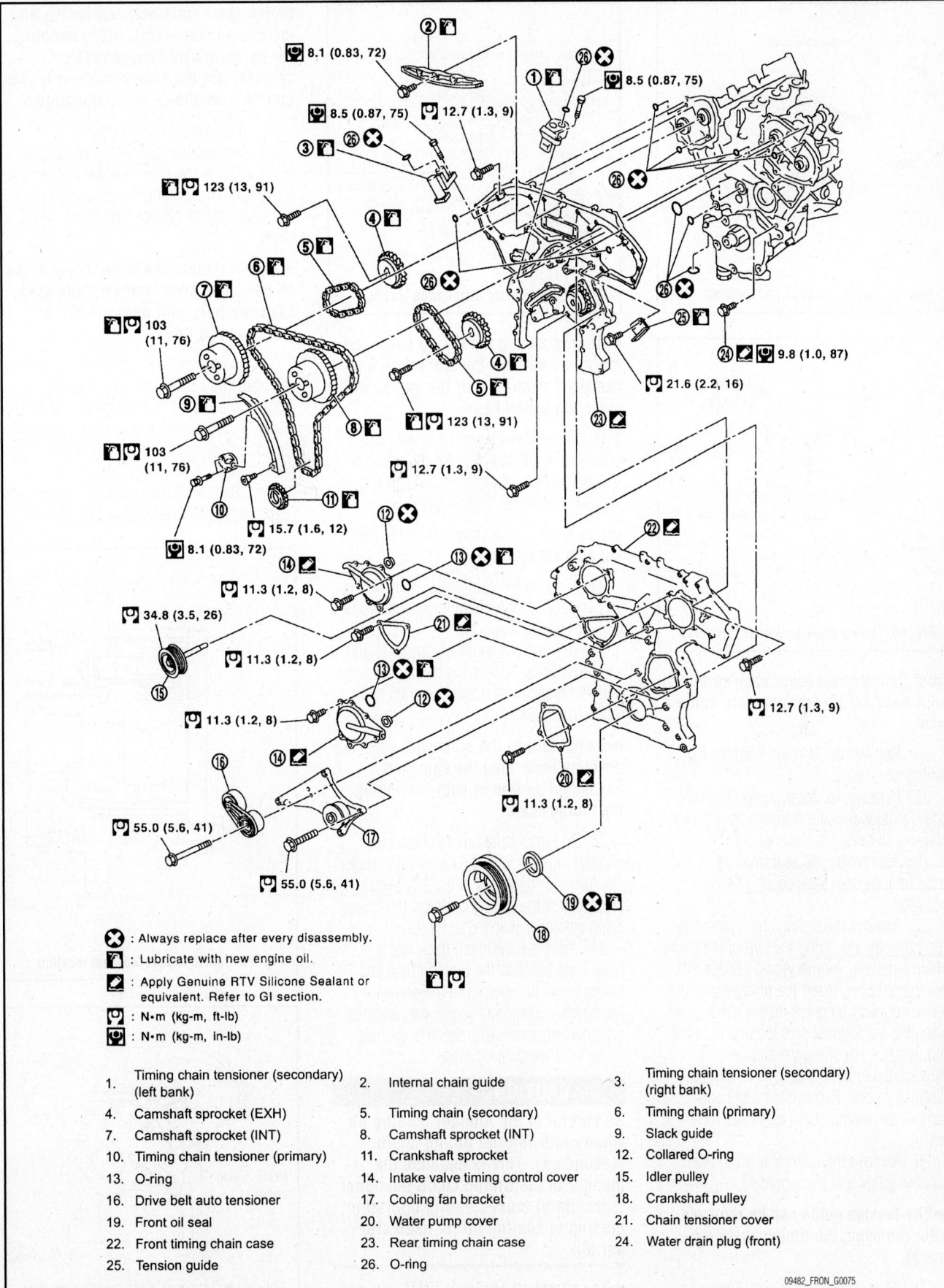

: Always replace after every disassembly.

: Lubricate with new engine oil.

: Apply Genuine RTV Silicone Sealant or equivalent. Refer to GI section.

: N·m (kg-m, ft-lb)

: N·m (kg-m, in-lb)

1.	Timing chain tensioner (secondary) (left bank)	
2.	Internal chain guide	
3.	Timing chain tensioner (secondary) (right bank)	
4.	Camshaft sprocket (EXH)	
5.	Timing chain (secondary)	
6.	Timing chain (primary)	
7.	Camshaft sprocket (INT)	
8.	Camshaft sprocket (INT)	
9.	Slack guide	
10.	Timing chain tensioner (primary)	
11.	Crankshaft sprocket	
12.	Collared O-ring	
13.	O-ring	
14.	Intake valve timing control cover	
15.	Idler pulley	
16.	Drive belt auto tensioner	
17.	Cooling fan bracket	
18.	Crankshaft pulley	
19.	Front oil seal	
20.	Water pump cover	
21.	Chain tensioner cover	
22.	Front timing chain case	
23.	Rear timing chain case	
24.	Water drain plug (front)	
25.	Tension guide	
26.	O-ring	

1. Timing chain tensioner (secondary) (left bank)
2. Internal chain guide
3. Timing chain tensioner (secondary) (right bank)
4. Camshaft sprocket (EXH)
5. Timing chain (secondary)
6. Timing chain (primary)
7. Camshaft sprocket (INT)
8. Camshaft sprocket (INT)
9. Slack guide
10. Timing chain tensioner (primary)
11. Crankshaft sprocket
12. Collared O-ring
13. O-ring
14. Intake valve timing control cover
15. Idler pulley
16. Drive belt auto tensioner
17. Cooling fan bracket
18. Crankshaft pulley
19. Front oil seal
20. Water pump cover
21. Chain tensioner cover
22. Front timing chain case
23. Rear timing chain case
24. Water drain plug (front)
25. Tension guide
26. O-ring

Fig. 44 Timing chain and related components

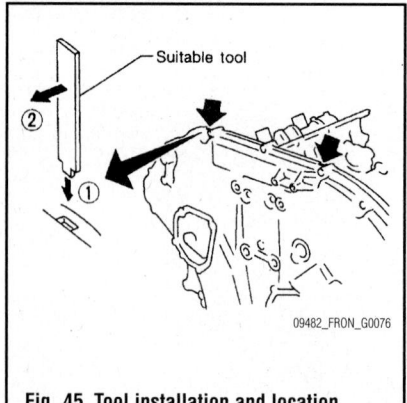

Fig. 45 Tool installation and location

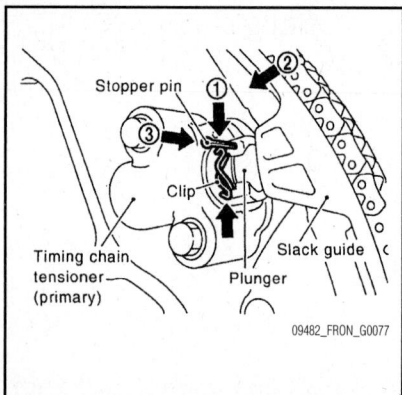

Fig. 46 Timing chain tensioner (primary)

front timing chain cover case carefully so it does not tilt, cant or warp under a load.

24. Remove the O-rings from the rear timing chain case.

25. Remove the water pump cover and chain tensioner cover from the front timing chain case cover, as required.

26. Remove the oil seal from the front timing chain case cover, as required.

27. Remove the timing chain tensioner (primary) by loosening the clip of the timing chain tensioner (primary) and release the plunger stopper. Insert the plunger into the tensioner body by pressing the slack guide. Keep the slack guide pressed and hold the plunger in by pushing the stopper pin through the tensioner body hole and the plunger groove. Remove the bolts and remove the timing chain tensioner (primary).

28. Remove the internal chain guide, tension guide and slack guide.

➡The tension guide can be removed after removing the timing chain (primary).

29. Remove the timing chain (primary) and the crankshaft sprocket.

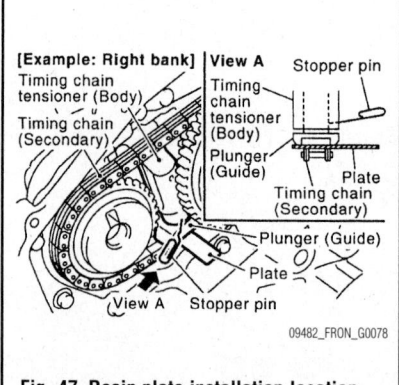

Fig. 47 Resin plate installation location

➡After removing the timing chain (primary), do not turn the crankshaft and camshaft separately or the valves will strike the piston heads.

30. To remove the timing chain (secondary) and camshaft sprockets, attach a suitable stopper pin to the right and left timing chain tensioner (secondary).

➡Use a 0.02 inch (approximate) metal pin as a stopper pin.

31. Remove the camshafts. Remove the valve lifters. Identify them for reinstallation in their original locations.

32. Remove the camshaft sprocket (INT and EXH) bolts. Secure the hexagonal portion of the camshaft using a wrench to loosen the bolts.

➡Do not loosen the bolts with securing anything other than the camshaft hexagonal portion or with tensioning the timing chain.

33. To remove the timing chain (secondary) together with the camshaft sprockets, turn the crankshaft slightly to secure slackness of the timing chain on the timing chain tensioner (secondary) side.

34. Insert a 0.020 inch thick metal or resin plate between the timing chain and timing chain plunger (guide). Remove the timing chain (secondary) together with the camshaft sprockets with the timing chain loose from the guide groove.

✳✳ CAUTION

Be careful of the plunger coming off when removing the timing chain (secondary). This is because the plunger of the timing chain tensioner (secondary) moves during operation, leading to coming off its fixed stopper pin.

➡The camshaft sprocket (INT) is a one piece integrated design sprocket for the

timing chain (primary) and for the timing chain (secondary). When handling the sprocket avoid shock to the sprocket. Do not disassemble or loosen bolt "A", as shown in the illustration.

35. Remove the water pump.

36. Remove the rear timing chain case cover bolts, in the reverse order of the tightening sequence. Using the proper tool, cut the liquid gasket sealant seal. Remove the cover.

➡Do not remove the metal cover of the oil passage. After removal, handle the case carefully so it does not tilt, or warp under a load.

37. Remove the O-rings from the cylinder head and No. 1 camshaft bracket. Remove the O-rings from the cylinder block.

38. If necessary, remove the timing chain tensioners (secondary) from the cylinder head by first removing the No. 1 camshaft bracket. Remove the timing chain tensioners (secondary) with the stopper pin attached.

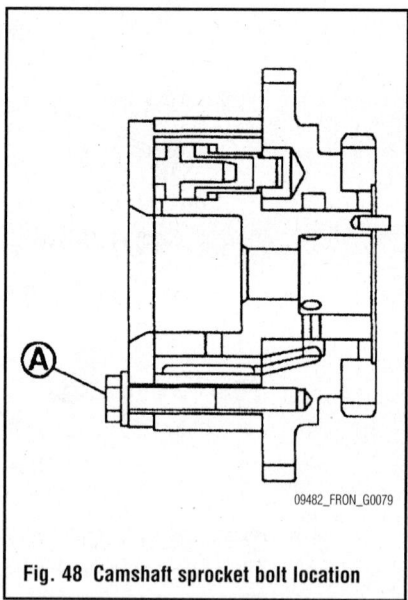

Fig. 48 Camshaft sprocket bolt location

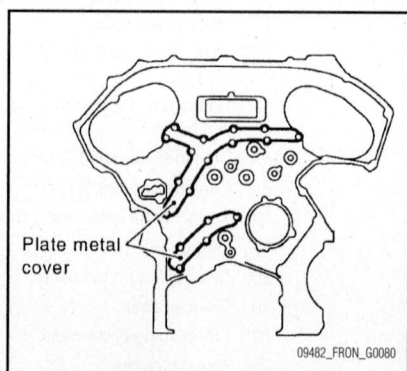

Fig. 49 Metal cover plate location on rear timing case cover

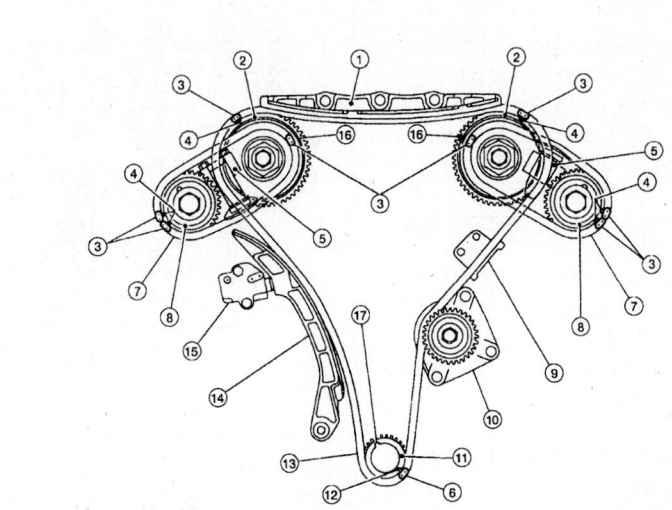

1. Internal chain guide
4. Mating mark (punched)
7. Secondary timing chain
10. Water pump
13. Primary timing chain
16. Mating mark (back side)

2. Camshaft sprocket (intake)
5. Secondary timing chain tensioner
8. Camshaft sprocket (exhaust)
11. Crankshaft sprocket
14. Slack guide
17. Crankshaft key

3. Mating mark (copper link)
6. Mating mark (yellow link)
9. Tensioner guide
12. Mating mark (notched)
15. Primary timing chain tensioner

09482_FRON_G0081

Fig. 50 Timing chain alignment

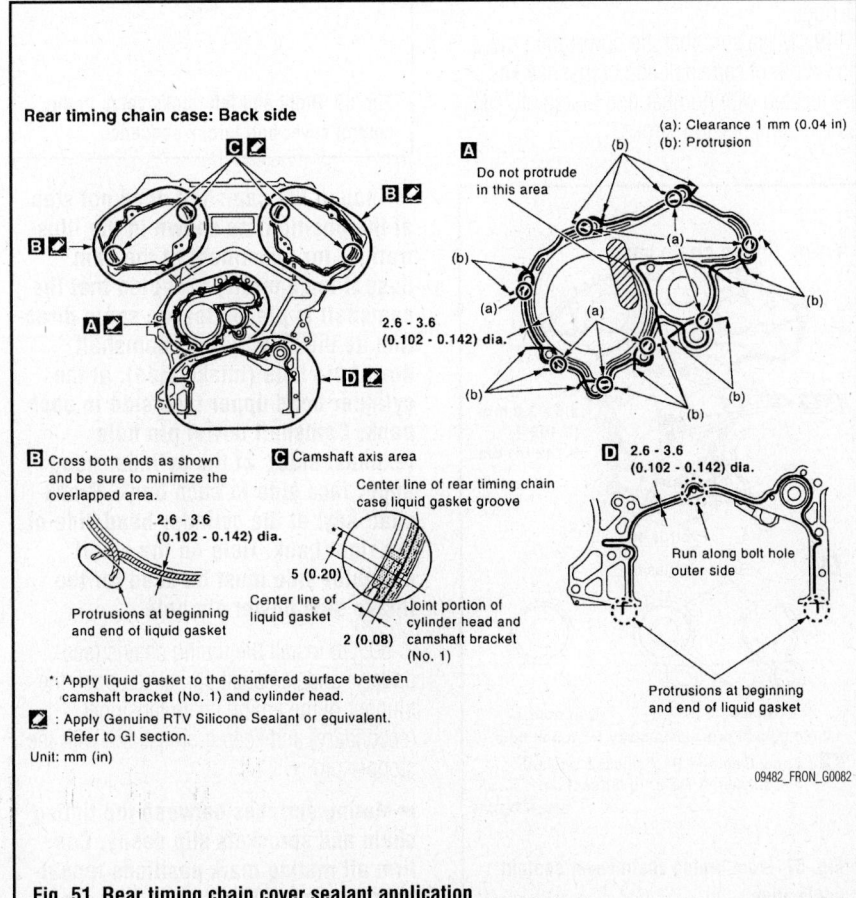

Rear timing chain case: Back side

(a): Clearance 1 mm (0.04 in)
(b): Protrusion

A Do not protrude in this area

2.6 - 3.6 (0.102 - 0.142) dia.

B Cross both ends as shown and be sure to minimize the overlapped area.

2.6 - 3.6 (0.102 - 0.142) dia.

Protrusions at beginning and end of liquid gasket

C Camshaft axis area

Center line of rear timing chain case liquid gasket groove

5 (0.20)

Center line of liquid gasket

2 (0.08)

Joint portion of cylinder head and camshaft bracket (No. 1)

D 2.6 - 3.6 (0.102 - 0.142) dia.

Run along bolt hole outer side

Protrusions at beginning and end of liquid gasket

*: Apply liquid gasket to the chamfered surface between camshaft bracket (No. 1) and cylinder head.

: Apply Genuine RTV Silicone Sealant or equivalent. Refer to GI section.

Unit: mm (in)

09482_FRON_G0082

Fig. 51 Rear timing chain cover sealant application

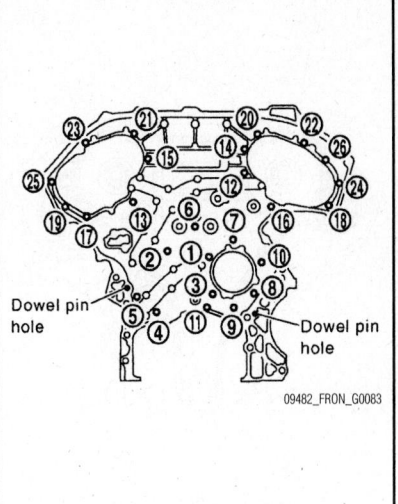

Dowel pin hole

Dowel pin hole

09482_FRON_G0083

Fig. 52 Rear timing chain cover bolt torque sequence

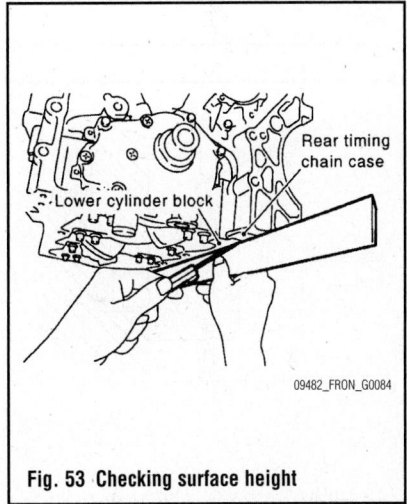

Rear timing chain case

Lower cylinder block

09482_FRON_G0084

Fig. 53 Checking surface height

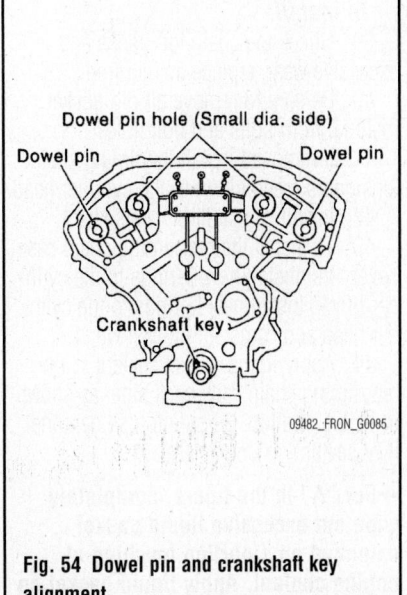

Dowel pin hole (Small dia. side)

Dowel pin

Dowel pin

Crankshaft key

09482_FRON_G0085

Fig. 54 Dowel pin and crankshaft key alignment

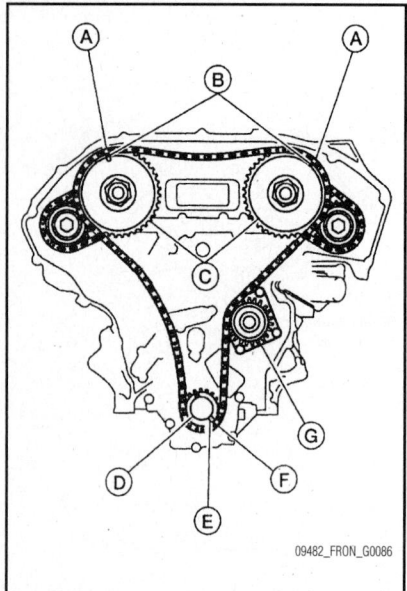

Fig. 55 Timing chain (primary) alignment

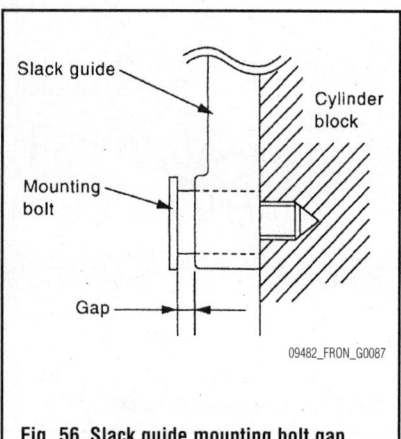

Fig. 56 Slack guide mounting bolt gap

To install:

39. Check the chain for cracks and excessive wear, replace as required.

40. Be sure to remove all old gasket material from bolts and bolt holes.

41. If removed install the timing chain tensioners (secondary) to the cylinder head.

42. Install camshaft brackets No. 1.

43. To install the rear timing chain case cover, first install new O-rings to the cylinder block, Install new O-rings to the cylinder head and camshaft bracket No. 1.

44. Apply liquid gasket sealant to the rear timing chain case back side, as shown in the illustration. Be sure to use genuine RTV sealant, or equivalent.

➡For "A" in the figure, completely wipe out excessive liquid gasket extended on a portion touching at engine coolant. Apply liquid gasket on the installation position of the water

pump and cylinder head very completely.

45. Align the rear timing case with dowel pins (right and left) on the cylinder block. Install the rear timing chain case. Make sure that the O-rings stay in place during installation to the cylinder block, cylinder head and camshaft bracket No. 1.

46. Tighten the bolts to specification and in the proper sequence.

    a. Bolt length: 0.79 inch. Bolt position: 1,2,3,6,7,8,9,and 10.

    b. Bolt length: 0.63 inch. Bolt position: except 1,2,3,6,7,8,9,and 10.

    c. Torque bolts to 9 ft. lbs.

    d. After all bolts are tightened, retighten them to specification and in the proper sequence

➡Be sure to wipe off any excess liquid gasket leaking to the surface for installing the oil pan.

47. After installing the rear timing case, check the surface height deference between the rear timing chain case and the lower cylinder block. Specification should be -0.0094–0.0055 inch. If not within specification, repeat the installation procedure.

48. Install the water pump, using new O-rings.

49. Make sure that the dowel pin hole, dowel pin of camshaft and crankshaft key are located with number one piston at TDC on the compression stroke.

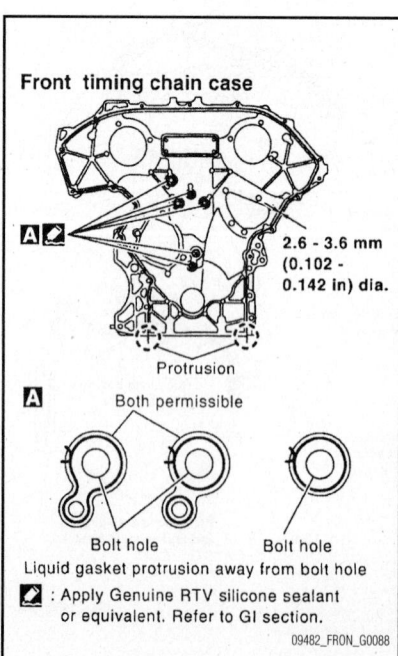

Fig. 57 Front timing chain cover sealant application

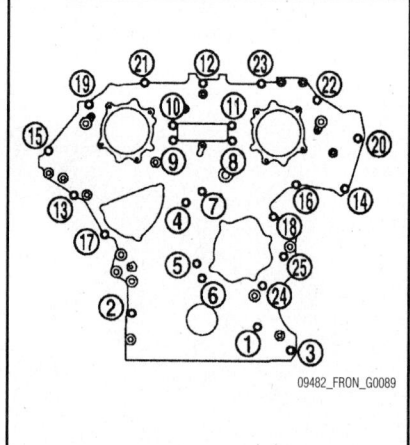

Fig. 58 Front timing chain cover bolt torque sequence

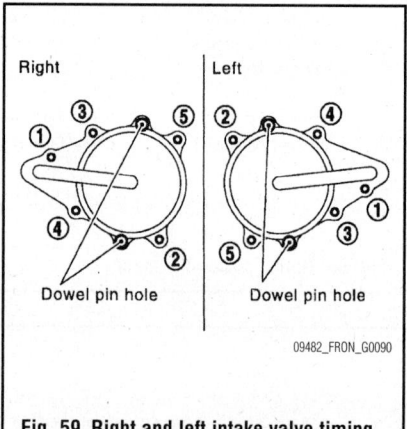

Fig. 59 Right and left intake valve timing control cover bolt torque sequence

➡Though the camshaft does not stop at the position, as shown in the illustration, for placement of the cam nose it is generally accepted that the camshaft is placed for the same direction as the illustration. Camshaft dowel pin hole (intake side): at the cylinder head upper face side in each bank. Camshaft dowel pin hole (exhaust side): at the cylinder head upper face side in each bank. Crankshaft key: at the cylinder head side of the right bank. Hole on the small diameter side must be used for the intake side dowel pin hole.

50. To install the timing chains (secondary) and camshaft sprockets, push the plunger of the timing chain tensioner (secondary) and keep it pressed in with the stopper pin.

➡Mating surfaces between the timing chain and sprockets slip easily. Confirm all mating mark positions repeatedly during the installation process.

51. Install the timing chains (secondary) and camshaft sprockets (INT and EXH).

52. Align the mating marks on the timing chain (secondary) cooper color link, with the ones on the camshaft sprockets (INT and EXH) punched and install them.

➡ **Mating marks for the camshaft sprocket (INT) are on the back side of the camshaft sprocket (secondary). There are two types of mating marks, circle and oval. They should be used for the right and the left banks, respectively. Right bank: circle type. Left bank: oval type.**

53. Align the dowel pin and pin hole on the camshafts with the groove and the dowel pin on the sprockets, and install them.

54. On the exhaust side, align the pin hole on the small diameter side of the camshaft front end with the dowel pin on the back side of the camshaft sprocket, and install them.

55. On the exhaust side, align the dowel pin on the camshaft front end with the pin groove on the camshaft sprocket, and install them.

➡ **In case that the positions of each mating mark and each dowel pin will not fit on the mating marks, make a fine adjustment to the position holding the hexagonal portion on the camshaft with a wrench, or equivalent.**

➡ **Bolts for the camshaft sprockets must be tightened. Tightening them by hand is enough to prevent the dislocation of the dowel pins. It may be difficult to visually check the dislocation of mating marks during and after installation. To make the matching easier, make a mating mark on the top of the sprocket teeth and its extended line in advance with paint.**

56. After confirming that the mating marks are aligned, tighten the camshaft sprocket bolts.

57. Pull the stopper pins out from the timing chain tensioners (secondary). Install the tension guide.

58. To install the timing chain (primary), install the crankshaft sprocket. Be sure that the mating marks on the crankshaft sprocket face the front of the engine.

59. Install the timing chain (primary).

➡ **Install the timing chain (primary) so that the mating mark punched on the camshaft sprocket is aligned with the copper link on the timing chain, while the mating mark notched on the crankshaft sprocket is aligned with the yellow link on the timing chain, as shown in the illustration. If it is difficult to align mating marks with and with of the timing chain (primary) with each sprocket, gradually turn the camshaft using a wrench on the hexagonal portion to align it with the timing marks. During alignment be careful to prevent dislocation of the mating marks alignments of the timing chains (secondary). Note indicates the water pump.**

60. Install the internal chain guide, slack guide and timing chain tensioner (primary).

➡ **Do not over tighten the slack guide bolts. It is normal for a gap to exist under the bolt seats when the bolts are tightened to specification.**

61. When installing the timing chain tensioner (primary), push in the plunger and keep it pressed in with the stopper pin. Remove any dirt on the surfaces. After installation, pull out the stopper pin by pressing the slack guide.

62. Make sure, again, that the mating marks on the camshaft sprockets and timing chain have not slipped out of alignment. Install new O-rings on the rear timing chain case.

63. Install a new front seal in the front timing chain case cover.

64. Install the water pump cover and chain tensioner cover to the front timing chain case cover. Apply a continuous bead of liquid gasket (0.091–0.130 inch diameter) to the front timing chain case cover before installing the water pump cover and chain tensioner cover. Be sure to use genuine RTV sealant, or equivalent.

65. Before installing the front timing chain case cover apply a continuous bead of liquid gasket (0.102–0.142 inch in diameter) to the front timing chain case back side, as shown in the illustration. Be sure to use genuine RTV sealant or equivalent.

66. Install new O-rings on the rear timing chain case. To assemble the front timing chain case cover, fit the lower end of the front timing chain case tightly onto the top face of the oil pan (upper). From the fitting point, make entire front timing chain case contact rear timing chain case completely.

➡ **Since the front timing chain case cover is offset for difference of holt holes; tighten the bolts temporarily while holding the front timing chain case cover from the front and the top. Now insert a dowel pin while holding the front timing chain case cover from the front and the top.**

67. Once the cover is installed, torque the retaining bolts to specification and in the proper sequence. There are four different types of bolts:

a. Bolt diameter: 0.39 inch. Bolt position: 1–5. Torque to 41 ft. lbs. (56 Nm).

b. Bolt diameter: 0.24 inch. Bolt position: 6–25. Torque to 9 ft. lbs. (12 Nm).

c. After all bolts are tightened, retighten them to specification and in the proper sequence.

68. Install the two bolts in the oil pan (upper). Torque to 16 ft. lbs.

69. Install new seal rings in the shaft grooves of the right and left intake valve timing control covers.

70. Apply a continuous bead of liquid gasket (0.083–0.122 inch in diameter) to the covers. Be sure to use genuine RTV sealant or equivalent.

71. Install new collared O-rings in the front timing chain case oil hole (left and right sides). Be careful not to move the seal ring from the installation groove, align the dowel pins on the front timing chain case with the holes to install the intake valve timing control covers.

72. Tighten the bolts in sequence and to specification.

73. Install the crankshaft pulley. Torque to specification.

74. Install the upper and lower oil pans.

75. Install the intake manifold collector.

76. Before installing the rocker cover, apply liquid gasket, be sure to use genuine RTV silicone sealant or equivalent, to the positions shown in the illustration. Refer to figure "a" to apply liquid gasket to joint part of camshaft bracket No. 1 and cylinder head. Refer to figure "b" to apply liquid gasket to the figure "a" squarely.

77. Install the rocker cover. Torque the retaining bolts to 17 inch lbs. and then to 74 inch lbs., in the proper sequence.

78. Continue the installation in the reverse order of the removal procedure.

➡ **If hydraulic pressure inside the timing chain tensioner drops after removal/installation, slack in the guide may generate a pounding noise during and just after engine start. This is normal the noise will stop after hydraulic pressure rises.**

VALVE LASH

ADJUSTMENT

See Figures 60 through 63.

1. Before servicing the vehicle, refer to the Precautions Section.

2. Disconnect the negative battery cable. Remove the engine under cover.

3. Remove the intake manifold collector.

4. Separate the engine harness and remove their brackets from the rocker covers. Remove the harness bracket from the cylinder head, if necessary.

5. Remove the ignition coil. Remove the PCV hoses. Remove the oil filler cap, if necessary.

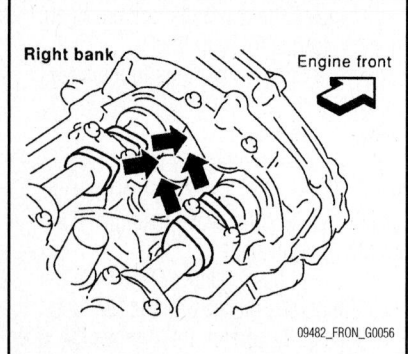

Fig. 60 No. 1 cylinder at TDC (compression stroke)—4.0L engine

6. Loosen the rocker cover retaining bolts, in the reverse order of the tightening sequence.

7. Remove the rocker covers from the engine.

8. Set the No. 1 cylinder at TDC of its compression stroke by rotating the crankshaft pulley clockwise to align the timing mark (grooved line without color) with the timing indicator. Make sure that the intake and exhaust cam noses on No. 1 cylinder (engine front side on right bank) are in alignment as shown in the illustration. If not, rotate the crankshaft in the clockwise direction 360 degrees.

9. Use a feeler gauge and measure the clearance between the valve lifter and the camshaft.

10. With the No. 1 piston at TDC, refer to the illustration and measure the valve clearances at the locations marked with an "X". The "X" locations are indicated in the illustration with an arrow.

11. Rotate the crankshaft pulley clockwise 240 degrees (when viewed from the engine front) to align No. 3 cylinder at TDC on the compression stroke.

➡**The crankshaft pulley bolt flange has a stamped line every 60 degrees, which can be used as a guide to rotation angle.**

12. With the No. 3 piston at TDC, refer to the illustration and measure the valve clearances at the locations marked with an "X". The "X" locations are indicated in the illustration with an arrow.

13. Rotate the crankshaft pulley clockwise 240 degrees (when viewed from the engine front) to align No. 5 cylinder at TDC on the compression stroke.

➡**The crankshaft pulley bolt flange has a stamped line every 60 degrees, which can be used as a guide to rotation angle.**

14. With the No. 5 piston at TDC, refer to the illustration and measure the valve clearances at the locations marked with an "X". The "X" locations are indicated in the illustration with an arrow.

15. If measurements are not within specification, proceed to the next step.

16. Remove the camshaft. Remove the valve lifters that are not within specification.

17. Measure the center thickness of the removed lifters, using a micrometer.

18. Use the equation $(t=t1+(C1-C2)$ to calculate valve lifter thickness for replacement.

➡**t= valve lifter thickness to be replaced. t1= removed valve lifter thickness. C1= measured valve clearance. C2= standard valve clearance.**

19. Intake valve lifter thickness of the new valve lifter can be identified by the stamp mark on the reverse side (inside the

Measuring position (right bank)		No. 1 CYL.	No. 3 CYL.	No. 5 CYL.
No. 1 cylinder at compression TDC	EXH		×	
	INT	×		
Measuring position (left bank)		No. 2 CYL.	No. 4 CYL.	No. 6 CYL.
No. 1 cylinder at compression TDC	INT			×
	EXH	×		

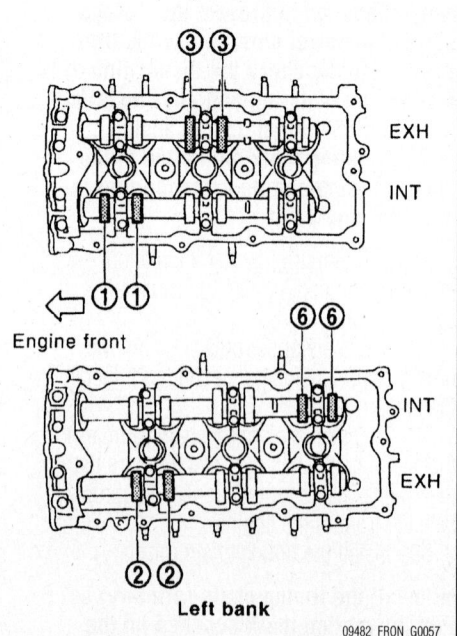

Fig. 61 Valve adjustment measurement No. 1 cylinder at TDC (compression stroke)—4.0L engine

Measuring position (right bank)		No. 1 CYL.	No. 3 CYL.	No. 5 CYL.
No. 3 cylinder at compression TDC	EXH			×
	INT		×	
Measuring position (left bank)		No. 2 CYL.	No. 4 CYL.	No. 6 CYL.
No. 3 cylinder at compression TDC	INT	×		
	EXH		×	

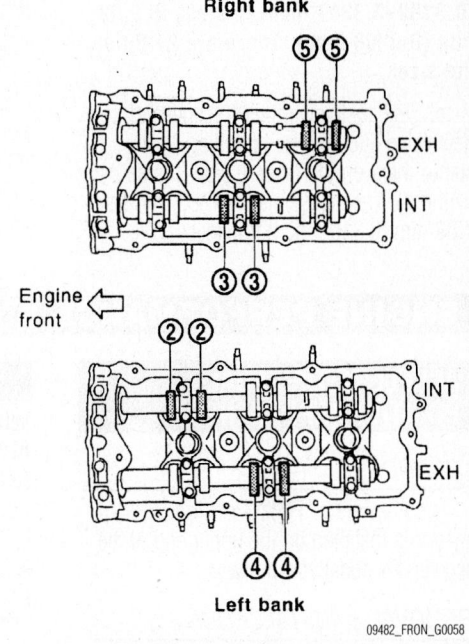

Fig. 62 Valve adjustment measurement No. 3 cylinder at TDC (compression stroke)—4.0L engine

Measuring position (right bank)		No. 1 CYL.	No. 3 CYL.	No. 5 CYL.
No. 5 cylinder at compression TDC	EXH	×		
	INT			×
Measuring position (left bank)		No. 2 CYL.	No. 4 CYL.	No. 6 CYL.
No. 5 cylinder at compression TDC	INT		×	
	EXH			×

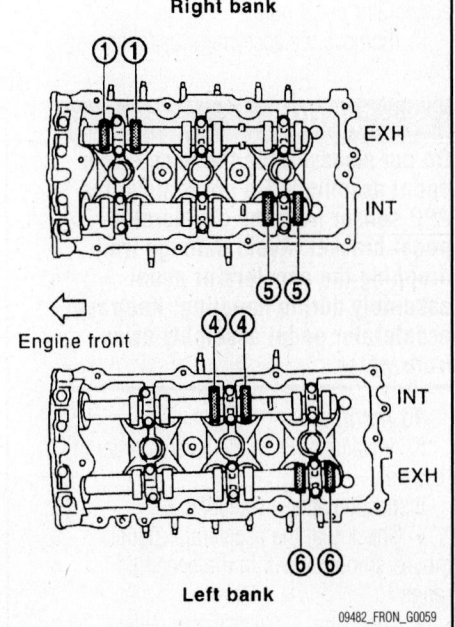

Fig. 63 Valve adjustment measurement No. 5 cylinder at TDC (compression stroke)—4.0L engine

cylinder). The stamp mark "788U" indicates 7.88 mm (0.3102 inch) thickness.

➡**Available thickness of a valve lifter ranges from 7.88–8.40 mm (0.3102–0.3307 inch) in steps of 0.02 mm (0.0008 inch). There are 27 different sizes.**

20. Exhaust valve lifter thickness of the new valve lifter can be identified by the stamp mark on the reverse side (inside the cylinder). The stamp mark "N788" indicates 7.88 mm (0.3102 inch) thickness.

➡**Available thickness of a valve lifter ranges from 7.88–8.36 mm (0.3102–0.3291 inch) in steps of 0.02 mm (0.0008 inch). There are 25 different sizes.**

21. Install the selected valve lifters.
22. Install the camshaft.
23. Manually rotate the crankshaft pulley in the clockwise direction a few rotations.
24. Check the valve clearance and be sure it is within specification.

25. When installing the rocker cover, apply liquid gasket, be sure to use genuine RTV silicone sealant or equivalent, to the positions shown in the illustration. Refer to figure "a" to apply liquid gasket to joint part of camshaft bracket No. 1 and cylinder head. Refer to figure "b" to apply liquid gasket to the figure "a" squarely.

26. Install the rocker cover. Torque the retaining bolts to 17 inch lbs. and then to 74 inch lbs., in the proper sequence.

27. Continue the installation in the reverse of the removal procedure.

ENGINE PERFORMANCE & EMISSION CONTROLS

ACCELERATOR PEDAL POSITION (APP) SENSOR

LOCATION

The Accelerator Pedal Position (APP) sensor is installed on the upper end of the accelerator pedal assembly.

REMOVAL & INSTALLATION

1. Disconnect the negative battery terminal.
2. Disconnect the accelerator position sensor electrical connector.
   • Pull the connector lock back to unlock the connector from the Accelerator Pedal Position sensor
   • Pull up on the connector to disconnect it from the APP sensor
3. Remove the two upper and one lower accelerator pedal nuts
4. Remove the accelerator pedal assembly

✳✳ CAUTION

Do not disassemble the accelerator pedal assembly. Do not remove the APP sensor from the accelerator pedal bracket. Avoid damage from dropping the accelerator pedal assembly during handling. Keep the accelerator pedal assembly away from water.

To install:

5. Installation is in the reverse order of removal.
Inspection after installation:
• Check that the accelerator pedal moves smoothly within the specified range
• Check that the accelerator pedal smoothly returns to the original position
• Perform an electrical inspection of the APP sensor

✳✳ CAUTION

When the harness connector of the APP sensor is disconnected, perform Accelerator Pedal Released Position Learning, Accelerator Pedal Released Position Learning and Accelerator Pedal Released Position Learning.

CAMSHAFT POSITION (CMP) SENSOR

LOCATION

See Figure 64.

Refer to the accompanying illustration.

REMOVAL & INSTALLATION

1. Loosen the fixing bolt of the sensor.
2. Disconnect CMP sensor (PHASE) harness connector.

3. Remove the sensor.

To install:

4. Installation is the reverse of the removal procedure.

CRANKSHAFT POSITION (CKP) SENSOR

LOCATION

See Figure 65.

Refer to the accompanying illustration.

REMOVAL & INSTALLATION

1. Loosen the fixing bolt of the sensor.
2. Disconnect CKP sensor (POS) harness connector.
3. Remove the sensor.

To install:

4. Installation is the reverse of the removal procedure.

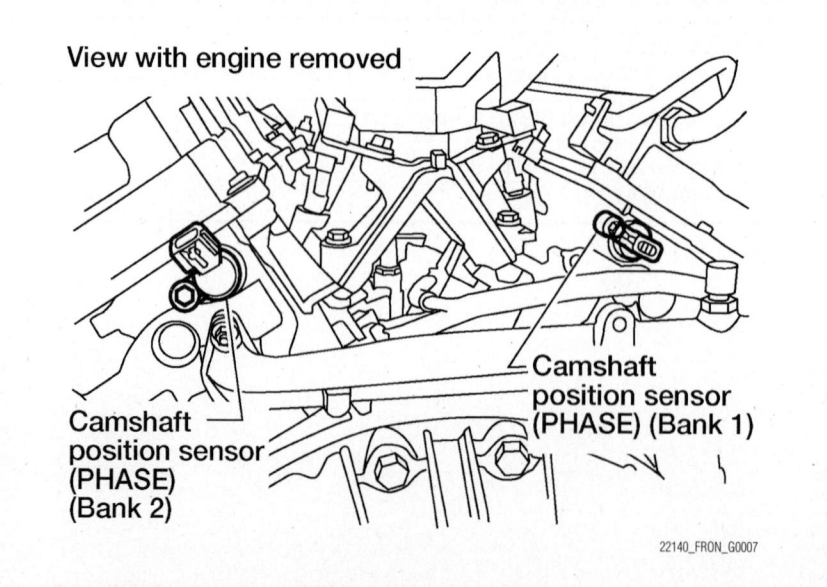

View with engine removed

Camshaft position sensor (PHASE) (Bank 2)

Camshaft position sensor (PHASE) (Bank 1)

22140_FRON_G0007

Fig. 64 Locating Camshaft Position (CMP) sensor—4.0L engine

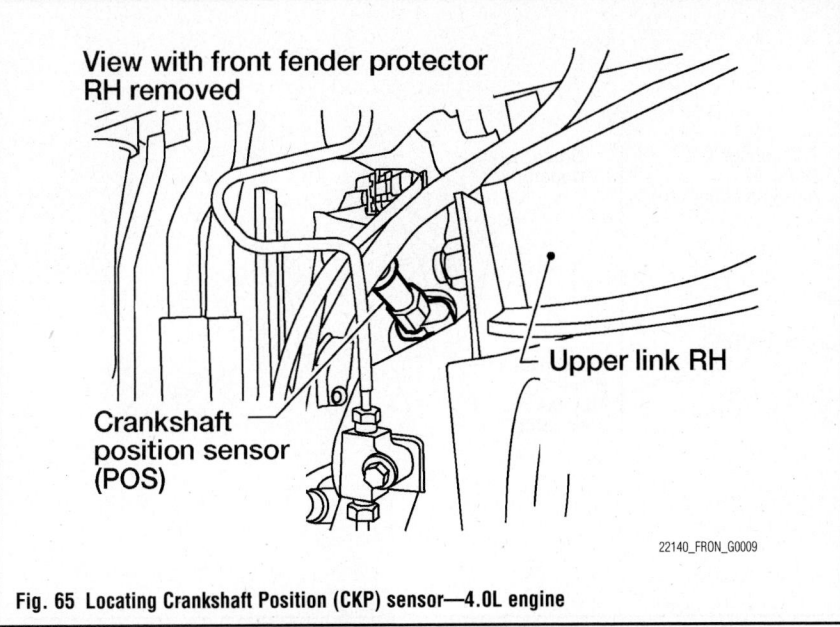

Fig. 65 Locating Crankshaft Position (CKP) sensor—4.0L engine

ELECTRONIC CONTROL MODULE (ECM)

LOCATION

See Figure 66.

Refer to the accompanying illustration.

HEATED OXYGEN (HO2S) SENSOR

LOCATION

See Figures 67 and 68.

Refer to the accompanying illustrations.

REMOVAL & INSTALLATION

> **※ CAUTION**
>
> **Perform the operation with the exhaust system fully cooled. The system will be hot just after the engine stops.**

1. Disconnect sensor harness connector.
2. Remove the sensor using heated oxygen sensor wrench KV10114400 (J-38365) or equivalent.

To install:
3. Installation is the reverse of removal procedure.

➡ Clean exhaust system threads before installing sensor.

INTAKE AIR TEMPERATURE (IAT) SENSOR

LOCATION

The Intake Air Temperature (IAT) sensor is built into Mass Air Flow (MAF) sensor.

REMOVAL & INSTALLATION

See Mass Air Flow (MAF) sensor.

KNOCK SENSOR (KS)

LOCATION

See Figure 69.

Refer to the accompanying illustration.

REMOVAL & INSTALLATION

1. Remove intake collector.
2. Remove sensor harness. Remove sensor.

To install:
3. Installation is the reverse of the removal procedure.

➡ Use care when installing sensor. Do not use any knock sensors that have been dropped or physically damaged. Use only new ones. Torque sensor properly.

MASS AIR FLOW (MAF) SENSOR

LOCATION

See Figure 70.

Refer to the accompanying illustration.

THROTTLE POSITION SENSOR (TPS)

LOCATION

Electric Throttle Control actuator consists of Throttle Control motor, Throttle Position (TPS) sensor, etc. The TPS responds to the throttle valve movement. The TPS has two sensors.

REMOVAL & INSTALLATION

The Electric Throttle Control actuator must be replaced as a unit. See Intake Manifold.

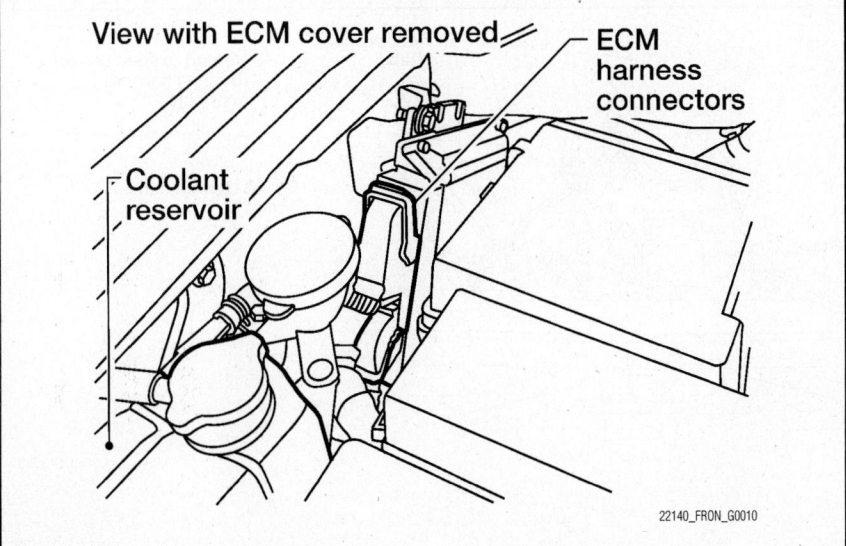

Fig. 66 Locating Electronic Control Module

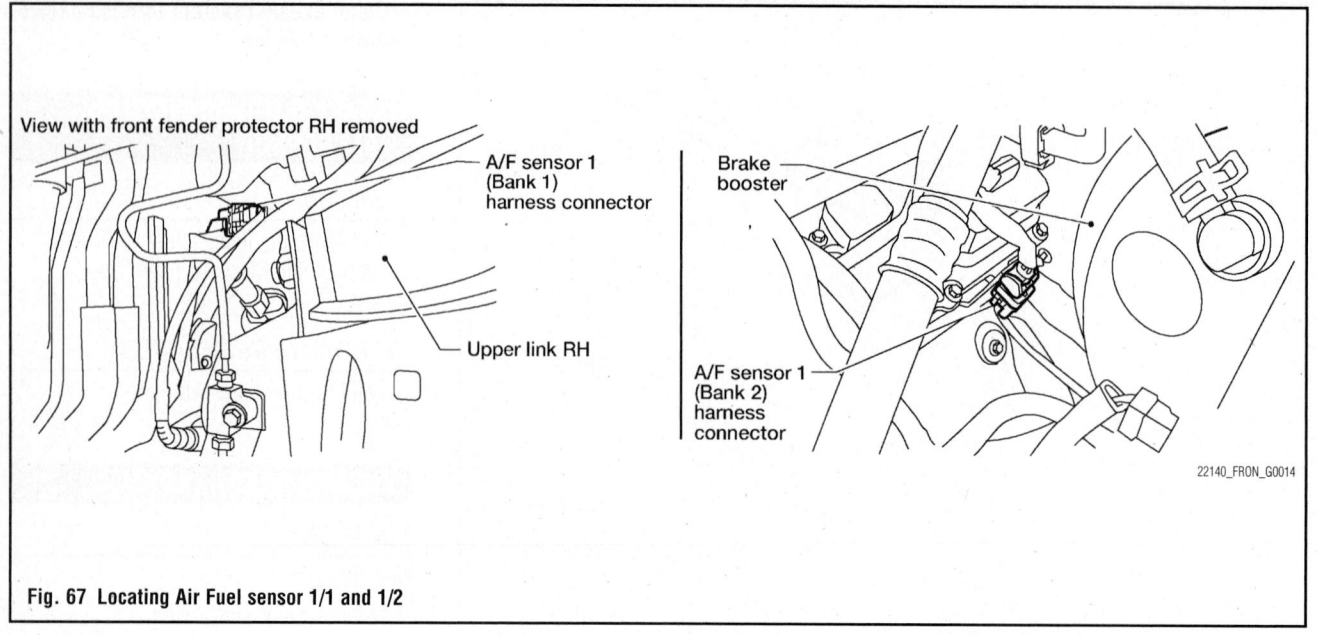

View with front fender protector RH removed

A/F sensor 1
(Bank 1)
harness connector

Upper link RH

Brake
booster

A/F sensor 1
(Bank 2)
harness
connector

22140_FRON_G0014

Fig. 67 Locating Air Fuel sensor 1/1 and 1/2

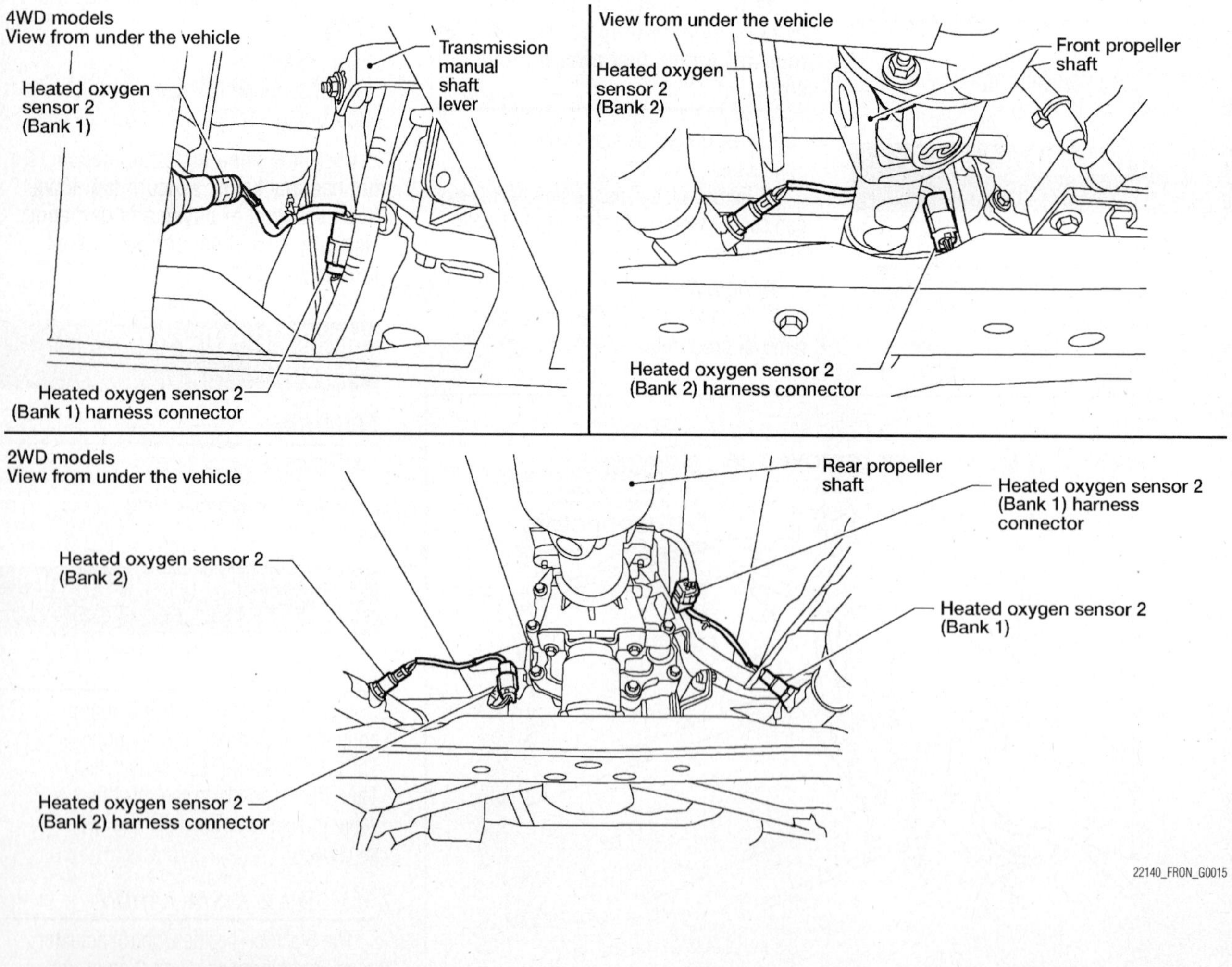

4WD models
View from under the vehicle

Heated oxygen
sensor 2
(Bank 1)

Transmission
manual
shaft
lever

Heated oxygen sensor 2
(Bank 1) harness connector

View from under the vehicle

Heated oxygen
sensor 2
(Bank 2)

Front propeller
shaft

Heated oxygen sensor 2
(Bank 2) harness connector

2WD models
View from under the vehicle

Rear propeller
shaft

Heated oxygen sensor 2
(Bank 1) harness
connector

Heated oxygen sensor 2
(Bank 2)

Heated oxygen sensor 2
(Bank 1)

Heated oxygen sensor 2
(Bank 2) harness connector

22140_FRON_G0015

Fig. 68 Locating Heated Oxygen Sensor (HO2S) 2/1 and 2/2

View with intake manifold collector removed

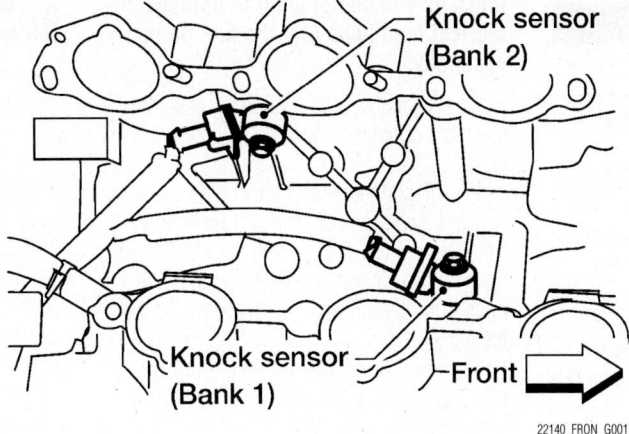

Knock sensor (Bank 2)

Knock sensor (Bank 1)

Front

22140_FRON_G0017

Fig. 69 Locating Knock Sensor (KS) bank 1 and bank 2

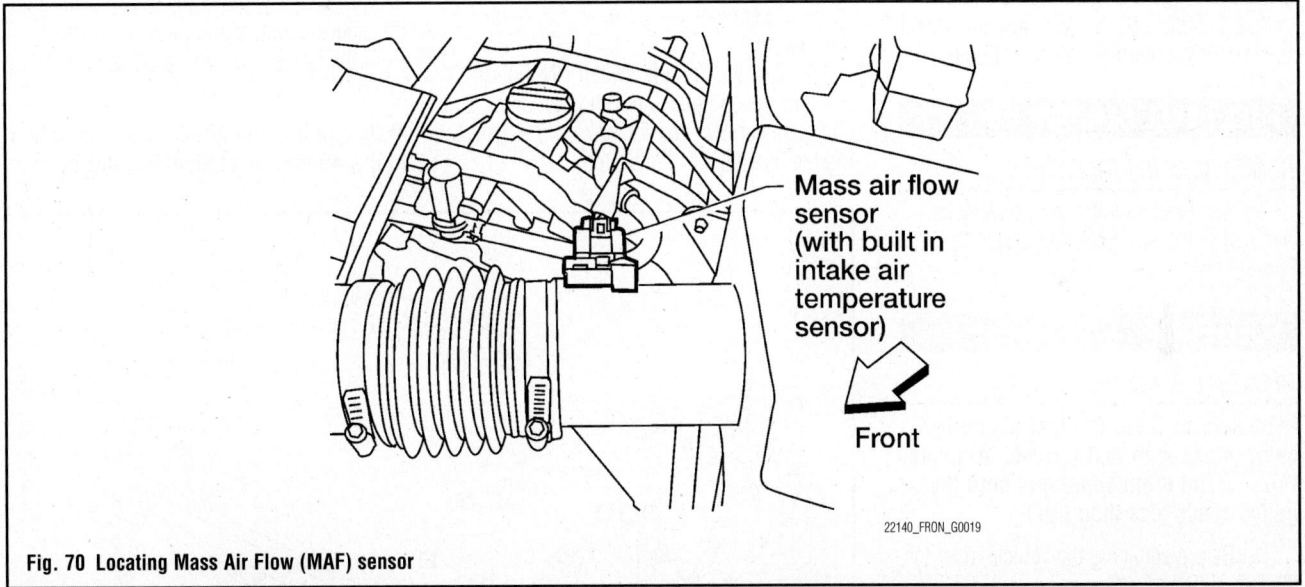

Mass air flow sensor (with built in intake air temperature sensor)

Front

22140_FRON_G0019

Fig. 70 Locating Mass Air Flow (MAF) sensor

FUEL   **GASOLINE FUEL INJECTION SYSTEM**

FUEL SYSTEM SERVICE PRECAUTIONS

Safety is the most important factor when performing not only fuel system maintenance but any type of maintenance. Failure to conduct maintenance and repairs in a safe manner may result in serious personal injury or death. Maintenance and testing of the vehicle's fuel system components can be accomplished safely and effectively by adhering to the following rules and guidelines.

• To avoid the possibility of fire and personal injury, always disconnect the negative battery cable unless the repair or test procedure requires that battery voltage be applied.

• Always relieve the fuel system pressure prior to disconnecting any fuel system component (injector, fuel rail, pressure regulator, etc.), fitting or fuel line connection. Exercise extreme caution whenever relieving fuel system pressure to avoid exposing skin, face and eyes to fuel spray. Please be advised that fuel under pressure may penetrate the skin or any part of the body that it contacts.

• Always place a shop towel or cloth around the fitting or connection prior to loosening to absorb any excess fuel due to spillage. Ensure that all fuel spillage (should it occur) is quickly removed from engine surfaces. Ensure that all fuel soaked cloths or towels are deposited into a suitable waste container.

• Always keep a dry chemical (Class B) fire extinguisher near the work area.

• Do not allow fuel spray or fuel vapors to come into contact with a spark or open flame.

• Always use a back-up wrench when loosening and tightening fuel line connection fittings. This will prevent unnecessary stress and torsion to fuel line piping.

• Always replace worn fuel fitting O-rings with new Do not substitute fuel hose or equivalent where fuel pipe is installed.

Before servicing the vehicle, make sure to also refer to the precautions in the beginning of this section as well.

FUEL SYSTEM PRESSURE

RELIEVING

1. Before servicing the vehicle, refer to the Precautions Section.
2. Remove the fuel pump fuse from the panel.
3. Start the engine and allow it to run until it stalls. Crank the engine for a few seconds to relieve additional fuel pressure.
4. Turn ignition switch off.
5. When repairs are complete, replace the fuel pump fuse and connect the negative battery cable.

FUEL FILTER

REMOVAL & INSTALLATION

Fuel Filter is serviced with fuel pump and sending unit assembly. See Fuel Pump.

FUEL LEVEL SENDING UNIT

REMOVAL & INSTALLATION

The fuel level sending unit is serviced with fuel pump and filter unit assembly. See Fuel Pump.

FUEL PUMP

REMOVAL & INSTALLATION

➡**Be sure to check the fuel gauge indicator. Make sure that it reads less than FULL. If not drain some fuel until the gauge reads less than FULL.**

1. Before servicing the vehicle, refer to the Precautions Section.
2. Properly relieve the fuel system pressure.
3. Disconnect the negative battery cable.
4. Remove the fuel filler cap. Remove the left rear tire and wheel assembly.
5. Disconnect the lower fuel filler hose from the fuel tank, the EVAP hose and the vent pipe quick connector.

➡**Disconnect the fuel feed hose from the molder clip in the side of the fuel tank.**

6. Remove the four tank shield retaining bolts. Remove the tank shield.
7. Remove the driveshaft.
8. Properly support the fuel tank. Remove the three fuel tank retaining strap bolts. Remove the fuel tank straps.

9. Lower the fuel tank to gain access to the top of the fuel pump assembly.

➡**Be careful not to lower the tank too much as you do not want to damage the fuel feed hose and the fuel pump assembly.**

10. Disconnect the fuel pump assembly electrical connector, and the fuel feed hose.
11. Disconnect the quick connector.
12. Lower the fuel tank and remove it from the vehicle. Disconnect the EVAP hose from the fuel pump and remove the EVAP hose from the molded clip in the top of the fuel tank.
13. Remove the fuel pump assembly lockring. Remove the fuel pump assembly. Discard the O-ring.

To install:

14. Installation is the reverse of the removal procedure.
15. Be sure to use a new O-ring upon installation.
16. Turn the ignition switch ON, but do not start the engine. Check the fuel lines and hose connections for leaks while applying fuel pressure to the system.

17. Start the engine and check for fuel leaks, correct as required.

FUEL RAIL (SUPPLY MANIFOLD) & INJECTOR

REMOVAL & INSTALLATION

See Figure 71.

1. Before servicing the vehicle, refer to the Precautions Section.
2. Properly relieve the fuel system pressure.
3. Disconnect the negative battery cable. Remove the fuel filler cap.
4. Remove the intake manifold collector.
5. Remove the quick connector cap (engine side). With the sleeve side of the quick connector release facing the quick connector, install the quick connector release on to the tube. Insert the quick connector release into the quick connector until the sleeve contacts and goes no further. Hold the quick connector release in that position.

➡**Disconnect the quick connector using tool J-45488, or equivalent, not by**

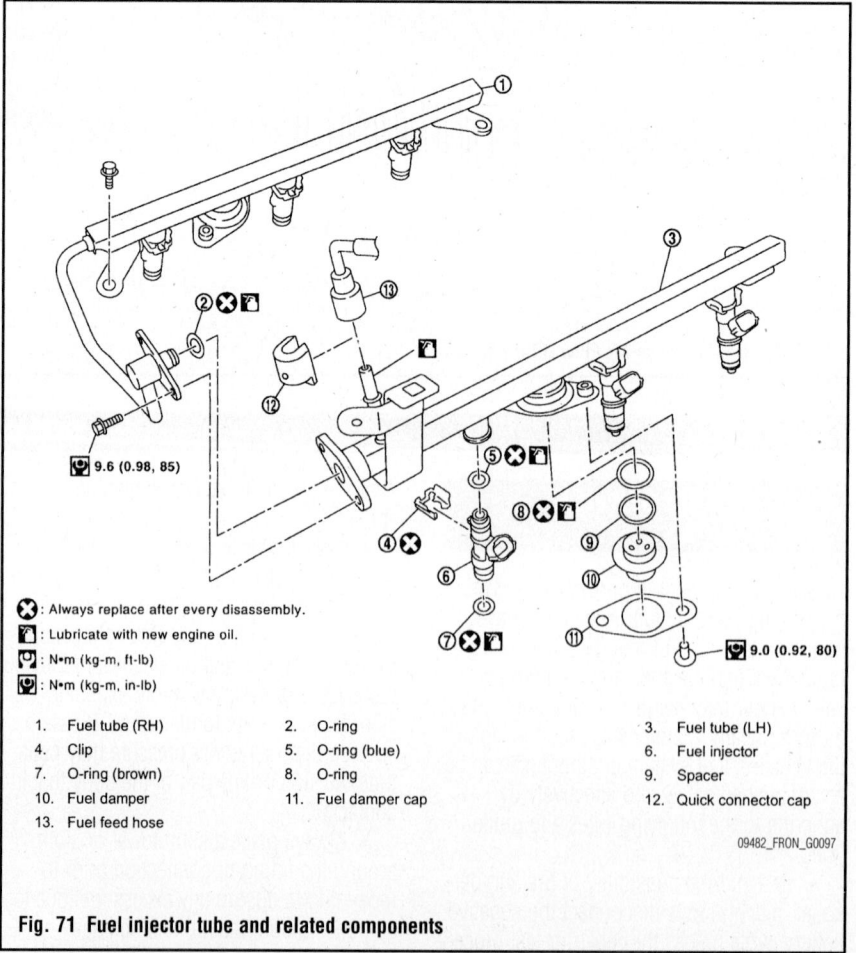

: Always replace after every disassembly.
: Lubricate with new engine oil.
: N•m (kg-m, ft-lb)
: N•m (kg-m, in-lb)

9.6 (0.98, 85)

9.0 (0.92, 80)

1. Fuel tube (RH)	2. O-ring	3. Fuel tube (LH)
4. Clip	5. O-ring (blue)	6. Fuel injector
7. O-ring (brown)	8. O-ring	9. Spacer
10. Fuel damper	11. Fuel damper cap	12. Quick connector cap
13. Fuel feed hose		

09482_FRON_G0097

Fig. 71 Fuel injector tube and related components

picking out the retainer tabs. Inserting the quick connector hard will not disconnect the quick connector. Hold the quick connector release where it contacts and goes no further.

6. Draw and pull out the quick connector straight from the fuel tube. Grasp the quick connector holding "A" in the illustration. Do not pull with lateral force applied and the O-ring inside the quick connector could be damaged.

➡️Have a cloth ready, as fuel will leak out. Avoid fire and sparks. Keep parts away from heat. Do not bend or twist the connection between the quick connector and the fuel feed hose. Cover the openings with a plastic bag.

7. Remove the PCV hose between the rocker covers.

8. Disconnect the harness for the fuel injector.

9. Loosen the retaining bolts. Remove the fuel tube and fuel injector assembly. Remove the bolts which connect the left and right fuel tubes.

10. To remove the fuel injectors from the fuel tube, open and remove the clip. Remove the injector by pulling it straight out.

11. Disconnect the right fuel tube from the left fuel tube. Loosen the bolts, to remove the fuel damper cap and fuel damper, if necessary.

To install:

➡️Use new O-ring seals for assembly. Note that the upper and lower O-rings are different. Do not confuse them. Fuel tube side: Blue. Nozzle side: Brown.

12. Installation is the reverse of the removal procedure.

13. When installing the fuel feed tube be sure to torque the retaining bolts to 7 ft. lbs

and then to 16 ft. lbs. in an alternating order.

14. Turn the ignition switch ON, but do not start the engine. Check the fuel lines and hose connections for leaks while applying fuel pressure to the system.

15. Start the engine and check for fuel leaks, correct as required.

FUEL TANK

REMOVAL & INSTALLATION

See Fuel Pump.

IDLE SPEED

ADJUSTMENT

Idle speed is maintained by the Powertrain Control Module (PCM). No adjustment is necessary or possible.

HEATING & AIR CONDITIONING SYSTEM

BLOWER MOTOR

REMOVAL & INSTALLATION

See Figure 72.

➡️Before servicing, or working around, the SRS system, turn the ignition switch OFF, disconnect both battery cables and wait at least three minutes. When servicing, or working around, the

SRS system do not work directly in front of the air bag module.

1. Before servicing the vehicle, refer to the Precautions Section.

2. Disconnect the negative battery cable. Disconnect the positive battery cable.

3. Remove the lower glove box assembly.

4. Disconnect the blower motor electrical connector.

5. Remove the blower motor retaining screws.

6. Remove the blower motor from its mounting.

To install:

7. Installation is the reverse of the removal procedure.

HEATER CORE

REMOVAL AND INSTALLATION

See Figures 73 through 79.

➡️Be sure to disarm the SRS system, prior to working on the vehicle. Turn the ignition switch OFF, disconnect both battery cables and wait at least three minutes before starting any work.

1. Before servicing the vehicle, refer to the Precautions Section.

2. Position the front wheels in the straight ahead direction.

3. Disconnect the negative battery cable. Disconnect the positive battery cable.

4. Drain the cooling system.

5. Properly discharge the air conditioning system.

6. If equipped with the 4.0L engine, remove the right side heater core pipe nuts.

7. Disconnect the heater core hoses from the heater core.

8. Disconnect the air conditioning refrigerant lines from the expansion valve.

9. Position the front seats in the rearmost position on the seat tracks.

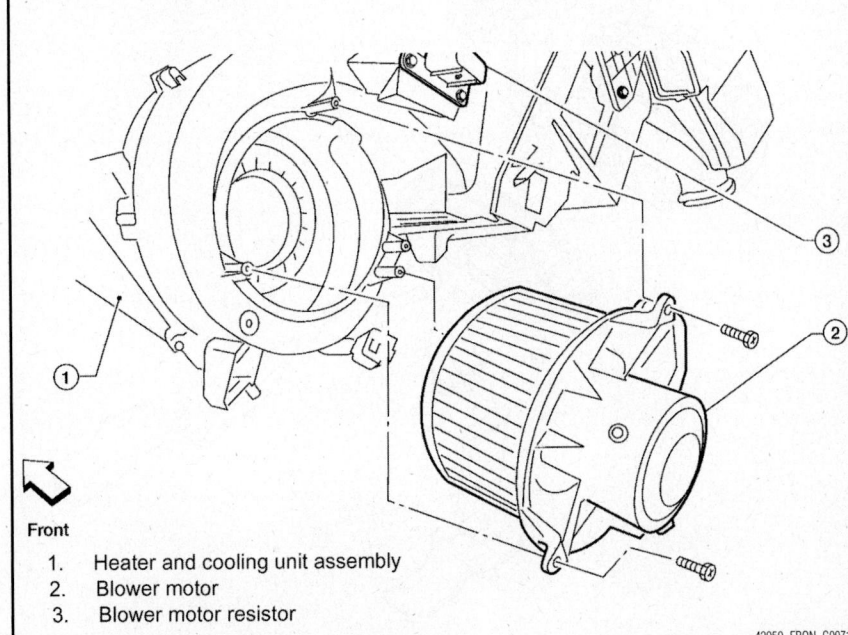

Front

1. Heater and cooling unit assembly
2. Blower motor
3. Blower motor resistor

42050_FRON_G0071

Fig. 72 Blower motor an related components

10. Remove the upper front pillar trim panel. Remove the steering lock escutcheon. Remove the cluster lid "A". Remove the combination meter. Disconnect the electrical connections.

11. Remove the optical sensor. Remove the audio unit. Remove the cluster lid "D".

12. Remove the glove box. Remove the two bolts, through the glove box opening, retaining the front passenger's side air bag module to the steering member. Disconnect the air bag module connectors.

13. Remove the instrument stay right side and left side bolts. Remove the instrument panel.

14. Remove the two front floor ducts.

15. To remove the driver's side air bag module, locate the retaining clip access hole under the steering wheel. Insert a suitable blunt tool (4 mm–6 mm in size).

➡ **Do not use sharp edged objects, such as a screwdriver, to release the driver's side airbag module from the steering wheel as SRS components may be unintentionally damaged.**

16. Press upward, toward the center of the steering wheel, on the retaining clip until the air bag module is released from the steering wheel.

17. Lift the air bag module from the steering wheel. Disconnect the electrical connectors. Remove the air bag module.

18. Disconnect the steering wheel switches. Remove the steering wheel center nut. Using a steering wheel removal tool, remove the steering wheel.

19. Remove the steering column upper and lower covers. Disconnect the wiper and washer switch connector. While pressing the tabs, pull the wiper and washer switch away from the spiral cable to remove it.

20. Disconnect the light and turn signal switch connector. While pressing the tabs, pull the light and turn signal switch toward the driver's door to remove it.

21. Remove the screws. While pressing the tab, pull the spiral cable away from the steering column assembly. Disconnect the electrical connectors.

➡ **With the steering linkage disconnected, the spiral cable may snap by turning the steering wheel beyond the limited number of turns. The spiral cable can be turned counterclockwise about 2.5 turns from the neutral position.**

22. Remove the lower knee protector.

23. Remove the locknut and bolt from the upper joint and then separate the upper joint from the upper shaft.

24. Remove the three nuts and bolt from the steering column and then remove the steering column assembly from the steering member.

25. Remove the hole cover seal and clamp. Remove the hole cover nuts, remove the hole cover from the dash panel.

26. Remove the bolt from the lower joint of the lower joint shaft and remove the lower joint shaft from the vehicle.

27. Disconnect the instrument panel wire harness at the right and left in-line connector brackets, and the fuse block (SMJ) electrical connectors.

28. Remove the covers and then remove the three steering member bolts from each side to disconnect the steering member from the vehicle body.

29. Remove the heater/evaporator case assembly with it attached to the steering member from the vehicle.

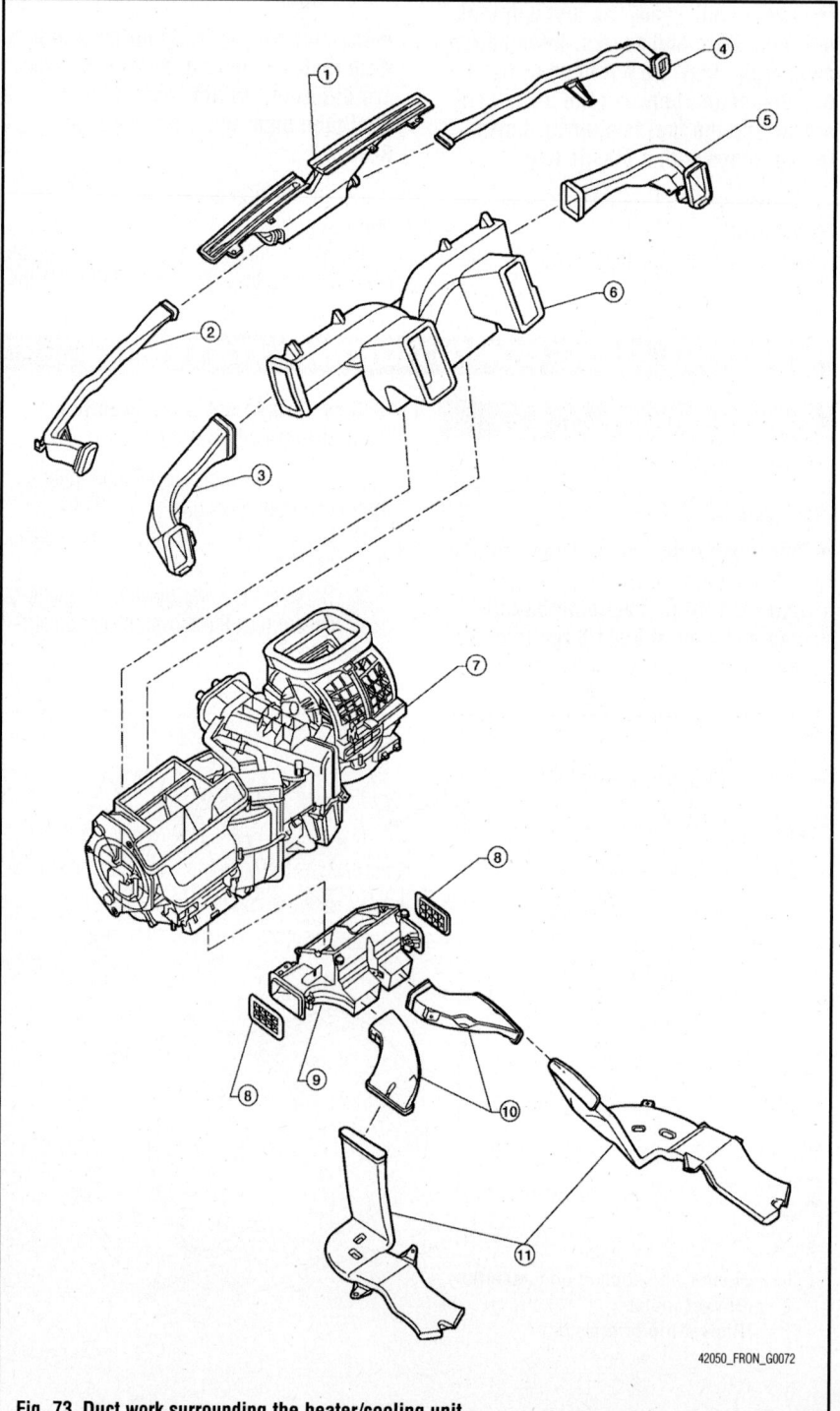

42050_FRON_G0072

Fig. 73 Duct work surrounding the heater/cooling unit

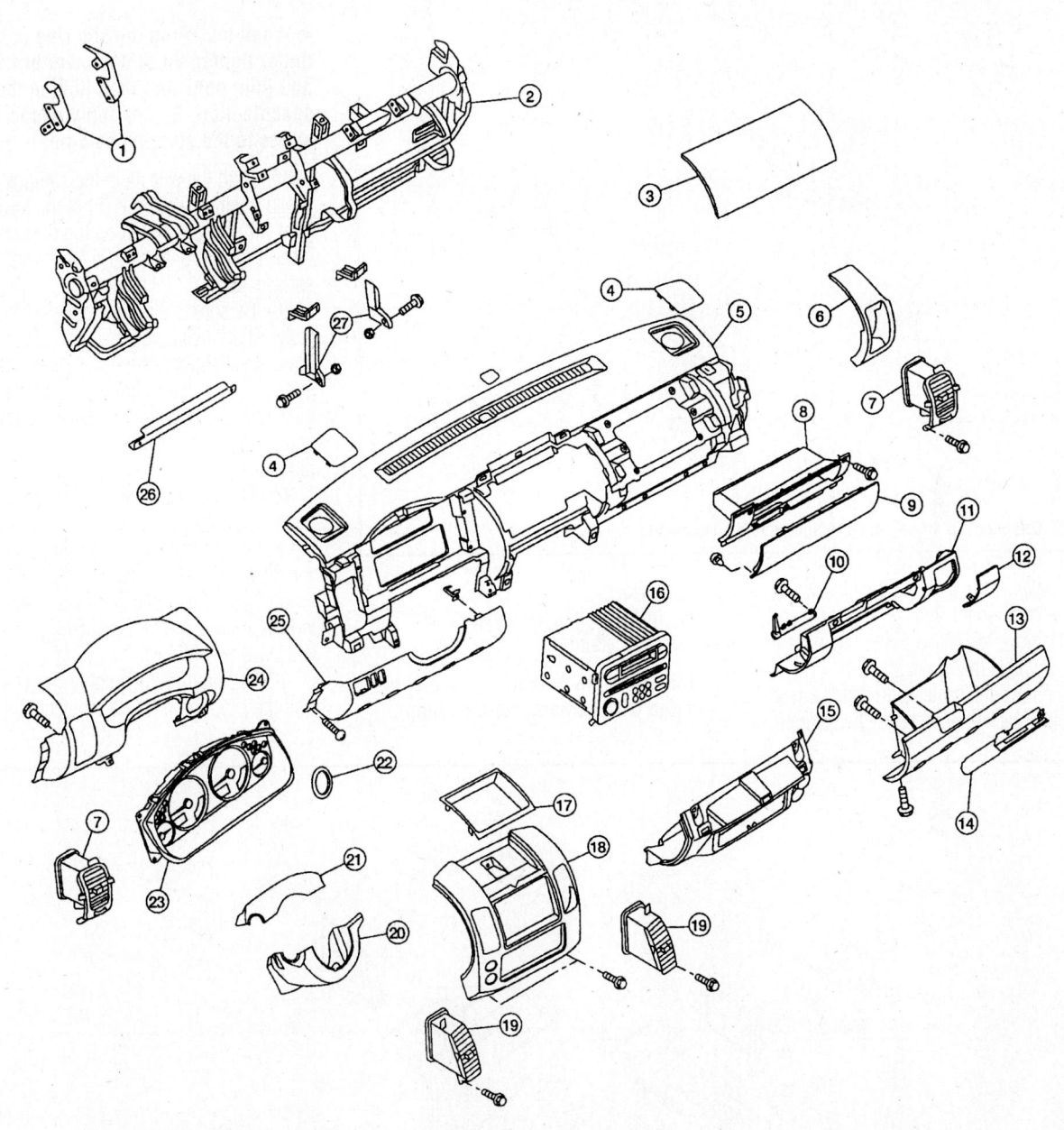

1.	Display unit bracket RH/LH	2.	Steering member assembly	3.	Passenger air bag module cover
4.	Speaker grille RH/LH	5.	Instrument panel and pad assembly	6.	Instrument side finisher
7.	Side ventilator assembly RH/LH	8.	Upper glove box bin	9.	Upper glove box door
10.	Lower glove box damper assembly	11.	Lower instrument panel RH	12.	Fuse block cover
13.	Lower glove box assembly	14.	Lower glove box latch assembly	15.	Cluster lid D
16.	Audio unit	17.	Storage tray	18.	Cluster lid C
19.	Center ventilator assembly RH/LH	20.	Steering column cover lower	21.	Steering column cover upper
22.	Steering lock escutcheon	23.	Combination meter	24.	Cluster lid A
25.	Lower instrument panel LH	26.	Knee protector brace	27.	Instrument stay RH/LH

09482_FRON_G0009

Fig. 74 Instrument panel and related components

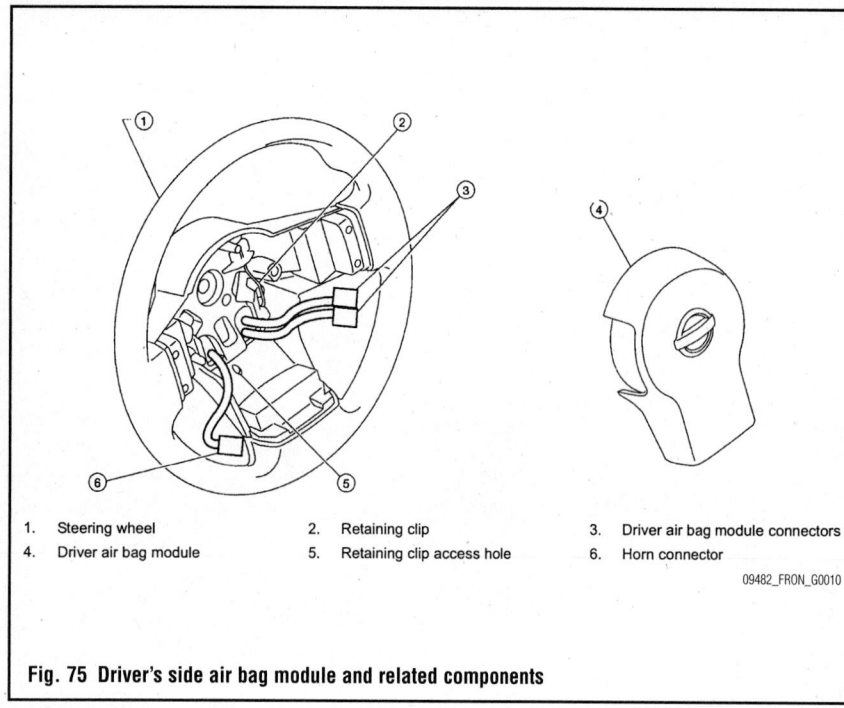

Fig. 75 Driver's side air bag module and related components

1.	Steering wheel	2.	Retaining clip	3.	Driver air bag module connectors
4.	Driver air bag module	5.	Retaining clip access hole	6.	Horn connector

09482_FRON_G0010

30. Separate the steering member from the heater/evaporator unit.

31. Remove the heater cover retaining screws. Remove the cover.

32. Remove the heater core and the evaporator pipe bracket. Remove the heater core.

To install:

33. Installation is the reverse of the removal procedure.

➡**If the in-cabin microfilters are contaminated with coolant, replace them.**

34. Be sure to use new steering column retaining bolts and pinch bolt, as required.

➡**When installing the steering column, finger tighten all of the lower bracket and joint bolts and then tighten them to specification. Do not apply undue stress to the steering column.**

35. With the wheels in the straight ahead position align the slit of the lower joint with the projection on the dust cover. Insert the joint until surface "A" contacts surface "B".

36. Be sure to align the spiral cable correctly when installing the steering wheel. Make sure that the cable is in the neutral position. The neutral position is detected by turning left 2.6 revolutions from the right end position and ending with the locating pin at the top.

37. To adjust the steering angle sensor neutral position, position the steering wheel in the straight ahead position and rive the vehicle at 10 mph or more for ten minutes. When the procedure is complete, the SLP indicator lamp and the VDC OFF indicator lamp will turn off.

38. Be sure to fill the cooling system with the proper grade and type coolant.

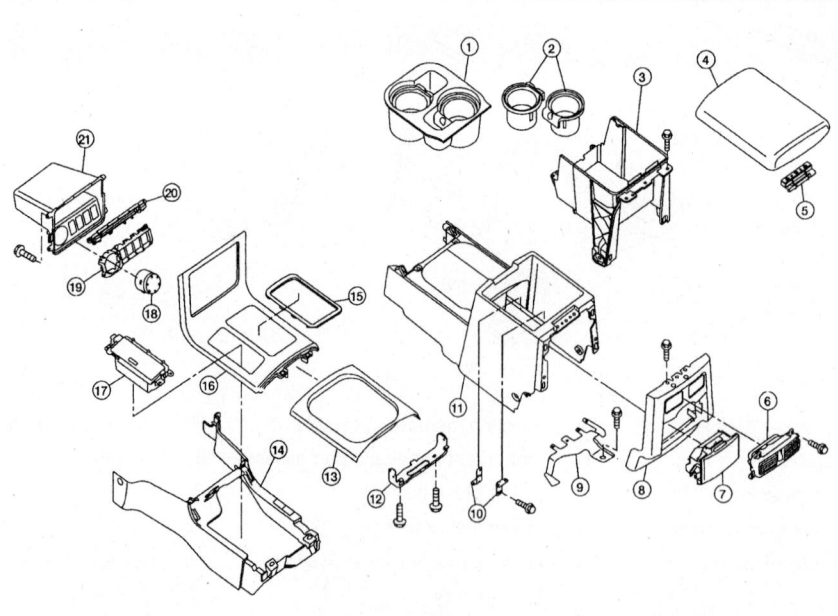

1.	Cup holder assembly	2.	Cup holder insert	3.	Center console bin
4.	Center console lid	5.	Hinge	6.	Ventilator console grille
7.	Rear cup holder assembly	8.	Rear finisher assembly	9.	Wire harness bracket
10.	Bracket DVD	11.	Center console rear base	12.	Bracket
13.	Cup holder finisher	14.	Center console front base	15.	A/T finisher bezel
16.	A/T finisher	17.	Ash tray	18.	Switch assembly
19.	Switch finisher	20.	CD changer door	21.	Console bin

09482_FRON_G0011

Fig. 76 Center console and related components

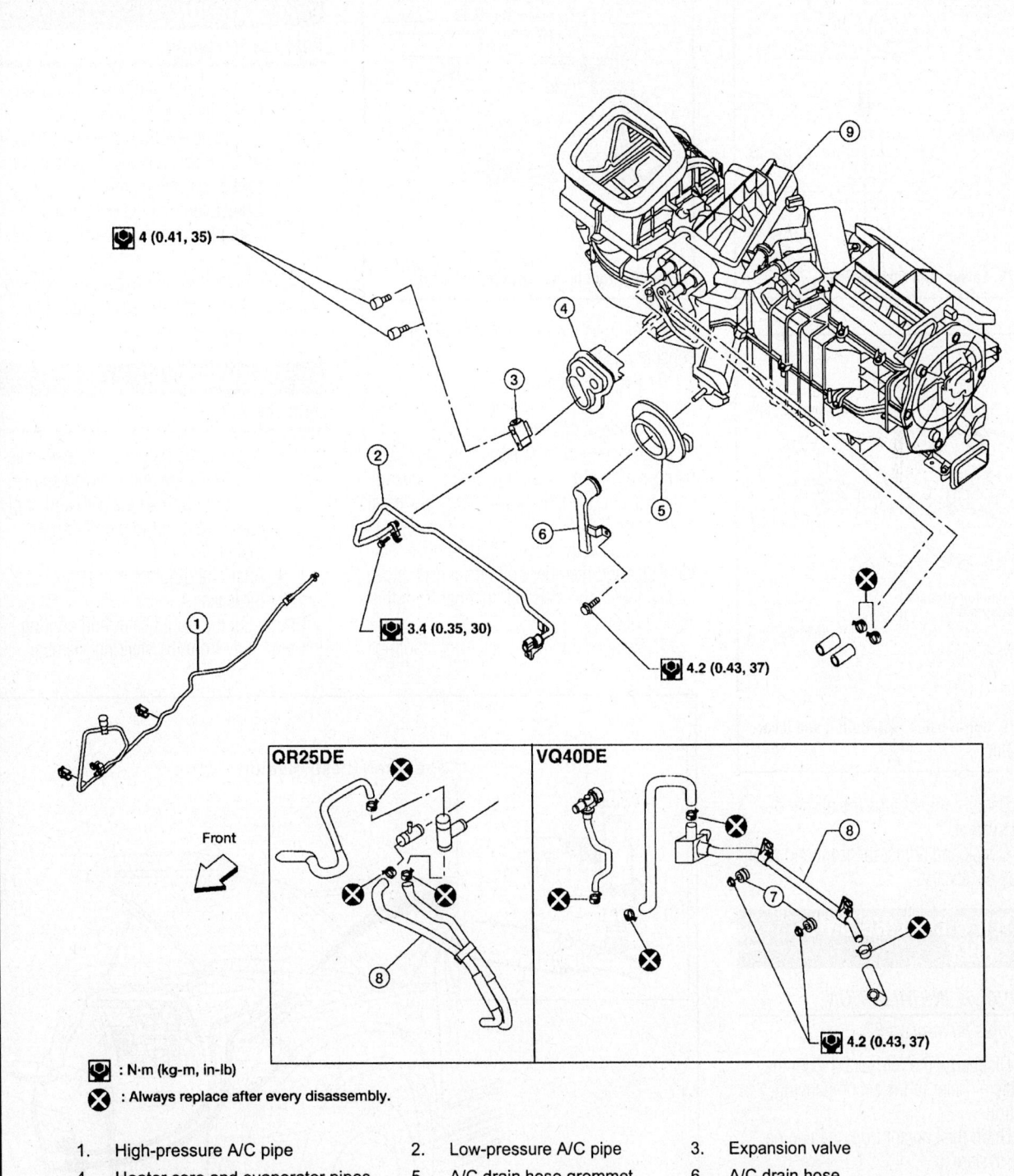

4 (0.41, 35)

3.4 (0.35, 30)

4.2 (0.43, 37)

QR25DE

VQ40DE

Front

4.2 (0.43, 37)

: N·m (kg-m, in-lb)

: Always replace after every disassembly.

1. High-pressure A/C pipe
2. Low-pressure A/C pipe
3. Expansion valve
4. Heater core and evaporator pipes grommet
5. A/C drain hose grommet
6. A/C drain hose
7. Heater core pipe mounts
8. Heater core pipes
9. Heater and cooling unit assembly

09482_FRON_G0012

Fig. 77 Heater/evaporator core and related components

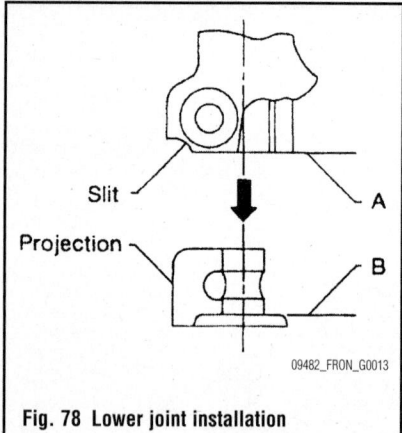

Slit

Projection

A

B

09482_FRON_G0013

Fig. 78 Lower joint installation

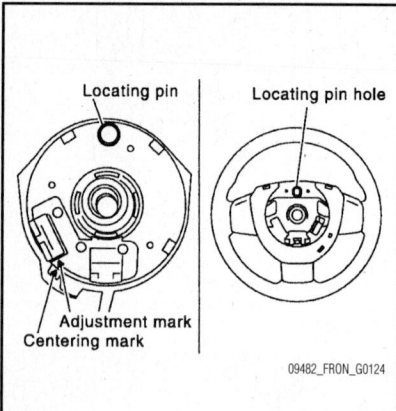

Locating pin

Locating pin hole

Adjustment mark
Centering mark

09482_FRON_G0124

Fig. 79 Spiral cable installation and locating point

39. Be sure to recharge the air conditioning system.

40. Check and adjust the front end alignment, as necessary.

HEATER & COOLING UNIT ASSEMBLY

REMOVAL & INSTALLATION

See Figures 80 through 82.

1. Discharge the refrigerant from the A/C system. Refer to the accompanying illustration.

2. Drain the coolant from the engine cooling system.

3. Disconnect the battery negative and positive terminals.

4. Remove the front heater core pipes RH nut.

5. Disconnect the front heater core hoses from the front heater core.

6. Disconnect the high- and low-pressure A/C pipes from the front expansion valve.

7. Move the two front seats to the rear most position on the seat track.

RH front heater core pipes nut

Front

37663_XTER_G0113

Fig. 80 Front heater core pipes nut RH

8. Remove the instrument panel and console panel.

9. Remove the two front floor ducts.

10. Remove the steering column.

11. Disconnect the instrument panel wire harness at the RH and LH in-line connector brackets, and the fuse block (SMJ) electrical connectors.

12. Remove the covers then remove the three steering member bolts from each side to disconnect the steering member from the vehicle body.

13. Remove the front heater and cooling

unit assembly with it attached to the steering member, from the vehicle.

✳✳ CAUTION

Note the following:

- Use care not to damage the seats and interior trim panels when removing the front heater and cooling unit assembly with it attached to the steering member.
- Use suitable plugs on the heater core pipes to prevent coolant leakage.

14. Remove the front heater and cooling unit assembly from the steering member.

To install:

✳✳ CAUTION

Note the following:

- Replace the O-ring of the low-pressure A/C pipe and high-pressure A/C pipe with a new one, and apply compressor oil to the O-ring for installation.
- After charging the refrigerant, check for leaks.

15. Install the front heater and cooling unit assembly from the steering member.

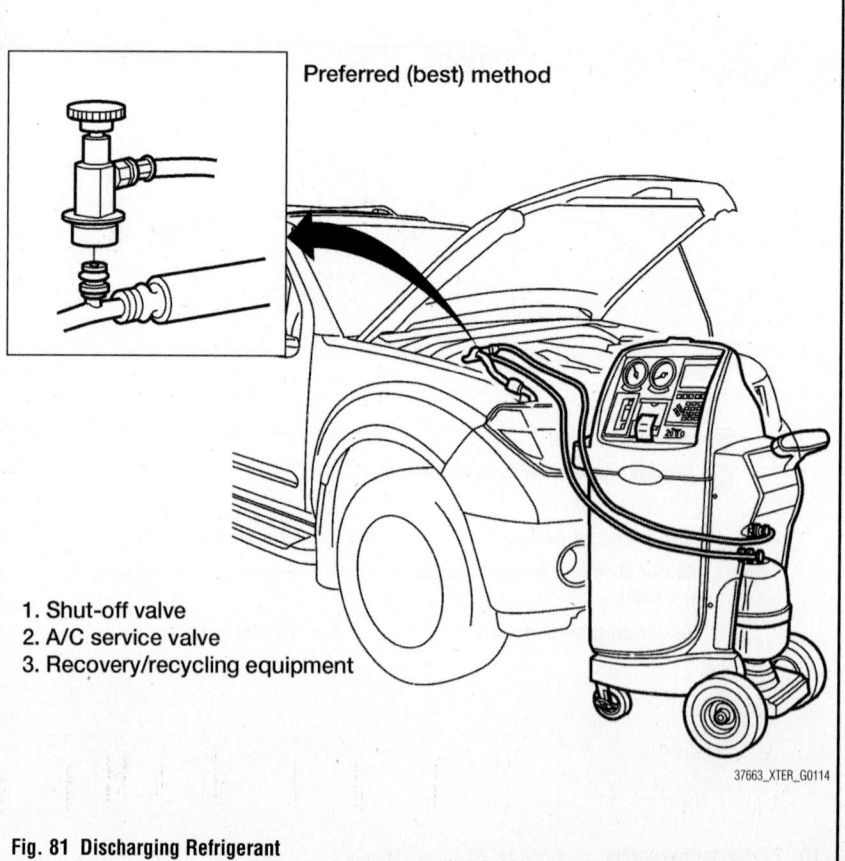

Preferred (best) method

1. Shut-off valve
2. A/C service valve
3. Recovery/recycling equipment

37663_XTER_G0114

Fig. 81 Discharging Refrigerant

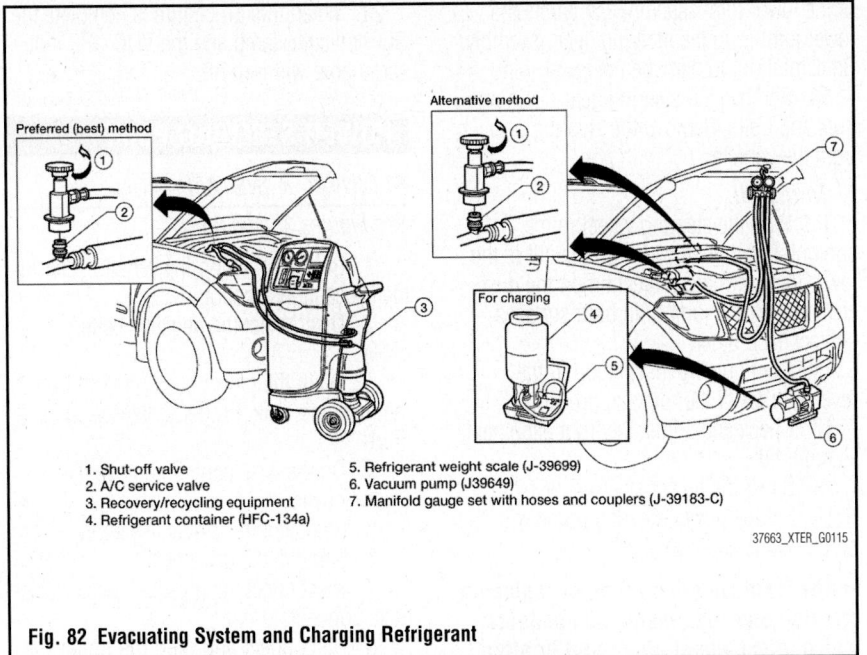

1. Shut-off valve
2. A/C service valve
3. Recovery/recycling equipment
4. Refrigerant container (HFC-134a)
5. Refrigerant weight scale (J-39699)
6. Vacuum pump (J39649)
7. Manifold gauge set with hoses and couplers (J-39183-C)

37663_XTER_G0115

Fig. 82 Evacuating System and Charging Refrigerant

16. Install the front heater and cooling unit assembly with it attached to the steering member, from the vehicle.

17. Install the covers then install the three steering member bolts from each side to disconnect the steering member from the vehicle body.

18. Connect the instrument panel wire harness at the RH and LH in-line connector brackets, and the fuse block (SMJ) electrical connectors.

19. Install the steering column.

20. Install the two front floor ducts.

21. Install the instrument panel and console panel.

22. Move the two front seats to the rear most position on the seat track.

23. Connect the high- and low-pressure A/C pipes from the front expansion valve.

24. Connect the front heater core hoses from the front heater core.

25. Install the front heater core pipes RH nut.

26. Connect the battery negative and positive terminals.

27. Fill the coolant to the engine cooling system.

28. Recharge the refrigerant to the A/C system. Refer to the accompanying illustrations.

✳✳ CAUTION
Note the following:

- Use care not to damage the seats and interior trim panels when removing the front heater and cooling unit assembly with it attached to the steering member.
- Use suitable plugs on the heater core pipes to prevent coolant leakage.

STEERING

POWER RACK & PINION STEERING GEAR

REMOVAL & INSTALLATION
See Figures 83 and 84.

➡The spiral cable may snap due to steering operation if the steering column is separated from the steering gear assembly. Be sure to secure the steering wheel to avoid turning.

1. Before servicing the vehicle, refer to the Precautions Section.
2. Position the front wheels in the straight ahead position.
3. Disarm the SRS system.
4. Disconnect the negative battery cable.
5. Drain the power steering fluid.
6. Raise and support the vehicle safely. Remove the tire and wheel assemblies.
7. Remove the undercover.
8. If equipped with 4WD, remove the final drive, then support the halfshafts, using wire.
9. Remove the stabilizer bar brackets, and position the stabilizer bar aside.

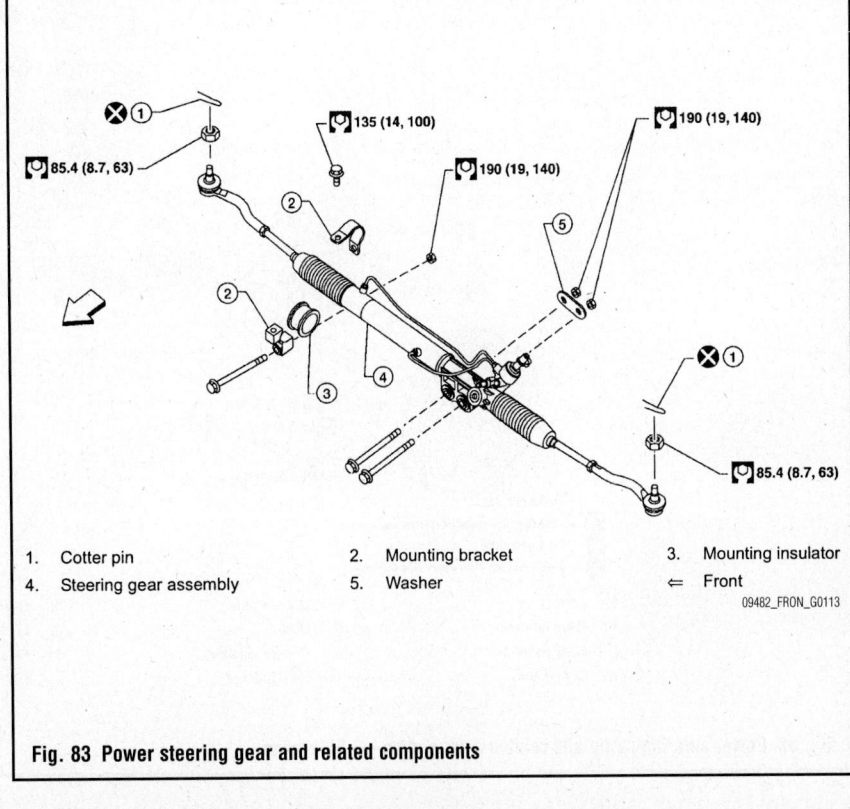

1. Cotter pin
2. Mounting bracket
3. Mounting insulator
4. Steering gear assembly
5. Washer
⇐ Front

09482_FRON_G0113

Fig. 83 Power steering gear and related components

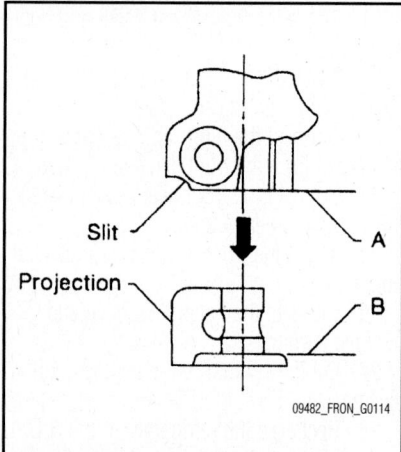

Fig. 84 Power steering gear lower joint installation alignment

10. Remove and discard the cotter pins at the steering outer sockets. Loosen the outer socket locknuts.

11. Remove the steering gear outer sockets from the steering knuckles, using tool HT72520000 (J-25730-A) or equivalent.

12. Disconnect and plug the power steering fluid lines at the steering gear.

13. Remove the bolt from the lower joint of the lower joint assembly. Separate the lower joint from the steering gear assembly. Be careful not to damage the lower joint.

14. Remove the steering gear retaining nuts and bolts. Remove the steering gear from the vehicle.

To install:

15. With the steering wheel in the straight ahead position, align the slit of the lower joint with the projection on the dust cover. Insert the joint until both surfaces contact each other.

16. Continue the installation in the reverse order of the removal procedure.

17. Check and adjust the front alignment, as required.

18. Bleed the power steering system.

19. Fill the power steering pump with the proper grade and type fluid.

➡ **After removing/installing or replacing steering and suspension components which effect wheel alignment or after adjusting wheel alignment, or the steering angle sensor or the ABS actuator electrical unit be sure to adjust the neutral position of the steering angle sensor before running the vehicle.**

20. Position the steering wheel in the straight ahead position.

21. When this procedure is complete the SLP indicator lamp and the VDC OFF indicator lamp will turn off.

POWER STEERING PUMP

REMOVAL & INSTALLATION

See Figures 85 and 86.

1. Before servicing the vehicle, refer to the Precautions section.

2. Disconnect the negative battery cable.

3. Drain the power steering fluid from the reservoir tank. Properly dispose of used fluid.

4. On the 4.0L engine, remove the engine cover.

5. Remove the air duct assembly.

6. Remove the drive belt.

7. Disconnect the pressure sensor electrical connector.

8. Disconnect and plug the fluid lines.

9. Remove the pump retaining bolts.

10. Remove the pump from the vehicle.

To install:

11. Installation is the reverse of the removal procedure.

12. Bleed the power steering system.

59.5 (6.1, 44)
61 (6.2, 45)
15.7 (1.6, 12)
61 (6.2, 45)
61 (6.2, 45)
48 (4.9, 35)
15.7 (1.6, 12)
15.7 (1.6, 12)
65 (6.6, 48)
61 (6.2, 45)
60.8 (6.2, 45)

: Apply Genuine NISSAN PSF or equivalent.
: N·m (kg-m, ft-lb)
: Always replace after every disassembly.

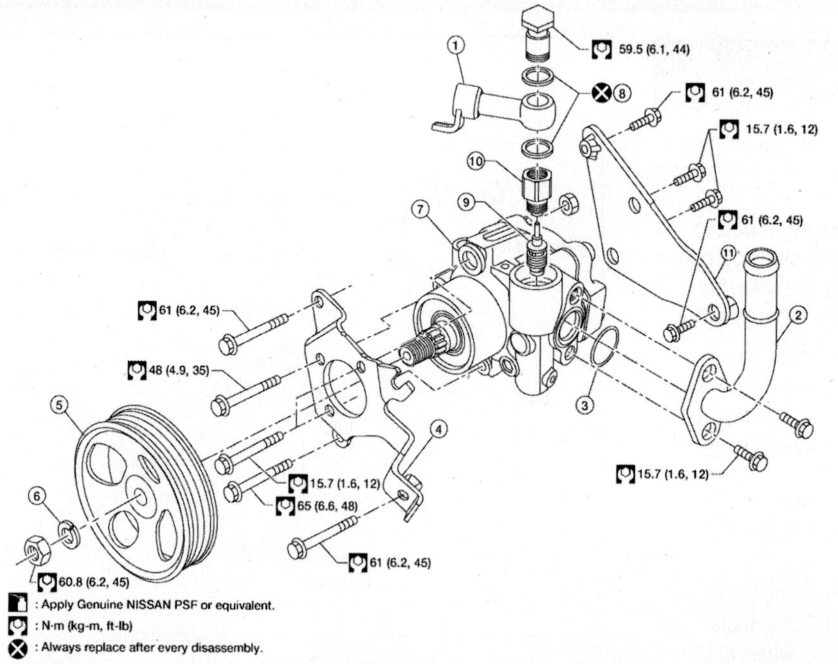

1. Joint
2. Suction pipe
3. O-ring
4. Front bracket
5. Pulley
6. Lock washer
7. Body assembly
8. Copper washers
9. Flow control valve and spring
10. Connector
11. Rear bracket

Fig. 85 Power steering pump and related components—2.5L engine

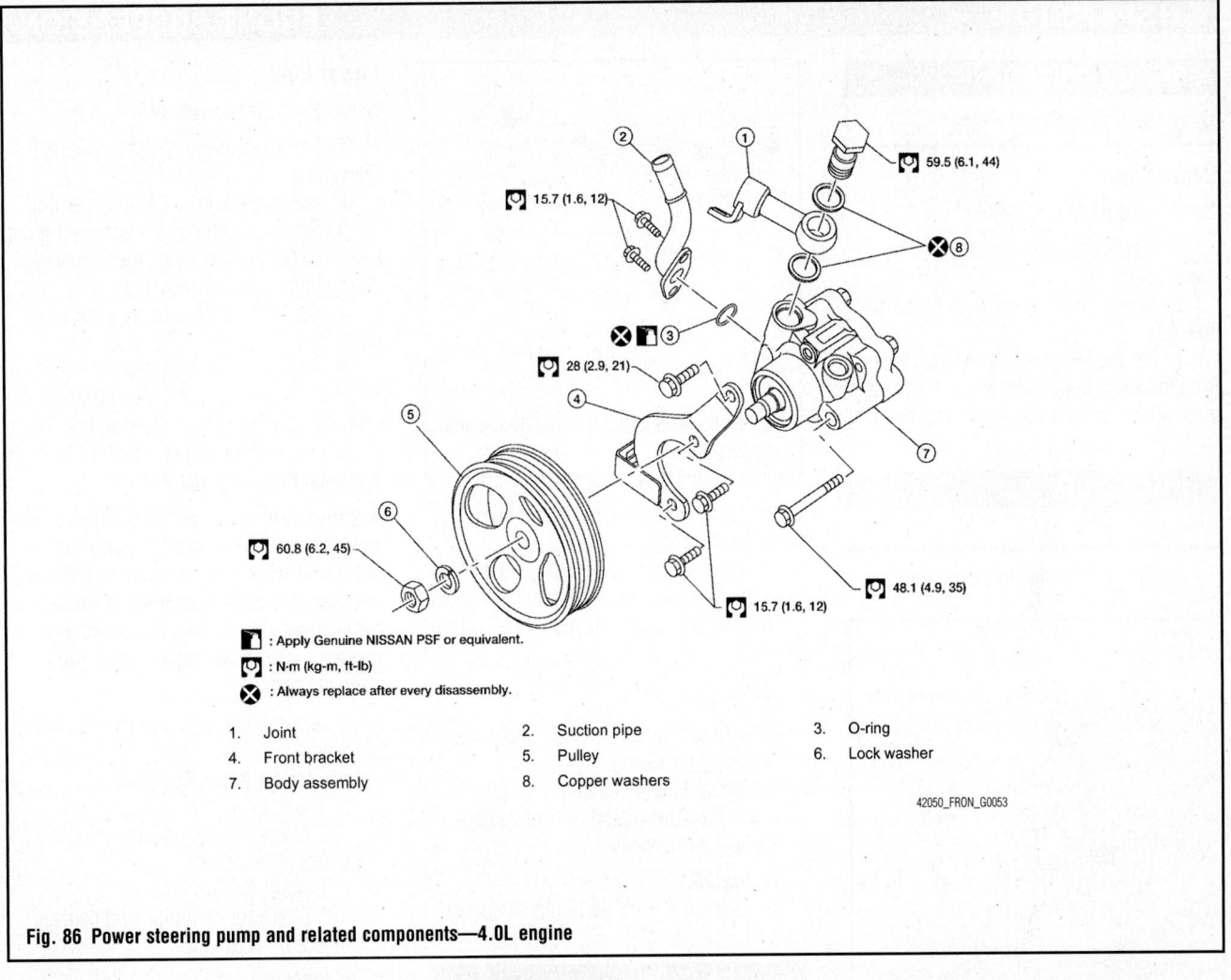

: Apply Genuine NISSAN PSF or equivalent.

: N·m (kg-m, ft-lb)

: Always replace after every disassembly.

1.	Joint	2.	Suction pipe	3.	O-ring
4.	Front bracket	5.	Pulley	8.	Copper washers
7.	Body assembly	6.	Lock washer		

42050_FRON_G0053

Fig. 86 Power steering pump and related components—4.0L engine

BLEEDING

1. Before servicing the vehicle, refer to the Precautions Section.

2. Fill the power steering system with the proper grade and type steering fluid.

➡**Do not allow the fluid level in the reservoir tank to go below the MIN level line. Check and add fluid as needed.**

3. Raise and safely support the vehicle.

4. Quickly turn the steering wheel to the full right and left detents and lightly touch the steering stoppers.

➡**Do not hold the steering wheel in the locked position for more than ten seconds.**

5. Repeat this operation until the fluid level no longer decreases.

6. Start the engine.

7. Quickly turn the steering wheel to the full right and left detents and lightly touch the steering stoppers.

➡**Do not hold the steering wheel in the locked position for more than ten seconds.**

8. Check for air bubbles or cloudy fluid. If found, repeat the bleeding procedure.

9. Stop the engine and check the fluid level. Correct as required.

SUSPENSION

CONTROL LINKS

REMOVAL & INSTALLATION

Upper Link

See Figures 87 through 89.

1. Remove the wheel and tire using power tool.

2. Support the lower link using a suitable jack.

3. For the LH side only, remove the bolt from the lower joint of the lower joint shaft, then reposition the lower joint shaft out of the way.

✳✳ CAUTION

Note the following:

- Do not damage the lower joint.

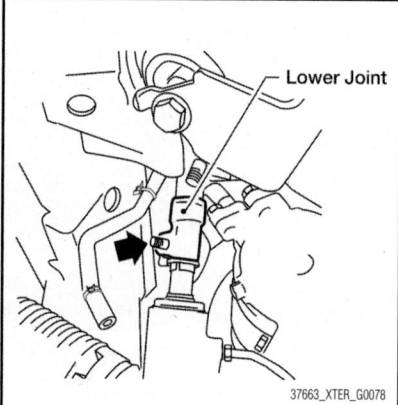

Fig. 87 For the left-hand side only, remove the bolt from the lower joint of the lower joint shaft

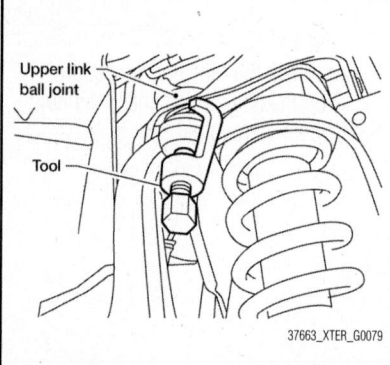

Fig. 88 Removal of the cotter pin and nut from upper link ball joint. Separate upper link ball joint stud from steering knuckle using tool number: ST290200001 (J-24319-01)

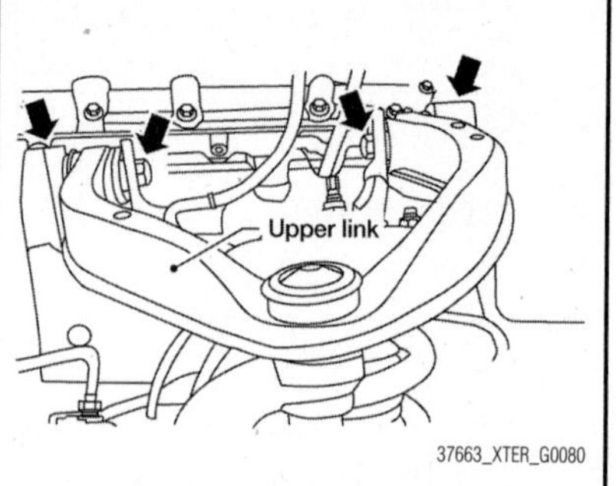

Fig. 89 Remove upper link mounting bolts and nuts

4. Remove cotter pin and nut from upper link ball joint.

5. Separate upper link ball joint stud from steering knuckle using a suitable tool.

6. Remove upper link mounting bolts and nuts.

Inspection After Removal

7. Upper Link

   a. Check for deformation and cracks. Replace if necessary.

8. Upper Link Ball Joint

   a. Check for distortion and damage. Replace if necessary.

To install:

9. Tighten all nuts and bolts to specification.

✳✳ CAUTION

Always replace drive shaft lock nut and cotter pin.

- After installation, check that the front wheel alignment is within specification.

10. Install upper link mounting bolts and nuts.

11. Install upper link ball joint stud to steering knuckle using
Tool.

12. Install cotter pin and nut from upper link ball joint.

✳✳ CAUTION

Note the following:

- Do not damage the lower joint.

13. For the LH side only, install the bolt from the lower joint of the lower joint shaft, then reposition the lower joint shaft.

14. Support the lower link using a suitable jack.

15. Install the wheel and tire using power tool.

FRONT SUSPENSION

Lower Link

See Figures 90 through 92.

1. Remove the wheel and tire using power tool.

2. Remove lower shock absorber bolt.

3. Remove stabilizer bar connecting rod lower nut using power tool, then separate connecting rod from lower link.

4. On 4WD models, remove the drive shaft. Refer to the Drive Train Section.

5. Remove pinch bolt from steering knuckle using power tool, then separate lower link ball joint from steering knuckle.

6. Remove lower link adjusting bolts and nuts, then the lower link.

➡**Some vehicles may be equipped with straight (non-adjustable) lower link bolts and washers. In order to adjust camber and caster on these vehicles, first replace the lower link bolts and washers with adjustable (cam) bolts and washers.**

7. Remove the jounce bumper from the lower link.

Inspection After Removal

8. Lower Link

   a. Check for deformation and cracks. Replace if necessary.

9. Lower Link Bushing

   a. Check for distortion and damage. Replace if necessary.

To install:

10. Tighten all nuts and bolts to specification.

11. When installing wheel and tire, refer to the Rotation information as follows: b. Remove wheels and tires.

   a. Rotate wheels and tires on each side from front to back as

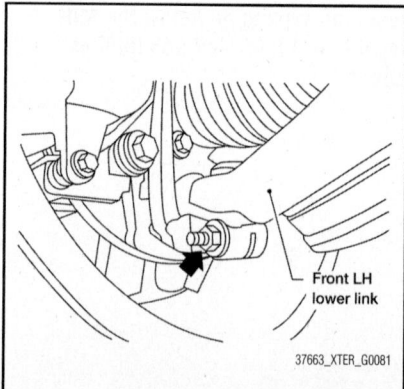

Fig. 90 Remove pinch bolt from steering knuckle using power tool, then separate lower link ball joint from steering knuckle

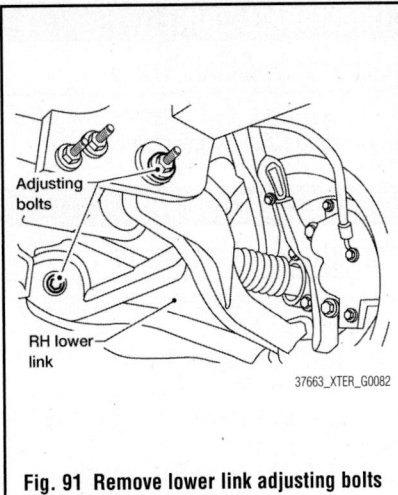

Fig. 91 Remove lower link adjusting bolts and nuts, then the lower link

shown. Do not include the spare wheel and tire when rotating the wheels and tires.

❊❊ CAUTION

When installing wheels and tires, tighten them diagonally by dividing the work two to three times in order to prevent the wheels from developing any distortion.

    b. Adjust the tire pressure to specification.

    c. After the wheel and tire rotation, retighten the wheel nuts after the vehicle has been driven for 1,000 km (600 miles), and also after any wheel and tire has been installed, such as after repairing a flat tire.

    12. After installation, check that the front wheel alignment is within specification.

    13. Install the jounce bumper from the lower link.

➡**Some vehicles may be equipped with straight (non-adjustable) lower link bolts and washers. In order to adjust camber and caster on these vehicles, first replace the lower link bolts and washers with adjustable (cam) bolts and washers.**

    14. Install lower link adjusting bolts and nuts, then the lower link.

    15. Install pinch bolt from steering knuckle using power tool, then separate lower link ball joint from steering knuckle.

    16. On 4WD models, install the drive shaft. Refer to the Drive Train Section.

    17. Install stabilizer bar connecting rod lower nut using power tool, then separate connecting rod from lower link.

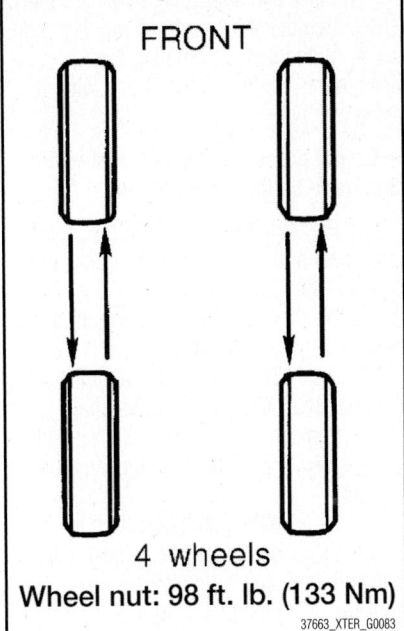

Fig. 92 Tire rotation and wheel nut tightening specifications

    18. Install lower shock absorber bolt.
    19. Install the wheel and tire using power tool.

LOWER BALL JOINT

REMOVAL & INSTALLATION

    The lower ball joint is serviced with the lower control arm as an assembly.

LOWER CONTROL ARM

REMOVAL AND & INSTALLATION

    1. Before servicing the vehicle, refer to the Precautions Section.
    2. Raise and support the vehicle safely.
    3. Remove the tire and wheel assembly.
    4. Remove the lower strut bolt.
    5. Remove the stabilizer bar connecting rod lower nut. Separate the connecting rod from the lower link.
    6. If equipped with 4WD, remove the halfshaft.
    7. Remove the pinch bolt from the steering knuckle. Separate the lower control arm ball joint stud from the steering knuckle, using the proper tool.
    8. Remove the lower control arm adjusting bolts and nuts. Lower the control arm and remove it from the vehicle.
    9. Remove the jounce bumper from the lower control arm.

To install:
    10. Installation is the reverse of the removal procedure.
    11. Be sure to replace all wearable components, as required.
    12. Check and adjust the front alignment, as required.

➡**After removing/installing or replacing steering and suspension components which effect wheel alignment or after adjusting wheel alignment, or the steering angle sensor or the ABS actuator electrical unit be sure to adjust the neutral position of the steering angle sensor before running the vehicle.**

    13. Position the steering wheel in the straight ahead position.
    14. Drive the vehicle at 10 mph for more than 10 minutes.
    15. When this procedure is complete the SLP indicator lamp and the VDC OFF indicator lamp will turn off.

MACPHERSON STRUT

REMOVAL & INSTALLATION

    1. Before servicing the vehicle, refer to the Precautions Section.
    2. Raise and support the vehicle safely.
    3. Remove the wheel and tire assembly.
    4. Support the lower link using a suitable jack.
    5. Remove connecting rod upper joints from stabilizer bar using power tool. Swing stabilizer bar down, repositioning it out of the way to access shock absorber lower mount.
    6. Remove the shock absorber lower bolt and nut.
    7. Remove the three shock absorber upper mounting nuts.
    8. Remove the coil spring and shock absorber assembly. Turn steering knuckle out to gain enough clearance for removal.

To install:
    9. Installation is the reverse of the removal procedure.
    10. The step in the strut assembly lower seat faces outside of vehicle.

OVERHAUL

Assembly

    1. When installing coil spring on strut, it must be positioned so spring ends align with proper spots on seats.
    2. Install the shock absorber mounting insulator in line with lower shock mount and step in lower seat. The step in the strut assembly lower seat faces outside of vehicle.

3. Tighten the new piston rod lock nut to 30 ft. lbs. (41 Nm).

4. Remove commercial service tool.

Disassembly

1. Set the shock absorber in a vise, then loosen (without removing) the piston rod lock nut.

✳✳ CAUTION

Do not remove piston rod lock nut at this time.

2. Compress the spring using commercial service tool until the shock absorber mounting insulator can be turned by hand.

✳✳ WARNING

Make sure that the pawls of the two spring compressors are firmly hooked on the spring. The spring compressors must be tightened alternately and evenly so as not to tilt the spring.

3. Remove the piston rod lock nut. Discard the piston rod lock nut, use a new nut for assembly.

4. Remove the components from the shock absorber. Keep the spring compressed in the commercial service tool if reusing it for assembly.

Inspection After Disassembly

1. Shock Absorber Assembly:
   - Check for smooth operation through a full stroke, both compression and extension
   - Check for oil leakage on welded or gland packing portions
   - Check piston rod for cracks, deformation or other damage and replace if necessary
2. Mounting Insulator and Rubber Parts:
   - Check cemented rubber-to-metal portion for separation or cracks. Check rubber parts for deterioration and replace if necessary
3. Coil Spring:
   - Check for cracks, deformation or other damage and replace if necessary
   - Check the free spring height. 2WD: 345.4 mm (13.6 inches), 4WD: 356 mm (14.0 inches).

STEERING KNUCKLE

REMOVAL & INSTALLATION

1. Before servicing the vehicle, refer to the Precautions Section.

2. Raise and support the vehicle safely.

3. Remove the wheel and tire assembly.

4. Without disassembling the hydraulic lines, remove brake caliper. Reposition it aside with wire.

➡ **Do not press the brake pedal while the brake caliper is removed.**

5. Put alignment marks on disc rotor and wheel hub and bearing assembly, then remove disc rotor.

6. Disconnect wheel sensor and remove bracket from steering knuckle.

7. On 4WD models, remove cotter pin, then remove lock nut from drive shaft.

8. Remove steering outer socket cotter pin at steering knuckle, then loosen mounting nut.

9. Remove the pinch bolt from the steering knuckle. Separate the lower control arm ball joint stud from the steering knuckle, using the proper tool.

10. Remove wheel hub and bearing assembly bolts.

11. Remove splash guard and wheel hub and bearing assembly from steering knuckle.

12. Remove cotter pin and nut from upper link ball joint.

13. Separate upper link ball joint from steering knuckle using Tool ST29020001 (J-24319-01).

14. Remove pinch bolt from steering knuckle, then separate lower link ball joint from steering knuckle.

15. Remove steering knuckle from vehicle.

To install:

16. Installation is the reverse of the removal procedure.

STABILIZER BAR (SWAY BAR) & LINKS

REMOVAL & INSTALLATION

1. Before servicing the vehicle, refer to the Precautions Section.

2. Remove the front valance center.

3. Raise and support the vehicle safely.

4. Remove the engine undercover.

5. Remove the connecting rod nuts.

6. Loosen the top bolts for the stabilizer bar mounting brackets. Remove the lower bolts from the mounting brackets.

7. Remove the stabilizer bar from the vehicle.

8. Remove the bushings from the stabilizer bar.

To install:

9. Installation is the reverse of the removal procedure.

UPPER BALL JOINT

REMOVAL & INSTALLATION

The upper ball joint is serviced with the upper control arm as an assembly.

UPPER CONTROL ARM

REMOVAL & INSTALLATION

1. Before servicing the vehicle, refer to the Precautions Section.

2. Raise and support the vehicle safely.

3. Remove the tire and wheel assembly.

4. Using a suitable jack, support the lower control arm.

5. If working on the left side, remove the bolt from the lower joint of the lower joint shaft, then reposition the lower joint shaft out of the way. Do not damage the lower joint.

6. Remove the cotter pin and nut from the upper control arm ball joint.

7. Separate the upper control arm ball joint stud from the steering knuckle, using tool ST29020001 (J-24319-01) or equivalent.

8. Remove the upper control arm retaining bolts and nuts.

9. Remove the upper control arm from the vehicle.

To install:

10. Installation is the reverse of the removal procedure.

11. Be sure to replace all wearable components, as required.

12. Check and adjust the front alignment, as required.

➡ **After removing/installing or replacing steering and suspension components which effect wheel alignment or after adjusting wheel alignment, or the steering angle sensor or the ABS actuator electrical unit be sure to adjust the neutral position of the steering angle sensor before running the vehicle.**

13. Position the steering wheel in the straight ahead position.

14. Drive the vehicle at 10 mph for more than 10 minutes.

15. When this procedure is complete the SLP indicator lamp and the VDC OFF indicator lamp will turn off.

WHEEL BEARINGS

REMOVAL & INSTALLATION

See Figure 93.

1. Before servicing the vehicle, refer to the Precautions Section.

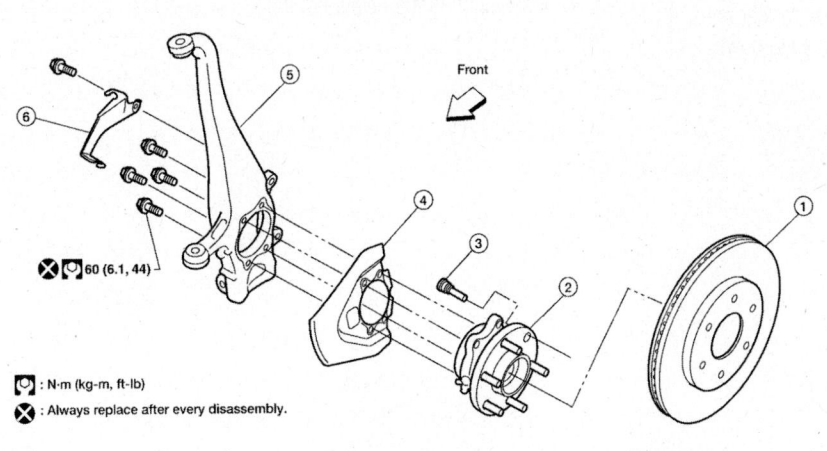

Fig. 93 Hub and bearing assembly

Labels:
1. Disc rotor
4. Splash guard
2. Wheel hub and bearing assembly
5. Steering knuckle
3. Wheel stud
6. Wheel sensor bracket

⊗ 🔧 60 (6.1, 44)

🔧 : N·m (kg-m, ft-lb)

⊗ : Always replace after every disassembly.

42050_FRON_G0104

2. Raise and support the vehicle safely.

3. Remove the tire and wheel assembly.

4. Remove the caliper and position it to the side with wire. Do not disconnect the brake fluid line.

➡**Do not press the brake pedal while the brake caliper is removed.**

5. Matchmark the brake rotor and the wheel hub. Remove the brake rotor.

6. Remove the cotter pin. Remove the lock nut.

7. Remove the halfshaft from the wheel hub and bearing assembly.

8. Remove the wheel sensor from the hub and bearing assembly. Do not pull on the wheel sensor harness.

9. Remove the wheel hub and bearing assembly bolts.

10. Remove the splash guard. Remove the wheel hub and bearing assembly from the steering knuckle.

➡**Carefully remove the wheel sensor and harness through the hole in the splash guard.**

To install:

11. Inspect the wheel sensor O-ring, replace the wheel speed sensor assembly, as required.

12. Installation is the reverse of the removal procedure.

13. Be sure to use new bolts when installing the wheel hub and bearing assembly.

SUSPENSION

LEAF SPRING

REMOVAL & INSTALLATION

1. Support rear axle assembly. Do not allow weight to be unsupported, but do not compress spring.

2. Remove lower end of shock absorber.

3. Remove u-bolt nuts and spring pad.

4. Remove rear shackle and bushings.

5. Remove front spring nut and bolt.

6. Remove spring.

To install:

7. Install front spring nut and bolt and rear shackle. Tighten finger tight.

8. Install U-bolts and bumper on axle. Install spring pad and nuts. Tighten nuts evenly to 53 ft. lbs. (72.5 Nm).

9. Install lower shock nut and bolt finger tight.

10. Lower vehicle and bounce to settle suspension. Tighten fasteners with vehicle on suspension.

11. Tighten lower shock absorber to 148 ft. lbs. (200 Nm).

12. Tighten front spring nut and bolt to 165 ft. lbs. (190 Nm).

REAR SUSPENSION

13. Tighten rear spring shackle nuts to 77 ft. lbs. (105 Nm)

SHOCK ABSORBER

REMOVAL & INSTALLATION

1. Support rear axle assembly.

2. Remove fasteners and shock.

To install:

3. Install shock absorber.

4. Install upper nut and tighten to 33 ft. lbs. (45 Nm).

5. Install lower nut and bolt. Tighten to 148 ft. lbs. (200 Nm).

6. Remove support.

STABILIZER BAR (SWAY BAR) & LINKS

REMOVAL & INSTALLATION

1. Disconnect the stabilizer bar ends from the connecting rods using power tool.

2. Remove the stabilizer bar clamps using power tool, and remove the bushings.
3. Remove the stabilizer bar.

To install:

4. Installation is the reverse of the removal procedure.

5. Install the stabilizer bar clamp and bushing so they are positioned outside of the crimp ring on the stabilizer bar.

NISSAN

Diagnostic Trouble Codes

DIAGNOSTIC TROUBLE CODES

OBD II VEHICLE APPLICATIONS

NISSAN

370
2011–2012
- 3.7L SMFI . . . . Engine Code: VQ37VHR

Altima
2011–2012
- 2.5L SMFI . . . . Engine Code: QR25DE
- 3.5L SMFI . . . . Engine Code: VQ35DE

Armada
2011-12
- 5.6L SMFI . . . . Engine Code: VK56DE

Cube
2011–2012
- 1.8L SMFI . . . . Engine Code: MR18DE

Frontier
2011–2012
- 2.5L SMFI . . . . Engine Code: QR25DE
- 4.0L SMFI . . . . Engine Code: VQ40DE

Juke
2011–2012
- 1.6L SMFI . . . . Engine Code: MR16DDT

Maxima
2011–2012
- 3.5L SFI . . . . . Engine Code: VQ35DE

Murano
2011–2012
- 3.5L MFI . . . . . Engine Code: VQ35DE

Pathfinder
2011–2012
- 4.0L MFI . . . . . Engine Code: VQ40DE
- 5.6L MFI . . . . . Engine Code: VK56DE

Quest
2011–2012
- 3.5L MFI . . . . . Engine Code: VQ35DE

Rogue
2011–2012
- 2.5L MPI . . . . . Engine Code: QR25DE

Sentra
2011–2012
- 2.0L SMFI . . . . Engine Code: MR20DE
- 2.5L SMFI . . . . Engine Code: QR25DE

Titan
2011–2012
- 5.6L SMFI . . . . Engine Code: VK56DE

Versa
2011–2012
- 1.6L MFI . . . . . Engine Code: HR16DE
- 1.8L MFI . . . . . Engine Code: MR18DE

Xterra
2011–2012
- 4.0L MFI . . . . . Engine Code: VQ40DE

OBD II Trouble Code List (P0XXX Codes)

DTC	Trouble Code Title and Conditions
DTC: P0010 **1T ECM, MIL: Yes** **Year:** 2011 **Model:** Altima, Quest **Engine:** 2.5L, 3.5L	**Camshaft Position "A" Actuator Circuit (Bank 1):** Monitor runs whenever following DTCs are not present: None All of the following conditions are met: A, B, C and D Starter: OFF Power switch: On (IG) Time after power switch off to on (IG): 0.5 seconds or more
DTC: P0011 **1T ECM, MIL: Yes** **Year:** 2011 **Model:** Altima, Quest **Engine:** 2.5L, 3.5L	**Camshaft Position "A" - Timing Over-Advanced or System Performance (Bank 1):** Monitor runs whenever following DTCs are not present: P0010 (Camshaft timing oil control valve assembly Bank 1) P0016 (VVT System Bank 1 - Misalignment) P0102, P0103 (Mass Air Flow Meter sub-assembly) P0107, P0108 (Manifold Absolute Pressure) P0115, P0117, P0118 (Engine Coolant Temperature Sensor) P0125 (Insufficient Engine Coolant Temperature for Closed Loop Fuel Control) P0335 (Crankshaft Position Sensor) P0340 (Camshaft Position Sensor) Battery voltage: 11 V or more Engine RPM: 800 to 4000 rpm Engine coolant temperature: 75 to 100°C (167 to 212°F)
DTC: P0011 **1T ECM, MIL: Yes** **Year:** 2011 **Model:** Titan **Engine:** 5.6L	**Intake Valve Timing Control Performance (Bank 1):** **NOTE: If DTC P0011 is displayed with DTC P0075, P0081, P1140 or P1145, first perform the trouble diagnosis for any other DTC before proceeding with P0011.** Condition A: The alignment of the intake valve timing control has been misregistered. Condition B: There is a gap between angle of target and phase-control angle degree.
DTC: P0011 **1T ECM, MIL: Yes** **Year:** 2011, 2012 **Model:** All (except Titan) **Engine:** 1.6L, 1.8L, 2.0L, 2.5L, 3.5L, 3.5L, 3.7L, 4.0L, 5.6L	**Intake Valve Timing Control Performance (Includes Hybrid Models):** HYBRID MODELS CAUTION: Hybrid systems use very high-voltage battery systems. Before starting any service work involving the battery system, turn the ignition switch OFF and then remove the service plug from pocket in the trunk. After removing the service plug, wait 10 minutes before touching any of the high-voltage connectors and terminals. **NOTE: When the malfunction is detected, the ECM enters fail-safe mode.** **NOTE: If DTC P0011 is displayed with DTC P0075, P0081, P0524, P1111, or P1136, perform the appropriate trouble diagnosis first.** * There is a gap between angle of target and phase-control angle degree.
DTC: P0012 **2T ECM, MIL: Yes** **Year:** 2011 **Model:** Altima, Quest **Engine:** 2.5L, 3.5L	**Camshaft Position "A" - Timing Over-Retarded (Bank 1):** Monitor runs whenever following DTCs are not present: P0010 (Camshaft timing oil control valve assembly Bank 1) P0016 (VVT System Bank 1 - Misalignment) P0102, P0103 (Mass Air Flow Meter sub-assembly) P0107, P0108 (Manifold Absolute Pressure) P0115, P0117, P0118 (Engine Coolant Temperature Sensor) P0125 (Insufficient Engine Coolant Temperature for Closed Loop Fuel Control) P0335 (Crankshaft Position Sensor) P0340 (Camshaft Position Sensor) Battery voltage: 11 V or more Engine RPM: 800 to 4000 rpm Engine coolant temperature: 75 to 100°C (167 to 212°F)
DTC: P0014 **1T ECM, MIL: Yes** **Year:** 2011 **Model:** Juke, Versa **Engine:** 1.6L	**Exhaust Valve Timing Control Performance:** There is a gap between angle of target and phase-control angle degree. ENG SPEED 1,200 - 3,175 rpm (A constant rotation is maintained.) COOLAN TEMP/S More than 60°C (140°F) Selector lever D position (CVT) 1st or 2nd position (M/T) Driving location Driving vehicle uphill (Increased engine load will help maintain the driving conditions required for this test.)

DTC	Trouble Code Title and Conditions
DTC: P0014 **1T ECM, MIL: Yes** **Year:** 2011, 2012 **Model:** Maxima **Engine:** 3.5L	**Exhaust Valve Timing (EVT) Control Performance (Bank 1):** **NOTE: If DTC P0014 or P0024 is displayed with DTC P0078, P0084, P1078 or P1084 first perform trouble diagnosis for respective DTC before proceeding with P0014.** There is a gap between angle of target and phase-control angle degree.
DTC: P0016 **2T ECM, MIL: Yes** **Year:** 2011 **Model:** Altima, Quest **Engine:** 2.5L, 3.5L	**Crankshaft Position - Camshaft Position Correlation (Bank 1 Sensor A):** Monitor runs whenever following DTCs are not present: P0010 (VVT Oil Control Valve Bank 1) P0102, P0103 (Mass Air Flow Meter Sub-assembly) P0107, P0108 (Manifold Absolute Pressure) P0115, P0117, P0118 (Engine Coolant Temperature Sensor) P0125 (Insufficient Engine Coolant Temperature for Closed Loop Fuel Control) P0335 (Crankshaft Position Sensor) P0340 (Camshaft Position Sensor) Engine RPM: 800 to 1000 rpm
DTC: P0021 **1T ECM, MIL: Yes** **Year:** 2011, 2012 **Model:** 370Z, Altima, Maxima, Murano, Pathfinder, Quest, Titan **Engine:** 2.5L, 3.5L, 3.5L, 3.7L, 4.0L, 5.6L	**Intake Valve Timing Control Performance (Bank 2):** * If DTC P0011 or P0021 is displayed with DTC P0075, P0081, P1111 or P1136, first perform trouble diagnosis for DTC P0075 or P0081. **NOTE: When the malfunction is detected, the ECM enters fail-safe mode.** * There is a gap between angle of target and phase-control angle degree. * When the malfunction is detected, the ECM enters fail-safe mode.
DTC: P0024 **1T ECM, MIL: Yes** **Year:** 2011, 2012 **Model:** Maxima **Engine:** 3.5L	**Exhaust Valve Timing (EVT) Control Performance (Bank 2):** **NOTE: If DTC P0014 or P0024 is displayed with DTC P0078, P0084, P1078 or P1084, first perform trouble diagnosis for the respective DTC before proceeding with P0024.** There is a gap between angle of target and phase-control angle degree.
DTC: P0031 **1T ECM, MIL: Yes** **Year:** 2011, 2012 **Model:** All **Engine:** 1.6L, 1.8L, 2.0L, 2.5L, 3.5L, 3.5L, 3.7L, 4.0L, 5.6L	**Air Fuel Ratio (A/F) Sensor 1 Heater Control Circuit Low (Includes Hybrid Models):** HYBRID MODELS CAUTION: Hybrid systems use very high-voltage battery systems. Before starting any service work involving the battery system, turn the ignition switch OFF and then remove the service plug from pocket in the trunk. After removing the service plug, wait 10 minutes before touching any of the high-voltage connectors and terminals. **NOTE: On and engines, this applies to Bank 1.** * The ECM performs ON/OFF duty control of the A/F sensor 1 heater corresponding to the engine operating condition to keep the temperature of A/F sensor 1 element at the specified range. * The current amperage in the A/F sensor 1 heater circuit is out of the normal range. (An excessively low voltage signal is sent to ECM through the A/F sensor heater.)
DTC: P0031 **1T ECM, MIL: Yes** **Year:** 2011 **Model:** Altima, Quest **Engine:** 2.5L, 3.5L	**Oxygen (A/F) Sensor Heater Control Circuit Low (Bank 1 Sensor 1):** Battery voltage: 10.5 V or more Time after engine start: 10 seconds or more Active heater OFF control: Not operating Active heater ON control: Not operating Heater output duty: 50% or more
DTC: P0032 **1T ECM, MIL: Yes** **Year:** 2011, 2012 **Model:** All **Engine:** 1.6L, 1.8L, 2.0L, 2.5L, 3.5L, 3.5L, 3.7L, 4.0L, 5.6L	**Air Fuel Ratio (A/F) Sensor 1 Heater (Bank 1) Control Circuit High (Includes Hybrid Models):** HYBRID MODELS CAUTION: Hybrid systems use very high-voltage battery systems. Before starting any service work involving the battery system, turn the ignition switch OFF and then remove the service plug from pocket in the trunk. After removing the service plug, wait 10 minutes before touching any of the high-voltage connectors and terminals. **Note: On and engines, this applies to Bank 1.** * The ECM performs ON/OFF duty control of the A/F sensor 1 heater corresponding to the engine operating condition to keep the temperature of A/F sensor 1 element at the specified range. * The current amperage in the A/F sensor 1 heater circuit is out of the normal range. (An excessively high voltage signal is sent to ECM through the A/F sensor heater.)
DTC: P0032 **1T ECM, MIL: Yes** **Year:** 2011 **Model:** Altima, Quest **Engine:** 2.5L, 3.5L	**Oxygen (A/F) Sensor Heater Control Circuit High (Bank 1 Sensor 1):** Battery voltage: 10.5 V or more Time after engine start: 10 seconds or more Active heater OFF control: Not operating Active heater ON control: Not operating Heater output duty: More than 0%

DTC	Trouble Code Title and Conditions
DTC: P0037 **1T ECM, MIL: Yes** **Year:** 2011, 2012 **Model:** All **Engine:** 1.6L, 1.8L, 2.0L, 2.5L, 3.5L, 3.5L, 3.7L, 4.0L, 5.6L	**Heated Oxygen Sensor 2 Heater Control Circuit Low (Includes Hybrid Models):** HYBRID MODEL CAUTION: Hybrid systems use very high-voltage battery systems. Before starting any service work involving the battery system, turn the ignition switch OFF and then remove the service plug from pocket in the trunk. After removing the service plug, wait 10 minutes before touching any of the high-voltage connectors and terminals. **Note: On and engines, this applies to Bank 1.** The current amperage in the heated oxygen sensor 2 heater circuit is out of the normal range. (An excessively low voltage signal is sent to ECM via the heated oxygen sensor 2 heater.)
DTC: P0037 **1T ECM, MIL: Yes** **Year:** 2011 **Model:** Altima, Quest **Engine:** 2.5L, 3.5L	**Oxygen Sensor Heater Control Circuit Low (Bank 1 Sensor 2):** Monitor runs whenever following DTCs are not present: None Case 1: Battery voltage: 10.5 V or more Engine: Running Starter: OFF Catalyst active air fuel ratio control: Not operating Time after heater ON: 10 seconds or more Learned heater OFF current operation completed flag: ON Learned heater OFF current get flag: ON Heated oxygen sensor heater high current fail (P0038): Not detected Case 2: Battery voltage: 10.5 V or more Engine: Running Starter: OFF Catalyst active air fuel ratio control: Not operating Time after heater ON: 10 seconds or more Learned heater OFF current operation completed flag: ON Heated oxygen sensor heater OFF current: More than 3.5 A Hybrid IC high current limiter monitor input: Fail Heated oxygen sensor heater high current fail (P0038): Not detected
DTC: P0038 **1T ECM, MIL: Yes** **Year:** 2011 **Model:** Altima, Quest **Engine:** 2.5L, 3.5L	**Oxygen Sensor Heater Control Circuit High (Bank 1 Sensor 2):** Monitor runs whenever following DTCs are not present: None Battery voltage: 10.5 V or more Engine: Running Starter: OFF Catalyst active air fuel ratio control: Not operating Time after heater ON: 10 seconds or more Learned heater OFF current operation completed flag: ON
DTC: P0038 **1T ECM, MIL: Yes** **Year:** 2011, 2012 **Model:** All **Engine:** 1.6L, 1.8L, 2.0L, 2.5L, 3.5L, 3.5L, 3.7L, 4.0L, 5.6L	**Heated Oxygen Sensor 2 Heater Control Circuit High (Includes Hybrid Models):** HYBRID MODELS CAUTION: Hybrid systems use very high-voltage battery systems. Before starting any service work involving the battery system, turn the ignition switch OFF and then remove the service plug from pocket in the trunk. After removing the service plug, wait 10 minutes before touching any of the high-voltage connectors and terminals. **Note: On and engines, this applies to Bank 1.** The current amperage in the heated oxygen sensor 2 heater circuit is out of the normal range. (An excessively high voltage signal is sent to ECM via the heated oxygen sensor 2 heater.)
DTC: P0043 **1T ECM, MIL: Yes** **Year:** 2011 **Model:** Altima, Rogue **Engine:** 2.5L	**Heated Oxygen Sensor 3 Heater Control Circuit Low (Includes Hybrid):** HYBRID MODELS CAUTION: Hybrid systems use very high-voltage battery systems. Before starting any service work involving the battery system, turn the ignition switch OFF and then remove the service plug from pocket in the trunk. After removing the service plug, wait 10 minutes before touching any of the high-voltage connectors and terminals. The current amperage in the heated oxygen sensor 3 heater circuit is out of the normal range. (An excessively low voltage signal is sent to ECM through the heated oxygen sensor 3 heater.)
DTC: P0044 **1T ECM, MIL: Yes** **Year:** 2011 **Model:** Altima, Rogue **Engine:** 2.5L	**Heated Oxygen Sensor 3 Heater Control Circuit High (Includes Hybrid Models):** HYBRID MODELS CAUTION: Hybrid systems use very high-voltage battery systems. Before starting any service work involving the battery system, turn the ignition switch OFF and then remove the service plug from pocket in the trunk. After removing the service plug, wait 10 minutes before touching any of the high-voltage connectors and terminals. The current amperage in the heated oxygen sensor 3 heater circuit is out of the normal range. (An excessively high voltage signal is sent to ECM through the heated oxygen sensor 3 heater.)

DTC	Trouble Code Title and Conditions
DTC: P0045 **1T ECM, MIL: Yes** **Year:** 2011 **Model:** Juke, Versa **Engine:** 1.6L	**TC BOOST CONTROL SOLENOID VALVE :** TC BOOST SOL/V (Turbocharger boost control solenoid valve circuit open) ECM detected the turbocharger boost control solenoid valve circuit is open.
DTC: P0047 **1T ECM, MIL: Yes** **Year:** 2011 **Model:** Juke, Versa **Engine:** 1.6L	**TC/SC BOOST CONTROL A SOLENOID VALVE CIRCUIT:** (Turbocharger boost control solenoid valve circuit low input) ECM detected the turbocharger boost control solenoid valve circuit is short to ground.
DTC: P0048 **1T ECM, MIL: Yes** **Year:** 2011 **Model:** Juke, Versa **Engine:** 1.6L	**TC/SC BOOST CONT A (Turbocharger boost control Circuit high input):** ECM detected the turbocharger boost control solenoid valve circuit s short to power.
DTC: P0051 **1T ECM, MIL: Yes** **Year:** 2011, 2012 **Model:** 370Z, Altima, Maxima, Murano, Pathfinder, Quest, Titan, Xterra **Engine:** 2.5L, 3.5L, 3.5L, 3.7L, 4.0L, 5.6L	**Air Fuel Ratio (A/F) Sensor 1 (Bank 2) Heater Control Circuit Low:** **Note: On and engines, this applies to Bank 2.** The ECM performs ON/OFF duty control of the A/F sensor 1 heater corresponding to the engine operating condition to keep the temperature of A/F sensor 1 element at the specified range. The current amperage in the A/F sensor 1 heater circuit is out of the normal range. (An excessively low voltage signal is sent to ECM through the A/F sensor heater.)
DTC: P0052 **1T ECM, MIL: Yes** **Year:** 2011, 2012 **Model:** 370Z, Altima, Maxima, Murano, Pathfinder, Quest, Titan, Xterra **Engine:** 3.5L, 3.5L, 3.7L, 4.0L, 5.6L	**Air Fuel Ratio (A/F) Sensor 1 (Bank 2) Heater Control Circuit High:** **Note: On and engines, this applies to Bank 2.** The ECM performs ON/OFF duty control of the A/F sensor 1 heater corresponding to the engine operating condition to keep the temperature of A/F sensor 1 element at the specified range. The current amperage in the A/F sensor 1 heater circuit is out of the normal range. (An excessively high voltage signal is sent to ECM through the A/F sensor heater.)
DTC: P0057 **1T ECM, MIL: Yes** **Year:** 2011, 2012 **Model:** 370Z, Altima, Maxima, Murano, Pathfinder, Quest, Titan, Xterra **Engine:** 2.5L, 3.5L, 3.5L, 3.7L, 4.0L, 5.6L	**Heated Oxygen Sensor 2 (Bank 2) Heater Control Circuit Low:** **Note: On and engines, this applies to Bank 2.** The current amperage in the heated oxygen sensor 2 heater circuit is out of the normal range. (An excessively low voltage signal is sent to ECM via the heated oxygen sensor 2 heater.)
DTC: P0058 **1T ECM, MIL: Yes** **Year:** 2011, 2012 **Model:** 370Z, Altima, Maxima, Murano, Pathfinder, Quest, Titan, Xterra **Engine:** 2.5L, 3.5L, 3.5L, 3.7L, 4.0L, 5.6L	**Heated Oxygen Sensor 2 (Bank 2) Heater Control Circuit High:** **Note: On and engines, this applies to Bank 2.** The current amperage in the heated oxygen sensor 2 heater circuit is out of the normal range. (An excessively high voltage signal is sent to ECM via the heated oxygen sensor 2 heater.)
DTC: P0075 **1T ECM, MIL: Yes** **Year:** 2011, 2012 **Model:** All **Engine:** 1.6L, 1.8L, 2.0L, 2.5L, 3.5L, 3.5L, 4.0L, 5.6L	**Intake Valve Timing Control Solenoid Valve Circuit (Includes Hybrid Models):** HYBRID MODELS CAUTION: Hybrid systems use very high-voltage battery systems. Before starting any service work involving the battery system, turn the ignition switch OFF and then remove the service plug from pocket in the trunk. After removing the service plug, wait 10 minutes before touching any of the high-voltage connectors and terminals. **Note: On and, this IVT is on bank 1.** * An improper voltage is sent to the ECM via the intake valve timing control solenoid valve.
DTC: P0078 **1T ECM, MIL: Yes** **Year:** 2011, 2012 **Model:** Juke, Maxima, Versa **Engine:** 1.6L, 3.5L	**Exhaust Valve Timing (EVT) Control Magnet Retarder (Bank 1) Circuit:** An improper voltage is sent to the ECM via the exhaust valve timing control magnet retarder.

DTC	Trouble Code Title and Conditions
DTC: P0081 **1T ECM, MIL: Yes** **Year:** 2011, 2012 **Model:** 370Z, Altima, Maxima, Murano, Pathfinder, Quest, Titan, Xterra **Engine:** 3.5L, 3.5L, 3.7L, 4.0L, 5.6L	**Intake Valve Timing Control Solenoid Valve Circuit (Bank 2):** **Note: Applies to and engines only.** * An improper voltage is sent to the ECM via the intake valve timing control solenoid valve. .
DTC: P0084 **1T ECM, MIL: Yes** **Year:** 2011, 2012 **Model:** Maxima **Engine:** 3.5L	**Exhaust Valve Timing (EVT) Timing Control (Magnet Retarder) Solenoid (Bank 2) Circuit:** An improper voltage is sent to the ECM via the exhaust valve timing control (magnet retarder).
DTC: P0087 **1T ECM, MIL: Yes** **Year:** 2011 **Model:** Juke, Versa **Engine:** 1.6L	**Fuel Rail Pressure (FRP) Too Low:** LOW FUEL PRES (High fuel pressure too low) Fuel rail pressure does not reach 1.3 MPa (13 bar, 13.3 kg/cm2, 188.5 psi) at engine cold start [water temperature 5°C (41°F) −40C° (104°F)]. Fuel rail pressure remains at 8.5 MPa (85 bar, 86.7 kg/cm2, 1232.8 psi) or less for 1 second or more during engine idle condition after cold start [water temperature 5°C (41°F) −40C° 104°F)].
DTC: P0088 **1T ECM, MIL: Yes** **Year:** 2011 **Model:** Juke, Versa **Engine:** 1.6L	**Fuel Rail Pressure (FRP) Too High:** (High fuel pressure too high) Fuel rail pressure remains at more than 12.5 MPa (125 bar, 127.5 kg/ cm2, 1813.0 psi) for 1 second or more during engine idle condition after cold start [water temperature 5°C (41°F) − 40C° (104°F)].
DTC: P0090 **1T ECM, MIL: Yes** **Year:** 2011 **Model:** Juke, Versa **Engine:** 1.6L	**FRP CONTROL SYSTEM CONTROL CIRCUIT/OPEN:** (High pressure fuel pump performance) Fuel rail pressure remains at 1.1 MPa (11 bar, 11.2 kg/ cm2, 159.5 psi) or less for 5 seconds or more during engine revFuel rail pressure remains at 18.5MPa (185 bar, 188.7 kg/cm2 , 2682.5 psi) or more for 0.3 seconds or more during engine rev.
DTC: P0096 **1T ECM, MIL: Yes** **Year:** 2011 **Model:** Juke, Versa **Engine:** 1.6L	**Intake Air Temperature (IAT) Sensor 2 Performance:** The comparison result of signals transmitted to ECM from each temperature sensor (IAT sensor 1, IAT sensor 2, ECT sensor, FTT sensor, and EOT sensor) shows that the voltage signal of the IAT sensor 2 is higher/lower than that of other temperature sensors when the engine is started with its cold state.
DTC: P0097 **1T ECM, MIL: Yes** **Year:** 2011 **Model:** Juke, Versa **Engine:** 1.6L	**IAT SENSOR CIRCUIT 2 B1 LOW INPUT:** ECM detects the following status continuously for 5 seconds or more: A voltage signal transmitted from the intake air temperature sensor 2 is 0.085 V or less.
DTC: P0098 **1T ECM, MIL: Yes** **Year:** 2011 **Model:** Juke, Versa **Engine:** 1.6L	**IAT SENSOR CIRCUIT 2 B1 HIGH INPUT:** ECM detects the following status continuously for 5 seconds or more: A voltage signal transmitted from the intake air temperature sensor 2 is 4.84 V or more.
DTC: P0101 **1T ECM, MIL: Yes** **Year:** 2011, 2012 **Model:** 370Z, Altima, Cube, Frontier, Juke, Maxima, Murano, Pathfinder, Quest, Rogue, Titan, Versa, Xterra **Engine:** 1.6L, 1.8L, 2.5L, 3.5L, 3.5L, 3.7L, 4.0L, 5.6L	**Mass Air Flow Sensor Circuit Range/Performance (Includes Hybrid Models):** HYBRID MODELS CAUTION: Hybrid systems use very high-voltage battery systems. Before starting any service work involving the battery system, turn the ignition switch OFF and then remove the service plug from pocket in the trunk. After removing the service plug, wait 10 minutes before touching any of the high-voltage connectors and terminals. * Condition A: A high voltage from the sensor is sent to ECM under light load driving condition. * Condition B: A low voltage from the sensor is sent to ECM under heavy load driving condition.
DTC: P0101 **1T ECM, MIL: Yes** **Year:** 2011 **Model:** Sentra **Engine:** 2.0L, 2.5L	**Mass Air Flow Sensor Circuit Range/Performance - Low Voltage:** * Engine: After warming up * Shift lever: P or N (CVT), Neutral (M/T) * Air conditioner switch: OFF * No load * A low voltage from the sensor is sent to ECM under heavy load driving condition.

DTC	Trouble Code Title and Conditions
DTC: P0101 **2T ECM, MIL: Yes** **Year:** 2011 **Model:** Altima, Quest **Engine:** 2.5L, 3.5L	**Mass Air Flow Circuit Range / Performance Problem:** Monitor runs whenever following DTCs are not present: None Throttle position (Throttle position sensor voltage): 0.2 to 2 V Time after engine start: 5 seconds or more Battery voltage: 10.5 V or more Engine coolant temperature: 70°C (158°F) or more Mass air flow sensor circuit fail (P0102, P0103): Not detected Intake air temperature sensor circuit fail (P0112, P0113): Not detected Engine coolant temperature sensor circuit fail (P0115, P0117, P0118): Not detected Crankshaft position sensor circuit fail (P0335, P0337, P0338): Not detected Throttle position sensor circuit fail (P0120, P0121, P0122, P0123, P0220, P0222, P0223, P2135): Not detected Canister pressure sensor circuit fail (P0452, P0453): Not detected EVAP leak detection pump fail (P2401, P2402): Not detected EVAP vent valve fail (P2419, P2420): Not detected
DTC: P0102 **T ECM, MIL: Yes** **Year:** 2011 **Model:** Altima, Quest **Engine:** 2.5L, 3.5L	**Mass or Volume Air Flow Circuit Low Input:** Monitor runs whenever following DTCs are not present: None
DTC: P0102 **1T ECM, MIL: Yes** **Year:** 2011, 2012 **Model:** All **Engine:** 1.6L, 1.8L, 2.0L, 2.5L, 3.5L, 3.5L, 3.7L, 4.0L, 5.6L	**Mass Air Flow Circuit Low Input (Includes Hybrid Models):** HYBRID MODELS CAUTION: Hybrid systems use very high-voltage battery systems. Before starting any service work involving the battery system, turn the ignition switch OFF and then remove the service plug from pocket in the trunk. After removing the service plug, wait 10 minutes before touching any of the high-voltage connectors and terminals. **NOTE: On and engines, this DTC is for Bank 1.** * An excessively low voltage from the sensor is sent to ECM.
DTC: P0103 **1T ECM, MIL: Yes** **Year:** 2011, 2012 **Model:** All **Engine:** 1.6L, 1.8L, 2.0L, 2.5L, 3.5L, 3.5L, 3.7L, 4.0L, 5.6L	**Mass Air Flow Sensor Circuit High Input (Includes Hybrid Models):** HYBRID MODELS CAUTION: Hybrid systems use very high-voltage battery systems. Before starting any service work involving the battery system, turn the ignition switch OFF and then remove the service plug from pocket in the trunk. After removing the service plug, wait 10 minutes before touching any of the high-voltage connectors and terminals. **NOTE: On and engines, this DTC is for Bank 1.** * An excessively high voltage from the sensor is sent to ECM.
DTC: P0103 **1T ECM, MIL: Yes** **Year:** 2011 **Model:** Altima, Quest **Engine:** 2.5L, 3.5L	**Mass or Volume Air Flow Circuit High Input:** Monitor runs whenever following DTCs are not present: None
DTC: P0106 **2T ECM, MIL: Yes** **Year:** 2011 **Model:** Altima, Quest **Engine:** 2.5L, 3.5L	**Manifold Absolute Pressure / Barometric Pressure Circuit Range / Performance Problem:** Monitor runs whenever following DTCs are not present: P0010 (Camshaft timing oil control valve assembly Bank 1) P0011 (VVT System Bank 1 - Advance) P0012 (VVT System Bank 1 - Retard) P0016 (VVT System Bank 1 - Misalignment) P0107, P0108 (Manifold Absolute Pressure) P0112, P0113 (Intake Air Temperature Sensor) P0115, P0117, P0118 (Engine Coolant Temperature Sensor) P0120, P0121, P0122, P0123, P0220, P0222, P0223, P2135 (Throttle Position Sensor) P0125 (Insufficient Engine Coolant Temperature for Closed Loop Fuel Control) P0335 (Crankshaft Position Sensor) P0340 (Camshaft Position Sensor) P106A (Evaporative Emission Control System Pressure Sensor - Manifold Pressure Sensor Correlation) Battery voltage: 10.5 V or more Engine coolant temperature: 10°C (50°F) or more Intake air temperature: 10°C (50°F) or more Engine speed: 1000 rpm or more Throttle position: Less than 2° Atmospheric pressure: 76 kPa-a (570 mmHg-a) or more

DTC	Trouble Code Title and Conditions
DTC: P0107 **1T ECM, MIL: Yes** **Year:** 2011 **Model:** Altima, Quest **Engine:** 2.5L, 3.5L	**Manifold Absolute Pressure / Barometric Pressure Circuit Low Input:** Monitor runs whenever following DTCs are not present: None Starter: Off Time after starter on to off: 2 seconds or more
DTC: P0108 **1T ECM, MIL: Yes** **Year:** 2011 **Model:** Altima, Quest **Engine:** 2.5L, 3.5L	**Manifold Absolute Pressure / Barometric Pressure Circuit High Input:** Monitor runs whenever following DTCs are not present: None Starter: Off Time after starter on to off: 2 seconds or more
DTC: P010A **1T ECM, MIL: Yes** **Year:** 2011, 2012 **Model:** 370Z **Engine:** 3.7L	**Manifold Absolute Pressure Sensor Circuit:** **NOTE: If DTC P010A is displayed with DTC P0643, first perform the trouble diagnosis for DTC P0643.** An excessively low voltage from the sensor is sent to ECM. An excessively high voltage from the sensor is sent to ECM.
DTC: P010B **1T ECM, MIL: Yes** **Year:** 2011, 2012 **Model:** 370Z **Engine:** 3.7L	**Mass Air Flow Sensor (Bank 2) Circuit/Range Performance:** Condition A: A high voltage from the sensor is sent to ECM under light load driving condition. Condition B: A low voltage from the sensor is sent to ECM under heavy load driving condition.
DTC: P010C **1T ECM, MIL: Yes** **Year:** 2011, 2012 **Model:** 370Z **Engine:** 3.7L	**Mass Air Flow Sensor (Bank 2) Circuit Low Input:** An excessively low voltage from the sensor is sent to ECM.
DTC: P010D **1T ECM, MIL: Yes** **Year:** 2011, 2012 **Model:** 370Z **Engine:** 3.7L	**Mass Air Flow Sensor (Bank 2) Circuit High Input:** An excessively high voltage from the sensor is sent to ECM.
DTC: P0111 **2T ECM, MIL: Yes** **Year:** 2011 **Model:** Altima, Quest **Engine:** 2.5L, 3.5L	**Intake Air Temperature Sensor Gradient Too High:** Monitor runs whenever following DTCs are not present: None After engine stop: Time after engine start: 10 seconds or more Battery voltage: 10.5 V or more Intake air temperature circuit fail (P0112, P0113): Not detected Engine coolant temperature sensor circuit fail (P0115, P0117, P0118): Not detected Mass air flow meter circuit fail (P0102, P0103): Not detected Accumulated mass air flow amount before engine stop: 2440 g or more Key-off duration: 30 minutes Engine coolant temperature when 30 minutes elapsed after engine stop: −40°C (−40°F) or more After cold engine start: Key-off duration: 5 hours Time after engine start: 10 seconds or more Intake air temperature circuit fail (P0112, P0113): Not detected Engine coolant temperature sensor circuit fail (P0115, P0117, P0118): Not detected Mass air flow meter circuit fail (P0102, P0103): Not detected Engine coolant temperature: 70°C (158°F) or more Accumulated mass air flow amount: 2440 g or more Either of the following conditions 1 or 2 is met: 1. Duration while engine load is low: 120 seconds or more 2. Duration while engine load is high: 10 seconds or more
DTC: P0111 **1T ECM, MIL: Yes** **Year:** 2011 **Model:** Juke, Versa **Engine:** 1.6L	**IAT SENSOR 1 B1 CIRCUIT RANGE/PERFORMANCE:** The comparison result of signals transmitted to ECM from each temperature sensor (IAT sensor 1, IAT sensor 2, ECT sensor, FTT sensor, and EOT sensor) shows that the voltage signal of the IAT sensor 1 is higher/lower than that of other temperature sensors when the engine is started with its cold state.

DTC	Trouble Code Title and Conditions
DTC: P0112 **1T ECM, MIL: Yes** **Year:** 2011 **Model:** Altima, Quest **Engine:** 2.5L, 3.5L	**Intake Air Temperature Circuit Low Input:** Monitor runs whenever following DTCs are not present: None Battery voltage: 8 V or more Power switch: On (IG)
DTC: P0112 **1T ECM, MIL: Yes** **Year:** 2011, 2012 **Model:** All **Engine:** 1.6L, 1.8L, 2.0L, 2.5L, 3.5L, 3.5L, 3.7L, 4.0L, 5.6L	**Intake Air Temperature Sensor Circuit Low Input (Includes Hybrid Models):** HYBRID MODELS CAUTION: Hybrid systems use very high-voltage battery systems. Before starting any service work involving the battery system, turn the ignition switch OFF and then remove the service plug from pocket in the trunk. After removing the service plug, wait 10 minutes before touching any of the high-voltage connectors and terminals. **NOTE: On and engines, this DTC is for Bank 1.** * An excessively low voltage from the sensor is sent to ECM.
DTC: P0113 **1T ECM, MIL: Yes** **Year:** 2011 **Model:** Altima, Quest **Engine:** 2.5L, 3.5L	**Intake Air Temperature Circuit High Input:** Monitor runs whenever following DTCs are not present: None Battery voltage: 8 V or more Power switch: On (IG)
DTC: P0113 **1T ECM, MIL: Yes** **Year:** 2011, 2012 **Model:** All **Engine:** 1.6L, 1.8L, 2.0L, 2.5L, 3.5L, 3.5L, 3.7L, 4.0L, 5.6L	**Intake Air Temperature Sensor Circuit High Input (Includes Hybrid Models):** HYBRID MODELS CAUTION: Hybrid systems use very high-voltage battery systems. Before starting any service work involving the battery system, turn the ignition switch OFF and then remove the service plug from pocket in the trunk. After removing the service plug, wait 10 minutes before touching any of the high-voltage connectors and terminals. **NOTE: On and engines, this DTC is for Bank 1.** * An excessively high voltage from the sensor is sent to ECM.
DTC: P0115 **1T ECM, MIL: Yes** **Year:** 2011 **Model:** Altima, Quest **Engine:** 2.5L, 3.5L	**Engine Coolant Temperature Circuit Malfunction:** Monitor runs whenever following DTCs are not present: None Engine coolant temperature sensor voltage: Less than 0.14 V, or more than 4.91 V
DTC: P0116 **2T ECM, MIL: Yes** **Year:** 2011 **Model:** Altima, Quest **Engine:** 2.5L, 3.5L	**Engine Coolant Temperature Circuit Range / Performance Problem:** Engine Coolant Temperature Sensor Cold Start Monitor: Monitor runs whenever following DTCs are not present: None Battery voltage: 10.5 V or more Time after engine start: 1 second or more Engine coolant temperature at engine start: Less than 60°C (140°F) Engine coolant temperature circuit fail (P0115, P0117, P0118, P0125): Not detected Intake air temperature sensor circuit fail (P0112, P0113): Not detected Soak time: 0 second or more Accumulated mass air flow: 828 g or more Fuel cut: OFF Difference between engine coolant temperature at engine start and intake air temperature: Less than 40°C (72°F) Engine Coolant Temperature Senosr Soak Monitor: Monitor runs whenever following DTCs are not present: None Battery voltage: 10.5 V or more Engine: Running Engine coolant temperature circuit fail (P0115, P0117, P0118, P0125): Not detected Intake air temperature circuit fail (P0112, P0113): Not detected Soak time: 5 hours or more Either of the following condition (a) or (b) is met: (a) Engine coolant temperature: 60°C (140°F) or more (b) Accumulated mass air flow: 1483 g or more
DTC: P0116 **1T ECM, MIL: Yes** **Year:** 2011, 2012 **Model:** 370Z, Altima, Frontier, Juke, Maxima, Murano, Pathfinder, Quest, Rogue, Titan, Versa, Xterra **Engine:** 1.6L, 1.8L, 2.5L, 3.5L, 3.5L, 3.7L, 4.0L, 5.6L	**Engine Coolant Temperature Sensor Circuit Range/Performance (Includes Hybrid Models):** HYBRID MODELS CAUTION: Hybrid systems use very high-voltage battery systems. Before starting any service work involving the battery system, turn the ignition switch OFF and then remove the service plug from pocket in the trunk. After removing the service plug, wait 10 minutes before touching any of the high-voltage connectors and terminals. **NOTE: If DTC P0116 is displayed with P0117 or P0118, first perform the trouble diagnosis for DTC P0117, P0118.** * Engine coolant temperature signal from engine coolant temperature sensor does not fluctuate, even when some time has passed after starting the engine with pre-warming up condition.

DTC	Trouble Code Title and Conditions
DTC: P0117 **1T ECM, MIL: Yes** **Year:** 2011, 2012 **Model:** All **Engine:** 1.6L, 1.8L, 2.0L, 2.5L, 3.5L, 3.5L, 3.7L, 4.0L, 5.6L	**Engine Coolant Temperature Circuit Low Input (Includes Hybrid Models):** HYBRID MODELS CAUTION: Hybrid systems use very high-voltage battery systems. Before starting any service work involving the battery system, turn the ignition switch OFF and then remove the service plug from pocket in the trunk. After removing the service plug, wait 10 minutes before touching any of the high-voltage connectors and terminals. * An excessively low voltage from the sensor is sent to ECM.
DTC: P0117 **1T ECM, MIL: Yes** **Year:** 2011 **Model:** Altima, Quest **Engine:** 2.5L, 3.5L	**Engine Coolant Temperature Circuit Low Input:** Monitor runs whenever following DTCs are not present: None Engine coolant temperature sensor voltage: Less than 0.14 V
DTC: P0118 **1T ECM, MIL: Yes** **Year:** 2011 **Model:** Altima, Quest **Engine:** 2.5L, 3.5L	**Engine Coolant Temperature Circuit High Input:** Monitor runs whenever following DTCs are not present: None Engine coolant temperature sensor voltage: More than 4.91 V
DTC: P0118 **1T ECM, MIL: Yes** **Year:** 2011, 2012 **Model:** All **Engine:** 1.6L, 1.8L, 2.0L, 2.5L, 3.5L, 3.5L, 3.7L, 4.0L, 5.6L	**Engine Coolant Temperature Sensor Circuit High Input (Includes Hybrid Models):** HYBRID MODELS CAUTION: Hybrid systems use very high-voltage battery systems. Before starting any service work involving the battery system, turn the ignition switch OFF and then remove the service plug from pocket in the trunk. After removing the service plug, wait 10 minutes before touching any of the high-voltage connectors and terminals. * An excessively high voltage from the sensor is sent to ECM.
DTC: P011B **2T ECM, MIL: Yes** **Year:** 2011 **Model:** Altima, Quest **Engine:** 2.5L, 3.5L	**Engine Coolant Temperature / Intake Air Temperature Correlation:** Monitor runs whenever following DTCs are not present: None All of the following conditions are met: Soak time: 7 hours or more Battery voltage: 10.5 V or more Time after engine start: 15 second or more Either of the following conditions is met: (a) or (b) (a) Minimum intake air temperature after engine start: -10°C (14°F) or more (b) Engine coolant temperature before engine start: -10°C (14°F) or more Engine coolant temperature sensor circuit fail (P0115, P0117, P0118, P0125): Not detected Intake air temperature sensor circuit fail (P0112, P0113): Not detected
DTC: P011C **1T ECM, MIL: Yes** **Year:** 2011 **Model:** Juke, Versa **Engine:** 1.6L	**CAT/IAT CRRLTN B1 (Charge air temperature/Intake air temperature correlation):** ECM detects a state that the temperature difference between intake air temperature sensor 1 and 2 remains 20°C (36°F) or less continuously for 5 seconds or more. ECM detects a state that the difference between the temperature of intake air temperature sensor 2 and its estimated temperature calculated by ECM from intake air temperature 1 and turbocharger boost sensor remains 106°C (191°F) or more continuously for 5 seconds or more.
DTC: P0120 **1T ECM, MIL: Yes** **Year:** 2011 **Model:** Altima, Quest **Engine:** 2.5L, 3.5L	**Throttle / Pedal Position Sensor / Switch "A" Circuit Malfunction:** Monitor runs whenever following DTCs are not present: None Either of the following conditions A or B is met: A. Power switch on (IG): 0.012 seconds or more B. Electronic throttle actuator power: ON
DTC: P0121 **1T ECM, MIL: Yes** **Year:** 2011 **Model:** Altima, Quest **Engine:** 2.5L, 3.5L	**Throttle / Pedal Position Sensor / Switch "A" Circuit Range / Performance Problem:** Monitor runs whenever following DTCs are not present: None Either of the following conditions A or B is set: A. Power switch: On (IG) B. Electric throttle motor power: ON Throttle position sensor malfunction (P0120, P0122, P0123, P0220, P0222, P0223, P2135): Not detected
DTC: P0121 **1T ECM, MIL: Yes** **Year:** 2011 **Model:** Xterra **Engine:** 4.0L	**Throttle Position (TP) Sensor 2 Circuit Range/Performance:** HYBRID MODELS CAUTION: Hybrid systems use very high-voltage battery systems. Before starting any service work involving the battery system, turn the ignition switch OFF and then remove the service plug from pocket in the trunk. After removing the service plug, wait 10 minutes before touching any of the high-voltage connectors and terminals. **NOTE:** If this DTC is displayed with DTC P0510, first perform the trouble diagnosis for DTC P0510. **NOTE:** On and engines, this DTC is for Bank 1. Condition A: A high voltage from the sensor is sent to ECM under light load driving condition. Condition B: A low voltage from the sensor is sent to ECM under heavy load driving condition.

DTC	Trouble Code Title and Conditions
DTC: P0122 **1T ECM, MIL: Yes** **Year:** 2011, 2012 **Model:** All **Engine:** 1.6L, 1.8L, 2.0L, 2.5L, 3.5L, 3.5L, 3.7L, 4.0L, 5.6L	**Throttle Position (TP) Sensor 2 Circuit Low Input (Includes Hybrid Models):** HYBRID MODELS CAUTION: Hybrid systems use very high-voltage battery systems. Before starting any service work involving the battery system, turn the ignition switch OFF and then remove the service plug from pocket in the trunk. After removing the service plug, wait 10 minutes before touching any of the high-voltage connectors and terminals. **NOTE: If this DTC is displayed with DTC P0510 or P0643, first perform the trouble diagnosis for DTC P0510 or P0643.** **NOTE: On and engines, this DTC is for Bank 1.** * An excessively low voltage from the TP sensor 2 is sent to ECM.
DTC: P0122 **1T ECM, MIL: Yes** **Year:** 2011 **Model:** Altima, Quest **Engine:** 2.5L, 3.5L	**Throttle / Pedal Position Sensor / Switch "A" Circuit Low Input:** Monitor runs whenever following DTCs are not present: None Either of the following conditions A or B is met: A. Power switch on (IG): 0.012 seconds or more B. Electronic throttle actuator power: ON
DTC: P0123 **1T ECM, MIL: Yes** **Year:** 2011 **Model:** Altima, Quest **Engine:** 2.5L, 3.5L	**Throttle / Pedal Position Sensor / Switch "A" Circuit High Input:** Monitor runs whenever following DTCs are not present: None Either of the following conditions A or B is met: A. Power switch on (IG): 0.012 seconds or more B. Electronic throttle actuator power: ON
DTC: P0123 **1T ECM, MIL: Yes** **Year:** 2011, 2012 **Model:** All **Engine:** 1.6L, 1.8L, 2.0L, 2.5L, 3.5L, 3.5L, 3.7L, 4.0L, 5.6L	**Throttle Position Sensor 2 Circuit High Input (Includes Hybrid Models):** HYBRID MODELS CAUTION: Hybrid systems use very high-voltage battery systems. Before starting any service work involving the battery system, turn the ignition switch OFF and then remove the service plug from pocket in the trunk. After removing the service plug, wait 10 minutes before touching any of the high-voltage connectors and terminals. **NOTE: If DTC P0122 or P0123 is displayed with DTC P0510 or P0643, first perform the trouble diagnosis for DTC P0510 or P0643.** **NOTE: On 3.7L engine, this DTC is for Bank 1.** * An excessively high voltage from the TP sensor 2 is sent to ECM. * When the malfunction is detected, ECM enters fail-safe mode and the MIL lights up.
DTC: P0125 **2T ECM, MIL: Yes** **Year:** 2011 **Model:** Altima, Quest **Engine:** 2.5L, 3.5L	**Insufficient Coolant Temperature for Closed Loop Fuel Control:** Monitor runs whenever following DTCs are not present: None Thermostat fail (P0128): Not detected Intake air temperature sensor circuit fail (P0112, P0113): Not detected Engine coolant temperature sensor circuit fail (P0115, P0117, P0118): Not detected Mass air flow meter circuit fail (P0102, P0103): Not detected
DTC: P0125 **1T ECM, MIL: Yes** **Year:** 2011, 2012 **Model:** All **Engine:** 1.6L, 1.8L, 2.0L, 2.5L, 3.5L, 3.5L, 3.7L, 4.0L, 5.6L	**Insufficient Engine Coolant Temperature for Closed Loop Fuel Control (Includes Hybrid Models):** HYBRID MODELS CAUTION: Hybrid systems use very high-voltage battery systems. Before starting any service work involving the battery system, turn the ignition switch OFF and then remove the service plug from pocket in the trunk. After removing the service plug, wait 10 minutes before touching any of the high-voltage connectors and terminals. * If DTC P0125 is displayed with P0116, P0117 or P0118, first perform the trouble diagnosis for the appropriate DTC, then proceed with P0125. * Voltage sent to ECM from the sensor is not practical, even when some time has passed after starting the engine. * Engine coolant temperature is insufficient for closed loop fuel control.
DTC: P0127 **1T ECM, MIL: Yes** **Year:** 2011, 2012 **Model:** All **Engine:** 1.6L, 1.8L, 2.0L, 2.5L, 3.5L, 3.5L, 3.7L, 4.0L, 5.6L	**Intake Air Temperature Too High (Includes Hybrid Models):** HYBRID MODELS CAUTION: Hybrid systems use very high-voltage battery systems. Before starting any service work involving the battery system, turn the ignition switch OFF and then remove the service plug from pocket in the trunk. After removing the service plug, wait 10 minutes before touching any of the high-voltage connectors and terminals. * Rationally incorrect voltage from the sensor is sent to ECM, compared with the voltage signal from engine coolant temperature sensor.

DTC	Trouble Code Title and Conditions
DTC: P0128 **2T ECM, MIL: Yes** **Year:** 2011 **Model:** Altima, Quest **Engine:** 2.5L, 3.5L	**Coolant Thermostat (Coolant Temperature Below Thermostat Regulating Temperature):** Monitor runs whenever following DTCs not present: P0010 (VVT Oil Control Valve Bank 1) P0011 (VVT System Bank 1 - Advance) P0012 (VVT System Bank 1 - Retard) P0016 (VVT System Bank 1 - Misalignment) P0031, P0032, P101D (Air Fuel Ratio Sensor Heater - Sensor 1) P0102, P0103 (Mass Air Flow Meter) P0107, P0108 (Manifold Absolute Pressure) P0112, P0113 (Intake Air Temperature Sensor) P0115, P0117, P0118 (Engine Coolant Temperature Sensor) P0120, P0121, P0122, P0123, P0220, P0222, P0223, P2135 (Throttle Position Sensor) P0171, P0172 (Fuel System) P0301, P0302, P0303, P0304 (Misfire) P0335 (Crankshaft Position Sensor) P0340 (Camshaft Position Sensor) P0351, P0352, P0353, P0354 (Igniter) P2195, P2196, P2237, P2238, P2239, P2252, P2253, P2A00 (Air Fuel Ratio Sensor - Sensor 1) Battery voltage: 11 V or more Either of the following conditions 1 or 2 is met: 1. All of following conditions are met: (a), (b) and (c) (a) Engine coolant temperature at engine start - Intake air temperature at engine start: -15 to 7°C (-27 to 12.6°F) (b) Engine coolant temperature at engine start: -10 to 56°C (14 to 133°F) (c) Intake air temperature at engine start: -10 to 56°C (14 to 133°F) 2. All of the following conditions are met: (d), (e) and (f) (d) Engine coolant temperature at engine start - Intake air temperature at engine start: More than 7°C (12.6°F) (e) Engine coolant temperature at engine start: 56°C (133°F) or less (f) Intake air temperature at engine start: -10°C (14°F) or more Accumulated time at 80 mph (128 km/h) or more: Less than 20 seconds
DTC: P0128 **1T ECM, MIL: Yes** **Year:** 2011, 2012 **Model:** 370Z, Altima, Cube, Frontier, Juke, Murano, Pathfinder, Quest, Rogue, Sentra, Titan, Versa, Xterra **Engine:** 1.6L, 1.8L, 2.0L, 2.5L, 3.5L, 3.5L, 3.7L, 4.0L, 5.6L	**Thermostat Function (includes Hybrid):** **NOTE: If DTC P0128 is displayed with DTC P0300, P0301, P0302, P0303, P0304, P0305, P0306, P0307 or P0308 first perform the trouble diagnosis for this DTC before continuing with P0128 diagnosis.** * The engine coolant temperature does not reach to specified temperature even though the engine has run long enough.
DTC: P0130 **1T ECM, MIL: Yes** **Year:** 2011, 2012 **Model:** All **Engine:** 1.6L, 1.8L, 2.0L, 2.5L, 3.5L, 3.5L, 3.7L, 4.0L, 5.6L	**Air Fuel Ratio (A/F) Sensor 1 Circuit (Includes Hybrid Models):** HYBRID MODELS CAUTION: Hybrid systems use very high-voltage battery systems. Before starting any service work involving the battery system, turn the ignition switch OFF and then remove the service plug from pocket in the trunk. After removing the service plug, wait 10 minutes before touching any of the high-voltage connectors and terminals. **NOTE: On and engines, this applies to Bank 1.** Condition A: The A/F signal computed by ECM from the A/F sensor 1 signal is constantly in the range other than approx. 2.2V. Condition B: The A/F signal computed by ECM from the A/F sensor 1 signal is constantly approx. 2.2V.
DTC: P0131 **1T ECM, MIL: Yes** **Year:** 2011, 2012 **Model:** All **Engine:** 1.6L, 1.8L, 2.0L, 2.5L, 3.5L, 3.5L, 3.7L, 4.0L, 5.6L	**Air Fuel Ratio (A/F) Sensor 1 Circuit Low Voltage (Includes Hybrid Models):** HYBRID MODELS CAUTION: Hybrid systems use very high-voltage battery systems. Before starting any service work involving the battery system, turn the ignition switch OFF and then remove the service plug from pocket in the trunk. After removing the service plug, wait 10 minutes before touching any of the high-voltage connectors and terminals. **NOTE: On and engines, this applies to Bank 1.** * To judge the malfunction, the diagnosis checks that the A/F signal computed by ECM from the A/F sensor 1 signal is not inordinately low. * The A/F signal computed by ECM from the A/F sensor 1 signal is constantly approx. 0V.

DTC	Trouble Code Title and Conditions
DTC: P0137 **2T ECM, MIL: Yes** **Year:** 2011 **Model:** Altima, Quest **Engine:** 2.5L, 3.5L	**Oxygen Sensor Circuit Low Voltage (Bank 1 Sensor 2):** Monitor runs whenever following DTCs not stored: P0016 (VVT System - Misalignment) P0031, P0032, P101D (Air Fuel Ratio Sensor Heater) P0037, P0038, P102D (Heated Oxygen Sensor Heater) P2195, P2196, P2237, P2238, P2239, P2252, P2253, P2A00 (Air Fuel Ratio Sensor) P0102, P0103 (Mass Air Flow Meter) P0107, P0108 (Manifold Absolute Pressure) P0112, P0113 (Intake Air Temperature Sensor) P0115, P0117, P0118 (Engine Coolant Temperature Sensor) P0125 (Insufficient Coolant Temperature for Closed Loop Fuel Control) P0120, P0121, P0122, P0123, P0220, P0222, P0223, P2135 (Throttle Position Sensor) P0128 (Thermostat) P0171, P0172 (Fuel System) P0301 - P0304 (Misfire) P0335 (Crankshaft Position Sensor) P0340 (Camshaft Position Sensor) P0451, P0452, P0453 (EVAP System) Heated Oxygen Sensor Output Voltage (Output Voltage, Hight Voltage and Low Voltage) Active air fuel ratio control: Performing Active air fuel ratio control begins when all of following conditions met: Battery voltage: 11 V or higher Engine coolant temperature: 75°C (167°F) or more Idling: OFF Engine speed: Less than 3200 rpm Air fuel ratio sensor status: Activated Fuel system status: Closed loop Fuel cut: OFF Engine load: 10 to 70% Heated Oxygen Sensor Impedance (Low Impedance): Battery voltage: 11 V or more Estimated sensor temperature: Below 700°C (1292°F) ECM (included in the hybrid vehicle control ECU) monitor: Completed DTC P0607: Not detected Heated Oxygen Sensor Impedance (High Impedance): Battery voltage: 11 V or more Estimated sensor temperature: 450 to 750°C (842 to 1382°F) DTC P0607: Not detected Heated Oxygen Sensor Output Voltage (Extremely High): Battery voltage: 11 V or more Time after engine start: 2 seconds or more Heated Oxygen Sensor Voltage During Fuel Cut: Engine coolant temperature: 70°C (158°F) or more Estimated catalyst temperature: 530°C (986°F) or more Fuel cut: ON
DTC: P0138 **1T ECM, MIL: Yes** **Year:** 2011, 2012 **Model:** All **Engine:** 1.6L, 1.8L, 2.0L, 2.5L, 3.5L, 3.5L, 3.7L, 4.0L, 5.6L	**Heated Oxygen Sensor 2 Circuit High Voltage (Includes Hybrid Models):** HYBRID MODELS CAUTION: Hybrid systems use very high-voltage battery systems. Before starting any service work involving the battery system, turn the ignition switch OFF and then remove the service plug from pocket in the trunk. After removing the service plug, wait 10 minutes before touching any of the high-voltage connectors and terminals. **NOTE: On and engines, this applies to Bank 1.** * Condition A: An excessively high voltage from the sensor is sent to ECM, or, * Condition B: The minimum voltage from the sensor is not reached to the specified voltage.

DTC	Trouble Code Title and Conditions
DTC: P0138 **2T ECM, MIL: Yes** **Year:** 2011 **Model:** Altima, Quest **Engine:** 2.5L, 3.5L	**Oxygen Sensor Circuit High Voltage (Bank 1 Sensor 2):** Monitor runs whenever following DTCs not stored: P0016 (VVT System - Misalignment) P0031, P0032, P101D (Air Fuel Ratio Sensor Heater) P0037, P0038, P102D (Heated Oxygen Sensor Heater) P2195, P2196, P2237, P2238, P2239, P2252, P2253, P2A00 (Air Fuel Ratio Sensor) P0102, P0103 (Mass Air Flow Meter) P0107, P0108 (Manifold Absolute Pressure) P0112, P0113 (Intake Air Temperature Sensor) P0115, P0117, P0118 (Engine Coolant Temperature Sensor) P0125 (Insufficient Coolant Temperature for Closed Loop Fuel Control) P0120, P0121, P0122, P0123, P0220, P0222, P0223, P2135 (Throttle Position Sensor) P0128 (Thermostat) P0171, P0172 (Fuel System) P0301 - P0304 (Misfire) P0335 (Crankshaft Position Sensor) P0340 (Camshaft Position Sensor) P0451, P0452, P0453 (EVAP System) Heated Oxygen Sensor Output Voltage (Output Voltage, Hight Voltage and Low Voltage) Active air fuel ratio control: Performing Active air fuel ratio control begins when all of following conditions met: Battery voltage: 11 V or higher Engine coolant temperature: 75°C (167°F) or more Idling: OFF Engine speed: Less than 3200 rpm Air fuel ratio sensor status: Activated Fuel system status: Closed loop Fuel cut: OFF Engine load: 10 to 70% Heated Oxygen Sensor Impedance (Low Impedance): Battery voltage: 11 V or more Estimated sensor temperature: Below 700°C (1292°F) ECM (included in the hybrid vehicle control ECU) monitor: Completed DTC P0607: Not detected Heated Oxygen Sensor Impedance (High Impedance): Battery voltage: 11 V or more Estimated sensor temperature: 450 to 750°C (842 to 1382°F) DTC P0607: Not detected Heated Oxygen Sensor Output Voltage (Extremely High): Battery voltage: 11 V or more Time after engine start: 2 seconds or more Heated Oxygen Sensor Voltage During Fuel Cut: Engine coolant temperature: 70°C (158°F) or more Estimated catalyst temperature: 530°C (986°F) or more Fuel cut: ON
DTC: P0139 **1T ECM, MIL: Yes** **Year:** 2011, 2012 **Model:** 370Z, Altima, Juke, Maxima, Pathfinder, Quest, Rogue, Sentra, Titan, Versa, Xterra **Engine:** 1.6L, 1.8L, 2.0L, 2.5L, 3.5L, 3.7L, 4.0L, 5.6L	**Heated Oxygen Sensor 2 Circuit Slow Response (Includes Hybrid Models):** HYBRID MODELS CAUTION: Hybrid systems use very high-voltage battery systems. Before starting any service work involving the battery system, turn the ignition switch OFF and then remove the service plug from pocket in the trunk. After removing the service plug, wait 10 minutes before touching any of the high-voltage connectors and terminals. **NOTE: On 4-cyl, sensor 2; on, sensor 2 bank 1** * It takes more time for the sensor to respond between rich and lean than the specified time.

DTC	Trouble Code Title and Conditions
DTC: P0139 **2T ECM, MIL:** Yes **Year:** 2011 **Model:** Altima, Quest **Engine:** 2.5L, 3.5L	**Oxygen Sensor Circuit Slow Response (Bank 1 Sensor 2):** Monitor runs whenever following DTCs not stored: P0016 (VVT System - Misalignment) P0031, P0032, P101D (Air Fuel Ratio Sensor Heater) P0037, P0038, P102D (Heated Oxygen Sensor Heater) P2195, P2196, P2237, P2238, P2239, P2252, P2253, P2A00 (Air Fuel Ratio Sensor) P0102, P0103 (Mass Air Flow Meter) P0107, P0108 (Manifold Absolute Pressure) P0112, P0113 (Intake Air Temperature Sensor) P0115, P0117, P0118 (Engine Coolant Temperature Sensor) P0125 (Insufficient Coolant Temperature for Closed Loop Fuel Control) P0120, P0121, P0122, P0123, P0220, P0222, P0223, P2135 (Throttle Position Sensor) P0128 (Thermostat) P0171, P0172 (Fuel System) P0301 - P0304 (Misfire) P0335 (Crankshaft Position Sensor) P0340 (Camshaft Position Sensor) P0451, P0452, P0453 (EVAP System) Heated Oxygen Sensor Output Voltage (Output Voltage, Hight Voltage and Low Voltage) Active air fuel ratio control: Performing Active air fuel ratio control begins when all of following conditions met: Battery voltage: 11 V or higher Engine coolant temperature: 75°C (167°F) or more Idling: OFF Engine speed: Less than 3200 rpm Air fuel ratio sensor status: Activated Fuel system status: Closed loop Fuel cut: OFF Engine load: 10 to 70% Heated Oxygen Sensor Impedance (Low Impedance): Battery voltage: 11 V or more Estimated sensor temperature: Below 700°C (1292°F) ECM (included in the hybrid vehicle control ECU) monitor: Completed DTC P0607: Not detected Heated Oxygen Sensor Impedance (High Impedance): Battery voltage: 11 V or more Estimated sensor temperature: 450 to 750°C (842 to 1382°F) DTC P0607: Not detected Heated Oxygen Sensor Output Voltage (Extremely High): Battery voltage: 11 V or more Time after engine start: 2 seconds or more Heated Oxygen Sensor Voltage During Fuel Cut: Engine coolant temperature: 70°C (158°F) or more Estimated catalyst temperature: 530°C (986°F) or more Fuel cut: ON
DTC: P0141 **2T ECM, MIL:** Yes **Year:** 2011 **Model:** Altima, Quest **Engine:** 2.5L, 3.5L	**Oxygen Sensor Heater Circuit Malfunction (Bank 1 Sensor 2):** Monitor runs whenever following DTCs are not present: None Case 1: Heated oxygen sensor heater circuit fail (P0037 and P0038): Not detected Battery voltage: 10.5 V or more Fuel cut: OFF Time after fuel cut ON to OFF: 30 seconds or more Accumulated heater ON time: 100 seconds or more Learned heater OFF current operation completed flag: ON Case 2: Duration that rear heated oxygen sensor impedance is less than 15 kΩ: 2 seconds or more

DTC	Trouble Code Title and Conditions
DTC: P0143 **1T ECM, MIL: Yes** **Year:** 2011, 2012 **Model:** Altima, Maxima, Murano, Pathfinder, Rogue, Sentra **Engine:** 2.5L, 3.5L, 3.5L, 4.0L, 5.6L	**Heated Oxygen Sensor 3 Circuit High Voltage (Includes Hybrid Models):** HYBRID MODELS CAUTION: Hybrid systems use very high-voltage battery systems. Before starting any service work involving the battery system, turn the ignition switch OFF and then remove the service plug from pocket in the trunk. After removing the service plug, wait 10 minutes before touching any of the high-voltage connectors and terminals. The minimum voltage from the sensor is not reached to the specified voltage.
DTC: P0144 **1T ECM, MIL: Yes** **Year:** 2011, 2012 **Model:** Altima, Frontier, Maxima, Murano, Pathfinder, Quest, Rogue **Engine:** 2.5L, 3.5L, 3.5L, 4.0L, 5.6L	**Heated Oxygen Sensor 3 Circuit Low Voltage (Includes Hybrid Models) :** HYBRID MODELS CAUTION: Hybrid systems use very high-voltage battery systems. Before starting any service work involving the battery system, turn the ignition switch OFF and then remove the service plug from pocket in the trunk. After removing the service plug, wait 10 minutes before touching any of the high-voltage connectors and terminals. The maximum voltage from the sensor is not reached to the specified voltage.
DTC: P0145 **1T ECM, MIL: Yes** **Year:** 2011, 2012 **Model:** Altima, Frontier, Maxima, Pathfinder, Quest, Rogue **Engine:** 2.5L, 3.5L, 4.0L, 5.6L	**Heated Oxygen Sensor 3 Circuit Slow Response (Includes Hybrid Models):** HYBRID MODELS CAUTION: Hybrid systems use very high-voltage battery systems. Before starting any service work involving the battery system, turn the ignition switch OFF and then remove the service plug from pocket in the trunk. After removing the service plug, wait 10 minutes before touching any of the high-voltage connectors and terminals. It takes more time for the sensor to respond between rich and lean than the specified time.
DTC: P0146 **1T ECM, MIL: Yes** **Year:** 2011, 2012 **Model:** Altima, Frontier, Maxima, Murano, Pathfinder, Quest, Rogue **Engine:** 2.5L, 3.5L, 3.5L, 4.0L, 5.6L	**Heated Oxygen Sensor 3 Circuit No Response Detected (Includes Hybrid Models):** HYBRID MODELS CAUTION: Hybrid systems use very high-voltage battery systems. Before starting any service work involving the battery system, turn the ignition switch OFF and then remove the service plug from pocket in the trunk. After removing the service plug, wait 10 minutes before touching any of the high-voltage connectors and terminals. An excessively high voltage from the sensor is sent to ECM.
DTC: P014C **1T ECM, MIL: Yes** **Year:** 2011 **Model:** Altima, Quest, Versa **Engine:** 1.8L, 2.5L, 3.5L	**Air Fuel Ratio (A/F) Sensor 1 Circuit Slow Response:** To judge the malfunction of A/F sensor 1, this diagnosis measures response time of the A/F signal computed by ECM from the A/F sensor 1 signal. The time is compensated by engine operating (speed and load), fuel feedback control constant, and the A/F sensor 1 temperature index. Judgment is based on whether the compensated time (the A/F signal cycling time index) is inordinately long or not. The response of the A/F signal computed by ECM from A/F sensor 1 signal takes more than the specified time.
DTC: P014D **1T ECM, MIL: Yes** **Year:** 2011 **Model:** Altima, Quest, Versa **Engine:** 1.8L, 2.5L, 3.5L	**Air Fuel Ratio (A/F) Sensor 1 Circuit Slow Response:** To judge the malfunction of A/F sensor 1, this diagnosis measures response time of the A/F signal computed by ECM from the A/F sensor 1 signal. The time is compensated by engine operating (speed and load), fuel feedback control constant, and the A/F sensor 1 temperature index. Judgment is based on whether the compensated time (the A/F signal cycling time index) is inordinately long or not. The response of the A/F signal computed by ECM from A/F sensor 1 signal takes more than the specified time.
DTC: P0150 **1T ECM, MIL: Yes** **Year:** 2011, 2012 **Model:** 370Z, Altima, Maxima, Murano, Pathfinder, Quest, Titan, Xterra **Engine:** 3.5L, 3.5L, 3.7L, 4.0L, 5.6L	**Air Fuel Ratio (A/F) Sensor 1 Bank 2 Circuit:** * The A/F signal computed by ECM from the A/F sensor 1 bank 2 signal is constantly in the range other than approx. 1.5V or 2.2V. * The A/F signal computed by ECM from the A/F sensor 1 bank 2 signal is constantly approx. 1.5V or 2.2V.
DTC: P0151 **1T ECM, MIL: Yes** **Year:** 2011, 2012 **Model:** 370Z, Altima, Maxima, Murano, Pathfinder, Quest, Titan, Xterra **Engine:** 3.5L, 3.5L, 3.7L, 4.0L, 5.6L	**Air Fuel Ratio (A/F) Sensor 1 Bank 2 Circuit Low Voltage:** * To judge the malfunction, the diagnosis checks that the A/F signal computed by ECM from the A/F sensor 1 signal is inordinately low. * The A/F signal computed by ECM from the A/F sensor 1 signal is constantly approx. 0V. * Maximum voltage is not reached.

DTC	Trouble Code Title and Conditions
DTC: P0152 **1T ECM, MIL: Yes** **Year:** 2011, 2012 **Model:** 370Z, Altima, Maxima, Murano, Pathfinder, Quest, Titan, Xterra **Engine:** 2.5L, 3.5L, 3.5L, 3.7L, 4.0L, 5.6L	**Air Fuel Ratio (A/F) Sensor 1 (Bank 2) Circuit High Voltage:** **NOTE: On and engines, this applies to Bank 2.** * Engine: After warming up. Maintaining engine speed at 2,000 rpm. * The A/F signal computed by ECM from the A/F sensor 1 signal is constantly approx. 5V. * An excessively high voltage from the sensor is sent to ECM.
DTC: P0153 **1T ECM, MIL: Yes** **Year:** 2011, 2012 **Model:** 370Z, Altima, Maxima, Murano, Pathfinder, Quest, Titan, Xterra **Engine:** 3.5L, 3.5L, 3.7L, 4.0L, 5.6L	**Air Fuel Ratio (A/F) Sensor 1 Bank 2 Circuit Slow Response:** The response of the A/F signal computed by ECM from A/F sensor 1 signal takes more than the specified time.
DTC: P0157 **1T ECM, MIL: Yes** **Year:** 2011, 2012 **Model:** 370Z, Altima, Maxima, Murano, Pathfinder, Quest, Titan, Xterra **Engine:** 3.5L, 3.5L, 3.7L, 4.0L, 5.6L	**Heated Oxygen Sensor 2 Bank 2 Circuit Low Voltage:** The maximum voltage from the sensor does not reach the specified voltage.
DTC: P0158 **1T ECM, MIL: Yes** **Year:** 2011, 2012 **Model:** 370Z, Altima, Maxima, Murano, Pathfinder, Quest, Titan, Xterra **Engine:** 3.5L, 3.5L, 3.7L, 4.0L, 5.6L	**Heated Oxygen Sensor 2 Bank 2 Circuit High Voltage:** * Condition A: An excessively high voltage from the sensor is sent to ECM, or, * Condition B: The minimum voltage from the sensor is not reached to the specified voltage.
DTC: P0159 **1T ECM, MIL: Yes** **Year:** 2011, 2012 **Model:** 370Z, Altima, Maxima, Murano, Pathfinder, Quest, Titan, Xterra **Engine:** 3.5L, 3.5L, 3.7L, 4.0L, 5.6L	**Heated Oxygen Sensor 2 Bank 2 Circuit Slow Response:** It takes more time for the sensor to respond between rich and lean than the specified time.
DTC: P015A **T ECM, MIL: Yes** **Year:** 2011 **Model:** Versa **Engine:** 1.8L	**Air Fuel (A/F) Ration Sensor 1 (Bank 2) Circuit Delayed Response:** The response time of the A/F sensor 1 signal delays more than the specified time computed by the ECM.
DTC: P015B **T ECM, MIL: Yes** **Year:** 2011 **Model:** Versa **Engine:** 1.8L	**Air Fuel (A/F) Ration Sensor 1 (Bank 2) Circuit Delayed Response:** The response time of the A/F sensor 1 signal delays more than the specified time computed by the ECM.

DTC	Trouble Code Title and Conditions
DTC: P0171 **2T ECM, MIL: Yes** **Year:** 2011 **Model:** Altima, Quest **Engine:** 2.5L, 3.5L	**System Too Lean (Bank 1):** Monitor runs whenever following DTCs not stored: P0010 (Camshaft Timing Oil Control Valve) P0011 (VVT System - Advance) P0012 (VVT System - Retard) P0016 (VVT System - Misalignment) P0031, P0032, P101D (Air Fuel Ratio Sensor Heater) P0102, P0103 (Mass Air Flow Meter) P0107, P0108 (Manifold Absolute Pressure) P0115, P0117, P0118 (Engine Coolant Temperature Sensor) P0125 (Insufficient Coolant Temperature for Closed Loop Fuel Control) P0120, P0121, P0122, P0123, P0220, P0222, P0223, P2135 (Throttle Position Sensor) P0335 (Crankshaft Position Sensor) P0340 (Camshaft Position Sensor) P0351 - P0354 (Igniter) Fuel system status: Closed loop Battery voltage: 11 V or higher Either of following conditions met: 1. Engine speed: Less than 1400 rpm 2. Engine load: 0.22 g/rev or more Catalyst monitor Not executed
DTC: P0171 **1T ECM, MIL: Yes** **Year:** 2011, 2012 **Model:** All **Engine:** 1.6L, 1.8L, 2.0L, 2.5L, 3.5L, 3.5L, 3.7L, 4.0L, 5.6L	**Fuel Injection System Too Lean (Includes Hybrid Models):** HYBRID MODELS CAUTION: Hybrid systems use very high-voltage battery systems. Before starting any service work involving the battery system, turn the ignition switch OFF and then remove the service plug from pocket in the trunk. After removing the service plug, wait 10 minutes before touching any of the high-voltage connectors and terminals. **NOTE: On and engines, this applies to Bank 1.** * Fuel injection system does not operate properly. * The amount of mixture ratio compensation is too large. (The mixture ratio is too lean.)
DTC: P0172 **2T ECM, MIL: Yes** **Year:** 2011 **Model:** Altima, Quest **Engine:** 2.5L, 3.5L	**System Too Rich (Bank 1):** Monitor runs whenever following DTCs not stored: P0010 (Camshaft Timing Oil Control Valve) P0011 (VVT System - Advance) P0012 (VVT System - Retard) P0016 (VVT System - Misalignment) P0031, P0032, P101D (Air Fuel Ratio Sensor Heater) P0102, P0103 (Mass Air Flow Meter) P0107, P0108 (Manifold Absolute Pressure) P0115, P0117, P0118 (Engine Coolant Temperature Sensor) P0125 (Insufficient Coolant Temperature for Closed Loop Fuel Control) P0120, P0121, P0122, P0123, P0220, P0222, P0223, P2135 (Throttle Position Sensor) P0335 (Crankshaft Position Sensor) P0340 (Camshaft Position Sensor) P0351 - P0354 (Igniter) Fuel system status: Closed loop Battery voltage: 11 V or higher Either of following conditions met: - 1. Engine speed: Less than 1400 rpm 2. Engine load: 0.22 g/rev or more Catalyst monitor: Not executed
DTC: P0172 **1T ECM, MIL: Yes** **Year:** 2011, 2012 **Model:** All **Engine:** 1.6L, 1.8L, 2.0L, 2.5L, 3.5L, 3.5L, 3.7L, 4.0L, 5.6L	**Fuel Injection System Too Rich (Includes Hybrid Models):** HYBRID MODELS CAUTION: Hybrid systems use very high-voltage battery systems. Before starting any service work involving the battery system, turn the ignition switch OFF and then remove the service plug from pocket in the trunk. After removing the service plug, wait 10 minutes before touching any of the high-voltage connectors and terminals. **NOTE: On and engines, this applies to Bank 1.** * Fuel injection system does not operate properly. * The amount of mixture ratio compensation is too large. (The mixture ratio is too rich.)

DTC	Trouble Code Title and Conditions
DTC: P0174 **1T ECM, MIL: Yes** **Year:** 2011, 2012 **Model:** 370Z, Altima, Maxima, Murano, Pathfinder, Quest, Titan, Xterra **Engine:** 3.5L, 3.5L, 3.7L, 4.0L, 5.6L	**Fuel Injection System Too Lean (Bank 2):** * Fuel injection system does not operate properly. * The amount of mixture ratio compensation is too large. (The mixture ratio is too lean.)
DTC: P0175 **1T ECM, MIL: Yes** **Year:** 2011, 2012 **Model:** 370Z, Altima, Maxima, Murano, Pathfinder, Quest, Titan, Xterra **Engine:** 3.5L, 3.5L, 3.7L, 4.0L, 5.6L	**Fuel Injection System Too Rich (Bank 2):** * Fuel injection system does not operate properly. * The amount of mixture ratio compensation is too large. (The mixture ratio is too rich.)
DTC: P0181 **1T ECM, MIL: Yes** **Year:** 2011, 2012 **Model:** All **Engine:** 1.6L, 1.8L, 2.0L, 2.5L, 3.5L, 3.5L, 3.7L, 4.0L, 5.6L	**Fuel Tank Temperature Sensor Circuit Range/Performance (Includes Hybrid Models):** HYBRID MODELS CAUTION: Hybrid systems use very high-voltage battery systems. Before starting any service work involving the battery system, turn the ignition switch OFF and then remove the service plug from pocket in the trunk. After removing the service plug, wait 10 minutes before touching any of the high-voltage connectors and terminals. * Rationally incorrect voltage from the sensor is sent to ECM, compared with the voltage signals from engine coolant temperature sensor and intake air temperature sensor.
DTC: P0182 **1T ECM, MIL: Yes** **Year:** 2011, 2012 **Model:** All **Engine:** 1.6L, 1.8L, 2.0L, 2.5L, 3.5L, 3.5L, 3.7L, 4.0L, 5.6L	**Fuel Tank Temperature Sensor Circuit Low Input (Includes Hybrid Models):** HYBRID MODELS CAUTION: Hybrid systems use very high-voltage battery systems. Before starting any service work involving the battery system, turn the ignition switch OFF and then remove the service plug from pocket in the trunk. After removing the service plug, wait 10 minutes before touching any of the high-voltage connectors and terminals. * An excessively low voltage from the sensor is sent to ECM.
DTC: P0183 **1T ECM, MIL: Yes** **Year:** 2011, 2012 **Model:** 370Z, Altima, Cube, Frontier, Maxima, Murano, Pathfinder, Quest, Rogue, Sentra, Titan, Xterra **Engine:** 1.8L, 2.0L, 2.5L, 3.5L, 3.5L, 3.7L, 4.0L, 5.6L	**Fuel Tank Temperature Sensor Circuit High Input (Includes Hybrid Models):** HYBRID MODELS CAUTION: Hybrid systems use very high-voltage battery systems. Before starting any service work involving the battery system, turn the ignition switch OFF and then remove the service plug from pocket in the trunk. After removing the service plug, wait 10 minutes before touching any of the high-voltage connectors and terminals. An excessively high voltage from the sensor is sent to ECM.
DTC: P0183 **1T ECM, MIL: Yes** **Year:** 2011 **Model:** Juke, Versa **Engine:** 1.6L	**Fuel Tank Temperature Sensor Circuit High Input:** An excessively high voltage from the sensor is sent to ECM.
DTC: P0190 **1T ECM, MIL: Yes** **Year:** 2011 **Model:** Juke, Versa **Engine:** 1.6L	**Fuel Rail Pressure Sensor Circuit:** Fuel rail pressure sensor circuit low input and high input Signal voltage from the fuel rail pressure sensor remains at more than 4.84 V / ess than 0.2 V for 5 seconds or more.
DTC: P0196 **1T ECM, MIL: Yes** **Year:** 2011, 2012 **Model:** 370Z, Altima, Juke, Maxima, Murano, Quest, Sentra, Versa **Engine:** 1.6L, 2.0L, 2.5L, 3.5L, 3.5L, 3.7L	**Engine Oil Temperature (EOT) Sensor Range/Performance:** **NOTE: If DTC P0196 is displayed with P0197 or P0198, first perform the trouble diagnosis for DTC P0197, P0198.** * Rationally incorrect voltage from the sensor is sent to ECM, compared with the voltage signals from engine coolant temperature sensor and intake air temperature sensor.

DTC	Trouble Code Title and Conditions
DTC: P0197 **1T ECM, MIL: Yes** **Year:** 2011, 2012 **Model:** 370Z, Altima, Juke, Maxima, Murano, Quest, Sentra, Versa **Engine:** 1.6L, 2.0L, 2.5L, 3.5L, 3.5L, 3.7L	**Engine Oil Temperature (EOT) Sensor Circuit Low Input:** An excessively low voltage from the sensor is sent to ECM.
DTC: P0198 **1T ECM, MIL: Yes** **Year:** 2011, 2012 **Model:** 370Z, Altima, Juke, Maxima, Murano, Quest, Sentra, Versa **Engine:** 1.6L, 2.0L, 2.5L, 3.5L, 3.5L, 3.7L	**Engine Oil Temperature (EOT) Sensor Circuit High Input:** An excessively high voltage from the sensor is sent to ECM.
DTC: P0201 **T ECM, MIL: Yes** **Year:** 2011 **Model:** Altima **Engine:** 2.5L	**No. 1 Cylinder Fuel Injector Circuit Open (Hybrid):** HYBRID MODELS CAUTION: Hybrid systems use very high-voltage battery systems. Before starting any service work involving the battery system, turn the ignition switch OFF and then remove the service plug from pocket in the trunk. After removing the service plug, wait 10 minutes before touching any of the high-voltage connectors and terminals. An excessively low voltage signal is sent to ECM through the No. 1 fuel injector
DTC: P0201 **1T ECM, MIL: Yes** **Year:** 2011 **Model:** Juke, Versa **Engine:** 1.6L	**Fuel Injector 1 Circuit Malfunction:** ECM detects No. 1 injector circuit is open or shorted.
DTC: P0202 **T ECM, MIL: Yes** **Year:** 2011 **Model:** Altima **Engine:** 2.5L	**No. 2 Cylinder Fuel Injector Circuit Open (Hybrid):** HYBRID MODELS CAUTION: Hybrid systems use very high-voltage battery systems. Before starting any service work involving the battery system, turn the ignition switch OFF and then remove the service plug from pocket in the trunk. After removing the service plug, wait 10 minutes before touching any of the high-voltage connectors and terminals. No. 2 cylinder fuel injector circuit open
DTC: P0202 **1T ECM, MIL: Yes** **Year:** 2011 **Model:** Juke, Versa **Engine:** 1.6L	**Fuel Injector 2 Circuit Malfunction:** ECM detects No. 2 injector circuit s open or shorted.
DTC: P0203 **1T ECM, MIL: Yes** **Year:** 2011 **Model:** Juke, Versa **Engine:** 1.6L	**Fuel Injector 3 Control Circuit:** ECM detects No. 3 injector circuit is open or shorted.
DTC: P0203 **T ECM, MIL: Yes** **Year:** 2011 **Model:** Altima **Engine:** 2.5L	**No. 3 Cylinder Fuel Injector Circuit Open (Hybrid):** HYBRID MODELS CAUTION: Hybrid systems use very high-voltage battery systems. Before starting any service work involving the battery system, turn the ignition switch OFF and then remove the service plug from pocket in the trunk. After removing the service plug, wait 10 minutes before touching any of the high-voltage connectors and terminals. An excessively low voltage signal is sent to ECM through the No. 3 fuel injector
DTC: P0204 **T ECM, MIL: Yes** **Year:** 2011 **Model:** Altima **Engine:** 2.5L	**No. 4 Cylinder Fuel Injector Circuit Open (Hybrid):** HYBRID MODELS CAUTION: Hybrid systems use very high-voltage battery systems. Before starting any service work involving the battery system, turn the ignition switch OFF and then remove the service plug from pocket in the trunk. After removing the service plug, wait 10 minutes before touching any of the high-voltage connectors and terminals. An excessively low voltage signal is sent to ECM through the No. 4 fuel injector
DTC: P0204 **1T ECM, MIL: Yes** **Year:** 2011 **Model:** Juke, Versa **Engine:** 1.6L	**Fuel Injector 4 Control Circuit:** ECM detects No. 4 injector circuit is open or shorted.

DTC	Trouble Code Title and Conditions
DTC: P02103 **1T ECM, MIL: Yes** **Year:** 2011 **Model:** Altima, Quest **Engine:** 2.5L, 3.5L	**Throttle Actuator Control Motor Circuit High:** * Short in throttle actuator circuit * Throttle actuator * Throttle valve * Throttle body assembly * Hybrid vehicle control ECU
DTC: P0220 **1T ECM, MIL: Yes** **Year:** 2011 **Model:** Altima, Quest **Engine:** 2.5L, 3.5L	**Throttle / Pedal Position Sensor / Switch "B" Circuit:** Monitor runs whenever following DTCs are not present: None Either of the following conditions A or B is met: A. Power switch on (IG): 0.012 seconds or more B. Electronic throttle actuator power: ON
DTC: P0222 **1T ECM, MIL: Yes** **Year:** 2011, 2012 **Model:** All **Engine:** 1.6L, 1.8L, 2.0L, 2.5L, 3.5L, 3.5L, 3.7L, 4.0L, 5.6L	**Throttle Position (TP) Sensor 1 Circuit Low Input (Includes Hybrid Models):** HYBRID MODELS CAUTION: Hybrid systems use very high-voltage battery systems. Before starting any service work involving the battery system, turn the ignition switch OFF and then remove the service plug from pocket in the trunk. After removing the service plug, wait 10 minutes before touching any of the high-voltage connectors and terminals. **NOTE: If DTC P0222 or P0223 is displayed with DTC P0643, first perform the trouble diagnosis for DTC P0643.** **NOTE: On and engines, this DTC is for Bank 1.** * An excessively low voltage from the TP sensor 1 is sent to ECM.
DTC: P0222 **1T ECM, MIL: Yes** **Year:** 2011 **Model:** Altima, Quest **Engine:** 2.5L, 3.5L	**Throttle / Pedal Position Sensor / Switch "B" Circuit Low Input:** Monitor runs whenever following DTCs are not present: None Either of the following conditions A or B is met: A. Power switch on (IG): 0.012 seconds or more B. Electronic throttle actuator power: ON
DTC: P0223 **1T ECM, MIL: Yes** **Year:** 2011, 2012 **Model:** All **Engine:** 1.6L, 1.8L, 2.0L, 2.5L, 3.5L, 3.5L, 3.7L, 4.0L, 5.6L	**Throttle Position (TP) Sensor 1 Circuit High Input (Includes Hybrid Models):** HYBRID MODELS CAUTION: Hybrid systems use very high-voltage battery systems. Before starting any service work involving the battery system, turn the ignition switch OFF and then remove the service plug from pocket in the trunk. After removing the service plug, wait 10 minutes before touching any of the high-voltage connectors and terminals. **NOTE: If DTC P0222 or P0223 is displayed with DTC P0643, first perform the trouble diagnosis for DTC P0643.** **NOTE: On and engines, this DTC is for Bank 1.** * An excessively high voltage from the TP sensor 1 is sent to ECM.
DTC: P0223 **1T ECM, MIL: Yes** **Year:** 2011 **Model:** Altima, Quest **Engine:** 2.5L, 3.5L	**Throttle / Pedal Position Sensor / Switch "B" Circuit High Input:** Monitor runs whenever following DTCs are not present: None Either of the following conditions A or B is met: A. Power switch on (IG): 0.012 seconds or more B. Electronic throttle actuator power: ON
DTC: P0227 **1T ECM, MIL: Yes** **Year:** 2011, 2012 **Model:** 370Z **Engine:** 3.7L	**Throttle Position Sensor 2 (Bank 2) Circuit Low Input:** **NOTE: If DTC P0122, P0123, P0227 or P0228 is displayed with DTC P0643, first perform the trouble diagnosis for DTC P0643.** An excessively low voltage from the TP sensor 2 is sent to ECM.
DTC: P0228 **1T ECM, MIL: Yes** **Year:** 2011, 2012 **Model:** 370Z **Engine:** 3.7L	**Throttle Position Sensor 2 (Bank 2) Circuit High Input:** **NOTE: If DTC P0122, P0123, P0227 or P0228 is displayed with DTC P0643, first perform the trouble diagnosis for DTC P0643.** An excessively high voltage from the TP sensor 2 is sent to ECM.
DTC: P0234 **1T ECM, MIL: Yes** **Year:** 2011 **Model:** Juke, Versa **Engine:** 1.6L	**Turbocharger Overboost Condition :** Turbocharger boost is higher than the target value.
DTC: P0237 **1T ECM, MIL: Yes** **Year:** 2011 **Model:** Juke, Versa **Engine:** 1.6L	**Turbocharger Boost Circuit Low Input:** An excessively low voltage from the turbocharger boost sensor is sent to ECM.

DTC	Trouble Code Title and Conditions
DTC: P0238 **1T ECM** **Year:** 2011 **Model:** Juke, Versa **Engine:** 1.6L	**Turbocharger Boost Circuit High Input:** An excessively high voltage from the turbocharger boost sensor is sent to ECM.
DTC: P0300 **2T ECM, MIL: Yes** **Year:** 2011 **Model:** Altima, Quest **Engine:** 2.5L, 3.5L	**Random / Multiple Cylinder Misfire Detected:** Misfire: Monitor runs whenever following DTCs not stored: P0016 (VVT System - Misalignment) P0102, P0103 (Mass Air Flow Meter) P0107, P0108 (Manifold Absolute Pressure) P0112, P0113 (Intake Air Temperature Sensor) P0115, P0117, P0118 (Engine Coolant Temperature Sensor) P0125 (Insufficient Coolant Temperature for Closed Loop Fuel Control) P0120, P0121, P0122, P0123, P0220, P0222, P0223, P2135 (Throttle Position Sensor) P0327, P0328 (Knock Control Sensor) P0335 (Crankshaft Position Sensor) P0340 (Camshaft Position Sensor) P0351 - P0354 (Igniter) Battery voltage: 8 V or higher VVT system: Not operated by scan tool Engine speed: 750 to 6400 rpm Either of following conditions (a) or (b) met: (a) Engine coolant temperature at engine start: Higher than -7°C (19°F) (b) Engine coolant temperature: Higher than 20°C (68°F) Fuel cut: OFF Monitor Period of Emission-relatd Misfire: First 1000 revolutions after engine start, or during check mode: Crankshaft 1000 revolutions Except above: Crankshaft 1000 revolutions x 4 Monitor Period of Catalyst-damaged Misfire (MIL Blinks): All of following conditions 1, 2 and 3 met: Crankshaft 200 revolutions x 3 1. Driving cycles: 1st 2. Check mode: OFF 3. Engine speed: Less than 3500 rpm Except above (MIL blinks immediately): Crankshaft 200 revolutions
DTC: P0300 **1T ECM, MIL: Yes** **Year:** 2011, 2012 **Model:** All **Engine:** 1.6L, 1.8L, 2.0L, 2.5L, 3.5L, 3.5L, 3.7L, 4.0L, 5.6L	**Multiple Cylinder Misfire Detected (Includes Hybrid Models):** HYBRID MODELS CAUTION: Hybrid systems use very high-voltage battery systems. Before starting any service work involving the battery system, turn the ignition switch OFF and then remove the service plug from pocket in the trunk. After removing the service plug, wait 10 minutes before touching any of the high-voltage connectors and terminals. * Multiple cylinder misfire. * One Trip Detection Logic (Three Way Catalyst Damage) **Note: On the 1st trip, when a misfire condition occurs that can damage the three way catalyst (TWC) due to overheating, the MIL will blink.** * When a misfire condition occurs, the ECM monitors the CKP sensor (POS) signal every 200 engine revolutions for a change. * When the misfire condition decreases to a level that will not damage the TWC, the MIL will turn off. * If another misfire condition occurs that can damage the TWC on a second trip, the MIL will blink. * Two Trip Detection Logic (Exhaust quality deterioration) **Note: For misfire conditions that will not damage the TWC (but will affect vehicle emissions), the MIL will only light when the misfire is detected on a second trip. During this condition, the ECM monitors the CKP sensor signal every 1,000 engine revolutions.** * A misfire malfunction can be detected in any one cylinder or in multiple cylinders.

DTC	Trouble Code Title and Conditions
DTC: P0301 **2T ECM, MIL: Yes** **Year:** 2011 **Model:** Altima, Quest **Engine:** 2.5L, 3.5L	**Cylinder 1 Misfire Detected:** Misfire: Monitor runs whenever following DTCs not stored: P0016 (VVT System - Misalignment) P0102, P0103 (Mass Air Flow Meter) P0107, P0108 (Manifold Absolute Pressure) P0112, P0113 (Intake Air Temperature Sensor) P0115, P0117, P0118 (Engine Coolant Temperature Sensor) P0125 (Insufficient Coolant Temperature for Closed Loop Fuel Control) P0120, P0121, P0122, P0123, P0220, P0222, P0223, P2135 (Throttle Position Sensor) P0327, P0328 (Knock Control Sensor) P0335 (Crankshaft Position Sensor) P0340 (Camshaft Position Sensor) P0351 - P0354 (Igniter) Battery voltage: 8 V or higher VVT system: Not operated by scan tool Engine speed: 750 to 6400 rpm Either of following conditions (a) or (b) met: (a) Engine coolant temperature at engine start: Higher than -7°C (19°F) (b) Engine coolant temperature: Higher than 20°C (68°F) Fuel cut: OFF Monitor Period of Emission-related Misfire: First 1000 revolutions after engine start, or during check mode: Crankshaft 1000 revolutions Except above: Crankshaft 1000 revolutions x 4 Monitor Period of Catalyst-damaging Misfire (MIL Blinks): All of following conditions 1, 2 and 3 met: Crankshaft 200 revolutions x 3 1. Driving cycles: 1st 2. Check mode: OFF 3. Engine speed: Less than 3500 rpm Except above (MIL blinks immediately): Crankshaft 200 revolutions
DTC: P0301 **1T ECM, MIL: Yes** **Year:** 2011, 2012 **Model:** All **Engine:** 1.6L, 1.8L, 2.0L, 2.5L, 3.5L, 3.5L, 3.7L, 4.0L, 5.6L	**No.1 Cylinder Misfire Detected (Includes Hybrid Models):** HYBRID MODELS CAUTION: Hybrid systems use very high-voltage battery systems. Before starting any service work involving the battery system, turn the ignition switch OFF and then remove the service plug from pocket in the trunk. After removing the service plug, wait 10 minutes before touching any of the high-voltage connectors and terminals. No. 1 cylinder misfires. 1. One Trip Detection Logic (Three Way Catalyst Damage) On the 1st trip, when a misfire condition occurs that can damage the three way catalyst (TWC) due to overheating, the MIL will blink. When a misfire condition occurs, the ECM monitors the CKP sensor (POS) signal every 200 engine revolutions for a change. When the misfire condition decreases to a level that will not damage the TWC, the MIL will turn off. 2. Two Trip Detection Logic (Exhaust quality deterioration) For misfire conditions that will not damage the TWC (but will affect vehicle emissions), the MIL will only light when the misfire is detected on a second trip. During this condition, the ECM monitors the CKP sensor signal every 1,000 engine revolutions. A misfire malfunction can be detected on any one cylinder or on multiple cylinders. If another misfire condition occurs that can damage the TWC on a second trip, the MIL will blink.
DTC: P03013-123 **T ECM, MIL: Yes** **Year:** 2011 **Model:** Altima, Quest **Engine:** 2.5L, 3.5L	**Battery Block 3 Becomes Weak:** Presence of a malfunctioning block is determined based on each battery block voltage (1 trip detection).

DTC	Trouble Code Title and Conditions
DTC: P0302 **2T ECM, MIL: Yes** **Year:** 2011 **Model:** Altima, Quest **Engine:** 2.5L, 3.5L	**Cylinder 2 Misfire Detected:** Misfire: Monitor runs whenever following DTCs not stored: P0016 (VVT System - Misalignment) P0102, P0103 (Mass Air Flow Meter) P0107, P0108 (Manifold Absolute Pressure) P0112, P0113 (Intake Air Temperature Sensor) P0115, P0117, P0118 (Engine Coolant Temperature Sensor) P0125 (Insufficient Coolant Temperature for Closed Loop Fuel Control) P0120, P0121, P0122, P0123, P0220, P0222, P0223, P2135 (Throttle Position Sensor) P0327, P0328 (Knock Control Sensor) P0335 (Crankshaft Position Sensor) P0340 (Camshaft Position Sensor) P0351 - P0354 (Igniter) Battery voltage: 8 V or higher VVT system: Not operated by scan tool Engine speed: 750 to 6400 rpm Either of following conditions (a) or (b) met: (a) Engine coolant temperature at engine start: Higher than -7°C (19°F) (b) Engine coolant temperature: Higher than 20°C (68°F) Fuel cut: OFF Monitor Period of Emission-related Misfire: First 1000 revolutions after engine start, or during check mode: Crankshaft 1000 revolutions Except above: Crankshaft 1000 revolutions x 4 Monitor Period of Catalyst-damaging Misfire (MIL Blinks): All of following conditions 1, 2 and 3 met: Crankshaft 200 revolutions x 3 1. Driving cycles: 1st 2. Check mode: OFF 3. Engine speed: Less than 3500 rpm Except above (MIL blinks immediately): Crankshaft 200 revolutions
DTC: P0302 **1T ECM, MIL: Yes** **Year:** 2011, 2012 **Model:** All **Engine:** 1.6L, 1.8L, 2.0L, 2.5L, 3.5L, 3.5L, 3.7L, 4.0L, 5.6L	**No. 2 Cylinder Misfire Detected (Includes Hybrid Models):** HYBRID MODELS CAUTION: Hybrid systems use very high-voltage battery systems. Before starting any service work involving the battery system, turn the ignition switch OFF and then remove the service plug from pocket in the trunk. After removing the service plug, wait 10 minutes before touching any of the high-voltage connectors and terminals. * No. 2 cylinder misfires. 1. One Trip Detection Logic (Three Way Catalyst Damage) - On the 1st trip, when a misfire condition occurs that can damage the three way catalyst (TWC) due to overheating, the MIL will blink. - When a misfire condition occurs, the ECM monitors the CKP sensor (POS) signal every 200 engine revolutions for a change. - When the misfire condition decreases to a level that will not damage the TWC, the MIL will turn off. 2. Two Trip Detection Logic (Exhaust quality deterioration) - For misfire conditions that will not damage the TWC (but will affect vehicle emissions), the MIL will only light when the misfire is detected on a second trip. - During this condition, the ECM monitors the CKP sensor signal every 1,000 engine revolutions. - A misfire malfunction can be detected on any one cylinder or on multiple cylinders. - If another misfire condition occurs that can damage the TWC on a second trip, the MIL will blink.

DTC	Trouble Code Title and Conditions
DTC: P0303 **1T ECM, MIL: Yes** **Year:** 2011, 2012 **Model:** All **Engine:** 1.6L, 1.8L, 2.0L, 2.5L, 3.5L, 3.5L, 3.7L, 4.0L, 5.6L	**No. 3 Cylinder Misfire Detected (Includes Hybrid Models):** HYBRID MODELS CAUTION: Hybrid systems use very high-voltage battery systems. Before starting any service work involving the battery system, turn the ignition switch OFF and then remove the service plug from pocket in the trunk. After removing the service plug, wait 10 minutes before touching any of the high-voltage connectors and terminals. * No. 3 cylinder misfires. 1. One Trip Detection Logic (Three Way Catalyst Damage) - On the 1st trip, when a misfire condition occurs that can damage the three way catalyst (TWC) due to overheating, the MIL will blink. - When a misfire condition occurs, the ECM monitors the CKP sensor (POS) signal every 200 engine revolutions for a change. - When the misfire condition decreases to a level that will not damage the TWC, the MIL will turn off. 2. Two Trip Detection Logic (Exhaust quality deterioration) - For misfire conditions that will not damage the TWC (but will affect vehicle emissions), the MIL will only light when the misfire is detected on a second trip. - During this condition, the ECM monitors the CKP sensor signal every 1,000 engine revolutions. - A misfire malfunction can be detected on any one cylinder or on multiple cylinders. - If another misfire condition occurs that can damage the TWC on a second trip, the MIL will blink.
DTC: P0303 **2T ECM, MIL: Yes** **Year:** 2011 **Model:** Altima, Quest **Engine:** 2.5L, 3.5L	**Cylinder 3 Misfire Detected:** Misfire: Monitor runs whenever following DTCs not stored: P0016 (VVT System - Misalignment) P0102, P0103 (Mass Air Flow Meter) P0107, P0108 (Manifold Absolute Pressure) P0112, P0113 (Intake Air Temperature Sensor) P0115, P0117, P0118 (Engine Coolant Temperature Sensor) P0125 (Insufficient Coolant Temperature for Closed Loop Fuel Control) P0120, P0121, P0122, P0123, P0220, P0222, P0223, P2135 (Throttle Position Sensor) P0327, P0328 (Knock Control Sensor) P0335 (Crankshaft Position Sensor) P0340 (Camshaft Position Sensor) P0351 - P0354 (Igniter) Battery voltage: 8 V or higher VVT system: Not operated by scan tool Engine speed: 750 to 6400 rpm Either of following conditions (a) or (b) met: (a) Engine coolant temperature at engine start: Higher than -7°C (19°F) (b) Engine coolant temperature: Higher than 20°C (68°F) Fuel cut: OFF Monitor Period of Emission-related Misfire: First 1000 revolutions after engine start, or during check mode: Crankshaft 1000 revolutions Except above: Crankshaft 1000 revolutions x 4 Monitor Period of Catalyst-damaging Misfire (MIL Blinks): All of following conditions 1, 2 and 3 met: Crankshaft 200 revolutions x 3 1. Driving cycles: 1st 2. Check mode: OFF 3. Engine speed: Less than 3500 rpm Except above (MIL blinks immediately): Crankshaft 200 revolutions

DTC	Trouble Code Title and Conditions
DTC: P0304 **1T ECM, MIL: Yes** **Year:** 2011, 2012 **Model:** All **Engine:** 1.6L, 1.8L, 2.0L, 2.5L, 3.5L, 3.5L, 3.7L, 4.0L, 5.6L	**No. 4 Cylinder Misfire Detected (Includes Hybrid Models):** HYBRID MODELS CAUTION: Hybrid systems use very high-voltage battery systems. Before starting any service work involving the battery system, turn the ignition switch OFF and then remove the service plug from pocket in the trunk. After removing the service plug, wait 10 minutes before touching any of the high-voltage connectors and terminals. * No. 4 cylinder misfires. * The misfire detection logic consists of the following two conditions. 1. One Trip Detection Logic (Three Way Catalyst Damage) - On the 1st trip, when a misfire condition occurs that can damage the three way catalyst (TWC) due to overheating, the MIL will blink. - When a misfire condition occurs, the ECM monitors the CKP sensor (POS) signal every 200 engine revolutions for a change. - When the misfire condition decreases to a level that will not damage the TWC, the MIL will turn off. - If another misfire condition occurs that can damage the TWC on a second trip, the MIL will blink. - When the misfire condition decreases to a level that will not damage the TWC, the MIL will remain on. - If another misfire condition occurs that can damage the TWC, the MIL will begin to blink again. 2. Two Trip Detection Logic (Exhaust quality deterioration) - For misfire conditions that will not damage the TWC (but will affect vehicle emissions), the MIL will only light when the misfire is detected on a second trip. - During this condition, the ECM monitors the CKP sensor signal every 1,000 engine revolutions. - A misfire malfunction can be detected on any one cylinder or on multiple cylinders.
DTC: P0304 **2T ECM, MIL: Yes** **Year:** 2011 **Model:** Altima, Quest **Engine:** 2.5L, 3.5L	**Cylinder 4 Misfire Detected:** Misfire: Monitor runs whenever following DTCs not stored: P0016 (VVT System - Misalignment) P0102, P0103 (Mass Air Flow Meter) P0107, P0108 (Manifold Absolute Pressure) P0112, P0113 (Intake Air Temperature Sensor) P0115, P0117, P0118 (Engine Coolant Temperature Sensor) P0125 (Insufficient Coolant Temperature for Closed Loop Fuel Control) P0120, P0121, P0122, P0123, P0220, P0222, P0223, P2135 (Throttle Position Sensor) P0327, P0328 (Knock Control Sensor) P0335 (Crankshaft Position Sensor) P0340 (Camshaft Position Sensor) P0351 - P0354 (Igniter) Battery voltage: 8 V or higher VVT system: Not operated by scan tool Engine speed: 750 to 6400 rpm Either of following conditions (a) or (b) met: (a) Engine coolant temperature at engine start: Higher than -7°C (19°F) (b) Engine coolant temperature: Higher than 20°C (68°F) Fuel cut: OFF Monitor Period of Emission-related Misfire: First 1000 revolutions after engine start, or during check mode: Crankshaft 1000 revolutions Except above: Crankshaft 1000 revolutions x 4 Monitor Period of Catalyst-damaging Misfire (MIL Blinks): All of following conditions 1, 2 and 3 met: Crankshaft 200 revolutions x 3 1. Driving cycles: 1st 2. Check mode: OFF 3. Engine speed: Less than 3500 rpm Except above (MIL blinks immediately): Crankshaft 200 revolutions

DTC	Trouble Code Title and Conditions
DTC: P0305 **1T ECM, MIL: Yes** **Year:** 2011, 2012 **Model:** 370Z, Altima, Frontier, Maxima, Murano, Pathfinder, Quest, Titan, Xterra **Engine:** 3.5L, 3.5L, 3.7L, 4.0L, 5.6L	**No. 5 Cylinder Misfire Detected:** * No. 5 cylinder misfires. * The misfire detection logic consists of the following two conditions. 1. One Trip Detection Logic (Three Way Catalyst Damage): - On the first trip, when a misfire condition occurs that can damage the three way catalyst (TWC) due to overheating, the MIL will blink. - When a misfire condition occurs, the ECM monitors the CKP sensor signal every 200 engine revolutions for a change. - When the misfire condition decreases to a level that will not damage the TWC, the MIL will turn off. - If another misfire condition occurs that can damage the TWC on a second trip, the MIL will blink. - When the misfire condition decreases to a level that will not damage the TWC, the MIL will remain on. - If another misfire condition occurs that can damage the TWC, the MIL will begin to blink again. 2. Two Trip Detection Logic (Exhaust quality deterioration): - For misfire conditions that will not damage the TWC (but will affect vehicle emissions), the MIL will only light when the misfire is detected on a second trip. - During this condition, the ECM monitors the CKP sensor signal every 1,000 engine revolutions. - A misfire malfunction can be detected in any one cylinder or in multiple cylinders.
DTC: P0306 **1T ECM, MIL: Yes** **Year:** 2011, 2012 **Model:** 370Z, Altima, Frontier, Maxima, Murano, Pathfinder, Quest, Titan, Xterra **Engine:** 3.5L, 3.5L, 3.7L, 4.0L, 5.6L	**No. 6 Cylinder Misfire Detected:** * No. 6 cylinder misfires. * The misfire detection logic consists of the following two conditions. 1. One Trip Detection Logic (Three Way Catalyst Damage): - On the first trip, when a misfire condition occurs that can damage the three way catalyst (TWC) due to overheating, the MIL will blink. - When a misfire condition occurs, the ECM monitors the CKP sensor signal every 200 engine revolutions for a change. - When the misfire condition decreases to a level that will not damage the TWC, the MIL will turn off. - If another misfire condition occurs that can damage the TWC on a second trip, the MIL will blink. - When the misfire condition decreases to a level that will not damage the TWC, the MIL will remain on. - If another misfire condition occurs that can damage the TWC, the MIL will begin to blink again. 2. Two Trip Detection Logic (Exhaust quality deterioration): - For misfire conditions that will not damage the TWC (but will affect vehicle emissions), the MIL will only light when the misfire is detected on a second trip. - During this condition, the ECM monitors the CKP sensor signal every 1,000 engine revolutions. - A misfire malfunction can be detected in any one cylinder or in multiple cylinders.No. 5 cylinder misfires.
DTC: P0307 **1T ECM, MIL: Yes** **Year:** 2011 **Model:** Pathfinder, Titan **Engine:** 5.6L	**No. 7 Cylinder Misfire Detected:** The misfire detection logic consists of the following two conditions: One Trip Detection Logic (Three Way Catalyst Damage) - On the 1st trip that a misfire condition occurs that can damage the three way catalyst (TWC) due to overheating, the MIL will blink. - When a misfire condition occurs, the ECM monitors the CKP sensor signal every 200 engine revolutions for a change. - When the misfire condition decreases to a level that will not damage the TWC, the MIL will turn off. - If another misfire condition occurs that can damage the TWC on a second trip, the MIL will blink. - When the misfire condition decreases to a level that will not damage the TWC, the MIL will remain on. - If another misfire condition occurs that can damage the TWC, the MIL will begin to blink again. Two Trip Detection Logic (Exhaust quality deterioration) - For misfire conditions that will not damage the TWC (but will affect vehicle emissions), the MIL will only light when the misfire is detected on a second trip. - During this condition, the ECM monitors the CKP sensor signal every 1,000 engine revolutions. - A misfire malfunction can be detected on any one cylinder or on multiple cylinders. No. 7 cylinder misfires.

DTC	Trouble Code Title and Conditions
DTC: P0308 **1T ECM, MIL: Yes** **Year:** 2011 **Model:** Pathfinder, Titan **Engine:** 5.6L	**No. 8 Cylinder Misfire Detected:** The misfire detection logic consists of the following two conditions: One Trip Detection Logic (Three Way Catalyst Damage) - On the 1st trip that a misfire condition occurs that can damage the three way catalyst (TWC) due to overheating, the MIL will blink. - When a misfire condition occurs, the ECM monitors the CKP sensor signal every 200 engine revolutions for a change. - When the misfire condition decreases to a level that will not damage the TWC, the MIL will turn off. - If another misfire condition occurs that can damage the TWC on a second trip, the MIL will blink. - When the misfire condition decreases to a level that will not damage the TWC, the MIL will remain on. - If another misfire condition occurs that can damage the TWC, the MIL will begin to blink again. Two Trip Detection Logic (Exhaust quality deterioration) - For misfire conditions that will not damage the TWC (but will affect vehicle emissions), the MIL will only light when the misfire is detected on a second trip. - During this condition, the ECM monitors the CKP sensor signal every 1,000 engine revolutions. - A misfire malfunction can be detected on any one cylinder or on multiple cylinders. No. 8 cylinder misfires.
DTC: P0308A-123 **T BCM** **Year:** 2011 **Model:** Altima, Quest **Engine:** 2.5L, 3.5L	**Hybrid Battery Voltage Sensor All Circuits Low:** Any of the battery block voltages become less than 2.0 V (open). (1 trip detection)
DTC: P0327 **1T ECM, MIL: Yes** **Year:** 2011 **Model:** Altima, Quest **Engine:** 2.5L, 3.5L	**Knock Sensor 1 Circuit Low Input (Bank 1 or Single Sensor):** Monitor runs whenever following DTCs are not present: None Battery voltage: 10.5 V or more Time after engine start: 5 seconds or more
DTC: P0327 **1T ECM, MIL: Yes** **Year:** 2011, 2012 **Model:** All **Engine:** 1.6L, 1.8L, 2.0L, 2.5L, 3.5L, 3.5L, 3.7L, 4.0L, 5.6L	**Knock Sensor Circuit Low Input (Includes Hybrid Models):** HYBRID MODELS CAUTION: Hybrid systems use very high-voltage battery systems. Before starting any service work involving the battery system, turn the ignition switch OFF and then remove the service plug from pocket in the trunk. After removing the service plug, wait 10 minutes before touching any of the high-voltage connectors and terminals. **NOTE: On and engines, this applies to Bank 1.** An excessively low voltage from the sensor is sent to ECM.
DTC: P0328 **1T ECM, MIL: Yes** **Year:** 2011, 2012 **Model:** All **Engine:** 1.6L, 1.8L, 2.0L, 2.5L, 3.5L, 3.5L, 3.7L, 4.0L, 5.6L	**Knock Sensor Circuit High Input (Includes Hybrid Models):** HYBRID MODELS CAUTION: Hybrid systems use very high-voltage battery systems. Before starting any service work involving the battery system, turn the ignition switch OFF and then remove the service plug from pocket in the trunk. After removing the service plug, wait 10 minutes before touching any of the high-voltage connectors and terminals. **NOTE: On and engines, this applies to Bank 1.** An excessively high voltage from the sensor is sent to ECM.
DTC: P0328 **1T ECM, MIL: Yes** **Year:** 2011 **Model:** Altima, Quest **Engine:** 2.5L, 3.5L	**Knock Sensor 1 Circuit High Input (Bank 1 or Single Sensor):** Monitor runs whenever following DTCs are not present: None Battery voltage: 10.5 V or more Time after engine start: 5 seconds or more
DTC: P0332 **1T ECM, MIL: Yes** **Year:** 2011, 2012 **Model:** 370Z, Altima, Maxima, Murano, Pathfinder, Quest, Titan, Xterra **Engine:** 3.5L, 3.5L, 3.7L, 4.0L, 5.6L	**Knock Sensor (KS) Bank 2 Sensor Circuit Low Input:** An excessively low voltage from the sensor is sent to ECM.
DTC: P0333 **1T ECM, MIL: Yes** **Year:** 2011, 2012 **Model:** 370Z, Altima, Frontier, Maxima, Murano, Pathfinder, Quest, Titan, Xterra **Engine:** 2.5L, 3.5L, 3.5L, 3.7L, 4.0L, 5.6L	**Knock Sensor (Bank 2) Circuit High Input:** An excessively high voltage from the sensor is sent to ECM.

DTC	Trouble Code Title and Conditions
DTC: P0335 **1T ECM, MIL: Yes** **Year:** 2011 **Model:** Altima, Quest **Engine:** 2.5L, 3.5L	**Crankshaft Position Sensor "A" Circuit:** Monitor runs whenever following DTCs are not present: None Case 1: Time after starter OFF to ON: 2.5 seconds or more Number of camshaft position sensor signal pulse: 6 times Battery voltage: 7 V or more Camshaft position sensor circuit fail (P0340): Not detected Power switch: On (IG) Case 2: Engine speed: 600 rpm or more Stater: Off Time after starter OFF to ON: 3 seconds or more
DTC: P0335 **1T ECM, MIL: Yes** **Year:** 2011, 2012 **Model:** All **Engine:** 1.6L, 1.8L, 2.0L, 2.5L, 3.5L, 3.5L, 3.7L, 4.0L, 5.6L	**Crankshaft Position Sensor (POS) Circuit (Includes Hybrid Models):** HYBRID MODELS CAUTION: Hybrid systems use very high-voltage battery systems. Before starting any service work involving the battery system, turn the ignition switch OFF and then remove the service plug from pocket in the trunk. After removing the service plug, wait 10 minutes before touching any of the high-voltage connectors and terminals. * The crankshaft position sensor (POS) signal is not detected by the ECM during the first few seconds of engine cranking. * The proper pulse signal from the crankshaft position sensor (POS) is not sent to ECM while the engine is running. * The crankshaft position sensor (POS) signal is not in the normal pattern during engine running.
DTC: P0338-885 **T ECM, MIL: Yes** **Year:** 2011 **Model:** Altima, Quest **Engine:** 2.5L, 3.5L	**Crankshaft Position Sensor "A" Circuit High:** NEI signal is not sent to the hybrid vehicle control ECU while the engine is running.
DTC: P0340 **1T ECM, MIL: Yes** **Year:** 2011 **Model:** Juke, Versa **Engine:** 1.6L	**Camshaft Position Sensor (Phase) Circuit :** The cylinder No. signal is not sent to ECM for the first few seconds during engine cranking. The cylinder No. signal is not sent to ECM during engine running. The cylinder No. signal is not in the normal pattern during engine running.
DTC: P0340 **1T ECM, MIL: Yes** **Year:** 2011, 2012 **Model:** 370Z, Altima, Cube, Frontier, Maxima, Murano, Pathfinder, Quest, Rogue, Sentra, Titan, Versa, Xterra **Engine:** 1.6L, 1.8L, 2.0L, 2.5L, 3.5L, 3.5L, 3.7L, 4.0L, 5.6L	**Camshaft Position Sensor Circuit (Includes Hybrid Models):** HYBRID MODELS CAUTION: Hybrid systems use very high-voltage battery systems. Before starting any service work involving the battery system, turn the ignition switch OFF and then remove the service plug from pocket in the trunk. After removing the service plug, wait 10 minutes before touching any of the high-voltage connectors and terminals. **NOTE: On and engines, this applies to Bank 1.** **NOTE: If DTC P0340 is displayed with DTC P0643, first perform the trouble diagnosis for DTC P0643.** * The cylinder No. signal or proper position signal is not sent to ECM for the first few seconds during engine cranking. * The cylinder No. signal or proper position signal is not set to ECM during engine running. * The cylinder No. signal or proper position signal is not in the normal pattern during engine running.
DTC: P0340 **1T ECM, MIL: Yes** **Year:** 2011 **Model:** Altima, Quest **Engine:** 2.5L, 3.5L	**Camshaft Position Sensor "A" Circuit (Bank 1 or Single Sensor):** Monitor runs whenever following DTCs are not present: None Engine speed: 600 rpm or more Starter: Off
DTC: P0341-747 **T ECM, MIL: Yes** **Year:** 2011 **Model:** Altima, Quest **Engine:** 2.5L, 3.5L	**Camshaft Position Sensor "A" Circuit High Input:** GI signal is not input for 2 sec. or more while the engine is running.
DTC: P0343-747 **T ECM, MIL: Yes** **Year:** 2011 **Model:** Altima, Quest **Engine:** 2.5L, 3.5L	**Camshaft Position Sensor "A" Circuit High Input:** GI pulse signal is not input for 2 sec. or more while the engine is running.
DTC: P0343-886 **T ECM, MIL: Yes** **Year:** 2011 **Model:** Altima, Quest **Engine:** 2.5L, 3.5L	**Camshaft Position Sensor "A" Circuit High Input:** GI signal is not sent to the hybrid vehicle control ECU while the engine is running.

DTC	Trouble Code Title and Conditions
DTC: P0345 **1T ECM, MIL: Yes** **Year:** 2011, 2012 **Model:** 370Z, Altima, Frontier, Maxima, Murano, Pathfinder, Quest, Xterra **Engine:** 2.5L, 3.5L, 3.5L, 3.7L, 4.0L, 5.6L	**Camshaft Position (CMP) Sensor Bank 2 Circuit:** **NOTE: if DTC P0340 or P0345 is displayed with DTC P0643, first perform the trouble diagnosis for DTC P0643.** * The cylinder No. signal is not sent to ECM for the first few seconds during engine cranking. * The cylinder No. signal is not sent to ECM during engine running. * The cylinder No. signal is not in the normal pattern during engine running.
DTC: P0351 **1T ECM, MIL: Yes** **Year:** 2011 **Model:** Altima, Quest **Engine:** 2.5L, 3.5L	**Ignition Coil "A" Primary / Secondary Circuit:** Monitor runs whenever following DTCs are not present: None Either of the following conditions A or B is met: A. Engine speed: 1500 rpm or less B. Starter: OFF Either of the following conditions C or D met: C. All of the following conditions (a) and (b) are met: (a) Engine speed: 500 rpm or less (b) Battery voltage: 6 V or more D. All of the following conditions (c), (d) and (e) are met: (c) Engine speed: More than 500 rpm (d) Battery voltage: 10 V or more (e) Number of sparks after CPU reset: 5 sparks or more Lost communication with hybrid vehicle control ECU (U0293): Not detected
DTC: P0352 **1T ECM, MIL: Yes** **Year:** 2011 **Model:** Altima, Quest **Engine:** 2.5L, 3.5L	**Ignition Coil "B" Primary / Secondary Circuit:** Monitor runs whenever following DTCs are not present: None Either of the following conditions A or B is met: A. Engine speed: 1500 rpm or less B. Starter: OFF Either of the following conditions C or D met: C. All of the following conditions (a) and (b) are met: (a) Engine speed: 500 rpm or less (b) Battery voltage: 6 V or more D. All of the following conditions (c), (d) and (e) are met: (c) Engine speed: More than 500 rpm (d) Battery voltage: 10 V or more (e) Number of sparks after CPU reset: 5 sparks or more Lost communication with hybrid vehicle control ECU (U0293): Not detected
DTC: P0353 **1T ECM, MIL: Yes** **Year:** 2011 **Model:** Altima, Quest **Engine:** 2.5L, 3.5L	**Ignition Coil "C" Primary / Secondary Circuit:** Monitor runs whenever following DTCs are not present: None Either of the following conditions A or B is met: A. Engine speed: 1500 rpm or less B. Starter: OFF Either of the following conditions C or D met: C. All of the following conditions (a) and (b) are met: (a) Engine speed: 500 rpm or less (b) Battery voltage: 6 V or more D. All of the following conditions (c), (d) and (e) are met: (c) Engine speed: More than 500 rpm (d) Battery voltage: 10 V or more (e) Number of sparks after CPU reset: 5 sparks or more Lost communication with hybrid vehicle control ECU (U0293): Not detected

DTC	Trouble Code Title and Conditions
DTC: P0354 **1T ECM, MIL: Yes** **Year:** 2011 **Model:** Altima, Quest **Engine:** 2.5L, 3.5L	**Ignition Coil "D" Primary / Secondary Circuit:** Monitor runs whenever following DTCs are not present: None Either of the following conditions A or B is met: A. Engine speed: 1500 rpm or less B. Starter: OFF Either of the following conditions C or D met: C. All of the following conditions (a) and (b) are met: (a) Engine speed: 500 rpm or less (b) Battery voltage: 6 V or more D. All of the following conditions (c), (d) and (e) are met: (c) Engine speed: More than 500 rpm (d) Battery voltage: 10 V or more (e) Number of sparks after CPU reset: 5 sparks or more Lost communication with hybrid vehicle control ECU (U0293): Not detected
DTC: P0400 **T ECM, MIL: Yes** **Year:** 2011 **Model:** Sentra **Engine:** 2.0L, 2.5L	**EGR System Function - No Flow:** No EGR flow is detected under conditions that call for EGR.
DTC: P0420 **2T ECM, MIL: Yes** **Year:** 2011 **Model:** Altima, Quest **Engine:** 2.5L, 3.5L	**Catalyst System Efficiency Below Threshold (Bank 1):** Monitor runs whenever following DTCs not stored: P0010 (Camshaft Timing Oil Control Valve) P0011 (VVT System - Advance) P0012 (VVT System - Retard) P0016 (VVT System - Misalignment) P0031, P0032, P101D (Air Fuel Ratio Sensor Heater) P0037, P0038, P102D (Heated Oxygen Sensor Heater) P2195, P2196, P2237, P2238, P2239, P2252, P2253, P2A00 (Air Fuel Ratio Sensor) P0136, P0137, P0138, P0139, P0607 (Rear Oxygen Sensor) P0102, P0103 (Mass Air Flow Meter) P0107, P0108 (Manifold Absolute Pressure) P0115, P0117, P0118 (Engine Coolant Temperature Sensor) P0125 (Insufficient Coolant Temperature for Closed Loop Fuel Control) P0120, P0121, P0122, P0123, P0220, P0222, P0223, P2135 (Throttle Position Sensor) P0171, P0172 (Fuel System) P0301 - P0304 (Misfire) P0335 (Crankshaft Position Sensor) P0340 (Camshaft Position Sensor) P0351 - P0354 (Igniter) Battery voltage: 11 V or more Intake air temperature: -10°C (14°F) or more Engine coolant temperature: 75°C (167°F) or more Atmospheric pressure: 76 kPa-a (570 mmHg-a) or more Idling: OFF Engine speed: Less than 3200 rpm Air fuel ratio sensor status: Activated Fuel system status: Closed loop Engine load: 10 to 70% All of the following conditions are met: Condition 1, 2 and 3: 1. Mass air flow rate: 10 to 28 gm/sec 2. Front catalyst temperature (estimated): 600 to 800°C (1112 to 1472°F) 3. Rear catalyst temperature (estimated): 370 to 675°C (698 to 1247°F)
DTC: P0420 **1T ECM, MIL: Yes** **Year:** 2011, 2012 **Model:** All **Engine:** 1.6L, 1.8L, 2.0L, 2.5L, 3.5L, 3.5L, 3.7L, 4.0L, 5.6L	**Catalyst System Efficiency Below Threshhold (Includes Hybrid Models):** HYBRID MODELS CAUTION: Hybrid systems use very high-voltage battery systems. Before starting any service work involving the battery system, turn the ignition switch OFF and then remove the service plug from pocket in the trunk. After removing the service plug, wait 10 minutes before touching any of the high-voltage connectors and terminals. **NOTE: On models with dual exhaust, this DTC refers to Bank 1.** * Three way catalyst (manifold) does not operate properly. * Three way catalyst (manifold) does not have enough oxygen storage capacity.

DTC	Trouble Code Title and Conditions
DTC: P0430 **1T ECM, MIL: Yes** **Year:** 2011, 2012 **Model:** 370Z, Altima, Maxima, Murano, Pathfinder, Quest, Titan, Xterra **Engine:** 3.5L, 3.5L, 3.7L, 4.0L, 5.6L	**Catalyst System Efficiency Below Threshhold (Bank 2):** * Three way catalyst (manifold) does not operate properly. * Three way catalyst (manifold) does not have enough oxygen storage capacity.
DTC: P043E **2T ECM, MIL: Yes** **Year:** 2011 **Model:** Altima, Quest **Engine:** 2.5L, 3.5L	**Evaporative Emission System Reference Orifice Clog Up:** Monitor runs whenever following DTCs are not present: None Atmospheric pressure: 70 to 110 kPa-a (525 to 825 mmHg-a) Battery voltage: 10.5 V or more Vehicle speed: Below 2.5 mph (4 km/h) Power switch: OFF Time after key off: 5, 7 or 9.5 hours Canister pressure sensor malfunction (P0452 and P0453): Not detected Purge VSV: Not operated by scan tool Vent valve: Not operated by scan tool Leak detection pump: Not operated by scan tool Both of the following conditions are met before key off: Conditions 1 and 2 1. Duration that vehicle driven: 5 minutes or more 2. EVAP purge operation: Performed Engine coolant temperature: 4.4 to 35°C (40 to 95°F) Intake air temperature: 4.4 to 35°C (40 to 95°F)
DTC: P043F **2T, MIL: Yes** **Year:** 2011 **Model:** Altima, Quest **Engine:** 2.5L, 3.5L	**Evaporative Emission System Reference Orifice High Flow:** Monitor runs whenever following DTCs are not present: None Atmospheric pressure: 70 to 110 kPa-a (525 to 825 mmHg-a) Battery voltage: 10.5 V or more Vehicle speed: Below 2.5 mph (4 km/h) Power switch: OFF Time after key off: 5, 7 or 9.5 hours Canister pressure sensor malfunction (P0452 and P0453): Not detected Purge VSV: Not operated by scan tool Vent valve: Not operated by scan tool Leak detection pump: Not operated by scan tool Both of the following conditions are met before key off: Conditions 1 and 2 1. Duration that vehicle driven: 5 minutes or more 2. EVAP purge operation: Performed Engine coolant temperature: 4.4 to 35°C (40 to 95°F) Intake air temperature: 4.4 to 35°C (40 to 95°F)

DTC	Trouble Code Title and Conditions
DTC: P0441 **2T, MIL: Yes** **Year:** 2011 **Model:** Altima, Quest **Engine:** 2.5L, 3.5L	**Evaporative Emission Control System Incorrect Purge Flow:** Monitor runs whenever following DTCs are not present: None Key Off Monitor: Atmospheric pressure: 70 to 110 kPa-a (525 to 825 mmHg-a) Battery voltage: 10.5 V or more Vehicle speed: Below 2.5 mph (4 km/h) Power switch: OFF Time after key off: 5, 7 or 9.5 hours Canister pressure sensor malfunction (P0452 and P0453): Not detected Purge VSV: Not operated by scan tool Vent valve: Not operated by scan tool Leak detection pump: Not operated by scan tool Both of the following conditions are met before key off Conditions 1 and 2: 1. Duration that vehicle driven: 5 minutes or more 2. EVAP purge operation: Performed Engine coolant temperature: 4.4 to 35°C (40 to 95°F) Intake air temperature: 4.4 to 35°C (40 to 95°F) Purge Flow Monitor: Engine: Running Engine coolant temperature: 4.4°C (40°F) or more Intake air temperature: 4.4°C (40°F) or more Canister pressure sensor malfunction: Not detected Purge VSV: Not operated by scan tool EVAP system check: Not operated by scan tool Atmospheric pressure: 70 to 110 kPa-a (525 to 825 mmHg-a) Battery voltage: 10 V or higher (Intake air temperature below 50°C [122°F]) 12 V or higher (Intake air temperature 50°C [122°F] or higher) Purge duty cycle: 8 % or more
DTC: P0441 **1T ECM, MIL: Yes** **Year:** 2011, 2012 **Model:** All **Engine:** 1.6L, 1.8L, 2.0L, 2.5L, 3.5L, 3.5L, 3.7L, 4.0L, 5.6L	**EVAP Control System Incorrect Purge Flow (Includes Hybrid Models):** HYBRID MODELS CAUTION: Hybrid systems use very high-voltage battery systems. Before starting any service work involving the battery system, turn the ignition switch OFF and then remove the service plug from pocket in the trunk. After removing the service plug, wait 10 minutes before touching any of the high-voltage connectors and terminals. **NOTE: If DTC P0441 is displayed with other DTC such as P2122, P2123 P2127, P2128, P2138, first perform trouble diagnosis for other DTC.** * Under normal conditions (non-closed throttle), sensor output voltage indicates if pressure drop and purge flow are adequate. If not, a malfunction is determined. * EVAP control system does not operate properly – EVAP control system has a leak between intake manifold and EVAP control system pressure sensor.
DTC: P0442 **1T ECM, MIL: Yes** **Year:** 2011, 2012 **Model:** All **Engine:** 1.6L, 1.8L, 2.0L, 2.5L, 3.5L, 3.5L, 3.7L, 4.0L, 5.6L	**EVAP Control System SmAll Leak Detected (Negative Pressure) (Includes Hybrid Models):** **NOTE: If DTC P0442 is displayed with DTC P0456, first perform the trouble diagnosis for DTC P0456.** * EVAP control system has a leak, EVAP control system does not operate properly.
DTC: P0443 **1T ECM, MIL: Yes** **Year:** 2011, 2012 **Model:** Frontier, Maxima, Murano, Rogue **Engine:** 2.5L, 3.5L, 3.5L, 4.0L	**EVAP Canister Purge Volume Control Solenoid Valve:** The canister purge flow is detected during the specified driving conditions, even when EVAP canister purge volume control solenoid valve is completely closed.
DTC: P0443 **1T ECM, MIL: Yes** **Year:** 2011, 2012 **Model:** 370Z, Altima, Cube, Juke, Pathfinder, Quest, Sentra, Titan, Versa, Xterra **Engine:** 1.6L, 1.8L, 2.0L, 2.5L, 3.5L, 3.7L, 4.0L, 5.6L	**EVAP Canister Purge Volume Control Solenoid Valve (Includes Hybrid Models):** HYBRID MODELS CAUTION: Hybrid systems use very high-voltage battery systems. Before starting any service work involving the battery system, turn the ignition switch OFF and then remove the service plug from pocket in the trunk. After removing the service plug, wait 10 minutes before touching any of the high-voltage connectors and terminals. Condition A: The canister purge flow is detected during the vehicle is stopped while the engine is running, even when EVAP canister purge volume control solenoid valve is completely closed. Condition B: The canister purge flow is detected during the specified driving conditions, even when EVAP canister purge volume control solenoid valve is completely closed.

DTC	Trouble Code Title and Conditions
DTC: P0444 **1T ECM, MIL: Yes** **Year:** 2011, 2012 **Model:** All **Engine:** 1.6L, 1.8L, 2.0L, 2.5L, 3.5L, 3.5L, 3.7L, 4.0L, 5.6L	**EVAP Canister Purge Volume Control Solenoid Valve Circuit Open (Includes Hybrid Models):** HYBRID MODELS CAUTION: Hybrid systems use very high-voltage battery systems. Before starting any service work involving the battery system, turn the ignition switch OFF and then remove the service plug from pocket in the trunk. After removing the service plug, wait 10 minutes before touching any of the high-voltage connectors and terminals. * An excessively low voltage signal is sent to ECM through the valve
DTC: P0445 **1T ECM, MIL: Yes** **Year:** 2011, 2012 **Model:** All **Engine:** 1.6L, 1.8L, 2.0L, 2.5L, 3.5L, 3.5L, 3.7L, 4.0L, 5.6L	**EVAP Canister Purge Volume Control Solenoid Valve Circuit Shorted (Includes Hybrid Models):** HYBRID MODELS CAUTION: Hybrid systems use very high-voltage battery systems. Before starting any service work involving the battery system, turn the ignition switch OFF and then remove the service plug from pocket in the trunk. After removing the service plug, wait 10 minutes before touching any of the high-voltage connectors and terminals. * An excessively high voltage signal is sent to ECM through the valve
DTC: P0447 **1T ECM, MIL: Yes** **Year:** 2011, 2012 **Model:** All **Engine:** 1.6L, 1.8L, 2.0L, 2.5L, 3.5L, 3.5L, 3.7L, 4.0L, 5.6L	**EVAP Canister Vent Control Valve Circuit Open (Includes Hybrid Models):** HYBRID MODELS CAUTION: Hybrid systems use very high-voltage battery systems. Before starting any service work involving the battery system, turn the ignition switch OFF and then remove the service plug from pocket in the trunk. After removing the service plug, wait 10 minutes before touching any of the high-voltage connectors and terminals. * An improper voltage signal is sent to ECM through EVAP canister vent control valve. * EVAP canister vent control valve remains open under specified driving conditions.
DTC: P0448 **1T ECM, MIL: Yes** **Year:** 2011, 2012 **Model:** All **Engine:** 1.6L, 1.8L, 2.0L, 2.5L, 3.5L, 3.5L, 3.7L, 4.0L, 5.6L	**EVAP Canister Vent Control Valve Closed (Includes Hybrid Models):** HYBRID MODELS CAUTION: Hybrid systems use very high-voltage battery systems. Before starting any service work involving the battery system, turn the ignition switch OFF and then remove the service plug from pocket in the trunk. After removing the service plug, wait 10 minutes before touching any of the high-voltage connectors and terminals. * EVAP canister vent control valve remains closed under specified driving conditions.
DTC: P0451 **2T ECM, MIL: Yes** **Year:** 2011 **Model:** Altima, Quest **Engine:** 2.5L, 3.5L	**Evaporative Emission Control System Pressure Sensor Range / Performance:** Monitor runs whenever following DTCs are not present: None Atmospheric pressure: 70 to 112 kPa-a (525 to 840 mmHg-a) Battery voltage: 10.5 V or more Intake air temperature: 4.4 to 50°C (40 to 122°F) Canister pressure sensor malfunction (P0452, P0453): Not detected Either of the following conditions 1 or 2 is met: 1. Engine condition: Running 2. Time after key-off: 5, 7 or 9.5 hours
DTC: P0451 **1T ECM, MIL: Yes** **Year:** 2011, 2012 **Model:** All **Engine:** 1.6L, 1.8L, 2.0L, 2.5L, 3.5L, 3.5L, 3.7L, 4.0L, 5.6L	**EVAP Control System Pressure Sensor Performance (Includes Hybrid Models):** HYBRID MODELS CAUTION: Hybrid systems use very high-voltage battery systems. Before starting any service work involving the battery system, turn the ignition switch OFF and then remove the service plug from pocket in the trunk. After removing the service plug, wait 10 minutes before touching any of the high-voltage connectors and terminals. * ECM detects a sloshing signal from the EVAP control system pressure sensor
DTC: P0452 **1T ECM, MIL: Yes** **Year:** 2011, 2012 **Model:** All **Engine:** 1.6L, 1.8L, 2.0L, 2.5L, 3.5L, 3.5L, 3.7L, 4.0L, 5.6L	**EVAP Control System Pressure Sensor Low Input (Includes Hybrid Models):** HYBRID MODELS CAUTION: Hybrid systems use very high-voltage battery systems. Before starting any service work involving the battery system, turn the ignition switch OFF and then remove the service plug from pocket in the trunk. After removing the service plug, wait 10 minutes before touching any of the high-voltage connectors and terminals. * An excessively low voltage from the sensor is sent to ECM.
DTC: P0452 **1T ECM, MIL: Yes** **Year:** 2011 **Model:** Altima, Quest **Engine:** 2.5L, 3.5L	**Evaporative Emission Control System Pressure Sensor / Switch Low Input:** * Canister pump module * Connector/wire harness (canister pump module - hybrid vehicle control ECU) * Hybrid vehicle control ECU
DTC: P0453 **1T ECM, MIL: Yes** **Year:** 2011, 2012 **Model:** All **Engine:** 1.6L, 1.8L, 2.0L, 2.5L, 3.5L, 3.5L, 3.7L, 4.0L, 5.6L	**EVAP Control System Pressure Sensor High Input (Includes Hybrid Models):** HYBRID MODELS CAUTION: Hybrid systems use very high-voltage battery systems. Before starting any service work involving the battery system, turn the ignition switch OFF and then remove the service plug from pocket in the trunk. After removing the service plug, wait 10 minutes before touching any of the high-voltage connectors and terminals. * An excessively high voltage from the sensor is sent to ECM.

DTC	Trouble Code Title and Conditions
DTC: P0453 **1T ECM, MIL: Yes** **Year:** 2011 **Model:** Altima, Quest **Engine:** 2.5L, 3.5L	**Evaporative Emission Control System Pressure Sensor / Switch High Input:** Monitor runs whenever following DTCs are not present: None Power switch: On (IG) Battery voltage: 8 V or more
DTC: P0455 **2T ECM, MIL: Yes** **Year:** 2011 **Model:** Altima, Quest **Engine:** 2.5L, 3.5L	**Evaporative Emission Control System Leak Detected (Gross Leak):** Monitor runs whenever following DTCs are not present: None Atmospheric pressure: 70 to 112 kPa-a (525 to 840 mmHg-a) Battery voltage: 10.5 V or more Vehicle speed: Below 2.5 mph (4 km/h) Power switch: OFF Time after key off: 5, 7 or 9.5 hours Canister pressure sensor malfunction (P0452 and P0453): Not detected Purge VSV: Not operated by scan tool Vent valve: Not operated by scan tool Leak detection pump: Not operated by scan tool Both of the following conditions are met before key off: Conditions 1 and 2 1. Duration that vehicle has been driven: 5 minutes or more 2. EVAP purge operation: Performed Engine coolant temperature: 4.4 to 35°C (40 to 95°F) Intake air temperature: 4.4 to 35°C (40 to 95°F)
DTC: P0455 **1T ECM, MIL: Yes** **Year:** 2011, 2012 **Model:** All **Engine:** 1.6L, 1.8L, 2.0L, 2.5L, 3.5L, 3.5L, 3.7L, 4.0L, 5.6L	**EVAP Control System Gross Leak Detected (Includes Hybrid Models):** * EVAP control system has a very large leak, such as fuel filler cap fell off. * EVAP control system does not operate properly. CAUTION: Never remove fuel filler cap during the DTC Confirmation Procedure.
DTC: P0456 **2T ECM, MIL: Yes** **Year:** 2011 **Model:** Altima, Quest **Engine:** 2.5L, 3.5L	**Evaporative Emission Control System Leak Detected (Very SmAll Leak):** Monitor runs whenever following DTCs are not present: None Atmospheric pressure: 70 to 112 kPa-a (525 to 840 mmHg-a) Battery voltage: 10.5 V or more Vehicle speed: Below 2.5 mph (4 km/h) Power switch: OFF Time after key off: 5, 7 or 9.5 hours Canister pressure sensor malfunction (P0452 and P0453): Not detected Purge VSV: Not operated by scan tool Vent valve: Not operated by scan tool Leak detection pump: Not operated by scan tool Both of the following conditions are met before key off: Conditions 1 and 2 1. Duration that vehicle has been driven: 5 minutes or more 2. EVAP purge operation: Performed Engine coolant temperature: 4.4 to 35°C (40 to 95°F) Intake air temperature: 4.4 to 35°C (40 to 95°F)
DTC: P0456 **1T ECM, MIL: Yes** **Year:** 2011, 2012 **Model:** All **Engine:** 1.6L, 1.8L, 2.0L, 2.5L, 3.5L, 3.5L, 3.7L, 4.0L, 5.6L	**Evaporative Emission Control System Very SmAll Leak (Negative Pressure Check) (Includes Hybrid Models):** HYBRID MODELS CAUTION: Hybrid systems use very high-voltage battery systems. Before starting any service work involving the battery system, turn the ignition switch OFF and then remove the service plug from pocket in the trunk. After removing the service plug, wait 10 minutes before touching any of the high-voltage connectors and terminals. **NOTE: If ECM judges a leak which corresponds to a very small leak, the very small leak P0456 will be detected.** **NOTE: If ECM judges a leak equivalent to a small leak, EVAP small leak P0442 will be detected.** **NOTE: If ECM judges there are no leaks, the diagnosis will be OK.** * If DTC P0456 is displayed with DTC P0442, first perform the trouble diagnosis for DTC P0456. * This diagnosis detects very small leakage in the EVAP line between fuel tank and EVAP canister purge volume control solenoid valve, using the intake manifold vacuum in the same way as conventional EVAP small leakage diagnosis. **NOTE: If ECM judges a leakage which corresponds to a very small leakage, the very small leakage P0456 will be detected.** * If ECM judges a leakage equivalent to a small leakage, EVAP small leakage P0442 will be detected. * If ECM judges that there are no leakage, the diagnosis will be OK. * EVAP system has a very small leak. * EVAP system does not operate properly.

DTC	Trouble Code Title and Conditions
DTC: P0460 **1T ECM, MIL: Yes** **Year:** 2011, 2012 **Model:** All **Engine:** 1.6L, 1.8L, 2.0L, 2.5L, 3.5L, 3.5L, 3.7L, 4.0L, 5.6L	**Fuel Level Sensor Circuit Noise (Includes Hybrid Models):** HYBRID MODELS CAUTION: Hybrid systems use very high-voltage battery systems. Before starting any service work involving the battery system, turn the ignition switch OFF and then remove the service plug from pocket in the trunk. After removing the service plug, wait 10 minutes before touching any of the high-voltage connectors and terminals. **NOTE: If DTC P0461 is displayed with DTC UXXXX, first perform the trouble diagnosis for DTC UXXXX.** **NOTE: If DTC P0460 is displayed with DTC P0607, first perform the trouble diagnosis for DTC P0607.** * When the vehicle is parked, naturally the fuel level in the fuel tank is stable. It means that output signal of the fuel level sensor does not change. If ECM senses sloshing signal from the sensor, fuel level sensor malfunction is detected. * Even though the vehicle is parked, a signal being varied is sent from the fuel level sensor to ECM.
DTC: P0461 **1T ECM, MIL: Yes** **Year:** 2011, 2012 **Model:** All **Engine:** 1.6L, 1.8L, 2.0L, 2.5L, 3.5L, 3.5L, 3.7L, 4.0L, 5.6L	**Fuel Level Sensor Circuit Range/Performance (Includes Hybrid Models):** HYBRID MODELS CAUTION: Hybrid systems use very high-voltage battery systems. Before starting any service work involving the battery system, turn the ignition switch OFF and then remove the service plug from pocket in the trunk. After removing the service plug, wait 10 minutes before touching any of the high-voltage connectors and terminals. **NOTE: If DTC P0461 is displayed with DTC U1000 or U1001, first perform the trouble diagnosis for appropriate "U" code.** **NOTE: If DTC P0461 is displayed with DTC P0607, first perform the trouble diagnosis for DTC P0607.** * This diagnosis detects the fuel gauge malfunction of the gauge not moving even after a long distance has been driven. Driving long distances naturally affect fuel gauge level. * The output signal of the fuel level sensor does not change within the specified range even though the vehicle has been driven a long distance.
DTC: P0462 **1T ECM, MIL: Yes** **Year:** 2011, 2012 **Model:** All **Engine:** 1.6L, 1.8L, 2.0L, 2.5L, 3.5L, 3.5L, 3.7L, 4.0L, 5.6L	**Fuel Level Sensor Circuit Low Input (Includes Hybrid Models):** HYBRID MODELS CAUTION: Hybrid systems use very high-voltage battery systems. Before starting any service work involving the battery system, turn the ignition switch OFF and then remove the service plug from pocket in the trunk. After removing the service plug, wait 10 minutes before touching any of the high-voltage connectors and terminals. **NOTE: If DTC P0462 or P0463 is displayed with DTC UXXXX, first perform the trouble diagnosis for DTC UXXXX.** **NOTE: If DTC P0462 or P0463 is displayed with DTC P0607, first perform the trouble diagnosis for DTC P0607.** * An excessively low voltage from the sensor is sent to ECM.
DTC: P0463 **1T ECM, MIL: Yes** **Year:** 2011, 2012 **Model:** All **Engine:** 1.6L, 1.8L, 2.0L, 2.5L, 3.5L, 3.5L, 3.7L, 4.0L, 5.6L	**Fuel Level Sensor Circuit High Input (Includes Hybrid Models):** HYBRID MODELS CAUTION: Hybrid systems use very high-voltage battery systems. Before starting any service work involving the battery system, turn the ignition switch OFF and then remove the service plug from pocket in the trunk. After removing the service plug, wait 10 minutes before touching any of the high-voltage connectors and terminals. **NOTE: If DTC P0462 or P0463 is displayed with DTC UXXXX, first perform the trouble diagnosis for DTC UXXXX.** **NOTE: If DTC P0462 or P0463 is displayed with DTC P0607, first perform the trouble diagnosis for DTC P0607.** * An excessively high voltage from the sensor is sent to ECM.
DTC: P0500 **2T ECM, MIL: Yes** **Year:** 2011, 2012 **Model:** Altima, Frontier, Quest, Sentra, Xterra **Engine:** 2.0L, 2.5L, 3.5L, 4.0L	**Vehicle Speed Sensor Circuit:** An almost 0 MPH signal from the vehicle speed sensor is sent to the ECM even when the vehicle is being driven.
DTC: P0500 **1T ECM, MIL: Yes** **Year:** 2011 **Model:** Juke **Engine:** 1.6L	**Vehicle Speed Sensor Circuit Malfunction:** At 20 km/h (13 MPH), ECM detects the following status continuously for 5 seconds or more: The difference between a vehicle speed calculated by a output speed sensor transmitted from TCM to ECM via CAN communication and the vehicle speed indicated on the combination meter exceeds 15km/h (10 MPH). **NOTE: If DTC P0500 is displayed with DTC UXXXX, first perform the trouble diagnosis for DTC UXXXX.**
DTC: P0500 **T, MIL: Yes** **Year:** 2011 **Model:** Altima, Quest **Engine:** 2.5L, 3.5L	**Vehicle Speed Sensor "A":** The vehicle speed signal from the vehicle speed sensor is cut while cruise control is operating.
DTC: P0500 **1T ECM, MIL: Yes** **Year:** 2011 **Model:** Juke, Versa **Engine:** 1.6L	**Vehicle Speed Sensor Circuit Malfunction:** The vehicle speed signal sent to ECM is almost 0 km/h (0 MPH) even when vehicle is being driven. **NOTE: If DTC P0500 is displayed with DTC UXXXX, first perform the trouble diagnosis for DTC UXXXX**
DTC: P0501 **1T ECM, MIL: Yes** **Year:** 2011 **Model:** Juke, Versa **Engine:** 1.6L	**Vehicle Speed Sensor "A" Range/Performance:** ECM detects a rear LH wheel sensor malfunction signal transmitted from the ABS actuator and electric unit (control unit) via CAN communication at least for 5 seconds in a row. **NOTE: If DTC P0501 or P2159 is displayed with DTC UXXXX, first perform the trouble diagnosis for DTC UXXXX.** Refer to EC-102, "DTC Index".

DTC	Trouble Code Title and Conditions
DTC: P0505 **2T ECM, MIL: Yes** **Year:** 2011 **Model:** Altima, Quest **Engine:** 2.5L, 3.5L	**Idle Control System Malfunction:** Monitor runs whenever following DTCs are not present: P0010 (Camshaft timing oil control valve assembly bank 1) P0011 (VVT system bank 1 - advance) P0012 (VVT system bank 1 - retard) P0016 (VVT system bank 1 - misalignment) P0031, P0032, P101D (Air fuel ratio sensor heater) P0102, P0103 (Mass air flow meter) P0107, P0108 (Manifold absolute pressure) P0115, P0117, P0118 (Engine coolant temperature sensor) P0120, P0121, P0122, P0123, P0220, P0222, P0223, P2135 (Throttle position sensor) P0125 (Insufficient engine coolant temperature for closed loop) P0171, P0172 (Fuel system) P0301, P0302, P0303, P0304 (Misfire) P0335 (Crankshaft position sensor) P0340 (Camshaft position sensor) P0351, P0352, P0353, P0354 (Igniter) P0451, P0452, P0453 (Evaporative emission control system) P2195, P2196 (Air fuel ratio sensor - rationality) P2237 (Air fuel ratio sensor - open) P2238, P2252 (Air fuel ratio sensor - low impedance) P2239, P2253 (Air fuel ratio sensor - high impedance) P2A00 (Air fuel ratio sensor - slow response) Engine: Running
DTC: P0506 **1T ECM, MIL: Yes** **Year:** 2011, 2012 **Model:** All **Engine:** 1.6L, 1.8L, 2.0L, 2.5L, 3.5L, 3.5L, 3.7L, 4.0L, 5.6L	**Idle Speed Control System RPM Lower Than Expected (Includes Hybrid Models):** HYBRID MODELS CAUTION: Hybrid systems use very high-voltage battery systems. Before starting any service work involving the battery system, turn the ignition switch OFF and then remove the service plug from pocket in the trunk. After removing the service plug, wait 10 minutes before touching any of the high-voltage connectors and terminals. **NOTE: If DTC P0506 is displayed with other DTC, first perform the trouble diagnosis for the other DTC.** * The idle speed is less than the target idle speed by 100 rpm or more.
DTC: P0507 **1T ECM, MIL: Yes** **Year:** 2011, 2012 **Model:** All **Engine:** 1.6L, 1.8L, 2.0L, 2.5L, 3.5L, 3.5L, 4.0L, 5.6L	**Idle Speed Control System RPM Higher Than Expected (Includes Hybrid Models):** HYBRID MODELS CAUTION: Hybrid systems use very high-voltage battery systems. Before starting any service work involving the battery system, turn the ignition switch OFF and then remove the service plug from pocket in the trunk. After removing the service plug, wait 10 minutes before touching any of the high-voltage connectors and terminals. **NOTE: If DTC P0507 is displayed with other DTC, first perform the trouble diagnosis for the other DTC.** * The idle speed is more than the target idle speed by 200 rpm or more.

DTC	Trouble Code Title and Conditions
DTC: P050A **2T ECM, MIL: Yes** **Year:** 2011 **Model:** Altima, Quest **Engine:** 2.5L, 3.5L	**Cold Start Idle Air Control System Performance:** Monitor runs whenever following DTCs are not present: P0010 (Camshaft timing oil control valve assembly bank 1) P0011 (VVT system bank 1 - advance) P0012 (VVT system bank 1 - retard) P0016 (VVT system bank 1 - misalignment) P0102, P0103 (Mass air flow meter) P0107, P0108 (Manifold absolute pressure) P0115, P0117, P0118 (Engine coolant temperature sensor) P0120, P0121, P0122, P0123, P0220, P0222, P0223, P2135 (Throttle position sensor) P0125 (Insufficient engine coolant temperature for closed loop) P0171, P0172 (Fuel system) P0301, P0302, P0303, P0304 (Misfire) P0335 (Crankshaft position sensor) P0340 (Camshaft position sensor) P0351, P0352, P0353, P0354 (Igniter) P2195, P2196 (Air fuel ratio sensor - rationality) P2237 (Air fuel ratio sensor - open) P2238, P2252 (Air fuel ratio sensor - low impedance) P2239, P2253 (Air fuel ratio sensor - high impedance) P2A00 (Air fuel ratio sensor - slow response) Battery voltage: 8 V or more Time after engine start: 3 seconds or more Starter: OFF Engine coolant temperature at engine start: -10°C (14°F) or more Engine coolant temperature: -10 to 50°C (14 to 122°F) Engine idling time: 0 seconds or more Fuel-cut: OFF Vehicle speed: Less than 37.5 mph (60.3 km/h) Atmospheric pressure: 76 kPa-a (570 mmHg-a) or more
DTC: P050A **1T ECM, MIL: Yes** **Year:** 2011 **Model:** Juke, Rogue, Versa **Engine:** 1.6L, 2.5L	**Cold Start Idle Air Control System Performance:** ECM does not control engine idle speed properly when engine is started with pre-warm up condition.

DTC	Trouble Code Title and Conditions
DTC: P050B **2T ECM, MIL: Yes** **Year:** 2011 **Model:** Altima, Quest **Engine:** 2.5L, 3.5L	**Cold Start Ignition Timing Performance:** Monitor runs whenever following DTCs are not present: P0010 (Camshaft timing oil control valve assembly bank 1) P0011 (VVT system bank 1 - advance) P0012 (VVT system bank 1 - retard) P0016 (VVT system bank 1 - misalignment) P0102, P0103 (Mass air flow meter) P0107, P0108 (Manifold absolute pressure) P0115, P0117, P0118 (Engine coolant temperature sensor) P0120, P0121, P0122, P0123, P0220, P0222, P0223, P2135 (Throttle position sensor) P0125 (Insufficient engine coolant temperature for closed loop) P0171, P0172 (Fuel system) P0301, P0302, P0303, P0304 (Misfire) P0335 (Crankshaft position sensor) P0340 (Camshaft position sensor) P0351, P0352, P0353, P0354 (Igniter) P2195, P2196 (Air fuel ratio sensor - rationality) P2237 (Air fuel ratio sensor - open) P2238, P2252 (Air fuel ratio sensor - low impedance) P2239, P2253 (Air fuel ratio sensor - high impedance) P2A00 (Air fuel ratio sensor - slow response) Battery voltage: 8 V or more Time after engine start: 3 seconds or more Starter: OFF Engine coolant temperature at engine start: -10°C (14°F) or more Engine coolant temperature: -10 to 50°C (14 to 122°F) Engine idling time: 0 seconds or more Fuel-cut: OFF Vehicle speed: Less than 37.5 mph (60.3 km/h) Atmospheric pressure: 76 kPa-a (570 mmHg-a) or more
DTC: P050B **1T ECM, MIL: Yes** **Year:** 2011 **Model:** Rogue **Engine:** 2.5L	**Cold Start Ignition Timing Performance:** **NOTE: If DTC P050B is displayed with other DTCs, perform the trouble diagnosis for the other DTCs first.** ECM does not control ignition timing properly when engine is started with pre-warming up condition.
DTC: P050E **1T ECM, MIL: Yes** **Year:** 2011 **Model:** Juke, Murano, Rogue, Versa **Engine:** 1.6L, 2.5L, 3.5L	**Cold Start Engine Exhaust Temperature Too Low:** The temperature of the catalyst inlet does not rise to the proper temperature when the engine is started with pre-warming up condition.
DTC: P0516-769 **T ECM, MIL: Yes** **Year:** 2011 **Model:** Altima, Quest **Engine:** 2.5L, 3.5L	**Battery Temperature Sensor Circuit Low:** Malfunction in the auxiliary battery thermometer sensor circuit (short to GND).
DTC: P0517-770 **T ECM, MIL: Yes** **Year:** 2011 **Model:** Altima, Quest **Engine:** 2.5L, 3.5L	**Battery Temperature Sensor Circuit High:** Malfunction in the auxiliary battery thermometer sensor circuit (open or +B short circuit).
DTC: P0520 **1T ECM, MIL: Yes** **Year:** 2011 **Model:** Juke, Versa **Engine:** 1.6L	**Engine Oil Pressure (EOP) Sensor Circuit:** Signal voltage from the EOP sensor remains at more than 4.9 V / less than 0.26 V for 5 seconds or more.

DTC	Trouble Code Title and Conditions
DTC: P0524 **1T ECM, MIL: Yes** **Year:** 2011 **Model:** Juke, Versa **Engine:** 1.6L	**Engine Oil Pressure Too Low:** An EOP sensor signal voltage applied to ECM remains lower than the specified value continuously for 10 seconds or more when the engine speed is 1,000 rpm or more.
DTC: P0524 **1T ECM, MIL: Yes** **Year:** 2011, 2012 **Model:** 370Z **Engine:** 3.7L	**Engine Oil Pressure Too Low:** Engine oil pressure is low because there is a gap between angle of target and phase-control angle.
DTC: P0550 **1T ECM** **Year:** 2011, 2012 **Model:** 370Z, Frontier, Maxima, Murano, Pathfinder, Titan, Xterra **Engine:** 2.5L, 3.5L, 3.5L, 3.7L, 4.0L, 5.6L	**Power Steering Pressure Sensor Circuit:** The MIL will not illuminate for this diagnosis. **NOTE: If DTC P0550 is displayed with DTC P0643, first perform the trouble diagnosis for DTC P0643.** * An excessively low or high voltage from the sensor is sent to ECM.
DTC: P0555 **T ECM, MIL: Yes** **Year:** 2011, 2012 **Model:** 370Z **Engine:** 3.7L	**Brake Boosere Pressure Sensor Circuit:** An excessively low voltage from the sensor is sent to ECM. An excessively high voltage from the sensor is sent to ECM.
DTC: P0560 **1T ECM, MIL: Yes** **Year:** 2011 **Model:** Altima, Quest **Engine:** 2.5L, 3.5L	**System Voltage:** Monitor runs whenever following DTCs are not present: None
DTC: P0560-117 **T, MIL: Yes** **Year:** 2011 **Model:** Altima, Quest **Engine:** 2.5L, 3.5L	**System Voltage:** The monitor will run whenever the following DTC is not present: TMC's intellectual property Other conditions belong to TMC's intellectual property: -
DTC: P0560-117 **T ECM, MIL: Yes** **Year:** 2011 **Model:** Altima, Quest **Engine:** 2.5L, 3.5L	**System Voltage:** Malfunction in the hybrid vehicle control ECU back-up power source circuit
DTC: P0571 **T, MIL: Yes** **Year:** 2011 **Model:** Altima, Quest **Engine:** 2.5L, 3.5L	**Brake Switch "A" Circuit:** Voltage of STP signal and that of ST1- signal of hybrid vehicle control ECU are less than 1 V for 0.5 seconds or more.
DTC: P0575 **T, MIL: Yes** **Year:** 2011 **Model:** Altima, Quest **Engine:** 2.5L, 3.5L	**Cruise Control Input Circuit:** When both of the following conditions are met: * STP signals input to the hybrid vehicle control ECU supervisory CPU and control ECU are different for 0.15 seconds or more * 0.4 seconds have passed after cruise cancel input signal (STP input) is input to the hybrid vehicle control ECU
DTC: P0603 **1T ECM, MIL: Yes** **Year:** 2011, 2012 **Model:** 370Z, Altima, Frontier, Juke, Maxima, Murano, Pathfinder, Quest, Rogue, Sentra, Versa, Xterra **Engine:** 1.6L, 2.0L, 2.5L, 3.5L, 3.5L, 3.7L, 4.0L, 5.6L	**ECM Power Supply Circuit (Includes Hybrid Models):** HYBRID MODELS CAUTION: Hybrid systems use very high-voltage battery systems. Before starting any service work involving the battery system, turn the ignition switch OFF and then remove the service plug from pocket in the trunk. After removing the service plug, wait 10 minutes before touching any of the high-voltage connectors and terminals. * ECM back-up RAM system does not function properly.

DTC	Trouble Code Title and Conditions
DTC: P0604 **T ECM, MIL: Yes** **Year:** 2011 **Model:** Altima, Quest **Engine:** 2.5L, 3.5L	**Random Access Memory (RAM):** Monitor runs whenever following DTCs are not present: None
DTC: P0605 **1T ECM, MIL: Yes** **Year:** 2011, 2012 **Model:** 370Z, Altima, Cube, Frontier, Juke, Maxima, Murano, Pathfinder, Quest, Rogue, Titan, Versa, Xterra **Engine:** 1.6L, 1.8L, 2.5L, 3.5L, 3.5L, 3.7L, 4.0L, 5.6L	**Engine Control Module (ECM) (Includes Hybrid Models):** HYBRID MODELS CAUTION: Hybrid systems use very high-voltage battery systems. Before starting any service work involving the battery system, turn the ignition switch OFF and then remove the service plug from pocket in the trunk. After removing the service plug, wait 10 minutes before touching any of the high-voltage connectors and terminals. A. ECM calculation function is malfunctioning. B. ECM EEP-ROM system is malfunctioning. C. ECM self shut-off function is malfunctioning.
DTC: P0606 **T ECM, MIL: Yes** **Year:** 2011 **Model:** Altima, Quest **Engine:** 2.5L, 3.5L	**ECM / PCM Processor:** Monitor runs whenever the following DTCs are not present: None
DTC: P0607 **T ECM, MIL: Yes** **Year:** 2011 **Model:** Altima, Quest **Engine:** 2.5L, 3.5L	**Control Module Performance:** Monitor runs whenever the following DTCs are not stored: None Engine: Running Estimated heated oxygen sensor temperature: 450 to 750°C (842 to 1382°F)
DTC: P0607 **1T ECM, MIL: Yes** **Year:** 2011, 2012 **Model:** All **Engine:** 1.6L, 1.8L, 2.0L, 2.5L, 3.5L, 3.5L, 3.7L, 4.0L, 5.6L	**CAN Communication Bus (Includes Hybrid Models):** HYBRID MODELS CAUTION: Hybrid systems use very high-voltage battery systems. Before starting any service work involving the battery system, turn the ignition switch OFF and then remove the service plug from pocket in the trunk. After removing the service plug, wait 10 minutes before touching any of the high-voltage connectors and terminals. When detecting error during the initial diagnosis of CAN controller of ECM.
DTC: P060A **T ECM, MIL: Yes** **Year:** 2011 **Model:** Altima, Quest **Engine:** 2.5L, 3.5L	**Internal Control Module Monitoring Processor Performance:** Monitor runs whenever following DTCs are not present: None
DTC: P060E **T ECM, MIL: Yes** **Year:** 2011 **Model:** Altima, Quest **Engine:** 2.5L, 3.5L	**Internal Control Module Throttle Position Performance:** Monitor runs whenever the following DTCs are not present: None DMA communication error: Not detected
DTC: P0611 **1T ECM, MIL: Yes** **Year:** 2011 **Model:** Juke, Versa **Engine:** 1.6L	**ECM Protection:** ECM overheat protection control is activated. ECM overheated
DTC: P0615 **T TCM, TCIL: Yes** **Year:** 2011, 2012 **Model:** 370Z, Altima, Frontier, Maxima, Murano, Pathfinder, Quest, Titan, Versa, Xterra **Engine:** 1.6L, 1.8L, 2.5L, 3.5L, 3.5L, 3.7L, 4.0L, 5.6L	**Starter Relay Circuit:** * This is not an OBD-II self-diagnostic item * This DTC will set if the starter monitor value is OFF when the ignition switch is ON at the" P" and "N" positions.
DTC: P0616-142 **T ECM, MIL: Yes** **Year:** 2011 **Model:** Altima, Quest **Engine:** 2.5L, 3.5L	**Starter Relay Circuit:** An ST signal from the hybrid vehicle control ECU is present when the ignition switch OFF.

DTC	Trouble Code Title and Conditions
DTC: P0617-142 **T, MIL: Yes** **Year:** 2011 **Model:** Altima, Quest **Engine:** 2.5L, 3.5L	**Starter Relay Circuit High:** An ST signal from the power management control ECU is present when the power switch is off.
DTC: P062B **1T ECM, MIL: Yes** **Year:** 2011 **Model:** Juke, Versa **Engine:** 1.6L	**ECM (Internal control module fuel injector control performance) :** Injector driver unit is malfunctioning.
DTC: P062F-143 **T ECM, MIL: Yes** **Year:** 2011 **Model:** Altima, Quest **Engine:** 2.5L, 3.5L	**EEPROM Malfunction:** The monitor will run whenever the following DTC is not present: TMC's intellectual property Other conditions belong to TMC's intellectual property: -
DTC: P062F-143 **T ECM, MIL: Yes** **Year:** 2011 **Model:** Altima, Quest **Engine:** 2.5L, 3.5L	**EEPROM Malfunction:** ECU internal error is detected.
DTC: P0630-804 **T ECM, MIL: Yes** **Year:** 2011 **Model:** Altima, Quest **Engine:** 2.5L, 3.5L	**VIN not Programmed or Mismatch-ECM / PCM:** * VIN not stored in hybrid vehicle control ECU * Input VIN in hybrid vehicle control ECU not accurate
DTC: P0643 **1T ECM, MIL: Yes** **Year:** 2011 **Model:** Juke, Versa **Engine:** 1.6L	**Sensor Power Supply Circuit Short:** The ECM detects a voltage of power source for sensor is excessively low or high.
DTC: P0643 **1T ECM, MIL: Yes** **Year:** 2011, 2012 **Model:** 370Z, Altima, Cube, Frontier, Maxima, Murano, Pathfinder, Quest, Rogue, Sentra, Titan, Versa, Xterra **Engine:** 1.6L, 1.8L, 2.0L, 2.5L, 3.5L, 3.5L, 3.7L, 4.0L, 5.6L	**Sensor Power Supply Circuit Short (Includes Hybrid Models):** HYBRID MODELS CAUTION: Hybrid systems use very high-voltage battery systems. Before starting any service work involving the battery system, turn the ignition switch OFF and then remove the service plug from pocket in the trunk. After removing the service plug, wait 10 minutes before touching any of the high-voltage connectors and terminals. * ECM detects a voltage of power source for sensor is excessively low or high. **NOTE: When the malfunction is detected, ECM enters fail-safe mode and the MIL illuminates.** * ECM stops the electric throttle control actuator control, throttle valve is maintained at a fixed opening (approx. 5 degrees) by the return spring.
DTC: P0657 **T ECM, MIL: Yes** **Year:** 2011 **Model:** Altima, Quest **Engine:** 2.5L, 3.5L	**Actuator Supply Voltage Circuit / Open:** Monitor runs whenever following DTCs are not present: None Throttle actuator power supply voltage: 7 V or more
DTC: P0700 **T TCM, MIL: Yes, TCIL: Yes** **Year:** 2011, 2012 **Model:** Frontier, Pathfinder, Titan, Xterra **Engine:** 2.5L, 4.0L, 5.6L	**TCM:** This is an OBD-II self-diagnostic item. Diagnostic trouble code P0700 is detected when the TCM is malfunctioning.
DTC: P0700 **T TCM, TCIL: Yes** **Year:** 2011, 2012 **Model:** Frontier **Engine:** 2.5L, 4.0L	**Transmission Range Switch A:** This is an OBD-II self-diagnostic item. Diagnostic trouble code "P0705" with CONSULT-III or 9th judgment flicker without CONSULT-III is detected under the following conditions: - When TCM does not receive the correct voltage signal from the transmission range switch 1, 2, 3, 4 based on the gear position. - When no other position but "P" position is detected from "N" positions.

DTC	Trouble Code Title and Conditions
DTC: P0703 **1T TCM, TCIL: Yes** **Year:** 2011 **Model:** Altima, Juke, Quest, Sentra, Versa **Engine:** 1.6L, 1.8L, 2.0L, 2.5L, 3.5L	**Brake Lamp Switch Circuit:** * ON, OFF status of the stop lamp switch is sent via the CAN communication from the combination meter to TCM using the signal. This is not an OBD-II self-diagnostic item. Diagnostic trouble code P0703 is detected when the stop lamp switch does not switch to ON and OFF. * The stop lamp switch does not switch to ON and OFF.
DTC: P0703 **T BCM, TCIL: Yes** **Year:** 2011, 2012 **Model:** Maxima, Murano, Rogue **Engine:** 2.5L, 3.5L, 3.5L	**Brake Switch B Circuit:** BCM detects ON/OFF state of the stop lamp switch and transmits the data to the TCM via CAN communication by converting the data to a signal. When the brake switch does not switch to ON or OFF, DTC is detected.
DTC: P0703 **T TCM, TCIL: Yes** **Year:** 2011 **Model:** Cube **Engine:** 1.8L	**Brake Switch B Circuit:** TCM detects malfunction in CAN communication between BCM TCM detects a state that ON/OFF of stop lamp switch signal is not switched
DTC: P0705 **T TCM, TCIL: Yes** **Year:** 2011 **Model:** Cube, Murano, Rogue, Versa **Engine:** 1.6L, 1.8L, 2.5L, 3.5L	**Transmission Range Sensor A Circuit (PRNDL Input):** TCM does not receive the correct voltage signal (based on the gear position) from the switch.
DTC: P0705 **1T TCM, TCIL: Yes** **Year:** 2011, 2012 **Model:** Altima, Maxima, Pathfinder, Quest, Sentra, Titan, Xterra **Engine:** 2.0L, 2.5L, 3.5L, 4.0L, 5.6L	**Park/Neutral Position Switch Circuit:** **NOTE: If DTC U1000 is displayed with this DTC, first perform the trouble diagnosis for DTC U1000.** * The PNP switch assembly includes a transaxle range switch. * The transaxle range switch detects the selector lever position and sends a signal to the TCM. * This is an OBD-II self-diagnostic item. * Diagnostic trouble code P0705 is detected under the following conditions: - When TCM does not receive the correct voltage signal from the PNP switches 1, 2, 3 and 4 based on the gear position. - When the signal from monitor terminal of PNP switch 3 is different from PNP switch 3. - When no other position but "P" position is detected from "N" positions.
DTC: P0705 **T TCM, TCIL: Yes** **Year:** 2011, 2012 **Model:** 370Z **Engine:** 3.7L	**Transmission Range Switch A:** * Transmission range switch 1 – 4 signals input with impossible pattern. * "P" position is detected from "N" position without any other position being detected in between.
DTC: P0705-757 **T ECM, MIL: Yes** **Year:** 2011 **Model:** Altima, Quest **Engine:** 2.5L, 3.5L	**Transmission Range Switch Circuit:** Transmission range switch pattern problem
DTC: P0705-758 **T ECM, MIL: Yes** **Year:** 2011 **Model:** Altima, Quest **Engine:** 2.5L, 3.5L	**Transmission Range Switch Circuit:** Shifting malfunction (open circuit in MJ)
DTC: P0710 **1T TCM, TCIL: Yes** **Year:** 2011 **Model:** Altima, Juke, Quest, Sentra, Versa **Engine:** 1.6L, 1.8L, 2.0L, 2.5L, 3.5L	**CVT Fluid Temperature Sensor Circuit:** The CVT fluid temperature sensor is included in the control valve assembly. The CVT fluid temperature sensor detects the CVT fluid temperature and sends a signal to the TCM. This is an OBD-II self-diagnostic item. Diagnostic trouble code P0710 is detected when TCM receives an excessively low or high voltage from the sensor.
DTC: P0710 **T TCM, TCIL: Yes** **Year:** 2011, 2012 **Model:** 370Z, Cube **Engine:** 1.8L, 3.7L	**Transmission Fluid Temperature Sensor A Circuit:** CVT fluid temperature does not rise to the specified temperature after driving for a certain period of time with the TCM-received oil temperature sensor value between $-39°C$ ($-38.2°F$) and $20°C$ ($-68°F$) CVT fluid temperature sensor value that TCM receives is more than $180°C$ ($356°F$) TCM-received CVT fluid temperature sensor value while driving is less than $-40°C$ ($-40°F$)

DTC	Trouble Code Title and Conditions
DTC: P0710 **T TCM, TCIL: Yes** **Year:** 2011, 2012 **Model:** Maxima, Murano, Rogue **Engine:** 2.5L, 3.5L, 3.5L	**CVT Transmission Fluid Temperature Sensor A Circuit:** During running, the CVT fluid temperature sensor signal voltage is excessively high or low.
DTC: P0715 **1T TCM, TCIL: Yes** **Year:** 2011 **Model:** Altima, Juke, Quest, Sentra, Versa **Engine:** 1.6L, 2.0L, 2.5L, 3.5L	**Input Speed Sensor Circuit (PRI Speed Sensor):** The input speed sensor (primary speed sensor) detects the primary pulley revolution speed and sends a signal to the TCM. This is an OBD-II self-diagnostic item. Diagnostic trouble code P0715 is detected when TCM does not receive the proper signal from the sensor. The engine speed signal is determined when the engine running at any speed. The scan tool display value should closely match the tachometer reading. The primary (PRI) speed signal is indicated during driving, with the Lock-Up ON. The display value should closely match engine speed.
DTC: P0715 **T TCM, TCIL: Yes** **Year:** 2011 **Model:** Juke, Versa **Engine:** 1.6L, 1.8L	**Input Speed Sensor A:** This is an OBD-II self-diagnostic item. Diagnostic trouble code P0715 is detected when TCM does not receive the proper signal from the sensor.
DTC: P0715 **T TCM, TCIL: Yes** **Year:** 2011, 2012 **Model:** Maxima, Murano, Rogue **Engine:** 2.5L, 3.5L, 3.5L	**Input/Turbine Speed Sensor A Circuit:** Input speed sensor (primary speed sensor) signal is not input due to an open circuit. An unexpected signal is input when vehicle is being driven.
DTC: P0715 **T TCM, TCIL: Yes** **Year:** 2011 **Model:** Cube **Engine:** 1.8L	**Input/Turbine Speed Sensor A Circuit:** Primary speed sensor signal is not transmitted to TCM Primary speed sensor value is less than 150 rpm while secondary pulley speed is more than 500 rpm
DTC: P0717 **T TCM, TCIL: Yes** **Year:** 2011, 2012 **Model:** 370Z **Engine:** 3.7L	**Input/Turbine Speed Sensor A Circuit No Signal:** * The revolution of input speed sensor 1 and/or 2 is 270 rpm or less. * This is an OBD-II self-diagnostic item. * Diagnostic trouble code "P0717" is detected under the following conditions: - When TCM does not receive the proper voltage signal from the sensor. - When TCM detects an irregularity only at position of 4GR for input speed sensor 2.
DTC: P0717 **1T TCM, TCIL: Yes** **Year:** 2011, 2012 **Model:** Frontier, Pathfinder, Titan, Xterra **Engine:** 2.5L, 4.0L, 5.6L	**Input Speed Sensor A (Turbine Revolution Sensor):** The input speed sensor detects input shaft rpm (revolutions per minute). It is located on the input side of the automatic transmission. Monitors revolution of sensor 1 and sensor 2 for non-standard conditions. This is an OBD-II self-diagnostic item. Diagnostic trouble code P0717 is detected under the following conditions: - When TCM does not receive the proper voltage signal from the sensor. - When TCM detects an irregularity only at position of 4th gear for input speed sensor 2.
DTC: P0720 **T TCM, TCIL: Yes** **Year:** 2011, 2012 **Model:** Maxima, Murano, Rogue **Engine:** 2.5L, 3.5L, 3.5L	**Output Speed Sensor Circuit:** * Signal from vehicle speed sensor CVT [output speed sensor (secondary speed sensor)] is not input due to open or short circuit. * An unexpected signal is input during running. * After ignition switch is turned ON, unexpected signal input from vehicle speed signal before the vehicle starts moving.
DTC: P0720 **T TCM, TCIL: Yes** **Year:** 2011 **Model:** Cube **Engine:** 1.8L	**Output Speed Sensor Circuit:** Secondary speed sensor signal is not transmitted to TCM Secondary speed sensor value is less than 150 rpm while primary pulley speed is more than 1,000 rpm
DTC: P0720 **1T TCM, TCIL: Yes** **Year:** 2011 **Model:** Altima, Quest **Engine:** 2.5L, 3.5L	**Vehicle Speed Sensor CVT (Secondary Speed Sensor):** * The vehicle speed sensor CVT [output speed sensor (secondary speed sensor)] detects the revolution of the CVT output shaft and emits a pulse signal. The pulse signal is sent to the TCM, which converts it into vehicle speed. * This is an OBD-II self-diagnostic item. * Diagnostic trouble code P0720 is detected TCM does not receive the proper signal from the sensor. * The VSS sensor signal during driving should closely match the speedometer.

DTC	Trouble Code Title and Conditions
DTC: P0720 **T TCM, TCIL: Yes** **Year:** 2011, 2012 **Model:** Maxima **Engine:** 3.5L	**Vehicle Speed Sensor (A/T Revolution Sensor):** A/T control unit (TCM) does not receive proper input signal from the sensor.
DTC: P0720 **1T TCM, TCIL: Yes** **Year:** 2011, 2012 **Model:** Frontier, Pathfinder, Titan, Versa, Xterra **Engine:** 1.6L, 1.8L, 2.5L, 4.0L, 5.6L	**Vehicle Speed Sensor A/T (Revolution Sensor/Output Speed Sensor):** * This is an OBD-II self-diagnostic item. * Diagnostic trouble code P0720 is detected under the following conditions: - When TCM does not receive the proper voltage signal from the sensor. - After ignition switch is turned "ON", irregular signal input from vehicle speed sensor MTR before the vehicle starts moving.
DTC: P0720 **T TCM, TCIL: Yes** **Year:** 2011, 2012 **Model:** 370Z **Engine:** 3.7L	**Output Speed Sensor Circuit:** * The output speed sensor recognizes that the vehicle speed is 5 km/h (3 MPH) or less even if the vehicle speed signal recognizes that the vehicle speed is 20 km/h (12 MPH) or more. (Only when starts after the ignition switch is turned ON.) * The vehicle speed recognized by the output speed sensor decelerates 36 km/h (23 MPH) or more during 60 msec when the output speed sensor recognizes that the vehicle speed is 36 km/h (23 MPH) or more and the vehicle speed signal recognizes that the vehicle speed is 24 km/h (15 MPH) or less. * The vehicle speed of output speed sensor decelerates 36 km/h (23 MPH) or more even if the vehicle speed of vehicle speed signal accelerates or decelerates 24 km/h (15 MPH) or less during 60 msec when the output speed sensor recognizes that the vehicle speed is 36 km/h (23 MPH) or more.
DTC: P0725 **T TCM, TCIL: Yes** **Year:** 2011, 2012 **Model:** 370Z, Xterra **Engine:** 3.7L, 4.0L	**Engine Speed Signal:** TCM does not receive proper voltage signal from ECM.
DTC: P0725 **T TCM, TCIL: Yes** **Year:** 2011 **Model:** Cube **Engine:** 1.8L	**Engine Speed Input Circuit:** * TCM detects a malfunction in CAN communication between TCM and ECM * When primary pulley speed is more than 1,000 rpm, engine speed (CAN signal) is less than 450 rpm
DTC: P0725 **1T TCM, TCIL: Yes** **Year:** 2011, 2012 **Model:** Altima, Frontier, Maxima, Murano, Pathfinder, Quest, Rogue, Sentra, Titan, Versa **Engine:** 1.6L, 1.8L, 2.0L, 2.5L, 3.5L, 3.5L, 4.0L, 5.6L	**Engine Speed Signal:** * The engine speed signal is sent from the ECM to the TCM. * Diagnostic trouble code P0725 is detected when TCM does not receive the engine speed signal or ignition signal (input by CAN communication) from ECM. * The engine speed signal is sent with the engine running and should closely match the tachometer reading.
DTC: P0729 **T TCM, TCIL: Yes** **Year:** 2011, 2012 **Model:** 370Z **Engine:** 3.7L	**Gear 6 Incorrect Ratio:** The gear ratio is: 0.914 or more 0.813 or less
DTC: P0730 **1T TCM, TCIL: Yes** **Year:** 2011 **Model:** Juke, Sentra, Versa **Engine:** 1.6L, 1.8L, 2.0L, 2.5L	**Belt Damage:** TCM selects the gear ratio using the engine load (throttle position), the primary pulley revolution speed, and the secondary pulley revolution speed as input signal. Then it changes the operating pressure of the primary pulley and the secondary pulley and changes the groove width of the pulley. This is not an OBD-II self-diagnostic item. TCM calculates the actual gear ratio with input speed sensor (primary speed sensor) and output speed sensor (secondary speed sensor). Diagnostic trouble code P0730 is detected, when TCM receives an unexpected gear ratio signal.
DTC: P0730 **T TCM, TCIL: Yes** **Year:** 2011, 2012 **Model:** 370Z **Engine:** 3.7L	**Incorrect Gear Ratio:** The revolution of under drive sun gear is 8,000 rpm or more. **NOTE: Not detected when in "P" or "N" position and during a shift to "P" or "N" position.**

DTC	Trouble Code Title and Conditions
DTC: P0730 **1T TCM, TCIL: Yes** **Year:** 2011, 2012 **Model:** Altima, Maxima, Murano, Quest, Rogue, Versa **Engine:** 1.6L, 1.8L, 2.5L, 3.5L, 3.5L	**Incorrect Gear Ratio:** This is not an OBD-II self-diagnostic item. TCM calculates the actual gear ratio with primary speed sensor and seconday speed sensor. TCM receives an unexpected gear ratio signal.
DTC: P0731 **T TCM, TCIL: Yes** **Year:** 2011, 2012 **Model:** 370Z, Frontier, Pathfinder, Titan, Xterra **Engine:** 2.5L, 3.7L, 4.0L, 5.6L	**A/T 1st Gear Function:** * This is an OBD-II self-diagnostic item. * Diagnostic trouble code P0731 is detected when TCM detects any inconsistency in the actual gear ratio.
DTC: P0732 **T TCM, TCIL: Yes** **Year:** 2011, 2012 **Model:** 370Z, Frontier, Pathfinder, Titan, Xterra **Engine:** 2.5L, 3.7L, 4.0L, 5.6L	**A/T 2nd Gear Function:** * This malfunction is detected when the A/T does not shift into 2GR position as instructed by TCM. This is not only caused by electrical malfunction (circuits open or shorted) but mechanical malfunction such as control valve sticking, improper solenoid valve operation. * This is an OBD-II self-diagnostic item. * Diagnostic trouble code P0732 is detected when TCM detects any inconsistency in the actual gear ratio.
DTC: P0733 **T TCM, TCIL: Yes** **Year:** 2011, 2012 **Model:** 370Z, Frontier, Pathfinder, Titan, Xterra **Engine:** 2.5L, 3.7L, 4.0L, 5.6L	**A/T 3rd Gear Function:** * This malfunction is detected when the A/T does not shift into 3GR position as instructed by TCM. This is not only caused by electrical malfunction (circuits open or shorted) but mechanical malfunction such as control valve sticking, improper solenoid valve operation. * This is an OBD-II self-diagnostic item. * Diagnostic trouble code P0733 is detected when TCM detects any inconsistency in the actual gear ratio.
DTC: P0734 **T TCM, TCIL: Yes** **Year:** 2011, 2012 **Model:** 370Z, Frontier, Pathfinder, Titan, Xterra **Engine:** 2.5L, 3.7L, 4.0L, 5.6L	**A/T 4th Gear Function:** * This malfunction is detected when the A/T does not shift into 4GR position as instructed by TCM. This is not only caused by electrical malfunction (circuits open or shorted) but mechanical malfunction such as control valve sticking, improper solenoid valve operation. * This is an OBD-II self-diagnostic item. * P0734 is detected when TCM detects any inconsistency in the actual gear ratio.
DTC: P0735 **T TCM, TCIL: Yes** **Year:** 2011, 2012 **Model:** 370Z, Frontier, Pathfinder, Titan, Xterra **Engine:** 2.5L, 3.7L, 4.0L, 5.6L	**A/T 5th Gear Function:** * This malfunction is detected when the A/T does not shift into 5GR position as instructed by TCM. This is not only caused by electrical malfunction (circuits open or shorted) but mechanical malfunction such as control valve sticking, improper solenoid valve operation. * This is an OBD-II self-diagnostic item. * Diagnostic trouble code P0735 is detected when TCM detects any inconsistency in the actual gear ratio.
DTC: P0740 **1T TCM, TCIL: Yes** **Year:** 2011, 2012 **Model:** Altima, Frontier, Maxima, Murano, Pathfinder, Quest, Rogue, Sentra, Titan, Versa, Xterra **Engine:** 1.6L, 1.8L, 2.0L, 2.5L, 3.5L, 3.5L, 4.0L, 5.6L	**Torque Converter Clutch Solenoid Valve:** * Diagnostic trouble code P0740 is detected under the following conditions: - TCM detects an improper voltage drop when it tries to operate the solenoid valve. - When TCM detects as irregular by comparing target value with monitor value.
DTC: P0740 **T TCM, TCIL: Yes** **Year:** 2011, 2012 **Model:** 370Z **Engine:** 3.7L	**Torque Converter Clutch Circuit - Open:** The torque converter clutch solenoid valve monitor value is 0.4 A or less when the torque converter clutch solenoid valve command value is more than 0.75 A.
DTC: P0740 **T TCM, TCIL: Yes** **Year:** 2011 **Model:** Cube **Engine:** 1.8L	**Torque Converter Clutch Circuit - Open:** * Torque converter clutch solenoid valve monitor voltage value of TCM is less than 70% of torque converter clutch solenoid valve target voltage value. * Torque converter clutch solenoid valve current command value of TCM and torque converter clutch solenoid valve current monitor value is deviated.

DTC	Trouble Code Title and Conditions
DTC: P0744 **T TCM, TCIL: Yes** **Year:** 2011, 2012 **Model:** 370Z, Altima, Frontier, Maxima, Murano, Pathfinder, Quest, Rogue, Sentra, Titan, Versa, Xterra **Engine:** 1.6L, 1.8L, 2.0L, 2.5L, 3.5L, 3.5L, 3.7L, 4.0L, 5.6L	**A/T TCC S/V Function (Lock-Up):** * This malfunction is detected when the A/T does not lock-up or does not shift to 5th gear. This is not only caused by electrical malfunction (circuits open or shorted) but also by mechanical malfunction such as control valve sticking, improper solenoid valve operation, etc. * This is an OBD-II self-diagnostic item. * Diagnostic trouble code P0744 is detected under the following conditions: - When A/T cannot perform lock-up even if electrical circuit is good. - When TCM detects as irregular by comparing difference value with slip rotation.
DTC: P0744 **T TCM, TCIL: Yes** **Year:** 2011 **Model:** Cube **Engine:** 1.8L	**Torque Converter Clutch Circuit Intermittent:** Torque converter slip speed is more than a certain value (40 rpm + vehicle speed/2) while TCM is in lock-up command state.
DTC: P0745 **T TCM, TCIL: Yes** **Year:** 2011 **Model:** Cube **Engine:** 1.8L	**Pressure Control Solenoid A:** * Monitor voltage value of TCM line pressure solenoid valve is less than 70% of the target voltage value of line pressure solenoid valve. * Current monitor value of the Line pressure solenoid valve differs from the TCM current command value of line pressure solenoid valve.
DTC: P0745 **T TCM, TCIL: Yes** **Year:** 2011, 2012 **Model:** Frontier, Maxima, Murano, Pathfinder, Rogue, Sentra, Titan, Versa, Xterra **Engine:** 1.6L, 1.8L, 2.0L, 2.5L, 3.5L, 3.5L, 4.0L, 5.6L	**Pressure Control Solenoid A:** * The line pressure solenoid valve regulates the oil pump discharge pressure to suit the driving condition in response to a signal sent from the TCM. * This is an OBD-II self-diagnostic item. * Diagnostic trouble code P0745 is detected under the following conditions: - When TCM detects an improper voltage drop when it tries to operate the solenoid valve. - When TCM detects as irregular by comparing target value with monitor value.
DTC: P0745 **T TCM, TCIL: Yes** **Year:** 2011, 2012 **Model:** 370Z **Engine:** 3.7L	**Pressure Control Solenoid A:** The line pressure solenoid valve monitor value is 0.4 A or less when the line pressure solenoid valve command value is more than 0.75 A.
DTC: P0746 **T TCM, TCIL: Yes** **Year:** 2011 **Model:** Cube **Engine:** 1.8L	**Pressure Control Solenoid A Performance - Stuck Off:** TCM detects a state that gear ratio is more than 2.9
DTC: P0746 **1T TCM, TCIL: Yes** **Year:** 2011, 2012 **Model:** Altima, Maxima, Murano, Quest, Rogue, Sentra, Versa **Engine:** 1.6L, 1.8L, 2.0L, 2.5L, 3.5L, 3.5L	**Pressure Control Solenoid A Performance (Stuck Off):** The pressure control solenoid valve A (line pressure solenoid valve) regulates the oil pump discharge pressure to suit the driving condition in response to a signal sent from the TCM. This is an OBD-II self-diagnostic item. Diagnostic trouble code P0746 is detected under the following conditions: - Unexpected gear ratio was detected in the LOW side due to excessively low line pressure.
DTC: P0750 **T TCM, TCIL: Yes** **Year:** 2011, 2012 **Model:** 370Z **Engine:** 3.7L	**Shift Solenoid A:** * The anti-interlock solenoid valve monitor value is ON when the anti-interlock solenoid valve command value is OFF. * The anti-interlock solenoid valve monitor value is OFF when the anti-interlock solenoid valve command value is ON.
DTC: P0775 **T TCM, TCIL: Yes** **Year:** 2011, 2012 **Model:** 370Z **Engine:** 3.7L	**Pressure Control Solenoid B:** The input clutch solenoid valve monitor value is 0.4 A or less when the input clutch solenoid valve command value is more than 0.75 A.

DTC	Trouble Code Title and Conditions
DTC: P0776 **1T TCM, TCIL: Yes** **Year:** 2011, 2012 **Model:** Altima, Juke, Maxima, Murano, Quest, Rogue, Sentra, Versa **Engine:** 1.6L, 1.8L, 2.0L, 2.5L, 3.5L, 3.5L	**Pressure Control Solenoid B Performance (Stuck Off):** The pressure control solenoid valve B (secondary pressure solenoid valve) regulates the secondary pressure to suit the driving condition in response to a signal sent from the TCM. This is an OBD-II self-diagnostic item. Diagnostic trouble code P0776 is detected when secondary pressure is too high or too low compared with the commanded value while driving.
DTC: P0776 **T TCM, TCIL: Yes** **Year:** 2011 **Model:** Cube **Engine:** 1.8L	**Pressure Control Solenoid B Performance - Stuck Off:** Difference of secondary pressure target value of TCM and secondary pressure actual value is more than 1.2 MPa
DTC: P0778 **T TCM, TCIL: Yes** **Year:** 2011 **Model:** Cube **Engine:** 1.8L	**Pressure Control Solenoid B Electrical:** Current monitor value of the secondary pressure solenoid valve differs from the TCM current command value of secondary pressure solenoid valve. Secondary pressure solenoid valve current command value of TCM and secondary pressure solenoid valve current monitor value is deviated.
DTC: P0778 **1T TCM, TCIL: Yes** **Year:** 2011, 2012 **Model:** Altima, Juke, Maxima, Murano, Quest, Rogue, Sentra, Versa **Engine:** 1.6L, 1.8L, 2.0L, 2.5L, 3.5L, 3.5L	**Pressure Control Solenoid B Electrical:** The pressure control solenoid valve B (secondary pressure solenoid valve) regulates the oil pump discharge pressure to suit the driving condition in response to a signal sent from the TCM. This is an OBD-II self-diagnostic item. Diagnostic trouble code P0778 is detected under the following conditions: - TCM detects an improper voltage drop when it tries to operate the solenoid valve. - When TCM compares target value with monitor value and detects an irregularity. Normal voltage not applied to solenoid due to cut line, short, or the like. TCM detects as irregular by comparing target value with monitor value.
DTC: P0780 **T TCM, TCIL: Yes** **Year:** 2011, 2012 **Model:** 370Z **Engine:** 3.7L	**Shift Error:** * When shifting from 3GR to 4GR with the selector lever in "D" position, the gear ratio does not shift to 1.412 (gear ratio of 4GR). * When shifting from 5GR to 6GR or 6GR to 7GR, the engine speed exceeds the prescribed speed. * The shift change time from 4GR to 3GR is 0.2 second or less.
DTC: P0795 **T TCM, TCIL: Yes** **Year:** 2011, 2012 **Model:** 370Z **Engine:** 3.7L	**Pressure Control Solenoid C:** The front brake solenoid valve monitor value is 0.4 A or less when the front brake solenoid valve command value is more than 0.75 A.
DTC: P0820 **T ECM, MIL: Yes** **Year:** 2011, 2012 **Model:** 370Z **Engine:** 3.7L	**Gear Lever Position Sensor Circuit:** Condition A: An excessively low voltage from the sensor is sent to ECM. An excessively high voltage from the sensor is sent to ECM. Condition B: There is a difference between target engine speed calculated by ECM and actual engine speed.
DTC: P0826 **T TCM, TCIL: Yes** **Year:** 2011, 2012 **Model:** Maxima, Murano, Rogue **Engine:** 2.5L, 3.5L, 3.5L	**Up & Down Shift Switch Circuit:** When an impossible pattern of switch signals is detected, a malfunction is detected. When shift up/down signal of paddle shifter continuously remains ON for 60 seconds.
DTC: P0826 **1T TCM, TCIL: Yes** **Year:** 2011 **Model:** Altima, Quest, Sentra **Engine:** 2.0L, 2.5L, 3.5L	**Manual Mode Switch (Up/Down Shift Switch) Circuit:** TCM sends the switch signals to combination meter via CAN communication line. Then manual mode switch position is indicated on the CVT position indicator. This is not an OBD-II self-diagnostic item. Diagnostic trouble code P0826 is detected when TCM monitors Manual mode, Non manual mode, Up or Down switch signal, and then detects irregular with impossible input pattern for 1 second or more. When an impossible pattern of switch signals is detected, a malfunction is detected.

DTC	Trouble Code Title and Conditions
DTC: P082B-575 **T ECM, MIL: Yes** **Year:** 2011 **Model:** Altima, Quest **Engine:** 2.5L, 3.5L	**Gear Lever X Position Circuit Low:** Open or GND short in select main sensor circuit.
DTC: P082C-576 **T ECM, MIL: Yes** **Year:** 2011 **Model:** Altima, Quest **Engine:** 2.5L, 3.5L	**Gear Lever X Position Circuit High:** +B short in select main sensor circuit
DTC: P082E-571 **T ECM, MIL: Yes** **Year:** 2011 **Model:** Altima, Quest **Engine:** 2.5L, 3.5L	**Gear Lever Y Position Circuit Low:** Open or GND short in shift main sensor circuit
DTC: P082F-572 **T ECM, MIL: Yes** **Year:** 2011 **Model:** Altima, Quest **Engine:** 2.5L, 3.5L	**Gear Lever Y Position Circuit High:** +B short in shift main sensor circuit
DTC: P0830 **T ECM, MIL: Yes** **Year:** 2011, 2012 **Model:** 370Z **Engine:** 3.7L	**Clutch Interlock Switch Circuit:** Condition A: ON signals from the clutch interlock switch and the clutch pedal position switch are sent to the ECM at the same time. Condition B: Clutch interlock switch ON signal is not sent to ECM for extremely long time.
DTC: P0833 **T ECM, MIL: Yes** **Year:** 2011, 2012 **Model:** 370Z **Engine:** 3.7L	**Clutch Pedal Position Switch Circuit:** Condition A: ON signals from the clutch pedal position switch and the clutch interlock switch are sent to the ECM at the same time. Condition B: Clutch pedal position switch ON signal is not sent to ECM for extremely long time.
DTC: P0840 **1T TCM, TCIL: Yes** **Year:** 2011, 2012 **Model:** Altima, Juke, Maxima, Murano, Quest, Rogue, Sentra, Versa **Engine:** 1.6L, 1.8L, 2.0L, 2.5L, 3.5L, 3.5L	**Transmission Fluid Pressure Sensor/Switch A Circuit:** The transmission fluid pressure sensor A (secondary pressure sensor) detects secondary pressure of CVT and sends TCM the signal. This is an OBD-II self-diagnostic item. Diagnostic trouble code P0840 is detected when TCM detects an improper voltage drop when it receives the sensor signal. Signal voltage of the secondary pressure sensor is too high or too low while driving.
DTC: P0840 **T TCM, TCIL: Yes** **Year:** 2011 **Model:** Cube **Engine:** 1.8L	**Transmission Fluid Pressure Sensor/Switch A Circuit:** Secondary pressure sensor voltage that TCM receives is more than 4.7 V Secondary pressure sensor voltage that TCM receives is less than 0.9 V
DTC: P0841 **1T TCM, TCIL: Yes** **Year:** 2011, 2012 **Model:** Altima, Juke, Maxima, Murano, Quest, Sentra, Versa **Engine:** 1.6L, 1.8L, 2.0L, 2.5L, 3.5L, 3.5L	**Transmission Fluid Pressure Sensor/Switch A Circuit Range/Performance:** Using the engine load (throttle position), the primary pulley revolution speed, and the secondary pulley revolution speed as input signal, TCM changes the operating pressure of the primary pulley and the secondary pulley and changes the groove width of the pulley to control the gear ratio. This is not an OBD-II self-diagnostic item. Diagnostic trouble code P0841 is detected when correlation between the values of the secondary pressure sensor and the primary pressure sensor is out of specification. Correlation between the values of the secondary pressure sensor and the primary pressure sensor is out of specification.
DTC: P0841 **T TCM, TCIL: Yes** **Year:** 2011 **Model:** Cube, Rogue **Engine:** 1.8L, 2.5L	**Transmission Fluid Pressure Sensor/Switch A Circuit Range/Performance:** Secondary pressure sensor value exceeds line pressure value.

DTC	Trouble Code Title and Conditions
DTC: P0845 **1T TCM, TCIL: Yes** **Year:** 2011, 2012 **Model:** Altima, Maxima, Murano, Quest **Engine:** 2.5L, 3.5L, 3.5L	**Transmission Fluid Pressure Sensor/Switch B Circuit:** Signal voltage of the primary pressure sensor is too high or too low while driving.
DTC: P0850 **1T ECM, MIL: Yes** **Year:** 2011, 2012 **Model:** All **Engine:** 1.6L, 1.8L, 2.0L, 2.5L, 3.5L, 3.5L, 3.7L, 4.0L, 5.6L	**Park/Neutral Position Switch:** * CVT & M/T: When the shift lever position is P or N (CVT), Neutral (M/T), park/neutral position (PNP) switch is ON. ECM detects the position because the continuity of the line (the ON signal) exists. * A/T: The signal of the park/neutral position (PNP) switch does not change in the process of engine starting and driving.
DTC: P0851-579 **T ECM, MIL: Yes** **Year:** 2011 **Model:** Altima, Quest **Engine:** 2.5L, 3.5L	**Park / Neutral Switch Input Circuit Low:** GND short in P position switch circuit
DTC: P0851-775 **T ECM, MIL: Yes** **Year:** 2011 **Model:** Altima, Quest **Engine:** 2.5L, 3.5L	**Neutral Signal Input Circuit Low:** N signal line malfunction
DTC: P0852-580 **T ECM** **Year:** 2011 **Model:** Altima, Quest **Engine:** 2.5L, 3.5L	**Park / Neutral Switch Input Circuit High:** Open or +B short in P position switch circuit
DTC: P085D-582 **T ECM, MIL: Yes** **Year:** 2011 **Model:** Altima, Quest **Engine:** 2.5L, 3.5L	**Gear Shift Control Module "A" Performance:** P position (PPOS) signal is logically inconsistent
DTC: P085D-599 **T ECM, MIL: Yes** **Year:** 2011 **Model:** Altima, Quest **Engine:** 2.5L, 3.5L	**Gear Shift Control Module "A" Performance:** P position (PPOS) signal malfunction (output pulse is abnormal)
DTC: P0861-597 **T ECM, MIL: Yes** **Year:** 2011 **Model:** Altima, Quest **Engine:** 2.5L, 3.5L	**Gear Shift Control Module "A" Communication Circuit Low:** GND short in P position (PPOS) signal circuit
DTC: P0862-598 **T ECM, MIL: Yes** **Year:** 2011 **Model:** Altima, Quest **Engine:** 2.5L, 3.5L	**Gear Shift Control Module "A" Communication Circuit High:** +B short in P position (PPOS) signal circuit
DTC: P0868 **T TCM, TCIL: Yes** **Year:** 2011, 2012 **Model:** Cube, Maxima, Murano, Rogue **Engine:** 1.8L, 2.5L, 3.5L, 3.5L	**Transmission Fluid Pressure Low:** Secondary fluid pressure is too low compared with the commanded value while driving.
DTC: P0868 **T TCM, TCIL: Yes** **Year:** 2011 **Model:** Juke, Versa **Engine:** 1.6L, 1.8L	**Transmission Fluid Secondary Pressure Down:** This is not an OBD-II self-diagnostic item. Diagnostic trouble code P0868 is detected when secondary fluid pressure is too low compared with the commanded value while driving.

DTC	Trouble Code Title and Conditions
DTC: P0868 1T TCM, TCIL: Yes Year: 2011 Model: Altima, Quest, Sentra Engine: 2.0L, 2.5L, 3.5L	**Secondary Pressure Down (Fluid Pressure Low):** The pressure control solenoid valve B (secondary pressure solenoid valve) regulates the secondary pressure to suit the driving condition in response to a signal sent from the TCM. This is not an OBD-II self-diagnostic item. Diagnostic trouble code P0868 is detected when secondary fluid pressure is too low compared with the commanded value while driving.
DTC: P0A01-725 T ECM, MIL: Yes Year: 2011 Model: Altima, Quest Engine: 2.5L, 3.5L	**Motor Electronics Coolant Temperature Sensor Circuit Range/Performance:** Sudden change in inverter coolant temperature sensor output
DTC: P0A01-725 T ECM, MIL: Yes Year: 2011 Model: Altima, Quest Engine: 2.5L, 3.5L	**Motor Electronics Coolant Temperature Sensor Circuit Range / Performance:** Unusual sudden change in HV coolant temperature sensor output occurs and the offset continues, or unusual sudden change in HV coolant temperature sensor output occurs repeatedly.
DTC: P0A01-726 T ECM, MIL: Yes Year: 2011 Model: Altima, Quest Engine: 2.5L, 3.5L	**Motor Electronics Coolant Temperature Sensor Circuit Range/Performance:** Inverter coolant temperature sensor output deviation
DTC: P0A01-726 T ECM, MIL: Yes Year: 2011 Model: Altima, Quest Engine: 2.5L, 3.5L	**Motor Electronics Coolant Temperature Sensor Circuit Range / Performance:** Temperature calculated by hybrid vehicle control ECU and actual temperature are different for 10 seconds or more.
DTC: P0A02-719 T ECM, MIL: Yes Year: 2011 Model: Altima, Quest Engine: 2.5L, 3.5L	**Motor Electronics Coolant Temperature Sensor Circuit Low:** Short to GND in the HV coolant temperature sensor circuit
DTC: P0A02-719 T ECM, MIL: Yes Year: 2011 Model: Altima, Quest Engine: 2.5L, 3.5L	**Motor Electronics Coolant Temperature Sensor Circuit Low:** Short to GND in the inverter coolant temperature sensor circuit
DTC: P0A02-720 T ECM, MIL: Yes Year: 2011 Model: Altima, Quest Engine: 2.5L, 3.5L	**Motor Electronics Coolant Temperature Sensor Circuit High:** Open or short to +B in the HV coolant temperature sensor circuit.
DTC: P0A03-720 T ECM, MIL: Yes Year: 2011 Model: Altima, Quest Engine: 2.5L, 3.5L	**Motor Electronics Coolant Temperature Sensor Circuit High:** Open or short to +B in the inverter coolant temperature sensor circuit.
DTC: P0A08/101 T Year: 2011 Model: Altima, Quest Engine: 2.5L, 3.5L	**DC / DC Converter Status Circuit:** Overheating of the hybrid vehicle converter (DC/DC converter)
DTC: P0A08/264 T Year: 2011 Model: Altima, Quest Engine: 2.5L, 3.5L	**DC / DC Converter Status Circuit:** Malfunction in the hybrid vehicle converter (DC/DC converter)

DTC	Trouble Code Title and Conditions
DTC: P0A08-101 **T** ECM, **MIL:** Yes **Year:** 2011 **Model:** Altima, Quest **Engine:** 2.5L, 3.5L	**DC / DC Converter Status Circuit:** Overheating of the hybrid vehicle converter (DC/DC converter).
DTC: P0A08-264 **T** ECM, **MIL:** Yes **Year:** 2011 **Model:** Altima, Quest **Engine:** 2.5L, 3.5L	**DC / DC Converter Status Circuit:** Malfunction in the hybrid vehicle converter (DC/DC converter)
DTC: P0A09-265 **T** ECM, **MIL:** Yes **Year:** 2011 **Model:** Altima, Quest **Engine:** 2.5L, 3.5L	**DC / DC Converter Status Circuit Low Input:** Open or short to GND in the hybrid vehicle converter (DC/DC converter) (NODD) signal line
DTC: P0A09-591 **T** ECM, **MIL:** Yes **Year:** 2011 **Model:** Altima, Quest **Engine:** 2.5L, 3.5L	**DC / DC Converter Status Circuit Low Input:** Hybrid vehicle converter (DC/DC converter) voltage switching (VLO) signal circuit malfunction (Open or short to GND)
DTC: P0A0D/350 **T** **Year:** 2011 **Model:** Altima, Quest **Engine:** 2.5L, 3.5L	**High Voltage System Inter-Lock Circuit High:** Operating any of the safety devices with the vehicle stopped (ILK signal is ON)
DTC: P0A0D-350 **T** ECM **Year:** 2011 **Model:** Altima, Quest **Engine:** 2.5L, 3.5L	**High Voltage System Inter-Lock Circuit High:** Operating any of the safety devices with the vehicle stopped (ILK signal is ON) and the power switch on (IG)
DTC: P0A0D-351 **T** ECM, **MIL:** Yes **Year:** 2011 **Model:** Altima, Quest **Engine:** 2.5L, 3.5L	**High Voltage System Inter-Lock Circuit High:** Interlock signal line opens (ILK signal is ON) while the vehicle is being driven
DTC: P0A0F/238 **T** **Year:** 2011 **Model:** Altima, Quest **Engine:** 2.5L, 3.5L	**Engine Failed to Start:** Engine does not start even though cranking it (transaxle input malfunction [engine system])
DTC: P0A0F-204 **T** ECM **Year:** 2011 **Model:** Altima, Quest **Engine:** 2.5L, 3.5L	**Engine Failed to Start:** Signal indicating abnormality input from the ECM portion of the hybrid vehicle control ECU (abnormal engine output)
DTC: P0A0F-205 **T** ECM **Year:** 2011 **Model:** Altima, Quest **Engine:** 2.5L, 3.5L	**Engine Failed to Start:** Signal indicating abnormality input from the ECM portion of the hybrid vehicle control ECU (engine is unable to start)
DTC: P0A0F-206 **T** ECM, **MIL:** Yes **Year:** 2011 **Model:** Altima, Quest **Engine:** 2.5L, 3.5L	**Engine Failed to Start:** Signal indicating abnormality input from the ECM portion of the hybrid vehicle control ECU (engine component malfunction)

DTC	Trouble Code Title and Conditions
DTC: P0A0F-238 **T ECM, MIL: Yes** **Year:** 2011 **Model:** Altima, Quest **Engine:** 2.5L, 3.5L	**Engine Failed to Start:** Engine does not start even though it is being cranked (transaxle input malfunction [engine system])
DTC: P0A0F-524 **T ECM, MIL: Yes** **Year:** 2011 **Model:** Altima, Quest **Engine:** 2.5L, 3.5L	**Engine Failed to Start:** Signal indicating abnormality input from the ECM portion of the hybrid vehicle control ECU (NE signal error)
DTC: P0A0F-525 **T ECM, MIL: Yes** **Year:** 2011 **Model:** Altima, Quest **Engine:** 2.5L, 3.5L	**Engine Failed to Start:** Signal indicating abnormality input from the ECM portion of the hybrid vehicle control ECU (GI signal error)
DTC: P0A10-263 **T ECM, MIL: Yes** **Year:** 2011 **Model:** Altima, Quest **Engine:** 2.5L, 3.5L	**DC / DC Converter Status Circuit High Input:** +B short in hybrid vehicle converter (DC/DC converter) NODD signal line
DTC: P0A10-592 **T ECM, MIL: Yes** **Year:** 2011 **Model:** Altima, Quest **Engine:** 2.5L, 3.5L	**DC / DC Converter Status Circuit High Input:** Hybrid vehicle converter (DC/DC converter) voltage switching (VLO) signal circuit malfunction (+B short)
DTC: P0A1A-151 **T ECM, MIL: Yes** **Year:** 2011 **Model:** Altima, Quest **Engine:** 2.5L, 3.5L	**Generator Control Module:** Run pulse signal cycle deviation or stop
DTC: P0A1A-155 **T ECM, MIL: Yes** **Year:** 2011 **Model:** Altima, Quest **Engine:** 2.5L, 3.5L	**Generator Control Module:** A/D converter error
DTC: P0A1A-156 **T ECM, MIL: Yes** **Year:** 2011 **Model:** Altima, Quest **Engine:** 2.5L, 3.5L	**Generator Control Module:** CPU ROM-RAM error
DTC: P0A1A-158 **T ECM, MIL: Yes** **Year:** 2011 **Model:** Altima, Quest **Engine:** 2.5L, 3.5L	**Generator Control Module:** CPU recognition error
DTC: P0A1A-166 **T ECM, MIL: Yes** **Year:** 2011 **Model:** Altima, Quest **Engine:** 2.5L, 3.5L	**Generator Control Module:** RD converter NM stop error
DTC: P0A1A-200 **T ECM, MIL: Yes** **Year:** 2011 **Model:** Altima, Quest **Engine:** 2.5L, 3.5L	**Generator Control Module:** The monitor will run whenever the following DTC is not present: TMC's intellectual property Other conditions belong to TMC's intellectual property: -

DTC	Trouble Code Title and Conditions
DTC: P0A1A-658 **T ECM, MIL: Yes** **Year:** 2011 **Model:** Altima, Quest **Engine:** 2.5L, 3.5L	**Generator Control Module:** ALU error
DTC: P0A1A-659 **T ECM, MIL: Yes** **Year:** 2011 **Model:** Altima, Quest **Engine:** 2.5L, 3.5L	**Generator Control Module:** TMC's intellectual property: -
DTC: P0A1A-791 **T ECM, MIL: Yes** **Year:** 2011 **Model:** Altima, Quest **Engine:** 2.5L, 3.5L	**Generator Control Module:** R/D converter communication error
DTC: P0A1A-792 **T ECM, MIL: Yes** **Year:** 2011 **Model:** Altima, Quest **Engine:** 2.5L, 3.5L	**Generator Control Module:** The monitor will run whenever the following DTC is not present: TMC's intellectual property Other conditions belong to TMC's intellectual property: -
DTC: P0A1A-793 **T ECM, MIL: Yes** **Year:** 2011 **Model:** Altima, Quest **Engine:** 2.5L, 3.5L	**Generator Control Module:** The monitor will run whenever the following DTC is not present: TMC's intellectual property Other conditions belong to TMC's intellectual property: -
DTC: P0A1A-796 **T ECM, MIL: Yes** **Year:** 2011 **Model:** Altima, Quest **Engine:** 2.5L, 3.5L	**Drive Motor "A" Control Module:** The monitor will run whenever the following DTC is not present: TMC's intellectual property Other conditions belong to TMC's intellectual property: -
DTC: P0A1B-163 **T ECM, MIL: Yes** **Year:** 2011 **Model:** Altima, Quest **Engine:** 2.5L, 3.5L	**Drive Motor "A" Control Module:** The monitor will run whenever the following DTC is not present: TMC's intellectual property Other conditions belong to TMC's intellectual property: -
DTC: P0A1B-164 **T ECM, MIL: Yes** **Year:** 2011 **Model:** Altima, Quest **Engine:** 2.5L, 3.5L	**Drive Motor "A" Control Module:** The monitor will run whenever the following DTC is not present: TMC's intellectual property Other conditions belong to TMC's intellectual property: -
DTC: P0A1B-168 **T ECM, MIL: Yes** **Year:** 2011 **Model:** Altima, Quest **Engine:** 2.5L, 3.5L	**Drive Motor "A" Control Module:** The monitor will run whenever the following DTC is not present: TMC's intellectual property Other conditions belong to TMC's intellectual property: -
DTC: P0A1B-192 **T ECM, MIL: Yes** **Year:** 2011 **Model:** Altima, Quest **Engine:** 2.5L, 3.5L	**Drive Motor "A" Control Module:** The monitor will run whenever the following DTC is not present: TMC's intellectual property Other conditions belong to TMC's intellectual property: -
DTC: P0A1B-193 **T ECM, MIL: Yes** **Year:** 2011 **Model:** Altima, Quest **Engine:** 2.5L, 3.5L	**Drive Motor "A" Control Module:** The monitor will run whenever the following DTC is not present: TMC's intellectual property Other conditions belong to TMC's intellectual property: -

DTC	Trouble Code Title and Conditions
DTC: P0A1B-195 **T ECM, MIL: Yes** **Year:** 2011 **Model:** Altima, Quest **Engine:** 2.5L, 3.5L	**Drive Motor "A" Control Module:** The monitor will run whenever the following DTC is not present: TMC's intellectual property Other conditions belong to TMC's intellectual property: -
DTC: P0A1B-198 **T ECM, MIL: Yes** **Year:** 2011 **Model:** Altima, Quest **Engine:** 2.5L, 3.5L	**Drive Motor "A" Control Module:** The monitor will run whenever the following DTC is not present: TMC's intellectual property Other conditions belong to TMC's intellectual property: -
DTC: P0A1B-511 **T ECM, MIL: Yes** **Year:** 2011 **Model:** Altima, Quest **Engine:** 2.5L, 3.5L	**Drive Motor "A" Control Module:** The monitor will run whenever the following DTC is not present: TMC's intellectual property Other conditions belong to TMC's intellectual property: -
DTC: P0A1B-512 **T ECM, MIL: Yes** **Year:** 2011 **Model:** Altima, Quest **Engine:** 2.5L, 3.5L	**Drive Motor "A" Control Module:** The monitor will run whenever the following DTC is not present: TMC's intellectual property Other conditions belong to TMC's intellectual property: -
DTC: P0A1B-661 **T ECM, MIL: Yes** **Year:** 2011 **Model:** Altima, Quest **Engine:** 2.5L, 3.5L	**Drive Motor "A" Control Module:** The monitor will run whenever the following DTC is not present: TMC's intellectual property Other conditions belong to TMC's intellectual property: -
DTC: P0A1B-786 **T ECM, MIL: Yes** **Year:** 2011 **Model:** Altima, Quest **Engine:** 2.5L, 3.5L	**Drive Motor "A" Control Module:** The monitor will run whenever the following DTC is not present: TMC's intellectual property Other conditions belong to TMC's intellectual property: -
DTC: P0A1B-788 **T ECM, MIL: Yes** **Year:** 2011 **Model:** Altima, Quest **Engine:** 2.5L, 3.5L	**Drive Motor "A" Control Module:** The monitor will run whenever the following DTC is not present: TMC's intellectual property Other conditions belong to TMC's intellectual property: -
DTC: P0A1B-791 **T ECM, MIL: Yes** **Year:** 2011 **Model:** Altima, Quest **Engine:** 2.5L, 3.5L	**Drive Motor "A" Control Module:** The monitor will run whenever the following DTC is not present: TMC's intellectual property Other conditions belong to TMC's intellectual property: -
DTC: P0A1B-795 **T ECM, MIL: Yes** **Year:** 2011 **Model:** Altima, Quest **Engine:** 2.5L, 3.5L	**Drive Motor "A" Control Module:** The monitor will run whenever the following DTC is not present: TMC's intellectual property Other conditions belong to TMC's intellectual property: -
DTC: P0A1D-103 **T ECM, MIL: Yes** **Year:** 2011 **Model:** Altima, Quest **Engine:** 2.5L, 3.5L	**Hybrid Powertrain Control Module:** The monitor will run whenever the following DTC is not present: TMC's intellectual property Other conditions belong to TMC's intellectual property: -
DTC: P0A1D-134 **T ECM, MIL: Yes** **Year:** 2011 **Model:** Altima, Quest **Engine:** 2.5L, 3.5L	**Hybrid Powertrain Control Module:** The monitor will run whenever the following DTC is not present: TMC's intellectual property Other conditions belong to TMC's intellectual property: -

DTC	Trouble Code Title and Conditions
DTC: P0A1D-135 **T ECM, MIL: Yes** **Year:** 2011 **Model:** Altima, Quest **Engine:** 2.5L, 3.5L	**Hybrid Powertrain Control Module:** The monitor will run whenever the following DTC is not present: TMC's intellectual property Other conditions belong to TMC's intellectual property: -
DTC: P0A1D-140 **T ECM, MIL: Yes** **Year:** 2011 **Model:** Altima, Quest **Engine:** 2.5L, 3.5L	**Hybrid Powertrain Control Module:** The monitor will run whenever the following DTC is not present: TMC's intellectual property Other conditions belong to TMC's intellectual property: -
DTC: P0A1D-141 **T ECM, MIL: Yes** **Year:** 2011 **Model:** Altima, Quest **Engine:** 2.5L, 3.5L	**Hybrid Powertrain Control Module:** The monitor will run whenever the following DTC is not present: TMC's intellectual property Other conditions belong to TMC's intellectual property: -
DTC: P0A1D-144 **T ECM, MIL: Yes** **Year:** 2011 **Model:** Altima, Quest **Engine:** 2.5L, 3.5L	**Hybrid Powertrain Control Module:** The monitor will run whenever the following DTC is not present: TMC's intellectual property Other conditions belong to TMC's intellectual property: -
DTC: P0A1D-145 **T ECM, MIL: Yes** **Year:** 2011 **Model:** Altima, Quest **Engine:** 2.5L, 3.5L	**Hybrid Powertrain Control Module:** The monitor will run whenever the following DTC is not present: TMC's intellectual property Other conditions belong to TMC's intellectual property: -
DTC: P0A1D-148 **T ECM, MIL: Yes** **Year:** 2011 **Model:** Altima, Quest **Engine:** 2.5L, 3.5L	**Hybrid Powertrain Control Module:** The monitor will run whenever the following DTC is not present: TMC's intellectual property Other conditions belong to TMC's intellectual property: -
DTC: P0A1D-162 **T ECM, MIL: Yes** **Year:** 2011 **Model:** Altima, Quest **Engine:** 2.5L, 3.5L	**Hybrid Powertrain Control Module:** The monitor will run whenever the following DTC is not present: TMC's intellectual property Other conditions belong to TMC's intellectual property: -
DTC: P0A1D-187 **T ECM, MIL: Yes** **Year:** 2011 **Model:** Altima, Quest **Engine:** 2.5L, 3.5L	**Hybrid Powertrain Control Module:** The monitor will run whenever the following DTC is not present: TMC's intellectual property Other conditions belong to TMC's intellectual property: -
DTC: P0A1D-393 **T ECM, MIL: Yes** **Year:** 2011 **Model:** Altima, Quest **Engine:** 2.5L, 3.5L	**Hybrid Powertrain Control Module:** The monitor will run whenever the following DTC is not present: TMC's intellectual property Other conditions belong to TMC's intellectual property: -
DTC: P0A1D-570 **T ECM, MIL: Yes** **Year:** 2011 **Model:** Altima, Quest **Engine:** 2.5L, 3.5L	**Hybrid Powertrain Control Module:** The monitor will run whenever the following DTC is not present: TMC's intellectual property Other conditions belong to TMC's intellectual property: -
DTC: P0A1D-721 **T ECM, MIL: Yes** **Year:** 2011 **Model:** Altima, Quest **Engine:** 2.5L, 3.5L	**Hybrid Powertrain Control Module:** The monitor will run whenever the following DTC is not present: TMC's intellectual property Other conditions belong to TMC's intellectual property: -

DTC	Trouble Code Title and Conditions
DTC: P0A1D-722 **T ECM, MIL: Yes** **Year:** 2011 **Model:** Altima, Quest **Engine:** 2.5L, 3.5L	**Hybrid Powertrain Control Module:** The monitor will run whenever the following DTC is not present: TMC's intellectual property Other conditions belong to TMC's intellectual property: -
DTC: P0A1D-723 **T ECM, MIL: Yes** **Year:** 2011 **Model:** Altima, Quest **Engine:** 2.5L, 3.5L	**Hybrid Powertrain Control Module:** The monitor will run whenever the following DTC is not present: TMC's intellectual property Other conditions belong to TMC's intellectual property: -
DTC: P0A1D-765 **T ECM, MIL: Yes** **Year:** 2011 **Model:** Altima, Quest **Engine:** 2.5L, 3.5L	**Hybrid Powertrain Control Module:** The monitor will run whenever the following DTC is not present: TMC's intellectual property Other conditions belong to TMC's intellectual property: -
DTC: P0A1D-787 **T ECM, MIL: Yes** **Year:** 2011 **Model:** Altima, Quest **Engine:** 2.5L, 3.5L	**Hybrid Powertrain Control Module:** The monitor will run whenever the following DTC is not present: TMC's intellectual property Other conditions belong to TMC's intellectual property: -
DTC: P0A1D-821 **T ECM, MIL: Yes** **Year:** 2011 **Model:** Altima, Quest **Engine:** 2.5L, 3.5L	**Hybrid Powertrain Control Module:** The monitor will run whenever the following DTC is not present: TMC's intellectual property Other conditions belong to TMC's intellectual property: -
DTC: P0A1D-822 **T ECM, MIL: Yes** **Year:** 2011 **Model:** Altima, Quest **Engine:** 2.5L, 3.5L	**Hybrid Powertrain Control Module:** The monitor will run whenever the following DTC is not present: TMC's intellectual property Other conditions belong to TMC's intellectual property: -
DTC: P0A1D-823 **T ECM, MIL: Yes** **Year:** 2011 **Model:** Altima, Quest **Engine:** 2.5L, 3.5L	**Hybrid Powertrain Control Module:** The monitor will run whenever the following DTC is not present: TMC's intellectual property Other conditions belong to TMC's intellectual property: -
DTC: P0A1F/123 **T BCM** **Year:** 2011 **Model:** Altima, Quest **Engine:** 2.5L, 3.5L	**Battery Energy Control Module (Hybrid):** Reception of an error signal from the battery smart unit
DTC: P0A1F-123 **T ECM, MIL: Yes** **Year:** 2011 **Model:** Altima, Quest **Engine:** 2.5L, 3.5L	**Battery Energy Control Module:** The monitor will run whenever the following DTCs are not present: TMC's intellectual property Other conditions belong to TMC's intellectual property: -
DTC: P0A1F-129 **T ECM, MIL: Yes** **Year:** 2011 **Model:** Altima, Quest **Engine:** 2.5L, 3.5L	**Battery Energy Control Module:** The monitor will run whenever the following DTC is not present: TMC's intellectual property Other conditions belong to TMC's intellectual property: -
DTC: P0A1F-150 **T ECM, MIL: Yes** **Year:** 2011 **Model:** Altima, Quest **Engine:** 2.5L, 3.5L	**Battery Energy Control Module:** The monitor will run whenever the following DTC is not present: TMC's intellectual property Other conditions belong to TMC's intellectual property: -

DTC	Trouble Code Title and Conditions
DTC: P0A2B-248 **T ECM, MIL: Yes** **Year:** 2011 **Model:** Altima, Quest **Engine:** 2.5L, 3.5L	**Drive Motor "A" Temperature Sensor Circuit Range / Performance:** Unusual sudden change in motor temperature sensor output occurs and the condition continues, or unusual sudden change in motor temperature sensor output occurs repeatedly.
DTC: P0A2B-250 **T ECM, MIL: Yes** **Year:** 2011 **Model:** Altima, Quest **Engine:** 2.5L, 3.5L	**Drive Motor "A" Temperature Sensor Circuit Range / Performance:** Motor temperature sensor output does not increase under conditions in which the value should increase, or output does not decrease under conditions in which the value should decrease.
DTC: P0A2C-247 **T ECM, MIL: Yes** **Year:** 2011 **Model:** Altima, Quest **Engine:** 2.5L, 3.5L	**Drive Motor "A" Temperature Sensor Circuit Low:** Short or short to GND in the motor temperature sensor circuit.
DTC: P0A2D-249 **T ECM, MIL: Yes** **Year:** 2011 **Model:** Altima, Quest **Engine:** 2.5L, 3.5L	**Drive Motor "A" Temperature Sensor Circuit High:** Open or short to +B in the motor temperature sensor circuit.
DTC: P0A37-258 **T ECM, MIL: Yes** **Year:** 2011 **Model:** Altima, Quest **Engine:** 2.5L, 3.5L	**Generator Temperature Sensor Circuit Range / Performance:** Unusual sudden change in generator temperature sensor output occurs and the condition continues, or unusual sudden change in generator temperature sensor output occurs repeatedly.
DTC: P0A37-260 **T ECM, MIL: Yes** **Year:** 2011 **Model:** Altima, Quest **Engine:** 2.5L, 3.5L	**Generator Temperature Sensor Circuit Range / Performance:** Generator temperature sensor output does not increase in which the value should increase, or output does not decrease under conditions in which the value should decrease.
DTC: P0A38-257 **T ECM, MIL: Yes** **Year:** 2011 **Model:** Altima, Quest **Engine:** 2.5L, 3.5L	**Generator Temperature Sensor Circuit Low:** Short to GND in the generator temperature sensor circuit.
DTC: P0A39-259 **T ECM, MIL: Yes** **Year:** 2011 **Model:** Altima, Quest **Engine:** 2.5L, 3.5L	**Generator Temperature Sensor Circuit High:** Open or short to +B in the generator temperature sensor circuit.
DTC: P0A3F-243 **T ECM, MIL: Yes** **Year:** 2011 **Model:** Altima, Quest **Engine:** 2.5L, 3.5L	**Drive Motor "A" Position Sensor Circuit:** The monitor will run whenever the following DTC is not present: TMC's intellectual property Other conditions belong to TMC's intellectual property: -
DTC: P0A40-500 **T ECM, MIL: Yes** **Year:** 2011 **Model:** Altima, Quest **Engine:** 2.5L, 3.5L	**Drive Motor "A" Position Sensor Circuit Range / Performance:** The monitor will run whenever the following DTC is not present: TMC's intellectual property Other conditions belong to TMC's intellectual property: -
DTC: P0A41-245 **T ECM, MIL: Yes** **Year:** 2011 **Model:** Altima, Quest **Engine:** 2.5L, 3.5L	**Drive Motor "A" Position Sensor Circuit Low:** The monitor will run whenever the following DTC is not present: TMC's intellectual property Other conditions belong to TMC's intellectual property: -

DTC	Trouble Code Title and Conditions
DTC: P0A4B-253 T ECM, MIL: Yes Year: 2011 Model: Altima, Quest Engine: 2.5L, 3.5L	**Generator Position Sensor Circuit:** The monitor will run whenever the following DTC is not present: TMC's intellectual property Other conditions belong to TMC's intellectual property: -
DTC: P0A4C-513 T ECM, MIL: Yes Year: 2011 Model: Altima, Quest Engine: 2.5L, 3.5L	**Generator Position Sensor Circuit Range / Performance:** The monitor will run whenever the following DTC is not present: TMC's intellectual property Other conditions belong to TMC's intellectual property: -
DTC: P0A4D-255 T ECM, MIL: Yes Year: 2011 Model: Altima, Quest Engine: 2.5L, 3.5L	**Generator Position Sensor Circuit Low:** The monitor will run whenever the following DTC is not present: TMC's intellectual property Other conditions belong to TMC's intellectual property: -
DTC: P0A51-174 T ECM, MIL: Yes Year: 2011 Model: Altima, Quest Engine: 2.5L, 3.5L	**Drive Motor "A" Current Sensor Circuit:** Motor current sensor high resolution circuit signal is out of range or there is a difference between it and the motor current sensor low resolution circuit current value.
DTC: P0A60-288 T ECM, MIL: Yes Year: 2011 Model: Altima, Quest Engine: 2.5L, 3.5L	**Drive Motor "A" Phase V Current:** The monitor will run whenever the following DTC is not present: TMC's intellectual property Other conditions belong to TMC's intellectual property: -
DTC: P0A60-290 T ECM, MIL: Yes Year: 2011 Model: Altima, Quest Engine: 2.5L, 3.5L	**Drive Motor "A" Phase V Current:** The monitor will run whenever the following DTC is not present: TMC's intellectual property Other conditions belong to TMC's intellectual property: -
DTC: P0A60-294 T ECM, MIL: Yes Year: 2011 Model: Altima, Quest Engine: 2.5L, 3.5L	**Drive Motor "A" Phase V Current:** The monitor will run whenever the following DTC is not present: TMC's intellectual property Other conditions belong to TMC's intellectual property: -
DTC: P0A60-501 T ECM, MIL: Yes Year: 2011 Model: Altima, Quest Engine: 2.5L, 3.5L	**Drive Motor "A" Phase V Current:** The monitor will run whenever the following DTC is not present: TMC's intellectual property Other conditions belong to TMC's intellectual property: -
DTC: P0A63-296 T ECM, MIL: Yes Year: 2011 Model: Altima, Quest Engine: 2.5L, 3.5L	**Drive Motor "A" Phase W Current:** The monitor will run whenever the following DTC is not present: TMC's intellectual property Other conditions belong to TMC's intellectual property: -
DTC: P0A63-298 T ECM, MIL: Yes Year: 2011 Model: Altima, Quest Engine: 2.5L, 3.5L	**Drive Motor "A" Phase W Current:** The monitor will run whenever the following DTC is not present: TMC's intellectual property Other conditions belong to TMC's intellectual property: -
DTC: P0A63-302 T ECM, MIL: Yes Year: 2011 Model: Altima, Quest Engine: 2.5L, 3.5L	**Drive Motor "A" Phase W Current:** The monitor will run whenever the following DTC is not present: TMC's intellectual property Other conditions belong to TMC's intellectual property: -

DTC	Trouble Code Title and Conditions
DTC: P0A63-502 **T ECM, MIL: Yes** **Year:** 2011 **Model:** Altima, Quest **Engine:** 2.5L, 3.5L	**Drive Motor "A" Phase W Current:** The monitor will run whenever the following DTC is not present: TMC's intellectual property Other conditions belong to TMC's intellectual property: -
DTC: P0A72-326 **T ECM, MIL: Yes** **Year:** 2011 **Model:** Altima, Quest **Engine:** 2.5L, 3.5L	**Generator Phase V Current:** The monitor will run whenever the following DTC is not present: TMC's intellectual property Other conditions belong to TMC's intellectual property: -
DTC: P0A72-328 **T ECM, MIL: Yes** **Year:** 2011 **Model:** Altima, Quest **Engine:** 2.5L, 3.5L	**Generator Phase V Current:** The monitor will run whenever the following DTC is not present: TMC's intellectual property Other conditions belong to TMC's intellectual property: -
DTC: P0A72-333 **T ECM, MIL: Yes** **Year:** 2011 **Model:** Altima, Quest **Engine:** 2.5L, 3.5L	**Generator Phase V Current:** The monitor will run whenever the following DTC is not present: TMC's intellectual property Other conditions belong to TMC's intellectual property: -
DTC: P0A72-515 **T ECM, MIL: Yes** **Year:** 2011 **Model:** Altima, Quest **Engine:** 2.5L, 3.5L	**Generator Phase V Current:** The monitor will run whenever the following DTC is not present: TMC's intellectual property Other conditions belong to TMC's intellectual property: -
DTC: P0A75-334 **T ECM, MIL: Yes** **Year:** 2011 **Model:** Altima, Quest **Engine:** 2.5L, 3.5L	**Generator Phase W Current:** The monitor will run whenever the following DTC is not present: TMC's intellectual property Other conditions belong to TMC's intellectual property: -
DTC: P0A75-336 **T ECM, MIL: Yes** **Year:** 2011 **Model:** Altima, Quest **Engine:** 2.5L, 3.5L	**Generator Phase W Current:** The monitor will run whenever the following DTC is not present: TMC's intellectual property Other conditions belong to TMC's intellectual property: -
DTC: P0A75-341 **T ECM, MIL: Yes** **Year:** 2011 **Model:** Altima, Quest **Engine:** 2.5L, 3.5L	**Generator Phase W Current:** The monitor will run whenever the following DTC is not present: TMC's intellectual property Other conditions belong to TMC's intellectual property: -
DTC: P0A75-516 **T ECM, MIL: Yes** **Year:** 2011 **Model:** Altima, Quest **Engine:** 2.5L, 3.5L	**Generator Phase W Current:** The monitor will run whenever the following DTC is not present: TMC's intellectual property Other conditions belong to TMC's intellectual property: -
DTC: P0A78-113 **T ECM, MIL: Yes** **Year:** 2011 **Model:** Altima, Quest **Engine:** 2.5L, 3.5L	**Drive Motor "A" Inverter Performance:** The monitor will run whenever the following DTC is not present: TMC's intellectual property Other conditions belong to TMC's intellectual property: -
DTC: P0A78-121 **T ECM, MIL: Yes** **Year:** 2011 **Model:** Altima, Quest **Engine:** 2.5L, 3.5L	**Drive Motor "A" Inverter Performance:** The monitor will run whenever the following DTC is not present: TMC's intellectual property Other conditions belong to TMC's intellectual property: -

DTC	Trouble Code Title and Conditions
DTC: P0A78-128 **T ECM, MIL: Yes** **Year:** 2011 **Model:** Altima, Quest **Engine:** 2.5L, 3.5L	**Drive Motor "A" Inverter Performance:** The monitor will run whenever the following DTC is not present: TMC's intellectual property Other conditions belong to TMC's intellectual property: -
DTC: P0A78-266 **T ECM, MIL: Yes** **Year:** 2011 **Model:** Altima, Quest **Engine:** 2.5L, 3.5L	**Drive Motor "A" Inverter Performance:** The monitor will run whenever the following DTC is not present: TMC's intellectual property Other conditions belong to TMC's intellectual property: -
DTC: P0A78-267 **T ECM, MIL: Yes** **Year:** 2011 **Model:** Altima, Quest **Engine:** 2.5L, 3.5L	**Drive Motor "A" Inverter Performance:** The monitor will run whenever the following DTC is not present: TMC's intellectual property Other conditions belong to TMC's intellectual property: -
DTC: P0A78-279 **T ECM, MIL: Yes** **Year:** 2011 **Model:** Altima, Quest **Engine:** 2.5L, 3.5L	**Drive Motor "A" Inverter Performance:** The monitor will run whenever the following DTC is not present: TMC's intellectual property Other conditions belong to TMC's intellectual property: -
DTC: P0A78-282 **T ECM, MIL: Yes** **Year:** 2011 **Model:** Altima, Quest **Engine:** 2.5L, 3.5L	**Drive Motor "A" Inverter Performance:** The monitor will run whenever the following DTC is not present: TMC's intellectual property Other conditions belong to TMC's intellectual property: - Motor inverter overvoltage signal detection (circuit malfunction).
DTC: P0A78-284 **T ECM, MIL: Yes** **Year:** 2011 **Model:** Altima, Quest **Engine:** 2.5L, 3.5L	**Drive Motor "A" Inverter Performance:** The monitor will run whenever the following DTC is not present: TMC's intellectual property Other conditions belong to TMC's intellectual property: - Motor inverter fail signal detection (overheat).
DTC: P0A78-286 **T ECM, MIL: Yes** **Year:** 2011 **Model:** Altima, Quest **Engine:** 2.5L, 3.5L	**Drive Motor "A" Inverter Performance:** The monitor will run whenever the following DTC is not present: TMC's intellectual property Other conditions belong to TMC's intellectual property: - Motor inverter fail signal detection (circuit malfunction)
DTC: P0A78-287 **T ECM, MIL: Yes** **Year:** 2011 **Model:** Altima, Quest **Engine:** 2.5L, 3.5L	**Drive Motor "A" Inverter Performance:** The monitor will run whenever the following DTC is not present: TMC's intellectual property Other conditions belong to TMC's intellectual property: - Motor inverter fail signal detection (over current due to inverter assembly malfunction)
DTC: P0A78-306 **T ECM, MIL: Yes** **Year:** 2011 **Model:** Altima, Quest **Engine:** 2.5L, 3.5L	**Drive Motor "A" Inverter Performance:** The monitor will run whenever the following DTC is not present: TMC's intellectual property Other conditions belong to TMC's intellectual property: - Motor torque execution monitoring malfunction.
DTC: P0A78-503 **T ECM, MIL: Yes** **Year:** 2011 **Model:** Altima, Quest **Engine:** 2.5L, 3.5L	**Drive Motor "A" Inverter Performance:** The monitor will run whenever the following DTC is not present: TMC's intellectual property Other conditions belong to TMC's intellectual property: - Motor inverter overvoltage signal detection (overvoltage due to MG ECU malfunction)
DTC: P0A78-504 **T ECM, MIL: Yes** **Year:** 2011 **Model:** Altima, Quest **Engine:** 2.5L, 3.5L	**Drive Motor "A" Inverter Performance:** The monitor will run whenever the following DTC is not present: TMC's intellectual property Other conditions belong to TMC's intellectual property: - Motor inverter overvoltage signal detection (overvoltage due to hybrid vehicle transaxle assembly malfunction)

DTC	Trouble Code Title and Conditions
DTC: P0A78-505 **T ECM, MIL: Yes** **Year:** 2011 **Model:** Altima, Quest **Engine:** 2.5L, 3.5L	**Drive Motor "A" Inverter Performance:** The monitor will run whenever the following DTC is not present: TMC's intellectual property Other conditions belong to TMC's intellectual property: - Motor inverter fail signal detection (over current due to MG ECU malfunction).
DTC: P0A78-506 **T ECM, MIL: Yes** **Year:** 2011 **Model:** Altima, Quest **Engine:** 2.5L, 3.5L	**Drive Motor "A" Inverter Performance:** The monitor will run whenever the following DTC is not present: TMC's intellectual property Other conditions belong to TMC's intellectual property: - Motor inverter fail signal detection (over current due to hybrid vehicle transaxle assembly malfunction)
DTC: P0A78-510 **T ECM, MIL: Yes** **Year:** 2011 **Model:** Altima, Quest **Engine:** 2.5L, 3.5L	**Drive Motor "A" Inverter Performance:** The monitor will run whenever the following DTC is not present: TMC's intellectual property Other conditions belong to TMC's intellectual property: - Motor inverter gate malfunction.
DTC: P0A78-586 **T ECM, MIL: Yes** **Year:** 2011 **Model:** Altima, Quest **Engine:** 2.5L, 3.5L	**Drive Motor "A" Inverter Performance:** The monitor will run whenever the following DTC is not present: TMC's intellectual property Other conditions belong to TMC's intellectual property: - Inverter voltage (VH) sensor performance problem.
DTC: P0A78-806 **T, MIL: Yes** **Year:** 2011 **Model:** Altima, Quest **Engine:** 2.5L, 3.5L	**Drive Motor "A" Inverter Performance:** The monitor will run whenever the following DTC is not present: TMC's intellectual property Other conditions belong to TMC's intellectual property: - Abnormal motor current value detection (MG ECU malfunction)3
DTC: P0A78-807 **T ECM, MIL: Yes** **Year:** 2011 **Model:** Altima, Quest **Engine:** 2.5L, 3.5L	**Drive Motor "A" Inverter Performance:** The monitor will run whenever the following DTC is not present: TMC's intellectual property Other conditions belong to TMC's intellectual property: - Abnormal motor current value detection (Inverter malfunction)
DTC: P0A78-808 **T ECM, MIL: Yes** **Year:** 2011 **Model:** Altima, Quest **Engine:** 2.5L, 3.5L	**Drive Motor "A" Inverter Performance:** The monitor will run whenever the following DTC is not present: TMC's intellectual property Other conditions belong to TMC's intellectual property: - Abnormal motor current value detection (Hybrid vehicle transaxle assembly malfunction)
DTC: P0A7A-122 **T ECM, MIL: Yes** **Year:** 2011 **Model:** Altima, Quest **Engine:** 2.5L, 3.5L	**Generator Inverter Performance:** The monitor will run whenever the following DTC is not present: TMC's intellectual property Other conditions belong to TMC's intellectual property: - Generator inverter fail signal detection (over current due to system malfunction)
DTC: P0A7A-130 **T ECM, MIL: Yes** **Year:** 2011 **Model:** Altima, Quest **Engine:** 2.5L, 3.5L	**Generator Inverter Performance:** The monitor will run whenever the following DTC is not present: TMC's intellectual property Other conditions belong to TMC's intellectual property: - Abnormal generator current value detection (System).
DTC: P0A7A-322 **T ECM, MIL: Yes** **Year:** 2011 **Model:** Altima, Quest **Engine:** 2.5L, 3.5L	**Generator Inverter Performance:** The monitor will run whenever the following DTC is not present: TMC's intellectual property Other conditions belong to TMC's intellectual property: - Generator inverter fail signal detection (overheating)
DTC: P0A7A-324 **T ECM, MIL: Yes** **Year:** 2011 **Model:** Altima, Quest **Engine:** 2.5L, 3.5L	**Generator Inverter Performance:** The monitor will run whenever the following DTC is not present: TMC's intellectual property Other conditions belong to TMC's intellectual property: - Generator inverter fail signal detection (circuit malfunction)

DTC	Trouble Code Title and Conditions
DTC: P0A7A-325 **T ECM, MIL: Yes** **Year:** 2011 **Model:** Altima, Quest **Engine:** 2.5L, 3.5L	**Generator Inverter Performance:** The monitor will run whenever the following DTC is not present: TMC's intellectual property Other conditions belong to TMC's intellectual property: - Generator inverter fail signal detection (overcurrent due to inverter assembly malfunction).
DTC: P0A7A-344 **T ECM, MIL: Yes** **Year:** 2011 **Model:** Altima, Quest **Engine:** 2.5L, 3.5L	**Generator Inverter Performance:** The monitor will run whenever the following DTC is not present: TMC's intellectual property Other conditions belong to TMC's intellectual property: - Generator torque execution monitoring malfunction
DTC: P0A7A-517 **T ECM, MIL: Yes** **Year:** 2011 **Model:** Altima, Quest **Engine:** 2.5L, 3.5L	**Generator Inverter Performance:** The monitor will run whenever the following DTC is not present: TMC's intellectual property Other conditions belong to TMC's intellectual property: - Generator inverter fail signal detection (overcurrent due to MG ECU malfunction)
DTC: P0A7A-518 **T ECM, MIL: Yes** **Year:** 2011 **Model:** Altima, Quest **Engine:** 2.5L, 3.5L	**Generator Inverter Performance:** The monitor will run whenever the following DTC is not present: TMC's intellectual property Other conditions belong to TMC's intellectual property: - Generator inverter fail signal detection (overcurrent due to hybrid vehicle transaxle assembly malfunction)
DTC: P0A7A-522 **T ECM, MIL: Yes** **Year:** 2011 **Model:** Altima, Quest **Engine:** 2.5L, 3.5L	**Generator Inverter Performance:** The monitor will run whenever the following DTC is not present: TMC's intellectual property Other conditions belong to TMC's intellectual property: - Generator inverter gate malfunction
DTC: P0A7A-809 **T ECM, MIL: Yes** **Year:** 2011 **Model:** Altima, Quest **Engine:** 2.5L, 3.5L	**Generator Inverter Performance:** The monitor will run whenever the following DTC is not present: TMC's intellectual property Other conditions belong to TMC's intellectual property: - Abnormal generator current value detection (MG ECU malfunction).
DTC: P0A7A-810 **T ECM, MIL: Yes** **Year:** 2011 **Model:** Altima, Quest **Engine:** 2.5L, 3.5L	**Generator Inverter Performance:** The monitor will run whenever the following DTC is not present: TMC's intellectual property Other conditions belong to TMC's intellectual property: - Abnormal generator current value detection (inverter malfunction)
DTC: P0A7A-811 **T ECM, MIL: Yes** **Year:** 2011 **Model:** Altima, Quest **Engine:** 2.5L, 3.5L	**Generator Inverter Performance:** The monitor will run whenever the following DTC is not present: TMC's intellectual property Other conditions belong to TMC's intellectual property: - Abnormal generator current value detection (hybrid vehicle transaxle assembly malfunction)
DTC: P0A7F/123 **Year:** 2011 **Model:** Altima, Quest **Engine:** 2.5L, 3.5L	**Hybrid Battery Pack Deterioration:** Internal resistance of HV battery is higher than the standard (1 trip detection) Difference in the capacity between battery blocks is larger than the standard (2 trip detection)
DTC: P0A7F-123 **1T ECM, MIL: Yes** **Year:** 2011 **Model:** Altima, Quest **Engine:** 2.5L, 3.5L	**Hybrid Battery Pack Deterioration:** The monitor will run whenever the following DTCs are not present: TMC's intellectual property Other conditions belong to TMC's intellectual property: -
DTC: P0A80/123 **Year:** 2011 **Model:** Altima, Quest **Engine:** 2.5L, 3.5L	**Replace Hybrid Battery Pack:** Difference in voltage between battery blocks is larger than the standard (2 trip detection)
DTC: P0A80-123 **2T ECM, MIL: Yes** **Year:** 2011 **Model:** Altima, Quest **Engine:** 2.5L, 3.5L	**Replace Hybrid Battery Pack:** The monitor will run whenever the following DTCs are not present: TMC's intellectual property Other conditions belong to TMC's intellectual property: -

DTC	Trouble Code Title and Conditions
DTC: P0A82/123 **Year:** 2011 **Model:** Altima, Quest **Engine:** 2.5L, 3.5L	**Hybrid Battery Pack Cooling Fan 1:** The speed of the battery cooling blower assembly is not within the specified range (1 trip detection)
DTC: P0A82-123 **1T ECM, MIL: Yes** **Year:** 2011 **Model:** Altima, Quest **Engine:** 2.5L, 3.5L	**Hybrid Battery Pack Cooling Fan 1:** The speed of the battery cooling blower assembly is not within the specified range.
DTC: P0A84/123 **Year:** 2011 **Model:** Altima, Quest **Engine:** 2.5L, 3.5L	**Hybrid Battery Pack Cooling Fan 1:** When the output voltage of the battery cooling blower assembly (VM) is too low compared to the target control voltage range (1 trip detection)
DTC: P0A84-123 **1T ECM, MIL: Yes** **Year:** 2011 **Model:** Altima, Quest **Engine:** 2.5L, 3.5L	**Hybrid Battery Pack Cooling Fan 1:** When the output voltage of the battery cooling blower assembly (VM) is too low compared to the target control voltage range.
DTC: P0A85/123 **Year:** 2011 **Model:** Altima, Quest **Engine:** 2.5L, 3.5L	**Hybrid Battery Pack Cooling Fan 1:** When the output voltage of the battery cooling blower assembly (VM) is too high compared to the target control voltage range (1 trip detection)
DTC: P0A85-123 **1T ECM, MIL: Yes** **Year:** 2011 **Model:** Altima, Quest **Engine:** 2.5L, 3.5L	**Hybrid Battery Pack Cooling Fan 1:** When the output voltage of the battery cooling blower assembly (VM) is too high compared to the target control voltage range.
DTC: P0A90-251 **T ECM, MIL: Yes** **Year:** 2011 **Model:** Altima, Quest **Engine:** 2.5L, 3.5L	**Drive Motor "A" Performance:** The monitor will run whenever the following DTC is not present: TMC's intellectual property Other conditions belong to TMC's intellectual property: - Motor magnetic force deterioration or same phase short circuit
DTC: P0A90-509 **T ECM, MIL: Yes** **Year:** 2011 **Model:** Altima, Quest **Engine:** 2.5L, 3.5L	**Drive Motor "A" Performance:** The monitor will run whenever the following DTC is not present: TMC's intellectual property Other conditions belong to TMC's intellectual property: - Motor system malfunction.
DTC: P0A92-261 **T ECM, MIL: Yes** **Year:** 2011 **Model:** Altima, Quest **Engine:** 2.5L, 3.5L	**Hybrid Generator Performance:** The monitor will run whenever the following DTC is not present: TMC's intellectual property Other conditions belong to TMC's intellectual property: - Generator magnetic force deterioration or same phase short circuit.
DTC: P0A92-521 **T ECM, MIL: Yes** **Year:** 2011 **Model:** Altima, Quest **Engine:** 2.5L, 3.5L	**Hybrid Generator Performance:** The monitor will run whenever the following DTC is not present: TMC's intellectual property Other conditions belong to TMC's intellectual property: - Generator system malfunction.
DTC: P0A93-346 **T ECM, MIL: Yes** **Year:** 2011 **Model:** Altima, Quest **Engine:** 2.5L, 3.5L	**Inverter Cooling System Performance:** The monitor will run whenever the following DTC is not present: TMC's intellectual property Other conditions belong to TMC's intellectual property: - Inverter cooling system malfunction (HV coolant malfunction)
DTC: P0A94-127 **T ECM, MIL: Yes** **Year:** 2011 **Model:** Altima, Quest **Engine:** 2.5L, 3.5L	**DC / DC Converter Performance:** The monitor will run whenever the following DTC is not present: TMC's intellectual property Other conditions belong to TMC's intellectual property: - Boost converter overvoltage signal detection (overvoltage due to system malfunction)

DTC	Trouble Code Title and Conditions
DTC: P0A94-172 **T ECM, MIL: Yes** **Year:** 2011 **Model:** Altima, Quest **Engine:** 2.5L, 3.5L	**DC / DC Converter Performance:** The monitor will run whenever the following DTC is not present: TMC's intellectual property Other conditions belong to TMC's intellectual property: - Boost converter fail signal detection (overcurrent due to system malfunction)
DTC: P0A94-442 **T ECM, MIL: Yes** **Year:** 2011 **Model:** Altima, Quest **Engine:** 2.5L, 3.5L	**DC / DC Converter Performance:** The monitor will run whenever the following DTC is not present: TMC's intellectual property Other conditions belong to TMC's intellectual property: - Abnormal voltage execution value.
DTC: P0A94-547 **T ECM, MIL: Yes** **Year:** 2011 **Model:** Altima, Quest **Engine:** 2.5L, 3.5L	**DC / DC Converter Performance:** The monitor will run whenever the following DTC is not present: TMC's intellectual property Other conditions belong to TMC's intellectual property: - Boost converter overvoltage signal detection (overvoltage due to MG ECU malfunction)
DTC: P0A94-548 **T ECM, MIL: Yes** **Year:** 2011 **Model:** Altima, Quest **Engine:** 2.5L, 3.5L	**DC / DC Converter Performance:** The monitor will run whenever the following DTC is not present: TMC's intellectual property Other conditions belong to TMC's intellectual property: - Boost converter overvoltage signal detection (overvoltage due to inverter malfunction).
DTC: P0A94-549 **T ECM, MIL: Yes** **Year:** 2011 **Model:** Altima, Quest **Engine:** 2.5L, 3.5L	**DC / DC Converter Performance:** The monitor will run whenever the following DTC is not present: TMC's intellectual property Other conditions belong to TMC's intellectual property: - Boost converter overvoltage signal detection (overvoltage due to hybrid vehicle transaxle assembly malfunction).
DTC: P0A94-550 **T ECM, MIL: Yes** **Year:** 2011 **Model:** Altima, Quest **Engine:** 2.5L, 3.5L	**DC / DC Converter Performance:** The monitor will run whenever the following DTC is not present: TMC's intellectual property Other conditions belong to TMC's intellectual property: - Boost converter overvoltage (OVL) signal detection (circuit malfunction)
DTC: P0A94-553 **T ECM, MIL: Yes** **Year:** 2011 **Model:** Altima, Quest **Engine:** 2.5L, 3.5L	**DC / DC Converter Performance:** The monitor will run whenever the following DTC is not present: TMC's intellectual property Other conditions belong to TMC's intellectual property: - Boost converter fail signal detection (boost converter overheating)
DTC: P0A94-554 **T ECM, MIL: Yes** **Year:** 2011 **Model:** Altima, Quest **Engine:** 2.5L, 3.5L	**DC / DC Converter Performance:** The monitor will run whenever the following DTC is not present: TMC's intellectual property Other conditions belong to TMC's intellectual property: - Boost converter fail signal detection (overcurrent due to MG ECU malfunction)
DTC: P0A94-555 **T ECM, MIL: Yes** **Year:** 2011 **Model:** Altima, Quest **Engine:** 2.5L, 3.5L	**DC / DC Converter Performance:** The monitor will run whenever the following DTC is not present: TMC's intellectual property Other conditions belong to TMC's intellectual property: - Boost converter fail signal detection (overcurrent due to inverter assembly malfunction).
DTC: P0A94-556 **T ECM, MIL: Yes** **Year:** 2011 **Model:** Altima, Quest **Engine:** 2.5L, 3.5L	**DC / DC Converter Performance:** The monitor will run whenever the following DTC is not present: TMC's intellectual property Other conditions belong to TMC's intellectual property: - Boost converter fail signal detection (overcurrent due to hybrid vehicle transaxle assembly malfunction)
DTC: P0A94-557 **T ECM, MIL: Yes** **Year:** 2011 **Model:** Altima, Quest **Engine:** 2.5L, 3.5L	**DC / DC Converter Performance:** The monitor will run whenever the following DTC is not present: TMC's intellectual property Other conditions belong to TMC's intellectual property: - Boost converter fail signal detection (circuit malfunction).

DTC	Trouble Code Title and Conditions
DTC: P0A94-585 **T ECM, MIL: Yes** **Year:** 2011 **Model:** Altima, Quest **Engine:** 2.5L, 3.5L	**DC / DC Converter Performance:** The monitor will run whenever the following DTC is not present: TMC's intellectual property Other conditions belong to TMC's intellectual property: - Boost converter voltage (VL) sensor performance problem.
DTC: P0A94-587 **T ECM, MIL: Yes** **Year:** 2011 **Model:** Altima, Quest **Engine:** 2.5L, 3.5L	**DC / DC Converter Performance:** The monitor will run whenever the following DTC is not present: TMC's intellectual property Other conditions belong to TMC's intellectual property: - Voltages from HV battery voltage (VB) sensor and boost converter voltage (VL) sensor deviate
DTC: P0A94-589 **T ECM, MIL: Yes** **Year:** 2011 **Model:** Altima, Quest **Engine:** 2.5L, 3.5L	**DC / DC Converter Performance:** The monitor will run whenever the following DTC is not present: TMC's intellectual property Other conditions belong to TMC's intellectual property: - Open or short to GND in the boost converter voltage (VL) sensor circuit
DTC: P0A94-590 **T ECM, MIL: Yes** **Year:** 2011 **Model:** Altima, Quest **Engine:** 2.5L, 3.5L	**DC / DC Converter Performance:** The monitor will run whenever the following DTC is not present: TMC's intellectual property Other conditions belong to TMC's intellectual property: - Short to +B in the boost converter voltage (VL) sensor circuit
DTC: P0A95/123 **Year:** 2011 **Model:** Altima, Quest **Engine:** 2.5L, 3.5L	**High Voltage Fuse:** Voltage between VB7 and VB8 terminals is below the standard despite the interlock switch being engaged (1 trip detection)
DTC: P0A95-123 **1T ECM, MIL: Yes** **Year:** 2011 **Model:** Altima, Quest **Engine:** 2.5L, 3.5L	**High Voltage Fuse:** Voltage between VC7 and VC8 terminals is below the standard despite the interlock switch being engaged.
DTC: P0A9C/123 **Year:** 2011 **Model:** Altima, Quest **Engine:** 2.5L, 3.5L	**Hybrid Battery Temperature Sensor "A":** When the battery temperature sensor performance is abnormal (1 trip detection/2 trip detection)
DTC: P0A9C-123 **1T ECM, MIL: Yes** **Year:** 2011 **Model:** Altima, Quest **Engine:** 2.5L, 3.5L	**Hybrid Battery Temperature Sensor "A" Range / Performance:** The monitor will run whenever the following DTCs are not present: TMC's intellectual property Other conditions belong to TMC's intellectual property: -
DTC: P0A9D/123 **Year:** 2011 **Model:** Altima, Quest **Engine:** 2.5L, 3.5L	**Hybrid Battery Temperature Sensor "A" Circuit Low:** When the temperature indicated by the battery temperature sensor is lower than a predetermined limit (open circuit) or is higher than a predetermined limit (short circuit) (1 trip detection)
DTC: P0A9D-123 **1T ECM, MIL: Yes** **Year:** 2011 **Model:** Altima, Quest **Engine:** 2.5L, 3.5L	**Hybrid Battery Temperature Sensor "A" Circuit Low:** The monitor will run whenever the following DTCs are not present: TMC's intellectual property Other conditions belong to TMC's intellectual property: -
DTC: P0A9E/123 **Year:** 2011 **Model:** Altima, Quest **Engine:** 2.5L, 3.5L	**Hybrid Battery Temperature Sensor "A" Circuit High:** When the temperature indicated by the battery temperature sensor is lower than a predetermined limit (open circuit) or is higher than a predetermined limit (short circuit) (1 trip detection)
DTC: P0A9E-123 **1T ECM, MIL: Yes** **Year:** 2011 **Model:** Altima, Quest **Engine:** 2.5L, 3.5L	**Hybrid Battery Temperature Sensor "A" Circuit High:** The monitor will run whenever the following DTCs are not present: TMC's intellectual property Other conditions belong to TMC's intellectual property: -

DTC	Trouble Code Title and Conditions
DTC: P0AA1-231 **T ECM, MIL: Yes** **Year:** 2011 **Model:** Altima, Quest **Engine:** 2.5L, 3.5L	**Hybrid Battery Positive Contactor Circuit Stuck Closed:** SMRB on the HV battery positive side is stuck closed.
DTC: P0AA1-233 **T ECM, MIL: Yes** **Year:** 2011 **Model:** Altima, Quest **Engine:** 2.5L, 3.5L	**Hybrid Battery Positive Contactor Circuit Stuck Closed:** SMRP, SMRB and SMRG on the HV battery positive and negative sides are stuck closed.
DTC: P0AA4-232 **T ECM, MIL: Yes** **Year:** 2011 **Model:** Altima, Quest **Engine:** 2.5L, 3.5L	**Hybrid Battery Negative Contactor Circuit Stuck Closed:** SMRG on the HV battery negative side stuck closed
DTC: P0AA6-526 **T ECM, MIL: Yes** **Year:** 2011 **Model:** Altima, Quest **Engine:** 2.5L, 3.5L	**Hybrid Battery Voltage System Isolation Fault:** Insulation resistance between the high-voltage circuit and the body has decreased.
DTC: P0AA6-611 **T BCM** **Year:** 2011 **Model:** Altima, Quest **Engine:** 2.5L, 3.5L	**Electric Compressor (Hybrid):** The DTC is output if there is insulation trouble with the high-voltage circuits in the air conditioning system. Possible causes are poor insulation in the compressor with motor assembly, or mixing of any oil other than ND-OIL 11 in the refrigerant cycle. The motor driven with high-voltage is built into the electrical compressor and is cooled directly with refrigerant. Compressor oil (ND-OIL 11) with high insulation performance is used because a leakage of electrical power may occur if regular compressor oil (DH-PS or DH-PR) is used. After READY and A/C ON High voltage system insulation malfunction - insulation resistance
DTC: P0AA6-611 **T ECM, MIL: Yes** **Year:** 2011 **Model:** Altima, Quest **Engine:** 2.5L, 3.5L	**Hybrid Battery Voltage System Isolation Fault:** Insulation resistance of the compressor with motor assembly has decreased.
DTC: P0AA6-612 **T ECM, MIL: Yes** **Year:** 2011 **Model:** Altima, Quest **Engine:** 2.5L, 3.5L	**Hybrid Battery Voltage System Isolation Fault:** Insulation resistance of the HV battery area has decreased.
DTC: P0AA6-613 **T ECM, MIL: Yes** **Year:** 2011 **Model:** Altima, Quest **Engine:** 2.5L, 3.5L	**Hybrid Battery Voltage System Isolation Fault:** Insulation resistance of the transaxle area has decreased.
DTC: P0AA6-614 **T ECM, MIL: Yes** **Year:** 2011 **Model:** Altima, Quest **Engine:** 2.5L, 3.5L	**Hybrid Battery Voltage System Isolation Fault:** Insulation resistance of the high-voltage DC area has decreased.
DTC: P0AA7-727 **T ECM, MIL: Yes** **Year:** 2011 **Model:** Altima, Quest **Engine:** 2.5L, 3.5L	**Hybrid Battery Voltage Isolation Sensor Circuit:** Malfunction in the insulation monitoring circuit in the battery smart unit
DTC: P0AAE-123 **T ECM, MIL: Yes** **Year:** 2011 **Model:** Altima, Quest **Engine:** 2.5L, 3.5L	**Hybrid Battery Pack Air Temperature Sensor "A" Circuit Low:** When the temperature indicated by the inlet air temperature sensor is lower than a predetermined limit (open circuit) or is higher than a predetermined limit (short circuit).

DTC	Trouble Code Title and Conditions
DTC: P0AAF-123 **T ECM, MIL: Yes** **Year:** 2011 **Model:** Altima, Quest **Engine:** 2.5L, 3.5L	**Hybrid Battery Pack Air Temperature Sensor "A" Circuit High:** When the temperature indicated by the inlet air temperature sensor is lower than a predetermined limit (open circuit) or is higher than a predetermined limit (short circuit).
DTC: P0ABF/123 **Year:** 2011 **Model:** Altima, Quest **Engine:** 2.5L, 3.5L	**Hybrid Battery Pack Current Sensor Circuit:** When the battery current sensor output is too low or high due to VIB/GIB failure
DTC: P0ABF-123 **1T ECM, MIL: Yes** **Year:** 2011 **Model:** Altima, Quest **Engine:** 2.5L, 3.5L	**Hybrid Battery Pack Current Sensor Circuit:** The monitor will run whenever the following DTCs are not present: TMC's intellectual property Other conditions belong to TMC's intellectual property: -
DTC: P0AC0-123 **T BCM** **Year:** 2011 **Model:** Altima, Quest **Engine:** 2.5L, 3.5L	**Hybrid Battery Pack Current Sensor Circuit Range/Performance:** The battery current sensor output characteristic is abnormal (offset/constant output) (1 trip detection/2 trip detection)
DTC: P0AC0-123 **1T ECM, MIL: Yes** **Year:** 2011 **Model:** Altima, Quest **Engine:** 2.5L, 3.5L	**Hybrid Battery Pack Current Sensor Circuit Range / Performance:** The monitor will run whenever the following DTCs are not present: TMC's intellectual property Other conditions belong to TMC's intellectual property: -
DTC: P0AC0-817 **T ECM, MIL: Yes** **Year:** 2011 **Model:** Altima, Quest **Engine:** 2.5L, 3.5L	**Hybrid Battery Pack Current Sensor Circuit Range / Performance:** The monitor will run whenever the following DTC is not present: TMC's intellectual property Other conditions belong to TMC's intellectual property: - HV battery current sensor performance problem.
DTC: P0AC1/123 **Year:** 2011 **Model:** Altima, Quest **Engine:** 2.5L, 3.5L	**Hybrid Battery Pack Current Sensor Circuit Low:** When the battery current sensor output is too low or high due to IB failure
DTC: P0AC1-123 **1T ECM, MIL: Yes** **Year:** 2011 **Model:** Altima, Quest **Engine:** 2.5L, 3.5L	**Hybrid Battery Pack Current Sensor "A" Circuit Low1:** The monitor will run whenever the following DTCs are not present: TMC's intellectual property Other conditions belong to TMC's intellectual property: -
DTC: P0AC2/123 **Year:** 2011 **Model:** Altima, Quest **Engine:** 2.5L, 3.5L	**Hybrid Battery Pack Current Sensor Circuit High:** When the battery current sensor output is too low or high due to IB failure
DTC: P0AC2-123 **1T ECM, MIL: Yes** **Year:** 2011 **Model:** Altima, Quest **Engine:** 2.5L, 3.5L	**Hybrid Battery Pack Current Sensor "A" Circuit High:** The monitor will run whenever the following DTCs are not present: TMC's intellectual property Other conditions belong to TMC's intellectual property: -
DTC: P0AC4 **T ECM, MIL: Yes** **Year:** 2011 **Model:** Altima **Engine:** 2.5L	**HV MIL On Request (Hybrid):** HYBRID MODELS CAUTION: Hybrid systems use very high-voltage battery systems. Before starting any service work involving the battery system, turn the ignition switch OFF and then remove the service plug from pocket in the trunk. After removing the service plug, wait 10 minutes before touching any of the high-voltage connectors and terminals. This DTC is displayed when a malfunction is detected by HV ECU. Check DTC for HV ECU and perform the trouble diagnosis.

DTC	Trouble Code Title and Conditions
DTC: P0AC6/123 **T BCM** **Year:** 2011 **Model:** Altima, Quest **Engine:** 2.5L, 3.5L	**Hybrid Battery Temperature Sensor "B" Range/Performance :** When the battery temperature sensor performance is abnormal (1 trip detection/2 trip detection)
DTC: P0AC6-123 **1T ECM, MIL: Yes** **Year:** 2011 **Model:** Altima, Quest **Engine:** 2.5L, 3.5L	**Hybrid Battery Temperature Sensor "B" Range / Performance:** The monitor will run whenever the following DTCs are not present: TMC's intellectual property Other conditions belong to TMC's intellectual property: -
DTC: P0AC7/123 **T BCM** **Year:** 2011 **Model:** Altima, Quest **Engine:** 2.5L, 3.5L	**Hybrid Battery Temperature Sensor "B" Circuit Low:** When the temperature indicated by the battery temperature sensor is lower than a predetermined limit (open circuit) or is higher than a predetermined limit (short circuit) (1 trip detection)
DTC: P0AC7-123 **1T ECM, MIL: Yes** **Year:** 2011 **Model:** Altima, Quest **Engine:** 2.5L, 3.5L	**Hybrid Battery Temperature Sensor "B" Circuit Low:** The monitor will run whenever the following DTCs are not present: TMC's intellectual property Other conditions belong to TMC's intellectual property: -
DTC: P0AC8/123 **T BCM** **Year:** 2011 **Model:** Altima, Quest **Engine:** 2.5L, 3.5L	**Hybrid Battery Temperature Sensor "B" Circuit High:** When the temperature indicated by the battery temperature sensor is lower than a predetermined limit (open circuit) or is higher than a predetermined limit (short circuit) (1 trip detection)
DTC: P0AC8-123 **1T ECM, MIL: Yes** **Year:** 2011 **Model:** Altima, Quest **Engine:** 2.5L, 3.5L	**Hybrid Battery Temperature Sensor "B" Circuit High:** The monitor will run whenever the following DTCs are not present: TMC's intellectual property Other conditions belong to TMC's intellectual property: -
DTC: P0ACB/123 **T BCM** **Year:** 2011 **Model:** Altima, Quest **Engine:** 2.5L, 3.5L	**Hybrid Battery Temperature Sensor "C" Range/Performance:** When the battery temperature sensor performance is abnormal (1 trip detection/2 trip detection)
DTC: P0ACB-123 **1T ECM, MIL: Yes** **Year:** 2011 **Model:** Altima, Quest **Engine:** 2.5L, 3.5L	**Hybrid Battery Temperature Sensor "C" Range / Performance:** The monitor will run whenever the following DTCs are not present: TMC's intellectual property Other conditions belong to TMC's intellectual property: -
DTC: P0ACC/123 **T BCM** **Year:** 2011 **Model:** Altima, Quest **Engine:** 2.5L, 3.5L	**Hybrid Battery Temperature Sensor "C" Circuit Low:** When the temperature indicated by the battery temperature sensor is lower than a predetermined limit (open circuit) or is higher than a predetermined limit (short circuit) (1 trip detection)
DTC: P0ACC-123 **1T ECM, MIL: Yes** **Year:** 2011 **Model:** Altima, Quest **Engine:** 2.5L, 3.5L	**Hybrid Battery Temperature Sensor "C" Circuit Low:** The monitor will run whenever the following DTCs are not present: TMC's intellectual property Other conditions belong to TMC's intellectual property: -
DTC: P0ACD/123 **T BCM** **Year:** 2011 **Model:** Altima, Quest **Engine:** 2.5L, 3.5L	**Hybrid Battery Temperature Sensor "C" Circuit High:** When the temperature indicated by the battery temperature sensor is lower than a predetermined limit (open circuit) or is higher than a predetermined limit (short circuit) (1 trip detection)

DTC	Trouble Code Title and Conditions
DTC: P0ACD-123 **1T ECM, MIL: Yes** **Year:** 2011 **Model:** Altima, Quest **Engine:** 2.5L, 3.5L	**Hybrid Battery Temperature Sensor "C" Circuit High:** The monitor will run whenever the following DTCs are not present: TMC's intellectual property Other conditions belong to TMC's intellectual property: -
DTC: P0ADB-227 **T ECM, MIL: Yes** **Year:** 2011 **Model:** Altima, Quest **Engine:** 2.5L, 3.5L	**Hybrid Battery Positive Contactor Control Circuit Low:** Short to GND in the SMRB circuit.
DTC: P0ADC-226 **T, MIL: Yes** **Year:** 2011 **Model:** Altima, Quest **Engine:** 2.5L, 3.5L	**Hybrid Battery Positive Contactor Control Circuit High:** Open or short to +B in the SMRB circuit.
DTC: P0ADF-229 **T ECM, MIL: Yes** **Year:** 2011 **Model:** Altima, Quest **Engine:** 2.5L, 3.5L	**Hybrid Battery Negative Contactor Control Circuit Low:** Short to GND in the SMRG circuit
DTC: P0AE0-228 **T ECM, MIL: Yes** **Year:** 2011 **Model:** Altima, Quest **Engine:** 2.5L, 3.5L	**Hybrid Battery Negative Contactor Control Circuit High:** Open or +B short in SMRG circuit
DTC: P0AE2-161 **T ECM, MIL: Yes** **Year:** 2011 **Model:** Altima, Quest **Engine:** 2.5L, 3.5L	**Hybrid Battery Precharge Contactor Circuit Stuck Closed:** When the power switch is on (READY) and regenerative braking is occurring, current is applied to SMRP (SMRG is turned off).
DTC: P0AE2-773 **T ECM, MIL: Yes** **Year:** 2011 **Model:** Altima, Quest **Engine:** 2.5L, 3.5L	**When the power switch is on (READY) and regenerative braking is occurring, current is applied to SMRP (SMRG is turned off).:** Current flows through SMRP when only SMRB is on during precharge (SMRP is stuck closed).
DTC: P0AE6-225 **T ECM, MIL: Yes** **Year:** 2011 **Model:** Altima, Quest **Engine:** 2.5L, 3.5L	**Hybrid Battery Precharge Contactor Control Circuit Low:** Open or short to GND in the SMRP circuit
DTC: P0AE7-224 **T ECM, MIL: Yes** **Year:** 2011 **Model:** Altima, Quest **Engine:** 2.5L, 3.5L	**Hybrid Battery Precharge Contactor Control Circuit High:** Short to +B in the SMRP circuit
DTC: P0AE9-123 **T BCM** **Year:** 2011 **Model:** Altima, Quest **Engine:** 2.5L, 3.5L	**Hybrid Battery Temperature Sensor "D" Range/Performance:** When the battery temperature sensor performance is abnormal (1 trip detection/2 trip detection)
DTC: P0AE9-123 **1T ECM, MIL: Yes** **Year:** 2011 **Model:** Altima, Quest **Engine:** 2.5L, 3.5L	**Hybrid Battery Temperature Sensor "D" Range / Performance:** The monitor will run whenever the following DTCs are not present: TMC's intellectual property Other conditions belong to TMC's intellectual property: -

DTC	Trouble Code Title and Conditions
DTC: P0AEA-123 **1T ECM, MIL: Yes** **Year:** 2011 **Model:** Altima, Quest **Engine:** 2.5L, 3.5L	**Hybrid Battery Temperature Sensor "D" Circuit Low:** The monitor will run whenever the following DTCs are not present: TMC's intellectual property Other conditions belong to TMC's intellectual property: -
DTC: P0AEA-123 **T BCM** **Year:** 2011 **Model:** Altima, Quest **Engine:** 2.5L, 3.5L	**Hybrid Battery Temperature Sensor "D" Circuit Low:** When the temperature indicated by the battery temperature sensor is lower than a predetermined limit (open circuit) or is higher than a predetermined limit (short circuit) (1 trip detection)
DTC: P0AEB-123 **T BCM** **Year:** 2011 **Model:** Altima, Quest **Engine:** 2.5L, 3.5L	**Hybrid Battery Temperature Sensor "D" Circuit High:** When the temperature indicated by the battery temperature sensor is lower than a predetermined limit (open circuit) or is higher than a predetermined limit (short circuit) (1 trip detection)
DTC: P0AEB-123 **1T ECM, MIL: Yes** **Year:** 2011 **Model:** Altima, Quest **Engine:** 2.5L, 3.5L	**Hybrid Battery Temperature Sensor "D" Circuit High:** The monitor will run whenever the following DTCs are not present: TMC's intellectual property Other conditions belong to TMC's intellectual property: -
DTC: P0AEE-276 **T ECM, MIL: Yes** **Year:** 2011 **Model:** Altima, Quest **Engine:** 2.5L, 3.5L	**Motor Inverter Temperature Sensor "A" Circuit Range / Performance:** Unusual sudden change in motor inverter temperature sensor output occurs and the offset continues, or unusual sudden change in motor inverter temperature sensor output occurs repeatedly.
DTC: P0AEE-277 **T ECM, MIL: Yes** **Year:** 2011 **Model:** Altima, Quest **Engine:** 2.5L, 3.5L	**Motor Inverter Temperature Sensor "A" Circuit Range / Performance:** Temperature calculated by power hybrid vehicle control ECU and actual temperature are different for 10 seconds or more.
DTC: P0AEF-275 **T ECM, MIL: Yes** **Year:** 2011 **Model:** Altima, Quest **Engine:** 2.5L, 3.5L	**Drive Motor Inverter Temperature Sensor "A" Circuit Low:** Open or short to GND in the motor inverter temperature sensor circuit
DTC: P0AF0-274 **2T ECM, MIL: Yes** **Year:** 2011 **Model:** Altima, Quest **Engine:** 2.5L, 3.5L	**Drive Motor Inverter Temperature Sensor "A" Circuit High:** Short to +B in motor inverter temperature sensor circuit.
DTC: P0B3D-123 **1T ECM, MIL: Yes** **Year:** 2011 **Model:** Altima, Quest **Engine:** 2.5L, 3.5L	**Hybrid Battery Voltage Sensor "A" Circuit Low:** The monitor will run whenever the following DTCs are not present: TMC's intellectual property Other conditions belong to TMC's intellectual property: -
DTC: P0B3D-123 **T BCM** **Year:** 2011 **Model:** Altima, Quest **Engine:** 2.5L, 3.5L	**Hybrid Battery Voltage Sensor "A" Circuit Low:** Any of the battery block voltages become less than 2.0 V (open). (1 trip detection)
DTC: P0B42-123 **1T ECM, MIL: Yes** **Year:** 2011 **Model:** Altima, Quest **Engine:** 2.5L, 3.5L	**Hybrid Battery Voltage Sensor "B" Circuit Low:** The monitor will run whenever the following DTCs are not present: TMC's intellectual property Other conditions belong to TMC's intellectual property: -

DTC	Trouble Code Title and Conditions
DTC: P0B42-123 **T BCM** **Year:** 2011 **Model:** Altima, Quest **Engine:** 2.5L, 3.5L	**Hybrid Battery Voltage Sensor "B" Circuit Low:** Any of the battery block voltages become less than 2.0 V (open). (1 trip detection)
DTC: P0B47-123 **1T ECM, MIL: Yes** **Year:** 2011 **Model:** Altima, Quest **Engine:** 2.5L, 3.5L	**Hybrid Battery Voltage Sensor "C" Circuit Low:** The monitor will run whenever the following DTCs are not present: TMC's intellectual property Other conditions belong to TMC's intellectual property: -
DTC: P0B47-123 **T BCM** **Year:** 2011 **Model:** Altima, Quest **Engine:** 2.5L, 3.5L	**Hybrid Battery Voltage Sensor "C" Circuit Low:** Any of the battery block voltages become less than 2.0 V (open). (1 trip detection)
DTC: P0B4C-123 **1T ECM, MIL: Yes** **Year:** 2011 **Model:** Altima, Quest **Engine:** 2.5L, 3.5L	**Hybrid Battery Voltage Sensor "D" Circuit Low:** The monitor will run whenever the following DTCs are not present: TMC's intellectual property Other conditions belong to TMC's intellectual property: -
DTC: P0B4C-123 **T BCM** **Year:** 2011 **Model:** Altima, Quest **Engine:** 2.5L, 3.5L	**Hybrid Battery Voltage Sensor "D" Circuit Low:** Any of the battery block voltages become less than 2.0 V (open). (1 trip detection)
DTC: P0B51-123 **T BCM** **Year:** 2011 **Model:** Altima, Quest **Engine:** 2.5L, 3.5L	**Hybrid Battery Voltage Sensor "E" Circuit Low:** Any of the battery block voltages become less than 2.0 V (open). (1 trip detection)
DTC: P0B51-123 **1T ECM, MIL: Yes** **Year:** 2011 **Model:** Altima, Quest **Engine:** 2.5L, 3.5L	**Hybrid Battery Voltage Sensor "E" Circuit Low:** The monitor will run whenever the following DTCs are not present: TMC's intellectual property Other conditions belong to TMC's intellectual property: -
DTC: P0B56-123 **1T ECM, MIL: Yes** **Year:** 2011 **Model:** Altima, Quest **Engine:** 2.5L, 3.5L	**Hybrid Battery Voltage Sensor "F" Circuit Low:** The monitor will run whenever the following DTCs are not present: TMC's intellectual property Other conditions belong to TMC's intellectual property: -
DTC: P0B56-123 **T BCM** **Year:** 2011 **Model:** Altima, Quest **Engine:** 2.5L, 3.5L	**Hybrid Battery Voltage Sensor "F" Circuit Low:** Any of the battery block voltages become less than 2.0 V (open). (1 trip detection)
DTC: P0B5B-123 **1T ECM, MIL: Yes** **Year:** 2011 **Model:** Altima, Quest **Engine:** 2.5L, 3.5L	**Hybrid Battery Voltage Sensor "G" Circuit Low:** The monitor will run whenever the following DTCs are not present: TMC's intellectual property Other conditions belong to TMC's intellectual property: -
DTC: P0B5B-123 **T BCM** **Year:** 2011 **Model:** Altima, Quest **Engine:** 2.5L, 3.5L	**Hybrid Battery Voltage Sensor "G" Circuit Low:** Any of the battery block voltages become less than 2.0 V (open). (1 trip detection)

DTC	Trouble Code Title and Conditions
DTC: P0B60-123 T BCM **Year:** 2011 **Model:** Altima, Quest **Engine:** 2.5L, 3.5L	**Hybrid Battery Voltage Sensor "H" Circuit Low:** Any of the battery block voltages become less than 2.0 V (open). (1 trip detection)
DTC: P0B60-123 1T ECM, MIL: Yes **Year:** 2011 **Model:** Altima, Quest **Engine:** 2.5L, 3.5L	**Hybrid Battery Voltage Sensor "H" Circuit Low:** The monitor will run whenever the following DTCs are not present: TMC's intellectual property Other conditions belong to TMC's intellectual property: -
DTC: P0B65-123 T BCM **Year:** 2011 **Model:** Altima, Quest **Engine:** 2.5L, 3.5L	**Hybrid Battery Voltage Sensor "I" Circuit Low:** Any of the battery block voltages become less than 2.0 V (open). (1 trip detection)
DTC: P0B65-123 1T ECM, MIL: Yes **Year:** 2011 **Model:** Altima, Quest **Engine:** 2.5L, 3.5L	**Hybrid Battery Voltage Sensor "I" Circuit Low:** The monitor will run whenever the following DTCs are not present: TMC's intellectual property Other conditions belong to TMC's intellectual property: -
DTC: P0B6A-123 1T ECM, MIL: Yes **Year:** 2011 **Model:** Altima, Quest **Engine:** 2.5L, 3.5L	**Hybrid Battery Voltage Sensor "J" Circuit Low:** The monitor will run whenever the following DTCs are not present: TMC's intellectual property Other conditions belong to TMC's intellectual property: -
DTC: P0B6A-123 T BCM **Year:** 2011 **Model:** Altima, Quest **Engine:** 2.5L, 3.5L	**Hybrid Battery Voltage Sensor "J" Circuit Low:** Any of the battery block voltages become less than 2.0 V (open). (1 trip detection)
DTC: P0B6F-123 T BCM **Year:** 2011 **Model:** Altima, Quest **Engine:** 2.5L, 3.5L	**Hybrid Battery Voltage Sensor "K" Circuit Low:** Any of the battery block voltages become less than 2.0 V (open). (1 trip detection)
DTC: P0B6F-123 1T ECM, MIL: Yes **Year:** 2011 **Model:** Altima, Quest **Engine:** 2.5L, 3.5L	**Hybrid Battery Voltage Sensor "K" Circuit Low:** The monitor will run whenever the following DTCs are not present: TMC's intellectual property Other conditions belong to TMC's intellectual property: -
DTC: P0B74-123 T BCM **Year:** 2011 **Model:** Altima, Quest **Engine:** 2.5L, 3.5L	**Hybrid Battery Voltage Sensor "L" Circuit Low:** Any of the battery block voltages become less than 2.0 V (open). (1 trip detection)
DTC: P0B74-123 1T ECM, MIL: Yes **Year:** 2011 **Model:** Altima, Quest **Engine:** 2.5L, 3.5L	**Hybrid Battery Voltage Sensor "L" Circuit Low:** The monitor will run whenever the following DTCs are not present: TMC's intellectual property Other conditions belong to TMC's intellectual property: -
DTC: P0B79-123 T BCM **Year:** 2011 **Model:** Altima, Quest **Engine:** 2.5L, 3.5L	**Hybrid Battery Voltage Sensor "M" Circuit Low:** Any of the battery block voltages become less than 2.0 V (open). (1 trip detection)

DTC	Trouble Code Title and Conditions
DTC: P0B79-123 **1T ECM, MIL: Yes** **Year:** 2011 **Model:** Altima, Quest **Engine:** 2.5L, 3.5L	**Hybrid Battery Voltage Sensor "M" Circuit Low:** The monitor will run whenever the following DTCs are not present: TMC's intellectual property Other conditions belong to TMC's intellectual property: -
DTC: P0B7E-123 **T BCM** **Year:** 2011 **Model:** Altima, Quest **Engine:** 2.5L, 3.5L	**Hybrid Battery Voltage Sensor "N" Circuit Low:** Any of the battery block voltages become less than 2.0 V (open). (1 trip detection)
DTC: P0B7E-123 **1T ECM, MIL: Yes** **Year:** 2011 **Model:** Altima, Quest **Engine:** 2.5L, 3.5L	**Hybrid Battery Voltage Sensor "N" Circuit Low:** The monitor will run whenever the following DTCs are not present: TMC's intellectual property Other conditions belong to TMC's intellectual property: -
DTC: P0B83-123 **1T ECM, MIL: Yes** **Year:** 2011 **Model:** Altima, Quest **Engine:** 2.5L, 3.5L	**Hybrid Battery Voltage Sensor "O" Circuit Low:** The monitor will run whenever the following DTCs are not present: TMC's intellectual property Other conditions belong to TMC's intellectual property: -
DTC: P0B83-123 **T BCM** **Year:** 2011 **Model:** Altima, Quest **Engine:** 2.5L, 3.5L	**Hybrid Battery Voltage Sensor "O" Circuit Low:** Any of the battery block voltages become less than 2.0 V (open). (1 trip detection)
DTC: P0B88-123 **T BCM** **Year:** 2011 **Model:** Altima, Quest **Engine:** 2.5L, 3.5L	**Hybrid Battery Voltage Sensor "P" Circuit Low:** Any of the battery block voltages become less than 2.0 V (open). (1 trip detection)
DTC: P0B88-123 **1T ECM, MIL: Yes** **Year:** 2011 **Model:** Altima, Quest **Engine:** 2.5L, 3.5L	**Hybrid Battery Voltage Sensor "P" Circuit Low:** The monitor will run whenever the following DTCs are not present: TMC's intellectual property Other conditions belong to TMC's intellectual property: -
DTC: P0B8D-123 **1T ECM, MIL: Yes** **Year:** 2011 **Model:** Altima, Quest **Engine:** 2.5L, 3.5L	**Hybrid Battery Voltage Sensor "Q" Circuit Low:** The monitor will run whenever the following DTCs are not present: TMC's intellectual property Other conditions belong to TMC's intellectual property: -
DTC: P0B8D-123 **T BCM** **Year:** 2011 **Model:** Altima, Quest **Engine:** 2.5L, 3.5L	**Hybrid Battery Voltage Sensor "Q" Circuit Low:** Any of the battery block voltages become less than 2.0 V (open). (1 trip detection)
DTC: P0B92-123 **T BCM** **Year:** 2011 **Model:** Altima, Quest **Engine:** 2.5L, 3.5L	**Hybrid Battery Voltage Sensor "R" Circuit Low:** Any of the battery block voltages become less than 2.0 V (open). (1 trip detection)
DTC: P0B92-123 **1T ECM, MIL: Yes** **Year:** 2011 **Model:** Altima, Quest **Engine:** 2.5L, 3.5L	**Hybrid Battery Voltage Sensor "R" Circuit Low:** The monitor will run whenever the following DTCs are not present: TMC's intellectual property Other conditions belong to TMC's intellectual property: -

DTC	Trouble Code Title and Conditions
DTC: P0C30-390 **T ECM, MIL: Yes** **Year:** 2011 **Model:** Altima, Quest **Engine:** 2.5L, 3.5L	**Hybrid Battery Pack State of Charge High:** Charge control error.
DTC: P0C76-523 **T ECM, MIL: Yes** **Year:** 2011 **Model:** Altima, Quest **Engine:** 2.5L, 3.5L	**Hybrid Battery System Discharge Time Too Long:** Inverter voltage (VH) sensor offset malfunction.

OBD II Trouble Code List (P1XXX Codes)

DTC	Trouble Code Title and Conditions
DTC: P100A **1T ECM, MIL: Yes** **Year:** 2011, 2012 **Model:** 370Z **Engine:** 3.7L	**Variable Valve Event & Lift (VVEL) Response Malfunction (Bank 1):** **NOTE: If DTC P100A or P100B is displayed with DTC P1090 or P1093, first perform the trouble diagnosis for DTC P1090 or P1093.** Actual event response to target is poor.
DTC: P100B **1T ECM, MIL: Yes** **Year:** 2011, 2012 **Model:** 370Z **Engine:** 3.7L	**Variable Valve Event & Lift (VVEL) Response Malfunction (Bank 2):** **NOTE: If DTC P100A or P100B is displayed with DTC P1090 or P1093, first perform the trouble diagnosis for DTC P1090 or P1093.** Actual event response to target is poor.
DTC: P101D **1T ECM, MIL: Yes** **Year:** 2011 **Model:** Altima, Quest **Engine:** 2.5L, 3.5L	**A/F Sensor Heater Circuit Performance Bank 1 Sensor 1 Stuck ON:** Battery voltage: 10.5 V or more Time after engine start: 10 seconds or more Air fuel ratio sensor heater duty-cycle ratio: 10 to 60% Air fuel ratio sensor heater ON current: 0.8 A or more Air fuel ratio sensor heater range check low current fail (P0031): Not detected Active heater OFF control: Not operating Active heater ON control: Not operating
DTC: P102D **1T ECM, MIL: Yes** **Year:** 2011 **Model:** Altima, Quest **Engine:** 2.5L, 3.5L	**O2 Sensor Heater Circuit Performance Bank 1 Sensor 2 Stuck ON:** Monitor runs whenever following DTCs are not present: None Battery voltage: 10.5 V or more Engine: Running Starter: OFF Catalyst active air fuel ratio control: Not operating Time after heater ON: 10 seconds or more Learned heater OFF current operation completed flag: ON Heated oxygen sensor heater OFF current: More than 3.5 A Hybrid IC high current limiter monitor input: Fail Heated oxygen sensor heater high current fail (P0038): Not detected
DTC: P106A **2T, MIL: Yes** **Year:** 2011 **Model:** Altima, Quest **Engine:** 2.5L, 3.5L	**Evaporative Emission System Pressure Sensor - Manifold Absolute Pressure Correlation:** Monitor runs whenever following DTCs are not present: P0106, P0107, P0108 (Manifold Absolute Pressure) P0452, P0453 (Evaporative Emission System Pressure Sensor) Arrive at after engine stop: 50 minutes Time after ECM (included in the hybrid vehicle control ECU) started by soak-timer: 60 seconds or more Battery voltage: 10.5 V or more Intake air temperature: -10°C (14°F) or more Engine coolant temperature sensor: -10°C (14°F) or more
DTC: P1078 **1T ECM, MIL: Yes** **Year:** 2011 **Model:** Juke, Versa **Engine:** 1.6L	**Exhaust Valve Timing Control Position Sensor (Bank 1) Circuit:** An excessively high or low voltage from the sensor is sent to ECM.

DTC	Trouble Code Title and Conditions
DTC: P1078 **1T ECM, MIL:** Yes **Year:** 2011, 2012 **Model:** Maxima **Engine:** 3.5L	**Exhaust Valve Timing Control Position Sensor (Bank 1) Circuit:** **NOTE: If this DTC is displayed with DTC P0643, first perform the trouble diagnosis for DTC P0643.** * An excessively high or low voltage from the sensor is sent to ECM.
DTC: P1084 **1T ECM, MIL:** Yes **Year:** 2011, 2012 **Model:** Maxima **Engine:** 3.5L	**Exhaust Valve Timing Control Position Sensor (Bank 2) Circuit:** **NOTE: If this DTC is displayed with DTC P0643, first perform the trouble diagnosis for DTC P0643.** * An excessively high or low voltage from the sensor is sent to ECM.
DTC: P1087 **T ECM, MIL:** Yes **Year:** 2011, 2012 **Model:** 370Z **Engine:** 3.7L	**Variable Valve Event & Lift (VVEL) SmAll Event Angle Malfunction (Bank 1):** **NOTE: If DTC P1087 or P1088 is displayed with DTC P1090 or P1093, perform the diagnosis for P1090 or P1093 first.** The event angle of VVEL control shaft is always small.
DTC: P1088 **T ECM, MIL:** Yes **Year:** 2011, 2012 **Model:** 370Z **Engine:** 3.7L	**Variable Valve Event & Lift (VVEL) SmAll Event Angle Malfunction (Bank 2):** **NOTE: If DTC P1087 or P1088 is displayed with DTC P1090 or P1093, perform the diagnosis for P1090 or P1093 first.** The event angle of VVEL control shaft is always small.
DTC: P1089 **T, MIL:** Yes **Year:** 2011, 2012 **Model:** 370Z **Engine:** 3.7L	**Variable Valve Event & Lift (VVEL) Control Shaft Position Sensor (Bank 1) Circuit:** **NOTE: If DTC P1089 or P1092 is displayed with DTC P1608, first perform the trouble diagnosis for DTC P1608.** An excessively low voltage from the sensor is sent to VVEL control module. An excessively high voltage from the sensor is sent to VVEL control module. Rationally incorrect voltage is sent to VVEL control module compared with the signals from VVEL control shaft position sensor 1 and VVEL control shaft position sensor 2.
DTC: P1090 **T, MIL:** Yes **Year:** 2011, 2012 **Model:** 370Z **Engine:** 3.7L	**Variable Valve Event & Lift (VVEL) System Performance (Bank 1) :** **NOTE: If DTC P1090 or P1093 is displayed with DTC P1091, first perform the trouble diagnosis for DTC P1091.** Event angle difference between the actual and the target is detected. Abnormal current is sent to VVEL actuator motor.
DTC: P1091 **T ECM, MIL:** Yes **Year:** 2011, 2012 **Model:** 370Z **Engine:** 3.7L	**Variable Valve Event & Lift (VVEL) Actuator Motor Relay Circuit:** VVEL control module detects the VVEL actuator motor relay is stuck OFF. VVEL control module detects the VVEL actuator motor relay is stuck ON.
DTC: P1092 **T ECM, MIL:** Yes **Year:** 2011, 2012 **Model:** 370Z **Engine:** 3.7L	**Variable Valve Event & Lift (VVEL) Control Shaft Position Sensor (Bank 2) Circuit:** **NOTE: If DTC P1089 or P1092 is displayed with DTC P1608, first perform the trouble diagnosis for DTC P1608.** An excessively low voltage from the sensor is sent to VVEL control module. An excessively high voltage from the sensor is sent to VVEL control module. Rationally incorrect voltage is sent to VVEL control module compared with the signals from VVEL control shaft position sensor 1 and VVEL control shaft position sensor 2.
DTC: P1093 **T ECM, MIL:** Yes **Year:** 2011, 2012 **Model:** 370Z **Engine:** 3.7L	**Variable Valve Event & Lift (VVEL) System Performance (Bank 2) :** **NOTE: If DTC P1090 or P1093 is displayed with DTC P1091, first perform the trouble diagnosis for DTC P1091** Event angle difference between the actual and the target is detected. Abnormal current is sent to VVEL actuator motor.
DTC: P1140 **1T ECM, MIL:** Yes **Year:** 2011 **Model:** Titan **Engine:** 5.6L	**Intake Valve Timing Control Position Sensor Circuit (Bank 1):** An excessively high or low voltage from the sensor is sent to ECM.

DTC	Trouble Code Title and Conditions
DTC: P1145 **1T ECM, MIL: Yes** **Year:** 2011 **Model:** Titan **Engine:** 5.6L	**Intake Valve Timing Control Position Sensor Circuit (Bank 2):** An excessively high or low voltage from the sensor is sent to ECM.
DTC: P1148 **1T ECM, MIL: Yes** **Year:** 2011 **Model:** Juke, Versa **Engine:** 1.6L	**Closed Loop Control Function (Bank 1):** The closed loop control function does not operate even when vehicle is being driven in the specified condition. **NOTE: DTC P1148 is displayed with DTC for A/F sensor 1. When the DTC is detected, perform the trouble diagnosis of DTC corresponding to A/F sensor 1.**
DTC: P1148 **1T ECM, MIL: Yes** **Year:** 2011, 2012 **Model:** 370Z, Altima, Cube, Frontier, Juke, Maxima, Murano, Pathfinder, Quest, Rogue, Titan, Versa, Xterra **Engine:** 1.6L, 1.8L, 2.5L, 3.5L, 3.5L, 3.7L, 4.0L, 5.6L	**Closed Loop Control Function (Includes Hybrid Models):** HYBRID MODELS CAUTION: Hybrid systems use very high-voltage battery systems. Before starting any service work involving the battery system, turn the ignition switch OFF and then remove the service plug from pocket in the trunk. After removing the service plug, wait 10 minutes before touching any of the high-voltage connectors and **NOTE: On and engines, this applies to Bank 1, except 3.7L which is Bank 2.** **NOTE: DTC P1148 or P1168 is displayed with another DTC for A/F sensor 1. Perform the trouble diagnosis for the corresponding DTC.** * The closed loop control function for bank 1 does not operate even when vehicle is being driven in the specified condition.
DTC: P1168 **1T ECM, MIL: Yes** **Year:** 2011, 2012 **Model:** Altima, Maxima, Murano, Pathfinder, Quest, Titan, Xterra **Engine:** 3.5L, 3.5L, 4.0L, 5.6L	**Closed Loop Control Function (Bank 2):** **NOTE: If DTC P1148 or P1168 is displayed with another DTC for air fuel ratio (A/F) sensor 2. Perform the trouble diagnosis for the corresponding DTC.** * The closed loop control function for bank 2 does not operate even when vehicle is being driven in the specified condition.
DTC: P1182 **T BCM, PATS: Yes** **Year:** 2011 **Model:** Rogue **Engine:** 2.5L	**LF Seat Belt Pre-Tensioner Circuit:** Front LH seat belt pre-tensioner 2 circuit is open.
DTC: P1186 **T BCM** **Year:** 2011 **Model:** Altima, Quest **Engine:** 2.5L, 3.5L	**LF Seat Belt Pre-Tensioner Circuit Open:** LH seat belt pre-tensioner circuit is open.
DTC: P1187 **T BCM** **Year:** 2011 **Model:** Altima, Quest **Engine:** 2.5L, 3.5L	**LF Seat Belt Pre-Tensioner Circuit Short:** LH seat belt pre-tensioner circuit is shorted to a power supply circuit.
DTC: P1188 **T BCM** **Year:** 2011 **Model:** Altima, Quest **Engine:** 2.5L, 3.5L	**LF Seat Belt Pre-Tensioner Circuit Short:** LH seat belt pre-tensioner circuit is shorted to ground.
DTC: P1189 **T BCM** **Year:** 2011 **Model:** Altima, Quest **Engine:** 2.5L, 3.5L	**LF Seat Belt Pre-Tensioner Circuit Short:** LH seat belt pre-tensioner circuits are shorted to each other.

DTC	Trouble Code Title and Conditions
DTC: P1195 **T ECM, MIL: Yes** **Year:** 2011 **Model:** Altima **Engine:** 2.5L	**Engine Does Not Start (Hybrid):** HYBRID MODELS CAUTION: Hybrid systems use very high-voltage battery systems. Before starting any service work involving the battery system, turn the ignition switch OFF and then remove the service plug from pocket in the trunk. After removing the service plug, wait 10 minutes before touching any of the high-voltage connectors and terminals. If DTC P1195 is displayed with DTC P0201, P0202, P0203, P0204, first perform the trouble diagnosis for DTC P0201, P0202, P0203, P0204. If DTC P1195 is displayed with DTC P0335, first perform the trouble diagnosis for DTC P0335. If DTC P1195 is displayed with DTC P0340, first perform the trouble diagnosis for DTC P0340. If DTC P1195 is displayed with DTC P0605, first perform the trouble diagnosis for DTC P0605. When the engine is abnormal, and the engine does not start.
DTC: P1196 **T ECM, MIL: Yes** **Year:** 2011 **Model:** Altima **Engine:** 2.5L	**Engine Lacks Power (Hybrid):** HYBRID MODELS CAUTION: Hybrid systems use very high-voltage battery systems. Before starting any service work involving the battery system, turn the ignition switch OFF and then remove the service plug from pocket in the trunk. After removing the service plug, wait 10 minutes before touching any of the high-voltage connectors and terminals. If DTC P1196 is displayed with DTC P0201, P0202, P0203, P0204, first perform the trouble diagnosis for DTC P0201, P0202, P0203, P0204. If DTC P1196 is displayed with DTC P0335, first perform the trouble diagnosis for DTC P0335. If DTC P1196 is displayed with DTC P0340, first perform the trouble diagnosis for DTC P0340. If DTC P1196 is displayed with DTC P0605, first perform the trouble diagnosis for DTC P0605. The estimated torque is excessively low compared with the target torque
DTC: P1197 **1T ECM, MIL: Yes** **Year:** 2011 **Model:** Juke, Versa **Engine:** 1.6L	**Fuel Run Out (Out of Gas):** Fuel rail pressure remains at 1.1 MPa (11 bar, 11.2 kg/cm2, 159.5 psi) or less for 5 seconds or more with the fuel level too low. Fuel rail pressure remains 2.7 MPa (27 bar, 27.5 kg/cm2, 392 psi) lower than a target fuel pressure for 5 seconds or more with the fuel level too low.
DTC: P1197 **T ECM, MIL: Yes** **Year:** 2011 **Model:** Altima **Engine:** 2.5L	**Fuel Run Out (Hybrid):** HYBRID MODELS CAUTION: Hybrid systems use very high-voltage battery systems. Before starting any service work involving the battery system, turn the ignition switch OFF and then remove the service plug from pocket in the trunk. After removing the service plug, wait 10 minutes before touching any of the high-voltage connectors and terminals. **Note:** This DTC may be detected if the vehicle continues turning counterclockwise over a certain speed for a length of time. Detecting condition for P1195 or P1196 is satisfied and low voltage from the fuel level sensor is sent to ECM.
DTC: P1211 **1T ECM** **Year:** 2011, 2012 **Model:** 370Z, Frontier, Maxima, Pathfinder, Titan, Versa, Xterra **Engine:** 1.8L, 3.5L, 3.7L, 4.0L, 5.6L	**TCS Control Unit:** * Freeze frame data is not stored in the ECM for this self-diagnosis. * The MIL will not illuminate for this self-diagnosis. * ECM receives malfunction information from "ABS actuator and electric unit (control unit)".
DTC: P1212 **1T ECM, MIL: Yes** **Year:** 2011, 2012 **Model:** 370Z, Cube, Maxima, Murano, Pathfinder, Rogue, Titan, Versa, Xterra **Engine:** 1.8L, 2.5L, 3.5L, 3.5L, 3.7L, 4.0L, 5.6L	**TCS Communication Line:** **NOTE: If DTC P1212 is displayed with DTC UXXXX, first perform the trouble diagnosis for DTC UXXXX.** **NOTE: If DTC P1212 is displayed with DTC P0607, first perform the trouble diagnosis for DTC P0607.** **NOTE: Be sure to erase the malfunction information such as DTC not only for "ABS actuator and electric unit (control unit)" but also for ECM after TCS related repair.** * Freeze frame data is not stored in the ECM for this self-diagnosis. * The MIL will not illuminate for this self-diagnosis. * ECM cannot receive the information from "ABS actuator and electric unit (control unit)".
DTC: P1212 **1T ECM, MIL: Yes** **Year:** 2011 **Model:** Juke, Versa **Engine:** 1.6L	**TCS Communication Line:** ECM can not receive the information from "ABS actuator and electric unit (control unit)" continuously. **NOTE: If DTC P1212 is displayed with DTC U1001, first perform the trouble diagnosis for DTC U1001.** If DTC P1212 is displayed with DTC P0607, first perform the trouble diagnosis for DTC P0607.
DTC: P1217 **1T ECM, MIL: Yes** **Year:** 2011 **Model:** Juke, Versa **Engine:** 1.6L	**Engine Over Temperature (Overheat) :** When the engine coolant temperature reaches an abnormally high temperature condition, a malfunction is indicated. CAUTION: When a malfunction is indicated, be sure to replace the coolant, and replace the engine oil.

DTC	Trouble Code Title and Conditions
DTC: P1217 **1T ECM, MIL: Yes** **Year:** 2011, 2012 **Model:** 370Z, Altima, Cube, Frontier, Maxima, Murano, Pathfinder, Quest, Rogue, Sentra, Titan, Xterra **Engine:** 1.8L, 2.0L, 2.5L, 3.5L, 3.5L, 3.7L, 4.0L, 5.6L	**Engine Over Temperature (Overheat) (Includes Hybrid Models):** HYBRID MODELS CAUTION: Hybrid systems use very high-voltage battery systems. Before starting any service work involving the battery system, turn the ignition switch OFF and then remove the service plug from pocket in the trunk. After removing the service plug, wait 10 minutes before touching any of the high-voltage connectors and terminals. **NOTE: If DTC P1217 is displayed with DTC UXXXX, first perform the trouble diagnosis for DTC UXXXX.** **NOTE: If DTC P1217 is displayed with DTC P0607, first perform the trouble diagnosis for DTC P0607.** * The ECM controls cooling fan relays through CAN communication line. * Cooling fan does not operate properly (overheat). * Cooling fan system does not operate properly (overheat). * Engine coolant was not added to the system using the proper filling method. * Engine coolant is not within the specified range.
DTC: P1220 **1T ECM, MIL: Yes** **Year:** 2011 **Model:** Titan **Engine:** 5.6L	**Fuel Pump Control Module (FPCM):** * An improper voltage signal from the FPCM, which is supplied to a point between the fuel pump and the dropping resistor, is detected by ECM. * During engine cranking, the signal voltage of the FPCM to the ECM is too low.
DTC: P1225 **1T ECM, MIL: Yes** **Year:** 2011, 2012 **Model:** All **Engine:** 1.6L, 1.8L, 2.0L, 2.5L, 3.5L, 3.5L, 3.7L, 4.0L, 5.6L	**Closed Throttle Position Learning Performance (Includes Hybrid Models):** HYBRID MODELS CAUTION: Hybrid systems use very high-voltage battery systems. Before starting any service work involving the battery system, turn the ignition switch OFF and then remove the service plug from pocket in the trunk. After removing the service plug, wait 10 minutes before touching any of the high-voltage connectors and terminals. **NOTE: For and, this DTC is for Bank 1.** * Closed throttle position learning value is excessively low.
DTC: P1226 **1T ECM, MIL: Yes** **Year:** 2011, 2012 **Model:** All **Engine:** 1.6L, 1.8L, 2.0L, 2.5L, 3.5L, 3.5L, 3.7L, 4.0L, 5.6L	**Closed Throttle Position Learning Performance (Includes Hybrid Models):** HYBRID MODELS CAUTION: Hybrid systems use very high-voltage battery systems. Before starting any service work involving the battery system, turn the ignition switch OFF and then remove the service plug from pocket in the trunk. After removing the service plug, wait 10 minutes before touching any of the high-voltage connectors and terminals. **NOTE: On abd, this DTC is for Bank 1.** * Closed throttle position learning is not performed successfully, repeatedly.
DTC: P1233 **T ECM, MIL: Yes** **Year:** 2011, 2012 **Model:** 370Z **Engine:** 3.7L	**Electric Throttle Control Performance (Bank 2):** **NOTE: If DTC P1233 or P2101 is displayed with DTC P1238, P1290, P2100 or 2119, first perform the trouble diagnosis for DTC P1238, P2119 or P1290, P2100.** Electric throttle control function does not operate properly
DTC: P1234 **1T ECM, MIL: Yes** **Year:** 2011, 2012 **Model:** 370Z **Engine:** 3.7L	**Closed Throttle Position Learning Performance (Bank 2):** Closed throttle position learning value is excessively low.
DTC: P1235 **1T ECM, MIL: Yes** **Year:** 2011, 2012 **Model:** 370Z **Engine:** 3.7L	**Closed Throttle Position Learning Performance (Bank 2):** Closed throttle position learning is not performed successfully, repeatedly.
DTC: P1236 **T ECM, MIL: Yes** **Year:** 2011, 2012 **Model:** 370Z **Engine:** 3.7L	**Throttle Control Motor (Bank 2) Circuit Short:** ECM detects short in both circuits between ECM and throttle control motor.
DTC: P1238 **T ECM, MIL: Yes** **Year:** 2011, 2012 **Model:** 370Z **Engine:** 3.7L	**Electrical Throttle Control Actuator (Bank 2):** Condition A: Electric throttle control actuator does not function properly due to the return spring malfunction. Condition B: Throttle valve opening angle in fail-safe mode is not in specified range. Condition CL ECM detect the throttle valve is stuck open.

DTC	Trouble Code Title and Conditions
DTC: P1239 **T ECM, MIL: Yes** **Year:** 2011, 2012 **Model:** 370Z **Engine:** 3.7L	**Throttle Position Sensor (Bank 2) Circuit Range/Performance:** Rationally incorrect voltage is sent to ECM compared with the signals from TP sensor 1 and TP sensor 2.
DTC: P1290 **T ECM, MIL: Yes** **Year:** 2011, 2012 **Model:** 370Z **Engine:** 3.7L	**Throttle Control Motor Relay Circuit Open (Bank 2):** ECM detects a voltage of power source for throttle control motor is excessively low.
DTC: P1421 **1T ECM, MIL: Yes** **Year:** 2011, 2012 **Model:** 370Z, Altima, Cube, Frontier, Juke, Maxima, Murano, Pathfinder, Quest, Sentra, Titan, Versa, Xterra **Engine:** 1.6L, 1.8L, 2.0L, 2.5L, 3.5L, 3.5L, 3.7L, 4.0L, 5.6L	**Cold Start Emission Reduction Strategy Monitoriing (Includes Hybrid Models):** HYBRID MODELS CAUTION: Hybrid systems use very high-voltage battery systems. Before starting any service work involving the battery system, turn the ignition switch OFF and then remove the service plug from pocket in the trunk. After removing the service plug, wait 10 minutes before touching any of the high-voltage connectors and terminals. **NOTE: If DTC P1421 is displayed with other DTC, first perform the trouble diagnosis for other DTC.** * ECM does not control ignition timing and engine idle speed properly when engine is started with pre-warming up condition.
DTC: P1423 **1T ECM, MIL: Yes** **Year:** 2011 **Model:** Juke, Versa **Engine:** 1.6L	**COLD START CONTROL (Cold start emission reduction strategy monitoring) :** ECM does not control fuel injection timing properly when engine is started with the engine cold. ECM
DTC: P1424 **1T ECM, MIL: Yes** **Year:** 2011 **Model:** Juke, Versa **Engine:** 1.6L	**COLD START CONTROL (Cold start emission reduction strategy monitoring) :** ECM does not control fuel injection quantity properly when engine is started with the engine cold.
DTC: P1451 **1T ECM, MIL: Yes** **Year:** 2011 **Model:** Juke, Versa **Engine:** 1.6L	**TC/SC PRES-EVAP PRES (EVAP control system pressure sensor/turbocharger boost sensor correlation):** ECM detects a state that the pressure difference remains -13.0 kPa (-98 mmHg, -3.83 inHg) or less/13.5 kPa (102 mmHg, 3.99 in-Hg) or more for continuously for 5 seconds or more under the condition that the pressure of the EVAP control system pressure sensor and that of the turbocharger boost sensor are equal.
DTC: P1550 **1T ECM, MIL: Yes** **Year:** 2011, 2012 **Model:** Cube, Frontier, Juke, Maxima, Murano, Pathfinder, Titan, Versa, Xterra **Engine:** 1.6L, 1.8L, 2.5L, 3.5L, 3.5L, 4.0L, 5.6L	**Battery Current Sensor Circuit Range/Performance:** * The MIL will not illuminate for this diagnosis. **NOTE: If DTC P1550 is displayed with DTC P0643, first perform the trouble diagnosis for DTC P0643.** * The output voltage of the battery current sensor remains within the specified range while engine is running.
DTC: P1551 **1T ECM, MIL: Yes** **Year:** 2011, 2012 **Model:** Cube, Frontier, Juke, Maxima, Murano, Pathfinder, Titan, Versa, Xterra **Engine:** 1.6L, 1.8L, 2.5L, 3.5L, 3.5L, 4.0L, 5.6L	**Battery Current Sensor Circuit Low Input:** * The MIL will not illuminate for this diagnosis. **NOTE: If DTC P1551 or P1552 is displayed with DTC P0643, first perform the trouble diagnosis for DTC P0643.** * An excessively low voltage from the sensor is sent to ECM.
DTC: P1551 **1T VCM** **Year:** 2011 **Model:** Leaf **Engine:** -L –	**BATTERY CURRENT SENSOR (Battery current sensor circuit low input):** An excessively low voltage from the sensor is sent to VCM.

DTC	Trouble Code Title and Conditions
DTC: P1552 **1T ECM, MIL: Yes** **Year:** 2011, 2012 **Model:** Cube, Frontier, Juke, Maxima, Murano, Pathfinder, Titan, Versa, Xterra **Engine:** 1.6L, 1.8L, 2.5L, 3.5L, 3.5L, 4.0L, 5.6L	**Battery Current Sensor Circuit High Input:** * The MIL will not illuminate for this diagnosis. **NOTE: If DTC P1551 or P1552 is displayed with DTC P0643, first perform the trouble diagnosis for DTC P0643.** * An excessively high voltage from the sensor is sent to ECM.
DTC: P1553 **1T ECM, MIL: Yes** **Year:** 2011, 2012 **Model:** Cube, Frontier, Juke, Murano, Pathfinder, Titan, Versa, Xterra **Engine:** 1.6L, 1.8L, 2.5L, 3.5L, 4.0L, 5.6L	**Battery Current Sensor Performance:** * The MIL will not illuminate for this diagnosis. **NOTE: If DTC P1553 is displayed with DTC P0643, first perform the trouble diagnosis for DTC P0643.** * The signal voltage transmitted from the sensor to ECM is higher than the amount of the maximum power generation.
DTC: P1554 **1T ECM, MIL: Yes** **Year:** 2011, 2012 **Model:** Cube, Frontier, Juke, Maxima, Murano, Pathfinder, Titan, Versa, Xterra **Engine:** 1.6L, 1.8L, 2.5L, 3.5L, 3.5L, 4.0L, 5.6L	**Battery Current Sensor Performance:** * The MIL will not illuminate for this diagnosis. **NOTE: If DTC P1554 is displayed with DTC P0643, first perform the trouble diagnosis for DTC P0643.** * The output voltage of the battery current sensor is lower than the specified value while the battery voltage is high enough.
DTC: P1555-181 **T ECM, MIL: Yes** **Year:** 2011 **Model:** Altima, Quest **Engine:** 2.5L, 3.5L	**Reactor Temperature Sensor Circuit Low:** Malfunction in the reactor temperature sensor wiring (short to GND).
DTC: P1556 **1T ECM, MIL: Yes** **Year:** 2011 **Model:** Juke, Versa **Engine:** 1.6L	**BATTERY TEMPERATURE SENSOR CIRCUIT (Low Input):** Signal voltage from Battery temperature sensor remains 0.16V or less for 5 seconds or more. **NOTE: Before measuring the terminal voltage, confirm that the battery is fully charged.**
DTC: P1556-182 **T ECM, MIL: Yes** **Year:** 2011 **Model:** Altima, Quest **Engine:** 2.5L, 3.5L	**Reactor Temperature Sensor Circuit High:** Malfunction in the reactor temperature sensor wiring (open or short to +B)
DTC: P1557 **1T ECM, MIL: Yes** **Year:** 2011 **Model:** Juke, Versa **Engine:** 1.6L	**BATTERY TEMPERATURE SENSOR CIRCUIT (High Input):** Signal voltage from Battery temperature sensor remains 4.84V or more for 5 seconds or more.
DTC: P1564 **1T ECM** **Year:** 2011, 2012 **Model:** 370Z, Altima, Cube, Frontier, Murano, Pathfinder, Quest, Rogue, Sentra, Titan, Versa, Xterra **Engine:** 1.8L, 2.0L, 2.5L, 3.5L, 3.5L, 3.7L, 4.0L, 5.6L	**ASCD Steering Switch Malfunction (Includes Hybrid Models):** HYBRID MODELS CAUTION: Hybrid systems use very high-voltage battery systems. Before starting any service work involving the battery system, turn the ignition switch OFF and then remove the service plug from pocket in the trunk. After removing the service plug, wait 10 minutes before touching any of the high-voltage connectors and terminals. * This self-diagnosis has the one trip detection logic. * The MIL will not illuminate for this self-diagnosis. **NOTE: If DTC P1564 is displayed with DTC P0605, first perform the trouble diagnosis for DTC P0605.** * An excessively high voltage signal from the ASCD steering switch is sent to ECM. * ECM detects that input signal from the ASCD steering switch is out of the specified range. * ECM detects that the ASCD steering switch is stuck ON.

DTC	Trouble Code Title and Conditions
DTC: P1564 **1T ECM, MIL: Yes** **Year:** 2011 **Model:** Juke, Versa **Engine:** 1.6L	**ASCD STEERING SWITCH :** An excessively high voltage signal from the ASCD steering switch is sent to ECM. ECM detects that input signal from the ASCD steering switch is out of the specified range. ECM detects that the ASCD steering switch s stuck ON.
DTC: P1571 **T, MIL: Yes** **Year:** 2011 **Model:** Altima, Quest **Engine:** 2.5L, 3.5L	**Radar Sensor Malfunction:** The hybrid vehicle control ECU detects a millimeter wave radar sensor assembly malfunction signal for 0.15 seconds or more while the dynamic radar cruise control is operating.
DTC: P1572 **2T ECM** **Year:** 2011 **Model:** Juke, Versa **Engine:** 1.6L	**ASCD BRAKE SWITCH :** Brake pedal position switch signal is not sent to ECM for extremely long time while the vehicle is driving. **NOTE: This self-diagnosis has the one trip detection logic. When malfunction A is detected, DTC is not stored in ECM memory. And in that case, 1st trip DTC and 1st trip freeze frame data are displayed. 1st trip DTC is erased when ignition switch OFF. And even when malfunction A is detected in two consecutive trips, DTC is not stored in ECM memory.**
DTC: P1572 **1T ECM, MIL: Yes** **Year:** 2011, 2012 **Model:** 370Z, Altima, Cube, Frontier, Maxima, Murano, Pathfinder, Quest, Rogue, Titan, Versa, Xterra **Engine:** 1.8L, 2.5L, 3.5L, 3.5L, 3.7L, 4.0L, 5.6L	**ACSD Brake Switch Malfunction (Includes Hybrid Models):** HYBRID MODELS CAUTION: Hybrid systems use very high-voltage battery systems. Before starting any service work involving the battery system, turn the ignition switch OFF and then remove the service plug from pocket in the trunk. After removing the service plug, wait 10 minutes before touching any of the high-voltage connectors and terminals. * This self-diagnosis has the one trip detection logic. * The MIL will not illuminate for this self-diagnosis. **NOTE: If DTC P1572 is displayed with DTC P0605, first perform the trouble diagnosis for DTC P0605.** * This self-diagnosis has the one trip detection logic. **NOTE: When malfunction A is detected, the DTC is not stored in ECM memory. And in that case, 1st trip DTC and 1st trip freeze frame data are displayed. 1st trip DTC is erased when ignition switch is turned OFF. And even when Malfunction A is detected in two consecutive trips, DTC is not stored in ECM memory.** * Malfunction A: When the vehicle speed is above 19 MPH, ON signals from the stop lamp switch and the ASCD brake switch are sent to ECM at the same time. * Malfunction B: ASCD brake switch signal is not sent to ECM for extremely long time while the vehicle is being driven.
DTC: P1572 **T, MIL: Yes** **Year:** 2011 **Model:** Altima, Quest **Engine:** 2.5L, 3.5L	**Improper Aiming of Radar Sensor Beam Axis:** The ECU detects that the millimeter wave radar sensor assembly beam axis is in an incorrect position (0.15 seconds or more) while the dynamic radar cruise control is operating.
DTC: P1574 **1T ECM, MIL: Yes** **Year:** 2011, 2012 **Model:** 370Z, Altima, Cube, Frontier, Maxima, Murano, Pathfinder, Quest, Rogue, Titan, Versa, Xterra **Engine:** 1.8L, 2.5L, 3.5L, 3.5L, 3.7L, 4.0L, 5.6L	**ASCD Vehicle Speed Sensor Malfunction (Includes Hybrid Models):** HYBRID MODELS CAUTION: Hybrid systems use very high-voltage battery systems. Before starting any service work involving the battery system, turn the ignition switch OFF and then remove the service plug from pocket in the trunk. After removing the service plug, wait 10 minutes before touching any of the high-voltage connectors and terminals. * The MIL will not illuminate for this self-diagnosis. **NOTE: If DTC P1574 is displayed with DTC UXXXX, first perform the trouble diagnosis for DTC UXXXX.** **NOTE: If DTC P1574 is displayed with DTC P0500, P0605 and/or P0607, first perform the trouble diagnosis for these DTCs before continuing with DTC P1574.** * ECM detects a difference between two vehicle speed signals is out of the specified range.
DTC: P1574 **2T ECM, MIL: Yes** **Year:** 2011 **Model:** Juke, Versa **Engine:** 1.6L	**ASCD VEHICLE SPEED SENSOR :** ECM detects a difference between two vehicle speed signals is out of the specified range. **NOTE: If DTC P1574 is displayed with DTC U1001, P0500, P0605 or P0607 perform the trouble diagnosis for them first.**
DTC: P1575 **T, MIL: Yes** **Year:** 2011 **Model:** Altima, Quest **Engine:** 2.5L, 3.5L	**Warning Buzzer Malfunction:** The hybrid vehicle control ECU receives a buzzer abnormal signal for 0.2 seconds or more while the dynamic radar cruise control is operating.
DTC: P1578 **T, MIL: Yes** **Year:** 2011 **Model:** Altima, Quest **Engine:** 2.5L, 3.5L	**Brake System Malfunction:** The hybrid vehicle control ECU receives a vehicle stability control system error signal for 0.2 seconds or more while the dynamic radar cruise control is operating.

DTC	Trouble Code Title and Conditions
DTC: P158A **T ECM** **Year:** 2011 **Model:** Juke, Versa **Engine:** 1.6L	**G SENSOR (G sensor calibration is incomplete) :** ECM detects a state that calibration of the G sensor is incomplete.
DTC: P159A **1T ECM, MIL: Yes** **Year:** 2011 **Model:** Juke, Versa **Engine:** 1.6L	**G SENSOR (G sensor circuit) :** The ECM detects the following status: A voltage signal transmitted from the G sensor s less than 0.5 V or more than 5.02 V continuously for 5 seconds or more.
DTC: P159B **1T ECM, MIL: Yes** **Year:** 2011 **Model:** Juke, Versa **Engine:** 1.6L	**G SENSOR (G sensor circuit range/performance) :** Every time when the vehicle is stopped, the ECM detects the following status 13 times in a row: A voltage signal transmitted from the G sensor is less than 2.275V or more than 2.725 V continuously for 5 seconds or more.
DTC: P159C **1T ECM, MIL: Yes** **Year:** 2011 **Model:** Juke, Versa **Engine:** 1.6L	**G SENSOR (G sensor circuit low input) :** When ECM detects the following status: A voltage signal transmitted from the G sensor is less than 0.5 V continuously for 5 seconds or more.
DTC: P159D **1T ECM, MIL: Yes** **Year:** 2011 **Model:** Juke, Versa **Engine:** 1.6L	**G SENSOR (G sensor circuit high input):** When ECM detects the following status: A voltage signal transmitted from the G sensor is more than 4.5 V continuously for 5 seconds or more.
DTC: P1604 **T BCM** **Year:** 2011 **Model:** Sentra **Engine:** 2.0L, 2.5L	**Steering Wheel Torque Sensor:** Perform self-diagnosis. Steering wheel: Not steering (there is no steering force).
DTC: P1604 **T BCM** **Year:** 2011 **Model:** Sentra **Engine:** 2.0L, 2.5L	**Steering Wheel (Column) Torque Sensor:** Steering wheel: Not steering (there is no steering force)
DTC: P1604 **1T BCM, MIL: Yes** **Year:** 2011 **Model:** Sentra **Engine:** 2.0L, 2.5L	**Steering Wheel/Column Torque Sensor:** Steering wheel: Not steering properly (lack of or too much steering force).
DTC: P1606 **T ECM, MIL: Yes** **Year:** 2011, 2012 **Model:** 370Z **Engine:** 3.7L	**Variable Valve Event & Lift (VVEL) Control Module:** VVEL control module calculation function is malfunctioning. VVEL EEPROM system is malfunctioning.
DTC: P1606-308 **T ECM, MIL: Yes** **Year:** 2011 **Model:** Altima, Quest **Engine:** 2.5L, 3.5L	**Collision Detection:** Shutoff signal from the airbag sensor assembly is determined.
DTC: P1606-317 **T ECM, MIL: Yes** **Year:** 2011 **Model:** Altima, Quest **Engine:** 2.5L, 3.5L	**Collision Detection:** A collision is determined due to a wiring malfunction.

DTC	Trouble Code Title and Conditions
DTC: P1607 **T** ECM, MIL: Yes **Year:** 2011, 2012 **Model:** 370Z **Engine:** 3.7L	**Variable Valve Event & Lift (VVEL) Control Module Circuit:** The internal circuit of the VVEL control module is malfunctioning.
DTC: P1608 **T** ECM, MIL: Yes **Year:** 2011, 2012 **Model:** 370Z **Engine:** 3.7L	**Variable Valve Event & Lift (VVEL) Sensor Power Supply Circuit:** VVEL control module detects a voltage of power source for sensor is excessively low or high.
DTC: P1610 **T** BCM, MIL: Yes **Year:** 2011, 2012 **Model:** 370Z, Cube, Frontier, Juke, Murano, Rogue, Versa, Xterra **Engine:** 1.6L, 1.8L, 2.5L, 3.5L, 3.7L, 4.0L	**Lock Mode:** * When the starting operation is carried out five or more times consecutively under the following conditions:
DTC: P1610 **T** BCM **Year:** 2011, 2012 **Model:** Altima, Maxima, Quest **Engine:** 2.5L, 3.5L	**NATS Antenna Amp. or Lock Mode:** Inactive communication between key slot and BCM.
DTC: P1611 **T** BCM **Year:** 2011, 2012 **Model:** 370Z, Altima, Cube, Frontier, Juke, Maxima, Murano, Quest, Rogue, Versa, Xterra **Engine:** 1.6L, 1.8L, 2.5L, 3.5L, 3.5L, 3.7L, 4.0L	**ID Discord – IMMU-ECM:** The ID verification results between BCM and ECM are NG. Registration is necessary. **NOTE: P1611 has the same meaning as B2192.**
DTC: P1612 **T** BCM **Year:** 2011, 2012 **Model:** 370Z, Altima, Cube, Frontier, Juke, Maxima, Murano, Quest, Rogue, Versa, Xterra **Engine:** 1.6L, 1.8L, 2.5L, 3.5L, 3.5L, 3.7L, 4.0L	**Chain of ECM-IMMU:** **NOTE: P1612 has the same meaning as B2193.** Inactive communication between ECM and BCM **NOTE: If DTC P1612 is displayed with DTC U1000 (for BCM), first perform the trouble diagnosis for DTC U1000**
DTC: P1614 **T** BCM **Year:** 2011, 2012 **Model:** 370Z, Frontier, Murano, Rogue, Xterra **Engine:** 2.5L, 3.5L, 3.7L, 4.0L	**Chain of IMMU-ECM:** Inactive communication between key slot or NATS antenna amp. and BCM. Mechanical key is malfunctioning.
DTC: P1615 **T** **Year:** 2011 **Model:** Altima, Quest **Engine:** 2.5L, 3.5L	**Communication Error from Distance Control ECU to ECM:** While the dynamic radar cruise control is either preparing for operation or operating, if communication data from the driving support ECU is logically inconsistent for a certain amount of time, the hybrid vehicle control ECU records this logical error code.
DTC: P1615 **T** BCM **Year:** 2011, 2012 **Model:** 370Z, Altima, Frontier, Maxima, Murano, Quest, Rogue, Xterra **Engine:** 2.5L, 3.5L, 3.5L, 3.7L, 4.0L	**Difference of Key:** The ID verification results between BCM and Intelligent Key or Mechanical Key are NG. Registration is necessary.

DTC	Trouble Code Title and Conditions
DTC: P1616 **T, MIL: Yes** **Year:** 2011 **Model:** Altima, Quest **Engine:** 2.5L, 3.5L	**Communication Error from ECM to Distance Control ECU:** While the dynamic radar cruise control is either preparing for operation or operating, if the hybrid vehicle control ECU continuously receives a logical error signal from the driving support ECU for more than a specific amount of time, the hybrid vehicle control ECU records this logical error code.
DTC: P1617 **T, MIL: Yes** **Year:** 2011 **Model:** Altima, Quest **Engine:** 2.5L, 3.5L	**Distance Control ECU Malfunction:** While the dynamic radar cruise control is either preparing for operation or operating, if a designation signal from the hybrid vehicle control ECU and a designation return signal from the driving support ECU do not match or the driving support ECU is malfunctioning for more than a specific amount of time, the hybrid vehicle control ECU records this DTC.
DTC: P1630 **T, MIL: Yes** **Year:** 2011 **Model:** Altima, Quest **Engine:** 2.5L, 3.5L	**Communication Error from VSC to ECM:** While the dynamic radar cruise control is either preparing for operation or operating, if communication data from the brake booster with master cylinder (skid control ECU) is logically inconsistent for a certain amount of time, the hybrid vehicle control ECU records this logical error code.
DTC: P1631 **T, MIL: Yes** **Year:** 2011 **Model:** Altima, Quest **Engine:** 2.5L, 3.5L	**Communication Error from ECM to VSC:** While the dynamic radar cruise control is either preparing for operation or operating, if the hybrid vehicle control ECU continuously receives a logical error signal from the brake booster with master cylinder (skid control ECU) for a certain amount of time, the hybrid vehicle control ECU records this logical error code.
DTC: P1650 **1T ECM, MIL: Yes** **Year:** 2011 **Model:** Juke, Versa **Engine:** 1.6L	**STARTER MOTOR RELAY 2 :** Starter relay is stuck ON. If DTC P1650 is displayed with DTC U1001,P0607, B209F, B20A0, B26F9 or B26FA perform the trouble diagnosis first. .
DTC: P1651 **1T ECM** **Year:** 2011 **Model:** Juke, Versa **Engine:** 1.6L	**STARTER MOTOR RELAY CIRCUIT:** A correlated error is detected for 2 seconds or more between a control signal transmitted from ECM and a feedback signal transmitted from IPDM E/R via CAN communication line.
DTC: P1652 **1T ECM, MIL: Yes** **Year:** 2011 **Model:** Juke, Versa **Engine:** 1.6L	**STARTER MOTOR SYSTEM (Starter motor communication line):** ECM detects malfunction in starter motor drive circuit of the IPDM E/R. **NOTE: If DTC P1650 is displayed with B209F or B20A0 of IPDM E/R, perform the trouble diagnosis for B209F, B20A0, B26F9 or B26FA first.**
DTC: P1700 **T TCM, TCIL: Yes** **Year:** 2011 **Model:** Cube, Murano, Sentra **Engine:** 1.8L, 2.0L, 2.5L, 3.5L	**CVT Control System:** This DTC is displayed with other DTC regarding TCM. Perform the trouble diagnosis for corresponding DTC.
DTC: P1701 **1T ECM** **Year:** 2011, 2012 **Model:** Maxima **Engine:** 3.5L	**VIAS Control Solenoid Valve 2 Circuit:** An excessively low or high voltage signal is sent to ECM via the VIAS control solenoid valve 2.
DTC: P1701 **T TCM, TCIL: Yes** **Year:** 2011, 2012 **Model:** Juke, Maxima, Murano, Rogue, Versa **Engine:** 1.6L, 1.8L, 2.5L, 3.5L, 3.5L	**Power Supply Circuit:** When the power supply to the TCM is cut off, for example because the battery is removed, and the self-diagnosis memory function stops. This is not a malfunction message (Whenever shutting off a power supply to the TCM, this message appears on the screen).

DTC	Trouble Code Title and Conditions
DTC: P1701 **T TCM, TCIL: Yes** **Year:** 2011 **Model:** Cube **Engine:** 1.8L	**Power Supply Circuit:** CAUTION: Immediately after TCM is replaced or after control valve or transaxle assembly is replaced (after TCM initialization is complete), self-diagnosis result of "P1701" may be displayed. In this case, erase selfdiagnosis result using CONSULT-III. After erasing self-diagnosis, perform reproduction procedures of DTC P1701 and check that a malfunction is not detected. Power supply (backup) of TCM is not supplied and learning function stops
DTC: P1701 **1T TCM, TCIL: Yes** **Year:** 2011 **Model:** Altima, Quest, Sentra **Engine:** 2.0L, 2.5L, 3.5L	**Transmission Control Module (Power Supply):** When the power supply to the TCM is cut OFF, for example because the battery is removed, and the self-diagnosis memory function stops, malfunction is detected. **NOTE: Since "P1701 TCM-POWER SUPPLY" will be indicated when replacing TCM, perform diagnosis after erasing "SELF-DIAG RESULTS"** This is not an OBD-II self-diagnostic item. Diagnostic trouble code "P1701 TCM-POWER SUPPLY" is detected when TCM does not receive the voltage signal from the battery power supply. This is not a malfunction message. (Whenever shutting OFF a power supply to the TCM, this message appears on the screen.)
DTC: P1705 **1T TCM, TCIL: Yes** **Year:** 2011, 2012 **Model:** Altima, Frontier, Maxima, Murano, Pathfinder, Quest, Rogue, Sentra, Titan, Versa, Xterra **Engine:** 1.6L, 1.8L, 2.0L, 2.5L, 3.5L, 3.5L, 4.0L, 5.6L	**Throttle Position/Accelerator Pedal Position Sensor Circuit:** Electric throttle control actuator consists of throttle control motor, accelerator pedal position sensor, throttle position sensor etc. The actuator sends a signal to the ECM, and ECM sends the signal to TCM with CAN communication. * This is not an OBD-II self-diagnostic item. * Diagnostic trouble code P1705 is detected when TCM does not receive the proper accelerator pedal position signals (input by CAN communication) from ECM.
DTC: P1705 **T TCM, TCIL: Yes** **Year:** 2011 **Model:** Cube **Engine:** 1.8L	**Accelerator Pedal Position Sensor Signal Circuit:** TCM detects that difference between 2 throttle opening signals (CAN communication) from ECM is 1/8 or more.
DTC: P1705 **T TCM, TCIL: Yes** **Year:** 2011, 2012 **Model:** 370Z **Engine:** 3.7L	**Accelerator Pedal Position (APP) Sensor Signal Circuit:** TCM detects improper accelerator pedal position signals received from ECM via CAN communication.
DTC: P1710 **T TCM, TCIL: Yes** **Year:** 2011, 2012 **Model:** Frontier, Pathfinder, Xterra **Engine:** 2.5L, 4.0L, 5.6L	**A/T Fluid Temperature Sensor Circuit:** * This is an OBD-II self-diagnostic item. * Diagnostic trouble code P1710 will be detected when TCM receives an excessively low or high voltage from the sensor. * A/T fluid temperature does not rise to the specified temperature while driving.
DTC: P1715 **1T ECM, TCIL: Yes** **Year:** 2011 **Model:** Pathfinder **Engine:** 4.0L, 5.6L	**Turbine Revolution Sensor:** This is an OBD-II self-diagnostic item. Diagnostic trouble code P0717 is detected under the following conditions: - When TCM does not receive the proper voltage signal from the sensor. - When TCM detects an irregularity only at position of 4th gear for turbine revolution sensor 2.
DTC: P1715 **1T TCM** **Year:** 2011, 2012 **Model:** 370Z, Altima, Cube, Frontier, Juke, Maxima, Murano, Pathfinder, Quest, Sentra, Versa, Xterra **Engine:** 1.6L, 1.8L, 2.0L, 2.5L, 3.5L, 3.5L, 3.7L, 4.0L, 5.6L	**Input Speed Sensor (Primary Speed Sensor/TCM Output):** * The MIL will not illuminate for this self-diagnosis. **NOTE: If DTC P1715 is displayed with DTC UXXXX, first perform the trouble diagnosis for DTC UXXXX.** **NOTE: If DTC P1715 is displayed with DTC P0335, P0340, P0605 and/or P0607, first perform the trouble diagnosis for the appropriate DTC before proceeding with P1715 diagnosis.** * Sensor signal is different from the theoretical value calculated by ECM from secondary speed sensor signal and engine rpm signal.
DTC: P1720 **1T ECM, MIL: Yes** **Year:** 2011, 2012 **Model:** Maxima, Murano **Engine:** 3.5L, 3.5L	**Vehicle Speed Sensor (TCM Output):** **NOTE: If DTC P1720 is displayed with DTC UXXXX first perform the trouble diagnosis for DTC UXXXX.** **NOTE: If DTC P1720 is displayed with DTC P0607, first perform the trouble diagnosis for DTC P0607.** The difference between two vehicle speed signals is out of the specified range.

DTC	Trouble Code Title and Conditions
DTC: P1721 **T TCM, TCIL: Yes** **Year:** 2011, 2012 **Model:** Frontier, Pathfinder, Titan, Xterra **Engine:** 2.5L, 4.0L, 5.6L	**Vehicle Speed Signal:** * This is not an OBD-II self-diagnostic item. * Diagnostic trouble code P1721 is detected when TCM does not receive the proper vehicle speed sensor MTR signal (input by CAN communication) from combination meter (unified meter and A/C amp).
DTC: P1721 **T TCM, TCIL: Yes** **Year:** 2011, 2012 **Model:** 370Z **Engine:** 3.7L	**Vehicle Speed Signal Circuit:** * The vehicle speed signal recognizes that the vehicle speed is 5 km/h (3 MPH) or less even if the output speed sensor recognizes that the vehicle speed is 20 km/h (12 MPH) or more. (Only when starts after the ignition switch is turned ON.) * The vehicle speed recognized by the vehicle speed signal decelerates 36 km/h (23 MPH) or more during 60 msec when the vehicle speed signal recognizes that the vehicle speed is 36 km/h (23 MPH) or more and the output speed sensor recognizes that the vehicle speed is 24 km/h (15 MPH) or less. * The vehicle speed of vehicle speed signal decelerates 36 km/h (23 MPH) or more even if the vehicle speed of output speed sensor accelerates or decelerates 24 km/h (15 MPH) or less during 60 msec when the vehicle speed sensor recognizes that the vehicle speed is 36 km/h (23 MPH) or more.
DTC: P1722 **1T TCM, TCIL: Yes** **Year:** 2011, 2012 **Model:** Altima, Juke, Maxima, Murano, Quest, Rogue, Versa **Engine:** 1.6L, 1.8L, 2.5L, 3.5L, 3.5L	**Vehicle Speed Signal Circuit:** CAN communication with the ABS actuator and the electric unit (control unit) is malfunctioning. There is a great difference between the vehicle speed signal from the ABS actuator and the electric unit (control unit), and the vehicle speed sensor signal.
DTC: P1722 **1T TCM, TCIL: Yes** **Year:** 2011 **Model:** Sentra **Engine:** 2.0L, 2.5L	**ESTM Vehicle Speed Signal:** The vehicle speed signal is transmitted from ABS actuator and electric unit (control unit) to TCM by CAN communication line. This is not an OBD-II self-diagnostic item. Diagnostic trouble code "P1722 ESTM VEH SPD SIG", with scan tool, is detected when TCM does not receive the proper vehicle speed signal (input by CAN communication) from ABS actuator and electric unit (control unit).
DTC: P1722 **T TCM, TCIL: Yes** **Year:** 2011 **Model:** Cube **Engine:** 1.8L	**Vehicle Speed Signal Circuit:** TCM detects a malfunction of CAN communication between ABS actuator and electric unit (control unit) When vehicle speed that TCM detects is 10 km/h (7 MPH) or more, vehicle speed signal (CAN signal) that is received from ABS actuator and electric unit (control unit) is less than 2 km/h (1 MPH) Change of vehicle speed signal (CAN communication) that TCM receives is large
DTC: P1723 **T TCM, TCIL: Yes** **Year:** 2011 **Model:** Cube **Engine:** 1.8L	**Speed Sensor:** When noise (pulse) that is generated because of connection malfunction caused by primary speed sensor and secondary speed sensor harness and others is detected, it is judged that a malfunction occurs. TCM detects that high frequency elements that are extracted from primary pulley speed and secondary pulley speed exceed a certain value
DTC: P1723 **1T TCM, TCIL: Yes** **Year:** 2011 **Model:** Altima, Juke, Quest, Rogue, Sentra, Versa **Engine:** 1.6L, 1.8L, 2.0L, 2.5L, 3.5L	**CVT Speed Sensor Circuit:** The vehicle speed sensor CVT [output speed sensor (secondary speed sensor)] detects the revolution of the parking gear and generates a pulse signal. The pulse signal is sent to the TCM, which converts it into vehicle speed. The input speed sensor (primary speed sensor) detects the primary pulley revolution speed and sends a signal to the TCM. Diagnostic trouble code P1723 is detected when there is a great difference between the vehicle speed signal and the secondary speed sensor signal. **NOTE:** P0720, P0715 or P0725 is displayed with the DTC at the same time.
DTC: P1726 **T TCM, TCIL: Yes** **Year:** 2011, 2012 **Model:** Altima, Juke, Maxima, Murano, Quest, Rogue, Sentra, Versa **Engine:** 1.6L, 1.8L, 2.0L, 2.5L, 3.5L, 3.5L	**Throttle Control Signal Circuit:** The electronically controlled throttle for ECM is malfunctioning.
DTC: P1726 **T TCM, TCIL: Yes** **Year:** 2011 **Model:** Cube **Engine:** 1.8L	**Throttle Control Signal Circuit:** TCM receives a malfunction signal of engine system from ECM

DTC	Trouble Code Title and Conditions
DTC: P1730 **T TCM, TCIL: Yes** **Year:** 2011, 2012 **Model:** Frontier, Pathfinder, Titan, Xterra **Engine:** 2.5L, 4.0L, 5.6L	**A/T Interlock:** * This is an OBD-II self-diagnostic item. * Diagnostic trouble code P1730 is detected when TCM does not receive the proper voltage signal from the sensor and switch. * TCM monitors and compares gear position and conditions of each ATF pressure switch when gear is steady.
DTC: P1730 **T TCM, TCIL: Yes** **Year:** 2011, 2012 **Model:** 370Z **Engine:** 3.7L	**Interlock:** The output sensor detects the deceleration of 12 km/h (7 MPH) or more for 1 second.
DTC: P1731 **T TCM, TCIL: Yes** **Year:** 2011, 2012 **Model:** Frontier, Pathfinder, Titan, Xterra **Engine:** 2.5L, 4.0L, 5.6L	**A/T 1st Engine Braking:** * This is not an OBD-II self-diagnostic item. * Diagnostic trouble code P1731 is detected under the following conditions. - When TCM does not receive the proper voltage signal from the sensor. - When TCM monitors each ATF pressure switch and solenoid monitor value, and detects as irregular when engine brake of 1st gear acts other than at "1" position.
DTC: P1734 **T TCM, TCIL: Yes** **Year:** 2011, 2012 **Model:** 370Z **Engine:** 3.7L	**7th Gear Incorrect Ratio:** DTC is set when any inconsistency is recognized in gear ratio.
DTC: P1740 **1T TCM, TCIL: Yes** **Year:** 2011, 2012 **Model:** Altima, Juke, Maxima, Murano, Quest, Rogue, Versa **Engine:** 1.6L, 1.8L, 2.5L, 3.5L, 3.5L	**Lock-Up Select Solenoid Valve:** Normal voltage not applied to solenoid due to cut line, short, or the like. TCM detects as irregular by comparing target value with monitor value.
DTC: P1740 **T TCM, TCIL: Yes** **Year:** 2011 **Model:** Cube **Engine:** 1.8L	**Lock-Up Select Solenoid Valve Circuit:** Lock-up select solenoid valve monitor value is OFF when lock-up select solenoid valve command value of TCM is ON Lock-up select solenoid valve monitor value is ON when lock-up select solenoid valve command value of TCM is OFF
DTC: P1740 **1T TCM, TCIL: Yes** **Year:** 2011 **Model:** Sentra **Engine:** 2.0L, 2.5L	**Lock-Up Select Solenoid Valve Circuit:** The lock-up select solenoid valve controls lock-up clutch pressure or forward clutch pressure (reverse brake pressure). When controlling lock-up clutch, the valve is turned OFF. When controlling forward clutch, it is turned ON. This is an OBD-II self-diagnostic item. Diagnostic trouble code P1740 is detected under the following conditions: - When TCM compares target value with monitor value and detects an irregularity.
DTC: P1745 **1T TCM, TCIL: Yes** **Year:** 2011, 2012 **Model:** Altima, Juke, Maxima, Quest, Rogue, Sentra, Versa **Engine:** 1.6L, 1.8L, 2.0L, 2.5L, 3.5L	**Line Pressure Control:** The pressure control solenoid valve A (line pressure solenoid valve) regulates the oil pump discharge pressure to suit the driving condition in response to a signal sent from the TCM. This is not an OBD-II self-diagnostic item. Diagnostic trouble code P1745 is detected when TCM detects the unexpected line pressure.
DTC: P1752 **T TCM, TCIL: Yes** **Year:** 2011, 2012 **Model:** Frontier, Pathfinder, Titan, Xterra **Engine:** 2.5L, 4.0L, 5.6L	**Input Clutch Solenoid Valve:** * This is an OBD-II self-diagnostic item. * Diagnostic trouble code P1752 is detected under the following conditions: - When TCM detects an improper voltage drop when it tries to operate the solenoid valve. - When TCM detects as irregular by comparing target value with monitor value.

DTC	Trouble Code Title and Conditions
DTC: P1754 **T TCM, TCIL: Yes** **Year:** 2011 **Model:** Pathfinder **Engine:** 4.0L, 5.6L	**Input Clutch Solenoid Valve Function:** * This is an OBD-II self-diagnostic item. * Diagnostic trouble code P1754 is detected under the following conditions: - When TCM detects that actual gear ratio is irregular, and relation between gear position and condition of pressure switch 3 is irregular during depressing accelerator pedal. (Other than during shift change) - When TCM detects that relation between gear position and condition of ATF pressure switch 3 is irregular during releasing accelerator pedal. (Other than during shift change)
DTC: P1757 **T TCM, TCIL: Yes** **Year:** 2011, 2012 **Model:** Frontier, Pathfinder, Titan, Xterra **Engine:** 2.5L, 4.0L, 5.6L	**Front Brake Solenoid Valve:** * This is an OBD-II self-diagnostic item. * Diagnostic trouble code P1757 is detected under the following conditions: - When TCM detects an improper voltage drop when it tries to operate the solenoid valve. - When TCM detects as irregular by comparing target value with monitor value.
DTC: P1759 **T TCM, TCIL: Yes** **Year:** 2011 **Model:** Pathfinder **Engine:** 4.0L, 5.6L	**Front Brake Solenoid Valve Function:** * This is an OBD-II self-diagnostic item. * Diagnostic trouble code P1759 is detected under the following conditions: - When TCM detects that actual gear ratio is irregular, and relation between gear position and condition of ATF pressure switch 1 is irregular during depressing accelerator pedal. (Other than during shift change) - When TCM detects that relation between gear position and condition of ATF pressure switch 1 is irregular during releasing accelerator pedal. (Other than during shift change)
DTC: P1762 **T TCM, TCIL: Yes** **Year:** 2011, 2012 **Model:** Frontier, Pathfinder, Titan, Xterra **Engine:** 2.5L, 4.0L, 5.6L	**Direct Clutch Solenoid Valve:** * This is an OBD-II self-diagnostic item. * Diagnostic trouble code P1762 will be detected under the following conditions: - When TCM detects an improper voltage drop when it tries to operate the solenoid valve. - When TCM detects as irregular by comparing target value with monitor value.
DTC: P1764 **T TCM, TCIL: Yes** **Year:** 2011 **Model:** Pathfinder **Engine:** 4.0L, 5.6L	**Direct Clutch Solenoid Valve Function:** * This is an OBD-II self-diagnostic item. * Diagnostic trouble code P1764 is detected under the following conditions: - When TCM detects that actual gear ratio is irregular, and relation between gear position and condition of ATF pressure switch 5 is irregular during depressing accelerator pedal. (Other than during shift change) - When TCM detects that relation between gear position and condition of ATF pressure switch 5 is irregular during releasing accelerator pedal. (Other than during shift change)
DTC: P1767 **T TCM, TCIL: Yes** **Year:** 2011, 2012 **Model:** Frontier, Pathfinder, Titan, Xterra **Engine:** 2.5L, 4.0L, 5.6L	**High & Low Reverse Clutch Solenoid Valve:** This is an OBD-II self-diagnostic item. Diagnostic trouble code P1767 will be detected under the following conditions: - When TCM detects an improper voltage drop when it tries to operate the solenoid valve. - When TCM detects as irregular by comparing target value with monitor value.
DTC: P1769 **T TCM, TCIL: Yes** **Year:** 2011 **Model:** Pathfinder **Engine:** 4.0L, 5.6L	**High/Low Reverse Clutch Solenoid Valve Function:** * This is an OBD-II self-diagnostic item. * Diagnostic trouble code P1769 is detected under the following conditions: - When TCM detects that actual gear ratio is irregular, and relation between gear position and condition of ATF pressure switch 6 is irregular during depressing accelerator pedal. (Other than during shift change) - When TCM detects that relation between gear position and condition of ATF pressure switch 6 is irregular during releasing accelerator pedal. (Other than during shift change)
DTC: P1772 **T TCM, TCIL: Yes** **Year:** 2011, 2012 **Model:** Frontier, Pathfinder, Titan, Xterra **Engine:** 2.5L, 4.0L, 5.6L	**Low Coast Brake Solenoid Valve:** * This is an OBD-II self-diagnostic item. * Diagnostic trouble code P1772 will be set when the TCM detects an improper voltage drop when it tries to operate the solenoid valve.

DTC	Trouble Code Title and Conditions
DTC: P1774 **T TCM, TCIL: Yes** **Year:** 2011, 2012 **Model:** Frontier, Pathfinder, Titan, Xterra **Engine:** 2.5L, 4.0L, 5.6L	**Low Coast Brake Solenoid Valve Function:** * This is an OBD-II self-diagnostic item. * Diagnostic trouble code P1774 will be detected under the following conditions: - TCM detects an improper voltage drop when it tries to operate the solenoid valve. - When TCM detects that actual gear ratio is irregular, and relation between gear position and condition of ATF pressure switch 2 is irregular during depressing accelerator pedal (other than during shift change). - When TCM detects that relation between gear position and condition of ATF pressure switch 2 is irregular during releasing accelerator pedal. (Other than during shift change)
DTC: P1777 **T TCM, TCIL: Yes** **Year:** 2011, 2012 **Model:** Maxima, Murano, Rogue **Engine:** 2.5L, 3.5L, 3.5L	**Step Motor Circuit:** Each coil of the step motor is not energized properly due to an open or a short.
DTC: P1777 **T TCM, TCIL: Yes** **Year:** 2011 **Model:** Cube **Engine:** 1.8L	**Step Motor Circuit:** Step motor monitor value is OFF when step motor command value of TCM is ON Step motor monitor value is ON when step motor command value of TCM is OFF
DTC: P1777 **1T TCM, TCIL: Yes** **Year:** 2011 **Model:** Altima, Juke, Quest, Sentra, Versa **Engine:** 1.6L, 1.8L, 2.0L, 2.5L, 3.5L	**Step Motor Circuit:** The step motor changes the step with turning 4 coils ON/OFF according to the signal from TCM. As a result, the flow of line pressure to primary pulley is changed and pulley ratio is controlled. This is an OBD-II self-diagnostic item. Diagnostic trouble code P1777 is detected under the following conditions: - When operating step motor ON and OFF, there is no proper change in the voltage of TCM terminal which corresponds to it.
DTC: P1778 **1T TCM, TCIL: Yes** **Year:** 2011, 2012 **Model:** Altima, Juke, Maxima, Murano, Quest, Rogue, Versa **Engine:** 1.6L, 1.8L, 2.5L, 3.5L, 3.5L	**Step Motor Circuit Malfunction:** There is a great difference between the number of steps for the stepping motor and for the actual gear ratio, when not changing the pulley ratio according to the instruction of TCM.
DTC: P1778 **T, TCIL: Yes** **Year:** 2011 **Model:** Cube **Engine:** 1.8L	**Step Motor Circuit Intermittent:** This DTC is not caused by an electrical malfunction (circuit open or short) but is caused by a mechanical malfunction (control valve clogging, solenoid valve sticking, and others). TCM detects that primary speed sensor value and primary pulley speed estimated from secondary speed sensor are in a deviated state, and target pulley ratio and actual pulley ratio are in a deviated state
DTC: P1778 **1T TCM, TCIL: Yes** **Year:** 2011 **Model:** Sentra **Engine:** 2.0L, 2.5L	**Step Motor Function:** The step motor's 4 aspects of ON/OFF change according to the signal from TCM. As a result, the flow of line pressure to primary pulley is changed and pulley ratio is controlled. This diagnosis item is detected when electrical system is OK, but mechanical system is NG. This diagnosis item is detected when the state of the changing the speed mechanism in unit does not operate normally. This is an OBD-II self-diagnostic item. Diagnostic trouble code P1778 is detected under the following conditions: - When not changing the pulley ratio according to the instruction of TCM.
DTC: P1800 **1T ECM** **Year:** 2011, 2012 **Model:** Frontier, Maxima, Murano, Pathfinder, Xterra **Engine:** 3.5L, 3.5L, 4.0L, 5.6L	**VIAS Control Solenoid Valve Circuit :** **NOTE: On engines with two valves, this DTC relates to Valve 1.** * The MIL will not illuminate for this self-diagnosis. * An excessively low or high voltage signal is sent to ECM through the valve.
DTC: P1801 **T TCM, TCIL: Yes** **Year:** 2011 **Model:** Murano **Engine:** 3.5L	**VIAS Control Solenoid Valve 2 Circuit:** An excessively low or high voltage signal is sent to ECM through the VIAS control solenoid valve 2.

DTC	Trouble Code Title and Conditions
DTC: P1805 **1T ECM** **Year:** 2011, 2012 **Model:** All **Engine:** 1.6L, 1.8L, 2.0L, 2.5L, 3.5L, 3.5L, 3.7L, 4.0L, 5.6L	**Brake Switch Signal Malfunction (Includes Hybrid Models):** HYBRID MODELS CAUTION: Hybrid systems use very high-voltage battery systems. Before starting any service work involving the battery system, turn the ignition switch OFF and then remove the service plug from pocket in the trunk. After removing the service plug, wait 10 minutes before touching any of the high-voltage connectors and terminals. * The MIL may not illuminate for this self-diagnosis. * A brake switch signal is not sent to ECM for extremely long time while the vehicle is being driven.
DTC: P1815 **T TCM, TCIL: Yes** **Year:** 2011, 2012 **Model:** 370Z, Pathfinder, Titan **Engine:** 3.7L, 4.0L, 5.6L	**Manual Mode Switch Circuit:** * TCM monitors manual mode, non manual mode, up or down switch signal, and detects as irregular when impossible input pattern occurs 2 seconds or more. * Shift up/down signal of paddle shifter continuously remains ON for 60 seconds.
DTC: P181A-596 **T ECM, MIL: Yes** **Year:** 2011 **Model:** Altima, Quest **Engine:** 2.5L, 3.5L	**Gear Lever X Position Circuit "A" / "B" Correlation:** Difference between select main sensor value and select sub sensor value is large.
DTC: P181B-595 **T ECM, MIL: Yes** **Year:** 2011 **Model:** Altima, Quest **Engine:** 2.5L, 3.5L	**Gear Lever Y Position Circuit "A" / "B" Correlation:** Difference between shift main sensor value and shift sub sensor value is large.
DTC: P182B-577 **T ECM, MIL: Yes** **Year:** 2011 **Model:** Altima, Quest **Engine:** 2.5L, 3.5L	**Gear Lever X Position "B" Circuit Low:** Open or GND short in select sub sensor circuit
DTC: P182C-578 **T ECM, MIL: Yes** **Year:** 2011 **Model:** Altima, Quest **Engine:** 2.5L, 3.5L	**Gear Lever X Position "B" Circuit High:** +B short in select sub sensor circuit
DTC: P182E-573 **T ECM, MIL: Yes** **Year:** 2011 **Model:** Altima, Quest **Engine:** 2.5L, 3.5L	**Gear Lever Y Position "B" Circuit Low:** Open or GND short in shift sub sensor circuit
DTC: P182F-574 **T ECM, MIL: Yes** **Year:** 2011 **Model:** Altima, Quest **Engine:** 2.5L, 3.5L	**Gear Lever Y Position "B" Circuit High:** +B short in shift sub sensor circuit
DTC: P1841 **T TCM, TCIL: Yes** **Year:** 2011 **Model:** Pathfinder **Engine:** 4.0L, 5.6L	**ATF Pressure Switch 1:** * This is not an OBD-II self-diagnostic item. * Diagnostic trouble code P1841 is detected when TCM detects that actual gear ratio is normal, and relation between gear position and condition of ATF pressure switch 1 is irregular during depressing accelerator pedal. (Other than during shift change)
DTC: P1843 **T TCM, TCIL: Yes** **Year:** 2011 **Model:** Pathfinder **Engine:** 4.0L, 5.6L	**ATF Pressure Switch 3:** * This is not an OBD-II self-diagnostic item. * Diagnostic trouble code P1843 is detected when TCM detects that actual gear ratio is normal, and relation between gear position and condition of ATF pressure switch 3 is irregular during depressing accelerator pedal. (Other than during shift change)
DTC: P1845 **T TCM, TCIL: Yes** **Year:** 2011 **Model:** Pathfinder **Engine:** 4.0L, 5.6L	**ATF Pressure Switch 5:** * This is not an OBD-II self-diagnostic item. * Diagnostic trouble code P1845 is detected when TCM detects that actual gear ratio is normal, and relation between gear position and condition of ATF pressure switch 5 is irregular during depressing accelerator pedal. (Other than during shift change)

DTC	Trouble Code Title and Conditions
DTC: P1846 **T TCM, TCIL:** Yes **Year:** 2011 **Model:** Pathfinder **Engine:** 4.0L, 5.6L	**ATF Pressure Switch 6:** * This is not an OBD-II self-diagnostic item. * Diagnostic trouble code P1846 is.detected when TCM detects that actual gear ratio is normal, and relation between gear position and condition of ATF pressure switch 6 is irregular during depressing accelerator pedal. (Other than during shift change)

OBD II Trouble Code List (P2XXX Codes)

DTC	Trouble Code Title and Conditions
DTC: P2004 **1T ECM, MIL:** Yes **Year:** 2011 **Model:** Rogue **Engine:** 2.5L	**Tumble Control Valve Stuck:** The target angle of tumble control valve controlled by ECM and the input signal from tumble control valve position sensor is not in the normal range.
DTC: P2014 **1T ECM, MIL:** Yes **Year:** 2011 **Model:** Rogue **Engine:** 2.5L	**Tumble Control Valve Position Sensor Circuit:** An excessively low or high voltage from the sensor is sent to ECM.
DTC: P2100 **1T ECM, MIL:** Yes **Year:** 2011, 2012 **Model:** 370Z, Altima, Cube, Frontier, Juke, Maxima, Murano, Pathfinder, Quest, Rogue, Sentra, Titan, Versa **Engine:** 1.6L, 1.8L, 2.0L, 2.5L, 3.5L, 3.5L, 3.7L, 4.0L, 5.6L	**Throttle Control Motor Relay Circuit is Open (Includes Hybrid Models):** HYBRID MODELS CAUTION: Hybrid systems use very high-voltage battery systems. Before starting any service work involving the battery system, turn the ignition switch OFF and then remove the service plug from pocket in the trunk. After removing the service plug, wait 10 minutes before touching any of the high-voltage connectors and terminals. **NOTE: On and, this DTC is for Bank 1.** * These self-diagnoses have the one trip detection logic. * ECM detects that the voltage of power source for throttle control motor is excessively low.
DTC: P2101 **1T ECM, MIL:** Yes **Year:** 2011, 2012 **Model:** All **Engine:** 1.6L, 1.8L, 2.0L, 2.5L, 3.5L, 3.5L, 3.7L, 4.0L, 5.6L	**Electric Throttle Control Performance (Includes Hybrid Models):** HYBRID MODELS CAUTION: Hybrid systems use very high-voltage battery systems. Before starting any service work involving the battery system, turn the ignition switch OFF and then remove the service plug from pocket in the trunk. After removing the service plug, wait 10 minutes before touching any of the high-voltage connectors and terminals. **NOTE: On and, this DTC refers to Bank 1.** **NOTE: If DTC P1233 or P2101 is displayed with DTC P1238, P1290, P2100 or 2119, first perform the trouble diagnosis for DTC P1238, P2119 or P1290, P2100.** * Electric throttle control function does not operate properly.
DTC: P2102 **1T ECM, MIL:** Yes **Year:** 2011 **Model:** Altima, Quest **Engine:** 2.5L, 3.5L	**Throttle Actuator Control Motor Circuit Low:** Monitor runs whenever following DTCs are not present: None Throttle actuator: ON Duty-cycle ratio to open throttle actuator: 80% or more Throttle actuator power supply: 8 V or higher Motor current change during latest 0.016 seconds: Less than 0.2 A
DTC: P2103 **1T ECM, MIL:** Yes **Year:** 2011, 2012 **Model:** All **Engine:** 1.6L, 1.8L, 2.0L, 2.5L, 3.5L, 3.5L, 3.7L, 4.0L, 5.6L	**Throttle Control Motor Relay Circuit is Short (Includes Hybrid Models):** HYBRID MODELS CAUTION: Hybrid systems use very high-voltage battery systems. Before starting any service work involving the battery system, turn the ignition switch OFF and then remove the service plug from pocket in the trunk. After removing the service plug, wait 10 minutes before touching any of the high-voltage connectors and terminals. * ECM detects the throttle control motor relay is stuck ON.
DTC: P2111 **1T ECM, MIL:** Yes **Year:** 2011 **Model:** Altima, Quest **Engine:** 2.5L, 3.5L	**Throttle Actuator Control System - Stuck Open:** Monitor runs whenever following DTCs are not present: None All of the following conditions are met: System guard* judge condition: ON Throttle actuator current: 2 A or more Duty-cycle to close throttle: 80% or more

DTC	Trouble Code Title and Conditions
DTC: P2112 **1T ECM, MIL: Yes** **Year:** 2011 **Model:** Altima, Quest **Engine:** 2.5L, 3.5L	**Throttle Actuator Control System - Stuck Closed:** Monitor runs whenever following DTCs are not present: None All of the following conditions are met: System guard* judge condition: ON Throttle actuator current: 2 A or more Duty-cycle to open throttle: 80% or more *: System guard is ON when the following conditions are set: Throttle actuator: ON Throttle actuator duty calculation: Executing Throttle position sensor fail: Not detected Throttle actuator current-cut operation: Not executing Throttle actuator power supply: 4 V or more Throttle actuator fail: Not detected
DTC: P2118 **1T ECM, MIL: Yes** **Year:** 2011 **Model:** Altima, Quest **Engine:** 2.5L, 3.5L	**Throttle Actuator Control Motor Current Range / Performance:** Monitor runs whenever following DTCs are not present: None Battery voltage: 8 V or higher Electronic throttle actuator power: ON
DTC: P2118 **1T ECM, MIL: Yes** **Year:** 2011, 2012 **Model:** All **Engine:** 1.6L, 1.8L, 2.0L, 2.5L, 3.5L, 3.5L, 3.7L, 4.0L, 5.6L	**Throttle Control Motor Circuit Short (Includes Hybrid Models):** HYBRID MODELS CAUTION: Hybrid systems use very high-voltage battery systems. Before starting any service work involving the battery system, turn the ignition switch OFF and then remove the service plug from pocket in the trunk. After removing the service plug, wait 10 minutes before touching any of the high-voltage connectors and terminals. **NOTE: On and, this DTC is for Bank 1.** * ECM detects short in both circuits between ECM and throttle control motor.
DTC: P2119 **1T ECM, MIL: Yes** **Year:** 2011, 2012 **Model:** All **Engine:** 1.6L, 1.8L, 2.0L, 2.5L, 3.5L, 3.5L, 3.7L, 4.0L, 5.6L	**Electric Throttle Control Actuator (Includes Hybrid Models):** HYBRID MODELS CAUTION: Hybrid systems use very high-voltage battery systems. Before starting any service work involving the battery system, turn the ignition switch OFF and then remove the service plug from pocket in the trunk. After removing the service plug, wait 10 minutes before touching any of the high-voltage connectors and terminals. **NOTE: When the malfunction is detected, ECM enters fail-safe mode and the MIL illuminates.** **NOTE: On and, this DTC is for Bank 1.** * Malfunction A: Electric throttle control actuator does not function properly due to the return spring malfunction. ECM controls the electric throttle actuator by regulating the throttle opening around the idle position. The engine speed will not rise more than 2,000 rpm. * Malfunction B: Throttle valve opening angle in fail-safe mode is not in specified range. ECM controls the electric throttle control actuator by regulating the throttle opening to 20 degrees or less. * Malfunction C: ECM detects the throttle valve is stuck open. While the vehicle is driving, it slows down gradually by fuel cut. After the vehicle stops, the engine stalls. * The engine can restart in N or P position (CVT), neutral (M/T), and engine speed will not exceed 1,000 rpm or more.
DTC: P2119 **1T ECM, MIL: Yes** **Year:** 2011 **Model:** Altima, Quest **Engine:** 2.5L, 3.5L	**Throttle Actuator Control Throttle Body Range / Performance:** Monitor runs whenever following DTCs are not present: None System guard* judge condition: ON *System guard is ON when the following conditions set: Throttle actuator: ON Throttle actuator duty calculation: Executing Throttle position sensor fail: Not detected Throttle actuator current-cut operation: Not executing Throttle actuator power supply: 4 V or more Throttle actuator fail: Not detected
DTC: P2120-152 **T ECM, MIL: Yes** **Year:** 2011 **Model:** Altima, Quest **Engine:** 2.5L, 3.5L	**Throttle / Pedal Position Sensor / Switch "D" Circuit:** Main sensor circuit wiring malfunction or level is not stable
DTC: P2121-160 **T ECM, MIL: Yes** **Year:** 2011 **Model:** Altima, Quest **Engine:** 2.5L, 3.5L	**Throttle / Pedal Position Sensor / Switch "D" Circuit Range / Performance:** Internal error of the main sensor.

DTC	Trouble Code Title and Conditions
DTC: P2122 **1T ECM, MIL: Yes** **Year:** 2011, 2012 **Model:** All **Engine:** 1.6L, 1.8L, 2.0L, 2.5L, 3.5L, 3.5L, 3.7L, 4.0L, 5.6L	**Accelerator Pedal Position Sensor 1 Circuit Low Input:** NOTE: If DTC P2122 or P2123 is displayed with DTC P0643, first perform the trouble diagnosis for DTC P0643. * An excessively low voltage from the APP sensor 1 is sent to ECM.
DTC: P2122-104 **T ECM, MIL: Yes** **Year:** 2011 **Model:** Altima, Quest **Engine:** 2.5L, 3.5L	**Throttle / Pedal Position Sensor / Switch "D" Circuit Low Input:** Open or short to GND in the main sensor circuit.
DTC: P2123 **1T ECM, MIL: Yes** **Year:** 2011, 2012 **Model:** All **Engine:** 1.6L, 1.8L, 2.0L, 2.5L, 3.5L, 3.5L, 3.7L, 4.0L, 5.6L	**Accelerator Pedal Position Sensor 1 Circuit High Input:** NOTE: If DTC P2122 or P2123 is displayed with DTC P0643, first perform the trouble diagnosis for DTC P0643. * An excessively high voltage from the APP sensor 1 is sent to ECM. * When the malfunction is detected, ECM enters fail-safe mode and the MIL illuminates.
DTC: P2123-105 **T ECM, MIL: Yes** **Year:** 2011 **Model:** Altima, Quest **Engine:** 2.5L, 3.5L	**Throttle / Pedal Position Sensor / Switch "D" Circuit High Input:** Short to +B in the main sensor circuit.
DTC: P2125-153 **T ECM, MIL: Yes** **Year:** 2011 **Model:** Altima, Quest **Engine:** 2.5L, 3.5L	**Throttle / Pedal Position Sensor / Switch "E" Circuit:** Sub sensor circuit wiring malfunction or level is not stable.
DTC: P2126-109 **T ECM, MIL: Yes** **Year:** 2011 **Model:** Altima, Quest **Engine:** 2.5L, 3.5L	**Throttle / Pedal Position Sensor / Switch "E" Circuit Range / Performance:** Internal error of the sub sensor.
DTC: P2127 **1T ECM, MIL: Yes** **Year:** 2011, 2012 **Model:** All **Engine:** 1.6L, 1.8L, 2.0L, 2.5L, 3.5L, 3.5L, 3.7L, 4.0L, 5.6L	**Accelerator Pedal Position Sensor 2 Circuit Low Input:** * An excessively low voltage from the APP sensor 2 is sent to ECM. * When the malfunction is detected, ECM enters fail-safe mode and the MIL illuminates.
DTC: P2127-109 **T ECM, MIL: Yes** **Year:** 2011 **Model:** Altima, Quest **Engine:** 2.5L, 3.5L	**Throttle / Pedal Position Sensor / Switch "E" Circuit Low Input:** Open or short to GND in the sub sensor circuit.
DTC: P2128 **1T ECM, MIL: Yes** **Year:** 2011, 2012 **Model:** 370Z, Altima, Cube, Frontier, Juke, Maxima, Murano, Quest, Rogue, Sentra, Titan, Versa, Xterra **Engine:** 1.6L, 1.8L, 2.0L, 2.5L, 3.5L, 3.5L, 3.7L, 4.0L, 5.6L	**Accelerator Pedal Position Sensor 2 Circuit High Input:** * An excessively high voltage from the APP sensor 2 is sent to ECM. * When the malfunction is detected, ECM enters fail-safe mode and the MIL illuminates.
DTC: P2128-108 **T ECM, MIL: Yes** **Year:** 2011 **Model:** Altima, Quest **Engine:** 2.5L, 3.5L	**Throttle / Pedal Position Sensor / Switch "E" Circuit High Input:** Short to +B in the sub sensor circuit.

DTC	Trouble Code Title and Conditions
DTC: P2132 **1T ECM, MIL: Yes** **Year:** 2011, 2012 **Model:** 370Z **Engine:** 3.7L	**Throttle Position Sensor 1 (Bank 2) Circuit Low Input:** An excessively low voltage from the TP sensor 1 is sent to ECM.
DTC: P2133 **1T ECM, MIL: Yes** **Year:** 2011, 2012 **Model:** 370Z **Engine:** 3.7L	**Throttle Position Sensor 1 (Bank 2) Circuit High Input:** An excessively high voltage from the TP sensor 1 is sent to ECM.
DTC: P2135 **1T ECM, MIL: Yes** **Year:** 2011, 2012 **Model:** All **Engine:** 1.6L, 1.8L, 2.0L, 2.5L, 3.5L, 3.5L, 3.7L, 4.0L, 5.6L	**Throttle Position Sensor Circuit Range/Performance (Includes Hybrid Models):** HYBRID MODELS CAUTION: Hybrid systems use very high-voltage battery systems. Before starting any service work involving the battery system, turn the ignition switch OFF and then remove the service plug from pocket in the trunk. After removing the service plug, wait 10 minutes before touching any of the high-voltage connectors and terminals. **NOTE: If DTC P2135 is displayed with DTC P0643, first perform the trouble diagnosis for DTC P0643.** **NOTE: On and, this DTC refers to Bank 1.** * Rationally incorrect voltage is sent to ECM compared with the signals from TP sensor 1 and TP sensor 2. * When the malfunction is detected, the ECM enters fail-safe mode and the MIL illuminates.
DTC: P2135 **1T ECM, MIL: Yes** **Year:** 2011 **Model:** Altima, Quest **Engine:** 2.5L, 3.5L	**Throttle / Pedal Position Sensor / Switch "A" / "B" Voltage Correlation:** Monitor runs whenever following DTCs are not present: None Either of the following conditions A or B is met: A. Power switch on (IG): 0.012 seconds or more B. Electronic throttle actuator power: ON
DTC: P2138 **1T ECM, MIL: Yes** **Year:** 2011, 2012 **Model:** 370Z, Altima, Frontier, Juke, Maxima, Murano, Pathfinder, Quest, Rogue, Sentra, Titan, Versa, Xterra **Engine:** 1.6L, 1.8L, 2.0L, 2.5L, 3.5L, 3.5L, 3.7L, 4.0L, 5.6L	**Accelerator Pedal Position Sensor Circuit Range/Performance:** **NOTE: If DTC P2138 is displayed with DTC P0643, first perform the trouble diagnosis for DTC P0643.** * Rationally incorrect voltage is sent to ECM compared with the signals from APP sensor 1 and APP sensor 2. * When the malfunction is detected, ECM enters fail-safe mode and the MIL illuminates.
DTC: P2138-110 **T ECM, MIL: Yes** **Year:** 2011 **Model:** Altima, Quest **Engine:** 2.5L, 3.5L	**Throttle / Pedal Position Sensor / Switch "D" / "E" Voltage Correlation:** Difference between the main sensor value and sub sensor value is large.
DTC: P2138-154 **T ECM, MIL: Yes** **Year:** 2011 **Model:** Altima, Quest **Engine:** 2.5L, 3.5L	**Throttle / Pedal Position Sensor / Switch "D" / "E" Voltage Correlation:** Main or sub sensor circuit wiring malfunction.
DTC: P2159 **1T ECM, MIL: Yes** **Year:** 2011 **Model:** Juke, Versa **Engine:** 1.6L	**Vehicle Speed Sensor "B" Range/Performance:** ECM detects a rear RH wheel sensor malfunction signal transmitted from the ABS actuator and electric unit (control unit) via CAN communication at least for 5 seconds in a row.
DTC: P2162 **1T ECM, MIL: Yes** **Year:** 2011 **Model:** Juke, Versa **Engine:** 1.6L	**VEHICLE SPEED SENSOR (A/B correlation) :** ECM detects a rear LH wheel sensor signal or a rear RH wheel sensor signal transmitted from the ABS actuator and electric unit (control unit) via CAN communication at least for 15 seconds in a row when the vehicle is in stopped condition.

DTC	Trouble Code Title and Conditions
DTC: P2195 **2T ECM, MIL: Yes** **Year:** 2011 **Model:** Altima, Quest **Engine:** 2.5L, 3.5L	**Oxygen (A/F) Sensor Signal Stuck Lean (Bank 1 Sensor 1):** Monitor runs whenever following DTCs not present: P0016 (VVT System Bank 1 - Misalignment) P0031, P0032, P101D (Air Fuel Ratio Sensor Heater - Sensor 1) P0102, P0103 (Mass Air Flow Meter) P0107, P0108 (Manifold Absolute Pressure) P0112, P0113 (Intake Air Temperature Sensor) P0115, P0117, P0118 (Engine Coolant Temperature Sensor) P0120, P0121, P0122, P0123, P0220, P0222, P0223, P2135 (Throttle Position Sensor) P0125 (Insufficient Engine Coolant Temperature for Closed Loop Fuel Control) P0128 (Thermostat) P0171, P0172 (Fuel System) P0301, P0302, P0303, P0304 (Misfire) P0335 (Crankshaft Position Sensor) P0340 (Camshaft Position Sensor) P0451, P0452, P0453 (EVAP System) P0505 (Idle speed control) Battery voltage: 11 V or more Atmospheric pressure: 76 kPa-a (570 mmHg-a) or higher Air fuel ratio sensor status: Activated Engine coolant temperature: 75°C (167°F) or more Continuous time of fuel cut: 3 to 10 seconds Lean Side Malfunction: Time after engine start: 30 seconds or more Fuel system status: Closed-loop
DTC: P2196 **2T ECM, MIL: Yes** **Year:** 2011 **Model:** Altima, Quest **Engine:** 2.5L, 3.5L	**Oxygen (A/F) Sensor Signal Stuck Rich (Bank 1 Sensor 1):** Monitor runs whenever following DTCs not present: P0016 (VVT System Bank 1 - Misalignment) P0031, P0032, P101D (Air Fuel Ratio Sensor Heater - Sensor 1) P0102, P0103 (Mass Air Flow Meter) P0107, P0108 (Manifold Absolute Pressure) P0112, P0113 (Intake Air Temperature Sensor) P0115, P0117, P0118 (Engine Coolant Temperature Sensor) P0120, P0121, P0122, P0123, P0220, P0222, P0223, P2135 (Throttle Position Sensor) P0125 (Insufficient Engine Coolant Temperature for Closed Loop Fuel Control) P0128 (Thermostat) P0171, P0172 (Fuel System) P0301, P0302, P0303, P0304 (Misfire) P0335 (Crankshaft Position Sensor) P0340 (Camshaft Position Sensor) P0451, P0452, P0453 (EVAP System) P0505 (Idle speed control) Battery voltage: 11 V or more Atmospheric pressure: 76 kPa-a (570 mmHg-a) or higher Air fuel ratio sensor status: Activated Engine coolant temperature: 75°C (167°F) or more Continuous time of fuel cut: 3 to 10 seconds Rich Side Malfunction: Time after engine start: 30 seconds or more Fuel system status: Closed-loop

DTC	Trouble Code Title and Conditions
DTC: P2237 **2T ECM, MIL: Yes** **Year:** 2011 **Model:** Altima, Quest **Engine:** 2.5L, 3.5L	**Oxygen (A/F) Sensor Pumping Current Circuit / Open (Bank 1 Sensor 1):** Monitor runs whenever following DTCs are not present: P0016 (VVT System Bank 1 - Misalignment) P0031, P0032, P101D (Air Fuel Ratio Sensor Heater - Sensor 1) P0102, P0103 (Mass Air Flow Meter) P0107, P0108 (Manifold Absolute Pressure) P0112, P0113 (Intake Air Temperature Sensor) P0115, P0117, P0118 (Engine Coolant Temperature Sensor) P0120, P0121, P0122, P0123, P0220, P0222, P0223, P2135 (Throttle Position Sensor) P0125 (Insufficient Engine Coolant Temperature for Closed Loop Fuel Control) P0128 (Thermostat) P0171, P0172 (Fuel System) P0301, P0302, P0303, P0304 (Misfire) P0335 (Crankshaft Position Sensor) P0401 (EGR System (Closed)) P0451, P0452, P0453 (EVAP System) P0505 (Idle speed control) Air Fuel Ratio Sensor Open Circuit Between A1A+ and A1A-: Estimated sensor temperature: 450 to 550°C (842 to 1022°F) Engine: Running Battery voltage: 11 V or more
DTC: P2238 **2T ECM, MIL: Yes** **Year:** 2011 **Model:** Altima, Quest **Engine:** 2.5L, 3.5L	**Oxygen (A/F) Sensor Pumping Current Circuit Low (Bank 1 Sensor 1):** Monitor runs whenever following DTCs are not present: P0016 (VVT System Bank 1 - Misalignment) P0031, P0032, P101D (Air Fuel Ratio Sensor Heater - Sensor 1) P0102, P0103 (Mass Air Flow Meter) P0107, P0108 (Manifold Absolute Pressure) P0112, P0113 (Intake Air Temperature Sensor) P0115, P0117, P0118 (Engine Coolant Temperature Sensor) P0120, P0121, P0122, P0123, P0220, P0222, P0223, P2135 (Throttle Position Sensor) P0125 (Insufficient Engine Coolant Temperature for Closed Loop Fuel Control) P0128 (Thermostat) P0171, P0172 (Fuel System) P0301, P0302, P0303, P0304 (Misfire) P0335 (Crankshaft Position Sensor) P0401 (EGR System (Closed)) P0451, P0452, P0453 (EVAP System) P0505 (Idle speed control) Air Fuel Ratio Sensor Low Impedance: Estimated sensor temperature: 700 to 800 °C (1292 to 1472°F) Engine coolant temperature: 10°C (50°F) or higher Fuel cut: No executed
DTC: P2239 **2T ECM, MIL: Yes** **Year:** 2011 **Model:** Altima, Quest **Engine:** 2.5L, 3.5L	**Oxygen (A/F) Sensor Pumping Current Circuit High (Bank 1 Sensor 1):** Monitor runs whenever following DTCs are not present: None Battery voltage: 11 V or more Power switch: On (IG) Time after power switch is off to on (IG): 5 seconds or more
DTC: P2252 **2T ECM, MIL: Yes** **Year:** 2011 **Model:** Altima, Quest **Engine:** 2.5L, 3.5L	**Oxygen (A/F) Sensor Reference Ground Circuit Low (Bank 1 Sensor 1):** Monitor runs whenever following DTCs are not present: None Battery voltage: 11 V or more Power switch: On (IG) Time after power switch is off to on (IG): 5 seconds or more
DTC: P2253 **2T ECM, MIL: Yes** **Year:** 2011 **Model:** Altima, Quest **Engine:** 2.5L, 3.5L	**Oxygen (A/F) Sensor Reference Ground Circuit High (Bank 1 Sensor 1):** Monitor runs whenever following DTCs are not present: None Battery voltage: 11 V or more Power switch: On (IG) Time after power switch is off to on (IG): 5 seconds or more

DTC	Trouble Code Title and Conditions
DTC: P2263 **1T ECM, MIL: Yes** **Year:** 2011 **Model:** Juke, Versa **Engine:** 1.6L	**TC SYSTEM B-1 (Turbocharger boost system performance) :** In spite of the boosting area, the boost does not increase.
DTC: P2401 **2T ECM, MIL: Yes** **Year:** 2011 **Model:** Altima, Quest **Engine:** 2.5L, 3.5L	**Evaporative Emission System Leak Detection Pump Control Circuit Low:** Monitor runs whenever following DTCs are not present: None Atmospheric pressure: 70 to 110 kPa-a (525 to 825 mmHg-a) Battery voltage: 10.5 V or more Vehicle speed: Below 2.5 mph (4 km/h) Power switch: OFF Time after key off: 5, 7 or 9.5 hours Canister pressure sensor malfunction (P0452 and P0453): Not detected Purge VSV: Not operated by scan tool Vent valve: Not operated by scan tool Leak detection pump: Not operated by scan tool Both of the following conditions are met before key off: Conditions 1 and 2 1. Duration that vehicle driven: 5 minutes or more 2. EVAP purge operation: Performed Engine coolant temperature: 4.4 to 35°C (40 to 95°F) Intake air temperature: 4.4 to 35°C (40 to 95°F)
DTC: P2402 **2T ECM, MIL: Yes** **Year:** 2011 **Model:** Altima, Quest **Engine:** 2.5L, 3.5L	**Evaporative Emission System Leak Detection Pump Control Circuit High:** Monitor runs whenever following DTCs are not present: None Atmospheric pressure: 70 to 110 kPa-a (525 to 825 mmHg-a) Battery voltage: 10.5 V or more Vehicle speed: Below 2.5 mph (4 km/h) Power switch: OFF Time after key off: 5, 7 or 9.5 hours Canister pressure sensor malfunction (P0452 and P0453): Not detected Purge VSV: Not operated by scan tool Vent valve: Not operated by scan tool Leak detection pump: Not operated by scan tool Both of the following conditions are met before key off: Conditions 1 and 2 1. Duration that vehicle driven: 5 minutes or more 2. EVAP purge operation: Performed Engine coolant temperature: 4.4 to 35°C (40 to 95°F) Intake air temperature: 4.4 to 35°C (40 to 95°F)
DTC: P2419 **2T ECM, MIL: Yes** **Year:** 2011 **Model:** Altima, Quest **Engine:** 2.5L, 3.5L	**Evaporative Emission System Switching Valve Control Circuit Low:** Monitor runs whenever following DTCs are not present: None Atmospheric pressure: 70 to 110 kPa-a (525 to 825 mmHg-a) Battery voltage: 10.5 V or more Vehicle speed: Below 2.5 mph (4 km/h) Power switch: OFF Time after key off: 5, 7 or 9.5 hours Canister pressure sensor malfunction (P0452 and P0453): Not detected Purge VSV: Not operated by scan tool Vent valve: Not operated by scan tool Leak detection pump: Not operated by scan tool Both of the following conditions are met before key off: Conditions 1 and 2 1. Duration that vehicle driven: 5 minutes or more 2. EVAP purge operation: Performed Engine coolant temperature: 4.4 to 35°C (40 to 95°F) Intake air temperature: 4.4 to 35°C (40 to 95°F)

DTC	Trouble Code Title and Conditions
DTC: P2420 **2T ECM, MIL: Yes** **Year:** 2011 **Model:** Altima, Quest **Engine:** 2.5L, 3.5L	**Evaporative Emission System Switching Valve Control Circuit High:** Monitor runs whenever following DTCs are not present: None Atmospheric pressure: 70 to 112 kPa-a (525 to 840 mmHg-a) Battery voltage: 10.5 V or more Vehicle speed: Below 2.5 mph (4 km/h) Power switch: OFF Time after key off: 5, 7 or 9.5 hours Canister pressure sensor malfunction (P0452 and P0453): Not detected Purge VSV: Not operated by scan tool Vent valve: Not operated by scan tool Leak detection pump: Not operated by scan tool Both of the following conditions are met before key off: Conditions 1 and 2 1. Duration that vehicle has been driven: 5 minutes or more 2. EVAP purge operation: Performed Engine coolant temperature: 4.4 to 35°C (40 to 95°F) Intake air temperature: 4.4 to 35°C (40 to 95°F)
DTC: P2423 **1T ECM, MIL: Yes** **Year:** 2011 **Model:** Altima, Rogue **Engine:** 2.5L	**HC Adsorption Catalyst Efficiency Below Threshhold (Hybrid):** HYBRID MODELS CAUTION: Hybrid systems use very high-voltage battery systems. Before starting any service work involving the battery system, turn the ignition switch OFF and then remove the service plug from pocket in the trunk. After removing the service plug, wait 10 minutes before touching any of the high-voltage connectors and terminals. HC adsorption catalyst (under floor) does not operate properly. HC adsorption catalyst (under floor) does not have enough oxygen storage capacity.
DTC: P2511-149 **T ECM, MIL: Yes** **Year:** 2011 **Model:** Altima, Quest **Engine:** 2.5L, 3.5L	**ECM/PCM Power Relay Sense Circuit Intermittent:** When the power switch is on (READY), the hybrid vehicle control ECU is reset.
DTC: P2519-766 **T ECM, MIL: Yes** **Year:** 2011 **Model:** Altima, Quest **Engine:** 2.5L, 3.5L	**A/C Request "A" Circuit:** Malfunction in the cooling fan operation condition signal circuit.
DTC: P2532-772 **T ECM, MIL: Yes** **Year:** 2011 **Model:** Altima, Quest **Engine:** 2.5L, 3.5L	**Ignition Switch Run Position Circuit High:** When no signals are received from the ECUs (skid control ECU and A/C amplifier) that are activated by the IG1 relay, but the hybrid vehicle control ECU is operating.
DTC: P2601-777 **T ECM, MIL: Yes** **Year:** 2011 **Model:** Altima, Quest **Engine:** 2.5L, 3.5L	**Oil Pump Control Circuit Range / Performance:** Command signal (OPM1) from hybrid vehicle control ECU is abnormal.
DTC: P2601-778 **T ECM, MIL: Yes** **Year:** 2011 **Model:** Altima, Quest **Engine:** 2.5L, 3.5L	**Oil Pump Control Circuit Range / Performance:** Oil pump fails to start more than set number of times and the temperature of the motor or generator exceeds load factor limit temperature when oil pump start request is accepted.
DTC: P2601-779 **T ECM, MIL: Yes** **Year:** 2011 **Model:** Altima, Quest **Engine:** 2.5L, 3.5L	**Oil Pump Control Circuit Range / Performance:** * Abnormal power source voltage is detected by oil pump with motor assembly * Power source voltage input to oil pump with motor assembly is 6.96 V or less for 2 seconds or more with OIL PMP relay on
DTC: P2602-767 **T ECM, MIL: Yes** **Year:** 2011 **Model:** Altima, Quest **Engine:** 2.5L, 3.5L	**Oil Pump Control Circuit Low:** * GND short in oil pump circuit * When OPM2 signal is on, OPST signal remains low

DTC	Trouble Code Title and Conditions
DTC: P2603-768 **T ECM, MIL: Yes** **Year:** 2011 **Model:** Altima, Quest **Engine:** 2.5L, 3.5L	**Oil Pump Control Circuit High:** * Open or +B short in oil pump circuit * When OPM2 signal is on, OPST signal remains high
DTC: P2610 **2T ECM, MIL: Yes** **Year:** 2011 **Model:** Altima, Quest **Engine:** 2.5L, 3.5L	**ECM / PCM Internal Engine Off Timer Performance:** Monitor runs whenever following DTCs are not present: None Case 1: Power switch: On (IG) Engine: Running Battery voltage: 8 V or more CPU clock elapsed time: 10 minutes or more Case 2: Internal engine OFF timer (elapsed time from engine stop): 30 minutes or more Battery voltage: 8 V or more Power switch: On (IG) Case 3: Internal engine OFF timer (elapsed time from engine stop): 40 minutes or more Battery voltage: 8 V or more Power switch: On (IG)
DTC: P2637 **T BCM** **Year:** 2011 **Model:** Altima, Quest **Engine:** 2.5L, 3.5L	**A/C Mode Door Actuator (Bi-Level Failure):** Mode door does not change even if auto amp. operates mode door motor.
DTC: P2713 **T TCM, TCIL: Yes** **Year:** 2011, 2012 **Model:** 370Z **Engine:** 3.7L	**Pressure Control Solenoid D:** The high and low reverse clutch solenoid valve monitor value is 0.4 A or less when the high and low reverse clutch solenoid valve command value is more than 0.75 A.
DTC: P2722 **T TCM, TCIL: Yes** **Year:** 2011, 2012 **Model:** 370Z **Engine:** 3.7L	**Pressure Control Solenoid E:** The low brake solenoid valve monitor value is 0.4 A or less when the low brake solenoid valve command value is more than 0.75 A.
DTC: P2731 **T TCM, TCIL: Yes** **Year:** 2011, 2012 **Model:** 370Z **Engine:** 3.7L	**Pressure Control Solenoid F:** The 2346 brake solenoid valve monitor value is 0.4 A or less when the 2346 brake solenoid valve command value is more than 0.75 A.
DTC: P2765 **T ECM, MIL: Yes** **Year:** 2011, 2012 **Model:** 370Z **Engine:** 3.7L	**Input Speed Sensor Circuit:** **NOTE: If DTC P2765 is displayed with DTC P0335, P0340 or P0345, first perform the trouble diagnosis for DTC P0335, P0340 or P0345.** There is a difference between engine speed signal calculated by ECM and input shaft speed sensor signal.
DTC: P2807 **T TCM, TCIL: Yes** **Year:** 2011, 2012 **Model:** 370Z **Engine:** 3.7L	**Pressure Control Solenoid G:** The direct clutch solenoid valve monitor value is 0.4 A or less when the direct clutch solenoid valve command value is more than 0.75 A.

DTC	Trouble Code Title and Conditions
DTC: P2A00 **1T ECM, MIL: Yes** **Year:** 2011, 2012 **Model:** All **Engine:** 1.6L, 1.8L, 2.0L, 2.5L, 3.5L, 3.5L, 3.7L, 4.0L, 5.6L	**Air Fuel (A/F) Sensor 1 Circuit Range/Performance (Includes Hybrid Models):** HYBRID MODELS CAUTION: Hybrid systems use very high-voltage battery systems. Before starting any service work involving the battery system, turn the ignition switch OFF and then remove the service plug from pocket in the trunk. After removing the service plug, wait 10 minutes before touching any of the high-voltage connectors and terminals. **NOTE: On and engines, this applies to Bank 1.** * To judge the malfunction, the A/F signal computed by ECM from the A/F sensor 1 signal is monitored not to be shifted to LEAN side or RICH side. * The output voltage computed by ECM from the A/F sensor 1 signal is shifted to the lean side for a specified period. * The A/F signal computed by ECM from the A/F sensor 1 signal is shifted to the rich side for a specified period.
DTC: P2A00 **2T ECM, MIL: Yes** **Year:** 2011 **Model:** Altima, Quest **Engine:** 2.5L, 3.5L	**A/F Sensor Circuit Slow Response (Bank 1 Sensor 1):** Monitor runs whenever following DTCs not stored: P0016 (VVT System - Misalignment) P0017 (Exhaust VVT System - Misalignment) P0031, P0032, P101D (Air Fuel Ratio Sensor Heater) P0102, P0103 (Mass Air Flow Sensor) P0112, P0113 (Intake Air Temperature Sensor) P0115, P0117, P0118 (Engine Coolant Temperature Sensor) P0125 (Insufficient Coolant Temperature for Closed Loop Fuel Control) P0120, P0121, P0122, P0123, P0220, P0222, P0223, P2135 (Throttle Position Sensor) P0128 (Thermostat) P0171, P0172 (Fuel System) P0301 - P0304 (Misfire) P0335 (Crankshaft Position Sensor) P0340 (Camshaft Position Sensor) P0451, P0452, P0453 (EVAP System) P0505 (IAC Valve) Active air fuel ratio control: Performing Active air fuel ratio control performed when following conditions met: Battery voltage: 11 V or higher Engine coolant temperature: 75°C (167°F) or higher Idling: OFF Engine speed: Less than 4000 rpm Air fuel ratio sensor status: Activated Fuel-cut: OFF Engine load: 10 to 70% Catalyst monitor: Not yet Intake air amount: 2.5 to 11 g/sec.
DTC: P2A00 **1T ECM, MIL: Yes** **Year:** 2011 **Model:** Juke, Versa **Engine:** 1.6L	**A/F SENSOR 1 B-1 (Air fuel ratio (A/F) sensor 1 circuit range/performance) :** Output voltage computed by ECM from the A/F sensor 1 signal is shifted to the lean side for a specified period. A/F signal computed by ECM from the A/F sensor 1 signal is shifted to the rich side for a specified period.
DTC: P2A03 **1T ECM, MIL: Yes** **Year:** 2011, 2012 **Model:** 370Z, Altima, Frontier, Maxima, Murano, Pathfinder, Quest, Sentra, Titan, Xterra **Engine:** 2.0L, 2.5L, 3.5L, 3.5L, 3.7L, 4.0L, 5.6L	**Air Fuel (A/F) Ratio Sensor 1 Circuit Range/Performance:** **NOTE: On and engines, this applies to Bank 2.** * The output voltage computed by ECM from the A/F sensor 1 signal is shifted to the lean side for a specified period. * The A/F signal computed by ECM from the A/F sensor 1 signal is shifted to the rich side for a specified period. * To judge the malfunction, the A/F signal computed by ECM from the A/F sensor 1 signal is monitored so it will not shift to LEAN side or RICH side.

OBD II Trouble Code List (P3XXX Codes)

DTC	Trouble Code Title and Conditions
DTC: P3000-388 **T ECM, MIL: Yes** **Year:** 2011 **Model:** Altima, Quest **Engine:** 2.5L, 3.5L	**HV Battery Malfunction:** Discharge inhibition control malfunction.

DTC	Trouble Code Title and Conditions
DTC: P3000-389 **T ECM, MIL: Yes** **Year:** 2011 **Model:** Altima, Quest **Engine:** 2.5L, 3.5L	**HV Battery Malfunction:** HV battery voltage drops
DTC: P3000-603 **T ECM, MIL: Yes** **Year:** 2011 **Model:** Altima, Quest **Engine:** 2.5L, 3.5L	**HV Battery Malfunction:** The hybrid vehicle control ECU detects a HV battery cooling system error signal.
DTC: P3004 **T ECM, MIL: Yes** **Year:** 2011 **Model:** Altima, Quest **Engine:** 2.5L, 3.5L	**Power Cable Malfunction:** The inverter voltage does not rise during precharge (time from when SMRP turns on until when SMRG turns on).
DTC: P3004-132 **T ECM, MIL: Yes** **Year:** 2011 **Model:** Altima, Quest **Engine:** 2.5L, 3.5L	**Power Cable Malfunction:** The inverter is not precharged.
DTC: P3004-133 **T ECM, MIL: Yes** **Year:** 2011 **Model:** Altima, Quest **Engine:** 2.5L, 3.5L	**Power Cable Malfunction:** A high-voltage wiring system error signal is detected in the hybrid vehicle control ECU.
DTC: P3004-801 **T ECM, MIL: Yes** **Year:** 2011 **Model:** Altima, Quest **Engine:** 2.5L, 3.5L	**Power Cable Malfunction:** Minimal overcurrent occurs during precharge (time from when SMRP turns on until when SMRG turns on).
DTC: P3004-801 **T ECM, MIL: Yes** **Year:** 2011 **Model:** Altima, Quest **Engine:** 2.5L, 3.5L	**Power Cable Malfunction:** Excessive overcurrent occurs during precharge (time from when SMRP turns on until when SMRG turns on).
DTC: P3004-803 **T ECM, MIL: Yes** **Year:** 2011 **Model:** Altima, Quest **Engine:** 2.5L, 3.5L	**Power Cable Malfunction:** While the power switch is on (READY), the electric battery fuse in the service plug grip is burned out, the service plug grip is removed, or SMRB or SMRG is open.
DTC: P3011-123 **T ECM, MIL: Yes** **Year:** 2011 **Model:** Altima, Quest **Engine:** 2.5L, 3.5L	**Battery Block 1 Becomes Weak:** Presence of a malfunctioning block is determined based on each battery block voltage (1 trip detection).
DTC: P3011-123 **1T ECM, MIL: Yes** **Year:** 2011 **Model:** Altima, Quest **Engine:** 2.5L, 3.5L	**Battery Block 1 Becomes Weak:** The monitor will run whenever the following DTCs are not present: TMC's intellectual property Other conditions belong to TMC's intellectual property: -
DTC: P3012-123 **T ECM, MIL: Yes** **Year:** 2011 **Model:** Altima, Quest **Engine:** 2.5L, 3.5L	**Battery Block 2 Becomes Weak:** Presence of a malfunctioning block is determined based on each battery block voltage (1 trip detection).

DTC	Trouble Code Title and Conditions
DTC: P3013-123 **T ECM, MIL: Yes** **Year:** 2011 **Model:** Altima, Quest **Engine:** 2.5L, 3.5L	**Battery Block 3 Becomes Weak:** Presence of a malfunctioning block is determined based on each battery block voltage (1 trip detection).
DTC: P3014-123 **T ECM, MIL: Yes** **Year:** 2011 **Model:** Altima, Quest **Engine:** 2.5L, 3.5L	**Battery Block 4 Becomes Weak:** Presence of a malfunctioning block is determined based on each battery block voltage (1 trip detection).
DTC: P3015-123 **T ECM, MIL: Yes** **Year:** 2011 **Model:** Altima, Quest **Engine:** 2.5L, 3.5L	**Battery Block 5 Becomes Weak:** Presence of a malfunctioning block is determined based on each battery block voltage (1 trip detection).
DTC: P3016-123 **T ECM, MIL: Yes** **Year:** 2011 **Model:** Altima, Quest **Engine:** 2.5L, 3.5L	**Battery Block 6 Becomes Weak:** Presence of a malfunctioning block is determined based on each battery block voltage (1 trip detection).
DTC: P3017-123 **T ECM, MIL: Yes** **Year:** 2011 **Model:** Altima, Quest **Engine:** 2.5L, 3.5L	**Battery Block 7 Becomes Weak:** Presence of a malfunctioning block is determined based on each battery block voltage (1 trip detection).
DTC: P3018-123 **T ECM, MIL: Yes** **Year:** 2011 **Model:** Altima, Quest **Engine:** 2.5L, 3.5L	**Battery Block 8 Becomes Weak:** Presence of a malfunctioning block is determined based on each battery block voltage (1 trip detection).
DTC: P3019-123 **T ECM, MIL: Yes** **Year:** 2011 **Model:** Altima, Quest **Engine:** 2.5L, 3.5L	**Battery Block 9 Becomes Weak:** Presence of a malfunctioning block is determined based on each battery block voltage (1 trip detection).
DTC: P3020-123 **T ECM, MIL: Yes** **Year:** 2011 **Model:** Altima, Quest **Engine:** 2.5L, 3.5L	**Battery Block 10 Becomes Weak:** Presence of a malfunctioning block is determined based on each battery block voltage (1 trip detection).
DTC: P3021-123 **T ECM, MIL: Yes** **Year:** 2011 **Model:** Altima, Quest **Engine:** 2.5L, 3.5L	**Battery Block 11 Becomes Weak:** Presence of a malfunctioning block is determined based on each battery block voltage (1 trip detection).
DTC: P3022-123 **T ECM, MIL: Yes** **Year:** 2011 **Model:** Altima, Quest **Engine:** 2.5L, 3.5L	**Battery Block 12 Becomes Weak:** Presence of a malfunctioning block is determined based on each battery block voltage (1 trip detection).
DTC: P3023-123 **T ECM, MIL: Yes** **Year:** 2011 **Model:** Altima, Quest **Engine:** 2.5L, 3.5L	**Battery Block 13 Becomes Weak:** Presence of a malfunctioning block is determined based on each battery block voltage (1 trip detection).

DTC	Trouble Code Title and Conditions
DTC: P3024-123 **T ECM, MIL: Yes** **Year:** 2011 **Model:** Altima, Quest **Engine:** 2.5L, 3.5L	**Battery Block 14 Becomes Weak:** Presence of a malfunctioning block is determined based on each battery block voltage (1 trip detection).
DTC: P3025-123 **T ECM, MIL: Yes** **Year:** 2011 **Model:** Altima, Quest **Engine:** 2.5L, 3.5L	**Battery Block 15 Becomes Weak:** Presence of a malfunctioning block is determined based on each battery block voltage (1 trip detection).
DTC: P3026-123 **T ECM, MIL: Yes** **Year:** 2011 **Model:** Altima, Quest **Engine:** 2.5L, 3.5L	**Battery Block 16 Becomes Weak:** Presence of a malfunctioning block is determined based on each battery block voltage (1 trip detection).
DTC: P3027-123 **T ECM, MIL: Yes** **Year:** 2011 **Model:** Altima, Quest **Engine:** 2.5L, 3.5L	**Battery Block 17 Becomes Weak:** Presence of a malfunctioning block is determined based on each battery block voltage (1 trip detection).
DTC: P3065/123 **T BCM** **Year:** 2011 **Model:** Altima, Quest **Engine:** 2.5L, 3.5L	**Hybrid Battery Temperature Sensor Range/Performance Stack A:** When the battery temperature sensor performance is abnormal (1 trip detection/2 trip detection)
DTC: P3065-123 **1T ECM, MIL: Yes** **Year:** 2011 **Model:** Altima, Quest **Engine:** 2.5L, 3.5L	**Hybrid Battery Temperature Sensor Range / Perfoemance Stack A:** The monitor will run whenever the following DTCs are not present: TMC's intellectual property Other conditions belong to TMC's intellectual property: -
DTC: P308A-123 **1T ECM, MIL: Yes** **Year:** 2011 **Model:** Altima, Quest **Engine:** 2.5L, 3.5L	**Hybrid Battery Voltage Sensor All Circuits Low:** The monitor will run whenever the following DTCs are not present: TMC's intellectual property Other conditions belong to TMC's intellectual property: -
DTC: P3107-213 **T ECM, MIL: Yes** **Year:** 2011 **Model:** Altima, Quest **Engine:** 2.5L, 3.5L	**Airbag ECU Communication Circuit Malfunction:** Short to GND in the communication circuit.
DTC: P3107-214 **T ECM, MIL: Yes** **Year:** 2011 **Model:** Altima, Quest **Engine:** 2.5L, 3.5L	**Airbag ECU Communication Circuit Malfunction:** Open or short to +B in the communication circuit
DTC: P3107-215 **T ECM, MIL: Yes** **Year:** 2011 **Model:** Altima, Quest **Engine:** 2.5L, 3.5L	**Airbag ECU Communication Circuit Malfunction:** Abnormal communication signal.
DTC: P3108-535 **T ECM, MIL: Yes** **Year:** 2011 **Model:** Altima, Quest **Engine:** 2.5L, 3.5L	**A/C Amplifier Communication Circuit Malfunction:** Serial communication malfunction.

DTC	Trouble Code Title and Conditions
DTC: P3108-536 **T ECM, MIL: Yes** **Year:** 2011 **Model:** Altima, Quest **Engine:** 2.5L, 3.5L	**A/C Amplifier Communication Circuit Malfunction:** A/C inverter malfunction.
DTC: P3108-538 **T ECM, MIL: Yes** **Year:** 2011 **Model:** Altima, Quest **Engine:** 2.5L, 3.5L	**A/C Amplifier Communication Circuit Malfunction:** Open in STB signal circuit.
DTC: P3110-139 **T ECM, MIL: Yes** **Year:** 2011 **Model:** Altima, Quest **Engine:** 2.5L, 3.5L	**IGCT Relay Malfunction:** There is a short to +B in the IGCT relay or the IGCT relay is stuck closed.
DTC: P3110-223 **T ECM, MIL: Yes** **Year:** 2011 **Model:** Altima, Quest **Engine:** 2.5L, 3.5L	**IGCT Relay Malfunction:** The IGCT relay remains stuck closed.
DTC: P3147-239 **T ECM, MIL: Yes** **Year:** 2011 **Model:** Altima, Quest **Engine:** 2.5L, 3.5L	**Transmission Malfunction:** Hybrid vehicle transaxle input malfunction (shaft damaged).
DTC: P3147-240 **T TCM, MIL: Yes** **Year:** 2011 **Model:** Altima, Quest **Engine:** 2.5L, 3.5L	**Transmission Malfunction:** MG1 locked
DTC: P3147-241 **T ECM, MIL: Yes** **Year:** 2011 **Model:** Altima, Quest **Engine:** 2.5L, 3.5L	**Transmission Malfunction:** Hybrid vehicle transaxle input malfunction (input system lock)
DTC: P3147-242 **T TCM, MIL: Yes** **Year:** 2011 **Model:** Altima, Quest **Engine:** 2.5L, 3.5L	**Transmission Malfunction:** Planetary gear locked
DTC: P3190 **1T ECM, MIL: Yes** **Year:** 2011 **Model:** Altima, Quest **Engine:** 2.5L, 3.5L	**Poor Engine Power:** Monitor runs whenever following DTCs are not present: None Fuel cut operation: Not operated Engine speed: 650 rpm or more (varies with engine coolant temperature) Communication with hybrid vehicle control control ECU: No malfunction
DTC: P3191 **1T VCM, MIL: Yes** **Year:** 2011 **Model:** Leaf **Engine:** -L –	**EV SYSTEM CAN COMMUNICATION ERROR:** When VCM detects an error signal that is received from traction motor inverter via CAN communication.
DTC: P3191 **1T ECM, MIL: Yes** **Year:** 2011 **Model:** Altima, Quest **Engine:** 2.5L, 3.5L	**Engine does not Start:** Monitor runs whenever following DTCs are not present: None Fuel cut operation: Not operated Engine speed: 650 rpm or more (varies with engine coolant temperature) Communication with hybrid vehicle control control ECU: No malfunction

DTC	Trouble Code Title and Conditions
DTC: P3193 **1T ECM, MIL: Yes** **Year:** 2011 **Model:** Altima, Quest **Engine:** 2.5L, 3.5L	**Fuel Run Out:** Monitor runs whenever following DTCs are not present: None Fuel cut operation: Not operated Engine speed: 650 rpm or more (varies with engine coolant temperature) Communication with hybrid vehicle control ECU: No malfunction
DTC: P3221-314 **T ECM, MIL: Yes** **Year:** 2011 **Model:** Altima, Quest **Engine:** 2.5L, 3.5L	**Generator Inverter Temperature Sensor Circuit Range / Performance:** Unusual sudden change in generator inverter temperature sensor output occurs and the offset continues, or unusual sudden change in generator inverter temperature sensor output occurs repeatedly.
DTC: P3221-315 **T ECM, MIL: Yes** **Year:** 2011 **Model:** Altima, Quest **Engine:** 2.5L, 3.5L	**Generator Inverter Temperature Sensor Circuit Range / Performance:** Temperature calculated by power hybrid vehicle control ECU and actual temperature are different for 10 seconds or more.
DTC: P3222-313 **T ECM, MIL: Yes** **Year:** 2011 **Model:** Altima, Quest **Engine:** 2.5L, 3.5L	**Generator Inverter Temperature Sensor Circuit High / Low:** Open or short to GND in the generator inverter temperature sensor signal circuit.
DTC: P3223-312 **T ECM, MIL: Yes** **Year:** 2011 **Model:** Altima, Quest **Engine:** 2.5L, 3.5L	**Generator Inverter Temperature Sensor Circuit High:** Short to +B in the generator inverter temperature sensor signal circuit.
DTC: P3226-562 **T ECM, MIL: Yes** **Year:** 2011 **Model:** Altima, Quest **Engine:** 2.5L, 3.5L	**DC/DC Boost Converter Temperature Sensor:** Unusual sudden change in boost converter temperature sensor output occurs and the offset continues, or unusual sudden change in boost converter temperature sensor output occurs repeatedly.
DTC: P3226-563 **T ECM, MIL: Yes** **Year:** 2011 **Model:** Altima, Quest **Engine:** 2.5L, 3.5L	**DC/DC Boost Converter Temperature Sensor:** Temperature calculated by power hybrid vehicle control ECU and actual temperature are different for 10 seconds or more.
DTC: P3227-583 **T ECM, MIL: Yes** **Year:** 2011 **Model:** Altima, Quest **Engine:** 2.5L, 3.5L	**Converter Temperature Sensor Circuit Low:** Open or short to GND in the boost converter temperature sensor signal circuit.
DTC: P3228-584 **T ECM, MIL: Yes** **Year:** 2011 **Model:** Altima, Quest **Engine:** 2.5L, 3.5L	**Converter Temperature Sensor Circuit High:** Short to +B in the boost converter temperature sensor signal circuit.
DTC: P3232-749 **T ECM, MIL: Yes** **Year:** 2011 **Model:** Altima, Quest **Engine:** 2.5L, 3.5L	**Open or Short to B+ in Blocking of HV Gate Connection:** Short to GND in the emergency shutdown signal line while the gate is shut down.
DTC: P3233-750 **T ECM, MIL: Yes** **Year:** 2011 **Model:** Altima, Quest **Engine:** 2.5L, 3.5L	**Short to B+ in Blocking of HV Gate Connection:** Open or short to +B in the emergency shutdown signal line when the gate is driving

GLOSSARY

ABS: Anti-lock braking system. An electro-mechanical braking system which is designed to minimize or prevent wheel lock-up during braking.

ABSOLUTE PRESSURE: Atmospheric (barometric) pressure plus the pressure gauge reading.

ACCELERATOR PUMP: A small pump located in the carburetor that feeds fuel into the air/fuel mixture during acceleration.

ACCUMULATOR: A device that controls shift quality by cushioning the shock of hydraulic oil pressure being applied to a clutch or band.

ACTUATING MECHANISM: The mechanical output devices of a hydraulic system, for example, clutch pistons and band servos.

ACTUATOR: The output component of a hydraulic or electronic system.

ADVANCE: Setting the ignition timing so that spark occurs earlier before the piston reaches top dead center (TDC).

ADAPTIVE MEMORY (ADAPTIVE STRATEGY): The learning ability of the TCM or PCM to redefine its decision-making process to provide optimum shift quality.

AFTER TOP DEAD CENTER (ATDC): The point after the piston reaches the top of its travel on the compression stroke.

AIR BAG: Device on the inside of the car designed to inflate on impact of crash, protecting the occupants of the car.

AIR CHARGE TEMPERATURE (ACT) SENSOR: The temperature of the airflow into the engine is measured by an ACT sensor, usually located in the lower intake manifold or air cleaner.

AIR CLEANER: An assembly consisting of a housing, filter and any connecting ductwork. The filter element is made up of a porous paper, sometimes with a wire mesh screening, and is designed to prevent airborne particles from entering the engine through the carburetor or throttle body.

AIR INJECTION: One method of reducing harmful exhaust emissions by injecting air into each of the exhaust ports of an engine. The fresh air entering the hot exhaust manifold causes any remaining fuel to be burned before it can exit the tailpipe.

AIR PUMP: An emission control device that supplies fresh air to the exhaust manifold to aid in more completely burning exhaust gases.

AIR/FUEL RATIO: The ratio of air-to-gasoline by weight in the fuel mixture drawn into the engine.

ALDL (assembly line diagnostic link): Electrical connector for scanning ECM/PCM/TCM input and output devices.

ALIGNMENT RACK: A special drive-on vehicle lift apparatus/measuring device used to adjust a vehicle's toe, caster and camber angles.

ALL WHEEL DRIVE: Term used to describe a full time four wheel drive system or any other vehicle drive system that continuously delivers power to all four wheels. This system is found primarily on station wagon vehicles and SUVs not utilized for significant off road use.

ALTERNATING CURRENT (AC): Electric current that flows first in one direction, then in the opposite direction, continually reversing flow.

ALTERNATOR: A device which produces AC (alternating current) which is converted to DC (direct current) to charge the car battery.

AMMETER: An instrument, calibrated in amperes, used to measure the flow of an electrical current in a circuit. Ammeters are always connected in series with the circuit being tested.

AMPERAGE: The total amount of current (amperes) flowing in a circuit.

AMPLIFIER: A device used in an electrical circuit to increase the voltage of an output signal.

AMP/HR. RATING (BATTERY): Measurement of the ability of a battery to deliver a stated amount of current for a stated period of time. The higher the amp/hr. rating, the better the battery.

AMPERE: The rate of flow of electrical current present when one volt of electrical pressure is applied against one ohm of electrical resistance.

ANALOG COMPUTER: Any microprocessor that uses similar (analogous) electrical signals to make its calculations.

ANODIZED: A special coating applied to the surface of aluminum valves for extended service life.

ANTIFREEZE: A substance (ethylene or propylene glycol) added to the coolant to prevent freezing in cold weather.

ANTI-FOAM AGENTS: Minimize fluid foaming from the whipping action encountered in the converter and planetary action.

ANTI-WEAR AGENTS: Zinc agents that control wear on the gears, bushings, and thrust washers.

ANTI-LOCK BRAKING SYSTEM: A supplementary system to the base hydraulic system that prevents sustained lock-up of the wheels during braking as well as automatically controlling wheel slip.

ANTI-ROLL BAR: See stabilizer bar.

ARC: A flow of electricity through the air between two electrodes or contact points that produces a spark.

ARMATURE: A laminated, soft iron core wrapped by a wire that converts electrical energy to mechanical energy as in a motor or relay. When rotated in a magnetic field, it changes mechanical energy into electrical energy as in a generator.

ATDC: After Top Dead Center.

ATF: Automatic transmission fluid.

ATMOSPHERIC PRESSURE: The pressure on the Earth's surface caused by the weight of the air in the atmosphere. At sea level, this pressure is 14.7 psi at 32°F (101 kPa at 0°C).

ATOMIZATION: The breaking down of a liquid into a fine mist that can be suspended in air.

AUXILIARY ADD-ON COOLER: A supplemental transmission fluid cooling device that is installed in series with the heat exchanger (cooler), located inside the radiator, to provide additional support to cool the hot fluid leaving the torque converter.

AUXILIARY PRESSURE: An added fluid pressure that is introduced into a regulator or balanced valve system to control valve movement. The auxiliary pressure itself can be either a fixed or a variable value. (See balanced valve; regulator valve.)

AWD: All wheel drive.

AXIAL FORCE: A side or end thrust force acting in or along the same plane as the power flow.

AXIAL PLAY: Movement parallel to a shaft or bearing bore.

AXLE CAPACITY: The maximum load-carrying capacity of the axle itself, as specified by the manufacturer. This is usually a higher number than the GAWR.

AXLE RATIO: This is a number (3.07:1, 4.56:1, for example) expressing the ratio between driveshaft revolutions and wheel revolutions. A low numerical ratio allows the engine to work easier because it doesn't have to turn as fast. A high numerical ratio means that the engine has to turn more rpm's to move the wheels through the same number of turns.

BACKFIRE: The sudden combustion of gases in the intake or exhaust system that results in a loud explosion.

BACKLASH: The clearance or play between two parts, such as meshed gears.

BACKPRESSURE: Restrictions in the exhaust system that slow the exit of exhaust gases from the combustion chamber.

BAKELITE®: A heat resistant, plastic insulator material commonly used in printed circuit boards and transistorized components.

BALANCED VALVE: A valve that is positioned by opposing auxiliary hydraulic pressures and/or spring force. Examples include mainline regulator, throttle, and governor valves. (See regulator valve.)

BAND: A flexible ring of steel with an inner lining of friction material. When tightened around the outside of a drum, a planetary member is held stationary to the transmission/transaxle case.

BALL BEARING: A bearing made up of hardened inner and outer races between which hardened steel balls roll.

BALL JOINT: A ball and matching socket connecting suspension components (steering knuckle to lower control arms). It permits rotating movement in any direction between the components that are joined.

BARO (BAROMETRIC PRESSURE SENSOR): Measures the change in the intake manifold pressure caused by changes in altitude.

BAROMETRIC MANIFOLD ABSOLUTE PRESSURE (BMAP) SENSOR: Operates similarly to a conventional MAP sensor; reads intake manifold pressure and is also responsible for determining altitude and barometric pressure prior to engine operation.

BAROMETRIC PRESSURE: (See atmospheric pressure.)

BALLAST RESISTOR: A resistor in the primary ignition circuit that lowers voltage after the engine is started to reduce wear on ignition components.

BATTERY: A direct current electrical storage unit, consisting of the basic active materials of lead and sulfuric acid, which converts chemical energy into electrical energy. Used to provide current for the operation of the starter as well as other equipment, such as the radio, lighting, etc.

BEAD: The portion of a tire that holds it on the rim.

BEARING: A friction reducing, supportive device usually located between a stationary part and a moving part.

BEFORE TOP DEAD CENTER (BTDC): The point just before the piston reaches the top of its travel on the compression stroke.

BELTED TIRE: Tire construction similar to bias-ply tires, but using two or more layers of reinforced belts between body plies and the tread.

BEZEL: Piece of metal surrounding radio, headlights, gauges or similar components; sometimes used to hold the glass face of a gauge in the dash.

BIAS-PLY TIRE: Tire construction, using body ply reinforcing cords which run at alternating angles to the center line of the tread.

BI-METAL TEMPERATURE SENSOR: Any sensor or switch made of two dissimilar types of metal that bend when heated or cooled due to the different expansion rates of the alloys. These types of sensors usually function as an on/off switch.

BLOCK: See Engine Block.

BLOW-BY: Combustion gases, composed of water vapor and unburned fuel, that leak past the piston rings into the crankcase during normal engine operation. These gases are removed by the PCV system to prevent the buildup of harmful acids in the crankcase.

BOOK TIME: See Labor Time.

BOOK VALUE: The average value of a car, widely used to determine trade-in and resale value.

BOOST VALVE: Used at the base of the regulator valve to increase mainline pressure.

BORE: Diameter of a cylinder.

BRAKE CALIPER: The housing that fits over the brake disc. The caliper holds the brake pads, which are pressed against the discs by the caliper pistons when the brake pedal is depressed.

BRAKE HORSEPOWER (BHP): The actual horsepower available at the engine flywheel as measured by a dynamometer.

BRAKE FADE: Loss of braking power, usually caused by excessive heat after repeated brake applications.

BRAKE HORSEPOWER: Usable horsepower of an engine measured at the crankshaft.

BRAKE PAD: A brake shoe and lining assembly used with disc brakes.

BRAKE PROPORTIONING VALVE: A valve on the master cylinder which restricts hydraulic brake pressure to the wheels to a specified amount, preventing wheel lock-up.

BREAKAWAY: Often used by Chrysler to identify first-gear operation in D and 2 ranges. In these ranges, first-gear operation depends on a one-way roller clutch that holds on acceleration and releases (breaks away) on deceleration, resulting in a freewheeling coast-down condition.

BRAKE SHOE: The backing for the brake lining. The term is, however, usually applied to the assembly of the brake backing and lining.

BREAKER POINTS: A set of points inside the distributor, operated by a cam, which make and break the ignition circuit.

BRINNELLING: A wear pattern identified by a series of indentations at regular intervals. This condition is caused by a lack of lube, overload situations, and/or vibrations.

BTDC: Before Top Dead Center.

BUMP: Sudden and forceful apply of a clutch or band.

BUSHING: A liner, usually removable, for a bearing; an anti-friction liner used in place of a bearing.

CALIFORNIA ENGINE: An engine certified by the EPA for use in California only; conforms to more stringent emission regulations than Federal engine.

CALIPER: A hydraulically activated device in a disc brake system, which is mounted straddling the brake rotor (disc). The caliper contains at least one piston and two brake pads. Hydraulic pressure on the piston(s) forces the pads against the rotor.

CAPACITY: The quantity of electricity that can be delivered from a unit, as from a battery in ampere-hours, or output, as from a generator.

CAMBER: One of the factors of wheel alignment. Viewed from the front of the car, it is the inward or outward tilt of the wheel. The top of the tire will lean outward (positive camber) or inward (negative camber).

CAMSHAFT: A shaft in the engine on which are the lobes (cams) which operate the valves. The camshaft is driven by the crankshaft, via a belt, chain or gears, at one half the crankshaft speed.

CAPACITOR: A device which stores an electrical charge.

CARBON MONOXIDE (CO): A colorless, odorless gas given off as a normal byproduct of combustion. It is poisonous and extremely dangerous in confined areas, building up slowly to toxic levels without warning if adequate ventilation is not available.

CARBURETOR: A device, usually mounted on the intake manifold of an engine, which mixes the air and fuel in the proper proportion to allow even combustion.

CASTER: The forward or rearward tilt of an imaginary line drawn through the upper ball joint and the center of the wheel. Viewed from the sides, positive caster (forward tilt) lends directional stability, while negative caster (rearward tilt) produces instability.

CATALYTIC CONVERTER: A device installed in the exhaust system, like a muffler, that converts harmful byproducts of combustion into carbon dioxide and water vapor by means of a heat-producing chemical reaction.

CENTRIFUGAL ADVANCE: A mechanical method of advancing the spark timing by using flyweights in the distributor that react to centrifugal force generated by the distributor shaft rotation.

CENTRIFUGAL FORCE: The outward pull of a revolving object, away from the center of revolution. Centrifugal force increases with the speed of rotation.

CETANE RATING: A measure of the ignition value of diesel fuel. The higher the cetane rating, the better the fuel. Diesel fuel cetane rating is roughly comparable to gasoline octane rating.

CHECK VALVE: Any one-way valve installed to permit the flow of air, fuel or vacuum in one direction only.

CHOKE: The valve/plate that restricts the amount of air entering an engine on the induction stroke, thereby enriching the air/fuel ratio.

CHUGGLE: Bucking or jerking condition that may be engine related and may be most noticeable when converter clutch is engaged; similar to the feel of towing a trailer.

CIRCLIP: A split steel snapring that fits into a groove to hold various parts in place.

CIRCUIT BREAKER: A switch which protects an electrical circuit from overload by opening the circuit when the current flow exceeds a pre-determined level. Some circuit breakers must be reset manually, while most reset automatically.

CIRCUIT: Any unbroken path through which an electrical current can flow. Also used to describe fuel flow in some instances.

CIRCUIT, BYPASS: Another circuit in parallel with the major circuit through which power is diverted.

CIRCUIT, CLOSED: An electrical circuit in which there is no interruption of current flow.

CIRCUIT, GROUND: The non-insulated portion of a complete circuit used as a common potential point. In automotive circuits, the ground is composed of metal parts, such as the engine, body sheet metal, and frame and is usually a negative potential.

CIRCUIT, HOT: That portion of a circuit not at ground potential. The hot circuit is usually insulated and is connected to the positive side of the battery.

CIRCUIT, OPEN: A break or lack of contact in an electrical circuit, either intentional (switch) or unintentional (bad connection or broken wire).

CIRCUIT, PARALLEL: A circuit having two or more paths for current flow with common positive and negative tie points. The same voltage is applied to each load device or parallel branch.

CIRCUIT, SERIES: An electrical system in which separate parts are connected end to end, using one wire, to form a single path for current to flow.

CIRCUIT, SHORT: A circuit that is accidentally completed in an electrical path for which it was not intended.

CLAMPING (ISOLATION) DIODES: Diodes positioned in a circuit to prevent self-induction from damaging electronic components.

CLEARCOAT: A transparent layer which, when sprayed over a vehicle's paint job, adds gloss and depth as well as an additional protective coating to the finish.

CLUTCH: Part of the power train used to connect/disconnect power to the rear wheels.

CLUTCH, FLUID: The same as a fluid coupling. A fluid clutch or coupling performs the same function as a friction clutch by utilizing fluid friction and inertia as opposed to solid friction used by a friction clutch. (See fluid coupling.)

CLUTCH, FRICTION: A coupling device that provides a means of smooth and positive engagement and disengagement of engine torque to the vehicle powertrain. Transmission of power through the clutch is accomplished by bringing one or more rotating drive members into contact with complementing driven members.

COAST: Vehicle deceleration caused by engine braking conditions.

COEFFICIENT OF FRICTION: The amount of surface tension between two contacting surfaces; identified by a scientifically calculated number.

COIL: Part of the ignition system that boosts the relatively low voltage supplied by the car's electrical system to the high voltage required to fire the spark plugs.

COMBINATION MANIFOLD: An assembly which includes both the intake and exhaust manifolds in one casting.

COMBINATION VALVE: A device used in some fuel systems that routes fuel vapors to a charcoal storage canister instead of venting them into the atmosphere. The valve relieves fuel tank pressure and allows fresh air into the tank as the fuel level drops to prevent a vapor lock situation.

COMBUSTION CHAMBER: The part of the engine in the cylinder head where combustion takes place.

COMPOUND GEAR: A gear consisting of two or more simple gears with a common shaft.

COMPOUND PLANETARY: A gearset that has more than the three elements found in a simple gearset and is constructed by combining members of two planetary gearsets to create additional gear ratio possibilities.

COMPRESSION CHECK: A test involving removing each spark plug and inserting a gauge. When the engine is cranked, the gauge will record a pressure reading in the individual cylinder. General operating condition can be determined from a compression check.

COMPRESSION RATIO: The ratio of the volume between the piston and cylinder head when the piston is at the bottom of its stroke (bottom dead center) and when the piston is at the top of its stroke (top dead center).

COMPUTER: An electronic control module that correlates input data according to prearranged engineered instructions; used for the management of an actuator system or systems.

CONDENSER: An electrical device which acts to store an electrical charge, preventing voltage surges.

2. A radiator-like device in the air conditioning system in which refrigerant gas condenses into a liquid, giving off heat.

CONDUCTOR: Any material through which an electrical current can be transmitted easily.

CONNECTING ROD: The connecting link between the crankshaft and piston.

CONSTANT VELOCITY JOINT: Type of universal joint in a halfshaft assembly in which the output shaft turns at a constant angular velocity without variation, provided that the speed of the input shaft is constant.

CONTINUITY: Continuous or complete circuit. Can be checked with an ohmmeter.

CONTROL ARM: The upper or lower suspension components which are mounted on the frame and support the ball joints and steering knuckles.

CONVENTIONAL IGNITION: Ignition system which uses breaker points.

CONVERTER: (See torque converter.)

CONVERTER LOCKUP: The switching from hydrodynamic to direct mechanical drive, usually through the application of a friction element called the converter clutch.

COOLANT: Mixture of water and anti-freeze circulated through the engine to carry off heat produced by the engine.

CORROSION INHIBITOR: An inhibitor in ATF that prevents corrosion of bushings, thrust washers, and oil cooler brazed joints.

COUNTERSHAFT: An intermediate shaft which is rotated by a mainshaft and transmits, in turn, that rotation to a working part.

COUPLING PHASE: Occurs when the torque converter is operating at its greatest hydraulic efficiency. The speed differential between the impeller and the turbine is at its minimum. At this point, the stator freewheels, and there is no torque multiplication.

CRANKCASE: The lower part of an engine in which the crankshaft and related parts operate.

CRANKSHAFT: Engine component (connected to pistons by connecting rods) which converts the reciprocating (up and down) motion of pistons to rotary motion used to turn the driveshaft.

CURB WEIGHT: The weight of a vehicle without passengers or payload, but including all fluids (oil, gas, coolant, etc.) and other equipment specified as standard.

CURRENT: The flow (or rate) of electrons moving through a circuit. Current is measured in amperes (amp).

CURRENT FLOW CONVENTIONAL: Current flows through a circuit from the positive terminal of the source to the negative terminal (plus to minus).

CURRENT FLOW, ELECTRON: Current or electrons flow from the negative terminal of the source, through the circuit, to the positive terminal (minus to plus).

CV-JOINT: Constant velocity joint.

CYCLIC VIBRATIONS: The off-center movement of a rotating object that is affected by its initial balance, speed of rotation, and working angles.

CYLINDER BLOCK: See engine block.

CYLINDER HEAD: The detachable portion of the engine, usually fastened to the top of the cylinder block and containing all or most of the combustion chambers. On overhead valve engines, it contains the valves and their operating parts. On overhead cam engines, it contains the camshaft as well.

CYLINDER: In an engine, the round hole in the engine block in which the piston(s) ride.

DATA LINK CONNECTOR (DLC): Current acronym/term applied to the federally mandated, diagnostic junction connector that is used to monitor ECM/PC/TCM inputs, processing strategies, and outputs including diagnostic trouble codes (DTCs).

DEAD CENTER: The extreme top or bottom of the piston stroke.

DECELERATION BUMP: When referring to a torque converter clutch in the applied position, a sudden release of the accelerator pedal causes a forceful reversal of power through the drivetrain (engine braking), just prior to the apply plate actually being released.

DELAYED (LATE OR EXTENDED): Condition where shift is expected but does not occur for a period of time, for example, where clutch or band engagement does not occur as quickly as expected during part throttle or wide open throttle apply of accelerator or when manually downshifting to a lower range.

DETENT: A spring-loaded plunger, pin, ball, or pawl used as a holding device on a ratchet wheel or shaft. In automatic transmissions, a detent mechanism is used for locking the manual valve in place.

DETENT DOWNSHIFT: (See kickdown.)

DETERGENT: An additive in engine oil to improve its operating characteristics.

DETONATION: An unwanted explosion of the air/fuel mixture in the combustion chamber caused by excess heat and compression, advanced timing, or an overly lean mixture. Also referred to as "ping".

DEXRON®: A brand of automatic transmission fluid.

DIAGNOSTIC TROUBLE CODES (DTCs): A digital display from the control module memory that identifies the input, processor, or output device circuit that is related to the powertrain emission/driveability malfunction detected. Diagnostic trouble codes can be read by the MIL to flash any codes or by using a handheld scanner.

DIAPHRAGM: A thin, flexible wall separating two cavities, such as in a vacuum advance unit.

DIESELING: The engine continues to run after the car is shut off; caused by fuel continuing to be burned in the combustion chamber.

DIFFERENTIAL: A geared assembly which allows the transmission of motion between drive axles, giving one axle the ability to rotate faster than the other, as in cornering.

DIFFERENTIAL AREAS: When opposing faces of a spool valve are acted upon by the same pressure but their areas differ in size, the face with the larger area produces the differential force and valve movement. (See spool valve.)

DIFFERENTIAL FORCE: (See differential areas)

DIGITAL READOUT: A display of numbers or a combination of numbers and letters.

DIGITAL VOLT OHMMETER: An electronic diagnostic tool used to measure voltage, ohms and amps as well as several other functions, with the readings displayed on a digital screen in tenths, hundredths and thousandths.

DIODE: An electrical device that will allow current to flow in one direction only.

DIRECT CURRENT (DC): Electrical current that flows in one direction only.

DIRECT DRIVE: The gear ratio is 1:1, with no change occurring in the torque and speed input/output relationship.

DISC BRAKE: A hydraulic braking assembly consisting of a brake disc, or rotor, mounted on an axle shaft, and a caliper assembly containing, usually two brake pads which are activated by hydraulic pressure. The pads are forced against the sides of the disc, creating friction which slows the vehicle.

DISPERSANTS: Suspend dirt and prevent sludge buildup in a liquid, such as engine oil.

DOUBLE BUMP (DOUBLE FEEL): Two sudden and forceful applies of a clutch or band.

DISPLACEMENT: The total volume of air that is displaced by all pistons as the engine turns through one complete revolution.

DISTRIBUTOR: A mechanically driven device on an engine which is responsible for electrically firing the spark plug at a pre-determined point of the piston stroke.

DOHC: Double overhead camshaft.

DOUBLE OVERHEAD CAMSHAFT: The engine utilizes two camshafts mounted in one cylinder head. One camshaft operates the exhaust valves, while the other operates the intake valves.

DOWEL PIN: A pin, inserted in mating holes in two different parts allowing those parts to maintain a fixed relationship.

DRIVELINE: The drive connection between the transmission and the drive wheels.

DRIVE TRAIN: The components that transmit the flow of power from the engine to the wheels. The components include the clutch, transmission, driveshafts (or axle shafts in front wheel drive), U-joints and differential.

DRUM BRAKE: A braking system which consists of two brake shoes and one or two wheel cylinders, mounted on a fixed backing plate, and a brake drum, mounted on an axle, which revolves around the assembly.

DRY CHARGED BATTERY: Battery to which electrolyte is added when the battery is placed in service.

DVOM: Digital volt ohmmeter

DWELL: The rate, measured in degrees of shaft rotation, at which an electrical circuit cycles on and off.

DYNAMIC: An application in which there is rotating or reciprocating motion between the parts.

EARLY: Condition where shift occurs before vehicle has reached proper speed, which tends to labor engine after upshift.

EBCM: See Electronic Control Unit (ECU).

ECM: See Electronic Control Unit (ECU).

ECU: Electronic control unit.

ELECTRODE: Conductor (positive or negative) of electric current.

ELECTROLYSIS: A surface etching or bonding of current conducting transmission/transaxle components that may occur when grounding straps are missing or in poor condition.

ELECTROLYTE: A solution of water and sulfuric acid used to activate the battery. Electrolyte is extremely corrosive.

ELECTROMAGNET: A coil that produces a magnetic field when current flows through its windings.

ELECTROMAGNETIC INDUCTION: A method to create (generate) current flow through the use of magnetism.

ELECTROMAGNETISM: The effects surrounding the relationship between electricity and magnetism.

ELECTROMOTIVE FORCE (EMF): The force or pressure (voltage) that causes current movement in an electrical circuit.

ELECTRONIC CONTROL UNIT: A digital computer that controls engine (and sometimes transmission, brake or other vehicle system) functions based on data received from various sensors. Examples used by some manufacturers include Electronic Brake Control Module (EBCM), Engine Control Module (ECM), Powertrain Control Module (PCM) or Vehicle Control Module (VCM).

ELECTRONIC IGNITION: A system in which the timing and firing of the spark plugs is controlled by an electronic control unit, usually called a module. These systems have no points or condenser.

ELECTRONIC PRESSURE CONTROL (EPC) SOLENOID: A specially designed solenoid containing a spool valve and spring assembly to control fluid mainline pressure. A variable current flow, controlled by the ECM/PCM, varies the internal force of the solenoid on the spool valve and resulting mainline pressure. (See variable force solenoid.)

ELECTRONICS: Miniaturized electrical circuits utilizing semiconductors, solid-state devices, and printed circuits. Electronic circuits utilize small amounts of power.

ELECTRONIFICATION: The application of electronic circuitry to a mechanical device. Regarding automatic transmissions, electrification is incorporated into converter clutch lockup, shift scheduling, and line pressure control systems.

ELECTROSTATIC DISCHARGE (ESD): An unwanted, high-voltage electrical current released by an individual who has taken on a static charge of electricity. Electronic components can be easily damaged by ESD.

ELEMENT: A device within a hydrodynamic drive unit designed with a set of blades to direct fluid flow.

ENAMEL: Type of paint that dries to a smooth, glossy finish.

END BUMP (END FEEL OR SLIP BUMP): Firmer feel at end of shift when compared with feel at start of shift.

END-PLAY: The clearance/gap between two components that allows for expansion of the parts as they warm up, to prevent binding and to allow space for lubrication.

ENERGY: The ability or capacity to do work.

ENGINE: The primary motor or power apparatus of a vehicle, which converts liquid or gas fuel into mechanical energy.

ENGINE BLOCK: The basic engine casting containing the cylinders, the crankshaft main bearings, as well as machined surfaces for the mounting of other components such as the cylinder head, oil pan, transmission, etc.

ENGINE BRAKING: Use of engine to slow vehicle by manually downshifting during zero-throttle coast down.

ENGINE CONTROL MODULE (ECM): Manages the engine and incorporates output control over the torque converter clutch solenoid. (Note: Current designation for the ECM in late model vehicles is PCM.)

ENGINE COOLANT TEMPERATURE (ECT) SENSOR: Prevents converter clutch engagement with a cold engine; also used for shift timing and shift quality.

EP LUBRICANT: EP (extreme pressure) lubricants are specially formulated for use with gears involving heavy loads (transmissions, differentials, etc.).

ETHYL: A substance added to gasoline to improve its resistance to knock, by slowing down the rate of combustion.

ETHYLENE GLYCOL: The base substance of antifreeze.

EXHAUST MANIFOLD: A set of cast passages or pipes which conduct exhaust gases from the engine.

FAIL-SAFE (BACKUP) CONTROL: A substitute value used by the PCM/TCM to replace a faulty signal from an input sensor. The temporary value allows the vehicle to continue to be operated.

FAST IDLE: The speed of the engine when the choke is on. Fast idle speeds engine warm-up.

FEDERAL ENGINE: An engine certified by the EPA for use in any of the 49 states (except California).

FEEDBACK: A circuit malfunction whereby current can find another path to feed load devices.

FEELER GAUGE: A blade, usually metal, of precisely predetermined thickness, used to measure the clearance between two parts.

FILAMENT: The part of a bulb that glows; the filament creates high resistance to current flow and actually glows from the resulting heat.

FINAL DRIVE: An essential part of the axle drive assembly where final gear reduction takes place in the powertrain. In RWD applications and north-south FWD applications, it must also change the power flow direction to the axle shaft by ninety degrees. (Also see axle ratio).

FIRING ORDER: The order in which combustion occurs in the cylinders of an engine. Also the order in which spark is distributed to the plugs by the distributor.

FIRM: A noticeable quick apply of a clutch or band that is considered normal with medium to heavy throttle shift; should not be confused with harsh or rough.

FLAME FRONT: The term used to describe certain aspects of the fuel explosion in the cylinders. The flame front should move in a controlled pattern across the cylinder, rather than simply exploding immediately.

FLARE (SLIPPING): A quick increase in engine rpm accompanied by momentary loss of torque; generally occurs during shift.

FLAT ENGINE: Engine design in which the pistons are horizontally opposed. Porsche, Subaru and some old VW are common examples of flat engines.

FLAT RATE: A dealership term referring to the amount of money paid to a technician for a repair or diagnostic service based on that particular service versus dealership's labor time (NOT based on the actual time the technician spent on the job).

FLAT SPOT: A point during acceleration when the engine seems to lose power for an instant.

FLOODING: The presence of too much fuel in the intake manifold and combustion chamber which prevents the air/fuel mixture from firing, thereby causing a no-start situation.

FLUID: A fluid can be either liquid or gas. In hydraulics, a liquid is used for transmitting force or motion.

FLUID COUPLING: The simplest form of hydrodynamic drive, the fluid coupling consists of two look-alike members with straight radial varies referred to as the impeller (pump) and the turbine. Input torque is always equal to the output torque.

FLUID DRIVE: Either a fluid coupling or a fluid torque converter. (See hydrodynamic drive units.)

FLUID TORQUE CONVERTER: A hydrodynamic drive that has the ability to act both as a torque multiplier and fluid coupling. (See hydrodynamic drive units; torque converter.)

FLUID VISCOSITY: The resistance of a liquid to flow. A cold fluid (oil) has greater viscosity and flows more slowly than a hot fluid (oil).

FLYWHEEL: A heavy disc of metal attached to the rear of the crankshaft. It smoothes the firing impulses of the engine and keeps the crankshaft turning during periods when no firing takes place. The starter also engages the flywheel to start the engine.

FOOT POUND (ft. lbs., lbs. ft. or sometimes, ft. lb.): The amount of energy or work needed to raise an item weighing one pound, a distance of one foot.

FREEZE PLUG: A plug in the engine block which will be pushed out if the coolant freezes. Sometimes called expansion plugs, they protect the block from cracking should the coolant freeze.

FRICTION: The resistance that occurs between contacting surfaces. This relationship is expressed by a ratio called the coefficient of friction (CL).

FRICTION, COEFFICIENT OF: The amount of surface tension between two contacting surfaces; expressed by a scientifically calculated number.

FRONT END ALIGNMENT: A service to set caster, camber and toe-in to the correct specifications. This will ensure that the car steers and handles properly and that the tires wear properly.

FRICTION MODIFIER: Changes the coefficient of friction of the fluid between the mating steel and composition clutch/band surfaces during the engagement process and allows for a certain amount of intentional slipping for a good "shift-feel".

FRONTAL AREA: The total frontal area of a vehicle exposed to air flow.

FUEL FILTER: A component of the fuel system containing a porous paper element used to prevent any impurities from entering the engine through the fuel system. It usually takes the form of a canister-like housing, mounted in-line with the fuel hose, located anywhere on a vehicle between the fuel tank and engine.

FUEL INJECTION: A system replacing the carburetor that sprays fuel into the cylinder through nozzles. The amount of fuel can be more precisely controlled with fuel injection.

FULL FLOATING AXLE: An axle in which the axle housing extends through the wheel giving bearing support on the outside of the housing. The front axle of a four-wheel drive vehicle is usually a full floating axle, as are the rear axles of many larger (1 ton and over) pick-ups and vans.

FULL-TIME FOUR-WHEEL DRIVE: A four-wheel drive system that continuously delivers power to all four wheels. A differential between the front and rear driveshafts permits variations in axle speeds to control gear wind-up without damage.

FULL THROTTLE DETENT DOWNSHIFT: A quick apply of accelerator pedal to its full travel, forcing a downshift.

FUSE: A protective device in a circuit which prevents circuit overload by breaking the circuit when a specific amperage is present. The device is constructed around a strip or wire of a lower amperage rating than the circuit it is designed to protect. When an amperage higher than that stamped on the fuse is present in the circuit, the strip or wire melts, opening the circuit.

FUSIBLE LINK: A piece of wire in a wiring harness that performs the same job as a fuse. If overloaded, the fusible link will melt and interrupt the circuit.

FWD: Front wheel drive.

GAWR: (Gross axle weight rating) the total maximum weight an axle is designed to carry.

GCW: (Gross combined weight) total combined weight of a tow vehicle and trailer.

GARAGE SHIFT: initial engagement feel of transmission, neutral to reverse or neutral to a forward drive.

GARAGE SHIFT FEEL: A quick check of the engagement quality and responsiveness of reverse and forward gears. This test is done with the vehicle stationary.

GEAR: A toothed mechanical device that acts as a rotating lever to transmit power or turning effort from one shaft to another. (See gear ratio.)

GEAR RATIO: A ratio expressing the number of turns a smaller gear will make to turn a larger gear through one revolution. The ratio is found by dividing the number of teeth on the smaller gear into the number of teeth on the larger gear.

GEARBOX: Transmission

GEAR REDUCTION: Torque is multiplied and speed decreased by the factor of the gear ratio. For example, a 3:1 gear ratio changes an input torque of 180 ft. lbs. and an input speed of 2700 rpm to 540 Ft. lbs. and 900 rpm, respectively. (No account is taken of frictional losses, which are always present.)

GEARTRAIN: A succession of intermeshing gears that form an assembly and provide for one or more torque changes as the power input is transmitted to the power output.

GEL COAT: A thin coat of plastic resin covering fiberglass body panels.

GENERATOR: A device which produces direct current (DC) necessary to charge the battery.

GOVERNOR: A device that senses vehicle speed and generates a hydraulic oil pressure. As vehicle speed increases, governor oil pressure rises.

GROUND CIRCUIT: (See circuit, ground.)

GROUND SIDE SWITCHING: The electrical/electronic circuit control switch is located after the circuit load.

GVWR: (Gross vehicle weight rating) total maximum weight a vehicle is designed to carry including the weight of the vehicle, passengers, equipment, gas, oil, etc.

HALOGEN: A special type of lamp known for its quality of brilliant white light. Originally used for fog lights and driving lights.

HARD CODES: DTCs that are present at the time of testing; also called continuous or current codes.

HARSH(ROUGH): An apply of a clutch or band that is more noticeable than a firm one; considered undesirable at any throttle position.

HEADER TANK: An expansion tank for the radiator coolant. It can be located remotely or built into the radiator.

HEAT RANGE: A term used to describe the ability of a spark plug to carry away heat. Plugs with longer nosed insulators take longer to carry heat off effectively.

HEAT RISER: A flapper in the exhaust manifold that is closed when the engine is cold, causing hot exhaust gases to heat the intake manifold providing better cold engine operation. A thermostatic spring opens the flapper when the engine warms up.

HEAVY THROTTLE: Approximately three-fourths of accelerator pedal travel.

HEMI: A name given an engine using hemispherical combustion chambers.

HERTZ (HZ): The international unit of frequency equal to one cycle per second (10,000 Hertz equals 10,000 cycles per second).

HIGH-IMPEDANCE DVOM (DIGITAL VOLT-OHMMETER): This styled device provides a built-in resistance value and is capable of limiting circuit current flow to safe milliamp levels.

HIGH RESISTANCE: Often refers to a circuit where there is an excessive amount of opposition to normal current flow.

HORSEPOWER: A measurement of the amount of work; one horsepower is the amount of work necessary to lift 33,000 lbs. one foot in one minute. Brake horsepower (bhp) is the horsepower delivered by an engine on a dynamometer. Net horsepower is the power remaining (measured at the flywheel of the engine) that can be used to turn the wheels after power is consumed through friction and running the engine accessories (water pump, alternator, air pump, fan etc.)

HOT CIRCUIT: (See circuit, hot; hot lead.)

HOT LEAD: A wire or conductor in the power side of the circuit. (See circuit, hot.)

HOT SIDE SWITCHING: The electrical/electronic circuit control switch is located before the circuit load.

HUB: The center part of a wheel or gear.

HUNTING (BUSYNESS): Repeating quick series of up-shifts and downshifts that causes noticeable change in engine rpm, for example, as in a 4-3-4 shift pattern.

HYDRAULICS: The use of liquid under pressure to transfer force of motion.

HYDROCARBON (HC): Any chemical compound made up of hydrogen and carbon. A major pollutant formed by the engine as a by-product of combustion.

HYDRODYNAMIC DRIVE UNITS: Devices that transmit power solely by the action of a kinetic fluid flow in a closed recirculating path. An impeller energizes the fluid and discharges the high-speed jet stream into the turbine for power output.

HYDROMETER: An instrument used to measure the specific gravity of a solution.

HYDROPLANING: A phenomenon of driving when water builds up under the tire tread, causing it to lose contact with the road. Slowing down will usually restore normal tire contact with the road.

HYPOID GEARSET: The drive pinion gear may be placed below or above the centerline of the driven gear; often used as a final drive gearset.

IDLE MIXTURE: The mixture of air and fuel (usually about 14:1) being fed to the cylinders. The idle mixture screw(s) are sometimes adjusted as part of a tune-up.

IDLER ARM: Component of the steering linkage which is a geometric duplicate of the steering gear arm. It supports the right side of the center steering link.

IMPELLER: Often called a pump, the impeller is the power input (drive) member of a hydrodynamic drive. As part of the torque converter cover, it acts as a centrifugal pump and puts the fluid in motion.

INCH POUND (inch lbs.; sometimes in. lb. or in. lbs.): One twelfth of a foot pound.

INDUCTANCE: The force that produces voltage when a conductor is passed through a magnetic field.

INDUCTION: A means of transferring electrical energy in the form of a magnetic field. Principle used in the ignition coil to increase voltage.

INITIAL FEEL: A distinct firmer feel at start of shift when compared with feel at finish of shift.

INJECTOR: A device which receives metered fuel under relatively low pressure and is activated to inject the fuel into the engine under relatively high pressure at a predetermined time.

INPUT: In an automatic transmission, the source of power from the engine is absorbed by the torque converter, which provides the power input into the transmission. The turbine drives the input(turbine)shaft.

INPUT SHAFT: The shaft to which torque is applied, usually carrying the driving gear or gears.

INTAKE MANIFOLD: A casting of passages or pipes used to conduct air or a fuel/air mixture to the cylinders.

INTERNAL GEAR: The ring-like outer gear of a planetary gearset with the gear teeth cut on the inside of the ring to provide a mesh with the planet pinions.

ISOLATION (CLAMPING) DIODES: Diodes positioned in a circuit to prevent self-induction from damaging electronic components.

IX ROTARY GEAR PUMP: Contains two rotating members, one shaped with internal gear teeth and the other with external gear teeth. As the gears separate, the fluid fills the gaps between gear teeth, is pulled across a crescent-shaped divider, and then is forced to flow through the outlet as the gears mesh.

IX ROTARY LOBE PUMP: Sometimes referred to as a gerotor type pump. Two rotating members, one shaped with internal lobes and the other with external lobes, separate and then mesh to cause fluid to flow.

JOURNAL: The bearing surface within which a shaft operates.

JUMPER CABLES: Two heavy duty wires with large alligator clips used to provide power from a charged battery to a discharged battery mounted in a vehicle.

JUMPSTART: Utilizing the sufficiently charged battery of one vehicle to start the engine of another vehicle with a discharged battery by the use of jumper cables.

KEY: A small block usually fitted in a notch between a shaft and a hub to prevent slippage of the two parts.

KICKDOWN: Detent downshift system; either linkage, cable, or electrically controlled.

KILO: A prefix used in the metric system to indicate one thousand.

KNOCK: Noise which results from the spontaneous ignition of a portion of the air-fuel mixture in the engine cylinder caused by overly advanced ignition timing or use of incorrectly low octane fuel for that engine.

KNOCK SENSOR: An input device that responds to spark knock, caused by over advanced ignition timing.

LABOR TIME: A specific amount of time required to perform a certain repair or diagnostic service as defined by a vehicle or after-market manufacturer.

LACQUER: A quick-drying automotive paint.

LATE: Shift that occurs when engine is at higher than normal rpm for given amount of throttle.

LIGHT-EMITTING DIODE (LED): A semiconductor diode that emits light as electrical current flows through it; used in some electronic display devices to emit a red or other color light.

LIGHT THROTTLE: Approximately one-fourth of accelerator pedal travel.

LIMITED SLIP: A type of differential which transfers driving force to the wheel with the best traction.

LIMP-IN MODE: Electrical shutdown of the transmission/ transaxle output solenoids, allowing only forward and reverse gears that are hydraulically energized by the manual valve. This permits the vehicle to be driven to a service facility for repair.

LIP SEAL: Molded synthetic rubber seal designed with an outer sealing edge (lip) that points into the fluid containing area to be sealed. This type of seal is used where rotational and axial forces are present.

LITHIUM-BASE GREASE: Chassis and wheel bearing grease using lithium as a base. Not compatible with sodium-base grease.

LOAD DEVICE: A circuit's resistance that converts the electrical energy into light, sound, heat, or mechanical movement.

LOAD RANGE: Indicates the number of plies at which a tire is rated. Load range B equals four-ply rating; C equals six-ply rating; and, D equals an eight-ply rating.

LOAD TORQUE: The amount of output torque needed from the transmission/transaxle to overcome the vehicle load.

LOCKING HUBS: Accessories used on part-time four-wheel drive systems that allow the front wheels to be disengaged from the drive train when four-wheel drive is not being used. When four-wheel drive is desired, the hubs are engaged, locking the wheels to the drive train.

LOCKUP CONVERTER: A torque converter that operates hydraulically and mechanically. When an internal apply plate (lockup plate) clamps to the torque converter cover, hydraulic slippage is eliminated.

LOCK RING: See Circlip or Snapring

MAGNET: Any body with the property of attracting iron or steel.

MAGNETIC FIELD: The area surrounding the poles of a magnet that is affected by its attraction or repulsion forces.

MAIN LINE PRESSURE: Often called control pressure or line pressure, it refers to the pressure of the oil leaving the pump and is controlled by the pressure regulator valve.

MALFUNCTION INDICATOR LAMP (MIL): Previously known as a check engine light, the dash-mounted MIL illuminates and signals the driver that an emission or driveability problem with the powertrain has been detected by the ECM/PCM. When this occurs, at least one diagnostic trouble code (DTC) has been stored into the control module memory.

MANIFOLD ABSOLUTE PRESSURE (MAP) SENSOR: Reads the amount of air pressure (vacuum) in the engine's intake manifold system; its signal is used to analyze engine load conditions.

MANIFOLD VACUUM: Low pressure in an engine intake manifold formed just below the throttle plates. Manifold vacuum is highest at idle and drops under acceleration.

MANIFOLD: A casting of passages or set of pipes which connect the cylinders to an inlet or outlet source.

MANUAL LEVER POSITION SWITCH (MLPS): A mechanical switching unit that is typically mounted externally to the transmission/transaxle to inform the PCM/ECM which gear range the driver has selected.

MANUAL VALVE: Located inside the transmission/transaxle, it is directly connected to the driver's shift lever. The position of the manual valve determines which hydraulic circuits will be charged with oil pressure and the operating mode of the transmission.

MANUAL VALVE LEVER POSITION SENSOR (MVLPS): The input from this device tells the TCM what gear range was selected.

MASS AIR FLOW (MAF) SENSOR: Measures the airflow into the engine.

MASTER CYLINDER: The primary fluid pressurizing device in a hydraulic system. In automotive use, it is found in brake and hydraulic clutch systems and is pedal activated, either directly or, in a power brake system, through the power booster.

MacPherson STRUT: A suspension component combining a shock absorber and spring in one unit.

MEDIUM THROTTLE: Approximately one-half of accelerator pedal travel.

MEGA: A metric prefix indicating one million.

MEMBER: An independent component of a hydrodynamic unit such as an impeller, a stator, or a turbine. It may have one or more elements.

MERCON: A fluid developed by Ford Motor Company in 1988. It contains a friction modifier and closely resembles operating characteristics of Dexron.

METAL SEALING RINGS: Made from cast iron or aluminum, their primary application is with dynamic components involving pressure sealing circuits of rotating members. These rings are designed with either butt or hook lock end joints.

METER (ANALOG): A linear-style meter representing data as lengths; a needle-style instrument interfacing with logical numerical increments. This style of electrical meter uses relatively low impedance internal resistance and cannot be used for testing electronic circuitry.

METER (DIGITAL): Uses numbers as a direct readout to show values. Most meters of this style use high impedance internal resistance and must be used for testing low current electronic circuitry.

MICRO: A metric prefix indicating one-millionth (0.000001).

MILLI: A metric prefix indicating one-thousandth (0.001).

MINIMUM THROTTLE: The least amount of throttle opening required for upshift; normally close to zero throttle.

MISFIRE: Condition occurring when the fuel mixture in a cylinder fails to ignite, causing the engine to run roughly.

MODULE: Electronic control unit, amplifier or igniter of solid state or integrated design which controls the current flow in the ignition primary circuit based on input from the pick-up coil. When the module opens the primary circuit, high secondary voltage is induced in the coil.

MODULATED: In an electronic-hydraulic converter clutch system (or shift valve system), the term modulated refers to the pulsing of a solenoid, at a variable rate. This action controls the buildup of oil pressure in the hydraulic circuit to allow a controlled amount of clutch slippage.

MODULATED CONVERTER CLUTCH CONTROL (MCCC): A pulse width duty cycle valve that controls the converter lockup apply pressure and maximizes smoother transitions between lock and unlock conditions.

MODULATOR PRESSURE (THROTTLE PRESSURE): A hydraulic signal oil pressure relating to the amount of engine load, based on either the amount of throttle plate opening or engine vacuum.

MODULATOR VALVE: A regulator valve that is controlled by engine vacuum, providing a hydraulic pressure that varies in relation to engine torque. The hydraulic torque signal functions to delay the shift pattern and provide a line pressure boost. (See throttle valve.)

MOTOR: An electromagnetic device used to convert electrical energy into mechanical energy.

MULTIPLE-DISC CLUTCH: A grouping of steel and friction lined plates that, when compressed together by hydraulic pressure acting upon a piston, lock or unlock a planetary member.

MULTI-WEIGHT: Type of oil that provides adequate lubrication at both high and low temperatures.

needed to move one amp through a resistance of one ohm.

MUSHY: Same as soft; slow and drawn out clutch apply with very little shift feel.

MUTUAL INDUCTION: The generation of current from one wire circuit to another by movement of the magnetic field surrounding a current-carrying circuit as its ampere flow increases or decreases.

NEEDLE BEARING: A bearing which consists of a number (usually a large number) of long, thin rollers.

NITROGEN OXIDE (NOx): One of the three basic pollutants found in the exhaust emission of an internal combustion engine. The amount of NOx usually varies in an inverse proportion to the amount of HC and CO.

NONPOSITIVE SEALING: A sealing method that allows some minor leakage, which normally assists in lubrication.

O2 SENSOR: Located in the engine's exhaust system, it is an input device to the ECM/PCM for managing the fuel delivery and ignition system. A scanner can be used to observe the fluctuating voltage readings produced by an O2 sensor as the oxygen content of the exhaust is analyzed.

O-RING SEAL: Molded synthetic rubber seal designed with a circular cross-section. This type of seal is used primarily in static applications.

OBD II (ON-BOARD DIAGNOSTICS, SECOND GENERATION): Refers to the federal law mandating tighter control of 1996 and newer vehicle emissions, active monitoring of related devices, and standardization of terminology, data link connectors, and other technician concerns.

OCTANE RATING: A number, indicating the quality of gasoline based on its ability to resist knock. The higher the number, the better the quality. Higher compression engines require higher octane gas.

OEM: Original Equipment Manufactured. OEM equipment is that furnished standard by the manufacturer.

OFFSET: The distance between the vertical center of the wheel and the mounting surface at the lugs. Offset is positive if the center is outside the lug circle; negative offset puts the center line inside the lug circle.

OHM'S LAW: A law of electricity that states the relationship between voltage, current, and resistance. Volts = amperes x ohms

OHM: The unit used to measure the resistance of conductor-to-electrical flow. One ohm is the amount of resistance that limits current flow to one ampere in a circuit with one volt of pressure.

OHMMETER: An instrument used for measuring the resistance, in ohms, in an electrical circuit.

ONE-WAY CLUTCH: A mechanical clutch of roller or sprag design that resists torque or transmits power in one direction only. It is used to either hold or drive a planetary member.

ONE-WAY ROLLER CLUTCH: A mechanical device that transmits or holds torque in one direction only.

OPEN CIRCUIT: A break or lack of contact in an electrical circuit, either intentional (switch) or unintentional (bad connection or broken wire).

ORIFICE: Located in hydraulic oil circuits, it acts as a restriction. It slows down fluid flow to either create back pressure or delay pressure buildup downstream.

OSCILLOSCOPE: A piece of test equipment that shows electric impulses as a pattern on a screen. Engine performance can be analyzed by interpreting these patterns.

OUTPUT SHAFT: The shaft which transmits torque from a device, such as a transmission.

OUTPUT SPEED SENSOR (OSS): Identifies transmission/transaxle output shaft speed for shift timing and may be used to calculate TCC slip; often functions as the VSS (vehicle speed sensor).

OVERDRIVE: (1.) A device attached to or incorporated in a transmission/transaxle that allows the engine to turn less than one full revolution for every complete revolution of the wheels. The net effect is to reduce engine rpm, thereby using less fuel. A typical overdrive gear ratio would be .87:1, instead of the normal 1:1 in high gear. (2.) A gear assembly which produces more shaft revolutions than that transmitted to it.

OVERDRIVE PLANETARY GEARSET: A single planetary gearset designed to provide a direct drive and overdrive ratio. When coupled to a three-speed transmission/transaxle configuration, a four-speed/overdrive unit is present.

OVERHEAD CAMSHAFT (OHC): An engine configuration in which the camshaft is mounted on top of the cylinder head and operates the valve either directly or by means of rocker arms.

OVERHEAD VALVE (OHV): An engine configuration in which all of the valves are located in the cylinder head and the camshaft is located in the cylinder block. The camshaft operates the valves via lifters and pushrods.

OVERRUNCLUTCH: Another name for a one-way mechanical clutch. Applies to both roller and sprag designs.

OVERSTEER: The tendency of some vehicles, when steering into a turn, to over-respond or steer more than required, which could result in excessive slip of the rear wheels. Opposite of under-steer.

OXIDATION STABILIZERS: Absorb and dissipate heat. Automatic transmission fluid has high resistance to varnish and sludge buildup that occurs from excessive heat that is generated primarily in the torque converter. Local temperatures as high as 6000F (3150C) can occur at the clutch plates during engagement, and this heat must be absorbed and dissipated. If the fluid cannot withstand the heat, it burns or oxidizes, resulting in an almost immediate destruction of friction materials, clogged filter screen and hydraulic passages, and sticky valves.

OXIDES OF NITROGEN: See nitrogen oxide (NOx).

OXYGEN SENSOR: Used with a feedback system to sense the presence of oxygen in the exhaust gas and signal the computer which can use the voltage signal to determine engine operating efficiency and adjust the air/fuel ratio.

PARALLEL CIRCUIT: (See circuit, parallel.)

PARTS WASHER: A basin or tub, usually with a built-in pump mechanism and hose used for circulating chemical solvent for the purpose of cleaning greasy, oily and dirty components.

PART-TIME FOUR WHEEL DRIVE: A system that is normally in the two wheel drive mode and only runs in four-wheel drive when the system is manually engaged because more traction is desired. Two or four wheel drive is normally selected by a lever to engage the front axle, but if locking hubs are used, these must also be manually engaged in the Lock position. Otherwise, the front axle will not drive the front wheels.

PASSIVE RESTRAINT: Safety systems such as air bags or automatic seat belts which operate with no action required on the part of the driver or passenger. Mandated by Federal regulations on all vehicles sold in the U.S. after 1990.

PAYLOAD: The weight the vehicle is capable of carrying in addition to its own weight. Payload includes weight of the driver, passengers and cargo, but not coolant, fuel, lubricant, spare tire, etc.

PCM: Powertrain control module.

PCV VALVE: A valve usually located in the rocker cover that vents crankcase vapors back into the engine to be reburned.

PERCOLATION: A condition in which the fuel actually "boils," due to excessive heat. Percolation prevents proper atomization of the fuel causing rough running.

PICK-UP COIL: The coil in which voltage is induced in an electronic ignition.

PING: A metallic rattling sound produced by the engine during acceleration. It is usually due to incorrect ignition timing or a poor grade of gasoline.

PINION: The smaller of two gears. The rear axle pinion drives the ring gear which transmits motion to the axle shafts.

PINION GEAR: The smallest gear in a drive gear assembly.

PISTON: A disc or cup that fits in a cylinder bore and is free to move. In hydraulics, it provides the means of converting hydraulic pressure into a usable force. Examples of piston applications are found in servo, clutch, and accumulator units.

PISTON RING: An open-ended ring which fits into a groove on the outer diameter of the piston. Its chief function is to form a seal between the piston and cylinder wall. Most automotive pistons have three rings: two for compression sealing; one for oil sealing.

PITMAN ARM: A lever which transmits steering force from the steering gear to the steering linkage.

PLANET CARRIER: A basic member of a planetary gear assembly that carries the pinion gears.

PLANET PINIONS: Gears housed in a planet carrier that are in constant mesh with the sun gear and internal gear. Because they have their own independent rotating centers, the pinions are capable of rotating around the sun gear or the inside of the internal gear.

PLANETARY GEAR RATIO: The reduction or overdrive ratio developed by a planetary gearset.

PLANETARY GEARSET: In its simplest form, it is made up of a basic assembly group containing a sun gear, internal gear, and planet carrier. The gears are always in constant mesh and offer a wide range of gear ratio possibilities.

PLANETARY GEARSET (COMPOUND): Two planetary gearsets combined together.

PLANETARY GEARSET (SIMPLE): An assembly of gears in constant mesh consisting of a sun gear, several pinion gears mounted in a carrier, and a ring gear. It provides gear ratio and direction changes, in addition to a direct drive and a neutral.

PLY RATING: A. rating given a tire which indicates strength (but not necessarily actual plies). A two-ply/four-ply rating has only two plies, but the strength of a four-ply tire.

POLARITY: Indication (positive or negative) of the two poles of a battery.

PORT: An opening for fluid intake or exhaust.

POSITIVE SEALING: A sealing method that completely prevents leakage.

POTENTIAL: Electrical force measured in volts; sometimes used interchangeably with voltage.

POWER: The ability to do work per unit of time, as expressed in horsepower; one horsepower equals 33,000 ft. lbs. of work per minute, or 550 ft. lbs. of work per second.

POWER FLOW: The systematic flow or transmission of power through the gears, from the input shaft to the output shaft.

POWER-TO-WEIGHT RATIO: Ratio of horsepower to weight of car.

POWERTRAIN: See Drivetrain.

POWERTRAIN CONTROL MODULE (PCM): Current designation for the engine control module (ECM). In many cases, late model vehicle control units manage the engine as well as the transmission. In other settings, the PCM controls the engine and is interfaced with a TCM to control transmission functions.

Ppm: Parts per million; unit used to measure exhaust emissions.

PREIGNITION: Early ignition of fuel in the cylinder, sometimes due to glowing carbon deposits in the combustion chamber. Preignition can be damaging since combustion takes place prematurely.

PRELOAD: A predetermined load placed on a bearing during assembly or by adjustment.

PRESS FIT: The mating of two parts under pressure, due to the inner diameter of one being smaller than the outer diameter of the other, or vice versa; an interference fit.

PRESSURE: The amount of force exerted upon a surface area.

PRESSURE CONTROL SOLENOID (PCS): An output device that provides a boost oil pressure to the mainline regulator valve to control line pressure. Its operation is determined by the amount of current sent from the PCM.

PRESSURE GAUGE: An instrument used for measuring the fluid pressure in a hydraulic circuit.

PRESSURE REGULATOR VALVE: In automatic transmissions, its purpose is to regulate the pressure of the pump output and supply the basic fluid pressure necessary to operate the transmission. The regulated fluid pressure may be referred to as mainline pressure, line pressure, or control pressure.

PRESSURE SWITCH ASSEMBLY (PSA): Mounted inside the transmission, it is a grouping of oil pressure switches that inputs to the PCM when certain hydraulic passages are charged with oil pressure.

PRESSURE PLATE: A spring-loaded plate (part of the clutch) that transmits power to the driven (friction) plate when the clutch is engaged.

PRIMARY CIRCUIT: The low voltage side of the ignition system which consists of the ignition switch, ballast resistor or resistance wire, bypass, coil, electronic control unit and pick-up coil as well as the connecting wires and harnesses.

PROFILE: Term used for tire measurement (tire series), which is the ratio of tire height to tread width.

PROM (PROGRAMMABLE READ-ONLY MEMORY): The heart of the computer that compares input data and makes the engineered program or strategy decisions about when to trigger the appropriate output based on stored computer instructions.

PULSE GENERATOR: A two-wire pickup sensor used to produce a fluctuating electrical signal. This changing signal is read by the controller to determine the speed of the object and can be used to measure transmission/transaxle input speed, output speed, and vehicle speed.

PSI: Pounds per square inch; a measurement of pressure.

PULSE WIDTH DUTY CYCLE SOLENOID (PULSE WIDTH MODULATED SOLENOID): A computer-controlled solenoid that turns on and off at a variable rate producing a modulated oil pressure; often referred to as a pulse width modulated (PWM) solenoid. Employed in many electronic automatic transmissions and transaxles, these solenoids are used to manage shift control and converter clutch hydraulic circuits.

PUSHROD: A steel rod between the hydraulic valve lifter and the valve rocker arm in overhead valve (OHV) engines.

PUMP: A mechanical device designed to create fluid flow and pressure buildup in a hydraulic system.

QUARTER PANEL: General term used to refer to a rear fender. Quarter panel is the area from the rear door opening to the tail light area and from rear wheel well to the base of the trunk and roof-line.

RACE: The surface on the inner or outer ring of a bearing on which the balls, needles or rollers move.

RACK AND PINION: A type of automotive steering system using a pinion gear attached to the end of the steering shaft. The pinion meshes with a long rack attached to the steering linkage.

RADIAL TIRE: Tire design which uses body cords running at right angles to the center line of the tire. Two or more belts are used to give tread strength. Radials can be identified by their characteristic sidewall bulge.

RADIATOR: Part of the cooling system for a water-cooled engine, mounted in the front of the vehicle and connected to the engine with rubber hoses. Through the radiator, excess combustion heat is dissipated into the atmosphere through forced convection using a water and glycol based mixture that circulates through, and cools, the engine.

RANGE REFERENCE AND CLUTCH/BAND APPLY CHART: A guide that shows the application of clutches and bands for each gear, within the selector range positions. These charts are extremely useful for understanding how the unit operates and for diagnosing malfunctions.

RAVIGNEAUX GEARSET: A compound planetary gearset that features matched dual planetary pinions (sets of two) mounted in a single planet carrier. Two sun gears and one ring mesh with the carrier pinions.

REACTION MEMBER: The stationary planetary member, in a planetary gearset, that is grounded to the transmission/transaxle case through the use of friction and wedging devices known as bands, disc clutches, and one-way clutches.

REACTION PRESSURE: The fluid pressure that moves a spool valve against an opposing force or forces; the area on which the opposing force acts. The opposing force can be a spring or a combination of spring force and auxiliary hydraulic force.

REACTOR, TORQUE CONVERTER: The reaction member of a fluid torque converter, more commonly called a stator. (See stator.)

REAR MAIN OIL SEAL: A synthetic or rope-type seal that prevents oil from leaking out of the engine past the rear main crankshaft bearing.

RECIRCULATING BALL: Type of steering system in which recirculating steel balls occupy the area between the nut and worm wheel, causing a reduction in friction.

RECTIFIER: A device (used primarily in alternators) that permits electrical current to flow in one direction only.

REDUCTION: (See gear reduction.)

REGULATOR VALVE: A valve that changes the pressure of the oil in a hydraulic circuit as the oil passes through the valve by bleeding off (or exhausting) some of the volume of oil supplied to the valve.

REFRIGERANT 12 (R-12) or 134 (R-134): The generic name of the refrigerant used in automotive air conditioning systems.

REGULATOR: A device which maintains the amperage and/or voltage levels of a circuit at predetermined values.

RELAY: A switch which automatically opens and/or closes a circuit.

RELAY VALVE: A valve that directs flow and pressure. Relay valves simply connect or disconnect interrelated passages without restricting the fluid flow or changing the pressure.

RELIEF VALVE: A spring-loaded, pressure-operated valve that limits oil pressure buildup in a hydraulic circuit to a predetermined maximum value.

RELUCTOR: A wheel that rotates inside the distributor and triggers the release of voltage in an electronic ignition.

RESERVOIR: The storage area for fluid in a hydraulic system; often called a sump.

RESIN: A liquid plastic used in body work.

RESIDUAL MAGNETISM: The magnetic strength stored in a material after a magnetizing field has been removed.

RESISTANCE: The opposition to the flow of current through a circuit or electrical device, and is measured in ohms. Resistance is equal to the voltage divided by the amperage.

RESISTOR SPARK PLUG: A spark plug using a resistor to shorten the spark duration. This suppresses radio interference and lengthens plug life.

RESISTOR: A device, usually made of wire, which offers a preset amount of resistance in an electrical circuit.

RESULTANT FORCE: The single effective directional thrust of the fluid force on the turbine produced by the vortex and rotary forces acting in different planes.

RETARD: Set the ignition timing so that spark occurs later (fewer degrees before TDC).

RHEOSTAT: A device for regulating a current by means of a variable resistance.

RING GEAR: The name given to a ring-shaped gear attached to a differential case, or affixed to a flywheel or as part of a planetary gear set.

ROADLOAD: grade.

ROCKER ARM: A lever which rotates around a shaft pushing down (opening) the valve with an end when the other end is pushed up by the pushrod. Spring pressure will later close the valve.

ROCKER PANEL: The body panel below the doors between the wheel opening.

ROLLER BEARING: A bearing made up of hardened inner and outer races between which hardened steel rollers move.

ROLLER CLUTCH: A type of one-way clutch design using rollers and springs mounted within an inner and outer cam race assembly.

ROTARY FLOW: The path of the fluid trapped between the blades of the members as they revolve with the rotation of the torque converter cover (rotational inertia).

ROTOR: (1.) The disc-shaped part of a disc brake assembly, upon which the brake pads bear; also called, brake disc. (2.) The device mounted atop the distributor shaft, which passes current to the distributor cap tower contacts.

ROTARY ENGINE: See Wankel engine.

RPM: Revolutions per minute (usually indicates engine speed).

RTV: A gasket making compound that cures as it is exposed to the atmosphere. It is used between surfaces that are not perfectly machined to one another, leaving a slight gap that the RTV fills and in which it hardens. The letters RTV represent room temperature vulcanizing.

RUN-ON: Condition when the engine continues to run, even when the key is turned off. See dieseling.

SEALED BEAM: A automotive headlight. The lens, reflector and filament from a single unit.

SEATBELT INTERLOCK: A system whereby the car cannot be started unless the seatbelt is buckled.

SECONDARY CIRCUIT: The high voltage side of the ignition system, usually above 20,000 volts. The secondary includes the ignition coil, coil wire, distributor cap and rotor, spark plug wires and spark plugs.

SELF-INDUCTION: The generation of voltage in a current-carrying wire by changing the amount of current flowing within that wire.

SEMI-CONDUCTOR: A material (silicon or germanium) that is neither a good conductor nor an insulator; used in diodes and transistors.

SEMI-FLOATING AXLE: In this design, a wheel is attached to the axle shaft, which takes both drive and cornering loads. Almost all solid axle passenger cars and light trucks use this design.

SENDING UNIT: A mechanical, electrical, hydraulic or electromagnetic device which transmits information to a gauge.

SENSOR: Any device designed to measure engine operating conditions or ambient pressures and temperatures. Usually electronic in nature and designed to send a voltage signal to an on-board computer, some sensors may operate as a simple on/off switch or they may provide a variable voltage signal (like a potentiometer) as conditions or measured parameters change.

SERIES CIRCUIT: (See circuit, series.)

SERPENTINE BELT: An accessory drive belt, with small multiple v-ribs, routed around most or all of the engine-powered accessories such as the alternator and power steering pump. Usually both the front and the back side of the belt comes into contact with various pulleys.

SERVO: In an automatic transmission, it is a piston in a cylinder assembly that converts hydraulic pressure into mechanical force and movement; used for the application of the bands and clutches.

SHIFT BUSYNESS: When referring to a torque converter clutch, it is the frequent apply and release of the clutch plate due to uncommon driving conditions.

SHIFT VALVE: Classified as a relay valve, it triggers the automatic shift in response to a governor and a throttle signal by directing fluid to the appropriate band and clutch apply combination to cause the shift to occur.

SHIM: Spacers of precise, predetermined thickness used between parts to establish a proper working relationship.

SHIMMY: Vibration (sometimes violent) in the front end caused by misaligned front end, out of balance tires or worn suspension components.

SHORT CIRCUIT: An electrical malfunction where current takes the path of least resistance to ground (usually through damaged insulation). Current flow is excessive from low resistance resulting in a blown fuse.

SHUDDER: Repeated jerking or stick-slip sensation, similar to chuggle but more severe and rapid in nature, that may be most noticeable during certain ranges of vehicle speed; also used to define condition after converter clutch engagement.

SIMPSON GEARSET: A compound planetary gear train that integrates two simple planetary gearsets referred to as the front planetary and the rear planetary.

SINGLE OVERHEAD CAMSHAFT: See overhead camshaft.

SKIDPLATE: A metal plate attached to the underside of the body to protect the fuel tank, transfer case or other vulnerable parts from damage.

SLAVE CYLINDER: In automotive use, a device in the hydraulic clutch system which is activated by hydraulic force, disengaging the clutch.

SLIPPING: Noticeable increase in engine rpm without vehicle speed increase; usually occurs during or after initial clutch or band engagement.

SLUDGE: Thick, black deposits in engine formed from dirt, oil, water, etc. It is usually formed in engines when oil changes are neglected.

SNAP RING: A circular retaining clip used inside or outside a shaft or part to secure a shaft, such as a floating wrist pin.

SOFT: Slow, almost unnoticeable clutch apply with very little shift feel.

SOFTCODES: DTCs that have been set into the PCM memory but are not present at the time of testing; often referred to as history or intermittent codes.

SOHC: Single overhead camshaft.

SOLENOID: An electrically operated, magnetic switching device.

SPALLING: A wear pattern identified by metal chips flaking off the hardened surface. This condition is caused by foreign particles, overloading situations, and/or normal wear.

SPARK PLUG: A device screwed into the combustion chamber of a spark ignition engine. The basic construction is a conductive core inside of a ceramic insulator, mounted in an outer conductive base. An electrical charge from the spark plug wire travels along the conductive core and jumps a preset air gap to a grounding point or points at the end of the conductive base. The resultant spark ignites the fuel/air mixture in the combustion chamber.

SPECIFIC GRAVITY (BATTERY): The relative weight of liquid (battery electrolyte) as compared to the weight of an equal volume of water.

SPLINES: Ridges machined or cast onto the outer diameter of a shaft or inner diameter of a bore to enable parts to mate without rotation.

SPLIT TORQUE DRIVE: In a torque converter, it refers to parallel paths of torque transmission, one of which is mechanical and the other hydraulic.

SPONGY PEDAL: A soft or spongy feeling when the brake pedal is depressed. It is usually due to air in the brake lines.

SPOOLVALVE: A precision-machined, cylindrically shaped valve made up of lands and grooves. Depending on its position in the valve bore, various interconnecting hydraulic circuit passages are either opened or closed.

SPRAG CLUTCH: A type of one-way clutch design using cams or contoured-shaped sprags between inner and outer races. (See one-way clutch.)

SPRUNG WEIGHT: The weight of a car supported by the springs.

SQUARE-CUT SEAL: Molded synthetic rubber seal designed with a square- or rectangular-shaped cross-section. This type of seal is used for both dynamic and static applications.

SRS: Supplemental restraint system

STABILIZER (SWAY) BAR: A bar linking both sides of the suspension. It resists sway on turns by taking some of added load from one wheel and putting it on the other.

STAGE: The number of turbine sets separated by a stator. A turbine set may be made up of one or more turbine members. A three-element converter is classified as a single stage.

STALL: In fluid drive transmission/transaxle applications, stall refers to engine rpm with the transmission/transaxle engaged and the vehicle stationary; throttle valve can be in any position between closed and wide open.

STALL SPEED: In fluid drive transmission/transaxle applications, stall speed refers to the maximum engine rpm with the transmission/transaxle engaged and vehicle stationary, when the throttle valve is wide open. (See stall; stall test.)

STALL TEST: A procedure recommended by many manufacturers to help determine the integrity of an engine, the torque converter stator, and certain clutch and band combinations. With the shift lever in each of the forward and reverse positions and with the brakes firmly applied, the accelerator pedal is momentarily pressed to the wide open throttle (WOT) position. The engine rpm reading at full throttle can provide clues for diagnosing the condition of the items listed above.

STALL TORQUE: The maximum design or engineered torque ratio of a fluid torque converter, produced under stall speed conditions. (See stall speed.)

STARTER: A high-torque electric motor used for the purpose of starting the engine, typically through a high ratio geared drive connected to the flywheel ring gear.

STATIC: A sealing application in which the parts being sealed do not move in relation to each other.

STATOR (REACTOR): The reaction member of a fluid torque converter that changes the direction of the fluid as it leaves the turbine to enter the impeller vanes. During the torque multiplication phase, this action assists the impeller's rotary force and results in an increase in torque.

STEERING GEOMETRY: Combination of various angles of suspension components (caster, camber, toe-in); roughly equivalent to front end alignment.

STRAIGHT WEIGHT: Term designating motor oil as suitable for use within a narrow range of temperatures. Outside the narrow temperature range its flow characteristics will not adequately lubricate.

STROKE: The distance the piston travels from bottom dead center to top dead center.

SUBSTITUTION: Replacing one part suspected of a defect with a like part of known quality.

SUMP: The storage vessel or reservoir that provides a ready source of fluid to the pump. In an automatic transmission, the sump is the oil pan. All fluid eventually returns to the sump for recycling into the hydraulic system.

SUN GEAR: In a planetary gearset, it is the center gear that meshes with a cluster of planet pinions.

SUPERCHARGER: An air pump driven mechanically by the engine through belts, chains, shafts or gears from the crankshaft. Two general types of supercharger are the positive displacement and centrifugal type, which pump air in direct relationship to the speed of the engine.

SUPPLEMENTAL RESTRAINT SYSTEM: See air bag.

SURGE: Repeating engine-related feeling of acceleration and deceleration that is less intense than chuggle.

SWITCH: A device used to open, close, or redirect the current in an electrical circuit.

SYNCHROMESH: A manual transmission/transaxle that is equipped with devices (synchronizers) that match the gear speeds so that the transmission/transaxle can be downshifted without clashing gears.

SYNTHETIC OIL: Non-petroleum based oil.

TACHOMETER: A device used to measure the rotary speed of an engine, shaft, gear, etc., usually in rotations per minute.

TDC: Top dead center. The exact top of the piston's stroke.

TEFLON SEALING RINGS: Teflon is a soft, durable, plastic-like material that is resistant to heat and provides excellent sealing. These rings are designed with either scarf-cut joints or as one-piece rings. Teflon sealing rings have replaced many metal ring applications.

TERMINAL: A device attached to the end of a wire or cable to make an electrical connection.

TEST LIGHT, CIRCUIT-POWERED: Uses available circuit voltage to test circuit continuity.

TEST LIGHT, SELF-POWERED: Uses its own battery source to test circuit continuity.

THERMISTOR: A special resistor used to measure fluid temperature; it decreases its resistance with increases in temperature.

THERMOSTAT: A valve, located in the cooling system of an engine, which is closed when cold and opens gradually in response to engine heating, controlling the temperature of the coolant and rate of coolant flow.

THERMOSTATIC ELEMENT: A heat-sensitive, spring-type device that controls a drain port from the upper sump area to the lower sump. When the transaxle fluid reaches operating temperature, the port is closed and the upper sump fills, thus reducing the fluid level in the lower sump.

THROTTLE POSITION (TP) SENSOR: Reads the degree of throttle opening; its signal is used to analyze engine load conditions. The ECM/PCM decides to apply the TCC, or to disengage it for coast or load conditions that need a converter torque boost.

THROTTLE PRESSURE/MODULATOR PRESSURE: A hydraulic signal oil pressure relating to the amount of engine load, based on either the amount of throttle plate opening or engine vacuum.

THROTTLE VALVE: A regulating or balanced valve that is controlled mechanically by throttle linkage or engine vacuum. It sends a hydraulic signal to the shift valve body to control shift timing and shift quality. (See balanced valve; modulator valve.)

THROW-OUT BEARING: As the clutch pedal is depressed, the throwout bearing moves against the spring fingers of the pressure plate, forcing the pressure plate to disengage from the driven disc.

TIE ROD: A rod connecting the steering arms. Tie rods have threaded ends that are used to adjust toe-in.

TIE-UP: Condition where two opposing clutches are attempting to apply at same time, causing engine to labor with noticeable loss of engine rpm.

TIMING BELT: A square-toothed, reinforced rubber belt that is driven by the crankshaft and operates the camshaft.

TIMING CHAIN: A roller chain that is driven by the crankshaft and operates the camshaft.

TIRE ROTATION: Moving the tires from one position to another to make the tires wear evenly.

TOE-IN (OUT): A term comparing the extreme front and rear of the front tires. Closer together at the front is toe-in; farther apart at the front is toe-out.

TOP DEAD CENTER (TDC): The point at which the piston reaches the top of its travel on the compression stroke.

TORQUE: Measurement of turning or twisting force, expressed as foot-pounds or inch-pounds.

TORQUE CONVERTER: A turbine used to transmit power from a driving member to a driven member via hydraulic action, providing changes in drive ratio and torque. In automotive use, it links the driveplate at the rear of the engine to the automatic transmission.

TORQUE CONVERTER CLUTCH: The apply plate (lockup plate) assembly used for mechanical power flow through the converter.

TORQUE PHASE: Sometimes referred to as slip phase or stall phase, torque multiplication occurs when the turbine is turning at a slower speed than the impeller, and the stator is reactionary (stationary). This sequence generates a boost in output torque.

TORQUE RATING (STALL TORQUE): The maximum torque multiplication that occurs during stall conditions, with the engine at wide open throttle (WOT) and zero turbine speed.

TORQUE RATIO: An expression of the gear ratio factor on torque effect. A 3:1 gear ratio or 3:1 torque ratio increases the torque input by the ratio factor of 3. Input torque (100 ft. lbs.) x 3 = output torque (300 ft. lbs.)

TRACTION: The amount of usable tractive effort before the drive wheels slip on the road contact surface.

TORSION BAR SUSPENSION: Long rods of spring steel which take the place of springs. One end of the bar is anchored and the other arm (attached to the suspension) is free to twist. The bars' resistance to twisting causes springing action.

TRACK: Distance between the centers of the tires where they contact the ground.

TRACTION CONTROL: A control system that prevents the spinning of a vehicle's drive wheels when excess power is applied.

TRACTIVE EFFORT: The amount of force available to the drive wheels, to move the vehicle.

TRANSAXLE: A single housing containing the transmission and differential. Transaxles are usually found on front engine/front wheel drive or rear engine/rear wheel drive cars.

TRANSDUCER: A device that changes energy from one form to another. For example, a transducer in a microphone changes sound energy to electrical energy. In automotive air-conditioning controls used in automatic temperature systems, a transducer changes an electrical signal to a vacuum signal, which operates mechanical doors.

TRANSMISSION: A powertrain component designed to modify torque and speed developed by the engine; also provides direct drive, reverse, and neutral.

TRANSMISSION CONTROL MODULE (TCM): Manages transmission functions. These vary according to the manufacturer's product design but may include converter clutch operation, electronic shift scheduling, and mainline pressure.

TRANSMISSION FLUID TEMPERATURE (TFT) SENSOR: Originally called a transmission oil temperature (TOT) sensor, this input device to the ECM/PCM senses the fluid temperature and provides a resistance value. It operates on the thermistor principle.

TRANSMISSION INPUT SPEED (TIS) SENSOR: Measures turbine shaft (input shaft) rpm's and compares to engine rpm's to determine torque converter slip. When compared to the transmission output speed sensor or VSS, gear ratio and clutch engagement timing can be determined.

TRANSMISSION OIL TEMPERATURE (TOT) SENSOR: (See transmission fluid temperature (TFT) sensor.)

TRANSMISSION RANGE SELECTOR (TRS) SWITCH: Tells the module which gear shift position the driver has chosen.

TRANSFER CASE: A gearbox driven from the transmission that delivers power to both front and rear driveshafts in a four-wheel drive system. Transfer cases usually have a high and low range set of gears, used depending on how much pulling power is needed.

TRANSISTOR: A semi-conductor component which can be actuated by a small voltage to perform an electrical switching function.

TREAD WEAR INDICATOR: Bars molded into the tire at right angles to the tread that appear as horizontal bars when 1/16 in. of tread remains.

TREAD WEAR PATTERN: The pattern of wear on tires which can be "read" to diagnose problems in the front suspension.

TUNE-UP: A regular maintenance function, usually associated with the replacement and adjustment of parts and components in the electrical and fuel systems of a vehicle for the purpose of attaining optimum performance.

TURBINE: The output (driven) member of a fluid coupling or fluid torque converter. It is splined to the input (turbine) shaft of the transmission.

TURBOCHARGER: An exhaust driven pump which compresses intake air and forces it into the combustion chambers at higher than atmospheric pressures. The increased air pressure allows more fuel to be burned and results in increased horsepower being produced.

TURBULENCE: The interference of molecules of a fluid (or vapor) with each other in a fluid flow.

TYPE F: Transmission fluid developed and used by Ford Motor Company up to 1982. This fluid type provides a high coefficient of friction.

TYPE 7176: The preferred choice of transmission fluid for Chrysler automatic transmissions and transaxles. Developed in 1986, it closely resembles Dexron and Mercon. Type 7176 is the recommended service fill fluid for all Chrysler products utilizing a lockup torque converter dating back to 1978.

U-JOINT (UNIVERSAL JOINT): A flexible coupling in the drive train that allows the driveshafts or axle shafts to operate at different angles and still transmit rotary power.

UNDERSTEER: The tendency of a car to continue straight ahead while negotiating a turn.

UNIT BODY: Design in which the car body acts as the frame.

UNLEADED FUEL: Fuel which contains no lead (a common gasoline additive). The presence of lead in fuel will destroy the functioning elements of a catalytic converter, making it useless.

UNSPRUNG WEIGHT: The weight of car components not supported by the springs (wheels, tires, brakes, rear axle, control arms, etc.).

UPSHIFT: A shift that results in a decrease in torque ratio and an increase in speed.

VACUUM: A negative pressure; any pressure less than atmospheric pressure.

VACUUM ADVANCE: A device which advances the ignition timing in response to increased engine vacuum.

VACUUM GAUGE: An instrument used for measuring the existing vacuum in a vacuum circuit or chamber. The unit of measure is inches (of mercury in a barometer).

VACUUM MODULATOR: Generates a hydraulic oil pressure in response to the amount of engine vacuum.

VALVES: Devices that can open or close fluid passages in a hydraulic system and are used for directing fluid flow and controlling pressure.

VALVE BODY ASSEMBLY: The main hydraulic control assembly of the transmission/transaxle that contains numerous valves, check balls, and other components to control the distribution of pressurized oil throughout the transmission.

VALVE CLEARANCE: The measured gap between the end of the valve stem and the rocker arm, cam lobe or follower that activates the valve.

VALVE GUIDES: The guide through which the stem of the valve passes.

The guide is designed to keep the valve in proper alignment.

VALVE LASH (clearance): The operating clearance in the valve train.

VALVE TRAIN: The system that operates intake and exhaust valves, consisting of camshaft, valves and springs, lifters, pushrods and rocker arms.

VAPOR LOCK: Boiling of the fuel in the fuel lines due to excess heat. This will interfere with the flow of fuel in the lines and can completely stop the flow. Vapor lock normally only occurs in hot weather.

VARIABLE DISPLACEMENT (VARIABLE CAPACITY) VANE PUMP: Slipper-type vanes, mounted in a revolving rotor and contained within the bore of a movable slide, capture and then force fluid to flow. Movement of the slide to various positions changes the size of the vane chambers and the amount of fluid flow. **Note:** GM refers to this pump design as variable displacement, and Ford terms it variable capacity.

VARIABLE FORCE SOLENOID (VFS): Commonly referred to as the electronic pressure control (EPC) solenoid, it replaces the cable/linkage style of TV system control and is integrated with a spool valve and spring assembly to control pressure. A variable computer-controlled current flow varies the internal force of the solenoid on the spool valve and resulting control pressure.

VARIABLE ORIFICE THERMAL VALVE: Temperature-sensitive hydraulic oil control device that adjusts the size of a circuit path opening. By altering the size of the opening, the oil flow rate is adapted for cold to hot oil viscosity changes.

VARNISH: Term applied to the residue formed when gasoline gets old and stale.

VCM: See Electronic Control Unit (ECU).

VEHICLE SPEED SENSOR (VSS): Provides an electrical signal to the computer module, measuring vehicle speed, and affects the torque converter clutch engagement and release.

VESPEL SEALING RINGS: Hard plastic material that produces excellent sealing in dynamic settings. These rings are found in late versions of the 4T60 and in all 4T60-E and 4T80-E transaxles.

VISCOSITY: The ability of a fluid to flow. The lower the viscosity rating, the easier the fluid will flow. 10 weight motor oil will flow much easier than 40 weight motor oil.

VISCOSITY INDEX IMPROVERS: Keeps the viscosity nearly constant with changes in temperature. This is especially important at low temperatures, when the oil needs to be thin to aid in shifting and for cold-weather starting. Yet it must not be so thin that at high temperatures it will cause excessive hydraulic leakage so that pumps are unable to maintain the proper pressures.

VISCOUS CLUTCH: A specially designed torque converter clutch apply plate that, through the use of a silicon fluid, clamps smoothly and absorbs torsional vibrations.

VOLT: Unit used to measure the force or pressure of electricity. It is defined as the pressure needed to move one amp through the resistance of one ohm.

VOLTAGE: The electrical pressure that causes current to flow. Voltage is measured in volts (V).

VOLTAGE, APPLIED: The actual voltage read at a given point in a circuit. It equals the available voltage of the power supply minus the losses in the circuit up to that point.

VOLTAGE DROP: The voltage lost or used in a circuit by normal loads such as a motor or lamp or by abnormal loads such as a poor (high-resistance) lead or terminal connection.

VOLTAGE REGULATOR: A device that controls the current output of the alternator or generator.

VOLTMETER: An instrument used for measuring electrical force in units called volts. Voltmeters are always connected parallel with the circuit being tested.

VORTEX FLOW: The crosswise or circulatory flow of oil between the blades of the members caused by the centrifugal pumping action of the impeller.

WANKEL ENGINE: An engine which uses no pistons. In place of pistons, triangular-shaped rotors revolve in specially shaped housings.

WATER PUMP: A belt driven component of the cooling system that mounts on the engine, circulating the coolant under pressure.

WATT: The unit for measuring electrical power. One watt is the product of one ampere and one volt (watts equals amps times volts). Wattage is the horsepower of electricity (746 watts equal one horsepower).

WHEEL ALIGNMENT: Inclusive term to describe the front end geometry (caster, camber, toe-in/out).

WHEEL CYLINDER: Found in the automotive drum brake assembly, it is a device, actuated by hydraulic pressure, which, through internal pistons, pushes the brake shoes outward against the drums.

WHEEL WEIGHT: Small weights attached to the wheel to balance the wheel and tire assembly. Out-of-balance tires quickly wear out and also give erratic handling when installed on the front.

WHEELBASE: Distance between the center of front wheels and the center of rear wheels.

WIDE OPEN THROTTLE (WOT): Full travel of accelerator pedal.

WORK: The force exerted to move a mass or object. Work involves motion; if a force is exerted and no motion takes place, no work is done. Work per unit of time is called power. Work = force x distance = ft. lbs. 33,000 ft. lbs. in one minute = 1 horsepower

ZERO-THROTTLE COAST DOWN: A full release of accelerator pedal while vehicle is in motion and in drive range.

Commonly Used Abbreviations

2

2WD	Two Wheel Drive

4

4WD	Four Wheel Drive

A

A/C	Air Conditioning
ABDC	After Bottom Dead Center
ABS	Anti-lock Brakes
AC	Alternating Current
ACL	Air cleaner
ACT	Air Charge Temperature
AIR	Secondary Air Injection
ALCL	Assembly Line Communications Link
ALDL	Assembly Line Diagnostic Link
AT	Automatic Transaxle/Transmission
ATDC	After Top Dead Center
ATF	Automatic Transmission Fluid
ATS	Air Temperature Sensor
AWD	All Wheel Drive

B

BAP	Barometric Absolute Pressure
BARO	Barometric Pressure
BBDC	Before Bottom Dead Center
BCM	Body Control Module
BDC	Bottom Dead Center
BPT	Backpressure Transducer
BTDC	Before Top Dead Center
BVSV	Bimetallic Vacuum Switching Valve

C

CAC	Charge Air Cooler
CARB	California Air Resources Board
CAT	Catalytic Converter
CCC	Computer Command Control
CCCC	Computer Controlled Catalytic Converter
CCCI	Computer Controlled Coil Ignition
CCD	Computer Controlled Dwell
CDI	Capacitor Discharge Ignition
CEC	Computerized Engine Control
CFI	Continuous Fuel Injection
CIS	Continuous Injection System
CIS-E	Continuous Injection System - Electronic
CKP	Crankshaft Position
CL	Closed Loop
CMP	Camshaft Position
CPP	Clutch Pedal Position
CTOX	Continuous Trap Oxidizer System
CTP	Closed Throttle Position
CVC	Constant Vacuum Control
CYL	Cylinder

D

DBC	Dual Bed Catalyst
DC	Direct Current
DFI	Direct Fuel Injection
DIS	Distributorless Ignition System
DLC	Data Link Connector
DMM	Digital Multimeter
DOHC	Double Overhead Camshaft
DRB	Diagnostic Readout Box
DTC	Diagnostic Trouble Code
DTM	Diagnostic Test Mode
DVOM	Digital Volt/Ohmmeter

E

EBCM	Electronic Brake Control Module
ECM	Engine Control Module
ECT	Engine Coolant Temperature
ECU	Engine Control Unit or Electronic Control Unit
EDIS	Electronic Distributorless Ignition System
EEC	Electronic Engine Control
EEPROM	Electrically Erasable Programmable Read Only Memory
EFE	Early Fuel Evaporation
EGR	Exhaust Gas Recirculation
EGRT	Exhaust Gas Recirculation Temperature
EGRVC	EGR Valve Control
EPROM	Erasable Programmable Read Only Memory
EVAP	Evaporative Emissions
EVP	EGR Valve Position

F

FBC	Feedback Carburetor
FEEPROM	Flash Electrically Erasable Programmable Read Only Memory
FF	Flexible Fuel
FI	Fuel Injection
FT	Fuel Trim
FWD	Front Wheel Drive

G

GND	Ground

H

HAC	High Altitude Compensation
HEGO	Heated Exhaust Gas Oxygen sensor
HEI	High Energy Ignition
HO2 Sensor	Heated Oxygen Sensor

I

IAC	Idle Air Control
IAT	Intake Air Temperature
ICM	Ignition Control Module
IFI	Indirect Fuel Injection
IFS	Inertia Fuel Shutoff
ISC	Idle Speed Control
IVSV	Idle Vacuum Switching Valve

Commonly Used Abbreviations

K

KOEO	Key On, Engine Off
KOER	Key ON, Engine Running
KS	Knock Sensor

M

MAF	Mass Air Flow
MAP	Manifold Absolute Pressure
MAT	Manifold Air Temperature
MC	Mixture Control
MDP	Manifold Differential Pressure
MFI	Multiport Fuel Injection
MIL	Malfunction Indicator Lamp or Maintenance
MST	Manifold Surface Temperature
MVZ	Manifold Vacuum Zone

N

NVRAM	Nonvolatile Random Access Memory

O

O2 Sensor	Oxygen Sensor
OBD	On-Board Diagnostic
OC	Oxidation Catalyst
OHC	Overhead Camshaft
OL	Open Loop

P

P/S	Power Steering
PAIR	Pulsed Secondary Air Injection
PCM	Powertrain Control Module
PCS	Purge Control Solenoid
PCV	Positive Crankcase Ventilation
PIP	Profile Ignition Pick-up
PNP	Park/Neutral Position
PROM	Programmable Read Only Memory
PSP	Power Steering Pressure
PTO	Power Take-Off
PTOX	Periodic Trap Oxidizer System

R

RABS	Rear Anti-lock Brake System
RAM	Random Access Memory
ROM	Read Only Memory
RPM	Revolutions Per Minute
RWAL	Rear Wheel Anti-lock Brakes
RWD	Rear Wheel Drive

S

SBC	Single Bed Converter
SBEC	Single Board Engine Controller
SC	Supercharger
SCB	Supercharger Bypass
SFI	Sequential Multiport Fuel Injection
SIR	Supplemental Inflatable Restraint
SOHC	Single Overhead Camshaft
SPL	Smoke Puff Limiter
SPOUT	Spark Output
SRI	Service Reminder Indicator
SRS	Supplemental Restraint System
SRT	System Readiness Test
SSI	Solid State Ignition
ST	Scan Tool
STO	Self-Test Output

T

TAC	Thermostatic Air Cleaner
TBI	Throttle Body Fuel Injection
TC	Turbocharger
TCC	Torque Converter Clutch
TCM	Transmission Control Module
TDC	Top Dead Center
TFI	Thick Film Ignition
TP	Throttle Position
TR Sensor	Transaxle/Transmission Range Sensor
TVV	Thermal Vacuum Valve
TWC	Three-way Catalytic Converter

V

VAF	Volume Air Flow, or Vane Air Flow
VAPS	Variable Assist Power Steering
VRV	Vacuum Regulator Valve
VSS	Vehicle Speed Sensor
VSV	Vacuum Switching Valve

W

WOT	Wide Open Throttle
WU-TWC	Warm Up Three-way Catalytic Converter

ENGLISH TO METRIC CONVERSION: TORQUE

To convert foot-pounds (ft. lbs.) to Newton-meters (Nm), multiply the number of ft. lbs. by 1.36
To convert Newton-meters (Nm) to foot-pounds (ft. lbs.), multiply the number of Nm by 0.7376

ft. lbs.	Nm	ft. lbs.	Nm	ft. lbs.	Nm	ft. lbs.	Nm
0.1	0.1	34	46.2	76	103.4	118	160.5
0.2	0.3	35	47.6	77	104.7	119	161.8
0.3	0.4	36	49.0	78	106.1	120	163.2
0.4	0.5	37	50.3	79	107.4	121	164.6
0.5	0.7	38	51.7	80	108.8	122	165.9
0.6	0.8	39	53.0	81	110.2	123	167.3
0.7	1.0	40	54.4	82	111.5	124	168.6
0.8	1.1	41	55.8	83	112.9	125	170.0
0.9	1.2	42	57.1	84	114.2	126	171.4
1	1.4	43	58.5	85	115.6	127	172.7
2	2.7	44	59.8	86	117.0	128	174.1
3	4.1	45	61.2	87	118.3	129	175.4
4	5.4	46	62.6	88	119.7	130	176.8
5	6.8	47	63.9	89	121.0	131	178.2
6	8.2	48	65.3	90	122.4	132	179.5
7	9.5	49	66.6	91	123.8	133	180.9
8	10.9	50	68.0	92	125.1	134	182.2
9	12.2	51	69.4	93	126.5	135	183.6
10	13.6	52	70.7	94	127.8	136	185.0
11	15.0	53	72.1	95	129.2	137	186.3
12	16.3	54	73.4	96	130.6	138	187.7
13	17.7	55	74.8	97	131.9	139	189.0
14	19.0	56	76.2	98	133.3	140	190.4
15	20.4	57	77.5	99	134.6	141	191.8
16	21.8	58	78.9	100	136.0	142	193.1
17	23.1	59	80.2	101	137.4	143	194.5
18	24.5	60	81.6	102	138.7	144	195.8
19	25.8	61	83.0	103	140.1	145	197.2
20	27.2	62	84.3	104	141.4	146	198.6
21	28.6	63	85.7	105	142.8	147	199.9
22	29.9	64	87.0	106	144.2	148	201.3
23	31.3	65	88.4	107	145.5	149	202.6
24	32.6	66	89.8	108	146.9	150	204.0
25	34.0	67	91.1	109	148.2	151	205.4
26	35.4	68	92.5	110	149.6	152	206.7
27	36.7	69	93.8	111	151.0	153	208.1
28	38.1	70	95.2	112	152.3	154	209.4
29	39.4	71	96.6	113	153.7	155	210.8
30	40.8	72	97.9	114	155.0	156	212.2
31	42.2	73	99.3	115	156.4	157	213.5
32	43.5	74	100.6	116	157.8	158	214.9
33	44.9	75	102.0	117	159.1	159	216.2

METRIC TO ENGLISH CONVERSION: TORQUE

To convert foot-pounds (ft. lbs.) to Newton-meters (Nm), multiply the number of ft. lbs. by 1.36

To convert Newton-meters (Nm) to foot-pounds (ft. lbs.), multiply the number of Nm by 0.7376

Nm	ft. lbs.	Nm	ft. lbs.	Nm	ft. lbs.	Nm	ft. lbs.	Nm	ft. lbs.
0.1	0.1	34	25.0	76	55.9	118	86.8	160	117.6
0.2	0.1	35	25.7	77	56.6	119	87.5	161	118.4
0.3	0.2	36	26.5	78	57.4	120	88.2	162	119.1
0.4	0.3	37	27.2	79	58.1	121	89.0	163	119.9
0.5	0.4	38	27.9	80	58.8	122	89.7	164	120.6
0.6	0.4	39	28.7	81	59.6	123	90.4	165	121.3
0.7	0.5	40	29.4	82	60.3	124	91.2	166	122.1
0.8	0.6	41	30.1	83	61.0	125	91.9	167	122.8
0.9	0.7	42	30.9	84	61.8	126	92.6	168	123.5
1	0.7	43	31.6	85	62.5	127	93.4	169	124.3
2	1.5	44	32.4	86	63.2	128	94.1	170	125.0
3	2.2	45	33.1	87	64.0	129	94.9	171	125.7
4	2.9	46	33.8	88	64.7	130	95.6	172	126.5
5	3.7	47	34.6	89	65.4	131	96.3	173	127.2
6	4.4	48	35.3	90	66.2	132	97.1	174	127.9
7	5.1	49	36.0	91	66.9	133	97.8	175	128.7
8	5.9	50	36.8	92	67.6	134	98.5	176	129.4
9	6.6	51	37.5	93	68.4	135	99.3	177	130.1
10	7.4	52	38.2	94	69.1	136	100.0	178	130.9
11	8.1	53	39.0	95	69.9	137	100.7	179	131.6
12	8.8	54	39.7	96	70.6	138	101.5	180	132.4
13	9.6	55	40.4	97	71.3	139	102.2	181	133.1
14	10.3	56	41.2	98	72.1	140	102.9	182	133.8
15	11.0	57	41.9	99	72.8	141	103.7	183	134.6
16	11.8	58	42.6	100	73.5	142	104.4	184	135.3
17	12.5	59	43.4	101	74.3	143	105.1	185	136.0
18	13.2	60	44.1	102	75.0	144	105.9	186	136.8
19	14.0	61	44.9	103	75.7	145	106.6	187	137.5
20	14.7	62	45.6	104	76.5	146	107.4	188	138.2
21	15.4	63	46.3	105	77.2	147	108.1	189	139.0
22	16.2	64	47.1	106	77.9	148	108.8	190	139.7
23	16.9	65	47.8	107	78.7	149	109.6	191	140.4
24	17.6	66	48.5	108	79.4	150	110.3	192	141.2
25	18.4	67	49.3	109	80.1	151	111.0	193	141.9
26	19.1	68	50.0	110	80.9	152	111.8	194	142.6
27	19.9	69	50.7	111	81.6	153	112.5	195	143.4
28	20.6	70	51.5	112	82.4	154	113.2	196	144.1
29	21.3	71	52.2	113	83.1	155	114.0	197	144.9
30	22.1	72	52.9	114	83.8	156	114.7	198	145.6
31	22.8	73	53.7	115	84.6	157	115.4	199	146.3
32	23.5	74	54.4	116	85.3	158	116.2	200	147.1
33	24.3	75	55.1	117	86.0	159	116.9	201	147.8

ENGLISH/METRIC CONVERSION: TEMPERATURE

To convert Fahrenheit (F°) to Celsius (C°), take F° temperature and subtract 32, multiply the result by 5 and divide the result by 9
To convert Celsius (C°) to Fahrenheit (F°), take C° temperature and multiply it by 9, divide the result by 5 and add 32

F°	C°	F°	C°	C°	F°	C°	F°
-40	-40.0	150	65.6	-38	-36.4	46	114.8
-35	-37.2	155	68.3	-36	-32.8	48	118.4
-30	-34.4	160	71.1	-34	-29.2	50	122
-25	-31.7	165	73.9	-32	-25.6	52	125.6
-20	-28.9	170	76.7	-30	-22	54	129.2
-15	-26.1	175	79.4	-28	-18.4	56	132.8
-10	-23.3	180	82.2	-26	-14.8	58	136.4
-5	-20.6	185	85.0	-24	-11.2	60	140
0	-17.8	190	87.8	-22	-7.6	62	143.6
1	-17.2	195	90.6	-20	-4	64	147.2
2	-16.7	200	93.3	-18	-0.4	66	150.8
3	-16.1	205	96.1	-16	3.2	68	154.4
4	-15.6	210	98.9	-14	6.8	70	158
5	-15.0	212	100.0	-12	10.4	72	161.6
10	-12.2	215	101.7	-10	14	74	165.2
15	-9.4	220	104.4	-8	17.6	76	168.8
20	-6.7	225	107.2	-6	21.2	78	172.4
25	-3.9	230	110.0	-4	24.8	80	176
30	-1.1	235	112.8	-2	28.4	82	179.6
35	1.7	240	115.6	0	32	84	183.2
40	4.4	245	118.3	2	35.6	86	186.8
45	7.2	250	121.1	4	39.2	88	190.4
50	10.0	255	123.9	6	42.8	90	194
55	12.8	260	126.7	8	46.4	92	197.6
60	15.6	265	129.4	10	50	94	201.2
65	18.3	270	132.2	12	53.6	96	204.8
70	21.1	275	135.0	14	57.2	98	208.4
75	23.9	280	137.8	16	60.8	100	212
80	26.7	285	140.6	18	64.4	102	215.6
85	29.4	290	143.3	20	68	104	219.2
90	32.2	295	146.1	22	71.6	106	222.8
95	35.0	300	148.9	24	75.2	108	226.4
100	37.8	305	151.7	26	78.8	110	230
105	40.6	310	154.4	28	82.4	112	233.6
110	43.3	315	157.2	30	86	114	237.2
115	46.1	320	160.0	32	89.6	116	240.8
120	48.9	325	162.8	34	93.2	118	244.4
125	51.7	330	165.6	36	96.8	120	248
130	54.4	335	168.3	38	100.4	122	251.6
135	57.2	340	171.1	40	104	124	255.2
140	60.0	345	173.9	42	107.6	126	258.8
145	62.8	350	176.7	44	111.2	128	262.4

LENGTH CONVERSION

To convert inches (in.) to millimeters (mm), multiply the number of inches by 25.4
To convert millimeters (mm) to inches (in.), multiply the number of millimeters by 0.04

Inches	Millimeters	Inches	Millimeters	Inches	Millimeters	Inches	Millimeters
0.0001	0.00254	0.005	0.1270	0.09	2.286	4	101.6
0.0002	0.00508	0.006	0.1524	0.1	2.54	5	127.0
0.0003	0.00762	0.007	0.1778	0.2	5.08	6	152.4
0.0004	0.01016	0.008	0.2032	0.3	7.62	7	177.8
0.0005	0.01270	0.009	0.2286	0.4	10.16	8	203.2
0.0006	0.01524	0.01	0.254	0.5	12.70	9	228.6
0.0007	0.01778	0.02	0.508	0.6	15.24	10	254.0
0.0008	0.02032	0.03	0.762	0.7	17.78	11	279.4
0.0009	0.02286	0.04	1.016	0.8	20.32	12	304.8
0.001	0.0254	0.05	1.270	0.9	22.86	13	330.2
0.002	0.0508	0.06	1.524	1	25.4	14	355.6
0.003	0.0762	0.07	1.778	2	50.8	15	381.0
0.004	0.1016	0.08	2.032	3	76.2	16	406.4

ENGLISH/METRIC CONVERSION: LENGTH

To convert inches (in.) to millimeters (mm), multiply the number of inches by 25.4
To convert millimeters (mm) to inches (in.), multiply the number of millimeters by 0.04

Inches		Millimeters	Inches		Millimeters	Inches		Millimeters
Fraction	Decimal	Decimal	Fraction	Decimal	Decimal	Fraction	Decimal	Decimal
1/64	0.016	0.397	11/32	0.344	8.731	11/16	0.688	17.463
1/32	0.031	0.794	23/64	0.359	9.128	45/64	0.703	17.859
3/64	0.047	1.191	3/8	0.375	9.525	23/32	0.719	18.256
1/16	0.063	1.588	25/64	0.391	9.922	47/64	0.734	18.653
5/64	0.078	1.984	13/32	0.406	10.319	3/4	0.750	19.050
3/32	0.094	2.381	27/64	0.422	10.716	49/64	0.766	19.447
7/64	0.109	2.778	7/16	0.438	11.113	25/32	0.781	19.844
1/8	0.125	3.175	29/64	0.453	11.509	51/64	0.797	20.241
9/64	0.141	3.572	15/32	0.469	11.906	13/16	0.813	20.638
5/32	0.156	3.969	31/64	0.484	12.303	53/64	0.828	21.034
11/64	0.172	4.366	1/2	0.500	12.700	27/32	0.844	21.431
3/16	0.188	4.763	33/64	0.516	13.097	55/64	0.859	21.828
13/64	0.203	5.159	17/32	0.531	13.494	7/8	0.875	22.225
7/32	0.219	5.556	35/64	0.547	13.891	57/64	0.891	22.622
15/64	0.234	5.953	9/16	0.563	14.288	29/32	0.906	23.019
1/4	0.250	6.350	37/64	0.578	14.684	59/64	0.922	23.416
17/64	0.266	6.747	19/32	0.594	15.081	15/16	0.938	23.813
9/32	0.281	7.144	39/64	0.609	15.478	61/64	0.953	24.209
19/64	0.297	7.541	5/8	0.625	15.875	31/32	0.969	24.606
5/16	0.313	7.938	41/64	0.641	16.272	63/64	0.984	25.003
21/64	0.328	8.334	21/32	0.656	16.669	1/1	1.000	25.400
			43/64	0.672	17.066			